Y0-BEE-992

American Men & Women of Science

1992-93 • 18th Edition

The 18th edition of *AMERICAN MEN & WOMEN OF SCIENCE* was prepared by the R.R. Bowker Database Publishing Group.

Stephen L. Torpie, Managing Editor
Judy Redel, Managing Editor, Research
Richard D. Lanam, Senior Editor
Tanya Hurst, Research Manager
Karen Hallard, Beth Tanis, Associate Editors

Peter Simon, Vice President, Database Publishing Group
Dean Hollister, Director, Database Planning
Edgar Adcock, Jr., Editorial Director, Directories

American Men & Women of Science

1992-93 • 18th Edition

A Biographical Directory of Today's Leaders in Physical, Biological and Related Sciences.

Volume 3 • G-I

R. R. BOWKER
New Providence, New Jersey

Published by R.R. Bowker, a division of Reed Publishing, (USA) Inc.

International Standard Book Number

Set:	0-8352-3074-0
Volume I:	0-8352-3075-9
Volume II:	0-8352-3076-7
Volume III:	0-8352-3077-5
Volume IV:	0-8352-3078-3
Volume V:	0-8352-3079-1
Volume VI:	0-8352-3080-5
Volume VII:	0-8352-3081-3
Volume VIII:	0-8352-3082-1

International Standard Serial Number: 0192-8570
Library of Congress Catalog Card Number: 6-7326
Printed and bound in the United States of America.

8 Volume Set

ISBN 0-8352-3074-0

9 780835 230742

Contents

Advisory Committee

Dr. Robert F. Barnes
Executive Vice President
American Society of Agronomy

Dr. John Kistler Crum
Executive Director
American Chemical Society

Dr. Charles Henderson Dickens
Section Head, Survey & Analysis Section
Division of Science Resource Studies
National Science Foundation

Mr. Alan Edward Fechter
Executive Director
Office of Scientific & Engineering Personnel
National Academy of Science

Dr. Oscar Nicolas Garcia
Prof Electrical Engineering
Electrical Engineering & Computer Science Department
George Washington University

Dr. Charles George Groat
Executive Director
American Geological Institute

Dr. Richard E. Hallgren
Executive Director
American Meteorological Society

Dr. Michael J. Jackson
Executive Director
Federation of American Societies for Experimental Biology

Dr. William Howard Jaco
Executive Director
American Mathematical Society

Dr. Shirley Mahaley Malcom
Head, Directorate for Education and Human Resources Programs
American Association for the Advancement of Science

Mr. Daniel Melnick
Sr Advisor Research Methodologies
Sciences Resources Directorate
National Science Foundation

Ms. Beverly Fearn Porter
Division Manager
Education & Employment Statistics Division
American Institute of Physics

Dr. Terrence R. Russell
Manager
Office of Professional Services
American Chemical Society

Dr. Irwin Walter Sandberg
Holder, Cockrell Family Regent Chair
Department of Electrical & Computer Engineering
University of Texas

Dr. William Eldon Splinter
Interim Vice Chancellor for Research,
Dean, Graduate Studies
University of Nebraska

Ms. Betty M. Vetter
Executive Director, Science Manpower Comission
Commission on Professionals in Science & Technology

Dr. Dael Lee Wolfe
Professor Emeritus
Graduate School of Public Affairs
University of Washington

Preface

American Men and Women Of Science remains without peer as a chronicle of North American scientific endeavor and achievement. The present work is the eighteenth edition since it was first compiled as *American Men of Science* by J. Mckeen Cattell in 1906. In its eighty-six year history *American Men & Women of Science* has profiled the careers of over 300,000 scientists and engineers. Since the first edition, the number of American scientists and the fields they pursue have grown immensely. This edition alone lists full biographies for 122,817 engineers and scientists, 7021 of which are listed for the first time. Although the book has grown, our stated purpose is the same as when Dr. Cattell first undertook the task of producing a biographical directory of active American scientists. It was his intention to record educational, personal and career data which would make "a contribution to the organization of science in America" and "make men [and women] of science acquainted with one another and with one another's work." It is our hope that this edition will fulfill these goals.

The biographies of engineers and scientists constitute seven of the eight volumes and provide birthdates, birthplaces, field of specialty, education, honorary degrees, professional and concurrent experience, awards, memberships, research information and adresses for each entrant when applicable. The eighth volume, the discipline index, organizes biographees by field of activity. This index, adapted from the National Science Foundation's Taxonomy of Degree and Employment Specialties, classifies entrants by 171 subject specialties listed in the table of contents of Volume 8. For the first time, the index classifies scientists and engineers by state within each subject specialty, allowing the user to more easily locate a scientist in a given area. Also new to this edition is the inclusion of statistical information and recipients of theNobel Prizes, the Craaford Prize, the Charles Stark Draper Prize, and the National Medals of Science and Technology received since the last edition.

While the scientific fields covered by *American Men and Women Of Science* are comprehensive, no attempt has been made to include all American scientists. Entrants are meant to be limited to those who have made significant contributions in their field. The names of new entrants were submitted for consideration at the editors' request by current entrants and by leaders of academic, government and private research programs and associations. Those included met the following criteria:

1. Distinguished achievement, by reason of experience, training or accomplishment, including contributions to the literature, coupled with continuing activity in scientific work;

or

2. Research activity of high quality in science as evidenced by publication in reputable scientific journals; or for those whose work cannot be published due to governmental or industrial security, research activity of high quality in science as evidenced by the judgement of the individual's peers;

or

3. Attainment of a position of substantial responsibility requiring scientific training and experience.

This edition profiles living scientists in the physical and biological fields, as well as public health scientists, engineers, mathematicians, statisticians, and computer scientists. The information is collected by means of direct communication whenever possible. All entrants receive forms for corroboration and updating. New entrants receive questionaires and verification proofs before publication. The information submitted by entrants is included as completely as possible within the boundaries of editorial and space restrictions. If an entrant does not return the form and his or her current location can be verified in secondary sources, the full entry is repeated. References to the previous edition are given for those who do not return forms and cannot be located, but who are presumed to be still active in science or engineering. Entrants known to be deceased are noted as such and a reference to the previous edition is given. Scientists and engineers who are not citizens of the United States or Canada are included if a significant portion of their work was performed in North America.

The information in AMWS is also available on CD-ROM as part of *SciTech Reference Plus.* In adition to the convenience of searching scientists and engineers, *SciTech Reference Plus* also includes *The Directory of American Research & Technology*, *Corporate Technology Directory*, sci-tech and medical books and serials from *Books in Print* and *Bowker International Series*. *American Men and Women Of Science* is available for online searching through the subscription services of DIALOG Information Services, Inc. (3460 Hillview Ave, Palo Alto, CA 94304) and ORBIT Search Service (800 Westpark Dr, McLean, VA 22102). Both CD-Rom and the on-line subscription services allow all elements of an entry, including field of interest, experience, and location, to be accessed by key word. Tapes and mailing lists are also available through the Cahners Direct Mail (John Panza, List Manager, Bowker Files 245 W 17th St, New York, NY, 10011, Tel: 800-537-7930).

A project as large as publishing *American Men and Women Of Science* involves the efforts of a great many people. The editors take this opportunity to thank the eighteenth edition advisory committee for their guidance, encouragement and support. Appreciation is also expressed to the many scientific societies who provided their membership lists for the purpose of locating former entrants whose addresses had changed, and to the tens of thousands of scientists across the country who took time to provide us with biographical information. We also wish to thank Bruce Glaunert, Bonnie Walton, Val Lowman, Debbie Wilson, Mervaine Ricks and all those whose care and devotion to accurate research and editing assured successful production of this edition.

Comments, suggestions and nominations for the nineteenth edition are encouraged and should be directed to The Editors, *American Men and Women Of Science*, R.R. Bowker, 121 Chanlon Road, New Providence, New Jersey, 07974.

Edgar H. Adcock, Jr.
Editorial Director

Major Honors & Awards

Nobel Prizes

Nobel Foundation

The Nobel Prizes were established in 1900 (and first awarded in 1901) to recognize those people who "have conferred the greatest benefit on mankind."

1990 Recipients

Chemistry:

Elias James Corey

Awarded for his work in retrosynthetic analysis, the synthesizing of complex substances patterned after the molecular structures of natural compounds.

Physics:

Jerome Isaac Friedman
Henry Way Kendall
Richard Edward Taylor

Awarded for their breakthroughs in the understanding of matter.

Physiology or Medicine:

Joseph E. Murray
Edward Donnall Thomas

Awarded to Murray for his kidney transplantation achievements and to Thomas for bone marrow transplantation advances.

1991 Recipients

Chemistry:

Richard R. Ernst

Awarded for refinements in nuclear magnetic resonance spectroscopy.

Physics:

Pierre-Gilles de Gennes*

Awarded for his research on liquid crystals.

Physiology or Medicine:

Erwin Neher
Bert Sakmann*

Awarded for their discoveries in basic cell function and particularly for the development of the patch clamp technique.

Crafoord Prize

Royal Swedish Academy of Sciences
(Kungl. Vetenskapsakademien)

The Crafoord Prize was introduced in 1982 to award scientists in disciplines not covered by the Nobel Prize, namely mathematics, astronomy, geosciences and biosciences.

1990 Recipients

Paul Ralph Ehrlich
Edward Osborne Wilson

Awarded for their fundamental contributions to population biology and the conservation of biological diversity.

1991 Recipient

Allan Rex Sandage

Awarded for his fundamental contributions to extragalactic astronomy, including observational cosmology.

Charles Stark Draper Prize

National Academy of Engineering

The Draper Prize was introduced in 1989 to recognize engineering achievement. It is awarded biennially.

1991 Recipients

Hans Joachim Von Ohain
Frank Whittle

Awarded for their invention and development of the jet aircraft engine.

National Medal of Science

National Science Foundation

The National Medals of Science have been awarded by the President of the United States since 1962 to leading scientists in all fields.

1990 Recipients:

Baruj Benacerraf
Elkan Rogers Blout
Herbert Wayne Boyer
George Francis Carrier
Allan MacLeod Cormack
Mildred S. Dresselhaus
Karl August Folkers
Nick Holonyak Jr.
Leonid Hurwicz
Stephen Cole Kleene
Daniel Edward Koshland Jr.
Edward B. Lewis
John McCarthy
Edwin Mattison McMillan**
David G. Nathan
Robert Vivian Pound
Roger Randall Dougan Revelle**
John D. Roberts
Patrick Suppes
Edward Donnall Thomas

1991 Recipients

Mary Ellen Avery
Ronald Breslow
Alberto Pedro Calderon
Gertrude Belle Elion
George Harry Heilmeier
Dudley Robert Herschbach
George Evelyn Hutchinson**
Elvin Abraham Kabat
Robert Kates
Luna Bergere Leopold
Salvador Edward Luria**
Paul A. Marks
George Armitage Miller
Arthur Leonard Schawlow
Glenn Theodore Seaborg
Folke Skoog
H. Guyford Stever
Edward Carroll Stone Jr
Steven Weinberg
Paul Charles Zamecnik

National Medal of Technology

U.S. Department of Commerce,
Technology Administration

The National Medals of Technology, first awarded in 1985, are bestowed by the President of the United States to recognize individuals and companies for their development or commercialization of technology or for their contributions to the establishment of a technologically-trained workforce.

1990 Recipients

John Vincent Atanasoff
Marvin Camras
The du Pont Company
Donald Nelson Frey
Frederick W. Garry
Wilson Greatbatch
Jack St. Clair Kilby
John S. Mayo
Gordon Earle Moore
David B. Pall
Chauncey Starr

1991 Recipients

Stephen D. Bechtel Jr
C. Gordon Bell
Geoffrey Boothroyd
John Cocke
Peter Dewhurst
Carl Djerassi
James Duderstadt
Antonio L. Elias
Robert W. Galvin
David S. Hollingsworth
Grace Murray Hopper
F. Kenneth Iverson
Frederick M. Jones**
Robert Roland Lovell
Joseph A. Numero**
Charles Eli Reed
John Paul Stapp
David Walker Thompson

*These scientists' biographies do not appear in *American Men & Women of Science* because their work has been conducted exclusively outside the US and Canada.

**Deceased [Note that Frederick Jones died in 1961 and Joseph Numero in May 1991. Neither was ever listed in *American Men and Women of Science*.]

Statistics

Statistical distribution of entrants in *American Men & Women of Science* is illustrated on the following five pages. The regional scheme for geographical analysis is diagrammed in the map below. A table enumerating the geographic distribution can be found on page xvi, following the charts. The statistics are compiled by tallying all occurrences of a major index subject. Each scientist may choose to be indexed under as many as four categories; thus, the total number of subject references is greater than the number of entrants in *AMWS*.

All Disciplines

	Number	Percent
Northeast	58,325	34.99
Southeast	39,769	23.86
North Central	19,846	11.91
South Central	12,156	7.29
Mountain	11,029	6.62
Pacific	25,550	15.33
TOTAL	**166,675**	**100.00**

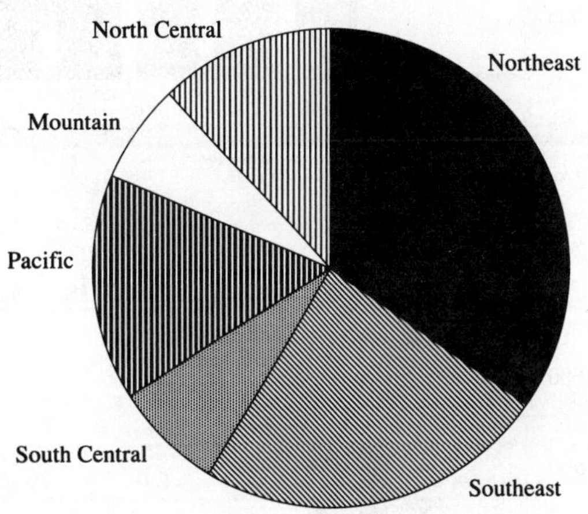

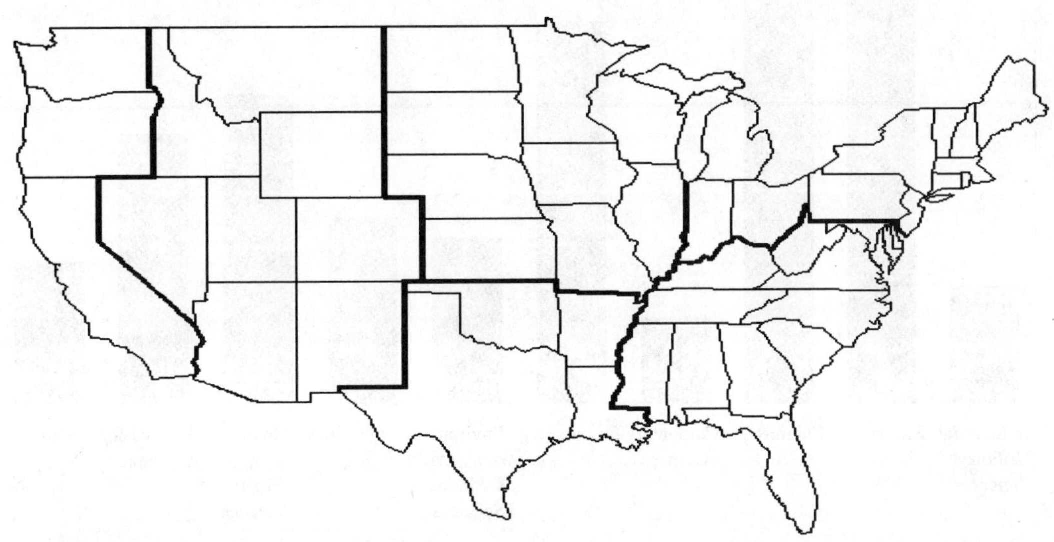

Age Distribution of American Men & Women of Science

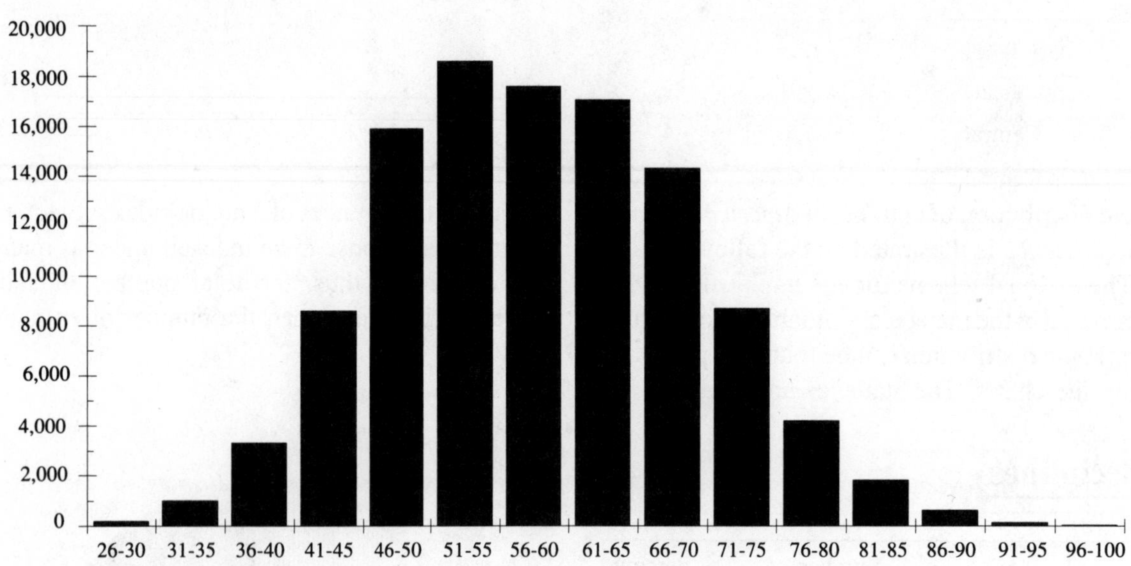

Number of Scientists in Each Discipline of Study

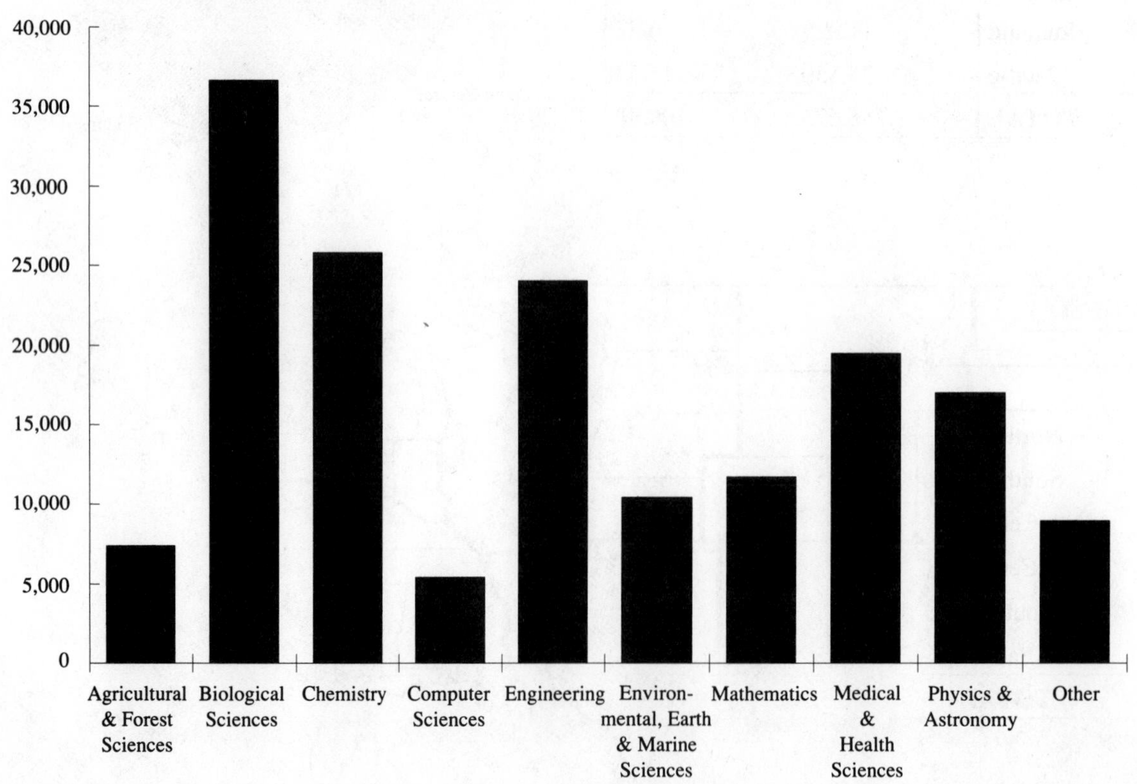

Agricultural & Forest Sciences

	Number	Percent
Northeast	1,574	21.39
Southeast	1,991	27.05
North Central	1,170	15.90
South Central	609	8.27
Mountain	719	9.77
Pacific	1,297	17.62
TOTAL	**7,360**	**100.00**

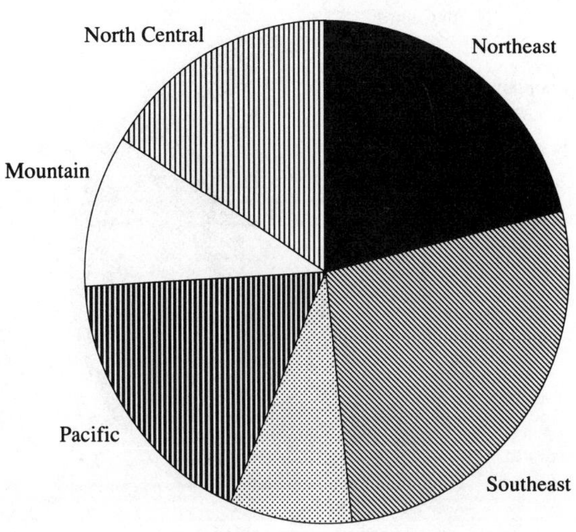

Biological Sciences

	Number	Percent
Northeast	12,162	33.23
Southeast	9,054	24.74
North Central	5,095	13.92
South Central	2,806	7.67
Mountain	2,038	5.57
Pacific	5,449	14.89
TOTAL	**36,604**	**100.00**

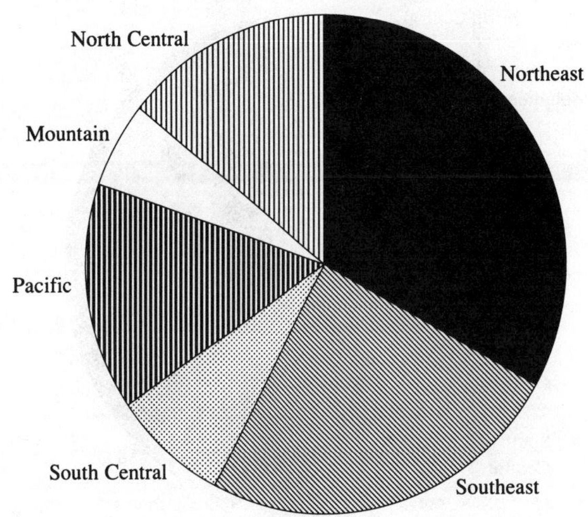

Chemistry

	Number	Percent
Northeast	10,343	40.15
Southeast	6,124	23.77
North Central	3,022	11.73
South Central	1,738	6.75
Mountain	1,300	5.05
Pacific	3,233	12.55
TOTAL	**25,760**	**100.00**

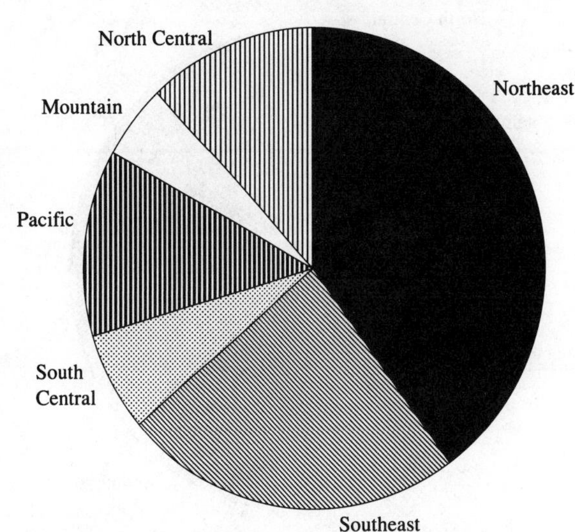

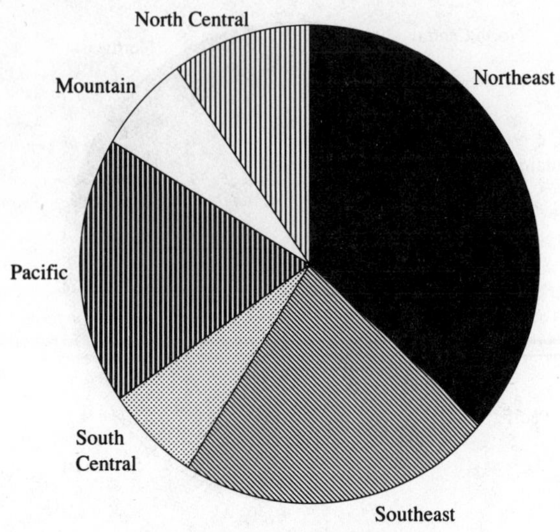

Computer Sciences

	Number	Percent
Northeast	1,987	36.76
Southeast	1,200	22.20
North Central	511	9.45
South Central	360	6.66
Mountain	372	6.88
Pacific	976	18.05
TOTAL	**5,406**	**100.00**

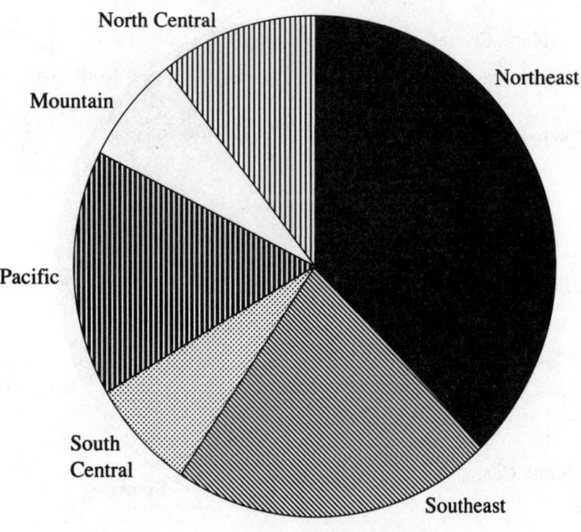

Engineering

	Number	Percent
Northeast	9,122	38.01
Southeast	5,202	21.68
North Central	2,510	10.46
South Central	1,710	7.13
Mountain	1,646	6.86
Pacific	3,807	15.86
TOTAL	**23,997**	**100.00**

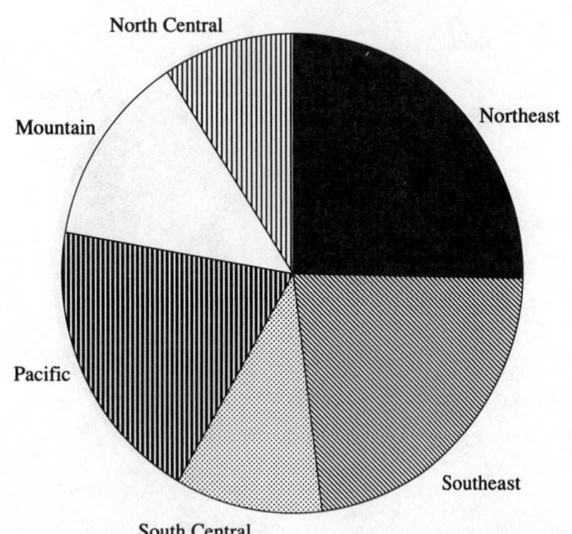

Environmental, Earth & Marine Sciences

	Number	Percent
Northeast	2,657	25.48
Southeast	2,361	22.64
North Central	953	9.14
South Central	1,075	10.31
Mountain	1,359	13.03
Pacific	2,022	19.39
TOTAL	**10,427**	**100.00**

Mathematics

	Number	Percent
Northeast	4,211	35.92
Southeast	2,609	22.26
North Central	1,511	12.89
South Central	884	7.54
Mountain	718	6.13
Pacific	1,789	15.26
TOTAL	**11,722**	**100.00**

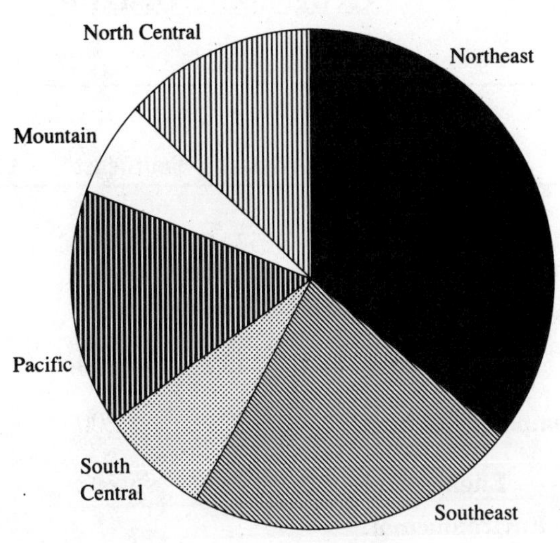

Medical & Health Sciences

	Number	Percent
Northeast	7,115	36.53
Southeast	5,004	25.69
North Central	2,577	13.23
South Central	1,516	7.78
Mountain	755	3.88
Pacific	2,509	12.88
TOTAL	**19,476**	**100.00**

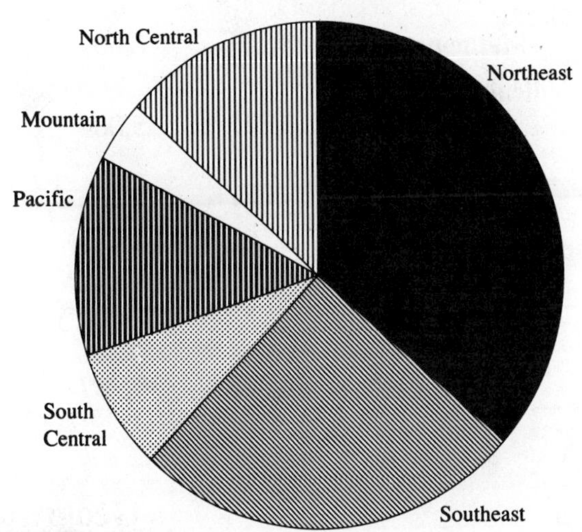

Physics & Astronomy

	Number	Percent
Northeast	5,961	35.12
Southeast	3,670	21.62
North Central	1,579	9.30
South Central	918	5.41
Mountain	1,607	9.47
Pacific	3,238	19.08
TOTAL	**16,973**	**100.00**

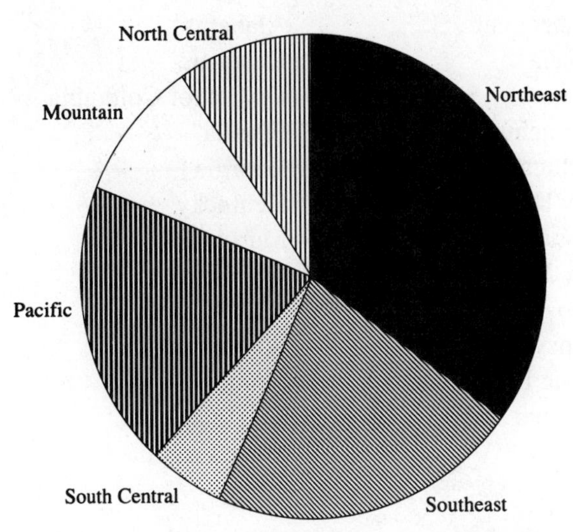

Geographic Distribution of Scientists by Discipline

	Northeast	Southeast	North Central	South Central	Mountain	Pacific	TOTAL
Agricultural & Forest Sciences	1,574	1,991	1,170	609	719	1,297	**7,360**
Biological Sciences	12,162	9,054	5,095	2,806	2,038	5,449	**36,604**
Chemistry	10,343	6,124	3,022	1,738	1,300	3,233	**25,760**
Computer Sciences	1,987	1,200	511	360	372	976	**5,406**
Engineering	9,122	5,202	2,510	1,710	1,646	3,807	**23,997**
Environmental, Earth & Marine Sciences	2,657	2,361	953	1,075	1,359	2,022	**10,427**
Mathematics	4,211	2,609	1,511	884	718	1,789	**11,722**
Medical & Health Sciences	7,115	5,004	2,577	1,516	755	2,509	**19,476**
Physics & Astronomy	5,961	3,670	1,579	918	1,607	3,238	**16,973**
Other Professional Fields	3,193	2,554	918	540	515	1,230	**8,950**
TOTAL	**58,325**	**39,769**	**19,846**	**12,156**	**11,029**	**25,550**	**166,675**

Geographic Definitions

Northeast
Connecticut
Indiana
Maine
Massachusetts
Michigan
New Hampshire
New Jersey
New York
Ohio
Pennsylvania
Rhode Island
Vermont

Southeast
Alabama
Delaware
District of Columbia
Florida
Georgia
Kentucky
Maryland
Mississippi
North Carolina
South Carolina
Tennessee
Virginia
West Virginia

North Central
Illinois
Iowa
Kansas
Minnesota
Missouri
Nebraska
North Dakota
South Dakota
Wisconsin

South Central
Arkansas
Louisiana
Texas
Oklahoma

Mountain
Arizona
Colorado
Idaho
Montana
Nevada
New Mexico
Utah
Wyoming

Pacific
Alaska
California
Hawaii
Oregon
Washington

Sample Entry

American Men & Women of Science (AMWS) is an extremely useful reference tool. The book is most often used in one of two ways: to find more information about a particular scientist or to locate a scientist in a specific field.

To locate information about an individual, the biographical section is most helpful. It encompasses the first seven volumes and lists scientists and engineers alphabetically by last name. The fictitious biographical listing shown below illustrates every type of information an entry may include.

The Discipline Index, volume 8, can be used to easily find a scientist in a specific subject specialty. This index is first classified by area of study, and within each specialty entrants are divided further by state of residence.

CARLETON, PHYLLIS B(ARBARA), b Glenham, SDak, April 1, 30. m 53, 69; c 2. ORGANIC CHEMISTRY. *Educ:* Univ Notre Dame, BSc, 52, MSc, 54, Vanderbilt Univ, PhD(chem), 57. *Hon Degrees:* DSc, Howard Univ, 79. *Prof Exp:* Res chemist, Acme Chem Corp, 54-59, sr res chemist, 59-60; from asst to assoc prof chem 60-63, prof chem, Kansas State Univ, 63-72; prof chem, Yale Univ, 73-89; CONSULT, CARLETON & ASSOCS, 89-. *Concurrent Pos:* Adj prof, Kansas State Univ 58-60; vis lect, Oxford Univ, 77, consult, Union Carbide, 74-80. *Honors & Awards:* Gold Medal, Am Chem Society, 81; *Mem:* AAAS, fel Am Chem Soc, Sigma Chi. *Res:* Organic synthesis, chemistry of natural products, water treatment and analysis. *Mailing Address:* Carleton & Assocs 21 E 34th St Boston MA 02108

Abbreviations

AAAS—American Association for the Advancement of Science
abnorm—abnormal
abstr—abstract
acad—academic, academy
acct—Account, accountant, accounting
acoust—acoustic(s), acoustical
ACTH—adrenocorticotrophic hormone
actg—acting
activ—activities, activity
addn—addition(s), additional
Add—Address
adj—adjunct, adjutant
adjust—adjustment
Adm—Admiral
admin—administration, administrative
adminr—administrator(s)
admis—admission(s)
adv—adviser(s), advisory
advan—advance(d), advancement
advert—advertisement, advertising
AEC—Atomic Energy Commission
aerodyn—aerodynamic
aeronaut—aeronautic(s), aeronautical
aerophys—aerophsical, aerophysics
aesthet—aesthetic
AFB—Air Force Base
affil—affiliate(s), affiliation
agr—agricultural, agriculture
agron—agronomic, agronomical, agronomy
agrost—agrostologic, agrostological, agrostology
agt—agent
AID—Agency for International Development
Ala—Alabama
allergol—allergological, allergology
alt—alternate
Alta—Alberta
Am—America, American
AMA—American Medical Association
anal—analysis, analytic, analytical
analog—analogue
anat—anatomic, anatomical, anatomy
anesthesiol—anesthesiology
angiol—angiology
Ann—Annal(s)
ann—annual
anthrop—anthropological, anthropology
anthropom—anthropometric, anthropometrical, anthropometry
antiq—antiquary, antiquities, antiquity
antiqn—antiquarian
apicult—apicultural, apiculture
APO—Army Post Office
app—appoint, appointed
appl—applied
appln—application
approx—approximate(ly)
Apr—April
apt—apartment(s)
aquacult—aquaculture
arbit—arbitration
arch—archives
archaeol—archaeological, archaeology
archit—architectural, architecture
Arg—Argentina, Argentine
Ariz—Arizona
Ark—Arkansas
artil—artillery
asn—association
assoc(s)—associate(s), associated
asst(s)—assistant(s), assistantship(s)
assyriol—Assyriology
astrodyn—astrodynamics
astron—astronomical, astronomy
astronaut—astonautical, astronautics
astronr—astronomer
astrophys—astrophysical, astrophysics
attend—attendant, attending
atty—attorney
audiol—audiology
Aug—August
auth—author
AV—audiovisual
Ave—Avenue
avicult—avicultural, aviculture

b—born
bact—bacterial, bacteriologic, bacteriological, bacteriology
BC—British Colombia
bd—board
behav—behavior(al)
Belg—Belgian, Belgium
Bibl—biblical
bibliog—bibliographic, bibliographical, bibliography
bibliogr—bibliographer
biochem—biochemical, biochemistry
biog—biographical, biography
biol—biological, biology
biomed—biomedical, biomedicine
biomet—biometric(s), biometrical, biometry
biophys—biophysical, biophysics
bk(s)—book(s)
bldg-building
Blvd—Boulevard
Bor—Borough
bot—botanical, botany
br—branch(es)
Brig—Brigadier
Brit—Britain, British
Bro(s)—Brother(s)
byrol—byrology
bull—Bulletin
bur—bureau
bus—business
BWI—British West Indies

c—children
Calif—California
Can—Canada, Canadian
cand—candidate
Capt—Captain
cardiol-cardiology
cardiovasc—cardiovascular
cartog—cartographic, cartographical, cartography
cartogr—cartographer
Cath—Catholic
CEngr—Corp of Engineers
cent—central
Cent Am—Central American
cert—certificate(s), certification, certified
chap—chapter
chem—chemical(s), chemistry
chemother—chemotherapy
chg—change
chmn—chairman
citricult—citriculture
class—classical
climat—climatological, climatology
clin(s)—clinic(s), clinical
cmndg—commanding
Co—County
co—Companies, Company
co-auth—coauthor
co-dir—co-director
co-ed—co-editor
co-educ—coeducation, coeducational
col(s)—college(s), collegiate, colonel
collab—collaboration, collaborative
collabr—collaborator
Colo—Colorado
com—commerce, commercial
Comdr—Commander

commun—communicable, communication(s)
comn(s)—commission(s), commissioned
comndg—commanding
comnr—commissioner
comp—comparitive
compos—composition
comput—computation, computer(s), computing
comt(s)—committee(s)
conchol—conchology
conf—conference
cong—congress, congressional
Conn—Connecticut
conserv—conservation, conservatory
consol—consolidated, consolidation
const—constitution, constitutional
construct—construction, constructive
consult(s)—consult, consultant(s), consultantship(s), consultation, consulting
contemp—contemporary
contrib—contribute, contributing, contribution(s)
contribr—contributor
conv—convention
coop—cooperating, cooperation, cooperative
coord—coordinate(d), coordinating, coordination
coordr—coordinator
corp—corporate, corporation(s)
corresp—correspondence, correspondent, corresponding
coun—council, counsel, counseling
counr—councilor, counselor
criminol—criminological, criminology
cryog—cryogenic(s)
crystallog—crystallographic, crystallographical, crystallography
crystallogr—crystallographer
Ct—Court
Ctr—Center
cult—cultural, culture
cur—curator
curric—curriculum
cybernet—cybernetic(s)
cytol—cytological, cytology
Czech—Czechoslovakia

DC—District of Columbia
Dec—December
Del—Delaware
deleg—delegate, delegation
delinq—delinquency, delinquent
dem—democrat(s), democratic
demog—demographic, demography
demogr—demographer
demonstr—demontrator
dendrol—dendrologic, dendrological, dendrology
dent—dental, dentistry
dep—deputy
dept—department
dermat—dermatologic, dermatological, dermatology
develop—developed, developing, development, developmental
diag—diagnosis, diagnostic
dialectol-dialectological, dialectology
dict—dictionaries, dictionary
Dig—Digest
dipl—diploma, diplomate
dir(s)—director(s), directories, directory
dis—disease(s), disorders
Diss Abst—Dissertation Abstracts
dist—district
distrib—distributed, distribution, distributive
distribr—distributor(s)
div—division, divisional, divorced
DNA—deoxyribonucleic acid
doc—document(s), documentary, documentation
Dom—Dominion
Dr—Drive

E—east
ecol—ecological, ecology
econ(s)—economic(s), economical, economy
economet—econometric(s)
ECT—electroconvulsive or electroshock therapy
ed—edition(s), editor(s), editorial
ed bd—editorial board
educ—education, educational
educr—educator(s)
EEG—electroencephalogram, electroencephalographic, electroencephalography
Egyptol—Egyptology
EKG—electrocardiogram
elec—elecvtric, electrical, electricity
electrochem-electrochemical, electrochemistry
electroph—electrophysical, electrophysics
elem—elementary
embryol—embryologic, embryological, embryology
emer—emeriti, emeritus
employ—employment
encour—encouragement
encycl—encyclopedia
endocrinol—endocrinologic, endocrinology
eng—engineering
Eng—England, English
engr(s)—engineer(s)
enol—enology
Ens—Ensign
entom—entomological, entomology
environ-environment(s), environmental
enzym—enzymology
epidemiol—epideiologic, epidemiological, epidemiology
equip—equipment
ERDA—Energy Research & Development Administration
ESEA—Elementary & Secondary Education Act
espec—especially
estab—established, establishment(s)
ethnog—ethnographic, ethnographical, ethnography
ethnogr—ethnographer
ethnol—ethnologic, ethnological, ethnology
Europ—European
eval—evaluation
Evangel—evangelical
eve—evening
exam—examination(s), examining
examr—examiner
except—exceptional
exec(s)—executive(s)
exeg—exegeses, exegesis, exegetic, exegetical
exhib(s)—exhibition(s), exhibit(s)
exp—experiment, experimental
exped(s)—expedition(s)
explor—exploration(s), exploratory
expos—exposition
exten—extension

fac—faculty
facil—facilities, facility
Feb—February
fed—federal
fedn—federation
fel(s)—fellow(s), fellowship(s)
fermentol—fermentology
fertil—fertility, fertilization
Fla—Florida
floricult—floricultural, floriculture
found—foundation
FPO—Fleet Post Office
Fr—French
Ft—Fort

Ga—Georgia
gastroenterol—gastroenterological, gastroenterology
gen—general
geneal—genealogical, genealogy
geod—geodesy, geodetic
geog—geographic, geographical, geography
geogr—geographer
geol—geologic, geological, geology
geom—geometric, geometrical, geometry
geomorphol—geomorphologic, geomorphology
geophys—geophysical, geophysics
Ger—German, Germanic, Germany
geriat—geriatric
geront—gerontological, gerontology
GES—Gesellschaft
glaciol—glaciology
gov—governing, governor(s)
govt—government, governmental
grad—graduate(d)
Gt Brit—Great Britain
guid—guidance
gym—gymnasium
gynec—gynecologic, gynecological, gynecology

handbk(s)—handbook(s)
helminth—helminthology
hemat—hematologic, hematological, hematology
herpet—herpetologic, herpetological, herpetology
HEW—Department of Health, Education & Welfare
Hisp—Hispanic, Hispania
hist—historic, historical, history
histol—histological, histology
HM—Her Majesty
hochsch—hochschule
homeop—homeopathic, homeopathy
hon(s)—honor(s), honorable, honorary
hort—horticultural, horticulture
hosp(s)—hospital(s), hospitalization
hq—headquarters

ABBREVIATIONS

HumRRO—Human Resources Research Office
husb—husbandry
Hwy—Highway
hydraul—hydraulic(s)
hydrodyn—hydrodynamic(s)
hydrol—hydrologic, hydrological, hydrologics
hyg—hygiene, hygienic(s)
hypn—hypnosis

ichthyol—ichthyological, ichthyology
Ill—Illinois
illum—illuminating, illumination
illus—illustrate, illustrated, illustration
illusr—illustrator
immunol—immunologic, immunological, immunology
Imp—Imperial
improv—improvement
Inc—Incorporated
in-chg—in charge
incl—include(s), including
Ind—Indiana
indust(s)—industrial, industries, industry
Inf—infantry
info—information
inorg—inorganic
ins—insurance
inst(s)—institute(s), institution(s)
instnl—institutional(ized)
instr(s)—instruct, instruction, instructor(s)
instrnl—instructional
int—international
intel—intellligence
introd—introduction
invert—invertebrate
invest(s)—investigation(s)
investr—investigator
irrig—irrigation
Ital—Italian

J—Journal
Jan—January
Jct—Junction
jour—journal, journalism
jr—junior
jurisp—jurisprudence
juv—juvenile

Kans—Kansas
Ky—Kentucky

La—Louisiana
lab(s)—laboratories, laboratory
lang—language(s)
laryngol—larygological, laryngology
lect—lecture(s)
lectr—lecturer(s)
legis—legislation, legislative, legislature
lett—letter(s)
lib—liberal
libr—libraries, library
librn—librarian
lic—license(d)
limnol—limnological, limnology
ling—linguistic(s), linguistical
lit—literary, literature
lithol—lithologic, lithological, lithology

Lt—Lieutenant
Ltd—Limited

m—married
mach—machine(s), machinery
mag—magazine(s)
maj—major
malacol—malacology
mammal—mammalogy
Man—Manitoba
Mar—March
Mariol—Mariology
Mass—Massechusetts
mat—material(s)
mat med—materia medica
math—mathematic(s), mathematical
Md—Maryland
mech—mechanic(s), mechanical
med—medical, medicinal, medicine
Mediter—Mediterranean
Mem—Memorial
mem—member(s), membership(s)
ment—mental(ly)
metab—metabolic, metabolism
metall—metallurgic, metallurgical, metallurgy
metallog—metallographic, metallography
metallogr—metallographer
metaphys—metaphysical, metaphysics
meteorol—meteorological, meteorology
metrol—metrological, metrology
metrop—metropolitan
Mex—Mexican, Mexico
mfg—manufacturing
mfr—manufacturer
mgr—manager
mgt—management
Mich—Michigan
microbiol—microbiological, microbiology
micros—microscopic, microscopical, microscopy
mid—middle
mil—military
mineral—mineralogical, mineralogy
Minn—Minnesota
Miss—Mississippi
mkt—market, marketing
Mo—Missouri
mod—modern
monogr—monograph
Mont—Montana
morphol—morphological, morphology
Mt—Mount
mult—multiple
munic—municipal, municipalities
mus—museum(s)
musicol—musicological, musicology
mycol—mycologic, mycology

N—north
NASA—National Aeronautics & Space Administration
nat—national, naturalized
NATO—North Atlantic Treaty Organization
navig—navigation(al)
NB—New Brunswick
NC—North Carolina
NDak—North Dakota
NDEA—National Defense Education Act
Nebr—Nebraska

nematol—nematological, nematology
nerv—nervous
Neth—Netherlands
neurol—neurological, neurology
neuropath—neuropathological, neuropathology
neuropsychiat—neuropsychiatric, neuropsychiatry
neurosurg—neurosurgical, neurosurgery
Nev—Nevada
New Eng—New England
New York—New York City
Nfld—Newfoundland
NH—New Hampshire
NIH—National Institute of Health
NIMH—National Institute of Mental Health
NJ—New Jersey
NMex—New Mexico
No—Number
nonres—nonresident
norm—normal
Norweg—Norwegian
Nov—November
NS—Nova Scotia
NSF—National Science Foundation
NSW—New South Wales
numis—numismatic(s)
nutrit—nutrition, nutritional
NY—New York State
NZ—New Zealand

observ—observatories, observatory
obstet—obstetric(s), obstetrical
occas—occasional(ly)
occup—occupation, occupational
oceanog—oceanographic, oceanographical, oceanography
oceanogr—oceanographer
Oct—October
odontol—odontology
OEEC—Organization for European Economic Cooperation
off—office, official
Okla—Oklahoma
olericult—olericulture
oncol—oncologic, oncology
Ont—Ontario
oper(s)—operation(s), operational, operative
ophthal—ophthalmologic, ophthalmological, ophthalmology
optom—optometric, optometrical, optometry
ord—ordnance
Ore—Oregon
org—organic
orgn—organization(s), organizational
orient—oriental
ornith—ornithological, ornithology
orthod—orthodontia, orthodontic(s)
orthop—orthopedic(s)
osteop—osteopathic, osteopathy
otol—otological, otology
otolaryngol—otolaryngological, otolaryngology
otorhinol—otorhinologic, otorhinology

Pa—Pennsylvania
Pac—Pacific
paleobot—paleobotanical, paleontology
paleont—paleontology

Pan-Am—Pan-American
parisitol—parasitology
partic—participant, participating
path—pathologic, pathological, pathology
pedag—pedagogic(s), pedagogical, pedagogy
pediat—pediatric(s)
PEI—Prince Edward Islands
penol—penological, penology
periodont—periodontal, periodontic(s)
petrog—petrographic, petrographical, petrography
petrogr—petrographer
petrol—petroleum, petrologic, petrological, petrology
pharm—pharmacy
pharmaceut—pharmaceutic(s), pharmaceutical(s)
pharmacog—pharmacognosy
pharamacol—pharmacologic, pharmacological, pharmacology
phenomenol—phenomenologic(al), phenomenology
philol—philological, philology
philos—philosophic, philosophical, philosophy
photog—photographic, photography
photogeog—photogeographic, photogeography
photogr—photographer(s)
photogram—photogrammetric, photogrammetry
photom—photometric, photometrical, photometry
phycol—phycology
phys—physical
physiog—physiographic, physiographical, physiography
physiol—physiological, phsysiology
Pkwy—Parkway
Pl—Place
polit—political, politics
polytech—polytechnic(s)
pomol—pomological, pomology
pontif—pontifical
pop—population
Port—Portugal, Portuguese
Pos:—Position
postgrad—postgraduate
PQ—Province of Quebec
PR—Puerto Rico
pract—practice
practr—practitioner
prehist—prehistoric, prehistory
prep—preparation, preparative, preparatory
pres—president
Presby—Presbyterian
preserv—preservation
prev—prevention, preventive
prin—principal
prob(s)—problem(s)
proc—proceedings
proctol—proctologic, proctological, proctology
prod—product(s), production, productive
prof—professional, professor, professorial
Prof Exp—Professional Experience
prog(s)—program(s), programmed, programming
proj—project(s), projection(al), projective
prom—promotion
protozool—protozoology
Prov—Province, Provincial
psychiat—psychiatric, psychiatry
psychoanal—psychoanalysis, psychoanalytic, psychoanalytical
psychol—psychological, psychology
psychomet—psychometric(s)
psychopath—psychopathologic, psycho pathology
psychophys—psychophysical, psychophysics
psychophysiol—psychophysiological, psychophysiology
psychosom—psychosomtic(s)
psychother—psychoterapeutic(s), psychotherapy
Pt—Point
pub—public
publ—publication(s), publish(ed), publisher, publishing
pvt—private

Qm—Quartermaster
Qm Gen—Quartermaster General
qual—qualitative, quality
quant—quantitative
quart—quarterly
Que—Quebec

radiol—radiological, radiology
RAF—Royal Air Force
RAFVR—Royal Air Force Volunteer Reserve
RAMC—Royal Army Medical Corps
RAMCR—Royal Army Medical Corps Reserve
RAOC—Royal Army Ornance Corps
RASC—Royal Army Service Corps
RASCR—Royal Army Service Corps Reserve
RCAF—Royal Canadian Air Force
RCAFR—Royal Canadian Air Force Reserve
RCAFVR—Royal Canadian Air Force Volunteer Reserve
RCAMC—Royal Canadian Army Medical Corps
RCAMCR—Royal Canadian Army Medical Corps Reserve
RCASC—Royal Canadian Army Service Corps
RCASCR—Royal Canadian Army Service Corps Reserve
RCEME—Royal Canadian Electrical & Mechanical Engineers
RCN—Royal Canadian Navy
RCNR—Royal Canadian Naval Reserve
RCNVR—Royal Canadian Naval Volunteer Reserve
Rd—Road
RD—Rural Delivery
rec—record(s), recording
redevelop—redevelopment
ref—reference(s)
refrig—refrigeration
regist—register(ed), registration
registr—registrar
regt—regiment(al)
rehab—rehabilitation
rel(s)—relation(s), relative
relig—religion, religious
REME—Royal Electrical & Mechanical Engineers
rep—represent, representative
Repub—Republic
req—requirements
res—research, reserve
rev—review, revised, revision
RFD—Rural Free Delivery
rhet-rhetoric, rhetorical
RI—Rhode Island
Rm—Room
RM—Royal Marines
RN—Royal Navy
RNA—ribonucleic acid
RNR—Royal Naval Reserve
RNVR—Royal Naval Volunteer Reserve
roentgenol—roentgenologic, roentgenological, roentgenology
RR—Railroad, Rural Route
Rte—Route
Russ—Russian
rwy—railway

S—south
SAfrica—South Africa
SAm—South America, South American
sanit—sanitary, sanitation
Sask—Saskatchewan
SC—South Carolina
Scand—Scandinavia(n)
sch(s)—school(s)
scholar—scholarship
sci—science(s), scientific
SDak—South Dakota
SEATO—Southeast Asia Treaty Organization
sec—secondary
sect—section
secy—secretary
seismog—seismograph, seismographic, seismography
seismogr—seismographer
seismol—seismological, seismology
sem—seminar, seminary
Sen—Senator, Senatorial
Sept—September
ser—serial, series
serol—serologic, serological, serology
serv—service(s), serving
silvicult—silvicultural, silviculture
soc(s)—societies, society
soc sci—social science
sociol—sociologic, sociological, sociology
Span—Spanish
spec—special
specif—specification(s)
spectrog—spectrograph, spectrographic, spectrography
spectrogr—spectrographer
spectrophotom—spectrophotometer, spectrophotometric, spectrophotometry
spectros—spectroscopic, spectroscopy
speleol—speleological, speleology
Sq—Square
sr—senior
St—Saint, Street(s)
sta(s)—station(s)
stand—standard(s), standardization
statist—statistical, statistics
Ste—Sainte
steril—sterility

stomatol—stomatology
stratig—stratigraphic, stratigraphy
stratigr—stratigrapher
struct—structural, structure(s)
stud—student(ship)
subcomt—subcommittee
subj—subject
subsid—subsidiary
substa—substation
super—superior
suppl—supplement(s), supplemental, supplementary
supt—superintendent
supv—supervising, supervision
supvr—supervisor
supvry—supervisory
surg—surgery, surgical
surv—survey, surveying
survr—surveyor
Swed—Swedish
Switz—Switzerland
symp—symposia, symposium(s)
syphil—syphilology
syst(s)—system(s), systematic(s), systematical

taxon—taxonomic, taxonomy
tech—technical, technique(s)
technol—technologic(al), technology
tel—telegraph(y), telephone
temp—temporary
Tenn—Tennessee
Terr—Terrace
Tex—Texas
textbk(s)—textbook(s)
text ed—text edition
theol—theological, theology
theoret—theoretic(al)
ther—therapy
therapeut—therapeutic(s)
thermodyn—thermodynamic(s)
topog—topographic, topographical, topography
topogr—topographer
toxicol—toxicologic, toxicological, toxicology
trans—transactions
transl—translated, translation(s)
translr—translator(s)
transp—transport, transportation
treas—treasurer, treasury
treat—treatment
trop—tropical
tuberc—tuberculosis
TV—television
Twp—Township

UAR—United Arab Republic
UK—United Kingdom
UN—United Nations
undergrad—undergraduate
unemploy—unemployment
UNESCO—United Nations Educational Scientific & Cultural Organization
UNICEF—United Nations International Childrens Fund
univ(s)—universities, university
UNRRA—United Nations Relief & Rehabilitation Administration
UNRWA—United Nations Relief & Works Agency
urol—urologic, urological, urology
US—United States
USAAF—US Army Air Force
USAAFR—US Army Air Force Reserve
USAF—US Air Force
USAFR—US Air Force Reserve
USAID—US Agency for International Development
USAR—US Army Reserve
USCG—US Coast Guard
USCGR—US Coast Guard Reserve
USDA—US Department of Agriculture
USMC—US Marine Corps
USMCR—US Marine Corps Reserve
USN—US Navy
USNAF—US Naval Air Force
USNAFR—US Naval Air Force Reserve
USNR—US Naval Reserve
USPHS—US Public Health Service
USPHSR—US Public Health Service Reserve
USSR—Union of Soviet Socialist Republics

Va—Virginia
var—various
veg—vegetable(s), vegetation
vent—ventilating, ventilation
vert—vertebrate
Vet—Veteran(s)
vet—veterinarian, veterinary
VI—Virgin Islands
vinicult—viniculture
virol—virological, virology
vis—visiting
voc—vocational
vocab—vocabulary
vol(s)—voluntary, volunteer(s), volume(s)
vpres—vice president
vs—versus
Vt—Vermont

W—west
Wash—Washington
WHO—World Health Organization
WI—West Indies
wid—widow, widowed, widower
Wis—Wisconsin
WVa—West Virginia
Wyo—Wyoming

Yearbk(s)—Yearbook(s)
YMCA—Young Men's Christian Association
YMHA—Young Men's Hebrew Association
Yr(s)—Year(s)
YT—Yukon Territory
YWCA—Young Women's Christian Association
YWHA—Young Women's Hebrew Association

zool—zoological, zoology

American Men & Women of Science

G

GAAFAR, SAYED MOHAMMED, b Tanta, Egypt, Jan 18, 24; nat US; m 49; c 4. VETERINARY PARASITOLOGY. *Educ:* Cairo Univ, BVSc, 44; Kans State Col, MS, 49, PhD, 50; Agr & Mech Col Tex, DVM, 55. *Prof Exp:* Veterinarian, Vet Serv, Egypt, 44-46; asst parasitologist, Vet Path Lab, Egypt, 46-47, parasitologist, 50-51; veterinarian, Rutheford Vet Hosp, 52-54; instr, Agr & Mech Col Tex, 55-56, asst prof, 56-58; asst prof vet sci, 58-60, assoc prof, 60-63, PROF PARASITOL, PURDUE UNIV, 63- *Concurrent Pos:* Vis prof, King Saud Univ, Saudi Arabia, Zagagig Univ, Egypt & Osaka Preferred Univ, Safai City, Osaka, Japan; consult & vis prof, Karestart Univ, Bangkok, Thailand; chief ed, Vet Parasitol. *Honors & Awards:* Fel, NSH, 64; Career Develop Award, NIH, 68; Japanese Soc Prom Sci Award, 87. *Mem:* Am Soc Trop Med & Hyg; Am Asn Pathologists; Am Asn Vet Parasitologists; Am Soc Parasitol; Am Vet Med Asn; World Fed Parasitologists; World Asn Advan Vet Parasitol; World Vet Asn; Coun Biol Eds. *Res:* Resistance against parasites as affected by dietary supplements and other organisms; surveys on parasitic infestations; immunopathology of helminth parasites in domestic animals; biology of ectoparasitism in domestic animals; immunology of ectoparasitism. *Mailing Add:* 2620 Newman Rd West Lafayette IN 47906

GAAL, ILSE LISL NOVAK, b Vienna, Austria, Jan 17, 24; nat US; m; c 2. SYMBOLIC LOGIC, ALGEBRA. *Educ:* Hunter Col, AB, 44; Radcliffe Col, MA, 44, PhD(math), 48. *Prof Exp:* Asst, Mass Eye & Ear Infirmary, 44-46; instr math, Wellesley Col, 48-50; Jewett fel, 50-52; instr, Cornell Univ, 52-53, asst prof, 54-59; res assoc, Yale Univ, 59-60; lectr, 64-78, ASSOC PROF MATH, UNIV MINN, 78-, RES ASSOC, 60- *Concurrent Pos:* Pres, NCent Sect, Math Asn Am, 85-86. *Honors & Awards:* C I Wilby Prize, Radcliffe, 48. *Mem:* Am Math Soc; Math Asn Am; Asn Symbolic Logic; Asn Women Math; Sigma Xi. *Res:* Mathematical logic; Galois theory. *Mailing Add:* 537 Vincent Hall Univ Minn Minneapolis MN 55455

GAAL, ROBERT A P, b Los Angeles, Calif, Mar 31, 29; m 53. PETROLEUM GEOLOGY, ENGINEERING GEOLOGY. *Educ:* Univ Calif, Los Angeles, BA, 53; Univ Southern Calif, MA, 58, PhD(geol), 66. *Prof Exp:* Regional petrol geologist, REDECO, Philippines, 59-61; marine geologist, Allan Hancock Found, 61-62; instr geol, Whittier Col, 62-64; cur mineral, Natural Hist Mus, Los Angeles County, 62-66; staff scientist & oceanogr, TRW Syst, 66-69, chief geologist, 69-73; res scientist gemology, Gemological Inst Am, 73-78; sr marine geologist, 78-82, GEOPHYSICIST, CALIF STATE LANDS COMN, 82- *Concurrent Pos:* Res assoc, Natural Hist Mus, Los Angeles County, 66-76; consult geol, various co, 73- *Mem:* Am Inst Mining & Metal; Sigma Xi; AAAS; Mineral Soc Am; Am Soc Oceanog (vpres, 69). *Res:* Instrumentation and techniques for detection and identification of properties of natural and synthetic gem materials, especially cathodoluminescence and thermoluminescence; marine geology; nearshore processes; oil seeps; shoreline erosion; marine minerals; offshore petroleum development and geothermal energy; petroleum geology. *Mailing Add:* 4408 Lucera Circle Palos Verde Estates CA 90274

GAAL, STEVEN ALEXANDER, b Budapest, Hungary, Feb 22, 24; US citizen; m 52. APPLIED MATHEMATICS. *Educ:* Budapest Univ, PhD(math), 47. *Prof Exp:* Instr math, Univ Szeged, 46-48; asst prof, Budapest Univ, 48; mem res, Nat Ctr Sci Res, Paris, 48-50; mem, Inst Adv Study, 50-52; instr, Cornell Univ, 53-54, asst prof math, 54-59; res assoc, Yale Univ, 59-60; vis assoc prof, 60-61, assoc prof, 61-63, PROF MATH, UNIV MINN, MINNEAPOLIS, 63- *Res:* Mathematical analysis; theory of numbers; topology; algebra; theoretical physics. *Mailing Add:* Dept Math Univ Minn 533 Vincent Hall Minneapolis MN 55455

GAALSWYK, ARIE, b Alvord, Iowa, June 14, 18; m 43; c 3. APPLIED MATHEMATICS. *Educ:* Luther Col, BA, 42; Univ Wis, MS, 47; Univ Minn, PhD(math), 63. *Prof Exp:* Instr math, Luther Col, 47-48; prin scientist, Mech Div, Gen Mills, Inc, Minn, 50-59; res assoc math, Univ Minn, 59-60; assoc prof, 60-70, chmn dept, 75-80, PROF MATH, AUGUSTANA COL, SDAK, 70- *Concurrent Pos:* Consult, Raven Indust, SDak, 60- *Mem:* Am Math Soc; Am Meteorol Soc. *Res:* Combustion shock waves; mathematical models for problems in mechanics and geophysics. *Mailing Add:* Dept Math Augustana Col 29 St S Sioux Falls SD 57102

GAAR, KERMIT ALBERT, JR, b Tyler, Tex, Jan 24, 34; m 58; c 4. MEDICAL PHYSIOLOGY. *Educ:* Univ Miss, BS(biol) & BS(pharm), 58; Univ Ark, MS, 65, PhD(physiol), 67. *Prof Exp:* From instr to asst prof physiol, Sch Med, Univ Miss, 67-69; from asst prof to assoc prof, 69-77, PROF PHYSIOL, SCH MED, LA STATE UNIV, SHREVEPORT, 77-, DIR ANIMAL CARE, 70- *Mem:* Am Physiol Soc. *Res:* Dynamics of control of the body fluids, particularly the microcirculatory and interstitial fluid dynamics as these relate to the pulmonary circulation and pulmonary edema formation. *Mailing Add:* Dept Physiol & Biophys Med Ctr LSU Med Ctr PO Box 33932 Shreveport LA 71130

GAARDER, NEWELL THOMAS, b La Crosse, Wis, Feb 17, 39; m 63; c 1. ELECTRICAL ENGINEERING. *Educ:* Univ Wis, BA, 61; Stanford Univ, MS, 62, PhD(elec eng), 65. *Prof Exp:* Res engr, Stanford Res Inst, 63-65; asst prof, Cornell Univ, 65-67; assoc prof, 67-76, PROF ELEC ENG, UNIV HAWAII, 76- *Mem:* AAAS; Inst Elec & Electronics Engrs. *Res:* Statistical decision theory; statistical communication theory; information sciences. *Mailing Add:* Dept Elec Eng Univ Hawaii Manoa 2540 Dole St Honolulu HI 96844

GAARENSTROOM, STEPHEN WILLIAM, b Minneapolis, Minn, Sept 20, 50; m 81; c 2. ANALYTICAL CHEMISTRY. *Educ:* Carleton Col, BA, 72; Purdue Univ, PhD(anal chem), 77. *Prof Exp:* SR STAFF RES SCIENTIST, GEN MOTORS RES LAB, 77- *Concurrent Pos:* Assoc ed, J Vac Sci Technol, 86-88; ed bd, Surface & Interface Analyst, 89-; div chmn, Am Vacuum Soc, 89-90. *Mem:* AAAS; Am Soc Testing & Mat; Am Chem Soc; Am Vacuum Soc; Asn Analytical Chemists; Am Soc Mat Int. *Res:* Surface analysis; electron spectroscopy; information content of spectra. *Mailing Add:* Dept Analytical Chem Gen Motors Res Lab Warren MI 48090-9055

GABALLAH, SAEED S, US citizen; m 60; c 3. BIOCHEMISTRY, MOLECULAR BIOLOGY. *Educ:* Cairo Univ, BS, 46; Univ Wis, PhD(biochem), 54. *Prof Exp:* Asst prof, Cairo Univ, 55-57; res assoc biochem, Col Med, Univ Ill, 57-61, asst prof, 62-63; asst prof, Chicago Col Osteop, 63-65; HEAD NEUROPSYCHIAT LAB, VET ADMIN HOSP, DOWNEY, 65- *Mem:* AAAS; Brit Biochem Soc; Am Chem Soc; NY Acad Sci. *Res:* Medical and biological research, especially central nervous system; organ and cell structure. *Mailing Add:* 221 W Sheridan Pl Lake Bluff IL 60044

GABAY, JONATHAN GLENN, b Queens, NY, Apr 10, 56; m 88. CUSTOM COMPUTER DESIGN, REAL TIME ACQUISITION & CONTROL. *Educ:* State Univ NY, Stony Brook, BSEE, 80. *Prof Exp:* Sr technol specialist, Tech Data Specialists, 78-81; sr design engr, Amecom Div, Litton Datalog, 81-84 & Add Div, NCR, 84-85; sr technol ed, Hearst Bus Publ, 85-88, VNU Publ, 88-89 & CMP Publ, 89-90; SR TECHNOL ED, MILLER FREEMAN PUBL, 90- *Concurrent Pos:* Consult, Computer Design, 84-, Eagle Telephonics, 88-89, Hydro Systs, Plus Logic & Viewlogic, 90- & Melard Systs & Hydro Systs, 91- *Res:* Specialty computer system design; tactical and efficient computer/human interfaces and specialty applications; evaluating state-of-the-art computer aided engineering tools and services for ASICS, PLDs, PCB and other engineering related design automation tools. *Mailing Add:* 124 N Bicycle Path Selden NY 11784

GABAY, SABIT, b Istanbul, Turkey, Mar 18, 22; nat US; div; c 3. BIOCHEMISTRY, PHARMACOLOGY. *Educ:* Galatasaray Col, Turkey, BA, 43; Istanbul Univ, BS, 46; Tex A&M Univ, MS, 54; Univ Madrid, PhD(biochem), 58. *Hon Degrees:* Dr, Mex Nat Inst Culture, Col Behav Sci, 76. *Prof Exp:* Res asst, Tex A&M Univ, 53-54; res biochemist, Univ Southern Calif, 55-59; sr res biochemist, Columbia Univ & Rockland State Hosp, 59-61; chief biochem res labs, Vet Admin Med Ctr, Brockton, Ma, 61-85; RETIRED. *Concurrent Pos:* Del Amo Found sr res fel, Postgrad Med Sch, Univ Madrid, 57-59; travel award, Sweden, Denmark, Spain, Italy, USSR, Romania, France, Hungary, Japan, Mexico, Arg, Venezuela, Turkey, Can & Nigeria, 58-85; secy-gen, Int Conf Phenothiazine Metab, France, 62; consult, Mo Psychiat Inst, 68-74 & Marcy State Hosp, 70-80; asst prof, Boston Univ, 62-70; asst prof pharmacol, Sch Dent Med, Harvard Univ, 70-77; Nat Acad Sci exchange prof to Romanian Univs, 71; secy-gen exec comt, World Fedn Biol Psychiat, 74-78; vis prof neuropharm, Ahmadu Bello Univ, Zaria, Nigeria, 77, Ramon & Cajal Ctr, Madrid, Spain (Sabbatical leave 83); adj prof psychiat, Brown Univ, 78-88; adj prof biochem, Univ VI, St Thomas, 86-

Honors & Awards: Gold Medal Award, Soc Biol Psych, 82. Mem: Soc Biol Psychiat (pres, 76); Int Soc Neurochem; Int Soc Biochem Pharmacol; Am Soc Neurochem; Asn Res Nerv & Ment Dis; fel Am Col Neuropsychopharmacol; Am Soc Exp Pharmacol & Therapeut; Am Soc Biochem & Molecular Biol. Res: Neurobiochemistry; biochemical psychopharmacology and neuropharmacology encompassing drug enzymology and their metabolic mechanisms; molecular pharmacology; drug-protein binding. Mailing Add: 1001 Cloverlea Rd Ruxton MD 21204-6812

GABB, TIMOTHY PAUL, b Thibodaux, La, Jan 22, 58. LOW CYCLE FATIGUE, THERMAL MECHANICAL FATIGUE. *Educ:* La State Univ, BS, 81, MS, 83; Case Western Reserve Univ, PhD(mat sci), 88. *Prof Exp:* RES METALLURGIST, LEWIS RES CTR, NASA, 83- *Mem:* Metall Soc. *Res:* The mechanical property/microstructural relationships of advanced high temperature materials; metallic and metal matrix composite materials. *Mailing Add:* 17599 Whitney Rd Apt 315 Strongsville OH 44136

GABBARD, FLETCHER, b Sand Gap, Ky, Sept 13, 30; m 57; c 2. NUCLEAR PHYSICS. *Educ:* Univ Ky, BS, 51; Rice Univ, MA, 57, PhD(physics), 59. *Prof Exp:* Physicist, US Naval Ord Lab, 51-52 & Nat Bur Standards, 52-53; from asst prof to assoc prof, 59-70, prof physics, 70-89, chmn dept physics & astron, 73-85, EMER PROF PHYSICS, UNIV KY, 89- *Mem:* AAAS; Am Phys Soc; Sigma Xi; Am Asn of Physics Teachers. *Res:* Energy levels in light nuclei; neutron induced reactions; neutron producing reactions in medium weight nuclei. *Mailing Add:* Dept Physics & Astron Univ Ky Lexington KY 40506-0055

GABBE, JOHN DANIEL, b Johannesburg, Rep S Africa, Jul 19, 29; m 56; c 2. PHYSICS. *Educ:* NY Univ, BA, 50, PhD(physics), 57; Univ ILL, MS, 51. *Prof Exp:* MEM TECH STAFF, AT&T BELL LABS, 56. *Mem:* Am Phys Soc; Am Asn Artificial Intelligence; Int Neurol Net Soc. *Res:* Machine learning and intelligent computerized design aids; geophysics, statistical data analysis; associative memory retrieval. *Mailing Add:* AT&T Bell Labs Holmdel NJ 07733

GABBERT, PAUL GEORGE, biochemistry, toxicology; deceased, see previous edition for last biography

GABBIANI, GIULIO, b Cremona, Italy, Mar 19, 37; m 63; c 3. CYTOSKELETON, WOUND HEALING. *Educ:* Univ Pavia, Italy, MD, 61; Inst Exp Med & Surg, Univ Montreal, Can, PhD(endocrinol), 65. *Prof Exp:* Asst prof, Inst Exp Med & Surg, Univ Montreal, Can, 65-67 & 68-69; res assoc, Dept Path, Harvard Med Sch, Boston, 67-68; asst prof, 69-75, PROF PATH, DEPT PATH, UNIV GENEVA, SWITZ, 75- *Mem:* Am Asn Pathologists; Am Soc Cell Biol; Int Acad Path; NY Acad Sci; Soc Exp Biol & Med; Swiss Soc Cellular & Molecular Biol. *Res:* Biology of cytoskeletal elements, acting in particular, in relation to wound healing and vascular pathology. *Mailing Add:* Dept Path Univ Geneva One ru Michel-Servet Geneva 4 1211 Switzerland

GABEL, ALBERT A, b Fremont, Ohio, Mar 3, 30; m 54; c 6. VETERINARY SURGERY. *Educ:* Ohio State Univ, DVM, 54, MSc, 59; dipl, Am Col Vet Surg. *Prof Exp:* Ambulatory clinician vet med, 54-55, instr, 57-60, from asst prof to assoc prof, 60-69, PROF VET SURG, COL VET MED, OHIO STATE UNIV, 69- *Concurrent Pos:* Univ develop fund grant, 58-59; Wyeth Lab res grant, 61-63; chmn eng res group, Ohio State Univ, Col Vet Med. *Honors & Awards:* Borden Award, 54. *Mem:* Am Col Vet Surg; Am Vet Med Asn. *Res:* Anesthesia and orthopedics of horses; exercise physiology in horses; evaluation of fitness; experimental training methods and effects of drugs on performance. *Mailing Add:* 1935 Coffey Rd Columbus OH 43210

GABEL, JAMES RUSSEL, b Pottstown, Pa, Aug 21, 18; m 44; c 1. PROTOZOOLOGY, NATURAL HISTORY. *Educ:* Pa State Col Lock Haven, BS, 47; Univ Pa, PhD(zool), 53. *Prof Exp:* Asst instr zool, Univ Pa, 47-53; asst prof, Fisk Univ, 53-56; assoc prof, WLiberty State Col, 56-59; from asst prof to assoc prof, San Francisco State Univ, 59-74, prof biol, 74-84; RETIRED. *Mem:* Nature Conserv; Mountain Lion Preserv Asn; Nat Wildlife Fed. *Res:* Protozoan parasitology; histology; desert biology; tarantula studies. *Mailing Add:* 440 Monticello St San Francisco CA 94127

GABEL, JOSEPH C, LYMPHATIC SYSTEMS. *Educ:* Ohio State Univ, MD, 64. *Prof Exp:* PROF & CHMN ANESTHESIOL, MED SCH, UNIV TEX, 80- *Res:* Mechanisms, causes & therapy of pulmonary edema. *Mailing Add:* Dept Anesthesiol Univ Tex Med Sch 6431 Fannin Houston TX 77030

GABEL, RICHARD ALLEN, b Sterling, Colo, Nov 4, 46; m 70; c 3. ORGANIC CHEMISTRY. *Educ:* Colo State Univ, BS, 69, PhD(org chem), 79; Univ Wis-Madison, MS, 71. *Prof Exp:* RES CHEMIST ORG CHEM, DIV SYNTEX, ARAPAHOE CHEM, INC, 78- *Mem:* Am Chem Soc. *Res:* Displacement of methoxyl group by carbon, nitrogen, oxygen and silicon nucleophiles on o-methoxyphenyloxazonlines; regio specific motalation and introduction of an electrophile or addition of a nucleaphile followed by oxidation of pyridyloxazolines. *Mailing Add:* 1477 Franklin Ct Louisville CO 80027

GABELMAN, IRVING J(ACOB), b Brooklyn, NY, Nov 12, 18; m 49; c 2. ELECTRONICS, COMPUTER SCIENCE. *Educ:* Brooklyn Col, BA, 38; City Col New York, BEE, 45; Polytech Inst Brooklyn, MEE, 48; Syracuse Univ, PhD(elec eng), 61. *Prof Exp:* Radio engr, Watson Labs, Red Bank, NJ, US Dept Air Force, 45-51; electronic scientist, Rome Air Develop Ctr, Griffiss AFB, 51-59, dir advan studies, 59-66, chief plans, 67-70, chief scientist, 71-76; PRES, TECH ASSOCS, 76- *Concurrent Pos:* US mem, avionics panel, Adv Group Aeronaut Res & Develop, NATO, 63-69, dep chmn, 69-71, chmn, 71- *Mem:* Fel AAAS; fel Inst Elec & Electronics Engrs. *Res:* Switching circuits, especially threshold element logical design; electronic computers. *Mailing Add:* 225 Dale Rd Rome NY 13440

GABELMAN, WARREN HENRY, b Tilden, Nebr, Apr 18, 21; m 45. PLANT GENETICS. *Educ:* Univ Nebr, BSc, 42; Yale Univ, PhD, 49. *Hon Degrees:* DSc, Univ Nebr, 76. *Prof Exp:* Asst hort, Exp Sta, Univ Nebr, 38-42; fel, Plant Genetics Chg Veg Breeding Prog, Conn Exp Sta, 46-49; from asst prof to assoc prof, 49-60, chmn dept, 65-73, PROF HORT, UNIV WIS-MADISON, 60- *Concurrent Pos:* Vchmn, Vegetable Systs Deleg China, Comt Scholarly Commun with People's Repub China, Nat Acad Sci, 77; mem, USDA Plant Genetic Resources Bd, 75-81. *Honors & Awards:* Marion Meadows Award, Am Soc Hort Sci, 66, 69, 73 & 79; Genetics & Plant Breeding Award, Nat Coun Com Plant Breeders, 80. *Mem:* Fel AAAS; fel Am Soc Hort Sci (pres, 78-79); Genetics Soc Am; Crop Sci Soc; Sigma Xi. *Res:* Cytoplasmic inheritance of male sterility in corn, onions, beets, and carrots; biosynthesis of carotenes in carrots; genetics of mineral nutrition efficiencies in higher plants. *Mailing Add:* Dept Hort Univ Wis Madison WI 53706

GABELNICK, HENRY LEWIS, b Boston, Mass, May 10, 40; m 87; c 2. CHEMICAL & BIOMEDICAL ENGINEERING. *Educ:* Mass Inst Technol, BS, 61, MS, 62; Princeton Univ, PhD(chem eng), 66. *Prof Exp:* Sr chem engr, Res Dept, Monsanto Co, Mass, 66-68; chem engr, Biomed Eng Br, NIH, 68-75, biomed engr, Contraceptive Develop Br, 75-86; DIR EXTRAMURAL PROGS, CONTRACEPTIVE RES & DEVELOP PROG, EASTERN VA MED SCH, 86-, PROF OBSTET-GYNEC, 86- *Concurrent Pos:* Expert consult, UN Develop Prog, Haifa, Israel, 73, World Health Orgn Spec Prog Human Reproduction, 78-; dir, Contraceptive Res & Develop Prog, Eastern Va Med Sch, 90- *Mem:* Am Inst Chem Engrs; Am Chem Soc; NY Acad Sci. *Res:* Kinetics of decomposition reactions; physical properties of polymers; blood rheology; kinetics of drug distribution in vivo; biomaterials; drug delivery systems; formulation of contraceptive drugs; development of barrier contraceptives. *Mailing Add:* Conrad Prog 1611 N Kent St Suite 806 Arlington VA 22209

GABER, BRUCE PAUL, b Chicago, Ill, Oct 15, 41; m 66, 85; c 2. PHYSICAL BIOCHEMISTRY. *Educ:* Hendrix Col, BA, 63; Univ Southern Calif, PhD(biochem), 68. *Prof Exp:* Fel, IBM Watson Res Lab, Columbia Univ, 68-70 & IBM T J Watson Res Ctr, 70-71; asst prof chem, Univ Mich, Dearborn, 71-75; sr res assoc chem, Univ Ore, 75-77; res assoc prof biochem, Sch Med, Univ Va, 77-80; consult, 77-80, staff scientist biophys optics sect, Naval Res Lab, Wash, DC, 80-84; dep head, Biomolecular Eng Br, 84-90, DEP HEAD, CTR BIOMOLECULAR SCI & ENG, 90- *Mem:* Biophys Soc; Am Soc Biochem & Molecular Biol; Molecular Graphics Soc; Sigma Xi. *Res:* Physical biochemistry of biomembranes; molecular modeling of membranes; protein engineering. *Mailing Add:* Ctr Biomolecular Sci & Eng Naval Res Lab Code 6090 Washington DC 20375

GABIS, DAMIEN ANTHONY, b Steubenville, Ohio, Feb 26, 42; m 65; c 3. FOOD MICROBIOLOGY. *Educ:* Col Steubenville, BA, 64; Univ Ky, MS, 67; NC State Univ, PhD(food sci), 70. *Prof Exp:* Assoc dir, 71-73, vpres, 73-87, PRES, SILLIKER LABS, 87-, CHIEF EXEC OFFICER, 90- *Concurrent Pos:* adj assoc prof biol, Ill Inst Technol, 78- *Mem:* Am Soc Microbiol; Inst Food Technologists; Can Soc Microbiol; Int Asn Milk, Food & Environ Sanit; Am Mgt Asn. *Res:* Analytical methods of detection of salmonella and all Enterobacteriaceae in foods; campylobacter. *Mailing Add:* Silliker Labs Group Inc 1304 Halsted St Chicago Heights IL 60411

GABLE, JAMES JACKSON, JR, b Oklahoma City, Okla, Apr 3, 18; m 41; c 5. INTERNAL MEDICINE. *Educ:* Univ Okla, BS, 40, MD, 42; Am Bd Internal Med, dipl, 53. *Prof Exp:* From asst prof to assoc prof, 50-72, CLIN PROF MED, SCH MED, UNIV OKLA, 72- *Concurrent Pos:* Attend physician, Vet Admin Hosp, 53-; chief med serv, Presby Hosp, 62-67, mem trustees, 64. *Mem:* Fel Am Col Physicians; NY Acad Sci. *Res:* Clinical aspects of internal medicine. *Mailing Add:* Okla Ma 2618 NW 65th St Oklahoma City OK 73116

GABLE, MICHAEL F, b Lancaster, Pa, June 1, 45. INVERTEBRATE BIOLOGY, AMPHIPOD TAXONOMY. *Educ:* Univ Fla, BS, 67; Univ NH, MS, 69, PhD(zool), 72. *Prof Exp:* From asst prof to assoc prof, 72-82, PROF BIOL, EASTERN CONN STATE UNIV, 82- *Concurrent Pos:* Consult; curtorial affil, Yale Peabody Mus, 88- *Mem:* Crustacean Soc; Am Soc Zoologists. *Res:* Systematics of Bermuda marine amphipod crustaceans. *Mailing Add:* Dept Biol Eastern Conn State Univ Willimantic CT 06226-2295

GABLE, RALPH WILLIAM, b San Antonio, Tex, Sept 27, 29; m 53; c 3. PHYSICAL CHEMISTRY. *Educ:* Univ Tex, BS, 50; Duke Univ, MA, 53, PhD(phys chem), 56. *Prof Exp:* From asst prof chem to assoc prof, Pfeiffer Col, 55-60, head dept, 55-60; from asst prof to assoc prof, 60-83, PROF CHEM, DAVIDSON COL, 83-; AT DEPT CHEM, EASTERN MICH UNIV. *Concurrent Pos:* Res assoc, Fla State Univ, 68-69; vis scientist, Oak Ridge Nat Lab, 78-79; proj seraphim, fel, Dept Chem, Eastern Mich Univ, 85-86. *Mem:* AAAS; Am Chem Soc; Am Asn Univ Prof; NY Acad Sci; AAAS. *Res:* Ion exchange; water structure; water pollution; chemical education; history of chemistry; computers. *Mailing Add:* Dept Chem Davidson Col Davidson NC 28036

GABLER, ROBERT EARL, b Lodi, Ohio, Nov 22, 27; m 50; c 3. PHYSICAL GEOGRAPHY. *Educ:* Univ Ohio, BS, 49; Pa State Univ, MS, 51; Columbia Univ, EdD(sci), 57. *Prof Exp:* High sch teacher, Ohio, 50-55; lectr geog, Hunter Col, 55-57; from asst prof to prof, Western Ill Univ, 57-90, chmn dept, 64-74, dir, NDEA Geog Inst, 65-68, dir int progs, 77-90; RETIRED. *Concurrent Pos:* Consult, Geog Inst Progs, US Off Educ, 65-66; dir coord, Nat Coun Geog Educ, 63-69; mem nominating comt, 68-69, const revision comt, 68-, mem exec bd, 69-, mem joint comt, Nat Coun Geog Educ-Asn Am Geog, 70- *Mem:* Asn Am Geogr. *Res:* Physical geography, especially physiography, with strong regional interests in the geography of Asia; application of geography in American education. *Mailing Add:* 711 E Piper St Macomb IL 61455

GABLER, WALTER LOUIS, b Chicago, Ill, May 30, 31; m 52; c 4. BIOCHEMISTRY. *Educ:* Northwestern Univ, DDS, 56, PhD(chem), 64. *Prof Exp:* Instr pedodont, Dent Sch, Northwestern Univ, 58-59, instr biochem, 60-64; asst prof biochem pedodont, Dent Sch, Univ Ore, 64-65, assoc prof, 65-67; assoc prof, Univ Ky, 67-69; assoc prof, 69-74, PROF ORAL BIOL & BIOCHEM, DENT SCH, UNIV ORE, 74-, AFFIL BIOCHEM MED SCH, 71-, CHMN BIOCHEM, DENT SCH. *Concurrent Pos:* Nat Heart Inst fel, 61-64; Pharmaceut Mfrs Asn grant, 70-71; Advan Inst Dent Res grant, 71; partic, Northwest Environ Health Conf, 71; Epilepsy Found Am grant, 71-72; vis scientist, Ore Regional Primate Res Ctr. *Mem:* AAAS; Int Asn Dent Res; NY Acad Sci. *Res:* Biochemistry of inflammation; metabolism of collagen; drug metabolism during pregnancy. *Mailing Add:* 611 SW Campus Univ Ore Portland OR 97201

GABLIKS, JĀNIS, b Nitaure, Latvia, Nov 1, 24; nat US; m 50; c 2. MICROBIOLOGY. *Educ:* Baltic Univ, Ger, DD, 49; Rutgers Univ, MS, 57, PhD, 63. *Prof Exp:* Dent surgeon, Int Refugee Orgn Eutin, Ger, 49-51; asst path & virol, E R Squibb & Sons, NJ, 52-55; asst biochem, Bur Biol Res, Rutgers Univ, 56-57; res assoc virol, Schering Co, 57-60; instr bact, Rutgers Univ, 60-63; asst prof cell biol, Mass Inst Technol, 63-67; assoc prof biol, 67-74, PROF BIOL, NORTHEASTERN UNIV, 74- *Mem:* Am Soc Microbiol; NY Acad Sci; Sigma Xi; AAAS. *Res:* Bacterial toxins; carcinogens; nutrition and viral infections; insecticides; fish viruses; application of cell cultures in toxicology, viral vaccines, antiviral agents. *Mailing Add:* 103 Cabot St Newton MA 02158

GABOR, ANDREW JOHN, b Budapest, Hungary, July 2, 35; m 61; c 3. NEUROLOGY, NEUROPHYSIOLOGY. *Educ:* George Washington Univ, BA, 56, MS, 58; Duke Univ, PhD(anat), 62, MD, 64. *Prof Exp:* Instr neuroanat, Med Ctr, Duke Univ, 62-64 & 66, resident neurol, 65-67; clin assoc, EEG & Clin Neurophysiol Br, NIH, 67-69; DIR EEG LAB, SACRAMENTO MED CTR & ASSOC PROF NEUROL, SCH MED, UNIV CALIF, DAVIS, 69- *Mem:* AAAS; Am Acad Neurol; Am Epilepsy Soc; Am EEG Soc. *Res:* Clinical neurology and clinical and electrographic manifestations of seizure disorders; cortical organization with special reference to the underlying mechanisms of epilepsy. *Mailing Add:* Dept Neurol Univ Calif Davis Med Ctr Sacramento CA 95817

GABOR, JOHN DEWAIN, b Chicago, Ill, Aug 8, 32; m 61; c 2. CHEMICAL ENGINEERING. *Educ:* Univ Ill, BS, 54; Cornell Univ, PhD(chem eng), 57. *Prof Exp:* Asst chem engr, 57-64, ASSOC CHEM ENGR, ARGONNE NAT LAB, 64- *Concurrent Pos:* Vis assoc prof, Univ Ill, Chicago Circle, 79-86. *Mem:* Am Inst Chem Engrs. *Res:* Fluidization; heat transfer; fluid mechanics; process development for nuclear fuel reprocessing; chemical reactors; nuclear reactor safety. *Mailing Add:* Argonne Nat Lab D206 9700 S Cass Ave Argonne IL 60439

GABOR, THOMAS, b Budapest, Hungary, June 28, 25; US citizen; m 63; c 2. PHYSICAL CHEMISTRY. *Educ:* Univ Sci, Hungary, BSc, 48; Univ London, PhD(phys chem), 59. *Prof Exp:* Res chemist org chem, Alco Gand, Belgium, 49-50, Taubmans Chems Co, Australia, 51-52 & Monsanto Chem Co, Australia, 53-54; chemist res labs, Westinghouse Elec Corp, 60-63 & Battelle Mem Inst, 64-69; res specialist, 69-81, SR RES SPECIALIST, CERAMIC TECHNOL CTR, INDUST & ELECTRONIC SECT & INDUST CONSUMER SECT, CENT RES LABS, 3M CO, 81- *Res:* Crystal growth; surface chemistry; chemical vapor deposition. *Mailing Add:* 3M Center Bldg 201-BE-15 St Paul MN 55144

GABOUREL, JOHN DUSTAN, b San Francisco, Calif, Oct 16, 28; m 51; c 5. PHARMACOLOGY. *Educ:* Univ Calif, BS, 50; Univ San Francisco, MS, 51; Univ Rochester, PhD(pharmacol), 57. *Prof Exp:* Res assoc, Atomic Energy Proj, Univ Rochester, 55-57; from instr to asst prof pharmacol, Med Sch, Stanford Univ, 57-64; assoc prof, 64-71, PROF PHARMACOL, MED SCH, UNIV ORE, 71- *Concurrent Pos:* Consult toxicol, Univ Rochester, 56-; vis prof, Walter & Eliza Hall, Inst Med Res, Melbourne, Australia, 73-74. *Mem:* Am Soc Pharmacol & Exp Therapeut. *Res:* Cancer research; immunosuppressants, drug effects on lymphocyte function; adrenal steroid effects on lymphoid tissue growth and metabolism; metabolism of atropine; anticonvulsants; ocular pharmacology; effects of vitamin A and other retinoids on cultured retinal pigment epthiliem; particular interest in differences between cultures derived from persons with inherited dystrophies such as retinitis pigmentosa. *Mailing Add:* Dept Pharmacol Univ Ore Med Sch Portland OR 97201

GABRIDGE, MICHAEL GREGORY, b Detroit, Mich, March 12, 43; m 65; c 2. MICROBIOLOGY. *Educ:* Mich State Univ,BS, 65, MS, 66; Univ Mich, PhD(microbiol), 71. *Prof Exp:* Res microbiologist, US Pub Health Serv, 67-68; asst prof microbiol, Med Sch, Univ Ill, 71-76, assoc prof, 76-80; SR SCIENTIST CELL BIOL, CELL SCI CTR, LAKE PLACID, NY, 80- *Concurrent Pos:* Vis scientist infectious dis, Clin Res Ctr, London, 77-78; curric coordr basic med sci, Med Sch, Univ Ill, 78-80; adj prof, Dept Path, Med Sch, Univ Vt, 80-83, Dept Biol, State Univ NY, Plattsburgh, 80-; pres, Bionique, Inc, 82-90; dir, Univ Fla Biotech Inst, 90- *Mem:* Am Soc Microbiol; Tissue Cult Asn; Am Asn Univ Professors; AAAS; Sigma Xi. *Res:* Pathogenesis of respiratory infections; trachea explant culture; mycoplasmas; in-vitro models for cytotoicity, instrumentation for tissue culture. *Mailing Add:* Univ Fla Biotech Inst One Progress Blvd Box 26 Alachua FL 32615

GABRIEL, BARBRA L, b Chicago, Ill, Dec 28, 53. ELECTRON MICROSCOPY, PROTEIN BIOPHYSICS. *Educ:* Elmhurst Col, BS, 75; Univ Chicago, PhD(biophys), 88. *Prof Exp:* Asst prof physics, Elmhurst Col, 77-88; res assoc, Univ Pa, 88-89 & Univ Va, 89-91; DIR ELECTRON MICROS, MUSCULOSKELETAL SCI RES INST, 91- *Concurrent Pos:* Dir electron micros, Packer Eng, 81-86; consult educ, Argonne Nat Lab, 77-82. *Mem:* Biophys Soc; Electron Micros Soc Am. *Res:* Structural analysis of proteins using electron microscopy. *Mailing Add:* Musculoskeletal Sci Res Inst 2190 Fox Mill Rd Herndon VA 22071

GABRIEL, CEDRIC JOHN, b Gustine, Calif, Mar 25, 35; m 64; c 2. SOLID STATE PHYSICS. *Educ:* Fresno State Col, AB, 56; Univ Calif, Berkeley, PhD(physics), 64. *Prof Exp:* Res physicist, Naval Ord Lab, Corona, 64-70; RES PHYSICIST, NAVAL OCEAN SYSTS CTR, 70- *Mem:* AAAS; Am Phys Soc. *Res:* Nuclear quadrupole resonance spectroscopy; magnetoptics in solids; photo detection; chemical dynamics. *Mailing Add:* 3778 Balboa Terr San Diego CA 92117

GABRIEL, EDWARD GEORGE, b Batavia, NY, Jan 26, 46; m 70; c 4. GENETICS, CELL BIOLOGY. *Educ:* Alfred Univ, BA, 68, MS, 69; Ohio State Univ, MS, 74, PhD(genetics), 77. *Prof Exp:* ASSOC PROF BIOL, LYCOMING COL, WILLIAMSPORT, PA, 77- *Concurrent Pos:* Genetic counr, Williamsport Hosp, 78- *Mem:* Genetics Soc Am; AAAS; Am Genetics Soc. *Res:* NOR staining of squirrel chromosomes; chromosome banding techniques in humans. *Mailing Add:* Dept Biol Lycoming Col Williamsport PA 17701

GABRIEL, EDWIN Z, b Union City, NJ, Aug 26, 13. COMPUTER SCIENCE, ENERGY TECHNOLOGY. *Educ:* Newark Col Eng, BSME, 36, ME, 39, MSEE, 52. *Prof Exp:* Heating engr, Webster Tallmadge & Co, NJ, 38-39; efficiency & mech engr, Prudential Ins Co Am, 39-41; asst mech engr, US Engr Off, NY, 41-42, Manhattan dist, Mass, 42-43; assoc mech engr, Watson Labs, USAF, NJ, 45-48, Cambridge Res Ctr, 48-50; electronic engr, Signal Corps Eng Labs, NJ, 50-52; proj engr, Curtiss-Wright Corp, 52-53; proj engr & consult, Kearfott Co, Inc, 53-55; asst prof elec eng, Univ Lehigh, 55-56; asst prof elec & mech eng, Villanova Univ, 56-60; asst prof elec eng, Fairleigh Dickinson Univ, 60-62; assoc prof, Weapons Dept, US Naval Acad, 62-64; writing textbk, 64-65; independent comput design, develop & mfr, 65; proj engr avionics lab, Electronics Command, US Army, 65-73; PRES & OWNER, GABRIEL COMPUT CO & GABRIEL ENG CONSULTS, 73- *Concurrent Pos:* Teacher physics, Monmouth Col, NJ, 78- *Mem:* Am Soc Mech Engrs; Am Soc Eng Educ; Inst Elec & Electronics Engrs; Am Helicopter Soc. *Res:* Study and design of automatic controls and analog and digital components and systems; use of teaching aids for effective presentation of the principles of physics; automatic control systems for helicopters; educational computers for high school and college teachers and students. *Mailing Add:* 91 Mt Tabor Way Ocean Grove NJ 07756

GABRIEL, GARABET J(ACOB), b Basrah, Iraq, Feb 15, 35; US citizen. ELECTRICAL ENGINEERING, PHYSICS. *Educ:* St Louis Univ, BS, 56; Ill Inst Technol, MS, 60; Northwestern Univ, PhD(elec eng, physics), 64. *Prof Exp:* Engr, Motorola Inc, Ill, 56-57; instr elec eng, Ill Inst Technol, 57-61; res asst microwave physics, Northwestern Univ, 61-64; staff scientist, Hallicrafters Co, 64-65; asst prof, 65-69, ASSOC PROF ELEC ENG, UNIV NOTRE DAME, 69-, STAFF MEM, RADIATION LAB, 65- *Concurrent Pos:* Res assoc, Res Inst, Chicago Med Sch, 60. *Mem:* Inst Elec & Electronics Engrs. *Res:* Electromagnetic and applied mathematics; theoretical and experimental work on microwaves and optics; statistical electrodynamics; high speed electromagnetic transients on superconducting coils and boundaries; statistical theoretical analysis of radiation scattering in fluctuating media. *Mailing Add:* Dept Elec Comp Eng Univ Notre Dame Notre Dame IN 46556

GABRIEL, HENRY, b Berlin, Ger, Apr 28, 14; nat US; m 38. INORGANIC CHEMISTRY, PHYSICAL CHEMISTRY. *Educ:* Long Island Univ, BS, 34; George Washington Univ, MS, 49; Stanford Univ, PhD(chem), 51. *Prof Exp:* Prof ger & asst dir, Berlitz Sch Langs, 36-43; assoc chem, George Washington Univ, 46-49; instr, Univ Santa Clara, 49-51; asst prof, Fisk Univ, 51-54; assoc prof, Siena Col, 54-56; mgr tech info, Phillips Petrol Co, 56-63, develop engr, Patent Div, 63-76; RETIRED. *Concurrent Pos:* Carnegie Found grant, Fisk Univ, 51-53. *Mem:* Am Chem Soc; Sigma Xi. *Res:* Rare earth elements, separation and compounds; preparation, properties and reactions of silicon compounds; ion exchange and adsorption; information storage and retrieval systems. *Mailing Add:* 1948 S Sante Fe Bartlesville OK 74003

GABRIEL, JOHN R, b Leeds, Eng, Apr 30, 31. APPLIED MATHEMATICS, COMPUTER SCIENCE. *Educ:* Univ Otago, NZ, BS(pure & appl math) & BS(physics & radio physics), 52, MS, 53. *Prof Exp:* Asst lectr physics, Univ Otago, NZ, 54-55, Imp Chem Industs fel, 55-57; theoret physicist, Inst Nuclear Sci, NZ, 57-58; sr spec res fel theoret physics, UK Atomic Energy Authority, Eng, 58-62; vis res fel, Solid Sci Div, 62-65, asst head comput ctr, Appl Math Div, 65-69, assoc comput scientist, 69-80, COMPUT SCIENTIST, APPL MATH DIV, ARGONNE NAT LAB, 80- *Mem:* Fel Brit Inst Physics & Phys Soc; Inst Elec & Electronics Engrs Comput Soc; Inst Elec & Electronics Engrs; Asn Comput Mach; NY Acad Sci. *Res:* Reliable computer systems in nuclear reactor control; automated inference systems for plant design and safety qualification; multiple instruction-multiple data computing, personal computers, data management for small businesses; group theory and quantum mechanics; numerical linear algebra. *Mailing Add:* Appl Math Div Argonne Nat Lab 9700 S Cass Ave Bldg 362 OACA Lemont IL 60439

GABRIEL, KARL LEONARD, b Philadelphia, Pa, July 27, 29; m 52; c 5. PHARMACOLOGY, TOXICOLOGY. *Educ:* Univ Pa, AB, 51, VMD, 56; Jefferson Med Col, PhD, 64; Drexel Inst Technol, MS, 65. *Prof Exp:* Asst dir, Indust Biol Res & Testing Lab, Inc, Pa, 56-61; vpres & dir res, AME Assocs, 65-66; assoc prof biol sci & environ eng, Drexel Inst Technol, 66-71; from assoc prof to prof pharmacol, Med Col Pa, 71-84, assoc prof community & preventive med & res assoc prof psychiat, 80-84; CONSULT. *Concurrent Pos:* From instr to asst prof, Sch Vet Med, Univ Pa, 60-65; vis asst prof, Med Col Pa, 65-71; pres, Biosearch, Inc, 68-; dipl, Am Bd Toxicol. *Mem:* Soc Toxicol; Am Soc Pharmacol & Exp Therapeut; NY Acad Sci; Am Asn Lab Animal Sci; Sigma Xi; fel Am Col Clin Pharm. *Res:* General pharmacology-toxicology; biomathematics. *Mailing Add:* PO Box 8598 Philadelphia PA 19101-8598

GABRIEL, LESTER H, b Brooklyn, NY, Mar 17, 28; m 50; c 2. CIVIL ENGINEERING. *Educ:* Cooper Union, BSCE, 49; Polytech Inst Brooklyn, MSCE, 56; Univ Calif, Berkeley, PhD, 71. *Prof Exp:* Civil engr, US Bur Reclamation, 49-51; struct engr, Farkas & Barron, 51-52 & Severud-Elstad-Krueger, 52-55; sr struct engr, Kaiser Engrs, 55-57; from asst prof to assoc prof civil eng, 57-67, chmn dept, 63-68, PROF CIVIL ENG, CALIF STATE UNIV, SACRAMENTO, 67- *Concurrent Pos:* Dir, Appl Res & Design Ctr, NSF sci fel, Calif, 61-; mem comt on soil-struct interaction, Transp Res Bd, Nat Acad Sci-Nat Res Coun, 64-76. *Mem:* Am Soc Civil Engrs; Am Concrete Inst; Am Soc Eng Educ. *Res:* Soil structure interaction for static and dynamic loads on buried structures; concrete materials and materials of manufacture. *Mailing Add:* 4841 Tono Way Sacramento CA 95841

GABRIEL, MORDECAI LIONEL, b New York, NY, Mar 18, 18; m 45; c 2. BIOLOGY, ACADEMIC ADMINISTRATION. *Educ:* Yeshiva Col, BA, 38; Columbia Univ, AM, 38, PhD(zool), 43. *Prof Exp:* Asst zool, Columbia Univ, 38-41, lectr, 41-42; instr genetics, Univ Conn, 43-45; instr, Brooklyn Col, 45-50, from asst prof to prof biol, 50-89, chmn dept, 65-71, dean, Sch Sci, 71-77, actg vpres acad affairs, 81-82, assoc provost acad affairs, 82-89, EMER PROF BIOL, BROOKLYN COL, 89- *Concurrent Pos:* Fac fel, Ford Found, 54-55; vis prof, Columbia Univ, 55; Fulbright lectr, Tel-Aviv Univ, 59-60. *Mem:* AAAS; Soc Vert Paleont; Am Soc Zool; Soc Study Evolution; Am Asn Anat; NY Acad Sci. *Res:* Comparative myology of Plectognath fishes; meristic variation of fishes; embryology of polydactylism in fowl. *Mailing Add:* Dept Biol Brooklyn Col Brooklyn NY 11210

GABRIEL, OSCAR V, b Manila, Philippines, Apr 22, 46; m 73; c 1. ATOMIC PHYSICS. *Educ:* Univ Philippines, BS, 66; Univ Pittsburgh, MS, 72, PhD(physics), 81. *Prof Exp:* Instr physics, Univ Philippines, 66-69; asst prof, Ateneo de Manila Univ, 73-78; asst prof physics, Univ Richmond, 81-83, Villanova Univ, 83-85; SR ENGR, GE AEROSPACE, 85- *Mem:* Am Phys Soc; Sigma Xi. *Res:* Theoretical study of inner-shell atomic processes at high energies by analytic perturbation theory and computer-assisted investigation of the same processes; radar systems engineering. *Mailing Add:* 202 Ramblewood Pkwy Mt Laurel NJ 08054

GABRIEL, R OTHMAR, b Vienna, Austria, Jan 10, 25; US citizen; m 49; c 2. BIOCHEMISTRY. *Educ:* Univ Vienna, PhD(chem), 54. *Prof Exp:* Res assoc biochem, Univ Vienna, 54-58; assoc, Med Ctr, Columbia Univ, 58-60; vis scientist, NIH, 60-64, res biochemist civil serv, 64-65; clin asst prof, Georgetown Univ, Med & Dent Sch, 61-64, prof lectr, 64-65, assoc prof, 65-70, actg chmn, 86-88, prof biochem, 70-90, EMER PROF BIOCHEM, GEORGETOWN UNIV, 90- *Mem:* Am Soc Biol Chemists. *Res:* Structure and function of carbohydrate containing cell surface components; precursors for the biosynthesis of polysaccharides; mechanism of enzyme action. *Mailing Add:* Georgetown Univ Med Ctr Four Research Ct Rockville MD 20850

GABRIEL, RICHARD FRANCIS, b New Rochelle, NY, Nov 30, 20; m 48; c 2. NUMERICAL ANALYSIS, DIFFERENTIAL EQUATIONS. *Educ:* Fordham Univ, AB, 43; Columbia Univ, MA, 47; Rutgers Univ, PhD(math), 55. *Prof Exp:* Instr math, Pub Sch, 46-47; educ therapist, Vet Admin, 47-48; instr math, St Francis Col, NY, 48-50; assoc prof & asst dir comput ctr, Rutgers Univ, 50-62; dir, Comput Ctr, 62-72, prof, 72-87, EMER PROF MATH, SETON HALL UNIV, 87- *Concurrent Pos:* NSF grant, 57; consult, Esso Res Eng Co & State Rehab Comput; chmn, NSF Numerical Analysis & Electronic Comput Conf, USAR, 58-62; prin investr, NIH Biomed Comput, Seton Hall Univ, 62-65; bd gov, Math Asn Am, 82-85. *Mem:* AAAS; Soc Indust & Appl Math; NY Acad Sci; Asn Comput Mach; Am Math Soc; Math Asn Am. *Res:* Complex variable; numerical analysis; electronic digital computers; evaluation of programming systems and equipment configurations. *Mailing Add:* 24 Gen Stanton Lane Charleston RI 02813

GABRIEL, ROBERT, synthetic organic chemistry, for more information see previous edition

GABRIEL, WILLIAM FRANCIS, b Sault Ste Marie, Mich, Oct 17, 25; m 48; c 2. ELECTRICAL ENGINEERING. *Educ:* Univ Wis, BS, 45, MS, 48, PhD(elec eng), 50. *Prof Exp:* Electronic scientist, Naval Res Lab, 50-59; sr res engr, Stanford Res Inst, 59-61; consult, Aero Geo Astro Corp, Va, 61-64; sr engr, NASA Goddard Space Flight Ctr, Md, 64-67; engr, Scanwell Labs, Inc, Va, 67-69; mem staff, Delex Systs, Inc, Arlington, Va, 69-70; head Antenna Systs Staff, Naval Res Lab, Washington, 70-86; CONSULT S/F ASSOC, LANDOVER, MD, 86- *Concurrent Pos:* AP-S distinguished lectr, Inst Elec & Electronics Engrs, 80-82. *Mem:* Fel Inst Elec & Electronics Engrs. *Res:* Adaptive array antenna systems; microwave components; instrumentation; superresolution spectral estimation techniques. *Mailing Add:* SF/Assoc Code 5318 Naval Res Lab Washington DC 20375

GABRIELE, ORLANDO FREDERICK, b North Providence, RI, June 6, 27; m 60; c 4. RADIOLOGY. *Educ:* Brown Univ, BA, 50; Yale Univ, MD, 54. *Prof Exp:* Instr radiol, Sch Med, Yale Univ, 57-59, lectr, 59-65, asst prof, 65-66; from assoc prof to prof, Sch Med, Univ NC, Chapel Hill, 66-72; PROF RADIOL & CHMN DEPT, WVA UNIV, 72- *Mem:* Radiol Soc NAm; Am Col Radiol; Am Roentgen Ray Soc; Asn Univ Radiologists. *Res:* Cardiovascular radiology. *Mailing Add:* Dept Radiol WVa Univ Med Ctr Morgantown WV 26506

GABRIELE, THOMAS L, b York, Pa, Sept 7, 40; m 65; c 3. ELECTRICAL ENGINEERING. *Educ:* Lehigh Univ, BSEE, 62; Johns Hopkins Univ, MS, 64, PhD(elec eng), 68. *Prof Exp:* Eng specialist, Martin Marietta Corp, 66-67; prin engr, Commun Div, Bendix Corp, Baltimore, 67-80, supvr, comput aided design, 80-84; dir software, 84-90, SR SYSTS ANALYST, BECTON DICKINSON, 90- *Mem:* Inst Elec & Electronics Engrs. *Res:* Pattern recognition; automated decision systems; computer assisted analysis; microprocessor based systems design. *Mailing Add:* Becton Dickinson Seven Loveton Circle PO Box 999 Sparks MD 21152-0999

GABRIELSE, HUBERT, b Golden, BC, Mar 1, 26; m 55; c 2. GEOLOGY. *Educ:* Univ BC, BASc, 48, MASc, 50; Columbia Univ, PhD(geol), 55. *Prof Exp:* Geologist, 53-70, head, Cordilleran Subdiv, 70-79, RES SCIENTIST, GEOL SURV CAN, 79- *Honors & Awards:* J Willis Ambrose Medal, Geol Asn Can. *Mem:* Geol Soc Am; NY Acad Sci; fel Geol Asn Can; mem Can Inst Mining & Metall; Royal Soc Can. *Res:* Regional stratigraphy, structure and tectonics. *Mailing Add:* 100 W Pender St Vancouver BC V6B 1R8 Can

GABRIELSEN, ANN EMILY, b Brooklyn, NY, Apr 18, 25. IMMUNOLOGY. *Educ:* Brooklyn Col, BA, 46; Univ Minn, MA, 53, PhD(microbiol), 70. *Prof Exp:* Res fel pediat, Univ Minn, 61-67, res specialist, 67-70; prin res scientist immunol, Kidney Dis Inst, NY State Dept Health, 70-83; RETIRED. *Mem:* Am Asn Immunologists. *Res:* Complement; autoimmune disease. *Mailing Add:* 1072 79th St Brooklyn NY 11228

GABRIELSEN, BERNARD L, b Woodland, Wash, May 29, 34; m 66; c 3. CIVIL & STRUCTURAL ENGINEERING. *Educ:* Ore State Univ, BS, 56; Stanford Univ, MS, 61, PhD(civil eng), 66. *Prof Exp:* Engr-scientist, Lockheed Missiles & Space Co, 58-61; sr engr, Western Develop Labs, Philco Corp, 61-63; acting asst prof civil eng, Stanford Univ, 63-65; from asst prof to assoc prof, 65-77, PROF CIVIL ENG, SAN JOSE STATE UNIV, 77-; CONSULT, 61- *Concurrent Pos:* Consult, Stanford Res Inst, 65-66 & Walter V Sterling, Inc, 65-; prin struct engr, URS Corp, 67-72; exec vpres, Sci Serv Inc, 74-84. *Mem:* Am Soc Civil Engrs; Am Concrete Inst; Am Metal Soc. *Res:* Structural research on the dynamic failure of building panels; application of statistics to structural problems. *Mailing Add:* Dept Civil Eng San Jose State Univ One Washington Sq San Jose CA 95192

GABRIELSEN, BJARNE, b Brooklyn, NY, Oct 22, 41. ORGANIC CHEMISTRY. *Educ:* Wagner Col, BS, 62; State Univ NY Stony Brook, MS, 64, PhD(phys org chem), 69. *Prof Exp:* From instr to assoc prof chem, Wagner Col, 70-80; MEM ASSOC, DEPT CHEM, UNIV FLA, GAINESVILLE, 80- *Mem:* Am Chem Soc. *Res:* Synthesis of labelled organic compounds of biological interest; pyrilium/pyridinium chemistry, specifically modification of amino groups; new experiments in teaching organic chemistry; organic chemistry under high pressure. *Mailing Add:* 4222 Garnet Dr Middletown MD 21769

GABRIELSEN, TRYGVE O, b Vest-Agder, Norway, Mar 27, 30; US citizen; m 55; c 4. RADIOLOGY. *Educ:* Univ Washington, BS, 53, MD, 56. *Prof Exp:* Asst chief diag radiol, Brooke Army Med Ctr, 60-62; from instr to assoc prof, 62-71, PROF RADIOL, MED SCH, UNIV MICH, ANN ARBOR, 71- *Honors & Awards:* Asn Univ Radiologists Travel Award, 68. *Mem:* Am Asn Neurol Surg; Radiol Soc NAm; Asn Univ Radiol; Am Soc Neuroradiol. *Res:* Diagnostic radiology, particularly neuroradiology. *Mailing Add:* Dept Radiol Univ Mich Ann Arbor MI 48109

GABRIELSON, RICHARD LEWIS, b Riverside, Calif, Feb 12, 31; div; c 4. PLANT PATHOLOGY. *Educ:* San Diego State Col, BA, 52; Univ Calif, Davis, PhD(plant path), 61. *Prof Exp:* From asst plant pathologist to assoc plant pathologist, 60-76, PLANT PATHOLOGIST, WESTERN WASH EXP STA, WASH STATE UNIV, 76- *Mem:* Am Phytopath Soc; Sigma Xi. *Res:* Etiology, epidemiology and control of vegetable diseases. *Mailing Add:* 1112 Nineth St SE D-1 Puyallup WA 98372

GABRUSEWYCZ-GARCIA, NATALIA, b Kiev, Ukraine, Nov 14, 34; US citizen; m 61. CELL BIOLOGY, CYTOLOGY. *Educ:* Univ Sao Paulo, BS, 58; Columbia Univ, PhD(zool), 64. *Prof Exp:* Instr histol & cytol, Univ Sao Paulo, 58-59; res assoc pharmacol, State Univ NY Upstate Med Ctr, 64-72; assoc prof, 72-80, PROF BIOL, ONONDAGA COMMUNITY COL, 80- *Concurrent Pos:* USPHS grant, 64-67. *Mem:* Am Soc Cell Biol. *Res:* Structure and nucleic acids metabolism of chromosomes, especially of polytene chromosomes. *Mailing Add:* Dept Biol Onondaga Community Col Syracuse NY 13215

GABUZDA, THOMAS GEORGE, b Freeland, Pa, Apr 27, 30; m 57; c 3. HEMATOLOGY. *Educ:* Lehigh Univ, BA, 51; Harvard Univ, MD, 55; Am Bd Internal Med, dipl, 63, cert hematol, 76, cert med oncol, 85. *Prof Exp:* From intern to asst resident med, Mass Gen Hosp, 55-57; USPHS fel, Med Dept B, Bispebjerg Hosp, Copenhagen, Denmark, 59-60; chief resident med, Lemuel Shattuck Hosp, Boston, 60-61; from res fel to res assoc, Med Sch, Harvard Univ, 61-64, instr, 64-65; from asst prof to assoc prof, 65-73, PROF MED, JEFFERSON MED COL, 73-, ASSOC MEM, CARDEZA FOUND HEMAT RES, 65-; CHIEF, DEPT HEMAT, LANKENAU HOSP, PHILADELPHIA, 70- *Concurrent Pos:* USPHS fel, Curtis Hemat Lab, Peter Bent Brigham Hosp, Boston, 61-63, asst med, Hosp, 61-64, jr assoc, 64-65; assoc med staff, Thomas Jefferson Univ Hosp, 75-; adj prof comp med, Sch Vet Med, Univ Pa, 75-81. *Mem:* Am Col Physicians; Am Soc Hemat; AMA; Am Fedn Clin Res; Sigma Xi. *Res:* Normal and abnormal hemoglobins; hemoglobin synthesis; ferritin; iron metabolism. *Mailing Add:* Lancaster/City Ave Lankenau Hosp Philadelphia PA 19151

GABY, WILLIAM LAWRENCE, b Hot Springs, NC, June 15, 17; m 40; c 2. BACTERIOLOGY. *Educ:* Univ Tenn, BA, 39, MS, 40; St Louis Univ, PhD(bact), 56; Am Bd Microbiol, dipl. *Prof Exp:* Sr bacteriologist, State Dept Pub Health, Tenn, 40-41; bacteriologist, Winthrop Chem Co, NY, 41-42; sr bacteriologist, Bristol Labs, Inc, 46-49; from asst prof to assoc prof bact, Hahnemann Med Col & Hosp, 49-64; chmn, Dept Health Sci, ETenn State Univ, 64-80, prof microbiol, 64-83; RETIRED. *Honors & Awards:* Commercial Solvents Award, 52. *Mem:* AAAS; Am Soc Microbiol; Am Soc Biol Chemists; Am Asn Immunologists; fel Am Acad Microbiologists. *Res:* Dissociation of bacteria; antigen; antibody response; penicillin production and antibiotic activity; mold mutation and metabolism; lipid and phospholipid role in metabolism. *Mailing Add:* 1408 College Heights Rd Johnson City TN 37604

GACS, PETER, b Budapest, Hungary, May 9, 47; m 72; c 2. INFORMATION THEORY. *Educ:* Roland Eotvos Univ, Budapest, MS, 70; J W Goeth Univ, WGer, PhD(math), 78. *Prof Exp:* Res assoc math, Math Inst Hungary Acad Sci, 70-77; asst prof, J W Goeth Univ, WGer, 77-78; res assoc comput sci, Stanford Univ, 79; asst prof math & computer sci, Univ Rochester, 80-84; ASSOC PROF COMPUTER SCI, BOSTON UNIV, 84- *Concurrent Pos:* Vis prof, Institut fur Statistik, G August Univ, WGer, 75; vis scientist, IBM Almaden Res Ctr, 88-89; various NSF grants. *Mem:* Asn Comput Mach; Am Math Soc; Inst Elec & Electronic Engrs. *Res:* Development of new techniques in Multi-terminal Information Theory (Shannon's Theory), including upper bound on common information; exact formulas connecting randomness tests with complexity; new ways to estimate descriptional complexity from below; fault-tolerant cellular automate. *Mailing Add:* Dept Comput Sci Boston Univ 111 Cummington St Boston MA 02215

GADAMER, ERNST OSCAR, b Berlin, Ger, Nov 1, 24; Can citizen; m 54; c 2. APPLIED MATHEMATICS. *Educ:* Univ Frankfurt, dipl physics, 51; Univ Toronto, MA, 56, PhD(aerophys), 62. *Prof Exp:* Data reduction engr, de Havilland of Can Co, Ont, 58-59; res fel, Inst Aerophys, Univ Toronto, 56-61; asst res officer hypersonics, Nat Res Coun Can, 61-62; asst prof eng analysis, 62-77, ASSOC PROF MATH SCI, MCMASTER UNIV, 77- *Res:* Low density aerodynamics; applied mathematics. *Mailing Add:* Dept Math & Statist McMaster Univ Hamilton ON L8S 4L8 Can

GADBERRY, HOWARD M(ILTON), b Kansas City, Mo, July 25, 22; m 52; c 2. TECHNOLOGY FORECASTING, CHEMICAL ENGINEERING. *Educ:* Univ Kans, BS, 43. *Prof Exp:* Chemist, Phillips Petrol Co, 43-46; assoc chem engr, 46-50, sr chem engr, 50-54, head indust chem sect, 54-58, asst mgr, Div Chem & Chem Eng, 58-61, asst dir, Econ Develop Div, 61-63, SR ADV TECH, MIDWEST RES INST, 63- *Mem:* Am Chem Soc; Am Inst Aeronaut & Astronaut; Sigma Xi; Am Asn Textile Chemists & Colorists; Am Ord Asn. *Res:* Advanced industrial technology; economic forecasting and modeling; energy resources and economics; technological forecasting; time-series analysis and forecasting; economic analysis of new technology; technology diffusion and history. *Mailing Add:* 4823 W 96th Terr Kansas City MO 64110

GADDIS, MONICA LOUISE, b Toledo, Ohio, Nov 25, 55; m 79. CARDIOVASCULAR PHYSIOLOGY, EXERCISE PHYSIOLOGY. *Educ:* Ind Univ, Bloomington, BS, 78, MS, 80, PhD(exercise physiol), 84. *Prof Exp:* RES ASSOC PHYSIOL, SCH MED, IND UNIV, 83- *Mem:* Am Physiol Soc. *Res:* Reflex control of venous capacitance; influences of chemoreceptor and baroreceptor control upon vascular capacitance. *Mailing Add:* Truman Med Ctr 2301 Holmes Kansas City MO 64108

GADDUM-ROSSE, PENELOPE, b Sept 4, 41; Brit citizen. BIOLOGICAL STRUCTURE. *Educ:* Univ Wales, BSc, 62; Univ Liverpool, PhD(reprod physiol), 65. *Prof Exp:* Fel biol struct, 66-68, from res assoc to res asst prof, 69-74, asst prof, 74-81, ASSOC PROF BIOL STRUCT, SCH MED, UNIV WASH, 81- *Mem:* Am Asn Anat; Soc Study Reprod; Soc Study Fertil. *Res:* Mammalian fertilization; mechanisms of gamete transport. *Mailing Add:* Dept Biol Struct Univ Wash Seattle WA 98195

GADDY, JAMES LEOMA, b Jacksonville, Fla, Aug 16, 32; m 52; c 2. CHEMICAL ENGINEERING. *Educ:* La Polytech Univ, BS, 55; Univ Ark, MS, 68; Univ Tenn, PhD(chem eng), 72. *Prof Exp:* Process engr, Ethyl Corp, Baton Rouge, La, 55-60; eng supvr, Ark, La Gas, 60-64, proj mgr, 64-67; from assoc prof to prof chem eng, Univ Mo, Rolla, 70-80; prof chem eng & head dept, 80-88, DISTINGUISHED PROF CHEM ENG, UNIV ARK, 88- *Concurrent Pos:* Consult, McDonnell Aircraft, 73-75, Aerojet Nuclear, 73-75, Monsanto Co, 73-75, UN, 77-79, Basler Hofmann, 77-78, Thermonetics Corp, 77-78, Banyon Eng, 77-78, Quaker Oats Co, 77-, Pritchard Co, 80-, Proctor & Gamble, 82-, Tenn Valley Authority, 83-, & Ethyl Corp, 83-; fac fel, Swiss Fed Inst Technol. *Honors & Awards:* Merck Lectr. *Mem:* Am Inst Chem Engrs; Am Chem Soc; AAAS; Am Soc Eng Educ; Sigma Xi. *Res:* 400 publication and presentations and $6,000,000 research contracts in chemical process optimization, synthesis, reliability and overdesign; biochemical engineering; biological conversion of biomass, coal and oil shale into energy and chemicals. *Mailing Add:* Dept Chem Eng Univ Ark Fayetteville AR 72701

GADDY, OSCAR, b Republic, Mo, July 18, 32; m 53; c 3. ELECTRICAL ENGINEERING. *Educ:* Univ Kans, BS, 57, MS, 59; Univ Ill, PhD(elec eng), 62. *Prof Exp:* Instr & res asst elec eng, Univ Kans, 57-59; res asst, 59-62, from asst prof to assoc prof, 62-69, assoc head dept,71-84, PROF ELEC ENG, UNIV ILL, URBANA-CHAMPAIGN, 69- *Mem:* Fel Inst Elec & Electronics Engrs; Sigma Xi. *Res:* Optical communication; quantum electronics; electron devices; infrared devices; subnanosecond optical pulse generation and detection. *Mailing Add:* 609 Evergreen Court E Urbana IL 61801

GADE, DANIEL W, b Niagara Falls, NY, Sept 28, 36; m 65; c 1. CULTURAL GEOGRAPHY, BIOGEOGRAPHY. *Educ:* Valparaiso Univ, BA, 59; Univ Ill, MA, 60, Univ Wis-Madison, MS, 62, PhD(geog), 67. *Prof Exp:* Vis instr geog, Univ Ore, 65-66; from asst prof to assoc prof, 66-79, dir Latin Am studies, 78-87, PROF GEOG, UNIV VT, 79- *Concurrent Pos:* Soc Sci Res Coun fel, Peru, 70; Nat Geog Soc res grant, Peru, 78; Fulbright res award, Madagascar, 83. *Mem:* Asn Am Geog; Conf Latin Am Geographers; Am Geog Soc; Soc Ethnobiol; Soc Econ Bot; Int Mountain Soc. *Res:* Past and present ecological relationships of primitive and folk societies; ethnobotany and ethnozoology; process of plant and animal domestication; Latin America; settlement geography; environmental conservation. *Mailing Add:* Dept Geog Old Mill Bldg Univ Vt Burlington VT 05405

GADE, EDWARD HERMAN HENRY, III, b St Joseph, Mo, Feb 19, 36; m 60; c 2. MATHEMATICS. *Educ:* Valparaiso Univ, BA, 58; Purdue Univ, MS, 60; Univ Pittsburgh, PhD(math), 65. *Prof Exp:* ASSOC PROF MATH, UNIV WIS, OSHKOSH, 66- *Res:* Summability methods analysis. *Mailing Add:* Dept Math Univ Wis Oshkosh WI 54901

GADE, SANDRA ANN, b Waterbury, Conn, Oct 27, 37; m 60; c 2. PHYSICS. *Educ:* Valparaiso Univ, BS, 59; Univ Pittsburgh, PhD(physics), 66. *Prof Exp:* Asst prof, 66-74, assoc prof, 74-80, PROF PHYSICS, UNIV WIS, OSHKOSH, 80 - *Mem:* Am Phys Soc. *Res:* Atomic physics; nuclear magnetic and electron paramagnetic resonance. *Mailing Add:* Dept Physics Univ Wis Oshkosh WI 54904

GADEBUSCH, HANS HENNING, b Charlottenburg, Ger, Jan 8, 24; nat US; m 49; c 7. MEDICAL MICROBIOLOGY. *Educ:* Univ Detroit, BS, 49; Univ Mich, MS, 51, PhD, 65. *Prof Exp:* Asst bact, Univ Mich, 49-51; res microbiologist, Irwin, Neisler & Co, Ill, 51-53; clin bacteriologist, Detroit Dept Health, 53; chief bacteriologist, Vet Admin Hosp, 53-57, coordr gen med res, 57-62, head microbiol-immunol res sect, 57-66; chief med microbiol, Squibb Inst Med Res, 66-71, asst dir dept microbiol, 71-79; mgr, New Antibiotic Res, Merck Sharp & Dohme Res Labs, 79-87; sr dir antibiotic eval & develop, 87-90, SR DIR MICROBIOL ADMIN, MERCK & CO, 90- *Mem:* AAAS; Am Soc Microbiol; Soc Exp Biol & Med; NY Acad Sci; fel Am Acad Microbiol; fel Infectious Dis Soc Am. *Res:* Microbiology; immunology; biochemistry; analytical chemistry; new drugs from nature research; animal models of human disease. *Mailing Add:* 711 Jade Rd Yardley PA 19067-3011

GADEK, FRANK JOSEPH, b Troy, NY, Oct 1, 41; m 70. ORGANIC CHEMISTRY. *Educ:* Siena Col, NY, BS, 63; Cath Univ Am, PhD(org chem), 69. *Prof Exp:* Asst, Cath Univ Am, 68; assoc prof, 68-80, PROF ORG CHEM, ALLENTOWN COL, 80- *Concurrent Pos:* Frederick Gardner Cottrell res grant-in-aid, 71-72. *Mem:* Am Soc Enol; Am Chem Soc; Sigma Xi; Am Asn Univ Professors. *Res:* Synthesis and the investigation of the properties of new heterocyclic aromatic compounds which are analogs of azulene and quinoline using organometallic reagents and novel dehydrogenating agents; french hybrid wines, deacidification and total phenols; chemical deacidification of wines; use of pectic enzymes in wine making. *Mailing Add:* Dept Chem Allentown Col Center Valley PA 18034

GAD-EL-HAK, MOHAMED, b Tanta, Egypt, Feb 11, 45; m 76; c 2. FLUID MECHANICS, TURBULENCE. *Educ:* Ain-Shams Univ, Egypt, BSc, 66; Johns Hopkins Univ, PhD(fluid mech), 73. *Prof Exp:* Instr mech eng, Ain-Shams Univ, 66-68; sr lectr fluid mech, Univ Southern Calif, 73-74; asst prof eng sci & systs, Univ Va, 74-76; sr res scientist, Flow Res Co, 76-86; PROF AEROSPACE & MECH ENG, UNIV NOTRE DAME, 86- *Concurrent Pos:* Vis prof, Univ Grenoble, France, 91-92; ed, J Am Inst Aeronaut & Astronaut & Appl Mech Reviews. *Mem:* Am Physical Soc; Am Soc Mech Engrs; Am Indust Aeronaut & Astronaut; Am Chem Soc. *Res:* Basic fluid mechanics; turbulence research especially experimental; development of unique flow visualization techniques. *Mailing Add:* Dept Aerospace & Mech Engr Univ Notre Dame Notre Dame IN 46556

GADEN, ELMER L(EWIS), JR, b Brooklyn, NY, Sept 26, 23; m 64; c 3. CHEMICAL ENGINEERING. *Educ:* Columbia Univ, BS, 44, MS, 47, PhD(chem eng), 49. *Hon Degrees:* DrEngr, Rensselaer Polytech Inst, 87. *Prof Exp:* Biochem eng head, Chas Pfizer & Co, 48-49; from asst prof to prof chem eng, Columbia Univ, 49-74, chmn dept, 60-69 & 71-74; dean eng, math & bus admin, Univ Vt, 75-79; chmn eng dept, 85-88, CHEM ENG DEPT, UNIV VA, 79- *Concurrent Pos:* Tech dir, Biochem Processes, Inc, 59-71. *Honors & Awards:* Egleston Medal, Columbia, 86; Founders Award, Am Inst Chem Engrs, 88. *Mem:* Nat Acad Eng; Am Chem Soc; Am Inst Chem Engrs; Am Soc Eng Educ. *Res:* Biochemical engineering; fermentation; renewable energy resources. *Mailing Add:* Chem Eng Dept Thornton Hall Univ Va Charlottesville VA 22903-2442

GADOL, NANCY, EXPERIMENTAL BIOLOGY. *Educ:* Carnegie-Mellon Univ, BA, 71; Univ Fla, MS, 74; Hahnemann Univ, PhD(immunol), 79. *Prof Exp:* Asst res immunologist, Sch Dent, Dept Oral Med, Univ Calif, San Francisco, 79-81; postdoctoral fel, Dept Neurol, Univ Calif & Vet Admin Med Ctr, San Francisco, 81-83; sr res assoc, Becton Dickinson Immunocytometry Systs, San Jose, Calif, 83-85, res scientist, 85-88, scientist, 88-89; SCIENTIST, IRWIN MEM BLOOD BANK, SAN FRANCISCO, CALIF, 89- *Concurrent Pos:* Adj & vis lectr, Dept Microbiol & Immunol, Univ Calif, San Francisco; lab instr microbiol, Pa Col Podiat Med, Philadelphia, lab instr immunol & microbiol, Hahnemann Univ. *Mem:* Am Asn Immunologists; AAAS. *Res:* Cellular immunology and biochemistry; autoimmune disorders and cancer; flow cytometry; sterile sorts; tissue culture; phenotypic characterization with monclonal antibodies; functional assays-cytotoxicity; ADDC proliferation; SDS-PAGE; cell separations; immunoprecipitation. *Mailing Add:* Irwin Mem Blood Ctrs 270 Masonic Ave PO Box 18718 San Francisco CA 94118

GADSBY, DWIGHT MAXON, b Dickens, Iowa, Oct 2, 32; m 61; c 2. AQUACULTURE, ECONOMICS OF RESOURCE CONSERVATION. *Educ:* Iowa State Univ, BS, 54, MS, 60. *Prof Exp:* Fulbright scholar, Univ Padova, 61-62; assoc agr econ, Dept Econ, Iowa State Univ, 62-63; agr economist, Div Trade Anal, 63-65, SUPVRY ECONOMIST, DIV ECON RES SERV, USDA, 65- *Concurrent Pos:* Coordr & lectr, USDA, 71-73 & 90. *Mem:* AAAS; Soil Conserv Soc Am. *Res:* Analysis of the economics of natural resource development; quantitative measurement of the impacts associated with rural economic development and associated community improvements. *Mailing Add:* 4711 Medora Dr Suitland MD 20746

GADSDEN, RICHARD HAMILTON, SR, b Denver, Colo, June 30, 25; m 53; c 6. BIOCHEMISTRY. *Educ:* Col Charleston, BS, 50; Med Col SC, MS, 52, PhD(biochem), 56. *Hon Degrees:* DLitt, Col Charleston, 87. *Prof Exp:* Instr org chem & biochem, 54-56, from asst prof to assoc prof biochem, 56-69, PROF BIOCHEM, PATH & LAB MED & DIR CLIN CHEM, MED UNIV SC, 69- *Honors & Awards:* Outstanding Contrib Educ, Am Asn Clin Chemists. *Mem:* fel AAAS; Am Chem Soc; fel Am Inst Chemists; Am Asn Clin Chemists; fel Asn Clin Scientists. *Res:* Clinical chemistry; toxicology; clinical analysis. *Mailing Add:* Dept Path & Lab Med Med Univ SC Charleston SC 29425

GADWOOD, ROBERT CHARLES, b Milwaukee, Wis, May 18, 52; m 75; c 2. MEDICINAL CHEMISTRY. *Educ:* Marquette Univ, BS, 74; Univ Wis, PhD(chem), 78. *Prof Exp:* NIH fel, Univ Pittsburgh, 78-80; asst prof chem, Northwestern Univ, 80-86; SR RES SCIENTIST, UPJOHN CO, 86- *Mem:* Am Chem Soc; AAAS. *Res:* Synthesis of new therapeutically useful agents. *Mailing Add:* Upjohn Co 301 Henrietta St Kalamazoo MI 49001

GADZUK, JOHN WILLIAM, b Philadelphia, Pa, Mar 28, 41; m 68; c 1. THEORETICAL SOLID STATE PHYSICS, SURFACE PHYSICS. *Educ:* Mass Inst Technol, BS, 63, MS, 65, PhD(solid state physics), 68. *Prof Exp:* Instr physics, Mass Inst Technol, 67-68; PHYSICIST, NAT BUR STANDARDS, 68- *Concurrent Pos:* Nordita vis prof, Chalmers Univ, Goteborg, Sweden, 78- *Honors & Awards:* Arthur S Fleming Award, 78. *Mem:* Am Phys Soc; Am Vacuum Soc. *Res:* Theory of solids; solid surfaces; chemisorption; photoemission; electron spectroscopies of solids; chemical dynamics at surfaces. *Mailing Add:* Surface Sci Div Nat Bur Standards Ctr Chem Phys Gaithersburg MD 20899

GAEDKE, RUDOLPH MEGGS, b Dallas, Tex, Apr 4, 38; m 60; c 3. NUCLEAR PHYSICS. *Educ:* Rice Univ, BA, 60; Univ SC, PhD(physics), 65. *Prof Exp:* Asst prof physics, Trinity Univ, Tex, 65-67; sr staff mem, Electro-Nuclear Div, Oak Ridge Nat Lab, 67; from asst prof to assoc prof, Trinity Univ, Tex, 67-75, chmn dept, 68-76 & 85-88, dir instnl res & planning, 76-80, assoc vpres acad affairs, 80-84, chmn dept, 85-88, PROF PHYSICS, TRINITY UNIV, TEX, 75- *Concurrent Pos:* Res scientist, Tex A&M Univ, 91. *Res:* Nuclear reaction induced by heavy ions; gamma spectroscopy; educational administration; elementary particles. *Mailing Add:* Dept Physics Trinity Univ 715 Stadium Dr San Antonio TX 78284

GAENSLER, EDWARD ARNOLD, b Vienna, Austria, Feb 5, 21; nat US; m 53; c 2. SURGERY, PHYSIOLOGY. *Educ:* Haverford Col, BS, 42; Harvard Univ, MD, 45. *Prof Exp:* From instr to assoc prof, 48-62, PROF SURG, SCH MED, BOSTON UNIV, 62- *Concurrent Pos:* USPHS fel, 48-50; Trudeau fel, 55-59; lectr, Harvard Med Sch & Sch Med, Tufts Univ, 60-; vis thoracic surgeon, Boston City Hosp & Univ Hosp; NIH res career award, 64- *Honors & Awards:* Edward Livingston Trudeau Gold Medal, 81. *Mem:* Fel Am Col Surg; fel Am Col Chest Physicians; Am Thoracic Soc; Am Soc Thoracic Surg; Am Fedn Clin Res. *Res:* Pulmonary physiology; physiology in thoracic surgery; chest radiology and pathology; interstitch and occupational lung disease. *Mailing Add:* Med Sch Boston Univ 80 E Concord St Boston MA 02118

GAER, MARVIN CHARLES, b Milwaukee, Wis, Apr 28, 35; m 60; c 3. PURE & APPLIED MATHEMATICAL ANALYSIS. *Educ:* Univ Wis, BS, 57, MS, 62; Univ Ill, PhD(math), 68; Univ Md, JD, 78. *Prof Exp:* Physicist, Allen-Bradley Co, Wis, 57-58; comput analyst, AC Electronics Div, Gen Motors Corp, 58-62; from instr to asst prof math, Univ Del, 67-75; math analyst, BDM Corp, Va, 77-79; PATENT ATTY, 79-; MATHEMATICIAN, US NAVAL AIR DEVELOP CTR, PA, 83- *Concurrent Pos:* Adj assoc prof math, Drexel Univ, 81-82. *Mem:* Am Math Soc; Math Asn Am; Soc Indust & Appl Math; Sigma Xi. *Res:* Mathematical analysis and its applications primarily in the area of signal processing in oceanographic acoustics. *Mailing Add:* 1756 Morris Dr Cherry Hill NJ 08003

GAERTNER, FRANK HERBERT, b San Rafael, Calif, July 12, 38; m 62; c 3. BIOCHEMISTRY, MICROBIOLOGY. *Educ:* Univ Ariz, BS, 61, MS, 62; Purdue Univ, PhD(microbiol), 66. *Prof Exp:* Res biologist biochem, Univ Calif, San Diego, 66-69; from asst prof to assoc prof biochem, Oak Ridge Grad Sch Biomed Sci, 69-81; DIR, NEW TECHNOLOGIES RES, MYCOGEN CORP. *Concurrent Pos:* USPHS fel, NIH, 66-68; NSF grant, 77-80; mem staff, Salk Inst, San Diego, 81- *Mem:* Am Soc Microbiol; Am Soc Biol Chemists. *Res:* Unique catalytic and structural properties of multienzyme clusters; regulation of the metabolism of aromatic compounds; plasmid gene expression in molds, yeast and bacteria; molecular genetics of bacillus thuringiensis toxins; delivery systems for biopesticides. *Mailing Add:* Dept Molecular Genetics Mycogen Corp 5451 Oberlin Dr San Diego CA 92109

GAERTNER, RICHARD F(RANCIS), b Pittsburgh, Pa, Aug 10, 33; m 62; c 4. MATERIALS SCIENCE, CHEMICAL ENGINEERING. *Educ:* Univ WVa, BS, 55, MS, 57; Univ Ill, PhD(chem eng), 59. *Prof Exp:* Res assoc, Gen Elec Res Lab, Gen Elec Co, 59-68, mgr tech mkt, Polymer Prod Oper, 68-69, mgr, Mfg Plastics Dept, 69-71, mgr eng, Laminating Metals Dept, 71-72, mgr tech planning, Chem & Metall Div, 72-75; res dir, Chem & Fiber Technol, 75-79, dir strategic tech planning, Owens-Corning Fiberglas corp, 79-86; DIR, CTR ADVAN TECHNOL DEVELOP, IOWA STATE UNIV, 86- *Mem:* Am Inst Chem Engrs; Am Chem Soc. *Res:* Heat transfer associated with change of phase; boiling phenomena; process engineering of polymers; glass reinforced plastic composites; glass reinforcements and yarns; research administration. *Mailing Add:* Appl Sci Complex II Iowa State Univ Ames IA 50011

GAERTNER, VAN RUSSELL, b Centralia, Ill, July 4, 24; m 52; c 3. ORGANIC CHEMISTRY. *Educ:* Bradley Univ, BS, 45; Univ Ill, PhD(org chem), 48. *Prof Exp:* Teacher inorg chem, Univ Ill, 45-47; NIH fel, Ohio State Univ, 48-49; asst prof chem, Univ Ore, 49-52; res chemist & group leader, Monsanto Agr Prod Co, 52-66, sr res specialist, 66-67, scientist, 67-70, unit leader, Res Dept, 71-76, sr res group leader, 76-81, sci fel, 81-87; RETIRED. *Mem:* Am Chem Soc; Am Inst Chemists. *Res:* Organometallic reactions; synthesis of carcinogenic hydrocarbons; heterocycles; Diels-Alder reactions of quinones; biological toxicants; surfactants; epoxyamines; paper chemicals; reactive polymers; small nitrogen heterocycles; amino phosphonates; herbicide and plant growth regulator synthesis. *Mailing Add:* 346 Sudbury Lane Ballwin MO 63011

GAERTNER, WOLFGANG WILHELM, b Vienna, Austria, July 5, 29; US citizen; m 55; c 3. SOLID STATE PHYSICS, ELECTRONICS. *Educ:* Univ Vienna, PhD(physics), 51; Vienna Tech Inst, dipl eng, 55. *Prof Exp:* Physicist, Siemens & Halske, Austria, 51-53; physicist solid state physics, US Army Signal Corps Res & Develop Labs, NJ, 53-59, chief scientist, Solid State Div, 59-60; vpres solid state physics & electronics, Columbia Broadcasting Syst, Inc, 60-65; PRES, W W GAERTNER RES INC, 65- *Mem:* Am Phys Soc; fel Inst Elec & Electronics Engrs. *Res:* Semiconductor devices; microelectronics; adaptive electronics; system design and software. *Mailing Add:* W W Gaertner Res Inc 140 Water SN Stamford CT 06903

GAERTTNER, MARTIN R, b Winchester, Mass, Apr 21, 44. ELECTRICAL ENGINEERING. *Educ:* Rensselaer Polytech Inst, BS, 65; Cornell Univ, PhD(exp physics), 70. *Prof Exp:* RES SCIENTIST, ENG & RES STAFF, FORD MOTOR CO, 73- *Mem:* Am Phys Soc. *Res:* Electromagnetic compatibility; high vacuum technology (radio frequency sputtering and ion bombardment evaporative deposition of thin films); electron tunneling; development of superconducting quantum interference detectors. *Mailing Add:* 2601 Detroit St Dearborn MI 48124

GAFARIAN, ANTRANIG VAUGHN, b Fresno, Calif, Dec 26, 24; c 2. MATHEMATICAL STATISTICS, PROBABILITY. *Educ:* Univ Mich, BS(eng math) & BS(mech eng), 47; Univ Calif, Los Angeles, PhD(math), 59. *Prof Exp:* Aerodynamicist, NAm Aviation, Inc, 47-51 & Northrop Aircraft, Inc, 51-52; aerodynamics engr, Aerophysics Develop Corp, 52-53; mem tech staff, Hughes Aircraft Co, 56-59; sr mathematician, Syst Develop Corp, 59-66, sr scientist, 66-69, head math & opers res prog staff, 67-69, mgr transp & telecommun dept, 69-73; PROF INDUST & SYSTS ENG, UNIV SOUTHERN CALIF, 73- *Concurrent Pos:* Teacher phys sci exten prog, Univ Calif, Los Angeles, 56-65; assoc prof, Calif State Univ, Northridge, 61-65. *Mem:* Opers Res Soc Am; Inst Indust Engrs; Sigma Xi. *Res:* Application of the mathematical theory of probability and mathematical statistics to the analysis of random phenomena. *Mailing Add:* 1033 Ocean Ave No 304 Santa Monica CA 90403

GAFFAR, ABDUL, b Dec 10, 39; US citizen. IMMUNOCHEMISTRY. *Educ:* Univ Karachi, BS, 56; Brigham Young Univ, MS, 65; Ohio State Univ, PhD(immunol), 67. *Prof Exp:* Res assoc biochem, Pakistan Res Coun, 60-63; res asst bact, Brigham Young Univ, 64-65; res asst immunol, Ohio State Univ, 65-67; res microbiologist, 67-70, sr res microbiologist, 70-73, res assoc, 73-75, sr assoc, 80, CORP RES FEL & ASSOC DIR, COLGATE-PALMOLIVE RES CTR, 81- *Mem:* AAAS; NY Acad Sci; Am Soc Microbiol. *Res:* Immunochemistry of tissue antigens and microbial products. *Mailing Add:* Colgate-Palmolive Res Ctr 909 River Rd Piscataway NJ 08854

GAFFEY, CORNELIUS THOMAS, b Philadelphia, Pa, Aug 24, 28; m 55. BIOPHYSICS. *Educ:* La Salle Col, BA, 49; Univ Tenn, MS, 51; Purdue Univ, PhD(biophys), 58. *Prof Exp:* Asst physiol, Purdue Univ, 51-52, instr physics, 58; BIOPHYSICIST PHYSICS, DONNER LAB, LAWRENCE BERKELEY LAB, UNIV CALIF, 58- *Mem:* Biophys Soc; Radiation Res Soc; Soc Neurosci; Tissue Cult Asn; Soc Exp Biol & Med. *Res:* Effects of irradiation on the nervous system; computer analysis of brain electrical activity; microelectrode techniques with central nervous system tissue cultures; excitation phenomenon in plant cell membranes; effects on non-ionizing radiation on biological systems. *Mailing Add:* 15 Roslyn Ct Berkeley CA 94670

GAFFEY, MICHAEL JAMES, b Meadville, Pa, Dec 1, 45; m 73. PLANETARY ASTRONOMY, REMOTE SENSING. *Educ:* Univ Iowa, BA, 68, MS, 70; Mass Inst Technol, PhD(planetary sci), 74. *Prof Exp:* Mem res staff planetary astron, Mass Inst Technol, 74-77; asst astronr planetary astron, Inst Astron, 77-80, asst researcher planetary geosciences, 80-81, assoc res planetary geosci, Hawaii Inst Geophys, Univ Hawaii, 81-84; PROF GEOL, RENSSELAER POLYTECH INST, 84- *Concurrent Pos:* Consult, Lunar & Planetary Rev Panel, NASA, 77-79. *Honors & Awards:* Asteroid named in honor, 3545 Gaffey. *Mem:* Fel Meteoritical Soc; Am Astron Soc; Am Geophys Union; AAAS. *Res:* Surface mineralogy of minor planets by remote sensing; early solar system processes from study of minor planets; utilization of extraterrestrial resources; remote sensing for terrestrial resources. *Mailing Add:* Dept Geol West Hall Rensselaer Polytech Inst Troy NY 12181

GAFFEY, WILLIAM ROBERT, b Jersey City, NJ, Feb 10, 24; m 49; c 3. BIOSTATISTICS, BIOMETRY. *Educ:* Univ Calif, AB, 48, PhD(math statist), 54. *Prof Exp:* Asst res biostatistician, Univ Calif, 54-55, from instr to asst prof biostatist, 55-60; statist consult, Calif Dept Pub Health, 60-69; sr biostatist consult, Inst Med Sci, Pac Med Ctr, 69-71; spec consult, Calif State Dept Pub Health, 71-72; dir health & epidemiol studies, Equitable Environ Health, Inc, 72-76; sr epidemiologist, SRI Int, 76-79; epidemiol dir, Monsanto Co, 79-89; RETIRED. *Mem:* Royal Soc Health; Am Statist Asn; Inst Math Statist. *Res:* Statistical problems in occupational health. *Mailing Add:* 11269 Pineside Dr St Louis MO 63146

GAFFIELD, WILLIAM, b Chicago, Ill, Dec 26, 35; m 61; c 2. STRUCTURE-TERATA RELATIONSHIPS OF PLANT ALKALOIDS. *Educ:* Univ Ill, Urbana, Champaign, BS, 57; Univ Iowa, PhD(chem), 63. *Prof Exp:* Res assoc org chem, Univ NH, Durham, 62-64; RES CHEMIST, WESTERN REGIONAL RES CTR, USDA, 64- *Concurrent Pos:* Vis res chemist, Nat Cancer Inst, NIH, Bethesda, 74-75. *Mem:* Fel Royal Soc Chem; Am Chem Soc; Am Soc Pharmacog. *Res:* Structure-terata relationships of plant alkaloids; chiroptical determination of natural product configuration and conformation. *Mailing Add:* Western Regional Res Ctr Agr Res Serv USDA Albany CA 94710

GAFFNEY, BARBARA LUNDY, b Houston, Tex, Dec 31, 44; m 66. ORGANIC CHEMISTRY, BIOCHEMISTRY. *Educ:* Rutgers Univ, BA, 66, PhD(org chem), 70. *Prof Exp:* Res assoc, Inst Microbiol, 69-70, lectr chem, Douglass Col, 70-84, ASSOC PROF, RUTGERS UNIV, 85- *Honors & Awards:* Joseph Hyman Award, Am Chem Soc, 89. *Mem:* Am Chem Soc; Biophysical Soc. *Res:* Synthesis and characterization of modified oligonucleotides. *Mailing Add:* Dept Chem Rutgers Univ New Brunswick NJ 08903

GAFFNEY, BETTY JEAN, biophysical chemistry, for more information see previous edition

GAFFNEY, EDWIN VINCENT, CANCER, IMMUNOLOGY. *Educ:* Cath Univ, PhD(cell biol), 68. *Prof Exp:* ASSOC PROF CELL BIOL, PA STATE UNIV, 71- *Mailing Add:* Dept Cell Biol & Microbiol Penn St Univ University Park PA 16802

GAFFNEY, F(RANCIS) J(OSEPH), b Middleton, Eng, June 27, 12; nat US; m 38; c 3. ENGINEERING. *Educ:* Northeastern Univ, BS, 35. *Prof Exp:* Chief engr, Browning Labs, Inc, 37-41; mem staff, Radiation Lab, Mass Inst Technol, 41-45; gen mgr, Polytech Res & Develop Co, 45-53; dir eng, Fairchild Guided Missiles Div, 53-55; vpres, Teleregister Corp, 55-59; vpres & gen mgr, FXR, Inc, 59; tech dir, NAm Philips Corp, 60-67, dir corp prod planning, 67-71, pres, Commun Systs Div, 71-75; RETIRED. *Concurrent Pos:* consult, Off Asst Secy Defense. *Mem:* Fel Inst Elec & Electronics Engrs. *Res:* Design of radio and radar equipment; measuring instruments; microwave equipment; guidance and control systems; engineering management. *Mailing Add:* 205 S Pebble Beach Blvd Sun City Center FL 33570

GAFFNEY, JEFFREY STEVEN, b San Bernardino, Calif, July 28, 49; m 90; c 3. PHYSICAL ORGANIC CHEMISTRY. *Educ:* Univ Calif, Riverside, BS, 71, MS, 73, PhD(chem), 75. *Prof Exp:* Res assoc, Dept Chem, 75-77, from assoc chemist to chemist, dept energy & environ, Brookhaven Nat Lab, 77-85; staff mem, Isotope & Nuclear Chem Div, Los Alamos Nat Lab, 85-88; CHEMIST, GROUP LEADER, ENVIRON RES DIV, ARGONNE NAT LAB, 89- *Mem:* Am Chem Soc; Am Asn Aerosol Res; Am Geophys Union. *Res:* Gas chromatography/mass spectrometry; isotope ratio mass spectrometry; gas phase kinetics; atmospheric chemistry; air pollution monitoring; mass spectroscopy of fragile molecules of biological interest; organic geochemistry; combustion chemistry; radiochem. *Mailing Add:* Bldg 203 Environ Res Div Argonne Nat Lab 9700 S Cass Ave Argonne IL 60439

GAFFNEY, PATRICK M, b New Orleans, La, Dec 3, 51; m 78; c 3. AQUACULTURE GENETICS, BIVALVE GENETICS. *Educ:* Univ Calif, Berkeley, AB, 73; State Univ NY, PhD(biol sci), 86. *Prof Exp:* Fel, State Univ NY, 86-87; ASST PROF MARINE BIOL, COL MARINE STUDIES, UNIV DEL, 87- *Concurrent Pos:* Fulbright fel, 73. *Mem:* Genetics Soc Am; Nat Shellfisheries Asn; Soc Study Evolution; World Aquacult Soc; Int Asn Genetics Aquacult. *Res:* Genetics of marine organisms, emphasizing population genetics of bivalves and applications of genetics to aquaculture. *Mailing Add:* Col Marine Studies Univ Del Lewes DE 19958

GAFFNEY, PAUL COTTER, b Du Bois, Pa, May 12, 17; m 44; c 7. MEDICINE. *Educ:* Univ Pittsburgh, BS, 40, MD, 42. *Prof Exp:* From instr to assoc prof, 51-62, dean admis sch med, 77-78, PROF PEDIAT, SCH MED, UNIV PITTSBURGH, 61-; EXEC DIR, MED ALUMNI ASN,81- *Concurrent Pos:* Sr staff, Children's Hosp, Philadelphia, Pa, 53-, med dir, 78-81. *Res:* Hematology. *Mailing Add:* Dept Pediat Scaife Hall M-250 Univ Pittsburgh 4200 Fifth Ave Pittsburgh PA 15261

GAFFNEY, PAUL G, II, b Attleboro, Mass, May 30, 46; m 74; c 1. OCEANOGRAPHY, ATMOSPHERIC SCIENCES. *Educ:* US Naval Acad, BS, 68; Cath Univ Am, MSE, 69; Naval War Col, dipl, 79; Jacksonville Univ, MBA, 86. *Prof Exp:* Res asst ocean eng, Cath Univ Am, 68-69; opers officer minesweeping, US Navy, 69-71; adv ocean hydrography, Vietnamese Navy Combat Hydro Surv Team, 71-72; ocean serv officer, Fleet Weather Cent, Rota, Spain, 72-75; exec asst to Oceanogr USN, 75-78; res fel Soviet naval affairs, Naval War Col, 78-79; cmndg officer hydrography, Oceano Unit Four, Indonesian Surv, 79-80; actg dir earth sci res, Off Naval Res, 80-81; mil asst int security affairs, Off Asst Secy Defense, 81-83; exec officer, Naval Res Lab, 83-84, cmndg officer, Naval Oceanog Command Facil, 84-86, dir resources & capt naval oceanog, Naval Res Lab, 86-89, asst chief naval res, 89-91, CMNDG OFFICER, NAVAL RES LAB, 91- *Mem:* Explorers Club; Oceanic Soc; Sigma Xi. *Res:* Field hydrographic survey operations in remote areas; mechanical properties of cables. *Mailing Add:* Cmndg Officer Naval Research Lab 4555 Overlook Ave SW Washington DC 20375-5000

GAFFNEY, PETER EDWARD, b Carbondale, Pa, Nov 24, 31; m 55; c 3. MICROBIOLOGY. *Educ:* Univ Scranton, BS, 53; Syracuse Univ, MS, 54; Rutgers Univ, PhD(environ sci), 58. *Prof Exp:* From asst prof to assoc prof appl biol res & eng exp sta, Ga Inst Technol, 58-65; ASSOC PROF MICROBIOL, GA STATE UNIV, 65- *Concurrent Pos:* Fulbright lectr, Trinity Col, Dublin, 67-68; consult, indust microbiol; pres, Southeast Labs, Inc, 71-78, Ga Environ Asn, Inc, 87; chmn, Analytical Serv, Inc, 78-83; exec secy, Ga Acad Sci, 88- *Mem:* AAAS; Am Soc Microbiol; Water Pollution Control Fedn; Sigma Xi; NY Acad Sci; fel Am Acad Microbiol. *Res:* Microbial physiology; growth kinetics; stream pollution; waste treatment; microbial deterioration of industrial products; industrial fermentations; effects of chlorobiphenyls on waste treatment processes. *Mailing Add:* Ga State Univ Plaza No 57 Atlanta GA 30302

GAFFNEY, THOMAS EDWARD, b East St Louis, Ill, Nov 5, 30; m 54; c 3. PHARMACOLOGY. *Educ:* Univ Mo, AB, 51, MS, 53; Univ Cincinnati, MD, 57. *Prof Exp:* Intern, Harvard Med Serv, Boston City Hosp, 58-59; asst resident medicine, Mass Gen Hosp, 59-60; instr pharmacol & asst med, Col Med, Univ Cincinnati, 60-61; clin assoc, Cardiol Br, Nat Heart Inst, 61-62; from asst prof to prof med, Col Med, Univ Cincinnati, 62-72; PROF PHARMACOL & MED & CHMN DEPT BASIC & CLIN PHARMACOL, MED UNIV SC, 72- *Concurrent Pos:* Clinician, Outpatient Dept & attend physician, Cincinnati Gen Hosp, 62-64, dir hypertension clin, 64-72; from assoc prof to prof pharmacol, Col Med, Univ Cincinnati, 62-72, dir div clin pharmacol, 62-72; res career develop award, Cincinnati Gen Hosp, 62-; mem med adv bd, Coun High Blood Pressure Res, Am Heart Asn; vis scientist toxicol, Mass Spectrometry Lab, Karolinska Inst, Stockholm, 69-70; mem & chmn prog rev comt pharmacol & toxicol, Nat Inst Gen Med Sci. *Honors & Awards:* Myrtle Wreath Achievement Award, Hadassah Res, 80. *Mem:* Soc Exp Biol & Med; Am Soc Pharmacol & Exp Therapeut; Am Fedn Clin Res; Protein Soc; Am Soc Clin Invest; Am Clin & Climat Assoc. *Res:* Clinical cardiovascular pharmacology; hypertension. *Mailing Add:* Dept Pharmacol Med Univ SC 171 Ashley Ave Charleston SC 29425

GAFFORD, LANELLE GUYTON, b Columbia, Miss, Feb 27, 30. VIROLOGY, MEDICAL MYCOLOGY. *Educ:* Univ Miss, BS, 51, MS, 55, PhD(microbiol), 69. *Prof Exp:* From instr to prof microbiol, Med Ctr, Univ Miss, 57-83; RETIRED. *Mem:* Am Soc Microbiol; Am Soc Exp Path; Med Mycol Soc Am; AAAS; Soc Gen Microbiol. *Res:* Biochemistry of avian poxvirus replicaton and hyperolasia; chemical characterization of fowlpox virus; physicochemical characterization of the viral DNA and mapping of the genome. *Mailing Add:* 1042 Meadowbrook Rd Jackson MS 39206

GAFFORD, WILLIAM R(ODGERS), civil engineering; deceased, see previous edition for last biography

GAGE, ANDREW ARTHUR, b Buffalo, NY, May 7, 22. SURGERY, ADMINISTRATION. *Educ:* Univ Buffalo, MD, 44. *Prof Exp:* Staff surgeon, Vet Admin Med Ctr, 53-68, chief surg serv, 68-84, chief staff, 71-86; PROF SURG, SCH MED, STATE UNIV NY, BUFFALO, 72-; DEP INST DIR, ROSWELL PARK MEM INST, BUFFALO, NY, 87- *Concurrent Pos:* Assoc dir clin affairs, Roswell Park Mem Inst, 86-87. *Mem:* Int Cardiovasc Soc; Soc Vascular Surg; Am Heart Asn; Soc Cryobiol; Am Col Surg. *Res:* Pacemaker work; cryosurgery; long term assistance of failing heart. *Mailing Add:* Roswell Park Mem Inst 666 Elm St Buffalo NY 14263

GAGE, CLARKE LYMAN, b Mason City, Iowa, Apr 20, 21; m 44; c 3. ORGANIC CHEMISTRY. *Educ:* Antioch Col, BS, 44; Ohio State Univ, PhD(chem), 51. *Prof Exp:* Res chemist, Polystyrene Copolymers, Monsanto Chem Co, 43-44; asst porphyrins, C F Kettering Found, 46; Vernay fel, 52; from asst prof to prof, 52-86, EMER PROF CHEM, ST LAWRENCE UNIV, 86- *Mem:* Am Chem Soc; Hist Sci Soc; Brit Soc Hist Sci; Sigma Xi. *Res:* Pyrrole chemistry; spirane synthesis; stereochemistry; history of science (chemistry). *Mailing Add:* Six Elm St Canton NY 13617

GAGE, DONALD S(HEPARD), b Evanston, Ill, June 10, 30; m 56; c 3. ELECTRICAL ENGINEERING. *Educ:* Northwestern Univ, BSEE, 53; Stanford Univ, MSEE, 54, PhD(elec eng), 58. *Prof Exp:* Asst elec eng, Stanford Univ, 55-58; from asst prof to assoc prof, Northwestern Univ, 58-62; assoc prof, Mich State Univ, 63-66; from assoc prof to prof elec eng, Colorado Springs, 66-77, PROF ELEC ENG & COMPUT SCI, UNIV COLO, DENVER, 77- *Concurrent Pos:* NSF fel, 53-55; consult, IBM, Boulder, Colo, 74, Hewlett Packard Corp, Colorado Springs, 84-85. *Mem:* Inst Elec & Electronics Engrs. *Res:* Explanation of semiconductor devices through an understanding of applied solid state physics with special interest in avalanche and transient radiation effects; automated material measurements at 94 gigahertz. *Mailing Add:* Elec Eng Univ Colo Campus Box 110 Denver CO 80217-3364

GAGE, FREDERICK WORTHINGTON, b Cleveland, Ohio, Dec 27, 12; m 40, 74; c 2. CHEMISTRY. *Educ:* Ill Wesleyan Univ, BS, 34; Northwestern Univ, MS, 37. *Prof Exp:* Res supvr, Columbia-Southern Chem Corp, 36-54; tech dir, Dayton Chem Prod Labs, Inc, Whittaker Corp, 54-58, tech dir & vpres, 58-70, gen mgr, Dayton Coatings & Chem Div, 70-80; RETIRED. *Mem:* Am Chem Soc. *Res:* Rubber reinforcing pigments; heavy chemicals; caustic soda; calcium bleach; chlorine; sodium bicarbonate; calcium chloride; general analytical methods; preparation of calcium carbonate and calcium silicate; use of rubber pigments; purification caustic soda; purification titanium tetrachloride; rubber to metal bonding agents. *Mailing Add:* 6650 Carinthia Dr Dayton OH 45459

GAGE, KENNETH SEAVER, b Boston, Mass, Nov 11, 42; m 69; c 2. GEOPHYSICS, METEOROLOGY. *Educ:* Brandeis Univ, AB, 64; Univ Chicago, MS, 66, PhD(geophys fluid dynamics), 68. *Prof Exp:* Asst prof meteorol, Univ Md, College Park, 68-72; vis assoc prof, Eng Exp Sta & Meteorol Dept, Univ Wis-Madison, 72-73; mem staff, Meteorol Dept, Control Data Corp, 73-76; PHYSICIST, AERONOMY LAB, ENVIRON RES LABS, NAT OCEANIC & ATMOSPHERIC ADMIN, 76- *Concurrent Pos:* Assoc ed, J Appl Meteorol, 82-90, Radio Sci, 87-; ed, J Geophys Res Atmospheres, 88- *Mem:* Am Meteorol Soc; Am Phys Soc; Am Geophys Union. *Res:* Geophysical fluid dynamics; hydrodynamic stability and turbulence theory; atmospheric turbulence and gravity waves; remote probing of the atmosphere; radar meteorology; atmospheric propagation of electromagnetic waves; mesoscale processes; tropical dynamics and climate variability. *Mailing Add:* Aeronomy Lab Environ Res Labs Boulder CO 80303

GAGE, L PATRICK, b Endicott, NY, May 4, 42; m 65, 85; c 4. MOLECULAR BIOLOGY, BIOCHEMISTRY. *Educ:* Mass Inst Technol, SB, 64; Univ Chicago, PhD(biophys), 69. *Prof Exp:* USPHS fel, Dept Embryol, Carnegie Inst, Washington, DC, 69-71; asst mem, Roche Inst Molecular Biol, 71-76, assoc mem cell biol, 77-80, head, Lab Recombinant DNA Res, 80-82; dir, Dept Molecular Genetics, 82-83, vpres & dir biol res & develop, 83-84, VPRES & DIR EXPLOR RES, ROCHE RES CTR, 84- *Concurrent Pos:* Adj assoc prof, Biochem PhD Prog, City Univ New York, 75-76; mem adv panel develop biol, NSF, 77-80; adj assoc prof cell biol, Sch Med, NY Univ, 77-; chmn biotechnology adv comt, Pharmaceut Mfrs Asn; bd mem, Life Sci Res Found. *Mem:* Soc Develop Biol; NY Acad Sci; Am Soc Biol Chemists; Am Soc Cell Biol; Sigma Xi; Indust Biotechnology Asn. *Res:* Responsible for all pharmaceutical discovery research, including synthetic chemistry, pharmacology, chemotherapy, immunology, virology, oncology, molecular genetics and biological chemistry; research in support of the fine chemicals, animal health and diagnostics businesses of the company; special interest in biotechnology, molecular immunology and virology. *Mailing Add:* Hoffmann-La Roche Bldg 76 Nutley NJ 07110

GAGE, TOMMY WILTON, b Stamford, Tex, Oct 6, 35; m 56; c 4. PHARMACOLOGY, PHYSIOLOGY. *Educ:* Univ Tex, Austin, BS, 57; Baylor Univ, DDS, 61, PhD(physiol), 69. *Prof Exp:* NIH spec fel, 66-69; PROF PHARMACOL & CHMN DEPT, COL DENT, BAYLOR UNIV, 69- *Concurrent Pos:* Mem, Coun on Dent Therapeut, Am Dent Asn, 85-90. *Honors & Awards:* Cooley Trophy, Tex Dent Asn. *Mem:* Am Asn Dent Sch;

Am Dent Asn; Int Asn Dent Res; Am Soc Pharmacol & Exp Therapeut; fel Am Col Dentists. *Res:* Cellular responses to oxygen and oxygenating agents, limited to tissues in the oral cavity; pain related to the oral cavity. *Mailing Add:* Dept Pharmacol Baylor Col Dent Dallas TX 75246

GAGEN, JAMES EDWIN, b Elyria, Ohio, Dec 27, 35; m 59; c 4. ORGANIC CHEMISTRY. *Educ:* Kent State Univ, BS, 57; Case Western Reserve Univ, MS, 64, PhD(org chem), 67. *Prof Exp:* Chemist, B F Goodrich Chem Co, 59-65; asst prof, 67-74, ASSOC PROF CHEM, UNIV TENN, MARTIN, 74- *Mem:* Am Chem Soc; Sigma Xi. *Res:* Aromatic substitution by sulfonyl peroxides. *Mailing Add:* Rte 4 Box 351 Tenn Martin TN 38237

GAGER, FORREST LEE, JR, b Philadelphia, Pa, Apr 23, 22; m 50, 86; c 2. ORGANO-ANALYTIC CHEMISTRY, NATURAL PRODUCTS CHEMISTRY. *Educ:* Haverford Col, BS, 49; Ind Univ, AM, 51. *Prof Exp:* Res chemist, Merck & Co, Inc, 51-55; res chemist, Philip Morris, Inc, 55-57, supvr org sect, 57-59, res assoc, 59-60, res scientist, 60-63, sr scientist, 63-85; RETIRED. *Concurrent Pos:* Vis fel, Univ Manchester, 65-66. *Mem:* AAAS; Am Chem Soc; Royal Soc Chem; Sigma Xi. *Res:* Chemistry of natural products; flavors and aromas; tobacco chemistry; chromatography; biosynthesis; isotope tracer techniques. *Mailing Add:* PO Box Z Sweet Briar VA 24595

GAGER, WILLIAM BALLANTINE, b Columbus, Ohio, April 18, 28; m 56; c 2. PHYSICS. *Educ:* Bowdoin Col, AB, 49; Ohio State Univ, MS, 51, PhD(physics), 56. *Prof Exp:* Physicist, Nat Bur Standards, 56-58 & Battelle Mem Inst, 58-67; ASSOC PROF PHYSICS, JACKSONVILLE UNIV, 67- *Mem:* Am Phys Soc. *Res:* Paramagnetic resonance; nuclear magnetic resonance; very low temperatures; solid state physics. *Mailing Add:* 10135 Lakeview Rd W Jacksonville FL 32211

GAGGE, ADOLF PHARO, b Columbus, Ohio, Jan 11, 08; m 36; c 4. BIOPHYSICS, PHYSIOLOGY. *Educ:* Univ Va, BA, 29, MA, 30; Yale Univ, PhD(physics), 33. *Prof Exp:* Instr math, Univ Va, 23-30; asst physics, Yale Univ, 30-31, from instr to asst prof biophys, 36-41; chief biophys br, Aero Med Lab, USAF, Wright Field, 42-45, chief aeromed res & opers, 45-50, chief med res div, Off Surgeon Gen, 50-51, chief, Human Factors & dir res & develop, 51-55, dep comdr sci, Off Sci Res, 55-60, comdr sci, Off Sci Res, 55-60, comdr, 60; prog mgr, Advan Res Proj Agency, Off Secy Defense, 60-63; assoc prof physiol, 63-69, prof environ health, 69-76, dep dir, John B Pierce Found Lab, 71-78, EMER PROF ENVIRON HEALTH, SCH MED, YALE UNIV, 76-, EMER FEL & CONSULT, JOHN B PIERCE FOUND LAB, 78- *Concurrent Pos:* Asst, Univ Va, 29-30; fel physics, Yale Univ, 33-35; res physicist, John B Pierce Found, Conn, 33-41, fel, 63-64; mem subcomt, Altitude Chambers, Nat Res Coun, 41; mem panel aviation med, Aeronaut Bd, 47; mem, Res & Develop Bd, 50-53; mem, Comt Hearing, Bioacoust & Biomech, Nat Acad Sci-Nat Res Coun, 66-67 & 72-75; sect ed, Am Physiol Soc J, 72-75; assoc ed, J Appl Physiol, 75-84. *Mem:* Nat Acad Eng; fel AAAS; Am Physics Soc; Am Physiol Soc; Biophys Soc; fel Am Soc Heating & Refrig Eng; fel Aerospace Med Asn. *Res:* Scattering of electrons in gases; atmospheric ionization; environmental physiology; exercise; partitional calorimetry; human temperature regulation and thermal comfort; aviation physiology; biometeorology; bio-astronautics. *Mailing Add:* John B Pierce Found Lab 290 Congress Ave New Haven CT 06519

GAGGIOLI, RICHARD A, b Lake Forest, Ill, Dec 3, 34; m 57; c 5. MECHANICAL ENGINEERING. *Educ:* Northwestern Univ, BSME, 57, MS, 58; Univ Wis, PhD(mech eng), 61. *Prof Exp:* NSF fel chem eng, Univ Wis, Madison, 61-62, from asst prof to assoc prof, mech eng, 62-69; res mem, US Army Math Res Ctr, 65-67; prof mech eng, Marquette Univ, 69-81, chmn dept, 69-72; dean eng & archit, Catholic Univ Am, 81-84; PROF MECH ENG, UNIV IOWELL, 84- *Concurrent Pos:* Vis fel, Battelle Mem Inst, 68-69. *Mem:* Am Soc Mech Engrs; Am Soc Heating, Refrig & Air Conditioning Engrs. *Res:* Applied thermodynamics; energy engineering; applied mathematics. *Mailing Add:* Dept Mech Eng Univ Lowell University Ave Lowell MA 01854

GAGLIANO, NICHOLAS CHARLES, b New Orleans, La, Apr 5, 27; m 51; c 2. PEDIATRICS, PREVENTIVE MEDICINE. *Educ:* Loyola Univ, La, BS, 48; La State Univ, MD, 52; Am Bd Pediat, dipl, 61. *Prof Exp:* From intern to resident, Charity Hosp La, New Orleans, 52-55; practicing pediatrician, 55-60; assoc prof pediat, La State Univ, 60-70, assoc prof prev med, 64-70, prof pediat & prev med, 70-, prof pub health, 77-, PROF FAMILY MED, 80-, EMER PROF PEDIAT, LA STATE UNIV, 88- *Concurrent Pos:* Asst proj dir collab child develop prog, Charity Hosp La, 61-64; med dir, Pediat Emergency Rm, Charity Hosp, New Orleans, 76-; vchmn & assoc dir, Residency Training Prog, Dept of Pediat, La State Univ, 84-88. *Mem:* Am Acad Pediat. *Res:* Preventive and ambulatory pediatrics. *Mailing Add:* Dept Pediat Sch Med La State Univ 1542 Tulane Ave New Orleans LA 70112

GAGLIANO, SHERWOOD MONEER, b New Orleans, La, Dec 10, 35; m 61. PHYSICAL GEOGRAPHY, GEOLOGY. *Educ:* La State Univ, BS, 59, MA, 63, PhD(geog), 67. *Prof Exp:* Supvr sediment lab, Coastal Studies Inst, La State Univ, 60-64, field investr, Coastal Geol, 64-65, instr coastal morphol, 65-67, asst prof coastal geog, 67-69, assoc prof, 69-73, asst prof marine sci, Univ, 69-73, instr geog & anthrop, 63-67, asst prof, 67-68; PRES, COASTAL ENVIRON, INC, 73- *Concurrent Pos:* Vis geologist, Am Geol Inst, 71; spec lectr, La State Univ, 73- *Mem:* Soc Am Archeol; Geol Soc Am; Soc Econ Paleontologists & Minerologists; US Quaternary Soc; Sigma Xi. *Res:* Coastal zone management; resource management; regional planning; environmental sciences; sedimentology; geomorphology and archeology of riverine and coastal areas; archeology of early man. *Mailing Add:* Coastal Environ Inc 929 E Lakeview Baton Rouge LA 70808

GAGLIARDI, GEORGE N(ICHOLAS), b Brooklyn, NY, May 17, 30; m 56; c 6. CHEMICAL ENGINEERING. *Educ:* Pratt Inst, BChE, 52; NY Univ, MChE, 61. *Prof Exp:* Chem engr process develop, Am Cyanamid Co, 52-56, res engr, 56-59, group leader process res agr chem, 59-66, mgr process develop, Agr Div, 66-68, PROD MGR PESTICIDES, AM CYANAMID CO, 68- *Mem:* Am Inst Chem Engrs; Am Chem Soc. *Res:* Process research on and manufacture of organic chemicals scaleup; techniques. *Mailing Add:* 16 Vander Veer Dr Lawrenceville NJ 08648

GAGLIARDI, L JOHN, b Camden, NJ, Dec 13, 35; m 61, 84; c 5. CHEMICAL PHYSICS, BIOPHYSICS. *Educ:* Villanova Univ, BS, 57; Temple Univ, MA, 65, PhD(physics), 71. *Prof Exp:* Asst prof physics, 66-74, ASSOC PROF PHYSICS, RUTGERS UNIV, 74- *Mem:* Am Phys Soc; Am Asn Physics Teachers. *Res:* Mechanism of protonic transfer and hydration in aqueous systems; protonic charge transfer mechanisms and cell division; membrane electrical stress and cell division. *Mailing Add:* Dept Physics Rutgers Univ 406 Penn Camden NJ 08102

GAGLIARDI, ROBERT M, b New Haven, Conn, Apr 7, 34; m 57; c 3. ELECTRICAL ENGINEERING. *Educ:* Univ Conn, BSEE, 56; Yale Univ, MS, 57, PhD(elec eng), 60. *Prof Exp:* Staff engr, Hughes Aircraft Co, 60-61; asst prof, 61-70, assoc prof, 70-80, PROF ENG, UNIV SOUTHERN CALIF, 80- *Concurrent Pos:* Consult, Hughes Aircraft Co, 61- *Mem:* Inst Elec & Electronics Engrs. *Res:* Problems of communication theory, information theory, telemetry and data handling system studies. *Mailing Add:* Dept Elec Eng Univ Southern Calif Los Angeles CA 90007

GAGLIARDI, UGO OSCAR, b Naples, Italy, July 23, 31; US citizen. COMPUTER SCIENCES. *Educ:* Univ Naples, dipl, 51, Dr Ing(elec eng), 54. *Prof Exp:* Dir comput lab, Elec Syst Div, US Air Force, 65-66; res fel comput eng, Harvard Univ, 66-68; vpres eng, Interactive Sci, Inc, 68-70; dir engr, Honeywell Inc, 70-75; PRES, GEN SYSTS GROUP, INC, 75-; CHMN BD, CTR SOFTWARE TECHNOL, 85- *Concurrent Pos:* Fulbright Scholar, Columbia Univ, 55-56; lectr, Harvard Univ, 66-74, prof prac comput eng, 74-83, Gordon McKay prof, 83-; mem, Nat Acad panel Inst Comput Sci & Technol, Nat Inst Standards & Technol, 85-, chmn, 89- *Res:* Software engineering; computer systems architecture; architecture of very large distributed computer systems; queuing theory; applications to computer systems design. *Mailing Add:* Dept Appl Scis Harvard Univ Cambridge MA 02138

GAGNA, CLAUDE EUGENE, b New York, NY, Sept 16, 56. IMMUNOLOGY, PATHOLOGY. *Educ:* St Peter's Col, BS, 79; Fairleigh Dickinson Univ, MS, 83; NY Univ, PhD(anat-biochem), 90. *Prof Exp:* Res asst biochem, 84-90, from instr to assoc prof anat, Basic Med Sci, 88-92, INSTR MOLECULAR BIOL, NY UNIV DENTAL CTR, 91- *Concurrent Pos:* Biomed consult, Herbert Law Firm, 90-; postdoctoral fel, NIH, 91- *Mem:* Fel Am Inst Chemists; Am Asn Anatomists; Asn Res Vision & Ophthal; NY Acad Sci; Genetics Soc Am; Protein Soc. *Res:* Examination of structure and function of left-handed Z-DNA, Z-RNA and Z-helical proteins in both normal and diseased human tissues; development of nonsurgical molecular biological techniques to turn off harmful genes. *Mailing Add:* 157 Morningside Lane Palisades Park NJ 07650

GAGNE, JEAN-MARIE, b Alma, Que, July 12, 32; m 59; c 1. SPECTROSCOPY, OPTICS. *Educ:* Univ Montreal, BSc, 58, MSc, 59; Univ Paris, DSc(physics), 65. *Prof Exp:* Asst prof physics, Polytech Sch, Univ Montreal, 59-63; fel spectros, Aime-Cotton Lab, 63-65; assoc prof, 65-80, PROF PHYSICS & ENG & HEAD OPTICAL LAB, POLYTECH SCH, UNIV MONTREAL, 80-, HEAD DEPT PHYS ENG, 70- *Mem:* AAAS; Can Asn Physicists; Optical Soc Am; Spectros Soc Can; Soc Photo-Optical Instrument Eng. *Res:* Hyperfine structure in line spectra and influence of isotopes on the hyperfine of atomic spectral lines; essentials of a problem os spectroscopy, especially Fabry-Perot spectrometers; detection of flux and spectrometers photon noise and detection noise. *Mailing Add:* Polytech Sch Univ Montreal CP 6079 succ A Montreal PQ H3C 3A7 Can

GAGNE, RAYMOND J, b Meriden, Conn, Aug 27, 35; m 63; c 2. SYSTEMATIC ENTOMOLOGY. *Educ:* Univ Conn, BA, 61; Iowa State Univ, MS, 63; Univ Minn, PhD(entom), 67. *Prof Exp:* RES ENTOMOLOGIST, SYST ENTOM LAB, USDA, 65- *Concurrent Pos:* Ed Proceedings Entom Soc Washington, 84-87. *Mem:* AAAS; Entom Soc Am. *Res:* Systematics of Diptera, particularly gall midges. *Mailing Add:* Syst Entom Lab USDA US Nat Mus Washington DC 20560

GAGNE, ROBERT RAYMOND, b Fitchburg, Mass, Mar 25, 48. ELECTROCATALYSIS. *Educ:* Worchester Polytech Inst, BS, 69; Stanford Univ, PhD(chem), 74. *Prof Exp:* Asst prof chem, Calif Inst Technol, 74-81; DIR CORP DEVELOP, HYPERION CATALYSIS INT, 81- *Mem:* Am Chem Soc; AAAS; Electrochem Soc; Royal Soc Chem. *Res:* Inorganic chemistry; homogeneous catalysis; electrocatalysis; electron transfer mechanisms; fuel cells and the conversion of solar energy to electricity and fuels. *Mailing Add:* 140 E Arrow Hwy San Dimas CA 91773

GAGNEBIN, ALBERT PAUL, b Torrington, Conn, Jan, 23, 09; m 35; c 2. METALLURGY. *Educ:* Yale Univ, BS, 30, MS, 32. *Prof Exp:* Res metallurgist, res lab, Int Nickel Co, 32-49, metallurgist in charge ductile iron, Res & Develop Div, 49-55, asst mgr, Nickel Sales Dept, 54-56, mgr, 56-58, asst vpres, 57-58, vpres & mgr, Primary Nickel Dept, 58-64, exec vpres, 64-88; RETIRED. *Concurrent Pos:* Vpres, Int Nickel Co Can, Ltd, 60-64, exec vpres, 64-67, pres, Inco Ltd, 67-72, chmn, 72-74, dir, 74-88. *Honors & Awards:* Simpson Gold Medal, Am Foundrymen's Soc, 52; Charles F Rand Mem Gold Medal, Am Inst Mining Metall & Petrol Engrs, 77. *Mem:* Nat Acad Eng; Am Foundrymen's Soc; Am Soc Mech Engrs; Am Inst Mining, Metall & Petrol Engrs; Am Soc Metals. *Res:* Development of nickel containing materials. *Mailing Add:* Inco Ltd One New York Plaza New York NY 10004

GAGNON, ANDRE, physiology, for more information see previous edition

GAGNON, CAMILIEN JOSEPH XAVIER, b Saint-Maurice, Que, Nov 11, 29; m 60; c 3. PLANT PATHOLOGY, PLANT BIOCHEMISTRY. *Educ:* Laval Univ, BA, 52, BScA, 56; Cornell Univ, MSc, 60, McGill Univ, PhD(plant path), 65. *Prof Exp:* Res officer, Forestry Lab, Agr Can, 56-65; res scientist, Forest Res Lab, Dept Forestry, 65-69; res scientist, Res Sta, Agr Can, 69-91, asst dir, 72-91; RETIRED. *Mem:* Can Soc Plant Pathologists. *Res:* Assessment of crop losses due to diseases; epidemiology of forage crop diseases; root rots of forage legumes; etiology and resistance. *Mailing Add:* 3155 Chambord St Ste-Foy PQ G1W 2Y2 Can

GAGNON, CLAUDE, b Montreal, Que, Jan 12, 50; m 73; c 3. REPRODUCTIVE BIOLOGY & MEDICINE, MALE INFERTILITY. *Educ:* Univ Montreal, BSc, 70, MSc, 71, PhD(biochem), 74. *Prof Exp:* Postdoctoral fel Neuropharmacol Bictr, Univ Basal, Switz, 74-76 & neuroendocrinol, NIH, Bethesda, 76-78; from asst prof to assoc prof pharmacol, Laval Univ, Que, 78-84; assoc prof, 84-89, PROF SURG UROL, MCGILL UNIV, 89- *Concurrent Pos:* Dir, Cell & Molecular Bioregulation Unit, CHUL, Laval Univ, 81-84 & Urol Res Lab, Royal Victoria Hosp, McGill Univ, 84-; chmn, Urol Res Comt, McGill Univ, 88-; pres, Biomed Res Consult, Can 160366, 88- *Mem:* Am Fertil Soc; Am Soc Cell Biol; Soc Study Reproduction; Am Soc Andrology; Can Fertil & Andrology Soc; Can Biochem Soc. *Res:* Regulation of male infertility; mechanisms controlling sperm motility and fertilizing capacity; new pathologies and molecular defects that may explain male infertility; new approaches to improve male infertility potential. *Mailing Add:* Urol Res Lab Royal Victoria Hosp Rm H6-46 McGill Univ 687 Pine Ave Montreal PQ H3A 1A1 Can

GAGNON, EUGENE GERALD, b Revere, Mass, Feb 20, 36; m 67; c 3. ELECTROCHEMISTRY, ELECTROCHEMICAL ENGINEERING. *Educ:* Loyola Col, Que, BSc, 57; Stevens Inst Technol, MS, 60; Pa State Univ, PhD, 70. *Prof Exp:* Asst, Stevens Inst Technol, 58-60; res chem eng, Socony Mobil Oil Co, 60-66; res asst, Pa State Univ, 66-69; asst res dir, C&D Batteries Div, Eltra Corp, 69-70; assoc sr res chemist, electrochem dept, 70-77, SR RES SCIENTIST, RES LAB, GEN MOTORS CORP, 77- *Concurrent Pos:* Instr, Rutgers Univ, 64-66 & Lawrence Inst Technol, 77; consult, Harry Diamond Labs, Army Mat Command, 68-69; vchmn, Electrochem Soc, 84-86, chmn, 86-88, mem bd dir, 86- 88; div ed, J Electrochem Soc, 87- *Mem:* Am Chem Soc; Electrochem Soc (secy-treas, 78-84); Sigma Xi. *Res:* Electrochemical studies on air cathodes; mathematical modeling in porous electrodes; fundamental studies on lead-acid batteries; kinetics of porous silver/silver oxide electrodes at low temperatures in KOH batteries; development of zinc-nickel oxide battery for electric vehicles. *Mailing Add:* Gen Motors Corp Res Labs 12 Mile & Mound Rds Warren MI 48090-9055

GAGNON, LEO PAUL, b Milford, NH, Oct 25, 29; m 57; c 6. PHARMACY. *Educ:* Mass Col Pharm, BS, 52; Purdue Univ, MS, 54, PhD, 57. *Prof Exp:* Res pharmacist, Pharm Res Lab, Miles Labs, Inc, 56-58, group leader, 58-59, sect head, 59-63; chief Norwich Prod develop sect, Norwich Pharm Co, 63-; AT PROD DEVELOP LAB, MENTHOLATUM CO. *Mem:* AAAS; Soc Cosmetic Chem; Am Pharmaceut Asn; Acad Pharmaceut Sci; NY Acad Sci; Sigma Xi. *Res:* Pharmaceutical product development; ethical and proprietary medicinals. *Mailing Add:* Product Develop Lab Mentholatum Co Inc 1360 Niagara St Buffalo NY 14213

GAGNON, MARCEL, b St Moise, Que, Jan 17, 28; m 54; c 3. FOOD TECHNOLOGY. *Educ:* Laval Univ, BA, 49, BS, 53; Univ Mass, MS, 55, PhD(food technol), 57. *Prof Exp:* Asst biologist, Que Dept Maritime Fisheries, Can, 50, technologist, 52-53, asst technologist, Econ Div, Can Dept Maritime Fisheries, Ont, 51; technologist, Fisheries Res Bd Can, 54-55; res instr food tech, Univ Mass, 55-56, asst res prof, 57; prod mgr & dir res & develop, Alphonse Raymond, Ltd, 58-61, vpres prod & res, 61-66, vpres & gen mgr, 66-68; vpres opers, R J Reynolds Foods, Ltd, 68-70; vrector, 70-71, dir food res, 71-82, DIR FOOD RES, INST ARMAND FRAPPIER, UNIV QUE, 82-; EXEC DIR, CTR IRRADIATION CAN, 86- *Concurrent Pos:* Lectr, Montreal Inst Technol, 60-; head, Food Mission to Europe, 64; Que Food Industs rep, Int Food Salon Paris, 64; mem bd dirs, Univ Que, Montreal, 70-71, Inst Armand Frappier, Univ Que, 87-89; exec vpres, Biopreserve Inc, 84. *Honors & Awards:* J Armand Bombardier Innovation Prize, 87; Gold Medal, Thailand Ministry Res, 89; Officer, Order of Quebec, 91. *Mem:* Am Chem Soc; AAAS; Inst Food Technologists; Can Inst Food Technologists (vpres, 62); Can Food Processors Asn (vpres, 66-67). *Res:* Food biochemistry; enzyme biochemistry; food management; environment; fisheries technology; food irradiation. *Mailing Add:* Cresala-Inst Armand-Frappier 531 Boul Des Prairies Cp 100 Laval PQ H7N 4Z3 Can

GAGNON, REAL, b Montreal, Que, May 9, 24; m 52; c 3. ANATOMY. *Educ:* Univ Montreal, BA, 45, MD, 51; Univ Mich, PhD(anat), 55. *Prof Exp:* From asst prof to assoc prof, 54-64, PROF ANAT, UNIV MONTREAL, 64- *Mem:* Can Asn Anat; French Asn Anat. *Res:* Gross anatomy; human embryology; anthropology. *Mailing Add:* Dept Anat Univ Montreal Cp 6128 Succursale A Montreal PQ H3C 3J7 Can

GAGOLA, STEPHEN MICHAEL, JR, b Buffalo, NY, June 25, 47; m 76; c 3. FINITE GROUPS. *Educ:* State Univ NY, Buffalo, BA & MA, 69; Univ Wis, PhD(math), 74. *Prof Exp:* Res fel, Mich State Univ, 74-76; asst prof, Tex A&M Univ, 76-; DEPT MATH, KENT STATE UNIV. *Mem:* Math Asn Am. *Res:* Characters of finite groups and in applying character theory to problems both within and outside of group theory. *Mailing Add:* Dept Math 302 AM Kent State Univ Kent OH 44242

GAHERTY, GEOFFREY GEORGE, b Montreal, PQ, Mar 7, 41; m 85; c 1. MACINTOSH CONSULTING & TRAINING. *Educ:* McGill Univ, BSc, 64; Univ Toronto, MA, 66, PhD(anthrop), 70. *Prof Exp:* Actg asst prof anthrop, Univ Calif, Santa Barbara, 68-70; asst prof, Univ Waterloo, 70-71; tutor sci & music, Cool Sch, Fac Med, McMaster Univ, 71-72, asst prof anthrop, 72-75; intern, Toronto Inst Human Rels, 76-79, resident, 79-81, supvr assoc, 81-85; assoc staff therapist, Bioenergetic Psychother Inst, 80-84; INSTR, UNIV TORONTO SCH CONTINUING STUDIES, 87- *Res:* Comparative osteology of contemporary and prehistoric populations of Africa, Europe and Southeast Asia. *Mailing Add:* 1365 Yonge St Suite 204 Toronto ON M4T 2P7 Can

GAHL, WILLIAM A, b Waukesha, Wis, Mar 11, 50; m 73; c 4. HUMAN GENETICS RESEARCH. *Educ:* Mass Inst Technol, BS, 72; Univ Wis, MD, 76, PhD(oncol res), 81; Am Bd Pediat, cert, 81; Am Bd Med Genetics, cert, 84. *Prof Exp:* Intern, Dept Pediat, Univ Wis Hosps, Madison, 76, resident, 77-78, chief resident, 79, instr, 80-81; med staff fel, Internist Genetics Prog, NIH, 81-84, sr staff fel, Sect Human Biochem Genetics, Human Genetics Br, Nat Inst Child Health & Human Develop, 84-86, actg chief, Human Genetics Br, 88, MED OFFICER & HEAD, SECT HUMAN BIOCHEM GENETICS, HUMAN GENETICS BR, NAT INST CHILD HEALTH & HUMAN DEVELOP, NIH, 86-, BR CHIEF, 89- *Concurrent Pos:* Fel, McArdle Labs Cancer Res, Madison, Wis, 80-81; mem, Nat Med Adv Bd, Cystinosis Found, 84-; dir, Interinst Genetics Prog, Clin Ctr, NIH, 89-; assoc ed, Am J Human Genetics, 90- *Mem:* Soc Inherited Metab Dis; Am Soc Human Genetics; Soc Pediat Res; Soc Study Inborn Errors Metab; Am Soc Clin Invest; Am Fedn Clin Res. *Mailing Add:* Human Genetics Br Bldg 10 Rm 95242 Nat Inst Child Health & Human Develop 9000 Rockville Pike Bethesda MD 20892

GAHLER, ARNOLD ROBERT, b Portland, Ore, Sept 6, 19; m 42; c 2. ANALYTICAL CHEMISTRY, ENVIRONMENTAL CHEMISTRY. *Educ:* Pac Univ, AB, 41; Purdue Univ, MS, 49. *Prof Exp:* Asst analytical chem, Purdue Univ, 46-48; analytical res chemist, Metal Res Labs, Electro Metall Co, Union Carbide & Carbon Corp, 50-59, sr develop chemist, Union Carbide Metals Co, 59-61, asst chief chemist, Tech Dept, Metals Div, Union Carbide Corp, 61-63, asst chief chemist, Alloy Plant, 63-64; asst biochem, Ore State Univ, 65; analytical chemist, Pac Northwest Water Lab, USPHS, 65-67, res chemist, 67-71, chief lab br, Region X, US Environ Protection Agency, 71-85; RETIRED. *Mem:* Am Chem Soc. *Res:* Analysis of environmental pollutants; quality assurance; laboratory design; laboratory certification. *Mailing Add:* 8152 SE Haida Dr Port Orchard WA 98366

GAIDIS, JAMES MICHAEL, b Baltimore, Md, Oct 17, 40; m 63; c 3. ORGANOMETALLIC CHEMISTRY. *Educ:* Harvard Univ, AB, 63; Univ Wis-Madison, PhD(inorg chem), 67. *Prof Exp:* Res chemist, Dow Chem Co, 67-70; res chemist, W R Grace & Co, Cambridge, 71-73, group leader, 73-74, CHEMIST, W R GRACE & CO, WASHINGTON RES CTR, COLUMBIA, MD, 74- *Mem:* Am Chem Soc; Am Soc Testing & Mat. *Res:* Silicon chemistry; industrial chemistry; materials science. *Mailing Add:* 7379 Rte 32 Columbia MD 21044

GAIDIS, MICHAEL CHRISTOPHER, b Madison, Wis, Mar 10, 67; m 90. SUPERCONDUCTIVITY & DEVICES. *Educ:* Mass Inst Technol, BS(elec eng), BS(physics) & MS, 89; Yale Univ, MS, 90. *Prof Exp:* Res asst III-V semiconductors, Lincoln Lab, Mass Inst Technol, 86-89, res asst, Elec Eng Lab, 88-89; RES ASST SUPERCONDUCTIVITY, YALE UNIV, 89- *Mem:* Am Phys Soc. *Res:* Superconducting single-photon X-ray detectors, primarily for astronomy applications (in conjunction with NASA); rapid thermal annealing of ion-implanted indium phosphide. *Mailing Add:* 1534 Ella T Grasso Blvd New Haven CT 06511

GAIDOS, JAMES A, b Clarksburg, WVa, July 15, 36; m 59, 88; c 2. PHYSICS. *Educ:* WVa Univ, AB, 58; Univ Wis, PhD(physics), 63. *Prof Exp:* Res assoc physics, Univ Wis, 63-65; asst prof, Hamilton Col, 65-68; from asst prof to assoc prof, 71-79, PROF PHYSICS, PURDUE UNIV, WEST LAFAYETTE, 79- *Res:* Elementary particles. *Mailing Add:* Dept Physics Purdue Univ West Lafayette IN 47907

GAIK, GERALDINE CATHERINE, b Chicago, Ill, July 6, 42; m 72. HEALTH SCIENCES. *Educ:* Mundelein Col, BS, 64; Loyola Univ, MS, 67; Northwestern Univ, PhD(anat), 72. *Prof Exp:* Asst prof gross anat & histol, Sch Dent Med, Univ Pittsburgh, 72-76; ASST PROF ANAT, SCH DENT, LOYOLA UNIV, 76- *Concurrent Pos:* Nat Inst Dent Res grant, 75; res grant, Univ Pittsburgh, 75; res grant, Nat Inst Dent Res, 78-80, Dept Health, Educ & Welfare, 78-80 & Pub Health Serv, 78-80; res comt grant, Sch Dent, Loyola Univ, 81. *Mem:* Sigma Xi; Asn Anatomists; Am Asn Anatomists; Soc Neurosci. *Res:* Morphological and experimental work on the development of the avian trigeminal ganglion and the synaptic morphology in the ventral horn gray matter of the spinal cord in paraplegic monkeys. *Mailing Add:* 2160 S First Ave Maywood IL 60153

GAILANI, SALMAN, b Baghdad, Iraq, May 25, 26; US citizen; m 56; c 3. MEDICINE, CANCER. *Educ:* Univ Baghdad, MB & ChB, 49. *Prof Exp:* Sr cancer res internist, Roswell Park Mem Inst, 61-65, assoc cancer res internist, 65-69, assoc chief med, 69-78; res asst prof med & res assoc prof pharmacol, Grad Sch, State Univ NY Buffalo, 70-78; DIR ONCOL, ST CATHERINE HOSP, 78-85; ONCOL, SANDRIDGE PROF CTR. *Concurrent Pos:* Clin assoc prof med, Chicago Med Sch, 78- & Ind Univ, 78- *Mem:* Am Soc Hemat; Am Fedn Clin Res; Am Asn Cancer Res; Am Soc Clin Oncol; Am Inst Nutrit. *Res:* Cancer chemotherapy; nutrition. *Mailing Add:* Sandridge Prof Ctr 9116 Columbia Ave Munster IN 46321

GAILAR, NORMAN MILTON, b Hornell, NY, Feb 20, 18; m 69. PHYSICS. *Educ:* Syracuse Univ, BA, 48, MS, 50; Ohio State Univ, PhD(physics), 58. *Prof Exp:* Physicist, Nat Bur Standards, 50-52 & Naval Res Lab, 52-53; from asst prof to prof physics, Univ Tenn, Knoxville, 58-85. *Mem:* Optical Soc Am. *Res:* Molecular and infrared spectroscopy; molecular structure. *Mailing Add:* 2709 Reagan Rd Knoxville TN 37931

GAILAR, OWEN H, b Rochester, NY, Nov 10, 25; m 52; c 2. PHYSICS. *Educ:* Univ Rochester, BS, 46; Purdue Univ. MS, 49, PhD(exp nuclear physics), 56. *Prof Exp:* Asst physics, Purdue Univ, 46-56; supvr reactor physics, statist, Combustion Eng, Inc, 57-62; ASSOC PROF NUCLEAR PHYSICS, PURDUE UNIV, 62- *Concurrent Pos:* Physicist, Nat Bur Standards, 51-52; consult, Pac Gas & Elec Co, Calif & Commonwealth Edison, 69-70; adj instr, Fresno City Col. *Res:* Reactor design; application of digital computers to reactor design and analysis; integration of power reactors into utility grids; nuclear fuel cycle; reactor physics; computer application. *Mailing Add:* 345 W San Carlos Fresno CA 93704

GAILEY, FRANKLIN BRYAN, b Atlanta, Ga, Oct 18, 18; m 48; c 4. BEHAVIOR-ETHOLOGY. *Educ:* Ga Tech, BS, 40; Univ Wis, MS, 42, PhD(biochem), 46. *Prof Exp:* Asst biochem, Univ Wis, 40-46; instr chem & biol, Lees Jr Col, 46-48; from asst prof to assoc prof biol, Berea Col, 48-57, chmn dept, 56-82, prof biol & chem, 57-89; RETIRED. *Concurrent Pos:* With Off Prod Res & Develop, 44. *Mem:* AAAS; Am Soc Plant Physiol. *Res:* Differentiation and separation of hog mucosa peptidases; production of penicillin by fermentation in pilot tanks; greening of dark-grown seedlings; protoplasmic streaming in slime mold; phospholipids in yeast. *Mailing Add:* CPO 795 Berea KY 40404-0795

GAILEY, KENNETH DURWOOD, chemistry, inorganic chemistry; deceased, see previous edition for last biography

GAILLARD, MARY KATHARINE, b New Brunswick, NJ, April 1, 39; c 3. GAUGE THEORIES, SUPER GRAVITY. *Educ:* Hollins Col, BA, 60; Columbia Univ, MA, 61; Univ Paris-sud, Orsay, France, DrSci(theory physics), 64 & 68. *Prof Exp:* Asst res, Nat Ctr Sci Res, 64-68, assoc res, 68-73, head res, 73-79, dir res, 80-81; PROF PHYSICS, UNIV CALIF, BERKELEY, 81- *Concurrent Pos:* Vis scientist, Cern, Geneva, 64-81, Nat Accelerater Lab firm, Batovia, Fl, 73-74, & 83, Inst Theory Physics, Univ Calif, Santa Barbara, 85; prin invest, NSF, 82-; faculty sr scientist, Lawrence Berkeley Lab, Univ Calif, Berkeley, 81-; chancellor's distinguished lectr, Univ Calif, Berkeley, 81; sci dir, Les Houches Summer Sch, 81; mem, Tech Assessment Comt Univ Progs, Dept Educ, 82; mem, Subpanel New Facil, HEPAP, 83; vis comt, Fermilab, 83-85, Astrophysics Adv Comt, 85-88, Physics Adv Comt, 86-90; adv comt, Theoretical Advan Study Inst Elem Particle Physics, 83-88; adv bd, Inst Advan Studies, Univ Calif, Santa Barbara, 85-88; mem, Subcomt Oversight Rev, NSF Theoret Physics Prog, 88; rev comt, Argonne Nat Lab High Energy Physics Div, 88-90; ser ed, Springer-Verlag, 88-; chair, Comt Status Women Physics, Am Phys Soc, 85, mem exec comt, Div Particles & Fields, 90-; mem, High Energy Physics Adv Panel, Dept Energy, 91- *Honors & Awards:* Prix Thibaud, 77; Loeb Lectr Physics, Harvard Univ, 80; Warner-Lambert Lectr, Univ Mich, 84; E O Lawrence Mem Award, 88. *Mem:* Nat Acad Sci; AAAS; fel Am Acad Arts & Sci; fel Am Phys Soc. *Res:* Elementary particle theory, phenomenology of gauge theories, physics of the early universe, unification of fundamental interactions, super collider physics, effective theories of particle physics based on superstring theories. *Mailing Add:* Lawrence Berkeley Lab 50A-3115 Berkeley CA 94720

GAIN, RONALD ELLSWORTH, b Glendale, Calif, Dec 8, 40; m 61; c 4. MYCOLOGY, BACTERIOLOGY. *Educ:* Wilmington Col, BS, 62; Miami Univ, MS, 65; WVa Univ, PhD(microbiol), 68. *Prof Exp:* Scientist microbiol, Worcester Found Exp Biol, 68-69; ASSOC PROF BACT & MYCOL, MARSHALL UNIV, 69- *Concurrent Pos:* Consult, Redi-Prod Corp, 74- *Mem:* Mycol Soc Am; Am Soc Microbiol; Sigma Xi. *Res:* Effect of iodine antiseptic formulations on bacterial endospores. *Mailing Add:* No Four Wildwood Lane Huntington WV 25701

GAINER, FRANK EDWARD, b Waynesboro, Ga, June 18, 38; m 61; c 3. ANALYTICAL CHEMISTRY. *Educ:* Morehouse Col, BS, 60; Tuskegee Inst, MS, 62; Iowa State Univ, MS, 64, PhD(chem), 67. *Prof Exp:* Sr analytical chemist, Analytical Develop Dept, 67-73, antibiotic assay coordr, Aniibiotic Assay Dept, 73-80, MGR ANTIBIOTIC ANALYTICAL AND QUAL CONTROL, ELI LILLY & CO, 80- *Mem:* Am Chem Soc; Sigma Xi. *Res:* Coordination of analytical chemical requirements associated with antibiotic development and production; chemical and microbiological assay development, lab automation, and computerization. *Mailing Add:* Eli Lilly & Co Indianapolis IN 46206

GAINER, GORDON CLEMENTS, organic chemistry, for more information see previous edition

GAINER, HAROLD, b New York, NY, Aug 6, 35; m 65; c 3. NEUROBIOLOGY. *Educ:* City Col New York, BS, 56; Univ Calif, Berkeley, PhD(physiol), 59. *Prof Exp:* Res fel physiol, Univ Calif, Berkeley, 59-60; USPHS res fels electrophysiol, Columbia Univ, 60-63; from asst prof to assoc prof zool, Univ Md, 63-69; res physiologist, Nat Inst Child Health & Human Develop, NIH, 69-73, chief sect functional neurochem, 73-80, lab chief, Lab Neurochem & Neuroimmunol, Nat Inst Child Health & Human Develop, 83-86, LAB CHIEF, LAB NEUROCHEM, NAT INST NEUROL DIS & COMMUN DIS, NIH, 87-, DIR, BASIC NEUROSCI PROG, 90- *Concurrent Pos:* Lectr, Sch Med, Univ Calif, San Francisco, 59-60; USPHS res grant, 63-; vis prof, Tel Aviv Univ, 69; prefector, George Washington Univ, Neurosci Training Prog, 75-83. *Mem:* AAAS; Am Physiol Soc; Am Soc Neurochem; Neurosci Soc; Sigma Xi. *Res:* Regulation of the biochemical morphological properties of neurons; neurobiology of neuronal cytoskeletal proteins and neuropeptides. *Mailing Add:* NIH Nat Inst Neurol Dis & Commun Dis Lab Neurochem Bldg 36 Rm 4D-20 Bethesda MD 20892

GAINER, JOHN LLOYD, b Grafton, WVa, July 19, 38; m 81; c 2. CHEMICAL ENGINEERING. *Educ:* Univ WVa, BSChE, 60; Mass Inst Technol, MS, 61; Univ Del, PhD(chem eng), 64. *Prof Exp:* Res chem engr, silicones div, Union Carbide Corp, 64-66; from asst prof to assoc prof, 66-82, PROF CHEM ENG, UNIV VA, 82- *Concurrent Pos:* Vis fel, Karolinska Inst, Sweden, 71-72; vis researcher, ICI Pharmaceut, Eng, 90. *Mem:* Am Inst Chem Engrs; Am Chem Soc; Am Soc Eng Educ; Sigma Xi. *Res:* Diffusion in liquids; surface chemistry; biological mass transfer; oxygen transport and atherosclerosis; treatment of diseases with carotenoids; biotechnology. *Mailing Add:* Dept of Chem Eng Univ Va Thornton Hall Charlottesville VA 22903-2442

GAINER, JOSEPH HENRY, b Atlanta, Ill, Oct 24, 24; m 54; c 6. VIROLOGY. *Educ:* Ohio State Univ, DVM, 46, MS, 47; Univ Mich, MS, 58. *Prof Exp:* Instr, Ohio State Univ, 47-49; res asst comp path, Mayo Clin, 49-50; res assoc med, Univ Chicago, 50-51; asst prof vet sci, Univ Ark, 51-53; vet virologist, Fla Dept Agr, 59-66; sr vet officer, Nat Cancer Inst, 66; virologist & vet dir, Nat Inst Environ Health Sci, USPHS, 67-73; VET MED OFFICER, FOOD & DRUG ADMIN, 73- *Concurrent Pos:* Adj prof animal sci, NC State Univ, 67-73. *Mem:* Am Soc Virol; Conf Res Workers Animal Dis; Am Soc Microbiol; Am Vet Med Asn; Int Acad Path; Am Asn Immunologists. *Res:* Ovine ketosis; experimental tuberculosis chemotherapy; radiation sickness; salmonellosis; chemical warfare; equine infectious anemia; metals and viral infections; encephalomyocarditis virus infections of animals; microbial resistance; interferons; experimental radiowave heating; iron deficient neutropenia in baby pigs; bioavailability of oxy tetracycline and amoxicillin in goats; gene control of caprine interferon. *Mailing Add:* Ctr for Vet Med HFV-520 Food & Drug Admin Beltsville MD 20705

GAINER, MICHAEL KIZINSKI, b St Louis, Mo, Feb 26, 33; m 55; c 8. STELLAR PHOTOMETRY, X-RAY DIFFRACTION ANALYSIS. *Educ:* WVa Univ, BS, 55, MS, 56. *Prof Exp:* Physicist, US Army Ballistics Res Lab, 56-62; asst prof, 62-66, ASSOC PROF PHYSICS, ST VINCENT COL, 66- *Concurrent Pos:* Consult, US Army Res Off, 62-74; guest lectr, Ind Univ Pa, 76; chairperson physics, St Vincent Col, 84-; secy, 84-86, vpres, 86-88, pres, Western Pa Sect, Am Asn Physics Teachers, 88- *Mem:* Am Asn Physics Teachers; Astron Soc Pac. *Res:* Stellar photometry; x-ray diffraction analysis; shock wave-metal interactions; author of various publications. *Mailing Add:* RD Three Box 415 Blairsville PA 15717

GAINES, ALAN MCCULLOCH, b Asheville, NC, Nov 13, 38; m; m 75; c 4. RESEARCH ADMINISTRATION, SCIENCE POLICY. *Educ:* Univ Chicago, BS, 60, MS, 63, PhD(geochem), 68. *Prof Exp:* Lectr chem, Univ Ill, Chicago, 60-62; lectr phys sci, Univ Chicago, 62-64, from res asst to res assoc geochem, 64-69; asst prof geochem, Univ Pa, 69-76; assoc prog dir, 76-79, prog dir geochem, 79-88, HEAD RES GRANTS SECT, NSF, 88- *Mem:* AAAS; Mineral Soc Am; Am Geophys Union; Geochem Soc; Int Asn Geochem & Cosmochem; Sigma Xi. *Res:* Kinetics and mechanisms of mineral reactions; geochemistry of carbonate minerals; crystal growth mechanisms. *Mailing Add:* Div Earth Sci NSF Washington DC 20550

GAINES, ALBERT L(OWERY), b Selma, Ala, Feb 28, 20; m 42; c 4. FISSION & FUSION POWER PLANTS. *Educ:* Auburn Univ, BME, 46; Univ Mo, MS, 50. *Prof Exp:* Mech engr, Phillips Petrol Co, Okla, 47; asst prof mech eng, Auburn Univ, 47-48; mech engr, Humble Oil & Refining Co, Tex, 48; instr mech eng, Univ Mo, 49-50; develop engr, Union Carbide Nuclear Co Div, Union Carbide Corp, Tenn, 50-56; supvr, Thermo Group, Combustion Eng, Inc, Chattanooga, 56-58, supvr, Spec Prod Eng Sect, 59-60, supvr, Proposition Eng Sect, 60-66, mgr desalination eng, Conn, 66-67, mgr mech design sect, Nuclear Power Dept, 67-69, exec asst to dir power systs eng, 69-70, proj mgr, Waterford Steam Elec Sta Nuclear Plant, 70-76, proj mgr, Advan Develop Dept, 76-84; PRES, A GAINES CO, CONSULT, 84- *Mem:* Am Soc Mech Engrs; Am Nuclear Soc; Am Soc Chem Engrs. *Res:* Design of large, heavy components for power systems; design of fuel assemblies and mechanisms for fission and fusion power plants; evaluation of project specifications and plans, including research and development. *Mailing Add:* 410 Leyswood Dr Greenville SC 29615

GAINES, DONALD FRANK, b Caldwell, Idaho. July 26, 36; m 60; c 2. INORGANIC CHEMISTRY. *Educ:* Col Idaho, BS, 58; Indiana Univ, PhD(inorg chem), 63. *Prof Exp:* Res assoc inorg chem, Indiana Univ, 63-64 & Univ Manchester, Eng, 64-65; from asst prof to assoc prof, 69-75, PROF INORG CHEM, UNIV WIS-MADISON, 75- *Mem:* Am Chem Soc; Royal Soc Chem. *Res:* Boron hydrides; organo-group III compounds; boron-group III and IV compounds; boron-nitrogen compounds; metallo-borane complexes; organo group II compounds. *Mailing Add:* Dept Chem Univ Wis 1101 University Ave Madison WI 53706

GAINES, EDWARD EVERETT, b Cleveland, Tenn, Jan 11, 37; m 60; c 1. ENERGETIC PARTICLE RADIATION IN SPACE, MAGNETOSPHERIC PHYSICS. *Educ:* Univ Chicago, BS, 58; Wash State Univ, MS, 61. *Prof Exp:* Physicist, US Naval Radiol Defense Lab, 60; scientist, Palo Alto Res Labs, 60-65, sr scientist, 65-77; RES SCIENTIST, LOCKHEED MISSILES & SPACE CO, 77- *Mem:* Am Geophys Union. *Res:* Measurement of penetrating particle radiations in space. *Mailing Add:* 737 Kendall Ave Palo Alto CA 94306

GAINES, EDWARD M(CCULLOCH), b Pullman, Wash, Sep 30, 13; m 34; c 2. FORESTRY. *Educ:* Wash State Col, BS, 34. *Prof Exp:* Forester, Nat Forest Admin, Forest Serv, USDA, 35-41, state & private forestry, 41-46, res ctr leader, 46-65, prog mgr res, 65-67, asst dir res Pac Southwest Forest Exp Sta, 67-74; RETIRED. *Mem:* fel AAAS; Soc Am Foresters. *Res:* Forestry and related resources (water, range, etc). *Mailing Add:* 1423 Blvd Park Dr Lacey WA 98503

GAINES, GEORGE LOWEREE, JR, b New Haven, Conn, Mar 7, 30; m 54; c 3. PHYSICAL CHEMISTRY. *Educ:* Yale Univ, BS, 50, MS, 52, PhD(chem), 54. *Prof Exp:* Chemist, E I du Pont de Nemours & Co, 50-51; asst chem res, Yale Univ, 52-54; PHYS CHEMIST, RES & DEVELOP CTR, GEN ELEC CO, 54- *Concurrent Pos:* Lectr, Rensselaer Polytech Inst, 70- *Mem:* Am Chem Soc. *Res:* Surface chemistry; ion exchange; silicate minerals; adsorption; monolayers; photosynthesis; polymer surface props. *Mailing Add:* 972 Charlton Rd Scotia NY 12302

GAINES, GORDON BRADFORD, b Fitzgerald, Ga, Aug 10, 23; m; c 1. SURFACE PHYSICS. *Educ:* Univ Ga, BS, 48, MS, 49; Capital Univ, MBA, 75. *Prof Exp:* Res engr phys electronics, Battelle Mem Inst, 50-52, asst div chief, solid state devices, 52-62, res assoc, Electronic Mat & Devices Div, 62-65, from fel to sr fel, 65-69, sr researcher, 73-78, chief, Electronic Mat & Devices Div, 69-87, res leader, 78-87; RETIRED. *Res:* Physical electronics; electron physics; solid state materials and devices; gas discharges; electron paramagnetic resonance, dielectrics. *Mailing Add:* 966 Faculty Dr Columbus OH 43221

GAINES, GREGORY, b Wiesbaden, Germany, Oct 14, 50; US citizen; div. MICROPLANKTON ECOLOGY, DINOFLAGELLATE CELL BIOLOGY. *Educ:* Grinnell Col, Iowa, AB, 72; Univ Southern Calif, Los Angeles, PhD(biol), 80. *Prof Exp:* Res assoc, dept bot, Southern Ill Univ, 80-84; fel, Dept Oceanog, Univ BC, 84-87; CONSULT SCIENTIST, US GOVT, 88- *Concurrent Pos:* Instr, Prog in Greece, Southern Ill Univ, 84- *Mem:* Am Soc Limnol & Oceanog; Phycol Soc Am; Soc Evolutionary Protistology; Soc Protozoologists. *Res:* Microplankton ecology, especially dinoflagellate heterotrophy, functional morphology, systematics and evolution in relation to other phytoplankton and microzooplankton, using techniques such as electron microscopy, fluorescence microscopy in both field and laboratory studies. *Mailing Add:* 2001 N Adams St No 903 Arlington VA 22201

GAINES, J H, b Luling, Tex, Mar 30, 31; m 52; c 3. ENGINEERING MECHANICS, AERONAUTICAL ENGINEERING. *Educ:* Univ Tex, Austin, BS, 57, MS, 59, PhD(eng mech), 66. *Prof Exp:* Instr eng mech, Univ Tex, Austin, 58-59; struct engr, Gen Dynamics, Ft Worth, 59-61; lectr civil eng, Grad Sch, Southern Methodist Univ, 61; instr eng mech, Univ Tex, Austin, 61-66; from asst prof to assoc prof aeronaut & mech eng, 66-72, PROF AEROSPACE ENG & ENG MECH, UNIV TEX, ARLINGTON, 72- *Concurrent Pos:* Consult, LTV Aerospace Corp, 68-70. *Mem:* Am Soc Civil Engrs; Am Inst Aeronaut & Astronaut; Am Soc Eng Educ. *Res:* Solid mechanics; structural analysis; vibrations; dynamics of structures. *Mailing Add:* Aerospace Eng Dept Univ Tex 1721 Ridgeview Ct Arlington TX 76012

GAINES, JACK RAYMOND, b Bozeman, Mont, May 9, 27; m 57; c 2. ORGANIC CHEMISTRY. *Educ:* Mont State Col, BS, 49, MS, 50, PhD(chem), 56. *Prof Exp:* Instr chem & physics, Western Mont Col Educ, 50-51; res chemist, Phillips Petrol Co, 56-57; assoc prof, 57-66, PROF ORG CHEM, SDAK SCH MINES & TECHNOL, 66- *Mem:* Am Chem Soc; Sigma Xi. *Res:* Basic organic chemistry research in heterocyclic compounds and medicinal chemistry; preparation of new plant growth regulators; production of a road de-icer from various biomass sources. *Mailing Add:* Dept Chem & Chem Eng SDak Sch Mines & Technol RR 1 Box 330 Rapid City SD 57702

GAINES, JAMES ABNER, b San Antonio, Tex, Aug 5, 27; m 57; c 9. ANIMAL SCIENCE & NUTRITION. *Educ:* Agr & Mech Col Tex, BS, 49, MS, 54; Iowa State Univ, PhD(animal breeding, genetics), 57. *Prof Exp:* ASSOC PROF ANIMAL SCI, VA POLYTECH INST & STATE UNIV, 56- *Concurrent Pos:* Mem staff, US AID, Argentina. *Mem:* Am Soc Animal Sci; Sigma Xi; Genetics Soc Am. *Res:* Animal genetics and husbandry; statistics; beef cattle crossbreeding. *Mailing Add:* Dept Animal Sci Va Polytech Inst & State Univ Blacksburg VA 24061

GAINES, JAMES R, b Cincinnati, Ohio, Sept 8, 35; m 67; c 2. SOLID STATE PHYSICS. *Educ:* Berea Col, AB, 56; Washington Univ, PhD(physics), 61. *Prof Exp:* From asst prof to assoc prof, 61-67, PROF PHYSICS, OHIO STATE UNIV, 67- *Concurrent Pos:* Consult, Avco Corp, 64-65; Alfred P Sloan fel, 64-67; vis scientist, AEC, Saclay, France, 65-66; lectr, Col France, 65-66; consult, Malaker Industs, 65-66; vis distinguished scholar, Univ Pa, 67; consult, Gardner Cryogenics Corp, 68-69, Lawrence Livermore Lab, 77- & Los Alamos, 77- *Mem:* Am Phys Soc. *Res:* Nuclear magnetic resonance on liquid and solid tritium, solid hydrogen, deuterium oxide, heavy water at temperatures below four degrees Kelvin; properties of semiconductors. *Mailing Add:* Dept Physics Univ Hawaii Honolulu HI 9682243210

GAINES, MICHAEL STEPHEN, b New York, NY, Jan 18, 43. POPULATION BIOLOGY. *Educ:* Tulane Univ, BS, 64; Ind Univ, MA, 65, PhD(zool), 70. *Prof Exp:* Asst prof systs & ecol, 70-75, ASSOC PROF SYST & ECOL, UNIV KANS, 75- *Concurrent Pos:* Res grant, NSF. *Mem:* Am Soc Study Evolution; Ecol Soc Am; Am Soc Mammalogists. *Res:* Population regulation of fluctuating vole populations; integrating genetics and ecology. *Mailing Add:* Dept Syst & Ecol Univ Kans 6007 Haw Lawrence KS 66045

GAINES, ROBERT D, b Bozeman, Mont, Sept 4, 33; m 57; c 2. BIOCHEMISTRY. *Educ:* Mont State Col, BS, 55, PhD(biochem), 60. *Prof Exp:* Res chemist, Minn Mining & Mfg Co, 54-57; sr chemist, Pillsbury Co, 60-61; from asst prof to assoc prof chem, 61-68, chmn dept, 62-66, PROF CHEM, CENT WASH STATE COL, 68-, CHMN DEPT, 77- *Res:* Carbohydrate chemistry; plant metabolism; microbial and animal physiology. *Mailing Add:* Dept Chem Cent Wash State Col Ellensburg WA 98926

GAINES, ROBERT EARL, b Champaign, Ill, Aug 16, 41; m 61; c 1. MATHEMATICS. *Educ:* Univ Ill, Urbana, BS, 62, MS, 63; Univ Colo, Boulder, PhD(math), 67. *Prof Exp:* Assoc prof, 67-77, PROF MATH & HEAD DEPT, COLO STATE UNIV, 77- *Concurrent Pos:* Vis Prof, Inst Math, Univ Catholique de Louvain, 75. *Mem:* Math Asn Am; Am Math Soc. *Res:* Boundary value problems for nonlinear ordinary differential equations; periodic solutions of nonlinear parabolic equations; mathematical economics. *Mailing Add:* 2533 Terry Lake Rd Ft Collins CO 80521

GAINES, TINSLEY POWELL, b Elberton, Ga, Feb 3, 39; m 68; c 2. ANALYTICAL CHEMISTRY, SOILS & SOIL SCIENCE. *Educ:* Univ Ga, BS, 61. *Prof Exp:* CHEMIST, COASTAL PLAIN EXP STA, UNIV GA, 65-; golf course consult, 75- *Concurrent Pos:* Pres, Tifton Phys Soil Testing Lab, Inc, 82. *Honors & Awards:* Outstanding Res Award, Sigma Xi, 83. *Mem:* Fel Asn Off Analytical Chemists; Am Soc Agron; Tobacco Chemists Res Conf; Sigma Xi; Am Peanut Res & Educ Soc; fel Am Inst Chemists. *Res:* Development of analytical chemical methods for soil and plant analysis; chemical analysis of soil and plant tissue. *Mailing Add:* 1412 Murray Ave Tifton GA 31793

GAINEY, LOUIS FRANKLIN, JR, b New Orleans, La, Nov 18, 47; c 4. PHYSIOLOGY, INVERTEBRATE ZOOLOGY. *Educ:* Fla State Univ, BS, 69, MS, 72, PhD(physiol), 76. *Prof Exp:* from asst prof to assoc prof, 76-89, chmn dept, 81-87, PROF DEPT BIOL, UNIV SOUTHERN MAINE, PORTLAND, 89- *Mem:* Am Soc Zoologists; AAAS; Sigma Xi. *Res:* Osmoregulation of bivalved molluscs and salt marsh insects; functional morphology of gastropod molluscs; effects of toxic dinoflagellates on bivalve mollusks. *Mailing Add:* Dept Biol Univ Southern Maine Portland ME 04103

GAINS, LAWRENCE HOWARD, b Brooklyn, NY, June 25, 48. ORGANIC CHEMISTRY, CHEMICAL EDUCATION. *Educ:* Univ Pa, BS, 70; Univ Ariz, MS, 77, PhD(chem), 80. *Prof Exp:* Res chemist, Nabisco Res Ctr, 70-72; instr organ chem, Univ Ariz, 78-79; res chemist, 79-85, SR RES CHEMIST, LORILLARD RES CTR, LORILLARD, INC, 85- *Concurrent Pos:* Lectr, Guilford Col, 81- *Mem:* Sigma Xi; Am Chem Soc; AAAS. *Res:* Synthetic chemistry, specializing in anaerobic and anhydrous techniques as applied to organo-metallic chemistry; transition metal complexes and natural product syntheses particularly syntheses of flavorants. *Mailing Add:* 602 Ashland Dr Greensboro NC 27403-1910

GAINTNER, JOHN RICHARD, b Lancaster, Pa, Feb 18, 36; m 61; c 3. MEDICINE. *Educ:* Lehigh Univ, BA, 58; Johns Hopkins Univ, MD, 62; Am Bd Internal Med, dipl, 71. *Prof Exp:* NIH fel hemat, Johns Hopkins Univ Hosp, 66-67; from asst prof to assoc prof med, Sch Med, Univ Conn Health Ctr, Farmington, 67-77, assoc dean clin affairs, 74-75, chief staff, Univ Hosp, 69-75; assoc dean admin, Sch Med, Johns Hopkins Univ, 77-80, assoc prof med, 77-83, vpres & dept dir, 81-83; PRES & CEO, ALBANY MED CTR & PROF MED, ALBANY MED COL, 83- *Honors & Awards:* Borden Res Award, 62. *Mem:* Fel Am Col Physicians; Am Med Asn; Med Admin Conf; fel Am Col Physician Exec; Soc Med Aminrs. *Res:* Health care delivery; medical administration; hematology, coagulation and blood platelets; malaria. *Mailing Add:* Albany Med Ctr 43 New Scotland Ave Albany NY 12208

GAIR, JACOB EUGENE, b Pittsfield, Mass, Apr 1, 22; m 44; c 3. GEOLOGY. *Educ:* Univ Rochester, AB, 46; Johns Hopkins Univ, PhD(geol), 49. *Prof Exp:* Asst prof geol, Univ Ore, 49-52; geologist, US Geol Surv, 52-87; RETIRED. *Concurrent Pos:* Ed Econ Geol. *Mem:* Geol Soc Am; Soc Econ Geol. *Res:* Petrography and stratigraphy of Precambrian iron ranges of northern Michigan, and of iron and gold region of Minas Gerais, Brazil; structural geology; central Appalachians; petrography; volcanic rocks; iron exploration of Turkey; tungsten exploration of North Carolina; massive sulfide deposits of southern Appalachians; mineral investigations in Eastern Desert of Egypt; mineral assessments of public and Indian lands and Piedmont to Blue Ridge belts in Georgia, North Carolina and South Carolina. *Mailing Add:* 9609 Hillridge Dr Kensington MD 20895

GAISSER, THOMAS KORFF, b Evansville, Ind, Mar 12, 40; m 64; c 1. PARTICLE PHYSICS, COSMIC RAY PHYSICS. *Educ:* Wabash Col, BA, 62; Univ Bristol, Eng, MSc, 65; Brown Univ, PhD(physics), 67. *Prof Exp:* Res assoc physics, Mass Inst Technol, 67-69; NATO fel, Cambridge Univ, Eng, 69-70; from asst prof to assoc prof, 70-79, PROF PHYSICS, BARTOL RES INST, UNIV DEL, 79- *Concurrent Pos:* Vis scientist, Brookhaven Nat Lab, 76 & Univ Rome, Italy, 90; vis prof, Univ Wis-Madison, 84. *Mem:* fel Am Phys Soc; Sigma Xi. *Res:* Cosmic rays and high energy physics and astrophysics; particle phenomenology. *Mailing Add:* Bartol Res Inst Univ Del Newark DE 19716

GAIT, ROBERT IRWIN, b Johannesburg, SAfrica, Sept 12, 38; Can citizen. MINERALOGY, CRYSTALLOGRAPHY. *Educ:* Univ Witwatersrand, BSc, 58, Hons, 59; Univ Man, MSc, 64, PhD(mineral), 67. *Prof Exp:* Geologist diamond prospecting, Williamson Diamonds Ltd, Tanzania, 60-62; asst cur, 67-71, assoc cur, 71-78, CUR MINERAL, ROYAL ONT MUS, TORONTO, 78- *Concurrent Pos:* Mineral res grant, Dept Univ Affairs, Ont, 69-70 & 70-71; hon dir, Can Gemmological Asn, Toronto. *Honors & Awards:* Mineral named in honor, Gaitite, 80. *Mem:* Mineral Asn Can (secy, 73); fel Mineral Soc Am; Geol Soc SAfrica; Mineral Asn SAfrica. *Res:* New minerals; mineral localities; pyrite and quartz crystallography. *Mailing Add:* Dept Mineralogy Royal Ont Museum 100 Queens Park Toronto ON M5S 2C6

GAITHER, ROBERT BARKER, b North Bay, Ont, Aug 12, 29; US citizen; m 54; c 2. MECHANICAL ENGINEERING. *Educ:* Auburn Univ, BME, 51; Univ Ill, MSME, 57, PhD(mech eng), 62. *Prof Exp:* Instr mech eng, Univ Ill, 57-62; assoc prof, 62-66, PROF MECH ENG, UNIV FLA, 66-, CHMN DEPT, 64- *Concurrent Pos:* Officer, USN, 51-54. *Honors & Awards:* ASME Centennial Medalion, Am Soc Mech Engrs, 80,. *Mem:* Am Soc Mech Engrs (pres, 81-82); Am Soc Eng Educ. *Res:* Statistical thermodynamics; gas dynamics; plasma diagnostics; transport properties; engineering education. *Mailing Add:* Dept Mech Eng Univ Fla 237 MEB Bldg Gainesville FL 32601

GAITHER, THOMAS WALTER, b Great Falls, SC, Nov 12, 38; m 68; c 1. BOTANY. *Educ:* Claflin Col, BS, 60; Atlanta Univ, MS, 64; Univ Iowa, PhD(bot), 68. *Prof Exp:* Teaching asst life sci, Univ Iowa, 64-66, gen bot, 66-67 & mycol, 67-68; assoc prof, 68-80, PROF GEN BIOL, GEN BOT & NONVASCULAR PLANT MORPHOL, SLIPPERY ROCK STATE COL, 80- *Mem:* AAAS; Mycol Soc Am; Bot Soc Am; Sigma Xi. *Res:* Ultrastructure of myxomycetes; morphology and development of capillitium; ultrastructural changes in angiosperm leaves throughout the growing season; mycology. *Mailing Add:* Dept Biol Slippery Rock State Col Slippery Rock PA 16057

GAITHER, WILLIAM SAMUEL, b Lafayette, Ind, Dec 3, 32; m 59; c 1. ACADEMIC ADMINISTRATION. *Educ:* Rose Polytech Inst, BScE, 56; Princeton Univ, MSE, 62, MA, 63, PhD(civil eng), 64. *Prof Exp:* Chief party, Ayrshire Collieries Corp, Ind, 54, construct inspector, 55; from field engr to res engr, Dravo Corp, Pa, 56-60; from field eng to field supt, Meyer Corp, Wis, 60-61; grad asst, Princeton Univ, 61-63, lectr, 64; assoc prof coastal eng, Univ Fla, 64-65; from supv engr to chief engr, port & coastal develop, Pipeline Div, Bechtel Corp, 65-67; from assoc prof to prof civil eng, Univ Del, 67-84, dean Col Marine Studies, 70-84; prof & pres, Drexel Univ, 84-87; vchmn, Roy F Weston Inc, 87-89; PRES & TRUSTEE, WESTON INST, 88- *Concurrent Pos:* Arthur Le Grand Doty fel, 61-62; Ford Found fel port planning, 62-64; dir, Roy F Weston Co, Inc, 74-90, Mutual Assurance Co, 85-, Philadelphia Elec Co, 85-90, Univ City Sci Ctr, 84- & Penjerdel Coun, 84-; mem, Marine Bd, Nat Res Coun, 75-81, Drexel Univ, 84-87 & Weston Inst, 88- *Honors & Awards:* Norman J Sollenberger Award, 83; William Chapin Award, 85. *Mem:* Am Soc Civil Engrs; Soc Naval Archit & Marine Engrs. *Res:* Marine structures; ocean engineering; marine transporation systems; marine and environmental organization and policy. *Mailing Add:* 3601 Baring St Philadelphia PA 19104

GAITZ, CHARLES M, b Victoria, Tex, May 7, 22; m 49; c 3. PSYCHIATRY, GERONTOLOGY. *Educ:* Rice Univ, BA, 42; Univ Tex Med Br, Galveston, MD, 46. *Prof Exp:* Resident psychiat, Johns Hopkins Hosp, 49-52; clin asst prof, Baylor Col Med, 52-78, dir residency training, 62-65; chief, Gerontol Res Sect, 66-78, dir, Clin Serv Div & head, Dept Appl Res, 78-80, head Gerontol Ctr, 80-85, HEAD MEM FAMILY PRACT CTR, TEX RES INST MENT SCI, 85- *Concurrent Pos:* Consult psychiat, Jewish Home for Aged, Houston, 56-; NIMH grants, 66-69 & 69-72, mem soc probs res rev comt, NIMH, 71-75; mem health sect, White House Conf Aging, 71; pres med staff, Houston Int Hosp, 72-73; mem coun res & develop, Am Psychiat Asn, 75-79, coun aging, 81-86; mem panel aging, President's Comn Mental Health, 77-78; clin prof psychiat, Baylor Col Med, 78- & Med Sch, Univ Tex, Houston, 83- *Mem:* Fel Am Psychiat Asn; fel Am Col Psychiatrists; Group Advan Psychiat; fel Geront Soc (pres, 76-77); fel Am Geriat Soc. *Res:* Delivery of health services to aged persons; relationship of leisure and mental health; treatment of senile dementia; issues faced by aging physicians. *Mailing Add:* 6550 Mapleridge Suite 208 Houston TX 77081

GAJAN, RAYMOND JOSEPH, b Missoula, Mont, Sept 30, 20; m 50; c 3. ANALYTICAL CHEMISTRY, ORGANIC CHEMISTRY. *Educ:* Univ Mont, BA, 43; Canisius Col, MS, 48. *Prof Exp:* Chemist, Nat Aniline Div, Allied Chem & Dye Corp, 43-49; inspection equip agency, US Army Chem Ctr, Md, 50-54, chem & radiol lab, 54-57; US Bur Mines, Md, 57-60; res chemist, Div Food Chem, Bur Sci, US Food & Drug Admn, 60-71, res chemist, 71-74, sr scientist, div chem & physics, 74-82, sr scientist, div chem technol, off sci, Bur Foods, 82-86; RETIRED. *Mem:* Sigma Xi; Am Chem Soc; fel Asn Off Anal Chem. *Res:* Development of electroanalytical procedures for the determination of pesticide residues, drugs, food additives and other contaminants when present in foods in less than microgram amounts. *Mailing Add:* 13109 Magellan Ave Rockville MD 20853

GAJDUSEK, DANIEL CARLETON, b Yonkers, NY, Sept 9, 23; m 36. PEDIATRICS, VIROLOGY. *Educ:* Univ Rochester, BA, 43; Harvard Med Sch, MD, 46. *Hon Degrees:* DSc, Univ Rochester, 77, Med Col Ohio, 77, Washington & Jefferson Col, Pa, 80, Harvard Med Sch, 82, Hahnemann Med Col & Hosp, 82; HHD, Hamilton Col, 77; Dr, Univ Aix-Marseille, 77; LLD, Univ Aberdeen, 80; Dr, Univ Sch Med & Dent, NJ, 87. *Prof Exp:* Intern pediat, Columbia-Presby Med Ctr, NY, 46-47; resident, Children's Hosp, Cincinnati, Ohio, 47-48; pediatrician, Med Mission, Ger, 48; sr resident pediat, Boston Children's Med Ctr, 49-51; sr investr infectious dis, Pasteur Inst Iran, 54-55; vis investr, Walter & Eliza Hall Inst Med Res, Australia, 55-57; CHIEF LAB SLOW LATENT & TEMPERATE VIRUS INFECTIONS & CHIEF LAB CENT NERVOUS SYST STUDIES, NAT INST NEUROL DISORDERS & STROKE, 58- *Concurrent Pos:* Nat Res Coun sr fel phys chem, Calif Inst Technol, 48-49; sr fel, Harvard Med Sch, 49-52 & 50-51; fel, Children's Med Ctr & Harvard Med Sch, Nat Found Infantile Paralysis, 50-52; Bicentnial lectr, Harvard Med Sch, 85. *Honors & Awards:* Nobel Prize in Physiol & Med, 76; Mead Johnson Award, Am Acad Pediat, 63; Health Clark Lectr, Univ London, Dyer Lectr, NIH, 74; Rachford Lectr, Children's Hosp, Cincinnati; Cotzias Prize, Am Acad Neurol, 78; Sterling Lectr, Yale Univ, 81; Hitchcock Lectr, Univ Calif, Berkeley, 82; Nelson Lectr, Univ Calif, Davis, 82; Harvey Lectr, NY Med Soc, 77; Eastmen Lectr, Univ Rochester, 85; Bayne Hones Lectr, Johns Hopkins Med Sch; Withering Lectr, Univ Burmingham, Eng; Cannon Elie Lectr, Children's Med Ctr, Boston; Lucian Dautreband Award Pathophysiol, Belg, 76. *Mem:* Nat Acad Sci; Am Pediat Soc; Soc Pediat Res; Am Anthrop Asn; Am Acad Neurol; Am Philos Soc; Am Acad Arts & Sci; Leopoldiana German Acad Natural Researchers; Am Neurol Asn; Nat Acad Med Belg (Antwerp & Brussels); Nat Acad Med Columbia; Nat Acad Med Mex; Nat Acad Med Slovakia. *Res:* Protein physical chemistry; mammalian virology; pathophysiology of autoimmune diseases and neurological degenerative disorders; human evolutionary studies in isolated populations; child behavior, development, nervous system patterning and learning and disease patterns in primitive cultures; theory of cyphers and notation for the coding of sensory and meter data in neurological information processing. *Mailing Add:* NIH Bldg 36 Rm SB-21 Bethesda MD 20892

GAJENDAR, NANDIGAM, b Nellore, India, Nov 29, 40; m 74; c 1. SOFTWARE SYSTEMS, COMPUTER SCIENCE. *Educ:* Sri Venkateswara Univ, India, BA, 59, MS, 61; Indian Inst Technol, Kharagpur, PhD(appl math), 65. *Prof Exp:* Lectr math, Indian Inst Technol, 65-70; asst prof math, 70-73, assoc prof, 73-80, PROF COMPUT SCI & SYSTS ANALYST, INFORM RESOURCE CTR, GRAMBLING STATE UNIV, 80- *Concurrent Pos:* Grant, Manned Spacecraft Ctr, NASA, 70-71. *Mem:* Asn Comput Mach; Am Asn Artifical Intel; Inst Elec & Electronics Engrs. *Res:* Computer software design and application development; numerical applications to structural problems. *Mailing Add:* Dept Math & Comput Sci Grambling State Univ Grambling LA 71245

GAJEWSKI, FRED JOHN, b New York, NY, May 2, 12; m 39; c 2. CHEMISTRY. *Educ:* NY Univ, BS, 34, PhD(chem), 38. *Prof Exp:* Asst chem, NY Univ, 34-37; res chemist, Calco Chem Co, 37-38; area supvr res & develop, GAF Corp, 38-55, mgr process res & develop, 55-60, tech dir, Antara Div, 60-62, asst to vpres & gen mgr dyestuff & chem div, 62-64, tech adminr res & develop, 64-68, dir shareholder rels, 68-70; dir tech opers, Azoplate Corp, 70-73, dir tech opers, 70-77, TECH CONSULT, AZOPLATE DIV, AM HOECHST CORP, 77- *Mem:* AAAS; Am Chem Soc; fel Am Inst Chem. *Res:* Surfactants; acetylene derivatives; dyestuffs; pigments; intermediates; metal carbonyls; metal powders; graphic arts. *Mailing Add:* 30 Westbrook Rd Westfield NJ 07090

GAJEWSKI, JOSEPH J, b Hammond, Ind, Nov 7, 39; m 62; c 2. ORGANIC CHEMISTRY. *Educ:* Loyola Univ Chicago, BS, 61; Univ Wis, PhD(org chem), 65. *Prof Exp:* NSF fel org chem, Columbia Univ, 65-66; from asst prof to assoc prof chem, 66-74, PROF CHEM, IND UNIV, BLOOMINGTON, 74- *Concurrent Pos:* Alfred P Sloan fel, 71-73. *Mem:* Am Chem Soc; Royal Soc Chem. *Res:* Energetics and stereochemistry of chemical reactions; application of molecular orbital theory to organic chemistry. *Mailing Add:* Dept Chem Ind Univ Bloomington IN 47401

GAJEWSKI, RYSZARD, b Warsaw, Poland, Feb 23, 30; US citizen; m 54; c 2. RESEARCH ADMINISTRATION, PLASMA PHYSICS. *Educ:* Warsaw Tech Univ, MS, 54; Inst Physics, Polish Acad Sci, PhD(physics), 58. *Prof Exp:* Var teaching pos, 50-67; prof physics, Warsaw Tech Univ, 67-68; DSR staff mem, Mass Inst Technol, 69-71; vis assoc prof, Brandeis Univ, 71-72; sr staff mem, Am Sci & Eng, 72-73, dir res, 73-76; dir, Div Advan Energy Projs, Dept Energy, 77-90; VPRES, RES & DEVELOP, PHYSICAL OPTICS CORP, 90- *Concurrent Pos:* Alfred P Sloan fel, Mass Inst Technol, 59-60; head lab plasma theory, Inst Nuclear Res, Warsaw, 58-68; fel, Case Inst Technol, 60-61; vis scientist, Boeing Sci Res Labs, 61 & 64; mem comt peaceful uses nuclear energy, Polish Acad Sci, 64-68; vis scientist, Inst Plasma Physics, Jutphaas, Neth, 69. *Mem:* AAAS; Am Phys Soc. *Res:* Plasma equilibrium and stability in laboratory and astrophysical magnetic fields; atoms in super strong magnetic fields; research administration. *Mailing Add:* Phys Optics Corp 20600 Gramercy Pl Torrance CA 90501

GAJEWSKI, W(ALTER) M(ICHAEL), b Hartford, Conn, Apr 4, 23; m 49; c 3. MECHANICAL & NUCLEAR ENGINEERING. *Educ:* Univ Conn, BEE, 49, MS, 51. *Prof Exp:* Engr, Bettis Atomic Power Lab, 51-55, sr engr & fel engr, 56, supvry engr, 56-59, sect mgr, 59-66, plant mgr, 66-68, asst proj mgr, Westinghouse-Bettis Atomic Power Plant, Pa, 68-70; eng mgr, Fast Flux Test Fac Proj, Hanford Eng Lab, Westinghouse Hanford Co, 70-72, mgr reactor opers eng, 72-85, asst mgr eng, 85-88; RETIRED. *Concurrent Pos:* Consult, Nuclear/Gen Eng Area. *Mem:* Inst Elec & Electronics Engrs; Am Nuclear Soc; fel Am Soc Mech Engrs; Nat Mgt Asn. *Res:* General engineering; reactor plant analysis and systems design; overall reactor plant design, development and evaluation; experimental/research reactor operations. *Mailing Add:* 3103 S Everett Pl Kennewick WA 99337

GAJJAR, JAGDISH T(RIKAMJI), b Bombay, India, May 23, 40; m 66; c 1. ELECTROOPTICS, ELECTRICAL ENGINEERING. *Educ:* Univ Bombay, BE, 60 & 61; Univ Okla, MEE, 63; Univ Houston, PhD(elec eng), 70. *Prof Exp:* Proj engr, Fischbach & Moore Systs Inc, Tex, 63-64; instr elec eng, Univ Tulsa, 66-67; asst prof to assoc prof, 70-84, PROF ELEC ENG & COMPUTER SCI, UNION COL, SCHENECTADY, NY, 84- *Concurrent Pos:* Prin investr, NSF eng res initiation grant, 71-72; consult, GE Corp Res & Develop Ctr, 77-86; Fulbright faculty fel, 85-86. *Mem:* Inst Elec & Electronics Engrs; Optical Soc Am. *Res:* Optical information processing; system theory and system diagnostics; applied instrumentation; atmospheric probing. *Mailing Add:* Dept Elec Eng Union Col Steinmetz 221 Schenectady NY 12308

GAKENHEIMER, WALTER CHRISTIAN, b Baltimore, Md, July 28, 16; m 41, 63. PHARMACEUTICAL CHEMISTRY. *Educ:* Univ Md, BS, 38, MS, 41, PhD(pharmaceut chem), 43. *Prof Exp:* Asst instr pharm, Univ Md, 38-42; synthetic org res chemist, Minerec Corp, Md, 42-45; sr chemist pharmaceut res, Merck & Co, Inc, 45-50, mgr tech serv, 50-53, mgr med mkt, 53-57; sci asst to pres, Stuart Co, 57-59, admin vpres, 59-63; tech dir int div, Atlas Chem Industs, Inc, 63-71; tech dir, 71-77, MGR RES QUAL ASSURANCE, LATIN AM DIV, ICI AMERICAS, 77- *Honors & Awards:* L S Williams Award. *Mem:* AAAS; Am Chem Soc; Drug Info Asn; NY Acad Sci. *Res:* Pharmaceutical research and development; synthesis of aminoalkanols of pharmacological interest. *Mailing Add:* 413 Stafford Rd Wilmington DE 19803

GAL, ANDREW EUGENE, b Budapest, Hungary, July 14, 18; nat US; m 51; c 3. ORGANIC CHEMISTRY. *Educ:* Swiss Fed Inst Technol, Chem Eng, 43, ScD(org chem), 47. *Prof Exp:* Res chemist, Cruet Labs, France, 48-50 & Roques, 50-53; sr chemist, Nat Drug Co, 53-62; supvr org chem, Hazleton Labs, 62-64; res chemist, Lab Neurochem, Nat Inst Neurol Dis & Stroke, 64-73, chief, Sect Neurochem Methodol, Nat Inst Neurol & Commun Dis & Stroke, 73-89; RETIRED. *Honors & Awards:* Spec Recognition Award, USPHS, 76. *Mem:* Am Chem Soc. *Res:* Synthesis of organic chemicals; synthetic estrogens; carbohydrates/alkaloids; barbituric acids; heterocyclic compounds; sphingolipids; radiochemicals; diagnostic reagents; analytical biochemistry; neurochemistry. *Mailing Add:* 707 Ware St SW Vienna VA 22180

GAL, GEORGE, b Pecel, Hungary, July 18, 21; m 45; c 1. ORGANIC CHEMISTRY. *Educ:* Pazmany Peter Univ, Hungary, PhD(chem), 44. *Prof Exp:* Res chemist explosives, Nitrochem Factory, Hungary, 44-46; pharmaceut, United Pharmaceut Factory, 46-52, mgr res, 52-56; sr res chemist, Merck Sharp & Dohme Labs, 57-80, sr res fel, 80-86; RETIRED. *Mem:* Am Chem Soc. *Res:* Reduction with complex metalhydrides; synthesis of amino alcohols, amino acids, heterocyclic compounds, steroids, antibiotics and peptides. *Mailing Add:* Five Timberline Way Watchung NJ 07060

GAL, JOSEPH, STEREOLSELECTIVE DRUG METABOLISM & ANALYSIS. *Educ:* Univ Calif, Davis, PhD(chem), 71. *Prof Exp:* ASSOC PROF MED & PHARMACOL, SCH MED, UNIV COLO, 83- *Mailing Add:* Box C237 UCHSC Denver CO 80262

GAL, SUSANNAH, b Battle Creek, Mich, Oct 29, 58; m 85. MOLECULAR BIOLOGY. *Educ:* John Hopkins Univ, PhD(biochem), 86. *Prof Exp:* Fel biotechnol, Nat Cancer Inst, NIH, 86-87; POSTDOC FEL, FRIEDRICH MIESCHER INST, 87. *Mem:* Am Soc Cell Biol; Sigma Xi; Mid-Atlantic Plant Molecular Biol Soc. *Res:* Plant molecular biology, studying the factors which affect genomic recombination in plants. *Mailing Add:* Friedrich Miescher Inst PO Box 2543 CH 4002 Basel Switzerland

GALA, RICHARD R, b Bayonne, NJ, July 2, 35; div; c 2. PHYSIOLOGY, ENDOCRINOLOGY. *Educ:* Rutgers Univ, BS, 57, PhD(neuro-endocrinol), 63. *Prof Exp:* Asst reproductive physiol, Univ Maine, 57-58; asst bact, Rutgers Univ, 58-60; res assoc endocrinol & biochem, Sch Med, Univ Louisville, 63-65; asst prof physiol, Sch Med, Boston Univ, 65-71; assoc prof, 71-76, PROF PHYSIOL, SCH MED, WAYNE STATE UNIV, 76- *Concurrent Pos:* Fel endocrinol & biochem, Sch Med, Univ Louisville, 63-65; vis scientist, Nat Inst Res Dairying, Shinfield, Reading, Eng, 79-80; vis distinguished scientist, Meiji Univ, Kanagawa, Japan, 88. *Mem:* AAAS; Soc Exp Biol & Med; Brit Endocrine Soc; Endocrine Soc; Am Physiol Soc; Int Soc Neuroendocrinology; Soc Neuroscience. *Res:* Relationship between nervous system and anterior pituitary; reproductive physiology; mammary gland physiology and tumorigensis. *Mailing Add:* Dept Physiol Wayne State Univ Sch Med Detroit MI 48201

GALABURDA, ALBERT MARK, b Santiago, Chile, July 20, 48; US citizen. NEUROLOGY, NEUROANATOMY. *Educ:* Boston Univ, AB & MD, 71. *Prof Exp:* Intern, Boston City Hosp, 71-72, resident internal med, 72-73, resident neurol, 73-76, asst vis neurologist, 76-77; ASST NEUROLOGIST, BETH ISRAEL HOSP, 76- *Concurrent Pos:* Mem, Am Bd Internal Med, 75 & Am Bd Psychiat & Neurol, 77. *Mem:* Am Acad Neurol; AAAS; Pan-Am Med Asn; NY Acad Sci; Soc Neurosci; Am Neurol Asn; Am Asn Anat; Behav Neurol Soc; Int Neurophysical Soc. *Res:* Neuroanatomical research in the fields of cytoarchitectonics, language and behavior. *Mailing Add:* Neurol Unit Beth Israel Hosp 330 Brookline Ave K-4 Boston MA 02215

GALAMBOS, JANOS, b Zirc, Hungary, Sept 1, 40; m 64. STATISTICS. *Educ:* Eotvos Lorand Univ, Budapest, MSc & PhD(probability), 63. *Prof Exp:* Asst prof porbability & statist, Eotvos Lorand Univ, 63-65; lectr math, Univ Ghana, 65-69 & Univ Ibadan, 69-70; from asst prof to assoc prof, 70-73, PROF MATH, TEMPLE UNIV, 73- *Concurrent Pos:* vis res fel statist lab, Cambridge Univ, 68; vis prof, Iowa State Univ, 73-74, Goethe Univ, Frankfurt, 74-75, Australian Nat Univ, 78, Keio Univ, Yokohama, 85 & Univ Beijing, 89. *Honors & Awards:* Fel, Humboldt Found, Bonn, 74-75. *Mem:* Am Math Soc; fel Inst Math Statist; Int Statist Inst. *Res:* Order statistics; distributions; probabilistic number theory; probabilistic inequalities; engineering applications of statistics. *Mailing Add:* Dept Math TU 038-16 Temple Univ Philadelphia PA 19122

GALAMBOS, JOHN THOMAS, b Budapest, Hungary, Oct 29, 21; nat US; c 3. MEDICINE. *Educ:* Univ Ga, BS, 48; Emory Univ, MD, 52. *Prof Exp:* Intern med, Barnes Hosp, St Louis, 52-53; resident, Billings Hosp, Chicago, 53-54; USPHS res fel, Chicago, 54-55; fel med, 55-57, assoc, 57-58, from asst prof to assoc prof, 58-67, PROF MED, SCH MED, EMORY UNIV, 67-,IN-CHG GASTROENTEROL TEACHING PROG, 58-, DIR DEPT MED, DIV DIGESTIVE DIS, 68- *Concurrent Pos:* Dir gastroenterol clin, Grady Mem Hosp, 57-; assoc physician, Emory Univ Hosp, 58-; consult, Vet Admin Hosp, Atlanta, Ga. *Mem:* AAAS; NY Acad Sci; Am Col Gastroenterol; Am Gastroenterol Asn; Am Col Physicians. *Res:* Liver function and alcoholic liver disease; portal hypertension and varices and nutrition in alcoholic liver disease; metabolism of adult human liver culture and glycosaminoglycan and collagen metabolism. *Mailing Add:* Dept Med Emory Univ 1364 Clifton Rd NE Atlanta GA 30322

GALAMBOS, ROBERT, b Lorain, Ohio, Apr 20, 14; m 39; c 3. PHYSIOLOGY, PSYCHOLOGY. *Educ:* Oberlin Col, AB, 35, MA, 36; Harvard Univ, AM, 38, PhD, 41; Univ Rochester, MD, 46; Yale Univ, MA, 62. *Hon Degrees:* MD, Univ Goteborg, Sweden, 71. *Prof Exp:* Instr & jr investr physiol, Harvard Med Sch, 42-43; intern, Emory Univ Hosp, 46, asst prof anat, Med Sch, 46-47; res fel, Psycho-Acoust Lab, Harvard Univ, 47-51; chief dept neurophysiol, Walter Reed Army Inst Res, 51-62; Higgins Prof psychol & physiol, Yale Univ, 62-68; prof, 68-81, EMER PROF NEUROSCI, SCH MED, UNIV CALIF, SAN DIEGO, 81-; RES SCIENTIST, CHILDREN'S HOSP RES CTR, SAN DIEGO, 81- *Mem:* Nat Acad Sci; Am Acad Arts & Sci; Am Physiol Soc; Acoust Soc Am. *Res:* Hearing; obstacle avoidance by bats; neurophysiology of learning. *Mailing Add:* Dept Neurosci 0608 Univ Calif San Diego 9500 Gilman Dr La Jolla CA 92093-0608

GALAMBOS, THEODORE V, b Budapest, Hungary, Apr 17, 29; US citizen; m 57; c 4. CIVIL ENGINEERING. *Educ:* Univ NDak, BS, 53, MS, 54; Lehigh Univ, PhD(civil eng), 59. *Hon Degrees:* DSc, Tech Univ Budapest, 72. *Prof Exp:* Stress analyst, Babcock & Wilcox Co, 54-56; from res asst to res assoc, Lehigh Univ, 56-59, from asst prof to assoc prof civil eng, 59-65; prof, Washington Univ, 65-68, chmn dept civil eng, 70-78, Harold R Jolley prof civil eng, 68-81; JAMES L RECORD PROF STRUCT ENG, UNIV MINN, 81- *Concurrent Pos:* Chmn Column Res Coun, 70-74. *Honors & Awards:* Norman Medal, Am Soc Civ Engrs, 83. *Mem:* Nat Acad Eng; Am Soc Civil Engrs; Column Res Coun; Int Asn Bridge & Struct Engrs; Nat Soc Prof Engrs. *Res:* Inelastic behavior of metal structures and structural elements; static instability of members and frames; probability-based structural design; testing of large concrete structure under dynamic load. *Mailing Add:* Civil & Mineral Eng Dept Univ Minn Minneapolis MN 55455

GALAN, LOUIS, b Pineres, Asturias, Spain, Aug 10, 28; nat US; m 59; c 2. AERONAUTICAL MANAGEMENT, ASTRONAUTICAL MANAGEMENT. *Educ:* Mass Inst Technol, BS, 51; Univ Mich, MS, 54 & 55. *Prof Exp:* Prog mgr, Apollo lunar exp, Mars Viking siesmometer, Toronto Zoo Ride Vehicle, Bendix Corp Aerospace Systs Div, 58-75, res engr, diesel fuel injection, Bendix Res Labs, 75-77; res engr energy systs, Environ Res Inst Mich, 77-81; dir eng, Photon Sources Inc, 81-84; DIR ENG, APPL INTELLIGENT SYSTS, 84- *Mem:* Am Soc Prof Engrs; Soc Mfg Engrs. *Mailing Add:* 4030 W Loch Alpine Dr Ann Arbor MI 48103

GALANE, IRMA B(ERESTON), b Baltimore, Md, Aug 23, 21; div; c 1. ELECTRONICS ENGINEERING. *Educ:* Goucher Col, BA, 40. *Prof Exp:* Asst math, Sch Eng, Johns Hopkins Univ, 41-42; physicist, Naval Ord Lab, 42-43; electronic engr, Bur Ships, USN, 43-49; electronic engr, Off Chief Signal Officer, US Army, 49-51; electronic engr, Bur Aeronaut, USN, 51-56; electronic scientist, Air Res & Develop Command, USAF, 56-57; electronic engr, Fed Commun Comn, 57-60; res engr, Goddard Space Flight Ctr, NASA, 60-62; supvry electronic engr, US Coast Guard Hqs, 62-64; eng specialist, Libr Cong, 64-66; gen engr, US Dept Navy, 66-71; SR RES ENGR, FED COMMUN COMN, 71- *Concurrent Pos:* Instr, Com Radio Inst, Md, 41-42; Navy mem & secy, Res & Develop Bd Radome Subpanel, Dept Defense, 51-56, mem int radio consult comt, Int Telecommun Union, 51-56; NASA Authorized Commun Technol Satellite experimenter, 77-; assoc, Smithsonian Inst. *Mem:* AAAS; sr mem Inst Elec & Electronics Engrs; Nat Soc Prof Engrs; Am Inst Aeronaut & Astronaut; Marine Technol Soc. *Res:* Satellite communication, radar and navigational systems; antennas; propagation; nuclear radiation detection instrumentation; infrared applications; missile radomes; oceanography. *Mailing Add:* 4201 Cathedral Ave NW Washington DC 20016

GALANIS, NICOLAS, b Athens, Greece, May 6, 39; Can citizen; m 66; c 2. HEAT & MASS TRANSFER. *Educ:* Nat Tech Univ, Greece, dipl eng, 64; Cornell Univ, MS, 67, PhD(thermal eng), 70. *Prof Exp:* Proj engr, Thermomech Consults, Athens, 64; from asst prof to assoc prof, 70-81, head dept, 81-84, assoc dean, 85-86, PROF THERMODYN HEAT TRANSFER, UNIV SHERBROOKE, 81- *Concurrent Pos:* Sci consult, Dept Econ Coord, Ctr Planning & Econ Res, Athens, Greece, 76; res assoc, Univ Queensland, Australia, 77; consult, BEAK Consult Ltd, Can, 78-79; res assoc, Ctr Nat de la Recherche Sci, France, 79-80; prin investr & adminr res contracts, Ministere de L'energie et des Ressources du Que, 80. *Honors & Awards:* R R Teetor Award, Soc Automotive Engrs, 72. *Mem:* Eng Inst Can; fel Can Soc Mech Eng (vpres, 80-81); Solar Energy Soc Can; Tech Chamber Greece. *Res:* Modelling, simulation and experimentation related to energy needs of buildings and to the utilization of heat pumps, solar and wind energy; author of over 25 technical publications. *Mailing Add:* Mech Eng Dept Fac Appl Sci Univ Sherbrooke 2500 University Blvd Sherbrooke PQ J1K 2R1 Can

GALARDY, RICHARD EDWARD, ENZYMOLOGY. *Educ:* Rockefeller Univ, PhD, 73. *Prof Exp:* ASSOC PROF BIOCHEM, UNIV KY, 80- *Mailing Add:* Dept Biochem Univ Ky 101 Sanders-Brown Bldg Lexington KY 40536

GALAS, DAVID JOHN, b St Petersburg, Fla, Feb 25, 44; m 67; c 2. MOLECULAR GENETICS. *Educ:* Univ Calif, Berkeley, AB, 67; Univ Calif, Davis, MS, 68, PhD(physics), 72. *Prof Exp:* Sr scientist, Biomed Div, Lawrence Livermore Lab, Univ Calif, Berkeley, 74-77; supvr res, Dept Molecular Biol, Univ Geneva, Switz, 77-81; from asst prof to assoc prof, 81-83, dir dept, 85-89, PROF, DEPT MOLECULAR BIOL, UNIV SOUTHERN CALIF, 88-; DIR HEALTH & ENVIRON RES, US DEPT ENERGY, 89- *Concurrent Pos:* Fel, Hertz Found. *Mem:* AAAS; Am Phys Soc; Genetics Soc Am. *Res:* Molecular genetics of transposition and the mechanisms and consequences of these recombination processes; molecular interactions of DNA with proteins and their consequences in gene control and recombination. *Mailing Add:* Dept Molecular Biol Univ Southern Calif Los Angeles CA 90089

GALASSO, FRANCIS SALVATORE, b Monson, Mass, Apr 26, 31; m 50; c 2. SOLID STATE CHEMISTRY, INORGANIC CHEMISTRY. *Educ:* Univ Mass, BS, 53; Univ Conn, MS, 57, PhD(chem), 60. *Prof Exp:* Res asst solid state chem, Univ Conn, 56-60; res scientist, Res Labs, United Aircraft Corp, 60-62, res supvr mat synthesis group, 62-67, chief mat synthesis sect, 67-74; prin scientist, 74-78, sr mat scientist, 78-85, MGR, MAT SYNTHESIS SECT, UNITED TECHNOL RES CTR, 85- *Concurrent Pos:* Sci & eng adv to US Rep Emilio Daddario, Conn; consult, NASA space exp; vis prof, Univ Conn, 85- *Honors & Awards:* Cert Recognition, NASA. *Mem:* Am Chem Soc; fel Am Ceramic Soc; Am Inst Mining, Metall & Petrol Eng. *Res:* X-ray crystallography; superconducting, pyrolytic, laser, ferroelectric and ferromagnetic, fiber and composite, thermoelectric and infrared optical materials; single crystal growth; perovskite type oxides; laser processing; ceramics; carbon-carbon composite coatings. *Mailing Add:* United Technol Res Ctr East Hartford CT 06108

GALASSO, GEORGE JOHN, b New York, NY, June 3, 32; m 58; c 3. MICROBIOLOGY, VIROLOGY. *Educ:* Manhattan Col, BS, 54; Univ NC, PhD(microbiol), 60. *Prof Exp:* Trainee virol, Univ NC, 60-62, res assoc, 62-63, res asst prof, 63-64; assoc prof, Sch Med, Univ Va, 64-68; mem grants assocs prog, NIH, 68-69, chief infectious dis br & head antiviral substances prog, 69-77, chief, Develop & Appln Br, Nat Inst Allergy & Infectious Dis, 77-83, ASSOC DIR, EXTRAMURAL AFFAIRS, NIH, 83- *Concurrent Pos:* Sect ed, Antiviral Res, ed, Antiviral Agents & Virus Dis of Man; mem ed adv bd, Encycl Virol. *Honors & Awards:* Super Serv Award, Pub Health Serv; Asst Secy Health's Award Except Achievement, Dept Health & Human Serv & Secy's Spec Citation; Cavaliere della Republica Italiana, Repub Italy. *Mem:* Am Soc Microbiol; Infectious Dis Soc; Am Acad Microbiol; Am Asn Immunologists; Am Soc Virol; Int Soc Antiviral Res; Int AIDS Soc. *Res:* Interferon and antiviral research; viral inhibition; antiviral substances; vaccine development; infectious diseases. *Mailing Add:* Bldg 1 Rm 152 Nat Insts Health Bethesda MD 20892

GALAWAY, RONALD ALVIN, b Oakland, Calif, June 23, 43; m 67. BIO-ORGANIC CHEMISTRY. *Educ:* Pac Union Col, BS, 66; Univ Calif, Riverside, MS, 72, PhD(chem), 75. *Prof Exp:* asst prof, 75-80, ASSOC PROF CHEM, LOMA LINDA UNIV, LA SIERRA CAMPUS, 80- *Mem:* Am Chem Soc. *Res:* Investigation into the mode of action of soybean lipoxygenase and prostaglandin synthetase complex from bovine source. *Mailing Add:* 4950 Via Campeche Riverside CA 92507

GALAYDA, JOHN NICOLAS, b Newark, NJ, Nov 29, 48; m 70; c 2. PHYSICS. *Educ:* Lehigh Univ, BA, 70; Rutgers Univ, PhD(physics), 77. *Prof Exp:* Asst physicist accelerator physics, Brookhaven Nat Lab, 77-80, assoc physicist, 80-90; DIR, APS ACCELERATOR SYSTS DIV, ARGONNE NAT LAB, 90- *Mem:* Am Phys Soc. *Res:* Accelerator physics. *Mailing Add:* 23 Center St Hinsdale IL 60521-3433

GALBIATI, LOUIS J, b Vineland, NJ, Feb 17, 25; m 53; c 4. ELECTRICAL ENGINEERING, MATHEMATICS. *Educ:* Johns Hopkins Univ, BE, 51; Cornell Univ, MS, 56, PhD(elec eng), 60, MEd, 67. *Prof Exp:* Engr, Gen Elec Co, Pa, 51-54; instr elec eng, Cornell Univ, 54-56; prin engr, Kimble Glass Co, Ohio, 56-58; scientist, Aero Lab, Cornell Univ, 59, asst 59-60; prof elec eng & head dept eng, Merrimack Col, 60-62; mgr space payloads, Radio Corp Am, Mass, 62; mem tech staff, Mitre Corp, Mass, 62-68; proj mgr & asst controller, Serv Technol Corp, 68-71; leader, RCA Corp, 72-74; dean instr,

Hartford State Tech Col, 74-76; dir eng technol, Univ Ark, 76-78; DEAN ENG TECHNOL, STATE UNIV NY COL TECHNOL, UTICA-ROME, 78- *Concurrent Pos:* Mem instrumentation eng staff, Avco Corp, Mass, 61-62; Merrimack Col rep, comt comput ctr, New Eng Col, 61-62; mem Andover Sch Comt, 65-68. *Mem:* Am Soc Eng Educ; Inst Elec & Electronics Engrs; Am Geophys Union; Instrument Soc Am. *Res:* Tropospheric control systems; arcs and magnetic fields; microprocessors; solar energy; machine vision. *Mailing Add:* Hartford State Tech Col 401 Flatbush Hartford CT 06106

GALBRAITH, DONALD BARRETT, b McDonald, Pa, Mar 10, 37; div; c 3. DEVELOPMENTAL GENETICS. *Educ:* Grove City Col, BS, 58; Brown Univ, ScM, 60, PhD(biol), & fel, 62. *Prof Exp:* Res assoc, Brown Univ, 63; NSF res grants genetics, 64-70, assoc prof, 70-77, PROF BIOL, TRINITY COL, CONN, 77-, CHAIR, BIOL DEPT, 87- *Concurrent Pos:* Vis assoc prof oral biol, Univ Conn Health Ctr, 73; vis prof oral biol, Univ Conn Health Ctr, 80, 87. *Mem:* AAAS; Am Genetic Asn; Am Soc Zool; Int Pigment Cell Soc; Sigma Xi; Pan Am Soc Pigment Cell Res. *Res:* Developmental genetics; experimental embryology; radiation biology; expression of color loci in mice; epithelial-mesenchymal interactions. *Mailing Add:* Dept Biol Trinity Col Hartford CT 06106

GALBRAITH, HARRY WILSON, b Detroit, Mich, Apr 8, 18. ORGANIC CHEMISTRY. *Educ:* Wayne State Col, BS, 40, MS, 42; Purdue Univ, PhD(org chem), 49. *Prof Exp:* Asst res chemist, Children's Fund, Mich, 40-43; asst, Cornell Univ & Purdue Univ, 43-46; analyst chem, Purdue Univ, 46-50; HEAD, GALBRAITH LABS, INC, 50- *Mem:* Am Chem Soc; Microchem Soc. *Res:* Organic microanalysis. *Mailing Add:* Galbraith Labs Inc PO Box 51610 Knoxville TN 37950

GALBRAITH, JAMES NELSON, JR, b Philadelphia, Pa, Apr 26, 36; m 83; c 2. GEOPHYSICS. *Educ:* Mass Inst Technol, SB, 58, PhD(geophys), 63. *Prof Exp:* Sr scientist, Geosci Inc, 63-67, vpres, 67-70; sr res geophysicist, Mobil Res & Develop Corp, 70-73, assoc geophys adv, 73-78, geophys coordr, 78-81, geophys mgr, 81-86, mgr seismic processing, 86-88, MGR GEOPHYS APPLICATIONS, MOBIL EXPLOR & PROD SERV, INC, 88- *Mem:* Am Geophys Union; Soc Explor Geophys; Seismol Soc Am; Inst Elec & Electronics Engrs; Europ Asn Explor Geophysicists. *Res:* Computer applications to geophysical problems, including time series analysis and optimum single and multichannel filtering of seismic and magnetic data and solution to boundry value problems. *Mailing Add:* 4366 Fawnhollow Dallas TX 75244

GALBRAITH, ROBERT MICHAEL, b Dec 1, 47. MICROBIOLOGY, IMMUNOLOGY. *Educ:* London Univ, MD, 71; Am Bd Internal Med, cert, 83. *Prof Exp:* From asst prof to assoc prof microbiol & immunol, Med Univ SC, Charleston, 77-80, from asst prof to assoc prof, Dept Path, 79-80, assoc prof, Dept Med, 81-85, vchmn clin affairs, Dept Microbiol & Immunol, 85-, PROF, DEPT MICROBIOL & IMMUNOL, MED UNIV SC, CHARLESTON, 84-, CHMN DEPT, 87-, PROF DEPT MED, 85- *Concurrent Pos:* Dir, Microbiol & Immunol Grad Prog, Med Univ SC; chmn, Multidisciplinary Task Force, Part II Hematopoietic & Immune Sect, Nat Bd Med Exams, 90-, chmn, 89-90; med dir, Liver Transplant Prog, Med Univ SC, Charleston, 90- *Honors & Awards:* Claude B Brown Mem Lectr, 79. *Mem:* Fel Am Col Physicians; Am Fedn Clin Res; Am Asn Immunologists; Am Soc Immunol Reprod; Am Soc Microbiologists; Asn Med Lab Immunologists; Clin Immunol Soc; AMA. *Res:* Microbiology; immunology. *Mailing Add:* Dept Microbiol & Immunol 203 Basic Sci Bldg Med Univ SC 171 Ashley Ave Charleston SC 29425

GALBRAITH, RUTH LEGG, b Lecompte, La, Nov 5, 23; m 50; c 1. TEXTILES. *Educ:* Purdue Univ, BS, 45, PhD(textile chem), 50. *Prof Exp:* Chemist orlon res, E I du Pont de Nemours & Co, 45-46; textile chemist detergent res, Gen Elec Co, 46-47; asst chem, Purdue Univ, 47-48; prof textiles, Univ Tenn, 50-55; from assoc prof to prof, Univ Ill, Urbana, 56-70; prof & head dept, consumer affairs, Auburn Univ, 70-73, dean Sch Home Econ, 73-85. *Concurrent Pos:* Mem nat adv comt, Fed Flammable Fabrics Act, 71-73; mem exec bd, Am Home Econ Asn, 75-76 & 78-80; mem, Comt of Nine, USDA, 81-83. *Mem:* Hon mem Am Soc Test & Mat; Am Asn Textile Chemists & Colorists; Am Chem Soc; Am Home Econ Asn. *Res:* Textile chemistry; detergents; textile fiber and fabric properties. *Mailing Add:* 368 Singleton Auburn AL 36830

GALBRAITH, WILLIAM, b Detroit, Mich, Mar 3, 45; m 67; c 2. BIOCHEMICAL PHARMACOLOGY. *Educ:* Western Reserve Univ, BA, 66; Univ Mich, Ann Arbor, MS, 68, PhD(biochem), 71. *Prof Exp:* Sr biochemist, Riker Labs, 3M Co, 71-75, res specialist biochem pharmacol, 75-79; SR RES PHARMACOLOGIST, E I DU PONT DE NEMOURS & CO, INC, 79- *Mem:* Am Chem Soc; Sigma Xi; NY Acad Sci; AAAS. *Res:* Biochemistry of the lung, mucus production and secretion; lysosomal enzymes and relation to drug effects; lectins, glycoproteins, carbohydrates and mucopolysaccharides; molecular basis of drug action. *Mailing Add:* 14 N Townview Lane Newark DE 19711

GALBREATH, EDWIN CARTER, vertebrate paleontology, geology; deceased, see previous edition for last biography

GALDES, ALPHONSE, b Malta, May 10, 52; m 76; c 2. COMPUTATIONAL CHEMISTRY, PHYSICAL CHEMISTRY. *Educ:* Univ Malta, BSc, 73, MSc, 75, Univ Oxford, PhD, 79. *Prof Exp:* Res fel, Harvard Med Sch, 79-81, res assoc, 81-84; sr chemist, 84-87, MGR, BOC GROUP TECH CTR, 87- *Concurrent Pos:* Rhodes scholar, 75. *Mem:* NY Acad Sci; AAAS; Am Soc Biochem Molecular Biol. *Res:* Computer assisted design; protein structure. *Mailing Add:* BOC Tech Ctr 100 Mountain Ave Murray Hill NJ 07974

GALDSTON, MORTON, b New York, NY, Nov 2, 12; m 41; c 3. MEDICINE. *Educ:* NY Univ, BS, 32, MD, 37. *Prof Exp:* NY Acad Med Bowen-Brooks scholar med, Johns Hopkins Hosp, 41-42; asst, 42-46, instr, 46-50, asst prof clin med, 50-54, asst prof med, 54-55, ASSOC PROF MED, NY UNIV MED CTR, 55-, RES ASSOC, 46- *Concurrent Pos:* From clin asst vis physician to vis physician, Goldwater Mem Hosp, 43-62; jr asst physician, Cardiac Clin, Lenox Hill Hosp, 48-52, sr asst physician, 52-58; asst vis physician, French Hosp, NY, 49-54, asst attend physician, 54-58; asst vis physician, NY Univ Hosp, 52-64, assoc vis physician, 64-; assoc vis physician, Bellevue Hosp, 55- *Mem:* Am Physiol Soc; Harvey Soc; Soc Clin Invest; Am Heart Asn; fel NY Acad Med. *Res:* Clinical research; physiology of respiration; circulation in general and of heart and kidneys; essential and experimental hypertension; humans under stress. *Mailing Add:* NY Univ Sch Med Dept Med 550 First Ave New York NY 10016

GALE, CHARLES, b Wyandotte, Mich, July 7, 26; wid; c 2. VETERINARY VIROLOGY. *Educ:* Mich State Univ, DVM, 52; Univ Minn, MPH, 55, PhD(virol), 58. *Prof Exp:* Instr vet bact & virol, Univ Minn, 52-56, asst head diag lab, 53-56; from asst res prof to assoc res prof, Ohio Agr Exp Sta, 56-62, assoc prof & adv grad sch, Ohio State Univ, 60-62; sr res virologist, 62-71, RES ASSOC, ELI LILLY & CO, 71- *Concurrent Pos:* Mem Conf Vet Diagnosticians; bd mem, Am Col Vet Microbiol; instr eve div, Indianapolis Univ. *Mem:* Am Vet Med Asn; Conf Res Workers Animal Dis; Am Asn Avian Path; US Animal Health Asn; Tissue Cult Asn; Am Soc Virol; AAAS. *Res:* Viral respiratory diseases, especially of cattle and companion animals; biological and pharmaceutical evaluations. *Mailing Add:* Eli Lilly & Co Box 708 Greenfield IN 46140

GALE, CHARLES C, JR, b Cleveland, Ohio, Sept 28, 26; m 58; c 2. PHYSIOLOGY. *Educ:* Ariz State Univ, BA, 51; Univ Pa, PhD(physiol), 60; Univ Stockholm, Fil Lic, 63, Fil Dr(physiol), 64. *Prof Exp:* From asst instr to instr physiol, Sch Med, Univ Pa, 56-61; NIH res fel, Royal Vet Col, Sweden, 61-64; res asst prof, Sch Med, Univ Wash, 64-65, from asst prof to assoc prof, 65-75, prof physiol, 75-82, prof biophys, 77-82; RETIRED. *Mem:* Am Physiol Soc; Int Soc Biometeorol; Endocrine Soc. *Res:* Neuroendocrinology, role of the central nervous system in regulation of the pituitary gland; thermoregulation, interaction of central nervous and endocrine systems; reproduction, gestation and lactation. *Mailing Add:* 3044 NW Market No 6 Seattle WA 98107

GALE, DAVID, b New York, NY, Dec 13, 21; m 54; c 2. MATHEMATICAL ECONOMICS. *Educ:* Swarthmore Col, BS, 43; Univ Mich, MA, 47; Princeton Univ, PhD(math), 49. *Prof Exp:* Mem staff, Radiation Lab, Mass Inst Technol, 43-45; instr math, Princeton Univ, 49-50; from asst prof to prof, Brown Univ, 50-66, chmn dept, 60-66; PROF MATH, ECON & OPER RES, UNIV CALIF, BERKELEY, 66- *Concurrent Pos:* NSF res grant, 52-53; Fulbright res scholar, Denmark, 53-54; consult, Rand Corp, 55-68; Guggenheim fel & vis prof, Univ Osaka, 62-63; NSF sr fel, Univ Copenhagen, 68-69. *Mem:* Nat Acad Sci; Am Math Soc; Mat Asn Am; Econometric Soc. *Res:* Mathematical economics; theory of games; geometry of convex sets; combinatorial problems. *Mailing Add:* Dept Math Univ Calif 909 Evans Hall Berkeley CA 94720

GALE, DOUGLAS SHANNON, II, b Kansas City, Mo, Aug 16, 42; m 65; c 2. INTERFACING, COMPUTING FACILITIES MANAGEMENT. *Educ:* Univ Kans, BS, 64; Univ Minn, MS, 66; Kans State Univ, PhD(physics), 72. *Prof Exp:* Instr physics, St Cloud State Univ, 66-69; asst prof physics, East Tex State Univ, 72-76, assoc prof, 76-79; dir, decentralized comput serv, Cornell Univ, 79-88; DIR, COMPUT RESOURCE CTR, UNIV NEBR, LINCOLN, 88- *Mem:* Am Phys Soc; Am Asn Physics Teachers; Inst Elec & Electronics Engrs. *Res:* Nuclear physics concerned with quasi-molecular resonances in heavy ion interactions, and heavy ion induced transfer reactions; microcomputer applications; experimental atomic physics, particularly optical and x-ray emission induced by low energy heavy ion bombardment. *Mailing Add:* Comput Resource Ctr Univ Nebr 326 Admin Lincoln NE 68588-0496

GALE, GEORGE OSBORNE, b Brooklyn, NY, May 3, 31; m 53; c 2. MICROBIOLOGY, SCIENCE ADMINISTRATION. *Educ:* Hofstra Univ, BA, 53; Purdue Univ, MS, 57. *Prof Exp:* Res asst microbiol, Purdue Univ, 54-56; biochemist, Lederle Labs, 56-59, res bacteriologist, Agr Div, 59-64, mgr Nutrit & Physiol, 70-73, GROUP LEADER, CLIN DEVELOP LAB, AM CYANAMID CO, 64-, MGR AGR DIV, 73- *Mem:* AAAS; Am Soc Microbiol; Am Soc Animal Sci; NY Acad Sci. *Res:* Reproductive physiology; bacterial resistance to antibiotics; chemotherapy of experimental infections; veterinary toxicology; herbicides; fungicides; insecticides; plant growth regulants. *Mailing Add:* Am Cyanamid Co PO Box 400 Princeton NJ 08543

GALE, GLEN ROY, b Florence, SC, Dec 1, 29; m 53; c 4. PHARMACOLOGY. *Educ:* Duke Univ, BA, 50, MA, 52, PhD(pharmacol), 54. *Prof Exp:* Physiologist, Vet Admin Hosp, Durham, 53-55; res assoc pharmacol, Duke Univ, 57-59, assoc, 59-65; pharmacologist, Vet Admin Hosp, Durham, NC, 59-65; from asst prof to assoc prof, 65-70, PROF PHARMACOL, MED UNIV SC, 70-; RES CAREER SCIENTIST, VET ADMIN MED CTR, CHARLESTON, SC, 80- *Concurrent Pos:* USPHS res grant, 61-; pharmacologist, Vet Admin Hosp, 65-80. *Honors & Awards:* Inventors Award, Dept Com, 80. *Mem:* AAAS; Sigma Xi; Am Soc Pharmacol & Exp Therapeut; Soc Exp Biol & Med; NY Acad Sci; Am Asn Univ Profs. *Res:* Mechanisms of action of antimicrobial agents and antitumor drugs; synthesis and evaluation of platinum complexes as antineoplastic agents; biochemical pharmacology; experimental chemotherapy; heavy metal toxicology. *Mailing Add:* Dept Pharmacol Med Univ SC 171 Ashley Ave Charleston SC 29425

GALE, HAROLD WALTER, b Syracuse, NY, Mar 19, 39. STRATEGIC COMMUNICATION. *Educ:* Syracuse Univ, BS, 60; Air Force Inst Technol, MS, 68; Purdue Univ, PhD(eng), 74. *Prof Exp:* Proj engr, USAF Rocket Propulsion Lab, 60-64, div chief develop eng, Plant Reproduction Off, 64-66, proj engr, Foreign Technol Div, 67-71, dep div chief, Ballistic Missile Off, 74-76, staff develop engr, Hq Syst Command, 76-80; SR STAFF ENGR, DEFENSE & SPACE SYST, COMMAND SUPPORT DIV, TRW CORP, 80- *Concurrent Pos:* Adj prof, Air Force Inst Technol, 69-70. *Mem:* Am Inst Aeronaut & Astronaut; Inst Elec & Electronics Engrs; Armed Forces Commun & Electronics Asn. *Res:* Interaction of electromagnetohydrodynamics processes with terrestrial systems and their utilization in dynamic multi-dimensional systems for national large-scale command and control; technical management. *Mailing Add:* 54630 Broadmoor St Alexandria VA 22310

GALE, HENRY H, MUSCLES. *Educ:* Univ Ill, Chicago, PhD(physiol), 66. *Prof Exp:* ASST PROF MED PHYSIOL, SCH MED, CREIGHTON UNIV, 66- *Mailing Add:* Dept Physiol Sch Med Creighton Univ 2500 Cal St Omaha NE 68178

GALE, JAMES LYMAN, b Boston, Mass, Dec 31, 34; m 66; c 4. EPIDEMIOLOGY, PUBLIC HEALTH. *Educ:* Harvard Univ, AB, 57; Columbia Univ, MD, 61; Univ Wash, MS, 69. *Prof Exp:* Intern, Bellevue Hosp, New York, 61-62; resident, Bellevue & Mem Hosps, 62-64; epidemic intel serv officer, USPHS, 64-67; fel prev med, 67-69, asst prof, 69-72, assoc prof epidemiol & int health, 72-78, assoc dean, 73-76, PROF EPIDEMIOL, SCH PUB HEALTH & COMMUNITY MED, UNIV WASH, 78- *Concurrent Pos:* Consult, Tri-Serv Gen Hosp, Taipei, Taiwan, 69-70; vis assoc prof, Sch Med, Nat Taiwan Univ, 69-72; USPHS career develop award, 69-74. *Mem:* Am Pub Health Asn; Infectious Dis Soc Am; Int Epidemiol Asn; Am Venereal Dis Asn; Soc Epidemiol Res. *Res:* Epidemiology of infectious disease; epilepsy; cerebal palsy; head injury. *Mailing Add:* Dept Epidem Univ Wash Seattle WA 98195

GALE, LAIRD HOUSEL, b San Francisco, Calif, Jan 24, 35; m 61; c 2. PHYSICAL ORGANIC CHEMISTRY. *Educ:* San Diego State Col, BS, 56; Univ Calif, PhD(org chem), 59. *Prof Exp:* Res technologist radiation chem, Martinez Res Lab, Shell Oil Co, 59-62, res chemist, Emeryville Res Ctr, Shell Develop Co, 62-66, chemist-exchange scientist, Shell Res Ltd, Thornton, Eng, 66-67, sr res chemist, Shell Develop Co, Tex, 62-74, staff res chemist, 74-80, STAFF RES CHEM, BIOL SCI RES LAB, SHELL DEVELOP CO, 80- *Mem:* Am Chem Soc; Am Inst Chemists; Sigma Xi. *Res:* Radiation chemistry and free radical chemistry; combustion chemistry; surface analysis techniques, including x-ray photoelectron spectroscopy, auger electron spectroscopy and secondary ion mass spectroscopy; heterogeneous catalysis; mass spectroscopy. *Mailing Add:* 10395 Old Placerville Rd Sacramento CA 95827

GALE, NORD LORAN, b Phoenix, Ariz, Feb 21, 38; m 60; c 3. MICROBIOLOGY, BIOCHEMISTRY. *Educ:* Brigham Young Univ, MS, 64, PhD(bact), 67. *Prof Exp:* Mem tech staff microbiol, TRW Inc, Calif, 66-68; from asst prof to assoc prof, 68-79, PROF LIFE SCI, UNIV MO, ROLLA, 80- *Mem:* Soc Environ Geochem & Health; Am Soc Microbiol. *Res:* Autotrophy; microbial membrane transport; distribution, physiology and toxicity of heavy metals; waste treatment. *Mailing Add:* Dept Life Sci Univ Mo Rolla MO 65401

GALE, PAULA JANE, b Joplin, Mo, July 26, 46. CHEMICAL PHYSICS. *Educ:* Randolph-Macon Woman's Col, BA, 68; Brandeis Univ, PhD(chem), 76. *Prof Exp:* Analytical chemist, Process Res, Inc, Mass, 69; RES ASSOC, MASON LAB, YALE UNIV, 75- *Mem:* Am Soc Mass Spectros. *Res:* Atomic and molecular scattering. *Mailing Add:* Bristol Myers Squibb PO Box 191 One Squibb Dr Bldg 101 New Brunswick NJ 08903-0191

GALE, ROBERT PETER, b New York, NY, Oct 11, 45. IMMUNOBIOLOGY, MEDICINE. *Educ:* Hobart Col, AB, 66; State Univ NY Buffalo, MD, 70; Am Bd Internal Med, dipl, 74; Univ Calif, Los Angeles, PhD, 78. *Prof Exp:* Intern med, 70-71, resident, 71-72, fel, 72-74, SCHOLAR IMMUNOL, UNIV CALIF, LOS ANGELES, 72-, ASST PROF MED, 74- *Concurrent Pos:* Scholar, Leukemia Soc Am, 77- *Mem:* Am Asn Cancer Res; Am Soc Clin Oncol; Am Soc Hematol; Sigma Xi; Soc Exp Biol & Med. *Res:* Bone marrow transplantation; leukemia research. *Mailing Add:* Dept Med Sch Med Div Hemat & Oncol Univ Calif Ctr Health Sci Los Angeles CA 90024

GALE, STEPHEN BRUCE, b Syracuse, NY, June 21, 40; m 63; c 3. ENVIRONMENTAL ENGINEERING, INDUSTRIAL WASTE. *Educ:* Syracuse Univ, BSChE, 62, MS, 64, PhD(sanit eng), 69. *Prof Exp:* Sanit engr, Air Pollution Eng Div, US Army Environ Hyg Agency, 69-72; tech dir & sales mgr, Andco Environ Processes Inc, 72-78; SALES MGR, NIAGARA ENVIRON ASSOCS, 78- *Mem:* Am Inst Chem Engrs. *Res:* Kinetics of oxidation of sulfites by dissolved oxygen; kinetics of photochemical oxidation of sulfur dioxide in dilute gas-air mixtures; air pollution instrumentation; electrochemical removal of chromates and other heavy metals from waste water effluents. *Mailing Add:* Niagra Environ Assn 88 Okell St Buffalo NY 14220

GALE, WILLIAM ARTHUR, b Houston, Tex, June 23, 39; c 2. COMPUTATIONAL LINGUISTICS, TEXT INTERPRETATION. *Educ:* Rice Univ, BA, 61, MA, 66, PhD(physics), 68. *Prof Exp:* Mem tech staff, Bell Commun Res, 69-73; MEM TECH STAFF, AT&T BELL LABS, 73- *Concurrent Pos:* Dir, Soc Artificial Intel & Statist, 87- *Mem:* Am Asn Artificial Intel; Am Statist Asn; Soc Artificial Intel & Statist. *Res:* Artificial intelligence; statistics; computational linguistics; economics. *Mailing Add:* AT&T Bell Labs Rm 2C278 600 Mountain Ave Murray Hill NJ 07974

GALEANO, CESAR, b Montevideo, Uruguay, Apr 8, 26; m 51; c 2. NEUROPHYSIOLOGY, NEURO-UROLOGY. *Educ:* Univ Republic, Uruguay, MD, 61. *Prof Exp:* Asst neurophysiol, Inst Res Biol Sci, Montevideo, Uruguay, 51-60, head electrophysiol, 61-65; int fel neurophysiol, Brain Res Inst, Mexico City, Mex, 65; NIH int fel, Sch Med, Stanford Univ, 65-67; from asst prof to assoc prof neurophysiol, 67-80, PROF PHYSIOL, SCH MED, UNIV SHERBROOKE, 80. *Mem:* Can Urol Asn; Uro-Dynamics Soc. *Res:* Physiology and physiopathology of micturition. *Mailing Add:* Dept Physiol Univ Sherbrooke Sch Med Sherbrooke PQ J1H 5N4 Can

GALEANO, SERGIO F(RANCIS), b Havana, Cuba, Apr 7, 34; US citizen; wid; c 3. ENVIRONMENTAL ENGINEERING. *Educ:* Univ Havana, MS, 57; Univ Fla, MSE, 64, PhD(bioenviron eng), 66. *Prof Exp:* Proj engr, Ingenieria Vame SAm, 56-58; chief design sect, Comision Nacional Acueductos, 59-60, head eng div, 60-61; process develop eng, Owens-Ill, Inc, 66-69, proj mgr new systs develop, 69-71, dir, Chem Div, 77-82, dir, Environ & Occup Health Div, Owens Health Div, 82; SR CONSULT, GA PAC CORP, 90- *Concurrent Pos:* Intersoc Bd Environ Eng. *Mem:* Am Inst Chem Engrs; Acad Environ Engrs; Sigma Xi; Nat Soc Prof Engrs; Am Soc Civil Engrs. *Res:* Deep well hydrology; water supply systems; environmental control; process development drying systems. *Mailing Add:* 133 Peachtree St NE Atlanta GA 30348

GALEENER, FRANK LEE, b Long Beach, Calif, July 31, 36; m 59; c 2. SOLID STATE PHYSICS. *Educ:* Mass Inst Technol, BS, 58, MS, 62; Purdue Univ, West Lafayette, PhD(physics), 70. *Prof Exp:* Physicist solid state physics, Lincoln Lab, Mass Inst Technol, 59-61 & Nat Magnet Lab, 61-64; scientist, 70-73, prin scientist solid state physics & mgr semiconductor res, 73-77, prin scientist solid state physics, Palo Alto Res Ctr, Xerox Corp, 79-87; PROF, DEPT PHYSICS, COLO STATE UNIV, 87- *Concurrent Pos:* co-chmn adv panel amorphous mat, Div Mat Sci, US Dept Energy, 80; co-ed, Structure & Excitations of Amorphous Solids, 76, The Physics of MOS Insulators, 80, Defects in Glasses, 86; mem, adv panel solid state physics, Off Naval Res, 80; vis scholar, Dept Theoretical Physics, Univ Oxford, 82; vis scientist, Cavendish Lab, Univ Cambridge, 83-84, Inst de Fisica, Univ Mex DF, 90; vis prof, Lab Elem Particle Physics, Univ Paris VI, 88. *Mem:* Fel Am Phys Soc; Optical Soc Am; Am Ceramic Soc; Mat Res Soc; Sigma Xi. *Res:* Optical properties, luminescence and light scattering in amorphous and crystalline semiconductors; solid state magnetoplasmas; classical electromagnetic scattering theory; artificial dielectrics; structure and vibrational properties of glasses; structure and growth of amorphous films; defects in microelectronic materials. *Mailing Add:* Dept Physics Colo State Univ Ft Collins CO 80523

GALEF, ARNOLD E, acoustics, fluids, for more information see previous edition

GALEHOUSE, JON SCOTT, b Doylestown, Ohio, Feb 16, 39; m 59; c 2. NEOTECTONICS, EARTHQUAKE STUDIES. *Educ:* Col Wooster, BA, 62; Univ Calif, Berkeley, PhD(geol), 66. *Prof Exp:* NSF fel, 66-67; from asst prof to assoc prof, 67-75, chmn dept, 73-76, PROF GEOL, SAN FRANCISCO STATE UNIV, 75- *Concurrent Pos:* Am Chem Soc Petrol Res Fund grant, 67-69; Personal Mobility grant, NSF, 72-73; res grants, US Geol Surv, 79-92. *Mem:* AAAS; Geol Soc Am; Soc Econ Paleont & Mineral; Am Geophys Union; Int Asn Sedimentologists; Sigma Xi. *Res:* Marine sedimentation and provenance of terriginous deposits in the ocean; creep rates on active faults. *Mailing Add:* Dept Geosci San Francisco State Univ San Francisco CA 94132

GALES, ROBERT SYDNEY, b Boston, Mass, Dec 12, 14; m 42; c 3. ACOUSTICS. *Educ:* Univ Calif, Los Angeles, AB, 38, MA, 42. *Prof Exp:* Asst physics, Univ Calif, Los Angeles, 40-42; assoc physicist & group leader, Div War Res, Univ Calif, 42-45; from physicist & leader, Psychol Physics Br to head, Listening Div, Naval Electronics Lab, 46-67; head listening div, Naval Undersea Ctr, 67-77; supvry physicist, Naval Ocean Systs Ctr, 77-80; staff scientist, Comput Sci Corp, 81-83; consult, SEACO Div Sci Appln Int Corp, 87-88; RETIRED. *Concurrent Pos:* Consult acoust, 48-; mem, Nat Res Coun Comt Hearing, Bioacoustics and Biomechanics, 52-80. *Mem:* Acoust Soc Am (pres, 75-76); Inst Noise Control Eng; Sigma Xi. *Res:* Hearing aids; audio masking; measurements and methods of detection of underwater sounds; noise measurement; voice communication; effects of noise on human performance. *Mailing Add:* 1645 Los Altos Rd San Diego CA 92109

GALETTO, WILLIAM GEORGE, b Grass Valley, Calif, Oct 5, 39; m 62; c 2. FOOD CHEMISTRY. *Educ:* Chico State Col, BS, 61; Univ Calif, Davis, PhD(agr chem), 67. *Prof Exp:* Res asst flavor chem, Dept Enol, Univ Calif, Davis, 62-67; res chemist, Western Regional Res Lab, USDA, 67-68; res chemist, Res & Develop Lab, 68-78, mgr tech dept, 78-80, RES MGR, FLAVOR DIV, McCORMICK & CO, 80- *Concurrent Pos:* Nat Acad Sci Res Coun assoc 67-68. *Mem:* AAAS; Am Chem Soc; Inst Food Technol; Am Soc Enol. *Res:* Stereochemistry of sesquiterpene lactones; synthesis of natural products; flavor chemistry; gas chromatographic separation of diastereoisomers; enzymatic flavor development. *Mailing Add:* 2712 Crystal Lane Baldwin MD 21013

GALEY, JOHN APT, b Oak Park, Ill, May 27, 28; m 52; c 5. REACTOR PHYSICS. *Educ:* Yale Univ, BS, 50; Univ Chicago, MS, 54, PhD(physics), 59. *Prof Exp:* Res assoc physics, Notre Dame Univ, 59-60; SR SCIENTIST, BETTIS ATOMIC POWER LAB, WESTINGHOUSE ELEC CORP, 60- *Res:* Photonuclear work; 100 million electron volt betatron; low energy physics; Van de Graff accelerators; experimental reactor physics. *Mailing Add:* 267 Toura Dr Pleasant Hills PA 15236

GALEY, WILLIAM RALEIGH, b Boise, Idaho, July 26, 43; m 83; c 4. PHYSIOLOGY, BIOPHYSICS. *Educ:* Lewis & Clark Col, BS, 65; Univ Ore, PhD(biochem), 69. *Prof Exp:* NIH & Mass Heart Asn fels biophys, Harvard Med Sch, 69-71; sr biophysicist, Alza Corp, 71-72; from asst prof to assoc prof, 72-84, PROF PHYSIOL, SCH MED, UNIV NMEX, 84- *Concurrent Pos:* Ed, IUPAB News, Quart Rev Biophys, 70-71; consult, Alza Corp, 72-73, Bell & Howell Corp, 73-74 & Vick Corp, 76-; Fulbright fel, Portugal, 83 & Turkey, 86. *Mem:* Am Physiol Soc; Biophys Soc; Sigma Xi; AAAS. *Res:* Effect of disease on water and nonelectrolyte transport in cells; electrolyte and water secretion by the exocrine pancreas; cell physiology; medical education. *Mailing Add:* Dept Physiol Univ NMex Sch Med Albuquerque NM 87131

GALGINAITIS, SIMEON VITIS, physics, for more information see previous edition

GALIBOIS, ANDRE, b Quebec, Que, Apr 21, 38; m 69; c 1. PHYSICAL METALLURGY, OPERATIONS RESEARCH. *Educ:* Laval Univ, BA, 57, BASc, 61, DSc(metall), 64. *Prof Exp:* Lectr eng mech, Laval Univ, 64; Ford Found res fel metall & mat sci, Univ Toronto, 64-65; from asst prof to assoc prof, 65-76, PROF MINES & METALL, LAVAL UNIV, 76-, ASST DIR DEPT, 70-, DIR ENG PHYSICS, 71- *Concurrent Pos:* Mem, Defence Res Bd Can, 69-, faculty coun, Laval Univ, 68-, Phys Metall Subcomt, Can Nat Adv Comt for Res in Mining & Metall, 69-; sr partner, Opers Res Consult Firm, 69- *Mem:* Am Soc Metals; Can Inst Mining & Metall. *Res:* Martensitic transformation and tempering mechanism in extra low carbon steels; grain boundaries migration and grain growth in high purity lead, tin and aluminum and in material doped with impurities; operations research applied to mining ventures. *Mailing Add:* Dept Mining & Metall Univ Laval Quebec PQ G1K 7P4 Can

GALIL, FAHMY, b Talkha, Egypt, Oct 20, 25; m 59; c 2. TEXTILE CHEMISTRY. *Educ:* Univ Cairo, BSc, 47, MSc, 54; Manchester Col Sci & Technol, Eng, PhD(polymer sci), 58. *Prof Exp:* Textile chemist, Misr Spinning & Weaving Co, Egypt, 47-50; asst eng, Univ Cairo, 50-58, lectr, 58-64, assoc prof, 64-68; res fel polymers, Univ Mainz, 68-69; technol mgr, 69-81, SR SPECIALIST, MONSANTO FIBERS CO, 81- *Concurrent Pos:* Humboldt Found fel, Reutlingen, Ger, 62-64. *Mem:* Am Chem Soc; sr mem Am Asn Textile Chemists & Colorists; corp mem Brit Soc Dyers & Colourists. *Res:* Flammability of textile materials; information management systems; theory of dyeing; polymer structure. *Mailing Add:* 2411 Circle Dr SE PO Box 2204 Decatur AL 35602

GALIL, KHADRY AHMED, b Egypt, 1942; Can citizen; m 75; c 1. PERIODONTAL RESEARCH, RESINS & ACID ETCHING. *Educ:* Alexandria Univ, DDS, 64, DrOralSurg, 67; Univ Western Ont, PhD(anat), 73, MSC Periodont, Univ Mich, 87. *Prof Exp:* Asst prof oral surg, Univ Alexandria, Egypt, 64-67; vis prof orthodont, Univ Mich, 69; asst prof anat, 73-76, ASSOC PROF ANAT & OEAL SURG, UNIV WESTERN ONT, CAN, 76- *Concurrent Pos:* Pvt pract periodontics, Univ Hosp London, Ont. *Mem:* Int Asn Dent Res; Can Dent Asn; fel Acad Gen Dent; Can Asn Dent Res; fel Int Acad Dent; Can Acad Periodont. *Res:* Wound healing. *Mailing Add:* Fac Med & Dent Univ Western Ont London ON N6A 5C1 Can

GALIN, DAVID, b New York, NY, Mar 6, 36; c 2. NEUROPSYCHOLOGY, PSYCHOPHYSIOLOGY. *Educ:* Antioch Col, BS, 57; Albert Einstein Col Med, MD, 61. *Prof Exp:* Res assoc, NIMH, 62-65; fel, Fels Res Inst, 65-68; asst prof in residence, 68-75, ASSOC PROF IN RESIDENCE, LANGLEY PORTER INST, UNIV CALIF, SAN FRANCISCO, 75- *Res:* Brain mechanisms related to behavior, particularly consciousness and subjective experience, neurodevelopment and specialization and interaction of the two cerebral hemispheres. *Mailing Add:* Dept Psychiat Univ Calif San Francisco CA 94143-0844

GALIN, MILES A, b New York, NY, Jan 6, 32; m 53; c 4. OPHTHALMOLOGY. *Educ:* NY Univ, AB, 51, MD, 55; Nat Bd Med Examrs, dipl, 56, Am Bd Ophthalmol, dipl, 60. *Prof Exp:* Intern, Mt Sinai Hosp, New York, 55-56; asst ophthalmol surg, Cornell Univ Med Col, 56-58, instr surg, 58-61, from clin asst prof to asst prof ophthalmol surg, 61-66; chmn, Dept Ophthalmol, NY Med Col, 66-73, prof, 66-79, dir res & planning, 73-79; ADJ PROF POLYMER SCI, UNIV LOWELL, 82- *Concurrent Pos:* From asst resident surgeon to resident surgeon ophthalmol, NY Hosp, 56-59, surgeon to out patients, 59-61, from asst attend surgeon to assoc attend surgeon, 61-66; consult ophthalmol, Mem Hosp, 60-66; attend ophthalmologist & chief, Flower & Fifth Ave Hosp, Metrop Hosp & Bird S Coler Hosp, 66-79, Blythedale Children's Hosp, 68-73, Westchester County Med Ctr, 75-79, Med Arts Ctr Hosp, 78- & Cabrini Med Ctr, 82-; consult ophthalmol, St Francis Hosp, 75- & Dept Surg, Catholic Med Ctr Brooklyn & Queens, 76-; lectr, Ramyer Found, 79; first Hermann Hosp lectr, Tex, 67; consult, Fed Aviation Admin, Regional Med Prog, Medicaid Adv Comt Qual Vision Care, Social Security Admin & Nat Multiple Sclerosis Found; hearing examr, Social Security Admin; mem bd dirs, Better Vision Inst; prin & coinvestr, Nat Soc Prev Blindness, Nat Soc Combat Blindness & US Pub Health Serv, Cornell Univ Med Col, 59-66, career scientist, Health Res Coun, 63-66, Nat Soc Prev Blindness & Nat Soc Combat Blindness, NY Med Col, 66-67, US Pub Health Serv, 66-, Susan Greenwall Found, 68-75, Nat Inst Allergy & Infections Dis, 68-, exchange scientist to USSR, US-Soviet Health Exchange, 69-71 & 74, United Health Found, 69, Am Cancer Found, 71-73 & Am Heart Found, 69-71; US ed, Annali de Ottalmologia; mem tech adv comt, Bur Handicapped Children; mem bd dirs, Pan Am Implant Asn, 85-; tech consult, Regional Med Prog. *Honors & Awards:* William Warner Hoppin Award, NY Acad Med, 59; Dr Ignacia Barraguer Mem Award, 68; Dr Henry Balconi Mem lectr, NY, 67; Binkhorst Award, Am Intra-Ocular Implant Soc, 78; Rayner Reddy Found lectr, UK Intraocular Implant Soc, 79; Dr P Siva Gold Medal, 82; Temoignage d' Honneur, Can Implant Asn, 75; Edward A Weisser Mem lectr, 76. *Mem:* AAAS; Am Acad Ophthal & Otolaryngol; AMA; Asn Res Vision & Ophthal; NY Acad Med; NY Acad Sci; Am Intraocular Implant Soc; Asn Univ Prof Ophthalmol; Israel Ophthalmol Soc; Royal Soc Med. *Mailing Add:* 345 E 37th St 3rd Fl New York NY 10016

GALINDO, ANIBAL H, b Buga, Colombia, Sept 11, 29; m 51; c 4. ANESTHESIOLOGY, NEUROPHYSIOLOGY. *Educ:* Rosary Col, Bogota, BSc, 46; Nat Univ Colombia, MD, 52; McGill Univ, PhD(physiol, neurophysiol), 68. *Prof Exp:* Internship, Univ Bogota Hosp, 52-53; pvt pract, 53-54; resident anesthesia, San Jose Hosp, Nat Univ Colombia, 54-56, instr, Univ, 58-59; vis scientist, Nat Inst Neurol Dis & Blindness, 59-63, head sect neuroanesthesia, 60-63; asst prof anesthesiol, Sch Med, McGill Univ, 64-68; from assoc prof to prof, Sch Med, Univ Wash, 68-74, chief div neurosurg anesthesiol, 72-74; PROF ANESTHESIOL, SCH MED, UNIV MIAMI, 74- *Mem:* Am Physiol Soc; Am Soc Pharmacol & Exp Therapeut. *Res:* Pharmacology of anesthetic drugs; effect of various anesthetics on cardiac excitability, cerebral and hepatic circulation; neuropharmacology, especially the effects of muscle relaxants and anesthetics on synaptic transmission. *Mailing Add:* 6500 Caballero Blvd Coral Galbes FL 33134

GALINSKY, ALVIN M, b Chicago, Ill, Oct 14, 31. PHARMACEUTICAL CHEMISTRY. *Educ:* Univ Ill, BS, 54, MS, 57, PhD(pharmaceut chem), 62. *Prof Exp:* From instr to asst prof org medicinals, Col Pharm, Columbia Univ, 60-63; from asst prof to assoc prof phys pharm, 63-68, PROF PHARM & ASSOC CHMN DEPT, SCH PHARM, DUQUESNE UNIV, 68- *Mem:* AAAS; Am Asn Cols Pharm; Am Pharmaceut Asn; Sigma Xi. *Res:* Structure-activity relationships; local anesthetics; dosage form, design and bioavailability; kinetics and stability of pharmaceutical dosage forms. *Mailing Add:* Dept Pharm Duquesne Univ Pittsburgh PA 15282

GALINSKY, IRVING, b New York, NY, June 15, 21; m 48; c 2. CYTOLOGY, GENETICS. *Educ:* McGill Univ, BSc, 44, MSc, 45; Univ Wis, PhD(cytogenetics), 48. *Prof Exp:* Asst, Univ Wis, 45-48; Hite cancer fel, Univ Tex, 48-49; Nat Cancer fel, Col Med, Baylor Univ, 49-51; res assoc bact genetics, Cold Spring Harbor Lab, 51-53; asst prof genetics & cytol, Univ Del, 53-54; res assoc, Biochem Res Found, 54-59; PROF BIOL, HOFSTRA UNIV, 59-, CHMN DEPT, 63- *Mem:* AAAS; Genetics Soc Am; Am Genetic Asn; Am Asn Cancer Res. *Res:* Cytology of meiosis and mitosis; bacterial genetics; cancer research; mitotic abnormalities in plant and animal cells. *Mailing Add:* Dept Biol Hofstra Univ 1000 Fulton Ave Hempstead NY 11550

GALINSKY, RAYMOND ETHAN, b Hartford, Conn, Jan 27, 48. PHARMACEUTICS, PHARMACOKINETICS. *Educ:* Univ Calif, Berkeley, BA, 70; Univ Calif, San Francisco, PharmD, 75. *Prof Exp:* Resident clin pharm, hosps & clins, Univ Calif, San Francisco, 75-76; instr, Philadelphia Col Pharm & Sci, 77-78; res fel pharmaceut, Sch Pharm, State Univ NY, Buffalo, 78-80, res asst prof, 80-83; ASST PROF PHARMACEUT, UNIV UTAH, 83-,. *Concurrent Pos:* NIH Grant & First Award, Nat Inst Aging, 87; adj asst prof pharmacol & toxicol, Col Pharm, Univ Utah, 87- *Mem:* AAAS; Am Soc Clin Pharmacol & Therapeut; Am Col Clin Pharm; NY Acad Sci; Am Soc Pharmacol Exp Therapeut; Am Asn Pharmaceut Scientists. *Res:* Effect of aging and hepatic disease on the pharmacokinetics and pharmacodynamics of drugs with emphasis on the factors controlling nonlinear conjugation reactions. *Mailing Add:* Dept Pharmaceut Col Pharm Univ Utah 301 Skaggs Hall Salt Lake City UT 84112

GALITZ, DONALD S, b Chicago, Ill, May 28, 35; m 60; c 2. PLANT PHYSIOLOGY. *Educ:* Monmouth Col, BA, 56; Univ Ill, MS, 60, PhD(agron), 61. *Prof Exp:* Asst prof biol sci, Western Ill Univ, 61-67; assoc prof, 68-75, PROF BOT, NDAK STATE UNIV, 75- *Concurrent Pos:* Res assoc, Dept Agron, Univ Ill, 67-68. *Mem:* Am Soc Plant Physiologists; Weed Sci Soc Am. *Res:* Plant physiology and biochemistry; plant metabolism, specifically nitrogen metabolism; stress physiology and physiology of seeds; weed physiology. *Mailing Add:* Dept Bot NDak State Univ St Univ Sta Fargo ND 58105

GALIVAN, JOHN H, b Albany, NY, June 19, 39; m 65; c 4. BIOCHEMISTRY. *Educ:* Union Univ, NY, BS, 60; State Univ NY Albany, MS, 63; Albany Med Col, PhD(biochem), 67. *Prof Exp:* Res asst biochem, Albany Med Col, 64-67; NIH fel, Scripps Clin & Res Found, 67-70; sr res scientist, Div Lab & Res, NY State Dept Health, 70-77, res scientist IV, 77-81; assoc prof biochem, Albany Med Col, 78-82; RES SCIENTIST VI, DIV LABS & RES, NY STATE DEPT HEALTH, 81-; PROF BIOCHEM, STATE UNIV NY, ALBANY, 85- *Concurrent Pos:* Exp Therapeut Study Sect, NIH, 85-89; ad hoc consult, Nat Cancer Inst, 83-93. *Mem:* AAAS; Am Chem Soc; Am Asn Cancer Res; Am Soc Biol Chem & Molecular Biol; Soc Pharmaceut Exp Therapeut. *Res:* Enzymology; folate and vitamin B12 metabolism; regulation of cell division; pyrimidine metabolism; cytoxic drugs. *Mailing Add:* Wadsworth Ctr Lab Res NY State Dept Health Empire State Plaza Albany NY 12201

GALKOWSKI, THEODORE THADDEUS, b Worcester, Mass, Nov 9, 21; m 49; c 2. ORGANIC CHEMISTRY. *Educ:* Col of the Holy Cross, BS, 47, MS, 48; Ohio State Univ, PhD(chem), 51. *Prof Exp:* Org chemist, Nat Bur Standards, 51-52; from asst prof to assoc prof, 52-60, PROF CHEM, PROVIDENCE COL, 60-, COORDR RES, 67- *Mem:* AAAS; Am Chem Soc; fel Am Inst Chemists. *Res:* Structure of starch; proof of structure of carbohydrates; synthesis of Carbon-14 labeled carbohydrates; configurations of branched carbohydrates; carbohydrates in natural products. *Mailing Add:* Dept Chem Providence Col River Ave Eaton St Providence RI 02918

GALL, CARL EVERT, b Burlington, Ont, Can, Dec 17, 31; m 55; c 4. CHEMICAL ENGINEERING. *Educ:* Univ Toronto, BASc, 55; Queen's Univ, Ont, MSc, 62; Univ Minn, PhD(chem eng), 66. *Prof Exp:* Lectr chem, Royal Mil Col, 53-54; tech rep, Imp Oil Ltd, 55-56; asst prof math, Haile Sellassie I Univ, 56-59; asst prof, Royal Mil Col, 59-61, asst prof chem eng, 61-62; res fel, Univ Minn, 62-64; asst prof, 64-66, ASSOC PROF CHEM ENG, UNIV WATERLOO, 66-, ASSOC CHMN UNDERGRAD CHEM ENG DEPT, 87- *Res:* Control science; applied mathematics; chemical reaction engineering. *Mailing Add:* Dept Chem Eng Univ Waterloo Waterloo ON N2L 3G1 Can

GALL, DONALD ALAN, b Reddick, Ill, Sept 13, 34; m 60; c 4. COMPUTER SCIENCE, BIOMEDICAL ENGINEERING. *Educ:* Univ Ill, BS, 56; Mass Inst Technol, SM, 58, MechE, 60, ScD, 64. *Prof Exp:* Exp res engr, Gen Motors Corp, 56-57; staff engr, Dynatech Corp, 59-60, mgr control systs, 62-63; asst prof automatic controls, Carnegie-Mellon Univ, 64-68, assoc prof mech eng, 68-69; res asst prof surg, Sch Med, Univ Pittsburgh, 68-69, res assoc prof surg & anesthesiol, 69-73; PRES, OMEGA COMPUT SYSTS, INC, 73- *Concurrent Pos:* Eng specialist, Thompson Ramo Wooldridge, Inc, 60-62. *Honors & Awards:* Taylor Medal, Int Conf Prod Res, 71. *Mem:* AAAS; Am Soc Mech Eng. *Res:* Medical computer systems; biological servomechanisms; computerized information systems; development of generalized data base management systems. *Mailing Add:* Omega Computer Systs Inc 4300 N Miller Rd No 136 Scottsdale AZ 85251

GALL, GRAHAM A E, b Moose Jaw, Sask, Sept 2, 36; m 60; c 3. FISH GENETICS. *Educ:* Univ Alta, BSc, 60, MSc, 63; Purdue Univ, PhD(animal genetics), 66. *Prof Exp:* Res asst genetics, Univ Alta, 63 & Purdue Univ, 63-66; from asst to assoc prof, 66-78, PROF ANIMAL SCI, UNIV CALIF, DAVIS, 78-, ANIMAL GENETICIST, EXP STA, 66- *Concurrent Pos:* Genetics consult, Calif Dept Fish & Game, 67-85; mem, Peer Rev Panel, Aquaculture, USDA, 82, 84, 85, 87-90; managing ed, Aquaculture, 85-; secy treas, Int Aquaculture Genetic Asn, 85- *Mem:* Am Fisheries Soc; AAAS; World Aquaculture Soc; Am Genetics Soc. *Res:* Genetics and breeding of fish; application of modern breeding methods to fish farming; population biology and management of Pacific salmon. *Mailing Add:* Dept Animal Sci Univ Calif Davis CA 95616-8521

GALL, JAMES WILLIAM, b Taylorville, Ill, Apr 22, 42; m 66; c 3. PHYSICAL CHEMISTRY, HAZARDOUS WASTE TREATMENT. *Educ:* Ohio State Univ, PhD(phys chem), 69. *Prof Exp:* Teaching asst chem, Ohio State Univ, 64-65; res scientist, Phillips Petrol Co, 69-80, sr res scientist chem, 80-85, supvr, 85-91, MGR, FUEL & LUBRICANT ANAL, PHILLIPS PETROL CO, 91- *Res:* Photochemistry; gas phase kinetics; chemistry of atmospheric pollutants; interaction of aqueous polymer solutions; thermodynamics and vapor-liquid equilibrium; oxidation; environmental services; hazardous waste treatment, incineration. *Mailing Add:* Phillips Res Ctr 139 Auto Lab Phillips Petrol Co Bartlesville OK 74004

GALL, JOSEPH GRAFTON, b Washington, DC, Apr 14, 28; m 55, 82; c 2. CELL BIOLOGY. *Educ:* Yale Univ, BS, 48, PhD(zool), 52. *Prof Exp:* From instr to assoc prof zool, Univ Minn, 52-64; prof biol, Yale Univ, 64-83; STAFF MEM, CARNEGIE INST, 83-, AM CANCER SOC PROF DEVELOP GENETICS, 84- *Mem:* Nat Acad Sci; Am Acad Arts & Sci; Am Soc Cell Biol (pres, 67-68); Genetics Soc Am; Soc Develop Biol (pres, 84-85). *Res:* Chromosome structure and function; nucleic acid metabolism; gene structure. *Mailing Add:* Carnegie Inst 115 W University Pkwy Baltimore MD 21210

GALL, MARTIN, b New York, NY, Oct 22, 44; m 68; c 3. MEDICINAL CHEMISTRY, ORGANIC CHEMISTRY. *Educ:* Trinity Col, Conn, BS, 66; Mass Inst Technol, PhD(org chem), 71. *Prof Exp:* Sr res scientist, Cent Nerv Syst Chem, Upjohn Co, 71-89; dir, Life Environ Sci Div, Syracuse Res Corp, 89-90; CHEM SECT MGR, NEUROMUSCULAR BLOCKING AGENT GI, ANAQUEST CORP, 90- *Mem:* Am Chem Soc; The Chem Soc; Sigma Xi. *Res:* Synthesize and evaluate new chemical substances which have clinical psychotherapeutic value; study new heterocyclic syntheses; metallation and reaction. *Mailing Add:* Anaquest Corp 100 Mountain Ave Murry Hill NJ 07974

GALL, WALTER GEORGE, b Passaic, NJ, Mar 11, 29; div; c 2. ORGANIC CHEMISTRY. *Educ:* Carnegie Inst Technol, BS, 50, MS, 50; Univ Rochester, PhD(chem), 53. *Prof Exp:* Res chemist org high polymer chem, Exp Sta, E I Du Pont de Nemours & Co, 53-62, sr res chemist, 63-69, res assoc, Plastics Prod & Resins Dept, 69-81; RETIRED. *Mem:* Sigma Xi; Am Chem Soc; fel Am Inst Chem. *Res:* Bridgehead nitrogen heterocycles; quinolizine; high polymer chemistry; stereospecific polymerizations; high temperature polymers; mineral-filled polymers; polymer flammability. *Mailing Add:* PO Box 436 El Cajon CA 92022

GALL, WILLIAM EINAR, b Warsaw, NY, Apr 12, 42; m 67. NEUROSCIENCES. *Educ:* Hamilton Col, AB, 63; Rockefeller Univ, PhD(life sci), 69. *Prof Exp:* Res assoc, 69-72, asst prof, 72-76, ASSOC PROF DEVELOP & MOLECULAR BIOL, ROCKEFELLER UNIV, 76-, RES DIR, NEUROSCI RES PROG, 81- *Mem:* Am Asn Immunologists; AAAS; Harvey Soc; Soc Neurosci. *Res:* Biochemistry of the cell surface; fertilization and early embryonic development; theoretical neurobiology. *Mailing Add:* Rockefeller Univ 1230 York Ave New York NY 10021-6399

GALLAGER, ROBERT G, b Philadelphia, Pa, May 29, 31. COMMUNICATIONS, DATA NETWORKS. *Educ:* Univ Pa, SB, 53; Mass Inst Technol, SM, 57, ScD, 60. *Prof Exp:* From asst prof to assoc prof, 60-67, FUJITSU PROF ELEC ENG, DEPT ELEC ENG & COMPUTER SCI, MASS INST TECHNOL, 67-, CO-DIR, LAB INFO & DECISION SYSTS, 86- *Concurrent Pos:* Co-chmn, Dept Area I, Control, Commun & Opers Res, Mass Inst Technol; consult, Codex Corp & Lincoln Labs. *Honors & Awards:* Medal of Honor, Inst Elec & Electronics Engrs, 90. *Mem:* Nat Acad Eng; fel Inst Elec & Electronics Engrs. *Res:* Data communication networks; information and communication theory. *Mailing Add:* Room 35-206 Mass Inst Technol Cambridge MA 02139

GALLAGHER, ALAN C, b Oak Park, Ill, June 14, 36; div; c 4. ATOMIC & MOLECULAR PHYSICS. *Educ:* Purdue Univ, BES, 58; Columbia Univ, PhD(physics), 64. *Prof Exp:* Res assoc physics, IBM Watson Lab, 64; JOINT INST FOR LAB ASTROPHYS, UNIV COLO, BOULDER, 64-; PHYSICIST, NAT BUR STANDARDS, 67- *Concurrent Pos:* Lectr, Univ Colo, 67-71, adj prof, 71- *Mem:* Fel Am Phys Soc. *Res:* Electron, atom and molecular collisions and radiation processes. *Mailing Add:* Dept Physics JILA A-709 Univ Colo Boulder CO 80309

GALLAGHER, BRENT S, b Greenfield, Mass, Nov 3, 39; m 58; c 2. PHYSICAL OCEANOGRAPHY. *Educ:* Univ Calif, Los Angeles, BS, 62; Scripps Inst, Univ Calif, PhD(oceanog), 65. *Prof Exp:* Asst res oceanogr, Inst Geophys & Planetary Physics, Univ Calif, San Diego, 66-67; asst prof, 67-71, ASSOC PROF PHYS OCEANOG, UNIV HAWAII, 71- *Concurrent Pos:* Consult, US Naval Radiological Defense Labs, 67-68. *Res:* Nonlinear aspects of ocean waves; fine-scale structure of physical variables in the ocean. *Mailing Add:* Box 407 Univ Hawaii Manoa 2500 Campus Rd Kortis Town HI 96760

GALLAGHER, BRIAN BORU, b Chicago, Ill, Sept 2, 34; m 58, 78; c 2. NEUROLOGY. *Educ:* Univ Notre Dame, BS, 56; Univ Chicago, PhD(biol psychol), 60, MD, 63. *Prof Exp:* Intern med, Univ Chicago, 63-64; resident neurol, Yale Univ-New Haven Hosp, 65-67; from instr to assoc prof, Sch Med, Yale Univ, 67-73; assoc prof pharmacol & neurol, Georgetown Univ, 73-77; PROF NEUROL, MED COL GA, 77- *Mem:* AAAS; Am Acad Neurol; Am Soc Neurochem; Soc Neurosci; NY Acad Sci; Sigma Xi. *Res:* Neuropharmacology and neurochemistry of epilepsy. *Mailing Add:* 3019 Bransford Rd Augusta GA 30909-3005

GALLAGHER, C(HARLES) E(DWARD), engineering, for more information see previous edition

GALLAGHER, CHARLES CLIFTON, b Boston, Mass, Feb 17, 37. ATMOSPHERIC SCIENCE, PHYSICS. *Educ:* Boston Col, BS, 58, MS, 60; Northeastern Univ, MS, 67. *Prof Exp:* Physicist plasma physics, Cambridge Res Lab, 60-74, PHYSICIST STRATOSPHERE, AIR FORCE GEOPHYSICS LAB, 74- *Mem:* Am Phys Soc; Sigma Xi. *Res:* Trace gas content of stratosphere studied as a function of altitude, latitude and time using chemiluminescence analysers and gas chromatography and with air samples obtained cryogenically, utilizing balloon-borne samplers. *Mailing Add:* PO Box 101 Concord MA 01742

GALLAGHER, DAVID ALDEN, b Chicago, Ill, April 20, 49; m 72; c 2. ELECTRON-TUBE PHYSICS. *Educ:* Univ Ill, Chicago, BS, 71; Univ Ill, Urbana, MS, 73, PhD(physics), 77. *Prof Exp:* ENG SPECIALIST, DEFENSE SYST DIV, NORTHROP CORP, 77- *Mem:* Am Physical Soc. *Res:* Fast-wave and millimeter- wave electron tubes, such as the peniotron, gyrotron, ortron and folded-waveguide traveling wave tube; development of both small and large signal theories of beam-wave interactions, their application to simulating tube performance and the conceptualization, development and evaluation of electron tubes. *Mailing Add:* 715 E Charles Arlington Heights IL 60004

GALLAGHER, GEORGE ARTHUR, b Paterson, NJ, Feb 11, 23; m 48; c 2. ORGANIC CHEMISTRY. *Educ:* Cornell Univ, AB, 43; Univ Pa, MS, 50, PhD(chem), 54. *Prof Exp:* Chemist, Socony Mobil Oil Co, Inc, 46-47; res chemist, Atlantic Refining Co, 47-53; chemist, Org Chem Dept, E I du Pont de Nemours & Co, 54-57, chemist, Elastomer Chem Dept, 57-67, personnel adminr, 67-72, lab adminr, 72-77, personnel mgr, 77-85, mgr, Spec Compensation & Benefits, 85; RETIRED. *Mem:* Am Chem Soc. *Res:* Elastomer research. *Mailing Add:* 832 Surrey Lane Media PA 19063

GALLAGHER, JAMES A, b Chicago, Ill, July 12, 26; m 57; c 4. POLYMER CHEMISTRY. *Educ:* St Louis Univ, BS, 49; Univ Mo, MA, 51, PhD(org chem), 55. *Prof Exp:* Chemist, Lubrizol Corp, Ohio, 51-52, Esso Res & Eng Co, 55-64; chemist, BASF Wyandotte Corp, 65-69, supvr polymer res, 69-71, mgr urethane appln, 72-73, asst to dir urethane chem res & develop, 74-79, asst to vpres, res & develop, 80-85, RES ASSOC, BASF WYANDOTTE CORP, 86- *Mem:* Am Chem Soc; Sigma Xi. *Res:* Synthetic resin development; synthesis and applications research on organic polymers, chiefly for surface coating, including cure mechanisms, surface chemistry. *Mailing Add:* 8439 Glengarry Rd Grosse Ile MI 48138

GALLAGHER, JAMES J, physics; deceased, see previous edition for last biography

GALLAGHER, JANE CHISPA, b Frankfurt, Ger, Jan 27, 50. PLANKTON ECOLOGY, POPULATION GENETICS. *Educ:* Stanford Univ, BS, 73; Univ RI, PhD(oceanog), 79. *Prof Exp:* Asst prof biol, 78-86, ASSOC PROF BIOL, CITY COL NEW YORK, 86- *Mem:* Am Soc Limnol & Oceanog; Phycological Soc Am; Sigma Xi; Soc Study Evolution; AAAS. *Res:* Ecology and physiology of marine plankton; species succession and population genetics of microalgae. *Mailing Add:* Biol Dept City Col New York Convent Ave 138 St New York NY 10031

GALLAGHER, JOEL PETER, b Chicago, Ill, Dec 12, 42; m 74; c 3. NEUROPHARMACOLOGY, NEUROPHYSIOLOGY. *Educ:* Col St Thomas, BS, 64; Univ Ill, BS & RPh, 67; Loyola Univ, Chicago, PhD(pharmacol), 72. *Prof Exp:* Community pharmacist, Westchester Apothecary, 67-71; res assoc pharmacol & NIH fel, Loyola Univ Chicago, 71-72, instr neurol & pharmacol, 72-73, NIH fel neurol & pharmacol, 73- 74, NIH fel neurophysiol, 74-75, adj asst prof, 75; from asst prof to assoc prof, 75-85, PROF PHARMACOL & TOXICOL, UNIV TEX MED BR, GALVESTON, 86- *Mem:* Soc Neurosci; Am Soc Pharmacol & Exp Therapeut; Acad Gen Pract Pharm; AAAS. *Res:* Analysis of the site and mechanism of action of drugs and the mechanism of normal and diseased physiological processes occurring at central synapses thought to be involved in the process of anxiety and space motion sickness. *Mailing Add:* Dept Pharmacol & Toxicol Univ Tex Med Br (MS-J-31) Galveston TX 77550

GALLAGHER, JOHN JOSEPH, JR, b Boston, Mass, Oct 7, 40; m 68; c 4. ROCK MECHANICS, BASIN ANALYSIS. *Educ:* Boston Col, BS, 62; Univ Mo, MA, 65; Texas A&M Univ, PhD(geol), 71. *Prof Exp:* Acting br chief, US Army CEngrs Sch, 66-68; res assoc tectonophysics, Cities Serv Exploration & Prod Res Lab, 71-75, mgr structural geol, 75-78; sr geol assoc, int exploration, Cities Serv Int Co, 78-80; mgr basin studies, Cities Serv Res & Technol, 80-83; mgr tectonics, Occidental Int Exploration & Prod Co, 83-85; mgr integr exploration res, 85-87, CONSULT EXPLORATION RES, ATLANTIC RICHFIELD RES & TECH SERV LAB, 87- *Concurrent Pos:* Mem, Marine Geol Comt, Am Assoc Petrol Geol, 87-, Geophys Comt, Int Comt; adj prof, Univ Tulsa, OK, 72-82; bd trustees, Int Basement Tectonics Assoc, 72-, chmn, 85-90. *Mem:* Am Geophysical Union; Int Basement Tectonics Asn; Geol Soc Am; Am Asn Petrol Geologists; Am Inst Prof Geologists; Sigma Xi. *Res:* Salt tectonics; application of dynamic structural geology and technophysics to oil and gas exploration; scanning electron microscopic fractographic study of fractured sand grains; photoelastic study of stress concentrations in granular-media and thrust-fault analog; oil and gas basin evaluation worldwide; computer applications to remote sensing interpretation. *Mailing Add:* AOGC Res & Tech Serv 2300 W Plano Pkwy Plano TX 75075-8427

GALLAGHER, JOHN LESLIE, b West Grove, Pa, Nov 4, 35; c 2. MARINE BOTANY. *Educ:* Univ Del, BSAg, 57, MS, 59, PhD(biol sci), 71. *Prof Exp:* Instr biol, West Nottingham Acad, 59-68; instr, Univ Del, 68-69; res assoc marine bot, Marine Inst, Univ Ga, 71-76, actg dir, 75-77, assoc marine scientist, 76-79; assoc prof marine studies, 80-85, PROF MARINE BIOL & BIOCHEM, UNIV DEL, 85- *Concurrent Pos:* Adj asst prof agron, Univ Ga, 75-79 & adj asst prof bot, 76-79. *Mem:* Ecol Soc Am; Am Soc Limnol & Oceanog; Estuarine Res Fedn (treas, 78-79); Botanical Soc Am; Brit Ecol Soc; Am Inst Biol Sci. *Res:* Ecology, physiology and genetics of salt marsh autotrophs; tissue culture of halophytes. *Mailing Add:* Col Marine Studies Univ Del Lewes DE 19958

GALLAGHER, JOHN M(ICHAEL), JR, b Charleston, WVa, Feb 26, 27; m 53; c 6. ELECTRICAL & NUCLEAR ENGINEERING. *Educ:* Rensselaer Polytech Inst, BEE, 51; Mass Inst Technol, MS, 54, EE, 56. *Prof Exp:* Asst & res engr, Dynamic Analytical & Control Lab, Mass Inst Technol, 51-56; sr engr comput analysis group, Atomic Power Dept, Westinghouse Elec Co, 56-59, supvry eng systs transient analysis group, 59-63, mgr systs analysis, 63-67, mgr control & elec systs, 67-69, consult engr, 70-75, mgr instrumentation & control develop, Nuclear Energy Systs, 75-88, consult engr, Control & Elec Systs, 88-90; NUCLEAR REGULATORY COMN, 91- *Concurrent Pos:* Mem int electro-tech comn, Am Nat Standards Inst. *Mem:* Am Nuclear Soc; Inst Elec & Electronics Engrs. *Res:* Development, design and testing of instrumentation; control electrical equipment and systems for nuclear power plants; distributed computer systems for control and for expert system advice for nuclear and fossil power plants. *Mailing Add:* 1450 Navahoe Dr Pittsburgh PA 15228

GALLAGHER, JOHN SILL, b Boston, Mass, Mar 26, 47; m 70; c 2. ASTROPHYSICS. *Educ:* Princeton Univ, AB, 69; Univ Wis-Madison, PhD(astron), 72. *Prof Exp:* Vis asst prof & res assoc astrophys, Univ Nebr-Lincoln, 72-74; asst prof astrophys, Univ Minn, Minneapolis, 74-77; assoc prof astron, Univ Ill, Urbana, 77-80, prof astron, 80-84, astronr, Kitt Peak Nat Observ, 84-86, dir, Lowell Observ, 86-89; PROF, ASTRON, UNIV WIS-MADISON, 91- *Concurrent Pos:* Assoc ed, Astrophys J, 86-; dir at large, AURA, Inc, 87- vpres, 89-; Aspen Theoret Astrophys Org Comt, 82- *Mem:* Int Astron Union; Am Astron Soc. *Res:* Observable properties of novae and related binary stars; stellar populations and evolution of extragalactic systems; gas content and structural form of galaxies. *Mailing Add:* Dept Astron Univ Wis 475 N Charter St Madison WI 53376

GALLAGHER, MARGIE LEE, EXPERIMENTAL BIOLOGY. *Educ:* Univ Tenn, BS, 69; Univ Fla, MS, 71; Univ Calif, PhD(nutrit), 76. *Prof Exp:* Postdoctoral res, Maine Heart Asn, 76; res asst prof, Univ Maine, Orono, 77-79; asst prof & asst scientist, 80-85, assoc prof & assoc scientist, 85-89, PROF & SCIENTIST, E CAROLINA UNIV, GREENVILLE, 89- *Concurrent Pos:* Vet Admin Ctr, Togus, Maine, 76; numerous grants, 81-; reviewer, Sea Grant Proposals, Nat Oceanic & Atmospheric Admin, proposals, USDA; vis sicentist, Israel & Italy, 84, Galilee Inst Technol, Kiryat Shomona, Israel, 85; consult, Charles Sockriter Farmes, Byrd Seafood Co, Inc, Int Nutrit, Inc, & Ospry Seafoods. *Mem:* Sigma Xi; Am Dietetic Asn; Am Fisheries Soc; Am Inst Nutrit; NY Acad Sci. *Res:* Nutrition; experimental biology; numerous technical publications. *Mailing Add:* East Carolina Univ Greenville NC 27858

GALLAGHER, MICHAEL TERRANCE, Gadsden, Ala, Nov 26, 43; m 73; c 2. TRANSPLANTATION IMMUNOLOGY, TUMOR IMMUNOLOGY. *Educ:* Univ Houston, BS, 66; Northwestern Univ, MS, 70; Baylor Col Med, PhD(immunol), 74. *Prof Exp:* Instr immunol, Div Exp Biol, Baylor Col Med, 74-75, asst prof, 75-80; asst res scientist, 80-82, ASSOC RES SCIENTIST IMMUNOL, DEPT CLIN & EXP, CITY OF HOPE NAT MED CTR, 82- *Mem:* Am Asn Cancer Res; Am Asn Immunologists; Int Soc Exp Hematol; The Transplantation Soc. *Res:* Transplantation immunology, particularly bone marrow transplantation and graft-vs-host disease; tumor immunology with emphasis on natural killer cells; differentiation of hemopoietic stem cells; the role of the reticuloendothelial system in autoimmune hemolytic anemia. *Mailing Add:* 1707 Colonial Ridge Rd Johnson City TN 37604

GALLAGHER, NEAL CHARLES, b Baltimore, Md, Jan 9, 49; m 71; c 2. ELECTRICAL ENGINEERING, OPTICS. *Educ:* Loyola Col, Md, BS, 71; Princeton Univ, MA & MSE, 73, PhD(elec eng), 74. *Prof Exp:* Asst prof, Case Western Reserve Univ, 74-76; asst prof, 76-79, ASSOC PROF ELEC ENG, PURDUE UNIV, 79- *Mem:* Inst Elec & Electronics Engrs; Optical Soc Am. *Res:* Statistical communication theory; digital holography. *Mailing Add:* Sch Elec Eng Purdue Univ West Lafayette IN 47907

GALLAGHER, NEIL IGNATIUS, b Cleveland, Ohio, Oct 12, 26; m 56; c 4. INTERNAL MEDICINE, HEMATOLOGY. *Educ:* Santa Clara Univ, BS, 47; Univ Ariz, MA, 51; St Louis Univ, MD, 54; Am Bd Internal Med, dipl, 62, cert hemat, 72. *Prof Exp:* Intern, St Louis City Hosp, Mo, 54-55; asst resident, St Louis Univ Hosp, 55-57; fel hemat, Vet Admin Hosp, St Louis, 57-59, clin investr, 59-61, chief radioisotope lab, 61-68, from assoc chief of staff to chief of staff res, 63-67, chief of med, 66-68; from instr to prof, internal med & vchmn dept, Sch Med, St Louis Univ, 59-73, staff mem, 59-87; STAFF MEM, INTERNAL MED, ST MARY'S HOSP, 87- *Concurrent Pos:* Staff physician, Univ Group Hosps, 59-; dir, St Louis Univ Med Serv, Vet Admin Hosp, 61-75. *Mem:* Am Fedn Clin Res; Am Soc Hemat; Int Soc Hemat; Soc Exp Biol & Med; fel Am Col Physicians. *Res:* Cell proliferation; humoral regulation of erythropoiesis; pathophysiology of dyserythropoiesis. *Mailing Add:* 869 Rampart Dr St Louis MO 63122

GALLAGHER, PATRICK KENT, b Waukegan, Ill, Mar 17, 31; m 53; c 2. INORGANIC CHEMISTRY. *Educ:* Univ Wis, BS, 52, MS, 54, PhD(inorg chem), 59. *Prof Exp:* mem tech staff inorg chem, Bell Labs, 59-89; DOW PROF MAT CHEM & ENG, DEPT CHEM & DEPT MAT SCI & ENG, OHIO STATE UNIV, 90- *Concurrent Pos:* Chmn comt E-37, Am Soc Testing & Mat, 74-76. *Honors & Awards:* Mettler Award, NAm Thermal Analysis Soc, 76; DuPont Award, Thermal Analysis, Int Confedn Thermal Analysis, 82; Kurnikov Medal, Acad Sci, USSR, 85; Semiconductor Int Res Award, 86. *Mem:* Fel NAm Thermal Analysis Soc (pres, 75, vpres, 74); Int Confedn Thermal Anal (vpres, pres-elect, 80, pres, 82, treas, 87); Am Chem Soc; fel Am Ceramic Soc; Am Soc Testing & Mat. *Res:* Synthesis, reactivity and characterization of inorganic materials, particularly oxides for electronic ceramics. *Mailing Add:* Dept Chem Ohio State Univ 120 W 18th Ave Columbus OH 43210-1173

GALLAGHER, PATRICK XIMENES, b Elizabeth, NJ, Jan 2, 35; m 61; c 2. MATHEMATICS. *Educ:* Harvard Univ, AB, 56; Princeton Univ, PhD(math), 59. *Prof Exp:* Asst math, Princeton Univ, 57-59; instr, Mass Inst Technol, 59-61; asst prof, Columbia Univ, 62-64; mem, Inst Advan Study, 64-65; from assoc prof to prof math, Barnard Col, 65-72, PROF MATH, COLUMBIA UNIV, 72- *Mem:* Am Math Soc. *Res:* Analytic number theory; finite groups. *Mailing Add:* Dept Math Columbia Univ Broadway & West 116th New York NY 10027

GALLAGHER, RICHARD HUGO, b New York, NY, Nov 17, 27; m 52; c 5. CIVIL ENGINEERING. *Educ:* NY Univ, BCE, 50, MCE, 55; State Univ NY, Buffalo, PhD(civil eng), 66. *Hon Degrees:* Dr Tech, Tech Univ Vienna, 87; Dr, Univ Col Swansea, Wales, 87. *Prof Exp:* Civil engr, US Dept Commerce, Civil Aeronaut Admin, 50-52; struct designer, Tex Co, 52-55; struct engr, Bell Aerosysts Co, 55-59, group leader struct res, 59-65, asst chief engr, 65-67; prof civil eng, Cornell Univ, 67-78, chmn dept struct eng, 70-78; dean, Col Eng, Univ Ariz, 78-84; vpres & dean fac, Worcester Polytech Inst, 84-88; PRES, CLARKSON UNIV, 88- *Concurrent Pos:* Consult, adv group aeronaut res & develop, NATO, 60-62; ed, Int J Numerical Methods Eng, 67- *Honors & Awards:* Worcester Reed Warner Medal, Am Soc Mech Engrs, 85; Structural Dynamics & Mat Award, Am Inst Aeronaut & Astronaut, 90; Benjamin Garver Lamme Award, Am Soc Eng Educ, 90. *Mem:* Nat Acad Eng; assoc fel Am Inst Aeronaut & Astronaut; Soc Exp Stress Analysis; fel Am Soc Mech Engrs; fel Am Soc Civil Engrs; Nat Soc Prof Engrs; Am Soc Eng Educ. *Res:* Solid mechanics, particularly the development of finite element techniques; development of methods of optimum design. *Mailing Add:* Clarkson Univ Potsdam NY 13699-5500

GALLAGHER, THOMAS FRANCIS, b Bronxville, NY, Nov 19, 44; m 74; c 1. RYABERG ATOMS, AUTOIONIZATION. *Educ:* Williams Col, AB, 66; Harvard Univ, AM, 68, PhD(physics), 71. *Prof Exp:* Res assoc physics, Dept Physics, Harvard Univ, 71, Univ Utah, 71-72; post doctoral physicist, Molecular Physics Lab, Stanford Res Inst, 72-74, physicist, 74-78, sr physicist, 78-83, prog mgr, 83-84; PROF PHYSICS, UNIV VA, 84- *Concurrent Pos:* Vis scientist, Atomic Physics Serv, CEN Saclay, 77, Inst Electronics Found, Univ Paris Sud, 80; prof assoc, Lab Aime Colton, Univ Paris Sud, 88. *Mem:* fel Am Phys Soc; Am Optical Soc. *Res:* Highly excited atoms to study quantitatively otherwise relatively inaccessible phenomena; examples are atoms in strong microwave fields; resonant collisional energy transfer and correlation of the electrons in autoionizing states. *Mailing Add:* Dept Physics Univ Va Charlottesville VA 22901

GALLAGHER, WILLIAM J(OSEPH), b Allentown, Pa, Jan 13, 31; m 56; c 1. MECHANICAL ENGINEERING. *Educ:* Drexel Inst, BS, 53; Lehigh Univ, MS, 55; Univ Pittsburgh, PhD(mech eng), 62. *Prof Exp:* Engr, Pa Power & Light Co, 53-54; assoc engr thermal & hydraul design, Westinghouse Elec Corp, 57-62, sr engr, 62-64; supvr, Bettis Atomic Power Lab, 64-66; mgr eng, NUS Corp, 66-69, dir tech opers Europe, Ger, 69-71; managing dir, Nuklear Ingenieur Serv GmbH, 71-74; vpres & gen mgr, Eng Consult Div, NUS Corp, 74-75, exec vpres, 82-87, pres, NUS Int, 87, SR VPRES ENVIRON SYSTS & CONSULT SERV GROUP, NUS CORP, 75-, ASST TO PRES, 78-; PRES & CHIEF EXEC OFFICER, NUMATEC, INC, 87- *Concurrent Pos:* Mem & chmn, Ger Govt Adv Comts Nuclear Power, 71-74. *Res:* Nuclear reactor heat transfer and fluid flow. *Mailing Add:* 2923 45th St NW Washington DC 20016

GALLAHER, DANIEL DAVID, b July 30, 52; m 79; c 2. DIETARY FIBER, ZINC ABSORPTION. *Educ:* Univ Calif, Davis, PhD(nutrit), 84. *Prof Exp:* ASSOC PROF FOOD & NUTRIT, NDAK STATE UNIV, 83- *Concurrent Pos:* Prin investr, Nat Cancer Inst grant. *Mem:* Am Inst Nutrit; Inst Food Technol; Am Diabetes Asn; Sigma Xi. *Res:* Influence of dietary fiber on gastro intestinal function; effects of fiber on bile acid metabolism and blood glucose control. *Mailing Add:* Dept Food & Nutrit Univ Minn 1334 Eckles Ave St Paul MN 55108

GALLAHER, LAWRENCE JOSEPH, b St Louis, Mo, May 31, 25; m 54; c 7. PHYSICS, COMPUTER SCIENCE. *Educ:* Rensselaer Polytech Inst, BS, 50; Washington Univ, PhD(physics), 55. *Prof Exp:* Lectr physics, Univ Minn, 55; from asst prof to assoc prof, Ohio Univ, 55-63; res scientist, Ga Inst Technol, 63-80, sr res scientist, 80-85; res scientist, Lockheed, 85-90; RETIRED. *Mem:* Am Phys Soc; Asn Comput Mach. *Res:* Applications of digital computers. *Mailing Add:* 3892 Brenton Way NE Atlanta GA 30319

GALLAHER, WILLIAM RICHARD, b Englewood, NJ, Dec 28, 44; div; c 4. VIROLOGY, CELL BIOLOGY. *Educ:* St Peters Col, NJ, BS, 66; Harvard Univ, PhD(microbiol & molecular genetics), 72. *Prof Exp:* Res assoc membrane biol, Scheie Eye Inst, Univ Pa, 71-73; asst prof microbiol, 73-80, ASSOC PROF MICROBIOL & IMMUNOL, LA STATE UNIV MED CTR, 80- *Concurrent Pos:* Grantee investr, Cancer Asn Greater New Orleans, 74-76, La Heart Asn, 78-80 & Nat Inst Allergy & Infectious Dis, 78- *Mem:* Am Soc Microbiol; Tissue Cult Asn; Am Soc Virol. *Res:* Biochemistry and genetics of mammalian cell membranes related to cell regulation and sensitivity to viral infection; identification of cellular receptors for animal viruses. *Mailing Add:* 1542 Tulane Ave Dept Microbiol La State Univ New Orleans LA 70112-1393

GALLANDER, JAMES FRANCIS, b Peoria, Ill, Apr 17, 37; m 63; c 1. FOOD SCIENCE. *Educ:* Ohio State Univ, BS, 60, PhD(food technol), 64. *Prof Exp:* From instr to assoc prof, 62-71, PROF FRUIT & VEG PROCESSING, OHIO AGR RES & DEVELOP CTR, 71- *Mem:* Inst Food Technologists; Am Soc Enol. *Res:* Freezing and canning of pomological crops; wine fermentation. *Mailing Add:* Dept Hort Ohio State Univ Ohio Agr Res & Develop Ctr Madison Ave Wooster OH 44691

GALLANT, DONALD, b Brooklyn, NY, Aug 9, 29; m 54; c 1. PSYCHIATRY, NEUROLOGY. *Educ:* Tulane Univ, BS, 51, MD, 55. *Prof Exp:* From instr to assoc prof, 61-68, PROF PSYCHIAT & NEUROL, SCH MED, TULANE UNIV, 69- *Concurrent Pos:* Prin investr, Tulane Psychopharmacol Serv Ctr grant, 62-; consult, La State Alcoholism Prog, 62- *Honors & Awards:* Robert C Lancaster Award for Humanitarian Efforts in Psychiat, 85. *Mem:* Am Col Neuropsychopharmacol; Am Psychiat Asn. *Res:* Investigation and evaluation of neuropsychopharmacologic compounds; alcoholism. *Mailing Add:* Dept Psychiat & Neurol Tulane Univ Sch Med 1415 Tulane Ave New Orleans LA 70112

GALLANT, ESTHER MAY, b Auburn, NY, 43. MALIGNANT HYPERTHERMIA, MUSCLE DISEASES. *Educ:* Ind Univ, PhD(bot), 70. *Prof Exp:* Asst prof vet anat, pharmacol & physiol, Iowa State Univ, 74-78; sr res fel & assoc consult, dept pharmacol, Mayo Clin, 79-82; asst prof, 82-85, ASSOC PROF VET BIOL, COL VET MED, UNIV MINN, 85- *Mem:* AAAS; Am Phys Soc; Biophys Soc; Am Asn Anatomists. *Res:* Skeletal muscle physiology; mechanical and electrophysiological studies of normal and diseased muscles. *Mailing Add:* Dept Vet Biol Univ Minn Col Vet Med St Paul MN 55108

GALLANT, JONATHAN A, b New York, NY, Sept 23, 37; m 61; c 3. MOLECULAR BIOLOGY. *Educ:* Haverford Col, BS, 57; Johns Hopkins Univ, PhD(biochem), 61. *Prof Exp:* From asst prof to assoc prof, 61-70, PROF GENETICS, UNIV WASH, 70- *Concurrent Pos:* NIH res grants, 62-65, 66-70 & 71-; NIH spec fel, Inst Biophys & Biochem, Paris, France, 64-65; dir sci & gas, Grosser Seattle, 69-; Guggenheim fel, 71. *Mem:* AAAS; Genetics Soc Am; Fedn Am Sci. *Res:* Accuracy in translation and transcription; regulation of enzyme synthesis; regulation of RNA synthesis; biology of aging. *Mailing Add:* Genetics Dept SK 50 Univ Wash Seattle WA 98195

GALLARDO-CARPENTIER, ADRIANA, b Sept 13, 30; m. PHARMACOLOGY. *Educ:* Univ Chile, DDS, 55. *Prof Exp:* Asst prof, 77-81, ASSOC PROF PHARMACOL, COL MED, HOWARD UNIV, 81- *Mem:* Am Soc Pharmacol & Exp Therapeut. *Res:* Effects of alcohol on the hypothalamus and on the heart. *Mailing Add:* 520 W St NW Washington DC 20059

GALLATI, WALTER WILLIAM, b Brooklyn, NY, Dec 7, 27; m 58; c 2. ZOOLOGY. *Educ:* Drew Univ, AB, 50; Univ Miami, MS, 52; Ohio State Univ, PhD(zool), 57. *Prof Exp:* Asst zool, Univ Miami, 50-51; asst zool, Ohio State Univ, 52-54, from asst instr to instr, 54-57; assoc prof sci, 57-59, PROF BIOL, INDIANA UNIV, PA, 59-, CHMN DEPT, 83- *Mem:* AAAS; Am Soc Parasitol; Am Micros Soc; Am Inst Biol Sci; Sigma Xi. *Res:* Parasitology. *Mailing Add:* 104 Shady Dr Indiana PA 15701

GALLAWAY, BOB MITCHEL, b Kosciusko, Miss, Oct 14, 16; m 41; c 3. CIVIL ENGINEERING. *Educ:* Tex A&M Univ, BS, 43, MS, 46, ME, 56. *Prof Exp:* Instr war training in eng, Tex A&M Univ, 43-44, instr eng drawing & descriptive geom, 44-46, asst prof, 46-48, from asst prof to assoc prof civil eng, 48-59, Brackett Prof, 78-80, PROF CIVIL ENG, TEX A&M UNIV, 59- *Concurrent Pos:* Res engr & head hwy mat dept, Tex Transp Inst; mem, Hwy Res Bd; pres, Consult & Res Serv, Inc, 67-; mem bd consults study 4-10, Nat Coop Hwy Res Prog; mem, Transp Ctr, Univ PR. *Mem:* Nat Soc Prof Engrs; Am Soc Testing & Mat; Am Soc Eng Educ; Am Concrete Inst; Asn Asphalt Paving Technol; Sigma Xi. *Res:* Materials research with particular interest in asphalt paving, portland cement, concrete and soil stabilization; skid resistance; waste solids utilization; synthetic aggregate. *Mailing Add:* Dept Civil Eng Tex A&M Univ College Station TX 77843

GALLE, KURT R(OBERT), b Newton, Kans, July 19, 25; m 46; c 2. MECHANICAL ENGINEERING. *Educ:* Purdue Univ, BS, 46, BS, 47, MS, 49, PhD(mech eng), 51. *Prof Exp:* Instr mech eng, Purdue Univ, 48-50; res engr, Boeing Co, Wash, 51-58, sr group engr, 58-60; ASSOC PROF MECH ENG, UNIV WASH, 60- *Mem:* Am Inst Aeronaut & Astronaut; Instrument Soc Am; Am Soc Eng Educ; Biomed Eng Soc; Simulation Coun; Sigma Xi. *Res:* Instrumentation research in bioengineering; systems analysis and simulation. *Mailing Add:* 8027 43rd Ave NE Seattle WA 98115

GALLEGLY, MANNON ELIHU, b Mineral Springs, Ark, Apr 11, 23; m 47; c 3. PLANT PATHOLOGY, PHYTOPHTHORA. *Educ:* Univ Ark, BS, 45; Univ Wis, MS, 46, PhD(plant path), 49. *Prof Exp:* From asst prof to prof plant sci, 49-86, dir div plant sci, 70-86, EMER PROF PLANT SCI, 86- *Concurrent Pos:* Consult, US/AID, EAfrica, 69-71. *Honors & Awards:* Campbell Award, AAAS, 60. *Mem:* Fel Am Phytopath Soc; Potato Asn Am; Sigma Xi. *Res:* The genus Phytophthora: taxonomy, sexuality, diseases; physiologic specialization and resistance to tomato and potato late blight. *Mailing Add:* 401 Brooks Hall Div Plant Sci WVa Univ Morgantown WV 26506-6057

GALLEGOS, EMILIO JUAN, b Del Norte, Colo, June 24, 32; m 63; c 1. PHYSICAL & ANALYTICAL CHEMISTRY. *Educ:* Regis Col, BS, 54; Kans State Univ, PhD(phys chem), 61. *Prof Exp:* Res chemist, 61-62, SR RES ASSOC, CHEVRON RES CORP, STANDARD OIL CO CALIF, 62- *Honors & Awards:* Outstanding Paper Award, Geochem Soc, 72. *Mem:* Am Chem Soc; Sigma Xi; Am Soc Mass Spectros. *Res:* Mass spectrometry; electron and optical microscopy; plasma; coal, shale and petroleum analysis; organic geochemistry; gas chromatography; mass spectrometry; computer. *Mailing Add:* 5618 Barrett Ave El Cerrito CA 94530

GALLELLI, JOSEPH F, b Brooklyn, NY, Mar 23, 36; m; c 5. PHARMACY. *Educ:* Long Island Univ, BS, 57; Temple Univ, MS, 59, PhD(phys pharm), 62. *Prof Exp:* Grad & teaching asst, Dept Pharm, Sch Pharm, Temple Univ, 57-58; jr investr, Pharmaceut Prod Develop Dept, Wyeth Inst Med Res, Radnor, Pa, 58-59; instr, Dept Pharm, Temple Univ, 59-60, Wyeth Labs teaching fel, Sch Pharm, 59-61; staff pharmacist, Pharmaceut Develop Serv, 61-62, chief, 62-70, CHIEF, PHARM DEPT, CLIN CTR, NIH, BETHESDA, MD, 70- *Concurrent Pos:* Chmn, Res Comt, Am Soc Hosp Pharmacists, 67-69 & Parenteral Prods Subcomt, Drug Standards Div, USP, 80-95; contrib ed, J Drug Intel, 67-70; clin instr pharm pract, Sch Pharm, Univ Md, 73-; assoc clin prof, Col Pharm & Pharmacol Sci, Howard Univ, 79-; mem, Drug Standards Consult Panel, 87-91. *Honors & Awards:* Abbott Award, Am Soc Hosp Pharmacists, 70. *Mem:* Am Pharmaceut Asn; Sigma Xi; Acad Pharmaceut Sci; Am Soc Hosp Pharmacists; Dist Columbia Soc Hosp Pharmacists; Int Pharmaceut Fedn; Am Asn Pharmaceut Scientists; Asn Mil Surgeons US. *Res:* Drug stability and assay development; formulation and stability of parenteral products; chemical kinetics and clinical pharmacokinetics; over 60 publications. *Mailing Add:* Warren Grant Magnuson Clin Ctr NIH Pharm Dept Bldg 10 Rm 1N257 Bethesda MD 20892

GALLEN, WILLIAM J, b Columbus, Ohio, July 4, 24; m 54; c 7. MEDICINE. *Educ:* Ohio State Univ, BA, 45, MD, 48; Am Bd Pediat, dipl, 57, cert cardiol, 61. *Prof Exp:* From intern to resident, Harper Hosp, Detroit, 48-51; resident pediat, Milwaukee Children's Hosp, 53-55; instr, Johns Hopkins Hosp, 55-56; Nat Heart Inst trainee, Karolinska Inst, Sweden, 56-57; from instr to prof pediat, Med Col Wis, 57-86; RETIRED. *Concurrent Pos:* Fel, Crippled Children's Bur, Johns Hopkins Univ Hosp, 55-56; dir, Cardiac Diag Unit, Milwaukee Children's Hosp, 57-86; dir, Fairchild Cardiac Study Ctr, 57-86, chief, Dept Pediat, 70-72 & med serv, 72-74; vchmn, dept pediat, Med Col Wis, 81. *Mem:* Am Col Cardiol; Am Acad Pediat; Am Heart Asn. *Res:* Pediatrics; pediatric cardiology. *Mailing Add:* PO Box 1997 Milwaukee WI 53201-1997

GALLENBERG, LORETTA A, b June 11, 57. PHARMACOLOGY, TOXICOLOGY. *Educ:* Univ Wis-Madison, BS, 79; Med Col Wis, MS, 85, PhD(pharmacol & Toxicol), 87. *Prof Exp:* Qual control microbiologist, Dairyland Food Labs, 79-81; sr res technician, 81-82, postdoctoral fel, Dept Anesthesiol, 87-91, ASST PROF PHARMACOL & TOXICOL ANESTHESIOL, MED COL WIS, 91- *Mem:* Am Soc Pharmacol & Exp Therapeut; Soc Toxicol; Am Physiol Soc. *Res:* Pharmacology; toxicology; author of 22 technical publications. *Mailing Add:* Anesthesia Res Vet Admin Med Ctr 151 National Ave Milwaukee WI 53295

GALLENT, JOHN BRYANT, b Wylam, Ala, Mar 28, 02; m 26. CHEMISTRY. *Educ:* Davidson Col, BS, 25; Univ NC, MS, 28, PhD(chem), 30. *Prof Exp:* Pub sch teacher, NC, 25-27; teacher, Univ NC, 27-30; prof chem, Jonesboro Agr & Mech Col, 30-31 & Brenau Col, 31-42; prof chem, 42-72, chmn dept, 62-72, EMER PROF CHEM, DAVIDSON COL, 72- *Concurrent Pos:* NSF fac fel, Univ Col, Univ London, 57-58; chem consult, 72- *Mem:* Am Chem Soc. *Res:* Sulfonation of vegetable oils; preparation of the ethers of p-di-hydroxybenzene and their sulfonates. *Mailing Add:* Chem Consult Serv Box 2 Davidson NC 28036

GALLER, BERNARD AARON, b Chicago, Ill, Oct 3, 28; m 51; c 4. INTELLIGENT VEHICLE-HIGHWAY SYSTEMS. *Educ:* Univ Chicago, PhB, 46, BS, 47, PhD(math), 55; Univ Calif, Los Angeles, AM, 49. *Prof Exp:* From instr to assoc prof math, Univ Mich, 55-66, assoc dean, Col Lit, Sci & Arts, 75-79, prof comput & commun sci, 66-84, assoc dir, Comput Ctr, 66-90, PROF ELECT ENG & COMPUT SCI, 85-, ASSOC DIR, RES SYSTS, INFO TECH DIV, UNIV MICH, ANN ARBOR, 90- *Concurrent Pos:* Ed-in-chief, Anns Hist Comput, 78-87. *Honors & Awards:* Distinguished Serv Award, AFIPS, 84; Distinguished Serv Award, Asn Comput Mach, 80. *Mem:* Asn Comput Mach (pres, 68-70). *Res:* Digital computers; intelligent vehicle-highway systems; automatic programming; computer analysis of musical sound; mathematical logic; linear programming; distributed systems. *Mailing Add:* 1056 Ferdon Rd Ann Arbor MI 48104

GALLER, JANINA REGINA, b Upsola, Sweden, US citizen; m 70; c 1. CHILD PSYCHIATRY. *Educ:* Sophie Newcomb Col, BS, 69; Albert Einstein Col Med, MD, 72. *Prof Exp:* Assoc chairperson, dept child psychiat, Sch Med, 80-85, dir, Residency Training Psychiat, 82-86, PROF PSYCHIAT, BOSTON UNIV, 82-, PROF PUB HEALTH, SCH PUB HEALTH, 84-, DIR, CTR BEHAV DEVELOP, SCH MED, 85- *Concurrent Pos:* Consult, UNICEF; mem steering comt, USAID. *Honors & Awards:* Blanche Ittleson Award, Am Psychiat Asn, 85. *Mem:* Am Inst Nutrit; Am Psychiat Asn; Am Acad Child Psychiat; Am Psychosomatic Soc. *Res:* Effects of nutritional deficiency on mental development in disadvantaged populations in animals, labaratory setting and developing countries especially Barbados and Mexico. *Mailing Add:* Dept Child Psychiat Boston Univ Med Sch 85 E Newton St Suite M921 Boston MA 02118

GALLER, SIDNEY ROLAND, b Baltimore, Md, Nov 9, 22; div; c 4. ECOLOGY, RESEARCH ADMINISTRATION. *Educ:* Univ Md, BS, 44, MS, 47, PhD(limnol), 48. *Prof Exp:* Asst agr exp sta, Univ Md, 42-43, asst zool, 46-48; consult human ecol & biophys & actg head biophys br, Off Naval Res, 48-50, head ecol sect, Biol Br, 50-51, head Biol Br, 51-65; asst secy for sci, 65-71, Smithsonian Inst, 65-71; dep asst secy, Com Environ Affairs, Dept Com, 71-80; PRES, SIDNEY R GALLER INC & ASSOCS, 80- *Concurrent Pos:* Collabr, US Fish & Wildlife Serv, 47; chmn US deleg, Proj-Harmonization Environ Standards, US-USSR Environ Agreement, 74-80. *Mem:* Fel AAAS; Marine Technol Soc; Sigma Xi; Am Soc Limnol & Oceanog. *Res:* Chemical, physical and biological investigations of acid ponds; resource conservation and recovery; designer of researchships of opportunity for oceanography and marine ecology; animal tracking systems; thermal pollution measurement systems; research policy and program planning. *Mailing Add:* 6242 Woodcrest Ave Baltimore MD 21209

GALLER, WILLIAM SYLVAN, b Chicago, Ill, June 12, 29; m 55. SANITARY ENGINEERING, OPERATIONS RESEARCH. *Educ:* Ill Inst Technol, BS, 51, MS, 61; Northwestern Univ, PhD(civil eng), 65. *Prof Exp:* Jr sanit engr, Water Purification Dept, City of Chicago, 51-53, sr sanit engr, 53-61; from asst prof to assoc prof civil eng, 64-75, PROF CIVIL ENG, NC STATE UNIV, 75- *Mem:* Am Water Works Asn; Am Water Resources Asn; Opers Res Soc Am; Water Pollution Control Fedn; Am Soc Civil Engrs; Sigma Xi. *Res:* Systems analysis approach to problems in sanitary engineering. *Mailing Add:* Dept Civil Eng NC State Univ PO Box 7908 Raleigh NC 27695-7908

GALLETTA, GENE JOHN, b Philadelphia, Pa, July 3, 29; m 57; c 2. CYTOLOGY, PLANT PHYSIOLOGY. *Educ:* Univ Md, BS, 51; Rutgers Univ, MS, 53; Univ Calif, PhD(genetics), 59. *Prof Exp:* Asst hort, Rutgers Univ, 51-53; asst pomologist, Univ Calif, 53-54, 56-59, pomologist, 56; from asst prof to prof hort sci, NC State Univ, 59-77, mem fac genetics, 70-77; RES PLANT GENETICIST, AGR RES SERV, USDA, 77- *Concurrent Pos:* G Darrow small fruit res award, Am Soc Hort Sci. *Mem:* Fel Am Soc Hort Sci; Am Genetic Asn; Am Pomol Soc; NAm Strawberry Growers Asn; Int Soc Hort Sci. *Res:* Small fruit crop breeding and genetics; plant and fruit morphology; inheritance of quantitative characters and disease resistance; plant propagation; fruit culture. *Mailing Add:* USDA Agr Res Serv BA PSI Fruit Lab BARC-W Beltsville MD 20705

GALLETTI, PIERRE MARIE, b Monthey, Switz, June 11, 27; US citizen; m 59; c 1. CARDIOPULMONARY PHYSIOLOGY. *Educ:* St Maurice Col, Switz, BA, 45; Univ Lausanne, MD, 51, PhD(physiol), 54. *Hon Degrees:* Roger Williams Col, 79, Univ Nancy, France, 82, Univ Ghent, Belg, 89. *Prof Exp:* Instr physiol, Univ Lausanne, 52-54; resident med, Zurich, 54-57; fel cardiovasc res, Cedars of Lebanon Hosp, Los Angeles, 57-58; from asst prof to prof physiol, Sch Med, Emory Univ, 58-67; chmn , Div Biol & Med Sci, 68-72, PROF MED SCI, BROWN UNIV, 67-, VPRES BIOL & MED, 72- *Concurrent Pos:* Int Union Against Cancer fel, Tumor Ctr, Palermo, 64-65; mem, Physiol Training Comt, NIH, 68-72; mem, Vet Admin Nephrol Rev Bd, 72-76; mem, cardiol adv comt, Nat Heart & Lung Inst, 76-79; chmn, Consensus Develop Conf Clin Appln Biomat, Biomed Eng Instrumentation Br, Div Res Serv & Off Med Appln Res & Devices & Technol Br Task Group, Devices & Technol Br, Div Heart & Vascular Dis, NIH, 81-82; Data Rev Bd, Clin Eval Ventricular Assist Devices, Nat Heart, Lung & Blood Inst, 81- & Sorin Biomed Bd Dirs, 84-; sci adv comt I Stat, Princeton, NJ, Cardiopulmonics, Salt Lake City, Utah & Cellular Transplants Inc, Providence, RI. *Honors & Awards:* Lilienthal lectr, Mt Sinai Hosp, NY, 64; Hastings lectr, NIH, 79; John H Gibbon Jr Award, Am Soc ExtraCorporeal Technol Inc, 80; Runzi Prize, 88. *Mem:* Am Physiol Soc; Am Soc Artificial Internal Organs; Am Col Cardiol. *Res:* Cardiorespiratory physiology; artificial heart, lung, kidney, liver and pancreas; physiology of death and resuscitation; bioengineering of artificial organs. *Mailing Add:* Div Biol & Med Brown Univ Providence RI 02912

GALLEZ, BERNARD, b Charleroi, Belg, June 17, 38; m 61; c 3. HYDRAULIC ENGINEERING. *Educ:* Cath Univ Louvain, CE, 61; Univ Liege, MScA, 63, DScA, 68. *Prof Exp:* Asst, Hydraulic Lab, Univ Liege, 61-63; lectr hydraulics, Sherbrooke Univ, 63-64, asst prof, 64-66; asst, Hydraulic Lab, Univ Liege, 66-68; from asst prof to assoc prof hydraulics, 68-71, head civil eng dept, 71-74, assoc prof, 71-76, PROF CIVIL ENG, UNIV SHERBROOKE, 76- *Concurrent Pos:* Hydraulic engr, SNC Montreal, 74-; vis prof & head civil eng dept, UNR, Rwanda, 90- *Honors & Awards:* Alexandre Galopin Found Laureat Award, 70. *Mem:* Int Asn Hydraul Res. *Res:* Hydraulic instabilities; sediment movement; pumping station design criteria; sanitary hydraulic engineering. *Mailing Add:* Dept Civil Eng Univ Sherbrooke 2500 Boulevard Sherbrooke PQ J1K 2R1 Can

GALLI, JOHN RONALD, b Salt Lake City, Utah, Oct 10, 36; m 60, 78; c 5. SOLID STATE PHYSICS. *Educ:* Univ Utah, BS, 58, MA, 60, PhD(physics), 63. *Prof Exp:* Physicist, Aerojet-Gen Corp Div, Gen Tire & Rubber Co, 63; from asst prof to assoc prof, 63-71, PROF PHYSICS, WEBER STATE COL, 71-, CHMN DEPT, 83- *Mem:* Am Asn Physics Teachers. *Res:* High pressure physics, mechanical properties of solids. *Mailing Add:* Dept Physics Weber State Univ Ogden UT 84408

GALLIAN, JOSEPH A, b Pennsylvania, Jan 5, 42; m; c 3. ALGEBRA, MATHEMATICAL SCIENCE. *Educ:* Slippery Rock State Univ, BA, 66; Notre Dame Univ, PhD(math sci), 71. *Prof Exp:* from asst prof to assoc prof, 72-80, PROF MATH SCI, UNIV MINN, DULUTH, 80- *Honors & Awards:* Allendoerfer Award, Math Asn Am. *Mem:* Am Math Asn; Math Asn Am. *Res:* Graph theory; finite groups. *Mailing Add:* Dept Math Sci Univ Minn Duluth MN 55812

GALLIE, THOMAS MUIR, b New York, NY, Aug 25, 25; m 45; c 4. COMPUTER SCIENCE. *Educ:* Harvard Univ, AB, 47; Univ Tex, MA, 49; Rice Univ, PhD(math), 54. *Prof Exp:* Pvt sch teacher, 49-50; asst, Rice Inst, 51-54; res instr math, Duke Univ, 54-55, from asst prof to prof, 56-71, dir comput lab, 58-64, dir comput sci prog, 71-72, prof comput sci, 71-89, EMER PROF COMPUT SCI, DUKE UNIV, 89- *Concurrent Pos:* Res engr, Humble Oil & Refining Co, 54 & 55-56; visitor, Swiss Fed Inst Technol, Zurich, 62-63 & 73-74; head educ res & training sect, Off Comput Activities, NSF, 68-69; vpres, Microelectronics Ctr NC, 80-81. *Mem:* Asn Comput Mach; Math Asn Am; Soc Indust & Appl Math; Inst Elec & Electronics Engrs; Sigma Xi. *Res:* Numerical analysis. *Mailing Add:* Dept Comput Sci Duke Univ Durham NC 27706

GALLIGAN, JAMES M, b Far Rockaway, NY, May 13, 31; m 58; c 2. PHYSICAL METALLURGY, SOLID STATE PHYSICS. *Educ:* Polytech Inst Brooklyn, BMetE, 55; Univ Ill, MS, 57; Univ Calif, Berkeley, PhD(metall), 63. *Prof Exp:* Res asst elec eng, Univ Ill, 55-57; res engr metall, E C Bain Lab, US Steel Corp, 57-58 & Lawrence Radiation Lab, Univ Calif, 58-62; from asst prof to assoc prof metall, Columbia Univ, 63-67; scientist, Brookhaven Nat Lab, 67-72; PROF METALL, UNIV CONN, 72- *Concurrent Pos:* Consult, Brookhaven Nat Lab, 65-67, 72-78, Stanley Tool Co & Plumb Tool Co, 73-77; vis scientist, Max Planck Inst Physics, 67; vis prof, Calif Inst Technol, 78-79. *Honors & Awards:* Von Humboldt Sr Sci Award, 86. *Mem:* Am Inst Min, Metall & Petrol Engrs; Am Phys Soc; Am Inst Mech Engrs; Sigma Xi; NY Acad Sci. *Res:* Defects in solids; hardening of crystals; superconductivity; metallurgical properties; nuclear materials. *Mailing Add:* Dept Metall Univ Conn U-136 Storrs CT 06268

GALLIGAN, JOHN D(ONALD), b Washington, DC, Oct 9, 32; m 58; c 5. PHYSICAL CHEMISTRY. *Educ:* Manhattan Col, BS, 55; Emory Univ, PhD, 58. *Prof Exp:* Chemist, Harris Res Labs, Washington, DC, 57-58, sr chemist, 58-62; proj chemist, Gillette Safety Razor Co, Boston, 62-65, vpres, Gillette Res Inst, Washington, DC, 65-72, vpres res & develop, personal care div, Chicago & Boston, 72-83, GROUP DIR, BOSTON RES & DEVELOP LABS, GILLETTE CO, 83- *Mem:* AAAS; Am Chem Soc. *Res:* Surface and polymer chemistry; lubrication; adsorption; electrochemistry; adhesion; chemical modification of textiles. *Mailing Add:* 30 Bakers Hill Rd Weston MA 02193-1761

GALLIN, JOHN I, b New York, NY, Mar 25, 43; m 66; c 2. EXPERIMENTAL BIOLOGY. *Educ:* Amherst Col, BA, 65; Cornell Univ, MD, 69. *Hon Degrees:* Dr, Amherst Col, 88. *Prof Exp:* Intern med, Bellevue Hosp, NY, 69-70, asst resident, 70-71, sr chief med res, 74-75; teaching asst, NY Univ Sch Med, 70-74, instr, 74-81; clin assoc, Lab Clin Invest, Nat Inst Allergy & Infectious Dis, NIH, 71-74, sr investr, 75-91, head, Bacterial Dis Sect, 78-86, DIR, DIV INTRAMURAL RES, NAT INST ALLERGY & INFECTIOUS DIS, NIH, 85-, CHIEF, LAB HOST DEFENSES, 91- *Concurrent Pos:* Consult, Infectious Dis, USN Med Ctr, Bethesda, Md, 72-, Allergy/Clin Immunol Serv, Walter Reed Army Med Ctr, Washington, DC, 82-; vchmn, Clin Res Subpanel, Nat Inst Allergy & Infectious Dis, 78-79, chmn, 79-80; co-chmn, Ann Phagocyte Workshop, 82-; assoc ed, J Immunol, 79-83, sect ed, 85-89; mem ad hoc working group, Immunocompetence in Space, NASA, 85, Infectious Dis Adv Coun, Merck, Sharp & Dohme Int, 85-90, numerous comt & panels, NIH, 81- *Honors & Awards:* Dean William Mecklenburg Polk Mem Prize in Res & Anthony Seth Werner Mem Prize in Infectious Dis, Cornell Med Col, 69; Squibb Award, Infectious Dis Soc Am, 84; Jeffrey Modell Found Lifetime Achievement Award, 90. *Mem:* Asn Am Physicians; Am Soc Clin Invest; Am Fedn Clin Res; fel Infectious Dis Soc Am; Am Asn Immunologists; Am Soc Cell Biol; Sigma Xi; Soc Leukocyte Biol. *Res:* Leukocyte (phagocyte) function. *Mailing Add:* Intramural Res Prog Nat Inst Allergy & Infectious Dis NIH 9000 Rockville Pike Bldg 10 Rm 11C103 Bethesda MD 20892

GALLINI, JOHN B(ATTISTA), b Detroit, Mich, June 5, 34; m 57; c 8. CHEMICAL ENGINEERING. *Educ:* Univ Detroit, BChE, 56; Univ Mich, MS, 58, PhD(chem, eng), 61. *Prof Exp:* Res engr, plastics dept, Yerkes Film Res & Develop, E I du Pont de Nemours & Co, Inc, 59-65, group mgr venture develop sect, film dept, 65-66, staff eng, 66-72, res assoc, Spruance Fibers Res & Develop, 72-85; RETIRED. *Concurrent Pos:* Consult. *Mem:* Am Chem Soc; Int Exec Sev Corp; Nat Exec Serv Corp. *Res:* High polymers; polymerization kinetics; physical and molecular properties of polymers; processing technology for thermoplastic materials; Kevlar polymerization, solvent recovery and spinning; coextrusion of film products; processes for nylon, Surlyn, Kapton, and polyacrylonintrile films; Vexar net products; Crofon fiber optics; flammability of nylon resins. *Mailing Add:* 2425 Triton Rd Richmond VA 23235

GALLISTEL, CHARLES RANSOM, b Indianapolis, Ind, May 18, 41; m 69; c 1. BEHAVIORAL PHYSIOLOGY, PHYSIOLOGICAL PSYCHOLOGY. *Educ:* Stanford Univ, BA, 63; Yale Univ, PhD(psychol), 66. *Prof Exp:* From asst prof to prof psychol, Univ Pa, 66-89, chmn dept, 81-84; PROF, UNIV CALIF-LOS ANGELES, 89- *Concurrent Pos:* Chair, publ bd, Psychonomic Soc, 87-89, Animal Res Comt, Fed of Behav, Psychol & Cognitive Sci. *Mem:* Soc Neurosci; fel Ctr Advan Study Behav Sci; fel Soc Exp Psychologists; fel AAAS; Psychonomic Soc. *Res:* Neural systems in the diencephalon that mediate motivation and reinforcement; electrical self-stimulation of the brain; organization of action; the theory of learning. *Mailing Add:* Dept Psychol Univ Calif Los Angeles 405 Hilgard Ave Los Angeles CA 90024-1563

GALLIVAN, JAMES BERNARD, b Sydney, NS, Aug 9, 38; m 61; c 4. PHYSICAL CHEMISTRY. *Educ:* St Francis Xavier Univ, BSc, 60, MA, 62; Notre Dame Univ, PhD(radiation chem), 65. *Prof Exp:* Res chemist, 65-70, sr res chemist, 70-73, proj leader, 73-77, mgr paper chem res & develop, 77-79, tech dir paper chem, 79-81, mgr new prod, mining chem, 81-83, MGR PHOSPHINE CHEM, AM CYANAMID CO, 84- *Mem:* AAAS; Am Chem Soc; Chem Inst Can; Tech Asn Pulp & Paper Indust; Am Mining Cong. *Res:* Radiation chemistry; electronic spectroscopy; chemicals for paper industry; analytical chemistry and instrumental analysis; mining chemicals; phosphing chemicals. *Mailing Add:* 45 Cranbury Rd Norwalk CT 06851-2616

GALLIZIOLI, STEVE, b Riva, Italy, July 25, 24; US citizen; m 49; c 3. WILDLIFE MANAGEMENT. *Educ:* Ore State Univ, BS, 50. *Prof Exp:* Dist wildlife biologist, Ariz Game & Fish Dept, 50-55, res biologist, 55-57, res supvr, 57-66, res chief, 66-78, chief, Wildlife Mgt Div, 79-83; RETIRED. *Concurrent Pos:* Partic, Food & Agr Orgn, UN, Assignment to Venezuela, 69; observ, Int Asn Fish & Wildlife Agencies, San Jose, Costa Rica, 78. *Honors & Awards:* Am Motors Conserv Award, 67; Conservationist of the Year Award, Ariz Wildlife Fedn, 78; Special Conserv Award, Nat Wildlife Fedn, 79; McCulloch Conserv of Yr, 86. *Mem:* Wildlife Soc; Int Asn Fish & Wildlife Agencies; Outdoor writers Asn Am. *Res:* Population dynamics of mule deer; investigation of factors controlling Gambel and scaled quail populations; development of inventory technique for Gambel quail, whitewing doves. *Mailing Add:* 4722 W Crittendon Lane Phoenix AZ 85031

GALLO, ANTHONY EDWARD, JR, b Apr 7, 31; US citizen; c 2. NEUROSURGERY. *Educ:* Tufts Univ, BS, 52; Harvard Univ, MD, 56. *Prof Exp:* Intern surg, Boston City Hosp, 56-57, sr asst resident neurosurg, 57-58; mem staff, Madigan Gen Hosp, Tacoma, Wash, 58-59; resident neurol & neurosurg, Letterman Gen Hosp, San Francisco, 59-61; chief resident pediat neurosurg & physiol, Hosps & Med Sch, Univ Pittsburgh, 61-63; dir intern training & chief neurosurg, Madigan Gen Hosp, Tacoma, Wash, 63-68; PROF NEUROSURG, MED SCH, UNIV ORE, 68-, PHYSICIAN & SURGEON, HEALTH SCI CTR, 77- *Concurrent Pos:* Asst & fel, Univ Calif, San Francisco, 59-60; teaching fel pediat neurosurg & physiol, Hosps & Med Sch, Univ Pittsburgh, 61-63; investr with Dr Paul Yakovlev, Boston, 63; prof pro-tem, Univ Saigon, 64-65; chief neurosurg training team, Cong H Hosp,

Saigon, 64-65; consult, Madigan Gen Hosp, 68-; consult, Myeomenigocele Clin, Crippled Children's Div, Univ Ore, 68-; liaison, US Army to Univ Ore Med Sch, 70- *Honors & Awards:* David W E Baird Award, Univ Ore Med Sch, 69. *Mem:* Cong Neurol Surg; Am Asn Neurol Surg. *Res:* Development of hydrocephalic beagles; surgical treatment of vascular lesions and complications of cerebrovascular disease; psychological evaluation of carotid occlusion; prefabricated cranioplasty. *Mailing Add:* Dept Surg Oregon Health Sci Univ 3181 SW Sam Jackson Park Rd Portland OR 97201

GALLO, AUGUST ANTHONY, b Rochester, NY, Feb 14, 51; m 83; c 1. ORGANIC CHEMISTRY, MEDICINAL CHEMISTRY. *Educ:* St John Fisher Col, BS, 73; Vanderbilt Univ, PhD(org chem), 78. *Prof Exp:* Res assoc org med chem, Univ Calif, San Francisco, 78-80; asst prof, 80-86, ASSOC PROF, DEPT CHEM, UNIV SOUTHWESTERN LA, 86- *Concurrent Pos:* Pres, Univ Southwestern La Sigma Xi Club, 85-86. *Mem:* Sigma Xi; Am Chem Soc; NY Acad Sci. *Res:* Synthesis of biologically important molecules and collaborative biological testing of these compounds; synthesis of novel heterocycles; organo-sulfur chemistry and new synthetic methodology; natural products chemistry. *Mailing Add:* Dept Chem Univ Southwestern La PO Box 44370 Lafayette LA 70504-4370

GALLO, CHARLES FRANCIS, b Mt Vernon, NY, July 22, 35; m 57; c 1. ENGINEERING PHYSICS, XEROGRAPHY. *Educ:* Rensselaer Polytech Inst, BS, 57. *Prof Exp:* Jr engr, Res Labs, Westinghouse Elec Corp, 57-62, eng specialist, 62-64; sr physicist, Xerox Corp, Rochester, 64-67, assoc scientist, 67-68, scientist, 68-71; res specialist, 3M Corp, 81-88; PRES & FOUNDER, SUPERCONIX INC, 88- *Concurrent Pos:* Sci consult, Inner City Schs, Rochester, NY; lectr, George Washington Univ, 75, Univ Western Ont, 74 & 76, World Electrotech Cong, Moscow, 77 & Purdue Univ, 80, Univ Southampton, 82, Univ Kyushu, 84. *Honors & Awards:* Hon Sci Award, Bausch & Lomb. *Mem:* Am Phys Soc; Japan Inst Electrostatics; Electrostatics Soc Am; Mat Res Soc. *Res:* Superconductivity; transport properties of solids; energy conversion; xerography; coronas; gas discharges; light sources; injection electroluminescence; optical radiation; work function; astrophysics; facsimile systems; alphanumeric printers; electrostatics; charge exchange; solid state physics. *Mailing Add:* 2440 Lisbon Ave Lake Elmo MN 55042

GALLO, DUANE GORDON, b Aberdeen, SDak, May 15, 26; m 52; c 10. ANALYTICAL CHEMISTRY, BIOCHEMISTRY. *Educ:* Univ NDak, BS, 51, MS, 53, PhD(biochem), 55. *Prof Exp:* Asst, Univ NDak, 52-55; from asst sr chemist to sr chemist, Mead Johnson & Co, 57-61, group leader, 61-68, sr res assoc, 68-70, prin investr, 70-73, sr prin investr, 73-75, prin res assoc, 75-78, prin res scientist, 78-87; RETIRED. *Mem:* AAAS; Am Soc Pharmacol & Exp Therapeut; Soc Exp Biol & Med. *Res:* Phosphatide and cholesterol synthesis and metabolism; lipid metabolism; endocrinology; reproductive physiology and biochemistry; drug analysis and metabolism. *Mailing Add:* Pharmacokinet & Metab Dept Bristol-Myers Pharm Res & Develop Evansville IN 47721

GALLO, FRANK J, structural engineering, engineering mechanics; deceased, see previous edition for last biography

GALLO, LINDA LOU, b Smithtown, WVa, Aug 20, 37; m 57; c 2. BIOCHEMISTRY, NUTRITION. *Educ:* WVa Univ, BS, 59; George Washington Univ, MS, 63, PhD(biochem), 69. *Prof Exp:* Chemist anal chem, Food & Drug Admin, 59-62; res asst biochem, 62-69, asst res prof, 69-75, from asst prof to assoc prof, 75-86, PROF BIOCHEM, GEORGE WASHINGTON UNIV, 86- *Concurrent Pos:* Sanders fel biochem, George Washington Univ, 62-64; prin investr, NIH grant, 75-88; consult, Cafritz Hosp, 72-75, Anheuser Busch, 75 & Lederle Labs, 84; mem, Am Heart Asn Arteriosclerosis Coun. *Mem:* Sigma Xi; Am Heart Asn; Grad Women Sci; NY Acad Sci; Soc Exp Biol & Med. *Res:* Cholesterol absorption; triglyceride absorption; atherosclerosis; dietary fiber; cholesterol; lipoprotein metabolism. *Mailing Add:* Dept Biochem & Molecular Biol George Washington Univ Med Ctr Washington DC 20037

GALLO, ROBERT C, b Waterbury, Conn, Mar 23, 37; m 61; c 2. CELL BIOLOGY, BIOCHEMISTRY. *Educ:* Providence Col, BA, 59; Jefferson Med Col, MD, 63. *Hon Degrees:* Numerous from US & foreign univs, 74-90. *Prof Exp:* Clin clerk, Metab Sect, Yale Univ Med Sch, 62-63; from intern to resident med, Univ Chicago, 63-65; clin assoc, Med Br, Nat Cancer Inst, NIH, 65-68, sr investr, 68-69, head, Sect Cellular Control Mech, Human Tumor Cell Biol Br, 69-72, CHIEF, LAB TUMOR CELL BIOL, NAT CANCER INST, NIH, 72- *Concurrent Pos:* Consult, M D Anderson Hosp & Tumor Inst, 70-71; consult virol, Roswell Park; consult microbiol, Univ SC; blood consult, Hahneman Med Sch Cancer Ctr; vis prof, Univ Minn, 71; lectr, numerous US & foreign univs, 71-90; adj prof, Dept Genetics, George Washington Univ, 72-, Dept Biol, Johns Hopkins Univ, 85-, Dept Microbiol, Immunol & Parasitol, Cornell Univ, 86-, Dept Molecular Genetics & Microbiol, Univ Med & Dent NJ, 87 & Dept Microbiol, Rutgers Univ, 88-; mem, numerous sci comts & panels, govt agencies & nat socs. *Honors & Awards:* Bryan Priestman Award, 72; Dameshek Award Res Hemat, 74; First Ciba-Geigy Award Biomed Sci, 76; Edward Rhodes Stitt Award & Lectr, Asn Mil Surgeons US, 82; Richard & Hinda Rosenthal Found Award & Lectr, Am Asn Cancer Res, 83; Griffuel Prize, Asn Res Cancer, 83; Medal of Honor, Am Cancer Soc, 83; Hermann Beerman Award, Soc Invest Dermat, 84; Lila Gruber Cancer Res Award, Am Acad Dermat, 84; NY Acad Sci Award, 85; First Sci Prize, Am Found AIDS Res, 87; David A Karnofsky Award, Am Soc Clin Oncol, 88; Lifetime Sci Medal, Inst Advan Studies Immunol & Aging, 90. *Mem:* Nat Acad Sci; Inst Med-Nat Acad Sci; Am Soc Hemat; Am Fedn Clin Res; Am Asn Cancer Res; AAAS; Am Soc Clin Invest. *Res:* Oncology, especially leukemia; tumor viruses; molecular biology; hematology; mechanisms involved in control of cell growth and differentiation and in particular how these apply in neoplasia; viral oncogenesis; human oncogenesis; role of inciting agents; virology. *Mailing Add:* NIH Bldg 37 Rm 6A11 Bethesda MD 20892

GALLO, ROBERT VINCENT, b New York, NY, Aug 18, 41. ENDOCRINOLOGY. *Educ:* Columbia Col, BA, 62; Purdue Univ, PhD(biol), 68. *Prof Exp:* NIH fel, Sch Med, Univ Calif, Los Angeles, 68-70; asst prof, 70-78, ADJ ASSOC PROF PHYSIOL, SCH MED, UNIV CALIF, SAN FRANCISCO, 78- *Concurrent Pos:* NIH grant, Sch Med, Univ Calif, San Francisco, 78-81. *Mem:* AAAS; Endocrine Soc; Int Soc Neuroendocrinol; Am Physiol Soc. *Res:* Reproductive neuroendocrinology. *Mailing Add:* Dept Biol Univ Conn Box U-42 Storrs CA 06268

GALLOP, PAUL MYRON, b New York, NY, Nov 24, 27; m 64; c 3. PROTEIN CHEMISTRY. *Educ:* Univ Pa, AB, 48; Mass Inst Technol, PhD(biophys), 53; Harvard Med Sch, MSc, 73. *Prof Exp:* Instr biophys, Mass Inst Technol, 53-54; biophysicist, L I Jewish Hosp, 54-59; vis asst prof, Albert Einstein Col Med, 57-59, from assoc prof to prof biochem, 59-72; CHMN DEPT ORAL BIOL, SCH DENT MED, HARVARD UNIV, 74-, PROF BIOL CHEM, DIV HEALTH SCI & TECHNOL, HARVARD-MASS INST TECHNOL, 76-, DEP DIR, MED SCIENTIST TRAINING PROG, SCH MED, 79-; BIOCHEMIST, LAB HUMAN BIOCHEM, CHILDREN'S HOSP MED CTR, 72-, DIR, LAB HUMAN BIOCHEM, DEPT ORTHOP SURG, 81- *Concurrent Pos:* Mem study sect, Adult Develop & Aging Res and Training Comt, NIH, Md, 73-77; mem, aging rev comt, Nat Inst Aging, 74- *Honors & Awards:* Career. *Mem:* AAAS; Am Chem Soc; Biophys Soc; Am Soc Biol Chem; Orthop Res Soc; Geront Soc. *Res:* Collagen structure; proteolytic enzymes; collagenase; mechanism of enzyme action; peptide synthesis; connective tissue structure; mass spectrometry; posttranslational modifications; glycohemoglobins; aging cells and tissues; vitamin K-dependent proteins; molecular basis of clinical disorders. *Mailing Add:* Children's Hosp Med Ctr 300 Longwood Ave Boston MA 02115

GALLOPO, ANDREW ROBERT, b Passaic, NJ, Mar 23, 40. ORGANIC CHEMISTRY, BIOCHEMISTRY. *Educ:* Rutgers Univ, AB, 62; Brown Univ, PhD(org chem), 67. *Prof Exp:* NIH fel, Univ Wis, 66-68; asst mgr, Philip Hunt Chem Corp, 68-69; from asst prof to assoc prof chem, Montclair State Col, 74-81; SR SCIENTIST, WARNER LAMBERT, 81- *Mem:* Am Chem Soc. *Res:* Organic chemical kinetics; enzymic kinetics; enzyme purification; drug delivery systems. *Mailing Add:* 133 Wessington Ave Garfield NJ 07026

GALLO-TORRES, HUGO E, US citizen; m; c 3. GASTROINTEROLOGY. *Educ:* San Marcos Univ, Lima, BSc, 54, MD, 63; Columbia Univ, MS, 64; Tulane Univ, PhD(biochem), 68; Am Bd Nutrit, cert, 78. *Prof Exp:* Resident med, Cent Army Hosp, Lima, 62-63; instr med, Charity Hosp La, 64-68; sr biochemist, Dept Biochem Nutrit, Hoffmann-La Roche Inc, 68-89 & 72-73, res group chief, 74-78; res sect head, Roche Res Ctr, 79-80, sr res physician, Dept Med Res, 81-83, asst med dir, Gastroenterol & Rheumatology, 83-85; dir gastroenterol, Serono Labs, Inc, 85-86; med officer, Div Cardio-Renal Drug Prod, 87, MED OFFICER, DIV GASTROENTEROL & COAGULATION DRUG PROD, FOOD & DRUG ADMIN, 87- *Concurrent Pos:* Physician, Specialized Clin Diag Neuro-endocrine Dis, Lima, 62-63; WHO fel nutrit, Incap, Guatemala, 64; NIH fel, Tulane Univ, 64-67; vis biochemist, Dept Vitamin & Nutrit Res, Hoffmann-La Roche & Co, Ltd, Basel, Switz, 70-72 & Dept Pharmacl, Roche Prod Ltd, Welwyn, Eng, 71; adj prof, Dept Home Econ, Hunter Col, 77-; adj prof med, Dept Med, Div Gastroenterol & Nutrit, Sch Med & Health Sci, George Washington Univ, 87- *Mem:* Fel Am Col Gastroenterol; Am Gastroenterol Asn; Am Physiol Soc; Am Acad Cert Med Nutritionists. *Res:* Internal medicine, especially gastrointestinal, cardiovascular and bronchopulmonary disorders; clinical nutrition; experimental physiology. *Mailing Add:* Div Gastroenterol & Coagulation Drug Prod Food & Drug Amin HFD-180 5600 Fishers Lane Rockville MD 20857

GALLOWAY, ETHAN CHARLES, b Howell, Mich, Oct 31, 30; c 3. ORGANIC CHEMISTRY. *Educ:* Mich State Univ, BSc, 51; Univ Calif, Berkeley, PhD(chem), 54. *Prof Exp:* Res chemist, Dow Chem Co, 54-58, mem staff, Tech Serv & Develop Dept, 58-60, head polymer intermediates, 60-61, head specialty chem, 61-62; dir res, Plastics Div, Nopco Chem Co, 62-65; mgr prod develop, 65-66, asst dir, Eastern Res Ctr, 66-67, dir, 67-69, dir res, 69-70, vpres & dir res, 70-78, vpres corp planning & develop, 78-81, EXEC VPRES TECHNOL, STAUFFER CHEM CO, 81- *Concurrent Pos:* Pres, Indust Res Inst, New York, 78-79. *Mem:* Am Chem Soc; Sci Res Soc Am; Indust Res Inst; Soc Chem Indust; Coun Chem Res. *Res:* Natural products; condensation polymers; specialty organics. *Mailing Add:* 6549 Thornbrook Circle Hudson OH 44236

GALLOWAY, GORDON LYNN, b Pottstown, Pa, Sept 25, 36; m 62; c 2. INORGANIC CHEMISTRY. *Educ:* Franklin & Marshall Col, BS, 57; Mich State Univ, PhD(inorg chem), 62. *Prof Exp:* Mem staff, Los Alamos Sci Lab, 61-63; asst prof chem & asst to dean col, St Lawrence Univ, 63-67; from asst prof to prof chem, Denison Univ, 67-89, chmn dept, 73-75; PROF & DIR PROG GEN CHEM, MICH STATE UNIV, 89- *Concurrent Pos:* Partic, NSF res prog col teachers, 64 & grant, 64-66; W B King vis prof gen chem, Iowa State Univ, 72-73; mem ed bd, J Chem Educ, 70- *Mem:* AAAS; Am Chem Soc; Sigma Xi; Am Asn Univ Profs. *Res:* Chemistry of the interaction of metal ions with organic molecules; analysis of the role of trace elements in health and disease. *Mailing Add:* Mich State Univ 119 Chem Bldg East Lansing MI 48824-1322

GALLOWAY, JAMES NEVILLE, b Annapolis, Md, Oct 26, 44; m 68; c 2. ENVIRONMENTAL CHEMISTRY. *Educ:* Whittier Col, BA, 66; Univ Calif, San Diego, PhD(chem), 72. *Prof Exp:* Consult chem oceanog, Southern Calif Coastal Water Res Proj, 72; pres, Shenandoah Crafts, Inc, 72-74; assoc environ chem, Cornell Univ, 74-76; asst prof environ chem, 76-80, assoc prof, 80-88, PROF ENVIRON SCI, UNIV VA, 88- *Concurrent Pos:* Pres, Bermuda Biol Stat Res Inc. *Mem:* Am Geophys Union. *Res:* The effect of acid deposition on aquatic ecosystems, the long range transport of atmospheric microcontaminants, the chemistry of natural waters, and air-sea exchange. *Mailing Add:* Environ Sci Dept Univ Va Charlottesville VA 22903

GALLOWAY, KENNETH FRANKLIN, b Columbia, Tenn, Apr 11, 41; m 59; c 2. ELECTRONICS ENGINEERING, RAD EFFECTS PHYSICS. *Educ:* Vanderbilt Univ, AB, 62; Univ SC, PhD(physics), 66. *Prof Exp:* Res assoc to assoc prof, Ind Univ, Bloomington, 66-72; res physicist, Naval Weapons Support Ctr, Crane, Ind, 72-74; proj leader, Electronic Tech Div, Nat Bur Standards, 74-77, chief, Semiconductor Devices & Circuits Div, 81-85, Semiconductor Electronics Div, 85-86; PROF & HEAD DEPT, ELECTRICAL & COMPUTER ENG, UNIV ARIZ, 86- *Concurrent Pos:* Prof elec eng, Univ Md, College Park, 80-86; gen chmn, Nuclear & Space Radiation Effects Conf & Hardened Electronics & Radiation Technol Conf, Inst Elec & Electronics Engrs, 85; prof optical sci, Univ Ariz, Tucson, 86-; mem, Nuclear & Plasma Sci Soc Admin Comt, Inst Elec & Electronics Engrs, 87-90, exec vchmn, Nuclear & Plasma Sci Soc Radiation Effects Comt, 88-91, vpres, Nuclear & Plasma Sci Soc, 89. *Honors & Awards:* Sci & Technol Fel, US Dept Com, 79, 80. *Mem:* Am Phys Soc; fel Inst Elec & Electronics Engrs; Electrochem Soc; AAAS; Am Soc Eng Educ; Sigma Xi. *Res:* Radiation effects on semiconductor devices; measurement techniques for semiconductor devices and circuits. *Mailing Add:* Electrical & Computer Eng Dept Univ Ariz Tucson AZ 85721

GALLOWAY, RAYMOND ALFRED, b Arbutus, Md, May 12, 28. PHYSIOLOGY. *Educ:* Univ Md, BS, 52, MS, 56, PhD(bot), 58. *Prof Exp:* From asst prof to assoc prof, 58-69, PROF PHYSIOL, UNIV MD, 69-, PROF BOT, 76- *Mem:* Bot Soc Am; Am Soc Plant Physiol; Phycol Soc Am. *Res:* Enzymology; heterotrophy in estuarine algae; physiology of algae; CO2 metabolism. *Mailing Add:* 330 N Central Ave Glendale CA 91203

GALLOWAY, SHEILA MARGARET, cytogenetics, for more information see previous edition

GALLOWAY, WILLIAM DON, b Dickson, Tenn, Dec 28, 39. PSYCHOPHYSICS. *Educ:* Univ Fla, BA, 61; La State Univ, MA, 63; Univ Md, PhD(exp psychol), 70. *Prof Exp:* Res asst, Human Resources Res Off, Ft Benning, Ga, 63-64; res asst animal perception, Inst Behav Res, 66-68; res asst psychol, Sch Med, Johns Hopkins Univ, 68-70; SUPVRY RES PSYCHOLOGIST, BUR RADIOL HEALTH, FOOD & DRUG ADMIN, 70- *Mem:* Am Psychol Asn. *Res:* Animal psychophysics, psychopharmacology, behavioral toxicology and teratology; behavioral effects on non-ionizing radiation. *Mailing Add:* 5600 Fishers Lane Rockville MD 20857

GALLOWAY, WILLIAM EDMOND, b Waco, Tex, Oct 8, 44; m 66; c 2. GEOLOGY. *Educ:* Tex A&M Univ, BS, 66; Univ Tex, Austin, MA, 68, PhD(geol), 71. *Prof Exp:* Res scientist assoc, Tex Bur Econ Geol, 70; res scientist, Explor Res Div, Continental Oil Co, 70-73, res group leader, 73-75, dir geol sect, 75; SR RES SCIENTIST, TEX BUR ECON GEOL, 75-; ELLIOTT PROF GEOL SCI, UNIV TEX, AUSTIN, 85- *Concurrent Pos:* Course leader, Clastic Sedimentation Sch, Continental Oil Co, 77-; consult energy supply, Off Technol Assessment, US Cong, 77-78; lectr, Continuing Educ Prog, Am Asn Petrol Geologists, 77-; vis prof, Univ Bergen, 80; Klabzuba vis prof, Univ Okla, 81; distinguished lectr, Am Asn Petrol Geologists, 85-86. *Honors & Awards:* A I Levorsen Award, Am Asn Petrol Geologists, 77 & 86; Wallace Pratt Mem Award, Am Asn Petrol Geologists, 83. *Mem:* Am Asn Petrol Geologists; Int Asn Sedimentologists; Sigma Xi; Am Geol Inst; Soc Econ Paleontologists & Mineralogists. *Res:* Clastic depositional systems; origin and geologic setting of petroleum deposits and sedimentary uranium ores; sedimentary deagenesis; basin hydrogeology; basin analysis; seismic and sequence stratigraphy. *Mailing Add:* Dept Geol Sci Univ Tex Austin TX 78713-7909

GALLOWAY, WILLIAM JOYCE, b Chicago, Ill, Sept 15, 24; m 46; c 2. PHYSICS, AERONAUTICAL ENGINEERING. *Educ:* Univ Calif, Los Angeles, BS, 49, MS, 50, PhD(physics), 53. *Prof Exp:* Physicist, Res Staff, Eng Labs, Signal Corps, 51-52; vpres, Bolt Beranek & Newman, Inc, 53-82; CONSULT, 82- *Concurrent Pos:* Chmn, Comt Hearing Bioacoustics & Biomechanics, Nat Res Coun, 79; designated eng rep aircraft acoustics, Fed Aviation Admin, 82- *Honors & Awards:* Arch T Colwell Merit Award, Soc Automotive Engrs, 73; Silver Medal in Noise, Acoust Soc Am, 88. *Mem:* Nat Acad Eng; Am Inst Aeronaut & Astronaut; Inst Noise Control Engrs (vpres, 78-80); fel Acoust Soc Am (vpres, 83-84); Sigma Xi. *Res:* Acoustics; cavitation in liquids; aircraft, rocket and traffic noise. *Mailing Add:* 19343 Olivos Dr Tarzana CA 91356

GALLUCCI, ROBERT RUSSELL, b New York, NY, Sept 26, 50; m; c 1. ORGANIC CHEMISTRY, POLYMER SCIENCE. *Educ:* City Col New York, BS, 72; Princeton Univ, MA, 74. *Prof Exp:* GE CRD, 76-82, GENERAL ELECTRIC PLASTICS, 82- *Concurrent Pos:* Instr polymer blends short course. *Mem:* Am Chem Soc; Soc Plastics Engrs. *Res:* Mechanisms; polymer synthesis; polymer blends. *Mailing Add:* Gen Elec Plastics One Lexan Lane Mt Vernon IN 47620

GALLUN, ROBERT LOUIS, b Milwaukee, Wis, Feb 21, 24; m 49; c 2. ENTOMOLOGY, GENETICS. *Educ:* Mich State Univ, BS, 48, MS, 50; Purdue Univ, PhD, 60. *Prof Exp:* Entomologist, Bur Entom & Plant Quarantine, 50-52, res entomologist, 53-72, RES LEADER & TECH ADV, AGR RES SERV, USDA, 72-; PROF ENTOM, AGR RES STA, PURDUE UNIV, 71- *Concurrent Pos:* Assoc prof entom, Purdue Univ, 70-71. *Mem:* Entom Soc Am; Am Soc Agron; Am Genetic Asn; Crop Sci Soc Am; Sigma Xi. *Res:* Host plant resistance to insects; biological races and genetics of the Hessian fly. *Mailing Add:* 215 Spring Valley Lane West Lafayette IN 47906

GALLUP, DONALD NOEL, limnology; deceased, see previous edition for last biography

GALMARINO, ALBERTO RAUL, b Buenos Aires, Arg, Sept 23, 28; m 61; c 1. MATHEMATICS. *Educ:* Univ Buenos Aires, Lic math, 55; Mass Inst Technol, PhD(math), 61. *Prof Exp:* Instr algebra & topology, Univ Buenos Aires, 57-58; res assoc math, North eastern Univ, 61-62; assoc prof, Univ Buenos Aires, 62-63; asst prof, 63, assoc prof, 68-78, PROF MATH, NORTHEASTERN UNIV, 78- *Concurrent Pos:* Math consult, Argentine Navy, 62. *Mem:* Am Math Soc; Argentine Math Soc. *Res:* Probability theory; stochastic processes; theory of games; mathematical analysis. *Mailing Add:* Dept Math Northeastern Univ 360 Huntington Ave Boston MA 02115

GALONSKY, AARON IRVING, b Brooklyn, NY, Apr 18, 29; m 51; c 3. NUCLEAR PHYSICS. *Educ:* Brooklyn Col, BA, 50; Univ Wis, MS, 51, PhD, 54. *Prof Exp:* Physicist, Oak Ridge Nat Lab, 54-59; group leader, Midwest Univs Res Asn, 59-64; assoc prof physics, 64-66, dir cyclotron lab, 67-69, assoc dir, 79-80, PROF PHYSICS, MICH STATE UNIV, 66- *Concurrent Pos:* Guest prof, Inst Nuclear Physics, Nuclear Res Lab, Julich, WGer, 75-76. *Mem:* Fel Am Phys Soc. *Res:* Nuclear reactions; scattering; radiation shielding; neutrons. *Mailing Add:* Dept Physics Mich State Univ East Lansing MI 48824

GALOSY, RICHARD ALLEN, b St Louis, Mo, Feb 6, 46. PSYCHOPHYSIOLOGY, NEUROBIOLOGY. *Educ:* Univ Mo-St Louis, BA, 68; Western Wash State Col, Bellingham, MS, 70; Univ NC, Chapel Hill, PhD(psychophysiol), 74, St Louis Univ; JD, 86. *Prof Exp:* From asst prof to assoc prof neurobiol, Univ Tex Health Sci Ctr, Dallas, 75-82; CHMN BD, ANAL DATA METHODS INC, 89- *Concurrent Pos:* Moss fel, Univ Tex Health Sci Ctr Dallas, 74-75, Tex Heart Asn fel, 75, prin investr cardiovasc adjust behav stress, 76-79, sympathetic stimulation & cardiovasc dynamics, 77-78, central nervous syst control hypertension, 79-81; atty, Schramm & Pines, 86- *Mem:* Soc Neurosci; Soc Psychophysiol Res; AAAS; Am Psychol Asn. *Res:* Neural control of cardiovascular function; effect of environmental stress on cardiovascular function. *Mailing Add:* Analysis Data Methods Inc PO Box 16621 Clayton MO 63105

GALPER, JONAS BERNARD, BIOCHEMISTRY, CELL BIOLOGY. *Educ:* Albert Einstein Col Med, PhD(biochem), 70, MD, 71. *Prof Exp:* ASST PROF CARDIOL, BRIGHAM & WOMEN'S HOSP, HARVARD MED SCH, 82- *Mailing Add:* Dept Med Brigham & Women's Hosp 75 Francis St Boston MA 02115

GALPERIN, IRVING, b Buffalo, NY, Dec 3, 26; m 58. PHYSICAL CHEMISTRY, MATERIALS SCIENCE. *Educ:* Alfred Univ, AB, 48; Canisius Col, MS, 52; Western Reserve Univ, PhD(phys chem), 57. *Prof Exp:* Chemist, Continental Can Corp, Ill, 58-60, Visking Div, Union Carbide Corp, 60-62, Plastics Div, NJ, 62-64; sr chemist, Interchem Corp, Clifton, 64-67; mem tech staff, Bell Tel Labs, 67-69; group mgr, Gen Cable Corp, 69-74; mat mgr, Hatfield Div, CCS Industs, NJ, 74-76; mat eng mgr, Cable Hydro Div, ITT, Calif, 76-78; SR STAFF ENGR, MAXWELL LABS, INC, 78- *Mem:* Am Chem Soc; Sigma Xi; Instrument Soc Am; Am Inst Chemists; Soc Advan Mat Process Eng. *Res:* Electrochemistry; structure and properties of polymers; materials engineering. *Mailing Add:* 3942 Country Trails Road Bonita CA 92002

GALPIN, DONALD R, b Butte, Mont, Apr 1, 33; m 57; c 2. MEDICINAL CHEMISTRY, PHARMACY. *Educ:* Univ Mont, BS, 57, MS, 62; Univ Wash, PhD(pharmaceut chem), 66. *Prof Exp:* Asst prof, 66-73, ASSOC PROF PHARM, COL PHARM, WASH STATE UNIV, 73- *Mem:* Am Pharmaceut Asn; Acad Pharmaceut Sci; Am Asn Col Pharm. *Res:* Drug metabolism; nuclear magnetic resonance; optical rotatory dispersion and circular dichroism; configurational and conformational analysis. *Mailing Add:* Col Pharm Wash State Univ Pullman WA 99163

GALSKY, ALAN GARY, b Chicago, Ill, Mar 6, 42; m 66. PLANT PHYSIOLOGY, MICROBIOLOGY. *Educ:* Roosevelt Univ, BS, 64; Northwestern Univ, MS, 67, PhD(biol), 69. *Prof Exp:* assoc prof, 69-76, PROF BIOL, BRADLEY UNIV, 76- *Mem:* Am Soc Plant Physiologists; Sigma Xi. *Res:* Role of 3', 5'-cyclic adenosine monophosphate in plants; mechanism of gibberellic acid action; production of plant hormones by phytopathogenic bacteria; biochemical aspects of crown-gall tumor formation. *Mailing Add:* Dept of Biol Bradley Univ Peoria IL 61625

GALSTAUN, LIONEL SAMUEL, b Kediri, Java, Dec 17, 13; US citizen; wid. CHEMICAL ENGINEERING, PHYSICAL CHEMISTRY. *Educ:* Univ Dayton, BS, 33; Mass Inst Technol, SM, 34, PhD(phys chem), 36. *Prof Exp:* Res chemist petrol, Tidewater Oil Co, 36-41, sr res engr refining, 41-45, supvr res, 46-58; sr investr chem, US Strategic Bombing Surv, 45-46; prin engr, Bechtel Corp, 58-62, mgr appl technol chem eng, 62-66, mgr process serv, Bechtel Assocs Prof Corp, 66-76, mgr process design, Bechtel Group Inc, 76-78, sr prin engr, 78-83; CONSULT, 83- *Mem:* Fel Am Inst Chem Engrs; Am Chem Soc; NY Acad Sci; Sigma Xi. *Res:* Fundamental properties of lubricants; processes for manufacture of high octane aviation gasoline components; optimization of refinery operations; desalting of sea water; conversion of coal to fluid fuels; design of prototype process plants from pilot data. *Mailing Add:* Four Lower Lake Rd Danbury CT 06811

GALSTER, RICHARD W, b Seattle, Wash, May 13, 30; m 51; c 2. ENGINEERING GEOLOGY. *Educ:* Univ Wash, Seattle, BS, 52, MS, 56. *Prof Exp:* Geologist, Grant County Pub Utility Dist, Wash, 54-55; geologist, US Corps Engrs, Seattle, 55-85, chief, Geol Sect, 73-85; CONSULT ENG GEOLOGIST, 85- *Concurrent Pos:* chmn, Eng Geol Div, Geol Soc Am, 77-78; mem, design rev team to People's Repub China, 80; mem, US Comn Large Dams; mem, US Comt Rock Mechs, Nat Res Coun, 86-89. *Mem:* Asn Eng Geol (secy, 80-81, vpres, 81-82, pres, 82-83); fel Geol Soc Am; Int Asn Eng Geol; Am Inst Prof Geologists. *Res:* Engineering geology; hydrogeology; rock mechanics; geotechnical foundation analysis; seismotectonics; rock excavation design. *Mailing Add:* 18233 13th NW Seattle WA 98177

GALSTER, WILLIAM ALLEN, b Kenosha, Wis, Apr 11, 32; m 55; c 5. TOXICOLOGY, ENVIRONMENTAL HEALTH. *Educ:* Univ Wis, Madison, BS, 58, MS, 61. *Prof Exp:* Chemist, Wis State Lab Hyg, 56-57; asst prof biol, St Benedict's Col, 61-64; asst prof comp physiol & coordr anal serv, Univ Alaska, 64-73, assoc zoochemist, Inst Arctic Biol, 73-76; biomed dir,

Flammability Res Ctr, 76-80, group leader, molec biol & biochem, Utah Biomed Test Lab, 80-82, res prof, clin chem, Med Technol Dept, 82-84, RES COORD, DEPT NEUROL, MED COL, UNIV UTAH, 84- *Concurrent Pos:* Comt combustion toxicol, Am Acad Sci, 79-80. *Mem:* AAAS; Am Soc Mammal; Am Zool Soc; Am Inst Chem; Int Soc Mammalian Hibernation. *Res:* Bioenergetics of natural populations of mammals; gluconeogenesis and lipolysis during cold stress; hibernation periodicity and energetics; acclimatization to cold and altitude; natural resistance to toxic substances; inhalation toxicology; carcinogenesis; immunosuppression; autoimunogenic diseases. *Mailing Add:* 4192 Holloway Dr Salt Lake City UT 84124

GALSTON, ARTHUR WILLIAM, b New York, NY, Apr 21, 20; m 41; c 2. PLANT PHYSIOLOGY. *Educ:* Cornell Univ, BS, 40; Univ Ill, MS, 42, PhD(bot), 43. *Hon Degrees:* LLD, Iona Col, New Rochelle, NY, 80. *Prof Exp:* Res fel biol, Calif Inst Technol, 43-44, sr res fel, 47-50, assoc prof biol, 51-55; prof plant physiol, Yale Univ, 55-65, chmn dept bot, 61-62, dir div biol sci, 65-66, chmn dept biol, 85-88, prof biol, 65-90, Eaton prof bot, 74-90, EMER PROF BIOL & BOT, YALE UNIV, 90- *Concurrent Pos:* Guggenheim fel, Med Nobel Inst, Stockholm, 50-51; consult, Cent Res Dept, E I du Pont de Nemours & Co, Inc, 55-77; chmn, Comt on Meetings, AAAS, 58-60; mem, metab biol panel, NSF, 59-60; Fulbright fel, Australia, 60-61; mem, Nat Res Coun, 64-76 & 84-87; NSF sr fac fel, Univ London, 68; Einstein fel, Hebrew Univ, Jerusalem, 80, fel, Wolfson Col, Cambridge, 83; Riken scientist, Japan, 88. *Mem:* Am Soc Plant Physiologists (secy, 55-57, vpres, 57-58, pres, 62-63); Bot Soc Am (pres, 67-68); Int Asn Plant Physiologists (secy-treas); Fedn Am Scientists Coun; Soc Social Responsibility in Sci (pres, 73-75); AAAS; Am Acad Arts & Sci. *Res:* Plant growth hormones; photobiology; polyamines; differentiation and morphogenesis; circadian rhythms as related to photoperiodism. *Mailing Add:* Dept Biol Yale Univ PO Box 6666 New Haven CT 06511-8112

GALSWORTHY, PETER ROBERT, b Montreal, Que, June 13, 39; m 63; c 2. BIOCHEMISTRY, GENETICS. *Educ:* Queen's Univ, BSc, 62; Univ Wis, Madison, PhD(biochem), 68. *Prof Exp:* NIH res fel biochem, Princeton Univ, 68-70; asst prof, 70-77, ASSOC PROF BIOCHEM, UNIV WESTERN ONT, 77- *Concurrent Pos:* Med Res Coun Can Res grant, 70-75. *Mem:* Can Biochem Soc; NY Acad Sci; Am Soc Microbiol. *Res:* Membrane structure and function; membrane transport proteins; energy-coupling mechanisms for active transport; phospholipids and membrane structure in relation to transport; role of transport systems in differentiation. *Mailing Add:* 41 Lonsdale Dr London ON N6G 1T4 Can

GALSWORTHY, SARA B, b Frankford, Ind, July 1, 38; Can citizen; m 63; c 1. IMMUNOLOGY. *Educ:* Pomona Col, BA, 60; Univ Wis, PhD(physiol chem), 65. *Prof Exp:* Asst prof, 74-81, ASSOC PROF MICROBIOL, UNIV WESTERN ONT, 81- *Mem:* Can Soc Microbiologists; Am Soc Microbiol; Can Soc Immunologists. *Res:* Listeria monogytogens infections; mechanisms of virulence; regulation of monocytopoiesis. *Mailing Add:* Dept Microbiol & Immunol Univ Western Ont London ON N6A 5C1 Can

GALT, CHARLES PARKER, JR, b Defiance, Ohio, Dec 27, 42. INVERTEBRATE ZOOLOGY, MARINE BIOLOGY. *Educ:* Univ Calif, Santa Barbara, BA, 65; Univ Wash, MS, 70, PhD(zool), 72. *Prof Exp:* Asst prof biol, Fla State Univ, 72-73; asst prof, 73-78, ASSOC PROF BIOL, CALIF STATE UNIV, 78- *Mem:* AAAS; Am Soc Limnol & Oceanog; Am Soc Zoologists; Marine Biol Asn UK; Sigma Xi. *Res:* Ecology; functional morphology; bioluminescence of pelagic tunicates of class larvacea; natural history and feeding biology of zooplankton. *Mailing Add:* Dept of Biol Calif State Univ Long Beach CA 90840

GALT, JOHN (ALEXANDER), b Toronto, Ont, Mar 8, 25; m 55; c 2. RADIO ASTRONOMY. *Educ:* Univ Toronto, BA, 49, MA, 52, PhD(physics), 56. *Prof Exp:* Geophysicist, Govt Can, 49-50; res physicist, Du Pont of Can, 56-57; res scientist, 63-81, ASTROPHYSICIST, DOM RADIO ASTROPHYS OBSERV, 59- *Concurrent Pos:* Res, Jodrell Bank, Univ Manchester, 57-59; guest prof, Chalmers Tech Univ, Gothenberg, Sweden, 72-73. *Mem:* AAAS; Am Astron Soc; Can Astron Soc; Can Asn Physicists; Inst Elec & Electronics Eng; Royal Astron Soc; Royal Astron Soc Can. *Res:* Physics; spectroscopy; high pressures; electronics; optics; radio astronomy; twenty-one centimeter research; long baseline interferometry, pulsars; radio telescope arrays; OH/IR stars, computers. *Mailing Add:* Dom Radio Astrophys Observ Box 248 Penticton BC V2A 6K3 Can

GALT, JOHN KIRTLAND, b Portland, Ore, Sept 1, 20; m 49; c 1. SOLID STATE SCIENCE. *Educ:* Reed Col, AB, 41; Mass Inst Technol, PhD(physics), 47. *Prof Exp:* Res assoc, Mass Inst Technol, 45-47; Nat Res Coun fel, Bristol, Eng, 47-48; mem tech staff, Bell Tel Labs, NJ, 48-61, dir solid state electronics res, 61-74; dir solid state sci res, Sandia Labs, 74-78, vpres, 78-85; prin scientist, Aero Space Corp, 85-90; RETIRED. *Concurrent Pos:* Mem, Nat Acad Comt Adv to Air Force Syst Command, 71-77, mem, Air Force Sci Adv Bd, 75- *Mem:* Fel Am Phys Soc; fel Inst Elec & Electronics Engrs; fel AAAS; Nat Acad Eng. *Res:* Mechanical and magnetic properties of solids; band structure of metals; lasers, nonlinear optics, luminescence and optical properties of solids. *Mailing Add:* 10501 Lagrima De Oro NE 807C Albuquerque NM 87111

GALTON, PETER MALCOLM, b London, Eng, Mar 14, 42, UK; div. VERTEBRATE ANATOMY, PALEONTOLOGY. *Educ:* Univ London, BSc, 64, PhD(zool), 67, DSc, 82. *Prof Exp:* Cur assoc vert paleont, Peabody Mus & mem res staff geol, Yale Univ, 67-70; from asst prof to assoc prof, 70-78, PROF BIOL, UNIV BRIDGEPORT, 78-; CUR AFFIL VERT PALEONT, PEABODY MUS, YALE UNIV, 85- *Mem:* Palaeont Soc (WGer); Sigma Xi; Soc Vert Paleont; Linnean Soc London. *Res:* Problems of interpreting dinosaurs as living animals; osteology, functional anatomy, systematics, classification, origins, evolution and interrelationships of different dinosaurs; Mesozoic zoogeography and continental drift. *Mailing Add:* Dept Biol Univ Bridgeport Bridgeport CT 06601

GALTON, VALERIE ANNE, b Louth, Eng, May 6, 34; m 77; c 2. ENDOCRINOLOGY. *Educ:* Univ London, BSc, 55, PhD(physiol), 58. *Prof Exp:* Res assoc endocrinol, Nat Inst Med Res, London, 55-58 & Thorndike Mem Lab, Harvard Med Sch, 59-61; from instr to assoc prof, 61-75, PROF PHYSIOL, DARTMOUTH MED SCH, 75- *Concurrent Pos:* Milton res fel med, Harvard Med Sch, 59-60; Life Ins med res fel, 61-63; USPHS res grants, 62-, res career develop award, 65-70. *Mem:* Endocrine Soc; Am Thyroid Asn. *Res:* Mode of action of the thyroid hormones, nature of their peripheral metabolism and relation between hormonal action and metabolism. *Mailing Add:* Dept Physiol Dartmouth Med Sch Hanover NH 03756

GALUSHA, JOSEPH G, JR, b Battle Creek, Mich, Feb 1, 45; m 68; c 2. ANIMAL BEHAVIOR, ETHOLOGY. *Educ:* Walla Walla Col, BA, 68; Andrews Univ, MA, 72; Oxford Univ, DPhil(ethology), 75. *Prof Exp:* Instr biol, Maplewood Acad, 68-71; res asst, Oxford Univ, 72-75; ASSOC PROF BIOL, WALLA WALLA COL, 75- *Mem:* Asn Study Animal Behavior; Brit Ornithologists Union; Am Ornithologists Union; Animal Behavior Soc. *Res:* Factors relating to density dependent behavior of social animals; psychobiology of group structure. *Mailing Add:* Dept Biol Sci Walla Walla Col College Place WA 99324

GALVIN, AARON A, b Brooklyn, NY, Apr 13, 32; m 56; c 3. ELECTRONICS ENGINEERING. *Educ:* Mass Inst Technol, BS & MS, 55. *Prof Exp:* Engr, Lincoln Lab, Mass Inst Technol, 55-59, group leader, 59-68; vpres advan develop, Aerospace Res, Inc, 68-78; DIR, ADT NEW ENG RES LAB, 78-, vpres, 82-87; CONSULT, 88- *Mem:* Fel Inst Elec & Electronics Engrs. *Res:* Radar; advanced radar system development, particularly optimal signal processing; radar and ultrasonic intrusion alarms; microprocessor security and energy management; mammography. *Mailing Add:* 130 Mt Auburn Cambridge MA 02138

GALVIN, CYRIL JEROME, JR, b Jersey City, NJ, June 16, 35. COASTAL ENGINEERING, COASTAL PROCESSES. *Educ:* St Louis Univ, BS, 57; Mass Inst Technol, SM, 59, PhD(geol), 63. *Prof Exp:* Res asst, Hydrodyn Lab, Mass Inst Technol, 59-63; phys oceanogr, US Army Coastal Eng Res Ctr, 63-70, chief, Coastal Processes Br, 70-78; COASTAL ENG CONSULT, 78- *Concurrent Pos:* US Geol Surv, 57, 58; res resident, Math Res Ctr, Univ Wis, 65-66; vis assoc prof geol sci, Northwestern Univ, 72. *Honors & Awards:* Huber Res Prize, Am Soc Civil Eng, 69, Norman Medal, 70. *Mem:* Am Shore & Beach Preserv Asn; Am Geophys Union; Am Soc Civil Engrs; Int Asn Hydraul Res; Soc Sedimentary Geol; Soc Hist Technol. *Res:* Longshore currents, marina entrance channels, coastal structures, sea level change; wave climate, sediment transport on beaches; inlets and barrier islands; Iridium anomaly at K/T boundary; history of 19th century science. *Mailing Add:* PO Box 623 Springfield VA 22150

GALVIN, FRED, b St Paul, Minn, Nov 10, 36; m 73; c 2. MATHEMATICS, COMBINATORICS. *Educ:* Univ Minn, BA, 58, MA, 61, PhD(math), 67. *Prof Exp:* Actg instr math, Univ Calif, Berkeley, 65-67, lectr, 67-68, asst prof, Los Angeles, 68-75; assoc prof, 75-78, PROF MATH, UNIV KANS, 78- *Mem:* Math Asn Am; Am Math Soc. *Res:* Elementary and recreational mathematics, combinatorics, classical set theory, general topology. *Mailing Add:* Dept Math Univ Kans 405 Snow Hall Lawrence KS 66045-2142

GALVIN, ROBERT W, b Marshfield, Wis, Oct, 1922. MANAGEMENT. *Prof Exp:* Mem staff, 40-50, sr officer, 59-90, CHMN EXEC COMT, MOTOROLA INC, 90- *Honors & Awards:* Nat Medal of Technol, 91. *Mailing Add:* Motorola Inc 1303 E Algonquin Rd Shaumburg IL 60196-1065

GALVIN, THOMAS JOSEPH, b Cisco, Tex, Mar 14, 34; m 52; c 3. PARASITOLOGY, VETERINARY MEDICINE. *Educ:* Tex A&M Univ, DVM, 57, BS & MS, 61; Tulane Univ, PhD(parasitol), 64. *Prof Exp:* Sta veterinarian, NC State Col, 57-58; from instr to assoc prof vet parasitol, Tex A&M Univ, 58-71, prof, 71-, prof vet microbiol, 77-; DEPT VET PARASITOL, WVA UNIV, MORGANTOWN. *Concurrent Pos:* NIH fel, 61-62; lectr, Univ EAfrica & Univ Col Nairobi, Kenya, 65-67. *Mem:* Wildlife Dis Asn; Am Vet Med Asn; Am Soc Vet Parasitol; Am Soc Parasitol; Am Soc Trop Med & Hyg. *Res:* Biology, control and treatment of helminth parasites of domestic and laboratory animals, especially those species which may affect human health. *Mailing Add:* Am Embassy AG-SEA PO Box 3087 Laredo TX 78044

GALYEAN, MICHAEL LEE, b Bentonville, Ark, Sept 2, 51; m 73; c 3. RUMINANT NUTRITION. *Educ:* NMex State Univ, BS, 73; Okla State Univ, MS, 75, PhD(animal nutrit), 77. *Prof Exp:* asst prof, 77-80, ASSOC PROF NUTRIT, 80-, PROF, ANIMAL NUTRIT, NMEX STATE UNIV, 86- *Mem:* Sigma Xi; Am Soc Animal Sci; Am Dairy Sci Asn; Am Inst Nutrit. *Res:* Manipulation of rumen fermentation; metabolic profile of stressed ruminants; nonprotein nitrogen utilization; grain processing for livestock; rate of passage of nutrients through the gastrointestinal tract of ruminants. *Mailing Add:* Dept Home Econ NMex State Univ Box 30003 Las Cruces NM 88003-0003

GAMBAL, DAVID, b Old Forge, Pa, Dec 16, 31; m 60; c 3. BIOCHEMISTRY, NUTRITION. *Educ:* Pa State Univ, BS, 53; Purdue Univ, MS, 56, PhD(biochem), 57. *Prof Exp:* Asst, Purdue Univ, 53-57; fel, McCollum-Pratt Inst, Johns Hopkins Univ, 57-59; from asst prof to assoc prof biochem, Vet Med Res Inst, Iowa State Univ, 59-65; assoc prof, 65-68, chmn dept, 76-79, PROF BIOCHEM, SCH MED, CREIGHTON UNIV, 68- *Concurrent Pos:* Vis res scientist, Nat Inst Health, 87-88. *Mem:* AAAS; Am Chem Soc; Soc Exp Biol & Med. *Res:* Isolation and characterization of proteins; hormonal control of cellular metabolism; arachidonic metabolism in brain; vitamin D and bone calcification. *Mailing Add:* Dept Biomed Sci Sch Med Creighton Univ 2500 California St Omaha NE 68178

GAMBERT, STEVEN ROSS, b New York, NY, Aug 22, 49; m 72; c 2. GERONTOLOGY, ENDOCRINOLOGY. *Educ:* Univ Col, NY Univ, BA, 71; Columbia Univ Col Physicians & Surgeons, MD, 75. *Prof Exp:* Resident med, Dartmouth Med Sch, 75-77; Fel endocrinol, Harvard Med Sch, 77-79; asst prof med, Med Col Wis, 79-81, assoc prof med & physiol, 81-83; PROF MED, NEW YORK MED COL, 83- *Concurrent Pos:* Consult, Res Planning Panel, Nat Inst Aging, 81 & Wis Dept Health & Social Service, 81-; co-chmn res comt, Milwaukee Admin on Aging, Long Term Care Geront Ctr, 81-; chief med serv, Ruth Taylor Geriat Inst; dir geriat, Westchester County Med Ctr. *Honors & Awards:* Geriat Med Acad Award, Nat Inst Aging. *Mem:* Fel Geront Soc Am; fel Am Geriat Soc; fel Am Col Physicians; Endocrine Soc; Am Aging Asn (pres). *Res:* Endocrine aspects of aging, primarily involving thyroid hormone economy; neuroendocrine changes; health care delivery. *Mailing Add:* Dept Med NY Med Col Dept Med Valhalla NY 10595

GAMBESCIA, JOSEPH MARION, b Philadelphia, Pa, June 10, 19; m 44; c 16. INTERNAL MEDICINE, GASTROENTEROLOGY. *Educ:* Philadelphia Col Pharm, BS, 39; Hahnemann Med Col, MD, 44. *Prof Exp:* Chief lab, US Army Hepatitis Res Ctr, 120th Sta Hosp, Ger, 46-48; assoc gastroenterol, 50-52, asst prof med, 52-55, dir res, Dept Gastroenterol, 51-61, clin prof med, 61-70, PROF MED, HAHNEMANN MED COL & HOSP, 70- *Concurrent Pos:* Instr, Div Grad Med, Univ Pa, 54-; asst, GI Clin, Philadelphia Gen Hosp, 56-60; pres, Philadelphia Nutrit Coun, 70. *Mem:* Am Col Gastroenterol; Am Soc Clin Invest; fel Am Col Physicians; Am Fedn Clin Res; fel Am Gastroenterol Asn. *Res:* Hepato pancreatic biliary disease. *Mailing Add:* 1811 S Broad St Philadelphia PA 19148

GAMBILL, JOHN DOUGLAS, b Rochester, Minn. PSYCHOPHARMACOLOGY, PSYCHIATRY. *Educ:* Northwestern Univ, BA, 63; Univ Minn, MD, 68. *Prof Exp:* Intern gen med, USPHS Hosp, San Francisco, 68-69; resident psychiat, Mass Gen Hosp, Harvard Univ, 69-72; res fel, 72-74, PSYCHIAT RESEARCHER, SCH MED, BOSTON UNIV, 75-, ASST PROF PSYCHIAT, 76- *Concurrent Pos:* Mem US-USSR health exchange psychiat, Inst Psychiat, Moscow, 74; consult, Boston City Hosp Alcoholism Clin, 75-; lectr psychobiol, Sch Med, Boston Univ, 76-, GRS grant, 77-; staff psychiatrist, Bedford Va Hosp, 78- *Res:* Biological basis of mental illness. *Mailing Add:* 29 Stults Rd Belmont MA 02178

GAMBILL, ROBERT ARNOLD, b Indianapolis, Ind, Feb 20, 27; m 50; c 2. MATHEMATICS. *Educ:* Butler Univ, AB, 50; Purdue Univ, MS, 52, PhD(math), 54. *Prof Exp:* Head theoret anal br, Math Div, US Naval Avionics Facil, 54-58; sr mathematician, Gen Motors Corp, 58-60; assoc prof math, 60-66, PROF MATH & ASST HEAD DEPT, 66- *Mem:* Am Math Soc; Math Asn Am; Soc Indust & Appl Math. *Res:* Ordinary and functional differential equations; calculus of variations; optimal control. *Mailing Add:* Div Math Purdue Univ West Lafayette IN 47907

GAMBINO, RICHARD JOSEPH, b New York, NY, May 17, 35; m 55; c 4. MAGNETIC RECORDING, MAGNETO-OPTICS. *Educ:* Univ Conn, BA, 57; Polytech Inst NY, MS, 76. *Prof Exp:* Phys scientist, US Army Signal Res Lab, Ft Monmouth, 56-60; Metallurgist, Pratt & Whitney Aircraft Div, United Aircraft Corp, 60-61; RES STAFF MEM, T J WATSON RES CTR, IBM, YORKTOWN HEIGHTS, 61- *Concurrent Pos:* Mem, Thin Film Div Bd, Am Vacuum Soc, 86-88. *Mem:* Am Vacuum Soc; Mat Res Soc; Inst Elec & Electronics Engrs; Inst Elec & Electronics Engrs Magnetic Soc; Sigma Xi. *Res:* Magnetic and superconducting properties of thin films of crystalline and amorphous metallic materials; magneto-optical and magneto-transport properties. *Mailing Add:* 2433 Hunterbrook Rd Yorktown Heights NY 10598

GAMBINO, S(ALVATORE) RAYMOND, b Brooklyn, NY, Oct 13, 26; m 53; c 2. CLINICAL PATHOLOGY, CLINICAL CHEMISTRY. *Educ:* Antioch Col, BS, 48; Univ Rochester, MD, 52. *Prof Exp:* Assoc pathologist, St Luke's Hosp, Milwaukee, Wis, 57-61; asst prof, 61-69, dir clin chem labs, 69-77, PROF PATH, COL PHYSICIANS & SURGEONS, COLUMBIA UNIV, 69- *Concurrent Pos:* Asst prof, Marquette Univ, 59-61; dir labs, Englewood Hosp, NJ, 61-69; attend pathologist, Presbyterian Hosp, New York, 69-77; chief pathologist & dir labs, St Luke's-Roosevelt Hosp Ctr, 78-82; founder & ed educ med letter, Lab Report for Physicians, 79- *Honors & Awards:* Ward Burdick Award, Am Soc Clin Path, 90. *Mem:* AAAS; Am Soc Clin Path; NY Acad Med; Am Asn Clin Chemists; Col Am Path. *Res:* Measurement and interpretation of blood pH, blood gases and serum bilirubin; diagnostic laboratory methodology. *Mailing Add:* Exec Vpres MetPath Inc One Malcolm Ave Teterboro NJ 07608

GAMBLE, DEAN FRANKLIN, b McDonald, Pa, Aug 6, 20; m 46; c 3. ORGANIC CHEMISTRY, INFORMATION SCIENCE. *Educ:* Pa State Univ, BS, 42, PhD(chem), 53. *Prof Exp:* Chem librn, NIH, 53-56, head sect doc, records & pubs, Cancer Chemother, Nat Serv Ctr, Nat Cancer Inst, 56-60; dir dept sci info, Miles Labs Inc, Ind, 60-66, actg dir, Info Dept, 66-71; head food & nutrit info & educ mat ctr, 71-72, dep dir libr serv, 72-76; mem dir's staff, Nat Agr Libr, 76-81; RETIRED. *Mem:* Am Chem Soc. *Res:* Chemical documentation; information processing, storage and retrieval. *Mailing Add:* Rte 4 Box 555 Luray VA 22835

GAMBLE, FRANCIS TREVOR, b Montpelier, Vt, July 10, 28; m 51; c 3. OPTICAL PATTERN RECOGNITION, LASER HOLOGRAPHY. *Educ:* Colgate Univ, AB, 58; Univ Conn, MA, 60, PhD(physics), 63. *Prof Exp:* Res asst physics, Univ Conn, 62-63; from asst prof to assoc prof, Denison Univ, 63-70, chmn dept, 69-70 & 83-89, dean students, 70-79, PROF PHYSICS, DENISON UNIV, 70- *Concurrent Pos:* Consult, Battelle Mem Inst, 64-70 & 80-; consult-evaluator, Comn Insts Higher Educ, N Central Asn, 74-; vis prof chem, Ohio State Univ, 84 & 85. *Mem:* Am Asn Physics Teachers; Am Phys Soc; NY Acad Sci; Optical Soc Am; Sigma Xi; Int Soc Opt Eng. *Res:* Laser holography; optical information processing and pattern recognition; optical system design. *Mailing Add:* Denison Univ Granville OH 43023

GAMBLE, FRED RIDLEY, JR, b Dallas, Tex, Apr 24, 41; m 64; c 2. SOLID STATE SCIENCE. *Educ:* Harvard Univ, BA, 64; Stanford Univ, PhD(chem physics), 68. *Prof Exp:* Sr chemist, Syva Corp, 68-71; group leader chem physics, Exxon Res & Eng Co, 71-75, dir, Phys & Mat Sci Lab, 75-81, mgr, mat technol div, 81-83; vpres & dir res, Schlumberger, 83-87; VPRES & CHIEF TECH OFFICER, SUPERCONDUCTOR TECHNOLOGIES, INC, 87- *Mem:* Am Chem Soc; fel Am Phys Soc. *Res:* New superconducting materials; intercalation compounds; interactions between metals and molecules. *Mailing Add:* Superconductor Technologies Inc 460 Wand Dr Santa Barbara CA 93111-2310

GAMBLE, JAMES LAWDER, JR, b Boston, Mass, Jan 8, 21; m 48; c 6. PHYSIOLOGY, BIOCHEMISTRY. *Educ:* Harvard Univ, BS, 43, MD, 45. *Prof Exp:* Intern pediat, Univ Hosp, Johns Hopkins Univ, 45-46, asst resident, Sch Med, 48-50; clin investr physiol, Brookhaven Nat Lab, 50-52; NIH fel pediat & biochem, 53-57, Am Heart Asn estab investr pediat & physiol, 57-62, asst prof physiol, 62-67, ASSOC PROF PHYSIOL, SCH MED, JOHNS HOPKINS UNIV, 67- *Mem:* Soc Pediat Res; Am Pediat Soc; Am Physiol Soc. *Res:* Electrolyte and fluid space physiology; mitochondrial electrolytes and ion transport. *Mailing Add:* Johns Hopkins Univ Sch Med Baltimore MD 21205

GAMBLE, JOHN ROBERT, internal medicine; deceased, see previous edition for last biography

GAMBLE, ROBERT OSCAR, b Greensboro, NC, Nov 20, 35; m 63; c 3. MATHEMATICS, COMPUTER SCIENCE. *Educ:* Duke Univ, BSME, 59; Clemson Univ, MS, 63, PhD(math), 71; Univ SC, MS, 88. *Prof Exp:* Develop engr, Celanese Fibers Co, SC, 59-62; instr math, Winthrop Col, 63-66, assoc prof, 70-90; ASST PROF COMPUTER SCI, USC-COASTAL CAROLINA COL, 90- *Mem:* Asn Comput Mach; Math Asn Am. *Res:* Investigation of ring-theoretic properties of matrices over finite local rings; public-key cryptosystems. *Mailing Add:* Dept Computer Sci USC-Coastal Carolina Col PO Box 1954 Conway SC 29526

GAMBLE, THOMAS DEAN, b Bellefonte, Pa, Aug 26, 47; m 69; c 3. PATTERN RECOGNITION, NEURAL NETS. *Educ:* Univ Mich, BS, 69; Univ Calif, Berkeley, PhD(physics), 78. *Prof Exp:* Physicist magneto tellurics, Earth Sci Div, Lawrence Berkeley Labs, 78-80; SR STAFF SCIENTIST, ENSCO INC, 80- *Mem:* Soc Explor Geophysicists. *Res:* Signal analysis; pattern recognition; magnetotellurics; statistics. *Mailing Add:* 5400 Port Royal Rd Springfield VA 22151-2388

GAMBLE, WILBERT, b Greenville, Ala, June 19, 32; m 57; c 1. BIOCHEMISTRY. *Educ:* Wayne State Univ, BS, 55, PhD, 60. *Prof Exp:* Asst physiol chem, Wayne State Univ, 55-59, NIH res fel, 59-62; from asst prof to assoc prof, 68-76, PROF BIOCHEM & BIOPHYS, ORE STATE UNIV, 76- *Concurrent Pos:* Vis assoc prof & NIH spec fel, Johnson Res Found, Univ Pa, 68-69; assoc, Danforth Found, 69; Fulbright fel, Univ Sci & Technol, Ghana, 71-72; vis res worker, Nat Heart, Lung & Blood Inst, 76-77, IPA investr, 83-84 & 90-91. *Honors & Awards:* Lehnard Fink Medal. *Mem:* AAAS; Am Chem Soc; Am Soc Biochem & Molecular Biol. *Res:* Enzymes and metabolism of vascular tissue; computer simulation; mechanism of action of polychlorinated biphenyls; atherosclerosis. *Mailing Add:* Dept Biochem & Biophys Ore State Univ Corvallis OR 97331

GAMBLE, WILLIAM LEO, b Elkhart, Kans, Nov 25, 36; m 62; c 3. STRUCTURAL ENGINEERING. *Educ:* Kans State Univ, BS, 59; Univ Ill, Urbana, MS, 61, PhD(civil eng), 62. *Prof Exp:* Res asst struct, Univ Ill, Urbana, 59-62; Fulbright res fel, div bldg res, Commonwealth Sci & Indust Res Orgn, Australia, 62-63; sr engr, Bechtel Corp, 64-65; from asst prof to assoc prof civil eng, 63-73, PROF CIVIL ENG, UNIV ILL, URBANA, 73- *Concurrent Pos:* Vis lectr, Univ Canterbury, Christchurch, NZ, 73-74; consult, distress concrete structures. *Mem:* Am Concrete Inst; Am Soc Civil Engrs; Am Soc Testing & Mat; Prestressed Concrete Inst. *Res:* Strength and behavior of reinforced concrete floor slab systems; long-term behavior of prestressed concrete bridges; prestressed concrete piling. *Mailing Add:* Dept Civil Eng Univ Ill Urbana-Champaign Urbana IL 61801-2397

GAMBOA, GEORGE JOHN, b Boise, Idaho, Jan 21, 46; m 67; c 1. BEHAVIORAL ECOLOGY. *Educ:* Idaho State Univ, BS, 68; Ariz State Univ, MS, 74; Univ Kans, PhD(biol), 79. *Prof Exp:* Asst & res asst zool, Ariz State Univ, 72-74; asst biol, Univ Iowa, 74-75; asst res fel social behav, Univ Kans, 75-78, res fel, 78-79; ASSOC PROF BIOL, OAKLAND UNIV, 80- *Mem:* AAAS; Entom Soc Can; Int Union Study Social Insects. *Res:* Evolution of insect social behavior; behavioral ecology of social hymenoptera. *Mailing Add:* Bio Sci Dept Oakland Univ Rochester MI 48309

GAMBORG, OLUF LIND, b Denmark, Nov 9, 24; Can citizen; m 53; c 3. PLANT CELL BIOLOGY & GENETICS. *Educ:* Univ Alta, BSc, 56, MSc, 58; Univ Sask, PhD(plant biochem), 62. *Prof Exp:* sr res officer plant biochem, Prairie Regional Lab, 58-79, res dir, Int Plant Res Inst, 79-81, SR RES FEL, NAT RES COUN CAN, 81- *Concurrent Pos:* Assoc ed, Environ & Exp Bot, 75-80; managing ed, Plant Cell Report, 80- *Honors & Awards:* Gold Medal, Can Soc Plant Physiologists, 77. *Mem:* Am Soc Plant Physiologists; Can Soc Plant Physiologists (pres, 75); Int Asn Plant Tissue Cult; Genetics Soc Can; Scand Soc Plant Physiol. *Res:* Plant cell culture; plant cell genetics, asexual production of plants from cultured cells and protoplasts; development of somatic hybrid plants by protoplast fusion; morphogenesis; plant biochemistry; plant developmental biology and genetics; plant molecular biology; genetic transformation. *Mailing Add:* 3200 Azalea Dr Ft Collins CO 80523

GAMBRELL, CARROLL B(LAKE), JR, b Birmingham, Ala, Dec 1, 24; m 44; c 2. INDUSTRIAL ENGINEERING, ENGINEERING MANAGEMENT. *Educ:* Clemson Univ, BS, 49; Univ Fla, MS, 52; Purdue Univ, PhD, 58; Fla Southern Col, BA, 77. *Prof Exp:* Instr, Clemson Univ, 49-51; asst prof indust eng, Lamar Univ, 52-55; from instr to asst prof, Purdue

Univ, 55-59; vpres acad affairs, 67-78, PROF, UNIV CENT FLA, 67- *Concurrent Pos:* Consult to var govt agencies & comts & var pvt industs, 46-; vis prof to var univs in US, 66-79; mem bd dirs, Regional Energy Training & Res Orgn, Inc, Cocoa, Fla, 71-76, Univ Cent Fla Found, Inc, 74-78, Winter Park Mem Hosp, 75- & Embry-Riddle Aeronaut Univ, 75-; distinguished lectr, Am Inst Indust Engrs, 73-; mem, Engrs Coun Prof Develop. *Mem:* Am Soc Eng Educ; fel Am Inst Indust Engrs; Sigma Xi. *Res:* Human engineering; system simulation; operations research; economic analysis. *Mailing Add:* 2050 Upper River Rd Macon GA 31211-1104

GAMBRELL, LYDIA JAHN, b Cleveland, Ohio, May 18, 04; m 32; c 2. BIOLOGY. *Educ:* Ohio State Univ, BA, 27, MA, 28, PhD(entom), 32. *Hon Degrees:* DSc, Keuka Col, 74. *Prof Exp:* Asst, Dept Zool & Entom, Ohio State Univ, 27-30; instr bact & physiol, Lindenwood Col, 30-32; tech asst entom, NY Agr Exp Sta, 35-38; field agt, Endicott Jr Col, 39-43; instr biol, Keuka Col, 45 & Hobart Col, 47; from instr to prof, Keuka Col, 47-74, head dept, 57-69, emer prof biol, 74-88; RETIRED. *Mem:* NY Acad Sci. *Res:* Insect anatomy and embryology. *Mailing Add:* Arbor at Lake 2020 W Lake Parker Dr Lakeland FL 13805

GAMBRELL, SAMUEL C, JR, b Owings, SC, Sept 15, 35; m 57; c 3. ENGINEERING MECHANICS. *Educ:* Clemson Univ, BS, 57, MS, 61; WVa Univ, PhD(eng mech), 65. *Prof Exp:* From instr to asst prof eng mech, Clemson Univ, 59-62; instr, WVa Univ, 62-63; from asst prof to assoc prof, Univ Ala, Tuscaloosa, 65-73, dir, Solid Mech Div, 69-76, asst dean eng & dir, Bur Eng Res, 76-83, dir eng placement, 78-83, PROF ENG MECH, UNIV ALA, TUSCALOOSA, 73- *Concurrent Pos:* Consult indust & legal firms. *Mem:* Am Soc Eng Educ; Soc Exp Mech; Nat Soc Prof Engrs; Sigma Xi. *Res:* Experimental stress analysis; photoelasticity; fatigue; experimental mechanics. *Mailing Add:* Eng Mech Dept Box 870278 Tuscaloosa AL 35487-0278

GAMBS, GERARD CHARLES, b Columbus, Ohio, May 2, 18; m 71; c 2. UTILIZATION OF FLY ASH IN CONCRETE & CONCRETE PRODUCTS. *Educ:* Ohio State Univ, BS, 40. *Prof Exp:* Jr mining engr, Pittsburgh Coal Co, 40-42; maj atomic bomb proj, US Army CEngrs, 42-46; asst prof, Eng Exp Sta, Ohio State Univ, 46-47; asst vpres, Consol Coal Co, 47-69; bus mgr, Gibbs & Hill, Inc, 69-70; vpres, Ford, Bacon & Davis Inc, 70-83; CONSULT ENGR, 83- *Concurrent Pos:* Mem bd dirs, Schott Energyy Corp, Tipperary Corp, 73-86, Onan Corp, 71-83, Ford, Bacon & Davis, Inc, 73-83. *Mem:* Am Inst Mining Metall & Petrol Engrs; Am Soc Mech Engrs; Am Nuclear Soc. *Res:* Energy; utilization of fly ash in concrete and concrete products. *Mailing Add:* 1725 York Ave Apt 33C New York NY 10128

GAMBS, ROGER DUANE, b Bozeman, Mont. ETHOLOGY, ANIMAL BEHAVIOR. *Educ:* Univ Idaho, BS, 63, MS, 65; Univ Mont, PhD(ethology), 73. *Prof Exp:* Instr biol, Wis State Univ, Whitewater, 65-67; prof zool, Ariz Western Col, 71-72; PROF BIOL, CALIF POLYTECH STATE UNIV, 74- *Mem:* Am Soc Ichthologists & Herpetologists; Animal Behavior Soc; Cooper Ornithological Soc; Am Ornithologists Union; Am Soc Mammalogists; Wildlife Soc. *Res:* Avian mobbing behavior; functional role of vertebrate vocalizations; evolutionaly significance of avian social organizations; social behavior in Tule Elk; effects of wildlife on Blue Oak regeneration; small mammal ecology. *Mailing Add:* Dept Biol Sci Calif Polytech State Univ San Luis Obispo CA 93407

GAME, JOHN CHARLES, b Tonbridge, Eng, Dec 14, 46. GENETICS OF DNA REPAIR, PULSED-FIELD GEL ELECTROPHORESIS. *Educ:* Oxford Univ, Eng, DPhil, 71. *Prof Exp:* Postdoctoral fel genetics, Biol Dept, York Univ, Ont, 71-72; res scientist, Nat Inst Med Res, London, 74-78; postdoctoral fel, 72-73, asst specialist, 78-82, STAFF SCIENTIST, LAWRENCE BERKELEY LAB, UNIV CALIF, BERKELEY, 82- *Mem:* Genetics Soc Am. *Res:* Processes of genetic recombination, meiosis, and the repair of DNA using the yeast Saccharomyces cerevisiae; botany including plants of California and Polynesia. *Mailing Add:* Donner Lab Lawrence Berkeley Lab Berkeley CA 94720

GAMELIN, THEODORE W, b Decorah, Iowa, Sept 24, 39; m 61; c 4. ANALYSIS & FUNCTIONAL ANALYSIS. *Educ:* Yale Univ, BS, 60; Univ Calif, Berkeley, PhD(math), 63. *Prof Exp:* Clemoore instr math, Mass Inst Technol, 63-65, asst prof, 67-68; prof, Nat Univ La Plata, Arg, 65-66; PROF MATH, UNIV CALIF, LOS ANGELES, 68- *Concurrent Pos:* Sloan Found fel, 69-71. *Mem:* Am Math Soc. *Res:* Function algebras; algebras of analytic functions. *Mailing Add:* Dept Math Univ Calif 405 Hilgard Ave Los Angeles CA 90024

GAMMAGE, RICHARD BERTRAM, b Whitby, Eng, Nov 5, 37; m 64. CHEMICAL HEALTH & SAFETY, HEALTH PHYSICS. *Educ:* Univ Exeter, BSc, 60, PhD(chem), 64. *Prof Exp:* NSF fel, Univ Fla, 64-67; res scientist oxide surfaces, 67-70, HEALTH PHYSICIST & GROUP LEADER, MONITORING TECHNOL & INSTRUMENTATION & SOLID STATE RADIATION DOSIMETRY, OAK RIDGE NAT LAB, 70- *Concurrent Pos:* Res assoc, Imp Col, Univ London, 64; mem, Comt Int Intercomparisons Environ Dosimeters, 76- *Mem:* Am Chem Soc; The Chem Soc; Am Indust Hyg Asn; Health Physics Soc. *Res:* Monitoring devices and measurements of indoor air pollutants, especially formaldehyge; real or near real-time monitoring devices for fugitive emissions from synthetic fossil fuels; solid state integrating radiation dosimeters; calcareous absorbents; ceramic oxides; lunar fines and their surface properties. *Mailing Add:* MS-6383 Bldg 7509 Oak Ridge Nat Lab PO Box 2008 Oak Ridge TN 37831-6113

GAMMAL, ELIAS BICHARA, b Cairo, Egypt, Nov 18, 30; Can citizen; m 61; c 5. ANATOMY, OBSTETRICS & GYNECOLOGY. *Educ:* Ain Shams Univ, Cairo, MB, ChB, 55; Univ Western Ont, PhD(med res), 66. *Prof Exp:* Intern, Ain Shams Univ Hosps, Egypt, 56-57; St Joseph's Infirmary, Atlanta, Ga, 57-58, resident obstet & gynec, 58-61; fel gynec cancer, Emory Univ, 61-62; fel med res, Collip Lab, 62-66, lectr obstet & gynec, 67, lectr anat, 68-69, asst prof, 69-73, assoc prof, 73-80, PROF ANAT, UNIV WESTERN ONT, 80-, ASST PROF OBSTET & GYNEC, 67 - *Concurrent Pos:* Vis prof anat & reproduction biol, Univ Hawaii, 80-81. *Mem:* Am Asn Anatomists; Can Asn Anatomists; Can Invests in Reprod; assoc Soc Obstetricians & Gynecologists Can. *Res:* placental trophoblast; endothelial differentiation and proliferation. *Mailing Add:* Dept Anat Univ Western Ont Health Sci Ctr London ON N6A 5C1 Can

GAMMEL, GEORGE MICHAEL, b Los Alamos, NMex, Sept 26, 52; m 78; c 1. PARTICLE BEAMS, ACCELERATOR DESIGN. *Educ:* Mich State Univ, BS & MS, 74; Cornell Univ, PhD(appl physics), 79. *Prof Exp:* Res asst appl physics, Lab Plasma Studies, Cornell Univ, 75-78; from asst scientist to assoc scientist, Accelerator Dept, Brookhaven Nat Lab, 80-83; staff mem, Princeton Plasma Physics Lab, 83-90; STAFF SCIENTIST, GRUMMAN CORP, 90- *Mem:* Am Physics Soc. *Res:* High current electron beam studies; particle beam fusion research; neutral beam development; plasma diagnostics; accelerator technology. *Mailing Add:* 2761 Covered Bridge Rd Merrick NJ 11566

GAMMEL, JOHN LEDEL, b Austin, Tex, July 9, 24; m 47; c 5. THEORETICAL PHYSICS. *Educ:* Univ Tex, BS, 44, MA, 46; Cornell Univ, PhD(theoret physics), 50. *Prof Exp:* Tutor physics, Univ Tex, 42-45, instr math, 45-46; asst physics, Cornell Univ, 46-50; mem staff, Los Alamos Sci Lab, 50-63; prof physics, Tex A&M Univ, 63-67; mem staff, Los Alamos Sci Lab, NMex, 67-74; chmn dept, 74-82, PROF PHYSICS, ST LOUIS UNIV, 74- *Concurrent Pos:* Instr, Univ Calif, Los Angeles, 52; Fulbright fel, 57-58; mem staff, Atomic Energy Res Estab, Harwell, Eng, 61-62; vis prof, Univ Western Ontario, Can, 79-80. *Mem:* Fel AAAS. *Res:* Theoretical nuclear physics. *Mailing Add:* Dept Physics St Louis Univ 221 N Grand St Louis MO 63103

GAMMELL, PAUL M, b Attleboro, Mass, Feb 27, 39; div. MEDICAL ULTRASOUND, NONDESTRUCTIVE EVALUATION. *Educ:* Univ Md, BS, 63; Am Univ, MS, 66; Catholic Univ Am, PhD(physics), 71. *Prof Exp:* Res assoc, Dept Chem, Purdue Univ, 71-74; Sci Res Coun sr vis fel, Dept Chem, Southampton Univ, 73; dir res, Med Div, Radionics, Ltd, 74-76; mem tech staff, Jet Propulsion Lab, Calif Inst Technol, 76-82; NAVAL SURFACE WARFARE CTR, 83- *Concurrent Pos:* Clin asst prof, Dept Radiol, Sch Med, Univ Southern Calif, 77-82. *Mem:* Sigma Xi; Am Soc Nondestructive Testing; Am Soc Testing Mat; Acoust Soc Am; Inst Elec & Electronics Engrs. *Res:* Ultrasonic physics and signal processing, particularly as applied to diagnostic medical ultrasound and to nondestructive evaluation. *Mailing Add:* Naval Surface Weapons Ctr Code R-34 White Oak Silver Spring MD 20410

GAMMILL, RONALD BRUCE, b Louisville, Miss, July 10, 48; m 68; c 2. ORGANIC CHEMISTRY. *Educ:* Millsaps Col, BS, 73; Univ SC, PhD(org chem), 76. *Prof Exp:* Am Cancer Soc fel, Univ Pittsburgh, 76-77; SCIENTIST ORG CHEM, UPJOHN CO, 77- *Mem:* Am Chem Soc. *Res:* New synthetic methods; alkylations; synthesis of novel molecules with biological activity. *Mailing Add:* 6704 Pleasant View Dr Kalamazoo MI 49002-1004

GAMMON, JAMES ROBERT, b Sparta, Wis, Apr 24, 30; m 53, 81; c 4. BIOLOGY, ECOLOGY. *Educ:* Wis State Col, Whitewater, BS, 56; Univ Wis, MS, 57, PhD(zool, bot), 61. *Prof Exp:* From asst prof to assoc prof, 61-73, PROF ZOOL, 73-, CHAIR, DEPAUW UNIV, 87- *Mem:* Am Fisheries Soc; Am Soc Limnol & Oceanog. *Res:* Fish ecology; river ecology; role of external metabolites in aquatic ecology. *Mailing Add:* Dept Biol Sci DePauw Univ Greencastle IN 46135

GAMMON, NATHAN, JR, b Cheyenne, Wyo, June 22, 14; m 41, 89; c 2. SOIL CHEMISTRY, PLANT NUTRITION. *Educ:* Univ Md, BS, 36, MS, 39; Ohio State Univ, PhD(soils), 41. *Prof Exp:* Asst soils, Univ Md, 36-39; asst agron, Ohio Exp Sta, 38-42; prof, 46-80, EMER PROF SOIL CHEM, INST FOOD & AGR SCI, UNIV FLA, 80- *Mem:* AAAS; Am Chem Soc; Soil Sci Soc Am; Am Soc Agron; Sigma Xi. *Res:* Ion exchange; plant nutrition; pasture soils and fertility maintenance; flame photometry; micronutrients; pecan and peach production; movement and availability of and factors influencing Cu, Zh, Fe, B, Mo and Mn in soils. *Mailing Add:* 1403 NW 11 Rd Gainesville FL 32605

GAMMON, RICHARD ANTHONY, b Lackawanna, NY, May 6, 37; m 63; c 2. MICROBIOLOGY. *Educ:* St Bonaventure Univ, BS, 59, PhD(biol), 65. *Prof Exp:* Asst biol, St Bonaventure Univ, 60-64; from asst prof to assoc prof, 64-77, PROF BIOL & COORDR NURSING PROG, GANNON COL, 77- *Concurrent Pos:* Res asst, Am Sterilizer Co, 66-; NIH res grant, 66-68. *Mem:* AAAS; Am Soc Microbiol. *Res:* Bacteriophages for the thermophile, Bacillus coagulans; kinetics of ethylene oxide sterilization. *Mailing Add:* Dept Biol Gannon Univ Erie PA 16541

GAMMON, RICHARD HARRISS, b Washington, DC, Apr 6, 43; m 73; c 1. PHYSICAL CHEMISTRY, ASTRONOMY. *Educ:* Princeton Univ, BA, 65; Harvard Univ, MA, 66, PhD(phys chem), 70. *Prof Exp:* NSF fel astron, Univ Calif, Berkeley, 70-71; res assoc astron & chem, Nat Radio Astron Observ, 71-73; vis prof astrochem, Radio Astron Ctr, Mackenzie Univ, Sao Paulo, Brazil, 74-76; vis fel radio astron, Battelle Observ, Pac Northwest Labs, 76-78; affil prof astron & chem, Univ Wash, 78; dir sci, Pac Sci Ctr, 78-80; Nat Oceanic & Atmospheric Admin, 80-89; PROF OCEANOG, UNIV WASH, 89- *Mem:* Am Astron Soc; AAAS; Brazilian Astron Soc. *Res:* Molecular spectroscopy; chemistry of interstellar space. *Mailing Add:* Dept Oceanog Univ Wash Seattle WA 98195

GAMMON, ROBERT WINSTON, b Washington, DC, Sept 1, 40; m 57; c 2. PHYSICS, PHOTON CORRELATION SPECTROSCOPY. *Educ:* Johns Hopkins Univ, AB, 61, PhD(physics), 67; Calif Inst Technol, MS, 63. *Prof Exp:* Mem tech staff, Hughes Res Labs, 62-63; res assoc physics, Johns Hopkins Univ, 66-67; from asst prof to assoc prof physics, Cath Univ Am, 67-72; asst prof, Inst Molecular Physics, 72-76, ASSOC PROF PHYSICS, INST PHYS SCI & TECHNOL, UNIV MD, 76- *Concurrent Pos:* Consult Naval Res Lab, 72-86, Nat Bur Stand, 74-78, TRACOR, 80-82, IBM Res Lab,

Zurich, 78-79; prin investr, critical fluid light scattering expon space shuttle, NASA Microgravity Sci & Appl Div, 84-; co-investr, critical point facil-thermal equilibrium exp, Int Mat Lab Space Shuttle Flight, 89- *Mem:* Am Phys Soc; Optical Soc Am. *Res:* Quantum optics, especially laser light scattering spectroscopy; experimental solid state physics, especially elasticity near phase transitions; thermodynamics, especially experimental critical phenomena in fluids generally and with reduced gravity. *Mailing Add:* Inst Phys Sci & Technol Univ Md College Park MD 20742

GAMMON, WALTER RAY, b Danville, Va, Oct 25, 42; m 66; c 2. DERMATOLOGY. *Educ:* NC State Univ, BS, 67; Univ NC, MD, 71; Am Bd Internal Med, cert, 74; Am Bd Dermat, cert, 79, cert dermat immunol & diag & lab immunol, 85. *Prof Exp:* Intern med, Barnes Hosp, St Louis, 71-72, jr asst resident, 72-73; sr asst resident med & dermat, NC Mem Hosp, Chapel Hill, 73-74; from instr to assoc prof, 76-86, PROF DERMAT & MED, SCH MED, UNIV NC, CHAPEL HILL, 86- *Concurrent Pos:* Vis prof & lectr, numerous univs, 76-91; NIH res grants, 77-94; mem, Task Force Immunopath, Am Acad Dermat, 85-88, Task Force Manpower, 86-89 & Spec Grants Rev Comt, Nat Inst Arthritis & Musculoskeletal & Skin Dis, 88. *Mem:* Am Acad Dermat; fel Am Col Physicians; Am Dermat Soc Allergy & Immunol; Am Fedn Clin Res; AMA; Am Soc Cell Biol; Am Soc Dermat Allergy & Immunol; Am Venereal Dis Asn; Soc Invest Dermat; Am Dermat Asn. *Res:* Author of more than 150 technical publications. *Mailing Add:* Dept Dermat Univ NC Mem Hosp CB No 7600 Chapel Hill NC 27514

GAMMON, WILLIAM HOWARD, information science, for more information see previous edition

GAMO, HIDEYA, b Ueda, Japan, Apr 1, 24; m 56; c 2. OPTICAL PHYSICS, ELECTRONICS ENGINEERING. *Educ:* Univ Tokyo, BS, 46, DSc(physics), 58. *Prof Exp:* Res assoc & lectr, Univ Tokyo, 48-58; res consult physicist, T J Watson Res Ctr, Int Bus Mach Corp, 58-59, res physicist, 59-63; vis prof elec eng, Univ Rochester, 63-64, prof, 64-68; PROF ELEC ENG, UNIV CALIF, IRVINE, 68- *Concurrent Pos:* Mem adv comt, Elec, Computer & Systs Div, NSF, 79-81; overseas ed, Japanese J Appl Physics, 82-; specialist, Chinese Univ Develop Proj, World Bank, 85; vis prof, Japanese Nat Defense Acad, Xian Jiaotong Univ, China, 85 & Tokyo Univ Sci, 89. *Mem:* Sr mem Inst Elec & Electronics Engrs; fel Optical Soc Am; Am Phys Soc; Phys Soc Japan; AAAS; fel Soc Photo-Optical Instrumentation Engrs; Japanese Soc Appl Physics; Inst Electronic Info & Commun Engrs, Japan. *Res:* Faraday rotation in a circular wave guide; intensity correlation interferometry; statistical properties of laser radiation; laser propagation through turbulent atmosphere; optical-electronic devices such as infrared isolator and photovoltaic detector; gas laser plasma; optical communication and information processing; passive undirectional system; matrix theory of partial coherence; unconventional imaging optics; nonlinear optics of organic materials; physical optics of electron waves. *Mailing Add:* 3812 Inlet Isle Dr Corona Del Mar CA 92625

GAMOTA, GEORGE, b Lviv, Western Ukraine, May 6, 39; US citizen; m 61; c 3. LOW TEMPERATURE PHYSICS. *Educ:* Univ Minn, BPhysics, 61, MS, 63; Univ Mich, PhD(physics), 66. *Prof Exp:* Res assoc & lectr, Univ Mich, 66-67; mem tech staff, Bell Labs, Murray Hill, NJ, 67-75; res specialist, Off Under Secy Defense, Res & Eng, Dept Defense, 75-77, dir res, 78-81; PROF PHYSICS & DIR, INST SCI & TECHNOL, UNIV MICH, ANN ARBOR, 81- *Concurrent Pos:* Mem, Adv Coun Res, 73-75; mem, NJ Gov's Comn to Evaluate Capital Needs of NJ, 75; chmn & founder, Sci & Technol Coun for Congressman M Rinaldo, 74-75; exec secy, Defense Sci Bd Study Fundamental Res in Univs, 76; mem, Pres Sci Adv Fed Coord Coun on Sci & Eng Technol, 76-77; exec secy, Defense Shale Oil Task Group, 78; chmn, Defense Comt Res, 78; mem, Panel Pub Affairs, Am Phys Soc, 81-84; sr corp consult, Sci Appln Int Corp, 81-; mem, Nat Res Coun Off Sci & Eng Personnel Adv Panel, 84; mem, NSF Policy Res & Anal/Sci Resource Studies Adv Comt, 85; mem, Adv Subcomt Electronics, Space Systs & Technol Adv Comt, NASA, 79-, Res Designee Adv Comt, 83 & Space Commercialization Comt, 84; bd adv, Nat Coalition Sci & Technol, 81. *Mem:* Fel AAAS; Am Phys Soc; Ukrainian Am Engrs Soc; Sigma Xi; Inst Elec & Electronics Engrs; Am Defense Preparedness Asn; NY Acad Sci. *Res:* Solid state physics; hydrodynamics; liquid crystals; quantum liquids and solids; soliton like waves in helium; charged particle beams. *Mailing Add:* Mitre Corp Burlington Rd Mail Stop E078 Bedford MA 01730

GAMOW, RUSTEM IGOR, b Washington, DC, Nov 4, 35; m 61. BIOENGINEERING, PHYSIOLOGY. *Educ:* Univ Colo, BA, 61, MBS, 63, PhD(microbiol), 67. *Prof Exp:* Teaching asst phys sci, Univ Colo, 61-63; NIH res fel biol, Calif Inst Technol, 67-68; asst prof bioeng, 68-73, assoc prof aerospace, 73-76, ASSOC PROF CHEM ENG, UNIV COLO, BOULDER, 76- *Concurrent Pos:* Vis prof, Dept Biochem, Univ Fla, 76; NSF res grant, 78-80; Food & Drug Admin res grant, 78-81. *Mem:* Am Soc Microbiol; Am Soc Plant Physiol; Am Inst Chem Engrs; Biomed Eng Soc. *Res:* Bacteriophage, kinetics; conformational changes in proteins; light response of Phycomyces; exercise physiology. *Mailing Add:* 186 Canon Park Boulder CO 80302

GAMSON, BERNARD W(ILLIAM), b Chicago, Ill, Aug 18, 17; m 14, 52; c 4. CHEMICAL ENGINEERING. *Educ:* Ill Inst Technol, BS, 38; Univ Mich, MS, 39; Univ Wis, PhD(chem eng), 43. *Prof Exp:* Develop engr, Socony Vacuum Oil Co, NJ, 39-41; res engr, War Prod Bd, Wis, 42-43; group leader, Great Lakes Carbon Corp, 43-49, chief process engr, 49-51, dir res & develop, 51-55; mgr nuclear fuel, Gen Elec Co, 55-56; assoc dir, Res Ctr, Borg-Warner Corp, Ill, 56-60, vpres res & eng, oil field prod, Byron Jackson Div, Calif, 60-65; consult chem engr & scientist, 65-70; VPRES, HARVEY ALUMINUM CO, TORRANCE, 70- *Mem:* Am Chem Soc; Am Inst Chem Engrs. *Res:* Petroleum refining; petro- and heavy chemicals; metallurgy of iron, aluminum and light metals; carbon technology; thermodynamics; sulfur recovery; nuclear reactor technology; oil well logging; thermoelectrics. *Mailing Add:* 1124 Hunt Club Dr Potomac MD 20854

GAMZU, ELKAN R, b Liverpool, Eng, Jan 27, 43; US citizen; m 65; c 2. DRUG DEVELOPMENT. *Educ:* Hebrew Univ, Jerusalem, Israel, BA, 67; Univ Pa, MA, 68, PhD(psychiat), 71. *Prof Exp:* Sr pharmacologist, Hoffman-LaRoche, Nutley, NJ, 71-77, res group chief, 78-85; assoc dir clin res, Warner-Lambert, Ann Arbor, Mich, 85-87; sr dir clin therapeut, Parke-Davis/Warner-Lambert, 87-89, vpres drug develop, 89; vpres develop, 89-90, PRES & CHIEF OPER OFFICER, CAMBRIDGE NEUROSCI, 90- *Concurrent Pos:* Mem bd gov, NY Acad Sci, 85-86. *Mem:* Fel Am Psychol Asn; fel Am Acad Neurol; Soc Neurosci; Am Soc Pharmacol & Exp Therapeut. *Res:* Techniques to permit rapid, efficient development of safe and efficacious treatment of psychiatric and neurological diseases. *Mailing Add:* Cambridge Neurosci One Kendall Sq Bldg 700 Cambridge MA 02139

GAN, JOSE CAJILIG, b Iloilo, Philippines, Nov 30, 33; m 60; c 2. BIOCHEMISTRY. *Educ:* Univ Wis, BS, 57; Univ Iowa, MS, 59; Univ Ill, PhD(biochem), 64. *Prof Exp:* Asst res physiologist, Univ Calif, Berkeley, 64-66; asst res biochemist, Hormone Res Lab, Med Ctr, Univ Calif, San Francisco, 66-68; from asst prof to assoc prof, 68-74, PROF BIOCHEM, UNIV TEX MED BR GALVESTON, 80- *Concurrent Pos:* San Francisco Heart Asn fel, 65-67. *Mem:* AAAS; Am Soc Biol Chem; Soc Exp Biol & Med. *Res:* Plasma protein and amino acid metabolism; biosynthesis, chemistry and biological activity of glycoproteins. *Mailing Add:* Dept Human Biol Chem & Genetics Univ Tex Med Br Galveston TX 77550

GANAPATHY, RAMACHANDRAN, b Tellicherry, India, Jan 16, 39; US citizen; m 71; c 1. NUCLEAR & COSMOCHEMISTRY, INORGANIC & ANALYTICAL CHEMISTRY. *Educ:* Univ Madras, BSc, 59; Univ Ark, PhD(chem), 67. *Prof Exp:* Sci officer nuclear chem, Atomic Energy Estt India, 60-63; sr res assoc cosmochem, Enrico Fermi Inst, Univ Chicago, 67-76; BAKER FEL, J T BAKER INC, 76- *Mem:* Geochem Soc. *Res:* Cosmochemistry: composition of meteorites; search for major meteorite impacts on the earth; high-purity chemicals. *Mailing Add:* Res Lab J T Baker Inc Phillipsburg NJ 08865

GANAPATHY, SEETHA N, clinical nutrition, nutritional anthropology; deceased, see previous edition for last biography

GANAPOL, BARRY DOUGLAS, b San Francisco, Calif, May 15, 44. NUCLEAR ENGINEERING. *Educ:* Univ Calif, Berkeley, BS, 66 & PhD(eng sci), 71; Columbia Univ, MS, 67. *Prof Exp:* Nuclear engr reactor physics, Swiss Fed Inst Reactor Res, 71-72; vis scientist, Ctr Nuclear Studies, Saclay, France, 72-74; nuclear engr reactor safety, Argonne Nat Lab, 74-76; from asst prof to assoc prof, 76-85, PROF NUCLEAR ENG, UNIV ARIZ, 85- *Concurrent Pos:* Consult, E G & G, 90, Argonne Nat Lab, 76-78, Los Alamos Nat Lab, 79-85, Swiss Fed Inst Reactor Res, 80-81, Sci Appln Inc, 81- 83; NASA summer fel, 87-89; fel, Air Force Off Sci Res, 84; Meyerhoff vis prof, Weigmann Inst Sci. *Mem:* Am Nuclear Soc; Soc Indust & Appl Math; AAAS; Sigma Xi; Math Asn Am. *Res:* Particle transport theory; fast reactor safety; applied mathematics; remote sensing. *Mailing Add:* Dept Nuclear Eng & Energy Univ Ariz Tucson AZ 85721

GANAS, PERRY S, b Brisbane, Australia, June 20, 37. ELECTRON-ATOM COLLISIONS, NUCLEAR SHELL MODEL CALCULATIONS. *Educ:* Univ Queensland, BSc, 61; Univ Sydney, PhD(physics), 68. *Prof Exp:* From asst prof to assoc prof, 70-82, PROF PHYSICS, CALIF STATE UNIV, LOS ANGELES, 82- *Concurrent Pos:* Vis prof physics, Univ Fla, 79-80 & Univ Calif, Los Angeles, 87; lectr physics, Univ Southern Calif, 85-86; instr physics, Santa Monica Col, 88 & East Los Angeles Col, 89- *Res:* Atomic physics; nuclear physics; electron-atom collisions; nuclear shell model calculations; author of numerous publications on theoretical atomic, molecular and nuclear physics. *Mailing Add:* Physics Dept Calif State Univ Los Angeles CA 90064

GANAWAY, JAMES RIVES, b East St Louis, Ill, Jan 2, 27; m 50; c 3. VETERINARY MEDICINE, MICROBIOLOGY. *Educ:* Univ Mo-Columbia, BS & DVM, 53; Johns Hopkins Univ, MPH, 58; Am Bd Vet Pub Health, dipl, 60. *Prof Exp:* Base veterinarian, USAF, 53-56, assoc virologist, Sch Hyg & Pub Health, Johns Hopkins Univ, 56-58, virologist, Armed Forces Inst Path, Walter Reed Army Med Ctr, 58-61; MICROBIOLOGIST, COMP PATH SECT, LAB AIDS BR, DIV RES SERV, NIH, 61- *Mem:* Am Vet Med Asn; Am Soc Microbiol; Am Asn Lab Animal Sci. *Res:* Investigations of the naturally occurring diseases of laboratory animals and their comparative aspects with human diseases, with emphasis on the infectious diseases, their etiology, pathogenesis, control and/or prevention. *Mailing Add:* 11620 Moorestown Pl Potomac MD 22309

GANCARZ, ALEXANDER JOHN, b Miami, Fla, Jan 15, 48. TECHNICAL MANAGEMENT. *Educ:* Princeton Univ, AB, 70; Calif Inst Technol, PhD, 76. *Prof Exp:* Staff mem, Los Alamos Nat Lab, 77-79, assoc group leader, Isotope Geochem Group, 80-81, dep group leader, 82-84, dep div leader, Isotope & Nuclear Chem Div, 85-89, DIV LEADER, ISOTOPE & NUCLEAR CHEM DIV, LOS ALAMOS NAT LAB, 90- *Mem:* Am Chem Soc; Geochem Soc; Am Geophys Union. *Res:* Isotope geochemistry; nuclear weapons physics and diagnostics; natural analogues for high-level nuclear waste repositories; separations chemistry; high precision mass spectrometry. *Mailing Add:* Los Alamos Nat Lab MS J515 Los Alamos NM 87545

GANCHOFF, JOHN CHRISTOPHER, b Wauwatosa, Wis, Aug 15, 33; m 66; c 2. ANALYTICAL CHEMISTRY, INORGANIC CHEMISTRY. *Educ:* Marquette Univ, BS, 55, MS, 57; Ga Inst Technol, PhD(chem), 63. *Prof Exp:* Asst prof chem, Rutgers Univ, 62-66; asst prof, 66-74, PROF CHEM, ELMHURST COL, 74- *Concurrent Pos:* Rutgers Univ & Res Coun grants, 63-64. *Mem:* Am Chem Soc. *Res:* Titrations, especially with spectrophotometric end point location, in both aqueous and non-aqueous solvents using complexing agents as titrants. *Mailing Add:* Dept Chem Elmhurst Col Elmhurst IL 60126

GANCHROW, DONALD, b Brooklyn, NY, Dec 30, 40; m 67; c 2. NEUROANATOMY, NEUROPSYCHOLOGY. *Educ:* Brooklyn Col, BS, 61; Duke Univ, PhD(psychol), 69. *Prof Exp:* Nat Acad Sci-Nat Res Coun vis scientist, Behav Sci Div, Pioneering Res Lab, US Army Natick Labs, Mass, 69-71; NINDS spec fel anat, Univ Wis-Madison & Univ Calif, San Francisco, 71-74; from lectr to sr lectr & fel anat, Hadassah Med Sch, Hebrew Univ, Jerusalem, 75-85; SR LECTR ANAT, SACKLER FAC MED, TEL AVIV UNIV, 85- *Concurrent Pos:* Nat Inst Neurol Dis & Stroke spec res fel, 72-74; vis asst prof, Dept Anat, Univ Conn Med Ctr, Farmington, 80-81. *Mem:* Israel Asn Anatomists; Am Asn Anatomists; Sigma Xi; NY Acad Sci; Am Chem Soc. *Res:* Evolution of sensory systems in mammals; trigeminal and gustatory and somesthetic nervous systems; regeneration in central nervous system of mammals; anatomy, electrophysiology, behavior. *Mailing Add:* Dept Anat & Anthropol Sackler Fac Med Tel Aviv Univ Ramat Aviv 69978 Tel Aviv Israel

GANCY, ALAN BRIAN, b New Haven, Conn, Jan 28, 32. PHYSICAL CHEMISTRY, INORGANIC CHEMISTRY. *Educ:* Trinity Col, BS, 53; Yale Univ, MS, 54, PhD(phys chem), 56; Syracuse Univ, MBA, 79. *Prof Exp:* Res chemist, Am Cyanamid Co, Conn, 56-57; from res chemist to sr res chemist, FMC Corp, NJ, 57-65; from res chemist to sr res chemist, Tyco Labs, Inc, 65-70; dir tech develop, Indust Chem Div, Allied Chem Corp, 70-74, dir advan technol, 74-79; consult chemist, 81; PRES, GANCY CHEM CORP, 82- *Concurrent Pos:* Adj prof orgn & mgt, Syracuse Univ, 79, vis prof, 80. *Mem:* Am Chem Soc; Electrochem Soc; AAAS; Asn Consult Chemists; Am Inst Chemists. *Res:* Development, testing, manufacture and sale of specialty chemicals, including environmentally safe proprietary deicing agents. *Mailing Add:* 8810 Wandering Way Baldwinsville NY 13027-1513

GANDELMAN, RONALD JAY, b New Haven, Conn, Feb 20, 44. PSYCHOBIOLOGY. *Educ:* Univ Pittsburgh, BS, 66; Univ Mass, MS, 67, PhD(psychol), 69. *Prof Exp:* Asst prof, 71-74, ASSOC PROF PSYCHOL, RUTGERS UNIV, 74- *Concurrent Pos:* Nat Inst Child Health & Human Develop fel, NIH, Univ Conn, 69-71. *Mem:* Int Soc Res Aggression; Int Soc Develop Psychobiol; Int Acad Sex Res; Int Soc Psychoneuroendocrinol. *Res:* Hormones and behavior; developmental psychobiology. *Mailing Add:* Dept Psychol Busch Campus Rutgers Univ New Brunswick NJ 08903

GANDER, FREDERICK W(ILLIAM), b New York, NY, Jan 14, 21; m 42; c 3. CHEMICAL ENGINEERING. *Educ:* Mass Inst Technol, SB, 42, SM, 46. *Prof Exp:* Res eng, Yerkes Res Lab, 46-53, res supvr, 53-55, res mgr, 55-57, lab dir, 57-62, res & develop mgr, 62-70, RES FEL, FILM DEPT, EXP STA, E I DU PONT DE NEMOURS & CO, INC, 70- *Mem:* Am Chem Soc; Am Inst Chem Engrs. *Res:* Polymer synthesis; chemical modification of polymers; polymer fabrication. *Mailing Add:* 1156 Mews Lane West Chester PA 19382-2077

GANDER, GEORGE WILLIAM, b Hamilton, Mont, June 27, 30; m 58; c 3. EXPERIMENTAL PATHOLOGY. *Educ:* Mont State Col, BS, 53; Cornell Univ, MS, 55, PhD(dairy chem), 59. *Prof Exp:* Res assoc dairy chem, Univ Conn, 59-61; res assoc path, Albany Med Col, 61-63; from asst prof to assoc prof, 63-74, PROF EXP PATH, MED COL, VA COMMONWEALTH UNIV, 74- *Mem:* Am Chem Soc; Am Soc Exp Path; NY Acad Sci; Reticuloendothelial Soc; Sigma Xi. *Res:* Mechanism of action of pyrogens; computers in clinical laboratory. *Mailing Add:* Dept Acad Path Med Col Va Commonwealth Univ Richmond VA 23298

GANDER, JOHN E, b Roundup, Mont, Mar 9, 25; m 51; c 3. BIOCHEMISTRY. *Educ:* Mont State Univ, BSc, 50; Univ Minn, MSc, 54, PhD(agr biochem), 56. *Prof Exp:* Asst prof chem, Mont State Univ, 55-58; from asst prof to assoc prof biochem, Ohio State Univ, 58-64; from assoc prof to prof biochem, Univ Minn, St Paul, 68-84; prof & chmn, 84-89, PROF MICROBIOL & CELL SCI, UNIV FLA, GAINESVILLE, 89- *Concurrent Pos:* Res career develop award, USPHS & NIH, 65-71. *Mem:* Am Soc Biol Chem & Molecular Biologists; Am Chem Soc; Am Soc Microbiologists; AAAS; Am Soc Plant Physiologists; Sigma Xi. *Res:* Structure, function and biosynthesis of galactofuranosyl-containing fungal glycoproteins and glycopeptides; mechanism of galactofuranosidase catalyzed reaction. *Mailing Add:* Microbiol & Cell Sci Dept Univ Fla 1059 McCarty Hall Gainesville FL 32611

GANDER, ROBERT JOHNS, b Eagle River, Wis, Sept 12, 18; m 48; c 2. POLYMER CHEMISTRY, ORGANIC CHEMISTRY. *Educ:* Univ Wis, BS, 40, MS, 42; Univ Ill, PhD(org chem), 44. *Prof Exp:* Asst Nat Defense Res Comt proj, Univ Wis, 40-42; Off Sci Res & Develop synthetic rubber proj, Univ Ill, 42-45; chemist, Firestone Plastics Co, 45-50; asst dir surg adhesives res, Johnson & Johnson Prod Inc, 50-66, mgr polymer res, Domestic Operating Co, 66-78, sr res fel, 78-87; RETIRED. *Mem:* Am Chem Soc. *Res:* Monomer and polymer synthesis; technology of vinyl chloride resins, plasticized vinyl films, pressure-sensitive adhesives, biomedical plastics, polymeric dental materials; retinoic acid derivatives. *Mailing Add:* PO Box 82 Whitehouse NJ 08888-0082

GANDERS, FRED RUSSELL, b Bremerton, Wash, Mar 10, 45. BIOSYSTEMATICS, POPULATION BIOLOGY. *Educ:* Wash State Univ, BA & BS, 67; Univ Calif, Berkeley, MA, 69, PhD(bot), 75. *Prof Exp:* From asst prof to assoc prof, 73-86, PROF BOT, UNIV BC, 86- *Concurrent Pos:* Vis prof, Bot, Univ Hawaii, 89-90. *Mem:* Soc Study Evolution; Bot Soc Am; Am Soc Plant Taxonomists (secy, 79-80). *Res:* Biosystematics and evolution of plants, effects of mating systems on genetic variability in natural plant populations, the function and evolution of mating systems in plant populations; adaptive radiation and evolution of Hawaiian Bidens. *Mailing Add:* Dept Bot Univ BC No 3529-6270 University Blvd Vancouver BC V6T 2B1 Can

GANDHI, HARENDRA SAKARLAL, b Calcutta, India, May 2, 41; m 66; c 2. PHYSICAL CHEMISTRY, CHEMICAL ENGINEERING. *Educ:* Univ Bombay, BChE, 63; Univ Detroit, MS, 66, DEng, 71. *Prof Exp:* Chem engr eng equipment design, Shwayder Chem, 64-65; specialty coating chemist, Specialty Coating Inc, 65-67; MGR CHEM ENG DEPT, SCI RES LABS, FORD MOTOR CO, 67- *Concurrent Pos:* Adv, UN/Govt India Proj on Emission Control from Indian vehicles; adj prof chem eng, Wayne State Univ. *Honors & Awards:* Ralph R Teetor Award, Soc Auto Engrs. *Mem:* Am Inst Chem Engrs; Am Chem Soc. *Res:* Heterogeneous catalysis; reaction kinetics; pollution prevention and coating chemistry; author of over 50 publications. *Mailing Add:* Sci Res Labs PO Box 2053 Dearborn MI 48121

GANDHI, OM P, b Multan, India, Sept 23, 34; US citizen; m 63; c 3. ELECTROMAGNETIC FIELDS, MICROWAVES. *Educ:* Univ Delhi, BS, 52; Indian Inst Sci, Bangalore, dipl elec commun eng, 55; Univ Mich, MSEE, 57, ScD(elec eng), 60. *Prof Exp:* Assoc res engr microwave tubes, Electron Physics Lab, Univ Mich, 58-60; res specialist solid-state devices, Philco Corp, 60-62; dep dir & head microwave devices, Cent Electronics Eng Res Inst, Pilani, India, 62-66; assoc prof, elec eng, 67-73 res prof bioeng, 75-78, PROF ELEC ENG, UNIV UTAH, 73- *Concurrent Pos:* Consult, Walter Reed Army Inst Res, 73-77 & 90, BSD Corp, 77-81, Equitable Environ Health, Inc, 78-80, Dosimeter Corp, 79-80 & 84-86, Remic Corp, 82, Hughes Aircraft Co & Univ Utah Res Inst, 83-, G E Milwaukee, 86-87, NIOSH, 86- & EG&G, Idaho, 86-88, SCEEE 82, 89-90, ERC Bioserv Corp, 90-91; mem, Int Union Radio Sci Comn B, 77-, Diag Radiol Sect, NIH, 78-81, Panel Eng Aspects of Pecision Acquisition of Vehicle Entry Phased Array Warning Syst, Air Force Radar Systs, Nat Acad Eng, 78-79, Panel Biol Effect Precision Acquisition of Vehicle Entry Phased Array Warning Syst, 78-79, chmn, Comt Man & Radiation, Inst Elec & Electronics Engrs, 80-82; mem, Panel Assessment Possible Health Effects Ground Wave Emergency Network (GWEN) Nat Res Coun, Nat Acad Sci 90-91; co-chmn, Inst Elec & Electronic Engrs SCC 28 Subcomt IV Safety Levels Radiation 3KHZ-3006HZ, 88- *Honors & Awards:* Distinguished Res Award, Univ Utah, 79. *Mem:* Bioelectromagnetic Soc; Int Microwave Power Inst; fel Inst Elec & Electronics Engrs. *Res:* Electromagnetic energy biological effects; biomedical and energy applications of microwaves, antennas and microwave communications; author over 200 journal articles. *Mailing Add:* Dept Elec Eng 3032 Merrill Eng Bldg Salt Lake City UT 84112

GANDLER, JOSEPH RUBIN, b Brooklyn, NY, Dec 2, 49; m 74; c 2. PHYSICAL ORGANIC CHEMISTRY. *Educ:* Brooklyn Col, BS, 71; Univ Calif, Santa Cruz, PhD(chem), 78. *Prof Exp:* NIH fel, State Univ NY Buffalo, 78-79, Brandeis Univ, 79-81; asst prof, 81-84, ASSOC PROF CHEM, CALIF STATE UNIV, FRESNO, 84- *Mem:* Am Chem Soc. *Res:* Acid-base catalysis; beta-elimination and proton transfer reactions. *Mailing Add:* Dept Chem Calif State Univ Fresno CA 93740

GANDOLFI, A JAY, b San Mateo, Calif, Dec 11, 46; m 68; c 3. TOXICOLOGY. *Educ:* Univ Calif, Davis, BA, 68; Ore State Univ, PhD(biochem), 72. *Prof Exp:* Trainee drug metab, Ore State Univ, 68-72; resident fel anesthetic toxicol, Mayo Clin Found, 72-75; sr res scientist inhalation toxicol, Pac Northwest Labs, Battelle Mem Inst, 75-77; res asst prof toxicol & anesthesiol, 78-85, assoc prof, 85-89, PROF ANESTHESIOL, PHARMACOL & TOXICOL, UNIV ARIZ, 89- *Concurrent Pos:* Res fel, Mayo Clin Found, 74-75. *Mem:* Soc Toxicol; Am Soc Pharmacol & Exp Therapeut; Am Soc Anesthesiol. *Res:* Biochemical pharmacology and toxicology of foreign compounds to man and animals with emphasis on the metabolism and disposition of the foreign agent and effect of agent on structure and function of endogenous biological components. *Mailing Add:* Univ Ariz Dept Anesthesiol Tucson AZ 85724

GANDOUR, RICHARD DAVID, b Sistersville, WVa, Feb 12, 45; m 71, 85; c 1. PHYSICAL ORGANIC CHEMISTRY, BIO-ORGANIC CHEMISTRY. *Educ:* Wheeling Jesuit Col, BS, 67; Rice Univ, PhD(org chem), 72. *Prof Exp:* From teaching assoc to res assoc chem, Univ Kans, 71-75; asst prof chem, 75-80, assoc prof 80-87, PROF CHEM, LA STATE UNIV, BATON ROUGE, 87- *Concurrent Pos:* Vis scientist, Ind Univ, 89-90; vis prof, Univ Calif, Los Angeles, 90. *Mem:* Am Chem Soc; fel AAAS; Sigma Xi; Am Soc Biochem & Molecular Biol. *Res:* Bio-organic chemical dynamics; design, synthesis and mechanistic studies of chemical models of enzymatic catalysis; structural bioorganic chemistry; molecular recognition; chemistry and biology of carnitine. *Mailing Add:* Dept Chem La State Univ Baton Rouge LA 70803-1804

GANEM, BRUCE, b Boston, Mass, Feb 7, 48; m. SYNTHETIC ORGANIC & NATURAL PRODUCTS CHEMISTRY. *Educ:* Harvard Univ, BA, 69; Columbia Univ, PhD(org chem), 72. *Prof Exp:* Fel chem, Stanford Univ, 73-74; from asst prof to assoc prof, 74-80, PROF CHEM, CORNELL UNIV, 80- *Mem:* Am Chem Soc; fel Royal Soc Chem. *Res:* Methods and reactions, including new organic reagents, for the synthesis of rare and natural products and biologically important molecules. *Mailing Add:* Dept Chem Cornell Univ Ithaca NY 14853

GANESAN, ADAYAPALAM T, b Madras, India, May 15; m 63. MOLECULAR BIOLOGY, GENETICS. *Educ:* Annamalai Univ, Madras, MA, 53; Stanford Univ, PhD(genetics), 63. *Prof Exp:* Res fel biochem, Indian Inst Sci, Bangalore, 53-55; res asst bot, Indian Agr Res Inst, New Delhi, 55-57; res assoc, 63-64, from asst prof to assoc prof, 65-76, PROF GENETICS, SCH MED, STANFORD UNIV, 77- *Concurrent Pos:* Res fel physiol, Carlesberg Lab, Denmark, 57-59; mem, Rask-Orstead Found, 57-58; NIH res grants, 66-92; NIH Career Develop Awards, 70-75. *Mem:* Genetics Soc Am; AAAS. *Res:* Cytogenetics; molecular biology of DNA replication and recombination; cell fusion; human biochemical genetics. *Mailing Add:* Genetics Dept Stanford Med Sch Stanford CA 94305

GANESAN, ANN K, b Denver, Colo, July 25, 33; m 63. MOLECULAR GENETICS. *Educ:* Wilson Col, BA, 54; Univ Wis, MS, 59; Stanford Univ, PhD(genetics), 61. *Prof Exp:* Res assoc microbiol, Palo Alto Med Res Found, 61-62; res assoc biochem, Syntex Inst Molecular Biol, 62-65; NIH fel radiol, Med Ctr, 65-66, res assoc, 66-71, RES ASSOC BIOL, STANFORD UNIV, 71- *Concurrent Pos:* Mem, NIH Study Sect, 86-90. *Mem:* Am Soc Microbiol; Genetics Soc Am. *Res:* Biochemical genetics of bacteria and cultured mammalian cells. *Mailing Add:* Dept Biol Stanford Univ Stanford CA 94305-5020

GANESAN, DEVAKI, b Bangalore, India, Feb 1, 40; m 67; c 2. INTERNAL MEDICINE, PHARMACOLOGY. *Educ:* Univ Mysore, India, MBBS, 63; Univ Poona, India, MD, 66. *Prof Exp:* Rotating intern med, surg, obstet & gynec, Bangalore Med Col, India, 62, asst surgeon, 63; lectr pharmacol, Armed Forces Med Col, 63-66; fel cardiovasc, Okla Med Res Found, 67-72, res assoc, Dept Biochem, 70-85, staff scientist, 72-74, asst mem & instr cardiovasc, 74-85; MED DIR, GEN MOTORS, 85- *Mem:* Am Fedn Clin Res; fel Am Heart Asn. *Res:* Lipid transport and lipoprotein metabolism in relation to atherogenesis, hyperlipidemias and physiology of fat absorption; characterization of lipolytic enzymes. *Mailing Add:* 7447 SE 74th Box 26527 Oklahoma City OK 73126

GANFIELD, DAVID JUDD, b Dubuque, Iowa, Jan 21, 41; m 64; c 2. IMMUNOLOGY, BIOCHEMISTRY. *Educ:* Parsons Col, BS, 63; Iowa State Univ, PhD(biochem), 71. *Prof Exp:* Chemist biochem, Nat Animal Dis Ctr, USDA, 63-71; from instr to asst prof biochem, Med Sch, Thomas Jefferson Univ, 71-77; res scientist immunol, Exp Sta, E I Du Pont De Nemours & Co Inc, 77-80, res scientist immunol, Du Pont Glenolden Lab, Pa, 80-82; dir labs, 83-87, PRES, DELMONT LABS, INC, SWARTHMORE, PA, 87- *Concurrent Pos:* HEW trainee immunol, Thomas Jefferson Univ, 71-74, adj asst prof biochem, 77-82. *Mem:* AAAS; Am Chem Soc; Am Soc Microbiol; Sigma Xi; Parenteral Drug Asn. *Res:* Mechanisms of lymphocyte regulation affecting the immune response; characterization of lymphocyte receptors; separation of lymphocyte subpopulations; genetic control of the immune response; characterization of immunomodulators therapeutic use of staphylocuccus aurens antigens. *Mailing Add:* 2233 Jamaica Dr Wilmington DE 19810

GANGAROSA, EUGENE J, b Rochester, NY, Aug 7, 26; m 50; c 4. EPIDEMIOLOGY. *Educ:* Univ Rochester, AB, 50, MD, 54, MS, 55. *Prof Exp:* Asst prof med & microbiol, Sch Med, Univ Md, 61-64; chief, Epidemic Intel Serv, Ctr Dis Control, 64-65, chief, Enteric Dis Sect, Bact Dis Br, Epidemiol Prog, 65-70, chief, Enteric Dis Br & dep dir, Bact Dis Div, Epidemiol Br, Ctr Dis Control, 70-78; dean fac, Health Sci, Am Univ Beirut, Lebanon, 78-81; prof & dir, MPH Prog, 82-89, PROF, INT HEALTH, EMORY UNIV, 89- *Concurrent Pos:* Vis assoc prof, Jefferson Med Col, 62; dir, Pakistan Med Res Ctr, Lahore, 62-64; consult, 81- *Mem:* Infectious Dis Soc; Am Epidemiol Soc. *Res:* Clinical investigation; cholera; shigellosis; salmonellosis; botulism; food poisoning; traveler's diarrhea, gastroenterology; hematology; drug toxicity. *Mailing Add:* Sch Pub Health Emory Univ 1599 Clifton Rd NE Atlanta GA 30322

GANGAROSA, LOUIS PAUL, SR, b Rochester, NY, June 8, 29; m 50; c 4. PHARMACOLOGY. *Educ:* Univ Rochester, BA, 52, MS, 61, PhD(pharmacol), 65; Univ Buffalo, DDS, 55. *Prof Exp:* Pvt pract, 55-61; asst prof dent res, Univ Rochester, 65-68, from instr to asst prof pharmacol, 65-68; assoc prof oral biol, 68-71, assoc prof pharmacol, 68-72, PROF ORAL BIOL, MED COL GA, 71-, PROF PHARMACOL, 72- *Concurrent Pos:* Lectr, Eastman Dent Ctr, 63-65, clin res assoc, 65-68; res grant, Nat Inst Dent Res, 67-71 & 81-83, consult training grant comt, 69-71, mem training comt, 71-73; chmn OTC panel on dentrices & dent care agents, Food & Drug Admin, 73-78. *Honors & Awards:* C V Mosby Award, 55. *Mem:* Am Soc Pharmacol & Exp Therapeut; Int Asn Dent Res; Am Dent Asn; Am Asn Dent Sch; Soc Exp Biol & Med; fel Am Col Dentists. *Res:* Dental research; local anesthetics; sympathomimets; dental drugs; Iontophoretic medication; antiviral chemotherapy; chronic pain. *Mailing Add:* Sch Dent Med Col Ga Augusta GA 30912- 1128

GANGE, RICHARD WILLIAM, b Hove, England, July 24, 45. PHOTOBIOLOGY. *Educ:* Cambridge Univ, Eng, MB, 69; Royal Col Physicians, London, MRCP, 72; Am Bd Dermat, cert, 87. *Prof Exp:* ASSOC PROF DERMAT, HARVARD UNIV, 86- *Mem:* Am Acad Dermat; Am Soc Dermat; Am Soc Photobiol; Soc Investigative Dermatopath. *Res:* Cutaneous effects of ultraviolet radiation; mechanisms of photoxicity. *Mailing Add:* Dept Dermat Wellman Lab Mass Gen Hosp Boston MA 02114

GANGEMI, FRANCIS A, b Syracuse, NY, Feb 20, 29; m 78; c 4. MOLECULAR PHYSICS, QUANTUM PHYSICS. *Educ:* Univ Notre Dame, BS, 54; Cath Univ Am, MS, 59, PhD(physics), 62. *Prof Exp:* Asst prof physics, Univ Portland, 62-65; assoc prof, Southern Ore Col, 65-67; assoc prof, 67-72, PROF PHYSICS & CHMN DEPT, OHIO NORTHERN UNIV, 72- *Concurrent Pos:* NSF grant, Cornell Conf Relativity, 63; NSF res grant, 63-64. *Mem:* Am Phys Soc; The Chem Soc. *Res:* Quantum chemistry; theoretical calculation of dipole movements of polyatomic molecules; particle physics; philosophy of science; improving teaching techniques. *Mailing Add:* Dept Physics Ohio Northern Univ Ada OH 45810

GANGI, ANTHONY FRANK, b Newark, NJ, Feb 19, 29; wid; c 4. GEOPHYSICS, SEISMOLOGY. *Educ:* Univ Calif, Los Angeles, BS, 53, MS, 54, PhD(physics, geophys), 60. *Prof Exp:* Sr electronics technician, Inst Geophys, Univ Calif, Los Angeles, 54-60; mem tech staff, Space Electronics Corp, Calif, 60-62; mgr antenna dept, Space-Gen Corp, 62-64; assoc prof geophys, Mass Inst Technol, 64-67; assoc prof, 67-70, actg head dept, 76-78, PROF GEOPHYS, TEX A&M UNIV, 70- *Concurrent Pos:* Shell Fel Geophys, 56-57; NATO vis prof, Instituto Nazionale de Geofiscia, Rome, 75-76; mgr, Tamu Seismic Studies Lab, 73-77; vis prof geophys, Univ Fed Bahia, Salvador, Brazil, 81-82. *Honors & Awards:* Basic Res Award Rock Mechanics, US Nat Comt, 79. *Mem:* Inst Elec & Electronics Engrs; Am Geophys Union; Seismol Soc Am; Soc Explor Geophys; Sigma Xi. *Res:* Seismology; elastic- and radio-wave propagation; antenna and antenna-array theory; boundary-value problems and mathematical physics; theoretical geophysics; theoretical tectonophysics and rock mechanics; data analysis. *Mailing Add:* Dept Geophys Tex A&M Univ College Station TX 77843-3114

GANGJEE, ALEEM, b Calcutta, India, July 30, 48; US citizen; m 72; c 2. MEDICINAL CHEMISTRY. *Educ:* Indian Inst Technol, BS, 69, MS, 71; Univ Iowa, PhD(med chem), 75. *Prof Exp:* Fel biochem, Univ Iowa, 75-76; trainee med chem, State Univ NY Buffalo, 76-79; from asst prof to assoc prof, 79-88, PROF MEDICINAL CHEM, SCH PHARM, DUQUESNE UNIV, PITTSBURGH, 88- *Mem:* Am Chem Soc; Am Asn Pharmaceut Scientists; Sigma Xi; NY Acad Sci. *Res:* Medicinal organic chemistry; structure-activity relationships; design of biologically active compounds; antitumor agents related to folates; cardiovascular agents and cholinergic agents; synthetic organic chemistry, stereochemistry and heterocyclic chemistry. *Mailing Add:* Sch Pharm Duquesne Univ Pittsburgh PA 15282

GANGLOFF, PIERRE, b Mulhouse, France, July 20, 41; Can citizen; m 65; c 5. GEOMORPHOLOGY. *Educ:* Univ Strasbourg, BA, 64, PhD(geomorphol), 70. *Prof Exp:* Chmn, dept geog, 77-78, PROF GEOMORPHOL, UNIV MONTREAL, 65- *Concurrent Pos:* Ed, J Geog Physique et Quaternaire, 79. *Res:* Holocene geomorphological evolution of landscapes; quaternary paleogeography; geomorphological mapping; ecological mapping. *Mailing Add:* Dept Geog Univ Montreal 520 Chaemim de La Cot-Ste-Cathrine Montreal PQ H2V 2B8 Can

GANGLOFF, RICHARD PAUL, b Pittsburgh, Pa, Aug 4, 48; m 70; c 1. MECHANICAL ENGINEERING. *Educ:* Lehigh Univ, BS, 70, MS, 72, PhD(mat sci), 74. *Prof Exp:* Metallurgist, Corp Res & Develop, Gen Elec Co, 74-80; metallurgist, Corp Sci Labs, Exxon Corp, 80-86; PROF MAT SCI & ENG, UNIV VA, 86- *Honors & Awards:* Henry Marion Howe Medal, Am Soc Metals; Award of Merit, Am Soc Testing & Mat. *Mem:* Am Soc Metals; Am Inst Mining, Metall & Petrol Engrs; fel Am Soc Testing & Mat; Sigma Xi. *Res:* Hydrogen embrittlement; fatigue; fracture mechanics; superalloy steels; nuclear materials; deformation and fracture; stress corrosion cracking; liquid metal embrittlement. *Mailing Add:* Dept Mat Sci Thornton Hall Univ VA Charlottesville VA 22903-2442

GANGSTAD, EDWARD OTIS, b Chippewa Falls, Wis, Dec 18, 17; m 45; c 4. ENVIRONMENTAL SCIENCES. *Educ:* Univ Wis, BS, 42, MS, 47; Rutgers Univ, PhD(agron), 50. *Prof Exp:* Asst biochem, Univ Wis, 46-47; asst agron, Rutgers Univ, 47-50; agronomist, Everglade Exp Sta, USDA, Fla, 50-51, res agronomist, 51-53, assoc agronomist, 53-54; from agronomist to prin agronomist, Tex Res Found, 54-66; mgt agronomist, Off Chief Engrs, US Army, Resource Mgt, Opers Div, Washington, DC, 66-69, chief, Aquatic Plant Control Prog, 69-75, botanist, 75-85; RETIRED. *Mem:* AAAS; Am Soc Agron; Weed Sci Soc Am; Am Inst Chem Engrs; Am Soc Civil Engrs. *Res:* Plant growth, breeding, production and utilization; forage crops for livestock production in the Southwest; management of military lands; aquatic plant control; pest control program. *Mailing Add:* 7909 Greeley Blvd Springfield VA 22152

GANGULEE, AMITAVA, b Rajshahi, India, Apr 26, 41. PHYSICAL METALLURGY, SOLID STATE PHYSICS. *Educ:* Univ Calcutta, BE, 61; Brown Univ, ScM, 64; Mass Inst Technol, ScD(metall, mat sci), 67. *Prof Exp:* Metallurgist, Steel & Allied Prod Ltd, India, 61-62; consult, Metal Eng & Treatment Co, 62-63; mem staff, div sponsored res, Mass Inst Technol, 67; sr assoc metallurgist, components div, 67-68, RES STAFF MEM, T J WATSON RES CTR, IBM CORP, 68- *Mem:* Am Phys Soc; Am Crystallog Asn; Am Inst Mining, Metall & Petrol Engrs; Am Vacuum Soc. *Res:* Magnetic and transport properties of solids; thermodynamics; x-ray diffraction; thin films; relation of structure with properties of solids. *Mailing Add:* 12 Hayden Lane Bedford MA 01730

GANGULI, AMAL(ENDU), metallurgy, for more information see previous edition

GANGULI, MUKUL CHANDRA, b Comilla, Bengal, India, Feb 28, 38; US citizen; m 70; c 3. HYPERTENSION, RENAL PHYSIOLOGY. *Educ:* Univ Calcutta, India, BSc, 56, BVSc & AH, 61; Agra Univ, India, MVSc, 63; Univ Minn, PhD(nutrit & biochem), 68. *Prof Exp:* Tech asst physiol, Nat Dairy Res Inst, India, 63-64; res asst nutrit, Univ Minn, 64-68; assoc nutrit, Iowa State Univ, 68-70; RES ASSOC NUTRIT, UNIV MINN, 70- *Mem:* Sigma Xi; Am Physiol Soc. *Res:* Sodium and prostaglandins and their relationship to the kidney and hypertension; protective role of dietary potassium and dietary fatty acids against sodium chloride hypertension; measurement of renal medullary blood flow and cardiac output in different types of experimental rat hypertension. *Mailing Add:* Dept Med Univ Minn Box 78 Mayo Minneapolis MN 55455

GANGULY, ASHIT K, b New Delhi, India, Aug 9, 34; m 66; c 1. RESOURCE MANAGEMENT. *Educ:* Delhi Univ, BSc(Hons), 53, MSc, 55, PhD(chem), 59; Univ London, PhD(chem), 62. *Prof Exp:* Sr scientist, Schering-Plough Corp, 68-70, prin scientist, 70-71, res fel, 71-79, assoc dir, 79-82, dir, 82-84, VPRES, SCHERING-PLOUGH CORP, 84- *Concurrent Pos:* Adj prof, Stevens Inst Technol, Hoboken, NJ, 74-; mem, Med Chem Study Sect, NIH, 86-; mem bd trustees, Bloomfield Col, NJ, 89- *Honors & Awards:* Khaira Lectr, Indian Asn Cultivation Sci, 75; Seshadri Mem Award, Delhi Univ, 82; Charles Sabat Lectr, Rutgers Univ, 87; Outstanding Scientist Award, Asn Scientists Indian Origin Am, 91. *Mem:* Fel Royal Soc Chem; Am Chem Soc; NY Acad Sci. *Res:* Structural elucidation and synthesis of natural products of biological importance; biosynthesis, mutasynthesis and synthesis of macrolide and penem antibodies; synthesis of dual antagonists of platelet aggregation factor and histamine; antiviral and antitumor research. *Mailing Add:* Schering-Plough Corp 60 Orange St Bloomfield NJ 07003-4795

GANGULY, BISHWA NATH, b Patna, India, Jan 2, 42; US citizen; m 66; c 2. PHYSICS. *Educ:* Patna Univ, BSc, 60, MSc, 62; Poona Univ, PhD(physics), 66. *Prof Exp:* Fel, Univ Toronto, 67-68; res assoc, La State Univ, Baton Rouge, 68-69; scientist solid state physics, Oak Ridge Nat Lab, 69-71; res asst prof physics, Univ Ill, Urbana, 71-73; SCIENTIST XEROGRAPHY, JOSEPH C WILSON CTR TECHNOL, XEROX CORP, 73- *Mem:* Am Phys Soc; Can Asn Physicists. *Res:* Theory of electronic states and conduction mechanisms in amorphous materials, particularly in organic crystals and polymers; superconductivity in metal hydrides; photo-generation and -conduction in amorphous materials. *Mailing Add:* 51 Parkhurst Rd Dayton OH 45440

GANGULY, JIBAMITRA, b Calcutta, India, Oct 24, 38; m 66; c 2. GEOLOGY, GEOCHEMISTRY. *Educ:* Univ Calcutta, BSc, 58; Jadavpur Univ, India, MSc, 60; Univ Chicago, PhD(geophys sci), 67. *Prof Exp:* Sci off, Atomic Energy Estab, India, 61-62; mem res staff, Yale Univ, 67-69, Jadavpur Univ, India, 69-71, & Birla Inst Technol & Sci, India, 71; mem res staff, Inst Geophys & Planetary Physics, Univ Calif, Los Angeles, 72-75; from asst prof to assoc prof, 75-86, PROF, DEPT GEOSCI, UNIV ARIZ, 86- *Concurrent Pos:* mem Comm on Thermodynamics, Int Mineral Asn. *Mem:* Am Geophys Union; fel Mineralogical Soc Am. *Res:* Phase equilibria and kinetic studies in geochemical systems; thermodynamics and crystal chemistry of rock forming minerals. *Mailing Add:* Dept Geosci Univ Ariz Tucson AZ 85721

GANGULY, PANKAJ, b Calcutta, India, Dec 31, 39; m 70; c 1. HEMATOLOGY, CELL PHYSIOLOGY. *Educ:* Univ Calcutta, BSc, 59, MSc, 61, PhD(biophys), 65. *Prof Exp:* Res assoc biophys, Blood Res Lab, Am Nat Red Cross, Washington, DC, 66-67; res scientist phys biochem, Am Red Cross, Bethesda, Md, 67-71; assoc mem hematol, Univ Tenn Ctr Health Sci, Memphis, 72-81, prof biochem, 82-84; prof biochem, Univ PR, San Juan, 84-87; PROG DIR, BLOOD DIS & RESOURCES, NIH, BETHESDA, MD, 87- *Concurrent Pos:* Fel Biophys Div, Saha Inst Nuclear Physics, Calcutta, 65-66; assoc prof biochem, Univ Tenn Ctr Health Sci, Memphis, 72- *Mem:* AAAS; Am Heart Asn; Biophys Soc; Am Soc Biol Chemists; Am Soc Hematol. *Res:* Blood platelets, their role in hemostasis and thrombosis blood coagulation factors; clinical conditions membrane receptors; protein structure. *Mailing Add:* Div Blood Dis & Resources NIH Fed Bldg 5A12 Bethesda MD 20892

GANGULY, RAMA, b Calcutta, India. BIOCHEMISTRY. *Educ:* Univ Calcutta, BSc, 57, MSc, 59, PhD(biochem), 64. *Prof Exp:* Res fel biochem, Cent Sci & Indust Res, Govt India, 59-64, from res officer to sr res officer immunol, Indian Coun Med Res, 64-71; assoc immunol, Dept Med & Med Microbiol, Col Med, Univ Fla, 71-74, instr, 75-76; asst prof to assoc prof infectious dis, Sch Med, WVa Univ, 76-81; ASSOC PROF MED, UNIV SOUTH FLA, 81- *Concurrent Pos:* Mem spec grants rev comt, Nat Inst Dent Res, NIH, 82-86. *Mem:* Am Soc Microbiol; Indian Sci Cong; Indian Soc Microbiol; Am Asn Immunologists; Am Acad Microbiol; Infectious Dis Soc Am. *Res:* Infectious diseases and immunology; cell-mediated immunity and secretory immunology; action of antibiotics on the immunologic functions of the host; development of vaccines; marrophage functions, nutrition and aging. *Mailing Add:* 11814 Raintree Lake Lane Apt C Tampa FL 33617-9413

GANGULY, SUMAN, b Dacca, India, Feb, 9, 42; US citizen. IONOSPHERIC RESEARCH, RADIO SCIENCE. *Educ:* Calcutta Univ, BS, 60, MS, 62, PhD(physics), 70. *Prof Exp:* Res fel ionospheric res, Bose Res Inst, Calcutta, 62-69, Delhi Univ, 70-72; sr res fel digital instruments, Inst Radiophysics & Electronics, Calcutta, 72-74; Leverhulme fel, Royal Soc London, satelite beaconats, Lancaster Univ, Eng, 74-76; sr res assoc ionospheric res, Arecibo Observ, PR, 76-79, Rice Univ, Houston, Tex, 79-85; PRES & SCIENTIST, PHYSICS & ELECTRONICS, CTR REMOTE SENSING, VA, 85- *Concurrent Pos:* Adj prof, Elec Eng Dept, Case Western Reserve Univ, Cleveland, 77-80; scientist, Computer Dept, Harvey Mudd Col, Claremont, Calif, 82-83; vis scientist, Max Planck Inst Aeronomie, WGer & Max Planck Inst Extraterrestrial Physics, Munchen, WGer, 83, Harvey Mudd Col, Claremont, Calif, 85-87. *Mem:* Am Geophys Union; Inst Elec & Electronics Engrs; Am Phys Soc; Int Soc Optical Eng; Union Radio Sci Int. *Res:* Ionospheric research using ground based and space sensors; electromagnetic sounding of the ionosphere covering ELF to microwave frequencies; high power radio frequency field interaction with ionosphere; generation of ELF signals; D-region studies. *Mailing Add:* Ctr Remote Sensing PO Box 9244 McLean VA 22102

GANGWAL, SANTOSH KUMAR, b Calcutta, India, Mar 26, 47; m 74; c 2. SYNTHETIC FUELS. *Educ:* Ind Inst Technol, Tech, 68; Univ Waterloo, MASc, 71, PhD(chem eng), 77. *Prof Exp:* Res assoc chem eng, Va Polytech Inst & State Univ, 76, fel, 77; SR CHEM ENGR, RES TRIANGLE INST, 77- *Concurrent Pos:* Adj prof, Appalachian State Univ, 79-, Duke Univ, 80- *Mem:* Am Chem Soc; Am Inst Chem Eng. *Res:* Coal gasification; synthetic fuels; heterogeneous catalysis; chromatography. *Mailing Add:* 807 Bacon St Durham NC 27703

GANGWER, THOMAS E, b Scranton, Pa, Oct 20, 46; m 69; c 3. PHYSICAL CHEMISTRY. *Educ:* Lebanon Valley Col, BS, 68; Univ Notre Dame, PhD(phys chem), 73. *Prof Exp:* Res assoc basic chem, 73-75, ASSOC CHEMIST, DEPT ENERGY & ENVIRON, BROOKHAVEN NAT LAB, 75- *Mem:* AAAS; Am Chem Soc. *Res:* Study of basic coal chemistry and of basic and applied radiation chemistry and photochemistry; microprocessor interfacing and machine language programming. *Mailing Add:* 739 Battle Front Trail Knoxville TN 37922-6611

GANGWERE, STANLEY KENNETH, b Canton, Ohio, Nov 12, 25; m 49; c 1. ZOOLOGY. *Educ:* Univ Mich, AB, 50, MS, 52, PhD(zool), 57. *Prof Exp:* Asst, Univ Mich, 51-55; from instr to assoc prof, 55-67, dir, Northwoods Biol Sta, 80-85, PROF BIOL, WAYNE STATE UNIV, 67-, ASSOC CHMN BIOL SCI, 85- *Concurrent Pos:* Fulbright sr lectr, Univ Valencia, 61; Fulbright res scholar, Span Entom Inst, Madrid, 62; sr lectr, La Plata Nat Univ, Argentine Fulbright Prog, 74-; mem Adv Screening Comt, Coun Int Exchange Scholars, Washington, 75- *Mem:* Asn d'Acridologie, Paris. *Res:* Ecology, biogeography and feeding behavior in orthopteroid insects, especially of Mediterranean subregion and related islands. *Mailing Add:* Dept Biol 210 Sci Hall Wayne State Univ 5950 Cass Ave Detroit MI 48202

GANIS, FRANK MICHAEL GANGAROSA, b Rochester, NY, Nov 26, 24; m 49; c 2. BIOCHEMISTRY. *Educ:* Univ Rochester, AB, 49, PhD(biochem), 56. *Prof Exp:* Asst, Univ Rochester, 50-51, res assoc, 51-56, instr biochem, 56-62, from instr to asst prof radiation biol, 59-62; dir labs, Clin Study Ctr, Univ Md, Baltimore, 62-66, asst prof biochem, Sch Med & chmn dept biochem, Sch Dent, 66-73; assoc dean sci, Univ Hartford, 73-75; dean, Sch Health Sci, Western Carolina Univ, 76-78; dean acad affairs, Ohio Col Podiatric Med, 78-85, vpres, 80-85, prof biochem, 78-84; RETIRED. *Concurrent Pos:* USPHS fel, Univ Rochester, 56-58; adj clin res assoc, Cleveland Clin Found, 84-; consult. *Res:* Intermediary metabolism of steroid hormones. *Mailing Add:* 8696 Camelot Dr Chesterland OH 44026

GANLEY, JAMES POWELL, b Altadina, Calif, Apr 25, 37; m 65; c 4. OPHTHALMOLOGY, UVEITIS. *Educ:* Mt St Mary's Col, Md, BS, 59; Georgetown Univ, MD, 63; Johns Hopkins Univ, MPH, 69, DrPH, 72. *Prof Exp:* Intern, Washington Hosp Ctr, Washington, DC, 63-64; resident ophthal, Upstate Med Ctr, State Univ NY, Syracuse, 65-68; resident prev med, Johns Hopkins Univ, 69-71; sr staff fel, Nat Eye Inst, NIH, 71-74; asst prof ophthal, Sch Med, Univ Ariz, 74-80; assoc prof & head ophthal dept, La State Univ Med Ctr, 80-82, PROF & HEAD DEPT OPHTHAL, LA STATE UNIV MED CTR, 82-; asst dean med affairs, 81-88, PROF OPHTHAL, MED CTR, LA STATE UNIV, SHREVEPORT, 82-, HEAD DEPT, 80- *Concurrent Pos:* Mem, Sci Adv Panel, Onchocerciasis Control Prog, WHO, 74-79, Ophthal Drugs Adv Comt, Food & Drug Admin, 76-83, adv comt, La Dept Health & Human Resources, 80-, Epidemiol & Dis Control Study Sect-1, NIH, 82-86, Comt Res, Regulatory Agencies & Fed Systs, Am Acad Ophthal, 86- & assoc secy, 90-95, Reviewers Reserve, NIH, 90-94; mem bd dirs, La Asn Blind, 80-, chmn, Human Serv Prog Comt, 85-88, first vchmn & secy, Exec Bd, 89-91; asst dean med affairs, La State Univ Med Ctr, 81-88; mem, bd dir, Northwest La Eye Bank, 87-88 & Shreveport Med Soc, 90-; asst prof ophthal, Med Ctr, La State Univ, Shreveport, 80-82. *Mem:* Int Soc Geog Ophthal (pres, 82-88, treas, 90-); Am Acad Ophthal; Am Col Epidemiol; Am Col Prev Med; Int Eye Found; Am Uveitis Soc. *Res:* Ophthalmic epidemiology; uveitis; onchocerciasis; ocular malignant melanoma. *Mailing Add:* Dept Ophthal La State Univ Med Ctr 1501 Kings Hwy Shreveport LA 71130

GANLEY, OSWALD HAROLD, b Amsterdam, Holland, Jan 28, 29; nat US; m 55; c 2. BACTERIOLOGY, PHYSIOLOGY. *Educ:* Hope Col, AB, 50; Univ Mich, MS, 51, PhD, 53; Harvard Univ, MPA, 65. *Prof Exp:* Asst med bact, Walter Reed Inst Res, 53-55; res assoc allergy & immunol, Merck Inst, 55-60, asst dir sci rels, Merck Sharp & Dohme Res Labs, 60-64; fel sci & pub policy, Harvard Univ, 64-65; spec asst to sci dir, AID, Md, 65-66; chief tech div, Int Sci & Technol Affairs, US Dept of State, Washington, DC, 66-69, sci counr, Am Embassy, Rome & Bucharest, 69-73, dir, Off Soviet & E Europ Sci & Technol Affairs, US Dept of State, 73-75, dep asst secy of state for sci & technol affairs, 75-78; res assoc, J F Kennedy Sch Govt, 78-80, EXEC DIR PROG INFO RESOURCES POLICY, HARVARD UNIV, 80-, LECTR PUB POLICY, J F KENNEDY SCH GOVT, 80- *Mem:* Asn Mil Surgeons US; Am Soc Microbiol; fel Am Acad Microbiol; Am Physiol Soc; Cosmos Club. *Res:* Shock, surgical infections, allergy and host resistance; foreign policy; computers/communication and information. *Mailing Add:* Harvard Univ 200 Aiken Cambridge MA 02138

GANLEY, W PAUL, b North Tonawanda, NY, Apr 1, 34; div. GALVANO LUMINESCENCE. *Educ:* State Univ NY, Buffalo, BA, 55, PhD(physics), 60. *Prof Exp:* Physicist, Cornell Aeornaut Lab, 60-61; asst prof physics, Bryn Mawr Col, 61-66; assoc prof & chmn, dept physics, Wilson Col, Pa, 66-72; res physicist, Aeronaut Lab, Cornell Univ, 72-74; PROF PHYSICS, ERIE COMMUNITY COL, 74- *Concurrent Pos:* Asst ed, Am J Physics, 61-66; consult, Calspan Corp, 77. *Mem:* Am Asn Physics Teachers. *Res:* Applied physics; solid state luminescence; acoustics and noise pollution; coherent optics and automatic target recognition; gamma-ray spectroscopy. *Mailing Add:* Dept Physics Erie Community Col S Campus 4140 Southwestern Blvd Orchard Park NY 14127

GANN, DONALD STUART, b Baltimore, Md, Feb 25, 32; m 59; c 4. PHYSIOLOGY, SURGERY. *Educ:* Dartmouth Col, AB, 52; Johns Hopkins Univ, MD, 56. *Hon Degrees:* MA, Brown Univ, 80. *Prof Exp:* Investr endocrinol, Nat Heart Inst, 58-60; instr surg, Med Col Va, 63-64; from sr instr to assoc prof surg, Western Reserve Univ, 64-70; assoc prof surg, Sch Med, Johns Hopkins Univ, 70-74, prof biomed eng, 70-79, prof emergency med, 74-79, dir dept emergency med, 73-79, dir div surg, 74-78, prof surg, 75-79,; dir div emergency med & trauma, 78-79; prof surg & chmn dept, RI Hosp, Brown Univ, 79-88; PROF SURG & ASSOC CHMN, DEPT SURG, UNIV MD SCH MED, 88-, PROF PHYSIOL, UNIV MD SCH MED, 89- *Concurrent Pos:* Spec fel, Nat Heart Inst, 63-64, res career develop award, 65-67; from sr instr to prof physiol, Western Reserve Univ, 64-70, prof & dir biomed eng, 67-70; asst surgeon, Univ Hosp Cleveland, 64-70; surgeon, Johns Hopkins Hosp, 70-79; chief surgeon, RI Hosp, 79-88; chief, Sect Endocrine Surg, Univ Md Hosp, 88- & Sect Trauma Surg, 90- *Mem:* Am Physiol Soc; Am Surg Asn; Am Col Surg; Am Asn Surg Trauma (secy, 81-86, pres, 87-88); Biomed Eng Soc (pres, 71-72); Endocrine Soc; Cent Soc Clin Res. *Res:* Endocrine physiology; electrolyte metabolism; mathematical models in biology and medicine; neuroendocrinology. *Mailing Add:* Dept Surg Sch Med Univ Md 225 S Greene St Baltimore MD 21201

GANN, RICHARD GEORGE, b Hartford, Conn, Sept 20, 44; div; c 2. PHYSICAL CHEMISTRY, COMBUSTION. *Educ:* Trinity Col, Conn, BS, 65; Mass Inst Technol, PhD(phys chem), 70. *Prof Exp:* Res assoc, Univ Pittsburgh, 70-72; res chemist, Naval Res Lab, 72-76; cheif prog chem, Ctr Fire Res, 76-79, head, explor fire res, 80-81, CHIEF, FIRE MEASUREMENT & RES DIV, NAT INST STANDARDS & TECHNOL, 81- *Concurrent Pos:* Postdoctoral res assoc, Space Res Coord Ctr, Univ Pittsburgh, 70-72; chair, Tech Study Group, Cigarette Safety Act, 84; sr exec fel, Harvard Univ, JFK Sch of Govt. *Mem:* Am Chem Soc; Combustion Inst. *Res:* Development of scientific understanding and engineering measurement method for fire research. *Mailing Add:* Bldg 224 Room B-250 Nat Inst Standards & Technol Gaithersburg MD 20899

GANNETT, E K, information science, for more information see previous edition

GANNON, MARY CAROL, US citizen. NUTRITIONAL BIOCHEMISTRY. *Educ:* Univ Minn, BS, 72, PhD(nutrit), 83. *Prof Exp:* Chemist, 73-83, NUTRIT BIOCHEMIST, DIV METAB RES, VET ADMIN MED CTR, MINNEAPOLIS, 83-; ASST PROF, UNIV MINN, 85- *Mem:* Am Inst Nutrit; Am Diabetes Asn; Inst Food Technologists; Europ Diabetes Asn. *Res:* Effect of ingestion of various foods and macronutrients on intermediary metabolism. *Mailing Add:* Metab Res 111G Vet Admin Med Ctr One Veterans Dr Minneapolis MN 55417

GANO, JAMES EDWARD, b Cleveland, Ohio, Sept 6, 41; m 65; c 2. ORGANIC CHEMISTRY, PHOTOCHEMISTRY. *Educ:* Miami Univ, AB, 63; Univ Ill, MS, 66, PhD(chem), 67. *Prof Exp:* From asst prof to assoc prof, 67-76, PROF CHEM, UNIV TOLEDO, 76- *Concurrent Pos:* Am Chem Soc res grant, 68-70; vis res assoc, Univ Calif, Los Angeles, 74-75; Res Corp res grant, 75-78. *Mem:* Am Chem Soc; Am Inst Chemists; Sigma Xi. *Res:* Novel new organic materials; short-lived reaction intermediates, their structures and properties; molecular recognition; inclusion complexes. *Mailing Add:* Dept Chem Univ Toledo Toledo OH 43606

GANO, RICHARD W, b Wilmington, Del, Jan 4, 56; m 80; c 2. CHEMICAL ENGINEERING. *Educ:* Univ Del, BS, 78; Univ Va, MS, 79. *Prof Exp:* Res eng, chem res dept, 78-79, res & develop eng, plastics technol eng, mfg eng, MFG SPECIALIST, ELASTOMER & POLYMER PROD DEPT, DUPONT, 79- *Mem:* Am Inst Chem Eng; Sigma Xi. *Res:* Catalytic hydrogemation; aerobic permentation. *Mailing Add:* 2582 Park Creek Dr Germantown TN 38139

GANO, ROBERT DANIEL, b Norfolk, Va, Jan 27, 22; m 53; c 4. ORGANIC CHEMISTRY. *Educ:* Univ Richmond, BS, 43; Univ NC, MA, 48. *Prof Exp:* From jr chemist to chemist, 43-75, sr process chemist, E I du Pont de Nemours & Co, Inc, 75-88; RETIRED. *Mem:* Am Chem Soc; Sigma Xi. *Res:* Aromatic nitro, amino and sulfonic acids; azo organic derivatives; new pharmaceutical CPDS (local anesthesics/antimalerials). *Mailing Add:* 1709 Shadybrook Rd Wilmington DE 19803

GANONG, WILLIAM FRANCIS, b Northampton, Mass, July 6, 24; m 48; c 4. NEUROENDOCRINOLOGY. *Educ:* Harvard Univ, AB, 45, MD, 49. *Prof Exp:* From intern to jr asst resident med, Peter Bent Brigham Hosp, 49-51; res fel surg, Harvard Med Sch, 52-55, dir, Surg Res Lab, 53-55; from asst prof to assoc prof, 55-64, fac res lectr, 68, chmn dept, 70-87, PROF PHYSIOL, SCH MED, UNIV CALIF, SAN FRANCISCO, 64-, LANGE PROF OF PHYSIOL, 82- *Concurrent Pos:* Mem, US Nat Comt, Int Union Physiol Sci, 75-81, chmn, 76-79; ed-in-chief, Neuroendocrinology, 79-84; mem, life sci adv comt, NASA, 80-86, comt on space biol & med, NAS, 86-90. *Honors & Awards:* Boylston Med Soc Prize, 49; IFI Golden Hippocrates Award, Italy, 70; Asn Chmn Departments Physiol Award, 78, 88; A A Berthold Medal, Ger, 85. *Mem:* Endocrine Soc; Asn Chmn Departments Physiol (pres, 76-77); Int Soc Neuroendocrinol (vpres, 76-80); fel AAAS; Am Soc Pharmacol & Exp Therapeut; Soc Exp Biol & Med; Am Physiol Soc (pres, 77-78); Sigma Xi; Soc Neurosci (treas, 84-85). *Res:* Neuroendocrinology; interrelation between endocrine and brain function. *Mailing Add:* 710 Hillside Ave Albany CA 94706

GANOTE, CHARLES EDGAR, b Blanchester, Ohio, Feb 9, 37. PATHOLOGY. *Educ:* Univ Cincinnati, BS, 60; Vanderbilt Univ, MD, 65. *Prof Exp:* Instr path, Sch Med, Vanderbilt Univ, 62-63, intern 65-66; staff assoc, Sect Path Anat, Lab Exp Path, Nat Inst Arthritis & Metab Dis, 66-69; consult pathologist, Baxter Labs, 69; from asst prof to prof path, Med Sch, Northwestern Univ, 69-89; PROF PATH, ETENN STATE UNIV, 89-, GRAD FAC MEM, 91- *Concurrent Pos:* Res fel, Sch Med, Vanderbilt Univ, 62-63; asst resident path, NIH, 66; chief, Electron Micros Labs & attend staff, Northwestern Mem Hosp, 69-89; res fel, Cardiothoracic Inst, Inst Cardiol, London, 73-74; vis prof, Nat Heart, Lung & Blood Inst, 79-80; anat path staff mem, Vet Admin Hosp, Mountain Home, Tenn, 89-; mem coun, Int Soc Heart Res, 91- *Mem:* Am Asn Pathologists & Bacteriologists; Am Soc Exp Path; Int Soc Heart Res. *Res:* Author of more than 80 technical publications. *Mailing Add:* Dept Path Vet Admin Grounds ETenn State Univ Bldg 1 PO Box 19540A Johnson City TN 37614

GANS, CARL, b Hamburg, Ger, Sept 7, 23; nat US; m 61. ZOOLOGY. *Educ:* NY Univ, BME, 44; Columbia Univ, MS, 50; Harvard Univ, PhD(biol), 57. *Hon Degrees:* Dr, Univ Antwerp, 85. *Prof Exp:* Asst proj mgr, Babcock & Wilcox Co, NY, 47-51, serv engr, 51-55, contract engr, 55; fel biol, Univ Fla, 57-58; from asst prof to prof biol, State Univ NY, Buffalo, 58-71, chmn dept, 70-71; prof zool & chmn dept, 71-75, PROF BIOL, UNIV MICH, 75- *Concurrent Pos:* Res assoc, Carnegie Mus, Pa, 52-53, Dept Amphibians & Reptiles, Am Mus Natural Hist, 59-, Mus Zool, Univ Kans, 82-; Guggenheim Mem fel, Dept Zool, Univ Brazil, 53-54, Field Asn Dept Physics, Univ Sydney & Australian Mus, 78-79; consult, Buffalo Mus Sci, 59-66, Buffalo Zool Park, 59-71, Time-Life, Ins, 65-67, Royal Soc Zool Anvers, 69-73; res fel, Dept Zool, Univ Leiden, Holland, vis prof, Dept Zool, Univ Tel Aviv, vis res, Dept Zool, Univ Bristol, sabbatical, 65-66; sr res sci, Mus Zool, Univ Mich, 71-; prof, Nat Mus Natural Hist, Paris, 85; vis prof, Dept Biol, Univ Instelling Antwerpen, Belg, 85-86. *Honors & Awards:* Gold Medal, Royal Soc Zool Anvers, 85. *Mem:* Asn Am Anatomists; Am Soc Mech Eng; Soc Study Evolution; Am Soc Ichthyologists & Herpetologists; Am Soc Biomech; Sigma Xi; fel AAAS; Am Inst Biol Sci; Am Soc Zoologists; Am Physiol Soc. *Res:* Behavioral aspects, functional morphology and systematics of reptiles and amphibians; respiration, feeding and burrowing adaptation; concepts of motor coordination; early evolution of vertebrates. *Mailing Add:* Div Biological Sci Univ Mich Natural Sci 2127 Ann Arbor MI 48109

GANS, DAVID, b New York, NY, Apr 10, 07; m 45. MATHEMATICS. *Educ:* NY Univ, BS, 28, PhD(math), 48; Harvard Univ, AM, 30. *Prof Exp:* Instr math, NY Univ, 28-42 & Univ Ill, 45; from instr to assoc prof, 46-74, EMER PROF MATH, WASH SQUARE COL, NY UNIV, 74- *Concurrent Pos:* Fac fel, Fund Adv Ed, 53-54. *Mem:* Math Asn Am; Am Math Soc. *Res:* Modern geometry, especially euclidean, non-euclidean, projective. *Mailing Add:* 190 Riverside Dr New York NY 10024

GANS, EUGENE HOWARD, b New York, NY, Dec 17, 29; m 53; c 2. PHARMACY, CHEMISTRY. *Educ:* Columbia Univ, BS, 51, MS, 53; Univ Wis, PhD(chem), 56. *Prof Exp:* Lab asst, Col Pharm, Columbia Univ, 51-53; Alumni Res Found fel pharmaceut res, Sch Pharm, Univ Wis, 55-56; sr scientist group leader, Hoffman-LaRoche, Inc, NJ, 56-60; head new prod develop sect, Vick Div Res & Develop Labs, Richardson-Merrell, NY, 60-64, asst dir develop, 64-67, dir, 67-71; assoc dir, Alza Inst Pharmaceut Chem, 71-72; dir invest res, 72-76, vpres & dir res & develop, Vicks Personal Care, Richardson-Vicks Div, Proctor & Gamble, Inc, 76-87, PRES, HASTINGS ASSOCS, 87- *Concurrent Pos:* Chmn proprietary drug task group, Antimicrobial II Rev Panel, Food & Drug Admin, 76-81; chmn sci adv comt, Cosmetics, Toiletries & Fragrance Asn. *Mem:* Am Pharmaceut Asn; Am Chem Soc; Sigma Xi; NY Acad Sci; Soc Investigative Dermat; Am Acad Dermat. *Res:* Optimizing pharmaceutical & consumer product research and development; corporate research and development management; development of superior drug, toiletry, cosmetic and proprietary products; advanced therapeutic and drug delivery systems; development and management of scientific personnel; synergizing research and development and marketing operations. *Mailing Add:* Five Fairview Dr Westport CT 06880

GANS, JOSEPH HERBERT, b Hartford, Conn, Dec 29, 22; m; c 3. PHARMACOLOGY. *Educ:* Univ Pa, VMD, 46; Jefferson Med Col, PhD(physiol), 58. *Prof Exp:* Prof pharmacol, State Univ NY Vet Col, Cornell Univ, 57-61; assoc prof, Sch Med, Ind Univ, 61-67; assoc prof, 67-69, PROF PHARMACOL, COL MED, UNIV VT, 69- *Mem:* AAAS; Am Physiol Soc; Am Soc Pharmacol & Exp Therapeut; Am Soc Nephrology; Soc Exp Biol & Med; Sigma Xi. *Res:* Chemically induced tissue injury and carcinogenesis; role of nucleic acids; kidney function and metabolism. *Mailing Add:* 410 N Civic Dr Apt 202 Walnut Creek CA 94596-3602

GANS, MANFRED, b Borken, Ger, Apr 27, 22; US citizen; m 48; c 2. OXIDATION PROCESSES, SYNTHETIC FIBERS. *Educ:* Univ Manchester, Eng, Hon, 50; Mass Inst Technol, SM & BSr Tech, 51. *Prof Exp:* Sr vpres technol, Halcon-Sci Design Co, 51-85; PRES, TECHNOL EVAL & DEVELOP ASN, INC, 86- *Concurrent Pos:* Consult, UN Develop Proj, 78- & UN Indust Develop Orgn, 81- *Mem:* Fel Am Inst Chem Engrs. *Res:* Development of oxidation processes for petrochemicals; creation of applied research institutes for chemical and refinery industries in developing countries. *Mailing Add:* Technol Eval & Develop Asn, Inc 313 First St Hoboken NJ 07030

GANS, PAUL JONATHAN, b Chicago, Ill, May 1, 33; m 59; c 2. CHEMICAL PHYSICS. *Educ:* Ohio State Univ, BSc, 54; Case Inst Technol, PhD(chem), 59. *Prof Exp:* Res assoc chem, Univ Ill, 59-62, instr, 61-62; from asst prof to assoc prof, 62-73, from asst chmn to chmn dept, 65-70, PROF CHEM, NY UNIV, 73- *Mem:* AAAS; Am Chem Soc; Am Phys Soc; Asn Comput Mach. *Res:* Statistical mechanics; phase transitions; polymeric systems; digital computation; Monte Carlo techniques. *Mailing Add:* Dept Chem New York Univ New York NY 10003

GANS, ROGER FREDERICK, b New York, NY, May 5, 41. FLUID MECHANICS, DYNAMICS. *Educ:* Mass Inst Technol, BS, 63; Univ Calif, Los Angeles, MS, 68, PhD(geol), 69. *Prof Exp:* Res assoc geophys & planetary sci, Calif Inst Technol, 69-71; instr appl math, Mass Inst Technol, 71-73, res assoc aeronaut & astronaut, 73-74; asst prof mech & aerospace sci, 74-78, assoc prof, 78-86, PROF, MECH ENG, UNIV ROCHESTER, 86-, CHMN DEPT, 84- *Concurrent Pos:* PI vanow NSF, 76-84; govt & indust consult; sabbatical Aberdeen Proving Ground & Marshall Space Flight Ctr, 82-83. *Mem:* Am Geophys Union; AAAS; Am Acad Mech; Am Soc Mech Eng; Am Phys Soc. *Res:* Boundary layer theory; lubrication; experimental fluid dynamics; nonlinear dynamics. *Mailing Add:* Dept Mech Eng Univ Rochester Rochester NY 14627

GANSCHOW, ROGER ELMER, b Buffalo, NY, Mar 21, 37; m 59; c 2. MOLECULAR GENETICS. *Educ:* Valparaiso Univ, BA, 59; State Univ NY Buffalo, PhD(biol), 67. *Prof Exp:* Am Cancer Soc fel biochem, Sch Med, Stanford Univ, 67-68; STAFF SCIENTIST, CINCINNATI CHILDREN'S HOSP RES FOUND, 68- *Concurrent Pos:* Asst prof, Med Sch, Univ Cincinnati, 68-71, assoc prof, 71-76, prof, 76-; Nat Inst Arthritis & Metab Dis res grant, 70-; assoc ed, Genetics, 78-; mem mammalian genetics study sect, NIH, 80-82. *Mem:* Am Soc Biol Chemists; Genetics Soc Am. *Res:* Regulation of mammalian gene expression. *Mailing Add:* Children's Hosp Res Found Cincinnati OH 45229-2899

GANT, FRED ALLAN, b Howard, Ala, Aug 7, 36; m 70; c 2. PHYSICAL CHEMISTRY, THERMODYNAMICS. *Educ:* Univ Ala, BS, 62, MS, 65, PhD(phys chem), 67. *Prof Exp:* Chemist, Cent Labs, Swift & Co, 60-61; asst prof chem, Mobile Col, 66-67; PROF CHEM, JACKSONVILLE STATE UNIV, 67- *Concurrent Pos:* Chmn, Health Careers Comt. *Mem:* Am Chem Soc (pres, Ala Sect, 78-79). *Res:* Thermodynamic studies of inorganic compounds using a copper block calorimeter; fused salt; reaction calorimetry; atomic absorption and gas chromatography. *Mailing Add:* Dept Chem Jacksonville State Univ Jacksonville AL 36265

GANT, KATHY SAVAGE, b Nashville, Tenn, Jan 5, 47; m 69; c 2. PHYSICS, HEALTH PHYSICS. *Educ:* Austin Peay State Univ, BA, 69; Univ Tenn, Knoxville, MS, 71, PhD(physics), 76. *Prof Exp:* Guest appointment, Atomic & Molecular Physics, 72-76, RES STAFF, CIVIL DEFENSE, RADIOL EMER PLANNING & TECHNOL, EMER PREPAREDNESS EXERCISES, OAK RIDGE NAT LAB, 76- *Concurrent Pos:* AEC radiol sci & protection fel, 69-71; USPHS radiol health fel, 71-74; Oak Ridge Assoc Univ grad res partic, 75-76. *Mem:* Sigma Xi; Health Physics Soc. *Res:* Radiation effects; radiological emergency planning; civil defense; emergency technology; electron mobilities and attachment rates in polyatomic gases; excimer formation; fluorescence; emergency preparedness exercises; design of federal radiological exercises. *Mailing Add:* Energy Div Oak Ridge Nat Lab, Bldg 4500 N PO Box 2008 Oak Ridge TN 37831-6190

GANT, NORMAN FERRELL, JR, b Wichita Falls, Tex, Feb 16, 39. MEDICAL SCIENCE. *Educ:* NTex State Univ, BA, 62; Univ Tex Southwestern Med Sch, MD, 64, MD(obstet & gynec), 68; Am Col Obstet & Gynec, dipl. *Prof Exp:* Fel reproductive physiol, 68-69 & 72-73, from asst prof to assoc prof, 68-76, chmn dept, 77-83, PROF OBSTET & GYNEC, UNIV TEX HEALTH SCI CTR DALLAS, 76- *Concurrent Pos:* Prin investr, NIH grant, 74-; co-investr, NIH grant, 74-79 & 77-82; prin investr, Robert Wood Johnson grant, 75-80. *Mem:* AMA; fel Am Col Obstetricians & Gynecologists; Soc Gynec Invest; Endocrine Soc. *Res:* Maternal-fetal medicine; pregnancy induced hypertension; endocrinology of pregnancy. *Mailing Add:* Ob-Gyn Dept Univ Tex Southwestern Med 5323 Harry Hines E Dallas TX 75235

GANTEN, DETLEV, b Lüneburg, Fed Repub Ger, Mar 28, 41; m; c 2. MOLECULAR BIOLOGY, PHARMACOLOGY. *Educ:* Univ Tübingen, Fed Repub Ger, MD, 68; McGill Univ, Montreal, Can, PhD, 73; Univ Heidelberg, Habilitation, 74. *Prof Exp:* Mem staff, Nephrological Sect, Hotel Dieu Hosp, Univ Montreal, Can, 69-73; PROF, DEPT PHARMACOL, UNIV HEIDELBERG, 75-, SCI DIR, GER INST HIGH BLOOD PRESSURE RES, 79-; COORDR, HYPERTENSION PROGS, FED MINISTRY SCI & TECHNOL, 79- *Concurrent Pos:* Deleg, Concerted Action Comt Biol Europ Community, Ger Res Found & Fed Govt; mem, Coun High Blood Pressure Res, Am Heart Asn; regional editor, Clin & Exp Hypertension; ed, Current Topics Neuroendocrinol; vis lectr, Univ Vancouver, Univ Wash & Univ Portland, 85. *Honors & Awards:* Chavez Award, Int Soc Hypertension, 81; Sechenev Mem Medal, USSR Acad Med Sci, 81; Wissenschaftspreis, Ger Hypertension League, 82. *Mem:* Am Heart Asn; Int Soc Hypertension; NY Acad Sci; Am Soc Pharmacol & Exp Therapeut; Europ Neurosci Asn; Europ Soc Clin Invest. *Res:* Molecular biology and pharmacology of neuropeptides; renin-angiotensin system; transgenic animals; pathophysiology of hypertension. *Mailing Add:* Dept Pharmacol Ger Inst High Blood Pressure Res Univ Heidelberg Im Neuenheimer Feld 366 Heidelberg D-6900 Germany

GANTHER, HOWARD EDWARD, b Adrian, Mo, Jan 17, 37; m 64; c 2. TRACE ELEMENTS, HEAVY METALS. *Educ:* Univ Mo, BS, 58; Univ Wis, MS, 61, PhD(biochem), 63. *Prof Exp:* Res assoc biol chem, Univ Mich, Ann Arbor, 63-64; assoc staff scientist, Jackson Lab, Bar Harbor, Maine, 64-65; asst prof biochem, Med Sch, Univ Louisville, 65-69; assoc prof, Univ Wis-Madison, 69-73, chmn dept, 82-85, prof nutrit sci, 73-90; PROF PREV MED & COMMUNITY HEALTH, UNIV TEX MED BR, GALVESTON, 91- *Honors & Awards:* Mead Johnson Award, Am Inst Nutrit, 75. *Mem:* Am Soc Biol Chemists; Am Inst Nutrit. *Res:* Selenium biochemistry; nutrition, metabolism, toxicity, interactions, functions and anticarcinogenicity; toxicology of heavy metals (mercury, cadmium); functions of glutathione and other sulfur compounds. *Mailing Add:* Dept Prev Med & Community Health Univ Tex Med Br 700 The Strand Galveston TX 77550

GANTI, VENKAT RAO, b Kakinada, India, July 13, 28; US citizen. ORGANIC CHEMISTRY. *Educ:* Univ Madras, BSc, 49; Andhra Univ, MSc, 51; Univ Wis, Madison, MS, 56; Univ Pa, PhD(chem), 59. *Prof Exp:* Res fel chem, Bristol Univ, 59-60 & Univ BC, 60-61; res assoc biochem, Banting & Best Inst, Toronto, 61-63; res fel chem, Univ Ottawa, 63-65; chemist, Nat Health & Welfare, Ottawa, 65-66; asst prof chem, Marianopolis Col, 66-67; sr chemist, Syracuse Univ Res Corp & Syracuse Univ, 67-73; sr scientist med chem, Endo Labs, Inc, 73-78, SR SCIENTIST, PHARMACEUT DEPT, E I DU PONT DE NEMOURS & CO, INC, 78- *Mem:* Am Chem Soc; AAAS; Sigma Xi. *Res:* Chemistry of alkaloids, carbohydrates, pesticides and phospholipids. *Mailing Add:* 115 Carlie Rd Wilmington DE 19803-3433

GANTNER, GEORGE E, JR, b St Louis, Mo, June 7, 27; m; c 7. MEDICINE, PATHOLOGY. *Educ:* St Louis Univ, BS, 49, MD, 53. *Prof Exp:* Pathologist & dir labs, Univ Hosps, 58-75, assoc prof, 62-69, PROF FORENSIC & ENVIRON PATH, SCH MED, ST LOUIS UNIV, 69-, ACTG CHMN DEPT PATH, 76- *Concurrent Pos:* Chief med exam, St Louis County, Mo, 69- *Mem:* Col Am Path; Nat Asn Med Examr (secy-treas); Am Acad Forensic Sci. *Res:* Chemistry of body after death; time of death; criteria of cell viability; data processing systems; drug effects on cytogenetics. *Mailing Add:* 233 Woodbourne Dr St Louis MO 63105

GANTT, ELISABETH, b Gakovo, Yugoslavia, Nov 26, 34; m 58; c 1. PLANT PHYSIOLOGY, BIOLOGICAL STRUCTURE. *Educ:* Blackburn Col, BA, 58; Northwestern Univ, MSc, 60, PhD(biol), 63. *Prof Exp:* NIH res assoc microbiol, Med Sch, Dartmouth Col, 63-66; NIH res assoc microbiol, Radiation Biol Lab, Smithsonian Inst, 66-88; PROF BOT, UNIV MD, 88- *Concurrent Pos:* Mem bd fels & assocs, Nat Res Coun, 73-76. *Honors & Awards:* Darbaker Prize, Bot Soc, 81. *Mem:* AAAS; Am Inst Biol Sci; Am Soc Photobiol; Am Soc Plant Physiologists; Phycol Soc Am (vpres, 77, pres, 78); Japan Soc Plant Physiologists. *Res:* Structure of photosynthetic apparatus; localization and characterization of phycobiliproteins; membrane structure. *Mailing Add:* Dept Bot Univ Md College Park MD 20742

GANTT, RALPH RAYMOND, b Chicago, Ill, Apr 2, 36; m 58. BIOCHEMISTRY. *Educ:* Blackburn Col, BA, 58; Univ Ill, Chicago, PhD(biochem, org chem), 64. *Prof Exp:* Am Cancer Soc fel, Dartmouth Med Sch, 63-66; CHEMIST, NAT CANCER INST, 66- *Mem:* Am Chem Soc; Tissue Cult Asn; Am Soc Biol Chemists. *Res:* Nucleic acid and protein synthesis; carcinogenesis. *Mailing Add:* Nat Cancer Inst Bldg 37 NIH Rm 4D05 Bethesda MD 20014

GANTZ, RALPH LEE, b Anthony, Kans, Sept 16, 32; m 61; c 2. WEED SCIENCE, ENTOMOLOGY. *Educ:* Kans State Univ, BS, 54; Univ Ill, MS, 56, PhD(agron), 58. *Prof Exp:* Asst agron, Univ Ill, 54-58, exten agronomist, 58-59; plant physiologist, Tex Div, Dow Chem Co, 59-63, field agriculturist, Plant Sci Res & Develop, Mich, 63-64, field agriculturist, Minn, 65-66, agr prod field res sta mgr, Davis, Calif, 66-72, agr develop specialist, Ag-Org Dept, Mich, 72-73, herbicide develop specialist, Agr Prod Dept, 73-78, prod develop mgr, 78-86, SR DEVELOP SPECIALIST, DOW CHEM USA, LUBBOCK, TEX, 86- *Concurrent Pos:* Agr consult, US Army CEngrs Res Proj. *Mem:* Am Soc Agron; Sigma Xi; Weed Sci Soc Am; Entomol Soc Am; Soc Nematologists. *Res:* Plant science; research and development. *Mailing Add:* 4009 94th St Lubbock TX 79423-3923

GANTZEL, PETER KELLOGG, b Pasadena, Calif, June 23, 34; m 56; c 3. PHYSICAL CHEMISTRY. *Educ:* Univ Colo, BA, 56; Univ Calif, Los Angeles, PhD(crystallog), 62. *Prof Exp:* Staff mem, Ga Technol, 62-86; RETIRED. *Mem:* Am Crystallog Asn. *Res:* X-ray diffraction and fluorescence analysis; electron diffraction and microscopy, computer programming for these areas; gamma-ray and optical emission spectroscopy. *Mailing Add:* 8308 Paseo del Ocaso La Jolla CA 92037

GANTZER, MARY LOU, b Minneapolis, Minn, Oct 3, 50. CLINICAL CHEMISTRY. *Educ:* Univ Minn, BChem, 72, MS, 76; Univ Va, PhD(chem), 80. *Prof Exp:* Instr, Biochem Lab, Dept Chem, Univ Va, 80-81; sr res scientist, Diag Div, 84-86, staff scientist, 86-87, SUPVR RES & DEVELOP, DIAG DIV, MILES INC, 87- *Concurrent Pos:* Mem, People to People Women Mgt Deleg, People's Repub China, 88- *Mem:* Am Chem Soc; fel Am Inst Chemists; Am Asn Clin Chem; NY Acad Sci. *Res:* Clinical chemistry; urinalysis; membrane biochemistry; enzymology; enzyme kinetics; cancer detection. *Mailing Add:* Diag Div Miles Inc PO Box 70 Elkhart IN 46515

GANZ, AARON, b New York, NY, Feb 4, 24; div; c 2. RESEARCH ADMINISTRATION. *Educ:* Univ Chicago, BS, 47, PhD(pharmacol), 50. *Prof Exp:* From instr to assoc prof pharmacol, Med Units, Univ Tenn, 50-61; exec secy, Res Career Award Comt, Nat Inst Gen Med Sci, NIH, 62-63, head res career sect, Res Fels Br, 63-64, training grants & fels officer, Off Dir, 64-68, assoc dir prog planning & eval, 68-70, chief gen oral sci prog, 70-72, CHIEF PAIN CONTROL & BEHAV STUDIES PROG, NAT INST DENT RES, 72- *Mem:* Fel AAAS; Int Asn Study Pain. *Res:* Oro-facial pain and behavioral research; biomedical research administration. *Mailing Add:* 9605 Shadow Oak Dr Gaithersberg MD 20879

GANZ, CHARLES ROBERT, b Brooklyn, NY, Dec 20, 42; m 63; c 2. ANALYTICAL CHEMISTRY, ORGANIC CHEMISTRY. *Educ:* NY Univ, BA, 63; Queen's Col, NY, MA, 66; Adelphi Univ, PhD(org chem), 69. *Prof Exp:* Res asst pharmaceut, Pfizer & Co, 63-66; asst chem, Adelphi Univ, 67-68, NASA res fel, 66-67, Petrol Res Fund fel, 68-69; group leader anal res, Ciba-Geigy Corp, 69-75; PRES, EN-CAS ANALYSIS LABS, 75- *Concurrent Pos:* Lectr, Penton Learning Systs, 77- *Mem:* Am Chem Soc; AAAS; Water Pollution Control Fedn; Am Asn Textile Chemists & Colorists. *Res:* Development and improvement of methods for analyzing trace organic and inorganic chemicals in various matrices; development of methods for environmental safety screening of chemical substances; pollutant measurement and detection. *Mailing Add:* 3702 Chiswell Ct Greensboro NC 27410

GANZA, KRESIMIR PETER, b Porec, Croatia, Feb 21, 48; Can citizen; m 73; c 3. AUGER SPECTROSCOPY, SURFACE RESEARCH. *Educ:* Uniz Zagreb, Croatia, BSc, 73. *Prof Exp:* Res asst elec mach, McMaster Univ, Ont, 79-81; SURFACE METHODS SPECIALIST, RES DIV, ONT HYRDO, 81- *Mem:* Can Micros Soc. *Res:* Auger surface spectrometry; residual gas analysis; ion etching methods; corrosion studies; semiconductors; surface phenomena. *Mailing Add:* 2497 Nikanna Rd Mississauga ON L5C 2W8 Can

GAO, JIALI, b Jixi, China, Jan 4, 62. COMPUTATIONAL CHEMISTRY, MOLECULAR MODELING. *Educ:* Beijing Univ, China, BS, 82; Purdue Univ, West Lafayette, PhD(org chem), 87. *Prof Exp:* Postdoctoral fel biophys, Dept Chem, Harvard Univ, 87-90; ASST PROF CHEM, DEPT CHEM, STATE UNIV NY, BUFFALO, 90- *Mem:* Am Chem Soc. *Res:* Quantum mechanical and statistical mechanical studies of organic reactions in solution; protein dynamics; mechanisms of enzyme-catalyzed reactions; molecular modeling. *Mailing Add:* Dept Chem State Univ NY Buffalo NY 14214

GAO, KUIXIONG, b Shanghai, China, May 3, 37; US citizen; m 67; c 2. IMMUNOHISTOCHEMISTRY, LIGHT & ELECTRON MICROSCOPY. *Educ:* Nanjing Univ, China, BS, 60, MS, 62; Chinese Acad Sci, Shanghai, PhD(immunol & embryol), 66. *Prof Exp:* Res assoc gas-liquid chromatog & steroids, Shanghai Inst Cell Biol, Chinese Acad Sci, 66-78, asst prof membrane biol, 78-84; asst prof liposome, Dept Biochem, 84-86, SR RES SCIENTIST, DEPT ANIMAL SCI, UNIV TENN, KNOXVILLE, 88- *Concurrent Pos:* Consult, J Longevity, China, 82-83 & Zymed Labs, Inc, 91-; vis scientist, Dept Anat, Univ NC, Chapel Hill, 86-87, vis prof neurosci, Dept Physiol, 87-88. *Mem:* NY Acad Sci; Am Soc Cell Biol; Am Soc Electron Micros. *Res:* Immunocytochemical study on antigenic sites in cells; water soluble embedding method for light and electron microscopy; membrane biogenesis; neuromuscular junction study; gas chromatographic analysis of steroids and prostoglanding; embryonic induction; solid core liposomes; one step stain and mount. *Mailing Add:* Dept Animal Sci Univ Tenn Knoxville TN 37901

GAPOSCHKIN, PETER JOHN ARTHUR, b Boston, Mass, Apr 5, 40. GROUP THEORY, QUANTUM MECHANICS. *Educ:* Mass Inst Technol, BSc, 61; Univ Calif, Berkeley, MA, 65, MA, 66, PhD(physics), 71. *Prof Exp:* Res asst astron, Nat Radio Astron Observ, 62 & Lick Observ, Mt Hamilton, 63; reader math & physics, Univ Calif, Berkeley, 64-71, res asst physics, Lawrence Berkeley Lab, 65-70, teaching asst physics & res asst math, 70; indust physicist, Naval Plant Rep Off, US Navy, Sunnyvale, Calif, 73-75, comput programmer, FNOC, Naval Postgrad Sch, 75-79; sr prog analyst, Informatics PMI, 79-80; instr physics & comput sci, Merritt Col, Oakland, 80-81; instr comput sci, City Col San Francisco, 81-82; PROGRAMMER ANALYST COMPUT SCI, BUR MGR INFO SYSTS, CITY OF SAN FRANCISCO, 83- *Concurrent Pos:* Asst programmer, Univ Calif, 76-77, programmer, Lawrence Berkeley Lab, 84; co-worker, physics res proj, Stanford Univ, 85- *Mem:* Am Astron Soc; Am Math Soc; Math Asn Am; Asn Comput Mach; Data Processing Mgt Asn; Sigma Xi. *Res:* Group theory. *Mailing Add:* 1442 A Walnut St Suite 371 Berkeley CA 94709-1496

GAPP, DAVID ALGER, b Washington, DC, Sept 22, 45; m 79; c 2. COMPARATIVE ENDOCRINOLOGY, IMMUNOHISTOCHEMISTRY. *Educ:* Col William & Mary, BS, 67, MA, 70; Boston Univ, PhD(biol), 77. *Prof Exp:* Postdoctoral endocrinol, Boston Univ, 77, Jackson Lab, 77-79; asst prof, 79-86, ASSOC PROF BIOL, HAMILTON COL, 86-, CHMN, DEPT BIOL, 90- *Concurrent Pos:* Lectr biol, Col William & Mary, 68-69; vis colleague, Histochem Dept, Hammersmith Hosp, Royal Postgraduate Med Sch, Univ London, 83. *Mem:* Sigma Xi; Endocrine Soc; Am Soc Zoologists; Am Diabetes Asn; Soc Study Amphibians & Reptiles; AAAS. *Res:* Comparative endocrinology of peptide hormones in the gastro- enteropancreatic endocrine system of reptiles; focus on the presence and distribution of regulatory peptides in turtles; the regulation of insulin secretion from turtle pancreas and gut. *Mailing Add:* Dept Biol Hamilton Col Clinton NY 13323

GARA, AARON DELANO, b Lamberton, Pa, Aug 9, 35; m 56, 85; c 4. RESEARCH ADMINISTRATION. *Educ:* Drexel Inst Technol, BS, 58; Wash Univ, PhD(physics), 65. *Prof Exp:* sr res physicist & proj mgr res labs, Gen Motors Corp, 65-83; TECH DIR, VPRES, NEWPORT CORP, 83- *Concurrent Pos:* Instr, Lawrence Inst Technol, 72- *Mem:* Fel Optical Soc Am; Soc Photo-Optical Instrument Eng; Am Phys Soc; Int Comn on Optics. *Res:* Optical information processing; optical data storage; electrooptic materials; laser physics; holography. *Mailing Add:* Newport Corp 18235 Mt Baldy Circle Fountain Valley CA 92708

GARA, ROBERT I, b Santiago, Chile, Dec 16, 31; m 58, 79; c 4. FOREST MANAGEMENT, FOREST ENTOMOLOGY. *Educ:* Utah State Univ, BS, 53; Oregon State Univ, MS, 62, PhD(entom), 64. *Prof Exp:* Forester, Kirby Lumber Corp, 57-60; res asst forest entom, Boyce Thompson Inst, Plant Res, 60-62, sr scientist, 62-63, proj leader, 63-66; asst prof forest entom, Col Forestry, Syracuse Univ, 66-68; PROF FOREST ENTOM, COL FOREST RESOURCES, UNIV WASH, 68- *Concurrent Pos:* Consult, Food & Agr Orgn, Turrialba, Costa Rica, 69-70, Santiago, Chile, 77-78 & Hanoi, VietNam, 90; vis prof, Univ Austral de Chile, Valdivia, 77-78; with USAID, Ecuador, 84-91; consult, Peace Corps, Chile, 89-90. *Mem:* Soc Am Foresters; Entom Soc Am. *Res:* Flight behavior of bark beetles; host selection behavior of forest insects; tropical forest entomology; tropical forestry. *Mailing Add:* Dept Forest Entom AR-10 Col Forest Resources Univ Wash Seattle WA 98195

GARABEDIAN, PAUL ROESEL, b Cincinnati, Ohio, Aug 2, 27; div. APPLIED MATHEMATICS. *Educ:* Brown Univ, AB, 46; Harvard Univ, AM, 47, PhD(math), 48. *Prof Exp:* Nat Res Coun fel, 48-49; asst prof math, Univ Calif, 49-50; from asst prof to prof, Stanford Univ, 50-59; PROF MATH, NY UNIV, 59- *Concurrent Pos:* Sloan fel, 60-62; Guggenheim fel, 66-67; Sherman Fairchild fel, Calif Inst Technol, 75. *Honors & Awards:* Birkhoff Prize, Am Math Soc, 83. *Mem:* Nat Acad Sci; Soc Indust & Appl Math; Am Math Soc; Am Acad Arts & Sci; Am Inst Aeronaut & Astronaut; Am Phys Soc. *Res:* Functions of a complex variable; hydrodynamics; partial differential equations. *Mailing Add:* Courant Inst NY Univ New York NY 10012

GARASCIA, RICHARD JOSEPH, b Detroit, Mich, Dec 25, 17; m 42; c 7. CHEMISTRY. *Educ:* Univ Detroit, BS, 40; Univ Mich, MS, 41; Univ Cincinnati, PhD(chem), 50. *Prof Exp:* Instr math, Univ Detroit, 41-42; from instr to assoc prof, 42-59, chmn dept, 61-66, PROF CHEM, XAVIER UNIV, OHIO, 59- *Concurrent Pos:* Res chemist, Cook Paint & Varnish Co, Detroit, 41-42. *Mem:* Am Chem Soc. *Res:* Organic synthesis; organic arsenic; phosphorus and antimony compounds; fluorene and acenaphthene chemistry. *Mailing Add:* Dept Chem Xavier Univ Evanston Sta Cincinnati OH 45207

GARAVELLI, JOHN STEPHEN, b Memphis, Tenn, Sept 7, 47. PROTEIN STRUCTURE, BIOMOLECULAR EVOLUTION. *Educ:* Duke Univ, BS, 69; Wash Univ, PhD(biochem), 75. *Prof Exp:* Res assoc, Marine Lab, Duke Univ, 75-76; res assoc, Chem Dept, Univ Del, 76-79, lectr gen chem, 79-80; lectr org chem, 80-81, res assoc, plant sci dept, Tex A&M Univ, 81-83; sr res assoc, Extraterrestrial Res Div, Nat Res Ctr, NASA-Ames Res Ctr, 83-85; res assoc & dir comput opers, Agouron Inst, La Jolla, Calif, 86; sr res specialist & syst mgr, Col Pharm, Univ Ill, Chicago, 86-88; SR RES SCIENTIST & DATABASE COORDR, NAT BIOMED RES FOUND, 89- *Mem:* AAAS; Am Chem Soc; Int Soc Study Origin of Life; Am Soc Gravitational & Space Biol; Comn Professionals Sci & Technol; Int Union Pure & Appl Chem. *Res:* Biochemical and biomedical science fields allied with phylogenetics, biogenesis and information theory; protein chemistry; structure and conformation prediction; computer modeling of biomolecules. *Mailing Add:* Box 3783 Washington DC 20007

GARAY, ANDREW STEVEN, b Pecs, Hungary, May 20, 26; nat US; m 68; c 5. MOLECULAR CHIRALITY, NITROGEN FIXATION. *Educ:* Eotvos Lorand Univ, Budapest, PhD(biochem), 52; Nat Acad, Hungary, DCS, 65. *Prof Exp:* Res scientist pharmaceut, Pharmaceut Res Inst, 53-56; prof plant physiol, Agr Res Sta, Fertod, 56-68; dir biophysics, Biol Res Ctr, Szeged, 68-75; PROF BIOPHYSICS, TEX A&M UNIV, 76- *Concurrent Pos:* Vis prof, Plant Virus Res Inst, Chiba, Japan, 71, Cairo Univ, 72 & Lab D'Optique Phys, Paris, 74; ed, Environ & Radiation Biophysics WGer, 76- *Mem:* Sigma Xi; Am Soc Biophysics; Am Soc Plant Physiol; Int Soc Origin Life; Nat Acad Hungary. *Res:* Origin and role of molecular chirality in biochemical processes; mechanism of nitrogen fixation. *Mailing Add:* Biochem & Biophys Dept Tex A&M Univ College Station TX 77843

GARAY, GUSTAV JOHN, b Carteret, NJ, May 24, 33; m 58; c 2. ENTOMOLOGY, MICROBIOLOGY. *Educ:* Columbia Univ, AB, 55; Rutgers Univ, MS, 62, PhD(med entom), 64. *Prof Exp:* Res asst immunol, Rockefeller Univ, 59; res asst entom, Rutgers Univ, 59-64; dir biol div, Ward's Natural Sci Estab, 64-74; PROF BIOL, MONROE COMMUNITY COL, 75- *Concurrent Pos:* Sabbatical, Gemencot Imt, 90. *Mem:* Am Soc Microbiol; Sigma Xi; Lepidopterist's Soc; Xerces Soc. *Res:* Mass rearing of insects; laboratory rearing of mosquitoes; Protozoan and Algae culture; photomicrography. *Mailing Add:* 3142 Rush Mendon Rd Honeoye Falls NY 14472

GARAY, LESLIE ANDREW, b Hosszuheteny, Hungary, Aug 6, 24; nat US. TAXONOMY. *Educ:* Tufts Univ, MSc, 61, PhD, 64. *Prof Exp:* Asst, Univ Toronto, 49-51, asst cur, 52-58; cur, orchid herbarium oakes ames, Bot Mus, Harvard Univ, 58-88; EMER PROF, HARVARD UNIV, 88. *Concurrent Pos:* Guggenheim fel, Can, 57-58; mem bd freshman adv, Harvard Univ, 60-63, mem fac arts & sci, 63-, chmn int orchid comn classification, nomenclature & registration, lectr biol, 65-70; mem, World Orchid Conf, London, 60, Singapore, 63; Int Bot Cong Montreal, Edinburgh, 64; field work in Colombia, Ecuador, Venezuela, Jamaica, Fiji, New Caledonia, New Guinea, Malaya, Ceylon. *Mem:* Int Asn Plant Taxon. *Res:* Taxonomy, phylogeny and evolution of the entire orchid family; orchids of Colombia, Ecuador, Haiti and Okinawa. *Mailing Add:* c/o Harvard Univ Herbaria 22 Divinity Ave Cambridge MA 02138

GARBACCIO, DONALD HOWARD, b Paterson, NJ, June 29, 30; m 52; c 4. MECHANICAL ENGINEERING. *Educ:* Newark Col Eng, BSME, 51; Princeton Univ, MSE, 53; Stanford Univ, PhD(eng mech), 55. *Prof Exp:* Res engr, Calif Res Corp, 55-58; sr engr, Rocketdyne Div, NAm Aviation, Inc, 58-60; sr staff mem, Nat Eng Sci Co, 60-64; sr res engr, Sci Eng Assocs Div, Kaman Aircraft Corp, 64-69; sr staff engr, Actron Industs, Inc, 69-77; MEM STAFF, C F BRAUN & CO, 77- *Mem:* Am Soc Mech Engrs. *Res:* Structural dynamics; engineering oceanography. *Mailing Add:* 425 Adams Sierra Madre CA 91024

GARBACZ, ROBERT J, b Buffalo, NY, Sept 12, 33. ELECTRICAL ENGINEERING. *Educ:* Univ Buffalo, BS, 55; Ohio State Univ, MS, 57, PhD(electro-magnetic scattering), 68. *Prof Exp:* Asst prof, 68-76, ASSOC PROF ELEC ENG, OHIO STATE UNIV, 76-, ASST SUPVR ELECTRO SCI LAB, 55- *Mem:* Inst Elec & Electronics Engrs; Sigma Xi. *Res:* Electromagnetic scattering by antennas; characteristic model expansions of fields scattered by obstacles of arbitrary shape. *Mailing Add:* Dept Elec Eng Ohio State Univ Columbus OH 43210

GARBARINI, EDGAR JOSEPH, b Jackson, Calif, Aug 1, 10; m 36; c 2. CIVIL ENGINEERING. *Educ:* Univ Calif, Berkeley, BS, 33. *Prof Exp:* Civil engr, Calif Comn, Golden Gate Int Exped, 38-39; engr, Pac Gas & Elec Co, 39-40; engr, 40-84, SR EXEC CONSULT, BECHTEL GROUP, INC, 84- *Mem:* Nat Acad Eng; fel Am Soc Civil Engrs; Mining & Metall Soc Am. *Mailing Add:* 1170 Sacramento St San Francisco CA 94108-1943

GARBARINI, GEORGE S, geology, for more information see previous edition

GARBARINI, VICTOR C, b New York, NY, May 24, 26; m 48; c 3. HIGH TECHNOLOGY SAFETY ENGINEERING, CHEMICAL CONTAMINATION CONTROL. *Educ:* Manhattan Col, BS, 44; NY Univ, PhD(chem), 56. *Prof Exp:* Engr & chemist, E I du Pont de Nemours & Co, Inc, Del, 51-55; engr, Esso Res & Eng Co, 55-59; mem tech staff, Bell Labs, 59-87; RETIRED. *Concurrent Pos:* Spec safety adv, silicon processing technol. *Mem:* Am Chem Soc; Am Soc Safety Engrs; Semiconductor Safety Asn. *Res:* Semiconductor device chemistry; process development and engineering; photolithography; analytical investigations and contamination controls; chemical safety engineering. *Mailing Add:* 201 Macada Rd Bethlehem PA 18017-2523

GARBER, ALAN J, b Philadelphia, Pa, Feb 27, 43; m; c 2. BIOCHEMISTRY, CELL BIOLOGY. *Educ:* Temple Univ, AB, 64, MD, 68; Fels Res Inst, PhD(physiol chem), 71. *Prof Exp:* Intern med, Temple Univ Hosp, 68-69, resident, 71-72; instr med, Sch Med, Wash Univ, St Louis, 72-74; from asst prof to assoc prof med, Baylor Col Med, 74-82, asst prof cell biol, 74-80, prog dir, Gen Clin Res Ctr, 77-87, assoc prof biochem & cell biol, 80-82, PROF MED & BIOCHEM & CELL BIOL, BAYLOR COL MED, 82-; CHIEF, DIABETES CLIN, BEN TAUB GEN HOSP, 74- & DIABETES-METAB UNIT, METHODIST HOSP, 85- *Concurrent Pos:* USPHS spec res fel, 72-74; attend physician, Ben Taub Gen Hosp, 74-; investr, Howard Hughes Med Inst, 74-78; counr, Southern Soc Clin Invest, 83-86. *Mem:* Am Fedn Clin Res; Am Diabetes Asn; Endocrine Soc; Southern Soc Clin Invest (pres, 88-89); Am Soc Biochem & Molecular Biol; Am Soc Clin Invest. *Res:* Author of 79 technical publications. *Mailing Add:* Baylor Col Med 6550 Fannin Suite 1045 Houston TX 77030

GARBER, CHARLES A, b Rock Island, Ill, May 23, 41; m 80. MATERIALS SCIENCE, POLYMER PHYSICS. *Educ:* Univ Ill, Urbana, BS, 63; Case Inst Technol, MS, 65, PhD(polymer solid state physics, morphol), 67. *Prof Exp:* Res fel, Case Inst Technol, 63-67; res physicist, Plastics Dept, E I du Pont de Nemours & Co, Inc, Del, 67-70; PRES, STRUCT PROBE, INC, WEST CHESTER, PA, 70- & METUCHEN, NJ, 73- *Concurrent Pos:* Instr, Cleveland State Univ, 66-67; adj prof, Drexel Inst Technol, 68-69 & Philadelphia Col Textiles & Sci, 69-70; dir, Asn Consult Chemists & Chem Engrs, 77- & Independent Labs Assurance Co, Ltd, 78- *Mem:* Am Chem Soc; Am Inst Chem Eng; Am Phys Soc; Electron Micros Soc Am; Sigma Xi; Soc Cosmetic Chemists; Am Soc Metals. *Res:* Characterization of solid materials with electron optical methods including scanning and transmission electron microscopy, electron probe microanalysis, Auger electron spectroscopy, electron spectroscopy for chemical analysis, thermal analysis and x-ray diffraction. *Mailing Add:* Steeplechase 60 Harrison Rd E West Chester PA 19380-6700

GARBER, DAVID H, b Norfolk, Va, June 18, 18; m 49; c 2. PHYSICS, EDUCATIONAL ADMINISTRATION. *Educ:* Univ Pa, AB, 40, AM, 41; Stanford Univ, PhD(physics), 53. *Prof Exp:* Asst physics, Univ Pa, 39-41 & Univ Rochester, 41-42; instr physics, Stanford Univ, 42-44; physicist, USN Electronics Lab, 44-45; res asst physics, Stanford Univ, 46-48; instr, Univ Wash, 49-51; res physicist, Cornell Aeronaut Lab, 51-55; sr staff scientist, Gen Dynamics Corp, 55-65; prog mgr, Philco-Ford Corp, 65-66; sci asst to dir, McDonnell Douglas Corp, 66-68, explor sci mgr, 69; consult, Claremont Univ Ctr, 69-70; exec secy Space Sci Bd & consult, Nat Acad Sci, 71-75; asst to vprovost acad admin, Univ Minn, Duluth, 76-80; prog dir, Univ Progs Div,

80-84, CONSULT EDUC & RES PROGS, OAK RIDGE ASSOC UNIV, 84- *Concurrent Pos:* Develop engr, Honeywell, Inc, 41-42; assoc ed, J Spacecraft & Rockets, 67-69; consult, Am Univ, Washington, DC, 75-76. *Mem:* AAAS; Am Asn Physics Teachers; assoc fel Am Inst Aeronaut & Astronaut; Am Phys Soc; Nat Coun Univ Res Adminrs; Sigma Xi. *Res:* Science policy; energy conversion and storage systems; space science and systems; atmospheric physics; atomic nuclear and radiation physics; nuclear magnetic resonance; physical oceanography. *Mailing Add:* 40 Brookside Dr Oak Ridge TN 37830-7670

GARBER, DONALD I, b Cleveland, Ohio, July 8, 36. NUCLEAR PHYSICS. *Educ:* Carnegie Inst Technol, BS, 58; Case Inst Technol, MS, 60, PhD(nuclear physics), 64. *Prof Exp:* From instr to asst prof physics, Case Western Reserve Univ, 64-67; from asst physicist to assoc physicist, 67-74, PHYSICIST, BROOKHAVEN NAT LAB, 74- *Mem:* Am Phys Soc. *Res:* Elastic and inelastic scattering of neutrons; polarization measurements of neutrons produced in deuteron reactions; radiative capture; automated nuclear physics publications; interactive computer graphics. *Mailing Add:* Three Wainscott Lane East Setauket NY 11733

GARBER, EDWARD DAVID, b New York, NY, Mar 22, 18; m 43; c 3. CYTOGENETICS. *Educ:* Cornell Univ, BS, 40; Univ Minn, MS, 42; Univ Calif, PhD(genetics), 49. *Prof Exp:* Asst res scientist & microbial geneticist, Naval Biol Lab, 49-53; from asst prof to prof, 53-88, EMER PROF BOT, UNIV CHICAGO, 88- *Mem:* Genetics Soc Am; Bot Soc Am; Am Soc Microbiol; Brit Soc Gen Microbiol; Sigma Xi. *Res:* Cytotaxonomy and cytogenetics of sorghum and collinsia; genetics of virulence; fungal genetics. *Mailing Add:* Dept Ecol & Evolution Univ Chicago 5801 Ellis Ave Chicago IL 60637

GARBER, FLOYD WAYNE, b Winfield, Kan, Aug 31, 41; m 62; c 1. HEALTH PHYSICS INSTRUMENTS, MINERALS ASSAY INSTRUMENTS. *Educ:* Southwestern Col, BA, 63; Univ Tenn, MS, 65, PhD(physics), 68. *Prof Exp:* Res & develop physicist, EG&G ORTEC, 68-69, detector mfg mgr, 69-80, prog mgr, 80-82; dir opers, Wemco I&C Div, 82-83; proj mgr, 83-85, DIR, DESIGN ENG & MFG DIV, TECHNOL FOR ENERGY CORP, 85-, CORP RADIATION SAFETY OFFICER, 85- *Mem:* Am Phys Soc; Health Physics Soc; Am Soc Mining Engrs. *Res:* Development, manufacturing and marketing of instruments that measure radiation. *Mailing Add:* 1113 W Outer Dr Oak Ridge TN 37830

GARBER, H(IRSH) NEWTON, b Philadelphia, Pa, Mar 16, 30; m 51; c 5. OPERATIONS RESEARCH, MANAGEMENT SCIENCE PRACTICE. *Educ:* Univ Pa, BS, 52; Mass Inst Technol, SM, 53, ScD(elec eng), 56. *Prof Exp:* Oper res analyst, RCA Corp, 56-65, sr scientist, 66-69, mgr opers res, 69-76, dir opers res, 76-86; VPRES & DIR MAN SCI, G P MERRILL LYNCH, 86- *Concurrent Pos:* Chmn, Col Practice Mgt Sci, Inst Mgr Sci, 76-78, pres, 83-84. *Mem:* Opers Res Soc Am; Soc Indust & Appl Math; Inst Elec & Electronics Engrs; Inst Mgr Sci (pres, 83). *Res:* Use of scientific methods in the problems of industrial management and services. *Mailing Add:* Three Point Woods Dr N Brunswick NJ 08902-1207

GARBER, HAROLD JEROME, b Cleveland, Ohio, Mar 12, 13; m 44; c 5. CHEMICAL ENGINEERING, NUCLEAR ENGINEERING. *Educ:* Univ Cincinnati, ChE, 35. *Prof Exp:* From instr to assoc prof chem eng, Univ Cincinnati, 36-47; prof, Univ Tenn, 47-55; mgr chem pressurized reactor develop, Westinghouse Nuclear Power Systs, 55-59; dir advan proj nuclear mat, Nuclear Mat & Equip Corp, 59-72; tech asst mgr fast breeder reactor design, Westinghouse Advan Reactors Div, 72-74, consult engr, Fusion Reactor Develop, Westinghouse Fusion Power Systs, 74-83; RETIRED. *Concurrent Pos:* Consult engr, USAF, 41-54, Gen Air Conditioning Corp, 42-54, Adler Co, 43-49 & Oak Ridge Nat Lab, 47-55. *Mem:* Am Inst Chem Engrs; Am Inst Chemists; Sigma Xi; AAAS. *Res:* Bubble mechanics; mass transfer; reaction kinetics; plutonium fuels; fusion reactor fueling; rocket propulsion fuels; moisture indication; gas dehydration; infrared radiant heating; fractional crystallization; radioisotope powered cardiac pacemakers; fractional distillation; gas absorption. *Mailing Add:* 6800 Ruh Rd Knoxville TN 37918

GARBER, JOHN DOUGLAS, b Minneapolis, Minn, May 12, 20; m 43; c 3. ORGANIC CHEMISTRY. *Educ:* Pa State Col, BS, 40; Univ Ill, PhD(chem), 43. *Prof Exp:* Res chemist, Standard Oil Develop Co, 43-47; res dir, E M Wanderman & Co, 47-48; res chemist, Merck & Co, Inc, 48-55, mgr agr & indust org chem develop, 55-64; asst to dir res & develop, Indust Chem Div, Am Cyanamid Co, 64-65, dir, 65-66; gen mgr basic & appl res, Moffett Tech Ctr, CPC Int, Inc, 66-69, sci vpres, Develop Div, 69-73, corp mgr indust res & develop, 73-76; CONSULT, 76- *Mem:* Am Chem Soc. *Res:* Conversion of carbohydrates to furans and their utilization as industrial organics; high temperature polymers and fluids; systemic approaches to plant agricultural chemicals; petrochemicals including specialty resins and organics; food components. *Mailing Add:* 43548 St Kitts Ct Palm Desert CA 92260

GARBER, LAWRENCE L, b Goshen, Ind, July 4, 42; m 65; c 1. TRANSITION METAL CHEMISTRY. *Educ:* Goshen Col, BA, 63; Mich State Univ, PhD(chem), 67. *Prof Exp:* Asst prof chem, Goshen Col, 68-69; from asst prof to assoc prof, 69-86, chmn, 79-83, PROF CHEM, IND UNIV, SOUTH BEND, 86- *Concurrent Pos:* Ind Univ Res Found grants, 70, 74, 78, 87. *Honors & Awards:* Sigma Xi Res Award, 67. *Mem:* Am Chem Soc; AAAS; Sigma Xi. *Res:* Synthesis and characterization of transition metal complexes; water quality studies. *Mailing Add:* Dept Chem Ind Univ South Bend IN 46634-7111

GARBER, MEYER, b Philadelphia, Pa, June 6, 28. PHYSICS. *Educ:* Univ Pa, BS, 49; Univ Ill, MS, 50, PhD(physics), 54. *Prof Exp:* Fulbright fel, Univ Leiden, Netherlands, 54-55; Nat Res Coun fel, Univ Ottawa, Can, 55-57; prof physics, Mich State Univ, 58-66; PHYSICIST, DEPT APPL SCI, BROOKHAVEN NAT LAB, 66- *Concurrent Pos:* Mem staff, Div Sponsored Res, Mass Inst Technol, 61-63; vis physicist, Brookhaven Nat Lab, 64. *Mem:* Am Phys Soc; Sigma Xi. *Res:* Low temperature physics. *Mailing Add:* Bldg 902B Brookhaven Nat Lab Upton NY 11973

GARBER, MORRIS JOSEPH, b New York, NY, Nov 6, 12; m 43; c 2. GENETICS, COMPUTER SCIENCES. *Educ:* Agr & Mech Col Tex, PhD(genetics), 51. *Prof Exp:* Asst prof genetics, Agr & Mech Col Tex, 47-56; biometrician & prof, 56-80, EMER PROF STATIST, UNIV CALIF, RIVERSIDE, 80- *Concurrent Pos:* Biometrician, Int Inst Trop Agr, Ibadan, Nigeria, 74-76; statistician, Int Ctr Res, Rice & Beans, Goiania, Goias, Brasil, 80-81. *Mem:* AAAS; Asn Comput Mach; Biomet Soc; Am Statist Asn; Am Genetic Asn. *Res:* Statistical design and analysis in agriculture; computer programming. *Mailing Add:* Dept Statist Univ Calif Riverside CA 92521

GARBER, MURRAY S, b New York, NY, Nov 10, 34; m 60; c 2. IMPACT OF COAL MINING ON HYDROLOGIC SYSTEM. *Educ:* Brooklyn Col, BS, 56; Univ Kans, MS, 62. *Prof Exp:* Hydrologist, US Geol Surv, 56-83; hydrologist, 83-86, CHIEF HYDROL BR, OFF SURFACE MINING, 86- *Mem:* Nat Water Well Asn; fel Geol Soc Am; Am Inst Prof Geologists. *Res:* Impact of coal mining on the hydrologic system. *Mailing Add:* 130 Foxcroft Rd Pittsburgh PA 15220-1704

GARBER, RICHARD HAMMERLE, b Beaver Falls, Pa, June 22, 21; m 45; c 4. PLANT PATHOLOGY. *Educ:* Geneva Col, BS, 47; Colo State Univ, MS, 50; Univ Calif, PhD(plant path), 60. *Prof Exp:* Asst bot & plant path, Colo State Univ, 48-50, resident supt, San Luis Valley Exp Sta, 50-52; plant pathologist, Agr Res Serv, USDA, Univ Calif, 54-86; RETIRED. *Concurrent Pos:* Mem, Cotton Dis Coun, secy, 65-66, chmn, 67-68; mem US team, Bilateral Conf Plant Protection, Tashkent, Ubekistan, USSR, 80; collabr, Agr Res Serv USDA; consult, Univ of Calif. *Mem:* Am Phytopath Soc; Sigma Xi. *Res:* Ecology of plant diseases; seed and soil treatments; fungicides-biological control. *Mailing Add:* US Cotton Res Sta 17053 Shafter Ave Shafter CA 93263

GARBER, RICHARD LINCOLN, b Seattle, Wash, May 8, 50; m 78; c 2. BIOLOGY, MOLECULAR BIOLOGY. *Educ:* Stanford Univ, BS, 72; Yale Univ, MPhil, 76, PhD(biol), 77. *Prof Exp:* fel biol, Roche Inst Molecular Biol, 77-79; AT GENETICS DEPT, UNIV WASH, SEATTLE, 83- *Concurrent Pos:* Univ Basel, Switz, 79-83. *Mem:* AAAS; Sigma Xi. *Res:* The genetic control of development is studied in fruitfly drosophila; our focus is on the homeotic genes loci which determine the fate of each cell in the organism; control of gene transcription; molecular cloning and nucleic acid sequencing; Japanese silkworm Bombyx mori. *Mailing Add:* Genetics Dept SK 50 Univ Wash Seattle WA 98195

GARBER, ROBERT WILLIAM, b Philadelphia, Pa, Aug 19, 43; m 66; c 2. ANALYTICAL CHEMISTRY, ENVIRONMENTAL CHEMISTRY. *Educ:* Philadelphia Col Pharm & Sci, BSc, 65; St Joseph's Col, Pa, MS, 67; Univ Pittsburgh, PhD(anal chem), 75. *Prof Exp:* Environ scientist, Brookhaven Nat Lab, 71-75, chemist air pollution studies, 75-80, res chemist environ studies, 80-88, RES CHEMIST, MATSEPARATION LAB, NAT FERTILIZER & ENVIRON RES CTR, TENN VALLEY AUTHORITY, 88-; ADJ PROF ENVIRON CHEM, ALA A&M UNIV, 88- *Res:* Field measurement and analytical methods development for environmental studies, especially air pollution studies. *Mailing Add:* Nat Fertilizer & Environ Res Ctr Muscle Shoals AL 35660

GARBERS, DAVID LORN, b La Crosse, Wis, Mar 17, 44; c 2. BIOLOGICAL CHEMISTRY. *Educ:* Univ Wis-Madison, BS, 66, MS, 70, PhD(biochem), 72. *Prof Exp:* Res assoc physiol, 72-74, asst prof, 74-76, ASSOC PROF PHARMACOL, VANDERBILT UNIV, 77- *Concurrent Pos:* Investr, Howard Hughes Med Inst, 76-; vis prof, Johns Hopkins Med Sch, 84-85; NIH study sect, 84-87. *Mem:* Am Soc Biol Chemists; Soc Study Reproduction. *Res:* Molecular biology of fertilization; signal transduction. *Mailing Add:* Rm 751 Light Hall Howard Hughes Med Inst Vanderbilt Univ Sch Med Nashville TN 37232

GARBRECHT, WILLIAM LEE, b Grand Rapids, Mich, Apr 18, 23; m 47; c 3. ORGANIC CHEMISTRY. *Educ:* Kalamazoo Col, AB, 48, MS, 49; Mich State Col, PhD(chem), 53. *Prof Exp:* Org chemist, 53-60, develop assoc, 60-68, res assoc, 68-74, RES ADV, ELI LILLY & CO, 74- *Mem:* Am Chem Soc; AAAS. *Res:* Chemistry of 5-amino tetrazole derivatives; chemistry of ergot alkaloids; chemistry of cephalosporin antibiotics; tetrazole sweeteners. *Mailing Add:* 7303 Wood Stream Dr Indianapolis IN 46254-9619

GARBUNY, MAX, b Koenigsberg, Ger, Nov 22, 12; nat US; m 47; c 3. OPTICAL PHYSICS, MOLECULAR PHYSICS. *Educ:* Tech Univ Berlin, Dipl Ing, 36, Dr Ing, 38. *Prof Exp:* Physicist, Allen-Bradley Co, 39-43; instr physics, Princeton Univ, 43-44; develop engr, Westinghouse Elec Corp, 44-46, sr res physicist, 46-52, mgr optical physics sect, 52-60, CONSULT OPTICAL PHYSICS, SCI & TECHNOL CTR, WESTINGHOUSE ELEC CORP, 60- *Concurrent Pos:* Mem nat exec comt, Infrared Info Symposia, 63-70; consult, Sci Adv Panel, US Army, 64-72, mem electronics adv group, Army Electronics Command, 69-74; NY State vis prof, Univ Rochester, 68-69; consult, Edgewood Arsenal, US Army, 72-76. *Mem:* Fel Am Phys Soc; fel Am Optical Soc. *Res:* Atomic physics; microwave devices; spectroscopy; solid state physics; infrared devices; low temperature physics; laser spectroscopy, particularly atomic and molecular resonance phenomena, laser chemistry, remote gas detection and laser engines; development of infrared parametric tunable oscillators; laser induced gas kinetics; laser cooling. *Mailing Add:* 2305 Marbury Rd Pittsburgh PA 15221

GARBUTT, JOHN THOMAS, b Janesville, Wis, Apr 19, 29; m 57; c 3. BIOCHEMISTRY. *Educ:* Beloit Col, BS, 51; Univ Wis-Madison, MS, 56, PhD(biochem), 58. *Prof Exp:* Sect leader, 58-72, MGR BIOCHEM, GRAIN PROCESSING CORP, 72- *Mem:* Am Chem Soc; AAAS. *Res:* Isolation and application of proteases and amylases for industrial use; industrial applications of carbohydrates and proteins. *Mailing Add:* Grain Processing Corp 1600 Oregon St Muscatine IA 52761-1476

GARCIA, ALBERT B, b Shirley, Mass, Nov 26, 44; m 68. COMPUTER DESIGN, NETWORK SIMULATION. *Educ:* WVa Univ, BSEE, 68, MSEE, 70; Farleigh Dickinson Univ, MBA; Univ Dayton, PHD(eng), 85. *Prof Exp:* asst prof comput eng, 85-89, PROD MGR, JOINT TACTICAL FUSION PROG, AIR FORCE INST TECHNOL, MCLEAN, VA, 89- *Concurrent Pos:* Adj asst prof, Univ Dayton, 85- *Mem:* Armed Forces Commun & Electronics Asn. *Res:* Digital logic design; computer hardware design; computer modeling and simulation; computer-communications networking; network simulations; network performance analysis. *Mailing Add:* 2150 Silentree Dr Vienna VA 22182-5169

GARCIA, ALFREDO MARIANO, b Itati-Corrientes, Arg, Sept 12, 27; US citizen; m 61. BIOLOGY, ANATOMY. *Educ:* Univ Buenos Aires, MD, 53, Dr Med, 58; Columbia Univ, PhD(zool), 62. *Prof Exp:* Res fel hemat, Mt Sinai Hosp, 57-59; instr biol, Columbia Univ, 59, lectr zool, 59-60, instr, 61-62; from asst prof to assoc prof, 62-74, PROF ANAT, STATE UNIV NY UPSTATE MED CTR, 74- *Concurrent Pos:* NIH fel, 60-62 & career develop award, 66-70. *Mem:* Histochem Soc. *Res:* Quantitative cytochemistry; fine structure and nucleic acid metabolism of mammalian blood cells. *Mailing Add:* 750 E Adams St Syracuse NY 13210

GARCIA, CARLOS E(RNESTO), b Las Vegas, Nev, May 14, 36; m 65; c 4. APPLIED MATHEMATICS. *Educ:* NMex State Univ, BS, 58, MS, 62, DSc(mech eng), 66, Indust Col Armed Forces, 84. *Prof Exp:* Assoc engr, Douglas Aircraft Co, 58-60; asst mech eng, NMex State Univ, 60-62, instr, 62-65; assoc res engr, NMex Inst Mining & Technol, 65-66, res engr & asst prof fluid dynamics, 66-67; sr scientist, Ling-Temco-Vought, 67-70; tech contract analyst, Atomic Energy Comn, US Dept Energy, 71-73, weapons develop engr, 73-77, sr prog engr, 77-81, chief, nuclear mat mgt br, 81-83, dir, Environ Safety & Health Div, 84- 86, dir energy technologies & waste mgt div, 86-89; PROG MGR, ENERGY & ENVIRON TECHNOLOGIES, LOS ALAMOS NAT LAB, 89- *Mem:* Nat Soc Prof Engrs; AAAS. *Res:* Hydromechanical missile control systems; shock wave phenomena; gas dynamics; thermodynamics; heat transfer; underground nuclear explosions; shock tubes; supersonic wind tunnels; subsonic and supersonic diffusers; boundary layer bleed; thermoelectrics; technical administration; science administration. *Mailing Add:* RR4 45 Sanchez Rd Belen NM 87002

GARCIA, CELSO-RAMON, b New York, NY, Oct 31, 21; m 50; c 2. OBSTETRICS & GYNECOLOGY, REPRODUCTIVE ENDOCRINOLOGY. *Educ:* Queens Col, NY, BS, 42; State Univ NY, MD, 45; Am Bd Obstet & Gynec, dipl. *Hon Degrees:* MA, Univ Pa, 72. *Prof Exp:* Intern Norweg Lutheran Hosp, Brooklyn, 45-46; resident path, Cumberland Hosp, 48-49, res fel obstet & gynec, 49-50, resident, 50-53; assoc, Sch Med & Trop Med, Univ PR, 53-54, asst prof, 54-55; res fel gynec, Harvard Med Sch, 55-56; asst dir, Rock Reproduction Clin, Inc, 56-58, co-dir, 58-61; dir training prog physiol of reproduction, Worcester Found Exp Biol, 60-62, sr scientist, 60-65; from assoc prof to prof obstet & gynec, 65-70, William Shippen Jr prof human reproduction, Sch Med & dir div, 70-87, vchmn dept obstet & gynec, Sch Med, 73-87, DIR REPRODUCTIVE SURG, UNIV PA HOSP, 87- *Concurrent Pos:* Res fel, Free Hosp Women, 55-57; asst, Harvard Med Sch, 58-59, asst obstet & gynec, 59-60, instr, 60-64, clin assoc, 64-65; assoc, San Juan Munic Hosp, 53-55; courtesy staff, Free Hosp Women, 57-58, from asst surgeon to assoc surgeon, 58-65; asst obstetrician & gynecologist, Boston Lying-in Hosp, 56-65; courtesy staff, Faulkner Hosp, 56-57, assoc staff, 57-65; courtesy staff, Glover Mem Hosp, 58-65 & Newton Wellesley Hosp, 63-65; consult, Worcester State Hosp, 61-65; asst surgeon & chief infertil clin, Mass Gen Hosp, 62-65; chief infertil & endocrine clin, Hosp Univ Pa, 65-70; consult, Philadelphia Gen Hosp, 65-70 & Pa Hosp, 73-; adj prof biol sci, Drexel Univ, 75-80. *Honors & Awards:* Carl Hartman Award, 61; Extraordinary Prof, Univ San Luis Potosi, Mex, 74; Pincus lectr, Sch Med, Wayne State Univ, 74. *Mem:* AAAS; Endocrine Soc; Am Fertil Soc (pres); AMA; Asn Planned Parenthood Physicians (pres, 74-75); Sigma Xi; Am Col Surgeons; Am Gynec & Obstet Soc; Soc Gynec Surgeons. *Res:* Reproductive physiology; infertility; menopause. *Mailing Add:* 109 Merion Rd Merion Station PA 19066

GARCIA, EUGENE N, b Guadalajara, Mex, Oct 27, 25; US citizen; wid; c 3. CLINICAL BIOCHEMISTRY, TOXICOLOGY. *Educ:* Gonzaga Univ, AB, 48; Univ San Francisco, MS, 51; Univ Calif, Los Angeles, PhD(physiol chem), 61. *Prof Exp:* Chemist org synthesis, Calif Corp Biochem Res, 54-55; res asst biochem, Med Ctr, Univ Calif, Los Angeles, 55-60; asst prof, Calif Col Med, Univ Calif, Irvine, 61-69; chmn dept health sci, 72-77, assoc prof, 72-79, PROF CHEM & HEALTH SCI, CALIF STATE UNIV, DOMINGUEZ HILLS, 80-; PARTNER, PROGRAMMEDIA ASSOCS, INC, 70- *Concurrent Pos:* Consult biomed educ, 69-; lectr, Univ Calif, Los Angeles, 70- *Mem:* AAAS; Am Chem Soc; Am Asn Clin Chem; Am Soc Qual Control; Nat Acad Clin Biochem; NY Acad Sci. *Res:* Ehrlich ascites carcinoma; irradiation effects and development of immunity in mice; lipids in Ehrlich ascites carcinoma; lipoprotein components. *Mailing Add:* ProgramMedia Assocs Inc 15608 Claretta Ave Norwalk CA 90650

GARCIA, HECTOR D, b Rio Grande City, Tex, Aug 26, 46; c 3. GENETIC TOXICOLOGY. *Educ:* Univ Tex, Houston, PhD(molecular biol), 79. *Prof Exp:* Res scientist, Phillip Morris USA, 81-87; consult, Functional Mgt Inst, 87-90; SR RES SCIENTIST, KRUG LIFE SCI, 90- *Mailing Add:* 5014 Camden Lane Pearland TX 77584

GARCIA, JOHN, b Santa Rosa, Calif, June 12, 17; m 43; c 3. CONDITIONED TASTE AVERSION. *Educ:* Univ Calif, Berkeley, BA, 48, MA, 49, PhD(psychol), 65. *Prof Exp:* Teaching asst psychol, Univ Calif, Berkeley, 49-51; psychologist, US Naval Radiol Defense Lab, San Francisco, 51-58; teacher biol sci, Oakland Pub Schs, Calif, 58-59; asst prof psychol, Calif State Col, Long Beach, 59-65; assoc biologist, Neurosurg Serv, Mass Gen Hosp, 65-68; prof psychol & chmn, Psychobiol Prog, State Univ NY, Stony Brook, 68-71, chmn, Dept, 71-72; prof, Univ Utah, Salt Lake City, 72-73; prof, 73-87, EMER PROF PSYCHOL, UNIV CALIF, LOS ANGELES, 87- *Concurrent Pos:* Consult neurophysiol, Long Beach Vet Admin Hosp, 59-65; lectr psychol, Dept Surg, Harvard Med Sch, Boston, 65-68 & Int Brain Res Sem, Kotor, Yugoslavia, 69; nat lectr, Sigma Xi, Southeastern US, 73-74; distinguished lectr, Univ Ill, 81; pres, Western Psychol Asn, 91-; William James fel, Am Psychol Asn. *Honors & Awards:* Howard Crosby Warren Medal, Soc Exp Psychologists, 78; Distinguished Sci Contrib Award, Am Psychol Asn, 79. *Mem:* Nat Acad Sci; fel Soc Exp Psychologists; AAAS; Am Psychol Soc. *Mailing Add:* 1950-A Chilberg Rd Mt Vernon WA 98273

GARCIA, JOSE DOLORES, JR, b Santa Fe, NMex, Jan 3, 36; m 60; c 2. ATOMIC PHYSICS. *Educ:* NMex State Univ, BS, 57; Univ Calif, Berkeley, MA, 59; Univ Wis, PhD(physics), 66. *Prof Exp:* NASA fel physics, Univ Pittsburgh, 66-67; from asst prof to assoc prof, 67-75, PROF PHYSICS, UNIV ARIZ, 75- *Concurrent Pos:* Consult, Lawrence Radiation Lab, Livermore, 70- & Air Force Weapons Lab, Kirtland AFB, 67; Fulbright grant, 57. *Mem:* Fel Am Phys Soc; Nat Physics Sci Consortium. *Res:* Atomic bound state theory; atomic scattering theory; use of super computers in theoretical physics. *Mailing Add:* Dept of Physics Univ of Ariz Tucson AZ 85721

GARCIA, JULIO H, b Armenia, Colombia, Dec 22, 33; nat US; m 66; c 2. NEUROPATHOLOGY, ELECTRON MICROSCOPY. *Educ:* Col St Bartholomew, Colombia, BS, 51; Nat Univ Colombia, MD, 58; Am Bd Path, dipl, 64. *Prof Exp:* Intern path, Hosp San Juan de Dios, Colombia, 58-59; resident physician, LI Jewish Hosp, 59-60; resident physician, Kings County Hosp, NY, 60-64; asst prof path, Med Col Va, 64-67; assoc prof path & neurol, Univ Tenn, Memphis, 67-70; assoc prof, Baylor Col Med, 70-71; prof path & head, Neuropath Div, Sch Med, Univ Md, Baltimore, 71-79, dir, Anat Path Labs, 77-79; PROF PATH & NEUROL, & DIR ANAT PATH, UNIV ALA, BIRMINGHAM MED CTR, 79- *Concurrent Pos:* Instr, State Univ NY, 62-64; consult, Eastern State Hosp, Va, 64-66, Off Chief Med Examr, Va, 64-66, Vet Admin Hosp, Richmond, Va, 64-66, Baptist Mem Hosp, Tenn, 67-, WTenn Psychiat Hosp & Inst, 67-, Vet Admin Hosp, Baltimore, 71- & Union Mem Hosp, 72-; USPHS grant, 69 & 73; fel coun cerebrovascular dis, Am Heart Asn, 70, mem res comt I, 71-73; mem stroke adv comt, Joint Comn Accreditation Hosps; mem, NSP-A, Nat Inst Health, 78-82. *Honors & Awards:* Humboldt Sr Scientist Award, Fed Rep Ger, 77-78. *Mem:* AAAS; NY Acad Sci; Am Asn Neuropath (vpres, 75-76); Am Asn Path & Bact; Asn Res Nerv & Ment Dis. *Res:* Ultrastructural changes of primates' brain after experimental production of regional ischemia; ultrastructural studies of mechanisms of nerve cell damage after ischemia, shock and trauma; effects of methanol poisoning on the brain; pathology. *Mailing Add:* Univ Ala Dept Path Univ Station Birmingham AL 35294

GARCIA, MANUEL MARIANO, b Dumaguete City, Philippines, Nov 25, 38; m 70. BACTERIOLOGY. *Educ:* Univ Philippines, BSA, 59, MSc, 62; Univ Guelph, PhD(microbiol, bact physiol), 67. *Prof Exp:* Res instr soil microbiol, Univ Philippines, 62-65; RES SCIENTIST & HEAD, BACT SECT, ANIMAL DIS RES INST, 67- *Mem:* AAAS; Am Soc Microbiol; Can Soc Microbiol; Sigma Xi. *Res:* Growth and survival of animal pathogens in soil; anaerobic bacteria and L-forms; microbial toxins; immunofluorescence; antigens and membranes of fusobacteria; phagocytosis and destruction of anaerobic bacteria by macrophages; microbial vaccines; pathogenesis of necrobacillosis. *Mailing Add:* Animal Dis Res Inst 801 Fallowfield Rd PO Box 11300 Nepean ON K2H 8P9 Can

GARCIA, MARCELO HORACIO, b Cordoba, Arg, Apr 22, 59; m 84. ENVIRONMENTAL HYDRODYNAMICS, HYDRAULIC MODELING. *Educ:* Nat Univ Litoral, Arg, Dipl Ing, 82; Univ Minn, MSc, 85, PhD(civil eng), 89. *Prof Exp:* Res docent hydraul, Nat Univ Litoral, Arg, 82-87; teaching asst fluid mech, Univ Minn, 84-86, res asst environ hydraul, 83-89; ASST PROF ENVIRON HYDRAUL, DEPT CIVIL ENG, UNIV ILL, 90- *Concurrent Pos:* Asst engr, Parano Medio Proj, AyEE, 82-83; vis engr, INCyTH, Arg, 83; consult, Northern States Power Co, Minn, 88-89; vis prof, Hokkaido River Res Found, Japan, 90, Hokkaido Lectr, 90. *Mem:* Am Geophys Union; Am Soc Civil Engrs; Int Asn Hydraul Res. *Res:* Environmental hydrodynamics; stratified flows; river mechanics; sediment transport; water-sediment interface phenomena and flows through porous media; man-made and natural phenomena. *Mailing Add:* 205 N Mathews Ave Urbana IL 61801

GARCIA, MARIANO, b Naguabo, PR, Sept 13, 18; m 40; c 2. MATHEMATICS. *Educ:* Washington & Jefferson Col, BS, 39, MA, 40; Univ Va, PhD(math), 44. *Prof Exp:* Instr math, Univ Richmond, 44; from asst prof to prof & head dept, Col Agr & Mech Arts, Univ PR, Mayaguez, 44-74; PROF MATH, HOSTOS COMMUNITY COL, 74- *Mem:* Am Math Soc; Math Asn Am. *Res:* Topology and theory of numbers. *Mailing Add:* Dept of Math Hostos Community Col 475 Grand Concourse Bronx NY 10451

GARCIA, MARIO LEOPOLDO, b Havana, Cuba, Sept 12, 40; US citizen. PHYSICAL CHEMISTRY, COSMETIC CHEMISTRY. *Educ:* Univ Havana, Dr Chem Sci, 64. *Prof Exp:* Mgr phys chem & anal chem dept, Cuban Inst Sugar Cane Derivatives, 64-66; res scientist phys chem, 66-68, sect head phys chem sect, 68-71, ASSOC DIR PHYS CHEM DEPT, RES LABS, CLAIROL INC, 71- *Concurrent Pos:* Assoc prof phys chem, Univ Havana, 65-66. *Mem:* Soc Rheol. *Res:* Physical chemistry of cosmetic products and their effects on human hair; mechanical, surface and optical properties of hair; structure of human hair; development of instrumental methods of chemical analysis. *Mailing Add:* 1459 Shippan Ave Stamford CT 06902

GARCIA, MEREDITH MASON, b Winchester, Mass; m; c 1. PHARMACOLOGY. *Educ:* St Mary's Dominican Col, BSc, 84; Tulane Univ, PhD(pharmacol), 90. *Prof Exp:* Dir, Radioimmunoassay Lab, Endocrine, Polypeptide & Cancer Inst, Vet Admin Med Ctr, New Orleans, 82-87; POSTDOCTORAL FEL, DEPT ANAT & PROG NEUROSCI, SCH MED, TULANE UNIV, 90- *Concurrent Pos:* Pharmaceut Mfrs Asn Found postdoctoral fel pharmacol- morphol, 90-92; Am Cancer Soc Inst res grant, 90-91; Nat Inst Drug Abuse res grant, 91-94; vis scientist, Transgenic Mouse Lab, Dept Molecular Biol, Squibb Inst Med Res, Princeton, 91. *Honors & Awards:* William G Carey Sci Award, AAAS, 90. *Mem:* Sigma Xi; Am Asn

Clin Chem; Am Asn Cancer Res; Am Soc Pharmacol & Exp Therapeut; Clin Ligand Assay Soc; Int Soc Differentiation; Soc Exp Biol Med; Soc Neurosci. *Res:* Molecular neurobiology of drug abuse; signal transduction at the opiate receptor; regulation of expression of immediate-early genes inn the brain; development of transgenic mouse models for the study of opiate actions; role of protein kinase C isoforms in neuronal plasticity and in the regulation of cellular proliferation and differentiation; author of numerous technical publications. *Mailing Add:* Dept Anat Tulane Med Sch, 1430 Tulane Ave New Orleans LA 70112

GARCIA, MICHAEL OMAR, b San Gabriel, Calif, Nov 18, 48; m 77; c 2. VOLCANOLOGY. *Educ:* Calif State Univ, Humboldt, BA, 71; Univ Calif, Los Angeles, PhD(geol), 76. *Prof Exp:* Asst prof, 76-81, ASSOC PROF GEOL, UNIV HAWAII, 81- *Mem:* Geol Soc Am; Am Geophys Union; Minerol Soc Am; Sigma Xi. *Res:* Volcanology; igneous petrology; field, microscopic and chemical analysis of the products of volcanoes including Kilauea and Mauna, Hawaii and Arenal, Costa Rica. *Mailing Add:* Dept Geol HIG 236 Univ Hawaii Manoa 2500 Campus Rd Honolulu HI 96822

GARCIA, OSCAR NICOLAS, b Havana, Cuba, Sept 10, 36; US citizen; m 62; c 2. ELECTRICAL ENGINEERING, COMPUTER SCIENCE. *Educ:* NC State Univ, BSEE, 61, MSEE, 64; Univ Md, PhD(elec eng), 69. *Prof Exp:* Asst elec eng, NC State Col, 61-62, instr, 62; elec engr, IBM Glendale Develop Labs, 62-63; asst prof elec eng, Old Dominion Col, 63-66; instr, Univ Md, 66-67, res asst, 67-68; assoc prof, Old Dominion Univ, 69-70; from assoc prof to prof elec eng, dept comput sci & eng, Univ SFla, Tampa, 70-78, chmn dept, 75-85; PROF ELEC ENG, DEPT ELEC ENG & COMPUT SCI, GEORGE WASHINGTON UNIV, 85- *Concurrent Pos:* Prin investr, NSF res initiation grant, 70-71; distinguished vis, Inst Elec & Electronics Engrs, 73-76, mem gov bd, 75-77; prog mgr, Instr Sci Equip Div, NSF, 77-78; prin investr, NSF grants, 80-83 & 90-91. *Honors & Awards:* Richard E Merwin Dist Serv Award, IEEE Computer Society, 88; Meritorious Serv Award, IEEE Computer Society, 91. *Mem:* Fel Inst Elec & Electronics Engrs; Am Soc Eng Educ; Asn Comput Mach; Sigma Xi; Inst Elec & Electronics Engrs Comput Soc (pres, 81-83); fel AAAS. *Res:* Application of coding theory to improved computer reliability and speed; computer architectural design and simulation of digital systems for design and diagnostic tests; artificial intelligence, expert systems, speech recognition. *Mailing Add:* Dept Elec Eng & Comput Sci George Washington Univ 801 22nd St NW Washington DC 20052

GARCIA, PILAR A, b Manila, Philippines, Nov 4, 26. NUTRITION. *Educ:* Univ Philippines, BS, 49; Univ Mich, MS, 50; Iowa State Univ, MS, 52, PhD(nutrit), 55. *Prof Exp:* Asst nutrit, 50-55, assoc & instr, 55-57, from asst prof to assoc prof food & nutrit, 57-74, PROF FOOD & NUTRIT, IOWA STATE UNIV, 74- *Mem:* Sigma Xi. *Res:* Human nutrition; energy expenditure of adult women; nutrition and aging of adult women; nutrition during adolescent pregnancy. *Mailing Add:* 34A Mackay Hall Iowa State Univ Ames IA 50011

GARCIA, RAUL, b Gijon, Spain, June 15, 35; Can citizen; m 60; c 3. HYPERTENSION, EXPERIMENTAL ENDOCRINOLOGY. *Educ:* Univ CHILE, MD, 60. *Prof Exp:* Asst prof int med, Univ CHILE, 63-68; res fel, Clin Res Inst Montreal, 68-70; assoc res prof, Inst Med Sci, Catholic Univ Chile, 70-73; res asst, 73-80, LAB DIR, CLIN RES INST MONTREAL, 80-; assoc prof, 83-90, PROF MED, UNIV MONTREAL, 90- *Concurrent Pos:* Assoc Mem Exp Med, Univ Mcgill, 84- *Mem:* Int Soc Hypertension; Can Soc Hypertension; Am Soc Hypertension; Am Heart Assoc. *Res:* Experimental hypertension endocrinology; atrial natriuretic factor, cardiovascular regulation. *Mailing Add:* Clin Res Inst Montreal 110 Pine Ave W Montreal PQ H2W 1R7 Can

GARCIA, RICHARD, b Sebastopol, Calif, Apr 26, 30; m 53; c 2. PARASITOLOGY, MEDICAL ENTOMOLOGY. *Educ:* Univ Calif, Berkeley, BS, 57, PhD(parasitol), 63. *Prof Exp:* USPHS fel arbovirus res, Rocky Mountain Lab, Mont, 63-65; asst res parasitologist, George Williams Hooper Found, Med Ctr, Univ Calif, San Francisco, 65-69; ASSOC ENTOMOLOGIST, UNIV CALIF, BERKELEY, 69- *Mem:* Entom Soc Am; Soc Insect Path; Am Mosquite Control Asn. *Res:* Behavior of bloodsucking arthropods to external stimuli and ecological studies of arboviruses in vector populations; biological control of mosquitoes. *Mailing Add:* Div Biol-1050 San Pablo Univ Calif Albany CA 94706

GARCIA-BENGOCHEA, FRANCISCO, b Havana, Cuba, Dec 15, 17; m 53; c 2. NEUROSURGERY. *Educ:* Univ Havana, MD, 41; Tulane Univ, MD, 49. *Prof Exp:* Res asst neurol, Col Physicians & Surgeons, Columbia Univ, 44-49, resident neurosurg, Presby Med Ctr, 48-49; instr neurol & neurosurg, Sch Med, Tulane Univ, 49-50, asst prof, 50-51; instr neurosurg, Col Med, Univ Fla, 60-61; asst prof, Sch Med, Univ Kans, 61-62; from assoc prof to prof, 62-76, distinguished serv prof, 70-83, EMER DISTINGUISHED SERV PROF NEUROSURG, COL MED, UNIV FLA, 83- *Concurrent Pos:* Traveling fel, Sch Med, Univ Havana, 43-45; surgeon, Greystone Brain Res Proj, Columbia Univ, 48-49; neurosurgeon, Charity Hosp New Orleans, Tulane Univ, 49-51; Markle scholar med sci, Sch Med, Tulane Univ, 50-51. *Mem:* Am Asn Neurosurg; Cong Neurol Surg; Neurosurg Soc Am. *Mailing Add:* Dept of Neurol Surg J 265 Univ of Fla Col of Med Gainesville FL 32610

GARCIA-CASTRO, IVETTE, b San Juan, PR, June 1, 37. ONCOLOGY, CELL CULTURE TECHNIQUES. *Educ:* Col Notre Dame Md, BA, 60; Cath Univ Am, MS, 70, PhD(cell biol), 73. *Prof Exp:* Chair, Sci Dept, Notre Dame High Sch, PR, 60-69; sr technician, Microbiol Assocs, Md, 71-73; asst dean, 82-87, chair, 85-87, PROF MICROBIOL, UNIV PR, RIO PIEDRAS, 73- *Concurrent Pos:* Guest scientist, Walter Reed Army Inst Res, Wash, DC, 81-82 & NIH, 91. *Mem:* Tissue Cult Asn; Am Soc Microbiol. *Res:* Determination of the rule of interleukin 1 and tumor neurosis factor and other cytohines on the systemic effects of tumors on the host. *Mailing Add:* Dept Biol Univ PR PO Box 22283 Rio Piedras PR 00931

GARCIA-COLIN, LEOPOLDO SCHERER, b Mex, Nov 27, 30; m 57; c 3. PHYSICS, THERMAL PHYSICS. *Educ:* Nat Univ Mex, BSc, 53 & 54; Univ Md, PhD(physics), 59. *Prof Exp:* Asst physics, Univ Md, 56-57, asst, Inst Fluid Dynamics & Appl Math, 57-59, res assoc, Inst Fluid Dynamics, 59-60; assoc prof, Nat Polytech Inst, Mex, 60-63; prof, Univ Puebla, 64-66; res prof physics, Nat Nuclear Energy Comn, 66-67; head appl res div, Mex Petrol Inst, 67-74; chmn dept physics & chem, 74-78, PROF PHYSICS & CHEM, METROP UNIV, IZTAPALAPA CAMPUS, MEX, 74- *Concurrent Pos:* Res asst, Nat Inst Sci Res, Mex, 53-62; consult, Nuclear Energy Comn, Mex, 60-76; lectr, Sch Mil Eng, 61-62; lectr, Nat Univ Mex, 61-62, part-time prof, 67-84; consult, Nat Inst Nuclear Energy, 79-82; Van der Waals Prof, Univ Amsterdam, 76; mem staff, El Colegio Nacional, Mex, 77. *Honors & Awards:* Sci Award, Acad Sci Res Mex, 65; Distinguished Prof, Univ Autonoma Metrop, 83; Fel Latin-Am Acad Sci, 87; Nat Prize Physics, Math & Natural Sci, 88. *Mem:* AAAS; Am Phys Soc; Am Asn Physics Teachers; Mex Physics Soc; El Colegio Nacional; fel Third World Acad Sci. *Res:* Statistical mechanics of equilibrium and non-equilibrium phenomena; superfluids; hydrodynamics of chemically reacting fluids; non-equilibrium thermodynamics; chemical kinetics. *Mailing Add:* Dept Physics Univ Autonoma Metrop-Iztapalapa Apdo Postal 55-534 Mexico 09340 DF Mexico

GARCIA-MUNOZ, MOISES, b Valencia, Spain, May 21, 22; m 55; c 1. ASTROPHYSICS. *Educ:* Univ Valencia, MS, 47; Univ Madrid, PhD(nuclear physics), 57. *Prof Exp:* Instr phys chem, Univ Valencia, 49-50, prof, Physics Div, Spanish Atomic Energy Comn, 51-56, investr, 56-59; res assoc atomic physics, 59-64, SR RES ASSOC, LAB ASTROPHYS & SPACE RES, ENRICO FERMI INST, UNIV CHICAGO, 64- *Mem:* Fel Am Phys Soc; Am Geophys Union. *Res:* Atomic and molecular processes; space physics. *Mailing Add:* Lab Astrophys & Space Res Univ Chicago 933 E 56th St Chicago IL 60637

GARCIA-PALMIERI, MARIO R, b Adjuntas, PR, Aug 2, 27; m 59; c 1. INTERNAL MEDICINE, CARDIOLOGY. *Educ:* Univ PR, BS, 47; Univ Md, MD, 51; Am Bd Internal Med, dipl, 58, cert cardiovasc dis, 62. *Prof Exp:* Intern, Fajardo Dist Hosp, 51-52; resident med, Bayamon Dist Hosp, 52-53; asst med, Sch Med, Univ PR, 53-54; Nat Heart Inst fel cardiol, 54-55; head dept med, Fajardo Dist Hosp, 55-56; from instr to prof, 55-84, assoc, 56-58, dir comprehensive med prog, 56-59, dir outpatient dept, 59-61, head dept med, 61-66, CHIEF SECT CARDIOL, SCH MED, UNIV PR, SAN JUAN, 61-, HEAD DEPT MED, 68-, DISTINGUISHED PROF MED, 84- *Concurrent Pos:* Resident med, San Patricio Vet Admin Hosp, 53-54; ed, Bull PR Med Asn, 60-66; head dept med, Univ Dist Hosp, 61; dir undergrad & postgrad cardiovasc training prog, Sch Med, Univ PR, 61-75; vis prof, Seton Hall Col Med, Sch Med, Univ Fla, 63 & Univ Ala, Birmingham, 81-82; vis lectr, Sch Med, Ind Univ, 63, Brooklyn Jewish Hosp, 64, Cent Univ Venezuela, 64 & Univs Barcelona & Madrid; lectr, Dominican Repub, 66 & 68; consult, San Patricio Vet Admin, Presby, San Jorge, San Juan City, Auxilio Muto & Doctor & Teachers Hosps; lectr cardiovasc epidemiol, Dept Prev Med & Pub Health, Univ PR; fel coun clin cardiol & coun epidemiol, Am Heart Asn; mem, presidential comn ethical aspects med, biomed & behav res, 79-82; mem, adv coun, Nat Inst on Aging, 79-83. *Mem:* AAAS; fel Am Col Physicians; fel Am Col Cardiol; Am Fedn Clin Res; Am Soc Trop Med & Hyg; Am Heart Asn; fel Royal Soc Health; fel Am Col Chest Physicians; Asn Am Med Cols; Asn Prof Med. *Res:* Tropical diseases; study of different electrocardiographical alternations, vectorcardiogram and coronary atherosclerosis. *Mailing Add:* MACP Box DG Caparra Heights San Juan PR 00922

GARCIA RAMOS, JUAN, b Queretaro, Mexico, June 26, 15; m 38; c 10. PHYSIOLOGY. *Educ:* Army Med Sch, Mex, MD, 36; Nat Polytech Inst, Mex, SciD, 64. *Prof Exp:* From asst prof to assoc prof physiol, Army Med Sch, Mex, 40-58, assoc prof pharmacol, 53, prof phys chem & gen physiol, 52-58, prof physiol, 58-61; assoc prof, Nat Polytech Inst, Mex, 42-44, prof physiol, Res Ctr Advan Study, 61-79; RETIRED. *Concurrent Pos:* Investr, Inst Cardiol, 44-53; vis investr, Rockefeller Inst, 48-49; Guggenheim fels, 48-49 & 51-52; sr res fel, Calif Inst Technol, 52; head dept physiol, Inst Neumol, 53-57; prof under contract, Res Ctr Advan Study, 80-87. *Honors & Awards:* Elias Sourasky Sci Award, 70. *Mem:* Soc Neurosci; Int Brain Res Org. *Mailing Add:* Dept Physiol & Biophys Nat Polytech Inst AD 14-740 Mexico 14 DF Mexico

GARCIA-RILL, EDGAR E, b Caracas, Venezuela, Oct 31, 48; Can citizen. NEUROPHYSIOLOGY. *Educ:* Loyola Col Montreal, BA, 69; McGill Univ, PhD(physiol), 73. *Prof Exp:* Res assoc neuropsychiat, Allan Mem Inst, Montreal, 72-73; fel res anat neurophysiol, Univ Calif, Los Angeles, 74-78; ASST PROF INTERDISCIPLINARY TOXICOL, UNIV ARK, 81-, ASSOC PROF ANAT, 82- *Concurrent Pos:* Fel Que Med Res Coun, Dept Psychiat, Univ Calif, Los Angeles, 73-74. *Mem:* Soc Neurosci; Can Physiol Soc; Am Asn Anat. *Res:* Neurophysiology of motor-sensory interactions; motor cortex; dorsal columns and basal ganglia; locomotion. *Mailing Add:* Dept Anat Univ Ark Med Sci 4301 Markham W Little Rock AR 72205

GARCIA-SAINZ, J ADOLFO, b Mexico, DF, Mar 10, 52; m 79; c 4. SIGNAL TRANSDUCTIONS, RECEPTORS & SECOND MESSENGERS. *Educ:* La Salle Col, BSc, 70; Nat Autonomous Univ Mexico, MD, 76, MSc, 78, PhD(biochem), 81. *Prof Exp:* Asst prof, 80-83, assoc prof, 80-82, PROF & CHMN BIOCHEM INST PHYSIOL, NAT AUTONOMOUS UNIV MEXICO, 83- *Concurrent Pos:* Ed, Europ J Pharmacol, 84-; fel John Guggenheim Found, 85; vpres & pres, Mexican Soc Biochem, 87-91. *Honors & Awards:* M Aleman Prize, Found Miguel Aleman, 85. *Mem:* Am Soc Biol Chemists; Am Soc Pharmacol & Exp Therapeut; Biochem Soc. *Res:* Processes involved in signal transduction; receptors, G-proteins, membrane effectors and second messengers are determined using currently available techniques; adrenergic receptors and protein kinase C. *Mailing Add:* Nat Autonomous Univ Mexico Apdo 70-248 Mexico City 04510 Mexico

GARCIA-SANTESMASES, JOSE MIGUEL, b Madrid, Spain, Dec 19, 52; m 82; c 2. DATA ANALYSIS METHODS & TECHNIQUES, EXPERT SYSTEMS. *Educ:* Univ Complutense, Bachelor, 75, PhD(math), 82. *Prof Exp:* Asst prof statist, 75-84, PROF TITULOR OPERS RES, UNIV COMPLUTENSE, 85- *Concurrent Pos:* Proj leader, Intel Decision Syst, SAm, 86-89; tech mgr, Grupo Apex, SAm, 90- *Mem:* Soc Expert Intel Opers. *Res:* Decision support systems; artificial intelligence techniques and its interface with operational research methods; mathematical modeling and its applications to medicine and ecology. *Mailing Add:* Dpto Ectaclistica Univ Complutense Madrid Spain

GARD, DAVID LYNN, CYTOSKELETON, MICROTUBULES. *Educ:* Calif Inst Technol, PhD(cell biol), 82. *Prof Exp:* SCHOLAR, MED SCH, UNIV CALIF, 82- *Mailing Add:* Dept Biochem and Biophysics Univ Calif Med Sch San Francisco CA 94143

GARD, DAVID RICHARD, b Connersville, Ind, Sept 28, 53; m 80; c 2. PHOSPHATE CHEMISTRY. *Educ:* Purdue Univ, BS, 75; Univ Ill, PhD(inorganic chem), 81. *Prof Exp:* Sr res chemist, 80-84, res specialist, 84-88, ASSOC FEL, MONSANTO CO, 88- *Mem:* Am Chem Soc. *Res:* Phosphate chemistry and technology; detergent ingredients; phase equilibria; kinetics; product and process development. *Mailing Add:* Monsanto Chem Co 800 N Lindbergh Blvd St Louis MO 63167

GARD, DON IRVIN, b Beaver Crossing, Nebr, June 18, 26. ANIMAL NUTRITION. *Educ:* Univ Nebr, BS, 50; Okla State Univ, MS, 52; Univ Ill, PhD(animal sci), 54. *Prof Exp:* Asst, Okla State Univ, 50-51 & Univ Ill, 52-54; dir res & nutrit, Crete Mills, Lauhoff Grain Co, Nebr, 54-57; sr scientist, 57-70, res scientist, 70-78, RES ASSOC, ELI LILLY & CO, 70- *Mem:* Am Soc Animal Sci; Poultry Sci Asn; Animal Nutrit Res Coun; World Poultry Sci Asn; Sigma Xi. *Res:* Research utilizing computers to develop new antibiotics and chemo-therapeutic agents for poultry. *Mailing Add:* 1735 Hickory Lane Greenfield IN 46140

GARD, GARY LEE, b Goodland, Kans, Nov 17, 37; m 72; c 4. FLUORINE CHEMISTRY. *Educ:* Univ Wash, BS, 60, PhD(chem), 64. *Prof Exp:* Sr res chemist, Allied Chem Co, 64-66; from asst prof to assoc prof, Portland State Univ, 66-75, chmn dept, 71-77, exec coun, Div Fluorine Chem, 77-80, actg dean, Col Sci & coordr, Environ Sci PhD Prog, 79-81, PROF CHEM, PORTLAND STATE UNIV, 75- *Concurrent Pos:* Mem staff, Marine Sci Comn, 79-81; consult, Col Chem Consults Serv, Am Chem Soc, 84-; adj prof, Ore Grad Ctr, 84- *Mem:* Am Chem Soc; Sigma Xi. *Res:* Preparation of new oxidizing agents; preparation of new dielectric materials, fuel cell electrolytes, surface active agents, polymers, and complexing systems that contain fluorine; preparative and physical chemicals studies of fluorine containing metals and non-metals systems. *Mailing Add:* Dept Chem Portland State Univ Portland OR 97207

GARD, JANICE KOLES, b Phoenix, Ariz, June 4, 54; m 80; c 2. NUCLEAR MAGNETIC RESONANCE SPECTROSCOPY. *Educ:* San Diego State Univ, BS, 77; Univ Ill, MS, 80, PhD(phys chem), 84. *Prof Exp:* Anal chemist, Am Testing Inst, 73-77; res asst, Los Alamos Nat Lab, 77; sr res chemist, 84-89, RES SPECIALIST, MONSANTO CO, 89- *Concurrent Pos:* Co-chair, St Louis Nuclear Magnetic Resonance Discussion Group, Am Chem Soc, 89-91; lab mgr, Phys Sci Ctr, Nuclear Magnetic Resonance Facil, Creve Coeur, Monsanto Co, 89-91. *Mem:* Am Chem Soc; Sigma Xi. *Res:* Multinuclear, multidimensional nuclear magnetic resonance structural characterization of bioinorganic, inorganic, and organic molecules. *Mailing Add:* 800 N Lindbergh St Louis MO 63167

GARD, O(LIVER) W(ILLIAM), b Berea, Ky, Oct 2, 21; m 43; c 5. MECHANICAL ENGINEERING. *Educ:* Univ Ky, BS, 48; Ga Inst Technol, MS, 51. *Prof Exp:* From instr to asst prof, 48-56, ASSOC PROF MECH ENG, UNIV KY, 56- *Mem:* Am Soc Mech Engrs. *Res:* Work simplification; time study; production engineering; statistical quality control; plant layout; machine design. *Mailing Add:* Keene Rd Lexington KY 40506

GARD, RICHARD, b Alhambra, Calif, July 6, 28; m 63; c 2. FISHERIES. *Educ:* Univ Calif, AB, 50, MA, 53, PhD(zool), 58. *Prof Exp:* Res zoologist, Univ Calif, 56-58, jr res zoologist, 58, lectr zool, 58-59, from jr res zoologist to asst res zoologist, 59-62; res biologist, US Bur Commercial Fisheries, 62-66; assoc res zoologist, Univ Calif, 69-70, Belvedere Sci Fund grant, 70 & 73-74; res biologist, Nat Marine Fisheries Serv, 71-73; assoc prof fishery & wildlife biol, Colo State Univ, 73-75; dir div fisheries & natural sci, 75-82, prof fisheries, 78-88, actg dean, Sch Fisheries & Sci, 82-83, EMER PROF FISHERIES, UNIV ALASKA SOUTHEAST, 88- *Concurrent Pos:* Prin investr, Belvedere Sci Fund, 69-74, Marine Mammal Comm, 74-76; consult, US Bur Reclamation, 74-75; prin investr, Alaska Sea Grant Prog, 76-79; prin investr, Univ Alaska Juneau Res Coun, 83-85, Alaska Dept Fish & Game, 85-86. *Mem:* Wildlife Soc; Am Fisheries Soc; Am Soc Mammalogists; Soc Marine Mammal; fel Am Inst Fishery Res Biologists. *Res:* Trout stream ecology; life history of sockeye salmon; life history and conservation of the gray whale; age and growth of Arctic char; stock separation of salmonids. *Mailing Add:* Juneau Ctr Fisheries & Ocean Sci Univ Alaska Fairbanks 11120 Glacier Hwy Juneau AK 99801

GARDELLA, JOSEPH AUGUSTUS, JR, b Detroit, Mich, Aug 22, 55; m 83. SURFACE SCIENCE & SPECTROSCOPY. *Educ:* Oakland Univ, BA(philos) & BS(chem), 77; Univ Pittsburgh, PhD(anal chem), 81. *Prof Exp:* Fac intern chem, Univ Utah, 81-82; asst prof, 82-87, ASSOC PROF CHEM, STATE UNIV NY, BUFFALO, 87-, RES ASST PROF STOMATOLOGY, 87- *Concurrent Pos:* Dir, Surface Sci Ctr, State Univ NY, Buffalo, 85-; consult, Air Prod & Chem Inc, 86-; secy, Niagara Frontier Sect, Soc Applied Spectros, 87-; mem, Joint Polymer Educ Comt-Am Chem Soc-Div Polymer Chem, Inc, 87-; co-dir, Indust Univ Ctr Biosurfaces, NSF Indust Univ Coop Res Ctr, 88-; prog officer, Chem Div, NSF, Wash, DC, 89-90; fel, Gelb Found, 86-89, Exxon Educ Found, 89-91. *Mem:* Am Chem Soc; Am Vacuum Soc; AAAS; Soc Applied Spectros; Mat Res Soc. *Res:* Study of structure, composition and function of macromolecular surfaces and interfaces; new methods of analysis of polymer surfaces and metal-organic interfaces and thin films; surface modification and characterization of multicomponent polymers; characterization of biological-material interfaces. *Mailing Add:* Dept Chem State Univ NY Buffalo Buffalo NY 14214

GARDELLA, LIBERO ANTHONY, b Chicago, Ill, July 24, 35; m 65; c 3. PHARMACEUTICAL CHEMISTRY. *Educ:* Univ Ill, BS, 59, PhD(pharmaceut chem), 62. *Prof Exp:* Fel org chem, Princeton Univ, 62-63; res pharmacist, Pharmaceut Div, Abbott Labs, 63-67, res pharmacist, Hosp Prod Div, 67-69, prog mgr prod develop, 69; dir qual affairs, Arnar-Stone Labs, Am Hosp Supply Corp, 69-80; dir qual control, 80-82, DIR PHARM DEVELOP, SMITH, KLINE & FRENCH LABS, 82- *Mem:* Am Pharmaceut Asn; Acad Pharmaceut Sci; Am Chem Soc; Am Asn Pharmaceut Scientists. *Res:* Medicinal chemistry; qual control of pharmaceutical dosage forms; formulation of pharmaceutical dosage forms. *Mailing Add:* Smith, Kline & French Labs PO Box 7929 Philadelphia PA 19101-7929

GARDENIER, JOHN STARK, b Portland, Maine, Apr 10, 37; m 77; c 6. MANAGEMENT SCIENCE, SIMULATOR SCIENCE. *Educ:* Yale Univ, BA, 59; George Washington Univ, DBusA, 73. *Prof Exp:* Mem tech staff mgt info syst, Comput Sci Corp, 68-69; sr analyst ship costs, CONSULTEC, 69-71; opers res analyst, Personnel Studies, Risk Anal & Marine Safety, USCG, 71-90; STATISTICIAN & SOFTWARE ENGR, NAT CTR HEALTH STATIST, 90- *Concurrent Pos:* Sponsor liaison, Maritime Transp Res Bd, Nat Res Coun, 73-83; rep, Int Marine Simulator Forum, 78-87; adj assoc prof, George Washington Univ, 81-82; prof lect, Am Univ, 82-84. *Honors & Awards:* Silver Medal, Dept Transp, 83. *Mem:* AAAS; Soc Comput Simulation; Am Statist Asn; Inst Mgt Sci. *Res:* Maritime transportation safety, especially relating to human factors, work demands, training and testing and simulators; alcohol and transportation safety; computer-based simulation, management science; modeling and statistical analysis in managing personnel productivity, effectiveness and economics; software engineering and management for developing computer assisted survey information collection software. *Mailing Add:* 115 St Andrews Dr NE Vienna VA 22180

GARDENIER, TURKAN KUMBARACI, b Istanbul, Turkey, Nov 10, 41; m; c 2. MEDICAL STATISTICS. *Educ:* Vassar Col, AB, 61; Columbia Univ, MA, 62, PhD(statist), 66. *Prof Exp:* Proj dir testing, Sci Res Assocs, 66-67; res scientist statist & opers res, IIT Res Inst, 67-69; asst prof statist & chmn dept, Mid East Tech Univ, 68-70; vis scientist appl math, Brookhaven Nat Lab, 70-71; assoc dir statist, Pfizer Pharmaceut, 71-73; asst prof marine transp, State Univ NY Maritime Col, 73-77; res assoc biostatist, Rockefeller Univ, 75-76; med statistician, US Environ Protection Agency, 78-81, PRES TKG CONSUTL LTD & PRO-RLE COMPUT INST, 82-; PRES TKG CONSULT LTD & PRO-RLE COMPUT INST, 82- *Concurrent Pos:* Cent Treaty Orgn res grant, 69; NSF res grant, 70; consult, Col Physicians & Surgeons, Columbia Univ, 73-76 & Dept Energy, 82-84; biostatistician, Ctr Prev Premature Arteriosclerosis, 75-76; asst presiding officer, State Univ NY Maritime Col, 75-77; consult Med Commun-Proj Aldactone, 75-76; systs consult, North Shore Univ Hosp, 75-76; adj asst prof biostatistics, Col Physicians & Surgeons, Columbia Univ, 78-80; adj assoc prof statist, George Washington Univ, 79-82; mem comt, Nat Acad Sci; assoc prof, Am Univ, 82-84; subcontractor, Anser Corp, 82- *Mem:* Opers Res Soc Am; Am Statist Asn; AAAS; Soc Risk Anal; Soc Comput Simulation; Biomet Soc. *Res:* Computer applications of medical research; large scale data base design; analysis of time series data, non-parametric statistics, index derivations for patient data; environmental risk estimation; metamodels and preprocessors for simulations. *Mailing Add:* 301 Maple Ave W Suite 100 Vienna VA 22180

GARDER, ARTHUR, b Kansas City, Mo, Dec 17, 25; m 53; c 2. NUMERICAL ANALYSIS. *Educ:* Univ Chicago, BS, 48; Wash Univ, MA, 50, PhD(math), 54. *Prof Exp:* Asst math, Wash Univ, 49-52; mathematician, United Gas Corp, 52-55; programmer utility routines, Int Bus Mach Corp, 55-56; mathematician numerical anal, Humble Oil & Refining Co, 56-64; assoc prof comput sci, Wash Univ, 64-66; assoc prof, 66-75, PROF MATH, SOUTHERN ILL UNIV, 75- *Mem:* Soc Indust & Appl Math; Math Asn Am. *Res:* Convolution transforms with totally positive kernels; solution of partial differential equations of elliptic and parabolic type by difference methods. *Mailing Add:* Fac of Math Studies Southern Ill Univ Edwardsville IL 62026

GARDIER, ROBERT WOODWARD, b Scranton, Pa, May 17, 27; m 51; c 3. PHARMACOLOGY. *Educ:* Univ Scranton, BS, 49; Univ Tenn, MS, 52, PhD(pharmacol), 54. *Prof Exp:* Pharmacologist, Pitman-Moore Co, 53-58; asst prof pharmacol & dir res anesthesiol, Sch Med, Ind Univ, 59-61; assoc prof pharmacol, Univ Tex Med Br, 61-63; assoc prof pharmacol & anesthesia res dir, Col Med, Ohio State Univ, 63-67, prof, 67-69; dir biol res, Bristol Labs, 69-71; prof pharmacol & anesthesiol, Col Med, Ohio State Univ, 71-77, actg chmn pharmacol, 73-74; assoc dean grad studies, 77-78, dir, Biomed Sci Doctoral Prog, 78-81, PROF & DIR BASIC PHARMACOL & TOXICOL, SCH MED, WRIGHT STATE UNIV, 77- *Concurrent Pos:* Mem med div, Inst Nuclear Studies, Oak Ridge, 51, consult, 51-52. *Mem:* AAAS; Soc Pharmacol & Exp Therapeut; Soc Exp Biol & Med; Soc Toxicol. *Res:* Drugs related to anesthesia; autonomic nervous system pharmacology; muscarinic receptor subtypes. *Mailing Add:* Dept of Pharmacol & Toxicol Sch of Med Wright State Univ Dayton OH 45435

GARDIN, JULIUS M, b Detroit, Mich, Jan 14, 49; m 82; c 2. CARDIOVASCULAR DISEASES, CARDIAC ULTRASOUND. *Educ:* Univ Mich, BS, 68, MD, 72. *Prof Exp:* Intern & resident internal med, Univ Mich Affil Hosps, 72-75; fel cardiol, Georgetown Univ Affil Hosps, 75-77; assoc med, Sch Med, Northwestern Univ, 77-78, asst prof, 78-79; from asst prof to assoc prof, 79-89, PROF MED & CARDIOL, UNIV CALIF, IRVINE MED CTR, 89- *Concurrent Pos:* Guest investr cardiol, Cardiol Br, Nat Heart, Lung & Blood Inst, NIH, 76-77; dir, Cardiopulmonary Resuscitation Comt & Cardiol Noninvasive Lab, Vet Admin, Lakeside Med Ctr, 77-79; assoc ed, Am J Cardiac Imaging, 85-; prin investr, Am Heart Asn, Calif Affil, 80-81,

82-84, Nat Inst Health, 88-; mem bd dirs, Am Soc Echocardiography, 85-, Int Cardiac Doppler Soc, 85-, Coun Geriatric Cardiol, 86-; fel, Coun Clin Cardiol & Coun Geriat Cardiol. *Mem:* Fel Am Col Physicians; fel Am Col Cardiol; fel Am Col Chest Physicians; Am Heart Asn; Int Cardiac Doppler Soc (vpres, 88-90, pres, 90-92); Am Soc Echocardiography (treas, 88-90, vpres, 90-92). *Res:* The application of ultrasound and magnetic resonance imaging to define anatomy and function of the heart (Doppler echocardiography) in experimental models and clinical situations; geriatric cardiology. *Mailing Add:* Div Cardiol Univ Calif Irvine Med Ctr 101 City Dr Rte 81 S Orange CA 92668

GARDINER, BARRY ALAN, electrification, entrainment, for more information see previous edition

GARDINER, DONALD ANDREW, b Buffalo, NY, Feb 2, 22; m 43; c 3. MATHEMATICAL STATISTICS. *Educ:* Univ Buffalo, BS, 43, MBA, 48; NC State Col, PhD(exp statist), 56. *Prof Exp:* Lectr statist, Univ Buffalo, 46-48; asst prof, Univ Tenn, 48-51; asst statistician, NC State Col, 55-56; statistician phys & eng sci, Oak Ridge Nat Lab, Tenn, 56-67, asst dir math div, 67-73; head math & statist res dept, Comput Sci Div, Nuclear Div, Union Carbide Corp, 73-; AT MARTIN MARIETTA ENERGY SYSTS. *Concurrent Pos:* Assoc prof, Univ Tenn, 65-73, prof, 73-; vis prof, Fla State Univ, 66-67; ed-elect, Technometrics, 71, ed, 72-74. *Mem:* Biomet Soc; fel Am Statist Asn; Inst Math Statist; Sigma Xi; fel AAAS. *Res:* Experimental statistics; design of experiments for physical sciences; statistical analysis of experiments in physical sciences; probability models; computational statistics. *Mailing Add:* 108 Mason Lane Oak Ridge TN 37830

GARDINER, JOHN ALDEN, b Providence, RI, Feb 9, 38; m 60; c 3. PESTICIDE CHEMISTRY, ANALYTICAL CHEMISTRY. *Educ:* NC Univ, BS, 60; Ohio State Univ, MS, 62, PhD(anal chem), 64. *Prof Exp:* Res chemist, Indust & Biochem Dept, 64-67, sr res chemist, 67-69, res supvr, 69-74, res mgr biochem dept, 74-76, asst mgr prod regist, 76-77, MGR REGIST & REGULATORY AFFAIRS, AGR PROD DEPT, E I DU PONT DE NEMOURS & CO, INC, 77- *Mem:* Am Chem Soc; Soc Chem Indust. *Res:* Metabolism/degradation of c-14 pesticides in soil, plants, animals, and water; pesticide residue analysis; assay methods; organic trace analysis. *Mailing Add:* Agr Prod Dept E I du Pont de Nemours & Co Inc Wilmington DE 19898

GARDINER, JOHN BROOKE, b Bryn Mawr, Pa, Nov 9, 29; m 56; c 2. ORGANIC CHEMISTRY. *Educ:* Haverford Col, AB, 51; Univ NC, PhD(org chem), 57. *Prof Exp:* Chemist prod res div, Exxon Chem Technol, 57-59 & high energy propellant proj, 59-64, sr chemist, 64-66, proj leader rubber adhesion, Enjay Polymer Labs, 66-68, staff planner prog & budget, 68-70, res assoc, 71, prog mgr, Enjay New Venture Additives Lab, 70-71, SECT HEAD, EXXON CHEM TECHNOL, LINDEN, 71-, PARAMINS, 72- *Mem:* AAAS; Am Chem Soc; Sigma Xi. *Res:* Optical isomerization; anthracene chemistry; synthetic motor oils; viscosity index improvers for motor oils; synthesis of high energy oxidizers-monomers-binders; scale-up formulation and micro rocket firing; rubber adhesion; microinterferometry; gear oils. *Mailing Add:* Exxon Chem Technol PO Box 536 Linden NJ 07036

GARDINER, KENNETH WILLIAM, b Chicago, Ill, Feb 10, 17; m 42; c 3. PHYSICAL CHEMISTRY, ANALYTICAL CHEMISTRY. *Educ:* Stanford Univ, AB, 39, MA, 40; Mass Inst Technol, PhD(instrumental anal chem), 52. *Prof Exp:* Sr res chemist, Lever Bros Co & Firestone Tire & Rubber Co, 41-49; asst, Mass Inst Technol, 50-52; dir res, Gardiner Instrument Res Lab, 52-56; dir res, Gen Chem Lab, Cent Res & Eng Div, Continental Can Co, 56-59; chief res chemist & asst dir, Cent Res Div, Consol Electrodynamics Corp Div, Bell & Howell Co, 59-60, dir chem res, Bell & Howell Res Ctr, 60-64; dir res & eng & gen mgr, Anal Systs Co, Teledyne Systs Corp, 64-70, pres & gen mgr, Teledyne Anal Instruments, Teledyne Inc, 70-72; prof appl sci, Col Phys Sci, 72-77, chmn, Appl Sci Prog, 73-77, prof admin, Grad Sch Admin, 72-82, prof, 77-87, PROF MGT, GRAD SCH MGT, UNIV CALIF, RIVERSIDE, 82-, EMER PROF ENVIRON SCI, COL NATURAL & AGR SCI, UNIV CALIF, 87- *Concurrent Pos:* Res assoc, Mass Inst Technol, 53; mem adv panel, Nat Res Coun-Nat Acad Sci for Anal Chem Div, Nat Bur Standards, 72-78, chmn, 77-78. *Mem:* Am Chem Soc; Am Phys Soc. *Res:* Management of technology, technology transfer, methods for measuring technology effectiveness and cost/benefit evaluations of research and development; instrumental analytical chemistry; physical and instrumental methods in research and chemical analysis. *Mailing Add:* 1125 Mesita Rd Pasadena CA 91107

GARDINER, LION FREDERICK, b Glen Cove, NY, June 21, 38; div. HIGHER EDUCATION DEVELOPMENT. *Educ:* Wheaton Col, Ill, BS, 60; Univ Mich, MS, 64; Univ RI, PhD, 72. *Prof Exp:* Instr biol, Delta Col, Mich, 64-65; lectr, 69-72, asst prof, 72-74, ASSOC PROF ZOOL, RUTGERS UNIV, NEWARK, 74- *Concurrent Pos:* Fac fel, NJ State Dept Higher Educ, 87-88. *Mem:* AAAS; Am Asn Higher Educ; Am Soc Study Higher Educ; Am Educ Res Asn. *Res:* Biology of the deep-sea fauna; systematics and biology of the Tanaidacea; marine benthic ecology; development of teaching skills of university faculty and graduate students; liberal education of undergraduates; academic management. *Mailing Add:* Dept Biol Sci Rutgers Univ 101 Warren St Newark NJ 07102-3192

GARDINER, WILLIAM CECIL, JR, b Niagara Falls, NY, Jan 14, 33; m 59, 91; c 3. PHYSICAL CHEMISTRY. *Educ:* Princeton Univ, AB, 54; Harvard Univ, PhD(chem), 60. *Prof Exp:* Res assoc chem, Max Planck Inst Phys Chem, Gottingen, 55-57; from instr to assoc prof, 60-72, PROF CHEM, UNIV TEX, AUSTIN, 72- *Concurrent Pos:* Guggenheim fel, Inst Phys Chem, Univ Gottingen, 75-76; Thyssen fel, 81; Lady Davis vis prof, Hebrew Univ Jerusalem, 85. *Honors & Awards:* Alexander von Humboldt Award, 79. *Mem:* AAAS; Am Phys Soc; Am Chem Soc; Combustion Inst. *Res:* Chemical kinetics; combustion; shock and detonation waves; molecular evolution. *Mailing Add:* Dept Chem Univ Tex Austin TX 78712

GARDLUND, ZACHARIAH GUST, b Lake City, Minn, Sept 12, 37; m 63; c 3. POLYMER CHEMISTRY, ORGANIC CHEMISTRY. *Educ:* Carleton Col, BA, 59; Univ Ariz, PhD, 64. *Prof Exp:* From assoc sr res chemist to sr res chemist, 64-69, sr res chemist & supvr org chem sect, 69-77, staff res scientist, 77-87, SUPVR ADVAN MAT GROUP, POLYMERS DEPT, RES LAB, GEN MOTORS TECH CTR, 77- , SR STAFF RES SCIENTIST, 88- *Mem:* Am Chem Soc. *Res:* Monomer-polymer synthesis; polymer structure-property relationship; block co-polymers; polymer blends; macromolecular composites. *Mailing Add:* Polymers Dept Res Labs Gen Motors Tech Ctr Warren MI 48090

GARDNER, ALVIN FREDERICK, b Chicago, Ill, Mar 22, 20; m 42, 82; c 1. PATHOLOGY. *Educ:* Emory Univ, DDS, 43; Univ Kansas City, cert, 46; Univ Ill, MS, 57; Georgetown Univ, PhD(path), 59. *Prof Exp:* Res assoc & instr oral path, Univ Ill, 57; resident oral path, Dent & Oral Br, Armed Forces Inst Path, 57-59; assoc prof path & oral path, Sch Dent, Univ Md, 59-63; pathologist, Bur Drugs, 63-67, dent officer, Off Drug Surveillance, 67-69, ORAL PATHOLOGIST, BUR DRUGS, FOOD & DRUG ADMIN, DEPT HEALTH & HUMAN SERV, 69- *Concurrent Pos:* Consult, Vet Admin, 60-; vis scientist, Nat Bur Standards; staff dentist, Kadlec Hosp, Hanford Works, Wash; mem dent serv, Stockton State Hosp, Calif; consult, Stedman's Med Dictionary; NIH, US Army Res & Develop Command, Am Cancer Soc & Sigma Xi res grants. *Mem:* AAAS; Am Nutrit Soc; fel Am Pub Health Asn; Am Med Writers' Asn; Am Dent Asn. *Res:* Experimental Lathyrism; disturbances in the metabolism of connective tissue; nutritional disturbance in mesoderm; oral pharmacology, effects of drugs on oral tissues. *Mailing Add:* 200 Hidden Valley Ln Silver Spring MD 20907

GARDNER, ANDREW LEROY, b Ogden, Utah, Feb 6, 19; m 41; c 5. PLASMA DIAGNOSTICS, MICROWAVE CIRCUITRY. *Educ:* Utah State Univ, BS, 40; Univ Calif, Berkeley, PhD(physics), 55. *Prof Exp:* Commun asst, Idaho Nat Forest, USDA, 41; radio engr, Off Chief Signal Officer, US War Dept, 42-44; mem staff, Radiation Lab, Mass Inst Technol, 44-45; res asst & assoc, Inst Eng Res, Univ Calif, 46-54, physicist, Lawrence Radiation lab, 54-64; from assoc prof to prof, 64-84, EMER PROF PHYSICS, BRIGHAM YOUNG UNIV, 85- *Concurrent Pos:* Consult, Inst Plasma Physics, Japan, 62 & Lawrence Radiation Lab, 68-75. *Mem:* Fel Am Phys Soc. *Res:* Experimental plasma physics; microwave circuitry; electronics; high voltage switching. *Mailing Add:* 555 E 2950 N Provo UT 84604

GARDNER, ARTHUR WENDEL, b Cedar City, Utah, Oct 23, 24. GENETICS. *Educ:* Utah State Univ, BS, 49; Kans State Univ, MS, 54, PhD(genetics), 56. *Prof Exp:* Instr genetics, Kans State Univ, 55; instr biol, Russell Sage Col, 61-64; asst prof, Washburn Univ, 64-67; assoc prof, 67-74, actg head dept, 68-69, PROF BIOL, W GA COL, 74- *Mem:* AAAS; Am Soc Animal Sci. *Res:* Genetics and physiology of the Syrian hamster. *Mailing Add:* Dept Biol WGa Col Carrollton GA 30118

GARDNER, BERNARD, b Brooklyn, NY, Oct 1, 31; m 54; c 3. SURGERY. *Educ:* NY Univ, AB, 52, MD, 56. *Prof Exp:* From asst prof to assoc prof, 65-72, prof surg, State Univ NY Downstate Med Ctr, 72-, dir surg oncol, Kings County Hosp, 70-; prof surg, Col Med & Dent NJ; CHIEF SURG, HACKENSACK MED CTR, 83- *Concurrent Pos:* Mem bd dirs, Res Found, State Univ NY, 68-; John & Mary R Markle scholar acad med, 68-; vis prof multiple insts; mem, bd dirs, Southern NY Res Found; mem study sect Cander Ed, Nat Cancer Inst; chmn training comt, Soc Surg Oncol, 88- *Mem:* Am Surg Asn; Asn Acad Surg; Soc Exp Biol & Med; Soc Univ Surg; Soc Surg Alimentary Tract; Soc Surg Oncol (secy, 88-91). *Res:* Tumor metastases; metabolic effects of tumor, particularly the relationship between tumor growth and calcium and phosphate metabolism; suspension stability of bile and its relationship to the electro-chemistry of the cholesterol-bile salt-lecithin micelle. *Mailing Add:* Dept Surg Hackensack Med Ctr 30 Prospect Ave Hackensack NJ 07601

GARDNER, BRYANT ROGERS, b McNary, Ariz, Sept 19, 30; m 53; c 8. SOIL CHEMISTRY, PLANT NUTRITION. *Educ:* Ariz State Univ, BS, 58; Univ Ariz, MS, 60, PhD(agr chem), 63. *Prof Exp:* Res assoc, 62-63, asst agr chemist, 63-68, assoc agr chemist, 68-73, AGR CHEMIST, UNIV ARIZ, 73- *Mem:* Am Soc Agron. *Res:* Soil fertility; plant physiology. *Mailing Add:* 21420 Via Del Palo Queen Creek AZ 85242

GARDNER, CHARLES OLDA, b Tecumseh, Nebr, Mar 15, 19; m 47; c 4. QUANTITATIVE GENETICS. *Educ:* Univ Nebr, BSc, 41, MS, 48; Harvard Univ, MBA, 43; NC State Univ, PhD(agron), 51. *Prof Exp:* Asst exten agronomist, Univ Nebr, 46-48, from assoc prof to prof, 52-70, chmn, Statist Lab, 57-68, found prof, 70-89, EMER PROF AGRON, UNIV NEBR, 89-, CONSULT, PLANT BREEDING, GENETICS & STATIST, 89- *Concurrent Pos:* Asst statistician, NC State Univ, 51-52; distinguished professorship, Univ Nebr, 70-89. *Honors & Awards:* Crop Sci Res Award, Crop Sci Soc Am, 78, DeKalb-Pfizer Distinguished Career Award, 84; Distinguished Serv Award, USDA, 88; Agron Serv Award, Am Soc Agron, 88. *Mem:* Fel AAAS (chmn, sect O, 87-88); Genetics Soc Am; fel Am Soc Agron (pres, 81-82); Am Genetic Asn; fel Crops Sci Soc Am (pres, 74-75); Biometric Soc; Sigma Xi. *Res:* Quantitative inheritance studies in plants; insecticide resistance management; biometrical genetics. *Mailing Add:* Dept Agron Univ Nebr Lincoln NE 68583-0915

GARDNER, CHARLES OLDA, JR, b Raleigh, NC, May 27, 49; m 73. PLANT VIROLOGY. *Educ:* Univ Nebr, BS, 71; Univ Fla, PhD(biochem & molecular biol), 76. *Prof Exp:* Res assoc, Dept Genetics, NC State Univ, 77; res assoc, 78-81, ASST PROF, DEPT BIOCHEM, OKLA STATE UNIV, 81- *Mem:* Am Phytopath Soc; Am Plant Physiologists; Sigma Xi. *Res:* Molecular biology of plant viruses; nucleic acid and protein synthesis in plants. *Mailing Add:* Dept Biochem Okla State Univ Stillwater OK 74074

GARDNER, CHESTER STONE, b Jamaica, NY, Mar 29, 47; m 68; c 2. ELECTRICAL ENGINEERING, ELECTRO-OPTICS. *Educ:* Mich State Univ, BS, 69; Northwestern Univ, MS, 71, PhD(elec eng), 73. *Prof Exp:* Mem tech staff, Bell Tel Labs, 69-71; teaching asst elec eng, Northwestern Univ, 71-73; from asst prof to assoc prof, 73-81, PROF ELEC ENG, UNIV ILL, URBANA, 81- *Concurrent Pos:* Consult, McGraw-Hill Bk Co, 76-77, Caterpillar Tractor Co, 78-, US Army Corp Engrs, 79- & Northern Ill Gas, 81- *Mem:* Optical Soc Am; Inst Elec & Electronics Engrs; Sigma Xi. *Res:* Optical communications; lidar; laser ranging; optical information processing and spread spectrum communications; optical metiology. *Mailing Add:* 1904 Trout Valley Rd Champaign IL 61820

GARDNER, CLIFFORD S, b Ft Smith, Ark, Jan 14, 24; m 67; c 2. APPLIED MATHEMATICS. *Educ:* Harvard Col, AB, 44; NY Univ, PhD(math), 52. *Prof Exp:* Physicist, Nat Adv Comt Aeronaut, 44-46; mathematician, Control Instrument Co, 47-48; physicist, Calif Res & Develop Co, 52-54; physicist, Radiation Lab, Univ Calif, 54-56; res scientist, Courant Inst, NY Univ, 56-62; physicist, Radio Corp Am, 62-64; vis res prof physics, Plasma Physics Lab, Princeton Univ, 64-68; PROF MATH, UNIV TEX, AUSTIN, 68- *Mem:* Am Math Soc. *Res:* Differential equations. *Mailing Add:* Dept Math Univ Tex Austin TX 78712

GARDNER, DANIEL, b New York, NY, Jan 23, 45; m 67; c 2. NEUROPHYSIOLOGY, BIOPHYSICS. *Educ:* Columbia Col, AB, 66; New York Univ, PhD(physiol), 71. *Prof Exp:* Programmer & comput math lectr, Goddard Inst Space Studies, NASA, 62-66; fel physiol, Sch Med, New York Univ, 66-71; sr fel neurophysiol, Sch Med, Univ Wash, 71-73; asst prof, 73-79, assoc prof, 79-89, PROF PHYSIOL, MED COL, CORNELL UNIV, 89- *Concurrent Pos:* Instr, Workshop Comput Math, Columbia Univ, 63-67; prin investr res grant, NIH, 73-, ONR, 86- *Mem:* Soc Neurosci; Biophys Soc; Am Physiol Soc; NY Acad Sci; Soc Gen Physiol; Am Asn Artificial Intel. *Res:* Biophysics of synaptic transmission between neurons in the nervous system of the marine mollusc aplysia; analysis of biological neural networks; development computerized teaching and simulations. *Mailing Add:* C-534 Dept Physiol Med Col Cornell Univ 1300 York Ave New York NY 10021

GARDNER, DAVID ARNOLD, b Ithaca, NY, June 19, 39; m 59; c 3. CLINICAL CHEMISTRY, BIOCHEMISTRY. *Educ:* Univ Rochester, BS, 61, PhD(biochem), 67; Ind Univ, South Bend, MBA, 83. *Prof Exp:* NIH res fel biochem, Brandeis Univ, 67-69; res scientist cell biol, Res Div, 69-76, tech prod mgr, 76-83, mgr, qual assurance develop, Ames Div, 83-87, MGR, QUAL ASSURANCE INFO SERV, DIAGNOSTICS DIV, MILES LABS, INC, ELKHART, 87- *Concurrent Pos:* Asst fac fel, Univ Notre Dame, 69-75. *Mem:* Am Asn Clin Chem; NY Acad Sci; Sigma Xi. *Res:* Regulation of cell division in tissue culture; use of cell cultures for the development of clinical diagnostic tests; glycolipid metabolism in cultured mammalian cells; new product development in clinical chemistry; management of technical support group for clinical diagnostic products; development of automated test systems for quality control of medical diagnostic products; automated laboratory information systems. *Mailing Add:* 16243 Marlowe Way Granger IN 46530

GARDNER, DAVID GODFREY, b Darlington, Eng, Feb 24, 36; Can citizen; m 60; c 3. ORAL PATHOLOGY. *Educ:* Univ Toronto, DDS, 58; Ind Univ, MSD, 65; Am Bd Oral Path, dipl, 69. *Prof Exp:* Dent officer, Can Army, 58-63; asst prof oral med, Univ BC, 65-66; from assoc prof to prof path, Univ Western Ont, 66-84, chmn div oral path, 70-84; PROF & CHMN DEPT PATH & RADIOL, UNIV TEX, HSC, HOUSTON DENT BR, 84- *Concurrent Pos:* Consult oral path & oral med, Children's Psychiat Res Inst, London, Ont, 66-84; mem consult panel, Can Tumour Reference Ctr, Nat Cancer Inst Can, 68-; mem active staff path & dent, Univ Hosp, London, Ont, 72-84; mem staff, Guy's Hosp, London, Eng, 73-74. *Mem:* Fel Am Acad Oral Path (pres, 85-86); Can Acad Oral Path (secy, 66-72, pres, 72-73). *Res:* Disturbances in odontegenesis; odontogenic tumors; disorders of maxillary sinus. *Mailing Add:* 455 Dexter St Denver CO 80220

GARDNER, DAVID MILTON, b Cleveland, Ohio, June 21, 28; m 55; c 5. PHYSICAL CHEMISTRY. *Educ:* Brown Univ, ScB, 50; Univ Pa, MS, 52, PhD(chem), 54. *Prof Exp:* Proj leader, Reaction Motors, Inc, 54-60, sr chemist, 56-57; group leader, 60-70, sr scientist, 70-82, RES MGR, PENNWALT CORP, 82- *Concurrent Pos:* Lectr, La Salle Col, 66-73. *Mem:* Am Chem Soc; Sigma Xi. *Res:* Thermodynamics; halogens; nitrogen; metallo-organic chemistry; process development; phase equilibria and kinetics; catalysis; fluorochemicals. *Mailing Add:* 727 W Valley Forge Rd King of Prussia PA 19406-1572

GARDNER, DAVID R, b London, Eng, Aug 10, 42; Can citizen. NEUROPHYSIOLOGY. *Educ:* Southampton Univ, Eng, BSc, 63, PhD(physiol-biochem & neurophys), 66. *Prof Exp:* SRC fel physiol-biochem, Southampton Univ, 66-67; from asst prof to assoc prof, 67-82, assoc undergrad chmn, 79-83, PROF BIOL, CARLETON UNIV, 82-, CHMN DEPT, 83- *Concurrent Pos:* Mem bd gov, Carleton Univ, 89- *Mem:* Soc Exp Biol; Soc Neurosci. *Res:* Modes of action of synthetic and natural product pesticides on excitable tissues of invertebrates. *Mailing Add:* Dept Biol Carleton Univ Colonel By Dr Ottawa ON K1S 5B6 Can

GARDNER, DONALD EUGENE, b Council Bluffs, Iowa, Nov 1, 31; m 60; c 4. INHALATION TOXICOLOGY, GENERAL TOXICOLOGY. *Educ:* Creighton Univ, BS, 55, MS, 58; Univ Cincinnati, PhD(environ health), 71. *Prof Exp:* Res immunologist, US Biol Labs, Ft Detrick, Frederick, Md, 60-62; res microbiologist, US Biol Proving Ground, Dugway, Utah, 62-64 & Nat Air Pollution Control Admin, 64-71; chief, Biomed Res Br, US Environ Protection Agency, 71-80, dir, Inhalation Toxicol Div, 80-82; dir res, environ sci, Northrop Serv, Inc, 82-89; VPRES & CHIEF SCIENTIST, MAN TECH ENVIRON TECHNOL, INC, 89- *Concurrent Pos:* Co-topic leader, Joint Coop Prog US-USSR, US Govt, 73-81, topic leader, US-Poland Joint Agreement, 76-82 & US-Yugoslavia Joint Res Agreement, US Environ Protection Agency, 81-82; adj prof physiol & pharmacol, Sch Med, Duke Univ, 75- & adj assoc prof, NC State Univ, 76-; mem, Task Force Environ Health, WHO, 76-77; mem sci adv comt, Environ Protection Agency, 80-; consult chem indust, pres, Inhalation Spec Sect, Soc Toxicol, 84-85, Immunotoxicol Spec Sect, 87-88 & Metal Spec Sect, 91; ed, J Inhalation Toxicol; grant & prog sci review comt, Food & Drug Admin, Environ Protection Agency & Nat Inst Environ Health Sci. *Mem:* Soc Toxicol; Am Col Toxicol; NY Acad Sci. *Res:* Environmental and occupational toxicology; immunotoxicology; inhalation toxicology; host defense mechanisms; short-term in vitro for predicting chronic health effects. *Mailing Add:* Man Tech Environ Technol Inc PO Box 12313 Research Triangle Park NC 27709

GARDNER, DONALD GLENN, b Chicago, Ill, Sept 25, 31; m 53, 69; c 2. NUCLEAR CHEMISTRY. *Educ:* Univ Ill, BS, 53; Univ Mich, MS, 54, PhD(nuclear chem), 57. *Prof Exp:* Chemist, Westinghouse Elec Corp, 57-59; asst prof chem, Univ Ark, 59-61; from asst prof to assoc prof, Ill Inst Technol, 61-68; mem staff, 65-67, SR CHEMIST, LAWRENCE LIVERMORE LAB, UNIV CALIF, 68- *Concurrent Pos:* Consult, Argonne Nat Lab, 62-64. *Mem:* Am Phys Soc; Am Chem Soc. *Res:* Nuclear reactions; cross sections; statistical theory. *Mailing Add:* L-234 Lawrence Livermore Nat Lab Livermore CA 94550

GARDNER, DONALD MURRAY, b Hartford, Conn, Dec 17, 28; m 51; c 3. COATINGS TECHNOLOGY. *Educ:* Hamilton Col, AB, 50; Columbia Univ, MA, 54, PhD(chem), 56; Am Int Col, MBA, 69. *Prof Exp:* Res chemist, Shawinigan Resins Corp, 55-61, group leader, 61-64; group leader, 64-81, com develop mgr, 81-90, PROJ LEADER, MONSANTO CO, 90- *Mem:* Am Chem Soc. *Res:* Instrumental analysis; water soluble polymers; surface chemistry; polymerization; adhesives; surface coatings. *Mailing Add:* Monsanto Co 730 Worcester St Springfield MA 01151-1099

GARDNER, EARL WILLIAM, JR, b Houston, Tex, July 31, 28; m 62; c 2. MICROBIOLOGY. *Educ:* Baylor Univ, BS, 50; Univ Tex, MA, 54, PhD(bact), 58. *Prof Exp:* From asst prof to prof biol, Tex Christian Univ, 58-75, chmn sci div, 72-75; scientist adminr, NIH, 75-77, DEP CHIEF BLOOD DIS BR, NAT HEART, LUNG & BLOOD INST, 77- *Concurrent Pos:* Vis asst prof, Univ Tex, 60. *Mem:* NY Acad Sci; Am Soc Microbiol; Am Inst Biol Sci. *Res:* Antigenicity and pathogenicity of vibrio comma. *Mailing Add:* Rte 1 Box 32A Santo TX 76472

GARDNER, EDWARD, JR, b Charleston, WVa, July 28, 25; m 49; c 3. IMMUNOLOGY. *Educ:* Morris Harvey Col, BS, 50; Ohio State Univ, MSc, 52, PhD(bact, immunol), 55; Am Bd Microbiol, dipl. *Prof Exp:* Asst bact, Ohio State Univ, 52-54, asst med, 54-55, res assoc, Div Hemat, Med Col Ga, 55-58, asst res prof, 58-62, assoc prof med & assoc prof med microbiol & pub health, 62-70; scientist/adminr, chief res & training br & exec secy, Environ Sci Training Comt, 70-74, scientist/adminr sci prog, 74-77, PROG DIR, REGULAR RES GRANTS DIV, NAT INST ENVIRON HEALTH SCI, 77- *Mem:* AAAS; Am Soc Microbiol; Soc Exp Biol & Med; fel Am Acad Microbiol; NY Acad Sci; Sigma Xi. *Res:* Autoimmune disease; erythropoietin; environmental health science. *Mailing Add:* Sci Prog Nat Inst Environ Health Sci PO Box 12233 Research Triangle Park NC 27709

GARDNER, EDWARD EUGENE, b Somerset, Pa, Aug 3, 23; m 48; c 4. SOLID STATE SCIENCE. *Educ:* Mass Inst Technol, SB, 48; Univ Minn, MA, 50; Cath Univ Am, PhD(physics), 55. *Prof Exp:* Mem res staff, Div Indust, Mass Inst Technol, 48-49; instr gen physics, Va Polytech Inst, 50-51; asst prof, US Naval Acad, 51-55; asst prof elec eng & physics, Lehigh Univ, 55-58; res physicist, Whirlpool Corp, 58-62; adv physicist, 62-69, sr physicist, Systs Prod Div, 69-76, sr physicist, Gen Technol Div, IBM Corp, 76-85; PROF, ELEC & COMPUT ENG DEPT, UNIV COLO, 85- *Concurrent Pos:* Prof, Gen Elec Co, 57; spec lectr, Mich State Univ, 59-; vis prof, Tuskegee Inst, 73-74. *Mem:* Electrochem Soc; Am Phys Soc; Inst Elec & Electronics Engrs; Am Soc Testing & Mat. *Res:* Solid state physics; semiconductor materials and process development; characterization of semiconductor materials; semiconductor materials measurement technique development. *Mailing Add:* 81 Benthaven Boulder CO 80303

GARDNER, ELDON JOHN, b Logan, Utah, June 5, 09; m 39; c 6. GENETICS. *Educ:* Utah State Univ, BS, 34, MS, 35; Univ Calif, PhD(zool), 39. *Hon Degrees:* DSc, Utah State Univ, 81. *Prof Exp:* Instr biol & dean lower div, Salinas Jr Col, 39-46; from asst prof to assoc prof biol, Univ Utah, 46-49, res prof, col med, 78-80; prof zool, 49-74, fac res lectr, 54, dean col sci, 62-67, dean sch grad studies, 67-74, EMER PROF BIOL, UTAH STATE UNIV, 74- *Concurrent Pos:* Vis investr, Univ Calif, 57-58 & Brit Mus & Univ London, 71; mem health res facil sci rev comt, Health Res Facil Br, Bur Educ Manpower Training, 67-71, chmn, 69-71. *Honors & Awards:* Willard Gardner Sci Award, 75. *Mem:* AAAS; Am Soc Naturalists; Genetics Soc Am; Am Soc Human Genetics (secy, 55-58, vpres, 61); Am Genetic Asn. *Res:* Genetics of abnormal growths and cancer. *Mailing Add:* Dept Biol Utah State Univ Logan UT 84322

GARDNER, ELIOT LAWRENCE, b Boston, Mass, Dec 31, 40; m 65; c 1. NEUROSCIENCE, PSYCHOBIOLOGY. *Educ:* Harvard Univ, AB, 62; McGill Univ, MA, 64, PhD(physiol psychol), 66. *Prof Exp:* Med officer, aeromed div, US Air Force, 66-69; res fel pharmacol, Albert Einstein Col Med, 69-72; assoc neurol, 72-73; assoc prof biopsychol, City Univ New York, 73-76; asst prof, 76-82, ASSOC PROF PSYCHIAT & NEUROSCI, ALBERT EINSTEIN COL MED, 82-, DIR, PROG BRAIN & BEHAV, 86- *Concurrent Pos:* USPHS fel, 69-72; adj asst prof psychobiol, NY Univ, 70-73; vis asst prof pharmacol, Albert Einstein Col Med, 73-76; USPHS res grants, 74- *Honors & Awards:* Career Scientist Award, Health Res Coun NY, 75. *Mem:* Soc Neurosci; Psychonomic Soc; Am Psychol Asn; Brit Brain Res Asn; NY Acad Sci. *Res:* Brain function and behavior; neuropsychology; neuropsychopharmacology. *Mailing Add:* Dept Psychiat SG 49 Albert Einstein Col Med 1300 Morris Park Ave Bronx NY 10461

GARDNER, ESTHER POLINSKY, b New York, NY, June 14, 41; m 67; c 2. NEUROPHYSIOLOGY. *Educ:* Smith Col, BA, 62; NY Univ, PhD(physiol), 71. *Prof Exp:* Fel physiol, Sch Med, NY Univ, 66-71; sr fel neurophysiol, Sch Med, Univ Wash, 71-73; asst prof, 73-81, ASSOC PROF PHYSIOL, SCH MED, NY UNIV, 81- *Concurrent Pos:* NIH fel, Sch Med, NY Univ, 67-71 & Univ Wash, 71-73; NIH res career develop award, 76-81; Career Scientist Award, Irma T Hirschl Found, 76-81. *Mem:* AAAS; Am Physiol Soc; Soc Neurosci; NY Acad Sci; Int Neural Networks Soc. *Res:* Sensory function of the hand; mechanisms used by single neurons in somatosensory areas of the primate brain to encode spatial and temporal properties of patterned cutaneous stimuli and their relation to sensation. *Mailing Add:* Dept Physiol NY Univ Sch Med New York NY 10016

GARDNER, FLORENCE HARROD, cellular physiology, biology symbiotic relationship, for more information see previous edition

GARDNER, FLOYD M, b Chicago, Ill, Oct 20, 29. COMMUNICATIONS, ELECTRONICS. *Educ:* Ill Inst Technol, BS, 50; Stanford Univ, MS, 51; Univ Ill, PhD(elec eng), 53. *Prof Exp:* CONSULT ENGR, 60- *Mem:* Fel Inst Elec & Electronics Engrs. *Res:* Synchronization of data communication; signal processing; author of book on phase lock techniques. *Mailing Add:* Gardner Res Co 1755 University Ave Palo Alto CA 94301

GARDNER, FRANK S(TREETER), b Baltimore, Md, Dec 9, 16; m 41; c 4. PHYSICAL METALLURGY. *Educ:* Mass Inst Technol, SB, 38, ScD(metall), 41. *Prof Exp:* Metallurgist, Am Brake Shoe Co, NJ, 41-46; metallurgist, Gen Elec Co, 46-53, supvr magnetic mat res, Transformer Div, 53-58; proj mgr, Nuclear Metals, Inc, 58-60; dep dir sci, Off Naval Res, 75-81, mat scientist, 60-81, res avd, 81-85; RETIRED. *Mem:* The Metall Soc; Am Soc Metals. *Res:* Metallography; pressure welding; metallurgy of soft magnetic materials; mechanical metallurgy; magnetic transformations; amorphous metals; research programming. *Mailing Add:* W Shore Pl Grantham NH 03753

GARDNER, FRANKLIN PIERCE, b Hillsville, Va, Mar 20, 24; m 43; c 3. AGRONOMY. *Educ:* Va Polytech Inst, BS, 49; Iowa State Univ, MS, 50, PhD(plant physiol, crop sci), 52. *Prof Exp:* Instr crop sci, Iowa State Univ; asst prof, Univ Ga, 52-54; asst prof, Ohio State Univ, 54-55; from assoc prof to prof, Iowa State Univ, 55-63; prof agron & head dept, Okla State Univ, 63-68; dean col appl sci, Western Ill Univ, 68-74, prof agron, 68-74, prof agron, 68-80; prof agron, Univ Fla, 80-90; RETIRED. *Mem:* Fel Am Soc Agron. *Res:* College administration; effect of photoperiod and temperature on development and flowering of perennial grasses and legumes; effect of cultural treatments on yield and quality of silage; effect of cultural treatments on photosynthetic efficiency and production. *Mailing Add:* Dept Agron Univ Fla Gainesville FL 32611

GARDNER, FRED MARVIN, b Kansas City, Kans, July 4, 22; m 63; c 4. ELECTRON PHYSICS. *Educ:* St Mary's Col, Minn, BS, 45; St Louis Univ, MS, 52; Univ Notre Dame, PhD(physics), 62. *Prof Exp:* Instr, Christian Bros Mil High Sch, Mo, 45-52; from asst prof to assoc prof physics, St Mary's Col, Minn, 52-62; from res scientist to sr res scientist, United Aircraft Res Labs, 62-69; vpres, High Tech Indust, 69-70; PROF PHYSICS, UNIV HARTFORD, 70- *Mem:* Am Phys Soc; Inst Elec & Electronics Engrs; Am Asn Physics Teachers; Sigma Xi (pres, Hartford Chap 73-74); Gynec Laser Soc; Laser Inst Am. *Res:* Medical lasers. *Mailing Add:* 111 Stockade Rd South Glastonbury CT 06073

GARDNER, FREDERICK ALBERT, b Middletown Springs, Vt, Nov 23, 27; m 49; c 6. FOOD SCIENCE & TECHNOLOGY, MICROBIOLOGY. *Educ:* Univ Vt, BS, 53; Agr & Mech Col, Tex, MS, 55; Univ Mo, PhD(poultry sci), 60. *Prof Exp:* Instr poultry sci, Agr & Mech Col, Tex, 54-55; asst, Iowa State Col, 55-56; asst, Univ Mo, 56-58, instr, 58-59; asst prof, 59-63, ASSOC PROF POULTRY SCI, TEX A&M UNIV, 63- *Mem:* Poultry Sci Asn; Inst Food Technologists; Sigma Xi. *Res:* Chemistry and microbiology of poultry and egg products with specific interest in fundamental product characterization and in the maintenance of product quality. *Mailing Add:* Dept Poultry Sci & Food Sci Tex A&M Univ College Station TX 77843

GARDNER, GERALD HENRY FRASER, b Ireland, Mar 2, 26; m 50. MATHEMATICS. *Educ:* Trinity Col, Dublin, BA, 47; Carnegie Inst Technol, MSc, 48; Princeton Univ, PhD(math), 53. *Prof Exp:* Lectr math, Trinity Col, Dublin, 49-54; instr, Cornell Univ, 54-55; asst prof, Carnegie Inst Technol, 55-56; sect head, Explor & Prod, Gulf Res & Develop Co, 56-66, sr scientist, 66-80; prof elec eng & prin investr, Seismic Acoust Lab, Univ Houston, 80-90; KECK PROF GEOPHYSICS, RICE UNIV, 90- *Mem:* Soc Explor Geophysicists; Math Asn Am; Europ Asn Explor Geophysicists; Soc Prof Well Log Analysts. *Res:* Relativity; mathematical physics and seismic exploration. *Mailing Add:* Keck Prof Geophysics Rice Univ Houston TX 77251

GARDNER, HAROLD WAYNE, b Carlisle, Pa, June 19, 35; m 81; c 5. BIOCHEMISTRY, LIPID CHEMISTRY. *Educ:* Pa State Univ, BS, 57, MS, 63, PhD(biochem), 65. *Prof Exp:* Assoc biochemist, Pineapple Res Inst, 65-66; asst res plant biochemist, Univ Calif, Los Angeles, 66-67; CHEMIST, NORTHERN REGIONAL LAB, US DEPT AGR, PEORIA, Ill, 67- *Mem:* Am Chem Soc; Am Oil Chemists' Soc; NY Acad Sci; Am Soc Plant Physiologists; Am Soc Biol Chemists. *Res:* Oxidation of unsaturated fatty acids by plant systems; lipid hydroperoxides and their reaction products. *Mailing Add:* Northern Regional Res Ctr US Dept Agr Peoria IL 61604

GARDNER, HOWARD SHAFER, b Brooklyn, NY, Sept 18, 08; m 31; c 2. PAPER CHEMISTRY, CHEMICAL ENGINEERING. *Educ:* Mass Inst Technol, SB, 30, SM, 31, ScD(chem eng), 46. *Prof Exp:* Staff chem engr, Eastman Kodak Co, NY, 31-36; asst prof chem eng & dir, Bangor Field Sta, Mass Inst Technol, 36-38; assoc prof chem eng & chmn dept, Univ Rochester, 38-47; dir res & develop, Fibreboard Corp, Calif, 47-61; sr res assoc & chmn eng & technol sect, Inst Paper Chem, Lawrence Univ, 62-64, admin staff, 64-65; PROF PULP & PAPER TECHNOL & CHEM ENG, UNIV WASH, 66- *Concurrent Pos:* Pvt consult, 38-47 & 66-; mem staff, Div Indust Coop & Res, Mass Inst Technol, 45; mem bd dirs, Tech Asn Pulp & Paper Indust, 72-75. *Mem:* Am Chem Soc; fel Tech Asn Pulp & Paper Indust; Am Inst Chem Engrs; Air Pollution Control Asn; Can Pulp & Paper Asn; Sigma Xi. *Res:* Pulp and paper technology; vapor explosions in kraft chemical recovery furnaces; air pollution; paper making properties of fibers. *Mailing Add:* 2373-C Via Mariposa W Laguna Hills CA 92653

GARDNER, JAMES VINCENT, b Topeka, Kans, May 28, 40; m 64, 81; c 2. MARINE GEOLOGY. *Educ:* Calif State Univ, San Diego, BS, 67; Columbia Univ, MA, 69, PhD(marine geol), 73. *Prof Exp:* Field geologist, Wm Ross Cabeen & Assoc, Peru, 67; geologist, Denver Res Ctr, Marathon Oil Co, 69-70; staff res assoc marine geol, Deep Sea Drilling Proj, Scripps Inst Oceanog, 73-75; MARINE GEOLOGIST, BR PAC MARINE GEOL, US GEOL SURV, 75- *Concurrent Pos:* Mem, Int Decade Ocean Explor Climap Proj, 70-73, corresp mem, 73-80; mem, Deep Sea Drilling Proj Rep, Joint Oceanog Inst Deep Earth Sampling Atlantic Panel, 74-75; vis res scientist, Inst Oceanographic Sci, Godalming, UK, 81-82. *Mem:* Geol Soc Am; Am Geophys Union; AAAS; Soc Econ Paleontologists & Mineralogists. *Res:* Marine geology and paleooceanography of equatorial Atlantic, Pacific and Bering Sea; deep-sea sedimentology, high-resolution geophysics and digital side-scan sonar processing and interpretation. *Mailing Add:* Br Pac Marine Geol US Geol Surv 345 Middlefield Rd Menlo Park CA 94025

GARDNER, JEFFERY FAY, GENETIC REGULATION, RECOMBINATION. *Educ:* Marquette Univ, PhD(biol), 75. *Prof Exp:* ASSOC PROF MICROBIOL, UNIV ILL, 78- *Mailing Add:* 131 Burrill Hall Univ Ill 407 S Goodwin Urbana IL 61801

GARDNER, JERRY DAVID, b Chanute, Kans, Jan 15, 41; m 62; c 2. GASTROENTEROLOGY, INTERNAL MEDICINE. *Educ:* Univ Kans, AB, 62; Univ Pa, MD, 66. *Prof Exp:* Clin assoc physiol & clin nutrit, 68-70, sr staff fel, Digestive Dis Br, 70-73, CHIEF, SECT GASTROENTEROL, NAT INST ARTHRITIS, METAB, DIGESTIVE & KIDNEY DIS, NIH, 73-, CHIEF, DIGESTIVE DIS BR, 77- *Honors & Awards:* Pollard Prize, Univ Mich, 79; Roche Award, Hoffman-LaRoche, 80. *Mem:* Am Soc Clin Invest; Am Fedn Clin Res; Am Gastroenterol Asn; Am Physiol Soc; AAAS. *Res:* Identification of the biochemical basis of action of gastrointestinal hormones. *Mailing Add:* Bldg 10 Rm 9C 103 Digestive Dis Br NIH Bethesda MD 20892

GARDNER, JOHN ARVY, JR, b Winona, Miss, Nov 5, 39; m 65; c 2. SOLID STATE PHYSICS. *Educ:* Rice Univ, BA, 61; Univ Ill, MS, 63, PhD(physics), 66. *Prof Exp:* Res assoc physics, Tech Univ, Munich, WGermany, 66-67; asst prof, Univ Pa, 67-73; assoc prof, 73-81, PROF PHYSICS, ORE STATE UNIV, 81- *Concurrent Pos:* Vis scientist, Univ Warwick, Eng, 79-80 & Max Planck Inst Solid State Res, Stuttgart, WGer, 80-81 & 85. *Mem:* Am Phys Soc; AAAS; Am Ceramic Soc; Mat Res Soc. *Res:* Experimental solid state physics; hyperfine interactions; magnetic and electronic properties of liquid and amorphous semiconductors; ceramics; high temperature semiconductors; high-Tc superconductors. *Mailing Add:* Dept Physics Ore State Univ Corvallis OR 97331

GARDNER, JOHN HALE, b Logan, Utah, Aug 24, 22; m 43; c 8. MAGNETIC RESONANCE, PLASMA PHYSICS. *Educ:* Utah State Agr Col, BS, 43; Harvard Univ, AM, 47, PhD(physics), 50. *Prof Exp:* Mem staff, Radiation Lab, Mass Inst Technol, 43-46; from asst prof to prof physics, Brigham Young Univ, 49-58; chmn dept, 61-63 & 64-71, PROF PHYSICS, BRIGHAM YOUNG UNIV, 58- *Concurrent Pos:* Mem tech staff, Thompson-Ramo-Woolidge Corp, 55-57 & 63-64, consult, Space Tech Labs, 57-63 & 64-68. *Mem:* Fel Am Phys Soc; Am Asn Physics Teachers; fel Brit Inst Physics Eng. *Res:* Soil moisture dynamics; microwave antennas; electromagnetic propagation; magnetic resonance; gaseous electronics; quantum electronics; plasma physics; theoretical physics; laws of physics from inevitable symmetries of nature. *Mailing Add:* Dept Physics 175 ESC Brigham Young Univ Provo UT 84602

GARDNER, JOHN OMEN, electron microscopy, for more information see previous edition

GARDNER, JOSEPH ARTHUR FREDERICK, b Nakusp, BC, Aug 17, 19; m 44; c 2. WOOD CHEMISTRY. *Educ:* Univ BC, BA, 40, MA, 42; McGill Univ, PhD, 44. *Prof Exp:* Res assoc, McGill Univ, 44-45; res chemist, Howard Smith Paper Mills, Ont, 45-47; head wood chem sect, Forest Prod Lab, 47-62, dir, 62-65; dean & prof, 65-83, EMER DEAN, FAC FORESTRY, UNIV BC, 84- *Concurrent Pos:* Hon lectr, Univ BC, 53-64, hon prof, 64; mem Can Environ Adv Coun, 84-87. *Mem:* Tech Asn Pulp & Paper Indust; Can Inst Forestry; fel Int Acad Wood Sci; fel Chem Inst Can; Can Pulp & Paper Asn; hon mem Asn BC Prof Foresters. *Res:* Chemistry of wood and its components, especially lignin and phenolic extractives. *Mailing Add:* 5537 Wallace St Vancouver BC V6N 2A1 Can

GARDNER, KENNETH DRAKE, JR, b San Francisco, Calif, Oct 3, 29; m 54; c 4. INTERNAL MEDICINE. *Educ:* Stanford Univ, BMS, 51, MD, 55. *Prof Exp:* Daland res fel clin med, 59-60; instr med, Sch Med, Univ Pa, 60-61; Daland res fel clin med, 61-63; instr med, Sch Med, Stanford Univ, 63-64, asst prof, 64-69; prof, Sch Med, Univ Hawaii, 69-73, assoc dean, 70-73, chmn dept, 71-73, chief, div renal dis, 73-90; PROF MED, SCH MED, UNIV NMEX, 73-, ASST DEAN GRAD MED EDUC, 74- *Concurrent Pos:* consult, Oversight Subcomt, US House Rep, 79-80; Fifth Cong Sci fel, Fedn Am Socs Exp Biol, 79-80; consult ed, Am J Kidney Dis, 80-; mem, Pub Affairs Comt, Federated Am Socs Exp Med; vchmn, Sci Adv Comt, Polycystic Kidney Res Found; secy-treas, Western Asn Physicians, 89- *Mem:* Am Fedn Clin Res; fel Am Col Physicians; Am Physiol Soc; Am Soc Nephrology. *Res:* Effect of hormones on molecular movement across collecting tubular walls of the mammalian kidney; clinical renal disease; renal cystic diseases and pathophysiology of human kidney disease. *Mailing Add:* Dept Med Univ NMex Sch Med Albuquerque NM 87131

GARDNER, LYTT IRVINE, b Reidsville, NC, Oct 1, 17; m 42; c 6. PEDIATRICS. *Educ:* Univ NC, AB, 38, MA, 40; Harvard Univ, MD, 43. *Prof Exp:* Res fel, Harvard Med Sch, 48-49; instr, Sch Med, Univ NC, 49-50; asst prof pediat, Sch Med, Johns Hopkins Univ, 50-52; assoc prof, Col Med, State Univ NY Upstate Med Ctr, 52-56; prof, Sch Med, Yale Univ, 56-57; PROF PEDIAT, COL MED, STATE UNIV NY UPSTATE MED CTR, 57- *Mem:* Soc Pediat Res (pres, 62-63); Am Pediat Soc; Endocrine Soc; Europ Soc Paediat Endocrinol; hon mem Latin-Am Soc Pediat Res. *Res:* Pediatric endocrinology; genetics; chromosome abnormalities in children; effects of maternal deprivation on endocrine system; steroid metabolism of fetal and infant adrenal cortex. *Mailing Add:* State Univ Hosp 750 E Adams St Syracuse NY 13210

GARDNER, MARIANNE LEPP, computer networks, for more information see previous edition

GARDNER, MARJORIE HYER, b Logan, Utah, Apr 25, 23; m 47; c 2. CHEMISTRY, SCIENCE EDUCATION. *Educ:* Utah State Univ, BS, 46; Ohio State Univ, MA, 58, PhD(sci educ, chem), 60. *Prof Exp:* High sch teacher, Utah, 47-49 & Nev, 49-53; instr chem & sci educ, Ohio State Univ, 58-60; asst exec secy & dir vistas sci, Nat Sci Teachers Asn, 61-64; from asst prof to assoc prof chem & sec educ, Univ Md, Col Park, 64-71, prof chem & sci educ, 71-84; dir & vis prof chem, 84-89, SR RES ASSOC, LAWRENCE HALL SCI, UNIV CALIF, BERKELEY, 90-; RES PROF, UTAH STATE UNIV, LOGAN, UTAH, 90- *Concurrent Pos:* Deleg, Int Conf & Cong, various countries, 65-; dir, NSF Inserv Insts Leadership Conf & Coop Col-Sch Sci Progs, 67-; dir, Interdisciplinary Approaches to Chem Proj, 70-; Fulbright award, 73-74 & 82-83; dir, Sci Educ Resource Improvement Div, NSF, 79-81. *Honors & Awards:* Robert H Carleton Award, 74; Catalyst Award, Chem Mfg Asn, 80; Nyholm Medal, Royal Chem Soc, 86; Pimentel Chem Educr Award, Am Chem Soc, 88. *Mem:* Fel AAAS; Am Chem Soc; Nat Sci Teachers Asn; Int Union Pure & Appl Chemists. *Res:* Curriculum development; evaluation; teacher preparation; instructional methods and materials; international chemical education. *Mailing Add:* Lawrence Hall Sci Univ Calif Berkeley CA 94720

GARDNER, MURRAY BRIGGS, b Lafayette, Ind, Oct 5, 29; m 61; c 4. PATHOLOGY, ONCOLOGY. *Educ:* Univ Calif, BA, 51, MD, 54. *Prof Exp:* From asst prof to prof path, Sch Med, Univ Southern Calif, 63-81; PROF PATH, SCH MED, UNIV CALIF, DAVIS, 81- *Concurrent Pos:* Pathologist, Univ Southern Calif-USPHS Air Pollution Proj, 63-68; prin investr, NAT Cancer Inst-Univ Southern Calif Res Contract, 68-; dep vet, Los Angeles County, 68- *Mem:* AAAS; Int Acad Path; Am Asn Path & Bact. *Res:* RNA tumor virology; etiology and epidemiology of cancer in humans and domestic pets; biological effects of urban air pollution; acquired immune deficiency in man and monkey. *Mailing Add:* 8313 Maxwell Lane Dixon CA 95620

GARDNER, PAUL JAY, b Wichita, Kans, May 25, 29; m 49; c 4. HEALTH SCIENCE ADMINISTRATION, HISTOLOGY. *Educ:* Univ Wichita, AB, 51, MS, 55; Univ Nebr, PhD(anat), 64. *Prof Exp:* Asst instr anat, Univ Kans, 55-56; prof biol & head dept, Vincennes Univ, 56-60; asst prof biol, Univ Nebr, Omaha, 60-63; USPHS trainee, 63-64, from instr to assoc prof, 64-77, vchmn anat dept, 77-83, interim dean grad studies & res, 83-84, PROF ANAT, UNIV NEBR MED CTR, OMAHA, 77-, ASST VCHANCELLOR ACAD AFFAIRS, 86- *Concurrent Pos:* Exec secy, cancer preclin prog proj rev comt, grants rev bd, div extramural activities, Nat Cancer Inst, NIH, 83-84. *Mem:* Am Soc Cell Biol; Am Asn Anat; Electron Micros Soc; Am Soc Andrology; AAAS. *Res:* Reproductive biology; electron microscopy; ultrastructure and function of male reproductive system. *Mailing Add:* Univ Nebr Med Ctr 600 S 42nd St-3022 Eppley Sci Hall Omaha NE 68198-6810

GARDNER, PETE D, organic chemistry; deceased, see previous edition for last biography

GARDNER, PHILLIP JOHN, b Pomona, Calif, July 28, 41; m 63; c 3. PHYSICAL CHEMISTRY. *Educ:* Univ Calif, Riverside, BA, 64; Fla State Univ, PhD(phys chem), 69. *Prof Exp:* Mem tech staff, GTE Labs, NY, 69-72, prog mgr, GTE Sylvania Lighting Ctr, Mass, 72-75, eng specialist, GTE Slyvania, Electro-Optics Orgn, 75-81; SR PRIN DEVELOP ENGR, RES & DEVELOP DEPT, COHERENT, INC, 81- *Mem:* Am Chem Soc; Am Phys Soc. *Res:* Experimental studies and development of lasers; developmental studies of atomic and molecular arc discharges including electrodeless systems; theoretical and experimental studies of radiative and non-radiative electronic transitions in molecules. *Mailing Add:* Coherent Inc 3210 Porter Dr Palo Alto CA 94304

GARDNER, REED MCARTHUR, b St George, Utah, Oct 24, 37; m 59; c 5. MEDICAL BIOPHYSICS, COMPUTING. *Educ:* Univ Utah, BSEE, 60, PhD(biophys, bioeng), 68. *Prof Exp:* Asst prof biophys & bioeng, 65-72, assoc prof med biophys & comput, 73-78, PROF MED BIOPHYS & COMPUT, UNIV UTAH, 78-; CO-DIR MED COMPUT, LDS HOSP, 85- *Concurrent Pos:* Consult, NASA Life Sci Adv Comt, 71 & Nat Libr Med. *Mem:* Inst Elec & Electronic Engrs; Biomed Eng Soc; Am Thoracic Soc; Asn Advan Med Instrumentation. *Res:* Application of bioengineering principles to solution of medical problems; computer applications in intensive care; cardiovascular laboratories and multiphasic screening centers; medical computing with emphasis on decisionmaking and intensive care technology. *Mailing Add:* Dept Med Informatics LDS Hosp 325 Eighth Ave Salt Lake City UT 84143

GARDNER, RICHARD A, b Oak Park, Ill, Dec 6, 41; m 64; c 3. FLUID MECHANICS, HEAT TRANSFER. *Educ:* Purdue Univ, BS, 63, MS, 65, PhD(fluid mech), 69. *Prof Exp:* From instr to asst prof aeronaut, astronaut & eng sci, Purdue Univ, 68-69; asst prof mech eng, 69-75, ASSOC PROF MECH ENG, WASHINGTON UNIV, 75- *Honors & Awards:* Ralph R Teetor Award, Soc Automotive Engrs, 71. *Mem:* Am Inst Aeronaut & Astronaut; Am Soc Mech Engrs; Biomagnetic Soc. *Res:* Magneto-fluid-mechanics; biomagnetic effects; heat transfer. *Mailing Add:* Dept Mech Eng Washington Univ St Louis MO 63130

GARDNER, RICHARD LYNN, b Clear Lake, Okla, June 4, 34; m 54; c 3. BIOCHEMISTRY. *Educ:* Panhandle Agr & Mech Col, BS, 57; Colo State Univ, PhD(chem), 63. *Prof Exp:* Instr math & chem, Panhandle Agr & Mech Col, 57-60; res asst chem, Colo State Univ, 60-61, NSF coop fel, 61-62, temporary instr, 62-63; PROF CHEM & HEAD DEPT, PANHANDLE STATE UNIV, 63-, CHMN, DIV MATH & SCI, 73- *Mem:* Am Chem Soc; Phytochem Soc NAm. *Res:* Phenols in plant disease resistance; plant sterols and cyclic triterpenes. *Mailing Add:* Dept Chem Panhandle State Univ PO Box 518 Goodwell OK 73939

GARDNER, ROBERT B, b Tarrytown, NY, Feb 27, 39; m 62; c 2. METHOD OF EQUIVALENCE, CONTROL THEORY. *Educ:* Princeton Univ, AB, 59; Columbia Univ, MA, 60; Univ Calif, Berkeley, PhD(math), 65. *Prof Exp:* Res asst math, Univ Calif, Berkeley, 62-65; vis mem, NSF & Sloan grants, Courant Inst Math Sci, NY Univ, 65-67; asst prof, Columbia Univ, 67-70; NSF grant, Inst Advan Study, 70-71; res mathematician & NSF grant, Univ Calif, Berkeley, 71; assoc prof, 71-77, assoc chmn, 77-82, PROF MATH, UNIV NC, CHAPEL HILL, 77- *Concurrent Pos:* Kenan vis prof, Univ Calif, Berkeley, 77-78; Kenan fel, 84; guest prof, Tech Univ, Berlin, 84, 89 & 90. *Mem:* Am Math Soc; Math Asn Am; Sigma Xi; Soc Indust & Appl Math; sr mem Inst Elec & Electronic Engrs. *Res:* Rigidity and uniqueness theorems for hypersurfaces; characterizations of isometries among volumes preserving diffeomorphisms; Cauchy problem; non linear wave equations; Backlund transformations; differential systems, method of equivalence and applications, geometric control theory, variational methods. *Mailing Add:* Dept Math Univ NC Chapel Hill NC 27599-3250

GARDNER, ROBERT WAYNE, b St George, Utah, July 24, 28; m 51; c 5. ANIMAL NUTRITION. *Educ:* Utah State Univ, BS, 59; Cornell Univ, MS, 62, PhD(animal nutrit), 64. *Prof Exp:* Res nutritionist, Cornell Univ, 63-64; asst prof dairy sci, Univ Ariz, 64-66; prof animal sci, 66-85, PROF BENSON AGR & FOOD INST, BRIGHAM YOUNG UNIV, 85- *Concurrent Pos:* Hon mem, Neuroallergy Comt, Am Col Allergists, 81-85; consult, Biomed & Environ Consults, Inc. *Mem:* Am Soc Animal Sci; Am Dairy Sci Asn; Am Inst Nutrit. *Res:* Energy requirements of animals for milk production; body chemical composition related to growth; interrelationships between selenium, vitamin E and muscular dystrophy; dairy calf nutrition; parturient paresis in dairy cattle; chemicals causing allergic reactions in humans and animals; medical profession in food allergies; completing writing a book on chemical causation and treatment of allergies. *Mailing Add:* Benson Agr & Food Inst Brigham Young Univ Provo UT 84602

GARDNER, ROBIN P(IERCE), b Charlotte, NC, Aug 17, 34; m 58. CHEMICAL ENGINEERING, NUCLEONICS. *Educ:* NC State Col, BChE, 56, MS, 58; Pa State Univ, PhD(fuel technol), 61. *Prof Exp:* Res asst fuel technol, Pa State Univ, 56-61; scientist, Oak Ridge Inst Nuclear Studies, 61-63; res engr & assoc dir, Measurement & Controls Lab, Res T riangle Inst, 63-67; assoc prof nuclear eng, 67-70, PROF NUCLEAR & CHEM ENG, NC STATE UNIV, 70-, DIR, CTR ENG APPLICATIONS RADIOISOTOPES, 80- *Concurrent Pos:* Consult, Oak Ridge Inst Nuclear Studies, 63-64; adj asst prof, NC State Univ, 64-; mem, Hwy Res Bd Comt Nuclear Principles & Applns, Nat Acad Sci-Nat Res Coun, 64- *Honors & Awards:* Radiation Indst Award, Am Nuclear Soc, 84. *Mem:* Am Nuclear Soc; Am Inst Chem Engrs; Am Chem Soc; Am Soc Testing & Mat; Fine Particle Soc. *Res:* Development of tracing and gauging techniques with radioisotopes; mathematical treatment of chemical engineering unit operations, particularly comminution. *Mailing Add:* Dept Nuclear Eng, Box 7909 NC State Univ Raleigh NC 27695-7909

GARDNER, RUSSELL, JR, b Granton, Wis, Mar 19, 38; m 60; c 3. PSYCHIATRIC EDUCATION, PSYCHOANALYSIS. *Educ:* MD, Univ Chicago, 62. *Prof Exp:* Instr psychiat, Albert Einstein Col Med, 68-71, asst prof, 71-74; PROF PSYCHIAT, DEPT NEUROSCI, SCH MED, UNIV NDAK, 74-, CHMN, DIV PSYCHIAT-BEHAV SCI, 74-, PROF, DEPT FAMILY MED, 80- *Concurrent Pos:* Consult, Psychiat Educ Br, NIMH, 76-, mem, Psychiat Educ Rev Comt, 79-82, chmn, 80-82; consult, Jamestown State Hosp, 78- *Mem:* Am Col Psychiatrists; Am Psychiat Asn; Am Psychosomatic Soc; Asn Acad Psychiat. *Res:* Stress effects on medical students, impaired physicians, adult women after childhood incest and confused institutionalized patients; sleep and dream psychophysiology; evolutionary model of manic-depressive disorder. *Mailing Add:* 1200 Graves Bldg Univ Tex Med Br Galveston TX 77550

GARDNER, SARA A, b San Diego, Calif, Mar 19, 38; m 74. PHARMACOLOGY, NEUROCHEMISTRY. *Educ:* San Diego State Col, AB, 60; Purdue Univ, West Lafayette, PhD(phys chem), 65. *Prof Exp:* Instr chem, Purdue Univ, 63-64; asst prof, Univ Iowa, 64-65, res assoc neurobiochem, 65-68, from asst prof to assoc prof, 68-73; grants assoc, NIH, 73-74, health scientist adminr, 74-78, dep prog dir, Pharmacol-Toxicol Prog, 78-79, dir, Pharmacol Sci Prog, Nat Inst Gen Med Sci, 79-; RETIRED. *Mem:* Int Soc Neurochem; Am Soc Neurochem; Soc Exp Biol & Med; Am Soc Biol Chemists; Am Soc Pharmacol & Exp Therapeut. *Res:* DNA synthesis in developing brain; indolealkylamines in brain. *Mailing Add:* 515 Country Aire Dr Grants Pass OR 97526

GARDNER, SHERWIN, b New York, NY, July 31, 28; m 54; c 2. MECHANICAL ENGINEERING, OPERATIONS RESEARCH. *Educ:* City Col New York, BME, 50. *Prof Exp:* From jr engr to engr, William L Gilbert Clock Corp, 50-51; engr res & develop labs, Bulova Watch Co, 54-56; prod engr, Kay Mfg Corp, 56; sr engr & prin engr, Ford Instrument Co, Sperry Rand Corp, 56-58, proj supvr, 58-62, stand mgr, 62-64; proj engr, Booz Allen Appl Res Inc, 64-66, res dir, 66-69, assoc, Booz Allen Systs, Inc, 69-70; asst comnr planning & eval, US Food & Drug Admin, 70-72, dep comnr, 72-79; SR VPRES SCI & TECHNOL, GROCERY MFG AM, 79- *Mem:* Inst Food Technologists. *Res:* System engineering analysis; management information systems; organizational and facility planning and evaluation; economic analysis. *Mailing Add:* Grocery Mfrs Am 1010 Wisconsin Ave NW Suite 800 Washington DC 20007

GARDNER, SYLVIA ALICE, b Pittsburgh, Pa, Nov 21, 47. ORGANOMETALLIC CHEMISTRY. *Educ:* Bucknell Univ, BS, 69; Univ Mass, MS, 71, PhD(inorg chem), 73. *Prof Exp:* Fel, Univ Mass, 73-74; res chemist, Res Labs, 74-81, sr chemist, Mfg Technol Div, 81-86, TECH ASSOC, EASTMAN KODAK CO, 86- *Concurrent Pos:* Sci secy, VI Int Conf Organometallic Chem, 72-73; adj fac, Rochester Inst Technol, 75- *Mem:* Am Chem Soc; Mat Res Soc. *Res:* Inorganic and organometallic synthetic chemistry; sol-gel chemistry. *Mailing Add:* Mat Sci & Eng Div B46 Eastman Kodak Co Rochester NY 14650

GARDNER, THOMAS WILLIAM, b East Stroudsburg, Pa, Jan 19, 49; m 73; c 2. FLUVIAL GEOMORPHOLOGY IN TECTONIC TERRAINS, REMOTE SENSING OF EARTH RESOURCES. *Educ:* Franklin & Marshall Col, BA, 71; Colo State Univ, MS, 73; Univ Cincinnati, PhD(geol), 77. *Prof Exp:* Geologist, Exxon Minerals Co, 78-79; asst prof, 79-85, ASSOC PROF GEOSCI, PA STATE UNIV, 85- *Mem:* Geol Soc Am; Am Geophys Union; Am Soc Photogram & Remote Sensing. *Res:* Fluvial hydrology, hydrology and geomorphology in disturbed and tectonically active terrains; Quaternary landscape evolution in semiarid climates; remote sensing and photogeology. *Mailing Add:* Dept Geosci Pa State Univ 507 Deike University Park PA 16802

GARDNER, WALTER HALE, b Beaver, Utah, Feb 24, 17; m 48; c 5. SOIL PHYSICS. *Educ:* Utah State Univ, BS, 39, MS, 47, PhD(soil physics), 50. *Prof Exp:* Asst soil physics, Cornell Univ, 40-41; spec instr math, Utah State Univ, 48-49; from asst prof to prof soils, 50-82, prof soils & biophys, 70-72, EMER PROF SOILS & BIOPHYS, WASH STATE UNIV, 82- *Concurrent Pos:* Guggenheim fel, 64-65; ed-in-chief, Proc of Soil Sci Soc Am, 65-69; mem, Am Soc Agron-NSF Vis Scientist Prog, 66-68; consult, Int Atomic Energy Agency, Vienna, 71-72; historian, Soil Sci Soc Am, 87- *Honors & Awards:* Distinguished Serv Award, Soil Sci Soc Am, 86- *Mem:* AAAS; fel Am Soc Agron; fel Soil Sci Soc Am (pres, 83-84); AAAS (pres, 85). *Res:* Physics of the soil with emphasis on soil water retention and flow and their effect on plant growth; published a history of soil physics in Advances in Soil Science. *Mailing Add:* NE 1505 Upper Dr Pullman WA 99163

GARDNER, WAYNE SCOTT, b Clifton, Colo, Jan 11, 20; m 44; c 4. PLANT PATHOLOGY, VIROLOGY. *Educ:* Utah State Univ, BS, 50, MS, 51; Univ Calif, Davis, PhD(plant path), 67. *Prof Exp:* Asst plant path, Utah State, 49-51; instr agr, Mesa Col, 51; plant pathologist, Lab Br, Crops Div, Tech Opers, Dugway Proving Ground, Utah, 51-52, crops div, Biol Warfare Facilities, 52-54; agr technician, Agr Div, Columbia-Geneva Steel Div, US Steel Corp, 54-61, chem unit, Fairless Works, Pa, 61-63; res asst plant path, Univ Calif, Davis, 63-65, lab technician, 65-67; from assoc prof to prof plant path sci, SDak State Univ, 67-85; RETIRED. *Mem:* Am Phytopath Soc. *Res:* Plant virology and electron microscopy; ultrastructure of plant pathogen, host and environment interaction; effects of air pollutants on plants; cereal and forage crop virus diseases; aerobiology and epiphytology. *Mailing Add:* 135 Leisureville Circle Woodland CA 95695

GARDNER, WAYNE STANLEY, b Granton, Wis, Sept 13, 41; m 64; c 2. WATER CHEMISTRY, MARINE CHEMISTRY. *Educ:* Univ Wis-Stephens Point, BS, 63; Univ Wis-Madison, MS, 64, PhD(water chem), 71. *Prof Exp:* Lt chem, USPHS, 64-66; asst prof marine chem, Skidaway Inst Oceanog, Univ Syst Ga, 70-77; mem staff, Columbia Nat Fishery Res Lab, 77-79; PHYS SCIENTIST, GREAT LAKES ENVIRON RES LABS, 79- *Mem:* AAAS; Am Soc Limnol & Oceanog. *Res:* Fluxes and biochemical transformations of organic compounds and trace metals through aquatic ecosystems and their effect on water quality. *Mailing Add:* Great Lakes Environ Res Labs 2205 Commonwealth Blvd Ann Arbor MI 48105

GARDNER, WESTON DEUAIN, b Cresco, Iowa, Jan 6, 17; m 44; c 2. GROSS ANATOMY. *Educ:* Pa State Col, BS, 38; Univ Pittsburgh, MD, 42. *Prof Exp:* Intern, St Francis Hosp, Pittsburgh, 42-43; practicing physician, 45-46; asst anat, Univ Wis, 46-47; from instr to prof, 47-79, EMER PROF ANAT, MED COL WIS, 79- *Concurrent Pos:* Dir med educ, Deaconess Hosp, 55-70, asst exec dir, Educ Coun Foreign Med Grads, 60-65; med dir, Curative Workshop Rehab Ctr, Milwaukee, 71-74. *Mem:* Am Asn Anatomists; Asn Med Illusrs; Asn Am Med Cols; Guild Natural Sci Illusrs. *Res:* Embryology of human ear; degeneration studies on the cingulum; repair of bone to experimental fractures; venous patterns of thorax and breast by infrared photographic methods; relation of superficial veins and thorax to breast tumors; development of broncho pulmonary segments in human lungs; segmental veins in human lungs; audiovisual communications; medical illustration. *Mailing Add:* 14905 Wisconsin Dr Elm Grove WI 53122

GARDNER, WILFORD ROBERT, b Logan, Utah, Oct 19, 25; m 49; c 3. SOIL PHYSICS. *Educ:* Utah State Agr Col, BS, 49; Iowa State Col, MS, 51, PhD(physics), 53. *Prof Exp:* Asst physics, Iowa State Col, 49-51, res assoc, Inst Atomic Res, 51-53; physicist soil physics, US Salinity Lab, USDA, Calif, 53-66; prof soil physics, Univ Wis-Madison, 66-80; head, Dept Soil & Water Sci, Univ Ariz, Tucson, 80-87; DEAN, COL NATURAL RESOURCES & DIR, AGR & NAT RESOURCES PROGS, UNIV CALIF, BERKELEY, 87- *Concurrent Pos:* NSF sr fel, Cambridge Univ, 59; Consult ed, Soil Sci, 61-; Fulbright lectr, Univ Ghent, 71-72; Haight traveling prof, Univ Wis, 78; mem, Sci Adv Bd, Environ Protection Agency, 82, Budget & Finance Comt, Am Soc Agron & Comt Irrigation Induced Water Qual Probs, Nat Acad Sci, 85; co-chair orgn comt, USDA-State Exp Sta Symp Plant Water Stress, 84. *Honors & Awards:* Am Soc Agron Award, 62. *Mem:* Nat Acad Sci; fel AAAS; Am Phys Soc; Soil Sci Soc Am; fel Am Soc Agron; Am Geophys Union. *Res:* Measurement of soil moisture by neutron scattering; soil physics; movement of fluids in porous media; soil-water-plant relations; soil salinity; plant biophysics; environmental physics. *Mailing Add:* Col Natural Resources Univ Calif 101 Giannini Hall Berkeley CA 94720

GARDNER, WILLIAM ALBERT, JR, b Sumter, SC, Aug 2, 39; m 60; c 3. GENITO-URINARY PATHOLOGY, LAB ADMINISTRATION. *Educ:* Wofford Col, BS, 60; Med Univ SC, MS, 65, MD, 65. *Prof Exp:* Teaching & res asst anat, Med Univ SC, 61-63; intern & path fel, Johns Hopkins Hosp, 65-67; asst & career resident, Duke Univ Med Ctr, 67-68, chief resident & instr path, 68-69; chief, lab serv, path, Vet Admin Med Ctr, Charleston & assoc prof path, Med Univ SC, 69-76, asst prof path, 72-76; chief lab serv path, Vet Admin Med Ctr, Nashville, 76-81; prof & vchmn path, Vanderbilt Univ Sch Med, 76-81; PROF & CHMN PATH, UNIV S ALA COL MED, 81- *Concurrent Pos:* Mem House Delegate, Col Am Pathologist, 74-77; vis prof path, Med Univ SC, 76-81; mem working cadre, NCI Nat Prostate Cancer Proj, 77-83; dist comnr, CAP Inspection & Accreditation Prog, 77-81; mem, US-Canadian, IAP educ comt, 78-82; CAP comn, govt rels, 83-85; chmn, Asn Path Chmn Vet Admin, 84-87; coun mem, US/Can Acad Path, 87-91; coun mem, Asn Path, chmn, 88-; ed-in-chief, Yearbook Path & Clin Path, 90-; pres, Ala Asn Pathologists, 91- *Mem:* Am Acad Path; US-Can Acad Path; Am Soc Clin Path; Asn Path Chmn. *Res:* Origin and pathogensis of prostate cancer; biology and host parasite relationships of trichomonads; etiological agents and natural history of prostatitis and prostate hyperplasia. *Mailing Add:* Dept Path 2451 Fillinghim St Mobile AL 36617

GARDNER, WILLIAM H, industrial chemistry, for more information see previous edition

GARDNER, WILLIAM HOWLETT, b New York, NY, Sept 25, 02; m 32; c 3. POLYMER CHEMISTRY. *Educ:* Cornell Univ, BChem, 23, PhD(inorg chem), 27; Dartmouth Col, MS, 25. *Prof Exp:* Instr chem, Dartmouth Col, 23-26; res chemist, Vitreous Enameling Co, Cleveland, 27; fel, Shellac Res Bur, Polytech Inst Brooklyn, 28-31, asst res prof chem eng, 31-34, res prof, 34-42, supvr shellac res bur, 34-42; chief specification sect, Conserv Div, War Prod Bd, Washington, DC, 42, chief, Mat Sect, 43, chief, Chem Mat Br, 43-44; mem staff, New Prods Div, Nat Aniline Div, Allied Chem Corp, 44-50, mgr chem lit, 50-67; consult forensic chem, 67-85; RETIRED. *Honors & Awards:* Paint Pioneer Award, NY Paint, Varnish & Lacquer Asn, 39; Roy Kienle Award, NY Soc Paint Technol, 60. *Mem:* Am Chem Soc; Am Soc Test & Mat; Am Inst Chem Eng; fel Am Inst Chem; Fedn Socs Paint Technol; Inst Food Technol. *Res:* Paint and varnish technology; resin chemistry; chemistry of high polymers; manufacture of pigments; maleic anhydride derivatives; aniline; food acidulants. *Mailing Add:* 29 Merriam Ave Bronxville NY 10708

GARDNER, WILLIAM LEE, b Carlisle, Pa, June 8, 40; m 65; c 3. PHYSICAL CHEMISTRY, INORGANIC CHEMISTRY. *Educ:* Pa State Univ, BS, 62; Purdue Univ, PhD(phys inorg chem), 68. *Prof Exp:* RES ASSOC, RES LABS, EASTMAN KODAK CO, 68- *Mem:* Soc Photog Scientists & Engrs. *Res:* Solution and titration calorimetry; adsorption and surface phenomena within an aqueous medium; properties of photographic gelatin; design and development of radiographic and color photographic materials. *Mailing Add:* Eight Keswick Way Fairport NY 14450

GARDNER-CHAVIS, RALPH ALEXANDER, b Cleveland, Ohio, Dec 3, 22; m 45; c 1. PHYSICAL CHEMISTRY. *Educ:* Univ Ill, BS, 43; Western Reserve Univ, MS, 52, PhD, 59. *Prof Exp:* Chemist, Argonne Nat Labs, 43-47 & Standard Oil Co, Ohio, 49-68; assoc prof, 68-87, EMER PROF CHEM, CLEVELAND STATE UNIV, 87-; DIR, MOLECULAR TECHNOL CORP, 83- *Concurrent Pos:* adj prof, Malone Col, 87- *Mem:* Am Chem Soc. *Res:* Infrared spectroscopy, heterogeneous catalysis. *Mailing Add:* Dept Chem Cleveland State Univ 1983 E 24th St Cleveland OH 44115

GARDOCKI, JOSEPH F, b Brooklyn, NY, Sept 5, 26; m. PHARMACOLOGY. *Educ:* Georgetown Univ, PhD, 51. *Prof Exp:* Pharmacologist, Food & Drug Admin, 47-48; pharmacologist, Hazelton Labs, 48-51; sr pharmacologist, Pfizer Therapeut Inst, 52-57; res assoc, Squibb Inst Med Res, 58-59; pharmacologist & head neuropsychopharmacol, 60-80, RES FEL, MCNEIL LABS, INC, 80- *Mem:* AAAS; Am Soc Pharmacol & Exp Therapeut; NY Acad Sci. *Res:* Intravenous anesthetics; tranquilizers; diuretics; analgesics; muscle relaxants; neurophysiological correlates of behavior; toxicology; blood chemistry; central nervous system stimulants; bioassay; operant conditioning. *Mailing Add:* 72 Meadow Lane Doylestown PA 18901

GARDON, JOHN LESLIE, b Budapest, Hungary, June 5, 28; US citizen; m 51; c 2. POLYMER CHEMISTRY, PHYSICAL CHEMISTRY. *Educ:* Swiss Fed Inst Technol, grad, 50; McGill Univ, PhD(phys chem), 52. *Prof Exp:* Chemist wood chem, Can Int Paper Co, Hawkesbury, Ont, 55-58; sr chemist textiles, Rohm & Haas Co, 58-60, group leader coatings, 60-62, res assoc polymers, 62-65, res mgr leather finishes, Springhouse, Pa, 65-68; dir, Corp Res Develop Div, M&T chem, Inc, Rahway, NJ, 68-73, dir res & develop, Coating & Ink Div, 73-80; vpres coatings res & develop, Sherwin Williams Co, Chicago, 80-; AT AKZO COATING AM. *Concurrent Pos:* Trustee, Paint Res Inst, 74-; chmn, Gordon Res Conf Adhesion, 76- *Honors & Awards:* Roon Award, Fed Soc Coating Technol, 66. *Mem:* Am Chem Soc (secy, 77-78)); Soc Plastics Engrs; Chem Inst Can; Adhesion Soc; NY Acad Sci; Sigma Xi. *Res:* Polymer synthesis; emulsion polymerization; thermodynamics of polymer solutions; coatings; finishes of textiles, paper and leather; adhesion; lignin; cellulose; polyelectrolytes; membrane separations. *Mailing Add:* Akzo Coatings Am-ICA Inc PO Box 7062 Troy MI 48007-7062

GARELICK, DAVID ARTHUR, b Woonsocket, RI, Nov 17, 37; m 60; c 1. PHYSICS. *Educ:* Mass Inst Technol, BS, 59, PhD(physics), 63. *Prof Exp:* Res staff, Lab Nuclear Sci, Mass Inst Technol, 63-67; assoc physicist, Brookhaven Nat Lab, 67-69; assoc prof physics, 69-74, PROF PHYSICS, NORTHEASTERN UNIV, 74- *Mem:* Am Physics Soc. *Res:* Elementary particles experimentation. *Mailing Add:* Dept Physics Northeastern Univ 360 Huntington Ave Boston MA 02115

GAREN, ALAN, b Brooklyn, NY, May 26, 26; m 59; c 5. DEVELOPMENTAL BIOLOGY. *Educ:* Univ Colo, BS, 45, PhD(biophysics), 53. *Prof Exp:* Res chemist, Oak Ridge Nat Lab, 46-48; fel, Nat Found Infantile Paralysis, Cold Spring Harbor, NY, 51-55; res assoc, Purdue Univ, 55-57; sr res assoc, Biol Dept, Mass Inst Technol, 57-60; from assoc prof to prof, Biol Div, Univ Pa, 60-63; PROF MOLECULAR BIOPHYS & BIOCHEM, YALE UNIV, 63-, PROF HUMAN GENETICS, 70- *Concurrent Pos:* Guggenheim fel, 70. *Honors & Awards:* Waksman Medal Microbiol, 62. *Mem:* Nat Acad Sci; fel Am Acad Arts & Sci. *Res:* Genetic and hormonal control of Drosophila development. *Mailing Add:* Dept Molecular Biophys & Biochem Yale Univ Kline Biol Tower New Haven CT 06511

GARETZ, BRUCE ALLEN, b St Paul, Minn, Nov 24, 49. PHYSICAL CHEMISTRY. *Educ:* Harvard Col, AB, 71; Mass Inst Technol, PhD(phys chem), 76. *Prof Exp:* Res assoc & lectr dept chem, Univ Toronto, 76-78; asst prof, 78-83, ASSOC PROF PHYS CHEM, POLYTECH UNIV, 83- *Concurrent Pos:* Ed, Advances in Laser Spectros, 82-; vis prof, dept chem, Indian Inst Technol, Kanpur, India, 86; vis scientist, Mass Inst Technol Laser Res Ctr, 84-; Alfred P Sloan res fel, 84-87. *Mem:* Am Phys Soc; Optical Soc Am. *Res:* Laser spectroscopy; nonlinear optics and multiphoton processes; polarization effects in molecular spectroscopy; molecular dynamics. *Mailing Add:* Dept Chem Polytech Univ Brooklyn NY 11201

GAREY, CARROLL LAVERNE, b Ft Collins, Colo, Nov 9, 17; m 41; c 3. CHEMISTRY. *Educ:* Univ Nebr, BSc, 39, MSc, 47; Purdue Univ, PhD(soils), 52. *Prof Exp:* Jr engr, State Hwy Dept, Nebr, 40-42; instr soils, Univ Nebr, 46-49; asst prof, Univ Ark, 51-57; res assoc & phys chemist, Inst Paper Chem, Wis, 57-75; mgr, polymer res & develop, Ralston Purina Co, 75-82; RETIRED. *Concurrent Pos:* Consult paper coating. *Honors & Awards:* Fulbright fel, Finland, 84. *Mem:* AAAS; Soc Rheol; Clay Minerals Soc; Am Chem Soc; Tech Asn Pulp & Paper Indust; Sigma Xi. *Res:* Evaluation of pigments for use as paper coatings and fillers; study of starches, proteins and polymers for use in paper manufacture, sizing and conversion; development of new natural polymers from protein type materials; solution of plant production problems in coatings and adhesives field; rheology and physical chemical problems of paper coating suspensions and on paper conversion processes; use of x-ray diffraction in study of cellulose, and structure of paper and coating. *Mailing Add:* 6310 Mesa Verde Dr Lincoln NE 68510

GAREY, MICHAEL RANDOLPH, b Manitowoc, Wis, Nov 19, 45; m 65. MATHEMATICS, COMPUTER SCIENCE. *Educ:* Univ Wis-Madison, BS, 67, MS, 69, PhD(comput sci), 70. *Prof Exp:* mem tech staff, Math Res Ctr, 70-81, dept head, Math Found Comput, 81-88, DIR, MATH SCI RES CTR, AT&T BELL LABS, 88- *Concurrent Pos:* Assoc ed, J Asn Comput Mach, 75-79, ed chief, 79-82, ed, MIT Press Series, 83- *Honors & Awards:* Lanchester Prize, Oper Res Soc Am, 79. *Mem:* Asn Comput Mach; Soc Indust & Appl Math; Oper Res Soc Am. *Res:* Design and analysis of combinatorial algorithms; graph theory; operations research. *Mailing Add:* AT&T Bell Labs 600 Mountain Ave Murray Hill NJ 07974

GAREY, WALTER FRANCIS, b Junction City, Ohio, Dec 13, 26; m 50; c 3. COMPARATIVE PHYSIOLOGY. *Educ:* Ohio State Univ, BS, 49, MA, 53; State Univ NY, Buffalo, PhD(physiol), 67. *Prof Exp:* Teacher biol & phys sci, Sec Sch, Ohio, Mich & Calif, 49-62; teaching fel, State Univ NY, Buffalo, 62-67; res physiologist, Scripps Inst Oceanog, 67-71, acad adminr, 71-77; dir grants prog, Kroc Found, 77-85; RETIRED. *Mem:* Am Col Rheumatology. *Res:* Comparative physiology of respiration, circulation and acid-base regulation; arthritis; diabetes; multiple sclerosis. *Mailing Add:* 1580 Monarch Dr Santa Ynez CA 93460

GARFIELD, ALAN J, b Utica, NY; m 79; c 3. RAY TRACING ALGORITHMS, REAL TIME ANIMATION. *Educ:* Univ Iowa, BA, 71; State Univ NY, Binghamton, MA, 75. *Prof Exp:* Asst prof art hist, Converse Col, 74-76 & Creighton Univ, 76-80; assoc prof computer graphics, Univ Wis-Madison, 90-91; CHAIR, ART & COMPUTER GRAPHICS DEPT, TEIKYO MARYCREST UNIV, 80- *Concurrent Pos:* Asst prof art hist, Southwest Mo State Univ, 75-76; pres, Digigraphic Systs, Inc, 84-; consult, McGladdey & Pullen Acct & Dynamic Graphics Inc, 86-90. *Mem:* Asn Comput Mach; Nat Computer Graphics Asn; Spec Interest Group Computer Graphics. *Res:* Exploring ray traced and animation algorithms; significant image research. *Mailing Add:* 34 Oak Lane Davenport IA 52804

GARFIELD, EUGENE, b New York, NY, Sept 16, 25; m; c 4. CHEMISTRY, INFORMATION SCIENCE. *Educ:* Columbia Univ, BS, 49, MS, 54; Univ Pa, PhD(struct ling), 61. *Prof Exp:* Res chemist, Evans Res & Develop Corp, 49-50; chem, Columbia Univ, 50-51; staff mem mach indexing proj, Johns Hopkins Univ, 51-53; PRES, INST SCI INFO, 56- *Concurrent Pos:* Lectr, Univ Pa, 63-, adj prof, 70-; consult, Smith, Kline & French Labs, Biol Abstracts, Nat Libr Med & Encycl Americana, 54-58; mem adv comt to cardiovasc lit proj, Nat Acad Sci; founder, publisher and ed, Current Contents, Index Chemicus & Sci Citation Index, Current Bibliog Directory, Chem Substruct Index, Social Sci Citation Index, Arts & Humanities Citation Index, Index to Sci Reviews, Index to Sci & Tech Proceedings, Jour Citation Reports, Automatic Subj Citation Alert; co-founder, former pres & chmn bd, Info Indust Asn, 68-; mem bd dirs, Ann Reviews; ed-in-chief, Scientometrics. *Honors & Awards:* Div Chem Info Sci Award, Am Chem Soc, 77; John Price Wetherill Medal, Franklin Inst; Derek J de Solla Price Mem Award, 84; John Scott Award, 87. *Mem:* Fel AAAS; Am Soc Info Sci; Nat Asn Sci Writers; NY Acad Sci; Am Chem Soc; Int Sci Policy Found. *Res:* Citation analysis; science policy. *Mailing Add:* Univ City Sci Ctr 3501 Market St Philadelphia PA 19104

GARFIELD, ROBERT EDWARD, b Douglas, Wyo, Jan 27, 39; m 64; c 2. REPRODUCTIVE PHYSIOLOGY. *Educ:* Univ Wyo, BSc, 67, MSc, 69; Univ Alberta, PhD(pharmacol), 73. *Prof Exp:* Fel, Dept Physiol, Univ Pa, 73-76; lectr, 76-78, from asst prof to assoc prof, 78-82, prof & chmn dept neurosci, 85-86, PROF DEPT BIOMED SCI & DEPT OBSTET & GYNEC, MCMASTER UNIV, CAN, 82- *Concurrent Pos:* Prin investr, Res Projs, Med Res Coun, Can, 76-; jr fel, Can & Ont Heart Found, 76-77; Can Heart Found scholar, 78-81; sr fel, Ont Heart Found, 81-87; Med Res Coun fel, Univ Paris, 85-86. *Mem:* Sigma Xi; Electron Micros Soc Am; Soc Obstetricians & Gynecologists Can; Soc Gynecol Invest; Biophys Soc; Am Fertil Soc. *Res:* Structure and function of muscle; uterine smooth muscle and the factors which bring about its activation for labor; vascular smooth muscle and changes during hypertension; author of 100 publications. *Mailing Add:* Dept Biomed Sci 1200 Main St W Hamilton ON L8N 3Z5 Can

GARFIELD, SANFORD ALLEN, b Feb 21, 43; m; c 1. CELL BIOLOGY. *Educ:* Univ Chicago, PhD(develop biol), 74. *Prof Exp:* Assoc prof anat & cell biol, Col Med, Univ Cincinnati, 79-86; DIR, DIABETES RES CTRS, NIH-NIDDK, 86- *Mem:* AAAS; Am Diabetes Asn. *Mailing Add:* NIH Nat Inst Diabetes Digestive & Kidney Dis Westwood Bldg 5333 Westbard Ave Rm 626 Bethesda MD 20816

GARFIN, DAVID EDWARD, b Minneapolis, Minn, July 7, 40; m 69; c 2. MOLECULAR BIOLOGY, ELECTRICAL ENGINEERING. *Educ:* Univ Minn, Minneapolis, BS, 62, MSEE, 64; Univ Calif, Berkeley, PhD(biophys), 72. *Prof Exp:* Res assoc, Univ Calif, San Francisco, 72-78, asst res biochemist, 78-79; mgr prod develop, Tago Inc, Burlingame, Calif, 79-80; head immunol develop & prod, Hana Biol Inc, Emeryville, Calif, 81-82; sr res biochemist, 83-84, group leader, Immunol Res, 84-86, GROUP LEADER, ELECTROPHORESIS & BIOCHEM RES, BIO-RAD LABS, HERCULES, CALIF, 86- *Mem:* AAAS; Am Chem Soc; Am Soc Microbiol; Biophys Soc; Electrophoresis Soc; IEEE; Inst Elec & Electronics Engrs. *Res:* Fractionation and analysis of biological molecules. *Mailing Add:* 112 Kenyon Ave Kensington CA 94708

GARFIN, LOUIS, b Mason City, Iowa, June 7, 17; m 43; c 2. ACTUARIAL SCIENCE. *Educ:* Univ Iowa, BA, 38, MS, 39, PhD(math), 42. *Prof Exp:* Asst math, Univ Iowa, 39-42; assoc instr radio operating, Training Sch, Army Air Force, Scott Field, 42-43; instr math in charge US Army radio prog, Ill Inst Technol, 43; instr, US Army Pre-Flight Prog, Univ Minn, 43-44; actuary, Ore State Dept Ins, 46-52; from assoc actuary to actuary, Pac Mutual Life Ins Co, 52-64, vpres & Chief Actuary, 64-82; CONSULT ACTUARY,82- *Mem:* Am Math Soc; fel Soc Actuaries; Am Acad Actuaries (vpres, 76-78); Int Actuarial Asn. *Res:* Pension funds; theoretical and practical bases and valuation; linear integral equations. *Mailing Add:* 371 Dartmoor St Laguna Beach CA 92651

GARFINKEL, ARTHUR FREDERICK, b New York, NY, Nov 13, 34; m 63; c 2. HIGH ENERGY PHYSICS. *Educ:* Columbia Univ, BA, 56, MS, 58, PhD(physics), 62. *Prof Exp:* Vis scientist, Res Estab Riso, Denmark, 62-64; proj assoc physics, Univ Wis-Madison, 64-67; asst prof, 67-73, assoc prof, 73-78, PROF PHYSICS, PURDUE UNIV, 79- *Concurrent Pos:* Exchange prof physics, Univ Hamburg, 77-78. *Mem:* Am Phys Soc. *Res:* Experimental particle physics research, primarily making use of counter techniques to study particle reactions; with special interest in proton and antiproton collisions at high energy. *Mailing Add:* Dept Physics Purdue Univ Lafayette IN 47907

GARFINKEL, BORIS, b Moscow, Russia, Nov 18, 04; nat US. ASTRONOMY. *Educ:* City Col New York, BS, Columbia Univ, MA , 27; Yale Univ, PhD(astron), 43. *Prof Exp:* Instr physics, Yale Univ, 43-46; res mathematician, Ballistic Res Lab, Aberdeen Proving Ground, 46-60, dep chief, Comput Lab, 60-63, chief res scientist, 63-67; sr res assoc & lectr astron, 67-73, emer sr res astronr, 73-83, EMER PROF ASTRON, YALE UNIV, 83- *Concurrent Pos:* Lectr, Univ Del, 48-56; res assoc, Yale Univ, 58-66, vis prof, 66-67; adv scientist, Lockheed Missile & Space Co, 61-62; Nat Acad Sci sr res assoc, 63-64. *Honors & Awards:* R H Kent Award, 59. *Mem:* Royal Astron Soc; Int Astron Union; Am Astron Soc. *Res:* Celestial mechanics; artificial satellite theory; the ideal resonance problem; astronomical refraction; calculus of variations; Trojan asteroids. *Mailing Add:* Dept Astron Yale Univ 260 Whitney Ave New Haven CT 06511

GARFINKEL, DAVID, biological & medical computing; deceased, see previous edition for last biography

GARFINKEL, HARMON MARK, b Brooklyn, NY, May 20, 33; m 56; c 2. MATERIALS SCIENCE. *Educ:* Brooklyn Col, BA, 57; Iowa State Univ, PhD(phys chem), 60; Harvard Univ, PMD, 73. *Prof Exp:* Sr chemist, Corning Glass Works, 60-64, mgr phys chem, 64-79, mgr bio org res, 69-73, dir biotech, 73-74, dir appl chem & biol, 74-78, dir res, 78-85; VPRES RES & DEVELOP, ENGELHARD CORP, 85- *Mem:* Am Chem Soc; Am Phys Soc; fel Am Inst Chemists; Am Ceramic Soc. *Res:* Developed techniques for improving the strength of ceramics via ion exchange techniques; the fundamentals (kineticsand thermodynamics) of the process. *Mailing Add:* Englehard Corp Menlo Park CN40 Edison NJ 08818

GARFINKEL, LAWRENCE, b New York, NY, Jan 11, 22; m 48; c 2. EPIDEMIOLOGY, BIOSTATISTICS. *Educ:* City Col New York, BBA, 47; Columbia Univ, MA, 49. *Prof Exp:* Biostatistician, Am Cancer Soc, 47-59, chief field & spec projs, 59-73, asst vpres, 73-79, vpres epidemiol & statist, 79-90; CONSULT, 90- *Concurrent Pos:* Asst res scientist, Col Dent, NY Univ, 62-69, from asst clin prof to assoc clin prof, 70-79; res consult to sr med investr, Vet Admin Hosp, East Orange, NJ, 65-80; mem epidemiol adv comt, Third Nat Cancer Surv, Nat Cancer Inst, 68-73, mem adv comt, Diet, Nutrit & Cancer Prog, 75-77, mem clearinghouse environ carcinogens, Nat Cancer Inst, 76-78; assoc prof, Mt Sinai Sch Med, 72; mem cancer com, Am Col Surgeons, 88; mem, Data Mgt Working Group, Nat Cancer Data Base, 89-; adv ed, Cancer, 78-, adv bd, 80- *Res:* Design and analysis of epidemiologic and pathologic studies in smoking and health; analysis of trends in cancer mortality, morbidity and survival. *Mailing Add:* Am Cancer Soc 1180 Ave of Americas New York NY 10036

GARFINKEL, PAUL EARL, b Winnipeg, Can, Jan 21, 46; m 67; c 3. PSYCHIATRY. *Educ:* Univ Manitoba, MD, 69; FRCP (C), 74; Univ Toronto, MSc, 77. *Prof Exp:* From assoc prof to prof psychiat & vchmn, Dept Psychiat, Univ Toronto, 82-90; phychiatrist-in-chief, Toronto Gen Hosp, 82-90; DIR & PSYCHIATRIST IN CHIEF, CLARKE INST PSYCHIAT, UNIV TORONTO, 90-, PROF & CHMN, DEPT PSYCHIAT, 90- *Concurrent Pos:* Vis prof, Nat Univ, Dublin, Ireland, 79 & dept psychiat, Univ Manitoba, 81; chief psychosomatic med unit, Clarke Inst Psychiat, 79-82, coordr res, 81-; mem, Health & Welfare Res Comt, Can, 80-82; coordr res, Dept Psychiat, Univ Toronto, 81-90. *Honors & Awards:* Paul Christie Mem Award, Ont Ment Health Found, 74; McNeil Award, Can Psychiat Asn, 75. *Mem:* Am Psychiat Asn; Psychiat Res Soc; Am Psychosomatic Soc; Soc Neurosci; fel Can Col Neuropharmacol. *Res:* Pathogenetic mechanisms in anorexia neurvosa; treatment of weight disorders; pharmacotherapies of depression. *Mailing Add:* Clarke Inst Psychiat 250 College St Toronto ON M5T 1R8 Can

GARFINKLE, BARRY DAVID, b Newark, NJ, Oct 3, 46; m 68; c 2. VIROLOGY, MICROBIOLOGY. *Educ:* Kans State Univ, BS, 68; Penn State Univ, MS, 70, PhD(microbiol), 72. *Prof Exp:* Res scientist virol, Philip Morris Res Ctr, 74-75; sr microbiologist, Merck Sharp & Dohme, 75-77, mgr biol qual tech serv, 77-81, mgr process validation, 81-83, mgr qual control pharmaceut prod, 83-85, mgr sterile process proj eng, 85-86, sr mgr biol prod technician serv, 86-87, dir quality assurance, 88-90, dir qual mgt validation, 90-91, SR DIR QUALITY MGT VALIDATION, MERCK SHARP & DOHME, 91- *Concurrent Pos:* NIH spec fel, Cold Spring Harbor, 71; fel, Roche Inst Molecular Biol, 72-74. *Mem:* Am Soc Microbiol; Am Tissue Culture Asn; Pharmaceut Mfg Asn; Parenteral Drug Asn. *Res:* All aspects of research and development of bacterial and viral vaccines; biochemistry. *Mailing Add:* 1274 Georgia Lane Hatfield PA 19440

GARFUNKEL, MYRON PAUL, b New York, NY, June 17, 23; m 46; c 3. PHYSICS. *Educ:* Rutgers Univ, BS, 47, PhD(physics), 51. *Prof Exp:* Physicist, Res Lab, Westinghouse Elec Corp, 51-59; PROF PHYSICS, UNIV PITTSBURGH, 59-, CHMN DEPT PHYSICS & ASTRON, 76- *Mem:* Am Phys Soc. *Res:* Solid state physics; cryogenics; superconductivity; microwave acoustics. *Mailing Add:* Dept Physics & Astron Allen Hall 100 Univ Pittsburgh 4200 Fifth Ave Pittsburgh PA 15260

GARG, ARUN, b Kanpur, India; US citizen; m 74; c 1. OCCUPATIONAL SAFETY & HEALTH. *Educ:* Inst Technol, Kanpur, India, BS, 69; Villanova Univ, MS, 71; Univ Mich, PE, 73, PhD(indust eng), 76. *Prof Exp:* Mgt, Sch Dent, Univ Mich, 73-76; asst prof indust eng, Univ Miami, 76-77; from asst prof to assoc prof, 77-88, PROF INDUST ENG, UNIV WIS, MILWAUKEE, 88- *Concurrent Pos:* Consult. Med Res Lab. Wright Patterson AFB, Ohio, 72 & Westinghouse Res & Develop Ctr, 77; vis asst prof, Tex Tech Univ, 78. *Mem:* sr mem Am Inst Indust Engrs; Human Factors Soc; Am Indust Hygiene Asn. *Res:* Occupational safety and health; ergonomics; occupational biomechanics and work physiology. *Mailing Add:* Indust & Systs Eng Univ Wis PO Box 784 Milwaukee WI 53201

GARG, ARUN K, b Agra, India, July 16, 46; Can citizen; m 70; c 1. MEDICAL BIOCHEMISTRY, LABORATORY MEDICINE. *Educ:* Agra Univ, BSC, 62, MSC, 64; Univ Sask, MSC, 68, PhD(biochem), 70; Univ BC, MD, 77; Royal Col Physicians & Surgeons, FRCP, 80. *Prof Exp:* DIR MED BIOCHEM, ROYAL COLUMBIAN & REGIONAL HOSP, 84- *Concurrent Pos:* Partner, C J Coody & Assocs, 80-, chmn, 89-; vis assoc prof, Dept Chem, Simon Fraser Univ, 80-84; clin assoc prof, Dept Path, Univ BC, 84-; count, Can Soc Clin Chemists, 81-83; secy & treas, Bd Dirs, BC Med Soc, 80-91; chmn, Sect Clin Path, Can Asn Path, 82-86; dir, med biochem, BC Biomed Lab, 89- *Mem:* Can Soc Clin Chemists; Am Col Physicians Execs; Col Am Pathologists; Asn Clin Biochemists UK; Can Med Asn; Can Asn Path. *Res:* Clinical applications of biochemical methods; role of ADH in Hyponatremia. *Mailing Add:* Dept Path Royal Columbian Hosp 330 E Columbia St New Westminster BC V3L 3W7 CAN

GARG, BHAGWAN D, b Dhuri Punjab, India, Aug 14, 40; US citizen; m 67; c 2. EXPERIMENTAL PATHOLOGY, MORPHOMETRY. *Educ:* Panjab Univ, India, BS, 61, MS, 62; McMaster Univ, PhD(biol), 66. *Prof Exp:* Res assoc, Case Western Reserve Univ Sch Med, 66-67; res assoc, Columbia Univ Col Physicians & Surgeons, 67-69; asst prof, Univ Montreal, 69-74; asst prof gross anat, Col Med, Howard Univ, 74-76; STAFF RES SCIENTIST, GM RES LABS, 76- *Concurrent Pos:* Lectr, Hunter Col, NY, NY, 68-69, Loyola Col, Montreal, Can, 72-74. *Mem:* Soc Quant Morphol; Int Soc Stereology. *Res:* Toxicologic pathology; health effects and carcinogenic potential of man made fibers. *Mailing Add:* Dept Biomed Sci GM Res Labs Warren MI 48090-9055

GARG, DEVENDRA PRAKASH, b Roorkee, India, Mar 22, 34; m 61; c 2. ROBOTICS & AUTOMATIC CONTROL SYSTEMS, TECHNOLOGY FORECASTING & ASSESSMENT. *Educ:* Agra Univ, BS, 54; Univ Roorkee, BEng, 57; Univ Wis, MS, 60; NY Univ, PhD(mech eng), 69. *Prof Exp:* Lectr mech eng, Univ Roorkee, 57-62, reader mech eng, 62-65; instr, NY Univ, 65-69; asst prof, Mass Inst Technol, 69-71, assoc prof & chmn eng projs lab, 71-72; dir undergrad studies, 77-86, PROF MECH ENG, DUKE UNIV, 72- *Concurrent Pos:* Vis lectr mech eng, Mass Inst Technol, 72-76; assoc ed, J Dynamic Systs, Measurement & Control, 71-73, J Interdisciplinary Modeling & Simulation, 77-80; mem organizing & prog comt, Int Symposium on Man Under Vibrations, Suffering & Protection, 79, 85; vis prof mech eng, Univ Roorkee, 78; fel, US Dept Transp, 79-80, Nat Aeronaut & Space Admin, 86- 87 & US Army Res Off, 88; chmn, Dynamic Systs & Control Div, Am Soc Mech Engrs, 85-86, tech panel, 87-; Fulbright sr scholar, 87-88. *Mem:* Fel Am Soc Mech Engrs; Inst Elec & Electronics Engrs; Sigma Xi. *Res:* Design of controllers for nonlinear systems using computational approach; dynamic modeling, simulation and control of socioeconomic systems; technology forecasting and assessment; high speed ground transportation; robot dynamics and control. *Mailing Add:* Sch Eng Duke Univ Durham NC 27706

GARG, DIWAKAR, b Kanpur, India, Sept 11, 52. MARKETING & FINANCE. *Educ:* H B T I, Kanpur, BS, 74; Okla State Univ, Ms, 75; Auburn Univ, PhD(chem eng), 79; Lehigh Univ, MBA, 85. *Prof Exp:* Sr res engr, Air Prod & Chems Inc, 79-82, prin res engr, 82- 85, sr prin res engr, 85-86, lead develop engr, 86-89, RES ASSOC, AIR PROD & CHEMS INC, 89- *Concurrent Pos:* New technol develop & mkt. *Mem:* Am Inst Chem Engrs; Am Chem Soc; Am Ceramic Soc; Am Soc Metals. *Res:* Advance coal liquefaction processes; coal products; hydrotreating and hydrocracking; heavy oils upgrading, abrasive, erosive and wear resistant coatings; chemical vapro deposition; reactor design; CVD diamond coatings; PVD coatings; PACVD coatings. *Mailing Add:* Air Prod & Chemicals Inc 7201 Hamilton Blvd Allentown PA 18195

GARG, HARI G, b Meerut, India, Oct 4, 31; m 60; c 2. BIOLOGICAL CHEMISTRY. *Educ:* Agra Univ, MSc, 52, PhD(chem), 60, DSc, 69. *Prof Exp:* Asst prof chem, Govt Uttar Pradesh, India, 52-56; res fel, Govt India, 56-59; fel & pool scientist peptide chem, CSIR, India, 60-62; vis res assoc carbohydrate chem, Ohio State Univ, 62-64; res fel peptide chem, Oxford Univ, 64-65; sci officer, Nucleic Acid Unit, Twyford Labs, London, 65-67; reader org chem, Univ Roorkee, 67-71; prin res assoc, 71-84, PRIN ASSOC BIOL CHEM, HARVARD MED SCH, 84-; ASSOC BIOCHEMIST SURG, MASS GEN HOSP, 90-; ASSOC BIOCHEMIST SURG, SHRINERS BURNS INST, 90- *Concurrent Pos:* Asst biochemist med, Mass Gen Hosp, 71-89, Shriners Burns Inst, 77-89. *Honors & Awards:* Khosla Res Prize, Univ Roorkee, 71. *Mem:* Am Chem Soc; Soc Complex Carbohydrates; Indian Chem Soc; The Chem Soc. *Res:* Synthesis of low molecular weight carbohydrate derivatives of biological significance; the chemistry and structural studies of glycoconjugates. *Mailing Add:* Shriners Burns Inst Boston MA 02114

GARG, JAGADISH BEHARI, b Kanpur, India, July 7, 29; m 55; c 2. NUCLEAR PHYSICS. *Educ:* Univ Allahabad, BSc, 48; Univ Lucknow, MSc, 51; Univ Paris, DSc(physics), 58. *Prof Exp:* Physicist, Indian Atomic Energy Dept, 51-55; fel physics, Lab Atomic & Molecular Physics, Col France, 55-58; Turner & Newall fel, Univ Manchester, 58-61; sr res assoc, Columbia Univ, 61-66; PROF PHYSICS, STATE UNIV NY ALBANY, 66- *Concurrent Pos:* Dir, Nuclear Accelerator Lab, State Univ NY Albany, 66-72; chmn, Int Conf Statist Properties of Nuclei, 71; vis scientist, Brookhaven Nat Labs, 66-; consult, Oak Ridge Nat Labs, 72-; vis scientist, Australian Atomic Energy Comn, 73, Nuclear Ctr, France, 80. *Mem:* Am Nuclear Soc; fel Am Phys Soc; NY Acad Sci; Sigma Xi; Fedn Am Scientists. *Res:* Reaction mechanism induced by charged particles at intermediate energies; high resolution neutron resonance spectroscopy. *Mailing Add:* Nuclear Accel 1400 Wash Ave Albany NY 12203

GARG, KRISHNA MURARI, b Lucknow, India, Aug 5, 32; m 66. MATHEMATICS. *Educ:* Lucknow Univ, BSc, 52, MSc, 55, PhD(math), 63. *Prof Exp:* Lectr math, Kanyakubja Col, Lucknow, 57-58, asst prof, 62-63, Govt India Coun Sci & Indust Res sr res fel, 63-64; asst prof, Univ Alta, 64-65; study & res, Inst Henri Poincare, Paris, France, 65-66; from asst prof to assoc prof, 66-77, PROF MATH, UNIV ALTA, 77- *Mem:* Am Math Soc; Math Asn Am; Can Math Cong; Math Soc France. *Res:* Nature of derivates and structure of level sets of real functions in general and nowhere monotone functions in particular. *Mailing Add:* Dept Math Univ Alberta Edmonton AB T6G 2E2 Can

GARG, LAL CHAND, b India, Jan 22, 33; m 59; c 2. PHARMACOLOGY, PHYSIOLOGY & MEDICINE. *Educ:* Punjab Univ, BS, 54, MS, 56; Univ Fla, PhD(pharmacol), 69. *Prof Exp:* From instr to asst prof pharmacol, Col Vet Med, India, 56-63; asst prof, Punjab Univ, 63-65; res assoc, 69-70, instr, 70-71, asst prof, 71-76, assoc prof, 76-86, PROF PHARMACOL, COL MED, UNIV FLA, 86- *Concurrent Pos:* NIH fac develop award biochem pharmacol, Univ Fla, 79-80. *Mem:* Sigma Xi; Am Soc Pharmacol & Exp Therapeut; NY Acad Sci; Am Soc Nephrol; Am Physiol Soc; Am Heart Asn. *Res:* Drug-enzyme interactions; renal and electrolyte pharmacology; hypertension. *Mailing Add:* Dept Pharmacol & Therapeut Univ Fla Col Med Box J-267 JHMHC Gainesville FL 32610

GARG, MOHAN LAL, medical statistics, for more information see previous edition

GARG, UMESH, b Bikaner, India, Mar 29, 53. NUCLEAR PHYSICS. *Educ:* Birla Inst Technol & Sci, Pilani, India, BSc, 72, MSc, 74; State Univ NY, Stony Brook, MA, 75, PhD(physics), 78. *Prof Exp:* Teaching asst, Dept Physics, State Univ NY, Stony Brook, 74-75, res asst, 75-78; RES ASSOC, CYCLOTRON INST, TEX A&M UNIV, 78- *Mem:* Am Phys Soc; Sigma Xi. *Res:* Gamma-ray spectroscopy following (HI,xn) reactions; giant resonances; heavy-ion reactions; collective properties of nuclei; spectroscopy following massive transfer reactions. *Mailing Add:* 18254 Westover Dr South Bend IN 46637

GARGARO, ANTHONY, b Chideock, Eng, Apr 11, 42. COMPUTER SCIENCE, SOFTWARE SYSTEMS. *Educ:* Brunel Univ, UK, Dipl, 63. *Prof Exp:* Computer scientist, 76-80, sr computer scientist, 80-84, LEAD SCIENTIST DEVELOP CONTROL SOFTWARE, COMPUTER SCI CORP, 84- *Concurrent Pos:* Area dir, Spec Interest Groups, Asn Comput Mach, 86- *Mem:* Asn Comput Mach; Brit Computer Soc. *Mailing Add:* Computer Sci Corp 200 Century Pkwy Mt Laurel NJ 08054

GARGES, SUSAN, b Chelsea, Mass, Jan 9, 53; m 82; c 2. GENE REGULATION, PROTEIN CONFORMATION. *Educ:* Univ Dayton, BA, 74, MS, 78; Univ Md, PhD(microbiol), 83. *Prof Exp:* Grad teaching asst biol, Univ Dayton, 74-76; fac res asst microbiol, Univ Md, 76-78, grad res asst, 78-79; MICROBIOLOGIST, NAT CANCER INST, NIH, 79- *Concurrent Pos:* Fac, NIH Grad Sch, Found Advan Educ in Sci, 86- *Mem:* Am Soc Microbiol. *Res:* Gene regulation and protein conformation, a genetic approach to the study of the subjects; camp receptor protein (CRP), a transcriptional activator, from Escherichia coli as the model system. *Mailing Add:* Nat Cancer Inst-NIH Bldg 37 Rm 2E20 Bethesda MD 20892

GARGUS, JAMES L, b Dalton, Ark, Oct 27, 22; m 43; c 5. TOXICOLOGY. *Educ:* George Washington Univ, BS, 50, MS, 54. *Prof Exp:* Res asst tissue culture, Warwick Clin, 50-54; instr microbiol, Rutgers Univ, 54-55; lab supvr, Hazleton Labs, Inc, 55-60, dir oncol dept, 58-75, dept chief, 61-65, res coordr, 65-68, dept dir appl biol, 69-87; RETIRED. *Mem:* AAAS; Soc Toxicol; Am Asn Cancer Res; Tissue Culture Asn; NY Acad Sci. *Res:* Supervision and research management in applied biology and toxicology; carcinogenesis, including bioassay procedures and ultrastructural changes induced by carcinogens; experimental cancer chemotherapy and toxicology. *Mailing Add:* 7108 Wayne Dr Annandale VA 22003

GARIBALDI, JOHN ATTILIO, b San Francisco, Calif, Apr 3, 16; m 59; c 2. BIOCHEMISTRY. *Educ:* Univ Calif, BS, 38, PhD(biochem), 58. *Prof Exp:* Chemist, 46-52, BIOCHEMIST, WESTERN REGIONAL RES LAB, USDA, 55- *Mem:* Am Soc Microbiol. *Res:* Microbial biochemistry; mineral nutrition of microorganisms; process development for the microbial production of antibiotics and vitamins; iron nutrition of microbes; shell egg microbiology. *Mailing Add:* 2311 Alva Ave El Cerrito CA 94530

GARIEPY, CLAUDE, b Quebec, Nov 3, 54. MEAT SCIENCE. *Educ:* Laval Univ, BScA, 79, MSc, 83, PhD(meat sci), 87. *Prof Exp:* Prof poultry sci, Inst Agr Technol, 79-80; postdoctoral fel meat sci, Lacombe Res Sta, 88-89; RES SCIENTIST, AGR CAN, 89- *Concurrent Pos:* Lectr, Laval Univ. *Mem:* Can Soc Animal Sci; Can Meat Sci Asn; Can Inst Food Sci Technol. *Res:* Effects of production factors on the keeping qualities and processing attributes of meat and meat products. *Mailing Add:* Food Res & Develop Ctr 3600 Casavant Blvd W St-Hyacinthe PQ J2S 8E3 Can

GARIEPY, RONALD F, b Mich, Apr 3, 40. MATHEMATICS. *Educ:* Wayne State Univ, PhD(math), 69. *Prof Exp:* PROF MATH, UNIV KY, 70- *Mem:* Am Math Asn. *Mailing Add:* Math Dept Univ Ky Lexington KY 40506

GARIK, VLADIMIR L, b Berlin, Ger, Mar 25, 13; nat US; m 44; c 1. PHYSICAL CHEMISTRY. *Educ:* Polytech Inst Brooklyn, BS, 44, MS, 47; Univ Conn, PhD, 53. *Prof Exp:* Assoc prof phys chem, Iona Col, 56-64; asst prof chem, 64-69, ASSOC PROF CHEM, MONTCLAIR STATE COL, 69- *Mem:* Am Chem Soc; Sigma Xi. *Res:* Hydrogenation of nitrocompounds with Raney nickel; kinetics of the decomposition of methyl in propyl ketone. *Mailing Add:* 71 Kensington Ave Clifton NJ 07014

GARIN, DAVID L, b New York, NY, June 14, 39. ORGANIC CHEMISTRY. *Educ:* City Col New York, BS, 60; Iowa State Univ, PhD(org chem), 64. *Prof Exp:* Van Leer fel, Weizmann Inst Sci, 64, NIH fel, Weizmann Inst Sci & Ind Univ, 64-66; asst prof, 66-71, ASSOC PROF CHEM, UNIV MO-ST LOUIS, 71- *Concurrent Pos:* Vis scientist, Environ Protection Agency, 79-80; comnr, Miss Hazardous Waste Mgt Comn, 78-81; cong res fel, Am Chem Soc, 80. *Mem:* Am Chem Soc; Nat Coalition Sci & Technol (treas). *Res:* Photochemistry; rearrangement reactions; reaction mechanisms; science and public policy. *Mailing Add:* Dept Chem Univ Mo 8001 Natural Bridge Rd St Louis MO 63121-4499

GARING, JOHN SEYMOUR, b Toledo, Ohio, Nov 6, 30; m 52; c 2. MOLECULAR SPECTROSCOPY. *Educ:* Ohio State Univ, BSc, 51, MSc, 54, PhD(physics), 58. *Prof Exp:* Asst physics, Ohio State Univ, 51-53; physicist, Geophys Res Directorate, Air Force Res Div, 58-61, chief, Infrared Physics Br, 61-63; dir, Optical Physics Div, Air Force Geophysics Lab, Hanscom AFB, Mass, 63-86; RETIRED. *Mem:* Fel AAAS; fel Optical Soc Am; Am Inst Aeronaut & Astronaut; Sigma Xi. *Res:* Infrared optics and spectroscopy; atmospheric transmission and emission; molecular structure and interactions; spectroscopic and interferometric instrumentation and techniques. *Mailing Add:* 157 Cedar St Lexington MA 02173-6507

GARITO, ANTHONY FRANK, b New Rochelle, NY; m 63; c 3. CONDENSED MATTER PHYSICS. *Educ:* Columbia Univ, BS, 62; Univ Pa, PhD, 68. *Prof Exp:* Res fel, Univ Pa, 64-68, res assoc physics, 68-70, from asst prof to assoc prof, 70-77, PROF PHYSICS, UNIV PA, 77-, TERM CHAIR PROF, 90- *Concurrent Pos:* DuPont Young fac fel, Univ Pa, 70-73; consult, 72-; vis scholar, Univ Paris, 77 & Soviet Acad Sci, 78; vis prof, Inst Phys & Chem Res, Japan, 85- *Mem:* Am Phys Soc; Optical Soc Am. *Res:* Many-body interactions and instabilities of lower dimensional systems; nonlinear optical phenomena in organic and polymeric systems. *Mailing Add:* Dept Physics 2n9d DRL/EL Univ Pa Philadelphia PA 19104

GARLAND, CARL WESLEY, b Bangor, Maine, Oct 1, 29; m 55; c 2. PHASE TRANSITIONS. *Educ:* Univ Rochester, BS, 50; Univ Calif, PhD(chem), 53. *Prof Exp:* Instr chem, Univ Calif, 53; from instr to assoc prof, 53-68, PROF CHEM, MASS INST TECHNOL, 68- *Concurrent Pos:* A P Sloan fel, 54-60; sci ed, Optics & Spectros, 60-81; Guggenheim fel, 63-64; vis prof, Univ Calif, 72, Univ Rome, 74, Kathol Univ Leuven, 77, Ben Gurion Univ, 80, Univ Paris, 81 & 82, Univ Bordeaux, 90. *Mem:* Am Phys Soc; fel Am Acad Arts & Sci. *Res:* Ultrasonic studies of phase transitions and critical points; bulk properties of solids at high pressures; dynamical aspects of cooperative phenomena; high-resolution calorimetry; phase transitions in liquid crystals. *Mailing Add:* 6-237 Mass Inst Technol Cambridge MA 02139

GARLAND, CEDRIC FRANK, b La Jolla, Calif, Nov 10, 46. EPIDEMIOLOGY, CANCER. *Educ:* Univ Southern Calif, Los Angeles, BA, 67; Univ Calif, Los Angeles, MPH, 70, PhD(pub health), 74. *Prof Exp:* Asst prof, Johns Hopkins Univ, Sch Hyg & Pub Health, 74-81; asst prof, 81-89, ASSOC PROF EPIDEMIOL, SCH MED LA JOLLA, UNIV CALIF, SAN DIEGO, 89-; ASSOC PROF EPIDEMIOL, GRAD SCH PUB HEALTH, SAN DIEGO STATE UNIV, 81- *Concurrent Pos:* Dir Epidemiol Prog, Univ Calif, San Diego, Cancer Ctr, 87- *Mem:* Soc Epidemiol Res; Am Pub Health Asn; fel Am Col Epidemiol; AAAS. *Res:* Epidemiologic studies of breast and colon cancer, melanoma, and diseases of the digestive system; epidemiology of environmentally-induced diseases, studies of effects of vitamin D and calcium on human health. *Mailing Add:* Dept Community & Family Med 0607 Sch Med Univ Calif San Diego La Jolla CA 92093-0607

GARLAND, CHARLES E, b Haverhill, Mass, June 5, 26; m 48; c 1. ORGANIC CHEMISTRY, COLOR SCIENCE. *Educ:* Colby Col, AB, 50; Univ NH, MS, 52. *Prof Exp:* Instr chem, Univ NH, 50-51; res chemist anthraquinone dyes, E I DuPont De Nemours & Co, 51-66, res chemist, 66-70, sr res chemist color measurement & control, 70-81, indust hyg assoc, 81-86; RETIRED. *Mem:* Am Chem Soc; Am Soc Testing & Mat; Am Indust Hygiene Asn. *Res:* Polarographic reduction of diazonium compounds; research in anthraquinone chemistry leading to new dyes for cotton and polyester fibers; measurement of color in solutions and dyed fabrics and computer shade matching; biological monitoring; permeation testing of safety fabrics; industrial hygiene. *Mailing Add:* Sharpley 314 Brockton Rd Wilmington DE 19803

GARLAND, DONITA L, b Emporia, Kans, 40; m 73; c 1. BIOCHEMISTRY. *Educ:* Washburn Univ, BA, 62; Univ Wash, PhD(biochem), 75. *Prof Exp:* Staff fel, Nat Heart, Lung & Blood Inst, NIH, 75-83, expert, Nat Eye Inst, 83-88, RES CHEMIST, LAB MECHANISMS OCULAR DIS, NAT EYE INST, NIH, 88- *Mem:* Am Soc Biochem & Molecular Biol; Asn Res Vision & Ophthal. *Res:* Regulation of metabolism of eye lens and mechanisms of cataract formation. *Mailing Add:* Nat Eye Inst NIH Bldg 6 Rm 235 Bethesda MD 20892

GARLAND, GEORGE DAVID, geophysics, for more information see previous edition

GARLAND, HEREFORD, b Lake Bluff, Ill, Jan 2, 05; m 40; c 3. FOREST PRODUCTS. *Educ:* Univ Calif, BS, 31, MS, 32; Wash Univ, PhD(plant physiol), 38. *Prof Exp:* Tech asst forestry, Univ Calif, 31-34; jr forester, Calif Forest Exp Sta, US Forest Serv, 34-35; instr forestry, Univ Ark, 39-40; asst conservationist, Calif Forest Exp Sta, US Forest Serv, 40-42, assoc forest prod technologist, 42-45; assoc prof forest prod res, Mich Technol Univ, 45-47, from assoc prof to prof, 47-74, dir, Inst Wood Res, 64-71, emer prof wood res, 74-81; RETIRED. *Mem:* Soc Am Foresters; Forest Prod Res Soc; Soc Wood Sci & Technol; Sigma Xi. *Res:* Mechanical properties of wood; anatomical studies of wood; coniferous wood in relation to its strength properties; forest products; wood production and economics. *Mailing Add:* 8570 Red Oak Dr NE Warren OH 44484-1630

GARLAND, HOWARD, b Detroit, Mich, Oct 27, 37; m 61; c 2. MATHEMATICS. *Educ:* Univ Chicago, BS, 57; Wayne State Univ, MS, 59; Univ Calif, Berkeley, PhD(math), 64. *Prof Exp:* Instr math, Yale Univ, 64-65; asst, Inst Advan Study, 65-66; asst prof, Yale Univ, 66-69; assoc prof, Cornell Univ, 69-71; prof, Columbia Univ, 71-72; prof, State Univ NY Stony Brook, 72-73; PROF MATH, YALE UNIV, 73- *Mem:* Am Math Soc. *Res:* Discrete subgroups of Lie groups. *Mailing Add:* Dept Math 431 DLLOM Yale Univ New Haven CT 06520

GARLAND, JAMES C, b Columbia, Mo, Aug 11, 42; c 2. EXPERIMENTAL CONDENSED MATTER PHYSICS. *Educ:* Princeton Univ, AB, 64; Cornell Univ, PhD(physics), 69. *Prof Exp:* NSF fel physics, Cambridge Univ, 69-70; from asst prof to assoc prof, 70-80, prof physics & dir Mat Res Lab, 80-85, CHMN PHYSICS DEPT, OHIO STATE UNIV, 86- *Mem:* Fel Am Phys Soc. *Res:* Electronic properties of metals, superconductors and composites; advanced materials. *Mailing Add:* Dept Physics Smith Lab Ohio State Univ Columbus OH 43210

GARLAND, JAMES W, b Washington, DC, Aug 1, 33; m 58; c 2. SOLID STATE PHYSICS, THEORETICAL PHYSICS. *Educ:* Univ Chicago, MS, 58, PhD(physics), 65. *Prof Exp:* Actg asst prof physics, Univ Calif, Berkeley, 63-65, asst prof, 66-67; assoc prof, 67-70, PROF PHYSICS, UNIV ILL, CHICAGO CIRCLE, 70- *Concurrent Pos:* Alfred P Sloan Found fel, 64-66; vis lectr, Cambridge Univ, 65; consult, Argonne Nat Lab, 67-, assoc physicist, 69-70. *Mem:* Am Phys Soc. *Res:* Theory of metals; semiconductor physics; magnetism; transition metals. *Mailing Add:* 331 S Peoria St, No 104 Chicago IL 60607-3526

GARLAND, JOHN KENNETH, b Cadillac, Mich, Dec 27, 35; m 58; c 5. CHEMISTRY. *Educ:* Univ Ill, BS, 57; Univ Kans, PhD(chem), 63. *Prof Exp:* Asst res chemist, Continental Oil Co, Okla, 57-59; asst prof chem, Univ Mo-Columbia, 63-70; ASSOC PROF CHEM, WASH STATE UNIV, 70- *Concurrent Pos:* Adj prof chem, Mont State Univ, 84-85. *Mem:* Am Chem Soc; AAAS. *Res:* Chemical education; hot atom chemistry, particularly recoil tritium; energy planning; X-ray crystallography. *Mailing Add:* Dept Chem Wash State Univ Pullman WA 99164

GARLAND, MICHAEL MCKEE, b Clarksville, Tenn, Jan 12, 39; m 58; c 2. SOLID STATE PHYSICS. *Educ:* Austin Peay State Col, BA, 61; Clemson Univ, PhD(physics), 65. *Prof Exp:* From asst prof to assoc prof, 65-77, PROF PHYSICS, MEMPHIS STATE UNIV, 77- *Mem:* Am Asn Physics Teachers; Am Vacuum Soc; Sigma Xi. *Res:* Superconducting behavior of thin films; superconductive tunneling between superimposed films. *Mailing Add:* Dept Physics Memphis State Univ Memphis TN 38152

GARLAND, ROBERT BRUCE, b Chicago, Ill, Nov 6, 32; m 55; c 4. ORGANIC CHEMISTRY. *Educ:* Univ Ill, BS, 53; Mass Inst Technol, PhD(chem), 57. *Prof Exp:* Sr res investr, Searle Labs, 59-80, RES SCIENTIST, RES & DEVELOP DIV, G D SEARLE & CO, 80- *Mem:* Am Chem Soc. *Res:* Synthetic organic chemistry; natural products. *Mailing Add:* Searle Res & Develop 4901 Searle Pkwy Skokie IL 60077

GARLAND, WILLIAM ARTHUR, b Washington, DC, Jan 28, 45; m 66; c 4. MEDICINAL CHEMISTRY. *Educ:* Univ San Francisco, BS, 68; Univ Wash, PhD(med chem), 74. *Prof Exp:* Sr scientist, Hoffman-La Roche Inc, 74-80, group chief, 80-85, res leader, 85-87, distinguished res leader, 87-90, ACTG DIR, DEPT DRUG METAB, HOFFMAN-LA ROCHE, INC, 90- *Res:* Mass spectrometry; drug metabolism; stable isotopes; pharmacokinetics. *Mailing Add:* Hoffmann-La Roche Inc Bldg 86 Nutley NJ 07110

GARLAND, WILLIAM JAMES, b St John's, Nfld, July 26, 48. EXPERT SYSTEMS. *Educ:* McMaster Univ, BEng, 70, MEng, 71, PhD(chem eng), 75. *Prof Exp:* Design engr, Ont Hydro, 75-79; supv design engr, Atomic Engegy Can Ltd, 79-83; ASSOC PROF NUCLEAR ENG, MCMASTER UNIV, 83- *Concurrent Pos:* Dept chair, McMaster Univ. *Mem:* Am Nuclear Soc; Can Nuclear Soc. *Res:* Integration of symbolic computation and reasoning into highly computerized numerically based real-time control systems for complex plant process management. *Mailing Add:* Dept Eng Physics McMaster Univ Hamilton ON L8S 4M1 Can

GARLICH, JIMMY DALE, b Okawville, Ill, Feb 18, 36; m 65; c 2. NUTRITION, BIOCHEMISTRY. *Educ:* Univ Ill, Urbana, BS, 58, MS, 59; Cornell Univ, PhD(animal nutrit), 64. *Prof Exp:* Nat Heart Inst fel human nutrit, Sch Pub Health, Univ Pittsburgh, 64-65 & Sch Med, Univ St Louis, 65-66; from asst prof to assoc prof, 66-85, PROF POULTRY SCI, NC STATE UNIV, 85- *Mem:* Poultry Sci Asn; Am Inst Nutrit. *Res:* Pharmacological effects of amino acids on lipid metabolism in human subjects; effect of plant proteolytic enzyme inhibitors on digestion by chicks; hormonal regulation of calcium metabolism in avian species. *Mailing Add:* Dept Poultry Sci NC State Univ Box 7608 Raleigh NC 27695

GARLICK, GEORGE DONALD, b Lusaka, Zambia, Nov 29, 34; m 64. GEOCHEMISTRY, PETROLOGY. *Educ:* Univ Witwatersrand, 56; Calif Inst Technol, PhD(geochem), 65. *Prof Exp:* Geologist, New Consol Gold Fields, SAfrica, 56-59; geochemist, Columbia Univ, 65-66, asst prof geol, 66-69; assoc prof, 69-74, PROF GEOL, HUMBOLDT STATE UNIV, 74- *Mem:* Am Geochem Soc; Geol Soc Am; Geophys Union; Sigma Xi. *Res:* Oxygen isotope geochemistry and petrology. *Mailing Add:* 500 Orchard Lane Arcata CA 95521

GARLICK, ROBERT L, b Cedar Rapids, Iowa, Feb 6, 49; m 79; c 1. PROTEIN CHEMISTRY, BIOTECHNOLOGY. *Educ:* Univ Pacific, AB, 70; Univ Ore, MS, 73; Univ Tex, PhD(zool), 79. *Prof Exp:* Postdoctoral fel, 79-82, from instr to asst prof med & biochem, Harvard Med Sch, 82-84; RES SCIENTIST, THE UPJOHN CO, 85- *Mem:* Am Chem Soc; Am Soc Biochem & Molecular Biol. *Mailing Add:* The Upjohn Co 1400-89-1 Kalamazoo MI 49001

GARLID, KEITH DAVID, b Albert Lea, Minn; m; c 1. PHARMACOLOGY. *Educ:* Gustavus Adolphus Col, BA, 56; Johns Hopkins Univ, MD, 61. *Prof Exp:* Fel biophys, Johns Hopkins Univ, 61-65, asst prof obstet & gynec, 65-71; assoc prof pharmacol, 71-84, PROF PHARMACOL, MED COL OHIO, 84- *Concurrent Pos:* Vis prof phys chem, Norwegian Inst Technol, 80-81, distinguished vis prof biol, Univ Miami, 87. *Mem:* Am Soc Biochem & Molecular Biol; Am Soc Pharmacol & Exp Therapeut; Biophys Soc. *Res:* Membrane biophysics and pharmacology; water structure in biological systems; membrane transport in mitochondria and its relationships to bioenergetics. *Mailing Add:* 422 W Broadway Maumee OH 43537

GARLID, KERMIT L(EROY), b Ellsworth, Wis, May 10, 29; m 54; c 4. NUCLEAR ENGINEERING, CHEMICAL ENGINEERING. *Educ:* Univ Wis, River Falls, BS, 50; Univ Minn, BChE, 56, PhD(chem eng), 61. *Prof Exp:* Engr qual control, Aero Div, Minneapolis-Honeywell Regulator Co, 53-54; instr chem eng, Univ Minn, 56-60; from asst prof to assoc prof, 60-71, assoc dean, Col Eng, 73-79, actg dean, 80-81, PROF NUCLEAR & CHEM ENG, UNIV WASH, 71-, vprovost, 82-86, CHMN, NUCLEAR ENG, 86- *Concurrent Pos:* Vis prof, Tech Univ, Munich, 68-69; consult, Hanford Atomic Power Oper, Gen Elec Co, 63-68, Pac Northwest Labs, Battelle Mem Inst, 65-68, Electronic Assoc, Inc, 68, Thermodyn, Inc, 70-73, Atlantic Richfield Hanford Co, 71-77, R W Beck & Assoc, 71-75, Adv Comt Reactor Safeguards, 73- & Rockwell Int-Hanford Opers, 77-; vis scientist, Battelle Human Affairs Res Ctr, 76-78b; vis prof, Norwegian Tech Univ, Trondheim, 80. *Mem:* AAAS; Am Inst Chem Engrs; Am Nuclear Soc; Sigma Xi. *Res:* control and optimization of chemical processes; nuclear reactor fuel cycle; nuclear reactor safety. *Mailing Add:* Nuclear Eng BF-20 Univ Wash Seattle WA 98195

GARMAISE, DAVID LYON, b Montreal, Que, Mar 26, 23; m 50; c 3. ORGANIC CHEMISTRY. *Educ:* McGill Univ, BSc, 42, PhD(chem), 45. *Prof Exp:* Assoc prof chem, Univ NB, 45-50; Brit Empire Cancer Soc res fel, Univ Wales, 50-52; res sect head, Monsanto Can Ltd, 52-63; MGR ORG CHEM RES, ABBOTT LABS LTD, 63-; SCI LITERATURE CONSULT, 86- *Mem:* Am Chem Soc; Chem Inst Can; AAAS. *Res:* Synthetic organic chemistry; medicinal chemistry; history of science. *Mailing Add:* Abbott Labs D-441 AP9A North Chicago IL 60064

GARMAN, BRIAN LEE, b Sturgis, Mich, July 6, 45. MATHEMATICS. *Educ:* Cornell Univ, AB, 67; Western Mich Univ, MA, 73, PhD(math), 76. *Prof Exp:* Instr math, Three Rivers High Sch, 67-71; mathematician, Univ Ky, 76-79 & Wesleyan Col, 79-80; mathematician, 80-85, ASSOC PROF MATH & CHMN DEPT, UNIV TAMPA, 85- *Mem:* Am Math Soc; Math Asn Am; Nat Coun Teachers Math. *Res:* Topological graph theory; mathematical education. *Mailing Add:* Dept Math Univ Tampa Tampa FL 33606

GARMAN, JOHN ANDREW, b Berlin, Pa, Aug 2, 21; m 45, 72; c 2. ORGANIC CHEMISTRY, RESEARCH ADMINISTRATION. *Educ:* Franklin & Marshall Col, BS, 43; Univ Md, PhD(org chem), 48. *Prof Exp:* Res chemist, Firestone Plastics Co Div, Firestone Tire & Rubber Co, 47-50; res chemist, US Indust Chem Co, 50-53, head org chem sect, 53-54; head org chem sect, Fairfield Chem Div, FMC Corp, 54-55, asst dir, Fairfield Br, Cent Chem Res, 56-57, mgr org chem res, Org Chem Div, 57-58, mgr org chem res, Chem & Plastics Div, 58-62, dir res & develop, Org Chem Div, 62-69; dir res & develop, Great Lakes Chem Corp, 70-83, asst to vpres corp technol, 83-86, patent consult, 86-88; RETIRED. *Mem:* Am Chem Soc. *Res:* Organic synthesis and process development; fine organic chemicals; pharmaceuticals; agricultural chemicals; flame retardants; polymer and plastic additives; bromine chemistry. *Mailing Add:* 1705 Beechwood Ave Baltimore MD 21228

GARMAN, ROBERT HARVEY, b New York City, NY, Aug 20, 41; m 63; c 2. TOXICOLOGY, COMPARATIVE PATHOLOGY. *Educ:* Cornell Univ, BA, 63, DVM, 66. *Prof Exp:* Vet clinician, Gen Vet Practice, Warrenton, Va, 66-67; Vet Officer, US Pub Health Serv, NIH, 67-69; fel path, Sch Med & Dent, Univ Rochester, 69-71, instr, 71-73, asst prof path & toxicol, 73-78; sr scientist path, Carnegie-Mellon Inst Res, 78-80; res scientist pathol, Bushy Run Res Ctr, 80-88; OWNER & PRES, CONSULTS IN VET PATH, 88- *Concurrent Pos:* Adj assoc prof, Univ Pittsburgh, 81- *Mem:* Am Vet Med Asn; Am Col Vet Pathologists; Int Acad Pathol; Am Asn Neuropath; Soc Toxicol Pathologists; NY Acad Sci. *Res:* Spontaneous diseases of animals and models of human disease; toxicologic pathology, including the testng of chemicals for acute, subchronic and chronic effects in animals. *Mailing Add:* PO Box 68 Murrysville PA 15668

GARMANY, CATHARINE DOREMUS, b New York, NY, Mar 6, 46; m 70; c 2. ASTRONOMY. *Educ:* Ind Univ, BS, 66; Univ Va, PhD(astron), 71. *Prof Exp:* Res assoc astron, Univ Va, 71-73; fel astron, 76-80, RES ASSOC, JOINT INST LAB ASTROPHYS, UNIV COLO, 80- *Honors & Awards:* Annie J Cannon Award, Am Astron Soc-Am Asn Univ Women, 76. *Mem:* Am Astron Soc; Int Astron Union. *Res:* O-type stars; binaries and stellar atmospheres; stellar kinematics of intermediate age stars. *Mailing Add:* Univ Colo JILA CB 440 Univ Colo Boulder CO 80309

GARMEZY, R(OBERT) H(ARPER), b Manila, Philippines, Dec 8, 23; c 4. MECHANICAL & ELECTRICAL ENGINEERING. *Educ:* Cornell Univ, BEE, 43, BME, 45; Chrysler Inst Eng, MAE, 47. *Prof Exp:* Instr, Elec Lab, Cornell Univ, 43-44; student engr, Chrysler Corp, 45-47, lab engr, 47-51; head radiator lab, 51-60, dir eng, 60-68, vpres eng, Automotive Div, Blackstone Corp, 68-81; RETIRED. *Concurrent Pos:* Instr, Night Sch, Lawrence Inst Technol, 46-51. *Mem:* Am Soc Mech Engrs; Soc Automotive Engrs; Inst Elec & Electronics Engrs. *Res:* Heat transfer; automotive radiators; heater cores. *Mailing Add:* 96 Gordon St Jamestown NY 14701

GARMIRE, ELSA MEINTS, b Buffalo, NY, Nov 9, 39; m 61; c 2. PHYSICS. *Educ:* Radcliffe Col, AB, 61; Mass Inst Technol, PhD(physics), 65. *Prof Exp:* Res fel physics, Mass Inst Technol, 65-66; scientist, Electronics Res Ctr, NASA, 66; res fel elec eng, Calif Inst Technol, 66-74; assoc dir, 74-81, PROF ELEC ENG, UNIV SOUTHERN CALIF, 81-, DIR, CTR LASER STUDIES, 84- *Concurrent Pos:* From part-time res fel to part-time sr res fel appl sci, Calif Inst Technol, 69-74; pres, Laser Images, Inc; consult, Stand Telecommun Labs, Eng, 74; mem tech staff, Aerospace Corp, 75-; consult, Northrop Corp, 75-; assoc ed, Optics Lett & Fiber & Integrated Optics; Optical Soc Am rep, Pub Policy Comt, Am Inst Physics. *Mem:* Nat Acad Eng; fel Inst Elec & Electronics Engrs; Optical Soc Am (vpres, 91); Am Phys Soc; Inst Elec & Electronics Engrs, Lasers & Electro-optics Soc. *Res:* Lasers; integrated optics; nonlinear optics; spectroscopy; quantum electronics; author of over 160 publications; awarded 8 patents. *Mailing Add:* 406 N Dianthus Manhattan Beach CA 90266

GARMIRE, GORDON PAUL, b Portland, Ore, Oct 3, 37; m 61, 76; c 4. X-RAY ASTRONOMY, ASTROPHYSICS. *Educ:* Harvard Univ, AB, 59; Mass Inst Technol, PhD(physics), 62. *Prof Exp:* Mem staff, Div Sponsored Res, Mass Inst Technol, 62-64; asst prof physics, 64-66, assoc prof, 67-68; sr fel, Calif Inst Technol, 66-67, from assoc prof to prof physics, 68-81; prof, 80-85, EVAN PUGH PROF ASTRON, PA STATE UNIV, 85- *Concurrent Pos:* Mem astron subcomt, NASA, 68-84; Fulbright sr fel & Guggenheim fel, 73-74; consult, Los Alamos Nat Lab, 80-88; chmn, high energy astrophys div, Am Astron Soc, 85; convener, Astron & Space Physics Coun, Univs Space Res Asn, 89- *Honors & Awards:* Except Sci Achievement Award, NASA, 78. *Mem:* Int Astron Union; Am Astron Soc. *Res:* X-ray and gamma ray astronomy; cosmic rays; high energy astrophysics. *Mailing Add:* Dept Astron 504 Davey Lab Pa State Univ University Park PA 16802

GARMON, LUCILLE BURNETT, b Johnstown, Pa, July 1, 36; m 56; c 2. HISTORY & PHILOSOPHY OF SCIENCE, CRYSTAL STRUCTURE. *Educ:* Univ Richmond, BS, 56, MS, 60; Univ Va, PhD(chem), 66. *Prof Exp:* Res chemist, Va Inst Sci Res, 57-61; assoc prof chem, ECarolina Col, 64-66; asst prof physics & res assoc, Auburn Univ, 66-68, fel solid state physics, 66-67; from asst prof to assoc prof physics, 68-74, assoc prof physics & chem, 74-78, actg chmn dept, 82-84, PROF CHEM & PHYSICS, WGA COL, 78-, CHMN DEPT, 84- *Concurrent Pos:* Consult, Freeport Kaolin Co, 67-68; vis prof, Chester Col, Eng, 77. *Mem:* AAAS; Am Chem Soc; Am Phys Soc; Am Asn Physics Teachers; Electron Micros Soc Am; Nat Sci Teachers Asn. *Res:* History of science as a means of enhancing science education; electron microscopy; teaching methods in physical and general chemistry. *Mailing Add:* Dept Chem WGa Col Carrollton GA 30118-0001

GARMON, RONALD GENE, b Charlotte, NC, Mar 13, 34; m 55; c 3. ANALYTICAL CHEMISTRY. *Educ:* Univ NC, BS, 58, PhD(chem), 61. *Prof Exp:* From res chemist to sr res chemist, Res Triangle Park Develop Ctr, 61-69, group leader, 68-81; GROUP LEADER, MONSANTO CO, DECATUR, ALA, 81- *Mem:* Am Chem Soc. *Res:* Analytical chemistry of polymers; radiochemistry; kinetic methods of analysis. *Mailing Add:* 2402 Burningtree Dr SE Decatur AL 35603

GARN, PAUL DONALD, b Fremont, Ohio, July 7, 20; m 45, 79; c 3. INSTRUMENTATION, THERMAL ANALYSIS. *Educ:* Ohio State Univ, BS, 48, MS, 49, PhD(chem), 52. *Prof Exp:* Mem tech staff, Bell Tel Labs, Inc, 52-63; assoc prof, 63-67, dir, Ctr Fire & Hazardous Mat Res, 84-87, PROF CHEM, UNIV AKRON, 67- *Concurrent Pos:* Consult, Apparatus Mfrs, Inc, 63-69; foreign lectr, Japanese Soc Calorimetry & Thermal Analysis, Japan, 70; res chemist, US Nat Bur Stand, 72-74; guest prof, Univ Cologne, Ger, 73-74; sr res fel US spec prog, Humboldt Found, 73; sr res assoc Nat Res Coun, Ballistic Res Lab, Aberdeen Proving Gound, Md, 87-88. *Honors & Awards:* Int Confederation Thermal Analysis Award, DuPont, 77. *Mem:* AAAS; Am Chem Soc; fel NAm Thermal Analysis Soc (pres, 69); Int Confedn Thermal Anal. *Res:* Thermoanalytical techniques; mechanisms of solid-solid transitions; chemistry of fire-related processes; kinetics of thermal decompositions; kinetics of reversible and irreversible high temperature reactions; chromatography. *Mailing Add:* 2241 Massillon Rd Akron OH 44312

GARN, STANLEY MARION, b New London, Conn, Oct 27, 22; m 50; c 2. PHYSICAL ANTHROPOLOGY, NUTRITION. *Educ:* Harvard Univ, AB, 42, AM, 47, PhD(phys anthrop), 48. *Prof Exp:* Res assoc, Mass Inst Technol, 42-44; tech ed, Polaroid Corp, 44-46, consult, 46; instr anthrop, Harvard Univ, 48-52; from assoc prof to prof, Antioch Col, 52-68, chmn dept growth & genetics, Fels Res Inst, 52-68; PROF NUTRIT SCH PUB HEALTH & FEL, CTR HUMAN GROWTH & DEVELOP, UNIV MICH, ANN ARBOR, 68- *Concurrent Pos:* Res fel, Mass Gen Hosp, 46-50, Forsyth Dent Infirmary, 47-52 & Int Univ Training Prog, 58; vis prof, Univ Chicago, 58, Southern Methodist Univ, 76; mem vis staff, Inst Nutrit Cent Am & Panama, 62; Walker-Ames vis prof, Univ Wash, 86. *Honors & Awards:* Neuhauser Lectr, Soc Pediat Radiol, 81; Raymond Pear Lectr, Human Biol Coun, 85. *Mem:* Nat Acad Sci; Am Soc Naturalists; Am Anthrop Asn; Am Asn Phys Anthrop; fel Am Acad Arts & Sci; fel Am Acad Pediat; Am Soc Clin Nutritionists; Int Asn Dent Res. *Res:* Growth and development of body tissues; applied physical anthropology; human evolution; interaction of nutrition and genetics on growth development and aging. *Mailing Add:* Ctr Human Growth 300 N Ingalls Bldg Ann Arbor MI 48109-0406

GARNAR, THOMAS E(DWARD), JR, b Vineland, NJ, June 24, 22; m 54. GEOLOGY, MINERAL ENGINEERING. *Educ:* WVa Univ, BS, 50, MS, 51. *Prof Exp:* Asst, WVa Univ, 48-51; res geologist res div, Int Minerals & Chem Corp, 51-56; mining engr, Humphreys Gold Corp, 56-57; develop engr, Pigments Dept, E I de Nemours & Co, Inc, 57-64, sr res geologist, 64-69, tech supvr, 69-71, res assoc, 71-85; PRIN CONSULT, TICON INC, 85- *Mem:* Am Inst Mining, Metall & Petrol Engrs; Soc Econ Geologists; Am Foundrymen's Soc; Sigma Xi. *Res:* Mineralogy and geology of industrial minerals; gravity, electrostatic, high tension, and magnetic concentration of titanium minerals, zircon, aluminum silicates; new uses for mineral products; foundry sand properties. *Mailing Add:* PO Box 417 Keystone Heights FL 32656

GARNEAU, FRANCOIS XAVIER, b Montreal, Que, May 6, 36; m 63; c 3. ORGANIC CHEMISTRY. *Educ:* Concordia Univ Montreal, BSc, 62; Col of the Holy Cross, MSc, 63; Univ Toronto, PhD (organic chem & photochem), 68. *Prof Exp:* Fels, Univ Alta, 68-70; PROF ORG CHEM, UNIV QUE, CHICOUTIMI, 70- *Mem:* Am Chem Soc; Fr-Can Asn Advan Sci; Chem Inst Can; Int Soc Chem Ecol. *Res:* Marine natural products and plant natural products. *Mailing Add:* Dept Chem Univ Que 555 Bl De L Universite Chicoutimi PQ G7H 2B1 Can

GARNER, ALBERT Y, b Washington, DC, May 8, 25; m 49; c 4. ORGANIC CHEMISTRY. *Educ:* Howard Univ, BS, 50, MS, 51; Pa State Univ, PhD(org chem), 56. *Prof Exp:* Instr, Howard Univ, 51-52, res assoc, 52-53; res chemist, Monsanto Chem Co, 56-62, res chemist, Monsanto Res Corp, 62-64, res specialist, 64-65, proj leader & group leader, 65-76, sr res specialist, Monsanto Res Corp, 76-79, sr res specialist, Monsanto Plastics & Resins Co, 79-82, fel, Monsanto Polymer Prod Co, 82-85, fel, Monsanto Chem Co, 86; RETIRED. *Mem:* Am Chem Soc; Am Inst Chem. *Res:* Organic and polymer synthesis; reaction mechanisms; organophosphorus chemistry; structure determination; reactive intermediates; reaction injection molding; nylon block copolymers. *Mailing Add:* 1463 Plumtree Rd Springfield MA 01119

GARNER, ANDREW, b Manchester, Eng, Oct 23, 45; Can citizen; m 69; c 3. CORROSION, METALLURGY. *Educ:* Liverpool Univ, BSc, 68; Univ BC, PhD(metall), 74. *Prof Exp:* Metallurgist aluminum, Alcan Industs Ltd, UK, 68-69; sr res engr corrosion, Molybdenum Res, Endako Mines, 75-78; ASSOC SCIENTIST CORROSION, PULP & PAPER RES INST CAN, 78- *Honors & Awards:* Weldon Medal, Can Paper & Pulp Asn, 85. *Mem:* Nat Asn Corrosion Engrs; Am Soc Metals; fel Brit Inst Metallurgists; Tech Asn Pulp & Paper Indust. *Res:* Corrosion in aqueous environments. *Mailing Add:* Pulp & Paper Res Inst Can 3800 Wesbrook Mall Vancouver BC V6S 2L9 Can

GARNER, CYRIL WILBUR LUTHER, b Scotland, Ont, Mar 20, 40; m 67. PURE MATHEMATICS. *Educ:* Univ NB, BSc, 61; Univ Toronto, MA, 62, PhD(math), 64. *Prof Exp:* From asst prof to assoc prof, 64-82, PROF MATH, CARLETON UNIV, OTTAWA, ONT, 82- *Mem:* Am Math Soc. *Res:* Projective geometry; finite geometries and the foundations of geometry with particular emphasis upon hyperbolic geometry; combinatorics. *Mailing Add:* Dept Math Carleton Univ Colonel By Dr Ottawa ON K1S 5B6 Can

GARNER, DANIEL DEE, b Minford, Ohio, Oct 7, 47; m 68; c 2. FORENSIC SCIENCE. *Educ:* David Lipscomb Col, BA, 69; Univ Tenn, PhD(med chem), 73. *Prof Exp:* Lab technologist, Baptist Mem Hosp, 66-69; forensic chemist, Bur Alcohol Tobacco & Firearms, 73-87; dir labs, 87-91, VPRES, CELLMARK DIAGNOSTICS, 91- *Concurrent Pos:* Med technologist, Howard County Gen Hosp, 73-75; assoc prof lectr, George Washington Univ, 75; assoc prof, Antioch Sch Law, 79- *Mem:* Am Acad Forensic Sci; Am Asn Blood Banks; Am Soc Human Genetics; Mid-Atlantic Asn Forensic Sci; Am Asn Clin Chemists. *Res:* Characterization of explosives and explosive residues of importance in forensic chemistry; development of identification and characterization techniques. *Mailing Add:* 20271 Goldenrod Lane Suite 120 Germantown MD 20876

GARNER, DUANE LEROY, b Madera, Calif, Feb 9, 36; m 60; c 3. REPRODUCTIVE PHYSIOLOGY, ANIMAL SCIENCE. *Educ:* Calif State Univ, Fresno, BS, 64; Wash State Univ, MS, 67, PhD(animal sci), 69. *Prof Exp:* Fel reproductive physiol, Dept Dairy Sci, Univ Ill, 69-70, res assoc, 70-72; prof reproductive physiol, Dept Physiol Sci, Okla State Univ, 72-85; PROF & CHAIR, DEPT ANIMAL SCI, UNIV NEV, RENO, 85- *Concurrent Pos:* Spec consult for NIH prog proj rev, NIH, 75, 80, 82, & 84, 86, 87, 88, 89, 90. *Mem:* Am Physiol Soc; Soc Study Reproduction; Am Soc Andrology; Am Soc Animal Sci; Soc Analytical Cytol. *Res:* Fluorescent staining of spermatozoa; fertilization, immunobiology of spermatozoa, flow cytometry of spermatozoa. *Mailing Add:* Dept Animal Sci Univ Nev-Reno Reno NV 89557

GARNER, GEORGE BERNARD, b Kirksville, Mo, Dec 7, 27; m 48; c 4. AGRICULTURAL BIOCHEMISTRY. *Educ:* Northeast Mo State Teachers Col, BS, 49; Univ Mo, MS, 51, PhD(agr chem), 57. *Prof Exp:* Teacher, Pub Sch, 51-53; from instr to assoc prof, 53-74, PROF AGR CHEM, UNIV MO, COLUMBIA, 74-, PROF BIOCHEM, 80- *Concurrent Pos:* Fulbright res scholar, 62-63. *Mem:* AAAS; Am Chem Soc; Am Soc Animal Sci; Sigma Xi. *Res:* Rumen nutrition and physiology; nitrate toxicity, fescue toxicosis and physiologically active constituents of forages; analytical methods applicable to agricultural products. *Mailing Add:* Animal Sci Ctr Univ Mo Columbia MO 65211

GARNER, HAROLD E, b Eldorado, Kans, May 8, 35; m 57; c 4. CARDIOVASCULAR DISEASE, ANESTHESIOLOGY. *Educ:* Kans State Univ, BS, 57, DVM, 62, MS, 64; Baylor Col Med, PhD(cardiovasc physiol, biomed eng), 71. *Prof Exp:* Pvt vet pract, 62-65; asst prof, Dept Clin Med, Univ Ill, 66-68; instr, asst prof & spec res fel, Dept Physiol & Surg, Baylor Col Med, 68-71; PROF VET MED & SURG & ASSOC INVESTR, DALTON RES CTR, UNIV MO-COLUMBIA, 71- *Mem:* Am Physiol Soc; NY Acad Sci; Am Soc Vet Anesthesiol; Am Acad Vet Cardiol; Am Vet Med Asn. *Res:* Specific and comparative aspects of equine hypertension-laminitis and human hypertension; automated quantitative administration of volatile anesthetics to animals; equine edotoxemia. *Mailing Add:* Dept Med & Surg Univ Mo Columbia MO 65211

GARNER, HARRY RICHARD, b East Liverpool, Ohio, Feb 4, 35; m 57; c 2. ELECTROCHEMISTRY. *Educ:* Kent State Univ, BS, 57; Western Reserve Univ, MS, 62. *Prof Exp:* Chemist, Harshaw Chem Co, 57-59, group leader, 59-63, unit mgr, Instrumental Analysis Lab, 63-66, from asst sect mgr to sect mgr, Cent Analysis Lab, 66-69, tech dir metal finishing res, 69-81, bus mgr, 82-83; vpres & gen mgr, metal finishing, Harshaw-Filtrol, 83-88; vpres & bus dir, Engelhard Corp, 88-90; CONSULT, M&T HARSHAW, 91- *Mem:* Am Electroplaters Soc. *Res:* Electrodeposition of metals; analytical chemistry; instrumental analysis; corrosion. *Mailing Add:* M&T Harshaw 1000 Harvard Ave Cleveland OH 44109

GARNER, HARVEY L(OUIS), b Lake, Colo, Dec 23, 26; m 50; c 2. COMPUTER ARITHMETIC, COMPUTER SYSTEM DESIGN. *Educ:* Univ Denver, BS, 49, MS, 51; Univ Mich, PhD(elec eng), 58. *Hon Degrees:* MA, Univ Pa. *Prof Exp:* Res assoc, Cosmic Radiation Lab, Univ Denver, 49-51; assoc res engr digital comput dept, Willow Run Labs, Univ Mich, 51-53, asst head dept, 53-55, from instr to prof elec eng, 55-70, dir info syst lab, 60-64; dir, Moore Sch Elec Eng, 70-77, PROF ELEC ENG, UNIV PA, 70-86; dir, MCC, 84-88; CONSULT, 88- *Concurrent Pos:* Nat lectr, Asn Comput Mach, 65; vis prof, Stanford Univ, 76-77. *Mem:* AAAS; Asn Comput Mach; fel Inst Elec & Electronics Engrs. *Res:* Digital computation; machine number systems and arithmetic; integration of hardware and software in digital systems. *Mailing Add:* 7400 Rockberry Cove Austin TX 78750

GARNER, HERSCHEL WHITAKER, b Jacksonville, Tex, June 25, 36; m 59; c 1. VERTEBRATE ZOOLOGY, ECOLOGY. *Educ:* Stephen F Austin State Univ, BS, 62; Tex Tech Univ, MS, 65, PhD(zool), 70. *Prof Exp:* Instr biol, Tex Tech Univ, 65-66; asst prof biol, 70-74, ASST PROF BIOL SCI, TARLETON STATE COL, 74- *Mem:* AAAS; Am Soc Mammal; Sigma Xi. *Res:* Biology and populations of small mammals. *Mailing Add:* Dept Biol Sci Tarleton State Col Stephenville TX 76401

GARNER, HESSLE FILMORE, b Creston, Iowa, Feb 24, 26; m 80; c 7. GEOMORPHOLOGY. *Educ:* Iowa State Col, BS, 50; Univ Iowa, MS, 51, PhD(geol), 53. *Prof Exp:* Asst phys geol, Univ Iowa, 50-51, asst paleont, 52-53; geologist, Richmond Petrol Co Calif, 53, geologist, Calif-Ecuador Petrol Co, 54-56; asst prof geol, Univ Ark, 56-60, assoc prof, 61-67; prof geol, Rutgers Univ, 67-76, prof II, 77-87; GEOL CONSULT & JUDICATION, 87- *Mem:* Fel Geol Soc Am; AAAS; Am Asn Petrol Geologists. *Res:* Invertebrate paleontology; paleozoic and mesozoic cephalopoda; tertiary stratigraphy and structure of Ecuador; Andes mountain geomorphology and climatic sedimentation; global drainage development; paleozoic stratigraphy of north Arkansas; global geomorphology; development of fluvial systems and continental clastic deposits; Quaternary paleoclimatology. *Mailing Add:* 202 Merrywood Dr Forest VA 24551

GARNER, JACKIE BASS, b Jonesboro, La, Aug 21, 34; m 57; c 2. MATHEMATICS. *Educ:* La Polytech Inst, BS, 55; Auburn Univ, MS, 57, PhD(math), 60. *Prof Exp:* From instr to prof math, La Tech Univ, 57-86; PROF MATH, MISS STATE UNIV, 86- *Concurrent Pos:* NSF res grant, 63-64; appointment, NIH, 76-79, grants, 72-75 & 80-86. *Mem:* Am Math Soc; Math Asn Am. *Res:* Ordinary and partial differential equations. *Mailing Add:* Dept Math Miss State Univ Drawer MA Miss State MS 39762

GARNER, JAMES G, b Astoria, NY, Dec 21, 38; m 64; c 2. CYTOGENETICS, MICROBIOLOGY. *Educ:* Providence Col, AB, 60; Long Island Univ, MS, 63; St John's Univ, NY, PhD(biol), 67. *Prof Exp:* Teaching fel biol, Long Island Univ, 60-62; teaching asst, St John's Univ, NY, 62-65; from instr to full prof, 65-76, PROF BIOL, C W POST COL, LONG ISLAND UNIV, 76- *Concurrent Pos:* Mem bd trustees & pub educ comt, Nassau-Suffolk County Chap, Nat Found-March of Dimes; biol consult, John Wiley & Sons, Inc, Publ, McGraw-Hill Book Co & Benjamin/Cummings Publ Co. *Mem:* Am Soc Cell Biol; Sigma Xi. *Res:* Cytogenetics, sex chromatin in vertebrates; Q- and G-banding in human and mammalian chromosomes; microbiology, action of plant hormones on bacteria. *Mailing Add:* Dept Biol C W Post Col Long Island Univ Greenvale NY 11548

GARNER, JASPER HENRY BARKDOLL, b Bulsar, India, Nov 7, 21; US citizen; m 54; c 1. MYCOLOGY, PLANT PATHOLOGY. *Educ:* Manchester Col, BA, 48; Ind Univ, MA, 53; Univ Iowa, PhD(bot), 55. *Prof Exp:* Asst prof bot, Univ Nebr, 55-56; from instr to asst prof, Univ Ky, 56-67; assoc prof biol, Eastern Tenn State Univ, 67-69; res assoc plant path, NC State Univ, 69-71; botanist, Off Air Progs, 71-72, botanist-microbiologist, spec studies staff, 72-78, biologist, 78-79, ECOLOGIST, ENVIRON CRITERIA

& ASSESSMENT OFF, ENVIRON PROTECTION AGENCY, 79- *Concurrent Pos:* Univ Ky-Int Coop Admin Prog vis prof, Col Agr & Vet Sci, Univ Indonesia, 57-61; chmn task force, Air Qual Criteria for Ozone and Other Photochem Oxidants, 75-78 & Health Asssessment of Cadmium, 77-81 & mem task force rewriting, Air Qual Criteria for Particulate Mat & Sulfur Oxides, 78-82; team mem, Air Qual Criteria for Oxides of Nitrogen, 88- *Mem:* Mycol Soc Am; Ecol Soc Am; Am Inst Biol Sci. *Res:* Gasteromycetes; forest pathology; interrelationships between natural gas pollution of soil and death of woody plants; vegetational effects of sulfur oxides and ozone; ecological effects of nitrogen oxides; particular nitrogen deposition and nutrient cycling; sulfur oxides and ozone particularly on forest ecosystems; author of several publications. *Mailing Add:* Environ Criteria & Assessment Off MD-52 Environ Protection Agency Research Triangle Park NC 27711

GARNER, LAFORREST D, b Muskogee, Okla, Aug 20, 33; m 64; c 3. ORTHODONTICS. *Educ:* Ind Univ, DDS, 57, MSD, 59, cert, 61. *Prof Exp:* Asst pedodontics, 57-58, res asst, 58-59, from instr to assoc prof, 59-71, PROF ORTHOD, SCH DENT, 71-, ASSOC DEAN, IND UNIV, INDIANAPOLIS, 88- *Concurrent Pos:* Chmn dept orthods, Sch Dent, Ind Univ, 70- *Mem:* Am Dent Asn; Am Soc Dent Children; Int Asn Dent Res; fel Am Col Dent; Sigma Xi; Am Asn Orthod. *Res:* Growth and development; posture of tongue in children with normal occlusions; cleft lip and palate rehabilitation; pedodontics; orthodontics. *Mailing Add:* Sch Dent Ind Univ Indianapolis IN 46202

GARNER, LYNN E, b Ontario, Ore, July 19, 41; m 60; c 5. MATHEMATICS. *Educ:* Brigham Young Univ, BS, 62; Univ Utah, MS, 64; Univ Ore, PhD(math), 68. *Prof Exp:* Instr, 64-66, asst prof, 68-74, assoc prof, 74-80, PROF MATH, BRIGHAM YOUNG UNIV, 81- *Mem:* Am Math Soc; Math Asn Am; Sigma Xi. *Res:* Algebra; commutative algebra; projective and algebraic geometry; number theory; philosophy of mathematics. *Mailing Add:* Dept Math Brigham Young Univ Provo UT 84602

GARNER, MERIDON VESTAL, b Belton, Tex, Sept 20, 28; m 56; c 3. MATHEMATICS. *Educ:* Sul Ross State Col, MEd, 56; NTex State Univ, EdD, 63. *Prof Exp:* Assoc prof math, Sul Ross State Col, 56-63; asst prof, 63-69, ASSOC PROF MATH, NTEX STATE UNIV, 69- *Mem:* Math Asn Am. *Res:* Preparation of subject-content material in mathematics for future elementary teachers. *Mailing Add:* Dept Math NTex State Univ PO Box 5116 Denton TX 76203

GARNER, REUBEN JOHN, b Oundle, Eng, Feb 4, 21; m 42; c 2. RADIATION BIOLOGY. *Educ:* Cambridge Univ, BA, 42, MA, 46; Univ Liverpool, MVSc, 52, DVSc, 61; MRCVS, 45, FRCVS, 52; ARIC, 56. *Prof Exp:* Vet res officer, Vet Res Lab, Nigeria, 46-50; lectr vet biochem, Univ Liverpool, 50-53; sr lectr chem path, Bristol Univ, 53-56; head radiobiol lab, Inst Res Animal Dis, Agr Res Coun, UK, 57-60; head pub health sect, Health & Safety Br, Radiol Protection Div, UK Atomic Energy Auth, 65; dir collab radiol health lab, Colo State Univ, 65-72; dir exp biol lab, Environ Protection Agency, Cincinnati, 72-75, dir health effects res lab, 75-81, sr res adv, exp biol div, 81-84; RETIRED. *Concurrent Pos:* Consult, Food & Agr Orgn & Int Atomic Energy Agency, 60-65. *Res:* Toxicology of biological, chemical and physical pollutants in the environment. *Mailing Add:* Four Lark Cr Chapel Hill NC 27514-2552

GARNER, RICHARD GORDON, food science & technology, for more information see previous edition

GARNER, ROBERT HENRY, b Mobile, Ala, Jan 7, 33; m 61; c 3. CHEMISTRY. *Educ:* Vanderbilt Univ, BA, 54; Rice Inst, PhD(chem), 58. *Prof Exp:* Fel chem, Yale Univ, 58-59; asst prof, 59-68, ASSOC PROF CHEM, UNIV ALA, 68-, ASST DEAN, COL SCI, 73- *Mem:* Am Chem Soc. *Res:* Organic chemistry. *Mailing Add:* Box 870336 Tuscaloosa AL 35487-0336

GARNET, HYMAN R, b Brooklyn, NY, Apr 4, 20; m 44; c 2. APPLIED MECHANICS, APPLIED MATHEMATICS. *Educ:* NY Univ, BS, 51; Columbia Univ, MA, 52; Polytech Inst Brooklyn, PhD(appl mech), 62. *Prof Exp:* Struct engr, Repub Aviation Corp, 52-57; mech engr, Mat Lab, Brooklyn Navy Yard, 57-58; struct engr, 58-59, SR RES SCIENTIST, RES DEPT, GRUMMAN AEROSPACE CORP, 59- *Concurrent Pos:* Lectr, Dept Eng Sci, Hofstra Univ, 65-72. *Mem:* Am Soc Mech Engrs; Am Acad Mech; Sigma Xi. *Res:* Solid mechanics: elasticity; plasticity; viscoelasticity; stress wave propagation; vibrations; statics and dynamics of shells; dynamic plasticity; development of nonlinear finite-element analyses; computational mechanics. *Mailing Add:* 2674 Janet Ave North Bellmore NY 11712

GARNETT, RICHARD WINGFIELD, JR, b Albemarle Co, Va, Mar 27, 15; m 41; c 5. PSYCHIATRY. *Educ:* Univ Va, BS, 36, MD, 40, MS, 49. *Prof Exp:* From asst prof to assoc prof, 49-57, actg chmn dept, 56-57 & 63-64, prof, 57-74, CLIN PROF PSYCHIAT, SCH MED, UNIV VA, 74-; SR STAFF PSYCHIATRIST, DAVID C WILSON PSYCHIAT HOSP, 81- *Concurrent Pos:* WHO fel community psychiat, Gt Brit & Western Europe, 63; mem sr fac sem, Harvard Med Sch, 65-67; consult, Vet Admin. *Mem:* Fel Am Psychiat Asn; AMA. *Res:* Psychosomatic medicine; academic and community psychiatry. *Mailing Add:* David C Wilson Neuropsy Hosp 2101 Arlington Blvd Charlottesville VA 22903

GARNSEY, STEPHEN MICHAEL, b Oceanside, Calif, Aug 3, 37; m 58; c 2. PLANT VIROLOGY. *Educ:* Univ Calif, Riverside, BA, 58, Univ Calif, Davis, PhD(plant path), 64. *Prof Exp:* Lab technician plant path, Univ Calif, Riverside, 58-59, res asst, Univ Calif, Davis, 59-63; RES PLANT PATHOLOGIST, US HORT RES LAB, USDA, 63-; assoc prof, 69-83, PROF PLANT PATH, INST FOOD & AGR SCI, UNIV FLA, 83- *Honors & Awards:* Lee M Hutchins Award, Am Phytopathol Soc, 81; Fel, Am Phytopath Soc, 87. *Mem:* AAAS; Am Phytopath Soc; Sigma Xi. *Res:* Mechanical transmission, purification and serology of plant viruses; identification and properties of citrus viruses and virus like pathogens; development of virus resistant citrus. *Mailing Add:* US Hort Res Lab USDA 2120 Camden Rd Orlando FL 32803

GARODNICK, JOSEPH, b Newark, NJ, Jan 29, 45; c 2. DATA COMMUNICATIONS. *Educ:* Rensselaer Polytech Inst, BEE, 66; Polytech Inst New York, MS, 69; City Univ New York, PhD(elec eng), 72. *Prof Exp:* Design engr radio frequency equip, Blonder Tongue Labs, 65-67, design engr digital equip, ITT Avionics, 67-69; res lab asst commun, City Col City New York, 69-72; mem tech staff, Bell Tel Labs, 72; dir commun systs, Goldmark Commun Corp, 72-75; exec vpres, Stern Telecommun Corp, 75-81; chief scientist, Phasecom Corp, 81-86; CORP VPRES, FIBRONICS INT INC, 86- *Concurrent Pos:* Adj prof Dept Elec Eng, City Col City Univ New York, 72-81; dir publ bd gov, Commun Soc, 79-83. *Mem:* Inst Elec & Electronics Engrs. *Res:* Digital processing of communication systems; digital phase locked loops; voice encoding; satellite communications; fiber optic data communications. *Mailing Add:* Fibronics Int Inc Communications Way Independence Park Hyannis MA 02601

GAROIAN, GEORGE, b Deadwood, SDak, July 23, 27; m 54; c 2. PARASITOLOGY. *Educ:* Wash Univ, AB, 49; Univ Ill, MS, 51, PhD(zool), 56. *Prof Exp:* Asst zool, Univ Ill, 49-55, asst vet med, 55-56; from asst prof to assoc prof, 56-84, asst chmn dept, 58-61, actg chmn dept, 71-72, chmn dept, 72-73, dir undergrad studies in dept, 73-80, prof, 84-87, EMER PROF ZOOL, SOUTHERN ILL UNIV, 87- *Mem:* Am Soc Parasitologists; Soc Protozoologists. *Res:* Protozoology; helminthology; animal diversity. *Mailing Add:* Dept Zool Southern Ill Univ Carbondale IL 62901-6501

GARON, CLAUDE FRANCIS, b Baton Rouge, La, Nov 5, 42; m 68; c 3. MICROBIOLOGY, BIOCHEMISTRY. *Educ:* La State Univ, BS, 64, MS, 66; Georgetown Univ, PhD(microbiol), 70. *Prof Exp:* RES MICROBIOLOGIST VIROL, NIH, 71-, CHIEF LAB VECTORS & PATHOGENS, NIH, MONT, 85- *Concurrent Pos:* Fel lab biol viruses, Nat Inst Allergy & Infectious Dis, NIH, 69-71, staff fel, 71-73, sr staff fel, 73-74. *Mem:* Am Soc Microbiol; Am Soc Virol; Am Soc Biochem & Molecular Biol. *Res:* Molecular biology; ultra structure of nucleic acids. *Mailing Add:* Lab Vectors & Pathogens Rocky Mountain Labs Hamilton MT 59840

GARON, OLIVIER, b Quebec, Que, Jan 24, 28; m 55; c 2. COMPARATIVE ANATOMY. *Educ:* Laval Univ, BA, 50; Univ Montreal, DVM, 55, MSc, 61, PhD, 64. *Prof Exp:* Head, dept basic sci, 61-72, dir animal anat & physiol, 72-80, PROF ANAT & BIOL, SCH VET MED, UNIV MONTREAL, 57- *Res:* Domestic and wild animals. *Mailing Add:* 2240 Carillon St Hyacinthe PQ J2S 7S9 Can

GARONE, JOHN EDWARD, b New York, NY. HUMAN BIOLOGY, SCIENCE WRITING. *Educ:* Columbia Col, BA, 48; Columbia Univ, MA, 49, PhD(natural sci), 58. *Prof Exp:* Instr natural sci, Columbia Univ, 55-57; PROF NATURAL SCI & CHMN DIV, JERSEY CITY STATE COL, 59- *Concurrent Pos:* Vis prof natural sci, Univ Southern Calif, 59-62; ed-in-chief, Natural Sci Bull, 68-72; pres & trustee, NJ Marine Sci Consortium, 70-73; chmn, Tommorrow's Scientists & Engrs, 70-73; adminr, Electronmicroscopy Ctr, 70- & Radiobiol Ctr, 73- *Mem:* AAAS; Am Inst Biol Sci; Fedn Am Socs Exp Biol; NY Acad Sci; Nat Sci Asn. *Mailing Add:* Dept Biol Jersey City State Col 2039 Kennedy Mem Blvd Jersey City NJ 07305

GAROUTTE, BILL CHARLES, b Absarokee, Mont, Mar 15, 21; m 48, 89; c 4. NEUROPHYSIOLOGY. *Educ:* Univ Calif, AB, 43, MD, 45, PhD(biophys), 54. *Prof Exp:* Intern, San Diego County Hosp, 45-46; asst resident neurol, Univ Hosp, 48 & 51-52, from lectr to prof, 49-86, EMER PROF ANAT & NEUROL, UNIV CALIF, SAN FRANCISCO, 86- *Concurrent Pos:* Fulbright scholar, Inst Neurol, London, 50-51; vis asst prof, Fac Med, Univ Indonesia, 56-57; vis investr, Inst Brain Res, Tokyo, 63, ext examr, Univ Malaysia, 78, Sci Univ Malaysia, 84. *Mem:* AAAS; Am Epilepsy Soc; Am Electroencephalog Soc; AMA; Am Asn Anat; Sigma Xi; Am Asn Electromyography & Electrodiagnosis. *Res:* Function of the nervous system. *Mailing Add:* 105 Molino Ave Mill Valley CA 94941

GARRARD, CHRISTOPHER S, b Wolverhantton, Eng, May 9, 45. PULMONARY DISEASE. *Educ:* Oxford Univ, PhD(physiol), 81. *Prof Exp:* ASSOC PROF MED & ANESTHESIOL, COL MED, UNIV ILL, 77- *Mailing Add:* John Radcliff Hosp Headington Oxford 0X3 9DU England

GARRARD, STERLING DAVIS, b Chicago, Ill, Nov 10, 19; m 45; c 3. PEDIATRICS. *Educ:* Univ Ill, MD, 45, BS, 48. *Prof Exp:* Instr pediat, Col Med, Univ Ill, 50-52; from asst prof to assoc prof, 55-65, actg chmn dept, 65-68; prof, State Univ NY Upstate Med Ctr, 65-72; prof pediat & dir, Riley Child Develop Ctr, Med Ctr, Ind Univ, 72-74; assoc prof pediat, Harvard Med Sch, 74-76; PROF FAMILY & COMMUNITY MED & PEDIAT, MED SCH, UNIV MASS, 76- *Concurrent Pos:* Mem ment hyg coun, NY State Dept Ment Hyg; chief physician, Walter E Fernald State Sch, 74-76; dir univ affil training prog, Eunice K Shriver Ctr Ment Retardation, 74-76; assoc pediatrician, Mass Gen Hosp, 74-76. *Mem:* AAAS; fel Am Asn Ment Deficiency; Am Acad Pediat; Soc Res Child Develop; Am Pediat Soc. *Res:* Mental retardation. *Mailing Add:* Dept Family Community Med Med Sch Univ Mass 55 Lake Ave N Worcester MA 01605

GARRARD, VERL GRADY, b Burley, Idaho, July 21, 23; m 62; c 3. PHYSICAL CHEMISTRY, ANALYTICAL CHEMISTRY. *Educ:* Univ Idaho, BS, 45, MS, 52; Univ Utah, PhD, 67. *Prof Exp:* From instr to asst prof, 47-70, assoc prof, 70-86, EMER ASSOC PROF CHEM, UNIV IDAHO, 86- *Mem:* AAAS; Am Chem Soc; Sigma Xi. *Res:* Chemical kinetics. *Mailing Add:* 422 N 700 E Provo UT 84606

GARRARD, WILLIAM LASH, JR, b Waco, Tex, Nov 7, 40; m 65. ENGINEERING MECHANICS. *Educ:* Univ Tex, BS, 62, PhD(eng mech), 68. *Prof Exp:* Asst prof, 68-76, ASSOC PROF AERONAUT & ENG MECH, UNIV MINN, MINNEAPOLIS, 76- *Concurrent Pos:* NSF grant, 68-69. *Mem:* Inst Elec & Electronics Engrs. *Res:* Control theory; stability and control of dynamical systems. *Mailing Add:* Dept Aeronaut Univ Minn 1107 Akerman Minneapolis MN 55455

GARRARD, WILLIAM T, b Seattle, Wash, Sept 16, 42. MOLECULAR BIOLOGY. *Educ:* Univ Calif, Los Angeles, PhD(microbiol), 71. *Prof Exp:* PROF BIOCHEM, UNIV TEX HEALTH SCI CTR, 74- *Mem:* Am Soc Biol Chemists; Am Soc Cell Biol; Sigma Xi. *Mailing Add:* Dept Biochem Health Sci Ctr Univ Tex 5323 Harry Hines Blvd Dallas TX 75235

GARRATTY, GEORGE, b London, Eng, July 2, 35; Brit citizen; m 69. IMMUNOHEMATOLOGY. *Educ:* AIMLS, 58; FIMLS, 60; MRCPath, 80; PhD, 85. *Prof Exp:* Chief med lab scientist hemat, Royal Postgrad Med Sch London, 63-68; res assoc immunohemat, Univ Calif, San Francisco, 68-70, assoc specialist & lectr, 71-78; res assoc, Inst Med Sci, San Francisco, 75-78; SCI DIR, AM RED CROSS BLOOD PROG, LOS ANGELES, 78-; CLIN PROF PATHOL, UNIV CALIF, LOS ANGELES, 85- *Concurrent Pos:* Lectr, Inner London Educ Authority, 59-68, Paddington Col, London, 59-68 & San Francisco State Univ, 71-78; consult, Irwin Mem Blood Bank, San Francisco, 73-78; mem comt, Tech Manual, Am Asn Blood Banks, 75-82, Sci Prog, 78-81, Standards, 82-89; mem comt, Adv Comt Antiglobulin Sera, Food & Drug Admin, 74-; Prog Chmn, Calif Blood Bank Syst, 77-78, chmn Bylaws Comt, 90-; pres, Calif Blood Bank Soc, 85-86. *Honors & Awards:* Ivor Dunsford Mem Award, Am Asn Blood Banks, 78, Emily Cooley Mem Award, 89; Charles R Drew Award, Am Red Cross, 87. *Mem:* Royal Col Pathologists & Inst Med Lab Sci, London; Am Asn Blood Banks; Am Fed Clin Res; Am Asn Immunologists; Am Soc Hematol; Int Soc Blood Tranfusion. *Res:* Immune red cell destruction (Alloimmune and Autoimmune); complement; antigen-antibody reactions in immunohematology. *Mailing Add:* Am Red Cross Blood Prog 1130 S Vermont Ave Los Angeles CA 90006

GARRAWAY, MICHAEL OLIVER, b Roseau, Dominica, Brit WI, Apr 29, 34; US citizen; m 66; c 3. PLANT PATHOLOGY. *Educ:* McGill Univ, BSc, 59, MSc, 62; Univ Calif, Berkeley, PhD(plant path), 66. *Prof Exp:* Asst res plant pathologist, Univ Calif, Berkeley, 66-68; asst prof, Ohio State Univ, 68-71, assoc prof, 71-78, PROF BOT PLANT PATH, OHIO STATE UNIV & OHIO AGR RES & DEVELOP CTR, 78- *Concurrent Pos:* Fred C Gloeckner Found Inc res grant dis of ornamentals. *Mem:* AAAS; Am Phytopath Soc; Am Inst Biol Sci; Can Phytopath Soc; Int Soc Plant Path; Mycol Soc Am. *Res:* Relation of nutrition to rhizomorph initiation and growth in Armillaria mellea (Vahl) Quel; nutritional and environmental factors affecting fungal growth and reproduction; influence of plant nutrition on disease susceptibility; response of maize leaves to biotic and abiotic stresses. *Mailing Add:* Rm 201 Dept Plant Path 201C Kottman Ohio State Univ 2021 Coffey Rd Columbus OH 43210

GARRELICK, JOEL MARC, b New York, NY, May 20, 41; m 63; c 3. STRUCTURAL DYNAMICS, PHYSICAL ACOUSTICS. *Educ:* City Univ New York, BCE, 63, ME, 65, PhD(mech), 69. *Prof Exp:* Lectr civil eng, City Col New York, 67-68; SR SCIENTIST VIBRATIONS & ACOUST, CAMBRIDGE ACOUST ASSOCS INC, 69- *Mem:* Am Soc Civil Engrs; fel Acoust Soc Am. *Res:* Theoretical studies in applied mechanics; structural dynamics; underwater acoustics; structure borne noise and vibration. *Mailing Add:* 46 Kenney Lane Concord MA 01742

GARRELL, MARTIN HENRY, b Brooklyn, NY, Jan 4, 39; m 65. ENVIRONMENTAL PHYSICS. *Educ:* Princeton Univ, AB, 60; Univ Ill, MS, 62, PhD(physics), 66. *Prof Exp:* Volkswagen Found res fel physics, Deutsches Elektronen Synchrotron, 66-68; vis asst prof, Univ Ill, Chicago Circle, 68-70; ASST PROF PHYSICS, ADELPHI UNIV, 70- *Mem:* Am Phys Soc; Sigma Xi. *Res:* Population models; air, water and pesticides pollution; fresh water and marine ecosystems. *Mailing Add:* Dept Physics Col Arts & Sci Adelphi Univ Garden City NY 11530

GARRELS, JAMES I, b Mt Pleasant, Iowa, Feb 14, 48. PROTEIN CHARACTERIZATION, CELLULR GROWTH REGULATION. *Educ:* Calif Inst Technol, BS, 71; Univ Calif, San Diego, PhD(biol), 78. *Prof Exp:* Am Cancer Soc postdoctoral fel, Cell Biol Group, Cold Spring Harbor Lab, 78-79, staff investr, 79-80, sr staff investr, 80-85, SR STAFF SCIENTIST & DIR, QUEST PROTEIN DATABASE CTR, COLD SPRING HARBOR LAB, 85- *Concurrent Pos:* NSF res grant, 78-80; NIH res grants, 79-82, 85-90 & 90-; Muscular Dystrophy Asn res grant, 80-84; Cystic Fibrosis Found res grant, 80-81; consult, Millipore Corp, 88- *Mem:* Int Electrophoresis Soc; Am Soc Biochem & Molecular Biol; Am Soc Microbiol. *Res:* Two-dimensional gel electrophoresis of proteins; quantitative computer analysis of two-dimensional gels; construction of protein databases; mammalian cell growth and transformation; author of numerous technical publications. *Mailing Add:* Quest Protein Database Ctr PO Box 100 Bungtown Rd Cold Spring Harbor NY 11724

GARRELS, ROBERT MINARD, geochemistry; deceased, see previous edition for last biography

GARRELTS, JEWELL MILAN, b McPherson, Kans, Oct 25, 03; m 24; c 3. CIVIL ENGINEERING. *Educ:* Valparaiso Univ, BS, 24; Columbia Univ, MS, 33. *Hon Degrees:* DSc, Doane Col, 78. *Prof Exp:* Draftsman, Ill Steel Co, 24-25; instr struct eng, Cornell Univ, 25-27; instr mech, Columbia Univ, 27-38, from asst prof to prof civil eng, 38-57, exec officer dept, 46-57, assoc dean col eng, 57-72, prof eng & chmn dept, 57-84, asst to dean, Sch Eng, 72-84; RETIRED. *Concurrent Pos:* Assoc engr, Hardesty & Hanover, NY, 36-51; consult engr, 51- *Mem:* AAAS; Am Soc Eng Educ; Am Soc Civil Engrs; Am Concrete Inst; Int Asn Bridge & Struct Eng. *Res:* Structural engineering and design. *Mailing Add:* 15 Brook Rd Tenafly NJ 07670

GARREN, HENRY WILBURN, b Hendersonville, NC, Apr 2, 25; m 46; c 2. PHYSIOLOGY, ENDOCRINOLOGY. *Educ:* Univ NC, AB, 47, BS, 49; Univ Md, MS, 51, PhD(physiol), 53. *Prof Exp:* Asst, Univ Md, 49-52; physiologist, Med Labs, Army Chem Ctr, Md, 52-53; assoc prof endocrinol, NC State Univ, 53-60, prof poultry sci & head dept, 60-68; DEAN & COORDR, COL AGR, UNIV GA, 68- *Honors & Awards:* Poultry Sci Res Award, 53. *Mem:* Am Physiol Soc; Poultry Sci Asn; Soc Exp Biol & Med; Sigma Xi. *Res:* Physiological response of animals to various stress stimuli, including disease, and conditions, nutritional and otherwise, affecting this response. *Mailing Add:* Ctr Improving Mountain Living Western Carolina Univ Cullowhee NC 28723

GARREN, RALPH, JR, b Rutland, Iowa, June 10, 21; m 42; c 3. HORTICULTURE. *Educ:* Ore State Col, BS, 50, MS, 54; Purdue Univ, PhD, 61. *Prof Exp:* Asst, Ore State Univ, 50-51, res asst & instr hort, 51-57, from asst prof to prof, 57-86, EMER PROF HORT, ORE STATE UNIV, 86- *Concurrent Pos:* Res found fel, Purdue Univ, 57-59. *Mem:* Am Soc Hort Sci; Int Soc Hort Sci. *Res:* Small fruits physiology and production problems. *Mailing Add:* Dept Hort Ore State Univ Corvallis OR 97331

GARRETSON, CRAIG MARTIN, b Glendale, NY, Sept 14, 24; m 49; c 3. ELECTROPHYSICS. *Educ:* Cooper Union, BEE, 49; Polytech Inst Brooklyn, MEE, 53, PhD(electrophys), 69. *Prof Exp:* Jr engr, Power Substa, New York City Bd Transp, 46-50; jr engr, Nat Union Radio Corp, 50-51; sr engr, Sylvania Elec Prod, 51-56; engr, Sperry Gyroscope Co, 56-63; from asst prof to assoc prof, 63-75, coordr, dept eng sci, 66-74, chmn dept physics 75-76 & 79-80, PROF PHYSICS, C W POST COL, LONG ISLAND UNIV, 75- *Mem:* Am Asn Physics Teachers; Inst Elec & Electronics Engrs. *Res:* Electron devices; electromagnetic theory. *Mailing Add:* Dept Physics C W Post Col Long Island Univ Brookville NY 11548

GARRETSON, HAROLD H, b Tacoma, Wash, Mar 3, 11; m 36; c 3. CHEMISTRY. *Educ:* Whitman Col, AB, 32; Univ Wash, MS, 33, PhD(chem), 40. *Prof Exp:* Asst prof chem, Agr & Mech Col, Tex, 38-43 & Univ NMex, 43-45; head dept, 64-74, PROF CHEM, LYNCHBURG COL, 45- *Concurrent Pos:* Consult, Oak Ridge Nat Labs. *Mem:* Am Chem Soc; Sigma Xi. *Res:* Thermodynamic ionization constants of sulfurous acid; electrometric titrations; stable isotope separations. *Mailing Add:* 3816 Faculty Dr Lynchburg VA 24501

GARRETT, ALFRED BENJAMIN, b Glencoe, Ohio, June 28, 06; m 34; c 3. CHEMISTRY. *Educ:* Muskingum Col, BS, 28; Ohio State Univ, MS, 31, PhD(chem), 32. *Hon Degrees:* DSc, Muskingum Col, 58, Ohio Wesleyan Univ, 62 & Denison Univ, 64. *Prof Exp:* Teacher pub sch, Pa, 28-29; asst chem, Ohio State Univ, 29-32; assoc prof, Kent State Col, 32-35; from instr to prof, 35-71, chmn dept chem, 58-62, vpres Univ, 62-69, EMER PROF CHEM, MCPHERSON LAB, OHIO STATE UNIV, 71- *Honors & Awards:* Am Chem Soc Award, 64; NASA Apollo 11 Award, 70. *Mem:* Nat Sci Teachers Asn (pres, 69-70); Sigma Xi; Am Chem Soc. *Res:* Photovoltaic cells; ionic equilibria in solution; physical and organic chemistry; low temperature studies of electrolytes; alkyl derivatives of boron hydrides. *Mailing Add:* Dept Chem Ohio State Univ, 140 W 18th Ave Columbus OH 43210

GARRETT, BARRY B, b Waco, Tex, Nov 10, 35; m 57; c 2. PHYSICAL CHEMISTRY, INORGANIC CHEMISTRY. *Educ:* Univ Tex, BS, 59, MA, 62, PhD(chem), 63. *Prof Exp:* Res assoc, Univ Ill, 63-64; from asst prof to prof chem, Fla State Univ, 64-84; SR SCIENTIST, OFF RES & DEVELOP (ORD), ARLINGTON, VA, 84- *Concurrent Pos:* Vis prof, Univ Wyo, 75. *Mem:* Am Chem Soc. Am Phys Soc. *Res:* Lattice effects and bonding in solids via electron and nuclear resonance studies. *Mailing Add:* 2100 N Military Rd Arlington VA 22207-3925

GARRETT, BENJAMIN CAYWOOD, b Richmond, Va, Jan 15, 49; m 71; c 2. OPERATIONS RESEARCH. *Educ:* Davidson Col, BS, 71; Emory Univ, MS & PhD(chem), 75. *Prof Exp:* Chemist, US Army Dugway Proving Ground, Utah, 76-78; asst prof chem, Davidson Col, 78-80; res scientist, Columbus Labs, Battelle Mem Inst, 80-91; CHIEF SCIENTIST, EAI CORP, 91- *Mem:* Am Chem Soc. *Res:* Analysis and detection of toxic substances; chemical and biological warfare. *Mailing Add:* EAI Corp 1308 Continental Dr Suite J Abingdon MD 21009

GARRETT, BOWMAN STAPLES, b Baton Rouge, La, July 8, 22; m 44; c 3. PHYSICAL CHEMISTRY. *Educ:* Univ La, BS, 40, MS, 42; Univ Ark, PhD, 54. *Prof Exp:* Shift supvr, Flintkote Co, 42; control chemist, Esso Standard Oil Co, 43-45; staff engr, Fuels Lab, Socony-Vacuum Co, 45-49; asst prof chem, Univ Ark, 50-52; sr phys chemist, Rohm & Haas, 53-57, lab head, 57-61, res supvr, 61-66, asst dir res, Res Div, 66-73, mkt mgr Polymers & Resins, Int Div, 73-76, Res & Develop Mgr, Latin Am, 76-78, consult, Latin Am, 78-82; RETIRED. *Mem:* Am Chem Soc. *Res:* Neutron diffraction; polymer physics; industrial polymer applications. *Mailing Add:* 385 Robin Ct Roswell GA 30076

GARRETT, CARLETON THEODORE, CARCINOGENESIS. *Educ:* Johns Hopkins Univ, MD, 66; Univ Wis-Madison, PhD(oncol), 77. *Prof Exp:* ASSOC PROF PATH, GEORGE WASHINGTON MED CTR, 77- *Mailing Add:* Dept Path George Washington Univ 2300 Eye St NW Washington DC 20037

GARRETT, CHARLES GEOFFREY BLYTHE, b Ashford, Kent, Eng, Sept 15, 25; nat US. PHYSICS. *Educ:* Cambridge Univ, BA, 46, MA, 50, PhD(physics), 50. *Prof Exp:* Asst low temperature physics, Royal Soc Mond Lab, Cambridge Univ, 46-50; instr physics, Harvard Univ, 50-52; mem tech staff, Bell Labs, 52-60, head, Optical Electronics Res Dept, 60-68 & Mat Sci Res Dept, 68-69, dir, Electron Device Process & Battery Lab, 69-73, dir, Integrated Circuit Lab, 73-79, dir, Common Subsystems Lab, 79-87; RETIRED. *Mem:* Fel Am Phys Soc; fel Inst Elec & Electronics Engrs. *Res:* Solid state; lasers; paramagnetics; surface physics; semiconductor device technology. *Mailing Add:* Seven Fithian Lane East Hampton NY 11837

GARRETT, DAVID L, (JR), b San Antonio, Tex, Sept 22, 44; m 75; c 4. PHYSICAL CHEMISTRY, POLYMER SCIENCE. *Educ:* Morehouse Col, BS, 67; Univ Iowa, MS, 69, PhD(phys chem), 72; Mich State Univ, MBA, 81. *Prof Exp:* Sr res chemist photog, Eastman Kodak Res Labs, 72-77; sr mem eng & res staff, dept polymer sci, Ford Motor Co, 77-80, prin eng & res scientist, 80-83; mgr, Eng Lab, 83-85, dir reliability & test, 85-87, DIR

TECHNOL TEST SERV, BOC GEN MOTORS, WARREN, MICH, 87- *Concurrent Pos:* Adj fac phys chem, Rochester Inst Technol, 74-77 & Wayne State Univ, 85; Merrill Europ Study travel fel, France, 65-66. *Mem:* Am Chem Soc; Soc Plastic Indust; Soc Automotive Engrs. *Res:* composites research; photo-oxidative degradation of paint enamel. *Mailing Add:* BOC Hq Technol Ctr 30009 Van Dyke Ave 2nd Floor Warren MI 48090

GARRETT, DONALD E(VERETT), b Long Beach, Calif, July 5, 23; m 46; c 4. CHEMICAL ENGINEERING, PHYSICAL CHEMISTRY. *Educ:* Univ Calif, BS, 47; Ohio State Univ, MS, 48, PhD(chem eng), 50. *Prof Exp:* Asst chem eng, Ohio State Univ, 47-50; res & develop engr & group leader, Dow Chem Co, Calif, 50-52 & Union Oil Co, 52-55; mgr res, Am Potash & Chem Co, 55-60; pres, Garrett Res & Develop Co, 60-75; exec vpres res & develop, Occidental Petrol Corp, 68-75; PRES, GARRETT ENERGY RES & ENG CO, 75-; PRES, SALINE PROCESSORS, 75-; PRES, LIQUID CHEM CORP, 75- *Concurrent Pos:* Pres, Assoc Chem Co, Calif, 60-64; mem gen tech adv comt, Off Coal Res, ERDA & Eng adv comt, Univ Calif, Ohio State Univ & US Dept Interior; adj prof, Univ Calif, Santa Barbara, 85- *Mem:* Am Chem Soc; Am Inst Chem Engrs. *Res:* Inorganic chemistry; research management; crystallization; evaporation; saline mineral processing. *Mailing Add:* PO Box 209 Ojai CA 93023-0209

GARRETT, EDGAR RAY, b Lordsburg, NMex, Aug 14, 21; m 42; c 4. SPEECH PATHOLOGY. *Educ:* Western NMex Univ, BS, 42; Univ NC, MA, 48; Univ Denver, PhD(speech path), 54. *Prof Exp:* From instr to assoc prof speech, 48-63, prof in charge speech & dir, Speech & Hearing Ctr, 63-65, asst dean Col Arts & Sci, 69-71, head dept speech, 65-85, DIR COMMUN DISORDERS, NMEX STATE UNIV, 85- *Concurrent Pos:* Res automated speech correction systs & prog lang for deaf, US Off Educ, 64-74. *Honors & Awards:* Sci Exhibit Award, Am Speech & Hearing Asn, 63 & 67. *Mem:* Fel AAAS; fel Am Speech & Hearing Asn; NY Acad Sci; Sigma Xi. *Res:* Audiology; psychology of learning; programmed instruction; systems analysis of therapy delivery. *Mailing Add:* Dept Spec Educ/Commun Dis NMex State Univ Box 3SPE Las Cruces NM 88003

GARRETT, EDWARD ROBERT, b New York, NY, Apr 9, 20; m 41; c 3. PHARMACEUTICAL CHEMISTRY, BIOPHARMACEUTICS. *Educ:* Mich State Univ, BS, 41, MS, 48; PhD(chem), 50. *Hon Degrees:* DSc, Mich State Univ, 74; Dr rer nat, Free Univ Berlin, WGer, 79; Dr, Acad Med, 89. *Prof Exp:* Asst foreman heavy chem prod, Gen Chem Co, 41; asst foreman acid prod & heavy chem, Keystone Ord Works, 42; chem process engr, Gen Tire & Rubber Co, 43-45; asst plant mgr sulfuric acid prod, Stauffer Chem Co, 45-46; asst, Mich State Univ, 46-49; sr res scientist, Upjohn Co, 50-61; grad res prof, 60-88, EMER GRAD RES PROF PHARM, UNIV FLA, 89- *Concurrent Pos:* Vis prof, Univ Wis, 59, Univ Calif, 63 & Univ Buenos Aires, 65; pres, Latin-Am Conf Biochem & Indust Pharm, 62; consult, Smith, Kline & French Co, 63-75; ed, Int J Clin Pharmacol. *Honors & Awards:* Upjohn Award, 59; Ebert Prize, 63; Res Achievement Award Phys Pharmaceut Chem, Am Pharmaceut Asn, 63, Res Achievement Award Pharmaceut Anal Chem, 70; J E Purkyne Medal, Czech Med Soc, 72; Indust Pharmaceut Technol Award, Acad Pharmaceut Sci, 76; Volwiler Award Res Excellence, 80; Stimulation Res Award, Acad Pharmaceut Sci, 81; Rho Chi lectr, 85; Res Achievement Award in Pharmacokinetics, Pharmacodynamics & Drug Distrib, 90. *Mem:* Fel AAAS; Am Chem Soc; NY Acad Sci; Am Pharmaceut Asn; fel Acad Pharmaceut Sci; Am Soc Clin Pharmacol & Ther; Am Col Clin Pharmacol; Am Asn Pharmaceut Sci. *Res:* Kinetics and mechanisms of reactions; prediction of stability of pharmaceuticals and antibiotics; pharmacokinetics of drug absorption, distribution and excretion in vivo, kinetic evalution of antibiotic activity. *Mailing Add:* 1826 NW 26th Way Gainesville FL 32605-3861

GARRETT, EPHRAIM SPENCER, III, b Binghamton, NY, July 20, 37; m 60; c 4. MICROBIOLOGY, FOOD TECHNOLOGY. *Educ:* Univ Southern Miss, BS, 62, MS, 64. *Prof Exp:* Res virologist poultry path, Charles Pfizer & Co, 63-65; res microbiologist food technol, US Bur Com Fisheries, 65-68; asst chief seafood inspection serv, US Dept Interior, 68-71; SPEC ASST, NAT MARINE FISHERIES SERV & LAB DIR, NAT SEAFOOD QUAL & INSPECTION LAB, US DEPT COM, 71- *Concurrent Pos:* US deleg, Food & Agr Orgn-WHO Codex Alimentarius Comn, Codex Comt Food Hyg, 69-; Food & Agr Orgn consult seafood harvesting & processing, 75-76; WHO-Food & Agr Orgn consult seafood microbiol, 77-79; chmn, Trop & Subtrop Technol Fisheries Conf of the Americas, 77. *Mem:* Am Soc Microbiol. *Res:* Presentations dealing with various product quality and safety issues relative to consumer protection in the consumption of fishery products. *Mailing Add:* Nat Seafood Inspect Lab US Dept Com PO Drawer 1207 Pascagoula MS 39568-1207

GARRETT, HENRY BERRY, b San Francisco, Calif, Feb 15, 48; m 76; c 1. SPACE PHYSICS. *Educ:* Rice Univ, BA, 70, MS, 73, PhD(space physics, astron), 74. *Prof Exp:* Res assoc space physics & astron, Rice Univ, 74; res physicist, Air Force Geophys Lab, 74-80; LEAD TECHNOLOGIST, RELIABILITY & QUAL ASSURANCE, JET PROPULSION LAB CALIF INST TECHNOL, 80- *Concurrent Pos:* Chmn, Int Asn Geomagnetism & Aeronomy Hist Comt, 78-86; assoc ed, J Spacecraft & Rockets, 87-89; consult, INTELSAT Corp, 89- *Mem:* Am Astron Soc; Am Phys Soc; Am Inst Aeronaut & Astronaut; Am Geophys Union. *Res:* Study of spacecraft changing; theoretical and experimental models of the upper atmosphere from 90 to 500 kilometers, as well as solar forecasting methods; interplanetary meteroid environment and earth space debris. *Mailing Add:* 2725 Brookhill Ave La Crescenta CA 91214

GARRETT, J MARSHALL, b Cleveland, Tenn, Nov 18, 32; m 54; c 2. GASTROENTEROLOGY. *Educ:* Univ Chattanooga, AB, 56; Univ Tenn, MD, 58; Univ Minn, MS, 64; Am Bd Internal Med, dipl, 67. *Prof Exp:* Intern med, John Gaston Hosp, Memphis, Tenn, 59; staff physician, United Mine Workers Hosp, Wise, Va, 60; resident internal med, Mayo Found, 60-63; chief gastroenterol, Vet Admin Hosp, 64-70, chief of staff, 69-70; DIR, GASTROENTEROL DEPT, ST VINCENT'S HOSP, 70- *Concurrent Pos:* Fel gastroenterol, Mayo Clin, 63-64; from instr to assoc prof med, Med Ctr, Univ Ala, Birmingham, 64-70, clin assoc prof, 70-; med dir, Montclair Ctr Digestive Dis, Baptist Med, Montclair, Birmingham. *Mem:* AAAS; AMA; Am Fedn Clin Res; Am Gastroenterol Asn; Sigma Xi. *Res:* Motility of the gastrointestinal tract. *Mailing Add:* 880 Montclair Rd No 577 Birmingham AL 35213-1980

GARRETT, JAMES M, b Magnolia, Ark, Aug 3, 41; m 66; c 3. ORGANIC CHEMISTRY. *Educ:* Arlington State Col, BS, 63; Univ Tex, PhD(org chem), 66. *Prof Exp:* Fel, Univ Fla, 66-67; from asst prof to assoc prof, 67-78, PROF CHEM, STEPHEN F AUSTIN STATE UNIV, 78- *Mem:* Am Chem Soc. *Res:* Synthetic organic photochemistry; chemistry of small ring systems; stable boron cations. *Mailing Add:* Carrizo Creek Box 3220 Nacogdoches TX 75961-9268

GARRETT, JAMES RICHARD, b Landrum, SC, May 4, 17; m 46; c 3. MATHEMATICS. *Educ:* Lenoir-Rhyne Col, AB, 40; Calif Inst Technol, BS, 41, MS, 52; Duke Univ, AM, 47, PhD(math), 50. *Hon Degrees:* DS, Lenoir-Rhyne Col, 72. *Prof Exp:* From asst prof to assoc prof math, Ga Inst Technol, 50-60; mgr math serv, Radio Corp Am Missile Test Proj, Patrick AFB, Fla, 60-67, mgr diag & test systs, Info Systs Div, RCA Corp, 67-69, mgr control systs, Comput Div, 69-71; mgr test systs, Univac Div, Sperry Rand Corp, 72; prof math & chmn dept, Lenoir-Rhyne Col, 72-80; CONSULT, 80- *Concurrent Pos:* Res assoc, Rich Comput Ctr, Ga Inst Technol, 58-60. *Mem:* Am Math Soc; Math Asn Am; Asn Comput Mach; Soc Indust & Appl Math. *Res:* Satellite and missile orbits and trajectories; mathematical problems. *Mailing Add:* 815 S Shannon Ave Indialantic FL 32903

GARRETT, JERRY DALE, b Springfield, Mo, Oct 1, 40; m 65; c 1. EXPERIMENTAL NUCLEAR PHYSICS. *Educ:* Univ Mo, Columbia, BS, 62; Univ Pa, MS, 67, PhD(physics), 70. *Prof Exp:* Res assoc physics, Tandem Accelerator Lab, Univ Pa, 70-71; res assoc nuclear physics, Los Alamos Sci Lab, 71-72; from asst physicist to assoc physicist, Brookhaven Nat Lab, 72-75; GROUP LEADER, OAK RIDGE NAT LAB, 88- *Concurrent Pos:* Assoc ed, Nuclear Physics J, 78-; guest scientist, Brookhaven Nat Lab, 78, Daresbury Lab, UK, 84, Joint Inst for Heavy Ion Res, Oak Ridge, 84, 85 & 87; Adv, Int Centre Theoret Physics, Trieste, 84-; lectr, Niels Bohr Inst, Univ Copenhagen, 74-; mem, NAm Steering Comt Radioactive Ion Beams, 90- *Mem:* AAAS; Am Phys Soc; NY Acad Sci; Europ Phys Soc; Danish Phys Soc. *Res:* Structure of rapidly-rotating nuclei. *Mailing Add:* Oak Ridge Nat Lab Bldg 6000 Mail Stop 6371 Oak Ridge TN 37831-6371

GARRETT, L(UTHER) W(EAVER), JR, b Corsicana, Tex, Apr 26, 25; m 47; c 2. CHEMICAL ENGINEERING. *Educ:* Univ Tex, BS, 47. *Prof Exp:* Chem engr, M W Kellogg Co, 47-50, process design engr, & asst to mgr synthol div, 50-55, actg mgr synthol div, 56-58, chief oper eng, 58-60, chief develop eng, 60-62, mgr iron & steel dept, 62-63; mgr, Proj Dept, Swindell-Dressler Pullman, Inc, 63-66, vpres equip opers, 66-69; eng mgr, Bechtel Corp, 69-74; sr vpres to pres, Fluor Utah Inc, 74-78; PRES, GARRETT ASSOC, INC, 78- *Mem:* Am Chem Soc; Am Inst Chem Engrs; Am Inst Mining Metall & Petrol Engrs; fel Am Inst Chemists. *Res:* Catalysis in the field of hydrocarbon synthesis; process metallurgy; processing synthetic hydrocarbon oils; ore reduction and metals processing. *Mailing Add:* 537 Virginia Ave San Mateo CA 94402

GARRETT, MICHAEL BENJAMIN, b San Jose, Calif, Mar 15, 41; m 70. PHYSICAL INORGANIC CHEMISTRY. *Educ:* Univ Calif, Berkeley, BA, 63; Univ Ill, Urbana, PhD(inorg chem), 70. *Prof Exp:* Instr chem, Drexel Univ, 69-70; ASST PROF CHEM, DEL VALLEY COL, 70- *Concurrent Pos:* Student assoc, Argonne Nat Lab, 66-68. *Mem:* Am Chem Soc; The Chem Soc. *Res:* Flash heating and kinetic spectroscopy; spectra of high multiplicity diatomic molecules. *Mailing Add:* Del Valley Col Doylestown PA 18901-2697

GARRETT, PAUL DANIEL, b Highland Park, Ill, Mar 15, 61; m 89. POLYMER MECHANICAL PROPERTIES, POLYMER INTERFACES. *Educ:* Univ Mich, BS, 83; Cornell Univ, MS, 85, PhD(mat sci), 88. *Prof Exp:* Sr res engr, 88-90, RES SPECIALIST, PHYS & ANALYTICAL SCI CTR, MONSANTO CHEM CO, MONSANTO CO, 90- *Mem:* Mat Res Soc; Am Chem Soc; Am Phys Soc; Am Soc Mat. *Res:* Deformation and fracture behavior of polymeric materials, including toughened plastics polymer-matrix composites, fibers and coatings; mechanical and thermodynamic behavior of polymeric thin films and interfaces. *Mailing Add:* Monsanto Chem Co 730 Worcester St Springfield MA 01151

GARRETT, PETER WAYNE, b Petoskey, Mich, Dec 20, 33; m 54; c 1. FOREST GENETICS. *Educ:* Mich State Univ, BS, 58; Univ Mich, MS, 62, PhD(forestry), 69. *Prof Exp:* Asst dist forester, State of Minn, 58-59; res forester, Union Camp Paper Corp, 59-61; asst prof forestry, Mich Technol Univ, 63-66; res forester genetics, 66-69, PROJ LEADER GENETICS, US FOREST SERV, 69- *Concurrent Pos:* Adj prof forest genetics, Univ NH, 70- *Mem:* Soc Am Foresters. *Res:* Supervising ecology research in Northeast with emphasis on northern hardwood species; regeneration and growth under different harvesting techniques; biodiversity of plant and animal species, migration rates under global change. *Mailing Add:* USDA Forest Serv Northeastern Forest Exp Sta Box 640 Durham NH 03824

GARRETT, REGINALD HOOKER, b Roanoke, Va, Sept 24, 39; m 89; c 3. BIOCHEMISTRY, MOLECULAR BIOLOGY. *Educ:* Johns Hopkins Univ, BS, 64, PhD(biol), 68. *Prof Exp:* Res asst biochem, McCollum-Pratt Inst, Johns Hopkins Univ, 56-64, fel, 68; from asst prof to assoc prof, 68-82, PROF BIOL, UNIV VA, 82- *Concurrent Pos:* NIH fel, 64-68; Fulbright fel, 75-76; vis fel, Univ Cambridge, 83. *Mem:* Am Soc Biol Chem; Am Soc Microbiol; Am Soc Plant Physiologists; Soc Gen Physiologists; AAAS. *Res:* Enzymology, genetics and regulation of nitrate assimilation; isolation and characterization of Neurospora crassa nit genes; regulation of metabolic potentiality and molecular biology of gene expression in lower eucaryotes. *Mailing Add:* Dept Biol Univ Va Charlottesville VA 22903

GARRETT, RICHARD E, b Chester, Pa, Sept 9, 33; m 55; c 2. MECHANICAL ENGINEERING. *Educ:* Univ Del, BSME, 56; Univ Fla, MSME, 63; Purdue Univ, PhD(mech eng), 67. *Prof Exp:* Develop engr, Hamilton Standard Div, United Aircraft Corp, 56-59; asst prof mech eng, Univ Fla, 59-67; from assoc prof to prof, Purdue Univ, 67-76; PROF MECH ENG & HEAD DEPT, UNIV CONN, STORRS, 76- *Concurrent Pos:* Eng consult, Midwest Appl Sci Corp, 66-68; consult & mem bd dir, TecTran, Inc, 68- *Mem:* Assoc mem Am Soc Mech Engrs; Soc Exp Stress Analysis. *Res:* Computer-aided design and computer graphics; optimization of mechanical devices; bio-medical engineering. *Mailing Add:* 78 Dunham Pond Rd Mansfield CT 06250

GARRETT, RICHARD EDWARD, b Roanoke, Va, Feb 17, 22; m 47. NUCLEAR PHYSICS. *Educ:* Roanoke Col, BS, 42; Ga Inst Technol, MS, 50; Univ Va, PhD(nuclear physics), 53. *Prof Exp:* Instr math & physics, Roanoke Col, 42-43 & 46-48; instr physics, Ga Inst Technol, 48-50 & Univ Va, 50-53; from asst prof to assoc prof, Hollins Col, 53-63; assoc prof physics, Univ Fla, 63-74, prof physics & phys sci, 74-89; RETIRED. *Concurrent Pos:* Consult, Radiation Lab, Univ Calif, 58-59; vis lectr, Va Polytech Inst & State Univ, 61; res assoc, Univ Va, 63. *Mem:* AAAS; Am Asn Physics Teachers. *Res:* Nature of nuclear emulsions, physical properties and their use in high energy particle physics; research in teaching development of new material; lecture demonstrations and laboratory experiments. *Mailing Add:* 4611 NW 32nd Pl Gainesville FL 32605

GARRETT, ROBERT AUSTIN, b Indianapolis, Ind, Jan 25, 19; m 46; c 4. UROLOGY. *Educ:* Miami Univ, AB, 40; Ind Univ, MD, 43. *Prof Exp:* From instr to prof urol, 48-87, chmn dept, 54-73, EMER PROF UROL, MED CTR, IND UNIV, INDIANAPOLIS, 87- *Concurrent Pos:* Mem staff, Vet Admin Hosp, Wishard Gen Hosp & Indianapolis Methodist Hosp. *Mem:* AMA; Am Urol Asn; Am Col Surgeons; Am Asn Genito-Urinary Surg; Soc Pediat Urol (pres, 69). *Res:* Pediatric urology. *Mailing Add:* Dept Urol Ind Univ Med Ctr Indianapolis IN 46273

GARRETT, ROBERT OGDEN, b Berkeley, Calif, Jan 11, 33; m 55; c 2. PHYSICS. *Educ:* Whitman Col, BA, 54; Cornell Univ, MS, 60; Univ Ore, PhD(physics), 64. *Prof Exp:* From asst prof to assoc prof, 64-76, PROF PHYSICS, BELOIT COL, 76-, CHMN DEPT & COORDR ENG PROG, 70- *Concurrent Pos:* Res assoc, Univ Ore, 69-70. *Mem:* Am Asn Physics Teachers. *Res:* Pressure broadening of spectral lines of atoms perturbed by foreign gases. *Mailing Add:* Dept Physics & Astron Beloit Col Beloit WI 53512

GARRETT, ROBERT ROTH, b Asheville, NC, Nov 29, 21; m 53; c 2. PHYSICAL CHEMISTRY. *Educ:* Duke Univ, BS, 44; Univ Louisville, MS, 51; Cornell Univ, PhD(chem), 57. *Prof Exp:* Asst, Univ Louisville, 50-51; asst, Cornell Univ, 52-56; res chemist, 57-80, res assoc, 80-85, CONSULT, E I DU PONT DE NEMOURS & CO INC, 85- *Res:* Phase transitions in high polymers; physical chemistry of elastomers; novel elastomeric adhesives, adhesion problems. *Mailing Add:* 1080 Wood Lane West Chester PA 19382

GARRETT, RUBY JOYCE BURRISS, b Greenville, SC, Apr 1, 46; m 69. PHYSIOLOGY, PHARMACOLOGY. *Educ:* Univ Tenn, BS, 68; Univ Ky, PhD(physiol, biophys), 71; Univ NC, Chapel Hill, JD, 81. *Prof Exp:* From instr to assoc prof pharmacodyn & toxicol, Univ Ky, 71-79; assoc, Vernon, Vernon, Wooten, Brown & Andrews, PA, 82-86; PARTNER, VERNON, VERNON, WOOTEN, BROWN, ANDREWS & GARRETT, PA, 87-; ADMIN LAW JUDGE, OCCUPATIONAL SAFETY & HEALTH, NC DEPT LABOR, RALEIGH, 84- *Concurrent Pos:* Vis investr, New York Blood Ctr, 73 & 74; grants, Sanders-Brown Res Ctr Aging, 76-78 & Tobacco & Health Inst, 76-79; staff mem, NC Law Rev, 80-81, note & comment ed, 81-82; consult/clerk, OSHA rev bd, NC Dept Labor, Raleigh, 81-83. *Mem:* Am Physiol Soc; Am Soc Cell Biol; Am Pharmaceut Asn; Am Soc Pharmacol & Exp Therapeut. *Res:* Physiological biochemistry; cell membrane structure and function; protein synthesis; author of numerous publications. *Mailing Add:* 407 James St Carrboro NC 27510

GARRETT, STEVEN LURIE, b Los Angeles, Calif, Apr 3, 49; m 74; c 1. FIBER-OPTIC SENSING, THERMOACOUSTICS ENGINES. *Educ:* Univ Calif, Los Angeles, BS, 70, MS, 72, PhD, 77. *Prof Exp:* Res assoc physics, Univ Calif, Los Angeles, 75-77, adj asst prof, 77-78; Hunt fel, Acoust Soc Am, MAPS, Univ Sussex, Eng, 78-79; res fel, Miller Inst, Univ Calif, Berkeley, 79-81; asst prof physics, 81-84, assoc prof, 84-88, PROF PHYSICS, NAVAL POSTGRAD SCH, MONTEREY, CALIF, 88- *Concurrent Pos:* Vis scientist, Los Alamos Nat Labs, 80-; consult, Sound Advise, 81-; Rosen prof solid state physics, Technion, Hiafa, Israel, 85. *Mem:* Fel Acoust Soc Am; Optical Soc Am; Am Phys Soc; Am Asn Physics Teachers; Soc Photo-Optical Instrument Engrs; Sigma Xi. *Res:* Acoustics and quantum fluids; transduction; fiber optics sensors; thermoacoustics. *Mailing Add:* Po Box 8716 Monterey CA 93943

GARRETT, THOMAS BOYD, b Pittsburgh, Pa, May 28, 41; m 67; c 1. THEORETICAL CHEMISTRY. *Educ:* Carnegie-Mellon Univ, BS, 63; Lehigh Univ, PhD(chem), 70. *Prof Exp:* Res chemist, Armstrong World Indust, Inc, 70-75, from res scientist to sr res scientist, 75-85, res assoc, 85-86, res unit mgr, 86-90, SR PRIN SCIENTIST, ARMSTRONG WORLD INDUST, INC, 90- *Mem:* Am Chem Soc; Sigma Xi; Am Asn Artificial Intel. *Res:* Decision support systems for research and development management. *Mailing Add:* 17 Buttonwood Dr Lititz PA 17543

GARRETT, WILLIAM NORBERT, b Cresson, Pa, June 8, 26; m 54; c 2. ANIMAL NUTRITION. *Educ:* Pa State Univ, BS, 50, MS, 51; Univ Calif, PhD(nutrit), 58. *Prof Exp:* Assoc animal husb, 53-56, from asst animal husbandman to assoc animal husbandman, 58-65, assoc prof, 65-69, PROF ANIMAL SCI, UNIV CALIF, DAVIS, 69- *Concurrent Pos:* Chmn, Dept Animal Sci, Univ Calif,. *Honors & Awards:* Am Feed Mfrs Asn Nutrit Res Award, Am Soc Animal Sci, 75, Morrison Award, 86. *Mem:* Am Soc Animal Sci (pres, 83-84); Am Inst Nutrit; Coun Agr Sci & Tech; Brit Soc. *Res:* Ruminant nutrition; energy metabolism. *Mailing Add:* Dept Animal Sci Univ Calif Davis CA 95616

GARRETT, WILLIAM RAY, b Warrior, Ala, Oct 17, 37; m 57; c 3. CHEMICAL PHYSICS. *Educ:* Univ Ala, BS, 60, MS, 62, PhD(physics), 63. *Prof Exp:* Res assoc physics, Res Inst, Univ Ala, 63-65, asst prof, 65-66; physicist, 66-74, CHIEF CHEM PHYSICS SECT, OAK RIDGE NAT LAB, 74- *Mem:* Am Phys Soc. *Res:* Theoretical atomic and molecular physics. *Mailing Add:* 101 Windham Rd Oak Ridge TN 37830

GARRETTSON, LORNE KEITH, b Pasadena, Calif, Mar 2, 34; m 63; c 3. CLINICAL PHARMACOLOGY, PEDIATRICS. *Educ:* Pomona Col, BA, 55; Johns Hopkins Univ, MD, 59. *Prof Exp:* Asst pediat, Emory Univ, 65-68; asst prof, State Univ NY Buffalo, 68-73; assoc prof pediat & pharm, Va Commonwealth Univ, 73-87; ASSOC PROF PEDIAT, EMORY UNIV, 88- *Concurrent Pos:* Vchmn, Accident & Poison Prev Comn, Am Acad Pediat, 74- & mem exec bd, Sect Clin Pharm & Therapeut, 76-79; mem exec bd, Am Asn Poison Control Ctrs, 76-79; consult, Glass Packaging Inst, Closure Comn, 77-79; prin investr, Gen Med Sci, NIH, 78-81. *Mem:* Am Soc Pharmacol & Exp Therapeut; Soc Pediat Res; Am Asn Poison Control Ctrs. *Res:* Age dependent elimination of drugs; pharmacokinetics of effects of drugs; pharmacokinetics of drug interactions. *Mailing Add:* Dept Pediat Emory Univ 69 Butler St SE Atlanta GA 30303

GARRICK, B(ERNELL) JOHN, b Eureka, Utah, Mar 5, 30; m 52; c 3. PHYSICS, ENGINEERING. *Educ:* Brigham Young Univ, BS, 52; Univ Calif, Los Angeles, MS, 62, PhD(eng), 68. *Prof Exp:* Physicist, Atomic Energy Div, Phillips Petrol Co, Idaho, 52-54; physicist & sr scientist, Reactor Hazards Eval Br, Atomic Energy Comn, Washington, DC, 55-57; vpres technol, Los Angeles, 57-76, GROUP PRES, NUCLEAR & SYST SCI GROUP, HOLMES & NARVER, INC, ANAHEIM, 76- *Concurrent Pos:* Tech adv & Atomic Energy Comn rep, Int Conf Peaceful Uses Atomic Energy, Switz, 58; mem reactor safety comt, Atomic Indust Forum, 60, steering comt pub understanding, 63 & ad hoc comt reactor regulation, 65; mem & chmn adv panel on reactor safety for Pakistan, Int Atomic Energy Agency, 62; consult, Adv Comt on Reactor Safeguards, 67. *Mem:* Am Nuclear Soc. *Res:* Systems analysis; nuclear engineering and applied physics. *Mailing Add:* 221 Crescent Bay Dr Laguna Beach CA 92651

GARRICK, ISADORE EDWARD, mathematics, physics; deceased, see previous edition for last biography

GARRICK, LAURA MORRIS, b Chicago, Ill, Sept 8, 45; m 70; c 1. MOLECULAR BIOLOGY. *Educ:* Marquette Univ, BS, 67; Univ Va, PhD(biol), 72. *Prof Exp:* Res asst instr med, 72-76, res instr med, 76-79, RES ASST PROF MED, 79-, STATE UNIV NY, BUFFALO,CLIN ASST PROF BIOCHEM, 86- *Concurrent Pos:* Richard E Whale fel, 72-73; fel, Maternal & Child Health Serv, State Univ NY, Buffalo, 74-75; NIH fel, State Univ NY, Buffalo, 75-77 & res assoc biochem, 78-86; Howard Hughes fel, Harvard Med Sch, 77-78; Prin investr, NSF, 87-90, NIH, 87-90. *Mem:* AAAS; Am Chem Soc; Genetics Soc Am; Am Soc Hematol. *Res:* Control of gene expression; structure and evolution of hemoglobins and their genes; animal models for genetic disease; iron metabolism. *Mailing Add:* Cary 25 State Univ NY Buffalo NY 14214

GARRICK, MICHAEL D, b Newport News, Va, July 25, 38; m 61, 70; c 1. BIOCHEMISTRY. *Educ:* Johns Hopkins Univ, BA, 59, PhD(biol), 63. *Prof Exp:* Asst prof genetics, McCoy Col, Johns Hopkins Univ, 63-64; asst prof biol, Univ Va, 64-70; res asst prof pediat, 70, res asst prof biochem, 70-72, from asst prof to assoc prof, 72-79, PROF BIOCHEM, STATE UNIV NY BUFFALO, 79-, RES ASSOC PROF PEDIAT, 76- *Concurrent Pos:* Fel med genetics, Hopkins Med Insts, 63-64. *Honors & Awards:* Special Award, Niagara Frontier Assoc Sickle Cell Dis, 75. *Mem:* AAAS; Genetics Soc; Am Soc Biochem & Molecular Biol; Am Soc Hemat; NY Acad Sci. *Res:* Biochemical and human genetics; gene action; protein biosynthesis; hemoglobinopathies; immunochemistry; molecular evolution. *Mailing Add:* Dept Biochem State Univ NY Buffalo NY 14214

GARRICK, RITA ANNE, CELL PHYSIOLOGY, MEMBRANE TRANSPORT. *Educ:* Dunbarton Col, BA, 63; Rutgers Univ, MS, 66, PhD(physiol), 71. *Prof Exp:* ASSOC PROF PHYSIOL, NJ MED SCH, UNIV MED & DENT, 76-; ASSOC PROF BIOL, COL LINCOLN CTR, FORDHAM UNIV, 84- *Concurrent Pos:* Adj prof, Rutgers Univ, 81 & 84. *Mem:* Am Physiol Soc; NY Acad Sci; Am Soc Zoologists; Sigma Xi; Asn Women in Sci. *Res:* Membrane transport. *Mailing Add:* 66B Lakeside Dr Millburn NJ 07041

GARRIGUS, UPSON STANLEY, b Willimantic, Conn, July 2, 17; m 42; c 2. ANIMAL SCIENCE. *Educ:* Univ Conn, BS, 40; Univ Ill, MS, 42, PhD(nutrit), 48. *Prof Exp:* Asst animal sci, Univ Ill, Urbana, 40-42 & 46-48, from instr to prof, 48-55, head ruminant div, 64-70, assoc head dept, 72-87, prof, 55-87, EMER PROF ANIMAL SCI & INT AGR, UNIV ILL, URBANA, 87- *Concurrent Pos:* Mem comt sheep nutrit, Nat Res Coun, 53- & comt use of non-protein nitrogen compounds as protein replacement for animals, 70-; Moorman res travel award, 70; consult agr, Indonesian Univ & consult animal prod, Thailand, 80; consult higher ed, Indonesian Univ, 89. *Mem:* Fel AAAS; fel Am Soc Animal Sci; Am Inst Nutrit; Am Inst Biol Sci; Sigma Xi. *Res:* Use of sulfur, antibiotics, arsenicals and non-protein nitrogen in lamb feeding; pelleting of ruminant rations; commercial ruminant feeding; farm flock feeding; breeding and management of ruminants; world animal agriculture. *Mailing Add:* 326 Mumford Univ Ill 1301 W Gregory Dr Urbana IL 61801

GARRINGTON, GEORGE EVERETT, dentistry, oral pathology, for more information see previous edition

GARRIOTT, JAMES CLARK, b Seymour, Ind, Sept 27, 38. TOXICOLOGY, PHARMACOLOGY. *Educ:* Univ Louisville, BA, 60; Ind Univ, PhD(toxicol), 67. *Prof Exp:* NIH fel, Inst Marine Sci, Univ Miami, 67-68; toxicologist, Conn State Dept Health, Hartford, 68-70; assoc toxicologist, Southwestern Inst Forensic Sci, 70-73, chief toxicologist, 73-81; from asst prof to assoc prof pharmacol & path, Dallas, 71-82, ASSOC PROF, PATH DEPT, HEALTH SCI CTR, UNIV TEX, SAN ANTONIO, 82-; CHIEF TOXICOLOGIST, BEXAR COUNTY MED EXAMR'S OFF & REGIONAL CRIME LAB, 82- *Concurrent Pos:* Consult, Vet Admin Hosp, Dallas, 73-80; affil mem, med-dent staff, Bexar County, Hosp Dist, San Antonio, Tex, 82- *Mem:* AAAS; fel Am Acad Forensic Sci; fel Am Acad Clin Toxicol; Am Chem Soc; Soc Toxicol. *Res:* Clinical and forensic toxicology; detection and determination of drugs and toxic substances and their metabolites in body specimens; clinical correlation of drug concentrations with their effects. *Mailing Add:* Bexar County Med Examr's Off 600 N Leona St San Antonio TX 78207

GARRIOTT, MICHAEL LEE, b Beech Grove, Ind, Jan 9, 51; m 72; c 1. CYTOGENETICS, TOXICOLOGY. *Educ:* Purdue Univ, BS, 73, MS, 76, PhD(mammalian & cytogenetics), 79. *Prof Exp:* Res asst mammalian & cytogenetics, Dept Animal Sci, Purdue Univ, 74-79; fel genetic toxicol, Argonne Nat Lab, 79-82; res investr, G D Searle & Co, 82-87; SR TOXICOLOGIST, ELI LILLY & CO, 87- *Mem:* Am Genetic Asn; Sigma Xi; Environ Mutagen Soc; Genetic Toxicol Asn. *Res:* Investigations of the effects of drugs, hormones, chemicals, physical agents and environmental stress on mammalian chromosomes; chromosomal aberrations as they adversely affect reproduction; mechanisms; improvement of genetic toxicology tests. *Mailing Add:* 9235 Indian Creek Rd S Indianapolis IN 46259

GARRIS, DAVID ROY, reproductive biology, developmental physiology, for more information see previous edition

GARRISON, ALLEN K, b Lake Wales, Fla, Oct 24, 31; m 57; c 2. PHYSICS. *Educ:* Davidson Col, BS, 53; Duke Univ, PhD(physics), 58. *Prof Exp:* From asst prof to assoc prof, 58-71, prof physics, Emory Univ, 71- 88; SR RES SCIENTIST, GA TECH RES INST, GA INST TECH. *Concurrent Pos:* Nat Res Coun fel, Naval Res Lab, 64-65. *Mem:* Am Phys Soc. *Res:* Microwave spectroscopy; solid state physics; electro-optics. *Mailing Add:* Ga Tech Res Inst Ga Tech EMC/EOD Atlanta GA 30332

GARRISON, ARTHUR WAYNE, b Greenville, SC, Sept 9, 34; m 56; c 5. ENVIRONMENTAL CHEMISTRY, ANALYTICAL CHEMISTRY. *Educ:* The Citadel, BS, 56; Clemson Univ, MS, 58; Emory Univ, PhD(org chem), 66. *Prof Exp:* Chemist, Dept Agr Chem, Clemson Univ, 61-62; anal chemist, Pesticide Pollution Lab, Div Water Supply & Pollution Control, USPHS, 62-65; res chemist, Southeast Water Lab, 65-73, supvry res chemist, Environ Res Lab, 73-85, CHIEF, CHEM BR, US ENVIRON PROTECTION AGENCY, 85- *Concurrent Pos:* Consult, WHO; mem adj fac, Clemson Univ, 83- *Honors & Awards:* Silver Medal, US Environ Protection Agency. *Mem:* Am Chem Soc; Sigma Xi; Int Asn Environ Analytical Chem. *Res:* Development of methods for analysis of organic pollutants in water; transformation and transport processes of pollutants in water. *Mailing Add:* Environ Res Lab US Environ Protection Agency Athens GA 30613

GARRISON, BARBARA JANE, b Big Rapids, Mich, Mar 7, 49; m 78. SURFACE CHEMISTRY. *Educ:* Ariz State Univ, BS, 71; Univ Calif, Berkeley, PhD(chem), 75. *Prof Exp:* Res fel chem, Purdue Univ, 75-77, vis asst prof, 78-79; lectr, Univ Calif, Berkeley, 77-78; from asst prof to assoc prof, 79-86, PROF CHEM, PA STATE UNIV, 86-, HEAD, DEPT CHEM, 89- *Concurrent Pos:* Alfred P Sloan Found res fel, 80; vis assoc chem, Calif Inst Technol, 85-86. *Mem:* Am Chem Soc; Am Phys Soc; Am Vacuum Soc. *Res:* Interaction of gases with solid surfaces; keV ion bombardment of solids; etching of semiconductors; molecular beam epitaxial growth of semiconductors and ionization phenomena near metal surfaces. *Mailing Add:* 152 Davey Lab Dept Chem Pa State Univ University Park PA 16802

GARRISON, BETTY BERNHARDT, b Danbury, Ohio, July 1, 32; m 68; c 1. MULTIPLICATIVE NUMBER THEORY, DISTRIBUTION OF PRIMES. *Educ:* Bowling Green State Univ, BSEd & BA, 54; Ohio State Univ, MA, 56; Ore State Univ, PhD(math), 62. *Prof Exp:* Instr math, Ohio Univ, 56-57; instr, 57-59, asst prof, 62-69, PROF MATH, SAN DIEGO STATE UNIV, 69- *Mem:* Am Math Soc; Math Asn Am. *Res:* Polynomials with large numbers of prime values; distributions of primes. *Mailing Add:* Dept Math San Diego State Univ San Diego CA 92182-0314

GARRISON, HAZEL JEANNE, b Washington, DC, Oct 8, 28; m 57; c 3. BOTANY. *Educ:* Howard Univ, BS, 50; Univ Mich, MS, 51; Pa State Univ, PhD(bot), 57. *Prof Exp:* Instr biol, Southern Univ, 51-53 & Grambling Col, 53-54; asst prof bot, Howard Univ, 57-58; res assoc, 65-66, assoc prof, 67-72, PROF BIOL, HAMPTON UNIV, 72-, DEAN, GRAD COL, 77-, ASST VPRES RES, 83- *Concurrent Pos:* Mem, Nat Coun Univ Res Admin. *Mem:* Bot Soc Am; Nat Inst Sci. *Res:* Floral morphology of Fagus grandifolia; studies on Zostera marina, especially growth and development. *Mailing Add:* The Graduate Col Hampton Univ Hampton VA 23668

GARRISON, JAMES C, b Detroit, Mich, Mar 29, 43; m 85. PHARMACY, BIOPHYSICS. *Educ:* Union Col, BS, 65; Univ Rochester, PhD(biophys), 71. *Prof Exp:* Postdoctoral fel, 71-75, from asst prof to prof, 75-91, PROF & ACTG CHAIR PHARMACOL, MED SCH, UNIV VA, 90- *Mem:* Sigma Xi; Am Soc Biochem & Molecular Biol; Am Soc Pharmacol & Exp Therapeut. *Res:* Mechanism of action of hormones, growth factors and drugs at the level of the plasma membrane. *Mailing Add:* Dept Pharmacol Univ Va Box 448 Charlottesville VA 22908

GARRISON, JOHN CARSON, b Lufkin, Tex, May 15, 35; m 57; c 2. QUANTUM OPTICS, STATISTICAL MECHANICS. *Educ:* Purdue Univ, BS, 57, MS, 59, PhD(physics), 61. *Prof Exp:* SR PHYSICIST, LAWRENCE LIVERMORE LAB, 61- *Concurrent Pos:* Lectr, Dept Appl Sci, Univ Calif, Davis, 69-; vis scientist, Max Planck Inst Quantum Optics, Garching, WGermany, 84-85. *Mem:* Am Phys Soc; Optical Soc Am; AAAS. *Res:* Quantum optics; topological phases in quantum mechanics and nonlinear optics; theory of amplified spontaneous emission; chaos in laser-active materials. *Mailing Add:* Lawrence Livermore Lab Livermore CA 94550

GARRISON, JOHN DRESSER, b Salt Lake City, Utah, Aug 9, 22; m 50, 68; c 4. PHYSICS. *Educ:* Univ Calif, Los Angeles, BA, 47, MA, 48; Univ Calif, PhD(physics), 54. *Prof Exp:* Instr physics, Yale Univ, 53-56; from asst prof to prof, 56-62, chmn dept, 66-69, PROF PHYSICS, SAN DIEGO STATE UNIV, 62- *Concurrent Pos:* Assoc physicist, Brookhaven Nat Lab, 62-63; consult, Atomic Co. *Mem:* Am Phys Soc. *Res:* Proton-proton scattering; solar energy; nuclear physics; neutron cross sections. *Mailing Add:* Dept Physics San Diego State Univ San Diego CA 92182

GARRISON, NORMAN EUGENE, b Asheville, NC, Mar 25, 43; m 65; c 2. BIOCHEMISTRY, DEVELOPMENTAL BIOLOGY. *Educ:* Mars Hill Col, BS, 65; Wake Forest Univ, MA, 67; Univ Mass, Amherst, PhD(biochem), 73. *Prof Exp:* Instr biol, Mars Hill Col, 66-68; asst prof, Madison Col, 68-70; PROF BIOL, JAMES MADISON UNIV, 73- *Concurrent Pos:* Southern Regional Educ Bd travel grant, 76-77 & 77-78; proj dir, Title VI-A sci equip prog grant, 78-79 & NSF instrnl sci equip prog grant, 78-80, 81-84. *Mem:* Sigma Xi. *Res:* Translational control of early embryonic development. *Mailing Add:* Dept Biol James Madison Univ Harrisonburg VA 22807

GARRISON, ROBERT EDWARD, b Dallas, Tex, Oct 25, 32; m 63; c 1. GEOLOGY. *Educ:* Stanford Univ, BS, 55, MS, 58; Princeton Univ, PhD(geol), 64. *Prof Exp:* Geologist petrol explor, Sunray D-X Oil Co, 59-61; res assoc geol, Princeton Univ, 64-65; asst prof, Univ Calif, Santa Barbara, 65-66; vis asst prof, Univ BC, 66-68; assoc prof, 68-73, chmn, Earth Sci Bd, 70-80, PROF EARTH SCI, UNIV CALIF, SANTA CRUZ, 73- *Concurrent Pos:* Guggenheim fel, 72-73; geologist, US Geol Surv, 73-83. *Mem:* AAAS; Geol Soc Am; Am Asn Petrol Geologists; Soc Econ Paleontologists & Mineralogists. *Res:* Sedimentology and stratigraphy; petrology of carbonate rocks and eugeosynclinal sedimentary rocks; electron microscopy of fine-grained sedimentary rocks. *Mailing Add:* Earth Sci Bd Univ Calif Santa Cruz CA 95064

GARRISON, ROBERT FREDERICK, b Aurora, Ill, May 9, 36; div; c 3. ASTRONOMY, ASTROPHYSICS. *Educ:* Earlham Col, BA, 60; Univ Chicago, PhD(astron & astrophys), 66. *Prof Exp:* Res asst astron, Yerkes Observ, 60-61; res assoc, Mt Wilson-Palomar Observ, 66-68; from asst prof to assoc prof, 68-78, PROF ASTRON, DAVID DUNLAP OBSERV, UNIV TORONTO, 78- *Concurrent Pos:* Pres, Stellar Classification Comn, Int Astron Union, 85-88; dir, Univ Toronto Southern Observ, Chile, 71-; Nat lectr, Sigma Xi, 88-90 & Shapley nat lectr, 88- *Honors & Awards:* Bronowski Mem lectr, 88. *Mem:* Am Astron Soc; Am Asn Variable Star Observers; Royal Astron Soc Can; Int Astron Union; Can Astron Soc. *Res:* Direct photography of galaxies and H II regions; stellar spectral classification; clusters and associations; Mira variables; galactic structure; stellar spectroscopy. *Mailing Add:* David Dunlap Observ Univ Toronto Richmond Hill ON L4C 4Y6 Can

GARRISON, ROBERT GENE, b Pittsburg, Kans, Aug 30, 25; m 58; c 2. MEDICAL MICROBIOLOGY. *Educ:* Kans State Col Pittsburg, BA, 49; Kans State Univ, MS, 51, PhD(bact), 54. *Prof Exp:* Asst antibiotics, Sch Med, La State Univ, 51-52; bacteriologist, Vet Admin Hosp, Grand Island, Nebr, 54-57, SCIENTIST, VET ADMIN HOSP, KANSAS CITY, 57- *Concurrent Pos:* Asst prof, Sch Med, Univ Kans, 61-67, assoc prof, 67-; clin asst prof, Sch Dent, Univ Mo-Kansas City, 70- *Mem:* Am Soc Microbiologists; Am Pub Health Asn; NY Acad Sci; fel Am Acad Microbiol. *Res:* Physiology of pathogenic fungi; fungal ultrastructure; electron cytochemistry. *Mailing Add:* Vet Admin Hosp 4801 Linwood Blvd Kansas City MO 64128

GARRISON, ROBERT J, b Washington, DC, Apr 5, 44. BIOMETRY RESEARCH. *Educ:* Univ Minn, BA, 66, MS, 68. *Prof Exp:* Statistician, Epidemiol Br, Epidemiol & Biomet Prog, Div Heart & Vascular Dis, 68-79, chief, Biomet Res Br, 79-84, CHIEF, FIELD STUDIES & BIOMET BR, EPIDEMIOL & BIOMET PROG, DIV EPIDEMIOL & CLIN APPLN, NAT HEART, LUNG & BLOOD INST, NIH, 84- *Concurrent Pos:* Fel, Coun Epidemiol, Am Heart Asn. *Res:* Relationship between cigarette smoking and HDL cholesterol levels; onset of obesity as the cause of atherogenic lipoprotein profile in young adults; enigmatic relationship between relative weight and mortality; non-invasive cardiological study protocol. *Mailing Add:* NIH Fed Bldg Rm 3A08 Nat Heart Lung Blood Inst 7550 Wisconsin Ave Bethesda MD 20892

GARRISON, WARREN MANFORD, b Seattle, Wash, June 6, 15; m 42; c 2. MOLECULAR RADIOLOGY. *Educ:* Univ Calif, Berkeley, BS, 37, MS, 39; NY Univ, PhD(photochem), 42. *Prof Exp:* Res assoc, Metall Lab, Univ Chicago, 42-43, assoc sect chief, Radiation Chem Sect, 43-44; res chemist, Hanford Eng Works, Gen Elec Co, Wash, 45-46; asst prof chem, Univ Wyo, 46-48; asst dir, Crocker Lab, 58-62, prin investr, Lawrence Berkeley Lab, 48-77, CONSULT, LAWRENCE BERKELEY LAB, UNIV CALIF, BERKELEY, 77- *Concurrent Pos:* Consult, E I du Pont de Nemours & Co, 58, Aerojet-Gen Nucleonics Div, Gen Tire & Rubber Co, 62-64, Battelle Sci Proj, US Army, 77-78 & Am Hosp Supply, 80; chmn, Gordon Res Conf Radiation Chem, 59; State Dept rep, Int Cong Nuclear Energy, Italy, 59. *Mem:* AAAS; Am Chem Soc; Radiation Res Soc; Am Inst Chem. *Res:* Reaction mechanism; radiation chemistry. *Mailing Add:* 1660 Ridgewood Rd Alamo CA 94507

GARRISON, WILLIAM EMMETT, JR, b Media, Pa, Nov 29, 33; m 57; c 6. ORGANIC CHEMISTRY, POLYMER CHEMISTRY. *Educ:* Juniata Col, BS, 55; Univ Ill, PhD(org chem), 59. *Prof Exp:* Res chemist, 58-66, sr res chemist, 66-72, res assoc, 72-85, res fel, 85-89, SR RES FEL, POLYMER PROD DEPT, EXP STA, E I DU PONT DE NEMOURS & CO, INC, 89- *Mem:* Am Chem Soc. *Res:* Polymers; product and process development. *Mailing Add:* Polymers Exp Sta E I du Pont de Nemours & Co Wilmington DE 19898

GARRITY, MICHAEL K, b Austin, Minn, Sept 1, 42; m 65, 85. BIOPHYSICS. *Educ:* St John's Univ, Minn, BS, 64; Ariz State Univ, MS, 65, PhD(math physics), 68. *Prof Exp:* From asst prof to assoc prof, 67-78, PROF PHYSICS, ST CLOUD STATE UNIV, 78- *Concurrent Pos:* Res fel, Am Soc Eng Educ, Johnson Space Ctr, NASA, 80 & 81; consult med physicist & radiol physics, 86. *Mem:* Am Asn Physics Teachers; Am Asn Med Physicist; Am Soc Therapeut Radiol & Oncol; Am Col Radiol. *Res:* Biomedical instrumentation; formulation of mathematical models; sensory information processing; physics of radiology. *Mailing Add:* Dept Physics St Cloud State Univ St Cloud MN 56301

GARRITY, THOMAS F, b Philadelphia, Pa, Nov 8, 43; m 69; c 3. MEDICAL BEHAVIORAL SCIENCE, HEALTH PSYCHOLOGY. *Educ:* Col Holy Cross, BS, 65; Duke Univ, MA, 67 & PhD(sociol), 71. *Prof Exp:* From instr to assoc prof, 70-81, PROF MED BEHAV SCI, UNIV KY, COL MED, 81- *Concurrent Pos:* Resident scholar, Nat Heart, Lung & Blood Inst, 76; grant reviewer, Behav Med Study Sect, NIH, 78-79, Vet Admin Merit Review Bd, 81-86; consult, Nat High Blood Pressure Educ Prog, 79-88, Nat Heart Lung & Blood Inst, 74-; reviewer, J Health & Social Behav, 75-, Social Sci & Med, 78-, J Am Dental Asn, 82- *Mem:* fel Soc Behav Med; fel Acad Behav Med Res; Am Psychosomatic Soc; Asn Behav Sci & Med Educ; Inst Soc, Ethics & Life Sci; Am Psychol Asn. *Res:* Studies of psychosocial factors that influence adherence of patients to preventive and therapeutic regimens; factors that influence translation of life stress into disease; factors that influence behav adjustment after heart attack. *Mailing Add:* Dept Behav Sci Univ Ky Col Med 800 Rose St Lexington KY 40536-0086

GARRO, ANTHONY JOSEPH, b New York, NY, Jan 13, 42; m 63; c 2. MICROBIOAL GENETICS. *Educ:* Manhattan Col, BS, 63; Columbia Univ, PhD(microbiol), 68. *Prof Exp:* Assoc, Mt Sinai Sch Med & City Col NY, 70-71, from asst prof to prof microbiol, 71-91; ASSOC DEAN ACAD AFFAIRS, NJ MED SCH, UNIV MD-NJ, 91- *Concurrent Pos:* NIH res fel microbial genetics, Albert Einstein Col Med, 68-70; res grants, Am Cancer Soc, NIH & March of Dimes. *Mem:* AAAS; Am Soc Microbiol; Am Asn Cancer Res; Environ Mutagen Soc; Am Asn Biol Chem; Sigma Xi. *Res:* Mechanisms of chemical carcinogenesis; fetal alcohol syndrome. *Mailing Add:* 11 Esmond Pl Tenafly NJ 07670

GARROD, CLAUDE, b New York, NY, Sept 25, 32; m 55; c 3. THEORETICAL PHYSICS. *Educ:* NY Univ, AB, 57, PhD(physics), 63. *Prof Exp:* Instr physics, Manhattan Col, 61-62; res scientist, Courant Inst Math Sci, NY Univ, 62-64; from asst prof to assoc prof, 62-76, PROF PHYSICS, UNIV CALIF, DAVIS, 76- *Mem:* Am Phys Soc. *Res:* Quantum theory of many-particle systems. *Mailing Add:* Dept Physics Univ Calif Davis CA 95616

GARROU, PHILIP ERNEST, b New York, NY, Apr 26, 49; m 69; c 2. CATALYSIS, ORGANOMETALLIC CHEMISTRY. *Educ:* NC State Univ, BS, 70; Ind Univ, PhD(inorg chem), 74. *Prof Exp:* TECHNOL DEVELOP MGR, DOW CHEM CO, 75- *Mem:* Am Chem Soc; Mat Res Soc; Am Ceramics Soc. *Res:* Currently involved in assessing internal and external technology and making recommendations for business opportunities. *Mailing Add:* Dow Chem Co 6100 Fairview Rd No 800 Charlotte NC 28210

GARROW, ROBERT JOSEPH, b Buffalo, NY, Dec 24, 29; m 64; c 3. MATHEMATICS, PHYSICS. *Educ:* Ohio State Univ, BS, 61; Xavier Univ, MEd, 69. *Prof Exp:* Engr, NAm Aviation, Inc, 57-60; instr math, Ohio State Univ, 60-61; chmn dept, 61-77, PROF MATH, FRANKLIN UNIV, 61- *Concurrent Pos:* Pres, Montessori Acad, Inc, 72-73, Montessori Child Develop Ctr, Inc, 73- & Franklin Educ Serv, Inc. *Mem:* Am Soc Eng Educ; Am Math Soc. *Res:* Evaluation of the reliability and validity of an electronics aptitude test; use of the computer as an aid in teaching mathematics and science; design and development of a Montessori school complex; open space school concept consisting of four classrooms and a learning center. *Mailing Add:* Dept Math Franklin Univ 201 S Grant Ave Columbus OH 43215

GARROWAY, ALLEN N, b Washington, DC, Oct 10, 43. PHYSICS. *Educ:* Rensselaer Polytech Inst, BS, 65; Cornell Univ, PhD(exp physics), 72. *Prof Exp:* Sci Res Coun fel nuclear magnetic resonance, Dept Physics, Univ Nottingham, 72-74; Nat Res Coun fel, 74-76, RES PHYSICIST, CODE 6120 CHEM DIV, NAVAL RES LAB, 76- *Mem:* Sigma Xi; Am Phys Soc; Fedn Am Scientists; AAAS; Am Chem Soc; Soc Magnetic Resonance Med. *Res:* Nuclear magnetic resonance in the solid state; polymers; NMR imaging. *Mailing Add:* 13404 Kris-Ran Ct Ft Washington MD 20744

GARRUTO, RALPH MICHAEL, b Binghamton, NY, Nov 20, 43; m 69; c 3. NEUROBIOLOGY, BIOMEDICAL ANTHROPOLOGY. *Educ:* Pa State Univ, BS, 66, MA, 69, PhD(human biol-anthrop), 73; Am Col Epidemiol, Cert, 83. *Prof Exp:* Fel, 72-73, staff fel, 73-75, sr staff fel neurobiol, 75-78, SR RES BIOLOGIST, NAT INST NEUROL & COMMUN DIS & STROKE, NIH, 78- *Concurrent Pos:* Adj prof med genetics, Univ SAla, 82-; pres, Am Dermatologic Asn, 87-88; exec secy, Int Union Anthropol & Ethnol Sci, Comn Aging, 85-; adj sr scientist, Penn State Univ, 85-; bd trustees, Nat Mus Health & Med Found, 90-92; sci adv comt, Int ALS MND Res Found, 90-; res comt neuroepidemiol, World Fedn Neurol. *Mem:* Soc Neurosci; Am Asn Phys Anthropologists; fel Human Biol Coun; Soc Epidemiol Res; NY Acad Sci; fel Am Col Epidemiol. *Res:* Epidemiology and molecular neurobiology of central nervous system disorders; biology of aging; experimental modelling; retrovirology; physiological and genetic adaptations of human populations to environmental stress. *Mailing Add:* Nat Inst Neurol Dis & Stroke NIH Bldg 36 Rm 58-21 Bethesda MD 20892

GARRY, FREDERICK W, b New Haven, Conn, July 12, 21. MECHANICAL ENGINEERING. *Educ:* Rose-Hulman Inst, BS, 51. *Hon Degrees:* DEng, Rose-Hulman Inst, 68; DSc, Clarkson Univ, 91. *Prof Exp:* Pres, Rohr Indust, Inc, 74-75, chmn & chief exec officer, 76-80; mgr eng progs, Gen Elec Co, 51-67, vpres, 68-73, vpres tech plans, Aircraft Eng Group, 73-74, vpres corp eng & mfg, 80-89; CONSULT, 89- *Concurrent Pos:* Mem bd, Elec Mutual Ins Co & Star Technologies, Inc; mem bd managers, Rose-Hulman Inst Technol; vchmn bd, Clarkson Univ; exec vpres, Nat Academies Corp, Nat Academies Eng & Sci; dir, Arnold & Mabel Beckman Ctr, Irvine, Calif. *Honors & Awards:* William Elgin Wickenden Award, Am Soc Eng Educ, 87; Nat Medal of Technol, 90. *Mem:* Nat Acad Eng; Air Force Asn; Aircraft Owners & Pilots Asn; Am Helicopter Soc; Asn US Army; AAAS; Am Soc Eng Educ; Soc Automotive Engrs. *Mailing Add:* 25 Lake Helix Dr La Mesa CA 91941-4434

GARRY, PHILIP J, b Bancroft, Iowa, Jan 19, 33. PATHOLOGY, NUTRITION. *Educ:* Univ Iowa, BA, 61, MS, 65; Ohio State Univ, PhD(biochem & nutrit), 74. *Prof Exp:* Res asst, Dept Pediat, Univ Iowa, 60-62, lab analyst, Dept Biochem, 62-63, asst, 64-65, res assoc, Dept Pediat, 65-66; res assoc, Dept Pediat, Col Med, Ohio State Univ, 66-71, instr, 71-74; from asst prof to assoc prof path, 74-87, DIR, CLIN NUTRIT LAB, UNIV NMEX, ALBUQUERQUE, 74-, PROF, DEPT PATH, SCH MED, 87- *Concurrent Pos:* Prog chmn, Arnold O Beckman Conf, Am Asn Clin Chem, 80; mem, Sci Rev Comt, Clin Nutrit & Early Develop Br, Nat Inst Child Health & Human Serv, 81, Ad Hoc Teaching Nursing Home Rev Comt, Nat Inst Aging, 82, Vitamin & Mineral Workshop, Nat Acad Sci, Nat Bur Standards Proj Session, NIH & Expert Panel Nutrit, Int Fedn Clin Chem, 84, US Dept Com Nat Bur Standards Adv Coun NIH Cancer Chemoprev Prog, 85 & Nutrit Study Sect, NIH, 90-94; corresp ed, Age & Nutrit J, 90. *Mem:* Am Asn Clin Chem; Am Inst Nutrit; Am Soc Clin Nutrit; assoc mem Am Dietetic Asn; fel Am Col Nutrit; Geront Soc Am. *Res:* Author of numerous technical publications. *Mailing Add:* Dept Path Sch Med Univ NMex 2701 Frontier Pl NE Albuquerque NM 87131

GARSIA, ADRIANO MARIO, b Tunis, Tunisia, Aug 20, 28; nat US; m 55; c 1. MATHEMATICS. *Educ:* Stanford Univ, PhD(math), 57. *Prof Exp:* C L E Moore instr math, Mass Inst Technol, 57-59; asst prof, Univ Minn, 59-61; from assoc prof to prof, Calif Inst Technol, 61-66; PROF MATH, UNIV CALIF, SAN DIEGO, 66- *Mem:* Am Math Soc; Math Asn Am. *Res:* Classical analysis; probability theory; classical differential geometry. *Mailing Add:* Dept Math Univ California San Diego Box 109 La Jolla CA 92093

GARSIDE, BRIAN K, b Ashton-Under-Lyne, Eng, Jan 29, 40; m 63, 83; c 7. OPTO-ELECTRONIC DEVICES & SYSTEMS, FIBER OPTICS. *Educ:* Oxford Univ Eng, BA, 62, MA & DPhil, 66. *Prof Exp:* Asst lectr physics, Oxford Univ, 65-66; mem tech staff coherent wave physics, Bell Labs NJ, 66-68; prof lasers & electro optics, McMaster Univ, Hamilton Ont, Can, 68-88; PRES, OPTO-ELECTRONICS INC, 78- *Concurrent Pos:* Vis prof, Clarendon Lab, Oxford, UK, 75; adv bd mem, Ryerson Polytechnic Univ Toronto, Inst Microstruct Sci, 88-90 & Nat Res Coun, Ottawa, Can, 90-; mem, Grant Selection Comt, Oper Strategic Panel, Nat Sci & Eng Res Coun, Ottawa, Can, 90- *Mem:* Optical Soc Am; Inst Elec & Electronics Engrs; Int Soc Optical Eng; Am Phys Soc; Can Asn Physicists. *Res:* Semiconductor diode laser sources and detectors; fiber optical testing and systems; fiber optical sensor systems. *Mailing Add:* Opto-Electronics Inc 2538 Speers Rd Oakville ON L6L 5K9 Can

GARSIDE, EDWARD THOMAS, b London, Ont, June 14, 30; m 75. ZOOLOGY. *Educ:* Queen's Univ, Ont, BA, 54; Univ Toronto, MA, 57, PhD(zool), 60. *Prof Exp:* Supvr fish culture, Ont Dept Lands & Forests, 61-62; asst prof zool, Univ Man, 62-65; asst prof biol, 65-67, assoc prof, 67-80, PROF BIOL, FAC ARTS & SCI, DALHOUSIE UNIV, 80- *Mem:* Am Fisheries Soc; Am Inst Fishery Res Biol; Am Soc Ichthyol & Herpet; Int Asn Theoret & Appl Limnol. *Res:* Embryogenesis and ecology of early stages of fish; limnological productivity; structural and physiologic responses of fish to pollutants. *Mailing Add:* Dept Biol Dalhousie Univ Halifax NS B3H 4H6 Can

GARSIDE, LARRY JOE, b Omaha, Nebr, May 2, 43; m 68. ECONOMIC GEOLOGY, VOLCANOLOGY. *Educ:* Iowa State Univ, BS, 65; Univ Nev-Reno, MS, 68. *Prof Exp:* Econ geologist, Nev Bur Mine & Geol, Univ Nev-Reno, 68-85, dep for res, 85-86, chief geologist, 86-87, actg dir & state geologist, 87-89, RES GEOLOGIST, UNIV NEV, RENO, 89- *Concurrent Pos:* Exec secy, Nev Oil & Gas Conserv Comn, 74-75; prin investr, Bendix Field Eng Contract, US Dept Energy, 82-, US Geol Serv, 90-, Nev Dept Minerals, 88. *Mem:* Geol Soc Am; Soc Econ Geologists; Am Asn Petrol Geologists. *Res:* Epithermal ore deposits; geologic mapping in mesozoic and tertiary volcanic terranes of Nevada; Nevada energy resources such as uranium, petroleum, and geothermal energy. *Mailing Add:* Nev Bur Mines & Geol Univ Nev-Reno Reno NV 89557-0088

GARSKE, DAVID HERMAN, b Kalamazoo, Mich, Mar 2, 37; div; c 1. MINERALOGY. *Educ:* Mich Technol Univ, BS, 59; Univ Mich, MS, 61, PhD(mineral), 70. *Prof Exp:* Asst prof mineral & geol, SDak Sch Mines & Technol, 65-76; REF MINERAL SUPPLIER, 70- *Mem:* Mineral Soc Am; Mineral Asn Can. *Mailing Add:* PO Box 83 Bisbee AZ 85603

GARST, JOHN ERIC, b Wichita, Kans, Oct 31, 46; m 71; c 2. TOXICOLOGY, MEDICINAL CHEMISTRY. *Educ:* Univ Kans, BA, 69; Univ Iowa, PhD(med chem), 74. *Prof Exp:* Fel biochem, Yale Univ, 73-74; fel physiol & pharmacol, Med Sch, Vanderbilt Univ, 74-75, fel toxicol, 75-78, res instr, 78; asst prof toxicol & asst prof animal sci, Univ Ill, Urbana, 78-85; assoc prof toxicol, Dept Surg & Med, Vet Med Col, Kans State Univ, 86-87; RES ASSOC PROF PHARM-TOXICOL, PRIMATE RES INST, NMEX STATE UNIV, HOLLOMAN AFB, 87- *Concurrent Pos:* Sect ed, J Animal Sci, 83-86. *Mem:* Am Chem Soc; Soc Toxicol; Am Soc Animal Sci; AAAS. *Res:* Structure-toxicity studies; pharmacology-medicinal chemistry; biochemical toxicology; author of several chapters and numerous articles. *Mailing Add:* 3008 Del Prado Alamogordo NM 88310-3960

GARST, JOHN FREDRIC, b Jackson, Miss, May 8, 32; m 55; c 1. CHEMISTRY. *Educ:* Miss State Col, BS, 54; Iowa State Col, PhD(chem), 57. *Prof Exp:* Instr chem, Yale Univ, 57-58; asst prof, Univ Calif, Riverside, 58-63; from asst prof to assoc prof, 63-78, Gen Sandy Beaver teaching prof chem, 82-85, PROF CHEM, UNIV GA, 78- *Mem:* Am Chem Soc. *Res:* Free radicals; solvent and metal effects on organoalkali systems; organo cobalt chemistry; Grignard reagent formation; fast reactions in solution. *Mailing Add:* Dept Chem Univ Ga Athens GA 30602

GARSTANG, MICHAEL, b Utrecht, Natal, SAfrica, Apr 4, 30; US citizen; m 53; c 2. METEOROLOGY, OCEANOGRAPHY. *Educ:* Univ Natal, BA, 52, MA, 58; Fla State Univ, MS, 61, PhD(meteorol), 64. *Prof Exp:* Asst geog, Univ Natal, 51; indust ed, African Explosives & Chem Industs, Ltd, 52; meteorologist, Brit Colonial Serv, 53-56; RES ASSOC MARINE METEOROL, WOODS HOLE OCEANOG INST, 57-; PROF ENVIRON SCI, UNIV VA, 71- *Concurrent Pos:* From asst prof to assoc prof, Fla State Univ, 65-70; mem, Adv Coun to Inst Trop Meteorol, UN Spec Fund Proj, Barbados, 68-70, Univ Corp Atmospheric Res Eval & Goals Comt, 69-70 & Adv Panel to Global Atmospheric Res Prog Anal Group, Nat Ctr Atmospheric Res & Adv Panel Global Atmospheric Res Prog Comt, Adv Comt to Nat Acad Sci, 75-78; vis scientist, Coun Sci & Indust Res, SAfrica, 70; consult, Nat Data Buoy Prog, Lockheed Aircraft Co & Southwest Res Inst, 70-71; prin investr, Trop Meteorol & Oceanog Progs, US Army Res Off, US Dept Defense & Environ Sci Serv Admin, US Forest Serv; chmn, Comt Hurricanes & Trop Meteorol, Am Meteorol Soc, 72-75, past chmn, 75-76; nat coordr, Planetary Boundary Layer Prog, Global Atmospheric Res Prog Atlantic Trop Exp, Nat Acad Sci, 75-78; pres, Simpson Weather Assoc, Inc. *Mem:* Fel Am Meteorol Soc; Sigma Xi; Univ Corp Atmospheric Res; Int Asn Meteorol & Atmoshpheric Physics. *Res:* Tropical meteorology and atmospheric chemistry, especially problems of trace gas exchanges between the surface layer, boundary layer and free atmosphere; rainfall distributions, convective cloud transports, wind energy potential and siting; thunderstorm analysis and short range prediction. *Mailing Add:* Dept Environ Sci Clark Hall Univ Va Charlottesville VA 22903

GARSTANG, ROY HENRY, b Southport, Eng, Sept 18, 25; m 59; c 2. ASTROPHYSICS, ATOMIC PHYSICS. *Educ:* Cambridge Univ, BA, 46, MA, 50, PhD(math), 54, ScD(physics & chem), 83. *Prof Exp:* Jr sci officer, Royal Aircraft Estab, Eng, 45-46; sci officer, Brit Ministry Works, 46-48; res assoc astrophys, Yerkes Observ, Univ Chicago, 51-52; lectr astron, Univ Col, Univ London, 52-60, reader, 60-64; asst dir univ observ, 59-64; chmn inst, 66-67, dir, Div Physics & Astrophys, 79-80, actg dir, Fiske Planetarium, 80-81, PROF PHYSICS, UNIV COLO, BOULDER, 64-, PROF ASTROPHYS, PLANETARY & ATMOSPHERIC SCI, 79-, FEL, JOINT INST LAB ASTROPHYS, 64- *Concurrent Pos:* Ed, Observ Mag, 53-60; astron adv ed, Chambers' Encycl, 60-66; guest worker, Nat Bur Standards, Washington, DC, 61-62, consult, 64-73; chmn comt transition probabilities, Int Astron Union, 64-76, vpres comn 14, 70-73, pres, 73-76; consult, Jet Propulsion Lab, 66-71; Erskine vis fels, Univ Canterbury & vis prof, Univ Calif, Santa Cruz, 71; chmn, Astron Educ Comt, Am Asn Physics Teachers, 87-89. *Mem:* Fel Am Phys Soc; Am Astron Soc; Int Astron Union; Royal Astron Soc; fel Brit Inst Physics; fel Optical Soc Am. *Res:* Spectroscopy; spectrum line intensities; forbidden transitions; spectroscopy of sun, stars and planetary nebulae; light pollution. *Mailing Add:* Joint Inst Lab Astrophys Univ Colo Boulder CO 80309-0440

GARSTENS, MARTIN AARON, b New York, NY, Mar 9, 11; m 36; c 1. SOLID STATE PHYSICS. *Educ:* City Col New York, BS, 32; Columbia Univ, MS, 34; Mass Inst Technol, ScD(fluid mech), 41. *Prof Exp:* Physicist, David Taylor Model Basin, 43-45 & Naval Res Lab, 46-61; physicist, Washington, DC, 61-74, PHYSICIST, OFF NAVAL RES, DEPT NAVY, ARLINGTON, 74- *Concurrent Pos:* Lectr, George Washington Univ, 47-48 & Howard Univ, 62-65. *Mem:* Am Phys Soc. *Res:* Nuclear and electron magnetic resonance; biophysics; foundations of physics. *Mailing Add:* 913 Buckingham Dr Silver Spring MD 20901

GARSTKA, WALTER U(RBAN), hydrology, for more information see previous edition

GART, JOHN JACOB, b Chicago, Ill, Apr 15, 31; m 61; c 4. STATISTICS. *Educ:* DePaul Univ, BSc, 53; Marquette Univ, MS, 55; Va Polytech Inst, PhD(statist), 58. *Prof Exp:* Asst prof biostatist, Sch Hyg & Pub Health, Johns Hopkins Univ, 58-62, assoc prof biostatist & statist, 62-65; mathematician, 65-67, HEAD MATH STATIST & APPL MATH SECT, BIOMET BR, NAT CANCER INST, 67- *Concurrent Pos:* Vis res fel, Univ London, 61-62, Va Polytech Inst, 71; adv, WHO, 66 & 80-86, FDA, 72 & 78-81. *Mem:* Fel Am Col Epidemiol; fel Am Statist Asn; fel Inst Math Statist; fel Royal Statist Soc. *Res:* Biometrics; mathematical statistics. *Mailing Add:* Math Statist & Appl Math Sect Nat Cancer Inst Bethesda MD 20205

GARTE, SEYMOUR JAY, b New York, NY, Oct 31, 47; m 69; c 2. CANCER RESEARCH, ENVIRONMENTAL ONCOLOGY. *Educ:* City Col New York, BS, 70; City Univ New York, PhD(biochem), 76. *Prof Exp:* Asst environ med, NY Univ Med Ctr, 75-76, assoc res scientist, 76-79, asst prof environ med, 79-87, ASSOC PROF ENVIRON MED, NY UNIV MED CTR, 87- *Concurrent Pos:* Prin investr, NIH grants, 83-; consult, Cahill, Gordon & Reindel, 84-85, Goldman, Hafetz, 87-88; mem, NIH rev panels, 86-; mem sci bd, Cancer Prev Res Inst, 90- *Honors & Awards:* Young Investr Res Award, Nat Cancer Inst, 79. *Mem:* Am Asn Cancer Res; Am Chem Soc. *Res:* Molecular biology and biochemistry of environmental carcinogenesis particularly in respect to oncogene activation; biochemical toxicology, chemical carcinogenesis, cell biology and risk assessment. *Mailing Add:* Dept Environ Med NY Univ Med Ctr 550 First Ave New York NY 10016

GARTEN, CHARLES THOMAS, JR, b Huntington, WVa, Mar 27, 48; m 70; c 2. RADIOECOLOGY. *Educ:* Washington & Lee Univ, BS, 70; Univ Ga, MS, 74. *Prof Exp:* Tech coordr, Mineral Cycling Studies, Savannah River Ecol Lab, Univ Ga, 73-76; res assoc, environ sci div, Union Carbide Corp, 76-83, RES STAFF MEM, OAK RIDGE NAT LAB, MARTIN MARIETTA ENERGY SYSTS, 84- *Concurrent Pos:* Sect ed, Nuclear Safety, 78-82. *Mem:* Am Chem Soc; AAAS. *Res:* Biogeochemistry of long-lived radionuclides; element cycling in terrestrial ecosystems; applications of stable isotopes in ecological studies. *Mailing Add:* Environ Sci Div Oak Ridge Nat Lab Oak Ridge TN 37831-6038

GARTENHAUS, SOLOMON, b Ger, Jan 3, 29; m; c 2. THEORETICAL PHYSICS. *Educ:* Univ Pa, BA, 51; Univ Ill, PhD(physics), 55. *Prof Exp:* Instr, Stanford Univ, 55-58; from asst prof to assoc prof, 58-64, asst dean grad sch, 72-77, PROF PHYSICS, PURDUE UNIV, 64-, SECY FAC, 80- *Concurrent Pos:* Distinguished vis prof, USAF Acad, 77-78; dir student progs, Univ Hamburg, 79-80. *Mem:* Fel Am Phys Soc; AAAS; NY Acad Sci; Am Asn of Physics Teachers. *Res:* Phase transitions. *Mailing Add:* Dept Physics Purdue Univ West Lafayette IN 47906

GARTH, JOHN CAMPBELL, b New York, NY, Sept 26, 34; m 60; c 2. HIGH ENERGY ELECTRON TRANSPORT IN SOLIDS. *Educ:* Princeton Univ, BSE, 56; Univ Ill, MS, 58, PhD(physics), 65. *Prof Exp:* Res asst physics, Univ Ill, 60-64; asst prof, Worcester Polytech Inst, 64-67; res physicist, Air Force Cambridge Res Labs, 67-76, PHYSICIST, ROME AIR DEVELOP CTR, SOLID STATE SCI DIV, RADIATION HARDENED ELECTRONICS TECHNOL BR, 76- *Mem:* Am Phys Soc; Sigma Xi; Am Nuclear Soc. *Res:* Theory of kilovolt electron transport in solids, including Boltzmann equation and Monte Carlo techniques; radiation dosimetry for microelectronic devices; electron spin resonance and optical properties of defects in solids; interaction of radiation with matter. *Mailing Add:* Rome Air Develop Ctr/ESR Hanscom AFB MA 01731

GARTH, JOHN SHRADER, b Los Angeles, Calif, Oct 3, 09; m 40; c 1. ZOOLOGY, ENTOMOLOGY. *Educ:* Univ Southern Calif, BMus, 32, MS, 35, PhD(zool), 41. *Prof Exp:* Asst zool, Univ Southern Calif, 35-37, res assoc, Allan Hancock Found, 37-42; civilian instr, Santa Ana Army Air Base, US War Dept, 42-44; res assoc, Allan Hancock Found, 46-52, cur, 63-70, chief cur, 70-75, assoc prof biol, 52-55, from adj assoc prof to assoc prof, 55-67, prof biol sci, 67-75, EMER PROF BIOL SCI, UNIV SOUTHERN CALIF, 75-, EMER CHIEF CUR, ALLAN HANCOCK FOUND, 75- *Concurrent Pos:* Entomologist & marine zoologist, Allan Hancock Found exped, Mex, Cent & SAm & Galapagos Islands, 31-41, exped leader, Ariz desert, 42, 46, 47 & 48; partic, US Prog Biol, Int Indian Ocean Exped, 64; mem sci adv comt, Charles Darwin Found, 65. *Honors & Awards:* John Adams Comstock Award, Lepidopterists Soc, 87. *Mem:* AAAS; Soc Syst Zool; Marine Biol Asn India; Carcinological Soc of Japan; Lepidopterists Soc. *Res:* Systematics, distribution and ecology of the brachyuran Crustacea, particularly of the Eastern and Indo-West Pacific regions; biological oceanography; zoogeography; butterflies of western National Parks; California butterflies. *Mailing Add:* Allan Hancock Found Univ Southern Calif Los Angeles CA 90089-0371

GARTH, RICHARD EDWIN, b Knoxville, Tenn, Mar 10, 26; m 50; c 6. GENETICS, BIOLOGY. *Educ:* Emory Univ, AB, 49, PhD(biol), 54; Univ Tenn, MS, 50. *Prof Exp:* Instr biol, Bloomfield Col, 50-51; asst prof, Mt Union Col, 54-55; from asst prof to assoc prof, E Tenn State Col, 55-58; from asst prof to assoc prof Northwestern State Col, La, 58-66; prof biol sci & head dept, Miss State Col Women, 66-69; head dept biol, 69-76, dir div sci & math, 71-73, PROF BIOL, UNIV TENN, CHATTANOOGA, 76- *Concurrent Pos:* Asst prog dir sec sch progs, NSF, 63-64, consult sci personnel & ed, Div, 64- *Res:* Autecology of Spanish moss; hypothalamus-gonadotrophin relationships; blastocyst implantation in rats; avian photoperiodism; physiology of development; molecular genetics. *Mailing Add:* Dept Biol Univ Tenn 615 McCallie Ave Chattanooga TN 37403

GARTHE, WILLIAM A, zoology; deceased, see previous edition for last biography

GARTHER, JOHN G, b Mar 3, 51; c 2. TRANSPLANTATION IMMUNOLOGY, BONE MARROW TRANSPLANTATION. *Educ:* McGill Univ, MD, 76; FRCP(C), 80. *Prof Exp:* ASSOC PROF PATH, CHILDREN'S HOSP OF WINNIPEG, 83-; ASSOC PROF PEDIAT & CHILD HEALTH & ADJ PROF IMMUNOL, UNIV MAN, 83- *Res:* Graft-versus host disease. *Mailing Add:* Dept Pathol Univ Man 770 Bannatyne Ave Winnipeg MB R3E 0W3 Can

GARTHWAITE, SUSAN MARIE, b Lancaster, WI, Jan 24, 50. CARDIAC DISEASE, ARRHYTHMIA. *Educ:* Alverno Col, Wis, BA, 72; Univ Mo, Columbia, PhD(physiol), 79. *Prof Exp:* Res scientist, 82-86, GROUP LEADER, G D SEARLE & CO, 86- *Mem:* Am Physiol Soc; Am Heart Asn. *Res:* Discovery and development of new antiarrhythmic agents, elucidation of mechanisms of cardiac arrhythmias and development of arrhythmia models. *Mailing Add:* G D Searle & Co 4901 Searle Pkwy Skokie IL 60077

GARTLAND, WILLIAM JOSEPH, b New York, NY, Apr 15, 41; m 81. GENETICS, MOLECULAR BIOLOGY. *Educ:* Holy Cross Col, BS, 62; Princeton Univ, MA, 64, PhD(biochem sci), 67. *Prof Exp:* Asst res scientist, NY Univ Med Ctr, 67-69; res biologist, Univ Calif, San Diego, 69-70; grants assoc, NIH, 70-71, prog admin genetics, 71-76, dir, Off Recombinant DNA Activ, 76-88, CHIEF RESOURCES & CTRS BR, NAT INST ALLERGY & INFECTIOUS DIS, NIH, 88- *Concurrent Pos:* Exec secy, recombinant DNA Adv Comt, NIH, 74-88, mem exec comt, 76-88; US rep, Comt Recombinant DNA Res, Europ Sci Found, 76-81; consult, NY State Adv Comt Recombinant DNA, 77-88; mem, USDA Agr Recombinant DNA Res Comt, 77-88; US rep, US-Japan Coop Prog Recombinant DNA Res, 81-88; US rep, Orgn Econ Coop & Develop Ad Hoc Comt Experts Safety & Reg Biotechnol, 82-88; US rep, Coun Europe Ad Hoc Comt Experts Genetic Eng, 83-88; US Head, Aids panel US-Japan Coop, Med Sci Prog, 88- *Honors & Awards:* NIH Director's Award, 78; Special Recognition Award, Pub Health Serv, 85. *Mem:* AAAS; Am Soc Human Genetics; Am Soc Microbiol. *Res:* Recombinant DNA research. *Mailing Add:* Div AIDS Allergy & Infectious Dis NIH Rm 243P 6003 Executive Blvd Bethesda MD 20892

GARTLER, STANLEY MICHAEL, b Los Angeles, Calif, June 9, 23; m 48. GENETICS. *Educ:* Univ Calif, Los Angeles, BS, 48; Univ Calif, PhD(genetics), 52. *Prof Exp:* Res assoc genetics, Columbia Univ, 52-57; res asst prof, 57-64, PROF MED & GENETICS, UNIV WASH, 64- *Concurrent Pos:* USPHS fel, 52-54, sr res fel, 59-, res career award, 64-; NIH merit scholar, 87. *Mem:* Genetics Soc Am (pres, 87); Am Soc Human Genetics; Am Soc Nat; Nat Acad Sci; Am Soc Human Genetics (pres, 87-88). *Res:* Human genetics; mammalian somatic cell genetics. *Mailing Add:* Dept Genetics SK-50 Univ Wash Seattle WA 98195

GARTNER, EDWARD A, b Milford, Conn, July 9, 28; m 54; c 3. ORGANIC CHEMISTRY. *Educ:* Univ Fla, BS, 51. *Prof Exp:* Process engr, Hercules Powder Co, 51-53, sr process engr, 55-57; develop chemist, Globe Mfg Co, 58-65, develop supvr, 65-68, spec projs coord, Fall River, 68-85, quality mgr, 74-85; RETIRED. *Mem:* Am Chem Soc; Am Soc Qual Control. *Res:* Manufacturing parameters of nitroglycerin; nitrocellulose and solid propellants; mechanism of cold injury; improvements in rubber thread; process-and-product patent for a spandex fiber; development of dyeing and finishings procedures and optimization of the thread; development and manufacture of fiber finishes and textile processibility. *Mailing Add:* 117 Mohawk Rd Somerset MA 02726

GARTNER, JOHN BERNARD, floriculture, ornamental horticulture, for more information see previous edition

GARTNER, LAWRENCE MITCHEL, b Brooklyn, NY, Apr 24, 33; m 56; c 2. MEDICINE, PHYSIOLOGY. *Educ:* Columbia Univ, AB, 54; Johns Hopkins Univ, MD, 58. *Prof Exp:* Intern pediat, Johns Hopkins Hosp, 58-59; resident, Bronx Munic Hosp Ctr, 59-61; from asst instr to instr, Albert Einstein Col Med, 61-63, assoc, 63-64, from asst prof to assoc prof, 64-74, prof pediat, 74-88, dir, div neonatology, 70-80, dir, Kennedy Ctr Clin Res Unit, 72-80; PROF PEDIAT, UNIV CHICAGO, 80-; CHMN, DEPT PEDIAT, UNIV CHICAGO, 80- *Concurrent Pos:* Chief resident, Bronx Munic Hosp Ctr, 61-62, asst dir premature Ctr, 62-67, dir, 67-; USPHS pediat res trainee, 62-64; Nat Inst Child Health & Human Develop spec fel & career develop award, 64-66; NIH career develop award, 66-74, res grant, 68-; United Health Found res grant, 67-68; Scripps Res fel, Inst Comp Biol, Univ Calif, San Diego, 67; chmn, Dept Pediat, Michael Reese Hosp, 86-89. *Mem:* Am Fedn Clin Res; Harvey Soc; Perinatal Res Soc; NAm Soc Pediat Gastroenterol (pres); Soc Pediat Res; Am Pediat Soc; Europ Soc Pediat Res. *Res:* Liver function and disease in premature and full-term newborn infants and children; bilirubin metabolism and physiology of bilirubin transport; management of premature and newborn infants; physiology of the newborn and premature infant. *Mailing Add:* Dept Pediat Univ Chicago Box 426 Chicago IL 60637

GARTNER, LESLIE PAUL, b Szolnok, Hungary, Mar 18, 43; US citizen; m 71; c 1. HISTOLOGY, DENTAL RESEARCH. *Educ:* Rutgers Univ, AB, 65, MS, 68, PhD(zool), 70. *Prof Exp:* From instr histol to asst prof anat, 70-75, ASSOC PROF ANAT, DENT SCH, UNIV MD, BALTIMORE, 75- *Concurrent Pos:* Bk reviewer, J Dent Educ, 75-; consult, US Army Dent Res, Walter Reed Army Hosp, Williams & Williams Publ Co; mem, Nat Bd Anat Sci Test Construct Comt, Am Dent Asn. *Mem:* Am Asn Anatomists; Pan Am Asn Anatomists; Am Asn Dent Schools. *Res:* Histo- and cytochemistry of odontogenesis and palate formation in rodents, as well as the teratogenic effects of radiation on oral embryology; aging and radiation-induced lifespan shortening in insects; development of trisomy 16 mouse. *Mailing Add:* Dept Anat Dent Sch Univ Md 666 W Baltimore St Baltimore MD 21201

GARTNER, NATHAN HART, b Cernauti, Romania, Aug 6, 39; US citizen; m 65; c 3. TRAFFIC CONTROL, ENGINEERING SYSTEMS ANALYSIS. *Educ:* Israel Inst Technol, BSc, 61, MSc, 67, DSc(eng), 70. *Prof Exp:* Lectr eng, Technion, Israel Inst Technol, 70-71; fel transp, Univ Toronto, 71-72; vis asst prof civil eng, Mass Inst Technol, 72-73; from asst prof to assoc prof, 73-81, PROF CIVIL ENG, UNIV LOWELL, MASS, 81- *Concurrent Pos:* Vis scientist, Opers Res Ctr, Mass Inst Technol, 73-76, Lab Info & Decision Systs, 79-85, Ctr Transp Studies, 86-; res fel, Fed Hwy Admin, Washington, DC, 76-78; chmn, Tranp Sci Sect, Opers Res Soc Am, 78-79; assoc ed, Transp Sci, 80-89; mem, Tranp Res Bd, Nat Acad Sci; guest ed, Transp Res, 90-91. *Mem:* Opers Res Soc Am; Inst Transp Engrs. *Res:* Urban traffic control; transportation systems analysis; optimization techniques. *Mailing Add:* Dept Civil Eng Univ Lowell One University Ave Lowell MA 01854

GARTNER, STEFAN, JR, b Hungary, Mar 28, 37; US citizen; m 63; c 2. MICROPALEONTOLOGY, MARINE GEOLOGY. *Educ:* Univ Conn, BA, 60; Univ Ill, MS, 62, PhD(geol), 65. *Prof Exp:* Res geologist, Esso Prod Res Co, 65-68; asst prof marine sci, Univ Miami, 68-72, assoc prof, 72-75; assoc prof geol, 75-81, PROF OCEANOG, TEX A&M UNIV, 80- *Mem:* Geol Soc Am; Sigma Xi. *Res:* Biostratigraphic and paleoecologic application of calcareous nannofossils in pelagic and hemipelagic sediments. *Mailing Add:* Dept Oceanog Tex A&M Univ College Station TX 77843

GARTNER, T KENT, DIRECT CELL-CELL INTERACTION, PLATELET AGGREGATION. *Educ:* Univ Calif, Davis, PhD(bacteriol), 65. *Prof Exp:* PROF BIOL, MEMPHIS STATE UNIV, 79- *Mem:* AAAS; Am Soc Cell Biol; Am Soc Hemat. *Res:* Role of adhesive glycoproteins in platelet aggregation. *Mailing Add:* Dept Biol Memphis State Univ Memphis TN 38152

GARTON, DAVID WENDELL, b Long Beach, Calif, July 14, 53. PHYSIOLOGICAL ECOLOGY, POPULATION GENETICS. *Educ:* Univ Ala, Huntsville, BS, 75; La State Univ, MS, 78, MApp Stat, 80, PhD(physiol), 83. *Prof Exp:* Instr biol, Univ Ala, Huntsville, 75-76; teaching & res asst zool & physiol, La State Univ, 76-83; postdoctoral fel, State Univ NY, Stony Brook, 83-85; ASST PROF, OHIO STATE UNIV, 85- *Mem:* Am Soc Zoologists; Am Soc Limnol & Oceanog; Am Malacological Union; AAAS. *Res:* Ecology, physiology and population genetics of invading invertebrate species in the Laurentian Great Lakes; physiology and genetics of adaptational responses to stressful environments by marine and freshwater invertebrates. *Mailing Add:* Dept Zool Ohio State Univ 1735 Neil Ave Columbus OH 43210

GARTON, RONALD RAY, b Billings, Mont, Feb 27, 35; m 59. RESEARCH ADMINISTRATION, GENERAL ENVIRONMENTAL SCIENCES. *Educ:* Univ Mont, BA, 58, BS, 63; Mich State Univ, MS, 67, PhD(fisheries, wildlife), 68. *Prof Exp:* Res aquatic biologist, Pac Northwest Water Lab, US Environ Protection Agency, 68-72, res aquatic biologist, Western Fish Toxicol Sta, 72-73, chief, 73-81; dir, Freshwater Div, Corvallis Environ Res Lab, 81-86; SR ENVIRON SCIENTIST, EA ENG, SCI & TECHNOL, INC, 86- *Concurrent Pos:* Mem stand comt, Am Nuclear Soc, 73-; consult, environ. *Honors & Awards:* Cert Fisheries Scientist, Am Fisheries Soc, 71; Gold Medal Except Serv for Res, Environ Protection Agency. *Mem:* Ecol Soc Am; Am Fisheries Soc; fel Am Inst Fish Res Biologists; Soc Environ Toxicol & Chem. *Res:* Effects of thermal pollution on freshwater organisms; effect of pollutants on aquatic species and aquatic ecosystems. *Mailing Add:* 1420 NW Ribier Pl Corvallis OR 97330

GARTRELL, CHARLES FREDERICK, b Baltimore, Md, Nov 4, 51; m 75; c 1. ORBITAL MECHANICS. *Educ:* Univ Md, BA, 73. *Prof Exp:* Task mgr & analyst, Syst Sci Div, Comput Sci Corp, 73-75; systems analyst, RCA Am Communications, 75-78; PRIN ENGR, GEN RES CORP, 78- *Concurrent Pos:* Lectr, Commun Technol, Georgetown Univ & Future Space Activ, Montgomery County Pub Sch; prin investr, AI appl for large space syst control. *Mem:* Am Inst Aeronaut & Astronaut; Optical Soc Am. *Res:* Advanced civilian and military space systems technology covering the next 20-30 years; design and studies of advanced earth observation; radar and communication satellites; orbital transfer vehicles, spacecraft control techniques and systems; spaceflight experimental concept design. *Mailing Add:* 10332 Ridgeline Dr Gaithersburg MD 20879

GARTSHORE, IAN STANLEY, b Calgary, Alta, Apr 27, 35; m 62; c 2. FLUID MECHANICS. *Educ:* Univ BC, BASc, 57; Univ London, MSc, 60; McGill Univ, PhD(mech eng), 65. *Prof Exp:* Sci officer, Nat Phys Lab, UK, 59-60; res officer, Nat Res Coun Can, 61-62; res dir, McGill Univ, 65-67; from asst prof to assoc prof mech eng, 67-79, PROF MECH ENG, UNIV BC, 79- *Mem:* Fel Can Aeronaut & Space Inst. *Res:* Swirling flow phenomena in laminar and turbulent motion; development of turbulent shear flows; wind effects on structures; wind energy. *Mailing Add:* Dept Mech Eng Univ BC 2075 Westbrook Pl Vancouver BC V6T 1W5 Can

GARTSIDE, PETER STUART, b Oldham, UK, Aug 12, 37; US citizen; m 63; c 4. BIOSTATISTICS. *Educ:* Brigham Young Univ, BS, 67, MS, 69; Univ Calif, Berkeley, PhD(biostatist), 76. *Prof Exp:* Instr, Dept Statist, Brigham Young Univ, 67-69; statistician, Sch Pub Health, Univ Calif, Berkeley, 72-73; asst prof, 73-78, ASSOC PROF BIOSTATIST, DEPT ENVIRON HEALTH, UNIV CINCINNATI, 78- *Mem:* Am Statist Asn; Biomet Soc; Am Math Soc; Soc Indust & Appl Math. *Res:* Environmental health research; occupational health research. *Mailing Add:* Biostatist Dept Environ Health 183G10 A WH Univ Cincinnati Med Ctr 231 Bethesda Ave Cincinnati OH 45267

GARTSIDE, ROBERT N(IFONG), b Fredericktown, Mo, Jan 1, 18; m 59; c 3. CHEMICAL ENGINEERING. *Educ:* Wash Univ, BS, 39. *Prof Exp:* Chem engr, Eastern Lab, Explosives Dept, E I du Pont de Nemours & Co, Inc, 39-42, 46-52, tech asst, Tech Div, 52-53, asst dir, Repauno Process Lab, 53-56, dir, Sales Develop Lab, 56-59, asst dir, Eastern Lab, 59-60, dir, Repauno Develop Lab, 60-67, admin asst, Eastern Lab, Gibbstown, 67-72, sr design consult, Eng Dept, 72-82; RETIRED. *Res:* Process development; administration. *Mailing Add:* 578 High St Woodbury NJ 08096

GARTY, KENNETH THOMAS, b Chicago, Ill, Nov 4, 16; m 50; c 4. ORGANIC CHEMISTRY. *Educ:* Purdue Univ, BS, 40; Stevens Inst Technol, MS, 47. *Prof Exp:* Chemist res, Bakelite Co, 40-48, group leader, 48-72, proj leader, 52-59, Union Carbide Corp, 59-65, sr chemist, Union Carbide Corp, 72-82; RETIRED. *Mem:* Am Chem Soc. *Res:* Vinyl polymerization and copolymerizations; organosulfur and epoxy polymers; polyethers. *Mailing Add:* 775 Hardgrove Rd Bridgewater NJ 08807

GARVEN, FLOYD CHARLES, b Baker, Minn, Mar 2, 22; m 45; c 3. ORGANIC CHEMISTRY. *Educ:* Moorhead State Teachers Col, BS, 46; NDak State Univ, MS, 48. *Prof Exp:* Org res chemist antituberculars and hypnotics, Abbott Labs, 49-54, & plant process, 54-59, org develop chemist process & prod develop, 59-62, mgr chem develop, 62-64, dir develop, 64-67, qual control, 67-69 & qual assurance, 69-73, dir qual assurance, Chem & Agr Prod Div, 74-84; RETIRED. *Concurrent Pos:* Tech mgt. *Mem:* Am Chem Soc; Am Soc Qual Control. *Res:* Antitubercular drugs; hypnotics; erythromycin structure work; chemotherapeutic drugs; process development. *Mailing Add:* 965 Norman St Gurnee IL 60031

GARVER, DAVID L, b Dayton, Ohio, Feb 3, 39; m 65; c 2. SCHIZOPHRENIA, PSYCHOPHARMACOLOGY. *Educ:* Oberlin Col, BA, 60; Western Reserve Univ Sch Med, MD, 65. *Prof Exp:* Resident psychiat, Univ Colo Med Ctr, 68-71; fel psychiat res, Ill State Psychiat Hosp, 71-72; instr psychiat, Pritzker Sch Med, Univ Chicago, 73-78; assoc prof, 78-81, PROF PSYCHIAT & PHARMACOL, COL MED, UNIV CINCINNATI, 81- *Concurrent Pos:* Asst prof psychiat, Rush Med Sch, 74-78; chmn, educ & training comt, Am Col Neuropsychopharmacol, 87; dir, Lab Psychobiol & Psychopharmacol, Med Univ Cincinnati, 78-, dept psychiat, 82- & Psychobiol Treat Unit, Univ Cincinnati Hosp, 78-; chmn Nat Inst Mental Health Treat Develop & Assessment, 86-; vis prof, dept psychiat, Wash Univ, St Louis, 88- *Mem:* Am Col Neuropsychopharmacol; Psychiat Res Soc (pres, 87-88); Soc Biol Psychiat; Am Psychiat Asn; Soc Neurosci; AAAS. *Res:* Resolution of the biologic heterogeneity within the group of psychotic disorders generally known as schizophrenia; differential drug-response patterns; differential neuroendocrinological response to specific probes; different neurotransmitter patterns in CSF and plasma are utilized to resolve such heterogeneity. *Mailing Add:* Dept Psychiat & Behav Neurobiol Univ Ala Birmingham AL 35294

GARVER, FREDERICK ALBERT, b Marion, Ohio, Oct 22, 36. CELL & MOLECULAR BIOLOGY. *Educ:* Ohio State Univ, BS, 59; Univ Colo, PhD(immunol), 69. *Prof Exp:* Postdoctoral fel, Microbiol Dept, Med Ctr, Univ Colo, 63-69 & Immunochem Dept, Max-Planck Inst Exp Med, Gottingen, Ger, 69-71; from asst prof to assoc prof, 71-81, PROF, DEPT CELL & MOLECULAR BIOL, MED COL GA, 81-, DIR, HYBRIDOMA FACIL, 89- *Concurrent Pos:* NIH res grants, 80-93; Nat Leukemia Asn res grant, 81-83; mem, Hemoglobin Variants Subcomt, Nat Comt Clin Lab Standards, 82- *Honors & Awards:* Merrell-Dow Award, Soc Perinatal Obstetricians, 82. *Mem:* Am Asn Immunologists; Am Soc Hemat; NY Acad Sci; Reticuloendothelial Soc; Am Soc Biol Chemists; AAAS; Am Asn Univ Professors. *Res:* Indentification of tumor antigens on human leukemia cells; immunochemical identification and characterization of variant hemoglobins; author of more than 100 technical publications. *Mailing Add:* Dept Cell & Molecular Biol Med Col Ga Augusta GA 30912-2100

GARVER, ROBERT VERNON, b Minneapolis, Minn, June 2, 32; m 57; c 5. MICROWAVE THEORY AND TECHNIQUES. *Educ:* Univ Md, BS, 56; George Washington Univ, MEA, 68. *Prof Exp:* Res physicist microwaves, 56-70, physicist fuze proj mgr, 70-75, SUPVR PHYSICIST, HARRY DIAMOND LABS, 75- *Concurrent Pos:* Assoc ed, Inst Elec & Electronics Engrs J Solid-State Circuits, 69-73; consult, Weinschel Eng Co, 70-75. *Mem:* Fel Inst Elec & Electronics Engrs. *Res:* Original ideas in microwave diode switches which served as the basis for all later development of diode switches, limiters, attenuators, and phase shifters. *Mailing Add:* 12205 Greenridge Dr Boyds MD 20841

GARVEY, GERALD THOMAS, b New York, NY, Jan 21, 35; m 59; c 3. NUCLEAR PHYSICS. *Educ:* Fairfield Univ, BS, 56; Yale Univ, PhD(physics), 62. *Prof Exp:* Res assoc physics, Yale Univ, 62-63; instr, Princeton Univ, 63-64; asst prof, Yale Univ, 64-66; from asst prof to prof physics, Princeton Univ, 66-76; dir, Physics Div, Argonne Nat Lab, 76-79, assoc dir physical res, 79-80, sr scientist, 76-84; prof physics, Univ Chicago, 78-84; DIR, LOS ALAMOS MESON PHYSICS FACIL, LOS ALAMOS NAT LAB, 85-, DEP ASSOC DIR, NUCLEAR & PARTICLE PHYSICS PROGS, 84- *Concurrent Pos:* Sloan Found fel, 67-69; consult, Brookhaven Nat Lab, 70-71. *Mem:* Am Phys Soc. *Res:* Experimental nuclear physics, particularly reactions and isoboric spin studies; weak interaction in nuclear systems; neutrino physics. *Mailing Add:* Los Alamos Meson Physics Facil MS H836 Los Alamos Nat Lab Los Alamos NM 87545

GARVEY, JAMES F, b Passail City, NJ, Feb 6, 57; m 87. MOLECULAR BEAMS, VAN DER WAALS CLUSTERS. *Educ:* Georgetown Univ, BS & MS, 78; Calif Inst Technol, PhD(chem), 85. *Prof Exp:* Postdoctoral fel, Univ Calif Los Angeles, 85-87; asst prof, 87-91, ASSOC PROF CHEM, STATE UNIV NY, BUFFALO, 91- *Concurrent Pos:* Res award, Alfred P Sloan Found, 91. *Honors & Awards:* Am Inst Chem Award, 78. *Mem:* Am Chem Soc; Am Phys Soc; Mat Res Soc; Am Soc Mass Spectrometry. *Res:* Chemical reactions within Van der Waals clusters; photochemistry within metal heyacarbonyl clusters thin film generation via cluster deposition. *Mailing Add:* Acheson Hall Dept Chem State Univ NY Buffalo NY 14214

GARVEY, JUSTINE SPRING, b Wellsville, Ohio, Mar 14, 22; m 46; c 2. IMMUNOCHEMISTRY. *Educ:* Ohio State Univ, BS, 44, MS, 48, PhD(microbiol, chem), 50. *Prof Exp:* Res fel chem, Calif Inst Technol, 51-57, sr res fel, 57-73, res assoc, 73-74; from assoc prof to prof immunochem, 74-89, prof, 80-89, EMER PROF BIOL, SYRACUSE UNIV, 90- *Concurrent Pos:* Vis prof microbiol, Univ Ill Med Ctr, 73; vis res assoc, Calif Inst Technol, 90- *Mem:* AAAS; Am Asn Immunol; NY Acad Sci. *Res:* Immunological methods; biology of aging; biological and chemical characterization of retained antigen; immunochemistry of metallothionein; role of antigen in immunity. *Mailing Add:* 698 Arden Rd Pasadena CA 91106-4408

GARVEY, R(OBERT) MICHAEL, b Winston-Salem, NC, Jan 4, 47; m 75. FREQUENCY AND TIME STANDARDS. *Educ:* Davidson Col, BS, 69; Duke Univ, PhD(physics), 75. *Prof Exp:* Res asst physics, Duke Univ, 72-77; mem staff, Nat Bur Standards, 77-79, scientist, Frequency & Time Systs, 79-82, dir, Eng, 82-90; CONSULT, 90- *Mem:* Sigma Xi; Inst Elec & Electronics Engrs. *Res:* Experimental aspects of molecular physics; emphasis on microwave spectroscopy, frequency and time standards, cesium frequency standards, quartz oscillators, frequency and time metrology. *Mailing Add:* 85 Monument Ave Swampscott MA 01907-1947

GARVEY, ROY GEORGE, b Pocatello, Idaho, Jan 19, 41; m 63; c 3. INORGANIC CHEMISTRY. *Educ:* Univ Utah, BA, 63, PhD(inorg chem), 66. *Prof Exp:* Asst prof, 66-69, ASSOC PROF INORG CHEM, NDAK STATE UNIV, 69- *Mem:* AAAS; Am Chem Soc; Am Crystallog Soc; Sigma Xi. *Res:* Coordination properties of M-D functional groups; kinetics of displacement reactions at transition metal sites; synthesis and physical properties of novel inorganic compounds. *Mailing Add:* Dept Chem NDak State Univ Fargo ND 58105-5516

GARVIN, ABBOTT JULIAN, PATHOLOGY. *Educ:* Med Univ SC, MD, 72, PhD(path), 75. *Prof Exp:* ASSOC PROF PATH, MED UNIV SC, 80-, DIR SURG PATH & CYTOPATH, 85- *Mailing Add:* Dept Path Med Univ SC 171 Ashley Ave Charleston SC 29403

GARVIN, DAVID, b Cleveland, Ohio, Aug 25, 23. PHYSICAL CHEMISTRY. *Educ:* Yale Univ, BS, 48; Harvard Univ, MA & PhD(chem), 51. *Prof Exp:* Instr chem, Princeton Univ, 51-55, asst prof, 55-61; chemist, Nat Bur Standards, 61-78, chief, Chem Thermodyn Div, 78-81, Chem Thermodyn Data Ctr, 81-86, chemist, 86-89; RETIRED. *Mem:* Am Chem Soc; Am Phys Soc. *Res:* Chemical thermodynamics; information retrieval; chemical kinetics; physical properties, data evaluation. *Mailing Add:* 18700 Walkers Choice Rd No 807 Gaithersburg MD 20879

GARVIN, DONALD FRANK, b Toledo, Ohio, Mar 14, 32; m 51; c 3. INDUSTRIAL MICROBIOLOGY. *Educ:* Wayne State Univ, BS, 64, MS, 66, PhD(biol), 75. *Prof Exp:* Clin lab technician, Woodward Gen Hosp, Highland Park, Mich, 55-59, asst supvr clin lab, 59-65; sr technologist, William Beaumont Hosp, Royal Oak, Mich, 65-66; res bacteriologist, Wyandotte Chem Corp, 66- 75, supvr, BASF Wyandotte Corp, 75-80, SUPVR REGULATORY AFFAIRS, DIVERSEY CORP, 80- *Mem:* Am Soc Microbiol; assoc mem Asn Off Anal Chem; Am Soc Testing & Mat; Am Chem Soc. *Res:* Identification and isolation of cell wall deficient varients of mycobacterium tuberculosis and clinical L-forms by gas chromotography, polyacrylamide electrophoresis and fluorescent antibody techniques; application research of germicides and fungicides. *Mailing Add:* Regulatory Affairs Diversey Corp Wyandotte MI 48192

GARVIN, JAMES BRIAN, b Poughkeepsie, NY, Mar 10, 56. PLANETARY GEOPHYSICS, RADAR & LASER REMOTE SENSING. *Educ:* Brown Univ, ScB, 78, MS, 81, PhD(geol sci), 84; Stanford Univ, MS, 79. *Prof Exp:* Fel planetary geol, Brown Univ, 84; GEOPHYSICIST GEOL, REMOTE SENSING, PLANETARY & COMETARY SCI, GODDARD SPACE FLIGHT CTR, NASA, 84- *Concurrent Pos:* Mem, topog sci working group & mars rover sample return working group, Washington, DC, 88, prin investr, topog profile and res proj grant & co-investr, shuttle laser altimeter exp, NASA, 86; prin investr, Iceland Volcanology, NASA Res proj grant, 88, co-investr, Mars Observer Laser Altimeter Exp, 86- *Mem:* Sigma Xi; Am Geophys Union; Asn Comput Mach. *Res:* Earth and planetary geophysics focusing on the surface processes on Venus, Earth and Mars, and involving radar(microwave), laser and image remote sensing; high resolution altimetry of the earth, moon and mars involving laser and radar systems; terrestrial impact craters from remote sensing datasets; quantitative volcanology of Iceland. *Mailing Add:* Goddard Space Flight Ctr NASA Code 921 Greenbelt MD 20771

GARVIN, JEFFREY LAWRENCE, b Pittsburgh, Pa, Dec 1, 57. PHYSIOLOGY. *Educ:* Univ Miami, BS, 79; Duke Univ, PhD(physiol), 84. *Prof Exp:* Instr med physiol, Dept Physiol, Duke Univ, 80; Nat Kidney Found fel, Lab Kidney & Electrolyte Metab, NIH, 84-85; ASSOC STAFF INVESTR, DEPT MED, HYPERTENSION & VASCULAR RES DIV, HENRY FORD HOSP, DETROIT, 88- *Concurrent Pos:* Guest worker, Lab Kidney & Electrolyte Metab, NIH, 85-88; Am Heart Asn res grant, 91-94. *Mem:* Am Soc Nephrology; Am Physiol Soc; AAAS. *Res:* Author of more than 30 technical publications. *Mailing Add:* Hypertension & Vascular Res Div Henry Ford Hosp 2799 W Grand Blvd Detroit MI 48202-2689

GARVIN, PAUL JOSEPH, JR, b Toledo, Ohio, Nov 16, 28; m 52; c 6. HEALTH RISK ASSESSMENT. *Educ:* St John's Univ, Minn, BA, 50; Univ Minn, MS, 59. *Prof Exp:* Res assoc, Sterling-Winthrop Res Inst, 54-58; sr res pharmacologist, Baxter Travenol Labs, Inc, 58-73, mgr safety eval, 73-77; dir toxicology, 77-78, SR CONSULT ENVIRON HEALTH, AMOCO CORP, 88- *Mem:* Soc Toxicol; Am Soc Pharmacol & Exp Therapeut; Am Indust Hyg Asn; Europ Soc Toxicol; AAAS; NY Acad Sci. *Res:* Health and environmental effects of petroleum products and chemicals; health risk assessment processes. *Mailing Add:* Amoco Corp 200 E Randolph Dr MC 4901 Chicago IL 60601

GARVIN, PAUL LAWRENCE, b Dec 5, 39; US citizen; m 64; c 2. MINERALOGY. *Educ:* Idaho State Univ, BS, 64; Univ Colo, Boulder, PhD(geol), 69. *Prof Exp:* Lectr geol, Idaho State Univ, 64-65; PROF GEOL, CORNELL COL, 69- *Concurrent Pos:* Consult, rock & mineral analysis. *Mem:* AAAS; Mineral Soc Am; Nat Asn Geol Teachers. *Res:* Sulfide phase relations; mineralogy and genesis of ore deposits. *Mailing Add:* Dept Geol Cornell Col Mt Vernon IA 52314

GARVINE, RICHARD WILLIAM, b Pottstown, Pa, Jan 7, 40; m 66; c 2. PHYSICAL OCEANOGRAPHY. *Educ:* Mass Inst Technol, BS, 61; Princeton Univ, PhD(aerodyn eng), 65. *Prof Exp:* Theoret aerodynamicist, Space Sci Lab, Gen Elec Co, 65-69; from asst prof to assoc prof mech eng, Marine Sci Inst, Univ Conn, 69-77; ASSOC PROF MARINE STUDIES, UNIV DEL, 77- *Mem:* AAAS; Am Geophys Union. *Res:* Estuarine dynamics; oceanic fronts; coastal upwelling. *Mailing Add:* Dept Marine Studies Univ Del Newark DE 19711

GARWIN, CHARLES A, b Savannah, Ga, Dec 23, 44. ACTUARIAL PROGRAMMING. *Educ:* Univ Chicago, BS, 64, MS, 65; Univ Calif-Berkeley, PhD(physics), 71. *Prof Exp:* Prog anal, 80-82, SR PROG ANALYSIS, NAT ASSOC CORP, 82- *Mem:* Am Phys Soc; Am Math Soc. *Mailing Add:* 1317A Franklin St Santa Monica CA 90404

GARWIN, EDWARD LEE, b Cleveland, Ohio, Mar 22, 33; m 54; c 3. APPLIED PHYSICS. *Educ:* Case Western Reserve Univ, BS, 54; Univ Chicago, MS, 55, PhD(physics), 58. *Prof Exp:* Res assoc, Univ Chicago, 58-59; res asst prof physics, Univ Ill, 59-60; prof mgr space simulation, Gen Tech Corp, 60-62; physicist, 62-68, HEAD APPL PHYSICS, STANFORD LINEAR ACCELERATOR CTR, 68-, PROF APPL PHYSICS, 75- *Concurrent Pos:* Consult, Space Technol Labs, Inc, 59-60, Thompson-Ramo-Wooldridge, Inc, 59-60, Gen Tech Corp, 62-64, Rand Corp, 69-74, Searle Cardio-Pulmonary Systs, Inc, 75-77 & Pacific Sierra Corp, 79-; sabbatical, Swiss Fed Inst Technol, Zurich, 85-86. *Mem:* Am Phys Soc; Sigma Xi; Am Vacuum Soc. *Res:* Secondary emission; surface and high energy physics; ultrahigh vacuum; medical electronics and instrumentation; superconductivity; solid state physics; polarized electron sources; high power flashlamps. *Mailing Add:* Stanford Linear Accelerator Ctr PO Box 4349 Bin 72 Stanford CA 94309

GARWIN, RICHARD LAWRENCE, b Cleveland, Ohio, Apr 19, 28; m 47; c 3. EXPERIMENTAL PHYSICS. *Educ:* Case Western Reserve Univ, BS, 47, DSc, 66; Univ Chicago, MS, 48, PhD(physics), 49. *Prof Exp:* From instr to asst prof physics, Univ Chicago, 49-52; physicist, IBM Watson Lab, Int Bus Machines Corp, 52-65, dir appl res, Thomas J Watson Res Ctr, 65-66, dir IBM

Watson Lab, 66-67, FEL, THOMAS J WATSON RES CTR, IBM CORP, 67- *Concurrent Pos:* Consult, Los Alamos Sci Lab, 49-; adj prof physics, Columbia Univ, 57-; vis scientist, Europ Orgn Nuclear Res, 59-60; consult, President's Sci Adv Comt, 58-62, mem, 62-65 & 69-72; mem, Defense Sci Bd, 66-68, exec comt, Assembly Math & Phys Sci, Nat Res Coun, 74-77, coun, Nat Acad Sci, coun, Inst Strategic Studies, London, 77-85; vis prof appl physics, Harvard Univ, 74; chmn, Solar Energy Res Inst Comt, Nat Res Coun, 75 & Panel Pub Affairs, Am Phys Soc, 78; prof pub policy, Harvard Univ, 79-81. *Mem:* Nat Acad Sci; Nat Acad Eng; Inst Med Nat Acad Sci; fel Am Phys Soc; fel Am Acad Arts & Sci; Coun Foreign Rels; Am Philos Soc. *Res:* Liquid and solid helium; general physics; electronics in communications and displays; avionics; strategic systems; author of over 200 technical publications; awarded 35 US patents. *Mailing Add:* T J Watson Res Ctr IBM Corp Box 218 Yorktown Heights NY 10598

GARWOOD, DOUGLAS LEON, b Taylorville, Ill, Feb 8, 44; m 73; c 2. PLANT BREEDING, BIOCHEMICAL GENETICS. *Educ:* Univ Ill, Urbana-Champaign, BS, 66, MS, 68; Pa State Univ, PhD(genetics), 73. *Prof Exp:* Asst prof to assoc prof plant breeding, Pa State Univ, 73-80; SECY-TREAS, GARWOOD SEED CO, 80- *Concurrent Pos:* Mem, Golden Harvest Seeds Res & Soybean Comts, 82-; mem, Golden Harvest Seeds Mkt Comt, 89- *Mem:* Am Soc Hort Sci; Am Soc Agron; Crop Sci Soc Am; Nat Sweet Corn Breeders Asn. *Res:* Genetic and biochemical analysis of sugar, starch and phytoglycogen biosynthesis in maize; performance evaluation of field corn hybrids; sweet corn quality evaluation. *Mailing Add:* Rte 1 Box 20 Stonington IL 62567

GARWOOD, MAURICE F, b Angola, Ind, July 31, 07. CHEMICAL ENGINEERING, METALLURGY. *Educ:* Ohio State Univ, BS, 33, BS, 47. *Prof Exp:* Exec engr, Chrysler Corp, 55-70; RETIRED. *Mem:* Fel Am Soc Metals; Soc Automotive Engrs; Am Standards Asn. *Res:* Author of one book on metallurgy. *Mailing Add:* 1000 Lely Palms Dr Apt 129E Naples FL 33962-8919

GARWOOD, ROLAND WILLIAM, JR, b Bogota, Colombia, Apr 13, 45; US citizen; m 67; c 3. AIR-SEA INTERACTION, OCEAN TURBULENCE. *Educ:* Bucknell Univ, BS, 67; Univ Wash, PhD(oceanog), 76. *Prof Exp:* Comn officer, Nat Oceanic & Atmospheric Admin Corps, 68-71; res & teaching asst, Univ Wash, 71-76; adj res prof, 76-79, PROF OCEANOG, NAVAL POSTGRAD SCH, 79- *Concurrent Pos:* Consult ocean modeling & air-sea interactions to govt & indust, 76-; prin investr res grants, Office Naval Res, 77-, NSF, 86-; contrib author, US Nat Report to Int Union Geol & Geophys, 79-; assoc ed, J Phys Oceanog, 83-88. *Mem:* Am Geophys Union; Am Meteorol Soc; Sigma Xi. *Res:* The ocean surface mixed layer; turbulence, coupled oceanic-atmospheric models; equatorial and tropical mixed layers; use of satellite remote sensing for ocean prediction and analysis; approximately two dozen archived publications on oceanic mixing and air-sea interaction. *Mailing Add:* Naval Postgrad Sch Code 68GD Monterey CA 93943

GARWOOD, VICTOR PAUL, b Detroit, Mich, Sept 13, 17; m 42; c 2. AUDIOLOGY. *Educ:* Univ Mich, BA, 39, MS, 48, PhD(speech path, exp phonetics), 52. *Prof Exp:* Clin asst, Speech Clin, Univ Mich, 46-48, chief exam div, 48-50; from instr to assoc prof, 50-59, asst dir, 52-59, co-dir, 59-77, chmn, Grad Prog Commun Dis, 67-71, res assoc, Geront Res Inst, 77-79, prof audiol, 59-88, prof, 60-88, EMER PROF AUDIOL & OTOLARYNGOL, SCH MED, UNIV SOUTHERN CALIF, 88- *Concurrent Pos:* Inst Neurol Dis & Blindness res fel, 57-58, spec res fel, 61-64; res assoc, Deafness Res Lab, Children's Hosp, Los Angeles, 58-61, consult, 64-; consult, Los Angeles County Hosp, 64-; mem hearing aid dispensers exam comt, Bd Med Examrs, State of Calif, 71-79, chmn, 77-79; med consult, Los Angeles Regional Off, Dept Health, 71-87; spec leave, sr audiologist, Audiol Resource Unit, Los Angeles Unified Sch Dist, 72-76. *Mem:* Emer mem Acoust Soc Am; emer mem Am Psychol Asn; fel Am Speech & Hearing Asn; emer mem Sigma Xi; Acad Rehab Audiol. *Res:* Medical audiology; audition; auditory neurophysiology. *Mailing Add:* 1240 Chautauqua Pacific Palisades CA 90272

GARWOOD, WILLIAM EVERETT, b Kirkwood, NJ, Oct 25, 19; m 46; c 3. ORGANIC CHEMISTRY. *Educ:* Univ NC, BA, 42. *Prof Exp:* Res chemist, Socony Vacuum Res Labs, Mobil Res & Develop Corp, 42-55, sr chemist, 55-71, res assoc, 71-81, res scientist, 81-87, CONSULT, MOBIL RES & DEVELOP CORP, 87- *Concurrent Pos:* Vis scientist, Univ Ill, 69; adj prof, Glassboro State Col, 90- *Mem:* Am Chem Soc; Catalysis Soc. *Res:* Chemical catalysis; lubricating oil additives; synthetic lubricants; organophosphorus chemistry; hydrodesulfurization; hydrocracking; zeolite-catalyzed reactions; carbonylation; alkylation. *Mailing Add:* Mobil Res & Develop Corp Paulsboro NJ 08066

GARY, JAMES H(UBERT), b Victoria, Va, Nov 18, 21; m 45; c 4. CHEMICAL ENGINEERING. *Educ:* Va Polytech Inst, BS, 42, MS, 46; Univ Fla, PhD(chem eng), 51. *Prof Exp:* Group engr chem eng, Standard Oil Co, 46-52; asst prof chem eng & res dir eng exp sta, Univ Va, 52-56; from assoc prof to prof, Univ Ala, 56-60; head dept, Colo Sch Mines, 60-72, dir & trustee, Res Inst, 70-72, vpres acad affairs, 72-79, dean fac, 77-79, dir & trustee, Res Inst, 81-84, prof, 60-91, EMER PROF, CHEM ENG & PETROL REFINING, COLO SCH MINES, 91- *Concurrent Pos:* Res assoc, Fla Eng Exp Sta, 49-51; consult, US Bur Mines, 57-60; chmn subcomt oil shale alt energy systs, Energy Resources Group, Nat Res Coun, 76-80; chmn oil shale adv comt, Off Technol Assessment, 77-80; mem, Govs Sci & Technol Coun, 80-82. *Honors & Awards:* Halliburton Award, 81; George R Brown Gold Medal, 87. *Mem:* Am Chem Soc; fel Am Inst Chem Engrs; Am Inst Mining, Metall & Petrol Engrs; Am Soc Eng Educ; Fel Am Assoc Adv Sci. *Res:* Desulfurization of petroleum and coal; organic nitrogen removal; distillation; shale oil; heavy oils processing. *Mailing Add:* 1021 18th St Golden CO 80401-1826

GARY, JULIA THOMAS, analytical chemistry, for more information see previous edition

GARY, NANCY E, b New York, NY. INTERNAL MEDICINE, NEPHROLOGY. *Educ:* Springfield Col, BS, 58; Med Col Pa, MD, 62. *Prof Exp:* Clin & res fel nephrology, Sch Med, Georgetown Univ, 65-67; chief nephrology, St Vincent's Hosp & Med Ctr, New York, 67-74; from asst prof to assoc prof, 74-81, PROF MED, RUTGERS MED SCH, UNIV MED & DENT NJ, 81-, ASSOC DEAN, 81- *Concurrent Pos:* Instr clin med, Sch Med, NY Univ, 68-74; consult ed, Am J Med, 72-; mem, grad fac physiol, Rutgers State Univ, 77-, grad fac toxicol, 82-; fel Robert Wood Johnson Health Policy, Inst Med, 87-88. *Mem:* Am Soc Nephrology; fel Am Col Physicians. *Res:* Clinical nephrology; toxic nephrology; medical education. *Mailing Add:* Dean Albany Med Col 47 New Scotland Ave Albany NY 12208

GARY, NORMAN ERWIN, b Ocala, Fla, Nov 1, 33; m 54; c 2. APICULTURE, ENTOMOLOGY. *Educ:* Univ Fla, BS, 55; Cornell Univ, PhD(apicult), 59. *Prof Exp:* Res assoc apicult, Cornell Univ, 59-62; from asst prof to assoc prof, 62-73, PROF ENTOM, UNIV CALIF, DAVIS, 73- *Concurrent Pos:* Consult, TV & film spec effects with insects. *Honors & Awards:* J I Hambleton Award. *Mem:* AAAS; Entom Soc Am; Int Bee Res Asn; Animal Behav Soc; Int Union Study Social Insects. *Res:* Behavior of insects, especially honey bees; economic entomology; foraging behavior of honey bees, flight range and distribution; stinging behavior; Africanized bee behavior. *Mailing Add:* Dept Entom Univ Calif Davis CA 95616

GARY, ROBERT, b Baltimore, Md, Apr 15, 28; div; c 2. SOFTWARE DOCUMENTATION, SCIENCE ADMINISTRATION. *Educ:* Loyola Col, Md, BS, 50; Yale Univ, MS, 51, PhD(chem), 54. *Prof Exp:* Res chemist, E I du Pont de Nemours & Co, 54-60; chemist, Nat Bur Standards, DC, 60-66, phys sci adminstr, 66-67; independent consult, Univ-Govt Rels, 67-70; mem prog planning staff, Nat Adv Comt Oceans & Atmosphere, US Dept Com, 73-77, asst to dir telecommun, 70-73, res appln analyst, 74-76, policy analyst & writer-ed, 77-82; staff leader, Software & Comput Syst Doc, Nat Weather Serv, 82-87; CONSULT, 87- *Res:* Synthetic fibers in non-garment uses; tire research and development; mixed electrolytes; solvent effects; pH standards in ordinary and heavy water; research planning and program development; telecommunications, teleconferencing, applications research, remote sensing, navigation systems, intelligibility of documentation; software and system documentation. *Mailing Add:* 132 Claybrook Dr Silver Spring MD 20902-3115

GARY, ROLAND THACHER, b Locker, Tex, Apr 29, 16; m 40; c 2. BIOLOGY. *Educ:* Southwest Tex State Teachers Col, BS, 40, MA, 46; George Peabody Col, PhD(higher educ), 53. *Prof Exp:* Teacher & adminr pub schs, Tex, 37-42; asst prof sci, Southwest Tex State Univ, 46-48; instr biol, George Peabody Col, 48-49; asst prof biol, Southwest Tex State Univ, 50-54, asst prof biol & gen sci, 55-56, prof, 56-77; RETIRED. *Concurrent Pos:* Conserv ed consult, 63-64. *Mem:* Nat Asn Biol Teachers. *Res:* Conservation education; science content of preservice courses for teachers of science in public schools at all levels; ecological studies on land snails. *Mailing Add:* 1204 Marlton St San Marcos TX 78666

GARY, STEPHEN PETER, b Cleveland, Ohio, Oct 3, 39; m 66; c 2. PLASMA PHYSICS. *Educ:* Case Western Reserve Univ, BS, 61; Washington Univ, St Louis, AM, 66, PhD(physics), 67. *Prof Exp:* Instr physics, Webster Col, 66-67; res assoc, Univ Iowa, 67-68; res asst, Univ St Andrews, 68-69; Leverhulme vis fel, Univ Col NWales, 69-70; from asst prof to assoc prof physics, Col William & Mary, 70-76; staff mem, 77-87, GROUP LEADER, LOS ALAMOS NAT LAB, 87- *Concurrent Pos:* Vis staff mem, Los Alamos Sci Lab, 74-75. *Mem:* Am Phys Soc; Am Geophys Union. *Res:* Linear, nonlinear plasma instabilities; plasma transport; space plasmas. *Mailing Add:* SST-8 MS D438 Los Alamos Nat Lab Los Alamos NM 87545

GARZA, CUTBERTO, b San Diego, Tex, Aug 26, 47; m 70; c 3. MATERNAL-INFANT HEALTH, PROTEIN & ENERGY METABOLISM. *Educ:* Baylor Univ, BS, 69; Baylor Col Med, MD, 73; Mass Inst Technol, PhD(nutrit biochem), 76. *Prof Exp:* From asst prof to assoc prof nutrit & gastrointestinal, dept pediat, 76-84, asst prof physiol, 77-85, assoc prof physiol, Baylor Col Med, 85-86; PROF, DEPT PEDIAT, CORNELL UNIV, 88-, DIR DIV NUTRIT SCI, 88- *Concurrent Pos:* Goldberger fel, 70; NIH trainee, nutrit biochem & metab, Mass Inst Technol, 72-75; mem consult staff, pediat serv, Harris County Hosp Dist, Houston, 76; adj prof nutrit, Sch Pub Health, Univ Tex, 76, adj asst prof, Sch Allied Health, 78; mem med staff, nutrit & gastrointestinal, Tex Children's Hosp, 82, internal med, St Luke's Episcopal Hosp, 84; mem food & nutrit bd, Inst Med, 90-93; mem bd Sci Counr, Nat Cancer Inst, 90-94. *Mem:* Am Soc Clin Nutrit; Am Col Nutrit; Am Inst Nutrit; AAAS; Soc Pediat Res; Am Pediat Soc. *Res:* Functional effects of feeding human milk or commercial formula to infants; regulation of lactation performance. *Mailing Add:* Div Nutrit Sci Savage Hall, Cornell Univ Ithaca NY 14853

GARZON, MAX, b Bogota, Columbia, Oct 1, 53; m 75; c 2. DISCRETE NEURAL NETWORKS, CELLULAR AUTOMATA. *Educ:* Nat Univ, Columbia, BS, 75; Univ Ill, MS, 80, PhD(computer sci), 84. *Prof Exp:* Asst prof math, Nat Univ, Columbia, 82; asst prof, 84-89, ASSOC PROF COMPUTER SCI, MEMPHIS STATE UNIV, 89- *Concurrent Pos:* Prin investr, NSF, 86-90; vis prof, UP-Ecole Normale, Lyon, France, 91-92. *Mem:* Am Math Soc; Inst Elec & Electronics Engrs; Asn Comput Mach. *Res:* Study models of computation by artificial and national intelligence, particularly complexity aspects of algorithms, turning machines, neural networks and automata networks. *Mailing Add:* Math Sci Memphis State Univ Memphis TN 38152

GARZON, RUBEN DARIO, b Quito, Ecuador, May 1, 37; US citizen; m 62; c 2. ARC INTERRUPTION, SOLID STATE SWITCHING. *Educ:* Calif State Univ, Los Angeles, BSEE, 68; Pa State Univ, MESc, 73. *Prof Exp:* Develop engr, ITE-Gould, 68-74; proj mgr, Gould-Brown Boueri, 74-79; sr proj mgr, Brown Boveri Elec, 80-85; DIR ENG, ASEA BROWN BOVERI, 85- *Concurrent Pos:* Prin investr, Energy Res & Develop Admin, US Dept

Energy, 74-76, Elec Power Res Inst, 78-80 & US Dept Energy, 80-81; mem adv panel, NSF Arc Res Proj grant, Univ Mich, 75-77. *Mem:* Sr mem Inst Elec & Electronics Engrs. *Res:* High current, high pressure arc interruption in sulfur hexaflouride circuit breakers; advanced development of switch gear devices using concepts for reduction and interruption of fault currents. *Mailing Add:* 240 Spring Valley Rd Columbia SC 29223

GASBARRE, LOUIS CHARLES, b Ridgway, Pa, May 15, 48; m 70; c 2. BOVINE IMMUNOLOGY. *Educ:* Ind Univ, Pa, BS, 70; Univ Md, MS, 74, PhD(zool), 78. *Prof Exp:* Postdoctoral fel immunol, Res & Training Ctr, WHO, 78-80; MICROBIOLOGIST IMMUNOPARASITOL, AGR RES SERV, USDA, 81- *Concurrent Pos:* Mem bd dirs, Am Asn Vet Immunologists, 89- *Mem:* Am Asn Immunologists; Am Asn Vet Immunologists; Am Asn Vet Parasitologists; AAAS. *Res:* Immunology of gastrointestinal nematodes infections of ruminants and definition of cellular immune responses of cattle. *Mailing Add:* LPSI HDL Agr Res Serv USDA Bldg 1040 Rm 2 Barc-E Beltsville MD 20705

GASCHO, GARY JOHN, b Bad Axe, Mich, Apr 9, 41; m 61; c 2. AGRONOMY. *Educ:* Mich State Univ, BS, 63, PhD(soil fertil & plant physiol), 68; Univ Ill, MS, 65. *Prof Exp:* from asst prof to prof plant nutrit, Agr Res & Educ Ctr, Univ Fla, 68-80; PROF SOIL FERTIL, COASTAL PLAIN EXP STA, UNIV GA, 80- *Concurrent Pos:* Actg dir, Agr Res & Educ Ctr, Belle Glade, Fla, 79-80; CONSULT. *Honors & Awards:* Outstanding Res Award, Sigma Xi, 83. *Mem:* Am Soc Agron; Soil Sci Soc Am; Am Peanut Res & Educ Soc; Coun Agr Sci & Technol; Am Soc Sugarcane Technol. *Res:* Peanut nutrition and fertility; sugarcane nutrition and physiology; alternate agronomic crop nutrition; fertility of irrigated crops; fertigation; tissue and soil analysis. *Mailing Add:* Dept Agron Coastal Plain Exp Sta Univ Ga Tifton GA 31793

GASCOIGNE, NICHOLAS ROBERT JOHN, b London, Eng, Jan 13, 58; m 86. T CELL RECEPTOR STRUCTURE & BIOLOGY, T CELL DEVELOPMENT. *Educ:* Univ Wales, UK, BSc Hons, 80; Univ London, PhD(immunol), 83. *Prof Exp:* Postdoctoral fel molecular immunol, Dept Med Microbiol, Stanford Univ Sch Med, 83-87; ASST MEM, DEPT IMMUNOL, SCRIPPS RES INST, 87- *Concurrent Pos:* Fac mem, Grad Prog Molecular & Cellular Biol, Scripps Res Inst, 90- *Mem:* Am Asn Immunologists; AAAS. *Res:* Various aspects of T-cell activation, particularly on the genes and structure of the T-cell receptor for antigen; produced soluble form of the receptor for biochemical analysis. *Mailing Add:* Dept Immunol IMMI Scripps Res Inst 10666 N Torrey Pines Rd La Jolla CA 92037

GASDORF, EDGAR CARL, b Decatur, Ind, Nov 27, 31; m 58; c 1. ANIMAL PHYSIOLOGY. *Educ:* Purdue Univ, BS, 53, MS, 56, PhD(biol sci), 59. *Prof Exp:* From instr to asst prof, 59-66, ASSOC PROF BIOL, BRADLEY UNIV, 66- *Mem:* AAAS; Ecol Soc Am; Am Inst Biol Sci. *Res:* Reptilian physiological ecology. *Mailing Add:* Dept Biol Bradley Univ 1501 W Bradley Ave Peoria IL 61625

GASH, KENNETH BLAINE, b Brooklyn, NY, Jan 2, 33; m 63; c 1. ORGANIC CHEMISTRY. *Educ:* Pratt Inst, BS, 60; Ariz State Univ, PhD(org chem), 68. *Prof Exp:* Lab asst chem, Chas Pfizer & Co, Inc, NY, 50-52, res asst, 56-60; teacher pub schs, NY, 60-63; assoc prof, 67-77, PROF CHEM, CALIF STATE COL, DOMINGUEZ HILLS, 77- *Mem:* Am Chem Soc; Sigma Xi. *Res:* Chemistry of carbonium ions, acyl and aroyl-oxonium ions; investigations of linear free energy relationships; synthesis of novel compounds; redesigning undergraduate organic laboratory programs. *Mailing Add:* Small Col Calif State Univ Dominguez Hills 1000 E Victoria Carson CA 90747

GASH, VIRGIL WALTER, b Rock Falls, Ill, June 28, 19; m 44; c 2. PROCESS TECHNOLOGY, AGRICULTURAL CHEMISTRY. *Educ:* Cornell Col, BA, 42; Univ Ill, MS, 47, PhD(chem), 52. *Prof Exp:* Res chemist org chem, William S Merrell Co, 47-50; sr res group leader, Monsanto Co, St Louis, 52-76, mgr res, Monsanto Agr Prod Co, 76-82; RETIRED. *Mem:* Am Chem Soc. *Res:* Fluorocarbons; nitrogen heterocyclics and unsaturated nitrogen compounds; steroids; oxidations; aromatic substitution; alicyclic stereochemistry; functional fluids; agricultural chemistry. *Mailing Add:* 350 Sudbury Lane Ballwin MO 63011

GASIC, GABRIEL J, b Punta Arenas, Chile, Mar 18, 12; m 53; c 3. PATHOLOGY. *Educ:* Univ Chile, MD, 38. *Prof Exp:* Prof biol, Sch Med, Univ Chile, 48-60, dir oncol, 60-65; res assoc, 60-63, res prof, 66-67, PROF PATH, UNIV PA, 67-; PROF, LAB EXP ONCOL, PA HOSP, PHILADELPHIA, PA. *Concurrent Pos:* John Simon Guggenheim Mem Found res fel, 43; Carnegie Inst Res Univ res fel, 44-46; res fel, Rockefeller Univ, 46-47; grants, Rockefeller Found, Damon Runyon Mem Fund, Am Cancer Soc/Nat Cancer Inst, NSF & Pop Coun. *Honors & Awards:* Paget-Ewing Award, Metastasis Res Soc, 90- *Mem:* AAAS; Am Soc Cell Biol; Histochem Soc; Am Asn Cancer Res; Chilean Acad Sci & Med; Int Soc Thrombosis & Hemostasis. *Res:* Cell surface histochemistry and biology; tumor invasiveness; tumor cell-platelet interactions; growth factors. *Mailing Add:* Lab Exp Oncol Pa Hosp Eighth & Spruce Sts Philadelphia PA 19107

GASICH, WELKO E, b Cupertino, Calif, Mar, 28, 22. AERONAUTICAL ENGINEERING. *Educ:* Stanford Univ, BA, 43, MS, 47. *Prof Exp:* Corp vpres & asst gen mgr technol, Northrop Corp, 61-66, Corp vpres, Ventura Div, 67-71, corp vpres, Aircraft Div, 71-76, corp vpres & group exec, Aircraft Group, 76-79, sr vpres advan projs, 79-85, exec vpres prog, 85-88; CONSULT, 88- *Mem:* Nat Acad Eng; Am Inst Aeronaut & Astronaut; Soc Automotive Engrs. *Mailing Add:* 3517 Caribeth Dr Encino CA 91436

GASIDLO, JOSEPH MICHAEL, b Detroit, Mich June 29, 35. REACTOR PHYSICS, NUCLEAR ENGINEERING. *Educ:* Wayne State Univ, BSE, 58, MSE, 61. *Prof Exp:* Teaching asst eng mech, Wayne State Univ, 58-59; resident assoc reactor eng, 59-60, physicist, fast reactors, 60-85, SPECIALIST NUCLEAR CRITICALITY SAFETY, ARGONNE NAT LAB, 65-, NUCLEAR SAFETY NON-REACTOR NUCLEAR FACIL, 87- *Mem:* Am Nuclear Soc. *Res:* Neutron physics of fast breeder reactors specializing in reactivity measurements and reaction rate measurements; criticality safety in handling and processing fissile materials. *Mailing Add:* Argonne Nat Lab PO Box 2528 Idaho Falls ID 83403-2528

GASIORKIEWICZ, EUGENE CONSTANTINE, b Grabiszew, Poland, Mar 11, 20; nat US; m 46; c 2. PLANT SCIENCE, PLANT PATHOLOGY. *Educ:* Marquette Univ, AB, 47, MS, 48; Univ Wis, PhD(plant path), 51. *Prof Exp:* Asst bot, Marquette Univ, 46-48; asst plant path, Univ Wis, 48-51, proj assoc, USDA, 51-52; asst res prof bot, Univ Mass, 52-58, asst prof plant path, Waltham Field Sta, 58-61; horticulturist & plant pathologist, Plant Sci Lab, Biol Res Ctr, S C Johnson & Son, Inc, 61-65; asst prof bot, Univ Wis-Parkside, Racine Campus, 66-67; dir res & tech serv, Can-Am Plant Co, Ltd, 67-68; from assoc prof to prof life sci, Univ Wis-Parkside, 68-70, chmn, Div Sci, 68-73, dir, natural sci areas, 68-77, EMER PROF LIFE SCI & ALLIED HEALTH, UNIV WIS-PARKSIDE, 88- *Concurrent Pos:* Med malpractice investr, 88- *Mem:* AAAS; Am Phytopath Soc; Am Inst Biol Sci. *Res:* Diseases of greenhouse florist crops; turf and ornamental plants; growth regulants and disease development; pathogen free plants and disease control; ecology of natural areas; Prairie restoration. *Mailing Add:* Dept Biol Scis Univ Wis-Parkside Box 2000 Kenosha WI 53141

GASIOROWICZ, STEPHEN G, b Gdansk, Poland, May 10, 28; nat US; m 53; c 3. PARTICLE PHYSICS. *Educ:* Univ Calif, Los Angeles, AB, 48, MA, 49, PhD(theoret physics), 52. *Prof Exp:* Physicist, Lawrence Radiation Lab, Univ Calif, 52-60; NSF fel, Inst Theoret Physics, Copenhagen & European Orgn Nuclear Res, 57-58; vis scientist, Max Planck Inst Physics & Astrophys, 59-60; assoc prof, 60-63, PROF PHYSICS, UNIV MINN, MINNEAPOLIS, 63- *Concurrent Pos:* Consult, Argonne Nat Lab, 61-70; vis scientist, Nordic Inst Theoret Atomic Physics, Copenhagen & Univ Marseille, 64 & Deutsches Elektron-Synchroton, Hamburg, 68-69 & 80; trustee, Aspen Ctr Physics, 80-86; vis prof, Univ Tokyo, 82. *Mem:* Fel Am Phys Soc. *Res:* Elementary particle physics. *Mailing Add:* Dept Physics Univ Minn Minneapolis MN 55455

GASKELL, DAVID R, b Glasgow, Scotland, Mar 11, 40; m 64; c 3. METALLURGY. *Educ:* Glasgow Univ, BSc, 62; McMaster Univ, PhD(metall), 67. *Hon Degrees:* MA, Univ Pa, 72. *Prof Exp:* Metallurgist, LaPorte Chem Ltd, Luton, Eng, 62-64; from asst prof to assoc prof metall, Univ Pa, 67-79, prof metall & geol, 79-82; PROF METALL ENG, PURDUE UNIV, 82- *Mem:* Am Inst Mining, Metall & Petrol Engrs; Can Inst Mining & Metall; Iron & Steel Inst Japan; Am Soc Metals. *Res:* Physical chemistry of liquid oxide systems; slag-metal chemistry; thermodynamics; chemical and extraction metallurgy. *Mailing Add:* Sch Mat Eng Mat & Elec Eng Bldg Purdue Univ West Lafayette IN 47907

GASKELL, PETER, b Lancashire, Eng, June 24, 17; m 40; c 3. PHYSIOLOGY. *Educ:* Univ Western Ont, MD, 50; Univ London, PhD(physiol), 55. *Prof Exp:* From asst prof to prof physiol, 55-85, EMER PROF, UNIV MAN, 86- *Concurrent Pos:* Med res assoc, Med Res Coun Can, 57-; mem clin invest unit, Winnipeg Gen Hosp, 57-85. *Res:* Peripheral circulation, especially in the limbs; hypertension. *Mailing Add:* 826 Campbell St Winnipeg MB R3N 1C6 Can

GASKELL, ROBERT EUGENE, b Grelton, Ohio, Jan 18, 12; m 40; c 2. APPLIED MATHEMATICS. *Educ:* Albion Col, AB, 33; Univ Mich, MS, 34, PhD(math), 40. *Prof Exp:* Instr math, Albion Col, 38-39 & Univ Ala, 40-42; res fel mech, Brown Univ, 42-43, res assoc, 43-46; asst prof math, Iowa State Col, 47-49, assoc prof, 49-51; supvr math servs unit, Boeing Airplane Co, 51-59; prof math, Ore State Univ, 59-66; chmn dept, 66-72, prof, 66-80, EMER PROF MATH, NAVAL POSTGRAD SCH, 80- *Mem:* AAAS; Soc Indust & Appl Math; Math Asn Am; Am Soc Eng Educ. *Res:* Industrial and engineering applications of mathematics. *Mailing Add:* 1207 Sylvan Rd Monterey CA 93940

GASKELL, ROBERT WEYAND, b Providence, RI, July 9, 45; m 67; c 3. PLANETARY PHYSICS. *Educ:* Brown Univ, ScB, 67; McGill Univ, PhD(physics), 72. *Prof Exp:* Fel physics, Carleton Univ, 72-74 & Univ Toronto, 74-75; res assoc physics, McGill Univ, 75-77; asst prof physics, Lafayette Col, 77-84; MEM TECH STAFF, JET PROPULSION LAB, 84- *Mem:* Am Phys Soc. *Res:* Satellite shape and topography; planetary surface simulations. *Mailing Add:* 1865 Sonoma Dr Altadena CA 91001

GASKILL, HERBERT STOCKTON, b Philadelphia, Pa, Jan 31, 09; m 38; c 4. PSYCHIATRY. *Educ:* Haverford Col, AB, 32; Univ Pa, MD, 37; Am Bd Psychiat & Neurol, dipl, 45. *Prof Exp:* Intern, Pa Hosp Philadelphia, 37-39; res neurol, Jefferson Med Col, 39-40; fel psychiat, Pa Hosp Philadelphia, 41-42; assoc psychiat & med, Univ Pa, 46-48, asst prof psychiat, 48-49; prof, Sch Med, Ind Univ, 49-53; prof psychiat & chmn dept, 53-73, EMER PROF PSYCHIAT, SCH MED, UNIV COLO, DENVER, 73- *Concurrent Pos:* Instr, Jefferson Med Col, 39-41, fel neuropath, 40-41; consult, Vet Admin Hosp, Lowry AFB, 53-57 & Fitzsimmons Army Gen Hosp, Denver, Colo, 53-75; mem selection comt for res scientist award, NIMH, 61-64, mem continuous ed training comt, 67-70, chmn, 67-68; chmn ad hoc comt, Nat Bd Med Exam, 61-64; mem adv comt residency training, Calif Dept Ment Hyg, chmn, 69-71; training & supv analyst, Denver Psychoanal Inst, 70-, chmn, 80-84. *Mem:* Am Psychoanal Asn; Am Psychiat Asn (vpres, 70-71); Am Psychosom Soc; Am Psychoanal Asn (pres, 75-77). *Mailing Add:* 4505 S Yosemite St No 364 Stoney Brook Denver CO 80237

GASKILL, IRVING E, b Mt Holly, NJ, Feb 24, 22; m 58; c 3. MATHEMATICS. *Educ:* Trenton State Teachers Col, BS, 43; Univ Pa, MA, 47. *Prof Exp:* Instr math, Bowling Green State Univ, 47-49; mathematician, Army Map Serv, 51-56 & US Army Engr Math Comput Agency, 56-70; dir math & comput lab, Nat Resource Anal Ctr, 70-75, Fed Preparedness Agency, 75-80; RETIRED. *Mem:* Am Math Soc. *Res:* Mathematical models for computation on electronic computer; economic models; war gaming vulnerability models; nuclear damage assessment models; geodetic network and datum adjustments. *Mailing Add:* 405 Belmont Dr SW Leesburg VA 22075-3510

GASKILL, JACK DONALD, b Ft Collins, Colo, Dec 9, 35; m 56; c 2. OPTICAL SCIENCES. *Educ:* Colo State Univ, BS, 57; Stanford Univ, MS, 65, PhD(elec eng), 68. *Prof Exp:* Electronics engr, Motorola, Inc, 57-58; res asst elec eng, Stanford Univ, 63-68; from asst prof to assoc prof, 68-75, PROF OPTICAL SCI, UNIV ARIZ, 75-, ADMINR ACAD AFFAIRS, 80- *Concurrent Pos:* Consult, US Army Res Off, 70-, Hughes Res Labs, 74- & Bendix Corp, 78- *Mem:* Inst Elec & Electronics Engrs; Optical Soc Am; fel Soc Photo-Optical Instrument Engrs. *Res:* Fourier optics; holography; medical optics. *Mailing Add:* Optical Sci Ctr Univ Ariz Tucson AZ 85721

GASKIN, DAVID EDWARD, b Croydon, Eng, June 21, 39; m 62; c 2. MARINE BIOLOGY, ENTOMOLOGY. *Educ:* Bristol Univ, BSc, 61; Massey Univ, NZ, PhD(entom), 68. *Prof Exp:* Whaling inspector & biologist, Brit Ministry Agr Fisheries & Food-Nat Inst Oceanog, Wormley, Eng, 61-62, asst exp officer fisheries res, Brit Ministry Agr Fisheries & Food, Lowestoft, 62; whale fisheries biologist, Fisheries Res Div, Wellington, NZ, 62-65; lectr zool, Massey Univ, NZ, 65-68; assoc prof, 68-80, PROF MARINE BIOL, DEPT ZOOL, UNIV GUELPH, 80- *Concurrent Pos:* Consult taxonomist, NZ Dept Agr, Wellington, 63-64; NZ sci rep, Int Whaling Comn, 63-68; NZ Univ Grants Comt res grant, 65-67; res grant, Nat Res Coun Can, 69-, mem grant comt, 79-82. *Mem:* Fel Royal Entom Soc London; NZ Entom Soc; Can Soc Zool. *Res:* Ecology and biology of Southern Hemisphere and North West Atlantic cetaceans; systematics and ecology of the family Crambinae; biogeography. *Mailing Add:* Dept Zool Univ Guelph Guelph ON N1G 2W1 Can

GASKIN, FELICIA, b Carlisle, Pa, Jan 17, 43; m 69; c 2. BIOCHEMISTRY, NEUROCHEMISTRY. *Educ:* Dickinson Col, AB, 65; Bryn Mawr Col, MA, 67; Univ Calif, San Francisco, PhD(biochem), 69. *Prof Exp:* Fel biochem, Stanford Univ, 69-71; res assoc phys biochem, Rockefeller Univ, 71-72 & Columbia Univ, 72-74; from asst prof to assoc prof path & biophys, Albert Einstein Col Med, 74-82; MEM & PROF, OKLA MED RES FOUND, 82- *Concurrent Pos:* Nat Inst Neurol Dis & Stroke spec res fel, 72-74; NIH res career develop award, 75-80. *Mem:* Am Chem Soc; Am Soc Biol Chem; Biophys Soc; Soc for Neurosci; Sigma Xi; Am Asn Neuropath. *Res:* Microtubule assembly; neurofibrous proteins in aging; immunological probes to study Alzheimer's disease. *Mailing Add:* Okla Med Res Found 825 N E 13th St Oklahoma City OK 73104

GASKIN, JACK MICHAEL, b Watertown, NY, Jan 8, 43; m; c 3. VETERINARY MICROBIOLOGY. *Educ:* Cornell Univ, DVM, 67, PhD(vet microbiol), 73. *Prof Exp:* Vet, NY, 67-73; asst prof vet sci & asst virologist, 73-80, ASSOC PROF, DEPT INFECTIOUS DIS, UNIV FLA, 80- *Mem:* Am Vet Med Asn. *Res:* Bovine syncytial virus isolations; canine distemper virus in domesticated cats and pigs; herpes virus infection of psittacine birds resembling Pacheio's parrot disease. *Mailing Add:* Dept Infectious Dis ISIS 633 Univ Fla Med Col Gainesville FL 32610

GASKINS, H REX, b Morehead City, NC, Jan 30, 58; m 87. DEVELOPMENTAL ENDOCRINOLOGY, LOCAL HORMONES & TISSUE DEVELOPMENT. *Educ:* NC State Univ, BS, 81, MS, 86; Univ Ga, PhD(cell biol), 89. *Prof Exp:* Postdoctoral fel, Am Diabetes Asn, 89-91, POSTDOCTORAL FEL, JUVENILE DIABETES FOUND INT, THE JACKSON LABORATORY, 91- *Mem:* Am Diabetes Asn; Am Asn Immunologists; Soc Exp Biol & Med; AAAS; Sigma Xi; Juvenile Diabetes Found Int. *Res:* Genetic mechanisms which underlie pancreatic beta cell destruction in mouse models of diabetes, utilize both classical and molecular genetics techniques to identify genes encoding diabetes susceptibility and determine how their expression in beta cells and macrophages contributes to disease. *Mailing Add:* The Jackson Laboratory Bar Harbor ME 04609

GASKINS, MURRAY HENDRICKS, b 1927. PLANT PHYSIOLOGY. *Educ:* Univ Ga, BS, 48; Univ Fla, MS, 56, PhD(agr), 58. *Prof Exp:* Horticulturist, Agr Res Serv, USDA, 58, officer in chg plant introd sta, Miami, 58-64, officer in chg, Fed Exp Sta, Mayaguez, PR, 64-71, res leader, Plant Sci Unit Agr Res Serv, 71-77; plant physiologist, USDA & prof plant physiol, Dept Agron, 77-86, EMER PROF, UNIV FLA, 86- *Mem:* Am Soc Hort Sci; Am Soc Plant Physiol; Soc Econ Bot. *Res:* Nitrogen fixation; fruit crops; tropical agriculture; plant growth substances. *Mailing Add:* 1259 NW 60th St Gainesville FL 32605

GASPAR, MAX RAYMOND, b Sioux City, Iowa, May 10, 15; m 38; c 5. SURGERY, VASCULAR SURGERY. *Educ:* Morningside Col, AB, 36; Univ SDak, BS, 38; Univ Southern Calif, MD, 41; Am Bd Surg, dipl; FACS. *Prof Exp:* SURGEON, HARRIMAN JONES CLIN HOSP, 48-; CHIEF OF STAFF, ST MARY'S HOSP, 67- *Concurrent Pos:* Clin instr, Univ Calif, Los Angeles, 52-54; assoc clin prof surg, Loma Linda Univ, 53-61, clin prof, 62-65; clin prof, Univ Southern Calif, 65-; dir peripheral vascular serv, Los Angeles County Hosp, 54-; consult vascular surg, US Naval Hosp, Long Beach. *Mem:* Int Soc Surg; Am Surg Asn; Soc Vascular Surg; Int Cardiovasc Soc; Soc Clin Vascular Surg (pres, 79-81). *Res:* Intestinal anastomosis; vascular surgery. *Mailing Add:* 1045 Atlantic Long Beach CA 90813

GASPAR, PETER PAUL, b Brussels, Belg, June 20, 35; US citizen; m 65; c 1. PHYSICAL ORGANIC CHEMISTRY, RADIOCHEMISTRY. *Educ:* Calif Inst Technol, BS, 57; Yale Univ, MS, 58, PhD(chem), 61. *Prof Exp:* NATO fel org chem, Univ Heidelberg, 61-62; res fel chem, Calif Inst Technol, 62-63; from asst prof to assoc prof, 63-73, PROF CHEM, WASHINGTON UNIV, 73- *Concurrent Pos:* Res collabr, Brookhaven Nat Lab, 69- *Honors & Awards:* Frederic Stanley Kipping Award, Organocilicon Chem, 86. *Mem:* Am Chem Soc; Royal Soc Chem. *Res:* Reaction mechanisms; reactive intermediates; photochemistry; gasphase reactions of free atoms and free radicals; organosilicon chemistry; main group organometalic chemistry; archaeometry. *Mailing Add:* Dept Chem Washington Univ Lindell-Skinker Blvd St Louis MO 63130

GASPARD, KATHRYN JANE, ENDOCRINOLOGY. *Educ:* Med Col Wis, PhD(physiol), 75. *Prof Exp:* ASST PROF PHYSIOL & PATHOPHYSIOL, SCH NURSING, MED COL WIS, 81- *Mailing Add:* 3745 N 97th St Milwaukee WI 53222

GASPARINI, FRANCIS MARINO, b Trieste, Italy, Oct 14, 41; US citizen; m 67; c 2. LOW TEMPERATURE PHYSICS. *Educ:* Villanova Univ, BS, 64; Univ Minn, MS, 66, PhD(physics), 70. *Prof Exp:* Instr physics, Villanova Univ, 66-67; fel, Ohio State Univ, 70-73; PROF PHYSICS, STATE UNIV NY, BUFFALO, 73- *Mem:* Fel Am Phys Soc; AAAS; Sigma Xi. *Res:* Critical aspects of the superfluid transition and surface properties of helium; finite size effects at phase transitions; two-dimensional behavior of He-3. *Mailing Add:* Dept Physics Fronczak Hall State Univ NY Buffalo Amherst NY 14260

GASPAROVIC, RICHARD FRANCIS, b Endicott, NY, Mar 9, 41; m 65; c 2. APPLIED PHYSICS, SYSTEMS ENGINEERING. *Educ:* Fordham Univ, BS, 63; Univ Pa, MS, 64; Rutgers Univ, PhD(physics), 69. *Prof Exp:* Engr electronics, RCA Corp, 69-71; PRIN STAFF PHYSICIST OCEAN REMOTE SENSING, APPL PHYSICS LAB, JOHNS HOPKINS UNIV, 71- *Mem:* Am Geophys Soc; Am Phys Soc; AAAS. *Res:* Remote sensing of ocean surface; air-sea interaction processes; infrared systems engineering; digital image processing. *Mailing Add:* Appl Physics Lab S1R Group MS 47-106 Johns Hopkins Univ Johns Hopkins Rd Laurel MD 20723-6099

GASPARRINI, CLAUDIA, MINERALOGY. *Educ:* Univ Rome, Italy, PhD(earth sci), 65. *Prof Exp:* Sr technician, res asst & res assoc, Univ Toronto, 66-72; phys scientist II, Geol Surv Can, Ottawa, 73; res scientist & mineralogist, Nat Inst Metall, Johannesburg, SAfrica, 74-75; PRES & OWNER, MINMET SCI LTD, TORONTO, CAN, 76-, TUCSON, ARIZ, 81- *Concurrent Pos:* Mem & vchmn, Process Mineral Comt, Am Inst Mining Engrs, 80; pres & owner, Space Eagle Publ Co Inc, Tucson, Ariz, 89-; mem, Res Bd Adv, Am Biog Inst, 90-; mem, Sci-by-Mail Prog, Mus Sci, Science Park, Boston, 90- *Mem:* Microbean Soc Am; Can Inst Mining & Metall; Metall Soc Am Inst Mining Engrs; Int Precious Metal Inst; Soc Geol Appl Mineral Deposits; Am Asn Women Sci; Am Soc Prof & Exec women; Nat Orgn Women; Nat Asn Prof & Exec Women. *Res:* Author of various publications; precious metals. *Mailing Add:* Minmet Sci Inc PO Box 41687 Tucson AZ 85717

GASPER, GEORGE, JR, b Hamtramck, Mich, Oct 10, 39; m 67; c 2. MATHEMATICAL ANALYSIS. *Educ:* Mich Technol Univ, BS, 62; Wayne State Univ, MA, 64, PhD(math), 67. *Prof Exp:* NSF vis lectr math, Univ Wis, Madison, 67-68; Nat Res Coun Can fel, 68-69; vis asst prof math, Univ Toronto, 69-70; from asst prof to assoc prof, 70-76, PROF MATH, NORTHWESTERN UNIV, 76- *Concurrent Pos:* NASA traineeship, 66-67; Alfred P Sloan fel, 73-75. *Mem:* Am Math Soc; Soc Indust & Appl Math. *Res:* Analysis; special functions; multipliers; orthogonal expansions; convolution structures; positive kernels; basic hypergeometric series. *Mailing Add:* Dept Math Northwestern Univ Evanston IL 60208

GASS, CLINTON BURKE, b Minn, Jan 9, 20; m 41; c 3. MATHEMATICS. *Educ:* Gustavus Adolphus Col, AB, 41; Univ Nebr, MA, 43, PhD(math), 54. *Prof Exp:* Instr math, Univ Nebr, 42-43; from assoc prof to prof, Nebr Wesleyan Univ, 43-47, prof & dean men, 47-53, chmn div natural sci, 53-54; from assoc prof to prof, 54-64, head dept math, astron & computer sci, 60-84, John T & Margaret Deal Prof Math, 64-86, sr prof, 86- 90, EMER PROF MATH, DEPAUW UNIV, 90- *Concurrent Pos:* Consult pub schs, Indianapolis, 58-59; chmn state math adv comt, Nat Defense Educ Act, 59-; assoc prog dir summer insts prog, NSF, Wash,DC, 65-66; math consult, US Dept Defense Overseas Schs, Europe, 69-70; resident dir, semester in Ger & France, DePauw Univ, 77; instr, Challenge Prog, US Dept Defense in Europe, 78-84, 87, 89 & 90. *Mem:* Am Math Soc; Math Asn Am. *Res:* Eigenfunction expansions; mathematical analysis. *Mailing Add:* 707 Highridge Ave Greencastle IN 46135

GASS, FREDERICK STUART, b Lincoln, Nebr, Apr 21, 43; m 66. MATHEMATICS. *Educ:* DePauw Univ, BA, 64; Dartmouth Col, AM, 66, PhD(math), 69. *Prof Exp:* Asst prof math, 68-75, ASSOC PROF MATH & STATIST, MIAMI UNIV, OXFORD, OHIO, 75- *Concurrent Pos:* Vis asst prof, Talladega Col, 69-70. *Mem:* Math Asn Am; Asn Symbolic Logic. *Res:* Mathematical logic; recursive function theory; ordinal notation theory. *Mailing Add:* Dept Math Miami Univ Oxford OH 45056

GASS, GEORGE HIRAM, b Sunbury, Pa, Sept 23, 24; m 48; c 3. ENDOCRINOLOGY, PHARMACOLOGY. *Educ:* Bucknell Univ, BS, 48; Univ NMex, MS, 52; Ohio State Univ, PhD(physiol), 55. *Prof Exp:* Exp biologist, Lederle Labs, Am Cyanamid Co, 48-51; asst chief, Endocrine Br, Div Pharmacol, Food & Drug Admin, 55-59; prof physiol & dir, Endocrinol Pharmacol Res Lab, Southern Ill Univ, 59-79; chmn, Dept Basic Sci, Okla Col Osteop Med & Surg, 79-84; CONSULT 84-, ADJ PROF, ORAL ROBERTS UNIV SCH MED, 87- *Concurrent Pos:* Mem biol sci comt, Grad Sch, USDA, 58-60; sr scientist fel, Alexander von Humboldt Found, WGer, 67-68 & fel; consult, A B Leo Pharmaceut Co, Sweden; vis prof, Nat Ctr Toxicol Res, 75, prin investr hormonal carcinogenesis; consult & ed, Endocrine Toxicol, 84-88; Fulbright fel, 72. *Mem:* Fel AAAS; Am Physiol Soc; Endocrine Soc; Am Soc Animal Sci; Soc Study Reprod. *Res:* Hormone assay; development of new assay methods; development of knowledge of the physiology and pharmacology of chemical compounds affecting the functions of the endocrine system; carcinogenicity of steroid hormones; mechanism of gastric function. *Mailing Add:* 6718 E 79th St Tulsa OK 74133

GASS, SAUL IRVING, b Chelsea, Mass, Feb 28, 26; m 46; c 2. MATHEMATICS. *Educ:* Boston Univ, BS & MA, 49; Univ Calif, PhD, 65. *Prof Exp:* Mathematician, Aberdeen Bombing Mission, 49-51; dir mgt anal, Hqs, USAF, 52-55; appl sci rep, Int Bus Mach Corp, 55-58; chief opers res br, C-E-I-R, Inc, 58-60; sr mathematician, Fed Systs Div, Int Bus Mach Corp, Md, 60-61, systs mgr, 61-66, mgr comput sci & oper res, 66-69; vpres, World Systs Labs, Inc, Md, 69-70; dir opers res, Mathematica, Inc, 70-75; chmn dept, 75-79, PROF MGT SCI & STATIST, COL BUS & MGT, UNIV MD, 75- *Mem:* AAAS; Am Fedn Info Processing Socs (secy, 63-65); Soc Indust & Appl Math; Opers Res Soc Am (vpres & pres, 75-77); Math Asn Am; Inst Mgt Sci; Am Inst Decision Sci. *Res:* Linear programming; game theory; operations research; digital computer applications. *Mailing Add:* Dept Mgt Sci & Statist Col Bus & Mgt Univ Md College Park MD 20742

GASSAWAY, JAMES D, b Fulton, Mass, Aug 17, 32. ELECTRICAL ENGINEERING. *Educ:* Univ Miss, BS, 57; Purdue Univ, PhD(elec eng), 64. *Prof Exp:* Design engr, Gen Dynamics Corp, Tex, 57-59; instr elec eng, Purdue Univ, 59-63, asst prof, 63-65; assoc prof, Univ Ala, 65-67; adj prof elec eng, Miss State Univ, 67-80, prof, 80-90; AT HOWARD INDUSTS, LAUREL, MISS, 90- *Concurrent Pos:* Res asst, Am Machine & Foundry, Purdue Univ, 62-63; consult, Wabash Magnetics, Ind, 64 & Radiation Inc, Fla, 68-69. *Mem:* Am Soc Eng Educ; Inst Elec & Electronics Engrs; Sigma Xi. *Res:* Modeling of semiconductor fabrication processes and semiconductor device behavior using digital computer; development of sensitive electronic instruments; thin magnetic films and magnetic properties of rolled sheets; techniques for reduction of limit cycle oscillations in non-linear feedback systems. *Mailing Add:* Howard Industs PO Box 1588 3225 Pendorf Rd Laurel MS 39441

GASSER, CHARLES SCOTT, b Ft Benning, Ga, July 26, 55; m 89. PLANT MOLECULAR BIOLOGY, MOLECULAR GENETICS. *Educ:* Univ Calif, Davis, BS, 78; Stanford Univ, PhD(biol), 85. *Prof Exp:* Sr res biologist, Monsanto Co, 85-87, res specialist, 87-89, sr res specialist, 89; ASST PROF, DEPT BIOCHEM & BIOPHYS, UNIV CALIF, DAVIS, 89- *Concurrent Pos:* NSF presidential young investr award, 90. *Mem:* Int Soc Plant Molecular Biol; Am Soc Plant Physiol; Tomato Genetics Soc; AAAS. *Res:* Immunophilins and prolyl isomerases of higher plants; molecular basis of flower development; flower-specific gene expression. *Mailing Add:* Dept Biochem & Biophys Univ Calif Davis CA 95616

GASSER, DAVID LLOYD, b Wadsworth, Ohio, Feb 12, 43; m 68; c 2. GENETICS, IMMUNOLOGY. *Educ:* Univ Akron, BS, 64; Univ Mich, MS, 66, PhD(zool), 70. *Prof Exp:* NIH fel, Univ Pa, 70-72, asst prof, 72-78, ASSOC PROF HUMAN GENETICS, UNIV PA, 78- *Mem:* AAAS; Am Soc Human Genetics; Genetics Soc Am; Am Asn Immunologists. *Res:* Genetic control of the immune response; genetics of susceptibility to teratogenic agents. *Mailing Add:* Dept Human Genet Univ Pa Sch Med Philadelphia PA 19104-6145

GASSER, HEINZ, b Trelex, Switz, Mar 23, 32; Can citizen; m 59; c 4. CROP PHYSIOLOGY, PLANT BREEDING. *Educ:* McGill Univ, BSc, 59, MSc, 62; Univ Nottingham, PhD(crop physiol), 65. *Prof Exp:* Actg officer-in-chg, Ft Chimo Substa, 59; plant protection officer, 59-60, res officer forage crops, Res Sta, 64-67, res scientist, Res Sta, Can Dept Agr, 67-76, sect head genetics & plant breeding, 70-76; dir res sta, Res & Teaching Br, Que Ministry Agr, 76-82; DIR TRAINING CTR, INT INST TROP AGR, NIGERIA, 82- *Concurrent Pos:* Res scientist, Can Dept Agr Res Br Transfer, Lusignan Plant Breeding Res Sta, France, 70-71. *Mem:* Crop Sci Soc Am; Am Soc Agron; Can Soc Agron; Inst Pub Admin Can. *Res:* Winter resistance and survival of alfalfa through plant breeding and management; Rhizobium specificity in varieties of forage legumes. *Mailing Add:* 2800 Ch St-Louis Ste-Foy PQ G1W 1P1 Can

GASSER, RAYMOND FRANK, b Cullman, Ala, Sept 13, 35; m 61; c 2. ANATOMY, PHYSIOLOGY. *Educ:* Spring Hill Col, BS, 59; Univ Ala, MS, 62, PhD(anat, physiol), 65. *Prof Exp:* Instr embryol, Med Ctr, Univ Ala, 65, from instr to assoc prof, 67-74, assoc dean sch med, 74-75, PROF ANAT, MED CTR, LA STATE UNIV, 74- *Mem:* Teratology Soc; Am Asn Clin Anatomists; Am Asn Anatomists. *Res:* Morphogenesis of the head and neck regions in man; morphometrics of neural tube and surrounding structures; effects of physical manipulation on cell differentiation, human embryology. *Mailing Add:* Dept Anat Med Ctr La State Univ New Orleans LA 70112-1393

GASSER, WILLIAM, b East Conemaugh, Pa, Nov 22, 23; m 60; c 3. ORGANIC CHEMISTRY. *Educ:* Waynesburg Col, BS, 47; Univ Mich, MS, 48; Univ Md, MS, 52, PhD(chem), 55. *Prof Exp:* Instr chem, Waynesburg Col, 48-49, from asst prof to assoc prof, 51-53; asst, Univ Md, 49-51 & 53-55; res chemist, Visking Co, Union Carbide Corp, 55-62; mem legal staff, Chas Pfizer & Co, 62-63; from asst prof to assoc prof, 63-71, PROF CHEM, QUINCY COL, 71- *Concurrent Pos:* Consult, Blessing Hosp. *Mem:* AAAS; Am Chem Soc; Sigma Xi. *Res:* Organic synthesis; polymers; patent law. *Mailing Add:* 1828 Maine St Quincy IL 62301

GASSIE, EDWARD WILLIAM, b Addis, La, Nov 29, 25; m 49; c 3. ANIMAL SCIENCE, INFORMATION SCIENCE. *Educ:* La State Univ, BS, 51, MS, 58, PhD(agr educ), 64. *Prof Exp:* From asst county agent to assoc county agent, 51-55, dist supvr, 55-64, assoc prof, 64-68, PROF EXTEN EDUC, LA STATE UNIV, BATON ROUGE, 68-, TRAINING SPECIALIST, COOP EXTEN SERV, 64-, HEAD, DEPT EXTEN & INT EDUC, 73- *Concurrent Pos:* Mem, Nat Exten Curric Develop Comt, 64- *Res:* Studies of behavioral changes in individuals and groups as a result of planned educational programs and the many variables associated with the extent of these changes. *Mailing Add:* 142 Old Forestry Bldg La State Univ Sch Voc Ed Baton Rouge LA 70803

GASSMAN, MERRILL LOREN, b Chicago, Ill, Feb 10, 43; m 67; c 3. PLANT PHYSIOLOGY, BIOCHEMISTRY. *Educ:* Univ Chicago, SB, 64, SM, 65, PhD(bot), 67. *Prof Exp:* Guest investr & USPHS res fel, Rockefeller Univ, 67-68; res plant physiologist, Int Minerals & Chem Corp, Ill, 68-69; res assoc, Argonne Nat Lab, 69; from asst prof to assoc prof, 69-82, PROF BIOL SCI, UNIV ILL, CHICAGO, 82- *Concurrent Pos:* Consult, Int Minerals & Chem Corp, Ill, 69-70; prin investr, NSF grants, 72-87; vis assoc res botanist, Univ Calif, Davis, 76; vis prof bot & microbiol, Ariz State Univ, Tempe, 83. *Mem:* AAAS; Am Soc Plant Physiol; Am Soc Photobiol; Am Soc Biochem & Molecular Biol; Sigma Xi. *Res:* Photo induction of chloroplast development; control systems of porphyrin and chlorophyll metabolism in plants. *Mailing Add:* Dept Biol Sci Box 4348 Univ Ill at Chicago Chicago IL 60680

GASSMAN, PAUL G, b Buffalo, NY, June 22, 35; m 57; c 7. ORGANIC CHEMISTRY. *Educ:* Canisius Col, BS, 57; Cornell Univ, PhD(org chem), 60. *Hon Degrees:* DSc, John Carroll Univ, 89. *Prof Exp:* From asst prof to prof chem, Ohio State Univ, 61-74; chmn dept, 75-79, prof chem, 74-88, REGENTS' PROF, UNIV MINN, 88- *Concurrent Pos:* Alfred P Sloan fel, 67-69. *Honors & Awards:* Petrol Chem Award, 72; James Flack Norris Award, Am Chem Soc, 85; Arthur C Cope Scholar Award, 86; Chem Pioneers Award, Am Inst Chemists, 90; Nat Catalyst Award, Chem Mfrs Asn, 90. *Mem:* Am Chem Soc; The Chem Soc; Am Inst Chemists; AAAS. *Res:* Mechanisms of catalysis; organoelectrochemistry; x-ray photoelectron spectroscopy; carbanion chemistry; chemistry of highly-strained molecules; synthesis of heterocyclic molecules; neighboring group participation in carbocation chemistry; oxidation of hydrocarbons; nitrenium ion chemistry; cycloaddition reactions. *Mailing Add:* Dept Chem Univ Minn Minneapolis MN 55455

GAST, JAMES AVERY, b New Rochelle, NY, Apr 28, 29; m 55; c 3. OCEANOGRAPHY. *Educ:* Amherst Col, AB, 53; Univ Wash, MS, 57, PhD(oceanog), 59. *Prof Exp:* Asst, Woods Hole Oceanog Inst, 47-48; asst Univ Wash, 53-58, assoc & actg instr, 58-59, sr oceanogr & exten asst prof, 59-60, res asst prof, 60-61; from asst prof to assoc prof, 61-70, coordr, 61-70, dir marine lab, 64-70, PROF OCEANOG, HUMBOLDT STATE COL, 70- *Concurrent Pos:* Oceanog consult, 62- *Res:* Oceanographic education; chemical oceanography. *Mailing Add:* Dept Oceanog Humboldt State Univ Arcata CA 95521

GAST, LYLE EVERETT, b Alden, Ill, Apr 9, 19; m 42; c 1. ORGANIC CHEMISTRY. *Educ:* Univ Ill, BS, 41; Univ Wis, PhD(org chem), 49. *Prof Exp:* Asst res chemist, Pa Salt Mfg Co, 41-44; res chemist, Northern Regional Labs, USDA, 49-60, head oil coatings invests, 60-77; RETIRED. *Mem:* Am Chem Soc; Am Oil Chemists' Soc. *Res:* Chlorination of organic compounds; fundamental investigations on higher unsaturated fatty acids. *Mailing Add:* 8606 Reese Rd Harvard IL 60033

GAST, ROBERT GALE, b Philadelphia, Mo, July 28, 31; m 54; c 3. SOIL CHEMISTRY. *Educ:* Univ Mo, BS, 53, MS, 56, PhD(soil chem), 59. *Prof Exp:* Asst scientist soil chem, Agr Res Lab, AEC, Tenn, 59-61, assoc prof, 61-68; res assoc soil sci, Mich State Univ, 68-69; prof soil sci, Univ Minn, St Paul, 70-77; head Dept Agron, Univ Nebr, 77-83; ASSOC DEAN & DIR, MICH AGR EXP STA, 83- *Concurrent Pos:* Lectr, Oak Ridge Inst Nuclear Studies, 63-70; ed, J Environ Qual, Am Soc Agron, 75-77. *Mem:* Fel Am Soc Agron (pres-elect, 86); Int Soil Sci Soc; Clay Minerals Soc; Soil Sci Soc Am (pres, 82). *Res:* Physical chemistry and mineralogy of soils; nutrient movement. *Mailing Add:* Dept Agron Univ Nebr Lincoln NE 68583

GASTEIGER, EDGAR LIONEL, neurophysiology, for more information see previous edition

GASTIL, R GORDON, b San Diego, Calif, June 25, 28; m 58; c 4. GEOLOGY. *Educ:* Univ Calif, Berkeley, AB, 50, PhD(geol), 54. *Prof Exp:* Geologist, Shell Oil Co, Alaska, 54 & Can Javelin Ltd, 56-58; lectr geol, Univ Calif, Los Angeles, 58-59; from asst prof to assoc prof, 59-65, chmn dept, 69-77, PROF GEOL, SAN DIEGO STATE UNIV, 65- *Concurrent Pos:* Res partic, NSF grants, 60-61, 62-69 & 71; NSF res grants, Baja, Calif, 68-70; mem int geol field conf, Am Geol Inst, EAfrica, 69; NSF res grants, Sonora, Mex, 71-72 & 82-88. *Mem:* AAAS; Geol Soc Am; Soc Econ Paleont & Mineral; Am Geophys Union; Chinese Acad Sci & Earth Sci. *Res:* Regional structural analysis; geochronology; origin and evolution of continents and ocean basins. *Mailing Add:* Dept Geol San Diego State Univ San Diego CA 92182

GASTINEAU, CLIFFORD FELIX, b Pawnee, Okla, Dec 18, 20; m 51; c 2. MEDICINE. *Educ:* Univ Okla, BA, 41, MD, 43; Univ Minn, PhD(med), 50. *Prof Exp:* CONSULT INTERNAL MED, MAYO CLIN, 50- *Mem:* AAAS; Endocrine Soc; Am Diabetes Asn; fel Am Col Physicians; Sigma Xi. *Res:* Clinical standpoint of diabetes and nutrition. *Mailing Add:* Mayo Med Sch Rochester MN 55901

GASTL, GEORGE CLIFFORD, b Shawnee, Kans, Feb 27, 38; m 67. TOPOLOGY, ALGEBRA. *Educ:* Univ Kans, AB, 60, MA, 62; Univ Wis, PhD(math), 66. *Prof Exp:* Asst prof, 66-70, ASSOC PROF MATH, UNIV WYO, 70- *Mem:* Math Asn Am; Am Math Soc. *Res:* Abstract topological spaces; extended topology; uniform spaces; proximity spaces. *Mailing Add:* Dept Math Univ Wyo 218 Ross Laramie WY 82071

GASTON, LYLE KENNETH, b Waterloo, Iowa, Nov 7, 30. PHEROMONES. *Educ:* Iowa State Univ, BS, 53; Univ Calif, Los Angeles, PhD(chem), 60. *Prof Exp:* Fel, Univ Colo, 60-62; CHEMIST & ADJ LECTR TOXICOL & PHYSIOL, UNIV CALIF, RIVERSIDE, 62- *Mem:* Am Chem Soc; Entom Soc Am; AAAS. *Res:* Isolation, identification and synthesis of insect sex pheromones; chemistry and analysis of pesticides and their residues. *Mailing Add:* Div Toxicol & Physiol Univ Calif Riverside CA 92521

GASTONY, GERALD JOSEPH, b Cleveland, Ohio, June 2, 40; m 67; c 1. SYSTEMATIC BOTANY. *Educ:* St Louis Univ, AB, 64; Tulane Univ, MS, 66; Harvard Univ, PhD(biol), 71. *Prof Exp:* Asst prof, 70-76, ASSOC PROF BIOL, IND UNIV, BLOOMINGTON, 76- *Concurrent Pos:* Assoc ed bot, J Rhodora, 72- & Am Fern J, 73- *Mem:* Bot Soc Am; Soc Study Evolution; Am Soc Plant Taxonomists; Int Asn Plant Taxon; Am Fern Soc (secy-treas, 74-77). *Res:* Palynology of tree ferns, both Cyatheaceae and Dicksoniaceae, and their allies; populational genetics of ferns; biosystematic studies of ferns and flowering plants. *Mailing Add:* Dept Plant Scis Ind Univ Bloomington IN 47401

GASTWIRTH, BART WAYNE, b Mineola, NY. PATHOMECHANICS OF THE FOOT. *Educ:* Ill Col Podiatric Med, DPM, 77; Am Bd Podiatric Orthop, cert, 85. *Prof Exp:* Instr, dept med sci, Ill Col Podiatric Med, 79-80, asst prof, dept podiatric med, 80-81, assoc prof, dept orthop sci, 81; assoc prof

& chmn, Dept Orthop Sci & adj assoc prof, Dept Community Health, 81-86, PROF & CHMN, DEPT ORTHOP, SCHOLL COL PODIATRIC MED, 81- *Concurrent Pos:* Consult, Vet Admin Hosp, Marion, IL, 83- *Mem:* Am Podiatric Med Asn; assoc mem Am Col Foot Surgeons; Am Pub Health Asn. *Res:* Biomechanics of the foot; diseases and surgery of the foot. *Mailing Add:* Dept Orthop Scholl Col Podiatric Med 1001 N Dearborn St Chicago IL 60610

GASTWIRTH, JOSEPH L, b New York, NY, Aug 31, 38. MATHEMATICAL STATISTICS. *Educ:* Yale Univ, BS, 58; Princeton Univ, MA, 60; Columbia Univ, PhD(math statist), 63. *Prof Exp:* Res assoc statist, Stanford Univ, 63-64; from asst prof to assoc prof, Johns Hopkins Univ, 64-72; PROF STATIST, GEORGE WASHINGTON UNIV, 72- *Concurrent Pos:* Vis assoc prof, Harvard Univ, 70-71; vis fac adv, Off Statist Policy, Exec Off of the President, 71-72, consult, 75-; assoc ed, J Statist Asn, 78-88; vis prof, Mass Inst Technol, 79; Guggenheim Found Fel, 85-86; mem, Am Statist Asn, Comn Law & Justice Statist, 88. *Mem:* Fel AAAS; fel Inst Math Statist; Int Statist Inst; fel Am Statist Asn; Royal Statist Soc; Indust Rels Res Asn. *Res:* Robust methods of inference; applied probability theory; economic statistics; statistics in law. *Mailing Add:* Dept Statist George Washington Univ Washington DC 20052

GASWICK, DENNIS C, b Hay Springs, Nebr, Feb 15, 42; m 64; c 2. INORGANIC CHEMISTRY. *Educ:* Nebr Wesleyan Univ, BA, 64; Ore State Univ, PhD(inorg chem), 68. *Prof Exp:* Res assoc, State Univ NY, Stony Brook, 68-69; PROF CHEM, ALBION COL, 69- *Concurrent Pos:* Vis scholar, Stanford Univ, 76-77; vis prof, Univ Nenchatel, Switz, 83-84, Northwestern Univ, 90-91. *Mem:* Am Chem Soc; Sigma Xi. *Res:* Mechanisms of chromium reactions and their intermediates; synthesis of binuclear chromium-cobalt complexes and the mechanisms of electron transfer reactions in coordination chemistry; kinetics and mechanistic studies of redox and substitution reactions involving complex ions in solution. *Mailing Add:* Dept Chem Albion Col 704 E Porter St Albion MI 49224

GAT, NAHUM, b Poland, Dec 19, 47; Israeli citizen; m 72; c 2. COMBUSTION, FLUID MECHANICS. *Educ:* Israel Inst Technol, BSc, 69; Univ Cincinnati, MS, 75, PhD(aerospace eng), 78. *Prof Exp:* Teaching asst exp aerospace eng, Univ Cincinnati, 76-78; mem tech staff, Defense & Space Systs Group, 78-80, sect mgr, Eng Sci Lab & Space Systs Group, 80-89, SR STAFF MEM & CONSULT, TRW INC, 89- *Concurrent Pos:* Lectr, Mech Eng Dept, Calif State Univ, Long Beach, 80- *Mem:* Am Inst Aeronaut & Astronaut; Combustion Inst; Am Soc Mech Engrs; Soc Automotive Engrs; Sigma Xi. *Res:* Automotive combustion; flame quenching and flame propagation; pulverized coal combustion and fluid mechanics of high swirl combustion; erosion of metals by airborne particles; two phase flow; combustion instrumentation and diagnostics. *Mailing Add:* TRW SNTG Bldg R1 Rm 1028 One Space Park Redondo Beach CA 90278

GAT, URI, b Jerusalem, Israel, June 28, 36; m 61; c 2. NUCLEAR ENGINEERING, THERMODYNAMICS. *Educ:* Israel Inst Technol, BSc, 63; Aachen Tech Univ, Dr Ing, 69; Ky Bd Regist, PE, 72. *Prof Exp:* Fel scientist nuclear reactor develop, Kernforschungsanlage, Ger, 63-69; asst prof mech eng, Univ Ky, 69-74; mgr, Gas Cooled Fast Reactor Prog, Oak Ridge Nat Lab, 74-80, mgr, Breeder Reactor Prog, 80-82, mgr aerols res, 83-87, reviewer, 87-88, mgr, Robotics Aircraft Paint Removal, 88-91, MGR, HIGH FLUX ISOTOPE REACTOR ENVIRON QUALIFICATION, OAK RIDGE NAT LAB, 89- *Concurrent Pos:* NSF res initiation grant, 71-72. *Honors & Awards:* Wilhelm Borchers Medaille. *Mem:* Am Nuclear Soc; Metric Asn; Sigma Xi. *Res:* Liquid metal fast breeder reactor; liquid fuel reactors; molten salt reactors; reactor evaluations; thermochemical equilibrium evaluation and calculation; phase diagrams; kinetics of chlorination of nuclear fuels; flow and heat transfer; metrication; gas cooled reactor; quality assurance; advanced inherently safe reactors; environmental qualification. *Mailing Add:* Oak Ridge Nat Lab PO Box 2009 Oak Ridge TN 37831-8088

GATELY, MAURICE KENT, b Omaha, Nebr, Feb 3, 46; m 72; c 2. CYTOKINES, IMMUNOPHARMACOLOGY. *Educ:* Johns Hopkins Univ, BA, 68, PhD(microbiol), 74, Johns Hopkins Sch Med, MD, 75. *Prof Exp:* Intern pediat, St Louis Children's Hosp 75-76; fel immunol, Harvard Med Sch, 76-79; sr staff fel, cellular Immunol Group Surg Neurol Branch, NIH, Bethesda, Md, 79-83; sr scientist immunopharmacol, 83-85, res investr, 85-88, RES LEADER, HOFFMANN-LA ROCHE, INC, 88- *Mem:* Am Asn Immunologists; AAAS; NY Acad Sci. *Res:* Identification of factors regulating cytolytic lymphocyte responses and the potential use of such factors in immunotherapy of tumors; author of numerous articles on lymphokines, on the mechanism of lymphocyte-mediated cytolysis and on mechanisms by which human brain tumors evade cellular immune attack. *Mailing Add:* Dept Immunopharmacol Hoffmann-La Roche Inc Nutley NJ 07110

GATENBY, ANTHONY ARTHUR, b London, Eng, Dec 25, 51; m 75; c 2. GENE EXPRESSION, PROTEIN ASSEMBLY. *Educ:* Portsmouth Polytechnic, UK, BSc, 73; Univ London, MSc, 74; Univ Nottingham, PhD(biochem), 77. *Prof Exp:* Res fel, Univ Edinburgh, UK, 77-79; higher sci officer, Plant Breeding Inst, UK, 79-82, sr sci officer, 82-85, prin sci officer, 85; res fel, Univ Wis, 85-86; PRIN INVESTR, E I DUPONT DE NEMOURS & CO INC, 86- *Concurrent Pos:* Hon res fel, Inst Enzyme Res, 84-85. *Mem:* Am Soc Biochem & Molecular Biol. *Res:* The folding and assembly of proteins in vitro and in vivo; protein translocation and secretion through membranes; production of foreign proteins in microorganisms, regulation of translation. *Mailing Add:* E I DuPont de Nemours & Co Experimental Sta PO Box 80402 Wilmington DE 19880-0402

GATES, ALLEN H(AZEN), JR, b Rockville, Conn, Nov 7, 29; m 65; c 2. REPRODUCTIVE BIOLOGY. *Educ:* La State Univ, BS, 51; Univ Edinburgh, Scotland, PhD(genetics), 59. *Prof Exp:* Res asst develop biol, Jackson Lab, Bar Harbor, Maine, 51-53, lab fel reproductive biol, 56-59, assoc staff scientist, 59-60; res assoc, Sch Med, Stanford Univ, 60-64, asst prof, 64-70; staff researcher, Syntex Res, Calif, 70-71; asst prof anat, Sch Med, Univ Rochester, 71-75, asst prof obstet & gynec, radiation biol & biophys, 75-82, asst prof toxicol in radiation biol & biophys, 82-88, ASST PROF GENETICS, SCH MED, UNIV ROCHESTER, 71-, ASST PROF TOXICOL, ENVIRON HEALTH SCI CTR, 88- *Concurrent Pos:* Nat Cancer Inst consult, 62-68; prin investr, 63-68 & 78-81. *Mem:* AAAS; Soc Study Reproduction. *Res:* Mammalian reproductive and developmental toxicology; effects of methylmercury and other environmental toxicants; effects of pulsed ultrasound (obstetrical diagnostic levels); teratology; ovum chromosomal aberrations; ovulation and preimplantation development; maintenance of unique mutant mouse strains. *Mailing Add:* Environ Health Sci Ctr Univ Rochester Med Ctr 575 Elmwood Ave Rochester NY 14642

GATES, BRUCE C(LARK), b Richmond, Calif, July 5, 40; m 67; c 2. CHEMICAL ENGINEERING. *Educ:* Univ Calif, Berkeley, BS, 61; Univ Wash, PhD(chem eng), 66. *Prof Exp:* Res engr, Chevron Res Co, Calif, 67-69; from asst prof to assoc prof, 69-77, assoc dir, Ctr Catalytic Sci & Technol, 77-81, PROF CHEM ENG, UNIV DEL, 77-, DIR, CATALYTIC CTR SCI & TECHNOL, 81-, H RODNEY SHARP PROF, 85-, PROF CHEM,88- *Concurrent Pos:* Fulbright res grant, Inst Phys Chem, Univ Munich, 66-67, 75-76 & 83-84. *Honors & Awards:* Del Sect Award, Am Chem Soc, 85. *Mem:* Am Inst Chem Engrs; Am Chem Soc. *Res:* Catalysis; surface chemistry and reaction kinetics; chemical reaction engineering; petroleum and petrochemical processes; catalysis by superacids, zeolites, soluble and supported transition-metal complexes and clusters; catalytic hydroprocessing. *Mailing Add:* Dept Chem Eng Univ Del Newark DE 19716

GATES, CHARLES EDGAR, b Rapid City, SDak, Mar 6, 26; m 51; c 3. AGRICULTURAL STATISTICS. *Educ:* Iowa State Univ, BS, 50; NC State Col, MS, 52, PhD(exp statist), 55. *Prof Exp:* Asst exp statist, NC State Col, 50-53, asst statistician, 53-54; asst prof statist, Univ Minn, St Paul, 56-60, assoc prof, 60-65, prof, 65-66, statistician, Agr Exp Sta, 56-66; PROF, INST STATIST, TEX A&M UNIV, 66- *Concurrent Pos:* Inst, Univ Louisville, 55-56. *Mem:* AAAS; Am Statist Asn; Royal Statist Soc; Biomet Soc (secy, E NAm Region, 75-78, pres-elect, 79, pres, 80); Sigma XI. *Res:* Wildlife population estimation; design of agricultural and biological experiments. *Mailing Add:* Statist/OE Teague Res Tex A&M Univ College Station TX 77843

GATES, D(ANIEL) W(ILLIAM), b Chicago, Ill, Oct 16, 21; m 45; c 2. CERAMICS. *Educ:* Univ Ill, BS, 48; Ga Inst Technol, MS, 51 & 53; Ala A&M Univ, MBA, 82. *Prof Exp:* Mem staff res & develop, Temco, Inc, 47-49; asst, Eng Exp Sta, Ga Inst Technol, 50-53; asst prof ceramics, Clemson Col, 53-54; aero materials res engr, Army Ballistic Missile Agency, NASA, 54-60, mat res engr, George G Marshall Space Flight Ctr, 60-77; INSTR COMPUT SCI, ALA A&M UNIV, 80- *Mem:* Am Ceramic Soc; Nat Inst Ceramic Engrs; Am Geophys Union; fel Am Inst Chem. *Res:* Geophysical exploration; porcelain enamels; graphite for rockets; materials for space environment; thermal-control coatings for space vehicles. *Mailing Add:* 3026 Crescent Circle SE Huntsville AL 35801

GATES, DAVID G(ORDON), b Kansas City, Mo, Mar 30, 31; m 50; c 2. INDUSTRIAL ENGINEERING. *Educ:* Univ Ark, BS, 56, MS, 59; Okla State Univ, PhD(eng), 62. *Prof Exp:* Indust engr, Corning Glass Works, 56-57; instr indust eng, Okla State Univ, 60-62; assoc prof, Auburn Univ, 62-63; PROF INDUST ENG, LAMAR UNIV, 63- *Mem:* Am Inst Indust Engrs; Am Soc Eng Educ. *Res:* Methods engineering; work measurement. *Mailing Add:* Dept Indust Eng Lamar Univ Box 10032 Beaumont TX 77710

GATES, DAVID MURRAY, b Manhattan, Kans, May 27, 21; m 44; c 4. PLANT PHYSIOLOGY, CLIMATOLOGY. *Educ:* Univ Mich, BS, 42, MS, 44, PhD(physics), 48. *Prof Exp:* Asst, Univ Mich, 41-44, res physicist, 42-44; res asst prof physics, Univ Denver, 47-54, assoc prof, 54-55; sci dir & liaison off, Off Naval Res, Eng, 55-57; asst chief radio propagation, Physics Div, Nat Bur Stand, 57-60, asst chief upper atmosphere & space physics div, 60-61, consult to dir, 61-64; prof natural hist, Univ & cur ecol, Univ Mus, Univ Colo, 65; prof biol, Wash Univ, & dir, Mo Bot Garden, 65-71; dir, Biol Sta, 71-86, prof bot, 71-91, EMER PROF BOT, UNIV MICH, ANN ARBOR, 91-; DISTINGUISHED VIS SCIENTIST, JET PROPULSION LAB, CALIF INST TECHNOL, 89- *Concurrent Pos:* Mem oper anal stand-by unit, Iowa State Col, 53-55 & Univ Denver, 55-; consult, Air Defense Command, Colo, 53-; ed, Radio Propagation & J of Res, 61-63; lectr, Univ Colo, 61-64 & vis prof, 65; chmn environ studies bd, Nat Acad Sci-Nat Acad Eng, 70-73; mem, Nat Sci Bd, 70-76; mem adv panel, Comn Sci & Astronaut, US House Rep, 71-75; mem bd, Conserv Found, 69-86, Nat Audubon Soc, 72-78, Cranbrook Inst Sci, 73-92 & World Wildlife Found, 86-87; mem biometeorol panel, US Nat Comt Int Biol Prog; chmn ad hoc comt environ, Environ Clearinghouse Inc, Washington, DC; mem nat air quality adv comt, USPHS, adv comt biol & med sci, NSF & ad hoc adv comt off ecol, Smithsonian Inst; bd dir, Detroit Edison Co, 79-92; mem bd, Acid Rain Found, 80-92 & L S B Leakey Found, 83-92. *Honors & Awards:* Gold Medal Ecol, Nat Coun State Garden Clubs, 71; Outstanding Bioclimatologist, Am Meteorol Soc, 71. *Mem:* Optical Soc Am; Sigma Xi; Ecol Soc Am; Bot Soc Am; Am Inst Biol Sci (vpres, 74, pres, 75). *Res:* Infrared spectroscopy in the near and far infrared; upper atmosphere research by infrared; geophysical exploration work; ecology; energy exchange for plants; transpiration; photosynthesis; ecological effects of energy use. *Mailing Add:* Biol Dept Univ Mich Ann Arbor MI 48109-1048

GATES, GEORGE O, geology; deceased, see previous edition for last biography

GATES, GERALD OTIS, b Brewerton, NY, Oct 18, 39. ECOLOGY. *Educ:* Univ Ariz, BA, 60, PhD(ecol), 63. *Prof Exp:* Asst prof biol, Univ of the Pac, 63-66; from asst prof to prof, 66-77, actg dean, 76-77, vpres, acad affairs, 79-82, ROBERTSON PROF BIOL, UNIV REDLANDS, 77- *Res:* Ecology and behavior of amphibians and reptiles. *Mailing Add:* Dept Biol Univ Redlands 1200 E Colton Ave Redlands CA 92373

GATES, HALBERT FREDERICK, b Milwaukee, Wis, Oct 30, 19; m 48; c 3. INTEGRATED PHYSICAL SCIENCE. *Educ:* Wis State Teachers Col, BS, 40; Univ Wis, PhM, 44; Mich State Col, PhD(physics), 54. *Prof Exp:* Asst prof physics, Berea Col, 48-50; asst prof & chmn dept, Cornell Col, 54-55; asst prof physics & phys sci, Univ Ill, 55-58; assoc prof physics, Northern Ariz Univ, 58-64, prof & chmn dept, 65-66; prof & chmn dept, Slippery Rock State Col, 67-68; chmn dept, 69-74, prof, 69-81, EMER PROF PHYSICS, BLOOMSBURG UNIV, PA, 81- *Mem:* Am Phys Soc; Am Asn Physics Teachers; Sigma Xi. *Res:* Ultrasonics and optics. *Mailing Add:* 12506 Amigo Dr Sun City West AZ 85375

GATES, HENRY STILLMAN, b Sharon, Pa, Oct 12, 29; m 53; c 5. APPLIED CHEMISTRY, SCIENTIFIC PROGRAMMING. *Educ:* Lehigh Univ, BS, 51; Univ Wis, PhD(inorg chem), 56. *Prof Exp:* Prof chem, Milton Col, 55-57; assoc prof, Wis State Univ, 57-58; asst prof, Utica Col, 58-62; prof chem & chmn div natural sci, Milton Col, 62-70; assoc prof chem, 70-72, PROF PHYS SCI, WESTMINSTER COL, UTAH, 72- *Mem:* AAAS; Am Chem Soc; Sigma Xi. *Res:* Microprocessor applications in chemical instrumentation; computer simulation of natural and artificial systems. *Mailing Add:* PO Box 17497 Salt Lake City UT 84117-0497

GATES, JAMES EDWARD, b Chicago, Ill, Jan 6, 43; m 68; c 2. PHYTOPATHOLOGY, MICROBIOLOGY. *Educ:* Northern Ill Univ, BSEd, 65, MS, 67; Univ Mo-Columbia, PhD(plant path), 72. *Prof Exp:* Instr biol, Northern Ill Univ, 67-69; asst prof biol, Randolph-Macon Col, 72-75; asst prof, 75-80, ASSOC PROF BIOL, VA COMMONWEALTH UNIV, 80- *Mem:* Sigma Xi; Am Phytopath Soc; Am Soc Microbiol; Mycol Soc Am. *Res:* Genetics of symbiotic development; role of bacteria in formation of acid mine drainage; nitrogen-fixation. *Mailing Add:* Dept Biol Va Commonwealth Univ 816 Park Richmond VA 23284

GATES, JOHN E(DWARD), b Parkersburg, WVa, June 22, 27; m 51; c 2. MATERIALS SCIENCE. *Educ:* Ohio State Univ, BSc, 53. *Prof Exp:* Lab technician, Battelle Mem Inst, 51-53, chemist radiochem, 53-54, proj leader radiation effects, 54-58, asst chief, 58-62, div chief, 62-68, assoc dept mgr, 68-76; pres, Aerospace Mat, Inc, 73-79; consult, 79-80; PRES, AEROSPACE LUBRICANTS, INC, 81- *Mem:* Inst Elec & Electronics Engrs. *Res:* Development of nuclear reactor fuels; radiation effects in fuels and other materials; development of radioisotope heat sources, turbine engine coatings and high temperature turbine blade materials. *Mailing Add:* Aerospace Lubricants Inc 1505 Delashmut Ave Columbus OH 43212

GATES, JOSEPH SPENCER, b Des Moines, Iowa, Jan 18, 35; m 64; c 2. HYDROGEOLOGY. *Educ:* Colo Sch Mines, Geol E, 56; Univ Utah, MS, 60; Univ Ariz, PhD, 72. *Prof Exp:* Hydraul engr, US Geol Surv, 56-58, geologist, 58-71, hydrologist, 71-74, hydrologist-in-charge, El Paso field hq, 74-77, proj chief, Rio Grande Environ Study, 71-77, CHIEF INVEST SECT, UTAH DIST, US GEOL SURV, 77- *Concurrent Pos:* Geologist, USAID, Cairo, Egypt, 65-67; grad teaching & res assoc, Dept Hydrol & Water Resources, Univ Ariz, 67-71. *Mem:* Geol Soc Am; Am Geophys Union. *Res:* Groundwater geology and hydrology of arid regions. *Mailing Add:* 2560 Cavalier Dr Salt Lake City UT 84121

GATES, LESLIE DEAN, JR, b Berwyn, Ill, Oct 10, 22; m 50; c 4. MATHEMATICS. *Educ:* Iowa State Col, BS, 47, MS, 50, PhD(math), 52. *Prof Exp:* Instr math, Iowa State Col, 52; mathematician, Armed Forces Spec Weapons Proj, US Dept Defense, 52-54; mathematician, US Naval Proving Grounds, 54-55, head appl math br, 55-58; mathematician, Atomic Energy Div, Babcock & Wilcox Co, 58-59, chief math sect, 59-61; ASSOC PROF MATH, SOUTHERN ILL UNIV, 61- *Mem:* Math Asn Am; Sigma Xi. *Res:* Numerical analysis. *Mailing Add:* Dept Math Southern Ill Univ Carbondale IL 62903

GATES, MARSHALL DEMOTTE, JR, b Boyne City, Mich, Sept 25, 15; m 41; c 4. ORGANIC CHEMISTRY. *Educ:* Rice Inst, BS, 36, MA, 38; Harvard Univ, PhD(org chem), 41. *Hon Degrees:* DSc, MacMurray Col, 63. *Prof Exp:* Asst prof chem, Bryn Mawr Col, 41-43; tech aid, Off Sci Res & Develop, Washington, DC, 43-46; assoc prof chem, Bryn Mawr Col, 46-49; lectr, 49-52, part-time prof, 52-60, prof chem, 60-81, HAUGHTON EMER PROF, UNIV ROCHESTER, 81- *Concurrent Pos:* Asst ed, J Am Chem Soc, 49-62, ed, 63-69; Tishler lectr, Harvard Univ, 53; mem, Comt Drug Addiction & Narcotics, Nat Res Coun, 58-70; Welch Found lectr, 60; mem, President's Comt, Nat Medal Sci, 68-70; vis prof, Dartmouth Col, 82 & 84-86. *Mem:* Nat Acad Sci; Am Chem Soc; fel Am Acad Arts & Sci; fel NY Acad Sci. *Res:* Chemistry of natural products and analgesics. *Mailing Add:* Dept Chem Univ Rochester Rochester NY 14627

GATES, MICHAEL ANDREW, b Corry, Pa, Dec 1, 46; Can citizen; m 74; c 4. PROTOZOOLOGY, SYSTEMATICS. *Educ:* Case Western Reserve Univ, BS, 68; Univ Toronto, MA, 73, PhD(zool), 76. *Prof Exp:* Fel protozool, Brit Mus Natural Hist, London, Eng, 76-77; fel, Cult Ctr Algae & Protozoa, Cambridge, Eng, 77-78; res asst prof zool, Dept Zool, Univ Toronto, 80-85; asst prof, 86-89, ASSOC PROF, DEPT BIOL, CLEVELAND STATE UNIV, 90- *Mem:* Am Math Soc; Soc Protozoologists; Am Soc Naturalists; Am Statist Asn; Soc Study Evolution. *Res:* Morphometric variation in ciliated protozoans; ciliate systematics and genetics; theoretical biology, especially the analysis of patterns. *Mailing Add:* Dept Biol Cleveland State Univ Cleveland OH 44115

GATES, OLCOTT, b New York, NY, Mar 19, 19; m 42; c 5. GEOLOGY. *Educ:* Harvard Univ, BS, 41; Univ Colo, MA, 50; Johns Hopkins Univ, PhD, 56. *Prof Exp:* Asst, Woods Hole Oceanog Inst, 46; geologist, US Geol Surv, 49-54; from asst prof geol to assoc prof, Johns Hopkins Univ, 54-63; mem, US Peace Corps, Ghana, 63-65; chmn dept, 66-77, PROF GEOL, STATE UNIV NY COL, FREDONIA, 66- *Concurrent Pos:* Geologist, Maine Geol Surv. *Mem:* Fel Geol Soc Am; Sigma Xi. *Res:* Volcanology; petrology; structural geology. *Mailing Add:* PO Box 234 Wiscasset ME 04578

GATES, RAYMOND DEE, b Akron, Ohio, Oct 10, 25; m 54; c 2. POLYMER CHEMISTRY. *Educ:* Univ Akron, BS, 49, MS, 51, PhD(chem), 61. *Prof Exp:* Res chemist, Firestone Tire & Rubber Co, 51-55, res chemist,Synthetic Rubber & Latex Div, 55-58; inst rubber res, Univ Akron, 58-60; sr res chemist, Chem Div, Int Latex Corp, 60-62, mgr appl polymer res, 62-66; res assoc,PPG Industs, 66-67, res supvr, 67-71; mgr chem servs, Oak Rubber Co, 71-72; mgr, SBR Adhesives Lab, Morgan Adhesives Co, 72-83; chemist, Algan, Inc, 83-91; RETIRED. *Mem:* Am Chem Soc. *Res:* Synthesis of elastomeric polymers and halogenated plastics; polymeric structures and the relationship of structural characteristics to physical properties; mechanism of polymeric network degradation; PVC plastisol compounding; adhesives compounding and characterization; water-based coating studies. *Mailing Add:* 3183 Silver Lake Blvd Cuyahoga Falls OH 44224

GATES, ROBERT LEROY, b Lincoln, Nebr, Dec 3, 17; m 43; c 2. BIOCHEMISTRY. *Educ:* Univ Nebr, BS, 39, PhD(chem), 52; Kans State Col, MS, 47. *Prof Exp:* Asst, Dept Milling Indust, Kans State Col, 46; res chemist, Farm Corp Processing Corp, 47; asst agr chem, Univ Nebr, 48-52; proj leader, Res Dept, Westvaco Chem Div, FMC Corp, 52-54, group leader, 54-58, dir res & develop dept, Niagara Chem Div, 58-77, mem staff, Agr Chem Div, 77-83; RETIRED. *Mem:* Am Chem Soc; Entom Soc Am. *Res:* Precipitation and concentration of amylase; action of amylases on raw starch; gelatinization and retrogradation of starches; role of starch in bread staling; preparation of herbicidally active organic compounds; formulation and toxicology of pesticides; synthesis and development of new pesticides. *Mailing Add:* 143 Elizabeth St Medina NY 14103

GATES, ROBERT MAYNARD, b Madison, Wis, June 26, 18; m 48; c 2. GEOLOGY. *Educ:* Univ Wis, BA, 41, MA, 41, PhD(geol), 49. *Prof Exp:* From instr to assoc prof, 49-60, PROF GEOL, UNIV WIS-MADISON, 60- *Concurrent Pos:* Geologist, Conn Geol Surv, 48-78; dir, Wasatch-Uinta Field Camp, 79-87. *Mem:* Fel Geol Soc Am; Mineral Soc Am; Geochem Soc. *Res:* Mineralogy and petrology of igneous and metamorphic rocks; geology of Western Connecticut. *Mailing Add:* Dept Geol 279 Weeks Hall Univ Wis Madison WI 53706

GATES, RONALD EUGENE, b Milwaukee, Wis, Sept 19, 41; m 69; c 3. BIOCHEMISTRY. *Educ:* St Mary's Col, Minn, BA, 63; Northwestern Univ, PhD(biochem), 68. *Prof Exp:* Fel biochem, St Jude Children's Res Hosp, 71-72, res consult, 73-76; res asst dept pediat, Med Ctr, Univ Tenn, 77-80; RES ASST PROF MED, MED SCH, VANDERBILT UNIV, 80- *Concurrent Pos:* Fel biochem, Vet Admin Hosp, Kansas City, Mo, 68-70. *Res:* Protein phosphorylation with emphasis on the function of hormones and growth factors; mechanism of action of vitamin A and its binding proteins. *Mailing Add:* Dept Dermat Med Vanderbilt Univ 21st Ave St & Garland Nashville TN 37232

GATES, SYLVESTER J, JR, b Tampa, Fla, Dec 15, 50. QUANTUM FIELD THEORY, SUPERSYMMETRICAL THEORIES. *Educ:* Mass Inst Technol, BS, 73, BS, 73, PhD(physics), 77. *Prof Exp:* Jr fel, Harvard Univ, 77-80; res assoc, Calif Inst Technol, 80-82; asst prof appl math, Mass Inst Technol, 82-84; assoc prof, 84-89, PROF PHYSICS, UNIV MD, 89-; PROF PHYSICS, HOWARD UNIV, 90- *Concurrent Pos:* Exec tech officer, Nat Soc Black Physicists, 90-92. *Mem:* Nat Soc Black Physicists. *Res:* Study of mathematical models as possible descriptions of the elementary particles and fundamental forces which occur in nature. *Mailing Add:* Dept Physics & Astron Univ Md College Park MD 20742

GATES, WILLIAM LAWRENCE, b South Pasadena, Calif, Sept 14, 28; m 51; c 3. METEOROLOGY. *Educ:* Mass Inst Technol, SB, 50, SM, 51, ScD, 55. *Prof Exp:* Asst, Mass Inst Technol, 50-53; res meteorologist, Air Force Cambridge Res Ctr, 53-57; asst prof meteorol, Univ Calif, Los Angeles, 57-59, assoc prof, 59-66; res scientist, Rand Corp, 66-76; PROF & CHMN DEPT ATMOSPHERIC SCI, ORE STATE UNIV, 76- *Mem:* Am Meteorol Soc; Am Geophys Union; Royal Meteorol Soc. *Res:* Dynamic meteorology; climate dynamics; numerical weather prediction; physical oceanography. *Mailing Add:* Dept Atmospheric Sci Ore State Univ Corvallis OR 97331

GATEWOOD, BUFORD ECHOLS, b Byhalia, Miss, Aug 23, 13; wid; c 1. AIRCRAFT STRUCTURES. *Educ:* La Polytech Inst, BS, 35; Univ Wis, MS, 37, PhD(math), 39. *Prof Exp:* Asst math, Univ Wis, 35-39; asst prof, La Polytech Inst, 39-42; asst stress analyst to asst chief struct engr, McDonnell Aircraft Corp, Mo, 42-46; struct res engr, Beech Aircraft, Kans, 46-47; from assoc prof to prof mech & head dept, Air Force Inst Tech, Wright Patterson Base, 47-55, res coordr & res prof, 55-60; prof aeronaut & astronaut eng, Ohio State Univ, 60-78; RETIRED. *Mem:* Am Inst Aeronaut & Astronaut; Am Soc Mech Eng; Math Asn Am; Sigma Xi. *Res:* Shear distribution in diagonal tension beams; thermal stresses in long cylindrical bodies; aircraft structures; buckling of tapered columns; fatigue and thermal stresses. *Mailing Add:* 2150 Waltham Rd Columbus OH 43221-4150

GATEWOOD, DEAN CHARLES, b Iowa City, Iowa, June 29, 25; m 53; c 4. BIOCHEMISTRY. *Educ:* Willamette Univ, BA, 50; Univ Ore, MA, 53. *Prof Exp:* Asst gen chem, 50-51, asst biochem, 52-53, asst endocrinol, Med Sch, 54-55, from instr to asst prof, 55-63, ASSOC PROF BIOCHEM, UNIV ORE HEALTH SCI CTR SCH DENT, 63- *Mem:* AAAS; Am Chem Soc. *Res:* Etiology of dental caries; biochemistry of selenium and other trace elements; metabolism of oral tissues. *Mailing Add:* Dept Biochem Univ Ore Den Sch Portland OR 97219

GATEWOOD, GEORGE DAVID, b St Petersburg, Fla, May 10, 40; m 59; c 1. ASTRONOMY. *Educ:* Univ SFla, BA, 65, MA, 68; Univ Pittsburgh, PhD(astron), 72. *Prof Exp:* From asst prof to assoc prof astron, 72-89, PROF ASTRON, ALLEGHENY OBSERV, UNIV PITTSBURGH, 89-, DIR OBSERV, 72- *Honors & Awards:* Nat Space Act Award, 87. *Mem:* Am Astron Soc; Royal Astron Soc; Int Astron Union. *Res:* The determination of stellar distances, luminosities and masses and the detection of extrasolar planetary systems. *Mailing Add:* Allegheny Observ Observ Sta Pittsburgh PA 15214

GATEWOOD, LAEL CRANMER, b Cleveland, Ohio, Nov 16, 38; m 61; c 2. HEALTH COMPUTER SCIENCE, BIOMETRY. *Educ:* Rockford Col, BA, 59; Univ Minn, Minneapolis, MS, 66, PhD(biomet), 71. *Prof Exp:* Technician biochem res, Mayo Clin, 59-61, technician biophys, 62-67; scientist health comput sci & biomet, 67-68, asst prof lab med & biomet & asst dir, Div Health Comput Sci, 71-74, sr assoc dir, Div Health Comput Sci, 74-79, ASSOC PROF LAB MED, PATH & BIOMET, UNIV MINN, MINNEAPOLIS, 74-, DIR, DIV HEALTH COMPUT SCI, 79- *Mem:* AAAS; Am Pub Health Asn; NY Acad Sci; Am Asn Med Systs Informatics; Asn Comput Mach; Sigma Xi. *Res:* Biomedical computation; simulation of dynamic physiological systems using both deterministic and stochastic modeling; techniques of quality assurance and data base management for clinical research; application of computers to health care services. *Mailing Add:* 2067 Goodrich Ave St Paul MN 55105-1019

GATH, CARL H(ENRY), chemical engineering; deceased, see previous edition for last biography

GATHERS, GEORGE ROGER, b Meridian, Okla, Feb 1, 36; m 69; c 1. EXPERIMENTAL SOLID STATE PHYSICS. *Educ:* Univ Southern Calif, BS, 60; Univ Calif, Berkeley, PhD(physics), 67. *Prof Exp:* Physicist, 67-69, group leader physics, 69-70, PHYSICIST, LAWRENCE LIVERMORE LAB, 70- *Mem:* Am Phys Soc. *Res:* Cyclotron resonance in lead; high pressure, high temperature equation of state of materials; shock hydrodynamics; the measurement of the equation of state and electrical properties of liquid metals at very high temperatures and under modest pressures using fast dynamic self-heating methods. *Mailing Add:* 1212 Harvest Rd Pleasanton CA 94566

GATHERUM, GORDON ELWOOD, b Salt Lake City, Utah, Oct 22, 23; m 47; c 2. FORESTRY PHYSIOLOGY, SOILS. *Educ:* Univ Wash, Seattle, BS, 49; Utah State Univ, MS 51; Iowa State Univ, PhD(silvicult, plant physiol), 59. *Prof Exp:* Asst range mgt, Utah State Univ, 49-51; asst prof, Tex Technol Col, 51-53; assoc prof forestry, Iowa State Univ, 53-64, prof, 64-69; prof & chmn dept, Ohio State Univ & Ohio Agr Res & Develop Ctr, 69-75, DIR SCH NATURAL RESOURCES, OHIO STATE UNIV, 75-, ASSOC DEAN, COL OF AGR & HOME ECON, 75- *Mem:* AAAS; Soc Am Foresters; Ecol Soc Am. *Res:* Tree physiology; forest soils. *Mailing Add:* 5710 Strathmore Lane Dublin OH 43017

GATHINGS, WILLIAM EDWARD, IMMUNOLOGY. *Educ:* University Ala, Birmingham, PhD(biol), 81. *Prof Exp:* PRES & CHIEF EXEC OFFICER, SOUTHERN BIOTECHNOL ASSOC, INC, 82- *Mailing Add:* South Biotechnol Assoc Inc PO Box 26221 Birmingham AL 35226

GATIPON, GLENN BLAISE, b New Orleans, La, July 10, 40; m 62; c 3. PSYCHOPHARMACOLOGY, NEUROPHYSIOLOGY. *Educ:* Tulane Univ, BS, 62; La State Univ, Baton Rouge, MS, 64; La State Univ, New Orleans, PhD(physiol), 70. *Prof Exp:* Instr neurophysiol, Sch Med, La State Univ, 70-72; ASST PROF ANESTHESIOL, PHARMACOL & NEUROSURG, MED CTR, UNIV MISS, 72-, RES ASSOC, TRAUMA CLIN, 73- *Concurrent Pos:* Nat Inst Dent Res spec dent res award, Med Ctr, Univ Miss, 73-76; consult pharmacol, Sch Nursing, Univ Southern Miss, 73- *Mem:* Int Asn Dent Res; Soc Neurosci. *Res:* Electrophysiol and computer analysis of single neuron activity. *Mailing Add:* Executive Park 95541 Atlanta GA 30347

GATLAND, IAN ROBERT, b London, Eng, Feb 17, 36; m 63; c 4. THEORETICAL PHYSICS. *Educ:* Univ London, BSc, 57, PhD(theoret physics), 60. *Prof Exp:* Fel theoret physics, European Orgn Nuclear Res, Geneva, 60-61; staff mem physics, Res Inst Advan Study, Md, 61-64; from asst prof to assoc prof, 64-74, PROF PHYSICS, GA INST TECHNOL, 74- *Mem:* Am Phys Soc; Sigma Xi. *Res:* Atomic collisions; ion-atom, ion-molecular interactions. *Mailing Add:* Sch Physics Ga Inst Technol Atlanta GA 30332-0430

GATLEY, IAN, b Runcorn, Eng, Feb 11, 50; m 76. INFRARED ASTRONOMY, ASTROPHYSICS. *Educ:* Univ London, BSc & ARCS, 72; Calif Inst Technol, PhD(physics), 78. *Prof Exp:* RES FEL PHYSICS, CALIF INST TECHNOL, 77- *Concurrent Pos:* Co-investr, Kuiper Airborne Observ, NASA, 78- *Mem:* Am Astron Soc. *Res:* Observational infrared astronomy. *Mailing Add:* 6340 N Placita De Eduardo Tucson AZ 85726

GATLEY, WILLIAM STUART, b Pueblo, Colo, Jan 24, 32; m 61; c 3. NOISE CONTROL, VIBRATIONS. *Educ:* Princeton Univ, AB, 54; Wash Univ, BS, 56, MS, 57; Purdue Univ, PhD, 67. *Prof Exp:* Res engr drilling, Jersey Prod Res Co, Standard Oil NJ, 57-62; prof mech eng, Univ Mo-Rolla, 66-72 & 73-79; prin acoust consult, Coffeen, Gatley & Assoc, 72-73; MGR MECH ENG RES, MOTOROLA INC, 79-, CHIEF MECH ENG. *Concurrent Pos:* Prin investr, NSF, 69-72; consult, Gulf Oil Corp, 74-77 & Gen Motors Corp, 76-; adj prof mech eng, Fla Atlantic Univ, 79-87; assoc, sci adv bd, Motorola, Inc, 84- *Honors & Awards:* Guerard Mackey Award, Am Soc Mech Engrs, 73. *Mem:* Am Soc Mech Engrs; Inst Noise Control Eng; Am Soc Eng Educ. *Res:* Noise control; vibrations; mechanical design. *Mailing Add:* Govt Electronics Group Motorola Inc 8201 E McDowell Rd Scottsdale AZ 85252

GATLIN, DELBERT MONROE, III, b Dallas, Tex, May 22, 58; m 82; c 1. FISH NUTRITION, NUTRITIONAL BIOCHEMISTRY. *Educ:* Miss State Univ, PhD(nutrit), 83. *Prof Exp:* Asst prof animal nutrit, Univ Ark, Pine Bluff, 85-87; ASST PROF, DEPT WILDLIFE & FISHERIES SCI, TEX A&M UNIV, 87- *Mem:* Am Inst Nutrit; Am Fisheries Soc; World Agr Soc. *Res:* Basic nutrition of fishes and shell fishes, with emphasis on species cultured for human consumption; determining dietary requirements for and metabolism of various nutrients; nutrition energetics, nutrition-diseases interactions, and nutritional biochemistry of fish as it relates to human health; comprehensive basic nutrition program targeted to improve production efficiency in aquaculture and enhance the quality of resulting products. *Mailing Add:* Dept Wildlife & Fisheries Sci Tex A&M Univ Col Sta TX 77843-2258

GATLIN, LILA L, b Hutchinson, Kans; m 47; c 4. THEORETICAL BIOLOGY, PHYSICAL CHEMISTRY. *Educ:* Univ Tulsa, BS, 57; Pa State Univ, MS, 59; Univ Tex, Austin, PhD(phys chem), 63. *Prof Exp:* NIH fel, Genet Found, Univ Tex, Austin, 63-64; asst prof phys chem, Drexel Univ, 64-66; vis lectr, Bryn Mawr Col, 66-67; res biophysicist, Space Sci Lab, Univ Calif, Berkeley, 70-74; ASSOC RES GENETICIST & LECTR GENETICS, UNIV CALIF, DAVIS, 74- *Mem:* AAAS. *Res:* Application of information theory to the living system; information theory and physical research. *Mailing Add:* 3790 El Camino No 338 Palo Alto CA 94306

GATLING, ROBERT RIDDICK, ANATOMIC PATHOLOGY. *Educ:* Tulane Univ, MD, 41. *Prof Exp:* CHIEF LAB SERV, VET ADMIN MED CTR, 70- *Mailing Add:* PO Box 4874 Fondren Sta Jackson MS 39216

GATOS, H(ARRY) C(ONSTANTINE), b Greece, Dec 27, 21; m 50; c 3. ELECTRONIC MATERIAL. *Educ:* Nat Univ Athens, dipl, 45; Ind Univ, AM, 48; Mass Inst Technol, PhD(chem), 50. *Hon Degrees:* DSc, Ind Univ, 83. *Prof Exp:* Instr inorg chem, Nat Univ Athens, 42-46; asst metall, Mass Inst Technol, 48-50, res assoc, 50-51, mem res staff, 51-52; res engr, E I du Pont de Nemours & Co, 52-55; leader chem & metall group, Lincoln Lab, 55-59, assoc head solid state div, 59-62, head, 62-66, PROF ELECTRONIC MAT & MOLECULAR ENG, MASS INST TECHNOL, 62- *Concurrent Pos:* Consult various indust & govt orgns; ed, Surface Sci & Surface Sci Letters. *Honors & Awards:* Solid State Sci & Technol Award, Electrochem Soc, 75, Acheson Medal, 82; Outstanding Sci Achievement Award, NASA, 75; Golden Cross Order Merit, Polish Peoples Repub, 80. *Mem:* Nat Acad Eng; fel AAAS; hon mem Electrochem Soc; Am Phys Soc; Am Inst Mining, Metall & Petrol Engrs; Am Acad Arts & Sci. *Res:* Electrochemistry; electronic materials; semiconductors; applications. *Mailing Add:* 20 Indian Hill Rd Weston MA 02193

GATROUSIS, CHRISTOPHER, b Norwich, Conn, Oct 8, 28; c 1. NUCLEAR CHEMISTRY, PHYSICAL CHEMISTRY. *Educ:* DePaul Univ, BS, 57; Univ Chicago, MS, 60; Clark Univ, PhD(nuclear chem), 65. *Prof Exp:* Res asst, Argonne Nat Lab, 56-61; asst scientist, Woods Hole Oceanog Inst, 64-66; chemist, 66-72, from asst div leader to div leader nuclear chem div, 72-85, ASSOC DIR, CHEM & MATS SCI, LAWRENCE LIVERMORE LAB, UNIV CALIF, 85- *Mem:* Am Phys Soc; Am Chem Soc. *Res:* Mechanisms of nuclear reactions; nuclear spectroscopy; radiochemistry and geochemistry of fallout radionuclides; detection of low level radioactivity. *Mailing Add:* Lawrence Livermore Lab Univ Calif Livermore CA 94550

GATSKI, THOMAS BERNARD, b Hazleton, Pa, Aug 30, 48; m 70; c 1. FLUID DYNAMICS, TURBULENCE. *Educ:* Pa State Univ, BS, 70, MS, 72, PhD(aerospace eng), 76. *Prof Exp:* Res assoc turbulence, Div Eng, Brown Univ, 75-77; res scientist aerocoust, 77-81, res scientist viscous flow br, 81-90, SR RES SCIENTIST THEORET FLOW PHYSICS BR, NASA LANGLEY RES CTR, 90- *Mem:* Am Phys Soc; Am Inst Aeronaut & Astronaut. *Res:* Aerodynamic drag reduction; sound generating mechanisms of turbulent shear flows and their suppression; flow of non-Newtonian fluids; turbulence closure modeling. *Mailing Add:* 721 E Tazewell Way Williamsburg VA 23185

GATTEN, ROBERT EDWARD, JR, b Lexington, Ky, Dec 21, 44; m 68; c 2. ENVIRONMENTAL PHYSIOLOGY, COMPARATIVE PHYSIOLOGY. *Educ:* Col William & Mary, BS, 66, MA, 68; Univ Mich, PhD(zool), 73. *Prof Exp:* Lectr biol, Col William & Mary, 66-67; asst prof biol, Col Wooster, 73-75; asst prof biol, Univ Toledo, 75-78; from asst prof to assoc prof, 78-86, PROF BIOL, UNIV NC, GREENSBORO, 86-, HEAD BIOL, 88- *Mem:* AAAS; Am Soc Zoologists; Am Soc Ichthyologists & Herpetologists'; Herpetologists League. *Res:* Aerobic and anerobic metabolism in amphibians and reptiles during exercise, diving, and hibernation. *Mailing Add:* Dept Biol Univ NC Greensboro NC 27412-5001

GATTERDAM, PAUL ESCH, b La Crosse, Wis, June 27, 29; m 57; c 3. AGRICULTURAL CHEMISTRY. *Educ:* Univ Wis, BS, 53, MS, 57, PhD(entom), 58. *Prof Exp:* Med entomologist, Agr Res Serv, USDA, 58-60; res asst prof entom, Pesticide Residue Lab, NC State Univ, 61-62; SR RES CHEMIST, AGR RES CTR, AM CYANAMID CO, 62- *Mem:* Am Entom Soc. *Res:* Toxicology, persistence, and metabolism of pesticides in or on plants, animals and soil. *Mailing Add:* Am Cyanamid Co PO Box 400 Princeton NJ 08540

GATTI, ANTHONY ROGER, b Buffalo, NY, Sept 18, 35; m 57; c 2. INORGANIC CHEMISTRY, RESEARCH ADMINISTRATION. *Educ:* Brown Univ, BS, 57; Pa State Univ, PhD(chem), 67. *Prof Exp:* Chemist, Arthur D Little, Inc, Mass, 60-63; from res chemist to sr res chemist & res supvr, Houston Res Lab, Shell Oil Co, 67-73, staff res chemist, 73-80, res supvr, 80-81, dept mgr res, Elastomers Dept, 81-90, MGR TECHNOL ACQUISITION, SHELL DEVELOP CO, 90- *Mem:* Catalysis Soc. *Res:* Catalysis; the inorganic chemistry and structure of transition metal and noble metal catalysts used in the processing of petroleum products; administration of catalysis and process research; process and product research in thermoplastic elastomers. *Mailing Add:* Shell Develop Co Westhollow Res Ctr PO Box 1380 Houston TX 77251-1380

GATTINGER, RICHARD LARRY, b Neudorf, Sask, June 28, 37; m 62; c 3. PHYSICS, AERONOMY. *Educ:* Univ Sask, BE, 60, MSc, 62, PhD(physics), 64. *Prof Exp:* Asst res off, 64-70, assoc res off physics, 70-77, SR RES OFFICER, NAT RES COUN CAN, 78- *Mem:* Am Geophys Union. *Res:* Observation and interpretation of the atmospheric phenomena classes as airglow and aurora, in near infrared, visible and ultraviolet. *Mailing Add:* 2143 Grafton Crescent Ottawa ON K1J 6K7 Can

GATTO, LOUIS ALBERT, b Montevideo, Repub of Uruguay, Feb 3, 50; nat US. PHYSIOLOGY. *Educ:* Fordham Univ, MS, 74, PhD(biol), 78. *Prof Exp:* PROF BIOL, STATE UNIV NY, 78- *Mem:* AAAS; Am Soc Zoologists; Sigma Xi; Am Physiol Soc; Am Micros Soc. *Res:* Histophysiology of the respiratory airways in mammals; the pharmacology of their mucosal glands is elucidated by histochemistry and correlated with the viscoelasticity of the mucus. *Mailing Add:* Dept Biol State Univ NY PO Box 2000 Cortland NY 13045

GATTONE, VINCENT H, II, b Philadelphia, Pa, June 8, 51; m 77; c 5. NEPHROLOGY, HYPERTENSION. *Educ:* Ursinus Col, BS, 73; George Washington Univ, MS, 75; Med Col Ohio, PhD(anat), 81. *Prof Exp:* Res fel renal med & hypertension, Sch Med, Ind Univ, 80-83; asst prof anat, M S Hershey Med Ctr, Pa State Univ, 83-86; asst prof, 86-89, ASSOC PROF ANAT, CELL BIOL, UNIV KANS MED CTR, 89- *Mem:* Am Asn Anatomists; Soc Neurosci; Am Soc Nephrology; Am Heart Asn; Int Soc Nephrology; Int Pediat Nephrology Asn. *Res:* Renal innervation and microvasculature, its development and how it relates to renal function and the development of hypertension; polycystic kidney disease; neuroimmunology. *Mailing Add:* Univ Kans Med Ctr Dept Anat Cell Biol 39th & Rainbow Blvd Kansas City KS 66103

GATTS, THOMAS F, b Kansas City, Mo, Oct 3, 33; m 54. ELECTRICAL ENGINEERING. *Educ:* Univ Md, BSEE, 62, MS, 66; Univ Iowa, PhD, 70. *Prof Exp:* Instr elec eng, Howard Univ, 62-64; engr, Sylvania Electronics Systs, NY, 64-65; asst prof elec eng, Howard Univ, 65-67; assoc prof, 69-72, dir, Comput Ctr, 72-76, DIR, COMPUT BASED HON PROG, UNIV ALA, 76- *Concurrent Pos:* Consult, Environ Protection Agency, Research Triangle Park, NC, 78- *Mem:* Inst Elec & Electronics Engrs; Am Soc Eng Educ; Sigma Xi. *Res:* Network theory; computer-aided design. *Mailing Add:* Comput Eng Dept Santa Clara Univ Santa Clara CA 95053

GATZ, ARTHUR JOHN, JR, b Oak Park, Ill, Aug 11, 47; m 86; c 2. ECOLOGY. *Educ:* Dickinson Col, AB, 69; Duke Univ, PhD(zool), 75. *Prof Exp:* From asst prof to assoc prof, 75-88, PROF ZOO, OHIO WESLEYAN UNIV, 88- *Concurrent Pos:* Vis scientist, Environ Sci Div, Oak Ridge Nat Lab, 83-85. *Mem:* AAAS; Ecol Soc Am; Sigma Xi; Soc Conserv Biol; Animal Behav Soc. *Res:* Population and community ecology of freshwater stream fishes; functional morphology of fishes; population biology of amphibians. *Mailing Add:* Dept Zool Ohio Wesleyan Univ Delaware OH 43015

GATZ, CAROLE R, b Omaha, Nebr, Jan 26, 33. PHYSICAL CHEMISTRY. *Educ:* Iowa State Univ, BS, 54; Univ Ill, PhD(phys chem), 60. *Prof Exp:* Phys chemist, Stanford Res Inst, 59-64; asst prof, 64-68, assoc prof, 68-78, PROF CHEM, PORTLAND STATE UNIV, 78- *Concurrent Pos:* Am Asn Univ Women fel, 70-71. *Mem:* Am Phys Soc; Sigma Xi. *Res:* Quantum chemistry; gas kinetics; atmospheric chemistry. *Mailing Add:* Chem Dept Portland State Univ PO Box 751 Portland OR 97207

GATZ, RANDALL NEAL, b Louisville, Ky, June 30, 43; m 76. PHYSIOLOGY. *Educ:* Univ Louisville, BS, 66; Univ Ky, MS, 69, PhD(physiol), 73. *Prof Exp:* Asst res, Max-Planck Inst Exp Med, 72-74; researcher physiol, Battelle Mem Inst, Geneva, Switz, 74-77; asst prof, Col Vet Med, Kans State Univ, 77-; assoc group head, Inhalation Toxicol & Physiol, Battelle Res Inst, Geneva, Switz; MEM STAFF, G W S CORP ASSOC SA. *Concurrent Pos:* Consult, Battelle Mem Inst, Geneva, Switz, 77- *Mem:* Am Physiol Soc; Soc Oxygen Transp Tissue; Sigma Xi; AAAS. *Res:* Gas exchange and pulmonary function in vertebrates; relationship between pulmonary blood flow and ventilation in simple lungs; phylogeny of lung function. *Mailing Add:* Chemin de la Montau Genolier CH-1261 Switzerland

GATZY, JOHN T, JR, b Philadelphia, Pa, June 14, 36; m 58; c 3. PHARMACOLOGY, TOXICOLOGY. *Educ:* Pa State Univ, BS, 58; Univ Rochester, PhD(pharmacol), 63. *Prof Exp:* From instr to asst prof pharmacol, Dartmouth Med Sch, 62-73; assoc prof pharmacol & toxicol, 73-83, dir pharmacol grad studies, 81-84, actg chair, 82-83, PROF PHARMACOL & TOXICOL, MED SCH, UNIV NC, CHAPEL HILL, 83- *Mem:* Am Soc Pharmacol & Exp Therapeut. *Res:* Bioelectric properties and solute transport of epithlial barriers; cell pharmacology and toxicology; cystic fibrosis. *Mailing Add:* Dept Pharmacol & Toxicol Univ NC Med Sch Chapel Hill NC 27514

GAU, JOHN N, b Boston, Mass, June 17, 45. ULTRASONICS, ATOMIC & MOLECULAR PHYSICS. *Educ:* Iowa State Univ, BS, 67; Univ Nebr, MS, 71, PhD(physics), 75. *Prof Exp:* Fel physics, 77-79, mem tech staff, 79-81, PROG & PROJ MGR, ANDERSON LAB, UNIV CONN, 81- *Concurrent Pos:* Vis assoc prof physics, Univ Conn, 85- *Mem:* Am Phys Soc; Inst Elec & Electronics Engrs. *Mailing Add:* Andersen Lab 45 Old Iron Ore Rd Bloomfield CT 06002

GAUCHER, GEORGE MAURICE, b Edmonton, Alta, Jan 6, 38; m 60; c 4. ANTIBIOTICS, BIOTECHNOLOGY. *Educ:* Univ Alta, BSc, 60; Univ Pa, PhD(biochem), 63. *Prof Exp:* Res assoc biochem, Univ Ill, 63-65; from asst prof to assoc prof, 65-77, PROF BIOCHEM, UNIV CALGARY, 77-, ADJ PROF CHEM, 87- *Concurrent Pos:* Smith Kline & French fel, 63-65; sect ed, Can J Microbiol, 79-82, 84-89, ed, 89-; mem, Biotechnol Comt, Nat Res Coun, 83-88; William Evans vis fel, Univ Otago, NZ, 87. *Mem:* Am Chem Soc; Can Biochem Soc; Can Soc Microbiologists; Am Soc Microbiol; Soc Indust Microbiol. *Res:* Enzymatic, metabolic, regulatory and genetic aspects of secondary metabolite antibiotic production by microscopic fungi; biotechnology and the development of immobilized cell bioreactors. *Mailing Add:* Biochem Div Dept Biol Sci Univ Calgary Calgary AB T2N 1N4 Can

GAUD, WILLIAM S, b Charleston, SC, June 17, 43. ECOLOGICAL MODELING. *Educ:* Col Charleston, BS, 65; Univ NC, Chapel Hill, PhD(zool), 70. *Prof Exp:* Asst prof, 70-81, ASSOC PROF BIOL, NORTHERN ARIZ UNIV, 81- *Concurrent Pos:* NSF fel syst ecol, Biol Sta, Univ Okla, 71; vis res, Swed Univ Agr Sci, 79-80. *Mem:* Ecol Soc Am; Soc Study Pop Ecol; Sigma Xi. *Res:* Mathematical modeling of ecosystems and of populations; environmental assessment of the effluent effects of coal-burning electric generating stations and of the transmission of electricity on big-horn sheep populations; nitrogen in pine trees; small game animal foraging behavior and its effects on forest nutrient cycles; insect herbivore-plant interactions. *Mailing Add:* 1968 N Crescent Dr Flagstaff AZ 86001

GAUDETTE, HENRI EUGENE, b Boston, Mass, Jan 26, 32; m 60; c 1. GEOLOGY, GEOCHEMISTRY. *Educ:* Univ NH, BA, 59; Univ Ill, MS, 62, PhD(geol), 63. *Prof Exp:* Res assoc clay mineral, Univ Ill, 63-65; from asst prof to assoc prof, 65-78, PROF GEOL, UNIV NH, 78- *Mem:* Fel Geol Soc Am; Mineral Soc Am; Clay Minerals Soc; Geochem Soc; Am Geophys Union. *Res:* Inorganic geochemistry; distribution of minor and trace elements in rocks and minerals; geochronology and isotope geology; marine geochemistry of estuarine and nearshore systems. *Mailing Add:* Dept Earth Sci Univ NH Durham NH 03824

GAUDETTE, LEO EWARD, b South Bellingham, Mass, Mar 29, 25; m 54; c 2. INDUSTRIAL TOXICOLOGY, CLINICAL TOXICOLOGY. *Educ:* Col of Holy Cross, BS, 49, MS, 50; Georgetown Univ, PhD(biochem), 57. *Prof Exp:* Chemist, Nat Heart Inst, 54-56; biochemist, Worcester Found Exp Biol, 56-58 & Nat Inst Allergy & Infectious Dis, 58-60; chief lab biochem pharmacol, Joseph E Seagram & Sons, 60-62; head neuropharmacol, Riker Labs, 62-63; tech dir, NEN Biomed Assay Labs, Inc, Mass, 63-75; lab mgr ref lab div, Diamond Shamrock Health Sci, Inc, 75-77; dir, Clin & Occup Med Toxicol, Tabershaw Occup Med Asn, 77-88; RETIRED. *Mem:* AAAS; Am Soc Pharmacol & Exp Therapeut; Am Chem Soc; NY Acad Sci; Am Asn Clin Chemists. *Res:* Drug metabolism, enzymes and mechanisms of action of central nervous system drugs; alcoholism; endocrinological function; isotope methodology, radioimmunoassay. *Mailing Add:* 26 Scribner Rd Tyngsboro MA 01879

GAUDIN, ANTHONY J, b New Orleans, La, Aug 11, 38; m 58; c 2. HERPETOLOGY, MORPHOLOGY. *Educ:* Univ Southern Calif, AB, 59, MS, 64, PhD(biol), 69. *Prof Exp:* Teacher high sch, 60-64; from instr to asst prof biol, Los Angeles Pierce Col, 64-70; asst prof, 70-74, assoc prof, 74-78, PROF BIOL, CALIF STATE UNIV, NORTHRIDGE, 78-, CHMN DEPT, 83- *Concurrent Pos:* NSF Sci fac fel, 68-69. *Mem:* Am Soc Ichthyol & Herpet; Soc Study Amphibians & Reptiles; Herpetologists League; Sigma Xi. *Res:* Amphibian embryology and osteology. *Mailing Add:* RR 3 Box 42A Morgantown IN 46160

GAUDINO, MARIO, b Buenos Aires, Arg, May 22, 18; nat US; m 47, 84; c 2. RENAL MEDICINE, PHYSIOLOGY. *Educ:* Univ Buenos Aires, BS, 34, MD, 44; NY Univ, PhD(physiol), 50. *Prof Exp:* Asst & chief lab & instr physiol, Sch Med, Univ Buenos Aires, 37-44, Arg Nat Cult Comn fel, 45; US Dept State fel, NY Univ, 46-48; asst prof physiol, Med Br, Univ Tex, 49; chmn dept biol physics, Sch Med, Nat Univ LaPlata, 50-51; res assoc surg, Col Med, NY Univ, 52-55, assoc prof, 55-57; med dir, Abbott Labs, Int Co & Abbott Universal, Ltd, 57-62; assoc med dir, Pfizer Int, Inc, 62-67; assoc dir & dir advan clin res int, Merck Sharp & Dohme Res Labs, 67-71, sr dir clin res int, 71-74; dir med compliance, 74-80, assoc dir, med serv, 80-89, DIR MED CONSULT SERV, CIBA-GEIGY PHARMACEUT CORP, 89- *Concurrent Pos:* Univ fel, NY Univ, 46-49, Dazian Found fel, 46-49; assoc dir med writing & advert, Lederle Labs Div, Am Cyanamid Co, 51-52; estab investr, Am Heart Asn, 54-57; clin asst prof, Med Col, Cornell Univ, 71-77; assoc dept med, Northwestern Univ Med Sch. *Mem:* Soc Exp Biol & Med; Am Physiol Soc; Microcirculatory Soc; fel NY Acad Sci; Am Soc Nephrol; AMA; Am Soc Clin Pharmacol & Therapeut; Am Fedn Clin Res. *Res:* Experimental hypertension; kidney; body and cellular water and electrolyte distribution, exchange and excretion; membrane permeability; biophysics; isotopes; hemodynamics; clinical research; pharmaceutical industry; management. *Mailing Add:* Ciba-Geigy Corp 556 Morris Ave Summit NJ 07901

GAUDIOSO, STEPHEN LAWRENCE, b Rochester, NY, Mar 27, 45; m 69; c 5. MANUFACTURING PROCESS & MATERIALS. *Educ:* Rensselaer Polytech Inst, BS, 67; Univ Mich, Ann Arbor, MS, 69, PhD(phys chem), 72. *Prof Exp:* Presidential intern, Argonne Nat Lab, 72-73; scientist, Xerox Corp, 73-80, tech specialist/proj mgr, Webster Res Ctr, 80-84, sr tech specialist/core mgr, advan prods & technol & advan mat & mfg processes, 84-86, mgr, technol develop, 86-88, MGR, MFG TECHNOL PLANNING, XEROX, 88- *Mem:* Am Chem Soc; Sigma Xi; Tech Asn Graphic Arts; NY Acad Sci. *Res:* Xerographic process and materials, magnetography, lithography, novel imaging; manufacturing materials and processes. *Mailing Add:* 1077 Everwild View Webster NY 14580

GAUDRY, ROGER, b Quebec, Que, Dec 15, 13; m 41; c 5. ORGANIC CHEMISTRY. *Educ:* Laval Univ, BA, 33, BSc, 37, DSc, 40. *Hon Degrees:* Eleven hon doctorates from various Univs. *Prof Exp:* Lectr chem fac med, Laval Univ, 40-45, from asst prof to prof, 45-54; asst dir res, Ayerst Labs, Inc & Ayerst, McKenna & Harrison, Ltd, 54-57, dir, 57-63, vpres & dir res, Ayerst, McKenna & Harrison, Ltd, 63-65; rector, Univ Montreal, 65-75; pres, Int Asn Univs, 75-80; RETIRED. *Concurrent Pos:* Vpres, Sci Coun Can, 66-72, pres, 72-75; mem, Acad Latin World, Paris, 67; pres bd, Asn Univs & Cols Can, 69; chmn, Coun UN Univ, 74-76. *Honors & Awards:* Pariseau Medal, French-Can Asn Advan Sci, 58. *Mem:* Am Chem Soc; fel Royal Soc Can; Chem Inst Can (pres, 55-56). *Res:* Amino acid synthesis and metabolism. *Mailing Add:* Univ Montreal PO Box 6128 Montreal PQ H3C 3J7 Can

GAUDY, ANTHONY F, JR, b Jamaica, NY, June 16, 25; m 55. BIOENGINEERING, CIVIL ENGINEERING. *Educ:* Univ Mass, BS, 51; Mass Inst Technol, MS, 55; Univ Ill, PhD(eng), 59. *Prof Exp:* Jr engr, Metcalf & Eddy & Alfred Hopkins Assoc, 51; engr, E F Carlson, Inc, 51-52 & Capuano Inc; res asst, Sedgwick Labs, Mass Inst Technol, 53-55; res engr, Nat Coun Stream Improv, 55-57; asst prof sanit eng, Univ Ill, 59-61; from assoc prof to prof, Okla State Univ, 61-68, actg head sch civil eng, 66-67, chmn bioeng & water resources prog, 67-68, Edward R Stapley prof civil eng & dir, Bioeng & Water Resources Prog, 68-79; H RODNEY SHARP PROF & CHMN, DEPT CIVIL ENG, UNIV DEL, 79- *Concurrent Pos:* USPHS grant develop grad prog bioenviron eng & res grants, 62-78; vpres, Thomas, Gaudy, McCaskill, Inc, Memphis, Tenn, 70- & Environ Eng Consult Inc, Stillwater, Okla, 74-; Sigma Xi lectr, 71-72. *Honors & Awards:* Harrison P Eddy Medal, Water Pollution Control Fedn, 67. *Mem:* Am Soc Civil Engrs; Water Pollution Control Fedn; Am Water Works Asn; Am Soc Microbiol; Am Chem

Soc; Sigma Xi. *Res:* Shock loading, kinetics and mechanism of activated sludge processes; response of biological systems to physical, chemical and biological environment and engineering control of such response. *Mailing Add:* 111 Bridleshire Ct Newark DE 19711

GAUFIN, ARDEN RUPERT, b Salt Lake City, Utah, Dec 25, 11; m 36; c 2. ZOOLOGY. *Educ:* Univ Utah, BS, 35, MS, 37; Iowa State Col, PhD(fisheries biol), 51. *Prof Exp:* Instr, Bd Ed, Salt Lake City, 36-43; instr, 46-49, assoc prof, 53-63, prof, 63-77, EMER PROF ZOOL, UNIV UTAH, 77- *Concurrent Pos:* Inspector, Salt Lake City Mosquito Abatement, 35-43 & Terminex of Utah, 39-43; in chg stream sanit res unit, Environ Health Ctr, Cincinnati, 50-53, Salt Lake City Metrop Water Dist, 54-64, NIH, 59-63 & USPHS, 63-65; asst dir biol sta, Univ Mont, 63-75, prof, 68-69. *Mem:* Am Soc Limnol & Oceanog; Am Fisheries Soc; Ecol Soc Am; Entom Soc Am. *Res:* Limnological surveys and fisheries investigations; ecology of stoneflies. *Mailing Add:* 5520 Breckenridge Dr Salt Lake City UT 84117

GAUGER, WENDELL LEE, b Eustis, Nebr, Nov 7, 27; m 52; c 4. MYCOLOGY. *Educ:* Univ Nebr, BS, 51; Univ Idaho, MS, 53; Purdue Univ, PhD(bot), 56. *Prof Exp:* Asst prof bot, Univ Nebr, 56-58; assoc prof biol, Nebr Wesleyan Univ, 58-59; from asst prof to assoc prof, 59-67, chmn dept, 65-71, PROF BOT, UNIV NEBR, LINCOLN, 67- *Mem:* Mycol Soc Am. *Res:* Genetics and variability of fungi. *Mailing Add:* Dept Biol Sci Univ Nebr Lincoln NE 68588-0118

GAUGHAN, RENATA RYSNIK, US citizen. PHYSICAL CHEMISTRY. *Educ:* Temple Univ, AB, 66; Univ Pa, PhD(phys chem), 71; Wharton Bus Sch, mgt cert, 80. *Prof Exp:* Res assoc phys chem, Brown Univ, 71-72; res assoc phys & inorg chem, Northwestern Univ, 73-74; sr chemist, Polymer Chem Res Labs, 74-78, sr analyst, Corp New Ventures, 78-80, MKT MGR MONOMERS, ROHM & HAAS CO, 80- *Mem:* Am Chem Soc; Tech Asn Pulp & Paper Indust; Chem Mkt Res Asn. *Res:* Physical chemistry of polymers; polymer synthesis. *Mailing Add:* 353 Evergreen Dr Norht Wales PA 19454-2701

GAUGHAN, ROGER GRANT, b Elizabeth, NJ, June 16, 51; m 76. POLYMER CHEMISTRY. *Educ:* NMex Inst Mining & Technol, BS, 74, MS, 76; Univ Utah, PhD(org chem), 80. *Prof Exp:* RES CHEMIST POLYMER SYNTHESIS, PHILLIPS PETROL CO, 80- *Mem:* Am Chem Soc; Soc Plastics Engrs. *Res:* Synthetic organic chemistry and synthesized terpenes; synthesized monomers and subsequently polymerized them; developing synthetic methods. *Mailing Add:* 501 Ellis Parkway Piscataway NJ 08854

GAUGHRAN, GEORGE RICHARD LAWRENCE, b Pa, Oct 19, 19; m 43; c 1. ANATOMY. *Educ:* Lehigh Univ, BA, 42; Univ Mich, MS, 47; PhD(zool), 52. *Prof Exp:* Res assoc, Mass Inst Technol, 43; from instr to asst prof anat, Med Sch, Univ Mich, 51-61; from assoc prof to prof, 61-85, vchmn dept, 64-74, EMER PROF ANAT, COL MED, OHIO STATE UNIV, 85- *Mem:* Am Asn Anatomists. *Res:* Gross human anatomy. *Mailing Add:* Dept Cell Biol Neurobiol & Anat Ohio State Univ Col Med Columbus OH 43210-1239

GAUGL, JOHN F, b Manchester, Conn, May 11, 37; c 3. CORONARY FLOW, RESPIRATORY PHYSIOLOGY. *Educ:* NTex State Univ, BA, 61, MA, 62; Univ Calif, Berkeley, PhD(physiol), 70. *Prof Exp:* Actg asst prof exercise physiol, Univ Calif, Berkeley, 70-71; chmn, Dept Physiol, 71-81, ASSOC PROF MED PHYSIOL, TEX COL OSTEOP MED, 71-; ASSOC PROF BASIC HEALTH SCI, NTEX STATE UNIV, 73- *Concurrent Pos:* Chmn, Div Physiol, Dept Basic Health Sci, NTex State Univ, 73-76. *Mem:* Am Physiol Soc; Am Col Sports Med. *Res:* Controlling factors of right coronary blood flow in dogs; interrelationships between the autonomic nervous system and endogenous opiates in cardiovascular function. *Mailing Add:* 2706 Chinquapin Oak Lane Arlington TX 76012

GAUGLER, ROBERT WALTER, b Paterson, NJ, Aug 12, 40; m 65; c 3. BIOCHEMISTRY. *Educ:* Hope Col, BA, 63; Pa State Univ, MS, 66; Georgetown Univ, PhD(biochem), 73; Roosevelt Univ, MBA, 80. *Prof Exp:* Res chemist biochem, Naval Med Res Inst, 66-68; chief, Lab Res Dept, Nat Naval Dent Ctr, 70-76, res biochemist dent biochem, Naval Dent Res Inst, 76-80; res asst prof, 80-83, asst res & develop, Naval Med Command, 83-87, EXEC OFFICER, NAVAL MED RES & DEVELOP COMMAND, UNIFORMED SERV UNIV, 87- *Mem:* Int Asn Dent Res; Am Chem Soc; Am Soc Microbiol. *Res:* Relationship of bacterial polysaccharides to dental disease; enzymatic degradation of insoluble glucans of streptococcus mutants; fluoride concentrations in dental plaque. *Mailing Add:* 9089 Wexford Vienna VA 22180

GAUHAR, ARUNA, b Nagpur, India, Jan 4, 40. NEUROLOGY, NEUROPATHOLOGY. *Educ:* Panjab Univ, India, ISc, 56, MB & BS, 61. *Prof Exp:* Resident path, Univ Chicago, 65-67, from instr to asst prof, 69-74; fel neuropath & electron micros, Northwestern Univ & Vet Admin Hosp, Hines, Ill, 67-69, resident neurol, Northwestern Univ, 74-76; assoc prof neurol & neuropath, Chicago Med Sch, 77-80, prof neurol & path, 80-84; consult & assoc prof, Chicago Osteop Col & Hosp, 78-83; STAFF NEUROPATHOLOGIST, VET ADMIN MED CTR, NORTH CHICAGO, 77- *Concurrent Pos:* House physician internal med, Irwin Hosp, New Delhi, India, 62; demonstr path, Christian Med Col, Ludhiana, India, 62, house physician internal med, 63; asst dir autopsy serv, Dept Path, Univ Chicago, 70-72, assoc dir neuropath, 70-74; pvt pract neurol, Baltimore, 84- *Mem:* Am Asn Neuropathologists; Int Acad Path; fel Am Acad Neurol; Sigma Xi. *Res:* Clincial neuropathology and electron microscopy; chromatolysis and axonal reaction in spinal cord motor neurons. *Mailing Add:* 11331 Manor Rd Glen Arm MD 21057

GAUL, RICHARD JOSEPH, b Pittsfield, Mass, Oct 3, 29; m 53; c 5. ORGANIC CHEMISTRY. *Educ:* Spring Hill Col, BS, 49; Mass Inst Technol, PhD(org chem), 54. *Prof Exp:* Res chemist, Am Cyanamid Co, 54-59; assoc prof chem & actg dir dept, 59-62, dir, 62-65, PROF CHEM, JOHN CARROLL UNIV, 62- *Mem:* Am Chem Soc. *Res:* Synthetic studies related to penicillin; products derived from acrylonitrile; chemistry of sulfur-nitrogen heterocycles; rocket propellants; structure of terreic acid; synthesis and polarographic behavior of polychlorinated propionitriles, acrylonitriles and derivatives. *Mailing Add:* Dept Chem John Carroll Univ Cleveland OH 44118-4582

GAULDEN, MARY ESTHER, b Rock Hill, SC, Apr 30, 21; m 56; c 2. RADIOLOGY, RADIOBIOLOGY. *Educ:* Winthrop Col, BS, 42; Univ Va, MA, 44, PhD(biol), 48. *Prof Exp:* Res asst cell biol, Univ Ala, 45-46; biologist radiation biol, NIH, 46-47; instr zool, Univ Tenn, 47-49; sr biologist radiation biol, Oak Ridge Nat Lab, 49-65; asst prof, 65-68, ASSOC PROF RADIOL, UNIV TEX HEALTH SCI CTR DALLAS, 68- *Concurrent Pos:* Vis prof, Univ NC, Chapel Hill, 54-55; consult, Oak Ridge Nat Lab, 65-70. *Mem:* Fel AAAS; Radiol Soc NAm; Am Soc Cell Biol; Radiation Res Soc; Am Soc Photobiol; Environ Mutagen Soc. *Res:* Effects of low doses of radiation, especially on chromosomes and on man; genotoxicity of chemicals. *Mailing Add:* Dept Radiol Univ Tex Health Sci Ctr 5323 Hines Blvd Dallas TX 75235-9071

GAULDIE, JACK, PATHOLOGY. *Educ:* Univ Col, London, PhD(biochem), 68. *Prof Exp:* PROF PATH, MCMASTER UNIV, 71- *Mailing Add:* Dept Path McMaster Univ 1200 Main St W Hamilton ON L8N 3Z5 Can

GAULL, GERALD E, b Sept 17, 30; m 75, 84; c 2. PEDIATRICS. *Educ:* Univ Mich, BA, 51; Boston Univ, MD, 55; Am Bd Pediat, dipl, 66. *Prof Exp:* Jr asst resident path, Peter Bent Brigham Hosp, Boston, Mass, 55-56, sr asst resident, 56-57; resident pre-med adv, John Winthrop House, Harvard Col, 57-60; jr asst resident pediat, Babies Hosp, Columbia-Presby Med Ctr, 60-61; instr, Sch Med, Emory Univ, 61-62; res assoc neurol, Neurol Clin Res Ctr, Col Physicians & Surgeons, Columbia Univ, 65-67; chief, dept human develop & nutrit, NY State Inst Basic Res Develop Disabilities, 67-84; prof pediat, Mt Sinai Sch Med, 74-84; VPRES, NUTRIT SCI, NUTRASWEET CO, 84- *Concurrent Pos:* NIH fel biochem & path, Harvard Med Sch, 57-60, teaching fel, 63-64; Nat Inst Neurol Dis & Blindness spec res fel human metab, Univ Col Hosp Med Sch, Univ London, 62-63 & Neuropsychiat. *Honors & Awards:* Borden Award, Am Acad Pediatrics, 78. Res Unit, Med Res Coun Labs, Eng, 64-65; sr resident, Children's Hosp Med Ctr, Boston, 63-64; asst attend pediatrician, Babies Hosp, Columbia-Presby Med Ctr, 65-67; attend, Mt Sinai Hosp, New York, 67- *Mem:* Soc Pediat Res; Am Pediat Soc; Int Soc Neurochem; Int Brain Res Orgn; Am Soc Neurochem. *Res:* Nutrition; neurochemistry. *Mailing Add:* Nutrasweet Co 1751 Lake Cook Rd Deerfield IL 60201

GAULT, DONALD E, b Chicago, Ill, Feb 12, 23; m 47; c 3. ASTROGEOLOGY, PLANETARY SCIENCES. *Educ:* Purdue Univ, BS, 44. *Prof Exp:* Res scientist aerodyn, NASA, 44-59, res scientist ballistics, 59-63, chief planetology br, 63-72, sr staff scientist, Ames Res Ctr, 72-76; CHIEF PLANETOLOGIST, MURPHYS CTR PLANETOLOGY, 76- *Concurrent Pos:* Guggenheim fel, Max Planck Inst für Kern Physik, 72; adj prof, Univ Ariz, 77; Fairchild distinguished scholar, Calif Inst Technol, 78. *Honors & Awards:* Barringer Medal, Meteoritical Soc, 86; G K Gilbert Award, Geol Soc Am, 87. *Mem:* Am Geophys Union; Meteoritical Soc. *Res:* Shock wave dynamics; meteoritics; selenology; impact cratering; comparative planetology. *Mailing Add:* Murphys Ctr Planetology Box 833 Murphys CA 95247

GAULT, FREDERICK PAUL, psychophysiology; deceased, see previous edition for last biography

GAULT, N(EAL) L, JR, b Austin, Tex, Aug 22, 20; m 47; c 3. MEDICINE. *Educ:* Univ Tex, BA, 50; Univ Minn, MB, 50, MD, 51. *Prof Exp:* From instr to assoc prof internal med, Univ Minn, Minneapolis, 53-67, from asst dean to assoc dean, Col Med Sci, 55-67; prof internal med & assoc dean, Sch Med, Univ Hawaii, 67-72; dean, 72-84, PROF INTERNAL MED, MED SCH, UNIV MINN, MINNEAPOLIS, 72- *Concurrent Pos:* Med adv, Seoul Nat Univ, 59-61; consult, Vet Admin Hosp, Minneapolis, 56-67, China Med Bd Ny, Inc, 63, 71 & AID, 64-67; dir postgrad med educ prog, Ryukyu Islands, 67-69; mem coun deans & exec coun, Asn Am Med Col, 74-80; pres, Minn Int Health Vol, 87-; mem, State Minn Bd Continuing Legal Educ, 86; sr consult, Minn Med Found, 86- *Honors & Awards:* Supreme Award, Japan Med Asn, 69. *Mem:* Asn Am Med Cols; AMA. *Res:* Rheumatology. *Mailing Add:* Univ Minn Mayo Mem Bldg Box 293 Minneapolis MN 55455

GAUM, CARL H, b New York, NY, July 29, 22; m 55; c 2. CIVIL ENGINEERING. *Educ:* Rutgers Univ, BSCE, 49, MSE, 62. *Prof Exp:* Hydraul engr, US Geol Surv, 49-53; consult engr, Gaum Prof Engrs, 53-55; supvry hydraul engr, Philadelphia Dist, US Army Corps Engrs, 55-60, asst chief, Hydrol Br, DC Dist, 60-61 & Basin Planning Br, Ohio River Div, 62-69, asst chief, Interagency & Spec Studies Br & chief tech assistance & int sect, Planning Div, Off Chief Engrs, 69-73, chief cent reports mgt br, Planning Div, Civil Works Directorate, Off Chief Engrs, 70-80; sr water resources engr, Greenhorne & O'Mara Inc, Riverdale, Md, 80-85; PRIN ENGR, GAUM & ASSOCS, KENSINGTON, MD, 85- *Concurrent Pos:* Planning assoc, Bd Engrs Rivers & Harbors, 61-62; mem, US Comt Large Dams; chmn US sect Publ Comt, Permanent Int Asn Navig Cong, Leningrad, 76, Edinborough, 81, Brussels, 85 & Osaka, 90. *Mem:* AAAS; fel Am Soc Civil Engrs; Nat Soc Prof Engrs; Permanent Int Asn Navig Cong. *Res:* Comprehensive water and related land resource planning, including surface and ground water supplies, water quality, navigation, flood control, sedimentation, drainage, irrigation, fish and wildlife, recreation, power and beach protection; environment; economic evaluation; resources management. *Mailing Add:* 9609 Carriage Rd Kensington MD 20895-3619

GAUMER, ALBERT EDWIN HELLICK, b Nazareth, Pa, Apr 27, 26. PHYSIOLOGICAL ECOLOGY. *Educ:* Moravian Col, BS, 50; Purdue Univ, MS, 52, PhD, 54. *Prof Exp:* Asst instr biol, Purdue Univ, 50; from asst prof to assoc prof, 54-56, chmn dept, 54-79, sr prof, 79-81, PROF, 56-, EMER SR PROF BIOL, MORAVIAN COL, 81- *Mem:* AAAS; Int Oceanog Found; Sigma Xi; Am Mus Natural Hist; Am Soc Zool. *Res:* Physiol ecology. *Mailing Add:* 5308 Bay Ave Ocean City NJ 08226

GAUMER, HERMAN RICHARD, b Wilmington, Del, Nov 17, 41; m 65; c 2. MICROBIOLOGY, IMMUNOLOGY. *Educ:* Dartmouth Col, AB, 64; Univ Del, MS, 67; Univ NC, Chapel Hill, PhD(microbiol), 71. *Prof Exp:* Fel immunol, Univ Colo Med Ctr, 71-73; asst prof microbiol, Sch Dent, Univ Minn, 73-76; asst prof clin immunol, Med Sch, Tulane Univ, 76-80; ASST PROF PATH, LA STATE UNIV MED CTR, 80-; DIR, SPEC IMMUNOL LAB, CHARITY HOSP, NEW ORLEANS, LA, 80- *Concurrent Pos:* Dir, Flow Cytometry Resource, Charity Hosp. *Mem:* AAAS; Am Soc Microbiol; Sigma Xi. *Res:* Clinical immunology; immunological classification of lymphoproliferative disorders; use of flow cytometry in clinical laboratory. *Mailing Add:* Dept Pathology La State Univ Med Ctr New Orleans LA 70112

GAUMNITZ, ERWIN ALFRED, applied statistics; deceased, see previous edition for last biography

GAUMOND, CHARLES FRANK, b Watertown, NY, Sept 8, 50. ACOUSTICS. *Educ:* Clarkson Univ, BS, 72; Univ Rochester, MA, 76, PhD(physics), 79. *Prof Exp:* RES PHYSICIST, NAVAL RES LAB, 79- *Concurrent Pos:* Adj prof physics, George Mason Univ, 85- *Mem:* Acoust Soc Am; Am Phys Soc; Inst Elec & Electronics Engrs; AAAS; Sigma Xi. *Res:* Acoustical scattering; acoustical imaging; target strength; sonar design and analysis. *Mailing Add:* Naval Research Lab Code 5132 Washington DC 20375-5000

GAUNAURD, GUILLERMO C, b Havana, Cuba, July 19, 40; US citizen; m 67. ACOUSTICS, APPLIED MECHANICS. *Educ:* Cath Univ Am, BA, 64, BSME, 66, MS, 67, PhD(acoust), 71. *Prof Exp:* Sr engr appl mech, McKiernan-Terry Div, Litton Indust, College Park, Md, 68-71; RES PHYSICIST/ACOUST, NAVAL SURFACE WARFARE CTR, WHITE OAK LAB, SILVER SPRING, MD, 71- *Concurrent Pos:* Consult engr, Ocean Systs Inc, Div Union Carbide, Arlington, Va, 66-67; res asst, Cath Univ Am, 69-71; lectr acoustics & dynamics, Univ Md Sch Eng, College Park, 84- *Honors & Awards:* Bernard Smith Award. *Mem:* Fel Acoust Soc Am; Sigma Xi; NY Acad Sci; Am Phys Soc; sr mem Inst Elect & Electronics Engrs; Am Soc Mech Engrs. *Res:* Radiation and scattering theory; absorption in materials; underwater acoustics; physical acoustics; mechanical vibrations and the general response of structures to acoustic and electromagnetic excitations. *Mailing Add:* 4807 Macon Rd Rockville MD 20852-2348

GAUNT, ABBOT STOTT, b Lawrence, Mass, July 4, 36; m 63. ZOOLOGY. *Educ:* Amherst Col, BA, 58; Univ Kans, PhD(zool), 63. *Prof Exp:* Instr biol, Middlebury Col, 63-66, asst prof, 66-67; fel, State Univ NY, Buffalo, 67-68, asst prof, 68-69; asst prof, 69-74, assoc prof zool, 74-86, PROF, OHIO STATE UNIV, 86- *Mem:* Fel AAAS; Soc Study Evolution; fel Am Ornith Union; Wilson Ornith Soc (2nd vpres, 77-79, 1st vpres, 79-81, pres, 81-); Am Soc Zool; Cooper Ornith Soc. *Res:* Avian vocal mechanisms; avian functional anatomy; structure and function of vertebrate muscle. *Mailing Add:* Dept Zool Ohio State Univ 1735 Neil Ave Columbus OH 43210

GAUNT, JOHN THIXTON, b Evansville, Ind, Feb 29, 36; m 58; c 1. STRUCTURAL ENGINEERING. *Educ:* Univ Cincinnati, CE, 58; Purdue Univ, MSCE, 59, PhD(struct), 66. *Prof Exp:* Design engr, Int Steel Co, Evansville, Ind, 59-62; from instr to asst prof struct eng, 64-76, ASSOC PROF STRUCT ENG, PURDUE UNIV, 76- *Concurrent Pos:* Lectr, Univ Evansville, 60-62. *Mem:* Am Soc Civil Engrs; Nat Soc Prof Engrs; Am Soc Eng Educ; Sigma Xi. *Res:* Design and behavior of steel structures; structural analysis; guyed towers. *Mailing Add:* 2906 Henderson West Lafayette IN 47906

GAUNT, PAUL, b Bootle, Eng, Feb 23, 32; m 64; c 2. SOLID STATE PHYSICS. *Educ:* Univ Sheffield, BSc, 53; Oxford Univ, DPhil(metall), 58. *Prof Exp:* Lectr physics, Univ Sheffield, 57-68; assoc prof, 68-75, PROF PHYSICS, UNIV MAN, 75- *Res:* Physical property and structural changes associated with magnetic and chemical ordering of metals and alloys; hard magnetic materials; electron microscopy; models of magnetic hardening. *Mailing Add:* Dept Physics Univ Man Winnipeg MB R3T 2N2 Can

GAUNT, ROBERT, b Macon, Mo, Apr 13, 07; m 33; c 1. ENDOCRINOLOGY. *Educ:* Univ Tulsa, BA, 29; Princeton Univ, MA, 30, PhD(biol), 32. *Prof Exp:* Asst biol, Princeton Univ, 29-31; prof & chmn dept, Col Charleston, 32-35; from asst prof to assoc prof, Washington Sq Col, NY Univ, 35-46; prof zool & chmn dept, Syracuse Univ, 46-51; dir endocrine res, Ciba Pharmaceut Co, 51-57, dir biol res, 58-66, dir basic biol sci, 66-72, consult, Ciba-Geigy Corp, 72-76; mem, Nat Prostatic Cancer Cadre, 72-80; RETIRED. *Concurrent Pos:* Guggenheim fel, Princeton Univ, 43; vis prof, Ohio State Univ, 55. *Mem:* Fel AAAS; Endocrine Soc; Am Physiol Soc; Am Asn Anat; fel NY Acad Sci. *Res:* Physiology of adrenal cortex and gonads; water and electrolyte metabolism; drugs affecting endocrine function and experimental hypertension; research administration. *Mailing Add:* 1150 Eighth Ave SW Apt 2410 Largo FL 34640

GAUNT, STANLEY NEWKIRK, b Elmer, NJ, June 20, 15; m 39; c 2. ANIMAL GENETICS. *Educ:* Rutgers Univ, BS, 38; NC State Col, PhD(animal genetics & physiol), 55. *Prof Exp:* Admin asst, Agr Adjust Admin, NJ, 38-39; from asst county agent to assoc county agent, Litchfield County exten serv, Conn, 39-45; prof dairy sci, 45-80, EMER PROF ANIMAL SCI & GENET, UNIV MASS, AMHERST, 80- *Concurrent Pos:* Researcher, NC State Col, 51-53; res grants, Nat Asn Artificial Breeders, 56, 57-65, Hood Found, 57-65; mem & chmn, Nat Res Comt Milk Compos, 60-67; res grants, Abelard Found, 62-74, George H Walker Res Fund, 73-79, Eastern Artificial Breeders Coop, 65-78 & Agway grant, 66-67; Fulbright res scholar, Denmark, 69; guest lectr univs, Aberdeen, Cambridge, Lisbon, Ljubljana & Thessalonikia, 69; dairy sci & genetics adv to Colanta, Medellin, Antioquia, Colombia, SAm, 79-84. *Honors & Awards:* DeLaval Exten Award, Am Dairy Sci, 77. *Mem:* Am Dairy Sci Asn (secy, 51-52); Am Soc Animal Sci; Am Genetic Asn; NY Acad Sci; Sigma Xi. *Res:* Measures of a dairy sire's genetic merit; programs to maximize genetic progress; genetic and environmental influences on non-fat components of milk; relationships between genetic markers to production and reproduction in dairy cattle and the genetics of mastitis. *Mailing Add:* 9441 SE Little Club Way N Tequesta FL 33469

GAUNTT, WILLIAM AMOR, b New Brunswick, NJ, Nov 21, 26; m 51; c 3. RESEARCH ADMINISTRATION, TECHNICAL MANAGEMENT. *Educ:* Rutgers Univ, BS, 51. *Prof Exp:* Dist mgr vitamins-minerals premixes, Nobco Chem Co, 52; nutritionist feed premixes & formulation, Hess & Clark Inc, 53-60; pres & gen mgr, Grain States Eastern Div Inc, 60-67; sales mgr & tech serv east region, Hoffman Taff Inc & Syntex, 67-75; VPRES NUTRIT & RES, ANIMAL FEEDS SUPPL & PREMIXES, PROVICO, INC, 75- *Concurrent Pos:* Reporter regional feed bus, Feedstuff Mag, 82-; mem, Coun Agr Sci Technol. *Mem:* Am Registry Prof Animal Scientist; Am Dairy Sci Asn; Am Animal Sci Asn; Am Poultry Sci Asn; NY Acad Sci. *Res:* Directs all nutrition and research for feeds, supplies, premixes, seeds and farm supplies and all personnel associated with same. *Mailing Add:* Provico Inc 104 Oak St PO Box 188 Botkins OH 45306

GAUNYA, WILLIAM STEPHEN, b Conn, Sept 28, 20; m 47; c 3. DAIRY HUSBANDRY. *Educ:* Univ Conn, BS, 47; Rutgers Univ, MS, 49, PhD, 61. *Prof Exp:* Res assoc dairy husb, Rutgers Univ, 48-49; asst prof dairy husb, Univ Conn, 49-62, from assoc prof to prof animal industs, 72-84; RETIRED; 203-429-4693. *Mem:* Am Dairy Sci Asn. *Mailing Add:* 12 Agron Rd Storrs CT 06268

GAURI, KHARAITI LAL, b Narowal, India, Oct 16, 33; m 64. PALEONTOLOGY, SEDIMENTOLOGY. *Educ:* Panjab Univ, India, BSc, 53, MA, 55; Univ Bonn, PhD(geol), 64. *Prof Exp:* Lectr geog, D A V Col, Jullundher, 55-58; sci assoc geol, Calif Inst Technol, 65-66; sci pool officer, Coun Sci & Indust Res, New Delhi, India, 66; from asst prof to assoc prof, 66-73, PROF GEOL, UNIV LOUISVILLE, 73-, CHMN DEPT, 76- *Concurrent Pos:* Res grants, Geol Soc Am, 67-68, Res Corp, 67-68, 70-71, Kress Found, 70-71, Am Res Ctr, Egypt, 80-84 & Ky Energy Cabinet, 85-86; deleg, Int Conf of Preservation of Stone Statuary, Bologna, Italy, 69, Int Inst Conserv, New York,70, 2nd Int Symp Deterioration Bldg Stones, Athens, 76, UNESCO/RILEM Int Symp Deterioration & Protection Stone Monuments, 78 & Fourth Int Cong Deterioration & Preserv of Stone Objects, Louisville, 82. *Mem:* Geol Soc Am; Int Inst Conserv. *Res:* Pollutan effects upon masonry materials; conservation of stone statuary. *Mailing Add:* 719 Fehr Rd Louisville KY 40206

GAUS, ARTHUR EDWARD, b Maplewood, Mo, Nov 30, 24; m 50; c 3. HORTICULTURE. *Educ:* Univ Mo, BS, 49, MS, 50, PhD, 57. *Prof Exp:* Asst instr hort, Univ Mo, 51-53; exten horticulturist, Kans State Col, 53-54; exten horticulturist, 54-66, PROF HORT, UNIV MO, COLUMBIA, 66- *Res:* Cost of production in vegetable crops. *Mailing Add:* 808 Hope Pl Columbia MO 65211

GAUS, PAUL LOUIS, b Athens, Ohio, Mar 26, 49; m 72; c 2. PHOTOCHEMISTRY AND ORGANOMETALLIC CHEMISTRY. *Educ:* Miami Univ, BS, 71; Duke Univ PhD(chem), 75. *Prof Exp:* Res assoc chem, State Univ NY Stony Brook, 75-77; ASST PROF CHEM, COL WOOSTER, 77- *Concurrent Pos:* Counr, Am Chem Soc, 77- *Mem:* Am Chem Soc; Sigma Xi. *Res:* Substitution and electron transfer reactions of transition metal ions; photochemical reactions of organometallic systems. *Mailing Add:* Dept Chem Col Wooster Wooster OH 44691

GAUSE, DONALD C, b Elkhart, Ind, May 4, 34; m 56; c 2. COMPUTER SCIENCE. *Educ:* Mich State Univ, BS, 56, MS, 57. *Prof Exp:* Assoc engr, Sperry Gyroscope Co, 57-59; proj engr, Gen Motors Truck & Coach, Gen Motors Corp, 59-60, sr proj engr, 60-61, sr res mathematician, Gen Motors Res Lab, 61-62; sr assoc programmer, IBM Corp, 62-63, staff engr, 63-68, mgr prog educ, 68-70; adj asst prof comput sci, 67-68, vis lectr, 68-70, assoc prof, 70-74, prof systs sci, Sch Advan Technol, 74-84, PROF SYSTS SCI, THOMAS J WATSON SCH ENG, STATE UNIV NY BINGH AMTON, 84-; ADJ STAFF, IBM CORP TECHNOL INST & THEIR SUCCESSORS, NY, 80- *Concurrent Pos:* Consult, various corps; pres, Ethnotech, Inc, 77-79; nat lectr, Asn Comput Mach; vis lectr, Sch Archit, Swiss Fed Inst, Zurich, Jung Inst, Norwegian univs & NZ univs. *Mem:* Asn Comput Mach; Soc Gen Systs Res. *Res:* Adaptive, self-organizing and general systems theory; creative processes; computer modelling and simulation; heuristic programming; decision theory; creative design; management of innovation; user oriented systems design; design methodology. *Mailing Add:* Three Kingsgate Lane Owego NY 13827

GAUSE, EVELYN PAULINE, b Detroit, Mich, Mar 11, 46. RADIOACTIVE WASTE MANAGEMENT, CORROSION. *Educ:* Univ Akron, BS, 67, MS, 73, PhD(inorg chem), 80. *Prof Exp:* Lit chemist polymers, Univ Akron, 67-68, mem staff, Grad Sch, 68-70; high sch teacher chem, Akron, 70-74; tech info specialist polymers, B F Goodrich, 74-76; proj mgr nuclear waste mgt, Brookhaven Nat Lab, 80-84; sr mat specialist, High Level Nuclear Waste Mgt, Roy F Weston Inc, 84-89; SR SCIENTIST, SRA TECHNOL INC, 89- *Concurrent Pos:* Lab supvr, Univ Akron, 79. *Mem:* Am Chem Soc; Am Nuclear Soc. *Res:* Evaluation of materials for packaging of nuclear waste for disposal in shallow land burial and in geologic repositories, involving package and site interactions over a period of thousands of years; chelation of radioisotopes. *Mailing Add:* 12850 Middlebrook Rd Suite 304 Germantown MD 20874

GAUSTAD, JOHN ELDON, b Minneapolis, Minn, May 23, 38; m 80; c 2. INFRARED ASTRONOMY. *Educ:* Harvard Col, AB, 59; Princeton Univ, PhD(astron), 62. *Prof Exp:* Res assoc astron, Princeton Univ, 62-63; res fel astron, Mt Wilson & Palomer Observ, 63-64; lectr math, Univ Nigeria, 64-67; from asst prof to prof astron, Univ Cal, Berkeley, 67-82; PROF ASTRON, SWARTHMORE COL, 82- *Concurrent Pos:* Chmn Astron Surv Comt, Nat Acad Sci, 70-72; chmn vis comt, Kitt Peak Nat Observ, 74-76; chmn Dept Astron, Univ Calif, Berkeley, 76-79; assoc dean Col Lett & Sci, 80-82; chmn Dept Astron, Swarthmore col, 82-86; dir, Sproul Observ,82-86; Edward Hicks McGill Prof, 90- *Mem:* AAAS; Am Astron Soc; Int Astron Union. *Res:* Stellar evolution; infrared astronomy; interstellar matter. *Mailing Add:* Dept Physics & Astron Swarthmore Col Swarthmore PA 19081

GAUSTER, WILHELM BELRUPT, b Vienna, Austria, Dec 25, 40; US citizen; m 69; c 2. SOLID STATE PHYSICS. *Educ:* Harvard Univ, AB, 61; Univ Tenn, PhD(physics), 66. *Prof Exp:* Res assoc, Oak Ridge Nat Lab, 66; supv, Phys Res Div, Sandia Nat Labs, 79-82, Fusion Technol Div, 82-87, mgr, Exploratory Nuclear Power Systs Dept, 87-89, MGR, NUCLEAR ENERGY SCI & MAT TECHNOL DEPT, SANDI NAT LABS, 89- *Concurrent Pos:* Mem tech staff, Sandia Nat Labs, 66-; adj prof, Univ NMex, 72-73; vis scientist, Kernforschungsanlage J lich, Ger, 74-75. *Mem:* Am Phys Soc. *Res:* Positron annihilation; radiation damage; muon spin rotation; plasma-wall interactions; thermomechanical effects; nuclear technology. *Mailing Add:* Dept 6420 Sandia Nat Lab Albuquerque NM 87185-5800

GAUSTER, WILHELM FRIEDRICH, b Vienna,Austria, Jan 6, 01; US citizen; m 40; c 2. SUPERCONDUCTIVITY, MAGNETISM. *Educ:* Tech Univ, Vienna, Dipl Ing, 22, DrSc(elec eng), 24. *Hon Degrees:* Dr, Tech Univ, Graz, Austria, 82. *Prof Exp:* Referent, Elin Ag, Vienna, 24-28, prokurist, 28-31, dir, 31-41, consult, 76-90; researcher submarine acoust, Ger Navy, 41-45; prof elec eng, Tech Univ, Vienna, 45, 48 & 49-50; prof, NC State Univ, 50-57; dir, Magnet Lab, Oak Ridge Nat Lab, 57-71, consult, 71-76; CONSULT, MINISTRY SCI & RES, AUSTRIAN GOVT, 76- *Concurrent Pos:* Privat dozent, Tech Univ, Vienna, 27-33, dozent, 33-45, univ prof, 83-; vis prof elec eng, Fordham Univ, NY, 48-49; consult, US Naval Res Lab, Gen Elec Co & Oak Ridge Nat Lab, 51-56. *Mem:* Fel Inst Elec & Electronics Engrs; fel Am Phys Soc. *Res:* Superconductivity; magnetism theory and technology; engineering problems in thermometer research. *Mailing Add:* Neuwaldegger Strasse 9 Vienna A-1170 Austria

GAUT, ZANE NOEL, b Nauvoo, Ala, Aug 29, 29; m 55; c 3. CLINICAL PHARMACOLOGY, NUTRITION. *Educ:* Birmingham-Southern Col, BS, 50; Tulane Univ, MD, 54, PhD(biochem), 64. *Prof Exp:* Intern med, St Thomas Hosp, Nashville, Tenn, 54-55; aerospace med specialist, Gen Dynamics Corp, 58-63; asst prof med & biochem, Tulane Univ, 64-66; clin pharmacologist, 66-72, dir clin nutrit, 72-77, asst med dir, 77-78, dir clin res, Endocrinol/Metab, Hoffmann-La Roche, Inc, 78-85; CONSULT, 85- *Concurrent Pos:* NIH res grant, 63-, fel nutrit, 64-; asst attend physician, St Luke's Med Ctr, NY; asst clin prof, Col Physicians & Surgeons, Columbia Univ; referee ed, Prof Soc Exp Biol & Med; attend, Newark Beth Israel Hosp, NJ. *Mem:* AAAS; Am Soc Pharmacol & Exp Therapeut; Am Inst Nutrit; Am Soc Clin Pharmacol & Therapeut; Am Chem Soc. *Res:* Metabolism in human blood platelets; phase I drug evaluation in man; bone disease; obesity. *Mailing Add:* Biomed-Pharmaceut Res Five Mountain Ave Warren NJ 07059

GAUTHIER, DIDIER, b Hauterive, Que, Can, Jan 5, 54. CELL BIOLOGY. *Educ:* Univ Sherbrooke, BSc, 76; Univ Laval, PhD(biochem), 87. *Prof Exp:* Res asst pharmacol, Univ Laval, 80-82, res asst biochem, 82-83; lectr, 84-87, ASST PROF BIOCHEM, UNIV MONCTON, 87- *Concurrent Pos:* Mem, Environ Sci Res Ctr, 88- *Res:* Interaction of cytoskeleton with other cell activities, mainly protein synthesis, glycolysis and ionic tunnels; biosynthesis of domoic acid by diatomea. *Mailing Add:* Dept de Chimie et Biochimie Univ Moncton Moncton NB E1A 3E9 Can

GAUTHIER, FERNAND MARCEL, b Soulanges, Que, May 4, 23; m 48; c 1. PLANT BREEDING. *Educ:* Laval Univ, BSA, 42; McGill Univ, MSc, 47; Univ Man, PhD, 67. *Prof Exp:* Cerealist, Exp Farm, Can Dept Agr, Que, 45-62; from assoc prof to prof plant breeding, Laval Univ, 62-74; asst dir gen res, Que Agr, 75-84; RETIRED. *Mem:* Genet Soc Can; Agr Inst Can. *Res:* Wheat and barley breeding; oat aneuploids. *Mailing Add:* 5286 Blvd St Joseph Lashine PQ H8T 1S2 Can

GAUTHIER, GEORGE JAMES, b Franklin, NH, July 22, 40; m 64; c 4. PHARMACEUTICAL CHEMISTRY. *Educ:* Univ Notre Dame, BSc, 62; Univ NH, PhD(org chem), 66; Univ New Haven, MBA, 75. *Prof Exp:* Res assoc organometallic chem, Frank J Seiler Lab, Off Aerospace Res, USAF Acad, 66-69; res chemist, Med Res Labs, 69-71, CHEMIST, PFIZER CHEM DIV, PFIZER, INC, 71- *Mem:* Am Chem Soc; Am Soc Brewing Chemists. *Res:* Investigations of nucleophilic additions to pyridinium salts; synthesis and reactions of metallocenes, ruthenocene, ferrocenes. *Mailing Add:* Pfizer Inc Eastern Point CT 06340

GAUTHIER, GERALDINE FLORENCE, b Haverhill, Mass, May 14, 31. CELL BIOLOGY. *Educ:* Mass Col Pharm, BS, 54, MS, 55; Radcliffe Col, AM, 56, PhD(anat), 62. *Prof Exp:* Res asst pharmacol, Harvard Med Sch, 56-58, teaching fel anat, 58-59, res fel, 59-62; from instr to asst prof biol, Brown Univ, 62-64; asst prof biol sci, 64-68, assoc prof biol, 68-73, prof biol, Wellesley Col, 73-; PROF ANAT, UNIV MASS MED SCH. *Concurrent Pos:* Fel, Am Found Pharmaceut Educ, 54-57; USPHS res award, 63-65, co-recipient, 64-74, prin investr grant, 75-80; Muscular Dystrophy Asns Am grant, 69-79. *Mem:* AAAS; Am Soc Cell Biol; Histochem Soc; Am Asn Anat; Int Soc Cell Biol. *Res:* Ultrastructural and cytochemical heterogeneity of skeletal muscle fibers; structural and functional relationships. *Mailing Add:* Dept Anat Med Sch Univ Mass 55 Lake Ave N Worcester MA 01655

GAUTHREAUX, SIDNEY ANTHONY, b Plaquemine, La, Oct 18, 40; m 63, 90; c 2. VERTEBRATE ZOOLOGY, ANIMAL BEHAVIOR. *Educ:* La State Univ, New Orleans, BS, 63; La State Univ, Baton Rouge, MS, 65, PhD(zool), 68. *Prof Exp:* Instr biol, La State Univ, Baton Rouge, 67-68; Stoddard-Sutton res fel zool, Univ Ga, 68-70; from asst prof to assoc prof, 70-77, PROF ZOOL, CLEMSON UNIV, 77- *Mem:* Ecol Soc Am; Am Soc Zoologists; Animal Behav Soc (secy, 81-84); Am Ornith Union; Am Soc Naturalists; Sigma Xi; fel AAAS. *Res:* Behavioral ecology; migratory behavior of animals, particularly birds and physiological mechanisms that underlie the behavior; behavioral dominance; radar and telescopic studies of bird migration; circadian rhythms. *Mailing Add:* Dept Biol Sci Clemson Univ Clemson SC 29631

GAUTIERI, RONALD FRANCIS, b Providence, RI, Oct 10, 33; m 62; c 1. PHARMACOLOGY. *Educ:* RI Col Pharm, BS, 55; Temple Univ, MS, 57, PhD(pharmacol), 60. *Prof Exp:* From asst prof to assoc prof, 60-70, PROF PHARMACOL, TEMPLE UNIV, 70-, CHMN DEPT, 71- *Concurrent Pos:* Vis lectr, Am Asn Cols Pharm, 68- *Mem:* AAAS; Acad Pharmaceut Sci; Am Soc Pharmacol & Exp Therapeut; Am Pharmaceut Asn; Am Asn Pharmaceut Scientists. *Res:* Cancer; human placental perfusions; toxicology; teratology; biochemistry; physiology; dental research; gastroenterology; mechanism of action of drugs; fetal pharmacology. *Mailing Add:* 418 Bolton Rd Glenside PA 19038

GAUTREAU, RONALD, b Newark, NJ, Jan 21, 40; m 61; c 3. PHYSICS. *Educ:* Lehigh Univ, BS, 61; Stevens Inst Technol, MS, 63, PhD(physics), 66. *Prof Exp:* PROF PHYSICS, NJ INST TECHNOL, 66- *Mem:* Am Phys Soc; Sigma Xi; NY Acad Sci. *Res:* Relativity physics with major emphasis on cosmology and the Schwarzschild field; science education. *Mailing Add:* Dept Physics NJ Inst Technol Univ Heights Newark NJ 07102

GAUTREAUX, MARCELIAN FRANCIS, b Nashville, Tenn, Jan 17, 30; m 52; c 4. RESEARCH ADMINISTRATION. *Educ:* La State Univ, BS, 50, MS, 51, PhD(chem eng), 58. *Prof Exp:* Asst supvr, Res & Develop Dept, Ethyl Corp, 51-55; from instr to asst prof chem eng, La State Univ, 55-58; mem staff, Res & Develop Dept, Ethyl Corp, 58-68, gen mgr, 68-69, vpres, 69-74, mem bd dir, 72-86, sr vpres, Res & Develop Dept, 74-86, sr vpres & sci adv, 81-86; CONSULT, 86- *Honors & Awards:* Ann Award, Chem Mkt Res Asn, 78. *Mem:* Nat Acad Eng; Am Inst Chem Engrs; Soc Eng Sci; Soc Chem Indust; Chem Mkt Res Asn. *Res:* Process development; commercial development. *Mailing Add:* PO Box 14799 Baton Rouge LA 70898

GAUTSCH, JAMES WILLARD, b Rockford, Ill, Oct 13, 41; m 68; c 2. RECOMBINANT DNA, GENETIC ENGINEERING. *Educ:* Univ Denver, BA, 63; Univ Wyo, MS, 68; Univ Calif, Irvine, PhD(molecular biol & biochem), 73. *Prof Exp:* Fel genetics, Jackson Lab, Bar Harbor, Maine, 74-76; fel immunol, Scripps Clin, La Jolla, Calif, 76-78, asst mem molecular biol, 78-86; CONSULT, 86- *Res:* Replication of retroviruses and their effect on host cell growth; mechanisms of gene expression in early mammalian embryo cells. *Mailing Add:* 451 S Granados Ave Solana CA 92075

GAUTSCHI, WALTER, b Basel, Switz, Dec 11, 27; US citizen; m 60; c 4. NUMERICAL ANALYSIS. *Educ:* Univ Basel, PhD, 53. *Prof Exp:* Fel Nat Inst Appl Calculus, Italy, 54-55 & comput lab, Harvard Univ, 55-56; res mathematician, Am Univ, 56-59; mathematician, Oak Ridge Nat Lab, 59-63; PROF MATH & COMPUT SCI, PURDUE UNIV, 63- *Concurrent Pos:* Res mathematician, Nat Bur Standards, 56-59; vis prof, Tech Univ, Munich, Ger, 70-71 & Univ Wis, 76-77. *Mem:* Am Math Soc; Math Asn Am; Soc Indust & Appl Math; Swiss Math Asn. *Res:* Numerical analysis; special functions; ordinary differential equations; orthogonal polynomials. *Mailing Add:* Dept Comput Sci Purdue Univ West Lafayette IN 47907

GAUVIN, J N LAURIE, b Shediac, NB, Oct 8, 29; m 64; c 2. THEORETICAL PHYSICS. *Educ:* Laval Univ, BSc, 54; Oxford Univ, PhD(theoret physics), 57. *Prof Exp:* From asst prof physics to assoc prof, Laval Univ, 57-70, prof 70-71; dir comn sci res, Ministry Educ, 71-81, sr adv, secretarial sci develop, 81-89; PRES, CGTS INC, 89- *Concurrent Pos:* Secy, Que Comt Sci Policy, 71-73. *Mem:* Can Asn Physicists. *Res:* Science policy; administration of grants for research. *Mailing Add:* 2650 Chemin St-Louis Quebec PQ G1W 1N3 Can

GAUVIN, WILLIAM H, b Paris, France, Mar 30, 13; Can citizen; m 36; c 1. CHEMICAL & METALLURGICAL ENGINEERING. *Educ:* McGill Univ, BEng, 41, MEng, 42, PhD(chem), 44. *Hon Degrees:* DEng, Univ Waterloo, 67, Univ McGill, 83, Queen's Univ, 84 & McMasters Univ, 86. *Prof Exp:* Lectr chem eng, McGill Univ, 42-44; plant supt, Frank W Horner Ltd, Montreal, 44-46; head chem eng div, Pulp & Paper Res Inst, 57-61; mgr, Noranda Res Centre, 61-70, dir res & develop, 70-83; sci adv to dir, HQ Res Inst, 83-90; assoc prof, 47-62, SR RES ASSOC, DEPT CHEM ENG, MCGILL UNIV, 62- *Concurrent Pos:* Mem, Nat Res Coun Can, 62-70, gen deleg, 70-71; mem, Sci Coun Can, 66-70; mem, Can Res Mgt Asn, 78-; pres, Interam Confedn Chem Eng, 78-81. *Honors & Awards:* Can Pulp & Paper Asn Weldon Medal Award, 58; Paper Award, Chem Inst Can, 60 & 61, Jane Mem Lect Award, 63; Sr Moulton Medal Award, Brit Inst Chem Engrs, 64; Medaille Archambault Award, Can Asn Adv Sci, 66; Chem Inst Can Medal, 66; Can Soc Chem Engrs Award, 68; Alcan Award, Can Inst Mining & Metall, 70; Gold Medal, Soc Invention, France, 74; Prix des Sci dú Quibec, 84; T W Eadic Medal, Royal Soc Can, 86; S W Killam Prize in Eng, 88. *Mem:* Foreign mem Nat Acad Engrs; Chem Inst Can (pres, 77-78); Can Soc Chem Engrs (pres, 66-67); hon mem Soc Indust Chem; fel Royal Soc Can; Sigma Xi; fel Can Acad Eng. *Res:* Electrochemistry; high temperature technology, particle dynamics and fluid dynamics, particularly chemical and metallurgical processes; plasma technology. *Mailing Add:* Seven Harrow Pl Beaconsfield PQ H9W 5C7 Can

GAVALAS, GEORGE R(OUSETOS), b Athens, Greece, Oct 7, 36; m 80. CHEMICAL REACTION ENGINEERING, COAL UTILIZATION. *Educ:* Nat Tech Univ Athens, dipl eng, 58; Univ Minn, MS, 62, PhD(chem eng), 64. *Prof Exp:* From asst prof to assoc prof, 64-75, PROF CHEM ENG, CALIF INST TECHNOL, 75- *Concurrent Pos:* Consult, Phillips Petrol Co, 67- & miscellaneous indust orgn. *Honors & Awards:* Wilhelm Award, Am Inst Chem Eng, 83; Wilhelm lectr, Princeton Univ, 87. *Mem:* Am Inst Chem Engrs; Am Chem Soc; Soc Petrol Engrs. *Res:* Chemical reaction engineering and catalysis; applied mathematics; coal combustion; ceramics processing. *Mailing Add:* Dept Chem Eng 206-41 Calif Inst Technol Pasadena CA 91125

GAVAN, JAMES ANDERSON, b Ludington, Mich, July 17, 16; m 45; c 2. ANTHROPOLOGY, ANATOMY. *Educ:* Univ Ariz, BA, 39; Univ Chicago, MA, 49, PhD(anthrop), 53. *Prof Exp:* Mem res staff, Yerkes Labs Primate Biol, 50-53; from asst prof anat to assoc prof, Med Univ SC, 53-62; assoc prof, Univ Fla, 62-67; chmn dept, 68-71 & 75-78, prof, 67-86, EMER PROF ANTHROP, UNIV MO, COLUMBIA, 86- *Concurrent Pos:* Mem, Dental Study Sect, NIH, 70-74, mem, Anthrop Rev Panel, NSF, 76-78. *Mem:* AAAS; Am Asn Phys Anthrop (secy-treas, 73-77, pres, 77-78); fel Am Anthrop Asn. *Res:* Growth, development and anatomy of primates; physical anthropology; gross anatomy. *Mailing Add:* Dept Anthrop Univ Mo Columbia MO 65211

GAVANDE, SAMPAT A, b India. SOIL PHYSICS, ENVIRONMENTAL QUALITY. *Educ:* Poona Univ, India, BS, 58; Kans State Univ, MS, 62; Utah State Univ, PhD(soil physics, irrig), 66. *Prof Exp:* From res asst to res assoc, Utah State Univ, 62-66; tech officer soil physics, UN Food & Agr Orgn-Inter-Am Inst, Orgn Am States proj, Turrialba, Costa Rica, 66-69; tech officer soil physics, trop soil & water mgt, UN Food & Agr Orgn proj, Nat Agr Ctr, Chapingo, Mex & Trop Agr Ctr, Tabasco, Mex, 69-72; prof soils & irrig & head dept, Grad Sch, Agrarian Autonomas Univ, Saltillo, Mex, 73-75; tech officer soil-water plant relations & co-dir, UN Food & Agr Orgn proj, Arid Lands Res Ctr, Saltillo, Mex, 75-77; sr scientist natural resources dept, Radian Corp, 77-82; proj mgr, Foreign Agr Orgn, UN Develop Prog, Paraguay, 87-79; CHIEF TECH SUPPORT BR, BUR SOLID WASTE, HEALTH DEPT AUSTIN, 89- *Concurrent Pos:* Soil & water conserv consult, Foreign Agr Orgn, Kenya, Chile, India, Iran, Peru & Indonesia. *Mem:* Sigma Xi; AAAS; Am Soc Agron; Soil Sci Soc Am; Int Soc Soil Sci; Am Soc Agr Engrs; Am Soc Mining Engrs; Am Sec Testing Mat. *Res:* Technological and environmental feasibility of waste disposal systems, and reclamation of surface mined lands; water conservation and tillage practices; landfill and surface improvements. *Mailing Add:* 3005 W Terrace Dr Austin TX 78731

GAVASCI, ANNA TERESA, b Siena, Italy, May 15, 24; US citizen; m 53; c 1. PETROLOGY. *Educ:* Univ Florence, DrLaurea(mineral), 48; Columbia Univ, PhD(geol), 69. *Prof Exp:* High sch teacher, Italy, 50-53; asst prof mineral, Univ Florence, 50-54; asst prof geol, 70-76, assoc prof geol & geog, Hunter Col, City Univ NY, 76-86; RETIRED. *Concurrent Pos:* Res award, City Univ NY, 72; res assoc geol, Lamont-Doherty Geol Observ, Columbia Univ, 70-76. *Mem:* AAAS; Geol Soc Am; Geochem Soc; Am Geophys Union; Int Asn Geochem & Cosmochem; Sigma Xi. *Res:* Physicochemical properties of Earth's mantle, ultramafic xenoliths, and deformation of amphiboles and pyroxenes. *Mailing Add:* 2001 Kimbrough Green Germantown TN 38138-4258

GAVENDA, JOHN DAVID, b Temple, Tex, Mar 25, 33; m 52; c 2. EXPERIMENTAL SOLID STATE PHYSICS, ELECTROMAGNETIC COMPATIBILITY. *Educ:* Univ Tex, BS, 54, MA, 56; Brown Univ, PhD(physics), 59. *Prof Exp:* Asst prof physics & res scientist, Defense Res Lab, 59-62, assoc prof, 62-67, PROF PHYSICS & EDUC, UNIV TEX, AUSTIN, 67- *Concurrent Pos:* Sr res fel, Inst Study Metals, Univ Chicago, 63; NATO sr fel sci, Univ Oslo, spring, 69; consult electromagnetic compatibility. *Mem:* AAAS; Am Asn Physics Teachers; fel Am Phys Soc; Inst Elec & Electronics Engrs. *Res:* Electronic properties of metals at low temperatures; physics education; measurement and reduction of electromagnetic emissions from computing equipment. *Mailing Add:* Dept Physics Univ Tex Austin TX 78712

GAVER, DONALD PAUL, b St Paul, Minn, Feb 16, 26. MATHEMATICS. *Educ:* Mass Inst Technol, SB, 50, SM, 51; Princeton Univ, PhD(math), 56. *Prof Exp:* Mem staff mil opers res, US Navy Opers Eval Group, 51-53; mem systs analysis res group, Princeton Univ, 53-56; res mathematician, Res Labs, Westinghouse Elec Corp, 56-60, supvry mathematician, Dept Math, 60, adv math, 62-64; assoc prof math & indust admin, Carnegie-Mellon Univ, 64-70, prof statist & indust admin, 70-74; PROF, DEPT OPERS RES, NAVAL POSTGRAD SCH, 74- *Mem:* Opers Res Soc Am; Am Statist Asn; Inst Math Statist. *Res:* Applications of probability and probability models; statistics; operational research. *Mailing Add:* Dept Opers Res Naval Postgrad Sch Code OR/GV Monterey CA 93943

GAVER, ROBERT CALVIN, b Chambersburg, Pa, Oct 2, 38; div; c 2. DRUG METABOLISM, PHARMACOKINETICS. *Educ:* Pa State Univ, BS, 60; Univ Pittsburgh, PhD(biochem), 64. *Prof Exp:* Res assoc, Univ Ill, 64-67; res scientist, Bristol Labs, Inc, 67-83; SR RES SCIENTIST, PHARMACEUT RES & DEVELOP DIV, DEPT METAB & PHARMACOKINETICS, BRISTOL-MYERS CO, 83- *Mem:* AAAS; Am Assoc Cancer Res. *Res:* Define the disposition of new therapeutic agents, particularly antihuman agents in animals and humans; development of analytical methodology; isolation and identification of metabolites; pharmacokinetics analyses and protein binding, tissue distribution and bio availability studies. *Mailing Add:* Pharmaceut Res & Develop Div Bristol-Myers Co PO Box 4755 Syracuse NY 13221

GAVIN, DAVID FRANCIS, b Indianapolis, Ind, Jan 4, 39; m 64; c 1. SYNTHETIC ORGANIC CHEMISTRY. *Educ:* Hofstra Univ, BS, 60; Southern Conn State Col, MS, 66. *Prof Exp:* SUPVR ORG CHEM RES & DEVELOP, OLIN CORP, 60- *Concurrent Pos:* Lectr, Southern Conn State Col, 73-; mgr govt contract R&D, Olin Chem Group. *Mem:* Am Chem Soc; Am Soc Lubrication Engrs; Sigma Xi. *Res:* High performance lubricants and functional fluids based on novel chemical principles. *Mailing Add:* 255 Sorghum Mill Cheshire CT 06410

GAVIN, DONALD ARTHUR, b Albany, NY, Mar 31, 31; m 55; c 7. NEUTRON & X-RAY SPECTROMETERS, NUCLEAR REACTOR PHYSICS. *Educ:* Holy Cross Col, BS, 53; Univ RI, MS, 55. *Prof Exp:* Supvr critical facil, 60-70, mgr non-destructive testing, 70-82, SR ENGR MAT ANALYSIS, KNOLLS ATOMIC POWER LAB, GEN ELEC CO, 86- *Concurrent Pos:* Instr physics, Schenectady Community Col, 78-81; prin investr inserv inspection naval nuclear reactors, 81-83. *Mem:* Am Soc Non Destructive Testing. *Res:* Analyses of materials used in nuclear reactors using neutron transmission; x-ray transmission; real time radiography; in service inspections of naval nuclear reactors. *Mailing Add:* 17 Rivercrest Dr Rexford NY 12148

GAVIN, GERARD BRENNAN, b Long Island City, NY, Aug 28, 23; m 47; c 5. REACTOR PHYSICS, NUCLEAR PHYSICS. *Educ:* Siena Col, Loudonville, NY, BS, 47. *Prof Exp:* Res asst physics, Knolls Atomic Power Lab, 51-54, physicist, 54-66, physicist reactor measurements, 66-74, proj engr design & construct, 74-75, sr engr nuclear 75-77, physicist prototype test, 77-81, sr physicist prototype test, 81-84; RETIRED. *Res:* Nuclear energy. *Mailing Add:* 51 Nicholas Dr Albany NY 12205

GAVIN, JOHN JOSEPH, b New Brunswick, NJ, Oct 21, 22; m 45; c 9. MICROBIOLOGY. *Educ:* Rutgers Univ, BS, 49, MS, 50, PhD, 64. *Prof Exp:* Head biol control, Smith, Kline & French Labs, Pa, 50-55; chief microbiologist, Food Res Labs, NY, 55-57; Fund Res Therapeut res fel biochem, Norristown State Hosp, 57-64; group leader bact res, Norwich Pharmacol Co & Eaton Labs Div, 64-66; sr scientist & head, Dept Allergy & Immunol, Dome Labs, 66-69; dir biol prod develop, Miles Labs, Inc, 69-71, dir molecular biol res, 71-75, dir allergy res affairs, 75-80, dir res, Hollister Stier Div, 80-84; sr med res & develop specialist, Eng & Econ Res Inc, 84-87, dir sci policy, 87-90; RETIRED. *Concurrent Pos:* Adj assoc prof, Univ Notre Dame, 79-81; adj prof, Eastern Washington Univ, 81-83. *Mem:* AAAS; fel Am Inst Chem; Am Soc Microbiol; Am Chem Soc; NY Acad Sci; Regulatory Affairs Prof Soc. *Res:* Immunology; nucleic acid metabolism; molecular biology. *Mailing Add:* 578 Tulip Poplar Crest Carmel IN 46032

GAVIN, JOSEPH GLEASON, JR, b Somerville, Mass, Sept 18, 20; m 43; c 3. AEROSPACE ENGINEERING & TECHNOLOGY. *Educ:* Mass Inst Technol, BS & MS, 42. *Hon Degrees:* ScD, Villanova, NY Regents. *Prof Exp:* Design engr, Grumman Aircraft Eng Corp, Bethpage, NY, 46-48, preliminary design group, 48-50, proj engr, 50-56, chief exp projs engr, 56-57, chief missile & space engr, 57-62, dir Lunar Module Prog Apollo, 62-72, vpres, 70-72, pres, 72-76, pres, chief operating officer & dir, 76-85, sr mgt consult, 85-90; RETIRED. *Concurrent Pos:* Chmn bd, Grumman Aircraft Eng Corp, 73-76; chmn, Int Coop Magnetic Fusion Comt, Nat Res Coun, 83-84. *Honors & Awards:* Distinguished Pub Serv Medal, NASA, 71. *Mem:* Nat Acad Eng; fel Am Inst Aeronaut & Astronaut; fel Am Astron Soc; Aerospace Industs Asn. *Mailing Add:* Six Endicott Dr Hunington NY 11743

GAVIN, LLOYD ALVIN, b New Orleans, La, Apr 20, 43; m 65; c 3. MATHEMATICS, STATISTICS. *Educ:* Xavier Univ La, BS, 64; Univ Kans, MA, 66; Ill Inst Technol, PhD(math), 73. *Prof Exp:* Asst prof math, Xavier Univ, 70-73; assoc prof, 73-84, PROF MATH, CALIF STATE UNIV, SACRAMENTO, 84- *Concurrent Pos:* Ford Found fel, Ill Inst Technol, 67-68, Whitney Young fel, 67 NDEA fel, 67, 68-70; distinguished dist gov, Toastmasters Int Dist 39, 87-88. *Mem:* Am Math Soc; Math Asn Am; Toastmasters Int. *Res:* Estimation; jacknife statistics. *Mailing Add:* 1213 Cedarbrook Way Sacramento CA 95831-4405

GAVIN, ROBERT M, JR, b Coatesville, Pa, Aug 16, 40; m 62; c 5. PHYSICAL CHEMISTRY. *Educ:* St John's Univ, Minn, BA, 62; Iowa State Univ, PhD(chem), 66. *Hon Degrees:* DSc, Haverford Col, 86. *Prof Exp:* From asst prof to prof chem, Haverford Col, 66-84, provost, 80-84; PRES, MACALESTER COL, 84. *Concurrent Pos:* Res assoc, Univ Mich, 66; fel, Univ Chicago, 69-70 & Univ Calif, Berkeley, 78-79. *Mem:* Am Chem Soc. *Res:* Molecular structure; spectroscopy; chemical bonding; photochemistry of vision. *Mailing Add:* Macalester Col 1600 Grand Ave St Paul MN 55105

GAVINI, MURALIDHARA B, b India, July 1, 47; m 75; c 1. RADIOCHEMISTRY, GEOCHEMISTRY. *Educ:* Andhra Univ, India, BSc, 70, MSc, 72; Univ Ark, Fayetteville, PhD(chem), 76. *Prof Exp:* investr aquatic geochem & atmospheric radiochem, Woods Hole Oceanog Inst, 76-79; VPRES & TECH DIR, US TESTING CO INC, 79- *Mem:* Am Chem Soc; Am Nuclear Soc. *Res:* Natural and artifically-produced radionuclides as potential atmospheric tracers; geochemical behavior of artificially produced radionuclides especially transuranium elements in the fresh water environments; technical management; analytical chemistry. *Mailing Add:* 58 Crane Circle New Providence NJ 07974-1107

GAVIS, JEROME, b Hartford, Conn, June 18, 28; m 54; c 2. ENVIRONMENTAL CHEMISTRY, ENVIRONMENTAL BIOLOGY. *Educ:* Polytech Inst Brooklyn, BChE, 49; Cornell Univ, PhD(chem), 53. *Prof Exp:* Asst prof, 56-60, ASSOC PROF ENVIRON ENG, JOHNS HOPKINS UNIV, 60- *Mem:* Am Inst Chem Engrs; Am Chem Soc; Am Soc Limnol & Oceanog; AAAS; Sigma Xi. *Res:* Natural water chemistry; transport phenomena in natural fluid systems; phytoplankton growth and uptake kinetics; interaction of phytoplankton and dissolved metal ions. *Mailing Add:* 2110 South Rd Baltimore MD 21209

GAVLIN, GILBERT, b Chicago, Ill, Jan 12, 20; m 47; c 3. ORGANIC CHEMISTRY, CHEMICAL ENGINEERING. *Educ:* Univ Ill, BS, 41; Cornell Univ, PhD(org chem), 48. *Prof Exp:* Res chemist, SAM Labs, Columbia Univ, 43-45 & Tenn Eastman Corp, 45-46; assoc org chemist, Armour Res Found, Ill Inst Tech, 46-47; asst, Cornell Univ, 47-48; res org chemist & dir, Nat Registry Rare Chems, Armour Res Found, 48-54; sr scientist, Richardson Co, 54-56, mgr res dept, 56-64; pres, Poly-Synthetix, Inc, 64-69, PRES, CUSTOM ORG, INC, 69- *Res:* Organic fluorine and chlorine compounds; polymer chemistry; mechanism of organic reactions; separation science and technology; mass transfer equipment design. *Mailing Add:* 6500 N Kenton Ave Lincolnwood IL 60646

GAVORA, JAN SAMUEL, b Brezova pod Bradlom, Czech, July 14, 33; Can citizen; m 59. BIOTECHNOLOGY APPLIED IN ANIMAL BREEDING. *Educ:* Agr Univ, Czech, Ing, 57, CSc, 67. *Prof Exp:* Dir artificial insemination animal breeding, State Breeding Bd, Nitra, Czech, 57-61; dir animal prod, Agr Co-op Farm, V Kostolany, Czech, 62; res scientist animal prod, Inst Sci Agr, Piestany, Czech, 63-66; res scientist poultry breeding, Poultry Res Inst, Ivanka pri Dunaji, Czech, 66-68; fel poultry breeding, Univ Man, 69-70; PRIN RES SCIENTIST & TEAM LEADER, MOLECULAR GENETICS, ANIMAL RES CTR, AGR CAN, OTTAWA, 71- *Honors & Awards:* Tom Newman Mem Award, 85; Merit Award, Pub Serv Can, 84; Travel Award, Japan Soc Prom Sci, 88; Cert of Merit, Can Soc Animal Sci, 89. *Mem:* World's Poultry Sci Asn; Poultry Sci Asn; Can Asn Animal Sci; Genetics Soc Can; Can Asn Advan Sci; Czech Soc Arts & Sci; Agr Inst Can. *Res:* Genetics of disease resistance; resistance against a neoplastic diseases of chickens, Marek's disease; breeding techniques for simultaneous improvement of production traits and disease resistance; lymphoid leukosis; biotechnology applied in livestock production. *Mailing Add:* Bldg No 34 Animal Res Ctr Ottawa ON K1A 0C6 Can

GAVURIN, LILLIAN, b New York, NY. CYTOGENETICS, DEVELOPMENTAL BIOLOGY. *Educ:* Brooklyn Col, BS, 51, MA, 58; NY Univ, PhD(biol), 71. *Prof Exp:* From asst prof to assoc prof, 71-80, PROF BIOL, FAIRLEIGH DICKINSON UNIV, 80- *Concurrent Pos:* Res grant, Sigma Xi, 72-73; res grants, Fairleigh Dickinson Univ, 72-74 & 77-78. *Mem:* Sigma Xi; AAAS; Am Soc Zoologists; Int Soc Differentiation. *Res:* Regeneration in blepharisma and planaria; macronuclear synthesis, macronuclear structure; cyclic AMP effects upon cell division, nuclear and chromosome structure, regeneration; brain differentiation in planaria. *Mailing Add:* Dept Biol Sci Fairleigh Dickinson Univ 1000 River Rd Teaneck NJ 07666

GAW, C VERNON, b Clinton, Mass, July 12, 42; m 66; c 2. BIOENGINEERING & BIOMEDICAL ENGINEERING. *Educ:* Univ Mass, BS, 68. *Prof Exp:* Electronics engr, Fenwal div, Walter Kidde, 68-74; sr proj engr, IItem Entry Control Div, Damon Corp, 74-79; sr engr, Valtec div, Int Tel & Tel, 79; RES & DEVELOP COORDR, INDUST MACH, BARRET CENTRIFUGALS INC, 79- *Mem:* Inst Elec & Electronics Engrs. *Res:* Liquid phase separation; liquid-solids separation; automatic controls; motion detection. *Mailing Add:* 37 Ford Rd RFD 3 Sterling MA 01564

GAWARECKI, STEPHEN JEROME, b Newark, NJ, July 31, 29; m 54; c 2. GEOLOGY. *Educ:* Rutgers Univ, BS, 51, MS, 52; Univ Colo, PhD(geol), 63. *Prof Exp:* Staff mining geologist, NJ Zinc Co, Colo & Pa, 52-53; geologist, Doeringsfeld, Amuedo & Ivey, Inc, Colo, 60-62; geologist, US Geol Surv, DC, 62-69, proj chief orbital photo anal, Br Regional Geophysics, 69-72, staff geologist remote sensing activ, 72-79, actg chief, Br Mideastern & Asian Geol & Br Europ & African Geol, Reston, Va, 79-84, chief, Br Resource Anal, 84-86; GEOL CONSULT, REMOTE SENSING DYNAMICS, 86- *Concurrent Pos:* Mem, UN Conf Sci & Technol for Develop, Vienna, 79; lectr geol, Northern Va Community Col, 89. *Mem:* AAAS; Am Soc Photogram & Remote Sensing; Sigma Xi. *Res:* Structural geology; photogeology; remote sensing of environment; regional tectonics; multispectral survey as an exploration tool for base metal deposits in humid tropical environments; tectonic map of North Thailand using ERTS satellite imagery; geologic hazard mitigation. *Mailing Add:* 7018 Vagabond Dr Falls Church VA 22042

GAWER, ALBERT HENRY, b New York, NY, July 22, 35. PHYSICAL CHEMISTRY. *Educ:* Rutgers Univ, BS, 57; Columbia Univ, AM, 58, PhD(chem), 63. *Prof Exp:* Instr chem, Brooklyn Col, 63-64; asst prof, Barnard Col, Columbia Univ, 64-68; ASST PROF CHEM, STATE UNIV NY, NEW PALTZ, 69- *Mem:* AAAS; Am Chem Soc; Am Phys Soc. *Res:* High resolution nuclear magnetic resonance spectroscopy; chemical applications of Mossbauer effect; instrumental methods of analysis. *Mailing Add:* Dept Chem State Univ NY Col New Paltz New Paltz NY 12561

GAWIENOWSKI, ANTHONY MICHAEL, b Newark, NJ, Oct 30, 24; m 55; c 5. BIOCHEMISTRY, ENDOCRINOLOGY. *Educ:* Villanova Col, BS, 48; Univ Mo, MS, 53, PhD(biochem), 56. *Prof Exp:* Control chemist, Merck & Co, 48; prof serv rep pharmaceut, Schering Corp, 49-52; res asst, Univ Mo, 53-56; res fel, Univ Tex, 56-57; asst prof, Kans State Univ, 57-63; PROF BIOCHEM, UNIV MASS, AMHERST, 63- *Concurrent Pos:* Guest worker, NIH, 69-70; guest scientist, Oak Ridge Asn Univ, 78. *Mem:* Fel AAAS; Am Chem Soc; Endocrine Soc; Sigma Xi; Phytochem Soc. *Res:* Steroid evolution; neurotransmitters; carotene precursors. *Mailing Add:* Dept Biochem Univ Mass Amherst MA 01003

GAWLEY, IRWIN H, JR, b Union City, NJ, Apr 20, 27; m 55; c 2. CHEMISTRY. *Educ:* Montclair State Col, AB, 49, AM 51; Columbia Univ, EdD, 57. *Prof Exp:* Teacher pub schs, NJ, 49-55; assoc prof, 55-60, PROF CHEM & DEAN SCH MATH & SCI, MONTCLAIR STATE COL, 60-, VPRES ACAD AFFAIRS, 73- *Mem:* AAAS; Am Chem Soc; Nat Asn Res Sci Teaching; Nat Sci Teachers Asn. *Res:* Chemical education; improvement of teaching of chemistry. *Mailing Add:* 177 McCosh Rd Upper Montclair NJ 07043

GAWLEY, ROBERT EDGAR, b Newark, NJ, Nov 7, 48; m; c 2. ORGANIC CHEMISTRY. *Educ:* Stetson Univ, BS, 70; Duke Univ, PhD(org chem), 75. *Prof Exp:* Res assoc med chem, Univ NC, Chapel Hill, 75-77; asst prof org chem, 77-82, ASSOC PROF, UNIV MIAMI, 82- *Mem:* Am Chem Soc; AAAS; Sigma Xi. *Res:* Synthetic methods; total syntheses; amino acid modification; nuclear magnetic resonance spectroscopy; asymmetric synthesis. *Mailing Add:* 6888 SW 59th St Miami FL 33143

GAWRON, OSCAR, b New York, NY, Aug 14, 14; m 47; c 2. BIOCHEMISTRY. *Educ:* Brooklyn Col, BS, 34; Columbia Univ, MA, 38; Polytech Inst Brooklyn, PhD(org chem), 45. *Prof Exp:* Chemist, Brooklyn Jewish Hosp, 35-40; chem supvr, Int Vitamin Corp, 40-45; res chemist, Am Home Prods Corp, NY, 45-46 & NY Quinine & Chem Works, Inc, 46-47; PROF BIOCHEM, DUQUESNE UNIV, 47-, DEAN GRAD SCH, 77- *Mem:* Am Chem Soc; Am Soc Biol Chem; NY Acad Sci. *Res:* Mechanisms of action of Krebs cycle enzymes; cis-aconitase and succinic dehydrogenase. *Mailing Add:* 14716D Canal View Dr Delray Beach FL 33484

GAY, BEN DOUGLAS, b Salt Lake City, Utah, Jan 11, 42; m 63; c 2. COMPUTER SCIENCE, ENGINEERING. *Educ:* Univ NMex, PhD(numerical gas dynamics), 69. *Prof Exp:* Dir comput sci prog, 73-75, asst prof comput sci & mech eng, Bucknell Univ, 69-, dir, off comput activ, 74-; AT MOTOROLA INC, ARLINGTON HEIGHTS, ILL. *Mem:* Asn Comput Mach. *Mailing Add:* Motorola Inc 1501 W Shure Dr Arlington Heights IL 60004

GAY, BRUCE WALLACE, JR, b Ludlow, Mass, July 23, 40; m 68; c 3. ATMOSPHERIC CHEMISTRY, KINETICS. *Educ:* Lowell Technol Inst, BS, 62; NMex Inst Mining & Technol, MS, 64. *Prof Exp:* Health serv officer, USPHS, 67-70; SR RES CHEMIST, US ENVIRON PROTECTION AGENCY, 70- *Concurrent Pos:* Teaching asst, Nat Univ Ireland, 64-66. *Honors & Awards:* Spec Sci Achievement Award, US Environ Protection Agency, 74, 76, 82, 84; Bronze Medal, 89. *Mem:* Am Chem Soc; AAAS; Sigma Xi. *Res:* Elucidation of atmospheric reactions to photochemical smog, long path infrared studies of the ambient urban atmosphere, atmospheric reactions of halogenated hydrocarbons and their effect on the earth's ozone layer; detection of naturally occuring hydrocarbons, their reactivity and overall burden to urban hydrocarbon pollution. *Mailing Add:* Mail Drop 84 Environ Protection Agency Research Triangle Park NC 27711

GAY, CAROL VIRGINIA LOVEJOY, b Belfast, Maine, Apr 8, 40; m 64; c 1. CELL BIOLOGY, SKELETAL METABOLISM. *Educ:* Univ Maine, BA, 62; Pa State Univ, MS, 67, PhD(physiol), 72. *Prof Exp:* Res assoc biochem & biophys, Pa State Univ, 72-80, sr proj assoc, 80-83, sr res assoc molecular & cell biol, 80-88, ASSOC PROF POULTRY SCI & CELL BIOL, PA STATE UNIV, 88- *Concurrent Pos:* NIH fel, Pa State Univ, 73-75; res grant, NIH, 77-94, res career develop award, 79-84; res grant, USDA, 81-85. *Mem:* AAAS; Am Soc Cell Biol; Am Physiol Soc; Am Soc Bone & Mineral Res; Royal Soc Med Affil. *Res:* Structure and metabolism of mineralized tissues; physiological roles and localization of carbonic anhydrase. *Mailing Add:* RD 1-Box 473 Bellefonte PA 16823

GAY, CHARLES FRANCIS, b Redlands, Calif, Oct 2, 46; m 69; c 1. PHYSICAL CHEMISTRY, ENGINEERING. *Educ:* Univ Calif, Riverside, BS, 68, PhD(chem), 78. *Prof Exp:* Engr, Spectrolab, 75-78; dir res, 78-80, vpres res & develop, 80-87, SR VPRES MFG, RES ENG, ARCO SOLAR, 87- *Concurrent Pos:* Nat Res Coun. *Mem:* Sigma Xi; Electrochemical Soc. *Res:* Solar energy; photovoltaics; silicon; critical region phase transitions; statistical thermodynamics. *Mailing Add:* Arco Solar 4650 Adohr Lane PO Box 6032 Camarillo CA 93010

GAY, DON DOUGLAS, b Oklahoma City, Okla, Jan 17, 44; m 66; c 3. PLANT PHYSIOLOGY. *Educ:* Augustana Col, Ill, BA, 66; Univ Iowa, MS, 71, PhD(plant physiol), 75. *Prof Exp:* Res plant physiologist, US Environ Protection Agency, Nev, 74-78; fel & assoc, Dept Prev Med & Environ Health, Univ Iowa, 78-79; sr scientist, Savannah River Lab, SC, 79-86; dir res & develop, Scientific Systs Inc, State Col Pa, 86; anal serv dept mgr, Environ Protection Systs, Jackson, Miss, 87; br mgr, Environ Protection Systs, Pensacola, Fla, 88-89; BR MGR, LAW ENVIRON, 89- *Concurrent Pos:* Vis asst prof biol, Univ Nev, Las Vegas, 75-78; pres, Symplectics, SC, 87. *Mem:* Sigma Xi. *Res:* Biochemical pathways in plant, animal and soil systems; analytical instrument development; new technologies and products research and development. *Mailing Add:* 2830 Hwy 297A Cantonment FL 32533

GAY, FRANK P, b Denoya, Okla, Jan 7, 25; m 57; c 2. POLYMER SCIENCE. *Educ:* Ind Univ, BS, 48; Univ Calif, PhD(chem), 51. *Prof Exp:* Asst chem, Univ Calif, 48-51; res chemist, Calif Spray Chem Corp, Standard Oil Co of Calif, 51-53; res chemist, 53-57, staff scientist, 57-59, res supvr, 59-65, res assoc film dept, 65-79, res assoc polymer prod dept, 79-86, RES FEL, E I DU PONT DE NEMOURS & CO, INC, 86- *Mem:* Am Chem Soc; Sigma Xi. *Res:* Organometallics; polymer physics and degradation; high temperature polymers; polyesters. *Mailing Add:* 527 Hemlocb Dr Hockessin DE 19707

GAY, HELEN, b Pittsfield, Mass, Aug 30, 18. BIOLOGY. *Educ:* Mt Holyoke Col, BA, 40; Mills Col, MA, 42; Univ Pa, PhD, 55. *Prof Exp:* Asst, Carnegie Inst, Dept Genetics, 42-43; jr prof asst, NIH, 43-45; asst, Dept Genetics, Carnegie Inst, 45-51, res assoc, 54-60, assoc cytogeneticist, 60-62, cytogeneticist, 62; PROF BIOL SCI, UNIV MICH, 62- *Concurrent Pos:* Lectr, Adelphi Col, 59-62; guest investr, Brookhaven Nat Lab; cytogeneticist, Carnegie Inst, 62-71. *Mem:* Fel AAAS; Am Soc Nat; Am Soc Zool; Genetics Soc Am; Soc Develop Biol; Am Soc Cell Biol. *Res:* Cytogenetics of Drosophila; cytochemistry; chromosome structure; electron microscopy. *Mailing Add:* Div Biol Sci Univ Mich Ann Arbor MI 48109

GAY, JACKSON GILBERT, b Selma, Ala, Dec 27, 32; m 55; c 2. THEORETICAL PHYSICS. *Educ:* Auburn Univ, BS, 55; Univ Fla, PhD(physics), 63. *Prof Exp:* Sr engr, Pratt & Whitney Aircraft Div, United Aircraft Co, 55-60; res assoc physics, Univ Fla, 63-64; sr res physicist, 64-85, SR STAFF SCIENTIST, GEN MOTORS RES LABS, 85- *Honors & Awards:* John M Campbell Award; Kettering Award. *Mem:* Am Phys Soc. *Res:* Theory of solid surfaces; Monte Carlo and molecular dynamics simulations of molecular systems. *Mailing Add:* Dept Physics Gen Motors Res Labs 12 Mile & Mounds Rds Warren MI 48090

GAY, LLOYD WESLEY, b Bryan, Tex, June 26, 33; m 63; c 4. WATERSHED MANAGEMENT. *Educ:* Colo State Univ, BS, 55; Australian Forestry Sch, Canberra, dipl, 59; Duke Univ, MF, 62, PhD(forest climat), 66. *Prof Exp:* Forester, Apache Nat Forest, Ariz, 55 & 57; res forester, Cent Sierra Snow Lab, Calif, 60-61; asst prof forest mgt, Ore State Univ, 66-70, from asst prof to assoc prof forest climatol, 70-75; assoc prof, 75-76, PROF WATERSHED MGT, SCH RENEWABLE NATURAL RESOURCES, UNIV ARIZ, 76-

Concurrent Pos: Nat Acad Sci-Polish Acad Sci exchangee, Inst Geog, Univ Warsaw, 70, vis scientist, 73; consult, Nat Cellulose & Paper Orgn, Cent Forest Exp Sta, Rome, 70 & 73; res off, Inst Hydrol, Eng, 73; vis scientist, Southwest Watershed Res Ctr, USDA/Agr Res Serv, Tucson, 82. *Mem:* Soc Am Foresters; Am Meteorol Soc. *Res:* Influence of forests on heat balance at earth's surface, especially snow and forest hydrology, forest meteorology and evapotranspiration processes. *Mailing Add:* Dept Watershed Mgmt Univ Ariz Tucson AZ 85721

GAY, MICHAEL HOWARD, b Tacoma, Wash, July 15, 43; m 65; c 1. PHARMACOLOGY. *Educ:* Univ Chicago, BA, 66; Wash Univ, PhD(pharmacol), 74. *Prof Exp:* Sr res asst pharmacol, Alcohol & Drug Abuse Res Ctr, McLean Hosp, 75-76; instr, dept pharmacol & med chem, Northeastern Univ, 76-80, asst prof, dept pharmacol, 80-83; SR SCIENTIST, BIOTEK INC, 83- *Concurrent Pos:* Res assoc psychiat, Harvard Med Sch, 75-76; consult, Biotek Inc, 80-83. *Mem:* AAAS. *Res:* Neurochemistry of central nervous system depressants; mechanisms of tolerance and physical dependence; pharmacology and sustained drug release technology. *Mailing Add:* 12 Spindle Hill Rd Apt 2-D Wolcott CT 06716

GAY, RENATE ERIKA, b Halle, EGer, Nov 11, 49; m 73; c 3. IMMUNOHISTOLOGY, COLLAGEN PATHOLOGY. *Educ:* Univ Med Sch Munich, MD, 75. *Prof Exp:* Intern med, Univ Med Sch Munich, 75-76; res specialist biochem, Rutgers Med Sch, 76; res assoc, Inst Dent Res, 76-81, RES ASST PROF DERMAT, UNIV ALA, 81-, ASSOC SCIENTIST, COMPREHENSIVE CANCER CTR & INVESTR, INST DENT RES, 81-, RES ASSOC PROF MED, 84- *Honors & Awards:* Caral-Nachman Prize, 84. *Mem:* Am Rheumatism Asn; NY Acad Sci; Am Soc Cell Biol. *Res:* Immunohistology and connective tissue pathology in rheumatology; dermatology and cancer research. *Mailing Add:* Dept Med Univ Med Ctr THT 433 Birmingham AL 35294

GAY, RICHARD LESLIE, b Redlands, Calif, Nov 17, 50. RADIOACTIVE & HAZARDOUS WASTE MANAGEMENT. *Educ:* Univ Calif, Los Angeles, BS & MS, 73, PhD(eng), 76. *Prof Exp:* Mem tech staff chem eng, Atomics Int, Rockwell Int, 76-86, MGR CHEM & PROCESS ENG, ROCKETDYNE DIV, ROCKWELL, 86- *Mem:* Sigma Xi; Combustion Inst; Am Chem Soc; Am Inst Chem Engrs. *Res:* Combustion technology; disposal of hazardous wastes; flue gas desulfurization; energy and resource recovery; volume reduction of radioactive waste; molten salt chemistry; high temperature materials compatibility; energy storage systems; high temperature superconductors; actinide partitioning and transmutation; seven patents awarded. *Mailing Add:* 10012 Hanna Ave Chatsworth CA 91311

GAY, STEFFEN, b Geyersdorf, EGe, Mar 22, 48; m 73; c 3. CONNECTIVE TISSUE PATHOLOGY, IMMUNOHISTOLOGY. *Educ:* Univ Med Sch Leipzig, EGer, MD, 72. *Prof Exp:* Fel path, Univ Med Sch Leipzig, 70-73; resident internal med, Poliklinik Leipzig, 72-73; res fel connective tissue biochem, Max Planck Inst, 73-76; vis investr, Inst Dent Res, 76-78, vis asst prof path, 77-78, assoc prof med, Dir Clin Immunol & Rheumat, 78-84, ASSOC PROF, DEPT DERMAT & SCIENTIST, INST DENT RES & COMPREHENSIVE CANCER CTR, 78-, SCIENTIST & PROF MATH, MULTIPURPOSE ARTHRITICS CTR, 84- *Concurrent Pos:* Ed-in-chief, J Collagen & Related Res; vis scientist, Dept Biochem, Rutgers Univ, 76; dir, Ctr for Rheumatology, WHO, 84-; assoc ed, Arthritis & Rheumat, 85-; dir, WHO Ctr Rheumatic Dis, 84. *Honors & Awards:* Alexander-Schmidt-Preis, Ger Soc Thrombosis Res, 75; Int Carol-Nachman Prize Rheumat, 78. *Mem:* Am Asn Pathologists; Am Rheumatism Asn; Ger Path Asn; Ger Rheumatology Asn; NY Acad Sci; Am Soc Cell Biol. *Res:* Biochemistry, immunology, histology and pathology of collagen types, glycoproteins and matrix degrading enzymes in normal and pathological connective tissues; cellular basis and molecular biology of joint destruction in rheumatic diseases. *Mailing Add:* Univ Ala UAB Sta THT 433 Birmingham AL 35294

GAY, STEVEN W, developmental biology, for more information see previous edition

GAY, THOMAS JOHN, b New York, NY, May 19, 40; m 64; c 2. ORAL BIOLOGY, PHYSIOLOGY. *Educ:* City Col, City Univ New York, BA, 62, PhD(speech sci), 67; Adelphi Univ, MS, 64. *Prof Exp:* Asst prof speech, Hunter Col, 67-69; res scientist speech, Haskins Labs, 69-70; from asst prof to assoc prof, 70-78, PROF ORAL BIOL, HEALTH CTR, UNIV CONN, FARMINGTON, 78- *Mem:* Acoust Soc Am; AAAS; Am Asn Phonetic Sci. *Res:* Articulatory and acoustic phonetics; physiological models of speech production; biological signal processing. *Mailing Add:* Dept BioStructure & Functions Univ Conn Health Ctr Farmington CT 06032

GAY, WILLIAM INGALLS, b Sussex, NJ, Jan 25, 26; m 48. MEDICAL ADMINISTRATION. *Educ:* Cornell Univ, DVM, 50. *Prof Exp:* Pvt pract, 50-52; chief dept animal husb, Walter Reed Army Med Serv, 52-54; chief animal hosp sect, NIH, 54-63, asst chief, Lab Aids Br, 61-63, lab animal specialist, Div Res Facil & Resources, 63-66; prog dir comp med, Nat Inst Gen Med, 66-67, chief res grants br, 67-70, assoc dir extramural progs, 71-77, dir extramural activ prog, 77-80, dir animal resources prog, Div Res Resources, Nat Inst Allergy & Infectious Dis, 80-88; CONSULT, R W SCI, 88- *Concurrent Pos:* Secy-treas, Am Col Lab Animal Med, 64-66. *Mem:* Am Vet Med Asn; Am Asn Lab Animal Sci (pres, 68). *Res:* Laboratory animal medicine; methods of animal experimentation; experimental surgery; medical research administration. *Mailing Add:* R W Sci 5515 Security Lane Suite 500 Rockville MD 20852

GAYED, SOBHY KAMEL, b Al-Minia, Egypt, Nov 13, 24; Can citizen; m 59; c 4. PLANT PATHOLOGY. *Educ:* Cairo Univ, BSc, 45, MSc, 51, PhD(phytopath), 55. *Prof Exp:* Demonstr bot, Fac Sci, Cairo Univ, 45-51, asst lectr, 51-55; lectr, Univ Cairo & Univ Khartoum, 59-63; res worker phytopath, Agrobiol Lab, Phillips Duphar, Holland, 64-65; res scientist phytopath, Agr Can Res Sta, 65-89; RETIRED. *Concurrent Pos:* Guest researcher, Phytopath Lab, Willie Commelin Scholten, Baarn, Neth, 57-58. *Mem:* Can Phytopath Soc. *Mailing Add:* 261 Markwell St Delhi ON N4B 2R1 Can

GAYER, KARL HERMAN, b Cleveland, Ohio, Aug 6, 13; m 50; c 3. CHEMISTRY. *Educ:* Case Western Res Univ, BS, 43, MS, 44; Ohio State Univ, PhD(chem), 48. *Prof Exp:* Instr chem, Ohio State Univ, 47-48; from asst prof to assoc prof, Wayne State Univ 48-59, head inorg div, 56-65, prof chem, 59-84; RETIRED. *Concurrent Pos:* Exec secy, dept chem, Wayne State Univ, 56-65. *Mem:* AAAS; Am Inst Chem; Calorimetry Conf; Am Chem Soc; Sigma Xi. *Res:* Thermochemistry; solution calorimetry; thermodynamics of solutions. *Mailing Add:* 7332 Mohansic Birmingham MI 48010

GAYLES, JOSEPH NATHAN, JR, b Birmingham, Ala, Aug 7, 37; c 2. PHYSICAL CHEMISTRY. *Educ:* Dillard Univ, AB, 58; Brown Univ, PhD(chem), 63. *Hon Degrees:* LLD, Dillard Univ, 83. *Prof Exp:* Res assoc phys chem, Ore State Univ, 63-64; Woodrow Wilson teaching assoc, Morehouse Col, 63-66, asst prof, 64-69, dir med prog, 72-75, prof phys chem, 69-80; pres, Talladega Col, 76-83; VPRES, MOREHOUSE SCH MED, 83-, PROF MED, 85- *Concurrent Pos:* Res assoc chem, Univ Iowa, 64 & Brown Univ, 66; res staff scientist, IBM Corp, 66-69; Dreyfus scholar, 70-; consult, NIH, 70-, EPA, 72- & NSF, 74-; premed adv, Morehouse Col, 70-77,; Mass Inst Technol bd vistors, 81-88; Woodrow Wilson Fel. *Mem:* Am Phys Soc; Am Chem Soc; Am Asn Polit & Social Scientists; Sigma Xi; AAAS. *Res:* Molecular structure and spectroscopy; laser phenomena; liquid crystals. *Mailing Add:* 1515 Austin Rd SW Atlanta GA 30331-2267

GAYLIN, WILLARD, b Cleveland, Ohio, 1925; m; c 2. PSYCHIATRY. *Educ:* Harvard Univ, AB, 47; Western Reserve Univ, MD, 51. *Prof Exp:* Intern, Cleveland City Hosp, Ohio, 51-52; resident psychiat, Bronx Vet Admin Hosp, NY, 52-54; prof psychiat & law, Sch Law, Columbia Col & Union Theol Sem, 70-80, CLIN PROF PSYCHIAT, COLUMBIA COL PHYSICIANS & SURGEONS, 80-; PRES, HASTINGS CTR, 70- *Concurrent Pos:* Pvt pract, 54-; from mem fac to clin prof psychiat, Psychoanal Sch, Columbia Univ, 56-; consult, AMA Judicial Coun & Inst of Life, Univ of Paris; mem bd dirs, Field Found, Helsinki Watch, comt Pub Justice & Nat Adv Bd Amnesty Int; lectr, Columbia Col Physicians & Surgeons, 1st commencement, 81; Chubb fel, Yale & vis prof, Harvard Med Sch; vis lectr, Sorbonne, Dartmouth Co, Hampshire Col, Princeton, Calif Inst Technol & Univ Hawaii; mem bd dirs, Planned Parenthood Fedn Am, Inc; mem, Human Rights Comt, Inst Med. *Honors & Awards:* George E Daniels Medal of Merit for Contrib Psychoanal Med; Elizabeth Cutter Morrow Lectr, Smith Col; Bloomfield Lectr, Med Sch, Case Western Reserve Univ. *Mem:* Inst Med-Nat Acad Sci; fel Am Psychiat Asn; Am Psychoanal Asn. *Res:* Judicial bias; ethics of behavior control; ethics, law and life sciences; author or editor of 15 books and about 70 articles. *Mailing Add:* Hastings Ctr 255 Elm Rd Briarcliff Manor NY 10510-9974

GAYLOR, DAVID WILLIAM, b Waterloo, Iowa, Apr 8, 30; m 54; c 4. BIOSTATISTICS. *Educ:* Iowa State Univ, BS, 51, MS, 53; NC State Univ, PhD(statist), 60. *Prof Exp:* Statistician, Hanford Atomic Prod Oper, Gen Elec Co, 53-55, Gen Dynamics/Convair, 55-57, Vallecitos Atomic Lab, Gen Elec Co, 60-62, Res Triangle Inst, 62-68 & Nat Inst Environ Health Sci, 68-72; CHIEF BIOMET, NAT CTR FOR TOXICOL RES, 72- *Concurrent Pos:* Adj assoc prof, NC State Univ, 67-72; adj prof, Med Sch, Univ Ark, 72- *Honors & Awards:* Shewell Award, 68 & Wilcoxin Prize, Chem Div, Am Soc Qual Control, 70. *Mem:* Fel Am Statist Asn; Biomet Soc. *Res:* Statistical design and analysis of experiments; quantitative risk assessment of chemicals. *Mailing Add:* Div Biometry Nat Ctr Toxicol Res Jefferson AR 72079

GAYLOR, JAMES LEROY, b Waterloo, Iowa, Oct 1, 34; m 56; c 4. MOLECULAR BIOLOGY, NUTRITION. *Educ:* Iowa State Univ, BS, 56; Univ Wis, MS, 58, PhD(biochem), 60. *Prof Exp:* From asst prof to assoc prof biochem, Grad Sch Nutrit, Cornell Univ, 60-69, prof biochem & molecular biol, 69-77, chmn sect, Div Biol Sci, 70-76; prof biochem & chmn dept, Univ Mo-Columbia, 77-80; dir health sci res, E I Du Pont de Nemours & Co, Inc, 81-85; dir biol res, 86-87, CORP DIR MOLECULAR BIOL, JOHNSON & JOHNSON, 87- *Concurrent Pos:* Vis lectr, Univ Ill, 64 & 65; vis mem staff, Dept Biochem, Sch Med, Univ Ore, 66-67; vis prof biochem, Sch Med, Osaka Univ, Japan, 73-74; Guggenheim fel; fel, Arteriosclerosis Coun, Am Heart Asn. *Mem:* Am Chem Soc; Am Soc Biochem & Molecular Biol; Am Inst Nutrit; Am Heart Asn; Am Asn Pharmaceut Scientists; AAAS. *Res:* Biosynthesis cholesterol and other sterols; microsomal electron transport of mixed function oxidases of biosynthetic processes; isolation and purification of microsomal enzymes; reconstitution of microsomal multienzymic systems. *Mailing Add:* Johnson & Johnson 410 George St Rm 1139 New Brunswick NJ 08901-2021

GAYLOR, MICHAEL JAMES, b Birmingham, Ala, May 16, 47; m 69; c 2. ENTOMOLOGY. *Educ:* Auburn Univ, BS, 69, MS, 71; Tex A&M Univ, PhD(entom), 75. *Prof Exp:* Asst prof urban entom, Tex Agr Exp Sta, Tex A&M Univ, 75-77; ASST PROF COTTON ENTOM, AUBURN UNIV, 78- *Mem:* Entom Soc Am; Sigma Xi. *Res:* Management of insect pests of cotton. *Mailing Add:* Dept Entom Auburn Univ Auburn AL 36849-5413

GAYLORD, EBER WILLIAM, b Pittsburgh, Pa, Nov 6, 22; m 62; c 1. MECHANICAL ENGINEERING. *Educ:* Carnegie Inst Technol, PhD(mech eng), 53. *Prof Exp:* Assoc prof mech eng, Carnegie Inst Technol, 53-59; RES ENGR, GULF RES & DEVELOP CO, 59- *Mem:* Am Soc Mech Engrs; Soc Automotive Engrs. *Res:* Experimental fluid mechanics; friction and wear dynamics; interface temperature between rubbing metals; momentum and mass transfer in jets; research and development in petroleum production engineering. *Mailing Add:* 1315 S Lake Rd Middlesex NY 14507

GAYLORD, NORMAN GRANT, b New York, NY, Feb 16, 23; m 45; c 4. POLYMER CHEMISTRY. *Educ:* City Col New York, BS, 43; Polytech Inst Brooklyn, MS, 49, PhD(chem), 50. *Prof Exp:* Chemist, Elko Chem Works, 43-44, Pa Salt Mfg Co, 45 & Merck & Co, Inc, 46-48; res assoc, Polytech Inst Brooklyn, 48-50; res chemist, Film Dept, E I du Pont de Nemours & Co, 50-55; group leader, Resin Dept, Interchem Corp, 55-56, asst dir, Org Chem Dept, 56-59; vpres, Polymer Div, Western Petrochem Corp, NY, 59-61; pres, Gaylord Res Inst Inc, 61-87; PRES, GAYLORD ASSOC, 87- *Concurrent*

Pos: Polymer consult, 61-; fel, Res Inst Scientists, Emer, Drew Univ, 87- *Honors & Awards:* Founders Award, Am Acad Optom, 85. *Mem:* Am Chem Soc; fel Soc Plastics Eng; Tech Asn Pulp & Paper Industs. *Res:* Polymer synthesis; polymerization kinetics; allyl polymerization; stereoregular polymers; organic chemistry of high polymers; block and graft copolymerization; charge transfer polymerization; photopolymerization; polymer modification; controlled release of pharmaceuticals; polymers for contact lenses. *Mailing Add:* 28 Newcomb Dr New Providence NJ 07974

GAYLORD, RICHARD J, b Plainfield, NJ, Dec 20, 47; m 70. THEORETICAL POLYMER PHYSICS, STATISTICAL MECHANICS. *Educ:* Polytech Inst Brooklyn, BS, 69; State Univ NY Syracuse, PhD(polymer sci), 73. *Prof Exp:* Res assoc, Polymer Res Inst, Univ Mass, 73-74; asst prof, 74-79, ASSOC PROF POLYMERS, DEPT METALL & MINING ENG, UNIV ILL, URBANA-CHAMPAIGN, 80- *Mem:* Am Phys Soc. *Res:* Statistical mechanics of polymeric systems; bulk polymer deformation; confined chain problems. *Mailing Add:* Dept Metal & Mining Engr 202a Mmb Univ Ill Urbana IL 61801

GAYLORD, THOMAS KEITH, b Casper, Wyo, Sept 22, 43; m 66; c 1. OPTICAL DATA PROCESSING. *Educ:* Univ Mo, Rolla, BS, 65, MS, 67; Rice Univ, PhD(elec eng), 70. *Prof Exp:* Res assoc, Rice Univ, 70-72; JULIUS BROWN CHAIR REGENTS PROF ELEC ENG, GA INST TECHNOL, 72- *Concurrent Pos:* Ed, Trans Educ, Inst Elec & Electronics Engrs, 78-82. *Honors & Awards:* Curtis W McGraw Award, Am Soc Elec Engrs, 79. *Mem:* Fel Inst Elec & Electronics Engrs; Sigma Xi; AAAS; fel Optical Soc Am. *Res:* Optical holographic data storage and processing; recording in electro-optic crystals; electromagnetic diffraction; volume gratings; semiconductor devices; instrumentation; optical computing. *Mailing Add:* Sch Elec Eng Ga Inst Technol Atlanta GA 30332

GAYNOR, JOHN JAMES, b Philadelphia, Pa, Dec 19, 53; m 76; c 4. GENE EXPRESSION, PLANT-PATHOGEN INTERACTIONS. *Educ:* St Joseph's Col, BS, 75; Rutgers Univ, MS, 78, PhD(plant physiol), 81. *Prof Exp:* Fel, Dept Biol, Yale Univ, 81-82; Lab Plant Molecular Biol, Rockefeller Univ, 82-84; ASST PROF BIOL, DEPT BIOL SCIS, RUTGERS UNIV, 84- *Concurrent Pos:* Res assoc, Life Sci Div, NASA, 81-82; Henry Rutgers fel, Rutgers Univ, 85-87. *Mem:* Am Soc Plant Physiologists; AAAS; Int Soc Plant Molecular Biol; Bot Soc Am; Sigma Xi. *Res:* Regulation of gene expressions in higher plants; the influence of small effector molecules (ie the phytohormone ethylene) on transcription and in the signal transduction pathway for this horm. *Mailing Add:* Dept Biol Sci Rutgers Univ 195 Univ Ave Newark NJ 07102

GAYNOR, JOSEPH, b New York, NY, Nov 15, 25; c 4. CHEMICAL ENGINEERING. *Educ:* Polytech Inst Brooklyn, BChE, 50; Case Inst Technol, MS, 52, PhD(chem eng), 55. *Prof Exp:* Asst unit opers & plastics lab, Case Inst Technol, 50-55; chem engr, Gen Elec Co, 55-59, chem process engr, 59-64, mgr, Info Mat Systs, 64-66; asst vpres res, Bus Equip Group, Bell & Howell Co, Ill, 66-68, vpres, 68, dir graphic media res, Res Labs, Calif, 68-71; mgr com develop group & mem pres off, Horizons Res Inc, 72-73; PRES, INNOVATIVE TECHNOL ASSOCS, 73- *Concurrent Pos:* Plenary lectr, Int Cong Photog Sci; gen chmn, Int Conf Electrophotog, Int Conf Bus Graphics, Int Cong Advan Non-Impact Printing Technol & Gordon Res Conf Phys & Chem Polymeric Films & Coatings. *Honors & Awards:* Indust Res IR-100 Awards, 63, 65. *Mem:* Fel AAAS; fel Am Inst Chem; sr mem Am Soc Photog Scientists & Engrs; Am Chem Soc; NY Acad Sci; Soc Photo-Optic Instrumentation Engrs; Soc Int Develop. *Res:* Solubility, diffusion, rheology, adhesion and lubrication of high polymers; engineering properties of high polymers; heat transfer; fluid flow; fluidization; physics and chemistry of the solid state; photochemistry; photoelectricity; unconventional image recording processes; electrophotography; photofabrication; polymeric films and coatings; optical and magnetic data storage discs; non-impact printing technologies; chemical processes. *Mailing Add:* Innovative Technol Assocs 3639 E Harbor Blvd #203E Ventura CA 93001-4277

GAZDA, I(RVING) W(ILLIAM), b Niagara Falls, NY, Nov 26, 41; m 66; c 1. MECHANICAL ENGINEERING, HEAT TRANSFER. *Educ:* Rensselaer Polytech Inst, BS, 63, MS, 65, PhD(heat transfer), 69. *Prof Exp:* Group leader electrode develop, 60-70, sect head electrode develop, 70-74, mgr tech control, 75-79, TECH DIR, GREAT LAKES CARBON CORP, 79- *Mem:* Am Soc Mech Engrs; Iron & Steel Soc, Am Inst Metall Engrs. *Res:* Heat transfer in fluids; physical and thermal property determinations for ceramics; thermal shock studies on ceramics. *Mailing Add:* Great Lakes Carbon Corp PO Box 667 Niagara Falls NY 14302

GAZDAR, ADI F, b Bombay, India, May 15, 37; US citizen; m 69. CANCER RESEARCH, CELL BIOLOGY. *Educ:* Univ London, BS & MB, 61. *Prof Exp:* SECT HEAD, NAT CANCER INST, 69- *Mem:* Am Asn Cancer Res; Inst Asn Study Lung Cancer; Am Soc Clin Oncol; Soc Exp Med & Biol. *Res:* Biology of human cancer, especially lung cancer and lymphomas; authored or co-authored approximately 150 publications on cancer research. *Mailing Add:* Nat Cancer Inst Div Cancer Treatment Clin Oncol Prog Bldg NNMC Rm 8-6108 Bethesda MD 20892

GAZIN, CHARLES LEWIS, b Colorado Springs, Colo, June 18, 04; m 27, 43; c 3. GEOLOGY. *Educ:* Calif Inst Technol, BS, 27, MS, 28, PhD(vert paleont), 30. *Prof Exp:* Asst vert paleont, Calif Inst Technol, 27-29; jr geologist, US Geol Surv, 30-32; from asst cur to cur in charge div vert paleont, US Nat Mus, 32-67; sr paleobiologist, 67-70, EMER PALEOBIOLOGIST, SMITHSONIAN INST, 70- *Concurrent Pos:* Mem, Nat Res Council, 48-51 & 57-60; dir, Am Geol Inst, 56-58; partic paleont expeds, Ore, Nev & Calif, Calif Inst Technol & Carnegie Inst; leader expeds var western states & Cent Am, Smithsonian Inst. *Honors & Awards:* Prize, Geol Soc Am, 30. *Mem:* Fel Geol Soc Am; fel Paleont Soc; Soc Vert Paleont (pres, 49); Am Soc Mammal; Soc Study Evolution. *Res:* Areal geology and stratigraphy of continental Cenozoic deposits; vertebrate paleontology; Tertiary and Quaternary mammals, particularly Paleocene and Eocene mammalian faunas. *Mailing Add:* 95-6600 Lucas Rd Richmond BC V7C 4T1

GAZIS, DENOS CONSTANTINOS, b Salonica, Greece, Sept 15, 30; US citizen; m 74; c 9. APPLIED MATHEMATICS. *Educ:* Tech Univ, Athens, BS, 52; Stanford Univ, MS, 54; Columbia Univ, PhD(eng mech), 57. *Prof Exp:* Designer engr civil eng, Tech Ministry, Athens, 52-53 & Tippetts & Assocs, NY, 55-57; sr res scientist mech, Gen Motors Res Labs, 57-61; res staff mem, 61-79, ASST DIR, IBM CTR, YORKTOWN HIGHTS, 82- *Concurrent Pos:* Instr, Tech Univ, Athens, 52-53; vis prof, Yale Univ, 69-70; dir gen sci, IBM Res Ctr, 71-74, tech adv to IBM chief scientist, IBM Corp Hq, Armonk, NY, 75-77, mem res rev bd, 77-78, 77-78, consult to dir res, IBM Res Ctr, 78-79 asst dir comput sci 79-82. *Honors & Awards:* Lanchester Prize, Opers Res Soc Am, 59. *Mem:* Opers Res Soc Am; Building Res Bd, Nat Res Coun; Int Platform Asn. *Res:* Operations research; computer science; applications of computers to social and environmental problems. *Mailing Add:* Lake Rd Katonah NY 10536

GAZLEY, CARL, JR, aerodynamics; deceased, see previous edition for last biography

GAZZANIGA, MICHAEL SAUNDERS, b Los Angeles, Calif, Dec 12, 39; m; c 6. PSYCHOBIOLOGY. *Educ:* Dartmouth Col, AB, 61; Calif Inst Technol, PhD(biol), 64. *Prof Exp:* USPHS res fel psychobiol, Calif Inst Technol, 64-66; from asst prof to assoc prof psychol & chmn dept, Univ Calif, Santa Barbara, 66-69; from assoc prof to prof, Grad Sch, NY Univ, 69-73; prof psychol, State Univ NY Stony Brook, 73-78, prof soc sci in med, 75-78; PROF PSYCHIAT & DIR DIV COGNITIVE NEUROSCI, CORNELL MED COL, 78- *Concurrent Pos:* NIMH res grant, 67-69; State Univ NY Univ-Wide Exchange scholar, 74- *Mem:* Am Physiol Soc; fel Am Psychol Asn; Am Acad Neurol; Int Neuropsychol Asn; NY Acad Sci. *Res:* Studies on human split-brain patient and other neurologic patients addressing the role of language in conscious experience and the cortical mechanisms involved in cognitive processing, visual-motor processing and memory. *Mailing Add:* Dartmouth Medical Sch Pike House Hanover NH 03756

GE, WEIKUN, b Beijing, China, Mar 25, 42; m 67; c 2. SEMICONDUCTING MICROSTRUCTURES, DEFECTS IN SEMICONDUCTORS. *Educ:* Peking Univ, BS, 65; Univ Manchester, UK, PhD(solid state electronics), 83. *Prof Exp:* Engr semiconductor mat devices, Beijing Inst Non-Ferrous Metals, 65-78; postdoctoral fel solid state electronics, Inst Sci Technol, Univ Manchester, 83; assoc prof physics, Inst Semiconductors, Chinese Acad Sci, 84-88; res assoc, 88-91, RES ASSOC PROF PHYSICS, DARTMOUTH COL, 91- *Honors & Awards:* Third Degree Prize for Progress in Sci & Technol, Chinese Acad Sci, 88, Second Degree Prize, 89. *Mem:* Am Phys Soc; fel Chinese Luminescence Soc. *Res:* Defects in GaAs and Si- for GaAs, local vibrational mode absorption induced by Ga-O-Ga- for Si, new donors and Pd-related defects; optical spectroscopic studies on semiconducting quantum wells and superlattices, especially on band structure of very short-period GaAs-AIAs superlattices. *Mailing Add:* Dept Physics & Astron Dartmouth Col Hanover NH 03755

GEACH, GEORGE ALWYN, b Eng, 1913. MATERIALS SCIENCE. *Educ:* Univ Sheffield, MSc, 36, PhD(metall), 39. *Prof Exp:* Res sect leader phys metall, Assoc Elec Industs Ltd, Eng, 41-64; mem fac, 64-80, emer prof eng sci, Univ Western Ont, 80-; RETIRED. *Mem:* Am Soc Metals; Brit Inst Metals; fel Brit Inst Metall. *Mailing Add:* 455 Dunedin Dr London ON N6H 3G9 Can

GEACINTOV, NICHOLAS, b Albi, France, Nov 9, 35; US citizen. CHEMICAL PHYSICS, BIOPHYSICS. *Educ:* State Univ NY Col Forestry, Syracuse, BS, 57, MS, 59, PhD(phys Chem), 61. *Prof Exp:* Res assoc photochem, Polytech Inst Brooklyn, 61-63; res scientist solid state physics, 63-69, from asst prof to assoc prof, 69-75, PROF CHEM, NY UNIV, 75- *Res:* Photophysics of aromatic molecules and crystals; spectroscopic properties of photosynthetic membranes; structure of carcinogen-nucleic acid complexes. *Mailing Add:* Dept Chem NY Univ Wash Sq New York NY 10003

GEADELMANN, JON LEE, b Anamosa, Iowa, June 4, 44; m 64; c 2. PLANT BREEDING. *Educ:* Iowa State Univ, BS, 66, PhD(plant breeding), 70. *Prof Exp:* Asst prof statist, Iowa State Univ, 70-72; asst prof plant breeding, 72-77, assoc prof, 77-80, PROF AGRON & PLANT GENETICS, UNIV MINN, 80- *Mem:* Am Soc Agron; Crop Sci Soc Am. *Res:* Population improvement; inbred line development; maize genetics; quantitative genetics. *Mailing Add:* 2030 W Skillman Roseville MN 55113

GEALER, ROY L(EE), b Detroit, Mich, Oct 23, 32; m 57; c 2. CHEMICAL ENGINEERING, BIO ENGINEERING & MATERIALS. *Educ:* Wayne State Univ, BS, 54; Univ Mich, MS, 55, PhD(chem eng), 58. *Prof Exp:* Asst combustion res, Univ Mich, 54-58; res engr, Ethyl Corp, 58-63; prin res eng assoc, Ford Motor Co, 63-68, prin staff engr, 68-78, prin res engr, 78-87; HARRINGTON ARTHRITIS RES CTR, 87- *Mem:* Am Chem Soc; Sigma Xi. *Res:* Automotive industry product and process research, including friction materials, industrial air pollution control, wastewater treatment, energy & fuels; development and evaluation of prosthetic materials and devices. *Mailing Add:* 514 W Marconi Ave Phoenix AZ 85023

GEALT, MICHAEL ALAN, b Philadelphia, Pa, Nov 27, 48; m 81. MOLECULAR BIOLOGY & MICROBIAL ECOLOGY. *Educ:* Temple Univ, Philadelphia, BA, 70; Rutgers Univ, PhD(microbiol), 74. *Prof Exp:* Res assoc, Col Med & Dent, Rutgers Univ, NJ, 74-76; trainee, Inst Cancer Res, Philadelphia, 76-78; asst prof biol sci, 78-84, assoc prof biosci & biotechnol, 84-90, PROF BIOSCI & BIOTECHNOL, DREXEL UNIV, PHILADELPHIA, 90- *Concurrent Pos:* Vis scientist, US Environ Protection Agency, Gulf Breeze, Fl, 90-91. *Mem:* AAAS; Am Soc Cell Biol; Am Soc Microbiol; Am Mycological Soc; Sigma Xi. *Res:* Microbial ecology; exchange of recombinant genetic material between micro-organisms in wastewater and other aqueous environments. *Mailing Add:* Dept Biosci & Biotechnol Drexel Univ 32nd & Chestnut St Philadelphia PA 19104

GEALY, ELIZABETH LEE, b Ft Worth, Tex, June 22, 23; div; c 4. GEOLOGY. *Educ:* Southern Methodist Univ, BS, 44, BA, 46; Radcliffe Col, MA, 51, PhD(geol), 53. *Prof Exp:* Private res, 53-57; assoc consult, D R McCord & Assocs, 58-61, dir geol serv, 61- 65; partner, Gealy & Gealy, Consult, 65-67; specialist, Scripps Inst Oceanog, Univ Calif, San Diego & exec staff geologist, deep sea drilling proj, 67-70; instr geol, San Diego State Col, 70-72; PRES, IVAN ALLEN INDUSTRIES, 72- *Mem:* AAAS; Am Asn Petrol Geol; Geol Soc Am; Sigma Xi. *Mailing Add:* 329 Vista De La Playa La Jolla CA 92037

GEALY, JOHN ROBERT, b Tokyo, Japan, Dec 4, 30; m 64. GEOLOGY. *Educ:* Southern Methodist Univ, BS, 51; Yale Univ, MS, PhD(geol), 55. *Prof Exp:* Geologist, Humble Oil & Refining Co, 55-74, GEOLOGIST, ESSO PRODS RES CO, 74- *Mem:* AAAS; Geol Soc Am; Am Asn Petrol Geol; Sigma Xi. *Res:* Regional geology. *Mailing Add:* 1416 Stanford St No 2 Houston TX 77019-4329

GEALY, WILLIAM JAMES, b Tokyo, Japan, Sept 7, 25; US citizen; m 45, 70; c 4. RESEARCH ADMINISTRATION. *Educ:* Univ Mich, BA, 46; Harvard Univ, MA, 51, PhD(geol), 53. *Prof Exp:* Geologist, Standard Oil Co of Calif, 53-56, consult geologist, 56-59; assoc dir develop, Ohio State Univ Res Found, 69-80; DIR OFF RES & SPONSORED PROG, NORTHWESTERN UNIV, 80. *Concurrent Pos:* lectr & asst prof, Southern Methodist Univ, 66-68; adj assoc prof geol, Ohio State Univ, 69-80; interim dir, Technol Commercialization Ctr, Northwestern Univ, 84-85; interim admin, Technol Transf Prog, Northwestern Univ, 89-90. *Mem:* Geol Soc Am; Am Asn Petrol Geol; Sigma Xi; Soc Res Adminrs; Asn Univ Technol Mgrs; Nat Coun Univ Res Admins. *Res:* Petroleum geology. *Mailing Add:* 803 Milburn St Evanston IL 60201

GEANANGEL, RUSSELL ALAN, b Cadiz, Ohio, Aug 2, 41; m 61; c 2. INORGANIC CHEMISTRY. *Educ:* Ohio State Univ, BS, 63, PhD(inorg chem), 68. *Prof Exp:* Asst prof, 68-71, assoc prof, 71-77, PROF CHEM, UNIV HOUSTON, 77- *Concurrent Pos:* Consult, 80- *Mem:* Am Chem Soc. *Res:* Synthesis and structural characterization of inorganic and organometallic compounds of structural, bonding and materials science interest, including compounds main group elements; multinuclei nuclear magnetic resonance and Mossbauer spectroscopy. *Mailing Add:* Dept Chem Univ Houston Houston TX 77204-5641

GEANKOPLIS, CHRISTIE J(OHN), b Minneapolis, Minn, June 18, 21. CHEMICAL ENGINEERING. *Educ:* Univ Minn, BChE, 43; Univ Pa, MS, 46, PhD(chem eng), 49. *Prof Exp:* Develop design & process engr, Atlantic Refining Co, 43-47; instr chem eng, Univ Pa, 47-48; from asst prof to prof chem eng, Ohio State Univ, 60-82; PROF CHEM ENG, UNIV MINN, MINNEAPOLIS, 82- *Concurrent Pos:* Consult chem engr, Battelle Mem Inst, 51-70 & Gen Mills Chem, Inc, 66-80, Henkel, Inc, 80-86, H B Fuller, Inc, 88-, Omega Source, Inc, 90- *Mem:* Am Chem Soc; Am Inst Chem Engrs. *Res:* Diffusion and mass transfer; transport processes; reaction kinetics; biochemical engineering. *Mailing Add:* 151 Amundson Hall Univ Minn 421 Washington Ave SE Minneapolis MN 55455

GEAR, ADRIAN R L, b Pretoria, SAfrica, Aug 31, 39; m 64; c 2. BIOCHEMISTRY. *Educ:* Oxford Univ, BA, 61, MA & DPhil(biochem), 65. *Prof Exp:* Fel biochem, Johns Hopkins Univ, 65-67; ASSOC PROF BIOCHEM, SCH MED, UNIV VA, 67- *Mem:* Am Soc Biol Chem; NY Acad Sci. *Res:* Mitochondria; blood platelets; survival, function and energy metabolism; mitochondrial biogenesis; ion transport; energy coupling; oxidative phosphorylation. *Mailing Add:* Dept Biochem Jordan Bldg Univ Va Sch Med Box 440 65A Charlottesville VA 22908

GEAR, CHARLES WILLIAM, b London, Eng, Feb 1, 35; div; c 2. COMPUTER SCIENCE, NUMERICAL SOFTWARE. *Educ:* Cambridge Univ, BA, 56; Univ Ill, MS, 57, PhD(math), 60. *Hon Degrees:* Doctorate, Royal Inst Technol, Stockholm, Sweden, 87. *Prof Exp:* Sr engr comput, IBM Corp, 60-62; prof computer sci, Univ Ill, Urbana, 62-90, dept head, 85-90; VPRES COMPUTER SCI, NEC RES INST, PRINCETON, 90- *Concurrent Pos:* Consult, Argonne Nat Lab, 66-71 & Inst Comput Appln Sci & Eng, 73-80. *Honors & Awards:* Forsythe Mem Award, Asn Comput Mach, 79. *Mem:* Asn Comput Mach; fel Inst Elec & Electronics Engrs; Soc Indust & Appl Math; fel AAAS. *Res:* Numerical analysis; computer graphics; computer software. *Mailing Add:* NEC Res Inst Four Independence Way Princeton NJ 08540

GEAR, JAMES RICHARD, b Kindersley, Sask, Apr 26, 35; m 61; c 1. ORGANIC CHEMISTRY, BIOCHEMISTRY. *Educ:* Univ Sask, BA, 56, MA, 58; McMaster Univ, PhD(biosynthesis), 62. *Prof Exp:* Fel chem, Univ Minn, 62-63; from asst prof to assoc prof, 63-73, PROF CHEM & BIOCHEM, UNIV REGINA, 73- *Concurrent Pos:* Coordr prof progs, Univ Regina, 71- *Mem:* Fel Chem Inst Can. *Res:* Organic synthesis; biosynthesis and chemistry of natural products. *Mailing Add:* Dept Chem Univ Regina Regina SK S4S 0A2 Can

GEARHART, PATRICIA JOHANNA, ANTIBODY DIVERSITY, B-CELL DEVELOPMENT. *Educ:* Univ Penn, PhD(immunol), 74. *Prof Exp:* ASST PROF BIOCHEM, Sch Hyg & Pub Health, Johns Hopkins Univ, 82- *Mailing Add:* Dept Biochem Sch Hyg & Pub Health Johns Hopkins Univ 615 N Wolfe St Baltimore MD 21205

GEARHART, ROGER A, b Chicago, Ill, Nov 27, 35; c 2. ENGINEERING PHYSICS, SCIENTIFIC MANAGEMENT. *Educ:* Univ Calif, Berkeley, BA, 64. *Prof Exp:* Physicist, Lawrence Berkeley Lab, 60-68; PHYSICIST, STANFORD LINEAR ACCELERATOR CTR, 68- *Concurrent Pos:* Vpres, Danmark Dania, 88-91. *Mem:* Inst Elec & Electronics Engrs. *Res:* High energy physics and nuclear physics using fixed targets at accelerators. *Mailing Add:* 614 Park Rd Redwood City CA 94062

GEARIEN, JAMES EDWARD, b Peoria, Ill, Aug 27, 19; m 48; c 2. MEDICINAL CHEMISTRY, ORGANIC CHEMISTRY. *Educ:* Univ Ill, BS, 41; Univ Mich, MS, 42, PhD(pharmaceut chem), 50. *Prof Exp:* Instr chem, 48-50, from asst prof to assoc prof, 50-60, PROF CHEM & CHMN DEPT, COL PHARM, UNIV ILL, CHICAGO, 60- *Mem:* Am Chem Soc. *Res:* Synthesis of organic medicinals; studies of organic structures and medicinal activity relationships; analgesics and angiotensin analogs. *Mailing Add:* 1301 Pennsylvania Des Plaines IL 60018-1118

GEARING, JUANITA NEWMAN, b Dallas, Tex, Dec 5, 45; m 73. ORGANIC GEOCHEMISTRY, MARINE ENVIRONMENTAL CHEMISTRY. *Educ:* Rice Univ, BA, 68; Univ Tex, Austin, PhD(chem), 73. *Prof Exp:* Fel org geochem, Marine Sci Lab, Univ Tex, 73-74; res assoc environ chem, Gulf Coast Res Lab, Ocean Spring, Miss, 74-76; res assoc oceanog & chem, 76-80, marine scientist oceanog, Univ RI, 80-85; RES SCIENTIST, CAN MINISTRY FISHERIES & OCEANS, 85- *Mem:* Geochem Soc Am; Europ Asn Org Geochemists; Am Soc Limnol & Oceanog. *Res:* Stable carbon isotope ratios; environmental measurements of hydrocarbons; rates of transport and fates of organic chemicals in the marine environment. *Mailing Add:* 40 Cole St Jamestown RI 02835

GEARY, JOHN CHARLES, b Chicago, Ill, Jan 2, 45; m 72. ASTRONOMY, ELECTRONICS. *Educ:* Mich State Univ, BA, 67; Univ Ariz, MS, 69, PhD(astron), 75. *Prof Exp:* Staff astronr instrumentation, Max Planck Inst Astron, Heidelberg, 75-78; STAFF ASTRONR INSTRUMENTATION, SMITHSONIAN ASTROPHYS OBSERV, 78- *Mem:* Am Astron Soc; Soc Photo-Optical Instrumentation Engrs. *Res:* Design and construction of advanced electronic detector systems for astronomical research. *Mailing Add:* Smithsonian Astrophys Observ 60 Garden St Cambridge MA 02138

GEARY, LEO CHARLES, b Pittsburgh, Pa, Nov 19, 42; m 65; c 2. ELECTRICAL & SYSTEMS ENGINEERING. *Educ:* Univ Pittsburgh, BS, 64, MS, 65 & PhD(elec eng), 68. *Prof Exp:* Group leader, Gulf Res Develop 68-71, systs engr & systs analyst, 68-72; cognitant engr, 72-77, prin engr, 77-80, MGR, WESTINGHOUSE 81- *Mem:* Inst Elec & Electronics Engrs. *Res:* Research and development of optimal control systems and pattern recognition techniques; design, development and implementation of computer and minor process computer system for data acquisition and process control. *Mailing Add:* Westinghouse Elec 37 Varden Dr Aiken SC 29801

GEARY, NORCROSS D, b Cambridge, Mass, Sept 6, 47. NEUROBIOLOGY, NEUROENDOCRINOLOGY. *Educ:* Harvard Col, BA, 69; Brown Univ, PhD(psychol), 78. *Prof Exp:* Res fel physiol chem & nutrit physiol, Inst Physiol, Ludwig-Maximilians Univ, Munich, Ger, 76-80; RES ASSOC, E W BOURNE LAB, NY HOSP, CORNELL MED CTR, WHITE PLAINS, 80- *Mem:* Soc Neurosci. *Res:* Neural and physiological control of appetitive behavior, in particular the neuroendocrine bases of food's potencies to reward and to satiate and the contribution of these mechansims to bodyweight regulation. *Mailing Add:* E W Bourne Lab NY Hosp Cornell Med Ctr 21 Bloomingdale Rd White Plains NY 10605

GEBALLE, GORDON THEODORE, b Berkeley, Calif, Sept 12, 47; m 71; c 3. URBAN ECOLOGY. *Educ:* Univ Calif, Berkeley, BA, 70; Yale Univ, PhD(biol), 81. *Prof Exp:* Assoc res scientist, 81-83, LECTR, SCH FORESTRY & ENVIRON STUDIES, YALE UNIV, 83-, ASST DEAN, 87- *Mem:* Am Soc Plant Physiologists; Am Phytopath Soc; AAAS; Sigma Xi; Int Fedn Syst Res. *Res:* Responses of plants to stress; physiological responses of oat leaves to wounding; effect of acid precipitation on host-pathogen interactions; soil ecosystems. *Mailing Add:* Sch Forestry & Environ Studies Yale Univ New Haven CT 06511

GEBALLE, RONALD, b Redding, Calif, Feb 7, 18; m 40; c 8. ATOMIC PHYSICS. *Educ:* Univ Calif, BS, 38, MA, 40, PhD(physics), 43. *Prof Exp:* Physicist, Radiation Lab, Univ Calif, 42-43; physicist, Appl Physics Lab, 43-46, from asst prof to assoc prof, 46-59, chmn dept, 57-73, actg dean, 75-76, assoc dean, Col Arts & Sci, 73-76, vprovost res & dean grad sch, 76-81, PROF PHYSICS, UNIV WASH, 59-, EMER PROF PHYSICS & EMER DEAN GRAD SCH, 85- *Concurrent Pos:* Consult, Army Res Off, NSF, 57-62; mem citizens comt educ, Wash State Legis, 60; guest scientist, FOM Inst Atomic & Molecular Physics, Neth, 64-65; chmn bd dirs, Pac Northwest Asn Col Physics, 65-70; mem comn, Col Physics, 66-71; secy, Int Conf Physics Electronic & Atomic Collisions, 67-77; mem adv comt grants, Res Corp, 67-73; mem adv panel physics, NSF, 68-73, chmn, 72-73; mem adv panel, Lab Astrophys Div, Inst Basic Stand, Nat Bur Stand, 70-75, chmn, 73-75; mem physics surv comt & chmn panel on educ, Nat Acad Sci-Nat Res Coun, 70-73; mem adv bd, Off Phys Sci, Assembly Math & Phys Sci, 75-; mem-at-large, US Nat Comt, Int Union Pure & Appl Physics, 74-; mem exec comt eval panels, Nat Bur Stand, Nat Acad Sci-Nat Res Coun, 75-; secy-treas, Asn Grad Sch, 78- *Mem:* Fel AAAS; fel Am Phys Soc; Am Asn Physics Teachers (pres, 69-70). *Res:* Atomic collision processes; physics education. *Mailing Add:* Dept Physics Univ Wash Seattle WA 98195

GEBALLE, THEODORE HENRY, b San Francisco, Calif, Jan 20, 20; m 41; c 5. LOW TEMPERATURE PHYSICS. *Educ:* Univ Calif, Berkeley, BS, 41, PhD(phys chem), 49. *Prof Exp:* Res assoc, Low Temperature Lab, Univ Calif, Berkeley, 50-52; mem tech staff, Bell Tel Labs, NJ, 52-68, head, Dept Low Temperature & Solid State Physics, Phys Res Lab, 57-68; chmn, Dept Appl Physics, 75-86, PROF, DEPT APPL PHYSICS & DEPT MAT SCI & ENG, STANFORD UNIV, 67-, THEODORE & SYDNEY ROSENBERG PROF APPL PHYSICS, 78- *Concurrent Pos:* Consult, Mat Sci Lab, Bell Labs, 68-; Guggenheim fel, Cavendish Lab, Cambridge, Eng, 75; mem, Solid State Physics Deleg, People's Repub China, 75; dir, Ctr Mat Res, Stanford Univ; mem, Nat Comn Superconductivity. *Honors & Awards:* Oliver E Buckley Solid State Physics Prize, Am Phys Soc, 70; First Bernd Matthias Mem Award, 89. *Mem:* Nat Acad Sci; fel Am Phys Soc; Am Acad Arts & Sci; Am Chem Soc. *Res:* Low temperature physics; superconductivity; materials science; experimental studies of superconductivity and magnetism in intermetallic compounds using heat capacity, transport and optical measurements. *Mailing Add:* Dept Appl Physics Stanford Univ Stanford CA 94305-4085

GEBALLE, THOMAS RONALD, b Seattle, Wash, Nov, 16, 44; m 67; c 2. INFRARED SPECTROSCOPY. *Educ:* Univ Calif, Berkeley, BA, 67, PhD(physics), 74. *Prof Exp:* Res fel physics, Univ Calif, Berkeley, 74-75; res scientist, Leiden Univ, 75-77; Carnegie fel, Mt Wilson & Las Campanas Observ, 77-81; astronr, 81-87, ASSOC DIR, UK INFRARED TELESCOPE, HILO, HAWAII, 87- *Mem:* Am Astron Soc; Int Astron Union. *Res:* Use of infrared spectroscopy to investigate the solar system, star formation, stellar evolution, interstellar medium and galactic nuclei; design and construction of infrared spectrometers for astronomical use. *Mailing Add:* Joint Astron Ctr 665 Komohana St Hilo HI 96720

GEBAUER, PETER ANTHONY, b Albany, Calif, Apr 15, 43; m 67; c 2. ORGANIC CHEMISTRY. *Educ:* Harvey Mudd Col, BS, 65; Univ Ill, PhD(org chem), 70. *Prof Exp:* Asst prof chem, Purdue Univ, Indianapolis, 70-75; from asst prof to assoc prof, 75-88, PROF CHEM, MONMOUTH COL, ILL, 88- , CHMN DEPT, 77- *Mem:* Am Chem Soc. *Res:* Elimination and substitution reactions employing phosphorus reagents; small ring chemistry. *Mailing Add:* Dept Chem Monmouth Col Monmouth IL 61462

GEBBEN, ALAN IRWIN, b Shelbyville, Mich, July 4, 31; m 53; c 4. PLANT ECOLOGY. *Educ:* Calvin Col, AB, 54; Vanderbilt Univ, MAT, 55; Univ Mich, MS, 59, PhD(commonragweed ecol), 65. *Prof Exp:* Asst biol, 55-58, from instr to assoc prof, 61-67, chmn dept, 80-85, PROF BIOL, CALVIN COL, 67- *Mailing Add:* Dept Biol Calvin Col 3201 Burton St Grand Rapids MI 49546

GEBBER, GERARD L, b New York, NY, Feb 12, 39; m 61; c 2. PHARMACOLOGY, NEUROPHYSIOLOGY. *Educ:* Long Island Univ, 60; Univ Mich, PhD(pharmacol), 64. *Prof Exp:* NIH res fel pharmacol, Univ Pa, 64-65; instr, Tulane Univ, 65-66; from asst prof to assoc prof pharmacol, 66-75, PROF PHARMACOL, MICH STATE UNIV, 75-, PROF PHYSIOL, 82- *Concurrent Pos:* Merit awardee, NIH, 87- *Mem:* AAAS; Soc Neurosci; Am Physiol Soc; Can Physiol Soc; Am Soc Pharmacol & Exp Therapeut. *Res:* Central autonomic reflex pathways; central nervous system control of circulation; nerve physiology. *Mailing Add:* 1835 N Harrison Rd East Lansing MI 48823

GEBBIE, KATHARINE BLODGETT, b Cambridge, Mass, July 4, 32; m 57. ASTROPHYSICS. *Educ:* Bryn Mawr Col, BA, 57; Univ London, BSc, 60, PhD(astrophys), 65. *Prof Exp:* Res assoc astrophys, Joint Inst Lab Astrophys, Univ Colo, 67-68, lectr physics & astrophys, 74-77; astrophysicist, Nat Bur Standards, 68-85, supvry physicist, 85-89; DIR, PHYSICS LAB, NAT INST STANDARDS & TECHNOL, 89- *Concurrent Pos:* Ed, The Observ, 65-67; adj prof astro-geophys, Univ Colo, 77-89. *Mem:* Int Astron Union; Am Astron Soc; fel Joint Inst Lab Astrophys; Am Phys Soc; Royal Astron Soc. *Res:* Planetary nebulae, stellar atmospheres; physics of solar atmosphere. *Mailing Add:* Physics Lab Nat Inst Standards & Technol Gaithersburg MD 20899

GEBELEIN, CHARLES G, b Philadelphia, Pa, July 16, 29; m 51; c 5. POLYMER CHEMISTRY, ORGANIC CHEMISTRY. *Educ:* Temple Univ, BA, 55, MA, 59, PhD(chem), 67. *Prof Exp:* Chemist, Res Labs, Rohm and Haas Co, Pa, 51-59 & Cent Res Lab, Borden Chem Co, 59-63; fels, Res Inst, Temple Univ, 63-67; from asst prof to assoc, 67-77, PROF CHEM, YOUNGSTOWN STATE UNIV, 77-; RES PROF, COL MED, NORTHEAST OHIO UNIV, 79- *Mem:* Am Chem Soc; AAAS; Sigma Xi; Soc Biomat; Controlled Release Soc. *Res:* Additions to carbon-carbon double bonds; polymer modification; polymerization; polymeric drugs; biomedical polymers; bioactive polymers; biomaterials. *Mailing Add:* Dept Chem Youngstown State Univ 265 Meadowbrook Ave Youngstown OH 44555-0001

GEBELT, ROBERT EUGENE, b Rockford, Ill, May 20, 37; m 62; c 3. INORGANIC CHEMISTRY. *Educ:* Univ Ill, BS, 59; Mich State Univ, PhD(chem), 65. *Prof Exp:* Res assoc, Okla State Univ, 64-66; asst prof, 66-70, assoc prof, 70-80, PROF CHEM & GEOL, MANKATO STATE UNIV, 80- *Mem:* AAAS; Am Chem Soc; Sigma Xi. *Res:* Thermodynamic properties of inorganic materials at high temperatures. *Mailing Add:* Dept Chem Mankato State Col Mankato MN 56001

GEBER, WILLIAM FREDERICK, b Rahway, NJ, Oct 26, 23; m 46; c 1. PHYSIOLOGY, PHARMACOLOGY. *Educ:* Dartmouth Col, AB, 47; Ind Univ, MS, 50, PhD(physiol), 54. *Prof Exp:* Teaching assoc physiol, Sch Med, Ind Univ, 53-54; res assoc phys med, Med Col, Univ Minn, 54; asst prof physiol, Sch Med, St Louis Univ, 54-58; assoc prof, Sch Med, Univ SDak, 58-65; assoc prof, 65-71, PROF PHARMACOL, MED COL GA, 71- *Mem:* AAAS; Am Soc Pharmacol & Exp Therapuet; Am Physiol Soc; Soc Toxicol; Am Chem Soc; Teratology Soc. *Res:* Quantitative measurement of blood flow, hyperemea causes, and drug effects in areas of kidney, intestine, muscle and spleen; vascular aging; cardiovascular responses to weightlessness; environmental causes of congenital malformations; drug induced teratogenesis. *Mailing Add:* Dept Pharmacol Med Col Ga Augusta GA 30912

GEBHARD, ROGER LEE, b Sioux City, Iowa, Jan 30, 45; m 66; c 2. GASTROENTEROLOGY, INTERNAL MEDICINE. *Educ:* Univ Minn, BA, 65, MD, 69. *Prof Exp:* Clin res assoc, NIH, 71-73; PROF, MED SCH, UNIV MINN, 77- *Concurrent Pos:* Staff physician, Vet Admin Med Ctr, Minneapolis, Minn, 77- *Mem:* Am Gastroenterol Asn; Am Asn Study Liver Dis; Am Physiol Soc; Am Soc Gastrointestinal Endoscopy. *Res:* Cholesterol metabolism; intestinal function and disease. *Mailing Add:* Minneapolis Vet Admin Med Ctr 111-D One Veteran Dr Minneapolis MN 55417

GEBHARDT, BRYAN MATTHEW, CELLULAR & OCULAR IMMUNOLOGY. *Educ:* Tulane Univ, PhD(biol), 67. *Prof Exp:* PROF OPHTHAL & MICROBIOL, LA STATE UNIV MED CTR, 78- *Mailing Add:* Dept Ophthal La State Univ Eye Ctr 2020 Gravier St Suite B New Orleans LA 70112-2234

GEBHARDT, JOSEPH JOHN, b New York, NY, Feb 28, 23; wid; c 2. REFRACTORY COMPOSITES DEVELOPMENT. *Educ:* Hobart Col, AB, 44; Carnegie Inst Technol, MSc, 50, DSc(chem), 51. *Prof Exp:* Instr chem, Manhattan Col, 46-47; instr, Carnegie Inst Tech, 47-49, asst, 49-51; sr res chemist, Titanium Div, Nat Lead Co, 51-57; sr res chemist, Am Potash & Chem Corp, 57-59, proj chemist, 59-60; res chemist, Missle & Space Div, Space Sci Lab, Gen Elec Co, 60-69, consult phys chem, Space Syst Div-RSO, 69-86; SR PHYS CHEM, GENERAL SCIENCES INC, 87- *Concurrent Pos:* Consult, Materials Applied Inc, 76-86; chmn, Ceramic Metal Systs Div, Am Ceramics Soc, 83. *Mem:* AAAS; fel Am Inst Chem; Am Chem Soc; fel Am Ceramics Soc; Sigma Xi. *Res:* High temperature materials; vapor deposition; solid state reactions; carbon and graphite; composite materials; vapor/solid reactions; formation of ceramic and metallic materials; optical structural properties; carbon-carbon composite processing and properties; inorganic materials processing. *Mailing Add:* 18 Beverly Ave Malvern PA 19355

GEBHART, BENJAMIN, b Cincinnati, Ohio, July 2, 23; m 68; c 2. ENGINEERING. *Educ:* Univ Mich, BS, 48, MS, 50; Cornell Univ, PhD, 54. *Hon Degrees:* MA, Univ Pa. *Prof Exp:* Instr mech eng, Univ Mich, 49-50 & Lehigh Univ, 50-51; from instr to prof, Cornell Univ, 51-75; prof mech eng & chmn dept, State Univ NY Buffalo, Amherst, 75-80; prof mech, 80, SAMUEL LANDIS GABEL PROF MECH ENG, UNIV PA, 80- *Concurrent Pos:* Preceptorship award, Cornell Univ, 55-57, Giordano Found res grant, 57-60, NSF res award, 59-; assoc prof, Aix Marseille, Nancy, 63, 66, 88; vis prof, Cornell Aeronaut Lab, 64-65, Univ Calif, Berkeley, 67, Ore State, 74, Naval Post Grad Sch, 80 & Ecole des Mines, France, 88; Freeman scholar, 78. *Honors & Awards:* Mem Award, Heat Transfer Div, Am Soc Mech Engrs, 72. *Mem:* AAAS; fel Am Soc Mech Engrs; Sigma Xi. *Res:* Fluid mechanics and heat transfer. *Mailing Add:* Towne Bldg Univ Pa Philadelphia PA 19104-6315

GECKLE, WILLIAM JUDE, b Baltimore, Md, May 26, 55; m 84; c 2. IMAGING PHYSICS, COMPUTER VISION. *Educ:* Loyola Col, Baltimore, BS, 77; Mich State Univ, MS, 79. *Prof Exp:* Grad asst, Mich State Univ, 77-79; assoc physicist, 79-85, PHYSICIST, APPL PHYSICS LAB, JOHNS HOPKINS UNIV, 85-, COORDR COLLAB PROGS IN IMAGE ANALYSIS, 85-, SECT SUPVR, 90- *Mem:* Inst Elec & Electronics Engrs. *Res:* Development of intelligent processing and modeling techniques applicable to medical imagery; algorithm development for MRI, single photon emission computed tomography, positron-emission tomography, computerized tomography and conventional imaging systems. *Mailing Add:* Appl Physics Lab Johns Hopkins Univ Johns Hopkins Rd Laurel MD 20707

GECZIK, RONALD JOSEPH, b Bronx, NY, Mar 22, 33; m 61; c 3. PHYSIOLOGY, PHARMACOLOGY. *Educ:* Fordham Univ, BS, 54, MS, 57, PhD(biol), 59; Seton Hall Univ, JD, 71. *Prof Exp:* Sr res scientist, Colgate-Palmolive Co, 59-62, sr res scientist, Pharmacol Sect, 62-64; asst to dir toxicol, Path Dept, 64-66, mgr biol data, 66-68, dir res admin, Squibb Inst Med Res, 68-73; vpres, Licensing Corp Develop, Carter Wallace, Inc, 73-84; vpres, SNW Inc, 84-89; PRES, PAUL MICHAEL ASSOC INC, 89- *Mem:* AAAS; NY Acad Sci; Am Soc Pharmacol & Exp Ther; Licensing Exec Soc; Sigma Xi. *Res:* Dental research, especially caries, periodontal disease and calculus; tissue culture; cytology; regeneration and wound healing; toxicology. *Mailing Add:* Eight Phyllis Place E Brunswick Township Milltown NJ 08850

GEDDES, DAVID DARWIN, biology, physiology; deceased, see previous edition for last biography

GEDDES, JOHN JOSEPH, b Columbus, Ohio, Oct 5, 40. ELECTRICAL ENGINEERING. *Educ:* Univ Minn, BSEE, 63; Purdue Univ, MSEE, 68, PhD(elec eng), 71. *Prof Exp:* Design engr circuits, Missle Div, Raytheon Co, 63-66; PRIN RES SCIENTIST, MAT SCI CTR, HONEYWELL CORP, 71- *Mem:* Sr mem Inst Elec & Electronics Engrs; Sigma Xi. *Res:* Magnetics; fiber optics; design and fabrication of Gallium Arsenide devices and monolithic circuits for microwave applications. *Mailing Add:* Systs & Res Ctr Honeywell Corp 10701 Lyndale Ave South Bloomington MN 55420

GEDDES, KEITH OLIVER, b North Battleford, Sask, Nov 4, 47; m 72; c 3. ALGEBRAIC ALGORITHMS, SYMBOLIC COMPUTATION. *Educ:* Univ Sask, BA, 68; Univ Toronto, MSc, 70, PhD(comput sci), 73. *Prof Exp:* Asst prof, 73-78, ASSOC PROF COMPUT SCI, UNIV WATERLOO, 78- *Mem:* Asn Comput Mach; Soc Indust & Appl Math; Can Appl Math Soc. *Res:* Algebraic algorithms; systems for symbolic computation, numerical approximation, scientific computation. *Mailing Add:* Dept Comput Sci Univ Waterloo Waterloo ON N2L 3G1 Can

GEDDES, LANELLE EVELYN, b Houston, Tex, Sept 15, 35; m 62. PHYSIOLOGY, NURSING. *Educ:* Univ Houston, BSN, 57, PhD(biophysics), 70. *Prof Exp:* Staff nurse, Houston Independent Sch Dist, 57-62; from instr to asst prof physiol, Baylor Col Med, 72-75; assoc prof, 75-78, PROF NURSING, PURDUE UNIV, 78-, HEAD, SCH NURSING, 80- *Concurrent Pos:* Fel physiol, Baylor Col Med, 70-72 & vis asst prof, 75-; coordr grad prog med & surg, Tex Women's Univ, 74-75. *Mem:* AAAS; NY Acad Sci. *Res:* Evaluation of cardiovascular dynamics and its clinical applications. *Mailing Add:* Sch Nursing Purdue Univ West Lafayette IN 47907

GEDDES, LESLIE ALEXANDER, b Scotland, May 24, 21; m 45, 61; c 1. CARDIOVASCULAR PHYSIOLOGY, BIOMEDICAL ENGINEERING. *Educ:* McGill Univ, BEng, 45, MEng, 53; Baylor Univ, PhD(physiol), 58. *Hon Degrees:* DSc, McGill Univ, 71. *Prof Exp:* Asst EEG & neurophysiol & demonstr elec eng, McGill Univ, 45-47; consult & supvr tech equip, Montreal Neurol Inst & Royal Victoria Hosp, 46-52; biophysicist, Baylor Col Med, 52, dir lab biophys, 53-57, asst prof physiol, 58-65, prof physiol & chief div biomed eng, 65-74; SHOWALTER DISTINGUISHED PROF BIOMED ENG, PURDUE UNIV, W LAFAYETTE, 74-, DIR BIOMED ENG CTR, 74- *Concurrent Pos:* Consult engr, 49-; partner & consult, Electro-Design Co,

50-52; consult, Nat Found Infantile Paralysis, Southwest Poliomyelitis Respiratory Ctr, 58-60 & US Air Force, 58-; prof physiol, Dent Col, Univ Tex, 57-74; prof vet physiol & pharmacol, Tex A&M Univ, 67-74, prof biomed eng, 68-74. *Honors & Awards:* Leadership Biomed Eng Award, Am Soc Advan Med Instrumentation, 85; Laufman-Greatbach Award, 87; Biomed Eng Award, 85. *Mem:* Nat Acad Eng; fel Am Col Cardiol; fel Inst Elec & Electronics Engrs; fel Australasian Col Physics & Eng Med Biol. *Res:* Electrophysiology; cardiology; biophysical instrumentation. *Mailing Add:* Biomed Eng Ctr Purdue Univ W Lafayette IN 47907

GEDDES, WILBURT HALE, b Eakly, Okla, Jan 10, 26; m 46; c 5. MARINE GEOPHYSICS. *Educ:* NC State Col, BS, 49. *Prof Exp:* Geophysicist airborne magnetics, 52-63, dir marine geophys surv, 63-68, dir acoust div, 70-78, dir oceanog off, 78-81, prin scientist, Marine Geophys Planning Syst Inc, 81-87; PRES, GEDDES GEOPHYSICAL ASSOCS, 87- *Honors & Awards:* Antartic Exploration Medal. *Mem:* Am Geophys Union; Soc Explor Geophysicists; Sigma Xi; Marine Technol Soc. *Res:* Underwater acoustics including propagation through the ocean floor, boundary reflection losses, volume scattering and background noise. *Mailing Add:* 765 E Scenic Dr Pass Christian MS 39571

GEDEON, GEZA S(CHOLCZ), b Miskolc, Hungary, June 12, 14; US citizen; c 3. ASTRODYNAMICS. *Educ:* Budapest Tech Univ, MS, 37, DSc(aero eng), 45. *Prof Exp:* Res asst aerodynamics, Aerotech Inst, 39-43; chief test pilot, Repulogepgyar, RT, 43-45; test engr, L Lang Mach & Eng Shop, 47-48; aircraft designer, L Breguet Ateliers d' Aviation, 48-50; lectr math & mech, Ind Tech Col, 50-55; design specialist, Int Harvester Co, 55-56; res scientist, Chance Vought Aircraft, 56-59 & Aeronutronic Div, Ford Motor Co, 59-60; head astrodynamics, Norair Labs, Northrop Corp, 60-62, chief flight mech, space labs, 62-65; staff engr, 65, mgr analysis mech dept, 65-68, sr scientist, TRW Systs, Inc, 68-; AT DEPT AEROSPACE ING, WEST COAST UNIV, LOS ANGELES, CALIF. *Concurrent Pos:* Lectr, Univ Calif, Los Angeles, 63-66 & 81- *Mem:* Celestial Mech Inst. *Res:* Astrodynamics related to trajectories, orbit determination and guidance. *Mailing Add:* 16055 Temecula Pacific Palisades CA 90272

GEDNEY, LARRY DANIEL, b Salt Lake City, Utah, Jan 22, 38; m 70; c 4. SEISMOLOGY. *Educ:* Univ Nev, Reno, BS, 60, MS, 66. *Prof Exp:* Asst geophysicist, 66-70, ASSOC PROF GEOPHYSICS, GEOPHYS INST, UNIV ALASKA, 71- *Concurrent Pos:* Geophysicist, US Earthquake Mechanism Lab, San Francisco, 70-71. *Mem:* Seismol Soc Am; Am Geophys Union; AAAS. *Res:* Seismology and tectonics of Alaska; tectonic mapping by remote sensing. *Mailing Add:* 546 Ester Loop Fairbanks AK 99701

GEDULDIG, DONALD, b New York, NY, Oct 27, 32; m 62; c 1. COMPUTER APPLICATIONS, DATA ACQUISITION & ANALYSIS. *Educ:* Cornell Univ, BEE, 55, MS, 57; Columbia Univ, PhD(biophys), 65. *Prof Exp:* Res engr med electronics, RCA Labs, 57-58; res engr, Rockefeller Univ, 58-61; asst res physiologist, Univ Calif, San Diego, 66-68; asst prof biophys, Sch Med, Univ Md, Baltimore, 68-75; assoc prof biophys, Fac Med, Mem Univ Nfld, 75-85; CONSULT, 85- *Concurrent Pos:* NIH fel, Physiol Lab, Cambridge Univ, 65-66; mem Brain Res Inst, Univ Calif, Los Angeles, 66-68; assoc ed, Can J Physiol & Pharmacol, 80- *Mem:* Biophys Soc; Soc Gen Physiol. *Res:* Membrane structure and function; transport of ions through membranes; excitation in nerve and muscle; electrophysiological instrumentation; biomedical engineering; computer systems for data acquisition and data analysis; system integration. *Mailing Add:* 6210 41st Pl Hyattsville MD 20781

GEE, ADRIAN PHILIP, b Whitehead, UK, Apr 30, 52. BONE MARROW TRANSPLANTATION, COMPLEMENT. *Educ:* Univ Birmingham, UK, BSc, 73; Univ Edinburgh, UK, PhD(immunol), 77. *Prof Exp:* Vis fel immunol, Nat Cancer Inst, 78-82; res assoc, dept path, Univ Toronto, 82-83; vis asst prof pediat, Univ Fla, 83-85, assoc prof immunol & med microbiol, 84-87, assoc prof pediat, 85-87, assoc dir res, Div Pediat Hemat, 85-87,; SR SCIENTIST, BAXTER HEALTHCARE CORP, 87- *Concurrent Pos:* Prin investr, Am Heart Asn, 84-86, Am Cancer Soc, 85-87 & Pardee Found, Mich, 85-87; co-prin investr, NIH, 84-87; adj assoc prof, Univ Fla, 87-; Hastilow res scholar; Lady Tata fel. *Mem:* Am Asn Cancer Res; Am Asn Immunologists; NY Acad Sci; Brit Soc Immunol; Inst Biol London. *Res:* Development of procedures to improve bone marrow transplantation; immunomagnetic depletion of subpopulations of cells from marrow to be used for autologous or haplotype-mismatched transplantation. *Mailing Add:* Baxter Healthcare Corp Fenwal BioMed Systs 3015 S Daimler Santa Ana CA 92705

GEE, ALLEN, b Patterson, Calif, Feb 23, 24; m 55; c 2. PHYSICAL CHEMISTRY. *Educ:* Univ Calif, BS, 47; Mass Inst Technol, PhD(phys chem), 51. *Prof Exp:* Asst chem, Univ Calif, 47; asst protein chem, Mass Inst Technol, 47-50; res assoc, Sugar Res Found, Nat Bur Standards, 51-57; sr process chemist, E I du Pont de Nemours & Co, 57-59; sr proj engr, Tex Instruments, Inc, 59-61; mem tech staff, Hughes Aircraft Co, 61-67, assoc mgr microelectronics lab, Newport Beach Div, 67-69; tech consult & regist investment adv, 70-74; RETIRED. *Mem:* Am Chem Soc; fel Am Inst Chem; Electrochem Soc; NY Acad Sci. *Res:* Serum proteins; bone, teeth and calcium phosphates; infrared and visible spectrophotometry; chromatography; waste treatment and water pollution; polymers, sugars, cellulose and synthetic fibers; semiconductors; surface effects; process development and control. *Mailing Add:* 2521 Sierra Vista Newport Beach CA 92660

GEE, CHARLES WILLIAM, b Des Moines, Iowa, Mar 19, 36; m 59; c 3. SCIENCE EDUCATION. *Educ:* Univ Wis, BS, 59; Okla State Univ, MS, 64; Mich State Univ, PhD(sci ed), 67. *Prof Exp:* Asst prof, 67-72, assoc prof, 72-80, PROF BIOL & SCI EDUC, MILLIGAN COL, 80- *Concurrent Pos:* Elem sci consult & staff mem, Eastern Tenn State Univ, 73, 74, 76, 78 & 81. *Mem:* Nat Sci Teachers Asn. *Res:* Experimentation with techniques of science presentation. *Mailing Add:* Dept Biol & Ed Milligan Col Milligan College TN 37682

GEE, DAVID EASTON, teaching, petroleum exploration; deceased, see previous edition for last biography

GEE, EDWIN AUSTIN, b Washington, DC, Feb 19, 20; m 44; c 3. CHEMICAL ENGINEERING. *Educ:* George Washington Univ, BS, 41, MS, 44; Univ Md, PhD(chem eng), 48. *Prof Exp:* Asst chemist, Naval Res Lab, Washington, DC, 41-42; chemist, US Bur Mines, Md, 42-43, phys chemist, 43-44, phys chemist, Ala, 44-45, chem engr, Md, 45-46, metallurgist, Washington, DC, 46-47, asst chief metall div, 47-48; metallurgist & chem engr, E I du Pont de Nemours & Co, 48-50, supvr res group, 50-51, mgr, Res Sect, 51-53, mgr plants technol, Pigments Dept, 53-57, asst dir sales, 57-60, asst dir develop dept, 57-63, dir, 63-66, mgr res & develop corp diversification, 66-68, gen mgr photo prod dept, 68-70, vpres, dir & mem exec comt, 70-80; pres, Int Paper Co, 78-81, chmn & chief exec officer, 81-85; RETIRED. *Concurrent Pos:* Mem bd, Buck Hills Falls Co, 77-85, Int Paper Co, 78-85, Am Home Prod Inc, 78-, Air Prod & Chem Inc, 80-91, Finger Matrix, 84-87, Bell & Howell Corp, 85-88, Oncogene Sci, Inc, 86-, Salomon Bros Fund, 87-, Bethlehem Steel, 87- & Clean Sites, 86- *Mem:* Nat Acad Eng; Am Chem Soc; Nat Soc Prof Engrs; Inst Mining, Metall & Petrol Engrs; Am Soc Metals. *Res:* Organic kinetics; rare metals; titanium and zirconium; extraction of ores; purification of salts; improved method for the acid decomposition of certain silicates; pigments, white and colored; semiconductor materials. *Mailing Add:* Box 362 Buck Hills Falls PA 18328

GEE, J BERNARD L, b Crewe, Eng, Mar 13, 27; m 57; c 2. MEDICINE, PHYSIOLOGY. *Educ:* Oxford Univ, BA, 48, MSc, 49, MA, 52, BM, ChB, 53. *Prof Exp:* Jr demonstr physiol, Oxford Univ, 48-49; intern surg, Guys Hosp, Univ London, 53-54; intern med, Postgrad Med Sch, Univ London, 56-57; resident, Radcliffe Infirmary, Oxford Univ, 57-59; fel, McGill Univ, 59-61; from instr to asst prof, Sch Med, Univ Wis, 61-64; from asst prof to assoc prof, Sch Med, Univ Pittsburgh, 64-69; assoc prof, 69-81, PROF MED, YALE UNIV, 81- *Concurrent Pos:* Consult, Vet Admin Hosps, 68- *Mem:* Am Thoracic Soc; Am Fedn Clin Res; fel Am Col Physicians; Am Physiol Soc; Reticuloendothelial Soc. *Res:* Pulmonary medicine and physiology; exercise and respiratory physiology; metabolic features of alveolar macrophages. *Mailing Add:* Yale Univ Sch Med 333 Cedar St PO Box 3333 New Haven CT 06510

GEE, JOHN HENRY, b New Westminster, BC, Can, July 25, 36; m 61. ECOLOGY. *Educ:* Univ BC, BCom, 59, MSc, 61; Univ Sydney, PhD(zool), 67. *Prof Exp:* Scientist, Fisheries Res Bd Can, 61-63; res fel, 66-67, from asst prof to assoc prof, 67-78, PROF ZOOL, UNIV MAN, 78-, HEAD, ZOOL DEPT, 81- *Concurrent Pos:* Res grants, Fisheries Res Bd Can, Nat Res Coun Can & Univ Manitoba, 67. *Mem:* Can Soc Zoologists. *Res:* Ecology of stream fish; buoyancy regulation in aquatic vertebrates. *Mailing Add:* Dept Zool Univ Man Winnipeg MB R3T 2N2 Can

GEE, LYNN LAMARR, b St Anthony Idaho, June 21, 12; m 33; c 2. MICROBIOLOGY. *Educ:* Brigham Young Univ, BS, 35; Colo A&M Col, MS, 37; Univ Wis, PhD(bacter), 41. *Prof Exp:* Asst, Brigham Young Univ, 35 & Univ Wis, 41; bacteriologist, Dept State Conserv, Wis, 39-41 & US Civil Serv, 46; soil microbiologist, Purdue Univ, 46-48; prof bact, Tex A&M Univ, 48-54; prof & head dept, Okla State Univ, 54-77; RETIRED. *Concurrent Pos:* Bacteriologist, US Dept Defense, 51-52. *Mem:* Am Soc Microbiol; Am Soc Prof Biol; Am Acad Microbiol; Soc Appl Bact. *Res:* Soil microbiology; aerobiology. *Mailing Add:* 116 S Orchard Lane Stillwater OK 74074

GEE, NORMAN, b Ottawa, Ont, Oct 17, 52; m 81; c 2. ELECTRON & ION TRANSPORT, RADIATION CHEM. *Educ:* Univ Alta, BSc, 74 & PhD(radiation chem), 79. *Prof Exp:* Technologist, 79-84, specialist tech, 84-89, FAC SERV OFFICER, UNIV ALTA, 89- *Mem:* Chem Inst Can. *Res:* Electron transport; ion transport; free ion yields; electron thermalization distances in dielectric gases, liquids and critical fluids; thermophysical properties of liquids. *Mailing Add:* Chem Dept Chem Bldg Radiation Res Ctr Edmonton AB T6G 2G2 Can

GEE, ROBERT WILLIAM, b Los Angeles, Calif, Jan 24, 36. BIOCHEMISTRY. *Educ:* Univ Southern Calif, PhD(biochem), 67. *Prof Exp:* Fel biochem, Scripps Inst Oceanog, 67-71; vis asst prof plant physiol, Ore State Univ, 71-72; RES ASSOC BIOCHEM, MICH STATE UNIV, 72- *Mem:* Am Soc Plant Physiologists; Am Chem Soc. *Res:* Studies of plant and animal micro body enzymes, including catalase glycerol phosphate dehydrogenase super oxide dismutase; role of micro bodies in cellular metabolism. *Mailing Add:* 531 Village Lansing MI 48910

GEE, SHERMAN, b Canton, China, July 18, 37; US citizen; m 65; c 2. ELECTRICAL ENGINEERING. *Educ:* Univ Calif, Berkeley, BS, 60; Mass Inst Technol, MS, 61; Stanford Univ, PhD(elec eng), 65. *Prof Exp:* Res asst radar astron, Stanford Univ, 63-65; adv develop engr, Sylvania Electronic Defense Labs, 61-63 & 65-66; electronic engr, Arnold Eng Develop Ctr, 66-67; electronic warfare engr, Air Force Avionics Lab, 67-69; mgr, Advan Technol Div, MB Assocs, 69-71; head, Technol Transfer Off, Naval Surface Weapons Ctr, 71-80, actg assoc dir, Navy Technol-Phys Sci & Electronics, 80-81, MGR, COMMAND, CONTROL & COMMUN, OFF NAVAL TECHNOL, 82- *Concurrent Pos:* Spec asst to dir, Nat Tech Info Serv, Dept Com, 76-77; staff specialist, Off Under Secy Defense Res & Eng, 79-80. *Mem:* Nat Security Indust Asn; Inst Elec & Electronics Engrs; Int Union Geod & Geophys. *Res:* Electromagnetics; propagation laser instrumentation; radiation and scattering; electronic systems analysis; technological innovation; technology transfer; communications. *Mailing Add:* Off Naval Technol 800 N Quincy St Arlington VA 22217

GEE, WILLIAM, b Riverhead, NY, Nov 4, 31; m 56; c 5. OCULAR PNEUMOPLETHYSMOGRAPHY, NONINVASIVE VASCULAR LABORATORY. *Educ:* State Univ NY, Brooklyn, MD, 61; Am Bd Surg, cert gen surg, 70, cert vascular surg, 84. *Prof Exp:* Internship & residency, Naval Hosp, St Albans, 61-67; fel, vascular surg, Univ Calif, San Francisco, 72-73; chief, vascular surg, Nat Naval Med Ctr, 73-77; DIR, VASCULAR LAB, LEHIGH VALLEY HOSP LAB, 77-, CLIN PROF SURG, HAHNEMANN UNIV, 86- *Concurrent Pos:* Consult, NIH Asymptomatic Carotid Atherosclerosis Study, 85; stroke coun, Am Heart Asn. *Mem:* Soc Vascular

Surg; Int Soc Cardiovas Surg; Am Col Surgeons. *Res:* Inventor, Ocular Pneumoplethysmograph; research in the design and development of this instrument from 1967 to present. Sixty four publications related to this and associated research. *Mailing Add:* Vascular Lab Box 689 Allentown PA 18105

GEEHERN, MARGARET KENNEDY, atmospheric physics, for more information see previous edition

GEELHOED, GLENN WILLIAM, b Grand Rapids, Mich, Jan 19, 42; c 2. SURGERY, PHYSIOLOGY. *Educ:* Calvin Col, AB, 64, BS, 65; Univ Mich, MD, 68; G W Univ, MA; Univ London, DTMH. *Prof Exp:* Asst surg, Peter Bent Brigham Hosp, Harvard Med Sch, 68-70; clin assoc cancer, Nat Cancer Inst, NIH, 70-72 & sr staff investr, 72-75, instr surg, 74-75; from asst prof to assoc prof, 75-86, DIR, SURG RES LABS & TRANSPLANT DIV, GEORGE WASHINGTON UNIV, 75-, CHIEF ENDOCRINE SURG, 75-, PROF SURG, 86- *Concurrent Pos:* Clin scholar health care res, Robert Wood Johnson Found, 75-77, assoc dir clin scholar prog, 77-; consult, Nat Cancer Inst, NIH, 75-, Walter Reed Army Hosp, 75-, US Senate Comt on the Judiciary, Mr Kennedy, 77- & WHO, Pan Am Health Orgn, World Bank, UN High Comnr on Refugees, 77-; James IV Traveling Surg Scholar, 86; prof, Int Med Educ; dir, Int Health Ctr. *Honors & Awards:* Gold Medal, Southeastern Surg Cong, 75. *Mem:* Fel Am Col Surgeons; Asn Acad Surg; Am Physiol Soc; Am Fedn Clin Res; Soc Univ Surgeons. *Res:* Surgical endocrinology; transplantation and tumor immunology; physiology especially renal, pulmonary and gastrointestinal; shock and trauma; oncology especially breast, melanoma, colon and sarcoma; pathophysiology especially thyroid, parathyroid, adrenal and pancreatic islets. *Mailing Add:* Dept Surg George Washington Univ 2150 Pennsylvania Ave NW Washington DC 20037

GEELS, EDWIN JAMES, b Hull, Iowa, Jan 24, 40; m 62; c 2. ORGANIC CHEMISTRY. *Educ:* Calvin Col, BS, 61; Iowa State Univ, PhD(chem), 65. *Prof Exp:* Fel, Iowa State Univ, 65; asst prof, 65-77, PROF CHEM, DORDT COL, 077- *Concurrent Pos:* Petrol Res Fund grant, 65-67. *Mem:* Am Chem Soc; Royal Soc Chem. *Res:* Electron transfer and free radical reactions of organic compounds. *Mailing Add:* Dept Chem Dordt College Sioux Center IA 51250

GEEN, GLEN HOWARD, zoology, aquatic biology; deceased, see previous edition for last biography

GEER, BILLY W, b Coin, Iowa, June 6, 35; m 57. GENETICS, NUTRITION. *Educ:* Northwest Mo State Col, BS, 57; Univ Nebr, MS, 60; Univ Calif, Davis, PhD(genetics), 63. *Prof Exp:* From asst prof to assoc prof, 63-75, chmn dept, 72-78, coordr, Know-Rush Med Prog, 79-85, PROF BIOL, KNOX COL, ILL, 75- *Concurrent Pos:* Vis assoc prof, Ore State Univ, 68-69; USPHS spec fel, 68-69; vis prof, Univ Calgary, 77-78 & 84-85; int sci exchange award, Nat Sci & Eng Res Coun Can, 84; vis sci award, Alta Heritage Found Med Res, 85; vis prof & Fulbright sr scholar, Monash Univ, 85. *Mem:* AAAS; Am Soc Zoologists; Am Inst Nutrit; Genetics Soc Am. *Res:* Biochemistry of reproduction; nutritional modification of enzymes; lipid metabolism. *Mailing Add:* Dept Biol Knox Col Galesburg IL 61401

GEER, IRA W, b Avoca, NY, Jan 18, 35; m 64; c 2. SCIENCE EDUCATION, METEOROLOGY. *Educ:* State Univ NY Col Brockport, BS, 57; Univ NC, MEd, 60; Pa State Univ, EdD(sci ed), 66. *Prof Exp:* Assoc prof, 61-70, chmn dept earth sci, 74-80, PROF METEOROL, STATE UNIV NY COL BROCKPORT, 70- *Concurrent Pos:* UNESCO sci educ adv, Ministry of Educ, Repub Korea, 69-70. *Mem:* Assoc mem Am Meteorol Soc. *Res:* Earth Sciences. *Mailing Add:* Dept Earth Sci State Univ NY Col Brockport NY 14420

GEER, JACK CHARLES, b Galesburg, Ill, Sept 19, 27; m 51; c 5. MEDICINE, PATHOLOGY. *Educ:* La State Univ, BS, 50, MD, 56. *Prof Exp:* Asst, Sch Med, La State Univ, 54-55, res assoc, 55-57, from instr to prof path, 57-66; prof, STex Med Sch, Univ Tex, 66-67; prof path, Ohio State Univ, 67-75, chmn dept, 67-72; assoc pathologist, Davidson Labs, 72-75; PROF PATH & CHMN DEPT, UNIV ALA, BIRMINGHAM, 75- *Concurrent Pos:* USPHS sr res career develop award, 59-66; vis investr, Rockefeller Inst, 60-61; consult, Spec Ctr Res Atherosclerosis, Bowman-Gray Sch Med, 75-80, NE Regional Primate Ctr, Harvard Univ, 79- *Mem:* Col Am Pathologists; Am Med Asn; Int Acad Path; Am Asn Pathologists & Bacteriologists. *Res:* Experimental pathology; electron microscopy. *Mailing Add:* 3744 Wimbleton Dr Birmingham AL 35223

GEER, JAMES FRANCIS, b Syracuse, NY, Oct 3, 40; m 65; c 4. APPLIED MATHEMATICS. *Educ:* Harpur Col, BA, 62; Univ Va, MA, 64; NY Univ, PhD(math), 67. *Prof Exp:* Mathematician, IBM, 67-69; PROF MATH, STATE UNIV NY, BINGHAMTON, 69- *Concurrent Pos:* Prin investr, res grants & contracts from various industs, NSF, 69-; consult, Math & Physics, IBM, Gen Elec & NASA, 69- *Mem:* Soc Indust & Appl Math; Am Acad Mech. *Res:* Asymptotic, perturbation and numerical methods to solve partial differential equations of mathematical physics, particularly in the areas of fluid mechanics, scattering theory and non-linear oscillations; slender body theory; free boundary value problems. *Mailing Add:* RD 3 Box 328 Endicott NY 13760

GEER, RICHARD P, b LaHarpe, Ill, Sept 23, 38. POLYMER CHEMISTRY, ORGANIC CHEMISTRY. *Educ:* Univ Ill, Urbana, BS, 60; Univ Rochester, PhD(org chem), 65. *Prof Exp:* RES CHEMIST, RES CTR, HERCULES INC, 64- *Mem:* Am Chem Soc. *Res:* Polymer and monomer synthesis; free radical chemistry. *Mailing Add:* 1115 Kelly Dr Newark DE 19711

GEER, RONALD L, b West Palm Beach, Fla, Sept 2, 26; m 51; c 2. OCEAN ENGINEERING. *Educ:* Ga Inst Technol, BME, 51. *Prof Exp:* Proj engr deepwater drilling prog, Shell Oil Co, New Orleans, 61-63, from staff mech engr to mgr marine tech group, Head Off Prod Dept, 63-66, chief, Pac Coast, 66-69, sr staff, 67-71, consult, 71-80, sr mech engr consult, Head Off Prod Dept, 80-86; RETIRED. *Concurrent Pos:* Mem vis comt, Dept Ocean Eng, MIT Corp, 78-; vchmn. Nat Res Coun-Assay Eng, Nat Acad Eng, 79-81, chmn, Marine Bd, 81, mem, Polar Res Bd, 84-86. *Mem:* Nat Acad Eng; Am Soc Mech Engrs; Am Petroleum Inst; Marine Technol Soc; hon mem ASME 85. *Res:* Offshore-marine systems engineering; petroleum drilling and production engineering. *Mailing Add:* PO Box 440335 Houston TX 77244-0335

GEERING, EMIL JOHN, b Yonkers, NY, Feb 8, 24; m 53; c 4. ORGANIC CHEMISTRY. *Educ:* Hobart Col, AM, 43; Polytech Inst Brooklyn, MS, 50, PhD(org chem), 54; State Univ NY Buffalo, MBA, 82. *Prof Exp:* Chem analyst, Best Foods, Inc, 46-47; chem analyst, Interchem Corp, 50-51; fel org chem, Polytech Inst Brooklyn, 52-53; TECH MGT, HOOKER CHEM & PLASTIC CORP, 54- *Mem:* Am Chem Soc; The Chem Soc; Sigma Xi. *Res:* Sulfur and chloro compounds; preparation, development and marketing of organic chemicals and polymers. *Mailing Add:* 4001 West River Pkwy Grand Island NY 14072

GEERS, THOMAS L, b El Paso, Tex, Nov 16, 39; m 61; c 2. STRUCTURAL DYNAMICS, WAVE PROPAGATION & SCATTERING. *Educ:* Mass Inst Technol, SB, 61, MS, 64, PhD(appl mech), 67. *Prof Exp:* Proj officer, David Taylor Model Basin, 61-63; res engr, Cambridge Acoust Assoc, 64-66; res scientist, Lockheed Palo Alto Res Lab, 66-74, staff scientist, 74-84, mgr, 84-85; chmn, 85-90, PROF MECH ENG, UNIV COLO, BOULDER, 90- *Concurrent Pos:* Consult, 86- *Honors & Awards:* Hess Award, Am Soc Mech Engrs, 70. *Mem:* fel Am Soc Mech Engrs; Acoust Soc Am; Am Acad Mech; Sigma Xi. *Res:* Structural dynamics and acoustics; analytical and computational methods in structure-medium interaction; asymptotic methods in wave propagation and scattering. *Mailing Add:* Dept Mech Eng Univ Colo Boulder CO 80302

GEESEMAN, GORDON E, b Fairview, Ill, Mar 18, 21; m 50; c 1. GENETICS, BOTANY. *Educ:* Univ Ill, BS, 43, MS, 46; Univ Wis, PhD(genetics, agron), 49. *Prof Exp:* With State Dept Agr, Wash, 53-57; asst agronomist, Mont State Col, 57-60; asst prof, 62-67, ASSOC PROF GENETICS & BOT, UNIV WIS, STEVENS POINT, 67- *Mem:* Nat Autobon Soc; Am Hort Soc; Am Soc Hort Sci. *Res:* Plant physiology, anatomy and genetics; selected economic plants. *Mailing Add:* Dept Biol Univ Wis Stevens Point WI 54481

GEESLIN, ROGER HAROLD, b Mt Healthy, Ohio, May 24, 31; m 54; c 3. MATHEMATICS. *Educ:* Kenyon Col, AB, 53; Yale Univ, MA, 57, PhD(math), 58. *Prof Exp:* Instr math, Int Christian Univ, Tokyo, 58-60; from asst prof to assoc prof, 60-68, from actg chmn to chmn dept, 69-73 & 74-79, actg dean, Univ Col, 73-74, PROF MATH, UNIV LOUISVILLE, 68- *Concurrent Pos:* NSF fel, 53-54. *Mem:* AAAS; Am Math Soc; Math Asn Am; Sigma Xi; Nat Coun Teachers Math. *Res:* Diffusion equations; functional analysis; real and complex variables; applications of computers in mathematics and the teaching of mathematics. *Mailing Add:* Shelby Campus Univ Louisville Louisville KY 40292

GEFFEN, ABRAHAM, b Atlanta, Ga, Sept 22, 16; m 48; c 3. RADIOLOGY. *Educ:* Emory Univ, BS, 37; Columbia Univ, MD, 41; Am Bd Radiol, dipl. *Prof Exp:* Asst radiotherapist, Mt Sinai Hosp, New York, 47-49; attend radiologist, 49-55, dir radiol, 55-75, sr attending radiologist, 75-87, CONSULT RADIOLOGIST, BETH ISRAEL MED CTR, 87-; PROF CLIN RADIOL, MT SINAI SCH MED, 68- *Mem:* Am Roentgen Ray Soc; Radiol Soc NAm; fel Am Col Radiol; NY Acad Med; AMA. *Mailing Add:* Beth Israel Med Ctr 10 Nathan D Perlman Pl New York NY 10003

GEFFEN, T(HEODORE) M(ORTON), b Calgary, Alta, Can, Feb 22, 22; nat US; m 85; c 3. PETROLEUM ENGINEERING. *Educ:* Univ Okla, BS, 43. *Prof Exp:* Res eng, Petrol & Natural Gas Conserv Bd, Alta, 44-45; petrol engr, Calif Stand Co, 45-46; jr engr, Pan Am Petrol Corp, 46-47, jr res engr, 47-49, res engr, 49-51, tech group leader, 51-53, tech group supvr, 53-55, res group supvr, 55-58, res sect supvr, 58-70; res sect mgr, Amoco Prod Co, 70-76, res consult, 76-81; PETROL CONSULT, 81- *Concurrent Pos:* Chmn, Oil Recovery Tech Domain Comt, Am Petrol Inst, 63-64 & Am Petrol Inst-Govt Res Liaison Comt, 64-65; mem bd dirs, Eng Socs Comn Energy, 81-83. *Honors & Awards:* John Franklin Carll Award, Soc Petrol Engrs, 74, Enhanced Oil Recovery Pioneer, 84. *Mem:* Am Inst Mining, Metall & Petrol Engrs; distinguished mem Soc Petrol Engrs; Am Asn Petrol Geologists. *Res:* Oil recovery; multiphase fluid flow in porous material; reservoir engineering; enhanced oil recovery. *Mailing Add:* 3314 E 51st St Tulsa OK 74135

GEFTER, MALCOLM LAWRENCE, b New York, NY, Mar 16, 42. MOLECULAR BIOLOGY, BIOCHEMISTRY. *Educ:* Univ Md, BS, 63; Albert Einstein Col Med, PhD(molecular biol), 67. *Prof Exp:* Asst prof biol sci, Columbia Univ, 69-72; assoc prof, 72-76, PROF BIOL, MASS INST TECHNOL, 76-, EXEC OFFICER, DEPT BIOL. *Honors & Awards:* Pfizer Award, Am Chem Soc, 75. *Mem:* Am Soc Biol Chemists. *Res:* Molecular biology of nucleic acids, including enzymology and genetics of DNA replication and RNA metabolism; genetics and biochemistry of cellular differentiation. *Mailing Add:* Dept Biochem Rm 56-705 Cambridge MA 02139

GEFTER, WILLIAM IRVIN, b Philadelphia, Pa, Jan 29, 15; m 39; c 4. INTERNAL MEDICINE. *Educ:* Univ Pa, AB, 35, MD, 39; Am Bd Internal Med, dipl, 47. *Prof Exp:* Intern, Philadelphia Gen Hosp, 39-41, resident med, 41-43; from instr to William J Mullen prof med, Woman's Med Col Pa, 43-66, pres staff, Hosp, 58-60, chief med, Col Div, Philadelphia Gen Hosp, 59-66; prof med, Sch Med, Temple Univ & dir dept med, Episcopal Hosp, Philadelphia, 66-74; DIR MED EDUC, ST JOSEPH HOSP, 74-; CLIN PROF MED, NY MED COL, 75- *Concurrent Pos:* Consult, Vet Admin Hosp, Philadelphia, 53-66; pres staff, Episcopal Hosp, Philadelphia, 70-72. *Mem:* Fel AMA; fel Am Col Physicians; fel Am Col Cardiol. *Res:* Clinical cardiology; internal medicine; medical education and administration. *Mailing Add:* 167 W Lane Revonah Woods Stamford CT 06905

GEGEL, HAROLD L(OUIS), b St Louis, Mo, Jan 30, 33; m 57; c 3. PHYSICAL METALLURGY. *Educ:* Univ Ill, Urbana, BS, 55; Ohio State Univ, MS, 62, PhD(metall eng), 65. *Prof Exp:* Develop engr, Frigidaire Div, Gen Motors Corp, 55-56; proj engr, 56-59, res metallurgist, Air Force Mat Lab, 59-88; DIR PROCESSING, UNIVERSAL ENERGY SYSTS (UES) AT BEAVER CREEK, 88- *Concurrent Pos:* Prof, Dept Mat Sci & Metall, Univ Cincinnati, 68- *Mem:* Am Soc Metals; Am Inst Mining, Metall & Petrol Engrs; Sigma Xi. *Res:* Alloy design; thermodynamic and electronic factors which control the phase stability of alpha and beta titanium alloys and the correlation of these factors with the important mechanical properties. *Mailing Add:* 4482 Blairgowrie Circle Dayton OH 45429

GEGENHEIMER, PETER ALBERT, b Tucson, Ariz, Mar 6, 50. NUCLEIC ACID BIOCHEMISTRY, RNA PROCESSING. *Educ:* Yale Univ, BS, 72; Wash Univ, St Louis, PhD(molecular biol), 79. *Prof Exp:* Res fel biochem & molecular biol, Univ Calif, San Diego, 80-83; vis res assoc, Univ Colo, Boulder, 83-85; ASST PROF BIOCHEM & BOT, UNIV KANSAS, LAWRENCE, 85- *Mem:* AAAS; Am Soc Biol Chemists; Int Soc Plant Molecular Biol. *Res:* Cellular roles and biochemical mechanisms of RNA processing reactions in bacteria, yeast, plant organelles and plant nuclei; RNA-catalysed reactions in organelles. *Mailing Add:* Dept Biochem & Bot Univ Kans 3038 Haworth Hall Lawrence KS 66045-2106

GEHA, ALEXANDER SALIM, b Beirut, Lebanon, June 18, 36, US citizen; m 67; c 3. CARDIOTHORACIC SURGERY, CARDIOPULMONARY PHYSIOLOGY. *Educ:* Am Univ Beirut, BS, 55, MD, 59; Univ Minn, MS, 67; Yale Univ, MA, 78. *Prof Exp:* Asst prof cardiothoracic surg, Univ Vt, Col Med, 67-69; from asst prof to assoc prof, Wash Univ, Sch Med, 69-75; from assoc prof to prof & chief, Yale Univ, Sch Med, 75-86; PROF & DIR CARDIOTHORACIC SURG, CASE WESTERN RESERVE UNIV, SCH MED, 86-, CHIEF, CARDIOTHORACIC SURG, UNIV HOSP CLEVELAND. *Concurrent Pos:* Consult cardiothoracic surg, Vet Admin Hosp, WHaven, Conn, 75-86, Waterbury Hosp, Conn, 76-86, Cleveland Vet Admin Hosp, 86-, Cleveland Metrop Gen Hosp, 86-, Mt Sinai Hosp, Cleveland, 90-; mem, Study Sect Surg, Anesthesiol & Trauma, NIH, 81-85; mem, bd dir, Conn Heart Asn, 81-86. *Mem:* Soc Thoracic Surg; Am Asn Thoracic Surg; Am Surg Asn; Am Col Surgeons; Soc Univ Surgeons; Soc Vascular Surg; Sigma Xi; Europ Surg Res Soc. *Res:* Clinical cardiothoracic surgery and cardiovascular research into cardiac hypertrophy and its physiologic effects and characteristics; cardiac and pulmonary transplantation and cardiac preservation and protection, both short-term and long term; cardiac neurophysiology. *Mailing Add:* Div Cardiothoracic Surg Case Western Reserve Univ & Univ Hosp Cleveland Cleveland OH 44106

GEHA, RAIF S, b Bechmezzine, Lebanon, Oct 12, 45; m 70; c 3. IMMUNOLOGY, ALLERGY. *Educ:* Am Univ, Beirut, BSc, 65, MD, 69. *Prof Exp:* Asst prof pediat, Am Univ, Beirut, 74-76; asst prof, 76-81, ASSOC PROF PEDIAT, HARVARD MED SCH, 81- *Mem:* Soc Pediat Res; Am Asn Immunologist; Am Thoracic Soc; Collegium Int Allergologicum. *Res:* Human TB cell interactions. *Mailing Add:* Dept Pediat Children's Hosp 300 Longwood Ave Boston MA 02115

GEHAN, EDMUND A, b Brooklyn, NY, Sept 2, 29; m 62; c 5. STATISTICS, CANCER. *Educ:* Manhattan Col, BA, 51; NC State Univ, MS, 53, PhD(exp statist), 57. *Prof Exp:* From instr to asst prof biostatist, Univ NC, 55-58; math statistician, Biomet Br, Nat Cancer Inst, 58-59, actg head, Biomet Sect, 59-61, head sect, 61-62, math statistician, 64-67; PROF BIOMET, UNIV TEX M D ANDERSON CANCER CTR, HOUSTON, 67- *Concurrent Pos:* Nat Cancer Inst spec fel statist, Birkbeck Col, Univ London, 62-64; mem cancer clin invest rev comt, Nat Cancer Inst, 71-74; fac develop leave, dept biomath & biophys, Univ Pierre & Marie-Curie, Paris, 81; Panelist, Consensus Dev Conf Limb Sparing Treat of Adult Soft Tissue Sarcomas & Osteosarconias, NIH, 84. *Honors & Awards:* Jeffrey A Gottleib Mem Award, 83. *Mem:* Biomet Soc (pres-elect, Eastern NAm Region, 71, pres, 72); fel Am Statist Asn; Am Soc Clin Oncol; Int Statist Inst; Soc For Clin Trials; Int Soc Clin Biostats (pres, 87-88). *Res:* Clinical trials in cancer research; statistical methodology with applications in cancer research. *Mailing Add:* Dept Biomath M D Anderson Cancer Ctr Univ Tex Houston TX 77030

GEHLBACH, FREDERICK RENNER, b Steubenville, Ohio, July 5, 35; m 60; c 2. ECOLOGY. *Educ:* Cornell Univ, AB, 57, MS, 59; Univ Mich, PhD(zool, conserv), 63. *Prof Exp:* From asst prof to assoc prof biol, 63-79, PROF BIOL & ENVIRON STUDIES, BAYLOR UNIV, 79- *Concurrent Pos:* Collabr, Nat Park Serv, 59-; Guggenheim fel, 70-71. *Honors & Awards:* Samuel T Dana Award, Univ Mich, 63. *Mem:* Am Soc Ichthyol & Herpet; Ecol Soc Am; Am Ornithol Union. *Res:* Ecology and behavior of southwestern vertebrates; conservation of natural communities and species. *Mailing Add:* Dept Biol Baylor Univ PO Box 6367 Waco TX 76798

GEHMAN, BRUCE LAWRENCE, b Akron, Ohio, Oct 22, 37; m 60; c 3. SOLID STATE PHYSICS, METALLURGY. *Educ:* Univ Mich, BS, 59; San Diego State Col, MS, 65; Univ Calif, San Diego, MS, 69, PhD(physics), 73. *Prof Exp:* Flight test engr instrumentation, Gen Dynamics Corp, 59-61; staff assoc physics, Gulf Energy & Environ Syst, 61-65; res asst, Univ Calif, San Diego, 65-71; mgr res & develop, Cominco Electronic Mat, 71-86, tech dir, deposition technol, 86-88; vpres bus develop, ISM Technologies, 88-89; TECH DIR, LEYBOLD MAT INC, 89- *Mem:* Am Soc Testing & Mat; Am Phys Soc; Am Soc Metals; Inst Elec & Electronic Engrs; Int Soc Hybrid Microelectronics. *Res:* Transmission electron spin resonance research in dilute localized magnetic moment systems; specialized metallurgical products for the electronics industry. *Mailing Add:* Leybold Mat Inc 16035 Vineyard Blvd Morgan Hill CA 95037

GEHO, WALTER BLAIR, b Wheeling, WVa, May 18, 39; m 62; c 5. PHARMACOLOGY. *Educ:* Bethany Col, BS, 60; Western Reserve Univ, PhD(pharmacol), 64, MD, 66. *Prof Exp:* Sr instr pharmacol, Case Western Reserve Univ, 66-67; sect head pharmaceut res, Procter & Gamble Co, 67-81; vpres & dir res, 81-89, PRES, TECHNOL UNLTD INC, 89- *Res:* Pharmacology of diphosphonates and metabolic bone diseases; targeted drug delivery, liposomes, hepatic directed insulin and the control of glucose metabolism. *Mailing Add:* Technol Unltd 146 S Bever St PO Box 723 Wooster OH 44691

GEHRELS, TOM, b Neth, Feb 21, 25; nat US; m 51; c 3. PLANETARY ASTRONOMY. *Educ:* Univ Leiden, BSc, 51; Univ Chicago, PhD(astron), 56. *Prof Exp:* Res assoc astron, Ind Univ, 56-61; assoc prof, 61-67, PROF, LUNAR & PLANETARY LAB, UNIV ARIZ, 67- *Concurrent Pos:* res assoc astron, Univ Chicago, 59-61; V A Sarabhai prof & hon fel, Phys Res Lab, Ahmedabad, India, 78-; gen ed, Space Sci Ser, 82-; corresp mem, Bataafsch Genootschap der Proefondervindelijke Wijsbegeerte, 84- *Honors & Awards:* Medal Except Sci Achievement, NASA, 74. *Mem:* Am Astron Soc; Int Astron Union; hon mem Hiroshima Astron Soc. *Res:* Minor planets; sky surveying; author of numerous publications regarding scannerscopy, ballooning and space programs, polarimetry, and minor planets and related objects. *Mailing Add:* Lunar & Planetary Lab Univ Ariz Tucson AZ 85721

GEHRENBECK, RICHARD KEITH, b St Paul, Minn, June 8, 34; m 58; c 3. HISTORY OF PHYSICS. *Educ:* Macalester Col, BA, 56; Univ Minn, MA, 66, PhD(hist of physics), 73. *Prof Exp:* Instr physics & math, Gerard Inst for Boys, Sidon, Lebanon, 57-60; from instr to asst prof physics, Park Col, 62-69; lab coordr physics, Univ Minn, 70-71; from asst prof to assoc prof, 72-88, PROF PHYS SCI, RHODE ISLAND COL, 88. *Concurrent Pos:* Area rep, Barnard-Columbia Hist of Physics Lab, 74-78. *Mem:* Am Asn Physics Teachers; Hist of Sci Soc; Sigma Xi. *Res:* History of physics; history of American science; science and society; history of astronomy. *Mailing Add:* 972 Smith St Providence RI 02908

GEHRI, DENNIS CLARK, b Beloit, Wis, Jan 28, 37; m 61; c 4. ENVIRONMENTAL CHEMISTRY. *Educ:* Univ Wis, BS, 59, PhD(phys chem), 68. *Prof Exp:* MEM SR STAFF, ATOMICS INT DIV, ROCKWELL INT CORP, 67- *Mem:* Am Chem Soc; Sigma Xi; Air Pollution Control Asn. *Res:* Experimental thermodynamics and molecular vibrations of gases; sodium chemistry, particularly interactions of carbon with liquid solium; air pollution control of stationary source and automotive emissions. *Mailing Add:* 5616 Mainmast Pl Agoura CA 91301

GEHRIG, JOHN D, b Watson, Minn, Feb 6, 24; m 53; c 7. ORAL SURGERY, ANATOMY. *Educ:* Univ Minn, DDS, 46, MS, 51; Am Bd Oral Surg, dipl. *Prof Exp:* Prof oral surg, Sch Dent, Univ Kansas City, 52-54; assoc prof, 54-67, chmn, Dept Oral Surg, 56-70, PROF ORAL SURG, SCH DENT, UNIV WASH, 67-, DIR, GRAD ORAL SURG TRAINING PROG, 70-, DIR ORAL SURG, UNDERGRAD CLIN, 72-, PROF, BIOL STRUCT DEPT, SCH MED, 88- *Concurrent Pos:* Attend, Children's Orthop Hosp, Providence Hosp, Vet Admin Hosp, Seattle & Vet Admin Hosp, Am Lake, Univ Hosp, Univ Wash, Harborview Hosp Med Ctr; consult, Madigan Gen Hosp, Tacoma & US Army Hq Dent Detachment, Ft Lewis; adj prof, Biol Struct Dept, Sch Med, Univ Wash, 75- *Mem:* Am Soc Oral Surg; Am Soc Oral Maxillofacial Surgeons; Am Asn Anatomists; International Asn Study Pain. *Res:* Inferior alveolar nerve regeneration on dogs through a Millipore filter sleeve; comparison of ATb sensitivities with ATb choice; blood and fluid volume changes in oral surgery; evaluation of concurrent methods of dental anesthesia; pain, chronic pain, dental sensory receptor (electron microscopy); vibratory analgesia for local anesthetic administration; topographic brain measures of human pain and pain responsivity; computerized imaging of human skull with measurements; human pain responsivity in a tonic pain model; psychological determinants. *Mailing Add:* Dept Biol Struct Univ Wash Seattle WA 98195

GEHRIG, ROBERT FRANK, b Manitowoc, Wis, Jan 3, 28; m 54; c 4. MICROBIOLOGY. *Educ:* Univ Wis, BS, 51, MS, 58, PhD(bact), 61. *Prof Exp:* Asst enzyme chem, Merck Inst Therapeut Res, NJ, 51-53; from asst prof to assoc prof, 62-72, chmn dept, 70-73 & 76-78& 84-87, PROF BIOL, RUSSELL SAGE COL, 72- *Mem:* AAAS; Am Soc Microbiol; Am Soc Parasitol. *Res:* Fatty acid metabolism of the filamentous fungi; physiology of penicillia and aspergilli. *Mailing Add:* Dept Biol Russell Sage Col Troy NY 12180

GEHRING, FREDERICK WILLIAM, b Ann Arbor, Mich, Aug 7, 25; m 53; c 2. MATHEMATICS. *Educ:* Univ Mich, BSE(elec eng) & BSE(math), 46, MA, 49; Cambridge Univ, PhD(math), 52, ScD, 76. *Hon Degrees:* PhD, Helsinki Univ, 77, Univ of Jyvaskyla. *Prof Exp:* Benjamin Pierce instr math, Harvard Univ, 52-55; from asst prof to assoc prof, 56-84, chmn dept, 73-75 & 77-84, T H HILDEBRANDT PROF MATH, UNIV MICH, 84- *Concurrent Pos:* Guggenheim & Fulbright fel, 58-59; NSF fel, 59-60; analytical ed, Duke Math J, 63-80; ed, D Van Nostrand Publ Co, 63-69, North Holland Publ Co, 70- & Springer-Verlag, 74-; vis prof, Stanford Univ, 64, Harvard Univ, 64-65, Univ Minn, 71, Mittag Leffler Inst, Sweden, 72 & 90 & Acad Finland, 89-; sr vis fel, UK Sci Res Coun, 81; Alexander von Humboldt Award, 81-82. *Honors & Awards:* Order White Rose, Finland, 86. *Mem:* Am Math Soc; Math Asn Am; Swiss Math Soc; Finnish Math Soc; Asn Women Math. *Res:* Analysis. *Mailing Add:* 2139 Melrose Rd Ann Arbor MI 48104

GEHRING, HARVEY THOMAS, b Chicago, Ill, Oct 27, 11; m 37; c 2. ORGANIC POLYMER CHEMISTRY. *Educ:* Univ Ill, BS, 34. *Prof Exp:* From asst dir to dir emulsion res, Sherwin Williams Co, 35-73; RETIRED. *Mem:* Am Chem Soc. *Res:* Emulsion and water thinned coatings. *Mailing Add:* 1164 E 169th St South Holland IL 60473

GEHRING, PERRY JAMES, b Yankton, SDak, Mar 15, 36; m 59; c 4. TOXICOLOGY, PHARMACOLOGY. *Educ:* Univ Minn, Minneapolis, BS & DVM, 60, PhD(pharmacol), 65. *Prof Exp:* Res assoc toxicol, Iowa State Univ, 60-61; USPHS fel pharmacol, Univ Minn, Minneapolis, 61-65; pharmacol toxicologist, Dow Chem Co, 65-68; assoc prof pharmacol, Mich State Univ, 68-70; from asst dir to dir toxicol, Dow Chem Co, 70-80, dir life sci, 80-81, vpres agr chem res & develop & dir health & environ res, 81-89;

VPRES, RES & DEVELOP, DOWELANCO. *Concurrent Pos:* Vis prof pharmacol, Mich State Univ, 74-; nonres lectr, Univ Mich Sch Pub Health, Ann Arbor. *Honors & Awards:* Founder's Award, Chem Indust Inst Toxicol, 83; Merit Award, Soc Toxicol, 83. *Mem:* AAAS; Int Union Toxicol(pres,86-89); Soc Toxicol(pres,80 & 81); Am Soc Pharmacol & Exp Therapeut. *Res:* Toxidynamics; chemical cataractogenesis; metal toxicology; pharmacokinetics; risk assessment. *Mailing Add:* DowElanco 9001 Purdue Rd Indianapolis IN 46268

GEHRIS, CLARENCE WINFRED, b Fleetwood, Pa, Oct 25, 17; m 41; c 2. BOTANY, PLANT ECOLOGY. *Educ:* Temple Univ, BS, 38, MEd, 45; Pa State Univ, DEd(biol sci), 64. *Prof Exp:* Instr pub sch, Pa, 45-62; assoc prof, 62-71, State Univ NY, Brockport, prof biol, 71-; AT DEPT LARYNGOLOGY OTOLOGY, JOHNS HOPKINS UNIV. *Concurrent Pos:* AAAS-Nat Sci Found fel, 59-61. *Mem:* Ecol Soc Am; Am Bot Soc; AAAS. *Res:* Bog ecology; peat development; pollen analysis. *Mailing Add:* 2112 Belair Rd Fallston MD 21047

GEHRKE, CHARLES WILLIAM, b New York, NY, July 18, 17; m 41; c 3. BIOCHEMISTRY. *Educ:* Ohio State Univ, BA, 39, MSc & BSc, 41, PhD(agr biochem), 47. *Prof Exp:* Asst bacteriologist, Ohio State Univ, 41; food & dairy inspector, Univ Ohio, 41-42; actg prof chem, Mo Valley Col, 42-43, prof & head dept, 43-45; instr, Ohio State Univ, 45-46; prof chem & head dept, Mo Valley Col, 46-49; assoc prof, 49-53, PROF BIOCHEM & MGR EXP STA CHEM LABS, COL AGR, UNIV MO, COLUMBIA, 54- *Concurrent Pos:* Co-investr, Apollo 11, 12, 14, 15, 16 & 17 lunar samples; mem, Am Chem Soc Adv Bd Chem Abstracts, 76; Mid Am State Univ Hon Lectr, 78; Sigma Xi res award, Univ Mo, 80. *Honors & Awards:* Harvey W Wiley Award, Asn Off Anal Chemists, 71; Chromatography Mem Medal, Sci Coun Chromatography, Acad Sci USSR, Moscow, 80. *Mem:* Am Chem Soc; NY Acad Sci; Int Soc Origin Life; fel Asn Off Anal Chemists; Int Asn Off Anal Chemists (pres, 84). *Res:* Physical biochemistry of proteins; gas chromatography of amino acids, genetic molecules and biological substances; analytical methods development. *Mailing Add:* 708 Edgewood Ave Columbia MO 65203

GEHRKE, HENRY, b Salina, Kans, May 11, 36; m 56; c 4. INORGANIC CHEMISTRY. *Educ:* Okla State Univ, BS, 58; Univ Iowa, MS, 62, PhD(inorg & org chem), 64. *Prof Exp:* From asst prof to assoc prof, 64-72, PROF CHEM, SDAK STATE UNIV, 72- *Mem:* AAAS; Am Chem Soc; Sigma Xi. *Res:* Coordination chemistry of rarer elements; bioinorganic chemistry of molybdenum and rhenium. *Mailing Add:* Dept Chem SDak State Univ Brookings SD 57007

GEHRKE, ROBERT JAMES, b Chicago, Ill, Nov 20, 40; m 63; c 3. NUCLEAR RADIATION METROLOGY, TECHNICAL MANAGEMENT. *Educ:* DePaul Univ, BS, 62; Univ Nev, Reno, MS, 66. *Prof Exp:* Physicist, Phillips Petrol Co-Idaho Nuclear Corp, Nat Reactor Testing Sta, 65-69, sr physicist, Aerojet Nuclear Co, 69-73; assoc scientist, Idaho Nat Eng Lab, EG&G Idaho, Inc, 73-79, sr scientist, 79-83, sci specialist, 83-85, unit mgr, 85-90, SCI SPECIALIST, IDAHO NAT ENG LAB, EG&G IDAHO, INC, 90- *Concurrent Pos:* Contrib ed, Radioactiv & Radiochem, J Appl Measurements. *Mem:* Am Phys Soc; Am Chem Soc; Am Nuclear Soc; Health Physics Soc. *Res:* Applied x-ray and gamma-ray spectroscopy, nuclear radiation metrology, applied radiation measurements, environmental radioanalysis, nuclear structure studies and radiation measurement instrumentation.; Technical management of a radiation measurement laboratory. *Mailing Add:* EG&G Idaho Inc PO Box 1625 Idaho Falls ID 83415

GEHRKE, WILLARD H, b Belmont, Wis, Jan 21, 20; m 44; c 3. CHEMICAL ENGINEERING. *Educ:* Univ Wis, BS, 42. *Prof Exp:* Chem engr, Monsanto Chem Co, Ohio, 42-47; process engr mfg eng, Marathon Div, Am Can Co, 47-51, supvr, 51, cent mfg eng, 51-53 & process eng, 53-57, mgr, 57-61, asst to vpres res & develop div, 61-62, dir paper prod res & develop, 62-64, dir rigid container res & develop, 64-66, dir fabric bus develop, 66-68; V PRES CO DEVELOP, CURWOOD INC, 68- *Concurrent Pos:* Mem, Indust Res Inst. *Mem:* Am Chem Soc; Tech Asn Pulp & Paper Indust. *Res:* Protective packaging films for foods and other products made from polyester, polyolefins, nylon, vinyl and aluminum foil. *Mailing Add:* 21 Meadowbrook Ct Appleton WI 54914

GEHRMANN, JOHN EDWARD, b Hoboken, NJ, June 25, 41; m 66; c 2. PHARMACOLOGY. *Educ:* Queens Col, NY, BS, 63; Stanford Univ, PhD(pharmacol), 69. *Prof Exp:* Res assoc psychopharmacol, Brain Res Labs, New York Med Col, 69-70; ASSOC RES PHARMACOLOGIST, UNIV CALIF, DAVIS, 70- *Mem:* AAAS; Am Chem Soc; NY Acad Sci; Soc Neurosci. *Res:* Computer based quantitative evaluation of the electroencephalogram; correlations with changes induced by conditioned behavioral responses and drugs acting on the central nervous system; electrophysiological assessment of acute spinal cord injury. *Mailing Add:* 738 Anderson Rd Davis CA 95616

GEHRS, CARL WILLIAM, b Elmhurst, Ill, May 4, 41; m 64; c 3. ECOLOGY, ENVIRONMENTAL TOXICOLOGY. *Educ:* Concordia Teacher's Col, BS, 63; Kans State Teacher's Col, MS, 67; Univ Okla, PhD(aquatic ecol), 72. *Prof Exp:* Biol teacher, Luther High Sch, South Chicago, Ill, 64-69, dean students, 67-69; life sci synthetic fuels prog coordr, 75-77, RES ECOLOGIST AQUATIC EFFECTS, ENVIRON SCI DIV, OAK RIDGE NAT LAB, 72-, PROG MGR ADV, FOSSIL ENERGY PROG, 75- *Mem:* AAAS; Ecol Soc Am; Sigma Xi; Soc Environ Toxicol & Chem; Soc Int Limnologists. *Res:* Environmental health concerns related to increased coal utilization; aquatic ecology; zooplantion population dynamics; environmental transport, fate and effects of organic contaminants; translation from laboratory to field; risk analysis. *Mailing Add:* 36 Montclair Rd Oak Ridge TN 37830

GEHRZ, ROBERT DOUGLAS, b Evanston, Ill, Dec 28, 44; m 70; c 2. ASTROPHYSICS. *Educ:* Univ Minn, BA, 67, PhD(physics), 71. *Prof Exp:* Res assoc physics, Univ Minn, 71-72; from asst prof to prof physics & astron, Univ Wyo, 72-85; PROF PHYSICS & ASTRON, UNIV MINN, 85-; DIR, MT LEMMON OBSERVING FACIL, 88- *Concurrent Pos:* Vis astronr, Kitt Peak Nat Observ, 69-; consult, Martin Marietta Aerospace, Colo, 74; dir-at-large, Aura, Inc, 76-; mem, NASA first team & NSF astron adv comt, 77-; mem, NASA dost team, 78-, Astron Surv Comt, Nat Acad Sci, 78-81; chmn, NSF Optical/Infrared subcomt, 78-; chmn, Observ Vis Comt, Asn Univ Res Astron, 82-84; adj prof physics & Astron, Univ Wyo, 85-; mem NSF adv comt, Astron, 76-79, 87-; NAS comt for planetary & lunar exploration; counr, Am Astron Soc, 86-89. *Mem:* Am Astron Soc; Royal Astron Soc; Sigma Xi; Int Astron Union; fel Explorers Club; AAAS. *Res:* Infrared astronomy; development of infrared instrumentation; construction of a 92 inch infrared telescope. *Mailing Add:* Dept Astron Univ Minn 116 Church St SE Minneapolis MN 55455

GEIBEL, JON FREDERICK, b White Plains, NY, June 3, 50; m 76. REACTION KINETICS. *Educ:* Harvey Mudd Col, BS, 72; Univ Calif, San Diego, MS, 74, PhD(chem), 76. *Prof Exp:* Scholar, Dept Med, Univ Calif, San Diego, 76-77; mem res staff high performance epoxy resins, Eng Res Ctr, Western Elec Co, 77-81; res chemist, High Temperature Thermoplastics, 81-85, SUPVR, ENG PLASTICS RES SECT, PHILLIPS RES CTR, 85- *Mem:* Am Chem Soc; Soc Plastics Eng. *Res:* Hemoglobin model compounds; structure-property relationships in aromatic polysulfide thermoplastics; material processing in microgravity. *Mailing Add:* 2600 Mountain Rd Bartlesville OK 74003

GEIDEL, GWENDELYN, b Lowville, NY, May 4, 53; m 76; c 2. ENVIRONMENTAL GEOCHEMISTRY. *Educ:* Univ SC, BS, 74, MS, 76, PhD, 82, JD, 89. *Prof Exp:* Res technologist I, Univ SC, 76-77, res technologist II, 77-80, res specialist I geochem, 80-82, res asst prof, 82-87; GEOLOGIST, 87- *Concurrent Pos:* Co-prin investr, Environ Protection Agency grant, Univ SC, 77-81; prin investr, US Bur Mines grant, Univ SC, 80-85, US Off Surface Mining & Develop Res Grants, 81-86, Hungarian Groundwater workshop, Environ Law Inst, 90-91; atty, 89- *Mem:* Sigma Xi; Can Land Reclamation Asn; AAAS; Am Soc Surface Mining & Reclamation; Int Asn Geochem & Cosmochem. *Res:* Environmental hydrogeological aspects of surface coal mining with primary emphasis on the prediction of coal mine drainage quality and ameliorative techniques and environmental legal issues. *Mailing Add:* 3823 Edinburgh Rd Columbia SC 29204

GEIDUSCHEK, ERNEST PETER, b Vienna, Austria, Apr 11, 28; nat US; m 55; c 2. MOLECULAR BIOLOGY, VIROLOGY. *Educ:* Columbia Univ, AB, 48; Harvard Univ, AM, 50, PhD(chem), 52. *Prof Exp:* Instr, Yale Univ, 52-53 & 55-57; asst prof chem, Univ Mich, 57-59; from asst prof to prof biophys, Univ Chicago, 59-70; chmn biol, 81-83, PROF BIOL, UNIV CALIF, SAN DIEGO, 70- *Concurrent Pos:* Guggenheim Found fel, 64-65; res career develop award, USPHS, 62-70; lectr, Europ Molecular Biol orgn, 77. *Honors & Awards:* Hilleman Lectr, Univ Chicago, 78. *Mem:* Nat Acad Sci; Am Soc Biol Chem; fel AAAS; fel Am Acad Arts & Sci; Am Soc Microbiol; Am Soc Virol. *Res:* Macromolecular structure; synthesis and function of nucleic acids; physico-chemical methods applicable to large molecules; synthesis and function of nucleic acids; genetic regulation; virus development; biochemistry. *Mailing Add:* Dept Biol & Ctr Molecular Genet Univ Calif San Diego 9500 Gilman Dr La Jolla CA 92093-0634

GEIGER, BENJAMIN, b Haifa, Israel, Jan 29, 47; m; c 3. CYTOLOGY, EMBRYOLOGY. *Educ:* Hebrew Univ, MSc, 72; Weizmann Inst Sci, PhD(chem immunol), 77. *Prof Exp:* Postdoctoral fel, Univ Calif, San Diego, 77-79, post-grad res assoc, 79; postdoctoral fel, Dept Chem Immunol, Weizmann Inst Sci, 79-80, sr scientist, 80-82, assoc prof, 83-88, PROF, DEPT CHEM IMMUNOL, WEIZMANN INST SCI, 88-; DEAN, FEINBERG GRAD SCH, 89- *Honors & Awards:* John F Kennedy Mem Prize, 77; Levinson Prize, 83; Fedn Europ Biochem Socs Prize, 84. *Mem:* Am Soc Cell Biol; Israeli Soc Biochem; Israeli Soc Cell Biol. *Res:* Molecular basis for cell communication and the involvement of the cytoskeleton in cellular interactions. *Mailing Add:* Dept Chem Immunol Weizmann Inst Sci Rehovot Israel

GEIGER, DAVID KENNETH, b Canton, Ohio, Apr 14, 56; m 81; c 3. BIOINORGANIC & STRUCTURAL CHEMISTRY. *Educ:* Franciscan Univ, BA, 78; Univ Notre Dame, PhD(inorg chem), 83. *Prof Exp:* Res assoc, dept energy, Notre Dame Radiation Lab, 83-85; ASST PROF CHEM, COL ARTS & SCI, STATE UNIV NY, GENESEO, 85- *Mem:* Am Chem Soc; Sigma Xi. *Res:* Structure-function relationships in coordination compounds which model the active site in metallo proteins, particularly iron porphyrins. *Mailing Add:* Dept Chem State Univ NY Geneseo NY 14454

GEIGER, DONALD R, b Dayton, Ohio, Feb 27, 33. PLANT PHYSIOLOGY. *Educ:* Univ Dayton, BS, 55; Ohio State Univ, MSc, 60, PhD(bot), 63. *Prof Exp:* Teacher, Cathedral Latin Sch, 55-60; from asst prof to assoc prof, 64-72, PROF BIOL, UNIV DAYTON, 72- *Concurrent Pos:* Prin investr, NSF res grant, 64-67 & 72-; res grant, US AEC, 68-71; prog dir strategies for responsible develop, 74-81. *Mem:* AAAS; Am Soc Plant Physiol; Can Soc Plant Physiol; Australian Soc Plant Physiol. *Res:* Translocation of sugar in higher plants using radioisotope tracer techniques; policy analysis for world development. *Mailing Add:* Dept Biol Univ Dayton Dayton OH 45469-2320

GEIGER, EDWIN OTTO, b Chicago, Ill, Apr 24, 39; m 63; c 4. BIOCHEMISTRY. *Educ:* DePaul Univ, BS, 61; Loyola Univ, MS, 64, PhD(biochem), 66, MBA, 75. *Prof Exp:* Fel biochem, Case Western Reserve Univ, 65-67; sr researcher, CPC Int, 67-75, team leader, 75-76, sect leader, 76-77; mgr corn res, Amstar Corp, 77-85; mgr microbiol devel, BW Biotec Inc, 85-86; SR SCIENTIST SEAGRAMS, 86- *Honors & Awards:* Medal, Am Inst Chem, 61. *Mem:* Am Chem Soc; Sigma Xi. *Res:* Enzyme chemistry; fermentation development; bound enzymes; process development. *Mailing Add:* 1 Maplewood Dr New Milford CT 06776-3830

GEIGER, GENE E(DWARD), b Pittsburgh, Pa, Oct 27, 28; m 59; c 2. MECHANICAL ENGINEERING. *Educ:* Carnegie Inst Technol, BSME, 50; Univ Pittsburgh, MSME, 55, PhD(mech eng), 64. *Prof Exp:* Res asst residual stress, Mellon Inst, 51; from instr to asst prof mech eng, Univ Pittsburgh, 51-64, assoc res prof, 64-66, assoc prof, 66-71, coordr grad prog, 66-83, PROF MECH ENG, UNIV PITTSBURGH, 71- *Concurrent Pos:* Consult, Pa Pub Utility Comn, 64, gas companies West Pa, 65-68, knowledge availability syst, Univ Pittsburgh, 65-, Anvil Prod, Inc, 67-68 & Mine Safety Appliances, Inc, 70-75 & 80-81; vis fel, Mellon Inst, 65-69; consult, Berner Int Corp, 75-81, Lectromelt, 79, Sunbeam Corp, 79-80, City of Pittsburgh, 81-90, Schneider Consult Engr, 85-86, Emec Consults, 89- *Mem:* Am Soc Eng Educ; Am Soc Mech Engrs; Nat Soc Prof Engrs. *Res:* Heat transfer; fluid dynamics and power; thermodynamics; energy conservation. *Mailing Add:* Dept Mech Eng Univ Pittsburgh Pittsburgh PA 15261

GEIGER, GORDON HAROLD, b Chicago, Ill, Apr 21, 37; m 60; c 2. METALLURGY, MATERIALS SCIENCE. *Educ:* Yale Univ, BE, 59; Northwestern Univ, MS, 61, PhD(mat sci), 64. *Prof Exp:* Process engr, Allis-Chalmers Mfg Co, 60-61; res asst, Northwestern Univ, 61-63; res engr, Jones & Laughlin Steel Corp, 64-65; from asst prof to assoc prof metall eng, Univ Wis-Madison, 65-68; from assoc prof to prof mat eng, Univ Ill, Chicago Circle, 69-73; prof metall eng & head dept, Univ Ariz, 73-80; sr proc consult, Inland Steel, 80-81; VPRES & TECH DIR, CHASE MANHATTAN BANK, 82- *Concurrent Pos:* Chmn adv comt steel technol, Off Technol Assessment, US Congress, 78-79. *Honors & Awards:* Campbell Award, Nat Asn Corrosion Engrs, 66; Stoughton Award, Am Soc Metals, 72. *Mem:* Am Inst Mining, Metall & Petrol Engrs; Nat Asn Corrosion Engrs; Am Soc Metals; Am Foundrymen's Soc; Am Rwy Eng Asn. *Res:* Thermodynamics and transport phenomena in process metallurgy; operations research as applied to metallurgical processes; steelmaking; copper production processes; mineral processing. *Mailing Add:* 4900 Emerson Ave S Minneapolis MN 55440

GEIGER, H JACK, b New York, NY, Nov 11, 25; m 51. EPIDEMIOLOGY, COMMUNITY MEDICINE. *Educ:* Case Western Reserve Univ, MD, 58; Harvard Univ, MSciHyg, 60. *Prof Exp:* Sci ed, Int News Serv, 49-54; intern med, Sch Med, Harvard Univ, 58-59, NIMH fel Joint Training Prog Soc Sci & Med, Dept Social Rel, 59-61, instr prev med, 61-62, asst med & resident, 62-64, asst prof pub health, Sch Pub Health, 64-65; assoc prev med, Sch Med, Tufts Univ, 65-66, prof, 66-68, prof community health & social med & chmn dept, 68-71; prof, Sch Med, State Univ NY Stony Brook, 71-80; ARTHUR C LOGAN PROF COMMUNITY MED, CITY COL NEW YORK, 80- *Concurrent Pos:* Mem bd gov, Inst Current World Affairs, 61-, chmn, 71; consult res policy comt, Peace Corps, 62-64, Rockefeller Found, 69-70 & Off Secy, US Dept Health, Educ & Welfare, 68-; res fel, Thorndike Lab, Boston City Hosp, 64; proj dir, Tufts Univ-Delta Health Ctr, Mound Bayou, Miss & Tufts Univ, Columbia Point Health Ctr, Boston, 65-71; Milbank Mem Fund fac fel, 66-71; mem, Nat Comn Hunger & Malnutrit US, 67-; mem bd, Nat Comn Inquiry Health Servs Americans, 69-71; mem planning comt, White House Conf Youth, 71-; consult, Inst Med-Nat Acad Sci, 80- *Mem:* Fel AAAS; fel Am Pub Health Asn; Soc Appl Anthrop; Asn Teachers Prev Med; Asn Am Med Cols. *Res:* Social medicine and change; community health; cultural factors in the epidemiology of hypertension, rheumatoid arthritis and schizophrenia. *Mailing Add:* 250 W 89th St New York NY 10024

GEIGER, JAMES STEPHEN, b Kitchener, Ont, Apr 5, 29; m 56; c 4. PHYSICS. *Educ:* McMaster Univ, BSc, 51, MSc, 52; Yale Univ, MS, 53, PhD(physics), 57. *Prof Exp:* Asst res off, 56-68, sr res off, 68-86, BR MGR NUCLEAR PHYSICS, ATOMIC ENERGY OF CAN, LTD, 86 - *Concurrent Pos:* Vis sr scientist, Lab Laser Energetics, Univ Rochester, 76-77. *Mem:* Fel Am Phys Soc; Can Asn Physicists. *Res:* Laser spectroscopy; nuclear physics. *Mailing Add:* PO Box 516 Deep River ON K0J 1P0 Can

GEIGER, JON ROSS, b Philadelphia, Pa, Nov 28, 43; m 66; c 2. GENETIC ENGINEERING, INDUSTRIAL MICROBIOLOGY. *Educ:* Pa State Univ, BS, 65; Univ Conn, PhD(genetics), 76. *Prof Exp:* Teaching asst genetics & biol, Univ Conn, 70-74, NIH cell biol trainee, 74-76, instr, 75; asst prof, genetics & microbiol, Smith Col, 76-80; sr res biologist, 81-83, res assoc biotechnol, 83-90, GROUP LEADER BIOTECHNOL & MICROBIOL, OLIN RES CTR, 90- *Concurrent Pos:* Proj dir, Genetics Prog, Smith Col, 77 & Med & Liberal Arts Prog, 78; vis scientist molecular genetics & prin investr spontaneous mutation res, Mass Inst Technol, 80-81; invited guest ed, Am Biol Teacher, 84. *Mem:* AAAS; Am Soc Microbiol; Nat Asn Biol Teachers; Sigma Xi. *Res:* Use of nucleic acid probes; use of recombinant DNA and transposons in the construction of microorganisms for biodegradation or bioproduction of specialty chemicals; industrial microbiology. *Mailing Add:* Olin Res Ctr PO Box 586 Cheshire CT 06410-0586

GEIGER, KLAUS WILHELM, b Berlin, Germany, Apr 26, 21; Can citizen; m 47; c 2. NUCLEAR PHYSICS. *Prof Exp:* Univ Tubingen, 49; Univ Mainz, PhD(physics), 51. Prof Exp: Res asst nuclear physics, Max Planck Inst Chem, Ger, 49-52; fel, Nat Res Coun Can, 52-54, res officer physics, 54-85; RETIRED. *Concurrent Pos:* Guest worker, Nat Res Coun Can, 85-86. *Mem:* Can Asn Physicists. *Res:* Radioactivity; cosmic rays; neutron standardization; neutron dosimetry; neutron spectra; photon dosimetry; electron dosimetry. *Mailing Add:* 865 Chapman Blvd Ottawa ON K1G 1V1 Can

GEIGER, MARION BRAXTON, industrial chemistry, for more information see previous edition

GEIGER, PAUL FRANK, b Meadville, Pa, July 12, 32; m 62; c 2. PHARMACOLOGY. *Educ:* Univ Tex, BA, 55, BS, 58, MS, 62, PhD(pharmacol), 66. *Prof Exp:* From asst prof to assoc prof, 67-75, PROF PHARMACOL, NORTHEAST LA UNIV, 75- *Mem:* Soc Neurosci; Soc Toxicol. *Res:* Effects of acetylcholine and biogenic amines on cholinesterase on drugs affecting the central nervous system; biochemical seizures; neuropharmacology and toxicology. *Mailing Add:* Div of Pharmacol & Toxicol Northeast La Univ Monroe LA 71209

GEIGER, PAUL JEROME, b Los Angeles, Calif, Jan 12, 30; m 55; c 4. BIOCHEMISTRY. *Educ:* Univ Calif, Berkeley, BS, 51; Johns Hopkins Univ, PhD(biochem), 62. *Prof Exp:* Res engr, Jet Propulsion Lab, Calif Inst Technol, 62, sr scientist, 62-68; res scientist, Biosci Lab, Dow Chem Co, 68-69; NIH res fel pharmacol, 69, asst prof, 69-73, ASSOC PROF PHARMACOL, UNIV SOUTHERN CALIF, 73- *Concurrent Pos:* Consult, Jet Propulsion Lab, Calif Inst Technol, 71. *Mem:* AAAS; Am Chem Soc; Sigma Xi. *Res:* Enzymology; kinetics of dehydrogenases; sulfhydryl structures; extraterrestrial biology and soil science; life detection on planets; detection of organic substances in arid soils; analytical biochemistry; intermediary metabolism of phosphate compounds; biochemical disorders in disease. *Mailing Add:* 1211 Marengo Ave South Pasadena CA 91030

GEIGER, RANDALL L, b Lexington, Nebr, May 17, 49; m 73; c 3. ELECTRICAL ENGINEERING, MATHEMATICAL STATISTICS. *Educ:* Univ Nebr, Lincoln, BS, 72, MS, 73; Colo State Univ, PhD(elec eng), 77. *Prof Exp:* Prof elec eng, Tex A&M Univ, 77-90; PROF ELEC ENG & CHMN, IOWA STATE UNIV, 90- *Concurrent Pos:* Pres, World Instr Inc, 82-85. *Mem:* Fel Inst Elec & Electronic Engrs; Audio Eng Soc. *Res:* Very-large-scale integration circuit design; analog methodologies; high frequency monolithic filter design; switched capacitor applications; biomedical applications; specialized amplifier design. *Mailing Add:* Dept Elec Eng & Computer Eng 201 Coover Hall Iowa State Univ Ames IA 50011

GEIGER, WILLIAM EBLING, JR, b Buffalo, NY, Feb 11, 44; m 65; c 2. ANALYTICAL CHEMISTRY. *Educ:* Canisius Col, BS, 65; Cornell Univ, PhD(anal chem), 69. *Prof Exp:* Res fel, Northwestern Univ, 69-70; asst prof chem, Southern Ill Univ, 70-74; assoc prof, 74-77, ASSOC PROF CHEM, UNIV VT, 77- *Mem:* Am Chem Soc. *Res:* Electroanalytical chemistry; organometallic and transition metal electrochemistry; elucidation of electronic structures by electron spin resonance; radical ion molecular complexes. *Mailing Add:* Dept Chem Univ Vt Burlington VT 05401

GEIGES, K S, b Providence, RI, Aug 24, 08. ELECTRICAL EQUIPMENT, SAFETY STANDARDS. *Educ:* NJ Inst Technol, BS, 28; Stevens Inst Technol, MS, 43. *Prof Exp:* Vpres, Underwriters Labs, 28-88; RETIRED. *Concurrent Pos:* US Rep, Int Electrotech Comn. *Honors & Awards:* Astin-Polk Medal, Am Nat Standards Inst, 78. *Mem:* Fel Inst Elec & Electronics Engrs. *Mailing Add:* 111 43rd St No A Newport Beach CA 48663-8927

GEIL, PHILLIP H, b Milwaukee, Wis, Sept 26, 30; m 77; c 5. POLYMER PHYSICS. *Educ:* Wis State Col, Milwaukee, BS, 52; Univ Wis, MS, 54, PhD(physics), 57. *Prof Exp:* Res physicist, E I du Pont de Nemours & Co, 56-62; sr res physicist, Res Triangle Inst, 62-63; from assoc prof to prof polymer sci & eng, Case Western Reserve Univ, 63-79; PROF POLYMER DIV, DEPT MAT SCI ENG, UNIV ILL, 79-, COORDR, POLYMER GROUP, 80-, CHMN DIV, 87- *Concurrent Pos:* Ed, J Macromolecular Sci (Physics), 63-; assoc dir, Nat Ctr Composite Mat Res, 86-90. *Mem:* Am Phys Soc; Am Chem Soc; Soc Plastics Eng. *Res:* Polymer Physics; relationship of polymer morphology to physical properties; mechanisms of polymer crystallization and deformation; polymer matrix composites. *Mailing Add:* Polymer Division Dept Mat Sci & Eng Univ Ill Urbana IL 61801

GEILKER, CHARLES DON, b Kingston, Mo, Dec 15, 33; m 58; c 2. ASTRONOMY, PHYSICS. *Educ:* William Jewell Col, AB, 55; Vanderbilt Univ, MA, 57; Case Western Reserve Univ, PhD(astron), 68. *Prof Exp:* Instr radiation physics, USPHS, 57-62; res assoc astron, Observ, Vanderbilt Univ, 62-64; assoc prof physics & astron, 68-77, PROF PHYSICS, WILLIAM JEWELL COL, 77-, CHMN DEPT, 80- *Mem:* Am Asn Physics Teachers; Health Physics Soc; Sigma Xi. *Res:* Electronic and optical instrumentation; astronomical photoelectric and photographic photometry; gamma spectrometry and radiation dosimetry; microcomputers. *Mailing Add:* 471 E Kansas Liberty MO 64068

GEINISMAN, YURI, b Kiev, Ukraine, USSR, Aug 7, 31; US citizen; m 71; c 1. NEUROBIOLOGY. *Educ:* First Moscow Med Univ, MD, 56; USSR Acad Sci, PhD(neuroanat), 62, DMS, 74. *Prof Exp:* Jr sci worker neuroanat, Inst Higher Nerv Activ & Neurophysiol, USSR Acad Sci, 62-66; guest scientist, Dept Anat, Semmelweis Univ Med Sch, Budapest, Hungary, 67-68; sr sci worker neuroanat, Inst Higher Nerv Activ & Neurophysiol, USSR Acad Sci, 68-74; asst prof, 75-77, ASSOC PROF ANAT, MED SCH, NORTHWESTERN UNIV, CHICAGO, 77- *Mem:* Am Soc Neurosci; Am Asn Anat; AAAS; Sigma Xi. *Res:* Structural and metabolic basis of manifestations of aging process and post-lesion plasticity in the central nervous system. *Mailing Add:* Dept Cell Biol & Anat Northwestern Univ 303 E Chicago Ave Chicago IL 60611

GEIS, AELRED DEAN, b Chicago, Ill, July 23, 29; m 51; c 1. BIOLOGY. *Educ:* Mich State Univ, BS, 51, MS, 52, PhD, 56. *Prof Exp:* Instr, Mich State Univ, 53-56; surv statistician, Bur Sport Fisheries & Wildlife, US Fish & Wildlife Serv, 56-58, wildlife res biologist & asst chief sect, Migratory Bird Pop & Distrib Studies, 58-61, chief sect waterfowl pop studies, Patuxent Wildlife Res Ctr, 61-62, asst dir Migratory Bird Populations Sta, 62-70, migratory bird specialist, Migratory Bird Pop Sta, 70-72, URBAN WILDLIFE SPECIALIST, FISH & WILDLIFE SERV, PATUXENT WILDLIFE RES CTR, 72- *Mem:* Am Ornithologist Union; Wildlife Soc. *Res:* Wildlife population dynamics, survey methods and biology. *Mailing Add:* Trotter Rd Clarksville MD 21029

GEISBUHLER, TIMOTHY PAUL, b Woodville, Ohio, July 16, 54; m 86; c 1. NUCLEOTIDE METABOLISM, CHROMATOGRAPHY. *Educ:* Bowling Green State Univ, BS, 76; Univ Cincinnati, MS, 79; Ohio State Univ, PhD(physiol chem), 83. *Prof Exp:* res assoc, 83-86, RES ASST PROF PHYSIOL, UNIV MO, 86- *Mem:* Am Physiol Soc; Int Soc Heart Res. *Res:* Biochemistry and physiology of diseased heart, with special emphasis on metabolic defects associated with myocardial ischemia; metabolism of nucleotides and nucleosides in healthy and diseased tissues. *Mailing Add:* Dept Physiol Univ Mo MA415 Columbia MO 65212

GEISE, MARIE CLABEAUX, b Buffalo, NY, July 5, 41. PHYSICAL ANTHROPOLOGY. *Educ:* State Univ NY Buffalo, BA, 63, MA, 66, PhD(phys anthrop), 67. *Prof Exp:* Lectr anthrop, Lehman Col, 66-67, from instr to assoc prof, 67-73; assoc prof anthrop & chmn dept, 73-80, PROF ANTHROP, STATE UNIV NY COL BUFFALO, 80- *Concurrent Pos:* Consult, WNED, Buffalo, NY, 65-66; dir, NSF Instructional Equip Grant, 69-71, recipient, NSF Sci Instructional Equip Grant, 74-76; grant reviewer, Univ Awards Comt, State Univ NY, 79- *Mem:* AAAS; fel Am Asn Phys Anthrop; Am Anthrop Asn; Human Biol Coun; Int Ref Orgn Forensic Med & Sci. *Res:* Palaeopathology of the pre-Columbian Indians of North America; physical and cultural environmental determinants of disease patterns in skeletal populations; methodology and analysis in osteology; forensic anthropology; biomedical anthropology. *Mailing Add:* Dept Anthrop State Univ NY Col Buffalo NY 14222

GEISELMAN, PAULA J, b Ironton, Ohio, June 30, 44; m 76. INGESTIVE BEHAVIOR, INGESTIVE PHYSIOLOGY. *Educ:* Ohio Univ, AB, 71, MS, 76; Univ Calif, Los Angeles, PhD(physiol psychol), 83. *Prof Exp:* Instr psychol, Ohio Univ, 74-76; fel physiol psychol, Univ Calif, Los Angeles, 77-79, res assoc, 78-79, asst exp psychol, 81, fel, 82, staff res assoc, multivariate statist, 83, vis asst prof physiol psychol & exp psychol, 83-86, ASST ADJ PROF, PHYSIOL PSYCHOL & EXP PSYCHOL, UNIV CALIF, LOS ANGELES, 86-, ASST RES PSYCHOLOGIST, 83- , DIR PSYCHOPHYSIOL RES, DIV CLIN NUTRIT, DEPT MED, SCH MED, 89- *Concurrent Pos:* Reviewer for numerous jour, 77-; prin investr, NSF grant, 85-87 & 87-91; career develop award, NIH, 89-94. *Mem:* Soc Neurosci; Am Psychol Asn; AAAS; Asn Advan Psychol; hon mem Brit Brain Res Soc; hon mem Europ Brain & Behav Soc; NY Acad Sci; Women Neurosci. *Res:* Behavioral, nutritional and physiological mechanisms that control energy, appetite and body weight; conditions under which carbohydrates suppress and stimulate appetite; published numerous articles in various journals. *Mailing Add:* Dept Psychol Univ Calif 405 Hilgard Ave Franz Hall Los Angeles CA 90024-1563

GEISLER, C(HRIS) D(ANIEL), b New York, NY, Jan 10, 33; m 61; c 2. BIOENGINEERING, NEUROPHYSIOLOGY. *Educ:* Mass Inst Technol, BS & MS, 56, ScD(elec eng), 61. *Prof Exp:* Assoc biocommun, Air Force Res & Develop Command, 60-61; mem tech staff, Bell Tel Labs, 61-62; from asst prof to assoc prof, 62-77, PROF ELEC & COMPUTER ENG & NEUROPHYSIOL, UNIV WIS-MADISON, 77- *Mem:* Fel Am Sci Affiliation; sr mem Inst Elec & Electronics Engrs; fel Acoust Soc; Am Soc Neurosci. *Res:* Quantitative study of the inner ear, including single-neuron recordings and mathematical models; auditory physiology and neurophysiology; sensory prosthesis. *Mailing Add:* Dept Elec & Computer Eng & Neurophysiol Univ Wis Madison WI 53706

GEISLER, FRED HARDEN, b Chicago, Ill, Mar 2, 47; m 68; c 1. MEDICAL PHYSIOLOGY. *Educ:* Case Inst Technol, BS, 67; Washington Univ, AM, 69, PhD(physics), 72; State Univ NY Buffalo, MO, 78. *Prof Exp:* Res assoc physics, Brookhaven Nat Lab, 72-74; surg resident, 78-79, RES ASSOC MED PHYSIOL, BUFFALO GEN HOSP, 75-, NEUROSURG RESIDENT, 79- *Mem:* Sigma Xi; Am Phys Soc; AAAS; AMA. *Res:* Measurement of cardio-pulmonary function in the intensive care hospital unit and patient management utilizing multidimensional computer analysis. *Mailing Add:* 7106 Long View Rd Columbia MD 21044

GEISLER, GRACE, b Rochester, NY, Dec 6, 12. BIOLOGY. *Educ:* Nazareth Col, BS, 39; Catholic Univ, MS, 41, PhD(biol), 44. *Prof Exp:* From instr to prof, Nazareth Col, NY, 43-78, chmn dept, 43-79, emer prof biol, 78-88; RETIRED. *Concurrent Pos:* Am Physiol Soc grant; Joseph Peabody fel, Catholic Univ Am. *Mem:* Am Inst Biol Sci; Nat Asn Biol Teachers. *Res:* Post-embryonic development of Hyalella, a crustacean; effects of humidity on Drosophila melanogaster pupae. *Mailing Add:* Sisters St Joseph Mother House 4095 East Ave Rochester NY 14618

GEISLER, JOHN EDMUND, b Pittsburgh, Pa, Apr 9, 34; m 74; c 2. DYNAMIC METEOROLOGY. *Educ:* Fla State Univ, BS, 59; Univ Ill, PhD(astron), 65. *Prof Exp:* Sr scientific officer aeronomy, Radio & Space Res Sta, Slough, Eng, 65-67; res assoc meteorol, Mass Inst Technol, 67-68; from asst prof to assoc prof, 68-75, prof meteorol, Univ Miami, 75-; HEAD OF DEPT METEOROL, UNIV UTAH. *Concurrent Pos:* Chief scientist meteorol, Off Climate Dynamics, NSF, 75- *Mem:* Am Meteorol Soc. *Res:* Numerical models of large-scale atmospheric waves; climate models. *Mailing Add:* Dept Meteorol Univ Utah Salt Lake City UT 84112

GEISMAN, RAYMOND AUGUST, SR, b St Louis, Mo, Oct 14, 21; m 52; c 3. ANALYTICAL CHEMISTRY. *Educ:* Wash Univ, BS, 43, MS, 44 & 54. *Prof Exp:* Analytical chemist, Scullin Steel Co, 42-44; prod supvr indust org chem, 46-49, asst chief chemist, 49-51, prod supvr indust org chem, 51-53, sales engr, 53-55, chief chemist, 55-77, PRIN ENGR, MONSANTO CO, 77- *Mem:* Am Chem Soc; Am Inst Chem Engrs; Sigma Xi; Instrument Soc Am. *Mailing Add:* 6231 Delor St Louis MO 63109

GEISON, RONALD LEON, BIOCHEMISTRY, NUTRITION. *Educ:* Univ Ill, PhD(nutrit biochem), 65. *Prof Exp:* CHEMIST, SIGMA CHEM CO, 75- *Res:* Preparation of synthetic and natural products, lipids. *Mailing Add:* 343 Nelda Kirkwood MO 63122

GEISS, GUNTHER R(ICHARD), b New York, NY, Oct 1, 38; m 62; c 2. SYSTEMS ANALYSIS, INFORMATION TECHNOLOGY. *Educ:* Polytech Inst Brooklyn, BEE, 59, MEE, 60, PhD(elec eng), 64. *Prof Exp:* Instr elec technol, Brooklyn Community Col, 60-61; res engr control theory res dept, Grumman Aircraft Eng Corp, 61-64, group leader, 64-69; dir systs sci div, Poseidon Sci Corp, NY, 69-71; mgr NY off, Skills Conversion Proj, Nat Soc Prof Engrs, 71-72; vis assoc prof, Inst Marine Sci, 72, from asst prof to assoc prof, 72-83, PROF, SCH SOCIAL WORK, ADELPHI UNIV, 83- *Concurrent Pos:* Lectr, Polytech Inst New York, 61-75, adj prof, dept elec eng, 75-, adj prof chem eng, 82; chmn, Oil Spillage Bd, Huntington, NY, 71-73; adj prof, Sch Bus, Adelphi Univ, 73-79 & Prog Training for Pub & Community Social Serv, State Univ NY Stonybrook. *Mem:* AAAS; Inst Elec & Electronics Engrs; Am Chem Soc; Coun Social Work Educ. *Res:* Application of systems theory and computer technology to problems of management and society; technology of oil spill control and clean up; planning of oil spill combat strategies and modeling, test and evaluation of spill control devices. *Mailing Add:* 8 Meadowlark Lane Huntington NY 11743

GEISS, ROY HOWARD, b Reading, Pa, Apr 17, 37; div; c 2. ELECTRON MICROSCOPY. *Educ:* Lafayette Col, BS, 59; Cornell Univ, PhD(appl physics), 68. *Prof Exp:* Res physicist, Carpenter Steel Corp, 59-63; sr scientist mat sci, Univ Va, 67-73; RES STAFF, IBM RES LAB, 73- *Concurrent Pos:* Mat analysis consult, US Army Foreign Sci & Technol Ctr, 72-73. *Mem:* Electron Micro Soc Am; Am Crystallog Asn; Am Soc Metals; Am Inst Mining, Metall & Petrol Engrs. *Res:* Use of transmission electron microscopy to study microstructural parameters of magnetic thin films; analytical electron microscopy including EDS and micro-diffraction; develop new techniques for use of instruments. *Mailing Add:* Almaden Res Ctr IBM Res Div 650 Harry Rd Dept K34 San Jose CA 95120-6099

GEISSER, SEYMOUR, b Bronx, NY, Oct 5, 29; m 55, 82; c 4. MATHEMATICAL STATISTICS. *Educ:* City Col New York, BA, 50; Univ NC, MA, 52, PhD(math statist), 55. *Prof Exp:* Sr asst scientist, USPHS, 55-57, mathematician, Nat Inst Mental Health, NIH, 57-61, chief biomet sect, Nat Inst Arthritis & Metab Dis, 61-65; prof math statist & chmn dept, State Univ NY Buffalo, 65-71; PROF STATIST & DIR SCH, UNIV MINN, MINNEAPOLIS, 71- *Concurrent Pos:* Lectr, USDA Grad Sch, 56-60; vis assoc prof, Iowa State Univ, 60; prof lectr, George Washington Univ, 60-65; vis prof, Univ Wis, 64; NSF vis lectr, 66-69; vis prof, Univ Tel-Aviv, Israel, 71, Carnegie-Mellon Univ, 76, Stanford Univ, 77, Univ Orange Free State, 78, Harvard Sch Pub Health, 81, Univ Chicago, 85, Univ Warrick 86 & Stanford Univ, 88; chmn, Panel Occup Safety & Health Statist, Nat Acad Sci, 86-87; Lady Davis vis prof, Hebrew Univ, Jerusalem, Israel, 91. *Mem:* Sigma Xi; fel Am Statist Asn; Biomet Soc; Math Asn Am; fel Inst Math Statist; fel Royal Statist Soc; Int Statist Inst; Bernoulli Soc. *Res:* Biometrics; statistics; research in multivariate analysis; Bayesian inference; predictivism; sample reuse, classification. *Mailing Add:* Sch Statist 270 Vincent Hall Univ Minn 206 Church St SE Minneapolis MN 55455

GEISSINGER, HANS DIETER, b Mannheim, Ger, Jan 15, 30; Can citizen; m 62; c 5. HISTOLOGY, MICROSCOPY. *Educ:* Univ Toronto, DVM, 60, MVSc, 62; Univ London, PhD(vet path), 66. *Prof Exp:* Res scientist, 62-63, assoc prof pathophysiol, 66-75, PROF MICROANATOMY, DEPT BIOMED SCI, ONT VET COL, UNIV GUELPH, 75- *Mem:* Can Vet Med Asn; Int Soc Stereology; Asn Sci, Eng & Technol Community Can; Micros Soc Can; Royal Micros Soc. *Res:* Pathology of atherosclerosis; pathophysiology and pathology of cardiovascular and neuromuscular diseases in man and animals; development of new microscopic methods for histology and histopathology. *Mailing Add:* Dept Biomed Sci Ont Vet Col Univ Guelph Guelph ON N1G 2W1 Can

GEISSINGER, LADNOR DALE, b Palm, Pa, Apr 23, 38; m 60; c 3. COMBINATORICS, APPLICATION GROUP CHARACTERS. *Educ:* Bluffton Col, BS, 59; Ind Univ, PhD(math), 63. *Prof Exp:* Asst prof math, Purdue Univ, 63-67; assoc prof, 67-77, PROF MATH, UNIV NC, CHAPEL HILL, 77- *Concurrent Pos:* Vis lectr, Univ of Ulm, Ger, 77-78. *Mem:* Am Math Soc; Math Asn Am. *Res:* Representation of groups; foundations of combinatorial theory; applications in chemistry and physics of finite symmetry groups. *Mailing Add:* Dept Math Phillips Hall Univ NC CB3250 Chapel Hill NC 27599-3250

GEISSLER, ERNST D(IETRICH), b Chemnitz, Ger, Aug 4, 15; nat US; m 41; c 2. AERONAUTICAL ENGINEERING. *Educ:* Dresden Tech Univ, BS, 36, MS, 39; Darmstadt Tech Univ, Dr Ing, 52. *Prof Exp:* Res engr control theory, Army Exp Inst, Peenemuende, Ger, 40-41, sect chief theory of flight & control, 41-45; aeronaut res engr, Ord Res & Develop Agency, Ft Bliss, Tex, 45-48, group leader aerodyn & flight mech, 48-50, chief aeroballistics sect, Redstone Arsenal, 50-56, dir aeroballistics lab, Army Ballistic Missile Agency, 56-59; mem res adv comt space vehicle aerodyn, NASA, 59-67, dir aero-astrodyn lab, Marshall Space Flight Ctr, 60-73; RETIRED. *Honors & Awards:* Except Civilian Serv Award, US Dept Army, 59; Except Sci Achievement Award, NASA, 63, Except Serv Medal, 68. *Mem:* Fel Am Inst Aeronaut & Astronaut; fel Am Astronaut Soc. *Res:* Stability and control; theory of flight; flight mechanics; guidance theory; dynamic analysis, especially flutter and vibration; missile aerodynamics; space mechanics. *Mailing Add:* 3604 Mae Dr SE Huntsville AL 35801

GEISSLER, PAUL ROBERT, b New York, NY, Jan 19, 32; m 60; c 3. CHEMICAL KINETICS. *Educ:* St Peter's Col, BS, 53; Fordham Univ, MS, 56; Univ Wis, PhD(phys chem), 62. *Prof Exp:* From res chemist to sr res chemist, Esso Res & Eng Co, Standard Oil Co, NJ, 62-68, res assoc, 68-76; MEM STAFF, EXXON CHEM CO, 76- *Mem:* Am Chem Soc. *Res:* Effect of neutron activation of organic liquids and solutions; radiation chemistry of hydrocarbon alkyl halide solutions; exploratory aromatic research; process optimization via mechanistic and kinetic studies; supercritical fluid chromatography. *Mailing Add:* 3848 Lake Latania Circle Baton Rouge LA 70816

GEISSMAN, JOHN WILLIAM, b Rockford, Ill, Oct 29, 52; m 75. PALEOMAGNETIST, TECTONOPHYSICIST. *Educ:* Univ Mich, BS, 73, MS, 76, PhD(geol), 80. *Prof Exp:* Student & lectr geol, Univ Mich, 77-79; fel & lectr, Univ Toronto, 79-80; ASST PROF GEOL, COLO SCH MINES, 80- *Concurrent Pos:* Consult, Anaconda Minerals Co, 81-; prin investr, Dept Energy grants, Colo Sch Mines, 81-, NSF grants, 82- *Mem:* Am Geophys Union; Sigma Xi. *Res:* Application of paleomagnetic research to problems in structural geology, economic geology and diagenesis. *Mailing Add:* Dept Geol Univ NMex Main Campus Albuquerque NM 87131

GEIST, J(OHN) C(HARLES), b Baltimore Co, Md, Sept 20, 15; m 38; c 2. COMMUNICATIONS. *Educ:* Univ Del, BEE, 37. *Prof Exp:* Test engr, Gen Elec Co, 37-38; jr engr, Hadley Transformer Co, 39; test engr, Westinghouse Mfg Corp, 40; radio engr, Signal Corps, US Army, 41; systs engr, Link Radio Corp, 46-47; commun engr, Appl Physics Lab, Johns Hopkins Univ, 48; missile systs engr, Vitro Corp Am, 48-58, mgr tech opers, Silver Spring Lab, 58-60, assoc dir, 60-70, vpres, Vitro Labs, 70-80; RETIRED. *Mem:* Sr mem Inst Elec & Electronics Engrs; Am Phys Soc. *Res:* Telemetering; missile guidance; computer applications. *Mailing Add:* 2205 Henderson Ave Silver Spring MD 20902

GEIST, JACOB M(YER), chemical engineering, cryogenics; deceased, see previous edition for last biography

GEIST, VALERIUS, b Nikolajew, Russia, Feb 2, 38; Can citizen; m 61; c 3. ZOOLOGY, ETHOLOGY. *Educ:* Univ BC, BSc, 60, PhD(zool), 67. *Prof Exp:* Asst prof, Environ Sci Ctr, Univ Calgary, 68-71, prog dir, Environ Sci, Fac Environ Design, 71-75, assoc prof zool, 71-76, assoc dean, Fac Environ Design, 77-80, PROF ENVIRON SCI, UNIV CALGARY, 76- *Concurrent Pos:* Nat Res Coun fel, Max Planck Inst Physiol Behav, 67-68; mem, Subcomt Conserv Terrestrial Communities, Can Comt Int Biol Prog, 69-76; Sci Adv Comt, World Wildlife Fund Can, 82-; coun, Can Soc Zoologists, 82-; dir, Alta Soc Prof Biologists; pres, Calgary Chap, Sigma Xi, 85-86; mem, JUCN Species Survival Comn. *Mem:* Wildlife Soc; Can Soc Zool; Can Soc Environ Biol; fel AAAS; Sigma Xi. *Res:* Behavior and evolution of large mammals, especially ungulates; relation between ecology and social behavior; evolution of Ice Age mammals; human biology in relation to environmental design. *Mailing Add:* Fac Environ Design Univ Calgary 2500 University Dr NW Calgary AB T2N 1N4 Can

GEISTERFER-LOWRANCE, ANJA A T, Can citizen. CARDIOLOGY. *Educ:* Univ Guelph, Can, BSc, 82; Univ Va, PhD(physiol), 89. *Prof Exp:* Res asst, Dept Chem, Univ Guelph, 81-83; POSTDOCTORAL FEL, CARDIOVASC DIV, BRIGHAM & WOMEN'S HOSP, HARVARD MED SCH, 83- *Concurrent Pos:* Paul Dudley White fel, 91. *Res:* Author of numerous publications. *Mailing Add:* Cardiovasc Div Brigham & Women's Hosp 75 Francis St Boston MA 02115

GEITZ, R(OBERT) C(HARLES), b McKeesport, Pa, Oct 23, 19; m 52; c 5. MECHANICAL ENGINEERING. *Educ:* Calif Inst Technol, BS, 41; Univ Pittsburgh, PhD(chem eng), 51. *Prof Exp:* Chem engr, E I du Pont de Nemours & Co, 41-43; partner, Calif Natural Prod, 48-49; sect chief org process develop, Res & Develop Div, Lever Bros, 49-56; group mgr res & develop, Am Mach & Foundry Co, 56-61; pres, Geitz Eng Co, 62-74; pres, 74-81, PROF CHEM ENG, SPRINGFIELD TECH COMMUNITY COL, 68-; PRES, GEITZ ENG CO, 81- *Mem:* Sigma Xi; Am Inst Chem Engrs; Am Soc Mech Engrs; Am Chem Soc; AAAS. *Res:* Organic chemical development and manufacture of detergents; product and process development; tobacco sheet and paper; application of mechanical engineering techniques to process industries; laboratory equipment; non-soap bars; process plant design and erection; closed die forging. *Mailing Add:* Box 207 Suffield CT 06078

GELATT, CHARLES DANIEL, JR, b La Crosse, Wis, Sept 5, 47; m 69; c 2. SOFTWARE SYSTEMS, SOLID STATE PHYSICS. *Educ:* Univ Wis-Madison, BA & MS, 69; Harvard Univ, PhD(physics), 74. *Prof Exp:* asst prof physics, Harvard Univ, 75-80; mem staff, Thomas J Watson Res Ctr, IBM Corp, 79-82; PRES, NORTHERN MICROGRAPHICS, INC, 83-; PRES, NMT CORP, 86- *Honors & Awards:* Indust Applns Physics Award, Am Inst Physics, 87. *Mem:* Am Phys Soc; Inst Elec & Electronics Engrs; Asn Comput Mach; AAAS. *Res:* Theoretical solid state physics; cohesion of transition metal elements and compounds; simulated annealing for optimization; computer graphics. *Mailing Add:* PO Box 2287 La Crosse WI 54602-2287

GELB, ARTHUR, b New York, NY, Sept 20, 37; m 58; c 3. NONLINEAR CONTROL SYSTEMS, ELECTRICAL ENGINEERING. *Educ:* City Col New York, BEE, 58; Harvard Univ, MS, 59; Mass Inst Technol, DSc(syst eng), 61. *Prof Exp:* Mgr, systs anal, Dynamics Res Corp, 61-66; PRES, THE ANALYTIC SCI CORP, 66- *Concurrent Pos:* Chmn, MIT Ctr Technol, 87- *Mem:* Fel Inst Elec & Electronics Engrs; fel Am Inst Aeronaut & Astronaut; Sigma Xi. *Res:* Optimal methods of navigation guidance and control; optimal estimation applications; nonlinear control systems. *Mailing Add:* 37 Meriam St Lexington MA 02173

GELB, LEONARD LOUIS, organic polymer chemistry; deceased, see previous edition for last biography

GELBAND, HENRY, b Austria, Aug 31, 36; US citizen; m 62; c 3. CARDIAL ELECTROPHYSIOLOGY. *Educ:* Washington & Jefferson Col, AB, 58; Jefferson Med Col, MD, 62. *Prof Exp:* Fel pediat cardiol, Col Physicians & Surgeons, Columbia Univ, 67-69, NIH spec res fel pharmacol, 69-71, vis fel pediat, 69-70, assoc pediat, 70-71; from asst prof to assoc prof pediat & pharmacol, 71-77, PROF PEDIAT & PHARMACOL, SCH MED, UNIV MIAMI, 77-, DIR, DIV PEDIAT CARDIOL, 76- *Concurrent Pos:* Instr pediat, Mt Sinai Sch Med, 67-68; consult, Children's Med Servs, Health & Rehab Servs, Fla, 71- & chmn, Cardiac Adv Coun, 85-; Basil O'Conner Res Award, Nat Found, March of Dimes, 74-76; prin investr, Nat Heart, Lung & Blood Inst, NIH, 76-; examr for bd cert, Sub-Bd Pediat Cardiol, Am Bd Pediat, 83. *Mem:* Am Heart Asn; fel Am Acad Pediat (treas, 79-80, vchmn, 81-83); fel Am Col Cardiol; emer mem Soc Pediat Res; Am Soc Pharmacol & Exp Therapeut; NAm Soc Pacing & Electrophysiol. *Res:* Measurement of calcium fluxes accross myocardial call in developing heart, its relationship to cardiac electro physiology & contractility as well as the use of calcium antagonists as antiarrhythmic agents. *Mailing Add:* Dept Pediat Sch Med Univ Miami PO Box 016960 (R-76) Miami FL 33101

GELBARD, ALAN STEWART, b Brooklyn, NY, Mar 28, 34; m 66; c 2. ENZYMOLOGY, NUCLEAR MEDICINE. *Educ:* Brooklyn Col, BS, 55; Univ Mass, MS, 56; Univ Wis, Madison, PhD(bact), 60. *Prof Exp:* Instr microbiol, State Univ NY Downstate Med Ctr, 59-62; res assoc, 62-74, assoc, Biophys Lab, 75-78, ASSOC MEM, BIOPHYS LAB, MEM SLOAN-KETTERING CANCER CTR, 79- *Concurrent Pos:* Instr, Sloan Kettering Div, Grad Sch Med Sci, Cornell Univ, 69-75, asst prof biophys, 75-81, assoc prof, 81- *Mem:* Am Soc Microbiol; Soc Nuclear Med; AAAS. *Res:* Enzymatic synthesis of amino acids labeled with positron-emitting, short-lived isotopes; development of organ and tumor scanning agents; metabolism and enzyme synthesis during cell cycle of cultured tumor cells; in-vivo studies of tumor and organ metabolism. *Mailing Add:* Mem Sloan-Kettering Cancer Ctr 1275 York Ave New York NY 10021

GELBART, ABE, b Paterson, NJ, Dec 22, 11; m 39; c 4. MATHEMATICS. *Educ:* Dalhousie Univ, BSc, 38; Mass Inst Technol, PhD(math), 40. *Hon Degrees:* LLD, Dalhousie Univ, 72; DSc, Bar-Ilan Univ, Israel, 85. *Prof Exp:* Asst, Mass Inst Technol, 39-40; instr math, NC State Col, 40-42; res assoc, Brown Univ, 42; assoc physicist, Nat Adv Comt Aeronaut, Langley Field, Va, 42-43; from asst prof to prof math, Syracuse Univ, 43-58; dir inst math, Yeshiva Univ, 58-59, dean, Belfer Grad Sch Sci, 59-70, DISTINGUISHED UNIV PROF MATH, YESHIVA UNIV, 68-, EMER DEAN, SCH, 70- *Concurrent Pos:* Mem, Inst Advan Study, 47-48; vis lectr, Sorbonne, Paris, 49; vis prof, Univ Southern Calif, 51; Fulbright lectr, Norway, 51-52; ed, Scripta Math, 57-; vis mem, Inst Advan Study, Princeton, 77-79; fel, Bard Col Ctr, 79-, vis prof math, Bard Col, 79-82; David & Rosalie Rose distinguished prof, Natural Sci & Math, Bard Col, 82-; mem adv bd, Ctr on Thinking & Learning, Pace Univ, 84-; mem, Bd Trustees, Bar-Ilan Univ, 83-, chmn, Sci Adv Comt, 83- *Mem:* Am Math Soc; Math Asn Am; assoc fel Am Inst Aeronaut & Astronaut. *Res:* Methods of generalizing complex function theory; nonlinear partial differential equations; functions of several complex varibles; theory of pseudo-analytic functions; existence theorems in integral equations; fluid dynamics. *Mailing Add:* 140 West End Ave New York NY 10023

GELBART, STEPHEN SAMUEL, b Syracuse, NY, June 12, 46; m 68; c 2. NUMBER THEORY. *Educ:* Cornell Univ, BA, 67; Princeton Univ, MA, 68, PhD(math), 70. *Prof Exp:* Instr math, Rutgers Univ, Newark, 68-70 & Princeton Univ, 70-71; asst prof, 71-75, assoc prof, 75-80, PROF MATH, CORNELL UNIV, 80- *Concurrent Pos:* Mem, Inst Advan Study, 72-73; Alfred P Sloan fel, 77-78; vis prof, Hebrew Univ, Jerusalem, 81-82. *Mem:* Am Math Soc. *Res:* Automorphic forms, group representations and applications to number theory. *Mailing Add:* Dept Theoret Math Weizmann Inst Math Rehovot 76100 Israel

GELBART, WILLIAM M, b Syracuse, NY, June 12, 46; m 68; c 2. CHEMICAL PHYSICS. *Educ:* Harvard Univ, BS, 67; Univ Chicago, MS, 68, PhD(chem physics), 70. *Prof Exp:* NSF-NATO fel, Univ Paris, 70-71; Miller Inst fel, Univ Calif, Berkeley, 71-72, asst prof chem, 72-75; assoc prof, 75-79, PROF CHEM, UNIV CALIF, LOS ANGELES, 79- *Concurrent Pos:* Alfred P Sloan Found fel, 74-78; Camille & Henry Dreyfus Found teacher-scholar, 76-81; vis prof, Univ Paris, 77-78 & 82-83; mem adv bd, J Statist Physics, 78-82, J Phys Chem, 80-84, Liquid Crystals, 85-, Chem Physics Lett, 86-, Langmuir, 86-, Molecular Physics, 87-; Brotherton vis prof, Univ Leeds, 88. *Mem:* Fel Am Phys Soc; Am Chem Soc. *Res:* Theory of gas-phase molecular relaxation processes and photochemistry; light scattering and optical properties of simple fluids near and away from their critical points; orientational order in liquid crystals, polymeric systems, interfacial films and surfactant solutions. *Mailing Add:* Dept Chem Univ Calif Los Angeles CA 90024

GELBAUM, BERNARD RUSSELL, b New York, NY, Feb 26, 22; m 42; c 4. MATHEMATICS. *Educ:* Columbia Univ, AB, 43; Princeton Univ, MA, 47, PhD(math), 48. *Prof Exp:* Instr math, Princeton Univ, 47-48; from asst prof to prof, Univ Minn, 48-64; prof & chmn, Univ Calif, Irvine, 64-68, assoc dean sch phys sci, 68-71; vpres acad affairs, 71-74, PROF MATH, STATE UNIV NY, BUFFALO, 74- *Mem:* Am Math Soc. *Res:* Linear spaces; topological algebra. *Mailing Add:* Dept Math State Univ NY Buffalo NY 14214-3093

GELBER, RICHARD DAVID, b Philadelphia, Pa, May 1, 47; m 69; c 2. BIOSTATISTICS. *Educ:* Cornell Univ, BS, 69, PhD(oper res), 75; Stanford Univ, MS, 70. *Prof Exp:* res asst prof statist, Sci & Statist Lab, State Univ NY Buffalo, 75-77; asst prof, Dept Biostatist, Sch Pub Health & Sidney Farber Cancer Inst, 77-83, ASSOC PROF DEPT BIOSTATIST & EPIDEMIOL, DANA FARBER CANCER INST, HARVARD UNIV, 83- *Concurrent Pos:* Statist & comput consult, Dean of Students Off, Cornell Univ, 73-75. *Honors & Awards:* Robert Wenner Prize, 87; Farmitalia Carlo Erba Prize, 87. *Mem:* Am Statist Asn; Biomet Soc; Am Soc Clin Oncol. *Res:* Application of statistical techniques to the conduct of clinical trials; special emphasis on sequential techniques; problems of ranking and selection; quality of life assessment; meta-analysis. *Mailing Add:* Div Biostatist & Epidemiol Mayer Bldg Dana Farber Cancer Inst 44 Binney St Boston MA 02115

GELBERG, ALAN, b New York, NY, May 28, 28; m 57; c 3. ORGANIC CHEMISTRY, INFORMATION SCIENCE. *Educ:* City Col New York, BS, 50; Univ Mo, MS, 53. *Prof Exp:* Asst chem, Univ Mo, 51-52, analytical chemist, Agr Exp Sta, 52-53; org chemist, US Army Chem Res & Develop Labs, Md, 55-57, group leader, 57-59, org chemist, Chem Struct Retrieval Prog, 59-61, asst chief, Indust Liaison Off, 61-63; document scientist, Diamond (Alkali) Shamrock Co, 64-67; chief chem info storage & retrieval staff, Sci Info Facil, FDA, 67-70, dir Mgt & Sci Info Systs Design Div, 70-72, dir, Drug Info Resources Div, Bur Drugs, 72-84, DEP CHIEF, SURVEILLANCE & DATA PROCESSING BR, CTR DRUG EVAL & RES, US FOOD & DRUG ADMIN, 85- *Concurrent Pos:* Mem, US Army Chem Corps Info Retrieval Comt, 60-63, liaison rep to mod methods comt, Div Chem Technol, Nat Res Coun, 61-63; mem tech adv bd, J Chem Doc, 69-71; prof lectr, Am Univ, 69-80, adj prof, Dept Chmn, 81-88; assoc ed,

Pesticides Index, 75- *Mem:* Am Chem Soc; Sigma Xi; Chem Notation Asn (pres, 70); AAAS. *Res:* Chemical information and structure retrieval; chemical line notations; automatic data processing equipment. *Mailing Add:* Ctr Drug Eval & Res HFD-737 US Food & Drug Admin 5600 Fishers Lane Rockville MD 20857

GELBKE, CLAUS-KONRAD, b Celle, WGer, May 31, 47. NUCLEAR STRUCTURE. *Educ:* Univ Heidelberg, dipl, 70, Dr rer nat, 73. *Prof Exp:* Vis asst nuclear physics, Max Planck Inst Nuclear Physics, 73-76; physicist nuclear physics, Lawrence Berkeley Lab, 76-77; assoc prof, 77-81, PROF PHYSICS, MICH STATE UNIV, 81-, UNIV DISTINGUISHED PROF, 90- *Concurrent Pos:* Vis, Brookhaven Nat Lab, 74, Univ Wash, 75; physicist, Lawrence Berkeley Lab, 78-79; Alfred P Sloan fel, 79-83. *Mem:* fel Am Phys Soc. *Res:* Nuclear reactions resulting from nucleus-nucleus collisions, including elastic and inelastic scattering, complete and incomplete fusion reactions, fragmentation reactions, fission and preequilibrium phenomena. *Mailing Add:* Cyclotron Lab & Dept Physics Mich State Univ East Lansing MI 48824

GELBOIN, HARRY VICTOR, b Chicago, Ill, Dec 21, 29; m 51; c 4. BIOCHEMISTRY, CANCER RESEARCH. *Educ:* Univ Ill, AB, 51; Univ Wis, MS, 56, PhD, 58. *Prof Exp:* Develop chemist, US Rubber Co, 51-54; asst biochem & cancer res, Univ Wis, 54-58; biochemist, NIH, 58-64, head chem sect, Carcinogenesis Studies Br, 64-66, CHIEF LAB MOLECULAR CARCINOGENESIS, NAT CANCER INST, NIH, 66- *Concurrent Pos:* Assoc ed, Cancer Res, 64; keynote speaker, Gordon Res Conf Cancer, 65; prin lectr, Franz Bielschowsky Mem, First Int Symp Molecular Biol, Carcino, NZ, 66; Claude Bernard award-vis prof, Univ Montreal, 70; lectr, Radiol Soc NAm, Ill, 70; prin lectr, US-Japan Coop Med Sci Prog, Charleston, SC, 73; Smith, Kline & French hon lectr, Univ Fla, Gainesville, 74, Univ Mich, 76; vis prof, Georgetown Univ, 78-81 & Hebrew Univ, Jerusalem, 85. *Mem:* Am Asn Cancer Res; Am Soc Biol Chemists; Am Soc Pharmacol & Exp Therapeut; AAAS; Int Soc Study Xenobiotics. *Res:* Biochemical mechanisms of carcinogenesis; drug and carcinogen metabolism; monoclonal antibody analyses of cytochromes P-450 and molecular biology of cytochromes P-450. *Mailing Add:* Rm 3E24 Bldg 37 Nat Cancer Inst NIH 9000 Rockville Pike Bethesda MD 20892

GELBWACHS, JERRY A, b New York, NY. QUANTUM OPTICS, LASERS. *Educ:* City Col NY, BE, 65; Stanford Univ, MS, 66, PhD(elec eng), 70. *Prof Exp:* NSF fel elec eng, 65-68, res asst lasers, Stanford Univ, 68-70; HEAD OPTICAL PHYS DEPT, AEROSPACE CORP, 70- *Honors & Awards:* Aerospace Corp Pres Sci Award. *Mem:* Am Phys Soc; Inst Elec & Electronics Engrs; fel Optical Soc Am. *Res:* Atomic resonance filters; trace vapor detection; remote sensing; laser spectroscopy. *Mailing Add:* Aerospace Corp PO Box 92957 MS M2-253 Los Angeles CA 90009

GELDERLOOS, ORIN GLENN, b Grand Rapids, Mich, July 28, 39; m 60; c 2. ENVIRONMENTAL PHYSIOLOGY. *Educ:* Calvin Col, BA, 61; Western Mich Univ, MA, 64; Northwestern Univ, PhD(environ biol), 70. *Prof Exp:* Instr high sch, Mich, 62-67; from asst prof to assoc prof biol, 70-77, dir, Sci Learning Ctr, 79-83, PROF BIOL SCI, UNIV MICH, DEARBORN, 77-, PROF ENVIRON STUDIES, 81- *Concurrent Pos:* NASA bio-space technol training prog, Wallops Island, 68; dir, environ study area, Univ Mich-Dearborn, 70-, dir environ studies prog, 75-; res assoc, Max-Planck Inst Physiol of Behav, Seewiesen, WGer; dir, comprehensive assistance to undergrad sci educ, NSF Grant, 78-82; dir, Rouge River Watershed Educ Proj, NSF grant, 89-93. *Mem:* Animal Behav Soc; Int Soc Chronobiol; AAAS; Sigma Xi; Natural Areas Asn. *Res:* Biological rhythms; microclimatology; mastery learning; urban ecology; bird migration; environmental physiology; environmental ethics and religion. *Mailing Add:* Dept Biol Sci Univ Mich Dearborn MI 48128-1491

GELDMACHER, R(OBERT) C(ARL), b Elgin, Ill, Apr 22, 17; m 41; c 3. ENGINEERING. *Educ:* Northern Ill Univ, BE, 42; Purdue Univ, MS, 46; Northwestern Univ, PhD(elec eng), 59. *Prof Exp:* Asst prof eng mech, Purdue Univ, 47-53, assoc prof eng sci, 53-60; prof eng sci & assoc dean eng, NY Univ, 60-66; head dept elec eng, 66-76, BURCHARD PROF ELEC ENG, STEVENS INST TECHNOL, 66- *Concurrent Pos:* Consult, Picatinny Arsenal, 52- *Mem:* Inst Elec & Electronics Engrs; Soc Indust & Appl Math. *Res:* Magnetomechanical solids; graph theory; network and system analysis and synthesis; applied elasticity. *Mailing Add:* PO Box 159 Erwinna PA 18920-0159

GELDREICH, EDWIN E(MERY), b Cincinnati, Ohio, May 9, 22; m 50; c 2. MICROBIOLOGY. *Educ:* Univ Cincinnati, AB, 47, MS, 48. *Prof Exp:* Res bacteriologist, Robert A Taft Sanit Eng Ctr, US Pub Health Serv, 48-67, res microbiologist, Nat Ctr Urban & Indust Health, 67-68, chief bact sect, Div Water Hyg, Nat Environ Res Ctr, 68-72; chief microbiol, 72-85, SR RES MICROBIOLOGIST, DRINKING WATER RES DIV WERL, US ENVIRON PROTECTION AGENCY, 85- *Concurrent Pos:* Dir water bacteriol lab eval serv; microbiol consult, Pan-Am Health Orgn, Brazil, spec microbiol consult, 69-88; spec microbiol consult, WHO, 72-; spec microbiol consult, Mediter Pilot study, UN Environ Prog, 76-87; chmn, Am Water Works Asn Organisms Water, 82-85, Joint Task Group, UN Environ Prog, 76-; vis prof, Am Soc Microbiol. *Honors & Awards:* Kimble Methodol Res Award, 55; Silver Medal, US Environ Protection Agency, 71; Regents' Water Microbiol Lectr, Univ Calif, 79; Res Award, Am Water Works Asn, 84; Bronze Medal, US Environ Prog, 84; Abel Wolman Award Excellence, 89. *Mem:* Am Soc Microbiol; Am Water Works Asn; Int Asn Water Pollution Res; fel Am Acad Microbiol. *Res:* Sanitary bacteriology; membrane filter techniques; rapid methods for the enumeration of bacterial indicators of pollution; biological tests for trace impurities in distilled water; microbiological contaminants in water supply; microbial criteria and standards; biofilms in water supply distribution. *Mailing Add:* 7330 Ticonderoga Ct Cincinnati OH 45230

GELEHRTER, THOMAS DAVID, b Liberec, Czech, Mar 11, 36; US citizen; m 59; c 2. MEDICAL GENETICS, CELL BIOLOGY. *Educ:* Oberlin Col, Ohio, BA, 57; Univ Oxford, Eng, MA, 59; Harvard Med Sch, MD, 63. *Prof Exp:* Intern & asst resident internal med, Mass Gen Hosp, Boston, 63-65; res assoc molecular biol, Nat Inst Arthritis & Metab Dis, NIH, Bethesda, Md, 65-69; fel med genetics, Univ Wash, Seattle, 69-70; asst prof human genetics, internal med & pediatrics, Sch Med, Yale Univ, New Haven, Conn, 70-73, assoc prof, 73-74; from assoc prof to prof internal med & human genetics, Med Sch, 74-87, dir, Div Med Genetics, 77-87, PROF & CHMN DEPT HUMAN GENETICS, PROF INTERN MED, MED SCH , UNIV MICH, ANN HARBOR, 87- *Concurrent Pos:* Mem, Bd Trustees, Oberlin Col, 70-75, Cell Biol Ad Hoc Rev Study Sect, NIH, 78, 83; vis scientist, Imperial Cancer Res Fund Labs, London, 79-80; mem, Med Knowledge Self-assessment Prog, Am Col Physicians, Genetics & Molecular Med Subcomt, chair, 89-91, mem med adv bd, Gene Screen; Chair; Macy Found fac scholar, Rhodes scholar. *Mem:* Am Soc Human Genetics; Am Soc Clin Invest; Am Soc Biol Chemists. *Res:* Hormonal regulation of fibrinolysis; molecular biology of plasminogen activator and plasminogen activator-inhibitor; biochemical and genetic analysis of the cellular actions of glucocorticoids and cyclic nucleotides. *Mailing Add:* Dept Human Genet Box 0618 Med Sch Univ Mich 1137 E Catherine St Ann Arbor MI 48109-0618

GELERINTER, EDWARD, b New York, NY, Oct 27, 36; m 63; c 3. CHEMICAL PHYSICS, LIQUID CRYSTAL PHYSICS. *Educ:* City Col New York, BEE, 58; Cornell Univ, PhD(physics), 66. *Prof Exp:* From asst prof to assoc prof, 71-78, PROF PHYSICS, KENT STATE UNIV, 78- *Concurrent Pos:* Vis prof, Ben Gurion Univ Negev, Israel, 80-81; prin investr grants & contracts, NSF, Binational Sci Found, NIH & NASA. *Mem:* Am Phys Soc. *Res:* Liquid crystal physics; spin label studies of glass formers: liquids, liquid crystals, polymers, etc; paramagnetic resonance of spin crossover systems. *Mailing Add:* Dept Physics Kent State Univ Kent OH 44242

GELERNTER, HERBERT LEO, b Brooklyn, NY, Dec 17, 29; m 52; c 3. PHYSICS, COMPUTER SCIENCE. *Educ:* Brooklyn Col, BS, 51; Univ Rochester, PhD(physics), 56. *Prof Exp:* Staff physicist, IBM Res, NY, 56-58, staff physicist & mgr theory automata group, 58-60, sr physicist, 60-64, sr physicist & mgr physics & comput appln res group, 64-66; PROF COMPUT SCI, STATE UNIV NY, STONY BROOK, 66- *Concurrent Pos:* Vis fel, Europ Orgn Nuclear Res, 60-61; mem ad hoc comt on-line data acquisition systs nuclear physics, Nat Res Coun-Nat Acad Sci, 68; Weizmann Mem Found fel, Weizmann Inst, Israel, 72-73. *Mem:* Am Phys Soc; Asn Comput Mach. *Res:* Simulation of intelligent behavior in machines; mathematical biophysics; computer applications in nuclear physics and chemistry; biomedical computer applications research. *Mailing Add:* Dept Comput Sci Div Math Sci State Univ of NY Stony Brook NY 11794

GELFAND, ALAN ENOCH, b New York, NY, Apr 17, 45; m 73; c 2. STATISTICS. *Educ:* City Col New York, BS, 65; Stanford Univ, MS, 67, PhD(statist), 69. *Prof Exp:* Asst prof, 69-74, assoc prof, 74-79, PROF STATIST, UNIV CONN, 80- *Mem:* Fel Am Statist Asn; Int Statist Inst; Inst Math Statist. *Res:* Bocyerian computation; statistical interference; statistical modeling. *Mailing Add:* Dept Statist U-120 Univ Conn Storrs CT 06269-3120

GELFAND, DAVID H, b New York, NY, June 9, 44; m 80; c 1. MOLECULAR BIOLOGY. *Educ:* Brandeis Univ, AB, 66; Univ Calif, San Diego, PhD(biol), 70. *Prof Exp:* Post grad res assoc biol, Univ Calif, San Diego, 70-72; lab mgr & asst res biochem, Univ Calif, San Francisco, 72-77; dir recombinant molecular res, 76-81, VPRES SCI AFFAIRS MOLECULAR GENETICS, CETUS CORP, 81- *Concurrent Pos:* Mem adv coun, NSF, 80-83. *Mem:* Am Soc Microbiol; AAAS; Am Soc Biol Chemists. *Res:* Isolation, characterization and expression of eukaryotic genes in bacteria, viral oncogenesis, mechanisms of RNA transcription and translation; study of mutations which increase plasmid copy number; isolation/characterization of thermostable DNA polymerases; in vitro DNA amplification. *Mailing Add:* Cetus Corp 1400 53rd St Emeryville CA 94608

GELFAND, ERWIN WILLIAM, b Montreal, Que, Mar 10, 41; c 2. IMMUNOLOGY, PEDIATRICS. *Educ:* McGill Univ, BSc, 62, MD, 66. *Prof Exp:* Res assoc, Max Planck Inst Immunobiol, 71-72; from asst prof to assoc prof pediat, Univ Toronto, 72-76; CHIEF DIV IMMUNOL & SR SCIENTIST, HOSP FOR SICK CHILDREN, 78- *Concurrent Pos:* Rotating internship, Montreal Children's Hosp, 66-67, jr asst resident, 67-68; sr resident, Children's Hosp Med Ctr, Boston, 68-69, fel med, 69-71; res fel, Harvard Med Sch, Boston, 69-71; Med Res Coun Can fel, 69-72; Queen Elizabeth II res scientist award, 75-81; Mead Johnson res award, 81. *Mem:* Am Fedn Clin Res; Am Asn Immunologists; Can Soc Immunol; Can Med Asn; Am Soc Clin Invest. *Res:* Cell differentiation; immunodeficiency diseases. *Mailing Add:* Jewish Ctr Immunol & Resp Med 1400 Jackson St Denver CO 80206

GELFAND, HENRY MORRIS, b New York, NY, Jan 7, 20; m 46; c 4. EPIDEMIOLOGY. *Educ:* Cornell Univ, BS, 40; Univ Chicago, MD, 50; Tulane Univ, MPH, 56. *Prof Exp:* Intern, USPHS Hosp, Staten Island, NY, 50-51; med entomologist, Liberian Inst, WAfrica, 51-53; assoc prof epidemiol, Sch Med, Tulane Univ, 53-59; chief, Enterovirus Unit, Commun Dis Ctr, 59-63, epidemiol adv, Nat Inst Commun Dis, Delhi, India, 63-65; spec asst res & eval, smallpox eradication prog, Commun Dis Ctr, Ga, 65-68; epidemiologist, Foreign Quarantine Prog, London, Eng, 68-70; chief eval br, Off Pop, US AID & mem staff, Off Int Health, USPHS, 70-72; prof & dir epidemiol prog, sch pub health, Univ Ill, 72-84; ADJ PROF, EPIDEMIOL, SCH PUB HEALTH, UNIV SC, 84- *Concurrent Pos:* Consult, Int Pub Health, World Health Orgn & US Agency for Int Develop. *Mem:* Am Epidemiol Soc; Am Pub Health Asn; Int Epidemiol Soc; AAAS. *Res:* Epidemiology of infectious diseases; smallpox eradication; foreign quarantine; health programs evaluation; family planning. *Mailing Add:* Sch Pub Health Univ SC Columbia SC 29208

GELFAND, JACK JACOB, b Newark, NJ, Aug 5, 44. CHEMICAL PHYSICS. *Educ:* Rutgers Univ, BA, 66, PhD(phys chem), 71. *Prof Exp:* Res staff metall, Solid State Div, RCA, 71-72; res assoc, 72-75, res staff astrophys, Dept Astrophys Sci, 75-77, RES STAFF PHYSICS, DEPT MECH & AEROSPACE ENG, PRINCETON UNIV, 77-; AT RCA. *Mem:* Am Chem Soc; Sigma Xi. *Res:* Laboratory chemical physics; vibrational spectroscopy and collisional energy transfer; ionospheric radio propagation. *Mailing Add:* RCA Labs Rte 1 Princeton NJ 08540

GELFAND, NORMAN MATHEW, b New York, NY, Jan 3, 39; m 79; c 3. EXPERIMENTAL HIGH ENERGY PHYSICS & ACCELERATOR PHYSICS. *Educ:* Columbia Univ, AB, 59, MA, 61, PhD(physics), 65. *Prof Exp:* Asst prof physics, Univ Chicago, 64-70, assoc prof, 70-80; MEM STAFF, FERMI NAT ACCELERATOR LAB, 79- *Concurrent Pos:* Sloan fel, 65-67; prog officer, NSF, 77-79. *Mem:* Am Phys Soc. *Res:* Particle physics; computer studies of accelerator lattices; lattice simulations. *Mailing Add:* Fermi Nat Accelerator Lab MS 306 PO Box 500 Batavia IL 60510

GELFANT, SEYMOUR, cell biology; deceased, see previous edition for last biography

GELIEBTER, ALLAN, b Frankfurt, Ger, Jan 22, 47; US citizen. OBESITY, BULIMIA NERVOSA. *Educ:* City Col NY, BS, 68; Columbia Univ, MA, 70, MPhil, 73, PhD(psychol), 76. *Prof Exp:* Adj lectr, Dept Psychol, Herbert H Lehman Col, City Univ NY, 74-76, adj asst prof, 76-78; from asst prof to assoc prof, 76-88, CHMN, DEPT PSYCHOL, TOURO COL, NEW YORK, 81-, PROF, 88- *Concurrent Pos:* Adj instr, Dept Psychol, Touro Col, New York, 75-76; res assoc, Obesity Res Ctr, Dept Med, St Luke's-Roosevelt Hosp Ctr, Columbia Univ, NY, 78-; prin investr grant, Stokely-Van Camp, 78-80, Obesity Core Feasibility Study, NIH, 81-83, biomed grant, St Lukes-Roosevelt Inst Health Sci, 84-85, R-Kane Prod, 89- & 90-; co-leader, Weight Control Unit, St Lukes-Roosevelt Hosp, 82-84, psychologist & instr clin psychol, Depts Psychiat & Med, 88-; investr biomed grant, Bleibtreu Found, 85-86; dir, Weight Control & Eating Disorders, 85-; coprin investr, NIH, 86-89 & St Lukes-Roosevelt Inst Health Sci, 89-90; co-chair, Joint Meeting Soc Study Ingestive Behav & Eastern Psychol Asn, NY, 91. *Mem:* Am Psychol Asn; AAAS; Am Asn Univ Professors; Sigma Xi; NAm Asn Study Obesity; NY Acad Sci; Soc Study Ingestive Behav; Am Soc Clin Nutrit; Am Inst Nutrit. *Res:* Regulation of food intake in obesity and bulimia nervosa; effect of type of exercise on body composition and resting metabolei during dieting in obesity. *Mailing Add:* Dept Med & Psychiat St Luke's-Roosevelt Hosp Amsterdam Ave 114th St WH-10 New York NY 10025

GELINAS, DOUGLAS ALFRED, b Nov 18, 40; US citizen. BOTANY, PHYSIOLOGY. *Educ:* Fitchburg State Col, BS, 63; Purdue Univ, MS, 66, PhD(biol), 68. *Prof Exp:* Asst prof bot, Univ Maine, 68-74, chair, bot & plant path, 75-82, assoc dean acad affairs, 82-89, ASSOC PROF BOT, UNIV MAINE, ORONO, 74-, CHAIR, DEPT PLANT BIOL & PATH, 90- *Mem:* AAAS. *Res:* Self-paced instruction in biology. *Mailing Add:* Univ Maine 216 Deering Hall Orono ME 04469

GELINAS, ROBERT JOSEPH, b Muskegon, Mich, Sept 25, 37; m 60; c 1. COMPUTATIONAL PHYSICS, CHEMICAL KINETICS. *Educ:* Univ Mich, BSE, 60, MSE, 61, PhD(nuclear eng), 65. *Prof Exp:* Physicist & group leader, Lawrence Livermore Lab, 66-75; SR SCIENTIST & MGR, BASIC SCI DIV, SCI APPLICATIONS, INC, 75- *Honors & Awards:* Mark Mills Award, Am Nuclear Soc, 65. *Mem:* AAAS; Am Phys Soc; Am Chem Soc; Combustion Inst Int; Sigma Xi. *Res:* Quantum statistical physics; radiation and plasma physics; non-equilibrium physics; chemical kinetics; combustion science; atmospheric physics; reactive fluid dynamics. *Mailing Add:* 1316 Canyon Side Ave San Ramon CA 94583

GELL, MAURICE L, b Brooklyn, NY, Dec 1, 37; m 60; c 2. METALLURGY. *Educ:* Columbia Univ, BA, 59, MA, 60, BS, 61; Yale Univ, MS, 63, PhD(metall), 65. *Prof Exp:* NSF fel, 65-66; res assoc, Adv Mat Res & Develop Lab, 66-67, sr res assoc & group leader, 67-71, group leader, 71-79, asst mgr, 79-81, MGR, MAT ENG LAB, PRATT & WHITNEY DIV, UNITED TECHNOL CORP, 81- *Honors & Awards:* Eng Mat Achievement Award, Am Soc Metals Int, 86. *Mem:* Fel Am Soc Metals. *Res:* Mechanical properties of metals including fatigue deformation and fracture, cleavage fracture, hydrogen embrittlement, deformation and fracture of gas turbine engine materials; development of gas turbine alloys. *Mailing Add:* Mat Eng Mail Stop 114-43 Pratt & Whitney East Hartford CT 06108

GELLAI, MIKLOS, b Enrod, Hungary, Dec 4, 30. CARDIOVASCULAR & RENAL PHYSIOLOGY. *Prof Exp:* MEM STAFF, SMITH KLINE & FRENCH LABS. *Mailing Add:* Smith Kline & French Labs PO Box 1539 L-521 King of Prussia PA 19406-0939

GELLER, ARTHUR MICHAEL, b New York, NY, Dec 18, 41; m; c 2. BIOCHEMISTRY. *Educ:* City Col New York, BS, 62; Duke Univ, PhD(biochem), 67. *Prof Exp:* Am Cancer Soc fel, Enzyme Inst, Univ Wis, 67-68; ASSOC PROF BIOCHEM, UNIV TENN, MEMPHIS, 68- *Concurrent Pos:* Vis scientist, Nat Inst Arthritis, Metab & Digestive Dis, NIH, 76-77; vis prof, Howard Hughes Med Inst, Duke Univ Med Ctr, 85; southeast regional dir & mem bd dirs, Sigma Xi, 84-91. *Mem:* AAAS; Sigma Xi; Am Soc Biol Chemists; Am Asn Univ Professors; Asn Res Vision & Ophthal. *Res:* Biochemistry of the lens; methionine adenosyltransferase; methylation; cataracts. *Mailing Add:* Dept Biochem Univ Tenn Memphis TN 38163

GELLER, DAVID MELVILLE, b Detroit, Mich, Dec 30, 30. BIOCHEMISTRY. *Educ:* Amherst Col, AB, 52; Harvard Univ, PhD(biochem), 57. *Prof Exp:* NSF fel, Oxford Univ, 57-58; instr biochem, Dept Chem, Univ Ill, 58-59; asst prof, 59-67, ASSOC PROF PHARMACOL, SCH MED, WASHINGTON UNIV, 67- *Res:* Oxidative and photophosphorylation; phosphorylation mechanisms; serum albumin biosynthesis and secretion. *Mailing Add:* Dept Pharmacol Wash Univ Sch Med 660 S Euclid Ave Box 8103 St Louis MO 63110

GELLER, EDWARD, b New York, NY, Dec 6, 28; m 52; c 3. NEUROCHEMISTRY. *Educ:* Univ Calif, Los Angeles, AB, 47, BS, 48, MS, 55, PhD(biochem), 56. *Prof Exp:* Asst biochem, 52-55, asst res biol chemist, Med Ctr, 59-65, asst prof, 65-70, ASSOC PROF PSYCHIAT, MED CTR, UNIV CALIF, LOS ANGELES, 70- *Concurrent Pos:* Res biochemist, Vet Admin, 57- *Mem:* Int Soc Neurochem; Am Col Neuropsychopharmacol; Endocrine Soc; Am Soc Neurochem; Int Soc Psychoneuroendocrinol; Sigma XI. *Res:* Relation of biochemistry to mental processes; biochemical correlates of behavior. *Mailing Add:* 22907 Gershwin Dr Woodland Hills CA 91364

GELLER, HARVEY, b New York, NY, July 6, 21; m 46; c 3. ANALYTICAL STATISTICS, RESEARCH MANAGEMENT. *Educ:* Brooklyn Col, AB, 43. *Prof Exp:* Statistician, Bur Rec & Statist, New York Health Dept, 46-49, Bur Cancer Control, DC Health Dept, 49-51, Div Civilian Health Req, USPHS, 51-53, Off Surgeon Gen, Dept Air Force, 53 & Sch Aviation Med, 54-55; supvry statistician, Chronic Dis Prog, USPHS, 55-59, chief opers studies, Cancer Control Prog, 59-68; head spec cancer surv sect, Nat Cancer Inst, NIH, 68-73, head field liaison sect, Biomet Br, 73-81; RETIRED. *Mem:* AAAS; fel Am Pub Health Asn; Am Statist Asn; Soc Epidemiol Res; Fedn Am Scientists. *Res:* Collection and analysis of cancer morbidity and mortality statistics; testing and evaluation of new cancer screening and diagnostic techniques and instruments; studies in the application of screening tests for cancer. *Mailing Add:* 116 Northway Greenbelt MD 20770

GELLER, HERBERT M, b New York, NY, Feb 20, 45. NEUROPHYSIOLOGY, NEUROPHARMACOLOGY. *Educ:* City Col New York, BEE, 65; Case Western Reserve Univ, PhD(bioeng), 70. *Prof Exp:* NIH fel physiol, Univ Rochester, 70-72; asst prof, 72-78, ASSOC PROF PHARMACOL, RUTGERS MED SCH, 78 - *Mem:* AAAS; Soc Neurosci; Biophys Soc; Am Physiol Soc; Am Soc Pharmacol & Exp Therapeut. *Res:* Physiology and pharmacology of neurotransmission; developmental neurobiology. *Mailing Add:* Dept Pharmacol Robert Wood Johnson Med Sch 675 Hoes Lane Piscataway NJ 08854-5635

GELLER, IRVING, b Boston, Mass, Oct 26, 25. EXPERIMENTAL PSYCHOLOGY. *Educ:* American Univ, PhD(psychol), 57. *Prof Exp:* DIR BEHAV PHARMACOL, SOUTHWESTERN FOUND BIOMED RES, 74- *Mem:* AAAS; Am Psychol Asn; Int Soc Biochem Pharmacol; Psychonomic Soc; Sigma Xi. *Mailing Add:* Nat Inst Drug Abuse Rm 10A-16 Parklawn Bldg 5600 Fishers Lane Rockville MD 20857

GELLER, KENNETH N, b Brooklyn, NY, Sept 22, 30; m 55; c 2. PHYSICS. *Educ:* Brooklyn Col, BS, 52; Univ Pa, PhD(physics), 60. *Prof Exp:* Res asst physics, Univ Pa, 55-60, asst prof, 60-66; assoc prof physics, Drexel Univ, 66-76, assoc prof atmospheric sci, 74-76, assoc dean grad sch, 74-83, dir, Ctr Multidisciplinary Study & Res, 76-83, dir, Off Sponsored proj, 83-87, PROF PHYSICS & ATMOSPHERIC SCI, DREXEL UNIV, 76-, ASST VPRES, RES & TECHNOL MGT, 87- *Mem:* AAAS; Am Phys Soc; Technol Transfer Soc; Licensing Exec Soc; Soc Res Adminrs; Asn Univ Tech Managers. *Res:* Study of low energy nuclear reactions; neutron detection techniques; lidar measurements of the atmosphere. *Mailing Add:* 7637 Brookhaven Rd Philadelphia PA 19151

GELLER, MARGARET JOAN, b Ithaca, NY, Dec 8, 47. COSMOLOGY. *Educ:* Univ Calif, Berkeley, AB, 70; Princeton Univ, MA, 72, PhD(physics), 75. *Prof Exp:* Fel theoret astrophys, Ctr Astrophys, 74-76; res assoc, 76-80, asst prof, 80-83, PROF ASTRON, HARVARD UNIV, 88-; ASTROPHYSICIST, SMITHSONIAN ASTROPHYS OBESERV, 83- *Concurrent Pos:* Mem Nat Comts, NSF, NASA & Am Astron Soc. *Honors & Awards:* MacArthur Award, 90; Newcomb-Cleveland Prize, Am Acad Arts & Sci, 90. *Mem:* Am Astron Soc; Int Astron Union; AAAS. *Res:* Nature and history of the galaxy distribution; the origin and evolution of galaxies; x-ray astronomy. *Mailing Add:* Ctr Astrophys 60 Garden St Cambridge MA 02138

GELLER, MARVIN ALAN, b Boston, Mass, Mar 19, 43; m 68; c 2. METEOROLOGY, PHYSICAL OCEANOGRAPHY. *Educ:* Mass Inst Technol, BS, 64, PhD(meteorol), 69. *Prof Exp:* From asst prof to assoc prof elec eng & meteorol, Univ Ill, Urbana-Champaign, 69-77; prof meteorol & phys oceanog, Univ Miami, 77-85; DIR, INST ATMOSPHERIC SCI, STATE UNIV NY, STONY BROOK. *Mem:* Am Meteorol Soc; Am Geophys Union. *Res:* Atmospheric dynamics with emphasis on atmospheric waves; climate variability studies; upper atmosphere; ocean dynamics. *Mailing Add:* Inst Atmospheric Sci State Univ NY B106 Physics Bldg Stony Brook NY 11794

GELLER, MILTON, b New York, NY, July 22, 22; m 49; c 2. ANALYTICAL & PHARMACEUTICAL CHEMISTRY. *Educ:* City Col New York, BS, 44; Brooklyn Col, MA, 60. *Prof Exp:* Anal chemist, Chas Pfizer & Co, Inc, 43-44 & 46-50; chemist, US Signal Corps, 50-51; anal chemist, Chase Chem Co, 51; anal chemist, Nepera Chem Co, 51-58, supvr qual control, 58; sr scientist, Warner Lambert Res Inst, 58-62, sr res assoc, 62-63, dir, Appl Anal Res, 63-66; sr scientist, Hoffmann La-Roche, Inc, Nutley, 66-78. *Mem:* Am Chem Soc; Am Pharmaceut Asn. *Res:* Methods development for pharmaceutical dosage forms. *Mailing Add:* 40-10 Kramer Pl Fairlawn NJ 07410

GELLER, MYER, b Winnipeg, Man, Can, Oct 24, 26; US citizen; m 54. ELECTROOPTICS. *Educ:* Univ Man, BS, 46; Univ Minn, MS, 48; Mass Inst Technol, PhD(physics), 55. *Prof Exp:* Mem tech staff semiconductors, Hughes Aircraft Co, Calif, 55-60, mem tech staff lasers, 62-64; sr scientist, Electro Optical Systs Inc, Calif, 60-62; RES PHYSICIST, US NAVY ELECTRONICS LAB, 64- *Mem:* Am Phys Soc; Optical Soc Am; Inst Elec & Electronics Eng. *Res:* Coherent optics, especially stimulated emission devices and non-linear optics; spectroscopy, atomic and plasma physics; optical atmospheric propagation. *Mailing Add:* 4009 Ventura Cyn Sherman Oaks CA 91423

GELLER, NANCY L, b New York, NY, Nov 3, 44. MATHEMATICAL STATISTICS, STATISTICS. *Educ:* City Col New York, BS, 65; Case Inst Technol, MS, 67; Case Western Reserve Univ, PhD(math), 72. *Prof Exp:* Asst prof statist, Univ Rochester, 70-72; asst prof statist, Univ Pa, 72-78; asst prof, 78-80, PROF COMMUNITY & PREV MED, MED COL PA, 80- *Mem:* Inst Math Statist; Am Statist Asn; Am Math Soc. *Res:* nonparametric statistics; use of citations in the study of science; health systems statistics. *Mailing Add:* Biostat Lab Sloan Kettering Cancer Ctr 1275 York Ave New York NY 10021

GELLER, ROBERT JAMES, b New York, NY, Feb 9, 52; m 78, 90. SEISMOLOGY. *Educ:* Calif Inst Technol, BS, 73, MS, 75, PhD(geophys), 77. *Prof Exp:* Res fel geophys, Calif Inst Technol, 77-78; asst prof geophys, Stanford Univ, 78-84; ASSOC PROF, TOKYO UNIV, 84- *Concurrent Pos:* Consult asst prof, Stanford Univ, 77-78; mem bd dir, Seismol Soc Japan, 88-; Guggenheim fel, 85. *Mem:* Am Geophys Union; Seismol Soc Am; Soc Explor Geophysicists; Royal Astron Soc; Sigma Xi. *Res:* Seismology; free oscillations; earthquake mechanisms and tectonics; earth structure. *Mailing Add:* Fac Sci Dept Geophys Tokyo Univ Bunkyo-Ku Tokyo 113 Japan

GELLER, RONALD G, b Peoria, Ill, Jan 15, 43; m 71; c 3. RESEARCH ADMINISTRATION, SCIENCE ADMINISTRATION. *Educ:* Univ Wis-Madison, BS, 64, PhD(physiol), 69. *Prof Exp:* Res assoc pharmacol, Nat Heart Inst, NIH, 69-71, sr staff fel, 71-72, grants assoc, 72-73, asst chief, Hypertension & Kidney Dis Br, Nat Heart & Lung Inst, 73-75, chief hypertension, 75-78, assoc dir vision res, Nat Eye Inst, 78-87, dir div prog anal, Off Sci Policy & Legis, 87-89, DIR DIV EXTEN AFFIL, NAT HEART LUNG & BLOOD INST, NIH, 89- *Concurrent Pos:* Mem, Coun High Blood Pressure Res, Am Heart Asn. *Mem:* Am Physiol Soc; Am Heart Asn; Soc Exp Biol & Med. *Res:* Reflex control of blood pressure; humoral control of blood pressure; physiology and pharmacology of the kallikrein-kinin system. *Mailing Add:* NIH Westwood Bldg Rm 7A17 Bethesda MD 20892

GELLER, SEYMOUR, b New York, NY, Mar 28, 21; m 42; c 2. SOLID STATE PHYSICS & CHEMISTRY. *Educ:* Cornell Univ, AB, 41, PhD(phys chem), 49. *Prof Exp:* Du Pont fel, Cornell Univ, 49-50; res chemist, Benger Lab, E I du Pont de Nemours & Co, 50-52; struct chemist, Bell Tel Labs, Inc, 52-64; struct chemist & group leader, Struct Properties Group, NAm Rockwell Sci Ctr, Calif, 64-71; prof, 71-90, EMER PROF ELEC ENG, UNIV COLO, BOULDER, 90- *Concurrent Pos:* Fac fel, Univ Colo Coun Res & Creative Work, 77-78; Croft res prof, Col Eng & Appl Sci, Univ Colo, 80; res award, Col Eng & Appl Sci, Univ Colo, 79, Croft res prof, 80, res lectr, 85, fac fel, 85-86; Creativity Award, NSF, 83-87. *Mem:* Fel Am Phys Soc; fel Mineral Soc Am; fel Inst Elec & Electronics Engrs; Sigma Xi. *Res:* Relations of properties to crystal structure; crystal chemistry; magnetic, superconducting, and semiconducting materials; solid electrolytes; high pressure phases; phase transitions; structures of inorganic and intermetallic compounds. *Mailing Add:* Dept Elec & Comput Eng Univ Colo Campus Box 425 Boulder CO 80309-0425

GELLER, STEPHEN ARTHUR, b Brooklyn, NY, Apr 26, 39; m 62; c 2. HEPATOPATHOLOGY, AUTOPSY PATHOLOGY. *Educ:* Brooklyn Col, NY, BA, 59; Howard Univ, Washington, DC, MD, 64. *Prof Exp:* From asst prof to assoc prof path, Mt Sinai Sch Med, NY, 71-78, actg chmn, 75-78, vchmn & prof, 78-84; CHMN PATH, CEDARS-SINAI MED CTR, LOS ANGELES, 84-; PROF PATH, UNIV CALIF, LOS ANGELES, 84- *Concurrent Pos:* Chmn, Anat Path Comt, Col Am Pathologists, 84-85; consult, Off Med Examr, Los Angeles County, 89-; prof lectr, Mt Sinai Sch Med, NY, 90- *Mem:* Am Asn Study Liver Dis; Col Am Pathologists; Am Soc Clin Pathologists; US-Can Acad Path; Hans Popper Hepatopath Soc (secy-treas, 89-); Am Asn Hist Med. *Res:* Mechanism of liver injury and hepatocarcinogenesis in a transgenic mouse model for anlpha-l-antitrypsin deficiency; tissue growth factors in liver tumors; manifestations of Epstein-Barr virus infection in human livers after transplantation. *Mailing Add:* 8700 Beverly Blvd Los Angeles CA 90048

GELLER, SUSAN CAROL, b Newark, NJ, Oct 27, 48. K-THEORY, CYCLIC HOMOLOGY. *Educ:* Case Inst Technol, BS, 70; Cornell Univ, MS, 72, PhD(math), 75. *Prof Exp:* Teaching asst math, Cornell Univ, 70-75; asst prof math, Purdue Univ, W Lafayette, 75-81; assoc prof, 81-89, PROF MATH, TEX A&M UNIV, 89- *Concurrent Pos:* Fac fel, Bunting Inst Radcliffe Col, Harvard Univ, 80-82; NSF vis prof women, Rutgers Univ, 87-88. *Mem:* Am Math Soc; Math Asn Am; Am Women Math. *Res:* Devise new techniques to compute K-theory of various rings with an emphasis on excision problems, negative K-theory and cyclic homology. *Mailing Add:* Dept Math Texas A&M Univ Col Sta TX 77843-3368

GELLERT, MARTIN FRANK, b Prague, Czech, June 5, 29; m 55, 74. BIOCHEMISTRY OF DNA. *Educ:* Harvard Univ, BA, 50; Columbia Univ, PhD(chem), 56. *Prof Exp:* Fel USPHS, Naval Med Res Inst, Md, 57-58; asst prof biochem, Dartmouth Med Sch, 58-59; res chemist, 59-69, CHIEF, SECT METAB ENZYMES, LAB MOLECULAR BIOL, NAT INST DIABETES, DIGESTIVE & KIDNEY DIS, NIH, 69- *Honors & Awards:* Merck Award Biochem, Am Soc Biol Chemists, 85; Richard Lounsbery Award, 85. *Mem:* Nat Acad Sci; AAAS; Am Soc Biochem & Molecular Biol. *Res:* Molecular genetics; enzymes of DNA synthesis and recombination. *Mailing Add:* Lab Molecular Biol NIDDK NIH Bethesda MD 20892

GELLERT, RONALD J, b New York, NY, July 24, 35; m 59. PHYSIOLOGY. *Educ:* NY Univ, BA, 57; Univ Calif, Berkeley, MA, 59, Univ Calif, San Francisco, PhD(physiol), 63. *Prof Exp:* NIH fel, Univ Calif, San Francisco, 63-64, Oxford Univ, 64-65 & Harvard Sch Dent Med, 65-67; sr investr, Pac Northwest Res Found, 67-70; res asst obstet & gynec, Sch Med, Univ Wash, 70-83; CLIN SCIENTIST, REPROD GENETICS, SWED HOSP MED CTR, 83- *Mem:* AAAS; Endocrine Soc; Soc Study Reproduction. *Res:* Neuroendocrinology of reproductive processes; control of onset of puberty; role of pars tuberalis of the pituitary in physiology of reproduction; pesticides and reproduction; hexachlorophene and neuroendocrine function; mechanism of action of certain anti-ovulatory drugs; vitamin A and reproductive function; xenobiotics and reproductive function, marihuana and reproductive function. *Mailing Add:* Reprod Genet Swed Hosp Med Ctr 747 Summit Ave Seattle WA 98195

GELLES, DAVID STEPHEN, b Leicester, UK, Mar 19, 45; US citizen; m 74; c 2. MATERIALS SCIENCE, METALLURGY. *Educ:* Harvey Mudd Col, BSc, 66; Mass Inst Technol, MSc, 68, ScD(phys metall), 71. *Prof Exp:* sr scientist mat develop, Hanford Eng Develop Lab, 74, prin engr, 82-87; sr engr, 87-88, STAFF ENGR, FUSION MAT, PAC NORTHWEST LAB, 88- *Concurrent Pos:* Sr res fel, Berkeley Nuclear Labs, Berkeley, Glos, UK, 71-74. *Mem:* Am Soc Metals; Am Inst Metall Engrs; Am Soc Testing & Mat. *Res:* Material response to radiation damage, microstructure/property relationships; ferritic alloy development for breeder reactor applications; ferritic alloy development for fusion reactor applications. *Mailing Add:* Pac Northwest Lab PO Box 999 P8-15 Richland WA 99352

GELLES, ISADORE LEO, b Philadelphia, Pa, Dec 15, 25; m 53; c 3. PHYSICS. *Educ:* Temple Univ, BA, 51, MA, 54. *Prof Exp:* Physicist, Pitman-Dunn Lab, Frankford Arsenal, 51-52; acoust engr, Radio Corp Am, 52-53; sr engr semiconductors, Res Div, Philco Corp, 53-54; mem staff microwave & optical spectros, Res Ctr, Int Bus Mach Corp, 56-61; physicist & group supvr, Res Lab, Am-Standard Corp, 61-63; from physicist to sr physicist, Ledgemont Lab, Kennecott Copper Corp, 63-73; RETIRED. *Concurrent Pos:* Consult, 73- *Mem:* Fel AAAS; Am Phys Soc; Sigma Xi; Metall Soc; Am Soc Metals. *Res:* Physical properties of refractory and nonferrous metals; audio spectrum analysis; semiconductor transport properties; magnetic resonance in solids; photoconductivity; ultrasonics in solids; crystal physics; holography. *Mailing Add:* 2423 Chilham Pl Potomac MD 20854

GELLES, S(TANLEY) H(AROLD), b Boston, Mass, Sept 12, 30; m 56; c 3. PHYSICAL METALLURGY. *Educ:* Mass Inst Technol, SB, 52, SM, 54, ScD, 57. *Prof Exp:* Asst metall, Mass Inst Technol, 53-57; proj leader phys metall, Nuclear Metals, Inc, Mass, 57-60, proj mgr, 60-63; phys res metallurgist, Ledgemont Lab, Kennecott Copper Corp, 63-68; assoc chief, Battelle Mem Inst, 68-75; dir, S H Gelles Assocs, 76-86, PRES, GELLES LABS, INC, 86- *Concurrent Pos:* Consult, Beryllium Comt, Mat Adv Bd, Nat Acad Sci, 64-67; consult, ad hoc comt high pressure technol, Mat Adv Bd, Nat Acad Sci, 72, Beryllium Metal Supply Options, 88-89. *Mem:* Am Inst Mining, Metall & Petrol Engrs; Am Soc Metals; Nat Asbestos Coun; Am Soc Testing Mat. *Res:* Metallurgy of beryllium; phase equilibria; metal deformation; high pressure technology; microstructure/property relations in metals; effect of gravity on microstructural development in alloys. *Mailing Add:* 2485 Wimbledon Rd Columbus OH 43220

GELLHORN, ALFRED, b St Louis, Mo, June 4, 13; m 39; c 4. MEDICAL EDUCATION, HEALTH POLICY. *Educ:* Washington Univ, MD, 37; Am Bd Internal Med, dipl, 56, cert, 77. *Hon Degrees:* ScD, Amherst Col, 69, City Col NY, 79, State Univ NY, 85, Albany Med Col, 86. *Prof Exp:* House officer surg, Barnes Hosp, St Louis, Mo, 37-39; house officer gynec, Passavant Hosp, Chicago, 39-40; res fel, Dept Embryol, Carnegie Inst & Johns Hopkins Univ, 40-43; from asst prof physiol to assoc prof pharmacol, Col Physicians & Surgeons, Columbia Univ, 43-46, from assoc prof pharmacol to assoc prof med, 46-58, prof med, 58-68, dir inst cancer res, 52-68; prof med & pharmacol, Dean Sch Med & dir med ctr, Univ Pa, 68-74; dir, Ctr Biomed Educ, City Col New York, 74-79, vpres health affairs, 74-79; vis prof, Sch Pub Health, Harvard Univ, 80-83; DIR MED AFFAIRS, NY STATE DEPT HEALTH, 83- *Concurrent Pos:* Chief med serv, Francis Delafield Hosp, 50-68; from assoc attend physician to attend physician, Presby Hosp, 56-68; vis clin prof, Albert Einstein Col Med, Yeshiva Univ, 56-68; pres, Coun Int Orgn Med Sci, 68-80; mem adv comt, USPHS, Am Cancer Soc & Nat Res Coun; vis prof, Sch Public Health, Harvard Univ, 80- *Mem:* Inst Med-Nat Acad Sci; Am Asn Cancer Res (pres, 63); Asn Am Physicians; Am Soc Pharmacol & Exp Therapeut; Am Soc Biol Chemists; fel Am Col Physicians. *Res:* Placental physiology; circulatory pharmacology; chemotherapy; infectious disease; clinical research in malignancy; mechanism of action of anti-tumor drugs on clinical cancer chemotherapy; medical aspects of neoplastic disease; lipid metabolism; medical oncology; medical education; medical ethics. *Mailing Add:* NY State Dept Health Tower Bldg Rm 910 Empire State Plaza Albany NY 11231

GELLIS, SYDNEY SAUL, b Claremont, NH, Mar 6, 14; m 39; c 2. MEDICINE. *Educ:* Harvard Univ, AB, 34, MD, 38. *Prof Exp:* Instr pediat, Johns Hopkins Univ, 43-46; asst prof, Harvard Univ, 46-56; prof & chmn dept, Sch Med, Boston Univ, 56-65; prof pediat & chmn dept, Sch Med, Tufts Univ, 65-81; RETIRED. *Concurrent Pos:* Consult, Surgeon Gen, 43-47; physician, Children's Med Ctr, Boston, 47-56, consult, 56-; pediatrician-in-chief, Beth Israel Hosp, Boston, 50-56 & New Eng Med Ctr Hosps, 65-81; ed, Year Book Pediat, 52-80; assoc ed, Am J Dis Children; lectr, Harvard Med Sch, 56-; dir pediat, Boston City Hosp, 56-65, consult pediat, 65-; lectr, Sch Med, Boston Univ, 65-; consult pediat, USAF, 69- *Honors & Awards:* Abraham Jacobi Award, Am Acad Pediat. *Mem:* Am Pediat Soc; Soc Pediat Res (secy, 52-58, pres, 58-59); cor mem Fr Soc Pediat; hon mem NZ Pediat Soc. *Res:* Liver disease and jaundice; gamma globulin; hepatitis. *Mailing Add:* 20 Ash St Boston MA 02111

GELLMAN, CHARLES, b New York, NY, Dec 18, 16; m 48; c 2. INDUSTRIAL ENGINEERING. *Hon Degrees:* LHD, New York Col Podiatric Med, 75. *Prof Exp:* Indust specialist, War Dept, 42-45; works mgr, Para Equip Co, 45-47 & Heppe Hudson Co, Inc, 47-50; exec purchasing off & tech adv supply mission, Israel, 50-53; tech dir, Am Technion Soc, 53-55; exec dir & consult, Grand Cent Hosp, New York, 55-63; pres & dir, Jewish Mem Hosp, 63-83; RETIRED. *Concurrent Pos:* Consult, Webb & Knapp Construct Co; pres, Greater New York Hosp Asn, 71-72; chmn, Hosp Asn NY State, 74-75. *Mem:* Am Soc Metals; Soc Automotive Engrs; fel Am Pub Health Asn; fel Royal Soc Health; NY Acad Sci. *Mailing Add:* 12 Birch Lane Green Acres Valley Stream NY 11581

GELLMAN, ISAIAH, b Akron, Ohio, Feb 19, 28; m 47; c 2. SANITATION. *Educ:* City Col New York, BChE, 47; Rutgers Univ, MS, 50, PhD(sanit), 52. *Prof Exp:* Res assoc sanit, Rutgers Univ, 48-52; process engr air & water, Abbott Labs, 52-56; regional engr pulp & paper-air & water pollution, Nat Coun Stream Improv, 56-67, asst tech dir, NY, 67-69, tech dir, 69-76, exec vpres, 76-87, PRES, NAT COUN PAPER INDUST FOR AIR & STREAM IMPROV, INC, 87- *Honors & Awards:* Tappi Fel & Environ Div Award. *Mem:* Water Pollution Control Fedn; Air Pollution Control Asn; Am Inst Chem Engrs; fel Tech Asn Pulp & Paper Indust. *Res:* Air and water pollution control; water resources development; treatment of gaseous and liquid effluents from pulp and paper production to prevent pollution problems. *Mailing Add:* NCASI 260 Madison Ave New York NY 10016

GELL-MANN, MURRAY, b New York, NY, Sept 15, 29; wid; c 2. THEORETICAL PHYSICS. *Educ:* Yale Univ, BS, 48; Mass Inst Technol, PhD(physics), 51. *Hon Degrees:* DSc, Yale Univ, 59, Univ Chicago, 67, Univ Ill, 68, Wesleyan Univ, 68, Univ Turin, 69, Univ Utah, 70, Columbia Univ, 77, Cambridge Univ, Eng, 80. *Prof Exp:* Mem sch math & physics, Inst Advan Study, 51-52; from instr to assoc prof physics, Inst Nuclear Studies Univ Chicago, 52-55; from assoc prof to prof physics, 55-67, MILLIKAN PROF THEORET PHYSICS, LAURITSEN LAB HIGH ENERGY PHYSICS, CALIF INST TECHNOL, 67- *Concurrent Pos:* Vis assoc prof, Columbia Univ, 54; mem, Inst Advan Study, 55 & 67-68; vis prof, Col France & Univ Paris, 59-60, Mass Inst Technol, 63 & Europ Orgn Nuclear Res, Geneva, Switz, 71-72 & 79-80; overseas fel, Churchill Col, Cambridge, Eng, 66; consult, President's Sci Adv Comt, 69-72; chmn, bd trustees, Aspen Ctr Physics, 73-79; citizen regent, Smithsonian Inst, 74-; mem, Sci & Grants Comt, Leakey Found, 77-; dir, J D & C T MacArthur Found, 79-; mem, Coun Foreign Relations; mem bd trustees, Santa Fe Inst, 82-85, co-chmn sci bd, 85- *Honors & Awards:* Nobel Prize Physics, 69; Heineman Prize, Am Inst Physics, 59; Ernest O Lawrence Award, Dept Energy, 66; Franklin Medal, Franklin Inst Philadelphia, 67; Joh J Carty Medal, Nat Acad Sci, 68; Res Corp Award, 69. *Mem:* Nat Acad Sci; Am Acad Arts & Sci; Am Phys Soc; foreign mem Royal Soc; hon mem French Phys Soc. *Res:* elementary particle theory; disperson or "S-matrix" theory; the renormalization group; strangeness; the weak interaction; broken symmetry as in the "eightfold way"; quarks; quantum chromodynamics. *Mailing Add:* Lauritsen Lab High Energy Phys Calif Inst of Technol Pasadena CA 91125

GELMAN, DONALD, b Brooklyn, NY, Sept 13, 38; m 62; c 2. FEYNMAN PATH INTEGRALS, VARIATIONAL BOUNDS. *Educ:* City Univ NY, BS, 59; NY Univ, MS, 64, PhD(physics), 69. *Prof Exp:* Teaching asst physics, Columbia Univ, 59-61; res physicist, Kollsman Instrument Corp, 61-64; instr sci, NY City Community Col, 64; ASSOC PROF PHYSICS, C W POST CAMPUS, LONG ISLAND UNIV, 64- *Mem:* Am Asn Physics Teachers; Sigma Xi. *Res:* feynman path integrals and variational bounds in scattering theory; astronomy. *Mailing Add:* Dept Physics CW Post Campus Long Island Univ Greenvale NY 11548

GELMAN, HARRY, b New York, NY, May 23, 35; m 57; c 5. ANTENNA DESIGN & ANALYSIS, DIGITAL SWITCHING SYSTEMS & NETWORKS. *Educ:* City Col New York, BS, 57; NY Univ, PhD(physics), 64. *Prof Exp:* Asst physics, Columbia Univ, 57-58 & NY Univ, 58-59; instr math & physics, US Merchant Marine Acad, 64; staff mem physics, Sandia Corp, 64-66; mem tech staff, Mitre Corp, 66-77; prin scientist, Physical Science Inc, 77-79; sr eng specialist, 79-82, eng mgr, 82-89, SR MEM TECH STAFF, GTE GOVT SYST CORP, 89- *Concurrent Pos:* Lectr, Northeastern Univ, 67-71. *Mem:* Am Asn Physics Teachers; Am Phys Soc. *Res:* Wave propagation; eikonal methods; theory of scattering; plasma physics; ionospheric physics; quantum theory of fields; special theory of relativity; general relativity; quantum theory of measurement. *Mailing Add:* Command Control & Commun Systs Commun Systs Div GTE Govt Syst Corp 77 A St Needham MA 02194-2892

GELMAN, SIMON, b May 26, 36; m; c 2. ANESTHESIOLOGY. *Educ:* First Leningrad Med Inst, USSR, MD, 59; 25 Oct-Hosp, Leningrad, PhD, 65; Leningrad Dzanelidze Inst First Aid, DSc, 73. *Prof Exp:* Head, Surg Off Polyclin, Siktivkar, USSR, 59-61; physician-resuscitationist, Ctr Treatment Patients Myocardial Infarction, Leningrad, USSR, 64-65; asst prof, Dept Anesthesiol, Leningrad Kirov Advan Training Inst Doctors, 65-73; sr anesthesiologist, Beilinson Med Ctr & Med Sch, Tel-Aviv Univ, Petah Tikva, Israel, 74-75; fel, Dept Anesthesiol, Case Western Reserve Univ, Cleveland, Ohio, 76-77, resident anesthesiol, Dept Anesthesiol, Univ Hosp, 77-78; assoc prof, Dept Anesthesiol, Sch Med, Univ Ala Birmingham, 78-81, dir clin res, 79-84, vchmn res, 84-89, PROF, DEPT ANESTHESIOL, SCH MED, UNIV ALA, BIRMINGHAM, 81-, CHMN DEPT, 89-, PROF, DEPT PHYSIOL & BIOPHYS, 89- *Concurrent Pos:* Mem res comt, Soc Cardiovasc Anesthesiologists, 81-82; vis prof, numerous insts, asns & univs, 81-91; mem sci adv bd, Asn Univ Anesthetists, 85-88 & ASA Comt Res, 88- *Mem:* Israel Soc Anesthesiol; Am Soc Anesthesiologists; Soc Cardiovasc Anesthesiologists; Asn Univ Anesthetists; Int Anesthesia Res Soc; Am Physiol Soc. *Mailing Add:* Dept Anesthesiol 845 Jefferson Tower Univ Ala 619 S 19th St Birmingham AL 35233-1924

GELMANN, EDWARD P, b New York, NY, May 31, 50. MOLECULAR BIOLOGY, GENETICS. *Educ:* Yale Univ, BS, 72; Stanford Univ, MD, 76. *Prof Exp:* Intern & resident internal med, Univ Chicago Hosps & Clins, 76-78; fel, 78-82, SR INVESTR ONCOL, NAT CANCER INST, NIH, 82- *Mem:* Am Col Physicians; Am Soc Microbiol. *Res:* Role of oncogenes in hormone dependent human neoplasma. *Mailing Add:* Lambardi Cancer Res Ctr 3800 Reservoir Rd NW Washington DC 20007

GELNOVATCH, V, ELECTRICAL ENGINEERING. *Educ:* Monmouth Col, BS, 63; NY Univ, MS, 66. *Prof Exp:* Signal Corps Sgt, US Army, 56-59; DIR MICROWAVE & SIGNAL PROCESSING DEV DIV, US ARMY ELECTRONICS TECHNOL & DEVICES LAB, 63- *Concurrent Pos:* Site chief, Holenstadt Microwave Radio Relay Sta, Ger; Mem exec & tech comt, Int Solid State Circuits Conf, 68-79; participant, Inst Elec & Electronics Engrs & USSR Popov Soc Exchange Prog, 74; army mem, Adv Group Electron Dev; chmn, Army Millimeter Wave Steering Group, Dept Defense; assoc ed, Microwave J; mem adv bd, Elec Eng Dept & vis prof elec eng, Univ Va; rep, Solid State Circuits Coun. *Mem:* Fel Inst Elec & Electronics Engrs. *Res:* Microwave solid state devices; microwave circuit design; optimization; measurement and synthesis; microwave transistor amplifiers; reflectometer modeling. *Mailing Add:* Microwave & Signal Processing Div US Army Eradcom Attn Delet-M Ft Monmouth NJ 07703

GELOPULOS, DEMOSTHENES PETER, b Valparaiso, Ind, Apr 24, 38; m 58; c 3. ELECTRICAL ENGINEERING. *Educ:* Valparaiso Univ, BS, 60; Univ Notre Dame, MS, 62; Univ Ariz, PhD(elec eng), 67. *Prof Exp:* Res engr comput, Northrop Nortronics Res Park, 62-63; asst prof elec eng, Univ Akron, 67-68; prof elec eng, Ariz State Univ, 68-; AT DEPT ENG, VALPARISO UNIV. *Concurrent Pos:* Consult, TRW, Inc, Euclid, Ohio, 67-68, Govt Electronics Div, Motorola, 68-69, Ariz Pub Serv Co, 68-78 & Boeing Comput Serv, 77-; prin investr, Res Prog at Ariz State Univ, Elec Power Res Inst, 76- *Honors & Awards:* Phoenix Outstanding Achievement Award, Inst Elec & Electronics Engrs, 73. *Mem:* Sr mem Inst Elec & Electronics Engrs. *Res:* Computer simulation of engineering systems; power systems dynamics. *Mailing Add:* Gellersen Ctr Valparaiso Univ Valparaiso IN 46383

GELPERIN, ALAN, b Cincinnati, Ohio, July 28, 41; m 82; c 5. NEUROPHYSIOLOGY, BEHAVIORAL BIOLOGY. *Educ:* Carleton Col, BA, 62; Univ Pa, PhD(zool), 66. *Prof Exp:* from asst prof to prof, Princeton Univ, 68-82; MEM TECH STAFF, BIOPHYS DEPT, AT&T BELL LABS, 82- *Concurrent Pos:* Dir, Neural Systs & Behav, Marine Biol Lab, Woods Hole, MA, 77-79; Guggenheim fel, 73. *Honors & Awards:* Newcomb-Cleveland Prize, AAAS, 71. *Mem:* Soc Neurosci; Biophys Soc; Am Physiol Soc. *Res:* Synaptic physiology and learning. *Mailing Add:* Biophysics Res Dept Bell Labs 600 Mountain Ave Murray Hill NJ 07974

GELTMAN, SYDNEY, b Philadelphia, Pa, May 23, 27; m 53; c 3. THEORETICAL PHYSICS. *Educ:* Yale Univ, BS, 48, MS, 49, PhD (physics), 52. *Prof Exp:* Physicist, Westinghouse Res Labs, 52-54; appl physics lab, Johns Hopkins Univ, 54-57; PHYSICIST, JOINT INST LAB ASTROPHYS, NAT BUR STANDARDS, 57- *Concurrent Pos:* Adj prof, Univ Colo, 63-; lectr, Univ Col, London, 66-67; consult Cen Saclay, 78-81, Livermore Nat Lab, 84-85. *Honors & Awards:* A V Humboldt US Sr Scientist Award, 82-83. *Mem:* Fel Am Phys Soc. *Res:* Theory of atomic scattering processes and ionization; atomic radiative processes. *Mailing Add:* Joint Inst Lab Astrophysics Nat Bur Standards & Univ Colo Boulder CO 80309-0440

GELUSO, KENNETH NICHOLAS, b New York, NY, Dec 10, 45. MAMMALOGY, ECOLOGY. *Educ:* Univ Vt, BA, 67; Univ Okla, MS, 70; Univ NMex, PhD(biol), 72. *Prof Exp:* Researcher, World Wildlife Fund & Nat Park Serv, 73-77; asst prof, 77-83, ASSOC PROF BIOL, UNIV NEBR, OMAHA, 83- *Mem:* Am Soc Mammalogists. *Res:* Mammalian biology. *Mailing Add:* Dept Biol Univ Nebr Omaha NE 68182

GELZER, JUSTUS, b Basel, Switz, Nov 8, 29; m 60; c 3. MICROBIOLOGY, IMMUNOLOGY. *Educ:* Univ Basel, MD, 55. *Prof Exp:* Fel microbiol, Univ Fla, 57-59; fel pediat, Childrens Hosp, Zurich, Switz, 59-62; fel immunol, Dept Microbiol, Sch Med, Columbia Univ, 62-64; sr microbiologist res labs, Pharmaceut Div, Ciba Ltd, Basel, 64-67, dir microbiol res, Ciba Pharmaceut Co, NJ, 67-70, dir biol res, Res Labs, 70-80, head strategy worldwide res, 80-84, HEAD MED DEPT, PHARMACEUT DIV, CIBA-GEIGY LTD, 85- *Mem:* Am Soc Microbiol; NY Acad Sci; Am Med Asn; Brit Pharm Soc; Swiss Soc Microbiol. *Res:* Host-parasite relationship in experimental infections and tumors; infectious immunity and resistance, chemotherapeutics and research management. *Mailing Add:* Pharmaceut Div Ciba-Geigy Med Dept Basel Switzerland

GEMAN, STAURT ALAN, b Chicago, Ill, Mar 23, 49. PHYSIOLOGY. *Educ:* Univ Mich, BS, 71; Dartmouth Col, MS, 73; Mass Inst Technol, PhD(math), 77. *Prof Exp:* PROF APPL MATH, BROWN UNIV, 77- *Mem:* AAAS. *Mailing Add:* Appl Math Dept Brown Univ Box F Providence RI 02912

GEMBICKI, STANLEY ARTHUR, b Davenport, Iowa, Dec 18, 41; m 70; c 2. CATALYST RESEARCH & DEVELOPMENT, SEPARATION PROCESS RESEARCH & DEVELOPMENT. *Educ:* Purdue Univ, BS, 64; Dartmouth Col, DE, 69. *Prof Exp:* Mgr catalyst develop, Exp Develop, UOP Inc, 75-78, catalyst res & develop, Process Develop, 78-80, sr mgr separation technol, Process Res & Develop, 80-86; DIR CATALYST RES, UOP RES CTR, 86- *Mem:* Am Inst Chem Eng; Am Chem Soc. *Res:* Discovery of new catalysts and research toward new petroleum and petrochemical processes. *Mailing Add:* Allied Signal Engineered MTRIS Box 5016 50 E Algonquin Rd Des Plaines IL 60017-5016

GEMIGNANI, MICHAEL C, b Baltimore, Md, Feb 23, 38; m 62; c 2. TOPOLOGY. *Educ:* Univ Rochester, BA, 62; Univ Notre Dame, MS, 64, PhD(math), 65; Ind Univ, JD, 80. *Prof Exp:* Instr math, St Mary's Col, Ind, 64-65; asst prof, State Univ NY, Buffalo, 65-68; assoc prof, Smith Col, 68-72; prof math sci & chmn dept, Ind Univ, Purdue Univ, Indianapolis, 72-81; prof math sci & dean col sci & humanities, Ball State Univ, Muncie, 81-86; dean, Col Arts & Sci, Univ Maine, 86-88; SR VPRES & PROVOST, UNIV HOUSTON, CLEAR LAKE, 88- *Concurrent Pos:* Consult, computer law; chmn, comt info processing, Ind Corp Sci & Technol. *Mem:* Math Asn Am; Am Bar Asn; AAAS; Sigma Xi; Computer Law Asn; Asn Comput Mach. *Res:* Computer related law. *Mailing Add:* Univ Houston Clear Lake Houston TX 77058

GEMINDER, ROBERT, b Wroclaw, Poland; Aug 3, 35; US citizen; m 59; c 3. INSTRUMENTATION, EXPERIMENTAL MECHANICS. *Educ:* Carnegie Mellon Univ, BS, 57. *Prof Exp:* Engr, Ling Temco Yought, 57-59; mgr, Aerojet Gen Corp, 59-65; asst dir, Mech Res Inc, 65-72; dir, Syst Develop Corp, 72-82; vpres, Hughes Tool Corp, 82-84; ENGR CONSULT, 84- *Concurrent Pos:* Instr, Citrus Col, 62-65; lectr, ocean seminars, 78. *Honors & Awards:* Seligman Award, Inst Environ Sci, 81. *Mem:* Fel Inst Environ Sci (pres, 75-76); sr mem Instrument Soc Am; Marine Tech Soc. *Res:* Underwater instrumentation; engineering analysis. *Mailing Add:* 710 Silver Spur Rd No 293 Rolling Hills Estates CA 90274

GEMMELL, GORDON D(OUGLAS), b Christchurch, NZ, Sept 18, 21; US citizen; m 58; c 2. PHYSICAL METALLURGY & CHEMISTRY. *Educ:* Univ NZ, BSc, 42, MSc, 43; Mass Inst Technol, MS, 53, ScD(metall), 56. *Prof Exp:* Physical chemist, Dept Sci & Indust Res, NZ, 43-51; res metallurgist, E I Du Pont de Nemours & Co, Inc, 56-86; CONSULT, 86- *Res:* Materials science. *Mailing Add:* 603 Weldin Rd Wilmington DE 19803-4941

GEMMELL, ROBERT S(TINSON), b Kenton, Ohio, Apr 14, 33; m 55; c 4. CIVIL ENGINEERING. *Educ:* Ohio State Univ, BCE, 56, MS, 57; Harvard Univ, PhD(eng), 63. *Prof Exp:* Instr civil eng, Rutgers Univ, 57; lectr & res fel div eng & appl physics & co-dir, USPHS res grant, Harvard Univ, 63-64; co-dir, USPHS, 64-65, res grants, 64-68, from asst prof to assoc prof, 64-75, chmn coun urban & regional planning, 75-79, PROF CIVIL ENG, TECHNOL INST, NORTHWESTERN UNIV, 75- *Concurrent Pos:* Off Water Resources res grants, 70-73; res contract, Forest Serv, USDA, 77-79. *Mem:* Am Soc Civil Engrs; Am Water Works Asn; Am Geophys Union; Am Water Resources Asn. *Res:* Coagulation; water treatment processes; water quality management; environmental health engineering; water resources systems; urban systems engineering. *Mailing Add:* Dept Civil Eng Technol Inst Northwestern Univ Evanston IL 60208

GEMMER, ROBERT VALENTINE, b Morristown, NJ, Nov 11, 46; m; c 2. PHYSICAL ORGANIC CHEMISTRY, COMBUSTION SCIENCE. *Educ:* Worcester Polytech Inst, BS, 68; Stanford Univ, PhD(org chem), 74. *Prof Exp:* Res asst org chem, Stanford Univ, 68-74; fel, Johns Hopkins Univ, 74-75; res assoc polymer chem, NASA, Ames Res Ctr, 75-77; res chemist polymer chem, Am Cyanamid Co, Stanford, Conn, 77-79; proj mgr, Duracell, Inc, 80-85; prog mgr, 85-89, SR PROJ MGR, COMBUSTION GAS RES INST, 89- *Concurrent Pos:* Resident res assoc, Nat Res Coun, 75-77. *Mem:* Am Chem Soc; Am Inst Chem Engrs, Combustion Inst. *Res:* Application of water soluble polymers to water treatment and purification; oxidation of polyhydrocarbons; chemistry of conducting organic materials; alkaline zinc batteries; combustion chemistry and physics. *Mailing Add:* GRI 8600 W Bryn Mawr Ave Chicago IL 60631

GEMPERLINE, MARGARET MARY CETERA, b Chicago, Ill, May 31, 53. ANALYTICAL CHEMISTRY. *Educ:* Ill Benedictine Col, BS, 75; ECarolina Univ, MS, 81. *Prof Exp:* Asst res chemist, CPC Int, Argonne, Ill, 75-77, asst plant chemist, 77-78; sci tech res asst, Ill Legis Coun, 78-79; RES ASSOC, EAST CAROLINA UNIV, 82- *Mem:* Am Chem Soc. *Res:* Instrument-computer interfacing; programmed experiments for physically impaired persons. *Mailing Add:* 1000 E Third St Greenville NC 27858

GEMPERLINE, PAUL JOSEPH, b Cleveland, Ohio; m 84. MULTICOMPONENT ANALYSIS, COMPUTERIZED DATA ANALYSIS. *Educ:* Cleveland State Univ, BS, 78, PhD(anal chem), 82. *Prof Exp:* ASST PROF CHEM, E CAROLINA UNIV, 82- *Concurrent Pos:* Consult, Burroughs Wellcome Co, 84-85. *Mem:* Sigma Xi; Am Chem Soc; Soc Appl Spectros. *Res:* Application of factor analysis to analytical chemistry, chromatographic curve resolution, multicomponent background correction; development of new methods of data analysis. *Mailing Add:* Dept Chem ECarolina Univ Greenville NC 27834

GEMSA, DIETHARD, b Berlin, Ger, Aug 9, 37; m 68; c 3. INFLAMMATION & CYTOKINES, IMMUNITY AGAINST VIRUSES. *Educ:* Univ Freiburg, Ger, MD, 64; Univ Heidelberg, Ger, Priv Doz(immunol), 74. *Prof Exp:* Res fel immunol, Med Sch, Univ Wash, Seattle, 65-67; resident internal med, Univ Mainz, Ger, 68-70; res assoc immunol, Dept Internal Med, Med Sch, Univ Calif, San Francisco, 70-73; asst prof immunol, Inst Immunol, Univ Heidelberg, Ger, 74-82; assoc prof immunopharmacol, Med Sch Hannover, Ger, 83-84; PROF & HEAD IMMUNOL, INST IMMUNOL, PHILIPPS UNIV, MARBURG, GER, 85- *Concurrent Pos:* Ed-in-chief, Immunobiol, 78- *Mem:* Am Asn Immunologists. *Res:* Role of activated monocytes/macrophages in the immune response; inflammation; antiviral defense systems; eicosanoid and cytokine release; tumor cytotoxicity. *Mailing Add:* Inst Immunol Philipps Univ Marburg D-3550 Germany

GEMSKI, PETER, b Bellingham, Mass, Oct 3, 36; div; c 2. IMMUNOLOGY. *Educ:* Brown Univ, AB, 58; Univ RI, MS, 60; Univ Pittsburgh, PhD(microbiol), 64. *Prof Exp:* NIH fels, Lister Inst Prev Med, London, Eng, 64-65 & Dept Med Microbiol, Sch Med, Stanford Univ, 66; res microbiologist, Dept Bact Immunol, Walter Reed Army Inst, 67-74, asst chief, Dept Appl Immunol, 74-80, Chief, Dept Biol Chem, 80-87, CHIEF, DEPT MOLECULAR PATH, WALTER REED ARMY INST RES, 88- *Mem:* AAAS; Am Soc Microbiol; Infectious Dis Soc. *Res:* Intergeneric bacterial hybridizations; genetic control of lipopolysaccharide biosynthesis; pathogenesis of enteric infections; virulence factors, molecular genetics; molecular and cell biology of bacterial toxins. *Mailing Add:* Dept Molecular Path Walter Reed Army Inst Res Washington DC 20307

GENAIDY, ASHRAF MOHAMED, b Cairo, Egypt, May 10, 57; m 82; c 1. WORK PHYSIOLOGY, BIOMECHANICS. *Educ:* Cairo Univ, BSc, 80; Univ Miami, MSc, 83, PhD(biomed & indust eng), 87. *Prof Exp:* ASST PROF ERGONOMICS, SAFETY & STATIST, WESTERN MICH UNIV, 87- *Concurrent Pos:* Lectr syst eng, Cairo Univ, 80-81; grad asst ergonomics, Univ Miami, 85-87. *Mem:* Sigma Xi; Human Factors Soc; Ergonomics; Inst Indust Engrs; Am Soc Biomech. *Res:* Ergonomics; safety; biomedical engineering; work methods and work measurement. *Mailing Add:* Dept Indust Eng Univ Miami Coral Gables FL 33124

GENAUX, CHARLES THOMAS, biochemistry, chemistry, for more information see previous edition

GENCO, JOSEPH MICHAEL, b Cleveland, Ohio, Apr 13, 39; m 67; c 1. CHEMICAL ENGINEERING, PHYSICAL CHEMISTRY. *Educ:* Case Western Reserve Univ, BS, 60; Ohio State Univ, MS, 62, PhD(chem eng), 65. *Prof Exp:* Assoc chief, Battelle Mem Inst, 65-74; CALDER PROF PULP & PAPER ENG, UNIV MAINE, ORONO, 74-, PROF CHEM ENG, 79- *Mem:* AAAS; Am Inst Chem Eng; Am Chem Soc; Tech Asn Pulp & Paper; Sigma Xi. *Res:* Fundamentals of pulping, bleaching and paper-making, pulp and paper technology. *Mailing Add:* 12 Winterhaven Dr Orono ME 04473-1118

GENCO, ROBERT J, b Silver Creek, NY, Oct 31, 38; m 57; c 3. IMMUNOCHEMISTRY, MICROBIOLOGY. *Educ:* State Univ NY Buffalo, DDS, 63; Univ Pa, PhD(microbiol), 67. *Hon Degrees:* DSc, Georgetown Univ, 90. *Prof Exp:* USPHS fel, 63-66; from asst prof to assoc prof, 67-74, dir, Grad Periodont, 68-85, PROF ORAL BIOL & PERIODONT, SCH DENT MED, STATE UNIV NY, BUFFALO, 74-, CHMN, DEPT ORAL BIOL, 77-; STAFF, BUFFALO GEN HOSP, 69- *Concurrent Pos:* Adv ed, Immunochem, 73-77; dir, Fel Prog Immunol & Periodont, State Univ NY, Buffalo, 74-, Periodont Dis Clin Res Ctr, 78-, assoc dean grad studies & res, Sch Dent Med, 85-, distinguished prof, 89-, interim chmn, Dept Microbiol, Sch Med & Biomed Sci, 90-; basic res in oral sci award, Int Asn Dent Res, 75, res periodont dis award, 81; chmn, Dent Sect, AAAS, 80; res forum, Am Acad Periodont, 84-; chmn, Ad Hoc Comt New Frontiers Oral Health Res, 85-89; mem, Coun Dent Res, Am Dent Asn, 87-90, chmn, 89-90; mem sci adv bd, Nat Inst Dent Res, 87-; ed, J Periodont, 88-; mem, Dent Prod Panel, Food & Drug Admin, 88- *Honors & Awards:* George Thorn Award, 77; William J Gies Found Award, Am Acad Periodont, 83; Seymour J Kreshover Lectr, NIH, 85. *Mem:* Inst Med-Nat Acad Sci; Am Acad Periodont; AAAS; Am Asn Dent Res (vpres, 83, pres-elect, 84, pres, 85); Am Asn Dent Schs; Am Asn Immunologists; Am Asn Microbiologists; Am Dent Asn; Int Asn Dent Res (vpres, 89, pres-elect, 90, pres, 91); NY Acad Sci. *Res:* Structure-function relationships of the various antibody molecules, especially those present in external secretions; host parasite interations in bacterial infectious, neutrophil function. *Mailing Add:* Dept Oral Biol Dent Sch State Univ NY Foster Hall Buffalo NY 14214

GENCSOY, HASAN TAHSIN, b Turkey, July 4, 24; US citizen; m 53; c 1. MECHANICAL ENGINEERING. *Educ:* Univ Calif, BS, 49; WVa Univ, MS, 51. *Prof Exp:* Customer engr, Int Bus Mach World Trade Corp, 51-52; prod engr, Bakir Sanayi Ltd, 53-55; from instr to prof, 56-85, EMER PROF MECH ENG, WVA UNIV, 85- *Concurrent Pos:* Consult, Chamber Indust, Turkey, 53-55; design eng consult, pvt cos, Forensic Eng Consult, 74-; consult, WVa Dept Nat Resources. *Honors & Awards:* Nat Adams Mem Award, Am Welding Soc, 66. *Mem:* Am Soc Mech Engrs; Am Soc Eng Educ; Sigma Xi. *Res:* Machine design; engineering analysis; engineering systems analysis; similitude in engineering; experimental stress analysis; machine design and stress analysis. *Mailing Add:* 3400 Galt Ocean Dr 902-S Ft Lauderdale FL 33308

GENDEL, STEVEN MICHAEL, MOLECULAR GENETICS, CYANOBACTERIA. *Educ:* Univ Calif, Irvine, PhD(cell biol), 77. *Prof Exp:* ASST PROF GENETICS, IOWA STATE UNIV, 83- *Res:* Plasmid molecular biology; nitrogen fixation. *Mailing Add:* Genet Dept Iowa State Univ Curtiss Hall Ames IA 50011

GENDERNALIK, SUE AYDELOTT, b Water Valley, Ky; m 46. CHEMISTRY. *Educ:* Marygrove Col, BS, 42. *Prof Exp:* Chemist org synthetic chem, Ethyl Corp, 42-44, res chemist anal method develop, 44-52, staff chemist engine fuel res, 52-72, sr res chemist, environ res, 73-81; RETIRED. *Honors & Awards:* Arch T Colwell Merit Award, Soc Automotive Engrs, 71. *Mem:* Am Chem Soc. *Res:* Fuel combustion; engine fuel relationships, particularly as related to exhaust emissions; public health impact of consumer use of lead and manganese; environmental research. *Mailing Add:* 682 N Rosedale Ct Grosse Pointe MI 48236

GENDLER, SANDRA J, b Minot, NDak, Sept 21, 44; m 66; c 2. MUCINGLYCOPROTEINS, BREAST CANCER BIOLOGY. *Educ:* Univ Minn, BA, 66; Univ Ill, MA, 73; Univ Southern Calif, PhD(biochem), 84. *Prof Exp:* Technician microbiol, Calif State Univ, Long Beach, 75-78; res asst biochem, Univ SCalif, 79-84; postdoctoral biochem & molecular biol, 84-87, HEAD, MOLECULAR EPITHELIAL CELL BIOL LAB, IMP CANCER RES FUND, 87- *Honors & Awards:* Harry J Deuel Award, Univ Southern Calif, 84. *Mem:* Am Soc Cell Biol; Int Soc Differentiation; Am Asn Cancer Res; Europ Asn Cancer Res; Brit Soc Cell Biol; Int Asn Breast Cancer Res; Sigma Xi. *Res:* Molecular studies of the function and regulation of molecules expressed during mammary gland differentiation and carcinogenesis. *Mailing Add:* Imp Cancer Res Fund PO Box 123 Lincoln's Inn Fields London WC2A 3PX England

GENEAUX, NANCY LYNNE, b Baltimore, Md, Dec 28, 42; m 68; c 2. ACOUSTICS, SIGNAL PROCESSING. *Educ:* Drexel Univ, BS, 65; George Washington Univ, MS, 73. *Prof Exp:* PHYSICIST ACOUSTICS, DAVID TAYLOR NAVAL SHIP RES & DEVELOP CTR, US NAVY, 60- *Mem:* AAAS. *Res:* New techniques in real time signal processing of ship acoustic data; special applications of specific data analysis procedures such as correlation, convolution, matched filtering, coherence and density. *Mailing Add:* 3300 Royale Glen Ave Davidsonville MD 21035

GENECIN, ABRAHAM, b Minneapolis, Minn, Aug 21, 18; m 41; c 2. INTERNAL MEDICINE, CARDIOLOGY. *Educ:* Columbia Univ, AB, 39; Johns Hopkins Univ, MD, 43. *Prof Exp:* From instr to asst prof, 44-66, ASSOC PROF MED, SCH MED, JOHNS HOPKINS UNIV, 66- *Concurrent Pos:* Pvt pract. *Mem:* AAAS; AMA; Am Heart Asn; Am Col Physicians. *Res:* Cardiology. *Mailing Add:* 611 Park Ave Baltimore MD 21201

GENEL, MYRON, b York, Pa, Jan 6, 36; m 68; c 3. PEDIATRICS, ENDOCRINOLOGY. *Educ:* Moravian Col, BS, 57; Univ Pa, MD, 61; Yale Univ, MA, 83. *Hon Degrees:* MA, Yale Univ, 83. *Prof Exp:* Intern med, Mt Sinai Hosp, 61-62; resident pediat, Children's Hosp Philadelphia, 62-64; pediatrician, US Army, TROP Res Med Lab, San Juan, PR, 64-66; teaching fel pediat endocrinol, Johns Hopkins Hosp, 66-67; fel genetics & metab dis, Children's Hosp Philadelphia, 67-69; assoc pediat, Sch Med, Univ Pa, 69-71; from asst prof to assoc prof pediat, Yale Univ Sch Med, 71-81, dir, Sect Pediat Endocrinol, 71-85, prog dir, Children's Clin Res Ctr, 71-86, PROF PEDIAT, YALE UNIV SCH MED, 81-, ASSOC DEAN GOVT & COMMUNITY AFFAIRS, 85- *Concurrent Pos:* Mem adv bd, Genetic Serv Prog, State of Conn, 79-82; Robert Wood Johnson Health Policy fel, Inst Med-Nat Acad

Sci, 82-83; consult health policy, subcomt on Invest & Oversight Comt Sci & Technol, US House Reps, 82-84; chmn, transplant adv comt, Off Comnr, Dept Income Maintenance, State of Conn, 84-; chmn & organizer Pub Policy Coun, Am Pediat Soc, Assn Med Sch Pediat Dept Chmn & Soc Pediat Res; organizer & moderator, Am Enterprise Inst Pub Policy Res, 84; mem, Health Policy Fel Bd, Inst Med-Nat Acad Sci, 89-; chmn, Coun Acad Socs, Asn Am Med Cols, 90-91. *Honors & Awards:* Jonathan May Award, Am Diabetes Asn, 79. *Mem:* Am Acad Pediat; Am Pediat Soc; Asn Am Med Col; Asn Prog Dirs (pres, 81-82); Endocrine Soc; Soc Pediat Res. *Res:* Disorders of growth and sexual development; genetics manken and detection of multiple endocrine neoplasia type 2A. *Mailing Add:* Off Govt & Community Affairs I-202 SHM Yale Univ Sch Med 333 Cedar St PO Box 3333 New Haven CT 06510-8056

GENENSKY, SAMUEL MILTON, b New Bedford, Mass, July 26, 27; m 53; c 2. VISUAL RESEARCH, APPLIED MATHEMATICS. *Educ:* Brown Univ, BS, 49, PhD(appl math), 58; Harvard Univ, MA, 51. *Prof Exp:* Mathematician, Nat Bur Standards, 51-54; assoc mathematician, Rand Corp, 58-59, res mathematician, 59-78; DIR, CTR FOR PARTIALLY SIGHTED, SANTA MONICA HOSP MED CTR, 78- *Concurrent Pos:* Pres, Coun Citizens With Low Vision, Am Coun for the Blind; mem adv comt on serv to the blind & partially sighted, Calif Voc Rehab Dept; mem adv comt on low vision serv, Am Found for the Blind. *Mem:* Sigma Xi; AAAS; Am Math Soc; fel Am Acad Optom. *Res:* Design of visual aids for disabled persons; closed circuit television system to aid the partially sighted; visual information transfer problems of the partially sighted; classical continuum mechanics, especially elasticity and viscous fluid theory. *Mailing Add:* 826 Jacon Way Pacific Palisades CA 90272

GENEROSO, WALDERICO MALINAWAN, b Bauan, Batangas, Philippines, Feb 16, 41; US citizen; m 63; c 2. GENETICS. *Educ:* Univ Philippines, BSA, 60, MS, 62; Univ Mo, Columbia, PhD(genetics), 67. *Prof Exp:* Instr genetics, Univ Philippines, 62-64; fel mutagenesis, 67-68, BIOLOGIST MUTAGENESIS, OAK RIDGE NAT LAB, 68- *Concurrent Pos:* Consult biol res, Third World Countries. *Mem:* Genet Soc Am; Environ Mutagen Soc; AAAS. *Res:* Chemical and radiation mutagenesis in mice; emphasis on transmissible genetic effects, expression of mutations and mechanisms for chromosome observation formation. *Mailing Add:* 112 Mohawk Rd Oak Ridge TN 37831

GENES, ANDREW NICHOLAS, b Boston, Mass, Oct 28, 32; m 63, 78; c 2. GEOLOGY, HYDROGEOLOGY. *Educ:* Boston Univ, AB, 55, AM, 60; Syracuse Univ, PhD(geol), 73. *Prof Exp:* Ed sci, McGraw-Hill Bk Co, 60-66; asst prof, 66-80, ASSOC PROF GEOL, UNIV MASS, BOSTON, 80- *Concurrent Pos:* Field mapper, Main State Geol Surv, 73-; chief geologist consult, Hidell-Eyster Assoc, Weytmouth, Mass. *Mem:* Norweg Geol Soc; Am Quaternary Asn. *Res:* Tillite genesis; cirque genesis and upland erosion surface denudation; deglaciation chronology of Northeastern Maine and Greece. *Mailing Add:* Dept Earth Sci & Geog Univ Mass/Boston Boston MA 02125

GENEST, JACQUES, b Montreal, Que, May 29, 19; m 53; c 5. NEPHROLOGY, ENDOCRINOLOGY. *Educ:* Jean de Brebeuf Col, BA, 37; Univ Montreal, MD, 42. *Hon Degrees:* LLD, Queen's Univ, Ont, 66 & Univ Toronto, 70; DSc, Laval Univ, 73, McGill Univ, 79 & Univ Ottawa, 80; DMedSci, Sherbrooke Univ, 74, Mem Univ Nfld, 78, St Francis Xavier Univ, Antigonish, NS, 83, State Univ NY, Buffalo, 84, Rockefeller Univ, 86, Concordia Univ, MTL, 86, Montpellier Univ, France, 89. *Prof Exp:* Sr intern, Hotel Dieu Hosp, 42, asst resident path, 43-44, chief resident med, 44-45; asst physician, Johns Hopkins Hosp, 45-48; with Rockefeller Inst Med Res, 48-51; med survr, Europ Med Res Ctrs for Que Govt, 51-52; chmn dept med, Fac Med, Univ Montreal & physician-in-chief, Hotel Dieu Hosp, 64-68; sci dir, 67-84, CONSULT, CLIN RES INST MONTREAL, 84- *Concurrent Pos:* Archbold fel med, Johns Hopkins Hosp, 45-46, Commonwealth Fund fel, 46-48; lectr, Royal Col, 61; chmn, Med Res Coun Quebec, 64-69; Sims Commonwealth Travelling Prof, 70; specialist & consult hypertension, nephrology & endocrinology; consult, dept nephrology-hypertension, Univ Montreal Hotel-Dieu Hosp, 84- *Honors & Awards:* Gairdner Award, Univ Toronto, 63; Companion of Can, 67; Flavelle Award, Royal Soc Can, 68; Stouffer Prize, 69; Killam Award, 86; Royal Bank Award, 80; F N G Starr Prize, 82; Izaak Walton Killam Mem Prize, Toronto, 86. *Mem:* Endocrine Soc; master Am Col Physicians; Am Clin & Climat Asn; Asn Am Physicians; fel Royal Soc Can; Peripatetic Club USA. *Res:* Human arterial hypertension; relationship of kidneys and adrenals to hypertension; electrolytes and renal function. *Mailing Add:* Clin Res Inst of Montreal 110 Pine Ave W Montreal PQ H2W 1R7 Can

GENET, RENE P H, organic chemistry; deceased, see previous edition for last biography

GENETELLI, EMIL J, b Brooklyn, NY, Feb 25, 37; m 60; c 3. SANITARY ENGINEERING, MICROBIOLOGY. *Educ:* Manhattan Col, BCE, 59; Rutgers Univ, MS, 62, PhD(environ sci), 65. *Prof Exp:* From instr to asst prof, 62-70, assoc prof, 70-80, PROF ENVIRON SCI, RUTGERS UNIV, 80- *Honors & Awards:* Heukelekian Award Indust Wastewater Treatment. *Mem:* Sigma Xi. *Res:* Microbiology of water and wastewater treatment; water resources; air pollution control; solid waste disposal; aquatic microbiology. *Mailing Add:* One Thrush Rd East Brunswick NJ 08816

GENETTI, WILLIAM ERNEST, chemical engineering; deceased, see previous edition for last biography

GENG, SHU, b China, Sept 3, 42; US citizen; m 68; c 2. APPLIED STATISTICS. *Educ:* Nat Taiwan Univ, BS, 64; Kans State Univ, MS, 69, PhD(statist), 72. *Prof Exp:* Statistician, Agr Res & Develop Div, Upjohn Co, 72-76; assoc prof, 76-84, PROF AGRON & RANGE SCI, UNIV CALIF, DAVIS, 84-, ASSOC DEAN, COL OF AGR & ENVIRON SCI. *Mem:* Biomet Soc; Am Statist Asn; Am Soc Agron. *Res:* Multivariate methods in design and data analysis; simulation models. *Mailing Add:* Dept Agron & Range Sci Univ Calif Davis CA 95616

GENGENBACH, BURLE GENE, b Grand Island, NB, Oct 25, 44; m 67; c 2. PLANT GENETICS. *Educ:* Univ Nebr, BS, 66, MS, 68; Univ Ill, PhD(genetics), 71. *Prof Exp:* Res assoc plant physiol, Dept Agron, Univ Ill, 71-72; from asst prof to assoc prof, 72-81, PROF PLANT GENETICS, DEPT AGRON, UNIV MINN, ST PAUL, 81- *Mem:* Fel Am Soc Agron (bd dir 85-88); fel Crop Sci Soc Am (bd dir 85-88); Genet Soc Am; Am Soc Plant Physiologists; Int Asn Plant Tissue Cell Cult; AAAS. *Res:* Mutant selection in maize cell cultures; selection and genetic studies of cytoplasmic traits of maize; enzymology of pathway regulating amino acid biosynthesis in crop plants; amyloplast differentation in developing maize kernels. *Mailing Add:* 1994 Beacon St Roseville MN 55113

GENGOZIAN, NAZARETH, b Racine, Wis, Feb 13, 29; m 48; c 3. IMMUNOLOGY, TRANSPLANTATION. *Educ:* Univ Wis, BS, 51, MA, 53, PhD(immunol), 55. *Prof Exp:* Res assoc, Oak Ridge Nat Lab, 55-56, biologist, 57-60, chief scientist, Oak Ridge Assoc Univs, 60-81; mem, Okla Med Res Found, 81-85; PROF, UNIV SFLA, 85- *Concurrent Pos:* USPHS fel, Nat Cancer Inst, 56-57; Ford Found prof, Univ Tenn, 68-76, prof, 72-81; adj prof, Wake Forest Univ, 69-77; transplantation, 71-86. *Mem:* Am Asn Immunologists; Radiation Res Soc; Soc Exp Biol & Med; Transplantation Soc. *Res:* Radiation immunology; primatology; bone marrow transplantation; monoclonal antibody; feline immunology. *Mailing Add:* Dept Pediat Univ SFla 140 Seventh Ave South St Petersburg FL 33701

GENIN, DENNIS JOSEPH, b Rockford, Ill, Sept 18, 38; m 60; c 3. SOLID STATE PHYSICS, ENGINEERING PHYSICS. *Educ:* Beloit Col, BS, 60; Iowa State Univ, PhD(physics), 66. *Prof Exp:* Assoc physics, Argonne Nat Lab, 66-68; sr assoc physicist, Thomas J Watson Res Ctr, 68-71, staff physicist, 71-75, develop engr, 75-76, proj engr, E Fishkill Facil, 76-84, sr engr, Syst Prod Div, 84-88, SR ENG, IBM, GEN TECHNOL DIV, 88- *Mem:* Am Phys Soc; Sigma Xi. *Res:* Electric and magnetic fields in solids using magnetic resonance techniques. *Mailing Add:* One Kellerhause Dr Poughkeepsie NY 12603-5441

GENIN, JOSEPH, b Norwalk, Conn, Sept 9, 34; m 64; c 3. SOLID MECHANICS. *Educ:* City Col New York, BCE, 54; Univ Ariz, MS, 57; Univ Minn, PhD(solid mech), 63. *Prof Exp:* Instr struct mech, Univ Ariz, 55-58; eng mech, Univ Minn, 58-62; sr engr, Gen Dynamics, Ft Worth, 63-64; from assoc prof to prof aeronaut, astronaut & eng sci, Purdue Univ, 64-72, prof mech eng, 72-81, chmn eng mech, 76-81; dean, Col Engr, 81-85, DIR, OPTICS & MAT SCI LAB, NMEX STATE UNIV, LAS CRUCES, 85- *Mem:* Am Soc Eng Educ; Am Inst Aeronaut & Astronaut; Am Soc Mech Engrs; Nat Soc Prof Engrs. *Res:* Solid mechanics; dynamic stability, vibrations, material damping, stress waves, structural mechanics; aeroelasticity, and viscoelasticity. *Mailing Add:* Box 3449 NMex State Univ Las Cruces NM 88003-0001

GENNARI, F JOHN, b Jersey City, NJ, May 18, 37; m 58; c 3. NEPHROLOGY, RENAL PHYSIOLOGY. *Educ:* Yale Univ, BS, 59, MD, 63. *Prof Exp:* Resident internal med; Univ Va Hosp, 63-66; captain, US Air Force, 66-68; fel nephrology, Tufts Univ-New Eng Med Ctr, 68-71; from asst prof to assoc prof med, Sch Med, Tufts Univ, 71-79; PROF MED & DIR NEPHROLOGY UNIT, COL MED, UNIV VT, 79- *Mem:* Am Fed Clin Res; Am Soc Clin Invest; Am Physiol Soc; Am Soc Nephrology. *Res:* Acid-base physiology; biocarbonate reabsorption patterns in the proximal tubule of the kidney in normal and disease states; carbon dioxide generation and diffusion across the tubular epithelium. *Mailing Add:* Dept Med D305 Given Bldg Univ Vt Col Med Burlington VT 05405

GENNARO, ALFONSO ROBERT, b Philadelphia, Pa, Dec 18, 25; m 49; c 5. MEDICINAL CHEMISTRY. *Educ:* Philadelphia Col Pharm, BSc, 48; Univ Pa, MSc, 51; Temple Univ, PhD, 56. *Prof Exp:* Chemist, E I du Pont de Nemours & Co, Inc, 51-53; chemist, Pa Salt Mfg Co, 53-55; from instr to assoc prof, 48-65, dir dept, 69-81, PROF CHEM, PHILADELPHIA COL PHARM & SCI, 65- *Concurrent Pos:* Mem, Comt Rev, US Pharmacopeia, 65-85, Pa Post-Sec Educ Planning Comn, 74-79 & Gould's Med Dict, 75; ed, Remin Pharm Sci, 80- *Mem:* Int Soc Heterocyclic Chem. *Res:* Synthetic medicinals; drug standards; electronic instrumentation. *Mailing Add:* Dept Chem Philadelphia Col Pharm & Sci 43rd St Wolnd Ave Philadelphia PA 19104

GENNARO, ANTONIO LOUIS, b Raton, NMex, Mar 18, 34; m 55; c 3. VERTEBRATE ECOLOGY. *Educ:* NMex State Univ, BS, 57; Univ NMex, MS, 61, PhD(vert ecol), 67. *Prof Exp:* Teacher high sch, NMex, 57-58; dir grad teaching asst biol, Univ NMex, 61-64; asst prof, St John's Univ, Minn, 65-66; from asst prof to assoc prof, 66-75, coordr biol sci, Div Natural Sci, 75-77, PROF BIOL, EASTERN NMEX UNIV, 75-, CUR, NATURAL HIST MUS, 67- *Concurrent Pos:* Consult, Nat Park Serv, 74-, NMex Dept Game & Fish, 74- & NMex Environ Inst, 75- *Mem:* Am Soc Mammal; Herpetologists League Inc; Sigma Xi. *Res:* Taxonomy and ecology in mammalogy and herpetology. *Mailing Add:* 816 W 19th Portales NM 88130

GENNARO, JOSEPH FRANCIS, b Brooklyn, NY, Apr 9, 24; m 44; c 5. CELL STRUCTURE. *Educ:* Fordham Univ, BS, 47; Univ Pittsburgh, MS, 49, PhD(zool), 52. *Prof Exp:* Asst prof biol, St John's Univ, Col Pharm, 51-52; instr anat, State Univ NY, Col Med, 53-56; asst prof, Univ Fla, Col Med, 56-64; assoc prof, Univ Louisville, Sch Med, 64-69; assoc prof biol & dir lab cellular biol, NY Univ, 69-88; prof anat, Nat Cheng Kung Univ, Col Med, 88-89; VIS PROF, DEPT ANAT, UNIV FLA, 89- *Concurrent Pos:* Res assoc, Brookhaven Nat Lab, 51-56; res fel, Harvard Univ, 64-65; lectr, Hunter Col, 78-80, NY Univ Med Ctr, 82; vis prof, Dept Anat, Col Med, Nat Cheng Kung Univ, Taiwan, 88-89. *Res:* Effect of membrane(plasma) hormone interaction on cell structure; visual information processing for new information storage and retrieved; G-protein switching mechanisms. *Mailing Add:* 4132 NW 13th Ave Gainesville FL 32605

GENNARO, JOSEPH J(OHN), b New York, NY, Apr 21, 19; m 46; c 4. STRUCTURAL ENGINEERING. *Educ:* City Col NY, BCE, 39; Columbia Univ, MS, 54. *Prof Exp:* Assoc prof civil eng, City Col NY, 46-50; PROF CIVIL ENG, STEVENS INST TECHNOL, 52- *Mem:* Fel Am Soc Civil Engrs. *Res:* Structures; civil engineering. *Mailing Add:* Dept Civil Eng Stevens Inst Technol Hoboken NJ 07030

GENNARO, ROBERT NASH, b Raton, NMex, Oct 14, 40; m 76; c 2. MICROBIOLOGY, BIOLOGY. *Educ:* NMex State Univ, BS, 63, MS, 65; Tex A&M Univ, PhD(microbiol), 71. *Prof Exp:* Res asst, NMex State Univ, 63-65; instr microbiol, Tex A&M Univ, 65-70; asst prof, 70-76, ASSOC PROF MICROBIOL, UNIV CENT FLA, 76- *Mem:* Sigma Xi; Am Soc Microbiol. *Res:* Environmental microbiology; resourse recovery; bioconversion of wastes to methane and alcohol. *Mailing Add:* Dept Molecular Biol & Microbiol Univ Cent Fla Orlando FL 32816

GENNIS, ROBERT BENNETT, b New York, NY, Oct 7, 44; m; c 2. BIOPHYSICAL CHEMISTRY. *Educ:* Univ Chicago, BS, 66; Columbia Univ, PhD(chem), 71. *Prof Exp:* Whitney fel, Harvard Univ, 71-73; asst prof chem, 73-79, assoc prof chem, 79-84, PROF BIOCHEM, UNIV ILL, URBANA, 84- *Concurrent Pos:* NIH-USPHS career develop grant, 75; Fulbright fel & Guggenheim fel, 89. *Mem:* Biophys Soc; Fed Am Soc Exp Biol; Am Soc Microbiol. *Res:* Protein-lipid interactions; structure and function of lipid requiring enzymes; physical probes of membrane structure and function; membrane-bond electron transport chains. *Mailing Add:* Dept Chem Box 15 Univ Ill Noyes Lab Urbana IL 61801

GENOVA, JAMES JOHN, b Bronx, NY, May 23, 46; m 66; c 6. SYSTEMS ANALYSIS, SIGNAL PROCESSING. *Educ:* Case Inst Technol, BS, 68; State Univ NY, Stony Brook, MA, 71, PhD(physics), 73. *Prof Exp:* Flight test engr, Grumman Aerospace, 69-73; scientist, Ensco, Inc, 73-76; staff engr, Tex Instruments, 76-81; systs engr, Norden Systs, 81-85; DIV MGR, DIGITAL SIGNAL CORP, 85- *Concurrent Pos:* Lectr, Richland Community Col, 78-80. *Mem:* Navy League; Asn Old Crows. *Res:* Signal progressing development and testing; feedback control theory and expert systems; electronics and anti-submarine warfare; radar systems. *Mailing Add:* Digital Signal Corp 80 Orville Dr Bohemia NY 11716

GENOWAYS, HUGH HOWARD, b Scottsbluff, Nebr, Dec 24, 40; m 63; c 2. SYSTEMATICS, MAMMALOGY. *Educ:* Hastings Col, Nebr, AB, 63; Univ Kans, PhD(zool), 71. *Prof Exp:* Res asst, Mus, Tex Tech Univ, 71-72, cur mammals, 72-76, lectr, Mus Sci Prog, 74-76, actg coordr res, 75-76; cur mammals, Carnegie Mus Natural Hist, Pittsburgh, 76-86; DIR, UNIV STATE MUS, UNIV NEBR, LINCOLN, 86- *Concurrent Pos:* Adj asst prof, dept vet & zool med, Sch Med, Tex Tech Univ, 73-75, dept biol sci, 73-76 & dept path, 75-76; managing ed, J Mammal, Am Soc Mammalogists, 74-78, publ ed, Carnegie Mus Natural Hist, 77-86; mem, Coun Biol Ed; courtesy prof biol sci, Univ Nebr, Lincoln, 87-, chair, Mus Studies Prog, 90- *Honors & Awards:* C Hart Merriam Award, Am Soc Mammalogist, 87. *Mem:* Am Soc Mammalogists(pres, 84-86); Am Asn Mus; Sigma Xi; Soc Syst Zoologists. *Res:* Systematics, biogeography and ecology of New World mammals, especially rodents and bats; application of computer to data analysis and data retrieval; technology of management and preservation of biological specimens. *Mailing Add:* Univ State Mus Univ Nebr Lincoln NE 68588-0338

GENS, RALPH S, b Ger, Nov 25, 24; m 52; c 2. ELECTRICAL ENGINEERING. *Educ:* Ore State Univ, BS, 49. *Prof Exp:* Asst adminr elec eng, Bonneville Power-Admin, 49-80; CONSULT, 80- *Honors & Awards:* Centennial Medal, Inst Elec & Electronic Engrs, Harbishaw, 82; Distinguished Serv Award, Dept Interior, 87. *Mem:* Nat Acad Eng; fel Inst Elec & Electronics Engrs. *Mailing Add:* PO Box 41 Mt Hood OR 97041

GENSAMER, MAXWELL, b Bradford, Pa, June 3, 02; m 50. PHYSICAL METALLURGY. *Educ:* Carnegie Inst Technol, BS, 24, MS, 31, DSc(metall), 33. *Prof Exp:* Plant metallurgist, Am Chain & Cable Co, Pa, 24-29; res metallurgist, Carnegie Inst Technol, 29-45, from asst prof to prof metall eng, 35-45; prof metall & head dept mineral technol, Pa State Univ, 45-47; asst dir res, Carnegie-Ill Steel Corp, 47-50; Howe prof metall, 50-71, EMER HOWE PROF METALL, COLUMBIA UNIV, 71- *Concurrent Pos:* Consult, Esso Res & Eng & US Steel Res Lab. *Honors & Awards:* Howe Medal, Am Soc Metals, 32. *Mem:* Fel Am Soc Metals; fel Am Inst Mining, Metall & Petrol Engrs. *Res:* Properties of alloys, especially steel as controlled by composition and microstructure; mechanical metallurgy, properties, fracture and failure analysis. *Mailing Add:* 1057 Forest Lakes Dr Apt 309 Naples FL 33942

GENSEL, PATRICIA GABBEY, b Buffalo, NY, Mar 18, 44; m 68; c 1. BOTANY, PALEOBOTANY. *Educ:* Hope Col, BA, 66; Univ Conn, MS, 69, PhD(bot), 72. *Prof Exp:* Res asst palynology, King's Col Univ, London, 66-67; res assoc paleobot, Univ Conn, 72-75; from asst prof to assoc prof bot, 75-90, PROF BOT, UNIV NC, CHAPEL HILL, 90- *Concurrent Pos:* NSF grants, 74, 77-79, 80-83 & 84-87, 88-90. *Mem:* Bot Soc Am; Am Asn Stratig Palynologists; Paleontol Asn; Am Fern Soc. *Res:* Morphology, diversification and evolution of early land vascular plants; studies of Devonian plants; fossil plants, fossil spores-pollen; plant morphology. *Mailing Add:* Dept Biol Coker 010-A CB 3280 Univ NC Chapel Hill NC 27599-3280

GENSHAW, MARVIN ALDEN, b Petoskey, Mich, Sept 24, 39; m 64; c 2. PHYSICAL CHEMISTRY. *Educ:* Mich Technol Univ, BS, 61; Univ Pa, PhD(phys chem), 66. *Prof Exp:* Res investr, Univ Pa, 66-69; PRIN STAFF SCIENTIST, AMES CO, DIV MILES LABS, INC, 69- *Concurrent Pos:* Adj asst prof, Notre Dame Univ, 77-78; mem, Coun Optical Radiation Measurement. *Mem:* Am Chem Soc; Electrochem Soc; Am Asn Clin Chem. *Res:* Physical chemical medical diagnostic tests; reflectance and color theory and measurement; electrochemistry. *Mailing Add:* Diagnostics Div Miles Inc 1127 Myrtle St Elkhart IN 46514

GENT, ALAN NEVILLE, b Leicester, Eng, Nov 11, 27; m 49; c 3. PHYSICS, MECHANICS. *Educ:* Univ London, BSc, 46 & 49, PhD(physics), 55. *Prof Exp:* Res asst, John Bull Rubber Co, 44-45, physicist, Brit Rubber Prod Res Asn, 49-58, prin physicist, 58-61; asst dir, Inst Polymer Sci, 64-78, dean Grad Studies & Res, 78-86, PROF POLYMER PHYSICS, UNIV AKRON, 61-, H A MORTON PROF POLYMER PHYSICS & ENG, 87- *Concurrent Pos:* Chmn, Gordon Res Conf Elastomers, 66, Gordon Res Conf Cellular Mat, 69, Gordon Res Conf Adhesion, 77 & Gordon Res Conf Composites, 91; vis prof, Queen Mary Col, Univ London, 69-70, Mc Gill Univ, 83 & Univ Minn, 85; consult, Gen motors & Goodyear Tire & Rubber Co. *Honors & Awards:* Corecipient, Mobay Award, 64; Bingham Medal, Soc Rheology, 75; Colwyn Medal, Plastics & Rubber Inst, 77; Int Res Award, Soc Plastics Engrs, 80; 3M Award, Adhesion Soc, 87; Goodyear Medal, Am Chem Soc Rubber Div, 90. *Mem:* Nat Acad Eng; Adhesion Soc (pres, 78); Soc Rheology (pres, 81); fel Am Phys Soc. *Res:* Mechanical behavior of elastomers; deformation; fracture; crystallization; adhesion; friction and wear; stress-cracking of polymers. *Mailing Add:* Polymer Eng Univ Akron Akron OH 44325-0301

GENT, MARTIN PAUL NEVILLE, b Luton, Eng, July 6, 50; m 72; c 2. BIOPHYSICAL CHEMISTRY. *Educ:* Oberlin Col, BA, 71; Yale Univ, PhD(chem), 75. *Prof Exp:* Res assoc biophys, Dept Life Sci, Univ Pittsburgh, 75-78; asst scientist, Dept Ecol & climat, 78-82, ASSOC SCIENTIST, DEPT FORESTRY & HORT, CONN AGR EXP STA, 83- *Mem:* AAAS; Crop Sci Soc Am; Plant Physiol Soc; Am Soc Hort Sci. *Res:* Carbohydrate metabolism in plants; translocation in plants; magnetic relaxation studies of biological membranes. *Mailing Add:* Dept Forestry & Hort PO Box 1106 New Haven CT 06504-1106

GENT, MICHAEL, b Durham, Eng, May 4, 34; m 57; c 2. MEDICAL STATISTICS. *Educ:* Univ Durham, BSc, 56, MSc, 60, DSc, 90. *Prof Exp:* Statistician, Imp Chem Indust, Eng, 57-60; sr lectr math, Bradford Inst Technol, Eng, 60-65; from lectr to sr lectr statist, Univ Bradford, 65-69; from assoc prof to prof, 69-79, chmn dept, 73-79, dean res, 79-84, PROF CLIN EPIDEMIOL & BIOSTATIST, McMASTER UNIV, 73- *Mem:* Royal Statist Soc; Am Heart Asn; Can Soc Clin Invest; fel Am Statist Asn; Int Epidemiol Asn. *Res:* Multicentre controlled clinical trials; cardiovascular research, particularly thrombosis. *Mailing Add:* Hamilton Civic Hosp Res Ctr 711 Concession St Hamilton ON L8V 1C3 Can

GENTILE, ANTHONY L, b New York, NY, Apr 23, 30; m 57; c 2. MATERIALS SCIENCE. *Educ:* City Col NY, BS, 59; NMex Inst Mining & Technol, MS, 57; Ohio State Univ, PhD(mineral), 60. *Prof Exp:* Ceramic & metall engr, Aerojet-Gen Corp, Calif, 60-61; head crystal chem group, Chem Physics Dept, 63-68, MEM TECH STAFF CRYSTAL GROWTH, HUGHES RES LABS, 61-, HEAD CHEM PHYSICS DEPT, ELECTRONIC MAT SECT, 68- *Mem:* Sigma Xi; Am Asn Crystal Growth; Mat Res Soc. *Res:* Crystal growth; application of chemical and physical principles to growth of single and poly-crystals; determination of crystal perfection and properties; evaluation of crystal defects; properties modifications by ionic substitution. *Mailing Add:* 11471 Woodbine Mar Vista CA 90066

GENTILE, ARTHUR CHRISTOPHER, b New York, NY, Nov 24, 26; m 49; c 1. PLANT PHYSIOLOGY. *Educ:* City Col NY, BS, 48; Brown Univ, ScM, 51; Univ Chicago, PhD(bot), 53. *Prof Exp:* Asst biol, City Col NY, 47-48 & Brown Univ, 49-51; asst bot, Univ Chicago, 51-53; univ fel, Duke Univ, 53, Nat Cancer Inst fel, 54-55; plant physiologist, US Forest Serv, 55-56; from asst prof to prof bot, Univ Mass, Amherst, 56-72, from asst dean to assoc dean grad sch, 65-72; dean grad col & vprovost res admin, Univ Okla, 72-74; vpres acad affairs, Univ Nev, Las Vegas, 74-79; exec dir, Am Inst Biol Sci, Arlington, Va, 79-83; dean acad affairs, 83-88, VICE CHANCELLOR ACAD AFFAIRS, IND UNIV, KOKOMO, 88- *Mem:* AAAS; Sigma Xi. *Res:* Growth and metabolism of neoplastic plant tissues. *Mailing Add:* Ind Univ-Kokomo 2300 S Washington St Kokomo IN 46902

GENTILE, DOMINICK E, b Asbury Park, NJ, Jan 12, 32; m 58; c 5. NEPHROLOGY. *Educ:* Univ Notre Dame, BS, 53; Georgetown Univ, MD, 57. *Prof Exp:* Fel nephrology, Georgetown Univ, 60-62; from instr to asst prof med, Sch Med, Univ Louisville, 64-68, assoc pediat, 67-68; asst prof biophys & physiol, Mt Sinai Sch Med, 68-69; from asst prof to assoc prof, 69-72, clin assoc prof, 72-80, CLIN PROF MED, UNIV CALIF, IRVINE, 80- *Mem:* Am Soc Nephrol; Int Soc Nephrol; Biophys Soc; NY Acad Sci. *Res:* Transport physiology and biophysics; renal physiology; kinetic and clinical hemodialysis. *Mailing Add:* 1310 W Stewart Dr Suite 606 Orange CA 92668

GENTILE, JAMES MICHAEL, b Chicago, Ill, Aug 31, 46; m 73; c 1. GENETICS, MOLECULAR BIOLOGY. *Educ:* St Mary's Col, Minn, BA, 68; Ill State Univ, MS, 70, PhD(genetics), 74. *Prof Exp:* Teaching asst biol, Ill State Univ, 69-72; researcher genet, Yale Univ Med Sch, 74-76; from asst prof to assoc prof, 76-84, HERRICK PROF BIOL, HOPE COL, 84- *Concurrent Pos:* Res contract, Environ Protection Agency, 76-80, Res Corp grant, 77-79; Nat Inst Environ Health Sci grant, 81-; Environ Protection Agency grant, 81-; WHO grant, 84-86. *Mem:* Am Soc Microbiol; Genet Soc Am; Environ Mutagen Soc; AAAS; Am Inst Biol Sci. *Res:* Environmental mutagenesis; in vivo and in vitro metabolism of chemicals to mutagens by both plant and animal systems. *Mailing Add:* Dept Biol Sci Hope Col Holland MI 49423

GENTILE, PHILIP, b New York, NY, Feb 27, 23. INORGANIC CHEMISTRY. *Educ:* City Col, BS, 43; Polytech Inst Brooklyn, MS, 48; Univ Tex, PhD(chem), 55. *Prof Exp:* Anal chemist, Ledoux & Co, NY, 43-44; res chemist, Nat Starch Prods, NY, 44-46; supvr, Nat Lead of Ohio, 55-57; from asst prof to assoc prof, 57-69, PROF CHEM, FORDHAM UNIV, 69- *Mem:* Am Chem Soc. *Res:* Nonaqueous solvents; kinetics; chelates; inorganic synthesis; solid-solid interactions. *Mailing Add:* Dept Chem Fordham Univ Bronx NY 10458

GENTILE, RALPH G, b Palermo, Italy, May 13, 14; US nat; m 53; c 4. ELECTRICAL ENGINEERING. *Educ:* Univ Rome, Italy, Dr Eng, 38 & 40. *Prof Exp:* Fel aeronaut eng, Rome Inst Advan Studies, 41; res engr, US Govt, 43-44; elec engr, Roger Williams Eng Co, 44-45; chief engr, TRM Elec Co, 45-49; fel nuclear eng, Argonne Nat Lab, 54; res group leader, Monsanto Chem Co, 60-63; mgr, Physics Sect, Babcock & Wilcox Co, 63-79; lectr, Calif Polytech Univ, 79-84; RETIRED. *Mem:* AAAS; Inst Elec & Electronics Engrs; Am Soc Eng Educ; Sigma Xi; AAAS. *Res:* Solid state electronics; industrial instrumentation; energy conversion. *Mailing Add:* 1426 Prefumo Canyon Rd San Luis Obispo CA 93405-6117

GENTILE, RICHARD J, b St Louis, Mo, June 25, 29. STRATIGRAPHY. *Educ:* Univ Mo-Columbia, BA, 56, MA, 58; Univ Mo-Rolla, PhD(geol), 65. *Prof Exp:* Geologist, Mo Geol Surv, Rolla, 58-65, chief geologist & head of coal geol div, 65-66; from asst prof to assoc prof, 66-75, PROF GEOL, UNIV MO, KANSAS CITY, 75- *Concurrent Pos:* NSF res grants, 68-69, 70 & 74. *Honors & Awards:* Cert Recognition, Sigma Xi, 81. *Mem:* Am Asn Petrol Geol; fel Geol Soc Am; Asn Eng Geol; Soc Econ Paleontologists & Mineralogists; Am Inst Prof Geologists; Sigma Xi. *Res:* Nonmetallic mineral resources studies of Missouri; stratigraphy of Pennsylvanian age strata of western Missouri and eastern Kansas. *Mailing Add:* Dept Geosci Univ Mo 5100 Rockhill Rd Kansas City MO 64110-2499

GENTILE, THOMAS JOSEPH, b New York, NY, Jan 17, 53. HIGH ENERGY PHYSICS EXPERIMENTATION. *Educ:* Columbia Univ, BS, 74; Cornell Univ, MS, 78, PhD(physics), 81. *Prof Exp:* Res assoc, Univ Rochester, 81-85; res assoc, Ohio State Univ, 85-86; STAFF SCIENTIST, MASS INST TECHNOL, LINCOLN LABS, 86- *Mem:* Am Phys Soc. *Res:* Remote sensing and radar imaging. *Mailing Add:* Mass Inst Technol Lincoln Labs PO Box 73 Lexington MA 02173

GENTLE, JAMES EDDIE, b Statesville, NC, May 31, 43; wid. STATISTICAL COMPUTING. *Educ:* Univ NC, Chapel Hill, BS, 66; La State Univ, Baton Rouge, MA, 69; Tex A&M Univ, MCS, 73, PhD(statist), 74. *Prof Exp:* Asst prof statist, Iowa State Univ, 74-78, assoc prof, 78-79; DIR RES & DESIGN, IMSL, INC, 79- *Concurrent Pos:* Adj prof math sci, Rice Univ, 80-; ed, Current Index Statist, Am Statist Asn & Inst Math Statist, 80-85; vis lectr, Am Statist Asn & Inst Math Statist, 81, vis prof math, Univ Tex, Arlington, 85; UN/FAO consult, Indian Agr Statist Res Inst, New Delhi, 85, 89; adj prof biostatist, Univ Texas Health Sci Ctr, 89-; chmn, Statist Comput Sect, Am Statist Asn, 80, mem bd dirs, 82-84; mem bd dirs, Am Fed Info Processing Socs, 86-88. *Mem:* Fel Am Statist Asn; Math Asn Am; Asn Comput Mach; AAAS; Int Statist Inst. *Res:* Statistical computing; robust procedures; simulation. *Mailing Add:* IMSL Inc 2500 City West Blvd Houston TX 77042-3020

GENTLE, KENNETH W, b Oak Park, Ill, Oct 27, 40; m 86. PLASMA PHYSICS. *Educ:* Mass Inst Technol, SB, 62, PhD(physics), 66. *Prof Exp:* Instr physics, Mass Inst Technol, 65-66; from asst prof to assoc prof, 66-77, PROF PHYSICS, UNIV TEX, AUSTIN, 77- *Concurrent Pos:* Fel, Alfred Sloan Found, 73-75. *Mem:* Am Phys Soc. *Res:* Linear and nonlinear wave phenomena in plasmas; plasma confinement and heating. *Mailing Add:* Dept Physics Univ Tex Austin TX 78712

GENTLEMAN, JANE FORER, b Washington, DC, Apr 8, 40; m 67; c 2. STATISTICAL ANALYSIS. *Educ:* Univ Chicago, BA, 62, MS, 65; Univ Waterloo, PhD(statist), 73. *Prof Exp:* Statist programmer, Univ Chicago, 62-65; assoc mem tech staff, Bell Tel Labs, 65-68; statist programmer, Imp Col, Univ London, 68-69; from instr to assoc prof, Univ Waterloo, Can, 69-82; sr res officer statist, Can, Social & Econ Div, 82-91, CHIEF HEALTH STATUS ANALYST, DIV CAN CTR HEALTH INFO, 91- *Mem:* Statist Soc Can; fel Am Statist Asn. *Res:* Data analysis; statistical computing; detection of outliers; analysis of mortality data; use of computers in teaching. *Mailing Add:* Statist Can Coats Bldg 18th Floor Tunney's Pasture Ottawa ON K1A 0T6 Can

GENTLEMAN, WILLIAM MORVEN, b Calgary, Alta, July 6, 42; div; c 2. COMPUTER SCIENCE, MATHEMATICS. *Educ:* McGill Univ, BSc, 63; Princeton Univ, MA, 64, PhD(math), 66. *Prof Exp:* Mem tech staff, Bell Tel Labs, 65-69; from asst prof to assoc prof appl anal & comput sci, Univ Waterloo, 69-74, prof comput sci, 74-84; res officer, 83-88, PRIN RES OFFICER, NAT RES COUN, 88-, HEAD, SOFTWARE ENG LAB, 90- *Concurrent Pos:* Sr res fel, Nat Phys Lab, 68-69, prin res fel, 75-76; vis res officer, Nat Res Coun, 82-83. *Honors & Awards:* Ross Medal, 89. *Mem:* Asn Comput Mach; Soc Indust & Appl Math. *Res:* Numerical algorithms and analysis; symbolic algebraic manipulation; software engineering; computer networks. *Mailing Add:* Inst Info Technol Nat Res Coun Montreal Rd M-50 Ottawa ON K1A 0R8 Can

GENTNER, NORMAN ELWOOD, b Carrot River, Sask, Apr 1, 43; div. BIOCHEMISTRY, MICROBIOLOGY. *Educ:* Univ Sask, BA, 63, Hons, 64; Univ Calif, Davis, PhD(biochem), 68. *Prof Exp:* lab demonstr, Univ Calif, Davis, 63-64, teaching asst, res asst & assoc biochem, 64-67, fel, 68; fel, Sch Med, Stanford Univ, 69-70; Res officer biol, Chalk River Nuclear Labs, Atomic Energy Can Ltd, 71-85, sr research officer, 85-88, sr research officer II, 88-89, HEAD, RADIATION BIOL BR, CHALK RIVER LABS, AECL RES, 89- *Concurrent Pos:* Nat Res Coun Can fel, 68; Jan Coffin Childs Mem Fund Med Res fel, 69-70; res student biochem, Atomic Energy Can Ltd, Chalk River, 63-64; mem biol sub-group, Effects of Atomic Radiation, UN Sci Comt; Regant's fel, Univ Calif, Davis, 64. *Mem:* Am Chem Soc; Am Soc Microbiol; Radiation Res Soc; Am Soc Photobiol; Environ Mutagen Soc. *Res:* DNA repair; survival, recovery, and mutation after radiation exposure; individual variation in radiation sensitivity and relation to cancer treatment; DNA repair inhibitors; recombinational repair of DNA lesions and consequences. *Mailing Add:* Radiation Biol Br Chalk River Labs AECL Res Chalk River ON K0J 1J0 Can

GENTNER, ROBERT F, b New York, NY, Oct 31, 38. PHYSICAL CHEMISTRY, ATOMIC PHYSICS. *Educ:* St John's Univ, BS, 60, MS, 62, PhD(phys chem), 68. *Prof Exp:* Res chemist, Picatinny Arsenal, Dover, NJ, 69-80; RES SCIENTIST, ARMAMENT RES & DEVELOP CTR, 80- *Mem:* Am Chem Soc; Am Phys Soc; AAAS. *Res:* Quantum mechanical calculations for atomic systems; detonation physics. *Mailing Add:* 13-05 135 St College Point NY 11356

GENTRY, ALWYN HOWARD, b Clay Center, Kans, Jan 6, 45; m 69; c 2. TAXONOMIC BOTANY. *Educ:* Kans State Univ, BS & BA, 67; Univ Wis, MS, 69; Wash Univ, PhD(biol), 72. *Prof Exp:* Cur, Summit Herbarium & Libr, 71-72; asst cur, Mo Bot Garden, 72-77, assoc cur, 77-86, cur, 86-90, SR CUR, MO BOT GARDEN, 90- *Concurrent Pos:* Instr phytogeog, Floristic Taxon Sem Trop Forests, 79- *Honors & Awards:* Distinguished Serv Award, Soc Conserv Biol, 90. *Mem:* Am Soc Plant Taxonomists; Int Asn Plant Taxonomists; Bot Soc Am; Asn Trop Biol; AAAS. *Res:* Tropical botany, pollination ecology, plant diversity, community ecology; systematics of bignoniaceae; neotropical floristics of Panama, Colombia, Ecuador and Peru. *Mailing Add:* Mo Bot Garden PO Box 299 St Louis MO 63166

GENTRY, CLAUDE EDWIN, b Oak Hill, WVa, Aug 3, 30; m 55; c 3. AGRONOMY. *Educ:* Univ Ky, BS, 58, MS, 60, PhD(plant path), 68. *Prof Exp:* Assoc agronomist, 60-69, assoc prof agron, 69-73, PROF AGRON & BIOL, BEREA COL, 73- *Concurrent Pos:* Fel, Univ Ky, 68. *Mem:* Am Phytopath Soc; Am Soc Agron. *Res:* Alkaloid content of tall fescue; plant pathology, including the interrelationship of Rhizoctonia solani, environment and genotype on the alkaloid content of tall fescue. *Mailing Add:* Dept Agron Berea Col Berea KY 40404

GENTRY, DONALD WILLIAM, b St Louis, Mo, Jan 18, 43; m 65; c 2. ENGINEERING EDUCATION. *Educ:* Univ Ill, BSE, 65; Univ Nevada, MS, 67; Univ Ar, PhD, 72. *Prof Exp:* PROF, DEAN ENG & DEPT HEAD MINING ENG, COLORADO SCHOOL MINES, 72- *Mem:* Sigma Xi. *Res:* Rock mass characterization. *Mailing Add:* 6590 Ridgeview Dr Morrison CO 80465

GENTRY, GLENN ADEN, b Athens, Ga, June 25, 31; m 58; c 2. VIROLOGY. *Educ:* Maryville Col, BA, 53; Vanderbilt Univ, MS, 56; Univ Miss, PhD(microbiol), 60. *Prof Exp:* Nat Cancer Inst fel, McArdle Mem Lab Cancer Res, 60-63; from asst prof to assoc prof, 63-68, PROF MICROBIOL, SCH MED, UNIV MISS, 68- *Concurrent Pos:* Nat Inst Allergy & Infectious Dis res career develop award, 63-73; Am Cancer Soc & USPHS grants, 63-73; vis prof virol, Glasgow Univ, 70-71. *Mem:* Am Soc Microbiol; Am Asn Exp Path; Sigma Xi. *Res:* Equine and human herpes viruses; pyrimidine metabolism in cell culture and in vivo in mammals; viral antimetabolites; viral physical chemistry; viral genetics. *Mailing Add:* Dept Microbiol Univ Miss Med Ctr Jackson MS 39216-4505

GENTRY, IVEY CLENTON, b Roxboro, NC, Apr 7, 19; m 43; c 3. MATHEMATICS. *Educ:* Wake Forest Col, BS, 40; Duke Univ, MA, 47, PhD(math), 49. *Prof Exp:* From asst prof to assoc prof, 49-56, PROF MATH, WAKE FOREST UNIV, 56- *Mem:* Am Math Soc; Am Math Asn. *Res:* Topology. *Mailing Add:* Dept Math Wake Forest Univ Box 7311 Reynolds Sta Winston-Salem NC 27109

GENTRY, JOHN TILMON, b St Louis, Mo, Dec 31, 21; m 49; c 5. PUBLIC HEALTH, HOSPITAL ADMINISTRATION. *Educ:* Washington Univ, AB, 44, BS & MD, 48; Harvard Univ, MPH, 51. *Prof Exp:* Intern, Clins, Univ Chicago, 48-49; resident physician, State Dept Health, NY, 49-50; asst to chief epidemiol br, Communicable Dis Ctr, 51-52, health officer, Great Anchorage Health Dist, Alaska, 52-53; dist state health officer, State Dept Health, NY, 54-57, regional health dir, 57-64; assoc prof pub health admin, 64-68, asst dean prog develop, Sch Pub Health, 64-68; prof pub health admin, Sch Pub Health & dir prog med care & health serv admin, Univ NC, Chapel Hill, 68-74, res prof inst res social sci, 70-74; dir bur health care serv & exec med dir, Medicaid Prog, New York Dept Health, 74-76; COMNR HEALTH, ERIE COUNTY DEPT HEALTH, BUFFALO, NY, 76- *Concurrent Pos:* From clin asst prof to clin assoc prof, State Univ NY Upstate Med Ctr, 55-64; dep chief health div & chief med educ br, US AID mission, India, 61-63; consult, Nat Comn Community Health Serv, 64-; adj prof, Grad Prog Hosp Admin, Duke Univ, 70-74; adj prof grad sch pub admin, NY Univ, 75- *Mem:* AAAS; fel Am Pub Health Asn; NY Acad Sci. *Res:* Identification of social, psychological and economic factors that enhance or impede the implementation of community health services. *Mailing Add:* 258 Atlantic Rd Gloucester MA 01930

GENTRY, KARL RAY, b Roxboro, NC, Apr 13, 38. MATHEMATICS. *Educ:* Wake Forest Univ, BA, 60; Univ Ga, MA, 62, PhD(math), 65. *Prof Exp:* ASSOC PROF MATH, UNIV NC, GREENSBORO, 65- *Mem:* Am Math Soc. *Res:* General topology. *Mailing Add:* Dept Math 383 Bus Econ Bldg Univ NC 1000 Spring Garden St Greensboro NC 27412

GENTRY, ROBERT CECIL, b Paducah, Ky, Nov 29, 16; m 48; c 4. METEOROLOGY. *Educ:* Murray State Col, BS, 37; Fla State Univ, PhD, 63. *Prof Exp:* Teacher pub sch, Ky, 37-40; res forecaster, US Weather Bur, 42-55, asst dir, Nat Hurricane Res Proj, 55-59, actg dir, 59-61, dir, 61-64, dir, Nat Hurricane Res Lab, 64-74, dir proj stormfury, 66-74; chief res meteorologist, Gen Elec Co, 75-78; adj prof, 78-79, res prof, 79-85, ADJ PROF ATMOSPHERIC PHYSICS, CLEMSON UNIV, 85- *Concurrent Pos:* Consult, World Meteorol Orgn, 60-75; cert consult meteorologist by Am Meteorol Soc, 77. *Honors & Awards:* Gold Medal Award Distinguished Achievement Fed Serv, US Dept Com, 70. *Mem:* Fel Am Meteorol Soc; Am Geophys Union; Sigma Xi. *Res:* Tropical meteorology, especially hurricanes; tornadoes, weather modification. *Mailing Add:* Rte 1 Box 632 Salem SC 29676

GENTRY, ROBERT FRANCIS, b Topeka, Kans, May 31, 21; m 42; c 3. POULTRY PATHOLOGY. *Educ:* Kans State Univ, DVM, 44; Univ Mo, MA, 47; Mich State Univ, PhD, 53. *Prof Exp:* Instr vet sci, Univ Mo, 44-47; vet, US Regional Poultry Res Lab, Mich, 47-53; prof vet sci, Vet Res Ctr, Pa State Univ, University Park, 54-81; RETIRED. *Mem:* Am Vet Med Asn; Poultry Sci Asn; US Animal Health Asn; Am Asn Avian Path; Conf Res Workers Animal Dis; Sigma Xi. *Res:* Avian disease research with emphasis on procedures for eradication of infectious diseases; detection of avian salmonella; measurement and control of egg and hatchery contamination; differential diagnosis of respiratory diseases; environmental control of poultry houses. *Mailing Add:* PO Box 482 Alva FL 33920

GENTRY, ROBERT VANCE, b Chattanooga, Tenn, July 9, 33; m 53; c 3. GEOCHEMISTRY, NUCLEAR GEOPHYSICS. *Educ:* Univ Fla, BS, 55, MS, 56; Columbia Union Col, DSc, 77. *Hon Degrees:* DSc, Columbia Union Col, 77. *Prof Exp:* Nuclear engr, Gen Dynamics/Convair, Tex, 56-57, aerophys engr, 57-58; sr engr, Martin Co, Fla, 58-59; instr math, Univ Fla, 59-61 & Walla Walla Col, 61-62; instr physics, Ga Inst Technol, 62-64; res physicist, Archaeol Res Found, 65-66; from asst prof to assoc prof physics, Columbia Union Col, 66-84; RES PHYSICIST, LECTR & AUTHOR, EARTH SCI ASSOCS, KNOXVILLE, TENN, 86- *Concurrent Pos:* Guest scientist, Chem Div, Oak Ridge Nat Lab, 69-82; grantee, 71-77, res fel grantee, 83- *Mem:* AAAS; Am Phys Soc; Am Geophys Union; fel NSF; NY Acad Sci; Sigma Xi. *Res:* Research, writing and lecturing relative to radioactive halos and their cosmological implications with respect to the mode of formation and age of the earth. *Mailing Add:* 6321 Cate Rd Powell TN 37849

GENTRY, ROGER LEE, b Bakersfield, Calif, Mar 19, 38; m 72; c 3. OCEANOGRAPHY. *Educ:* Calif State Univ, San Francisco, BA, 63, MA, 66; Univ Calif, Santa Cruz, PhD(biol), 70. *Prof Exp:* Proj leader biol, Stanford Res Inst, 69-70; fel ethol, Univ Adelaide, S Australia, 70-71; res assoc biol, Univ Calif, Santa Cruz, 71-74; WILDLIFE BIOLOGIST, MARINE MAMMAL DIV, NAT MARINE FISHERIES SERV, DEPT COM, 74- *Concurrent Pos:* Sabbatical to Coastal Marine Studies, Univ Calif, Santa Cruz, 78-80. *Mem:* AAAS; Soc Marine Mammal; Oceanog Soc. *Res:* Social behavior; social organization; and behavioral ecology of vertebrates, especially pinnipeds; diving and foraging behavior of marine vertebrates. *Mailing Add:* Nat Marine Fishery Service NOAAMarine Mammal Div NWFC 7600 Sand Pt Way NE C15700 Seattle WA 98115-0070

GENTRY, WILLARD MAX, JR, b Omaha, Nebr, May 2, 23; m 49; c 4. ORGANIC CHEMISTRY. *Educ:* Harvard Univ, AB, 43, AM, 48; Boston Univ, PhD(chem), 51. *Prof Exp:* Chemist, Dow Chem Co, 51-63, group leader, 63-70, anal coordr, 70-71, mgr res planning, 71-83; RETIRED. *Mem:* Am Chem Soc; Sigma Xi. *Res:* Budgeting and project planning for development of new agricultural products and organic chemicals. *Mailing Add:* 19415 Indian Summer Lane Monument CO 80132-9422

GENTRY, WILLIAM RONALD, b Texarkana, Tex, July 7, 42; m 64. PHYSICAL CHEMISTRY, CHEMICAL PHYSICS. *Educ:* Univ Redlands, BS, 64; Univ Calif, Berkeley, PhD(phys chem), 67. *Prof Exp:* Res assoc chem, Mass Inst Technol, 67-68, NSF fel, 68-70; from asst prof to assoc prof, 70-79, PROF CHEM & CHEM PHYSICS, UNIV MINN, 79-, CHAIR, DEPT CHEM, 89- *Concurrent Pos:* Sloan Found res fel, 75-77; pres, Beam Dynamics, 75-; mem, Chem Rev Panel, Air Force Off Sci Res, 80-84; adv ed, Chem Physics Letters, 80-; assoc ed, J Am Chem Soc, 82-; fel, Yamada Sci Found, 86. *Mem:* AAAS; Am Phys Soc; Sigma Xi; Am Chem Soc; Fel Am Phys Soc, 87. *Res:* Chemical dynamics of molecular collisions, including inelastic and reactive processes in the gas phase. *Mailing Add:* Dept Chem Univ Minn Minneapolis MN 55455

GENTZLER, ROBERT E, b York, Pa, Aug 24, 43; m 65; c 2. CHEMICAL KINETICS. *Educ:* Dartmouth Col, MA, 67; Univ Mass, PhD(chem), 70. *Prof Exp:* res chemist, Plastics Dept, Wash Lab, 70-80, RES SUPVR, PLASTIC PROD DIV, EXP STA, E I DU PONT DE NEMOURS & CO, INC, 80- *Mem:* Am Chem Soc. *Res:* Nuclear magnetic resonance studies of probe nuclei in solvation and inorganic complex environments; thermoplastic polymer properties and organic reaction kinetics/catalysis. *Mailing Add:* 126 Weldin Park Dr Weldin Park DE 19850

GENUTH, SAUL M, b South Norwalk, Conn, Mar 13, 31; m 53; c 2. ENDOCRINOLOGY. *Educ:* Harvard Col, AB, 53; Case Western Reserve Univ, MD, 57. *Prof Exp:* Clin instr med, 64-70, asst prof, 70-73, assoc prof med, 73-78, PROF MED, SCH MED, CASE WESTERN RESERVE UNIV, 78-; from asst to assoc med, Mt Sinai Hosp, 64-68, DIR ISOTOPE LAB MED & DIR SALTZMAN INST CLIN INVEST, MT SINAI HOSP, CLEVELAND, 66- *Mem:* Am Fedn Clin Res; Am Diabetes Asn; Endocrine Soc; Am Col Physicians. *Res:* Insulin delivery in diabetes; microvascular complications of diabetes; obesity. *Mailing Add:* Mt Sinai Med Ctr One Mt Sinai Dr Cleveland OH 44106

GENYS, JOHN B, b Lithuania, Aug 12, 23; US citizen; m 65; c 2. FOREST GENETICS, FORESTRY. *Educ:* Univ Göttingen, dipl, 49; Mich State Univ, PhD(forestry, genetics), 60. *Prof Exp:* Forest res aide, Lake States Forest Exp Sta, USDA, 55-57; res asst forest genetics, Mich State Univ, 57-60; instr soil sci, Univ Wis, 60-61; biologist, Natural Resources Inst, 61-62, res asst prof natural resources, 62-66, res assoc prof, 66-76, chmn, Inland Resources lab, 78-79, PROF DENDROGENETICS & FOREST SCI, CTR ENVIRON & ESTUARINE STUDIES, UNIV MD, 76- *Concurrent Pos:* Assoc ed, Chesapeake Sci, 63-77; app, Nat Plant Genetic Resources Bd, 82-88. *Honors & Awards:* Cert Appreciation, US Dept Agr, 87. *Res:* Genetic variations in Pinus, Larix, Picea, Prunus and Liriodendron species; selections for productivity, pest and pollution resistance; interspecific pinus hybrids. *Mailing Add:* Appalachian Environ Lab Univ Md Frostburg State Col Frostburg MD 21532

GENZER, JEROME DANIEL, b New York, NY, July 23, 25; m 50; c 2. ORGANIC CHEMISTRY. *Educ:* NY Univ, BA, 47; Ind Univ Bloomington, MA, 48. *Prof Exp:* Jr chemist, Warner-Lambert Co, 48-53, from scientist to sr scientist, 53-63, sr res assoc, 63-75, assoc dir & dir chem develop, 77-80; RETIRED. *Mem:* Am Chem Soc. *Res:* Organic synthesis; process research and development of organic chemicals for medicinal use from laboratory through pilot plant in preparation for production, including chemical and equipment design and evaluation and cost evaluation. *Mailing Add:* 27 Washington Ct Livingston NJ 07039

GEOFFROY, GREGORY LYNN, b Honolulu, Hawaii, July 8, 46; m 71; c 4. ORGANOMETALLIC CHEMISTRY. *Educ:* Univ Louisville, BS, 68; Calif Inst Technol, PhD(chem), 74. *Prof Exp:* From asst prof to assoc prof, 74-82, prof, 82-88, PROF & HEAD CHEM, PA STATE UNIV, UNIVERSITY PARK, 88- *Concurrent Pos:* Alfred P Sloan Found res fel, 78-80; Camille & Henry Dreyfus teacher scholar award, 77-82; John Simon Guggenheim Found fel, 82-83. *Mem:* Am Chem Soc. *Res:* Photochemistry of transition metal organometallic compounds; homogeneous catalysis; synthetic organometallic chemistry; heterogeneous catalysis. *Mailing Add:* 813 W Foster Ave State College PA 16801-3938

GEOGHEGAN, ROSS, b Dublin, Ireland, July 2, 43; US citizen; m 69; c 2. GEOMETRIC TOPOLOGY, ALGEBRAIC TOPOLOGY. *Educ:* Univ Col Dublin, Ireland, BSc, 63, MSc, 64; Cornell Univ, PhD(math), 70. *Prof Exp:* Mem staff, Inst Advan Study, 70-72; from asst prof to assoc prof, 72-84, PROF STATE UNIV NY, BINGHAMTON, 84- *Concurrent Pos:* Vis asst prof math, Univ Ga, 74-75; mem, Inst Advan Study, 78-79; vis, Institut des Hautes Etudes Scientifiques, Bures-sur-Yvette, France, 85-86. *Mem:* Am Math Soc; Math Asn Am; Irish Math Soc. *Res:* Geometric topology; homology of groups; fixed point theory; shape theory; infinite-dimensional topology. *Mailing Add:* Dept Math Sci State Univ Ny Binghamton NY 13901

GEOGHEGAN, THOMAS EDWARD, M-RNA STRUCTURE, GENE EXPRESSION. *Educ:* Pa State Univ, PhD(biochem), 75. *Prof Exp:* ASSOC PROF BIOCHEM, UNIV LOUISVILLE, 79- *Mailing Add:* Dept Biochem Univ Louisville Box 35260 Louisville KY 40292

GEOGHEGAN, WILLIAM DAVID, b Milford, Conn, July 2, 43; m 70; c 1. CELL BIOLOGY, IMMUNOCHEMISTRY. *Educ:* Univ Bridgeport, BA, 70, MS, 72; Ohio State Univ, PhD(anat), 77. *Prof Exp:* Fel cell biol, 77-80, asst prof, Med Col Wis, 80-; AT DEPT DERMAT, UNIV TEX HEALTH SCI CTR. *Concurrent Pos:* Res chemist, Vet Admin, 80- *Mem:* AAAS; Reticuloendothelial Soc; Histochem Soc; Soc Invest Dermat. *Res:* Developed methods for the controlled absorption of proteins to colloidal gold for use in immunochemical assays and immunocytochemistry in light and electron microscopy; immunology of the lung and skin. *Mailing Add:* Dept Dermat Univ Tex Health Sci Ctr 6431 Fannin Houston TX 77030

GEOKAS, MICHAEL C, b Villia-Attiha, Greece, Aug 25, 24. PANCREATIC ENZYMES, PROTEINMATIC ENZYMES. *Educ:* Athens Univ, MD, 51; McGill Univ, MSc, 64, PhD(invest med), 66. *Prof Exp:* CHIEF MED SERV, VET ADMIN MED CTR, MARTINEZ, CALIF, 74- *Mem:* NY Acad Sci; Am Physiol Soc; fel Am Col Physicians; Am Gastroenterol Asn. *Mailing Add:* Vet Admin Med Ctr 150 Muir Rd Martinez CA 94553

GEOKEZAS, MELETIOS, b Erythrai, Greece, June 10, 36; US citizen; m 63; c 2. INFORMATION SCIENCE, ELECTRICAL ENGINEERING. *Educ:* Univ Wash, BS, 60, MS, 63, PhD(elec eng), 68. *Prof Exp:* Res engr, Boeing Co, 61-63 & 65-68; res engr, Honeywell Inc, 68-77, sect chief, 77-79, mgr, 79-80, dir, 80-87, dir systs, 87-90, DIR ADVAN PROG, 90- *Mem:* Inst Elec & Electronic Engrs; Am Defense Preparedness Asn. *Res:* Signal processing; pattern recognition. *Mailing Add:* Precision Armament Systs Alliant Techsysts Inc 10400 Yellow Circle Dr Minnetonka MN 55343

GEORGAKAKOS, KONSTANTINE P, b Athens, Greece, Sept 12, 54; m 84; c 1. ESTIMATION THEORY, FLASH-FLOOD FORECASTING. *Educ:* Nat Tech Univ Athens, Dipl, 77; Mass Inst Technol, MS, 80, ScD(hydrol & water resources), 82. *Prof Exp:* Res asst hydrol & water resources, Mass Inst Technol, 77-82; Nat Res Coun-Nat Oceanic & Atmospheric Admin res assoc hydrol water resources, Hydrol Res Lab, Nat Weather Serv, 82-85, res hydrologist, 85-86; asst prof hydrol & water resources, 86-89, ASSOC PROF, HYDROMETEOROL, UNIV IOWA, 89- *Concurrent Pos:* Presidential young investr, NSF, 87-92; Mem, Comt Large Scale Exp Hydrol, Am Geophys Union, 84-86, chair precipitation, 90-92 & chair hydrol, Am Meteorol Soc, 91-93. *Mem:* Am Geophys Union; Inst Elec & Electronic Engrs; AAAS; Sigma Xi; Am Meteorol Soc; Am Soc Civil Engrs. *Res:* Design, development and testing of flood and flash-flood forecasting systems that are based on hydrometeorological principles and capable for real-time quantitative probabilistic forecasts; research in precipitation prediction; hydrologic effects of expotential climate change. *Mailing Add:* 44 Tucson Pl Iowa City IA 52240-9484

GEORGAKIS, CHRISTOS, US citizen. PROCESS CONTROL, CHEMICAL REACTORS. *Educ:* Nat Tech Univ, Athens, ChE dipl, 70; Univ Ill, MS, 72, Univ Minn, PhD(chem eng), 75. *Prof Exp:* From asst prof to assoc prof chem eng, Mass Inst Technol, 75-83; prof measurement & control, Univ Thessaloniki, 80-83; assoc prof chem eng, 83-87, CTR DIR, LEIHIGH UNIV, 85-, PROF CHEM ENG & PROCESS CONTROL, 87- *Concurrent Pos:* DuPont prof, Mass Inst Technol, 75-76, Edgerton prof, 78-80; Dreyfus Found teacher-scholar, 79; vis prof, Rhone Poulenc Industrialization, France, 91. *Mem:* Am Inst Chem Engrs; Am Chem Soc. *Res:* Chemical process modeling; optimization and control; polymerization processes; bioreactors; intelligent measurement; state estimation; statistical process control. *Mailing Add:* Lehigh Univ 111 Research Dr Bethlehem PA 18015

GEORGAKIS, CONSTANTINE, b Vonitsa, Greece, Mar 14, 37; US citizen. MATHEMATICS, STATISTICS. *Educ:* DePaul Univ, BS, 61, MS, 63; Ill Inst Technol, PhD(math), 69. *Prof Exp:* From instr to asst prof, 62-77, ASSOC PROF MATH, DePAUL UNIV, 77- *Concurrent Pos:* Dir undergrad res prog, NSF, 68-70. *Mem:* Am Math Soc; Math Soc Am. *Res:* Fourier analysis on groups, probability theory and mathematical statistics. *Mailing Add:* Dept Bus Math & Statis De Paul Univ 2323 N Sem Ave Chicago IL 60614-3298

GEORGANAS, NICOLAS D, b Athens, Greece, June 15, 43; Can citizen; m 72; c 2. COMPUTER-COMMUNICATIONS. *Educ:* Nat Tech Univ Athens, Dipl Eng, 66; Univ Ottawa, PhD(control theory), 70. *Prof Exp:* From lectr to assoc prof, 70-80, chmn dept, 81-84, PROF ELEC ENG, UNIV OTTAWA, 80-, DEAN ENG, 86- *Concurrent Pos:* Vis prof commun syst archit, IBM CER, La Gaude, France, 77-78 & advan studies, Bull Transac & Inst Nat Res Informatics & Autom, France, 84-85; consult, Dept Commun, Govt Can, 78-83. *Mem:* Fel Inst Elec & Electronics Engrs. *Res:* Multimedia broadband communications; performance modeling and evaluation of computer systems. *Mailing Add:* Dean's Off Fac Eng Univ Ottawa Ottawa ON K1N 6N5 Can

GEORGE, A(LBERT) R(ICHARD), b New York, NY, Mar 12, 38; m 59; c 3. AEROACOUSTICS, MANUFACTURING ENGINEERING. *Educ:* Princeton Univ, BSE, 59, MA, 61, PhD(aerospace & mech sci), 64. *Prof Exp:* Res assoc aerospace eng, Princeton Univ, 64; from asst prof to assoc prof, Grad Sch Aerospace Eng, Cornell Univ, 64-72, assoc prof aerospace eng, 72-77, dir, Sibley Sch Mech & Aerospace Eng, 77-87, PROF MECH & AEROSPACE ENG, CORNELL UNIV, 77-, DIR, CORNELL MFG ENG & PRODUCTIVITY PROG, 91- *Concurrent Pos:* Vis asst prof, Univ Wash, 64-65; vis sr res fel, Southampton Univ, UK, 71-72, consult, 65-, prin investr, 68-; head sect, BMW Act, Munich Ger, 87-88; NRC sr assoc, NASA Am Res Ctr, 88; chmn, Wind Noise Comt, Soc Automotive Engrs. *Mem:* Am Inst Aeronaut & Astronaut; Am Phys Soc; Am Soc Mech Engrs; Soc Automotive Engrs; Am Helicopter Soc. *Res:* Aerodynamics of aircraft and ground vehicles; aerodynamic noise mechanisms; noise control; design and performance of automobiles; sonic boom. *Mailing Add:* Dept Mech & Aerospace Eng 105 Upson Hall Cornell Univ Ithaca NY 14853-7501

GEORGE, ALBERT EL DEEB, b Alexandria, Egypt, May 1, 36. PETROLEUM CHEMISTRY, HEAVY OIL RECOVERY. *Educ:* Univ Cairo, BSc, 57, MSc, 62, PhD(petrol chem), 67; Enrico Mattei Inst, Milan, Italy, dipl petrol ref, 69. *Prof Exp:* Res asst, Petrol Chem Sect, Nat Res Ctr, Cairo, Egypt, 58-67, res supvr petrol technol sect, 67-69; nat res coun can fel, Fuels Res Ctr, 69-72; res scientist, Energy Res Labs, 72-80, sr res scientist, 80-88, HEAD, BITUMEN & OIL RECOVERY SECTION, DEPT ENERGY, MINES & RESOURCES, OTTAWA, 88- *Concurrent Pos:* Ital Govt training grant, ENI Petrol Corp, Italy, 68-69. *Mem:* Am Chem Soc; Soc Petrol Engrs; Petrol Soc of Can Inst Mining & Metall. *Res:* Chemistry of bitumens, heavy oils and synthetic fuels; separation and analysis of hydrocarbons nitrogenous and sulfur compounds in petroleum distillates; processing of bitumens and heavy crude oils and their different fractions using high pressure catalytic processes; developing analytical methods for bitumens heavy oils and synthetic fuels; geochemistry of petroleum chemistry of enhanced oil recovery. *Mailing Add:* Energy Res Labs c/o 555 Booth St Ottawa ON K1A 0G1 Can

GEORGE, ANNE DENISE, b Leeds, Eng, Nov 19, 41; m 68; c 2. ORGANIC CHEMISTRY, PHYSICAL ORGANIC CHEMISTRY. *Educ:* Univ Manchester, BSc Hons, 63, PhD(org chem), 66. *Prof Exp:* Fel organometallic chem, Univ Calif, 66-68; res assoc & temp asst prof org chem, Univ Nebr, 68-78; from asst prof to assoc prof, 78-90, PROF ORG CHEM, NEBR WESLEYAN UNIV, 90- *Mem:* Am Chem Soc; The Chem Soc; Asn Women Sci. *Res:* Mechanisms of nucleophilic aromatic substitutions and of abnormal nucleophilic substitutions in allylic systems; physical organic and synthetic studies of nitrogen heterocyclic compounds. *Mailing Add:* Nebr Wesleyan Univ 5000 Saint Paul Lincoln NE 68504

GEORGE, BOYD WINSTON, b Burlington, Iowa, Mar 17, 25; m 50; c 3. ENTOMOLOGY. *Educ:* Univ Iowa, BA, 48, MS, 50; Iowa State Col, PhD(entom), 57. *Prof Exp:* Instr zool, Univ SDak, 50-54; from instr to asst prof, Iowa State Univ, 56-61; entomologist, Northern Grain Insects Res Lab, Entom Res Div, Agr Res Serv, USDA, 61-65, dir & invests leader lab, 65-68, asst chief, Veg & Specialty Crops Insect Res Br, Plant Indust Sta, 68-73, asst area dir, Ore-Wash Area, 73-75, res entomologist, Agr Res Serv, Veg & Sugarbeet Insects Lab, 75-77; res entomologist, Western Cotton Res Lab, Sci & Educ Admin-Agr Res, USDA, 77-82; RETIRED. *Mem:* Entom Soc Am. *Res:* Control of arthropod pests of vegetables and sugarbeets in the southwest; insect resistance in cotton. *Mailing Add:* 906 E Laguna Dr Tempe AZ 85282

GEORGE, CARL JOSEPH WINDER, b Cincinnati, Ohio, Oct 24, 30; m 62. AQUATIC ECOLOGY. *Educ:* Univ Mich, BS, 56; Harvard Univ, PhD(biol), 60. *Prof Exp:* Asst prof biol, San Fernando Valley State Col, 60-61; asst prof, American Univ Beirut, 61-67, Rockefeller Found grants, 63-67; assoc prof, 67-82, PROF BIOL, UNION COL, NY, 82- *Concurrent Pos:* Partic fel, UNESCO Int Cong, Moscow, 66; Smithsonian Inst res support grant fisheries in UAR, 68-72; prin investr lake restoration studies, Environ Protection Agency, 75-79; vis prof, Univ RI, Kingston, 80-81 & Marine Lab, Aberdeen, Scotland, 81; Am field serv, liaison tour ecology of El Nino, 84; sci exhib develop, Schinectody Mus, 86; chmn comn fish, Standard Methods, Am Water Fowl Asn, 88- *Mem:* Am Soc Ichthyol & Herpet; Ecol Soc Am; Am Soc Limnol & Oceanog. *Res:* Vertebrate ecology and general limnology with particular attention to vertebrates of northeastern NY state; ecology of regional and world agriculture; seasonal regional moments of water birds, Upper Hudson Valley. *Mailing Add:* Dept Biol Union Col Schenectady NY 12308

GEORGE, CHARLES REDGENAL, b Faison, NC, July 21, 38; m 61; c 2. ENTOMOLOGY, PARASITOLOGY. *Educ:* A&T Col NC, BS, 60; Okla State Univ, MS, 65; Cornell Univ, PhD(entom), 70. *Prof Exp:* Teacher high sch, NC, 60-64; partic, NSF Acad Year Inst, Okla State Univ, 64-65; instr biol, Fayetteville State Col, 65-66; res asst insect path, Cornell Univ, 66-67; assoc prof, 70-74, PROF BIOL, NC CENT UNIV, 74- *Mem:* Am Soc Parasitol; Entom Soc Am; Soc Invert Path. *Res:* Effects of malnutrition on growth and mortality of the red rust flour beetle, Tribolium castaneum parasitized by Nosema whitei Weiser. *Mailing Add:* Dept Biol NC Cent Univ Durham NC 27707

GEORGE, CLIFFORD EUGENE, b Wyandotte, Mich, Mar 16, 42; m 65; c 3. ENERGY FROM BIOMASS, ELECTROMAGNETIC POWER APPLICATIONS. *Educ:* Miss State Univ, BSChE, 66, MS, 76, PhD(chem eng), 85. *Prof Exp:* Develop engr, Copolymer Rubber Co, Inc, 66-70; chief process engr, Calumet Industs, 70-78; gen supt, Crosby Chem Co, 78-81; ASST PROF CHEM ENG, MISS STATE UNIV, 81- *Mem:* Air & Waste Mgt Asn; Am Inst Chem Engrs. *Res:* Improved techniques for the application of electromagnetic energy to industrial applications; energy from biomass processes; contaminated soil remediation. *Mailing Add:* Dept Chem Eng Miss State Univ PO Drawer CN Mississippi State MS 39762

GEORGE, DICK LEON, b Oklahoma City, Okla, Mar 10, 36; m 57; c 2. MATHEMATICS. *Educ:* Okla State Univ, BS, 58; Duke Univ, PhD(math), 62. *Prof Exp:* From instr to asst prof math, NC State Col, 61-63; assoc prof, Charlotte Col, 63-65; assoc prof, Trinity Univ, Tex, 65-68; PROF MATH & CHMN DEPT, GA COL MILLEDGEVILLE, 68- *Mem:* Am Math Soc; Math Asn Am; Soc Indust & Appl Math. *Res:* Mixed boundary value problems in thermoelasticity. *Mailing Add:* Dept Math Ga Col Milledgeville GA 31061

GEORGE, DONALD WAYNE, b Topeka, Kans, May 1, 21; m 43; c 2. PLANT BREEDING. *Educ:* Kans State Col, BS, 48, MS, 49. *Prof Exp:* Asst agronomist, Tex Agr Exp Sta, 49-51; asst plant breeder, Univ Ariz, 51-54; agronomist, Pendleton Br Exp Sta, Wash State Univ, 54-65, agronomist, sci & educ admin, USDA & plant sci res div, 65-82; RETIRED. *Res:* Cold and frost hardiness; vernalization, photoperiod and temperature influence on development; crown placement; coleoptile tiller and secondary crown development; adaptation to early seeding for erosion-pollution control; post harvest dormancy of wheat. *Mailing Add:* NW 1610 Deane Pullman WA 99163

GEORGE, EDWARD THOMAS, b North Adams, Mass, Dec 27, 25; m 55; c 4. CHEMICAL ENGINEERING. *Educ:* Worcester Polytech Inst, BS, 47, MS, 49; Yale Univ, DEng(chem eng), 53. *Prof Exp:* Res scientist, B F Goodrich Chem Co, Ohio, 53-57; eng develop engr, Sci Design Co, NY, 57-59, comput dept mgr, 59-60; res develop mgr & comput dept dir, Quantum, Inc, Conn, 60-62; pres, founder & owner, Conn Sci Ctr, Inc, 62-69; assoc prof indust eng, 70-77, PROF INDUST ENG, UNIV NEW HAVEN, 77- *Concurrent Pos:* Teaching assignments, New Haven Col, 64-70; chmn air & water conserv comt, Chamber of Commerce, 66- *Mem:* NY Acad Sci; fel Am Inst Chemists. *Mailing Add:* Dept Indust Eng Univ New Haven 300 Orange Ave West Haven CT 06516

GEORGE, ELMER, JR, b Kanosh, Utah, Apr 15, 28; m 55; c 4. MICROBIOLOGY, BIOCHEMISTRY. *Educ:* Utah State Univ, BS, 54, MS, 56; Univ Minn, PhD(dairy indust), 64. *Prof Exp:* Creamery foreman, instr & res, Utah State Univ, 54-55; res asst, Univ Minn, 55-58, res & teaching fel, 58-60; lab dir, Qual Control Comt, St Paul, Minn, 60-64; LAB DIR, NY STATE FOOD LAB, 64- *Mem:* Am Dairy Sci Asn; Am Soc Microbiol; fel Asn Off Anal Chem; Int Asn Milk & Environ Sanitarians; Sigma Xi. *Res:* Applied microbiology; chemistry of foods. *Mailing Add:* Dir Food Labs Bldg 7 State Campus Albany NY 12235

GEORGE, FREDRICK WILLIAM, b Fairfield, Iowa, Dec 16, 46; m 68; c 2. SEXUAL DIFFERENTATION. *Educ:* Iowa State Univ, BS, 69, MS, 72, PhD(zool), 75. *Prof Exp:* Fel endocrinol, 75-77, instr, 77-79, ASST PROF INTERNAL MED, MED SCH, UNIV TEX SOUTHWESTERN, 79-, ASST PROF CELL BIOL, 83- *Mem:* Endocrine Soc; Sigma Xi; Soc Study of Reproduction. *Res:* Fetal endocrinology; sexual differentiation. *Mailing Add:* Dept Cell Biol & Anat Univ Tex Southwestern Med Ctr 5323 Harry Hines Blvd Dallas TX 75235

GEORGE, HARVEY, b New York, NY, Apr 28, 35; div; c 3. BIOCHEMISTRY, CLINICAL PATHOLOGY. *Educ:* Cornell Univ, BA, 57; Univ Tenn, MS, 61, PhD(biochem), 63. *Prof Exp:* Res fel biochem, Sch Med, Tufts Univ, 64-66; sr biochemist, Collab Res, Inc, Mass, 67-68, dir biochem res, 68-70; dir clin labs, Lahey Clin Found, 70-85; DIR DIAG LABS, MASS CTR DIS CONTROL, DEPT PUB HEALTH, COMMONWEALTH OF MASS. *Concurrent Pos:* Asst res prof, Med Sch, Boston Univ, 67-69; assoc prof allied health, Northeastern Univ, Boston. *Mem:* Am Asn Clin Chem; Am Chem Soc; NY Acad Sci; Am Soc Microbiol; Am Soc Clin Path; Am Public Health Asn; Asn Off Anal Chemists. *Res:* Clinical chemistry; protein and nucleic acid biosynthesis. *Mailing Add:* State Lab Inst 305 South St Jamaica Plain MA 02130

GEORGE, JAMES E, b Pittsburgh, Pa, July 25, 38; m 63; c 4. INORGANIC CHEMISTRY. *Educ:* Allegheny Col, BS, 60; Univ Ill, PhD(chem), 64. *Prof Exp:* Asst prof chem, Oberlin Col, 63-65; asst prof, 65-69, ASSOC PROF CHEM, DePAUW UNIV, 69- *Concurrent Pos:* Asst prof chem, Univ Marshal. *Mem:* AAAS; Am Chem Soc. *Res:* Chemical education; structure of coordination compounds; coordination chemistry of biological compounds. *Mailing Add:* Dept Chem DePauw Univ Locust St Greencastle IN 46135

GEORGE, JAMES FRANCIS, b Kansas City, Mo, Dec 18, 29; m 57; c 1. TESTING & EVALUATION OF DEVELOPMENT SYSTEMS. *Educ:* Univ Ill, BS, 51; George Washington Univ, MS, 60. *Prof Exp:* Engr bridge design, Howard, Needles, Tummen & Bergendorff, 54-55; bridge engr,

Michael Baker Jr, Inc, 55-60; assoc engr, missle res & develop technol & energy, 61-63, sr engr, 63-70, PROJ ENGR, ENERGY & MIL SYSTS DEVELOP, APPL PHYSICS LAB, JOHNS HOPKINS UNIV, 70- *Mem:* Am Inst Aeronaut & Astronaut. *Res:* Flywheel energy storage systems; surface effect ships for Navy use; ocean thermal energy conversion (OTEC) applications; safety system to prevent damaging fires in shipboard electrical systems. *Mailing Add:* Environ Affairs Dept El Paso Natural Gas Co PO Box 1492 El Paso TX 79978

GEORGE, JAMES HENRY BRYN, b Swansea, Wales, Feb 5, 29; nat US; m 63; c 2. PHYSICAL CHEMISTRY, CHEMICAL ENGINEERING. *Educ:* Oxford Univ, BA, 49, MA & PhD(phys chem), 52; Mass Inst Technol, SM, 53. *Prof Exp:* Instr chem eng, Mass Inst Technol, 53; chem engr, Ionics, Inc, 53-54; sr phys chemist, Arthur D Little, Inc, 54-60, group leader, 60-66, sect head phys chem, 66-69, sect head chem systs, 69-74, vpres chem systs, 72-80; PRES, GEORGE CONSULT INT INC, 80- *Mem:* Electrochem Soc; Am Chem Soc. *Res:* Thermodynamics; electrochemistry; ion exchange; technology, economics and applications of batteries and fuel cells; electric vehicle; world battery industry studies. *Mailing Add:* 53 Spring Road Concord MA 01742

GEORGE, JAMES Z, b Lynn, Mass, Dec 29, 22; m 52; c 2. ATOMIC HYPERFINE STRUCTURE, ATOMIC QUANTUM PHYSICS. *Educ:* Northeastern Univ, SB, 48. *Prof Exp:* Physicist mass spectrometry, nuclear radiation physics, NIH, Bethesda, Md, 49-51 & chem div, US Naval Med Res Inst, 51-55; sr physicist atomic beam clock, physics dept, Nat Co, Melrose, Mass, 55-58, sr res physicist tech staff, Malden, Mass, 58-60, mgr atomic beam resonators, dept phys electronics, Melrose, 60-66; pres, Frequency Control Corp, Topsfield, Mass, 66-69; tech consult atomic beam physics, JG Res & Technol, Frequency Electronics Inc, NY & US Navy, 69-74; dir, atomic res, atomic beam resonators, Frequency Electronics, New Hyde Park, NY, 74-78; res physicist, Salem, Mass, 78-82, res physicist, Melrose, Mass, 82-86, RES PHYSICIST ATOMIC BEAM RESONATORS, JG RES & TECHNOL, DANVERS, MASS, 87- *Concurrent Pos:* Consult, US Navy, 86- *Mem:* Am Phys Soc; Am Inst Physics; NY Acad Sci. *Res:* Hyperfine structure of the "hydrogenic" alkalai atoms; including cesium & rubidium; atomic beam physics. *Mailing Add:* JG Res & Technol One Southside Rd Danvers MA 01923-1408

GEORGE, JOHN ALLEN, entomology, for more information see previous edition

GEORGE, JOHN ANGELOS, b Sault Ste Marie, Mich, Sept 10, 34; m 59; c 3. AEROSPACE ENGINEERING. *Educ:* St Louis Univ, BS, 55, PhD(physics), 67; Calif Inst Technol, MS, 56. *Prof Exp:* Flight test engr, McDonnell Aircraft Corp, 58-59; from asst prof to assoc prof aerospace eng, 59-70, actg chmn dept gen sci, 68-70, PROF AEROSPACE ENG, PARKS COL, ST LOUIS UNIV, 70-, CHMN AEROSPACE ENG & ENG SCI DEPT, 77- *Concurrent Pos:* Res assoc, St Louis Univ, 65-70. *Mem:* Am Inst Aeronaut & Astronaut; Am Soc Eng Educ. *Res:* Teaching in areas of gas dynamics, aerodynamics and flight mechanics; systems studies for aircraft condition monitoring. *Mailing Add:* Dept Aerospace Eng Parks Col St Louis Univ Cahokia IL 62206

GEORGE, JOHN CALEEKAL, b Kerala, India, June 16, 21; m 50; c 3. ZOOLOGY, PHYSIOLOGY. *Educ:* Univ Bombay, BSc, 42, PhD(zool), 48. *Prof Exp:* Demonstr zool, IY Col, Bombay, India, 45-48; lectr, Inst Sci, Bombay, India, 48; zoologist, Dept Anthrop, Govt India, 48-50; reader & head zool, Univ Baroda, 50-56, prof & head, 57-67; from assoc prof to prof, 67-86, ACTG CHMN DEPT, 74-, EMER PROF ZOOL, UNIV GUELPH, 86- *Concurrent Pos:* Fulbright-Smith-Mundt fel, Univ Pa, 53-54; ed, J Animal Morphol & Physiol, 56-; mem int teamwork embryol & Dutch Govt scholar, Holland, 58; Dorabji Tata travel grant to Holland, 58; Fulbright res scholar & lectr, Wash State Univ, 61-62; ed & founder, Pavo, Indian J Ornith, 63-; res grants, Muscular Dystrophy Asn Am, 64-67 & USDA, 65-67; mem biol res comt, Coun Sci & Indust Res, Govt India, 65-67; food & agr comt, Govt India Atomic Energy Comn, 67; negotiated develop grant, Nat Res Coun Can, 70-74; mem animal biol comt, Nat Res Coun Can, 75-77, chmn, 77, adv, 78; Guelph Res Award, Sigma Xi, 79. *Mem:* Fel NY Acad Sci; fel Zool Soc India; Sigma Xi, (pres, 83). *Res:* Vertebrate anatomy; herpetology; ornithology; environmental physiology; comparative physiology; muscle physiology; avian and insect physiology endocrinology; histochemistry; embryology; pineal physiology; thermal physiology. *Mailing Add:* Dept Zool Univ Guelph Guelph ON N1G 2W1 Can

GEORGE, JOHN HAROLD, b Bucyrus, Ohio, Nov 29, 35; m; c 3. MATHEMATICS. *Educ:* Ohio State Univ, BS, 57; Univ Ala, MA, 61, PhD(math), 66. *Prof Exp:* Aerospace scientist, Guid & Control Lab, US Army Ballistic Missile Agency, Ala, 57-60; aerospace scientist & sect chief missile guid & control, Astrionics Lab, Marshall Space Flight Ctr, NASA, 60-66, spec asst to chief res math, Aero-Astrodynamics Lab, Ala, 66-67; assoc prof, 67-74, PROF MATH, UNIV WYO, 74- *Concurrent Pos:* Sr Humboldt award, 73-74. *Mem:* Math Model Soc; Math Asn Am; Soc Indust & Appl Math. *Res:* Oil shale kinetics and chemical process control; modeling of large scale processes; free boundary problems; statistical forces on ocean structures. *Mailing Add:* Dept Math Univ Wyo Box 3036 Laramie WY 82070

GEORGE, JOHN LOTHAR, b Milwaukee, Wis, Apr 17, 16; m 44, 66; c 5. BIOLOGY. *Educ:* Univ Mich, BS, 39, MS, 41, PhD(zool), 52. *Prof Exp:* 4H county camp dir, State Agr Exten Serv, WVa, 39; asst, Univ Mich, 39-42 & 46-47; asst prof zool, Vassar Col, 50-57; assoc cur mammals, NY Zool Soc, 57-58; biologist, US Fish & Wildlife Serv, 58-63; assoc prof forestry, Pa State Univ, 63-70, wildlife mgt, 70-75, chmn wildlife planning group, 70-81, prof, 75-81, EMER PROF WILDLIFE MGT, PA STATE UNIV, 81- *Concurrent Pos:* Ranger naturalist, Great Smokies Nat Park, 40-41; biologist, US Fish & Wildlife Serv, 46-47; chmn comt ecol effects of chem controls, Int Union Conserv Nature & Natural Resources, Switz, 63-74; trustee, Nat Parks Asn, 69-80; Cong Adv ad hoc comt on environ, US Cong, 70-80; mem steering comt, Symp Continuing Educ, Soc Am Foresters, 73; area chmn conserv award comt, Wildlife Soc, 75. *Honors & Awards:* Creative Res Award, Nat Asn Univ Exten Associations, 72. *Mem:* Wildlife Soc; Wilderness Soc; Am Soc Mammal; Soc Am Foresters; Am Inst Biol Sci. *Res:* Avian ecology; effects of pesticides on wild life; avian and mammal populations and life histories; endangered species; urban wildlife; continuing education. *Mailing Add:* 685 Westerly Rkwy State College PA 16801

GEORGE, JOHN RONALD, b Pasco, Wash, Apr 28, 40; m 62; c 2. CROP PHYSIOLOGY, FORAGE CROP PHYSIOLOGY. *Educ:* Washington State Univ, BS, 62; Purdue Univ, MS, 64, PhD(crop physiol), 67. *Prof Exp:* Agronomist, Am Potash Inst, 66-69; asst prof, 69-72, assoc prof, 72-75, PROF AGRON, IOWA STATE UNIV, 75- *Honors & Awards:* Merit Cert, Am Forage & Grassland Coun, 85; Fel Award, Nat Asn Col Teachers Agriculture, 90. *Mem:* Am Forage & Grassland Coun (pres, 79-80); fel Am Soc Agron; fel Crop Sci Soc Am; Nat Asn Cols & Teachers Agr. *Res:* Nitrate accumulation in forage crops; forage production; management and physiology research; establishment, renovation, fertilization and stand dynamics; defoliation management of cool-season and warm-season forage species; the role of warm-season prairie grasses in modern forage-livestock systems. *Mailing Add:* Dept Agron Iowa State Univ Ames IA 50011

GEORGE, JOHN WARREN, inorganic chemistry, for more information see previous edition

GEORGE, KALANKAMARY PILY, b Kerala, India, June 13, 33; m 58; c 2. SOIL & STRUCTURAL ENGINEERING. *Educ:* Nat Inst Eng, India, BE, 56; Iowa State Univ, MS, 61, PhD(civil eng), 63. *Prof Exp:* Dist asst engr, Pub Works Dept, Kerala, India, 56-59; res assoc soils eng, Eng Exp Sta, Iowa State Univ, 59-63; from asst prof to assoc prof struct eng & soils & res engr, 63-68, PROF CIVIL ENG, UNIV MISS, 68- *Concurrent Pos:* State Hwy Dept res grant, 63-78, proj dir hwy res, 63-; NSF grant, 71-74. *Mem:* Am Soc Civil Engrs. *Res:* Soil stabilization; pavement mgt; design of soil-cement bases; pavement analysis; soil structure interaction. *Mailing Add:* Dept Civil Eng Univ Miss University MS 38677

GEORGE, KENNETH DUDLEY, b Eltham, NZ, Oct 31, 16; US citizen; m 43; c 2. HEALTH PHYSICS, RADIATION DOSIMETRY. *Educ:* Univ Auckland, BSc, 39, MSc, 40. *Prof Exp:* With NZ Dept Sci & Indust Res, 40-44, UK-Can Atomic Energy Proj, Can, 44-48 & 49-50 & US Army Munitions Command, Picatinny Arsenal, 51-64; reactor supvr, Res Ctr, Union Carbide Corp, Tuxedo, NY, 64-66, supt nuclear opers, 66-68, sr res scientist, 68-81; RETIRED. *Mem:* Am Nuclear Soc. *Res:* Medical and industrial applications of radioisotopes and nucleonics; nuclear reactor science. *Mailing Add:* RD 3 Box 278 Boonton NJ 07005-0278

GEORGE, M COLLEEN, b Austin, Tex, Aug 16, 38; m 70. EXERCISE PHYSIOLOGY, CARDIOVASCULAR PHYSIOLOGY. *Educ:* Univ Tex, BS, 60, MEd, 63, EdD, 68; Tex Tech Univ, PhD(physiol), 82. *Prof Exp:* From asst prof to assoc prof phys educ, Tex Tech Univ, 64-70, res asst physiol, Sch Med, 77-81; assoc prof phys educ, N Tex State Univ, 70-76; res dir, Spinal Injury Res & Rehab Group, 85-86; CONSULT, PHYSIOL DATA SYSTS LTD, 87- *Concurrent Pos:* Vis prof exercise sci, Ariz State Univ, 84-85. *Mem:* AAAS; Sigma Xi; Am Physiol Soc. *Res:* Motor integration and development of children and the mentally retarded; direct effects of exercise and high altitude on rat myocardium; body composition and exercise tolerance of spinal cord injured humans. *Mailing Add:* 8009 E Del Tiburon Dr Scottsdale AZ 85258

GEORGE, MELVIN DOUGLAS, b Washington, DC, Feb 13, 36; m 58; c 2. MATHEMATICS. *Educ:* Northwestern Univ, BA, 56; Princeton Univ, PhD(math), 59. *Prof Exp:* Res assoc, Inst Fluid Dynamics & Appl Math, Univ Md, 59-60; from asst prof to prof math, Univ Mo, 60-70, assoc dean grad sch, 67-69; prof math & dean col arts & sci, Univ Nebr, Lincoln, 70-75; PROF MATH & VPRES ACAD AFFAIRS, UNIV MO, 75- *Mem:* Am Math Soc; Math Asn Am. *Res:* Mathematical economics; partial differential equations; functional analysis. *Mailing Add:* Off of the Pres St Olaf Col Northfield MN 55057

GEORGE, MICHAEL JAMES, b Staten Island, NY, Nov 13, 41. PHYSICS, SHOCK HYDRODYNAMICS. *Educ:* Univ NC, BS, 63; Calif Inst Technol, PhD(physics), 69. *Prof Exp:* Res fel physics, Calif Inst Technol, 69-70; MEM STAFF PHYSICS, LOS ALAMOS NAT LAB, UNIV CALIF, 70- *Honors & Awards:* Award of Excellence, Dept of Energy. *Mem:* Sigma Xi. *Res:* Shock wave and detonation physics; cosmic rays and space physics. *Mailing Add:* 214 Los Pueblos Los Alamos NM 87544

GEORGE, MILON FRED, b St Cloud, Minn, Jan 19, 44; m 66; c 2. PLANT PHYSIOLOGY. *Educ:* Univ Minn, BS, 66, MS, 73, PhD(plant physiol), 75. *Prof Exp:* Instrumentation engr electronics, McDonnell-Douglas, Santa Monica, Calif, 66-69; asst prof plant physiol, Va Polytech Inst, 75-78; ASST PROF PLANT PHYSIOL, UNIV MO, COLUMBIA, 78- *Mem:* Am Soc Plant Physiologists; AAAS; Am Soc Hort Sci. *Res:* Physiology of plant survival at low temperature. *Mailing Add:* 382 Crown Point Columbia MO 65211

GEORGE, NICHOLAS, b Council Bluffs, Iowa, Oct 29, 37; m 66. OPTICS, ELECTRICAL ENGINEERING. *Educ:* Univ Calif, Berkeley, BS; Univ Md, MS; Calif Inst Technol, PhD(elec eng & physics). *Prof Exp:* Physicist to sect chief, Nat Bur Standards, Washington, DC; chief, physics sect, Emerson Res Labs, Washington, DC; sr staff physicist, Hughes Aircraft Co, El Segundo, Calif; prof elec eng & appl physics, Calif Inst Technol; dir, 77-81, PROF OPTICS, INST OPTICS, UNIV ROCHESTER, 77- *Concurrent Pos:* Consult var Fortune 500 Co; dir-at-large, Optical Soc Am; founding dir, Ctr Optoelectronic Systs Res, 86- & Rochester Imaging Consortium, 91- *Mem:* Fel Optical Soc Am; fel Soc Photo-Optical Instrumentatiom Engrs; Am Phys Soc; Am Asn Univ Profs; Sigma Xi; fel Inst Elec & Electronics Engr. *Res:* Optical systems; opto-electronic systems emphasizing automatic pattern recognition, remote sensing and precision meterology; speckle in optical noise with an emphasis on its wavelength dependence and sub-resolution; x-ray diffraction, interference and holography. *Mailing Add:* Inst Optics Univ Rochester Wilson Blvd Rochester NY 14627

GEORGE, PATRICIA MARGARET, b Wales, Nov 23, 48. INTERFACE CHEMISTRY & PHYSICS. *Educ:* Calif State Univ, BS(chem), 76 & BA(math), 76; Calif Inst Technol, PhD(chem), 81. *Prof Exp:* Fel chem, Calif Inst Technol, 81-82; sr mem tech staff, 82-85, ENG SUPVR, AEROJET ELECTROSYSTS CO, 85- *Mem:* Am Phys Soc; Am Chem Soc; Am Vacuum Soc. *Mailing Add:* Aerojet Elec Systs Co PO Box 296 Mail Stop 538244 Azusa CA 91702

GEORGE, PAUL JOHN, organic chemistry, for more information see previous edition

GEORGE, PETER KURT, b Schenectady, NY, Dec 1, 42; m 64; c 2. MAGNETIC HEADS, MAGNETIC MATERIALS. *Educ:* Univ Mich, BS, 64; Case Inst Technol, MS, 66; Case Western Reserve Univ, PhD(elec sci appl phys), 69. *Prof Exp:* Postdoctoral elec mat, Delft Tech Univ, 70-71; mem tech staff mat res, Rockwell Int, 71-75; mem tech staff mag res, IBM, 76-77; sr prof design mgr magnetic bubbles, Nat Semiconductor, 77-82; dir peripheral comp, Control Data, 82-90; DIR PERIPHERAL COMP, SEAGATE TECHNOL, 90- *Mem:* Inst Elec & Electronic Engrs; Am Inst Physics. *Res:* Developing advanced components for rigid disk drives; magnetoresistive/thin film heads; thin film metal disks. *Mailing Add:* 8741 Sandro Rd Bloomington MN 55438

GEORGE, PHILIP, b Maidstone, Eng, Jan 30, 20; m 46; c 6. BIOPHYSICAL CHEMISTRY. *Educ:* Cambridge Univ, BA, 41, MA, 44, PhD, 45. *Prof Exp:* Asst tutor, Christ's Col, Cambridge Univ, 45-47; Brotherton res lectr phys chem, Univ Leeds, 47-49; asst dir res, Dept Colloid Sci, Cambridge Univ, 49-55; res prof biophys chem, 55-73, prof biol chem, 73-77, PROF BIOL, UNIV PA, 77- *Concurrent Pos:* Fel, Christ's Col, Cambridge Univ, 53-55. *Mem:* Biophys Soc; Am Chem Soc; Am Soc Biol Chem; Hist Sci Soc; Royal Soc Chem; Sigma Xi. *Res:* Coordination chemistry; hemoprotein reactions; biochemical thermodynamics; history of chemistry. *Mailing Add:* Four Herford Pl Lansdowne PA 19050

GEORGE, PHILIP DONALD, b Baltimore, Md, June 8, 21; m 49; c 2. ELECTRICAL INSULATION, CAPACITORS. *Educ:* Mt St Mary's Col, Md, BS, 43; Pa State Col, MS, 44, PhD(org chem), 48. *Prof Exp:* Asst chem, Pa State Col, 42-45; res assoc, 46-56, mgr insulation & plastics behav, 56-59, mgr plastics, 59-61 & advan develop lab, 61-69, mgr eng & mfg film prod, 69-76, MGR FILM LICENSING, GEN ELEC CO, 76- *Mem:* AAAS; Am Chem Soc; Soc Plastics Engrs; Am Mgt Asn; fel Am Inst Chemists. *Res:* Penicillin; boron hydrides; silicones, materials behavior; liquid and solid dilectrics; plastic film and capacitors. *Mailing Add:* 22 Coolidge Ave Glens Falls NY 12801

GEORGE, RAYMOND S, b San Bernardino, Calif, Sept 4, 36; m 63; c 2. NUCLEAR CHEMISTRY. *Educ:* Univ Calif, Riverside, BA, 58; Northwestern Univ, PhD(anal chem), 62. *Prof Exp:* Fulbright fel, H C Orsteds Inst, Copenhagen Univ, 62-63; staff mem, Los Alamos Sci Lab, 63-70; sr safeguards engr, Albuquerque Opers Off, Nuclear Mat Mgt Div, US AEC, 71-75; br chief, Nuclear Safeguards & Accountability, US Energy Res & Develop Admin, 75-78, PROG MGR NUCLEAR WEAPONS, US DEPT ENERGY, 78- *Concurrent Pos:* Consult solar energy anal & appln. *Res:* Boric acid complexes; inorganic complexation chemistry; analytical research and development; analytical instrumentation; explosives chemistry; crime laboratory establishment and techniques; radiation chemistry; nuclear materials safeguards and accountability; factory automation. *Mailing Add:* 1432 Honeysuckle NE Albuquerque NM 87122

GEORGE, ROBERT, b Turlock, Calif, Feb 10, 23; m 58; c 2. PHARMACOLOGY. *Educ:* Univ Ore, AB, 49; Univ Calif, Berkeley, PhD(physiol), 53. *Prof Exp:* Asst physiol, Univ Calif, Berkeley, 50-53; jr res pharmacologist, Univ Calif Med Ctr, San Francisco, 53-56; USPHS fel neuroendocrinol, Maudsley Hosp, Univ London, 56-58; asst res pharmacologist & asst prof pharmacol, 58-61; assoc prof, 61-67, PROF PHARMACOL, CTR HEALTH SCI, UNIV CALIF, LOS ANGELES, 67- *Concurrent Pos:* Co-ed, Ann Rev Pharmacol & Toxicol; consult, Nat Inst Drug Abuse & Nat Inst Neurol & Commun Disorders & Stroke. *Mem:* AAAS; Am Soc Pharmacol & Exp Therapeut; Am Physiol Soc; Int Soc Neuroendocrinol. *Res:* Analgesics and neuroendocrine function; neural control of anterior pituitary; diabetes mellitus; pharmacology and pathology of tremor. *Mailing Add:* Dept Pharmacol Ctr Health Sci Univ Calif Los Angeles CA 90024-1735

GEORGE, ROBERT EUGENE, b Bowling Green, Ohio, Nov, 24, 29; m 52; c 7. RADIOLOGICAL PHYSICS, HEALTH PHYSICS. *Educ:* Ohio State Univ, BSc, 52; Univ Rochester, MSc, 61; Purdue Univ, PhD(bionucleonics), 66. *Prof Exp:* Chief pharm serv, Naval Hosp, Med Serv Corps, US Navy, Quantico, Va, 53-56, chief pharm serv, Naval Hosp, Newport, RI, 56-60, instr nuclear, chem & biol weapons, Naval Unit, Army Chem Sch, 61-64, res projs group dir, Armed Forces Radiobiol Res Inst, 66-67, chmn dept radiation biol, 67-72, radiol physicist, Naval Hosp, Bethesda, Md, 72-73; ASSOC PROF BIONUCLEONICS DEPT, PURDUE UNIV, 73-76; assoc prof & chief radiol physicist, Dept Radiation Oncol, Ind Univ, Indianapolis, 75-89; RETIRED. *Honors & Awards:* Meritorious Serv Medal, Dept Defense, 72. *Mem:* Radiation Res Soc; Health Physics Soc; Am Asn Physicist in Med; Am Col Radiol. *Res:* Applications of high energy electrons and photons in radiation therapy. *Mailing Add:* Dept Radiation Oncol Ind Univ 1100 W Michigan St Indianapolis IN 46223

GEORGE, ROBERT PORTER, b San Rafael, Calif, June 18, 37; m 67; c 3. DEVELOPMENTAL BIOLOGY, INVERTEBRATE ZOOLOGY. *Educ:* Univ Calif, Berkeley, BA, 61; Univ Hawaii, MS, 67, PhD(microbiol), 68. *Prof Exp:* Res fel bacteriol, Univ Wis-Madison, 68-70; asst prof, 70-75, ASSOC PROF ZOOL, UNIV WYO, 75- *Mem:* Soc Develop Biol; AAAS; Soc Free Radical Res. *Res:* Transmission and scanning electron microscopy of cellular slime molds; autoradiographic study of stalk formation in Dictyostelium; effect of inhibitors and chemotactic substances on morphogenesis in Dictyostelium; isolation of cell receptors for cyclic-adenosine monophosphate; study of human carnosinase and its inhibitors; protozoan defenses against free radicals. *Mailing Add:* Dept Zool & Physiol Univ Wyo Univ Sta 3166 Laramie WY 82070

GEORGE, RONALD BAYLIS, b Zwolle, La, Nov 17, 32. INTERNAL MEDICINE, PULMONARY DISEASES. *Educ:* Univ Ala, BA, 54; Tulane Univ, MD, 58; Am Bd Internal Med, dipl, 65. *Prof Exp:* From asst prof internal med to assoc prof med, Sch Med, Tulane Univ, 66-72, PROF MED & HEAD PULMONARY DIS SECT, SCH MED, MED CTR, LA STATE UNIV, SHREVEPORT, 72- *Concurrent Pos:* Ed, Am J Med Sci, Respiratory Care. *Mem:* Am Thoracic Soc; Am Col Chest Physicians. *Res:* Pulmonary physiology; pneumonias and their x-ray appearance; pulmonary mycoses; lung cancer; obstructive airways disease. *Mailing Add:* 1501 Kings Hwy Shreveport LA 71130

GEORGE, RONALD EDISON, b Winnipeg, Man, Can, Oct 1, 37; m 60; c 3. COMPUTER SCIENCE. *Educ:* Univ Sask, BEng, 59, MSc, 61; Univ Waterloo, PhD(mgt sci), 78. *Prof Exp:* Lectr mech eng, Univ Sask, 59-61; from anal programmer comput sci to exec dir, Govt of Sask, 61-71; from asst prof to assoc prof comput sci, Univ Guelph, Ont, 74-80; prof dept comput sci, Univ Calgary, 80-89; PRES & CHIEF EXEC OFFICER, ACTC TECHNOL, 89- *Concurrent Pos:* Consult, Atomic Energy Can, 60; vchmn & bd mem, Sask Govt Comput Ctr, 65-71; adv, S Sask Hosps Bd, 68-71; consult minister health, Study of Psychiat Serv, Govt of Sask, 69. *Mem:* Can Info Processing Soc; Asn Comput Mach; Asn Prof Engrs. *Res:* Management information systems; factors effecting the use of information by senior management. *Mailing Add:* ACTC Technol 350 6715 Eighth Ave NE Calgary AB T2N 1N4 Can

GEORGE, SARAH B(REWSTER), b Tacoma, Wash, Nov 16, 56; m 85; c 1. EVOLUTION, MUSEUM SCIENCE. *Educ:* Univ Puget Sound, BS, 78; Ft Hays State Univ, MS, 80; Univ NMex, PhD(biol), 84. *Prof Exp:* Asst cur, 85-90, ASSOC CUR MAMMAL, NATURAL HIST MUS LOS ANGELES COUNTY, 90- *Concurrent Pos:* Adj asst prof biol, Univ Southern Calif, 85- & Univ Calif, Los Angeles, 89-; mem bd dirs, Am Soc Mammalogists, 85-88, 88-91 & 91-94. *Mem:* Am Soc Mammalogists; Soc Study Evolution; Soc Syst Zool; Asn Women Sci. *Res:* Evolution, systematics, population genetics, conservation biology and taxonomy of small mammals, particularly shrews; museum science and vertebrate collection management. *Mailing Add:* Natural His Mus 900 Exposition Blvd Los Angeles CA 90007

GEORGE, SIMON, b India, May 10, 31; nat US. ATOMIC SPECTROSCOPY. *Educ:* Univ Travancore, BS, 51; Univ Saugar, MS, 54; Univ BC, PhD(physics), 62. *Prof Exp:* Lectr physics, Hislop Col, Nagpur, 54-55, Gauhati Univ, India, 55-57 & Univ BC, 59-60; from asst prof to assoc prof, 61-69, PROF PHYSICS, CALIF STATE UNIV, LONG BEACH, 69- *Mem:* Optical Soc Am; Am Asn Physics Teachers; Am Phys Soc. *Res:* Atomic spectroscopy; Fabry Perot interferometry; energy levels in spectra and laser spectroscopy; and physics education. *Mailing Add:* Dept Physics-Astron Calif State Univ 1250 Bellflower Blvd Long Beach CA 90840

GEORGE, STEPHEN ANTHONY, b Seattle, Wash, May 31, 43; m 69; c 2. BIOLOGY, NEUROSCIENCE. *Educ:* Univ BC, BSc, 64; Johns Hopkins Univ, PhD(biophysics), 70. *Hon Degrees:* AM, Amherst Col, 84. *Prof Exp:* Asst prof biol sci, Univ Md, Baltimore County, 70-73; assoc prof biol, 73-83, PROF BIOL, AMHERST COL, 84- *Concurrent Pos:* Prin investr, NIH grant, 75-; vis scientist develop biol, Nat Inst Med Res, Eng, 77; vis scientist, Univ Col London, 82, Inst Biophys, Beijing, China, 89. *Mem:* Soc Neurosci; Asn Res Vision & Ophthal. *Res:* Visual system development; nerve cell excitability. *Mailing Add:* Dept Biol Amherst Col Amherst MA 01002

GEORGE, STEPHEN L, b Lubbock, Tex, Dec 11, 43; m 65; c 1. BIOSTATISTICS. *Educ:* Tex Tech Univ, BA, 65; NC State Univ, MS, 67; Southern Methodist Univ, PhD(statist), 69. *Prof Exp:* Asst prof biomet, Univ Tex, M D Anderson Hosp & Tumor Inst, 69-74; dir data ctr, Europ Orgn Res Treat of Cancer, 74-75; assoc prof biomet, Univ Tex, M D Anderson Hosp & Tumor Inst, 75-76; dir, Sect Biostatist, St Jude Children's Res Hosp, 76-88; PROF BIOSTATIST, DUKE UNIV MED CTR, 88- *Concurrent Pos:* Adj asst prof math sci, Rice Univ, 70-75; group statistician, Cancer & Leukemia Group B, Duke Univ Med Ctr, 90- *Mem:* Sigma Xi; fel Am Statist Asn; Biomet Soc; Inst Math Statist; Soc Clin Trials; Am Soc Clin Oncol. *Res:* Statistical methods in clinical research; selection problems; sequential analysis; survival studies. *Mailing Add:* Cancer Ctr Biostatist Duke Univ Med Ctr Box 3958 Durham NC 27710

GEORGE, T ADRIAN, b Darlington, Eng, Feb 1, 42; m 68; c 2. INORGANIC CHEMISTRY, ORGANOMETALLIC CHEMISTRY. *Educ:* Manchester Col Sci & Technol, BS, 63; Univ Sussex, PhD(chem), 66. *Prof Exp:* Res assoc, Univ Calif, Riverside, 66-68; asst prof, 68-74, assoc prof, 74-78, PROF CHEM, UNIV NEBR, LINCOLN, 78- *Mem:* Royal Soc Chem; Am Chem Soc; Sigma Xi. *Res:* Fixation of dinitrogen using molybdenum complexes as the active specie; preparation of ammonia, hydrazine and organonitrogen compounds from molecular nitrogen. *Mailing Add:* Dept Chem Univ Nebr Lincoln NE 68588-0304

GEORGE, TED MASON, b Lynnville, Tenn, Sept 22, 22; m 60. NUCLEAR PHYSICS, ELECTRONICS. *Educ:* Vanderbilt Univ, BA, 49, MA, 57, PhD(physics), 64. *Prof Exp:* Asst prof physics, Murray State Univ, 56-59 & Furman Univ, 63-64; prof physics & chmn dept, Eastern Ky Univ, 64-88; RETIRED. *Mem:* Am Asn Physics Teachers. *Res:* Gamma and beta ray spectroscopy. *Mailing Add:* 125 Allen Douglas Dr Richmond KY 40475-0950

GEORGE, THOMAS D, b Robstown, Tex, Mar 3, 40; m 70. MATERIALS SCIENCE, ANALYTICAL CHEMISTRY. *Educ:* Tex Tech Univ, BS, 62, MS, 65; Northwestern Univ, PhD(mat sci), 68. *Prof Exp:* Mem tech staff, Cent Res Labs, 68-69, mgr anal labs, Qual & Reliability Assurance, Components Group, 69-77, opers mgr calculator prod, 77-78, MGR TIME PROD, US CONSUMER GROUP, TEX INSTRUMENTS, 78- *Mem:* Electrochem Soc; Sigma Xi. *Mailing Add:* 8009 Del Tiburon Dr Scottsdale AZ 85258

GEORGE, THOMAS FREDERICK, b Philadelphia, Pa, Mar 18, 47; m 70. THEORETICAL CHEMISTRY. *Educ:* Gettysburg Col, BA, 67; Yale Univ, MS, 68, PhD(chem), 70. *Prof Exp:* Res assoc chem, Mass Inst Technol, 70-71 & Univ Calif, Berkeley, 71-72; from asst prof to prof chem, Univ Rochester, 72-85; PROF CHEM & PHYSICS & DEAN FAC NATURAL SCI & MATH, STATE UNIV NY, BUFFALO, 85- *Concurrent Pos:* Dreyfus Found teacher-scholar, 75-82; Sloan Found res fel, 76-80; comt mem, US Army Basic Sci Res, 78-81; distinguished vis lectr, Dept Chem, Univ Tex, Austin, 78; vchmn, Sixth Int Conf on Molecular Energy Transfer, Rodez, France, 79; lectr semiclassical methods in molecular scattering & spectroscopy, Adv Study Inst, NATO, Cambridge, Eng, 79; distinguished speaker, Dept Chem, Univ Utah, 80; adv bd mem, J Phys Chem, 80-; distinguished lectr, Air Force Weapons Lab, Kirtland AFB, 80; chmn, Gordon Res Conf, NH, 81; mem prog comt, Fourth-Sixth Int Conf on Lasers & Appl, New Orleans, 81-82, San Francisco, 83 & Int Laser Sci Conf, Dallas, 85; organizer, NSF Workshop on Theoret Aspects of Laser Radiation & Its Interaction with Atomic & Molecular Systs, 77; mem exec comt, Phys Div, Am Chem Soc, 79-82, 85-88, vchmn, 85-86, chmn, 86-88; Guggenheim Mem Found fel, 83-84. *Honors & Awards:* Marlow Medal & Prize, Faraday Div, Royal Soc Chem, 79. *Mem:* Am Chem Soc; fel Am Phys Soc; Royal Soc Chem; fel NY Acad Sci; Sigma Xi; Soc Photo-Optical Instrumentation Engrs; Coun Cols Arts & Sci; Europ Phys Soc; AAAS. *Res:* Theory of laser-induced chemical physics; nonlinear optics; molecular collision dynamics; chemical reactions; energy transfer; molecular clusters; surface and solid state chemistry/physics; high temperature superconductivity; polymers; 400 major publications and 7 edited books. *Mailing Add:* 239 Fronzak Hall State Univ NY Buffalo NY 14260

GEORGE, TIMOTHY GORDON, b Pittsburgh, Pa, Dec 12, 58; m 81; c 2. FAILURE ANALYSIS, REFRACTORY METALS. *Educ:* Univ Pittsburgh, BS, 80; Univ Phoenix, MA, 90. *Prof Exp:* Staff metallurgist, Kennametal Inc, 80-83; SECT LEADER, LOS ALAMOS NAT LAB, 83- *Concurrent Pos:* Consult, 85- *Mem:* Am Soc Metals; Am Welding Soc. *Res:* Failure analyses of components used in radioisotope space power systems; evaluate materials through investigation of properties at elevated temperatures at high strain rates and in complex strain states. *Mailing Add:* 430 Ridgecrest Ave Los Alamos NM 87544

GEORGE, WILLIAM, b Santa Cruz, Calif, Feb 2, 25. ORNITHOLOGY, ZOOLOGY. *Educ:* Univ Ariz, BA, 57, MS, 58, PhD(zool), 61. *Prof Exp:* Chapman fel, Am Mus Natural Hist, 61-62; NSF grant, 62-64; asst prof, 64-72, PROF ZOOL, SOUTHERN ILL UNIV, CARBONDALE, 78- *Mem:* Am Ornith Union; Wilson Ornith Soc; Cooper Ornith Soc. *Res:* Raptor biology; evolution of bird wing; domestic cat ecology. *Mailing Add:* Dept Zool & Life Sci 0355B Southern Ill Univ Carbondale IL 62901

GEORGE, WILLIAM JACOB, b Houtzdale, Pa, June 19, 38; m 64; c 2. BIOCHEMICAL PHARMACOLOGY. *Educ:* Pa State Univ, BS, 60; Univ Pittsburgh, BS, 64; Univ Mich, PhD(pharmacol), 68. *Prof Exp:* Fel pharmacol, Univ Minn, 68-70; asst prof, 70-73, assoc prof, 73-78, PROF PHARMACOL, SCH MED, TULANE UNIV, 78- *Concurrent Pos:* Mem, Int Study Group Res Cardiac Metab. *Honors & Awards:* Merck Found Award, 70. *Mem:* AAAS; Am Soc Pharmacol & Exp Therapeut; Am Soc Hemat; NY Acad Sci. *Res:* Drug metabolism; intermediary metabolism; cyclic nucleotides; hormonal control of metabolic processes. *Mailing Add:* Dept Pharmacol Tulane Univ Med Sch New Orleans LA 70112

GEORGE, WILLIAM KENNETH, JR, b Camp Shelby, MS, Apr 19, 45; m 66; c 2. FLUID MECHANICS, TURBULENCE. *Educ:* Johns Hopkin Univ, BSE, 67, PhD(mech), 71. *Prof Exp:* Instr aerospace eng, 68-71, res assoc, App Res Lab, Pa State Univ, 71-74; from asst prof to assoc prof mech eng, 74-80, PROF MECH & AEROSPACE ENG, STATE UNIV NY, BUFFALO, 80- *Concurrent Pos:* Vis scientist, Factory Mutual Res Corp, 74; Ctr Nuclear Studies, Grenoble, 76; Danish Nat Lab, 80-81; vis scientist, Calspan Corp, 87-88; vis prof, DTH, Danish Tech Univ, 88. *Mem:* Am Soc Mech Eng; Am Phys Soc; Sigma Xi. *Res:* Fluid mechanics and heat transfer especially turbulent fluids; development experimental techniques, laser Doppler anemometer, measurement and theoretical understanding of turbulent shear flows. *Mailing Add:* Dept Mech & Aerospace Eng 339 Jarvis Hall Buffalo NY 14260

GEORGE, WILLIAM LEO, JR, b Riverside, NJ, June 1, 38; m 60; c 2. PLANT GENETICS, PLANT BREEDING. *Educ:* Del Valley Col, BS, 60; Rutgers Univ, MS, 62, PhD(horticulture), 66. *Prof Exp:* Res asst genetics, Rutgers Univ, 60-66; asst geneticist, Conn Agr Exp Sta, 66-71; from assoc prof, to prof hort, Ohio State Univ & Ohio Agr Res & Develop Ctr, 71-77; prof hort & head dept, 77-84, ASSOC DEAN & DIR RES INSTR, COL AGR, UNIV ILL, 84- *Concurrent Pos:* Res dir, Am Soc Hort Sci, 77-78; vis prof, Univ Fla, Bradenton, 83. *Mem:* AAAS; Am Genetic Asn; fel Am Soc Hort Sci; Sigma Xi; Coun Agr Sci & Technol. *Res:* Genetics and developmental genetics of sexuality and unstable gene systems in higher plants; development of new breeding systems in vegetable crops; genetics and physiology of fruiting in vegetable crops. *Mailing Add:* Col Agr Univ Ill 1301 W Gregory Dr Urbana IL 61801

GEORGE-NASCIMENTO, CARLOS, b Santiago, Chile, Aug 28, 45; US citizen; m; c 2. PROTEIN CHEMICALS. *Educ:* Univ Chile, MS, 67; Baylor Col Med, PhD(biochem), 75. *Prof Exp:* Instr biochem, Sch Chem & Pharm, Univ Chile, 68-70; postdoctoral res fel, Dept Microbiol, Sch Med, Univ Conn, 76-78, instr, 79-80; asst prof, Dept Biochem, Sch Med, Univ PR, 80-83; prin scientist & mgr, 83-87, SR SCIENTIST, PROTEIN CHEM SECT, CHIRON CORP, EMERYVILLE, CALIF, 87- *Concurrent Pos:* Lectr biochem, Med Sci Campus, Univ PR, 81-83; prin investr, Am Heart Asn & NSF, 81-84; invited prof, Cath Univ, Santiago, Chile, 86, 90 & Univ Chile, Santiago, 88. *Res:* Author of numerous publications. *Mailing Add:* Protein Chem Dept Chiron Corp 4560 Horton St Emeryville CA 94608-2916

GEORGE-WEINSTEIN, MINDY, b Philadelphia, Pa, Feb 5, 53. ANATOMY. *Educ:* Thomas Jefferson Univ, BS, 78, PhD(anat), 84. *Prof Exp:* Postdoctoral fel, Dept Biochem & Biophys, Univ Pa, 84-87 & Dept Microbiol, 87-88; PROF ANAT, PHILADELPHIA COL OSTEOP MED, 88- *Concurrent Pos:* NIH fel, 84-86 & grant, 90-92; Muscular Dystrophy Asn fel, 86-87. *Mem:* Develop Biol Soc; Am Soc Cell Biol. *Res:* Author of numerous publications. *Mailing Add:* Dept Anat Philadelphia Col Osteop Med 4150 City Ave Philadelphia PA 19131

GEORGHIOU, GEORGE PAUL, b Famagusta, Cyprus, Nov 23, 25; wid; c 2. ENTOMOLOGY, TOXICOLOGY. *Educ:* Cornell Univ, BS, 52, MS, 53; Univ Calif, Berkeley, PhD(entom), 60. *Hon Degrees:* PhD, Univ Thessaloniki, Greece, 81. *Prof Exp:* Govt entomologist, Dept Agr, Cyprus, 54-58; jr specialist entom, 58-60, lectr, 60-69, from asst entomologist to assoc, 60-69, head, div toxicol & physiol, 75-83, chmn, dept entom, 83-84, PROF ENTOM, UNIV CALIF, RIVERSIDE, 69- *Concurrent Pos:* Res grants, NIH, 64-67 & 85-, WHO, 65-85, USDA, 66-69 & DAMD, 85-88; consult, Food & Agr Orgn, 73, 74, 75 & 78; mem WHO expert adv panel on vector biol & control, 75-, Food & Agr Orgn, 76- & AID, 83, 85 & 89; Guggenheim fel, 67-68; fac res lectr, Univ Calif, 87. *Honors & Awards:* Bussart Award, Entom Soc Am, 87; Superior Serv Award, USDA, 89. *Mem:* Fel AAAS; Entom Soc Am; Am Chem Soc. *Res:* Insect resistance to insecticides; genetics of resistance; insect toxicology. *Mailing Add:* Dept Entom Univ Calif Riverside CA 92521

GEORGHIOU, PARIS ELIAS, b Cairo, Egypt, July 20, 46; Can citizen; m 71; c 2. ORGANIC CHEMISTRY, ANALYTICAL CHEMISTRY. *Educ:* Witwatersrand Univ, SAfrica, BSc Hons, 67; McGill Univ, PhD(chem), 73. *Prof Exp:* Instr chem, Dawson Col, Montreal, 72-73; fel, Univ Alta, Can, 73-75; ASSOC PROF CHEM, MEM UNIV NFLD, 75- *Concurrent Pos:* Chem Consult to Nat Health & Welfare, Can, Nat Res Coun, Dept Consumer & Corp Affairs. *Mem:* Chem Inst Can; Am Chem Soc. *Res:* Synthesis and mechanisms in organic chemistry; chemical characterization of petroleum and its products; analytical chemistry of trace pollutant gases and respirable particulate matter in indoor air; mutagenicity and carcinogenicity of environmental pollutants and selected organic molecules such as the sterols. *Mailing Add:* Dept Chem Mem Univ Nfld St John's NF A1B 3X7 Can

GEORGHIOU, SOLON, b Kato Zodia, Cyprus, Aug 11, 39; US citizen; m 68; c 2. PHOTOPHYSICS. *Educ:* Univ Athens, Greece, BSc, 62; Univ Manchester, MSc, 65, PhD(physics), 68. *Prof Exp:* Res assoc, Univ Minn, 67-69 & Johns Hopkins Univ, 70-73; from asst prof to assoc prof physics, 73-83, PROF PHYSICS, UNIV TENN, 83-, ADJ PROF, BIOCHEM, 87- *Mem:* Biophys Soc; Am Phys Soc; Am Chem Soc; Am Soc Photobiol. *Res:* Intrinsic and extrinsic fluorescent probes for studying the dynamics and structure of biomolecules and biological assemblies; mechanisms of deexcitation of molecular electronic states; optical, Raman and infrared spectroscopies. *Mailing Add:* Dept Physics Biophysics Lab Physics Lab Univ Tenn Knoxville TN 37996-1200

GEORGI, DANIEL TAYLAN, b Zurich, Switz, June 29, 48; US citizen. RESERVOIR GEOPHYSICS, PHYSICAL OCEANOGRAPHY. *Educ:* Univ Calif, San Diego, BA, 71; Columbia Univ, MA, 73, PhD(geol), 77. *Prof Exp:* Scholar phys oceanog, Woods Hole Oceanog Inst, 77-78, asst scientist, 78-81; SR RES GEOPHYSICIST, EXXON PROD RES CORP, 81- *Mem:* Am Geophys Union; Soc Prof Well Log Analyst; Soc Exploration Geophysicists. *Res:* Reservoir geophysics and geotechnical properties of sediments; antarctic physical oceanography; water mass formation; finistructure and mixing; fractured reservoir analysis. *Mailing Add:* Eff Resources Can Ltd 237 Fourth Ave SW Calgary AB T2P 0H6 Can

GEORGI, HOWARD, b San Bernardino, Calif, Jan 6, 47; c 2. THEORETICAL PHYSICS. *Educ:* Harvard Col, BA, 67; Yale Univ, PhD, 71. *Prof Exp:* Res fel, 71-73, jr fel, Soc Fels, 73-76, Alfred P Sloan Found fel, 76-80, PROF PHYSICS, HARVARD UNIV, 80- *Concurrent Pos:* Sr fel, Soc Fels, Harvard Univ, 81- *Mem:* Am Phys Soc. *Res:* Particle theory particularly unified theories of particle interactions at short distances and su(2)xu(1) breaking. *Mailing Add:* Dept Physics Harvard Univ Cambridge MA 02138

GEORGI, JAY R, b New York, NY, Nov 9, 28; m 52; c 4. VETERINARY MEDICINE. *Educ:* Cornell Univ, DVM, 51, PhD, 62. *Prof Exp:* Asst prof phys biol, NY State Col Vet Med, Cornell Univ, 65-66, from assoc prof to prof, 66-, EMER PROF PARASITOL, NY STATE COL VET MED, CORNELL UNIV. *Mem:* AAAS; Am Vet Med Asn. *Res:* Veterinary clinical parasitology and nematode taxonomy, morphology and bionomics. *Mailing Add:* Diag Lab c/o Col Vet Med PO Box 786 Ithaca NY 14851

GEORGIADE, NICHOLAS GEORGE, b Lowell, Mass, Dec 25, 18; m 42; c 3. PLASTIC SURGERY. *Educ:* Columbia Univ, DDS, 44; Duke Univ, MD & BS, 49; Am Bd Plastic Surg, dipl; Am Bd Oral Surg, dipl. *Prof Exp:* Intern oral surg, Kings County Hosp, 44; intern & asst res gen surg, 49-52, asst res & res plastic surg, 52-54, from instr to assoc prof, 54-64, PROF PLASTIC, MAXILLOFACIAL & ORAL SURG, SCH MED, DUKE UNIV, 64- & CHMN DIV, 75- *Concurrent Pos:* Nat Cancer Inst clin fel, 52-54; consult plastic, maxillofacial & oral surgeon, Vet Admin Hosp, Durham; consult, US Army & US Air Force; res proj, NIH; mem, Plastic Surg Res Coun; ed, Cleft Palate J, 70-76; vchmn, Am Bd Plastic Surg, 74-75; vchmn, Plastic Surg Residency Rev Comt. *Mem:* Am Soc Plastic & Reconstruct Surg; Am Soc Oral Surg; Am Asn Plastic Surg (secy, 72-75, vpres, 76, pres elect, 77, pres, 78); fel Am Col Surg; Am Soc Maxillofacial Surg (pres, 63); Sigma Xi; Int Soc Aesthetic Plastic Surg. *Res:* Plastic, maxillofacial and oral surgery; tissue preservation; burns; maxillofacial growth and development; numerous publications. *Mailing Add:* Box 3098 Duke Univ Hosp Durham NC 27710

GEORGIADIS, JOHN G, b Filiates, Greece, Sept 14, 59; m 87; c 1. HEAT TRANSFER, FLUID MECHANICS. *Educ:* Nat Tech Univ, Athens, Greece, dipl mech eng, 83; Univ Calif, Los Angeles, MS, 84, PhD(mech eng), 87. *Prof Exp:* ASST PROF, DEPT MECH ENG & MAT SCI, DUKE UNIV, 87- *Concurrent Pos:* Investr, Duke/NSF Eng Res Ctr Emerging Cardiovascular Technologies, 89-; presidential young investr award, NSF, 91. *Honors & Awards:* Eng Res Initiation Award, Am Soc Mech Engrs & Eng Found, 88. *Mem:* Am Soc Mech Engrs; Am Phys Soc. *Res:* Computational fluid mechanics and heat transfer; experimental heat transfer; magnetic resonance imaging flow diagnostics; transport phenomena in random media; bifurcation analysis; cardiovascular fluid mechanics. *Mailing Add:* Dept Mech Eng & Mat Sci Duke Univ Durham NC 27706

GEORGIAN, VLASIOS, b Quincy, Mass, Sept 5, 19; m 49; c 2. ORGANIC CHEMISTRY. *Educ:* Harvard Univ, SB, 41, MA, 43, PhD(chem), 50. *Prof Exp:* Res chemist, Polaroid Corp, 42-44; asst, Harvard Univ, 45-46; from instr to asst prof chem, Northwestern Univ, 51-57, res assoc, 57-60; from asst prof to assoc prof, 60-90, EMER PROF CHEM, TUFTS UNIV, 90- *Concurrent Pos:* USPHS fel, Harvard Univ, 50-51; consult, Smith Kline & French Labs, 60-; vis prof, Exten Progs, Harvard Univ, 68-80. *Mem:* Am Chem Soc; Am Acad Arts & Sci; NY Acad Sci; Royal Soc Chem. *Res:* Synthesis of small and strained ring systems; additions to cyclooctatetraene; molecular rearrangements; steroid syntheses and transformations; syntheses of natural products; syntheses of beta lactam antibiotics; tetracycline modifications. *Mailing Add:* 139 Clifton St Belmont MA 02178

GEORGIEV, VASSIL ST, b Sofia, Bulgaria, May 29, 36; US citizen; c 1. CHEMOTHERAPY. *Educ:* Higher Med Inst, Sofia, MSc, 61; Higher Inst Chem Technol, Sofia, MSc, 65; Bulgarian Acad Sci, PhD(org chem), 71. *Prof Exp:* Sr res assoc, dept chem, Pa State Univ, 72-74; sr res chemist, Immunol & Inflammatory Dis Res Team, USV Pharmaceut Corp, NY, 74-75, dept med chem, 75-78; sr res scientist, 79-81, prin investr & sect head, chemother & immunother res, dept org chem, 81-87; SR PRIN INVESTR & SECT HEAD IMMUNOL, INFECTIOUS & METAB DIS RES, DEPT OF ORG CHEM, PENNWALT PHARMACEUT DIV, 87- *Mem:* Am Chem Soc; fel NY Acad Sci; Swiss Chem Soc; German Chem Soc; Chem Soc Japan; fel Royal Soc Chem. *Res:* Design and synthesis of drugs with potential therapeutic importance in areas of antifungal, antiviral, antimicrobial, anti-inflammatory and allergy-immunology activities; structural elucidation and synthesis of natural products; application of physical organic methods in the chemistry of natural products. *Mailing Add:* PO Box 1032 Penfield NY 14526-0632

GEORGIOU, GEORGE, b Athens, Greece, May 18, 59; US citizen; m 84. GENETIC ENGINEERING, ENVIRONMENTAL BIOTECHNOLOGY. *Educ:* Univ Manchester Inst Sci & Technol BSc, 81; Cornell Univ, MS, 83, PhD(chem eng), 87. *Prof Exp:* Asst prof, 86-91, ASSOC PROF CHEM ENG, UNIV TEX, AUSTIN, 91- *Concurrent Pos:* NSF presidential young investr, 87; Dow Chem Co young fac award, 88. *Mem:* Am Chem Soc; AAAS; Am Soc Microbiol; Am Inst Chem Engrs. *Mailing Add:* Dept Chem Eng Univ Tex Austin TX 78712

GEORGOPAPADAKOU, NAFSIKA ELENI, b Thessaloniki, Greece, Jan 6, 50; nat US. MICROBIAL BIOCHEMISTRY, ENZYMOLOGY. *Educ:* Mills Col, BA, 71; Yale Univ, PhD(chem), 75. *Prof Exp:* Res fel, Harvard Univ, 76-77; res investr biochem, Squibb Inst Med Res, 77-81, sr res investr microbiol, 81-84; res group chief chemother, 84-85, res investr chemother & pharmacol, 85-88 RES LEADER DEPT CHEMOTHER, ROCHE RES CTR, 88- *Mem:* AAAS; Am Chem Soc; Am Soc Microbiol; Am Acad Microbiol; Am Soc Biol Chemists. *Res:* Microbial cell wall biosynthesis; proteolytic enzymes; mechanism of enzyme action; antibiotic action and design. *Mailing Add:* Hoffmann-LaRoche Inc 340 Kingsland St Nutley NJ 07110-1199

GEORGOPOULOS, APOSTOLOS P, NEUROPHYSIOLOGY, BEHAVIOR. *Educ:* Univ Athens, Greece, MD, 68. *Prof Exp:* ASSOC PROF NEUROSCI, JOHNS HOPKINS UNIV, 83- *Mailing Add:* Dept Neurosci Phillip Barn Lab Neurophys Johns Hopkins Univ 725 N Wolfe St Baltimore MD 21205

GEORGOPOULOS, CONSTANTINE PANOS, b Skouruchori, Pyrgos, Greece, Jan 27, 42; m 66. BACTERIOPHAGE BIOLOGY, HEAT SHOCK REGULATION. *Educ:* Amherst Col, Mass, BS, 64; Mass Inst Technol, PhD(microbiol), 69. *Prof Exp:* Fel biochem, Stanford Univ, 70-71; res assoc molecular biol, Univ Geneva, 72-76; from asst prof to assoc prof, 77-82, PROF MOLECULAR BIOL, UNIV UTAH MED CTR, SALT LAKE CITY, 82- *Concurrent Pos:* Prin investr, NIH Grant, 77-, 84-, NSF Int Travel Award, 86, NSF Intern Exchange grant, 90-; mem, Microbiol, Physiol & Genetics Study Sect, NIH, 85-; vis prof, Dept Molecular Biol, Univ Geneva, 86. *Mem:* Sigma Xi; Am Soc Microbiol; Genetics Am Soc; Am Soc Biochem & Molecular Biol. *Res:* Mechanisms of DNA replication; RNA transcription and morphogenics in bacteria and bacteriophages; regulation and function of the highly conserved stress response in escherichia coli and its role in bacteriophage lambda growth. *Mailing Add:* Dept Cellular & Viral & Molecular Biol Univ Utah Med Ctr Salt Lake City UT 84132

GEORGOPULOS, PETER DEMETRIOS, b Bronx, NY, Oct 1, 44; m 67; c 1. NUCLEAR PHYSICS. *Educ:* Long Island Univ, BS, 66; Pa State Univ, PhD(physics), 71. *Prof Exp:* ASSOC PROF PHYSICS, PA STATE UNIV, DEL COUNTY CAMPUS, 71- *Concurrent Pos:* Regional fac affil, Bartol Res Found, 71-79; vis scientist, Lawrence Livermore Lab, 78 & 79. *Mem:* Am Phys Soc; Am Asn Physics Teachers; Sigma Xi. *Res:* Nuclear structure information obtained by gamma ray studies; statistical properties of matter. *Mailing Add:* Del County Campus Penn State Univ Media PA 19063

GEPHART, LANDIS STEPHEN, operations research, applied mathematics; deceased, see previous edition for last biography

GEPNER, IVAN ALAN, b Newark, NJ, Sept 10, 45. DEVELOPMENTAL BIOLOGY, GENETICS. *Educ:* Rutgers Univ, BA, 67; Princeton Univ, MS, 69, PhD(biol), 72. *Prof Exp:* Asst prof, 73-80, ASSOC PROF BIOL, MONMOUTH COL, 80- *Mem:* Soc Develop Biol; Am Soc Zoologists. *Res:* Cell interactions in development; intercellular adhesion. *Mailing Add:* Dept Biol Monmouth Col Cedar Ave West Long Branch NJ 07764

GEPPERT, GERARD ALLEN, b Belleville, Ill, Nov 26, 32; m 56; c 5. EXTRACTIVE METALLURGY, ANALYTICAL CHEMISTRY. *Educ:* St Mary's Univ, Tex, BS, 54. *Prof Exp:* Res chemist extractive metall, Alcoa Res Labs, East St Louis, Ill, 54-72, sr res scientist, 73-75; chief chemist, anal chem, mobile works, Aluminum CoAm, 75-82, Point Comfort Opers, 82-85; consult. *Mem:* Am Chem Soc; Metall Soc. *Res:* Refining of bauxite, alumina and alumina chemicals; fire retardant alumina hydrate; quality control and applications of alumina hydrate; alumina extraction from non-bauxitic ores; improvements of the Bayer process. *Mailing Add:* 3720 Claridge Rd S Mobile AL 36608-1752

GERACE, LARRY R, b West Carthage, NY, Oct 27, 51. MOLECULAR BIOLOGY. *Educ:* Johns Hopkins Univ, BA, 73; Rockefeller Univ, PhD, 79. *Prof Exp:* Postdoctoral fel, Rockefeller Univ, 79-80; from asst prof to assoc prof cell biol & anat, Sch Med, Johns Hopkins Univ, 80-87, ASSOC MEM, DEPT MOLECULAR BIOL, RES INST, SCRIPPS CLIN, 87- *Concurrent Pos:* NIH res career develop award, 84; managing ed, J Cell Biol, 87-90; sect ed, Current Opinion Cell Biol, 88-90; co-chair, prog comt, Am Soc Cell Biol, 90. *Honors & Awards:* R R Bensley Mem Award, 87. *Res:* Author of numerous publications. *Mailing Add:* Dept Molecular Biol Res Inst Scripps Clin 10666 N Torrey Pines Rd La Jolla CA 92037

GERACE, MICHAEL JOSEPH, b Brooklyn, NY, Jan 11, 44; m 68, 85; c 3. ADHESIVE & SEALANT FORMULATION, NEW PRODUCT DEVELOPMENT & COMMERCIALIZATION. *Educ:* Long Island Univ, BS, 65; Tufts Univ, PhD(chem), 70. *Prof Exp:* Sr chemist, W R Grace & Co, 75-78; res & develop mgr, Norton Co, 79-81; vpres res & develop, Protective Treatments, 81-86; PRES, ASTER INC, 86- *Mem:* Am Chem Soc; Adhesives & Sealant Coun; Soc Automotive Engrs; Soc Mfg Engrs. *Res:* Commercialization of automotive and construction adhesives and sealants; plastisols; epoxys; acrylics; hot melts; pressure sensitives; butyls and butyl tapes. *Mailing Add:* 320 E Peach Orchard Ave Oakwood OH 45419

GERACE, PAUL LOUIS, b Batavia, NY, Aug 20, 34; m 65; c 2. INORGANIC CHEMISTRY, ELECTROPHOTOGRAPHY. *Educ:* Univ Notre Dame, BS, 56, PhD(chem), 61. *Prof Exp:* Teaching fel chem, Univ Notre Dame, 56-58; res chemist, Solvay Process Div, Allied Chem Corp, NY, 61-66; scientist, Xerox Corp, 66-75, tech specialist & proj mgr, 75-83; res assoc, Amoco Corp, 85-86; RETIRED. *Mem:* Am Chem Soc; Sigma Xi; NY Acad Sci. *Res:* Special materials development and specification, encompassing transfer of product and process technology from research to development to manufacturing; photoconductors and photoreceptors; surface coatings; chemical safety. *Mailing Add:* One Pickthorn Dr Batavia NY 14020

GERACI, JOSEPH E, b Newark, NJ, Feb 24, 16; m 49; c 1. INTERNAL MEDICINE. *Educ:* Marquette Univ, MD, 40; Univ Minn, MS, 49. *Prof Exp:* PROF MED, MAYO GRAD SCH MED, UNIV MINN, ROCHESTER, 62-, CONSULT, MAYO CLIN, 51- *Mem:* Am Col Physicians; Am Col Chest Physicians; Am Fedn Clin Res; Infectious Dis Soc Am; Am Thoracic Soc. *Res:* Infectious disease. *Mailing Add:* 1760 Eighth St SW Rochester MN 55902

GERAGHTY, JAMES JOSEPH, b New York, NY, Nov 20, 20; m 42; c 2. GEOLOGY. *Educ:* City Col, BS, 49; NY Univ, MS, 53. *Prof Exp:* Geologist, Water Resources Div, Ground Water Br, US Geol Surv, 49-55; ground water geologist, Leggette, Brashears & Graham, 55-57; ground water geologist, Port Washington, 57-74 & Tampa, 75-78, Annapolis, 78-82, GROUND WATER GEOLOGIST, GERAGHTY & MILLER INC, TAMPA, 78- *Concurrent Pos:* Lectr, Hofstra Col, 52-55, State Univ NY, 58 & NY Univ, 62-; tech adv, UN, 62-74. *Mem:* Geol Soc Am; Am Water Works Asn; Am Asn Petrol Geol; Am Geophys Union; Int Asn Hydrogeol; Am Soc Civil Engrs. *Res:* Ground water geology; hydrology. *Mailing Add:* 4641 Westford Circle Tampa FL 33688-3630

GERAGHTY, MICHAEL A, b Chicago, Ill, May 15, 30; m 60; c 4. MATHEMATICS. *Educ:* Univ Notre Dame, BSc, 52, PhD(math), 59. *Prof Exp:* Off Naval Res res assoc math, Northwestern Univ, 59-60, asst prof, 60-62; vis assoc prof, Res Inst, Univ Ala, 62-64; asst prof, 64-65, ASSOC PROF MATH, UNIV IOWA, 65- *Mem:* Am Math Soc; Math Asn Am. *Res:* Topology. *Mailing Add:* Dept Math Univ Iowa Iowa City IA 52240

GERALD, MICHAEL CHARLES, b New York, NY, Nov 20, 39; m 65; c 2. PHARMACOLOGY, EDUCATION ADMINISTRATION. *Educ:* Fordham Univ, BS, 61; Ind Univ, PhD(pharmacol), 68. *Prof Exp:* USPHS fel psychiat & pharmacol, Univ Chicago, 68-69; from asst prof to assoc prof, 69-80, actg chmn, 80-81, PROF PHARMACOL, COL PHARM, OHIO STATE UNIV, 80-, ASSOC DEAN, 84- *Concurrent Pos:* Consult, WHO, 83-84; USP comt rev, 80-85; Gustavus A Pfeiffer Mem Res fel, 83-84. *Mem:* NY Acad Sci; AAAS; Am Soc Pharmacol & Exp Therapeut; fel Acad Pharmaceut Sci (secy, 75-77, vchmn, 78-79); Soc Neurosci; Am Asn Cols Pharm; Am Found Pharmaceut Educ. *Res:* Neuropharmacology; drug abuse; phychopharmacology; national drug policy; author of textbooks on general and nursing pharmacology; drugs and poisons of Agatha Christie. *Mailing Add:* Col Pharm Ohio State Univ Columbus OH 43210-1291

GERALD, PARK S, b Omaha, Nebr, June 30, 21; m 56; c 3. PEDIATRICS. *Educ:* Iowa State Col, BS, 43; Creighton Univ, MD, 47. *Prof Exp:* From instr to assoc prof pediat, Chilren's Hosp Med Ctr, 58-70, assoc, 59-62, assoc clin prof, 65-67, prof pediat, 70-84, chief clin genetics div, 66-82; RETIRED. *Concurrent Pos:* Nat Heart Inst spec fel, 55-59; mem, Joint Comn Ment Health of Children, Inc, 67-69; chmn ment retardation res & training comt,

Nat Inst Child Health & Human Develop, 69-72. *Honors & Awards:* Mead-Johnson Award, Am Acad Pediat, 62. *Mem:* Soc Pediat Res; Am Soc Clin Invest; Am Pediat Soc; Am Soc Human Genetics. *Res:* Clinical genetics; cytogenetics; inherited variants of proteins. *Mailing Add:* 71 Hundreds Rd Wellesley MA 02181

GERALDSON, CARROLL MORTON, b Manitowoc, Wis, Apr 8, 18; m 49; c 7. SOIL CHEMISTRY. *Educ:* St Olaf Col, BA, 40; Univ Wis, PhD(soils, plant physiol), 51. *Prof Exp:* From asst prof to prof soil chem, Inst Food & Agr Sci, Agr Res & Educ Ctr, 51-90, EMER PROF SOIL CHEM, GULF COAST RES CTR, UNIV FLA, 91- *Honors & Awards:* Vaughan Award, Am Soc Hort Sci, 55; Fla Fruit & Vegetable Award, 56. *Mem:* Am Soc Agron; Am Soc Hort Sci. *Res:* Nutrition of vegetable crops; control of physiological disorders such as black heart of celery and blossom-end rot of tomatoes and peppers; development and utilization of the intensity and balance soil solution testing procedure for evaluating the ionic root environment for optimal production; maintenance of 3-dimensional gradients; establishment in the soil profile by surface application of nutrients with a mulch covering and constant source of water. *Mailing Add:* Gulf Coast Res Ctr 5007 60th St E Bradenton FL 34203-9324

GERARD, CLEVELAND JOSEPH, b Milton, La, Sept 25, 24; m 63; c 2. CONSERVATION, SOIL FERTILITY. *Educ:* Southwestern La Inst, BS, 48; Kans State Univ, MS, 50; Agr & Mech Col Tex, PhD(soil physics), 55. *Prof Exp:* Soil scientist, Agr Res Serv, USDA, Ore State Col, 54-57; soil physicist, Tex Agr Exp Sta, 57-89, EMER PROF SOIL PHYSICSIST, 89- *Concurrent Pos:* Res award, Tex A&M Univ, 80-81. *Mem:* Soil Sci Soc Am; Int Soil Sci Soc; Int Soil Tillage Res Orgn. *Res:* Soil chemistry; soil physics and soil-plant relationships. *Mailing Add:* Tex Agr Exp Sta Box 1658 Vernon TX 76384

GERARD, GARY FLOYD, b Saginaw, Mich, June 1, 44; m 66; c 3. MOLECULAR BIOLOGY, VIROLOGY. *Educ:* Pa State Univ, BS, 66; Mich State Univ, PhD(biochem), 72. *Prof Exp:* Res assoc molecular virol, Inst Molecular Virol, Sch Med, St Louis Univ, 71-73, asst prof, 73-77, assoc prof, 77-82; sr scientist, Bethesda Res Lab, 82-85; DIR RES & DEVELOP, LIFE TECHNOL INC, 85- *Res:* Retro viral and eukaryotic DNA replication. *Mailing Add:* Life Technol Inc 8717 Grovemont Circle Gaithersburg MD 20877

GERARD, JESSE THOMAS, b Windsor, Ont, Mar 19, 41; m 62; c 2. ANALYTICAL CHEMISTRY, NUCLEAR CHEMISTRY. *Educ:* Univ Windsor, BSc, 64, PhD(anal chem), 68. *Prof Exp:* Res fac mem & co-investr Apollo lunar samples, Dept Chem, Cornell Univ, 68-71; res chemist & Nat Acad Sci-Nat Res Coun inhouse resident res assoc theoret studies br, Lab Space Physics, Goddard Space Flight Ctr, NASA, 71-72; SR RES CHEMIST, ANALYTICAL SCI DIV, RES LABS, EASTMAN KODAK CO, 72- *Mem:* Fel AAAS; fel Am Chem Soc; fel Geochem Soc. *Res:* Neutron activation analysis; radiochemistry; x-ray fluorescence analysis; classical chemical analysis; separations; ultra trace, trace, minor and major element analysis. *Mailing Add:* Four Horizons Apt 108C 1450 E Harmon Ave Las Vegas NV 89119-5944

GERARD, VALRIE ANN, b Amityville, NY, Feb 21, 48. SEAWEED ECOLOGY, AQUACULTURE. *Educ:* State Univ NY Buffalo, BA, 70; Univ Calif, Santa Cruz, MA, 74, PhD(biol), 76. *Prof Exp:* Killam fel biol, Dalhousie Univ, 76-78; res fel environ eng, Calif Inst Technol, 78-83; asst prof, 83-88, ASSOC PROF, STATE UNIV NY STONYBROOK, 88- *Concurrent Pos:* Lectr, Univ Calif, Santa Cruz, 78, Univ Southern Calif, 79 & 82; vis assoc prof, Univ Calif, Davis, 90. *Res:* Physiological ecology of marine macroalgae; seaweed aquaculture. *Mailing Add:* Marine Sci Res Ctr State Univ NY Stony Brook NY 11794

GERARDO, JAMES BERNARD, b Toluca, Ill, Oct 18, 36; m 63. LASERS, ATOMIC PHYSICS. *Educ:* Univ Ill, BS, 59, MS, 60, PhD(elec eng), 63. *Prof Exp:* From res asst to res assoc elec eng, Univ Ill, Urbana, 59-67; mem res staff, Org 5100, 66-67, supvr plasma physics, 67-74, MGR LASER RES, SANDIA LABS, SANDIA CORP, 74- *Mem:* Fel Am Phys Soc; Optical Soc Am; Mat Res Soc; Inst Elec & Electronic Engrs. *Res:* Laser research and development; molecular physics; gas kinetics. *Mailing Add:* Sandia Natl Labs Dept 1120 Albuquerque NM 87185

GERASIMOWICZ, WALTER VLADIMIR, b Phoenixville, Pa, July 23, 52. SPECTROSCOPY, THERMODYNAMICS. *Educ:* Ursinus Col, BS, 74; Villanova Univ, MS, 77, PhD(physical chem), 81. *Prof Exp:* Res fel, 81-83, STAFF SCIENTIST, EASTERN REGIONAL RES CTR, USDA, 83- *Concurrent Pos:* Instr chem, Pa State Univ, Del County, 79-80, lectr, 80. *Mem:* Am Chem Soc; AAAS; Royal Soc Chem. *Res:* Experimental physical chemistry; metal-ligand interactions; fourier transform infrared; visible and nuclear magnetic resonance spectroscopy; thermodynamic and equilibria studies; vibrational or normal mode analysis. *Mailing Add:* 822 Cherry St Phoenixville PA 19460

GERATZ, JOACHIM DIETER, b Gloethe, Ger, July 3, 29; US citizen; m 61; c 3. PATHOLOGY. *Educ:* Univ Frankfurt, MD, 53. *Prof Exp:* Asst path, Univ Frankfurt, 54-55; intern, Jefferson Hosp, Roanoke, Va, 55-56; res path, 56-60, from instr to assoc prof, 60-72, PROF PATH, UNIV NC, CHAPEL HILL, 72- *Mem:* AAAS; Int Acad Path; Am Asn Path Am Soc Clin Path; Int Soc Thrombosis & Haemostasis. *Res:* Blood coagulation; inhibitors of proteolytic enzymes; physiology and pathology of the pancreas; antiviral, antiarthritic and anticancer agents. *Mailing Add:* Dept Path Univ NC at Chapel Hill Chapel Hill NC 27599-7525

GERBA, CHARLES PETER, b Blue Island, Ill, Sept 10, 45; m 70; c 2. ENVIRONMENTAL MICROBIOLOGY. *Educ:* Ariz State Univ, BS, 69; Univ Miami, PhD(microbiol), 73. *Prof Exp:* Postdoctoral fel virol, Baylor Col Med, 73-74, asst prof, 74-81; assoc prof, 81-84, PROF MICROBIOL, UNIV ARIZ, 84- *Concurrent Pos:* AAAS fel environ sci, 84; chmn, Div Appl & Environ Microbiol, Am Soc Microbiol, 83-84 & 87-88; consult, Sci Adv Bd, US Environ Protection Agency, 86- *Mem:* Am Soc Microbiol; Int Asn Water Pollution Control & Res; AAAS; Am Water Works Asn; Inst Food Technologists. *Res:* Fate, transport, and removal of pathogenic microorganisms in the environment; new methods for the detection of pathogens in the environment and their removal by water and wastewater treatment. *Mailing Add:* 1980 W Paseo Monserrat Tucson AZ 85704

GERBER, BERNARD ROBERT, b New York, NY, May 31, 35; m 57, 87; c 3. POLYMER CHEMISTRY. *Educ:* Hunter Col, AB, 55; NY Univ, MS, 58, PhD(phys chem), 64. *Prof Exp:* Operator comput, AEC facil, Inst Math Sci, 55-58; res asst, Rheum Dis Study Group, NY Univ Med Ctr, 58-64; Posidoc res fel, Helen Whitney Found, Inst Molecular Biol, Nagoya Univ, Japan, 64-67; res assoc chem contractile proteins, State Univ NY Downstate Med Ctr, 67-68; asst prof biol, Univ Pa, 68-75; assoc prof phys & org chem & dir sch grad studies, State Univ NY Downstate Med Ctr, 75-79, assoc prof biochem, Sch Med, 75-79; mgr, Databit, 79-80; vpres res, Reseal Cont Corp, 80-85; SCI TECH MGT CONSULT, 80- *Concurrent Pos:* John Polachek Found Med Res fel, 68; prin investr, Biomed Sci support grant, 69-70; NSF instnl grant, 69-71, res grant, 71-78; NY State Instnl grant, 75-80; consult, several cos, 80- *Mem:* AAAS; NY Acad Sci; Biophys Soc; Soc Gen Physiol; Am Inst Chemists. *Res:* Invention and research management of multidose dispensing systems for preservative-free foods, pharmaceuticals and other products; static and dynamic aspects of three-dimensional structure of polymers; self-assembly processes; connective tissue chemistry; bacterial flagella and bacterial motility. *Mailing Add:* 583 Mello Ln Santa Cruz CA 95062-2707

GERBER, DONALD ALBERT, b New York, NY, Apr 10, 32; m 64; c 2. INTERNAL MEDICINE. *Educ:* Columbia Univ, AB, 53, MD, 57; Am Bd Internal Med, dipl. *Prof Exp:* Intern, Osler Med Serv, Johns Hopkins Hosp, Baltimore, Md, 57-58, asst resident & asst med, 58-59; asst resident, Presby Hosp, New York, 59-60; from instr to asst prof, 63-69, ASSOC PROF MED, STATE UNIV NY DOWNSTATE MED CTR, 69- *Concurrent Pos:* Arthritis Found vis fel, Col Physicians & Surgeons, Columbia Univ, 60-63; spec investr, Arthritis Found, 63-66; career scientist award, Health Res Coun, City of New York, 65-75; prin investr res grant, Nat Inst Arthritis, Metab & Digestive Dis, 66-79 & 82-85. *Mem:* Am Col Rheumatology; Am Fedn Clin Res; Harvey Soc; fel Am Col Physicians. *Res:* Biochemical abnormalities in rheumatoid arthritis and systemic lupus erythematosus; mechanism of action of anti-rheumatic drugs, especially gold thiomalate and D-penicillamine; the role of L-histidine, copper, sulfhydryl groups, and hypochlorite in rheumatic disease; aggregation of immunoglobulin G; mechanism of action of lupus-inducing drugs. *Mailing Add:* Dept Med Box 42 State Univ NY Health Sci Ctr Brooklyn 450 Clarkson Ave Brooklyn NY 11203-2098

GERBER, EDUARD A, physical electronics; deceased, see previous edition for last biography

GERBER, GEORGE HILTON, b St Walburg, Sask, Mar 19, 42. ENTOMOLOGY, PHYSIOLOGY. *Educ:* Univ Sask, BSA, 64, PhD(entom), 69. *Prof Exp:* RES SCIENTIST ENTOM, AGR CAN, 69-, SECT HEAD, 89- *Concurrent Pos:* Assoc ed, Int J Insect Morphol & Embryol, 91- *Honors & Awards:* C Gordon Hewitt Award, Entom Soc Can. *Mem:* Entom Soc Am; Entom Soc Can; Can Soc Zoologists; Sigma Xi. *Res:* Reproductive biology and physiology of insects; pest management system for rape crop insects; pest management system for cutworms of cereals. *Mailing Add:* Agr Can Res Sta Winnipeg MB R3T 2M9 Can

GERBER, H JOSEPH, b Vienna, Austria, Apr 17, 24; m 53; c 2. PRODUCT DEVELOPMENT. *Educ:* Rensselaer Polytech Inst, BA, 46. *Hon Degrees:* PhD (Eng), Rensselaer Polytech Inst, Univ New Haven, 90. *Prof Exp:* Dir, Gerber Sci Inst Ltd, UK, NV Gerber Sci Instrument S A Belgium, Gerber Sci, Italy, pres, Gerber Sci Inst Co, 48-78, PRES CHMN BD & CHIEF EXEC OFFICER, GERBER SCI, INC, 78- *Concurrent Pos:* Trustee, Rensselaer Polytech Inst, Troy NY,. *Honors & Awards:* Holden Medal, 83; ORT Sci & Technol Award, 88. *Mem:* Nat Acad Eng. *Mailing Add:* Gerber Sci Inc 83 Gerber Rd W South Windsor CT 06074

GERBER, JAY DEAN, VACCINE RESEARCH & DEVELOPMENT, IMMUNOMODULATIONS. *Educ:* Univ Kans, PhD(microbiol immunol), 68. *Prof Exp:* PROJ DIR, NORDEN LABS, 85- *Res:* Immunomodulations. *Mailing Add:* Smith Kline Beecham Animal Health 601 W Cornhusker Hwy Lincoln NE 68508

GERBER, JOHN FRANCIS, b Versailles, Mo, Dec 13, 30; m 55; c 3. AGRICULTURAL METEOROLOGY, SATELLITE METEOROLOGY. *Educ:* Univ Mo, BS, 56, MS, 57, PhD(soils), 60. *Prof Exp:* Instr soils, Univ Mo, 59-60; from asst prof to assoc prof, 60-69, asst dean res, Agr Exp Sta, 71-72, dir Ctr Environ Progs, 73-77, dir inst food & agr sci grants prog, 77-87, PROF FRUIT CROPS & CLIMATOLOGIST, UNIV FLA, 69- *Concurrent Pos:* Consult, United Fruit Co, 63, Corps Eng, US Army, 65-68, Walt Disney World, Aikins Tech Inc, 67; fac develop grant from Univ Fla & vis prof, Dept Hort, Pa State Univ, 69-70; rev ed, Jour, Am Soc Hort Sci; prin investr, Dept Educ, 79-, NASA, 78-80; mem, Nat Res Coun Study Panels, 77, 78, 79, bd atmospheric sci & climat; consult, Org Trop Studies, Environ Protection Agency, USDA, 70-81, NIH, 88; dir, Biotechnol Inst Res & Develop, 87-89, Univ Indust Relations, USDA Off Agr Biotechnol, 89-90. *Mem:* Am Soc Hort Sci; Am Meteorol Soc; fel Am Soc Hort Sci; Sigma Xi; AAAS. *Res:* climatology; cold protection of plants; heat budget studies; agricultural climatology; biotechnology; research administration and science policy. *Mailing Add:* Dept Fruit Crops Univ Fla Gainesville FL 32611

GERBER, JOHN GEORGE, CLINICAL PHARMACOLOGY, INTERNAL MEDICINE. *Educ:* Med Col Va, MD, 72. *Prof Exp:* ASSOC PROF, UNIV COLO, 78- *Mailing Add:* Health Sci Ctr Univ Colo 4200 E 9th Ave Denver CO 80220

GERBER, LEON E, b Brooklyn, NY, Sept 5, 41; m 64; c 7. GEOMETRY. *Educ:* Brooklyn Col, BS, 60; Yeshiva Univ, MA, 62, PhD(math), 68. *Prof Exp:* Programmer, Brookhaven Nat Lab, 60; from instr to asst prof math, 66-76, ASSOC PROF MATH, ST JOHN'S UNIV, NY, 76- *Mem:* Math Asn Am. *Res:* Geometry of n-dimensions; asymptotic relations between solutions of differential equations and difference equations. *Mailing Add:* 1488 E 18th St Brooklyn NY 11230-6706

GERBER, LINDA M, b New York, NY, April 12, 53; m 82; c 2. EPIDEMIOLOGY, MEDICAL ANTHROPOLOGY. *Educ:* State Univ NY, Binghamton, BA, 73; Univ Colo, MA, 76, PhD(anthrop), 78. *Prof Exp:* Res asst psychol, Inst Behav Sci, Univ Colo, 74-75 & instr anthrop, 76; res intern pop, EW Pop Inst, EW Ctr, Hawaii, 76-77; res asst anthrop, Inst Behav Sci, Univ Colo, 75-78, res assoc, 78; fel, Cornell Univ Med Col, 79-81, res assoc pub health, 79- 81, asst prof, 82-84, clin asst prof, 84-86, ASST PROF PUB HEALTH, CORNELL UNIV MED COL, 87-, ASST PROF EPIDEMIOL MED, 87- *Concurrent Pos:* Res training prog culture change, NIMH, 75-78; postdoctoral fel, Pub Health Serv, 79-81; preceptor, Dept Pub Health, Med Col, Cornell Univ, 80-; res scientist & epidemiologist, Nassau County Dept Health, 84-86; asst prof clin community & prev med (epidemiol), Sch Med, State Univ NY, Stony Brook 85-87; mem, Human Biol Coun & Epidemiol Coun, Am Heart Asn. *Mem:* Human Biol Coun; Am Asn Phys Anthropologists; Am Heart Asn; Coun Epidemiol. *Res:* Epidemiology of hypertension, with particular interest in body fat distribution as a risk factor for both hypertension and cardiovascular disease. *Mailing Add:* Cardiovasc Ctr 525 E 68th St Starr 4 New York NY 10021

GERBER, LOUIS P, biochemistry, for more information see previous edition

GERBER, MICHAEL A, b Kessel, WGer, Oct 18, 39; m 72; c 1. PATHOLOGY. *Educ:* Gutenberg Gymn, Weisbaden, WGer, MS, 60; Gutenberg Univ, Mainz, WGer, MD, 66; Am Bd Clin & Anat Path, dipl, 72. *Prof Exp:* Internship, Gutenberg Univ, Mainz & Kreiskrankenhaus Bad Harzburg, WGer, 66-67; internship, Middlesex Gen Hosp, New Brunswick, NJ, 67-68, resident path, 68-69; resident path, Mt Sinai Hosp, NY, 69-70, trainee exp path, 70-72; assoc path, Mt Sinai Sch Med & City Hosp Ctr Elmhurst, 72-73; chief electron micros, Vet Admin Hosp, Bronx, NY, 73-74; from asst prof to prof path, Mt Sinai Sch Med, 73-87; PROF & CHMN, DEPT PATH, SCH MED, TULANE UNIV, NEW ORLEANS, 87-, DIR, GRAD PROG MOLECULAR & CELL BIOL, 91- *Concurrent Pos:* Mem, Gastrointestinal Drugs Adv Comt, Food & Drug Admin, 82-85; dir path, City Hosp Ctr Elmhurst, NY, 82-87; co-dir cellular & molecular path, Biomed Sci Doctoral Prog, Mt Sinai Sch Med, City Univ New York, 82-87; mem rev comt, Nat Liver Transplant Data Base, NIH, 86-89 & rev comt, Liver Ctr, 90; bk ed, Hepatol, 88-, assoc ed, 89- *Mailing Add:* Dept Path Sch Med Tulane Univ 1430 Tulane Ave New Orleans LA 70112

GERBER, NAOMI LYNN HURWITZ, REHABILITATION MEDICINE. *Educ:* Smith Col, AB, 65; Harvard Univ, MA, 66; Tufts Univ, MD, 71; Am Bd Internal Med, dipl, 75; Am Bd Phys Med & Rehab, dipl, 79. *Prof Exp:* Res assoc, Med Sch, Tufts Univ, 66-67; intern med, New Eng Med Ctr, Boston, 71-72, resident, 72-73; clin assoc, Arthritis & Rheumatism Br, Nat Inst Arthritis, Metab & Digestive Dis, 73-75, CHIEF, DEPT REHAB MED, CLIN CTR, NIH, BETHESDA, 76-, PHYSICIAN, ARTHRITIS & RHEUMATISM BR, NAT INST ARTHRITIS, DIABETES & DIGESTIVE & KIDNEY DIS, 76- *Concurrent Pos:* Resident phys med & rehab, George Washington Univ, 75-77, adj assoc prof internal med, 75-; mem, Coun Rehab Rheumatology, Am Rheumatism Asn, 82-, secy, 82-84, pres, 84-86; panel chief orthop surg, NIH, 84-; clin prof phys med & rehab, Georgetown Univ, 88-90; mem, Prof Educ Comt, Arthritis Found, 88-; Kovacs vis fel, Royal Soc Med, Eng, 91. *Honors & Awards:* Women in Sci & Eng Award, 86; Sidney Licht Mem Lectr, 89. *Mem:* Am Col Rheumatology; Am Acad Phys Med & Rehab; Asn Acad Psychiatrists. *Mailing Add:* Warren Grant Magnuson Clin Ctr NIH Dept Rehab Med Bldg 10 Rm 65235 Bethesda MD 20892

GERBER, SAMUEL MICHAEL, b New York, NY, June 21, 20; wid; c 2. ORGANIC CHEMISTRY, TEXTILE CONSERVATION. *Educ:* City Col, BS, 42; Columbia Univ, MA, 48, PhD(chem), 52. *Prof Exp:* Asst chem, Columbia Univ, 47-51; chemist, Am Cyanamid Co, 51-55, asst to tech dir, Org Chem Div, 56-58, group leader, 58-68, chief chemist, Org Chem Div, 68-74, mgr dyes & chem res & develop, Chem Res Div, 74-80; mgr dyes & chem res & develop, Atlantic Chem Corp, 80-81; CONSULT, 82-; RES ASSOC, RUTGERS UNIV, NEW BRUNSWICK, 85-; ADJ PROF, FASHION INST TECHNOL, NY, 87- *Mem:* Am Chem Soc; Brit Soc Dyers & Colourists; The Chem Soc; Am Asn Textile Chemists & Colorists. *Res:* Development and use of dyes, textile chemicals and their intermediates. *Mailing Add:* 70 Hillcrest Rd Martinsville NJ 08836

GERBERG, EUGENE JORDAN, b New York, NY, June 1, 19; m 41; c 5. ENTOMOLOGY. *Educ:* Cornell Univ, BS, 39, MS, 41; Univ Md, PhD, 54. *Prof Exp:* Entomol technician, State Conserv Dept NY, 40-41; asst entomologist, USPHS, 41-43; pres, Cornell Chem & Equip Co, 48-79; pres, Am Biochem Lab Inc, 48-81; PRES & DIR, INSECT CONTROL & RES, INC, 46- *Concurrent Pos:* Spec surv, Ministry Agr, Venezuela, 50; coop scientist, USDA, 54-; trop agr specialist, US Mkt & Bus Develop Mission, Nigeria, 61 & Pakistan, 68; consult, WHO, 72, Pan Am Health Orgn, Neth Antilles, 74 & Agency for Int Develop, US Dept of State, Sri Lanka, 77, Thailand, 78; proj leader, EAfrica Aedes Res Unit, Tanzania, WHO, 69 & 70; res assoc, Fla Dept Agr, 72-; pres, Biol Res Inst Am, Inc, 73-; adj prof, Univ Serv, Univ Health Sci, 86- *Honors & Awards:* Outstanding Award, Med & Vet Entom, Am Regist Prof Entom, 83; Meritorious Servive Award, Am Mosquito Control Asn. *Mem:* Entom Soc Am; Am Mosquito Control Asn; Inst Food Tech; Am Soc Trop Med & Hyg; Royal Soc Trop Med & Hyg. *Res:* Tropical and medical entomology; insect ecology; coleoptera; biolgical control of mosquitoes. *Mailing Add:* 6603 Johnnycake Rd Woodlawn MD 21207

GERBERICH, JOHN BARNES, b Wooster, Ohio, Apr 21, 16; m 48; c 1. MICROBIOLOGY. *Educ:* Kent State Univ, BS, 39, MA, 41; Ohio State Univ, PhD, 51. *Prof Exp:* Instr zool, Ohio State Univ, 44-46; instr biol, Mich State Col, 46-47; instr zool, Univ Minn, Duluth, 47-53, insect control res prods, 53-54; from asst prof to prof zool, 54-67, PROF BIOL & DIR ALLIED HEALTH PROGS, UNIV WIS-EAU CLAIRE, 67- *Mem:* Entom Soc Am; Sigma Xi. *Res:* Biological control methods of insects; immature insects; insect microbiology. *Mailing Add:* 1810 Hoover Eau Claire WI 54701

GERBERICH, WILLIAM WARREN, b Wooster, Ohio, Dec 30, 35; m 59; c 3. METALLURGY, MATERIALS SCIENCE. *Educ:* Case Inst Technol, BS, 57; Syracuse Univ, MS, 59; Univ Calif, Berkeley, PhD(mat sci & eng), 71. *Prof Exp:* Proj eng metall, Jet Propulsion Lab, Pasadena, Calif, 59-62; sr res scientist, Aeronutronic, Newport Beach, Calif, 62-65; eng specialist, Aerojet Gen Corp, Sacramento, Calif, 65-67; res scientist mat sci, Lawrence Radiation Lab, Univ Calif, Berkeley, 67-71; PROF MAT SCI, UNIV MINN, MINNEAPOLIS, 71-, ASSOC HEAD DEPT, 80- *Concurrent Pos:* Lectr mat sci, Univ Calif, Berkeley, 67-71; consult, Standard Oil Calif, 68-69 & Meyer Indust, Int Tel & Tel Corp, 76-; mem, nat Mat Adv Bd, Nat Res Coun, 73-76 & Bd Publ, Metall Trans, 77-; adv, Minn Pollution Control Agency, 74-76; chmn fracture mech comt, Am Soc Metals, 78-; vchmn fatigue res comt, Am Soc Testing & Mat, 78-; mem mat adv bd, Argonne Labs, 80-; bd dir, Acta Metall, 83-86, chmn bd, 86-89. *Honors & Awards:* William Spraragen Award, Am Welding Soc, 68. *Mem:* Am Inst Mining, Metall & Petrol Engrs; fel Am Soc Metals; Am Soc Testing & Mat; Sigma Xi; Mat Res Soc. *Res:* Strengthening mechanisms and fracture phenomena including hydrogen embrittlement, ductile-brittle transition phenomena, fatigue crack growth; thin film mechanics including epitaxial interface studies and electron channeling of strain gradients. *Mailing Add:* Dept Chem Eng & Mat Sci Univ Minn Minneapolis MN 55455

GERBI, SUSAN ALEXANDRA, b New York, NY, Mar 13, 44. CELL BIOLOGY. *Educ:* Barnard Col, BA, 65; Yale Univ, MPhil, 68, PhD(biol), 70. *Prof Exp:* Fel biol, Max Planck Inst Biol Tübingen, WGer, 70-72; from asst prof to assoc prof, 72-82, PROF BIOL & DIR GRAD PROG MOLECULAR BIOL, CELL BIOL & BIOCHEM, BROWN UNIV, 82- *Concurrent Pos:* Childs res grant, 73; USPHS res grant, 74-, career develop award, 75-80; Am Cancer Soc res grant, 77-83 & 86-; NSF res grant, 78, 81 & 82-; vis assoc prof, Duke Univ, 81-82. *Mem:* Am Soc Cell Biologists; Soc Develop Biol; Genetics Soc Am; Sigma Xi. *Res:* Structure and function of ribosomal RNA regions conserved during evolution; fine structure and replication of sciarid amplified DNA puff DNA; control of chromosome movement. *Mailing Add:* 48 Cynthia Rd Seekonk MA 02771

GERBIE, ALBERT B, b Toledo, Ohio, Nov 20, 27. OBSTETRICS & GYNECOLOGY. *Educ:* George Washington Univ, MD, 51. *Prof Exp:* Intern, Michael Reese Hosp, 51-52; assoc staff, 52-60, ATTEND STAFF OBSTET & GYNEC, NORTHWESTERN MEM HOSP, HOSP & PRENTICE WOMEN'S 60-; PROF OBSTET & GYNEC, MED SCH, NORTHWESTERN UNIV, 72-, DIR GRAD EDUC PROG, 74- *Concurrent Pos:* Chief, Div Obstet-Gynec, Children's Mem Hosp; assoc ed & bk ed, Am J Obstet & Gynec; assoc ed, Surg, Gynec & Obstet; vpres, Am Bd Obstet & Gynec. *Mem:* Fel Am Col Obstetricians & Gynecologists; fel Am Col Surgeons; fel Am Asn Obstetricians & Gynecologists; fel Am Gynec Soc; AMA. *Res:* Prenatal genetics; trophoblasic diseases. *Mailing Add:* Soc Hume Gene 707 Fairbanks Ct Suite 500 Chicago IL 60611

GERDEEN, JAMES C, b Escanaba, Mich, July 16, 37; m 60; c 3. ENGINEERING MECHANICS. *Educ:* Mich Technol Univ, BSME, 59; Ohio State Univ, MSEM, 62; Stanford Univ, PhD(eng mech), 66. *Prof Exp:* Res engr, Appl Mech Div, Battelle Mem Inst, Ohio, 59-63, sr res engr, Adv Solid Mech Div, 65-68; assoc prof eng mech, Mich Technol Univ, 68-76, prof eng mech, 76-89; PROF MECH ENG, UNIV COLO, 89- *Concurrent Pos:* Guest lectr, Ohio State Univ, 67; mem subcomt shells, pressure vessel res comt, Welding Res Coun, 67- *Mem:* Soc Exp Stress Anal; Am Soc Mech Engrs. *Res:* Stress analysis of thin plate and shell structures, particularly pressure vessels; plastic deformation in structures and metal working operations; inelastic behavior and rock mechanics. *Mailing Add:* Dept Mech Eng Campus Box 112 Mich Technol Univ PO Box 173364 Denver CO 80217-3364

GERDES, CHARLES FREDERICK, b Keokuk, Iowa, June 25, 45; m 80; c 1. AQUATIC INSECTS, THYSANOPTERA. *Educ:* Western Ill Univ, BS, 67, MS, 74; Univ Ill, PhD(entom), 79. *Prof Exp:* Res asst entom, Ill Natural Hist Surv, 77-80; vis asst prof biol, Ill State Univ, 80-81; res assoc biol, Northeast Mo State Univ, 81-84, consult aquatic entom, 82-85; LAB MGR, DEPT PHARMACOL, KIRKSMITH COL OF OSTEOPATHOLOGY, 82- *Mem:* AAAS; Am Registry Prof Entomologists. *Res:* Identification of aquatic insects for environmental impact studies; taxonomic studies of thysanoptera. *Mailing Add:* 808 Kings Ct Apt D Kirksville Col at Osteopathology Kirksville MO 63501

GERDING, DALE NICHOLAS, b Belgrade, Minn, May 16, 40; m 64; c 3. INFECTIOUS DISEASES, HOSPITAL EPIDEMIOLOGY. *Educ:* St Johns Univ, Minn, BS, 62; Univ Minn, MD, 68. *Prof Exp:* Space syst analyst, Hughes Aircraft, 62-64; resident med, Peter Bent Brigham Hosp, Boston, 68-69; clin assoc, comput, US Pub Health Serv, NIH, 69-71; resident med, Dept Med & Lab Med & Path, Univ Minn, 71-73, fel infectious dis, 73-75, instr dept med, 74-75, from asst prof to assoc prof, 75-86; staff physician infectious dis, Vet Admin Med Ctr, 74-80; SECT CHIEF, VET AFFAIRS MED CTR, 80-; PROF, DEPT MED & LAB MED & PATH, UNIV MINN, 86- *Mem:* Fel Infectious Dis Soc Am; Am Soc Microbiol. *Res:* The use of molecular biology techniques to characterize hospital infection epidemiology by using plasmid and chromosomal restriction techniques for clostridium difficile; s aureus and gentamicin-resistant gram-negatives. *Mailing Add:* Infectious Dis Sect Vet Affairs Ctr Minneapolis MN 55417

GERDING, THOMAS G, b Evanston, Ill, Feb 11, 30; m 55; c 4. TECHNICAL MANAGEMENT, MEDICAL SCIENCES. *Educ:* Purdue Univ, BS, 52, MS, 54, PhD(pharm), 60. *Prof Exp:* Instr pharm, Purdue Univ, 57-60, asst prof, 60-61; sr res pharmacist, Pitman-Moore Co Div, Dow Chem Co, 62-63, asst dir prod develop, 63, head prod develop, 63-65; tech dir, Glenbrook Labs Div, Sterling Drug, Inc, 65-67, dir prod develop div, Sterling-Winthrop Res Inst Div, 67-71; vpres res & develop, Calgon Consumer Prod Co, Merck & Co, 71-77; vpres & dir res & develop, Johnson & Johnson Prod Inc, 77-88; ADJ PROF, COL PHARM & DIR, DRUG DYNAMICS, UNIV TEX, AUSTIN, 88- *Concurrent Pos:* Consult, Health Care Indust. *Mem:* Sigma Xi; Am Chem Soc; Soc Chem Indust. *Res:* Pharmaceutics and wound care and healing. *Mailing Add:* 30105 Oakmont Dr Georgetown TX 78628

GERDY, JAMES ROBERT, b Chicago, Ill, Oct 10, 43; m 67; c 1. DEVELOPMENTAL BIOLOGY, GENETICS. *Educ:* Lake Forest Col, BS, 65; Northern Ill Univ, MS, 66; Southern Ill Univ, PhD(zool), 75. *Prof Exp:* Instr biol, Col St Francis, 66-69; vis lectr, Valparaiso Univ, 74-75; asst prof biol, Denison Univ, 75-78; mem staff, Smith Kline & French, 78-87; PRES, MEDCORE ASSOCS, 87- *Mem:* AAAS; Am Inst Biol Sci; Sigma Xi; Am Soc Zoologists; Soc Develop Biol. *Res:* Cell surface studies as related to ontogenic and phylogenetic development; development of genetic regulatory mechanisms; evolutionary embryology. *Mailing Add:* 658 Maple Ct Frankfort IL 60423

GERE, JAMES MONROE, b Syracuse, NY, June 14, 25; m 46; c 3. STRUCTURAL ENGINEERING, ENGINEERING MECHANICS. *Educ:* Rensselaer Polytech Inst, BCE, 49, MCE, 51; Stanford Univ, PhD(eng mech), 54. *Prof Exp:* Instr struct eng, Rensselaer Polytech Inst, 49-50, res assoc, 50-52; from asst prof to assoc prof, 54-62, exec head, Dept Civil Eng, 67-72, dir, Blume Earthquake Eng Ctr, 74-87, PROF STRUCT ENG, STANFORD UNIV, 62- *Mem:* Am Soc Civil Engrs; Am Soc Eng Educ; Earthquake Eng Res Inst; Sigma Xi. *Res:* Vibrations and buckling of structural components; analysis of framed structures; earthquake engineering. *Mailing Add:* 932 Valdez Pl Stanford CA 94305-1008

GEREBEN, ISTVAN B, b Sopron, Hungary, Jan 17, 33; US citizen; m 56; c 4. OCEANOGRAPHY. *Educ:* Budapest Tech Univ, MS, 56. *Prof Exp:* Res asst geophys, Hungarian Acad Sci, 55-56; field engr, Nat Admin Geophys & Geod Hungary, 56-57; res asst, Lamont Geol Observ, 59-62; res engr, Underseas Div, Westinghouse Elec Corp, 62-66 & Hydrospace Res Corp, 66-67, mgr acoust dept, 67-71; proj dir, Tracor Inc, 71-72; mem tech staff, Systs Group, 72-80, SR OCEANOGR, TRW FEDERAL SYSTS GROUP, 80- *Mem:* Am Geophys Union; NY Acad Sci; US Naval Inst. *Res:* Response of fixed or free floating bodies or structures to ocean wave forces; bearing capacity of ocean bottom sediments; hydrodynamical studies and experimental tests of deep towed bodies; underwater acoustics; submarine silencing; sonar technology; acoustic systems management; towed and fixed surveillance array performance analysis; ocean acoustics; military effectiveness of force structures. *Mailing Add:* 4101 Blackpool Rd Rockville MD 20853

GERECHT, J FRED, b New York, NY, Nov 8, 15; m 44; c 1. CHEMISTRY. *Educ:* Polytech Inst Brooklyn, BS, 42, MS, 45, PhD(org chem), 48. *Prof Exp:* Lab helper, E R Squibb & Sons, NY, 37-39; technician, Rockefeller Inst, 39-40; chemist, Schering Corp, NJ, 40-41; res chemist, Colgate-Palmolive-Peet Co, 41-49, group leader, Colgate-Palmolive Co, 49-57, res assoc, 57-79; RETIRED. *Concurrent Pos:* Instr, Fairleigh Dickenson Univ, 60-63; instr, Rutgers Univ, 63-73, adj prof, 73-79; adj prof, Somerset County Col, 79-86. *Mem:* Am Chem Soc; Am Oil Chemists Soc. *Res:* Synthesis, polymerization and oxidation of fatty acids; thermal polymerization of olefinic materials; sulfation, sulfonation and carbonium ion reactions. *Mailing Add:* 1236 Crim Rd Bridgewater NJ 08807

GEREN, COLLIS ROSS, b Miami, Okla, Mar 28, 45; m 67; c 2. PROTEIN CHEMISTRY. *Educ:* Northeastern Okla State Univ, BS, 67; Kans State Col Pittsburg, MS, 72; Okla State Univ, PhD(biochem), 74. *Prof Exp:* Teacher high sch sci, Picher, Okla Public Schs, 67-70; res assoc, Dept Biochem, Med Ctr, Univ Kans, 74-76; from asst prof to assoc prof, 76-83, PROF CHEM, UNIV ARK, 83- *Concurrent Pos:* Chair & coordr of Ark Biotechnol Ctr, 87. *Mem:* Am Soc Biol Chemists; Am Chem Soc; Sigma Xi; Am Asn Univ Prof; Int Soc Toxinology. *Res:* Venoms of North American spiders and snakes: venoms are fractionated, their major toxic components identified, and characterized both as to structure and function, and possible uses for isolated venom components are examined. *Mailing Add:* Dept Chem & Biochem Univ Ark Fayetteville AR 72701

GERENCSER, GEORGE A, EPITHELIAL TRANSPORT CENTER, CHLORIDE PUMP. *Educ:* Ind Univ, PhD(physiol), 71. *Prof Exp:* Assoc prof, 80-85, PROF PHYSIOL, SCH MED, UNIV FLA, 85- *Res:* Anion-ATP asc. *Mailing Add:* Dept Physiol J Hillis Miller Health Ctr Univ Fla Med Coll Gainesville FL 32610

GERENCSER, MARY ANN (AIKEN), b Macon, Ga, Mar 24, 27; m 57. BACTERIOLOGY. *Educ:* Ga Col Milledgeville, BS, 48; Smith Col, MA, 50; Univ Ky, PhD(bact), 58. *Prof Exp:* Technician, Ga State Dept Health, 50-51; res technician, Commun Dis Ctr, USPHS, 51-54; res asst bact, Univ Ky, 54-57; dir, Consol Labs, 57-59; supvr clin lab microbiol, Michael Reese Hosp, Chicago, Ill, 59-61; RES ASSOC MICROBIOL, SCH MED, W VA UNIV, 62- *Mem:* Am Soc Microbiol. *Res:* Clinical microbiology; genus Actinomyces; taxonomy; serology; dental microbiology. *Mailing Add:* 1264 Colonial Dr Morgantown WV 26505

GERENCSER, VINCENT FREDERIC, b New Brunswick, NJ, Jan 17, 27; m 57. BACTERIOLOGY. *Educ:* Fordham Univ, BS, 51; Univ Ky, PhD(bact), 58. *Prof Exp:* Asst bact, Univ Ky, 52-57; res assoc microbiol, Col Med, Univ Ill, 57-59, instr, 59-61; asst prof, 61-66, ASSOC PROF MICROBIOL, SCH MED, W VA UNIV, 66- *Mem:* AAAS; Am Soc Microbiol; Brit Soc Gen Microbiol; Sigma Xi. *Res:* Aquatic microbiology; microbial ecology; taxonomy; genetics; morphology; cytology. *Mailing Add:* Dept Microbiol WVa Univ Med Morgantown WV 26506

GERETY, ROBERT JOHN, b Jersey City, NJ, Oct 16, 39; m 67; c 3. IMMUNOLOGY, PEDIATRICS. *Educ:* Rutgers Univ, BA, 62; George Washington Univ, MD, 70; Stanford Univ, MA & PhD(immunol), 71. *Prof Exp:* Res assoc med microbiol, Stanford Univ, 70-71; res assoc viral immunol, Div Virol, NIH, 71-73; CHIEF HEPATITIS BR, BUR BIOLOGICS, FOOD & DRUG ADMIN, 73- *Concurrent Pos:* Chmn, Comt License Biologics US, 72-; contracts consult, Nat Heart, Lung & Blood Inst, NIH, 73-; consult, WHO, 75-; chmn, Hepatitis Coord Comt, NIH, 75-80; chmn, Comt Assess Primate Usage Infectious Dis Res, 78. *Mem:* Am Asn Immunologists; Am Acad Pediat; Sigma Xi; Am Asn Microbiologists. *Res:* Viral hepatitis; immunology; in vitro serologic tests; pediatrics; vaccine production, testing and usage policies; biological standardization. *Mailing Add:* 14 Cambridge Ctr Cambridge MA 02142

GEREZ, VICTOR, b Santander, Spain, Apr 11, 34; Mexican citizen; m 84; c 2. ELECTRICAL ENGINEERING, MECHANICAL ENGINEERING. *Educ:* Nat Univ Mexico, EE, 57; Univ Calif, MEE, 69, PhD(elect eng), 72. *Prof Exp:* Res eng, Fed Elect Bd, 65-73; dept head & prof elect eng, Col Eng Nat Univ Mex, 73-76; dept dir power eng, Inst Elect Res, 77-83; assoc prof dept head elect eng, 83-84, PROF ELECT ENG, MONT STATE UNIV, 84- *Concurrent Pos:* Lect, Ibero Am Univ, 59-65; col eng, 59-68, 77-83, dept head control eng, Col eng, Nat Univ Mex 72-73. *Mem:* Inst Elec & Electronics Engrs; Am Asn Eng Educ; Int Asn Sci Technol Develop. *Res:* Power system transmissin Planning and operation; Pollution Flashover Phenomenon in Transmission systems; electrical engineering education. *Mailing Add:* Dept Elec Eng Mont State Univ Bozeman MT 59717

GERFEN, CHARLES OTTO, b Breese, Ill, Apr 10, 20; m 48; c 4. PHYSICAL CHEMISTRY. *Educ:* Ill State Univ, normal, BS; Univ Mo, PhD(chem), 51. *Prof Exp:* Chemist, Atmospheric Nitrogen Corp, Ky, 42-43; res chemist, Inorg Res Dept, Mallinckrodt Chem Works, 51-55, group leader, 55-57, asst dir, 57-60, dir gen res, 60-63, spec prod dept, 63-71 & narcotics & dangerous drugs, 71-73, dir govt rels, 73-76, DIR SPEC CORP PURCHASES, MALLINCKRODT, INC, 76- *Mem:* AAAS; Am Chem Soc; Am Inst Chem. *Res:* Chemistry of group IV and V elements, niobium, tantalum; titanium; zirconium and hafnium; ore beneficiation; chemistry of opium and opiates. *Mailing Add:* 37 Portland Dr St Louis MO 63131

GERGELY, JOHN, b Budapest, Hungary, May 15, 19; nat US; m 45; c 8. BIOCHEMISTRY. *Educ:* Univ Budapest, MD, 42; Leeds Univ, PhD(phys chem), 48. *Prof Exp:* From asst prof pharmacol to asst prof biochem, Univ Budapest, 42-48; asst prof biochem, New Sch Soc Res, 48-50; Nat Heart Inst sr trainee, Univ Wis, 50-51; res assoc med, 51-62, asst prof, 62-71, assoc prof, 71-80, PROF BIOL CHEM, HARVARD MED SCH, 80-; BIOCHEMIST, MASS GEN HOSP, 69-; DIR DEPT MUSCLE RES, BOSTON BIOMED RES INST, 70- *Concurrent Pos:* NIH spec res fel, 48-50; estab investr, Am Heart Asn, 51-58; from asst biochemist to assoc biochemist, Mass Gen Hosp, 54-69; tutor biochem sci, Harvard Med Sch, 57-72; dir dept muscle res, Retina Found, 61-70. *Mem:* Am Soc Biol Chem; Biophys Soc; Am Chem Soc; NY Acad Sci; Brit Biochem Soc. *Res:* Biochemistry of muscle contraction; enzymes; nuclear magnetic resonance; physical chemistry of proteins; electron spin resonance. *Mailing Add:* Dept Biol Chem Harvard Med Sch 25 Shattuck St Boston MA 02115

GERGELY, PETER, b Budapest, Hungary, Feb 12, 36; US citizen; m 64; c 2. STRUCTURAL & EARTHQUAKE ENGINEERING. *Educ:* McGill Univ, BEng, 60; Univ Ill, Urbana, MS, 62, PhD(civil eng), 63. *Prof Exp:* Struct engr, Dominion Bridge Co, Montreal, 60; struct engr steel design, Pittsburgh-Des Moines Steel Co, 69-70; from asst prof to assoc prof, 63-75, chmn dept, 83-88, dir, sch civil & environ eng, 85-88, PROF STRUCT ENG, CORNELL UNIV, 75- *Concurrent Pos:* Consult, Pittsburgh-Des Moines Steel Co, 64-75; vis prof, Univ Toronto, 76; Nat Acad Sci grant, 76; Lawrence Livermore Lab, Am Concrete Inst, 83-, mem bd dirs, 85-88. *Honors & Awards:* Am Soc Civil Engrs Award, 74, Res Prze, 76. *Mem:* Fel Am Concrete Inst; fel Am Soc Civil Engrs; Earthquake Eng Res Inst. *Res:* Reinforced concrete; structural dynamics; nuclear reactor structures; earthquake engineering. *Mailing Add:* Dept Struct Eng Cornell Univ Hollister Hall Ithaca NY 14853

GERGIS, SAMIR D, US citizen. ANESTHESIOLOGY. *Educ:* Cairo Univ, Egypt, MB, 54, MD(anesthesia), 62; FACA, 71. *Prof Exp:* House officer, Cairo Univ Hosps, Egypt, 55, resident, Dept Anesthesia, 57-58; fel, Anesthesiol Ctr, WHO & Univ Copenhagen, Denmark, 63; res fel, Col Med, Univ Iowa, 68, from instr to assoc prof, Dept Anesthesia, 69-76, actg head dept, 77-78, dir, Clin Serv, Hosps & Clins, 78-80, PROF, DEPT ANESTHESIA, COL MED, UNIV IOWA, IOWA CITY, 76-, VCHMN CLIN AFFAIRS, DEPT ANESTHESIA, UNIV IOWA HOSPS & CLINS, 85- *Concurrent Pos:* Mem, Subcomt Neuromuscular Transmission, Am Soc Anesthesiologists, 86- & Comt Surg Anesthesia, 91; bd mem, Asn Anesthesia Clin Dirs. *Mem:* Int Anesthesia Res Soc; AMA; Am Soc Anesthesiologists; NY Acad Sci; Am Soc Pharmacol & Exp Therapeut; Soc Exp Biol & Med; AAAS; Nat Soc Med Res; Am Soc Clin Pharmacol & Therapeut; Asn Anesthesia Clin Dirs. *Res:* Neuromuscular physiology and pharmacology; clincial research in anesthesia. *Mailing Add:* Dept Anesthesia Univ Iowa Col Med Iowa City IA 52242

GERHARD, EARL R(OBERT), b Louisville, Ky, Aug 9, 22; m 47; c 5. CHEMICAL ENGINEERING, ENVIRONMENTAL ENGINEERING. *Educ:* Univ Louisville, BChE, 43, MChE, 47; Univ Ill, PhD(chem eng), 53. *Prof Exp:* From asst prof to assoc prof, 51-64, chmn dept, 69-73, assoc dean, 73-80, PROF CHEM ENG, UNIV LOUISVILLE, 64-, DEAN, 80- *Mem:* Am Chem Soc; Am Soc Eng Educ; Am Inst Chem Engrs; Nat Soc Prof Engrs. *Res:* Transport phenomenon; reaction kinetics; coal research; kinetics & reactor design; environmental impacts and assessment. *Mailing Add:* 5723 Prince William St Louisville KY 40207

GERHARD, GLEN CARL, b Albion, NY, Mar 1, 35; m 57; c 4. BIOENGINEERING & BIOMEDICAL ENGINEERING. *Educ:* Syracuse Univ, BEE, 56; Ohio State Univ, MSc, 58, PhD(elec eng), 63. *Prof Exp:* Engr, Gen Elec Co, 56; asst elec eng, Ohio State Univ, 56-57; develop engr, Eastman Kodak Co, 57; res asst elec eng, Ohio State Univ, 57-58, instr elec eng & res assoc electron devices, 58-62; res engr, Electronics Lab, Gen Elec Co, 62-63 & 64-67; from asst prof to prof elec eng, Univ NH, 67-90, dir, Biomed Ear Nose & Throat Ctr, 81-90; VIS PROF, UNIV ARIZ, 90- *Concurrent Pos:* Consult, Gen Elec Co, 67-68, Kidder Press Div, Moore Bus Forms, 68-70, Hewlett-Packard, 87-89; assoc dir, Clin Eng Ctr, Univ NH, 73-81, dir, 81-; vis assoc prof elec eng & surg, Univ Ariz, 81-82; legal expert witness, 76- *Mem:* Inst Elec & Electronics Engrs; Eng Med & Biol Soc; Sigma Xi; Asn Advan Med Instrumentation; Am Soc Eng Educ. *Res:* Biomedical instrumentation; physiological transducers and sensors; electrosurgery; neurosurgical instrumentation; prosthetic devices; clinical engineering; analog VLSI circuits. *Mailing Add:* Dept Elect & Comp Eng Univ Ariz Tucson AZ 85721

GERHARD, LEE C, b Albion, NY, May 30, 37; m 64; c 1. STRATIGRAPHY, RESEARCH ADMINISTRATION. *Educ:* Syracuse Univ, BS, 58; Univ Kans, MS, 61, PhD(geol), 64. *Prof Exp:* Explor geologist, Sinclair Oil & Gas Co, 64-65, region stratigrapher, 65-66; from asst prof to assoc prof geol, Southern Colo State Col, 66-72; asst dir & actg dir, WIndies Lab & assoc prof, Dickinson Univ, St Croix, 72-75; asst state geologist, NDak Geol Surv, 75-78, state geologist & dir, 78-81; explor mgr, Rocky Mountain div, Supron Energy Corp, 81-82; getty prof geol eng, Colo Sch Mines, 82-87; STATE GEOLOGIST & DIR, KANS GEOL SURV, UNIV KANS, 87- *Concurrent Pos:* Assoc prof geol, Univ NDak, 75-78, prof, 78-81, chmn dept, 78-81. *Mem:* Am Asn Petrol Geol; Geol Soc Am; Soc Econ Paleont & Mineral; Sigma Xi. *Res:* Structural geologic history; carbonate petrography; stratigraphy; sedimentary petrology. *Mailing Add:* Univ Kans 1930 Constant Ave Campus W Lawrence KS 66046

GERHARD, WALTER ULRICH, b Zurich, Switz. VIROLOGY, IMMUNOLOGY. *Educ:* Univ Zurich, Med Prac, 69, Dr med, 71. *Prof Exp:* Fel virol, Inst Immunol, Basel, 71-72; fel immunol, Dept Path, Univ Pa, 73-74; res assoc viral immunol, 74-77, ASSOC PROF VIRAL IMMUNOL, WISTAR INST, 78- *Mem:* Am Asn Immunologists. *Res:* Antigenic drift of influenza virus; pathogenesis of multiple sclerosis. *Mailing Add:* Wistar Inst 36th St & Spruce St Philadelphia PA 19104

GERHARDT, DON JOHN, b Evansville, Ind, Oct 30, 43; m 68; c 3. HIGH PRESSURE FLUIDS, COMPRESSORS. *Educ:* Purdue Univ, BSME, 65; Univ Detroit, MS, 67; Univ Mich, PhD(bioeng), 75, MBA, 82. *Prof Exp:* Test engr, Gen Motors Proving Grounds, 65, design engr, Pontiac Div, 66-67, res engr, Res Labs, 68; eng supvr, Ford Motor Co, 72-83; eng mgr, 84-88, CHIEF ENGR, INGERSOLL-RAND, 89- *Concurrent Pos:* Researcher, Hwy Safety Inst, Univ Mich, 69-70, lectr, Med Sch, 70-71; mem, Noise Reduction Comt, Soc Automotive Engrs, 79-80, Chassis Design Comt, 82-83 & Electronics Comt, 85-88; appointee, Exhaust & Emissions, Motor Vehicle Mfrs Asn, 79-80. *Mem:* Am Soc Mech Engrs; Inst Elec & Electronics Engrs; Soc Automotive Engrs; Nat Soc Prof Engrs; Am Soc Metals; Vol Tech Assistance. *Res:* Impact of science on people and environment; product safety; advanced industrial equipment; expert computer systems; microprocessor control theory; advanced compressors; sound and vibration analysis; high pressure fluids; electroencephalogram computer analysis; neuroscience; holder of three US patents. *Mailing Add:* 3474 Tanglebrook Trail Clemmons NC 27102

GERHARDT, GEORGE WILLIAM, b Shaler Twp, Pa, Nov 7, 15. ORGANIC CHEMISTRY. *Educ:* Univ Pittsburgh, BS, 36, PhD(chem), 51. *Prof Exp:* From fel to adv fel, Mellon Inst, 36-59; asst tech dir, Stoner-Mudge Div, Mobil Finishes Co, Inc, 59-66; asst tech dir, Mobil Chem Co, 66-71, lab dir, 71-73, mgr com develop & customer serv, Packaging Coatings Dept, 73-81. *Concurrent Pos:* Mem comt on packaging, Nat Res Coun, 75-78; consult, 85- *Mem:* Am Chem Soc; Inst Food Technol; Fedn Soc Paint Technol; Master Brewers Asn Am. *Res:* Organic coatings; packaging; resin manufacture. *Mailing Add:* 2733 Cole Rd Wexford PA 15090

GERHARDT, H CARL, JR, b Newport News, Va, May 23, 45; m 66; c 2. ACOUSTIC COMMUNICATION, NEUROETHOLOGY. *Educ:* Univ Ga, BS, 66; Univ Tex, Austin, MA, 68, PhD(zool), 70. *Prof Exp:* Res assoc animal commun, Cornell Univ, 70-71; asst prof, 71-77, assoc prof, 77-80, PROF, DIV BIOL SCI, UNIV MO, COLUMBIA, 81- *Concurrent Pos:* NIH res career develop award, 76-81; vis prof, Gesamthochschule, WGer, 78, Dept Zool, Univ Melbourne, 81-82 & 84, Univ Vienna, 84. *Mem:* AAAS; fel Animal Behav Soc; Int Soc Neuroethology (secy, 84-87). *Res:* Sound pattern recognition in animals, using synthetic acoustic signals to identify pertinent properties of the vocalizations of animals; neurobiological mechanisms and evolution of acoustic communication. *Mailing Add:* Div Biol Sci Tucker Hall Univ Mo Columbia MO 65211

GERHARDT, JON STUART, b Springfield, Ohio, June 5, 43; m 70; c 2. MECHANICAL ENGINEERING. *Educ:* Univ Cincinnati, BScME, 66, MS, 68, PhD(mech eng-math), 71. *Prof Exp:* Coop engr, Whirlpool Corp, 61-66; res assoc & instr, Univ Cincinnati, 69-71; asst prof eng technol, Univ NC, Charlotte, 71-73; proj engr, Duff-Norton Co, 73; sr develop engr, Gencorp, Inc, 73-77, group leader rubber prod eng, 77-79, mgr tech staff develop, 79-84, mgr prod eng & res ctr admin, 84-87, DIR RES, GEN TIRE, INC, 87- *Concurrent Pos:* Mem bd dirs, Am Soc Mech Engrs; lectr, Univ Akron, 77-; chmn bd, Chapel Hill Christian Schs. *Mem:* Am Soc Mech Engrs; Soc Automotive Engrs; Am Chem Soc. *Res:* Measurement and analysis of complex dynamic systems; interaction of tire dynamics and vehicles that produce ride quality attributes. *Mailing Add:* Gen Tire Inc One General St Akron OH 44329-0026

GERHARDT, KLAUS OTTO, b Drengfurt, Germany, Aug 6, 35. ORGANIC CHEMISTRY, ANALYTICAL CHEMISTRY. *Educ:* Tech Univ Berlin, Diplom Chemiker, 64, Dr rer nat(anal chem), 67. *Prof Exp:* Fel, Univ Mo-Columbia, 67-68; asst prof phys & gen chem, Lincoln Univ, Mo, 68-69; res anal chemist, 69-77, sr res chemist, 77-88, RES ASSOC PROF BIOCHEM, UNIV MO-COLUMBIA, 88- *Mem:* Am Chem Soc. *Res:* Analytical methods of gas-liquid chromatography and high pressure liquid chromatography for fatty acids, amino acids, nucleosides, and biological markers for the detection of cancer; mass spectrometry. *Mailing Add:* Rm 4 Agr Bldg Univ Mo Columbia MO 65211

GERHARDT, LESTER A, b Bronx, NY, Jan 28, 40; m 61; c 2. ELECTRICAL & SYSTEMS ENGINEERING. *Educ:* City Col New York, BEE, 61; State Univ NY Buffalo, MS, 64, PhD(commun systs), 69. *Prof Exp:* Sr elec engr avionics systs, Bell Aerospace Corp, 61-64, sect head signal & info processing res, 64-69; assoc prof, Systs Div, Rensselaer Polytech Inst, 69-74, chmn dept, 75-86, dir computer integrated mfg, 86-91, PROF ELEC SYSTS ENG, RENSSELAER POLYTECH INST, 74-, ASSOC DEAN ENG, 91- *Concurrent Pos:* Asst to dir advan res & consult, Bell Aerospace Corp, 69-75; consult, USAF-Rome Air Develop Ctr, 72- & Gen Elec Co, 78-; appointed panel mem, Adv Group Aerospace Res & Develop, NATO, US rep, Collab Res Panel. *Mem:* Sigma Xi; fel Inst Elec & Electronics Engrs. *Res:* Adaptive systems research with applications to communications, control and pattern recognition; voice and image processing; digital, signal and information processing. *Mailing Add:* Three Bear Brook Ct Clifton Park NY 12065

GERHARDT, MARK S, b Bronx, NY, Mar 22, 46. COMPUTER SCIENCE, REAL-TIME SYSTEMS ARCHITECTURE. *Educ:* City Col NY, BS, 67; Princeton Univ, MS, 68. *Prof Exp:* Sr staff engr, 87-89, CHIEF TECHNOLOGIST SOFTWARE DIRECTORATE, ENG DIV, ESL INC, 89- *Concurrent Pos:* Vchair, Spec Interest Group Ada, Asn Comput Mach, 87-89, distinguished reviewer, Ada 9X, 88-, chair, 89- *Mem:* Asn Comput Mach. *Mailing Add:* ESL Inc MS J5-3 495 Java Dr PO Box 3510 Sunnyvale CA 94088-3510

GERHARDT, PAUL DONALD, b Riverside, Calif, July 10, 17; m 68; c 1. ENTOMOLOGY. *Educ:* Univ Calif, BS, 40, MS, 41, PhD(entom), 49. *Prof Exp:* Asst entom, Univ Calif, 40-41, assoc, Exp Sta, Univ Calif, Riverside, 46-49, from jr entomologist to asst entomologist, 49-54; from asst entomologist to assoc entomologist, 55-62, ENTOMOLOGIST, UNIV ARIZ, 63-, PROF ENTOM, 68- *Mem:* Entom Soc Am; Lepidop Soc. *Res:* Biology and control of insects attacking vegetable crops; field crops and citrus; use of insecticides; Hymenoptera; Mutillidae; Lepidoptera; Papillionidae. *Mailing Add:* 838 E Tenth St Mesa AZ 85203

GERHARDT, PHILIPP, b Milwaukee, Wis, Dec 30, 21; m 45; c 3. MICROBIOLOGY. *Educ:* Univ Wis, PhB, 43, MS, 47, PhD(bact), 49; Am Bd Microbiol, dipl. *Prof Exp:* Asst prof bact, Ore State Univ, 49-51; chief lab div biol develop, Pine Bluff Arsenal, 51-52; from asst prof to prof microbiol, Med Sch, Univ Mich, 53-65; chmn dept, 65-75, assoc dean res & grad study, osteop med, 75-87, PROF MICROBIOL & PUB HEALTH, MICH STATE UNIV, 65- *Concurrent Pos:* Consult, Nat Ctr Toxicol Res, 75-78; dir, Ribi Immuno Chem Res Inc, 85-; adj sr scientist, Mich Biotech Inst, 85- *Mem:* Int Union Microbiol Soc (pres, 82-86); Am Soc Microbiol (secy, 61-67, vpres, 73-74, pres, 74-75); fel Am Acad Microbiol; Soc Gen Microbiol; fel AAAS; Sigma Xi; hon Polish Med Soc. *Res:* Spore resistance mechanisms, fermentations; membranes and permeability; spores; dialysis culture. *Mailing Add:* Dept Microbiol & Pub Health Mich State Univ East Lansing MI 48824-1101

GERHARDT, REID RICHARD, b Monmouth, Ill, May 2, 41; m 65; c 2. ENTOMOLOGY. *Educ:* Va Polytech Inst, BS, 63, MS, 65; NC State Univ, PhD(entom), 72. *Prof Exp:* Instr biol, Radford Col, 64-66; asst prof, 73-78, ASSOC PROF AGR BIOL, UNIV TENN, KNOXVILLE, 78- *Mem:* Entom Soc Am. *Res:* Effect of insect pests on livestock; biology and ecology of biting diptera. *Mailing Add:* Path Dept Univ Tenn Knoxville TN 37901

GERHART, JAMES BASIL, b Pasadena, Calif, Dec 15, 28; m 58; c 2. NUCLEAR PHYSICS, HISTORY OF SCIENCE. *Educ:* Calif Inst Technol, BS, 50; Princeton Univ, MA, 52, PhD(physics), 54. *Prof Exp:* Asst physics, Princeton Univ, 50-54, instr, 54-56; from asst prof to assoc prof, 56-65, PROF PHYSICS, UNIV WASH, 65- *Concurrent Pos:* Chmn, Pac NW Asn Col Physics, 70-72 & exec officer, 72-; gov bd, Am Inst Physics, 72-78 & 80-82. *Honors & Awards:* Millikan Medal, Am Asn Physics Teachers, 85 & Distinguished Serv Award, 83. *Mem:* Fel AAAS; fel Am Phys Soc; Am Asn Physics Teachers (secy, 71-77, vpres, 77, pres-elect, 78, pres, 79); Am Inst Physics; Pac NW Asn Col Physics (chmn, 70-72); Am Inst Physics. *Res:* Beta and gamma ray spectroscopy; nuclear scattering and reactions. *Mailing Add:* Dept Physics Univ Wash Seattle WA 98195

GERHART, JOHN C, b Cincinnati, Ohio, Mar 27, 36; m 64. BIOCHEMISTRY. *Educ:* Harvard Univ, AB, 58; Univ Calif, Berkeley, PhD(biochem), 62. *Prof Exp:* Asst prof molecular biol & virol, 62-67, assoc prof, 67-73, PROF MOLECULAR BIOL & RES BIOCHEMIST, VIRUS LAB, UNIV CALIF, BERKELEY, 73- *Res:* Cell growth and regulation; control of enzyme activity; metazoan development. *Mailing Add:* Dept Molecular Biol Univ Calif 2120 Oxford St Berkeley CA 94720

GERHOLD, GEORGE A, b Kewanee, Ill, Mar 17, 37; m 58. PHYSICAL CHEMISTRY. *Educ:* Univ Ill, BS, 58; Univ Wash, Seattle, PhD(chem), 63. *Prof Exp:* NSF fel chem, Univ Col, Univ London, 63-65; asst prof, Univ Calif, Davis, 65-70; from asst prof to assoc prof, 70-76, PROF CHEM, WESTERN WASH UNIV, 76- *Res:* Electronic and crystal spectra; excitons. *Mailing Add:* Dept Chem Western Wash Univ 516 High St Bellingham WA 98225

GERHOLD, HENRY DIETRICH, b Mahwah, NJ, Feb 1, 31; m 56; c 3. FOREST GENETICS. *Educ:* Pa State Univ, BS, 52, MF, 54; Yale Univ, PhD(forest genetics), 59. *Prof Exp:* Soil conserv aid, Soil Conserv Serv, USDA, 49, fire control aid, Forest Serv, Mont, 51, forester, Idaho, 52-53; res forester, Northeastern Forest Exp Sta, 54-55; from instr to assoc prof, 56-69, PROF FOREST GENETICS, PA STATE UNIV, 69-, ASST DIR RES, SCH FOREST RESOURCES, 85- *Concurrent Pos:* NSF travel grant, World Consult Forest Genetics, UN Food & Agr Orgn, 63; NATO sr sci fel, 70; chmn, Northeastern Forest Tree Improvement Conf, 70-72; chmn genetic resistance to dis & insects subject group, Int Union Forest Res Orgns, 66-74, deputy coodr, division 2, 74-79. *Mem:* Soc Am Foresters; Sigma Xi; Int Soc Arboricult; Am Forestry Asn. *Res:* Forest genetics; genetics of Christmas trees and landscape trees; resistance to diseases, insects and air pollutants. *Mailing Add:* 109 Ferguson Bldg Pa State Univ University Park PA 16802

GERICH, JOHN EDWARD, DIABETES, COUNTER-REGULATION. *Educ:* Georgetown Univ, MD, 69. *Prof Exp:* PROF MED & PHYSIOL, MAYO MED SCH, ROCHESTER, MINN, 79. *Res:* Insulin resistance. *Mailing Add:* Clin Res Ctr 3488 Presbyterian Univ Hosp 230 Lothrop Pittsburgh PA 15261

GERICKE, OTTO LUKE, b San Francisco, Calif, July 16, 07; m 34, 84; c 4. PSYCHIATRY. *Educ:* Univ Calif, AB, 29, MD, 33; Univ Pa, cert, 39; Sch Mil Neuropsychiat, 43; Am Bd Psychiat & Neurol, dipl, 50. *Prof Exp:* Intern, San Francisco Hosp, 32-33; from asst resident to resident, St Joseph's Hosp, San Francisco, 33-35; pvt pract med, Calif, 35-36; asst surgeon, Dist 36, Ft Douglas Civilian Conserv Corps, 36; physician & psychiatrist, Mendocino State Hosp, Talmage, Calif, 39-42; asst supt, Stockton State Hosp, 45-46; med dir, Patton State Hosp, 46-72; from asst clin prof to clin prof, Loma Linda Univ, 47-75; CONSULT, 75- *Concurrent Pos:* Med examr criminal cases & lunacy cases, 51- *Mem:* AMA; fel Am Psychiat Asn; Pan-Am Med Asn; Asn Mil Surgeons. *Mailing Add:* 6987 Palm Ave Highland CA 92346

GERICKE, OTTO REINHARD, b Detmold, Ger, Sept 7, 21; US citizen; m 52; c 1. PHYSICS. *Educ:* Univ Gottingen, BS, 44, MS, 50. *Prof Exp:* Physicist, Siemens-Reiniger Co, Ger, 51-54 & Siemens & Halske Co, 54-58; physicist, US Army Mat Res Agency, 58-63, res physicist, 63-66, chief appl physics br, 66-69, head nondestructive testing br, 69-78, SUPVRY RES PHYSICIST, US ARMY MAT TECHNOL LAB, 78- *Honors & Awards:* Cert Outstanding Achievement, Sci Conf, US Army, 62. *Mem:* Soc Nondestructive Test; Acoust Soc Am. *Res:* Nondestructive testing; studies of various types of physical phenomena to determine their potential usefulness for nondestructive evaluation of materials including ultrasonic spectroscopy; holography and imaging, image enhancement; computer-based ultrasonic signal processing. *Mailing Add:* Seven Crest Circle Medfield MA 02052

GERIG, JOHN THOMAS, b Windham, Ohio, Nov 7, 38; m 61; c 2. BIO-ORGANIC CHEMISTRY. *Educ:* Col Wooster, BA, 60; Brown Univ, PhD(chem), 64. *Prof Exp:* Asst org chem, Brown Univ, 60-64; res fel, Calif Inst Technol, 64-66; assoc prof, 66-77, PROF CHEM, UNIV CALIF, SANTA BARBARA, 77- *Honors & Awards:* Res Career Develop Award, Pub Health Serv, 73-78. *Mem:* Am Chem Soc; AAAS; Int Soc Magnetic Res; Am Soc Biol Chemists; Molecular Biol & Protein Soc. *Res:* Nuclear magnetic resonance spectroscopy; protein chemistry; enzymatic reactions. *Mailing Add:* Dept Chem Univ Calif Santa Barbara CA 93106

GERIG, THOMAS MICHAEL, b Washington, DC, May 25, 42; m 66. MATHEMATICAL STATISTICS. *Educ:* George Washington Univ, AB, 65; Univ NC, Chapel Hill, PhD(statist), 71. *Prof Exp:* Assoc prof, 75-80, PROF STATIST, NC STATE UNIV, 80- *Mem:* Am Statist Asn; Inst Math Statist. *Res:* Multivariate analysis; categonial data; sampling; statistical computing. *Mailing Add:* Inst Statist Box 8203 NC State Univ Raleigh NC 27695

GERIK, JAMES STEPHEN, b Waco, Tex, Sept 24, 56; m 88. SOILBORNE FUNGI, FUNGAL VECTORS OF VIRUSES. *Educ:* Tex A&M, BS, 77, MS, 79; Univ Calif, Berkeley, PhD(plant path), 84. *Prof Exp:* Res assoc, Univ Wis-Madison, 84-85; postdoctoral, Agr Res Serv, USDA, Salinas, Calif, 85-87, res plant pathologist, 87-91; RES PLANT PATHOLOGIST, HOLLY SUGAR CORP, TRACY, CALIF, 91- *Mem:* Am Phytopath Soc; Mycol Soc Am; Am Soc Sugarbeet Technologists; AAAS. *Res:* Ecological and epidemological aspects of soilborne sugarbeet diseases and fungal vectors of soilborne viruses. *Mailing Add:* PO Box 60 Tracy CA 95378

GERIN, JOHN LOUIS, b St Paul, Minn, Sept 28, 37; m 60; c 1. VIROLOGY, BIOCHEMISTRY. *Educ:* Georgetown Univ, BS, 59; Univ Tenn, Knoxville, MS, 61, PhD(zool), 64. *Prof Exp:* Res scientist, Sci Div, Abbott Lab, 64-66, group leader biochem & biophys virol, 66-67; head, Rockville Lab, Molecular Anat Prog, Oak Ridge Nat Lab, 67-78; PROF MICROBIOL & HEAD DIV MOLECULAR VIROL & IMMUNOL, GEORGETOWN UNIV MED CTR, 78- *Mem:* AAAS; Am Soc Microbiol; Am Asn Immunol; Am Soc Microbiol; Infect Dis Soc Am. *Res:* Biochemical and biophysical characteristics of animal viruses, especially human respiratory viruses and hepatitis-viruses; purification of viral antigens for vaccine production; antivirals; viral diagnostic. *Mailing Add:* Div Molec Virol & Immunol/ Georgetown Univ 5640 Fishers Lane Rockville MD 20852

GERING, ROBERT LEE, b Parker, SDak, Feb 18, 20; m 45; c 2. ZOOLOGY, SCIENCE EDUCATION. *Educ:* Univ Utah, AB, 47, MA, 48, PhD(invert zool), 50. *Prof Exp:* Chmn nat sci div & biol dept, Bethel Col, 48-53; asst dir ecol res proj, Univ Utah, 53-54; assoc prof biol, Wells Col, 54-59, prof & chmn dept, 59-65; INDEPENDENT RESEARCHER & DEVELOP EDUC PROD, 65-; PRES, INFO APPLNS, INC, 66- *Concurrent Pos:* Vis prof, Ward's Natural Sci Estab, 63-64 & Univ Rochester, 65-66; prof biol, Rochester Inst Technol, 66-68; coord computerized multi-media instr, Nat Technol Inst for Deaf, 68-69. *Honors & Awards:* Res Award, Am Cancer Soc, 51. *Res:* Advanced audio-visual systems; morphology and behavior study of spiders; genetics. *Mailing Add:* Info Applns Inc 2169 Baird Rd Penfield NY 14526-2419

GERJUOY, EDWARD, b Brooklyn, NY, May 19, 18; m 40; c 2. THEORETICAL PHYSICS, ENVIRONMENTAL LAW. *Educ:* City Col, New York, BS, 37; Univ Calif, MA, 40, PhD(physics), 42; Univ Pittsburgh, JD, 77. *Prof Exp:* Res physicist, Off Sci Res & Develop, Columbia Univ, 42-46; asst dir sonar anal group, 46; asst prof physics, Univ Southern Calif, 46-49, assoc prof, 49-52; from assoc prof to prof, Univ Pittsburgh, 52-58; mem res staff, Gen Atomic Div, Gen Dynamics Corp, 58-61 & E H Plesset Assocs, 61-62; dir plasma & space appl physics, Defense Electronics Prod Div, Radio Corp Am, 62-64; prof, 64-82, hearing examr, Pa Environ Hearing Bd, 80-81, EMER PROF PHYSICS, UNIV PITTSBURGH, 82- *Concurrent Pos:* Mem adv comt, Army Res Off Nat Acad Sci, 65-68; mem health physics vis comt, Oak Ridge Nat Lab, 68-74; vis fel, Joint Inst Lab Astrophys, Colo, 70; consult, Westinghouse Res Labs, Rand Corp, Inst Defense Anal, Lockheed Electronics Div, Inst Energy Anal, Oak Ridge, Tenn & Environ Protection Agency, 77-; ed, Comments on Atomic and Molecular Physics, 72-74; consult & counsr legal-tech issues, 76-; assoc of law firm, 78-80; ed-in-chief, Jurimetrics, 81-87; mem & chmn, Panel Public Affairs, Am Phys Soc, 78-81, Comt Int Freedom Scientists, 82-85; mem coun, Sect Sci & Technol, Am Bar Asn, 78-80, 84-86 & 87-; mem coun, Rose Schmidt, Hasley & DiSalle, Law Firm, 81-86, mem firm, 87-; sr ed, Jurimetrics, 87-; mem, Nat Conf Lawyers & Scientists, Am Bar Asn, 86- *Mem:* Fel AAAS; fel Am Phys Soc; Am Bar Asn. *Res:* Atomic collision theory; technical-legal subjects. *Mailing Add:* Dept Physics Univ Pittsburgh Pittsburgh PA 15260

GERKE, JOHN ROYAL, b New York, NY, May 29, 27; m 48; c 4. DRINKING AND WASTEWATER, MARINE AQUACULTURE. *Educ:* Duke Univ, BA, 47; Univ Ill, MS, 49; Rutgers Univ, PhD, 64. *Prof Exp:* Res asst, State Water Surv, Ill, 48-51; sr scienist, E R Squibb & Sons, 51-64, Squibb fel, 62-64; sr microbiologist, Hoffman La Roche, NJ, 64-67; assoc prof microbiol, Pac Univ, 67-76; mem staff, World Book-Childcraft Int, Inc, 76-82; INSTR, TILLAMOOK BAY COMMUNITY COL, 82-; OWNER/MGR, JOHN R GERKE PHD APPL SCI LAB, 82- *Mem:* Am Chem Soc; Am Soc Microbiol; Inst Food Technologists; Int Asn Milk, Food & Environ sanitarians. *Res:* Ocular diseases, pseudomonas diseases; analytical microbiology; microbiological chemistry; microbiological transformation of steroids; antibiotics; water bacteriology and chemistry; marine aquaculture. *Mailing Add:* 1775 Vista View Dr Tillamook OR 97141

GERKEN, GEORGE MANZ, b Hackensack, NJ, July 12, 33; m 56; c 2. HEARING, BRAIN FUNCTION. *Educ:* Mass Inst Technol, BS, 55; Univ Chicago, PhD(physiol psychol), 59. *Prof Exp:* Asst prof psychol, Univ Va, 59-67; res assoc, Callier Ctr Commun Disorders, 67-70, dir grad studies, 74-77; assoc prof, 73-83, PROF COMMUN DISORDERS, UNIV TEX, DALLAS, 84-; RES SCIENTIST, CALLIER CTR COMMUN DISORDERS, 70- *Concurrent Pos:* Vis prof otolaryngol, Univ Tex Southwestern Med Ctr, 88-90. *Mem:* Am Speech-Lang Hearing Asn; Psychonomic Soc; Soc Neurosci; Am Audiol Soc; Acoustical Soc Am; Asn Res Otolaryngol. *Res:* Brain function; neuropsychology; hearing. *Mailing Add:* Callier Ctr Commun Disorders 1966 Inwood Rd Dallas TX 75235

GERKING, SHELBY DELOS, b Elkhart, Ind, Nov 16, 18; m 43; c 3. ECOLOGY, FISH BIOLOGY. *Educ:* DePauw Univ, AB, 40; Ind Univ, PhD(zool), 44. *Prof Exp:* Res assoc physiol, Ind Univ, Bloomington, 44-46, instr zool, 46-49, from asst prof to prof, 49-67, dir biol sta, 59-67, assoc dir water resources res ctr, 63-67; prof, 67-83, chmn dept, 67-74, EMER PROF ZOOL, ARIZ STATE UNIV, 83- *Concurrent Pos:* Mem, Off Sci Res & Develop, 44; res assoc lake & stream surv, State Dept Conserv, Ind, 46-53; Ciba grant, 59; NSF grant & sci fac fel, 59; coordr, mem & dep convenor, Int Biol Prog Biol Basis of Freshwater Fish Prod, Nat Res Coun, Eng, 66; mem adv panel environ biol, NSF, 66-68; Ariz State Univ rep founding insts, Inst Ecol, 64-78; staff mem, jury panel for limnetic prize, inst limnol, Czech Acad Sci & Dept Agron, France, 72; mem Ecol Adv Comt, Sci Adv Bd, Environ Protection Agency, 75-78; consult, Commonwealth Edison, Chicago, 75-78 & Elec Power Res Inst, 84 & 85; vis scientist biol, Euratom Ispra, Italy, 74-75 & Dept Zool, Univ Cape Town, SAfrica, 82; consult, Lake Ohrid Yugoslavia Proj, Smithsonian Inst, 75-78; partic, Symp Lake Metab/Lake Mgt, Univ Uppsala, Sweden, 77; assoc ed, Environ Biol Fishes, J Fis Biol, Marine Ecol Progress Series. *Honors & Awards:* Mercer Award, Ecol Soc Am, 55; Silver Medal, Am Fisheries Soc; Hickman Lectr, DePauw Univ, 88. *Mem:* Fel AAAS; fel Am Inst Fishery Res Biol; Am Soc Zool; Am Fisheries Soc (vpres, 82-86, pres, 87-88); Ecol Soc Am (treas, 69-71); Int Soc Limnol; Am Soc Ichthy & Herp; Am Soc Limnol & Oceanog; Int Soc Limnol; Sigma Xi. *Res:* Fish populations in lakes and streams; fish nutrition; fish production; temperature tolerance; fish reproduction and stress. *Mailing Add:* Dept Zool Ariz State Univ Tempe AZ 85287-1501

GERLACH, A(LBERT) A(UGUST), b Columbus, Ohio, May 22, 20; m 43; c 5. APPLIED MATHEMATICS. *Educ:* Ohio State Univ, BS, 42; Ill Inst Technol, MS, 48 & 50, PhD, 58. *Prof Exp:* Sr engr & asst to chief engr, Rowe Eng Corp, 46-48; sr engr, Motorola, Inc, 48; res engr, Armour Res Found, Ill Inst Technol, 48-53; exec engr & mgr, Res Sect, Cook Res Labs Div, Cook Elec Co, 53-61, from asst dir labs to assoc dir labs, Tech-Ctr Div, 61-69, dir res, 69-70; dir res, 70-71, head signal processing br, 71-85, SUPVRY RES PHYSICIST, US NAVAL RES LAB, WASHINGTON, DC, 85- *Mem:* Inst Elec & Electronics Engrs; Acoust Soc Am. *Res:* Circuit and network theory; theory of modulation; integral transforms; information theory and computers; electronic instrumentation; underwater acoustics and signal processing. *Mailing Add:* 123 Quay St Alexandria VA 22314

GERLACH, EBERHARD, b Berlin, Ger, Mar 10, 34. MATHEMATICS. *Educ:* Ind Univ, AM, 59; Univ Kans, PhD(math), 64. *Prof Exp:* Res asst math, Univ Kans, 62-64; asst prof, Univ BC, 64-71; ASSOC ED, MATH REVIEWS, UNIV MICH, ANN ARBOR, 71- *Concurrent Pos:* Fel, Univ Edinburgh, 69-70. *Mem:* Am Math Soc; Math Asn Am; Can Math Cong; Sigma Xi. *Res:* Functional analysis, in particular, Hilbert spaces, linear operators and applications to differential problems. *Mailing Add:* 904 Bruce St Ann Arbor MI 48103

GERLACH, EDWARD RUDOLPH, b McConnelsville, Ohio, June 27, 31; m 51; c 5. ANALYTICAL CHEMISTRY. *Educ:* Muskingum Col, BS, 56; Ohio State Univ, MS, 62; Walden Univ, PhD, 81. *Prof Exp:* Fac mem chem, biol & math, Batavia, Ohio schs, 56-57; chmn dept, 77-80, MEM FAC CHEM, MUSKINGUM COL, 57- *Concurrent Pos:* Res grants, Ohio Acad Sci, 65-66 & 68-69; NSF grants, 73-79; partic dept energy, Citizen's Energy Workshops, 77-79; consult hazardous chem safety, 80- *Mem:* Am Chem Soc; Coblentz Soc; Nat Sci Teachers Asn. *Res:* Trace elements in strip mine soils; zinc and copper in body fluids; laboratory and home air pollution. *Mailing Add:* Dept of Chem Muskingum Col New Concord OH 43762

GERLACH, HOWARD G, JR, b Cheektowaga, NY, Nov 30, 40; m 64; c 2. CLINICAL CHEMISTRY. *Educ:* Cleveland State Univ, BS, 63; Case Western Reserve Univ, PhD(org chem), 66. *Prof Exp:* Res chemist, Exp Sta, 66-73, nat training supvr, 73-78, training mgr, Automatic Clin Anal Div, 78-80, TECH SUPPORT MGR, PROD MGT, MED PROD DEPT, E I DU PONT DE NEMOURS & CO, INC, WILMINGTON, 81- *Mem:* Am Chem Soc; Royal Soc Chem; Am Soc Training & Develop; Am Asn Clin Chem. *Res:* Identification and synthesis of alkaloids; photochromic agents; photopolymerization systems; non-silver photographic systems; international sales & divestitures of businesses. *Mailing Add:* 2603 Pennington Dr Brandywood DE 19810

GERLACH, JOHN LOUIS, b Quincy, Ill; c 3. DEVELOPED & IMPROVED ANALYTICAL CHEMICAL METHODS, DEVELOPED CORROSION TESTS. *Educ:* Univ Ill, BS, 51; Univ Iowa, MS, 56; Univ Cincinnati, MS, 67. *Prof Exp:* Analytical chemist, Corn Prod Co, 51-53; sr res chemist & proj leader, Nalco Chem Co, 57-66; MGR RES & DEVELOP, NCH CORP, 67- *Mem:* Am Chem Soc; Nat Asn Corrosion Engrs; Am Water Works Asn. *Res:* Innovate and develop products to prevent scale and corrosion in boilers, cooling tower-condenser systems, chill water, and automotive cooling systems; develop new products for the clarification of water and industrial wastes; develop new inhibited acid cleaners; holder of three United States patents. *Mailing Add:* 1011 Sam Hill Irving TX 75062

GERLACH, JOHN NORMAN, b Portland, Ore, Aug 24, 47. HYDROMETALLURGY. *Educ:* Portland State Univ, BS, 69; Univ Calif, Riverside, PhD(chem), 74. *Prof Exp:* Phys scientist, US Army, 70-71; sr chemist, Kennecott Copper Corp, 74-78; res chemist, Phelps Dodge Corp, 79-87; LEAD DESIGN CHEMIST, CHEM DIV, 87- *Mem:* Am Chem Soc; AAAS; Metall Soc, Am Inst Mining, Metall & Petrol Engrs. *Res:* Copper, silver, gold, nickel, cobalt, vanadium and molybdenum extractive metallurgy; solvent extraction, electrochemical processing, leaching, mineral flotation, smelter pollution control, in-situ mining and engineering economics; facilities of all scales, from laboratories to commercial plants. *Mailing Add:* Unocal Corp 1511 E Orangethorpe Ave Fullerton CA 92631-5204

GERLACH, ROBERT LOUIS, b Guthrie, Okla, Nov 16, 40; m 62; c 4. ELECTRON OPTICS, SURFACE SCIENCE. *Educ:* Northwestern Univ, BS, 64; Cornell Univ, PhD(appl physics), 69. *Prof Exp:* Mem tech staff surface anal, Sandia Corp, 68-73; sr engr instrument develop, Varian Assoc, 73-74; proj scientist, 74-85, DIR, RES & DEVELOP, PHYS ELECTRONICS DIV, PERKIN-ELMER CORP, 85- *Honors & Awards:* IR 100 Award, Indust Res Mag, 80 & 81. *Mem:* Am Phys Soc; Am Vacuum Soc; Electron Micros Soc Am. *Res:* Development of surface analysis instruments such as the scanning Auger microprobe, x-ray photoelectron spectrometer and secondary ion spectrometer. *Mailing Add:* Phys Elec Ind Inc 6509 Flying Cloud Dr Eden Prairie MN 55344

GERLACH, TERRENCE MELVIN, b New London, Wis. GEOLOGY, GEOCHEMISTRY. *Educ:* Univ Wis, BS, 64, MS, 67; Univ Ariz, PhD(geol), 74. *Prof Exp:* Field geologist, Oliver Mining Co, 65; consult geochem, Anaconda Corp, 73-74; res geologist & geochemist, Sandia Labs, 76-89; RES GEOLOGIST & GEOCHEMIST, US GEOL SURVEY, 89- *Mem:* Am Geophys Union; Geol Soc Am. *Res:* Chemical thermodynamics of igneous processes. *Mailing Add:* US Geol Surv Box 25046 DFC MS 903 Denver CO 80225

GERLOFF, GERALD CARL, b Aurora, Nebr, Jan 26, 20; m 49, 61; c 2. BOTANY. *Educ:* Univ Nebr, BS, 41; Univ Wis, PhD(soils), 48. *Prof Exp:* Proj assoc, 48-49, from asst prof to prof, 49-86, EMER PROF BOT, UNIV WIS-MADISON, 86- *Mem:* AAAS; Am Soc Plant Physiol. *Res:* Mineral nutrition of plants, plant physiology and nutritional ecology. *Mailing Add:* Dept Bot Birge Hall B119 Univ Wis-Madison Madison WI 53706

GERLT, JOHN ALAN, b Sycamore, Ill, July 28, 47; m 86. BIOCHEMISTRY. *Educ:* Mich State Univ, BS, 69; Harvard Univ, AM, 70, PhD(biochem), 74. *Prof Exp:* asst prof chem, Yale Univ, 75-81, assoc prof chem, 81-84; PROF CHEM & BIOCHEM, UNIV MARYLAND, 84- *Concurrent Pos:* Jane Coffin Childs Mem Fund fel med res, 74-75; Career Development Award, NIH, 78-83; Alfred Sloan fel, 81-85. *Mem:* Am chem Soc; Am Soc Biochem & Molecular Biol; Protein Soc. *Res:* Mutagenic and structural studies of the mechanism of the Staphylococcal nuclease reaction; mutagenic and mechanistic studies of the enzymes of mandelate metabolism; enzymology of dark DNA repair. *Mailing Add:* Dept Chem & Biochem Univ Maryland College Park MD 20742

GERMAIN, RONALD N, b New York, NY, Oct 29, 48. IMMUNOLOGY. *Educ:* Brown Univ, ScB & ScM, 70; Harvard Univ, PhD & MD, 76. *Prof Exp:* From instr to assoc prof path, Harvard Med Sch, 76-82; sr investr, 82-87, CHIEF, LYMPHOCYTE BIOL SECT, LAB IMMUNOL, NAT INST ALLERGY & INFECTIOUS DIS, NIH, 87- *Concurrent Pos:* Milton Fund Award, 77-78; assoc ed, J Imunol & Jour Reticuloendothial Sos, 80-84, J Molecular & Cellular Immunol, 81-83; scholar grant, Am Cancer Soc, 81-82; guest investr, Lab Molecular Genetics, Nat Inst Child Health & Human Develop, NIH, 81-82; mem, Adv Comt Clin Invest Immunol & Immunother, Am Cancer Soc, 85-87 & Clin Rev Subpanel, Nat Inst Allergy & Infectious Dis, NIH, 87-88; dep ed, J Immunol, 87- *Mem:* Am Asn Immunologists; Sigma Xi. *Mailing Add:* Lymphocyte Biol Sect Lab Immunol NIAID NIH Bldg 10 Room 11N311 Bethesda MD 20892

GERMAN, DWIGHT CHARLES, b Elmhurst, Ill, May 28, 44; m; c 1. NEUROSCIENCE. *Educ:* Southern Methodist Univ, BA, 66; Univ Okla, MS, 67, Health Sci Ctr, PhD(biol psychol), 72. *Prof Exp:* Fel neurophysiol, dept physiol, Univ Wash Med Sch, 72-75; asst prof, dept physiol, 75-79, asst prof, dept physiol & psychiat, 76-79, ASSOC PROF, DEPTS PHYSIOL & PSYCHIAT, UNIV TEX HEALTH SCI CTR, DALLAS, 79- *Concurrent Pos:* Prin investr, NIMH res grants, 76-; NIH res grant, 83-; res affil, Regional Primate Res Ctr, Med Sch, Univ Wash, 84; ed adv bd, Life Sciences, 85- *Mem:* Soc Neurosci; AAAS; Sigma Xi; Int Brain Res Orgn. *Res:* Structure and function of brain catecholamine-containing neurons; role of catecholamine containing neurons in normal and pathological behavior; neuropathology of Parkinsons disease and Alzheimers disease. *Mailing Add:* 5931 Encore Dr Univ Tex Health Sci Ctr Dallas TX 75240

GERMAN, JAMES LAFAYETTE, III, b Grayson County, Tex, Jan 2, 26; m 56; c 2. HUMAN CYTOGENETICS. *Educ:* La Polytech Inst, BS, 45; Southwestern Med Col, MD, 49; Am Bd Internal Med, dipl, 58. *Prof Exp:* Intern, Cook County Hosp, Chicago, Ill, 49-51; res physician internal med, Vet Admin Hosp, McKinney, Tex, 52-55; clin assoc, NIH, 56-58; res assoc & asst physician, Rockefeller Inst, 58-62, asst prof & assoc physician, 63; assoc prof anat & pediat, Med Col, Cornell Univ, 65-75; SR INVESTR & DIR HUMAN GENETICS, NY BLOOD CTR, 68- *Concurrent Pos:* Mem, Corp Marine Biol Lab, Woods Hole, Mass; clin prof pediat, Cornell Univ Med Col, 75-; adj prof, Rockefeller Univ, 75-86; consult genetics, Dept Path, Mem Hosp Cancer & Allied Disorders, NY, 76-; prof genetics, Cornell Univ Grad Sch Med Sci, 63-; vis prof, Univ Geneva, 82. *Mem:* Am Soc Human Genetics; Europ Soc Human Genetics; Genetics Soc Am; Am Soc Cell Biol; Sigma Xi; Harvey Soc; Am Soc Clin Invest; fel Japan Soc Promotion Sci. *Res:* Genetics of human cancer; human genetics. *Mailing Add:* NY Blood Ctr 310 E 67th St New York NY 10021

GERMAN, VICTOR FREDERICK, US citizen. PEDIATRICS. *Educ:* Univ Richmond, BS 58; Univ Ill, PhD(org chem), 63. *Hon Degrees:* MD, Univ Chicago, 75. *Prof Exp:* NIH fel, Univ Calif, Berkeley, 63-64; staff mem, Robins rep, Math & Sci Ctr, Richmond Pub Sch, Va Commonwealth Univ, 64-72; intern & resident, Duke Univ, 75-77, asst prof, Div Allergy, 77-81; asst prof, 81-86, ASSOC PROF, DEPT PEDIAT, UNIV TEX HEALTH SCI CTR, SAN ANTONIO, 86- *Concurrent Pos:* Sr res chemist, A H Robins, 64-72; assoc staff physician, Infant Transport Serv, Univ Calif, 77-79 & Berkeley Health Dept, 80; emergency physician, Med Ctr, Children's Hosp, Oakland Calif, 77-81; fel cardiovasc res & perinatal & pulmonary med, Univ Calif, San Francisco, 77-81; physician, Berkeley Pub Health, 79-81; assoc staff mem, Dept Pediat, Kaiser Permanent Med Ctr, 80-81. *Mem:* Sigma Xi; Am Physiol Soc; Am Soc Microbiol; Am Thoracic Soc; Am Acad Pediat. *Res:* Mechanisms of gram bacterial adherence to respiratory tract epithelia; alveolar marcophage function (1) in vivo (2) in vitro culture; airway secretions; biosynthesis and release; tracheal submucosal gland physiology gram negative. *Mailing Add:* Dept Pediat Univ Tex Health Sci Ctr 7703 Floyd Curl Dr San Antonio TX 78284

GERMANE, GEOFFREY JAMES, b Cleveland Heights, Ohio, July 3, 50; m 76; c 5. MECHANICAL ENGINEERING. *Educ:* Rose Hulman Inst Technol, BS, 72, MS, 75; Brigham Young Univ, PhD(mech eng), 78. *Prof Exp:* Asst prof, 76-79, ASSOC PROF MECH ENG, BRIGHAM YOUNG UNIV, 79- *Concurrent Pos:* Prin investr, Brigham Young Univ, Utah Power & Light, 79-, US Dept Energy, 80-; consult, Hercules, Inc, 80-81, Utah Power & Light Co, 81-, Collision Safety Eng, 81-, Carvern Petrochem, 80-, UHI Corp, 80- *Honors & Awards:* Teetor Award, Soc Automotive Engrs, 81. *Mem:* Soc Automotive Engrs; Sigma Xi. *Res:* Internal combustion engine computer control; alcohol fuels and lean limit combustion; coal-fired power plant fires and explosions prevention; basic combustion and pollutant formation studies of coal-water mixtures. *Mailing Add:* Brigham Young Univ Dept Mech Eng CB 242 Provo UT 84602

GERMANN, ALBERT FREDERICK OTTOMAR, II, b Cleveland, Ohio, Jan 4, 29; m 54; c 3. ANIMAL NUTRITION. *Educ:* Purdue Univ, BS, 51, MS, 56, PhD, 58. *Prof Exp:* Vpres, 58-77, PRES NUTRIT RES ASSOCS, INC, 77- *Mem:* AAAS; Am Soc Animal Sci. *Res:* Relationship of lysine requirement to protein level of weanling swine; nutrient requirements of the chinchilla and rabbit. *Mailing Add:* 307 W Columbia St South Whitley IN 46787

GERMANN, RICHARD P(AUL), b Ithaca, NY, Apr 3, 18; m 42; c 1. ORGANIC CHEMISTRY, APPLICATION RESEARCH. *Educ:* Univ Colo, BA, 39. *Hon Degrees:* PhD, Hamilton State Univ, 73. *Prof Exp:* Chief anal chemist, Taylor Refining Co, Tex, 43-44; res develop chemist, Alrose Chem Co Div, Geigy Chem Corp, 52-55; new prod develop chemist, Res Div, W R Grace & Co, Md, 55-60; chief chemist, Soap & Cosmetic Div, G H Packwood Mfg Co, 60-61; coordr chem prod develop, Abbott Labs, Ill, 61-71; consult chemist, 71-72; pres, Germann Int Ltd, 73-82; PRES, RAMTEK INT LTD, 73- *Concurrent Pos:* Consult, Chem Dept, Bowling Green State Univ, 88-89. *Mem:* fel AAAS; Am Chem Soc; Am Pharmaceut Asn; Am Asn Textile Chemists & Colorists; Chem Soc London; Com Develop Asn; Chem Mkt Res Asn; Sigma Xi. *Res:* Vitamin nutrition; trace element use in growth of agricultural crops; biocides; pollution control; organic, pharmaceutical, polymer, petroleum, leather, textile, agricultural, dye, paint, detergent, insecticides and biocides, by-product recovery. *Mailing Add:* Six Vinewood Dr Norwalk OH 44857

GERMANY, ARCHIE HERMAN, b Dixon, Miss, Nov 18, 17; m 43; c 3. ORGANIC CHEMISTRY. *Educ:* Miss Col, BA, 39; Univ NC, PhD(org chem), 43. *Prof Exp:* Lab asst chem, NC, 39-41; res assoc metall lab, Chicago, 43-44; chemist, Clinton Labs, Oak Ridge, 44-46; from assoc prof chem to prof, Miss Col, 46-60, head dept, 60-61, chmn div sci & math, 61-; RETIRED. *Res:* Synthetic organic chemistry. *Mailing Add:* 803 E Leak Clinton MS 39056

GERMESHAUSEN, KENNETH J(OSEPH), electronics; deceased, see previous edition for last biography

GERMINARIO, RALPH JOSEPH, b Jersey City, NJ, Oct 27, 43; m 76; c 2. BIOLOGICAL CHEMISTRY. *Educ:* Seton Hall Univ, BA, 65, MS, 67; Univ NDak, PhD(microbiol), 70. *Prof Exp:* Fel cell genetics, 70-74, STAFF INVESTR, CELL BIOL LAB, LADY DAVIS INST MED RES, 74- *Concurrent Pos:* Instr, Dept Biol, Concordia Univ, 75-79; assoc mem, Dept Med, McGill Univ, 78-89; adj prof, Dept Biol, Concordia Univ, 79-, Dept Med, McGill Univ, 90- *Honors & Awards:* Can Asn Geront. *Mem:* Am Soc Microbiol; Am Soc Cell Biol; NY Acad Sci; Tissue Cult Asn; Can Soc Clin Invest; Cell Cycle Soc; Can Asn Geront, Chmn Div Biol Sci. *Res:* Insulin action genetic control; regulation of sugar transport; modulation of insulin action in cultured cells; diabetes and aging. *Mailing Add:* Lady Davis Inst for Med Res 3755 Cote St Catherine Rd Montreal PQ H3T 1E2 Can

GERMINO, FELIX JOSEPH, b New York, NY, July 14, 30; m 52; c 7. FOOD CHEMISTRY. *Educ:* Fordham Univ, BS, 52; Univ of Chicago, MBA, 74. *Prof Exp:* Food inspector, Food Inspection Serv, USDA, NY, 52; asst chemist, Gen Foods Tech Ctr, 54-59; assoc chemist, Morehead Patterson Res Ctr, Am Mach & Foundry Co, 59-64; proj leader carbohydrate chem, Corn Prod Co, Argo, 64-71, dir, com dev, CPC Int, Inc, 64-72, sect leader foods & paper textiles, 65-71, asst mgr, 71-72; assoc dir, Quaker Oats Co, 72-74, dir pet foods, 74-76, vpres res & develop pet foods, 76-78, vpres human foods, 78-82; PRES, F GERMINO & ASSOCS INC, 83- *Mem:* Am Chem Soc; Inst Food Technol; Indust Res Inst; Coun Foreign Relations; Am Asn Cereal Chem. *Res:* Physical structure of starch; synthesis of polysaccharide derivatives for use in textiles, paper and food related fields; physical chemistry of starch and starch fractions; environmental sciences; food sugar. *Mailing Add:* 12414 83rd Ave Palos Park IL 60464

GERMROTH, TED CALVIN, b Latrobe, Pa, Feb 21, 52; m 75; c 2. CELLULOSE & WOOD CHEMISTRY. *Educ:* Col William & Mary, Va, BS, 74; Univ Calif, Berkeley, MS, 76, PhD(org chem), 79. *Prof Exp:* SR RES CHEMIST, EASTMAN CHEM DIV, EASTMAN KODAK, 79- *Mem:* Am Chem Soc. *Res:* Derivativization of cellulose and wood products to prepare commercially useful products such as cellulose esters and ethers; wood; pulping; synthetic organic and natural product chemistry. *Mailing Add:* 4538 Grace Dr Kingsport TN 37664

GERNANT, ROBERT EVERETT, b Geneseo, Ill, Dec 3, 41; m 67; c 2. PALEOECOLOGY OF MICROFOSSILS, PALEOENVIRONMENTS OF SEDIMENTARY ROCKS. *Educ:* Univ Ill, BS, 63; Univ Mich, MS, 65, PhD(geol), 69. *Prof Exp:* Geologist, Shell Oil Co, 63, Humble Oil & Refining Co, 65; chmn dept geol, 76-80, ASSOC PROF GEOL, UNIV WIS-MILWAUKEE, 68-, DIR, CTR IMPROV INSTR, 83- *Concurrent Pos:* Field dir, Md Acad Sci, 69-70; ed consult, Kendall Hunt Publ Co, 79-88. *Mem:* Sigma Xi; Paleont Soc; Geol Soc Am. *Res:* Paleoecology of marine protist and invertebrate fossils; freshwater Ostracoda; general geology of Colorado Plateu and vicinity; evolutionary patterns of fossil Ostracoda. *Mailing Add:* Dept Geol & Geophys Sci Univ Wis Milwaukee WI 53201

GERNER, EUGENE WILLARD, b Sheboygan, Wis, Aug 8, 47; m 82; c 4. CELL BIOLOGY, RADIOBIOLOGY. *Educ:* Univ Wis-Madison, BA, 69, MS, 70; Univ Tex Grad Sch Biomed Sci, Houston, PhD(biophys), 74. *Prof Exp:* Comput programmer, High Energy Physics Dept, Univ Wis-Madison, 67-69, med physicist, Radiol Dept, 69-71; fel biophys, Univ Tex Grad Sch Biomed Sci, Houston, 71-74; from asst prof to assoc prof, 74-83, PROF, DEPT RADIATION ONCOL & BIOCHEM, HEALTH SCI CTR, UNIV ARIZ, 83- *Honors & Awards:* Res Award, Radiation Res Soc, 88. *Mem:* Radiation Res Soc; Cell Biol Soc; Am Soc Biol Chemists; Am Asn Cancer Res; Am Soc Therapeut Radiol & Oncol. *Res:* Cellular and molecular radiation biology, emphasizing studies on the biochemical controls of mammalian cell cycle kinetics and the effects of anticancer agents, such as radiations, hyperthermia and drugs on cell proliferation. *Mailing Add:* Radiation Oncol Dept Health Sci Ctr Univ Ariz 1501 N Campbell Ave Tucson AZ 85724

GERNS, FRED RUDOLPH, b Mannheim, Ger, Nov 28, 25; nat US; m 56; c 2. POLYMER CHEMISTRY. *Educ:* Ohio State Univ, BA, 48; Syracuse Univ, MS, 51; Univ Va, PhD(chem), 59. *Prof Exp:* Jr med chemist, Smith, Kline & French Labs, 52-56; sr org chemist, Burroughs Wellcome & Co, Inc, 59-63; chemist, Chas Pfizer & Co, Conn, 63-65; sr res chemist, 65-82, RES ASSOC, GREAT LAKES CHEM CORP, 82- *Mem:* Am Chem Soc. *Res:* Nitrogen and sulfur heterocycles; halogen chemistry; polymers; photoresists. *Mailing Add:* 248 Connolly St West Lafayette IN 47906-2724

GERO, ALEXANDER, b Budapest, Hungary, May 26, 07; nat US; m 35; c 2. CHEMISTRY. *Educ:* Univ Vienna, PhD(chem), 30. *Prof Exp:* Asst org chem, Inst Tech, Berlin, 32-34; res chemist, Syngala, Inc, Vienna, 35-36, Dr Wander, Inc, Budapest, 36-38, Dr Roussel Labs, Paris, 38-39 & Silva-Araujo-Roussel Labs, Rio de Janeiro, 41-42; chief chemist, Berkeley Chem Corp, NJ, 42-45; chem dir, Nat Foam System & Purocaine, Inc, Pa, 45-47; from asst prof to assoc prof chem, Villanova Col, 47-52; from assoc prof to prof, 52-82, EMER PROF PHARMACOL, HAHNEMANN MED COL, 82- *Concurrent Pos:* Fulbright prof, Valladolid & Oviedo, 62-63. *Mem:* Int Narcotics Res Conf; Am Soc Pharmacol & Exp Therapeut. *Res:* Mechanisms of drug action; opiates; receptor theory. *Mailing Add:* 143 Kendal Kennett Square PA 19348

GEROCH, ROBERT PAUL, b Akron, Ohio, June 1, 42. THEORETICAL PHYSICS. *Educ:* Mass Inst Technol, BS, 63; Princeton Univ, PhD(physics), 67. *Prof Exp:* Air Force Off Sci Res fel, Birkbeck Col, 67-68; NSF fel, 68-69; fel physics, Syracuse Univ, 69-70; assoc prof, Univ Tex, Austin, 70-71; assoc prof, 71-75, PROF PHYSICS & MATH, ENRICO FERMI INST, UNIV CHICAGO, 75- *Res:* General relativity. *Mailing Add:* Dept Physics-Math RI 365 Univ Chicago 5801 Ellis Ave Chicago IL 60637

GEROLA, HUMBERTO CAYETANO, b Buenos Aires, Arg, Feb 7, 43; m 66; c 2. THEORETICAL ASTROPHYSICS. *Educ:* Univ Rome, PhD(physics), 69. *Prof Exp:* Asst prof astrophysics, Univ Buenos Aires, 69-72; res assoc, Univ Colo, 72-74 & NY Univ, 74-76; MEM STAFF, THOMAS J WATSON RES CTR, IBM CORP, 76- *Concurrent Pos:* Dir, Nat Ctr Cosmic Radiation, Arg, 70-71. *Honors & Awards:* Outstanding Innovation Award, IBM Corp, 78. *Mem:* Int Astron Union; Am Astron Soc. *Res:* Structure and evolution of galaxies; processes of star formation; physics of the interstellar medium. *Mailing Add:* 20390 Knollwood Dr Saratoga CA 95070

GEROLIMATOS, BARBARA, b New York, NY, July 13, 50. ENDOCRINOLOGY. *Educ:* Fordham Univ, BS, 72; Columbia Univ, MPhil, 76, PhD(biochem), 79. *Prof Exp:* Fel, dept surg, Col Physicians & Surgeons, Columbia Univ, 78, NIH fel trainee biochem & endocrinol, dept obstet & gynec, 79-83; group leader, Ayerst Int, 84-86, clin monitor, Ayerst Labs, 86-88; med writer, Ayerst Int, 86-88; mgr, Boehringer Engelheim, 88-89; ASSOC DIR, CLIN & SCI AFFAIRS, PFIZER CONSUMER HEALTH CARE, 89- *Concurrent Pos:* Mem & cofounder, Rape Crisis Intervention Prog, Columbia Med Ctr, 78-83; secy, Metrop NY Chap Asn Women Sci, 82-84, vpres, 84-85, pres, 86-89; chairperson, scholarship winners alumni asn career coun comt, IBEW Educ & Cult Fund, 85-90; mem, Women in Sci Comt, NY Acad Sci, 86- *Mem:* Asn Women Sci; NY Acad Sci. *Res:* Drug development and clinical trials in areas of consumer health care products. *Mailing Add:* 90 Park Terrace East Apt 2B New York NY 10034

GERONE, PETER JOHN, b Oakfield, NY, Apr 11, 28; m 51; c 7. VIROLOGY. *Educ:* State Univ NY, Buffalo, BA, 49, MA, 51; Johns Hopkins Univ, ScD(microbiol), 54. *Prof Exp:* Instr biol, State Univ NY, Buffalo, 50-51; supvry microbiologist, US Army Biol Labs, 54-71; DIR, DELTA REGIONAL PRIMATE RES CTR, TULANE UNIV, LA, 71-, ADJ PROF IMMUNOL & MICROBIOL, 73-, ADJ PROF TROP MED, 77- *Concurrent Pos:* Dir, Gulf South Res Inst, 78-89. *Honors & Awards:* Dept Army Meritorious Civilian Serv Award, 68. *Mem:* Int Primatol Soc; Sigma Xi; Am Soc Primatologists; Am Soc Microbiol. *Res:* Infectious diseases in nonhuman primates. *Mailing Add:* Delta Regional Primate Res Ctr Tulane Univ Covington LA 70433

GEROW, CLARE WILLIAM, b Detroit, Mich, Oct 8, 27; m 51; c 13. ORGANIC POLYMER CHEMISTRY. *Educ:* Univ Detroit, BS, 51; Iowa State Univ, PhD(org chem), 56. *Prof Exp:* Res polymer chemist, Film Dept, E I du Pont de Nemours & Co, 56-61, tech mkt eval investr, 61-63, res polymer chemist, 63-67, staff scientist, Yerkes Res & Develop Lab, 67-71, staff scientist, Spruance Film Plant, 71-76, res assoc, Spruance Polymer Prod & Res Plant, 76-81, res assoc, 81-84, sr res assoc, Spruance Fibers Plant, 84-90; RETIRED. *Concurrent Pos:* Mem adj fac, Va Commonwealth Univ, 71-89. *Mem:* Am Chem Soc. *Res:* Organogermanium and organosilicon chemistry; polymer chemistry, including polyolefins, polythio ketones and aldehydes and polyimides; modification, formulation and application of barrier coatings to cellulosic and polyester films; polyaramid chemistry. *Mailing Add:* RTE 640 Waterview VA 23180

GERPHEIDE, JOHN H, b Manitowoc, Wis, Sept 17, 25; m 51; c 3. MECHANICAL ENGINEERING, AERONAUTICS. *Educ:* Calif Inst Tech, BS, 45, MS, 48. *Prof Exp:* Engr, US Naval Ord Test Sta, Inyokern, 46-47; engr, Calif Inst Technol, 48-53, proj engr, 53-59, flight test engr, 54-55, sect mgr spacecraft develop, 60-62, consult Apollo support, 62-63, staff specialist proj planning, 64, proj mgr develop Mars Landing Craft, 67-68, sect mgr syst design & integration, 64-73, proj mgr solar elec propulsion, 73-74, Sea Satellite syst mgr, Jet Propulsion Lab, 74-79, chief engr, flight projs, 79-80, PROJ MGR, VENUS ORBITING IMAGING RADAR, JET PROPULSION LAB, CALIF INST TECHNOL, 80-; PROJ MGR, MAGELLAN. *Mem:* Sigma Xi; Planetary Soc. *Res:* Development of advanced designs, new technologies and system approaches to exploration of space. *Mailing Add:* 2165 Queensberry Rd Pasadena CA 91104

GERRARD, JOHN WATSON, b Kasenga, Rhodesia, Apr 14, 16; m 41; c 3. PEDIATRICS. *Educ:* Oxford Univ, BA, 38, BM, BCh, 41, DM, 51; FRCP(C), 56; FRCP, 68. *Prof Exp:* Lectr, Univ Birmingham, 48-51, chief asst, 51-55; head, Dept Pediat, Univ Sask, 55-71, prof, 55-83, emer prof pediat, Univ Hosp, 83-87; RETIRED. *Honors & Awards:* John Scott Award, Philadelphia City Trusts, 62; Enuresis Found Award, 68; Ross Award, Can Pediat Soc, 85. *Mem:* Can Pediat Soc; Am Acad Pediat; Am Acad Allergy; Am Col Allergists; Soc Clin Ecol. *Res:* Gastrointestinal and genito-urinary allergies. *Mailing Add:* Dept Pediat Univ Sask Hosp Saskatoon SK S7N 0X0 Can

GERRARD, JONATHAN M, b Oct 13, 47; Can citizen. HEMOSTASIS, THROMBOSIS. *Educ:* Univ Sask, BA, 67; McGill Univ, Montreal, MMCM, 71; Univ Minn, PhD(med), 78; Am Acad Pediat, cert, 76; FRCP(C), 82. *Prof Exp:* Intern pediat, Univ Minn, 71-72, resident, 72-73, fel, 73-76, from instr to asst prof, 76-80; from asst prof to assoc prof pediat, 80-87, DIR, RES & CLIN INVEST, DEPT PEDIAT & CHILD HEALTH, UNIV MAN, WINNIPEG, 84-, HEAD, SECT PEDIAT HEMAT, 85-, PROF PEDIAT, 87- *Concurrent Pos:* Estab investr, Am Heart Asn, 78; adj prof zool, Univ Man, Winnipeg, 85- *Honors & Awards:* Scientist Award, Med Res Coun Can, 85. *Mem:* Soc Pediat Res; Am Pediat Soc; Am Soc Hemat; Int Soc Thrombosis & Hemostasis; NY Acad Sci. *Res:* Blood platelet and a role in bleeding and clotting disorders; basic mechanism of platelet cell biology; applied studies of patients with bleeding disorders; published papers on other blood cells including neutrophils and natural killer cells; bald eagles; published papers and books on bald eagles. *Mailing Add:* Box 113 RR 1 Headingly MB R0H 0J0 Can

GERRARD, THERESA LEE, MONOCYTES, ANTIGEN PRESENTATION. *Educ:* Med Col Va, PhD(microbiol & immunol), 80. *Prof Exp:* SR STAFF FEL, US FOOD & DRUG ADMIN, 84- *Mailing Add:* Div Virol Bldg 29A Rm 2A- 21 US Food & Drug Admin 8800 Rockville Pike Bethesda MD 20892

GERRARD, THOMAS AQUINAS, b La Crosse, Wis, Feb 7, 33; m 58; c 3. GEOLOGY. *Educ:* Univ Cincinnati, BS, 56; Miami Univ, Ohio, MS, 59; Univ Ariz, PhD(geol), 64. *Prof Exp:* Explor geologist, Chevron Oil Co, 64-66; from asst prof to assoc prof, 66-76, PROF GEOL, WITTENBURG UNIV, 76-CHMN DEPT, 73- *Concurrent Pos:* Lectr, Tulane Univ, 65-66; geol consult, Minerals Dept, Exxon Co, USA, 69-79. *Mem:* Am Asn Petrol Geol; Soc Econ Paleont & Mineral; Nat Asn Geol Teachers; Sigma Xi; Am Inst Prof Geologists. *Res:* Sedimentary petrology and geochemistry; regional stratigraphy; marine geology. *Mailing Add:* Dept Geol Wittenberg Univ Springfield OH 45501

GERRATH, JOSEPH FREDRICK, b Saskatoon, Sask, June 25, 36; m 66; c 2. BOTANY, PHYCOLOGY. *Educ:* Univ BC, BA, 59, BSc, 63, MSc, 65, PhD(bot), 68. *Prof Exp:* Asst prof, 68-80, ASSOC PROF BOT, UNIV GUELPH, 80- *Mem:* Can Bot Asn (treas, 75-77); Int Phycol Soc; Phycol Soc Am; Brit Phycol Soc; Sigma Xi. *Res:* Taxonomy and cytology of desmids. *Mailing Add:* Dept Bot Univ Guelph Guelph ON N1G 2W1 Can

GERRAUGHTY, ROBERT JOSEPH, b Newton, Mass, Aug 30, 28; m 53; c 5. PHARMACY. *Educ:* Mass Col Pharm, BS, 50, MS, 52; Univ Conn, PhD(pharm), 59. *Prof Exp:* Asst pharm, Univ Conn, 55-58; asst prof, Rutgers Univ, 58-60; from assoc prof to prof, Univ RI, 60-72; dean, Sch Pharm, Creighton Univ, 72-77, assoc vpres health sci, 77-79, vpres admin, 79-90; RETIRED. *Concurrent Pos:* consult, US Food & Drug Admin, various Pharmaceut Co & AID. *Mem:* Am Pharmaceut Asn; Am Asn Col Pharm. *Res:* Ultrasonics; pharmaceutical formulation; educational research on learning modules; drug inspection. *Mailing Add:* Dept Pharm Creighton Univ Omaha NE 68178

GERRING, IRVING, b Bridgeport, Conn, Apr 16, 09; m 44; c 3. ENVIRONMENTAL SCIENCES, PUBLIC HEALTH ADMINISTRATION. *Educ:* Univ Conn, BS, 31; Columbia Univ, MSPH, 35. *Prof Exp:* Teacher pub sch, Conn, 32; dir, Div Environ Health, Bridgeport, 35-39; assoc pub health engr, USPHS, 41-43, sanitarian, 44-46, exec secy, Pub Health & Med Res Admin, 47-56, exec secy, Parasitol, Radiation & Pop Res Study Sects, 56-71, health scientist adminr, Div Res Grants, NIH, 71-73; RETIRED. *Concurrent Pos:* Spec asst, War Assets Admin, 47. *Mem:* Fel Am Pub Health Asn. *Res:* Bacteriology; public health; air and water pollution; occupational health; food technology. *Mailing Add:* 7 Pinecrest Ct Greenbelt MD 20770

GERRISH, JAMES RAMSAY, b Centralia, Ill, Feb 27, 56; m 78; c 3. GRASSLAND ECOLOGY. *Educ:* Univ Ill, BS, 78; Univ Ky, MS, 81. *Prof Exp:* Res assoc, 81-84, res assoc & supt, 84-87, RES ASST PROF & SUPT, FORAGE SYST RES CTR, UNIV MO, COLUMBIA, 88- *Concurrent Pos:* Mem, Am Forage & Grassland Coun. *Mem:* Am Soc Agron. *Res:* Systems analysis of grassland ecosystems grazed by domestic ruminants; plant community dynamics as impacted by grazing behavior. *Mailing Add:* UMC-Forage Syst Res Ctr Rte 1 Box 80 Linneus MO 64653

GERRITSEN, ALEXANDER NICOLAAS, b The Hague, Netherlands, Nov 29, 13; m 43; c 2. METAL PHYSICS, LOW TEMPERATURE PHYSICS. *Educ:* State Univ, Leiden, Drs, 37, Dr, 48. *Prof Exp:* Govt asst physics, State Univ, Leiden, 37, head & instr physics lab, 43, scientist, Found Fundamental Res Matter, 47, sr scientist, 47-56; vis prof, 54, assoc prof, 56, prof, 60, EMER PROF PHYSICS, PURDUE UNIV, 79- *Mem:* Fel Am Phys Soc; Netherlands Asn Sci Invest (secy, 48-53). *Res:* Transport phenomena in metals; low temperatures. *Mailing Add:* 100 Wheeler Lane West Lafayette IN 47906

GERRITSEN, FRANCISCUS, b Dordrecht, Netherlands, Apr 23, 23; m 48; c 4. COASTAL & OCEAN ENGINEERING. *Educ:* Delft Univ Technol, MS, 50, Univ Trondheim, Norway, Dr Tech, 81. *Prof Exp:* Hydraul engr, Delft Univ Technol, 48-51; res engr, Royal Netherlands Bd Rd & Waterways, 51-56; from asst prof to assoc prof coastal eng, Univ Fla, 56-62; chief engr, Royal Netherlands Bd Rd & Waterways, 62-69; chmn dept ocean eng, 72-75, PROF OCEAN ENG, UNIV HAWAII, 69- *Concurrent Pos:* Consult for foreign govts & pvt pract, 56-; lectr, Int Course Hydraul Eng, 67-69; vis prof, Delft Univ Technol, 75-76. *Mem:* Netherlands Royal Inst Eng; Am Soc Civil Engrs; Int Asn Hydraul Res; Permanent Int Asn Navig Cong. *Res:* Harbor engineering, coastal processes, estuaries, construction, transportation; wave mechanics, ocean environment, air-sea interaction. *Mailing Add:* Dept Ocean Eng-Hol 401 Univ Hawaii at Manoa 2500 Campus Rd Honolulu HI 96822

GERRITSEN, GEORGE CONTANT, b Passaic, NJ, Dec 28, 26; m 50; c 2. PHYSIOLOGY. *Educ:* Hope Col, AB, 50; Mich State Univ, MS, 55, PhD(physiol), 60. *Prof Exp:* Res instr agr biochem, Mich State Univ, 50-55; sr scientist, Mead Johnson & Co, 60-62; res scientist, Diabetes Res Dept, Upjohn Co, 62-90; RETIRED. *Mem:* AAAS; Am Diabetes Asn; Soc Exp Biol & Med. *Res:* Etiology and pathogenesis of diabetes; hypoglycemic agents; metabolic interrelationships between carbohydrate and lipid metabolism; genetics of diabetic animals and prediabets. *Mailing Add:* 114 Walnut Wood Court Plainwell MI 49080

GERRITSEN, HENDRIK JURJEN, b The Hague, Netherlands, Jan 19, 27; m 72; c 4. PHYSICS, CHEMISTRY. *Educ:* State Univ Leiden, BS, 48, PhD(physics, chem), 55. *Prof Exp:* Asst prof low temperature res, State Univ Leiden, 52-55; res physicist, Res Labs, Radio Corp Am, Zurich, 55-57, res physicist & mem tech staff low temperature magnetism, 57-67; assoc prof, 67-72, PROF PHYSICS, BROWN UNIV, 72- *Concurrent Pos:* Assoc prof, Chalmers Univ Technol, Sweden, 61-62; vis prof physics, State Univ Utrecht, Neth, 74 & State Univ Karlsruhe, WGer, 81-82; Int Res & Exchanges Bd Scholar, Lithuanian Soviet Socialist Repub, 84. *Mem:* Swiss Phys Soc; Sigma Xi; Am Optical Soc; Fedn Am Scientists; Am Friends Serv Comt. *Res:* Radio astronomy; microwave properties of solids, particularly at low temperatures and in magnetic materials; gaseous electronics; optics, spectroscopy; nonlinear optics and holography; optical properties of ternary III and V semiconductor layers; plasma dynamics in semiconductors on picosecond time scale; holographic optical elements. *Mailing Add:* Dept Physics Brown Univ Providence RI 02912

GERRITSEN, JEROEN, b Leiden, Neth, Nov 6, 51; m 79; c 2. PLANKTON ECOLOGY, THEORETICAL ECOLOGY. *Educ:* Antioch Co, BS, 74; Johns Hopkins Univ, MA, 76, PhD(ecol & evolution), 78. *Prof Exp:* Res fel, dept zool, Univ Ga, 78-80, proj mgr, Okefenokee Ecosyst Invest, 81-84, res assoc & asst ecologist, Inst Ecol, Univ Ga, 80-87; SR SCIENTIST ESM OPERS, VERSAR INC, 87-, PROG MGR, 89- *Concurrent Pos:* Prin investr, NSF grant, 85-87; consult, Nat Geog Educ films, 84-85; peer reviewer, Nat Acid Precipitation Assessment Prog, 84; dir, Flow Cytometry Facil, Univ Ga, 85-87; prin investr, contracts, 87- *Mem:* Estuarine Res Fedn; Am Geophys Union; Am Soc Limnol & Oceanog; Ecol Soc Am; Atlantic Estuarine Res Soc. *Res:* Mechanisms of ecological processes; aquatic feeding ecology; environmental effects on trophic interactions; ecosystem response to environmental variability; wetland ecology; long-term ecological monitoring; acid deposition. *Mailing Add:* Versar Inc ESM Operations 9200 RumseyRd Columbia MD 21045

GERRITSEN, MARY ELLEN, b Calgary, Alta, Can, Sept 20, 53; m 81; c 3. ALLERGY & INFLAMMATION, CARDIOVASCULAR PHYSIOLOGY. *Educ:* Univ Calgary, BSc, 75, PhD(pharmacol), 78. *Prof Exp:* Teaching fel pharmacol, Univ Calif, San Diego, 78-80; from asst prof to assoc prof physiol, NY Med Col, 81-90; SR STAFF SCIENTIST, MILES INC, 90- *Honors & Awards:* Pharm Award, Microcirculatory Soc, 83, Mary Weideman Award, 84. *Mem:* Microcirculatory Soc; Am Physiol Soc; Am Soc Pharmacol & Exp Therapeut. *Res:* Hormonal regulation of microvascular endothelial cell function; roles of inflammatory mediators and arachidonic acid metabolites; glucocorticoid and insulin actions on microvascular endothelial cell metabolism; leukocyte adhesion. *Mailing Add:* Inst Arthritis & Autoimmunity Miles Inc 400 Morgan Lane West Haven CT 06516

GERRITY, ROSS GORDON, b Regina, Sask, May 27, 45; m 69; c 3. EXPERIMENTAL PATHOLOGY, CARDIOVASCULAR DISEASES. *Educ:* Univ Sask, BA, 67, MA, 69; Australian Nat Univ, PhD(exp path), 72. *Prof Exp:* Fel path, McMaster Univ, 73-75, asst prof, 75-76; SR STAFF MEM CARDIOVASC DIS RES, CLEVELAND CLIN FOUND, 76- *Concurrent Pos:* Fel, Med Res Coun Can, 73-74; consult electron microscopist, McMaster Univ, 73-74, lectr path, 74; res fel, Ont Heart Found, 75-76; dir electron micros facil & microscopist, Fac Health Sci, McMaster Univ, 75-76. *Mem:* Australian Soc Exp Path; Electron Micros Soc Can; fel Am Heart Asn. *Res:* Structure and function of the vascular wall, particularly the relationship between endothelial permeability and vascular disease and the biology of the smooth muscle cell in vivo and in vitro. *Mailing Add:* CLeveland Res Inst 2351 E 22nd St Cleveland OH 44040

GERRODETTE, TIMOTHY, b May 3, 46. CONSERVATION BIOLOGY. *Educ:* Carleton Col, BA, 68; Univ Calif San Diego, PhD(oceanog), 79. *Prof Exp:* Fishery biologist, Honolulu, 83-89, OPERS RES ANALYST, NAT MARINE FISHERIES SERV, LA JOLLA, 89- *Mem:* Soc Conserv Biol; Soc Marine Mammal; Wildlife Soc. *Res:* Conservation and management of marine mammals; population dynamics of rare species, especially the Hawaiian monk seal. *Mailing Add:* SW Fisheries Sci Ctr F-SWC PO Box 271 La Jolla CA 92038-0271

GERRY, EDWARD T, b Boston, Mass, Sept 7, 38; m 60; c 2. QUANTUM ELECTRONICS. *Educ:* Col William & Mary, BS, 59; Cornell Univ, MS, 62; Mass Inst Technol, PhD(nuclear eng), 65. *Prof Exp:* Chmn laser res comt, Avco Everett Res Lab, 61-70, dir laser prog off, 70-71; chief laser technol div, Advan Res Proj Agency, Dept Defense, 71-75; PRES & CHIEF OPER OFFICER, W J SCHAFER ASSOCS, 75- *Mem:* AAAS; Am Phys Soc. *Res:* Physics of fully ionized plasmas; physics of gas lasers; chemical kinetics. *Mailing Add:* 1901 N Ft Meyer Dr Suite 800 Arlington VA 22209

GERRY, MICHAEL CHARLES LEWIS, b Victoria, BC, Nov 8, 39; m 67. CHEMICAL PHYSICS, SPECTROCHEMISTRY. *Educ:* Univ BC, BA, 60, MSc, 62; Cambridge Univ, PhD(phys chem), 65. *Prof Exp:* Fel physics, Duke Univ, 65-67; from asst prof to assoc prof, 67-73, PROF CHEM, UNIV BC, 81- *Concurrent Pos:* DFG guest prof, Univ Kiel & Munich, 88-89. *Honors & Awards:* Union Carbide Award for Chem Educ, Chem Inst, Can, 87; Herzberg Award, Spectros Soc, Can. *Mem:* Royal Soc Chem; Chem Inst Can; Can Asn Physicists. *Res:* Electron spin resonance of irradiated solids; microwave spectroscopy of gases; nuclear quadrupole resonance spectroscopy; high resolution infrared spectroscopy of gases. *Mailing Add:* Dept Chem Univ BC 2036 Main Mall Vancouver BC V6T 1Y6 Can

GERRY, RICHARD WOODMAN, b Lewiston, Maine, Nov 23, 14; wid; c 3. POULTRY NUTRITION. *Educ:* Univ Maine, BS, 38; Purdue Univ, MS, 46, PhD(animal nutrit), 48. *Prof Exp:* Asst poultry husb, Purdue Univ, 43-48; from assoc prof to prof, 48-84, EMER PROF POULTRY SCI, UNIV MAINE, ORONO 85- *Mem:* Poultry Sci Asn; World Poultry Sci Asn; Sigma Xi. *Res:* Nutritive value of forest and industrial waste products, potato products, marine products and cereal grains in poultry feed; restricted feeding and watering of growing and laying hens; mineral and vitamin nutrition of poultry. *Mailing Add:* Dept Animal & Vet Sci Univ Maine 136 Hitchner Hall Orono ME 04473

GERSBACHER, WILLARD MARION, b Springerton, Ill, Mar 25, 06; m 38; c 5. ZOOLOGY, ECOLOGY. *Educ:* Southern Ill State Norm Univ, EdB, 26; Univ Ill, MA, 28, PhD(zool), 32. *Prof Exp:* Asst zool, Univ Ill, 27-29; instr biol, Southern Ill State Norm Univ, 29-30; prof sci & math & head of dept, Eastern NMex Jr Col, 34-35; instr zool, Eastern Ill Col, 35-36; from asst prof to prof, Southern Ill Univ, 36-66, head dept, 38-55; prof, Southeast Mo State Col, 66-72; EMER PROF ZOOL, SOUTHERN ILL UNIV, 66- *Mem:* AAAS; fel Ecol Soc Am; Am Soc Limnol & Oceanog; Am Micros Soc; Wildlife Soc. *Res:* Ecology of plants and animals; fresh water biology; development of stream and lake communities; methods in biology teaching; nature conservancy. *Mailing Add:* 1709 Colonial Dr Cape Girardeau MO 63701

GERSCH, HAROLD ARTHUR, b New York, NY, Jan 8, 22; m 47; c 3. PHYSICS. *Educ:* Ga Inst Technol, BS, 48; Johns Hopkins Univ, PhD(physics), 53. *Prof Exp:* From asst prof to prof, 54-62, REGENTS' PROF PHYSICS, GA INST TECHNOL, 70- *Concurrent Pos:* Vis lectr, Johns Hopkins Univ, 56-57; consult, Oak Ridge Nat Lab, 58-; fel, Ford Found, 66; sr fel, NATO, 73; consult prof, Univ New Orleans, 78; vis prof, Oglethorpe Univ, 87, US Mil Acad, 89. *Mem:* AAAS; fel Am Phys Soc. *Res:* Statistical mechanics; quantum mechanics. *Mailing Add:* Sch Physics Ga Inst Technol Atlanta GA 30332

GERSCH, WILL, b New York, NY, Jan 24, 29. ENGINEERING SCIENCE. *Educ:* City Col NY, BEE, 50; NY Univ, MS, 56; Columbia Univ, DrEngSc, 61. *Prof Exp:* Jr engr, Res Div, Philco Corp, Ford Motor Co, Pa, 50-51; engr, W L Maxson Corp, NY, 51-53 & Math Inst, NY Univ, 53-56; sr res engr, Electronic Res Lab, Columbia Univ, 56-61; Nat Acad Sci-Nat Res Coun fel eng, Imp Col, Univ London, 61-62; sr res engr, Electronic Res Lab, Columbia Univ, 62-63; assoc prof stochastic processes, Purdue Univ, 63-70; PROF STOCHASTIC PROCESSES, UNIV HAWAII, 70- *Concurrent Pos:* Vis prof, Dept Eng Mech, Stanford Univ, 66-67, NIH spec training & res fel neurol, 68-70, consult, 69; res fel, Am Statist Asn, 81-82. *Mem:* AAAS; Soc Indust & Appl Math; Inst Elec & Electronics Engrs. *Res:* Analysis and modeling of time series; applications to dynamical systems; neurophysiology; electroencephalogram and electrocardiogram analysis and modeling; modeling and analysis of econometric data. *Mailing Add:* Info Sci Prog Univ Hawaii 2500 Campus Rd Honolulu HI 96844

GERSCHENSON, LAZARO E, b Buenos Aires, Arg, Apr 25, 36; m 62, 90; c 3. CELL BIOLOGY, PATHOLOGY. *Educ:* Univ Buenos Aires, MD, 59, PhD(physiol), 63. *Prof Exp:* Asst res path, Inst Cardiol, Nat Acad Med, Arg, 62-63; fel biochem, Lab Nuclear Med & Radiation Biol, Univ Calif, Los Angeles, 63-66; chief instr biol chem dept, Univ Buenos Aires, 67-68; asst res biochemist, Lab Nuclear Med & Radiation Biol, Univ Calif, Los Angeles, assoc prof path, Med Sch & assoc res biologist, Lab Nuclear Med & Radiation Biol, 71-77; PROF PATH, actg chmn dept, 84-87, PROF PATH, UNIV COLO MED CTR, 77-, CHMN DEPT, 87- *Concurrent Pos:* Nat Coun Sci Res, Arg foreign res fel, 63-65, career res award, 67-68. *Mem:* NY Acad Sci; Am Asn Path; Soc Study Reproduction; Am Soc Cell Biol; Endocrine Soc; Int Acad Path. *Res:* Atherosclerosis; metabolism of the arterial wall; biology of cells in culture; hormonal mechanism of action; cancer research. *Mailing Add:* Dept Path Univ Colo Med Ctr Denver CO 80262

GERSH, EILEEN SUTTON, b Bishop's Stortford, Eng, July 8, 13; nat US; m 44; c 2. CYTOGENETICS. *Educ:* Oxford Univ, BA, 34; Univ London, PhD(genetics), 39. *Prof Exp:* Res worker genetics, John Innes Hort Inst, Eng, 35-38; asst Drosophila cytogenetics, Carnegie Inst, 38-42; lectr cytogenetics, McGill Univ, 42-43; instr biol, Johns Hopkins Univ, 43-46; res assoc zool, Univ Chicago, 54-63; from res assoc to res asst prof, Univ Pa, 63-71, res assoc prof animal biol, Sch Vet Med, 71-76, lectr, 76-85; RETIRED. *Concurrent Pos:* Vis lectr, Swarthmore Col, 67-68; adj assoc prof biol, Univ Pa, 74-76. *Mem:* Genetics Soc Am. *Res:* Cytogenetics of Drosophila. *Mailing Add:* Biol-212 G1-G5 Univ Pa Philadelphia PA 19104

GERSH, MICHAEL ELLIOT, b Brooklyn, New York, June 3, 43. GAS SENSOR, SIGNATURE ANALYSIS. *Educ:* Yale Univ, BS, 65; Univ Wisconsin, PhD, 71. *Prof Exp:* Res assoc, Univ Fl, 71-73; prin scientist, Aerodyne Res Inc, 73-81; VPRES, SPECTRAL SCIENCES INC, 81- *Mem:* Am Physical Soc; Am Chem Soc; AAAS; Sigma Xi. *Res:* Research and development of sensors to characterize gasses and surfaces for industrial processes; aerospace target and background signature phenomenology development and analysis. *Mailing Add:* Spectral Sci Inc 99 S Bedford St Burlington MA 01803-5169

GERSHBEIN, LEON LEE, b Chicago, Ill, Dec 22, 17. ORGANIC CHEMISTRY, BIOCHEMISTRY. *Educ:* Univ Chicago, SB, 38, SM, 39; Northwestern Univ, PhD(chem), 44. *Prof Exp:* Res assoc org chem, Northwestern Univ, 45-47; asst prof biochem, Col Med, Univ Ill, 47-53; assoc prof, Ill Inst Technol, 53-59; DIR, NORTHWEST INST MED RES, 58- *Concurrent Pos:* Adj prof, Ill Inst Technol, 59- *Mem:* AAAS; Am Asn Cancer Res; Am Oil Chem Soc; Asn Clin Scientists; Am Fedn Clin Res; Intern Soc Study Xenobiotics; Nat Acad Clin Biochem; NY Acad Sci; Soc Appl Spectroscopy; Soc Cosmetic Chem; Soc Exp Biol Med. *Res:* Chemistry of natural products; biochemical pharmacology; sulfur compounds; endocrine metabolism; growth promotors and deccelerators; liver regeneration; electron transfer systems; tumorigenesis; lipid metabolism; sebaceous glandular secretions. *Mailing Add:* Northwest Inst Med Res 5645 W Addison Chicago IL 60634

GERSHBERG, HERBERT, b New York, NY, Dec 1, 17; m 64. INTERNAL MEDICINE. *Educ:* City Col New York, BS, 37; Univ Md, MS, 37; Med Col Va, MD, 41. *Prof Exp:* Fel physiol chem, Sch Med, Yale Univ, 46-48; fel internal med, Col Med, NY Univ, 48-50; from instr to asst prof physiol & med, 50-65, ASSOC PROF MED, COL MED, NY UNIV MED CTR, 65- *Concurrent Pos:* Dir diabetes & endocrine clin, Bellevue Hosp, 61-, assoc attend, 65-; dir training grant endocrinol, NIH, 61-67; consult, WHO, Geneva, 64; assoc attend, Univ Hosps, 65- *Mem:* Soc Exp Biol & Med; Endocrine Soc; Am Physiol Soc; Am Diabetes Asn; Am Col Physicians. *Res:* Metabolism, nutrition and endocrinology; pituitary, adrenal cortex and renal function; diabetes; parathyroid; growth hormone. *Mailing Add:* Dept Clin Med NY Univ Sch Med 614 Second Ave New York NY 10016

GERSHENGORN, MARVIN CARL, b New York, NY, May 26, 46; m 69; c 1. MOLECULAR MECHANISMS. *Educ:* City Col City Univ NY, BS, 67; New York Univ Sch Med, MD, 71. *Prof Exp:* Resident med, Strong Mem Hosp, Univ Rochester, 71-73; clin assoc endocrinol, Nat Inst Health, 73-76; from asst prof to assoc prof med, New York Univ Sch Med, 76-83; PROF MED, CORNELL UNIV MED COL, 83- *Honors & Awards:* Van Meter Prize, Am Thyroid Asn, 85. *Mem:* Am Soc Clin Invest; Endocrine Soc; Am Fedn Clin Res; Am Thyroid Asn; Am Soc Biol Chemists; Am Physiol Soc. *Res:* To understand the molecular mechanisms of hormone action. *Mailing Add:* 1300 York Ave New York NY 10021

GERSHENOWITZ, HARRY, b New York, NY, Mar 27, 26. HISTORY OF SCIENCE, PHILOSOPHY OF SCIENCE. *Educ:* St John's Univ, NY, BS, 48; Long Island Univ, BA, 54, MS, 57; Columbia Univ, EDD(sci educ), 67; Am Bd Pharm, dipl. *Prof Exp:* Instr biol, Panzer Col, 55-58; instr sci, Fairleigh Dickinson Univ, 58-62; asst prof biol, Wilkes Col, 62-65; PROF SCI, GLASSBORO STATE COL, 65- *Mem:* Am Technion Soc; Nat Sci Teachers Asn; Am Inst Aeronaut & Astronaut; Nat Hist Soc. *Res:* The Darwinian Age. *Mailing Add:* Dept Life Sci Glassboro State Col Glassboro NJ 08028

GERSHENSON, HILLEL HALKIN, b New York, NY, Mar 27, 35; m 62; c 3. MATHEMATICS. *Educ:* Univ Wis, BA, 55, MA, 57; Univ Chicago, PhD(math), 61. *Prof Exp:* Instr math, Princeton Univ, 61-63; asst prof, Cornell Univ, 63-68; ASSOC PROF MATH, UNIV MINN, MINNEAPOLIS, 68- *Concurrent Pos:* Vis lectr, Aarhus Univ, 66-67. *Mem:* Am Math Soc. *Res:* Algebraic topology; homological algebra. *Mailing Add:* Dept Math 127 Vincent Hall Univ Minn 206 Church St SE Minneapolis MN 55455

GERSHENZON, M(URRAY), b Brooklyn, NY, Nov 17, 28; m 52; c 3. SOLID STATE ELECTRONICS. *Educ:* City Col New York, BS, 49; Columbia Univ, AM, 53, PhD(chem), 57. *Prof Exp:* Mem tech staff solid state physics, Bell Tel Labs, NJ, 57-66; PROF MAT SCI & ELEC ENG, UNIV SOUTHERN CALIF, 66- *Mem:* Am Phys Soc; Inst Elec & Electronics Engrs; Mat Res Soc. *Res:* Radiative recombination in semiconductors; molecular beam epitaxy. *Mailing Add:* Dept Mat Sci Univ Southern Calif Univ Park Los Angeles CA 90089-0241

GERSHINOWITZ, HAROLD, b New York, NY, Aug 31, 10; m 35. PHYSICAL CHEMISTRY. *Educ:* City Col New York, BS, 31; Harvard Univ, AM, 32, PhD(chem), 34. *Prof Exp:* Harvard Parker traveling fel, Princeton Univ, 34-35; res assoc chem, Columbia Univ, 35-36 & Harvard Univ, 36-38; petrol technologist, Shell Oil Co Inc, Mo, 38-39, dir res lab, Houston, 39-42, dir mfg res east of Rockies, 42-45, dir explor & prod res div, 45-51, vpres explor & prod tech div, 51-53, pres, Shell Develop Co, 53-62, chmn res coun & res coordr, Royal Dutch/Shell, 62-65; CONSULT, 66- *Concurrent Pos:* Consult, Orgn Econ Coop & Develop, Paris, 66-70; chmn environ studies bd, Nat Acad Sci-Nat Acad Eng, 67-70; affil, Rockefeller Univ, 67-78, adj prof, 78-81; dir, Bataafse Petrol Maatschappij NV & Shell Int Res Maatschappij NV, Shell Res, Ltd & Shell Res NV, 62-65. *Mem:* Fel AAAS; Am Chem Soc; Sigma Xi. *Res:* Economics and administration of research; reaction kinetics from standpoint of quantum and statistical mechanics. *Mailing Add:* 25 Sutton Pl S Apt 9-G New York NY 10022

GERSHMAN, LEWIS C, b Cleveland, Ohio, Mar 11, 38; m 64; c 2. CARDIOLOGY, BIOPHYSICS. *Educ:* Princeton Univ, AB, 62; State Univ NY Downstate Med Ctr, MD & PhD(physiol, biophys), 68. *Prof Exp:* Intern med & pediat, Kings County Hosp & State Univ NY Downstate Med Ctr, 68-69, resident internal med, 69-70; instr med, Col Med, Univ Rochester, 72-73; ASSOC PROF MED & PHYSIOL, ALBANY MED COL, 73- *Concurrent Pos:* Staff physician med, Vet Admin Hosp, Bath, NY, 72-73; staff physician, Vet Admin Med Cntr, Albany, 73- *Mem:* Biophys Soc; Am Fedn Clin Res; NY Acad Sci; AAAS. *Res:* Biochemistry of muscle proteins. *Mailing Add:* Vet Admin Med Cntr Holland Ave Albany NY 12208

GERSHMAN, LOUIS LEO, b Nov 5, 20; US citizen; m 48; c 1. ANALYTICAL CHEMISTRY. *Educ:* City Col, New York, BS, 41; Polytech Inst, Brooklyn, MS, 57. *Prof Exp:* Chemist, 48-58, supv chemist, 60-66, LAB DIR, US FOOD & DRUG ADMIN, 67- *Mem:* Am Chem Soc; Asn Off Anal Chem. *Res:* Infrared determination of endrin residues. *Mailing Add:* 151 Coolidge Ave Watertown MA 02172

GERSHMAN, MELVIN, b Hartford, Conn, Aug 24, 27; m 50; c 2. BACTERIOLOGY. *Educ:* Ohio State Univ, BSc, 54; Univ Mass, MSc, 57. *Prof Exp:* Instr bact, Smith Col, 56-57 & Springfield Hosp Nursing Sch, Mass, 57; asst prof bact, 58-63, assoc prof microbiol & animal path, 63-77, PROF MICROBIOL, UNIV MAINE, 77- *Mem:* Am Soc Microbiol; Sigma Xi. *Res:* Diagnostic bacteriology; enteric diseases; mycoplasma; phage typing. *Mailing Add:* Six Frost Lane Orono ME 04473

GERSHOFF, STANLEY NORTON, m. NUTRITION. *Educ:* Univ Wis, BA, 43, MS, 48, PhD(biochem), 51. *Prof Exp:* Res asst, Dept Biochem, Univ Wis-Madison, 47-51, res assoc, 51-52; res assoc, Dept Nutrit, Sch Pub Health, Harvard Univ, 52-56, from asst prof to assoc prof, 56-77; prof nutrit & chmn, Grad Dept Nutrit, Tufts Univ, 77-81, dir, Nutrit Inst, 77-81, prin investr, USDA-Human Nutrit Res Ctr, 77-84, DEAN, SCH NUTRIT, TUFTS UNIV, 81-, PRIN INVESTR, USDA- HUMAN NUTRIT RES CTR, 84- *Concurrent Pos:* Consult, Unicef, Thailand, Pakistan & Indonesia, 74-80; vis lectr, Sch Pub Health, Harvard Univ, 77-85; counr, Am Soc Clin Nutrit, 79-82; hon prof nutrit & food hyg, WChina Univ Med Sci, 87; mem, numerous comts, coun & bds. *Honors & Awards:* Borden Award Nutrit, Am Inst Nutrit, 72; Martha F Trulson Lectr, Am Diabetic Asn, 77. *Mem:* Am Inst Nutrit; Am Soc Clin Nutrit; Brit Nutrit Soc; Sigma Xi. *Mailing Add:* Sch Nutrit Tufts Univ 132 Curtis St Medford MA 02155

GERSHON, ANNE A, b Pa, Aug 30, 38; m 61; c 3. MEDICINE, INFECTIOUS DISEASE. *Educ:* Smith Col, AB, 60; Cornell Univ, MD, 64. *Prof Exp:* Intern pediat, New York Hosp, 64-65; NIH fel bacteriol, Dunn Sch Path, Oxford, Eng, 65-66; resident pediat, New York Hosp, 66-68; fel infectious dis, 68-70, instr & asst prof pediat, 70-74, assoc prof pediat, 74-80, PROF PEDIAT, NY UNIV MED CTR, 80- *Mem:* Am Acad Pediat; Soc Pediat Res; Infectious Dis Soc; Am Soc Microbiol. *Res:* Viral infections of man and immunologic responses to these infections; viral vaccines; herpesvirus infections. *Mailing Add:* Colum Univ Col P & S 650 W 168th St New York NY 10032

GERSHON, ELLIOT SHELDON, b Brooklyn, NY, June 5, 40; m 67; c 2. GENETICS. *Educ:* Harvard Univ, AB, 61, Harvard Med Sch, MD, 65. *Prof Exp:* Intern med, Mt Sinai Hosp, 65-66; resident psychiat, Mass Ment Health Ctr, 66-69; clin assoc, Lab Clin Sci, NIMH, 69-71; dir res, Jerusalem Ment Health Ctr, 71-74; unit chief, 74-78, chief sect psychogenetics, Biol Psychiat Br, 78-84, CHIEF, CLIN NEUROGENETICS BR, NIMH, 84-; MED DIR, USPHS, 80- *Concurrent Pos:* Teaching fel, Harvard Med Sch, 66-69; consult, Peter Bent Brigham Hosp, 68-69 & Prince George's County Health Dept, Md, 69-70; mem sci adv bd, Israel Ctr Psychobiol, 72-74; mem fac psychiat, Wash Sch Psychiat, 76- & NIMH Staff Col, 77-; sci adv bd, Found Depression & Manic-Depression, 78-, prof adv bd, Jerusalem Ment Health Ctr, 78-; dir, Off Sci, Alcohol Drug Abuse & Mental Health Admin, 86-87. *Honors & Awards:* Anna Morika Found Prize, 79. *Mem:* Am Psychiat Asn; Am Psychopath Asn; Am Col Neuropsychopharmacol; AAAS. *Res:* Manic-depressive illness; genetics; psychopharmacology; psychobiology. *Mailing Add:* NIH Bldg 10 Rm 3N218 9000 Rockville Pike Bethesda MD 20892

GERSHON, HERMAN, b Brooklyn, NY, Jan 27, 21; m 55; c 2. BIOCHEMISTRY. *Educ:* Brooklyn Col, BA, 42; Fordham Univ, MS, 47; Univ Colo, PhD(biochem), 50. *Prof Exp:* Chemist, Food Res Labs, Inc, 42-43 & Dr R J Block Lab, 46-47; biochemist, Vet Admin Hosp, Northport, NY, 50-51; dir res & develop, Pharmaceut & Fine Chem, United Org Corp, 51-55; biochem res, Pfister Chem Works, Inc, 55-62; sr org chemist, Boyce Thompson Inst Plant Res, Inc, 62-84; collabr, USDA, Ithaca, NY, 85; SCIENTIST, DEPT CHEM, FORDHAM UNIV, 86- *Concurrent Pos:* Adj scientist, NY Bot Garden, 91- *Mem:* Fel AAAS; Am Chem Soc; fel NY Acad Sci; Sigma Xi. *Res:* Amino acid antagonists; synthesis of pyrimidines, quinolines, and fluorinated metabolite analogues; mode of action of antifungal agents. *Mailing Add:* 60 Brewer Rd Monsey NY 10952

GERSHON, MICHAEL DAVID, b New York City, NY, Mar 3, 38; m 61; c 3. NEUROBIOLOGY. *Educ:* Cornell Univ, AB, 58, MD, 63. *Prof Exp:* USPHS res fel, Cornell Univ, 63-64, instr anat, Med Col, 64-65; USPHS res assoc pharmacol, Oxford Univ, 65-66; from asst prof to prof anat, Cornell Univ, 66-75; PROF & CHMN ANAT, COL PHYSICIANS & SURGEONS, COLUMBIA UNIV, 75- *Concurrent Pos:* Mem, Neurol Disorders Prog Proj, Gastrointestinal Drug Adv Comn, Food & Drug Admin, 73-75; prin investr, Neural Control Gastrointestinal Activ, 76, Neurol A Study; vis prof, Grass Found, 81. *Honors & Awards:* Jacob Javits Award, Nat Inst Neurol Commun Disorders & Stroke. *Mem:* Am Asn Anatomists; Am Soc Cell Biol; Am Physiol Soc; Am Soc Pharmacol & Exp Therapeut; Soc Neurosci; Sigma Xi; Asn Anat Chmn (pres); AAAS; Am Gastroenterol Asn; Am Soc Pharmacol & Exp Therapeut. *Res:* Cellular biology patterns of neural organization and factors that govern the development of the intrinsic nervous system of the gut (enteric nervous system). *Mailing Add:* Col Physicians & Surgeons Columbia Univ 630 W 168th St New York NY 10032

GERSHON, SAMUEL, b Poland, Dec 13, 27; US citizen; c 2. PSYCHOPHARMACOLOGY. *Educ:* Univ Sydney, BBS, 50; Univ Melbourne, DPM, 56; Royal Col Psychiatrists, FRC, 73. *Prof Exp:* Dep psychiat supt, Ballaret Ment Hosp, 54-60; chief psychopharmacol sect, Schizophrenia & Psychopharmacol Joint Res Proj, Univ Mich, 60-63; prin res scientist, Inst Psychiat, Univ Miss, 63-65; dir neuropsychopharmacol res unit, Dept Psychiat, NY Med Sch, 65-79; PROF & CHMN PSYCHIAT, DEPT PSYCHIAT, SCH MED, WAYNE STATE UNIV, 79-; DIR ADMIN, LAFAYETTE CLIN, MICH DEPT MENT HEALTH, 79- *Concurrent Pos:* Assoc prof physiol, pharmacol & psychiat, Miss Inst Psychiat, Univ Miss, 63-65, vis prof, 75-; distinguished prof, Lakeland AFB, US Air Force Med Ctr, 75-; res assoc prof, Dept Psychiat, NY Univ Med Sch, 65-68, res prof, 68-70, prof, 71-79; res collaborator, Med Dept, Brookhaven Nat Lab Assoc Univ, 78-79. *Mem:* Fel Am Col Neuropsychopharmacol; Am Soc Clin Pharmacol & Therapeut; Am Col Clin Pharmacol; Psychiat Res Soc. *Res:* Biological psychiatry; gerontology; psychopharmacol. *Mailing Add:* Dept Psychiat-1464 Lafayette Wayne State Univ Sch Med 540 E Canfield Detroit MI 48201

GERSHON, SOL D, b Chicago, Ill, Oct 18, 10; m 34; c 2. CHEMISTRY. *Educ:* Univ Ill, PhC, 30; Univ Chicago, BS, 34, MS, 35, PhD(chem), 38. *Prof Exp:* Asst chem, Col Pharm, Univ Ill, 30-38, from instr to assoc, 38-42, asst prof, 42-43; res chemist, Pepsodent Co, 43-47, dir new prods develop, Pepsodent div, Lever Bros Co, 47-48, res mgr, 49-52, asst dir res prod improv & develop, 52-60, develop mgr household prod, 60-63, assoc res dir, 63-65, asst dir develop, 65-68, tech planning dir, Edgewater, 68-74; exec dir, Soc Cosmetic Chemists, 74-77, CONSULT, 74- *Honors & Awards:* Medal Award, Soc Cosmetic Chemists; Fairchild Scholar. *Mem:* Am Chem Soc; Am Oil Chemists Soc; hon mem Soc Cosmetic Chemists (vpres, 51, pres, 52); Am Pharmaceut Asn; Int Asn Dent Res; fel Am Inst Chem; hon pres, Int Fedn Soc Cosmetic Chem. *Res:* Chemotherapy; synthesis of organic medicinals; drug assay; carbohydrates; cosmetic, dentifrice, detergent and edible product development. *Mailing Add:* 1363 Mercedes St Teaneck NJ 07666

GERSHOWITZ, HENRY, b New York, NY, Sept 22, 24; m 49; c 4. HUMAN GENETICS. *Educ:* Brooklyn Col, BA, 49; Calif Inst Technol, PhD(genetics), 54. *Prof Exp:* Res fel immunogenetics, Univ Wis, 54-56; res assoc, 57-61, from asst prof to assoc prof, 61-70, prof human genetics, Univ Mich-Ann Arbor, 70-88; RETIRED. *Concurrent Pos:* Consult, Blood Bank, Univ Hosp, Mich, 57-75; consult, Nat Legal Labs, 79-89, dir, 89- *Mem:* AAAS; Am Soc Human Genetics; Am Asn Immunol; Genetics Soc Am; Am Asn Blood Banks; Sigma Xi. *Res:* Blood group inheritance and correlations; genetic control of immunoglobulin structure and antibody specificity; paternity tests. *Mailing Add:* 2019 Winsted Blvd Ann Arbor MI 48103

GERSHUN, THEODORE LEONARD, mechanical engineering, nuclear engineering; deceased, see previous edition for last biography

GERSON, LOWELL WALTER, b New York, NY, Sept 26, 42; m 64; c 2. EPIDEMIOLOGY, GERONTOLOGY. *Educ:* Case Western Reserve Univ, BA, 64, MA, 66, PhD(med sociol), 70. *Prof Exp:* Asst prof sociol, John Carroll Univ, 68-70; asst prof med sociol, Mem Univ Nfld, 70-74, assoc prof, 74-75; assoc prof clin epidemiol, McMaster Univ Med Ctr, 75-78; assoc prof epidemiol, 78-82, ASSOC DIR, DIV COMMUNITY HEALTH SCI, NORTHEASTERN OHIO UNIV COL MED, 81-, PROF EPIDEMIOL, 82- *Concurrent Pos:* Mem, Non Med Use of Drug Sci Rev Comt, Can, 74-76, Nat Health Welfare Can Sci Rev Comn, 74-81 & Exec Comt, Can Liver Found Liver Epidemiol, 79-; prof, Kent State Univ, 80-; adj assoc prof, Grad Sch Univ Pittsburgh, 81-88; adj prof, Univ Akron, 84-; adj grad fac, Kent State Univ, 80-; mem, Ohio Acad Family Pract Res, CHC, 82; vis fel, South Australia Health Comn, 86; chair, Task Force on Acad Base Pub Health, Ohio Pub Health Asn, 88-89; mem rev panel, Health Care Financing Admin, 89. *Honors & Awards:* Hon Paramedic Akron. *Mem:* Am Pub Health Asn; Asn Teachers Prev Med; Int Epidemiol Asn; Soc Epidemiol Res; Geront Soc Am; Asn Med Educ & Res Alcoholism. *Res:* Health services research, primarily ambulatory care; geriatrics; alcoholic treatment. *Mailing Add:* 7385 La Costa Dr Hudson OH 44236

GERSON, ROBERT, b New York, NY, Dec 5, 23; m 48; c 3. PHYSICS. *Educ:* City Col NY, BChE, 43; NY Univ, PhD(physics), 54. *Prof Exp:* Physicist, Erie Resistor Corp, Pa, 53-56; sr physicist, Clevite Res Ctr, Ohio, 56-62; PROF PHYSICS, UNIV MO, ROLLA, 62- *Mem:* Am Phys Soc; AAAS; Am Asn Physics Teachers; Sigma Xi. *Res:* Dielectric and semiconducting materials; ferroelectricity; ferroelectric-ferromagnetic interaction; electrets; Mossbauer effect; ion implantation; development of novel educational methods for mathematical concepts. *Mailing Add:* Dept Physics Univ Mo PO Box 249 Rolla MO 65401

GERSPER, PAUL LOGAN, b Columbus, Ohio, Oct 12, 36; m 56; c 4. SOIL MORPHOLOGY, SOIL MINERALOGY. *Educ:* Ohio State Univ, BSc, 61, MSc, 63, PhD(soil sci), 68. *Prof Exp:* Res assoc soil sci, Ohio State Univ, 61-68; asst prof pedology & asst pedologist, 68-75, asst dean, Col Agr Sci, 72-74, chmn, Dept Conserv & Resource Studies, 76-80, ASSOC PROF PEDOLOGY, UNIV CALIF, BERKELEY, 75-, ASSOC PEDOLOGIST, AGR EXP STA, 75-, ASSOC PROF, DEPT PLANT & SOIL BIOL, 80- *Mem:* Am Soc Agron; Soil Sci Soc Am; Am Polar Soc; Ecol Soc Am; Nat Geog Soc; Am Land Res Asn; Nature Conserv. *Res:* Soil morphology, development and classification; interactions of soil forming factors and their effects on soil development; soil-plant relationships; soil resource evaluation; land use planning. *Mailing Add:* Dept Plant & Soil Biol Univ Calif 2120 Oxford St Berkeley CA 94720

GERST, IRVING, b New York, NY, May 30, 12; wid; c 2. COMBINATORIAL IDENTITIES. *Educ:* City Col NY, BS, 31; Columbia Univ, MA, 32, PhD(math), 47. *Prof Exp:* Teacher, Bd Educ, NY, 38-42; instr, US Air Force Tech Sch, Keesler Field, Miss, 42-44; tech consult, Transportation Corps, US Army Serv Forces, NY, 44-46; mathematician & head appl anal group, Control Instrument Co Div, Burroughs Corp, 46-58; sr proj mem & leader networks group, Radio Corp Am, 58-61; prof, 61-82, EMER PROF APPL MATH STATE UNIV NY, STONY BROOK, 82- *Concurrent Pos:* Lectr, City Univ NY, 58-61; consult, Sperry-Rand, Inc, 61-63. *Mem:* Am Math Soc; Math Asn Am. *Res:* Network theory; complex variable; functional equations; operational methods; number theory; combinatorics. *Mailing Add:* Dept Appl Math & Statist State Univ NY Stony Brook NY 11794

GERST, JEFFERY WILLIAM, b San Francisco, Calif, Apr 23, 44; m 66; c 2. ZOOLOGY, PHYSIOLOGY. *Educ:* Chico State Col, BA, 66, MA, 68; Univ Nebr-Lincoln, PhD(zool), 73. *Prof Exp:* Asst prof physiol, 73-77, ASST PROF ZOOL, N DAK STATE UNIV, 77- *Mem:* Am Soc Zoologists; Sigma Xi. *Res:* Compensatory changes in physiology elicited by changes in the external environment; transport enzyme systems; water-mineral balance. *Mailing Add:* Dept Zool NDak State Univ Fargo ND 58102

GERST, PAUL HOWARD, b Sept 24, 27; US citizen; m 57; c 3. SURGERY, PHYSIOLOGY. *Educ:* Columbia Univ, AB, 48, MD, 52. *Prof Exp:* Instr physiol, Univ Pa, 55-56; instr surg, 62-64, ASST PROF SURG, COL PHYSICIANS & SURGEONS, COLUMBIA UNIV, 64-; DIR SURG, BRONX-LEBANON HOSP CTR, 64-; PROF SURG, ALBERT EINSTEIN COL MED, 72- *Concurrent Pos:* USPHS res fel, 55-56 & res career develop award, 63-64; Am Col Surgeons award, 60-63. *Mem:* Am Col Surgeons; Am Asn Thoracic Surg; Am Physiol Soc; Biophys Soc; Am Col Chest Physicians. *Res:* Thoracic and cardiovascular diseases. *Mailing Add:* Dept Surg Bronx Lebabon Hosp Ctr 1650 Grand Concourse Bronx NY 10457

GERSTEIN, BERNARD CLEMENCE, b Monticello, NY, Oct 18, 32; m 58; c 5. PHYSICAL CHEMISTRY. *Educ:* Purdue Univ, BS, 53; Iowa State Univ, PhD(phys chem), 60. *Prof Exp:* Res assoc phys chem, 60-61, from asst prof to assoc prof, 61-74, PROF CHEM, IOWA STATE UNIV, 75- *Concurrent Pos:* Vis assoc prof, Calif Inst Technol, 72-73; vis lectr, Univ Paris, 81. *Mem:* AAAS; Am Chem Soc; Am Phys Soc. *Res:* Heterogeneous catalysis in insulators and semiconductors; applications of pulsed NMR to the study of electronic structures of molecules and solids. *Mailing Add:* Dept Chem Iowa State Univ Ames IA 50011

GERSTEIN, GEORGE LEONARD, b Berlin, Ger, Apr 12, 33; div; c 2. NEUROBIOLOGY. *Educ:* Harvard Univ, BA, 52, MA, 54, PhD(physics), 58. *Prof Exp:* NIH fel biophys, Mass Inst Technol, 58-60, instr physics, 60-61, res assoc commun sci, 61-63; mem staff, Ctr Comput Technol, 63-64; asst prof biophys, 64-66, assoc prof biophys & physiol, 66-69, PROF BIOPHYS & PHYSIOL, UNIV PA, 69- *Concurrent Pos:* Mem neurol A study sect, 67-71 & Computer & Biomath Sci Study Sect, NIH, 72-76. *Mem:* AAAS; Physiol Soc; Soc for Neurosci. *Res:* Electrical activity of the nervous system; auditory system; mathematical analysis and computer simulation in neurophysiology. *Mailing Add:* Dept Physiol/G4 Univ Pa Philadelphia PA 19104-6085

GERSTEIN, LARRY J, b Leavenworth, Kans, Aug 28, 40; m 69; c 2. MATHEMATICS. *Educ:* Columbia Col, AB, 62; Univ Notre Dame, MS, 63, PhD(math), 67. *Prof Exp:* Asst prof, 67-72, assoc prof, 72-78, PROF MATH, UNIV CALIF, SANTA BARBARA, 78- *Concurrent Pos:* Vis asst prof, Mass

Inst Technol, 70-71; vis assoc prof, Univ Notre Dame, 72-73; hon res fel, Harvard Univ, 78-79; vis prof, Dartmouth Col, 85-86. *Mem:* Am Math Soc; Math Asn Am. *Res:* Number theory; quadratic and hermitian forms. *Mailing Add:* Dept Math Univ Calif Santa Barbara CA 93106

GERSTEIN, MELVIN, b Chicago, Ill, May 8, 22; m 44; c 1. CHEMISTRY. *Educ:* Univ Chicago, BS, 42, PhD(chem), 45. *Prof Exp:* Asst instr, Army Specialized Training Prog, Univ Chicago, 42-44, jr chemist, Metall Proj, 44, res assoc, 44-46; chemist, NASA, 46-49, head combustion fundamentals sect, 49-54, chief chem br, 54-57, asst chief propulsion chem div, 57-59; chief phys sci div, Jet Propulsion Lab, Calif Inst Technol, 59-60; from vpres to pres, Dynamic Sci Corp, 60-66; assoc dean, Sch Eng, Univ Southern Calif, 70-84, prof mech eng, 66-88; RETIRED. *Concurrent Pos:* Lectr, Fenn Col, 48-49; sci rep, Adv Group Aeronaut Res & Develop, NATO, Rome, 52, Cambridge, 53, mem combustion panel, 54-57, mem combustion & propulsion panel, 57-64; ed, Isotopics, 55-58; sr lectr, Calif Inst Technol, 62-; vpres res & develop, Ginter Corp, 69-; Adv Group Aerospace Res & Develop, Energy & Propulsion Panel, Working Group on Aircraft Fires, NATO, 77-; chmn comt fire resistant hydraul fluids, Nat Res Coun, 77- *Honors & Awards:* Outstanding Achievement Award, Cleveland Tech Soc Coun, 57. *Mem:* Am Soc Mech Engrs; Combustion Inst; Sigma Xi. *Res:* Chemistry of fuels and propellants; hydride chemistry; acid-base theory; combustion theory and applications; propulsion. *Mailing Add:* 1141 Via Romero Palos Verdes Estates CA 90274

GERSTEL, DAN ULRICH, b Berlin-Dahlem, Ger, Oct 23, 14; nat US; m 38; c 2. BOTANY, CYTOLOGY. *Educ:* Univ Calif, Davis, BS, 40; Univ Calif, Berkeley, MS, 42, PhD(genetics), 45. *Prof Exp:* Asst bot, Univ Calif, 42-44, assoc genetics, 44-46; assoc geneticist, Res Inst, Stanford Univ, 47 & Natural Rubber Plant Res Sta, USDA, 47-49; res fel, Calif Inst Technol, 49-50; asst prof agron, 50-53, assoc prof, 53-56, prof field crops, 56-63, Reynolds prof, 64-80, EMER REYNOLDS PROF CROP SCI & GENETICS, NC STATE UNIV, 80- *Concurrent Pos:* Vis prof, Weizmann Inst, 61-62; NC-Israel exchange prof, Bet Dagan, Israel, 79. *Mem:* Genetics Soc Am. *Res:* Cytogenetics and speciation of Nicotiana, Gossypium and Parthenium; interspecific hybridization and genetic instability; origin and breeding of cultivated crops. *Mailing Add:* Dept Crop Sci NC State Univ Raleigh NC 27695-7620

GERSTEN, JEROME WILLIAM, b New York, NY, Apr 20, 17; m 41; c 5. PHYSICAL MEDICINE & REHABILITATION. *Educ:* City Col New York, BS, 35; NY Univ, MD, 39; Univ Minn, MS, 49. *Prof Exp:* Asst, Mayo Clin, 47-49; from asst prof to assoc prof, 49-57, head dept, 57-81, PROF PHYS MED & REHAB, SCH MED, UNIV COLO, DENVER, 57- *Mem:* Am Physiol Soc; Soc Exp Biol & Med: Am Asn Electromyography & Electrodiag; Am Cong Rehab Med (pres, 69); Am Heart Asn; Am Pain Soc. *Res:* Stroke; learning disability; delivery of health care. *Mailing Add:* Univ Colo Sch Med Denver CO 80262

GERSTEN, JOEL IRWIN, b New York, NY, Mar 18, 42; m 64; c 4. THEORETICAL SOLID STATE PHYSICS. *Educ:* City Col NY, BS, 62; Columbia Univ, MA, 63, PhD(physics), 68. *Prof Exp:* Mem tech staff physics, Bell Labs, 68-70; from asst prof to assoc prof, 70-77, PROF PHYSICS, CITY COL NY, 77- *Concurrent Pos:* Consult, Bell Labs, 70-73; fel, Inst Advan Studies, Hebrew Univ Jerusalem, 78-79. *Mem:* Am Phys Soc. *Res:* Theoretic atomic physics; radiation processes; low energy electron diffraction; nonlinear optical processes; solid state physics; surface physics. *Mailing Add:* Dept Physics City Coll NY Convent Ave 138th St New York NY 10031

GERSTEN, STEPHEN M, b Utica, NY, Dec 2, 40. TOPOLOGY, ALGEBRA. *Educ:* Princeton Univ, AB, 61; Cambridge Univ, PhD(math), 65. *Prof Exp:* Instr math, Princeton Univ, 63-64 & Rice Univ, 64-65; NSF fel, Oxford Univ, 65-66; from asst prof to assoc prof, Rice Univ, 66-74; mem staff, 74-76, PROF MATH, UNIV UTAH, 76- *Mem:* Am Math Soc. *Res:* Algebraic topology; homological algebra; projective class groups and Whitehead groups of algebras and geometric applications. *Mailing Add:* Dept Math Univ Utah Salt Lake City UT 84112

GERSTENHABER, MURRAY, b Brooklyn, NY, May 6, 27; m 56; c 3. MATHEMATICS. *Educ:* Yale Univ, BS, 48; Univ Chicago, MS, 49, PhD(math), 51; Univ Pa, JD, 73. *Prof Exp:* Jewett fel, 51-53; from asst prof to assoc prof, 53-61, PROF MATH, UNIV PA, 61- *Concurrent Pos:* Mem, Inst Advan Study, 57-59, 62, 65-66 & 81-82; mem staff, Inst Defense Anal, 61-62; ed, Bull Am Math Soc, 65-71. *Mem:* Am Math Soc; Math Asn Am; Soc Indust & Appl Math. *Res:* Algebra; deformation of algebras; inseparable fields; probabilistic inference in law. *Mailing Add:* 237 Hamilton Rd Merion Station PA 19066

GERSTER, ROBERT ARNOLD, industrial filtrations, process control systems; deceased, see previous edition for last biography

GERSTING, JOHN MARSHALL, JR, b Cincinnati, Ohio, Nov 1, 40; m 62; c 2. APPLIED MATHEMATICS, COMPUTER SCIENCE. *Educ:* Purdue Univ, BS, 62; Ariz State Univ, MS, 64, PhD(eng sci), 70. *Prof Exp:* Analyst math modeling, Lewis Res Ctr, NASA, 62; analyst trajectory simulation, US Govt, Washington, DC, 64-66; fac assoc eng, Ariz State Univ, 66-70; asst prof eng & comput sci, 70-74, assoc prof 75-79, PROF COMPUT SCI, IND UNIV-PURDUE UNIV, INDIANAPOLIS, 79- *Mem:* Asn Comput Mach; Am Phys Soc. *Res:* Computational methods; stiff differential systems; hydrodynamic stability; database management systems; relational and hierarchical databases; search techniques. *Mailing Add:* Comput Sci Ind Univ-Purdue Univ 1201 E 38th St Indianapolis IN 46202

GERSTING, JUDITH LEE, b Springfield, Vt, Aug 20, 40; m 62; c 2. COMPUTER SCIENCE EDUCATION. *Educ:* Stetson Univ, BS, 62; Ariz State Univ, MA, 64, PhD(math), 69. *Prof Exp:* From asst prof to assoc prof, 70-80, PROF COMPUT SCI, IND UNIV-PURDUE UNIV, INDIANAPOLIS, 81- *Concurrent Pos:* Assoc prof comput sci, Univ Cent Fla, 80-81; vis prof computer sci, Univ Hawaii, Hilo, 90-91. *Mem:* Asn Comput Mach; Inst Elec & Electronics Engrs Comput Soc. *Res:* Fault-tolerant computing systems. *Mailing Add:* Comput Sci Ind Univ-Purdue Univ 1201 E 38th St Indianapolis IN 46205

GERSTL, BRUNO, NEUROCHEMISTRY. *Educ:* Univ Vienna, MD, 27. *Prof Exp:* Prof path, Stanford Univ, 61-72; RETIRED. *Mailing Add:* 824 Mayfield Ave Stanford CA 94305

GERSTL, SIEGFRIED ADOLF WILHELM, b Ger, Aug 5, 39; US citizen; c 2. REMOTE SENSING, SPACE SCIENCE. *Educ:* Univ Stuttgart, Dipl, Phys, 64; Univ Karlsruhe, PhD(physics), 67. *Prof Exp:* Scientist physics, Univ Stuttgart, 64-65; scientist nuclear physics, Univ Karlsruhe, 65-68; sr engr appl physics, Westinghouse Advan Reactors Div, 68-71; mem staff physics, Argonne Nat Lab, 71-74; MEM STAFF PHYSICS, LOS ALAMOS NAT LAB, 74- *Concurrent Pos:* Mem, nat prog comt mem, Am Nuclear Soc, 72-82, NASA satellite instrument panels, 86-; prin investr, NASA & Dept Energy. *Mem:* Am Nuclear Soc; Am Geophys Union; AAAS; Am Soc Photog. *Res:* Physics of satellite remote sensing; modeling of atmospheric optics and imaging; atmospheric correction algorithms; spectral and angular signatures of earth reflectance features; radiation beam propagation for solar and laser radiation; computational solutions to radiative transfer equation. *Mailing Add:* Los Alamos Nat Lab ADAL-A104 Los Alamos NM 87545

GERSTLE, FRANCIS PETER, JR, b Louisville, Ky, June 23, 42; m 67, 87; c 2. MATERIALS SCIENCE. *Educ:* St Joseph's Col, Ind, BA, 64; Mass Inst Technol, SB, 65, SM, 66; Duke Univ, PhD(mech eng & mat sci), 72. *Prof Exp:* Tech staff mem test facil design, Sandia Labs, 66-68; instr & res asst mech eng & mat sci, Duke Univ, 68-71; mem tech staff composite mat, Sandia Labs, 72-73, mem tech staff mech mat, 73-74, supvr composite mat, 74-83, supvr ceramics develop, 84-89, SUPVR CERAMICS & GLASS PROCESSING, SANDIA LABS, 89- *Concurrent Pos:* Comt mem characterization org matrix composites, Nat Mat Adv Bd, 78-80; chmn, Gordon Res Conf Composites, 81; comt mem, Lightweight Mil Combat Vehicles, Nat Mat Adv Bd, 81-83; mem panel, Nat Acad Sci, 84. *Mem:* Soc Eng Sci; Am Soc Mech Engrs; Soc Advan Mat & Process Eng; Soc Exp Stress Anal. *Res:* Analysis and characterization of filamentary composite materials; design of composite structures, especially pressure vessels; interfacial and residual viscoelastic effects in seals. *Mailing Add:* Org 1845 Bldg T47 Rm 4 Sandia Labs Albuquerque NM 87185

GERSTLE, KURT H, b Munich, Ger, Nov 11, 23; US citizen; m 51; c 4. STRUCTURAL ENGINEERING. *Educ:* Univ Calif, Berkeley, BS, 49, MS, 52; Univ Colo, PhD(civil eng), 56. *Prof Exp:* From instr to assoc prof, 57-70, PROF CIVIL ENG, UNIV COLO, BOULDER, 70- *Concurrent Pos:* NSF sci fac fel, Brown Univ, 59-60; vis prof struct eng, SEATO Grad Sch Eng, Thailand, 63-64; Fulbright lectr, Munich Tech Univ, 70-71; vis prof, Norwegian Tech Univ, Trondheim, 78; Alexander von Humboldt sr scientist award, 78-79; Erskine fel, Canterbury Univ, Christchurch, NZ, 86. *Honors & Awards:* Wason Medal, Am Concrete Inst, 64; Moisseiff Award, Am Soc Civil Engr, 84; Higgins Lectureship, Inst Steel Construct, 89. *Mem:* Am Soc Civil Engrs; Am Concrete Inst; Sigma Xi. *Res:* Elastic and inelastic analysis and behavior of structures; behavior of reinforced concrete structures. *Mailing Add:* Campus Box 428 Univ Colo Boulder CO 80309-0428

GERSTMAN, HUBERT LOUIS, b Buffalo, NY, Feb 20, 34; m 59; c 3. AUDIOLOGY, SPEECH SCIENCES. *Educ:* State Univ NY Col Geneseo, BS, 55; Pa State Univ, MEd, 60, DEd(audiol), 62. *Prof Exp:* Asst prof Speech, Univ Akron, 63-65; instr oral biol, Sch Dent Med, 72-87, ASSOC PROF OTOLARYNGOL & PHYS MED, SCH MED, TUFTS UNIV, 65-; ASSOC PROF OTOLARYNGOL, DEPT SURG, STATE UNIV NY, STONYBROOK, 89- *Concurrent Pos:* HEW fel, St Elizabeth Hosp, 62-63; chief, Speech, Hearing & Lang Ctr, New Eng Med Ctr Hosp, 65-; pres, Mass Speech-Lang-Hearing Asn, 68; pres, Acoust Corp Am, 71-83; ed-treas, Assoc Serv Prog Commun Disorders, 77- & pres, 83-; chmn, Bd Prof Advisors, VNA, Boston, 78-; mem, Prof Serv Bd, Am Speech-Lang-Hearing Asn, 82-84 & Standards Coun, 86-; vpres & pres elec, Nat Alliance Stuttering; mgt consult, 87-89. *Mem:* Fel Am Speech & Hearing Asn; Acoust Soc Am; Am Cleft Palate Asn; Am Auditory Soc; Sigma Xi. *Res:* Speech perception and hearing aids; management of hearing, speech and language impairment; use of artificial speech devices for non-vocal; test and measurement in psycho-acoustic experiments; administration and management of practices and clinical programs. *Mailing Add:* Speech Lang & Hearing Prog 33 Research Way East Stauket NY 11733

GERSTNER, ROBERT W(ILLIAM), b Chicago, Ill, Nov 10, 34; m 58; c 2. STRUCTURAL ENGINEERING. *Educ:* Northwestern Univ, BS, 56, MS, 57, PhD(civil eng), 60. *Prof Exp:* From asst prof to assoc prof, 60-69, PROF STRUCT MECH, UNIV ILL, 69- *Concurrent Pos:* NSF res grant, 64-66. *Mem:* AAAS; Am Soc Civil Engrs; Am Concrete Inst; Am Soc Eng Educ. *Res:* Interface adjustment techniques and numerical methods in solution of problems in structural mechanics and elasticity; stress distributions in layered and sandwich systems. *Mailing Add:* 2628 W Agatite Ave Chicago IL 60625

GERTEIS, ROBERT LOUIS, b San Diego, Calif, Sept 1, 36; m 62; c 2. INORGANIC CHEMISTRY. *Educ:* Univ Wichita, BS, 58; Univ Ill, MS, 61, PhD(inorg chem), 63. *Prof Exp:* Res chemist, Esso Res & Eng Co, 63-68; res scientist, NJ Zinc Co, Palmerton, 68-76, sr res scientist, 76-77; sr res scientist, G&W Natural Resource Group, Bethlehem, 77-81; chief chemist, 81-83, MGR ANAL SERV, GOLD FIELDS MINING CORP, GOLDEN, CO, 83- *Mem:* Am Chem Soc. *Res:* Analysis of gold, titanium and zinc ores. *Mailing Add:* 28806 Clover Lane Evergreen CO 80439

GERTEISEN, THOMAS JACOB, b Owensboro, Ky, Oct 21, 43; m 74; c 2. ORGANIC CHEMISTRY. *Educ:* Brescia Col, BA, 66; Univ Tenn, Knoxville, PhD(chem), 70. *Prof Exp:* Asst prof chem, Brescia Col, 70-75; chem res, Sci Adv Bd, US Environ Protection Agency, Chicago, 75-76; asst prof chem, Univ Pittsburgh, Bradford, 76-78; asst prof chem, Columbus Col, 78-81;

CHEM MGR QUAL ASSURANCE, CUTTER LABS, 81- *Mem:* Am Chem Soc; Am Soc Qual Control; Am Inst Chemists; Asn Off Anal Chemists. *Res:* Physical properties of fused ring organic compounds; instrumental analysis--infrared hydrogen-bonding studies, mass spectroscopy, nuclear magnetic resonance; gas chromatographic analysis of trace organic contaminants. *Mailing Add:* Dept Biogen 14 Cambridge Ctr Cambridge MA 02142

GERTH, FRANK E, III, b San Antonio, Tex, Oct 8, 45. ALGEBRA, NUMBER THEORY. *Educ:* Rice Univ, BA, 67; Princeton Univ, MA, 71, PhD(math), 72. *Prof Exp:* Mem prof staff eng, TRW Systs, 68-70; asst math, Princeton Univ, 70-72; instr, Univ Pa, 72-74; from asst prof to assoc prof, 74-87, PROF MATH, UNIV TEX, AUSTIN, 87- *Mem:* Am Math Soc. *Res:* Ideal class groups of algebraic number fields. *Mailing Add:* Dept Math Univ Tex Austin TX 78712

GERTJEJANSEN, ROLAND O, b Vesta, Minn, May 7, 36; m 57; c 3. FOREST PRODUCTS. *Educ:* Univ Minn, BS, 61, MS, 62, PhD(wood sci), 66. *Prof Exp:* Technologist, US Forest Prod Lab, 62-63; from insr to assoc prof, 63-75, PROF FOREST PROD, COL NATURAL RESOURCES, UNIV MINN, ST PAUL, 75- *Concurrent Pos:* Forest prod consult. *Mem:* Tech Asn Pulp & Paper Indust; Forest Prod Res Soc; Soc Wood Sci & Technol. *Res:* Design and development of new wood base composites; physical and mechanical properties of wood fiber and particle products; evaluation of alternative raw materials for particleboard. *Mailing Add:* Dept Forest Prod Col Natural Resourses Univ Minn 2004 Folwell Ave St Paul MN 55108-1011

GERTLER, MENARD M, b Saskatoon, Sask, May 21, 19; nat US; m 43; c 3. INTERNAL MEDICINE. *Educ:* Univ Sask, BA, 40; McGill Univ, MD, CM, 43, MSc, 46; NY Univ, DSc, 58. *Prof Exp:* Demonstr physiol, Med Sch, McGill Univ, 45-47; resident cardiol & exec dir coronary res proj, Mass Gen Hosp, 47-50; instr med, Col Physicians & Surgeons, Columbia Univ, 50-54; assoc prof, 58-66, dir cardiovasc res, 58-86, PROF MED, RUSH INST, NY UNIV, 66- *Concurrent Pos:* Asst, Presby Hosp, NY, 50-54; asst attend physician & physician-in-chg cardiovasc dis, Francis Delafield Div, Columbia Presby Med Ctr, 51-54; consult & lectr, St Albans Naval Hosp; attend physician med, NY Univ Med Ctr, 74- *Mem:* Am Chem Soc; fel Am Col Physicians; Am Fedn Clin Res; NY Acad Sci; NY Acad Med. *Res:* Cardiovascular disease and biochemistry; atherosclerosis and epidemiology of heart disease; biochemistry of congestive heart failure. *Mailing Add:* Rush Inst 400 E 34th St New York NY 10016

GERTNER, SHELDON BERNARD, b New York, NY, Feb 16, 27; m 60; c 3. PHARMACOLOGY. *Educ:* Brooklyn Col, BS, 48; Yale Univ, PhD, 53. *Prof Exp:* Rockefeller fel, Nat Inst Med Res, London, Eng, 53-55; fel, Inst Super di Sanita, Rome, Italy, 55; assoc, Col Physicians & Surgeons, Columbia Univ, 55-57; from asst prof to assoc prof, 57-67, PROF PHARMACOL, NJ MED SCH-UMDNJ, NEWARK, 67- *Mem:* AAAS; Am Soc Pharmacol & Exp Therapeut; Pharmacol Soc Can; Soc Exp Biol & Med; Am Soc Hypertens. *Res:* Autonomic pharmacology; histaminergic neurotransmission; fat embolism. *Mailing Add:* Dept Pharmacol NJ Med Sch-UMDNJ Newark NJ 07103

GERTZ, SAMUEL DAVID, b Baltimore, Md, June 19, 47; m 73; c 3. PATHOLOGIC ANATOMY. *Educ:* Yeshiva Univ, BA, 68; Univ Md, MS, 71, PhD(anat), 75; Hebrew Univ, MD, 87. *Prof Exp:* Grad instr, Dept Anat, Sch Dent, Univ Md, 69-71, grad instr, Dept Anat, Sch Med, 71-75, res asst, 72-75, res assoc vascular path, Dept Neurol, 75-76, from instr to asst prof, 76-78; lectr, 76-78, SR LECTR, DEPT ANAT, HADASSAH MED SCH, HEBREW UNIV, JERUSALEM, 78-, CHMN DEPT, 90- *Mem:* Soc Neurosci; AAAS; Israel Soc Electron Micros; Israel Soc Anat Sci (secy, 77-78); Am Asn Anatomists. *Res:* Clinical and cardiac morphological findings in patients treated with tissue plasminogen activator during acute myocardial infarction; hemodynamic factors in rupture of atherosclerotic plague; role of coronary vasospasm in the pathogenesis of myocardial infarction. *Mailing Add:* Dept Anat Hadassah Med Sch PO Box 1172 Jerusalem Israel

GERTZ, STEVEN MICHAEL, b Philadelphia, Pa, Feb 15, 43; m 67; c 2. ENVIRONMENTAL SCIENCES, WASTE MANAGEMENT. *Educ:* Philadelphia Col Pharm & Sci, BS, 65; Drexel Univ, MS, 68, PhD(environ eng), 73. *Prof Exp:* Environ sanitarian, Philadelphia Dept Pub Health, 65-67, indust sanitarian, 68-69; sr biophysicist, Radiation Mgt Corp, 72-74; partner, Porte-Gertz Consults, Inc, 74-80; vpres, Environ Consult & Testing Serv, Inc, 80-81; vpres & assoc div mgr, Weston Inc, 81-89; VPRES & GEN MGR, BURNS & ROE ENVIRON SERV INC, 89- *Mem:* Am Soc Testing & Mat; Hazardous Mat Control Res Inst; Health Physics Soc; Acad Cert Hazard Control Mgrs; Acad Cert Hazardous Mat Mgrs. *Res:* Management of interdisciplinary environmental programs; environmental monitoring; hazardous waste management; emergency environmental response; health physics; health and safety (chemical); statistics. *Mailing Add:* 306 Bangor Rd Bala Cynwyd PA 19004

GERUGHTY, RONALD MILLS, b San Francisco, Calif, Aug 9, 32; m 55; c 5. PATHOLOGY. *Educ:* Univ Calif, DDS, 61, PhD(path), 65; Am Bd Oral Path, dipl, 67. *Prof Exp:* Asst dir clins, Sch Dent, Univ Calif, 61-62; training grant, NIH, 62-66; chmn dept oral path, Col Dent Med, Univ SC, 66-74, assoc prof path, Col Med, 66-85, prof oral path, Col Dent Med, 71-85, chief, Div Head & Neck Path, 68-85; dean, Col Health, Univ Cent Fla, 85-88. *Concurrent Pos:* Consult, Vet Admin Hosp, Charleston & Columbia, SC, 66-, US Army, Ft Jackson, 66-, Roper Hosp, 70, St Francis Xavier Hosp, Charleston, 70 & US Navy, Beaufort, 73, consult. *Mem:* Am Dent Asn; Am Acad Oral Path; Am Bronchoesophagol Asn; Int Acad Path. *Res:* Immunopathology; study of cellular immune mechanisms, including transplantation and tumor immunity; head and neck pathology, particularly cancer and its biological behavior. *Mailing Add:* Dean Health Scis Univ Cent Fla PO Box 25000 Orlando FL 32816

GERVAIS, FRANCINE, b Montreal, Que, Oct 10, 51; m 76; c 1. INFLAMMATION, AMYLOIDOSIS. *Educ:* Univ Montreal, BSc, MSc,77, PhD(microimmunol), 80-82. *Prof Exp:* Fel hemat, Res Ctr, Sacre-Coeur Hosp, Montreal, 80-82; RES ASSOC, MONTREAL GEN HOSP RES INST, 82- *Concurrent Pos:* Asst prof, McGill Univ, Dept Exp Med, 83-88; fel immunol, Bristol-Myers, 82; chmn, Animal Care Comt, 89-; assoc prof, McGill Univ, Dept Exp Med, 90- *Mem:* Am Asn Immunol; Reticulocredotheliol Soc; Can Soc Immunologists. *Res:* Role of the anyloid enhancing factor in the pathogenesis of amyloidosis (secondary type and Alzheimer's disease); genetic control of susceptibility to murine AIDS (MAIDS). *Mailing Add:* Montreal Gen Hosp Res Inst 1650 Cedar Ave Montreal PQ H3G 1A4 Can

GERVAIS, PAUL, b St Barthelemy, Que, Aug 27, 15; m 55; c 4. AGRONOMY. *Educ:* Laval Univ, BSA, 37; McGill Univ, MSc, 48; Univ Wis, PhD, 58. *Prof Exp:* Res officer cereals & forages, Exp Farm, Can Dept Agr, Lennoxville, 39-60 & legumes & pastures, 60-62; prof forages, 62-84, res officer, 84-86, EMER PROF, LAVAL UNIV, 86- *Honors & Awards:* Agron Merit Award, Quebec, 80. *Mem:* Am Soc Agron; Can Soc Agron (pres, 63-64); Crop Sci Soc Am. *Res:* Management and physiology of forage crops. *Mailing Add:* Dept Plant Sci Laval Univ Quebec PQ G1K 7P4 Can

GERVAY, JOSEPH EDMUND, b Dec 29, 31; US citizen; m 56; c 1. ORGANIC CHEMISTRY. *Educ:* Univ Montreal, BS, 61; Univ BC, MS, 63, PhD(org chem), 65. *Prof Exp:* From res chemist to sr res chemist, Res & Develop Div, Electronics Dept, 66-84, RES ASSOC, E I DU PONT DE NEMOURS & CO, INC, 84- *Mem:* Am Chem Soc; Soc Photog Sci & Eng. *Res:* Oxo reaction, Fischer-Tropsch synthesis, phenol oxidations and biosynthesis of alkaloids and steroids in vivo and in vitro using isotopic labels; photopolymers; principles, processes and materials; photoresists, unconventional photographic systems; photographic chemistry; organic reaction mechanism; photopolymer systems; electronic materials; electroless plating processes; printed circuit board manufacture; permanent protective coatings; polymer chemistry; solder mask coating; flex printed circuits. *Mailing Add:* Dupont Exp Sta 336/106 Wilmington DE 19880-0336

GERWE, RAYMOND DANIEL, b Cincinnati, Ohio, May 28, 04; m 33; c 2. CHEMISTRY. *Educ:* Univ Miami, Ohio, BS, 27; Univ Cincinnati, MA, 29, PhD(org chem), 32. *Prof Exp:* Instr chem, Oxford Col, 27-28; head res & develop sect, Kroger Food Found, 32-38; dir res, Fla Div, FMC Corp, 38-70; RETIRED. *Concurrent Pos:* Consult, Citrus Processing, 73-; past chmn-Fla sect, Inst Food Technol. *Mem:* Am Chem Soc; Inst Food Tech; Sigma Xi. *Res:* Food technology, processing and chemistry; chemical education; photochemistry as applied to organic chemistry. *Mailing Add:* 2131 Reaney Rd Lakeland FL 33803

GERWICK, BEN CLIFFORD, JR, b Berkeley, Calif, Feb 22, 19; m 41; c 4. CIVIL ENGINEERING. *Educ:* Univ Calif, BS, 40. *Prof Exp:* US Navy 40-45; mem staff, Ben C Gerwick, Inc, San Francisco, 46-52, pres, 52-70; exec vpres, Santa Fe-Pomeroy, Inc, 68-71; prof, 71-89, EMER PROF CIVIL ENG, UNIV CALIF, BERKELEY, 89-; CHMN, BEN C GERWICK INC, CONSULT ENGRS, SAN FRANCISCO, 88- *Concurrent Pos:* Sponsoring mgr, Richmond-San Rafael Bridge substruct, 53-56 & San Mateo-Hayward Bridge, 64-66; lectr construct eng, Stanford Univ, 62-68; consult major bridge & marine construct projs; consult construct engr ocean struct & concrete offshore struct in North Sea, Japan, MidEast, Australia, Southeast Asia & Arctic; chmn marine bd, Nat Res Coun, 78-80. *Honors & Awards:* Turner Award & Corbetta Award, Am Concrete Inst, 74; Karp Award, Am Soc Civil Engrs, 76; Lockheed Award, Marine Technol Soc, 77; Blakeley Smith Medal, Soc Naval Architects & Marine Engrs, Franklin Inst, Frank P Brown Medal, 81; Freyssinet Medal, Int Fedn Prestressing, 82; Emil Morsch Medal, Deutsche Beton Verein, 82; Swedish Concrete Award, 85; Presidents Award & Peurifoy Award, Am Soc Civil Engrs, 89. *Mem:* Nat Acad Eng; hon mem Am Soc Civil Engrs; fel Am Concrete Inst; Fedn Int Precontrainte (pres, 74-78); French Concrete Soc; Prestressed Concrete Inst (pres, 57-58); Soc Naval Architects & Marine Engrs; Royal Swedish Acad Tech Sci; Norweg Acad Tech Sci; Ger Engrs Soc. *Res:* Marine, ocean and arctic structures; concrete structures. *Mailing Add:* 601 Montgomery St Rm 1400 San Francisco CA 94111

GERWICK, BEN CLIFFORD, III, b Berkeley, Calif, Mar 2, 53; m 74; c 2. PLANT PHYSIOLOGY, WEED SCIENCE. *Educ:* Alaska Methodist Univ, BA, 74; Wash State Univ, PhD(bot), 78. *Prof Exp:* Instr phys educ, Alaska Methodist Univ, 74; res assoc plant physiol, Univ Ga, 78-79; Sr res biologist, Dow Chemical, 79-86, leader, 86-90; TECH LEADER, HERBICIDE RES, DOW ELANCO, 90- *Concurrent Pos:* NSF fel, 78-79. *Mem:* Am Soc Plant Physiologists; Sigma Xi; Weed Sci Soc. *Res:* Discovery of new herbicides and herbicide; mode-of-actions. *Mailing Add:* Dow Elanco 2001 W Main St Greenfield IN 46140

GERWIN, BRENDA ISEN, b Boston, Mass, May 2, 39; m 60; c 3. MOLECULAR BIOLOGY. *Educ:* Radcliffe Col, BA, 60; Univ Chicago, PhD(biochem), 64. *Prof Exp:* Res assoc biochem, Rockefeller Inst, 64-66; instr, Sch Med, Case Western Reserve Univ, 66-69; biochemist, Molecular Anat Prog, Oak Ridge Nat Lab, 69-71; sr staff fel, Nat Cancer Inst, 71-73, CHEMIST, NAT CANCER INST, 73- *Mem:* Am Soc Biochem & Molecular Biol; Am Soc Microbiol; Sigma Xi; AAAS; Am Chem Soc. *Res:* cancer research; enzymology; cell biology. *Mailing Add:* Bldg 37 Rm 2C15 Nat Cancer Inst Bethesda MD 20892

GERWIN, RICHARD A, b Chicago, Ill, March 13, 34; m 60; c 2. PLASMA INSTABILITIES, PLASMA TRANSPORT. *Educ:* Univ Chicago, BA, 54, BS, 56, MS, 57; Eindhoven Tech Univ, DSc, 66. *Prof Exp:* Staff mem plasma physics, Boeing Sci Res Lab, 59-71; staff mem, Plasma Focus Group, Los Alamos Nat Lab, 71-73, Magnetic Fusion Theory Group, 73-79, group leader, 79-89, STAFF MEM, MAGNETIC FUSION THEORY GROUP, LOS ALAMOS NAT LAB, 89- *Concurrent Pos:* Vis scientist, Inst Fundamental Res on Matter, Neth, 64-66; vis prof, dept nuclear eng, Univ Wash, Seattle,

85. *Mem:* Fel Am Phys Soc. *Res:* Theoretical plasma physics within the context of the magnetic fusion energy goals; magnetic confinement of hot plasmas sufficient to obtain useful energy from fusion reactions; plasma thrusters. *Mailing Add:* Los Alamos Nat Lab MS F647 Los Alamos NM 87545

GERY, IGAL, IMMUNOLOGY. *Educ:* Hebrew Univ, Jerusalem, PhD, (immunol), 63. *Prof Exp:* SECT HEAD EXP IMMUNOL, NAT EYE INST, NIH, 81- *Res:* autoimmunity in the eye. *Mailing Add:* Nat Eye Inst NIH Bldg 10 Rm 10N208 Bethesda MD 20892

GESCHKE, CHARLES MATTHEW, b Cleveland, Ohio, Sept 11, 39; m 64; c 3. COMPUTER SCIENCE. *Educ:* Xavier Univ, Ohio, AB, 62, MS, 63; Carnegie-Mellon Univ, PhD(comput sci), 72. *Prof Exp:* Instr math, John Carroll Univ, 63-68; res scientist comput sci, Palo Alto Res Ctr, Xerox Corp, 72-80, mgr, Imaging Sci Lab, 80-87; PRES, ADOBE SYSTS, 87- *Mem:* Asn Comput Mach; Math Asn Am. *Res:* Programming languages; machine design for efficient emulation of higher level languages; computer imaging and graphics. *Mailing Add:* Adobe Systs 1585 Charleston Rd PO Box 7900 Mountain View CA 94043

GESCHWIND, STANLEY, b Brooklyn, NY, Nov 22, 21; m 57; c 3. PHYSICS. *Educ:* City Col New York, BS, 43; Univ Ill, MS, 47; Columbia Univ, PhD(physics), 51. *Prof Exp:* Res physicist, Columbia Univ, 51-52; physicist, Bell Labs, Murray Hill, 52-66, head, Quantum & Solid State Physics Dept, 66-83; RETIRED. *Mem:* Fel Am Phys Soc; Sigma Xi. *Res:* Microwave spectroscopy; millimeter wave generation; magnetism; paramagnetic resonance; optical spectra of solids; microwave optical double resonance in solids; light scattering. *Mailing Add:* PO Box 883 New Providence NJ 07974

GESELL, THOMAS FREDERICK, b East Cleveland, Ohio, Apr 28, 40; m 64; c 3. HEALTH PHYSICS, RADIATION DOSIMETRY. *Educ:* San Diego State Univ, BS, 65; Univ Tenn, MS, 68, PhD(physics), 71. *Prof Exp:* From asst prof to assoc prof health physics, Sch Pub Health, Univ Tex Health Sci Ctr, Houston, 71-81. *Concurrent Pos:* Health physics fel, 65-68, radiol health fel, 68-71, consult, Tex State dept, Health Resources, 73-81, Houston Lighting & Power Co, 73-76 & US Environ Protection Agency, 74-81; adj assoc prof, Rice Univ, 75-81; staff mem, President's Comn on the accident at Three Mile Island, 79-80; mem, sci comts, Nat Coun Radiation Protection & Measurements, 80-; Health Physics Soc Standards Comt, 81-87; assoc ed, Health Physics J, 85-90, Dept Energy Liason to Idaho State Gov, 89, consult. *Mem:* Health Physics Soc. *Res:* Measurement of ionizing radiation, especially beta dosimetry, and personnel and environmental dosimetry; evaluation of human exposure to radionuclides in the environment, especially radon and its decay products. *Mailing Add:* Radiol & Environ Sci Lab 785 Doe Pl Idaho Falls ID 83402

GESELOWITZ, DAVID B(ERYL), b Philadelphia, Pa, May 18, 30; m 53; c 3. BIOMEDICAL ENGINEERING. *Educ:* Univ Pa, BS, 51, MS, 54, PhD(elec eng), 58. *Prof Exp:* From asst prof to assoc prof, Univ Pa, 58-71; PROF BIOENG, PA STATE UNIV, 71-, PROF MED, 82- *Concurrent Pos:* Consult, Provident Mutual Life Ins Co, 59-71, Burroughs Corp, 61-64 & Vet Admin Hosp, Washington, DC, 62-74; NIH fel & vis assoc prof, Mass Inst Technol, 65-66; ed, Trans Biomed Eng, Inst Elec & Electronics Engrs, 67-71; chmn electrocardiog comt, Am Heart Asn, 76-81; Guggenheim fel & vis prof, Duke Univ, 78-79; circulatory syst devices panel, Food & Drug Admin, 83-87. *Honors & Awards:* Centennial Medal, Inst Elec & Electronics Engrs, 84; Career Achievement Award, Inst Elec & Electronic Engrs/EMRS Soc. *Mem:* Nat Acad Eng; Biomed Eng Soc; fel Inst Elec & Electronics Engrs; Int Soc Comput Electrocardiogrpahy; fel Am Col Cardiol; AAAS. *Res:* Electrocardiography; cardiac electrophysiology; artificial hearts. *Mailing Add:* 232 Hallowell Bldg Pa State Univ University Park PA 16802

GESHNER, ROBERT ANDREW, b Chicago, Ill, Feb 8, 28; m 52; c 2. PHYSICS, MATHEMATICS. *Educ:* Cornell Univ, AB, 52. *Prof Exp:* Engr, Western Elec Co, 52-56; dept head photog & printed wire develop, Gen Dynamics Corp, 55-61; prin engr, Govt & Commercial Systs Div, RCA Corp, 61-64, head, Microphotolithography Lab, 74-80, mgr, Solid State Technol Ctr, Advan Mask Technol, 80-83, MGR, SOLID STATE DIV, PHOTOMASK TECHNOL OPERS, RCA CORP, 83- *Honors & Awards:* Inst Printed Circuits President's Award, 66. *Mem:* Sr mem Inst Elec & Electronics Engrs. *Res:* Printed wiring manufacturing and artwork; microelectronic packaging; solid state device microphotolithography involving optics, metrology, microphotographic processes, laser and electron beam lithography and inspection sampling plans. *Mailing Add:* Rockaway Rd Lebanon NJ 08833

GESINSKI, RAYMOND MARION, b Monessen, Pa, July 16, 32; m 66. CELL BIOLOGY, HEMATOLOGY. *Educ:* Kent State Univ, BS, 60, MA, 62, PhD(biol sci), 68. *Prof Exp:* From instr to prof, 62-85, EMER PROF BIOL SCI, KENT STATE UNIV, 85- *Concurrent Pos:* Vis scholar, Bayside Lab, Univ Del, Lewes, 65; Tuscarawas County Univ Found res grant, 70; res fel, Kent State Univ, 70-; vis scientist, Manned Space Ctr, 71-; pres, Midwest Biol Corp; chmn bd, Geltech Inc, 87. *Mem:* AAAS; Exp Hemat Soc; Am Soc Cell Biol; Sigma Xi. *Res:* Influence of micro-environment modifications on hemic cellular metabolism; homoestasis of cell differentiation, maturation, and molecular determinants of neoplasms. *Mailing Add:* 6000 Frank NW Canton OH 44720

GESKIN, ERNEST S, b Dnepropetrovsk, USSR, June 4, 35; US citizen; m 61; c 1. MECHANISM OF WATERJET MACHINING, WATERJET MACHINING TECHNOLOGY. *Educ:* Dnepropetrovsk Inst Metall, USSR, MS, 57; Moscow Inst Steel & Alloys, PhD(metall), 67. *Prof Exp:* Engr, Inst Automation, USSR, 56-67, mgr lab, 67-74; assoc res prof thermal sci, George Washington Univ, 77-78; assoc prof, Clarkson Col Technol, 79-80; res scientist, Revere Copper & Bass, 81-83; spec lectr, 84-85, assoc prof, 86-89, PROF THERMAL SCI & MFG, NJ INST TECHNOL, 89- *Honors & Awards:* Cert Recognition, Minerals, Metals & Mat Soc, 89; Cert Recognition, Am Soc Mech Engrs, 90 & 91. *Mem:* Metall Soc; Waterjet Technol Asn; Sigma Xi. *Res:* Development of the formalism of nin-equilibrium thermodynamics; second law analysis and variational methods; application of the second law techniques to development of manufacturing technologies; technology and fundamentals of waterjet machining. *Mailing Add:* Mech Eng Dept NJ Inst Technol 323 King Blvd Newark NJ 07102

GESNER, BRUCE D, b Fall River, Mass, May 7, 38; m 59; c 4. ORGANIC POLYMER CHEMISTRY. *Educ:* Bradford Durfee Col Technol, BS, 60; Univ Idaho, PhD(org chem), 63. *Prof Exp:* Polymer chemist, Bell Tel Lab, 63-69, supvr mat & microconnections group, 69-71, supvr mat chem group, 71-78, field rep, Bell Labs, 78-83, div mgr field liaison, Bellcore, 84; DIV MGR TECHNOL INTROD & SUPPORT, PAC BELL, 84- *Concurrent Pos:* Adj prof, Morris Brown Col, Atlanta, GA, 73; nat defense act Fel, 60-63. *Mem:* Am Chem Soc; AAAS; Sigma Xi. *Res:* Organic reactions mechanisms, composition and structure of polyblends; time dependent relationship of conformationally mobile structures. *Mailing Add:* 1300 Fountain Springs Circle Danville CA 94526-5625

GESSAMAN, JAMES A, b Dayton, Ohio, Dec 10, 39; c 3. PHYSIOLOGICAL ECOLOGY, AVIAN PHYSIOLOGY. *Educ:* Univ Ill, Urbana, MS, 64, PhD(zool), 68. *Prof Exp:* Asst prof physiol, 68-73, assoc prof biol, 73-89, PROF BIOL, UTAH STATE UNIV, 90- *Res:* Bioenergetics; thermoregulation; hawk migration. *Mailing Add:* Dept Biol Utah State Univ Logan UT 84322

GESSAMAN, MARGARET PALMER, b Florence, Ariz, Oct 7, 34; m 65. NONPARAMETRIC STATISTICS, DISCRIMINATION. *Educ:* Mont State Univ, BS, 56, MS, 65, PhD(statist), 66. *Prof Exp:* Asst prof math, Ithaca Col, 67-70; from asst prof to assoc prof, 70-76, head dept, 73-80, PROF MATH, UNIV NEBR, OMAHA, 76-,. *Concurrent Pos:* Mem, Math Achievement Comt, Col Level Exam Bd, Educ Testing Serv, 75-80. *Mem:* Am Math Soc; Am Statist Asn; Asn Comput Mach; Inst Math Statist. *Res:* Nonparametric statistics, discrimination and density estimation and statistical decision theory. *Mailing Add:* Dean Grad Studies & Res Univ of Nebr Omaha NE 68182

GESSEL, IRA MARTIN, b Philadelphia, Pa, Apr 9, 51; m 89. ENUMERATIVE COMBINATORICS. *Educ:* Harvard Univ, AB, 73; Mass Inst Technol, PhD(math), 77. *Prof Exp:* Fel, T J Watson Res Ctr, IBM, 77-78; instr appl math, Mass Inst Technol, 78-80, asst prof, 80-84; from asst prof to assoc prof, 84-90, PROF MATH & COMPUT SCI, BRANDEIS UNIV, 90- *Concurrent Pos:* Vis asst prof math, Univ Calif, San Diego, 78. *Mem:* Am Math Soc; Math Asn Am; Asn Comput Mach. *Res:* Enumerative combinatorics; problems involving counting permutations, paths and graphs. *Mailing Add:* Dept Math Brandeis Univ Waltham MA 02254-9110

GESSEL, STANLEY PAUL, b Utah, Oct 14, 16; m 47, 74; c 5. FORESTRY, SOILS. *Educ:* Utah State Agr Col, BS, 39; Univ Calif, Berkeley, PhD(soils), 50. *Prof Exp:* Asst, Univ Calif, 46-48; instr, Univ Wash, 48-51, from asst prof to assoc prof, 51-62, prof forest soils, 62-84, assoc dean, Col Forest Resources, 64-84, EMER PROF, UNIV WASH, 84- *Concurrent Pos:* NSF lectr, 63; dir western coniferous forest biome, Int Biol Prog; consult, Forestry & Timber Bur, Australia, 71 & Forestry Bur, Taiwan, 74; Forestry Comm NSW, 83-84, 87; Forest Res Inst, New Zealand, 87; consult, Indonesia, 90. *Honors & Awards:* Distinguished Serv Award, Int Union Forestry Res Orgn, 89. *Mem:* Fel AAAS; fel Soil Sci Soc Am; fel Soc Am Foresters; Soc Range Mgt; Am Geophys Union. *Res:* Forest soils, forest growth, forest ecology; mineral cycling and tree nutrition; tropical forestry, especially soils. *Mailing Add:* Col Forest Resources Univ Wash Seattle WA 98195

GESSER, HYMAN DAVIDSON, b Montreal, Que, Apr 24, 29; m 52; c 3. PHYSICAL CHEMISTRY. *Educ:* Loyola Col, Can, BSc, 49; McGill Univ, PhD(phys chem), 52. *Prof Exp:* Fel photochem, Univ Rochester, 52-54; Nat Res Coun Can fel, Ottawa, Can, 54-55; assoc prof, 55-67, PROF CHEM, UNIV MAN, 67- *Concurrent Pos:* Res fel, Israel Inst Technol, 61-62; UN Develop Prog tech asst expert, Ctr Indust Res, Israel, 68-69. *Honors & Awards:* Borden Award, 83. *Mem:* Am Chem Soc; fel Chem Inst Can; Chem Soc; Sigma Xi. *Res:* Atomic, thermal and photochemical kinetics; gas chromatography; surface reactions; liquid fuels from natural gas. *Mailing Add:* Dept Chem Univ Man Winnipeg MB R3T 2N2 Can

GESSERT, CARL F, b St Louis, Mo, Apr 14, 23; m 58; c 2. PHARMACOLOGY. *Educ:* Wash Univ, AB, 49; Univ Wis, PhD(biochem), 55. *Prof Exp:* Res asst, Sch Med, Wash Univ, 55-58, res instr, 58-62; asst prof pharmacol, 62-67, actg chmn dept, 67-69, ASSOC PROF PHARMACOL, COL MED, UNIV NEBR, OMAHA, 67- *Mem:* Affil Am Soc Pharmacol & Exp Therapeut; Sigma Xi. *Res:* Pharmacology of neurotransmitters. *Mailing Add:* Dept Pharmacol Univ Nebr Col Med Omaha NE 68198-6260

GESSERT, WALTER LOUIS, b Detroit, Mich, May 26, 19; m 44; c 3. PHYSICS. *Educ:* Eastern Mich Univ, BS, 44; Wayne State Univ, MS, 47; Mich State Univ, PhD(physics), 54. *Prof Exp:* Instr gen physics, Wayne State Univ, 46-50; asst ultrasonics res, Mich State Univ, 51-54; physicist tire noise, US Rubber Co, 54-57, group leader in chg passenger tire design, 57-61; from assoc prof to prof, 61-86, EMER PROF PHYSICS & ASTRON, EASTERN MICH UNIV, 86- *Concurrent Pos:* Teacher musical acoust, Eastern Mich Univ, 86- *Mem:* Acoust Soc Am; Am Asn Physics Teachers; Optical Soc Am; Sigma Xi. *Res:* Ultrasonics; instrumentation; optics. *Mailing Add:* Dept Physics-Astron Eastern Mich Univ Ypsilanti MI 48197

GESSLER, ALBERT MURRAY, b Staten Island, NY, Nov 10, 18; m 43; c 2. RUBBER CHEMISTRY. *Educ:* Cornell Univ, BS, 42. *Prof Exp:* Chemist synthetic rubber, Standard Oil Develop Co, 42-58, res assoc, Esso Res & Eng Co, 58-66, SR RES ASSOC SYNTHETIC RUBBER, CHEM RES DIV, EXXON CHEM CO, 66- *Concurrent Pos:* Chmn, NY Rubber Group, 65; chmn elastomers, Gordon Res Conf, 71. *Mem:* Brit Inst Rubber Indust; Am Chem Soc. *Res:* Elastomer and high polymer chemistry and technology; reinforcement with carbon blacks. *Mailing Add:* 448 Orchard Cranford NJ 07016

GESSLER, JOHANNES, b Basel, Switz, Oct 19, 36; m 62; c 4. FLUID MECHANICS, HYDRAULICS. *Educ:* Swiss Fed Inst Technol, BS, 60, PhD(hydraul), 65. *Prof Exp:* Jr engr, Swisselectra, Basel, 60-61; res engr, Lab Hydraul Res, Swiss Fed Inst Technol, 61-66; asst prof civil eng, 66-70, assoc prof, 70-88, PROF CIVIL ENG, COLO STATE UNIV, 88- *Mem:* Am Soc Civil Engrs; Swiss Eng & Archit Soc; Int Asn Hydraul Res. *Res:* River mechanics, especially sediment transport; ophmization techniques of water supply systems. *Mailing Add:* 3205 Shore Rd Ft Collins CO 80524

GESSNER, ADOLF WILHELM, b Berlin, Ger, Aug 26, 28; m 53; c 2. CHEMICAL ENGINEERING. *Educ:* Williams Col, BA, 52; Mass Inst Technol, ScD(chem eng), 54. *Prof Exp:* Process engr, Chemische Werks Huls, WGer, 54-55; develop engr, Jones & Laughlin Steel Corp, Pa, 55-57; process engr & mgr chem eng sect, Sci Design Co, NY, 57-61; sr develop engr, Lummus Co, NJ, 61-67; supvr process develop, Foster Wheeler Corp, 67-70; dir eng & develop, Givaudan Corp, Clifton, 70-80; sr process coordr, Lurgi Corp, River Edge, NJ, 80-82; sr engr, Synthetic Fuels Corp, Washington DC, 82-86; chief, tech assessment br, US Dept Energy, Morgantown, WVa, 86-87; ASSOC, CHENPATENTS, SILVER SPRING, MD, 85-; MGR ENVIRON ENG & DEVELOP LAB, VERSAR INC, SPRINGFIELD, VA, 87- *Concurrent Pos:* Lectr, Polytech Inst Brooklyn, 61-63, adj prof, 63- *Mem:* Am Chem Soc; Am Inst Chem Engrs; Sigma Xi. *Res:* Thermodynamics; unit operations. *Mailing Add:* 3013 Birchtree Lane Silver Spring MD 20906-3035

GESSNER, FREDERICK B(ENEDICT), b Newark, NJ, June 11, 37; m 62; c 4. MECHANICAL ENGINEERING. *Educ:* Lehigh Univ, BS, 59; Purdue Univ, MS, 60, PhD(mech eng), 64. *Prof Exp:* Res scientist, Res Div, Am Radiator & Standard Sanit Corp, 63-65; asst prof mech eng, Va Polytech Inst, 65-67; from asst prof to assoc prof, 67-76, PROF MECH ENG, UNIV WASH, 77- *Concurrent Pos:* NASA res grants, 66-67 & 76-; consult res div, Am Radiator & Standard Sanit Corp, 66-67; NSF res grants, 68-71 & 73-79; assoc ed, J Fluids Eng, 75-78; Off Naval Res grant, 76- *Mem:* Am Soc Mech Engrs; Sigma Xi; Am Soc Eng Educ; Am Inst Aeronaut & Astronaut. *Res:* Fluid mechanics; heat transfer; turbulence; secondary flow in non-circular ducts; subsonic flow in diffusers; pressure-flow behavior in distensible tubes; fluid dynamic studies of flow in the lower urinary tract; Reynolds stress modeling of corner flows; supersonic corner flows; hot-wire measurement techniques for complex flows. *Mailing Add:* Dept Mech Eng Univ Wash Seattle WA 98195

GESSNER, IRA HAROLD, b Rockville Center, NY, June 23, 31; m 59; c 3. PEDIATRIC CARDIOLOGY. *Educ:* State Univ Iowa, AB, 52; Univ Vt, MD, 56; Am Bd Pediat, dipl, 64, cert cardiol, 66. *Prof Exp:* Intern pediat, Ohio State Univ, 56-57; resident, 60-61, chief resident, 62, from instr to assoc prof, 62-70, PROF PEDIAT, COL MED, UNIV FLA, 70- *Concurrent Pos:* Mem coun cardiovasc dis in the young, Am Heart Asn; Am Heart Asn advan res fel, Univ Fla, 62-64 & Wenner-Gren Inst, Stockholm, 64-65; NIH career develop award, 67; mem sub-bd pediat cardiol, Am Bd Pediat, 73-79. *Mem:* Soc Pediat Res; Teratology Soc; Am Acad Pediat; fel Am Col Cardiol; Am Heart Asn; Am Pediat Soc. *Mailing Add:* Div Pediat Cardiol Med Col Univ Fla J Hillis Health Ctr Gainesville FL 32610

GESSNER, PETER K, b Warsaw, Poland, May 3, 31; m 59; c 2. PHARMACOLOGY. *Educ:* Univ London, BSc, 55, PhD(biochem), 58. *Prof Exp:* Res assoc, Res Div, Cleveland Clin, 58-61, asst staff mem, 61-62; asst prof pharmacol, 62-67, assoc prof, 67-75, PROF PHARMACOL, STATE UNIV NY BUFFALO, 75- *Concurrent Pos:* Career teacher alcoholism & substance abuse, Nat Inst Alcohol Abuse & Alcoholism/Nat Inst Drug Abuse, 78-82. *Mem:* Am Soc Pharmacol & Exp Therapeut. *Res:* Characterization of drug interactions including investigation of concurrent changes in the metabolic fate of the drugs and other biochemical events to determine mechanism; study of interaction of drugs and ethanol; study of withdrawal phenomena; evaluation of instruction; phenomenology and mechanism of action of hallucinogens; study of absorption on activated charcoal; pharmacological action and metabolic fate of disulfiram and diethyldithiocarbamate. *Mailing Add:* Dept of Pharmacol Sch of Med SUNY Health Sci Ctr 102 Faber Hall 3435 Main St Buffalo NY 14214

GESSNER, ROBERT V, b Elizabeth, NJ, Apr 17, 48; m 81. MYCOLOGY, MICROBIOLOGY. *Educ:* Rutgers Univ, BS, 70; Univ RI, MS, 72, PhD(biol sci), 75. *Prof Exp:* Fel marine mycol, Inst Marine Sci, Univ NC, Chapel Hill, 75-76, res assoc & instr, 76-78; from asst prof to assoc prof, 78-88, PROF, DEPT BIOL SCI, WESTERN ILL UNIV, 88- *Concurrent Pos:* Res assoc, Morton Arboretum, 81- *Mem:* Mycol Soc Am; Brit Mycol Soc; Am Soc Microbiol; Am Phytopath Soc; Int Soc Arboricult. *Res:* Ecology, physiology and morphology of fungi; morels; arboriculture; salt-marsh and marine fungi. *Mailing Add:* Dept Biol Sci Western Ill Univ Macomb IL 61455

GESSNER, TERESA, b Stanislawow, Poland, Sept 28, 33; US citizen. BIOCHEMICAL PHARMACOLOGY, TOXICOLOGY. *Educ:* Univ London, BSc, 56, PhD(org chem), 59. *Prof Exp:* Res chemist, B F Goodrich Co, Cleveland, 59-62; res assoc med chem, State Univ NY Buffalo, 62-64, from instr to asst prof biochem pharmacol, 64-71; sr cancer res scientist, 72-75, ASSOC CANCER RES SCIENTIST, DEPT EXP THERAPEUT, ROSWELL PARK MEM INST, 75- *Concurrent Pos:* From res asst prof to res assoc prof pharmacol, Roswell Park Grad Div, State Univ NY Buffalo, 71-77, res prof, 77-, dir grad studies, 74-76; mem, Nat Bladder Cancer Proj Study Sect, 77- *Mem:* Biochem Soc Gt Brit; Am Chem Soc; Am Soc Pharmacol & Exp Therapeut. *Res:* Metabolism of drugs, carcinogens and other xenobiotics; metabolic conjugations; drug interactions. *Mailing Add:* Dept Pharm Roswell Park Mem Inst 666 Elm St Buffalo NY 14263

GESSOW, ALFRED, b Jersey City, NJ, Oct 13, 22; m 47; c 4. PHYSICS, APPLIED MATHEMATICS. *Educ:* City Col New York, BCE, 43; NY Univ, MAero Eng, 44. *Prof Exp:* Aeronaut res scientist, Langley Res Ctr, Nat Adv Cmt Aeronaut, Va, 44-49, head fluid & space physics sect, NASA HQ, 59-61, chief fluid physics br, 61-67, asst dir res div, 67-71, chief areodyn & fluid mech, Aerodyne & Vehicle Systs Div, Off Advan Res & Technol, 71-80; PROF & DIR, CTR ROTORCRAFT EDUC RES, UNIV MD, 80- *Concurrent Pos:* Lectr, Grad Exten Ctr, Univ Va, Hampton, 45-58; Adv Group Aeronaut Res & Develop helicopter consult to France & Ger, 59; exec secy adv comt fluid mech, NASA, 59-63, mem res & tech subcomt fluid mech, 67-, mem subcomt electrophys & adv coun basic res, 68-; adj prof, NY Univ, 68 & Cath Univ Am, 70-; chmn indust & prof adv coun, Dept Aerospace Eng, Pa State Univ, 70-; invited prof, Korean Inst Advan Sci, Seoul, Korea, 80; mem, bd army sci & technol, Nat Acad Sci, 80-86; mem, Army Sci Bd, 86-; chmn dept aerospace eng, Univ Md. *Honors & Awards:* Nikolsky Hon Lectureship, Am Helicopter Soc, 85. *Mem:* Fel Am Inst Aeronaut & Astronaut; hon fel Am Helicopter Soc(founding ed of jour, 55); Am Soc Eng Educ. *Res:* Rotating-wing aerodynamics; fluid physics; electrophysics; aerodynamics; computational aerodynamics. *Mailing Add:* 7308 Durbin Terr Bethesda MD 20817

GEST, HOWARD, b London, Eng, Oct 15, 21; nat US; m 41; c 3. MICROBIOLOGY. *Educ:* Univ Calif, Los Angeles, BA, 42; Wash Univ, PhD(microbiol), 49. *Prof Exp:* Asst metall lab, Univ Chicago, 43; from jr chemist to assoc chemist, Clinton Labs, Tenn, 43-46; asst radiol, Sch Med, Wash Univ, 46-49; from instr to assoc prof microbiol, Sch Med, Western Reserve Univ, 49-59; prof microbiol, Henry Shaw Sch Bot, Wash Univ, 59-64, dept zool, 64-66; chmn dept, 66-70, prof microbiol, 66-78, DISTINGUISHED PROF MICROBIOL, DEPT BIOL, IND UNIV, BLOOMINGTON, 78-, ADJ PROF, DEPT HIST PHILOS SCI, 83- *Concurrent Pos:* USPHS spec res fel, Calif Inst Technol, 56-57; mem adv panel metab biol, NSF, 63-66; NSF sr fel, Nat Inst Med Res, London, 65-66; mem, Study Sect Bact & Mycol, NIH, 66-68, chmn, Study Sect Microbiol Chem, 68-69 & Study Sect Microbiol Physiol & Genetics, 88-90; mem comt microbiol probs man in extended space flight, Nat Acad Sci-Nat Res Coun, 67-69; Guggenheim fel, Imp Col, Univ London, Univ Stockholm & Univ Tokyo, 70; vis prof, Univ Tokyo & Japanese Soc Promotion Sci, 70; Guggenheim fel, Univ Calif, Los Angeles & Imp Col, Univ London, 80; distinguished fac lect award, Ind Univ, 87. *Mem:* Am Soc Microbiol; Am Soc Biochem & Molecular Biol; Brit Soc Gen Microbiol; Am Acad Microbiol. *Res:* Physiology and intermediary metabolism of microorganisms; photosynthesis; metabolism of molecular hydrogen and nitrogen; electron transport mechanisms; metabolic regulatory mechanisms; biogeochemistry. *Mailing Add:* Photosynthetic Bact Group Biol Dept Ind Univ Jordan Hall 469Hall Bloomington IN 47405

GESTELAND, RAYMOND FREDERICK, b Madison, Wis, April 2, 38; m 60; c 4. GENE REGULATION, TRANSLATION. *Educ:* Univ Wis, BS, 60, MS, 61; Harvard Univ, PhD(biochem), 65. *Prof Exp:* NSF fel, Inst Molecular Biol, Univ Geneve, 62-66; asst dir res, Cold Spring Harbor, NY, 67-78; INVESTR & PROF HUMAN GENETICS, HOWARD HUGHES MED INST, UNIV UTAH,78- *Res:* Regulation of gene expression and the mechanism of decoding, including nonsense suppressors, tRNA structure and context effects. *Mailing Add:* Dept Human Genetics Howard Hughes Med Inst Univ Utah 6160 Eccles Genetics Bldg Salt Lake City UT 84112

GESTELAND, ROBERT CHARLES, b Madison, Wis, July 1, 30; m 61; c 3. NEUROPHYSIOLOGY, ELECTRICAL ENGINEERING. *Educ:* Univ Wis, BS, 53; Mass Inst Technol, SM, 57, PhD(neurophysiol), 61. *Prof Exp:* Engr, Gen Radio Co, 53-54; mem staff, Harvard-Peabody-Smithsonian Exped, Kalahari Desert, 57-58; mem res staff, Electronics Res Lab, Mass Inst Technol, 61-62, res assoc biol, 62-65; mem res staff life sci, Sci Eng Inst, 62-65; assoc prof elec eng & biol, Northwestern Univ, 65-67, assoc prof to prof biol sci, 67-80, prof neurobiology & physiol, 80-85; PROF ANAT & CELL BIOL, UNIV CINCINNATI MED CTR, 85-, ASSOC DEAN, BIOMED SCI, 89- *Concurrent Pos:* Consult, Arthur D Little Co, Inc, Mass Ment Health Ctr, Invention Group Inc & Unilever NV; pres, Taste & Smell Consult Group, Inc, Evanston, Ill, 79-87; NIH Javits, Pepper Investr. *Mem:* AAAS; Am Soc Zool; Am Phys Soc; Soc Neurosci; Am Chemoreception Sci. *Res:* Sensory neurophysiology; electrode processes; electronic circuit theory. *Mailing Add:* 318 Chenora Ct Cincinnati OH 45215

GESUND, HANS, b Vienna, Austria, Sept 18, 28; nat US; m; c 2. STRUCTURAL ENGINEERING. *Educ:* Yale Univ, BEng, 50, MEng, 53, DEng, 58. *Prof Exp:* Instr civil eng, Yale Univ, 53-58; dir grad studies civil eng, 84-88, PROF STRUCT ENG, UNIV KY, 58-, CHMN, DEPT CIVIL ENG, 87- *Mem:* Fel Am Soc Civil Engrs; Am Soc Eng Educ; Int Asn Bridge & Struct Engrs; fel Am Concrete Inst. *Res:* Structural design; reinforced concrete; structural mechanics; limit and ultimate load design; design of concrete slabs; author of more than 50 publications.. *Mailing Add:* Dept Civil Eng Univ Ky Lexington KY 40506-0046

GETCHELL, THOMAS V, b Erie, Pa, Nov 16, 39; m; c 2. PHYSIOLOGY, BIOPHYSICS. *Educ:* Gannon Col, BA, 63; Villanova Univ, MS, 66; Northwestern Univ, PhD(neurosci), 69. *Prof Exp:* Instr, Biol Dept, Gannon Col, 63-64; assoc physiol, Med Sch & assoc mem, Monell Chem Senses Ctr, Univ Pa, 69-72; NIH spec fel neurophysiol, Yale Univ, 73-74, asst prof, Dept Physiol, 74-78; from assoc prof to prof, Dept Anat & Cell Biol, Wayne State Univ, 78-89, chairperson, Neurosci Prog, 81-85, assoc dean, Grad Sch, 86-89; PROF, DEPT PHYSIOL & BIOPHYS, UNIV KY, 89-, ASSOC DEAN RES & BASIC SCI, UNIV KY, 89- *Concurrent Pos:* Vis scientist, Roche Inst Molecular Biol, 76, 88-89, NIH sr fel, 88-89; assoc prog dir, Sensory Physiol & Perception Prog, Directorate Biol, Behav & Social Sci, NSF, 78-79; adj prof, Sch Med, Wayne State Univ, 80-89; mem, Commun Sci Study Sect, NIH, 81-82, mem & chmn, Sensory Dis & Lang Study Sect, 82-86, mem, Bd Friends, Nat Inst Deafness & Other Commun Dis, NIH, 91- & Nat Adv Bd, 91-95; assoc ed, Behav & Brain Sci, 84-; exec chairperson, Asn Chemoreception Sci, 89-90. *Mem:* AAAS; Am Physiol Soc; Asn Chemoreception Sci; Asn Neurosci Depts & Progs; Asn Res Otolaryngol; Europ Chemoreception Res Orgn; NY Acad Sci; Soc Neurosci; Sigma Xi. *Res:* Author of numerous publications. *Mailing Add:* A B Chandler Med Ctr Col Med Univ Ky 800 Rose St Lexington KY 40536-0084

GETCHELL, THOMAS VINCENT, CELL BIOLOGY, NEURAL SCIENCE. *Educ:* Northwestern Univ, PhD(neural sci), 69. *Prof Exp:* PROF NEURAL SCI & ANAT, WAYNE STATE UNIV, 82-, ASSOC DEAN GRAD SCH, 86- *Concurrent Pos:* Sr Fel NRC-NIH. *Honors & Awards:* Javits Neurosci Investr Award. *Mailing Add:* Dept Neuro-Sci Wayne State Univ Sch Med 540 E Canfield Detroit MI 48201

GETHMANN, RICHARD CHARLES, b Yakima, Wash, June 8, 41; m 74; c 2. GENETICS. *Educ:* Wash State Univ, BS, 64; Ore State Univ, MS, 66; Univ Chicago, PhD(biol), 70. *Prof Exp:* NSF fel genetics, Univ Calif, San Diego, 70-71; asst prof, 71-76, ASSOC PROF BIOL, UNIV MD, BALTIMORE COUNTY, 77- *Mem:* Genetics Soc Am. *Res:* Chromosome behavior in Drosophila. *Mailing Add:* Dept Biol Sci Univ Md Baltimore County 5401 Wilkens Ave Catonsville MD 21228

GETHNER, JON STEVEN, b Chicago, Ill, July 12, 46; m 83. ANALYTICAL INSTRUMENTATION, PROCESS ANALYTICAL CHEMISTRY. *Educ:* Univ Chicago, SB, 68; Columbia Univ, MPhil, 73, PhD(chem), 76. *Prof Exp:* Systs engr, IBM Corp, 68-70; staff fel biophysics, NIH, 76-78; res chemist, Corp Res Lab, Exxon Res & Eng Corp, 78-80, staff chemist, 80-91; PRES, ADAPTIVE ANALYZER TECHNOLOGIES, 91- *Mem:* Am Phys Soc; Biophys Soc; Am Chem Soc. *Res:* Hydrodynamics of particles in solution; nucleation phenomena; biological self-assembly systems; intermolecular interactions; laser light scattering; structure and properties of gel and polymer networks; coal physical chemistry; FTIR spectroscopy; NIR and FTIR analyzer systems; real-time chemical process measurement. *Mailing Add:* Adaptive Analyzer Technologies 1985 Winding Brook Way Scotch Plains NJ 07090

GETOOR, RONALD KAY, b Royal Oak, Mich, Feb 9, 29; m 59; c 1. MATHEMATICS. *Educ:* Univ Mich, AB, 50, MS, 51, PhD(math), 54. *Prof Exp:* Instr math, Princeton Univ, 54-56; from asst prof to prof, Univ Wash, 56-66; PROF MATH, UNIV CALIF, SAN DIEGO, 66- *Concurrent Pos:* NSF fel, 59-60; vis prof, Stanford Univ, 64-65. *Mem:* Am Math Soc; fel Inst Math Statist. *Res:* Probability theory, especially general theory of Markov processes and their associated potential theory. *Mailing Add:* Dept Math C-012 Univ Calif San Diego La Jolla CA 92093

GETSINGER, WILLIAM J, b Waterbury, Conn, Jan 24, 24. MICROWAVES, TRANSMISSION LINE ANALYSIS. *Educ:* Univ Conn, BS, 49, Stanford Univ, MS, 59, EE, 62. *Prof Exp:* Design engr, Technicraft Labs, 49-50 & 52-57, Westinghouse Elec Corp, 50-52; sr res engr, Stanford Res Inst, 57-62; staff mem, Lincoln Labs, Mass Inst Technol, 62-69; sr scientist, Comsat Labs, 69-84; CONSULT, 84- *Mem:* Fel Inst Elec & Electronics Engrs. *Mailing Add:* Star Route Box 36B Bozman MD 21612

GETTING, IVAN ALEXANDER, b New York, NY, Jan 18, 12; m 37; c 3. PHYSICS. *Educ:* Mass Inst Technol, BS, 33; Oxford Univ, DPhil(astrophys), 35. *Hon Degrees:* DSc, Northeastern Univ, 54, Univ Southern Calif, 86. *Prof Exp:* Jr fel, Harvard Univ, 35-40; mem radiation lab, Mass Inst Technol, 40-45, prof, 45-50; asst develop planning, Dep Chief Staff Develop, US Air Force, 50-51; vpres eng & res, Raytheon Co, 51-60; pres & trustee, The Aerospace Corp, 60-77; CONSULT. *Concurrent Pos:* Mem sci adv bd, US Air Force, 45-; mem, Res & Develop Adv Coun, Sig Corps, 52-60; consult, var panels of the President's Sci Adv Comt, 61-75; chmn, Naval Warfare Panel, 71-75; consult, Nat Security Coun, 75-77; mem undersea warfare comn, Nat Acad Sci. *Honors & Awards:* Naval Ord Develop Award, 45; President's Medal of Merit, 48; Except Serv Award, US Air Force, 60; Kitty Hawk Award, Los Angeles Chamber com, 75; Pioneer Award, Inst Elec & Electronics Engrs, 75, Founder's Medal, 89. *Mem:* Nat Acad Eng; fel Am Phys Soc; fel Inst Elec & Electronics Engrs (pres, 78); fel Am Acad Arts & Sci. *Res:* Particle accelerators; nuclear physics; radar; fire control; gaseous discharges; astrophysics; multivibrator high speed router; automatic tracking of targets by radar; rapid scanning radar antennas. *Mailing Add:* 312 Chadbourne Ave Los Angeles CA 90049

GETTING, PETER ALEXANDER, b Boston, Mass, Nov 28, 44; m 69; c 2. NEUROSCIENCE, ELECTROPHYSIOLOGY. *Educ:* Mass Inst Technol, BS, 67; Univ Calif, Berkeley, PhD(biophys), 71. *Prof Exp:* Fel neurophysiol, Univ Wash, 71-73; asst prof biol, Stanford Univ, 73-80; from asst prof to assoc prof, 80-85, PROF PHYSIOL, UNIV IOWA, 85- *Concurrent Pos:* Mem, adv comt, Friday Harbor Labs, 82-83; mem, Neurobiol II Study Sect, NIH, 83- *Mem:* Soc Neurosci; AAAS; Am Physiol Soc; Int Soc Neuroethology. *Res:* Description of neuronal circuitry mediating behavior using intra and extra-cellular recording techniques; biophysical approaches applied to mechanisms of neuronal excitability and synaptic transmission; neuronal control of mammalian respiration. *Mailing Add:* Dept Physiol & Biophys Univ Iowa Iowa City IA 52242

GETTINS, PETER, b Sunderland, UK, Nov 8, 53. MULTINUCLEAR MAGNETIC RESONANCE, PROTEASE INHIBITORS. *Educ:* Oxford Univ, BA, 76 & DPhil(biochem), 79. *Prof Exp:* Postdoc asst molecular biophys, Yale Univ, 79-83, res scientist, 83-84; ASST PROF BIOCHEM, VANDERBILT UNIV SCH MED, 84-88, ASSOC PROF BIOCHEM, 88- *Mem:* Am Soc Biochem & Molecular Biol; Am Chem Soc; Am Asn Advan Sci; Int Soc Magnetic Resonance. *Res:* Structure and function relationships of proteins with particular emphasis on human plasma protease inhibitors; use of spectroscopic methods; nuclear magnetic resonance; electron paramagnetic resonance. *Mailing Add:* Dept Biochem Vanderbilt Univ Sch Med Light Hall Nashville TN 37232-0146

GETTLER, JOSEPH DANIEL, b Brooklyn, NY, Mar 5, 16; m 55; c 4. PHYSICAL ORGANIC CHEMISTRY. *Educ:* Columbia Univ, BA, 37, MA, 39, PhD(chem), 43. *Prof Exp:* Res chemist, Air Reduction Co, 43-46; from instr to asst prof, 46-54, ASSOC PROF CHEM, NY UNIV, 54- *Concurrent Pos:* Adj assoc prof chem, Columbia Univ, 74-; tech adv, Syn-Zyme Labs. *Mem:* Am Chem Soc; Royal Soc Chem; Sigma Xi. *Res:* Solution kinetics; condensation reactions; molecular rearrangements; statistical analysis. *Mailing Add:* 209 Jennifer Lane Yonkers NY 10710

GETTNER, MARVIN, b Rochester, NY, July 21, 34; m 59; c 3. PHYSICS. *Educ:* Univ Rochester, BS, 56; Univ Pa, PhD(physics), 61. *Prof Exp:* Res assoc physics, Univ Pa, 60-61; from asst prof to assoc prof, 61-70, PROF PHYSICS, NORTHEASTERN UNIV, 70- *Concurrent Pos:* Vis scientist, Rutherford Lab, Eng, 68-69, vis scientist, Max Planck Inst, Cern, Geneva, 76-77; prin investr, high energy exp group, NSF, 72- *Mem:* Fel Am Phys Soc. *Res:* Experimental study of high energy electron-position collision to reveal the fundamental nature of matter, forces and energy; techniques used include radiation and nuclear particle detectors, real time data acquisitions, digital electronics and large scale data analysis. *Mailing Add:* Dept Physics Northeastern Univ 360 Huntington Ave Boston MA 02115

GETTY, ROBERT J(OHN), b St Louis, Mo, Dec 30, 22; m 50. CHEMICAL ENGINEERING. *Educ:* Wash Univ, BS, 48, MS, 49, DSc, 55. *Prof Exp:* Jr engr piping layouts, Midwest Piping & Supply Co, 42; res chem engr, Res Lab, 49-59, sr res engr, 59-63, asst chief, 63-66, tech mgr chem prods, 66-70, mgr chem prod, tech mgr chem prods, 80-83, CONSULT, ALUMINUM CO AM, 80- *Mem:* Am Soc Heat, Refrig & Air Conditioning Engrs; Sigma Xi; Am Inst Chem Engrs. *Res:* Adsorption; chromatography; dehydration of liquids and gases; alumina processes; manufacture and application of alumina chemicals, refrigerants, binders for ceramics and refractories. *Mailing Add:* 526 N Kirkwood Rd/Unit 3B Kirkwood MO 63122

GETTY, WARD DOUGLAS, b Detroit, Mich, Aug 8, 33; m 55; c 4. PLASMA WAVE PROPAGATION, EXPERIMENTAL PLASMA PHYSICS. *Educ:* Univ Mich, BS(math) & BS(elec eng), 55, MS, 56; Mass Inst Technol, ScD(elec eng), 62. *Prof Exp:* Asst prof elec eng, Mass Inst Technol, 62-66; assoc prof, 66-77, PROF ELEC ENG, UNIV MICH, ANN ARBOR, 77- *Concurrent Pos:* Ford Found fel, 62-64; consult, Raytheon Co, 63-66; vis scientist, Princeton Plasma Physics Lab, 73-74, Lawrence Livermore Lab, 79. *Mem:* Inst Elec & Electronics Engrs; Am Phys Soc. *Res:* Energy conversion; plasma physics in application to controlled thermonuclear fusion; electron beam sources. *Mailing Add:* Dept Electrical-Comp Eng Univ Mich Ann Arbor MI 48109

GETZ, GODFREY S, b Johannesburg, SAfrica, June 18, 30; m 55; c 4. BIOCHEMISTRY, PATHOLOGY. *Educ:* Univ Witwatersrand, MB, BCh, 54, BSc, 55; Oxford Univ, DPhil(biochem), 63. *Prof Exp:* Lectr chem path, Witwatersrand Univ, 56; demonstr biochem, Oxford Univ, 56-59; lectr chem path, Witwaterstrand, 59-63; res assoc, Harvard Med Sch, 63-64; asst prof path & res assoc biochem, 64-67, assoc prof, 67-72, assoc dean col & div biol sci, 74-77, PROF PATH & BIOCHEM, UNIV CHICAGO, 72-, CHMN, DEPT PATH, 88- *Mem:* Brit Biochem Soc; Am Asn Pathologists; Am Soc Biochem & Molecular Biol; Sigma Xi. *Res:* Lipids and membrane biogenesis, especially interaction of lipid biosynthesis and mitochondrial formation; assembly and secretion of plasma lipoproteins. *Mailing Add:* Dept Path Fac Exc Univ Chicago Chicago IL 60637

GETZ, LOWELL LEE, b Chesterfield, Ill, Sept 21, 31; m 53; c 2. ECOLOGY. *Educ:* Univ Ill, BS, 53; Univ Mich, MS, 59, PhD(zool), 60. *Prof Exp:* Res assoc ecol, Univ Mich, 59-61; instr zool, Univ Conn, 61-62, from asst prof to assoc prof, 62-69; head, Dept Ecol, Ethology & Evolution, 75-80, PROF, UNIV ILL, URBANA, 69- *Mem:* Am Soc Mammal; Brit Ecol Soc; Ecol Soc Am; Soc Study Evolution; Sigma Xi; fel AAAS. *Res:* Ecology of mammals and mollusks. *Mailing Add:* 606 E Healey St Champaign IL 61820

GETZ, MICHAEL JOHN, b Peoria, Ill, Apr 28, 44; m 64; c 2. CELL BIOLOGY, MOLECULAR BIOLOGY. *Educ:* WTex State Univ, BSc, 67, MSc, 68; Univ Tex Grad Sch Biomed Sci, PhD(molecular biol), 72. *Prof Exp:* Nat Cancer Inst fel molecular biol, Beatson Inst Cancer Res, Glasgow, Scotland, 72-74; Mayo Found fel exp path, Mayo Clin & Mayo Found, 74-75, assoc consult path, anat & molecular med, 75-77; from instr to asst prof biochem, Mayo Med Sch, 76-81, assoc prof cell biol, 81-85, consult path, anat & molecular med, Mayo Clin & Found, 77-80, consult cell biol, 80-85, PROF BIOCHEM & MOLECULAR BIOL, MAYO MED SCH, 87-, CONSULT BIOCHEM & MOLECULAR BIOL, MAYO CLIN & FOUND, 85- *Concurrent Pos:* Am Cancer Soc res grant, 76-78; res grant, NIH, Pub Health Serv, 78-94; biol sci study sect, Pub Health Serv, NIH, 90- *Mem:* AAAS; Am Soc Cell Biol. *Res:* Regulation of gene expression by peptide growth factors; mechanisms of cell proliferation and chemical carcinogenesis. *Mailing Add:* Dept Biochem & Molecular Biol Mayo Clin & Found 200 SW First St Rochester MN 55905

GETZEN, FORREST WILLIAM, b Stuart, Fla, Feb 28, 28; m 56; c 3. PHYSICAL CHEMISTRY. *Educ:* Va Mil Inst, BS, 50; Mass Inst Technol, PhD(chem), 56. *Prof Exp:* Instr chem, Va Mil Inst, 50-51; asst, Mass Inst Technol, 53-55; res engr, Humble Oil & Refining Co, 56-61; assoc prof, 61-80, PROF CHEM, NC STATE UNIV, 80- *Concurrent Pos:* Mem US Eng Team, Kabul, Afghanistan, 65-67. *Mem:* AAAS; Am Chem Soc; Am Soc Qual Control; Am Inst Mining Metall & Petrol Eng; fel Am Inst Chem; Sigma Xi. *Res:* Petroleum reservoir fluid behavior; phase behavior of hydrocarbon mixtures; compressibility of gaseous argon; interfacial phenomena; surface and colloid chemistry; solution thermodynamics. *Mailing Add:* 2009 Banbury Rd Raleigh NC 27608

GETZEN, RUFUS THOMAS, b Columbia, SC, Apr 11, 44; m 68. GROUNDWATER HYDROLOGY. *Educ:* Wake Forest Col, AB, 65; Univ SC, MS, 69; Univ Ill, PhD(geol), 74. *Prof Exp:* HYDROLOGIST, US DEPT INTERIOR, GEOL SURV, 71- *Mem:* Nat Water Well Asn; AAAS; Soc Econ Paleontologists & Mineralogists. *Res:* Hydrologic effects of petrofabrics. *Mailing Add:* 345 Middlefield Rd MS 496 US Geol Surv Menlo Park CA 94025

GETZENDANER, MILTON EDMOND, b Grandview, Wash, Apr 17, 18; m 41; c 2. PESTICIDE CHEMISTRY. *Educ:* Whitman Col, AB, 40; Univ Wash, MS, 45; Univ Tex, PhD(biochem), 49. *Prof Exp:* Res chemist, Univ Wash, 44-46; res scientist concentration growth factor, Univ Tex, 48-49; biochemist, Hanford Works, Gen Elec Co, 49-53; agr chem research, Dow Chem Co, 53-

59, group leader residue res, Bioprod Dept, 59-69, res dir residue res, Agr Organics Dept, 69-75, assoc scientist govt regist, Health & Environ Res, 76-81; RETIRED. *Concurrent Pos:* Jr chemist, Chem Warfare Serv, US Army, 43-46. *Mem:* Am Chem Soc; Am Soc Testing & Mat; Asn Off Analytical Chem. *Res:* Effect of radiation on bacterial growth; metabolite destruction by ionizing radiations; concentration of growth factor; lignin research; methods for and analysis of agricultural chemical residues; protocols for studies on crops, animals, soil, water, elucidating residues of agricultural chemicals to establish safety; registration and tolerances for agricultural products through United States Environmental Protection Agency; global registration coordinator. *Mailing Add:* 6008 Melbourne Indianapolis IN 42608

GETZIN, LOUIS WILLIAM, entomology; deceased, see previous edition for last biography

GETZIN, PAULA MAYER, b New York, NY, Oct 6, 41; c 2. GENERAL COMPUTER SCIENCES. *Educ:* Radcliffe Col, BA, 61; Columbia Univ, MA, 62, PhD(chem), 67; Stevens Inst Technol, MS(computer sci), 86. *Prof Exp:* Fel chem, Rutgers Univ, 67-69; asst prof, 69-77, ASSOC PROF CHEM, KEAN COL, 77- *Concurrent Pos:* mem, Highland Park Bd Educ, 80-87. *Mem:* Am Chem Soc; Sigma Xi; Asn Comput Mach. *Res:* Theoretical chemical calculations. *Mailing Add:* 423 Lincoln Ave Highland Park NJ 08904

GEUMEI, AIDA M, CLINICAL PHARMACOLOGY, INTERNAL MEDICINE. *Educ:* Univ Alexandria, Egypt, PhD(pharmacol), 65. *Prof Exp:* CLIN ASSOC PROF INTERNAL MED, SOUTHWESTERN MED SCH, DALLAS, 78-; DIR PULMONARY DEPT, HENDERSON MEM HOSP, 78- *Mailing Add:* Henderson Mem Hosp 701 N High St Suite No 3 Henderson Mem Hosp 701 N High St Suite No 3 Henderson TX 75652

GEURTS, MARIE ANNE H L, b Leopoldville, Congo, June 4, 47. PALYNOLOGY. *Educ:* Cath Univ Louvain, SC, 69, DcSc, 75. *Prof Exp:* Asst geog & physics, Cath Univ Lorrain, 69-76; researcher, Nat Ctr Geomorphol Res, Belg, 76-78; adj prof, 78-82, PROF GEOG & PALYNOLOGY, UNIV OTTAWA, 82- *Concurrent Pos:* Traveling fel, Belg Ministry Educ, 76. *Mem:* Belg Asn Geog; French Asn Geog; Asn French Speaking Palynologists; Can Palynologists. *Res:* Contemporary pollen spectra and palynostratigraphy in Yukon territory and Northwest territories (Canada); formation and stratigraphy of continental limestone in Belgium, France and Spain. *Mailing Add:* Dept Geog Univ Ottawa Ottawa ON K1N 6N5 Can

GEVANTMAN, LEWIS HERMAN, b New York, NY, Sept 12, 21; m 48; c 2. PHYSICAL CHEMISTRY. *Educ:* Johns Hopkins Univ, BE, 42; Univ Notre Dame, PhD(chem), 51. *Prof Exp:* Chem operator, Bethlehem Steel Co, 42 & Johns Hopkins Univ, 42-43; assoc chemist, Clinton Labs, Oak Ridge Nat Labs, 43-46; supvry chemist, Radiation Chem, US Naval Radiol Defense Lab, 51-64; sr sci adv, US Mission, UN Int Atomic Energy Agency, 64-67; sr prog analyst, 72-73, prog mgr off standard ref data, Nat Inst Standards & Technol, 67-83, CONSULT, 83- *Concurrent Pos:* Consult, Nuclear Sci & Eng Corp, 56-60 & mem staff, Coun Environ Qual, 79-80; secy, Solubility Data Proj, Int Union Pure & Appld Chem, 80- *Mem:* Sigma Xi; Am Chem Soc; Int Union Pure & Appl Chem. *Res:* Effect of ionizing radiation on the rates of chemical reactions; distribution of absorbed energy in matter and energy transfer mechanisms; radiation dosimetry; nuclear weapons effects; chemical kinetics; solid state; scientific data evaluation and management. *Mailing Add:* 11608 Toulone Dr Potomac MD 20854

GEVARTER, WILLIAM BRADLEY, b Brooklyn, NY, July 19, 27; m 71; c 5. SIMULATION OF MOTIVATED HUMAN DECISION MAKING. *Educ:* Univ Mich, BS, 51; Univ Calif, Los Angeles, MS, 55; Stanford Univ, PhD(aeronaut & astronaut eng), 66. *Prof Exp:* Staff engr, Lockheed Missiles & Space Co, 57-66; adv engr, Int Bus Mach, 66-67; mem tech staff, Bellcom, Div Bell Labs, 67-71; opers res analyst, NASA Hq, Wash, DC, 70-75, mgr automation res & technol, 75-81, res assoc artificial intel, 81-83, computer scientist, Ames Res Ctr, 84-90; CONSULT COGNITIVE SCI & ARTIFICIAL INTEL, GEVARTER & ASSOCS, 90- *Concurrent Pos:* Mem, Robotics & Automation Coun, Inst Elec & Electronics Engrs, 84-85. *Mem:* Inst Elec & Electronics Engrs. *Res:* Artificial intelligence; cognitive science; robotics; expert systems; simulation of human decision making considering emotions; author of publications on artificial intelligence and robotics. *Mailing Add:* 9550 St Helena Rd Santa Rosa CA 95404

GEVECKER, VERNON A(RTHUR) C(HARLES), b St Louis, Mo, Jan 19, 09; m 32; c 2. CIVIL ENGINEERING. *Educ:* Univ Mo, BSCE, 31, CE, 50; Calif Inst Technol, MSCE, 37. *Prof Exp:* Surveyman & student engr, US Eng Dept, Mo, 31-34; asst, Mo Sch Mines, 34-35 & Calif Inst Technol, 35-37; plant eng trainee, Procter & Gamble Mfg Co, 37-38; from instr to prof civil eng, Univ Mo-Rolla, 38-74, asst dean of fac, 53-59; RETIRED. *Concurrent Pos:* Part time consult. *Mem:* Am Soc Civil Engrs; Nat Soc Prof Engrs. *Res:* Open channel flows; particle size effects on certain phenomena. *Mailing Add:* Rte 2 Box 39 Rolla MO 65401

GEWANTER, HERMAN LOUIS, b Bronx, NY, May 26, 27; m 66; c 2. ORGANIC CHEMISTRY. *Educ:* Long Island Univ, BS, 52; Univ Fla, PhD(fluorocarbons), 62. *Prof Exp:* Chemist, Standard Chem Prod, Inc, 52-56; res asst, Univ Fla, 56-61; proj leader org Synthesis, Res & Develop Chem Div, Union Carbide Corp, 62-66; staff scientist tech serv & prod develop, 66-69, MGR TECH SERV, TECH SERV CTR, PFIZER INC, 69- *Mem:* AAAS; Sigma Xi; Sci Res Soc Am; Am Chem Soc; Am Asn Textile Chemists & Colorists; Steel Structures Painting Coun; NY Acad Sci; Am Oil Chemists Soc; Am Soc Metals. *Res:* Fluorocarbon chemistry; alkyl amines and polyalkylene amines and derivatives; polymer intermediates; detergent additives; chelation; citrates, gluconates, itaconates, erythorbates, 2-ketogluconates, fumarates, sorbitol and derivatives and applications. *Mailing Add:* 33 Greentree Dr Waterford CT 06385

GEWARTOWSKI, J(AMES) W(ALTER), b Chicago, Ill, Nov 10, 30; m 56; c 5. ELECTRONICS ENGINEERING. *Educ:* Ill Inst Technol, BS, 52; Mass Inst Technol, SM, 53; Stanford Univ, PhD(elec eng), 58. *Prof Exp:* Res asst, Electronics Lab, Stanford Univ, 54-57; mem tech staff, 57-62, supvr microwave source group, 62-71, supvr microwave integrated circuits, Amplifiers & Dielec Resonators Group, 81-84, SUPVR SL RECEIVER & OPTICAL RELAY GROUP, AT&T BELL LABS, 84- *Honors & Awards:* Thompson Mem Award, Inst Radio Eng, 60. *Mem:* Inst Elec & Electronics Engrs; Sigma Xi. *Res:* Microwave solid state circuits and lightweight receivers. *Mailing Add:* 2908 Edgemont Dr Allentown PA 18103

GEWERTZ, BRUCE LABE, b Philadelphia, Pa, Aug 27, 49; m 79; c 3. MEDICINE. *Educ:* Pa State Univ, BS, 69; Jefferson Med Sch, MD, 72. *Prof Exp:* Surg res, Univ Mich, 72-77; asst prof surg, Southwestern Med Sch, 77-81; assoc prof, 81-88, PROF SURG, 88-, FAC DEAN, MED EDUC, UNIV CHICAGO, 89- *Concurrent Pos:* Ed, J Surg Res, 90-; Teaching Scholar, Am Heart Asn, 78-81. *Mem:* Soc Vascular Surg; Soc Univ Surgeons; Asn Surg Educ(pres, 82-83); Am Col Surgeons; Am Physiol Soc; Soc Clin Surg. *Res:* Regulation of blood flow in the intestine, kidneys and spinal cord. *Mailing Add:* Dept Surg Box 129 5841 S Maryland Ave Chicago IL 60637

GEWIRTZ, ALLAN, b Brooklyn, NY, May 30, 31; m 54; c 4. MATHEMATICS, COMPUTER SCIENCES. *Educ:* Brooklyn Col, BS, 59, MA, 64; City Univ New York, PhD(math), 67. *Prof Exp:* Standards engr, Western Elec Corp, 59-61; electronics proj leader, Veeco Instruments Corp, 61-64; assoc prof math, Pace Col, 64-68; asst dean, Sch Gen Studies, 71-74, dean, Continuing Higher Educ, 81-84, PROF MATH, BROOKLYN COL, 68- *Concurrent Pos:* Consult, microcomputers, 82-; Treas Freehold Area Hosp, 88-89; vchmn, Centrastate Med Ctr, 90- *Mem:* Math Asn Am; fel NY Acad Sci; AAAS. *Res:* Combinatorial mathematics, including graph theory, block designs and game theory; low & direct current detection, including mass spectrometry and ion gauges; microcomputer applications in mathematics; medical quality assurance. *Mailing Add:* 63 Blenheim Rd Englishtown NJ 07726

GEWIRTZ, DAVID A, b Ger, Aug 28, 48; US citizen; m 74; c 3. EXPERIMENTAL CHEMOTHERAPY. *Educ:* Brooklyn Col, BS, 70; City Univ NY, Mt Sinai, PhD(physiol), 77. *Prof Exp:* Fel, Dept Med, 77-80, from instr to asst prof, 80-88, ASSOC PROF PHARMACOL, MED COL VA, 88- *Mem:* Am Asn Cancer Res; Am Soc Pharmacol Exp Ther. *Res:* Analysis of the biochemical factors conferring intrinsic resistence to chemotherapy in the hepatoma cell; studies of the biochemical pharmacology of antineoplastic drugs in MCF7 breast tumor cells. *Mailing Add:* Dept Med Med Col Va Box 230 MCV Sta Richmond VA 23298

GEWURZ, HENRY, b Barcelona, Spain, June 1, 36. IMMUNOLOGY. *Educ:* Johns Hopkins Univ, MD, 62. *Prof Exp:* PROF IMMUNOL & CHMN, DEPT IMMUNOL & MICROBIOL, ST LUKE'S MED CTR, 71- *Mem:* Am Fedn Clin Res; Am Asn Physicians; Am Asn Immunologists; Am Soc Clin Invest. *Mailing Add:* Dept Immunol & Microbiol Rush Presby-St Luke's Med Ctr 1753 W Congress Pkwy Chicago IL 60612

GEYER, JOHN CHARLES, b Neosho, Mo, Aug 11, 06; m 33; c 1. SANITARY ENGINEERING. *Educ:* Univ Mich, BS, 31; Harvard Univ, MS, 33; Johns Hopkins Univ, DEng, 43. *Prof Exp:* From instr to asst prof sanit eng, Univ NC, 34-37; assoc civil eng, Johns Hopkins Univ, 37-42; asst chief engr, Health & Sanit Div, Off Inter-Am Affairs, Washington, DC, 42-43; assoc prof sanit eng, Johns Hopkins Univ, 46-48, chmn, Dept Geog & Environ Eng, 57-69, prof, 48-80, emer prof sanitary eng, 80-88; RETIRED. *Concurrent Pos:* Engr, Md Water Resources Comn, 38; prin investr, Water Filtration Proj, 47-51, waste disposal projs, AEC, 48-, Storm Drainage Res Proj, 49-, low flow augmentation proj, NIH, 57-63, ground water measurement res, 62-, sanit sewerage res proj, Fed Housing Admin, 59-63, residential water use proj, 59- & cooling water res proj, Edison Elec Inst, 63-; consult indust & govt, 48-; mem tech comt water supply, Interstate Comn Potomac River Basin, 48-, subcomt water supply, Nat Res Coun, 49-, adv comt small sewers, bldg res, adv bd, 56-57, comt educ objectives, 57; Adv Comt Spec Weapons Defense; Eng Joint Coun-Am Water Works Asn rep, Comt, Nat Water Policy, 57-; study sect environ sci & eng, NIH, 59-63; comnr, Md Dept Geol, Mines & Water Resources, 55-; mem adv comt reactor safeguards, AEC, 64, vpres, 64. *Mem:* Nat Acad Eng; AAAS; Am Soc Civil Engrs; Am Water Works Asn; Water Pollution Control Fedn. *Res:* Textile waste treatment and recovery; industrial wastes; ground water in Baltimore industrial area; water supply and wastewater disposal. *Mailing Add:* 3811 Canterbury Rd No 204 Baltimore MD 94305

GEYER, MARK ALLEN, b Portland, Ore, Dec 19, 44. NEUROSCIENCES, PSYCHOPHARMACOLOGY. *Educ:* Univ Ore, BA, 66; Univ Iowa, MA, 68; Univ Calif, San Diego, PhD(psychol), 72. *Prof Exp:* Res asst psychol, Univ Iowa, 66-68, res asst pharmacol, 68 & 68-70; staff res assoc psychol, 70-72, asst res psychobiologist, dept psychiat, 73-78, from asst prof to assoc prof psychiat, 79-89, PROF PSYCHIAT, SCH MED, UNIV CALIF, SAN DIEGO, 89- *Concurrent Pos:* Founding partner, San Diego Instruments; NIMH res sci develop awardee, 83- *Mem:* Soc Biol Psychiat; Soc Neurosci; Am Col Neuropsychopharmacol. *Res:* Behavioral, histochemical and pharmacological studies of noradrenergic, dopaminergic and serotonergic neurons in mammalian brain; studies of drugs of abuse. *Mailing Add:* Dept Psychiat Univ Calif San Diego 0804 La Jolla CA 92093-0804

GEYER, RICHARD ADAM, b New York, NY, Oct 27, 14; m 40; c 2. OCEANOGRAPHY. *Educ:* NY Univ, BS, 37, MS, 40; Princeton Univ, MA, 50, PhD(geophys), 51. *Prof Exp:* Res geophys & geol, Standard Oil Co, NJ, 38-42; physicist-in-chg degaussing range, Bur Ord, US Navy Dept, RI, 42-44, sr field instr, Oceanog Inst, Woods Hole, 44-45; sr res geophysicist, Humble Oil & Refining Co, 45-48, head oceanog sect, 49-54; chief geophysicist, Gravity Dept, Geophys Serv, Inc, Tex Instruments, Inc, 54-59, mgr, 59-66; head, Dept Oceanog, 66-76, dir geosci develop progs, 76-80, prof, 66-80, EMER PROF OCEANOG, TEX A&M UNIV, 80-; PRES, GEOPHYS

ASSOCS, INC, 81- *Concurrent Pos:* Instr, Princeton Univ, 39-42; adj asst prof, Univ Houston, 48-61; ed, Soc Explor Geophys, 49-51; mem oceanwide surv panel, Comt Oceanog, Nat Acad Sci, 61-; tech dir oceanog, Geosci Dept, Geophys Serv, Inc, Tex Instruments, Inc, 63-66; vchmn, President's Comn Marine Sci Eng & Resources, 67-69; chmn adv comt, Int Decade Ocean Explor, NSF, 70-; consult, US Coast & Geodetic Surv; ed, Submersibles & Their Use in Oceanog Eng & Marine Pollution; adj prof, Offshore Technol Res Ctr, Tex A&M Univ, 89- *Mem:* Hon mem Soc Explor Geophys; Nat Ocean Indust Asn; Marine Technol Soc. *Res:* Exploration geophysics; application of oceanographic science to marine engineering operations and to military aspects of underwater sound. *Mailing Add:* 300 Greenway Dr Bryan TX 77801

GEYER, ROBERT PERSHING, b Racine, Wis, Sept 28, 18; m 45; c 2. BIOCHEMISTRY. *Educ:* Univ Wis, BS, 41, MS, 43, PhD(biochem), 46. *Hon Degrees:* MA, Harvard. *Prof Exp:* Fel, 46-48, from asst prof to assoc prof, 49-71, PROF NUTRIT, SCH PUB HEALTH, HARVARD UNIV, 71-, CHMN DEPT, 77- *Concurrent Pos:* Mem, Coun Thrombosis, Am Heart Asn & Int Conf Biochem of Lipids; mem blood dis & resources adv bd, Nat Heart & Lung Inst, 74-76 & 79-83; consult, 84-88; co-chair 2nd & 3rd Int Symp on Artificial Blood Substitutes. *Mem:* Fel AAAS; Am Chem Soc; Am Asn Cancer Res; Am Inst Nutrit; Tissue Cult Asn; Sigma Xi; Soc Exp Biol Med; Mat Res Soc. *Res:* Development and study of artificial blood substitutes; nutritional, biochemical and chemical studies on oils and fats; parenteral nutrition and the metabolic fate of parenteral nutrients; tissue culture biochemistry. *Mailing Add:* Harvard Sch Pub Health Nutrit Bldg 2 665 Huntington Ave Boston MA 02115

GEYER, STANLEY J, b Pittsburgh, Pa, July 25, 49. LABORATORY MEDICINE. *Educ:* Jefferson Med Col, MD, 74; Am Bd Path, cert, 78. *Prof Exp:* Resident path, Med Ctr, NY Univ, 74-75 & Univ Health Ctr Pittsburgh, 75-76; postdoctoral fel exp path & immunol, Dept Path, Sch Med, Univ Pittsburgh, 76-77, from asst prof to assoc prof path, 77-84, asst dean vet affairs, 82-84; staff pathologist & chief, Autopsy Sect, Pittsburgh Vet Admin Med Ctr, 77-79, chief, Lab Serv, 79-82, transfusion officer & chief, Blood Bank, 81-83, chief staff, 82-84; assoc prof path & assoc dean, Sch Med, Univ Wash, 84-90; PROF PATH, SCH MED, GEORGETOWN UNIV, 90-, DIR, DEPT LAB MED, HOSP, 90-, DIR, AUTOPSY PATH, 91- *Concurrent Pos:* Chief staff, Seattle Vet Admin Med Ctr, 84-90; fac mem, Health Systs Specialist Training Prog, Vet Admin Cent Off, 84-90, chmn, Data Mgt Task Force Support Peer Rev, 85-86 & Nat Qual Care Task Force, 88-89; mem, Nat Adv Comt Peer Rev Orgn Vet Admin, 85-90 & Task Force Physician Staffing Needs Vet Admin, Inst Med, 88- *Mem:* Am Asn Neuropathologists; AAAS; Am Asn Immunologists; fel Col Am Pathologists; NY Acad Sci; Soc Med Decision Making; Group Res Path Educ; Acad Clin Lab Physicians & Scientists; Am Soc Clin Pathologists. *Mailing Add:* Dept Lab Med Georgetown Univ Hosp 3800 Reservoir Rd NW Washington DC 20007-2197

GEYLING, F(RANZ) TH(OMAS), b Tientsin, China, Sept 7, 26; nat US; m 61; c 4. COMPUTER MODELING PHYSICAL DESIGN. *Educ:* Stanford Univ, BS, 50, MS, 51, PhD(eng mech), 54. *Prof Exp:* Instr, Stanford Univ, 50-52; mem tech staff, mech res, 54-60, head analytical mech & eng physics, 60-70, DISTINGUISHED MEM TECH STAFF, MAT RES & ENG, BELL TEL LABS, 70- *Mem:* Am Soc Mech Engrs; Am Phys Soc; AAAS. *Res:* Computer fluid mechanics; modelling of material processes; structural mechanics and stress analysis; process modelling and manufacturability studies relating to material science and physical devices. *Mailing Add:* 6706 Cypress Pt Austin TX 78746

GEYMAN, JOHN PAYNE, b Santa Barbara, Calif, Feb 9, 31; m 56; c 3. FAMILY MEDICINE. *Educ:* Princeton Univ, AB, 52; Univ Calif, San Francisco, MD, 60; Am Bd Family Pract, dipl. *Prof Exp:* Intern, Los Angeles County Gen Hosp, 60-61; resident, Sonoma County Hosp, Santa Rosa, 61-63; pvt pract, Mt Shasta, 69-71; dir, Family Pract Residency Prog, Community Hosp Sonoma County, 69-71 & Davis Family Residency Network Prog, Univ Calif, 72-; chmn, div family pract & assoc prof community & family med, Univ Utah, 71-72; prof family pract & vchmn dept, Sch Med, Univ Calif, Davis, 72-76; chmn dept, 76-90, PROF FAMILY MED, SCH MED, UNIV WASH, 76- *Concurrent Pos:* Asst clin prof ambulatory & community med, Univ Calif, 69-71; ed, J Family Pract, 73-90, J Am Bd Family Pract, 90-; chmn, pictorial comt, Am Bd Family Pract, 74-77; consult, Am Acad Family Physicians, 72-, Family Med, NIH, 72-77 & Residency Assistance Prog, 75-80. *Honors & Awards:* Thomas Johnson Award, Am Acad Family Physicians, 80; Curtis Hames Res Award, 90. *Mem:* Fel Am Acad Family Physicians; Soc Teachers Family Med; AMA; Inst Med-Nat Acad Sci. *Res:* Family practice curriculum development and evaluation; role of family physician; changing trends in medical education and clinical practice. *Mailing Add:* Dept Family Med Univ Wash Seattle WA 98195

GEZON, HORACE MARTIN, b Grand Rapids, Mich, Nov 12, 14; m; c 2. MICROBIOLOGY, EPIDEMIOLOGY. *Educ:* Calvin Col, AB, 38; Univ Chicago, MD, 40. *Prof Exp:* From instr to asst prof pediat, Univ Chicago, 47-52; from assoc prof epidemiol to prof epidemiol & microbiol, Univ Pittsburgh, 52-66; prof pediat, Sch Med, Boston Univ, 66-75, chmn dept pediat, 66-70; EPIDEMIOLOGIST & MICROBIOLOGIST, ATLANTIC ANTIBODIES, 75- *Concurrent Pos:* Res fel pediat path, Univ Chicago, 46-47; vis prof, Am Univ, Beirut, 50-51; consult to Surgeon Gen, US Army; dir comn enteric infections, Armed Forces Epidemiol Bd, 63-70; dir pediat serv, Boston City Hosp, 66-70; lectr pediat, Harvard Med Sch, 67-; lectr microbiol, Harvard Sch Pub Health, 67-69; consult pediat, Children's Hosp, 67-; clin prof pediat, Sch Med, Tufts Univ, 76-81; consult pediat. *Mem:* Soc Pediat Res; Am Epidemiol Soc; Am Pediat Soc. *Res:* Action of antibiotics on bacteria; epidemiology of streptococcal, staphylococcal, enteric diseases of human and chlamydial, corynebacterial and mycobacterial diseases of goats; bacterial metabolism; pediatrics. *Mailing Add:* 15 Western Ave Gorham ME 04038

GFELLER, EDUARD, b Aarburg, Switz, Sept 11, 37; nat US; m 59, 74; c 2. NEUROANATOMY, PSYCHIATRY. *Educ:* Univ Berne, DrMed, 64. *Prof Exp:* From instr to asst prof neuroanat, Johns Hopkins Univ, 65-72; from asst prof to assoc prof psychiat, Sch Med, Univ Ala, Birmingham, 74-78; prof psychiat & neurosci, Col Med, Univ Fla, 78-87; PVT PRACT, 87- *Concurrent Pos:* Fel endocrinol, Univ Berne, 64-65; chief, Psychiat Serv, Vet Admin Hosp, Birmingham, 75-77; chief staff, Vet Admin Hosp, Tuscaloosa, 77-78; chief psychiat serv, Vet Admin Hosp, Gainesville, 78- *Mem:* AAAS; Am Asn; Sigma Xi. *Res:* Catecholamine metabolism in brain; etiology and genetics of schizophrenia. *Mailing Add:* 201 N Lakemont Ave Suite 2100 Winter Park FL 32792

GHAFFAR, ABDUL, b Macchli Shaher, India, July 6, 42; m 66; c 3. CELLULAR IMMUNOBIOLOGY, CANCER IMMUNOLOGY. *Educ:* Univ Karachi, BSc, 62; Univ London, MPhil, 70; Univ Edinburgh, PhD(immunol), 73. *Prof Exp:* Res asst, Univ Edinburgh, 70-72, res fel, 73-75; res asst prof, Univ Miami, 75-78; staff investr, Comprehensive Cancer Ctr, State Fla, 75-78; ASSOC PROF IMMUNOL, SCH MED, UNIV SC, 78- *Concurrent Pos:* Asst ed, J Reticuloendothelial Soc. *Mem:* Brit Soc Immunol; Am Asn Immunologists; Reticuloendothelial Soc; Am Soc Microbiol; Royal Col Pathologists. *Res:* Cellular immunology; cancer immunology; immunosuppression and immunopotentiation; macrophage functions and immunoregulation. *Mailing Add:* Dept Microbiol & Immunol Sch Med Univ SC Columbia SC 29208

GHAI, GEETHA R, b Madras, India, Jan 7, 46; m 68; c 2. BIOCHEMICAL PHARMACOLOGY, CARDIOVASCULAR PHARMACOLOGY. *Educ:* Gujarat Univ, India, BSc Hons, 65; Calcutta Univ, India, MSc, 68; Baroda Univ, India, PhD(microbiol), 72. *Prof Exp:* Res assoc pharmacol, Dept Pharmacol & Exp Therapeut, State Univ NY, Buffalo, 75-78; res assoc pharmacol, Dept Pharmacol, Univ Southern Ala, Mobile, 78-80, asst prof, 80-83; res scientist, 83-87, SR RES SCIENTIST, RES DEPT, PHARMACOL DIV, CIBA-GEIGY, SUMMIT, NJ, 87- *Concurrent Pos:* Reviewer, NJ Affil Peer Rev Comt, Am Heart Asn, 89-91. *Mem:* Am Soc Pharmacol & Exp Therapeut. *Res:* Biotechnology and molecular biology-related advances. *Mailing Add:* Pharmacol Div Ciba-Geigy Corp 556 Morris Ave Summit NJ 07901

GHALY, THARWAT SHAHATA, b Cairo, Egypt, Oct 12, 39; m 65; c 2. GEOLOGY, PETROLOGY. *Educ:* Ain Shams Univ, Cairo, BSc, 59; Glasgow Univ, PhD(igneous & metamorphic petrol), 65. *Prof Exp:* Instr geol, Fac Sci, Ain Shams Univ, Cairo, 59-61; res asst, Glasgow Univ, 62-65; mineralogist, Thermal Syndicate Ltd, Wallsend, Eng, 66-67; PROF GEOL, E TEX STATE UNIV, 67- *Mem:* Geol Soc Am; Geol Soc London; Geol Soc Glasgow. *Res:* Igneous and metamorphic petrology with mineralogy and geochemistry; structures, petrology and metamorphic differentiation of Precambrian rocks. *Mailing Add:* Dept Earth Sci E Tex State Univ Commerce TX 75428

GHANAYEM, BURHAN I, b Kufrzibad, Jordan, Mar 11, 52; US citizen; m 79; c 3. TOXICOLOGY. *Educ:* Cairo Univ, Egypt, BS, 75; NTex State Univ, MS, 79; Univ Tex Med Br, PhD(pharmacol-toxicol), 83. *Prof Exp:* Pharmacist, Kuwait, 75-77; res technician, Tex Col Osteop Med, Ft Worth, Tex, 77-79; res assoc, Univ Tex Med Br, Galveston, 79-80, asst lab supvr, 80-81, res assoc, 81-83; vis postdoctoral fel, Nat Inst Environ Health Sci, NIH, 83-84, staff fel, 84-86, sr staff fel, 86-88, TOXICOLOGIST, NAT INST ENVIRON HEALTH SCI, NIH, RESEARCH TRIANGLE PARK, NC, 88- *Concurrent Pos:* Instr, Durham Tech Community Col, NC, 85-90. *Mem:* Soc Toxicol; Am Soc Pharmacol & Exp Ther. *Res:* Metabolism of environmental chemicals and their toxicity; mechanisms of chemical-induced carcinogenicity. *Mailing Add:* 800 Carpenter Fletcher Rd Durham NC 27713

GHANDAKLY, ADEL AHMAD, b Alexandria, Egypt, Mar 15, 45; Can citizen; m 81. ELECTRIC ENERGY SYSTEMS. *Educ:* Univ Alexandria, BSc, 67; Univ Calgary, MSc, 73, PhD(elec eng), 75. *Prof Exp:* Instr elec eng, Univ Alexandria, 67-69; design engr, Pub Utility, Egypt, 69-71; res asst, Univ Calgary, 71-75; sr design engr, Montreal Energy Co, 75-77; asst prof, Univ New Orleans, 77-79; ASSOC PROF ELEC ENG, UNIV TOLEDO, 79- *Concurrent Pos:* Consult, Siemens-Allis Co, 77-79, La Power & Light, 78 & Detroit Edison Co, 79- *Mem:* Inst Elec & Electronics Engrs; Asn Prof Engrs. *Res:* Computer applications to power system studies and control; power system design; control of large motor drives; study and design of solar and wind energy systems. *Mailing Add:* Dept Elec Eng Univ Toledo 2801 W Bancroft Toledo OH 43606

GHANDEHARI, MOHAMMAD HOSSEIN, b Tehran, Iran, Sept 29, 43; m 64; c 2. MATERIALS SCIENCE & ENGINEERING METALLURGY. *Educ:* Weber State Col, BS, 68; Univ Utah, PhD(chem), 70. *Prof Exp:* Asst prof chem, Tehran Univ Technol, Iran, 70-73; instr & res assoc, Univ Utah, 73-75; res chemist, Fansteel Res Ctr, 75-80; RES SCIENTIST, SCI & TECHNOL DIV, UNION OIL CALIF, 80- *Mem:* Inst Elec & Electronics Engrs; Mat Res Soc; Am Chem Soc; Electrocham Soc. *Res:* Powder metallurgy of rare earth magnets; electrometallurgy of rare earth metals in molten salts; ceramics processing; rare earth metals and alloys; high temperature superconductors; fuel cells. *Mailing Add:* Unocal Sci & Tech Div Unocal Corp Brea CA 92621

GHANDHI, SORAB KHUSHRO, b Allahabad, India, Jan 1, 28; US citizen; m 81; c 3. ELECTRICAL ENGINEERING. *Educ:* Benares Hindu Univ, BSc, 47; Univ Ill, MS, 48, PhD, 51. *Prof Exp:* Mem tech staff, Electronics Lab, Gen Elec Co, 51-60; res mgr, Philco Corp, Ford Motor Co, 60-63; chmn div electrophys, 67-74, PROF ELECTROPHYS, RENSSELAER POLYTECH INST, 63- *Concurrent Pos:* Consult, Sprague Elec Co, Mass, Analog Devices, Mass, RCA, Pa & Stauffer Chem Co, Am Cyanamid. *Mem:* Sigma Xi; fel Inst Elec & Electronics Engrs. *Res:* Solid state devices, processes and materials; microelectronics. *Mailing Add:* Seven Linda LAne Niskayuna Schenectady NY 12309

GHARRETT, ANTHONY JOHN, b Seattle, Washington, Dec 23, 45; m 76. FISH BIOCHEMICAL AND POPULATION GENETICS. *Educ:* Calif Inst Technol, BS, 67; Ore State Univ, MS, 73, PhD(genetics), 75. *Prof Exp:* NIH fel trainee genetics & cell biol, Univ Minn, 74-76; from asst prof to assoc prof, 76-88, PROF FISHERIES, UNIV ALASKA FAIRBANKS, 88- *Concurrent Pos:* Prin investr, Alaska Sea Grant Col Prog, Univ Alaska, 77-; vis assoc prof, Univ Mich, 85-86; tech comt mem, Western Regional Aquacult Consortium, 86- *Mem:* AAAS; Sigma Xi; Pac Fishery Biologists; Am Inst Fishery Res Biologists; Am Fisheries Soc. *Res:* Biochemical genetics of salmonid; population genetics, stock separation and identification; genetic marking; mitochondrial DNA studies of salmonids. *Mailing Add:* Univ Alaska Fairbanks 11120 Glacier Hwy Juneau AK 99801

GHASSEMI, MASOOD, b Tehran, Iran, Mar 7, 40; m 68. ENVIRONMENTAL ENGINEERING, CHEMICAL ENGINEERING. *Educ:* Univ Wash, BS, 61, MS, 63, PhD(environ eng), 67. *Prof Exp:* Proj engr, Havens & Emerson Engrs, Ohio, 67-68; mem tech staff, Atomics Int, N Am Rockwell Corp, 68-73; mem tech staff, TRW Chem Eng Div, 73-75, sr proj engr, TRW Environ Eng Div, 75-83; prin & tech dir, Meesa, San Pedro, Calif, 83-85; chief hazardous waste engr, Chem Hill, Santa Ana, Calif, 85-87; dir, Solid & Hazardous Waste Mgt Prog, URS Consults Inc, Long Beach, Calif, 87-89; SUPVR, ENVIRON COMPLIANCE, LOCKHEED MISSILES & SPACE CO, SUNNYVALE, CALIF, 89- *Concurrent Pos:* Sr lectr, Univ Southern Calif, 72-76. *Mem:* Hazardous Mat Control Res Inst; Am Chem Soc; Sigma Xi. *Res:* Hazardous waste management; environmental assessment of new energy technologies; water and wastewater treatment technologies; environmental impact assessment. *Mailing Add:* 962 El Cajon Way Palo Alto CA 94303

GHATE, SUHAS RAMKRISHNA, b Nasik, India, Oct 2, 46; m 73; c 2. COMPUTER SIMULATION & MODELING. *Educ:* Univ Poona, BE, 67; NC State Univ, Raleigh, MS, 71, PhD(agr eng), 74. *Prof Exp:* Asst lectr eng, Tech Inst, Roona, India, 68-69; res technician, NC State Univ, Raleigh, 74-75; res assoc mech noise control, Univ Guelph, 75-78, asst prof mech, 79; from asst prof to assoc prof, 79-90, PROF, COASTAL PLAIN EXP STA, UNIV GA, TIFTON, 90- *Concurrent Pos:* Reviewer, Am Soc Agr Engrs journals. *Mem:* Am Soc Agr Engrs; Am Soc Mech Engrs. *Res:* Mechanization of horticultural crop production; gel planter for vegetable crops; drying of pecans using desiccants; machine design; computer simulation and modeling. *Mailing Add:* Dept Agr Eng Coastal Plain Exp Sta PO Box 748 Tifton GA 31793-0748

GHAUSI, MOHAMMED SHUAIB, b Kabul, Afghanistan, Feb 16, 30; m 61; c 2. ELECTRICAL ENGINEERING. *Educ:* Univ Calif, Berkeley, BS, 56, MS, 57, PhD(elec eng), 60. *Prof Exp:* From asst prof to prof elec eng, NY Univ, 60-74; prof elec & comput eng & chmn dept, Wayne State Univ, 74-78; dean eng & John F Dodge prof eng, Oakland Univ, 78-; DEAN, COL ENG, UNIV CALIF, DAVIS. *Concurrent Pos:* Sect head, Elec Sci & Analysis, NSF, 72-74. *Honors & Awards:* Inst Elec & Electronics Engrs Medal. *Mem:* Fel Inst Elec & Electronics Engrs, circuits & systs soc (pres, 76-77); NY Acad Sci; Inst Elec & Electronics Engrs. *Res:* Electronic circuits and systems; author of numerous publications. *Mailing Add:* Sch Eng Univ Calif Davis CA 95616

GHAZARIAN, JACOB G, b Baghdad, Iraq, Nov 28, 37; US citizen; m 65; c 1. BIOCHEMISTRY, NUTRITION. *Educ:* Murray State Univ, BSc, 63; Memphis State Univ, MSc, 67; Univ Neb, Lincoln, PhD(biochem), 71. *Prof Exp:* Vis scientist biochem, Univ Tenn, Memphis, 70-71; fel, Univ Wis, Madison, 71-74, proj assoc, 74-75; asst prof, 75-80, ASSOC PROF BIOCHEM, MED COL WIS, 80- *Concurrent Pos:* Prin investr, NSF grants, 66-78, NIH fel, 71-74 & NIH grant, 76-79; NIH career develop award, 76-81. *Mem:* Sigma Xi; Am Soc Bone & Mineral Res; Int Soc Supramolecular Biol; Am Chem Soc; AAAS. *Res:* Physiochemical and immunological characterization of kidney and liver mitochondrial and microsomal hydroxylases (monooxygenases) associated with the orderly control of the vitamin D endocrine system. *Mailing Add:* Dept Biochem Med Col Wis 8701 Watertown Plank Rd Milwaukee WI 53226

GHEBREHIWET, BERHANE, b Asmara, Ethiopia, Sept 28, 46. MOLECULAR IMMUNOLOGY, BIOCHEMISTRY. *Educ:* Sch Vet Med, Warsaw, Poland, DVM, 71; Vet Sch, Alfort, France, MSc, 73; Univ Paris, DSc(immunol), 74. *Prof Exp:* Res assoc molecular immunol, Scripps Clin & Res Found, 74-80; asst prof dept med, 79-85, ASSOC PROF DEPT MED & PATH, SCH MED, STATE UNIV NY STONY BROOK, 85- *Honors & Awards:* Sigma Xi. *Mem:* AAAS; NY Acad Sci; Am Asn Immunol; Am Chem Soc; Am Asn Vet Immunol; Am Fedn Clin Res; Soc Leukocyte Biol. *Res:* Complement and the interaction of complement with cell surface receptors; role of complement in inflammation. *Mailing Add:* Dept Med Health Sci Ctr State Univ NY Stony Brook NY 11794-8161

GHEITH, MOHAMED A, b Kherbeta, Kom Hamada, Egypt, Feb 11, 25; m; c 2. GEOLOGY, MINERALOGY. *Educ:* Cairo Univ, BSc, 45; Univ Minn, MS, 50, PhD(geol), 51. *Prof Exp:* Demonstr geol, Cairo Univ, 45-46; lectr, Ain-Shams, Cairo, 52-57; lectr, 58-59, from asst prof to assoc prof, 59-69, chmn dept, 64-75, dir, Spec Int Health Progs, 78-85, dir, Spec External Progs Mid East, 78-86, PROF GEOL, BOSTON UNIV, 69- *Concurrent Pos:* Consult, Bur Mining & Com, Egypt, 54-57; fel, Sch Advan Studies, Mass Inst Technol, 57-59; distinguished vis prof, Kuwait Univ, 73, Am Univ, Cairo, Cairo Univ, Alexandria Univ, Tanta & Mansoura Univ, 73-74 & Univ Qatar, Egypt, 83; deleg, Nubian Arabian Shield Geol & Mineralization Conf, Saudi Arabia, 78, Morocco, 82. *Mem:* Fel Geol Soc Am; Am Crystallog Asn; Nat Asn Geol Teachers; Mineral Soc Am; Geol Soc Egypt; AAAS. *Res:* Geochemistry of iron oxides and oxide hydrates; geochronology of Northeast Africa; crystallography, genesis, stability relations and synthesis of some phosphate minerals and silicates; science education; archeological geology. *Mailing Add:* Boston Univ 271 Forest St Arlington MA 02174

GHEN, DAVID C, b Pittsburgh, Pa, Feb 7, 39; m 60; c 2. SONAX UNDERWATER ACOUSTICS, NAVAL ASW ANALYSIS. *Educ:* Muskingum Col, BS, 61; Penn State Univ, MS, 62. *Prof Exp:* Sr analyst, Analysis & Technol Inc, 74-76, group mgr, 76-77, br mgr, 77-78, prin sci, 78-79, chief scientist, 79-80; res specialist, Shearwater Inc, 80-84; corp scientist, 84-87, DEPT MGR, ANALYSIS & TECHNOL INC, 87- *Mem:* Am Acoust Soc. *Res:* Development and evaluation of submarine tactics and associated system performance requirements. *Mailing Add:* 84 Ocean View Ave Mystic CT 06355

GHENT, ARTHUR W, b Toronto, Ont, Sept 8, 27; m 61; c 3. ECOLOGY, BIOMETRICS. *Educ:* Univ Toronto, BScF, 50; MA, 54; Univ Chicago, PhD(zool), 60. *Prof Exp:* Res officer, Forest Insect Lab, Can Dept Agr, 50-59; asst prof zool, Univ Okla, 60-64; asst prof, 64-65, assoc prof, 65-69, prof quant biol, 70-76, PROF ENTOM & ZOOL, UNIV ILL, URBANA, 76- *Res:* Insect and molluscan ecology and behavior; forest ecology; nonparametric methods. *Mailing Add:* Dept Ecol Ethology & Evolution Univ Ill 515 Morrill Hall Urbana IL 61801

GHENT, EDWARD DALE, b Little Rock, Ark, Oct 4, 37; m 62. GEOLOGY. *Educ:* Yale Univ, BS, 59; Univ Calif, Berkeley, PhD(geol), 64. *Prof Exp:* Asst prof geol, San Jose State Col, 64; lectr, Victoria Univ Wellington, 64-67; from asst prof to assoc prof, 67-75, PROF GEOL, UNIV CALGARY, 75-, HEAD DEPT, 87- *Mem:* Mineral Asn Can; Geol Soc Am; Mineral Soc Am; Geol Asn Can; Geochem Soc. *Res:* Metamorphic petrology; geochemistry; electron microprobe analysis. *Mailing Add:* Dept Geol Univ Calgary Calgary AB T2N 1N4 Can

GHENT, KENNETH SMITH, b Hamilton, Ont, June 29, 11; nat US; m 42; c 3. MATHEMATICS. *Educ:* McMaster Univ, BA, 32; Univ Chicago, SM, 33, PhD(math), 35. *Prof Exp:* From instr to asst prof math, Univ Ore, 35-42; physicist, 11th Naval Dist, USN, 42-45; from assoc prof to prof math, 47-77, EMER PROF MATH, UNIV ORE, 77-, DIR INT STUDENT SERV, 74- *Mem:* Am Math Soc; Math Asn Am. *Res:* Algebra and number theory; sums of values of polynomials multiplied by constants. *Mailing Add:* Dept Math 1925 E 26 Ave Eugene OR 97403

GHENT, WILLIAM ROBERT, b Hamilton, Ont, Apr 25, 22; c 3. SURGERY. *Educ:* Queen's Univ, Ont, MD, CM, 47; FRCS(C), 54, FACS 58. *Prof Exp:* Clin asst, 55-60, from asst prof to assoc prof, 60-71, PROF SURG, QUEEN'S UNIV, ONT, 71- *Concurrent Pos:* Chief of surg, Hotel Dieu Hosp, 57-73. *Mem:* Fel Am Col Surg; fel Int Col Surg. *Mailing Add:* Ten Montreal St Queens Univ Kingston ON K7L 3G6 Can

GHEORGHIU, PAUL, b Rumania, June 27, 16; US citizen; m 46; c 1. PHYSICS, INSTRUMENTATION. *Educ:* Signal Corps Mil Acad, BSEE, 37; Polytech Sch, lic es sc, 40; Advan Prof Sch, dipl life sci, 55. *Prof Exp:* Asst prof physics & electronics, Signal Corps Mil Acad, 42-43; ed, Asn Free Press, 48-56; proj engr electronics, Manson Labs, 57-59 & Bulova Res & Develop Electronics, Bulova Watch Co, 59-60; sr systs engr res & develop, Transitron Electronic Corp, 60-62; dir bio-med & electronic res, Advan Res Ctr, Hi-G, Inc, Conn, 62-66; vpres, Med Electrosci, Inc, NY, 66-70; mgr indust controls div, Opto Mechanisms, Inc, NJ, 70-72; pres, Today's Technol Assoc, 72-74; mgr advan res & develop, Scott Electronics Div, NCR Corp, 74-78; PRES, PAUL SIGMA INC, 78- *Concurrent Pos:* Secy, Joint Mil Comt Defense Res, 40-42; sr staff scientist, Frequency Electronics; consult, Lorad Electronics & Math Assocs Inc. *Mem:* Sr mem Inst Elec & Electronics Engr; Int Fedn Med Electronics & Biol Eng; NY Acad Sci; Int Soc Cybernet Med. *Res:* Metrology of standards and traceability; medical electronics, electro-chemical and solid state switching techniques; life sciences; bio-engineering; environmental technology; applied physics; communication in their design; production and marketing; management of advanced electronics; international marketing; cybernetics; operations research; computer sciences. *Mailing Add:* 2402 Loyal Lane Bel Air MD 21014-6308

GHERARDI, GHERARDO JOSEPH, b Lucca, Italy, July 1, 21; nat US; m 57; c 4. PATHOLOGY. *Educ:* Princeton Univ, AB, 42; Columbia Univ, MD, 45. *Prof Exp:* From instr to asst prof path, Col Med, State Univ NY Downstate Med Ctr, 50-54; assoc prof, Sch Med, Tufts Univ, 54-70; ASSOC PROF PATH, SCH MED, BOSTON UNIV, 70-; PATHOLOGIST, FRAMINGHAM UNION HOSP, 70- *Concurrent Pos:* Assoc vis pathologist, Kings County Hosp, 54; pathologist, Tufts-New Eng Med Ctr, 54-70; lectr path, Northeastern Univ, 60-78; consult, Cushing Hosp, Framingham & Medfield State Hosp, 70- *Mem:* AMA; Col Am Path. *Res:* Human diagnostic pathology. *Mailing Add:* 25 Kenilworth Rd Wellesley MA 02181

GHERING, MARY VIRGIL, b Grand Rapids, Mich, July 18, 10. ENZYMOLOGY, CHEMISTRY GENERAL. *Educ:* Cent Mich Univ, AB, 35; Marquette Univ, MS, 48; St Thomas Inst, PhD, 68. *Prof Exp:* High sch teacher, Mich, 36-38, Nottawa Twp Sch Unit, 41-42 & Marywood Acad, 42-43; pub sch teacher, 43-49; from asst prof to prof chem, Aquinas Col, 49-68, chmn dept phys sci, 59-63; librn, St Thomas Inst, 68-87; RETIRED. *Mem:* Am Chem Soc; Am Inst Chemists. *Res:* Effect of six vitamins of B complex upon the growth of Fusarium solani in a synthetic medium; effect of low velocity electrons upon the activity of trypsin. *Mailing Add:* 2025 E Fulton St Grand Rapids MI 49503-3895

GHERING, WALTER L, b Edinboro, Pa, Nov 3, 30. ACOUSTICS. *Educ:* US Naval Acad, BS, 56; Pa State Univ, BS, 61, MS, 63, PhD(physics), 68. *Prof Exp:* Res physicist, Babcock and Wilcox Res Ctr, 68-73, res specialist, 73-87; RETIRED. *Mem:* Inst Noise Control Eng. *Res:* Acoustic instrumentation, noise and vibration control; engineering physics (sonic cleaning and structural vibrational response prediction); industrial instrumentation (two-phase flow measurement and high-temperature pressure sensors). *Mailing Add:* 2460 S Linden Ave Alliance OH 44601

GHERNA, ROBERT LARRY, b Los Angeles, Calif, Nov 15, 37; m 62; c 3. BACTERIOLOGY, BIOCHEMISTRY. *Educ:* Univ Southern Calif, AB, 60, PhD(bact), 64. *Prof Exp:* Sr scientist bact, 66-67, asst cur, 73-74, from actg head to head, Dept Comput Sci, 76-85, HEAD, DEPT BACT, AM TYPE CULT COLLECTION, 74- *Concurrent Pos:* Fel, Stanford Univ, 64-66; adj assoc prof microbiol, Univ Md, 76-79. *Mem:* Am Soc Microbiol; Sigma Xi; US Fedn Cult Collections (vpres, 78); Am Inst Biol Sci; NY Acad Sci. *Res:* Bacterial physiology; general microbiology; bacterial systematics; sulfur and heterocyclic metabolism. *Mailing Add:* Am Type Cult Collection 12301 Parklawn Dr Rockville MD 20852

GHEZZO, MARIO, b Trieste, Italy, Nov 20, 37; US citizen; m 73. INTEGRATED CIRCUITS. *Educ:* Univ Trieste, Dr physics, 62. *Prof Exp:* Staff, Res Lab, Sprague Elec Co, Mass, 66-69; STAFF PHYSICS, CORP RES DEVELOP CTR, GEN ELEC CO, 69- *Concurrent Pos:* Lectr, integrated circuit yield enhancement, Rensselaer Polytechnic Inst. *Mem:* Electrochem Soc; Inst Elec & Electronics Engrs. *Res:* Development of a 1 micron complementary metal-oxide semiconductor bulk process for very large scale integration circuit applications; investigation of advanced metal oxide semiconductor field effect transistor structures and integration in a viable integrated circuit process; yield and productivity enhancement of integrated circuits. *Mailing Add:* Gen Elec Co Corp Res Develop Bldg KW-B1309 PO Box 8 Schenectady NY 12301

GHIA, KIRTI N, b Bombay, India. FLUID DYNAMICS, APPLIED MECHANICS. *Educ:* Gujarat Univ, India, BS, 60; Ill Inst Technol, MS, 65, PhD(mech & aerospace eng), 69. *Prof Exp:* Res engr, Premier Automobiles Ltd, India, 60-61; res asst fluid dynamics, Ill Inst Technol, 61-62, instr, 62, asst, 62-69; from asst prof to assoc prof, 69-78, PROF FLUID DYNAMICS, UNIV CINCINNATI, 78- *Concurrent Pos:* Consult, Huyck Corp, 65-67, Kenner Prod Co, Cincinnati, 72-76, Gen Elec Co, Cincinnati, 73-, Air Force Flight Dynamics Lab, Wright Patterson AFB, 76-, Naval Ship Res & Develop Ctr, Bethesda, 77- & Reynolds Metal Co, Ala, 78-; co-prin ininvestr, NSF grants, 72-79, Aerospace Res Lab, 72-75, Gen Elec Co grants, 74-75 & 76-77 & Off Sci Res grants, 78-79. *Mem:* Am Inst Aeronaut & Astronaut; Am Soc Mech Engrs; Am Soc Eng Educ; Sigma Xi. *Res:* Analysis and numerical solutions of three-dimensional viscous internal flow problems; use of numerical coordinate transformations and higher-order spline techniques and direct solvers in the solution of navier-stokes equations. *Mailing Add:* 8640 Sturbridge Dr Cincinnati OH 45236

GHIARA, PAOLO, b Naples, Italy, Apr 18, 58; m 84; c 2. IMMUNOLOGY. *Educ:* Classic Lyceum A Genovesi, Naples, Bachelor, 76; Nat Music Conserv, Livorno, dipl, 80; Univ Naples, MSc, 82. *Prof Exp:* Internal student pharmacol, Inst Exp Pharmacol, Univ Naples, 79-82, fel, 82-83; res asst immunopharmacol, 83-87, SR INVESTR IMMUNOPHARMACOL, SCLAVO, RES CTR, 87- *Concurrent Pos:* Fel, Wellcome Res Labs, Beckenham, UK, 85, Ludwig Inst, Epalinges, 87. *Mem:* Gruppo di Coop Immunol; Am Asn Immunologists. *Res:* Immunopharmacology of the inflammatory process; cDNA cloning and expression of interleuken-1 beta; structure-activity relationships of interleuken-1; identification and characterization of membrane receptors for interleukin-1. *Mailing Add:* Lab Immunopharmacol Sclavo Res Ctr Via Fiorentina 1 Siena I-53100 Italy

GHIDONI, JOHN JOSEPH, b Yonkers, NY, Feb 22, 31; m 56; c 6. PATHOLOGY. *Educ:* Fordham Univ, BSc, 53; State Univ NY, MD, 57. *Prof Exp:* Intern, Brooklyn Hosp, 57-58; resident path, Bronx Munic Hosp Ctr, NY, 58-62; instr, Albert Einstein Col Med, 62-63; asst prof, Baylor Col Med, 65-69, dir lab exp path, 66-69; PROF PATH, UNIV TEX MED SCH SAN ANTONIO, 69- *Concurrent Pos:* USPHS fel, Albert Einstein Col Med, 61-62, spec fel, 62-63; consult path, Vet Admin Hosp, Houston, Tex, 67-69. *Mem:* Electron Micros Soc Am; Int Acad Path; Am Soc Cell Biol; Am Asn Path. *Res:* Amyloidosis; epithelial metaplasia; cell injury by proton particles; effects of isoproteronal on salivary gland; cardiovascular prostheses; cellular dedifferentiation and differentiation. *Mailing Add:* Dept Path-Cell Bio Univ Tex Med Sch 7703 Floyd Curl Dr San Antonio TX 78284-7750

GHIGO, FRANK DUNNINGTON, b Richmond, Va, Jan 18, 45; m 72. RADIO ASTRONOMY. *Educ:* Haverford Col, BA, 66; Univ Tex at Austin, PhD(astron), 76. *Prof Exp:* Res assoc, Physics Dept, Brandeis Univ, 76-79, instr, 78; RES ASSOC, ASTRON DEPT, UNIV MINN, 79- *Mem:* Int Astron Union; Am Astron Soc; Royal Astron Soc; Astron Soc Pac. *Res:* Detailed radio imaging of quasars, radio galaxies, and interacting galaxies; development of software for processing of data from radio interferometer systems, and for processing of images from automated photograph digitizers. *Mailing Add:* Nat Radio Astron Observ PO Box 2 Green Bank WV 24944

GHIORSO, ALBERT, b Vallejo, Calif, July 15, 15; m 42; c 2. PHYSICS. *Educ:* Univ Calif, BS, 37. *Hon Degrees:* PhD, Gustavus Adolphus Col, 66. *Prof Exp:* Mem staff, Metall Lab, Univ Chicago, 42-46; physicist, 46-69, DIR HEAVY ION LINEAR ACCELERATOR, DEPT CHEM, LAWRENCE BERKELEY LAB, UNIV CALIF, BERKELEY, 69- *Mem:* Am Phys Soc. *Res:* Transuranium elements; co-discoverer of elements 95-106, inclusive; nuclear properties of heavy element isotopes; fission counters; electronic apparatus for measurement of nuclear radiations; reactions induced by heavy ions; systematics of radioactive decay. *Mailing Add:* 687 Vincente Ave Berkeley CA 94707

GHIORSO, MARK STEFAN, b San Francisco, Calif, Oct 21, 54. AQUEOUS GEOCHEMISTRY, IGNEOUS PETROLOGY. *Educ:* Univ Calif, Berkeley, AB, 76, MA, 78, PhD(geol), 80. *Prof Exp:* From asst prof to assoc prof, 80-88, PROF GEOL, UNIV WASH, 88- *Concurrent Pos:* Presidential young investr award, NSF, 85-90. *Mem:* Geol Soc Am; Mineral Soc Am; Am Geophys Union; Geochem Soc; Am Math Asn. *Res:* Thermodynamic modelling and properties of silicate liquids; mass transfer in hydrothermal and geothermal systems; experimental determination of the kinetics of mineral dissolution; diffusion in silicate liquids. *Mailing Add:* Dept Geol Sci AJ-20 Univ Wash Seattle WA 98195

GHIRARDELLI, ROBERT GEORGE, b San Francisco, Calif, Nov 12, 30; m 57; c 5. ORGANIC CHEMISTRY. *Educ:* Univ San Francisco, BS, 52; Calif Inst Technol, PhD(chem), 56. *Prof Exp:* Asst, Calif Inst Technol, 56-57 & Ga Inst Technol, 57-58; asst prof chem, Robert Col, Istanbul, 58-60; chief org chem br, 60-67, assoc dir chem div, 67-74, assoc dir chem & biol sci div, 74-77, DIR CHEM & BIOL SCI DIV, US ARMY RES OFF, 77- *Concurrent Pos:* Vis asst prof, Duke Univ, 62-, sr res assoc chem, 68-71, adj assoc prof, 72-81, adj prof, 82- *Mem:* AAAS; Am Chem Soc. *Res:* Stereochemistry; reaction mechanisms; macrocyclic polyethers; circular dichroism. *Mailing Add:* Chem Biol Sci Div US Army Res Off Box 12211 Research Triangle Park NC 27709

GHIRON, CAMILLO A, b Turin, Italy, Nov 11, 32; US citizen; m 59; c 3. BIOPHYSICS. *Educ:* Mass Inst Technol, SB, 54; Univ Utah, PhD(molecular biol), 64. *Prof Exp:* Res asst neurosurg, Sch Med, NY Univ, 56-57; res asst biophys, Brookhaven Nat Lab, 57-60; res assoc chem, Univ Minn, 63-64; asst prof physiol & res assoc biochem, 64-66, from asst prof to assoc prof, 66-80, PROF BIOCHEM, SCH MED & SCH AGR, UNIV MO-COLUMBIA, 80- *Concurrent Pos:* Partic, AEC training prog radiation phys chem, Univ Minn, 63-64; Nat Heart Inst sr fel, Biol Div, Oak Ridge Nat Lab, 71-72. *Mem:* AAAS; Radiation Res Soc; Biophys Soc; Am Chem Soc; Brit Biochem Soc; Sigma Xi. *Res:* Radiation inactivation kinetics of enzymes; photodynamic action; photophysics; dynamic topography of proteins. *Mailing Add:* Dept Biochem 322 Chem Univ Mo Columbia MO 65211

GHISELIN, MICHAEL TENANT, b Salt Lake City, Utah, May 13, 39. ZOOLOGY. *Educ:* Univ Utah, BA, 60; Stanford Univ, PhD(biol), 65. *Prof Exp:* Fel systs, Marine Biol Lab, 65-67; from asst prof to assoc prof zool, Univ Calif, Berkeley, 67-78; Guggenheim fel, 78-79, fel Mac Arthur Prize, 81-86; res prof biol, Univ Utah, 80-83, RES FEL, CALIF ACAD SCI, 83 - *Honors & Awards:* Pfizer Prize, Hist Sci Soc, 70. *Mem:* AAAS; Soc Syst Zool; Am Soc Nat; Paleont Soc; Soc Study Evolution. *Res:* Comparative invertebrate anatomy; evolutionary biology; history, methodology and philosophy of biology. *Mailing Add:* Calif Acad Sci Golden Gate Park San Francisco CA 94118-9961

GHISTA, DHANJOO NOSHIR, b Bombay, India, Jan 10, 40; m 67; c 3. BIOMEDICAL ENGINEERING, SURGERY. *Educ:* Univ Bombay, BEng, 60; Stanford Univ, PhD(eng mech), 64. *Prof Exp:* Res assoc eng mech & aerospace eng, Nat Acad Sci-Nat Res Coun, 64-67; res scientist, Ames Res Ctr, NASA, 67-69; assoc prof biomed eng, mech eng & surg, Washington Univ, 69-71; prof & head biomed eng, Indian Inst Technol, Madras, 71-75; sr biomed res scientist, Ames Res Ctr, NASA, 75-77; rehab eng scientist spinal cord injury ctr, Vet Admin, Palo Alto, 77-78; prof biomed eng & eng mech, Mich Technol Univ, 78-80; prof med, mech eng & eng physics & dir biomed eng, McMaster Univ, 80-87; VPRES, CORP MED DEVICES & INDUST DEVELOP, 88- *Concurrent Pos:* Mem Int Panel Biomed Eng, Am Soc Eng Educ, 75; ed, Automedica J & Renaissance Universal J, 76-; dir, NATO Advan Study Inst Spinal Cord Med Eng, 81, Int Spinal Injury Trust; reviewer, res grants, NSERC, Med Res Coun, Coalition Health Funding, NSF. *Honors & Awards:* Rotary Prize, 60. *Mem:* Am Soc Mech Engrs; Int Soc Biomech; Soc Biomed Eng; Soc Math Biol; Soc Biomat; Am Acad Mech; Sigma Xi. *Res:* Cardiovascular engineering physics of physiological mechanisms, monitoring, diagnostic, prosthetic, assist devices and procedures; orthopaedic biomechanics of joints orthoplasty, fracture fixation, surgical simulation and correction of spinal deformities and fracture stabilization; urological mechanics of dysynergic bladder assessment and transurethral surgery; rehabilitation engineering, limb prosthesis design, assessment of back and joint stiffness; societal science, self-reliant community development, economic democracy; author or coauthor of over 220 publications; GI surgical engineering of artificial pacing for reflux disorders and short-bowel syndrome; analyses of evoked potentials for neurological assessment; neo-humanism philosophy; integrated theory of mind, matter and consciousness. *Mailing Add:* Dept Med McMaster Univ Hamilton ON L8S 4L7 Can

GHOLSON, LARRY ESTIE, b San Luis Obispo, Calif, July 13, 49; m. ENTOMOLOGY. *Educ:* US Merchant Marine Acad, BS, 71; Calif Polytech State Univ, MS, 75; Iowa State Univ, PhD(entom), 78. *Prof Exp:* Lab instr entom, zool & natural hist, Calif Polytech State Univ, 74-75; res assoc entom, Iowa State Univ, 76-78; asst prof entom, NMex State Univ, 78-80; ext specialist, crop sci, NC State Univ, Raleigh, 80-81; ext specialist, Integrated Pest Mgt, Univ Wyo, 81-87; MGR AGR DEVELOP, WESTERN SUGAR CO, 87- *Concurrent Pos:* Mem comt pest detection & surv Entom Soc Am, 78-80; tech adv, major cereal grain prod, Egypt, 79; mem, Int Agr Liason Off, State of Wyo, Trade Mission, Egypt, Saudi Arabia & Jordan, 85, Saudi Arabia & Jordan, 86; dir, Beet Sugar Develop Found, 87-91. *Mem:* Entom Soc Am; Sigma Xi; Am Phytopath Soc; AAAS; Am Soc Sugar Beet Technologists. *Res:* Intergrated crop production strategies for forages and small grains in Wyoming and the Middle East; development of sugar beet production technology. *Mailing Add:* 1700 Broadway No 1600 Denver CO 80290

GHOLSON, ROBERT KARL, b McLeansboro, Ill, Feb 13, 30; m 52; c 3. BIOCHEMISTRY. *Educ:* Univ Chicago, BA, 50; Univ Ill, BS, 55, PhD(biochem), 58. *Prof Exp:* Res assoc biochem, Okla State Univ, 58-59; res assoc med sch Univ , Univ Mich, 59-61 & Kyoto Univ, 61-62; res assoc, 62-66, assoc prof, 66-69, PROF BIOCHEM, OKLA STATE UNIV, 69- *Mem:* Am Soc Biol Chem; Am Chem Soc; Sigma Xi; NY Acad Sci. *Res:* Biosynthesis of pyridine nucleotides; plant-pathogen interactions. *Mailing Add:* Dept Biochem Okla State Univ Stillwater OK 74075

GHONEIM, MOHAMED MANSOUR, b Egypt, Feb 6, 34; US citizen; m 66. ANESTHESIA, PSYCHOPHARMACOLOGY. *Educ:* Ain Shams Univ, Cairo, Egypt, MB, ChB, 57; Fac Anaesthetists, Royal Col Surg Eng, 63. *Prof Exp:* Sr house officer, anesthesia Nuffield Dept Anaesthetics, Oxford, Eng, 63-64; lectr, Assiut Univ, Egypt, 64-65; from asst prof to assoc prof, 67-72, PROF ANESTHESIA, UNIV IOWA, 76- *Mem:* Am Soc Anesthesiologists; Am Soc Pharmacol & Exp Therapeut; Asn Univ Anesthetists; Fac

Anaesthetists, Royal Col Surgeons, Eng. *Res:* Psychopharmacology, particularly the effects of drugs on human cognition and memory; psychological outcomes of anesthesia and surgery; effects of general anesthetics on human memory. *Mailing Add:* Dept Anesthesia Iowa Univ Hosps & Clins Iowa City IA 52242

GHONEIM, YOUSSEF AHMED, b Cairo, Egypt. CONTROL SYSTEM DESIGN, AUTOMOTIVE APPLICATIONS. *Educ:* Cairo Univ, Egypt, BSc, 76, MSc, 78; McGill Univ, Montreal, Can, MEng, 80, PhD(elec eng), 85. *Prof Exp:* Res engr teaching, Cairo Univ, 76-78; teaching asst, McGill Univ, 79-85; sr res engr, 85-90, STAFF RES ENGR AUTOMOTIVE INDUST, GEN MOTORS RES LABS, 90- *Mem:* Inst Elec & Electronics Engrs. *Res:* Vehicle traction; chassis control; analytical and experimental investigation. *Mailing Add:* Elec Eng No 40 Gen Motors Res Lab 12 Mile & Mound Rd Warren MI 48090

GHOSE, HIRENDRA M, b Patna, India; US citizen; m 60; c 3. PHYSICAL CHEMISTRY, INORGANIC CHEMISTRY SURFACE PHYSICS & TRIBOLOGY. *Educ:* Bihar Nat Col, Patna, BSc, 49; Sci Col Patna, MSc, 53; Mont State Col, PhD(phys chem), 60. *Prof Exp:* Demonstr physics, Bihar Nat Col, Patna, 49-51, prof chem, 53-55; instr, Skidmore Col, 59-60; sr res scientist, Glidden Co, 60-67; res supvr, Addressograph Multigraph Corp, 67-70; PROF CHEM & CHMN DEPT PHYS SCI, CUYAHOGA COMMUNITY COL, 70-; PRES, GHOSE & ASSOCS, 78-; CONSULT, FED ENIVRON PROTECTION AGENCY, 72-; CONSULT, NSF, 76- *Concurrent Pos:* Vis lectr, State Univ NY Col Plattsburgh, 59-60; consult, Friends Psychiat Res Inst, Spring Grove Ment Hosp, Baltimore, Md, 66-67; consult environ, chem & educ areas; pres,Ghose Industs Inc, 72-78. *Mem:* Am Chem Soc; Am Vacuum Soc; Soc Tribologists & Lubrication Engrs. *Res:* Surface reactivity, fluid chemistry, Auger Emission Spectroscopy; interaction of lubricant antiwear additives with alloy and ceramic surfaces; published books and technical papers in the fields of physics, chemistry, mathematics, high technology science and engineering. *Mailing Add:* Dept Phys Sci 2900 Community College Ave Cleveland OH 44115

GHOSE, RABINDRA NATH, b Howrah, India, Sept 1, 25; US citizen; m 64; c 1. BIOPHYSICS, PHYSICS. *Educ:* Univ Jadavpur, Calcutta, BEE, 46; Indian Inst Sci, Bangalore, dipl, 48; Univ Wash, MS, 52; Univ Ill, MA & PhD(elec eng), 54, EE, 56, LLB, 70; Golden Gate Univ, MBA, 86. *Prof Exp:* Instr elec eng, Jadavpur Univ, India, 46-49; tech officer & chief tech instr, Hq Western Command Indian Signals, 49-51; mem tech staff, Radio Corp Am, 54-56 & Space Tech Lab, 56-59; dir res & adv develop, Space-Gen Corp, 59-63; chmn bd, Am Nucleonics Corp, 63-83; PRES & CHMNN BD, TECHNOL RES INT, 83- *Concurrent Pos:* Reviewer sci proposals, NSF, 62; session chmn, Int Conf Microwaves, Commun & Info Theory, 64; mem res & tech adv bd, NASA, 74-76; mem, div adv group, Air Force Sci Adv Bd, 80-84; mem, adv comt & panel chmn, Defense Intel. *Mem:* Fel AAAS; fel Inst Elec & Electronics Engrs; fel Am Phys Soc; fel Inst Elec Engrs London; fel Inst Phys; fel Inst Engrs India. *Res:* Electromagnetic field theories; microwaves; nuclear science antennas. *Mailing Add:* 8167 Mulholland Terr Los Angeles CA 90046

GHOSE, SUBRATA, b Jamshedpur, India; m 67; c 2. MINERALOGY, CRYSTALLOGRAPHY. *Educ:* Univ Calcutta, MSc, 55; Univ Chicago, MS & PhD(mineral), 59. *Prof Exp:* Res assoc, Univ Pittsburgh, 59-61 & Univ Bern, 61-63; sr res assoc, Swiss Fed Inst Technol, 63-65; asst prof chem, Worcester Polytech Inst, 65-67; Nat Res Coun-Nat Acad Sci sr res fel, Goddard Space Flight Ctr, NASA, Md, 67-71; vis lectr crystallog, Univ Calif, Berkeley, 71-72; res assoc prof, Univ Wash, 72-75, res prof mineral, 75-77, actg prof geol sci, 77-78, PROF MINERAL PHYSICS & CRYSTALLOG, UNIV WASH, 78- *Concurrent Pos:* Vis prof, Univ Sci & Med, Grenoble, France, 80-81; Rollin D Salisbury fel, Univ Chicago, 56; vis prof, Univ Kyoto, Kyoto, Japan, 86; adj prof mats sci & eng & geophys, Univ Wash. *Mem:* Fel Mineral Soc Am; Mineral Asn Can; Am Crystallog Asn; Am Geophys Union. *Res:* Crystal chemistry of rock-forming silicates, phosphates and borates; cation order-disorder and thermodynamics and kinetics of crystalline solutions; physics of minerals; lattice dynamics; structural and magnetic phase transitions. *Mailing Add:* Dept Geol Sci Univ Wash Seattle WA 98195

GHOSE, TARUNENDU, b Begusarai, India, May 25, 28. PATHOLOGY. *Educ:* Univ Calcutta, BS, 45, MB, BS, 50; Indian Cancer Res Ctr, Bombay, PhD(exp path & oncol), 59; FRCP, 74. *Prof Exp:* House surgeon & registr, R G Kar Med Col Hosp & Chittaranjan Cancer Hosp, Calcutta, 50-54; asst med officer & res officer, Indian Cancer Res Ctr & Tata Mem Hosp, Bombay, 55-58, 59-61; med officer, Nat Acad Sci, US, 58-59; univ fel, Univ Aberdeen, 61-62, lectr path, 63-65; univ fel, Univ Leeds, 63; sr lectr, Monash Univ, Australia, 65-68; assoc prof, 69-77, PROF PATH & ASSOC PROF MICROBIOL, DALHOUSIE UNIV, 77- *Concurrent Pos:* Sr registr, Aberdeen Royal Infirmary, 63-65; assoc pathologist, Prov NS; head, Sect Immunopath, Victoria Gen Hosp, Halifax, NS, 69-; mem, Grant Rev Panels, Med Res Coun, Can & Cancer Res Soc, Montreal. *Honors & Awards:* Med Res Coun Prize, India, 59. *Mem:* Path Soc Gt Brit & Ireland; Am Asn Cancer Res. *Res:* Cancer immunology; immunopathology of respiratory and joint diseases. *Mailing Add:* Dept Path Dalhousie Univ Halifax NS B3H 4H7 Can

GHOSH, AMAL KUMAR, b Thaton, Burma, June 21, 31; m 61; c 3. SOLID STATE PHYSICS. *Educ:* Univ Calcutta, BS, 51, MS, 54, PhD(physics), 61. *Prof Exp:* Res assoc, Univ Notre Dame, 58-60; resident res assoc, Argonne Nat Lab, 60-62, asst physicist, 62-65; mem sci staff, Itek Corp, 66-70; res assoc, Esso Res & Eng Co, 70-74, res assoc, Exxon Res Ctr, 74-75, SR RES ASSOC, EXXON RES CTR, 75- *Mem:* Am Phys Soc. *Res:* Optical properties of solids; radiation effects; photoimaging and semiconductor devices. *Mailing Add:* 75 Twin Oaks Rd Bridgewater NJ 08807

GHOSH, AMAL KUMAR, b Berhampore, WBengal, India, June 1, 26; m 65; c 2. BIOCHEMISTRY. *Educ:* Univ Calcutta, BSc, 46, MSc, 48, DSc, 59. *Prof Exp:* Asst res officer biochem, Univ Calcutta, 56-58; fel biochem, Univ Pa, 59-63; vis asst prof biochem, Johnson Found, Univ Pa, 67-69; asst prof biochem, Harrison Dept Surg Res, 69-77, sr res specialist, 77-79, CORE DIR, DIABETES RES CTR, DEPT BIOCHEM & BIOPHYS, UNIV PA, 80-; RES SPECIALIST DIABETES RES CTR, DEPT BIOCHEM & BIOPHYS, UNIV PA. *Mem:* Am Soc Biol Chemists; NY Acad Sci. *Res:* Regulation and synchronization of biological oscillations; regulation of energy and redox states in normal and cancer cells and in the brain. *Mailing Add:* Dept Biochem & Biophys 410 Anat Chem Bldg G3 Univ Pa Diabetes Res Ctr Philadelphia PA 19104

GHOSH, AMIT KUMAR, mechanical metallurgy, materials science, for more information see previous edition

GHOSH, AMITAVA, b Calcutta, W Bengal, India, Oct 20, 57. BRITTLE FRACTURE FORMATION & PROPAGATION MATHEMATICAL MODELING. *Educ:* Indian Inst Technol, Kharagpur, BTech, 78; Univ Ariz, MS, 83, PhD(mining eng), 90. *Prof Exp:* Asst mgr res & serv, I D L Chemicals Ltd, India, 78-81; grad assoc res, Univ Ariz, Tucson, 82-84, grad assoc teaching, 85-90; POSTDOCTORAL FEL RES, UNIV NEV, RENO, 90- *Concurrent Pos:* Consult, Hargis & Assocs, Inc, Phoenix, 89. *Honors & Awards:* Chandrakala Div Medal, Mining, Metall & Geol Inst, India, 79. *Mem:* Int Asn Math Geol. *Res:* Fracture of rock water static and dynamic loads; fractal fracture phenomena; geomechanics; fragmentation and ground vibration from rock blasting; knowledge-based systems in geotechnology; mathematical, numerical and stochastic modeling. *Mailing Add:* 3865 E Leonesio Dr Apt No 32 Reno NV 89512

GHOSH, ANIL CHANDRA, b Kamargaon, India, Sept 1, 36; m 68; c 1. ORGANIC CHEMISTRY, MEDICINAL CHEMISTRY. *Educ:* Gauhati Univ, India, BS, 56, MS, 58; Poona Univ, PhD(chem), 63; FRIC. *Prof Exp:* Lectr chem, JB Col, India, 59; res fel chem, Nat Chem Lab, Poona, 60-63, sr res fel, 63-64; res assoc, Univ Neb, Lincoln, 64-66; res assoc, NC State Univ, Raleigh, 66-67; res chemist, Swiss Fed Inst Technol, 68-69; sr res assoc, Roswell Park Mem Inst, Buffalo, 69-71; sr res assoc, John C Sheehan Inst Res, Inc, 71-72; sr res scientist, SISA Inc, 72-; MEM STAFF AT RANBAXY LABS LTD, INDIA. *Concurrent Pos:* Adj sr res scientist, John C Sheehan Inst for Res Inc, 72- *Mem:* Royal Inst Chem London; Chem Soc; Am Chem Soc. *Res:* Synthesis of new drugs; steroids; alkaloids; CNS agents; fermentation products; natural products chemistry; photochemistry; mold metabolites. *Mailing Add:* Glaxo India Ltd Second Pokhran Rd Thane 400601 Maharash TRA India

GHOSH, ARATI, b Raipvr, MP, India, Mar 1, 36; US citizen; m 60; c 1. MEMBRANE TRANSPORT-PROTEIN, ELECTRON MICROSCOPY. *Educ:* Calcutta Univ, India, BSc, 55, MSc, 57, PhD(biochem), 63. *Prof Exp:* Sr man power fel microbiol, Calcutta Univ, India, 58-63, sr UGC fel, 63-64; postdoctoral fel neurochem, Univ Western Ont, London, Can, 64-66; res assoc microbial genetics, Waksman Inst Microbiol, Rutgers Univ, NJ, 67-73; res specialist microbial cell physiol, Rutgers Med Sch, Univ Med & Dent NJ, Piscataway, 73-85; RES SCIENTIST TOXICOL/ENVIRON POLLUTION, NJ STATE DEPT HEALTH, TRENTON, 85- *Concurrent Pos:* Vis scientist, Nat Inst Med Res, Mill Hill, London, UK, 77. *Mem:* Am Soc Cell Biol; Electron Micros Soc Am. *Res:* membrane structure and membrane regulation of microbial protein secretion; mechanism of action of antibiotics on cell membrane; electron microscopic analysis of environmental pollution (asbestos). *Mailing Add:* Eight Constitution Ct East Brunswick NJ 08816

GHOSH, ARUP KUMAR, b Calcutta, India, May 31, 50; m 72; c 2. PHYSICS, MATERIAL SCIENCE. *Educ:* Univ Delhi, BSc, 69, MSc, 71; Univ Rochester, PhD(physics), 76. *Prof Exp:* Res assoc physics, Univ Rochester, 76; res assoc, 76-78, asst physicist, 78-79, ASSOC PHYSICIST, BROOKHAVEN NAT LAB, 80- *Mem:* Am Phys Soc. *Res:* Superconductivity; A-15 material preparation; transport properties of superconductors and amorphous material; superconductors and amorphous material; amorphous silicon. *Mailing Add:* Brookhaven Nat Lab Bldg No 902 Upton NY 11973

GHOSH, ASOKE KUMAR, b Dacca, India, June 17, 38; Can citizen; m 64; c 3. PHYSICS, ELECTRONICS. *Educ:* Calcutta Univ, BSc, 56, MSc, 59; Jadavpur Univ, PhD(plasma physics), 64. *Prof Exp:* Sr lectr physics, N Bengal Univ, 62-63; mem sci staff plasmas, laser electronics, RCA Ltd, Can, 65-77; DIR PROG DEVELOP PLASMAS, ENERGY & ELECTRONICS, MPB TECHNOL INC, 77- *Res:* Plasma physics; gas discharge; lasers; microwave; solar energy; pollution monitoring devices; ARC technology; heat pipe application; devices; laser applications (zone crystallization); iodine laser. *Mailing Add:* 151 Hymus Pt Claire PQ H9R 1E9 Can

GHOSH, BHASKAR KUMAR, b Dibrugarh, India, Feb 10, 36; m 60; c 3. MATHEMATICAL STATISTICS. *Educ:* Univ Calcutta, BSc, 55; Univ London, PhD(statist), 59. *Prof Exp:* Res asst statist, Univ Col, Univ London, 58-59; statistician, Atomic Power Construct, Ltd, Eng, 59-60; asst lectr statist & math, Chelsea Col, Univ London, 60-61; from asst prof to assoc prof math, 61-68, chmn dept, 81-85, PROF MATH, LEHIGH UNIV, 68- *Concurrent Pos:* Consult, Pa Power & Light Co, 62-63, Int Tel & Tel Corp, 64, Beryllium Corp, 65 & Howmet Aluminum Corp, 79, BOC Group, 84-; vis prof, Mass Inst Technol, 68, Va Polytech Inst, 78-80; vis prof, Univ Münster, WGer, 86-87; Humboldt Sr US Scientist, 86-87. *Mem:* Inst Math Statist; Royal Statist Soc. *Res:* Probability; sequential analysis; statistical inference. *Mailing Add:* Dept Math Lehigh Univ Bethlehem PA 18015

GHOSH, BIJAN K, b India, 35; US citizen; m 60; c 1. PHYSIOLOGY, ELECTRON MICROSCOPY. *Educ:* Calcutta Univ, India, BSc, 55, MSc, 57, DSc(microbiol & physiol), 63. *Prof Exp:* Asst res prof, Waksman Inst, 68-73, assoc prof, 73-79, PROF PHYSIOL, ROBERT WOOD JOHNSON MED SCH, UNIV MED & DENT NJ & HON PROF, WAKSMAN INST MICROBIOL, RUTGERS UNIV, 79- *Concurrent Pos:* Vis prof, Univ Amsterdam, 73; vis lectr, Chinese Acad Sci, 85. *Mem:* Can Soc Biochem; Am Soc Chem Biol; Am Soc Microbiol; Electron Micros Soc Am. *Res:* The mechanism of regulation of protein secretion by membrane; nature of bacterial membrane compartments; application of quantitative electron microscopy to study microbial physiology and molecular biology. *Mailing Add:* Dept Physiol & Biophys Robert Wood Johnson Med Sch 675 Hoes Lane Piscataway NJ 08854-5635

GHOSH, CHITTA RANJAN, b Mulghar, India, Mar 1, 36. IMMUNOLOGY, MICROBIOLOGY. *Educ:* Univ Calcutta, BVS, 57; Univ Ky, PhD(microbiol), 69. *Prof Exp:* Vet surgeon, Govt WBengal, India, 58-63; instr obstet & gynec, Sch Med, Tufts Univ, 69-74; immunologist, St Joseph's Hosp, 73-75; immunologist, dept path, Sch Med, Creighton Univ, 75-80; LAB SCIENTIST, OMAHA-DOUGLAS COUNTY HEALTH DEPT, 80- *Concurrent Pos:* Immunologist, St Margaret's Hosp, 69-74. *Mem:* Am Soc Microbiol. *Res:* Autoimmune diseases; clinical immunology and serology; immunology of cancer. *Mailing Add:* 1308 N 40th Omaha NE 68131

GHOSH, DIPAK K, b Kalupur, WBengal, India, Feb 21, 47; m 76; c 2. SKIN CARE RESEARCH, DERMATOLOGICAL RESEARCH. *Educ:* Jadavpur Univ, India, BS, 67, PhD(pharm), 73; Banaras Hindu Univ, India, MS, 69. *Prof Exp:* Sr res chemist pharmaceut, Tata-Fison Labs, India, 69-70; res fel pharm, Rutgers Univ, 73-74; sr scientist pharmaceut, Block Drug Co, NJ, 74-78; DIR RES & DEVELOP, CHESEBROUGH-POND'S RES LAB, 78- *Mem:* Am Pharmaceut Asn; Soc Cosmetic Chemists; Royal Soc Chem. *Res:* Development, evaluation and clinical testing of dermatological semisolid formulations; cough and cold medicine; antacid, analgesic and other over the counter products; several US patents. *Mailing Add:* Chesebrough-Pond's Res Labs Trumbull Indust Park Trumbull CT 06611

GHOSH, HARA PRASAD, b India, June 1, 37; m 68; c 2. GENETIC ENGINEERING & BIOTECHNOLOGY. *Educ:* Calcutta Univ, BSc Hon, 57, MSc, 59, PhD(biochem), 63. *Prof Exp:* Fel biochem, Univ Calif, Davis, 64-66 & Univ Wis-Madison, 66-67; from asst prof to assoc prof, 69-76, PROF BIOCHEM, MCMASTER UNIV, 76-, CHMN BIOCHEM, 83- *Concurrent Pos:* Med Res Coun Can scholar, 69-74; mem, Molecular Biol & Virol Grants Panel, Nat Cancer Inst Can, 82-84, Molecular Biol Grants Panel, Med Res Coun Can, 83-85; vis prof biol, Mass Inst Technol, 84-85. *Mem:* Am Soc Biol Chemists; Can Biochem Soc; Am Soc Microbiol; Am Soc Virol. *Res:* Targeting of viral membrane glycoproteins: cloning, construction and expression of chimeric genes and engineered mutagenesis; introduction and expression of foreign genes into mammalian cells and tissues using highly transmissible recombinant retrovirus vectors with extended host-range. *Mailing Add:* Dept Biochem McMaster Univ 1200 Main St W Hamilton ON L8N 3Z5 Can

GHOSH, MRIGANKA M(OULI), b Calcutta, India, Nov 5, 35; m 67; c 1. WATER CHEMISTRY, ENVIRONMENTAL ENGINEERING. *Educ:* Indian Inst Technol, Kharagpur, BTech, 58; Univ Ill, Urbana, MS, 62, PhD(sanit eng), 65. *Prof Exp:* Jr engr, Hindusthan Steel Ltd, India, 58-59, asst engr, 59-60; grad asst civil eng, Univ Ill, Urbana, 61-65, asst prof, 65-66; reader, Jadavpur Univ, India, 66-68; from asst prof to prof, Univ Maine, 68-76; prof civil eng, Univ Mo, 76-; PROF ENVIRON ENG, PA STATE UNIV. *Concurrent Pos:* Consult, Calcutta Metrop Planning Orgn, 66-68, Edward C Jordan & Co, 69-70, city of Somersworth, 69- & city of Dover, 71-; ed, J Environ Engr Div, Am Soc Civil Engrs, 78-80. *Mem:* Am Water Works Asn; Am Soc Civil Engrs (exec secy, 80-84); Water Pollution Control Fedn; Am Chem Soc. *Res:* Chemistry of iron and manganese in natural waters; cultural eutrophication of lakes; effect of mercury and cadmium on stream biota; role of yellow organic acids in natural waters; physical chemical parameters affecting the removal of collods by porous media; removing trace organics from drinking waters. *Mailing Add:* Dept Civil Eng Pa State Univ Main Campus University Park PA 16802

GHOSH, NIMAI KUMAR, b Kandi, India, Feb 19, 37; m 58; c 2. EXPERIMENTAL MEDICINE. *Educ:* Calcutta Univ, BSc, 56, MSc, 58, DPhil(biochem), 64; FRSC, 69. *Prof Exp:* Lectr chem, Kandi Raj Col, Calcutta Univ, 59-60; res chemist, Calcutta Sch Trop Med, 60-64; res assoc & asst prof path, Sch Med, Tufts Univ, 64-67; asst prof med & biochem pharmacol, Brown Univ, 68-70; res scientist, 70-75, asst prof, 75-77, ASSOC PROF EXP MED, NY UNIV MED CTR, 77-; assoc prof, Long Island Univ, 81-82, prof, Arnold Marie Schwart* Col, Pharm & Health Sci, 82-87; prof, Pace Univ, 85-87; sr pharmaceut sci, Pharmacol, Danbury Univ, Conn, 86; supvr, Biochem, New York City Health & Hosp Corp, 87; EXEC VPRES, ADMIN SCI & TECH DIR, TRI COUNTY DIAGNOSTIC LABS CORP, 87- *Concurrent Pos:* Vis investr, Montreal Cancer Inst, Can, 68; consult, Centralized Lab Serv, Inc, Health Ins Plan Greater New York, 70-76; Int Union Biochem cong award, Am Soc Biol Chemists, 71-78; sci reviewer, Biochimica Biophysica & Biophys, J of Nat Cancer, 71-87 Arch Biochem & Biophys, J Nat Cancer Inst & NSF, 79; dir clin biochem, Path Dept, Catholic Med Ctr, 78- *Honors & Awards:* Am Inst Chemists. *Mem:* Am Chem Soc; Am Soc Biol Chemists; Can Biochem Soc; Brit Biochem Soc; Indian Sci Cong Asn. *Res:* Isolation and characterization of oncofetal isoenzymes of phosphohydrolase in human placenta and cervical cancer cells; identification of ectopic peptide hormones and regulation of their biosynthesis in cancer cells by antineoplastic inhibitors of DNA replication; drug-induced biochemical markers of cancer oncofetal (placental) antigens in cancer and pregnancy endorphins; alfa fetoprotein, angiotensin and renin in placenta and pregnancy serum. *Mailing Add:* 76-08 45th Ave Queens NY 11369

GHOSH, SAKTI P, b Calcutta, India, Feb 6, 35. MATHEMATICAL STATISTICS, COMPUTER SCIENCE. *Educ:* Univ Calcutta, BSc, 55, MSc, 57; Univ Calif, Berkeley, PhD(statist), 62. *Prof Exp:* Statistician, State Statist Bur, Govt WBengal, India, 58-59; asst statist, Univ Calif, Berkeley, 59-62; mem res staff, T J Watson Res Ctr, 62-68, MEM RES STAFF, RES LAB, INT BUS MACH CORP, 68- *Concurrent Pos:* Adj asst prof, NY Univ, 64-; co-chmn, Int Symposium Very Large Data Bases, 76; prog co-chmn & mem, Steering Comt, Nat Comput Conf, 78; invited US rep, South East Asia Comput Conf, Manila, 78; adj prof, San Jose State Univ, 82-; ed, Software Eng, Inst Elec & Electronics Engrs, 84- *Honors & Awards:* Spec Award For Model Curric, Comput Socs Group, Inst Elec & Electronics Engrs, 77; IBM Res Div Award, 78; Invention Achievement Award, IBM, 82. *Mem:* Fel Inst Elec & Electronics Engr, 90. *Res:* Sampling theory; information retrieval theory; information sciences; computer language; coding theory; information and data conversion; manufacturing research; statistical database management. *Mailing Add:* IBM Almaden Res Ctr San Jose CA 95120

GHOSH, SAMBHUNATH, b Calcutta, India, Aug 1, 35; m 61; c 2. ENVIRONMENTAL ENGINEERING, BIOENGINEERING. *Educ:* Univ Calcutta, BS, 56; Univ Ill, Urbana, MS, 63; Ga Inst Technol, PhD(civil & environ eng), 70. *Prof Exp:* Jr engr civil eng, Pub Works Dept, State of Assam, India, 56-57; asst engr struct & sanit eng & asst construct engr, Durgapur Steel Proj, Hindustan Steel, Govt India, 58-61; grad res asst environ eng, Univ Ill, Urbana, 61-63; res assoc, Univ NC, Chapel Hill, 63-64; engr sanit eng, Wiedeman & Singleton, Ga, 64-65; trainee environ & civil eng, Ga Inst Technol, 65-69, fel civil eng, 70; mgr bioeng res, Inst Gas Technol, Ill Inst Technol, 70-85; CHMN, CIVIL ENG DEPT, UNIV UTAH, 85- *Concurrent Pos:* Consult, Aqua Tech, Atlanta, Ga, 69-70; vis lectr, Dept Environ Eng, Ill Inst Technol, 71-72; US deleg, sci exchange visit People's Repub China, US Nat Acad Sci & Chinese Asn Sci & Technol, 82; mem, Nat Prog Vis Scientists People's Repub China, US Nat Acad Sci, 86-87; Fulbright res & lectr award, Nat Environ Eng Res Inst, Nagpur, India, 90-91. *Honors & Awards:* Monie A Ferst Mem Res Award, Sigma Xi. *Mem:* Am Soc Civil Engrs; Int Asn Water Pollution Res; Water Pollution Control Fedn; Soc Indust Microbiol; Sigma Xi. *Res:* Methanogenic and nonmethanogenic fermentation of wastes and biomass for fuel production; resource recovery; water and waste treatment; biotransformation of fossil fuels; process development through bench and pilot scale investigation; solid and hazardous waste. *Mailing Add:* Dept Civil Eng Univ Utah Merrill Eng Bldg 3220 Salt Lake City UT 84112

GHOSH, SANJIB KUMAR, b Calcutta, India, Sept 9, 25; m 51; c 2. PHOTOGRAMMETRY, GEODESY. *Educ:* Univ Calcutta, BSc, 45; Int Training Ctr Aerial Surv, Delft, Netherlands, Photog Engr, 57; Ohio State Univ, PhD(photogram), 64. *Prof Exp:* Surveyor, Surv India, 48-60; from res asst to res assoc photogram, 60-61; from instr to assoc prof photogram & geod, Ohio State Univ, 62-79; PROF PHOTOGRAM, LAVAL UNIV, 79- *Concurrent Pos:* Mem panel experts photogram, UNESCO, 66-; vis prof, Fed Univ Parana, Brazil, 76-77; invited prof, Hosei Univ, Japan, 77-78; UN consult, 80-; external examr, Univ Teknologi Malaysia, Lagos Univ, Nigeria, Univ Delhi, India, Univ Rajasthan, India; invited prof, Inst Agr & Vet, Hassan II, Morocco, 87-88, Univ Sao, Paolo, Brazil, 89; UN fel, Commendation Ohio House Represenatives, 78. *Honors & Awards:* Presidential Award, Am Soc Photogram, 71 & 79. *Mem:* Am Soc Photogram; fel Am Cong Surv & Mapping; Int Soc Photogram; Geog Soc India; Brit Photogram Soc; Can Inst Surv; Soc Photo-optical Instrumentation Engrs. *Res:* Geometric and physical aspects of photogrammetry; orientation of photogrammetric models; aerial triangulation; photogrammetric system calibration; un-conventional photogrammetry; remote sensing. *Mailing Add:* Dept Photogram Laval Univ Quebec PQ G1K 7P4 Can

GHOSH, SATYENDRA KUMAR, b Berhampore, India, Sept 17, 45; m 73; c 2. STRUCTURAL ENGINEERING, EARTHQUAKE ENGINEERING. *Educ:* Univ Calcutta, BE, 66; Univ Waterloo, MASc, 69, PhD(struct eng), 72. *Prof Exp:* Res & teaching asst civil eng, Univ Waterloo, 67-69, 70-72 & Univ Pittsburgh, 69-70; fel, Univ Waterloo, 72-73, res assoc, 73, adj prof, 73-74; struct engr, 74-75, sr struct engr, 75-79, PRIN STRUCT ENGR, PORTLAND CEMENT ASN, 79- *Concurrent Pos:* Partner, Elan Assocs, Waterloo, Ont, 72-73; consult, Peter Sheffield & Assocs, Toronto, 73; partic earthquake eng workshop, NSF, Univ Calif, Berkeley, 77; mem joint comts, Am Concrete Inst & Am Soc Civil Eng, 77- *Mem:* Am Concrete Inst; Am Soc Civil Engrs. *Res:* Analysis and design of earthquake resistant reinforced concrete building structures.. *Mailing Add:* Dept Civil Eng-MC 246 Univ Ill Box 4348 Chicago IL 60680

GHOSH, SUBIR, b Calcutta, India, Aug 26, 50; m 78. STATISTICAL PLANNING OF EXPERIMENTS. *Educ:* Univ Calcutta, BS, 68, MS, 70, Colo State Univ, PhD(statist), 76. *Prof Exp:* Res asst statist, Colo State Univ, 72-75 & fel, 76; researcher math statist, Indian Statist Inst, India, 71, vis fel, 76-77 & lectr, 77-80; asst prof, 80-85, ASSOC PROF STATIST, UNIV CALIF, RIVERSIDE, 85- *Mem:* Inst Math Statist; Am Statist Asn. *Res:* Statistical planning or design of experiments; search design; robustness of design against the unavailability of data; linear models; planning with correlated data; optimum designs; sampling methods. *Mailing Add:* Dept Statist Univ Calif 900 Univ Ave Riverside CA 92521

GHOSH, SWAPAN KUMAR, b Calcutta, India, Jan 1, 42; m 70; c 2. IMMUNOLOGY, BIOCHEMISTRY. *Educ:* Univ Calcutta, India, BSc, 62, MSc, 64, PhD(biochem), 69. *Prof Exp:* Lectr biochem, Med Col Calcutta Univ, 66-68; asst prof biochem, Med Col, NBengal Univ, 68-70; USPHS fel pharmacol, Med Ctr, Univ Ill, 71-72; sr res assoc neurochem, State Univ NY, Buffalo, 73-75; Nat Cancer Inst fel immunol, 76-78, CANCER RES SCIENTIST III IMMUNOL, ROSWELL PARK MEM INST, 79- *Concurrent Pos:* Adj asst prof immunol, State Univ NY, Buffalo, 85. *Mem:* NY Acad Sci; Am Asn Immunologists. *Res:* Immunological aspects of tumor metastasis; biochemical and immunological characterization of mammalian cell surfaces; mechanism of generation of tumor variants. *Mailing Add:* Dept Life Sci Ind State Univ Sixth & Chestnut Terre Haute IN 47803

GHOSH, VINITA JOHRI, b New Delhi, India, Sept 5, 49; m 72; c 2. COMPUTER MODELLING. *Educ:* Univ Dehli, BSc, 69, MSc, 71; Univ Rochester, PhD(physics), 79. *Prof Exp:* PHYSICS ASSOC, MULTILAYER SYSTS, BROOKHAVEN NAT LAB, 80- *Concurrent Pos:* Tech ed, Phys Rev, 80-81. *Mem:* Am Women Sci. *Res:* Use of computer modelling to study the problem of flux pinning in type II superconductors; characteristics of defect migration in A15 compounds and light-induced photoconductivity in amorphous silicon; modelling of positron implantation profiles. *Mailing Add:* Bldg 480 Brookhaven Nat Lab Upton NY 11971

GHOSHAL, NANI GOPAL, b Dacca, India, Dec 1, 34; m 71; c 1. COMPARATIVE ANATOMY, NEUROANATOMY. *Educ:* Bengal Vet Col, Calcutta, GVSc, 55; Royal (Dick) Sch Vet Studies, Edinburgh, DTVM, 61; Fac Vet Med, Hannover, Dr med vet, 62; Iowa State Univ, PhD(anat), 66. *Prof Exp:* Vet asst surgeon, Civil Vet Dept, Govt W Bengal, India, 55; demonstr comp vet anat, Bengal Vet Col, Calcutta, 55-56; res asst, MB Govt Col Vet Sci & Animal Husb, Mhow, 56-59; pool scientist, Indian Coun Agr Res, New Delhi, 63; instr, 63-66, from asst prof to assoc prof, 67-74, PROF VET ANAT, IOWA STATE UNIV, 74- *Concurrent Pos:* Tollygunj Calcutta Scholarship, 54, Ger Acad Exchange Serv Scholar, Govt WGer, 61; adj prof, Inst Agron & Vet, Hassan II, Rabat, Morocco, 84-88; consult, Minn-Morocco Proj, USAID, 84-88. *Honors & Awards:* Raymond Star Gold Medal, Govt W Bengal, India, 55. *Mem:* AAAS; Am Asn Vet Anat; fel Royal Zool Soc Scotland; World Asn Vet Anat; Pan Am Asn Anat; Am Asn Anat; NY Acad Sci; Sigma Xi. *Res:* Gross anatomy; functional anatomy; brain temperature regulation; author and co-author of textbooks. *Mailing Add:* Dept Vet Anat Col Vet Med Iowa State Univ Ames IA 50011

GHOSHTAGORE, RATHINDRA NATH, b Sribari, Bangladesh, July 8, 37; m 67; c 2. MATERIALS SCIENCE, SEMICONDUCTOR INTEGRATED CIRCUITS. *Educ:* Univ Calcutta, BS, 57, MS, 60; Mass Inst Technol, ScD(ceramics), 65. *Prof Exp:* Sr physicist, Fundamental Res Lab, Xerox Corp, NY, 65-66; mem tech staff mat & processes, Fairchild Res & Develop Lab, Calif, 66-67; sr engr, 68-74, fel scientist, 74-79, adv scientist, Westinghouse Res Labs, 79-80, ADV ENGR & PROG MGR, WESTINGHOUSE ADVAN TECHNOL LAB, 80- *Mem:* Fel Am Phys Soc; Electrochem Soc; sr mem Inst Elec & Electronics Engrs. *Res:* Insulators and semiconductors; chemical vapor deposition of thin solid films in amorphous, polycrystalline, epitaxial and glassy state; diffusion in semiconductors; process development and fabrication of advanced silicon integrated circuits for defense and space systems. *Mailing Add:* Westinghouse Advan Technol Lab PO Box 1521 Baltimore MD 21203

GHOWSI, KIUMARS, b Esfahan, Iran, July 28, 60; m 86. SOLID LIQUID INTERFACES PHENOMENA, HIGH ELECTRIC FIELD BREAKDOWN IN SOLIDS & LIQUIDS. *Educ:* La State Univ, BS, 82, MS, 84, PhD(chem), 90. *Prof Exp:* Postdoctoral chem, La State Univ, 90-91; ASST PROF, TEX TECH UNIV, 91- *Mem:* Electrochem Soc; Am Chem Soc. *Res:* Insulators electrochemistry; ionic and electronic transport phenomena in solids and liquids; colloid and interface phenomena. *Mailing Add:* Dept Chem & Biochem Tex Tech Univ Lubbock TX 79409-1061

GHOZATI, SEYED-ALI, b Tehran, Iran, Oct 19, 44. ELECTRICAL ENGINEERING, COMPUTER SCIENCE. *Educ:* Tehran Univ, BS, 67; Columbia Univ, MS, 72, MPH & PhD(elec eng, comput sci), 76. *Prof Exp:* Teaching asst, Dept Elec Eng & Comput Sci, Columbia Univ, 71-76, assoc consult work, 76; asst prof, 76-81, ASSOC PROF COMPUT SCI, QUEENS COL, 81- *Concurrent Pos:* City Univ grants, 78-79, 80-81, 81-82. *Mem:* Inst Elec & Electronics Engrs; Sigma Xi. *Res:* Stochastic systems; simulation; microprocessors. *Mailing Add:* Two Jackie Dr Westbury NY 11590

GHUMAN, GIAN SINGH, b Barchuhi, India, July 7, 29; m 48; c 3. GEOENVIRONMENTAL SCIENCE. *Educ:* Punjab Univ, India, BS, 52, MS, 55; Univ Calif, Davis, PhD(soil sci), 67. *Prof Exp:* Res asst fertilizer exp, Dept Agr, Punjab Univ, India, 55-57; asst prof soil chem, Shri Karan Narinder Col Agr, India, 57-62; assoc prof, 67-71, PROF EARTH SCI, SAVANNAH STATE COL, 71- *Concurrent Pos:* Dir, NASA Res Proj Kennedy Space Ctr, 74-77; dir, Marshland Res Proj, SE Atlantic Coast, Environ Protection Agency, 79-83; dir, heavy metal compos, Munic Wastewate Res Proj, 84-85; dir, Release rates of Toxic Metals from Coastal Soils Res Proj, 88-89. *Honors & Awards:* Presidential Achievement Award, 82. *Mem:* AAAS; Am Soc Agron; Soil Sci Soc Am; Int Soc Soil Sci; Clay Minerals Soc; Environ Geochem & Health Soc. *Res:* Availability of iron and manganese as affected by soil treatments; investigations on mineral manganocalcite; clay minerals and mineralization in surface and ground waters; water quality and marine sediments; heavy metals in estuaries. *Mailing Add:* Dept Earth Sci Savannah State Col Savannah GA 31404

GHURYE, SUDHISH G, b Bombay, India, Nov 10, 24; m 53. STATISTICS. *Educ:* Univ Bombay, MSc, 47; Univ NC, PhD(math statist), 52. *Prof Exp:* Asst prof math, Univ Ore, 53-54; reader statist, Univ Lucknow, 54-56; asst prof, Univ Chicago, 56-58; assoc prof math, Northwestern Univ, 59-61; assoc prof, Univ Minn, 61-62; prof, Ind Univ, Bloomington, 62-68, chmn dept, 64-67; chmn dept, 71-75, PROF MATH, UNIV ALTA, 68-, PROF STATIST, 81- *Concurrent Pos:* Assoc ed, Ann Math Statist, 70-73; ed, Can J Statist, 80- *Mem:* Fel Inst Math Statist; Statist Soc Can; Int Statist Inst. *Res:* Applied probability; mathematical statistics. *Mailing Add:* 1333 Lincoln St No 385 Bellingham WA 98226

GIACCHETTI, ATHOS, b Florence, Italy, Jan 1, 21; US citizen; m 48; c 2. SPECTROSCOPY. *Educ:* Univ Florence, Dr(physics), 47. *Prof Exp:* Asst prof physics, Nat Univ South, Arg, 48-53; head spectros sect, Nat AEC, 53-58; assoc prof atomic spectros, La Plata Nat Univ, 58-60; Orgn Am States fel & vis prof, Purdue Univ, 60-61; assoc physicist, Argonne Nat Lab, 61-71; specialist, Dept Sci Affairs, Orgn Am States, 71-86; RETIRED. *Concurrent Pos:* Vis scientist, Aimè Cotton Lab, Nat Ctr Sci Res, France, 69-70; hon prof, Univ La Plata, Arg, 80. *Mem:* Optical Soc Am. *Res:* Atomic spectroscopy; spectrochemistry; standard wave lengths; term analysis. *Mailing Add:* 1211 Woodside Pkw Silver Spring MD 20910

GIACCO, ALEXANDER FORTUNATUS, b San Giovanni di Gerace, Italy. SCIENCE ADMINISTRATION. *Educ:* Va Polytech Inst, BS, 42. *Hon Degrees:* DBus, William Carey Col, 80; LLD, Widener Univ, 84, Cath Univ Am, 90; DBus Admin, Goldey Beacom Col, 84; LHD, Mt St Marys Col, 88. *Prof Exp:* Mgt, prod, mkt & planning positions, Hercules Inc, 42-71, mem bd dirs, 71-73, gen mgr, Hercules Europe, 73-74, vpres, 74-76, exec vpres, 76-77, chief exec officer & 6th pres, 77-80, chmn bd, 80-87; chief exec officer, 87-90, CHMN BD, HIMONT INC, 83- *Concurrent Pos:* Chmn, econ develop comt, Del Round Table; dep chmn, Propulsion Comt Guided Missiles & Jet-Assisted Take-Off, Am Defense Preparedness Asn; mem, Adv Coun Japan-US Econ Rel, adv bd, New Ctr Hist Chem; mem bd dirs, Montedison, SpA, 83-90, Ferruzzi Finanziaria, 88- *Honors & Awards:* Order of Merit, Ital Repub, 85. *Mem:* Nat Acad Eng; Am Asn Soverign Mil Order; Soc Chem Indust; Soc Plastics Indust; Soc Automotive Engrs. *Res:* One patent on the design of solid rocket propellant grains. *Mailing Add:* Himont Inc 2801 Centerville Rd PO Box 15439 Wilmington DE 19850-5439

GIACCONI, RICCARDO, b Genoa, Italy, Oct 6, 31; nat US; m 57; c 3. ASTROPHYSICS. *Educ:* Univ Milan, PhD(physics), 54. *Hon Degrees:* DSc, Univ Chicago, 83; Laurea ad Honoreum Astron, Univ Padua, 84. *Prof Exp:* Asst prof physics, Univ Milan, 54-56; Fulbright fel, Ind Univ, 56-58; res assoc, Princeton Univ, 58-59; exec vpres & mem bd dirs, Am Sci & Eng, Inc, Mass, 59-73; prof astron & assoc dir, High Energy Astrophys Div, Ctr Astrophys, Harvard Univ, 73-81; DIR, SPACE TELESCOPE SCI INST, BALTIMORE, 81-; PROF PHYSICS & ASTRON, JOHNS HOPKINS UNIV, 82- *Concurrent Pos:* Vis comt, Asn Univ Res Astron, 71-75, Univ Padova & Univ Chicago; counr, Am Astron Soc, 79-82; pres, Comt 48, Int Astron Union, 82-86; mem adv comt, Max-Planck Inst Physics & Astrophys; vis comt, Asn Univ Res Astron, 71-75; prof physics, Univ Milan, Italy, 91- *Honors & Awards:* Helen B Warner Prize, Am Astron Soc, 66; Como Prize, Ital Phys Soc, 67; Rontgen Prize Astrophys, Physikalisch-Medizinische Gessellschaft, Germany, 71; Medal for Except Sci Achievement, NASA, 71 & 80, Distinguished Pub Ser Award, 72; Richtmyer Mem Lectr, Am Asn Physics Teachers, 75; Space Sci Award, Am Inst Aeronaut & Astronaut, 76; Elliot Cresson Medal, Franklin Inst, 80; Catherine Wolfe Bruce Gold Medal, Astron Soc Pac, 81; Dannie Heineman Prize Astrophys, Am Astron Soc/Am Inst Physics, 81; Henry Norris Russell lectr, Am Astron Soc, 81; Gold Medal, Royal Astron Soc, 82; A Cressy Morrison Award, NY Acad Sci, 82; Wolf Prize in Physics, 87. *Mem:* Nat Acad Sci; Am Acad Arts & Sci; fel AAAS; Am Phys Soc; Int Astron Union; Comt Space Res; Am Astron Soc; fel Royal Astron Soc. *Res:* X-ray astronomy; fields and particles. *Mailing Add:* Space Telescope Sci Inst Homewood Campus 3700 San Martin Dr Baltimore MD 21218

GIACOBBE, THOMAS JOSEPH, b Newark, NJ, June 25, 41; m 65; c 3. ORGANIC CHEMISTRY, BIOORGANIC CHEMISTRY. *Educ:* Bowdoin Col, BA, 63; Univ Vt, PhD(chem), 68. *Prof Exp:* NIH fel, Univ Wis-Madison, 68-69; spec assignments dept, Dow Chem Co USA, 69-71, res biol chemist, 71-72, herbicide proj mgr, 72-76, mgr agr chem process develop, 76-78; group leader agr chem process develop, 78-81, group leader polyethylene pilot plant, 81-84, GROUP LEADER NEW PROD RES, MOBIL CHEM CO, 84- *Concurrent Pos:* Fel, Univ Wis, 68-69; instr, Univ Calif Exten, 76. *Mem:* Am Chem Soc. *Res:* Synthetic organic chemistry; structure-activity correlations; physical organic and process chemistry; lubrication chemistry; agricultural chemistry. *Mailing Add:* 17 Nassau Ct Skillman NJ 08558

GIACOBINI, EZIO, AGING & DEVELOPMENT OF NERVOUS SYSTEM. *Educ:* Univ Turin, Italy, MD, 53; Karolinska Inst, Sweden, PhD(physiol), 59. *Prof Exp:* PROF PHARMACOL & CHMN DEPT, SCH MED, ILL UNIV, 82- *Mailing Add:* Dept Pharm Southern Ill Univ PO Box 19230 Springfield IL 62794-9230

GIACOLETTO, L(AWRENCE) J(OSEPH), b Clinton, Ind, Nov 14, 16; m 41; c 1. ELECTRONICS. *Educ:* Rose-Hulman Inst Technol, BS, 38; Univ Iowa, MS, 39; Univ Mich, PhD(electronics), 52. *Prof Exp:* Asst, Univ Iowa, 38-39; asst elec mach, Univ Mich, 39-41; res engr, Labs Div, Radio Corp of Am, NJ, 46-56; res mgr, Electronics Dept, Sci Lab, Ford Motor Co, 56-60; prof, 60-87, EMER PROF ELEC ENG, MICH STATE UNIV, 87- *Concurrent Pos:* Pres, CoRes Inst, Okemos, Mich. *Mem:* Fel AAAS; fel Inst Elec & Electronics Engrs. *Res:* Theory, design and application of solid state devices; electronics; power electronics. *Mailing Add:* Dept Elec Eng Mich State Univ East Lansing MI 48824

GIACOMELLI, FILIBERTO, b Pisa, Italy, Nov 18, 28; m 58. EXPERIMENTAL PATHOLOGY. *Educ:* Univ Pisa, MD, 54. *Prof Exp:* Asst med path, Univ Pisa, 56-57 & Univ Rome, 57-63; asst gen path, Univ Pisa, 63-66; asst prof path, Ind Univ, Indianapolis, 66-67; assoc, Columbia Univ, 67-68; from asst prof to assoc prof path, New York Med Col, 68-78; PROF PATH, SCH MED, WAYNE STATE UNIV, 78- *Concurrent Pos:* NIH fel, Columbia Univ, 61-63. *Mem:* Am Asn Path; Am Heart Asn; Am Soc Cell Biol; Electron Micros Soc Am; AAAS; NY Acad Sci. *Res:* Light and electron microscopy; cytochemistry of normal and abnormal tissues; ultrastructure and biology of hypertensive and diabetic cardiovascular disease in experimental animals; metabolic and morphological correlates. *Mailing Add:* Dept Path 540 E Canfield Ave Detroit MI 48201

GIACOMETTI, LUIGI, b Gubbio, Italy, Jan 21, 26; US citizen; m 55; c 2. BIOLOGY. *Educ:* Brown Univ, MSc, 62, PhD(biol), 64. *Prof Exp:* Asst scientist, Ore Regional Primate Res Ctr, 64-67, assoc scientist, 67-69, sci dir, Ore Zool Res Ctr, 69-73; prog dir corneal dis & cataract, Nat Eye Inst, 73-77, EXEC SECY VISUAL SCI BR, DIV RES GRANTS, NIH, 77- *Concurrent Pos:* Asst prof, Johns Hopkins Univ, 74. *Mem:* AAAS; Am Asn Anat. *Res:* Anatomy. *Mailing Add:* Div Res Grants Referral & Rev Br NIH Bldg WB Rm 303 Bethesda MD 20892

GIAEVER, IVAR, b Norway, Apr 5, 29; nat US; m 52; c 4. BIOPHYSICS, PHYSICS. *Educ:* Norweg Inst Technol, Siv Ing, 52; Rensselaer Polytech Inst, PhD(theoret physics), 64. *Hon Degrees:* Numerous honorary doctorate degrees. *Prof Exp:* Maintenance engr, Norweg Army, 53; patent examr, Norweg Govt, 54; engr, Advan Eng Prog, Gen Elec Co, Can, 54-56; appl mathematician, Gen Elec Co, NY, 56-58, staff mem, Res & Develop Ctr, 58-88; INST PROF SCI, RENSSELAER POLYTECH INST, 89- *Concurrent Pos:* Guggenheim fel, Cambridge Univ, 69-70; adj prof, Univ Calif, San Diego, 75; vis prof, Salk Inst Biol Studies, La Jolla, 75; mem, Comt Scholarly Commun with People's Repub China, Nat Acad Sci, 76; prof-at-large, Univ Oslo, 88. *Honors & Awards:* Nobel Prize in Physics, 73; Oliver

E Buckley Prize, Am Phys Soc, 65; Vladimir K Zworkin Award, Nat Acad Eng, 74. *Mem:* Nat Acad Sci; Nat Acad Eng; Norweg Prof Engrs; fel Am Phys Soc; Norweg Acad Sci; Inst Elec & Electronics Engrs; Am Acad Arts & Sci; hon mem Norweg Acad Technol; hon mem Am Soc Mech Engrs; Swed Acad Eng. *Res:* Application of the tools of physics to solve biological problems; determine how mammalian cells recognize each other, and to find out why cancer cells metastasize; understand the process of protein absorption on various materials; develop a microscopic process that can image DNA and protein without damaging the molecules; author of 75 papers; recipient of 34 patents; tissue culture. *Mailing Add:* Rensselaer Polytech Inst Troy NY 12180-3590

GIALLORENZI, THOMAS GAETANO, b New York, NY, Feb 28, 43; m 66; c 2. QUANTUM OPTICS. *Educ:* Cornell Univ, BS, 65, MS, 66, PhD(appl physics), 69. *Prof Exp:* Res physicist, Gen Tel & Electronics Labs, 69-70; head, Optical Tech Br, 70-78, SUPT OPT SCI DIV, NAVAL RES LAB, 78- *Concurrent Pos:* Consult to var govt agencies. *Honors & Awards:* Award Appl Sci, Sci Res Soc Am, 73. *Mem:* Fel Optical Soc Am; Am Phys Soc; Inst Elec & Electronics Engrs. *Res:* Fiber and integrated optics; Raman and parametric scattering; laser physics. *Mailing Add:* Off Naval Res Lab Code 6500 4555 Overlook Ave SW Washington DC 20375-5000

GIAM, CHOO-SENG, b Singapore, Apr 2, 31; m 56; c 3. PHYSICAL ORGANIC CHEMISTRY, ANALYTICAL CHEMISTRY. *Educ:* Univ Malaya, BSc, 54, Hons, 55; Univ Sask, MSc, 61, PhD(chem), 63. *Prof Exp:* Govt analyst, Chem Dept, Govt Singapore, 55-58; lectr chem, Univ Malaya, 58-59; res chemist, Imp Oil, Can, 63-64; res assoc chem, Univ Calif, 64-66; prof chem & oceanog, Tex A&M Univ, 66-81; prof chem & geol sci, Univ Tex El Paso, 81-83, dean, Col Sci, 81-83; PROF, TEX A&M UNIV, GALVESTON, 88- *Concurrent Pos:* Analyst, Munic Coun, Malaya, 58-59; fels, Nat Res Coun Can, 63-64 & Pa State Univ, 64-65. *Mem:* Am Chem Soc; sr mem Chem Inst Can; NY Acad Sci. *Res:* Chemistry of heterocycles, effects of structures of organic compounds on their reactivities; environmental chemistry; nucleophilic reactions and mechanisms of these reactions. *Mailing Add:* Coastal Zone Lab PO Box 1675 Galveston TX 77553-1675

GIAMATI, CHARLES C, JR, b Akron, Ohio, Aug 26, 27; m 54; c 2. COMPUTER SCIENCE. *Educ:* Oberlin Col, AB, 50; Univ Mich, AM, 52; Case Inst Technol, PhD(physics), 62. *Prof Exp:* Aeronaut res scientist, Nat Adv Comt Aeronaut, 52-55, physicist, Lewis Res Ctr, 55-85, SR SYSTS ENGR, LOCKHEED MISSILES & SPACE CTR, NASA, 86- *Concurrent Pos:* Lectr, Fenn Col, 61-65, Oberlin Col, 65-66 & Cleveland State Univ, 67-70; Fulbright-Hayes lectr grant, Istanbul Tech Univ, 66-67. *Mem:* AAAS; Am Phys Soc. *Res:* Cosmic ray physics; large scintillation and Cerenkov counters; scattering of nucleons from complex nuclei; computer control systems; computer image processing of remote sensor data; color coded analysis of fluid flow research data; computer aided software engineering. *Mailing Add:* 2760 Wagar Rd Cleveland OH 44116

GIAMBRONE, JOSEPH JAMES, b Norristown, Pa, June 24, 50; m; c 2. MICROBIOLOGY, IMMUNOLOGY. *Educ:* Univ Del, BS, 72, MS, 74; Univ Ga, PhD(microbiol), 77. *Prof Exp:* ASSOC PROF POULTRY PATH & VET PATHOBIOL, AUBURN UNIV, 77- *Concurrent Pos:* Vis scientist, CSIRO Animal Health Div, Parkville, Australia. *Honors & Awards:* Res Award, Poultry Sci Assoc. *Mem:* Poultry Sci Asn; Am Soc Microbiol; World Vet Poultry Sci Asn; Am Asn Avian Pathologists; World Poultry Sci. *Res:* Determination of various immune mechanisms which render poultry resistant to specific disease; molecular biology of RNA viruses of poultry. *Mailing Add:* Dept Poultry Sci Auburn Univ Auburn AL 36849-5416

GIAMEI, ANTHONY FRANCIS, b Corning, NY, Oct 14, 40; m 62; c 2. METALLURGY, MATERIALS SCIENCE. *Educ:* Yale Univ, BE, 62; Northwestern Univ, PhD(mat sci), 67. *Prof Exp:* Res assoc alloy studies, Adv Mat Res & Develop Lab, 66-68, sr res assoc, 68-69, group leader, 69-71, group leader alloy res, Mat Eng & Res Lab, 71-77, sr staff scientist, Pratt & Whitney Aircraft Group, 77-81, PRIN SCIENTIST, UNITED TECHNOL RES CTR, 81- *Honors & Awards:* George Mead Gold Medal, 81. *Mem:* Am Soc Metals; Am Inst Mining Metall & Petrol Engrs; Sigma Xi. *Res:* Phase transformation morphology and kinetics; quantitative phase analysis by x-ray diffraction; stacking faults in ordered lattices; temperature dependence of strength in intermetallics; influence of structural changes on strength; undirectional solidification; crystal growth; superalloy forming; rapid solidification. *Mailing Add:* 54 Virginia Dr Middletown CT 06457

GIAMMARA, BERVERLY L TURNER SITES, b Gove City, Kans, Feb 4, 38; m 56; c 3. ELECTRON MICROSCOPY, HISTOCHEMISTRY. *Educ:* Univ Louisville, BLS, 76, MS, 81. *Prof Exp:* Lab asst, Col Med, Univ Fla, 60-64; res asst, Sch Med, Univ Louisville, 64-65; electron microscopist, Am Standard Develop & Eng Lab, 65-71; dir, Electron Micros Lab, Grad Prog & Res, Univ Louisville, 78-73; CONSULT, 85- *Concurrent Pos:* Res asst prof, Univ NC, 83-85. *Honors & Awards:* Patent on Silver Methenamine Stain. *Mem:* Electron Micros Soc Am (chmn); Int Asn Dent Res; Mat Res Soc; Sigma Xi. *Res:* Biological electron microscopy; ultrastructural and histochemical studies of normal and pathologic tissues; material-tissue interactions; multiple grid staining device and epoxy slide embedment. *Mailing Add:* 2205 Weber Ave Louisville KY 40205

GIAMMONA, CHARLES P, JR, b Chicago, Ill, Aug 22, 48; m 70; c 3. OCEANOGRAPHY. *Educ:* St Mary's Col, BA, 70; Tex A&M Univ, PhD(oceanog), 78. *Prof Exp:* Lectr, Dept Biol, Univ Wis, 70-72; appl oceanographer & res assoc, Tex A&M Univ, 72-79; prog mgr, Oceanog Div, Univ Petrol & Minerals, Saudi Arabia, 79; asst dir, NJ Sea Grant Prog, 79-80; DEP DIR, STRATEGIC PETROL RESERVE PROG, TEX A&M UNIV, 80-, ASSOC HEAD, ENVIRON ENG, CIVIL ENG DEPT, 80- *Concurrent Pos:* Expert witness pollution study, Mississippi River, 72; aquanaut, Hydrolab Underwater Habitat, 73; consult, Hess Oil Co, LGL Environ Consults, 74-78, Cultural Resource Serv, US Army Corps Engrs, 78-; res award, Southern Regional Educ Biol, 77. *Mem:* Sigma Xi; Soc Limnol & Oceanog; Am Soc Civil Engrs. *Res:* Scientific diving and underwater photo interpretation; strategic petroleum reserve sites monitoring studies. *Mailing Add:* Civil Eng Dept Tex A&M Univ College Station TX 77843

GIAMMONA, SAMUEL T, b Grand Rapids, Mich, Dec 17, 30. PEDIATRICS. *Educ:* Mich State Univ, BS, 51; Yale Univ, MD, 54. *Prof Exp:* Jr instr pediat, Univ Mich, 58-59; asst prof, Med Ctr, Ind Univ, 62-65; assoc prof, Sch Med, Univ Miami, 65-69, sci dir ment retardation prog, 67-69; prof pediat, 69-71, ADJ PROF PEDIAT, UNIV CALIF, SAN DIEGO, 71-; CHMN DEPT, DIS CHEST SECT, CHILDREN'S HOSP, 71- *Mem:* Am Acad Pediat; Am Fedn Clin Res. *Res:* Pulmonary physiology in infants and other children. *Mailing Add:* Dept Pediat Childrens Hosp SF 3700 California St San Francisco CA 94119

GIAMPAPA, MARK STEVEN, b Cincinnati, Ohio, Oct 4, 54. RADIATIVE TRANSFER, DATA ANALYSIS & SPECTRAL DIAGNOSTICS. *Educ:* Univ Southern Calif, BS, 76; Univ Ariz, PhD(astron), 80. *Prof Exp:* Teaching asst astron, Univ Southern Calif, 74-76; teaching asst astron, Steward Observ, Univ Ariz, 76-80, res assoc, 80; postdoctoral fel, Harvard-Smithsonian Ctr Astrophys, 80-82; STAFF SCIENTIST, NAT OPTICAL ASTRON OBSERV, 82- *Concurrent Pos:* Prin investr, Int Untraviolet Explor Prog, 80-; ed, Second Cambridge Workshop Sun & Cool Stars, 81-82 & Astron Quart, 82-; mem or chair sci organizing comt, Cambridge Workshops Sun & Cool Stars, 81, 85 & 91; vis fel, Smithsonian Astrophys Observ, 82-83; vis scientist, Sch Physics, Univ Sydney, Australia, 87; mem, proposal peer review panel, NASA Sci Progs, 87-90, subcomt space physics, US NASA, 88. *Honors & Awards:* George Van Biesbroeck Award, 85. *Mem:* Am Astron Soc; Int Astron Union. *Res:* Solar-stellar physics; analogs of solar magnetic phenomena such as spots and flares as they occur on other stars; the sun as a star; measurement of stellar magnetic fields, detection of surface features; the delineation of the properties of stellar dynamos; pre-main sequence stars and star formation; high resolution spectroscopy. *Mailing Add:* Nat Solar Observ 950 N Cherry Ave PO Box 26732 Tucson AZ 85726-6732

GIANELLY, ANTHONY ALFRED, b Boston, Mass, Aug 19, 36; m 60; c 2. ORTHODONTICS. *Educ:* Harvard Univ, AB, 57, DMD, 61, cert orthodont, 63; Boston Univ, PhD(biochem), 67, MD, 74. *Prof Exp:* Res assoc orthodont, Sch Dent Med, Harvard Univ, 63-64; Nat Inst Dent Res fel, Boston Univ, 64-67; assoc prof, 67-69, PROF ORTHODONT & CHMN DEPT, GRAD SCH DENT, BOSTON UNIV, 69-, RES PROF BIOCHEM & HEAD ORTHODONT SECT, UNIV HOSP, 76- *Concurrent Pos:* Consult, Mass Medicaid Prog, 72- *Mem:* Am Asn Orthod. *Res:* Growth, development and modification of craniofacial region. *Mailing Add:* 92 Windsor Rd Waban MA 02168

GIANETTO, ROBERT, b Montreal, Que, Aug 7, 27; m 52; c 4. BIOCHEMISTRY. *Educ:* Univ Montreal, BSc, 49, MSc, 51, PhD(biochem), 53. *Prof Exp:* Nat Res Coun Can fel, Cath Univ Louvain, 52-53; from assoc prof to prof biochem, Univ Montreal, 53-88; RETIRED. *Mem:* AAAS; NY Acad Sci; Can Biochem Soc. *Res:* Enzymology; lysosomes. *Mailing Add:* Dept Biochem Fac Med Univ Montreal Montreal PQ H3C 3J7 Can

GIANINO, PETER DOMINIC, b East Boston, Mass, May 8, 32; m 54; c 3. OPTICAL PHYSICS, SOLID STATE PHYSICS. *Educ:* Boston Col, BS, 53; Northeastern Univ, MS, 59. *Prof Exp:* Physicist, Sylvania Elec Prod Div, Gen Tel & Electronics Corp, 55-58 & Ewen-Knight Corp, 58-59; res physicist, Cambridge Res Labs, 59-76, RES PHYSICIST, ROME AIR DEVELOP CTR, USAF, 76- *Concurrent Pos:* Lectr, Northeastern Univ. *Honors & Awards:* Marcus O'Day Annual Award, USAF Cambridge Res Labs, 74. *Mem:* Optical Soc Am; Am Asn Physics Teachers; Soc Photo-Optical Instrumentation Engrs; Sigma Xi. *Res:* Antenna pattern synthesis of linear arrays; missile systems; three-level solid state masers; magnetic anisotropy and resonance at low temperatures; secondary electron emission; laser windows; fiber optics; optical signal processing. *Mailing Add:* Rome Air Develop Ctr ESO Hanscom AFB Bedford MA 01731

GIANNELIS, EMMANUEL P, b Rhodes, Greece, Sept 5, 57; m 87. MATERIALS CHEMISTRY, SYNTHESIS OF NEW MATERIALS. *Educ:* Univ Athens, Greece, BS, 80; Mich State Univ, PhD(inorg chem), 85. *Prof Exp:* Res assoc chem, Mich State Univ, 85-86, res assoc chem eng, 86-87; ASST PROF, MAT SCI & ENG, CORNELL UNIV, 87- *Concurrent Pos:* Consult, Therm, Inc, 88-90, Corning, Inc, 90- & S Adelman & Assoc, 91- *Mem:* Sigma Xi; AAAS; Am Chem Soc; Am Ceramic Soc; Mat Res Soc; Clay Minerals Soc. *Res:* Physics and chemistry of intercalation; molecular assemblies of electroactive polymers; polymer-ceramic nanocomposites; ceramic thin films; materials for optoelectronic packaging; sensing devices; materials for optical waveguides and photonics. *Mailing Add:* Cornell Univ Bard Hall Ithaca NY 14853-1501

GIANNETTI, RONALD A, b Chicago, Ill, May 21, 46; m 75; c 1. BEHAVIORAL MEDICINE, COMPUTER SOFTWARE DEVELOPMENT. *Educ:* Univ Calif, Berkeley, AB, 67, PhD(psychol), 73. *Prof Exp:* Treat team leader, 73-74, patient evaluator, 74-75, eval coordr, Vet Admin Hosp, Salt Lake City, 76-78; from asst prof to assoc prof, 78-85, prof, dept psychiat, Eastern VA Med Sch, 85-88; CHAIR FAC, FIELDING INST, SANTA BARBARA, 88- *Concurrent Pos:* Instr, Col Med Univ Utah, 74-77, res asst prof, 77-78; dir, internship training, Eastern Va Med Sch, 78-81; chair, Va Consortium Prof Psychol, 79-88; consult, Vet Admin Med Ctr, Hampton, Va, 82-88. *Mem:* Am Psychol Asn; fel Soc Personality Assessment; Biofeedback Soc Am; Soc Comput in Psychol. *Res:* Applications of computer technology to problems of mental health care delivery, particularly computer based patient evaluation; evaluation and treatment of patients with psychophysiological disorders. *Mailing Add:* Fielding Inst 2112 Santa Barbara St Santa Barbara CA 93105

GIANNINI, A JAMES, b Youngstown, Ohio, June 11, 47; m 75; c 2. CLINICAL TOXICOLOGY, BIOPSYCHIATRY. *Educ:* Youngstown State Univ, BS, 70; Univ Pittsburgh, MD, 74. *Prof Exp:* Chief res consult, Pa Justice Comn, 74-75; fel psychiat, Yale Univ, 75-78; assoc prof, 78-84, PROF & VCHMN PSYCHIAT, NORTHEAST OHIO MED COL, 84-; PROF PSYCHIAT, OHIO STATE UNIV, 84- *Concurrent Pos:* Co-dir, Wed Clin Yale New Haven Hosp, 76-78; sr consult, Fair Oaks Hosp, 70-, Regent Hosp,

79-, Smith Kline-Bechman Laboratories, 82; chmn, Natl Adv Comm on Rape, NIH, 83-84, Mahoning County Mental Health Bd, 83-87, psychiat & toxicol Western Reserve Care Syst, 84-,; mem, Adv Comn NIDA, 84-; dir, clin res, Princeton Diag Labs, 87-; clin assoc prof psychiat, Ohio State Univ, 82-84; consult, sci adv bd, Neurodata Inc, Los Angeles, 88-; forensic psychiatrist, Mahoning County Prosecutor's Off, Youngstown, OH, 89-; med dir, Chem Abuse Ctrs Inc, 89-; examnr, LaTrohs Univ, Bundvora, Australia, 89-91; Am partic, Drug Abuse Prog, US Unfo Agency, Cyprus, Italy & Yugoslavia, 90- *Honors & Awards:* Bronze Award, Brit Med Asn, 82. *Mem:* Soc Neurosci; Sigma Xi; fel Am Col Clin Pharmacol; Am Acad Clin Psychiat; Acad Psychosom Med; fel, Am Psychiat Asn. *Res:* Physiology of psychopathological processes; cocaine addiction; psychosis; premenstrual syndrome; anorexia nervosa. *Mailing Add:* Col Med NE Ohio Univ Youngstown OH 44504

GIANNINI, GABRIEL MARIA, b Rome, Italy, Oct 21, 05; nat US; m 31; c 3. PHYSICS. *Educ:* Univ Rome, Dr(physics), 29. *Prof Exp:* Student engr, Radio Corp of Am, Victor Mfg Co, Camden, 30-31; res engr, Curtis Inst Music, Philadelphia, 31-34, John D Rockefeller, Jr & Riverside Church, NY, 33-35; res engr & co-exec, Transducer Corp, 36-39; cost control engr, Vultee Aircraft Corp, Calif, 39-41; staff asst coordr, Off Vpres In Chg Mfg, Lockheed Aircraft Corp, 41-43; pres, G M Giannini & Co, Inc, 44-57, dir res lab & pres, Giannini Sci Corp, 57-65, PRES, GIANNINI INST, 65- *Concurrent Pos:* Instr, Eve Sch, Univ Calif, 41-43; pres & chief engr, Autoflight Corp, 44-47; mem, Nat Air Coun, 48- *Mem:* Acoust Soc Am; Am Phys Soc; Soc Automotive Engrs; assoc fel Am Inst Aeronaut & Astronaut; sr mem Inst Elec & Electronics Engrs; assoc fel Royal Aeronaut Soc UK; Am Soc Mech Engrs; Am Soc Naval Engrs; Marine Technol Soc; US Naval Inst. *Res:* Reaction power plants and automatic flight equipment; plasma technology; underwater electrical equipment. *Mailing Add:* 51555 Madison St Indio CA 92201-9740

GIANNINI, MARGARET JOAN, b Camden, NJ, May 27, 21; m 48; c 4. PEDIATRICS. *Educ:* Hahnemann Med Col, MD, 45; Am Bd Pediat, dipl, 50. *Prof Exp:* From assoc prof to prof pediat, NY Med Col, 48-79, dir, Univ Affil Ment Retardation Inst, 50-79; dir, Nat Inst Handicapped Res, Washington, DC, 79-81; dir rehab develop, Vet Admin, Washington, DC, 81-91, dep asst chief med dir prosthetics & rehab, 88-91; RETIRED. *Concurrent Pos:* Consult, Bur Handicapped Children, NY Health Dept, 60-; mem state wide planning comt ment retardation & NY State Dept Ment Hyg, 64; mem bd dirs, Avard Learning Ctr; mem adv bd, Ment Retardation Sect, Headstart Proj, Massive Econ Neighborhood Develop; mem adv coun, Asn Help Retarded Children; chmn, Int Sem Ment Retardation; chmn ment retardation task force State Wide Planning Voc Rehab Serv, NY State Dept Educ. *Mem:* Inst Med-Nat Acad Sci; fel Am Acad Pediat; Asn Univ Affil Facil. *Res:* Mental retardation. *Mailing Add:* 2301 E St NW Apt A1110 Washington DC 20037

GIANNOTTI, RALPH ALFRED, b Long Island City, NY, May 12, 42; m 64; c 4. ORGANIC CHEMISTRY. *Educ:* St John's Univ, BS, 63, MS, 65, PhD(org chem), 69. *Prof Exp:* Res asst org chem, St John's Univ, 64-69; res assoc, Mass Inst Technol, 69-70; from asst prof to assoc prof, 70-84, PROF CHEM, STATE UNIV NY AGR & TECH COL FARMINGDALE, 84- *Concurrent Pos:* Adj prof chem, Nassau Community Col, 72- *Mem:* Am Chem Soc; Coblentz Soc; NY Acad Sci. *Res:* Synthesis and characterization of polypeptides with known repeating sequence of amino acids; solid phase peptide synthesis; proteins. *Mailing Add:* Dept Chem State Univ NY Agr & Tech Col Farmingdale NY 11735

GIANNOVARIO, JOSEPH ANTHONY, b Brooklyn, NY, Mar 18, 48; m 70; c 2. ANALYTICAL CHEMISTRY, NON-METALLIC MATERIALS. *Educ:* Villanova Univ, BS, 70, MS, 74, PhD(anal chem), 75. *Prof Exp:* Res chemist, Space Div, 74-77, proj scientist life sci, Space Systs, 77-79, sr scientist life sci, 79-81, TEAM LEADER SYSTS ENG, SPACE LAB 4, MATSCO DIV, GEN ELEC CO, 81- *Mem:* Am Chem Soc. *Res:* Air pollution analysis by gas chromatography; free flow, continuous particle electrophoresis-space applications; life sciences support-space shuttle environmental systems. *Mailing Add:* Ten Sycamore Ct Paoli PA 19301

GIANOLA, UMBERTO FERDINANDO, b Birmingham, Eng, Oct 29, 27; m 52; c 3. SYSTEMS DESIGN, SYSTEMS SCIENCE. *Educ:* Univ Birmingham, Eng, BSc, 48, PhD, 51. *Prof Exp:* Consult, Royal Aircraft Estab, Eng, 51; res fel, Univ BC, 51-53; mem tech staff, Bell Labs, 53-63, head solid state digital device dept, 63-69 & ocean res dept, 69-71, dir Ocean Systs Studies Ctr, 71-84, Defense Systs Ctr, 84-85, Govt Systs Planning Ctr, 85-88; CONSULT, 88- *Mem:* Sci Res Soc Am; fel Inst Elec & Electronics Engrs; Sigma Xi. *Res:* Electron optics; nuclear radiation detectors; solid state devices for memory, logic and communication systems; ocean acoustics and antisubmarine warfare surveillance systems. *Mailing Add:* 1112 Indian Mound Trail Vero Beach FL 32963

GIANTS, THOMAS W, b Lowell, Mass, Jan 3, 40. POLYMERIC OPTICAL MATERIALS, CONDUCTIVE POLYMERS. *Educ:* Lowell Technol Inst, BS, 61, MS, 66; Tufts Univ, PhD(org-anal chem), 71. *Prof Exp:* Instr chem, Boston State Col, 70-72; res assoc org polymer chem, Univ Ariz, 72-73; res chemist, Hughes Aircraft Co, 73-78, head mat chem sect, 78-86, sr staff engr, 86-88; MEM TECH STAFF, MAT SCI LAB, AEROSPACE CORP, 88- *Mem:* Am Chem Soc; Sigma Xi. *Res:* Structure-property relationships in polymer systems; polymeric thin films; synthesis and characterization of high temperature materials involving organic and organometallic polymers; nuclear magnetic resonance spectrometry; conformational analysis. *Mailing Add:* Aerospace Corp MS M2/250 PO Box 92957 Los Angeles CA 90009-2957

GIANTURCO, MAURIZIO, b Potenza, Italy, Dec 2, 28; m 54; c 2. ORGANIC CHEMISTRY. *Educ:* Univ Rome, DrChem, 51. *Prof Exp:* Instr org chem, Univ Rome, 51-52; sr res assoc, Univ Ill, 53-56; res chemist, Tenco, Inc, 56-61, head fundamental res sect, 61-63; head tech res & develop sect, 63-68, dir corp res dept, 68-73, asst to vpres, 73-75, asst to sr vpres, 75-76, vpres sci, 76-81, SR VPRES SCI, COCA-COLA CO, 81- *Concurrent Pos:* Donegani res fel, 51-52; Fulbright fel, 52-53. *Mem:* Am Chem Soc; Inst Food Technol; NY Acad Sci; AAAS; Am Inst Chemists; Am Inst Food Technol. *Res:* Organic natural products; infrared and mass spectrometry. *Mailing Add:* Coca Cola Co PO Drawer 1734 Atlanta GA 30301

GIAQUINTA, ROBERT T, b Lawrence, Mass, Feb 22, 47; m 70; c 2. PLANT PHYSIOLOGY, BIOCHEMISTRY & BIOTECHNOLOGY. *Educ:* Merrimack Col, BA, 68; Univ Dayton, MS, 70, PhD(plant physiol), 72. *Prof Exp:* Res assoc fel plant biochem, Purdue Univ, 72-74; res scientist, E I Du Pont de Nemours & Co, Inc, 75-79, res supvr plant physiol, 79-81, res mgr, Cent Res & Develop Dept, 81-84, res mgr, 85-87, MGR BIOTECH BUS DEVELOP, AGR PROD DEPT, E I DU PONT DE NEMOURS & CO, INC, 87- *Concurrent Pos:* Mem exec comt, Am Soc Plant Physiol, 83-86; mem ed bd, J Plant Physiol, 83-; chmn, Plant Growth Regulator Soc Am, 85-86. *Honors & Awards:* Charles A Shull Award, Am Soc Plant Physiologists, 85. *Mem:* Am Soc Plant Physiologists; Plant Growth Regulation Soc Am. *Res:* Mechanism and control of photosynthate translocation in crop plants; loading and unloading of assimilates in the phloem and the cellular events governing assimilate partitioning within the plant. *Mailing Add:* Agr Prod Dept DuPont Co Barley Mill Wilmington DE 19880

GIARDINI, ARMANDO ALFONZO, b Salamanca, NY, June 5, 25; m 50; c 4. MINERALOGY. *Educ:* Univ Mich, BS, 51, MS, 53, PhD(mineral), 56. *Prof Exp:* Res scientist, Electrotech Lab, US Bur Mines, Tenn, 52; res engr, Res & Develop Lab, Carborundum Co, NY, 53-55; res assoc & teaching fel mineral, Univ Mich, 56-57; phys scientist & proj leader, Res Prog, US Army Signal Res & Develop Lab, 57-65; prof geol, Univ Ga, 65-88; RETIRED. *Res:* Geological and geophysical studies, evolution of the earth's mantle, crust, oceans, atmosphere, and petroleum deposits; genesis of natural diamond and diamond synthesis. *Mailing Add:* City Hall VFW Dr PO Box 27 Watkinsville GA 30677

GIAROLA, ATTILIO JOSE, b Jundiai, Brazil, Oct 26, 30; m 55; c 2. MICROWAVE TECHNOLOGY ANTENNAS. *Educ:* Univ Sao Paulo, BS, 54; Univ Wash, MS, 59, PhD(elec eng), 63. *Prof Exp:* Instr elec eng, Aeronaut Inst Technol, Brazil, 55-57, assoc prof, 63-65; instr, Seattle Univ, 57-60 & Univ Wash, 60-62; res engr, Boeing Co, 62-63, res scientist, 65-68; assoc prof elec eng, Tex A&M Univ, 68-74; dean grad sch, 75-86, acad vpres, 80-82, PROF ELEC ENG, STATE UNIV CAMPINAS, 75- *Concurrent Pos:* Chmn, Nat Electronics Conf, Brazil, 64, Intern Symp Microwave Tech Indust Develop, Brazil, 85; vis prof, Univ Sao Paulo, 65; consult, Capes, Brazil, 79-, FAPESP, 85-, CNPQ, 87-; Unicamp prof, 87- *Mem:* Sr mem Inst Elec & Electronics Engrs; Brazilian Microwave Soc (vpres, 85-87, pres, 87-); Brazilian Telecommunication Soc. *Res:* Solid state and microwave devices; parametric devices; optical devices; traveling wave tubes; frequency selective limiters; elastic-, spin-, and magnetoelastic-delay lines; bioeffects of electromagnetic radiation; electromagnetic wave propagation; antennas; planar structures such as strip, microstrip and fin lines; patch antennas; dispersion in optical fibers. *Mailing Add:* Faculdade de Eng Eletrica Unicamp CP 6101 Campinas SP 13081 Brazil

GIAROLI, JOHN NELLO, b Memphis, Tenn, Feb 14, 28; m 58; c 6. ORAL SURGERY. *Educ:* Memphis State Univ, BS, 50; Univ Tenn, DDS, 53. *Prof Exp:* Instr, Loyola Univ, New Orleans, 55-56; intern, Charity Hosp, New Orleans, La, 56-57; resident, Confederate Mem Hosp, Shreveport, 57-59; from asst prof to assoc prof, 60-76, PROF ORAL & MAXILLOFACIAL SURG, COL DENT, UNIV TENN, MEMPHIS, 76- *Concurrent Pos:* Chief, Dent Serv, St Francis Hosp; consult, LeBonheur Children's Hosp, Baptist Mem Hosp Cent, Baptist Mem Hosp E, St Joseph Hosp, Methodist Hosp N & Methodist Hosp Cent. *Mem:* Am Dent Asn; Am Soc Dent Children; Am Acad Oral Med; Am Soc Advan Gen Anesthesia Dent; Int Dent Fedn; Am Soc Oral & Maxillofacial Surg; Am Dent Soc Anesthesiol; Int Fedn Dent Anesthesiol; Am Asn Univ Prof; Am Col Oral & Maxillofacial Surgeons. *Mailing Add:* Dept Oral Surg Col Dent Univ Tenn Memphis TN 38101

GIARRUSSO, FREDERICK FRANK, b Little Falls, NY, May 23, 36; m 62. CHEMISTRY. *Educ:* Ariz State Univ, BS, 58; Univ Mich, PhD(chem), 66. *Prof Exp:* Jr res chemist, Merck & Co, Inc, 58-61; fel natural prod, Calif Inst Technol, 65-66; res chemist, Gen Elec Res & Develop Ctr, NY, 66-68; sr res scientist, Squibb Inst Med Res, 68-71, sect head, Chem & Biol Res Admin, 71-74, head, Preclinical Res Admin, 74-76; mgr, Sci Admin Res & Develop, 76-80, dir, comput lab systs, res & develop, Revlon Health Care Group, 80-; mem staff, Tripos Assoc, Mo; MGR BUS DEVELOP, RES & DEVELOP SYSTS, DIGITAL EQUIP CORP, 88. *Mem:* AAAS; Am Chem Soc; Chem Soc; Drug Info Asn (vpres, 80-81). *Res:* Synthetic organic and natural product chemistry; steroids; the application of computers to biological and chemical research. *Mailing Add:* Five Lincoln Dr Acton MA 01720-3110

GIBALA, RONALD, b New Castle, Pa, Oct 3, 38; c 4. METALLURGY. *Educ:* Carnegie Inst Technol, BS, 60; Univ Ill, MS, 62, PhD(metall eng), 64. *Prof Exp:* From asst prof to prof metall, Case Western Reserve Univ, 64-84, prof macromolecular sci, 77-84, co-dir, Mat Res Lab, 81-84; prog dir metall, NSF, 82-83; PROF & CHMN DEPT MAT SCI & ENG, UNIV MICH, ANN ARBOR, 84- *Concurrent Pos:* NSF res grant, 66-71 & 76-; Dept Energy res contract, 67-89; USAF Off Sci Res grant, 71-75, 90-; vis prof, Centre d'Etudes Nucleaires de Grenoble, 73-74; res consult, Gen Motors Corp Res Labs, 77-82. *Honors & Awards:* Alfred Noble Prize, Am Soc Civil Engrs, 69. *Mem:* Fel Am Soc Metals; Am Inst Mining Metall & Petrol Engrs; Sigma Xi; AAAS; Mat Res Soc; Am Ceramics Soc. *Res:* Physical metallurgy; defects in solids; internal friction; mechanical properties of solids. *Mailing Add:* Mat Sci & Eng Dept Univ Mich Ann Arbor MI 48109

GIBALDI, MILO, b New York, NY, Dec 17, 38; m 60; c 1. PHARMACOLOGY. *Educ:* Columbia Univ, BS, 60, PhD(pharmaceut), 63. *Hon Degrees:* DSc, Col Pharmaceut Sci, Columbia Univ, 76. *Prof Exp:* Asst prof pharm, Columbia Univ, 63-66; from asst prof to assoc prof, State Univ

NY Buffalo, 66-69, prof pharmaceut, 69-78; PROF PHARMACEUT & DEAN SCH PHARM, UNIV WASH, 78-, ASSOC VPRES HEALTH SCI, 82- *Concurrent Pos:* NIH res grant, 67-70; consult, Hoffman-La Roche, Ciba Geigy, Beohringen Ingelheim, Ricker/3M, Ortho, Searle, 66-; investr, NIGNIS/NIH Prog Proj Grant, 83-; mem, FDA Panel on Generic Drugs, 86. *Mem:* Inst Med-Nat Acad Sci; fel AAAS; Am Pharmaceut Asn; fel Acad Pharmaceut Sci; Am Asn Cols Pharm; Am Chem Soc. *Res:* Drug absorption; physical-chemical and biological properties of bile salts; dissolution phenomena; pharmacokinetics; Author of series of textbooks. *Mailing Add:* Sch Pharm Health Sci Ctr SC-69 Univ Wash Rm T-341 Seattle WA 98195

GIBB, JAMES WOOLLEY, b Magrath, Alta, Apr 19, 33; m 56; c 2. BIOCHEMICAL PHARMACOLOGY. *Educ:* Univ Alta, BS, 58, MS, 61; Univ Mich, PhD(pharmacol), 65. *Prof Exp:* res assoc pharmacol, Nat Heart Inst, 65-67; from asst prof to assoc prof, 67-75, PROF PHARMACOL, UNIV UTAH, 75-, CHMN PHARMACOL & TOXICOL, 74- *Concurrent Pos:* Vis res prof, Pharmakologisches Inst, Univ Innsbruck, 73-74. *Mem:* AAAS; NY Acad Sci; Am Soc Pharmacol & Exp Therapeut; Soc Neurosci. *Res:* Biosynthesis of catecholamines and indoleomines; neurochemistry; drug abuse; neuropeptides. *Mailing Add:* Dept Pharmacol & Toxicol Univ Utah Salt Lake City UT 84112

GIBB, RICHARD A, b Fraserburgh, Scotland, Feb 16, 36; m 60; c 3. GEOPHYSICS, GEOLOGY. *Educ:* Aberdeen Univ, BSc, 58; Univ Birmingham, MSc, 59, PhD(geol), 61; Carleton Univ, Ottawa, BA, 75. *Prof Exp:* Geophysicist, Bur Mineral Resources, Geol & Geophys, Australia, 62-65; res scientist, 65-82, chief scientist gravity & geodynamics, Earth Physics Br, Energy Mines & Resources Can, 83-86, CHIEF AEROMAGNETICS, GRAVITY & GEODYNAMICS, GEOPHYS DIV, GEOL SURV CAN, 86- *Concurrent Pos:* Secy subcomt gravity, Assoc Comt Geod & Geophys, Can, 67-72; prog mgr, geophysics activ, Can Nuclear Fuel Waste Mgt Prog, 83-87; ed, Can Geophys Bull. *Mem:* Fel Geol Asn Can; Can Geophys Union; Soc Explor Geophysicists; Am Geophys Union. *Res:* Solid-earth geophysics, tectonophysics, potential field methods and interpretation; application of geophysicsl methods to research areas of Canadian Nuclear Fuel Waste Management Program. *Mailing Add:* Dept Energy Mines & Resources Geophys Div Geol Surv Can One Observatory Crescent Ottawa ON K1A 0Y3 Can

GIBB, THOMAS ROBINSON PIRIE, b Belmont, Mass, Feb 10, 16; m 39, 83; c 2. ANALYTICAL CHEMISTRY, MARINE CHEMISTRY. *Educ:* Bowdoin Col, BS, 36; Mass Inst Technol, PhD(org-metallic chem), 40. *Prof Exp:* Instr, Mass Inst Technol, 40-43, asst prof, 43-46; dir chem res, Metal Hydrides, Inc, 46-51; assoc prof & dir sponsored res, 52-58, prof, 58-80, EMER PROF CHEM, TUFTS UNIV, 81- *Concurrent Pos:* Consult, Chem Warfare Serv Develop Lab, Mass Inst Technol, 44-46; vis prof, Univ Fla, 63; NSF fel, 63-64; hon res assoc, Univ Col, Univ London, 64; guest investr, Woods Hole Oceanog Inst, 70. *Mem:* Am Chem Soc. *Res:* Instrumental analysis; inorganic hydrides; structural inorganic; trace-constituents of sea water; foam separation; exotic air pollutants. *Mailing Add:* 115 Wellesley Ave Needham Heights MA 02194

GIBBARD, BRUCE, b Detroit, Mich, Oct 18, 42; m 66. EXPERIMENTAL HIGH ENERGY PHYSICS. *Educ:* Univ Mich, BS, 64, MS, 66, PhD(physics), 70. *Prof Exp:* Jr vis scientist physics, Europ Orgn for Nuclear Res, 70-72; sr res assoc & instr lab nuclear studies & physics dept, Cornell Univ, 72-78; assoc physicist, Brookhaven Nat Lab, 78-80, physicist head, Data Acquisition Group, Accelerator Dept, 80-81, physicist, 81-90, PHYSICIST & HEAD, HIGH ENERGY & NUCLEAR PHYSICS COMPUT GROUP, PHYSICS DEPT, BROOKHAVEN NAT LAB, 90- *Mem:* Am Phys Soc. *Res:* Strong interactions involving neutrons in the initial and/or final states; electroproduced multiparticle final states; neutrino-induced reactions, high energy pp interactions; techniques and instrumentation for data acquisition and processing. *Mailing Add:* Physics Dept Bldg 510C Brookhaven Nat Lab Upton NY 11973

GIBBARD, H FRANK, b Norman, Okla, Dec 27, 40; m 57; c 4. PHYSICAL CHEMISTRY, ELECTROCHEMISTRY. *Educ:* Univ Okla, BS, 62; Mass Inst Technol, SM, 64, PhD(phys chem), 67. *Prof Exp:* Res assoc, Mass Inst Technol, 66-67; from asst prof to assoc prof chem, Southern Ill Univ, Carbondale, 67-76; sr staff scientist, Gould Labs, 76-78, prin scientist, 78-80, Gould phys chem fel, 80-86; DIR RES, POWER CONVERSION INC, ELMWOOD PARK, NJ, 86- *Mem:* Electrochem Soc; AAAS; Am Chem Soc. *Res:* Electrochemistry; thermal properties of battery systems; physical chemistry of solutions of electrolytes and nonelectrolytes; primary and secondary batteries; calorimetry. *Mailing Add:* 237 Farmingdale Rd Wayne NJ 07470

GIBBENS, ROBERT PARKER, b Ness City, Kans, Nov 15, 28. RANGE MANAGEMENT. *Educ:* Ft Hays Kans State Col, BS, 50, MS, 52; Univ Wyo, PhD(range mgt), 72. *Prof Exp:* Asst specialist range mgt, Sch Forestry, Univ Calif, Berkeley, 55-67; specialist, Plant Sci Div, Univ Wyo, 67-72; fel animal sci, SDak State Univ, 72-74; RANGE SCIENTIST, AGR RES SERV, USDA, 74- *Mem:* AAAS; Soc Range Mgt; Ecol Soc Am. *Res:* Structure and function of rangeland ecosystems and rangeland ecosystem modelling; grazing management strategies; rangeland revegetation; rangeland plant phenology. *Mailing Add:* 2701 Fairway Dr Las Cruces NM 88001

GIBBINS, BETTY JANE, b Canton, Ohio, May 7, 23. CHEMISTRY, SCIENCE EDUCATION. *Educ:* Mt Union Col, BA, 45; Ohio State Univ, MS, 47, PhD(chem), 53. *Prof Exp:* Asst, Res Lab, Goodyear Tire & Rubber Co, 45-46; instr chem, Col Wooster, 48-51; asst prof, Franklin & Marshall Col, 53-58; from assoc prof to prof, 58-84, EMER PROF CHEM, LAKE ERIE COL, 84- *Mem:* AAAS; Sigma Xi; Am Chem Soc. *Res:* Molecular additive compounds and phase diagrams. *Mailing Add:* 405 Bank St Apt 5 Painesville OH 44077

GIBBINS, SIDNEY GORE, b Mt Vernon, NY, Feb 24, 26; m 53; c 3. INORGANIC CHEMISTRY. *Educ:* Calif Inst Technol, BS, 49; Univ Wash, Seattle, PhD(chem), 55. *Prof Exp:* Res chemist, E I du Pont de Nemours & Co, NY, 55-57, Olin Mathieson Chem Corp, Calif, 57-59, Nat Eng Sci Co, 59-61 & Aerospace Corp, 61-64; asst prof chem, NMex Highlands Univ, 64-65; asst prof chem, Univ Victoria, BC, 65-69, assoc prof, 69-87; RETIRED. *Mem:* Am Chem Soc. *Res:* Boron, silicon and transition metal hydrides; hydrogen peroxide; hydrazine; instrumental analysis; mass and infrared spectrometry; nuclear magnetic resonance; gas chromatography; high vacuum techniques. *Mailing Add:* 701 Fieldston Rd Bellingham WA 98225

GIBBON, NORMAN CHARLES, b Buffalo, NY, June 16, 29; m 60; c 2. MECHANICAL ENGINEERING. *Educ:* Iowa State Univ, BS, 57; State Univ NY, Buffalo, MBA, 72. *Prof Exp:* Engr, Union Carbide, 57-62, proj engr, 62-64, sect engr, 64-67, div engr, 67-70, supvr, 70-77, proj mgr, Linde Div, 77-91; RETIRED. *Res:* Thermal insulation; vacuum technology; heat transfer; cryogenic equipment development. *Mailing Add:* 4342 E Riner Rd Grand Island Tonawanda NY 14150

GIBBONS, ASHTON FRANK ELEAZER, b Mahaicony, Guyana, Apr 20, 35; US citizen; m 63; c 2. REPRODUCTIVE PHYSIOLOGY, ENDOCRINOLOGY. *Educ:* Atlantic Union Col, BA, 62; Boston Univ, MA, 67, PhD(reproductive physiol), 70. *Prof Exp:* Res asst endocrinol, Worcester Found Exp Biol, 63-65, 66-67, sr res asst, 68-70, res assoc, 70-72, dir summer res training prog, 73-78, staff scientist, 73-78; CHMN & PROF, DEPT BIOL SCI, OAKWOOD COL, 82- *Concurrent Pos:* Vis prof biol, Atlantic Union Col, 73-74; NIH res fel, 79, 80 & 85; extramural assoc, NIH, 89. *Mem:* NY Acad Sci; Soc Study Fertility; Soc Study Reproduction; Radiation Res Soc; Am Soc Zoologists; Am Physiol Soc; Sigma Xi; Int Soc Hypertension Blacks. *Res:* Early development; radiation biology of early mammal development; physiology of hypertension; cardiovascular physiology. *Mailing Add:* Dept Biol Oakwood Col Huntsville AL 35896

GIBBONS, BARBARA HOLLINGWORTH, b Newark, Del, Jan 17, 32; m 61; c 2. BIOCHEMISTRY. *Educ:* Mt Holyoke Col, AB, 53; Harvard Univ, PhD(biochem), 63. *Prof Exp:* Asst researcher, 67-75, assoc res cytol, 75-82, RESEARCHER, UNIV HAWAII, 82- *Mem:* Am Soc Cell Biol. *Res:* Kinetics of the non-enzymatic and enzymatic hydration of carbon dioxide; molecular mechanism of motility in sperm flagella. *Mailing Add:* Pac Biomed Res Ctr Univ Hawaii Honolulu HI 96822

GIBBONS, DAVID LOUIS, b Cleveland, Ohio, May 11, 46; m 70; c 2. PHYSICAL CHEMISTRY, FUEL TECHNOLOGY. *Educ:* Univ Chicago, BS, 69; Univ Ill, Urbana, PhD(chem), 74. *Prof Exp:* Asst prof chem, Colgate Univ, 74-76; RES SCIENTIST, MARATHON OIL CO, 76- *Mem:* Am Chem Soc; Soc Petrol Engrs; Sigma Xi; Soc Core Analysis. *Res:* Rock and fluid properties. *Mailing Add:* Marathon Oil Co 7400 S Broadway Littleton CO 80122

GIBBONS, DONALD FRANK, b Birmingham, Eng, July 23, 26; m 50; c 2. BIOMEDICAL ENGINEERING. *Educ:* Univ Birmingham, BSc, 47, PhD, 50, DSc, 73. *Prof Exp:* Res fel, Univ Chicago, 50-52; res assoc, Royal Mil Col, Can, 52-54; mem tech staff, Bell Tel Labs, 54-62; prof metall, Case Western Reserve Univ, 62-67, dir, Ctr Study Mat, 62-74, prof biomed eng & exp path, 68-81; STAFF SCIENTIST BIOSCI LAB, 3M CTR, ST PAUL, MINN, 82- *Mem:* AAAS; Soc Biomat; Am Soc Test & Mat; Biol Eng Soc UK; NY Acad Sci. *Res:* Biomaterials; histopathology; improvement of present and development of new materials suitable for biological implantation, including encapsulation of microelectronic circuits and electrode materials; examination of ancient works of art. *Mailing Add:* Six Blue Goose Rd North Oaks MN 55114

GIBBONS, IAN READ, b Hastings, Eng, Oct 30, 31; m 61; c 2. MOLECULAR BIOLOGY. *Educ:* Cambridge Univ, BA, 54, PhD(biophys), 57. *Prof Exp:* Res fel biophys, Univ Pa, 57-58; res fel biol, Harvard Univ, 58-62, lectr, 62-63, asst prof, 64-67; assoc prof, 67-69, PROF BIOPHYS, PAC BIOMED RES CTR, UNIV HAWAII, 69- *Mem:* Am Soc Cell Biol; Royal Soc, London; Am Soc Biol Chem; Biophys Soc. *Res:* Cell motility; molecular organization of subcellular organelles, especially cilia and flagella. *Mailing Add:* Pac Biomed Res Ctr Univ Hawaii Honolulu HI 96822

GIBBONS, J WHITFIELD, b Montgomery, Ala, Oct 5, 39; m 63; c 4. POPULATION ECOLOGY, HERPETOLOGY. *Educ:* Univ Ala, BS, 61, MS, 63; Mich State Univ, PhD(zool), 67. *Prof Exp:* NIH fel ecol, res assoc, 68-73, SR ECOLOGIST, SAVANNAH RIVER ECOL LAB, 73-; PROF ZOOL, UNIV GA, 85- *Concurrent Pos:* Dir NSF Undergrad Res Participation Proj, 69 & 70; adj prof, Wake Forest Univ, 86; adj prof, Mich State Univ, 88. *Mem:* Am Soc Ichthyologists & Herpetologists; Ecol Soc Am; Sigma Xi; Herpetologists' League (pres, 89-90). *Res:* Population dynamics and ecology of fish, amphibians and reptiles; reproductive ecology and evolution of reptiles; effects of thermal effluents on natural populations of animals; the effects of artifical elevation of environmental temperatures on animal populations, particularly vertebrates; life history phenomena in reptile and amphibian populations. *Mailing Add:* Savannah River Ecol Lab Drawer E Aiken SC 29802

GIBBONS, JAMES F, b Leavenworth, Kans, Sept 19, 31; m 54; c 3. ELECTRICAL ENGINEERING. *Educ:* Northwestern Univ, BS, 53; Stanford Univ, MS, 54, PhD(elec eng), 56. *Prof Exp:* Fulbright fel, Cambridge Univ, 56-57; from asst prof to prof, 57-83, dean, Sch Eng, 84, REID WEAVER DENNIS PROF ELEC ENG, STANFORD UNIV, 83-, FREDERICK EMMONS TERMAN DEAN, SCH ENG, 84- *Concurrent Pos:* Mem tech staff, Bell Tel Labs, Inc, 56; consult, Shockley Transistor Corp, 57-63, Fairchild Camera & Instrument Co, 64-71, Avantek, Inc, 64-, Electronics Br, Atomic Energy Res Estab, Harwell, Eng, 70-71, Acad Educ Develop, 73-75, Technol Aids Basic Educ Develop, Alaska Natives Found, 73-75, Gen Elec Co, United Technols, Inc & Micropower Systs, Inc, 77-; NSF sr fel, 63-64; Fulbright lectr, 63-64; assoc ed, Transactions Electron Devices, Inst Elec & Electronics Engrs, 64-70; mem grad fel panel, NSF, 64-70, chmn eng fel panel, 68-70; mem bd dirs, Avantek, Inc, 67- & comt higher educ, HEW, 69-74; vis prof, Nuclear Physics Dept, Oxford Univ, 70-71 & Univ Tokyo, 71; mem, US Sci Team Exchanges on Ion Implantation & Beam Processing, Japan, 71, People's Repub China, 76, USSR, 77 & 79, Australia, 81; founder & chmn, Solar Energy Res Assocs, 75-; bd dirs, Lockheed Corp, Raychem Inc & Technol Strategies & Alliances; sci adv bd, KRI, Int. *Honors & Awards:* Jack A Morton Award, Inst Elec & Electronics Engrs, 80; Award in Solid State Sci & Technol, Electrochem Soc, 89. *Mem:* Nat Acad Sci; Nat Acad Eng; Royal Swed Acad Eng Sci; fel Inst Elec & Electronics Engrs; Sigma Xi; Norweg Acad Tech Sci; fel Am Acad Arts & Sci. *Res:* Transistor circuits; solid state devices; ion implantation in semiconductors; semiconductor device analysis; process physics and solar energy. *Mailing Add:* Terman Bldg Rm 214 Stanford Univ Stanford CA 94305

GIBBONS, JAMES JOSEPH, b Springfield, Mo, Oct 31, 46. ANALYTICAL CHEMISTRY, INORGANIC CHEMISTRY. *Educ:* Drury Col, AB, 68; La State Univ, Baton Rouge, PhD(analytical & inorg chem), 74. *Hon Degrees:* DLitt, Xavier Univ, Calcutta, India, 76. *Prof Exp:* Res chemist, Hoffman-Taff Div, Syntex Pharmaceut, 69-70; teaching asst gen chem, La State Univ, 71-74; dir, Drury Res Inst, 74-84, lectr gen chem, Drury Col, 74-75, from asst prof to assoc prof chem, 76-84, coordr, dept, chem, math & physics, 78-79; MGR, ANALYTICAL SERV, DAYCO TECH CTR, 85- *Concurrent Pos:* Vis scientist lectr chem, Jadavpur Univ & distinguished scientist exchange prog, Kalyani Univ, India, 75-76 & 81-82; vis assoc prof chem, Univ Okla, 83; adj prof chem, Drury Col, 86- *Mem:* AAAS; Am Chem Soc; Am Soc Testing & Mat; fel Am Inst Chemists; Sigma Xi; NY Acad Sci. *Res:* Solution chemistry with materials testing and polymer synthesis. *Mailing Add:* Mgr Analytical Serv Dayco Tech Ctr PO Box 3258 Springfield MO 65808

GIBBONS, JEAN DICKINSON, b St Petersburg, Fla, Mar 14, 38; m 58, 74. STATISTICAL ANALYSIS. *Educ:* Duke Univ, AB, 58, MA, 59; Va Polytech Inst, PhD(statist), 63. *Prof Exp:* Asst prof math, Mercer Univ, 58-60; asst prof math & statist, Univ Cincinnati, 61-63; from asst prof to assoc prof statist, Univ Pa, 63-70; PROF STATIST, UNIV ALA, 70- *Concurrent Pos:* Res prof; consult; Fulbright scholar. *Mem:* Fel Am Statist Asn; Int Statist Inst. *Res:* Performance of nonparametric tests based on ranks of observations; ranking and selection procedures. *Mailing Add:* Dept Statist Univ Ala Box 870226 Tuscaloosa AL 35487-0226

GIBBONS, JOHN HOWARD, b Harrisonburg, Va, Jan 15, 29; m 55; c 3. SCIENCE, TECHNOLOGY POLICY. *Educ:* Randolph-Macon Col, BS, 49; Duke Univ, PhD(physics), 54. *Prof Exp:* Res assoc nuclear physics, Duke Univ, 53-54; physicist, Oak Ridge Nat Lab, 54-75, group leader, Geophys Lab, 65-69, dir, Environ Prog, 70-74; dir, Off Energy Conserv, Fed Energy Admin, 73-74; prof physics & dir energy environ & resources, Univ Tenn, 74-79; DIR, OFF TECHNOL ASSESSMENT, US CONG, 79- *Concurrent Pos:* Bd dirs, Resources Future & AAAS, Coun Foreign Rels. *Honors & Awards:* Pub Serv Award, Fedn Am Scientists, 90; Szilard Award, Am Inst Physics, 91. *Mem:* Fel AAAS; fel Am Phys Soc; Sigma Xi. *Res:* Energy and environment; population problems; socio-technical problems and environmental quality; technology assessment; energy policy; energy conservation; solar utilization; solid waste reduction; resource and energy recovery; origins of the solar system; nuclear structure; energy conservation. *Mailing Add:* PO Box 497 The Plains VA 22171

GIBBONS, JOSEPH H(ARRISON), b Turbeville, SC, Sept 4, 34; m 56; c 2. CHEMICAL ENGINEERING. *Educ:* Univ SC, BS, 56; Univ Pittsburgh, MS, 58, PhD(heat transfer), 61. *Prof Exp:* From jr engr to sr engr, Westinghouse Elec Corp, 56-63; assoc prof chem eng, 63-74, PROF CHEM ENG, UNIV SC, 74-, CHMN DEPT, 77- *Mem:* Am Inst Chem Engrs; Am Soc Eng Educ; Am Chem Soc; Nat Soc Prof Engrs. *Res:* Heat transfer and fluid dynamics. *Mailing Add:* Dept Chem Eng Univ of SC Columbia SC 29208

GIBBONS, LARRY V, b Harrisburg, Ill, Mar 18, 32; m 54; c 4. INDUSTRIAL HYGIENE, HAZARDOUS WASTE ANALYSIS. *Educ:* Wash Univ, AB, 54; Southern Ill Univ, MS, 58, PhD(physiol), 70. *Prof Exp:* Res supvr, Biol Lab, UMC Industs, 58-63, res dir, Unidynamics, 70-71; sr physiologist, adv space craft design, McDonnell Douglas, 63-68; assoc dir res, Intersci Res Inst, 68-70; PRES & LAB DIR, APPL RES & DEVELOP LAB, ARDL, INC, 71- *Honors & Awards:* b Harrisburg, Ill, Mar 18, 32. *Mem:* Sigma Xi; Am Chem Soc; Am Physiol Soc; Air Pollution Control Asn. *Mailing Add:* RR 5 Box 316 Mt Vernon IL 62864

GIBBONS, LOREN KENNETH, b Cheboygan, Mich, Dec 13, 38; m 61; c 2. ORGANIC CHEMISTRY. *Educ:* Albion Col, AB, 60; Univ Kans, PhD(org chem), 64. *Prof Exp:* Res chemist, Niagara Chem Div, FMC Corp, 64-76, mgr org synthesis, 76-80, CONSULT, 80- *Mem:* Am Chem Soc. *Res:* Synthesizing potential herbicides; heterocyclic compounds. *Mailing Add:* 130 Clayton Ct Vallejo CA 94590

GIBBONS, LOUIS CHARLES, b Lost Springs, Wyo, Aug 8, 14; m 40; c 1. PETROLEUM CHEMISTRY. *Educ:* Univ Ohio, Athens, BS, 36; Ohio State Univ, MS, 38, PhD(org chem), 40. *Prof Exp:* Chemist, Nat Adv Comt Aeronaut, Va, 41, head org synthesis sect, Ohio, 42-45, chief fuels br, 45-50, assoc chief fuels & combustion res div, 50-55; res chem supvr, Ohio Oil Co, 55-61, assoc res dir, Marathon Oil Co, Colo, 61-70, res adv, Ohio, 70-72, assoc res dir, 72-75, res assoc, Colo, 75-79; RETIRED. *Mem:* Am Chem Soc. *Res:* Synthesis and combustion of hydrocarbons; paraffinic hydrocarbons derived from tetrahydrofuryl alcohol; tertiary oil recovery; aircraft fuels; petroleum chemistry. *Mailing Add:* 133 Sherbrook Rd Mansfield OH 44907

GIBBONS, MATHEW GERALD, b Oakland, Calif, Jan 21, 19. ATMOSPHERIC PHYSICS, NUCLEAR PHYSICS. *Educ:* St Mary's Col, Calif, BS, 40; Univ Calif, Berkeley, MA, 45, PhD(physics), 53. *Prof Exp:* Instr chem, St Mary's Col, Calif, 40-42, instr physics, 42-45, asst prof & dept head, 45-56; res physicist, US Naval Radiol Defense Lab, San Francisco, 56-64, phys sci adminr, 64-69, sci ed optics & spectros, 60-76; RETIRED. *Res:* Nuclear magnetic resonance; atmospheric transmission of visible and near infrared radiation; nuclear reactor hazards; nuclear weapon effects. *Mailing Add:* 3535 Coolidge Ave Apt 45 Oakland CA 94602-3304

GIBBONS, MICHAEL FRANCIS, JR, b Laconia, NH, Mar 20, 41. BIOLOGICAL ANTHROPOLOGY, ANATOMY. *Educ:* Yale Univ, BA, 63, MPhil, 70, PhD(anthrop), 74. *Prof Exp:* Teaching fel anat, Yale Univ, 71-72; instr, 72-74, asst prof, 74-79, ASSOC PROF ANTHROP, UNIV MASS, 79- *Concurrent Pos:* Vis prof anthrop & biol, Univ Alaska, 77-78; vis scientist, Mass Audubon Soc, 78-79; Ford prof, Univ Mass, 86-87. *Mem:* Soc Vert Paleont; fel Am Anthrop Asn; Sigma Xi; Am Geol Inst; AAAS; Am Asn Anatomists. *Res:* Time and evolution; evolution of speech sound generation potential; regional anatomy of the head and neck in primates; human and primate paleontology; mammalian reproductive systems; forensic osteology. *Mailing Add:* Anthrop Harbor Campus Univ Mass Boston MA 02125-3393

GIBBONS, PATRICK C(HANDLER), b Washington, DC, Dec 18, 43; m 68; c 4. QUASICRYSTALS, NANOCLUSTERS. *Educ:* Georgetown Univ, BS, 65; Harvard Univ, PhD(physics), 71. *Prof Exp:* From instr to asst prof physics, Princeton Univ, 71-76; from asst prof to assoc prof, 76-89, PROF PHYSICS, WASH UNIV, 89- *Mem:* Am Phys Soc. *Res:* Electron microscopy and electron energy loss studies of quasicrystals, nanoclusters and crystalline C60. *Mailing Add:* Dept Physics Campus Box 1105 Wash Univ One Brookings Dr St Louis MO 63130-4899

GIBBONS, RONALD J, b New York, NY, Dec 10, 32; m 59; c 3. MICROBIOLOGY, BACTERIAL PHYSIOLOGY. *Educ:* Wagner Col, BS, 54; Univ Md, MS, 56, PhD(microbiol), 58. *Hon Degrees:* Dr Odontol, Univ Goteborg, Sweden, 77; MD, Univ Utrecht, Holland, 80. *Prof Exp:* Res assoc microbiol, 59-61, assoc bact, 61-64, asst prof, 64-65, assoc staff mem, 65-67, SR STAFF MEM, FORSYTH DENT CTR, 67-, CLIN PROF, HARVARD UNIV, 74- *Concurrent Pos:* Res fel bact, Forsyth Dent Ctr, Harvard Univ, 58-59. *Honors & Awards:* Res Award, Int Asn Dent Res, 67; Dent Caries Award, Int Asn Dent Res, 90. *Mem:* Am Soc Microbiol; Int Asn Dent Res; Am Dent Asn. *Res:* Microbiological ecology; physiology and ecology of bacteria indigenous to mucous membranes of man; microbiology of mixed anaerobic infections, dental caries and periodontal disease; anaerobic metabolism. *Mailing Add:* Forsyth Dent Ctr 140 Fenway Boston MA 02115

GIBBS, ALAN GREGORY, b Weaverville, Calif, Feb 23, 39; m 69. APPLIED MATHEMATICS. *Educ:* Stanford Univ, BS, 60, MS, 61, PhD(nuclear eng), 65. *Prof Exp:* SR RES SCIENTIST, PAC NORTHWEST LABS, BATTELLE MEM INST, 65- *Concurrent Pos:* Lectr, Ctr Grad Study Richland, Wash, 67-68; Battelle vis prof, Univ Wash, 68-69. *Mem:* Sigma Xi. *Res:* Mathematics of transport theory, neutron scattering theory, nuclear reactor theory; statistical mechanics; stochastic processes; atmospheric physics. *Mailing Add:* 1920 Mahan Richland WA 99352

GIBBS, ANN, b Corpus Christi, Tex, May 19, 40. NUCLEAR CHEMISTRY. *Educ:* Univ Tex, BA, 62; Univ Ark, MS, 64, PhD(nuclear chem), 66. *Prof Exp:* Res assoc marine chem, Inst Marine Sci, Univ Tex, 60-63; res asst accelerators, Univ Ark, 62-66; CHEMIST NUCLEAR CHEM, SAVANNAH RIVER PROJ, 66- *Mem:* Am Nuclear Soc; Am Chem Soc; Am Phys Soc; AAAS; Am Soc Testing & Mat. *Res:* Spectroscopy of transuranium elements and non-destructive analysis of plutonium; analytical chemistry of heavy water and transuranium elements. *Mailing Add:* Savannah River Site Westinghouse Savannah River Co Aiken SC 29808

GIBBS, CHARLES HOWARD, b Salt Lake City, Utah, Dec 24, 40; m 64; c 2. DENTAL RESEARCH, BIOMEDICAL ENGINEERING. *Educ:* Univ Utah, BSME, 64; Case Western Reserve Univ, MS, 66, PhD(eng), 69. *Prof Exp:* Proj engr, Eng Design Ctr, Case Western Reserve Univ, 64-71, asst clin prof & sr res assoc, 71-74; asst prof, 74-80, ASSOC PROF BASIC DENT SCI, COL DENT, UNIV FLA, GAINESVILLE, 80- *Concurrent Pos:* Prin investr, Nat Inst Dent Res res grants, 71- *Mem:* Am Inst Dent Res; Am Dent Asn; Sigma Xi. *Res:* Jaw movements, occlusion, forces on teeth and jaw; tooth mobility; biomedical engineering applied to dentistry. *Mailing Add:* 1918 SW 48th Ave Gainesville FL 32608

GIBBS, CLARENCE JOSEPH, JR, b Washington, DC, Dec 10, 24. VIROLOGY. *Educ:* Cath Univ Am, AB, 50, MS, 52, PhD, 62; Univ Mass, MD. *Hon Degrees:* DSc, Cath Univ Am. *Prof Exp:* Med bacteriologist clin path, Div Vet Med, Walter Reed Army Inst Res, 52-55; virologist, Dept Hazardous Opers, Div Commun Dis, 55-59; virologist arbovirus sect, Lab Trop Virol, Nat Inst Allergy & Infectious Dis, 59-63, DEP CHIEF LAB CNS STUDIES & CHIEF LAB SLOW, LATENT & TEMPERATE VIRUS INFECTIONS, NAT INST NEUROL & COMMUN DIS & STROKE, 63-, NEUROVIROL COORDR, INTRAMURAL RES, 74- *Concurrent Pos:* Assoc prof epidemiol, Sch Pub Health & Hyg & assoc prof neurol, Sch Med, Johns Hopkins Univ; Nat Soc Med Sci fel comn; mem comt virol, Nat Comn Med Sci. *Honors & Awards:* Nat Inst Allergy & Infectious Dis. *Mem:* Am Asn Immunol; World Fedn Neurol; AAAS; Am Soc Trop Med & Hyg; Am Asn Path; Sigma Xi. *Res:* Virus induced immunopathology; zoology, epidemiology and immunogenicity of arboviruses; aging process in man; behavioral changes in man and animals associated with infectious processes; oncogenic and tumor viruses; disease patterns in primitive populations; hemorrhagic fevers; ecology of infectious diseases; viral epidemiology. *Mailing Add:* 326 E St NE Nat Inst Neurol Commun Dis & Stroke Washington DC 20002

GIBBS, DANIEL, b Ft Jackson, SC, Feb 4, 43; m 79; c 1. NEUROBIOLOGY, INSECT PHYSIOLOGY. *Educ:* Wesleyan Univ, BA, 65; Stanford Univ, MA, 70, PhD(biol), 74. *Prof Exp:* Res assoc zool, Univ Washington, 74-76; from asst prof to assoc prof biol, De Paul Univ, 76-87; SOFTWARE DEVELOPER, ODESTA CORP, 87- *Concurrent Pos:* Prin investr, NIH grants, De Paul Univ, 79-81. *Mem:* Soc Neurosci; Am Soc Zoologists; Entom Soc Am; AAAS; Sigma Xi. *Res:* Neuroanatomy and electrophysiology of prothoracic tropic hormone-secreting cells in the brain of an insect, Manduca Sexta; these cells initiate larval molting and adult development in insects. *Mailing Add:* Odesta Corp 4084 Commercial Ave Northbrook IL 60062

GIBBS, DAVID LEE, b San Francisco, Calif, July 5, 48. CLINICAL MICROBIOLOGY, PUBLIC HEALTH. *Educ:* Cornell Univ Med Ctr, PhD(microbiol), 74; Am Bd Med Microbiol, dipl, 80. *Prof Exp:* Fel public health & med microbiol, Ctr Dis Control, 74-76; consult smallpox prog, WHO, India, 75; instr med, Med Col, Cornell Univ, 77-78, asst prof, 78-79; dir clin microbiol, Santa Clara Valley Med Ctr, 79-80; asst dir new drug develop, 80-81, assoc dir clin microbiol, 81-86, SR ASSOC DIR CLIN RES, PFIZER PHARMACEUT, INC, 86- *Concurrent Pos:* Vis prof, Fed Univ Bahia, Brazil, 77-79; US dir, Commun Dis Ctr, Brazil, 77-79. *Mem:* Am Soc Microbiol. *Res:* Antibiotics; clinical microbiology and immunology methods. *Mailing Add:* 235 E 42nd St New York NY 10017

GIBBS, DOON LAURENCE, b Urbana, Ill, Feb 5, 54. SOLID STATE PHYSICS. *Educ:* Univ Utah, BS, 77; Univ Ill, MS, 79; PhD(physics), 82. *Prof Exp:* Res asst, Univ Ill, 78-82; asst physicist, 83-84, ASSOC PHYSICIST, BROOKHAVEN NAT LAB, 84- *Mem:* NY Acad Sci. *Res:* Magnetic x-ray scattering studies of the magnetic structure and phase transitions of solids using synchrotron radiation; surface x-ray scattering studies of solids. *Mailing Add:* 127 Senix Ave Center Moriches NY 11933

GIBBS, FINLEY P, b Washington, DC, Aug 15, 40; m 63; c 3. PHYSIOLOGY, ANATOMY. *Educ:* Univ Calif, Berkeley, AB, 63; Univ Ore, PhD(physiol), 68. *Prof Exp:* Asst prof, Sch Med & Dent, Univ Rochester, 68-75, assoc prof anat & physiol, 75-81; ASSOC PROF ANAT, SCH MED, UNIV, MO COLUMBIA, 81- *Mem:* Am Physiol Soc; Endocrine Soc; Am Asn Anatomists. *Res:* Control of ACTH secretion, the control of metatonin secretion and mechanisms involved with the biological clock (circadian rhythms); neuroendocrinology. *Mailing Add:* Dept Anat Sch Med Univ Mo Columbia MO 65212

GIBBS, FREDERIC ANDREWS, b Baltimore, Md, Feb 9, 03; m 30; c 2. MEDICAL SCIENCE, CEREBRAL BLOOD FLOW. *Educ:* Yale Univ, AB, 25; Johns Hopkins Univ, MD, 29. *Hon Degrees:* DSc, Univ Montpellier, 65, Univ Ill, 77. *Prof Exp:* Res fel neuropath, Harvard Med Sch, 29-30; res fel, Johnson Found Med Physics, Univ Pa, 30-32; res fel, Harvard Med Sch, 33-34, res fel physiol, 34-35, instr neurol, 36-44; assoc prof psychiat, 44-49, PROF NEUROL, UNIV ILL, CHICAGO, 50- *Honors & Awards:* Lasker Award, 51. *Mem:* Am Acad Cerebral Palsy; Am Acad Neurol; Am EEG Soc (pres, 48-49); Am Epilepsy Soc (pres, 49); Am Med EEG Soc (pres, 78-79); Norweg Acad Sci. *Res:* Cerebral blood flow; epilepsy and anti-epilepsy therapy; electroencephalography; brain tumors and brain trauma; encephalitis; psychosis and behavior disorders; intellectual defects; psychopharmacology. *Mailing Add:* Data Bank & Conf Ctr 618 N 400 E Valparaiso IN 46383

GIBBS, GERALD V, b Hanover, NH, June 28, 29; m 59; c 4. MINERALOGY. *Educ:* Univ NH, BA, 55; Univ Tenn, MS, 57; Pa State Univ, PhD(mineral), 62. *Prof Exp:* Res assoc mineral, Univ Chicago, 61-62; res mineralogist, Linde Co, Union Carbide Corp, NY, 62-63; from asst prof to assoc prof mineral, Pa State Univ, 63-67; prof mineral, 67-81, UNIV DISTINGUISHED PROF, VA POLYTECH INST & STATE UNIV, 81- *Concurrent Pos:* NSF fel, 60-61 & 64-82, NASA grants, 71-73; vis distinguished prof, Ariz State Univ, 79, 83; vis res sci, Sch Higher Educ, 86. *Honors & Awards:* Roebling Gold Medalist, 86. *Mem:* Mineral Soc Am (pres, 81); Am Crystallog Asn; fel Am Geophys Union. *Res:* Silicate mineralogy; bonding in minerals; mathematical crystallography. *Mailing Add:* Dept Mineral Va Polytech Inst & State Univ Blacksburg VA 24060

GIBBS, GORDON EVERETT, b Cordova, Ill, Sept 25, 11; m 41; c 4. PEDIATRICS. *Educ:* Univ Redlands, AB, 32; Univ Calif, MA, 35, PhD(physiol), 39, MD, 42. *Prof Exp:* From intern to resident, Univ Hosp, Univ Calif, 42-47; assoc prof pediat res, Univ Md, 50-54; assoc prof pediat, 54-56, prof & chmn dept, 56-66, res prof pediat, 66-81, EMER PROF PEDIAT, COL MED, UNIV NEBR, 81- *Concurrent Pos:* Sr fel pediat, Nat Res Coun, Calif, 47-49 & Univ Ill, 49-50; prin investr, NIH grants. *Honors & Awards:* Bronze Star Medal, US Army, 44; Meritorious Serv Medal, USAF, 71. *Mem:* Am Pediat Soc; Soc Exp Biol & Med; Am Diabetes Asn; Am Acad Pediat; Sigma Xi. *Res:* Juvenile diabetes; experimental diabetes in monkeys; effect of diet, insulin and hypophysectomy on retinal and renal complications; cystic fibrosis of the pancreas; clinical research; numerous published articles. *Mailing Add:* 88120 Ave 73 Thermal CA 92274-9683

GIBBS, HAROLD CUTHBERT, b Barbados, BWI, Apr 29, 28; m 54; c 4. PARASITOLOGY. *Educ:* McGill Univ, BSc, 51, MSc, 56, PhD, 58; Ont Vet Col, DVM, 55. *Prof Exp:* Pathologist, Can Wildlife Serv, 57-58; hon asst prof parasitol, Macdonald Col, McGill Univ, 58-63, assoc prof, 63-71; PROF ANIMAL PATH & WILDLIFE RESOURCES, UNIV MAINE, ORONO, 71-, CO-OP PROF FOREST RESOURCES, 77- *Concurrent Pos:* Lectr, Univ Ottawa, 57-58; res officer, Animal Path Labs, Can Dept Agr, 58-62; sr parasitologist, Averst, McKenna & Harrison, 62-63. *Mem:* Am Soc Parasitol; Wildlife Dis Asn; Am Vet Med Asn; Sigma Xi. *Res:* Helminthology. *Mailing Add:* RFD 1-Box 333 Hampden ME 04444

GIBBS, HUGH HARPER, b Can, Oct 22, 30; m 56; c 1. ORGANIC CHEMISTRY. *Educ:* Queen's Univ, Can, BSc, 52, MSc, 53; Univ Ill, PhD(chem), 56. *Prof Exp:* Res chemist, E I Du Pont de Nemours & Co, Inc, 56-67, sr res chemist, 67-70, res assoc polymer prods dept, 70-85, sr res assoc, Textile Fibers Dept, 85-90, RES FEL, FIBERS DEPT, E I DU PONT DE NEMOURS & CO, INC, 90- *Mem:* Am Chem Soc; Soc Advan Mat & Proc Eng. *Res:* Synthetic organic chemistry; mechanisms of organic reactions; synthesis and reactions of fluorocarbon compounds; polymer chemistry; high performance composites; polyimides. *Mailing Add:* Fibers Dept Chestnut Run Bldg 702 E I Du Pont de Nemours & Co Inc Wilmington DE 19880-0702

GIBBS, HYATT MCDONALD, b Hendersonville, NC, Aug 6, 38. NONLINEAR OPTICS, OPTICAL BISTABILITY. *Educ:* NC State Univ, BS(elec eng) & BS(eng physics), 60; Univ Calif, Berkeley, PhD(physics), 65. *Prof Exp:* Actg asst prof physics, Univ Calif, Berkeley, 65-67; mem tech staff, Bell Labs, 67-80; PROF OPTICAL SCI, UNIV ARIZ, 80-, DIR OPTICAL CIRCUITRY COOP, 84- *Concurrent Pos:* Exchange scientist, Philips Res Labs, Eindhoven, Netherlands, 75-76; vis lectr, Princeton Univ, 78-79; Franklin Inst fel. *Honors & Awards:* Michelson Medal, Franklin Inst, 84. *Mem:* Fel Am Phys Soc; fel Optical Soc Am; sr mem Inst Elec & Electronics Engrs; fel AAAS. *Res:* Spin exchange; time reversal; polarization of lead and thallium; self-induced transparency pulse breakup, peak amplification, pulse compression, degeneracy, Faraday rotation, self focusing, pulse collisions; neoclassical theory tests; subnatural linewidth fluorescence; optical transistor, optical bistability; superfluorescence; energy transfer and Anderson localization; laser spectroscopy; coherent optics; optical turbulence; nonlinear optical signal processing. *Mailing Add:* Optical Sci Ctr Univ Ariz Tucson AZ 85721

GIBBS, JAMES ALBERT, b Montgomery, Ala, Oct 1, 17; m 43; c 4. ORGANIC CHEMISTRY. *Educ:* Fisk Univ, AB, 38, MA, 40; Harvard Univ, MA, 47. *Prof Exp:* Asst chem, Fisk Univ, 40-41 & US Civil Serv, Washington, DC, 42-43; instr chem, Fisk Univ, 43-44; asst prof, Hampton Inst, 45-46; res chemist physiol, Med Sch, Tufts Univ, 47-49; chemist, Tracerlab, Inc, 49-57; tech dir & treas, Volk Radiochem Co, Ill, 58-60; mgr chem labs, Packard Instrument Co, Inc, 60-67, mgr chem supplies, 67-74, regulatory affairs mgr, 74-85; RETIRED. *Mem:* AAAS; Am Chem Soc; Am Inst Chem; NY Acad Sci. *Res:* Synthesis of isotopically labeled compounds, carbon-14, hydrogen-2, hydrogen-3, sulfur-35, iodine-131, phosphorus-32; radioassay techniques and instrumentation. *Mailing Add:* 8242 S Langley Ave Chicago IL 60619

GIBBS, JAMES GENDRON, JR, b Charleston, SC, Dec 28, 38; m 64; c 2. NEUROBIOLOGY, PSYCHIATRY. *Educ:* Trinity Col, BS, 60; Med Col SC, MD, 64. *Prof Exp:* Intern, Univ Hosp, Univ Mich, 64-65; med officer, US Army, 65-67; resident psychiat, NY Hosp-Cornell Med Ctr, 68-72; res assoc, 72-73, asst prof, 73-78, ASSOC PROF PSYCHIAT, CORNELL UNIV MED COL, 78- *Concurrent Pos:* Fel, Found Fund Res Psychiat, 72-73; Glorney-Raisbeck fel, NY Acad Med, 73-74; NIMH res scientist develop award, 74-80; Irma T Hirschl career scientist award, 79-84. *Mem:* Am Physiol Soc; Am Psychiat Asn; Asn Res Nervous & Ment Dis; Soc Neurosci. *Res:* Physiological mechanisms of motivated behaviors. *Mailing Add:* Dept Psychol 21 Bloomingdale Rd White Plains NY 10605

GIBBS, MARTIN, b Philadelphia, Pa, Nov 11, 22; m 50; c 5. BIOLOGY. *Educ:* Philadelphia Col Pharm, BS, 43; Univ Ill, PhD(plant physiol), 47. *Prof Exp:* Asst chem, Univ Ill, 43-44, asst bot, 44-45, asst agron, 45-47; mem dept biol, Brookhaven Nat Lab, 47-56; assoc prof biochem, Cornell Univ, 56-60, prof, 60-64; chmn dept, 65-68, prof biol, 64-78, PROF PHOTOBIOL, BRANDEIS UNIV, 78-, ABRAHAM & GERTRUDE BURG PROF LIFE SCI, 85- *Concurrent Pos:* Vis prof, Univ Pa, 54 & Queens Col, 58; vis scientist, Res Inst Adv Study, Md, 59 & 60; consult, NSF, 61-64 & 67-70; consult, NIH, 64-67; ed, Physiologie Vegetale, 68-; Sci Res Soc Am-Sigma Xi nat lectr, 69; consult, NATO Fel Bd; ed in chief, Plant Physiol, 63-, ed, Annual Rev, 67-70; vis prof, Univ Calif, Riverside, 77- *Mem:* Nat Acad Sci; AAAS; Am Soc Plant Physiol; Am Soc Biol Chem; Am Acad Arts & Sci. *Res:* Photosynthesis; carbohydrate metabolism of higher plants and algae. *Mailing Add:* Inst Photobiol Brandeis Univ Waltham MA 02254

GIBBS, MARVIN E, b St Louis, Mo, Dec 30, 34; m 60; c 2. CHEMICAL ENGINEERING. *Educ:* Washington Univ, St Louis, BS, 56, PhD(chem eng), 60. *Prof Exp:* Sr res engr, Monsanto Co, St Louis, 60-66, res specialisr, 66-67, res group leader, 67-70, mfg supt, Tex, 70-77, res mgr, 77-78, dir planning & control, 78-80, dir res & develop, 80-; MEM STAFF SOFTWARE ENG INST, CARNEGIE-MELLON UNIV. *Mem:* Am Inst Chem Engrs. *Res:* Process research, using special techniques, including and reaction kinetics, to provide all necessary information for large scale production. *Mailing Add:* 13445 Kings Glen Dr St Louis MO 63131

GIBBS, NORMAN EDGAR, b Keyport, NJ, Nov 27, 41; m 67; c 2. COMPUTER SCIENCE. *Educ:* Ursinus Col, BS, 64; Purdue Univ, MS, 66, PhD(comput sci), 69. *Prof Exp:* Asst prof math, Col William & Mary, 69-77, assoc prof math & comput sci, 77-81; asst chmn & prof, comput sci dept, Ariz State Univ, Tempe, Az, 81-83; chmn & prof, Comput Sci & Info Studies, Bowdain Col, 83-85; DIR SOFTWARE EDUC, SOFTWARE ENG INST, CARNEGIE-MELLON UNIV, 85- & DIR PROD & SERV DIV, 90- *Concurrent Pos:* Asst dir comput ctr, Col William & Mary, 69- *Mem:* Asn Comput Mach. *Res:* Graph theoretic algorithms; programming systems and languages; data structures; computer science education. *Mailing Add:* Software Eng Inst Carnegie-Mellon Univ Pittsburgh PA 15213

GIBBS, PETER (GODBE), b Salt Lake City, Utah, Dec 7, 24; m 53; c 3. PHYSICS. *Educ:* Univ Utah, BS, 47, MS, 49, PhD(physics), 51. *Prof Exp:* Res assoc physics, Univ Ill, 51-52, instr, 52-54; Fulbright lectr theoret physics, Univ Ceylon, 54-55; assoc prof physics, 57-62, from actg chmn to chmn dept, 67-76, PROF PHYSICS, UNIV UTAH, 62-, ASSOC RES PROF CERAMIC ENG, 56- *Concurrent Pos:* Fulbright lectr, Univ Sao Paulo, 63; consult, Atomics Int Div, NAm Aviation, Inc, 58-63, Dept Sci Affairs, Orgn Am States, 63 & Stanford Res Inst, 65-67; res assoc physics, Univ Calif, Berkeley, 77-78. *Honors & Awards:* Purdy Prize, Am Ceramic Soc, 62. *Mem:* Fel AAAS; NY Acad Sci; Biophys Soc; fel Am Phys Soc; Am Asn Physics Teachers; Sigma Xi. *Res:* Solid state and biological physics. *Mailing Add:* 201 James Fletcher Bldg Univ Utah Salt Lake City UT 84112

GIBBS, R DARNLEY, b Ryde, Eng, June 30, 04; m 61; c 1. BOTANY. *Educ:* Univ London, BSc, 25, PhD, 33; McGill Univ, MSc, 26. *Prof Exp:* Biochemist, Am Rubber Producers, 27-28; demonstr bot, 25-26 & 28-29, lectr, 29-37, from asst prof to prof, 37-65, Macdonald prof, 65-71, EMER PROF BOT, McGILL UNIV, 71- *Mem:* Am Soc Plant Physiol; Royal Soc Can; Linnean Soc London. *Res:* Comparative chemistry of higher plants as applied to problems of taxonomy. *Mailing Add:* 32 Orchards Way Southampton S02 1RE England

GIBBS, RICHARD LYNN, b Buffalo, NY, May 12, 39; m 59; c 3. ATOMIC PHYSICS, PLASMA PHYSICS. *Educ:* Univ of the South, BA, 61; Clarkson Col Technol, MS, 63, PhD(physics), 66. *Prof Exp:* From asst prof to assoc prof, 66-74, PROF PHYSICS, LA TECH UNIV, 74- *Mem:* AAAS; Am Phys Soc; Am Asn Physics Teachers. *Res:* Atomic and molecular structure calculations; mathematical physics; experimental plasma measurements. *Mailing Add:* Dept Physics La Tech Univ PO Box 3187 Tech Sta Ruston LA 71272

GIBBS, ROBERT JOHN, b New York, NY, Aug 14, 26; m 52; c 5. PHYSICAL CHEMISTRY, SCIENCE ADMINISTRATION. *Educ:* Fordham Univ, BS, 48, MS, 49, PhD(phys org chem), 52. *Prof Exp:* Res assoc phys chem, Sch Med, Univ Va, 52-54; mem staff, Mass Inst Technol, 54-56; chemist, Eastern Utilization Res & Develop Div, USDA, 56-64; grants assoc, Div Res Grants, USPHS, 64, asst to extramural opers & procedures officer, Off Dir, NIH, 64-69, opers anal off, 69-71, chief gen res support br, Div Res Resources, 71-74; asst vpres planning & develop, Univ DC, Washington, 74-78, admin & serv, 78-83; RETIRED. *Mem:* Fel AAAS; Am Chem Soc; Sigma Xi. *Res:* Science administration; kinetics of protein denaturation; physical properties of proteins; structural proteins of muscle. *Mailing Add:* 19630 White Rock Dr Sun City West AZ 85375

GIBBS, RONALD JOHN, b Joliet, Ill, Dec 26, 33. GEOCHEMISTRY. *Educ:* Northwestern Univ, Ill, BS, 57, MS, 60; Univ Calif, San Diego, PhD(oceanog), 65. *Prof Exp:* Asst prof geol, Univ NMex, 65-66 & Univ Calif, Los Angeles, 66-70; assoc prof, Northwestern Univ, Ill, 70-74; PROF, COL MARINE STUDIES, UNIV DEL, 74-. DIR, COLLOIDAL CTR, 85- *Mem:* AAAS; Geol Soc Am; Clay Minerals Soc. *Res:* Environmental studies, pollution, sedimentology; clay mineralogy; geochemistry of dissolved and suspended loads of river systems; oceanic processes affecting discharged materials of river systems & pollution transport. *Mailing Add:* Col Marine Studies Univ Del Newark DE 19716

GIBBS, SAMUEL JULIAN, b Amory, Miss, Apr 1, 32; m 58; c 3. DENTAL RADIOLOGY, RADIOBIOLOGY. *Educ:* Emory Univ, DDS, 56; Univ Rochester, PhD(radiation biol), 69; Am Bd Oral & Maxillofacial Radio, dipl, 81. *Prof Exp:* Pvt pract, Ala, 59-63; asst prof radiol & dent res, Univ Rochester, 68-70; asst prof, 70-76, ASSOC PROF DENT & RADIOL, VANDERBILT UNIV, 76- *Concurrent Pos:* Fel radiation biol, Univ Rochester, 63-68; consult, Vet Admin Hosp, Nashville, Tenn, 70-; dir, Am Bd Oral & Maxillofacial Radiol, 83-, vpres, 84-87, pres, 88- *Mem:* Fel Am Acad Dental Radiol (vpres, 79-80, pres, 80-81); Radiation Res Soc; Int Asn Dent Res; Am Dent Asn; Am Col Radiol; Radiol Soc NAm. *Res:* Experimental dental radiology; hazards of diagnostic radiology. *Mailing Add:* CTR Radiation Oncology E-1200 Med Ctr North Vanderbilt Univ Hosp Nashville TN 37232

GIBBS, SARAH PREBLE, b Boston, Mass, May 25, 30; div; c 2. CELL BIOLOGY, PHYCOLOGY. *Educ:* Cornell Univ, AB, 52, MS, 54; Radcliffe Col, PhD(cell biol), 62. *Prof Exp:* NIH fel biol, Harvard Univ, 61-62; NIH fel bact, Edinburgh Univ, 62-63, res assoc animal genetics, 63-65; from asst prof to assoc prof, 66-74, PROF BIOL, MCGILL UNIV, 74- *Concurrent Pos:* Mem fel comt, Nat Res Coun Can, 75-78; mem Cell & Genetics Grant Comt, Natural Sci & Eng Res Coun Can, 82-85. *Honors & Awards:* Darbaker Prize, Bot Soc Am, 75. *Mem:* Can Soc Cell Biol (pres, 72-73); Am Soc Cell Biol; Phycol Soc Am; Can Soc Plant Molecular Biol; AAAS. *Res:* Cell biology, molecular organization of chloroplasts, evolution of algal chloroplasts from eukaryotic endosymbionts. *Mailing Add:* Dept Biol McGill Univ 1205 Docteur Penfield Ave Montreal PQ H3A 1B1 Can

GIBBS, TERRY RALPH, b St Joseph, Mo, Aug 7, 37; m 58; c 3. MECHANICAL ENGINEERING, NUCLEAR ENGINEERING. *Educ:* Univ Mo, Rolla, BSME, 59; Univ NMex, MM, 79. *Prof Exp:* Staff mem detonator design & eng, Los Alamos Nat Lab, 59-66, staff mem nuclear weapons effects, 66-73, asst nuclear weapons, 73-75, alt group leader detonation physics, 75-79, asst div leader appl physics, 79-84, lab budget officer, 84-88, finance oficer, 88-89, DEP CONTROLLER, LOS ALAMOS NAT LAB, 89- *Res:* Phenomenology associated with nuclear weapons vulnerability of nuclear weapon components to antiballistic missile-produced radiation, electromagnetic environments and explosives. *Mailing Add:* Los Alamos Nat Lab PO Box 1663 CONT MS A119 Los Alamos NM 87545

GIBBS, THOMAS W(ATSON), b Alexandria, Va, Sept 27, 32; m 54; c 3. METALLURGY, MATERIALS ENGINEERING. *Educ:* Mass Inst Technol, SB, 54, SM, 55, ScD(metall), 64. *Prof Exp:* Assoc scientist, Avco Corp, 57-59, sr scientist, 59-64, staff scientist, 64-65; res metallurgist, 65-68, res supvr, 68-71, venture specialist, 71-75, consult mgr mat engr, 75-79, MGR CONSULT SERVICES, E I DU PONT DE NEMOURS & CO, INC, 79- *Mem:* Am Soc Metals; Nat Asn Corrosion Engrs. *Res:* Mechanical properties; high temperature metallurgy; high intensity arcs; mechanical testing; materials engineering; fiber metallurgy; composite materials; semiconductor preparation and properties; dimensional stability; wear; foundry metallurgy. *Mailing Add:* 2527 Deepwood Dr Wilmington DE 19810

GIBBS, WILLIAM EUGENE, b Akron, Ohio, Sept 23, 30; m 48; c 5. PHYSICAL CHEMISTRY. *Educ:* Univ Akron, BS & MS, 54, PhD(chem), 59. *Prof Exp:* Res chemist, Inst Rubber Res, Univ Akron, 53-55 & Goodyear Tire & Rubber Co, 55-58; res chemist, Air Force Materiel Lab, 58-59, group leader, 59-62, chief polymer br, 62-66, dir, 66-70; vpres & dir res & develop, Foster Grant Co, Inc, 70-78; VPRES & DIR RES & DEVELOP, PLASTICS DIV, AM HOECHST CORP, 78- *Concurrent Pos:* Mem comt fire toxicol, Nat Acad Sci. *Mem:* AAAS; Am Chem Soc; Soc Plastics Eng; Faraday Soc; Chem Soc London. *Res:* Synthesis and properties of polymers; physical chemistry of polymers; mechanisms of polymerizations and degradation. *Mailing Add:* 317 Jule Dr Chesapeake VA 23320

GIBBS, WILLIAM ROYAL, b Dublin, Tex, July 6, 34; m 55; c 2. PHYSICS. *Educ:* Univ Tex, BS, 55, MA, 57; Rice Univ, PhD(physics), 61. *Prof Exp:* Res assoc physics, Univ Neuchatel, 61-62; group leader T-5, 73-75 & 89-90, mem res staff, 62-90, LAB ASSOC, LOS ALAMOS NAT LAB, 90- *Mem:* Fel Am Phys Soc. *Res:* Nuclear reaction mechanisms; hadron-nucleus. *Mailing Add:* Los Alamos Nat Lab T-5 MS-B283 Los Alamos NM 87545

GIBBY, IRVIN WELCH, medical microbiology, for more information see previous edition

GIBEAULT, VICTOR ANDREW, b Pawtucket, RI, Oct 21, 41; m 65; c 2. ORNAMENTAL HORTICULTURE. *Educ:* Univ RI, BS, 63, MS, 65; Ore State Univ, PhD(farm crops), 71. *Prof Exp:* COOP EXTEN SPECIALIST TURF & LANDSCAPE, COOP EXTEN SERV, UNIV CALIF, RIVERSIDE, 69- *Mem:* Am Soc Agron; Weed Sci Soc Am. *Res:* Adaptation of turf grass species and varieties to various habitats; herbicide influence on growth and development of turf grasses. *Mailing Add:* 5675 Via Mensabe Riverside CA 92506

GIBIAN, GARY LEE, b Englewood, NJ, April 17, 49. SIGNAL PROCESSING, PSYCHOACOUSTICS. *Educ:* Mass Inst Technol, BS, 71; Washington Univ, MA, 73, PhD(physics), 80. *Prof Exp:* Teaching asst, Washington Univ, 71-73, res asst, Cent Inst Deaf, 73-76, res asst physiol & biophysics, 76-79; assoc sr res scientist, Gen Motors Res Labs, 79-81, sr res scientist, 81-82; asst prof, Am Univ, Physics Dept, 82-88; SR SCIENTIST, PLANNING SYSTS INC, 89- *Mem:* Acoustical Soc Am; Audio Engr Soc; Comput Music Assoc. *Res:* Physiological acoustics, cochlear nonlinearites; psychoacoustics, subjective response to transportation noise. *Mailing Add:* Planning Systs Inc 7923 Jones Br Dr McClean VA 22102

GIBIAN, MORTON J, organic chemistry, enzymology; deceased, see previous edition for last biography

GIBIAN, THOMAS GEORGE, b Prague, Czech, Mar 20, 22; nat US; m 49; c 4. CHEMISTRY. *Educ:* Univ NC, BSc, 42; Carnegie-Mellon Univ, PhD(chem), 48. *Prof Exp:* Petrol res, Atlantic Ref Co, 48-51; develop engr, Dewey & Almy Chem Div, W R Grace & Co, Md, 51-52, plant mgr, 53-56, gen mgr, Battery Separator Div, 56-57, vpres, Org Chem Div, 57-62 & Chem Group, 62-63, pres, Res Div, 63-66, vpres & tech group exec, 66-74; pres, Chem Construct Corp, NY, 74-76; chmn, Henkel Corp, 80-85; PRES, TECH GUID INT, INC, GREENWICH, CONN, 76- *Concurrent Pos:* Lectr, Drexel Inst Technol, 49-51. *Mem:* Am Chem Soc; Indust Res Inst; Soc Chem Indust; Soc Chimie Industrielle. *Res:* Administration and management. *Mailing Add:* PO Box 219 Sandy Spring MD 20860

GIBILISCO, JOSEPH, b Omaha, Nebr, Feb 6, 24; m 51; c 3. DENTISTRY. *Educ:* Univ Minn, DDS, 48, MSD, 51. *Prof Exp:* Chmn dept oper dent, Creighton Univ, 49-51; assoc prof dent & anat, Univ Ala, 51-54; consult, Dept Dent, 56-62, chmn dept dent, Mayo Clin, 62-76, PROF DENT, MAYO MED SCH, 75- *Concurrent Pos:* Chmn, Southeastern Minn Health Systs Agency, 76-80 & Minn State Coord Coun, 76- *Mem:* AAAS; Am Dent Asn; Int Asn Dent Res; fel Am Col Dent; Am Acad Oral Med; Sigma Xi. *Res:* Clinical dentistry, particularly oral diagnosis and management of complex craniofacial pain problems. *Mailing Add:* Mayo Clin Rochester MN 55901

GIBLETT, ELOISE ROSALIE, b Tacoma, Washington, Jan 17, 21. HEMATOLOGY, IMMUNOGENETICS. *Educ:* Univ Wash, BS, 42, MS, 47, MD, 51. *Prof Exp:* Clin assoc med, Sch Med, Univ Wash, 55-57, from clin instr to clin prof, 58-67, res prof med, 67-87, emer prof med & emer exec dir, Puget Sound Blood Ctr, 87-; RETIRED. *Concurrent Pos:* USPHS fel hemat, Univ Wash & Postgrad Med Sch, London, 53-55; USPHS trainee genetics, Case Western Reserve Univ, 60; mem NIH genetics study sect, Nat Heart, Lung & Blood Res Rev Comt, Nat Blood Resources Comt, Food & Drug Admin Toxicology Adv Comt; head immunogeneticss & assoc dir, Puget Sound Blood Ctr, 55-79, exec dir, 79-87. *Honors & Awards:* Emily Cooley Award, 75; Karl Landsteiner Award, 76; Philip Levine Award, 78. *Mem:* Nat Acad Sci; Am Soc Hemat; Am Soc Human Genetics (pres, 73); Am Asn Immunol; Asn Am Physicians. *Res:* Erythrokinetics; genetic polymorphisms of all blood components; blood group antibodies; inherited enzyme defects in immunodeficiency diseases. *Mailing Add:* Puget Sound Blood Ctr 921 Terry Ave Seattle WA 98104

GIBLEY, CHARLES W, JR, b Philadelphia, Pa, Oct 28, 34; m 56; c 4. ZOOLOGY, EMBRYOLOGY. *Educ:* Villanova Univ, BS, 56; Iowa State Univ, MS, 59, PhD(zool), 61. *Prof Exp:* Lab instr biol, Villanova Univ, 57; res asst zool, Iowa State Univ, 57-61; asst prof biol, Villanova Univ, 61-65; prof histo-embryol, Pa Col Podiatric Med, 63-70, acad dean, 67-74, prof anat, 70-84, vpres acad affairs, 74-84; CONSULT, 84- *Concurrent Pos:* Asst biologist & asst prof, Univ Tex M D Anderson Hosp & Tumor Inst, 65-67. *Mem:* Am Soc Zool; Soc Develop Biol; AAAS; Am Asn Anatomists; Am Soc Cell Biol. *Res:* Experimental embryology; histochemistry and developmental causes underlying the morphogenesis of embryonic and adult kidney in vertebrates. *Mailing Add:* Dean Arts-Sci Phila Col Pharm-Sci Woodland Ave at 43rd St Philadelphia PA 19104

GIBLIN, DENIS RICHARD, b Montreal, Que, Mar 14, 27; m 66; c 2. NEUROLOGY. *Educ:* McGill Univ, BSc, 48, MD, 52. *Prof Exp:* Intern, Royal Victoria Hosp, Montreal, 52-53; intern med, Duke Univ Hosp, Durham, NC. 53-54; resident neurol, Montreal Neurol Inst, 54-55 & Peter Bent Brigham Inst, Boston, 55-56; Nuffield traveling fel, London, 56-57; ASSOC PROF NEUROL, ALBERT EINSTEIN COL MED, 73- *Honors &*

Awards: A Cressy Morrison Award Nat Sci, NY Acad Sci, 63. *Res:* Neurophysiology, clinical, evoked somatosensory and visual responses in healthy subjects and patients with neurological disorders; neurophysiology, sensory coding by individual neurons. *Mailing Add:* Dept Neurol Albert Einstein Col Med Bronx NY 10461

GIBLIN, FRANK JOSEPH, b St Petersburg, Fla, Feb 24, 42; m 65; c 2. BIOCHEMISTRY. *Educ:* Univ Windsor, BASc, 65; State Univ NY Buffalo, PhD(biochem), 74. *Prof Exp:* Engr chem, Procter & Gamble Co, 65-69; assoc ocular res, 74-77, asst prof ocular res, inst biol sci, 77-83, ASSOC PROF BIOMED RES, EYE RES INST, OAKLAND UNIV, 83- *Honors & Awards:* Rohto Cataract Res Award, 81; Alcon Res Recognition Award, 85. *Mem:* AAAS; Sigma Xi; Asn Res Vision & Ophthal. *Res:* Mechanism of formation of cataract in the ocular lens and the possible role of glutathione in maintaining the transparency of this tissue by preventing oxidative damage. *Mailing Add:* 1651 Bretton Dr N Rochester MI 48063

GIBLIN-DAVIS, ROBIN MICHAEL, b Berkeley, Calif, Mar 6, 55; m 82; c 3. PHYTOPATHOLOGY, ENTOMOLOGY. *Educ:* Univ Calif, Davis, BS, 77, PhD(entom), 82. *Prof Exp:* Postdoctoral assoc, Univ Calif, Riverside, 82-85; asst prof, 85-90, ASSOC PROF ENTOM, UNIV FLA, FT LAUDERDALE, 90- *Mem:* Soc Nematologists; Soc Invert Pathologists; Entom Soc Am; Orgn Trop Am Nematologists. *Res:* The taxonomy, ecology and physiology of nematode parasites and associates of plants (turfgrasses and ornamentals) and insects. *Mailing Add:* Univ Fla 3205 College Ave Ft Lauderdale FL 33314

GIBOFSKY, ALLAN, b New York, NY, Sept 7, 49; m 82; c 3. RHEUMATOLOGY, FORENSIC MEDICINE. *Educ:* Brooklyn Col, NY, 69; Cornell Univ Med Col, MD, 73; Fordham Univ, NY, JD, 85. *Prof Exp:* Intern pathol, NY Hosp, 73-74, res med, 74-77; res fel immunol, Rockefeller Univ, 77-79; asst prof med, 79-85, ASSOC PROF MED, CORNELL MED COL, NY, 85-, ASST PROF PUB HEALTH, 89- *Concurrent Pos:* Adj assoc prof immunol, Rockefeller Univ, 85-; pres, med sci comt, NY Arthritis Found, 88-; chair by laws comt, Am Soc Histocompatibility & Immunogenetics, 86-; adj assoc prof law, Fordham Univ, 87-; mem, Data & Technol Assessment Prog, Am Med Asn, 87-; Jonas Salk scholar, City NY, 69. *Honors & Awards:* Nat Res Serv Award, Nat Inst Health, 77. *Mem:* Fel Am Col Rheumatism (secy-treas, 85-); fel Am Col Legal Med; Am Col Physicians; Am Asn Immunologists; Am Fed Clin Res. *Mailing Add:* 535 E 70th St New York NY 10021

GIBOR, AHRON, b Jaffa, Israel, Sept 16, 25; nat US; m 50; c 2. PHYSIOLOGY. *Educ:* Univ Calif, BA, 50, MA, 52; Stanford Univ, PhD(biol), 56. *Prof Exp:* Res biologist, Alaska Dept Fish & Game, 57-59; chief biol sect, Resources Res, Inc, 59-60; res assoc physiol, Rockefeller Inst, 60-63, asst prof, 64-66; assoc prof, 66-68, PROF PHYSIOL, UNIV CALIF, SANTA BARBARA, 68- *Concurrent Pos:* Guggenheim fel, 78-79; vis prof, Kyoto Univ, 85. *Mem:* Am Soc Cell Biol; AAAS. *Res:* General and algal physiology. *Mailing Add:* Dept Biol Univ Calif Santa Barbara CA 93106

GIBORI, GEULA, b Beirut, Lebanon, Aug 8, 45; m 70; c 3. ENDOCRINOLOGY, REPRODUCTIVE PHYSIOLOGY. *Educ:* Lebanese Univ, BS, 67; Sorbonne, MS, 68; Tel Aviv Univ, PhD(physiol), 73. *Prof Exp:* Lectr physiol, Univ Tel Aviv, 71-73; fel reproduction, Case Western Reserve Univ, 73-75 & Univ Mich, 75-76; asst prof, 76-80, assoc prof, 80-86, PROF PHYSIOL, UNIV ILL, 86- *Concurrent Pos:* Fulbright Found fel, 73-76; NSF grant, 78-84, NIH grants, 78-91; Merit Award, NIH, 88- *Mem:* Endocrine Soc; Soc Study Reproduction; Soc Study Fertility; Am Physiol Soc; Soc Exp Biol Med; fel AAAS. *Res:* Placental derived regulators and the molecular control of ovarian function. *Mailing Add:* Dept Physiol & Biophys Col Med Univ Ill PO Box 6998 m/c 901 Chicago IL 60680

GIBSON, ATHOLL ALLEN VEAR, b Eshowe, S Africa, Nov 25, 40; m 70; c 2. SOLID STATE PHYSICS, LOW TEMPERATURE PHYSICS. *Educ:* Univ Natal, BSc, 61, PhD(physics), 71. *Prof Exp:* Lectr physics, Univ Natal, 63-68 & 69-70; lectr, Nottingham Univ, 70-71; fel, Univ Fla, 71-73, asst prof, 73-76; res assoc, Northwestern Univ, 76-78; asst prof physics, Tex A&M Univ, 78-84; VPRES RES & DEVELOP, NALORAC CRYOGENICS CORP, 84- *Concurrent Pos:* Consult, Los Alamos Nat Lab, 82-84. *Mem:* Am Phys Soc. *Res:* Nuclear magnetic resonance as a probe of the solid state and of surfaces at low and very low temperatures and high pressures; nuclear magnetic resonance spectrometer design. *Mailing Add:* 837 Arnold Dr No 600 Martinez CA 94553

GIBSON, AUDREY JANE, b Paris, France, Oct 5, 24; m 51. MICROBIOLOGY, BIOCHEMISTRY. *Educ:* Cambridge Univ, BA, 46; Univ London, PhD(biochem), 49. *Prof Exp:* Commonwealth Fund fel microbiol, Hopkins Marine Sta, Stanford Univ, 49-50; res assoc, Agr Res Coun Unit Microbiol, Univ Sheffield, 50-51 & 53-63; asst prof microbiol & phys biochem, Johnson Res Found, Pa, 63-65; asst prof microbiol, 66-70, assoc prof, 70-79, PROF BIOCHEM, MOLECULAR & CELL BIOL, CORNELL UNIV, 79- *Concurrent Pos:* Res assoc, Univ Ill, 61. *Mem:* Soc Biol Chemists; Am Soc Microbiol. *Res:* Growth regulation in photosynthetic prokaryotes; characterization and regulation of transport processes; carbon, phosphate and nitrogen metabolism; membrane giogenesis. *Mailing Add:* Biochem Sect Div Biol Sci Cornell Univ Ithaca NY 14853

GIBSON, BENJAMIN FRANKLIN, V, b Madisonville, Tex, Sept 3, 38; m 68; c 3. NUCLEAR PHYSICS, FEW-BODY PHYSICS. *Educ:* Rice Univ, BA, 61; Stanford Univ, PhD(physics), 66. *Prof Exp:* Fel, Lawrence Radiation Lab, Univ Calif, 66-68; Nat Res Coun res assoc, Nat Bur Standards, DC, 68-70; res assoc physics, Brooklyn Col, 70-72; STAFF PHYSICIST, THEORET PHYSICS DIV, LOS ALAMOS NAT LAB, 72- *Concurrent Pos:* Detailee, Depart Energy, 80-81; vis scholar, Univ Melbourne, 86-87, Flinders Univ, 87; assoc ed, Phys Review C, 88-; vchmn, Few-Body Topical Group, Am Phys Soc, 90-91, chmn, 91-92. *Mem:* Fel Am Phys Soc; fel Jap Soc Prom Sci. *Res:* Theoretical nuclear, hypernuclear and elementary particle physics; medium energy physics; few-body physics. *Mailing Add:* MS-B283 Los Alamos Nat Lab Los Alamos NM 87545

GIBSON, CARL H, b Springfield, Ill, Sept 26, 34; div; c 2. FLUID DYNAMICS, OCEANOGRAPHY. *Educ:* Univ Wis, BS, 56, MS, 57; Stanford Univ, PhD(chem eng), 62. *Prof Exp:* Chem engr, Oak Ridge Nat Lab, 57-58; Peace Corps teacher, Osmania Univ, India, 62-64; asst res engr, 65-66, ASSOC PROF ENG PHYSICS & OCEANOG, UNIV CALIF, SAN DIEGO, 65- *Concurrent Pos:* Guggenheim fel, 73. *Mem:* Fel Am Phys Soc; Am Inst Chem Eng; Sigma Xi. *Res:* Turbulence; turbulent mixing of passive and reacting scalar properties; transport phenomena; nuclear reactor engineering; oceanography; atmospheric and oceanic diffusive phenomena. *Mailing Add:* Appl Mech B-010 Univ Calif San Diego 5246 Urey Hall La Jolla CA 92037

GIBSON, COLVIN LEE, b Detroit, Mich, Apr 12, 18; m 41; c 3. PARASITOLOGY. *Educ:* Univ Mich, AB, 40, AM, 41, PhD(zool), 51. *Prof Exp:* Parasitologist, Onchocerciasis Res Proj, Pan-Am Sanit Bur & USPHS, 48-52, parasitologist, Lab Trop Dis, Nat Inst Allergy & Infectious Dis, NIH, 52-57, spec asst to chief extramural progs br, 57-61, chief virus reagents prog, 61-62, chief res reference br, 62-63, sci commun officer, 63, chief res grants br, 63-65, chief parasitol & med entom br, 65-68, asst to dir, NIH, 68-75; CONSULT, WHO, 74- *Concurrent Pos:* Ed, Trop Med & Hyg News, 66-87. *Mem:* Am Soc Parasitol; Am Soc Trop Med & Hyg. *Res:* Filarial diseases, especially onchocerciasis; epidemiology of toxoplasmosis. *Mailing Add:* 3307 Harrell St Wheaton MD 20906-4148

GIBSON, COUNT DILLON, JR, b Covington, Ga, July 10, 21; m; m 50; c 4. MEDICINE. *Educ:* Emory Univ, BS, 42, MD, 44. *Prof Exp:* Asst med, Col Physicians & Surgeons, Columbia Univ, 50-51; from asst prof to assoc prof, Med Col Va, 51-57; prof prev med & chmn dept, Med Sch, Tufts Univ, 58-69; assoc dean, Community Health Progs, 69-80, prof community & prev med & chmn dept, 69-87, PROF HEALTH RES & POLICY, MED CTR, STANFORD UNIV, 88- *Mem:* Fel Am Col Physicians; fel Am Pub Health Asn. *Res:* Infectious diseases; medical care. *Mailing Add:* Dept Health Res & Policy Stanford Univ Med Ctr Stanford CA 94305

GIBSON, DAVID F(REDERIC), b West Newton, Mass, Jan 10, 42; m 63; c 2. INDUSTRIAL ENGINEERING. *Educ:* Purdue Univ, BSIE, 63, MSIE, 64, PhD(indust eng), 69. *Prof Exp:* Indust engr, Naval Ord Plant, Ill, 63; res asst, Purdue Univ, 63-64; proj indust engr & chief methods & stand br, Sacramento Army Depot, 65-66; instr indust eng, Purdue Univ, 68-69; asst prof indust & mgr eng, Mont State Univ, 69-71; dean, Sch Syst Sci, Ark Tech Univ, 71-72; from assoc prof to prof indust eng, 72-76, asst dean, 77-82, DEAN, COL ENG, MONT STATE UNIV, 83- *Concurrent Pos:* Asst dir eng exp sta, Mont State Univ, 77-83. *Mem:* Inst Indust Engrs; Am Soc Eng Educ; Nat Soc Prof Eng; Nat Coun Examrs Eng & Surveyors. *Res:* Management systems analysis and design; computerized planning systems; surface coal mining operations; forest engineering. *Mailing Add:* Col Eng Mont State Univ Bozeman MT 59717

GIBSON, DAVID MARK, b Kokomo, Ind, Aug 7, 23; m 51; c 5. BIOCHEMISTRY. *Educ:* Wabash Col, AB, 44; Harvard Univ, MD, 48. *Prof Exp:* Asst prof, Enzyme Inst, Univ Wis, 55-58; assoc prof, 58-61, actg chmn dept biochem, 65-67, prof, 61-75, chmn dept, 67-88 GRACE M SHOWALTER PROF BIOCHEM, SCH MED, IND UNIV, 75- *Concurrent Pos:* Res fel biochem, Univ Ill, 50-53 & Enzyme Inst, Univ Wis, 53-58; estab investr, Am Heart Asn, 57-62; Career Develop Award, NIH, 62-67. *Mem:* AAAS; Am Soc Cell Biol; Am Soc Biol Chemists; Biochem Soc Gt Brit; Am Heart Asn. *Res:* Control of fatty acid and cholesterol biosynthesis; regulation of hydroxymethylglutaryl co-enzyme A reductase by reversibile phosphorylation. *Mailing Add:* Dept Biochem Sch Med Ind Univ Indianapolis IN 46202-5122

GIBSON, DAVID MICHAEL, b Joplin, Mo, Sept 1, 45; m 75; c 2. ORGANIC CHEMISTRY. *Educ:* Southwest Mo Univ, BS, 67; Univ New Orleans, PhD(org chem), 75. *Prof Exp:* Res assoc, Int Paper Co, 75-79; sr chemist, 79-86, staff chemist, 86-99, STAFF QUAL CONTROL CHEMIST, EXXON CHEM AMERICAS, 89- *Mem:* Am Soc Testing & Mat. *Res:* Analytical techniques for analysis of light hydrocarbons and intermediate alcohols. *Mailing Add:* Exxon Chem Americas Chem Plant Lab PO Box 241 Baton Rouge LA 70821

GIBSON, DAVID THOMAS, b Wakefield, Eng, Feb 16, 38; m 63; c 2. BIOCHEMISTRY, MICROBIOLOGY. *Educ:* Univ Leeds, BSc, 61, PhD(biochem), 64. *Prof Exp:* From asst prof to prof microbiol, Univ Tex, Austin, 69-88, PROF BIOCATALYSIS & MICROBIOL, UNIV IOWA, 88- *Concurrent Pos:* Res grants, NIH, 70-, ENVIRON PROTECTION AGENCY, 86-; NIH grant, 71; consult, Gen Elec, 75-; assoc ed, Soc Indust Microbiol, 76-79; study sect mem microbial chem, NIH, 77-81. *Honors & Awards:* Minnie F Piper Stevens Award, 83. *Mem:* Am Soc Microbiol; Am Chem Soc; AAAS; Soc Indust Microbiol; Sigma Xi. *Res:* Mechanisms used by micro-organisms to oxidize polycyclic aromatic hydrocarbons; mechanisms of enzymatic oxygen fixation. *Mailing Add:* Dept Microbiol Univ Iowa Iowa City IA 52240

GIBSON, DOROTHY HINDS, b Italy, Tex, July 19, 33. ORGANIC CHEMISTRY, ORGANOMETALLIC CHEMISTRY. *Educ:* Tex Christian Univ, BA, 54, MA, 56; Univ Tex, Austin, PhD(chem), 65. *Prof Exp:* Instr chem, Tex Christian Univ, 56-61; res assoc, Univ Tex, Austin, 64-65 & Univ Colo, 65-69; from asst prof to assoc prof, 69-75, PROF CHEM, UNIV LOUISVILLE, 75- *Concurrent Pos:* Vchair, chem, Univ Louisville, 78-87, actg chair, 82-83. *Mem:* Sigma Xi; Am Chem Soc. *Res:* Synthesis and properties of transition metal carbonyl complexes; reactive organometallic intermediates; transition metal hydrides as reducing agents. *Mailing Add:* Dept Chem Univ Louisville Louisville KY 40292

GIBSON, EARL DOYLE, b Putnam, Okla, July 20, 23; m 57; c 4. COMMUNICATIONS ENGINEERING. *Educ:* Okla Inst Technol, BSEE, 49; Univ Md, MSEE, 60. *Prof Exp:* Engr, Gen Elec Co, 49-51 & US Naval Ord Lab, Md, 51-56; chief analysis sect, ACF Industs, Inc, 56-63; sr staff engr,

Aerospace Corp, Calif, 63-66; SR STAFF SCIENTIST, ELECTRONICS RES CTR, ROCKWELL INT, 66- *Mem:* Inst Elec & Electronics Engrs. *Res:* Data communications; adaptive equalization; adaptive receivers. *Mailing Add:* 171 Sudbury Dr Milpitas CA 95035

GIBSON, EDWARD F, b Colorado Springs, Colo, Apr 2, 37; m 63; c 4. NUCLEAR PHYSICS. *Educ:* Univ Colo, Boulder, BA, 59, MA, 64, PhD(physics), 66. *Prof Exp:* Physicist, Cryogenic Eng Div, Nat Bur Standards, 58-64; res asst nuclear physics, Univ Colo, 64-66, res assoc, 66; res assoc, Univ Ore, 66-68, scientist in residence, US Naval Radiol Defense Lab, 68-69; from asst prof to assoc prof, 69-78, chmn dept, 79-88, PROF PHYSICS, CALIF STATE UNIV, SACRAMENTO, 78- *Concurrent Pos:* Consult, Calif Energy Comn, 77-79 & Control Data Corp, 81-86; vis prof, Univ Colo, 80. *Mem:* Am Phys Soc; Sigma Xi. *Res:* Nuclear spectroscopy; nuclear reactions; elastic and inelastic 3He scattering; gamma-ray spectroscopy; nuclear life times; low temperature electrical and thermal conductivity; meson scattering; intermediate energy meson physics; computer assisted instruction. *Mailing Add:* Dept Physics Calif State Univ 6000 J St Sacramento CA 95819

GIBSON, EDWARD GEORGE, b Buffalo, NY, Nov 8, 36; m 59; c 2. ATMOSPHERIC PHYSICS, ENGINEERING. *Educ:* Univ Rochester, BS, 59; Calif Inst Technol, MS, 60, PhD(eng & physics), 64. *Prof Exp:* With Aeronutronic Res Lab, Philco Corp, Calif, 64-65; scientist-astronaut, NASA, 65-75 & Skylab Prog, 72-75; mem staff, Aerospace Corp, 75-76; consult, ERNO Raumfahrttechnik, Bremen, W Ger, 76-77; ASTRONAUT, NASA-JOHNSON SPACE CTR, 77- *Concurrent Pos:* Mem solar physics subcomt, NASA. *Mem:* Am Astron Soc; Am Inst Physics; Am Inst Aeronaut & Astronaut. *Mailing Add:* 3809 Dina Terrace No 9 Cincinnati OH 45211-6558

GIBSON, EVERETT KAY, JR, b Seagraves, Tex, May 13, 40; m 73; c 1. GEOCHEMISTRY, ANALYTICAL CHEMISTRY. *Educ:* Tex Tech Univ, Bs, 63, MS, 65; Ariz State Univ, PhD(Geochem), 69. *Prof Exp:* Teaching asst chem, Tex Tech Univ, 63-65, instr, 65; aerospace technician, 70-74, PLANETARY SCIENTIST SPACE SCI, JOHNSON SPACE CTR, NASA, 74- *Concurrent Pos:* Res assoc, Nat Acad Sci-Nat Res Coun, Johnson Space Ctr, NASA, 69-70, lunar sample prelim exam team, 69-73, lunar sample analysis & planning team, 74-77; consult, Economist, London, 70-75 & Brit Broadcasting Corp, London, 75-80; lunar sample prin investr, NASA, Washington, DC, 71-90; dir, Clear Creek Basin Authority, Harris County, Tex, 74-76; assoc ed, Proc 5th to 9th & 12th Lunar & Planetary Sci Conf, 74-78; adj prof geol, Univ Houston, 74-90, lectr, Sch Continuing Educ, 76-; prin investr planetary geol, 78-88 & NASA Planetary Biol Prog, 83-; Leverhulme sr fel, Open Univ, Eng, 84-85. *Honors & Awards:* Sustained Performance Award, Johnson Space Ctr, NASA, 71, 72, & 76; Outstanding Lectr Award, Am Astron Soc, 76. *Mem:* Meteoritical Soc (secy, 74-80); Am Chem Soc; AAAS; Int Asn Geochem & Cosmochem; Sigma Xi. *Res:* Geochemistry of meteorites and lunar samples; nature of volatiles in terrestrial and extraterrestrial materials; development of analytical methods of analysis for volatile elements; Martian geochemical processes; nature of Archean atmosphere and ocean. *Mailing Add:* SN2 Planetary Sci Br Johnson Space Ctr NASA Houston TX 77058

GIBSON, FLASH, b Spokane, Wash, Apr 3, 44; m 63; c 2. AQUATIC ENTOMOLOGY. *Educ:* Eastern Wash Univ, BA, 66; Ore State Univ, MA, 69, PhD(zool), 71. *Prof Exp:* ASST PROF BIOL, EASTERN WASH UNIV, 71- *Mem:* Ecol Soc Am; Am Soc Zoologists; Northwest Sci Asn; Sigma Xi. *Res:* Ecological physiology of aquatic invertebrates of northwestern United States wetlands. *Mailing Add:* Dept Biol Eastern Wash Univ Cheney WA 99004

GIBSON, GARY EUGENE, b Greeley, Colo, Oct 3, 45; m 73. BIOCHEMISTRY. *Educ:* Univ Wyo, BS, 68; Cornell Univ, PhD(physiol), 73. *Prof Exp:* NIH trainee, 73-76, asst res biochemist, Neuropsychiat Inst, Univ Calif, Los Angeles, 76-78; from asst prof to assoc prof biochem & neurol, 78-90, PROF NEUROSCI, CORNELL MED SCH, 90- *Honors & Awards:* Jordi Folch-Pi Award, Am Soc Neurochem, 82. *Mem:* Sigma Xi; Am Soc Neurochem; AAAS; Soc Neurosci; Int Soc Neurochem; Fedn Am Soc Exp Biol & Med. *Res:* Examination of interactions in brain and other tissues of calcum, neurotransmitters and oxidative metabolism during aging, vitamin deficiencies and Alzheimer's disease; vitamin deficiencies and Alzheimer's disease. *Mailing Add:* 60 Fernwood Rd Larchmont NY 10538

GIBSON, GEORGE, b Windsor, Ont, July 11, 31; m 63; c 2. WILDLIFE PATHOLOGY. *Educ:* Univ Toronto, BA, 55, MA, 57; Univ BC, PhD(zool), 65. *Prof Exp:* Asst prof biol, Notre Dame Univ, BC, 65-66; res scientist, Can Wildlife Serv, 66-84; CUR PARASITES, CAN MUS NATURE, 85- *Mem:* Am Soc Parasitol; Wildlife Dis Asn; Can Soc Zool. *Res:* Surveys, identification, epizootiology of helminths of wildlife, principally migratory birds. *Mailing Add:* Can Mus Nature Box 3443 Sta D Ottawa ON K1P 6P4 Can

GIBSON, GEORGE, b Yonkers, NY, Mar 20, 09; m 40; c 2. CHEMISTRY. *Educ:* Polytech Inst Brooklyn, BS, 32, MS, 35, PhD(inorg chem), 42. *Prof Exp:* Chemist, Chemco Photo Prod, Inc, NY, 32-33; asst chem, Polytech Inst Brooklyn, 32-42; from asst prof to prof chem, Ill Inst Technol, 42-60, actg chmn dept, 53; prof, 60-74, chmn dept, 62-72, EMER PROF CHEM, BROOKLYN COL, CITY UNIV, 74- *Concurrent Pos:* Teacher, Pub Sch, NY, 35-36; analytical chemist, Charles Pfizer Co, NY, 36-37; res chemist, E I Du Pont de Nemours & Co, NJ, 37-42; mem bd trustees, Cape Cod Mus Natural Hist, 78-81. *Mem:* Am Chem Soc; Sigma Xi. *Res:* Chemistry of uranium; nonaqueous solvents; chemistry of hydrazine; coordination compounds. *Mailing Add:* 284 Scudder Rd PO Box 573B Osterville MA 02655

GIBSON, GEORGE R, b Kendaia, NY, Oct 2, 05. EXPLORATION GEOLOGY, PETROLEUM GEOLOGY. *Educ:* Univ Minn, BA, 30, PhD(geol), 34. *Prof Exp:* Instr geol, Univ Minn, 33-34, Ohio State Univ, 40-41; assoc prof geol, Carleton Col, 34-38; geologist, Socony-Vacuum Co, 38-40, Magnalia Oil Co, 41-43; dist mgr, Richfield Oil, 43-48; mgr explor, Seaboard Petrol, 48-52; GEOL CONSULT, 52- *Mem:* Sigma Xi; Earth Scientist; fel Geol Soc Am; Soc Independent Professional Earth Scientists; Am Asn Petrol Geologists; Soc Econ Paleontologists & Mineralogists. *Mailing Add:* PO Box 2296 Midland TX 79702

GIBSON, GERALD W, b Saluda Co, SC, Oct 27, 37; m 68; c 3. ORGANIC CHEMISTRY. *Educ:* Wofford Col, BS, 59; Univ Tenn, PhD(chem), 63. *Prof Exp:* From asst prof to prof chem, Col Charleston, 65-84, chmn dept, 68-82; DEAN, ROANOKE COL, 84- *Mem:* Am Chem Soc. *Res:* Organolithium compounds; general and organic chemistry problem-solving; Wiswesser line notation; reactions of organolithium compounds with alkyl halides. *Mailing Add:* Roanoke Col Salem VA 24153

GIBSON, GORDON, b McKeesport, Pa, Jan 9, 26; m 55; c 4. THEORETICAL PHYSICS. *Educ:* Univ Pittsburgh, BS, 49, MS, 52, PhD(physics), 55. *Prof Exp:* Instr physics & res asst, Univ Pittsburgh, 50-55; sr scientist, Atomic Power Div, Westinghouse Elec Corp, 55-64, fel scientist, Astronuclear Lab, 64-67, adv scientist, 67-74, mgr plasma & nuclear eng dept, 74-84, consult scientist, Advan Energy Systs Div, 84-86, CONSULT SCIENTIST, SOURCE & TECH CTR, WESTINGHOUSE ELEC CORP, 86- *Concurrent Pos:* Consult, Lawrence Radiation Lab, Univ Calif, 55-64. *Mem:* Am Phys Soc. *Res:* Neutron cross sections; nuclear rocket reactor; plasma and reactor physics; controlled fusion. *Mailing Add:* 120 Lamar Rd Pittsburgh PA 15241

GIBSON, HAROLD F(LOYD), b Retrop, Okla, Feb 24, 21; m 41; c 2. MILITARY OPTICAL SYSTEMS, LASER USING WEAPONS. *Educ:* Univ Okla, BS, 43, MS, 48. *Prof Exp:* Physicist radiation lab, Nat Bur Standards, 48-53; physicist, Harry Diamond Labs, 53-56, supvr physicist, 56-62, supvr res & develop, 62-69, chief, Appl Physics Br, 69-76, chief, Components & Mat Lab, 76-80; RETIRED. *Concurrent Pos:* Consult, 80- *Mem:* AAAS; Am Phys Soc. *Res:* Solid state physics; energy sources and conversion techniques; laser materials; applied infrared and laser technology. *Mailing Add:* 10705 Jamaica Dr Silver Spring MD 20902

GIBSON, HAROLD J(AMES), b Wixom, Mich, June 18, 09; m 32; c 3. MECHANICAL ENGINEERING. *Educ:* Univ Mich, BS, 29, MS, 30. *Prof Exp:* Res engr, Ethyl Corp, 30-44, res coordr, 44-47, res supvr, 47-54, dir prod appln, 54-57, tech dir res labs, 57-63, mgr labs, 63-76; RETIRED. *Mem:* AAAS; fel Soc Automotive Engrs. *Res:* Mechanical and chemical research on automotive emissions and petroleum fuels, lubricants and additives. *Mailing Add:* 28505 Inkster Rd Farmington Hills MI 48334-5258

GIBSON, HARRY WILLIAM, b Syracuse, NY, May 2, 41; m 62; c 3. PHYSICAL ORGANIC CHEMISTRY. *Educ:* Clarkson Col Technol, BS, 62, PhD(chem), 66. *Prof Exp:* Investr chem with Prof Ernest L Eliel, Univ Notre Dame, 65-66; res chemist, Res & Develop Lab, Chem Div, Union Carbide Corp, 66-69; scientist, Res Labs, Xerox Corp, 69-74, sr scientist, 74-82, sr mem res staff, 82-84, sr res scientist & dir allied signal, Eng Mat Res Ctr, 84-86; PROF DEPT CHEM, VA POLYTECH INST & STATE UNIV, 86- *Concurrent Pos:* Jones Laughlin scholar, 58-62; fel Nat Defense Educ Act, 62-65; vis lectr, Dept Chem, Univ Ill, 78; mem joint educ comt, Div Polymer Chem & Org Coatings & Plastics, Am Chem Soc, 79-84; co-ed, Polymer Educ Newsletter, Am Chem Soc, 81-83; adj prof, Dept Chem, Univ Rochester, 82-84; chmn, Rochester Sect, Am Chem Soc, 84; vis lectr, dept chem, Univ Ottawa, 85; participant, Mat Sci eng study, Nat Res Coun, Nat Acad Sci, 85. *Mem:* Am Chem Soc; Sigma Xi; Chem Inst Can. *Res:* Chemistry of Reissert compounds; liquid crystals; chemical modification of polymers; electrical properties of organics; electrically conductive polymers; polymer synthesis and characterization; asymmetric synthesis; novel polymer architectures. *Mailing Add:* 4789 Susannah Dr Blacksburg VA 24060-6939

GIBSON, HENRY CLAY, JR, b Philadelphia, Pa, Mar 15, 22; m 52; c 3. ATOMIC PHYSICS. *Educ:* Princeton Univ, AB, 50. *Prof Exp:* Pres, Radiation Res Corp, Fla, 50-56; PRES, FRANKLIN GNO CORP, 56-, PUB RELS OFFICER, 74- *Concurrent Pos:* Dir, John Dusenbery Co, NJ, 58- *Mem:* AAAS; Am Inst Physics; Am Phys Soc; Inst Elec & Electronics Engr. *Res:* Plasma chromatograph, a gaseous electrophoresis instrument for rapid chemical analysis. *Mailing Add:* 400 S Ocean Blvd Palm Beach FL 33480

GIBSON, J MURRAY, b Forres, Scotland, Mar 8, 54; m 80. ELECTRON MICROSCOPY, SEMICONDUCTOR PHYSICS. *Educ:* Univ Aberdsen, Scotland, BSc, 75; Univ Cambridge, Eng, PhD(physics), 78. *Prof Exp:* Res fel, Dept Physics, Univ Cambridge, 78 & Res Div, IBM, 78-80; mem tech staff, Iterface Physics Dept, 80-87, DISTINGUISHED MEM TECH STAFF & HEAD ELECTRONICS & PHOTONICS MAT RES DEPT, BELL LABS, 87- *Honors & Awards:* Burton Medal, Electron Micros Soc Am. *Mem:* Am Phys Soc; Inst Physics; Electron Micros Soc Am. *Res:* Ultra-high resolution electron microscopy applied to problems in solid-state physics and materials science. *Mailing Add:* Bell Labs 1E 234 600 Mountain Ave Murray Hill NJ 07974

GIBSON, JAMES (BENJAMIN), b Ellensburg, Wash, June 9, 28; m 68; c 1. ASTROMETRY, PHOTOMETRY. *Educ:* Univ Calif, Berkeley, AB, 52. *Prof Exp:* Physicist, Lawrence Livermore Lab, Univ Calif, 53-57; math analyst, Missiles & Space Div, Lockheed Aircraft Corp, 57-58; asst, Lick Observ, Univ Calif, 58-60; astronr, Flagstaff Sta, US Naval Observ, Ariz, 60-61; res asst astron, Van Vleck Observ, Wesleyan Univ, 61-63, res assoc, 64-67; res asst, Observ, Yale Univ, 68-71; prin observer, Yale-Columbia Southern Observ, 71-74; tech officer astron, Univ of the Orange Free State, 74-76; consult, 76-78; mem tech staff, Jet Propulsion Lab, 78-86, optical astro engr, ITT, Fed Elec Corp & Jet Propulsion Lab, 86-88, OPTICAL ASTRO ENGR, OAO CORP & JET PROPULSION LAB, 88- *Concurrent Pos:* Res assoc, Inst

Advan Study, 66-67; NASA res grant for eclipse exped, 66-67; vis astronr, Kitt Peak Nat Observ, 66-67; guest investr, Palomar Observ, 81- *Mem:* Fel AAAS; Am Astron Soc; fel Royal Astron Soc; Royal Astron Soc Can; Sigma Xi; Int Astron Union. *Res:* Observational astronomy; astrometry; comet and minor planet positions from photographic plates and charge coupled device frames; intermediate bandwidth photoelectric photometry; systematic errors of trigonometric parallaxes; spectroscopy and spectral classification; airborne optical instrumentation; variable stars in galactic and globular clusters; flare stars. *Mailing Add:* 6838 Greeley St Tujunga CA 91042

GIBSON, JAMES DARRELL, b South Bend, Ind, Apr 8, 34; m; c 3. MECHANICAL ENGINEERING. *Educ:* Purdue Univ, BS, 57, MS, 59; Univ NMex, PhD, 68. *Prof Exp:* Sr engr, Gen Dynamics Corp, 59-63; instr mech eng, Univ NMex, 63-68; assoc prof, Univ Wyo, 68-72; dir grad studies, 79-87, PROF MECH ENG, ROSE-HULMAN INST TECHNOL, 72- *Concurrent Pos:* NASA-Am Soc Eng Educ fel, Langley Res Ctr, 69, proj dir systems design fel prog, 71, 72 & 73. *Mem:* Am Soc Eng Educ; Inst Noise Control Eng; Am Soc Mech Engrs. *Res:* Vibrations and noise; mechanical design. *Mailing Add:* Rose-Hulman Inst Technol 5500 Wabash Ave Terre Haute IN 47803

GIBSON, JAMES DONALD, organic chemistry; deceased, see previous edition for last biography

GIBSON, JAMES EDWIN, b Des Moines, Iowa, Aug 22, 41; m 61; c 3. PHARMACOLOGY, TOXICOLOGY. *Educ:* Drake Univ, BA, 64; Univ Iowa, MS, 67, PhD(pharmacol), 69. *Prof Exp:* Chemist, Chem Labs, Iowa State Dept Agr, 62-64; jr res scientist pharmacol, Neuropharmacol Div, Abbott Labs, Ill, 64-65; asst prof, Mich State Univ, 69-72, assoc prof pharmacol, 72-76; VPRES & DIR RES, CHEM INDUST INST TOXICOL, 76- *Concurrent Pos:* Vis prof pharmacol, Univ Mainz, 75-76; Alexander von Humboldt US Sr Scientist Award, 75-76. *Mem:* AAAS; Soc Develop Biol; Am Soc Pharmacol & Exp Therapeut; Soc Toxicol; Teratology Soc. *Res:* Determination and description of drug or chemical induced teratogenicity, embryotoxicity and perinatal toxicity in mammals with particular emphasis on molecular mechanisms of action; teratology. *Mailing Add:* Chem Indust Inst Toxicol PO Box 12137 Research Triangle Park NC 27709

GIBSON, JAMES H, b Morgantown, WVa, May 3, 30; m; c 2. ENVIRONMENTAL SCIENCES. *Educ:* WVa Univ, BS, 52; Cornell Univ, PhD(anal chem), 57. *Prof Exp:* Chemist, Eastman Kodak Co, NY, 57-60; fel & res assoc, Cornell Univ, 60-61; chief chem div, US Army Chem Corps, Dugway Proving Grounds, Utah, 61-63; assoc prof anal chem, 63-80, dir, natural resource ecol lab, 73-83, COORDR, NAT ATMOSPHERIC DEPOSITION PROG, COLO STATE UNIV, 78- *Concurrent Pos:* Consult, US Army Chem Corps, Dugway Proving Ground, 63-65; NSF grants, 64-70 & 74-78; dir admin & serv, Natural Resource Ecol Lab, Colo State Univ, 71-73; dir grassland biome studies, US/Int Biol Prog, 74-78; chmn US grazinglands comt & mem US nat comt, Man & Biosphere Prog, UNESCO, 75-76 & 79-, mem comt 14, environ pollution, 82-; chmn, comt on monitoring and assessment of trends in acid deposition, Nat Res Coun, 83-85; prin investr, atmospheric deposition, Bur Land Mgt, 79-, USDA, 79- Nat Park Serv, 79-, Nat Oceanog & Atmospheric Admin, 81-, Environ Protection Agency, 83-, US Geol Survey, 83-, US Forest Serv, 85- *Mem:* AAAS; Am Chem Soc. *Res:* Measurements and monitoring of atmospheric deposition; effects in natural ecosystems; 00039538xadministration in areas of systems ecology and environmental sciences. *Mailing Add:* Dept Mining Southwest Va Community Col Box SVCC Richlands VA 24641

GIBSON, JAMES JOHN, b St Albans, Eng, Mar 12, 23; US citizen; m 57; c 3. ELECTRONICS ENGINEERING. *Educ:* Royal Inst Technol, Sweden, CivIng, 47; Chalmers Inst Technol, Sweden, TekLic, 71. *Hon Degrees:* DSc, Chalmers Inst Technol, Sweden, 85. *Prof Exp:* Res engr electronics, Res Inst Nat Defense, Sweden, 46-52 & RCA Labs, 52-54; group head, Royal Inst Technol, Sweden, 54-56; mem tech staff, RCA LABS, 56-59, fel tech staff electronics, 69-87; CONSULT, SIGNAL SYST RES, 87- *Concurrent Pos:* Fac mem, LaSalle Col, 71-73; mem, Cable Television Adv Comt, 72-75, Nat Quadrophonic Radio Comt, 72-75 & Digital Audio Stand Comt, 78-81; vis prof elec eng, Rutgers Univ, 78-; mem, Fed Commun Comn Adv Comt, tv receiver noise figures, 79-80; mem, Electronic Indust Asn Comt, tv multichannel sound, 82-85; mem admin comt, Inst Elec & Electronics Engrs, Consumer Electronics Soc, 87-89; chmn, Inst Elec & Electronics Engrs Masarulbuka Award subcomt, 87-89. *Honors & Awards:* David Sarnoff Award, RCA Labs, 85. *Mem:* Fel Inst Elec & Electronics Engrs; fel Audio Eng Soc; Soc Motion Picture & Television Engrs; AAAS; Sigma Xi. *Res:* Consumer electronic systems; solid state circuits; communication systems television and frequency modulation broadcast systems; antennas; computer memories. *Mailing Add:* Signal Syst Res PO Box 7636 Princeton NJ 08543-7636

GIBSON, JOHN E(GAN), b Providence, RI, June 11, 26; m 50; c 4. ENGINEERING MANAGEMENT. *Educ:* Univ RI, BS, 50; Yale Univ, ME, 52, PhD(elec eng), 56. *Prof Exp:* From instr to asst prof elec eng, Yale Univ, 52-57; from assoc prof to prof, Purdue Univ, 57-65, dir, Control & Info Syst Labs, 61-65; dean eng, Oakland Univ, 65-73, John Dodge Prof eng, 70-73; COMMONWEALTH PROF & DEAN ENG & APPL SCI, UNIV VA, 73- *Concurrent Pos:* Consult, various aerospace & electronics firms & Dept Com, 66-67; adv, Electronics Res Ctr, NASA, Mass, 66-70 & NSF, 81. *Mem:* Fel Inst Elec & Electronics Engrs; Am Soc Elec Engrs; Nat Soc Prof Engrs. *Res:* Nonlinear automatic control; large scale systems; transportation systems; urban system studies, long range planning and management of research and development. *Mailing Add:* Systs Engr-AM Bldg 117E Univ Va Charlottesville VA 22901

GIBSON, JOHN KNIGHT, Springfield, Mass, Oct 22, 57; m 82; c 1. HIGH TEMPERATURE CHEMISTRY, SOLID STATE CHEMISTRY. *Educ:* Boston Univ, BA, 79; Univ Calif, Berkeley, PhD(phys chem), 83. *Prof Exp:* Res asst, Lawrence Berkeley Lab, 79-83; RES STAFF, OAK RIDGE NAT LAB, 83- *Mem:* Am Chem Soc. *Res:* Thermochemistry; solid state chemistry; high temperature chemistry; high temperature vaporization; physicochemical properties of actinide and lanthanide elements and compounds; transplutonium element chemistry. *Mailing Add:* Chem Div Oak Ridge Nat Lab Oak Ridge TN 37831-6375

GIBSON, JOHN MICHAEL, b Franklin, Ky, Aug 11, 40; m 64. NEUROPHYSIOLOGY, BIOENGINEERING. *Educ:* Univ Ky, BSEE, 63, PhD(physiol & biophys), 70. *Prof Exp:* Fel neurophysiol, Univ Wis-Madison, 70-73, asst scientist, 73-80, assoc scientist neurophysiol, 80-84, asst prof biometry, 84-88; asst prof otolaryngology, Med Univ SC, 88-90; INSTR, ENG TECHNOL, SAVANNAH TECH, 91- *Mem:* Soc Neurosci; Inst Elec & Electronics Engrs; Inst Elec & Electronics Engrs Comput Soc; Inst Elec & Electronics Engrs Med & Biol Soc; Int Neural Network Soc; Asn Comput Soc. *Res:* Computer simulation studies of artificial neural networks and their enhancement through simulated evolution and selection; application of engineering and computer techniques to the biomedical sciences. *Mailing Add:* Dept Electronics Eng Technol Savannah Tech 5717 White Bluff Rd Savannah GA 31499

GIBSON, JOHN PHILLIPS, b Pittsburg, Kans, Sept 18, 30; m 53; c 2. VETINARY MEDICINE. *Educ:* Kans State Univ, BS, 53, DVM & MS, 59; Ohio State Univ, PhD(vet path), 64. *Prof Exp:* Instr vet bact, Purdue Univ, 59-60; res assoc vet path, Ohio State Univ, 60-64; pathologist toxicol, 64-70, head Dept Path & Toxicol, 70-90, DIR DRUG SAFETY ASSESSMENT, MERRELL-DOW PHARMACEUT, INC, 90- *Concurrent Pos:* Adj asst prof, Col Med, Univ Cincinnati, 65-, adj assoc prof Lab Animal Med, 72- *Mem:* Am Col Vet Path; Soc Toxicol; Int Acad Path; Soc Toxicol Pathologists; Teratol Soc; Am Vet Med Asn. *Res:* Eperythrozoonsis in swine; canine ascariasis; canine distemper; gnotobiotic dog; toxicity testing of therapeutic agents. *Mailing Add:* Drug Safety Assessment Marion Merrell Dow Inc 2110 E Galbraith Rd Cincinnati OH 45215

GIBSON, JOSEPH W(HITTON), JR, b Norristown, Pa, Feb 24, 22; m 46; c 3. CHEMICAL ENGINEERING. *Educ:* Worcester Polytech Inst, BS, 44. *Prof Exp:* Res chem engr fiber & dyeing, E I Du Pont De Nemours & Co, 46-51, res chem engr clothing comfort, 51-52, 53-57, res chem engr leather-like prods, 52-53, res chem engr end use textiles, 57-67, sr res engr textile res, 67-79, sr tech specialist, Imaging Systs Mkt, 79-90; RETIRED. *Concurrent Pos:* Postgrad, Princeton Univ & Mass Inst Technol, 44-45. *Honors & Awards:* Olney Medal, Am Asn Textile Chemists & Colorists, 79. *Mem:* Am Chem Soc; hon mem Fiber Soc; Sigma Xi; Am Asn Textile Chemists & Colorists; Int Platform Asn. *Res:* Technical evaluations of photopolymer printing plates; preparation and evaluation of end use textiles; clothing comfort with human subjects; dyeing of synthetic fibers. *Mailing Add:* 1215 Hillside Blvd Carrcroft Wilmington DE 19803

GIBSON, JOSEPH WOODWARD, water chemistry, corrosion; deceased, see previous edition for last biography

GIBSON, JOYCE C, lipoprotein metabolism, for more information see previous edition

GIBSON, JOYCE CORREY, b Jan 5, 48; m; c 2. BIOCHEMISTRY. *Educ:* Mt Holyoke Col, AB, 70; Harvard Univ, MSc, 72, DSc(nutrit), 74. *Prof Exp:* Res asst, Dept Nutrit, Harvard Univ, 70, Dept Cell Cult, Strangeways Lab, Eng, 71; NIH postdoctoral fel, Dept Nutrit, Cornell Univ, 74-76; sci officer, Med Prof Unit, St Vincent's Hosp, Univ New South Wales, 76-79; asst prof med, Div Arteriosclerosis & Metab, Mt Sinai Sch Med, 79-83; res assoc prof, Dept Med, Sch Med, Univ Miami, 83-86; MGR ATHEROSCLEROSIS BIOCHEM, CIBA-GEIGY CORP, NJ, 86- *Mem:* Assoc mem Sigma Xi; Am Heart Asn; AAAS; Am Inst Nutrit; Am Inst Clin Nutrit. *Res:* Author of numerous scientific journal articles. *Mailing Add:* Dept Atherosclerosis-Biochem Ciba-Geigy Corp 556 Morris Ave Summit NJ 07901

GIBSON, KATHLEEN RITA, b Philadelphia, Pa, Oct 9, 42. PHYSICAL ANTHROPOLOGY, NEUROANATOMY. *Educ:* Univ Mich, Ann Arbor, BA, 63; Univ Calif, Berkeley, PhD(anthrop), 70. *Prof Exp:* From asst prof to assoc prof, 70-80, PROF ANAT, DENT BR, UNIV TEX HEALTH SCI CTR, HOUSTON, 80- *Concurrent Pos:* Mem, Grad Sch Biomed Sci, Univ Tex, Houston, 71-; lectr anthrop, Rice Univ, 72-76, adj assoc prof, 81- *Mem:* AAAS; Am Anthrop Asn; Am Soc Primatologists; Am Asn Phys Anthrop; Am Anat Asn; Int Primatological Asn; Am Asn Dent Schs. *Res:* Comparative neurology; primate behavior; brain maturation; evolution of language, intelligence and tool use; dental evolution. *Mailing Add:* Dept Anat Univ Tex Health Sci Ctr Houston TX 77225

GIBSON, KENNETH DAVID, b Kuala Lumpur, Malaya, May 9, 26; m 52; c 3. BIOCHEMISTRY. *Educ:* Cambridge Univ, MA, 50; Univ London, PhD(biochem), 56. *Prof Exp:* Leverhulme res fel, Univ London, 56-59; vis scientist, NIH, 59-60; lectr chem path, St Mary's Hosp, London, Eng, 60-65; vis prof chem, Cornell Univ, 65-68; asst mem physiol chem, 68-71, ASSOC MEM PHYSIOL CHEM, ROCHE INST MOLECULAR BIOL, 71- *Mem:* AAAS; Am Soc Biol Chemists. *Res:* Biosynthesis of porphyrins and chlorophyll; biosynthesis of choline; theoretical studies of polypeptide structure; biogenesis of bacterial membranes; mechanism of action of growth-promoting hormones. *Mailing Add:* Dept Chem Cornell Univ Baker Lab Ithaca NY 14850

GIBSON, LEE B, b Chicago, Ill, Mar 3, 26; m 53; c 1. MICROPALEONTOLOGY. *Educ:* Wash Univ, BA, 49, MA, 52; Univ Okla, PhD(geol, paleobot), 61. *Prof Exp:* Instr geol, Univ NH, 53-54; paleontologist, Creole Petrol Corp, Venezuela, 54-57, sr paleontologist, 57-59; sr res geologist, Mobil Oil Corp, 61-70, res assoc, Field Res Lab, 70-72, chief stratigrapher, Mobil Explor & Producing, 72-84; CONSULT, 85- *Concurrent Pos:* Consult, 85- *Mem:* Am Asn Petrol Geol; Sigma Xi. *Res:* Paleoecology; application of animal population characteristics to paleoenvironmental interpretation in conjunction with the development of new biostatigraphic methodology; regional and applied biostratigraphy; eustatics. *Mailing Add:* 10215 Epping Dallas TX 75229

GIBSON, LUTHER RALPH, applied physics, space physics; deceased, see previous edition for last biography

GIBSON, MARY MORTON, b Bardstown, Ky, Nov 20, 39; m 64. NEUROPHYSIOLOGY. *Educ:* Univ Ky, BSME, 63, PhD(physiol & biophys), 70. *Prof Exp:* Fel, Univ Wis-Madison, 70-73, ASST SCIENTIST RES AUDITORY & VISUAL NEUROPHYSIOL, DEPT NEUROPHYSIOL & OPHTHAL, 73- *Mem:* AAAS; Acoust Soc Am; Soc Neurosci. *Res:* Single unit and intracranial and extracranial evoked potential studies of the auditory and visual pathways of animals. *Mailing Add:* 1109 Hartsbluff Rd Wadmalaw Island SC 29487

GIBSON, MAURICE HENRY LINDSAY, anatomy, for more information see previous edition

GIBSON, MELVIN ROY, b St Paul, Nebr, June 11, 20. PHARMACOGNOSY. *Educ:* Univ Nebr, BS, 42, MS, 47; Univ Ill, PhD(pharmacog), 49;. *Hon Degrees:* DSc, Univ Nebr, 85. *Prof Exp:* From asst prof to prof, 49-85, EMER PROF, PHARMACOG, WASH STATE UNIV, 85. *Concurrent Pos:* Ed, Am J Pharmaceut Educ, 56-61; sr vis fel sci, Orgn Econ Coop & Develop, Neth & Sweden, 62; pres, Am Soc Pharmacog, 64-65; mem, US Pharmacopeia Rev Comt, 70-75; bd dir, Am Found Pharmaceut Educ, 80-85. *Honors & Awards:* Nat Kappa Psi Citation for Serv to the Prof of Pharm, 61; R A Lyman Award, Am Asn Cols Pharm, 73. *Mem:* Fel AAAS; NY Acad Sci; Int Pharmaceut Fedn; Am Inst Hist Pharm; Acad Pharmaceut Sci; Sigma Xi; hon mem Am Found Pharmaceut Educ. *Res:* Public health education for pharmacy students; enzyme-alkaloid relations in plants; sterile plant tissue culture; plant biosynthesis. *Mailing Add:* W 707 Sixth Ave 4 Spokane WA 99204

GIBSON, PETER MURRAY, b Laurel Hill, NC, Sept 1, 39; m 70. MATHEMATICS. *Educ:* NC State Univ, BS, 61, MS, 63, PhD(math), 66. *Prof Exp:* Instr math, NC State Univ, 66-67; from asst prof to assoc prof, 67-74, PROF MATH, UNIV ALA, HUNTSVILLE, 74- *Concurrent Pos:* Hon fel, Univ Wis-Madison, 74-75. *Mem:* Am Math Soc; Math Asn Am. *Res:* Combinatorial mathematics; linear algebra. *Mailing Add:* Dept Math Univ Ala Box 1247 Huntsville AL 35899

GIBSON, QUENTIN HOWIESON, b Aberdeen, Scotland, Dec 9, 18; m 51; c 4. PHYSIOLOGY, BIOPHYSICAL CHEMISTRY. *Educ:* Queen's Univ Belfast, MB, BCh, 41, MD, 44, PhD(biochem), 46, DSc, 51. *Prof Exp:* Demonstr physiol, Queen's Univ Belfast, 41-46, lectr, 46-47; lectr, Sch Med, Univ Sheffield, 47-55, prof biochem & chmn dept, 55-63; prof phys biochem & physiol, Sch Med, Univ Pa, 63-65; GREATER PHILADELPHIA PROF BIOCHEM, MOLECULAR & CELL BIOL, CORNELL UNIV, 65- *Concurrent Pos:* Fogarty Fel, NIH 88; assoc ed, J Biol Chem. *Mem:* Nat Acad Sci; fel Royal Soc; Brit Physiol Soc; Am Acad Arts & Sci; Am Soc Biol Chem. *Res:* Etiology and biochemistry of idiopathic methemoglobinemia; hemoglobinometry; measurement of rapid reactions; mechanisms of enzyme reactions. *Mailing Add:* Dept Biochem & Molecular Biol 225 Biotech Bldg Cornell Univ Ithaca NY 14853

GIBSON, RAYMOND EDWARD, b San Juan, PR, Oct 27, 46; m 80; c 2. RADIOPHARMACEUTICAL CHEMISTRY, RECEPTOR PHARMACOLOGY. *Educ:* Univ Calif, Santa Cruz, BA, 69; Univ Calif, Santa Barbara, PhD(chem), 72. *Prof Exp:* NIH postdoctoral fel, Dept Biol, Cornell Univ, 72-74, res assoc, 74-76; from asst res prof to assoc res prof, Dept Radiol, George Wash Univ Med Ctr, 76-83, sr res scientist, 83-87, sr staff scientist, 87-88, ADJ ASSOC PROF, DEPT RADIOL, GEORGE WASHINGTON UNIV MED CTR, 88-; RES FEL, MERCK, SHARP & DOHME RES LAB, 88- *Concurrent Pos:* Bd dir, Radiopharmaceut Sci Coun, Soc Nuclear Med, 88-90. *Mem:* Am Chem Soc; Soc Nuclear Med; AAAS. *Res:* Development of receptor-binding radiopharmaceuticals for clinical diagnosis and for basic pharmaceutical research. *Mailing Add:* Merck Sharp & Dohme Res Lab West Point PA 19486

GIBSON, RICHARD C(USHING), b Cambridge, Mass, Dec 31, 19; m 42; c 4. INSTRUMENTATION. *Educ:* Mass Inst Technol, BS, 42, MS, 46, ScD(instrumentation), 53. *Prof Exp:* Radar & commun officer, USAF, 42-45, asst prof elec eng, Air Force Inst Technol, 46-51, chief missile guid, Res & Develop Command, 53-56, dep chief of staff, Missile Develop Ctr, 56-58, dir exp vehicles & instrumentation, Res & Develop Command, 59-60, prof astronaut, Air Force Acad, 60-65, head dept, 60-62, vcomdr nat range div, 65-67; head, Elec Eng Dept, Univ Maine, Orono, 67-77, prof elec eng, 67-80; consult, 81-82; RETIRED. *Concurrent Pos:* Lectr, Univ Conn, 60-66; dir, F J Seiler Res Lab, 62-65; mem, US Air Force Sci Adv Bd, 70-81. *Mem:* Am Soc Eng Educ. *Res:* Human response characteristics; missiles; range instrumentation inertial guidance; astronautics. *Mailing Add:* Box 105 Surry ME 04684

GIBSON, ROBERT HARRY, b Clover, SC, Jan 25, 38; m 60; c 2. ANALYTICAL CHEMISTRY. *Educ:* Erskine Col, AB, 60; Columbia Univ, MA, 61, PhD(chem), 65. *Prof Exp:* Assoc prof, 65-75, PROF CHEM & CHMN DEPT, UNIV NC, CHARLOTTE, 75- *Mem:* Am Chem Soc; AAAS; Electrochem Soc; Sigma Xi. *Res:* Electroanalytical chemistry; polarography; electrochemistry of organic compounds. *Mailing Add:* Dept Chem Univ NC UNCC Sta Charlotte NC 28223

GIBSON, ROBERT JOHN, JR, biophysics, for more information see previous edition

GIBSON, ROBERT WILDER, b Greensburg, Kans, Dec 19, 17. ALGEBRA. *Educ:* Ft Hays Kans State Col, AB, 38; Univ Ill, Urbana, AM, 39, PhD(math), 43. *Prof Exp:* Assoc prof math, Kans State Col, 43; prof physics, William Penn Col, 47-49; eng aide, Kans State Hwy Dept, 58-59; from asst prof to assoc prof math, Okla State Univ, 59-69; assoc prof civil eng, 69-80, ASSOC PROF MECH ENG, AUBURN UNIV, 80- *Mem:* AAAS; Math Asn Am; Soc Indust & Appl Math; Inst Mgt Sci; Sigma Xi. *Res:* Mathematical programming; abstract algebra; the transportation problem of linear programming; six- and eight-bar linkage cognates; mathematical theory of flexagons. *Mailing Add:* 263 Payne St Auburn AL 36830

GIBSON, ROSALIND SUSAN, b Northumberland, UK, Nov 20, 40; m 63; c 2. NUTRITIONAL ASSESSMENT, TRACE ELEMENTS. *Educ:* Univ London, BSc, 62 & PhD(nutrit), 79; Univ Calif Los Angeles, MS, 65. *Prof Exp:* Res biochemist, Inst Orthop, Univ London, 62-63; nutrit biochemist, Ethio-Swed Children's Nutrit Unit, Addis Ababa, Ethiopia, 65-68; lectr nutrit, Trinity & All Saints Col, 68-71 & Polytech N London, 71-78; res assoc nutrit, dept pediat, Dalhousie Univ, 77-78; PROF APPL HUMAN NUTRIT, DEPT FAMILY STUDIES, UNIV GUELPH, 79- *Concurrent Pos:* Mem health & welfare, Expert Comt Dietary Fibre, 84-85; vis prof, dept human nutrit, Univ Otago, Dunedin, NZ & Inst of Med Res, Madang, Papua New Guinea, 86-87; rep nutrit sci, Int Union Nutrit Sci, 87- *Mem:* Can Soc Nutrit Sci (treas, 82-85); Can Dietetic Asn; Brit Nutrit Soc; Am Inst Nutrit. *Res:* Trace elements essential in human nutrition; assessment of trace element status of groups at risk for trace element deficiencies; use of hair as index of trace element status; nutritional assessment with emphasis on pre-school children in less industrialized countries. *Mailing Add:* Appl Human Nutrit Dept Family Studies Univ Guelph Guelph ON N1G 2W1 Can

GIBSON, SAM THOMPSON, b Covington, Ga, Jan 1, 16; m 42, 86; c 4. INTERNAL MEDICINE, GOVERNMENT & REGULATORY. *Educ:* Ga Inst Technol, BS, 36; Emory Univ, MD, 40. *Prof Exp:* Spec res assoc, Harvard Med Sch, 43; asst med dir, Blood Prog, Am Nat Red Cross, 49-51, assoc med dir, 51-53, assoc dir, 53-56, dir, 56-66, sr med officer, 57-67; asst dir, Div Biol Stand, NIH, 67-72; asst dir, Bur Biologics, FDA, 72-74, asst to dir, 74-77, dir, Div Biologics Eval, 77-83, dir, Div Biol Prod Compliance, 83-85, assoc dir sci & tech, Off Compliance, Ctr Drug Eval & Res, 85-88; RETIRED. *Concurrent Pos:* Res fel med, Harvard Med Sch, 41-42, Milton fel, 47-49; med house officer, Peter Bent Brigham Hosp, 40-41, asst resident, 46-47, asst, 47-49; assoc, Sch Med, George Washington Univ, 49-63, asst clin prof, 63-, asst, Univ Hosp, 49-, clin asst prof med, Uniformed Serv, Univ Health Sci, 80-; consult, Nat Naval Med Ctr, Md, 50-63; adv blood transfusion serv, League Red Cross Socs, 55-66. *Mem:* Int Soc Blood Tranfusion; Am Soc Hemat; Int Soc Hemat; AMA; Am Fedn Clin Res. *Res:* Effect of serum albumin on kidney and liver function in disease; blood banking and plasma fractionation. *Mailing Add:* 5801 Rossmore Dr Bethesda MD 20814

GIBSON, THOMAS ALVIN, JR, b Joplin, Mo, June 17, 19; m 45; c 5. HEALTH PHYSICS. *Educ:* The Citadel, BS, 40; Univ Va, MS, 48; Univ Calif, Berkeley, MS, 50. *Prof Exp:* Asst prof chem & physics, NGa Col, 62-64; res chemist, Lawrence Livermore Nat Lab, 64-86; RETIRED. *Mem:* Am Chem Soc; Health Physics Soc. *Res:* Bioradiology; laboratory research and nuclear weapons tests to determine weapons effects; nuclear radiation, residual radiation and fallout; industrial and engineering applications of nuclear explosions; radiation health physics. *Mailing Add:* 40 Mariposa Ct Danville CA 94526

GIBSON, THOMAS CHOMETON, b Burnley, Eng, Apr 30, 21; US citizen; m 59; c 5. INTERNAL MEDICINE, CARDIOLOGY. *Educ:* Cambridge Univ, BA, 42, MA, 45, MB, BCh, 46; MRCP, 55, FRCP, 75. *Prof Exp:* House physician med, London Hosp, Eng, 46-47; resident physician, London & Bath Hosps, 53-55; pvt pract, 56-57; from instr to asst prof med, Univ NC, 59-62; asst prof, 62-66, assoc prof, 66-76, PROF MED, COL MED, UNIV VT, 76- *Concurrent Pos:* Fel cardiol, Sch Med, Univ NC, 57-59; chmn subcomt congestive heart failure, Sect Community Serv, Nat Conf Cardiovascular Dis, 63-64; mem adv comt clin criteria for congestive heart failure, USPHS Heart Dis Control Prog, 64; attend physician, Med Ctr Hosp, Vt, Burlington. *Mem:* AMA; Am Heart Asn; Am Geriat Soc; fel Am Col Physicians; fel Am Col Cardiol. *Res:* Clinical cardiology; phonocardiography; electrocardiography; echocardiography; cardiovascular epidemiology. *Mailing Add:* Shelburne Pt Shelburne VT 05482

GIBSON, THOMAS GEORGE, b Milwaukee, Wis, Aug 10, 34; m 56; c 3. INVERTEBRATE PALEONTOLOGY. *Educ:* Univ Wis, BS, 56, MS, 59; Princeton Univ, PhD(geol), 62. *Prof Exp:* Geologist, Shell Oil Co, 57; GEOLOGIST, US GEOL SURV, PALEONT & STRATIG BR, US NAT MUS, 62- *Mem:* Paleont Soc; Marine Biol Asn UK. *Res:* Distribution and taxonomy of marine Mollusca and Foraminifera in the Tertiary deposits of the Atlantic and Gulf Coastal Plains and the Recent of the Atlantic Shelf. *Mailing Add:* 970 Nat Ctr US Geol Surv Reston VA 22092

GIBSON, THOMAS RICHARD, b Orange, CA, June 26, 51. POST-TRANSLATIONAL PROCESSING, BIOSYNTHESIS. *Educ:* Univ Calif, BS, 73, PhD(biochem), 79. *Prof Exp:* Res assoc, Harbor, Univ Calif, Los Angeles, Med Ctr, 79-82; RES ASST PROF, SAN DIEGO STATE UNIV, 86-; PRIN INVESTR, AM HEART ASN, 87- *Mem:* Am Soc Biochem & Molecular Biol; Am Soc Hypertension. *Res:* Atrial Natrionetic Factor is produced in the heart and acts on the kidney; biosynthesis and processing of Atrial Natrionetic Fact0r. *Mailing Add:* Dept Biol San Diego State Univ 5300 Campanile Dr San Diego CA 92182

GIBSON, THOMAS WILLIAM, b Petoskey, Mich, Oct 11, 35; m 60; c 5. ORGANIC CHEMISTRY. *Educ:* Aquinas Col, BS, 59; Purdue Univ, PhD(chem), 63. *Prof Exp:* RES CHEMIST, MIAMI VALLEY LABS, PROCTER & GAMBLE CO, 63- *Mem:* Am Chem Soc. *Res:* Organic synthesis, especially as applied to naturally occurring terpenoids; photochemistry of various classes of organic compounds. *Mailing Add:* PO Box 599 Procter & Gamble Co 11810 E Miami River Rd Cincinnati OH 45239

GIBSON, WALTER MAXWELL, b Enoch, Utah, Nov 11, 30; m 53, 67; c 7. EXPERIMENTAL PHYSICS. *Educ:* Univ Utah, BS, 54; Univ Calif, PhD(nuclear chem), 56. *Prof Exp:* Mem tech staff, Bell Tel Labs, 59-76; prof physics & chmn dept, 76-84, vpres res & dean grad studies, 84-86, DISTINGUISHED PROF PHYSICS, STATE UNIV NY, ALBANY, 88- *Concurrent Pos:* Res collabr, Brookhaven Nat Labs, 60-; adj prof, Rensselaer Polytech Inst, 63-76 & Rutgers Univ, 63-76; ed, Radiation Effects, 84-88. *Mem:* Fel Am Phys Soc; Am Vacuum Soc; Sigma Xi; AAAS. *Res:* Nuclear and solid state physics, principally nuclear fission mechanism studies;

principles and application of solid state detectors and interaction of charged particles with crystalline and amorphous media, surfaces, structure and dynamics. *Mailing Add:* Dept Physics State Univ NY 1400 Washington Ave Albany NY 12222

GIBSON, WILLIAM ANDREW, HISTOCHEMISTRY, CYTOCHEMISTRY. *Educ:* Georgetown Univ, PhD(path), 76. *Prof Exp:* ASST DEAN, GRAD STUDY & RES, BAYLOR COL DENT, 86- *Mailing Add:* Grad Studies & Res Baylor Col Dent 3302 Gaston Ave Dallas TX 75246

GIBSON, WILLIAM LOANE, b Corpus Christi, Tex, July 20, 44; m 68; c 1. MATHEMATICAL ANALYSIS. *Educ:* Univ Tex, Austin, BA, 66, MA, 68; Univ Houston, PhD(math), 74. *Prof Exp:* Res assoc integral equations, Math Physics Br, NASA-Johnson Space Ctr, 73-74; sr engr space shuttle, McDonnell Douglas Tech Serv Co, 74-78; ENG PROJ MGR SOFTWARE, SPACE SYSTS CO, MARTIN MARIETTA ASTRONAUT GROUP, 78- *Res:* Product integral solutions for Stieltjes and Stieltjes-Volterra integral equations; nonlinear integral equations; hereditary systems. *Mailing Add:* 6665 S Webster St Littleton CO 80123

GIBSON, WILLIAM RAYMOND, b Murphysboro, Ill, Oct 21, 23; m 46; c 2. PHARMACOLOGY. *Educ:* Butler Univ, BS, 48, MS, 56. *Prof Exp:* Pharmacologist, Eli Lilly & Co, 56-63, asst head biol qual control, 63-65, head, Toxicity Dept, 65-77, dir toxicol planning, 77-79; consult toxicol, 79-90; RETIRED. *Concurrent Pos:* Lectr, Butler Univ, 58-67, res assoc toxicol, 80-87. *Mem:* Soc Toxicol; Sigma Xi. *Res:* Pharmacology of anesthetics, sedatives, antihistamines and anticholinergics; toxicology of drugs and agricultural chemicals. *Mailing Add:* 12026 Castle Row Overlook Carmel IN 46032

GIBSON, WILLIAM WALLACE, b Philadelphia, Pa, Sept 26, 28; m 51; c 3. ENTOMOLOGY. *Educ:* Univ RI, BS, 51; Kans State Univ, MS, 55, PhD(entom), 57. *Prof Exp:* Jr asst entom & plant path, Univ Exp Sta, Univ RI, 52, asst zool, 52-53; asst entom, Kans State Univ, 53-56; cur insect collection, Rockefeller Found Mex Agr Prog, 56-58; entomologist, Assoc Seed Growers, Inc, 58-61; temporary asst prof biol, Northeast La State Col, 62; from asst prof to assoc prof, 62-68, PROF BIOL, STEPHEN F AUSTIN STATE UNIV, 68- *Mem:* Entom Soc Am; Mex Soc Entom; Sigma Xi. *Res:* Biological and ecological investigations of insects; taxonomy of scarab dung beetles. *Mailing Add:* Dept Biol Stephen F Austin State Univ Nacogdoches TX 75961

GICLAS, HENRY LEE, b Flagstaff, Ariz, Dec 9, 10; m 36; c 1. ASTRONOMY. *Educ:* Univ Ariz, BS, 37. *Hon Degrees:* DSc, Northern Ariz Univ, 80. *Prof Exp:* Res asst astron, Lowell Observ, 31-42, astronr, 42-81, exec officer, 53-75; RETIRED. *Concurrent Pos:* Adj prof, Ohio State Univ, 68-81; exec vpres, Raymond Educ Found, Ariz, 71-77, pres, 77-; adj prof, Northern Ariz Univ, 72- *Mem:* Fel AAAS; Int Astron Union; Am Astron Soc. *Res:* Photographic and photoelectric photometry; positional astrometry, comets and minor planets; proper motion survey. *Mailing Add:* 120 E Elm Ave Flagstaff AZ 86001

GICLAS, PATRICIA C, b Albuquerque, NMex. BIOCHEMISTRY, IMMUNOPATHOLOGY. *Educ:* NMex Inst Mining & Technol, BS, 70; Univ Ariz, PhD(molecular biol), 76. *Prof Exp:* Fel, Scripps Clin & Res Found, 75-77; fel, Nat Jewish Hosp & Res Ctr, 77-79, res assoc, 79-80; sr investr, Dept Pediat, Nat Jewish Hosp & Res Ctr, 80-89, CO-DIR, DIAG IMMUNOL LAB, NAT JEWISH CTR IMMUNOL RESPIRATORY MED, 89- *Concurrent Pos:* Instr, Dept Med, Health Sci Ctr, Univ Colo, 79-80, asst prof, Pulmonary Div, 80- *Mem:* AAAS; Am Soc Microbiol; Am Asn Immunologists; NY Acad Sci; Int Soc Develop & Comp Immunol. *Res:* Biochemistry of complement activation and complement-mediated aspects of acute inflammation particularly as related to pulmonary disease; mechanisms of protein synthesis and turnover in acute inflammation. *Mailing Add:* Dept Pediat & Allergy-Immunol Nat Jewish Ctr 1400 N Jackson Denver CO 80206

GIDAL, GEORGE, b Munich, Ger, Sept 8, 34; US citizen; m 62; c 2. EXPERIMENTAL HIGH ENERGY PHYSICS. *Educ:* Columbia Univ, AB, 55, MA, 57, PhD(physics), 60. *Prof Exp:* Res assoc physics, Lawrence Radiation Lab, 60-61; res physicist, Inst Physics, Torino, Italy, 64-65; SR STAFF PHYSICIST, LAWRENCE BERKELEY LAB, 66- *Concurrent Pos:* Vis assoc prof, Tel Aviv Univ, 69-70. *Mem:* Am Phys Soc. *Res:* Strong interactions of particles; measurements of interaction mechanisms via details of production and decay distributions; backward inelastic scattering; electron-positron interactions; photon-photon interactions. *Mailing Add:* Physics Div Lawrence Berkeley Lab Berkeley CA 94720

GIDARI, ANTHONY SALVATORE, b Cambridge, Mass, Sept 20, 43. PHARMACOLOGY, HEMATOLOGY. *Educ:* Tufts Univ, BS, 65; NY Univ, MS, 70, PhD(physiol, hemat), 72. *Prof Exp:* Asst res scientist hemat & physiol, Lab Exp Hemat, NY Univ, 71-72; from instr to asst prof med, 72-77, asst prof, 77-80, ASSOC PROF PHARMACOL, STATE UNIV NY DOWNSTATE MED CTR, 80- *Mem:* Am Soc Pharmacol & Exp Therapeut. *Res:* Biochemistry and physiology of blood cell formation including the regulation of erythropoiesis by the hormone erythropoietin; mechanism of action of vasodilator agents. *Mailing Add:* Dept Pharmacol Box 29 State Univ NY Downstate Med Ctr 450 Clarkson Ave Brooklyn NY 11203

GIDDA, JASWANT SINGH, b Hoshiarpur, Panjab, India, Oc 1, 46; US citizen; m 75; c 2. AUTONOMIC PHARMACOLOGY, GASTROENTEROLOGY. *Educ:* Panjab Univ, India, BSc, 65 & 67, MSc, 68, PhD(biol), 73. *Prof Exp:* Res scientist neurobiol, Univ Tex, Dallas, 74-78, instr gastroenterol, Health Sci Ctr, San Antonio, 78-81; instr, Harvard Med Sch, 81-84; sr res scientist pharmacol, Bristol-Myers Co, Syracuse, 84-87; RES SCIENTIST PHARMACOL, LILLY RES LAB, ELI LILLY & CO, 87- *Concurrent Pos:* Lectr pharmacol, Harvard Med Sch, 84-87; asst res prof physiol, State Univ NY, New York, 85-87. *Mem:* Am Motility Soc; Am Gastroenterol Asn; Am Physiol Soc. *Res:* Understanding the neuromuscular mechanisms which control motility & secretionin the gastrointestinal tract. *Mailing Add:* Eli Lilly & Co Lilly Res Labs MC304 Lilly Corporate Ctr Indianapolis IN 46285

GIDDENS, DON P(EYTON), b Augusta, Ga, Oct 24, 40; m 57; c 4. FLUID MECHANICS, GAS DYNAMICS. *Educ:* Ga Inst Technol, BAE, 63, MSAE, 65, PhD(aerospace eng), 66. *Prof Exp:* Assoc aircraft engr, Lockheed-Ga Co, 63; mem tech staff, Aerospace Corp, Calif, 65 & 66-67; asst prof gas dynamics, 68-70, from assoc prof to prof aerospace eng, 70-83, REGENTS' PROF MECH ENG, GA INST TECHNOL, 83-, DIR AERO ENG, 88- *Mem:* Sigma Xi; Am Asn Mech Engrs; Am Inst Aeronaut & Astronaut; Am Soc Eng Educ; fel Am Heart Asn. *Res:* Fluid mechanics of the cardiovascular system and the application to medical problems; rarefied gas dynamics; turbulent flows. *Mailing Add:* Sch Aero Eng Ga Inst Technol Atlanta GA 30332-0150

GIDDENS, JOEL EDWIN, b Eastman, Ga, Feb 11, 17; m 42; c 3. SOIL MICROBIOLOGY. *Educ:* Univ Ga, BS, 40, MS, 42; Rutgers Univ, PhD(soils), 50. *Prof Exp:* Asst, Univ Ga, 40-42; jr chemist, Southern Regional Res Lab, USDA, 42-45; asst agronomist, Univ Ga, 46-48, 50-52, assoc prof, 52-59, prof agron, 59-85; RETIRED. *Mem:* Fel Am Soc Agron; Soil Sci Soc Am; Am Chem Soc; Am Soc Microbiol; Soil Conserv Soc Am. *Res:* Soil fertility; nitrogen fixation by plants; waste management. *Mailing Add:* 315 Parkway Dr Athens GA 30606

GIDDENS, WILLIAM ELLIS, JR, b Dublin, Ga, Oct 8, 37; m 61; c 1. VETERINARY PATHOLOGY. *Educ:* Iowa State Univ, DVM, 61; Mich State Univ, PhD(path), 68. *Prof Exp:* NIH fel path, Mich State Univ, 65-67, res instr, 67-68; asst prof, Univ Wash, 68-78, assoc prof path, Div Animal Med, Sch Med, 78-87; RETIRED. *Mem:* Am Col Vet Path; Am Vet Med Asn; Int Acad Path. *Res:* Comparative pathology; respiratory diseases of animals and man; pathogenesis of virus diseases of nonhuman primates. *Mailing Add:* Div Animal Med Univ Wash Sch Med Seattle WA 98195

GIDDINGS, GEORGE GOSSELIN, b Worcester, Mass, Mar 26, 37; m 66; c 2. FOOD PROCESSING, FOOD ENGINEERING. *Educ:* Univ Mass, BSc, 63; Mich State Univ, MSc, 69, PhD(food sci & technol), 72. *Prof Exp:* Res food technologist, US Army, Res & Develop Command, Natick, Mass, 63-66; res assoc, Mich State Univ, 66-72; prof food sci, NC State Univ, 72-77; sr tech adv & mgr food technol, Fundacion Chile, 77-81; dir food appl, Radiation Technol, Inc, 81-83; dir food irradiation serv, Isomedix Inc, NJ, 83-87; RADIATION PROCESSING CONSULT, 87- *Mem:* Am Chem Soc; Inst Food Technologists. *Res:* Meat pigment chemistry and biochemistry; radiation preservation of foods. *Mailing Add:* 61 Beech Rd Randolph NJ 07869

GIDDINGS, JOHN CALVIN, b American Fork, Utah, Sept 26, 30; c 2. PHYSICAL CHEMISTRY. *Educ:* Brigham Young Univ, BS, 52; Univ Utah, PhD(chem), 54. *Prof Exp:* Res assoc, Univ Wis, 55-56; res assoc, 56-57, from asst prof to assoc prof, 57-61, from assoc res prof to res prof, 61-66, PROF CHEM, UNIV UTAH, 66- *Concurrent Pos:* Mem adv bd anal chem, US Air Force Off Sci Res, 62-64, mem chem res eval panel, 64-69; Foster lectr, State Univ NY Buffalo, 71; Fulbright grant, Cayetano Heredia Univ, Lima, Peru, 74; ed, Separation Sci & Technol, 66-; adv ed, Chromatography, 65-; mem adv bd, J Liquid Chromatography, J High Resolution Chromatography, 78- *Res:* Separation methods; field-flow fractionation; chromatography; diffusion; environmental science. *Mailing Add:* Dept Chem Univ Utah Salt Lake City UT 84112

GIDDINGS, SYDNEY ARTHUR, b Mildura, Victoria, Australia, Sept 5, 29; US citizen; m 54; c 3. POLYMER CHEMISTRY. *Educ:* Univ Melbourne, Australia, BS, 52; Ohio State Univ, PhD(inorg chem), 59. *Prof Exp:* Sr scientist, Am Cyanamid, 59-67; group leader, Formica, 67-71, plant mgr, 71-73, dir res, 73-86; VPRES TECH OPER, CONGOLEUM, 87- *Mem:* Am Chem Soc. *Res:* Product and proces research of surfacing products based on polyvinyl chloride, melamine and phenolic resins. *Mailing Add:* 861 Sloan Ave Trenton NJ 08619

GIDDINGS, THOMAS H, JR, ELECTRON MICROSCOPY, MICROBIOLOGY. *Educ:* Univ Colo, PhD(biol). *Prof Exp:* MEM STAFF, DEPT MOLECULAR BIOL, UNIV COLO. *Mailing Add:* Dept Molecular & Cellular Develop Biol Univ Colo Boulder CO 80309-0347

GIDDINGS, WILLIAM PAUL, b Indianapolis, Ind, May 7, 33; m 61; c 3. PHYSICAL ORGANIC CHEMISTRY. *Educ:* DePauw Univ, BA, 54; Harvard Univ, MA, 56, PhD(phys org chem), 59. *Prof Exp:* Res instr chem, Univ Wash, 59-60; asst prof, Albion Col, 60-62; from asst prof to assoc prof chem, Pac Lutheran Univ, 62-68, chmn dept, 66-70 & 83-89, chmn, Div Natural Sci, 69-75, PROF CHEM, PAC LUTHERAN UNIV, 68- *Concurrent Pos:* NSF res grant, 64-66; Petrol Res Fund res grant, 66-69; vis scholar, Dept Civil Eng & Atmospheric Sci, Univ Wash, 76 & Dept Environ Health, 89-90. *Mem:* Am Chem Soc; Sigma Xi; AAAS; Am Asn Univ Prof. *Res:* Nature of carbonium ion intermediates in solvolysis reactions; atmospheric chemistry of organic compounds. *Mailing Add:* Dept Chem Pac Lutheran Univ Tacoma WA 98447

GIDEZ, LEWIS IRWIN, b Boston, Mass, Jan 27, 27; m 55; c 5. BIOCHEMISTRY. *Educ:* Iowa State Col, BS, 48; Harvard Univ, PhD(biochem), 53. *Prof Exp:* Asst med biochemist, Med Dept, Brookhaven Nat Lab, 52-58; asst prof, 58-69, ASSOC PROF BIOCHEM, ALBERT EINSTEIN COL MED, 69- *Concurrent Pos:* Estab investr, Am Heart Asn, 58-63; career scientist, Health Res Coun, NY, 64-70; exec ed, J Lipid Res, 70- *Mem:* Am Chem Soc; Am Soc Biol Chem. *Res:* Chemistry and metabolism of lipids and lipoproteins. *Mailing Add:* 9650 Rockville Pike Bethesda MD 20814

GIDLEY, J(OHN) L(YNN), b Lytle, Tex, Dec 30, 24; m 59; c 7. CHEMICAL & PETROLEUM ENGINEERING. *Educ:* Univ Tex, BS, 50, MS, 52, PhD(chem eng), 55. *Prof Exp:* Sr res engr, Prod Res Div, Humble Oil & Refining Co, 54-63; new uses adv, Standard Oil Co, NJ, 63-64, res assoc, Prod Res Div, Esso Prod Res Co, 64-68; hq supv engr, 68-69, tech adv, 69-81, SR TECH ADV, EXXON CO USA, 81- *Concurrent Pos:* Distinguished lectr,

Soc Petrol Engrs, 80-81, emer distinguished lectr, 89-90. *Honors & Awards:* Distinguished Serv Award, Soc Petrol Engrs, 90. *Mem:* Am Chem Soc; Am Inst Mining, Metall & Petrol Engrs; Soc Petrol Engrs. *Res:* Well stimulation methods; sand control, cementing, perforating; engineering training. *Mailing Add:* 5211 Caversham Dr Houston TX 77096

GIDWANI, RAM N, b India, Mar 11, 36; US citizen; m 73; c 2. BIOPHARMACEUTICS, PHYSICAL PHARMACY. *Educ:* L M Col Pharm, India, BS, 58; St John's Univ, MS, 59; Univ Alta, PhD(biopharmaceut), 75. *Prof Exp:* Res pharmacist res & develop, Warner Lambert, NJ, 69-70; res pharmacist, Miles Labs, Ind, 70-71; asst prof pharmaceut, Arnold & Marie Schwartz Col Pharm, Brooklyn, NY, 75-77; sr scientist res & develop, Block Drug Co, NJ, 78; sr scientist res & develop, USV Pharmaceut, 78-80; sr scientist, FMC, Maine, 80-82; sr scientist, Sandoz Labs, Neb, 82-86; sr scientist, NIH, Md, 86-88; SR SCIENTIST, IMMUNOBIOL RES INST, JOHNSON & JOHNSON, 88- *Concurrent Pos:* Fel Nuclear Med Res Labs, Univ Calif, Los Angeles, 75. *Mem:* Am Pharmaceut Asn; Acad Pharmaceut Sci; Soc Cosmetic Chemists; Fedn Am Soc Exp Biol. *Res:* Evaluation of solid dispersion systems; differential thermal analysis; solubility; dissolution; bioavailability and pharmacokinetic studies. *Mailing Add:* 43 Gridley Circle Milford NJ 08848

GIEBISCH, GERHARD HANS, b Vienna, Austria, Jan 17, 27; US citizen; m 52; c 2. PHYSIOLOGY. *Educ:* Univ Vienna, MD, 51. *Hon Degrees:* DLitt, Univ Uppsala, 77 & Univ Bern, 79. *Prof Exp:* Instr pharmacol, Med Sch, Univ Vienna, 51, asst prof, 56-57; intern, Milwaukee Hosp, 52-53; fel physiol, Med Col, Cornell Univ, 53-54, instr, 55-56, from asst prof to prof, 57-68; prof & chmn, Dept Physiol, 68-73, STERLING PROF CELLULAR & MOLECULAR PHYSIOL, SCH MED, YALE UNIV, 70- *Concurrent Pos:* Estab investr, Am Heart & Asn, 57-62; pub health res career award, 62-68; mem, Physiol Study Sect, NIH, 64-69 & 82-84; NIH career award, 65-68; sect ed, Kidney & Electrolyte Metab, Am J Physiol, Am J Appl Physiol, 67-69; mem, Renal Dis & Urol Training Grant Study Sect, 70-73; hon prof, Univ Lausanne, 74-75; fac scholar award, Josiah Macy, Jr Found, 74-75; mem coun, Soc Gen Physiol, 80-82, Am Physiol Soc, 88-90 & Nat Inst Diabetes, Digestive Dis & Kidney Dis, 88-; Wellcome vis prof, 83; ed, Physiol Rev, 85-90. *Honors & Awards:* Homer Smith Award, 71; Johannes Müller Medal, Ger Physiol Soc, 80; Alexander von Humboldt Prize, 87; Volhard Medal, Ger Nephrological Soc, 88; Ernst Jung Prize Med, 90. *Mem:* Nat Acad Sci; Soc Gen Physiologists (pres, 86-87); Am Physiol Soc; Biophys Soc; Am Soc Nephrology (pres, 71-72); Soc Clin Res; Am Acad Arts & Sci. *Res:* Electrolyte metabolism; renal physiology, particularly studies on single nephrons, employing methods of micropuncture. *Mailing Add:* Dept Cellular & Molecular Physiol Yale Univ Sch Med 333 Cedars St New Haven CT 06510-8026

GIEDT, W(ARREN) H(ARDING), b Leola, SDak, Nov 1, 20; m 50. MECHANICAL ENGINEERING, MATERIALS SCIENCE. *Educ:* Univ Calif, Berkeley, BAS, 44, MS, 46, PhD(mech eng), 50. *Prof Exp:* From instr to prof mech eng, Univ Calif, Berkeley, 47-65; prof, 65-83, head dept, 65-69, assoc dean grad study, Col Eng, 72-80, EMER PROF MECH ENG, UNIV CALIF, DAVIS, 83- *Concurrent Pos:* Serv engr, Babcock & Wilcox Co, 50; consult, Bechtel Corp, 51; proj leader, Detroit Controls Corp, 52-56; consult, Am Standard Corp, 56-65, Lawrence Livermore Nat Lab, 60-, Boeing Co, 62-66 & Sandia Nat Labs, 83-; Fulbright prof, Univ Tokyo, 63; consult, NASA, 64-67; ed, J Heat Transfer, 67-72; fel, Japan Soc Prom Sci, 80. *Honors & Awards:* Jennings Award, Am Welding Soc, 70; Western Elec Award, Am Soc Eng Educ, 71, G Edwin Burks Award, 74; Heat Transfer Mem Award, Am Soc Mech Engrs, 76, James Harry Potter Gold Medal, 85; Thermal Eng Mem Award, Japan Soc Mech Eng, 90. *Mem:* Fel Am Soc Mech Eng; Am Welding Soc; Am Soc Eng Educ. *Res:* Heat transfer; thermodynamics; welding. *Mailing Add:* Dept Mech Eng Univ Calif Davis CA 95616

GIEGEL, JOSEPH LESTER, b New York, NY, June 27, 38; m 64; c 3. CLINICAL BIOCHEMISTRY. *Educ:* Univ Miami, BS, 62, PhD(microbiol), 68. *Prof Exp:* Instr dermat & microbiol, Sch Med, Univ Miami, 68-70; dir clin chem res & develop, Dade Div, Am Hosp Supply Corp, 70-85; PRES, DIAMEDIX CORP, 85- *Concurrent Pos:* Tech dir, La Huis Clin Labs, Miami, Fla, 68-70. *Mem:* Am Chem Soc; Am Asn Clin Chemists; Sigma Xi. *Res:* Diagnostic procedures of research and development in clinical chemistry and immunology. *Mailing Add:* Diamedix Corp 2140 N Miami Ave Miami FL 33127

GIELISSE, PETER JACOB MARIA, b 's-Hertogenbosch, Netherlands, Mar 7, 34; US citizen; m 56; c 2. MINERALOGY, CRYSTALLOGRAPHY. *Educ:* Boston Col, MS, 59; Ohio State Univ, PhD(mineral), 61. *Prof Exp:* Res engr physics, Comstock & Wescott, Inc, Mass, 57-59; Res Found fel, Ohio State Univ, 59-61; res physicist, Air Force Cambridge Res Labs, Mass, 61-63; res engr solid state physics, Metall Prod Dept, Gen Elec Co, Detroit, 63-68; PROF MAT & CHMN ENG, UNIV RI, 68- *Concurrent Pos:* Res asst geophys, Boston Col, 58-59, instr geol & geophys, 61-63. *Mem:* Am Geophys Union; Am Mineral Soc. *Res:* Thermochemical mineralogy; physical, structural and optical properties of materials; phase equilibria in multi-component systems; crystal synthesis; very high pressure-high temperature materials; semiconductor devices. *Mailing Add:* Capitol Technol PO Box 4227 Tallahassee FL 32315

GIERASCH, LILA MARY, b Needham, Mass, Sept 18, 48; div. BIOPHYSICAL CHEMISTRY. *Educ:* Mt Holyoke Col, AB, 70; Harvard Univ, PhD(biophys), 75. *Prof Exp:* Teaching fel, Harvard Univ, 72-73; asst prof chem, Amherst Col, 74-79 & Univ Del, 79-81; from assoc prof to prof chem, Univ Del, 81-87; ROBERT A WELCH PROF BIOCHEM & PHARMACOL, UNIV TEX SOUTHWESTERN, DALLAS, 88- *Concurrent Pos:* A P Sloan fel, 84-86; Guggenheim fel, 86; assoc mem, Int Union Pure Appl Chem, Comm Biotechnol, 86-89. *Honors & Awards:* Vincent Du Vigneaud Award, 84; Mary Lyon Award, 85. *Mem:* Am Chem Soc; fel AAAS; NY Acad Sci; Biophys Soc; Am Soc Biol Chemists; Am Soc Microbiol. *Res:* Studies of polypeptide conformations and interactions with membranes by physical chemical methods, including nuclear magnetic resonance and circular dichroism; protein folding and localization. *Mailing Add:* Dept Pharmacol Univ Tex Southwestern Med Ctr Dallas TX 75235-9041

GIERASCH, PETER JAY, b Washington, DC, Dec 19, 40; m 64; c 2. ASTRONOMY. *Educ:* Harvard Univ, BA, 62, PhD(appl math), 68. *Prof Exp:* Asst prof meteorol, Fla State Univ, 69-72; assoc prof, 72-80, PROF ASTRON, CORNELL UNIV, 80- *Concurrent Pos:* Res fel, Alfred P Sloan Found, 75. *Mem:* Am Meteorol Soc; Am Astron Soc; Int Astron Union. *Res:* Atmospheric motions on planets; solar convection. *Mailing Add:* Dept Astron Cornell Univ Ithaca NY 14853

GIERE, FREDERIC ARTHUR, b Galesville, Wis, Dec 10, 23; m 55; c 3. MAMMALIAN PHYSIOLOGY. *Educ:* Luther Col, AB, 47; Syracuse Univ, MS, 51; Univ NMex, PhD(physiol), 53. *Prof Exp:* Instr biol, Luther Col, 47-49; asst zool, Syracuse Univ, 49-51; asst physiol, Univ NMex, 51-53, asst prof, 53-55; assoc prof biol, Luther Col, 55-62; prof & chmn, 62-88, EMER PROF & CHMN BIOL DEPT, LAKE FOREST COL, 88- *Concurrent Pos:* Lectr, Univ SDak, 61; consult, Argonne Nat Lab, 67-84; USPHS fel, Inst Work Physiol, Oslo, Norway, 68-69; vis scientist molecular anat prog, Argonne Nat Lab, 77-78 & 84-85; consult, Abbott Labs, 78-82, Nat Bureau Standards, 83; adj prof, Northwestern Univ, 88- *Mem:* Fel AAAS; Am Soc Zool; Am Physiol Soc; Soc Exp Biol & Med; NY Acad Sci; Electrophoresis Soc. *Res:* Water and electrolyte metabolism; 2-D SDS electrophoresis of tissues and urine metabolites, work physiology. *Mailing Add:* 321 E Washington Lake Bluff IL 60044

GIERER, PAUL L, b New York, NY. ORGANIC CHEMISTRY. *Educ:* City Univ NY, BA, 59; Southern Ill Univ, PhD(chem), 72. *Prof Exp:* Chemist, GAF Corp, 63-66; res chemist, Mallinckrodt, Inc, 73-81; TECH MGR, WITCO CHEM CO, 81- *Mem:* Am Chem Soc. *Res:* Drug chemicals; organometallics; heterocycles. *Mailing Add:* 806 Morris Turnpike Short Hills NJ 07078

GIERING, JOHN EDGAR, b Easton, Pa, Feb 8, 29; m 54; c 3. PHARMACOLOGY, PHYSIOLOGY. *Educ:* Moravian Col, BS, 51; Purdue Univ, MS, 53, PhD(endocrinol), 57. *Prof Exp:* Sr res scientist, Astra Pharmaceut Prod, Inc, 56-65, head hemat sect, 65-66; head, Dept Develop Pharmacol, 66-73, mgr int sci opers, 73-76, assoc med dir, Pharmaceut div, 76-80, DIR SCI & COORDR LICENSING, PHARMACEUT DIV, PENNWALT CORP, 80- *Concurrent Pos:* Mem coun thrombosis, Am Heart Asn. *Mem:* AAAS; affil Am Soc Pharmacol & Exp Therapeut; assoc Pharmacol Soc Can; NY Acad Sci. *Res:* Reproductive physiology; action of sex steroids on uterine enzymes and morphology; blood coagulation; pharmacology of fibrinolytic agents and platelet aggregation; drug screening and evaluation; clinical pharmacology; analgesic drug development. *Mailing Add:* Pennwalt Corp Pharmaceut Div PO Box 1710 Rochester NY 14603

GIERING, WARREN PERCIVAL, b Troy, NY, Aug 17, 41; m 67; c 2. ORGANOMETALLIC CHEMISTRY. *Educ:* Rensselaer Polytech Inst, BS, 63; State Univ NY Stony Brook, PhD(chem), 69. *Prof Exp:* Res assoc chem, Brandeis Univ, 69-71; asst prof chem, 71-77, ASSOC PROF CHEM, BOSTON UNIV, 77- *Mem:* Am Chem Soc. *Res:* Synthesis and chemistry of reactive pi-complexes of alkenes, alkynes, carbenes, and cyclobutadienes. *Mailing Add:* Dept Chem Boston Univ 25 Buick St Boston MA 02215

GIERKE, TIMOTHY DEE, b Spokane, Wash, Dec 27, 46; m 70; c 4. PHYSICAL CHEMISTRY, POLYMER PHYSICS. *Educ:* Gonzaga Univ, BS, 69; Univ Ill, PhD(phys chem), 74. *Prof Exp:* Res asst phys chem, Univ Ill, 71-74; res chemist, 74-80, SR SUPVR, CENT RES & DEVELOP, E I DU PONT DE NEMOURS & CO INC, 80- *Mem:* Am Phys Soc. *Res:* Physical chemistry of liquid crystals and polymers; optical microscopy; ion containing polymers; Rayleigh light scattering; molecular theories of fluids. *Mailing Add:* 2405 Granby Rd Wilmington DE 19810

GIESBRECHT, JOHN, plant breeding, genetics; deceased, see previous edition for last biography

GIESE, ARTHUR CHARLES, b Chicago, Ill, Dec 19, 04; m 28; c 1. BIOLOGY. *Educ:* Univ Chicago, BS, 27; Stanford Univ, PhD(biol), 33. *Prof Exp:* Asst, 29-30, actg instr, 30-33, from instr to prof, 33-46, EMER PROF BIOL, STANFORD UNIV, 70- *Concurrent Pos:* Rockefeller Found fel, Princeton & Woods Hole, 39-40; Guggenheim fel, Calif Inst Technol & Northwestern Univ, 47-48 & European Marine Biol Stas, 59. *Mem:* AAAS; Soc Protozool; Am Soc Zool; Soc Gen Physiol; Am Soc Photobiol. *Res:* Cell physiology; effect of ultraviolet light on cell viability and functions; bioluminescence; photodynamic action; reproduction of marine invertebrates. *Mailing Add:* 792 Santa Maria Ave SW Palo Alto CA 94305-2493

GIESE, CLAYTON, b Minneapolis, Minn, July 19, 31; m 85. CHEMICAL PHYSICS, LASER SPECTROSCOPY. *Educ:* Univ Minn, BS, 53, PhD(physics), 57. *Prof Exp:* From instr to asst prof physics, Univ Chicago, 57-65; assoc prof, 65-74, PROF PHYSICS, UNIV MINN, MINNEAPOLIS, 74- *Mem:* Am Phys Soc. *Res:* Dynamics of molecular and ionic collisions; spectroscopy of molecular beams. *Mailing Add:* Sch Physics & Astron Univ Minn Minneapolis MN 55455

GIESE, DAVID LYLE, b Wells, Minn, Aug 18, 33. APPLIED STATISTICS, ACADEMIC ADMINISTRATION. *Educ:* Univ Minn, BS, 55, MA, 58, PhD(educ psychol), 65. *Prof Exp:* Instr math, 57-63, coordr res, 63-74, from asst prof to assoc prof math, 65-74, PROF MATH & ASST DEAN GEN COL, UNIV MINN, MINNEAPOLIS, 74- *Mem:* AAAS; Am Educ Res Asn; Psychomet Soc; Asn Instnl Res; Am Statist Asn. *Res:* Application of statistical methods to problems of educational curriculum development and evaluation with emphasis on program evaluation rather than individual achievement. *Mailing Add:* 106 Nicholson Univ Minn Minneapolis MN 55455

GIESE, GRAHAM SHERWOOD, b Newport News, Va, Oct 13, 31; m 58; c 5. COASTAL OCEANOGRAPHY. *Educ:* Trinity Col, Conn, BS, 53; Univ RI, MS, 64; Univ Chicago, PhD(geophys sci), 66. *Prof Exp:* Res asst coastal processes, Woods Hole Oceanog Inst, 56-62, asst scientist, 67; from asst prof to assoc prof oceanog, Univ PR, 67-72; assoc scientist oceanog, Marine Consult Assocs, Inc, 72-76; from assoc scientist to sr scientist oceanog, Provincetown Ctr Coastal Studies, 76-83; assoc dir, Marine Sci Res Ctr, State Univ NY, Stonybrook, 83-85; from guest investr to RES SPECIALIST, WOODS HOLE OCEANOG INST, 85- *Mem:* AAAS; Am Geophys Union; Am Meteorol Soc. *Res:* Coastal and near-shore oceanography. *Mailing Add:* PO Box 154 Provincetown MA 02657-0154

GIESE, JOHN H, b Chicago, Ill, Mar 10, 15; m 46; c 2. MATHEMATICS. *Educ:* Univ Chicago, BS, 36; Princeton Univ, PhD(math), 40. *Prof Exp:* Instr math, Princeton Univ, 39-40; instr, Rutgers Univ, 40-42; instr, Purdue Univ, 42-44; aerodynamicist, Bell Aircraft Corp, NY, 44-46; mathematician, Ballistic Res Labs, 46-58, chief comput lab, 59-68, chief appl math div, 68-74; RETIRED. *Concurrent Pos:* Instr math, Princeton Univ, 41-42; lectr, Univ Mich, 54-55; prof math, statist & comput sci, Univ Del, 63-77. *Res:* Isohedral polyhedra; elementary geometrical examples for computer graphics; personal computing; numerical analysis. *Mailing Add:* 2123 Sherwood Lane Havre de Grace MD 21078

GIESE, ROBERT FREDERICK, b Milwaukee, Wis, Apr 2, 43; m 66; c 1. PHYSICS. *Educ:* Univ Wis-Madison, BS, 65; Stanford Univ, PhD(physics), 73. *Prof Exp:* Fel physics, Argonne Nat Lab, 72-75, res asst systs analysis, 75-77, asst environ engr systs analysis, 77-78, SYSTS ANALYST, ENERGY SYSTS DIV, ARGONNE NAT LAB, 78- *Concurrent Pos:* Woodrow Wilson fel, Stanford Univ, 65-66, teaching asst, 66-67, res asst, 67-72; systs analysis, Off Dir, Argonne Nat Lab, 85- *Res:* Systems analysis in energy related fields; simulation of storage within electric utility supply systems. *Mailing Add:* Energy Systs Div 9700 S Cass Ave Argonne IL 60439

GIESE, ROBERT PAUL, b Greenbay, Wis, June 23, 36; m 75. PHYSICAL GEODESY, GEOGRAPHICAL INFORMATION SYSTEMS. *Educ:* Univ Wis-Madison, BS, 61; Univ Mo, Columbia, MA, 68; Univ Houston, PhD(math), 74. *Prof Exp:* Cartogr, Aeronaut Chart & Info Ctr, USAF, 62-65; computer programmer, Cancer Res Ctr, Columbia, 65-66; instr math, Univ Mo, Columbia, 66-67; computer consult, Robert P Giese Consults, 67-82; VPRES, G-TECH CORP, HOUSTON, 82- *Concurrent Pos:* Systs analyst, Calif Computer Prods, 68-69; sr res staff mem, Ray Geophys, Houston, 69-71; computer programmer, Dept Biomath, Univ Tex, Houston, 71-73; lectr math, Univ St Thomas, Houston, 75-80, adj prof bus, 80-88; vis lectr, Rice Univ, Houston, 83-84; adj prof, Grad Sch Mgt Systs, Univ Houston, 84-85, Dept Technol, 87-89 & Dept Indust Eng, 90- *Mem:* Math Asn Am. *Res:* Applied mathematics; computer science. *Mailing Add:* G-Tech Corp 6102 Queenswood Lane Houston TX 77008-6341

GIESE, ROGER WALLACE, b St Paul, Minn, Jan 26, 43; m 70; c 3. CLINICAL CHEMISTRY, CHROMATOGRAPHY. *Educ:* Hamline Univ, BS, 65; Mass Inst Technol, PhD(org chem), 69; Am Bd Clin Chem, dipl, 77. *Prof Exp:* Assoc on staff med, Peter Bent Brigham Hosp, 69-73; assoc prof, 77-81, PROF, NORTHEASTERN UNIV, 81- *Concurrent Pos:* Fel, Woodrow Wilson Soc, 65-66 & Am Cancer Soc, 69-71; res fel biol chem, Harvard Med Sch, 69-73, trainee human biochem training prog, 70-73; fac fel clin chem, Barnett Inst Chem Anal, Northeastern Univ, 74-; assoc ed, J Chromatogr, 88- *Mem:* Am Asn Clin Chem; AAAS; Am Chem Soc; Clin Radioassay Soc. *Res:* Ultratrace organic and biological analysis. *Mailing Add:* 110 Mugar Hall Northeastern Univ Boston MA 02115

GIESE, RONALD LAWRENCE, b Milwaukee, Wis, June 28, 34; m 54; c 2. FOREST ENTOMOLOGY. *Educ:* Wis State Col, Milwaukee, BS, 56; Univ Wis, MS, 58, PhD(entom, plant ecol), 60. *Prof Exp:* From asst prof to prof entom, Purdue Univ, West Lafayette, 60-75, dir natural resources & environ sci prog, 70-75; PROF FORESTRY & CHMN DEPT, UNIV WIS-MADISON, 75- *Mem:* Fel AAAS; Soc Am Foresters; Ecol Soc Am; Entom Soc Am; Entom Soc Can. *Res:* Population dynamics; animal and plant ecology; computer science; radiology; bioclimatology; symbioses; forest stand structure and species composition related to insect fauna; mycangia; periodicity; pest management; simulation. *Mailing Add:* Dept Forestry Russell Labs 1630 Linden Univ Wis Madison WI 53706

GIESE, ROSSMAN FREDERICK, JR, b New York, NY, Jan 7, 36; m 60; c 3. CRYSTALLOGRAPHY. *Educ:* Columbia Univ, BA, 56, MA, 59, PhD(mineral), 62. *Prof Exp:* Sr physicist, Carborundum Co, 61-66; asst prof geol sci, 66-68, assoc prof, 68-77, actg chmn dept, 70-72, PROF GEOL SCI, STATE UNIV NY BUFFALO, 77- *Concurrent Pos:* Sr cancer res scientist, Ctr Crystallog Res, Roswell Park Mem Inst, 66-68; sr res assoc, Nat Res Coun, 74-75; res assoc, Nat Ctr Sci Res, France, 75-76. *Honors & Awards:* Ralph Grim lectr, Univ Ill, 79. *Mem:* AAAS; Am Crystallog Asn; Mineral Soc Am. *Res:* Crystal structure and crystal chemistry of minerals, particularly clays and micas; interatomic forces in silicate minerals; crystal chemistry of water and hydroxyl in minerals. *Mailing Add:* Dept Geol 4240 Ridge Lea Amherst NY 14226

GIESECKE, ADOLPH H, b Oklahoma City, Okla, Apr 19, 32; m 54; c 4. MEDICINE, ANESTHESIOLOGY. *Educ:* Univ Tex, MD, 57. *Prof Exp:* Intern, William Beaumont Army Hosp, El Paso, Tex, 57-58; resident, Parkland Mem Hosp, Dallas, 60-63; from asst prof to assoc prof, 63-69, PROF ANESTHESIOL, UNIV TEX HEALTH SCI CTR, 69-; CHMN, DEPT ANESTHESIOL, SOUTHWESTERN MED SCH, UNIV TEX, 81- *Concurrent Pos:* Attend anesthesiologist, Parkland Mem Hosp, Children's Med Ctr & Vet Admin Hosp, Dallas, Tex, 63- & Presby Hosp, 67-; Fulbright lectr & guest prof, Johannes Gutenberg Univ, Ger, 70. *Mem:* AMA; Am Soc Anesthesiol; Int Anesthesia Res Soc. *Res:* Anesthesia for trauma and for obstetrics. *Mailing Add:* Dept Anesthesiol Univ Tex Health Sci Ctr 5323 Harry Hines Blvd Dallas TX 75235

GIESEKE, JAMES ARNOLD, b Granite City, Ill, Oct 16, 36; m 82; c 3. CHEMICAL ENGINEERING, AEROSOL SCIENCE. *Educ:* Univ Ill, BS, 59; Univ Wash, MS, 63, PhD(chem eng), 64. *Prof Exp:* From res chem engr to sr chem engr, 63-70, assoc fel, 70-74, res leader, 74-87, SECT MGR, BATTELLE COLUMBUS LABS, 87- *Mem:* Sigma Xi; Air Pollution Control Asn; Am Inst Chem Engrs; Am Chem Soc; Am Asn Aerosol Res. *Res:* Mechanics and physics of aerosols; dust collection problems; fission product transport analyses related to nuclear reactor safety. *Mailing Add:* 3930 Smiley Rd 505 King Ave Hilliard OH 43026

GIESEKING, JOHN ELDON, b Altamont, Ill, Oct 1, 05; m 36; c 1. SOIL SCIENCE. *Educ:* Univ Ill, BS, 26, MS, 27, PhD(soils), 34. *Prof Exp:* Asst soils, Univ Ill, 27-32; asst soil chem, Univ Mo, 32-33; from asst prof to prof soil physics, 34-74, EMER PROF SOIL CHEM, UNIV ILL, URBANA, 74- *Concurrent Pos:* Lectr, Univ Nebr, 58; ed, Monogr on Soil Components. *Mem:* AAAS; Am Chem Soc; Am Soil Sci Soc; Int Soc Soil Sci; fel Am Inst Chem. *Res:* Cation and anion exchange studies; mutual flocculation between positive and negative colloids; use of x-rays for diffraction studies; electron microscopy; petroleum cracking catalysts; use of radioactive potassium and phosphorus in cation and anion exchange and fixation studies. *Mailing Add:* 1221 W William Champaign IL 61821-4504

GIESEL, JAMES THEODORE, b Toledo, Ohio, Nov 17, 41; m 64; c 1. POPULATION BIOLOGY, ECOLOGY. *Educ:* Mich State Univ, BS, 63; Univ Ore, PhD(biol), 68. *Prof Exp:* Ford Found fel pop biol, Univ Chicago, 69-70; asst prof, 70-76, ASSOC PROF ZOOL, UNIV FLA, 76- *Mem:* Ecol Soc Am; Genetics Soc Am; Soc Study Evolution. *Res:* Analysis of the temporal aspects of interspecific competition; isozymic analysis of the genetics of natural populations; theoretical analysis of the effects of age distribution in populations with overlapping generations on effective number and selection. *Mailing Add:* Dept Zool Univ Fla Gainesville FL 32610

GIESLER, GREGG CARL, b Chicago, Ill, Sept 11, 44; m 68; c 4. NUCLEAR CHEMISTRY, COMPUTER PROGRAMMING. *Educ:* Univ Ill, Urbana, BS, 66; Mich State Univ, PhD(chem physics), 71. *Prof Exp:* Staff mem nuclear chem, Los Alamos Nat Lab, Univ Calif, 72-86 & Jomar Systs, 86-88; STAFF MEM NUCLEAR CHEM, SALEM TECH SERV, 88- *Mem:* Am Chem Soc; Am Phys Soc. *Res:* Pion interactions with complex nuclei; gamma-ray spectroscopy; computer-controlled data acquisition and processing; radiation dosimetry. *Mailing Add:* G Cubed 660 Navajo Los Alamos NM 87544

GIESS, EDWARD AUGUST, b Mineola, NY, Sept 12, 29; m 53; c 3. SOLID STATE PHYSICS, CERAMICS. *Educ:* State Univ NY Col Ceramics, Alfred, BS, 51, MS, 52, PhD(ceramics), 58. *Prof Exp:* Ceramic engr, Res Div, Nat Lead Co, 52-55; res staff mem crystal chem, Res Div, 58-84, MFG RES, INT BUS MACH CORP, 84- *Concurrent Pos:* Exec Comt, Am Asn Crystal Growth, 90- *Mem:* AAAS; fel Am Inst Chem; fel Am Ceramic Soc; Am Phys Soc; Am Asn Crystal Growth (vpres, 81-84). *Res:* Flux melt crystal growth; solid state reactions and sintering; electrooptic and magnetic materials; magnetic bubble garnet liquid phase epitaxy, glass ceramics, superconducting oxides. *Mailing Add:* Four Cotswold Dr Purdy NY 10578-9801

GIESSEN, BILL C(ORMANN), b Pittsburgh, Pa, June 8, 32; m 60; c 1. SOLID STATE CHEMISTRY, MATERIALS SCIENCE. *Educ:* Univ Gottingen, DSc(metall), 58. *Prof Exp:* Res assoc metall, Mass Inst Technol, 59-68; assoc prof chem, 68-83, PROF CHEM & MECH ENG, NORTHEASTERN UNIV, 73-; ASSOC DIR, BARNETT INST CHEM ANALYSIS & MAT SCI, 74- *Concurrent Pos:* Vis prof, Mass Inst Technol, 75, Harvard Univ, 76; dir, Encogy Mats Corp, 78-88; secy alloy phase comt, Am Inst Mining, Metall & Petrol Engrs, 70-72; dir & chmn, Cambridge Anal Asn, Inc, 79-89; dir, Marko Mat, Inc, 78-, Cambridge Mkt Analysis Corp, 89- *Honors & Awards:* Hume-Rothery Award, Metall Soc, Am Indst Mech Eng, 90. *Mem:* Am Chem Soc; Am Crystallog Soc; Am Inst Mining, Metall & Petrol Engrs; Mat Res Soc (secy, 79-82); Am Soc Metals. *Res:* Physical metallurgy; x-ray crystallography; structural and alloy chemistry, rapid solidification processing, mechanical, electronic and magnetic properties of alloys, metallic glasses; ceramic superconductors; ion nitriding. *Mailing Add:* Barnett Inst Northeastern Univ Boston MA 02115

GIESY, JOHN PAUL, JR, b Youngstown, Ohio, Aug 9, 48; m 70; c 1. LIMNOLOGY. *Educ:* Alma Col, BS, 70; Mich State Univ, MS, 72, PhD(limnol), 74. *Prof Exp:* Res assoc limnol, Savannah River Ecol Lab, Univ Ga, 74-80; PROF FISHERIES & WILDLIFE & COORDR, ENVIRON EFFECTS RES, PESTICIDE RES CTR, MICH STATE UNIV, 81- *Concurrent Pos:* NSF res fel, 69-70; fel, Woodrow Wilson Found, 70; instr ecol, Alma Col, 72; fel, North Cent Res Found, 72-74; instr ecol, Univ SC, 75, Univ Fla & Emory Univ, 78-; Fulbright fel, Univ Bayreuth, WGer, 87-88; mem bd dirs, Soc Environ Toxicol Chem, 87- *Mem:* Ecol Soc Am; Am Soc Limnol Oceanog; Int Soc Theoret Appl Limnol; Soc Environ Toxicol Chem (pres, 90-91). *Res:* Cycling of heavy metals; uptake and availability of heavy metals in aquatic systems; aquatic toxicology; pesticides; dioxins; PLBs; author of over 130 books, book chapters and journal publications. *Mailing Add:* Fisheries & Wildlife Mich State Univ East Lansing MI 48824

GIESY, ROBERT, b Columbus, Ohio, July 16, 22; m 50; c 4. PLANT MORPHOLOGY. *Educ:* Ohio State Univ, AB, 46, BSc, 54, PhD(plant morphol), 57. *Prof Exp:* Fulbright scholar, Univ Col NWales, 57-58; from instr to aasoc prof bot, Ohio State Univ, 58-78; RETIRED. *Mem:* Bot Soc Am; Electron Micros Soc Am. *Mailing Add:* 5318 N High St Columbus OH 43214

GIETZEN, DOROTHY WINTER, PHYSIOLOGICAL SCIENCES. *Educ:* Calif State Univ, Sacramento, BS, 70; Univ Calif, Davis, MS, 78, PhD(physiol), 83. *Prof Exp:* Sch health consult, Placer County Off Educ, Auburn, Calif, 71-76; grad student, Dept Nutrit, Univ Calif, Davis, 76-78, res assoc & teaching asst, Dept Animal Physiol, 78-83, postdoctoral fel, Food Intake Lab & Dept Physiol Sci, Sch Vet Med, 83-86, dir res, dir, Neurochem

Lab & asst res neurophysiologist, Dept Psychiat, Sch Med, 86-91, asst res neurophysiologist, Dept Physiol Sci, Sch Vet Med, 87-90, assoc res neurophysiologist, Univ, 90-91, EXEC DIR, FOOD INTAKE LAB & SR SCIENTIST, CLIN NUTRIT RES UNIT, UNIV CALIF, DAVIS, 88-, ASST PROF, DEPT PHYSIOL SCI, SCH VET MED, 91- *Concurrent Pos:* Sch health consult, Rocklin Sch Dist, Calif, 76-78; mem, Animal Resources Comt, Dept Nutrit, Univ Calif, Davis, 89, Space Allocation Comt, 89-90, Dean's Res Coord Coun, Sch Med, 89-91 & Biomed Res Support Grant Selection Adv Comt, 90. *Mem:* Am Inst Nutrit; Am Physiol Soc; Sigma Xi; Soc Neurosci; Soc Study Ingestive Behav; Int Brain Res Orgn; Women Neurosci; AAAS; Am Asn Univ Women. *Mailing Add:* Dept Physiol Sci Univ Calif Sch Vet Med Davis CA 95616

GIEVER, JOHN BERTRAM, b Omaha, Nebr, Sept 18, 19; m 43; c 3. MATHEMATICS. *Educ:* Creighton Univ, BS, 42; Mass Inst Technol, PhD(math), 48. *Prof Exp:* Instr math, Boston Univ, 48-51, asst prof, 51-52; mem staff, Instrumentation Lab, Mass Inst Technol, 52-53; from asst prof to assoc prof math, Univ Okla, 53-59; PROF MATH SCI, NMEX STATE UNIV, 59- *Mem:* Am Math Soc; Asn Symbolic Logic; Math Asn Am. *Res:* Algebraic topology. *Mailing Add:* 475 Milton Ave Las Cruces NM 88005

GIFFEN, MARTIN BRENER, b Pittsburgh, Pa, Sept 13, 19; c 2. PSYCHIATRY, PSYCHOLOGY. *Educ:* Univ Mich, AB, 41; Univ Pittsburgh, MD, 45. *Prof Exp:* Chief psychiat, USAF Hosp, Fla, 54-57, Wiesbaden, Ger, 57-61, chmn dept psychiat, Wilford Hall USAF Hosp, Tex, 61-68, dir psychiat residency prog & hosp serv, 65-68, vcomdr, 68; prof psychiat, Univ Tex Med Sch, San Antonio, 68-85, coordr psychiat residency training, Univ Tex Health Sci Ctr, 72-85, clin prof psychiat, 85; RETIRED. *Concurrent Pos:* Consult psychiat, Surgeon Gen, USAF Europe, 57-61, Surgeon Gen, US Air Force, 61-68, US Attorney Western Dist Tex, 63-69 & Wilford Hall USAF Hosp, Tex, 68-; examr, Am Bd Psychiat & Neurol, 71-72. *Mem:* Fel AAAS; fel Am Psychiat Asn; fel Am Col Psychiat; Asn Mil Surgeons US; Am Acad Psychiat & Law. *Mailing Add:* Dept Psychiat Univ Tex Med Sch San Antonio TX 78284

GIFFEN, ROBERT H(ENRY), b Pottsville, Pa, Feb 10, 22; m 49; c 3. ENVIRONMENTAL ASSESSMENT, CORROSION. *Educ:* Newark Col Eng, BS, 43; Iowa State Col, MS, 47, PhD(chem eng), 51. *Prof Exp:* Control chemist, Gen Chem Co, NJ & Ill, 43-44; chem engr, Los Alamos Sci Lab, Univ Calif, 46; jr res asst, Inst Atomic Res, Iowa State Col, 47-51; sr engr, 51-57, supvr engr, 57-64, fel engr, 64-66, supvr, 66-69, mgr radiation technol, 69-72, ADV ENGR, BETTIS ATOMIC POWER LAB, WESTINGHOUSE ELEC CORP, 72- *Mem:* Am Inst Chem Engrs. *Res:* Design and development of fluid systems for pressurized water nuclear power plants; environmental assessments. *Mailing Add:* Bettis Atomic Power Lab Westinghouse Elec Corp Box 79 West Mifflin PA 15122-0079

GIFFEN, WILLIAM MARTIN, JR, b Akron, Ohio, June 23, 33; m 61; c 3. POLYMER CHEMISTRY. *Educ:* Univ Akron, BS, 55, MS, 56, PhD(polymer chem), 61. *Prof Exp:* Asst proj chemist, Amoco Chem Corp, Ind, 61-62; sr res chemist, Gen Tire & Rubber Co, 62-64; sr chemist, Tech Dept, Marbon Chem Div, Borg-Warner Corp, WVa, 64-68, res assoc, Develop Div, 68-70; mem fac, Wash Tech Col, 71-72; sr chemist, Addressograph-Multigraph Corp, Ohio, 72-74; sr chemist, Standard Oil Co, Ohio, 74-85; RETIRED. *Mem:* Am Chem Soc. *Res:* Emulsion, graft, and alkylene oxide polymerizations; cationic and anionic solution polymerizations; vinyl polymerizations and copolymerizations; latex formulation; high impact resins; photo-conductive polymers; latex can and metal coatings; food packaging plastics; reinforced and filled plastics composites; composites for metal replacement; membrane separation of liquid mixtures; polyvinyl chloride blends & alloys. *Mailing Add:* 7250 Darien Dr Hudson OH 44236

GIFFIN, EMILY BUCHHOLTZ, b Madison, Wis, Sept 9, 47; m 69; c 2. VERTEBRATE PALEONTOLOGY, PALEONEUROLOGY. *Educ:* Col of Wooster, BA, 69; Univ Wis, MS, 71; George Washington Univ, PhD(biol sci), 74. *Prof Exp:* Cur paleont, Pa State Mus, 74-76; asst prof geol, Univ Wis-Milwaukee, 80-88; asst prof geol, 77-80, ASST PROF BIOL, WELLESLEY COL, 88- *Mem:* Soc Vert Paleont; Am Soc Zoologists; Paleont Soc; Geol Soc Am. *Res:* Paleontology. *Mailing Add:* Dept Biol Sci Wellesley MA 02181

GIFFIN, WALTER C(HARLES), b Walhonding, Ohio, Apr 22, 36; m 56; c 2. INDUSTRIAL ENGINEERING. *Educ:* Ohio State Univ, BIndustEng & MSc, 60, PhD(mass transp), 64. *Prof Exp:* Res engr, Gen Motors Res Labs, 60-61; res assoc opers res, Eng Exp Sta, 61-62, from instr to prof, 62-87, EMER PROF INDUST ENG, OHIO STATE UNIV, 87-; PROF & CHAIR ENG, UNIV SOUTHERN COLO, 87- *Mem:* Am Inst Indust Engrs; Am Soc Eng Educ; Opers Res Soc Am; Inst Mgt Sci. *Res:* Air traffic control, inventory control and transportation systems; queueing phenomena. *Mailing Add:* Dept Eng Univ Southern Colo 2200 Bonforte Blvd Pueblo CO 81001-4901

GIFFORD, CAMERON EDWARD, b New Bedford, Mass, Sept 23, 31; m 52; c 2. ECOLOGY, PHYSIOLOGY. *Educ:* Earlham Col, BA, 55; Harvard Univ, MA, 59; Univ Ga, PhD(zool), 64. *Prof Exp:* Assoc prof biol & chmn dept, Earlham Col, 61-72, dir, David Worth Dennis Biol Sta, 62-72; res specialist environ systs lab, Woods Hole Oceanog Inst, 72-76, oil spill res coordr ecosystems ctr, Marine Biol Lab, 76-78; OWNER & VPRES, GIFFORD & GIFFORD INC, DBA H V LAWRENCE, 77- *Mem:* Int Ornith Cong; Sigma Xi. *Res:* Avian physiology; homing behavior in bats; chlorophyll determinations in various marine algae; lipid determinations and migratory behavior in the bobolink; mariculture, investigations for the biological requirements and development of technical systems for mass culturing various phycocolloid producing marcoscopic marine algae. *Mailing Add:* 14 Beechwood Dr Falmouth MA 02540-2337

GIFFORD, DAVID STEVENS, b Glens Falls, NY, Nov 14, 24; m 51; c 4. ORGANIC CHEMISTRY. *Educ:* Dartmouth Col, AB, 49; Univ Conn, PhD(org chem), 60. *Prof Exp:* Res chemist, Naugatuck Chem Div, US Rubber Co, 50-54; asst col instr chem, Univ Conn, 55-56, asst, 56-57, asst col instr, 57-59; res asst, Purdue Univ, 59-61; from asst ed to assoc ed org indexing, 61-69, sr assoc indexer, 69-73, SR ASSOC ED, CHEM ABSTR SERV, OHIO STATE UNIV, 73- *Mem:* AAAS; Sigma Xi. *Res:* Abstracting and indexing terpenes and carbohydrates from Russian and Japanese literature. *Mailing Add:* Chem Abstr Serv Ohio State Univ Columbus OH 43210

GIFFORD, ERNEST MILTON, b Riverside, Calif, Jan 17, 20; m 42; c 1. BOTANY. *Educ:* Univ Calif, Berkeley, AB, 42, PhD(bot), 49. *Prof Exp:* From instr to assoc prof bot, 49-57, jr botanist, Agr Exp Sta, 49-51, from asst botanist to assoc botanist, 51-57, chmn dept, 63-68, 70-71 & 73-78, PROF BOT & BOTANIST, AGR EXP STA, UNIV CALIF DAVIS, 62- *Concurrent Pos:* Merck sr res fel, Harvard Univ, 56-57; NSF res grant, Univ Calif, 58-66, & 79-82; John Simon Guggenheim Mem Found fel, 66-67; Fulbright res scholar, Nat Ctr Sci Res, France, 66-67; NATO sr fel, France, 74; ed-in-chief, Am J Bot, 75-79. *Mem:* Am Inst Biol Sci; Bot Soc Am (vpres, 81, pres, 82); Sigma Xi; Linnean Soc London; Int Soc Plant Morphologists (vpres, 80-84). *Res:* Developmental anatomy and ultrastructure of vascular plants; quantitative studies of lower vascular plant meristems; spermatogenesis of Ginkgo biloba. *Mailing Add:* Dept Bot Univ Calif Davis CA 95616-8537

GIFFORD, FRANKLIN ANDREW, JR, b Union City, NJ, May 7, 22; m; c 2. METEOROLOGY. *Educ:* NY Univ, BS, 47; Pa State Univ, MS, 54, PhD(meteorol), 55. *Prof Exp:* Meteorologist, Northwest Airlines, Inc, 45-50; res meteorologist, US Weather Bur, Nat Oceanic & Atmospheric Admin, 50-66, dir, Atmospheric Turbulence & Diffusion Lab, 66-80; CONSULT, 80- *Concurrent Pos:* Mem adv comt reactor safeguards, US AEC, 58-68, consult, Adv Comt Reactor Safety, 68-80; consult, Int Atomic Energy Agency, 66-83. *Honors & Awards:* Gold Medal, US Dept Com, 63; Outstanding Contrib to Advanc Appl Meteorol Award, Am Meteor Soc, 90. *Mem:* Fel Am Meteorol Soc; fel AAAS; Sigma Xi. *Res:* Atmospheric turbulence and diffusion; air pollution; reactor hazards; meteorology of other planets. *Mailing Add:* 109 Gorgas Lane Oak Ridge TN 37831

GIFFORD, GEORGE EDWIN, b Minneapolis, Minn, Dec 6, 24; m 56; c 2. VIROLOGY, MICROBIOLOGY. *Educ:* Univ Minn, Minneapolis, BA, 49, MS, 53, PhD(bact, immunol), 55. *Prof Exp:* Instr immunol & bact, Sch Med, Univ Minn, Minneapolis, 55-56; from asst prof to assoc prof microbiol, 57-68, actg chmn dept, 65-66, 70-73 & 82-83, grad coord dept, 70-82, PROF IMMUNOL & MED MICROBIOL, COL MED, UNIV FLA, 68-, ASSOC DEAN GRAD EDUC, 90- *Concurrent Pos:* USPHS spec fel, Nat Inst Med Res, London, 62-63; vis prof, Hadassah Med Sch, Hebrew Univ, Israel, 84 & Fraunhofer Inst, WGer, 85. *Mem:* Fel AAAS; fel Am Acad Microbiol; Am Soc Microbiol; Am Asn Immunol; Soc Gen Microbiol; Am Soc Virol. *Res:* Natural resistance to viral disease at the cellular level; macrophage killing of tumor cells; role of interferon in cell-mediated immunity; production and action of tumor necrosis factor; effect of interferon, vitamin A and tumor necrosis factor on tumor cells. *Mailing Add:* Dept Immunol & Med Microbiol Univ Fla Col Med Gainesville FL 32610-0266

GIFFORD, GERALD F, b Chanute, Kans, Oct 24, 39; m 82. RANGELAND HYDROLOGY. *Educ:* Utah State Univ, BS, 62, MS, 64, PhD(watershed sci), 68. *Prof Exp:* Lectr watershed mgt, Univ Nev, Reno, 65-67; from asst prof to prof range watershed sci & chmn, Watershed Sci Unit, Utah State Univ, 67-84, dir, Inst Land Rehab, 82-84; HEAD, DEPT RANGE, WILDLIFE & FORESTRY, UNIV NEV, RENO, 84- *Concurrent Pos:* Consult, Nat Park Serv, Nat Comn Water Qual, Bur Land Mgt, Amax Coal Co, Smithsonian Inst, Tex Tech Found, Mountain Fuel Supply, Univ Minn Morocco Proj & Off Tech Assessment, 67-; vis prof, Div Land Resources Mgt, Commonwealth Sci & Indust Res Orgn, Alice Springs & Canberra, Australia, 73-74; mem, Wild & Free-Roaming Horses & Burro Comt, Nat Acad Sci, 79-80, Sci Comt Rangeland Hydrol, Soc Range Mgt; assoc ed, J Range Mgt, 82-87, 91- & Arid Soil Res & Rehab, 84-90; AAAS Comt, Arid Lands, 87-90. *Mem:* Soc Range Mgt; Am Water Resources Asn; Soil Conserv Soc Am; Soc Wetlands Scientists. *Res:* Man's impact on the hydrologic cycle in arid and semiarid rangeland environments; runoff; erosion; infiltration; evapotranspiration; groundwater recharge. *Mailing Add:* Dept Range Wildlife & Forestry Univ Nev 1000 Valley Rd Reno NV 89512

GIFFORD, HAROLD, b Omaha, Nebr, Jan 25, 06; m 36; c 3. OPHTHALMOLOGY. *Educ:* Univ Nebr, BSc, 30, MD, 31. *Prof Exp:* From instr to assoc prof, 36-64, prof, 64-80, EMER PROF OPHTHAL, COL MED, UNIV NEBR, 80- *Concurrent Pos:* Practicing physician. *Mem:* Am Ophthal Soc; fel Am Acad Ophthal & Otolaryngol. *Res:* Clinical research. *Mailing Add:* 3636 Burt St Omaha NE 68131

GIFFORD, JAMES FERGUS, b Lynn, Mass, Mar 3, 40; m 63. HISTORY OF MEDICINE. *Educ:* Dartmouth Col, BA, 61; Andover Newton Theol Sch, BD, 64, STM, 65; Duke Univ, PhD(social & intellectual hist), 69. *Prof Exp:* Assoc prof hist, Guilford Col, 69-77; ASSOC PROF HIST MED, DUKE UNIV, 77- *Concurrent Pos:* Josiah Macy Jr Found fel, 75-76. *Mem:* Am Asn Hist Med; Orgn Am Historians. *Mailing Add:* Dept Commun & Family Med Duke Univ Box 3702 Durham NC 27706

GIFFORD, JOHN A, b Strasbourg, France, Feb 17, 47; US citizen. ARCHAEOLOGICAL GEOLOGY, GEOARCHAEOLOGY. *Educ:* Univ Mass, Amherst, BS, 69; Univ Miami, MS, 73; Univ Minn, Minneapolis, PhD(archaeol geol), 78. *Prof Exp:* Assoc dir, Archaeometry Lab, Univ Minn, 78-82; asst prof archaeology, 83-86, ASSOC PROF ARCHAEOL, UNIV MIAMI, CORAL GABLES, 86- *Mem:* Archaeol Inst Am; Geol Soc Am; Soc Prof Archeol; Soc Archaeol Sci. *Res:* Physical environmental changes affecting coastal cultures, primarily prehistoric in the Mediterranean and Caribbean. *Mailing Add:* Dept Anthrop PO Box 248106 Coral Gables FL 33124

GIFFORD, RAY WALLACE, JR, b Westerville, Ohio, Aug 13, 23; m 47, 73; c 4. MEDICINE. *Educ:* Otterbein Col, BS, 44; Ohio State Univ, MD, 47; Univ Minn, MS, 52. *Hon Degrees:* DSc, Otterbein Col, 86. *Prof Exp:* Intern, Colo Gen Hosp, Denver, 47-48; resident physician internal med, Univ Hosp, Ohio State Univ, 48-49; fel, Mayo Found, 49-52, consult sect internal med, Mayo Clinic, 52-61, instr med, Mayo Found, Univ Minn, 53-58, asst prof, 58-61; mem staff, 61-67, head dept hypertension & nephrology, 67-85, VCHMN DIV MED, CLEVELAND CLIN FOUND, 78- *Concurrent Pos:* Mem, Adv Comt to Dir NIH, 82-86; mem bd trustees, AMA, 86-90. *Honors & Awards:* Oscar B Hunter Award, Am Soc Clin Pharmacol & Therapeut, 79; Simon Rodbard Award, Am Col Chest Physicians, 82. *Mem:* AMA; Am Heart Asn; Am Fedn Clin Res; fel Am Col Physicians; fel Am Col Cardiol; fel Am Col Chest Physicians. *Res:* Hypertension and renal disease. *Mailing Add:* 9500 Euclid Ave Cleveland OH 44195-5042

GIGER, ADOLF J, b Solothurn, Switz, Jan 4, 27; US citizen; m 58; c 3. RADIO COMMUNICATIONS, RADIO PROPAGATION. *Educ:* Swiss Fed Inst Technol, dipl elec eng, 50, Dr sc tech, 56. *Prof Exp:* Res engr, Inst High Frequency Tech, Swiss Fed Inst Technol, 50-56; mem tech staff, AT&T Bell Labs, 56-58, supvr, 58-71, dept head, 71-89; CONSULT, GIGER RADIO CONSULT, 90- *Concurrent Pos:* US deleg, Study Group 9, Int Consultative Radio Comt, Geneva, Switz, 77-89. *Mem:* Fel Inst Elec & Electronics Engrs. *Res:* Design and analysis of digital radio relay systems; physics and modeling of microwave propagation; microwave interference analysis and prediction; development of antennas and circuits for interference reduction; consulting on air navigation systems. *Mailing Add:* 27 Olde Farms Rd Boxford MA 01921

GIGLI, IRMA, b Cordoba, Arg, Dec 22, 31. DERMATOLOGY, IMMUNOLOGY. *Educ:* Nat Univ Cordoba, MD, 57. *Prof Exp:* Intern med, Cook County Hosp, Chicago, 57-58, resident dermat, 58-60; fel, NY Univ, 60-61; vis investr immunol, Howard Hughes Med Inst, Miami, Fla, 61-64; investr, Univ Frankfurt, 65-67; res assoc, Harvard Med Sch, 67-69, from asst prof to assoc prof dermat, 69-75, assoc immunol, 69-76, sr assoc dermat, 71-75; prof dermat & exp med, NY Univ, 76-, Dir, Asthma & Allergic Dis Ctr Immunodermat Studies, 80-; CHIEF DIV DERMAT, SAN DIEGO MED CTR. *Concurrent Pos:* Med Found award, 68-69; Am Cancer Soc fac res award, 70-72; NIH res career develop award, 72-76; Guggenheim award, 74-75; vis scientist biochem, Univ Oxford, 74-75; chief dermat, Peter Bent Brigham Hosp, 71-75. *Mem:* Am Soc Clin Invest; Am Asn Immunologists; Am Acad Dermat; Soc Invest Dermat; Asn Am Physicians. *Res:* Immunochemistry and immunobiology or the complement system; studies of skin diseases which may have immunological basis. *Mailing Add:* Chief Div Dermat Univ Calif San Diego Med Ctr 225 Dickinson St San Diego CA 92103

GIGLIO, RICHARD JOHN, b Hartford, Conn, Aug 27, 37; m 60; c 2. OPERATIONS RESEARCH, INDUSTRIAL ENGINEERING. *Educ:* Mass Inst Technol, BS, 59; Stanford Univ, MS, 62, PhD(opers res), 65. *Prof Exp:* Engr aerospace eng, 59-60, sr proj engr indust eng & opers res, 63-67, prof, 67-77, HEAD INDUST ENG & OPERS RES, UNIV MASS, 78- *Concurrent Pos:* Prin investr grants, New Eng Elec Co, 72-73, HEW, 73-76, Army Res & Develop Command, 76-77 & Mass Dept Health, 77-78. *Mem:* Opers Res Soc Am; Inst Mgt Sci. *Res:* Design and analysis of large scale industrial and public systems, especially health related systems. *Mailing Add:* Dept Indust Eng & Opers Res Univ Mass Amherst MA 01003

GIGLIOTTI, HELEN JEAN, b Rochester, NY, July 27, 36. BIOCHEMISTRY. *Educ:* Vassar Col, BA, 58; Univ Mich, PhD(biochem), 63. *Prof Exp:* Res assoc biochem, Scripps Clin & Res Found, 63-66; asst prof, 66-69, assoc prof, 69-73, chmn dept, 77-80, PROF CHEM, CALIF STATE UNIV, FRESNO, 74-, ASST VPRES ACAD AFFAIRS, BUDGET & INSTRUCT RESOURCES, 81- *Concurrent Pos:* Consult, Cent Calif Med Labs. *Mem:* AAAS; Am Chem Soc; Am Asn Clin Chem. *Res:* Clinical chemistry methods, especially chromatographic and enzymological; plant biochemistry. *Mailing Add:* Off VPres Acad Affairs Calif State Univ Fresno CA 93726

GIGUERE, JOSEPH CHARLES, b Ottawa, Ont, Aug 10, 39; m 63. ELECTRICAL ENGINEERING. *Educ:* McGill Univ, BEng, 60; NS Tech Col, MEng, 65, PhD(elec eng), 69. *Prof Exp:* Lectr elec eng, NS Tech Col, 65-69; asst dean, 72-80, ASSOC PROF ELEC ENG, CONCORDIA UNIV, 69-, ASSOC DEAN ENG, 80- *Concurrent Pos:* Consult, EMI Electronics Ltd, NS, 65-69; dir, Ctr Res & Info, Montreal, 83- *Mem:* Inst Elec & Electronics Engrs. *Res:* Electrical network theory. *Mailing Add:* Dept Elec Eng Concordia Univ 1455 De Maisonneuve W Rm GM1100 Montreal PQ H3G 1M8 Can

GIKAS, PAUL WILLIAM, b Lansing, Mich, July 23, 28; m 52; c 3. PATHOLOGY. *Educ:* Univ Mich, BA, 50, MD, 54. *Prof Exp:* From instr to assoc prof, 60-69, PROF PATH & ASST DEAN ADMIN, UNIV MICH MED SCH, ANN ARBOR, 69- *Concurrent Pos:* Chief lab serv, Vet Admin Hosp, Ann Arbor, 60-68; mem adv comt traffic safety, Dept HEW, 66-68; consult, Armed Forces Inst Path & USPHS, 67-68 & 71. *Mem:* AMA; US-Can Acad Path. *Res:* Pathogenesis of injuries in highway accidents; long term preservation of blood by freezing; correlation of morphologic findings in prostatic carcinoma with imaging techniques. *Mailing Add:* Dept Path Univ Mich Ann Arbor MI 48109-0054

GIL, JOAN, b Barcelona, Catalonia, pain, June 26, 40; m 70; c 2. ASBESTOS PATHOLOGY. *Educ:* Univ Barcelona Med Sch, Med, 64, Doctorate Med, 68; Univ Berne Med Sch, Habilitation, 74. *Hon Degrees:* MA, Univ Pa, 83. *Prof Exp:* Mem staff, anat, Med Sch, Univ Berne, 66-76; assoc prof med, Univ Miami, 76-77; assoc prof med, Univ Pa, 77-84; PROF PATH, MT SINAI SCH MED, 84- *Mem:* Int Soc Stereology; Am Thoracic Soc; Am Physiol Soc; NY Acad Sci. *Res:* Image analysis in histology and cytology, computerized diagnosis, correlations between structure and function of lung pathology. *Mailing Add:* Dept Path Box 1194 Mt Sinai Med Ctr 100th St & Fifth Ave New York NY 10029-6574

GIL, SALVADOR, b Salta, Arg, Sept 8, 50; c 2. NUCLEAR ASTROPHYSICS. *Educ:* Univ Nat Tucuman, Argentina, BS, 75; Univ Wash Seattle, MS, 80 & PhD(physics), 84. *Prof Exp:* Teaching asst physics, Univ Nat Tucuman, 71-77; res asst nuclear physics, Tandar Cnea, Buenos Aires, 78-79 & 84-86; res asst, 79-84, RES ASSOC, NUCLEAR PHYSICS LAB, UNIV WASH, 86- *Concurrent Pos:* Exchange visitor, Am Field Serv, 68-69; fel Orgn Am States, 82-84; assoc prof, Univ Buenos Aires, 88- *Mem:* Am Phys Soc; Am Teacher Asn; Argentine Physics Asn. *Res:* Heavy ion nuclear physics; sub-barrier fusion, in particular, the spin distribution of the compound nucleus form in heavy ion reactions. *Mailing Add:* Nuclear Physics Lab Univ Wash Seattle GL 10 Seattle WA 98198

GILANI, SHAMSHAD H, b Lahore, WPakistan, Feb 4, 37; m 70; c 2. TERATOLOGY, CARDIOLOGY. *Educ:* Univ Punjab, WPakistan, BS, 55, MS, 58; State Univ NY Buffalo, PhD(exp embryol), 67. *Prof Exp:* From instr to assoc prof, 67-81, PROF ANAT, NJ MED SCH, COL MED & DENT NJ, 81- *Honors & Awards:* Golden Apple Award, AMA, 73, 82, 84, 85, 86. *Mem:* Am Asn Anat; Teratology Soc; Europ Teratology Soc; Soc Develop Biol; Sigma Xi. *Res:* Analyzing the mechanism of abnormal development of the heart. *Mailing Add:* Anat Dept NJ Med Sch 100 Bergen St Newark NJ 07103

GILARDI, EDWARD FRANCIS, b Brooklyn, NY, Apr 26, 36; m 63; c 4. ENVIRONMENTAL ENGINEERING, ENVIRONMENTAL SCIENCE. *Educ:* Manhattan Col, BCE, 57; Rutgers Univ, MS, 63, PhD(environ sci), 66. *Prof Exp:* Engr, NY State Conserv Dept, 57-59; res assoc, Rutgers Univ, 59-65; PROJ MGR, ROY F WESTON INC, 65- *Mem:* Am Soc Civil Engrs; Air Pollution Control Asn; Am Acad Environ Engrs; Water Pollution Control Asn; Sigma Xi. *Res:* Biological waste treatment; application of wastes to land environment; reaction of air pollutants with building materials. *Mailing Add:* 12 Long Lane Malvern PA 19355

GILARDI, GERALD LELAND, medical bacteriology, for more information see previous edition

GILARDI, RICHARD DEAN, b Wisconsin Rapids, Wis, Feb 23, 40; m 60; c 1. STRUCTURAL CHEMISTRY. *Educ:* Mass Inst Technol, BS, 61; Univ Md, PhD(phys chem), 66. *Prof Exp:* Chemist, Am Instrument Co, Md, 62-63; res asst, Univ Md, 63-66; chemist, Inst Defense Analysis, Va, 66; res assoc x-ray diffraction, 66-68, RES CHEMIST, NAVAL RES LAB, 68- *Concurrent Pos:* Consult, Inst Defense Analysis, 65-66. *Mem:* Am Crystallog Asn; Am Chem Soc; Sigma Xi. *Res:* Techniques of diffraction analysis; molecular structure determination by x-ray diffraction; correlation of molecular structure with biological activity; biopolymer structure. *Mailing Add:* Naval Res Lab Code 6030 Washington DC 20375

GILBARG, DAVID, b Brooklyn, NY, Sept 17, 18; m 41; c 1. APPLIED MATHEMATICS. *Educ:* City Col New York, BS, 38; Ind Univ, PhD(math), 41. *Prof Exp:* Asst math, Ind Univ, 39-41; physicist, Nat Bur Standards, 41-42; chief, Fluid Dynamics Test Sect, Naval Ord Lab, 42-45, chief, Theoret Mech Subdiv, 45-46; from asst prof to assoc prof math, Ind Univ, 46-57; exec head dept, 59-69, PROF MATH, STANFORD UNIV, 57- *Mem:* Am Math Soc; Math Asn Am; Ger Soc Appl Math & Mech. *Res:* Fluid dynamics; partial differential equations. *Mailing Add:* Dept Math Stanford Univ Stanford CA 94305

GILBERT, ALLAN HENRY, b Liverpool, Eng, Oct 20, 29; m 46; c 2. SYNTHETIC ORGANIC CHEMISTRY. *Educ:* Univ Liverpool, BSc, 51, PhD(org chem), 54. *Prof Exp:* Can Res fel, Univ NB, 54-56; res chemist, 56-65, sect chief detergent solids sect, 65-72, mgr tech serv, 72-75, dir sci res, 75-83, VPRES, SCI AFFAIRS LEVER BROS CO, EDGEWATER, 83- *Mem:* Am Chem Soc; Asn Res Dirs. *Mailing Add:* Lever Bros 45 River Rd Edgewater NJ 07020

GILBERT, ALTON LEE, b Elmira, NY, Apr 13, 42; m 62; c 2. INFORMATION SCIENCES, RESEARCH MANAGEMENT. *Educ:* NMex State Univ, BSEE, 70, MS, 71, ScD(info sci), 73. *Prof Exp:* Res engr commun, NMex State Univ, 69-73; res electronics engr info sci, White Sands Missile Range, 73-82; TECH DIR, TECH SOLUTIONS INC, MESILLA PARK, NMEX. *Concurrent Pos:* Adj assoc prof, Dept Elec Eng, NMex State Univ, 73-84; evaluator res prog, US Army Res Off, 73-82; partner & engr, Gilbert-Roman Eng Asn, 74-77; mem tech review comt, Joint Serv Electronics Prog, 75-82; adv, Tex Tech Control Theory Prog, Off Naval Res, US Navy, 76-82. *Honors & Awards:* Edward Gamble Award, 79. *Mem:* Inst Elec & Electronics Engrs; Sigma Xi. *Res:* Real-time applications of pattern recognition and machine intelligence; improvements in communications systems for data transmission and processing; novel computer architectures for data handling in real-time; robotics and artificial intelligence; test and evaluation instrumentation; real-time video tracking. *Mailing Add:* Tech Solutions Inc PO Box 1148 Mesilla Park NM 88047

GILBERT, ARTHUR CHARLES, b New York, NY, Sept 23, 26; div; c 2. VIBRATION ANALYSIS & DESIGN, PRODUCT APPLICATIONS. *Educ:* NY Univ, BAeroE, 46 MAeroE, 47 ScD(eng), 56. *Prof Exp:* Sr tech exec res & develop & vpres eng, Cent Intel Agency, Bendix & United Aircraft Corp, 61-70; sci adv, Chief Naval Opers, USN, Dept Defense, 70-75; vpres, Data Solutions Corp, 75-77 & Unified Industs, Inc, 77-78; dir, OAO Corp, 78-81; construct engr, Daughters Am Revolution, 82-87; CONSULT, FMC CORP & LITTON CORP, 82-, USN DEPT DEFENSE, 88- *Concurrent Pos:* Founder & dir, Auto-Train Corp, 68-79; consult, NIH, 69-70; consult & lectr, Naval War Col, USN, 75-76; expert witness, numerous atty partnerships, 85- *Res:* System analysis and design; structural dynamics; advanced technical planning; engineering management; weapon system requirements analyses and design; aircraft structural failures and patent infringements. *Mailing Add:* 1201 S Eads St Arlington VA 22202

GILBERT, ARTHUR DONALD, b Niagara Falls, NY, Aug 12, 16; m 41; c 4. INDUSTRIAL CHEMISTRY. *Educ:* Middlebury Col, AB, 38; Cornell Univ, PhD(org chem), 42. *Prof Exp:* Asst instr chem, Cornell Univ, 39-41; res chemist, Eastern Lab, E I du Pont de Nemours & Co, Inc, 42-50, lab sect head, 50-52, tech specialist, Explosives Dept, Tech Div, 52-55, asst dir, Burnside Lab, 55-56, mgr tech div, Foreign Rels Dept, 56-58, mgr patent & licensing sect, Int Dept, 58-60, asst dir, Eastern Lab, 60-63, dir, Explosives Exp Sta Lab, 63-68, tech specialist, Res & Develop Div Staff, 68-73, patents & licensing specialist, Polymer Intermediates Dept, 73-77; RETIRED. *Mem:* Am Chem Soc; Soc Chem Indust. *Res:* Detonator research; organic polymer intermediate research; fundamental chemistry of organic reactions of nitric acid. *Mailing Add:* Gardenside M-2 Shelburne VT 05482

GILBERT, BARRIE, b Bournemouth, Eng, June 5, 37. SEMICONDUCTOR DESIGN, DEVICE DESIGN. *Educ:* Bournemouth Municipal Col, HNC (appl physics), 62. *Prof Exp:* Scientific asst, Signal & Develop Estab, UK Ministry Defense, 54-55; engr, Mullard, 58-64; design engr, Vickers-Armstrong, 64; group leader Techtronics, 64-70 & Analog Devices, 77-80, DIV FEL ANALOG DEVICES, PLESSY RES LABS, 81- *Mem:* Fel Inst Elec & Electronics Engrs. *Res:* Circuit design; high performance non-linear analog circuits; translinear principle; translinear multiplier. *Mailing Add:* 1100 NW Compton Dr Suite 301 Beaverton OR 97006

GILBERT, BARRY JAY, b Brooklyn, NY, Feb 2, 43; m 65; c 2. PHYSICS. *Educ:* Polytech Inst Brooklyn, BS, 63; Lehigh Univ, MS, 65, PhD(physics), 68. *Prof Exp:* Asst prof physics, 68-77, ASSOC PROF PHYSICS, RI COL, 77- *Mem:* Am Asn Physics Teachers. *Res:* Theoretical plasma physics. *Mailing Add:* Seven Ducarl Dr Lincoln RI 02865

GILBERT, BARRY KENT, b Chicago, Ill, Sept 22, 44; m 73; c 1. PHYSIOLOGY, ELECTRICAL ENGINEERING. *Educ:* Purdue Univ, BS, 65; Univ Minn, PhD(physiol, biophys), 72. *Prof Exp:* Res asst physiol & biophys, Mayo Clin, Mayo Found, 71-73, res assoc, 73-75, instr physiol, 75-77, asst prof physiol & biophys, Mayo Med Sch, 78-80, STAFF SCIENTIST, DEPT PHYSIOL & BIOPHYS, MAYO CLIN/MAYO FOUND, 80- *Concurrent Pos:* NIH fel, Nat Heart, Lung & Blood Inst, 72-74; assoc staff scientist physiol & biophys, Mayo Clin/Mayo Found, 75-77, staff scientist, 78-, assoc prof biophys, 80-85, prof, Physiol/Biophysics, 86- *Mem:* Am Physiol Soc; Inst Elec & Electronics Engrs; AAAS; Mat Res Coun, Defense Adv Res Proj Agency. *Res:* Application of engineering and computational methods to the solution of computation bound problems in biomedical research and clinical medicine; design of high performance signal processors using gallium arsenide integrated circuits; development of computer aided design (CAD) software for signal processor and integrated circuit design. *Mailing Add:* Dept Physiol & Biophys Mayo Clin 200 First St SW Rochester MN 55901

GILBERT, BRIAN E, b Hollywood, Calif, Jan 31, 42; m 70; c 2. VIROLOGY. *Educ:* Univ Calif, Berkeley, AB, 64; Univ Calif, Los Angeles, PhD(med microbiol & immunol), 70. *Prof Exp:* Postdoctoral Fel, microbiol, Sch Med, Northwestern Univ, 70-73; MEM FAC, DEPT MICROBIOL & IMMUNOL, BAYLOR COL MED, TEX MED CTR, 73- *Mem:* Am Soc Microbiol; Am Soc Virol; Int Soc Antiviral Res; AAAS. *Res:* Role of cystine in the morphogenesis of Histoplasma capsulatum; regulation of protein synthesis at the translational level during mouse brain development; antiviral chemotherapy; aerosol therapy for viral respiratory disease. *Mailing Add:* Dept Microbiol & Immunol Baylor Col Med Tex Med Ctr Houston TX 77030

GILBERT, CARTER ROWELL, b Huntington, WVa, May 23, 30; m 58; c 2. ICHTHYOLOGY. *Educ:* Ohio State Univ, BSc, 51, MSc, 53; Univ Mich, PhD(zool, ichthyol), 60. *Prof Exp:* Asst zool, Ohio State Univ, 52; asst ichthyol, Div Fishes, Mus Zool, Univ Mich, 58-59; res assoc, US Nat Mus, 59-61; asst cur ichthyol, Fla State Mus & asst prof zool, Univ Fla, 61-72, assoc cur & joint assoc prof & Latin Am Studies, 72-87, CUR ICHTHYOL, FLA MUS NATURAL HIST & PROF ZOOL, UNIV FLA, 87- *Concurrent Pos:* Numerous grants, NSF, Am Physiol Soc. *Mem:* Am Soc Ichthyol & Herpet (hon secy, 81-82, secy, 82-90); Japanese Soc Ichthyol; Soc Systematic Zool; Sigma Xi. *Res:* Eastern North American freshwater fishes; sharks; western Altantic marine fishes; Over 65 publications. *Mailing Add:* Fla Mus Natural Hist Univ Fla 276 Mus Gainesville FL 32611

GILBERT, CHARLES MERWIN, physical geology; deceased, see previous edition for last biography

GILBERT, DANIEL LEE, b Brooklyn, NY, July 2, 25; m 64; c 1. PHYSIOLOGY. *Educ:* Drew Univ, AB, 48; Univ Iowa, MS, 50; Univ Rochester, PhD(physiol), 55. *Prof Exp:* Instr physiol, Sch Med & Dent, Univ Rochester, 55-56; from instr to asst prof, Albany Med Col, 56-60; from asst prof to assoc prof, Jefferson Med Col, 60-63; head sect cellular biophys, Lab Biophys, Nat Inst Neurol Dis & Stroke, 63-71, PHYSIOLOGIST, NIH, 62- *Concurrent Pos:* Consult, Grad Coun, George Washington Univ, 65-70; mem corp, Marine Biol Lab, Woods Hole; counr, Oxygen Soc, 87-89. *Honors & Awards:* Bowditch Lectr, Am Physiol Soc, 64. *Mem:* Fel AAAS; Biophys Soc; Am Physiol Soc; Oxygen Soc; Soc Neurosci; Sigma Xi. *Res:* Oxygen poisoning; oxygen in biological evolution; cell permeability; neurophysiology. *Mailing Add:* NIH Bldg 9 Rm 1E-124 Bethesda MD 20892

GILBERT, DAVID ERWIN, b Fresno, Calif, June 23, 39; m 60; c 2. ATOMIC SPECTROSCOPY. *Educ:* Univ Calif, Berkeley, AB, 62; Univ Ore, MA, 64, PhD(physics), 68. *Prof Exp:* Teaching asst physics, Univ Ore, 62-65, res asst, 65-68; assoc prof physics & chmn dept, 68-74, dean acad affairs, 77-83, PRES, EASTERN ORE STATE COL, 83- *Concurrent Pos:* Res assoc, Univ Ore, 69-; vis scientist, Sect Astrophys, Paris Observ, Meudon, France, 75-76. *Mem:* Sigma Xi; Am Asn Physics Teachers. *Res:* Studies of the effects of foreign gas on atomic absorption lines, total line shape studies. *Mailing Add:* Eastern Ore State Col La Grande OR 97850

GILBERT, DEWAYNE EVERETT, b Dixon, Ill, Oct 18, 24; m 50; c 3. CEREAL CROPS, FORAGE CROPS. *Educ:* Iowa State Col, BS, 50, MS, 56, PhD(argon), 59. *Prof Exp:* Instr agron, Iowa State Col, 55-59; asst prof agron, Ohio State Uinv, 59-63; farm adv bioclimatology, Univ Calif, 63-64 & 73-76, exten specialist, 64-73; assoc prof, 76-79, head dept, 78-83, PROF AGRON, UNIV NEV, 79-, EXTEN AGRONOMIST, 76- *Concurrent Pos:* Head dept, 78-83, exten argonomist, Univ Nev, Reno, 76- *Mem:* Am Soc Agron; Crop Sci Soc Am; Sigma Xi. *Res:* Crop weather interactions; alfalfa. *Mailing Add:* 3358 Dana Way Sparks NV 89431

GILBERT, DON DALE, b Ponca City, Okla, June 5, 34; m 63; c 2. ANALYTICAL CHEMISTRY. *Educ:* Univ Calif, Berkeley, BS, 56; Univ Minn, PhD(analytical chem), 59. *Prof Exp:* Res chemist, Analytical & Phys Measurement Div, Calif Res Corp, Stand Oil Co Calif, 59-65; from asst prof to assoc prof, 65-75, chair, Chem Dept, 82-87, PROF CHEM, NORTHERN ARIZ UNIV, 75- *Mem:* Am Chem Soc; fel AAAS; Sigma Xi. *Res:* Digital computers in chemical instrumentation; colorimetry; trace analysis; electroanalytical chemistry. *Mailing Add:* Dept Chem PO Box 5698 Northern Ariz Univ Flagstaff AZ 86011

GILBERT, DOUGLAS L, b La Veta, Colo, June 28, 25; m 49; c 2. ECOLOGY. *Educ:* Colo State Univ, BS, 50, MS, 51; Univ Mich, PhD(wildlife mgt), 62. *Prof Exp:* Res biologist, Colo Game & Fish Dept, 51-52; instr wildlife mgt, Colo Agr & Mech Col, 52-53; pub rels specialist, Colo Game & Fish Dept, 53-55; asst prof forestry, Univ Mont, 55-56; assoc prof major wildlife mgt, Colo State Univ, 57-66, prof wildlife biol, 66-69; prof wildlife sci, Cornell Univ, 69-71; asst dean col forestry & natural resources, 71-75, PROF WILDLIFE BIOL & CHMN DEPT FISHERY & WILDLIFE BIOL, COLO STATE UNIV, 71- *Honors & Awards:* Outstanding Wildlife Bi. *Mem:* Nat Wildlife Fedn; AAAS; Wildlife Soc; Am Soc Range Mgt; Wildlife Dis Asn. *Res:* Natural resource ecology; public relations in natural resource management; wildlife management. *Mailing Add:* 1205 Ellis St Ft Collins CO 80524

GILBERT, E(DWARD) O(TIS), b Joliet, Ill, Mar 29, 30; m 54; c 3. ENGINEERING. *Educ:* Univ Mich, BSE, 52, MSE, 53, PhD(instrumentation eng), 57. *Prof Exp:* From instr to assoc prof aeronaut eng, Univ Mich, 53-64; vpres res & eng, Appl Dynamics, Inc, 63-69; SR VPRES TECHNOL, APPL DYNAMICS COMPUT SYSTS DIV, RELIANCE ELEC, 69- *Concurrent Pos:* Mem tech staff, Space Tech Labs, Inc, Calif, 59-60; consult. *Mem:* AAAS; Inst Elec & Electronics Engrs. *Res:* Design and application of analog and hybrid computers; automatic control; space and flight mechanics; instrumentation engineering; industrial control computer design and applications. *Mailing Add:* 2969 Hickory Lane Ann Arbor MI 48104

GILBERT, EDGAR NELSON, b Woodhaven, NY, July 25, 23; m 48; c 3. MATHEMATICS. *Educ:* City Col New York, BS, 43; Mass Inst Technol, PhD(math), 48. *Prof Exp:* Asst physics, Univ Ill, 43; staff mem, Radiation Lab, Mass Inst Technol, 44-46; MEM TECH STAFF, BELL TEL LABS, 48- *Mem:* Inst Elec & Electronics Engrs; Soc Indust Appl Math. *Res:* Electromagnetic theory; differential equations; information theory; probability. *Mailing Add:* AT&T Bell Labs Rm 2C381 Murray Hill NJ 07974

GILBERT, EDWARD E, b New York, NY, May 1, 25; m 61; c 5. ENTOMOLOGY. *Educ:* Southern Methodist Univ, BS & MS, 50; Univ Calif, Berkeley, PhD, 61. *Prof Exp:* Asst prof biol, State Univ NY Stony Brook, 58-65; prof, Northeast Mo State Teachers Col, 65-69; HEAD DEPT BIOL, W GA COL, 69- *Concurrent Pos:* NSF fel radioecol inst, Oak Ridge Inst Nuclear Studies, 63, NSF grant, 67-69. *Mem:* AAAS; Ecol Soc Am; Am Inst Biol Sci. *Res:* Life history biology, particularly Tribolium; taxonomy of Curculionidae. *Mailing Add:* Dept Biol WGa Col Carrollton GA 30118

GILBERT, ELMER G(RANT), b Joliet, Ill, Mar 29, 30. ENGINEERING. *Educ:* Univ Mich, BSE, 52, MSE, 53, PhD(instrumentation eng), 57. *Prof Exp:* From instr to assoc prof, 53-63, PROF AERONAUT ENG, DEPT AEROSPACE ENG, UNIV MICH, ANN ARBOR, 63-, PROF INFO & MEM COMPUT, INFO & CONTROL ENG PROG, DEPT AEROSPACE ENG, 71- *Mem:* Inst Elec & Electronics Engrs; Soc Indust & Appl Math. *Res:* Automatic control; systems theory; optimization. *Mailing Add:* Dept Aerospace Eng Univ Mich Ann Arbor MI 48109

GILBERT, ETHEL SCHAEFER, b Toledo, Ohio, Apr 10, 39; m 68; c 2. PUBLIC HEALTH & EPIDEMIOLOGY. *Educ:* Oberlin Col, BA, 61; Univ Mich, MPH, 64, PhD(biostatist), 66. *Prof Exp:* From instr to asst prof biostatist, Univ Wash, 67-68; sr res scientist, 73-80, staff scientist, 80-90, SR STAFF SCIENTIST, PAC NORTHWEST LABS, BATTELLE MEM INST, 80- *Honors & Awards:* Schnedecor Award, 80. *Mem:* Fel Am Statist Asn; Biometrics Soc; Soc Epidemiol Res. *Res:* Occupational and environmental epidemiology; effects of radiation exposure. *Mailing Add:* Pac Northwest Div Battelle Mem Inst Richland WA 99352

GILBERT, EUGENE CHARLES, b Manchester, NH, Nov 16, 42; m 65; c 2. SYNTHETIC ORGANIC CHEMISTRY, ORGANIC POLYMER CHEMISTRY. *Educ:* St Anselm's Col, AB, 65; Univ Notre Dame, PhD(phys org chem), 69. *Prof Exp:* Res assoc org chem, Johns Hopkins Univ, 69-70; from group leader to sr group leader res & develop, Chem Div, Quaker Oats Co, 70-75; prog mgr develop, Chem Div, Chemetron Corp, 75-77; res & develop mgr, Chem Div, Thiokol Corp, 77-79; sr res & develop assoc, PVC Div, BF Goodrich, 79, mfr mgr, 80-83, res & develop mgr, 84-85, bus mgr, Med Prod, 85-90, MGR, POWER MAT, BF GOODRICH, 91- *Mem:* Am Chem Soc. *Res:* Conformational analysis; thermodynamics; polymer stabilizers; radiation curable coatings; thermoplastic and thermosetting plastics and elastomers; polymer films; synthetic wound dressings; polymer based medical disposables. *Mailing Add:* 9921 Brecksville Rd Brecksville OH 44141

GILBERT, FRANCIS CHARLES, nuclear physics, for more information see previous edition

GILBERT, FRANCIS EVALO, b Mattoon, Ill, June 8, 16; m 42; c 2. CHEMISTRY. *Educ:* Univ Ill, BS, 39. *Prof Exp:* Chemist, Cuneo Press, Ill, 39-41 & Eversharp, Inc, 46; CHIEF CHEMIST & DIR RES, SANFORD INK CO, 46-, VPRES, 67- *Mem:* Am Chem Soc. *Res:* Specialty inks, writing, stamping, marking; adhesives, dextrine, gum and rubber types; packaging of above items for retail distribution. *Mailing Add:* 1080 Chestnut San Francisco CA 94109

GILBERT, FRANK ALBERT, biology; deceased, see previous edition for last biography

GILBERT, FRANKLIN ANDREW, SR, b Burlington, NJ, June 8, 19; m 44; c 3. HORTICULTURE. *Educ:* Rutgers Univ, BS, 42, MS, 48, PhD, 52. *Prof Exp:* Asst exten specialist pomol, Rutgers Univ, 44-46, instr & res assoc, 46-50; horticulturist, Univ Exp Sta, Peninsula Br Exp Sta, 50-82, from asst prof to prof, 50-82, emer prof horticult, 82-83; RETIRED. *Honors & Awards:* Int Dwarf Tree Asn Award, 81. *Mem:* Am Soc Hort Sci; Am Pomol Soc. *Res:* Strawberry breeding; apple rootstock studies. *Mailing Add:* 403 W Pine Sturgeon Bay WI 54235

GILBERT, FRED, b Brooklyn, NY, Nov 24, 41. HUMAN GENETICS. *Educ:* Mass Inst Technol, BS, 62; Albert Einstein Col Med, MD, 66. *Prof Exp:* Intern & asst resident internal med, Barnes Hosp, Wash Univ Sch Med, 66-68; clin assoc biochem genetics, Nat Heart & Lung Inst, 68-71; fel human genetics & asst biol, Yale Univ & Sch Med, 71-74; asst prof human genetics & assoc pedt, Univ Pa Sch Med, 74-83; assoc prof pediat & human genetics, Mt Sinai Med Sch, NY, 83-88; ASSOC PROF PEDIAT & CO-DIR HUMAN GENETICS, CORNELL UNIV MED COL, 88- *Mem:* Am Soc Human Genetics; AAAS. *Res:* Cell biology; differentiation and neurobiology. *Mailing Add:* Cornell Univ Med Col 1300 York Ave New York NY 10021

GILBERT, FRED IVAN, JR, b Newark, NJ, 1920. INTERNAL MEDICINE, PUBLIC HEALTH. *Educ:* Univ Calif, Berkeley, BS, 42; Stanford Univ, MD, 45; Am Bd Internal Med, dipl, 53. *Prof Exp:* Intern, Stanford Univ, 45-46; spec worker med, 47; mem staff, VA Hosp, Ft Miley, 48-50; PROF PUB HEALTH, UNIV HAWAII, MANOA, 76-; MED DIR, PAC HEALTH RES INST, 77- *Concurrent Pos:* Attend staff & chief med, Queen's Hosp & St Francis Hosp, Honolulu; clin instr med, Stanford Univ, 48-51. *Mem:* Inst Med-Nat Acad Sci. *Mailing Add:* Straub Clin & Hosp 888 S King St Honolulu HI 96813

GILBERT, FREDERICK EMERSON, JR, b Birmingham, Ala, June 1, 41; m 62; c 2. PATHOLOGY, CLINICAL CHEMISTRY. *Educ:* Birmingham Southern Col, BS, 63; Univ Ala, MS, 65, MD, 68. *Prof Exp:* Res assoc biochem, Mem Inst Path, 65-71; chief cytopath, Ctr Dis Control, 71-73; LAB DIR, COWETA GEN HOSP, NEWNAN HOSP, GILBERT LAB, 73- *Concurrent Pos:* Southern Med Asn res fel, 69-70; mem, Ad Hoc Adv Comt Cytol, 71-77. *Honors & Awards:* Med Col Ala Res Award, 68. *Mem:* Am Chem Soc; Am Asn Clin Chemists; Am Soc Clin Path; Col Am Path; Am Soc Hist Med. *Res:* Multiphasic screening to study the diseased codon; inorganic pyrophosphatase and its role in nucleotide metabolism; quality control cytology. *Mailing Add:* Two Pinehollow Dr Newnan GA 30263

GILBERT, FREDERICK FRANKLIN, b Toronto, Can, Aug 5, 41; US citizen; m 64, 81; c 3. FISH & WILDLIFE SCIENCES. *Educ:* Acadia Univ, BS, 65; Univ Guelph, MS, 66, PhD(zool), 68. *Prof Exp:* Big game prog leader, Maine Inland Fish & Game, 68-72; from asst prof to assoc prof zool, Univ Guelph, 72-81; dir wildlife biol, Wash State Univ, 81-85, prof zool & wildlife biol, 85-88, interim chair natural resource sci, 88-91, PROF NATURAL RESOURCE SCI, WASH STATE UNIV, 88- *Concurrent Pos:* Asst prof, Univ Maine, 68-72; dir, Ecol Serv Planning, 75-77; chair, Fac Senate, Wash State Univ, 86-88; chair, US Tech Adv Group ISO/TC191, 87- *Mem:* AAAS; Am Soc Mammalogists; Wildlife Soc; Soc Am Foresters; Soc Northwestern Vertebrate Biol. *Res:* Forest management practices in relation to wildlife populations; bioenergetic relationships in natural and perturbed ecosystems; role of environmental factors in modifying population density, behavior and physiology of wildlife species. *Mailing Add:* Dept Natural Resource Sci Wash State Univ Pullman WA 99164-6410

GILBERT, GARETH E, b Fall River, Mass, Sept 30, 21; m 49; c 2. BOTANY. *Educ:* Ohio State Univ, BSc, 48, MSc, 49, PhD(plant ecol), 53. *Prof Exp:* From instr to asst prof, 52-61, ASSOC PROF BOT, OHIO STATE UNIV, 61- *Mem:* Ecol Soc Am; Sigma Xi. *Res:* Plant ecology. *Mailing Add:* Dept Bot Ohio State Univ 1735 Neil Ave Columbus OH 43210

GILBERT, GEORGE LEWIS, b Abington, Mass, Sept 10, 33; m 62; c 3. INORGANIC CHEMISTRY. *Educ:* Antioch Col, BS, 58; Mich State Univ, PhD(inorg chem), 63. *Prof Exp:* Res chemist, Lawrence Radiation Lab, Univ Calif, 63-64; chmn, chem dept, Denison Univ, 60-74, from asst prof to assoc prof chem, 64-80, actg chmn dept, 67-68, sci coordr, 70-74, coordr Learning Resources Ctr, 80-83. *Mem:* AAAS; Sigma Xi; Royal Soc Chem. *Res:* Reactions of noble gas compounds; coordination compounds and their structures. *Mailing Add:* Dept Chem Denison Univ Granville OH 43023

GILBERT, HARRIET S, b Philadelphia, Pa, June 22, 30; m 57; c 3. HEMATOLOGY. *Educ:* Bryn Mawr Col, AB, 51; Columbia Univ, MD, 55; Am Bd Internal Med, dipl, 65; Am Bd Hemat, dipl, 72. *Prof Exp:* From intern to asst resident internal med, Mt Sinai Hosp, 55-58, clin asst hemat, 58-63, asst attend hematologist, 63-69; from asst prof to assoc prof med, Mt Sinai Sch Med, 66-81, asst dean res, 77-81 prof med & assoc dean res, 81-86; PROF MED & DIR GEN CLIN RES CTR, ALBERT EINSTEIN COL MED, YESHIVA UNIV, BRONX, NY, 86- *Concurrent Pos:* Fel, Mt Sinai Hosp, 58-61, from res asst to res assoc, 61-64; Am Cancer Soc fel, 59-61, grant, 65; consult, Elmhurst Hosp, New York, 64-; attend hematologist, Bronx Vet Hosp, 71-73, consult, 73-; asst dir, Res Activities Comt & Off & asst prog dir, Clin Res Ctr, Mt Sinai Med Ctr, 75-79, assoc dir, 79-; assoc attend hematologist, Mt Sinai Hosp, 69-81; lectr, Mt Sinai Sch Med, 86-; attend Weiler Hosp, Albert Einstein Col Med, 86-, Bronx Munic Hosp Ctr, 86-, Montefiore Med Ctr, 86-; assoc ed, Mt Sinai J Med. *Mem:* Biophys Soc; Harvey Soc; Am Soc Hemat; Am Fedn Clin Res; Am Col Physicians; Am Inst Nutrit. *Res:* Hematology, especially myeloproliferative diseases; biochemical changes, including lipid metabolism, erythrocyte metabolism, histamine and serotonin metabolism and leukocyte membrane development; clinical problems, particularly surgical and neurological complications; chemotherapy in control of myeloproliferative disease; vitamin B-12 binding proteins; automation in hematology. *Mailing Add:* Albert Einstein Col Med 1300 Morris Park Ave Bldg VDE Rm 230 Bronx NY 10461

GILBERT, JACK PITTARD, b Lenior County, NC, Nov 8, 25; m 44; c 2. ANALYTICAL CHEMISTRY, GOOD MANUFACTURING PRACTICES. *Educ:* Wagner Col, BS, 52. *Prof Exp:* Chemist, Radioactive Lab, Merck Sharp & Dohme Res Labs Div, Merck & Co, Inc, 51-56, suprv analytical chem, Pilot Plant Control Lab, 56-67, sr res chemist, 67-73, group leader, 73-76, sect leader analytical chem, Microanal & Pilot Plant Control Lab, 76-90; RETIRED. *Mem:* Am Chem Soc; Am Microchem Soc. *Res:* Functional group analysis; micro and macro; monaqueous titrations; acid base analysis; spectrophotometer analysis; specific ion analysis and potentiometric titrations; good manufacturing practices, good laboratory practices, and specification. *Mailing Add:* 4072 Richmond Ave Staten Island NY 10312-5634

GILBERT, JAMES ALAN LONGMORE, b Grantown-on-Spey, Scotland, Jan 28, 18; m 44; c 8. INTERNAL MEDICINE. *Educ:* Univ Edinburgh, MB & ChB, 41, MD, 47; FRCP, 47 & 65; FRCP(E), 47; FRCPS(C), 50. *Prof Exp:* Assoc prof, 57-63, prof med, univ alta, 63-; dir, Clin Teaching Unit, 70-86, EMER PROF, ROYAL ALEXANDRA HOSP, EDMONTON, 86- *Honors & Awards:* I Provincial Award, Excellence in Med, 78. *Mem:* Fel Am Col Physicians; fel Royal Soc Arts; Am Gastroenterol Asn; Can Asn Gastroenterol (pres, 70); Int Soc Res Med Educ (pres, 74). *Res:* Prediabetic syndromes; physiological basis of the dumping syndrome following partial gastrectomy for peptic ulcer; medical education, evaluation in medical education; ethanol metabolism and its complications. *Mailing Add:* 204 Hys Ctr 11010 101st St Edmonton AB T5H 4B9 Can

GILBERT, JAMES FREEMAN, b Vincennes, Ind, Aug 9, 31; m 59; c 3. GEOPHYSICS, SEISMOLOGY. *Educ:* Mass Inst Technol, BS, 53, PhD(geophys), 56. *Prof Exp:* Res assoc, Mass Inst Technol, 56-57; asst prof, Inst Geophys, Univ Calif, Los Angeles, 57-59, assoc prof, 60; sr res geophysicist, Tex Instruments, Inc, 60-61; assoc dir, 76-88, PROF GEOPHYS, INST GEOPHYS & PLANETARY PHYSICS, SCRIPPS INST OCEANOG, UNIV CALIF, SAN DIEGO, 61- *Concurrent Pos:* Guggenheim fel, Cambridge Univ, 64-65, Guggenheim overseas fel, Churchill Col, Cambridge, 72-73; guest lectr, Veining-Meinesz Geophys Inst, Univ Utrecht, 73 & Acad Sci, USSR, 76; distinguished lectr, Dept Physics, Univ Alta, 80; Sherman Fairchild distinguished scholar, Div Geol & Planetary Sci, Calif Inst Technol, 87; chmn, Grad Dept, Scripps Inst Oceanog, Univ Calif, San Diego, 88-91, fac res lectr, Univ, 91-92. *Honors & Awards:* Gold Medal, Royal Astron Soc, 81; Arthur L Day Medal, Geol Soc Am, 85; Balzan Prize, 90. *Mem:* Nat Acad Sci; Am Phys Soc; Am Math Soc; fel Am Geophys Union; Seismol Soc Am; Sigma Xi; fel Am Acad Arts & Sci; foreign hon fel Europ Union Geosci. *Res:* Elastodynamics; normal mode theory; geophysical problems of inversion and inference; structure of the earth and of earthquake sources; earthquake mechanism; computational geophysics. *Mailing Add:* Inst Geophys Planetary Physics Univ Calif San Diego La Jolla CA 92093

GILBERT, JAMES ROBERT, b Jefferson, Mo, Feb 16, 46; m 69; c 2. WILDLIFE SCIENCE, BIOSTATISTICS. *Educ:* Colo State Univ, BS, 68; Univ Minn, MS, 70; Univ Idaho, PhD(wildlife), 74. *Prof Exp:* Res assoc wildlife, Univ Idaho, 71-73; instr, Cornell Univ, 73-74; res assoc, Univ Wash, 74-75; asst prof, 75-80, ASSOC PROF WILDLIFE, UNIV MAINE, 80- *Concurrent Pos:* Consult, US Fish & Wildlife Serv, 76-78, Bur Land Mgt, 78- & Fed Hwy Admin, 77-81; mem working group grey seals, Int Coun Explor of Sea, 77-79. *Mem:* Wildlife Soc; Am Soc Mammalogists; Biometr Soc; Bear Biologists Asn. *Res:* Population dynamics and censusing, especially of large terrestrial and marine mammals. *Mailing Add:* Dept Wildlife Resources Univ Maine Orono ME 04469

GILBERT, JEROME B, RESEARCH ADMINISTRATION. *Prof Exp:* RETIRED. *Mem:* Nat Acad Eng. *Mailing Add:* 324 Tappen Terr Orinda CA 94563

GILBERT, JIMMIE D, b Quitman, La, July 12, 34; m 53, 74; c 6. ALGEBRA. *Educ:* La Polytech Inst, BS, 56; Auburn Univ, MS, 57, PhD(math), 60. *Prof Exp:* Instr math, Auburn Univ, 57-58; from asst prof to assoc prof, 58-62, prof math, La Tech Univ, 65-86; PROF MATH, UNIV SC, Spartanburg, 86- *Mem:* Am Math Soc; Math Asn Am. *Res:* Linear algebra, author of ten college mathematics textbooks. *Mailing Add:* Dept Math Univ SC Spartanburg SC 29303

GILBERT, JOEL STERLING, b Wichita, Kans, Aug 29, 35; m 59; c 3. MECHANICAL ENGINEERING. *Educ:* Univ Okla, BS, 58, PhD(mech eng), 65; Okla State Univ, MS, 60. *Prof Exp:* Assoc engr, Tex Instruments, Inc, 58-59; res asst fluid contamination, Okla State Univ, 59-60; systs engr, Sandia Labs, 60-61; instr mech eng, Univ Okla, 61-62; res asst radiant head transfer, 64-65; asst prof mech eng, Univ Fla, 65-70; staff mem, Theoret Design Div, 70-75, STAFF MEM, NUCLEAR & ENG TECHNOL DIV, LOS ALAMOS NAT LAB, 75- *Concurrent Pos:* Consult, Sandia Labs & Martin Marietta Corp, Fla, 68-69. *Mem:* Am Soc Mech Engrs; Am Inst Aeronaut & Astronaut; Am Soc Eng Educ; Nat Soc Prof Engrs. *Res:* Thermodynamics; heat transfer; fluid mechanics; electromagnetic propagation. *Mailing Add:* 208 Canada Way Los Alamos NM 87544

GILBERT, JOHN ANDREW, b Ger, Sept 3, 48; US citizen; m 85; c 5. EXPERIMENTAL STRESS ANALYSIS, APPLIED OPTICS. *Educ:* Polytech Inst Brooklyn, BS, 71, MS, 73; Ill Inst Technol, PhD(solid mech), 75. *Prof Exp:* From lectr to assoc prof eng, Univ Wis-Milwaukee, 75-85;

PROF MECH ENG, UNIV ALA-HUNTSVILLE, 85- *Concurrent Pos:* Consult, Allen Bradley Co, 76-77, 82-85, Appl Power, 79-81, Gen Elec Med Systs Div, 80-81, Waukesha Engine, 81 & AT&T Bell Lab, 81-89; prin investr, Army Res Off Contracts, 80-87 & NSF, 83-85, NASA, 87-; adj prof mat sci, Univ Ala, Tuscaloosa, 88-, Univ Ala, Birmingham, 89-; pres, Optechnol Inc, 89-; res fel, Am Soc Nondestructive Testing, 87. *Mem:* Soc Exp Mech; Am Acad Mech; Sigma Xi; Brit Soc Strain Measurement; Am Soc Eng Educ; Soc Photo-Optical Instrumentation Engrs; Am Soc Civil Engrs. *Res:* Holographic interferometry and speckle metrology; fiber optic sensing; radial metrology. *Mailing Add:* Dept Mech Eng Univ Ala Huntsville AL 35899

GILBERT, JOHN BARRY, b Hull, Eng, Jan 21, 37; m 59; c 2. PHYSICAL CHEMISTRY. *Educ:* Univ Hull, BSc, 57, PhD(chem), 60. *Prof Exp:* Res fel photochem, Univ Alta, 60-62; from res assoc to sr res assoc, Esso Petrol Can, 62-83, res adv, 83-91; RETIRED. *Mem:* Fel Chem Inst Can. *Res:* Petroleum refining; catalytic processes, especially catalytic hydrogenation and hydrocracking; catalytic theory and reactor design theory; environmental protection; petroleum products quality. *Mailing Add:* 1675 Winton Rd Sarnia ON N7V 4C1 Can

GILBERT, JOHN CARL, b Laramie, Wyo, Jan 30, 39; m 65; c 1. PHYSICAL ORGANIC CHEMISTRY. *Educ:* Univ Wyo, BS, 61; Yale Univ, MS, 62, PhD(chem), 66. *Prof Exp:* From asst prof to assoc prof, 65-84, chmn dept, 87-91, PROF CHEM, UNIV TEX, AUSTIN, 84- *Concurrent Pos:* NIH prin investr, Welch Found. *Mem:* Am Chem Soc; Sigma Xi. *Res:* Thermal and photochemical isomerization of hydrocarbons; organometallics; carbenes; synthesis; reaction mechanisms. *Mailing Add:* Dept Chem Univ Tex Austin TX 78712

GILBERT, JOHN JOUETT, b Southampton, NY, July 18, 37; m 59; c 2. FRESHWATER BIOLOGY, INVERTEBRATE ZOOLOGY. *Educ:* Williams Col, BA, 59; Yale Univ, PhD(biol), 63. *Prof Exp:* NIH fel, Univ Wash, 63-64; asst prof biol, Princeton Univ, 64-67; from asst prof to assoc prof, 67-74, PROF BIOL, DARTMOUTH COL, 74- *Mem:* Am Soc Zool; Am Soc Limnol & Oceanog; Ecol Soc Am; Int Asn Theoret & Appl Limnol. *Res:* Biology of rotifers; ecology; invertebrate biology; sexuality and form-change in ploimate rotifers; biology of freshwater sponges; trophic interactions in and competition among freshwater zooplankton. *Mailing Add:* Dept Biol Sci Dartmouth Col Hanover NH 03755

GILBERT, LAWRENCE IRWIN, b New York, NY, Jan 24, 29; m 52; c 3. ZOOLOGY. *Educ:* Long Island Univ, BS, 50; NY Univ, MS, 55; Cornell Univ, PhD(zool), 58. *Prof Exp:* From asst prof to prof biol sci, Northwestern Univ, Evanston, 58-80; William Rand Kenan prof zool, 80-82, KENAN PROF & CHAIR BIOL, UNIV NC, CHAPEL HILL, 82- *Concurrent Pos:* NSF sr fel, Univ Berne, 64-65; vis scientist, Am Physiol Soc, 63-; mem, Presidential Task Force Pest Mgt, 71-72; ed, Insect Biochem, 84-; fel, Am Acad Arts & Sci, 90. *Mem:* Soc Growth & Develop; Soc Exp Biol; Am Soc Cell Biol; Entom Soc Am; Sigma Xi. *Res:* Invertebrate endocrinology; biochemical effect of insect hormones; insect physiology; lipid metabolism and transport in insects; endocrine gland ultrastructure; sterols and terpenes in insects. *Mailing Add:* Dept Biol Univ NC CB 3280 Coker Hall Chapel Hill NC 27599-3280

GILBERT, MARGARET LOIS, b Wakefield, RI, June 9, 28. BOTANY, ECOLOGY. *Educ:* Univ RI, BS, 49; Univ Wis, PhD(bot), 53. *Prof Exp:* Asst prof biol, Northwestern State Col, La, 53-54; from asst prof to assoc prof, 54-61, Nelson C White chair, Natural Sci, 81, PROF BIOL & CHMN, DEPT BIOL SCI & DIV NATURAL SCI, FLA SOUTHERN COL, 61- *Concurrent Pos:* Consult. *Mem:* Bot Soc Am; Ecol Soc Am; Sigma Xi. *Res:* Plant ecology. *Mailing Add:* Dept Biol Sci Fla Southern Col Lakeland FL 33801

GILBERT, MURRAY CHARLES, b Lawton, Okla, Jan 21, 36; m 58; c 3. PETROLOGY. *Educ:* Univ Okla, BS, 58, MS, 61; Univ Calif, Los Angeles, PhD(geol), 65. *Prof Exp:* Asst res geologist, Univ Calif, Los Angeles, 65; fel, Carnegie Inst Geophys Lab, 65-68; from asst prof to prof petrol, Va Polytech Inst & State Univ, 68-83, chmn dept, 75-80; head dept, Tex A&M Univ, 85-90; DIR GEOL & GEOPHYS, UNIV OKLA, 90- *Concurrent Pos:* Geologist, Okla Geol Surv, 77-78. *Mem:* Am Geophys Union; Am Asn Petrol Geologists; Geol Soc Am; Mineral Soc Am; Mineral Soc Gt Brit & Ireland. *Res:* Experimental mineralogy and petrology; stability relations of amphiboles, pyroxenes, olivine, and aluminum silicates; sulfide-silicate relations; geology and petrology of Wichita Mountains, Oklahoma and Southern Oklahoma Anlacogen; planetology. *Mailing Add:* Dept Geol & Geophys Sch Geol & Geophys Univ Okla 100 E Boyd Norman OK 73019

GILBERT, MYRON B, b Rochester, NY, Sept 3, 21; m 45; c 3. COMMUNICATIONS SCIENCE, SCIENCE ADMINISTRATION. *Educ:* Cornell Univ, AB, 47; Mass Inst Technol, MS, 49. *Prof Exp:* Meteorologist, Pan Am Grace Airways, 49-57; chief, Systs & Applns Br, Geophys Res Directorate, 57-60, chief, Eval Div, Air Force Cambridge Res Labs, 60-67, dir, Air Force Environ Consult Serv, 67-69; mgr tech commun, Honeywell Info Systs, 69-74; managing ed, Johns Hopkins Appl Tech Digest & ed supvr, Appl Phys Lab, Johns Hopkins Univ, 75-88; RETIRED. *Mem:* Am Meteorol Soc; Am Geophys Union; AAAS; Inst Elec & Electronics Engrs. *Res:* Dissemination of scientific and technical information; writing and publication of scientific reports in the areas of geophysics, mathematics, physics and engineering. *Mailing Add:* 700 Seventh St SW Apt 143 Washington DC 20024

GILBERT, PAUL WILNER, b Rochester, NY, Feb 14, 16; m 41, 89; c 2. MATHEMATICS. *Educ:* Univ Rochester, AB, 36, AM, 37; Duke Univ, PhD(math), 40. *Prof Exp:* Instr math, Tex Technol Col, 40-42 & US Mil Acad, 45-46; from instr to prof math, Syracuse Univ, 46-81; RETIRED. *Mem:* Math Asn Am. *Res:* Numerical analysis. *Mailing Add:* 202 Wellington Rd De Witt NY 13214

GILBERT, PERRY WEBSTER, b North Branford, Conn, Dec 1, 12; m 38; c 8. MARINE BIOLOGY, VERTEBRATE MORPHOLOGY. *Educ:* Dartmouth Col, AB, 34; Cornell Univ, PhD(zool), 40. *Hon Degrees:* DHumL, York Col, Pa, 78. *Prof Exp:* Instr zool, Dartmouth Col, 34-36; asst, Cornell Univ, 37-40, from instr to prof zool, 40-68; exec dir, 67-78, EMER DIR, MOTE MARINE LAB, 78-; PROF NEUROBIOL & BEHAV, CORNELL UNIV, 68- *Concurrent Pos:* Instr, Marine Biol Lab, Woods Hole, 41; Carnegie fel, 49-50; Guggenheim fel, 57, 64; chmn shark res panel, Am Inst Biol Sci, 58-; consult comt polar res, Nat Acad Sci, 59-; leader, Tahiti-Tikehau Exped, 64; chief scientist, Brit Honduras Exped Shark, 69. *Mem:* Fel Am Inst Fishery Res Biologists; Am Soc Zool; Am Soc Ichthyol & Herpet; Am Soc Mammal; Soc Study Evolution. *Res:* Morphology of birds, fish and mammals; development of vertebrate eyeball musculature; structural and functional adaptations of aquatic birds; biology of elasmobranch fishes. *Mailing Add:* Mote Marine Lab 1600 City Island Park Sarasota FL 34236

GILBERT, R(OBERT) J(AMES), b Morristown, Tenn, Dec 14, 23; m 47; c 8. MECHANICAL ENGINEERING. *Educ:* Univ Tenn, BS, 48. *Prof Exp:* Engr, Nylon Tech, E I du Pont de Nemours & Co, 49-52, res engr, Dacron Tech, 52-53, group supvr, Tech & Process, 53-56, res supvr, Dacron Res, 56-57, tech supvr, 57-62, sr supvr, Spunbonded Res & Develop, 62-69, prod planning supvr, 69-74, venture planner, 74-85; RETIRED. *Concurrent Pos:* Financial consult. *Mem:* Am Inst Chem Engrs. *Res:* Polymer gear meter pumps; spinning and drawing of nylon continuous filament yarn; polymerization; spinning and drawing of Dacron; staple and continuous filment yarn; processes for spunbonded webs; ecomomics of spunbonded processes. *Mailing Add:* 802 Nella Dr Goodlettsville TN 37072

GILBERT, RICHARD CARL, b Ft Wayne, Ind, Sept 15, 27; m 56; c 3. MATHEMATICAL ANALYSIS. *Educ:* Harvard Univ, AB, 51; Univ Calif, Los Angeles, PhD(math), 58. *Prof Exp:* Asst, Univ Calif, Los Angeles, 53-55; actg instr math, Univ Calif, Riverside, 55-57, actg asst prof, 57-58, asst prof, 58-63; assoc prof, 63-67, PROF MATH, CALIF STATE UNIV, FULLERTON, 67- *Concurrent Pos:* Vis asst prof, Univ Chicago, 61-62; vis assoc prof, Math Res Ctr, Univ Wis, 62-63. *Mem:* Am Math Soc; Math Asn Am; Am Sci Affil. *Res:* Spectral theory of linear operators; ordinary differential equations; functional analysis. *Mailing Add:* Dept Math Calif State Univ Fullerton CA 92634

GILBERT, RICHARD DEAN, b Winnipeg, Man, Mar 14, 20; nat US; m 44, 82; c 1. POLYMER CHEMISTRY. *Educ:* Univ Man, BSc, 42, MSc, 43; Univ Notre Dame, PhD, 50. *Prof Exp:* Jr res chemist, Polymer Corp, 46-47, res chemist, 50-51; res chemist, Ky Synthetic Rubber Corp, 51-55; group leader, Uniroyal Chem Co, 55-60, sect mgr synthetic rubber & latex, 60-66; assoc prof, 66-68, prof textile chem, 68-90, PROF WOOD & PAPER SCI, NC STATE UNIV, 90- *Concurrent Pos:* Consult, Borg-Warner Chem, 69-77, Monsanto, 69-78, Gen Tire, 79 & Catawba-Char Lab, 81-, Lord Corp, 85-87, 90-92, Prod Int, 88. *Mem:* Am Chem Soc; NY Acad Sci; Fiber Soc; Sigma Xi; AAAS. *Res:* High polymers; biodegradable polymers; biopolymers elastomeric fibers; photodegradation of polymers; spin labelling studies of dye diffusion; amisotropic solutions of cellulose; epoxy-graphite fiber composites, polymer blends; block copolymers. *Mailing Add:* Box 8302 NC State Univ Raleigh NC 27695

GILBERT, RICHARD E(ARLE), b Brooklyn, NY, Jan 24, 33; m 57; c 5. CHEMICAL ENGINEERING. *Educ:* Worcester Polytech Inst, BS, 54; Princeton Univ, PhD(chem eng), 59. *Prof Exp:* From asst prof to assoc prof, 58-69, PROF CHEM ENG, UNIV NEBR, 69- *Mem:* Am Chem Soc; Am Inst Chem Engrs; Am Soc Eng Educ. *Mailing Add:* Dept Chem Eng Univ Nebr Lincoln NE 68588-0126

GILBERT, RICHARD GENE, b Holdenville, Okla, Dec 3, 35; c 2. PLANT PATHOLOGY, SOIL MICROBIOLOGY. *Educ:* Colo State Univ, BS, 61, MS, 63, PhD(plant path), 64. *Prof Exp:* Res scientist, Soils Lab, USDA, Md, 64-71, res microbiologist, US Water Conserv Lab, Ariz, 71-82, res plant pathologist, Agr Res Serv, Prosser, Wash, 82-89, NAT AGRICHEM SPECIALIST, SOIL CONSERV SERV, USDA, WASHINGTON, DC, 89- *Mem:* Am Phytopath Soc; Soil Sci Soc Am; Am Soc Agron. *Res:* Plant pathology; soil-borne diseases; environmental ecology; soil biochemistry; nitrogen transformations in soil; environmental biology. *Mailing Add:* USDA-Soil Conserv Serv PO Box 2890 Washington DC 20013

GILBERT, RICHARD LAPHAM, JR, b Schenectady, NY, Oct 5, 16; m 39, 55; c 4. CHEMISTRY. *Educ:* Cornell Univ, BChem, 38. *Prof Exp:* Chem microscopist, Am Cyanamid Co, 38-44, res chemist, 44-48; res chemist, Lion Oil Co, 48-50; res chemist, 50-55, group leader process develop, 55-62, group leader phosphorus & nitrogen res, 62-73, group leader prep agr chem, 73-77, SR RES CHEM, AM CYANAMID CO, 77- *Mem:* Am Chem Soc; Am Inst Chem Eng. *Res:* Process analysis; economic analyses; fermentation products as agricultural chemicals; laboratory construction and ventilation. *Mailing Add:* 59 Shady Brooklane Princeton NJ 08540

GILBERT, ROBERT, b Chicago, Ill, Dec 5, 24; m; c 1. MEDICINE. *Educ:* Univ Ill, BS, 50; Am Bd Internal Med, dipl. *Prof Exp:* Intern, Cook County Hosp, 52-53; resident, Upstate Med Ctr, 53-56; from instr to assoc prof, 56-72, PROF, DEPT MED, HEALTH SCI CTR, STATE UNIV NY, SYRACUSE, 72- *Mem:* Fel Am Col Physicians; Am Physiol Soc; Fedn Clin Res; Am Heart Asn; Am Thoracic Soc. *Res:* Author of numerous publications. *Mailing Add:* Dept Med Health Sci Ctr State Univ NY 750 E Adams St Syracuse NY 13210

GILBERT, ROBERT L, b Chicago, Ill, Jan 28, 31; m 55; c 2. SOLID STATE PHYSICS. *Educ:* Ill Inst Technol, BS, 55, MS, 58, PhD(physics), 67. *Prof Exp:* Physicist, Admiral Corp, 55-57; proj physicist, Nuclear-Chicago Corp, 58-62; from asst prof to assoc prof physics, 66-77, chmn dept, 70-77, PROF PHYSICS, NORTHEASTERN ILL UNIV, 77- *Concurrent Pos:* Consult, Solid State Div, Oak Ridge Nat Lab, 66, fel, 67-68. *Mem:* Am Inst Physics; Sigma Xi. *Res:* Theory of imperfections of solids with emphasis upon optical properties of F-centers in alkali-halides. *Mailing Add:* 1574 Anderson Lane Buffalo Grove IL 60089

GILBERT, ROBERT PERTSCH, b New York, NY, Jan 8, 32; m 55; c 2. MATHEMATICS. *Educ:* Brooklyn Col, BS, 52; Carnegie Inst Technol, MS(math) & MS(physics), 55, PhD(math), 58. *Prof Exp:* From instr to asst prof math, Univ Pittsburgh, 57-60; asst prof, Mich State Univ, 60-61; asst prof, Univ Md, 61-64, res assoc prof, 64-65; prof, Georgetown Univ, 65-66; prof, Ind Univ, Bloomington, 66-75; UNIDEL PROF MATH & DIR, MATH SCI INST, UNIV DEL, 75- *Concurrent Pos:* Ed-in-chief, Applicable Analysis, main ed, Complex Variables; ed, J Math Analysis; nonlinear anal consult, Hahn-Meitner Inst, Berlin, 74-75; consult ed, Pitman Press, Eng, 75- *Honors & Awards:* Alexander von Humboldt Sr Scientist Award, 75 & 85. *Mem:* Am Math Soc; Soc Indust & Appl Math. *Res:* Classical analysis; functions of several complex variables; partial differential equations; complex analysis; physical mathematics. *Mailing Add:* Dept Math Univ Del Newark DE 19716

GILBERT, ROBERT PETTIBONE, b Chicago, Ill, Sept 29, 17; m 43; c 6. INTERNAL MEDICINE. *Educ:* Haverford Col, BA, 38; Northwestern Univ, MD, 43. *Prof Exp:* Resident, Stanford Univ Hosp, 46-48; clin asst, instr & assoc med, Sch Med, Northwestern Univ, 50-55, from asst prof to assoc prof, 55-65; assoc dean, 65-72, CLIN PROF MED, JEFFERSON MED COL, THOMAS JEFFERSON UNIV, 65-, DIR, STUDENT EMPLOYEE HEALTH, 80- *Concurrent Pos:* Fel, Stanford Univ Hosp, 46-48; Markle scholar, 52-57; res assoc, Univ Minn, 56-57; sr staff physician, Vet Admin Res Hosp, 54-56; dir educ & res, Evanston Hosp, 57-65. *Mem:* Soc Exp Biol & Med; AMA; Am Heart Asn; Cent Soc Clin Res; Sigma Xi. *Res:* Myocardial infarction; cardiogenic shock; bacteremic shock; microcirculation. *Mailing Add:* Jefferson Med Col 1025 Walnut St Philadelphia PA 19107

GILBERT, SCOTT F, b New York, NY, April 13, 49; m 71; c 2. DEVELOPMENTAL BIOLOGY. *Educ:* Wesleyan Univ, BA, 71; Johns Hopkins Univ, MA, 76, PhD(biol), 76. *Prof Exp:* Fel molecular biol, Univ Wis, 76-78 & immunol, 78-80; MEM STAFF, SWARTHMORE COL, 80- *Honors & Awards:* Dwight J Ingle Prize, 84. *Mem:* Soc Develop Biol; Am Soc Human Genetics; Int Soc Develop Commun Immunol; Am Zool Soc; Hist Sci Soc. *Res:* Investigation of developmentally important compounds on mammalian cell surfaces using monoclonal antibodies and cell culture. *Mailing Add:* Dept Biol Swarthmore Col Swarthmore PA 19081

GILBERT, SEYMOUR GEORGE, b Orange, NJ, Mar 24, 14; m 39; c 4. FOOD SCIENCE. *Educ:* Rutgers Univ, BS, 35, MS, 38, PhD(plant physiol), 41. *Prof Exp:* Asst bot, Rutgers Univ, 36-39, plant physiol, 39-41, res assoc pomol, 41-42; assoc plant physiologist, Field Lab, Tung Invest, US Dept Agr, 42-51; sr chemist in charge malt res sect, Pabst Brewing Co, 51-58; lab mgr & staff asst to res & develop dir, Milprint, Inc, 58-60, res mgr, 60-63, corp tech dir, 63-65; dir, Packaging Inst, 75-78; prof packaging sci, 65-75, PROF II, RUTGERS UNIV, 75- *Concurrent Pos:* Instr, Univ Fla, 46-51; consult, Nat Res Coun, 74-78. *Honors & Awards:* Ebert Award, Am Acad Pharm Sci, 75; Prof Award, Packaging Inst, 75. *Mem:* Am Chem Soc; Inst Food Technol; Packaging Inst; Am Soc Testing & Mat. *Res:* Packaging; enzyme technology; plant biochemistry; migration to foods; physical chemistry of foods. *Mailing Add:* Dept Food Sci Rutgers Univ New Brunswick NJ 08903

GILBERT, STEPHEN MARC, b Long Island, NY, Mar 31, 41; m 62; c 2. ELECTRICAL ENGINEERING, MATHEMATICS. *Educ:* NJ Inst Technol, BS, 62; Univ Southern Calif, MS, 64; Cornell Univ, PhD(elec eng), 67. *Prof Exp:* Engr design, Hughes Aircraft Co, 62-64; instr teaching, Cornell Univ, 64-67; supvr res & develop, Bell Tel Labs, 67-71; dept mgr develop, Teledyne Brown Eng Co, 71-74; vpres res & develop, 74-89, DIR NEW PROD DEVELOP, DYNETICS, INC, 89- *Concurrent Pos:* Hughes masters fel, Hughes Aircraft Co, 62-64; Ford fel, Cornell Univ, 64-65, Faraday fel, 65-66, NSF fel, 66-67; instr, Fairleigh Dickenson Univ, 66-67; mem bd dir & chmn acad coun, Southeastern Inst Technol, 75-; assoc prof, Univ Ala, Huntsville, 72-75. *Mem:* Sigma Xi. *Res:* Application of engineering and mathematical principles to radar and communications systems; electronic countermeasures; electronic counter-countermeasures; waveform design; signal processing; system evaluation. *Mailing Add:* Dynetics Inc PO Drawer 1000 Explorer Blvd Huntsville AL 35814-5050

GILBERT, SUSAN POND, AXONAL TRANSPORT, CELL MOTILITY. *Educ:* Dartmouth Col, PhD(cell biol), 86. *Prof Exp:* RES ASSOC, PA STATE UNIV, 86- *Mailing Add:* Dept Cell & Molecular Biol Pa State Univ 301 Althouse Lab University Park PA 16802

GILBERT, THEODORE WILLIAM, JR, b Attleboro, Mass, Nov 4, 29. ANALYTICAL CHEMISTRY. *Educ:* Mass Inst Technol, BS, 51; Univ Minn, PhD(chem), 56. *Prof Exp:* Res chemist, Oak Ridge Nat Lab, 56-57; asst prof chem, Pa State Univ, 57-60; prof, 60-88, EMER PROF ANALYTICAL CHEM, UNIV CINCINNATI, 88- *Concurrent Pos:* Mem staff, Brookhaven Nat Lab, 66-67. *Mem:* Am Chem Soc. *Res:* Liquid chromatography; ion exchange; trace analysis; analytical solvent extraction. *Mailing Add:* 46 N Liberty St Nantucket MA 02554

GILBERT, THOMAS LEWIS, b Topeka, Kans, Nov 24, 22; m 46; c 3. ENVIRONMENTAL SCIENCE, THEORETICAL CHEMICAL PHYSICS. *Educ:* Calif Inst Technol, BS, 44, MS, 49; Ill Inst Technol, PhD(theoret physics), 56. *Prof Exp:* Asst, Armour Res Found, 44-46, from asst physicist to assoc physicist, 47-56, res physicist, 56; from asst physicist to physicist, 56-76, SR PHYSICIST, ARGONNE NAT LAB, 76-; ASSOC DIR, CHICAGO CTR RELIG & SCI, 88- *Mem:* Am Phys Soc; Sigma Xi; AAAS; Health Physics Soc; Soc Risk Analysis; Soc Sci Study Relig; Inst Relig & Sci. *Res:* Electronic structure of atoms, molecules and solids; interatomic forces; environmental analyses and project management; risk assessment; philosophy of science; relationship of science and religion. *Mailing Add:* 11919 Ford Rd Palos Park IL 60464

GILBERT, THOMAS REXFORD, b Rochester, NY, Oct 29, 46; m 83; c 3. ANALYTICAL CHEMISTRY. *Educ:* Clarkson Col Technol, BS, 68; Mass Inst Technol, PhD(anal chem), 71. *Prof Exp:* Res assoc, New Eng Aquarium, 71-77, assoc dir res, 77-81; asst prof, 81-85, ASSOC PROF CHEM, NORTHWESTERN UNIV, 86- *Concurrent Pos:* Res assoc fel chem, Brandeis Univ, 73-77. *Mem:* Am Chem Soc; Soc Appl Spectros; Sigma Xi. *Res:* Environmental analytical chemistry of trace metals; plasma emission spectroscopy; bioanalytical separations. *Mailing Add:* 22 Castle Rd Norfolk MA 02056

GILBERT, W(ILLIAM) D(OUGLAS), b Kingston, Ont, Feb 13, 10; m 40; c 3. MECHANICAL ENGINEERING. *Educ:* Queen's Univ, BSc, 32; Mass Inst Technol, SM, 35. *Prof Exp:* Lectr mech eng, Queen's Univ, 46-49, from asst prof to assoc prof, 49-59, head dept, 56-69, PROF MECH ENG, QUEEN'S UNIV, 59- *Concurrent Pos:* Mem, Civilian Atomic Power Dept, Can Gen Elec Co, 56-59; mem, Atomic Energy Can Ltd, Ont, 69-70. *Res:* Two phase fluid flow; thermodynamics. *Mailing Add:* Dept Mech Eng Queen's Univ Kingston ON K7L 3N6 Can

GILBERT, WALTER, b Boston, Mass, Mar 21, 32; m 53; c 2. MOLECULAR BIOLOGY. *Educ:* Harvard Univ, AB, 53, Am, 54; Cambridge Univ, PhD(math), 57. *Hon Degrees:* DSc, Univ Chicago & Columbia Univ, 78, Univ Rochester, 79 & Yeshiva Univ, 81. *Prof Exp:* Chmn & prin exec officer, Biogen-NV, 81-84; NSF fel physics, Harvard Univ, 57-58, lectr & res fel, 58-59, asst prof, 59-64, assoc proof biophys, 64-68, prof biochem, 68-72, Am Cancer Soc prof molecular biol, 72-81, sr assoc biochem & molecular biol, 82-84, prof biol, 85-86, H H Timken prof sci, 86-87, CARL M LOEB UNIV PROF & CHMN, DEPT CELLULAR & DEVELOP BIOL, HARVARD UNIV, 87- *Concurrent Pos:* Guggenheim fel, Paris, 68-69. *Honors & Awards:* Nobel Prize in Chem, 80; US Steel Found Award Molecular Biol, Nat Acad Sci, 68; V D Mattia Lectr, Roche Inst Molecular Biol, 76; Smith Kline & French Lectr, Univ Calif, Berkeley, 77; Louis & Bert Freedman Award, NY Acad Sci, 77; Charles-Leopold Mayer Prize, Acad Sci, Inst France, 77; Harrison Howe Award, Am Chem Soc, 78; Gairdner Found Award, 79; Albert Lasker Basic Med Res Award, Albert & Mary Laskr Found, 79; Prize Biochem Anal, German Soc Clin Chem, 80; Sober Award, Am Soc Biol Chemists, 80. *Mem:* Nat Acad Sci; Am Acad Arts & Sci; Am Phys Soc; Am Soc Biol Chem; foreign mem Royal Soc. *Res:* Gene evolution; genetic control mechanisms; repressors; protein-DNA interactions; author/co-author of 98 articles on theoretical physics and molecular biology. *Mailing Add:* Biol Labs Harvard Univ 16 Divinity Ave Cambridge MA 02138

GILBERT, WALTER WILSON, b Johnson City, Tenn, July 10, 22; m 50; c 2. INDUSTRIAL CHEMISTRY. *Educ:* Ga Inst Technol, BS, 44, MS, 47; Univ Wis, PhD(chem), 50. *Prof Exp:* Res chemist, Cent Res Dept, E I Du Pont de Nemours & Co, Inc, 50-63, supvr mat res, 63-67, res mgr, Develop Dept, 67-70, mgr chem res sect, Electrochem Dept, 70-72, lab adminr, Indust Chem Dept, 72, res supvr, Indust Chem Dept, 73-77, res supvr Chem & Pigments Dept, Exp Sta, 78-85; RETIRED. *Mem:* Am Chem Soc; Sigma Xi. *Res:* Reactions under high pressure; catalysis; inorganic crystal growth; tropolone; magnetic materials; compositions for hybrid circuits, industrial chemicals; amines and polyols for polymer intermediates; THF polymerization. *Mailing Add:* 519 Summit Dr Berkeley Ridge Hockessin DE 19707-9647

GILBERT, WILLIAM BEST, b Berea, Ky, Feb 13, 21; m 48; c 2. AGRONOMY. *Educ:* Berea Col, BS, 42; Univ Ky, MS, 52; NC State Univ, PhD(physiol, ecol), 56. *Prof Exp:* Teacher, Pub Schs, Ky, 45-50; asst, Univ Ky, 50-52; asst, NC State Univ, 52-55, from asst prof to assoc prof, 58-76, prof crop sci, 76-; RETIRED. *Concurrent Pos:* Consult, Univ NC Mission, Peru, 55-58. *Mem:* Am Soc Agron. *Mailing Add:* Dept Crop Sci NC State Univ Raleigh NC 27650

GILBERT, WILLIAM HENRY, III, b Glen Ridge, NJ, Nov 30, 39; m 68; c 2. ECOLOGY, ENVIRONMENTAL SCIENCES. *Educ:* Yale Univ, BA, 62; Univ Mass, Amherst, PhD(ecol), 73. *Prof Exp:* Head sci dept, Ethiopian Ministry Educ, US Peace Corps, 64-66; res trainee ecol, Marine Biol Lab, 69-70; asst prof biol, Colby Col, 70-76; asst prof environ studies, Ottawa Univ, 76-78; from asst prof to assoc prof, 78-86, PROF, SIMPSON COL, 86- *Mem:* Ecol Soc Am; Nat Audubon Soc; Nature Conservancy. *Res:* Marine molluscan ecology and systematics; mathematical models for dispersion patterns and species diversity; bird distribution; environmental studies. *Mailing Add:* Dept of Biol Simpson Col Indianola IA 50125

GILBERT, WILLIAM IRWIN, b Philadelphia, Pa, Mar 6, 15; m 46; c 2. PETROLEUM CHEMISTRY. *Educ:* Pa State Col, BS, 36; Princeton Univ, AM, 38, PhD(org chem), 39. *Prof Exp:* Instr chem, Western Md Col, 39-40; res chemist, Gulf Res & Develop Co, 40-45, asst head sect new processes, 45-52, head petrochem sect, 52-55, asst dir process div, 55-61, dir petrochem div, 61-68, mgr, Kansas City Lab, 68-71, dir petrochem div, 71-75, SR POLICY ADV, GULF OIL CORP, 75- *Mem:* AAAS; Am Chem Soc. *Res:* Petroleum refining; petrochemicals. *Mailing Add:* 31 Oakglen Dr Oakmont PA 15139

GILBERT, WILLIAM JAMES, b Shelton, Wash, Feb 10, 16; m 42; c 4. ALGOLOGY. *Educ:* Univ Wash, BS, 38; Univ Mich, MS, 39, PhD(bot), 42. *Prof Exp:* Asst bot, Univ Mich, 38-40; res biologist, Com Solvents Corp, Ind, 43-46; from asst prof to assoc prof biol, 46-57, chmn dept, 57-71, chmn div sci & math, 62-65, assoc acad dean, 71-73, PROF BIOL, ALBION COL, 57- *Mem:* Phycol Soc Am; Int Phycol Soc. *Res:* Morphology and taxonomy of cryptograms and marine algae. *Mailing Add:* Dept Biol Albion Col Albion MI 49224

GILBERT, WILLIAM SPENCER, b New York, NY, May 25, 27; m 56; c 2. NUCLEAR PHYSICS. *Educ:* Univ Calif, AB, 48, PhD(physics), 52. *Prof Exp:* Physicist, Radiation Lab, Univ Calif, 49-52; res engr, Atomic Energy Res Dept, NAm Aviation, Inc, 52-54; physicist, Lawrence Livermore Lab, 54-63, PHYSICIST, LAWRENCE BERKELEY LAB, UNIV CALIF, 63- *Mem:* Am Phys Soc. *Res:* Accelerator design; radioactivity problems; shielding; experimental areas; superconductivity; cryogenics; superconducting magnets. *Mailing Add:* 2952 Claremont Blvd Berkeley CA 94720

GILBERT-BARNESS, ENID F, b Sydney, Australia, May 31, 27; nat US; m 54, 87; c 5. PEDIATRICS, DEVELOPMENTAL PATHOLOGY. *Educ:* Univ Sydney, MS, 50, MD, 83; Am Col Pathol, FACP, 63; Royal Col Path, Australia, FRAC Path. *Prof Exp:* From asst prof to assoc prof path, WVa Univ, 63-70; PROF PATH, UNIV WIS, 71-, PROF PEDIAT, 72- *Concurrent Pos:* Dir Pediat Path, Univ Wis, 71- & surg path, 75- med alumni prof, 86-; Vis prof, Univ SFla, 78-; mem, Cancer Educ Comt, 80-83; mem, Study Sect, Nat Heart Lung & Blood Inst, NIH, 80-83; pres, Soc Pediat Path, 86-87. *Mem:* Am Soc Clin Path; Soc Pediat Path (pres, 86-87); Terarology Soc; Arthur Purdy Stout Soc; Am Acad Pediat; Int Acad Path; Am Pediat Soc; Int Pediat Path Asn (pres, 90-92). *Res:* Genetic and developmental pathology; birth defects and malformation syndromes; sudden infant death syndrome. *Mailing Add:* Dept Surgery Pathology Unv Wis Hosp 600 Highland Ave Madison WI 53792

GILBERTSEN, RICHARD B, US citizen; m 73; c 2. IMMUNOPATHOLOGY. *Educ:* Beloit Col, BS, 69; Duke Univ, PhD(immunol), 74. *Prof Exp:* Res assoc, Med Ctr, Duke Univ, 74-75; proj assoc, Immunobiol Res Ctr, Univ Wis-Madison, 75-76, postdoctoral fel, 76-77; immunopharmacologist, Dept Pharmacol, Anti-Arthritis Proj, Abbott Labs, North Chicago, 77-80; res assoc & group leader, Dept Pharmacol, Immunopharmacol Sect, Parke-Davis Pharmaceut Res, Warner-Lambert Co, 80-85, sr res assoc & group leader, 85-89, sr res assoc & group leader, Dept Exp Ther, Immunopath Sect, 89-90, ASSOC RES FEL, DEPT EXP THER, IMMUNOPATH SECT, PARKE-DAVIS PHARMACEUT RES, WARNER-LAMBERT CO, 90- *Mem:* Am Col Rheumatology; Inflammation Res Asn; Am Soc Pharmacol & Exp Therapeut. *Mailing Add:* Pharmacol Dept Div Warner-Lambert Co Parke-Davis Pharmaceut Res 2800 Plymouth Rd Ann Arbor MI 48105

GILBERTSEN, VICTOR ADOLPH, b Winona, Minn, Nov 4, 24; m 48. SURGERY, CANCER. *Educ:* Hamline Univ, BA, 48; Univ Minn, Minneapolis, BS, 50, MB, 52, MD, 53, MS, 57; Am Bd Surg, dipl. *Prof Exp:* From instr to asst prof, Sch Med, 58-71, asst prof surg, Grad Sch, Univ Minn, Minneapolis, 61-77, assoc prof, Health Sci Ctr, 71-77, dir, Cancer Detection Ctr, 60-77, CHIEF INVESTR CANCER CONTROL, UNIV MINN HEALTH SCI CTR, 77-; MEM STAFF MINN DEPT HEALTH. *Mem:* Am Col Surg; Asn Acad Surg; Int Union Against Cancer; Accademia Tiberina; Asn Am Med Cols. *Res:* Cancer detection, control and treatment; medical education. *Mailing Add:* Dept Surgery Univ Minn Box 147 Mayo Minneapolis MN 55455

GILBERTSON, DONALD EDMUND, b Whitehall, Wis, Oct 22, 34; m 57; c 2. PARASITOLOGY, MALACOLOGY. *Educ:* Wis State Univ, BS, 59; SDak State Univ, MS, 62; Univ Cincinnati, PhD(zool), 66. *Prof Exp:* Asst zool, SDak State Univ, 60-62 & Univ Cincinnati, 62-65; res fel med parasitol, Sch Pub Health, Harvard Univ, 66-68; from asst prof to assoc prof zool, 68-84, PROF ECOL & BEHAV BIOL, UNIV MINN, MINNEAPOLIS, 84-; DIR, JAMES FORD BELL MUS NATURAL HIST, 83- *Mem:* AAAS; Am Soc Trop Med & Hyg; Am Soc Parasitol; Sigma Xi. *Res:* Host parasite relationships; protein and nucleic acid metabolism of mollusks; control of disease bearing snails; biochemistry of mollusk body fluids. *Mailing Add:* Ecol 109 Zool Bldg Univ Minn 318 Church St SE Minneapolis MN 55455

GILBERTSON, JOHN R, biochemistry; deceased, see previous edition for last biography

GILBERTSON, MICHAEL, b Durham, Eng, Feb 4, 45; Can citizen. ECOTOXICOLOGY. *Educ:* Queen's Univ, Belfast, BD, 67, MSc, 69. *Prof Exp:* Biologist limnol, Can Ctr Inland Waters, 69-71; biologist pesticides, Can Wildlife Serv, 71-74; biologist contaminants, Environ Protection Serv, Environ Can, 75-81; biologist contaminants, Fish Habitats Mgt, Dept Fisheries & Oceans, Ottawa, 81-86; SECY, WATER QUAL BD & BIOLOGIST, INT JOINT COMN, 86- *Concurrent Pos:* Leader, Can Task Force Polychlorobiphenyls, 75-76 & Can Task Force Mirex, 76-77. *Honors & Awards:* Chandler-Misener Award, Int Asn Great Lakes Res, 72. *Res:* Effects and movement of chemical contaminants in ecosystems; chick edema disease; polychlorinated dibenzo-p-dioxins in the Great Lakes; transboundary pollution and United States/Canada relations. *Mailing Add:* Int Joint Comn PO Box 32869 Detroit MI 48232-2869

GILBERTSON, ROBERT LEE, b Hamilton, Mont, Jan 15, 25; m 48; c 2. MYCOLOGY, FOREST PATHOLOGY. *Educ:* Univ Mont, BA, 49; Univ Wash, MS, 51; State Univ NY, PhD(mycol), 54. *Prof Exp:* Asst, State Univ NY Col Forestry, Syracuse, 51-54; asst prof forestry, Univ Idaho, 54-59; from asst prof to assoc prof forest bot, State Univ NY Col Forestry, Syracuse, 59-67; PROF PLANT PATH, UNIV ARIZ, 67- *Concurrent Pos:* Consult, US Forest Serv, 57-; vis assoc prof, Biol Sta, Univ Mont, 64, 66 & Univ Minn, 70; chmn, Western Int Forest Dis Work Conf, 80. *Mem:* Mycol Soc Am (pres, 79); Soc Am Foresters; Phytopath Soc Am; Brit Mycol Soc. *Res:* Taxonomy of woodrotting fungi of North America; biology of wood-rotting basidiomycetes of North America, including taxonomy, floristics, genetics of sexuality and vegetative incompatibility, pathological relationships, and ecological significance. *Mailing Add:* Dept Plant Path Univ Ariz Tucson AZ 85721

GILBERTSON, TERRY JOEL, b La Crosse, Wis, May 18, 39; m 67; c 2. CLINICAL BIOCHEMISTRY, CLINICAL CHEMISTRY. *Educ:* Wis State Univ, La Crosse, BS, 62; Univ Minn, Minneapolis, PhD(org chem), 67. *Prof Exp:* Trainee biochem, Univ Minn, St Paul, 67-68; asst prof, SDak State Univ, 68-72; res scientist, 72-84, DIR, UPJOHN CO, 85- *Concurrent Pos:* Sabbatical, Mayo Clinic, 82-83. *Mem:* Am Chem Soc; Am Asn Clin Chemists; NY Acad Sci; Asn Official Anal Chem. *Res:* Development of clinical laboratory assays and synthesis of labeled compounds and the application of these to clinical studies and agriculture. *Mailing Add:* Upjohn Co Kalamazoo MI 49007

GILBOE, DANIEL PIERRE, b Amboy, Ill, Dec 29, 34; m 61. MICROBIOLOGY, ORGANIC CHEMISTRY. *Educ:* Univ Wis-Madison, BSc, 56, MSc, 59; Univ Minn, PhD(biochem), 67. *Prof Exp:* Chemist, Minn Mining & Mfg Co, 56; res chemist, Archer-Daniels-Midland Co, Minn, 60-63; res assoc, Dept Dairy Sci, Univ Ill, Urbana, 67-68; RES BIOCHEMIST, VET ADMIN HOSP, 68- *Concurrent Pos:* Instr biochem, Univ Minn, 71-74, asst prof, 74-; ed, J Minn Acad Sci. *Mem:* AAAS; Am Chem Soc; Am Soc Biochem & Molecular Biol; Sigma Xi. *Res:* Hormonal and metabolite regulation of liver glycogen metabolism with particular reference to control of glycogen synthase phosphatase; initiation of liver glycogen synthesis. *Mailing Add:* Vet Affairs Med Ctr One Veterans Dr Minneapolis MN 55417

GILBOE, DAVID DOUGHERTY, b Richland Center, Wis, July 13, 29; m 51; c 2. BIOCHEMISTRY, ANIMAL PHYSIOLOGY. *Educ:* Miami Univ, BA, 51; Univ Wis, MS, 55, PhD(biochem), 58. *Prof Exp:* Res asst biochem, 55-58, from instr physiol chem & surg to prof physiol & surg, 58-89, PROF PHYSIOL & NEUROSURG, UNIV WIS-MADISON, 89- *Concurrent Pos:* Fulbright lectr, Univ Chile, 70; Neurol & Study Sect, 80-84; ed bd, J Neurochem & J Neurochem Res, 86- *Mem:* Am Chem Soc; Am Physiol Soc; Int Soc Neurochem; Am Soc Biol Chem; Int Soc Cerebral Blood Flow & Metabolism; Am Soc Neurochem. *Res:* Biochemistry of brain function; transport of metabolites, intermediary metabolism of carbohydrates and amino acids in the central nervous system during hypoxia and recovery; pharmacologic treatment of post ischemia brain injury. *Mailing Add:* Dept Neurosurg Univ Wis Med Sch 4630 Med Sci Ctr Madison WI 53706-1532

GILBREATH, MICHAEL JOYE, cellular immunology, immune suppression, for more information see previous edition

GILBREATH, SIDNEY GORDON, III, b Atlanta, Ga, Aug 11, 31; m 57; c 3. INDUSTRIAL ENGINEERING. *Educ:* Univ Tenn, BS, 58, MS, 62; Ga Inst Technol, PhD(indust eng), 67. *Prof Exp:* Indust engr, Robertshaw Corp, 59-60; sales engr, Wallace & Tiernan Inc, 60-61; civil engr, Tenn Valley Authority, 61-62; instr indust eng, Ga Inst Technol, 62-66; asst prof, Va Polytech Inst, 67-68; dir mat mgt, Westmoreland Coal Co, Inc, 78-79, gen mgr, Va Oper, 79-80; from assoc prof to prof, Tenn Technol Univ, 68-78, chmn dept, 69-78 & 80-82, prof, 82-91 INTERIM CHMN & DEPT INDUST ENG, TENN TECHNOL UNIV, 91- *Concurrent Pos:* Consult indust energy & mgt, 68-78. *Mem:* Am Soc Qual Control; Am Inst Indust Engrs (pres, 78-79); Am Soc Eng Educ. *Res:* Statistical sampling; methods engineering; engineering economy; production; inventory control and management; facilities design and material handling. *Mailing Add:* Indust Eng Dept Tenn Tech PO Box 5011 Cookeville TN 38505

GILBREATH, WILLIAM POLLOCK, b Portland, Ore, Nov 10, 36; m 65; c 2. SPACE BIOLOGY. *Educ:* Reed Col, BA, 58; Univ Wash, PhD(inorg chem), 62. *Prof Exp:* Res scientist, NASA, 62-75 & 77-81, prog mgr hq, 75-76 & 82-83, proj mgt, 82-85, FLIGHT PROG MGT, AMES RES CTR, NASA HQ, 86- *Mem:* Am Chem Soc; sr mem Am Inst Aeronaut & Astronaut; Sigma Xi. *Res:* Sponsor the development of biological and medical research and technology investigations for spaceflight. *Mailing Add:* 700 Seventh St SW Apt 507 Washington DC 20024

GILBRECH, DONALD ALBERT, b Holly Grove, Ark, Apr 12, 27; m 49; c 3. ENGINEERING MECHANICS. *Educ:* Univ Ark, BSIE, 53, MS, 54; Purdue Univ, PhD(mech eng), 58. *Prof Exp:* From instr to assoc prof, 53-63, PROF ENG MECH, UNIV ARK, FAYETTEVILLE, 63- *Mem:* Am Soc Mech Engrs. *Res:* Fluid mechanics, especially instrumentation, acoustics, pulsating flow and gas dynamics; experimental stress analysis; engineering science; computer interfacing. *Mailing Add:* Dept Mech Eng Univ Ark Fayetteville AR 72701

GILBY, STEPHEN WARNER, b Dayton, Ohio, Sept 22, 39; m 62; c 3. STEELMAKING. *Educ:* Univ Cincinnati, BS, 62; Ohio State Univ, PhD(metall eng), 66. *Prof Exp:* Res engr steelmaking, Youngstown Steel Co, 66-76; res engr, Armco Steel Co, 67-69, sr res engr, 69-72, res assoc, 72-75, mgr steelmaking res, 75-82, DIR PROCESS RES, ARMCO STEEL CO, 82- *Concurrent Pos:* Chmn, external adv comn, Mat Sci & Eng Dept, Ohio State Univ, 88- *Mem:* Am Iron & Steel Soc; Am Soc Metals Int. *Res:* Steelmaking and continuous casting process development. *Mailing Add:* 121 Sunset Ct Monroe OH 45050

GILCHRIST, BRUCE, b Pontefract, Eng, Aug 4, 30; nat US; wid; c 3. COMPUTER SCIENCES. *Educ:* Univ London, BSc, 50, PhD(meteorol), 52. *Prof Exp:* Vis meteorologist, Inst Advan Study, 52-54, mem staff, 54-56; asst prof math & dir comput ctr, Univ Syracuse, 56-59; mgr prog & comput, Res Ctr, Int Bus Mach, 59-61, dir systs eng prog, Corp Staff, 61-63; dir planning, Serv Bur Corp, 63-65; mgr sci opers, Int Bus Mach Data Processing Div, 65-68; exec dir, Am Fedn Info Processing Socs, Inc, 68-73; dir comput activities, 73-85, SR ADV INFO STRATEGY, COLUMBIA UNIV, 86- *Concurrent Pos:* Consult, US Weather Bur, 55-59; mem panel weather & river serv, Nat Acad Sci/Nat Acad Engrs comt adv Environ Sci Serv Admin, 68-69; panel nat progs comput sci & eng bd, Nat Acad Sci, 68-69; consult, US Off Educ, 71-72 & US Gen Acctg Off, 76-90. *Mem:* AAAS; Asn Comput Mach (secy, 60-62, vpres, 62-64); Am Meteorol Soc; Inst Elec & Electronics Engrs; Am Fedn Info Processing Soc (pres, 66-68). *Res:* Methods and applications of high speed computation; manpower and regulatory aspects of the computer industry; electronic funds transfer systems. *Mailing Add:* PO Box 656 Chappaqua NY 10514

GILCHRIST, RALPH E(DWARD), b Milwaukee, Wis, Dec 17, 26; m 55; c 3. PETROLEUM ENGINEERING. *Educ:* Univ Denver, BA, 47; Univ Tex, BS, 50, MS, 51; Pa State Univ, PhD(petrol & natural gas eng), 58. *Prof Exp:* Res engr petrol prod res, Texaco, Inc, 51-52; res assoc petrol & natural gas eng, Pa State Univ, 54-57; asst proj engr petrol prod res, Sinclair Res Labs, Inc, 57-59; res engr, Phillips Petrol Co, 59-66; sr res engr prods, Tenneco Oil Co, 66-69, dir prod res, 69-71; mgr explor & prod res, Southwest Res Inst, 71-73; dean & prof technol, Corpus Christi State Univ, 73-79; PRES, RALPH

E GILCHRIST INC, 79- *Concurrent Pos:* Lectr, Univ Tulsa, 59-60. *Mem:* Am Inst Mining, Metall & Petrol Engrs; Sigma Xi; Am Soc Oceanog. *Res:* Drilling; production; reservoir engineering; oil and gas recovery. *Mailing Add:* 12502 Taylorcrest Houston TX 77024

GILDE, HANS-GEORG, b Ger, July 8, 33; US citizen; m 57; c 2. ORGANIC CHEMISTRY. *Educ:* Albright Col, BS, 57; Ohio Univ, PhD(org chem), 61. *Prof Exp:* Instr chem, Ohio Univ, 60-61; from asst prof to assoc prof org chem, 61-70, prof chem, 70-84, EBENEZER BALDWIN ANDREWS CHAIR CHEM & NATURAL SCI, MARIETTA COL, 84- *Concurrent Pos:* Assoc, Danforth Found, 78-84; Harness Fel, 86-89. *Mem:* Am Chem Soc; Sigma Xi. *Res:* Molecular spectroscopy. *Mailing Add:* Dept Chem Marietta Col Marietta OH 45750

GIL DE LAMADRID, JESÚS, b San Juan, PR, Aug 20, 26; m 50; c 3. MATHEMATICS. *Educ:* Univ Chicago, BS, 48, MS, 49; Univ Mich, PhD(math), 55. *Prof Exp:* Res assoc-Navy Logistics, George Washington Univ, Wash, 50-52; instr math, Ohio State Univ, 55-57; from asst prof to assoc prof, 57-67, PROF MATH, UNIV MINN, MINNEAPOLIS, 67- *Concurrent Pos:* Res assoc & lectr, Yale Univ, 61-62; residence, Centre Univ Int, Paris, 64-65; vis prof, Univ Rennes, 71-72 & Ludwig-Maximilians-Univ, Munich, 74-75 & 78-79. *Res:* Harmonic analysis; representation theory of topological groups and algebras. *Mailing Add:* Dept Math Univ Minn Minneapolis MN 55455

GILDEN, DONALD HARVEY, b Baltimore, Md, Aug 30, 37; m 66; c 2. NEUROLOGY, NEUROBIOLOGY. *Educ:* Dartmouth Col, BA, 59; Univ Md, MD, 63. *Prof Exp:* Intern, Univ Ill Res & Educ Hosp, 63-64; resident neurol, Univ Chicago Hosps, 64-67; staff neurologist, Walter Reed Army Med Ctr, US Army, 67-69; fel neurovirol, Sch Hyg & Pub Health, Johns Hopkins Univ, 69-71; res assoc, Wistar Inst, 71-76, asst prof, 71-76, ASSOC PROF NEUROL, SCH MED, UNIV PA & WISTAR INST, 76- *Mem:* Am Acad Neurol; Am Asn Immunologists; Am Asn Neuropathologists; Am Soc Microbiol; AAAS. *Res:* Fast and slow virus infection of the central nervous system; virus-induced immunopathology; multiple sclerosis. *Mailing Add:* Univ Col HTH Sci Ctr Dept Neurol, 4200 E 9th Ave Box B182 Denver CO 80262

GILDEN, RAYMOND VICTOR, b Chicago, Ill, Aug 4, 35; m 57; c 3. IMMUNOLOGY, GENETICS. *Educ:* Univ Calif, Los Angeles, AB, 57, MA, 59, PhD(zool), 62. *Prof Exp:* USPHS fel, Calif Inst Technol, 62-63; assoc mem, Wistar Inst, 63-65; vpres res div, Flow Labs, Inc, 65-76; DIR, MOLECULAR ONCOL PROG, FREDERICK CANCER RES CTR, 76- *Mem:* Sigma Xi. *Res:* Tumor viruses. *Mailing Add:* Frederick Cancer Res Inst Nat Cancer Inst Bldg 427 PO Box B Frederick MD 21701

GILDENBERG, PHILIP LEON, b Hazleton, Pa, Mar 15, 35; m 55; c 4. NEUROSURGERY, NEUROPHYSIOLOGY. *Educ:* Univ Pa, AB, 55; Temple Univ, MD & MS, 59, PhD(neurophysiol), 70. *Prof Exp:* Res assoc neurophysiol, Max Planck Inst Brain Res, 68; staff neurosurgeon, Cleveland Clin Found, 68-72; assoc prof neurosurg, Col Med, Univ Ariz, 72-75; PROF & CHIEF DIV NEUROSURG, UNIV TEX MED SCH, HOUSTON, 75- *Concurrent Pos:* Fel, Max Planck Inst Brain Res, 68; asst ed, Progress Neurol & Psychiat, 69-74; adj asst prof, Case Western Reserve Univ, 70-72; assoc ed, Confinia Neurologica, 72-75; ed, Appl Neurophysiol, 74-; mem bd dirs, World Soc Stereotactic & Functional Neurosurg, 72-76, secy-treas, 76- *Mem:* Fel Am Col Surg; Cong Neurol Surg; Am Asn Neurol Surg; Soc Univ Neurosurg; Am Soc Stereotactic & Functional Neurosurg (secy-treas, 72-). *Res:* Physiology and stereotactic treatment of involuntary movements; pain and its treatment and relation to narcotics. *Mailing Add:* Div Neurosurg Univ Tex Med Sch 6560 Fannin No 1530 Houston TX 77030

GILDENHORN, HYMAN L, b Cleveland, Ohio, May 27, 21; m 55. MEDICINE, RADIOLOGY. *Educ:* Ohio State Univ, BS, 43, MS, 47; Cornell Univ, MD, 51; Am Bd Radiol, dipl, 56. *Prof Exp:* From intern med to resident radiol, Michael Reese Hosp, Chicago, Ill, 51-55; assoc radiologist, 56-57, dir dept, 57-75, CHMN DIV DIAG RADIOL, CITY OF HOPE MED CTR, 77- *Concurrent Pos:* Asst clin prof, Univ Southern Calif, 69-75; mem coun cardiovascular radiol, Am Heart Asn. *Mem:* AMA; Radiol Soc NAm; NY Acad Sci; Int Col Radiol; Sigma Xi. *Res:* Physical factors related to diagnostic roentgenology; clinical research in diagnostic roentgenology; synergism of irradiation and cholesterol diet in production of arteriosclerotic lesions in animals. *Mailing Add:* City of Hope Nat Med Ctr 1500 E Duarte Rd Duarte CA 91010

GILDERSLEEVE, BENJAMIN, b Damascus, Va, June 7, 07. ECONOMIC GEOLOGY, GEOLOGICAL MAPPING. *Educ:* Univ Va, BS, 30, MS, 31; Johns Hopkings Univ, PhD(geol), 39. *Prof Exp:* Field asst, Va Geol Sur, 30-31; jr geol, US Geol Sur, 31-34; from asst geol to assoc geol, 34-51, Tenn Valley Authority, group chief, commodity geologist & head, Bowling Green Field Off, 51, 77; RETIRED. *Mem:* Sigma Xi; fel Geol Soc Am. *Mailing Add:* 837 Ridgecrest Way Bowling Green KY 42104-3822

GILDERSLEEVE, RICHARD E, b Flushing, NY, Aug 17, 14; m 48; c 2. ELECTRICAL ENGINEERING. *Educ:* Rensselaer Polytech Inst, BEE, 48; Univ Syracuse, PhD(elec eng), 58. *Prof Exp:* Eng asst, Consol Edison Co, 34-41; from instr to assoc prof elec eng, 48-77, ASST DEAN, COL ENG, SYRACUSE UNIV, 77- *Mem:* Am Soc Eng Educ; Inst Elec & Electronics Engrs. *Res:* Microwave antennas. *Mailing Add:* 718 Maple Dr Fayetteville NY 13066

GILDSETH, WAYNE, b Sioux Falls, SDak, July 10, 35; m 60; c 2. PHYSICAL CHEMISTRY. *Educ:* Augustana Col, SDak, BA, 57; Iowa State Univ, PhD(phys chem), 64. *Prof Exp:* Prof chem, Pac Lutheran Univ, 64-66; prof chem, Augustana Col, SDak, 66-79; at Univ SDak, Sch Med, 79-83; DEAN, ST CLOUD STATE UNIV, 83- *Concurrent Pos:* SDak Statehouse fel, 72-73. *Res:* Electrolytic solution chemistry; ionic complexes. *Mailing Add:* Southern Ark Univ Box 1397 Magnolia AR 71753

GILE, LELAND HENRY, b Alfred, Maine, Feb 23, 20; m 47. SOIL GENESIS, SOIL CLASSIFICATION. *Educ:* Univ Maine, BS, 53; Univ Wis, MS, 54. *Prof Exp:* Conserv aide, Soil Conserv Serv, USDA, 46-50, soil scientist, 55-57, res soil scientist, 57-76; INDEPENDENT RES, 76- *Concurrent Pos:* Cert Merit, Arid Zone Res, AAAS, 87. *Honors & Awards:* Kirk Bryan Award, Geol Soc Am, 83. *Mem:* Fel AAAS; Soil Sci Soc Am; Geol Soc Am; Am Quaternary Asn; Int Soc Soil Sci; Sigma Xi. *Res:* Soil-geomorphic studies in arid and semiarid lands. *Mailing Add:* 2600 Desert Dr Las Cruces NM 88001

GILES, EUGENE, b Salt Lake City, Utah, June 30, 33; m 64; c 2. PHYSICAL & BIOLOGICAL ANTHROPOLOGY, FORENSIC ANTHROPOLOGY. *Educ:* Harvard Univ, AB, 55, AM, 60, PhD(anthrop), 66; Univ Calif, Berkeley, MA, 56. *Prof Exp:* Instr anthrop, Univ Ill, Urbana, 64-66; asst prof, Harvard Univ, 66-70; assoc prof, 70-73, head dept, 75-80 & 82-83, assoc dean, Grad Col, 86-89, PROF ANTHROP, UNIV ILL, URBANA-CHAMPAIGN, 73- *Concurrent Pos:* NSF fel demog, Australian Nat Univ, 67-68; vis asst prof anthrop, Univ Utah, 69; vis fel prehist, Australian Nat Univ, 78. *Mem:* Fel Am Anthrop Asn; Am Soc Human Genetics; Am Asn Phys Anthropologists (vpres, 79-80, pres, 81-83); Human Biol Coun; fel Am Acad Forensic Sci; fel AAAS. *Res:* Analysis of morphological variation in crania; demography, physical variation and genetic structure of noncosmopolitan human populations in oceania; forensic anthropology; history of physical anthropology. *Mailing Add:* Dept Anthrop 109 Davenport Hall Univ Ill 607 S Mathews Urbana IL 61801

GILES, JESSE ALBION, III, b New Kensington, Pa, June 2, 31; m 53; c 3. ORGANIC CHEMISTRY. *Educ:* Univ NC, BS, 53; Univ Ala, MS, 55. *Prof Exp:* Res chemist, 54-67, sect head res dept, 67-76, mgr analytical res div, 76-80, DIR TECH SERV, R J REYNOLDS TOBACCO CO, 80- *Concurrent Pos:* Abstractor, Chem Abstr, Am Chem Soc, 55-66. *Mem:* Am Chem Soc; Sigma Xi. *Res:* Chemistry of plant natural products; diterpenes; aliphatic sulfur and boron compounds; radiocarbon and tritium tracer techniques. *Mailing Add:* 2626 Village Trail Winston-Salem NC 27106

GILES, JOHN CRUTCHLOW, b London, Eng, Jan 15, 34; m 56; c 2. PHYSICS. *Educ:* Sheffield Univ, BSc, 55; Univ Exeter, PhD(physics), 58. *Prof Exp:* Nat Res Coun Can fel, Univ BC, 58-60, from instr to asst prof physics, 60-63; lectr, Aberdeen Univ, 63-64; from asst prof to assoc prof, 64-70, PROF PHYSICS, CALIF STATE UNIV, HAYWARD, 70- *Mem:* Am Asn Physics Teachers. *Res:* Solid state physics; semiconducting substances; optical and transport properties. *Mailing Add:* Dept Physics Calif State Univ Hayward CA 94542

GILES, MICHAEL ARTHUR, b Toronto, Ont, Oct 6, 43; m 74; c 4. PHYSIOLOGY, AQUATIC TOXICOLOGY. *Educ:* Univ Man, BSc, 65; Univ BC, MSc, 69, PhD(zool), 73. *Prof Exp:* Fisheries biologist salmon enhancement, Dept Fisheries, Can, 65-67; res scientist herring biol, Fisheries & Marine Serv Environ Can, 73-74; res assoc & lectr environ biol, Sir George Williams Univ, Montreal, 75-76; RES SCIENTIST AQUATIC TOXICOL, FRESHWATER INST FISH & MARINE SERV ENVIRON CAN, 76- *Concurrent Pos:* Teaching fel, Univ BC, 72-73 & 74-75. *Mem:* Can Soc Zoologists. *Res:* Respiratory physiology of fish; physiological toxicology of fish; ontogenetic variation in physiology of fish; aquaculture. *Mailing Add:* Freshwater Inst 501 University Crescent Winnipeg MB R3T 2N6 Can

GILES, MICHAEL KENT, b Logan, Utah, Oct 24, 45; m 68; c 7. OPTICAL ENGINEERING, ELECTRICAL ENGINEERING. *Educ:* Brigham Young Univ, BES, 71, MS, 71; Univ Ariz, MS, 76, PhD(optical sci), 76. *Prof Exp:* Electronics engr electro-optics, US Naval Weapons Ctr, China Lake, Calif, 71-77; res electronics engr, Electro-optics & Imaging Processing, White Sands Missile Range, 77-80; PROF, DEPT ELEC & COMPUT ENG, NMEX STATE UNIV, LAS CRUCES, 82- *Concurrent Pos:* Res physicist optics, USAF Weapons Lab, Albuquerque, NMex, 80-82. *Honors & Awards:* Cert Outstanding Achievement, US Army Dept Chief of Staff for Res Develop & Acquisition, 78. *Mem:* Optical Soc Am; Soc Photo-Optical Instrumentation Engrs. *Res:* Image processing (digital and optical); lasers; sensors; visual optics. *Mailing Add:* Dept Elec & Comput Eng Box 3-0 NMex State Univ Las Cruces NM 88003

GILES, NORMAN HENRY, b Atlanta, Ga, Aug 6, 15; m 39, 69; c 2. GENETICS. *Educ:* Emory Univ, AB, 37; Harvard Univ, MA, 38, PhD(biol), 40. *Hon Degrees:* DSc, Emory Univ, 80. *Prof Exp:* Parker fel, Harvard Univ, 40-41; from instr to prof bot, Yale Univ, 41-61, Eugene Higgins prof genetics, 61-72; Fuller E Callaway prof 72-86, EMER PROF GENETICS, UNIV GA, 86- *Concurrent Pos:* Prin biologist, Oak Ridge Nat Lab, 47-50; consult biol, Oak Ridge Nat Lab, 51-64 & Brookhaven Nat Lab, 56-64; Fulbright & Guggenheim fels, Univ Genetics Inst, Copenhagen, 59-60; mem genetics study sect, NIH, 60-64, mem genetics training comt, 66-70; Guggenheim fel genetics, Australian Nat Univ, 66. *Honors & Awards:* Thomas Hunt Morgan Medal, Genetics Soc of Am 88. *Mem:* Nat Acad Sci (chmn, Genetics Sect, 76-79); AAAS; fel Am Acad Arts & Sci; Am Soc Naturalists (pres, 77); Am Inst Biol Sci; Genetics Soc Am (treas, 54-56, vpres, 69, pres, 70); foreign mem Royal Danish Acad Sci & Letters. *Res:* Molecular organization function and regulation of gene clusters and multienzyme complexes, especially in lower eucaryotes. *Mailing Add:* Dept Genetics Univ Ga Athens GA 30602

GILES, PETER COBB, b Albany, Calif, Nov 21, 29; div; c 4. PHYSICS. *Educ:* Univ Calif, BS, 52, MS, 53, PhD(physics), 58. *Prof Exp:* SR PHYSICIST, LAWRENCE LIVERMORE LAB, UNIV CALIF, 58- *Mem:* Am Phys Soc; Sigma Xi. *Res:* Nuclear physics; application of computers and numerical techniques to physical problems. *Mailing Add:* 1881 De Lean Livermore CA 94550

GILES, RALPH E, b Rahway, NJ, Mar 26, 41; m 63; c 5. PHARMACOLOGY, BIOCHEMISTRY. *Educ:* Fordham Univ, BS, 62; Univ Minn, PhD(pharmacol), 66. *Prof Exp:* Asst prof pharmacol, Fordham Univ, 66-68; scientist pharmacol, Warner-Lambert Res Inst, 68-69, sr scientist, 69-72, sr res assoc, 72-75; mgr pharmacol, 75-79, DIR PHARMACOL, ICI-AMERICAS, 79- *Mem:* NY Acad Sci; Am Soc Pharmacol & Exp Therapeut; Soc Exp Biol & Med; fel Royal Soc Med; Sigma Xi. *Res:* Respiratory pharmacology; ens pharmacology; bronchodilators; experimental production of emphysema; renal pharmacology; catechol-o-methyl transferase. *Mailing Add:* 1303 Grayson Rd E Welshire Wilmington DE 19803

GILES, ROBERT H, JR, b Lynchburg, Va, May 25, 33; m 56; c 2. ECOLOGY, WILDLIFE MANAGEMENT. *Educ:* Va Polytech Inst, BS, 55, MS, 58; Ohio State Univ, PhD(wildlife mgt), 64. *Prof Exp:* Dist biologist, Va Comn Game & Inland Fisheries, 58-60; asst prof wildlife mgt, Univ Idaho, 63-67; from assoc prof to prof forestry, 67-74, PROF WILDLIFE MGT & ENVIRON & URBAN SYST, VA POLYTECH INST & STATE UNIV, 74- *Mem:* Wildlife Soc. *Res:* Conservation education evaluation; insecticide-ecology; forest ecology; computer simulation of ecological systems. *Mailing Add:* Dept Wildlife Mgt Va Polytech Inst & State Univ Blacksburg VA 24061

GILES, ROBIN, b London, Eng, Jan 30, 26; m 54; c 3. FUZZY REASONING, FOUNDATIONS OF PHYSICS. *Educ:* Glasgow Univ, BSc, 46, DSc(found of thermodyn), 66. *Prof Exp:* From asst lectr to sr lectr physics, Glasgow Univ, 46-66; assoc prof, 66-69, PROF MATH, QUEEN'S UNIV, ONT, 69- *Concurrent Pos:* Carnegie fel physics, Inst Theoret Physics, Copenhagen, Denmark, 54-55; res assoc math, Tulane Univ, 63-64. *Honors & Awards:* Kelvin Medal, Glasgow Univ, 68. *Mem:* Math Asn Am; Soc Exact Philos; Can Math Soc; Asn Symbolic Logic; Int Fuzzy Sets Asn. *Res:* Use of formal languages and nonclassical logic for exact reasoning under uncertainty; application of this to artificial intelligence and to foundations of physics. *Mailing Add:* Dept Math & Statist Queen's Univ Kingston ON K7L 3N6 Can

GILES, ROBIN ARTHUR, b London, Eng, Oct 4, 36; m 61; c 4. PHYSICS. *Educ:* Univ London, BSc, 58, PhD(physics), 62. *Prof Exp:* Lectr physics, King's Col, Univ London, 60-68; from lectr to sr lectr, Queen's Univ Belfast, 68-76; PROF PHYSICS & DEAN SCI, BRANDON UNIV, 76- *Concurrent Pos:* UK Sci Res Coun grants, 69-76. *Mem:* Fel Inst Physics, London. *Res:* Elastic scattering of thermal energy alkali metal atoms; experimental and theoretical studies of total and differential cross-sections. *Mailing Add:* Dean Sci Brandon Univ Brandon MB R7A 6A9 Can

GILES, THOMAS DAVIS, b Greenwood, Miss, Feb 24, 38; m; c 3. MEDICINE. *Educ:* Tulane Univ, MD, 62; Am Bd Internal Med, dipl, 69. *Prof Exp:* Intern med, Charity Hosp, New Orleans, 62-63, resident internal med, 63-66, chief resident, 65-66; fel cardiol, Sch Med, Tulane Univ, 68-70, from instr to assoc prof med, 64-74, clin assoc prof, 74-76, assoc prof, 76-77, PROF MED, DEPT & SCH MED, TULANE UNIV, 77-, DIR, CARDIOVASC RES LAB, SCH MED, 81- *Concurrent Pos:* Vis physician & cardiologist, Charity Hosp La, 69-78, dir, Qual Assurance Prog, 74-76, sr med vis physician, Tulane Div, 78-; vis physician med staff, Hotel Dieu Hosp, 73-; chief, Med Serv, Vet Admin Med Ctr, New Orleans, 76-; mem, Cardiovasc Task Force, Joint Comn Accreditation Health Care Orgn, 88-; adj prof, Dept Physiol, Sch Med, Tulane Univ, 89-; fel, Coun Circulation, Coun Basic Sci & Coun High Blood Pressure Res, Am Heart Asn. *Mem:* Fel Am Col Angiol; fel Am Col Cardiol; fel Am Col Chest Physicians; Am Col Clin Pharmacol; fel Am Col Physicians; Am Fedn Clin Res; Am Soc Clin Pharmacol & Therapeut; Am Soc Hypertension; Am Soc Pharmacol & Exp Therapeut. *Res:* Cardiomyopathies; congestive heart failure; hypertension; polypeptides and the cardiovascular system; regulation of CNS neurotransmitters and drugs in cardiovascular control. *Mailing Add:* Dept Med Sch Med Tulane Univ 1430 Tulane Ave New Orleans LA 70112-2699

GILETTI, BRUNO JOHN, b New York, NY, Dec 6, 29; m 84; c 2. GEOCHEMISTRY. *Educ:* Columbia Univ, AB, 51, BS, 52, MA, 54, PhD(geol), 57. *Prof Exp:* Res fel, Lamont Geol Observ, Columbia Univ, 57-58; res fel geol & mineral, Oxford Univ, 58-60; from asst prof to assoc prof geol, 60-67, chmn dept, 77-80, PROF GEOL, BROWN UNIV, 67- *Concurrent Pos:* Assoc physician, Univ Paris, VI, 85. *Mem:* Am Geol Soc; Geochem Soc; Am Geophys Union. *Res:* Kinetics of geological processes, especially diffusion; absolute age determination of geological materials by isotopic analysis; distribution and abundance of radioactive and stable nuclides in the earth and their significance in geology. *Mailing Add:* Dept Geol Sci Brown Univ Brown Sta Providence RI 02912

GILFEATHER, FRANK L, b Great Falls, Mont, Sept 29, 42; m; m; c 2. MATHEMATICS. *Educ:* Univ Mont, BA, 64, MA, 66; Univ Calif, Irvine, PhD(math), 69. *Prof Exp:* Assoc, Off Naval Res, 69-70; asst prof math, Univ Hawaii, 70-74; from assoc prof math to prof math, Univ Nebr, Lincoln, 74-88; PROF & CHMN, DEPT MATH & STATIST, UNIV NMEX, 88- *Concurrent Pos:* Prog dir modern analysis, NSF, 83-85; staff dir, bd math sci, Nat Res Coun, 85-87. *Mem:* Am Math Soc; Math Asn Am. *Res:* Operator theory on Hilbert spaces; non self-adjoint operator algebras. *Mailing Add:* Dept Math & Statist Univ NMex Albuquerque NM 87131

GILFERT, JAMES C(LARE), b Tamaqua, Pa, June 21, 27; m 49; c 3. ELECTRICAL ENGINEERING. *Educ:* Antioch Col, BS, 50; Ohio State Univ, MSc, 51, PhD(physics), 57. *Prof Exp:* Res assoc antenna lab, Ohio State Univ, 53-56, from instr to assoc prof elec eng, 57-67; from assoc prof to prof, 67-83, chmn dept, 69-83, EMER PROF ELEC ENG, OHIO UNIV, 83-; FOUNDER & PRES, ATHENS TECH SPECIALISTS, INC, 82- *Concurrent Pos:* Sr tech specialist, NAm Aviation, Inc, 61-67; Am Coun Educ fel, 71-72; vis prof, Chubu Inst Technol, Nagoya, Japan, 73, 80; chmn eng res coun prog, Am Soc Eng Educ, 75-76; design consult, Ohio State Dept Transp, 77- *Mem:* Am Soc Eng Educ; Inst Elec & Electronics Engrs. *Res:* Microprocessor applications; instrumentation; communications. *Mailing Add:* Dept Elec Eng Ohio Univ Athens OH 45701

GILFILLAN, ALASDAIR MITCHELL, b Perth, Scotland, May 2, 56; m 81. PHOSPHOLIPID METABOLISM IN MAST CELL-BASOPHIL SIGNALLING, PROTEIN KINASE C ACTIVATION IN MAST CELL SIGNALLING. *Educ:* Univ Strathclyde, BSc Hons, 77; Univ Manchester, MSc, 79, PhD(pharmacol), 81. *Prof Exp:* Post doctoral assoc biochem, Dept Pediat, Yale Univ, 82-85, assoc res scientist, 85-87; sr scientist, 87-89, ASSOC RES INVESTR BIOCHEM, DEPT PHARMACOL, HOFFMANN-LA ROCHE, 89- *Concurrent Pos:* Prin investr, Dept Pharmacol, Hoffmann-La Roche, 87- *Mem:* Royal Pharmaceut Soc Gt Brit; Am Soc Biochem & Molecular Biol; Am Soc Cell Biol. *Res:* Signal transduction mechanisms leading to inflammatory mediator release from mast cells; mechanisms of pulmonary surfactant secretion. *Mailing Add:* 340 Kingsland St Nutley NJ 07110

GILFILLAN, ROBERT FREDERICK, b Roanoke, Va, Oct 30, 23; m 51; c 3. MICROBIOLOGY. *Educ:* Univ Tenn, BA, 49, MS, 50, PhD(microbiol, biochem), 56. *Prof Exp:* Asst bact, Univ Tenn, 49-50 & 54-55; asst bact metab, Univ Minn, 50-53; microbiologist, US Dept Defense, Gen Mills, Minn, 53-54; res assoc biol, Oak Ridge Nat Lab, Tenn, 55-56; asst prof bact & virol, Med Col SC, 56-59; asst prof obstet & gynec, Med Sch, Tufts Univ, 59-61; chief virol unit, Found Res Nerv Syst, 61-64; microbiologist, St Margaret's Hosp, Boston, 59-61, chief virologist, 64-70; asst prof obstet & gynec, Sch Med, Tufts Univ, 69-80; chief, Virol Lab, State Lab Inst, Dept Pub Health, Mass, 70-82; assoc prof Vet Med & Pathol, Tufts Sch Vet Med, 82-90; CONSULT, 90- *Concurrent Pos:* Soc Am Bact fel, Sch Med, Yale Univ, 57; res assoc, Sch Med, Univ Boston, 61-64; lectr appl microbiol, Harvard Sch Pub Health, 73- *Mem:* AAAS; Am Soc Microbiol; Sigma Xi. *Res:* Virology; experimental pathology. *Mailing Add:* 2533 Buena Vista Rd Winston-Salem NC 27104

GILFIX, EDWARD LEON, b Cambridge, Mass, Mar 14, 23; m 52; c 3. OPERATIONS ANALYSIS & SYSTEMS REQUIREMENTS, MANAGEMENT INFORMATION SYSTEMS DESIGN. *Educ:* Univ Mass, BS, 50; Univ Mich, MS, 51. *Prof Exp:* Mem tech staff, Willow Run Res Ctr, Univ Mich, 51-53; staff mem, Chrysler Corp, 53-55; mgr customer training, Datamatic Corp, 55-62; mem tech staff, Mitre Corp, 62-67; PRIN ENGR, MISSILE SYSTS DIV, RAYTHEON CO, 67- *Res:* Multi-discipline systems analysis, design, and documentation in programs involving air defense missile systems; command and control systems; centralized product assurance data analysis systems; electric utility distribution automation systems. *Mailing Add:* 42 Peacock Farm Rd Lexington MA 02173

GILFORD, LEON, b Warsaw, Poland, Feb 14, 17; nat US; m 50. MATHEMATICAL STATISTICS, ENGINEERING STATISTICS. *Educ:* Brooklyn Col, AB, 39; George Wash Univ, AM, 49. *Prof Exp:* Math statistician, US Bur Census, 46-55, chief opers res & qual control br, 55-60; sr scientist, Opers Res, Inc, 60-69, prin scientist, 69-71; chief statistician & dir ADP, US Tariff Comn, 71-74; spec asst reliability, energy res & develop admin, 74-76; consult, 76-77; spec asst statist res, US Bur Census, 77-78, actg assoc dir statist standards & methodology, 81; vpres res & develop, Corbo Corp, 81-90; CONSULT, 90- *Concurrent Pos:* Mem, Coun Am Statist Asn, 68-70; mem, Am Nat Standards Comt Z-1, Subcomt Statist Methods, 76-83; mem adv coun, US Dept Educ, 79; mem adv comt, Dept Energy, Energy Info Admin, 81-84; consult, 81-; mem bd dirs, Cobro Corp, 83-; mem, Panel on Qual Control, Nat Welfare Progs, Nat Acad Sci, 86-87; mem bd dirs, Planning Sci Inc, 88-89. *Honors & Awards:* Silver Medal, US Dept Com, 56. *Mem:* Fel AAAS; fel Am Statist Asn; Inst Statist Inst. *Res:* Mathematical models of physical and social systems, data collection and processing. *Mailing Add:* 6602 Rivercrest Ct Bethesda MD 20816

GILFRICH, JOHN VALENTINE, b Springfield, Mass, Sept 14, 27; m 54; c 5. ANALYTICAL CHEMISTRY. *Educ:* Am Int Col, BA, 49. *Prof Exp:* Analytical chemist, Nat Bur Stand, 48-50, phys chemist, 50-52; analytical chemist, US Naval Ord Lab, 52-60, phys chemist, 60-66; res chemist, Naval Res Lab, 66-71, head, Spectrochem Analytical Sect, 71-77, consult, X-ray Optics Br, 77-81, assoc head, Condensed Matter Physics Br, 81-82; RETIRED. *Concurrent Pos:* Consult, 87-; consult, Condensed Matter Physics Br, Naval Res Lab, 83-87. *Mem:* Am Chem Soc; fel Am Inst Chem; Am Crystallog Asn; Sigma Xi; Microbeam Analytical Soc; Soc Appl Spectros. *Res:* General analytical chemistry, including application of x-ray diffraction and spectroscopy to analytical problems; application and study of x-ray physics. *Mailing Add:* 8710 Lowell St Bethesda MD 20817-3218

GILGAN, MICHAEL WILSON, b Burns Lake, BC, Feb 26, 38; m 60; c 3. SHELLFISH TOXINS, FISH PRODUCT QUALITY. *Educ:* Univ BC, BSc, 59, MA, 62; Univ Wis, PhD(biochem), 65. *Prof Exp:* RES SCIENTIST, DEPT FISHERIES & OCEANS, 65- *Concurrent Pos:* Assoc referee, Asn Off Analytical Chemists. *Mem:* AAAS; Sigma Xi; Am Chem Soc; Asn Official Analytical Chemists. *Res:* Brain and smooth muscle phosphorylases; paralytic shellfish poison; toxins of the mold Fusarium tricinctum; biochemistry and physiology of molt and sex hormones of crustaceans; natural products from starfish; bioactive amines in fish products; chemical indicators of fish product quality; marine toxins in fish and shellfish. *Mailing Add:* Sci & Tech Serv Lab Dept Fisheries & Oceans PO Box 550 Halifax NS B3J 2S7 Can

GILHAM, PETER THOMAS, b Sydney, Australia, Nov 12, 30; div; c 3. ORGANIC CHEMISTRY, BIOCHEMISTRY. *Educ:* Univ Sydney, BSc, 51, MSc, 53, DSc, 75; Univ NSW, PhD(org chem), 56. *Prof Exp:* BC Res Coun fel, Univ BC, 56-58; Imp Chem Indust res fel org chem, Imp Col, London, 58-59; lectr, Univ Adelaide, 59-60; asst prof enzyme inst, Univ Wis, 60-62; assoc prof, 62-69, PROF BIOL SCI, PURDUE UNIV, WEST LAFAYETTE, 69- *Concurrent Pos:* Mem subcomt biol chem div chem & chem technol, Nat Acad Sci-Nat Res Coun, 64-66. *Mem:* Am Chem Soc; Am Soc Biol Chem. *Res:* Stereochemistry; nucleic acid structure and synthesis. *Mailing Add:* Dept Biol Sci Purdue Univ West Lafayette IN 47907

GILINSKY, VICTOR, b Warsaw, Poland, May 28, 34; US citizen; div; c 2. THEORETICAL PHYSICS. *Educ:* Cornell Univ, BEngPhys, 56; Calif Inst Technol, PhD(theoret physics), 61. *Prof Exp:* Mem tech staff, Aerospace Corp, 61-71; phys scientist, Rand Corp, 61-71, head phys sci dept, 74-75; comnr, US Nuclear Regulatory Comn, 75-84; CONSULT, 84- *Concurrent Pos:* Asst dir policy & prog rev, AEC, 71-73. *Mem:* Am Phys Soc; Inst Strategic Studies. *Res:* Many-body aspects of plasma physics; electromagnetic waves in the atmosphere; nuclear power technology; arms control; science and technology policy. *Mailing Add:* 48 Wellesley Circle Glen Echo MD 20812

GILINSON, PHILIP J(ULIUS), JR, b Lowell, Mass, July 28, 14; m 43; c 2. ELECTRICAL ENGINEERING, AEROSPACE ENGINEERING. *Educ:* Mass Inst Technol, BS, 36, MS, 52. *Prof Exp:* Engr, Heinze Elec Co, Mass, 36-38, Pac Mills, Inc, 38-40 & Doelcam Corp, 46-47; engr electromagnetics, Instrumentation Lab, Mass Inst Technol, 47-54, asst dir, 54-62, dep assoc dir, 62-73; staff consult, Charles Stark Draper Lab, Inc, 73-80; RETIRED. *Mem:* Sigma Xi; Nat Soc Prof Engrs. *Res:* Development and design of electromagnetic devices in aeronautics and astronautics, particularly inertial guidance in instrumentation systems; development of new precision measurement techniques in electromechanics and viscometer used in polymer chemistry and blood rheology research. *Mailing Add:* Eight Fuller Rd Chelmsford MA 01824

GILJE, JOHN, b Elkader, Iowa, Feb 24, 39; m 70. INORGANIC CHEMISTRY. *Educ:* Univ Minn, BChem, 61; Univ Mich, PhD(inorg chem), 65. *Prof Exp:* From asst prof to assoc prof, 65-75, prof hons prog, 76-79, PROF CHEM, UNIV HAWAII, 75- *Concurrent Pos:* Vis prof, Univ Tex, Austin, 72-73; mem organizing comt, US-Japan Organotransition Element Symp, 74, Am Chem Soc/Chem Soc, Japan Chem Cong, 79, Cong Pac Basin Chem Soc, 84; Alexander von Humboldt fel, Tech Univ Braunschweig, WGer, 80-81; acad vis, Imperial Col Sci & Technol, London, 79-80; vis prof, Univ Goettingen, Ger, 90. *Honors & Awards:* Honeywell Prize in Sci & Eng, 61; Alexander von Humboldt Distinguished Scientist Award, 90. *Mem:* Am Chem Soc. *Res:* Organometallic and main group chemistry; synthesis of new materials; F-element chemistry. *Mailing Add:* Dept Chem Univ Hawaii Honolulu HI 96822

GILKERSON, WILLIAM RICHARD, b Greenville, SC, June 5, 26; m 56; c 3. PHYSICAL CHEMISTRY. *Educ:* Univ SC, BS, 49; Univ Kans, PhD(chem), 53. *Prof Exp:* Res fel chem, Univ SC, 53 & Calif Inst Technol, 53-54; res fel, 54-55, from asst prof to assoc prof, 55-66, PROF CHEM, UNIV SC, 66- *Mem:* Am Chem Soc. *Res:* Conductance and dielectric properties of electrolytes in solution; ion pairing. *Mailing Add:* Dept Chem Univ SC Columbia SC 29208

GILKESON, M(URRAY) MACK, b Augusta, Kans, Feb 8, 22; m 44; c 4. CHEMICAL ENGINEERING. *Educ:* Univ Southern Calif, BE, 44; Kans State Univ, MS(chem eng), 47; Univ Mich, MSE, 51, PhD(chem eng), 52; Claremont Grad Sch, PhD(govt), 77. *Prof Exp:* Cost acct, El Dorado Foundry, Inc, 46; asst, Eng Res Inst, Univ Mich, 49-51; from asst prof to assoc prof, Tulane Univ, 52-61; assoc prof, 61-64, PROF ENG, HARVEY MUDD COL, 64- *Concurrent Pos:* Dir, Prog Pub Policy Studies, Claremont Cols, 73-75; vis coordr, Engenharia Clinica-Fed Univ, Brazil, 76-77; prog assoc, NSF Int Prog, 78-80; assoc dir, Am Soc Eng Educ, 84-86; vis prof, Papua, New Guinea, Univ Tech, 87. *Mem:* Am Inst Chem Engrs; Am Soc Eng Educ. *Res:* Contact catalysis; engineering design; overseas development; public policy studies; entrepreneurship. *Mailing Add:* Dept Eng Harvey Mudd Col Claremont CA 91711

GILKESON, RAYMOND ALLEN, b Sutherland, Nebr, Mar 29, 21; m 45; c 3. SOIL SCIENCE. *Educ:* Univ Nebr, BSc, 49; Wash State Univ, MSc, 51. *Prof Exp:* Soil scientist, Soil Conserv Serv, 51-53; from asst soil scientist to soil scientist, 53-73, from asst prof to assoc prof, 53-70, PROF SOILS, WASH STATE UNIV, 70- *Mem:* Soil Sci Soc Am. *Res:* Soil survey and classification; methods of soil survey of forest and range lands; airphoto interpretation. *Mailing Add:* 102 Island View Dr Sequim WA 98382

GILKESON, ROBERT FAIRBAIRN, b Philadelphia, Pa, June 26, 17; m 41; c 5. ELECTRICAL ENGINEERING. *Educ:* Cornell Univ, EE, 39. *Prof Exp:* Engr, Philadelphia Elec Co, 39-40, mem staff, Westinghouse Atomic Power Div, 51-53, exec vpres, 62-65, pres, 65-71, chief exec officer, 70-78, dir, 62-88, chmn bd, 71-88; RETIRED. *Concurrent Pos:* Pres & dir, Philadelphia Elec Power Co & Susquehanna Power Co; dir, Penn Mutual Life Ins Co; mem, Nat Sci Bd, 82- *Mem:* Nat Acad Eng; Inst Elec & Electronics Engrs. *Res:* Development and commercial application of advanced steam cycles and nuclear power generation. *Mailing Add:* 10313 Desert Forrest Circle Sun City AZ 85351

GILKEY, JOHN CLARK, CRYOBIOLOGY, INSTRUMENT DEVELOPMENT. *Educ:* Purdue Univ, PhD(biol), 77. *Prof Exp:* DIR RES, RESEARCH MFG CO, 85- *Concurrent Pos:* Asst res scientist & dir, Biotechnol Div, Life Sci Core Facil Micros, Univ Ariz. *Res:* Physiology and biophysics of fertilization and early development. *Mailing Add:* Biotechnol Univ Ariz Life Sci S Tucson AZ 85721

GILKEY, PETER BELDEN, b Utica, NY, Feb 27, 46; m 78; c 2. MATHEMATICS. *Educ:* Yale Univ, BS & MA, 67; Harvard Univ, PhD(math), 72. *Prof Exp:* Instr comput sci, NY Univ, 71-72; lectr math, Univ Calif, Berkeley, 72-74; asst prof math, Princeton Univ, 74-80; assoc prof math, 81-85, PROF MATH, UNIV ORE, 85- *Concurrent Pos:* Sloan Found grant, 75; NSF grant, 72- *Mem:* Am Math Soc. *Res:* Differential geometry and global analysis. *Mailing Add:* Dept Math Univ Ore Eugene OR 97403

GILKEY, RUSSELL, b Hopkinsville, Ky, Nov 19, 20; m 44; c 2. POLYMER CHEMISTRY. *Educ:* Univ Ky, BS, 43; Univ Ill, PhD(chem), 49. *Prof Exp:* Res chemist, B F Goodrich Co, 43-45; sr res chemist, Tenn Eastman Co, 49-67, res assoc, 67-73, sr res assoc, 73-79; RETIRED. *Mem:* Am Chem Soc; Sigma Xi. *Res:* Polymers; fibers; plastics; adhesives; protective coatings. *Mailing Add:* 1704 Springfield Ave Kingsport TN 37664

GILL, ARTHUR, electrical engineering, computer science, for more information see previous edition

GILL, AYESHA ELENIN, b Fresno, Calif, Oct 31, 33; div; c 2. POPULATION GENETICS, MEDICAL GENETICS. *Educ:* Univ Calif, Berkeley, BA, 57, BA, 61, PhD(genetics), 72; Am Bd Med Genetics, dipl, 84. *Prof Exp:* Asst prof biol, Univ Calif, Los Angeles, 72-79; from asst prof to assoc prof, Univ Nev, 79-84; res scientist, NY State Dept Health, 84-87; dir followup, Newborn Screening, 87-88; RES SCIENTIST & STATISTICIAN, 89- *Concurrent Pos:* Fel pop biol, Univ Chicago, 71-72; human & med geneticist, State Nev Genetics Prog, 79-81, dir, 81-82. *Mem:* Soc Study Evolution; Genetics Soc Am; Am Soc Mammalogists; Am Soc Human Genetics; Soc Syst Zool; Am Pub Health Asn. *Res:* Human genetic diseases, care and prevention; newborn screening; population genetics and evolutionary dynamics of natural populations; emphasis on speciation in rodents, using a multivariate approach. *Mailing Add:* Inst Health Policy Studies Univ Calif 1388 Sutters St 11th Floor San Francisco CA 94109

GILL, BIKRAM SINGH, b Dhudike, India; US citizen; m; c 4. PLANT GENETICS, AGRICULTURE. *Educ:* Khalsa Col Amritsar, BS, 63; Punjab Univ Chandigarh, MS, 66; Univ Calif, Davis, PhD(genetics), 73. *Prof Exp:* D F Jones fel genetics, Univ Mo, Columbia, 73-74; res assoc cell biol, Wash Univ, 74-75; res geneticist, Univ Calif,75-78; prof genetics, Agr Res & Educ Ctr, Univ Fla, 78-79; from asst prof to assoc prof, 79-87, PROF PLANT PATH, KANS STATE UNIV,87- *Concurrent Pos:* Lectr, Univ Calif, Riverside, 77; vis scientist, Coun Sci & Res Orgn, Div of Plant Indust, Canberra, Australia, 86-87; vis prof, US Nat Acad Sci, Ger Dem Repub, 87 & Sci Sem, Peoples Repub China, 87; assoc ed, J Heredity, 84- *Mem:* Genetics Soc Am; Agron Soc Am; Crop Sci Soc Am. *Res:* Cytogenetics of crop plants such as wheat, tomato and sugarcane; molecular biology of mouse cells in tissue culture; giemsa banding and chromosome identification work in plants. *Mailing Add:* Dept Plant Path Throckmorton Hall Kans State Univ Manhattan KS 66506

GILL, C(HARLES) BURROUGHS, b Sudbury, Ont, Apr 8, 21; m 55; c 2. METALLURGICAL ENGINEERING. *Educ:* Univ Toronto, Ont, BASc, 45; Mo Sch Mines, MS, 47, PhD(metall), 52. *Prof Exp:* Res metallurgist, Mo Sch Mines, 52-55; tech supt, Deloro Smelting & Refining Co, 55-57; prof metall eng, Lafayette Col, 57-89; RETIRED. *Mem:* Am Inst Mining, Metall & Petrol Engrs; Sigma Xi. *Res:* Extractive metallurgy and environmental engineering. *Mailing Add:* 405 Monroe St Easton PA 18042

GILL, CLIFFORD CRESSEY, b Cape Town, SAfrica, Oct 7, 21; Can citizen; m 72. PLANT VIROLOGY, PHYTOPATHOLOGY. *Educ:* Univ Cape Town, BSc, 57; Univ Calif, Berkeley, PhD(plant path), 64. *Prof Exp:* Chem analyst, Fuel Res Inst, SAfrica, 45; control chem analyst, Petersen, Ltd, 46-48; tech adv, Kodak, Ltd, 48-56; control engr, Ont Paper Co, Can, 56-57; analytical chemist, Gooch, Ltd, Calif, 57-58; RES SCIENTIST, CAN AGR RES STA, 64- *Concurrent Pos:* Assoc ed phytopath, SAfrican Bot Soc. *Mem:* Am Phytopath Soc; Can Phytopath Soc; SAfrican Bot Soc; Sigma Xi. *Res:* Study and control of virus diseases of cereal crops; control of barley yellow dwarf virus; gel electrophoresis; oat necrotic mottle virus. *Mailing Add:* 92 Minnetonka St Winnipeg MB R2M 3Y3 Can

GILL, DAVID MICHAEL, b London, Eng, Sept 17, 40; m 65; c 2. BIOCHEMISTRY, TOXINOLOGY. *Educ:* Cambridge Univ, MA, 64, PhD(biochem), 66. *Prof Exp:* Fel biochem, Cambridge Univ, 66-67 & Mass Inst Technol, 67-68; fel biol, 68, from instr to assoc prof biol, Harvard Univ, 68-78; assoc prof, 78-81, PROF MOLECULAR BIOL & MICROBIOL, SCH MED, TUFTS UNIV, 81- *Concurrent Pos:* Vis assoc prof, Sch Med, Yale Univ, 78. *Res:* Biochemical basis of the action of bacterial toxins, especially diphtheria, cholera and clostridial toxins ADP ribosylate; biological role of poly (adenosine diphosphate ribose). *Mailing Add:* Dept Molecular Biol & Microbiol 136 Harrison Ave Tufts Univ Sch Med Boston MA 02111

GILL, DHANWANT SINGH, b India, Dec 1, 41; m 70; c 4. STATISTICS, MATHEMATICS. *Educ:* Punjab Univ, India, BA, 60, MA, 63; Miami Univ, MSc, 73; Kans State Univ, PhD(statist), 77. *Prof Exp:* Lectr math, SN Col, Banga, India, 63-66 & SGGS Col, Chandigarh, India, 66-71; asst prof statist, 77-82, dir statist, 86-88, ASSOC PROF STATIST, NDAK STATE UNIV, 82-, DIR CONSULT CTR, 86-, CHMN, DEPT STATIST, 88- *Mem:* Am Statist Asn. *Res:* Multivariate statistical analysis; distribution theory. *Mailing Add:* Dept Statist NDak State Univ St Univ Sta Fargo ND 58105

GILL, DOUGLAS EDWARD, b New York, NY, Jan 25, 44; m 67. POPULATION BIOLOGY, ECOLOGY. *Educ:* Marietta Col, BS, 65; Univ Mich, Ann Arbor, MA, 67, PhD(zool), 71. *Prof Exp:* Asst prof, 71-76, ASSOC PROF ZOOL, UNIV MD, 76- *Concurrent Pos:* Mem fac, Orgn Trop Studies, 71, coordr, 72 & 79. *Mem:* Am Soc Zoologists; Ecol Soc Am; Soc Study Evolution; Brit Ecol Soc; Am Soc Naturalists. *Res:* Population dynamics of natural populations; dynamics of natural metapopulations of the red-spotted newt; clinal variation in life history traits of frogs; co-evolution of parasites and hosts; evolution of trees. *Mailing Add:* Dept Zool Univ Md College Park MD 20742

GILL, FRANK BENNINGTON, b New York, NY, Oct 2, 41; m 65; c 1. ORNITHOLOGY. *Educ:* Univ Mich, BS, 63, PhD(zool), 69. *Prof Exp:* Asst cur ornith, 69-74, ASSOC CUR ORNITH, ACAD NATURAL SCI PHILADELPHIA, 74-, DIR SYSTS, 73- *Concurrent Pos:* Adj assoc prof, Univ Pa, 74-; dir Hawk Mountain Sanctuary, 75-; assoc ed, Am Midland Naturalist, 75-78. *Mem:* Am Ornithologists Union; Am Soc Naturalists; Soc Study Evolution; Soc Syst Zool; Ecol Soc Am. *Res:* Birds; foraging ecology; energetics; speciation and hybridization; geographic variation and evolutionary flexibility; vocal communication. *Mailing Add:* Acad Natural Sci 19th & Pkwy Philadelphia PA 19103

GILL, GEORGE WILHELM, b Sterling, Kans, June 28, 41; m 62, 75; c 4. PHYSICAL ANTHROPOLOGY. *Educ:* Univ Kans, BA, 63, MPhil, 70, PhD(anthrop), 71; Am Bd Forensic Anthrop, dipl, 78. *Prof Exp:* Instr, 71, from asst prof to assoc prof, 71-85, PROF ANTHROP, UNIV WYO, 85- *Concurrent Pos:* Sci consult for skeletal identification, Wyo State Law

Enforcement Agencies, 72-; sci leader, Easter Island Anthrop Exped, 81; bd dirs, Am Bd Forensic Anthrop, 85-; secy, Phys Anthrop Sect, Am Acad Forensic Sci, 85-87, chmn, 87-88. *Mem:* Am Asn Phys Anthrop; fel Am Acad Forensic Sci. *Res:* Human evolution and skeletal biology of late prehistoric and modern populations, especially North America, Mesoamerica and Polynesia; human osteology and skeletal identification (forensic anthropology), especially racial identification. *Mailing Add:* Dept Anthrop Univ Wyo Laramie WY 82071

GILL, GORDON NELSON, b Montgomery, Ala, Dec 19, 37; m 74; c 4. ENDOCRINOLOGY. *Educ:* Vanderbilt Univ, BA, 60, MD, 63; FACP. *Prof Exp:* Internship internal med, Vanderbilt Univ Hosp, 63-64; resident internal med, Yale-New Haven Hosp, 64-66; fel, Sch Med, Yale Univ, 66-68; NIH res fel, 68-69, asst prof, 69-73, assoc prof, 73-78, PROF MED, SCH MED, UNIV CALIF, SAN DIEGO, 78-, CHIEF DIV ENDOCRINOL, 74- *Concurrent Pos:* Sect ed, Int Encycl Pharmacol & Therapeut, 72-; mem endocrinol study sect, NIH, 76-80. *Mem:* Am Fedn Clin Res; Am Soc Clin Invest; Am Soc Biol Chemists; Endocrine Soc; Asn Am Physicians. *Res:* Mechanisms through which polypeptide hormones and cyclic nucleotides control cell growth and differentiated function. *Mailing Add:* Dept Med 0650 Univ Calif San Diego La Jolla CA 92093-0650

GILL, GURCHARAN S, b Moga, Punjab, India, Mar 26, 35; US citizen; m 58; c 4. MATHEMATICS. *Educ:* Brigham Young Univ, BS, 58; Univ Utah, MS, 60, PhD(math), 65. *Prof Exp:* From asst prof to assoc prof, 65-72, PROF MATH, BRIGHAM YOUNG UNIV, 72- *Mem:* Am Math Soc; Math Asn Am; Soc Indust & Appl Math; Nat Coun Teachers Math. *Res:* Functional analysis; topology. *Mailing Add:* 3292 N Canyon Rd Provo UT 84604

GILL, HARMOHINDAR SINGH, b Ferozpur, India, Apr 7, 33; m 59; c 3. PLANT PATHOLOGY, MYCOLOGY. *Educ:* Punjab Univ, India, BSc, 53, MSc, 55; Univ Ill, Urbana, PhD(plant path), 65. *Prof Exp:* Res asst mycol, Indian Agr Res Inst, 56-61; exp officer plant path, Dept Agr, Tanzania, 61-62; res assoc, Univ Ill, Urbana, 65-67; plant pathologist-nematologist, Riverside County Dept Agr, 67-75, PLANT PATHOLOGIST, UNIV CALIF, RIVERSIDE, 75- *Mem:* Am Phytopath Soc; Sigma Xi. *Res:* General plant pathology especially diagnostic investigations of phytopathogenic organisms and nematodes; phytophthora diseases of ornamentals and taxomony; regulatory and extension; nematology. *Mailing Add:* 2282 Kentwood Dr Riverside CA 92507

GILL, HAROLD HATFIELD, b Hays, Kans, Sept 26, 21; m 40; c 3. ANALYTICAL CHEMISTRY. *Educ:* Kans State Col, BS, 49, MS, 50. *Prof Exp:* Chemist, 51-54, analytical specialist, 54-62, analytical res specialist, 63-68, sect leader, 69-70, RES MGR, DOW CHEM CO, 71- *Mem:* Am Chem Soc; fel Am Inst Chemists; AAAS; Sigma Xi. *Res:* Molecular spectroscopy; thermal chemistry; analysis of environmental samples for trace components. *Mailing Add:* 2004 Rapanos Dr Midland MI 48640

GILL, JAMES EDWARD, b Berkeley, Calif, Apr 12, 31; m 64, 79; c 3. CYTOCHEMISTRY, CYTOLOGY. *Educ:* Univ Calif, Berkeley, AB, 55; Stanford Univ, MS, 58, PhD(physics), 63. *Prof Exp:* Physicist, Film Dept, E I Du Pont de Nemours & Co, 62-65; biophysicist, Biomed Div, Lawrence Livermore Lab, Univ Calif, 65-75; asst prof path, Sch Med & Dent, Univ Rochester, 75-79; Becton Dickson Facs Syst, Mt View, Calif, 79-89; MEM STAFF, UNIPATH CO, MT VIEW, CA, 89- *Mem:* Int Soc Analytical Cytol. *Res:* Fluorescence cytochemistry for cell identification; flow cytomietry for clinical diagnosis and monitoring; molecular physics; cytology; light microscopy. *Mailing Add:* 1452 Gilmore St Mountain View CA 94040

GILL, JOHN LESLIE, b La Harpe, Ill, May 25, 35; m 82; c 2. BIOMETRICS. *Educ:* Univ Ill, BS, 56; Iowa State Univ, MS, 61, PhD, 63. *Prof Exp:* Res assoc statist, Iowa State Univ, 61-62; assoc prof, Va Polytech Inst, 63-64; assoc prof, 64-70, PROF ANIMAL SCI, MICH STATE UNIV, 72- *Concurrent Pos:* Vis fel math, Univ NSW, 71; vis prof, Univ Reading, 78, Swiss Fed Inst Technol, 78, Agr Univ, Norway, 84, Tech Univ, Munich, 85, 91 & Agr Univ Wageningen, Neth, 85. *Mem:* Bernoulli Soc; Am Soc Animal Sci; Biomet Soc; Am Statist Asn. *Res:* Statistical genetics; design and analysis of experiments in biological research. *Mailing Add:* Dept Animal Sci Mich State Univ East Lansing MI 48824-1225

GILL, JOHN PAUL, JR, b Tuscaloosa, Ala, Feb 16, 37. MATHEMATICAL ANALYSIS. *Educ:* Univ Ga, BS, 58; Univ Ala, MA, 64; Colo State Univ, PhD(math), 71. *Prof Exp:* Instr math, Murray State Univ, 64-67; assoc prof 71-81, prof math & chmn dept, Univ Southern Colo, 81-83; PROF DEPT MATH, UNIV S COLO, 83- *Concurrent Pos:* Gov, Rocky Mountain Sect Math Asn Am, 87-90. *Mem:* Am Math Soc; Sigma Xi; Math Asn Am. *Res:* Complex analysis; continued fractions and infinite series. *Mailing Add:* Math Dept Univ S Colo Pueblo CO 81005

GILL, JOHN RUSSELL, JR, b Richmond, Va, July 22, 29; m 64; c 3. HYPERTENSION. *Educ:* Univ Va, BA, 50, MD, 54; Am Bd Internal Med, cert, 62. *Prof Exp:* Med intern, St Louis City Hosp, Wash Univ Serv, 54-55; jr asst resident, Duke Hosp, Durham, NC, 55-56, sr asst resident, 56-57; clin assoc, Nat Heart Inst, NIH, 57-59; Nat Found fel, Inst Biol Chem, Univ Copenhagen, Denmark, 59-60; sr investr, 60-88, chief clin serv, 78-88, EMER SCIENTIST, HYPERTENSION-ENDOCRINE BR, NAT HEART, LUNG & BLOOD INST, NIH, BETHESDA, MD, 88-; CLIN ASSOC PROF MED, GEORGETOWN UNIV, WASH, DC, 72- *Concurrent Pos:* Clin asst prof med, Georgetown Univ, 66-72; chmn, Med Records Comt, Clin Ctr, NIH, 74-75, Animal Care & Use Comt, Nat Heart, Lung & Blood Inst, 85-88, mem, Intramural Rev Bd, 78-88; mem, Coun High Blood Pressure Res, Am Heart Asn & Coun Circulation, Sect Renal Dis. *Mem:* Am Fedn Clin Res; Am Soc Clin Invest; Am Physiol Soc; Am Soc Nephrology; Am Soc Hypertension; Endocrine Soc; fel Am Col Physicians. *Res:* Author of numerous publications. *Mailing Add:* Hypertension Endocrine Br Nat Heart Lung & Blood Inst NIH Bldg 10 Rm 8C104 Bethesda MD 20892

GILL, MERTON, b Chicago, Ill, July 26, 14. PSYCHOANALYSIS. *Educ:* Univ Chicago, MD, 38. *Prof Exp:* Prof psychiat, State Univ NY Downstate Med Ctr, 63-68; PROF PSYCHIAT, UNIV ILL MED CTR, 71- *Concurrent Pos:* Fel, Res Ctr Ment Health, NY Univ, 68-71. *Honors & Awards:* Menninger Prize, Am Psychoanal Asn, 61. *Mem:* Am Psychiat Asn; Am Psychoanal Asn; Am Psychol Asn. *Res:* Audio-recorded psychoanalysis. *Mailing Add:* Dept Psychiat Univ Ill Col Med PO Box 6998 Chicago IL 60680

GILL, PIARA SINGH, b Bassuwal, India, Feb 15, 40; m 67; c 2. PHYSICAL CHEMISTRY, PHOTO CHEMISTRY. *Educ:* Panjab Univ, India, BSc, 61, MSc, 62; Kans State Univ, MS, 65, PhD(chem), 67. *Prof Exp:* Fel chem, Univ Houston, 67-68; res assoc, Wright-Patterson AFB, Ohio, 68-69; from asst prof to assoc prof, 69-77, PROF CHEM, TUSKEGEE UNIV, 77- *Res:* Chemical reactions induced by radiation, specifically the collisional energy transfer from excited ions to neutral molecules; mass-spectrometric studies of ion-molecule reactions. *Mailing Add:* Dept Chem Tuskegee Univ Tuskegee AL 36088

GILL, ROBERT ANTHONY, b Darlington, Eng, Apr 29, 28; m 57; c 3. PHYSICAL CHEMISTRY. *Educ:* Univ Durham, BSc, 49, PhD(chem), 53. *Prof Exp:* Sr sci officer analytical chem, Royal Naval Sci Serv, Eng, 53-57 & Atomic Energy Res Estab, Eng, 57-58; sr phys chemist, 58-64, group leader, 64-67, head paper & nonwovens chem res dept, 67-79, HEAD TEXTILE PAPER & NONWOVENS CHEM RES DEPT, ROHM & HAAS CO, 80- *Mem:* Am Chem Soc; Fiber Soc; Tech Asn Pulp & Paper Indust. *Res:* Physical chemistry of polymers, adhesives, elastomers and textile fibers. *Mailing Add:* 900 Midland Ave Apt 2F Yonkers NY 10704

GILL, ROBERT WAGER, b Waterbury, Conn, Jan 19, 40; m 77; c 2. ECOLOGY. *Educ:* Oberlin Col, BA, 61; Univ Mich, MS, 63, PhD(zool), 67. *Prof Exp:* Asst prof biol, 67-71, asst prof biol & statist, 71-74, asst to vchancellor, 74-87, EXEC ASST TO CHANCELLOR, UNIV CALIF, RIVERSIDE, 87- *Mem:* AAAS. *Res:* Population and community ecology; population genetics; theoretical ecology. *Mailing Add:* Off Chancellor Univ Calif Riverside CA 92502

GILL, RONALD LEE, b Mechanicsburg, Pa, Nov 30, 49; m 71; c 2. ON-LINE ISOTOPE SEPARATION, G-FACTOR MEASUREMENTS. *Educ:* Pa State Univ, BS, 71; Iowa State Univ, MS, 74, PhD(nuclear chem), 77. *Prof Exp:* Teaching fel, Iowa State Univ, 77-79; PHYSICIST, BROOKHAVEN NAT LAB, 79- *Mem:* Am Phys Soc; Am Chem Soc. *Res:* Nuclear structure studies of short lived isotopically separated fission products. *Mailing Add:* Dept Physics Bldg 510 A Brookhaven Nat Lab Upton NY 11973

GILL, STANLEY JENSEN, b Salt Lake City, Utah, Aug 21, 29; m 52; c 2. BIOPHYSICAL CHEMISTRY. *Educ:* Harvard Univ, AB, 51; Univ Ill, PhD(phys chem), 54. *Prof Exp:* Asst chem, Univ Ill, 51, 52; res assoc, Cornell Univ, 54; from asst prof to assoc prof, 56-64, PROF CHEM, UNIV COLO, BOULDER, 64- *Concurrent Pos:* Du Pont fel, 57; NSF sr fel, 66-67. *Mem:* Am Chem Soc; Am Phys Soc; The Chem Soc; Soc Rheology. *Res:* Physical chemistry of high polymers; application of physical techniques to study biological materials; new experimental methods; thermodynamic properties of solutions. *Mailing Add:* Dept Chem Univ Colo Campus Box 215 Boulder CO 80309

GILL, STEPHEN PASCHALL, b Baltimore, Md, Nov 13, 38; m 61. GAS DYNAMICS, SIGNAL PROCESSING. *Educ:* Mass Inst Technol, BS, 60; Harvard Univ, BA, 61, PhD(appl physics), 64. *Prof Exp:* Physicist, Stanford Res Inst, 64-66, head high energy gas dynamics, 66-68; head high energy gas dynamics, Physics Int Co, 68-70, mgr, Shock Dynamics Dept, 70-72; pres & founder, 72-77, CHIEF SCIENTIST & CHMN BD, ARTEC ASSOCS INC, 77-; FOUNDER & CHIEF SCIENTIST, MAGNETIC PULSE INC, 85- *Concurrent Pos:* Pres & founder, Votan, 79-81, cheif scientist & chmn bd, 81- *Mem:* Am Math Soc; Am Phys Soc; Inst Elec & Electronics Engrs; AAAS; Soc Indust & Appl Math. *Res:* Applied mathematics; shock wave physics; signal processing. *Mailing Add:* 32 Flood Circle Atherton CA 94025

GILL, THOMAS JAMES, III, b Malden, Mass, July 2, 32; m 61; c 3. PATHOLOGY, IMMUNOLOGY. *Educ:* Harvard Univ, AB, 53, MA, 57; Harvard Med Sch, MD, 57; Am Bd Path, dipl, 65. *Prof Exp:* Asst path, Peter Bent Brigham Hosp, 57-58; med intern, New York Hosp-Cornell Med Ctr, 58-59; from asst to sr assoc path, Peter Bent Brigham Hosp, 59-71; PROF PATH & CHMN DEPT, SCH MED, UNIV PITTSBURGH & PATHOLOGIST IN CHIEF, UNIV HEALTH CTR, 71-, PROF HUMAN GENETICS, 84- *Concurrent Pos:* From res fel to assoc, Harvard Med Sch, 59-65, from asst prof to assoc prof, 70-72; Lederle med fac award, 62-65; NIH res career develop award, 65-71; consult govt & indust, 66-; assoc mem comn immunization, Armed Forces Epidemiol Bd, 66-70, mem, 70-72; consult, Surgeon Gen, US Army, 70-76; mem sci adv bd, St Jude Children's Res Hosp, 69-77; mem allergy & immunol res comt, Nat Inst Allergy & Infectious Dis, 73-76; mem, Merit Rev Bd Immuniol, Med Res Serv, Vet Admin, 76-79, bd dirs, Allegheny County Chap Easter Seal Soc, 72-77, Sci Adv Comt, Damon Runyon-Walter Winchell Cancer Fund, 78-, Comt Animal Models & Genetic Stocks, Nat Res Coun, 78-, chmn, 83-86, Comt Rabbit Genetic Resources, 79-80 & Comt Preserv Lab Animal Resources, 85-; trustee, Am Bd Path, 81-; jr fel, Soc fels, Harvard Univ, 59-62. *Honors & Awards:* Smith, Kline & French Distinguished Lectr, Hahneman Med Col, 84; Whipple Lectr, Univ Rochester, 84. *Mem:* AAAS; Am Chem Soc; Am Asn Immunol; fel Am Soc Clin Path; Transplantation Soc (vpres, 82-84). *Res:* Immunology; genetics; reproductive immunology; immunogenetics and its application to tissue transplantation and reproduction in inbred rats and humans. *Mailing Add:* Dept Path Univ Pittsburgh Sch Med Pittsburgh PA 15261

GILL, WILLIAM D(ELAHAYE), b Portugal, Jan 18, 35; Can citizen; m 62; c 3. SOLID STATE PHYSICS, TECHNICAL MANAGEMENT. *Educ:* Univ BC, BASc, 60, MASc, 62; Stanford Univ, PhD(mat sci), 69. *Prof Exp:* Res physicist, Res Div, 62-79, MGR & ADVAN PRINTER TECHNICIAN,

INFO PROD DIV, INT BUS MACH CORP, 79- *Mem:* Am Phys Soc. *Res:* Electronic properties of metallic polymers; transport properties in insulating solids; photoconductivity; photovoltaic effects in heterojunctions; advanced printer technology. *Mailing Add:* 1928 Cowper St Palo Alto CA 94301

GILL, WILLIAM JOSEPH, b Romulus, Mich, July 17, 44; m 67; c 2. FOOD SCIENCE. *Educ:* Mich State Univ, BS, 66, MS, 68; Pa State Univ, PhD(food sci), 76. *Prof Exp:* Food scientist, Mead Johnson & Co, 68-69; sr mgr prod res & develop, H J Heinz & Co, 69-76; corp vpres res & tech servs, Pepsico, Inc, 76-83; corp exec vpres menu & opers develop, Pillsburg Co Burger King Corp, 84-89; OWNER & PRES, SEASONSHIELD & CROWN WINDOW CO, 89- *Concurrent Pos:* Mgmt consult, Triangle Group, 83-84. *Mem:* AAAS; Inst Food Technologists; Food Safety Coun; Am Chem Soc; Int Life Sci Inst. *Res:* Processes, materials, packages, quality assurance, scientific and regulatory affairs relative to food and beverage products. *Mailing Add:* Seasonshield & Crown Window Co 355 Center Ct Venice FL 34292

GILL, WILLIAM N(ELSON), b New York, NY, Sept 13, 28; m 54, 82; c 4. CHEMICAL ENGINEERING. *Educ:* Syracuse Univ, BS, 51, MA, 55, PhD(chem eng), 60. *Prof Exp:* Field engr, Am Blower Corp, 51-55; res assoc, Syracuse Univ, 55-57, from instr to assoc prof chem eng, 57-65; prof & chmn dept, Clarkson Col Technol, 65-71; dean & prof chem eng, State Univ NY, Buffalo, 71-86; PROF & CHMN DEPT, RENSSELAER POLYTECH INST, 86- *Concurrent Pos:* Consult, Corning Glass Corp, 75, Carborundum Co, Niagara Falls, 76- & NASA; Fulbright-Hays sr scholar, US-UK Educ Comn, 77; ed, Chem Eng Commun, 79-; Glenn Murphy Distinguished Prof, Iowa State Univ, 80-82; Fulbright Sr Res Scholar, US-Australia Educ Comt, 86-87. *Mem:* Am Inst Chem Engrs. *Res:* Turbulence and transport phenomena; role of transport phenomena in chemical reactions; reverse osmosis studies; dispersion phenomena; crystal growth; semiconductor processing; ultrafiltration of proteins. *Mailing Add:* Chem Eng Dept Rensselaer Polytech Inst Troy NY 12180-3590

GILL, WILLIAM ROBERT, b McDonald, Pa, July 21, 20; m 47; c 5. AGRONOMY. *Educ:* Pa State Univ, BS, 42; Univ Hawaii, MS, 49; Cornell Univ, PhD(agron, soils), 55. *Prof Exp:* Soil scientist, Pineapple Res Inst, Univ Hawaii, 49-50; asst, Cornell Univ, 52-55; res soil scientist, 55-71, dir, 71-80, COLLABR, NAT SOIL DYNAMICS LAB, USDA, 80- *Concurrent Pos:* Adj prof, Grad Fac, Auburn Univ, 57-87; US exchange scientist, USSR, 70. *Honors & Awards:* John Deere Gold Medal, Am Soc Agr Engrs, 90. *Mem:* Am Soc Agron; Soil Sci Soc Am; Int Soil Sci Soc; Am Soc Agr Eng. *Res:* Dynamic relations of soil-machine systems with emphasis on tillage and traction in agricultural soils as they influence the efficiency and production of crops and soil physical conditions. *Mailing Add:* Nat Soil Dynamics Lab USDA Box 792 Auburn AL 36831-0792

GILLAM, BASIL EARLY, b Wellsville, Mo, Oct 24, 13; m 39; c 1. MATHEMATICS. *Educ:* Univ Mo, AB, 35, MA, 36, PhD(math), 40. *Prof Exp:* Instr math, Univ Mo, 37-44; PROF MATH & HEAD DEPT, DRAKE UNIV, 44- *Mem:* Am Math Soc; Math Asn Am. *Res:* Metric geometry; new set of postulates for Euclidean geometry. *Mailing Add:* Dept Math & Comput Sci Drake Univ 25th St & Univ Ave Des Moines IA 50311

GILLARD, BAIBA KURINS, b Ger, May 14, 46; m 67. CARBOHYDRATE CHEMISTRY, AUTOIMMUNITY. *Educ:* Purdue Univ, BS, 67; Washington Univ, MA, 69, PhD(chem), 72. *Prof Exp:* Robert A Welch Found fel biochem, Baylor Col Med, 72-74; Cystic Fibrosis Found res fel, 74-76, adj asst prof pediat, Sch Med, Univ Calif, Los Angeles, 76-83; asst prof exp med, 83-90, RES ASSOC PROF, BAYLOR COL MED, HOUSTON, TEX, 90- *Mem:* Am Chem Soc; Am Soc Biol Chem & Molecular Biol; Carbohydrate Soc. *Res:* Glycosphingolipids; endothelium; immune function. *Mailing Add:* Dept Med Baylor Col One Baylor Plaza Houston TX 77030

GILLARY, HOWARD L, b New York, NY, Feb 6, 40. NEUROPHYSIOLOGY. *Educ:* Oberlin Col, AB, 61; Johns Hopkins Univ, PhD(biol), 66. *Prof Exp:* Nat Acad Sci-Nat Res Coun res assoc biophys, Naval Med Res Inst, Md, 65-67; res assoc & USPHS fel biol, Stanford Univ, 67-69; asst prof, 69-74, assoc prof, 74-79, PROF PHYSIOL, UNIV HAWAII, 79- *Mem:* Soc Neurosci. *Res:* Sensory physiology; quantitative electrophysiological studies on labellar taste receptors of the blowfly, eyes of a terrestrial gastropod, and neural control of behavior in the crayfish; photoreceptors in a marine gastropod. *Mailing Add:* Dept Physiol Univ Hawaii 1960 East-West Rd Honolulu HI 96822

GILLASPY, JAMES EDWARD, b Bartlett, Tex, Oct 15, 17; m 48; c 2. ENTOMOLOGY. *Educ:* Agr & Mech Col, Tex, BS, 40; Ohio State Univ, BS, 41; Univ Calif, PhD(entom), 54. *Prof Exp:* Entomologist, Bur Entom & Plant Quarantine, USDA, 46-48; asst, Univ Calif, 49-51 & Univ Tex, 53; salesman, Calif Spray Chem Corp, Stand Oil Co Calif, Idaho, 54-55; res grasshopper ecol, Univ Idaho, 55; instr life sci, San Bernardino Valley Union Jr Col, 56-57; instr chem, Mt San Antonio Jr Col, 57-58; teacher pub sch, Calif, 59; asst prof natural sci & math, Tex Lutheran Col, 59-60; asst prof biol, Sul Ross State Col, 60-61; res assoc, Harvard Mus Comp Zool, 61-63; asst prof biol, Mankato State Col, 63-66; assoc prof, 66-75, PROF BIOL, TEX A&I UNIV, 75- *Mem:* AAAS; Am Entom Soc; Am Inst Biol Sci; Am Registry Prof Entomologists. *Res:* Biology and taxonomy of aculeate Hymenoptera; organic evolution; animal behavior; economic entomology; management of Polistes wasps for venom production and for caterpillar predation. *Mailing Add:* Dept Biol Box 158 Tex A&I Univ Kingsville TX 78363

GILLE, JOHN CHARLES, b Akron, Ohio, Oct 12, 34; m 63; c 2. ATMOSPHERIC SOUNDING, MIDDLE ATMOSPHERE. *Educ:* Yale Univ, BS, 56; Cambridge Univ, BA, 58, MA, 66; Mass Inst Technol, PhD(geophys), 64. *Prof Exp:* Res asst meteorol, Harvard Univ, 60-64; from asst prof to assoc prof, Fla State Univ, 64-72; prog scientist, 72-73, leader, Upper Atmosphere Proj, 73-77, HEAD, GLOBAL ATMOSPHERIC CHANGE SECT, NAT CTR ATMOSPHERIC RES, 77- *Concurrent Pos:* Consult, Honeywell, Inc, 64-71, Int Bus Mach Corp, 67-68, Barnes Eng, 71, Arthur D Little, 83 & TRW Corp, 89-; mem, var adv groups on earth & planetary atmospheres, NASA, 69- & var nat comns, 70-; vis prof astrogeophysics, Univ Colo, 70, lectr, 72-; assoc ed, J Atmospheric Sci, 74-80; assoc ed, J Geophys Res, 80-82; vis prof, Geophysics Dept, Kyoto Univ, 88. *Honors & Awards:* Except Sci Achievement Award, NASA, 82. *Mem:* Fel, AAAS; Am Geophys Union; fel Am Meteorol Soc. *Res:* Inversion of satellite measurements, limb infrared emission to obtain stratosphere & mesosphere temperature and composition; calculation, and measurements of atmospheric infrared, visible radiation; middle atmospheric dynamics, chemistry; infrared spectroscopy; radiative effects on fluid dynamics. *Mailing Add:* Nat Ctr Atmospheric Res PO Box 3000 Boulder CO 80307-3000

GILLEAN, MARFRED ELWOOD, electrical engineering, for more information see previous edition

GILLELAND, MARTHA JANE, b Monroe, La, Sept 9, 40. BIOORGANIC CHEMISTRY. *Educ:* La Polytech Inst, BS, 62; La State Univ, Baton Rouge, PhD(org chem, biochem), 68. *Prof Exp:* Res assoc biochem, Edsel B Ford Inst Med Res, Mich, 68-69; fel, Northwestern Univ, Evanston, 70-71; assoc prof, 72-77, chairperson dept, 74-81, PROF CHEM, CALIF STATE UNIV, BAKERSFIELD, 77- *Mem:* Sigma Xi; AAAS; Am Chem Soc. *Res:* Mechanism of enzyme action. *Mailing Add:* 2016 Klemer St Bakersfield CA 93312-9310

GILLEN, KEITH THOMAS, b Cleveland, Ohio, Aug 11, 42. ATOMIC & MOLECULAR COLLISIONS, BEAM-SURFACE INTERACTIONS. *Educ:* Calif Inst Technol, BS, 64; Univ Wis, Madison, PhD(phys chem), 70. *Prof Exp:* Postdoctoral fel, Chem Dept, Univ Calif, Berkeley, 70-73; postdoc fel, 73-75, from chem physicist to sr chem physicist, 75-83, PROG MGR, SRI INT, 83- *Mem:* AAAS; Am Chem Soc; Am Phys Soc; Am Soc Mass Spectrometry. *Res:* Atomic, molecular and ionic collisions both in the gas phase and on surfaces; mass spectrometric surface analysis techniques. *Mailing Add:* 134 Carmelita Dr Mountain View CA 94040-3255

GILLEN, KENNETH TODD, b Cleveland, Ohio, Aug 11, 42; m 64; c 2. POLYMER SCIENCE, NUCLEAR MAGNETIC RESONANCE. *Educ:* Univ Calif, Berkeley, BS, 64; Univ Wis-Madison, PhD(chem), 70. *Prof Exp:* Mem staff chem physics, Bell Tel Labs, NJ, 70-72; mem staff electronics, Jeol, Inc, NJ, 72-74; MEM STAFF POLYMER SCI, SANDIA LABS, 74- *Mem:* Am Chem Soc; Am Phys Soc. *Res:* Polymers; mechanical properties; viscoelasticity; aging; radiation effects on polymers. *Mailing Add:* 8104 Northridge Ave NE Albuquerque NM 87109

GILLENWATER, JAY YOUNG, b Kingsport, Tenn, July 27, 33; m 55; c 3. UROLOGY. *Educ:* Univ Tenn, BS, 54, MD, 57. *Prof Exp:* Instr histol & gross anat, Sch Med, Univ Tenn, 56; intern med, Sch Med, Univ Louisville, 60-62; asst prof urol, 65-67, PROF UROL & CHMN DEPT, SCH MED, UNIV VA, 67- *Concurrent Pos:* USPHS fel renal & cardiovasc dis, Univ Pa Grad Hosp, 64-65; NIH urol training grant, 65-69, study grants, 66-68 & 66-71; lectr, Sch Med, Univ Louisville, 61-62. *Mem:* AAAS; Am Physiol Soc; Am Fedn Clin Res; Am Col Surg; AMA. *Res:* Renal physiology. *Mailing Add:* Dept Urol Univ Va Sch Med Box 422 Charlottesville VA 22908

GILLER, E(DWARD) B(ONFOY), b Jacksonville, Ill, July 8, 18; m 43; c 5. NUCLEAR ARMS CONTROL, SCIENCE POLICY. *Educ:* Univ Ill, BS, 40, MS, 48, PhD(chem eng), 50. *Prof Exp:* Jr engr petrol ref, Sinclair Oil Co, 40-41; US Air Force, 41-, chief radiation br, Armed Forces Spec Weapons Proj, 50-54, dir res, Spec Weapons Ctr, 54-59, spec asst to comdr, Off Aerospace Res, 59-64, dir sci & technol, Hq, 64-67; dir mil appln, US AEC, Washington, DC, 67-72, asst gen mgr nat sec, 72-75, dep asst adminr nat sec, ERDA, 75-77, joint chiefs of staff, 77-84, CONSULT, JOINT CHIEFS STAFF, DEPT DEFENSE, 84- *Concurrent Pos:* Prog mgr, arms control, Pac-Sierra Res Corp, 84- *Mem:* AAAS; Sigma Xi; fel Am Inst Chem; Am Inst chem Engrs. *Res:* Low temperature viscosity of hydrocarbons; heat transfer; gaseous thermal diffusion; atmospheric transmission optics; effects of high intensity radiant energy; nuclear weapons effects; high energy particle physics; government research; arms control treaty negotiation. *Mailing Add:* 216 Wapiti Dr Bagfield CO 81122-9243

GILLES, FLOYD HARRY, b Elgin, Ill, Oct 18, 30; div; c 5. NEUROPATHOLOGY, NEUROLOGY. *Educ:* Univ Chicago, BA, 51, BS & MD, 55; Am Bd Psychiat & Neurol, dipl, 62; Am Bd Path, dipl & cert neuropath, 74. *Prof Exp:* Intern, Johns Hopkins Hosp, 55-56, asst neurol, Sch Med, 56-59; instr clin neurol, Sch Med, Georgetown Univ, 59-61; neuropathologist, 62-71, assoc prof, 71-80, PROF NEUROPATH, CHILDREN'S HOSP MED CTR, HARVARD MED SCH, 80- *Concurrent Pos:* Nat Insts Neurol Dis & Blindness spec fel neuropath, Cent Anatomic Lab, Md, 61-62; asst resident, Baltimore City Hosps, 56-58 & Johns Hopkins Hosp, 58-59; from instr clin neurol to asst prof neuropath, Harvard Med Sch, 62-71; mem path task force, Perinatal Res Br, Nat Inst Neurol Dis & Stroke, 70; assoc neuropath, Beth Israel Hosp, Boston, 71-, New Eng Deaconess Hosp & New Eng Baptist Hosp, 81- *Mem:* Am Asn Neuropath (asst secy-treas, 70). *Res:* Pediatric neuropathology; reaction of fetal brain to insult. *Mailing Add:* 300 Longwood Ave Boston MA 02115

GILLES, KENNETH ALBERT, b Minneapolis, Minn, Mar 6, 22; m 44; c 2. AGRICULTURAL BIOCHEMISTRY. *Educ:* Univ Minn, BS, 44, PhD(biochem), 52. *Prof Exp:* Chem engr, Pillsbury Mills, Inc, 46-49; instr agr biochem, Univ Minn, 49-51; proj leader cereal biochem, Gen Mills, Inc, 52-60 & basic milling res, 60-61; prof cereal technol & chmn dept, NDak State Univ, 61-70, prof cereal chem & vpres agr, 69-81; ADMINR, FED GRAIN INSPECTION SERV, USDA, 81- *Concurrent Pos:* Ed, Cereal Chem, 61-68. *Mem:* AAAS; Am Chem Soc; Am Asn Cereal Chem (pres, 71-72); Inst Food Technol; Asn Oper Millers; Sigma Xi. *Res:* Cereal pentosans; amylolytic enzymes; lipids; synthetic condiments and food technology; wheat, semolina and flour quality and utilization. *Mailing Add:* 5415 Thetford Pl Alexandria VA 22310

GILLES, PAUL WILSON, b Kansas City, Kans, Jan 13, 21; m 44; c 3. PHYSICAL CHEMISTRY. *Educ:* Univ Kans, AB, 43; Univ Calif, PhD(phys chem), 47. *Prof Exp:* Asst chem, Univ Calif, 43-44 & Manhattan Proj, 44-47; from asst prof to prof, 47-63, UNIV DISTINGUISHED PROF CHEM, UNIV KANS, 63- *Mem:* AAAS; Am Chem Soc; Am Phys Soc; Sigma Xi. *Res:* High temperature chemistry; thermodynamics; vaporization processes; vapor pressures; properties of refractory borides, carbides and oxides; dissociation energies and stabilities of high temperature gases; x-ray crystallography; mass spectrometray; high molecular weight of boron sulfides. *Mailing Add:* Dept Chem Univ Kansas 20 Mallott Lawrence KS 66045

GILLESPIE, ARTHUR SAMUEL, JR, b Peking, China, Nov 21, 31; US citizen; m 53; c 3. INDUSTRIAL CHEMISTRY. *Educ:* Wake Forest Univ, BS, 53; Duke Univ, MA, 55. *Prof Exp:* Staff mem, Battery Lab, Sania Corp, 55-56; res engr, Phys Chem Dept, Aluminum Co Am, 56-61; res chemist, Measurements & Controls Lab, Res Triangle Inst, 61-66; sr chemist, Tex Gulf Sulphur Co, 66-67; res chemist, 67-77, dir corp serv, 77-88, ENVIRON MGR, LITHIUM DIV, FMC CORP, 88- *Mem:* Am Chem Soc. *Res:* Commercial utilization of mineral wastes; nuclear chemistry and radiochemistry; instrumentation design and development; water resources; water, air and hazardous wastes methods and management; aluminum and lithium chemistry; general corporation management matters; governmental affairs; environmental management. *Mailing Add:* PO Box 795 Bessemer City NC 28016

GILLESPIE, CLAUDE MILTON, b Huntsville, Ala, Dec 13, 32; c 2. NUCLEAR PHYSICS, RADIATION HYDRODYNAMICS. *Educ:* Ga Inst Technol, BS, 55; Ohio State Univ, PhD(physics), 66. *Prof Exp:* MEM STAFF NUCLEAR WEAPON EFFECTS & PHENOMENOL, NUCLEAR WEAPON DESIGN, INERTIAL FUSION, LOS ALAMOS NAT LAB, 66- *Mem:* Am Phys Soc; Am Geophys Union. *Res:* Beta and gamma ray spectroscopy; phenomenology and hydrodynamics of explosions; theoretical and experimental nuclear weapon effects; laser fusion; design and testing of nuclear weapons; weapons system analysis. *Mailing Add:* 427 Estante Way Los Alamos NM 87544

GILLESPIE, DANIEL THOMAS, b Springfield, Mo, Aug 15, 38; m 76; c 2. STATISTICAL PHYSICS, CHEMICAL PHYSICS. *Educ:* Rice Inst, BA, 60; Johns Hopkins Univ, PhD(physics), 68. *Prof Exp:* Res asst physics, Johns Hopkins Univ, 64-68; res assoc, Univ Md, 68-71; res physicist, 71-80, MATHEMATICIAN, NAVAL WEAPONS CTR, 80 - *Concurrent Pos:* Instr, Johns Hopkins Univ, 66-68 & Univ Md, 71. *Mem:* Am Phys Soc; Sigma Xi. *Res:* Kinetic theory; Monte Carlo methods; theory and simulation of stochastic processes, especially Markov processes; chemical kinetics of nonequilibrium systems. *Mailing Add:* 812 West Vicki Ave Ridgecrest CA 93555

GILLESPIE, ELIZABETH, b Montreal, Que, May 7, 36. PHARMACOLOGY, BIOCHEMISTRY. *Educ:* McGill Univ, BSc, 57, PhD(biochem), 66. *Prof Exp:* Res assoc pharmacol, Yale Univ, 68-71; from instr to asst prof med, Johns Hopkins Univ, 71-81; mgr, Respiratory Sect, Mead Johnson & Co, 81-82; sr res scientist cardiovasc biol, Bristol Myers Pharmaceut Res & Develop, 83-89; RETIRED. *Concurrent Pos:* USPHS grant pharmacol, Yale Univ, 66-68. *Mem:* Am Asn Immunol; Am Soc Pharmacol & Exp Therapeut. *Res:* Biochemical pharmacology; immediate hypersensitivity; atherosclerosis cardiovascular disease. *Mailing Add:* 2116 Hobbs Rd C-8 Nashville TN 37215

GILLESPIE, G(EORGE) RICHARD, chemical engineering; deceased, see previous edition for last biography

GILLESPIE, GEORGE H, b Dallas, Tex, Sept 9, 45; c 3. ATOMIC COLLISIONS AT HIGH ENERGIES, PHYSICS. *Educ:* Rice Univ, Houston, Tex, BA & MEE, 68, Univ Calif, San Diego, MS, 69, PhD(physics), 74. *Prof Exp:* Engr, Int Bus Mach, 67; res asst, Los Alamos Sci Lab, 68 & Univ Calif, San Diego, 68-74; STAFF SCIENTIST, PHYS DYNAMICS, INC, 75- *Concurrent Pos:* Assoc, La Jolla Inst, 76-; consult, Sci Appln Int Corp, 85- *Mem:* Am Phys Soc; AAAS. *Res:* Atomic, electromagnetic and nuclear interactions at intermediate energies; superconducting instruments; physics of particle accelerators; mathematical physics; technology assessment. *Mailing Add:* 12520 High Bluff Dr Suite 330 San Diego CA 92130

GILLESPIE, GEORGE YANCEY, b Greenwood, Miss, Oct 29, 43; m 66; c 1. IMMUNOBIOLOGY OF NEOPLASMS, MOLECULAR BIOLOGY OF MACROPHAGES. *Educ:* Univ Miss, BA, 65, MS, 68, PhD(immunol & immunochem), 71. *Prof Exp:* Res asst immunol, dept biol, Univ Miss, 68-71; res fel immunopathol, dept path, Med Ctr, Univ Kans, Kansas City, 71-73, instr path, 73-75; res assoc immunopathol, Scripps Clin & Res Found, La Jolla, Calif, 75-77; res asst prof path, Univ NC, Chapel Hill, 77-81, from res asst prof to res assoc prof surg, dept surg, 81-87; ASSOC PROF, DEPT SURG, UNIV ALA, 88- *Concurrent Pos:* Mem core fac & dir, Immuno-Neurooncol Res Prog, Lineberger Cancer, Res Ctr, Univ NC, Chapel Hill, 79-, dir, tissue cult facil, 84-; topic ed, Surv Immunol Res, 81- *Mem:* AAAS; Am Asn Immunologists; Am Asn Pathologists; Am Soc Microbiol; NY Acad Sci; Tissue Cult Asn. *Res:* The role of mononuclear phagocyte-derived growth factors in wound healing, neovascularization and neoplastic cell proliferation; immunobiology and immunotherapy of malignant brain tumors; central nervous system regulation of immune function. *Mailing Add:* Div Neurosurg Dept Surg PO Box 65 Tinsely Harrison Birmingham AL 35294

GILLESPIE, JAMES HOWARD, b Bethlehem, Pa, Nov 26, 17; m 41; c 3. VETERINARY SCIENCE, VIROLOGY. *Educ:* Univ Pa, VMD, 39; Am Col Vet Microbiologists, dipl. *Prof Exp:* Instr poultry dis & asst poultry pathologist, Univ NH, 40; asst prof poultry dis, 46-48, asst prof bact, 48-50, assoc prof vet bact, 50-56, asst dir lab dis of dogs, 51-61, chmn dept, 72-81, PROF VET BACT, NY STATE COL VET MED, CORNELL UNIV, 56- *Concurrent Pos:* USPHS fels, State Vet Res Inst, Amsterdam & Univ Calif, Berkeley, 60-61; exec secy adv comt foot & mouth dis, Nat Acad Sci, 61-; chmn, Nat Acad Sci-Nat Res Coun Foot & Mouth Dis Mission, Arg, 62; chmn animal resources adv comt, Div Res Resources, NIH, 74-75; chmn bd comp virol, WHO, 75-80; consult, Plum Island Animal Dis Ctr, USDA, 75-81; bd mem, Am Col Vet Microbiologists. *Honors & Awards:* Gaines Award, Am Vet Med Asn. *Mem:* AAAS; Am Soc Microbiol; Am Vet Med Asn; Conf Res Workers Animal Dis. *Res:* Animal virology and bacteriology; viral diseases of domesticated animals, fin-fish and shellfish. *Mailing Add:* 616-A Vet Res Cornell Univ Dept Microbiol Ithaca NY 14853

GILLESPIE, JERRY RAY, b Lincoln, Nebr, Jan 25, 37. PHYSIOLOGY, PATHOLOGY. *Educ:* Okla State Univ, BS & DVM, 61; Univ Calif, Davis, PhD(comp path), 65. *Prof Exp:* Asst anat, Okla State Univ, 57-61, supvr animal care, Univ Dairy Barn, 60-61; gen practitioner, Gotherburg Animal Hosp, Nebr, 61-62; asst specialist anat, 62-65, from asst prof clin sci to assoc prof, 66-71, assoc dean student serv, 70-72, assoc prof med physiol, 69-75 & vet med physiol, 71-75, PROF PHYSIOL, SCH VET MED, UNIV CALIF, DAVIS, & PROF PHYSIOL, SCH MED, 75- *Concurrent Pos:* Fel, Cardiovasc Res Inst, San Francisco, 65-66. *Mem:* AAAS; Am Physiol Soc; Am Vet Med Asn; Am Col Cardiol; Am Soc Anesthesiol. *Res:* Investigations of respiratory mechanics in healthy animals and those with chronic respiratory diseases. *Mailing Add:* Dept Clin Sci Col Vet Med Kans State Univ Manhattan KS 66506-5606

GILLESPIE, JESSE SAMUEL, JR, b Lynchburg, Va, Dec 20, 21; m 50; c 4. ORGANIC CHEMISTRY. *Educ:* Va Mil Inst, BS, 43; Univ Va, PhD(chem), 49. *Prof Exp:* Asst prof chem, Univ Richmond, 49-51; sr chemist, Va-Carolina Chem Corp, 51-53, group leader, 53-54, asst div mgr, 54-56, mgr org & agr chem, 56-58; partner, Cox & Gillespie Chemists & Chem Engrs, 58-62; sr chemist, 62-68, actg dir, 68-69, dir, Va Inst Sci Res, Univ Richmond, 69-82, prof chem, 72-82; ADV, THOMAS F & KATE MILLER JEFFRESS MEM TRUST, 82- *Concurrent Pos:* Dir sponsored progs, Univ Richmond, 72-82. *Mem:* Am Chem Soc; Sigma Xi. *Res:* Mechanisms of organic reactions; synthetic organic and medicinal chemistry. *Mailing Add:* 303 Hillwood Rd Richmond VA 23226

GILLESPIE, JOHN, b Buffalo, NY, Oct 30, 36. THEORETICAL PHYSICS. *Educ:* Univ Rochester, BS, 58; Univ Calif, Berkeley, PhD(theoret physics), 63. *Prof Exp:* Physicist, Lawrence Radiation Lab, Univ Calif, 63; res assoc, Columbia Univ, 63-65; res assoc, Stanford Linear Accelerator Ctr, 65-67; mem staff, Theoret Physics Ctr, Polytech Sch, Paris, 67-68; res physicist, AEC, Saclay, France, 68-69; asst prof physics, Boston Univ, 69-72; prof, Inst Nuclear Sci, Univ Grenoble, France, 72-75; instr res, Polytech High Sch, Paris, 75-78; ASSOC PROF PHYSICS, LEHMAN COL, CITY UNIV NY, 78- *Concurrent Pos:* Asst prof, Univ Calif, Santa Cruz, 65-66; vis investr biophysics, Sloan-Kettering Inst, NY, 78- *Mem:* AAAS; Am Phys Soc; NY Acad Sci; Europ Phys Soc; Am Asn Physicists Med; Sigma Xi. *Res:* Nuclear theory; elementary particles; medical physics. *Mailing Add:* Dept Physics & Astron Herbert H Lehman Col Bronx NY 10468

GILLESPIE, JOHN PAUL, organic chemistry, analytical chemistry, for more information see previous edition

GILLESPIE, LAROUX KING, b Colorado Springs, Colo, Nov 11, 42; m 66; c 5. PRECISION MINIATURE MACHINING, DEBURRING & BURR TECHNOLOGY. *Educ:* Kansas Univ, BS, 65, MS, 68; Utah State Univ, MS, 73. *Prof Exp:* Process engr, Bendix Corp, 66-73, sr engr, 73-78, staff engr, 78-81; SR PROJ ENGR, ALLIED-SIGNAL AEROSPACE CO, 81- *Concurrent Pos:* Dir, Soc Mfg Engrs, 77-81. *Honors & Awards:* Nat Award Merit, Soc Mfg Engrs, 74, Albert Sargent Progress Award, 84. *Mem:* Soc Mfg Engrs; Am Soc Mech Engrs. *Res:* Metal cutting; precision miniature machining; burr formation and deburring. *Mailing Add:* Allied-Signal Aerospace Co PO Box 419159 Kansas City MO 64141-6159

GILLESPIE, ROBERT HOWARD, b Richmond, Ind, Jan 31, 16; m 44; c 3. ORGANIC CHEMISTRY. *Educ:* Ind Univ, BS, 38; Univ Wis, PhD(org chem), 44. *Prof Exp:* Asst, Nat Defense Res Comt Projs, Univ Wis & Kendall Mills, 40-44, textile chemist, Res Dept, Kendall Mills, 44-55, res chemist, Theodore Clark Lab, 55-59; res assoc, Inst Paper Chem, 59-60; chemist, Forest Prod Lab, Forest Serv, USDA, 60-66, supvr res chemist, 66-87; RETIRED. *Mem:* Fel AAAS; fel Am Inst Chemists; Fiber Soc; fel Am Soc Testing & Mat; Forest Prod Res Soc. *Res:* Organic synthesis; textile development; cellulose chemistry; pressure-sensitive adhesives; rubbers, resins and plastics; non-woven fabrics; chemistry of wood; wood adhesives. *Mailing Add:* Three Down Wind Hilton Head Island SC 29928

GILLESPIE, RONALD JAMES, b London, Eng, Aug 21, 24; m 50; c 2. INORGANIC CHEMISTRY & GENERAL CHEMISTRY. *Educ:* Univ London, BSc, 45, PhD, 49, DSc, 57. *Hon Degrees:* LLD, Concordia Univ, Montreal, 88. *Prof Exp:* Lectr chem, Univ Col, London, 49-58; assoc prof, 58-60, PROF CHEM, MCMASTER UNIV, 60- *Concurrent Pos:* Commonwealth Fund fel, Brown Univ, 53-54; vis prof, Univ Manchester, Eng, Univ Sci Tech Languedoc, Montpellier, France, Univ Geneva, Switz, Univ Gottingen, WGer, Australian Nat Univ, Canberra, Univ Melbourne, Australia, Univ Auckland, New Zealand, Panjab Univ, Chandigarh, India, 65-83; mem Comt Teaching Chem, Int Union of Pure & Appl Chem. *Honors & Awards:* Harrison Mem Prize, Chem Soc, 53; Distinguished Serv Award Advan Inorg Chem, Am Chem Soc, 73; Chem Inst Can, Union Carbide Award for Chem Educ, 76; Chem Inst Can Medal, 77; Silver Jubilee Medal, 78; Nyholm lectr, Royal Soc Chem, 78-79; Fluorine Chem Award, Am Chem Soc, 81; Henry Marshall Tory Medal, Royal Soc of Canada, 83; Izaak Walter Killam Memorial Prize, Canada Coun, 87. *Mem:* Am Chem Soc; Chem Inst Can; Royal Soc Chem; fel Royal Inst Chem; fel Royal Soc Can; fel Royal Soc London. *Res:* Inorganic and physical chemistry of nonaqueous solvents; structural inorganic chemistry; fluorine chemistry; chemistry of the nonmetallic elements; chemical education. *Mailing Add:* Dept Chem McMaster Univ 1280 Main St W Hamilton ON L8S 4L8 Can

GILLESPIE, TERRY JAMES, b Vancouver, BC, Jan 5, 41; m 66; c 1. AGRICULTURAL METEOROLOGY. *Educ:* Univ BC, BSc, 62; Univ Toronto, MA, 63; Univ Guelph, PhD(meteorol), 68. *Prof Exp:* Meteorologist, Can Govt Serv, 63-66; ASSOC PROF METEOROL, UNIV GUELPH, 68- *Mem:* Am Meteorol Soc. *Res:* Meteorology as applied to agriculture. *Mailing Add:* Dept Agrometeorology Univ Guelph Guelph ON N1G 2W1 Can

GILLESPIE, THOMAS DAVID, b New Brighton, Pa, Dec 3, 39; m 63; c 2. AUTOMOTIVE ENGINEERING. *Educ:* Carnegie Inst Technol, BS, 61; Pa State Univ, MS, 65, PhD(mech eng), 70. *Prof Exp:* Engr glass res, Glass Res Ctr, PPG Indust, 63-64; proj officer res & develop, Armor & Eng Bd, US Army, 64-66; res assoc, Pa State Univ, 70-73; sr develop engr automotive mfg, Ford Motor Co, 73-76; RES SCIENTIST VEHICLE RES, TRANSP RES INST, UNIV MICH, 76- *Concurrent Pos:* Pvt consult, 77- *Honors & Awards:* L Ray Buckendale Award, Soc Automotive Engrs, 84. *Mem:* Soc Automotive Engrs; Am Soc Testing & Mat; Am Soc Mech Engrs; Sigma Xi. *Res:* Heavy vehicle braking, handling and ride performances; highway roughness and frictional charcteristics. *Mailing Add:* 1083 Bandera Ann Arbor MI 48103

GILLESPIE, WALTER LEE, b Hamilton, Ohio, Jan 6, 30; m 53; c 2. SCIENCE EDUCATION. *Educ:* Miami Univ, AB, 52, MA, 54; Univ Ill, PhD(zool), 60. *Prof Exp:* Asst prof zool, Butler Univ, 59-60; from instr to assoc prof biol, Wells Col, 60-67; assoc prog dir, Teacher Educ Sect, 67-71, prog dir, Div Pre-Col Educ Sci, 71-73, head, Instrnl Improv Implementation Sect, Pre-Col Educ, 73-75, dir, Div Sci Educ Resources Improv, 75-78, dep asst dir sci educ, 78-80, ACTG ASST DIR SCI & ENG EDUC, NSF, 80- *Mem:* AAAS; Ecol Soc Am; Am Soc Zool; Am Ornith Union; Arctic Inst NAm; Sigma Xi. *Res:* Bird populations of the subarctic; improvement of science and engineering education programs. *Mailing Add:* 1800 Kimberly Rd Silver Spring MD 20903

GILLESPIE, WILLIAM HARRY, b Webster Springs, WVa, Jan 8, 31; m 50; c 5. PALEOBOTANY, FORESTRY ADMINISTRATION. *Educ:* Univ WVa, BS, 52, MS, 54. *Prof Exp:* Forest biologist, Plant Pest Control Div, 56-66, from asst dir to dir, 66-69, admin asst, 69-82, asst comnr, 82-86, DIR WVA DIV FORESTRY, 86-; BOTANIST, COAL RESOURCES BR, US GEOL SURV, 74- *Concurrent Pos:* Instr, WVa Univ, 58-74, adj assoc prof, 74-87, adj prof, 87- *Mem:* Soc Am Foresters; Bot Soc Am; Int Asn Plant Taxon; Am Asn Petrol Geologists; Am Inst Biol Sci; Geol Soc Am. *Res:* Paleobotany; economic botany of the Appalachian area, particularly edible and poisonous plants; paleobotany of the Upper Paleozoic of the United States, especially compression floras; control of forest tree diseases, general agricultural and forestry administration. *Mailing Add:* 916 Churchill Circle Charleston WV 25314

GILLETT, JAMES WARREN, biochemistry, for more information see previous edition

GILLETT, LAWRENCE B, b Montreal, Que, Aug 22, 31; m 61. GEOLOGY. *Educ:* McGill Univ, BS, 53, MS, 56; Princeton Univ, MA, 56, PhD(geol), 62. *Prof Exp:* Asst prof geol, Univ NDak, 59-62; assoc prof, 62-73, PROF GEOL, STATE UNIV NY COL PLATTSBURGH, 73- *Mem:* Am Geophys Union; Nat Asn Geol Teachers; Can Inst Mining & Metall; Geol Asn Can; Sigma Xi. *Res:* Structures of diabase dikes; petrology of syenites; base metal exploration in Canada; glacial geomorphology; liquid immiscibility. *Mailing Add:* Two Adirondack Lane Plattsburgh NY 12901

GILLETT, TEDFORD A, b Burley, Idaho, Aug 18, 35; m 60; c 5. FOOD SCIENCE, MEAT SCIENCE. *Educ:* Univ Idaho, BS, 60, MS, 62; Mich State Univ, PhD(food sci), 66. *Prof Exp:* Res assoc biochem, Mich State Univ, 66-68; asst prof food & animal sci, Utah State Univ, 69-76; RES GROUP LEADER, UNION CARBIDES FOOD SCI INST, 76- *Mem:* Inst Food Technol; Am Meat Sci Asn; Am Soc Animal Sci. *Res:* Body composition; enzyme purification and characterization; meat processing; effect of nitrite upon bioavailability of iron; ham processing and emulsion technology. *Mailing Add:* 9400 Keswick Ave Stillwater MN 55082

GILLETTE, DAVID DUANE, b South Bend, Ind, Mar 20, 46; m 71; c 1. VERTEBRATE PALEONTOLOGY, EVOLUTIONARY BIOLOGY. *Educ:* Mich State Univ, BS, 67; Southern Methodist Univ, PhD(geol), 74. *Prof Exp:* Vis prof geol, Bryn Mawr Col, 74-75; asst prof, Sul Ross State Univ, 75-76; assoc prof biol, Col Idaho, 76-81; CUR, SHULER MUS PALEONTOL & ASST PROF GEOL, SOUTHERN METHODIST UNIV, 81- *Concurrent Pos:* Gerald Beadle fel, Tall Timbers Res Sta, Tallahassee, 74-76; res assoc, Acad Natural Sci, Philadelphia, 75-76; adj cur, Shuler Mus Paleont, Southern Methodist Univ, 75-81; dir, Mus Natural Hist, Col Idaho, 76-81; Inst Study Earth & Man fel, Southern Methodist Univ, 77-; adj prof geol, Boise St Univ, 79-81. *Mem:* Soc Vert Paleont; Geol Soc Am; Am Soc Mammalogists; Am Soc Ichthyologists & Herpetologists. *Res:* Evolution of edentates in North America; pleistocene vertebrate paleontology; mesozoic mammals; history of geology in North America; evolution of small cats; biology of the pigmy killer whale; stratigraphy; oceanography; mammalogy. *Mailing Add:* 4601 Fairfax Ave Dallas TX 75209

GILLETTE, DEAN, b Chicago, Ill, Aug 11, 25; m 49; c 1. MATHEMATICS. *Educ:* Ore State Col, BS, 48; Univ Calif, AM, 50, PhD(math), 53. *Prof Exp:* Asst math, Univ Calif, 50-53; mem tech staff, Bell Labs, 56-62, dir, Mil Anal Ctr, 62-66, exec dir, 66-71, exec exec dir systs res, 71-79, exec dir corp studies, Transmission Systs Eng Div, 79-; PROF, ENG DEPT, HARVEY MUDD COL. *Mem:* AAAS; Am Math Soc; Soc Indust & Appl Math; Inst Elec & Electronics Engrs; Sigma Xi. *Res:* Systems analysis; operations research. *Mailing Add:* Info Sci Claremont Grad Sch Claremont CA 91711

GILLETTE, EDWARD LEROY, b Coffeyville, Kans, May 21, 32; m; c 4. EXPERIMENTAL RADIATION THERAPY, COMPARATIVE ONCOLOGY. *Educ:* Kans State Univ, BS & DVM, 56; Colo State Univ, MS, 61, PhD(physiol, radiation biol), 65. *Prof Exp:* From instr to assoc prof, 59-71, PROF RADIOL, COLO STATE UNIV, 71-, DIR COMP ONCOL, COL VET MED & BIOMED SCI, 74- CHMN, DEPT RADIOL HEALTH SCI, 89- *Concurrent Pos:* Advan fel sect exp radiother, Univ Tex MD Anderson Hosp & Tumor Inst, 68-69; vis scientist, Los Alamos Sci Lab, Univ Calif, 72-76; assoc ed, Radiation Res, 79-82, 86-89; Cancer Res Manpower Rev Comt, NIH, 80-83; counr, Radiation Res Soc, 88-91. *Honors & Awards:* Ralston Purina Res Award, 88. *Mem:* Am Col Vet Radiol (pres, 72-73); Am Vet Med Asn; Radiation Res Soc; Am Soc Therapeut Radiol & Oncol; Am Asn Cancer Res; Vet Cancer Soc (pres, 82-84); Am Col Vet Internal Med. *Res:* Experimental radiotherapy; hyperthermia and radiation response of spontaneous tumors and normal tissue response. *Mailing Add:* Comp Oncol Unit Vet Teaching Hosp Colo State Univ Ft Collins CO 80523

GILLETTE, JAMES ROBERT, b Hammond, Ind, Feb 9, 28; m 53; c 2. BIOCHEMICAL PHARMACOLOGY. *Educ:* Cornell Col, AB, 47; Univ Iowa, MS, 49, PhD(biochem), 54. *Hon Degrees:* DSc, Cornell Col, 79. *Prof Exp:* Asst prof biol & chem, Jamestown Col, 49-51; biochemist, 54-58, dep chief, 67-71, actg chief, 71-72, CHIEF LAB CHEM PHARMACOL, NAT HEART, LUNG & BLOOD INST, 72-, HEAD SECT DRUG ENZYME INTERACTION, 58- *Honors & Awards:* Roland T Lakey Hon Lect Award, 67; Claude Bernard Vis Prof Award, Univ Montreal, 71; Bernard B Brodie Award Drug Metab, 78; Alan D Bass lectr, 79. *Mem:* AAAS; Am Soc Biol Chem; Am Chem Soc; Am Soc Pharmacol; Int Soc Study Xenobiotics. *Res:* Metabolism of drugs and other foreign compounds. *Mailing Add:* Lab Chem Pharmacol Nat Heart Lung & Blood Inst NIH Bldg 10 Rm 8N117 Bethesda MD 20892

GILLETTE, KEVIN KEITH, b Walla Walla, Wash, July 5, 61; m 83; c 1. MATHEMATICS, OPERATIONS RESEARCH. *Educ:* Stanford Univ, BS & MS, 82. *Prof Exp:* Sr res analyst, mgt sci, Bank Am, 82-83; res analyst, opers res, 84-86, SR SYSTS ANALYST, OPERS ENG, AM AIRLINES, 86- *Concurrent Pos:* Lectr, Sch Mgt, Univ Tex, Dallas, 85-88, Computer Sci Dept, 88 & 91. *Mem:* Soc Indust & Appl Math. *Res:* Improving accuracy of airline forecasting systems related to flight planning and weather. *Mailing Add:* Dept Opers Eng Dallas-Ft Worth Airport Am Airlines PO Box 619616 MD 5423 Dallas TX 75261-9616

GILLETTE, MARTHA ULBRICK, b Lincoln, Nebr, Nov 5, 45; m 69; c 2. NEUROPHYSIOLOGY, NEUROCHEMISTRY. *Educ:* Grinnell Col, Iowa, BA, 67; Univ Hawaii, MS, 69; Univ Toronto, PhD(zool), 75. *Prof Exp:* Fel neurophysiol, Univ Calif, Santa Cruz, 75-78; vis asst prof physiol & biol, 78-85, vis assoc prof physiol, 85-88, assoc prof, Dept Physiol & Biophys, 88-90, ASSOC PROF CELL & STRUCT BIOL, PHYSIOL & COL MED, UNIV ILL, 90- *Concurrent Pos:* Lectr neurophysiol, Div Nat Sci, Univ Calif, Santa Cruz, 78. *Mem:* Soc Neurosci; Int Soc Neuroethol; Found Biomed Res; Soc Res Biol Rhythms. *Res:* Electrophysiology and biochemistry of circadian time-keeping neurons in the mammalian brain. *Mailing Add:* Dept Cell & Struct Biol 506 Morrill Hall Univ Ill Urbana IL 61801

GILLETTE, NORMAN JOHN, b Cicero, NY, Mar 17, 11; m 37; c 3. PALEOBOTANY. *Educ:* Syracuse Univ, AB, 32, MA, 33; Univ Chicago, PhD(paleobot), 37. *Prof Exp:* Asst bot, Univ Chicago, 35-37; asst prof, Univ Idaho, 37-47; from asst prof to assoc prof, Syracuse Univ, 47-64; dir trop biol prog, 71-78, prof, 64-76, EMER PROF BOT, STATE UNIV NY COL OSWEGO, 76- *Mem:* Sigma Xi. *Res:* Morphology of fossil plants; Tertiary flora of Idaho; tropical botany of Jamaica. *Mailing Add:* RD 3 Box 138 Brown Dr Oswego NY 13126

GILLETTE, PAUL CRAWFORD, b 1942; c 2. PEDIATRIC CARDIOLOGY, CARDIAC ELECTROPHYSIOLOGY. *Educ:* Univ NC, BA, 65; Med Col SC, MD, 69. *Prof Exp:* Intern pediat, Baylor Col Med, 69-70, resident, 70-72, fel pediat cardiol, 72-74, fel cell biophys, 74-75, from asst prof to assoc prof pediat cardiol & cell biophys, 75-80, prof pediat cardiol, 80-84; PROF & DIR PEDIAT CARDIOL, MED UNIV SC, 84- *Concurrent Pos:* Consult, NIH, 77-78, 81 & 82, AMA Residency Rev Comt, 79-, AMA Diag Therapeut; prin investr, Nat Res & Demonstration Ctr, NIH, 77-79, res career develop award, 79-84; educ grant, Am Acad Pediat, 70. *Mem:* Am Col Cardiol; Am Acad Pediat; NAm Soc Pacing & Electrophysiol (pres); Am Heart Asn; Soc Pediat Res. *Res:* Developmental cardiac electrophysiology and pharmacology; developed techniques for electrophysiological study of cardiac arrhythmia in children and treatment with medical, pacing and surgical techniques; studied developmental cardiac physiology and pharmacology and the DNA metabolism of the heart. *Mailing Add:* 171 Ashley Ave Charleston SC 29425

GILLETTE, PHILIP ROGER, b Mt Vernon, Iowa, May 12, 17; wid; c 2. PHYSICS. *Educ:* Cornell Col, BA, 37; Univ Ill, BS, 38, MS, 39, PhD(physics), 42. *Prof Exp:* Mem staff, Radiation Lab, Mass Inst Technol, 42-45; proj engr, Sperry Gyroscope Co, NY, 45-48; physicist, Gen Elec Co, 48-50; sr res engr, SRI Int, 50-57, sr res physicist, 57-86; RETIRED. *Concurrent Pos:* Instr philos, Col Notre Dame, 87-88. *Mem:* Am Phys Soc; Inst Elec & Electronics Engrs; AAAS; Sigma Xi. *Res:* Molecular absorption and fluorescence spectroscopy; characteristics and design of pulseforming networks and pulse transformers; weapons, information and training systems design and operations analysis. *Mailing Add:* 151 Canada Cove Ave Half Moon Bay CA 94019

GILLETTE, RHANOR, b Bushnell, Fla, July 7, 43; m 69; c 2. NEUROPHYSIOLOGY. *Educ:* Univ Miami, BS, 67; Univ Hawaii, MS, 69; Univ Toronto, PhD(zool), 74. *Prof Exp:* Fel, Univ Calif at Santa Cruz, 74-78; ASST PROF NEUROPHYSIOL, DEPT PHYSIOL & BIOPHYS, UNIV ILL, URBANA, 78- *Mem:* Soc Neurosci. *Res:* Mechanisms of plasticity in behavior; mechanisms of plasticity in neuron function; evolution of nervous systems. *Mailing Add:* Dept Physiol & Biophys Univ Ill Urbana Campus 407 S Goodwin Ave Urbana IL 61801

GILLETTE, RICHARD F, b Chicago, Ill, July 2, 34; m 54; c 4. AVIONIC SYSTEMS, ELECTRONIC COUNTER MEASURES. *Educ:* Ill Inst Technol, BS, 66; Loyola Univ, MBA, 69. *Prof Exp:* Engr, Beltone Electronics, 55-62; mgr sustaining engr, Knight Div Allied Radio, 62-64; eng specialist, 64-71, prod line mgr, 71-73, prog mgr, 73-75, mgr advan systs, 75-77, dir eng, 77-79, VPRES ENG, NORTHROP CORP DEFENSE SYSTS, 79- *Mem:* Inst Elec & Electronics Engrs; Nat Soc Prof Engrs; Asn Old Crows. *Res:* Microwave signal detection and analysis; broad band microwave power generation; integerated systems for electronic counter measures. *Mailing Add:* North Crop Defense Systs 600 Hicks Rd Rolling Meadows IL 60008

GILLETTE, RONALD WILLIAM, VIROLOGY, MOLECULAR BIOLOGY. *Educ:* Univ Pittsburgh, PhD(biol sci), 55. *Prof Exp:* Dir cellular immunol, Meloy Labs, Inc, 81-87; MEM STAFF, DEPT MOLECULAR, & CELL BIOL, CYTOGEN CORP, 88- *Res:* Cellular immunology. *Mailing Add:* Cytogen Corp 201 College Rd E Princeton NJ 08540

GILLHAM, JOHN K, b London, Eng, Aug 7, 30; m 61; c 3. POLYMER SCIENCE. *Educ:* Cambridge Univ, BA, 53, MA, 57; McGill Univ, PhD(chem), 59. *Prof Exp:* Res chemist, Stamford Labs, Am Cyanamid Co, 58-65; assoc prof, 65-75, PROF CHEM ENG, PRINCETON UNIV, 75- *Concurrent Pos:* Vis res chemist, Plastics Prog, Princeton Univ, 64-65; mem, Nat Mat Adv Bd Comts, Washington, 81, 84; vis fel, Japan Soc Prom Sci, 83; vis scholar, Chinese Acad Sci, People's Repub China, 84. *Honors & Awards:* Chem of Plastics & Coatings Award, Am Chem Soc, 78; Award in Thermal Analysis, NAm Thermal Analysis Soc, 78; Doolittle Award, Am Chem Soc, 80; Roon Award, Fedn Socs Coatings Technol, 83 & 89; Ann Int Res Award, Soc Plastics Engrs, 88. *Mem:* Am Chem Soc; Soc Plastics Eng; Am Inst Chem Eng; Am Phys Soc; NY Acad Sci; Plastics & Rubber Inst; NAm Thermal Analysis Soc. *Res:* Molecular structure and temperature-dependent properties of polymeric materials; thermosetting polymers; spiral fractures; dynamic mechanical analysis; development of torsional braid analysis (TBA) technique; development of time-temperature-transformation (TTT) cure diagram. *Mailing Add:* Dept Chem Eng Princeton Univ Princeton NJ 08544-5263

GILLHAM, NICHOLAS WRIGHT, b New York, NY, May 14, 32; m 56. MOLECULAR BIOLOGY. *Educ:* Harvard Univ, AB, 54, AM, 55, PhD(biol), 62. *Prof Exp:* USPHS fel, Yale Univ, 62-63; from instr to asst prof biol, Harvard Univ, 63-68; from assoc prof to prof, 68-82, JAMES B DUKE PROF ZOOL, DUKE UNIV, 82- *Concurrent Pos:* Vis prof, Rockefeller Univ, 74-75; mem, President's Panel Med Res, 75 & gov task force on sci & technol, 82; mem study sect genetics, NIH, 76-80; Guggenheim fel, Cold Spring Harbor, 84-85, Duke Univ, 85; vchmn bd dirs & Chmn Sci comt, Am Type Cult Collection, 90- *Mem:* Soc Cell Biol; Genetics Soc Am; Sigma Xi. *Res:* Organelle heredity, molecular biology and transformation; genetics and molecular biology of chloroplasts and mitochondria. *Mailing Add:* Dept Zool Duke Univ Durham NC 27706

GILLHAM, ROBERT WINSTON, b Newmarket, Ont, Dec 29, 40; m 63; c 1. GROUNDWATER RESOURCE & CONTAMINATION, REMEDIATION OF CONTAMINATED GROUNDWATER. *Educ:* Univ Toronto, BSA, 63; Univ Guelph, MSc, 68; Univ Ill, PhD(soil physics), 73. *Prof Exp:* Fel soils, Univ Guelph, 73-74; res asst prof groundwater, 74-77, from asst prof to assoc prof, 77-86, PROF GROUNDWATER, UNIV WATERLOO, 86-; DIR, WATERLOO CTR GROUNDWATER RES, 87- *Concurrent Pos:* Mem, Working Group Contrib Contaminants Great Lakes by Groundwater, Internal Joint Comt, 84-88, Groundwater Contamination Working Group, Scope, 86-, Comt Hydrol, Nat Res Coun Asn, 88-89; ed-in-chief, J Contaminant Hydrol, 87- *Honors & Awards:* Thomas Roy Award, Can Geotech Soc, 88. *Mem:* Am Geophys Union; Am Soc Agron. *Res:* Response of aquifers to pumping, groundwater-surface water interactions and migration of contaminants in groundwater; remediation of contaminated aquifers. *Mailing Add:* 11 Crawford St Guelph ON N1G 1Y9

GILLIAM, CHARLES HOMER, b Lexington, Tenn, Oct 14, 52; m 71; c 2. HORTICULTURE, PLANT PHYSIOLOGY. *Educ:* Univ Tenn, Martin, BS, 74; Va Polytech Inst & State Univ, MS, 76, PhD(hort), 77. *Prof Exp:* Asst prof, Ohio State Univ, 77-80; asst prof, 80-83, ASSOC PROF HORT, AUBURN UNIV, 83- *Honors & Awards:* R P White Award, Hort Res Inst, 78. *Mem:* Am Soc Hort Sci; Int Plant Propagators Soc; Sigma Xi. *Res:* Weed control in nursery crops; improving production efficiency of ornamentals. *Mailing Add:* Dept Hort Auburn Univ Auburn AL 36849

GILLIAM, JAMES M, aging, electromicroscopy instrumentation; deceased, see previous edition for last biography

GILLIAM, JAMES WENDELL, b Chicota, Tex, July 18, 38; m 58; c 3. SOIL CHEMISTRY. *Educ:* Okla State Univ, BS, 60; Miss State Univ, MS, 63, PhD(soil chem), 65. *Prof Exp:* From asst prof to assoc prof, 65-76, PROF SOIL CHEM, NC STATE UNIV, 76- *Mem:* Fel Am Soc Agron; AAAS; fel Soil Sci Soc Am; Sigma Xi. *Res:* Fertilizer reactions in soils; uptake of nutrients by plants and plant analysis; cation exchange reactions of soil organic matter; contribution of fertilizers to contamination of surface waters; nonpoint source pollution; agricultural water management; wetlands and agriculture; effect of buffers and riparian areas on water quality. *Mailing Add:* Dept Soil Sci NC State Univ Raleigh NC 27650

GILLIAM, OTIS RANDOLPH, b Waverly, Va, Sept 19, 24; m 53; c 3. PHYSICS. *Educ:* Randolph-Macon Col, BS, 43; Duke Univ, PhD(physics), 50. *Prof Exp:* Instr, 50-51 & 53-55, from asst prof to assoc prof, 56-67, PROF PHYSICS, UNIV CONN, 67- *Concurrent Pos:* Consult, Am Optical Co, 66-67, Brookhaven Nat Lab & Picatinny Arsenal. *Mem:* Am Phys Soc; Am Asn Physics Teachers. *Res:* Determination of the structure of matter by microwave methods, paramagnetism and radiation damage. *Mailing Add:* Dept Physics U-46 Univ Conn Storrs CT 06268

GILLICH, WILLIAM JOHN, b Washington, DC, Jan 8, 35; m 59; c 3. PENETRATION MECHANICS, ARMOR CONCEPT DESIGN & DEMONSTRATION. *Educ:* Johns Hopkins Univ, BES, 57, MS, 60, PhD(mech), 64. *Prof Exp:* From jr instr to instr mech eng, Johns Hopkins Univ, 57-60, res asst, 61-64; res physicist, Aberdeen Proving Ground, 64-71, supv res physicist & chief wave propagation & mat sect, Solid Mech Br, 71, actg chief, 71-73, supv res physicist & chief, Penetration Mech Br, 73-82, SUPV RES PHYSICIST & CHIEF, ARMOR MECH BR, TERMINAL BALLISTICS DIV, US ARMY BALLISTIC RES LAB, ABERDEEN PROVING GROUND, 82-; Supv Res physicist & chief, Armor Mech Br, 82-86. *Concurrent Pos:* Res asst, Johns Hopkins Univ, 57-64; instr, Ballistic Inst, Univ Del Exten, 69-70. *Honors & Awards:* US Army Res & Develop Award, 77, 79, 83; R H Kent Award, 81; Am Defense Preparedness Asn Awards. *Mem:* Soc Natural Philos; Am Soc Mech Engrs; Soc Exp Stress Analysis. *Res:* Plastic and elastic wave propagation; dynamic continuum mechanics; shock wave reflections; Moire technology. *Mailing Add:* Armor Mech Br Terminal Ballistics Div Aberdeen Proving Ground MD 21005-5066

GILLIES, ALASTAIR J, b Halifax, Eng, Oct 7, 24; US citizen; m 52; c 4. ANESTHESIOLOGY, PHARMACOLOGY. *Educ:* Univ Edinburgh, BSc, 47, MB, ChB, 48. *Prof Exp:* Resident house surgeon & resident anesthetist, Royal Infirmary, Edinburgh, 48-49; asst res anesthesiol, Mass Gen Hosp, 52; asst anesthetist, Strong Mem Hosp, 54-55; asst prof anesthesiol, Sch Med, Yale Univ, 55-59; prof pharmacol & toxicol, 59-77, PROF ANESTHESIOL & CHMN DEPT PHARMACOL & TOXICOL, MED CTR, UNIV ROCHESTER, 59- *Concurrent Pos:* Fel, Mayo Found, 53; anesthesiologist, Grace-New Haven Community Hosp, 55-59; anesthetist in chief, Strong Mem Hosp, 59-77; consult, Bur Heart Dis, NY State Dept Health. *Mem:* Am Soc Anesthesiol; Asn Univ Anesthetists; NY Acad Sci; Pan-Am Med Asn; Int Anesthesia Res Soc; Brit Med Asn. *Mailing Add:* Dept Anesthesiol Univ Rochester Med Ctr Rochester NY 14642

GILLIES, CHARLES WESLEY, b Wilmington, Del, Dec 1, 46; m 67, 87; c 3. PHYSICAL & PHYSICAL ORGANIC CHEMISTRY. *Educ:* West Chester State Col, BA, 68; Univ Mich, PhD(phys chem), 72. *Prof Exp:* Fel phys chem, Harvard Univ, 73-76; PROF PHYS CHEM, RENSSELAER POLYTECH INST, 76- *Concurrent Pos:* Referee, J Am Chem Soc, 72-, J Phys Chem, 78- & J Molecular Spectros; proj dir, Grant Petrol Res Found, 77-80 & Res Corp, US Dept Energy, 81; reviewer, NSF, 78- & Nat Res Coun, 79. *Honors & Awards:* Fajans Award, 72. *Mem:* Am Chem Soc. *Res:* Applications of microwave spectroscopy to structures and mechanisms; Fourier transform microwave spectroscopy. *Mailing Add:* Dept Chem Rensselaer Polytech Inst Troy NY 12181

GILLIES, GEORGE THOMAS, b Rugby, NDak, Aug 20, 52; m 79; c 1. MEDICAL PHYSICS, BIBLIOGRAPHIC STUDIES. *Educ:* NDak State Univ, BSc, 74; Univ Va, MSc, 76, PhD(eng physics), 80. *Prof Exp:* Physicist, Int Bur Weights & Measures, 81-83; develop staff mem, Martin Marietta Energy Systs, Inc, 83-85; postdoctoral fel, 80-81, res asst prof, 85-88, RES ASSOC PROF ENG PHYSICS, UNIV VA, 88- *Concurrent Pos:* Vis scientist, US Nat Bur Standards, 81, Cavendish Lab, Cambridge Univ, 81-82 & Univ Trieste, 82; prin investr, Video Tumor Fighter Proj, 87- *Mem:* Sr mem Inst Elec & Electronics Engrs; Am Phys Soc; Optical Soc Am. *Res:* Experimental tests of Newtonian gravitation; design of magnetic suspension systems; development of the video tumor fighter which is a method of nonlinear stereotaxis that can deliver hyperthermia to brain tumors. *Mailing Add:* Dept Physics Univ Va Charlottesville VA 22901

GILLIGAN, DIANA MARY, CELL BIOLOGY. *Educ:* Albert Einstein Col Med, MD & PhD(anat & cell struct), 85. *Prof Exp:* resident, Internal Med, NC Mem Hosp, 85-90; RES ASSOC, DEPT MED, DUKE UNIV MED CTR, 90- *Mailing Add:* Howard Hughes Med Inst Box 3892 Duke Univ Med Ctr Durham NC 27710

GILLIGAN, LAWRENCE G, b New York, NY, June 19, 48; m; c 2. MATHEMATICS. *Educ:* NY Univ, BA, 69, MA, 73. *Prof Exp:* PROF MATH, UNIV CINCINNATI, 84- *Mem:* Am Asn Univ Prof; Am Soc Eng Educ; Am Math Asn Two Yr Col (vpres, 84); Math Asn Am. *Mailing Add:* Univ Cincinnati 2220 Victory Pkwy Cincinnati OH 45206

GILLIKIN, JESSE EDWARD, JR, b Morehead City, NC, Jan 22, 52; m 70; c 2. ANALYTICAL CHEMISTRY, INORGANIC CHEMISTRY. *Educ:* East Carolina Univ, BA, 74, MS, 76. *Prof Exp:* Qual assurance control scientist, 78, qual assurance supvr, 79-81, qual assurance specialist, 81-82, QUAL ASSURANCE SUPVR, BURROUGHS WELLCOME CO, 82- *Mem:* Am Chem Soc. *Res:* Electroanalytical chemistry, with an emphasis towards computerized laboratory equipment. *Mailing Add:* 123 Forest Acres Dr Greenville NC 27834

GILLILAN, JAMES HORACE, b Kansas City, Mo, Dec 21, 32; m 70. MATHEMATICS. *Educ:* Univ Mo, BA, 54, MA, 55; Univ Ill, PhD(math), 61. *Prof Exp:* Instr math, Univ Mich, 61-63; ASST PROF MATH, UNIV MO-KANSAS CITY, 66- *Mem:* Am Math Soc. *Res:* Integration theory. *Mailing Add:* Dept Math Univ Mo Kansas City MO 64110

GILLILAN, LOIS ADELL, b Mapleton, Utah, June 15, 11. ANATOMY, PHYSIOLOGY. *Educ:* Mt Holyoke Col, BA, 35; Vassar Col, AM, 37; Univ Mich, PhD(anat), 40; Univ Pittsburgh, MD, 47. *Prof Exp:* Asst physiol, Vassar Col, 35-38; vol asst neuroanat, Univ Mich, 39-42, asst physiol, 41; instr anat, Univ Pittsburgh, 42-45; mem med staff health serv, Univ Ill, 48-49; assoc prof anat, Grad Sch Med, Univ Pa, 49-60; from assoc prof to prof, 60-76, EMER PROF ANAT, MED CTR, UNIV KY, 76- *Concurrent Pos:* Gelston fel, Univ Mich, 41-42. *Mem:* Am Asn Anat; Am Neurol Asn; assoc Am Acad Neurol; fel Coun Cerebrovasc Dis; hon mem Royal Soc Health; hon mem Am Heart Asn. *Res:* Physiology of lung lymphatics; anatomy of thalamus and brainstem of mammals; autonomic nervous system; blood supply of the central nervous system. *Mailing Add:* Mayfair Village No 608 3310 Tates Creek Rd Lexington KY 40502

GILLILAND, BOBBY EUGENE, b Epps, La, Aug 6, 36; m 59; c 1. ELECTRICAL ENGINEERING, INSTRUMENTATION. *Educ:* La Polytech Inst, BS, 58; Univ Ark, MS, 64, PhD(electronics, instrumentation), 68. *Prof Exp:* Res asst electronics & instrumentation, Univ Ark, Little Rock, 62-67; from asst prof to assoc prof elec eng, 67-76, from asst dean to assoc dean eng, 73-86, PROF ELEC & COMPUT ENG, CLEMSON UNIV, 76-, SPEC ASST TO PRES, 86- *Concurrent Pos:* Mem, SC Nuclear Adv Coun, 73-80, chmn, 75-76 & SC Energy Exten Serv Task Force, 78-81. *Mem:* Inst Elec & Electronics Engrs; Am Soc Eng Educ; Nat Soc Prof Engrs; Instrument Soc Am; Sigma Xi. *Res:* instrumentation including digital, textile and biomedical; digital computer control systems; modeling and simulation; innovative graduate engineering education. *Mailing Add:* 411 Shorecrest Dr Clemson SC 29631

GILLILAND, DENNIS CRIPPEN, b Warren, Pa, July 23, 38; m 57; c 3. STATISTICS, MATHEMATICS. *Educ:* Kent State Univ, BA, 59; Mich State Univ, MS, 63, PhD(statist), 66. *Prof Exp:* Develop engr, Goodyear Aerospace Corp, 59-66; from asst prof to assoc prof, 66-74, dept chairperson, 85-89, PROF STATIST & PROBABILITY, MICH STATE UNIV, 74- *Concurrent Pos:* Lectr, Univ Calif, Berkeley, 66-67; vis scholar, Univ Chicago, 82. *Mem:* Inst Math Statist; Am Statist Asn; Am Soc Qual Control. *Res:* Applied statistics; decision theory; statistics and law. *Mailing Add:* Dept Statist & Probability Mich State Univ East Lansing MI 48824-1027

GILLILAND, FLOYD RAY, JR, b Cotter, Ark, Dec 18, 39; m 62; c 1. ENTOMOLOGY, ZOOLOGY. *Educ:* Ark Polytech Col, BS, 62; Univ Ark, MS, 64; Miss State Univ, PhD(entom), 67. *Prof Exp:* From asst prof to prof entom, Auburn Univ, 67-77; CONTRACT RES & CONSULT, 77- *Mem:* AAAS; Entom Soc Am. *Res:* Basic and applied studies of cotton insects, especially methods of biological control of cotton insects. *Mailing Add:* PO Box 65 Pike Road AL 36064

GILLILAND, HAROLD EUGENE, b Duncan, Okla, Sept 9, 37. CHEMICAL ENGINEERING, PHYSICAL CHEMISTRY. *Educ:* Mass Inst Technol, SB, 59, SM, 61, PhD(chem eng), 65. *Prof Exp:* Res scientist, 64-67, res group supvr, Continental Oil Co, 67-80, sect dir, 80-88, MGR RESERVOIR ENG, CONOCO INC, 88- *Mem:* Am Inst Mining, Metall & Petrol Engrs. *Res:* Heterogeneous catalysis; thermal methods of recovering hydrocarbon resources; petroleum recovery processes; reservoir engineering; drilling and completions. *Mailing Add:* Prod Eng & Res Conoco Inc PO Drawer 2197 Houston TX 77079

GILLILAND, JOE E(DWARD), b Alhambra, Calif, Dec 4, 27; m 52; c 2. CHEMICAL ENGINEERING. *Educ:* Tex Tech Col, BS, 49; Okla State Univ, MS, 60. *Prof Exp:* Chem engr, Tex Brine Corp, 49-51; chem engr, 51-57, pilot plant supvr, 57-63, SUPT, SPEC CHEM DEPT, OZARK-MAHONING CO, 63- *Mem:* Inst Chem Engrs. *Res:* Inorganic fluorine compounds; crystallization. *Mailing Add:* 5512 S Yorktown Pl Tulsa OK 74105

GILLILAND, JOHN L(AWRENCE), JR, b Clearfield, Pa, Oct 4, 10; m 35; c 3. CHEMICAL ENGINEERING. *Educ:* Univ Colo, BS, 33. *Prof Exp:* From jr chem engr to chem engr, US Bur Reclamation, 33-51; gen chemist, Ideal Cement Co, 51-64, dir qual control, 64-67, tech dir, 67-70, dir environ qual, Cement Div, Ideal Basic Indust, Inc, 70-77; RETIRED. *Concurrent Pos:* Mem nat adv comt control tech, Nat Air Pollution Control Admin. *Mem:* Am Chem Soc; Am Soc Testing & Mat; Am Acad Environ Engrs. *Res:* Air pollution control; quality control of Portland cement manufacture; environmental controls. *Mailing Add:* 3753 S Granby Way Aurora CO 80014

GILLILAND, ROBERT MCMURTRY, b Galveston, Tex, Jan 16, 21; m 44; c 2. PSYCHIATRY, PSYCHOANALYSIS. *Educ:* Univ Tex, BA, 41; Univ Tex Med Br Galveston, MD, 45. *Prof Exp:* Resident psychiat, Vet Admin Hosp, Waco, Tex, 46-48 & Univ Tex Med Br Galveston, 48-49; pvt pract psychiat & psychoanalysis, 49-64; PROF PSYCHIAT, BAYLOR COL MED, 64- *Concurrent Pos:* Supr analyst, Psychoanalysis Inst, Houston-Galveston, 64- *Mem:* Am Psychoanalysis Asn; Am Psychiat Asn; Int Psychoanalysis Asn. *Mailing Add:* 2201 W Holcombe Suite 320 Houston TX 77030

GILLILAND, RONALD LYNN, b Emporia, Kans, July 16, 52. SOLAR & STELLAR VARIABILITY. *Educ:* Univ Kans, Lawrence, BA, 74; Univ Calif, Santa Cruz, PhD(astrophys), 79. *Prof Exp:* Fel, Advan Study Prog, High Altitude Observ, Nat Ctr Atmospheric Res, 79-81, staff scientist, 81-88; ASSOC ASTROMR, SPACE TELESCOPE SCI INST, 88- *Mem:* Am Astron Soc; Int Astron Union. *Res:* Solar and stellar variability from both theoretical and obervational approaches. *Mailing Add:* Space Telescope Sci Inst 3700 San Martin Dr Baltimore MD 21218

GILLILAND, STANLEY EUGENE, b Minco, Okla, June 24, 40; m 60, 90; c 4. DAIRY & FOOD MICROBIOLOGY. *Educ:* Okla State Univ, BS, 62, MS, 63; NC State Univ, PhD(food sci), 66. *Prof Exp:* Assoc prof food sci, NC State Univ, 65-76; from assoc prof to prof, 76-86, REGENTS PROF ANIMAL SCI, OKLA STATE UNIV, 86- *Honors & Awards:* Pfizer Award, Am Dairy Sci Asn, 79, Dairy Res Award, 87. *Mem:* Fel Am Acad Microbiol; Inst Food Technol; Am Dairy Sci Asn; Coun Agr Sci & Technol; Sigma Xi; Am Soc Microbiol. *Res:* Nutrition of lactic starter cultures; growth of high population bacterial starter cultures; antagonisms of starter cultures toward psychrotrophic bacteria & food-borne pathogens; factors which limit the growth and action of starter cultures; microbial dietary supplements; intestinal microecology; health and/or nutritional benefits from lactobacilli as dietary adjuncts. *Mailing Add:* Animal Sci Dept Okla State Univ Stillwater OK 74078-0425

GILLILAND, WILLIAM NATHAN, b Portsmouth, Ohio, May 23, 19; m 44; c 2. GEOLOGY. *Educ:* Ohio State Univ, BA, 41, PhD(geol), 48. *Prof Exp:* Asst geol, Ohio State Univ, 41-42 & 46-47; geologist, US Geol Surv, 48-49; from instr to prof geol, Univ Nebr, 49-65, chmn dept, 51-64; dean, Newark Col Arts & Sci, 65-68, PROF GEOL, RUTGERS UNIV, NEWARK, 65- *Concurrent Pos:* Geologist, Tenn Valley Authority, 42; asst prof, Utah Field Sta, Ohio State Univ, 49-50; consult, Calif Co, 51-53, Shell Oil Co, 55 & Creole Petrol Corp, 56-57. *Mem:* Fel Geol Soc Am; Am Asn Petrol Geol; Am Inst Prof Geologists. *Res:* Stratigraphy; structural geology; petroleum geology. *Mailing Add:* Dept Geol Rutgers Univ Newark NJ 07102

GILLIN, JAMES, b Floral Park, NY, Sept 16, 25; m 49; c 2. CHEMICAL ENGINEERING, PROCESS CHEMISTRY. *Educ:* Cornell Univ, BChE, 47, PhD(chem eng), 51. *Prof Exp:* Sr engr chem eng process develop, Merck & Co Inc, 49-52, group leader, 52-56, sect mgr pilot plant, 56-59, mgr, 59-62, dir, 62-64, chem engr res & develop, 64-69, exec dir new drug develop, 69-70, exec dir planning & admin, 70-71, vpres develop, Merck, Sharp & Dohme Res Labs, 71-79, pres, MSD AGVET Div, 79-87; RETIRED. *Mem:* Nat Acad Eng; Am Chem Soc; Am Inst Chem Engrs; NY Acad Sci; fel Am Inst Chem; AAAS. *Res:* Ion exchange; vapor phase catalytic oxidation. *Mailing Add:* 11354 Golf View Lane North Palm Beach FL 33408

GILLINGHAM, JAMES CLARK, b Waukesha, Wis, Oct 15, 44; m 67; c 2. HERPETOLOGY. *Educ:* Wis State Univ, Oshkosh, BS, 67, MS, 72; Univ Okla, PhD(zool), 76. *Prof Exp:* Fac asst biol, Univ Wis, Oshkosh, 67-72; vis instr zool, Univ Okla, 75-76; from asst prof to assoc prof, 76-84, PROF BIOL, CENT MICH UNIV, 84- *Concurrent Pos:* Dir, Cent Mich Biol Sta, 85- *Honors & Awards:* Ortenberger Award & Arthur N Bragg Natural Hist Award, Univ Okla, 76. *Mem:* Am Soc Zoologists; Animal Behav Soc; Am Soc Icthyologists & Herpetologists; Herpetologists League; Soc Study Amphibians & Reptiles; Sigma Xi. *Res:* Behavioral ecology of amphibians and reptiles; ritualized behavior of snakes; behavioral morphology of reptiles. *Mailing Add:* Dept Biol Cent Mich Univ Mt Pleasant MI 48859

GILLINGHAM, JAMES MORRIS, b Gallipolis, Ohio, Mar 29, 24; m 48; c 2. PHARMACEUTICAL CHEMISTRY. *Educ:* Denison Univ, BA, 49; Northwestern Univ, 50. *Prof Exp:* Asst res chemist, Parke, Davis & Co, 50-55; asst sci dir, Warren Teed Prods, 55-59; prod mgr, Diamond Labs, 59-60, dir pharmaceut prods develop, 60-61; vpres, Vale Chem Co, 61-70; mgr pharmaceut prod, Animal Health Div, Am Hoechst Corp, 70-72; dir tech serv & qual assurance, Marion Labs, Inc, 72-76, sr tech adv, regulatory affairs, 76-89; GILLINGHAM CONSULT, 89- *Honors & Awards:* E Baugh Award Chem, 49. *Mem:* Am Soc Qual Control. *Res:* Research and products development in pharmaceutical and analytical chemistry; production and processing of veterinary and human pharmaceutical products; quality assurance and regulatory compliance of pharmaceutical products and processes. *Mailing Add:* 9706 Overbrook Rd Leawood KS 66206

GILLINGHAM, ROBERT J, b Galvin, Wash, Oct 29, 23. SOLID STATE PHYSICS, ELECTRONICS. *Educ:* Gonzaga Univ, MA, 49; La State Univ, MS, 60, PhD(physics), 63. *Prof Exp:* From asst prof to assoc prof physics, 62-76, assoc prof elec eng, 74-80, ASSOC PROF PHYSICS, GONZAGA UNIV, 80- *Concurrent Pos:* Head dept sci & dir seismol observ, Mt St Michael's Sem, 62-67. *Mem:* Am Phys Soc; Am Asn Physics Teachers; Seismol Soc Am. *Res:* Electronic instrumentation. *Mailing Add:* Dept Physics Gonzaga Univ Spokane WA 99258

GILLIOM, RICHARD D, b Bluffton, Ind, June 25, 34; m 58; c 3. ORGANIC CHEMISTRY. *Educ:* Southwestern at Memphis, BS, 56; Mass Inst Technol, PhD(org chem), 60. *Prof Exp:* Res chemist, Esso Res Labs, Humble Oil & Refining Co, 60-61; from asst prof to prof chem, Southwestern, Memphis, 61-84; prof chem, Rhodes Col, 84-90; RETIRED. *Concurrent Pos:* Fulbright lectr, Univ Skoplje, 68-69; assoc, Drug Design Div, Univ Tenn Ctr Health Sci, 75; consult, Molecular Design Int, 76-88; res investr, VA Hosp, Memphis, 82. *Mem:* Am Chem Soc. *Res:* Effects of structure upon reactivity; quantitative structure-activity relationship; computational chemistry. *Mailing Add:* 1667 Miller Farms Rd Germantown TN 38138

GILLIS, BERNARD THOMAS, b Pierre, SDak, Mar 7, 31; m 53; c 4. ORGANIC CHEMISTRY. *Educ:* Loras Col, BS, 52; Wayne State Univ, PhD(org chem), 56. *Prof Exp:* Asst org chem, Wayne State Univ, 52-53; fel, Mass Inst Technol, 56-57; from asst prof to prof chem, Duquesne Univ, 57-70, assoc chmn dept, 65-68, dean grad sch, 68-70; acad vpres & provost, Indiana Univ, 70-80; ACAD VPRES, YOUNGSTOWN STATE UNIV, 80- *Mem:* AAAS; Sigma Xi. *Res:* Organic synthesis, structure elucidation of natural products; divinyl ethers; oxidation of hydrazides and hydrazones; azo dienophiles; educational administration. *Mailing Add:* Acad VPres Youngstown State Univ Youngstown OH 44555-0001

GILLIS, CHARLES NORMAN, b Glasgow, Scotland, Feb 3, 33; US citizen; m 60; c 2. PHARMACOLOGY. *Educ:* Glasgow Univ, BSC, 54, PhD(pharmacol), 57. *Prof Exp:* Asst lectr exp pharmacol, Glasgow Univ, 54-57; asst prof pharmacol, Univ Alta, 57-61; from asst prof to assoc prof pharmacol, Yale Univ, 61-68; head cardiovasc pharmacol, Squibb Inst Med Res, 68-69; assoc prof anesthesiol & pharmacol, 69-73, PROF ANESTHESIOL PHARMACOL, SCH MED, YALE UNIV, 73- *Concurrent Pos:* Estab investr, Am Heart Asn, 64-68; vis assoc prof, Med Sch, Rutgers Univ, 68-69; assoc ed, Biochem Pharmacol, 81- *Mem:* Am Soc Pharmacol & Exp Therapeut; Pharmacol Soc Can. *Res:* Cardiopulmonary and autonomic pharmacology; neuropharmacology. *Mailing Add:* Dept Anesthesiol Yale Univ Sch Med 333 Cedar St New Haven CT 06510

GILLIS, HUGH ANDREW, b Sydney, NS, Aug 11, 35; Can citizen; m 77; c 2. MASS SPECTROMETRY, RADIATION CHEMISTRY. *Educ:* St Francis Xavier Univ, BSc, 54; Univ Notre Dame, PhD(chem), 57. *Prof Exp:* Gen Elec Co res fel phys chem, Univ Leeds, 57-59; assoc res officer, Nat Res Coun Can, Ottawa, 67-74, sr res officer, Physics Div, 74-81, liaison officer, Off Atlantic Regional Dir, 81-85, sr res officer, Atlantic Res Lab, Nat Res Coun Can, Halifax, NS, 85-87; DEAN SCI, ST FRANCIS XAVIER UNIV, 87- *Concurrent Pos:* vis fel, Mellon Inst, Carnegie-Mellon Univ, 60, 65, 67 & 68, Centre D'Etudes Nucleaire De Saclay, France, 73. *Mem:* Can Asn Physicists; Am Chem Soc; fel Chem Inst Can. *Res:* Trapped and solvated electrons in irradiated aqueous, alcoholic and hydrocarbon liquids and solids. *Mailing Add:* Dean Sci St Francis Xavier Univ Antigonish NS B2G 1C0 Can

GILLIS, JAMES E, JR, b Lewiston, Maine, Apr 22, 20; c 1. GEOLOGY, CARTOGRAPHY. *Educ:* Cath Univ Am, BS, 41; George Washington Univ, MSBA, 66; Indust Col Armed Forces, dipl, 66. *Prof Exp:* Tech secy, Comt Geophys Sci, Joint Res & Develop Bd, US Dept Army, 46-47; secy, Comt Geophys & Geog, Res & Develop Bd, US Dept Defense, 47-49, res analyst, Planning Div, 49-51, panel dir, 51-52; chief, Coordination Br, Snow, Ice & Permafrost Res Estab, Corps Eng, US Dept Army, 52-53, actg dir, 53-58, adminr, 58-59, phys sci adminr in charge res satellite geod & mapping, Eng Res & Develop Labs, Ft Belvoir, 59-60, chief, Intel Div, US Army Engr Geod Intel & Mapping Res & Develop Agency, 60-67, assoc dir mapping & geog sci lab, US Army Eng Topog Labs, 67-69, tech dir, Directorate Advan Systs, US Army Topog Command, Corps Engrs, 69-72, asst dep dir plans requirements & technol, Defense Mapping Topographic Ctr, 72-74, asst chief, Dept Cartog, 74-76, chief, 76-78, dep dir, 79-80; CONSULT, 80- *Concurrent Pos:* Consult, NASA; Nat Acad Sci-Nat Res Coun deleg, Int Geophys Year, Moscow, Russia, 59. *Mem:* Geol Soc Am; Am Soc Photogram; Soc Photog Sci & Eng; Am Geophys Union; Optical Soc Am. *Res:* Field and laboratory research and development programs in the physical sciences; 00039919x geodesy; mapping satellites; optical and electronic tracking techniques and equipment; methods, techniques and systems for collection and processing of military geographic intelligence; production of topographic maps and digital topographic data. *Mailing Add:* 1132 E Paseo Pavon Tucson AZ 85718

GILLIS, JAMES THOMPSON, b Boston, Mass, Aug 30, 56; m 88. SIGNAL PROCESSING & WAVELET WAVEPOCKET ANALYSIS, PRECISION POINTING & CONTROL OF SPACECRAFT. *Educ:* Wash Univ, St Louis, BS, 79; Univ Calif Los Angeles, MS, 85, PhD(elec eng), 88. *Prof Exp:* Mem tech staff, Space & Commun Group, Hughes Aircraft, 79-83; mem tech staff, Controls Analysis Dept, 83-88, ENG SPECIALIST, PRECISION POINTING OFF, AEROSPACE CORP, 88- *Concurrent Pos:* Res assoc, Systs Res Ctr, Univ Md, 90-91. *Mem:* Inst Elec & Electronics Engrs; Soc Indust & Appl Math; Am Math Soc; Math Asn Am; Am Inst Aeronaut & Astronaut. *Res:* Precision pointing and vibration control of spacecraft and launch vehicles; signal processing including random fields, wavelet and wavepacket analysis and Armax identification. *Mailing Add:* PO Box 0196 Seabrook MD 20703

GILLIS, JOHN ERICSEN, b White Plains, NY, May 16, 43. EVOLUTIONARY BIOLOGY. *Educ:* Univ Pa, BA, 65; Colo State Univ, MS, 68, PhD(zool), 75. *Prof Exp:* Teaching asst zool, Colo State Univ, 68 & 70; asst prof biol, Univ Maine, Ft Kent, 75-76, Allegheny Col, 76-83 & Univ Nebr-Lincoln, 84; ASSOC PROF BIOL, UNIV WIS-LA CROSSE, 84- *Mem:* AAAS; Soc Study Evolution; Herpetologists League; Soc Study Amphibians & Reptiles; Sigma Xi. *Res:* Amphibian water economy; evolutionary relationships in the Rana pipiens complex; color polymorphisms in grasshoppers; thermal ecology of grasshoppers. *Mailing Add:* Dept Biol Univ Wis La Crosse 1725 State St La Crosse WI 54601

GILLIS, MARINA N, b Harrisburg, Pa, Jan 30, 34; m 61. POLYMER CHEMISTRY. *Educ:* Wilson Col, BA, 55. *Prof Exp:* res chemist, chem div, Thiokol Corp, 57-81, SR RES CHEMIST, CONGOLEUM CORP, TRENTON, 81- *Mem:* Am Chem Soc. *Res:* Polymer synthesis; polymer structure and properties; polyethylene sulfide degradation and stabilization; polyacrylates; polyurethanes; polymeric coatings; radiation polymerization; synthesis and reactions of urethane curatives; polysulfides, rotogravure inks, vinyl foams and chemical embossing. *Mailing Add:* 12 Lawndale Rd Yardley PA 19067

GILLIS, MURLIN FERN, b Santa Cruz, Calif, Apr 26, 35; m 71; c 3. VETERINARY MEDICINE, PHYSICS. *Educ:* Univ Wash, BS, 56; Calif Inst Technol, MS, 59; Wash State Univ, DVM, 63. *Prof Exp:* From res scientist to sr res scientist, Pac Northwest Labs, Battelle Mem Inst, 66-71, res assoc, 71-72, res & develop mgr, 72-75, mgr proj develop, 75-80, assoc mgr, Biol Dept, 80-86, RES OPERS MGR, LIFE SCI CTR, PAC NORTHEST LABS, BATTELLE MEM INST, 86- *Mem:* Am Vet Med Asn. *Res:* Experimental surgery; bioengineering; biomaterials; implantology; fertility control; hyperbaric medicine; bioelectromagnetism; research administration. *Mailing Add:* 2330 Camas Ave Richland WA 99352

GILLIS, PETER PAUL, b Newport, RI, Dec 23, 30; m 53, 80; c 6. SOLID MECHANICS, MATERIALS SCIENCE. *Educ:* Brown Univ, ScB, 53, ScM, 61, PhD(eng), 64. *Prof Exp:* Prod engr, Fram Corp, RI, 56-57; develop engr, Leesona Corp, 57-58; instr mach design, RI Sch Design, 58-59; res asst eng, Brown Univ, 59-64; asst prof eng mech, 64-68, assoc prof mat sci, 68-75, PROF MAT SCI, UNIV KY, 75- *Concurrent Pos:* Consult, Spindletop Res Inc, Ky, 64-66, Lawrence Livermore Lab, 65-73, Sandia Corp, NMex, 67 & Los Alamos Sci Lab, 68-79; Fulbright-Hays res scholar, Physics & Eng Lab, Dept Sci & Indust Res, NZ, 70-71; consult div compliance, Atomic Energy Comn, 70-73; vis scientist, State Univ NY Col Forestry, Syracuse, 71. *Mem:* Am Inst Mining, Metall & Petrol Engrs; fel Am Soc Metals Int; Am Soc Mech Engrs; Fedn Am Scientists; Volunteers Tech Assistance. *Res:* Flow and fracture properties of materials, particularly relations between these charactersitics and the microstructure of the material in terms of the concepts of crystal physics. *Mailing Add:* Dept Mat Sci & Eng Univ Ky Lexington KY 40506-0046

GILLIS, RICHARD A, b Rochester, NY, Mar 20, 38; m 60; c 1. PHARMACOLOGY. *Educ:* Miami Univ, BA, 60; McGill Univ, PhD(pharmacol), 65. *Prof Exp:* Instr, Harvard Med Sch, 67; from instr to assoc prof, 67-77, PROF PHARMACOL, SCH MED, GEORGETOWN UNIV, 77- *Concurrent Pos:* Res fel pharmacol, Harvard Med Sch, 65-66. *Mem:* AAAS; Am Soc Pharmacol & Exp Therapeut. *Res:* Cardiovascular and autonomic nervous system pharmacology. *Mailing Add:* Dept Pharmacol NW 408 Med-Dent Georgetown Univ Sch Med 3900 Reservoir Rd NW Washington DC 20007

GILLISPIE, GREGORY DAVID, b Painesville, Ohio, June 23, 49; m 72; c 3. PHYSICAL CHEMISTRY, CHEMICAL PHYSICS. *Educ:* Mich State Univ, BS, 71, PhD(chem), 75. *Prof Exp:* Fel chem, Wayne State Univ, 75-77; asst prof, State Univ NY, Albany, 77-82; from asst prof to assoc prof, chem, 83-90; NDAK STATE UNIV, CHAIR, 89-, PROF CHEM, 90- *Mem:* Am Phys Soc; Am Chem Soc; Int Soc Optical Eng. *Res:* Laser spectroscopy; radiationless transitions; fluorescence; phosphorescence; intramolecular hydrogen bonds. *Mailing Add:* Dept Chem NDak State Univ Fargo ND 58105

GILLMAN, DAVID, b New York, NY, Sept 6, 38. MATHEMATICS. *Educ:* Univ Wis, BS, 58, MS, 59, PhD(math), 62. *Prof Exp:* Res instr math, Cornell Univ, 62; mem, Inst Advan Study, 62-64; asst prof, 64-69, ASSOC PROF MATH, UNIV CALIF, LOS ANGELES, 69- *Res:* Topology. *Mailing Add:* Dept Math 6356 Math Scis Univ Calif 405 Hilgard Ave Los Angeles CA 90024

GILLMAN, HYMAN DAVID, b Brooklyn, NY, Dec 21, 41; m 67; c 3. INORGANIC CHEMISTRY. *Educ:* Long Island Univ, BS, 63; Tufts Univ, PhD(inorg chem), 68. *Prof Exp:* Sr res chemist, Pennwalt Chem Corp, 68-81; SR RES CHEMIST, ARCO CHEM CO, 81- *Mem:* Am Chem Soc. *Res:* Preparation and study of coordination compounds; synthesis of coordination polymers with inorganic backbones; UV resins and coatings. *Mailing Add:* Rd 1 Box 213B Spring City PA 19475-9720

GILLMAN, LEONARD, b Cleveland, Ohio, Jan 8, 17; m 38; c 2. MATHEMATICS. *Educ:* Columbia Univ, BS, 41, MA, 45, PhD(math), 53. *Prof Exp:* Asst math, Columbia Univ, 41-42, lectr, 43; assoc, Tufts Col, 43-45; mem staff, Opers Eval Group, Mass Inst Technol, 45-51; from instr to assoc prof math, Purdue Univ, 52-60; prof & chmn dept, Univ Rochester, 60-69; chmn dept, 69-73, prof, 69-87, EMER PROF MATH, UNIV TEX, AUSTIN, 87- *Concurrent Pos:* Guggenheim Mem fel, 58-59; mem, Inst Advan Study, 58-60; NSF sr fel, 59-60; vis lectr, Math Asn Am, 61-69; mem comt on regional develop, Nat Acad Sci-Nat Res Coun, 63-65, chmn, 65-66; mem, US Comn on Math Instr, 65-66, chmn, 66-69; US deleg, Int Comn Math Instr, 66-69, organizing comt, First Int Cong Math Educ, 69; math ed, W W Norton Co, 67-80; mem ed bd, Gen Topology & Its Appln, 71- *Mem:* Am Math Soc (assoc secy, 69-71); Math Asn Am (treas, 73-86, pres-elect, 86-87, pres, 87-89); Nat Coun Teachers Math. *Res:* Theory of sets; topology; rings of continuous functions. *Mailing Add:* 1606 The High Rd Austin TX 78746-2236

GILLMOR, R(OBERT) N(ILES), b Marion, Iowa, Dec 25, 06; m 30; c 2. PHYSICAL METALLURGY, PHYSICAL CHEMISTRY. *Educ:* SDak Sch Mines & Technol, BS, 28. *Prof Exp:* Develop engr process metall, Int Smelting Co, 28-31; consult & develop engr phys metall, Gen Elec Co, 31-42, mgr works lab, 42-45, mgr mat & process lab, 45-71; RETIRED. *Mem:* Fel Am Soc Metals. *Res:* Physical metallurgy in both ferrous and nonferrous metallurgy, especially austenitic, ferritic and martensitic stainless steels and their passive characteristics in various environmental media. *Mailing Add:* 7903 Oak Hollow Lane Fairfax VA 22039

GILLMORE, DONALD W(OOD), b Lorain, Ohio, June 24, 19; m 44; c 3. FUEL ENGINEERING, CHEMICAL ENGINEERING. *Educ:* Williams Col, BA, 41; Pa State Univ, PhD(fuel tech), 54. *Prof Exp:* Asst, Titanium Div, Nat Lead Co, 41-43; res assoc, Inst Gas Tech, 43-44; sr chemist, Butadiene Div, Koppers Co, Inc, 44-45; res chemist, Houdry Process Corp, 45-48; asst & assoc fuel tech div, Pa State Univ, 48-51; supvr carbon res, Pittsburgh Coke & Chem Co, 53-59; mgr tech br, Electro Minerals Div, Carborundum Co, Niagara Falls, 60-67, plant mgr, 67-70; res supvr, US Bur Mines, 70-75; res supvr, Energy Res & Develop Admin, Morgantown Energy Res Ctr, WVa, 75-77; sect leader, Morgantown Energy Technol Ctr, WVa, 77-80, br chief, Dept Energy, 80-85; RETIRED. *Mem:* Am Chem Soc. *Res:* Fuel and coal technology; physical chemistry; activated carbon; electron microscopy; Fischer-Tropsch synthesis; abrasives; electric furnacing; refractories; coal gasification, fluid-bed combustion and gasification; fly ash utilization; strip mine and coal refuse reclamation; coal preparation; oil and gas recovery; in-situ coal gasification. *Mailing Add:* 661 South View St Morgantown WV 26505

GILLOOLY, GEORGE RICE, b West Chester, Pa, June 6, 30; m 55; c 2. LUMINESCENT MATERIALS RESEARCH. *Educ:* Univ Mich, BS, 53; Dartmouth Col, MA, 57. *Prof Exp:* Teacher sci, Flat Rock High Sch, 53-54; instr chem, Flint Jr Col, 54-55; teaching fel chem, Dartmouth Col, 55-57; prod engr halophosphors, Gen Elec Co, Cleveland, 57-58, chemist phosphors, 58-64, res chemist, 64-78, sr develop chemist phosphors, 78-90, sr res chemist phosphors, 90, CONSULT PHOSPHORS, GEN ELEC CO, 90- *Concurrent Pos:* Client instr, Kepner-Tregoe, 73-83. *Honors & Awards:* Saul Dushman Award, Gen Elec Corp Res & Develop Ctr, 80. *Mem:* Electrochem Soc; Am Chem Soc; fel Am Inst Chemists. *Res:* Luminescence materials research including identifying developing and implementing new phosphors and phosphor improvements; halophosphate lamp phosphors; halophosphate phosphor synthesis. *Mailing Add:* 1375 E Colonial Dr Salisbury NC 28144

GILLOTEAUX, JACQUES JEAN-MARIE A, b Mons, July 9, 44; Belg citizen; m 72; c 1. HISTO-CYTOCHEMISTRY, HISTOLOGY. *Educ:* Univ Louvain, BSc, 64, MSc, 66, ScD, 74. *Prof Exp:* Instr biol & chem, Col St Pierre, 66; prof histol, Univ Louvain, 70-71; lectr cell biol, State Univ NY, Stony Brook, 76; asst prof physiol, Upstate Med Ctr, State Univ NY Syracuse, 77-79; adj prof biol, Onondaga Community Col, 79; ASSOC PROF MICROS ANAT, COL MED, NORTHEASTERN OHIO UNIV, 79- *Concurrent Pos:* Vis prof, Med Sch, St Georges Univ, 79. *Mem:* Royal Zool Soc Belg; Am Soc Cell Biol; AAAS; Am Asn Anatomists; Histochem Soc. *Res:* Structure and function of contractile cells and tissues; comparative histochemistry and cytochemistry; cell biology related to cardiac tissues and to hormone-induced carcinogenesis. *Mailing Add:* Anat Dept Col Med Northeastern Ohio Univ PO Box 95 Rootstown OH 44272

GILLOTT, DONALD H, b Connellsville, Pa, Aug 25, 31; m 70; c 3. ELECTRICAL ENGINEERING. *Educ:* Univ Pittsburgh, BS, 56, MS, 59, PhD(elec eng), 63. *Prof Exp:* Jr engr, Latrobe Steel Co, 54-56; from instr to assoc prof elec eng, Univ Pittsburgh, 56-68; prof elec eng & chmn dept, 68-76, DEAN SCH ENG, CALIF STATE UNIV, SACRAMENTO, 76- *Concurrent Pos:* NSF grant, 64-, travel grant, Europe, 66; design engr, Latrobe Steel Co, 56-58; consult, Power Supply Div, Int Bus Mach Corp, 59-63 & Nat Acad Sci, 63- *Mem:* Am Soc Eng Educ; sr mem Inst Elec & Electronics Engrs. *Res:* Flux penetration in magnetic materials, including effects of saturation and hysteresis. *Mailing Add:* 3077 Emerald Ct Cameron Park CA 95682-8536

GILLOW, EDWARD WILLIAM, b Scranton, Pa, Apr 14, 39; m 62; c 3. PHYSICAL CHEMISTRY, ANALYTICAL CHEMISTRY. *Educ:* Pa State Univ, BS, 60; State Univ NY Buffalo, PhD(phys chem), 66. *Prof Exp:* Fel chem, Univ Tex, Austin, 66-68; res chemist, Jackson Lab, E I du Pont de Nemours & Co, Inc, 78-88, Pigments Dept, Exp Sta Lab, 68-70, tech serv chemist, Chestnut Run Lab, 70-72, res chemist, Pigments Dept, Exp Sta, 72-78, res chemist, Chem & Pigments Dept, 78-88; CONSULT, 88- *Res:* Chemical kinetics; analytical methods development; solution chemistry with nonaqueous solvents; chromatography; porosimetry. *Mailing Add:* 1811 Bybrook Rd Wilmington DE 19803

GILLUM, AMANDA MCKEE, b Rochester, NY, Apr 13, 47; m 68; c 1. BIOCHEMISTRY, GENETIC ENGINEERING. *Educ:* Ind Univ, AB, 69; Mass Inst Technol, PhD(biochem), 76. *Prof Exp:* Scholar, Med Sch, Stanford Univ, 76-78; RES INVESTR, E R SQUIBB & SONS, INC, 78- *Mem:* Am Soc Microbiol; AAAS; NY Acad Sci. *Res:* Application of molecular genetics and nucleic acid biochemistry to develoment of pharmaceutical compounds, particularly anti-fungal agents; development of new techniques for nucleic acid sequencing and analysis; role of plasmids in antibiotics, biosynthesis and resistance. *Mailing Add:* 192 Penn View Dr Pennington NJ 08534

GILLUM, RONALD LEE, b Decatur, Ill, Feb 7, 38; m 60; c 3. CLINICAL PATHOLOGY. *Educ:* DePauw Univ, BA, 60; Univ Ill Col Med, MD, 64; Am Bd Path, dipl & cert anat path & clin path, 69 & dipl & cert radioisotopic path, 74. *Prof Exp:* Intern, Ill Cent Hosp, Chicago, 64-65; resident path, Univ Ill Res & Educ Hosps, 65-69, teaching asst, Col Med, 65-67, instr, 67-69; asst prof, Univ Tex Med Br Galveston, 72-75; assoc prof, Univ Tex Med Sch, Houston, 75-77, Col Med, Univ Okla, 77-85; chief path serv, 80-84, DIR CLIN LABS, OKLA MEM HOSP, 84-; PROF PATH, COL MED, UNIV OKLA, 85- *Mem:* Am Asn Clin Chemists; Am Soc Clin Pathologists; Acad Clin Lab Physicians & Scientists; AMA; Col Am Pathologists; NY Acad Sci; AAAS. *Res:* Selection and utilization of instrumentation and methods in the clinical laboratory; the place of continuing education in quality control in the clinical laboratory; selection and interpretation of lab tests; nutritional disorders and trace elements. *Mailing Add:* Okla Mem Hosp Path Serv PO Box 26307 E B 402 Oklahoma City OK 73126

GILMAN, ALBERT F, III, b Chicago, Ill, June 25, 31; m 64; c 10. MATHEMATICS. *Educ:* Northwestern Univ, BS, 52; Univ Mont, MA, 58; Ind Univ, MA, 62, PhD(math), 63. *Prof Exp:* From instr to asst prof math, Bowdoin Col, 63-66; from assoc prof to prof math & chmn div sci & math, Col VI, 66-69; asst vpres acad affairs, 69-73, dir comput ctr, 70-72, actg vice chancellor acad affairs, 72, PROF MATH, WESTERN CAROLINA UNIV, 69- *Concurrent Pos:* Exec dir, Republican Study Group, US House of Rep, 73-74; NEH Fel, Univ Chi, 82. *Honors & Awards:* NAS Vis Exchange Prof, Bulgarian Acad Sci, 78. *Mem:* Am Math Soc; Math Asn Am. *Res:* Noncommutative schemes; economic models; psephology; politicometrics. *Mailing Add:* Dept Math Western Carolina Univ Cullowhee NC 28723

GILMAN, ALFRED G, b New Haven, Conn, July 1, 41. BIOCHEMISTRY, PHARMACOLOGY. *Educ:* Yale Univ, BA, 62; Case Western Univ, MD & PhD(pharmacol), 69. *Prof Exp:* Pharmacol res assoc, Lab Biochem Genetics, Nat Heart & Lung Inst, NIH, 69-71; from asst prof to prof pharmacol, Univ Va, 71-81; PROF & CHMN DEPT PHARMACOL, UNIV TEX SOUTHWESTERN MED CTR, 81-, RAYMOND & ELLEN WILLIE DISTINGUISHED CHAIR MOLECULAR NEUROPHARMACOL, 87- *Concurrent Pos:* Mem bd sci counrs, Nat Health, Lung & Blood Inst, NIH, 82-86; ad hoc mem pharmacol study sect, NIH, 73-75, mem, 77-81; mem sci adv comt biochem & chem carcinogenesis, Am Cancer Soc, 82-86; mem sci rev bd, Howard Hughes Med Inst, 86- *Honors & Awards:* John J Abel Award Pharmacol, Am Soc Pharmacol & Exp Therapeut, 75; Poul Edvard Poulsson Award, Norweg Pharmacol Soc, 82; Gairdner Found Int Award, 84; Richard Lounsbery Award, Nat Acad Sci, 87; Distinguished Res Award Biomed Sci, Am Asn Med Col, 88; Albert Lasker Basic Med Res Award, 89; Passano Found Award, 90; Louis S Goodman & Alfred Gilman Award, Drug Receptor Pharmacol, Am Soc Pharmacol & Exp Therapeut, 90; Basic Sci Res Prize, Am Heart Asn, 90. *Mem:* Nat Acad Sci; Inst Med-Nat Acad Sci; Am Soc Pharmacol & Exp Therapeut; Am Soc Biol Chemist; fel Am Acad Arts & Sci; fel AAAS. *Res:* Author of over 100 technical journal articles. *Mailing Add:* Dept Pharmacol Univ Tex Health Southwestern Med Ctr 5323 Harry Hines Blvd Dallas TX 75235-9041

GILMAN, DONALD LAWRENCE, b Hartford, Conn, Oct 15, 31; m 61, 87; c 3. METEOROLOGY. *Educ:* Harvard Univ, AB, 52; Mass Inst Technol, MS, 54, PhD(meteorol), 57. *Prof Exp:* Mem res staff, Meteorol Dept, Mass Inst Technol, 55-58; res meteorologist, Extended Forecast Div, Nat Weather Serv, Nat Oceanic & Atmospheric Admin, 58-64, chief, Develop & Testing Sect, 64-71, chief, Long Range Prediction Group, Nat Meteorol Ctr, 72-79, chief, Prediction Br, Climate Analysis Ctr, 79-89; RETIRED. *Mem:* AAAS; fel Am Meteorol Soc; Am Geophys Union; Sigma Xi. *Res:* Objective techniques of extended and long range forecasting; forecast verification; meteorological statistics. *Mailing Add:* 4335 Cedarlake Ct Alexandria VA 22309

GILMAN, FREDERICK JOSEPH, b Lansing, Mich, Oct 9, 40; m 67; c 4. THEORETICAL HIGH ENERGY PHYSICS. *Educ:* Mich State Univ, BS, 62; Princeton Univ, PhD(physics), 65. *Prof Exp:* NSF res fel, Calif Inst Technol, 65-66, res fel theoret physics, 65-67; res assoc, 67-69, assoc prof, 69-73, PROF, STANFORD LINEAR ACCELERATOR CTR, 73-; ASSOC DIR PHYSICS RES DIV, SSC LAB, 90- *Concurrent Pos:* Vis, Fermilab, 72, Caltech, 73, Inst Advan Study, 74, Weizmann Inst, 82; mem, SLAC Prog Adv Comt, 69-72, Beratron Scheduling Comt, 72-74, Exec Comt, Div Particles & Fields, APS, 72-74, Exp Prog Adv Comt, SLAC, 78-80, Fermilab Prog Adv Comt, 78-81, Cornell Prog Adv Comt, 80-85, Rev Comt, High Energy Physics Div, LBL, 81-84, SSC Detector Cost/Model Adv Comt, 85, Adv Bd, ITP, Santa Barbara, 88-91; ed, Int J Modern Physics A, J Modern Physics Letters A; vchmn, Sakurai Prize Comt, APS, 86, chmn, 87; vchmn, div Particles & Fields, APS, 88, chmn, 89; chmn, steering comt, High Energy Physics in 90's, Snowmass, 88. *Mem:* Fel Am Phys Soc. *Res:* Theoretical elementary particle physics. *Mailing Add:* Res Div SSC Lab 2550 Beckleymeade Ave Dallas TX 75237

GILMAN, JANE P, b Washington, DC, April 17, 45. MATHEMATICS. *Educ:* Univ Chicago, BS, 65; Columbia Univ, PhD, 71. *Prof Exp:* Instr, State Univ NY, Stony Brook, 71-72; from asst prof to assoc prof, Newark Col Arts & Sci, 72-84, PROF, DEPT MATH & COMPUT SCI, RUTGERS UNIV, NEWARK, NJ, 84-, CHAIR DEPT, 82- *Concurrent Pos:* Mem, Inst Advan Study, Sch Math, Princeton, NJ, 79-80; mem, Math Sci Res Inst, Berkeley, Calif, 86; vis res mathematician, Princeton Univ, 88-; mem, Coun Am Math Soc, 86-88. *Mem:* Am Math Soc; Asn Comput Mach; Am Women Math. *Res:* Riemann surfaces; fuchsian groups teichmuller theory; geometric topology; combinatorial group theory; author of 17 technical research articles. *Mailing Add:* Dept Math & Comput Sci Rutgers Univ Smith Hall Newark NJ 07102

GILMAN, JOHN JOSEPH, b St Paul, Minn, Dec 22, 25; m 50, 76; c 7. PHYSICS, METALLURGY. *Educ:* Ill Inst Technol, MS, 48; Columbia Univ, PhD(metall), 52. *Prof Exp:* Res engr metall, Crucible Steel Co, 48-52; res assoc, Gen Elec Co, 52-60; prof eng, Brown Univ, 60-63; prof physics & metall, Univ Ill, Urbana-Champaign, 63-68; dir, Mat Res Ctr, Allied Chem Corp, 68-78, dir, Corp Develop Ctr, 78-80; mgr corp res, Standard Oil Co, Ind, 80-85; assoc dir, 85-87, SR SCIENTIST, LAWRENCE BERKELEY LAB, 87- *Concurrent Pos:* Adj prof, Columbia Univ, 68-74; consult, Lawrence Radiation Lab; mem mat res coun, Advan Res Proj Agency, Dept Defense, 68- *Honors & Awards:* Mathewson Medal, Am Inst Mining, Metall & Petrol Eng, 59; Campbell lectr, Am Soc Metals, 66; Reduction to Pract Award, Metall Soc, 86; IR-100 Awards, 74, 76, 80. *Mem:* Nat Acad Eng; fel Am Soc Metals; Am Inst Mining, Metall & Petrol Eng; fel Am Phys Soc; Mat Res Soc. *Res:* Mechanical behavior of solids; solid state physics; behavior of shock and detonation waves in solids; research management; technical management. *Mailing Add:* Lawrence Berkeley Lab 1 Cyclotron Rd Berkeley CA 94720

GILMAN, JOHN RICHARD, JR, b Malden, Mass, July 6, 25; m 60; c 2. SCIENCE ADMINISTRATION. *Educ:* Harvard Univ, AB, 46; NY Univ, MSW, 83. *Prof Exp:* At John H Breck, Inc, 47-48, dir publicity, 49-53, asst adv mgr, 50-53, dir new prod, Mkt Res, 55-58, tech dir, 56-62; dir new prod, Acco Labs, Am Cyanamid Corp, 63; exec vpres, treas, pres & dir, August Sauter Am, Inc, New York, 65-78; PRES CONSULT, JOHN R GILMAN & ASSOCS, 78- *Concurrent Pos:* Lectr, Niagara Univ, Niagara Falls, NY, 85 & 86, Newport Col, Newport, RI, 86 & 87. *Mem:* NY Acad Sci; fel Am Orthopsychiat Asn. *Res:* Psychological impediments to creative investigation and to effective administration in science and technology; effective communications within the research and development function, and between research and development and other organizational functions. *Mailing Add:* 395 Punkateest Neck Tiverton RI 02878

GILMAN, LAUREN CUNDIFF, b Bozeman, Mont, Nov 24, 14. GENETICS. *Educ:* Baker Univ, AB, 36; Johns Hopkins Univ, PhD(genetics), 40. *Prof Exp:* Asst biol, Johns Hopkins Univ, 37-39, Brooks Fund res grant, 40-41; asst prof zool, Univ SDak, 46-47; assoc prof, 47-80, EMER PROF ZOOL, UNIV MIAMI, 80- *Mem:* Fel AAAS; Soc Protozool; Am Genetic Asn; Am Micros Soc; Am Soc Zool; Sigma Xi. *Res:* Occurrence distribution and interrelationship of mating types, inheritance of mating type and factors involved in the mating reaction of Paramecium caudatum. *Mailing Add:* Dept Biol Univ Miami PO Box 249118 Coral Gables FL 33124

GILMAN, MARTIN ROBERT, b Brooklyn, NY, Mar 28, 37; m 63; c 1. TOXICOLOGY. *Educ:* Del Valley Col, BS, 63; Drexel Univ, MS, 68, PhD(toxicol), 71. *Prof Exp:* Consult toxicologist, Gilmar Assocs, 71-72; dir, Cannon Labs, 72-74; Prod Safety Toxicologist, Colgate-Palmolive Co, 74-83; TOXICOLOGIST, HAZELTON LABS, 83- *Mem:* AAAS; NY Acad Sci; Soc Toxicol; Teratol Soc. *Res:* Consumer product safety; Federal Hazardous Substances Act evaluations; food products safety; animal toxicologic study evaluations; human irritation/sensitization; inhalation teratology; animal reproduction. *Mailing Add:* Hazleton Res Prod 6321 S 6th St Kalamazoo MI 49009

GILMAN, NORMAN WASHBURN, b Augusta, Maine, June 6, 38; m 58; c 2. ORGANIC CHEMISTRY. *Educ:* Occidental Col, BA, 63; Princeton Univ, MA, 65, PhD(org chem), 67. *Prof Exp:* NIH fel, Harvard Univ, 67-68; sr res chemist, 68-80, res group chief, 80-85, RES INVESTR, HOFFMANN-LA ROCHE INC, 85- *Mem:* AAAS; Am Chem Soc; NY Acad Sci. *Res:* Structure-activity relationships; heterocyclic chemistry for application to medicinal chemistry; bronchiopulmonary research. *Mailing Add:* Five Normandy Dr Wayne NJ 07470

GILMAN, PAUL BREWSTER, JR, b Havana, Cuba, Nov 21, 29; US citizen; m 52; c 4. PHYSICAL CHEMISTRY. *Educ:* Univ Mass, BS, 51; Northeastern Univ, MS, 54; Rutgers Univ, PhD(phys chem), 57. *Prof Exp:* From res chemist to res assoc emulsion res, Eastman Kodak Co, 57-70, sr res assoc, 70-89, CONSULT, 89- *Mem:* Fel Soc Photog Sci & Eng. *Res:* Photochemistry; photographic science; spectral sensitization; luminescence; photothermographic materials. *Mailing Add:* Eastman Kodak Co Res Labs B-82 Floor 6 Rochester NY 14650-2111

GILMAN, PETER A, b Hartford, Conn; m 66, 76; c 4. CONVECTION THEORY, DYNAMO THEORY. *Educ:* Harvard Col, BA, 62; Mass Inst Technol, SM, 64, PhD(meteorol), 66. *Prof Exp:* Asst prof astro-geophys, Univ Colo, Boulder, 66-69; sci vis, Advan Study Prog, 69-70, chmn, Advan Study Prog, 71-75, mem staff, 70-73, sr scientist, Nat Ctr Atmospheric Res, 73-, head, Solar Variability Sect, 77-87, SR SCIENTIST, 75- DIR, HIGH ALTITUDE OBSERV, 87- *Concurrent Pos:* Lectr, Univ Colo, Boulder, 70-77, adj prof astrophys, planetary & atmospheric sci, 79-; mem solar & space physics comt, Space Sci Bd, Nat Acad Sci, 80-83. *Mem:* AAAS; Am Meteorol Soc; Am Astron Soc; Am Geophys Union; Int Astron Union. *Res:* Fluid dynamics and magneto hydrodynamics of the sun and planets. *Mailing Add:* Nat Ctr Atmospheric Res Box 3000 Boulder CO 80307

GILMAN, RICHARD ATWOOD, b Concord, NH, Jan 22, 35; m 58; c 1. GEOLOGY. *Educ:* Dartmouth Col, AB, 57; Univ Ill, MS, 59, PhD(geol), 61. *Prof Exp:* Instr phys sci, Univ Ill, 61-63; from asst prof to prof geol, 63-74, dept chmn, 75-87, DISTINGUISHED TEACHING PROF GEOL, STATE UNIV NY COL FREDONIA, 74- *Concurrent Pos:* Consult, Maine Geol Surv, 62-; res found grant-in-aid, State Univ NY Col Fredonia, 63; NSF fel, 71-72. *Mem:* AAAS; Geol Soc Am; Nat Asn Geol Teachers; Sigma Xi. *Res:* Structural analysis of igneous and metamorphic rocks; field mapping in Southern Maine. *Mailing Add:* Dept Geosci State Univ NY Fredonia NY 14063

GILMAN, ROBERT EDWARD, b Concord, NH, Jan 19, 32; m 54; c 2. ORGANIC CHEMISTRY. *Educ:* Dartmouth Col, AB, 54; Univ Mich, MS, 57, PhD(chem), 59. *Prof Exp:* Res chemist, W R Grace & Co, 58-60; Nat Res Coun Can fel, 60-62; vis asst prof, Williams Col, 62-64; from asst prof to assoc prof, 64-72, PROF CHEM, ROCHESTER INST TECHNOL, 73- *Concurrent Pos:* NSF sci fac fel, Univ Calif, Los Angeles, 70-71. *Mem:* Am Chem Soc. *Res:* Natural products; stereochemistry; reaction mechanisms; organic synthesis; cyclophane chemistry. *Mailing Add:* Dept Chem Rochester Inst Technol Lomb Memorial Dr Rochester NY 14623

GILMAN, ROBERT HUGH, b Utica, NY, July 28, 42; m 69; c 2. DYNAMICAL SYSTEMS, GROUP THEORY. *Educ:* Princeton Univ, AB, 64; Columbia Univ, PhD(math), 69. *Hon Degrees:* Hon M Eng, Stevens Inst technol, 87. *Prof Exp:* From asst prof to assoc prof, 74-82, PROF MATH, STEVENS INST TECHNOL, 82- *Concurrent Pos:* Vis mem, Courant Inst Math Sci, 71-72; mem, Inst Advan Study, 84-85. *Mem:* Am Math Soc; Math Asn Am; Sigma Xi. *Res:* Dynamical systems and computational group theory. *Mailing Add:* Dept Math Stevens Inst Technol Hoboken NJ 07030

GILMAN, S(ISTER) JOHN FRANCES, b Malden, Mass, July 29, 30. COMPUTER SCIENCE, MATHEMATICS. *Educ:* Manhattanville Col, BA, 52; Cath Univ Am, MA, 61; St Louis Univ, PhD(math), 67. *Prof Exp:* Assoc prof math, St Joseph Col, Md, 66-73; PROF MATH, NIAGARA UNIV, 73-, PROF COMPUT SCI, 81- *Concurrent Pos:* Consult, Div Instrnl TV, Md State Dept Educ, 72-74; mem, NY State Adv Coun Postsec Continuing Educ Serv, 76-78 & Adult Learning Serv, 78, consult & workshop dir, math, 77-81; mem adv bd, Orleans-Niagara Teacher Ctr, 85-89; mem adv bd, Niagara-Falls Teacher Ctr, 87-91. *Mem:* Am Math Soc; Math Asn Am; Nat Coun Teachers Math; Asn Comput Mach; Data Processing Mgt Asn; Asn Systs Managers. *Res:* Mathematics education in secondary and elementary schools; number theory; evaluation of large projects. *Mailing Add:* 1129 Niagra Ave Niagara Falls NY 14305-2743

GILMAN, SID, b Los Angeles, Calif, Oct 19, 32; m 62, 84; c 1. CLINICAL NEUROLOGY, NEUROPHYSIOLOGY & NEUROPHARMACOLOGY. *Educ:* Univ Calif, Los Angeles, BA, 54, MD, 57. *Prof Exp:* Intern, Univ Calif Hosp, Los Angeles, 57-58; res assoc, NIH, 58-60; resident, Neurol Unit, Boston City Hosp, Mass, 60-63; instr neurol, Harvard Med Sch, 65-66, assoc, 66-68; from asst prof to assoc prof, Col Physicians & Surgeons, Columbia Univ, 68-72, prof neurol, 72-76, H Houston Merritt prof neurol, 76-77; PROF & CHMN DEPT NEUROL, UNIV MICH, 77- *Concurrent Pos:* Ambrose & Gladys Bowyer Found fel med, 57-58; res fel, Harvard Med Sch at Boston City Hosp, 62-65. *Honors & Awards:* Weinstein Goldenson Award, 81. *Mem:* AAAS; Am Acad Neurol; Am Neurol Asn (pres 88-89); Am Soc Clin Invest; Soc Neurosci; Sigma Xi. *Res:* Disorders of movement and behavior in humans and animals with lesions of the central nervous system; positron emission tomography studies of cerebellar degenerations; dementia; and epilepsy. *Mailing Add:* Dept Neurol TC 1914 Univ Mich 1500 E Medical Center Dr Ann Arbor MI 48109-0316

GILMAN, STEVEN CHRISTOPHER, Urbana, Ill, Dec 25, 52; m; c 4. IMMUNOPHARMACOLOGY, INFLAMMATION. *Educ:* Miami Univ, Ohio, BS, 75; Pa State Univ, MS, 77, PhD(microbiol), 79. *Prof Exp:* Res assoc immunol, Scripps Clin & Res Found, 79-82; res scientist immunol, Wyeth Labs, Inc, 82-87; DIR, BIOL RES, CYTOGEN CORP, PRINCETON, NJ, 87- *Concurrent Pos:* Adj asst prof, Sch Med, Temple Univ, 83-; assoc ed, Int Arch Pharmacodynamics & Ther, 85-; fel, NIH & Coun Tobacco Res. *Mem:* Am Asn Immunol; Am Soc Microbiol; Inflammation Res Asn; Soc Nuclear Med; Am Rheumatism Asn. *Res:* Regulation of immune responses in disease states such as arthritis and aging; role of hormones and other soluble mediators in bone and joint destruction; monoclonal antibody immunotherapy. *Mailing Add:* Cytogen Corp 201 College Rd E Princeton NJ 08540

GILMARTIN, AMY JEAN, botany; deceased, see previous edition for last biography

GILMARTIN, MALVERN, b Los Angeles, Calif, Nov 14, 26; m 76; c 5. BIOLOGICAL OCEANOGRAPHY. *Educ:* Pomona Col, BA, 54; Univ Hawaii, MS, 56; Univ BC, PhD(oceanog), 60. *Prof Exp:* Sr scientist, Inter-Am Trop Tuna Comt, 60-65; assoc prof, Univ Hawaii, 65-67; prof biol oceanog & dir oceanog, Stanford Univ, 67-69 & 70-74; prog dir biol oceanog, NSF, 69-70; dir, Australian Inst Marine Sci, 74-78; prof biol oceanog & dir, Ctr Marine Studies & Sea Grant, 78-85, PROF ZOOL, DEPT ZOOL, UNIV MAINE, 85- *Concurrent Pos:* Consult, Nat Fishing Inst, Ecuador, 61-64 & Empresa Puertos de Colombia, 63; coordr, Inter-Am El Nino Proj, 62-64; mem, NSF biol oceanog panel, 70-74; NSF R/V Alpha Helix adv bd, 71-74; sr sci adv, Ctr Marine Res, Yugoslavia, 72-, Plank Comt, Int Comn, Sci Explor Mediter Sea, 74-; mem, Comt Oceanog Res, Australia; consult, Comt Great Bar Reef Mar Park Authority; adv, Fish Inst, Makerere Univ, Uganda, 79-81; Fulbright scholar, Yugoslavia, 83-84 & 90. *Mem:* AAAS; Am Soc Limnol & Oceanog; Am Fisheries Soc; Int Phycol Soc; Phycol Soc Am; Sigma Xi. *Res:* Primary and secondary aquatic production; fisheries; oceanography. *Mailing Add:* Zool Dept 306 Murray Hall Univ Maine Orono ME 04469

GILMARTIN, THOMAS JOSEPH, b Rochester, NY, Aug 5, 40; m 67; c 2. ELECTRO-OPTICAL SYSTEMS. *Educ:* Georgetown Univ, BS, 62; Purdue Univ, MS, 64, PhD(elec eng), 68. *Prof Exp:* Staff, Optics Div, Lincoln Lab, Mass Inst Technol, 67-74; assoc proj leader, Lawrence Livermore Nat Lab, 74-75, dep div leader, 75-79, asst prog leader, 79-81, syst group leader, Laser Prog, 81-86, opers mgr, 86-91, DIR STAFF, INSTNL PLANNING, LAWRENCE LIVERMORE NAT LAB, 91- *Res:* Applications of lasers for laser radar, inertial confinement fusion, and isotope separation; design and integration of systems; performance and cost optimization; program planning. *Mailing Add:* Lawrence Livermore Nat Lab PO Box 808 L1 Livermore CA 94550

GILMER, DAVID SEELEY, b Manila, Philippines, Dec 1, 37; US citizen; m 63; c 2. WILDLIFE POPULATIONS & HABITATS. *Educ:* US Naval Acad, BS, 59; Univ Mich, MWM, 67; Univ Minn, PhD(ecol), 71. *Prof Exp:* Wildlife res biologist, Northern Prairie Wildlife Res Ctr, 72-79, biologist-in-charge, Wildlife Res Field Sta, 79-85, CHIEF SECT PAC STATES ECOL, US FISH & WILDLIFE SERV, DIXON, CALIF, 85- *Mem:* Wildlife Soc; Wilson Ornith Soc; Am Ornithologist Union; Ecol Soc Am. *Res:* Waterfowl ecology; raptor ecology; wetland ecology; remote sensing for wildlife habitat and population assessment; biotelemetry. *Mailing Add:* Northern Prairie Wildlife Res Ctr Res Field Sta US Fish & Wildlife Serv 6924 Tremont Rd Dixon CA 95620-9603

GILMER, GEORGE HUDSON, b Hampden-Sydney, Va, Sept 3, 37. STATISTICAL MECHANICS. *Educ:* Davidson Col, BS, 58; Univ Va, PhD(physics), 62. *Prof Exp:* Res assoc, Cornell Univ, 62-64; from asst prof to assoc prof physics, Washington & Lee Univ, 64-72; MEM TECH STAFF, BELL LABS, 72- *Concurrent Pos:* Vis, Delft Technol Univ, 70-71. *Mem:* Am Phys Soc. *Res:* Crystal growth theory; grain boundary diffusion; critical point phenomena. *Mailing Add:* 1D-447 Bell Labs Murray Hill NJ 07974

GILMER, PENNY JANE, b Hackensack, NJ, Aug 19, 43; m 80; c 2. BIOPHYSICAL CHEMISTRY, IMMUNOCHEMISTRY. *Educ:* Douglass Col, BA, 65; Bryn Mawr Col, MA, 67; Univ Calif, Berkeley, PhD(biochem), 72. *Prof Exp:* Fel biophys, Stanford Univ, 73-75, fel immunochem, 75-77; asst prof chem, 77-84, interim assoc dean arts & sci, 90-91, ASSOC PROF CHEM, FLA STATE UNIV, 84- *Concurrent Pos:* Actg asst prof human biol, Stanford Univ, 76; mem comt fac res, Fla State Univ, 78 & 84. *Mem:* Am Chem Soc; AAAS. *Res:* Fast reaction kinetics; enzyme mechanisms; membrane-mediated phenomena; prostaglandins; cell-cell recognition; biodegradation of hazardous waste. *Mailing Add:* Arts & Sci Off Dean Fla State Univ Tallahassee FL 32306-1022

GILMER, ROBERT, b Pontotoc, Miss, July 3, 38; m 63; c 2. ALGEBRA. *Educ:* Miss State Univ, BS, 58; La State Univ, MS, 60, PhD(math), 61. *Prof Exp:* Res instr math, La State Univ, 61-62; vis lectr, Univ Wis, 62-63; from asst prof to prof math, 63-81, ROBERT O LAWTON DISTINGUISHED PROF, 81- *Concurrent Pos:* Res assoc, Off Naval Res, 62-63; vis prof, Miss State Univ, 62; Alfred P Sloan Found fel, 65-67; prin investr, NSF grants, 65-; assoc ed, Am Math Monthly, 71-73; Fulbright sr scholar res award, Australian-Am Educ Found, 73-74; vis prof, Latrobe Univ, 74, Univ Tex, Austin, 76-77, vis res prof, Univ Conn, 82; mem, ed bd, J Commun in Algebra, 74-85. *Mem:* Am Math Soc; Math Asn Am. *Res:* Commutative ring theory; multiplicative ideal theory; polynomial and power series rings; dimension theory; monoid rings; field theory. *Mailing Add:* Dept Math Fla State Univ Tallahassee FL 32306-3027

GILMER, ROBERT MCCULLOUGH, b Lawrence, Kans, Dec 10, 20; m 55. PHYTOPATHOLOGY. *Educ:* Univ Wis, BS, 47, MS, 48, PhD(phytopath), 50. *Prof Exp:* From asst prof to prof, 50-76, head dept, 69-72, EMER PROF PLANT PATH, NY EXP STA, CORNELL UNIV, 76- *Mem:* Am Phytopath Soc. *Res:* Virology of deciduous fruit trees. *Mailing Add:* 2042 Ted Rd Royal Oaks Brooksville FL 33512

GILMER, THOMAS EDWARD, JR, b Draper, Va, Mar 4, 25; m 51; c 5. SOLID STATE PHYSICS, SEMICONDUCTORS. *Educ:* Hampden-Sydney Col, BS, 48; Univ NC, MS, 53, PhD(physics), 56. *Prof Exp:* Instr math, Ky Mil Inst, 48-50; asst, Duke Univ, 53 & Univ NC, 53-54; sr scientist, Exp, Inc, Va, 55-56, head solid state physics lab, 57-58; assoc prof, 58-62, actg head dept, 59-60, assoc dean, Col Arts & Sci, 69-79, PROF PHYSICS, VA POLYTECH INST & STATE UNIV, 62-, HEAD DEPT, 82- *Mem:* Am Phys Soc. *Res:* Neutron spectra and cross-sections; ionizing energy loss; magnetic resonance; electrical properties of semiconductors; infrared absorption in solids; photovoltaics. *Mailing Add:* Physics Dept Va Polytech Inst & State Univ Blacksburg VA 24061

GILMONT, ERNEST RICH, b Newton, Mass, July 1, 29; m 65. ORGANIC CHEMISTRY. *Educ:* Middlebury Col, AB, 51, MSc, 52; Mass Inst Technol, PhD(org chem), 56. *Prof Exp:* Sr res chemist, FMC Corp, NJ, 56-58, group leader, 58-61; dir res, US Peroxygen Corp, Calif, 61-62; sr scientist, Millmaster Onyx Corp, Newark, NJ, 62-66, dir res & develop, A Gross & Co Div, 66-70 & tech dir, 70-78, gen mgr copygraphics, 78-84; CONSULT, ARTHUR D LITTLE, 84- *Concurrent Pos:* Comnr, Sci Manpower Comn, 74-; lectr, Robert A Welch Found, 75; chmn, Coun Scientific Soc Pres, 64-69; dir, Asn Res Dirs, 84-86. *Mem:* Fel AAAS; Am Chem Soc; fel Am Inst Chem (pres, 73-75); Am Oil Chem Soc; Am Inst Chem Engrs. *Res:* Fatty acids; hydrogenation; organic peroxides; polymerization catalysis; halogenation. *Mailing Add:* 146 Central Park W New York NY 10023

GILMONT, ROGER, b New York, NY, Dec 30, 16; m 41; c 1. CHEMICAL ENGINEERING. *Educ:* Cooper Union, BChE, 39; Polytech Inst Brooklyn, MChE, 43, DrChE, 47. *Prof Exp:* Draftsman, US Coast & Geod Surv, Washington, DC, 40-41; jr chem engr, QM Corps, US War Dept, 41; jr mat engr, Nat Bur Standards, 41-42, jr phys chemist, 42-43; teacher chem eng, City Col New York, 43-44; phys chemist, Gen Foods Corp, NJ, 44-47; tech dir, Emil Greiner Co, 47-55, pres, Manostat Corp, 55-61; ADJ PROF CHEM ENG, POLYTECH INST BROOKLYN, 51-, PRES, GILMONT INSTRUMENTS, INC, 61- *Mem:* AAAS; Am Chem Soc; Am Technion Soc; Instrument Soc Am; Am Inst Chem; Am Inst Chem Engrs; Sigma Xi. *Res:* Thermodynamics; instrument design; food spoilage; correlation of vapor liquid equilibria. *Mailing Add:* 3841 240th St Douglas Manor NY 11363

GILMORE, ALVAN RAY, b Pensacola, Fla, Nov 6, 21; m 43; c 2. FORESTRY. *Educ:* Univ Fla, BSF, 49; Duke Univ, MF, 50, DF, 61. *Prof Exp:* Instr forestry, Univ Fla, 50-52; asst prof, Ala Polytech Inst, 53-58; assoc prof forestry, Univ Ill, Urbana-Champaign, 58-68, prof, 68-; RETIRED. *Res:* Forest tree physiology; forest soils; forest ecology. *Mailing Add:* PO Box 865 Dadeville AL 36853

GILMORE, ARTHUR W, b Louisville, Ky, Oct 5, 20; m 48; c 4. RESOURCE MANAGEMENT, OPERATIONS RESEARCH. *Educ:* Rensselaer Polytech Inst, BAE, 42; Univ Colo, MS, 56. *Prof Exp:* Instr aerodyn eng, Rensselaer Polytech Inst, 42-43; aerodynamicist A, aircraft design, Consol-Vultee Corp, 43-44; asst prod eng autopilot design, Sperry Gyroscope Co, 45-48; section head, dir eng design, Grumman Aerospace Corp, 48-77; asst prof aerodyn eng, Univ Colo, 55-57; dir, Grad Studies Indust Mgt, 78-89, EMER LECTR, STATE UNIV NY, STONY BROOK, 89- *Concurrent Pos:* Mem, res adv comt aerodyn, NASA, 60-63; mem, Nat Comn Eng Manpower, 66-68; consult, Am Asn Eng Socs, 77-84, Long Island Lighting Co, 78-79 & AIL Div, Eaton Corp, 84-85. *Mem:* Assoc Sigma Xi. *Res:* Aerodynamics; engineering economics; production and operations analysis; engineering manpower. *Mailing Add:* PO Box 179 East Setauket NY 11733

GILMORE, EARL C, b Whitney, Tex, Sept 7, 30; m 54; c 3. GENETICS. *Educ:* Tex A&M Univ, BS, 52, MS, 57; Univ Minn, PhD(genetics), 67. *Prof Exp:* Instr agron, Tex A&M Univ, 57-59; res agronomist, USDA, 59-65 & Dekalb Agr Asn, Inc, 65-66; from asst prof to prof plant breeding, 66-79, RES DIR RESIDENT & PROF, TEX AGR EXP STA, TEX A&M UNIV, VERNON, 79- *Concurrent Pos:* Consult hybrid wheat, Rohm & Haas Co, 73-74. *Mem:* Am Soc Agron; Crop Sci Soc Am. *Res:* Development of hybrid wheat, breeding wheat and flax; quantitative genetics of agronomic crops. *Mailing Add:* Tex A&M Agr Res Ctr PO Box 1658 Vernon TX 76384

GILMORE, EARL HOWARD, b Turkey, Tex, July 9, 23; m 46; c 2. MATHEMATICS. *Educ:* Tex Tech Col, BS, 43, MS, 47; Univ Calif, PhD(chem), 51. *Prof Exp:* Phys chemist, US Naval Ord Test Sta, 50-53; from asst prof to assoc prof chem, Okla State Univ, 53-58; from asst prof to assoc prof math, Tex Tech Col, 58-68; RES CHEMIST, HELIUM RES CTR, US BUR MINES, 68- *Concurrent Pos:* Res assoc, Res Found, Okla State Univ, 53-57. *Mem:* Am Chem Soc; Am Phys Soc; Math Asn Am. *Res:* Photoelectric photometry; energy level systems for excited molecule species. *Mailing Add:* Rte 1 Box 430 Amarillo TX 79106-9801

GILMORE, FORREST RICHARD, b Cisco, Tex, Aug 25, 22; div; c 2. NUCLEAR WEAPONS EFFECTS. *Educ:* Calif Inst Technol, BS, 44, PhD(physics), 51. *Prof Exp:* Res analyst, Douglas Aircraft Co, 44-45; instr appl mech & res engr, Calif Inst Technol, 50-53; physicist, Rand Corp, 53-71; SR STAFF SCIENTIST, R&D ASSOCS, 71- *Mem:* Fel Am Phys Soc; Am Optical Soc; Am Geophys Union. *Res:* Atmospheric physics; atomic and molecular physics; radiation; hydrodynamics; nuclear weapons effects. *Mailing Add:* R&D Assocs Box 9695 Marina del Rey CA 90295-2095

GILMORE, JAMES EUGENE, b Tokay, NMex, July 13, 27; m 53; c 2. ENTOMOLOGY, TEPHRITID FRUIT FLY RESEARCH. *Educ:* NMex State Univ, BS, 50; Ore State Univ, MS, 57. *Prof Exp:* County office mgr prod & mkt admin, USDA, 50-54; asst pests of pome & stone fruits, Ore State Univ, 55-57; entomologist, Citrus Insect Invests, Entom Res Div, Calif, 57-64, entomologist & asst to chief, Fruit & Veg Insects Res Br, USDA, Honolulu, 64-69, asst to dir, Entom Res Div, 69-72, asst dir, Calif-Hawaii-Nev Area, Western Region, 72-82, dir, Trop Fruit & Veg Res Lab, Agr Res Serv, 82-90; RETIRED. *Mem:* Entom Soc Am. *Res:* Biology and control of insects and mites attacking citrus and deciduous fruits; insect and mite pathogens for biological control; resistance of mites to chemicals; control, quarantine and eradication of tropical fruit flies. *Mailing Add:* 2841 Holly Ave Clovis CA 93612

GILMORE, JOHN T, b Oakes, NDak, Sept 1, 31; m 59; c 2. NUCLEAR CHEMISTRY. *Educ:* NDak State Univ, BS, 53; Univ Calif, Berkeley, PhD(nuclear chem), 60. *Prof Exp:* From res chemist to sr res chemist, 60-75, RES SCIENTIST, CHEVRON RES & TECHNOL CO, STANDARD OIL CO CALIF, 75- *Mem:* Am Chem Soc. *Res:* X-ray and microprobe analysis; scanning electron microscopy; activation analysis. *Mailing Add:* Chevron Res & Technol Co 100 Chevron Way Richmond CA 94802-0627

GILMORE, JOSEPH PATRICK, b Brooklyn, NY, Sept 30, 28; m 50; c 4. PHYSIOLOGY. *Educ:* St John's Univ, NY, BS, 51, MS, 52; George Washington Univ, PhD(physiol), 62. *Prof Exp:* Asst zool & physiol, St John's Univ, NY, 50-52; physiologist, Naval Med Field Res Lab, Camp Lejeune, 52-55, head dept physiol, 55-58; physiologist, Lab Cardiovasc Physiol, Nat Heart Inst, 58-64, dep chief, 64-66; from assoc prof to prof physiol, Med Sch, Univ Va, 66-70; prof physiol & biophys & chmn dept, Col Med, Univ Nebr, Omaha, 70-88; RETIRED. *Concurrent Pos:* NIH career develop award, 67; assoc, George Washington Univ, 62-66; mem study sect cardio B, NIH, 68-72; mem Cardiovascular A Res Study Comt, Am Heart Asn, 64-76; mem coun, Soc Exp Biol & Med, 78- mem coun high blood pressure res, Med Adv Bd. *Mem:* Am Physiol Soc; Am Heart Asn; Am Soc Nephrology; Am Soc Pharmacol & Exp Therapeut; fel Am Col Cardiol. *Res:* Electrophysiology of cardiac nerve; renal physiology with particular reference to body salt and water control; renin-angiotensin. *Mailing Add:* 1389 Ventnor Ave Tarpon Springs FL 34689

GILMORE, MAURICE EUGENE, b New York, NY, Jan 2, 38; m 64; c 3. MATHEMATICS, ALGEBRAIC TOPOLOGY. *Educ:* Georgetown Univ, AB, 59; Syracuse Univ, BS, 61; Univ Calif, Berkeley, PhD(math), 67. *Prof Exp:* Teaching asst, Syracuse Univ, 59-61; asst, Univ Calif, Berkeley, 61-66; from instr to assoc prof, 66-78, chmn dept, 75-88, PROF MATH, NORTHEASTERN UNIV, 78- *Concurrent Pos:* Prof, State Tech Univ, Chile, 68-69. *Mem:* Am Math Soc; Math Asn Am. *Res:* Algebraic topology; vector fields; homotopy groups; obstruction theory; differential topology; homotopy theory; three-manifolds; knot theory. *Mailing Add:* Dept Math Northeastern Univ Boston MA 02115

GILMORE, PAUL CARL, b Can, Dec 5, 25; nat US; m 54; c 2. COMPUTER SCIENCE. *Educ:* Univ BC, BA, 49; Cambridge Univ, MA, 51; Univ Amsterdam, PhD(math), 53. *Prof Exp:* Univ res assoc math & Nat Res Coun Can res fel, Univ Toronto, 53-55; asst prof, Pa State Univ, 55-58; staff mathematician, Thomas J Watson Res Ctr, IBM Corp, 58-77, mgr combinatorial math, 64-67, asst to chief scientist, 74-75; PROF & DEPT HEAD COMPUTER SCI, UNIV BC, 77-; STAFF MATHEMATICIAN, THOMAS J WATSON RES CTR, INT BUS MACH CORP, 58- *Concurrent Pos:* Vis prof, Univ BC, 71-72; Asn for Symbolic Logic rep, Div Math Sci, Nat Res Coun, 69-75; adj prof math statist, Columbia Univ, 72-77; mem coun Can Math Soc; mem Grant Selection Comt Comput & Info Sci of Natural Sci & Eng Res Coun Can. *Honors & Awards:* Lanchester Prize, Opers Res Soc Am, 64. *Mem:* Asn Comput Mach; Opers Res Soc Am; Can Info Processing Soc; Can Math Soc. *Res:* Programming languages. *Mailing Add:* Dept Computer Sci Univ BC 2075 Wesbrook Mall Vancouver BC V6T 1W5 Can

GILMORE, ROBERT BEATTIE, b 1913; m 40; c 4. FUEL TECHNOLOGY & PETROLEUM ENGINEERING, ENVIRONMENTAL SCIENCES. *Educ:* Univ Tulsa, BS, 39. *Prof Exp:* Engr, Shell Oil Corp, 34-41; vpres & dir, De Golyer & MacNaughton, 49-61, pres, 62-66, chmn bd, 67-71, sr chmn, 72-77, vchmn exec comt, 78-88; RETIRED. *Concurrent Pos:* Petrol Eng lectr, Univ Tex, Dallas, Oil & Gas Financial Mgt, 79-87. *Honors & Awards:* DeGolyer Medal, Soc Petrol Engrs, 75. *Mem:* Soc Petrol Engrs (pres, 55). *Mailing Add:* S 400 One Energy Sq 4925 Greenville Ave Dallas TX 75206-4001

GILMORE, ROBERT SNEE, b Pittsburg, Pa, Apr 20, 38; m 63; c 2. NONDESTRUCTIVE TESTING, QUALITY ASSURANCE. *Educ:* Rensselaer Polytech Inst, BCE, 60, MS, 64, PhD(geophysics), 67. *Prof Exp:* Fel Biophys, Rensselaer Polytech Inst, 66-68; group leader ultrasonics, NASA Electronics Res Ctr, 68-70; mem staff, Bioeng Dept, Forsyth Dent Ctr, 70-71; develop engr, 72-77, RES PHYSICIST, RES & DEVELOP CTR, GEN ELEC CO, 77- *Concurrent Pos:* Res affil, Mass Inst Technol, 69-71; adj prof mat eng, Rensselaer Polytech Inst, 78-80, bioeng, 82- *Honors & Awards:* IR-100 Award for Comput Assisted Ultrasonic Microscope; fel, Am Soc Nondestructive Testing. *Mem:* Fel Am Soc Nondestructive Testing; Sigma Xi. *Res:* Ultrasonic characterization of materials with respect to elastic constants, grain size, composition and suitability for service at high temperatures and stresses; ultrasonic imaging; ultrasonic microscopy; nondestructive testing; ultrasonic spectroscopy and microspectroscopy. *Mailing Add:* Bdlg KWD Rm 251 Gen Elec Co PO Box 43 Schenectady NY 12301

GILMORE, SHIRLEY ANN, b Connellsville, Pa, Jan 1, 35. NEUROBIOLOGY, SPINAL CORD. *Educ:* Thiel Col, BA, 57; Univ Cincinnati, PhD(anat), 61. *Prof Exp:* From instr to assoc prof, 62-75, PROF ANAT, MED CTR, UNIV ARK, LITTLE ROCK, 75-, CHMN DEPT ANAT, UNIV ARK mED SCI, 84- *Concurrent Pos:* USPHS res fel, Uppsala, 61-62; mem panel postdoctoral fels, NSF, 71; consult, Associateship Off, Comn Human Rescources, Nat Acad Sci, 81; mem Life Sci Panel, Nat Res Coun Assoc Prog; Neurol A study sect, NIH, 88- *Honors & Awards:* Jacob Javits Neurosci Investr Award, NIH. *Mem:* AAAS; Am Asn Anat; Am Inst Biol Sci; Am Asn Hist Med; Soc Neurosci. *Res:* Effects of radiation on maturing nervous system; normal maturation and myelin formation in spinal cord; regenerative and reparative capacities of immature spinal cord. *Mailing Add:* Dept Anat 510 Univ Ark Med Sci Little Rock AR 72205-7199

GILMORE, STUART IRBY, b New York, NY, July 24, 30; c 6. SPEECH PATHOLOGY, AUDIOLOGY. *Educ:* State Univ NY, Albany, BA, 50, MA, 51; Univ Wis, PhD(speech), 62. *Prof Exp:* Instr commun, Talladega Col, 51-53; dir speech path & audiol, Jr League Speech & Hearing Clin, Columbia, 54-60; asst prof speech path, Peabody Col, Vanderbilt Univ, & dir, Bill Wilkerson Speech & Hearing Ctr, 60-62; asst prof speech, Univ Wis, 62-65; PROF SPEECH, LA STATE UNIV, 65- *Concurrent Pos:* Consult, 65-, Belle Chasse State Sch, 70-; lectr & supvr, Mayo Clin Sem on Laryngectomy Rehab, 81-85, Int Asn Laryngectomees Voice Inst, 82-85; adj prof, dept otorhinolaryngology, La State Univ Med Sch & Earl K Long Hosp. *Mem:* Sigma Xi; fel Am Speech-Lang & Hearing Asn; Comput Users in Speech; Am Cleft Palate Asn. *Res:* Communicative disorders of the aged; measures of speech and language proficiency; psychosocial concomitants laryngectomy. *Mailing Add:* 429 Bayside Rd Apt B Arcata CA 95521

GILMORE, THOMAS MEYER, b Milheim, Pa, Mar 10, 42; m 72; c 2. ORGANIC CHEMISTRY. *Educ:* Lock Haven Univ, BS, 65; Univ Del, MS, 70; Pa State Univ, PhD(food sci), 76. *Prof Exp:* Instr chem/physics, Univ Del, 67-72; prof chem, Del Tech & Com Col, 75-78; asst prof dairy chem, SDak State Univ, 78-82; sr food technician, Hershey Foods Corp, 82-85; TECH DIR, DAIRY & FOOD INDUST ASN, INC, 85- *Concurrent Pos:* Expert comt mem, Int Dairy Fedn, 85-; consult, USDA/ERS Team USSR-Gosagriprom, 89-90. *Mem:* Am Dairy Sci Asn; Inst Food Technologists; Int Asn Milk Food & Environ Sanitarians; Int Dairy Fedn; Am Soc Testing & Mat; Am Soc Agr Engrs. *Res:* organic chemistry; photochemical reactions in milk; chemical, physical and organoleptic properties of dairy products; new product development; research and writing sanitary design standards for dairy food processing equipment. *Mailing Add:* 6245 Executive Blvd Rockville MD 20852-3938

GILMORE, WILLIAM FRANKLIN, b Bailey, Miss, Mar 26, 35; m 63; c 2. MEDICINAL CHEMISTRY. *Educ:* Va Mil Inst, BS, 57; Mass Inst Technol, PhD(org chem), 61. *Prof Exp:* Chemist antiradiation drugs, Walter Reed Army Inst Res, 62-63; res assoc peptides, Inst Molecular Biophys, Fla State Univ, 63-64; assoc chemist org synthesis, Midwest Res Inst, 64-65, sr chemist, 65-67; from asst prof to assoc prof pharmaceut chem, 67-71, chmn dept, 69-71, PROF MED CHEM, SCH PHARM, UNIV MISS, 71-, CHMN DEPT, 71-80 & 88- *Concurrent Pos:* Mem pharmacol & toxicol training comt, Nat Inst Gen Med Sci, 67-70. *Mem:* AAAS; Am Chem Soc; Am Asn Cols Pharm; Sigma Xi. *Res:* Organic synthesis; mechanism of Favorskii rearrangement; synthesis of radioprotective compounds; synthesis of polypeptides; peptido minetics. *Mailing Add:* Sch Pharm Univ Miss University MS 38677

GILMORE, WILLIAM STEVEN, astronomy, physics, for more information see previous edition

GILMOUR, CAMPBELL MORRISON, b Scotland, July 2, 16; nat US; m 43; c 3. BACTERIOLOGY. *Educ:* Univ BC, BSA, 41, MSA, 45; Univ Wis, PhD(bact), 49. *Prof Exp:* Assoc prof bact, Okla Agr & Mech Col, 49-51; from assoc prof to prof, Ore State Univ, 51-67; dir ctr environ biol, Univ Utah, 67-70; HEAD DEPT BACT, UNIV IDAHO, 70- *Concurrent Pos:* Study-lectureship, Rockefeller Found; consult water pollution. *Mem:* Am Soc Microbiol. *Res:* Classification and metabolism of soil microorganisms. *Mailing Add:* 2335 15th St Lewiston ID 83501

GILMOUR, ERNEST HENRY, b Adin, Calif, Aug 17, 36; m 82; c 4. PALEONTOLOGY. *Educ:* Univ Southern Calif, BS, 60; Univ Mont, MS, 64, PhD(geol), 67. *Prof Exp:* Eng geologist, Calif Dept Water Resources, 56-61; geologist, Cominco Am, Inc, 63; instr, Univ Mont, 64; geologist, US Geol Surv, 64-65; asst prof mineral fuels & geologist, Mont Bur Mines & Geol, 65-67; PROF GEOL, EASTERN WASH UNiV, 67-, VPROVOST, GRAD STUDIES & RES, 87- *Concurrent Pos:* Teaching asst, Univ Southern Calif, 60-61; mem staff, Goudkoff & Hughes, Calif, 67; consult. *Honors & Awards:* Trustees Medal, Nat Acad Sci; Fulbright lectr, Pakistan. *Mem:* Geol Soc Am; Am Asn Petrol Geol; AAAS; Paleont Soc; Soc Econ Paleont Mineral; Sigma Xi. *Res:* Carbonate petrology, paleontology and paleoecology; Paleozoic Bryozoa, Coelenterata, Protozoa and Brachiopoda; stratigraphic problems dealing with stratabound ore deposits; taxonomic and biostratigraphic studies of Permian bryozoa in western US, Pakistan, USSR, Poland, and Canada. *Mailing Add:* Dept Geol Eastern Wash Univ Cheney WA 99004

GILMOUR, HUGH STEWART ALLEN, b Alta, Can, Apr 25, 26; m 51; c 2. PHYSICAL CHEMISTRY, PHOTOGRAPHY. *Educ:* Univ BC, BA, 49; Univ Utah, PhD(chem), 53. *Prof Exp:* Fel chem, Univ Ill, 53-55; res chemist, 55-68, res assoc, Eastman Kodak Co, 68-89; CONSULT, 90- *Mem:* Soc Photog Scientists & Engrs. *Res:* Photosynthesis; photochemistry; color photographic systems; color thermographic systems and media. *Mailing Add:* 183 Oakdale Dr Rochester NY 14618-1148

GILMOUR, MARION NYHOLM H, b Vancouver, BC, Jan 29, 28; US citizen; m 51; c 2. MICROBIOLOGY, DENTAL RESEARCH. *Educ:* Univ BC, BA, 49; Univ Ill, Urbana, MSc, 52, PhD(microbiol), 57. *Prof Exp:* Jr res officer microbial genetics, Atomic Energy Can, Ltd, 49-51; head dept microbiol & res assoc, Eastman Dent Ctr, 57-77; clin instr, Univ Rochester, 60-69, from asst prof to assoc prof dent, 74-88; RETIRED. *Concurrent Pos:* Consult, Nat Inst Dent Res, 71-, mem study sect, 73-75; consult microbiol core curriculum comt, Am Asn Dent Schs. *Mem:* AAAS; Am Inst Biol Sci; Am Soc Microbiol; Int Asn Dent Res; NY Acad Sci. *Res:* Oral microorganism, their physiological amd metabolic capabilities and factors affecting them both in vitro and in vivo. *Mailing Add:* 183 Oakdale Dr Rochester NY 14618

GILMOUR, THOMAS HENRY JOHNSTONE, b Dunoon, Scotland, Sept 27, 36; m 62; c 2. INVERTEBRATE ZOOLOGY. *Educ:* Univ Glasgow, BSc, 60, PhD(zool), 63. *Prof Exp:* Asst lectr zool, Univ Exeter, 63-64; from asst prof to assoc prof, 64-80, PROF INVERT ZOOL, UNIV SASK, 80- *Mem:* Can Soc Zool; Am Soc Zool. *Res:* Functional morphology of invertebrates. *Mailing Add:* Dept Biol Univ Sask Saskatoon SK S7N 0W0 Can

GILMOUR - STALLSWORTH, LISA K, b Rochester, NY, Jan 9, 59; m 83. CARDIOVASCULAR MODELLING, CHEMICAL ANALYSIS-HAZARDOUS WASTE. *Educ:* Cornell Univ, BS, 80, AB, 81; Univ Tex, Austin, MS, 83. *Prof Exp:* Engr, Sun Oil Co, 79 & Univ Tex, 81-90. *Res:* Pharmacologic manipulation of peritoneal mass transfer; engineering analysis of cardiovascular function. *Mailing Add:* 7630 Woodhollow No 157 Austin TX 78731

GILOW, HELMUTH MARTIN, b Cedarburg, Wis, Sept 11, 33; m 67; c 1. ORGANIC CHEMISTRY. *Educ:* Wartburg Col, BA, 55; Univ Iowa, MS, 57, PhD, 59. *Prof Exp:* Assoc prof, 59-69, chmn dept, 76-78, prof chem, Southwestern Memphis, 69-; MEM STAFF CHEM DEPT, RHODES COL, MEMPHIS. *Mem:* Am Chem Soc; Sigma Xi. *Res:* Pyrimidine and purine type derivatives; reactions and properties of beta diketones; substituent effects of positive poles in aromatic substitution; pyrrole chemistry; photochemistry. *Mailing Add:* Dept Chem Southwestern Memphis Memphis TN 38112

GILPIN, MICHAEL JAMES, b Washington, Iowa, Mar 4, 41; m 66; c 1. PURE MATHEMATICS. *Educ:* Parsons Col, BS, 63; Univ Ariz, MA, 65; Univ Ore, PhD(math), 72. *Prof Exp:* Instr math, Gonzaga Univ, 65-68; asst prof, 72-80, ASSOC PROF MATH, MICH TECHNOL UNIV, 80- *Mem:* Am Math Soc; Math Asn Am. *Res:* Symmetric spaces and symmetric forms of degree greater than three over fields. *Mailing Add:* Dept Math Mich Technol Univ Houghton MI 49931

GILPIN, ROGER KEITH, b Middlesboro, Ky, Apr 29, 47; m 67; c 4. CHROMATOGRAPHY, SPECTROSCOPY. *Educ:* Ind State Univ, BS, 69; Univ Ariz, PhD(anal chem), 73. *Prof Exp:* Sr scientist anal chem, McNeil Labs Inc, 73-76, group leader, 76-78; From to asst prof, to assoc prof 78-86, PROF ANALYTICAL CHEM, DEPT CHMN, KENT STATE UNIV, 86- *Concurrent Pos:* Sr tech adv, IBM Instruments, 81-83. *Mem:* Am Chem Soc; Sigma Xi; Soc Appl Spectros; AAAS. *Res:* Fundamental and applied thin layer, gas, and high pressure liquid chromatography; organometallic surface reactions; infrared radiation, electronic spin resonance and nuclear magnetic resonance of chemically modified surfaces; pharmaceutical analysis. *Mailing Add:* Dept Chem Kent State Univ Kent OH 44242

GILREATH, JAMES PRESTON, b Greenville, SC, Sept 20, 47; m 75. WEED CONTROL, HERBICIDE DEGRADATION. *Educ:* Clemson Univ, BS, 74, MS, 76; Univ Fla, PhD(hort), 81. *Prof Exp:* Asst prof, 81-86, ASSOC PROF WEED SCI, UNIV FLA, 86- *Mem:* Weed Sci Soc Am; Am Soc Hort Sci. *Res:* Development of chemical weed control programs for floricultural and vegetable crops and investigations into herbicide movement and in soil; herbicide degradation; weed biology. *Mailing Add:* Inst Food & Agr Sci Univ Fla Gulf Coast Res & Educ Ctr 3604 38th Ave S Bradenton FL 34203

GILROY, JAMES JOSEPH, b Scranton, Pa, June 5, 26; m 58; c 4. BACTERIOLOGY. *Educ:* Univ Scranton, BS, 49; Cath Univ Am, MS, 51; Univ Md, PhD(bact), 58. *Prof Exp:* Clin bacteriologist, Children's Hosp, Washington, DC, 52-53; bacteriologist, Army Chem Ctr, Ft Detrick, Md, 53-54; asst bact, Univ Md, 54-57; from asst prof to assoc prof, Colo State Univ, 57-64; from asst chmn to chmn dept, 67-70, ASSOC PROF BACT, BOSTON COL, 64- *Mem:* Am Soc Microbiol. *Res:* Bacterial metabolism; nutritional requirements for growth of rumen bacteria; nutritional requirements for toxin production by Clostridium perfringens; induction of fermentation enzymes in bacteria; bacterial growth inhibition by d-camphor. *Mailing Add:* Dept Biol Boston Col 140 Commonwealth Ave Chestnut Hill MA 02167

GILROY, JOHN, b Eng, March 29, 25; US citizen; m 52, 75; c 3. NEUROAUDIOLOGY. *Educ:* Univ Durham, MB & BS, 48, MD, 57, FRCP(C), 65; FACP, 71. *Prof Exp:* Instr neurol, 63-65, asst prof, 65-66, assoc prof, 66-67, PROF NEUROL, WAYNE STATE UNIV, 67, CHMN, 68- *Concurrent Pos:* Chief neurol, Harper-Crace Hosp, Detroit, 68- & Detroit Receiving Hosp, 68- *Mem:* Am Neurol Asn; Am Acad Neurol; Am Soc Neuroimaging. *Res:* Cerebrovascular disease and neuroaudiology; author and co-author of 80 publications. *Mailing Add:* 27207 Lahser No 102 South Field MI 48034

GILRUTH, ROBERT R(OWE), b Nashwauk, Minn, Oct 8, 13; m 37; c 1. AERONAUTICAL ENGINEERING, AEROSPACE ENGINEERING. *Educ:* Univ Minn, BS, 35, MS, 36. *Hon Degrees:* DSc, Univ Minn, 62, George Washington Univ, 62, Ind Inst Technol, 62; DEng, Mich Technol Univ, 63; LLD, NMex State Univ, 70. *Prof Exp:* Mem, Nat Adv Comt Aeronaut, NASA, 37-43, asst chief, Flight Res Div, 43-45, chief, Pilotless Aircraft Res Div, 45-51, asst dir, Langley Res Ctr, 51-58, dir, Proj Mercury, Space Task Group, NASA, 58-61, dir, Manned Spacecraft Ctr, 61-73, consult to admir, Johnson Space Ctr, 74-76; RETIRED. *Concurrent Pos:* Consult, Res & Develop Bd, 46-58; mem, Sci Adv Bd, US Air Force, 51-57 & Ballistic Missile Defense Comt, 55; mem bd dir, Bunker-Ramo Corp, 74-78; consult aerospace, 78- *Honors & Awards:* Reed Award, Am Inst Aeronaut & Astronaut, 50; Louis Hill Space Transportation Award, 62; Goddard Mem Trophy, 62; Spirit of St Louis Medal, Am Soc Mech Engrs, 65; Guggenheim Int Astronaut Award, Inst Aeronaut Sci, 66; Space Flight Award, Am Astronaut Soc, 67; James Watts Int Medal, 71; Robert J Collier Trophy, 72; Nat Aviation Club Award for Achievement, 71. *Mem:* Nat Acad Sci; Nat Acad Eng; hon fel Am Inst Aeronaut & Astronaut; hon fel Royal Aeronaut Soc. *Res:* Airplane stability and structures; hydrofoil craft; aerodynamics; rocket propulsion; manned space flight; high temperature facilites. *Mailing Add:* Rte 1 Box 1486 Kilmarnock VA 22482

GILSTEIN, JACOB BURRILL, b NY, Feb 5, 23; wid; c 5. PHYSICS, RESEARCH ADMINISTRATION. *Educ:* City Col New York, BS, 43; NY Univ, MS, 50, PhD, 58. *Prof Exp:* Physicist pilotless aircraft instrumentation, Nat Adv Comt Aeronaut, 46-47; res assoc combustion, rocket propulsion, detonation, Res Div, NY Univ, 47-56; physicist, Missiles & Space Vehicles Dept, Gen Elec Corp, Pa, 56-59, physicist systs appln, Defense Systs Dept, 59-61, systs engr, Missile & Space Div, 61-66, mgr aerospace physics lab, Reentry Systs Dept, 66-68, mgr aerospace physics & systs analysis, 68; Dept Ballistic Missile Defense, Off Asst Secy Army & dir, Army Advan Ballistic Missile Defense Agency, 68-74; mgr advan syst eng, 74-79, prog gen mgr advan systs, Gen Elec Reentry Systs Div, 79-84, mgr Penetration & Defense Systs, 84-87, CHIEF SCIENTIST, STRATEGIC SYST DEPT, 87- *Mem:* Am Phys Soc; assoc fel Am Inst Aeronaut & Astronaut; Sigma Xi; Am Defense Preparedness Asn; Old Crows. *Res:* Instrumentation; combustion studies of jets and rockets; boundary layer theory; detonation in solid explosives; hypersonic aerodynamics; plasma physics; vulnerability and hardening; penetration aids; strategic defense systems; management of high energy laser developments; electronic counter measures. *Mailing Add:* 666 W Germantown Pike Apt 611N Plymouth Meeting PA 19462

GILSTRAP, FRANKLIN EPHRIAM, b Clovis, Calif, Apr 24, 44; m 64; c 1. ENTOMOLOGY. *Educ:* Fresno State Col, BA, 68; Univ Calif, Riverside, MS, 71, PhD(entom), 74. *Prof Exp:* Asst prof, 74-80, ASSOC PROF ENTOM, TEX A&M UNIV, 80- *Mem:* Entom Soc Am; Can Entom Soc; Int Orgn Biol Control; Acarological Soc Am. *Res:* Biological control of arthropod pests of grain sorghum, wheat, corn and citrus; to obtain, release, establish and evaluate exotic natural enemies developing bio-ecological information for both pests and beneficials. *Mailing Add:* Dept Entom Tex A&M Univ College Station TX 77843

GILTINAN, DAVID ANTHONY, b Jamestown, NY, Dec 7, 36; m 67; c 2. PHYSICS. *Educ:* Case Inst Technol, BS, 59, MS, 63, PhD(physics), 68. *Prof Exp:* Assoc prof, 68-70, head dept, 69-72, PROF PHYSICS, EDINBORO UNIV PA, 70- *Mem:* Am Asn Physics Teachers. *Res:* Theoretical and nuclear physics; elementary particle reactions; phenomenological computer analysis. *Mailing Add:* Dept Physics Edinboro Univ Pa Edinboro PA 16444

GILULA, LOUIS ARNOLD, b Lubbock, Tex, Oct 21, 42; m 70; c 2. RADIOLOGY, MAMMOGRAPHY. *Educ:* Univ Ill, Chicago, MD, 67; Am Bd Radiol, dipl, 73. *Prof Exp:* From instr to assoc prof, 73-82, PROF RADIOL, MALLINCKRODT INST RADIO, WASH UNIV SCH MED, 82- *Concurrent Pos:* Co-dir musculoskeletal sect, Mallinckrodt Inst Radiol, Wash Univ Sch Med, 75-, dir vis fel prog. *Mem:* Fel Am Col Radiol; Int Skeletal Soc; AMA; Radiol Soc NAm; Asn Univ Radiol; Am Roeutgen Ray Soc; hon mem Am Soc Surg Hand. *Res:* Musculoskeletal radiology; three dimensional imaging of the musculoskeletal system; mammography; author of more than 250 technical articles & publications; wrist radiology and anatomy. *Mailing Add:* 250 Dielman Rd St Louis MO 63124

GILULA, NORTON BERNARD, b Dec 9, 44. MOLECULAR & CELL BIOLOGY. *Educ:* Southern Ill Univ, BA, 66, MA, 68; Univ Calif, Berkeley, PhD(physiol), 71. *Prof Exp:* Res fel anat, Harvard Med Sch, 72; postdoctoral fel cell biol, Rockefeller Univ, 72-73, from asst prof to assoc prof, 73-81; prof cell biol, Baylor Col Med, 81-86; MEM MOLECULAR BIOL, SCRIPPS RES INST, 86-, DEAN GRAD STUDIES, 88- & CHAIR, DEPT CELL BIOL, 91- *Concurrent Pos:* Assoc ed, Develop Biol, 76-78, J Neurocytol, 77-78; chmn, Biol Sci Sect, NY Acad Sci, 77 & Molecular Biol Study Sect, NIH, 80-81; mem, Sci Adv Bd, Wills Found, 82-, Nat Adv Coun, Nat Inst Gen Med Sci, NIH, 84-87 & Panel Cell Biol, Nat Res Coun, 85-86; ed, Current Opinion Cell Biol, 89- *Honors & Awards:* Burton Award, Electron Micros Soc Am, 79. *Mem:* Fel AAAS. *Mailing Add:* Dept Cell Biol MB6 Scripps Res Inst 10666 N Torrey Pines Rd La Jolla CA 92037

GILVARG, CHARLES, b New York, NY, June 13, 25; m 49; c 4. BIOCHEMISTRY. *Educ:* Cooper Union, BChE, 48; Univ Chicago, PhD(biochem), 51. *Prof Exp:* Nat Found Infantile Paralysis fel, 52-54; from instr to prof biochem, Sch Med, NY Univ, 54-64; prof chem, Frick Chem Lab, Princeton Univ, 64-70, chmn dept biochem sci, 70-73, prof, 70-89, PROF MOLECULAR BIOL, PRINCETON UNIV, 90- *Concurrent Pos:* Cancer res scholar, 55-58; USPHS sr res fel, 58-61, career res develop award, 61-64, consult, 62-; Guggenheim Fel, 64-66; LaRoche Fel, 73-74; ed, J Biol Chem, 65-70, 72-78 & Biochem et Biophysica Acta, 78-81. *Honors & Awards:* Paul-Lewis Award, Am Chem Soc, 62. *Mem:* AAAS; Am Soc Biol Chem; Am Chem Soc; Harvey Soc. *Res:* Intermediary metabolism, particularly in microorganisms. *Mailing Add:* Dept Molecular Biol Princeton Univ Princeton NJ 08544

GILVARRY, JOHN JAMES, b Manchester, Eng, July 15, 17; nat US. SOLID STATE SCIENCE, PLANETARY SCIENCES. *Educ:* City Col New York, BS, 40; Princeton Univ, MA & PhD(physics), 43. *Prof Exp:* Res physicist, Off Sci Res & Develop Proj, Princeton Univ, 42-43, instr physics, 43; res physicist, Naval Ord Lab, Washington, DC, 43-46, NAm Aviation, Inc, Calif, 46-48, Douglas Aircraft Co, 48-49 & Rand Corp, 49-56; res physicist res labs, Allis-Chalmers Mfg Co, Wis, 56-61; consult res physicist, Gen Dynamics/Convair, 61 & Gen Dynamics/Astronaut, 61-64; Nat Acad Sci-Nat Res Coun sr resident res assoc, Space Sci Div, Theoret Studies Br, Ames Res Ctr, NASA, 64-67; resident consult, Phys Sci Dept, Rand Corp, 68-74; RES PHYSICIST, PHYSICS DIV, TEX INST, INC, 74- *Concurrent Pos:* Consult, Inst Plasma Physics, Stanford Univ, Inst Defense Anal, Washington, DC, Res Div, Plas-Tech Equip Corp, Mass, Benson Lehner Corp, Calif & Jet Propulsion Lab, Calif Inst Technol, 73-; prin investr, Planetol Prog, NASA, 73- *Mem:* Fel AAAS; fel Am Phys Soc; Am Math Soc; Am Astron Soc; Math Asn Am. *Res:* Nuclear and atomic physics; solid state physics; electronics; astronomy; astrophysics; geophysics; gravitational physics. *Mailing Add:* Physics Div Tex Instruments Inc 1240 Harvard St Suite 5 Santa Monica CA 90404

GIMARC, BENJAMIN M, b Nogales, Ariz, Dec 5, 34; m 59. THEORETICAL CHEMISTRY. *Educ:* Rice Univ, BA, 56; Northwestern Univ, PhD(phys chem), 63. *Prof Exp:* Instr chem, Northwestern Univ, 61-62; res assoc, Johns Hopkins Univ, 62-63, USPHS fel & lectr chem, 63-64; asst prof, Ga Inst Technol, 64-66; from asst prof to assoc prof, 66-78, head dept, 73-76 & 82-85, PROF CHEM, UNIV SC, 78- *Concurrent Pos:* Vis scholar, Princeton Univ, 76-77, Univ Calif, Berkeley, 85; Nat Acad Sci exchange scientist, Zagreb, Yugoslavia, 78, 82, 84. *Mem:* Am Chem Soc; Am Phys Soc. *Res:* Development of qualitative molecular orbital theory and applications to problems of molecular shapes; relative molecular stabilities; barriers to inversion and rotation; reaction mechanisms; chemical applications of graph theory. *Mailing Add:* Dept Chem Univ SC Columbia SC 29208

GIMBLE, JEFFREY M, b Oct 14, 55. IMMUNOBIOLOGY. *Educ:* Dartmouth Col, BA, 76; Yale Univ, MA, 80, PhD(cell biol), 81, MD, 82. *Prof Exp:* Intern, Dept Internal Med, Barnes Hosp, Wash Univ, St Louis, Mo, 82-83, jr resident, 83-84; med staff fel res, Lab Immunogenetics, Nat Inst Allergy & Infectious Dis, NIH, 84-87; ASST MEM, OKLA MED RES FOUND, OKLAHOMA CITY, 87- *Concurrent Pos:* Adj asst prof, Dept Microbiol & Immunol, Health Sci Ctr, Univ Okla, Oklahoma City, 88- *Res:* Author of numerous publications. *Mailing Add:* Dept Immunobiol & Cancer Oklahoma Med Res Found 825 NE 13th St Oklahoma City OK 73104

GIMBRONE, MICHAEL ANTHONY, JR, b Buffalo, NY, Nov 16, 43; m 71; c 3. EXPERIMENTAL PATHOLOGY, CELL BIOLOGY. *Educ:* Cornell Univ, AB, 65; Harvard Med Sch, MD, 70. *Prof Exp:* Intern, resident & res fel, Mass Gen Hosp, 70-72; staff assoc pathophysiol, Nat Cancer Inst, 72-74; res assoc, 74-76, from asst prof to assoc prof, 79-85, PROF PATH, HARVARD MED SCH, 85- *Concurrent Pos:* Mem vis fac, W Alton Jones Cell Sci Ctr, 76-78; consult, Nat Heart, Lung & Blood Inst, NIH, 76-; estab investr, Am Heart Asn, 77-82; head, Vascular Pathophysiol Res Lab, 77-85, dir, vascular res div, Brigham & Women's Hosp, 85- *Honors & Awards:* Exp Path Award, Warner-Lambert Parke-Davis, 82. *Mem:* Am Heart Asn; Am Soc Cell Biol; AAAS; Tissue Cult Asn; Am Soc Hematol; Am Asn Pathologists. *Res:* Cardiovascular pathophysiology, especially atherosclerosis, thrombosis and hypertension; vascular cell biology. *Mailing Add:* Dept Path Brigham & Women's Hosp 75 Francis St Boston MA 02115

GIMELLI, SALVATORE PAUL, b New York, NY, Feb 22, 19; m 44; c 1. PHARMACEUTICAL CHEMISTRY. *Educ:* St John's Univ, NY, BS, 41; Fordham Univ, MS, 47; Rutgers Univ, PhD, 66. *Prof Exp:* Chemist, Philco Corp, 42-45; asst chem, Fordham Univ, 45-47; instr, Lafayette Col, 47-50; res chemist pharmaceut, Reed & Carnrick, 50-54; from asst prof to assoc prof, 54-70, PROF CHEM, FAIRLEIGH DICKINSON UNIV, 70-, CHMN DEPT, 77- *Mem:* Am Chem Soc. *Res:* Synthetic medicinal products. *Mailing Add:* 42 Terrace Ave Rocehelle Park NJ 07662

GIMLETT, JAMES I, b Salt Lake City, Utah, Dec 10, 29; m 54; c 5. OPTICS, GEOPHYSICS. *Educ:* Stanford Univ, BS, 50, MS, 52 & 61, PhD, 65. *Prof Exp:* From geophysicist to chief geophysicist, Hycon Aerial Surv, Inc, 55-58; asst prof geophys, Univ Nev, 58-63; sr res scientist, Actron, 63-66, mgr optics eng, 66-72, chief scientist, 72-79; CHIEF TECHNOL ENGR, MCDONNELL DOUGLAS ASTRONAUT CO, 79- *Concurrent Pos:* Air Force Off Sci Res grant geophys, 62-63. *Mem:* AAAS; Am Soc Photogram; Soc Explor Geophys; Europ Asn Explor Geophys. *Res:* Information science; photogrammetry. *Mailing Add:* 1324 Sierra Madre Villa Pasadena CA 91107

GIMPLE, GLENN EDWARD, b Lovewell, Kans, Sept 15, 40; m 62; c 2. INORGANIC CHEMISTRY. *Educ:* Kans State Teachers Col, BA, 62; Univ Kans, PhD(chem), 69. *Prof Exp:* From asst prof to assoc prof chem, Emporia State Univ, 65-86; INSTRNL COORDR, WOLF CREEK NUCLEAR OPER CORP, 86- *Mem:* Am Chem Soc. *Res:* Coordination compounds; use of substituted 2, 2-bipyrimidines as ligands in complexes of transition metals; kinetics and mechanisms of inorganic reactions; electron transfer reactions. *Mailing Add:* RR 2 No 120C Emporia KS 66801

GIN, JERRY BEN, b Tucson, Ariz, June 29, 43; m 65; c 2. PATHOLOGY, PHARMACOLOGY. *Educ:* Univ Ariz, BS, 64; Univ Calif, Berkeley, PhD(biochem), 68; Loyola Col, MBA, 78. *Prof Exp:* Res scientist, Div Arthritis & Metab Dis, NIH, 68-70; asst dir clin chem, Van Nuys Br, 70-73 & New York Br, 73-75; dir clin chem, Biosci Labs, Baltimore-Washington Br, 75-78; oper mgr, Corp Prod Dept, 78-81, DIR INT OPERS, DIAG & INSTRUMENTATION, DOW CHEM CO, 81- *Mem:* Fel Am Asn Clin Chemists; Am Chem Soc. *Res:* All areas of clinical chemistry with special emphasis on radioimmunoassays and clinical enzymology; management. *Mailing Add:* 2023 N Talbet Indianapolis IN 46202

GIN, W(INSTON), b San Francisco, Calif, Aug 24, 28; m; c 1. ENGINEERING PHYSICS. *Educ:* Univ Ariz, BA, 50, BS, 51; Univ Calif, Los Angeles, MA, 55, MS, 59. *Prof Exp:* Mem tech staff, Hughes Aircraft Co, 53-57; res engr, 57-59, res group supvr, 59-61, asst sect chief, Solid Propellant Eng Sect, 61-63, sect chief, 63-75, sect mgr, Solid Propulsion & Environ Systs Sect, 75-78, sect mgr, propulsion systs sect, Jet Propulsion Lab, 78-82, DIV STAFF ENGR, JET PROPULSION LAB, CALTECH, CALIF INST TECHNOL, 78- *Honors & Awards:* Solid Propellant Rocketry Team Award, Nat Aeronaut & Space Admin, 76. *Mem:* Am Inst Aeronaut & Astronaut; Sigma Xi. *Res:* Rocket propulsion; gas dynamics; pyrotechnics; chemical processes. *Mailing Add:* 1418 Star Ridge Dr Monterey Park CA 91754

GINDLER, JAMES EDWARD, b Highland, Ill, Jan 4, 25; m 49; c 4. NUCLEAR CHEMISTRY. *Educ:* Eastern Ill State Col, BS, 50; Univ Ill, MS, 51, PhD(chem), 54. *Prof Exp:* Assoc chemist, Argonne Nat Lab, 53-72, chemist, 72-90, asst div dir, 83-90; RETIRED. *Concurrent Pos:* Vis chemist, Nuclear Res Ctr, Karlsruhe, 64-66. *Honors & Awards:* Co-recipient, Genia Czerniak Prize, Ahavat Zion Found, 74. *Mem:* Sigma Xi; Am Chem Soc; AAAS. *Res:* Heavy element chemistry and nuclear properties; fission product chemistry and spectroscopy; fission process. *Mailing Add:* 625 62nd St Downers Grove IL 60516

GINDSBERG, JOSEPH, b Leipzig, Ger, Sept 3, 20; US citizen; m 49; c 3. ELECTROMAGNETIC PROPAGATION, NOISE MODULATION. *Educ:* City Col New York, BS, 44; Northeastern Univ, MS, 61. *Prof Exp:* Electronics engr, Loral Electronics Corp, NY, 50-51, proj engr, 51-55; sr engr, Missile Systs Div, Raytheon Electronics, Inc, 61-63, mgr microwave tech lab, 63-65; prin engr, Missile Systs Div, Raytheon Co, Mass, 65-70; sr staff engr, Eljim Div, KMS Industs, Inc, 70-75, mgr qual assurance, Eljim Div, Elbit Ltd, 75-77; RES ENGR, MBT DIV, ISRAEL AIRCRAFT INDUSTS, 77- *Mem:* Sr mem Inst Elec & Electronics Engr. *Res:* Generation and processing of microwave signals, particularly for airborne radar; microwave antennas; parametric amplification; modulation theory; noise theory and measurement; radiation hazards. *Mailing Add:* 20/5 Hen St Petah Tikua 49 520 Israel

GINELL, ROBERT, b New York, NY, Apr 27, 12; m 44; c 2. CHEMICAL PHYSICS. *Educ:* Polytech Inst Brooklyn, BS, 36, MS, 40, PhD(chem), 43. *Prof Exp:* Asst chem, Polytech Inst Brooklyn, 37-43; res assoc, Nat Defense Res Comt Proj, Univ NC, 43-45; asst prof, NJ State Teachers Col, Jersey City, 45; from asst prof to prof, 46-78, EMER PROF CHEM, BROOKLYN COL, 78- *Concurrent Pos:* Vis prof, Univ Utah, 60-61; exec officer, PhD Chem Prog, City Univ New York Syst, 62-65; res adv, Food & Drug Admin, Dept HEW, New York Dist, 65-69, Environ Protection Agency, Edison, NJ, 84. *Honors & Awards:* Morrison Prize, NY Acad Sci, 53. *Mem:* Fel AAAS; Am Chem Soc; Am Phys Soc; fel Am Inst Chem; fel NY Acad Sci. *Res:* Theoretical chemistry; association theory; sorption; virial coefficients; theory of liquids and solids; thermonuclear reactions; basic concepts of thermodynamics. *Mailing Add:* Dept Chem Brooklyn Col Brooklyn NY 11210

GINELL, WILLIAM SEAMAN, b New York, NY, Aug 24, 23; m 46; c 3. ART CONSERVATION SCIENCE, MATERIAL SCIENCE. *Educ:* Polytech Inst Brooklyn, BS, 43; Univ Wis, PhD(chem), 49. *Prof Exp:* Asst chem, Substitute Alloy Material Labs, Columbia Univ, 43-44 & Univ Wis, 46-49; from assoc chemist to chemist, Brookhaven Nat Lab, 49-58; sr res specialist, Atomics Int Div, NAm Aviation, Inc, 58-61; head inorg chem sect, Aerospace Corp, 61-63; prin scientist, Douglas Aircraft Co, Calif, 63-67; sect chief, McDonnell Douglas Astronaut Co, Huntington Beach, 67-78, prin staff scientist, 78-84; HEAD MAT SCI, J PAUL GETTY CONSERV INST, 84- *Concurrent Pos:* Consult, J Paul Getty Mus. *Mem:* Fel AAAS; Am Chem Soc; Am Inst Conserv; Int Inst Conserv; Am Phys Soc. *Res:* Photochemistry; thermodynamics of high temperature reactions; nuclear reactor chemistry; crystal growth; liquid metal chemistry; nuclear and laser radiation damage; semiconductors; infrared materials. *Mailing Add:* Getty Conserv Inst 4503B Glencoe Ave Marina Del Rey CA 90292-6372

GINÉ-MASDÉU, EVARIST, b Catalonia, Spain, July 31, 44; m 66; c 2. LIMIT THEOREMS IN PROBABILITY THEORY, INFINITE DIMENSIONAL SPACES. *Educ:* Univ Barcelona, dipl, 66, DSc, 76; Mass Inst Technol, PhD(math), 73. *Prof Exp:* Asst prof math, Univ Carabobo, Venezuela, 66-70; lectr math & statist, Univ Calif, Berkeley, 74-75; res assoc, Venezuelan Inst Sci Res, 73-80, head dept, 75-79; assoc prof, Autonomous Univ Barcelona, Spain, 77-80; from assoc prof to prof math, Tex A&M Univ, 83-90; PROF MATH, UNIV CONN, 90- *Concurrent Pos:* Vis prof, Louis Pasteur Univ, France, 78; prin investr, NSF, 80- *Honors & Awards:* Venezuelan & Catalan Nat Prizes in Math, 77. *Mem:* Fel Inst Math Statist; Am Math Soc; Bernoulli Soc; Catalan Sci Soc; Int Statist Inst. *Res:* Statistics of directions; central limit theorem in infinite dimensional spaces; limit theorems for empirical processes; stochastic process. *Mailing Add:* Dept Math Univ Conn Storrs CT 06269

GINER, JOSE DOMINGO, b Teruel, Spain, Aug 4, 28; US citizen; c 5. ELECTROCHEMISTRY, PHYSICAL CHEMISTRY. *Educ:* Univ Valencia, BS & MS, 51; Univ Madrid, Dr(chem), 55. *Prof Exp:* Staff scientist electrochem res, CSIC, Madrid, 55-56; res fel, Univ Erlangen, 56-58 & Univ Bonn, 58-61; proj engr fuel cell res, Pratt & Whitney Aircraft, 61-65; asst dir electrochem res, Tyco Labs, 65-73; PRES, GINER INC, 73- *Mem:* Am Chem Soc; Am Inst Chem Engrs; Int Soc Electrochem; Electrochem Soc; Am Soc Testing & Mat. *Res:* Fuel cells; electrocatalysis; batteries; electrochemical sensors; corrosion; general electrode kinetics; biomedical devices. *Mailing Add:* PO Box 9147 Waltham MA 02254-9147

GINER-SOROLLA, ALFREDO, b Vinaros, Spain, Sept 23, 19; m 64; c 1. ORGANIC CHEMISTRY, BIOCHEMISTRY. *Educ:* Univ Madrid, MS, 47, Dr Pharm, 54; Cornell Univ, PhD, 58. *Prof Exp:* Asst tuberc study, Andreu Labs, Barcelona, 47-52; sci adv, Span Div, Farbwerke Hoechst, 52-54; res assoc, Sloan-Kettering Inst, 57-60; vis scientist, Univ Chem Lab, Cambridge Univ, 60-63; assoc, Sloan-Kettering Inst, 63-67; asst prof, 65-74, assoc prof biochem, Sloan-Kettering Div, Med Col, Cornell Univ, 74-82, ASSOC MEM, SLOAN-KETTERING INST, 67- *Concurrent Pos:* Res fel, Sloan-Kettering Inst, 54-57; asst, Sch Pharm, Univ Barcelona, 51-54; vis prof, sr investr, Dept Pharmacol, Immunopharmacol Prog, Col Med, Univ SFLA, 83- *Honors & Awards:* Valencia Sci Prize, 77. *Mem:* Am Chem Soc; Am Asn Cancer Res; AAAS. *Res:* Synthesis of new purine antimetabolites and immunomodulators as antiviral and anticancer agents; carcinogenesis studies of nitrosated DNA and RNA bases and psychotropic and their prevention. *Mailing Add:* Col Med Univ SFla 12901 N 30th St Box 9 Tampa FL 33612-4799

GINEVAN, MICHAEL EDWARD, b Amsterdam, NY, Oct 22, 46; m 70; c 1. RISK ANALYSIS, ENVIRONMENTAL MONITORING. *Educ:* State Univ NY Albany, BS, 68; Univ Mass, Amherst, MS, 71; Univ Kans, PhD(mat biol), 76. *Prof Exp:* Res assoc, Environ Res Lab, Univ Kans Ctr Res Inc, 76-78; res assoc, Argonne Nat Lab, 78-79, asst statistician, Div Biol & Med Res, 79-82; biostatistician, Health Effects Br, US Nuclear Regulatory Comn, 82-86, biostatistical consult, 86-87; sr sci adv statist, Environ Corp, 87-88; PRIN SCIENTIST BIOSTATIST & EPIDEMIOL, VERSAR, INC, 88- *Concurrent Pos:* Courtesy asst prof, Dept Entom, Univ Kans, 77-78; consult environ sci, Terrestial Biomonitoring Prog, Environ Protection Agency, 79-82; mem, Nuclear Regulatory Res Comt, Am Statist Asn, 84-; vchmn, Am Statist Asn, Comt on Statist & the Environ, 87-; chmn, organizing comt, Eight Annual Am Statist Asn, Conf on Radiation & Health, 88-89; mem, Radon Mitigation Subcomt US Environ Protection Agency, Radiation Adv Comt, 87; secy, Am Statist Asn Sect on Statist & the Environ, 89-90; mem, Nat Res Coun Comt Assessment of Possible Health Effects of the Ground Wave Emergency Network, 90-91. *Mem:* Am Statist Asn; Biomet Soc; Soc Epidemiol Res; AAAS; Soc Risk Analysis; NY Acad Sci. *Res:* Applications of statistics to public health and environmental problems; theoretical studies of meta-analysis and environmental sampling methodologies, risk analysis and environmental epidemiology. *Mailing Add:* 307 Hamilton Ave Silver Spring MD 20901

GING, ROSALIE J, b Detroit, Mich, June 26, 21; m; c 2. PSYCHIATRY. *Educ:* Univ Mich, BA, 42, MD, 45; Am Bd Psychiat & Neurol, dipl, 57. *Prof Exp:* Intern, New Eng Hosp Women & Children, Boston, 46; resident, Pottstown Hosp, Pa, 48; pvt pract, 49; resident psychiat, Philadelphia State Hosp, 50-51, Conn State Hosp, Middletown, 52-53 & Univ Kans Hosp, Kansas City, 53-54; clin instr psychiat, Univ Kans, Kansas City, 55-56; instr, 56-65, ASST PROF PSYCHIAT, MED SCH, UNIV MICH, ANN ARBOR, 65-; CHIEF PSYCHIAT SERV, VET ADMIN HOSP, 60- *Concurrent Pos:* Staff psychiatrist, Vet Admin Hosp, Kansas City, 54-56; staff psychiatrist, Vet Admin Hosp, Ann Arbor, 56-58, asst chief psychiat serv, 58-60; mem clin psychopharmacol res rev comt, NIMH, 73-77; mem exec comt coop chemother studies in psychiat, Vet Admin. *Mem:* AMA; fel Am Psychiat Asn. *Res:* Consistency of electroencephalograms of normal subjects and psychiatric patients; correlation of electroencephalograms and multiple physical symptoms; clinical electroencephalography. *Mailing Add:* 2215 Fuller Rd Ann Arbor MI 48105

GINGELL, RALPH, b Pontypool, Gt Brit, July 26, 45. DRUG METABOLISM. *Educ:* Univ Birmingham, BSc, 66; Univ London, PhD(biochem), 70; Am Bd Toxicol, dipl. *Prof Exp:* Lectr biochem, Brooklands Tech Col, Weybridge, 70-71; assoc prof biochem, Eppley Cancer Res Inst & Dept Biochem, Univ Nebr Med Ctr, Omaha, 71-80; SR TOXICOLOGIST, DEPT TOXICOL, SHELL OIL CO, 80- *Concurrent Pos:* Sainsbury res fel, Nat Col Food Technol, Weybridge, Gt Brit, 70-71. *Mem:* Soc Toxicol; Int Soc Study Xenobiotics. *Res:* Metabolism of xenobiotics, including toxicants and carcinogens in mammals in relation to their modes of action; industrial toxicology. *Mailing Add:* Dept Toxicol Shell Oil Co PO Box 4320 Houston TX 77210

GINGER, LEONARD GEORGE, b Chicago, Ill, Oct 1, 18; m 43; c 6. ORGANIC CHEMISTRY. *Educ:* Northwestern Univ, BS, 39; Univ Chicago, MS, 41; Yale Univ, PhD(org chem), 43. *Prof Exp:* Jr chemist, Merck & Co, NJ, 40-41; Pittsburgh Plate Glass fel, Northwestern Univ, 44, Int, Inc fel, 45-49; asst sci dir, Baxter Int, Inc, 49-53, dir org chem res, 53-57, dir res, 57-59, vpres res & develop, 59-69, sr vpres, 69-81; PRES, L G GINGER & ASSOCS, 81- *Concurrent Pos:* Assoc chemist, Manhattan Proj, Univ Chicago, 44 & Off Sci Res & Develop, Northwestern Univ, 46; consult, Baxter Travenol Labs Inc, 82-84, BW-Biotec, 85-87, Energy Consults, Inc, 87- *Mem:* AAAS; Am Chem Soc; NY Acad Sci; Indust Res Inst. *Res:* Bacterial chemistry; pyrogens; organic synthesis; micro-analytical methods; chemistry of tubercle bacillus; pharmaceuticals; thyroid chemistry; parenteral nutrition; biomedical engineering; medical electronics; blood preservation; microbial enzymes. *Mailing Add:* 2100 Burr Oak Dr L G Ginger & Assocs Glenview IL 60025

GINGERICH, KARL ANDREAS, b Lahr, Germany, Oct 8, 27. INORGANIC CHEMISTRY, PHYSICAL CHEMISTRY. *Educ:* Univ Freiburg, dipl, 54, Dr rer nat, 57. *Prof Exp:* Asst, Univ Freiburg, 53-57; res assoc, Univ Ill, 57-58; asst prof chem, Pa State Univ, 58-64; sr chemist, Battelle Mem Inst, 64-68; PROF CHEM, TEX A&M UNIV, 68- *Mem:* Am Chem Soc. *Res:* Knudsen cell mass spectrometric and matrix isolation spectroscopic studies of high temperature systems and derived thermodynamic and structural properties. *Mailing Add:* Dept Chem Tex A&M Univ College Station TX 77843

GINGERICH, OWEN (JAY), b Washington, Iowa, Mar 24, 30; m 54; c 3. HISTORY OF ASTRONOMY. *Educ:* Goshen Col, BA, 51; Harvard Univ, MA, 53, PhD(astron), 62. *Prof Exp:* From instr to asst prof astron & dir observ, Am Univ, Beirut, 55-58; lectr astron, Wellesley Col, 58-59; lectr, 60-69, assoc prof, 68-69; astrophysicist, 62-86, SR ASTRON, SMITHSONIAN ASTROPHYS OBSERV, 86-; PROF ASTRON & HIST SCI, HARVARD UNIV, 69- *Concurrent Pos:* Dir, Cent Bur Astron Telegrams, Int Astron Union, 65-68, assoc dir, 68-79, pres, Comn on Hist of Astron, 71-76; counr, Am Astron Soc, 73-76, chmn, Educ Adv Comt, 75-77, chmn, Hist Astron Div, 80 & 83-84; assoc ed, J Hist Astron, 75-; fel, Churchill Col, Cambridge, Eng, 85-86. *Honors & Awards:* Sigma Xi Nat Lectr, 71; George Darwin Lectr, Royal Astron Soc, 71. *Mem:* AAAS; Hist Sci Soc; Am Astron Soc; Am Philos Soc (vpres, 82-85); Int Acad Hist Sci; Am Acad Arts & Sci. *Res:* History of astronomy, especially Kepler and Copernicus. *Mailing Add:* Ctr Astrophys 60 Garden St Cambridge MA 02138

GINGERICH, PHILIP DERSTINE, b Goshen, Ind, Mar 23, 46. VERTEBRATE PALEONTOLOGY. *Educ:* Princeton Univ, AB, 68; Yale Univ, MPhil, 72, PhD(paleont), 74. *Prof Exp:* Asst prof paleont, Univ Mich, 74-75; NATO fel, Univ Montpellier, 75-76; from asst prof to assoc prof, 76-83, dir mus paleont, 81-87, PROF PALEONT, UNIV MICH, 83- *Concurrent Pos:* J S Guggenheim fel, 83. *Honors & Awards:* Shuchert Award, Paleont Soc, 81. *Mem:* AAAS; Am Soc Mammalogists; Sigma Xi; Soc Study Evolution; Soc Vert Paleont; Paleont Soc. *Res:* Primate evolution; origin and radiation of mammals; fossil record and evolution at the species level. *Mailing Add:* Mus Paleont Univ Mich Ann Arbor MI 48109-1079

GINGERY, ROY EVANS, b Lodi, Ohio, June 3, 42; m 64; c 2. BIOCHEMISTRY, PLANT PATHOLOGY. *Educ:* Carnegie-Mellon Univ, BS, 64; Univ Wis, MS, 67, PhD(biochem), 68. *Prof Exp:* Asst prof, Ohio Agr Res & Develop Ctr, 68-69, RES CHEMIST PLANT PATH, USDA & OHIO AGR RES & DEVELOP CTR, OHIO STATE UNIV, 69- *Mem:* Am Phytopath Soc. *Res:* Nucleic acid metabolism in virus infected plants; biochemical mechanisms of disease resistance in plants. *Mailing Add:* Plant Path Ohio State Univ Main Campus Kottman Hall Columbus OH 43210

GINGLE, ALAN RAYMOND, b Chicago, Ill, Jan 22, 49. DEVELOPMENTAL BIOLOGY, BIOPHYSICS. *Educ:* Ill Inst Technol, BS, 71; Univ Chicago, PhD(physics), 75. *Prof Exp:* Res asst solid state physics, Zenith Radio Corp, 67-72; RES ASSOC BIOPHYS, UNIV CHICAGO, 75- *Mem:* Sigma Xi. *Res:* Development in the cellular slime moulds; early chick embryonic development; growth properties of drosophila cell lines. *Mailing Add:* 5040 N Sayre Ave Chicago IL 60656

GINGOLD, KURT, b Vienna, Austria, Aug 7, 29; nat US; m 57; c 2. CHEMISTRY. *Educ:* Tulane Univ, BS, 50; Harvard Univ, MA, 52, PhD(chem), 53. *Prof Exp:* Lit chemist, Ethyl Corp, Mich, 53-54; sr info scientist, Am Cyanamid Co, 56-86; CONSULT, 86- *Concurrent Pos:* Vpres, Int Fedn Translr, 63-66, mem coun, 66-77 & 84-90. *Mem:* Am Chem Soc; fel Inst Linguists, London; Am Translr Asn (pres, 63-65). *Res:* Scientific literature; technical writing, editing and translation; organometallic and inorganic chemistry; information science. *Mailing Add:* 35 Windsor Ln Cos Cob CT 06807

GINGRAS, BERNARD ARTHUR, b Montreal, Que, Jan 23, 27; m 52; c 4. ORGANIC CHEMISTRY. *Educ:* Univ Montreal, BSc, 48, MSc, 49, PhD, 52; Oxford Univ, DPhil, 54. *Hon Degrees:* Dr Sci, York Univ, 79. *Prof Exp:* Lectr org chem, Univ Montreal, 51-52; Nat Res Coun Can overseas fel, Oxford Univ, 52 & Merck & Co, Inc, 53; from assoc res officer to sr res officer, Div Appl Chem, Nat Res Coun Can, 54-67, assoc awards officer, 67-70, dir off grants & scholar, 70-72, asst vpres univ res, 72-74, vpres univ grants & scholar, 74-78, vpres external rel, 78-84, exec vpres, 85-87, vpres externternal rel, 87-90; RETIRED. *Mem:* Fel Chem Inst Can; Fr-Can Asn Advan Sci (pres). *Res:* Textiles; fungicides for prevention of deterioration of fibers by microorganisms; photochemical degradation of cellulose; coordination chemistry. *Mailing Add:* 919 Bermuda Ave Ottawa ON K1K 0V8 Can

GINGRASS, RUEDI PETER, b Milwaukee, Wis, May 10, 32; m 58, 83; c 5. PLASTIC SURGERY. *Educ:* Univ Mich, AB, 54, MD, 58; Marquette Univ, MS, 63. *Prof Exp:* Asst prof plastic surg & exec dir, Sch Med, Marquette Univ, 67-71; assoc prof plastic surg & chmn dept, Med Col Wis, 71-83; CONSULT, 83- *Concurrent Pos:* NIH head & neck cancer trainee, Duke Univ Med Ctr,

63-66. *Mem:* Am Soc Plastic & Reconstructive Surg; Soc Head & Neck Surg; Am Asn Plastic Surg; Plastic Surg Res Coun. *Res:* Wound healing and wound infection; skin grafting. *Mailing Add:* 9800 W Blue Mound Rd Milwaukee WI 53226

GINGRICH, NEWELL SHIFFER, b Orwigsburg, Pa, Jan 29, 06; m 28; c 2. PHYSICS. *Educ:* NCent Col, AB, 26; Lafayette Col, MA, 27; Univ Chicago, PhD(physics), 30. *Prof Exp:* Lab instr physics, Lafayette Col, 26-27, instr, 27-28; asst prof, Mt Allison Univ, 30-31; instr, Mass Inst Technol, 31-36; from asst prof to prof, 37-73, EMER PROF PHYSICS, UNIV MO-COLUMBIA, 73- *Concurrent Pos:* Grants-in-aid, AAAS, Rumford Fund, Am Acad Sci & Elizabeth Thompson Sci Fund, Res Corp, NSF; physicist, Naval Ord Lab, 41-43; tech aide, Liaison Off, Off Sci Res & Develop, Washington, DC, 43-44, Oak Ridge Nat Lab, 52-53 & Argonne Nat Lab, 56-58; NSF fel, 59-60. *Mem:* AAAS; fel Am Phys Soc; Am Asn Physics Teachers; Am Crystallog Asn; Sigma Xi. *Res:* Compton effect in x-rays; circuits used with geiger counter work; diffraction of x-rays by elements in liquid state; cosmic rays; atomic distribution in liquids; magnetic structure in crystals; neutron diffraction. *Mailing Add:* 320 Physics Bldg Univ Mo Columbia MO 65211

GINI, MARIA LUIGIA, b Milano, Italy, July 30, 46; m 82. ARTIFICIAL INTELLIGENCE, ROBOTICS. *Educ:* Univ Milano, Italy, Doctor(physics), 72. *Prof Exp:* Res assoc computer sci, Politecnico Milano, Italy, 73-76, sr res assoc, 78-82; vis res assoc Artificial Intel, Stanford Univ, 76-77; asst prof, 82-88, ASSOC PROF COMPUTER SCI, UNIV MINN, MINNEAPOLIS, 88- *Concurrent Pos:* Vis prof, Univ Salerno, Italy, 88-89 & Stanford Univ, 89. *Mem:* Asn Comput Mach; Am Asn Artificial Intel; Computer Prof Social Responsibility; Inst Elec & Electronic Engrs; Nat Serv Robot Asn. *Res:* Integration of artificial intelligence with robotics for navigation, exploration and task execution in changing environments; planning and sensing to detect and recover from errors in robotic assembly. *Mailing Add:* 4-192 Elec Eng & Computer Sci Bldg Univ Minn 200 Union St SE Minneapolis MN 55455-0191

GINIS, ASTERIOS MICHAEL, b Thessaloniki, Greece, Feb 16, 45; m 71; c 1. FOOD SCIENCE, BIOCHEMISTRY. *Educ:* Univ Thessaloniki, BS, 68; Univ Wis-Madison, MS, 73, PhD(food sci), 76. *Prof Exp:* Res asst plant breeding, Cereals Inst & Cotton Inst Greece, 66-67; res asst, Univ Wis-Madison, 71-76; sr res food scientist res & develop, 76-80, develop leader, 80-89, DEPT HEAD, GEN MILLS INC, 89-, DIR, CEREAL PARTNERS WORLDWIDE. *Mem:* Inst Food Technologists; Am Meat Sci Asn; Am Asn Cereal Chemists. *Res:* Basic and exploratory food; product and process development; frozen foods; storage life prediction of foods; microwave technology. *Mailing Add:* Romanel-sur-Morges 1122 Switzerland

GINN, H EARL, b Tylertown, Miss, July 7, 31; m; c 2. INTERNAL MEDICINE, NEPHROLOGY. *Educ:* Baylor Univ, BS, 53; Emory Univ, MD, 57. *Prof Exp:* Intern, Med Ctr, Univ Okla, 57-58, resident internal med, 58-60; instr med, New York Hosp, Med Ctr, Cornell Univ, 60-61; instr med & physiol, Med Ctr, Univ Okla, 62, asst prof med, physiol & urol, 62-65; from asst prof to prof, Sch Med, Vanderbilt Univ, 65-73, prof med, 73-89, prof urol, 65-89; PARTNER & PHYSICIAN, NASHVILLE NEPHROLOGY ASSOC, 89-; SR VPRES, REN CORP, 90- *Concurrent Pos:* USPHS fel renal & electrolyte physiol, Cornell Univ, 60-61; NIH career res develop award, Med Ctr, Univ Okla, 64-65 & Univ Hosp, Vanderbilt Univ, 65-; clin investr, Vet Admin Hosp, Med Ctr, Okla, 61-64, chief renal & electrolyte sect, Okla City Vet Admin Hosp & chief kidney sect, Med Ctr, Univ Okla, 63-65; chief nephrology div & chief renal dialysis units, Vanderbilt Univ Hosp, 65-; gen consult, Kidney Dis Control Prog, 67-68; mem coun circulation, Am Heart Asn. *Mem:* Am Physiol Soc; Am Soc Artificial Internal Organs; Am Soc Nephrology; Harvey Soc. *Res:* Renal transplantation; dynamics of hemodialysis; biochemistry of uremia. *Mailing Add:* Nashville Nephrol Asn 4515 Harding Rd Suite 308 Nashville TN 37205

GINN, ROBERT FORD, b Alamosa, Colo, Aug 13, 31. CHEMICAL ENGINEERING, MATHEMATICS. *Educ:* Univ Colo, Boulder, BS, 53; Lawrence Univ, MS, 57; Univ Del, MChE, 64, PhD(chem eng), 68. *Prof Exp:* Chem engr, Albert E Reed Paper Group, Eng, 57-58; res engr, Nat Vulcanized Fibre Co, Del, 59-60 & Westvaco Corp, Md, 65-69; sr scientist, Am Can Co, Wis, 69-71; res fel, Inst Paper Chem, 71-77; ASST PROF CHEM ENG, MICH TECHNOL UNIV, 77- *Mem:* Am Chem Soc; Soc Rheology; Brit Soc Rheology; Am Inst Chem Eng; Sigma Xi. *Res:* Polymer science and engineering. *Mailing Add:* Dept Chem Eng Mich Technol Univ Houghton MI 49931

GINN, THOMAS CLIFFORD, b Medford, Ore, July 18, 46. AQUATIC ECOLOGY, FISHERY BIOLOGY. *Educ:* Ore State Univ, BS, 68, MS, 70; NY Univ, PhD(aquatic ecol), 77. *Prof Exp:* Res asst radiol physics, X-ray Sci & Eng Lab, Ore State Univ, 69-71; asst res scientist aquatic ecol, Inst Environ Med, NY Univ Med Ctr, 71-77; mgr biol serv, 77-80, dir environ consult, Tetra Tech, Inc, 80-87; VPRES ENVIRON CONSULT, PTI ENVIRON SERV, 87- *Mem:* Am Fisheries Soc; AAAS; Sigma Xi; Am Inst Fishery Res Biologists; Estuarine Res Fedn; Water Pollution Control Fedn. *Res:* Effects of toxic substances on marine and estuarine biota; response of freshwater ecosystems to lake restoration techniques; toxic responses of freshwater and marine organisms to chlorinated effluents; sewage discharge effects on marine ecosystems. *Mailing Add:* PTI Environ Serv 15375 SE 30th Pl Suite 250 Bellevue WA 98007-6500

GINNARD, CHARLES RAYMOND, b Detroit, Mich, Oct 2, 47; m 69; c 3. ANALYTICAL CHEMISTRY. *Educ:* Wayne State Univ, BS, 69. *Prof Exp:* Analytical chemist surface analysis & polymer characterization, 69-78, sect supvr chem analysis, 78-83, RES ADMINR, CENT RES & DEVELOP DEPT, E I DU PONT DE NEMOURS & CO, INC, 83- *Mem:* Am Chem Soc; Sigma Xi. *Res:* Analytical intrumentation and techniques for polymer, chemical and biophysical characterization. *Mailing Add:* Du Pont Cent Res & Develop Exp Sta E328/414 Wilmington DE 19880-0328

GINNING, P(AUL) R(OLL), b Shreveport, La, June 16, 23; m 47; c 2. CHEMICAL ENGINEERING. *Educ:* NC State Col, BS, 48. *Prof Exp:* Jr chemist, Viscose Control Lab, Indust Rayon Corp, 48-49, jr chem engr, Prod Supt Staff, 49-52; res engr, Tire Cords & Fabrics, Res Div, Goodyear Tire & Rubber Co, 52-53, chem engr, De-Icing Res, 53-55, group leader ice-guard res, 55-62, group leader new prod res, 62-65, sect head mat eng res, 65-68, sect head flexible films & sheeting, 68-77, sect head plastics res, Res Div, 77-80, sect head, Polyester Applications Dept, 80-81, PRIN ENGR, POLYESTER APPLICATIONS DEPT, GOODYEAR TIRE & RUBBER CO, 81- *Mem:* Am Chem Soc; Soc Plastics Engrs; Am Inst Chem. *Res:* Viscose preparation and spinning, textiles as related to tire cords; conductive rubber; de-icing; rain erosion resistance of aircraft materials at high speeds; adhesives; structural laminates for aircraft ice-guards; rubber compounding; plastics processing; acoustical rubbers; solution cast and extruded polyvinyl chloride films; processing of thermoplastic polymers; investigation of polymer blends; research to develop copolyesters suitable for processing into films, sheeting and other articles suitable for industrial food packaging applications; patents issued. *Mailing Add:* 668 Highland Ave Wadsworth OH 44281

GINNINGS, GERALD KEITH, b Greensboro, NC, Sept 16, 28; m 58; c 3. MATHEMATICS, STATISTICS. *Educ:* Elon Col, BA, 50; Appalachian State Teachers Col, MA, 62; Auburn Univ, EdD(math educ), 66. *Prof Exp:* Teacher electronics, H L Yoh Co Commun, Ft Sill, Okla, 59; field engr, Inter-Continental Ballistic Missile Launching Sites, Radio Corp Am Serv Co, NJ, 59-61; design engr, Martin Co, Colo, 61; asst prof math, Berry Col, 62-65; assoc prof, 66-68, PROF MATH, E TENN STATE UNIV, 68- *Res:* Modern and traditional algebra; breeching the barrier in communications which exists between the applied mathematics areas and the pure areas. *Mailing Add:* Dept Math E Tenn State Univ Johnson City TN 37614

GINNS, JAMES HERBERT, b New Britain, Conn, June 21, 38; m 67. MYCOLOGY, FOREST PATHOLOGY. *Educ:* Univ Conn, BS, 60; WVa Univ, MS, 62; NY State Univ, Syracuse, PhD(mycol), 67. *Prof Exp:* Res plant path, WVa Agr Exp Sta, 60-62; res forest path, Southlands Exp Forest, Int Paper Co, 62-64; teacher bot, NY State Univ Col Forestry, 65-67; res forest path, Can Forest Serv, 67-69; RES MYCOL, CAN DEPT AGR, 69- *Concurrent Pos:* Counr, Mycol Soc Am, 77-79; Friends of Farlow Herbarium, 82-; mem, Int Comn Taxon Fungi, 86-, Species Survival Comn, Fungi Group, 89-92, exec coun, Mycol Soc Am 90-93; cur, Can Collection Fungus Cultures, 75-82, Nat Mycol Herbarium Can, 88- *Mem:* Mycol Soc Am; Sigma Xi; Asn Syst Collections; US Fedn Cul Collections. *Res:* Taxonomy, classical and biochemical of wood; decaying fungi, especially polyporaceae and corticiaceae of the boreal forest. *Mailing Add:* Saunders Bldg Cent Exp Farm Ottawa ON K1A 0C6 Can

GINOCCHIO, JOSEPH NATALE, b Summit, NJ, Dec 25, 36; m 64; c 2. THEORETICAL NUCLEAR PHYSICS. *Educ:* Lehigh Univ, BS, 58; Univ Rochester, PhD(physics), 64. *Prof Exp:* Res assoc nuclear physics, Rutgers Univ, 64-66 & Mass Inst Technol, 66-68; asst prof nuclear physics, Yale Univ, 68-75; STAFF PHYSICIST, LOS ALAMOS NAT LAB, 75- *Mem:* Fel Am Phys Soc. *Res:* Study of nuclear structure with shell model; random phase approximation applied to nuclear bound and scattering states; group theory as used in nuclear physics; four nucleon correlations in nuclei, pion and intermediate energy proton scattering from nuclei. *Mailing Add:* Los Alamos Nat Lab Mail Stop B283 PO Box 1663 Los Alamos NM 87545

GINOS, JAMES ZISSIS, b Hillsboro, Ill, Feb 1, 23; m 47; c 2. MEDICINAL CHEMISTRY, ORGANIC CHEMISTRY. *Educ:* Columbia Col, BA, 54; Stevens Inst Technol, MS, 57, PhD(org chem), 64. *Prof Exp:* Chemist, Colgate-Palmolive Co, NJ, 52-57; chief chemist, Diamond Alkali, NJ, 57-58; chemist, Nopco Chem Co, 59-64; from asst scientist to scientist, Brookhaven Nat Labs, NY, 64-75; sr res assoc, 80-86, ASSOC LAB MEM, MEM SLOAN-KETTERING CANCER CTR, 86-; RES ASSOC PROF CHEM, MED COL, CORNELL UNIV, 75. *Concurrent Pos:* Asst instr, Newark Col Eng, 59-60; res asst prof, Mt Sinai Sch Med, 68-70. *Mem:* AAAS; Am Chem Soc; Am Soc Pharmacol & Exp Therapeut; Soc Nuclear Med; NY Acad Sci; Am Soc Clin Pharmacol & Therapeut. *Res:* Synthesis of radiopharmaceuticals for use in position emission tomography. *Mailing Add:* Sloan-Kettering Inst 1275 York Ave New York NY 10021

GINSBERG, ALVIN PAUL, b Brooklyn, NY, Jan 7, 32. INORGANIC CHEMISTRY. *Educ:* NY Univ, AB, 54; Columbia Univ, AM, 55, PhD(chem), 59. *Prof Exp:* Instr chem, Brown Univ, 59-60; MEM TECH STAFF, BELL TEL LABS, 60- *Mem:* AAAS; Am Chem Soc; Am Phys Soc; Royal Soc Chem. *Res:* Preparation, spectroscopic and magnetic properties of transition metal compounds; applications of valence theory; hydride complexes. *Mailing Add:* 119 Spring Ridge Dr Berkeley Heights NJ 07922

GINSBERG, BARRY HOWARD, b Brooklyn, NY, May 9 45; m 67; c 2. DIABETES MELLITUS, ENDOCRINOLOGY. *Educ:* Harpur Col, State Univ NY, Binghamton, BA, 65; Einstein Col Med, PhD(molecular biol), 71, MD, 72. *Prof Exp:* Resident med, Harvard Univ, 74; NIH fel endocrinol, 77; asst prof med, Univ Iowa, 77-82, asst prof biochem, 78-82, assoc prof, 82-88, PROF MED & BIOCHEM, UNIV IOWA, 88- *Concurrent Pos:* Mem bd dirs, Am Diabetes Asn, 82-85. *Mem:* Endocrine Soc; Am Diabetes Asn; Am Fedn Clin Res. *Res:* Effect of membrane biophysical properties on hormone action; develoment of a glucose sensor; relationship of diabetes control to complications; computers in medicine. *Mailing Add:* Becton Dickinson & Co One Becton Dr Franklin Lakes NJ 07417-1883

GINSBERG, DONALD MAURICE, b Chicago, Ill, Nov 19, 33; m 57; c 2. SUPERCONDUCTIVITY, LOW TEMPERATURE PHYSICS. *Educ:* Univ Chicago, BA, 52, BS, 55, MS, 56; Univ Calif, PhD(physics), 60. *Prof Exp:* From res assoc to assoc prof, 59-66, PROF PHYSICS, UNIV ILL, URBANA-CHAMPAIGN, 66- *Concurrent Pos:* Alfred P Sloan res fel, 60-64, NSF fel, 66-67; mem, Educ Rev Comt Solid State Div, 77-82, chmn, 79-80, Mat Sci & Technol Div, Argonne Nat Lab, 82-83; vis scientist, physics, Am Assoc Physics Teachers & Am Inst Physics, 65-71. *Mem:* Fel Am Phys Soc; AAAS; Sigma Xi. *Res:* Superconductivity. *Mailing Add:* Dept Physics Loomis Lab Univ Ill Urbana Champaign 1110 W Green St Urbana IL 61801

GINSBERG, EDWARD S, b New York, NY, Oct 4, 38; div; c 3. HIGH ENERGY PHYSICS. *Educ:* Brown Univ, AB & ScB, 59; Stanford Univ, MS, 61, PhD(physics), 65. *Prof Exp:* Res assoc physics, Univ Pa, 64-66; asst prof, 66-71, ASSOC PROF PHYSICS, UNIV MASS, BOSTON, 71- *Concurrent Pos:* NSF res grant, 67-73. *Mem:* Am Phys Soc. *Res:* High energy theoretical physics; weak and electromagnetic interactions; radiative convections; planetary science. *Mailing Add:* Dept Physics Univ Mass Harbor Campus Boston MA 02125-3393

GINSBERG, HAROLD SAMUEL, b Daytona Beach, Fla, May 27, 17; m 49; c 4. VIROLOGY, MICROBIOLOGY. *Educ:* Duke Univ, AB, 37; Tulane Univ, MD, 41. *Prof Exp:* Resident, Mallory Inst Path, Boston City Hosp, 41, asst, 42, intern, 4th Med Serv, Harvard Univ, 42-43, asst resident, Thorndike Mem Lab, 43; asst physician, Hosp & asst, Rockefeller Inst, 46-49, res physician hosp & assoc, 49-51; assoc prof prev med & asst prof med, Sch Med, Western Reserve Univ, 51-60; prof microbiol & chmn dept, Univ Pa, 60-73; prof & chmn dept, 73-85, EUGENE HIGGINS PROF MED & MICROBIOL, COL PHYSICIANS & SURGEONS, COLUMBIA UNIV, 85- *Concurrent Pos:* Consult, Nat Inst Allergy & Infectious Dis, Surgeon Gen, USPHS, NASA, 69-73; mem comn acute respiratory dis, Armed Forces Epidemiol Bd, 59-71, adv bd, Am Cancer Soc, 70-74, 75-; mem microbiol comn, Nat Bd Med Examrs, 72-, chmn, 75-84; ed, J Bact, J Virol, J Exp Med, J Infectious Dis & Intervirol; actg dir, Comprehensive Cancer Ctr, Columbia Univ, 83-85. *Honors & Awards:* Alexander von Humboldt Found Sr US Scientist Award, 85. *Mem:* Nat Acad Sci; Inst Med-Nat Acad Sci; Am Soc Clin Invest; Asn Am Physicians; Harvey Soc; Am Asn Immunol; Am Acad Microbiol (vpres, 70-71, pres, 71-72). *Res:* Biochemistry and genetics of viral infections. *Mailing Add:* Dept Med Col Physicians & Surgeons Columbia Univ 650 W 168th St Rm 1014 New York NY 10032

GINSBERG, JERRY HAL, b New York, NY, Sept 18, 44. STRUCTURAL ACOUSTICS & VIBRATION, NONLINEAR ACOUSTICS. *Educ:* Cooper Union, BSCE, 65; Columbia Univ, MS, 66, EScD(eng mech), 70. *Prof Exp:* Asst prof eng mech, Sch Aero, Astro & Eng Sci, Purdue Univ, 69-74, assoc prof, Sch Mech Eng, 74-80; prof, 80-89, GEORGE W WOODRUFF CHAIR, SCH MECH ENG, GA INST TECHNOL, 89- *Concurrent Pos:* Assoc prof, Nat Super Sch Elec Mech, Nancy, France, 75-76; Fulbright Hays advan res fel, US State Dept, 75; consult, Lawrence Livermore Nat Lab, 79- & Gen Motors Res Labs, 86-89. *Mem:* Fel Acoust Soc Am; fel Am Soc Mech Engrs; Sigma Xi; Am Acad Mech. *Res:* Acoustics and vibrations involving dynamic response of submerged structures; nonlinear effects in propagation of acoustic waves; high-intensity sound beams; wave propagation in heterogeneous elastic media. *Mailing Add:* Sch Mech Eng Ga Inst Technol Atlanta GA 30332

GINSBERG, JONATHAN I, b Brooklyn, NY, Mar 18, 41; m 63; c 5. MATHEMATICAL ANALYSIS. *Educ:* Yeshiva Univ, BA, 61, MA, 64, PhD(math), 69. *Prof Exp:* From instr to assoc prof math, Yeshiva Col, 68-86; PRIN, YESHIVA CHOFETZ CHAIN, 86- *Mem:* Am Math Soc. *Res:* Completeness theorems in various Banach spaces and algebras. *Mailing Add:* 25 Dover Terr Monsey NY 10952

GINSBERG, MARK HOWARD, b New York, NY, Aug 30, 45; m 68; c 2. IMMUNOLOGY, CELL BIOLOGY. *Educ:* State Univ NY Downstate Med Ctr, MD, 70. *Prof Exp:* Intern med, Univ Chicago, 70-71, resident, 71-73, fel rheumatol, 73-75; fel, 75-78, ASST MEM IMMUNOPATH, SCRIPPS CLIN & RES FOUND, 78-, ASSOC MEM, 82- *Concurrent Pos:* Clin instr, Sch Med, Univ Calif, San Diego, 76-; clin investr, Nat Inst Arthritis, Metab & Digestive Dis, NIH, 77- *Mem:* Am Asn Immunologists; Am Fedn Clin Res; Am Rheumatism Asn; Am Soc Clin Res. *Res:* Role of platelets in inflammation; mechanisms of platelet secretion and adhesion; biology of platelet factor four. *Mailing Add:* Dept Immunopath Scripps Found 10666 N Torrey Pines Rd La Jolla CA 92037

GINSBERG, MURRY B(ENJAMIN), b New York, NY, Oct 26, 28; m 55; c 2. ELECTRONICS ENGINEERING. *Educ:* City Col New York, BEE, 49; Univ Md, MS, 58. *Prof Exp:* Electronics engr, Bur Ships, US Navy Dept, 50-57; electronics engr, systs anal off, Harry Diamond Labs, Dept Army, 57-84; sr engr, Westinghouse Electric, 84-85; PRIN ELEC ENG, PRC INC, 86- *Mem:* Inst Elec & Electronics Engrs. *Res:* Systems analysis and engineering of electronic weapon systems; radar and information theory. *Mailing Add:* PRC Inc 2121 Crystal Dr Suite 300 Arlington VA 22202

GINSBERG, MYRON DAVID, b Denver, Colo, Aug 26, 39; m 69; c 2. NEUROLOGY. *Educ:* Wesleyan Univ, BA, 61; Harvard Med Sch, MD, 66. *Prof Exp:* Clin fel, Harvard Med Sch & Boston City Hosp, 66-68; resident & fel neurol, Harvard Med Sch & Mass Gen Hosp, 68-70; staff assoc physiol Lab Perinatal Physiol, NIH, 70-72; res fel neuropath, Mass Gen Hosp, Boston, 72-73; asst to assoc prof neurol, Sch Med, Univ Pa, 73-79; ASSOC PROF TO PROF NEUROL & RADIOL, SCH MED, UNIV MIAMI, 79-, DIR, CEREBROVASCULAR RES CTR, 81- *Concurrent Pos:* estab investr, Am Heart Asn, 76-81; attend neurologist, Jackson Mem Hosp; Am Heart Asn, Nat Res Comt, 86; Javits Neurosci Invstr Award, NIH, 85-92. *Mem:* Am Acad Neurol; Am Physiol Soc; John Morgan Soc; Am Neurol Asn; Int Soc Cerebral Blood Flow & Metab; Soc Neurosci. *Res:* Physiology and neuropathology of energy-deprivation states in the central nervous system; cerebral blood flow and metabolism in hypoxic-ischemic encephalopathy. *Mailing Add:* Dept Neurol Sch Med Univ Miami PO Box 016960 Miami FL 33101

GINSBERG, THEODORE, b Brooklyn, NY, Aug 28, 41; m 67; c 2. MECHANICAL ENGINEERING. *Educ:* Pratt Inst, BChE, 63; Pa State Univ, MS, 66, PhD(nuclear eng), 70. *Prof Exp:* Assoc mech engr, Argonne Nat Lab, 70-72; sr lectr nuclear eng, Technion-Israel Inst Technol, 72-74; SR SCIENTIST MECH ENG, BROOKHAVEN NAT LAB, 74- *Mem:* Am Nuclear Soc; Am Soc Mech Engrs. *Res:* Heat transfer; fluid mechanics; two-phase flow and heat transfer; nuclear reactor safety; experimental thermofluid sciences. *Mailing Add:* Brookhaven Nat Lab Bldg 820-M Upton NY 11973

GINSBERG, THEODORE, b New York, NY, May 30, 33; m 55; c 3. VIRAL IMMUNOLOGY, CELL MEDIATED IMMUNITY. *Educ:* City Col New York, BS, 55; Rutgers Univ, PhD(biochem), 68. *Prof Exp:* Biochemist, Ciba Pharmaceut, Summit, NJ, 56-60; bio-org chemist, Carter-Wallace Labs, Cranbury, NJ, 60-64; fel, Ciba Ltd, Basel, Switz, 68-69; dir biochem, 71-72, asst dir res, 72-75, DIR SCI AFFAIRS, NEWPORT PHARMACEUT, 76- *Mem:* Am Chem Soc; Am Soc Microbiol; AAAS. *Res:* Nucleotide, nucleic acid metabolism; biosynthesis of unusual ribonucleotides; nucleic acid sequencing; structural determinations; transfer RNA biosynthesis and structure; viral immunology; cell-mediated immunity; T-cell function. *Mailing Add:* 424 Los Robles Laguna Beach CA 92651

GINSBERG-FELLNER, FREDDA VITA, b New York, NY, Apr 21, 37; m 61; c 2. MEDICINE, ENDOCRINOLOGY. *Educ:* Cornell Univ, AB, 57; NY Univ, MD, 61. *Prof Exp:* Intern pediat, Albert Einstein Col Med, 61-62, fel pediat endocrinol, 62-65, asst resident pediat, 63-64, sr resident, 65-66; assoc, 67-69, from asst prof to assoc prof, 69-81, PROF PEDIAT, MT SINAI SCH MED, 81-, DIR PEDIAT ENDOCRINOL, 77- *Concurrent Pos:* Sr clin pediatrician, Mt Sinai Hosp, 67-69, asst attend pediatrician, 69-74, assoc attend pediatrician, 74-81, attend pediatrician, 81-, dir, Carole & Michael Friedman & Family Young Peoples Diabetes Treat Unit, 84- *Honors & Awards:* Paul Lacy Award, Nat Diabetes Res Interchange, 82; Mary Jane Kugel Award, Juv Diabetes Found, 88. *Mem:* Am Diabetes Asn; Endocrine Soc; Soc Pediat Res; Am Pediat Soc; Am Fedn Clin Res; Am Acad Pediat. *Res:* Mechanisms of development of type I insulin dependent diabetes mellitus; genetic and immune factors; gestational diabetes; childhood obesity. *Mailing Add:* Dept Pediat Mt Sinai Sch Med One Gustave L Levy Pl Box 1198 New York NY 10029-6574

GINSBURG, ANN, b 1932; m 55; c 2. BIOCHEMISTRY, PHYSICAL BIOCHEMISTRY. *Educ:* Univ Calif, Berkeley, BA, 54, MA, 55; George Washington Univ, PhD(biochem), 64. *Prof Exp:* CHIEF, SECT PROTEIN & CHEM, NAT HEART, LUNG & BLOOD INST, NIH, 66- *Mem:* Am Soc Biochem & Molecular Biol; Calorimetry Conf. *Res:* Enzyme structure/regulation; thermodynamics of ligand-protein & protein-protein infractions. *Mailing Add:* Nat Lung & Blood Inst NIH Bldg 3 Rm 208 Bethesda MD 20892

GINSBURG, BENSON EARL, b Detroit, Mich, July 16, 18; m 41; c 3. GENETICS. *Educ:* Wayne State Univ, BS, 39, MS, 41; Univ Chicago, PhD(zool), 43. *Prof Exp:* Instr zool & physiol, Univ Chicago, 43-44, res assoc pharmacol, 44-46, asst prof biol sci, 46-49, from assoc prof to prof natural sci, 49-59, chmn dept, 50-57, assoc dean univ, 57-59, prof biol & head sect, 59-63, William Rainey Harper prof, 63-69; PROF BIOBEHAV SCI & CHMN DEPT, UNIV CONN, 69- *Concurrent Pos:* Sci assoc, Jackson Lab; fel, Ctr Advan Study Behav Sci, 57-58; consult ed, Encycl Britannica Films, 59-; mem adv comt animal resources, NIH, 65-; mem panel behav biol, Nat Acad Sci-Nat Res Coun, 66-67. *Mem:* Genetics Soc Am; Am Soc Naturalists; Biomet Soc; Am Soc Human Genetics; Am Genetic Asn. *Res:* Gene action in mammalian nervous system; inheritance of emotionality; biological education; zoology. *Mailing Add:* 149 Hillyndale Rd Storrs CT 06268

GINSBURG, CHARLES P, b 1920; US citizen. RADIO ENGINEERING, ELECTRICAL ENGINEERING. *Educ:* San Jose State Col, BA, 48. *Prof Exp:* Studio & transmitter engr radio stas, San Francisco Bay Area, 42-52; mem staff, Ampex Corp, 52-75, vpres advan develop, 75-86; RETIRED. *Honors & Awards:* David Sarnoff Gold Medal, Soc Motion Picture & TV Engrs, 57; Vladimir K Zworykin TV Prize, Inst Radio Engrs, 58; Valdemar Poulsen Gold Medal, Danish Acad Tech Sci, 60; Howard N Potts Medal, Franklin Inst, 69. *Mem:* Nat Acad Eng; fel Soc Motion Picture & TV Engrs; fel Inst Elec & Electronics Engrs; Franklin Inst. *Mailing Add:* 2311 W 29th Ave Eugene OR 97405

GINSBURG, DAVID, b New York, NY, Sept 5, 20; m 40; c 2. ORGANIC CHEMISTRY. *Educ:* City Col New York, BS, 41; Columbia Univ, MA, 42; NY Univ, PhD(org chem), 47. *Prof Exp:* Prod chemist, US Rubber Co, Pa, 42-43; res group leader, NY Quinine & Chem Works, Brooklyn, 43-48; res chemist, Daniel Sieff Res Inst, Weizmann Inst Sci, 48-50, sr res chemist, 50-54; vpres res, 59-60, actg pres, 61-62, PROF CHEM, ISRAEL INST TECHNOL, 54- *Concurrent Pos:* Fel, USPHS & Harvard Corp, Harvard Univ, 52-53, Guggenheim fel, 60-61; Lipsky fel, Oxford Univ, 53, Royal Soc vis prof, 70-71, vis fel, Merton Col, 70-71; chmn, Israel Coun Res & Develop, 61-62; vis prof, Brandeis Univ, 61 & 68, Univ Zurich, 61, Univ Sask, 64, NY Univ, 64 & 65, Boston Univ, 67, Weizmann Inst, 68, McGill Univ, 68 & 69, Oxford Univ, 70-71, Univ Basel, 75, Tex Christian Univ, 76, Hebrew Univ, Jerusalem & Max Planck Inst, Heidelberg, 77 & 82, Stanford Univ & Univ de Paris, Sud, 78, ETH, Zürich, 79, Univ Leiden, Univ Lorrvain-la-Nure, 81, Max Planck Inst, Müheim/Ruhr, 84 & Univ Alberta, 84; mem bur, Int Union Pure & Appl Chem, 63-69; vis res prof, Royal Soc, Cambridge Univ, 83; vis fel, Robinson Col, 83. *Honors & Awards:* Weizmann Prize, 54; Rothschild Prize, 65; Israel Prize, 73; August Wilhelm von Hofmann Medal, Soc German Chemists, 83. *Mem:* Am Chem Soc; Royal Soc Chem; Swiss Chem Soc; Israel Acad Sci & Humanities; Israel Chem Soc (pres, 55-57 & 66-67). *Res:* Natural products; sterochemistry of alicyclic systems; propellanes. *Mailing Add:* Internal Med/Kresge Univ Mich Ann Arbor MI 48109

GINSBURG, HERBERT, b Schenectady, NY, May 2, 28; m 51; c 3. EXPERIMENTAL STATISTICS. *Educ:* State Univ NY, BA, 50, MA, 51; NC State Univ, MS, 54. *Prof Exp:* Eng statistician, 54-61, sr res statistician, Res Labs, 61-63, FEL STATISTICIAN, BETTIS ATOMIC POWER LAB, WESTINGHOUSE ELEC CORP, 63- *Concurrent Pos:* Adj mem grad faculty, Univ Pittsburgh, 62- *Mem:* Am Statist Asn; Royal Statist Soc; Sigma Xi. *Res:* Experimental design; empirical model building; teaching. *Mailing Add:* Westinghouse Elec Co PO Box 79 West Mifflin PA 15122

GINSBURG, JACK MARTIN, b Philadelphia, Pa, May 31, 28; m 55; c 1. PHYSIOLOGY. *Educ:* Univ Pa, BA, 49; Tulane Univ, PhD(physiol), 53. *Prof Exp:* Lab asst physiol, Tulane Univ, 50-53; res participant, Biol Div, Oak Ridge Nat Lab, 53-54; instr pharmacol, Tulane Univ, 54-55; instr physiol, Univ Cincinnati, 57-59; asst prof, Col Med, Univ Rochester, 59-68; assoc prof, 68-73, PROF PHYSIOL & ASSOC PROF MED, MED COL GA, 73- *Mem:* AAAS; Am Physiol Soc; Am Soc Nephrology; Soc Exp Biol & Med. *Res:* Biological transport; body fluid; electrolytes. *Mailing Add:* Dept Physiol Med Col Ga Augusta GA 30912

GINSBURG, MERRILL STUART, b Chicago, Ill, July 20, 35; m 71; c 2. APPLIED SEISMOLOGY, SIGNAL PROCESSING. *Educ:* Mass Inst Technol, BS, 59, MS, 60; Univ Utah, PhD(geophys), 63. *Prof Exp:* Sr geophys engr, Geophys Serv Ctr, Mobil Oil Corp, 63-65, sr geophys interpreter, 65-70, assoc geophysicist, Explor Develop Dept, Explor Serv Ctr, 71-72, assoc geophysicist, Gravity-Magnetic Sect, 72-74, geophys specialist, 74-77, staff geophysicist, Gravity Magnetic Sect, Explor Serv Ctr, 71-82; staff geophysicist, Velocity Refraction Modeling Unit, 82-83, sr staff geophysicist, 83-85, geophys adv, 85-89, geophys adv, Appl Geophys Dept, 89-90, GEOPHYS ADV, SIGNAL PROCESSING DEPT, SEISMIC GRAVITY MAGNETIC SERV UNIT MOBIL EXPLOR & PROD SERV, INC, 90- *Mem:* Am Geophys Union; Soc Explor Geophys; Europ Asn Explor Geophys; NY Acad Sci; Sigma Xi. *Res:* Gravity and magnetics; applied mathematics; seismic modeling; seismic signal processing; issued one patent. *Mailing Add:* Mobil Oil Corp-MEPSI Box 650232 Dallas TX 75265-0232

GINSBURG, NATHAN, b Casey, Ill, Aug 25, 10; m 42; c 1. BIOPHYSICS. *Educ:* Ohio State Univ, BA, 31, MS, 32; Univ Mich, PhD(physics), 35. *Prof Exp:* Res assoc physics, Univ Mich, 35-36; res assoc physics, Johns Hopkins Univ, 38-40, Johnston fel physics, 39, res assoc, 39-40; res assoc physics, Dept Embryol, Carnegie Inst, 40-42; asst prof physics, Univ Tex, Austin, 41-46; from asst prof to prof, Syracuse Univ, 46-76, dept chmn, 66-76; RETIRED. *Concurrent Pos:* Vis scientist, Naval Res Lab, Washington, 76-79; dean, Col Arts & Sci, Syracuse Univ, 64-65. *Mem:* Fel Am Phys Soc; fel Optical Soc Am. *Res:* Author of more than 30 publications in spectroscopy. *Mailing Add:* 1111 W 12th St No 114 Austin TX 78703

GINSBURG, ROBERT NATHAN, b Wichita Falls, Tex, Apr 26, 25; m 56. SEDIMENTOLOGY. *Educ:* Univ Ill, AB, 48; Univ Chicago, MA, 50, PhD(geol), 53. *Prof Exp:* Asst marine geol, Marine Lab, Univ Miami, 50-54; res geologist, Shell Develop Co, 54-60, sr res assoc geol, 60-65; prof geol & oceanog, Johns Hopkins Univ, 65-70; PROF SEDIMENTOLOGY, COMP SEDIMENTOLOGY LAB, ROSENSTIEL SCH MARINE & ATMOSPHERIC SCI, UNIV MIAMI, 70- *Concurrent Pos:* Queen's fel, Australia, 78; distinguished lectr, Am Asn Petrol Geologists, 84 & 85; pres, IUGS Global Sedimentary Geol Comn, 87. *Honors & Awards:* Cloos Lectr, Johns Hopkins, 80; Twenhofel Medal, Soc Econ Paleont & Mineral, 85; Gold Medal, Fla Acad Sci, 76. *Mem:* Fel Geol Soc Am; fel AAAS; Int Asn Sedimentol (vpres, 82-); hon Soc Econ Paleont & Mineral (pres, 68); Am Asn Petrol Geol; Am Geophys Union. *Res:* Recent sediments; coral reefs; ancient and modern algal structures; marine geology. *Mailing Add:* Sch Marine & Atmospheric Sci Univ Miami Comput Sedimentology Lab 4600 Rickenbacker Causeway Miami Beach FL 33139

GINSBURG, SEYMOUR, b Brooklyn, NY, Dec 12, 27; div; c 2. COMPUTER SCIENCE. *Educ:* City Col New York, BS, 48; Univ Mich, MS, 49, PhD(math), 52. *Prof Exp:* Asst prof math, Univ Miami, Fla, 51-55; mathematician, Nat Cash Register Co, 56-59; head systs synthesis & orgn sect, Hughes Res Labs, 59-60 & sr mathematician, Syst Develop Corp, 60-71; prof, 66-77, FLETCHER JONES PROF COMPUTER SCI, UNIV SOUTHERN CALIF, 77- *Concurrent Pos:* Guggenheim Fel, 74-75. *Mem:* Am Math Soc; Math Asn Am; Asn Comput Mach; fel Inst Elec & Electronics Engrs; Soc Indust & Appl Math. *Res:* Automata; formal language theory; grammar theory; theory of data bases. *Mailing Add:* Comput Sci Dept Univ Southern Calif Los Angeles CA 90089-0782

GINSBURG, VICTOR, b Singapore, Mar 22, 30; nat US; m 55; c 2. BIOCHEMISTRY. *Educ:* Univ Calif, BA, 52, PhD, 55. *Prof Exp:* CHIEF BIOCHEM SECT, NAT INST ARTHRITIS, METAB & DIGESTIVE DIS, 56- *Concurrent Pos:* Fel, Nat Found, 58-59. *Mem:* Am Chem Soc; Am Soc Biol Chem. *Res:* Carbohydrate biochemistry. *Mailing Add:* NIDDK Lab Struct Biol NIH Bldg 8 Rm 110 Bethesda MD 20892

GINSBURGH, IRWIN, b Brooklyn, NY, Apr 15, 26; m 46; c 4. ENGINEERING PHYSICS. *Educ:* City Col New York, BS, 47; Rutgers Univ, PhD(physics), 51. *Prof Exp:* Asst, Rutgers Univ, 47-50; res physicist, 51-57, sr proj physicist, 57-61, sr res supvr, 61-66, res assoc, 66-76, sr res assoc, Amoco Oil Co, Standard Oil Co Inc, 76-84. *Concurrent Pos:* Consult & legal expert, fire & explosion, 84- *Honors & Awards:* R & D 100 Award, Indust Res Mag, 68, 69, 71, 72 & 87. *Mem:* Am Phys Soc; Combustion Inst. *Res:* Engineering physics; fuel technology and petroleum engineering; acoustics; atmospheric chemistry and physics; hydrogen explosions; static electricity; automation; explosions; vapor recovery; air and water conservation; new energy sources. *Mailing Add:* 24125 Clearbank Newhall CA 91321

GINSKI, JOHN MARTIN, b Chicago, Ill, May 26, 26; m 56; c 3. PHYSIOLOGY. *Educ:* Loyola Univ, Ill, BS, 49, MS, 53, PhD(physiol), 56. *Prof Exp:* Res assoc pharmacol, Univ Chicago, 55-56; res assoc physiol & pharmacol, Col Med, Univ Sask, 57; from instr to asst prof, Col Med, Univ Nebr, 57-60; asst prof, Med Units, 60-67, ASSOC PROF PHYSIOL & BIOPHYS, UNIV TENN CTR HEALTH SCI, MEMPHIS, 67- *Concurrent Pos:* Vis prof, Univ Valle, Columbia, 63-64. *Mem:* AAAS; Am Physiol Soc; Soc Exp Biol & Med. *Res:* Electrolyte exchange across cellular membranes and the correlation of these mechanisms with physiological activity of the cells. *Mailing Add:* Dept Physiol & Biophys Univ Tenn Ctr Health Sci 894 Union Memphis TN 38163

GINSLER, VICTOR WILLIAM, biochemistry, for more information see previous edition

GINSPARG, PAUL H, b Chicago, Ill. QUANTUM FIELD THEORY, STATISTICAL PHYSICS. *Educ:* Harvard Univ, BA, 77; Cornell Univ, PhD(physics), 81. *Prof Exp:* Jr fel, Soc Fels, Harvard Univ, 81-84, from asst prof to assoc prof physics, 84-90; STAFF MEM, LOS ALAMOS NAT LAB, 90- *Concurrent Pos:* Vis, Prof CEN Saclay, 79-80, Princeton Univ, 82-, Stanford Linear Accelerator Ctr, 87- *Res:* Quantum field theory; string theory; confromal field theory; statistical mechanics. *Mailing Add:* Los Alamos Nat Lab MS-B285 Los Alamos NM 87545

GINTAUTAS, JONAS, b Justinava, Lithuania, Oct 3, 38; US citizen; m 70; c 2. RESEARCH ADMINISTRATION, SCIENCE EDUCATION. *Educ:* Moscow Pedag Inst, BS & MS, 65; Northwestern Univ, PhD(appl neurosci & speech path); Autonomy Univ Juarez, MD, 84. *Prof Exp:* Assoc prof speech path, psychol & neurosci, Texas Tech Univ Health Sci Ctr, 75-77, fel physiol, 77-79, assoc prof anesthesiol & dir res labs, 79-82; DIR BASIC & CLIN RES & PROF NEUROL, BROOKDALE HOSP MED CTR, BROOKLYN, NY, 85- *Concurrent Pos:* Vpres med affairs, Albert Int Corp, NY, 86-; ed consult, J Aphasia, Apraxia & Agnosia, 79- *Mem:* AAAS; Int Anesthesia Res Soc; NY Acad Sci. *Res:* Cardiovascular and smooth muscle physiology; pharmacology; action of calcium channel blocks and local anesthetics; higher cortical function disorders. *Mailing Add:* Basic & Clin Res Dept Brookdale Hosp Med Ctr Brooklyn NY 11212

GINTER, MARSHALL L, b Chico, Calif, Aug 24, 35; m 57; c 2. ATOMIC & MOLECULAR STRUCTURE, VUV INSTRUMENTATION & RADIATION. *Educ:* Chico State Col, BS, 58; Vanderbilt Univ, PhD(phys chem), 61. *Prof Exp:* Res assoc phys chem, Vanderbilt Univ, 59-60, dir spectros lab, 61-62; res assoc physics, Univ Chicago, 62-66; from asst prof to assoc prof, 66-74, PROF MOLECULAR PHYSICS, UNIV MD, COLLEGE PARK, 74- *Concurrent Pos:* Consult, Naval Res Lab, Washington, DC, 67-; guest worker, Nat Bur Standards, Gaithersburg, Md, 84- *Mem:* Fel Am Phys Soc; Am Chem Soc; fel Optical Soc Am; Sigma Xi. *Res:* Atomic and molecular structure; high resolution electronic spectroscopy; electronic structure of small molecules; atomic and molecular Rydberg states; VUV light sources and instrumentation. *Mailing Add:* Inst Phys Sci & Technol Univ Md College Park MD 20742

GINTHER, ROBERT J, b Lewiston, Maine, July 19, 17; m 43; c 3. INORGANIC CHEMISTRY. *Educ:* Northeastern Univ, BS, 40. *Prof Exp:* Chemist, Sylvania Elec Prod, Inc, 40-46; chemist & head res group, struct & compos luminescent mat, 46-69, head cent mat res activ, 69-73, RES PHYSICIST, MAT SCI DIV, NAVAL RES LAB, 73- *Mem:* Am Phys Soc; Electrochem Soc; Sigma Xi; Am Ceramic Soc (vchmn glass div, 75). *Res:* Luminescence of inorganic crystals and glasses. *Mailing Add:* 3107 Myrtle Ave Temple Hills MD 20748

GINZBERG, ELI, b New York, NY. RESOURCE MANAGEMENT, ECONOMICS. *Educ:* Columbia Univ, AB, 31, AM, 32, PhD, 34. *Hon Degrees:* DLitt, Jewish Theol Seminary Am, 66 & Columbia Univ, 82; LLD, Loyola Univ, 69. *Prof Exp:* A Barton Hepburn prof, 35-79, A BARTON HEPBURN EMER PROF ECON & SPEC LECTR, GRAD SCH BUS, COLUMBIA UNIV, 79- *Concurrent Pos:* Spec asst to chief statistician, US War Dept, 42-44; dir, Resources Anal Div, Surgeon Gen's Off, 44-46; mem, Med Adv Bd, Secy War, 46-48; dir, NY State Hosp Study, 48-49, med consult, Hoover Comn, 52; bd gov, Hebrew Univ, Jerusalem, 53-59; chmn, Nat Manpower Adv Comt, 62-73 & Nat Comn Manpower; consult, Depts State & Labor & Gen Acctg Off; co-chair adv comt, Job Creation Proj, Nat Comt Full Employment, 84- *Honors & Awards:* Martin Mem lectr, Am Col Surgeons, 80. *Mem:* Inst Med-Nat Acad Sci; Am Acad Arts & Sci; fel AAAS; Am Econ Asn; assoc mem Soc Med Consult Armed Forces. *Res:* Author of 100 books in economics. *Mailing Add:* Eisenhower Ctr Conserv Human Resources Columbia Univ New York NY 10027

GINZBURG, LEV R, b Moscow, USSR, Jan 11, 45; m 70; c 2. THEORETICAL POPULATION BIOLOGY. *Educ:* Univ Leningrad, MS, 67; Agrophys Inst Leningrad, PhD(math biol), 70. *Prof Exp:* Res assoc, Agrophys Inst Leningrad, 67-70, sr res assoc, 70-75; asst prof math, Northeastern Univ, Boston, Mass, 76-77; assoc prof, 77-83, PROF ECOL & EVOLUTION, STATE UNIV NY, STONY BROOK, 83- *Concurrent Pos:* Pres, Appl Biomath, 82- *Res:* Theoretical population biology, including macroevolutionary theory, population genetics, mathematical models in ecology and ecological genetics; applied ecology relating to ecological risk analysis as a methodology for environmental impact assessment. *Mailing Add:* Dept Ecol & Evolution State Univ NY Stony Brook NY 11794

GINZEL, KARL-HEINZ, b Reichenberg, Czech, June 1, 21; US citizen; m 58; c 2. PHARMACOLOGY. *Educ:* Univ Vienna, MD, 48. *Prof Exp:* Res asst pharmacol, Vienna, 48-53; sr sci officer, Biol Sect, Glaxo Labs, Eng, 54-55; sr lectr neuro-pharmacol, Inst Neurol, Nat Hosp Nerv Dis, London, 57-60; head neuropharmacol sect, Riker Labs, Calif, 60-70; PROF PHARMACOL, UNIV ARK MED SCI, LITTLE ROCK, 71- *Concurrent Pos:* WHO res fel pharmacol, Oxford Univ, 52-53; sr res fel neurophysiol & pharmacol, Univ Birmingham, 55-57; res assoc anatomist, Univ Calif, Los Angeles, 69-71; vis prof, Cardiovasc Res Inst, Univ Calif, San Francisco, 76. *Mem:* Am Soc Pharmacol & Exp Therapeut; Brit Physiol Soc; Ger Pharmacol Soc. *Res:* Pharmacology of neuromuscular and ganglionic transmission; introduction to clinical use of succinylcholine; central muscle relaxant agents; respiratory stimulant vanillic acid diethyamide; psychotomimetic drugs; 5-hydroxytryptamine; autonomic and cardiovascular pharmacology; neuropharmacology; sensory stimulants: prostaglandins, nicotine; educational programs on tobacco; expert reviewer of the Surgeon General's Report 87, Nicotine Addiction. *Mailing Add:* Dept Pharmacol Univ Ark Med Sci 4301 W Markham Little Rock AR 72205

GINZTON, EDWARD LEONARD, b Dnepropetrovsk, Ukraine, Dec 27, 15; US citizen; m 39; c 4. APPLIED PHYSICS. *Educ:* Univ Calif, BS, 36, MS, 37; Stanford Univ, EE, 38, PhD, 40. *Prof Exp:* Res engr, Sperry Gyroscope Co, NY, 40-46; asst prof appl physics & elec eng, Stanford Univ, 46-47; chmn bd dirs, Varian Assocs, 59-84, chief exec officer, 59-72, pres, 64-68, MEM BD DIRS, VARIAN ASSOCS, 48-, CHMN, EXEC COMT, 84- *Concurrent Pos:* From assoc prof to prof elec eng, Stanford Univ, 47-68, dir microwave lab, 49-59, dir Proj M, 57-60; co-chmn, Stanford Mid-Peninsula Urban Coalition, 68-72; mem, bd dir Mid-Peninsula Coalition Housing Develop Corp, 70-; chmn, Comt Motor Vehicle Emissions, Nat Acad Sci, 71-74, co-chmn, Comt Nuclear Energy Study, 75-80, mem, Comt Sci & Nat Security, 82-84 & Comt Lab Animals Biomed & Behav Res, 85-88; mem coun, Nat Acad Eng, 74-80; mem bd dir, Stanford Hosp, 75; mem bd trustees, Stanford Univ, 77- *Honors & Awards:* Liebmann Mem Prize, Inst Elec & Electronics Eng, 58, Medal of Honor, 69. *Mem:* Nat Acad Sci; Nat Acad Eng; AAAS; Am Acad Arts & Sci; fel Inst Elec & Electronics Eng; Nat Acad Econ Res. *Res:* Microwave tube and measurements; linear electron accelerators; circuits. *Mailing Add:* Varian Assoc Inc 3100 Hansen Way PO Box 10800 Palo Alto CA 94303

GIOANNINI, THERESA LEE, b Galesburg, Ill, Nov 21, 49; m 79; c 2. BIOORGANIC & ORGANIC CHEMISTRY. *Educ:* St Mary-of-the-Woods Col, BS, 71; NY Univ, MS, 76, PhD(chem), 78. *Prof Exp:* Biochem technician spec chem clin lab, Mercy Hosp, 71-73; asst res scientist chem opiate receptor, 77-84, RES ASST PROF, DEPT PSYCHOL, NY UNIV MED CTR, 84-; ASSOC PROF, DEPT NATURAL SCI, BARUCH COL, CITY UNIV NEW YORK, 90- *Mem:* Am Chem Soc; AAAS; NY Acad Sci. *Res:* Use of photo affinity labels to investigate biological systems; chemical characterization and isolation of the opiate receptors; examination of biological systems with photophysical probes. *Mailing Add:* Dept Natural Sci Baruch Col City Univ New York 17 Lexington Ave New York NY 10010

GIOGGIA, ROBERT STEPHEN, b Bronx, NY, Sept 29, 43. OPTICS. *Educ:* NY Univ, BS, 65; Trinity Col, MS, 67; Bryn Mawr Col, PhD(physics), 75. *Prof Exp:* From instr to assoc prof, 67-84, PROF PHYSICS, WIDENER COL, 84- *Mem:* Am Asn Physics Teachers; Am Phys Soc; Sigma Xi; Optical Soc Am. *Res:* Instabilities and chaos in lasers. *Mailing Add:* Dept Physics Widener Col Chester PA 19013

GIOIA, ANTHONY ALFRED, b Torrington, Conn, Apr 7, 34; m 78; c 3. ARITHMETIC FUNCTIONS. *Educ:* Univ Conn, BA, 55; Univ Mo, MA, 61, PhD(math), 64. *Prof Exp:* Instr math, Univ Mo, 59-64; asst prof, Tex Tech Univ, 64-66; assoc prof, 66-74, PROF MATH, WESTERN MICH UNIV, 74- *Mem:* Am Math Soc; Math Asn Am; Fibonacci Asn. *Res:* Theory of numbers; arithmetic functions. *Mailing Add:* Dept Math Western Mich Univ Kalamazoo MI 49008-3899

GIOLLI, ROLAND A, b San Vito, Italy, Feb 22, 34; US citizen; m 63; c 3. NEUROANATOMY, HISTOLOGY. *Educ:* Univ Calif, Davis, AB, 56; Univ Calif, Berkeley, PhD(anat), 60. *Prof Exp:* Asst anat, Univ Calif, Berkeley, 57-59, res anatomist, Sch Optom, 60-63; from asst prof anat to assoc prof, 64-70, PROF ANAT, NEUROBIOL & BIOL SCI, UNIV CALIF, IRVINE-CALIF COL MED, 74- *Concurrent Pos:* Ment Health Training Prog fel, Dept Anat & Brain Res Inst, Univ Calif, Los Angeles, 63-64; Alexander von Humboldt Sr Scientist Award, 82-83. *Mem:* Am Asn Anat; Soc Neurosci. *Res:* Visual system; visual-vestibular mechanisms; oculomotor system; neurotransmitters. *Mailing Add:* Dept Anat & Neurobiol Col Med Univ Calif Irvine CA 92717

GIOMETTI, CAROL SMITH, b Milford, Conn, Aug 10, 50; m 79; c 2. TWO DIMENSIONAL ELECTROPHORESIS, PROTEIN BIOCHEMISTRY. *Educ:* Knox Col, Galesburg, Ill, BA, 72; Rush Med Col, Rush-Presby-St Luke's Med Ctr, Chicago, Ill, MT, 73; Univ Ill, Chicago, PhD(biochem), 78. *Prof Exp:* Teaching asst biochem, Univ Ill, Chicago, 73-; fel, 78-82, asst biologist, 82-83, ASST BIOCHEMIST, ARGONNE NAT LAB, ILL, 83- *Honors & Awards:* Young Clin Chemist Res Award, Am Asn Clin Chem, 84. *Mem:* Am Soc Cell Biol; Electrophoresis Soc; Am Asn Clin Chem; Am Soc Clin Pathologists. *Res:* Two dimensional gel electrophoresis used to detect protein alterations related to genetic variation and-or pathological processes in human cells and animal model systems; proteins found to be markers for specific abnormalities are characterizied biochemically and specific antisera are developed. *Mailing Add:* Div Biol & Med Res Argonne Nat Lab 9700 S Cass Ave Argonne IL 60439

GION, EDMUND, b Altheimer, Ark, Sept 27, 29; m 60; c 3. PHYSICS. *Educ:* Reed Col, BA, 59; Lehigh Univ, MS, 61, PhD(physics), 65. *Prof Exp:* RES PHYSICIST, BALLISTIC RES LAB, ABERDEEN PROVING GROUND, 65- *Mem:* Am Phys Soc; Am Defense Preparedness Asn. *Res:* Shock tubes; blast waves; non-equilibrium flow phenomena; fluid dynamics. *Mailing Add:* 809 S Washington St Havre De Grace MD 21078

GIORDANO, ANTHONY B(RUNO), b New York, NY, Feb 1, 15; m 39; c 1. ELECTRONICS ENGINEERING, RESEARCH ADMINISTRATION. *Educ:* Polytech Inst Brooklyn, BEE, 37, MEE, 39, DEE, 46. *Prof Exp:* From instr to prof, 39-85, from assoc dean to dean grad studies, 57-85, assoc provost, 83-85, EMER PROF ELEC ENG & EMER DEAN, POLYTECH UNIV, 85- *Concurrent Pos:* Asst. Off Sci Res & Develop Proj, 42-45; sr res assoc, Microwave Res Inst, 45-49, res supvr, 49-60; assoc ed, Radio Sci, 72-75; vpres, Eng Found, 77-79, pres, 80. *Honors & Awards:* Western Elec Fund Award, Am Soc Eng Educ, 81, Collins Achievement Award, 86; Centennial medal, Inst Elec & Electronics Engrs, 84, Outstanding Leadership Award, 90. *Mem:* Fel Am Soc Eng Educ (vpres, 74-75, 84-86, pres-elect, 88-89, pres, 89-90); fel Inst Elec & Electronics Engrs; fel AAAS; Int Sci Radio Union; Sigma Xi. *Res:* Lightning research; precision waveguide attenuators utilizing thin metallic films; radar systems. *Mailing Add:* Off Acad Affairs Polytech Univ NY 333 Jay St Brooklyn NY 11201

GIORDANO, ARTHUR ANTHONY, b Boston, Mass, Sept 28, 41; m 64; c 2. DIGITAL COMMUNICATIONS, DIGITAL SIGNAL PROCESSING. *Educ:* Northeastern Univ, BS, 64, MS, 66; Univ Pa, PhD(elec eng), 70. *Prof Exp:* Eng specialist, GTE Sylvania, 70-72; sr tech staff, Stein Assocs, 72-74; eng specialist, GTE Sylvania, 70-72, sr eng specialist, 78-80, radio systs dept mgr, 80-81, survivable systs dept mgr, 81-85; VPRES SYSTS, CNR, INC, 85- *Concurrent Pos:* Instr, Northeastern Univ, 73-75 & 78-79, Tufts Univ, 80-81 & Northeastern Univ, 81-; vchmn, Comn E, Int Union Radio Sci, 78-81, chmn, 81-84. *Mem:* Int Union Radio Sci; Inst Elec & Electronics Engrs. *Res:* Design and performance evaluation of various radio systems, including waveform design (modulation and coding), channel modeling, receiver design, network design, propagation analysis, digital signal processing, spread spectrum and message handling systems. *Mailing Add:* VP Systs CNR 220 Reservoir St Needham MA 02194

GIORDANO, NICHOLAS J, b Frankfort, Ind, Dec 21, 51; m 73; c 1. THERMAL & SOLID STATE PHYSICS. *Educ:* Purdue Univ, BS, 73; Yale Univ, PhD(eng & appl sci). *Prof Exp:* Asst prof eng & appl sci, Yale Univ, 77-79; ASST PROF PHYSICS, PURDUE UNIV, 79- *Concurrent Pos:* Vis scientist, Hahn-Meitner Inst, Berlin, 77. *Mem:* Am Phys Soc. *Res:* Fabrication and properties of microstructures; low frequency noise in solids. *Mailing Add:* Physics Dept Purdue Univ West Lafayette IN 47907

GIORDANO, PAUL M, b Providence, RI, Apr 4, 36. AGRICULTURAL RESEARCH. *Educ:* Univ RI, BS, 59, MS, 60; Univ Conn, PhD(soil chem), 64. *Prof Exp:* MEM STAFF TENN VALLEY AUTHORITY. *Mem:* Soil Sci Soc Am; Am Soc Agron; Nat Mgt Asn. *Mailing Add:* Tenn Valley Authority F 137N FERC Bldg Muscle Shoals AL 35660-1010

GIORDANO, THOMAS HENRY, b Philadelphia, Pa, Feb 13, 50. GEOCHEMISTRY. *Educ:* Millersville State Col, BA, 72; Pa State Univ, PhD(geochem), 78. *Prof Exp:* Fel geochem, Depts Physics & Earth Sci, 77-79, ASST PROF GEOL, NMEX STATE UNIV, 80- *Mem:* Am Chem Soc; Am Geophys Union; Soc Econ Geologists; Geochem Soc. *Res:* Chemical hydrothermal systems; chemical speciation in natural waters; experimental determination of mineral solubilities; transport and deposition of ore minerals. *Mailing Add:* Dept Earth Sci NMex State Univ Box 3AB Las Cruces NM 88003

GIORDANO, TONY, b Cleveland, OH, May 8, 60; m 82; c 2. GENE REGULATION, MRNA PROTEIN INTERACTIONS. *Educ:* New Col Univ SFla, BA, 83; Ohio State Univ, MSc, 84, PhD(molecular genetics), 88. *Prof Exp:* Fel biotechnol, Nat Cancer Inst, NIH, 88-90, staff fel, Nat Inst Aging, 90-91; RES SCIENTIST, ABBOTT LABS, 91- *Mem:* Am Soc Cell Biol; Soc Neurosci. *Res:* Gene regulation in Alzheimer's disease; clone genes novel to the disease. *Mailing Add:* Abbott Lab Dept 47W AP-10 Abbott Park IL 60064

GIORDMAINE, JOSEPH ANTHONY, b Toronto, Ont, Apr 10, 33; m 58; c 3. OPTICAL PHYSICS, SOLID STATE PHYSICS. *Educ:* Univ Toronto, BA, 55; Columbia Univ, AM, 57, PhD(physics), 60. *Prof Exp:* Instr physics, Columbia Univ, 59-61; dir chem physics res, Bell Labs, 71-76, Solid State Electronics, 74-81, Electronic & Photonic Technol, 81-83, III & V Electronic & Display, 83-85, dir, Advan Technol Develop, 85-87, mem tech staff, 61-89; VPRES, NEC RES INST, 89- *Concurrent Pos:* Consult, Fairchild Camera & Instrument Corp, 61; lectr, Univ Calif Exten, 64-68; vis prof, Munich Tech Univ, 66; mem comt basic res advan, Army Res Off, Nat Res Coun, 67-72, panel atomic, molecular & electronics physics, Physics Surv Comt, 70-71; adv ed, Optics Commun, 70-75 & J Non-Metals 72-80; mem comt on atomic & molecular physics, Nat Res Coun, 71-74; vis comt, Mass Inst Technol Mat Magnet Lab, 71-73; nomination comt, Am Physical Soc, 73, Frack Isakson Prize Award Comt, 81; prog chmn, Int Quantum Electronics Conf, 76, chmn, 78; assoc ed, Optics Lett, 77-79; mem steering comt, Conf Laser Eng & Appln, 77-79; mem comt employment foreign scientist to US, Nat Res Coun, 79-80, US liason comt, IUPAP, 81-87; ed comt, Ann Rev Mats Sci Asn, 79-, assoc ed, 83-; mem, Inst Imaging Sci Adv Comt, Polytechnic Inst, NY, 81; mem Comn Quantum Electronics, Int Union Rive & Appl Physics, 81-87, vchmn, 86-87. *Mem:* AAAS; fel Am Phys Soc; sr mem Inst Elec & Electronics Eng; Am Astron Soc; Optical Soc Am; Sigma Xi. *Res:* Molecular beams; paramagnetic resonance and relaxation; masers and lasers; nonlinear optical effects; optical properties of solids; radio astronomy; solid state electronics. *Mailing Add:* Four Independence Way Princeton NJ 08540

GIORGI, ANGELO LOUIS, b Syracuse, NY, July 18, 17; m 46; c 2. RADIOCHEMISTRY. *Educ:* Syracuse Univ, BS, 39; Univ NMex, MS, 54, PhD, 57. *Prof Exp:* Chemist, Gen Motors Corp, 40-42; jr chemist, US Bur Mines, 42-44; asst chemist, Naval Res Lab, 44-46; MEM STAFF RADIOCHEM, LOS ALAMOS NAT LAB, 46- *Mem:* Am Phys Soc; Mat Res Soc. *Res:* Superconductivity; study of superconducting properties of refractory carbides and nitrides; magnetism and superconductivity in actinide metals. *Mailing Add:* 151 El Gancho Los Alamos NM 87544

GIORGI, ELSIE A, b New York, NY. HEALTH CARE PLANNING. *Educ:* Hunter Col, BA, 31; Col Physicians & Surgeons, Columbia Univ, MD, 49; Am Bd Internal Med, dipl, 56. *Prof Exp:* Asst prof clin med, Cornell Univ Med Col, 57-62; asst prof, Univ Calif, Los Angeles, 62-66; asst prof med, Univ Southern Calif, Sch Med, 66-69; assoc clin prof community & family med, Univ Calif, Col Med, Irvine, 69-72; ASSOC CLIN PROF MED, SCH MED, UNIV CALIF, LOS ANGELES, 72- *Concurrent Pos:* Dir med, NY Infirmary, 60-61; attend physician, Dept Med, Cedars-Sinai Hosp, Los Angeles, 62-; guest lectr, Sch Social Welfare, Univ Calif, Los Angeles, 64-81; attend physician, Los Angeles Co Hospital, Univ Southern Calif Med Ctr, 66-71; assoc mem, Dept Internal Med, Orange Co Med Ctr, 69-; mem staff, St John's Hosp, Calif, 70- *Mem:* Inst Med-Nat Acad Sci; Geront Soc; Am Pub Health Asn. *Mailing Add:* 153 S Laskey Dr Suite 3 Beverly Hills CA 90212

GIORGI, JANIS V(INCENT), b Meadville, Pa, May 23, 47. FLOW CYTOMETRY, LYMPHOCYTE DIFFERENTIATION. *Educ:* Univ Fla, Gainesville, BA, 69, Univ NMex, PhD(microbiol & immunol), 77. *Prof Exp:* Fel, 77-80, asst resident prof, path, Univ NMex Sch Med, 80-81; instr surg, immunol, Harvard Med Sch, 81-83; ASST PROF, IMMUNOL, UNIV CALIF SCH MED, LOS ANGELES, 84- *Concurrent Pos:* Asst surg, immunol, Mass Gen Hosp, 81-83. *Mem:* Am Asn Immunologists; Asn Women in Sci; Clin Immunol Soc; Soc Analytic Cytol; Sigma Xi. *Res:* Basic laboratory research on human immunology, especially immune deficiencies, virus-host interactions, and transplantation biology. *Mailing Add:* Dept Med/CIA/CIC Univ Calif Sch Med 10833 Le Conte Ave Los Angeles CA 90024-1745

GIORGIO, ANTHONY JOSEPH, b Hartford, Conn, Feb 8, 30; m 61; c 3. HEMATOLOGY, BIOCHEMISTRY. *Educ:* Boston Univ, AB, 52, MD, 57; Columbia Univ, MSPH, 53; Am Bd Internal Med, dipl; Am Bd Hematol, dipl. *Prof Exp:* Instr med, Univ Utah, 62-66; asst prof pediat, NY Med Col, 67-70; asst prof med, Sch Med, Univ Pittsburgh, 70-72; chief, Hemat Div, Martin Luther King, Jr Gen Hosp, 72-78; PROF MED, CHARLES R DREW POSTGRAD MED SCH, UNIV CALIF, LOS ANGELES, 72-; ASSOC PROF, UNIV SOUTHERN CALIF, 72- *Concurrent Pos:* NIH fel biochem, Univ Utah, 64-66; NIH res grant, 69- *Mem:* Am Inst Nutrit; Am Soc Clin Nutrit; Am Fedn Clin Res; Am Soc Hematology; fel Am Col Physicians. *Res:* Nutritional anemias; vitamin B-12 metabolism; clinical laboratory methodology. *Mailing Add:* 3440 W Lomita Blvd Torrance CA 90505

GIORI, CLAUDIO, b Milano, Italy, June 2, 38; US citizen; m 63; c 2. POLYMER CHEMISTRY, MEDICAL APPLICATIONS OF PLASTICS. *Educ:* Univ Milano, Dr(polymer chem), 62. *Prof Exp:* Res scientist, ENI Res Labs, 63-65 & Fibers Div, Allied Chem Corp, 65-68; res scientist, IIT Res Inst, 68-74, sr scientist, 74-81; sr res scientist, Borg-Warner Corp, 81-87; POLYMER SPECIALIST, HOLLISTER INC, 87- *Concurrent Pos:* Nat dir, Soc Aerospace Mat & Process Eng, 84-85. *Mem:* AAAS; Am Chem Soc; Soc Aerospace Mat & Process Eng. *Res:* Synthesis of high temperature resistant polymers; kinetics and mechanisms of polymerization; stereospecific polymerization; thermal degradation and stabilization of polymers; polymers for reverse osmosis processes; effect of radiation on polymers; polymers for medical devices. *Mailing Add:* 2975 Orange Brace Rd Riverwoods IL 60015

GIOVACCHINI, PETER L, b New York, NY, Apr 12, 22; m 45; c 3. PSYCHIATRY, PSYCHOSOMATIC MEDICINE. *Educ:* Univ Chicago, BS, 41, MD, 44. *Prof Exp:* Res fel psychiat, Univ Chicago Clins, 48-50; from clin asst prof to clin assoc prof, 54-62, CLIN PROF PSYCHIAT, UNIV ILL COL MED, 62- *Concurrent Pos:* Consult, Michael Reese Hosp, Chicago, 51-57, Wil Wilmette Family Serv Bur, Ill, 54-, United Charities, Chicago, 55-64 & Res & Educ Hosp, Chicago, 58-; vis prof psychiat, Univ Southern Calif, 68; ed-in-chief, Tactics & Tech Psychoanal Treat, 72-; co-ed, Adolescent Psychiat, 72-81; lectr, Am Psychother Inst, 73- *Mem:* Am Psychoanalytic Asn; fel Am Psychiat Soc; fel Am Orthopsychiat Asn. *Res:* Psychotherapeutic process; characterological disorders; psychosomatic medicine, particularly multiple psychosomatic conditions in the same patsents; psychotic states. *Mailing Add:* 505 N Lakeshore Dr Chicago IL 60611

GIOVACCHINI, RUBERT PETER, b Fresno, Calif, June 2, 28; m 49; c 3. MEDICINE. *Educ:* Creighton Univ, BS, 48, MSc, 54; Univ Nebr, PhD, 58. *Hon Degrees:* ScD, Univ Nebr, 80. *Prof Exp:* Instr anat, Col Med, Univ Nebr, 57-58; res histopathologist, Toni Co, Ill, 58-59, asst med dir, 60-64; dir med eval, Gillette Med Res Inst, Washington, DC, 64-67; dir med eval div & vpres, Gillette Res Inst, 67-70, dir med eval dept & vpres, Gillette Co Res Inst, 70-71, pres, Gillette Med Eval Labs, 71-74, VPRES CORP PROD INTEGRITY, GILLETTE CO, 74- *Concurrent Pos:* Assoc anesthesia dept, Bishop Clarkson Mem Hosp, 57-58. *Honors & Awards:* Gold Metal Award, Soc Cosmetic Chemists, 76. *Mem:* Soc Toxicol; Am Acad Dermat; Am Acad Clin Toxicol; Acad Toxicol Sci; Europ Soc Toxicol; Sigma Xi. *Res:* Toxicological and clinical evaluation of foods, drugs, chemicals and cosmetics. *Mailing Add:* 401 Professional Dr Gaithersburg MD 20879

GIOVANELLA, BEPPINO C, b Merano, Italy, June 12, 32; m 71; c 2. CANCER. *Educ:* Univ Rome, Laurea(biol), 56, Libera Docenza (gen path), 62. *Prof Exp:* Asst cancer res, Regina Elena Inst, Rome, 56-60; res assoc, McArdle Lab, Univ Wis-Madison, 62-70; LAB DIR, CANCER RES LAB, ST JOSEPH'S HOSP, 70- *Concurrent Pos:* Clin asst prof oncol, Col Med Baylor Univ, 71-75, adj asst prof oncol, 75-77, adj assoc prof, Baylor Col Med, 77- *Mem:* Am Asn Cancer Res; Am Asn Pathologists; NY Acad Sci. *Res:* Radiobiology of tumors; skin carcinogenesis; effects of supranormal temperatures on normal and neoplastic cells; chemotherapy of malignant tumors; immunobiology of neoplastic cells; heterotransplantation of human tumors in nude thymusless mice. *Mailing Add:* Cancer Res Lab St Joseph's Hosp Houston TX 77002

GIOVANELLI, RICCARDO, b Reggio Emilia, Italy, Aug 30, 46; m 77. RADIO ASTRONOMY. *Educ:* Univ Bologna, Laurea, 68; Ind Univ, Bloomington, MA, 71, PhD(astron), 76. *Prof Exp:* Res asst astron, Univ Bologna, 68-69; vis prof physics, Nat Univ El Salvador, 73-75; staff scientist astron, Univ Bologna, 76-77; vis prof, Univ Okla, 77-78; sr res assoc astron, Nat Astron & Ionosphere Ctr, 78-91, head, Radio Astron Div, 84-91; dir, Areciho Observ, 87-88; PROF ASTRON, CORNELL UNIV, 91- *Honors & Awards:* H Draper Medal Astron Physics, Nat Acad Sci, 89. *Mem:* Soc Ital Astron; Am Astron Soc; Int Astron Union; AAAS. *Res:* Structure of the interstellar medium; supernova remnants; high velocity hydrogen clouds; interacting galaxies; clusters of galaxies; cosmology. *Mailing Add:* Arecibo Observ PO Box 995 Arecibo PR 00613

GIOVANIELLI, DAMON V, b Teaneck, NJ, May 8, 43; m 68; c 2. PLASMA PHYSICS. *Educ:* Princeton Univ, AB, 65; Dartmouth Col, PhD(physics), 70. *Prof Exp:* Instr appl sci, Yale Univ, 70-72; mem staff, Los Alamos Nat Lab, 72-77, alt group leader, 77-79, assoc div leader, 79-80, prog mgr exp inertial fusion, 80-83, dep assoc dir, 83-87, PHYSICS DIV LEADER, LOS ALAMOS NAT LAB, 87- *Concurrent Pos:* Consult, Sci Appl, Inc, 80-84. *Mem:* Am Phys Soc; Am Asn Physics Teachers; Sigma Xi. *Res:* Inertial fusion. *Mailing Add:* Los Alamos Nat Lab MS D434 Los Alamos NM 87545

GIOVINETTO, MARIO BARTOLOME, b La Plata, Arg, May 5, 33; m 66; c 1. CLIMATOLOGY, SEA ICE. *Educ:* Univ Wis-Madison, MS, 66, PhD(geog), 68. *Prof Exp:* Res asst glaciol, Ohio State Univ, 59-61; res assoc geol, Univ Mich, 59; proj asst glaciol, Univ Wis-Madison, 61-67, actg instr climat, 67-68; from asst prof to assoc prof, Univ Calif, Berkeley, 68-73; prof & head dept, 73-83, PROF CLIMAT, UNIV CALGARY, 83- *Concurrent Pos:* Mem, Nat Res Coun, 71-74. *Mem:* Am Meteorol Soc; Am Geophys Union; Asn Am Geog; Am Water Resources Asn; Glaciol Soc. *Res:* Arctic and Antarctic glaciology; physical climatology of glaciers and its implications on paleoclimates; precipitation phenomena in arid and semi-arid lands. *Mailing Add:* Dept Geog Univ Calgary Calgary AB T2N 1N4 Can

GIPPIN, MORRIS, b New York, NY, Apr 4, 08; m 44; c 1. ORGANIC CHEMISTRY. *Educ:* Univ NC, AB, 35; Brooklyn Col, MA, 38; Western Reserve Univ, PhD(chem), 51. *Prof Exp:* Sr res chemist, Firestone Tire & Rubber Co, 51-65, sr res scientist, 65-70; RETIRED. *Mem:* Am Chem Soc. *Res:* Catalyst systems in transition metal solution polymerization of dienes; Ziegler-Natta catalysis; studies in polymer stereoregularity, macro structure, and polymer properties; aqueous free radical polymerization. *Mailing Add:* 220 Kenridge Rd Akron OH 44313

GIPSON, ILENE KAY, b Hoberg, Mo, Oct 13, 44; m. CELL BIOLOGY. *Educ:* Drury Col, BA, 66; Univ Ark, MS, 68, PhD(zool), 73. *Prof Exp:* Teaching asst zool, Univ Ark, 66-68, res asst plant path, 68-70, asst zool, 70-73; res assoc & instr cell biol cornea, Univ Ore, 74-76, asst prof, ophthal, Health Sci Ctr, 76-79; asst prof ophthal, Harvard Med Sch, Boston, 79-85; assoc scientist & head Morphol Unit, 79-84, SR SCIENTIST & HEAD CORNEA UNIT, EYE RES INST RETINA FOUND, BOSTON, 84-; ASSOC PROF OPHTHAL ANAT & CELL BIOL, HARVARD MED SCH, BOSTON, 85- *Concurrent Pos:* Consult vision res planning comt, Nat Eye Inst, 76, prin investr NIH Grant, 77-, res career develop award, 78-83, mem vision res prog comt, study sect, 79-84. *Honors & Awards:* Alcon Res Award, 85; Merit Award, Nat Eye Inst, 90. *Mem:* Am Soc Cell Biologists; Asn Res Vision & Ophthal. *Res:* Motility and adhesion of corneal epithelium. *Mailing Add:* Eye Res Inst 20 Staniford St Boston MA 02114

GIPSON, MACK, JR, b Trenton, SC, Sept 15, 31. GEOLOGY. *Educ:* Paine Col, BA, 53; Univ Chicago MS, 61, PhD(geol), 63. *Prof Exp:* Prof & chmn, Dept Geol Sci, Va State Univ, 64-75; res assoc, Exxon Prod Res Co, 75-82; mgr plastic geol & sr plastic geologist, NLER Co, 82-84; explorationist, Aminoil Inc, 84; sr proj specialist, Phillips Petrol, 84-86; PROF GEOL, UNIV SC, 86- *Mem:* Sigma Xi; fel Geol Soc Am; Asn Petrol Geologists; Nat Asn Black Geologists & Geophysicists; Am Geophys Union; Nat Tech Asn; Nat Asn Geol Teachers. *Mailing Add:* Dept Geol Univ SC Main Campus Columbia SC 29208

GIPSON, PHILIP, b Ft Smith, Ark, Jan 2, 43. WILDLIFE ECOLOGY. *Educ:* State Col Ark, BS, 64; Univ Ark, MS, 67, PhD(zool), 71. *Prof Exp:* Teacher sci, Juneau Pub Sch Syst, Alaska, 65-66; res assoc wildlife ecol, Univ Nebr, Lincoln, 71-74, asst prof wildlife ecol & exten wildlife specialist, 74-76; ASSOC PROF WILDLIFE MGT & ASST LEADER ALASKA COOP WILDLIFE RES UNIT, UNIV ALASKA, 76-; WILDLIFE BIOLOGIST, POLICY & PROG DEVELOP, PLANNING & RISK ANALYSIS SYSTS, USDA. *Concurrent Pos:* Consult, Ark Dept Planning, 72-73. *Mem:* Wildlife Soc; Am Soc Mammal; Sigma Xi. *Res:* Interrelationships of wildlife and agriculture, emphasizing damage assessment and control; predator and prey interactions; wildlife and habitat relationships. *Mailing Add:* USDA-APHIS/PPD/PRAS 6505 Belcrest Rd Fed Bldg Rm 806 Hyattsville MD 20782

GIPSON, ROBERT MALONE, b Odessa, Tex, Apr 9, 39; m 61; c 2. INDUSTRIAL ORGANIC CHEMISTRY. *Educ:* Abilene Christian Col, BS, 61; Univ Tex, MA, 63, PhD(org chem), 65. *Prof Exp:* Res chemist, Texaco Chem Co Austin Labs, 65-76, supvr applns res, 76-82, mgr develop, 82-86, mgr tech servs, 86-88; DIR, TEXACO PORT ARTHUR RES LABS, 88- *Mem:* Am Chem Soc; Catalysis Soc. *Res:* Catalysis; process research; surface active agents. *Mailing Add:* Texaco Port Arthur Res Labs PO Box 1608 Port Arthur TX 77641

GIRALDO, ALVARO A, b Colombia, SAm, Jan 7, 45; nat US; m; c 2. IMMUNOPATHOLOGY. *Prof Exp:* Internship, St Ignacio Univ Hosp, SAm, 67; resident path, Cent Mil Hosp, SAm, 69; internship, St John Hosp, Mich, 70, resident path, 71-74; fel path cancer, Univ Tex Syst Cancer Ctr, M D Anderson Hosp & Tumor Inst, 75; ASSOC PATHOLOGIST, ST JOHN HOSP, DETROIT, 76-, HEAD, DIV IMMUNOPATH, 78- *Concurrent Pos:* Fel immunol, Wayne State Univ, 76-78, adj asst prof, Dept Immunol & Microbiol, Sch Med, 78-88, Dept Path, 80-88, adj assoc prof, Dept Immunol, Microbiol & Path, 88-, vis fel histocompatability & molecular biol, 89-; vis fel cardiovasc path, NIH, 79; vis scholar hematopath, Univ Southern Calif, Los Angeles, 84. *Mem:* Am Soc Clin Pathologists; Int Acad Path; Am Asn Pathologists; Soc Hematopath; NY Acad Sci; AAAS; Soc Cardiovasc Path; Soc Med Lab Immunol; Am Asn Immunologists; Clin Immunol Soc. *Mailing Add:* Dept Path St John Hosp 22101 Moross Rd Detroit MI 48236

GIRARD, DENNIS MICHAEL, b Detroit, Mich, July 26, 39; m 62; c 3. MATHEMATICS, STATISTICS. *Educ:* Univ Detroit, BS, 61, MA, 62; Ohio State Univ, PhD(math), 68. *Prof Exp:* Aerospace engr, Lewis Res Ctr, NASA, 62-63; asst prof math, Ohio State Univ, 68-69; asst prof ecosyst anal, 69-73, assoc prof math & statist, 73-90, PROF INFO & COMPUT SCI, UNIV WIS-GREEN BAY, 90- *Mem:* AAAS; Am Math Soc; Math Asn Am; Am Statist Asn; Biometr Soc; Inst Math Statist; Inst Elec & Electronics Engrs. *Res:* Biomathematics; biostatistics; image analysis; statistical methodology in toxicology. *Mailing Add:* Col Environ Sci Univ Wis Green Bay WI 54301

GIRARD, FRANCIS HENRY, b Chicago, Ill, Dec 28, 35; m 58; c 3. TECHNOLOGY COMMERCIALIZATION, TECHNOLOGY DEVELOPMENT & IDENTIFICATION. *Educ:* De Paul Univ, BS, 57 & MS, 59; Northwestern Univ, PhD(chem), 64. *Prof Exp:* Instr chem, Evanston Hosp Sch Nursing, 60; res chemist, Toni Co, 64-67, sr res chemist, 67-69, res supvr 69-71; from asst res dir to res dir, Personal Care Div, Gillette Co, 72-78, dir prod develop, 78-90; dir, New Technol Develop, Toiletries Technol Labs, 90. *Concurrent Pos:* Ed, J Soc Cosmetic Chemists, 83-86. *Mem:* Am Chem Soc; Soc Cosmetic Chemists; Sigma Xi. *Res:* Cosmetic, toiletries and personal care products research and development; cosmetic science particularly chemistry and physics of hair and skin; polymer chemistry; phosphoramidic acid chemistry; guanidine chemistry; intermediary metabolism. *Mailing Add:* Ten Ivy Lane Sherborn MA 01770

GIRARD, G TANNER, b Jacksonville, Fla, May 15, 52; m 76; c 2. NATURAL RESOURCE POLICY, ORNITHOLOGY. *Educ:* Principia Col, BS, 74; Univ Cent Fla, MS, 76; Fla State Univ, PhD(sci educ), 79. *Prof Exp:* ASSOC PROF BIOL & ENVIRON SCI, BIOL DEPT, PRINCIPIA COL, ELSAH, ILL, 77- *Concurrent Pos:* Mem, Ill Nature Preserves Comn, 87-, chmn, 89-; vis prof, Univ del Valle, Guatemala, 88, consult, Surv Proposed Nat Parks, 89; consult, Ill River's Proj, 90- *Mem:* Am Inst Biol Sci; Am Ornithologists Union. *Res:* Ecology; natural resource policy; ornithology; science education; wildlife management. *Mailing Add:* Principia Col Elsan IL 62028

GIRARD, JAMES EMERY, b Joliet, Ill, July 1, 45; m 70; c 6. ANALYTICAL CHEMISTRY. *Educ:* Lewis Col, BA, 67; Pa State Univ, PhD(org chem), 71. *Prof Exp:* Scholar chem, Pa State Univ, 71-72; fel, Univ Calif, San Diego, 72-73; asst prof chem, Col Holy Cross, 73-76; staff scientist, Gen Elec Corp Res & Develop, 77-79; assoc prof, 79-84, PROF & CHMN DEPT CHEM, AMERICAN UNIV, 85- *Concurrent Pos:* consult & expert witness. *Mem:* AAAS; Am Chem Soc; Sigma Xi. *Res:* High pressure liquid chromatography, gel permeation chromatography, gas chromatography; ion chromatography; cancerostatic agents; pharmaceuticals; environment. *Mailing Add:* Chem Dept Am Univ Washington DC 20016

GIRARD, KENNETH FRANCIS, b Cohoes, NY, Sept 20, 24; m 53; c 5. BACTERIOLOGY, IMMUNOLOGY. *Educ:* Siena Col, BS, 48; McGill Univ, MSc, 50, PhD(bact), 52. *Prof Exp:* Mgr educ serv, Lederle Labs, Am Cyanamid Co, 52-54; asst prof bact & immunol, Med Sch, Dalhousie Univ, 54-56; asst prof, McGill Univ, 56-58; asst prof, Sch Med, Tufts Univ, 59-62; tech dir, Clin Labs, Boston Dispensary, New Eng Med Ctr, 59-61; ASST DIR HEALTH SERV, MASS DEPT PUB HEALTH, 61- *Concurrent Pos:* Res assoc, Sch Pub Health, Harvard Univ, 62-82; lectr, Simmons Col, 64-70. *Mem:* Am Soc Microbiol; Am Asn Path; emer mem Am Bd Med Microbiol. *Res:* Listeriosis; rabies; syphilis serology. *Mailing Add:* Mass State Dept Pub Health 305 S St Jamaica Plain MA 02130

GIRARDEAU, MARVIN DENHAM, JR, b Lakewood, Ohio, Oct 3, 30; m 56; c 3. STATISTICAL MECHANICS, QUANTUM MANY-BODY THEORY. *Educ:* Case Inst Technol, BS, 52; Univ Ill, MS, 54; Syracuse Univ, PhD(physics), 58. *Prof Exp:* NSF fel physics, Inst Advan Study, Princeton Univ, 58-59; res assoc, Brandeis Univ, 59-60; mem staff, Boeing Sci Res Labs, Wash, 60-61; res assoc, Enrico Fermi Inst Nuclear Studies, Univ Chicago, 61-63; assoc prof physics, 63-67, dir, Inst TheoretSci, 67-69, chmn dept, 74-76, PROF PHYSICS, UNIV ORE, 67-, MEM, INST THEORET SCI, 63-, MEM, CHEM PHYS INST, 81- *Concurrent Pos:* Prin investr, NSF, 65-78 & Off Naval Res, 81-87; vis prof Nordic Inst Theoret Atomic Physics, Norges Tekniske Hogskole, Trondheim, Norway, 69-70, Univ Colo, 78, Univ Paul Sahatier, Toulouse, France, 79 & Univ Nice, France, 80; vis scientist, Univ Libre, Bruxelles, Belgium, 70, Max Planck Inst, Fed Repub Germany, 84-86; mem, physics rev panel, Nat Res Coun, 79, 81, 84, 85. *Honors & Awards:* Alexander von Humboldt Prize, Fed Repub Germany, 84. *Mem:* Fel Am Phys Soc; Sigma Xi. *Res:* Quantum many-body problems; statistical mechanics; atomic molecular and chemical physics; construction of second-quantization representations for composite-particle systems and application to theory of reactive collisions. *Mailing Add:* Inst Theoret Sci Univ Ore Eugene OR 97403-1274

GIRARDI, ANTHONY JOSEPH, b Philadelphia, Pa, Mar 19, 26; m 58; c 2. MICROBIOLOGY. *Educ:* Pa State Univ, BS, 49; Univ Pa, PhD(med microbiol), 52. *Prof Exp:* Instr prev med, Sch Med, Yale Univ, 52-53; instr virol, Univ Pa, 53-56; chief res, Microbiol Res Found, 56-59; res assoc, Merck Inst Therapeut Res, 59-63; res assoc, Wistar Inst, 63-70, assoc mem, 70-74; dir, E Tenn Cancer Res Ctr, 74-77; SR MEM & HEAD DEPT VIROL & IMMUNOL, INST MED RES, 77- *Concurrent Pos:* NIH res fel, Sch Med, Yale Univ, 52-53, Nat Cancer Inst grant, 57-; res assoc, Children's Hosp, Philadelphia, 53-56; consult, Microbiol Assocs, Inc, 53-56. *Mem:* Tissue Cult Asn; NY Acad Sci; Am Asn Cancer Res; Am Asn Immunol; Sigma Xi. *Res:* Virology; cancer virus; tumor immunology. *Mailing Add:* 14 E Wayne Terr Village Pine Run Blackwood NJ 08012

GIRARDOT, JEAN MARIE DENIS, b Lons Le Saunier, France, Nov 10, 44; m 66; c 2. ROLE OF VITAMIN K IN CARBOXYLATION OF BLOOD PROTEIN, CALMODULIN IN BRAIN HIPPOCAMPUS. *Educ:* Univ Louis Pasteur, France, BS, 66; Univ Okla, MS, 76, PhD(biochem & molecular biol), 77. *Prof Exp:* Res tech, Neurochem Ctr, Strasbourg, France, 66-70; sr res tech, Merrell Toraude Strasbourg, France, 70-72, Okla Med Res Found, Oklahoma City, 72-74; post-doctorate, Univ Louis Pasteur, France, 77-80; asst prof biochem, Found & Dept Biochem, Univ Okla Health Sci Ctr, 80-84; sr res sci, Kimberly Clark Corp, 84-87; PRES, BIOMED DESIGN INC, 87- *Concurrent Pos:* Asst mem biochem, Okla Med Res Found, 84; grant referee, NSF & NIH, 83-84; prin investr, Small Bus Innovation Res Prog, Nat Heart Blood & Lung Inst, NIH, 89; consult, Medtronic Heart Valve Div, 88- *Mem:* Soc Biochem & Molecular Biol; Soc Exp Biol & Med; Soc Biomat. *Res:* Fixation of tissue heart valve replacement; anticalcification of tissue heart valve replacement and other bioprostheses. *Mailing Add:* Biomed Design Inc 430 Tenth St NW S-204 Atlanta GA 30318

GIRARDOT, PETER RAYMOND, b Detroit, Mich, Aug 15, 22; m 53; c 7. GENERAL ENVIRONMENTAL SCIENCES. *Educ:* Univ Detroit, BS, 44; Univ Mich, MS, 48, PhD(chem), 52. *Prof Exp:* Group leader, Metall Lab, Univ Chicago, 44-45; jet instrumentation, Appl Physics Lab, Johns Hopkins Univ, 45; instr chem, Univ Mich, 50-51; asst, Eng Res Inst, 51-52; proj leader phys & inorg chem, Bjorksten Res Labs, Inc, 52-55; sr res supvr explor inorg chem, Pittsburgh Plate Glass Co, 55-66; dean sci, 66-73, PROF CHEM, UNIV TEX, ARLINGTON, 73- *Concurrent Pos:* prof chem, Inst Teknologi, Mara, Malaysia, 86-87. *Mem:* Am Chem Soc; Int Union Pure & Appl Chem; Am Soc Enol & Viticult. *Res:* Inorganic chlorinations; halides, subhalides and polyhalides; extractive metallurgy; high temperature reactions; heavy industrial chemicals; metal cluster compounds; fermentation. *Mailing Add:* Dept Chem Univ Tex Arlington PO Box 19065 Arlington TX 76019-0065

GIRAUDI, CARLO, b Turin, Italy, Mar 28, 26; US citizen; m 50; c 2. CHEMICAL ENGINEERING. *Educ:* Milan Polytech Inst, DSc(chem eng), 49. *Prof Exp:* Chem engr, Ultra Chem Works Div, Witco Chem Co, 50-57, plant engr, 57-58, tech dir & vpres, 58-61, vpres, res & develop 61-70, group vpres, 70-80; RETIRED. *Mem:* Am Chem Soc; Am Inst Chem Engrs. *Res:* Organic chemistry. *Mailing Add:* 413 Devonshire Dr Venice FL 34293

GIRD, STEVEN RICHARD, b San Diego, Calif, June 15, 43; m 68; c 2. PRODUCTION EFFICIENCY OPTIMIZATION, QUALITY CONTROL. *Educ:* San Diego State Univ, BS, 67, MS, 71. *Prof Exp:* Microanal chemist, RocketDyne-Div NAm Rockwell, 67-69; sr res chemist, Caligrapo, Inc-Div Southern Comfort, 71-79; res chemist & prod mgr, Brown Forman, 79-81; tech dir, Mohawk Liqueur Corp-McKesson Corp, 81-88; DIR, NEW PROD DEVELOP, JIM BEAM BRANDS CO, 88- *Mem:* Inst Food Technologists. *Res:* Develop and refine flavor systems; use new technologies to create new alcoholic beverages, reduce costs and improve product stability. *Mailing Add:* Jim Beam Brands Co 7324 Paddock Rd Cincinnati OH 45216

GIRERD, RENE JEAN, b Le Creusot, France, Mar 1, 20; m 49; c 1. PATHOLOGY. *Educ:* Univ Lyon, MD, 53, PhD, 60. *Prof Exp:* Asst, Univ Montreal, 52-53; biochemist, F W Horner Ltd, 53-54; instr med, Univ Southern Calif, 54-55; biologist, Nepera Chem Co, 55-57; sr scientist, Warner-Lambert Res Inst, 57-62; path res, NY Med Col, 62-63 & Albert Einstein Col Med, 63-65; asst prof path, NY Med Col, 66-67; pathologist, Morristown Mem Hosp, 67-68; lab dir, Dover Gen Hosp, NJ, 69-87; FORENSIC PATHOLOGIST, MORRIS CO, 78- *Mem:* Am Physiol Soc. *Mailing Add:* Four Gunston Ct Morris Plains NJ 07950

GIRI, JAGANNATH, b Bharahopur, India, Jan 16, 33; US citizen; m 58; c 4. STRUCTURAL ANALYSIS, FLUTTER ANALYSIS. *Educ:* Bihar Inst Technol, India, BS, 57; Univ Maine, Orono, MS, 72; Ga Inst Technol, PhD(struct optimization), 76. *Prof Exp:* Foreman, Prod Planning, Indian Tube Co, 58-62; asst prof mech eng, Regional Inst Technol, 62-70; res engr, Ga Inst Technol, 76-79; sr engr, Cessna Aircraft Co, Wichita, 79-82; sr group engr, 82-89, STAFF ENGR STRUCT DYNAMICS, BEECH AIRCRAFT CORP, WICHITA, 89- *Mem:* Am Inst Aeronaut & Astronaut; Am Soc Mech Engrs. *Res:* Structural analysis; optimization. *Mailing Add:* 2524 Fox Run Wichita KS 67226

GIRI, LALLAN, m; c 3. EXPERIMENTAL BIOLOGY. *Educ:* Univ Gorakhpur, India, BS, 65; Banaras Hindu Univ, India, MS, 67; Univ Mont, Missoula, PhD(biol sci), 74. *Prof Exp:* Res assoc, Dept Chem, Univ Mont, Missoula, 74-75; group leader protein biochem, Max Planck Inst Molecular Genetics, WGer, 75-78; res biochemist, Dept Biol Chem, Sch Med, Univ Md, 78-79; asst prof, Dept Biochem & Microbiol, Univ Mass, Amherst, 79-83; mgr chem res & develop, Pharmacia LKB Biotechnol Inc, NJ, 83-89; DIR QUAL CONTROL, CONNAUGHT LABS, INC, 89- *Concurrent Pos:* Consult, NIH, 84-87 & NSF, 90-91. *Mem:* Am Soc Biol Chemists; Am Soc Biochem & Molecular Biol; NY Acad Sci; AAAS; Sigma Xi. *Res:* Proteins and peptides; process development; fermentation; molecular biology. *Mailing Add:* Connaught Labs Inc PO Box 187 Swiftwater PA 18370

GIRI, NARAYAN C, b India, May 1, 28; m 62; c 2. MATHEMATICAL STATISTICS. *Educ:* Midnapur Col, BSc, 51; Univ Calcutta, MSc, 53; Stanford Univ, PhD(math statist), 61. *Prof Exp:* Statist asst, Jute Agr Res Inst, India, 53-55; res investr biomet, Indian Coun Agr Res, 55-58; asst prof math, Univ Ariz, 61-62; asst prof, Cornell Univ, 62-64; assoc prof, 64-70, PROF MATH, UNIV MONTREAL, 70- *Concurrent Pos:* Grants, US Off Naval Res, 61-63 & 64-70, Nat Res Coun Can, 65-78, & Nat Sci Cong Asn; Quebec Res Grant, 70-78; mem bd dir, Ctr Res Math, 70-72; ed, J, Can Statist Asn, 72-74. *Mem:* Hon fel Inst Math Statist; hon fel Am Statist Asn; fel Royal Statist Soc; Can Statist Asn; Int Statist Inst. *Res:* Multivariate analysis; biometry; design of experiments; sample survey; mathematics. *Mailing Add:* Dept Math Univ Montreal Montreal PQ H3C 3J7 Can

GIRI, SHRI N, b India, Jan 30, 34; US citizen; m 66; c 2. PULMONARY TOXICOLOGY, BIOCHEMICAL PHARMACOLOGY. *Educ:* Univ Allahabad, India, BSc, 55; Mathura, India, BVSc/AH, 59; Mich State Univ, MS, 61; Univ Calif, Davis, PhD(comp pharmacol & toxicol), 65. *Prof Exp:* Fel, Dept Pharmacol, Med Ctr, Univ Calif, San Francisco, 65-67; res assoc, Dept Pharmacol, Stanford Univ, 67-68; asst prof pharmacol, Dept Physiol Sci, Univ Calif, Davis, 68-74; vis scientist, Lab Chem Pharmacol, NIH, 76-77; assoc prof, 74-80, actg chmn, 80-81, PROF PHARMACOL, DEPT PHYSIOL SCI, UNIV CALIF, DAVIS, 80- *Concurrent Pos:* Prin investr, NIH grants, 81- *Mem:* AAAS; Am Soc Pharmacol & Exp Therapeut; Am Soc Toxicol. *Res:* Biochemical mechanism of lung damage in response to lung toxicants; roles of prostaglandins and cyclic nucleotides in health and disease. *Mailing Add:* Dept Vet Pharm/Toxicol Univ Calif Davis CA 95616

GIRIFALCO, LOUIS A(NTHONY), b Brooklyn, NY, July 3, 28; m 50; c 8. SOLID STATE PHYSICS, METALLURGY. *Educ:* Rutgers Univ, BS, 50; Univ Cincinnati, MS, 52, PhD(phys chem), 54. *Prof Exp:* Res chemist, E I du Pont de Nemours & Co, 54-55; solid state physicist, NASA, 55-59, head, Solid State Physics Sect, Lewis Res Ctr, Ohio, 59-61; assoc prof metall eng, 61-65, dir lab struct of matter, 67-69, chmn, Dept Metall & Mat Sci, 72-74, assoc dean, Col Eng & Appl Sci, 74-79, vprovost res, 79-81, actg provost, 81, PROF METALL & MAT SCI, UNIV PA, 65- *Concurrent Pos:* Pres, Cara Corp, 69-70. *Mem:* AAAS; Am Phys Soc; Am Inst Mining, Metall & Petrol Engrs. *Res:* Surface chemistry; diffusion in solids; statistical mechanics of solids; lattice vibrations; imperfections in metals; cohesion in metals; theory of alloys. *Mailing Add:* Dept Mat Sci & Eng 403 LRSM-K1 Univ Pa 3231 Walnut St K1 Philadelphia PA 19104

GIRIJAVALLABHAN, VIYYOOR MOOPIL, b Kerala, India, May 31, 42; m; c 2. ORGANIC CHEMISTRY, MEDICINAL CHEMISTRY. *Educ:* Univ Kerala, MSc, 65, PhD(org chem), 70. *Prof Exp:* Res fel org chem, Regional Res Labs, Hyderabad, India, 65-69; fel, Imp Col, Univ London, 70-74; sr scientist & fel med res, 75-77, sect leader, 81, assoc dir, infectious dis, 82-87, prin scientist med res, 78-81, DIR, INFECTIOUS DIS & TUMOR BIOL, SCHERING CORP, 88- *Honors & Awards:* President's Award, Schering-Plough, 83. *Mem:* Am Chem Soc; AAAS; NY Acad Sci. *Res:* Chemical and biological sciences; enzymology and related pharmaceutical sciences such as antibiotics, antivirals, cardiovascular agents; antitumor agents. *Mailing Add:* Schering-Plough Corp Res Ctr Bloomfield NJ 07003

GIRIT, IBRAHIM CEM, b Ankara, Turkey, Dec 21, 51; m 76; c 2. LOW TEMPERATURE PHYSICS. *Educ:* Hacettepe Univ, Turkey, BSc, 74; Sussex Univ, Eng, PhD(nuclear physics), 80. *Prof Exp:* Res fel physics, Sussex Univ, 80-82; from asst prof to assoc prof physics, Hacettepe Univ, 82-85; res assoc, Vanderbilt Univ-Univ Tenn, 85-87, Vanderbilt Univ-Orau, 87-89; ASST PROF PHYSICS, PRINCETON UNIV, 89- *Mem:* Am Phys Soc. *Res:* Low temperature nuclear orientation; parity and time reversal violation; mixed symmetry states in vibrational nuclei; bolometric detector development and measurement of 17 KeV neutrino mass. *Mailing Add:* Dept Physics Jadwin Hall PO Box 708 Princeton NJ 08544-0708

GIROLAMI, ROLAND LOUIS, b Milwaukee, Wis, Sept 19, 24; m 49. MICROBIOLOGY. *Educ:* Univ Wis, BS, 50, MS, 52, PhD(bact), 55. *Prof Exp:* Asst bact, Univ Wis, 51-55; res assoc, Oak Ridge Nat Lab, 55-56; res microbiologist, Res Div, Nat Dairy Prod Corp, 56-57; res microbiologist, 57-69, dir qual assurance, Abbott Sci Prod Div, Calif, 69-73, SR MICROBIOLOGIST, ABBOTT LABS, 73- *Mem:* Am Soc Microbiol; Sigma Xi. *Res:* Clinical microbiology; antimicrobial agents; secondary metabolism. *Mailing Add:* 538 Ridgewood Lane Libertyville IL 60048

GIROTTI, ALBERT WILLIAM, b Springfield, Mass, Aug 9, 37; m 69; c 1. BIOCHEMISTRY. *Educ:* Mass Inst Technol, SB, 59; Univ Mass, MS & PhD(protein chem), 65. *Prof Exp:* Res assoc, Med Col, Cornell Univ, 65-68; from asst prof to assoc prof, 68-89, PROF BIOCHEM, MED COL WIS, 89- *Concurrent Pos:* NSF grants, 71-73 & 82-, NIH grants, 74-79 & 88- *Mem:* Am Chem Soc; Am Soc Photobiol; Am Soc Biol Chemists; Sigma Xi; Oxygen Soc. *Res:* Mechanisms of lipid peroxidation in cell membranes; antineoplastic effects of phototherapeutic sensitizing agents; oxygen radical biochemistry. *Mailing Add:* Dept Biochem Med Col Wis Milwaukee WI 53226

GIROU, MICHAEL L, b St Louis, Mo, July 2, 47; div; c 2. DISTRIBUTED COMPUTING, PARALLEL ALGORITHMS. *Educ:* Univ Mo-Columbia, AB, 68, PhD(math), 89. *Prof Exp:* Mgr tech develop, Honeywell Inc, 69-72; pres & chief exec officer, MFD Inc, 73-84; div dir, Presearch Inc, 84-87; PRES, MT SYSTS CO, 87- *Concurrent Pos:* Adj prof, Univ Tex, Dallas, 90- *Mem:* Am Math Soc; Asn Comput Mach; Soc Indust & Appl Math. *Res:* Algorithms for parallel and distributed computers; general topology. *Mailing Add:* 17250 Knoll Trail No 1205 Dallas TX 75248

GIROUARD, FERNAND E, b St Mary, NB, Sept 19, 38; m 60; c 3. SOLID STATE PHYSICS. *Educ:* St Joseph Univ, BSc, 59; Univ Notre Dame, PhD(physics), 65. *Prof Exp:* From asst prof to assoc prof, 64-72, head dept math, Physics & Comput Sci, 67-73 & 85-91, PROF PHYSICS, UNIV MONCTON, 72- *Mem:* Can Asn Physicists; Am Asn Physics Teachers. *Res:* Optical properties of thin films and solids (vacuum ultraviolet to the infrared); solar selective surfaces; electrochromic systems. *Mailing Add:* Dept Math, Physics & Comput Sci Univ Moncton 98 Chapman Moncton NB E1A 3E9 Can

GIROUARD, RONALD MAURICE, b Sudbury, Ont, Mar 22, 36; m 67; c 2. PLANT MORPHOLOGY, PLANT PHYSIOLOGY. *Educ:* Univ Toronto, BSA, 60; Ohio State Univ, MSc, 62; Purdue Univ, PhD(hort), 67. *Prof Exp:* RES SCIENTIST, LAURENTIAN FORESTRY CENTRE, FORESTRY CAN, 66- *Mem:* Int Plant Propagators Soc; Am Soc Hort Sci. *Res:* Root form, morphological and physiological qualities of container-grown forest trees seedlings. *Mailing Add:* 3349 Place-Radison Ste-Foy PQ G1X 2K2 Can

GIROUX, GUY, b Levis, Que, July 13, 26; m 62; c 2. OPTICS. *Educ:* Laval Univ, BA & BPh, 45, BS, 50, MS, 53, DSc(physics), 55. *Prof Exp:* Defence serv sci officer, Can Res & Develop Estab, Can, 55-63; Liaison Off, Mass, 63-66; sci consult, Can Res & Develop Estab, 66-67; sect head optical & infrared surveillance, Defence Res Estab, 67-78, DIR COMMAND & CONTROL DIV, RES ESTAB, VALCARTIER, 78- *Honors & Awards:* Queen's Jubilee Medal, 78. *Res:* Study of semiconductors used as photodetectors; infrared physics; nuclear physics; beta-rays spectroscopy; use of Fourier analysis in optics and noise problems; reentry physics; lasers; optical and infrared surveillance equipment and systems; optical and infrared counter-surveillance; computer simulation; army and navy command & control. *Mailing Add:* 699 Rue Dalquier Ste-Foy PQ G1V 3H4 Can

GIROUX, VINCENT A(RTHUR), b Los Angeles, Calif, Nov 26, 21; m 40; c 8. ELECTRICAL ENGINEERING. *Educ:* Univ Calif, Los Angeles, BS, 49; Univ Southern Calif, MSEE, 56. *Prof Exp:* Eng asst, Southern Calif Edison Co, 49-51; eng assoc, Los Angeles Dept Water & Power, 51-57; from asst prof to assoc prof, 57-66, PROF ENG, CALIF STATE UNIV, LOS ANGELES, 66- *Concurrent Pos:* Lectr, Univ Southern Calif, 60- *Mem:* Inst Elec & Electronics Engrs; Am Soc Eng Educ. *Res:* Power system analysis and economics; interruption of large magnitude currents. *Mailing Add:* 322 14th St Santa Monica CA 90402

GIROUX, YVES M(ARIE), b Quebec, Que, June 15, 35; m 58; c 3. STRUCTURAL ENGINEERING. *Educ:* Laval Univ, BA, 55, BAS, 59; Mass Inst Technol, MS, 60, DSc(struct eng), 66. *Prof Exp:* Asst civil eng, 60-62, from asst prof to assoc prof, 64-72, head dept, 67-72, PROF CIVIL ENG, LAVAL UNIV, 72-, ASSOC VRECTOR ACAD & RES, 77- *Concurrent Pos:* Assoc ed, Can J Civil Engrs, 74-82; Nat Coun Univ Res Admin. *Mem:* Can Soc Civil Engrs; Can Asn Union Res Admin (pres, 83-84); Am Soc Eng Educ. *Res:* Numerical analysis of structures; connections for steel structures; research management. *Mailing Add:* 1236 Petitclerc Cap-R Quebec PQ G1Y 3G2 Can

GIRSCH, STEPHEN JOHN, b Roanoke, Va, May 13, 46; m 71. PHOTOBIOLOGY. *Educ:* Pa State Univ, BS, 68; Univ Okla, PhD(biophys), 72. *Prof Exp:* Res fel bioluminescence, Biol Labs, Harvard Univ, 72-74; res assoc, Univ Sussex, 74-75; fel vision, Heinz Steinitz Marine Lab, Israel, 75-76; res fel vision, dept molecular biol, Univ Ore, 76-; MEM STAFF DEPT OPHTHAL, UNIV ROCHESTER MED CTR. *Concurrent Pos:* NIH Eye Inst fel, 72; res assoc bioluminescence, Hebrew Univ Jerusalem, 75; vis res fel, Tobias Landau Found, 75. *Mem:* AAAS; Am Chem Soc; Am Soc Photobiol. *Res:* Characterization and comparison of various bioluminescent systems with special emphasis on the chemical mechanisms of light emission; additional studies on visual systems and their interfacing to bioluminescent systems. *Mailing Add:* Dept Ophthal Univ Rochester Med Ctr 601 Elmwood Rochester NY 14642

GIRSE, ROBERT DONALD, graph theory, generating functions; deceased, see previous edition for last biography

GIRVIN, EB CARL, b Georgetown, Tex, Dec 27, 17; m 44; c 3. GENETICS. *Educ:* Univ Tex, BA, 40, MA, 41, PhD, 48. *Prof Exp:* Prof biol & zool, Millsaps Col, 48-53; chmn div natural & appl sci, Southwestern, Univ, 67-70, head dept, 53- 81, prof biol, 53-88; RETIRED. *Concurrent Pos:* Mem, Tex State Bd Examr Basic Sci, 60-75. *Mem:* AAAS. *Mailing Add:* 1256 Main St Georgetown TX 78626

GIRVIN, JOHN PATTERSON, b Detroit, Mich, Feb 5, 34; Can citizen. NEUROSURGERY, SURGICAL TREATMENT OF EPILEPSY. *Educ:* Univ Western Ont, MD, 58; McGill Univ, PhD(physiol), 65; FRCS(C), 68; Am Bd Neurol Surg, dipl, 85. *Prof Exp:* From asst prof to prof, 68-84, PROF NEUROSURG & PHYSIOL & CHMN, DEPT CLIN NEUROL SCI, UNIV WESTERN ONT, LONDON, ONT, 84- *Concurrent Pos:* Prof & chmn, dept clin neurol sci, Univ Western Ont, 84-89. *Mem:* Am Asn Neurol Surgeons; Res Soc Neurol Surgeons; Soc Neurosci; Am Epilepsy Asn; AAAS; Royal Col Physicians & Surgeons. *Res:* Neurophysiology of epilepsy, both in laboratory animals and humans; neuroprostheses. *Mailing Add:* Univ Hosp 339 Windermere London ON N6A 5A5 Can

GIRVIN, STEVEN M, b Austin, Tex, Apr 5, 50; m 72; c 2. THEORETICAL CONDENSED MATTER PHYSICS, MANY-BODY THEORY. *Educ:* Bates Col, BS, 71; Univ Maine, MS, 73; Princeton Univ, MS, 74, PhD(physics), 77. *Prof Exp:* Res assoc, Ind Univ & Chalmers Univ, Gothenburg, Sweden, 77-79; physicist, Nat Bureau Standards, 79-87; PROF PHYSICS, IND UNIV, 87- *Mem:* Am Phys Soc. *Res:* Theoretical condensed-matter physics and many-body theory; the two dimensional electron gas inversion layers and superconductivity. *Mailing Add:* Dept Physics Ind Univ Swain Hall W 117 Bloomington IN 47405

GISH, DUANE TOLBERT, b White City, Kans, Feb 17, 21; m 46; c 4. ORIGINS-MOLECULAR BIOLOGY & PALEONTOLOGY. *Educ:* Univ Calif, Los Angeles, BS, 49; Univ Calif, Berkeley, PhD(biochem), 53. *Prof Exp:* Lilly postdoctoral fel, Med Col, Cornell Univ, New York, 53-55, asst prof biochem, 55-56; asst res assoc biochem, Virus Lab, Univ Calif, Berkeley, 56-60; res assoc biochem, Upjohn Co, Kalamazoo, Mich, 60-71; assoc dir, 71-81, VPRES, INST CREATION RES, 81- *Concurrent Pos:* Prof, Christian Heritage Col, 71-81. *Mem:* Am Chem Soc; fel Am Inst Chemists; AAAS. *Res:* Writing on all aspects of the subject of origins; lecture and debate extensively. *Mailing Add:* 4300 Summit Dr La Mesa CA 91941

GISLASON, ERIC ARNI, b Oak Park, Ill, Sept 9, 40; m 62; c 2. ION-MOLECULE COLLISIONS & REACTIONS. *Educ:* Oberlin Col, AB, 62; Harvard Univ, PhD(chem physics), 67. *Prof Exp:* Nat Ctr Air Pollution Control spec fel chem, Univ Calif, Berkeley, 67-69; from asst prof to assoc prof, 69-77, PROF CHEM, UNIV ILL, CHICAGO CIRCLE, 77- *Concurrent Pos:* Vis scientist, FOM Inst Atomic & Molecular Physics, Amsterdam, The Netherlands, 77-78; assoc prof, Univ Paris at Orsay, 85. *Mem:* Am Phys Soc; Am Chem Soc. *Res:* Molecular beam studies of chemical reactions and intermolecular potentials; theoretical studies of molecular collisions. *Mailing Add:* Dept Chem Univ Ill Chicago IL 60680

GISLASON, I(RVING) LEE, b Nanaimo, BC, July 21, 43; m 69; c 2. CHILD & ADOLESCENT PSYCHIATRY. *Educ:* Univ BC, MD, 69. *Prof Exp:* Gen practice med, 70-72; from clin asst prof to assoc prof, 77-87, CLIN PROF PSYCHIAT, UNIV CALIF, IRVINE, 87- *Mem:* Fel Am Acad Child Psychiat; fel Royal Col Physicians & Surgeons Can. *Res:* Infant, adolescent, adult and group psychiatry. *Mailing Add:* Med Ctr Rt 88 Univ Calif 101 City Dr S Orange CA 92668

GISLER, GALEN ROSS, b Clovis, NMex, June 21, 50; m 84; c 2. ASTRONOMY, PLASMA & SPACE PHYSICS. *Educ:* Yale Univ, BS, 72; Cambridge Univ, PhD(astrophysics), 76. *Prof Exp:* Res assoc astron, Leiden Univ Observ, 76-77; res assoc, Kitt Peak Nat Observ, 77-79; asst scientist astron, Nat Radio Astron Observ, 79-81; MEM STAFF, LOS ALAMOS NAT LAB, 81- *Mem:* Royal Astron Soc; Am Astron Soc. *Res:* Evolution of galaxies and clusters of galaxies; gas dynamics inside and outside galaxies; computational plasma physics; particle acceleration. *Mailing Add:* Los Alamos Natl Lab MS D438 Space Plasma Physics Los Alamos NM 87545

GISOLFI, CARL VINCENT, b New York, NY, Nov 15, 42; m 65; c 3. PHYSIOLOGY. *Educ:* Manhattan Col, BS, 64; Ind Univ, PhD(physiol), 69. *Prof Exp:* From instr to assoc prof, 69-81, PROF EXERCISE SCI, PHYSIOL & BIOPHYS, UNIV IOWA, 81- *Mem:* Am Physiol Soc; Soc Neurosci; fel Am Col Sports Med (pres). *Res:* Mechanisms enabling man to adapt to stress and the regulation of body temperature during exercise; Work-heat tolerance, heat acclimatization; circulatory response to hyperthermia; body fluid homeostasis; gastrointestinal function during exercise. *Mailing Add:* Dept Physiol & Biophys Univ Iowa Iowa City IA 52240

GISSEN, AARON J, anesthesiology, physiology; deceased, see previous edition for last biography

GISSER, DAVID G, electrical engineering, for more information see previous edition

GIST, LEWIS ALEXANDER, JR, b Richmond, Va, Nov 17, 21; m 48; c 2. ORGANIC CHEMISTRY, SCIENCE ADMINISTRATION. *Educ:* Va Union Univ, BS, 47; Howard Univ, MS, 49; Iowa State Univ, PhD(organometallic chem), 56. *Prof Exp:* Asst chem, Howard Univ, 47-49, asst instr, 49; res asst, George Washington Carver Found, Tuskegee Inst, 49-52, res assoc & asst prof, 56; assoc prof & head dept, Va Union Univ, 56-58; prof & chmn dept, Va State Col, 58-64; assoc prog dir, Summer Study Prog & coordr foreign activ, NSF, 64-73, prog mgr, Div Higher Educ in Sci, 73-74, dir, Equal Employ Opportunity, 74-76, dir, Div Sci Personnel Improv, 76-82; RETIRED. *Mem:* Am Chem Soc; Am Inst Chemists; Nat Inst Sci (vpres, 58-62, pres, 62-63 & 71-72). *Res:* Chemistry of organometallic compounds of lithium, sodium, magnesium, tin and lead; chemistry of dienes, diketones and heterocyclic compounds. *Mailing Add:* 1336 Locust Rd NW Washington DC 20012

GISZCZAK, THADDEUS, b Detroit, Mich, Oct 6, 16. METALLURGICAL ENGINEERING. *Educ:* Wayne State Univ, BS, 38. *Prof Exp:* Mem staff, Gen Motors Corp, 51-60, div chief metallurgist, 60-81; RETIRED. *Concurrent Pos:* Mem var comts, Am Soc Testing & Mat. *Honors & Awards:* Award of Sci Merit, Am Foundrymen's Soc, 72; Indust Award, Iron Casting Soc, 80. *Mem:* Fel Am Soc Metals. *Mailing Add:* 4410 Dicker Rd Saginaw MI 48603

GITAITIS, RONALD DAVID, b Wilmington, Del, Apr 25, 50; m 77; c 2. PHYTOBACTERIOLOGY. *Educ:* Univ Del, BS, 74; Univ Fla, MS, 76, PhD(plant path), 79. *Prof Exp:* Asst prof, 80-85, ASSOC PROF PLANT PATH, UNIV GA, 85- *Mem:* Am Phytopath Soc; Sigma Xi. *Res:* Epidemiologic studies of bacterial plant diseases that occur in southern United States; investigate seed treatments for control of seed-borne bacteria; screen plant germplasm for resistance to plant diseases. *Mailing Add:* Dept Plant Path Univ Ga Coastal Plain Exp Sta Tifton GA 31793

GITELMAN, HILLEL J, b Rochester, NY, Dec 23, 32; m; c 4. MEDICINE. *Educ:* Princeton Univ, AB, 54; Univ Rochester, MD, 58; Am Bd Internal Med, cert, 66. *Prof Exp:* Intern med, Duke Hosp, Durham, NC, 58-59, jr asst resident, 59-60, resident, 62-63; clin assoc, Nat Inst Arthritis & Metab Dis, NIH, 60-62; trainee & instr med, Dept Med, Med Ctr, Duke Univ, 63; res fel metab, Dept Med, Sch Med, Univ NC, Chapel Hill, 64-66, from asst prof to assoc prof med, 66-73, med dir, Clin Chem Lab, 67-71, PROF MED, DEPT & SCH MED, UNIV NC, 73- *Concurrent Pos:* Counr, Am Fedn Clin Res, 70-73. *Mem:* Sigma Xi; Southern Soc Clin Invest; Am Fedn Clin Res; Am Physiol Soc. *Mailing Add:* Dept Med Sch Med Univ NC CB 7155 3034 Old Clin Bldg Chapel Hill NC 27599-7155

GITHENS, JOHN HORACE, JR, b Woodbury, NJ, Jan 2, 22; m 45; c 2. PEDIATRICS, PEDIATRIC HEMATOLOGY. *Educ:* Swarthmore Col, BA, 44; Temple Univ, MD, 45; Am Bd Pediat, dipl, 52, dipl pediat hem/onc, 74. *Prof Exp:* From instr to assoc prof pediat, Univ Colo, 51-60; prof & chmn dept, Col Med, Univ Ky, 60-63; assoc dean, 64-73, prof 64-84, EMER PROF PEDIAT, UNIV COLO MED CTR, DENVER, 84- *Concurrent Pos:* Mem consult staff, Denver Childrens Hosp; dir, Colo Sickle Cell Ctr, 74-88. *Honors & Awards:* Ross Award in Pediat Educ, 57. *Mem:* Am Acad Pediat; Soc Pediat Res; Am Pediat Soc. *Res:* Pediatric hematology; medical education. *Mailing Add:* 7100 E Severn Pl Denver CO 80220

GITLIN, HARRIS MARTLIN, b Columbus, Ohio, Feb 20, 14; m 45; c 1. DRIP IRRIGATION SYSTEMS. *Educ:* Ohio State Univ, BSc, 40, BAE, 41; Mich State Univ, MSc, 62. *Prof Exp:* Asst prof, dept agr eng, Ohio State Univ, 46-49; sr res engr, agr res dept, Ford Tractor & Implement Div, 49-59; exten specialist teaching & res, 62-83, EMER EXTEN SPECIALIST, COL TROP AGR, UNIV HAWAII, 83- *Mem:* Am Soc Agr Engrs. *Res:* Crop drying methods of water loss by grain & grass, and methods of accelerating loss; methods and limitations of minimum tillage; haymaking processes and equipment; hydraulics of drip irrigation systems. *Mailing Add:* 10735 Crosby Dr Sun City AZ 85351

GITLITZ, MELVIN H, b Montreal, Que, Feb 28, 40; m 64; c 2. ORGANOMETALLIC CHEMISTRY, TOXICOLOGY. *Educ:* McGill Univ, BSc, 61; Univ Western Ont, PhD(chem), 65. *Prof Exp:* Sr res chemist, Corp Res Lab, 65-72, res assoc develop div, 72-75, res mgr, 75-76, res mgr chem div, 76-88, SR STAFF SCIENTIST, M&T CHEM INC, RAHWAY, 88- *Mem:* NY Acad Sci; Am Chem Soc. *Res:* Organometallic chemistry and industrial applications of organo metallics; organometallic pesticides and biocides; metal-organic chemical vapor deposition (MOCVD); toxicology of organometallic compounds. *Mailing Add:* Two Vauxhall Ct Edison NJ 08820-1809

GITNICK, GARY L, b Omaha, Nebr, Mar 13, 39; c 4. GASTROENTEROLOGY. *Educ:* Univ Chicago, BS, 60, MD, 63. *Prof Exp:* Internship & fel internal med, Johns Hopkins Hosp, 63-64; residency internal med, Mayo Clinic, 64-65; res assoc, Sect Infectious Dis & head, Univ Vaccine Develop, NIH, 67-69; from asst prof to assoc prof, 69-79, chief staff, Med Ctr, 90-92, PROF MED GASTROENTEROL, UNIV CALIF, LOS ANGELES, 79- *Concurrent Pos:* Residency gastroenterol, Mayo Clin, 67-69; asst proj dir, Clin Res Ctr, Univ Calif, Los Angeles, 69-71. *Mem:* AAAS; Am Asn Study Liver Dis; Am Col Physicians; Am Fed Clin Res; Am Gastroenterol Asn. *Res:* Basic studies regarding the causation of Crohn's disease and ulcerative colitis and a major effort investigating the forms of acute viral hepatitis and the progression of acute hepatitis to chronic liver disease. *Mailing Add:* Dept Med Univ Calif Ctr Health Sci Los Angeles CA 90024

GITTELMAN, BERNARD, b Philadelphia, Pa, Oct 28, 32; m 58; c 3. ELEMENTARY PARTICLE PHYSICS. *Educ:* Mass Inst Technol, BS, 54, PhD(physics), 58. *Prof Exp:* Res assoc physics, Princeton Univ, 58-66 & Stanford Univ, 66-69; PROF PHYSICS, CORNELL UNIV, 69- *Mem:* Am Phys Soc. *Res:* Electromagnetic interactions of elementary particles. *Mailing Add:* 109 Tudor Rd Cornell Univ Ithaca NY 14850

GITTELMAN, DONALD HENRY, b Reading, Pa, Feb 16, 29; m 80; c 1. POLYMER CHEMISTRY, BIOCHEMISTRY. *Educ:* Albright Col, Reading Pa, BS, 48; Univ Pa, MSChE, 53; Temple Univ, MBA, 66. *Prof Exp:* Polymer chemist, Glidden Co, Reading, Pa, 52-54, polymer chemist, 56-58, sr polymer chemist, 58-59, sect head, Polymer Res Lab, 59-64 & Wood Finish & Lacquer Lab, 64-65, lab mgr, 65-68; chem engr, Walter Reed Army Med Ctr, Washington, DC, 54-56; sr develop engr intergrated circuit photochem, Western Elec Co, AT&T Inc, Reading, Pa, 68-71, sr engr indust hyg, 71-80 & indust hyg & safety, 80-82, sr engr & chem eng consult, 82-88; PVT CONSULT, ENVIRON, SAFETY & HEALTH ENG, JUPITER, FL, 89- *Concurrent Pos:* Lectr chem, Albright Col, 61-65, lectr econ, 66-70; consult, Lincoln Ind Chem Co, 71-82; pres, Berksiana Found, Reading, Pa, 75-78. *Mem:* Fel Am Inst Chemists; sr mem Am Inst Chem Engrs; Nat Soc Prof Engrs; Am Chem Soc; Am Indust Hyg Asn; Soc Plastics Engrs; Semiconductor Equip & Mat Inst; Chem Safety Task Force. *Res:* Developed procedures for environmental control, industrial hygiene and safety in processing semiconductor devices of both silicon and III-V substrates through material preparatives steps. *Mailing Add:* 131 Palm Ave Unit 36 Jupiter FL 33477-5195

GITTES, RUBEN FOSTER, b Majorca, Spain, Aug 4, 34; US citizen; m 55; c 3. GENITOURINARY SURGERY, ENDOCRINOLOGY. *Educ:* Harvard Univ, AB, 56, MD, 60. *Prof Exp:* Clin assoc surg, Nat Cancer Inst, 63-65; clin asst urol, Mass Gen Hosp, 66-67; asst prof surg & urol, Univ Calif, Los Angeles, 68-69; from assoc prof to prof surg & urol, Univ Calif, San Diego, 71-76; PROF SURG, HARVARD MED SCH & CHIEF UROL, PETER BENT BRIGHAM HOSP, 76- *Concurrent Pos:* Vis asst, Inst Urol, Hosp de Santa Cruz y San Pablo, Barcelona, Spain, 67; mem surg training comt, NIH, 69- *Mem:* AAAS; Asn Acad Surg; Soc Univ Urol; Soc Univ Surg; fel Am Col Surg. *Res:* Experimental hyperparathyroidism; thyrocalcitonin and urinary calcium homeostasis; control of gonadal development; renal transplantation. *Mailing Add:* Scripps Clin & Res Found 10666 N Torrey Pines La Jolla CA 92037

GITTINS, ARTHUR RICHARD, b Edmonton, Alta, May 26, 26; m 49; c 2. ENTOMOLOGY. *Educ:* Univ Alta, BSc, 52; Univ Idaho, MS, 55; Mont State Col, PhD, 63. *Prof Exp:* Entomologist, City of Edmonton, Alta, 49-52; from instr to prof entom, Univ Idaho, 55-78, head dept, 69-78, dean grad sch, 78-88, assoc vpres res, 84-88; RETIRED. *Concurrent Pos:* Ed, Idaho Acad Sci J, 89-; chmn, Comn Health & Environ, 89- *Mem:* Am Entom Soc; Soc Syst Zool; Entom Soc Am; Entom Soc Can; fel Royal Entom Soc London. *Res:* Systematic entomology; insect anatomy; insect systematics, biology with special emphasis on solitary wasps. *Mailing Add:* 624 N Lincoln Moscow ID 83843

GITTINS, JOHN, b Eng, Aug 12, 32; Can citizen; m 58; c 3. GEOLOGY, GEOCHEMISTRY. *Educ:* McMaster Univ, BSc, 55, MSc, 56; Cambridge Univ, PhD(petrol), 59, ScD, 83. *Prof Exp:* Vis res assoc geochem, Pa State Univ, 59-60, asst prof mineral, 60-61; lectr, 61-63, assoc prof, 63-75, PROF GEOL, UNIV TORONTO, 75- *Concurrent Pos:* Vis prof, Cambridge Univ, 68-69, 77 & 83-84; Bye fel, Robinson Col, Cambridge, 83-84, 89-90. *Mem:* Royal Hort Soc; Geol Asn Can; Mineral Asn Can; Mineral Soc Gt Brit; Mineral Soc Am. *Res:* Igneous and metamorphic petrology; phase equilibrium; experimental mineralogy and petrology; carbonatites and alkaline rocks; ore genesis. *Mailing Add:* Dept Geol Univ Toronto Toronto ON M5S 3B1 Can

GITTLEMAN, ARTHUR P, b Brooklyn, NY, Oct 7, 41; m 86; c 1. COMPUTER SCIENCES. *Educ:* Univ Calif, Los Angeles, AB, 62, MA, 65, PhD(math), 69. *Prof Exp:* Tech aide math, E H Plesset Assocs, 62; asst, Univ Calif, Los Angeles, 62-65, asst, Inst Geophys, 65-66; from asst prof to assoc prof, 66-75, PROF MATH, CALIF STATE UNIV, LONG BEACH, 75- *Mem:* Math Asn Am; Asn Comput Mach; Inst Elec & Electronics Engrs Comput Soc. *Res:* Programming languages, artificial intelligence. *Mailing Add:* Dept Math & Comput Sci Calif State Univ Long Beach CA 90840

GITTLEMAN, JONATHAN I, b Newark, NJ, Feb 5, 26; m 47; c 3. MAGNETISM, SUPERCONDUCTIVITY. *Educ:* Rutgers Univ, BS, 48, PhD(physics), 52. *Prof Exp:* Instr physics, Rutgers Univ, 52-53; scientist, Franklin Inst, 53-55; mem tech staff, RCA Labs, 55-83, sr mem tech staff, 83-87; RETIRED. *Mem:* Am Phys Soc; Mat Res Soc. *Res:* Electrical, magnetic, thermal and optical properties of metals; superconductors, semiconductors and composites. *Mailing Add:* 15 Bearfort Way Lawrenceville NJ 08648

GITTLER, FRANZ LUDWIG, b Breslau, Germany, Mar 12, 24; US citizen; m 51; c 4. PHYSICAL CHEMISTRY. *Educ:* Syracuse Univ, BA, 48; Univ Buffalo, MA, 50; Pa State Univ, PhD(phys chem), 54. *Prof Exp:* Sr chemist, Sylvania Elec Prod, Inc, 54-57; scientist, Linde Lab, Union Carbide Corp, 57-58; MEM RES STAFF, BELL TEL LABS, 58- *Concurrent Pos:* Lectr, Pa State Univ, Fogelsville. *Mem:* Am Chem Soc; Electrochem Soc; Inst Elec & Electronics Engr. *Res:* Statistical thermodynamics and thermodynamic properties of substances; chemistry and physics of semiconductor materials, processing and devices. *Mailing Add:* 1141 N Broad St Allentown PA 18104

GITTLESON, STEPHEN MARK, b Washington, DC, July 6, 38; m 63. PROTOZOOLOGY, MACINTOSH HARDWARE & SOFTWARE. *Educ:* Tulane Univ, BS, 60, MS, 62; Univ Calif, Los Angeles, PhD(zool), 66. *Prof Exp:* Res scholar, Univ Calif, Los Angeles, 66; asst prof zool, Univ Ky, 66-70; res assoc, Stevens Inst Technol, 71-72; from asst prof to assoc prof, 72-78, PROF BIOL, FAIRLEIGH DICKINSON UNIV, 78- *Concurrent Pos:* Partic exec ed training prog, Rockefeller Univ, 70-71 & Heart & Lung Inst, NIH, Stevens Inst Technol, 71-72. *Res:* Cellular effects of narcotic gases; motile behavior of individual and aggregate swimming Protozoa; ecology of underground water Protozoa; applications of computer in education. *Mailing Add:* Dept Biol 1000 River Rd Teaneck NJ 07666

GITZENDANNER, L G, b New York, NY, Mar 27, 19; m 43; c 4. MECHANICAL & ELECTRICAL ENGINEERING. *Educ:* Lehigh Univ, BS, 41. *Prof Exp:* Develop engr elec mech, Gen Elec Co, 41-46, mgr invest sect, Gen Eng Lab, 46-47, design sect, 47-50, mech develop, 50-55 & mech equip eng, 55-61, consult engr mech eng, Adv Tech Labs, 61-67; mgr disc design eng, Honeywell Info Systs Inc, 67-75; proj mgr, Diablo Systs Inc, 75-77; MGR ADVAN CONCEPTS LAB, MAGNETIC PERIPHERALS INC, 77- *Honors & Awards:* C A Coffin Award, 46. *Mem:* Inst Elec & Electronics Engrs; Am Soc Mech Engrs; Nat Soc Prof Engrs. *Res:* Electromechanical development engineering; automation; remotely operated tools for nuclear work; underwater sound; mechanical design; computer peripherals. *Mailing Add:* 4949 NW 31 Terr Oklahoma City OK 73122

GIUFFRIDA, ROBERT EUGENE, b New York, NY, Aug 23, 28; m 55; c 2. ORGANIC CHEMISTRY. *Educ:* Columbia Univ, BS, 52. *Prof Exp:* Chemist, US Testing Co, 54-55 & Ciba Pharmaceut Prod, 55-60; res dir, Org Prod Inc, 60-64; res dir, 64-80, QUAL CONTROL MGR, ELAN CHEM CO, 80- *Mem:* AAAS; Am Chem Soc; Am Inst Chemists; NY Acad Sci. *Res:* Synthetic organic chemistry; synthetic methods for manufacture of aroma and flavor chemicals. *Mailing Add:* 73 W Trail Stamford CT 06903

GIUFFRIDA, THOMAS SALVATORE, b Meriden, Conn, Sept 23, 49; m 72. COMMUNICATIONS ENGINEERING, RADIO ASTRONOMY. *Educ:* Wesleyan Univ, BA, 71; Mass Inst Technol, PhD(physics), 78. *Prof Exp:* MEM TECH STAFF COMMUN ENG, BELL TEL LABS, 77- *Concurrent Pos:* Comt mem, Radio Commun Comt, Commun Soc Inst Elec & Electronics Engrs, 78- *Mem:* Am Astron Soc; Sigma Xi. *Res:* System and planning studies for radio facilities to be used in the national communications network; radio interferometry and its application to atmospheric and astronomical measurements. *Mailing Add:* 61 Carnegie Ct Middletown NJ 07748

GIULIANELLI, JAMES LOUIS, b Beverly, Mass, Aug 7, 40; m 66; c 2. SOLAR ENERGY, FUELS. *Educ:* Univ Mass, BS, 62; Univ Wis-Madison, PhD(chem), 69. *Prof Exp:* Res fel, Univ Tex, 69-71; contract prof chem, Univ Los Andes, Venezuela, 71-77; vis asst prof chem, 77-79, asst prof chem, Colo Sch Mines, 79-85; dir, Inst Chem Educ, 90-91, ASSOC PROF CHEM, REGIS COL, 85- *Concurrent Pos:* Res fel, Jet Propulsion Lab, Calif Inst Technol, Pasadena, 80, consult, 80-81, res fel, 81. *Mem:* Am Chem Soc. *Res:* Separation of sulfur forms in coal; rates and mechanisms by electron spin resonance; technology and applications of solar ponds; transmission of light in solar ponds. *Mailing Add:* Dept Chem Regis Col Denver CO 80221

GIULIANO, ROBERT MICHAEL, b Altoona, Pa, June 4, 54; m 81; c 3. SYNTHETIC ORGANIC & NATURAL PRODUCTS CHEMISTRY. *Educ:* Pa State Univ, BS, 76; Univ Va, PhD(chem), 81. *Prof Exp:* Postdoctoral assoc chem, Univ Md, 81-82; asst prof, 82-88, ASSOC PROF CHEM, VILLANOVA UNIV, 88- *Concurrent Pos:* Regional ed, J Carbohydrate Chem, 89-; mem chmn, Carbohydrate Div, Am Chem Soc, 89-; vis assoc prof res, Brown Univ, 90-91. *Mem:* Am Chem Soc; Sigma Xi. *Res:* Synthetic organic and carbohydrate chemistry; synthesis of carbohydrate components of antibiotics; cycloaddition reactions; electrophilic amination. *Mailing Add:* 60 Jefferson St Bala Cynwyd PA 19004

GIULIANO, VINCENT E, b Detroit, Mich, Nov 17, 29; m 54; c 8. APPLIED MATHEMATICS. *Educ:* Univ Mich, AB, 52, MS, 56; Harvard Univ, PhD(appl math), 59. *Prof Exp:* Staff mathematician, Gen Motors Res Ctr, 53-54; res assoc appl math, Comput Lab, Wayne State Univ, 56 & Harvard Univ, 57-59; mem sr prof staff opers res, 59-67 & 71-83, SR STAFF MEM, ARTHUR D LITTLE, 71-; CHIEF SCIENTIST & VPRES, MIRROR SYSTS INC, 83- *Concurrent Pos:* Res fel math ling & Gordon McKay vis lectr, Harvard Univ, 60-63; lectr, NATO Advan Study Inst, 63; prof & dean, Sch Info & Libr Studies, State Univ NY Buffalo, 67-71. *Mem:* Inst Elec & Electronics Eng. *Res:* Information processing research; studies concerning diverse aspects of processing; storage retrieval and use of scientific, biomedical and natural-language information; visual and language pattern-processing problems; evaluatory studies of information systems. *Mailing Add:* Mirror Systs Inc 2067 Mass Ave 5th Floor Cambridge MA 02140

GIUS, JOHN ARMES, b Fairbanks, Alaska, June 2, 08; m 36; c 1. SURGERY. *Educ:* Univ Ore, BA, 31, MD, 34; Columbia Univ, DSc(med), 39. *Prof Exp:* From instr to assoc prof surg, Med Sch, Univ Ore, 39-52; from assoc prof to prof, 52-72, EMER PROF SURG, UNIV IOWA, 72-; ATTEND SURGEON & DIR MED EDUC, POMONA VALLEY COMMUNITY HOSP, 71- *Mem:* Soc Univ Surg; Am Col Surg. *Res:* Venous circulation; bilary tract diseases. *Mailing Add:* 1798 N Garey St Pomona CA 91767

GIVEN, KERRY WADE, inorganic chemistry, polymer chemistry, for more information see previous edition

GIVEN, PETER HERVEY, fuel science, organic geochemistry; deceased, see previous edition for last biography

GIVEN, ROBERT R, b Los Angeles, Calif, July 20, 32; m 59; c 3. ENVIRONMENTAL BIOLOGY. *Educ:* Chico State Col, AB, 53; Univ Southern Calif, MS, 63, PhD(biol), 70. *Prof Exp:* Diver-biologist marine zool, Pomona Col, 57-59; marine biologist, Univ Southern Calif, 59-61, 62-63; diver-biologist, Calif Dept Fish & Game, 63-65; staff biologist, 65-69, asst dir, 69-70, asst prof & dir res, Marine Lab, 70-76, RES SCIENTIST, INST MARINE & COASTAL STUDIES & DIR, CATALINA MARINE SCI CTR, UNIV SOUTHERN CALIF, 76- *Mem:* Soc Syst Zool; Sigma Xi; Asn Syst Collections. *Res:* Ecology, taxonomy and distribution of southern California marine invertebrates; training of scientist-divers; manned submersibles on continental shelf; biological effects of marine pollution (benthic, in shore). *Mailing Add:* Box 651 Avalon CA 90704

GIVENS, EDWIN NEIL, b St Louis, Mo, Dec 8, 35; m 57; c 2. ORGANIC CHEMISTRY. *Educ:* William Jewell Col, AB, 57; Univ Del, MS, 63, PhD(org chem), 65. *Prof Exp:* Sr res chemist petrol, Mobil Res & Develop Corp, 64-76; sr res chemist coal, 76-78, sect mgr coal, 78-80, mgr, coal liquid res & develop, 80-84, MGR, RES & DEVELOP SAFETY & SPEC SERV, AIR PROD & CHEM INC, 84- *Mem:* Am Chem Soc. *Res:* Coal liquefaction; coal liquids upgrading and hydrotreating; petroleum processing; naphtha reforming; petrochemicals; zeolite synthesis; zeolite catalysis. *Mailing Add:* 3425 Gail Ln Bethlehem PA 18017

GIVENS, JAMES ROBERT, b Huntsville, Ala, Aug 6, 30. REPRODUCTIVE ENDOCRINOLOGY. *Educ:* David Lipscomb Col, BS, 52; Vanderbilt Univ, MS, 53; Univ Tenn, Memphis, MD, 56. *Prof Exp:* Internship, Memphis Hosps, 56-57, residency internal med, 59-61; instr med, Tufts Univ, 61-62 & Vanderbilt Univ, 62-64; from instr to asst prof med, 64-68, asst dir clin res ctr, 65-69, dep chief, 70-71, co-dir, 71-73, assoc prof med, 68-76, assoc prof obstet & gynec, 70-76, PROF MED & OBSTET & GYNEC, UNIV TENN, MEMPHIS, 76-, DIR DIV REPRODUCTIVE MED, 70- *Concurrent Pos:* USPHS fel, Tufts Univ, 61-62. *Mem:* Soc Gynec Invests; Endocrine Soc; Am Fedn Clin Res; Am Fertil Soc; Am Col Physicians. *Res:* Understanding the inheritance and characterization of the pathophysiology of polycystic ovaries. *Mailing Add:* 1831 Georgia Hwy 112 Sylvester GA 31791

GIVENS, PAUL EDWARD, b Pawhuska, Okla, Aug 12, 34; m 57; c 2. PACKAGING ENGINEERING, MANPOWER-MACHINE SYSTEMS. *Educ:* Univ Ark, BS, 57; Creighton Univ, MBA, 68; Univ Tex, Arlington, PhD(indust eng), 74. *Prof Exp:* Div indust engr pipeline opers, Northern Natural Gas Co, 65-67, mgr employee rels, 67-69; mgr personnel, Western Co NAm, 69-72; grad teaching asst indust eng, Univ Tex, Arlington, 72-74, instr mgt, 74; assoc prof, Miss State Univ, 74-80; mgt specialist bus opers, Miss State Coop Exten Serv, 77-80; vpres opers, cotton mkt, Staplcotn, Greenwood, Miss, 80-83; dir technol, Ctr Technol Develop, Univ Mo, Rolla, 84-85, assoc prof eng mgt, 83-87, dir, Small Bus Inst, 84-86, assoc dir, Ctr Technol Develop, 85-86; CHMN & PROF, INDUST & MGT SYSTS ENG DEPT, UNIV SFLA, TAMPA, 87- *Concurrent Pos:* Consult, manpower planning & develop, LTV Aerospace, 71 & Tech Eng-Archit Mgt Proj, State Mo, 84-85; dir, Mgt Div, Inst Indust Engrs, 78 & 79; pres, Soc Eng & Mgt Systs, Inst Indust Eng, 91-92. *Honors & Awards:* Edward A Smith Res Award, Edward A Smith Found, 85. *Mem:* Inst Indust Engrs; Am Soc Eng Mgt; Am Soc Eng Educ; Acad Mgt; Inst Packaging Profs. *Mailing Add:* Eng 118 Col Eng Tampa FL 33620-5350

GIVENS, RICHARD SPENCER, b Buffalo, NY, May 19, 40; m 66; c 4. ORGANIC CHEMISTRY, BIOANALYTICAL CHEMISTRY. *Educ:* Marietta Col, BS, 62; Univ Wis, PhD(org chem), 67. *Prof Exp:* NIH fel org chem, Iowa State Univ, 66-67; from asst prof to assoc prof, 67-76, PROF CHEM, UNIV KANS, 76- *Concurrent Pos:* Assoc dir, Ctr Bioanalytical Res, Univ Kans; vis foreign scientist, Sci Res Coun (England), Univ Sheffield; hon lectr, Mid-Am State Univ, 84-85. *Mem:* Am Chem Soc; Royal Soc Chem; NY Acad Sci; Inter-Am Photochem Soc. *Res:* Organic photochemistry, both synthetic and mechanistic investigations; applications of chemiluminescence and fluorescence to trace analysis of bioactive substances. *Mailing Add:* Dept Chem 5023 MAL Univ Kans Lawrence KS 66045

GIVENS, SAMUEL VIRTUE, b Roanoke, Va, Nov 22, 46; m 68; c 2. BIOSTATISTICS. *Educ:* Va Polytech Inst & State Univ, BS, 68, MS, 70, PhD(statist), 73. *Prof Exp:* Consult statist, Stochastics Inc, 71-72; asst dir res statist, 72-, INT DIR BIOSTATIST, HOFFMAN-LA ROCHE, INC. *Concurrent Pos:* Teaching asst statist, Va Polytech Inst & State Univ, 72; adj prof statist, Seton Hall Univ, 74- *Mem:* Am Statist Asn; Biomet Soc. *Res:* General problems in the bioassay area and multivariate design of experiments restricted by a cost criterion. *Mailing Add:* 33 Seneca Pl Upper Montclair NJ 07043

GIVENS, WILLIAM GEARY, b Camden, Ark, Sept 10, 32; m 58. PHYSICAL CHEMISTRY. *Educ:* Rice Inst, BA, 54; Univ Wis, PhD, 59. *Prof Exp:* Res chemist, Jersey Prod Res Co, 59; asst prof chem, Norwich Univ, 59-61 & Exten Div, Univ Wis, 61-62; res chemist, Gen Dynamics/Astronaut, 62-64 & Rocketdyne Div, NAm Aviation, Inc, 64-66; MEM FAC, DEPT CHEM, GROSSMONT COL, 66- *Mem:* Am Chem Soc. *Mailing Add:* Dept Chem Grossmont Col El Cajon CA 92020

GIVENS, WYATT WENDELL, b Forestburg, Tex, Aug 1, 32; m 57; c 2. NUCLEAR PHYSICS. *Educ:* NTex State Univ, BA, 56, MA, 57; Rice Univ, MA, 60, PhD(nuclear physics), 63. *Prof Exp:* Res asst, Socony Mobil Oil Co, Inc, 57-58, sr res technologist, 62-71, RES ASSOC, FIELD RES LAB, MOBIL RES & DEVELOP CORP, 71- *Mem:* AAAS; Am Phys Soc; Sigma Xi. *Res:* Basic and applied nuclear physics, especially oil well logging and oil exploration. *Mailing Add:* 10338 Carry Back Dallas TX 75229

GIVI, PEYMAN, b Tehran, Iran, June 26, 58; US citizen; m 83; c 2. COMBUSTION, COMPUTATIONAL FLUID DYNAMICS. *Educ:* Youngstown State Univ, BE, 80; Carnegie Mellon Univ, ME, 82, PhD(mech eng), 84. *Prof Exp:* Res scientist, Flow Industs Inc, 85-88; asst prof, 88-91, ASSOC PROF MECH & AEROSPACE ENG, STATE UNIV NY, BUFFALO, 91- *Concurrent Pos:* Prin investr, NASA, NSF, Off Naval Res & Air Force Off Sci Res, 85-; mem panel, Workshops Combustion & Fluid Mech, 85-, several comts, NSF, 89-; vis scholar, NASA Lewis Res Ctr, 88; vis scientist, NASA Langley Res Ctr, 89 & 90; Off Naval Res young investr award, 90-93; NSF presidential young investr award, 90-95; fac adv, Soc Women Engrs, 90- *Mem:* Combustion Inst; Am Inst Aeronaut & Astronaut. *Res:* Computational fluid mechanics; numerical combustion; use of advanced computational methods and modern supercomputer to understand the complex phenomena of turbulent combustion; turbulence. *Mailing Add:* Dept Mech & Aerospace Eng State Univ NY Buffalo NY 14260

GIVLER, ROBERT L, b Mason City, Iowa, May 8, 31; m 56; c 1. MEDICINE, PATHOLOGY. *Educ:* Univ Iowa, MD, 56. *Prof Exp:* Intern, King County Hosps, Seattle, Wash, 56-57; instr path, Univ Iowa, 57-58, asst, 58-61, assoc, 61, from asst prof to assoc prof, 63-72; PATHOLOGIST, RES HOSP & MED CTR, KANS CITY, 72-; CLIN ASSOC PROF, UNIV MO, KANSAS CITY, 84- *Mem:* Col Am Path; Am Soc Clin Path; Int Acad Path. *Res:* Leukemia; lymphoma. *Mailing Add:* Dept Path Res Med Ctr Kansas City MO 64132

GIVNER, MORRIS LINCOLN, biochemistry, for more information see previous edition

GIVONE, DONALD DANIEL, b Paterson, NJ, July 10, 36; m 72; c 2. ELECTRICAL ENGINEERING. *Educ:* Rensselaer Polytech Inst, BSEE, 58; Cornell Univ, MS, 61, PhD(elec eng), 63. *Prof Exp:* PROF ELEC ENG, STATE UNIV NY BUFFALO, 63- *Mem:* Inst Elec & Electronics Engrs; Sigma Xi. *Res:* Switching circuit theory and logic design; computer technology. *Mailing Add:* Dept Elec & Comput Eng State Univ NY Buffalo NY 14260

GIZA, CHESTER ANTHONY, b Three Rivers, Mass, May 18, 30; m 61; c 2. SYNTHETIC ORGANIC CHEMISTRY, ORGANIC STEREOCHEMISTRY. *Educ:* Univ Mass, Amherst, BS, 55, MS, 58; Univ Notre Dame, PhD(org chem), 68. *Prof Exp:* Asst chem, Univ Mass, Amherst, 54-57; chemist, Union Carbide Res Inst, NY, 59-63; asst chem, Univ Notre Dame, 63-67; asst prof, 67-70, chmn dept, 74-75, ASSOC PROF ORG CHEM, WHEELING COL, 70- *Concurrent Pos:* Chem consult, 74-; vis scientist, WVa Univ, 75, Carnegie-Mellon Univ, Pa, 81. *Mem:* Sigma Xi; Am Chem Soc. *Res:* Synthesis and evaluation of biologically active organic compounds, with special attention to stereochemical aspects of structure; growth-promoting substances. *Mailing Add:* 1501 Walnut Grove Wheeling WV 26003-9632

GIZA, YUEH-HUA CHEN, b Taipei, Taiwan, China, Mar 18, 29; m 61; c 2. ORGANIC BIOCHEMISTRY. *Educ:* Nat Univ Taiwan, BSc, 52; Tufts Univ, MS, 55; Univ Mass, PhD, 59. *Prof Exp:* Asst chem, Yale Univ, 58-61; res assoc, Inst Muscle Dis, NY, 61-63; res assoc chem, Univ Notre Dame, 64-68; from asst prof to assoc prof, 69-74, adj assoc prof, 74-76, ASSOC PROF CHEM, WHEELING COL, 76- *Concurrent Pos:* Sigma Xi Res grant; WVa Heart Assn Res grant; fac res fel (NSF), Univ Pittsburg, 78 & Carnegie-Mellon Univ, 79; Pew res fel, Univ Ky, 86-87. *Mem:* Am Chem Soc; Sigma Xi. *Res:* Polymerization of sulfur containing organic compound; thermal stable polymers; metabolism of amino acids in plants; effect of synthetic hydroxyamines on ATP-ase activity; a model for transport of L-histidine across cytoplasmic membrane; synthesis and structural studies of Cu-chelating polymers. *Mailing Add:* 1501 Walnut Grove Wheeling WV 26003

GIZIS, EVANGELOS JOHN, b Tinos, Greece, Apr 1, 34; m 67; c 2. BIOCHEMISTRY, FOOD SCIENCE. *Educ:* Nat Univ Athens, BS; Ore State Univ, PhD(food sci & biochem), 63. *Prof Exp:* Fel food sci, Mich State Univ, 64-65; fel enzymes & pectins, Mellon Inst, 65-66; biochemist, L I Jewish Hosp, Queens Hosp Ctr Affil, Jamaica, NY, 66-70; assoc prof natural sci, City Univ NY, 70-72; vpres, 77-85, ACTG PRES, BOR MANHATTAN COMMUNITY COL, CITY UNIV NY, 85-; PROF NATURAL SCI & ASSOC DEAN FAC, HOSTOS COMMUNITY COL, CITY UNIV NY, 72-, DEAN ARTS & SCI, 74- *Concurrent Pos:* Res collab, Med Dept, Brookhaven Nat Lab, 66-70; res chemist, Vet Admin Hosp, Brooklyn, NY, 70-72. *Mem:* Soc Exp Biol & Med; Am Chem Soc; Inst Food Technol; NY Acad Sci; Am Inst Nutrit. *Res:* Pectinolytic enzymes and pectins; vitamin B12 binders in milk, in normal human and pernicious anemia serum; chromatographic techniques. *Mailing Add:* 427 Ryder Manhasset NY 11030-2761

GJEDDE, ALBERT, b Copenhagen, Denmark, Jan 10, 46; m 83; c 3. NEUROPHARMACOLOGY, NEUROLOGY. *Educ:* Rungsted Acad, Copenhagen, Artium, 64; Univ Copenhagen, Cand Med, 73, Dr med(physiol), 83. *Prof Exp:* Postdoctoral fel, Dept Neurol, Cornell Univ Med Col, 73-76; postdoctoral fel, Dept Neurosura, Rigshospitalet, Univ Copenhagen, 76-79, from asst prof to assoc prof physiol, Panum Inst, 79-86; assoc prof, 86-89, PROF NEUROL & NEUROSURG, MONTREAL NEUROL INST, MCGILL UNIV, 89- *Concurrent Pos:* Assoc dir, McConnell Brain Imaging Ctr, Montreal Neurol Inst, 87-, dir, Positron Imaging Labs, 89-; dir, Int Soc Cerebral Blood Flow & Metab, 87-; dep chief ed J Cerebral Blood Flow Metab, 87-; coordr & prin investr, Med Res Coun Spec Proj Positron Imaging, 89- *Mem:* Int Soc Neurochem; Am Physiol Soc; Soc Neurosci. *Res:* Cerebral blood flow and metabolism; blood-brain transfer of nutrients and water, relationship between blood flow; glucose metabolism and oxygen consumption in the brain of mammals and man; dopaminergic neurotransmission and metabolism of amino acids; disorders of brain metabolism and dopaminergic neurotransmission in humans, including mitochondrial encephalopathiesm; Parkinson's disease; schizophrenia; temporal lobe epilepsy and related disorders; positron emission tomography. *Mailing Add:* Montreal Neurol Inst 3801 University St Montreal PQ H3A 2B4 Can

GJESSING, HELEN WITTON, b Boston, Mass, Nov 22, 27; m 55; c 4. MICROBIOLOGY. *Educ:* Beloit Col, BA, 50; Univ Mass, MA, 52. *Prof Exp:* Res technician virus, Res Infectious Dis Div, Children's Med Ctr & Sch Med, Harvard Univ, 52-53, res asst, 53-55; res & teaching asst parasitol res & lab instr, Sch Trop Med, San Juan, PR, 55-57; from instr to assoc prof, 63-77, PROF BIOL, UNIV VI, 77- *Concurrent Pos:* Actg dean, Univ VI, 68-69, chmn sci & math div, 69-72 & 76-78; dir minority biomed support prog res grant, NIH, 76-77; dir minority inst sci improvement prog grant, NSF, 78-81. *Mem:* Am Soc Microbiol; Sigma Xi. *Res:* Antibiotic marine bacteria, particularly those in sponges; antibiotic marine sponges. *Mailing Add:* Box 1844 St Thomas VI 00803

GJOSTEIN, NORMAN A, b Chicago, Ill, May 26, 31; m 59; c 2. MATERIALS SCIENCE, ELECTRONIC SYSTEMS. *Educ:* Ill Inst Technol, BS, 53, MS, 54; Carnegie-Mellon Univ, PhD(metall eng), 58. *Prof Exp:* Res engr, Thompson-Ramo-Wooldridge, Inc, 58-60; sr res scientist, Ford Motor Co, 60-61, prin res scientist assoc, 61-64, staff scientist, 64-69, prin res scientist, 69-73, mgr metall dept, 73-76, mgr Europ res liaison, 76-78, mgr res planning, 78-79, dir long-range & systs res, 79-81, dir systs res lab, 81-86, DIR POWER TRAIN & MAT, FORD MOTOR CO, 86- *Mem:* Sigma Xi; Inst Elec & Electronics Engrs; Am Inst Mech Engrs; Nat Acad Eng; fel Am Soc Metals Int. *Res:* Physics and chemistry of interfaces and surfaces; composite materials; applications of advanced automotive materials; leed and auger spectroscopy. *Mailing Add:* 544 S Claremont Dearborn MI 48124

GLABE, CHARLES G, b Columbus, Ohio, Jan 16, 52. CELL BIOLOGY, BIOCHEMISTRY. *Educ:* Calif State Univ, Sacramento, BA, 73; Univ Calif, Davis, PhD(zool), 77. *Prof Exp:* Res asst, Univ Calif, Davis, 73-77; res assoc, Sch Med, Johns Hopkins Univ, 78-; MEM STAFF WORCESTER FEDN EXP BIOL. *Concurrent Pos:* Rockefeller Found fel, 78- *Mem:* Am Soc Cell Biol. *Res:* Biochemistry of fertilization, cell adhesion and cell surface interactions. *Mailing Add:* Dept Molecular Biol & Biochem Univ Calif Irvine CA 92717

GLABERSON, WILLIAM I, b Chicago, Ill, Nov 8, 44; m 66; c 1. LOW TEMPERATURE PHYSICS. *Educ:* Univ Chicago, SB, 64, PhD(physics), 69. *Prof Exp:* Asst prof, 68-73, ASSOC PROF PHYSICS, RUTGERS UNIV, 73-, ASSOC CHMN DEPT PHYSICS & DIR GRAD PROG, 75- *Concurrent Pos:* Alfred P Sloan fel, 71-73. *Mem:* Am Phys Soc. *Res:* Superfluidity; hydrodynamics; superfluidity of He4; quantized vorticity; ionic impurities. *Mailing Add:* Dept Physics & Astron Rutgers Univ New Brunswick NJ 08903

GLABMAN, SHELDON, b New York, NY, Apr 13, 32; m 54; c 3. INTERNAL MEDICINE, NEPHROLOGY. *Educ:* Univ Pa, BA, 53; Chicago Med Sch, MD, 57. *Prof Exp:* Div dialysis unit, Mt Sinai Hosp, 68-86; asst prof, 68-73, ASSOC PROF MED, MT SINAI SCH MED, 73- *Concurrent Pos:* USPHS fel, Med Col, Cornell Univ, 60-62 & Mt Sinai Hosp, 62-65; asst attend med, Mt Sinai Hosp, 68-73, assoc attend, 73-; consult, Lincoln Hosp, New York, 70-72. *Mem:* Am Fedn Clin Res; Am Soc Nephrology; Int Soc Nephrology. *Res:* Renal physiology; hemodialysis; renal transplantation. *Mailing Add:* 1175 Park Ave New York NY 10128

GLADDEN, BRUCE, b Hohenwald, Tenn, Jan 27, 51; m 73; c 2. EXERCISE PHYSIOLOGY, MUSCLE FATIGUE. *Educ:* Univ Tenn, Knoxville, PhD(zool), 76. *Prof Exp:* ASSOC PROF EXERCISE PHYSIOL, DIV ALLIED HEALTH, UNIV LOUISVILLE, 78- *Mem:* Am Physiol Soc; Am Col Sport Med. *Res:* Muscle fatigue, lactate metabolism; the role of oxygen in muscle performance; substrate utilization. *Mailing Add:* Exercise Physiol Lab Univ Louisville Crawford Gym Louisville KY 40292

GLADDING, GARY EARLE, b Brownwood, Tex, Apr 16, 44; m 66, 84; c 2. EXPERIMENTAL HIGH ENERGY PHYSICS. *Educ:* Univ Ill, BS, 65; Harvard Univ, AM, 68, PhD(physics), 71. *Prof Exp:* Res assoc physics, 71-72, from asst prof to assoc prof, 73-85, PROF PHYSICS, UNIV ILL, 86- *Concurrent Pos:* Vis scientist physics, Europ Orgn for Nuclear Res, 72; NSF fel, 72. Mem; Am Phys Soc. *Res:* Counter and proportional wire chamber experiments studying the production of new particles with large transverse momentum. *Mailing Add:* Dept Physics Univ Ill 1110 W Green St Urbana IL 61801

GLADE, RICHARD WILLIAM, embryology; deceased, see previous edition for last biography

GLADFELTER, WAYNE LEWIS, b Bryn Mawr, Pa, Feb 16, 53; m 78; c 2. INORGANIC CHEMISTRY. *Educ:* Colo Sch Mines, BS, 75; Pa State Univ, PhD(chem), 78. *Prof Exp:* NSF fel chem, Calif Inst Technol, 78-79; from asst prof to assoc prof, 79-88, PROF CHEM, UNIV MINN, 88- *Honors & Awards:* Nobel Laureate Signature Award, Am Chem Soc, 80. *Mem:* Am Chem Soc; Sigma Xi; Mat Res Soc. *Res:* Synthesis and characterization of metal carbonyls, nitrosyls and organometallic cluster compounds; catalysis organometallic precursors to solid state materials; chemical vapor deposition. *Mailing Add:* Dept Chem Univ Minn Minneapolis MN 55455

GLADFELTER, WILBERT EUGENE, b York, Pa, April 29, 28; m 52; c 3. PHYSIOLOGY, NEUROBIOLOGY. *Educ:* Gettysburg Col, AB, 52; Univ Pa, PhD(physiol), 60. *Prof Exp:* Asst instr physiol, Univ Pa, 54-56, 58-59; from instr to asst prof, 59-69, ASSOC PROF PHYSIOL, WVA UNIV, 69- *Mem:* Am Physiol Soc; Soc Neurosci; Sigma Xi; Am Soc Zoologists. *Res:* Neurophysiology; regulation of energy exchange; hypothalamus and motor activity. *Mailing Add:* Dept Physiol WVa Univ Morgantown WV 26506

GLADMAN, CHARLES HERMAN, b Harrison Co, Ohio, May 24, 17; m 47; c 1. MATHEMATICS. *Educ:* Ohio State Univ, BS, 38, MA, 48. *Prof Exp:* From instr to assoc prof, 48-83, chmn dept, 65-68, EMER ASSOC PROF MATH, UNIV TEX, EL PASO, 83- *Concurrent Pos:* Mathematician, Schellenger Res Labs, 58-60. *Mem:* AAAS; Math Asn Am; Soc Indust & Appl Math; Sigma Xi. *Res:* Mathematical analysis. *Mailing Add:* 436 Granada Ave El Paso TX 79912

GLADNEY, HENRY M, b Prague, Czech, Feb 8, 38; Can citizen; m 69; c 2. COMPUTER SCIENCE. *Educ:* Univ Toronto, BA, 60; Princeton Univ, MA, 62, PhD(chem), 63. *Prof Exp:* Res staff mem, 63-68, tech adv to vpres & chief scientist, 68-70, mgr res comput facility, 70-72, mgr, Molecular Dynamics Dept, 72-76, mgr res comput facility, 77-79, RES STAFF MEM COMPUTER SCI, ALMADEN RES CTR, INT BUS MACH CORP, 79- *Concurrent Pos:* Res sabbatical, 76-77. *Mem:* Asn Comput Mach; fel Am Phys Soc. *Res:* Laboratory and factory automation; software security; resource control in computer facilities; programming languages; distributed data systems. *Mailing Add:* IBM Almaden Res Ctr 650 Harry Rd Dept K51/802 San Jose CA 95120-6099

GLADNEY, WILLIAM JESS, entomology, acarology, for more information see previous edition

GLADROW, ELROY MERLE, b Cleveland, Ohio, Sept 2, 15; m 47; c 2. PHYSICAL CHEMISTRY. *Educ:* Heidelberg Col, BS, 38; Iowa State Col, PhD(chem), 48. *Prof Exp:* Res assoc chemist, Nat Defense Res Comt, Manhattan Proj, AEC, 42-47; res chemist, Standard Oil Develop Co, 47-58 & Esso Standard Oil Co, 58-63; res assoc, Exxon Res & Eng Co, 63-70, sr res assoc, 70-79; RETIRED. *Concurrent Pos:* Consult, Exxon Res & Eng Co, 79-80, Ethyl Corp, 83-84. *Res:* Hydrocarbon catalysis; cracking, reforming, hydrocracking, hydrodesulfurization, isomerization, fuel cell; separation and preparation of pure rare earths; uranium fission products chemistry; radiation chemistry. *Mailing Add:* 9884 Spanish Moss Ct Sun City AZ 85373

GLADSTONE, HAROLD MAURICE, b Brooklyn, NY, Jan 23, 32; div; c 2. MATHEMATICS, COMPUTER SCIENCES. *Educ:* Rensselaer Polytech Inst, BS, 52; Adelphi Univ, MS, 56; Polytech Inst NY, PhD(org chem), 61. *Prof Exp:* Res chemist, Armstrong Cork Co, 56-57; eve instr, Adelphi Univ, 58-60; res chemist, Esso Res & Eng Co, 61-62; proj mgr, Quantum Inc, 62-63, dir lab, 63-65; mem tech staff, Bell Tel Labs, 65-69; asst chmn, 70-71, chmn dept, 71-74, from asst prof to assoc prof, 69-75, PROF CHEM, MIDDLESEX COUNTY COL, 75- *Mem:* Am Chem Soc. *Res:* Organic synthesis; reaction mechanisms; monomers; polymers. *Mailing Add:* Dept Chem Middlesex County Col Woodbridge Ave & Mill Rd Edison NJ 08818

GLADSTONE, MATTHEW THEODORE, b Manchester, NH, Apr 25, 19; m 46; c 3. ORGANIC CHEMISTRY. *Educ:* Univ Chicago, BS, 40, PhD(org chem), 48. *Prof Exp:* Lab asst chem, Univ Chicago, 41-42 & 46-47; res assoc, Gen Elec Co, 47-50; res assoc, Abrasives Div, Tech Dept, Behr-Manning Corp, 51-54, group leader, 54-64; asst dir res, Coated Abrasives Div, Norton Co, NY, 64-72, res assoc, 74-83; tech consult, Riken-Norton Co, Ltd, Tokyo, 72-74; RETIRED. *Mem:* AAAS; Am Chem Soc; Am Inst Chem. *Res:* Alkyd and phenolic resins; plastics; free radical reactions; reactions of free radicals in solution; decomposition of acetyl peroxide in acids, nitroalkanes and halogenated esters; fluorocarbon polymers; abrasives, epoxy resins; cloth finishing; polyurethanes. *Mailing Add:* 761 Cent Parkway Schenectady NY 12309

GLADSTONE, WILLIAM TURNBULL, b Syracuse, NY, May 5, 31; m 56; c 2. FOREST GENETICS, WOOD SCIENCE. *Educ:* State Univ NY Col Forestry, Syracuse Univ, BS, 53; Yale Univ, MF, 65; NC State Univ, PhD(forest genetics, wood sci), 69. *Prof Exp:* Asst pulp mill supt, Union Bag Camp Paper Corp, Va, 53-63; asst prof forest genetics, State Univ NY Col Forestry, Syracuse Univ, 68-72; forest geneticist, Weyerhaeuser Co, 72-74, mgr, Southern Forestry Res Ctr, 74-76, mgr trop forestry res, 76-89; ED, SOUTHERN J APPL FORESTRY, 89- *Mem:* Tech Asn Pulp & Paper Indust; Forest Prod Res Soc; Soc Am Foresters. *Res:* Variability and heritability of wood properties; relationships between wood fiber properties and products manufactured from wood; environmental influences on wood properties. *Mailing Add:* RD 2 Box 140AA Avoca NY 14809

GLADUE, BRIAN ANTHONY, b Norwich, Conn, Nov 30, 50. PSYCHOBIOLOGY, NEUROSCIENCES. *Educ:* Northeastern Univ, BA & BS, 73; Mich State Univ, PhD(zool), 79. *Prof Exp:* Instr animal behav, Dept Zool, Mich State Univ, 79; fel psychobiol, Human Sexuality Training Grant, Dept Psychiat, State Univ NY, Stony Brook, 79-81; res fel psychoendocrinol, Long Island Res Inst, 80-81, res scientist, 81-84; PROG DIR HUMAN SEXUALITY & ASST PROF PSYCHOL, NDAK STATE UNIV, 84- *Concurrent Pos:* Nat res serv award, NIMH, 79-81. *Mem:* Int Soc Human Ethology; Endocrine Soc; Animal Behav Soc; NY Acad Sci; AAAS; Int Acad Sex Res. *Res:* Gonadal hormone influences on the development of sexual behavior (sexual differentiation); interactions of hormone and physiology of human sexual arousal and sexual orientation (psychoendocrinology). *Mailing Add:* NDak State Univ State Univ Sta Fargo ND 58105

GLADWELL, GRAHAM M L, b Otford, Eng, Feb 21, 34; Can citizen; m 58; c 3. INVERSE PROBLEMS, CONTACT PROBLEMS. *Educ:* Univ London, BSc, 54, PhD(math), 57, DSc, 69. *Prof Exp:* Asst lectr math, Univ Col London, 56-59, lectr, 59-60; lectr, Univ WI, 60-62; lectr mech eng, Mass Inst Technol, 62; lectr aeronaut, Univ Southampton, 62-63, sr lectr vibration theory, Inst Sound & Vibration Res, 63-69; PROF CIVIL ENG, UNIV WATERLOO, 69-, PROF APPL MATH, 79- *Concurrent Pos:* Dir, Am Acad Mech, 80-83; ser ed, solid mech & applications, Kluwer Acad Publ, 89- *Honors & Awards:* Cancam Medal, Can Cong Appl Mech, 91. *Mem:* Fel Inst Math & Its Applications; fel Am Acad Mech; Soc Indust & Appl Math. *Res:* Contact problems in the classical theory of elasticity; theory of vibration, particularly inverse problems in vibration; inverse scattering theory. *Mailing Add:* Dept Civil Eng Univ Waterloo Waterloo ON N2L 3G1 Can

GLADWELL, IAN, b Bolton, Eng, Oct 20, 44; m 79. MATHEMATICAL SOFTWARE, NUMERICAL ANALYSIS. *Educ:* Oxford Univ, BA, 66, PhD(math), 70; Univ Manchester, MS, 67. *Prof Exp:* Teaching asst math, Univ Manchester, 67-69, lectr, 69-80, sr lectr, 80-87; assoc prof, 87-88, PROF & CHMN MATH, SOUTHERN METHODIST UNIV, 88- *Concurrent Pos:* Res fel, Dept Computer Sci, Univ Toronto, 75; Royal Soc indust fel, NAG Ltd, Oxford, 86; assoc ed, J Numerical Analysis, Inst Math & Applications, 87- *Mem:* Fel Inst Math & Applications; Soc Indust & Appl Math; Am Math Soc. *Res:* Numerical analysis; mathematical software; parallel computation; knowledge based systems; scientific computation. *Mailing Add:* Math Dept Southern Methodist Univ Dallas TX 75275

GLADYSZ, JOHN A, b Kalamazoo, Mich, Aug 13, 52. ORGANIC CHEMISTRY, ORGANOMETALLIC CHEMISTRY. *Educ:* Univ Mich. BSChem, 71; Stanford Univ, PhD(org chem), 74. *Prof Exp:* Asst prof org chem, Univ Calif, Los Angeles, 74-82; assoc prof, 82-85, PROF, UNIV UTAH, 85- *Concurrent Pos:* Alfred P Sloan Foundation fel, 80-84; Camille & Henry Dreyfus teacher-scholar grant, 80-85; assoc ed, Chem Rev, 84- *Honors & Awards:* Arthur C Cope Scholar Award, 88. *Mem:* Am Chem Soc; Chem Soc; AAAS. *Res:* Transition metal and main-group element reagents for organic synthesis; reactions of coordinated ligands; heterogeneous and homogeneous catalysis; chemistry of metal atoms; high pressure synthesis. *Mailing Add:* Dept Chem Univ Utah Salt Lake City UT 84112

GLAENZER, RICHARD H, b St Louis, Mo, Nov 29, 33; m 63; c 4. ELECTRICAL ENGINEERING, PHYSICS. *Educ:* Wash Univ, St Louis, BS, 60, MS, 64; Carnegie-Mellon Univ, PhD(elec engr), 68. *Prof Exp:* Engr, McDonnell Douglas Corp, 60-62, res assoc, 62-64, res scientist, 67-71; staff engr, McDonnell Douglas Astronaut Co, 71-77; sect mgr, McDonnell Douglas Electronics Co, 77-83; vis prof, Inst Nat de Astrofisica, Optica, & Electronics, 83-84; sr res specialist, Monsanto Co, 84-88; SR PROC ENGR SPECIALIST, MEMC ELECTRONIC MATS, 88- *Mem:* Inst Elec & Electronics Engrs; AAAS. *Res:* Integrated circuits; photodetectors; infrared to visible image conversion; electrical behavior of defects in semiconductor materials and devices. *Mailing Add:* 7112 Westmoreland University City MO 63130

GLAESER, HANS HELLMUT, b Chemnitz, Ger, June 30, 34; nat US; m 62; c 2. INORGANIC CHEMISTRY. *Educ:* Karlsruhe Tech Univ, BS, 55, MS, 58, Dr rer nat, 61. *Prof Exp:* Res fel inorg chem, Wash State Univ, 62-66; from res chemist to sr res chemist, 66-73, res assoc, 73-83, RES FEL, C&P DEPT, E I DU PONT DE NEMOURS & CO, INC, 83- *Mem:* Assoc Inst Mech Engrs. *Res:* Extractive metallurgy, Ti02 pigment manufacture, thermochemical equilibrium calculations; reaction mechanisms; chlorination technology of ilmenite and other ores. *Mailing Add:* Meadows 11 Meadow Lane Wilmington DE 19807

GLAESER, ROBERT M, b Kenosha, Wis, July 20, 37; m 60; c 3. BIOPHYSICS. *Educ:* Univ Wis-Madison, BS, 59; Univ Calif, Berkeley, PhD(biophys), 64. *Prof Exp:* NSF fel, Math Inst, Oxford Univ, 63-64; NIH traineeship biophys, Univ Chicago, 64-65; lectr med physics, 65-66, from asst prof to assoc prof, 66-76, PROF BIOPHYS, UNIV CALIF, BERKELEY, 76- *Mem:* AAAS; Biophys Soc; Electron Micros Soc Am (pres, 86-); Am Crystallog Asn; Am Soc Cell Biol; Royal Micros Soc. *Res:* Molecular structure and molecular organization of cell membranes; electron optics and the interpretation of electron microscopic images. *Mailing Add:* Univ Calif 363 Donner Labs Berkeley CA 94720

GLAESER, WILLIAM A(LFRED), b Utica, NY, Aug 25, 23; m 51; c 3. METALLURGY, TRIBOLOGY. *Educ:* Cornell Univ, BME, 49; Ohio State Univ, MS, 59. *Prof Exp:* Develop engr, Clark Bros, Inc, NY, 49-51; prin engr, 51-57, proj leader, 57-59, asst chief, Eng Mech Div, 59-64, assoc chief, Exp Physics Div, 64-69, fel lubrication mech, Div Mech & Systs Eng Dept, 69-72, RES LEADER, STRUCT MAT & TRIBOLOGY SECT, COLUMBUS LABS, BATTELLE MEM INST, 72- *Concurrent Pos:* Adj asst prof, Ohio State Univ; chmn, Gordon Conf Friction Lubrication & Wear, 82. *Mem:* Am Soc Metals; Sigma Xi. *Res:* Bearings lubrication and wear phenomena, especially as concerned with unusual environments; metallurgical aspects of wear and friction; solid state physics aspects of wear and friction; near surface microsctructures of worn metals and alloys by electron microscopy. *Mailing Add:* Battelle Mem Inst 505 King Ave Columbus OH 43201

GLAGOV, SEYMOUR, b New York, NY, Aug 8, 25; m 46; c 1. PATHOLOGY. *Educ:* Brooklyn Col, BA, 46; Univ Geneva, MD, 53; Am Bd Path, dipl, 60. *Prof Exp:* From instr to assoc prof, 58-70, PROF PATH, UNIV CHICAGO, 70- *Concurrent Pos:* Res fel, Am Heart Asn, 58-60; estab investr, Am Heart Asn, 62- *Mem:* Sigma Xi. *Res:* Experimental pathology; pathophysiology and biology of blood vessels; diseases of the liver; human pathology. *Mailing Add:* 5233 S Univ Ave Chicago IL 60615

GLAHN, HARRY ROBERT, b Shelbyville, Mo, July 28, 28; m 49; c 2. METEOROLOGY. *Educ:* Northeast Mo State Teachers Col, BS & BSEd, 53; Mass Inst Technol, MS, 58; Pa State Univ, PhD(meteorol), 63. *Prof Exp:* Elem teacher, Duncan Sch, Mo, 47-51; from res meteorologist to supvry res meteorologist, US Weather Bur, 58-67; supvry res meteorologist, 67-76, DIR, TECH DEVELOP LAB, NAT WEATHER SERV, 76- *Honors & Awards:* Silver Medal, Dept of Com, 68 & Gold Medal, 75; Nat Atmospheric & Oceanic Admin Eng & Applns Award, 75. *Mem:* Fel Am Meteorol Soc. *Res:* Application of statistics to meteorology. *Mailing Add:* Tech Develop Lab Nat Weather Serv 1325 Eastwest Hwy Rm 10214 MC W-OSD2 Silverspring Metro Ctr Silver Spring MD 20910

GLAID, ANDREW JOSEPH, III, b Pittsburgh, Pa, July 14, 23; m 55; c 5. BIOCHEMISTRY. *Educ:* Duquesne Univ, BS, 49, MS, 50; Duke Univ, PhD(biochem), 55. *Prof Exp:* From asst prof to assoc prof, 54-61, PROF CHEM, DUQUESNE UNIV, 61-, CHMN DEPT, 75- *Mem:* Am Chem Soc; Sigma Xi; Am Asn Univ Prof. *Res:* Kinetics of enzymatic reactions; stereochemistry of biologically active compounds. *Mailing Add:* Dept Chem Duquesne Univ Pittsburgh PA 15282

GLAMKOWSKI, EDWARD JOSEPH, b Brooklyn, NY, May 20, 36; m 63; c 3. ORGANIC CHEMISTRY. *Educ:* Fordham Univ, BS, 58; Ohio State Univ, PhD(chem), 63. *Prof Exp:* Sr res chemist, Merck, Sharp & Dohme Res Labs, Rahway, 63-71; res assoc, 71-77, RES GROUP MGR, HOECHST-ROUSSEL PHARMACEUTICALS, INC, SOMERVILLE, 77- *Mem:* Am Chem Soc. *Res:* Indoles and other heterocyclic compounds; analgesic, anti-inflammatory and antipsychotic agents; drugs for the treatment of alzheimer's disease. *Mailing Add:* Seven Owens Dr Warren NJ 07060

GLANCEY, BURNETT MICHAEL, b New Orleans, La, May 31, 30; m 56, 81. MEDICAL ENTOMOLOGY, ECONOMIC ENTOMOLOGY. *Educ:* La State Univ, BS, 55, MS, 58; Cornell Univ, PhD(entomol), 63. *Prof Exp:* Res entomologist mosquito biol, Area Control, Agr Res Serv, 63-68; MED RES ENTOMOLOGIST FIRE ANT PROJ, SCI & EDUC ADMIN-FED RES, USDA, 68- *Mem:* Sigma Xi. *Res:* Pheromones of fire ants; ant physiology. *Mailing Add:* 1126 St Johns Ave Green Cove Springs FL 32043

GLANCY, DAVID L, b Cincinnati, Ohio, Oct 17, 34; m 57; c 3. CARDIOLOGY. *Educ:* Emory Univ, BA, 55; Johns Hopkins Univ, MD, 61. *Prof Exp:* Intern & asst resident med, Johns Hopkins Hosp, 61-63; asst & chief resident, Grady Mem Hosp, Atlanta, Ga, 63-65; instr, Emory Univ, 65-66; staff assoc cardiol, Nat Heart & Lung Inst, La State Univ Med Ctr, 66-68, sr investr, 68-69, chief cardiac diag, 69-72; chief cardiol & prof med, 72-74, CLIN PROF MED, LA STATE UNIV MED CTR, NEW ORLEANS, 74-; DIR CARDIOL DEPT, HOTEL DIEU HOSP, 74- *Concurrent Pos:* Fel coun clin cardiol, Am Heart Asn; founding mem Cardiac Angiography & Interactions. *Mem:* Fel Am Col Cardiol; fel Am Col Physicians; fel Am Col Chest Physicians; Am Fedn Clin Res. *Res:* Clinical cardiology; cardiovascular physiology; cardiac pathology; interventional cardiology. *Mailing Add:* Cardiol Dept Hotel Dieu Hosp 2021 Perdido St New Orleans LA 70112

GLAND, JOHN LOUIS, b Valparaiso, Ind, Feb 22, 47; m 67; c 2. SURFACE CHEMISTRY. *Educ:* Wittenberg Univ, BA, 69; Univ Calif, Berkeley, PhD(phys chem), 73. *Prof Exp:* Staff res scientist, Gen Motors Res Labs, 73-82; assoc corp res, Exxon Res & Eng Co, 82-88; PROF CHEM & CHEM ENG, UNIV MICH, 88- *Concurrent Pos:* John M Campbell basic res award, Gen Motors Res. *Honors & Awards:* Giuseppe Parravano Award for Excellence in Catalysis Res. *Mem:* Am Chem Soc; Am Phys Soc; Am Vacuum Soc; Sigma Xi. *Res:* Reactions with and over solid surfaces; effects of the geometry and electronic properties of the surface on reactivity; soft x-ray absorption studies of absorbed species and surface reactions using synchrotron radiation; development of fluorescence yield near edge spectroscopy (FYNES) in the soft x-ray region; high resolution electron energy loss spectroscopy; low-energy electron diffraction; auger and electron spectroscopic chemical analysis applied to mechanistic studies of catalysis. *Mailing Add:* Chem Dept Univ Mich Ann Arbor MI 48109

GLANDT, EDUARDO DANIEL, b Buenos Aires, Arg, Mar 4, 45; US citizen. THERMODYNAMICS. *Educ:* Univ Buenos Aires, BS, 68; Univ Pa, MS, 75, PhD(chem eng), 77. *Prof Exp:* From asst prof to prof, 77-90, CARL V S PATTERSON PROF CHEM ENG, UNIV PA, 90- *Concurrent Pos:* Gulf vis prof, Dept Chem Eng, Carnegie-Mellon Univ, 89-90. *Honors & Awards:* Victor K Lamer Award, Am Chem Soc, 79. *Mem:* Am Inst Chem Engrs; Am Chem Soc; Am Phys Soc; AAAS. *Res:* Classical and statistical thermodynamics: liquids and their mixtures, computer simulations; interfacial phenomena: adsorption, colloids, membranes; heterogeneous media: percolation and gelation phenomena. *Mailing Add:* Dept Chem Eng Univ Pa Philadelphia PA 19104-6393

GLANTZ, RAYMON M, b Brooklyn, NY, July 1, 41. NEUROPHYSIOLOGY. *Educ:* Brooklyn Col, BA, 63; Syracuse Univ, MS, 64, PhD(physiol psychol), 66. *Prof Exp:* Instr physiol, Sch Med, NY Univ, 66-67; res fel neurophysiol, Calif Inst Technol, 67-69; asst prof biol, 69-74, assoc prof biol & elec eng, 74-80, PROF BIOL & ELEC ENG, RICE UNIV, 80-, CHMN BIOL, 87- *Concurrent Pos:* Grass Found fel neurophysiol, Woods Hole Marine Biol Lab, 67; vis res assoc, Med Fac, Rotterdam Univ, summers 70-; investr, Woods Hole Marine Biol Lab, 79. *Mem:* AAAS; Am Soc Zool; Soc Neurosci; Int Soc Neuroethnol. *Res:* Information processing in the invertebrate visual system; neural control of behavior; mechanism of phototransduction; biophysical properties of dendrites; synaptic physiology of neural networks. *Mailing Add:* Dept Biochem Rice Univ PO Box 1892 Houston TX 77251

GLANTZ, STANTON ARNOLD, b Cleveland, Ohio, May 3, 46; m 72; c 2. CARDIOVASCULAR PHYSIOLOGY, INVOLUNTARY SMOKING & TOBACCO CONTROL. *Educ:* Univ Cincinnati, BS, 69; Stanford Univ, MS, 70, PhD(appl mech), 73. *Prof Exp:* Sr fel cardiovasc res, 75-77, from asst prof to assoc prof, 77-86, PROF MED, UNIV CALIF, SAN FRANCISCO, 86- *Concurrent Pos:* Aerospace engr, NASA Manned Spacecraft Ctr, 69; mem, Cardiovasc Res Inst, Univ Calif, San Francisco, 77-; vis prof med, Univ Vt, 88- *Mem:* Am Heart Asn; Am Physiol Soc; AAAS. *Res:* Mechanics of cardiac function; applied biostatistics; involuntary smoking; bioengineering; statistics; public health policy. *Mailing Add:* Div Cardiol Univ Calif Box 0124 San Francisco CA 94143-0124

GLANVILLE, JAMES OLIVER, b London, Eng, July 24, 41; m 65; c 2. INORGANIC CHEMISTRY, ANALYTICAL CHEMISTRY. *Educ:* Univ London, BSc & ARCS, 62; Univ Md, PhD(chem), 67. *Prof Exp:* Res chemist, Res & Develop Dept, Inorg Div, FMC Corp, 68-69; assoc prof chem, Va Western Community Col, 69-76; vis prof chem, Va Tech, 77-78; VPRES RES & DEVELOP, WEN-DON CORP, 78- *Mem:* Am Chem Soc; Am Inst Chemists; Am Soc Testing & Mat; Sigma Xi. *Mailing Add:* Chem Dept Va Polytech Inst State Univ Blacksburg VA 24061

GLANZ, FILSON H, b Los Angeles, Calif, Aug 7, 34; m 67; c 1. ELECTRICAL ENGINEERING. *Educ:* Stanford Univ, BS, 56, MS, 57, PhD(elec eng), 65. *Prof Exp:* Engr, Librascope, Inc, 57-59 & Stanford Res Inst, 61-63; asst prof, 65-70, ASSOC PROF ELEC ENG, UNIV NH, 70- *Mem:* Inst Elec & Electronics Engrs; Sigma Xi. *Res:* Adaptive pattern recognition; information theory; digital signal processing; non-uniform sampling. *Mailing Add:* 25 Orchard Dr Durham NH 03824

GLANZ, PETER K, b Portsmouth, Va, Oct 23, 41; m 63. PHYSICS. *Educ:* Bates Col, BS, 63; Bucknell Univ, MS, 65; Univ Conn, PhD(physics), 71. *Prof Exp:* Asst prof, 71-76, ASSOC PROF PHYSICS, RI COL, 76- *Concurrent Pos:* Mem, Mus Holography. *Mem:* Nat Asn Sci Teachers; Am Asn Physics Teachers. *Res:* Interferometric holography. *Mailing Add:* Dept Physics RI Col Providence RI 02908

GLANZ, WILLIAM EDWARD, b Ypsilanti, Mich, Jan 27, 49; m 80. MAMMALOGY, COMMUNITY ECOLOGY OF VERTEBRATES. *Educ:* Dartmouth Col, AB, 70; Univ Calif, Berkeley, PhD(zool), 77. *Prof Exp:* Teaching fel trop biol, Smithsonian Trop Res Inst, 77-78; lectr environ studies, Univ Calif, Santa Cruz, 78-79, vis lectr biol, Univ Calif, Los Angeles, 79; asst prof zool, 79-85, coop asst prof wildlife, 83-85, ASSOC PROF ZOOL & COOP ASSOC PROF, UNIV MAINE, ORONO, 85- *Mem:* Am Soc Mammalogists; Am Ornith Union; AAAS; Asn Trop Biol; Cooper Ornith Soc; Ecol Soc Am. *Res:* Community ecology of mammals and birds in North and South America; seed predation, caching behavior and territoriality in squirrels; population dynamics of mammals in tropical forests. *Mailing Add:* Dept Zool Univ Maine Orono ME 04469-0146

GLAROS, GEORGE RAYMOND, b Minneapolis, Minn, Nov 23, 41; m 67; c 2. ORGANIC CHEMISTRY. *Educ:* Univ Minn, BChem, 65; Univ Nebr, PhD(org chem), 71. *Prof Exp:* Peace Corps lectr chem, Cameroon Protestant Col, 65-67; instr chem, Univ Conn, 71-72; ASSOC PROF CHEM, RUSSELL SAGE COL, 72- *Mem:* Am Chem Soc. *Res:* Chemistry of sulfoxides; allylic rearrangements. *Mailing Add:* Dept Chem Russell Sage Col 45 Ferry St Troy NY 12180

GLARUM, SIVERT HERTH, b Providence, RI, June 6, 33; m 59; c 2. PHYSICAL CHEMISTRY. *Educ:* Kalamazoo Col, BA, 55; Brown Univ, PhD(chem), 60. *Prof Exp:* MEM TECH STAFF, BELL LABS, 59- *Mem:* Am Chem Soc; Am Phys Soc; Sigma Xi. *Res:* Dielectrics; solid state; paramagnetic resonance. *Mailing Add:* Bell Labs Murray Hill NJ 07971

GLASCOCK, HOMER HOPSON, II, b Hannibal, Mo, Apr 10, 29; m 58; c 3. PHYSICS. *Educ:* Univ Mo, BS, 51, MS, 56, PhD(solid state physics), 61. *Prof Exp:* res physicist, Gen Elec Co, 60-90; RES PHYSICIST, HARRIS CORP, 90- *Honors & Awards:* IR 100 Award, 79. *Mem:* Am Phys Soc; Inst Elec & Electronics Engrs. *Res:* Solid state and surface physics; physical electronics. *Mailing Add:* Harris Corp PO Box 804 Schenectady NY 12301

GLASCOCK, MICHAEL DEAN, b Hannibal, Mo, June 27, 49. ARCHAEOLOGICAL CHEMISTRY, GEOCHEMISTRY. *Educ:* Univ Mo-Rolla, BS, 71; Iowa State Univ, PhD(nuclear physics), 75. *Prof Exp:* Res asst nuclear physics, Ames Lab, USAEC, 73-75; res assoc nuclear physics, Univ MD, College Park, 75-78; SR RES SCIENTIST, UNIV MO, 79- *Mem:* Am Phys Soc; Sigma Xi; Am Chem Soc; Am Nuclear Soc; Soc Archaeol Scientists. *Res:* Development of applications for capture gamma-ray spectroscopy as an analytical technique for various disciplines including agriculture, archaeology, biology, chemistry, nuclear engineering, geology and physics. *Mailing Add:* Univ Mo Res Reactor Facil Columbia MO 65211

GLASEL, JAY ARTHUR, b New York, NY, Apr 30, 34; m 62. MOLECULAR IMMUNOLOGY, BIOCHEMISTRY. *Educ:* Calif Inst Technol, BS, 55; Univ Chicago, PhD(chem physics), 59. *Prof Exp:* Asst prof biochem, Columbia Univ, 64-70; vis scientist, Oxford Univ, 70-71; assoc prof, 71-75, PROF BIOCHEM, UNIV CONN HEALTH CTR, 75- *Concurrent Pos:* NSF fel, Univ Calif, San Diego, 60-61 & Imp Col, Univ London, 61-62. *Mem:* Am Physical Soc; Am Chem Soc; Am Soc Biol Chemists. *Res:* characterization of opiate receptors; nuclear magnetic resonance spectroscopy. *Mailing Add:* Dept Biochem Univ Conn Health Ctr Farmington CT 06030

GLASER, CHARLES BARRY, PROTEIN CHEMISTRY, PURIFICATION. *Educ:* Polytech Inst New York, PhD(org chem), 69. *Prof Exp:* SR SCIENTIST, CODON, 86- *Mailing Add:* 284 Wawona St S San Francisco CA 94080

GLASER, DONALD ARTHUR, b Cleveland, Ohio, Sept 21, 26; m; c 2. PHYSICS, MOLECULAR BIOLOGY. *Educ:* Case Inst Technol, BS, 46, Calif Inst Technol, PhD(physics), 50. *Hon Degrees:* DSc, Case Inst Technol, 59. *Prof Exp:* From instr to prof physics, Univ Mich, 49-59; vis prof, Univ Calif Berkeley 59-60, prof physics, 60-64, Miller Res Biophysicist, 62-64, prof Physics & Molecular Biol, 64-89, PROF PHYSICS & PROF MOLECULAR & CELL BIOL, DIV NEUROBIOL, UNIV CALIF, BERKELEY, 89- *Concurrent Pos:* distinguished res fel, Smith-Kettlewell Inst for Vision Res, 83-84. *Honors & Awards:* Nobel Prize, 60; Charles Vernon Boys Prize, Brit Inst Physics & Phys Soc, 58; Prize, Am Phys Soc, 59; Golden Plate Award, Am Acad Achievement, 89. *Mem:* Nat Acad Sci; fel Am Phys Soc; Neurosci Inst; Int Acad Sci; Asn Res Vision & Opthal; Ny Acad Sci; AAAS; Am Asn Artificial Intel; Am Soc Microbiol; Fedn Am Scientists; Sigma Xi. *Res:* Nuclear physics; cosmic rays; molecular genetics; neurobiology. *Mailing Add:* Dept Molecular & Cell Biol Univ Calif 337 Stanley Hall Berkeley CA 94720

GLASER, EDMUND M, b New York, NY, Oct 17, 27; m 59; c 3. COMPUTER SCIENCE, PHYSIOLOGY. *Educ:* Cooper Union, BEE, 49; Johns Hopkins Univ, MSE, 54, DEng, 60. *Prof Exp:* Electromech engr control systs, Glenn L Martin Co, 50-52; res assoc, Radiation Lab, Johns Hopkins Univ, 52-60; assoc prof physiol, 62-70; PROF PHYSIOL & COMPUT SCI, SCH MED, UNIV MD, BALTIMORE, 70- *Concurrent Pos:* Fel physiol, Sch Med, Johns Hopkins Univ, 62; consult, Hoover Electronics, 58-59, Electronic Commun, Inc, 60-62, Westinghouse Elec Corp, 62-63, Johns Hopkins Univ, 62-64. *Mem:* AAAS; Soc Neurosci; Inst Elec & Electronics Engrs; Acoust Soc Am; Sigma Xi. *Res:* Sensory neurophysiology, especially auditive, and its relationship to information theory; biological control systems; application of computers to the neurosciences; biomedical engineering. *Mailing Add:* 2411 Brambleton Rd Baltimore MD 21209

GLASER, EDWARD L(EWIS), computer design, hardware & software; deceased, see previous edition for last biography

GLASER, FREDERIC M, b Toledo, Ohio, Dec 7, 35; m 61; c 1. QUANTUM PHYSICS. *Educ:* Purdue Univ, BS, 57; Ohio State Univ, PhD(phys chem), 63. *Prof Exp:* Summer res asst, Am Cyanamid Co, 56-59; NSF fel & res assoc physics, Univ Chicago, 63-64; asst prof, Bowling Green State Univ, 64-66; mem staff, Houston Opers, TRW Systs Inc, 66-69; ASSOC PROF PHYSICS, PAN AM UNIV, 69- *Mem:* AAAS; Am Phys Soc; Am Chem Soc; Am Math Soc; Soc Indust & Appl Math. *Res:* Theoretical solid state physics; bonding theory of small molecules and ultraviolet spectra; experimental reflectance spectroscopy. *Mailing Add:* Dept Phys & Geol Univ Tex Pan Am Edinburg TX 78539-2999

GLASER, FREDERICK BERNARD, b Rochester, NY, Nov 8, 35; div; c 1. PSYCHIATRY. *Educ:* Univ Wis, BS, 55; Harvard Univ, MD, 59. *Prof Exp:* Assoc prof psychiat & chief, Sect Drug & Alcohol Abuse, Med Col Pa, 72-75; cong fel health policy, Robert Wood Johnson Found, 74-75; PROF PSYCHIAT, FAC MED, UNIV TORONTO, 75-; HEAD PSYCHIAT, ADDICTION RES FOUND CLIN INST, 75- *Concurrent Pos:* John & Mary

R Markle Found scholar, 68-73; consult treatment, Gov's Coun Drug & Alcohol Abuse, Commonwealth Pa, 72-75; chmn, Demonstration Rev Comt, Nat Inst Drug Abuse, 75-77; vis lectr, Felton Bequest, Melbourne, Australia, 79; assoc ed, Res Advan Alcohol & Drug Problems, 75- *Mem:* Am Psychiat Asn; Can Psychiat Asn; Royal Col Physicians & Surgeons Can. *Res:* Development and implementation of a systems approach to health care delivery, including alcohol and drug dependence. *Mailing Add:* 44 Charles W Toronto ON M4Y 1R5 Can

GLASER, GILBERT HERBERT, b New York, NY, Nov 10, 20; m 46; c 2. NEUROLOGY. *Educ:* Columbia Univ, AB, 40, MD, 43, ScD(med), 51. *Hon Degrees:* MA, Yale Univ, 63. *Prof Exp:* Intern, Mt Sinai Hosp, NY, 43-44; asst resident neurol, Neurol Inst, 44-45, chief resident, 45-46; res asst, Columbia Univ, 48-50, instr, 50-51, assoc, 51-52; from asst prof to prof neurol, 52-91, head sect, 52-71, chmn dept, 71-86, EMER PROF NEUROL, SCH MED, YALE UNIV, 91- *Concurrent Pos:* Resident psychiat, NY Psychiat Inst, 48-49, sr res scientist, 49-50; consult to Surgeon Gen, Neurol Res Training Grant Comt, USPHS, 56-60, mem neurol res progs grant comt, 68-72, mem epilepsy adv comt, 74-77; vis prof neurol, Hosp for Sick Children & Univ Col, London, 65-66; Hunan Med Col, Changsha, China, 86; vis prof, Nat Hosp, London, 72 & Park Hosp, Oxford, 73, 74, 75, 78, 81, 84, & 87, Brain Res Inst, Univ Niigato, Japan, 89; mem neuropharmacol adv comt, Food & Drug Admin, Dept Health, Educ & Welfare, 70-72; ed, Epilepsia, J Neurol Sci, J Nervous & Mental Dis; Fulbright Distinguished Prof, Univ Zagreb, Yugoslavia, 81; vis scholar, Green Col, Oxford Univ, Eng, 87, 88. *Honors & Awards:* Lennox Lectr Award, Am Epilepsy Soc, 85; Distinguished Scholar Award, Med Col Va, 86. *Mem:* Am EEG Soc; Am Epilepsy Soc (pres, 63); Am Neurol Asn (1st vpres, 77-78); fel Am Col Physicians; Am Acad Neurol (pres, 73-75); hon mem Asn Brit Neurologists. *Res:* Clinical neurology and neurophysiology; electroencephalography; epilepsy; neuromuscular disorders; metabolic disorders of the nervous system; developmental neurology. *Mailing Add:* 205 Millbrook Rd Yale Univ North Haven CT 06473

GLASER, HAROLD, b Kurseni, Lithuania, Aug 28, 24; US citizen; m 45; c 3. THEORETICAL PHYSICS. *Educ:* Roosevelt Univ, BS, 48; Northwestern Univ, MS, 49, PhD(physics), 53. *Prof Exp:* Instr, Roosevelt Univ, 49-51; sr physicist, Appl Physics Lab, Johns Hopkins Univ, 52-54; head theoret analysis sect, Syst Analysis Br, Naval Res Lab, 54-57, physicist, Off Naval Res, 57-64, head nuclear physics br, 64-66; dep chief solar physics, Off Space Sci & Appln, NASA, 66-67, chief, 67-70; chief planning, Nat Bur Standards, 70-71; Off Sci & Technol, Exec Off of President, 71-72; dep & actg dir exp technol incentives prog, Nat Bur Standards, 72; dep dir sci & technol, Off Nat Res & Develop Assessment, NSF, 72-75; dir solar terrestrial progs, Nat Aeronaut & Space Admin, 75-80; consult, Off Mgt & Budget, 80-81; asst to pres, Lab Affairs, Univ Calif, Berkeley, 81-84; CONSULT, 84- *Concurrent Pos:* Res assoc, Univ Southern Calif, 59; consult, Lawrence Livermore Nat Lab, 84. *Honors & Awards:* Meritorious Civilian Serv Award, Dept Navy, 65. *Mem:* Int Astron Union; Am Phys Soc; Am Geophys Union; Am Astron Soc; Sigma Xi. *Res:* Electromagnetic propagation; noise theory; radio astronomy; solar physics. *Mailing Add:* 1346 Bonita Ave Berkeley CA 94709

GLASER, HERMAN, b Brooklyn, NY, Nov 11, 23; m 55; c 1. PHYSICS. *Educ:* Brooklyn Col, BA, 43; Johns Hopkins Univ, PhD(physics), 50. *Prof Exp:* Instr physics, Univ NC, 43-44; physicist, Naval Ord Lab, 44-45; jr instr physics, Johns Hopkins Univ, 45-49; assoc prof, Tex Tech Col, 50-53; from asst prof to assoc prof, 53-69, PROF PHYSICS, HOFSTRA UNIV, 69-, CHMN DEPT, 75- *Mem:* Am Phys Soc; Am Asn Physics Teachers; Sigma Xi. *Res:* X-ray spectroscopy; low energy nuclear physics. *Mailing Add:* Dept Physics Hofstra Univ Hempstead NY 11550

GLASER, JANET H, b Eugene, Ore, Mar 9, 44; m 68; c 2. CELL CULTURE, CARBOHYDRATE BIOCHEMISTRY. *Educ:* Univ Wash, BS, 66; Univ Calif, San Diego, PhD(biol), 71. *Prof Exp:* Asst human genetics, Sch Med, Wash Univ, St Louis, 71-74; vis asst prof microbiol, 74-75, res asst biochem, 75-85, asst dir, 85-87, ASSOC DIR BIOTECHNOL CTR, UNIV ILL, 87- *Mem:* Am Soc Biol Chemists; AAAS. *Res:* Regulation of synthesis of proteoglycans and collagen by chondrocytes in serum-free tissue culture and factor controlling expression of differentiated phenotype. *Mailing Add:* Univ Ill 901 S Mathews Urbana IL 61801

GLASER, KEITH BRIAN, b Munich, WGer, May 27, 60; US citizen; m; c 3. IMMUNOPHARMACOLOGY. *Educ:* Tex A&M Univ, Galveston, BS, 82; Univ Calif, Santa Barbara, PhD(marine pharmacol), 87. *Prof Exp:* Teaching asst pharmacol, Dept Biol Sci, Univ Calif, Santa Barbara, 83, trainee marine pharmacol, Marine Sci Inst, 83-87; NIH postdoctoral fel, Sch Med, Wash Univ, St Louis, 87; res chemist, Univ Calif, San Diego, 87-88, NIH postdoctoral fel, 88-89; RES SCIENTIST, DIV IMMUNOPHARMACOL, WYETH-AYERST RES, 89- *Concurrent Pos:* Mem, Gordon Res Conf Marine Natural Prod Chem, 84 & Int Conf Therapeut Control Inflammatory Dis, Inflammation Res Asn, 86 & 90. *Mem:* Assoc mem Am Soc Biochem & Molecular Biol. *Res:* Site and mechanism of action of anti-inflammatory and immunomodulatory agents; pharmacological mode of action of anti-inflammatory marine natural products; author of more than 20 technical publications. *Mailing Add:* Inflammation & Bone Metab Div Wyeth-Ayerth Res CN8000 Princeton NJ 08543-8000

GLASER, KURT, b Vienna, Austria, Feb 16, 15; nat US; m 46; c 4. CHILD PSYCHIATRY. *Educ:* Univ Vienna, 33-38; Univ Lausanne, MD, 39; Univ Ill, MSc, 48. *Prof Exp:* Instr pediat, Col Med, Univ Ill, 45-50; asst chief physician, Med Sch, Hebrew Univ, Israel, 50-54; from clin instr to clin assoc prof pediat & psychiat, Sch Med, Univ Md, Baltimore, 54-81; asst prof, 72-81, EMER ASST PROF PEDIAT & PSYCHIAT, MED SCH, JOHNS HOPKINS UNIV, BALTIMORE, 81-; EMER CLIN ASSOC PROF PSYCHIAT, SCH MED, UNIV MD, BALTIMORE, 81- *Concurrent Pos:* Consult, Dept Pediat & Psychiat, Sinai Hosp, Baltimore, 59-, Ctr Eval Clin Children, Univ Md, 59-75, John F Kennedy Inst, 76-, Social Security Admin, 85- & Harbel Ment Health Clin, 86-; clin dir, Rosewood State Hosp, Owings Mills, Md, 61-72; dir, Adolescent Unit, Springfield Hosp Ctr, Sykesville, Md, 72-81; staff psychiatrist, Sheppard Pratt Hosp, Baltimore, 84-85. *Mem:* Emer fel Am Acad Pediat; fel Am Psychiat Asn; Am Soc Adolescent Psychiat. *Res:* Growth and development of premature infants; cellular composition of the bone marrow in normal infants and children; problems of child psychiatry, including school phobia, learning disorders, depression and suicide in children and adolescents, maternal deprivation in children; mental retardation. *Mailing Add:* 6114 Biltmore Ave Baltimore MD 21215-3602

GLASER, LESLIE, b DeKalb, Ill, Sept 4, 37; m 58; c 4. TOPOLOGY. *Educ:* DePauw Univ, BA, 59; Univ Wis, MS, 61, PhD(topol), 64. *Prof Exp:* Asst prof math, Rice Univ, 64-68; assoc prof, 68-71, PROF MATH, UNIV UTAH, 71- *Concurrent Pos:* Vis mem, Inst Advan Study, 69-70; Sloan fel, 69-71. *Res:* Point set topology; combinatorial topology. *Mailing Add:* Dept Math Univ Utah Salt Lake City UT 84112

GLASER, LUIS, b Vienna, Austria, Mar 30, 32; US citizen; m 61; c 2. BIOCHEMISTRY. *Educ:* Univ Toronto, BA, 53; Washington Univ, PhD(biochem), 56. *Prof Exp:* From instr to prof biochem, Sch Med, Wash Univ, 56-86, chmn dept, 75-86; EXEC VPRES & PROVOST, UNIV MIAMI, FLA, 86- *Mem:* Am Soc Biol Chemists; Am Soc Cell Biol; Am Chem Soc; Am Soc Neurosci. *Res:* Mechanisms of sugar synthesis; control of cell growth; bacterial cell wall components; neuronal development; embryonal development. *Mailing Add:* Univ Miami 240 Ash Bldg Coral Gables FL 33124

GLASER, MICHAEL, b Cleveland, Ohio, Mar 11, 45; m 68; c 2. BIOCHEMISTRY. *Educ:* Univ Calif, Los Angeles, BS, 66; Univ Calif, San Diego, PhD(chem), 71. *Prof Exp:* Fel biochem, Med Sch, Washington Univ, 71-74; from asst prof to assoc prof, 74-83, PROF BIOCHEM, UNIV ILL, 83- *Concurrent Pos:* NIH fel, 71-73; mem, NIH Molecular Cytology Study Sect, 81-85; mem, Ill Heart Asn Grant Rev Comt, 76-77 & Am Heart Asn Physiol Chem B Res Study Comt, 79-81; Ctr Advan Study fel, Univ Ill, 78; NIH res career develop award, 76-81. *Mem:* Am Chem Soc; Am Soc Microbiol; Am Soc Biochem & Molecular Biol; Biophys Soc; Am Soc Cell Biol. *Res:* Membrane structure, function and biosis; lipid synthesis, macromolecular synthesis and cell growth; fluorescene microscopy and spectroscopy. *Mailing Add:* Dept Biochem Univ Ill 1209 W California St Urbana IL 61801

GLASER, MILTON ARTHUR, b New York, NY, Sept 4, 12; m 37; c 3. ORGANIC POLYMER CHEMISTRY. *Educ:* Tufts Univ, BS, 34. *Prof Exp:* Engr, Mass, 31-32 & 34-36; chief chemist & dir res, Standard Varnish Works, 36-45; vpres & tech dir, Midland Div, Dexter Corp, Waukegan, 45-64, exec vpres res & develop, 64-70, div vpres res & develop, 70-77; CONSULT, 78- *Concurrent Pos:* Lectr, NDak State Univ, Ill Inst Technol & Univ Mo; pres, Fedn Socs Paint Technol, 56-57; vpres & trustee, Paint Res Inst, 58, pres, 65-66; titular mem org coating sect, Int Union Pure & Appl Chem, 67-; Mattiello lectr, 74. *Honors & Awards:* Outstanding Serv Award, 57; Distinguished Serv Award, 59; Heckel Award, 63; Merit Award, 68; Pregl Award, NY Acad Sci, 86. *Mem:* Am Chem Soc; Am Oil Chemists' Soc; Nat Asn Corrosion Eng; fel Am Inst Chemists; Inst Food Technol. *Res:* Organic coatings including alkyd, silicone, epoxy, urethane and phenolic polymers; high-temperature resistant coatings; corrosion resistant coatings for container linings; specialty coatings for severe exposure environments; consult on reasearch and development to improve innovative productivity in laboratories. *Mailing Add:* 171 Wentworth Ave Glencoe IL 60022

GLASER, MYRON B(ARNARD), b New York, NY, Dec 31, 27; m 85. ELECTRICAL ENGINEERING, ELECTRONICS ENGINEERING. *Educ:* City Col New York, BEE, 50; Yale Univ, MEng, 51. *Prof Exp:* Assoc proj engr, Sperry Gyroscope Co, 51-53, proj engr, 53-57, sr engr, 57-58, prin engr, Sperry Phoenix Co, 58-62; sr res engr, SRI Int, 62-89; PRIN, GLASER ASSOCS, 89- *Concurrent Pos:* Alt deleg, Radio Technol Comn Aeronaut, 64, partic spec comt 117, 68- *Mem:* Inst Elec & Electronics Engrs; Inst Navig. *Res:* Telecommunications and information systems; public safety communications systems; electronic voting systems; geographic information systems; systems planning and implementation assistance; user; systems design & systems science; two United States patents. *Mailing Add:* Glaser Assocs PO Box 4067 Mountain View CA 94040-0067

GLASER, PETER E(DWARD), b Czech, Sept 5, 23; nat US; m 55; c 3. ENGINEERING, SOLAR ENERGY & SPACE TECHNOLOGY. *Educ:* Leeds Col Tech, Eng, dipl, 43; Charles Univ, Prague, dipl, 47; Columbia Univ, MS, 51, PhD, 55. *Prof Exp:* Head design dept, Werner Textile Consults, 49-53; sr res engr, 55-61, group leader, 61-66, sect head eng sci, 66-73, vpres eng sci, 73-78, CORP VPRES, ARTHUR D LITTLE, INC, 78- *Concurrent Pos:* Mem mat adv bd, Nat Acad Sci, 58, mem study group solar energy, 71; mem, Int Inst Refrig, 59-72; ed-in-chief, J Solar Energy, Int Solar Energy Soc, 71-85; pres, Sunsat Energy Coun, 78-; co-investr, Shuttle, Initial Blood Storage Exp, 78-86; mem Solar Power Satellite Adv Panel, Off Technol Assessment, 80-81; chmn, Space Power comts, Int Astronaut Fedn, 84-89, NASA Task Force on space goals, 86-87, Lunar Energy Enterprise case study task force, NASA, 88-89; adv bd, Ctr Space Power, Tex A&M Univ & Space Studies Inst. *Honors & Awards:* Carl F Kayan Medal, Columbia Sch Eng, 74; Farrington Daniels Award, Int Solar Energy Soc, 85. *Mem:* Fel AAAS; Am Soc Mech Engrs; Int Solar Energy Soc (pres, 68-70); assoc fel Am Inst Aeronaut & Astronaut; Int Acad Astronaut. *Res:* Thermal insulation developments for extreme temperatures; design of solar furnace and arc imagining furnace; high temperature research with imagining furnaces; lunar surface research; solar power satellite development; solar heating and cooling; photovoltaic conversion; rural electrification systems using renewable resources; lunar surface missions; commercial space power utility; remote sensing applications; extravehicular activity on the lunar surface; selection of launch sites in Hawaii; space station habitability module appliances; advanced space transportation systems for operation in cis-lunar space; space-based sensor systems to identify CO2-induced climate changes; space station portable contamination detector; space suit gloves and boot soles for moon and Mars surface missions; dust protection during extravehicular activities on the moon and mars. *Mailing Add:* 62 Turning Mill Rd Lexington MA 02173

GLASER, ROBERT, b Providence, RI, Jan 18, 21; c 2. LEARNING THEORY, COGNITIVE PSYCHOLOGY INSTRUCTION. *Educ:* City Col NY, BS, 42; Ind Univ, MA, 47, PhD(psychol measurement & learning theory), 49. *Hon Degrees:* DSc, Univ Leuven, 80, Ind Univ, 84 & Univ Gateborg, 85. *Prof Exp:* Instr psychol, Ind Univ, 48-49; asst prof, Univ Ky, 49-50; res asst prof, Univ Ill, 50-52; sr res scientist, Am Inst Res, 52-56; prof psychol, 57-72, prof educ, 64-72, UNIV PROF PSYCHOL & EDUC, UNIV PITTSBURGH, 72-, DIR LEARNING RES & DEVELOP CTR, 63- *Concurrent Pos:* Consult, Ford Found Proj Indust Literacy, 81; vis prof, Japan Soc Prom Sci, 82; mem Comt Res Math, Sci & Technol Educ, Nat Res Coun, 84, Math Sci Educ Bd, 85-87; mem, Forum Res Mgt, Fedn Behav, Psychol & Cognitive Sci, 87-89, bd sci affairs, Am Psychol Asn, 88-, Design & Analysis Comt, Nat Asn Educ Publishers, 83- *Honors & Awards:* Outstanding Res Field Instrnl Mats Award, Am Educ Res Asn/Am Educ Publ Inst, 70; Distinguished Res Educ Award, Am Educ Res Asn, 76; E L Thorndike Award, Distinguished Psychol Contribs, Am Psychol Asn, 81; Distinguished Sci Award Applns Psychol, 87. *Mem:* Am Psychol Asn; Nat Acad Educ (pres, 81-85); Psychonomic Soc; Am Educ Res Asn; AAAS. *Res:* Cognitive analysis of the acquisition of knowledge, the nature of expertise; inference and discovery in science; cognition, learning and instructional systems. *Mailing Add:* 833 LRDC Bldg Univ Pittsburgh Pittsburgh PA 15260

GLASER, ROBERT J, b Brooklyn, NY, Mar 17, 42; m 62; c 2. ORGANIC CHEMISTRY, BIO-ORGANIC CHEMISTRY. *Educ:* Univ Pa, BA, 63; Rutgers Univ, PhD(org chem), 70. *Prof Exp:* Develop chemist, Indust Adhesives Div, PPG Industs, Inc, 63-66; res assoc org chem, Princeton Univ, 69-70, res assoc x-ray crystallog, 70-71; lectr, Ben Gurion Univ of the Negev, Israel, 71-77, sr lectr org chem, 77; LP MARKLEY CHARITABLE TRUST. *Concurrent Pos:* NIH res fel, 69-71; vis res staff mem stereochem, Princeton Univ, 78-79; sci adv, Bromine Compounds Ltd, 80-81. *Mem:* Am Chem Soc; Royal Soc Chem; Israel Chem Soc. *Res:* Asymmetric synthesis via homogeneous chiral catalysts; heterogenization of homogeneous catalysts via polymer attachment; stereochemistry of transition-metal complexes. *Mailing Add:* LP Markley Charitable Trust 525 Middlefield Rd Suite 130 Menlo Park CA 94025

GLASER, ROBERT JOY, b St Louis, Mo, Sept 11, 18; m 49; c 3. MEDICINE. *Educ:* Harvard Univ, SB, 40, MD, 43; Am Bd Internal Med, dipl, 51. *Hon Degrees:* ScD, Temple Univ, Chicago Med Sch, Univ Colo, Univ NH, Mt Sinai Med Sch, Wash Univ; LHD, Rush Med Col. *Prof Exp:* Intern med, Barnes Hosp, St Louis, Mo, 44; resident, Peter Bent Brigham Hosp, Boston, Mass, 44-45; resident, Barnes Hosp, St Louis, Mo, 45-47; from instr to assoc prof, Med Sch, Washington Univ, 49-57, asst & assoc dean, 53-57, chief div immunol, 54-57; prof & dean, Univ Colo, 57-63, vpres med affairs, 59-63; prof social med, Harvard Med Sch, 63-65; prof med, dean sch med & vpres med affairs, Stanford Univ, 65-70; vpres, Commonwealth Fund, 70-72; pres, Henry J Kaiser Family Found, 72-83; TRUSTEE & DIR MED SCI, LUCILLE P MARKEY CHARITABLE TRUST, 84- *Concurrent Pos:* Nat Res Coun fel, Med Sch, Washington Univ, 47-49; asst med, Washington Univ, 45-47; asst physician, Barnes Hosp, St Louis, Mo, 49-57; chief rheumatic fever clin, Washington Univ Clins, 49-57; vis physician, Univ I Med Serv, St Louis City Hosp, 50, chief serv, 50-53; attend physician, Colo Gen Hosp, 57-63; pres, Affiliated Hosps Ctr, Inc, Boston, 63-65; chief of staff & attend physician, Stanford Univ Hosp, 65-70; consult, State of Mo Crippled Children's Serv, 49-55, div of serv for crippled children, Univ Ill, 50-57, Vet Admin Hosps, Denver & Grand Junction, Colo & Albuquerque, NMex, 57-63, consult, Fitzsimons Army Hosp, Denver, Colo, 57-63, Lowry AFB Hosp, 57-63, Nat Health Res Facilities Adv Coun, 58-61, Vet Admin, Mass, 63-65, Harvard Med Servs, Boston City Hosp, 63-65 & Peter Bent Brigham Hosp, Boston, 63-65; consult prof med, Stanford Univ, 73-; mem sci adv coun, Rheumatic Fever Res Inst, 56-59; mem grad training grant comt, Nat Inst Allergy & Infectious Dis, 58-61; mem, Nat Health Res Facilities Adv Coun, 61-65 & adv coun, Nat Inst Dent Res, 65-69; mem adv comt higher educ, Dept Health, Educ & Welfare, 67-68, ad hoc adv comt, 67-68, nat adv ment health coun, 70-; mem bd dir, Kaiser Found Hosps & Health Plan, Inc, 67-79; mem bd, Commonwealth Fund, 69-88, Kaiser Family Found, 70-83, Hewlett-Packard, 71-91, Calif Water Serv Co, 72-, Dynapol, 72-83, Equitable Life Ins Soc, 79-86, First Boston Inc, 82-88, Resonex, Inc, 84-86, Packard Found, 84- & Alza Corp, 87-; mem comt med affairs, Yale Univ, 69-82, mem adv bd, Yale Sch Org & Mgt, 76-84; Alpha Omega Alpha lectr, Univ Alta, 63, Boston Univ, 63, Univ Colo, 64, Univ WVa, 64, Univ Louisville, 65, Univ Cincinnati, 66, Univ Calif, San Francisco, 67, NY Med Col, 69, Univ Ind, 70. *Honors & Awards:* Lowell Lectr, Mass Gen Hosp, 65; Stockton Kimball Lectr, 72; Abraham Flexner Award, Assoc Am Med Col, 84; Hubert Humphrey Cancer Res Ctr Award, 85; Merrimon Lectr, 85. *Mem:* Inst Med, Nat Acad Sci; Am Fedn Clin Res (secy, 52-53, chmn, 54-55); Am Med Cols (asst secy, 56-60, vpres, 63-64, pres-elect, 67-68, chmn, 68-69); Am Acad Arts & Sci (vpres, 72-76); Am Clin & Climat Asn (vpres, 74-75, pres, 82-83). *Res:* Experimental streptococcal infections; rheumatic fever. *Mailing Add:* 525 Middlefield Rd Suite 130 Menlo Park CA 94025

GLASER, ROGER MICHAEL, b Brooklyn, NY, Feb 6, 44; m 74. EXERCISE PHYSIOLOGY. *Educ:* Queens Col, NY, BA, 68, MS, 69; Ohio State Univ, PhD(exercise physiol), 71. *Prof Exp:* Asst res & develop, United Sci Labs, 61-63; electronic engr, GEDCO, 64; asst prof biol sci, 72-74, from asst prof to assoc prof physiol, 74-81, PROF PHYSIOL, SCH MED, WRIGHT STATE UNIV, 81- *Concurrent Pos:* Fel physiol, Col Med, Ohio State Univ, 71-72; consult pulmonary dis, Wright Patterson AFB Hosp & cardiol, Dayton Vet Admin Hosp, 75- *Mem:* Aerospace Med Asn; Am Physiol Soc; fel Am Col Sports Med; Biomed Eng Soc; Inst Elec & Electronics Engrs; Sigma Xi. *Res:* Examining energy cost, power efficiency, cardiovascular and pulmonary stresses involved with wheelchair locomotion to reduce physiological risks and increase physical work capacity of wheelchair patients for their improved rehabilitation. *Mailing Add:* Dept Physiol Wright State Univ Sch Med 3171 Research Blvd Dayton OH 45420

GLASER, RONALD, b New York, NY, Feb 27, 39; m; c 2. VIROLOGY, CYTOLOGY. *Educ:* Univ Bridgeport, BA, 62; Univ RI, MS, 64; Univ Conn, PhD(virol), 68. *Prof Exp:* Asst prof virol, Life Sci Dept, Ind State Univ, 69-70; asst prof microbiol, Hershey Med Ctr, Pa State Univ, 70-73, from assoc prof to prof, 73-78; PROF MED & MICROBIOL & CHMN DEPT MED MICROBIOL & IMMUNOL OHIO STATE UNIV MED CTR, 78- *Concurrent Pos:* NIH postdoctoral fel, Baylor Col Med, 68-69; United Cancer Coun grant, Hershey Med Ctr, Pa State Univ, 71-73, USPHS grant, 74-77 & 77-81; sr mem grad sch, Pa State Univ, University Park, 74-78; fels, Franco-Am Exchange Prog-Fogarty Int Nat Ctr, 75 & 77; contract NCI, 78-81. *Honors & Awards:* Leukemia Soc Scholar Award, Leukemia Soc Am, 74. *Mem:* AAAS; Am Soc Microbiol. *Res:* Association of oncogenic herpes viruses and cancer; regulation of the Epstein-Barr virus in human cells and its association with Nasopharyngeal Carcinoma; neuroimmunology, stress and health. *Mailing Add:* Dept Med Microbiol & Immunol Ohio State Univ Med Ctr 333 W Tenth Ave Columbus OH 43210

GLASER, WARREN, b Brooklyn, NY, Apr 8, 28; m 61; c 2. MEDICINE, PHYSIOLOGY. *Educ:* Columbia Univ, AB, 46, MD, 50. *Prof Exp:* Resident exp med, Oak Ridge Inst Nuclear Studies, 55-56; from instr to assoc prof med, State Univ NY Downstate Med Ctr, 63-72; CLIN PROF MED, SCH MED & DENT, UNIV ROCHESTER, 72- *Concurrent Pos:* Consult, Med Div, Oak Ridge Inst Nuclear Studies; attend physician, Strong Mem Hosp. *Mem:* Fel Am Col Physicians; Am Fedn Clin Res; fel NY Acad sci; fel NY Acad Med; Sigma Xi. *Mailing Add:* 601 Elmwood Ave Univ Rochester Rochester NY 14642

GLASER DE LUGO, FRANK, US citizen. APPLIED MATHEMATICS. *Educ:* Loyola Univ, Los Angeles, BA, 63; Calif State Col, Los Angeles, MA, 66; Univ Calif, Riverside, PhD(math), 70. *Prof Exp:* Teaching asst math, Calif State Col, Los Angeles, 65-66 & Univ Calif, Riverside, 66-70; from lectr to assoc prof, 70-79, PROF MATH, CALIF STATE POLYTECH UNIV, POMONA, 79- *Concurrent Pos:* Lectr, Univ Madrid, 80, Univ Fes, Morocco, 82 & Univ Aalborg, Denmark, 87. *Mem:* Asn Turkish Am Scientists. *Res:* Developing a theory of means of function relative to a weight function in N-dimensional space, consisting of a generalization of the work of E D Cashwell and C J Everett; translation of Whitney's Singularity Theory and Catastrope Theory by Zeeman into Spanish; non-euclidean geometry; developing a form of non-euclidean geometry called "Hyper- polar-geometry". *Mailing Add:* Calif State Polytech Univ 3801 W Temple Ave Pomona CA 91768-4033

GLASFORD, GLENN M(ILTON), b Arcola, Tex, Nov 4, 18; m 48; c 2. ELECTRICAL ENGINEERING. *Educ:* Univ Tex, BS, 40; Iowa State Col, MS, 42. *Prof Exp:* Mem staff, Radiation Lab, Mass Inst Technol, 42-45; head, Adv Develop Group, TV Transmitter Dept, Allen B DuMont Labs, 45-47; from instr to prof, 47-86, EMER PROF ELEC ENG, SYRACUSE UNIV, 86- *Concurrent Pos:* Consult, Gen Elec Co, 48-53 & Int Bus Mach Corp, 57-59. *Mem:* Fel AAAS; fel Inst Elec & Electronics Engrs; Soc Motion Picture & TV Engrs; Am Soc Eng Educ. *Res:* Electronics; physics and application of semiconductor devices; television; radar. *Mailing Add:* Dept Elec & Comput Eng Syracuse Univ Syracuse NY 13244-1240

GLASGOW, DALE WILLIAM, b Sandyville, Iowa, Jan 29, 25; m 53; c 2. NUCLEAR PHYSICS. *Educ:* Simpson Col, BA, 49; Ore State Univ, MS, 56, PhD(nuclear physics), 61. *Prof Exp:* Physicist, Gen Elec Co, 49-53; sr physicist, Hanford Labs, Gen Elec Co, 61-65; sr physicist, Pac Northwest Lab, Battelle Mem Inst, 65, sr res scientist, 65-69; Ohio State Univ vis res assoc, Aerospace Res Lab, Wright Patterson AFB, 69-71, sr nuclear physicist, Adena Corp, 71-73; sr res scientist, Dept of Physics, Duke Univ, 73-75; STAFF MEM, LOS ALAMOS NAT LAB, UNIV CALIF, 75- *Mem:* Am Phys Soc; Am Asn Physics Teachers. *Res:* Beta and gamma ray spectroscopy; fast neutron physics; fast pulse techniques and time of flight techniques; particle accelerators; environmental science; nuclear reactors; controlled thermonuclear reaction. *Mailing Add:* 85 San Juan Los Alamos NM 87544

GLASGOW, DAVID GERALD, b Apple Creek, Ohio, Aug 25, 36; m 58; c 3. POLYMER CHEMISTRY. *Educ:* Col Wooster, AB, 58; Univ Cincinnati, PhD(org chem), 63. *Prof Exp:* Sr res chemist, Org Sect, 62-65, Polymer Synthesis Sect, 65-68, res group leader, 68-75, sr res group leader, 75-78, SR RES SPECIALIST, MONSANTO RES CORP, 78- *Mem:* Am Chem Soc. *Res:* Polymer synthesis; polyurethanes; polyesters; polyamides; adhesives; elastomers; thermoplastics; organic synthesis; chlorinated and brominated aromatics. *Mailing Add:* 361 S Village Dr Dayton OH 45459

GLASGOW, JOHN CHARLES, b Nashville, Tenn, Dec 14, 32; m 58; c 3. APPLIED MECHANICS, STRUCTURAL DYNAMICS. *Educ:* Univ Notre Dame, BS, 54; St Louis Univ, MS, 60. *Prof Exp:* Struct engr, McDonnell Aircraft Corp, Mo, 54-56, aero loads engr, 56-57, dynamics engr, 57-61; sr engr, Dynamics, NAm Aviation Inc, Ohio, 61-63; head, Dynamics Sect, 63-68, res engr, Short Takeoffs & Landing Propulsion Br, 68-72, sr proj engr, 72-75, SR RES ENGR, WIND ENERGY SYSTS, LEWIS RES CTR, NASA, 63- *Concurrent Pos:* Instr, Wash Univ, 59-61. *Res:* Large horizontal axis wind turbines associated primarily with structural dynamics and controls aspects of the machines. *Mailing Add:* 5862 Gareau Dr North Olmsted OH 44070

GLASGOW, LOUIS CHARLES, b Chicago, Ill, Sept 28, 43; m 67; c 2. PHYSICAL CHEMISTRY, CHEMICAL PROCESS TECHNOLOGY. *Educ:* Univ Ill, BS, 65; Univ Wis, MS, 70, PhD(chem), 71. *Prof Exp:* Fel photochem, Radiation Chem Div, Max Planck Inst, Mulheim, Ger, 70-72; asst prof phys chem, Northeastern Univ, 72-73; res chemist textile fibers, Lycra Res & Develop, E I du Pont de Nemours, 73-75, res chemist atmospheric sci, Freon Prod Div, 75-77, sr res chemist, 77-78, sr suprvr, 78-79, supt, Kel-Chlor Plant, 79-80, dir, Sabine River Lab, 80-82, prin consult corp res & develop planning, 82-84, planning mgr, Atomic Energy Div, 84-86, prog mgr nylon, 87-89, LAB DIR, DU PONT CHEM, E I DU PONT DE NEMOURS, 89- *Mem:* Sigma Xi; Am Chem Soc; AAAS. *Res:* Photochemistry; gas phase reactions; resource management; research management. *Mailing Add:* Du Pont Chem E I Du Pont de Nemours & Co Inc Exp Sta262/225 Wilimington DE 19898

GLASGOW, LOWELL ALAN, microbiology, pediatrics; deceased, see previous edition for last biography

GLASHAUSSER, CHARLES MICHAEL, b Newark, NJ, Dec 7, 39; m 65; c 2. NUCLEAR PHYSICS. *Educ:* Boston Col, BS, 61; Princeton Univ, PhD(physics), 66. *Prof Exp:* Physicist, Saclay Nuclear Res Ctr, France, 65-67 & Lawrence Radiation Lab, Univ Calif, Berkeley, 67-69; from asst prof to assoc prof, 69-79, PROF PHYSICS, RUTGERS UNIV, NEW BRUNSWICK, 79- *Concurrent Pos:* Guest prof, Univ Munich, 75-76; physicist, Saclay Nuclear Res Ctr, France, 81 & 85; chmn, Los Alamos Meson Physics Facil Users Group, Inc, 84; prog adv comt, Ind Univ Cyclotron, 85-87, nuclear sci adv comt, 87-; mem, prog adv comt, Tri-Univ Meson Facil, Univ BC, 89-; vis prof, Univ Paris, Orsay, 91. *Mem:* Fel Am Phys Soc; AAAS. *Res:* Spin dependence in inelastic scattering at intermediate energy; polarization phenomena in nuclear reactions; the delta resonance in nuclei. *Mailing Add:* Dept Physics Rutgers Univ New Brunswick NJ 08903

GLASHOW, SHELDON LEE, b New York, NY, Dec 5, 32; m 71; c 4. ELEMENTARY PARTICLE PHYSICS. *Educ:* Cornell Univ, BA, 54; Harvard Univ, AM, 55, PhD(physics), 58. *Hon Degrees:* DSc, Yeshiva Univ, 78, Univ Aix-Marseille, 82, Adelphi Univ, Bar Ilan Univ & Gustavus Adolphus, 89. *Prof Exp:* NSF fel, Inst Theoret Phys, Copenhagen Univ, 58-60; res fel physics, Calif Inst Technol, 60-61; asst prof, Stanford Univ, 61-62; assoc prof, Univ Calif, Berkeley, 62-66; prof, 66-78, HIGGINS PROF PHYSICS, HARVARD UNIV, 79-, MELLON PROF SCI, 87- *Concurrent Pos:* Sloan fel, 62-66; vis scientist, Niels Bohr Inst, 64 & Europ Orgn Nuclear Res, 68; consult, Brookhaven Nat Lab, 68-; vis prof, Univ Aix-Marseille, 71 & Mass Inst Technol, 74-75 & 79-80; affil sr scientist, Houston, 83-; distinguished univ prof & vis prof physics, Boston Univ, 83-84, distinguished vis scientist, 84; pres, Int Sakharov Comt, 85-88; coun mem, Am Acad Achievement. *Honors & Awards:* Nobel Prize physics, 79; Oppenheimer Mem Prize, 77; George Ledlie Prize, 78; Castiglione de Sicilia Prize, 83. *Mem:* Nat Acad Sci; fel AAAS; Am Acad Arts & Sci; Sigma Xi; fel Am Phys Soc. *Res:* Theory of elementary particles and their interactions; unified picture of strong, weak and electrodynamic interactions; identification of the fundamental constituents of matter. *Mailing Add:* Lyman Lab Physics Harvard Univ Cambridge MA 02138

GLASKY, ALVIN JERALD, b Chicago, Ill, June 16, 33; m 57; c 4. BIOCHEMISTRY. *Educ:* Univ Ill, BS, 54, PhD(biochem), 58. *Prof Exp:* Asst biochem, Univ Ill, 54-56; NSF fel Biochem Inst, Univ Lund, 58-59 & Wenner-Grens Inst, Univ Stockholm, 59; dir biochem res labs, Michael Reese Hosp, Chicago, Ill, 59-62; group leader drug enzym, Dept Pharmacol, Abbott Labs, 62-66; dir res, Int Chem Nuclear Corp, 66-68; pres, chief exec officer & chmn, Newport Pharmaceut Int, Inc, 68-84; exec dir, Am Soc Health Asn, 84-85; PRES, GLASKY ASSOC, 80-; PRES, CHIEF EXEC OFFICER & CHMN, ADVAN IMMUNOTHERAPEUT, INC, 87- *Concurrent Pos:* Asst prof, Col Med, Univ Ill, 59-68; assoc prof, Chicago Med Sch, 68- *Mem:* AAAS; Am Chem Soc; Am Soc Microbiol; Am Pharmaceut Asn; NY Acad Sci; Sigma Xi. *Res:* Mechanisms of drug action; biochemical pharmacology; viral chemotherapy; immunopharmacology; biochemistry of learning and memory. *Mailing Add:* Advan Immunotherapeut 1582 Deere Ave Irvine CA 92714

GLASNER, MOSES, b Cluj, Rumania, June 29, 42; US citizen. MATHEMATICS. *Educ:* Univ Calif, Los Angeles, BA, 63, MA, 65, PhD(math), 66. *Prof Exp:* Actg asst prof math, Univ Calif, Los Angeles, 66-67; asst prof, Calif Inst Technol, 67-75; ASSOC PROF MATH, PA STATE UNIV, UNIVERSITY PARK, 75- *Mem:* Am Math Soc; Optical Soc Am. *Res:* Global aspects of potential theory; beam propagation methods. *Mailing Add:* Dept Math Pa State Univ University Park PA 16802

GLASOE, PAUL KIRKWOLD, b Northfield, Minn, Nov 22, 13; m 35, 85; c 3. CHEMISTRY. *Educ:* St Olaf Col, AB, 34; Univ Wis, PhD(inorg chem), 38. *Prof Exp:* Asst chem, Univ Wis, 34-38, instr gen chem, 38; instr, Univ Ill, 38-40; res chemist, Eastman Kodak Co, 40-47; from assoc prof to prof chem, Wittenberg Col, 47-51; prof, Carthage Col, 51-52; prof chem, Wittenberg Univ, 52-79, chmn dept, 60-68; RETIRED. *Concurrent Pos:* NSF fac fel, Cornell Univ, 58-59 & King's Col, Univ London, 68-69. *Mem:* Am Chem Soc; Sigma Xi. *Res:* Colloids; chemical analysis of photographic developers; non-aqueous solvents; acidity in heavy water; exchange of deuterium for hydrogen in hydrogen-bonded systems. *Mailing Add:* Rte 1 Box 140 Sarona WI 54870

GLASPEY, JOHN WARREN, b Glens Falls, NY, July 18, 44; m 69; c 2. ASTRONOMY. *Educ:* Case Inst Technol, BSc, 66; Univ Ariz, PhD(astron), 71. *Prof Exp:* Res assoc astron, Univ BC, 71-73, lectr & instr, 73-75, res assoc, 75-76; astronomer & engr astron, Univ Montreal, 76-88; CONSULT, 88- *Mem:* Int Astron Union; Am Astron Soc; Can Astron Soc; Astron Soc Pac. *Mailing Add:* Can-France Hawaii Telescope Corp PO Box 1597 Kamuela HI 96743

GLASPIE, DONALD LEE, b Oxford, Ind, Feb 16, 22; m 45; c 1. THERMODYNAMICS, HEAT TRANSFER. *Educ:* Purdue Univ, BSME, 47; Wichita State Univ, MS, 61. *Prof Exp:* Engr air conditioning, Shelby-Skipwith, Inc, Tenn, 47-52; co-owner, Ozark-York, Inc, Mo, 52-55; engr, Boeing Airplane Co, Kans, 55-63; design engr, New Orleans, 63-66; specialist, Vought Aeronaut Div, LTV, Inc, Dallas, 66-71; owner mech, Glaspie Consult Engr, 71-73; mgr res & develop mech, Glitsch, Inc, Subsid Foster Wheeler, Dallas, 71-83; mgr, eng serv dept, CACI, Inc-Fed, Dallas, 83-84; OWNER, GLASPIE CONSULT ENGR, 84- *Mem:* Am Soc Mech Engrs; Am Soc Heating Refrig & Air Conditioning Engrs. *Res:* Air conditioning systems design for aircraft and buildings; petrochemical tower packing and other internal equipment design; fabrication systems and mechanical equipment design; engineering computer software development. *Mailing Add:* 3768 Northaven Rd Dallas TX 75229-2752

GLASPIE, PEYTON SCOTT, b Commerce, Tex, Dec 16, 46; m 72. PHYSICAL ORGANIC CHEMISTRY, SURFACE CHEMISTRY. *Educ:* ETex State Univ, Commerce, BS, 69; Rice Univ, PhD(chem), 74. *Prof Exp:* Asst chem, Rice Univ, 69-74; sr chemist process develop-surface chem, Monsanto Polymers & Petrochem Co, 74-78, catalysis group supvr, Monsanto Chem Intermed, Co, 78-81, process technol mgr anal chem, 81-82, mgr personnel planning, Monsanto Corp Res Co, 83-85, mgr, 86-89, MGR EXTERNAL RES & DEVELOP FUNDING, MONSANTO CORP RES CO, 90-, MGR INFO SYSTS, 91- *Mem:* Am Chem Soc; Catalysis Soc; Sigma Xi. *Res:* Surface chemistry explorations of currently used industrial processes by novel instrument applications; applying the results to better understand the chemical kinetics of the reactions and thereby improve the system; laboratory automation and computer applications. *Mailing Add:* Monsanto Co 800 N Lindbergh Blvd St Louis MO 63167

GLASS, ALASTAIR MALCOLM, c Harrogate, Yorkshire, Eng, Aug 2, 40; m 66; c 4. ELECTRONIC & OPTOELECTRONIC APPLICATIONS. *Educ:* Univ Col, London, BSc; Univ Brit Columbia, PhD(elec eng), 64. *Prof Exp:* Res fel, Univ London, Kings Col, 65-67; mem tech staff, AT&T Bell Labs, 67-78, supvr, 78-83, head device mat & res, 83-87, Mat Res Lab, 87-89, ASST DIR, PASSIVE COMPONENTS RES LAB, AT&T BELL LABS, 89- *Concurrent Pos:* adv comt, Piezoelec, Off Naval Res, 79-85, mat res, 85, 87; adv comt, Basic Scientific Res, 80-83; mat res adv comt, Pa State Univ, 83-; electronic mat working group, NASA, 84- *Mem:* Nat Acad Eng; Mat Res Soc; Inst Elec & Electronics Engrs. *Res:* Development and understanding of new materials for electronic and optoelectronic applications, materials include ferroelectrics, compound semiconductors and organics. *Mailing Add:* AT&T Bell Labs 600 Mountain Ave Murray Hill NJ 07974

GLASS, ALEXANDER JACOB, b Pittsfield Twp, NY, Jan 4, 33; m 59; c 3. PHYSICS, OPTICS. *Educ:* Rensselaer Polytech Inst, BS, 54; Yale Univ, MS, 55, PhD(physics), 63. *Prof Exp:* Res staff mem, Res Inst Advan Study, Md, 57-58; res assoc, Yale Univ, 63-64; staff scientist Inst Defense Analysis, 64-66; chief laser physics br, Naval Res Lab, 66-68; prof elec eng, Wayne State Univ, 68-70, chmn dept, 70-73; head theory & design analysis, Lawrence Livermore Lab, 73-78, asst assoc dir lasers, 78-81; PRES & COORDR, KMS FUSION, INC, 81- *Concurrent Pos:* Fel Yale Univ, 60; lectr, space sci, Cath Univ Am, 65-68 & appl sci, Univ Calif, Davis, 73-; mem, Inst Res Eng Sci, 68-70. *Honors & Awards:* Oustanding teacher, Col Eng, Wayne State Univ, 69; Res Publ Award, US Naval Res Lab, 69. *Mem:* Fel Optical Soc Am; Inst Elec & Electronics Engrs; Am Phys Soc. *Res:* Non-linear optics; high power laser design; laser damage; quantum electronics; numerical methods. *Mailing Add:* 206 Hillcrest Rd Berkeley CA 94705

GLASS, ARTHUR WARREN, b Flint, Mich, Mar 14, 21; m 50; c 4. GENETICS. *Educ:* Gustavus Adolphus Col, AB, 43; Univ Minn, MA, 48, PhD, 54. *Prof Exp:* Assoc prof, 50-59, head dept, 50-72, PROF BIOL, GUSTAVUS ADOLPHUS COL, 59- *Mem:* Sigma Xi; Soc Study Evolution. *Res:* Population genetics. *Mailing Add:* 510 Capitol Dr St Peter MN 56082

GLASS, BILLY PRICE, b Memphis, Tenn, Sept 9, 40; m 66; c 2. MARINE GEOLOGY, ASTROGEOLOGY. *Educ:* Univ Tenn, BS, 63; Columbia Univ, PhD(marine geol), 68. *Prof Exp:* Res asst marine geol, Lamont Geol Observ, 65-67; res scientist, Goddard Space Flight Ctr, US Army Corps Engrs, 68-70, Nat Res Coun resident res assoc, 70; from asst prof to assoc prof, 70-83, PROF GEOL, UNIV DEL, 83-, DEPT CHMN, 86- *Concurrent Pos:* Prin investr, Apollo 12-17 Lunar Samples; mem, US-Japan Antarctic Search for Meteorites, 77-78; prin investr, NSF, 72-88; assoc ed, meteoritics. *Honors & Awards:* Nininger Meteorite Award, 66-67. *Mem:* Fel Geol Soc Am; fel Meteoritical Soc; Sigma Xi; Am Geophys Union; Int Astron Union; Nat Asn Geol Teachers. *Res:* Correlation and dating of marine sediments; tektites and microtektites; effects of large impacts. *Mailing Add:* 387 Hobart Dr Greenbridge Newark DE 19713

GLASS, BRYAN PETTIGREW, b Mandeville, La, Aug 21, 19; m 46; c 2. MAMMALOGY. *Educ:* Baylor Univ, AB, 40; Agr & Mech Col Tex, MS, 46; Okla State Univ, PhD, 52. *Prof Exp:* Asst zool, Agr & Mech Col Tex, 40-42 & 45-46; from asst prof to prof, 46-85, EMER PROF ZOOL, OKLA STATE UNIV, 85- *Concurrent Pos:* Dir mus, Okla State Univ, 66-85. *Mem:* Am Soc Ichthyologists & Herpetologists; Am Soc Mammal. *Res:* Systematics of Oklahoma rodents; ecology of muskrats in Oklahoma; biology; taxonomy; migration of bats; Ethiopian mammals; Brazilian mammals; author and co-author of several publications. *Mailing Add:* 517 S Willis Stillwater OK 74074

GLASS, DAVID BANKES, b Indianapolis, Ind, Apr 30, 47; m 73; c 1. PHARMACOLOGY, BIOCHEMISTRY. *Educ:* Purdue Univ, BS, 70; Univ Minn, PhD(pharmacol), 76. *Prof Exp:* Fel biochem, Univ Calif, Davis, 76-77; fel pharmacol, Univ Wash, 77-78; asst prof, 78-84, ASSOC PROF PHARMACOL, EMORY UNIV, 84- *Concurrent Pos:* Med res fel, Bank of Am-Giannini Found, 76-77; res investr, Am Heart Asn, Ga affil, 80-83. *Mem:* Protein Soc; AAAS; NY Acad Sci; Am Chem Soc; Am Soc Pharmacol & Exp Therapeut; Am Soc Biol Chemists. *Res:* Biochemical pharmacology; biochemistry of protein kinases; peptide synthesis; peptide structure-function studies. *Mailing Add:* Dept Pharmacol Emory Univ Sch Med Atlanta GA 30322

GLASS, DAVID CARTER, b New York, NY, Sept 17, 30; m 82. PSYCHOLOGICAL STRESS, BEHAVIORAL MEDICINE. *Educ:* NY Univ, AB, 52, MA, 54, PhD(social psychol), 59. *Prof Exp:* Postdoctoral fel psychophysiol, NY Univ, 59-62; asst prof social psychol, Ohio State Univ, 62-63; staff social psychologist, Russell Sage Found, 63-66; assoc prof psychol, Rockefeller Univ, 66-68; prof psychol, NY Univ, 68-72; prof & chair psychol, Univ Tex, Austin, 72-76; prof psychol, Grad Ctr, City Univ New York, 76-82; vprovost, 82-86, spec adv to the provost, 86-89, PROF, STATE UNIV NY, STONY BROOK, 82-, VPROVOST, 89- *Concurrent Pos:* Adj assoc prof psychol, Columbia Univ, 65-66, 67-68; staff social psychol, Russell Sage Found, 66-71, vis scholar, 75-76; dir, Lab Biobehav, Grad Ctr, City Univ New York, 77-82. *Honors & Awards:* Socio Psychol Prize, AAAS, 71; Am Psychol Asn Award, 88. *Mem:* Sigma Xi; fel Am Psychol Asn; fel AAAS; Acad Behav Med Res, (pres, 80-81); Soc Psychophysiol Res; Am Psychosomatic Soc. *Res:* Effects of psychosocial factors on physical illness. *Mailing Add:* Provost Off State Univ NY Admin Bldg Stony Brook NY 11794

GLASS, EDWARD HADLEY, b Waltham, Mass, Feb 19, 17; m 42; c 2. ENTOMOLOGY. *Educ:* Mass State Col, BS, 38; Va Polytech Inst, MS, 40; Ohio State Univ, PhD(entom), 43. *Prof Exp:* Asst entomologist, Exp Sta, Va Polytech Inst, 40-41; entomologist, Am Cyanamid Co, NY, 43-48; assoc prof, 48-55, head dept, 69-82, PROF ENTOM, NY STATE AGR EXP STA, 55- *Concurrent Pos:* Vis prof, Univ Philippines, 66-67. *Mem:* Fel AAAS; Entom Soc Am; Sigma Xi. *Res:* Biology and control of apple orchard insect pests; insect photoperiodism; tropical rice insects and their control; use of sex pheromones for insect detection and control. *Mailing Add:* 377 White Springs Rd Geneva NY 14456

GLASS, EDWARD NATHAN, b New York, NY. THEORETICAL PHYSICS. *Educ:* Carnegie-Mellon Univ, BS, 59; Syracuse Univ, MS & PhD(physics), 69. *Prof Exp:* Math analyst, Guid Systs, Norden Div United Aircraft Corp, 59-62; vis asst prof physics, Univ Cincinnati & Trenton State Col, 71-73; vis res fel, Princeton Univ, 73-74; assoc prof, 74-81, PROF PHYSICS, UNIV WINDSOR, 81- *Concurrent Pos:* Fel, Aeronaut Res Lab, Nat Acad Sci, 69-71; vis scholar, Stanford Univ, 81-82. *Mem:* Am Phys Soc; Am Asn Physics Teachers. *Res:* General relativity; relativistic astrophysics. *Mailing Add:* Dept Physics Univ Windsor Windsor ON N9B 3P4 Can

GLASS, GEORGE, b Vienna, Austria, June 15, 36; US citizen; m 64; c 4. HIGH ENERGY PHYSICS. *Educ:* Mass Inst Technol, SB, 59, PhD(physics), 64. *Prof Exp:* Mem res staff physics, Mass Inst Technol, 64-65; res assoc, Northeastern Univ, 65-66; asst prof, Univ Wash, 66-74; MEM STAFF, TEX A&M UNIV, 74- *Mem:* Am Phys Soc; Am Asn Physics Teachers. *Res:* High energy nuclear physics; particles; electronics; light. *Mailing Add:* Los Alamos Nat Lab LAMPF MS H831 Los Alamos NM 87545

GLASS, GEORGE B JERZY, b Warsaw, Poland, Jan 9, 03; nat US; m 33; c 2. GASTROENTEROLOGY. *Educ:* Univ Warsaw, MD, 27; Am Bd Internal Med, dipl. *Prof Exp:* Intern, Hosp Wolski, Warsaw, 27-28, resident, 28-29; resident Med Div, Allgemeines Hosp, Univ Vienna, 31-32; resident, Hosp Hotel Dieu, Paris, France, 32-33; asst physician, Hosp Child Jesus, 33-38; from asst clin prof to assoc clin prof med, 48-54, assoc prof, 54-62, chief sect gastroenterol, 61-69, PROF MED & DIR GASTROENTEROL LAB, METROP MED CTR, NEW YORK MED COL, 62- *Concurrent Pos:* Fel, Krankenaus am Urban, Berlin, Ger, 30-31; assoc physician, Metrop Hosp, New York, 50-61, vis physician, 62-; Bird S Coler Mem Hosp, 52-; assoc attend physician, Flower & Fifth Ave Hosps, 54-61, attend physician, 62-; assoc clinician, Sloan-Kettering Inst, 60-66; assoc vis physician, James Ewing Hosp, 61-66; consult gastroenterol, New York Infirmary, 67-; assoc consult, Mem Hosp. *Mem:* Asn Am Physicians; Am Fedn Clin Res; Am Physiol Soc; Am Gastroenterol Asn; Am Soc Hemat. *Res:* Physiology and biochemistry of the stomach; intrinsic factor and vitamin B-12 metabolism; chemistry and biological significance of large molecular substances in gastric juices; immunology of atrophic gastritis and pernicious anemia. *Mailing Add:* 120 E 79th St New York NY 10022

GLASS, GRAHAM PERCY, b Birmingham, Eng, July 21, 38; m 65. PHYSICAL CHEMISTRY. *Educ:* Univ Birmingham, BSc, 59; Cambridge Univ, PhD(chem), 63. *Prof Exp:* Fel chem, Harvard Univ, 63-65; lectr, Univ Essex, 65-67; asst prof, 67-73, ASSOC PROF CHEM, RICE UNIV, 73- *Mem:* Am Phys Soc; Am Chem Soc. *Res:* Reaction kinetics; molecular energy transfer; shock tube chemistry. *Mailing Add:* 2340 Robin Hood Houston TX 77005-2606

GLASS, HERBERT DAVID, b New York, NY, Oct 14, 15; m 44; c 2. MINERALOGY. *Educ:* NY Univ, BA, 37; Columbia Univ, MA, 47, PhD(geol), 51. *Prof Exp:* GEOLOGIST, STATE GEOL SURV, ILL, 48- *Mem:* Mineral Soc Am; Geol Soc Am; Mineral Soc Gt Brit & Ireland. *Res:* Clay mineralogy and its application to stratigraphic correlation, interpretation and classification. *Mailing Add:* State Geol Surv Univ Ill 131 Nat Res Bldg Champaign IL 61820

GLASS, HIRAM BENTLEY, b Laichowfu, Shantung, China, Jan 17, 06; m 34; c 2. GENETICS. *Educ:* Baylor Univ, AB, 26, MA, 29, LLD, 58; Univ Tex, PhD(genetics), 32. *Hon Degrees:* Nine from US Univs, 57-77. *Prof Exp:* Teacher pub sch, Tex, 26-28; fel genetics, Nat Res Coun, Univ Oslo, Kaiser-Wilhelm Inst & Univ Mo, 32-34; instr zool, Stephens Col, 34-38; from asst prof to prof biol, Goucher Col, 38-48; from assoc prof to prof, Johns Hopkins Univ, 48-65; acad vpres, 65-71, distinguished prof, 65-76, EMER DISTINGUISHED PROF BIOL, STATE UNIV NY, STONY BROOK, 76- *Concurrent Pos:* Res assoc, Teachers Col, Columbia, 36-37; asst ed, Quart Rev Biol, 44-48, assoc ed, 49-57, ed, 58-86; res assoc, Baltimore RH Blood Typing Lab, 47-52; consult, US Dept State, Ger, 50-51; dir surv biol abstracting, Biol Abstracts, 52-54, mem bd trustees, 54-60, pres, 58-60; del, Int Union Biol Scis, 53, 55; ed, Surv Biol Progress, 54-62; mem bd sch comnrs, Baltimore, Md, 54-58; adv comt biol & med, AEC, 55-63, chmn, 62-63; mem comt genetic effects of atomic radiation, genetics panel, Nat Acad Sci, 55-64; lectr, Davis Washington Mitchell, Tulane Univ, 58; adv, Continuing Comt Conf Sci & World Affairs, 58-64; democratic adv, Coun Comt Sci & Tech, 59-60; adv, Governor's Adv Comt Nuclear Energy, Md, 59-65; chmn, Biol Sci Curriculum Study, 59-65; nat lectr, Sigma Xi, 58-59 & bicentennial lect, 75-76; lectr, Stoneburner, Va Med Col & John Calvin McNair, Univ NC, 63; chmn, bd trustees, Cold Spring Harbor Lab, 67-73; chmn, Life Sci Prog, Space Sci Bd Rev, NASA, 70; vis prof, Univ Calif, Santa Cruz, 71-72; dir, Hist & Genetics Proj, Am Philos Soc, 78-86. *Mem:* Nat Acad Sci; AAAS (pres, 69); Genetics Soc Am (vpres, 60); Am Soc Naturalists (secy, 50-53, pres, 65); Nat Asn Biol Teachers (pres, 71); Am Soc Human Genetics (pres, 67); Czechoslovak Acad Sci. *Res:* Genetics of Drosophila; human genetics; history of genetics; suppressor genes; Rh blood types. *Mailing Add:* Box 65 East Setauket NY 11733

GLASS, HOWARD GEORGE, b Chicago, Ill, May 25, 09; m 42; c 2. PHARMACOLOGY. *Educ:* Univ Ill, BS, 32; Northwestern Univ, MS, 35; Univ Chicago, PhD(pharmacol), 42; Marquette Univ, MD, 49. *Prof Exp:* Chemist, Bauer & Balk, Ill, 35-38; res assoc toxicity lab, Off Sci Res & Develop, Chicago, 41-44; instr pharmacol, Sch Med, Univ Okla, 44-46; instr, Marquette Univ, 46-48; pharmacologist, Armour Labs, Ill, 49-53; pharmacologist, Miles-Ames Res Lab, Ind, 53-62; ASST DIR, DEPT DRUGS, AMA, 62- *Concurrent Pos:* Med consult, 77- *Mem:* Am Soc Pharmacol & Exp Therapeut; Soc Exp Biol & Med. *Res:* Anoxia; respiratory physiology; toxicology of drugs; sedatives; clinical pharmacology of enzymes; antihypertensives; diuretics; antineoplastic agents. *Mailing Add:* 1330 Meadowlake Way Monument CO 80132

GLASS, I(RVINE) I(SRAEL), b Poland, Feb 23, 18; nat Can; m 42, 83; c 3. AEROSPACE ENGINEERING, FLUIDS. *Educ:* Univ Toronto, BASc, 47, MASc, 48, PhD(aerophysics), 50. *Hon Degrees:* Hon Prof Nanjing Aeronaut Inst, China. *Prof Exp:* Aeronaut engr, Canadair Ltd, Que, 46, Can Car & Foundry, 47 & A V Roe, Ont, 48; instr aeronaut eng, 48-49, res assoc, Inst Aerophysics, 50-54, from asst prof to assoc prof, 54-60, asst dir, 68-74, chmn dept aerospace eng sci, Grad Studies, 61-66, PROF AEROSPACE SCI & ENG, INST AEROSPACE STUDIES, UNIV TORONTO, 60- *Concurrent Pos:* Aerodynamicist, Can Armament Res & Develop Estab, Que, 52; invited lectr, USSR & Siberian Acad Sci, 61 & 69; mem assoc comt space res, Nat Res Coun Can, 62-65, standing subcomt on high speed flow, 67- & assoc comt aerodynamics, 68-71; res adv comt on fluid mech, NASA, 65-70; consult, Am & Can indust; lectr, Royal Can Inst, 65 & 75; W Rupert Turnbull lectr, Can Aeronaut & Space Inst, 67; chmn, Int Shock Tube Symp, 69; assoc ed, Physics of Fluids, 67-70 & Progress in Aerospace Sci, 74-; Lady Davis fel, Dept Aerospace Eng, Israel Inst Technol, 74-75; vis prof, Japan Soc Promotion Sci, Kyoto Univ, 75. *Honors & Awards:* Distinguished Dryden lectr, Aerospace Industs Asn Am, 86. *Mem:* Fel AAAS; fel Am Phys Soc; fel Am Inst Aeronaut & Astronaut; fel Can Aeronaut & Space Inst; fel Royal Soc Can. *Res:* Hypersonic gas dynamics; aerophysics; shock-wave phenomena; sonic boom; dusty-gas dynamics; fusion from explosive-driven implosions. *Mailing Add:* Inst Aerospace Studies 4925 Dufferin St Downsview ON M3H 5T6 Can

GLASS, JAMES CLIFFORD, b Los Angeles, Calif, Sept 20, 37; m 59; c 4. MACROMOLECULAR STRUCTURE, HIGH T SUPERCONDUCTORS. *Educ:* Univ Calif, Berkeley, BA, 60; Calif State Univ, MS, 65; Univ Nev, PhD(physics). *Prof Exp:* Physicist, Rocketdyne, 60-64; sr physicist, Electro-Optical, 64-65; from asst prof to prof physics, NDak State Univ, 68-88, chair physics, 73-85, chair eng sci, 85-88; DEAN SCI, MATH & TECHNOL, EASTERN WASH UNIV, 88-, PROF PHYSICS, 88- *Concurrent Pos:* Vis scientist, Kernforschungsanlage Jülich, BRD, 77-78. *Mem:* Am Phys Soc; Am Asn Physics Teachers; Sigma Xi. *Res:* Hyperfine interactions: gamma-gamma correlations, Mössbauer in ferroelectrics, superconductors and macromolecules; molecular biophysics: positron annihilation applied to conformational free volume; active transport; enzyme dynamics. *Mailing Add:* S 4423 Hogan St Spokane WA 99203

GLASS, JOHN RICHARD, b Lancaster, Pa, Aug 31, 17; m 40; c 2. ANALYTICAL CHEMISTRY. *Educ:* Elizabethtown Col, BS, 38. *Prof Exp:* Res chemist water purification, Wallace & Tiernan Co, 39-45; sr res chemist anal res, Mobil Oil Corp, 45-61, res assoc, 61-82; RETIRED. *Concurrent Pos:* Consult, 82-83. *Honors & Awards:* John D Goodell Award, Am Water Works Asn, 43; IR 100 Award, Indust Res Mag, 76. *Mem:* Am Chem Soc. *Res:* Colorimetric and trace metal analysis; analytical instruments. *Mailing Add:* 189 Cedar Rd Mickleton NJ 08056

GLASS, LAUREL ELLEN, b Selma, Calif, Oct 1, 23. DEVELOPMENTAL BIOLOGY, GERIATRIC MEDICINE. *Educ:* Univ Calif, BA, 51; Duke Univ, PhD(exp embryol), 58; Univ Calif, San Francisco, MD, 74. *Prof Exp:* Asst zool, Duke Univ, 53-56; res assoc, Path Res Lab, Vet Admin Hosp, Durham, NC, 57-58; from instr to prof anat, Sch Med, Univ Calif, San Francisco, 59-89, exec dir, Ctr Deafness, 84-89, prof psychiat, 84-89, EMER PROF ANAT & PSYCHIAT, SCH MED, UNIV CALIF, SAN FRANSCISCO, 89- *Concurrent Pos:* Instr anat, Duke Univ, 58. *Mem:* Am Asn Anat; Soc Develop Biol; Am Deafness & Rehab Asn; Self-Help Hard Hearing People; Asn Learning Disabled Adults. *Res:* Cell interactions during gonadogenesis; macromolecular transfer between maternal body and oocyte or embryo; affective and behavioral effects of acquired hearing loss. *Mailing Add:* Langley Porter Psychiat Inst Univ Calif 401 Parnassus Box GMO San Francisco CA 94143-0984

GLASS, NATHANIEL E, b Philadelphia, Pa, Mar 2, 49. CONDENSED MATTER PHYSICS. *Educ:* Univ Pa, BA, 70; State Univ NY, Stony Brook, PhD(physics), 76. *Prof Exp:* Res asst, Swiss Fed Inst Technol, Lausanne, 77-80, fel physics dept, Univ Calif, Irvine, 80-83, adj prof, physics dept, Naval Postgrad Sch, 84-87; MEM TECH STAFF, ROCKWELL INT SCI CTR, 87- *Mem:* Am Phys Soc; Optical Soc Am. *Res:* Condensed matter theory; lattice dynamics theories of imperfect crystals and of dislocation dynamics; surface electromagnetic waves; acoustic waves; nonlinear optics and superconductivity. *Mailing Add:* Rockwell Sci Ctr 1049 Camino Dos Rios Thousand Oaks CA 91360

GLASS, RICHARD STEVEN, b New York, NY, Mar 5, 43; m 70; c 2. ORGANIC CHEMISTRY. *Educ:* NY Univ, BA, 63; Harvard Univ, PhD(chem) 67. *Prof Exp:* NIH fel, Stanford Univ, 66-67; sr res chemist org chem, Hoffmann La Roche, Inc, 67-70; from asst prof to assoc prof, 70-82, PROF CHEM, UNIV ARIZ, 82- *Mem:* Fel AAAS; Am Chem Soc; Royal Soc Chem; Sigma Xi. *Res:* Total synthesis of natural products; synthetic methods; organosulfur chemistry; biorganic mechanisms; organometallic chemistry. *Mailing Add:* Dept Chem Univ Ariz Tucson AZ 85721

GLASS, RICHARD THOMAS, b Jacksonville, Fla, March 9, 41; c 3. ORAL PATHOLOGY, FORENSIC ODONTOLOGY. *Educ:* Emory Univ, DDS, 65; Univ Chicago, PhD(path), 72; Am Bd Oral Path, dipl. *Prof Exp:* From asst prof to assoc prof path, 72-75, CHIEF SECT ORAL PATH, UNIV OKLA, 76-, PROF ORAL PATH & CHMN DEPT, COL DENT, 80- *Concurrent Pos:* Consult, Vet Admin Hosp, Muskogee, Oklahoma City, 72, Fed Aviation Admin & Okla State Med Examr, 73-; staff, Childrens Mem Hosp, Oklahoma City, 72-, Presby Hosp, 80-; oral pathologist, Health Sci Ctr Hosp, Univ Okla,

72- *Mem:* Fel Am Acad Oral Path; Am Dent Asn; Int Asn Dent Res; fel Sigma Xi. *Res:* Cancer of the head and neck; oral mucosal inflammatory diseases; forensic odontology; transmission of disease (infected toothbrush denture). *Mailing Add:* Dept Oral Path Col Dent Okla Univ PO Box 26901 Oklahoma City OK 73190

GLASS, ROBERT LOUIS, b Chicago, Ill, May 15, 23; m 46; c 4. BIOCHEMISTRY. *Educ:* Univ Ill, BS, 50; Univ Minn, MS, 54, PhD(agr biochem), 56. *Prof Exp:* From instr to assoc prof, 56-72, PROF BIOCHEM, UNIV MINN, ST PAUL, 72- *Mem:* Am Chem Soc; Am Oil Chem Soc. *Res:* Lipid biochemistry. *Mailing Add:* 28 Oakwood Dr New Brighton MN 55112-3327

GLASS, ROGER I M, b Somerville, NJ, Jan 10, 46; c 3. MEDICINE, EPIDEMIOLOGY. *Educ:* Harvard Col, AB, 67; Harvard Med Sch, MD, 72; Harvard Univ, MPH, 72; PhD, Goteboz, Swed, 84. *Prof Exp:* Mem staff environ epidemiol, Nat Inst Environ Health Sci, 73-74; lectr med, Radcliffe Infirmary, 74; instr med & fel clin epidemiol, Mt Sinai Sch Med, 76-77; EPIDEMIOLOGIST, CTR DIS CONTROL, 77- *Concurrent Pos:* Scientist, Winstand Ctr Diarrheal Dis Res, Bangladesh, 75-83; med officer, NIH. *Mem:* Soc Epidemiol Res; Am Fedn Clin Res; AAAS; Am Pub Health Asn; Am Epidemiol Soc. *Mailing Add:* Ctr Dis Control 1600 Clifton Rd NE Atlanta GA 30333

GLASS, WERNER, b Berlin, Ger, May 27, 27; US citizen; m 52; c 3. CHEMICAL ENGINEERING. *Educ:* Syracuse Univ, BChE, 50; Mass Inst Technol, SM, 51, ScD(chem eng), 56. *Prof Exp:* Res engr, Esso Res & Eng Co, Standard Oil Co (NJ), 56-58, group head, 58-59, sr engr, 59-63; asst dir res, Ionics, Inc, 63-64, dep dir res, 64-67; sr engr, Esso Res & Eng Co, 67-68, eng assoc, 68-71; sr staff adv, Exxon Res & Eng Co, 71-86; CONSULT, 86- *Honors & Awards:* Winquist Medal, Swed Acad Sci. *Mem:* Am Inst Chem Engrs. *Res:* Planning and evaluation; communication skills. *Mailing Add:* 913 Boulevard Westfield NJ 07090

GLASS, WILLIAM A, b Winfield, Kans, Dec 9, 31; m 56; c 3. RADIOLOGICAL PHYSICS. *Educ:* Southwestern Col, BA, 53; Univ Kans, MA & MS, 58. *Prof Exp:* Asst prof physics, Kans State Teachers Col, 58-62; sr res scientist radiol physics, Pac Northwest Labs, Battelle Mem Inst, 63-69; res collabr, Brookhaven Nat Lab, 69-70; mgr, radiol physics, 72-80, dept mgr, 80-85, ASST DIR RES, PAC NORTHWEST LABS, BATTELLE MEM INST, 85- *Concurrent Pos:* Clin lectr, Univ Wash, 64-69; adj prof radiol, Univ Wash, 74- *Mem:* AAAS; Am Phys Soc; Radiation Res Soc. *Res:* Interaction of charged particles in matter; radiation dosimetry; microdosimetry; ionization phenomena; biological effects of ionizing radiation. *Mailing Add:* 2113 Beech Ave Richland WA 99352

GLASSBRENNER, CHARLES J, b Albany, NY, Apr 9, 28; m 59; c 3. PHYSICS. *Educ:* St Bernardine of Siena Col, BS, 50, MS, 56; Univ Conn, PhD(physics), 63; Northeastern Univ, MS, 77. *Prof Exp:* Elec engr, NY State Architects Off, 51-55; physicist, Watervliet Arsenal, NY, 56-57; asst physics, Univ Conn, 57-60; physicist, Res Lab & Knolls Atomic Power Lab, Gen Elec Co, 60-64; sr physicist, Controls for Radiation, Inc, Mass, 64-66 & Ion Physics Corp, 66-67; PROF COMPUT SCI, WORCESTER STATE COL, 67- *Mem:* Asn Comput Mach; Am Phys Soc; Sigma Xi. *Res:* Radiation effects on solids; thermal conductivity; mechanical strength of solids; thermodynamics; computer science. *Mailing Add:* 380 Grove St Paxton MA 01612-1143

GLASSER, ALAN HERBERT, b Brooklyn, NY, Sept 8, 43; m 66; c 2. THEORETICAL PLASMA PHYSICS. *Educ:* Columbia Col, BA, 65; Univ Calif, San Diego, MS, 67, PhD(physics), 72. *Prof Exp:* Res assoc mem physics, Plasma Physics Lab, Princeton Univ, 72-75, res staff, 75-80; mem fac, Phys Dept, Auburn Univ, 80-84; MEM STAFF, LOS ALAMOS NAT LAB, 84- *Mem:* Sigma Xi; AAAS; Am Phys Soc. *Res:* Equilibrium, stability and transport of magnetically confined plasmas, with an emphasis on the influence of realistic geometry; numerical simulation of fluids. *Mailing Add:* T-DOT Group LANL MS-647 PO Box 1663 Los Alamos NM 87545

GLASSER, ARTHUR CHARLES, pharmaceutical chemistry; deceased, see previous edition for last biography

GLASSER, JAY HOWARD, b New Haven, Conn, May 6, 35. BIOSTATISTICS, COMMUNITY HEALTH. *Educ:* Univ Conn, BS, 57; Columbia Univ, MS, 60; NC State Univ, PhD(exp statist), 67. *Prof Exp:* From instr to asst prof biostatist, Univ NC, 64-69; ASSOC PROF BIOMET & COMPUT SCI, UNIV TEX SCH PUB HEALTH HOUSTON, 69- *Concurrent Pos:* Consult, HEW, 67-; sr lectr community med, St Thomas's Hosp Sch, Eng, 74-; HEW career develop award, 70. *Mem:* Am Statist Asn; Biomet Soc; fel Am Pub Health Asn; fel Royal Statist Soc; Int Union Sci Study Pop. *Res:* Health services. *Mailing Add:* Dept Biometry Univ Tex Health Sci Ctr PO Box 20036 Houston TX 77225

GLASSER, JOHN WEAKLEY, b Baltimore, Md, Oct 3, 44; m 81; c 1. MARINE COMMUNITY ECOLOGY. *Educ:* Princeton Univ, AB, 66; Duke Univ, PhD(zool), 76. *Prof Exp:* Teaching asst, Duke Univ, 71-76; asst prof zool, 76-83, ASSOC SCI ECOL, UNIV GA, 83- *Mem:* Ecol Soc Am; Am Soc Limnol & Oceanog; Sigma Xi; Am Soc Naturalists. *Res:* Individual and population phenomena such as strategies of resource aquisition and life histories, which evolve in a community context; community phenomena such as colonization and succession, which result from interactions among populations and between them and their abiotic environment; theoretical population and community ecology. *Mailing Add:* Div Immunization Ctrs Dis Control 1600 Clifton Rd Atlanta GA 30333

GLASSER, JULIAN, b Chicago, Ill, May 23, 12; m 42; c 2. CHEMICAL METALLURGY. *Educ:* Univ Ill, BS, 33, MS, 35; Pa State Univ, PhD(phys chem), 39. *Prof Exp:* Asst, Univ Ill, 34 & Pa State Univ, 35-38; res engr, Battelle Mem Inst, 38-42; chief chemist & chief metallurgist, Aluminum Div, Olin Industs, Inc, Wash, 42-45; dir res, Gen Abrasive Co, NY, 45-47; supvr & res engr, Armour Res Found, Ill, 47-53; tech dir, Cramet, Inc, Tenn, 53-58; consult, 58-60; PRES, CHEM & METALL RES, INC, 60- *Concurrent Pos:* Consult & staff metallurgist, Mat Adv Bd, Nat Res Coun, 51-53 & 58-61; consult to head metall br, Navy Dept, Off Naval Res, DC, 52-53; vis prof, Vanderbilt Univ, 58-59. *Mem:* Am Chem Soc; Am Soc Metals; Electrochem Soc; Am Inst Mining, Metall & Petrol Eng; Am Ceramics Soc. *Res:* Extractive metallurgy; refractory metals; light metals; high temperature materials; vacuum and high temperature metallurgy; electrothermics; electrochemistry; electrometallurgy. *Mailing Add:* 3308 Alta Vista Dr Chattanooga TN 37411

GLASSER, LEO GEORGE, b Wilkes-Barre, Pa, July 20, 16; m 41; c 3. PHYSICAL OPTICS. *Educ:* Cornell Univ, AB, 38, MA, 40. *Prof Exp:* Asst, Eastman Kodak Co, NY, 38; asst, Nat Geog Soc, Cornell Univ, 38-40; physicist, US Navy Bur Ord, Washington, DC, 40-44; sr physicist, Tenn Eastman Corp, Oak Ridge, 44-45; res engr & supvr eng res lab, E I Du Pont de Nemours, 45-55, res mgr, 56-59, elec res, Mech Develop Lab, 59-60, asst dir eng res lab, 60-63, dir eng physics lab, 63-70, mgr div progs, 70-73, dir eng physics lab, 73-80; CONSULT 80- *Concurrent Pos:* Dir, Mt Cuba Astron Observ, 71- *Mem:* Optical Soc Am; Sigma Xi; Am Astron Soc. *Res:* Development and application of colorimetric instruments and optical and other analytical instruments for process control; light measurement; refractometry; ultraviolet and infrared spectrophotometry; lasers, holography. *Mailing Add:* 726 Loveville Rd No 84 Hockessin DE 19707

GLASSER, M LAWRENCE, b Crookston, Minn, Oct 5, 33; m 56; c 4. SOLID STATE PHYSICS, APPLIED MATHEMATICS. *Educ:* Univ Chicago, BA, 53, MS, 55; Univ Miami, MS, 57; Carnegie Inst Technol, PhD(physics), 62. *Prof Exp:* Instr physics, Univ Miami, 57-58; sr physicist, Battelle Mem Inst, 62-64; asst prof physics, Univ Wis, 64-66; sr physicist, Battelle Mem Inst, 66-67, staff scientist, 67-74; prof appl math, Univ Waterloo, 74-77; PROF MATH, CLARKSON UNIV, 77- *Concurrent Pos:* Consult, Battelle Mem Inst, 64. *Honors & Awards:* J Graham Res Award, 78. *Mem:* Am Phys Soc; Math Asn Am; Soc Indust & Appl Math. *Res:* Theoretical solid state physics; classical mathematical analysis; statistical mechanics. *Mailing Add:* Dept Math Clarkson Univ Potsdam NY 13699-9999

GLASSER, RICHARD LEE, b Baltimore, Md, Mar 26, 27; m 50; c 3. NEUROBIOLOGY. *Educ:* Johns Hopkins Univ, AB, 49; Univ Md, PhD(physiol), 57. *Prof Exp:* From instr to asst prof, 57-65, ASSOC PROF PHYSIOL, SCH MED, UNIV NC, CHAPEL HILL, 65- *Concurrent Pos:* Vis asst prof, Sch Med, Duke Univ, 59-61. *Mem:* AAAS; Am Physiol Soc; Animal Behav Soc; Soc Neurosci; Soc Values Higher Educ. *Res:* Neurophysiology; neural control of respiration; neural control of cardiovascular activity; neural control of sweat gland activity; memory consolidation mechanisms. *Mailing Add:* Dept Physiol Univ NC Sch Med Res Wing CB No 7545 Chapel Hill NC 27599-7545

GLASSER, ROBERT GENE, b Chicago, Ill, Apr 14, 29; div; c 4. PHYSICS. *Educ:* Univ Chicago, AB, 48, BS, 50, MS, 52, PhD(physics), 54. *Prof Exp:* Asst, Univ Chicago, 52-54, res assoc, 54-55; physicist, US Naval Res Lab, 55-65; from assoc prof physics to prof physics & comput sci, 65-75, PROF PHYSICS & ASTRON, UNIV MD, COLLEGE PARK, 75- *Concurrent Pos:* Lectr, Am Univ, 57 & Univ Md, College Park, 58-65. *Mem:* AAAS; Am Phys Soc; Fedn Am Scientists; Asn Comput Mach. *Res:* Elementary particle physics; mathematical physics; computers. *Mailing Add:* Dept Physics Univ Md College Park MD 20742

GLASSER, STANLEY RICHARD, b New York, NY, Dec 2, 26; m 50; c 2. REPRODUCTIVE BIOLOGY, ENDOCRINOLOGY. *Educ:* Cornell Univ, BA, 48; Rutgers Univ, PhD(zool), 52. *Prof Exp:* From instr to asst prof radiation biol, Med Sch, Univ Rochester, 52-62; assoc prof obstet & gynec, Med Sch, Vanderbilt Univ, 62-72; ASSOC PROF CELL BIOL, BAYLOR COL MED, 72-, DEPT DIR GRAD STUDIES, 75- *Concurrent Pos:* Vis asst prof, Ill Wesleyan Univ, 58; lectr, Am Inst Biol Sci-Med Educ Nat Defense, 58-65; vis prof, Baylor Univ, 59; prof obstet & gynec, Maternal Health & Family Planning Prog, Meharry Med Sch, 64-72; prof obstet & gynec, Hahnemann Med Col, 68-77; fel cancer res, Weizmann Inst Rehovat, Israel, 78. *Honors & Awards:* Medal Pop Res, Franklin Inst of Philadelphia, 72. *Mem:* Endocrine Soc; Soc Study Reproduction; Brit Soc Study Fertil; Am Physiol Soc; Soc Develop Biol. *Res:* Mammalian reproductive biology; hormone action; biochemistry of decidua and implantation; steroid receptors; uterus, corpus luteum, steroidogenesis; biochemistry of trophoblast growth and differentiation. *Mailing Add:* Dept Cell Biol & Ctr Pop Res Baylor Col Med One Baylor Plaza Houston TX 77030

GLASSER, WOLFGANG GERHARD, b Zwickau, Ger, Oct 9, 41; m 69; c 2. LIGNIN CHEMISTRY, PULP & PAPER CHEMISTY. *Educ:* Univ Hamburg, Ger, dipl-Holzwirt, 66, Dr rer nat (wood chem), 69. *Prof Exp:* Postdoctoral res assoc & asst prof chem eng, Univ Wash, Seattle, 69-71; from asst prof to assoc prof, 72-80, PROF WOOD CHEM, VA POLYTECH INST & STATE UNIV, 80-, DIR, BIOBASED MAT CTR, 88- *Concurrent Pos:* Panel chmn, Nat Acad Sci, 74-76; dir, Pulp & Paper Res Inst, Sao Paulo, Brazil, 76; sabbatical, Inst Resquisas Technol, Sao Paulo, Brazil, 76 & Centre de Recherches sur les Macromolecules Vegetales, 85; publicity chair, Am Chem Soc, 86-89, chmn, 90. *Honors & Awards:* George Olmsted Award, Am Paper Inst, 74; Sci Achievement Award, Int Union Forestry Res Inst, 86. *Mem:* Am Chem Soc; Tech Asn Pulp & Paper Indust; Soc Wood Sci Technol; Soc Plastics Engrs; Sigma Xi. *Res:* Chemistry research with wood; separation of constitutive biopolymers from lignocellulosic resources; use of wood-derived polymers for structural materials and plastics; chemical modification of lignin, xylan and cellulose; author of numerous technical publications. *Mailing Add:* Dept Wood Sci & Forest Prod Va Polytech Inst & State Univ Blacksburg VA 24061-0503

GLASSGOLD, ALFRED EMANUEL, b Philadelphia, Pa, July 20, 29; m 53; c 2. THEORETICAL PHYSICS, ASTROPHYSICS. *Educ:* Univ Pa, BA, 50; Mass Inst Technol, PhD(physics), 54. *Prof Exp:* Physicist, Oak Ridge Nat Lab, 54-55; lectr theoret physics, Univ Minn, 55-57; res physicist, Univ Calif, 57-63; assoc prof, 63-65, head dept, 69-74, PROF PHYSICS, NY UNIV, 65- *Mem:* AAAS; Am Astron Soc; Int Astron Union; Am Phys Soc. *Mailing Add:* Dept Physics NY Univ Four Wash Pl New York NY 10003

GLASSICK, CHARLES ETZWEILER, b Columbia, Pa, Apr 6, 31; m 52; c 5. ORGANIC CHEMISTRY. *Educ:* Franklin & Marshall Col, BS, 53; Princeton Univ, MA, 55, PhD(chem), 57. *Hon Degrees:* DSc, Univ Richmond, 77; LLD, Dickinson Law Sch, 86. *Prof Exp:* Res chemist, Rohm & Haas Co, 57-62; assoc prof chem, Adrian Col, 62-67; Am Coun Educ fel acad admin, Fresno State Col, 67-68; vpres, Great Lakes Cols Asn, 68-69; assoc dean, Albion Col, 69-71; vpres acad affairs, Univ Richmond, 71-72, vpres & provost, 72-77; PRES, GETTYSBURG COL, 77- *Concurrent Pos:* Instr, Temple Univ, 58-62; consult, Nat Endowment for the Humanities, 73-74 & State Coun Higher Educ in Va, 74-76. *Mem:* Am Chem Soc. *Res:* Synthesis of polycyclic aromatics and carcinogenics; plant growth regulatory chemicals and alkaloids. *Mailing Add:* 243 W Broadway Gettysburg PA 17325

GLASSLEY, WILLIAM EDWARD, b Ventura, Calif, July 1, 47; c 1. PETROLOGY, TECTONICS. *Educ:* Univ Calif, San Diego, BA, 69; Univ Wash, MS, 71, PhD(tectonics, petrol), 73. *Prof Exp:* Fel, Mineral-Geol Mus, Univ Oslo, 73-74; asst prof tectonics & petrol, dept oceanog, Univ Wash, 74-76; asst prof geol, Middlebury Col, 76-85; TECH AREA LEADER GEOCHEM, LAWRENCE LIVERMORE LAB, 85- *Mem:* Am Geophys Union; AAAS. *Res:* Structural and petrologic evolution of early Precambrian continental crust and the role of fluids in deep crustal processes. *Mailing Add:* Dept Earth Sci Lawrence Livermore Nat Lab Livermore CA 94550

GLASSMAN, ARMAND BARRY, b Paterson, NJ, Sept 9, 38; m 58; c 3. LABORATORY MEDICINE, NUCLEAR MEDICINE. *Educ:* Rutgers Univ, BA, 60; Georgetown Univ, MD, 64. *Prof Exp:* Assoc prof & dir clin labs path, Med Col Ga, 71-74, prof & dir, 74-76; PROF & MED DIR MED TECHNOL & MED LAB TECHNOL & PROF & CHMN LAB MED, MED UNIV SC, 76-, ASSOC DEAN & PROF RADIOL & NUCLEAR MED, COL MED, 79- *Concurrent Pos:* Consult, Vet Admin Hosp, Augusta, Ga, 71-76, Lab Nuclear Med, Charleston, 76-; prof, Col Grad Studies, Med Univ SC, 76- *Mem:* Soc Cryobiol; Acad Clin Lab Physicians & Scientists; Asn Clin Scientists; Am Asn Blood Banks; Int Soc Path Coun; Sigma Xi. *Res:* Quantitation and functional characterization of lymphocytes using surface markers, mitogenic responsiveness, carbohydrate metabolism and biomathematical modeling. *Mailing Add:* Sr Vpres Med Affairs Montefiore Med Ctr 111 E 210th St Bronx NY 10467

GLASSMAN, EDWARD, b New York, NY, Mar 18, 29; m 56; c 4. GENETICS, ALCOHOL METABOLISM. *Educ:* NY Univ, AB, 49, MS, 51; Johns Hopkins Univ, PhD(biol), 55. *Prof Exp:* Adam T Bruce predoctoral fel, Johns Hopkins Univ, 54-55; res assoc biochem, City of Hope Med Res Ctr, 57-58, staff mem genetics, 59-60; from asst prof to assoc prof, 60-67, dir neurobiol prog, 65-72, dir div chem neurobiol, 71-78, PROF BIOCHEM & GENETICS, SCH MED, UNIV NC, CHAPEL HILL, 67-, HEAD, PROG TEAM EFFECTIVENESS & CREATIVITY, 82- *Concurrent Pos:* Am Cancer Soc fel cancer res, Calif Inst Technol, 55-57; NIH fel, Inst Animal Genetics, Edinburgh, Scotland & Zool Inst, Zurich, 57-58; NIH career develop award, 61-71; Guggenheim Mem Found fel, 68-69; vis prof, Sch Med, Stanford Univ, 68-69; pres, NC Chapter of the Soc Neurosci, 74; mem bd adv, Neurochem Res, 75-78; mem, Univ NC Group & Orgn Develop Team, 77-78; vis prof, Sch Biol Sci, Univ Calif, Irvine, 78; vis fel, Ctr Creative Leadership, Greensboro, NC, 83; vis scientist, Stanford Res Ctr, Menlo Park, Calif, 86; creativity & leadership consult, 80-; head, Prog Team Effectiveness & Creativity, Sch Med, Univ NC, Chapel Hill, 82- *Mem:* Am Soc Neurochem; Int Soc Neurochem; Am Soc Biol Chemists; fel AAAS; fel Royal Soc Edinburgh. *Res:* Genetic control of alcohol metabolism. *Mailing Add:* Dept Biochem Univ NC Sch Med Chapel Hill NC 27516

GLASSMAN, HAROLD NELSON, b Waterbury, Conn, Sept 25, 12. BIOCHEMISTRY. *Educ:* Univ Pa, AB, 33, MS, 36; Georgetown Univ, PhD(chem), 49. *Prof Exp:* Asst physiol, Univ Pa & Marine Biol Lab, Woods Hole, 35-38; asst biochem, Wistar Inst, Philadelphia, 38-39; biochemist, Chem Corps, US Army, Camp Detrick, 44-56, asst sci dir, Ft Detrick, 56-72; dir admin, 72-74, Frederick Cancer Res Ctr, asst to gen mgr, 74-79; RETIRED. *Honors & Awards:* Barnett L Cohen Award, Am Soc Microbiol, 79. *Mem:* Am Chem Soc; Am Soc Microbiol; Sigma Xi; Am Acad Microbiol. *Res:* Surface active agents; erythrocyte permeability; toxins; spray drying; experimental airborne infection; toxoids; vaccines; pathogenesis of infectious disease; epidemiology; planning, coordination, and evaluation of large-scale research and development programs. *Mailing Add:* 1701 W 7th St Frederick MD 21702

GLASSMAN, IRVIN, b Baltimore, Md, Sept 19, 23; m 51; c 3. ENGINEERING, PHYSICAL CHEMISTRY. *Educ:* Johns Hopkins Univ, BE, 43, DrEng, 50. *Prof Exp:* Asst chem eng res, Substitute Alloy Mat Labs, Columbia, 43-46; res assoc mech & aerospace eng, 50-55, from asst prof to assoc prof, 55-64, PROF MECH & AEROSPACE ENG, PRINCETON UNIV, 64- *Concurrent Pos:* Nat Sci Found sr fel & vis prof, Univ Naples, 66-67, vis prof, 78-79; chmn, Propulsion & Energetics Panel, Adv Group Aerospace Res & Develop, NATO, 66-79; ed, Combustion Sci & Technol, 68-; mem comt on motor vehicle emissions, Nat Res Coun; Standard Oil Calif vis prof combustion, Stanford Univ, 75- *Honors & Awards:* Egerton Gold Medal, Combustion Inst, 82; Roe Award, Am Soc Eng Educ, 84. *Mem:* Am Asn Univ Profs; Am Chem Soc; Combustion Inst. *Res:* Combustion problems in energy, environment, propulsion, fire safety. *Mailing Add:* PO Box 14 Princeton NJ 08542-0014

GLASSMAN, JEROME MARTIN, b Philadelphia, Pa, Mar 2, 19; m 52; c 3. TOXICOLOGY, BIOSTATISTICS. *Educ:* Univ Pa, AB, 39, MA, 42; Yale Univ, PhD(pharmacol, pub health), 50. *Prof Exp:* Statistician, Bur Census, US Dept Com, Washington, DC, 40-43; hematologist, Div Pharmacol, Food & Drug Admin, 43-45; instr microanat & physiol, Essex Col Med & Surg, 45-46; res assoc, Biol Div, Schering Corp, 46; instr biol sci, Hunter Col, 46-47; asst, Lab Appl Physiol, Yale Univ, 50-51; sr res pharmacologist, Wyeth Labs, 51-59, head dept pharmacodynamics, 59-62; dir div pharmacol, USV Pharmaceut Corp, NY, 62-68, dir biol res, 68-70; dir clin res & pharmacol, Wampole Labs Div, Denver Chem Mfg Co, 70-75; dir clin invest, Wallace Labs Div, Carter Wallace Inc, 75-88; PVT CONSULT, 88- *Concurrent Pos:* Assoc prof, NY Med Col, 63-85. *Honors & Awards:* Scouter's Award, 61. *Mem:* Fel AAAS; Soc Toxicol; fel Am Col Clin Pharmacol Chemother; fel Am Col Clin Pharmacol; fel NY Acad Sci; Am Soc Pharmacol & Exp Therapeut; Sigma Xi. *Res:* Nature of fibrillae of cells; toxicology of dichloro-diphenyl-trichloroethane; toxicology of benzidine and its congeners; adsorbents and anticholinergic drugs; pharmacology of hyaluronidase; cardiotonicity and cardiotoxicity of sympathomimetic amines; inflammation; analgesic agents; local anesthetics; antidiabetic agents; otitis externa; immunologics; anticoagulants; venereal disease; headache; antibiotics, muscle relaxants, mucolytics; awarded patents in field. *Mailing Add:* 280 Sleepy Hollow Rd Briarcliff Manor NY 10510-2139

GLASSMAN, SIDNEY FREDERICK, b Chicago, Ill, July 30, 19; m 50; c 2. PLANT TAXONOMY. *Educ:* Univ Ill, BS, 42, MA, 47; Univ Okla, PhD(bot), 50. *Prof Exp:* Instr bot, Univ Wyo, 51-52; asst botanist, Field Mus, Univ Ill, Chicago, 51, from asst prof to assoc prof biol sci, 52-63, res fel, 57, PROF BIOL SCI, UNIV ILL, CHICAGO, 63-, CUR, HERBARIUM, 55-, RES ASSOC, FIELD MUS, 63- *Concurrent Pos:* Fel, Nat Res Coun, Caroline Islands, 49; instr, Wilson Jr Col, 53; NSF award, 65, 69 & 76, res trips, Europe, 66, Utrecht & Berlin, 74, Univ Tex & Ill, 80, Kew Gardens, Eng, Univ strathclyde, Glasgow, 81-, palm collecting trips, Brazil, 65, 69, 76 & 88. *Mem:* Am Soc Plant Taxon; Int Asn Plant Taxon; Int Palm Soc. *Res:* Revision of index of American palms and palm genus Copernicia; grass flora of Chicago region and revision of genus Syagrus; revision of palm genera Attalea, Scheelea, Orbignya and Maximiliana. *Mailing Add:* 8942 Bellefort Ave Morton Grove IL 60053

GLASSOCK, RICHARD JAMES, NEPHROLOGY, TRANSPLANTATION. *Educ:* Univ Calif, Los Angeles, MD, 60. *Prof Exp:* PROF INTERNAL MED & NEPHROLOGY & CHMN DEPT MED, SCH MED, UNIV CALIF, LOS ANGELES, 80- *Mailing Add:* Dept Med Univ Calif 1000 W Carson Torrance CA 90509

GLASSTONE, SAMUEL, b London, Eng, Mar 5, 97; nat; m 29, 69. NUCLEAR SCIENCE. *Educ:* Univ London, BSc, 16, MSc, 20, PhD(phys chem), 22, DSc, 26. *Prof Exp:* Lectr phys chem, Univ London, 19-21, Univ Col of Southwest, 21-28 & Univ Sheffield, 28-39; res assoc chem, Princeton Univ, 39-41 & sci educ, Univ Press, 41-42; prof chem, Univ Okla, 42-43; consult electrochem, Zenith Radio Corp, Ill, 43-46; prof chem, Boston Col, 47-48; consult, Dept of Energy, 48-82 & Los Alamos Nat Lab, 52-70; RETIRED. *Honors & Awards:* Worcester Reed Warner Medal, Am Soc Mech Eng, 59; Arthur Holly Compton Award, Am Nuclear Soc, 68. *Mem:* AAAS; Royal Soc Chem; Am Nuclear Soc. *Res:* Physical chemistry; electrochemistry; atomic energy; space science. *Mailing Add:* 103 Wiltshire Dr Oak Ridge TN 37830

GLATSTEIN, ELI J, b Muscatine, Iowa, Feb 20, 38. RADIATION ONCOLOGY. *Educ:* State Univ Iowa, BA, 60; Stanford Univ, MD, 64; Am Bd Radiol, cert, 72. *Prof Exp:* Intern, NY Hosp, Med Ctr, Cornell Univ, 64-65; Picker Found fel radiobiol, Hammersmith Hosp, London, UK, 70-71, fel, Gray Lab, Mt Vernon Hosp, Northwood, Middlesex, UK, 71-72; resident & fel radiation ther, Sch Med, Stanford Univ, 67-70, asst prof radiol, 72-77, univ adv, 73-77, fac sen, Med Sch Senate, 75-77; cancer expert, Radiol Oncol Br, 77-78, actg dir, Clin Oncol Prog, 88-89, CHIEF, RADIATION ONCOL BR, CLIN ONCOL PROG, DIV CANCER TREATMENT, NAT CANCER INST, BETHESDA, 78-, ACTG DIR, RADIATION RES PROG, 91- *Concurrent Pos:* Consult radiation biol, Lawrence Radiation Lab, Berkeley, 74-76; mem, Cancer Clin Invest Rev Comt, Nat Cancer Inst, 76-78; assoc prof radiol, Uniformed Serv Univ Health Sci, Bethesda, 79-81, prof, 81-; Nat Cancer Inst rep, Am Joint Comt Cancer, 90- *Honors & Awards:* Presidential Meritorious Exec Rank Award, 85. *Mem:* Am Soc Therapeut Radiologists; Int Soc Lymphology; Brit Inst Radiol; Am Radium Soc; Am Soc Clin Oncol; Int Asn Study Lung Cancer; Radiation Res Soc; Asn Univ Professors; Am Asn Cancer Res; AAAS. *Res:* Radiation effects; normal tissue tolerance; lymphomas-Hodgkin's Disease; genitourinary cancer; sarcomas; pediatric neoplasia; lung cancer; radiation therapy treatment planning; photodynamic therapy; radioimmunoglobin therapy; intraoperative radiotherapy. *Mailing Add:* Radiation Oncol Br Bldg 10 Rm B3B69 NIH Nat Cancer Inst 9000 Rockville Pike Bethesda MD 20892

GLATZER, LOUIS, b New York, NY, Aug 7, 40; m 65; c 1. MICROBIAL GENETICS, MOLECULAR BIOLOGY. *Educ:* Dartmouth Col, AB, 63; NC State Univ, MS, 65; Univ Tex, Austin, PhD(zool, biochem), 70. *Prof Exp:* ASST PROF BIOL, UNIV TOLEDO, 73- *Concurrent Pos:* Nat Inst Gen Med Sci fel, Oak Ridge Nat Lab, 70-73. *Mem:* Am Soc Photobiol; Am Soc Microbiol. *Res:* Plasmid genetics; effects of near ultraviolet light irradiation on bacterial systems; molecular biology of recombination; homologies between related and unrelated DNA's. *Mailing Add:* Dept Biol Univ Toledo 2801 W Bancroft Toledo OH 43606

GLAUBER, ROY JAY, b New York, NY, Sept 1, 25; m 60; c 2. PHYSICS. *Educ:* Harvard Univ, SB, 46, AM, 47, PhD(physics), 49. *Prof Exp:* Asst theoret physics, Los Alamos Sci Lab, 44-46; AEC fel, Inst Advan Study, 49-50, mem, 49-51, Jewett fel, 50-51; lectr, theoret physics, Calif Inst Technol, 51-52; lectr & Bayard Cutting fel, 52-54, from asst prof to prof, 54-76, MALLINCKRODT PROF PHYSICS, HARVARD UNIV, 76- *Concurrent Pos:* AEC fel, Swiss Fed Inst Technol, 50; Fulbright lectr, Grenoble, France,

54; Guggenheim Mem Found fel, 59; mem bd & ed J Math & Physics, 61-63, J Nuclear Physics, 72; US-Soviet exchange lectr, Leningrad, 64; consult, Radiation Lab, Univ Calif, Bell Tel Labs, Am Tel & Tel Co & Lewis Res Ctr, NASA, Ohio, Los Alamos Nat Lab; NSF sr fel, 66-67; vis scientist, CERN, Geneva, 67 & 83, guest prof, 72-73; Lorentz prof, Univ Leidan, Holland, 74; vis prof, Nordita, Copenhagen, 74 & Col de France, 83; adj prof physics, Univ Ariz, Tucson, 88-; A von Humboldt res award, Bonn, 89. *Honors & Awards:* A A Michelson Medal, Franklin Inst, 85; Max Born Medal & Award, Am Optical Soc, 85. *Mem:* Fel Nat Acad Sci; fel Am Phys Soc; fel Am Acad Arts & Sci; fel Am Optical Soc. *Res:* Nuclear physics; particle diffraction and diffusion problems; elementary particle theory and high energy physics; quantum optics; statistical mechanics; quantum theory of measurement. *Mailing Add:* Lyman Lab Physics Harvard Univ Cambridge MA 02138-2901

GLAUBERMAN, GEORGE ISAAC, b New York, NY, Mar 3, 41; m 65; c 2. FINITE GROUP THEORY. *Educ:* Polytech Inst Brooklyn, BS, 61; Harvard Univ, MA, 62; Univ Wis, PhD(math), 65. *Prof Exp:* Asst prof math, Univ Wis, 65; from instr to assoc prof, 65-70, PROF MATH, UNIV CHICAGO, 70- *Concurrent Pos:* Sloan Found res fel, 67-69; NSF fel, Univ Oxford, Eng, 71-72; Guggenheim fel & vis sr res fel, Jesus Col, Oxford Univ, Eng, 78-79; vis, Math Inst, Univ Oxford, Eng, 79-80. *Mem:* Am Math Soc; Math Asn Am. *Res:* Algebra, principally the theory of finite groups. *Mailing Add:* Dept Math Univ Chicago 5734 S University Ave Chicago IL 60637

GLAUBMAN, MICHAEL JUDA, b Baltimore, Md, Dec 31, 24; m; c 4. PARTICLE PHYSICS, NUCLEAR PHYSICS. *Educ:* Hebrew Univ, BS, 47; Univ Ill, MA, 50, PhD(physics), 53. *Prof Exp:* Asst physics, Univ Ill, 49-53 & Princeton Univ, 53-55; res assoc, Columbia Univ, 55-56; sr res physicist, Atomics Int Div, NAm Aviation, Inc, 56-59; from asst prof to assoc prof, 59-70, PROF PHYSICS, NORTHEASTERN UNIV, 70- *Mem:* Am Phys Soc; AAAS. *Res:* Experimental nuclear structure; high energy experimental physics. *Mailing Add:* Dept Physics Northeastern Univ Boston MA 02115

GLAUDEMANS, CORNELIS P, b Semarang, Java, Apr 16, 32; US citizen; m 56; c 6. MEDICINE. *Educ:* Fed Univ, Utrecht, Netherlands, BS, 54; McGill Univ, Can, PhD(chem), 58. *Prof Exp:* Vis res scientist, NIH, 62-65; prof chem, Yale Univ, 65-67; CHIEF, NIH, 67- *Mailing Add:* 8/BIA-23 NIH Bethesda MD 20892

GLAUERT, HOWARD PERRY, b St Louis, Mo, June 19, 52. NUTRITION & CANCER. *Educ:* Univ Mo, Columbia, AB, 74; Mich St Univ, PhD(nutrit), 82. *Prof Exp:* Lab technician food sci, Univ Mo, 73-76; grad asst nutrit, Mich State Univ, 76-82; fel environ toxicol, Univ Wis, 82-85; ASST PROF NUTRIT, UNIV KY, 85- *Mem:* Am Inst Nutrit; Sigma Xi; NY Acad Sci; AAAS; Soc Toxicol; Am Asn Cancer Res. *Res:* Effect of nutrition on chemical carcinogenesis; natural history of colon and liver carcinogenesis; hypolipidemic peroxisome proliferators and their role in carcinogenesis. *Mailing Add:* Dept Nutrit & Food Sci Univ Ky 212 Funkhouser Bldg Lexington KY 40506-0054

GLAUNSINGER, WILLIAM STANLEY, b Newark, Ohio, May 10, 45; m 67; c 1. SOLID STATE CHEMISTRY, PHYSICAL CHEMISTRY. *Educ:* Miami Univ, BS, 67; Cornell Univ, PhD(chem), 72. *Prof Exp:* Teaching asst, Cornell Univ, 67-69, res asst, 69-72; asst prof chem, 72-76, ASSOC PROF CHEM, ARIZ STATE UNIV, 76- *Concurrent Pos:* Res Corp res grant, 74; NSF grant, 76; Phys Res Found grant, 77. *Mem:* Am Chem Soc; AAAS; Am Phys Soc; Int Soc Magnetic Resonance; Sigma Xi. *Res:* Magnetic resonance spectroscopy using electron paramagnetic resonance, nuclear magnetic resonance and electron-nuclear double resonance techniques; theoretical and experimental investigation of metal-ammonia systems; transition metal oxides; rare earth hexaborides; II-VI compounds and alloys; metallic microcrystals; spin-labeled biomolecules. *Mailing Add:* 128 W Harrison Chandler AZ 85224-3724

GLAUSER, ELINOR MIKELBERG, b Philadelphia, Pa, Aug 24, 31; m 52; c 3. PHARMACOLOGY, PHYSIOLOGY. *Educ:* Univ Pa, BA, 52; Woman's Med Col of Pa, MD, 57. *Prof Exp:* Assoc res, Dept Med, Woman's Med Col Pa, 61-63; from instr to asst prof, 63-70, ASSOC PROF PHARM & MED, SCH MED, TEMPLE UNIV, 70- *Concurrent Pos:* Fel physiol, Trudeau Soc, Sch Med, Univ Pa, 58-59; Kate Hurdmead fel exp med, Cambridge Univ, 59-60; fel med, Harvard Med Sch, 60-61; NIH fel, 61-63; Am Thoracic Soc fel, 63-64. *Mem:* AAAS; Am Thoracic Soc; Am Chem Soc; AMA; NY Acad Med. *Res:* Pulmonary physiology, pulmonary pharmacology and chronic pulmonary disease in man. *Mailing Add:* 630 Richards Rd Wayne PA 19087

GLAUSER, FREDERICK LOUIS, b Philadelphia, Pa, June 15, 37; m; c 2. PULMONARY MEDICINE. *Educ:* Ursinus Col, BS, 59; Hahnemann Med Col, MD, 63. *Prof Exp:* Resident internal med, Vet Admin Hosp, Univ Ore, 66-68, resident pulmonary med, 68-69; fel pulmonary dis, Univ Pittsburgh Hosp, 69-70 & Stanford Univ, 70-71; clin pract internal med & pulmonary dis, San Jose Med Clin, Calif, 71-72; asst prof med in line, Pulmonary Div, Sch Med, Univ Calif, Irvine, 72-75, asst prof med in residence, Univ, 75-77; chief, Pulmonary Sect, 84-88, CHIEF, PULMONARY DIV, MCGUIRE VET ADMIN MED CTR, MED COL VA, RICHMOND, 87-, PROF, 89- *Concurrent Pos:* Actg dir respiratory ther, Orange County Med Ctr, 72-75; consult, Long Beach Vet Admin Med Ctr, 73-75, chief, Pulmonary Sect B, 75-77, co-dir MICU, 75-76, dir, 76-77; co-dir, MICU, Med Col Va & McGuire Vet Admin Med Ctr, 78-; Nat Heart, Lung & Blood Inst consult, site visits, 80 & 86; chmn, Multidisciplinary Crit Care Comt, Med Col Va, 81-86, mem, 86-; chmn, Crit Care Comt, McGuire Vet Admin Med Ctr, 85-88. *Mem:* Am Physiol Soc; Fedn Am Socs Exp Biol; Microcirculatory Soc. *Mailing Add:* Pulmonary Div Med Col Va 1200 E Broad St Box 50 Richmond VA 23298-0050

GLAUSER, STANLEY CHARLES, b Philadelphia, Pa, June 19, 31; m 52; c 3. PHARMACOLOGY. *Educ:* Univ Pa, BA, 51, MS & MD, 55, PhD(phys chem), 59. *Prof Exp:* Asst phys chem, Sch Med, Univ Pa, 52-54; res assoc, NIH, 57-59; NSF res fel molecular biol, Med Res Coun Unit, Cavendish Lab, Cambridge Univ, 59-60; res assoc biol, Mass Inst Technol, 60-61; asst prof molecular biol & physiol, Sch Med, Univ Pa, 61-65; from assoc prof to prof molecular pharmacol, Sch Med, Temple Univ, 65-79; CHIEF SECT PHARMACOL, DEPT MED, GRAD HOSP, PHILADELPHIA, PA, 79- *Concurrent Pos:* Instr, USDA Grad Sch, 58-59; prof, Hadassah Med Sch, Hebrew Univ, Jerusalem, Israel, 83- *Mem:* AAAS; Am Chem Soc; Am Med Asn; NY Acad Sci; Am Soc Pharmacol & Exp Therapeut. *Res:* Molecular biology; physical chemistry of hemoproteins and other porphyrin-like compounds of biological interest; thermodynamics; kinetics; genetics. *Mailing Add:* 630 Richards Rd Wayne PA 19087

GLAUZ, ROBERT DORAN, b Detroit, Mich, Aug 13, 27; m 53; c 3. APPLIED MATHEMATICS. *Educ:* Univ Mich, BSE, 48, MS, 49; Brown Univ, PhD(appl math), 53. *Prof Exp:* Instr math, Mont State Col, 49-50; asst, Brown Univ, 51-53; mem staff, Los Alamos Sci Lab, Univ Calif, 53-57; eng anal specialist, Aircraft Gas Turbine Div, Gen Elec Corp, 57-58; tech specialist, Aerojet-Gen Corp Div, Gen Tire & Rubber Co, 58-66, mgr eng anal & prog, 66-68; PROF MATH, UNIV CALIF, DAVIS, 68- *Mem:* Am Math Soc; Math Asn Am; Asn Comput Mach. *Res:* Viscoelasticity; numerical analysis. *Mailing Add:* Dept Math Univ Calif Davis CA 95616

GLAUZ, WILLIAM DONALD, b Grand Rapids, Mich, Oct 26, 33; m 56; c 4. ENGINEERING SCIENCE, DATA ACQUISITION AND ANALYSIS. *Educ:* Mich State Univ, BS, 56; Purdue Univ, MS, 59, PhD(eng sci), 64. *Prof Exp:* Instr, Purdue Univ, 56-63; assoc analyst, 63-64, sr analyst, 64-66, prin analyst, 66-69, head anal & appl math, 69-71, mgr hwy & traffic systs eng, 71-80, DIR SAFETY & ENG ANALYSIS, MIDWEST RES INST, 80- *Concurrent Pos:* Consult, Midwest Appl Sci Corp, 59-63; lectr, Univ Mo-Kansas City, 64-79; comt chmn, Hwy Res Bd, Nat Acad Sci-Nat Res Coun. *Mem:* Inst Transp Eng. *Res:* Research management; highway safety; traffic analysis; simulation; digital computation; mechanics; applied mathematics; aerodynamics. *Mailing Add:* 11600 Minor Dr Kansas City MO 64114

GLAVIANO, VINCENT VALENTINO, b Frankfurt, NY, July 19, 20; m 45; c 2. PHYSIOLOGY. *Educ:* City Col New York, BS, 50; Columbia Univ, PhD(physiol), 54; Chicago Med Sch, MD, 82. *Prof Exp:* Asst physiol, Col Physicians & Surgeons, Columbia Univ, 51-53, instr, 53-54; asst prof, Col Med, Univ Ill, 56-60; from assoc prof to prof, Stritch Sch Med & Grad Sch, Loyola Univ, Ill, 60-70; prof physiol & chmn dept, Chicago Med Sch, Univ Health Sci, 70-85; EMER PROF & DIR, BIOTECHNOL RES ASSOCS, 86- *Concurrent Pos:* Consult clin physics, Hines Vet Admin Hosp. *Mem:* AAAS; Am Physiol Soc; Soc Exp Biol & Med; Harvey Soc; NY Acad Sci; Sigma Xi. *Res:* Cardiovascular physiology; catecholamines; cardiac metabolism; electrolytes and shock. *Mailing Add:* 517 Carlisle Ct Glen Ellyn IL 60137

GLAVIN, GARY BERTRUN, b Winnipeg, Man, Feb 2, 49; m 73; c 1. NEURO-GASTROENTEROLOGY, BRAIN-GUT RELATIONSHIPS. *Educ:* Univ Man, BA, 71, MSc, 73 & PhD(psychopharmacol), 75. *Prof Exp:* Assoc prof pharmacol, 84-88, ASSOC PROF NEUROSURG, UNIV MAN, 87-, PROF PHARMACOL, 88- *Mem:* Am Soc Pharmacol & Exp Therapeut; Soc Neurosci; AAAS; NY Acad Sci; Am Gastroenterol Asn. *Res:* Role of central and peripheral dopamine in experimental gastroduodinal ulcer disease; monoamine neurochemistry of stress and of focal seizure disorder. *Mailing Add:* Dept Pharmacol Fac Med Univ Man Winnipeg MB R3E 0W3 Can

GLAWE, LLOYD NEIL, b Des Moines, Iowa, Aug 21, 32; m 70; c 2. INVERTEBRATE PALEONTOLOGY, STRATIGRAPHY-SEDIMENTATION. *Educ:* Univ Ill, BS, 54; La State Univ, MS, 60, PhD(paleont), 66. *Prof Exp:* From asst prof to assoc prof, 64-77 PROF GEOL, NORTHEAST LA UNIV, 77- *Concurrent Pos:* Res assoc, La Geol Survey, 61, Ala Geol Survey, 64 & Oil & Gas Consult & Contract Geol, 80- *Mem:* Paleont Res Inst; Soc Econ Paleontologists & Mineralogists; Sigma Xi. *Res:* Classification and distribution of Gulf Coast Tertiary Pectinidae; foraminiferal biostratigraphy and paleoenvironment. *Mailing Add:* Dept Geosci NE Louisiana Univ 700 University Ave Monroe LA 71209-0550

GLAZE, ROBERT P, b Birmingham, Ala, Apr 14, 33; m 58; c 2. BIOCHEMISTRY. *Educ:* Univ of the South, BS, 55; Univ Rochester, PhD(biochem), 61. *Prof Exp:* Asst prof biochem, Univ Ala, 64-72, from asst dean to assoc dean, Schs Med & Dent, 67-72, dir instnl study prog, 72-73, coordr res grants, 67-76, asst to pres, Univ Ala, 73-76, dean admin, 76-77, vpres res & grad studies, 77-81, actg vpres instnl advan, 80-81, vpres res & instnl advan, 81-88, exec dir, UAB Res Found, 88, vpres res develop, 88-90, ASST TO PRES SPECIAL PROG, UNIV ALA, 91- *Concurrent Pos:* Fel physiol chem, Sch Med, Johns Hopkins Univ, 61-64; USPHS fel, 62-63 & trainee, 62-64; NSF res grant, 66-68. *Mem:* AAAS; Am Chem Soc; Am Inst Chemists. *Res:* Mitochondrial electron transport; oxidative phosphorylation; phosphate transfer enzymes; uremia. *Mailing Add:* Univ Ala UAB Sta Birmingham AL 35294

GLAZE, WILLIAM H, b Sherman, Tex, Nov 21, 34; m 76; c 4. ENVIRONMENTAL CHEMISTRY. *Educ:* Southwestern Univ Tex, BS, 56; Univ Wis, MS, 58, PhD(chem), 61. *Prof Exp:* Robert A Welch fel, Rice Univ, 60-61; from asst prof to assoc prof, NTex State Univ, 61-65, assoc dean arts & sci & dir, Inst Environ Studies, 73-75, prof chem, 65-80, dir, Inst Appl Sci, 75-80; prof chem & environ sci, head, Grad Prog Environ Sci & dir, Ctr Energy & Environ Sci, Univ Tex, Dallas, 80-84; PROF & PROG DIR, ENVIRON SCI & ENG PROG, SCH PUBL HEALTH, UNIV CALIF, LOS ANGELES, 84- *Concurrent Pos:* Robert A Welch res grant organometallic compounds, 63-74; NSF undergrad equip fel consult, 64 & res grant organometallic photochem, 69-71; res grants, Environ Protection Agency, 74- & mem, Sci Adv Bd; Tex Water Develop Bd grant, Tex Dept Health grant & NSF sponsored appointee res exchange prog, Shell Develop Co, 76; James M Montgomery & Metrop Waste Dist Contract, 86-88; res grants, US-Mex Policy Rels, Hewlett Found, Advan Oxidation Processes, La Dept Water & Power, US Environ Protection Agency, San Diego Wastewater Reuse, Western Consortium Pub Health, Hazardous Substances Control, NSF & Eng

Res Ctr, 87- *Honors & Awards:* F J Zimmerman Award, Environ Sci, 86. *Mem:* Am Chem Soc; Environ Protection Agency. *Res:* Chemistry of water disinfection compounds; reaction kinetics and mechanisms; water purification by activated carbon; trace organics in the environment; mass spectrometry and gas chromatography. *Mailing Add:* Environ Sci Univ Tex Dallas Grad Prog Richardson TX 75080

GLAZENER, EDWARD WALKER, b Raleigh, NC, Feb 3, 22; m 47; c 2. POULTRY HUSBANDRY, HIGHER EDUCATION. *Educ:* NC State Col, BS, 43; Univ Md, MS, 45, PhD(poultry genetics), 49. *Prof Exp:* Poultry supvr, Green Pastures, NC, 39-41; asst county agent, Exten Serv, NC State Col, 44; asst prof poultry husb, Univ Md, 46; assoc prof poultry genetics, NC State Univ, 46-49, head dept poultry sci, 55-60, dir instr, Sch Agr & Life Sci, 60-70, actg dean, 70-71, prof poultry sci & genetics, 49-86, dir acad affairs, Sch Agr & Life Sci, 60-86, assoc dean & dir, 74-86, spec asst, 86-88, EMER DIR SPEC ASSIGNMENTS, JEFFERSON SCHOLAR HONORS PROG-COMMUNITY COL PROG DEVELOP, NC STATE UNIV, 88- *Concurrent Pos:* Vis scholar, Pa State Univ, 72; chmn, Coun Higher Educ, Agr Sci, Southern Regional Educ Bd, 74-, Manpower Adv Assessment Comt, USDA, 80-86 & Higher Educ Comt, USDA, 82-84; mem, Adv Comt, Health Sci, NC State Univ. *Honors & Awards:* Martin Litwack Vet Progression Award, 86. *Mem:* Fel AAAS; Am Genetic Asn; Poultry Sci Asn; Am Inst Biol Sci; Am Asn Higher Educ. *Res:* Genetics and endocrinology of poultry; thyroid activity as related to strain differences in growing chickens; curriculum development, particularly in agricultural and biological sciences; pre-professional medical sciences. *Mailing Add:* 115 Patterson Hall NC State Univ Raleigh NC 27607

GLAZER, ALEXANDER NAMIOT, b July 7, 35. BIOCHEMISTRY, MOLECULAR BIOLOGY. *Educ:* Univ Sydney, Australia, BSc(Hon), 56, MSc, 57; Univ Utah, PhD(biochem), 60. *Prof Exp:* Postdoctoral, biophys, Weizmann Inst Sci, Israel, 61-62; postdoctoral, Molecular Biol Lab, Med Res Ctr, Cambridge, Eng, 62-63; from asst prof to prof biochem, Sch Med, Univ Calif, Los Angeles, 64-76; prof microbiol, Dept Microbiol & Immumnl, 76-89, PROF BIOCHEM & MOLECULAR BIOL, DEPT MOLECULAR & CELL BIOL, UNIV CALIF, BERKELEY, 89- *Concurrent Pos:* Assoc mem, Inst Molecular Biol, Univ Calif, Los Angeles, 65-76; mem, NIH Biophys & Biophys Chem A Study Sect, 69-73; chmn, Dept Microbiol & Immunol, Univ Calif, Berkeley, 77-82; expert analyst, Chemtracts in Biochem & Molecular Biol, 89-; sr staff, Div Chem Biodynamics, Lawrence Berkeley Lab, 90- *Honors & Awards:* Endeavour Prize, British Asn Advan Sci, 56; Darbaker Prize, Bot Soc Am, 80; Sci Reviewing Prize, Nat Acad Sci, 91. *Mem:* Am Soc Biochem & Molecular Biol; AAAS; Am Soc Microbiol; Protein Soc. *Res:* Protein structure-function relationships, enzymology, biophysical chemistry-Examples: assembly of macromolecular complexes (flagella, phycobilisomes), development of fluorescent reagents for cell sorting, cell analyses, and detection of reactive oxygen species and physiologically important antioxidants, methods for DNA detection. *Mailing Add:* 229 Stanley Hall Univ Calif Berkeley CA 94720

GLAZER, ROBERT IRWIN, b New York, NY, May 14, 42; m 65; c 2. PHARMACOLOGY, ONCOLOGY. *Educ:* Columbia Univ, BS, 65, MS, 67; Ind Univ, PhD(pharmacol), 70. *Prof Exp:* Asst prof pharmacol, Emory Univ, 72-77; head appl pharmacol sect, Med Chem & Biol Lab, 77-85, LAB BIOL CHEM, NAT CANCER INST, 85- *Concurrent Pos:* Nat Cancer Inst fel pharmacol, Sch Med, Yale Univ, 70-72 & res grant, Emory Univ, 73-76; Pharmaceut Mfrs Asn found fac develop award, 73-75; Bristol-Myers Multidrug Resistance Grant, 88- *Mem:* AAAS; Am Soc Pharmacol & Exp Therapeut; Am Asn Cancer Res; Am Soc Biol Chem. *Res:* Studies the role of protein phosphorylation and serine and tyrosine protein kinases in differentiation; multidrug resistance and human immunodeficiency virus (HIV) replication. *Mailing Add:* Dept Pharmacol Georgetown Univ Sch Med Four Research Ct Rockville MD 20850

GLAZIER, ROBERT HENRY, b Amherst, Mass, Oct 1, 26. ORGANIC CHEMISTRY. *Educ:* Amherst Col, AB, 48; Univ NH, MS, 50; Univ Kans, PhD(chem), 52. *Prof Exp:* Prof chem & chmn dept, Alderson-Broaddus Col, 52-54 & Nebr Wesleyan Univ, 54-61; vis prof, Kalamazoo Col, 61-62; assoc prof, 62-70, PROF CHEM, WASHBURN UNIV, 70- *Mailing Add:* Dept Chem Washburn Univ Topeka KS 66621

GLAZKO, ANTHONY JOACHIM, b San Francisco, Calif, Aug 15, 14; m 59; c 3. CHEMICAL PHARMACOLOGY. *Educ:* Univ Calif, AB, 35, PhD(biochem), 39. *Prof Exp:* Asst biochem, Univ Calif, 36-39; res assoc, Univ Mich, 39-41; asst prof biochem, Sch Med, Emory Univ, 46-47; lab dir chem pharmacol, Parke-Davis & Co, Div Warner-Lambert & Co, 47-80; CONSULT, 80- *Concurrent Pos:* Mem comn antiepileptic drugs, Int League Against Epilepsy, 75-78. *Mem:* Am Chem Soc; Am Soc Pharmacol & Exp Therapeut; Am Soc Clin Pharmacol & Therapeut; NY Acad Sci; Soc Exp Biol & Med. *Res:* Drug metabolism; species differences in metabolism; radioisotopes; analytical procedures. *Mailing Add:* 1245 Fair Oaks Pkwy Ann Arbor MI 48104

GLAZMAN, YULI M, b Kiev, Russia, Aug 7, 11; m 50; c 1. SURFACE CHEMISTRY, COLLOID CHEMISTRY. *Educ:* Kiev State Univ, MChem, 34; Ministry Higher Educ, USSR, PhD(chem), 39; USSR Acad Sci, DrChem, 59. *Hon Degrees:* Ministry Higher Educ, USSR, prof chem, 60. *Prof Exp:* From asst prof to assoc prof phys chem, Kiev State Univ, 37-41; assoc prof phys chem & colloid chem, Technol Inst Light Indust, Kiev, USSR, 45-60, prof & chmn dept, 60-74; vis prof chem, Northeastern Univ, 75-77; PROF CHEM ENG, TUFTS UNIV, 75- *Concurrent Pos:* Mem acad coun colloid chem, USSR Acad Sci, 60-74; mem acad coun disperse systs stability, Ukrainian SSR Acad Sci, 64-74; chmn, Precipitation & Crystallization Sect, USSR Mendeleyev Chem Soc, 71-75; consult, Polaroid Corp, 76-87 & Drew Chem Corp, 77-80; lectr, Ctr Prof Advan, 79- 88. *Mem:* Am Chem Soc. *Res:* Disperse systems stability; adsorption from solution; surfactants and their micelle formation in solution; surface forces in disperse systems; stabilization of coal-oil slurries; over 110 publications. *Mailing Add:* 1788 Beacon St Apt 3A Brookline MA 02146

GLEASON, ANDREW MATTEI, b Fresno, Calif, Nov 4, 21; m 59; c 3. MATHEMATICS. *Educ:* Yale Univ, BS, 42. *Hon Degrees:* AM, Harvard Univ, 53. *Prof Exp:* From asst prof to prof math, 50-69, HOLLIS PROF MATH & NATURAL PHILOS, HARVARD UNIV, 69- *Honors & Awards:* Cleveland Prize, 52. *Mem:* Nat Acad Sci; Am Math Soc (vpres, 62-63, pres, 81-82); Math Asn Am; Am Acad Arts & Sci; Am Philos Soc. *Res:* Topological groups; Banach algebras. *Mailing Add:* 110 Larchwood Dr Cambridge MA 02138

GLEASON, CLARENCE HENRY, b Montreal, Que, May 19, 22; m 50; c 3. MEDICINAL CHEMISTRY, RESEARCH ADMINISTRATION. *Educ:* McGill Univ, BSc, 44, PhD(chem), 47. *Prof Exp:* Lectr, McGill Univ, 44-45; res chemist, Charles E Frosst & Co, 47-62, chief res chemist, 62-69, exec asst to dir res, 69-82, dir res admin & planning, Merck Frosst Labs, 83-87; RETIRED. *Concurrent Pos:* Lectr, Sir George Williams Col, 47-48; dir, Belair Chem Ltd, Can, 64-65. *Mem:* Fel Chem Inst Can; Am Chem Soc; Soc Res Adminr; NY Acad Sci; Can Res Mgt Asn. *Res:* Novel and improved medicinal agents. *Mailing Add:* 7514 Mountbatten Rd Montreal PQ H4W 1J8 Can

GLEASON, EDWARD HINSDALE, JR, b North Adams, Mass, May 20, 27; m 50; c 4. ORGANIC CHEMISTRY. *Educ:* Northeastern Univ, BS, 53; State Univ NY, PhD(chem), 59. *Prof Exp:* Res chemist, Am Cyanamid Co, 53-56; res chemist, Koppers Co, Inc, 59-67, mgr latices res, 67-74; sr proj scientist, Arco/Polymers Inc, 74-80; sci adv, Polysar Latex, 80-88; SR RES ASSOC, BASF CORP, 88- *Mem:* Am Chem Soc; Tech Asn Pulp & Paper Indust. *Res:* Polymer chemistry; vinyl polymerization; copolymerization; emulsion polymerization. *Mailing Add:* BASF Corp 2200 Polymer Dr Chattanooga TN 37421

GLEASON, GALE R, JR, b Battle Creek, Mich, Oct 8, 27; m 49; c 4. ENVIRONMENTAL SCIENCE. *Educ:* Cent Mich Univ, BS, 50; Mich State Univ, MS, 51, PhD(fisheries, wildlife), 61. *Prof Exp:* From instr to assoc prof biol, Cent Mich Univ, 54-65; prof biol & chmn dept, Lake Superior State Col, 65-86, chmn div natural sci, 68-86; RETIRED. *Concurrent Pos:* AEC res fel, 55-57; consult, Mich State Dept Hwy & Transp, 75; proj dir, US Dept Interior Bur Outdoor Recreation, 75-76. *Mem:* Am Soc Limnol & Oceanog. *Res:* Trophodynamics of the St Marys River System, interconnecting waters of Lake Superior and Lake Huron. *Mailing Add:* 3989 Nicolet Rd Sault Ste Marie MI 49783

GLEASON, GEOFFREY IRVING, b Los Angeles, Calif, Apr 23, 23; m 50; c 2. RADIOCHEMISTRY. *Educ:* Univ Southern Calif, BChE, 47. *Prof Exp:* Asst, Am Potash & Chem Co, 48-49; res assoc, Abbott Labs, 49-52, mgr, Oak Ridge Div, 52-58, res scientist, 58-64, sr scientist, Oak Ridge Assoc Univs, 64-79, analytical chemist, Oak Ridge Nat Lab, 79-83, analytical chemist, Oak Ridge Assoc Univs, 83-87; RETIRED. *Concurrent Pos:* Consult, 87- *Mem:* Am Chem Soc; Sigma Xi. *Res:* Radiochemical separations; analytical chemistry; radiation measurement. *Mailing Add:* 127 Cumberland View Dr Oak Ridge TN 37830

GLEASON, JAMES GORDON, b Hammondsport, NY, Mar 24, 15; m 40; c 1. MECHANICAL ENGINEERING. *Educ:* Ala Polytech Inst, BS, 38; Univ Ark, MS, 54. *Prof Exp:* Asst instr, Ala Polytech Inst, 37-38, lab technician & instr, 38-40; from instr to asst prof mech eng, 40-45, assoc prof mech eng & head aeronaut div, 45-54, PROF MECH & AERONAUT ENG, UNIV ARK, FAYETTEVILLE, 54- *Concurrent Pos:* Plant engr, Forrest Park Canning Co, 70; vis res engr, Power Plant Unit, Boeing Airplane Co, Pratt & Whitney Aircraft & Main Burner Group. *Mem:* Am Soc Eng Educ; Am Soc Mech Engrs; Soc Automotive Engrs; Am Soc Heating, Refrig & Airconditioning Engrs; Nat Soc Prof Engrs. *Res:* Theory of aeronautics; internal combustion engines; aircraft engines; thermodynamics; refrigeration; airconditioning; machine design; compressible fluid flow. *Mailing Add:* 1139 Sunset Dr Fayetteville AR 72701

GLEASON, LARRY NEIL, b Independence, Ore, June 3, 39; div; c 2. PARASITOLOGY. *Educ:* Chico State Col, AB, 64; Univ NC, Chapel Hill, MSPH, 65, PhD(parasitol, zool), 69. *Prof Exp:* Res assoc pharmacol, Sch Med, Univ Fla, 69-70; from asst prof to assoc prof, 70-81, PROF BIOL & PARASITOL, WESTERN KY UNIV, 81- *Mem:* Am Soc Parasitol; Wildlife Dis Asn. *Res:* Taxonomy, life history, ecology and host-parasite relationship of intestinal helminths. *Mailing Add:* Dept Biol Western Ky Univ Bowling Green KY 42101

GLEASON, RAY EDWARD, b Burlington, Vt, Dec 17, 31; m 64; c 4. BIOSTATISTICS, GENETICS. *Educ:* Univ Vt, BSc, 54; Univ Mass, MS, 60; Tex A&M Univ, PhD(genetics, statist), 63. *Prof Exp:* Geneticist, Nedlar Farms, Inc, NH, 54-58; trainee, Div Math Biol, Harvard Med Sch & Peter Bent Brigham Hosp, 63-65; res assoc, 69-70, sr investr, Elliott P Joslin Res Lab, Harvard Med Sch & prin assoc med, Med Sch, 70-85; BIOSTATISTICIAN, CLIN RES CTRS, BRIGHAM & WOMEN'S HOSP, 85-, MASS INST TECHNOL, 87- *Concurrent Pos:* Res fel, Elliott P Joslin Res Lab, Harvard Med Sch, 65-68; res assoc, Dartmouth Med Sch, Norris Cotton Cancer Res Ctr, 85-86. *Mem:* Biomet Soc; Boston chap, Am Statist Asn. *Res:* Endocrinology-hypertension, nutrition and biostatistical analysis. *Mailing Add:* Endocrinol-Hypertension Div Brigham & Women's Hosp 221 Longwood Ave Boston MA 02115

GLEASON, ROBERT WILLARD, b Santiago, Chile, Feb 9, 32; US citizen; m 58; c 3. ORGANIC CHEMISTRY. *Educ:* Middlebury Col, BA, 54, MS, 56; Mass Inst Technol, PhD(org chem), 60. *Prof Exp:* From instr to assoc prof, Middlebury Col, 60-72, from actg chmn to chmn dept, 69-77, dean sci, 82-84, chmn, Div Natural Sci, 77-82, dean fac, 84-88, PROF CHEM, MIDDLEBURY COL, 72- *Concurrent Pos:* Vis fac res assoc, Univ Colo, 67-68; vis prof, Univ Cologne, 74-75; C Nuirmiu, VT-EPSCOR, 84- *Mem:* Am Chem Soc; Sigma Xi. *Res:* Oxidation of 1, 1-disubstituted hydrazines; reductions of N-nitrosamines. *Mailing Add:* Dept Chem Middlebury Col Middlebury VT 05753

GLEASON, THOMAS JAMES, b Louisville, Ky, Sept 23, 41; m 69; c 2. ELECTRO-OPTICS, SOLID STATE PHYSICS. *Educ:* Johns Hopkins Univ, BA, 63, PhD(physics), 68. *Prof Exp:* Res asst low energy nuclear physics, Johns Hopkins Univ, 66-68; physicist basic res, Harry Diamond Labs, US Army, 69-71, res physicist appl physics, 71-74, supvry physicist, electro-optic systs, 74-77, br chief, 77-80; prog mgr, Syst Planning Corp, 80-82; PRES, GLEASON RES ASSOCS, INC, 82- *Mem:* Am Phys Soc; Sigma Xi. *Res:* Military electro-optic systems; non-linear optical systems; laser devices. *Mailing Add:* 10375 Launcelot Lane Columbia MD 21044

GLEATON, HARRIET ELIZABETH, b Altoona, Pa, Aug 25, 37. ANESTHESIOLOGY. *Educ:* Temple Univ, MD, 62. *Prof Exp:* Instr anesthesia, Sch Med, Univ Pa, 67-69; asst prof, Pritzker Sch Med, Univ Chicago, 69-71; ASSOC PROF ANESTHESIA, SCH MED, UNIV OKLA, 71-, CHIEF ANESTHESIA SECT & MED DIR RESPIRATORY THER, UNIV HOSP, 71-; ANESTHESIOLOGIST, BARTLESVILLE MED PART CTR, 85- *Concurrent Pos:* USPHS fel anesthesia res, Sch Med, Univ Pa, 65-67; staff anesthesiologist, Hosp Univ Pa, 67-69; asst attend physician, Michael Reese Hosp & Med Ctr, 69-71, assoc attend physician, 71; consult anesthesia, Vet Admin Hosp, Oklahoma City, 71-; med dir respiratory ther, Oscar Rose Jr Col, 78-85. *Mailing Add:* 3400 E Frank Phillips Ste 308 Bartlesville Med Parks Ctr Bartlesville OK 74006

GLEAVES, EARL WILLIAM, b Miami, Okla, Apr 3, 30; m 50; c 3. ANIMAL NUTRITION. *Educ:* Okla State Univ, BS, 53, MS, 61, PhD(animal nutrit), 65. *Prof Exp:* Instr poultry nutrit, Okla State Univ, 63-64; from asst prof to assoc prof, 64-73, PROF POULTRY EXTEN & NUTRIT, UNIV NEBR, LINCOLN, 73- *Concurrent Pos:* Eval team, Syrian Animal Sci Prog, 78-79; consult, US Feed Grains Coun, Italy, Japan, South Korea, Taiwan & Poland, 82, 85 & 87. *Mem:* Poultry Sci Asn; World Poultry Sci Asn. *Res:* Interrelationships of nutrition and physiology in the domestic fowl. *Mailing Add:* C206 Animal Sci Univ Nebr Lincoln NE 68583-0908

GLEAVES, JOHN THOMPSON, b Louisville, Ky, May 2, 46; m 68; c 2. PHYSICAL CHEMISTRY. *Educ:* Univ Louisville, BS, 68; Univ Ill, MS, 73, PhD(phys chem), 75. *Prof Exp:* SR RES CHEMIST, MONSANTO POLYMERS & PETROCHEMICALS, 75- *Mem:* Am Chem Soc. *Res:* Molecular dynamics; application of multiphoton spectroscopic techniques to the study of molecular dynamics; reactions at gas-solid interfaces; infrared and visible chemiluminescence. *Mailing Add:* RR 1 Box 171-A2 Foley MO 63347-9728

GLEBE, BRIAN DOUGLAS, b Kitchener, Ont, Jan 14, 48; m 73; c 3. BIOLOGY. *Educ:* Univ Guelph, BSc, 71; McGill Univ, PhD(biol), 77. *Prof Exp:* Nat Res Coun fel biol, 77-78, researcher & fish culturist, 78-85, AQUACULTURE COORDR, HUNTSMAN MARINE LAB, N AM SALMON RES LAB, 85- *Mem:* Can Soc Zoologists; Am Fisheries Soc. *Res:* Salmon ecology, genetics and aquaculture. *Mailing Add:* Huntsman Marine Lab PO Box 488 St Andrews NB E0G 2X0 Can

GLECKMAN, PHILIP LANDON, b Boston, Mass, Nov 16, 61; m 87. SOLID STATE LASERS, SOLAR ENERGY. *Educ:* Mass Inst Technol, BS, 84; Univ Chicago, MS, 86, PhD(physics), 88. *Prof Exp:* Res asst, 85-88, RES ASSOC, ENRICO FERMI INST, UNIV CHICAGO, 88- *Concurrent Pos:* Consult, Philips Corp, 87-88. *Mem:* Optical Soc Am; Sigma Xi. *Res:* Novel concepts in optical design called nonimaging optics to establish a record high concentration of sunlight. *Mailing Add:* 5308 S Blackstone Apt 3 Chicago IL 60615

GLEDHILL, BARTON L, b Philadelphia, Pa, Sept 29, 36; m 59; c 2. VETERINARY MEDICINE, CYTOLOGY. *Educ:* Pa State Univ, BS, 58; Univ Pa, VMD, 61; FRVCS, Royal Vet Col Sweden, 62, PhD(vet reproduction), 66; Am Col Theriogenologists, dipl, 73. *Prof Exp:* Am Vet Med Res Trust res fel, 61-64; NIH fel, 64-66; from asst prof to assoc prof reproduction, Sch Vet Med, Univ Pa, 69-72; sr staff scientist, 73-80, dep div leader, 80-82, DIV LEADER, BIOMED SCI DIV, LAWRENCE LIVERMORE NAT LAB, UNIV CALIF, 82- *Concurrent Pos:* Docent, Royal Vet Col Sweden, 66-; NIH spec fel, 68-70; ad hoc mem, Study Sect, NIH, 85- & Study Sect Rev Bd, USDA, 86-; consult, Cytogam, 88-90. *Honors & Awards:* Ortho Lectr, Mc Master Univ, 84; Tap lectureship, Am Fertil Soc, 90. *Mem:* Am Vet Med Asn; Soc Study Reproduction; Am Col Theriogenology (pres, 75-76); Environ Mutagen Soc; Soc Analytical Cytol (pres elect, 90-91). *Res:* Male gametogenesis and fertilization; DNA adduct detection; gender preselection; accelerator mass spectrometry. *Mailing Add:* Div Biomed Sci L-452 Lawrence Livermore Nat Lab Univ Calif Livermore CA 94550

GLEDHILL, ROBERT HAMOR, b South Kingstown, RI, Feb 2, 31; m 54; c 3. POULTRY NUTRITION, MANAGEMENT. *Educ:* Univ RI, BS, 52, MS, 54; Purdue Univ, PhD(poultry nutrit), 57. *Prof Exp:* Nutritionist, Archer-Daniels-Midland Co, 57-60; res assoc, Corn Prod Co, 60-61; res nutritionist, Hales & Hunter Co, Ill, 61-66; sr poultry nutritionist, Cent Soya Co, Inc, 66-89; RETIRED. *Mem:* Poultry Sci Asn. *Res:* Nutritional research on broilers, replacement birds, layers and turkeys. *Mailing Add:* Rte 4 Box 158 Oakwood Decatur IN 46733

GLEDHILL, RONALD JAMES, b Cleveland, Ohio, Nov 23, 14; m 42; c 4. CHEMISTRY, INFORMATION SCIENCE. *Educ:* Ohio State Univ, BA, 39; Univ Minn, PhD(phys chem), 46. *Prof Exp:* Asst chem, Univ Minn, 39-43; res chemist, Eastman Kodak Co, 46-67, res assoc, 67-76; RETIRED. *Mem:* Assoc mem Am Chem Soc. *Res:* Color and constitution of dyes; color photography research; particle-size measurement; technical information retrieval systems. *Mailing Add:* 166 Dake Ave Rochester NY 14617

GLEDHILL, WILLIAM EMERSON, b Baltimore, Md, Aug 1, 41; c 2. MICROBIOLOGY, BIOCHEMISTRY. *Educ:* Univ Del, BA, 63, MS, 66; Pa State Univ, PhD(microbiol, biochem), 69. *Prof Exp:* Res microbiologist, Tenneco Chem, Inc, 69-70; SCI FEL, MONSANTO CO, 70- *Mem:* Am Soc Microbiol; Soc Indust Microbiol; Soc Environ Toxic Chem; Soap & Detergent Asn; Chem Mfg Asn. *Res:* Microbial physiology and ecology; industrial application of novel microorganisms; environmental impact of synthetic organic chemicals; environmental sciences. *Mailing Add:* 2041 King Arthur Ct St Louis MO 63146-6016

GLEESON, AUSTIN M, b Philadelphia, Pa, Apr 5, 38; m 60; c 2. ELEMENTARY PARTICLE PHYSICS. *Educ:* Drexel Inst, BSc, 60; Univ Pa, MSc, 63, PhD(physics), 65. *Prof Exp:* Comput designer, Radio Corp Am, 56-58 & 60-62; tech rep, Burroughs Corp, 58-59; instr physics, Drexel Inst, 59-60 & 62-63; res asst, Univ Pa, 63-65; res assoc, Syracuse Univ, 65-67, asst prof, 67-69; from asst prof to assoc prof, 69-77, PROF PHYSICS, UNIV TEX AUSTIN, 77- *Concurrent Pos:* Assoc dean, Col Nat Sci, 80-85. *Mem:* Am Inst Physics. *Res:* Strong interaction elementary particle phenomenology. *Mailing Add:* Dept Physics Univ Tex Austin TX 78712

GLEESON, JAMES NEWMAN, b Louisville, Ky, Mar 30, 40; m 74. PHYSICAL ORGANIC CHEMISTRY, POLYMER CHEMISTRY. *Educ:* Bellarmine Col, AB, 62; Xavier Univ, MS, 64; Univ Cincinnati, PhD(phys org chem), 69. *Prof Exp:* Res chemist polymer chem, 68-77, RES CHEMIST PHYS ORG & POLYMER CHEM, AMOCO CHEM CORP, STANDARD OIL CO, IND, 77- *Mem:* Am Chem Soc; Soc Advan Mat & Process Eng. *Res:* Advanced structural materials; non-biological solar photochemistry; technology forecasting; research planning. *Mailing Add:* 1441 Trilium Lane Naperville IL 60540

GLEESON, RICHARD ALAN, b Pittsburgh, Pa, Sept 26, 47; m 69; c 1. CHEMORECEPTION. *Educ:* Franklin & Marshall Col, BA, 69; Pa State Univ, MS, 72; Col William & Mary, PhD(marine sci), 78. *Prof Exp:* Postdoctoral fel, Monell Chem Senses Ctr, 78-80, asst mem, 80-83; RES SCIENTIST, WHITNEY LAB, UNIV FLA, 83- *Concurrent Pos:* Vis asst prof, Dept Zool, Univ Fla, 80-81. *Mem:* Am Soc Zoologists; Soc Neurosci; Asn Chemoreception Sci. *Res:* Chemosensory neurobiology and behavior of aquatic organisms, particularly crustaceans; pheromonal communication systems; morphology, neurophysiology and biochemistry/cytochemistry of chemoreceptors. *Mailing Add:* Whitney Lab Univ Fla 9505 Ocean Shore Blvd St Augustine FL 32086-8623

GLEESON, THOMAS ALEXANDER, b New York, NY, Aug 11, 20; m 42; c 2. METEOROLOGY. *Educ:* Harvard Univ, BS, 46; NY Univ, MS, 47, PhD(meteorol), 50. *Prof Exp:* Asst, NY Univ, 47-49; from asst prof to assoc prof, 49-59, PROF METEOROL, FLA STATE UNIV, 59- *Concurrent Pos:* State climatologist, Florida, 84- *Mem:* AAAS; fel Am Meteorol Soc; Am Geophys Union. *Res:* General meteorology and climatology; probability and statistical relations in meteorology. *Mailing Add:* Dept Meteorol Fla State Univ Tallahassee FL 32306

GLEGHORN, G(EORGE) J(AY), (JR), b San Francisco, Calif, May 27, 27; m 48; c 3. ELECTRICAL ENGINEERING, APPLIED MATHEMATICS. *Educ:* Univ Colo, BS, 47; Calif Inst Technol, MS, 48, PhD(elec eng), 55. *Prof Exp:* Mem tech staff, TRW Space Technol Labs, 54-59, head launch opers, Pioneer I & V, Explorer VI Satellites, 59-60, head controls dept, 58-60, prog dir, Able 5 Lunar Probe Satellite, 60-61, prog dir, Orbit Geophys Observ, 61-65, asst dir opers, Space Systs Prog Mgt, 65-66, mgr spacecraft opers, Space Vehicles Div, 66-70, mgr, Defense Space Systs Opers, 70-73, asst gen mgr eng, Space Syst Div, TRW Defense & Space Systs Group, 74-81, VPRES & CHIEF ENGR, TRW SPACE & TECHNOL GROUP, 81- *Mem:* Fel Am Inst Aeronaut & Astronaut; Inst Elec & Electronics Engrs; Sci Res Soc Am; Sigma Xi; Nat Acad Eng. *Res:* Control system and digital computer analysis and engineering; spacecraft and ballistic missile system engineering. *Mailing Add:* 28850 Crestridge Rd Rancho Palos Verdes CA 90274

GLEICH, CAROL S, b Kewanee, Ill, Jan 18, 35. HEALTH SCIENCES EDUCATION. *Educ:* Univ Iowa, BA, 58, MS, 67, PhD(health sci educ), 72. *Prof Exp:* From instr to asst prof health, Col Med, Univ Iowa, 71-77, dir med technol prog, 72-77; CHIEF, SPEC PROJ & DATA ANALYSIS, DIV MED & EXEC SECY COUN GRAD MED ED, BUR HEALTH PROFESSIONS, HEALTH RESOURCES & SERV ADMIN, HEALTH & HUMAN SERV, 77- *Concurrent Pos:* Adj assoc prof, Dept Path, Sch Med, Univ Md, 80-; assoc ed, Am J Allied Health; chief, Res Develop Div Med, Health Res & Serv Admin. *Mem:* Am Soc Clin Path; Am Soc Allied Health Prof; Am Soc Med Technol; Nat Coun Int Health. *Res:* Clinical laboratory manpower; international consulting on manpower and allied health education; physician manpower and medical education. *Mailing Add:* Bur Health Professions Parklawn Bldg R-4C-16 Rockville MD 20857

GLEICH, GERALD J, b Escanaba, Mich, May 14, 31; m 56, 76; c 7. IMMUNOLOGY, INTERNAL MEDICINE. *Educ:* Univ Mich, BA, 53, MD, 56; Am Bd Internal Med, dipl, 65; Am Bd Allergy, dipl, 66. *Prof Exp:* Intern, Philadelphia Gen Hosp, Pa, 56-57; resident internal med, Jackson Mem Hosp, Miami, Fla, 59-61; trainee allergy & immunol, Med Ctr, Univ Rochester, 61-63, instr med & microbiol, Sch Med & Dent, 63-65; from instr to asst prof med & microbiol, 65-73, assoc prof med, microbiol & immunol, Mayo Grad Sch Med, 73-77, chmn, Dept Immunol, 82-90, PROF MED & IMMUNOL, MAYO MED SCH, UNIV MINN, 77- *Concurrent Pos:* Consult, Methodist Hosp & St Mary's Hosp, Rochester, Minn, 65-; NIH res grants, 66-; bd sci counrs, Nat Int Allergy & Infectious Dis, 81-83. *Mem:* Am Acad Allergy; Am Fedn Clin Res; Am Asn Immunol; fel Am Col Physicians; Am Soc Clin Invest. *Res:* Antibody formation; hypersensitivity; eosinophils. *Mailing Add:* Dept Immunol Mayo Found Rochester MN 55905

GLEICHER, GERALD JAY, b Brooklyn, NY, Jan 31, 39; m 66; c 3. ORGANIC CHEMISTRY. *Educ:* Brooklyn Col, BS, 59; Univ Mich, MS, 61, PhD(chem), 63. *Prof Exp:* Instr chem, Univ Mich, 63; res assoc, Univ Tex, 64-65 & Princeton Univ, 65-66; asst prof, 66-70, assoc prof, 70-80, PROF CHEM, ORE STATE UNIV, 80- *Mem:* Am Chem Soc; Royal Soc Chem. *Res:* Free radical reaction mechanisms; quantum chemistry; linear free energy relationships; steric effects. *Mailing Add:* Dept Chem Ore State Univ Corvallis OR 97331

GLEIM, CLYDE EDGAR, b Wheelersburg, Ohio, July 22, 13; m 49; c 3. ORGANIC CHEMISTRY, OPERATIONS RESEARCH. *Educ:* Ohio Univ, BS, 35; Ohio State Univ, MS, 38; Pa State Univ, PhD(org chem), 41. *Prof Exp:* Asst inorg chem, Ohio Univ, 35; asst shift chemist, Sharple Solvents Co, Mich, 35-36, res chemist, 36; asst Ohio State Univ, 37-38 & Pa State Univ, 39-41; from res chemist to sr res chemist, Goodyear Tire & Rubber Co, 41-58, head plastics polymerization sect, 58-65 & polymer characterization & specialty polymers sect, 65-67, info analyst, Polyester Div, Fiber Tech Ctr, 67-78; CONSULT 78- *Concurrent Pos:* Instr night col, Univ Akron, 48-50. *Mem:* Am Chem Soc. *Res:* Amyl naphthalenes and amines; pure hydrocarbons; urethans and derivatives; vinyl, allyl, urea-formaldehyde and phenolformaldehyde resins and rubber adhesives; molding and laminating resins; high molecular weight addition and condensation polymers; mostly saturated polyester polymers and copolymers. *Mailing Add:* 2501 Kensington Rd Akron OH 44333

GLEIM, PAUL STANLEY, b Wheelersburg, Ohio, Dec 20, 23; m 50; c 2. SOLID STATE CHEMISTRY, SEMICONDUCTORS. *Educ:* Ohio Univ, BS, 49. *Prof Exp:* Chemist, Res Div, Armco Steel Corp, 51-57; engr, Semiconductor Components Div, Tex Instruments Inc, 57-59, group leader, Germanium Develop Dept, 59-62 & Semiconductor Res & Develop Lab, 62-69, mgr advan prod, Chem Mat Div, 69-72, mgr eng silicon mat, 73-77, mgr eng, Gadolinium Gallium Garnet Components Group, 77-81; CONSULT, SEMICONDUCTOR INDUST, 82- *Mem:* Electrochem Soc. *Res:* Semiconductor crystal growth; vapor phase deposition of semiconductors and insulators; processing of semiconductors, especially technology of cutting, lapping, mechanical and chemical polishing; semiconductor fabrication; engineering and manufacturing of advanced silicon materials systems for integrated circuits; processing of gadolinium gallium garnet, especially technology of slicing, lapping, edge beveling, stress relieving and polishing for magnetic bubble memories. *Mailing Add:* Eight Cumberland Pl Richardson TX 75080

GLEIM, ROBERT DAVID, b Philadelphia, Pa, Nov 11, 46; m 70. BIO-ORGANIC CHEMISTRY. *Educ:* Elizabethtown Col, BS, 68; Brown Univ, PhD(org chem), 73. *Prof Exp:* Fel chem, Case Western Reserve Univ, 74-75; fel pharm, Univ Wis-Madison, 75-77; RES CHEMIST, CORP NEW VENTURES, ROHM & HAAS CO, 77- *Mem:* Am Chem Soc; AAAS; Sigma Xi. *Res:* Application of synthetic techniques to the total synthesis of natural products development of electrically conductive polymers (polyacetylene) with improved stability and enhanced transport properties particularly adriamycin, pre-prostaglandin endoperoxide and bicyclic alkaloids. *Mailing Add:* 1986 Street Rd New Hope PA 18938

GLEISER, CHESTER ALEXANDER, VETERINARY & COMPARATIVE PATHOLOGY. *Educ:* Univ Pa, VMD, 40. *Prof Exp:* PROF PATH & VET PATHOLOGIST, DEPT PATH, UNIV TEX HEALTH SCI CTR, 80- *Mailing Add:* 5403 Laneashire San Antonio TX 78230

GLEISSNER, GENE HEIDEN, b Brooklyn, NY, Feb 1, 28; m 67. COMPUTER APPLICATION, SYSTEMS DEVELOPMENT. *Educ:* Columbia Univ, AB, 47, MA, 48. *Prof Exp:* Lectr math, Columbia Univ, 48-50; dep head programming & coding br, US Naval Weapons Lab, 50-54, head comput div, 55-63, asst dir comput, Comput & Analysis Lab, 63-65, assoc tech dir comput, Math & Logistics, David Taylor Naval Ship Res & Develop Ctr, 65-86, spec asst to dir navy Labs Interlab Comput, 74-86; vpres, Integrated Microcomput Systs, Inc, 86-87; PRES, GLEISSNER ASSOCS, 87- *Mem:* Fel AAAS; Math Asn Am; Inst Elec & Electronics Eng, Comput Soc; Soc Indust & Appl Math; Asn Comput Mach. *Res:* Computer and information systems technology and applications; development and management of computer networks; logistics systems analysis; office automation; systems design. *Mailing Add:* 10532 Farnham Dr Bethesda MD 20814-2222

GLEIT, CHESTER EUGENE, analytical chemistry, physical chemistry; deceased, see previous edition for last biography

GLEITER, MELVIN EARL, b Alma, Wis, June 9, 26; m 55; c 2. BIOCHEMISTRY, ENVIRONMENTAL CHEMISTRY. *Educ:* Wartburg Col, BA, 51; Purdue Univ, MS, 56, PhD(biochem), 58. *Prof Exp:* Asst biochem, Purdue Univ, 51-52 & 54-58; sr res biochemist, Monsanto Co, 58-64; from asst prof to assoc prof chem, 64-72, PROF CHEM, UNIV WIS-EAU CLAIRE, 72-, CHMN DEPT, 88- *Mem:* Am Chem Soc; AAAS; Sigma Xi. *Res:* Effects of pesticides on environment; air and water pollution analysis; energy and environmental effects; nature conservancy. *Mailing Add:* 2816 Irene Dr Eau Claire WI 54701

GLEMBOTSKI, CHRISTOPHER CHARLES, b Sommerville, NJ, Dec 13, 52; m 76. ENZYMOLOGY, TISSUE CULTURE. *Educ:* Calif Polytecch State Univ, San Luis Obispo, BS, 73; Univ Calif, Los Angeles, PhD(biochem), 79. *Prof Exp:* Instr biochem & physiol, Univ Colo Med Sch, 80-83; asst prof, pharmacol, Univ Pa Sch Med, 83-86; PROF BIOL, SAN DIEGO STATE UNIV, 86- *Concurrent Pos:* Res fel, Univ Colo Med Sch, 79-83. *Mem:* Sigma Xi; NY Acad Sci. *Res:* Post-translational processing of pro-adrenocorticotropic hormone/endorphin in the pituitary. *Mailing Add:* Dept Biol San Diego State Univ San Diego CA 92182

GLEN, ROBERT, entomology; deceased, see previous edition for last biography

GLENDE, ERIC A, JR, b Fergus Falls, Minn, Nov 12, 38; m 69, 79; c 4. TOXICOLOGY. *Educ:* Concordia Col, Moorhead, Minn, BA, 60; Univ NDak, MS, 62, PhD(biochem), 66. *Prof Exp:* Res assoc path, 66-67, physiol, 67-68, asst prof, 68-75, ASSOC PROF PHYSIOL, SCH MED, CASE WESTERN RESERVE UNIV, 75- *Mem:* Am Physiol Soc; Oxygen Soc. *Res:* Toxic and nutritional liver injury. *Mailing Add:* Dept Physiol Sch Med Case Western Reserve Univ 2109 Adelbert Rd Cleveland OH 44106

GLENDENING, NORMAN WILLARD, b Chicago, Ill, Nov 1, 13; m 40; c 2. APPLIED CHEMISTRY, SURFACE COATINGS. *Educ:* St Olaf Col, BA, 36; NDak State Univ, MS, 40. *Prof Exp:* Asst, Univ Wis, 37-38; asst tech dir, Albron Pigments Div, Aluminum Co Am, 40-46; res assoc alkyd resins, Miner Labs, Ill, 47-48; formulator, Appliance Finishes Res & Develop, Midwest Indust Div, Glidden Co, Glidden-Durkee Div, SCM Corp, 48-53, group leader, 53-62, sect supvr, 62-64, mgr qual control, Midwest Chem Coatings, 64-76; RETIRED. *Mem:* Am Chem Soc; Fedn Socs Coatings Technol. *Res:* Volatile by-products of air-dried linseed oil films; organic finishing systems; protective coatings. *Mailing Add:* 1310 Lynn St Heber Springs AR 72543-9545

GLENDENNING, NORMAN KEITH, b Galt, Can, Jan 17, 31; m 63; c 2. THEORETICAL NUCLEAR PHYSICS. *Educ:* McMaster Univ, BSc, 54, MSc, 55; Ind Univ, PhD(physics), 59. *Prof Exp:* Physicist, 58-63, head nuclear theory group, 65-83, SR STAFF SCIENTIST, NUCLEAR SCI DIV, LAWRENCE BERKELEY LAB, UNIV CALIF, BERKELEY, 63-, CHMN DIV STAFF COMT, 78- *Concurrent Pos:* Physicist, Lab Nuclear Physics, Orsay, France, 62-63 & Saclay Nuclear Res Ctr, 62-63; mem, Int Conf Nuclear Struct, Kingston, Ont, 60, Padua, Italy, 62, Gatlinberg, Tenn, 66, Varenna, Italy, 67, Dubna, Russia, 68, Montreal, Que, 69, Argonne, 72, Nashville, 74, Tokyo, Bombay & Calcutta, 77, Hirschegg, Austria, Madison, Wis & Bangalore, India, 78, Roscoff, France, 79, Vancouver, Can, 79, Hirschegy, Austria, 80, 81, 85, 86 & 88, Balaton, Hungary, Fontevraud, France, & Florence, Italy, 83, Darmstadt, WGermany & Helsinki, 84, Kyoto, Japan, 87; lectr, Radium Inst, Orsay, France, 62-63; guest prof, Univ Frankfurt, 69, Lab Physique Theorique, France, 67, 73, 76, 79, Univ Pierre & Marie Curie, 80, 81, 83, 85 & 87; chmn, Sci Prog Comt Conf Intermediate Energy Heavy Ion Symp, Berkeley, Calif, 75 & Third Summer Study High Energy Heavy Ion Physics, 76. *Mem:* Fel Am Phys Soc. *Res:* Theoretical research in the field of direct nuclear reactions; two-nucleon transfer reactions; microscopic theory of inelastic scattering; nuclear structure; nuclear collisions at relativistic energies; properties and phase transitions in super dense nuclear matter, neutron stars. *Mailing Add:* Lawrence Berkeley Lab Univ Calif Berkeley CA 94704

GLENISTER, BRIAN FREDERICK, b Albany, Australia, Sept 28, 28; nat; m 56; c 3. PALEOBIOLOGY. *Educ:* Univ Western Australia, BSc, 49; Univ Melbourne, MSc, 53; Univ Iowa, PhD(geol), 56. *Prof Exp:* Asst, Univ Melbourne, Australia, 50-54 & Univ Iowa, 54-56; sr lectr geol, Univ Western Australia, 56-59; from asst prof to prof, 59-74, chmn dept, 68-74, A K MILLER PROF GEOL, UNIV IOWA, 74- *Mem:* AAAS; Paleont Soc; Soc Econ Paleontologists & Mineralogists; Geol Soc Am; Brit Paleont Asn; Sigma Xi. *Res:* Fossil cephalopods; biostratigraphy. *Mailing Add:* Box 148B Rte 2 North Liberty IA 52317

GLENISTER, PAUL ROBSON, b Chicago, Ill, June 18, 18. FOOD SCIENCE. *Educ:* Chicago Teachers Col, BEd, 39; Univ Chicago, SM, 41, PhD(bot), 43. *Prof Exp:* Teacher biol & bot, Wilson Jr Col, 43-50; researcher, J E Siebel Sons' Co, 50-82; RETIRED. *Mem:* AAAS; Am Soc Brewing Chemists. *Res:* Beer and brewing. *Mailing Add:* 6432 St Calif Ave Chicago IL 60629-2836

GLENN, ALAN HOLTON, b Brigham City, Utah, May 9, 50; m 74; c 3. INSTRUMENTATION, VALVE NOISE. *Educ:* Brigham Young Univ, BS, 74, ME, 75, PhD, 87. *Prof Exp:* PROD MGR RES & DEVELOP, VALTEK, INC, 75- *Mem:* Assoc mem Am Soc Mech Eng. *Res:* Noise generated in control valves. *Mailing Add:* PO Box 577 Salem UT 84653

GLENN, ALFRED HILL, b Yonkers, NY, June 3, 21; m 47. MARINE METEOROLOGY, PHYSICAL OCEANOGRAPHY. *Educ:* Univ Wis, BS, 42; NY Univ, MS, 43. *Prof Exp:* Jr engr, Holway & Cochrane, Tulsa, 39-41; mem staff, Chicago Bridge & Iron Co, 42; meteorologist, Air Weather Serv, Washington, DC, 45-46; CONSULT METEOROLOGIST & PRES, A H GLENN & ASSOCS, 46- *Concurrent Pos:* Captain, US Army Air Force, 43-45; panel mem oceanog, Res & Develop Bd, 47-48, mem subcomt on oceanog, joint meteorol comt; consult, Am Meteorol Soc. *Honors & Awards:* Appl Meteorol Award, Am Meteorol Soc, 62; Outstanding Serv to Meteorol Award, Am Meteorol Soc, 71. *Mem:* Fel Am Meteorol Soc; fel Royal Meteorol Soc. *Res:* Engineering meteorology and oceanography; hurricane and severe storm forecasting; decision theory applications in meteorological-oceanographic forecasting. *Mailing Add:* New Orleans Lakefront Airport PO Box 26337 New Orleans LA 70126

GLENN, BARBARA PETERSON, b Lincoln, Nebr; m 75; c 1. RUMINANT NUTRITION, FORAGE UTILIZATION. *Educ:* Univ Ky, BS, 75, PhD(animal sci), 80. *Prof Exp:* Fac res assoc, Univ Md & USDA, 80-82; RES ANIMAL SCIENTIST, USDA, 82- *Concurrent Pos:* Secy, Northeast Sect, Am Soc Animal Sci, dairy livestock comt, 89. *Mem:* Am Soc Animal Sci; Am Dairy Sci Asn; Am Forage & Grassland Coun; Coun Agr Sci & Technol. *Res:* Evaluation of nitrogen and forage utilization by ruminants, specifically dairy cattle; intake, nutrient digestibility and metabolism, rates of passage and nutrient flows to the lower gut have been measured in ruminants consuming harvested forages. *Mailing Add:* Agr Res Serv USDA Ruminant Nutrit Lab Bldg 200 Rm 101A BARC-E Beltsville MD 20705

GLENN, BERTIS LAMON, b Duncan, Okla, Sept 9, 22; m 54. VETERINARY PATHOLOGY, COMPARATIVE PATHOLOGY. *Educ:* Okla State Univ, DVM, 52, MS, 61; Univ Okla, PhD(comp path), 63; Am Col Vet Path, dipl. *Prof Exp:* From asst prof to assoc prof, 53-67, PROF VET PATH, OKLA STATE UNIV, 67- *Concurrent Pos:* NSF sci fac fel, 61-62; NIH spec fel, 62-63. *Mem:* AAAS; Am Vet Med Asn; NY Acad Sci; Sigma Xi. *Res:* Photosensitivity diseases; liver function; congenital porphyria; genital diseases of the dog caused by hormonal disturbances; cytogenetics of domestic animals. *Mailing Add:* Dept Vet Path Okla State Univ Stillwater OK 74075

GLENN, FURMAN EUGENE, b Gastonia, NC, Nov 3, 44; m 68; c 3. ORGANIC POLYMER CHEMISTRY. *Educ:* NC Col, Durham, BS, 66; Wayne State Univ, PhD(org chem), 73. *Prof Exp:* Res chemist, 72-81, area supt, 81-89, SR RES ASSOC, E I DU PONT DE NEMOURS & CO, INC, 89- *Mem:* Am Chem Soc; Sigma Xi. *Res:* Development and study of nonconventional methods for synthesis of elastomeric block copolymers containing polyurethane segments; development of new methods for control of free radical emulsion polymerizations leading to hydrocarbon synthetic elastomers; study of colloid properties of latices employed in binder and adhesive applications; development of new heat resistant hydrocarbon elastomers; management of diene monomer synthesis and polymerization. *Mailing Add:* 4713 Nottinghamshire Dr Louisville KY 40299

GLENN, GEORGE R(EMBERT), b Anderson, SC, Sept 2, 23; m 44; c 3. CIVIL & GEOTECHNICAL ENGINEERING. *Educ:* Clemson Univ, BCE, 43, MSCE, 57; Southern Baptist Theol Sem, MDiv, 55; Iowa State Univ, PhD, 63. *Prof Exp:* Field engr, Daniel Construct Co, 46-47; resident engr, J E Sirrine Co, 48-50; consult engr, 50-51; instr eng, Speed Sci Sch, Louisville, 51-55; prof eng & head dept, Bluefield Col, 55-57 & Wingate Col, 57-58; asst prof appl sci, Southern Ill Univ, 58-64; asst dean, Col Eng, Rutgers Univ, 64-69, prof civil & environ eng, 64-87, assoc dean, Col Eng, 69-70; vis prof eng, Clemson Univ, 88-90; RETIRED. *Concurrent Pos:* Design engr, 54, 56, 58; with NSF res participation prog, Iowa State Univ, 60, Ford Found grant, 61-63; res civil engr, US Naval Civil Eng Lab, Calif, 64; pres, Am Field Serv, NJ; Mem, Lime-Pozzolan Stabilization Comn, Soils & Geol Div & Physicochem Phenomena In Soils Comn, Transp Res Bd, Nat Res Coun, 80-86 & Stabilization of Soil & Rock with Chemicals Coun, 87-; on leave, head Civil Eng Technol Dept, Greenville, SC, 80-81. *Mem:* Am Soc Civil Engrs; Am Soc Eng Educ; Sigma Xi. *Res:* Advanced geotechnical engineering; evaluation of soil properties; physico-chemical phenomena in soil stabilization; effect of soil properties upon use as an engineering material in foundations; structures. *Mailing Add:* 134 Chestnut Pl Arden NC 28704

GLENN, JAMES FRANCIS, b Lexington, Ky, May 10, 28; m 48; c 4. UROLOGY. *Educ:* Univ Rochester, BA, 50; Duke Univ, MD, 53; Nat Bd Med Examrs, dipl, 54; Am Bd Urol, cert, 62; FRCS, 87. *Prof Exp:* Intern gen surg, Peter Bent Brigham Hosp, Boston, 52-54; resident urol surg, Med Ctr, Duke Univ, 56-59, instr, Sch Med, 58-59; asst prof, Yale Univ, 59-61; assoc prof, Bowman Gray Sch Med, 61-63; prof & chief urol, Med Ctr, Duke Univ, 63-80; prof surg & dean, Sch Med, Emory Univ, 80-83; actg dean, Mt Sinai Sch Med, 83-84, prof urol, 83-87, pres, Med Ctr, Sch Med & Hosp, 83-87; PROF SURG, COL MED, UNIV KY, 87- *Concurrent Pos:* Assoc surg, Yale-New Haven Hosp, 59-61; attend urologist, Vet Admin Hosp, W Haven, Conn, 59-61, NC Baptist Hosp, 61-63, Duke Univ Med Ctr, 63-80 & Mt Sinai Hosp, New York, 83-87; consult urologist, Conn State Hosp, 60-61, Vet Admin Med Ctr, Durham, 59-61, Lincoln Hosp, Babies Hosp, Watts Hosp & Durham County Gen Hosp, 63-80, Womack Army Hosp, Ft Bragg, NC, 67-80, Vet Admin Med Ctr, Oteen, NC, 69-80, Atlanta, Ga, 80-83, Bronx, NY, 83-87 & Lexington, Ky, 87-, Comn Handicapped Children, Commonwealth of Ky, 87- & Shriners' Hosp for Crippled Children, Lexington, Ky, 88-; prin investr, Cancer Chemotherapy Prog, NIH, 61-63, Adjuvant Bladder Cancer Study, 64-70; consult, Div Med Sci, Nat Acad Sci-Nat Res Coun, 67-68, mem, Comt Genito-Urinary Syst, 68-71; vis prof, Great Ormond St Hosp for Sick Children & Inst Child Health, Univ London, 72-73; consult, Agency Int Develop, Vietnam Med Sch, AMA, 67-71; consult to surgeon gen, USAF, 75-80; attend surgeon, Emory Univ Hosp, Grady Mem Hosp, Crawford W Long Hosp & Henrietta Egleston Children's Hosp, Atlanta, 80-83 & Univ Ky Hosp, 87-; dir, New England Critical Care Inc, 86- *Honors & Awards:* Robert V Day Mem lectr, Western Sect, Am Urol Asn, 69; William T Belfield lectr, Chicago Urol Soc, 77. *Mem:* Am Col Surgeons; Am Acad Pediat; Am Urol Asn; Am Asn Clin Urologists; Soc Pediat Urol; AMA; Am Fedn Clin Res; Int Soc Urol; Am Soc Nephrology; Am Asn Genito-Urinary Surgeons; Am Surg Soc; NY Acad Sci; NY Acad Med; Can Urol Asn; Colombian Urol Soc; German Urol Asn; Pan Am Med Asn; Royal Soc Med; Int Soc Surg; Soc Univ Surgeons; Soc Univ Urologists. *Res:* Adrenal surgery; pediatric urology; genitourinary malignancies. *Mailing Add:* Dept Surg Univ Ky Med Ctr Lexington KY 40536

GLENN, JOSEPH LEONARD, b Albany, NY, Jan 20, 25; m 50; c 4. BIOCHEMISTRY. *Educ:* St Lawrence Univ, BS, 50; Syracuse Univ, MS, 52, PhD, 54. *Prof Exp:* From asst prof to assoc prof, 56-66, chmn dept, 75-79, PROF BIOCHEM, ALBANY MED COL, 66- *Concurrent Pos:* Nat Heart Inst fel, Enzyme Inst, Univ Wis, 54-56; Lederle Med Fac award, 58-61; USPHS res career develop award, 62-72. *Mem:* Am Soc Biol Chem. *Res:* Cellular metabolism; enzymology. *Mailing Add:* 24 David Ave Troy NY 12180

GLENN, KEVIN CHALLON, b Albuquerque, NMex, Mar 29, 54; m; c 2. BIOLOGY SCIENCE. *Educ:* Calif State Polytech Univ, BS, 75; Univ Calif, Irvine, PhD(biol), 79. *Prof Exp:* Am Heart Asn postdoctoral fel, Univ Wash, Seattle, 79-81; sect leader cell biol, Bethesda Res Labs, Inc, 81-82; staff res scientist, Flow Labs, Inc, Va, 82; sr res biologist, Monsanto Co, 83-85, res specialist, 85-87, res group leader I, 87-88, ASSOC SCI FEL, MONSANTO CO, ST LOUIS, MO, 88- *Concurrent Pos:* Adj asst prof, In Vitro Cell Biol & Biotechnol Prog, State Univ NY, Plattsburgh & Miner Inst, Chazy, 87-; course instr, Molecular Biol Eukaryotic Cells Biotechnol Prog, Univ Col, Wash Univ, St Louis, Mo, 90- *Mem:* AAAS; Am Soc Cell Biol; Endocrine Soc. *Mailing Add:* Dept Biol Sci Monsanto Co 700 Chesterfield Village Pkwy St Louis MO 63198

GLENN, LEWIS ALAN, b Brooklyn, NY, Nov 17, 36; m 58; c 2. MECHANICAL ENGINEERING, COMPUTATIONAL PHYSICS. *Educ:* Polytech Inst NY, Brooklyn, BChE, 58; NJ Inst Technol, ME, 61; Univ Southern Calif, PhD(mech eng), 68. *Prof Exp:* Proj engr, US Naval Air Rocket Test Sta, 59-62 & NAm Aviation, Inc, 62-68; staff scientist, Appl Theory, Inc, Los Angeles, 68-71; sr staff scientist, Inst CERAC, Ecublens, Switz, 71-77; assoc div leader, Physics Dept, Lawrence Livermore Lab, Univ Calif, 77-87; SHOCK PHYSICS PROG MGR, EARTH SCI DEPT, LAWRENCE LIVERMORE NAT LAB, 87- *Concurrent Pos:* Consult & fel, NAm Rockwell, 66-68; pres, Glenn Technol Consults. *Mem:* Am Phys Soc; assoc fel Am Inst Aeronaut & Astronaut; Am Acad Mech. *Res:* Continuum mechanics; fracture; dynamic properties of materials. *Mailing Add:* Lawrence Livermore Lab PO Box 808 Livermore CA 94550

GLENN, LOYD LEE, b Monterey, Calif, Dec 10, 52; m 78; c 2. NEUROPHYSIOLOGY, NEUROANATOMY. *Educ:* Univ Calif, Santa Cruz, AB, 74; Stanford Univ, PhD(neurobiol), 79. *Prof Exp:* NSF fel, Laval Univ, 79-80; res fel, NIH, 80-82; asst prof, 82-86, ASSOC PROF NEUROBIOL, OHIO COL PODIATRIC MED, 86- *Concurrent Pos:* Prin investr, 85- *Honors & Awards:* Donald B Lindsley Award, Soc Neurosci, 80. *Mem:* Soc Neurosci. *Res:* Morphology and physiology of spinal motoneurons and interneurons. *Mailing Add:* Dept Physiol Ohio Col Podiatric Med Cleveland OH 44106-3073

GLENN, RICHARD A(LLEN), b Brooklyn, NY, Mar 16, 25; m 47; c 4. CIVIL ENGINEERING. *Educ:* Purdue Univ, BS, 49; Wash Univ, PhD(geol), 53. *Prof Exp:* Geologist & geophysicist, Standard Oil Co, Calif, 53-61; instr eng, 61-67, dean instr, 67-76, pres col, 76-85, EMER PRES, VENTURA COL, 86- *Mem:* Am Asn Petrol Geol; Sigma Xi. *Res:* Geology and its relationship to the location of economic mineral deposits and to engineering; seismology. *Mailing Add:* 1880 Rice Rd Ojai CA 93023

GLENN, ROLLIN COPPER, b Nov 25, 27; m 53; c 3. SOILS. *Educ:* Va Polytech Inst & State Univ, BS, 55, MS, 57; Univ Wis, PhD(soils), 59. *Prof Exp:* From asst prof to prof agron, Miss State Univ, 59-73; chmn dept, 73-77, PROF PLANT SCI, UNIV MAINE, 73- *Mem:* Am Soc Agron; Soil Sci Soc Am; Soil Conserv Soc Am; Clay Minerals Soc; Int Soc Soil Sci; Sigma Xi. *Res:* Soil chemistry, mineralogy and fertility; soil conservation; turf soils; production agronomy. *Mailing Add:* Dept Plant Soil Sci Univ Maine Deering Hall Orono ME 04473

GLENN, THOMAS M, b Detroit, Mich, July 20, 40; m 62; c 3. PHARMACOLOGY, PHYSIOLOGY. *Educ:* Rockhurst Col, AB, 62; Univ Mo-Kansas City, MS, 65, PhD(pharmacol), 68. *Prof Exp:* Assoc prof pharmacol, NDak State Univ, 66-68; assoc prof, Fla A&M Univ, 68-69; asst prof, Med Col Pa, 71-73; prof pharmacol & chmn dept, Univ S Ala, 73-82; exec dir biol res, Ciba-Geigy Corp, 82-84, sr vpres & dir res, 84-88; sr dir pharmacol sci, Genentech Inc, 88, vpres pharmacol sci, 88-89; pres & chief exec officer, Biocrust, Inc, 89-90; ASSOC VCHANCELLOR HEALTH AFFAIRS, DUKE UNIV, 91- *Concurrent Pos:* Nat Heart & Lung Inst fel physiol, Sch Med, Univ Va, 69-71; mem study sect, Nat Heart, Lung & Blood Inst, 78-81; co-dir, Clin Pharmacol Res Unit, 81-82. *Mem:* AAAS; Am Soc Pharmacol & Exp Therapeut; NY Acad Sci; Am Heart Asn; Am Physiol Soc; Am Soc Clin Pharmacol. *Res:* Biochemistry; cardiovascular pharmacology, especially pathogenesis of circulatory shock and the therapeutics of shock, special emphasis on toxic factors in shock and the role of leucocytes, complement, free radicals and lysosomal enzymes in the pathogenesis of circulatory shock. *Mailing Add:* Duke Univ Med Ctr Box 3701 Durham NC 27705

GLENN, WILLIAM ALEXANDER, statistics, for more information see previous edition

GLENN, WILLIAM GRANT, b Lansdowne, Pa, Dec 21, 16; m 39. ZOOLOGY. *Educ:* Trinity Univ, Tex, BSc, 48, MA, 52; Rutgers Univ, MSc, 55, PhD(zool), 56; Am Bd Microbiol, dipl, 65. *Prof Exp:* Asst biol & genetics, Rutgers Univ, 48; instr biol, Trinity Univ, Tex, 49; teacher pub sch, Tex, 49-53; asst biol & genetics, Rutgers Univ, 53-55; head immunol sect, USAF Sch Aviation Med, 56-63, chief immunobiol unit, USAF Sch Aerospace Med, 63-70; dir automated biosysts ctr & chief immunol serv, Cybertek, Inc, 70-74; mgr biol sci, Bioquest, 74-75; tech dir, Q C Monitors, Inc, 75-77; consult microbiol, 77-79; sci lectr, Incarnate Word Col, 83-87; VIS SCIENTIST, SAN ANTONIO ISD, 87- *Concurrent Pos:* Publ free-lance author in var sci subj for children, 83- *Mem:* Fel Am Acad Microbiol. *Res:* Immunochemistry; antigen-antibody diffusion; plasma proteins; comparative serology; biomedical instrumentation. *Mailing Add:* 8543 Braewick San Antonio TX 78239

GLENN, WILLIAM HENRY, JR, b Philadelphia, Pa, Dec 22, 37; m 66; c 2. OPTICAL PHYSICS, APPLIED MATHEMATICS. *Educ:* Mass Inst Technol, SB & MS, 60, PhD(physics), 66. *Prof Exp:* Instr physics, Southeastern Mass Tech Inst, 65; sr res scientist, 66-68, prin scientist, 68-71, sr prin scientist, 71-90, SR CONSULT SCIENTIST, UNITED TECHNOL RES CTR, 90- *Concurrent Pos:* Lectr, Hartford Grad Ctr, Rensselaer Polytech Inst, 68; assoc ed, Inst Elec & Electronics Engrs J Quantum Electronics, 82-85. *Mem:* Fel Optical Soc Am; Inst Elec & Electronics Eng. *Res:* Quantum & nonlinear and fiber optics; laser and optical physics; photonics; mathematical modeling. *Mailing Add:* United Technologies Res Ctr 400 Main St East Hartford CT 06108

GLENN, WILLIAM WALLACE LUMPKIN, b Asheville, NC, Aug 12, 14; m 43; c 2. SURGERY. *Educ:* Univ SC, BS, 34; Jefferson Med Col, MD, 38. *Hon Degrees:* MA, Yale Univ, 6l, Dr, Univ Cadiz, Spain. *Prof Exp:* Intern, Pa Hosp, 38-40; resident surg, Mass Gen Hosp, 40-41; asst physiol, Sch Pub Health, Harvard Univ, 41-43; resident surg, Mass Gen Hosp, 45-46; assoc, Jefferson Med Col, 46-48; from instr to assoc prof, 48-61, chief cardiovascular surg, 48-68, chief cardiothoracic surg, 68-75, Charles W Ohse prof surg, 61-85, sr res scientist, 85-87, EMER CHARLES W OHSE PROF SURG, SCH MED, YALE UNIV, 85- *Concurrent Pos:* Pres, Am Heart Asn, 70-71; bd dirs, Charles E Culpepper Found, 78-85; ed, Thoracic & Cardiovasc Surgeon. *Honors & Awards:* Alton Oschner lectr; Samuel Harvey lectr; George Rosen Mem Award; Gold Heart Award, Am Heart Asn. *Mem:* Am Surg Asn; Am Col Surgeons; AMA; Am Thoracic Asn; Am Soc Artifical Internal Organs. *Res:* Cardiovascular physiology and surgery; physiology of the lymphatics; medical education; cardiothoracic surgery; author or co-author of over 190 publications. *Mailing Add:* Dept Surg Yale Univ Sch Med New Haven CT 06510

GLENNAN, T(HOMAS) KEITH, b Enderlin, NDak, Sept 8, 05; m 31; c 4. EDUCATION ADMINISTRATION, PUBLIC SERVICE. *Educ:* Yale Univ, BS, 27. *Hon Degrees:* MA, Yale Univ, 61; DSc, numerous US univs; DE, Fenn Col, 53, Stevens Inst Technol, 54, Case Inst Technol, 60; LLD, var US univs. *Prof Exp:* Mem staff, Elec Res Prod, Inc, NY, 27-35; opers mgr, Paramount Pictures, Inc, Calif, 35-39, studio mgr, 39-41; exec, Vega Airplane Corp, 61; studio mgr, Samuel Goldwyn Studios, 41-42; dir US Navy underwater sound lab, Div War Res, Columbia, 42-45; exec, Ansco Corp, 45-47; pres, Case Western Univ, 47-66, pres, Case Western Inst Technol, 65-68; pres, Assoc Univs, Inc, 65-68; asst to John W Gardner, Nat Urban Coalition, 68-70; US rep, Int AEC, Vienna, Austria, 70-73; consult, US State Dept, 73-77; RETIRED. *Concurrent Pos:* Comnr, Atomic Energy Comn, 50-52; first adminr, NASA, 58-61; trustee, Rand Corp, 62-72 & Case Western Reserve Univ, 70-78; pres, Assoc Univs, Inc, 65-68; asst to chmn, Urban Coalition, 68-69; US ambassador, Int Atomic Energy Agency, 70-73; alumni fel, Yale Corp, 62-68; former dir, Repub Steel Corp, Standard Oil Co, Ohio, Air Prod & Chem, Inc & Avco Corp. *Honors & Awards:* Medal for Merit, US Govt; Distinguished Serv Medal, NASA, 67; Distinguished Honor Award, US Dept State, 73. *Mem:* Nat Acad Eng; fel Am Acad Arts & Sci; Benjamin Franklin fel Royal Soc Arts; fel AAAS. *Res:* Academic and research administration. *Mailing Add:* 10450 Lottsford Rd Unit 4107 Mitchellville MD 20721

GLENNE, BARD, b Oslo, Norway, May 2, 35; m 67; c 1. CIVIL & ENVIRONMENTAL ENGINEERING. *Educ:* Wash State Univ, BS, 57; Mass Inst Technol, MS, 59; Univ Calif, Berkeley, PhD(civil eng), 66. *Prof Exp:* Res engr, Norwegian Inst Water Res, 60-61; res engr, Swiss Fed Inst Technol, 62-63; asst prof civil eng, Univ Nev, 63-64; res assoc, Univ Calif, Berkeley, 64-66; hydraul engr, Norconsult A S, Oslo, Norway, 66-68; asst prof civil eng, Ore State Univ, 68-70; assoc prof, 70-75, prof civil eng, Univ Utah, 75-81; LECTR, DEPT MECH ENG, UNIV WASH, SEATTLE, 81- *Mem:* Am Soc Civil Engrs; Int Asn Hydraul Res; Norweg Soc Prof Engrs. *Res:* Utilization and management of fresh and salt water resources. *Mailing Add:* Dept Mech Eng Univ Wash Seattle WA 98195

GLENNER, E J, b Chicago, Ill, July 7, 26; US citizen; m; c 3. TELECOMMUNICATIONS, SWITCHING SYSTEMS. *Educ:* Northwestern Univ, BS, 48. *Prof Exp:* Dir, indust liaison, exec dir, switching systs & other positions, GTE Commun Systs, 50-87; PRES, RENOIR ENG, 87- *Concurrent Pos:* Chmn, Commun Switch & Commun Soc, Inst Elec & Electronics Engrs, 72, dir, meetings & conf, 80-90; head, Strategic Planning & mem Adv Group, Comt TI Standards Orgn. *Honors & Awards:* McClellan Award, Inst Elec & Electronic Engrs. *Mem:* Fel Inst Elec & Electronics Engrs. *Mailing Add:* Renoir Eng 4022 E Greenway Rd Phoenix AZ 85032

GLENNER, GEORGE GEIGER, b Brooklyn, NY, Sept 17, 27; m; c 4. PATHOLOGY. *Educ:* Johns Hopkins Univ, AB, 49, MD, 53; Am Bd Path, dipl, 58. *Prof Exp:* Intern surg, Mt Sinai Hosp NY, 53-54; resident path, Mallory Inst Path, Boston City Hosp, 54-55; res pathologist, Sect Histochem, Path Lab, NIH, 55-57; asst pathologist, Hosp, Johns Hopkins Univ, 57-58; med dir, USPHS, 64-82; chief sect histochem, Path Lab, 59-70, CHIEF SECT MOLECULAR PATH, LAB EXP PATH, NAT INST ARTHRITIS, METAB & DIGESTIVE DIS, 71-; RES PATHOLOGIST, SCH MED, UNIV CALIF, SAN DIEGO, 82- *Concurrent Pos:* Asst pathologist, Boston Univ, 54-55 & Harvard Sch Legal Med, 55; asst, Sch Med, Johns Hopkins Univ, 57-58; guest instr, Sch Med, Univ Kans, 60, 61, 64, 65; assoc prof, Sch Med, Georgetown Univ, 65-; vis prof, Heller Inst, Tel-Hashomer Hosp, Tel Aviv, Israel, 67; chmn dept med & physiol, NIH Grad Sch Found Advan Educ Sci, 68. *Honors & Awards:* Merit Award, Nat Res Coun Aging, NIH, 88; Med Res Award Metrop Life Found, 89; Potamkin Award, Alzheimers Dis Res, Am Acad Neurol, 89. *Mem:* Am Asn Mil Surgeons; NY Acad Sci; Am Asn Pathologists; Ger Nat Acad Sci; Histochem Soc; Am Fedn Clin Invest. *Res:* Disturbances of protein metabolism; pathology of endocrine tumors; histochemistry and biochemistry of amino acids and proteolytic enzymes; histochemical enzyme kinetics; nature and pathogenesis of amyloidosis and Alzheimer's disease. *Mailing Add:* Dept Path M-012 Univ Calif Sch Med La Jolla CA 92093

GLENN-LEWIN, DAVID CARL, b Chicago, Ill, Sept 22, 43; m 74; c 3. PLANT ECOLOGY. *Educ:* Knox Col, AB, 65; Cornell Univ, PhD(ecol, evolutionary biol), 72. *Prof Exp:* From asst prof to assoc prof, Iowa State Univ, 72-87, chair bot, 87-90, PROF BOT, IOWA STATE UNIV, 87-, ASSOC DEAN, COL LIB ARTS & SCI, 90- *Concurrent Pos:* Consult, Black & Veatch Eng, 75, 78 & 81; vis prof, York, UK, 80, Utrecht, Neth, 81-82. *Mem:* Ecol Soc Am; Brit Ecol Soc; Int Asn Ecol; Int Assoc Vegetation Sci. *Res:* Numerical ecology; vegefution science; species diversity and ecosystem structure. *Mailing Add:* Dept Bot Iowa State Univ Ames IA 50011-1020

GLENNON, JOSEPH ANTHONY, b New York, NY, Feb 21, 31. THYROID DISEASES. *Educ:* State Univ NY, MD, 57. *Prof Exp:* PROF MED & CHIEF DIV ENDOCRINOL & METAB, THOMAS JEFFERSON UNIV, 77- *Res:* Metabolic diseases. *Mailing Add:* 5149 N Ninth Ave No 202 Pensacola FL 32504-8733

GLESER, LEON JAY, b St Louis, Mo, Dec 17, 39; m 69; c 1. MATHEMATICAL STATISTICS,. *Educ:* Univ Chicago, BS, 60; Stanford Univ, MS, 62, PhD(statist), 63. *Prof Exp:* Asst prof math statist, Columbia Univ, 63-65; from asst prof to assoc prof statist & biostatist, Johns Hopkins Univ, 65-72; from assoc prof to prof statist, Purdue Univ, 72-89; PROF MATH & STATIST, UNIV PITTSBURGH, 89- *Concurrent Pos:* Vis fel, Educ Test Serv, 71-72; assoc ed, J Am Statist Asn, 71-74 & 86- & Psychometrika, Psychometric Soc, 72-79; vis prof, Sidney Farber Cancer Inst & Sch Pub Health, Harvard Univ, 79-80; vis fac, Nat Bur Standards, 88- *Mem:* Fel Am Statist Asn; fel Inst Math Statist; Psychometric Soc. *Res:* Statistical inference; multivariate statistical analysis; large sample theory; model building and goodness of fit tests; psychometrics; biometrics. *Mailing Add:* Dept Math & Statist Univ Pittsburgh Pittsburgh PA 15260

GLESS, ELMER E, b Rogers, Nebr, Feb 3, 28; m 62; c 2. ACAROLOGY. *Educ:* Ariz State Univ, BAEd, 55; Iowa State Univ, MS, 57, PhD(acarology), 68. *Prof Exp:* Med entomologist, USDA, Tex, 58-59 & William Cooper & Nephews, Inc, 59-62; instr biol, Northern Ill Univ, 62-64; field entomologist, Bernice P Bishop Mus, Hawaii, 65-67; res assoc, Iowa State Univ, 67-68; asst prof biol, 68-70, assoc prof biol sci, 70-72, head dept, 70-88, PROF BIOL SCI, MONT COL MINERAL SCI & TECHNOL, 70- *Mem:* Am Entom Soc; Sigma Xi; Sci Soc. *Res:* Prostigmatid mite biology and taxonomy; water microbiology including bacteria and invertebrate insects. *Mailing Add:* Dept Biol Sci Mont Col Mineral Sci & Technol Butte MT 59701

GLESS, GEORGE E, b Schuyler, Nebr, Jan 15, 17; m 40; c 2. ELECTRICAL ENGINEERING, ELECTRIC VEHICLES. *Educ:* Univ Colo, BS, 40, MS, 48; Iowa State Univ, PhD(elec eng), 63. *Prof Exp:* Student engr, Commonwealth Edison Co, Ill, 40-41, electronic & control engr, 41-46; from instr to assoc prof, 46-64, in charge hybrid comput facil, 66-69, prof, 64-, EMER PROF ELEC ENG, UNIV COLO, BOULDER. *Concurrent Pos:* Consult, Elec Vehicle. *Mem:* Am Soc Eng Educ; Inst Elec & Electronics Engrs. *Res:* Computers and control systems; electric vehicles. *Mailing Add:* Dept Elec Eng Campus Box 425 Univ Colo Boulder CO 80309-0425

GLESSNER, ALFRED JOSEPH, b Chester, Pa, Jan 19, 43; m 66; c 2. REGULATORY AFFAIRS. *Educ:* Villanova Univ, BChE, 64; Univ Pa, MS, 67, PhD(chem eng), 69. *Prof Exp:* Res engr, Res & Develop, Sun Oil Co, 69-77, mgr bus develop, 77-83, DIR, ADMIN & REGULATORY AFFAIRS, SUN TECH. *Mem:* Am Inst Chem Engrs; Am Chem Soc; Regulatory Affairs Prof Soc. *Res:* Physical adsorption; thermodynamics; statistical mechanics; molecular sieves; separation processes; solid-liquid phase equilibrium; petroleum refining processes; catalytic reforming; catalytic cracking. *Mailing Add:* 16 Hawthorne Lane Glen Mills PA 19342

GLETSOS, CONSTANTINE, b Stylis, Greece, Aug 5, 34; m 67; c 4. ORGANIC CHEMISTRY, MEDICINAL CHEMISTRY. *Educ:* Nat Univ Athens, dipl chem, 58; Univ BC, MSc, 65, PhD(org chem), 68. *Prof Exp:* Res chemist, Gen Develop Co, Greece, 61-63; analytical chemist, Gen Chem Labs, Greece, 63; SR RES SCIENTIST ORG CHEM, AM HOME PROD CORP, 68- *Concurrent Pos:* Lab asst, Chropei Co, 57; lab dir & instr, Army, 59-61; teaching asst, Univ BC, 63-68; grants, Nat Cancer Inst, Can, 63, Nat Res Coun, Can, 65 & 68. *Mem:* Can Inst Chemists; Am Chem Soc. *Res:* Isolation of diphosphopyridine nucleotide; steroids; alkaloids; terpenoids; penicillins; cephalosporins; peptides; heterocyclic drugs and intermediates. *Mailing Add:* Wyeth Ayerst Labs Chem Develop Dept 64 Maple St Rouses Point NY 12979

GLEW, DAVID NEVILLE, b Stamford, Eng, Sept 27, 28; m 52; c 4. PHYSICAL CHEMISTRY. *Educ:* Cambridge Univ, BA, 49, PhD, 52. *Prof Exp:* Fel, Nat Res Coun Can, Ottawa, 52-54; AEC contract with Prof J H Hildebrand, Univ Calif, 54-55; sr lectr chem & dept head,Univ Natal, Durban, SAfrica, 55-57; res chemist, Dow Chem Can, Inc, 57-61, assoc scientist, 61-67, res scientist, 67-91; RETIRED. *Mem:* Royal Soc Chem; fel Chem Inst Can. *Res:* Water and aqueous solution structure, properties and H-bonding; reaction kinetics; thermodynamics and phase equilibria of aqueous systems; materials science and thermoplastic composites. *Mailing Add:* 536 Highbury Park Sarnia ON N7V 2J9 Can

GLEWWE, CARL W(ILLIAM), b St Paul, Minn, May 1, 27; m 53; c 3. ALTERNATIVE ENERGY, COMPUTER SCIENCES. *Educ:* Univ Minn, BS, 50, MS, 55, PhD(elec eng), 58. *Prof Exp:* Instr elec eng, Univ Minn, 53-58; syst design engr digital comput, Univac Div, Sperry Rand Corp, 58-61, mgr 61-69, group mgr digital comput, 69-79; dir prog mgt, Mid-Am Solar Energy Ctr, 79-81; vpres eng, Burdick Corp, 82-83; dir eng, Sentech Med Corp, 84-85; OWNER & VPRES, LYNN SERVS, INC, 86- *Concurrent Pos:* Sr engr, Gen Mills, Inc, 57-58; dir res, Minn Energy Agency, 77-78. *Mem:* Inst Elec & Electronics Engrs. *Res:* Computer design; alternative energy sources. *Mailing Add:* 1821 University Ave St Paul MN 55104

GLEYSTEEN, JOHN JACOB, b Sioux City, Iowa, Jan 3, 41; m 71; c 2. SURGERY, GASTROINTESTINAL PHYSIOLOGY. *Educ:* Grinnell Col, BA, 63; Univ Iowa, MD, 67; Am Bd Surg, cert, 76 & 86. *Prof Exp:* Resident gen surg, Univ Ala Med Ctr, Birmingham, 70-75; staff physician surg serv, 75-80, asst chief surg, 80-87, CHIEF SURG, VET ADMIN MED CTR, 87-; ASSOC PROF SURG, UNIV ALA, BIRMINGHAM, 87- *Concurrent Pos:* Asst prof surg, Med Col Wis, 75-87. *Mem:* Asn Acad Surg; Am Col Surgeons; Am Gastroenterol Asn; Soc Surg Alimentary Tract. *Res:* Gastrointestinal physiology; gastric motility and emptying; portal hypertension. *Mailing Add:* Dept Surg Univ Ala 700 S 19th St Birmingham AL 35223

GLEYZAL, ANDRE, b New Orleans, La, Nov 23, 08. MATHEMATICS. *Educ:* Ohio State Univ, BA, 31, MA, 33, PhD(math), 36. *Prof Exp:* Instr physics, Boston Col, 37-38; prof math, St Michael's Col, 38-41; mathematician, US Dept Navy, 41-51 & Nat Bur Standards, 51-53; mathematician, US Naval Surface Weapons Lab, 53-72; RETIRED. *Mem:* Am Math Soc. *Res:* General flow of metals; vibration of ships; non-steady viscous flow of fluids; aerodynamical trajectories; tenser analysis; relativity, unified field theory, absolute Newtonian calculus; analytic space, time, matter; analytical energy-electric charge universes; square root vector calculus and enerel wave mechanics in a Riemannian geometry of several complex variables; encrel mathematical physics; mathematical electronics; new n-dimensional geometric physics and quantized geometric electronic charge. *Mailing Add:* 300 NE 44th St Boca Raton FL 33431

GLEZEN, WILLIAM PAUL, b Oblong, Ill, Mar 15, 31; m 53; c 2. MEDICINE, INFECTIOUS DISEASES. *Educ:* Purdue Univ, BS, 53; Univ Ill, BS, 54, MD, 56. *Prof Exp:* Instr pediat, Sch Med, Univ NC, 61-62; epidemiologist, Kansas City Field Sta, Commun Dis Ctr, USPHS, 62-64, chief respiratory & enteric virus dis unit, 64-65; from asst prof to assoc prof pediat, Sch Med, Univ NC, Chapel Hill, 65-75; assoc prof, 75-77, PROF MICROBIOL & PEDIAT & EPIDEMIOLOGIST, INFLUENZA RES

CTR, DEPT MICROBIOL & IMMUNOL, BAYLOR COL MED, HOUSTON, 77-, HEAD, PREV MED SECT, DEPARTMENTS MICROBIOL, IMMUNOL & PEDIAT, 89- *Concurrent Pos:* Instr, Sch Med, Univ Kans, 63-65; adj prof epidemiol, Univ Tex Sch Pub Health, 83- *Res:* Epidemiology and pathogenesis of acute respiratory disease. *Mailing Add:* Dept Microbiol & Immunol Baylor Col Med One Baylor Plaza Houston TX 77030

GLICK, ARNOLD J, b Brooklyn, NY, Nov 7, 31; m 53, 79; c 3. THEORETICAL SOLID STATE PHYSICS. *Educ:* Brooklyn Col, BA, 55; Univ Md, PhD(physics), 61. *Prof Exp:* NSF fel, Weizmann Inst, 59-61; from asst prof to assoc prof, 61-78, PROF PHYSICS, UNIV MD, COLLEGE PARK, 78- *Concurrent Pos:* NSF fel, Univ Paris, Orsay & Ctr for Nuclear Study, Saclay, France, 67-68; vis prof, Hebrew Univ, 78 & Weizmann Inst, 81-82; Lady Davis fel, Israel Inst Technol, 78. *Mem:* Am Phys Soc. *Res:* Quantum mechanical many body problem; theory of solids and condensed matter. *Mailing Add:* Dept Physics Univ Md College Park MD 20740

GLICK, BERNARD ROBERT, b New York, NY, April 9, 45; Can citizen; m 68; c 1. PHOTOSYNTHESIS, NITROGEN FIXATION. *Educ:* City Col New York, BS, 69; Univ Waterloo, MSc, 72, PhD(chem), 74. *Prof Exp:* Fel molecular biol, Univ Toronto, 74-78; res assoc, Nat Res Coun Can, 78-79; group leader, ENS Bio-Logicals, Inc, 79-82; assoc prof molecular biol, 82-89, PROF MOLECULAR BIOL & BIOTECHNOL, UNIV WATERLOO, 89- *Mem:* Am Soc Microbiol; Can Soc Microbiol. *Res:* Regulation of PEP carboxylase gene expression in sorghum bicolor and anabaena variabilis; genetic modification of free living diazotrophs; introduction of cellulase genes; chromosomal integration; development of a two-stage airlift fermenter for recombinant microorganisms. *Mailing Add:* Dept Biol Univ Waterloo Waterloo ON N2L 3G1 Can

GLICK, BRUCE, b Pittsburgh, Pa, May 5, 27; m 50; c 3. PHYSIOLOGY, IMMUNOLOGY. *Educ:* Rutgers Univ, BSc, 51; Univ Mass, MSc, 53; Ohio State Univ, PhD(genetics, physiol), 55. *Prof Exp:* Asst, Univ Mass, 51-52 & Ohio State Univ, 52-55; from asst to prof poultry immunol & physiol, 55-86, coordr animal physiol prog, 73-86, PROF IMMUNOL & HEAD POULTRY SCI DEPT, MISS STATE UNIV, 86- *Concurrent Pos:* Merck res award, 78. *Mem:* Fel AAAS; Am Physiol Soc; Am Asn Immunol; fel Poultry Sci Asn; Am Ornith Union; Reticuloendothelial Soc. *Res:* Physiology of lymphoid tissue, especially bursa of Fabricius; cellular interactions in the chickens immune system; kinetics of the lymphomatoid complex; ruiandendritic cells. *Mailing Add:* Dept Poultry Sci 129 Poole Agr Ctr Clemson Univ Clemson SC 29634

GLICK, CHARLES FREY, b Allentown, Pa, Apr 21, 17; m 46; c 1. ANALYTICAL CHEMISTRY. *Educ:* Lehigh Univ, BS, 38, MS, 40; Mass Inst Technol, PhD(phys chem), 43. *Prof Exp:* Res chemist, E I du Pont de Nemours & Co, NJ, 43-46; supvr phys res lab, Barrett Div, Allied Chem & Dye Corp,; asst head physics lab, Rohm & Haas Co, 53-56, res technologist, 56-59; sect supvr, Physics & Analytical Chem Div, US Steel Corp, 59-82; RETIRED. *Mem:* Am Chem Soc. *Res:* Chemical physics; instrumental methods of organic and inorganic analysis. *Mailing Add:* Rte 1 Box 90 Maysville WV 26833-9443

GLICK, DAVID, b Homestead, Pa, May 3, 08; m 29, 41, 45, 81; c 4. ANALYTICAL CHEMISTRY. *Educ:* Univ Pittsburgh, BS, 29, PhD(biochem), 32; Am Bd Clin Chem, dipl, 53. *Hon Degrees:* LLD, Univ Glasgow, 80. *Prof Exp:* Asst chem, Univ Pittsburgh, 29-32; Hernsheim fel chem, Mt Sinai Hosp, NY, 32-34; chief chemist, Mt Zion Hosp, San Francisco, 34-36; Rockefeller fel, Carlsberg Lab, Denmark, 36-37; chief chemist, Newark Beth Israel Hosp, 37-42; head vitamin & enzyme res, Russell-Miller Milling Co, Minn, 43-46; from assoc prof to prof physiol chem, Univ Minn, 46-61; prof & head div,histochem, Med Sch, Stanford Univ, 61-73; dir, Ctr Histochem Res, 73-78, STAFF SCIENTIST, SRI INT, 78-; EMER PROF PATH, MED SCH, STANFORD UNIV, 73-, MEM STAFF, CANCER BIOL RES LAB, 78- *Concurrent Pos:* Vis res chemist, Carlsberg Lab, Denmark, 33; fel, Commonwealth Fund, Carlsberg Lab, Denmark, Inst Cell Res, Stockholm, 49, 58-59 & Zool Sta, Naples, 59; consult, Toxicity Lab, Univ Chicago, 45-46, Vet Admin Hosp, Minneapolis, 46-47, Ft Detrick, 57-67 & Vet Admin Hosp, Palo Alto, 61-89; mem cytochem panel, Comt Growth, Nat Res Coun, 51-55; MacFarlane prof exp med, Univ Glasgow, 70-71; ed, Methods Biochem Analysis, 53-88; co-ed, Tech Biochem & Biophys Morphol, 72-77. *Honors & Awards:* Van Slyke Award, 77; Ames Award, 80; Evans Lectr, 81; Glick Lectr, 84- *Mem:* Fel AAAS; Am Chem Soc; Soc Exp Biol & Med; Am Soc Biol Chem; hon mem Histochem Soc (vpres, 50, 69, pres, 51, 70); Am Soc Cell Biol; hon mem Royal Danish Acad Sci; hon mem Finnish Histochem Soc; hon assoc Int Fedn Socs Histochem & Cytochem (vpres, 68-72, pres 72-76). *Res:* Histochemistry quantitative techniques and applications; biochemistry, enzymes, vitamins. *Mailing Add:* 680 Junipero Serra Blvd Stanford CA 94305-8444

GLICK, DAVID M, b San Francisco, Calif, Jan 4, 36; m 61; c 3. BIOCHEMISTRY, PROTEIN CHEMISTRY. *Educ:* Oberlin Col, BA, 57; Case Western Reserve Univ, PhD(biochem), 62. *Prof Exp:* Res assoc biochem, Weizmann Inst, 62-63; res assoc, Brookhaven Nat Lab, 64-66; asst prof, 66-72, ASSOC PROF, MED COL WIS, 72- *Concurrent Pos:* Fel, Arthritis & Rheumatism Found, 62-64; res assoc, King's Col, London, 63 & State Univ NY Buffalo, 64; vis prof, Free Univ Brussels, 73-74; vis assoc prof technion, Haifa, 87. *Mem:* Am Chem Soc; Am Soc Biochem Molecular Biol; London Biochem Soc. *Res:* Chemistry, physiology and mechanism of activation of pepsinogens and other aspartate proteinases. *Mailing Add:* Makor Chems LTD PO Box 6570 Jerusalem 91064 Israel

GLICK, FORREST IRVING, b Glasgow, Mont, May 24, 34; m 58, 64; c 3. LOW TEMPERATURE PHYSICS. *Educ:* St Olaf Col, BA, 56; Univ Minn, MS, 59, PhD(physics), 66. *Prof Exp:* Instr physics, Univ Minn, 63-64; from asst prof to assoc prof, 64-71, PROF PHYSICS, MANKATO STATE UNIV, 71-, ASST TO DEAN, SCH PHYSICS, ENG & TECHNOL, 85- *Mem:* Am Asn Physics Teachers; Am Phys Soc. *Res:* Liquid helium; low temperature physics. *Mailing Add:* Sch Physics, Eng & Technol Mankato State Univ Mankato MN 56002

GLICK, GERALD, b Brooklyn, NY, Apr 3, 34; m 63; c 4. INTERNAL MEDICINE, CARDIOLOGY. *Educ:* Cornell Univ, AB, 55, MD, 58; Am Bd Internal Med, dipl, 66; Am Bd Cardiovasc Dis, dipl, 68. *Prof Exp:* Intern med, Sch Med, Rochester & Strong Mem Hosp, 58-59, resident, 59-60; sr investr, Cardiol Br, Nat Heart Inst, 62-68; assoc prof pharmacol & med & dir div clin cardiovasc pharmacol, Baylor Col Med, 68-71; dir, Cardiovasc Inst, Michael Reese Hosp & Med Ctr, 71-78; prof med, 74-78, clin prof Med, Pritzker Sch Med, Univ Chicago, 78-90; CONSULT, 90- *Concurrent Pos:* USPHS fel cardiol, 60-62; Am Heart Asn estab investr, 63-69; attend physician, Nat Heart Inst, 63-69; hon med asst, Postgrad Med Sch & Hammersmith Hosp, London, Eng, 64-65; chief electrocardiographer, NIH, 66-68. *Mem:* Fel Am Col Physicians; Am Physiol Soc; fel Am Col Cardiol; Am Soc Clin Invest; Am Soc Pharmacol & Exp Therapeut. *Res:* Clinical cardiology and investigations relating to cardiovascular pharmacology, the autonomic nervous control of the cardiovascular system and the determinants of ventricular function. *Mailing Add:* 55 E Washington Chicago IL 60602

GLICK, HAROLD ALAN, b US citizen. BIOPHYSICS. *Educ:* Univ Calif, Los Angeles, BS, 64; Univ Chicago, MS, 66; Univ Calif, Riverside, PhD(physics), 76. *Prof Exp:* Res asst physics, Gen Motors Res Lab, 66; instr, Calif State Univ, San Bernadino, 71; NIH fel, Univ Chicago, 76-77; asst prof, Riverside City Col, 78; asst prof physics, Claremont Men's Col, 78-79; MEM STAFF, FLEET ANALYSIS CTR, US NAVY, 80- *Mem:* Am Phys Soc; Sigma Xi. *Mailing Add:* 8013 Rockford Cr Riverside CA 92509

GLICK, J LESLIE, b New York, NY, Mar 2, 40; c 2. BIOTECHNOLOGY. *Educ:* Columbia Univ, AB, 61, PhD(zool), 64. *Prof Exp:* USPHS vis res fel biochem, Princeton Univ, 64-65; sr cancer res scientist, Roswell Park Mem Inst, 65-67, assoc cancer res scientist, 67-69; exec vpres, Assoc Biomedic Systs, Inc, 69-72, pres, 72-75, chmn bd, 72-77; pres, Genex Corp, 77-84, chief exec officer, 77-87, chmn bd, 83-85, pres, 85-87; CHIEF EXEC OFF, BIONEX CORP, 87- *Concurrent Pos:* Asst res prof, Roswell Park Div, State Univ NY Buffalo, 67-68, assoc res prof & chmn, 68-70, mem exec comt, Grad Sch, 68-70; res prof, Canisius Col, 68-70, Niagara Univ, 68-70; pres & chmn bd, HTI Corp, 72-75; vpres, Nat Asn Life Sci Indust, 75-76, dir, 75-77; pres, Inst Sci & Social Accountability, Inc, 75-78; pres & dir, Indust Biotechnol Asn, 81-83, dir, 83-84; overseer, Simon's Rock of Bard Col, 84-85, first chmn bd overseers, 85; mem, Biotechnol Tech Adv Comt, US Dept Com, 85-87, trustee, Nat Fac Humanities, Arts & Sci, 85-87; mem, bd ed adv, Strategic Directions, 84-87 & adv coun, High Technol Mkt Rev, 86-87; adj prof, Grad Sch Technol & Mgt, Univ Col, Univ Md, 88-, mem adv panel, 88- *Honors & Awards:* John S Newberry Prize. *Mem:* Am Physiol Soc; NY Acad Sci; Sigma Xi. *Res:* Pathogenesis and treatment of Alzheimer's disease and other dementias; technology assessments pertaining to biotechnology; bioethical problems; evaluation systems on performance of research. *Mailing Add:* 10899 Deborah Dr Potomac MD 20854

GLICK, JANE MILLS, b Memphis, Tenn. Nov 26, 43; m 68; c 2. MOLECULAR BIOLOGY. *Educ:* Randolph-Macon Woman's Col, AB, 65; Columbia Univ, PhD(biochem), 71. *Prof Exp:* NIH fel, Lab Biochem, Nat Cancer Inst, 72-73; Cancer Res Found Chicago fel, Radiobiol Res Div, Sch Med, Stanford Univ, 73-74; res assoc, Sch Dent Med, Univ Pa, 74-75, res asst prof & nat res serv fel, Dept Biochem, 75-77; from asst prof to assoc prof, 77-90, PROF BIOCHEM, MED COL PA, 90- *Concurrent Pos:* Consult, Lab Chem Pharmacol, Nat Heart & Lung Inst, NIH, 71. *Mem:* AAAS; Am Soc Biol Chemists; Am Heart Asn. *Res:* Enzymology; lipoprotein metabolism. *Mailing Add:* Dept Physiol & Biochem Med Col Pa 3300 Henry Ave Philadelphia PA 19129

GLICK, JOHN H, b New York, NY, May 9, 43; m 68; c 2. INTERNAL MEDICINE, MEDICAL ONCOLOGY. *Educ:* Princeton Univ, AB, 65; Columbia Univ, MD, 69. *Prof Exp:* From asst prof to assoc prof, 74-83, PROF MED, UNIV PA, 83-, DIR, CANCER CTR, 85-, MADLYN & LEONARD ABRAMSON PROF CLIN ONCOL, 88- *Concurrent Pos:* mem, COCERT, Am Bd Med Specialties, 85-, bd gov, Am Bd Internal Med, 87-89, Instnl Res Grants Review Comn Am Chem Soc, 89-; prin investr, Ctr Support Grant, Univ Pa, NCI, 85-94, Eastern Coop Oncol Group Grant, 91-; chmn, Am Bd Internal Med, Subspec Bd Med Oncol, 87-89. *Mem:* Fel Am Col Physicians; Am Soc Clin Oncol; Am Asn Cancer Res; Am Fedn Clin Res; Am Soc Hemat. *Res:* Development of improved treatment for patients with Hodgkin's disease and breast cancer, including the use of adjuvant therapy in early stage breast cancer; founded the Philadelphia Bone Marrow Transplant Program, which is now testing the efficacy of ABMT in combination with high-dose chemotherapy in the treatment of metastatic breast cancer. *Mailing Add:* Cancer Ctr Univ Pa 6 Penn Tower 3400 Spruce St Philadelphia PA 19104-4283

GLICK, JOHN HENRY, JR, b Cumberland, Md, June 16, 24; m 54; c 5. CLINICAL CHEMISTRY. *Educ:* Cath Univ, AB, 49; St Louis Univ, PhD(biochem), 56. *Prof Exp:* From instr to asst prof biochem, Col Med & Dent NJ, Newark, 56-66; from asst prof to assoc prof biochem in path, 66-73, ASSOC PROF PATH & ONCOL, MED CTR, UNIV KANS, 73- *Mem:* AAAS. *Res:* Lipoproteins; patient test data analysis. *Mailing Add:* Depts Path & Oncol Univ Kans Med Ctr 39th St & Rainbow Blvd Kansas City KS 66103

GLICK, MARY CATHERINE, PEDIATRICS. *Educ:* Pa State Univ, BS, 48; Univ Pa, MS, 53, PhD(microbiol), 58. *Prof Exp:* Assoc, Dept Anat & Dept Therapeut Res, Sch Med, Univ Pa, 60-65, from asst prof to assoc prof, Dept Therapeut Res, 65-72, assoc prof, Dept Pediat, 72-75, PROF PEDIAT RES, SCH MED, UNIV PA, PHILADELPHIA, 75- *Concurrent Pos:* Res fel, Lab Carbohydrate Res, Harvard Med Sch, Mass Gen Hosp, 63-64; mem, Grad Group Molecular Biol, Sch Med, Univ Pa, 69-, Exp Biol Sect, breast Cancer Task Force, NIH, 73-76 & Physiol Chem Study Sect, NIH, 77-82; vis prof, Dept Genetics & Dept Biochem, Weizmann Inst Sci, Israel, 71-73; admin chmn, Sci Adv Comt, Daman Runyon-Walter Winchell Cancer Fund, NY, 78-79; assoc ed, Cancer Res, 82-85. *Mem:* Am Soc Biol Chemists; Nat Soc Complex Carbohydrates (pres-elect & pres, 81-82). *Mailing Add:* Dept Pediat Children's Hosp Univ Pa Med Sch 34th St & Civic Ctr Blvd Philadelphia PA 19104

GLICK, MILTON DON, b Memphis, Tenn, July 30, 37; m 65; c 2. PHYSICAL CHEMISTRY, INORGANIC CHEMISTRY. *Educ:* Augustana Col Ill, AB, 59; Univ Wis, PhD(phys chem), 65. *Prof Exp:* Fel x-ray crystallog, Cornell Univ, 64-66; from asst prof to prof, 66-83, chmn chem dept, Wayne State Univ, 78-83; dean, Col Arts & Sci, Univ Mo, 83-88; PROVOST & INTERIM PRES, IOWA STATE UNIV, 88- *Mem:* Am Chem Soc; Am Crystallog Asn. *Res:* X-ray crystallographic studies on compounds of chemical and biological interest; computer applications in chemistry; computer graphics; structure-magnetic correlations; structure-relativity correlations. *Mailing Add:* Off Provost Iowa State Univ Beardshear Hall Ames IA 50011

GLICK, RICHARD EDWIN, b Chicago, Ill, Apr 7, 27; div; c 4. PHYSICAL CHEMISTRY. *Educ:* Univ Ill, BS, 51; Univ Calif, Los Angeles, PhD(chem), 54. *Prof Exp:* Res chemist, Brookhaven Nat Lab, 54-55; Milton fel, Harvard Univ, 55-56; asst prof chem, Pa State Univ, 56-59; from asst prof to assoc prof, 59-66, PROF CHEM, FLA STATE UNIV, 66-, DIR INST FUTURE RESOURCES, 82- *Concurrent Pos:* Vis scientist, Peymeinade, France, 70-71; pres, Corp Future Resources, UNLTD, Inc, 86-, CFR Bio-Gen Corp, 86. *Mem:* Am Soc Mass Spectrometry; Sigma Xi. *Res:* Bio-mass energy processes and by-product utilization; cosmochemistry; chemical kinetics. *Mailing Add:* Dept Chem Fla State Univ Tallahassee FL 32306-3006

GLICK, ROBERT L, b Ashland, Ky, Jan 26, 34; m 58; c 2. ROCKET PROPULSION. *Educ:* Purdue Univ, BSME, 55, MSME, 59, PhD(mech eng), 66. *Prof Exp:* Prin engr, Huntsville Div, Thiokol Corp, 63-66; tech specialist, Cummins Engine Co, Inc, Ind, 66-69; assoc prof mech eng, Univ Kans, 69-72; prin engr, Thiokol Corp, 72-78; res engr, Res Inst, Univ Dayton, 78-81; SR RESEARCHER, DEPT AERONAUT & ASTRONAUT, PURDUE UNIV, 81- *Mem:* Assoc fel Am Inst Aeronaut & Astronaut; Am Soc Mech Engrs. *Res:* Internal ballistics of solid propellant rockets; thrust modulation and combustion of solid propellants; swirl and heat rejection in diesel engines; temperature measurement in nonsteady nonisothermal flows; turbulent entrance region flow; surface vehicle drag reduction. *Mailing Add:* 525 N 101st Pl Mesa AZ 85207

GLICK, SAMUEL SHIPLEY, b Baltimore, Md, Dec 18, 00; m 27; c 2. PEDIATRICS, PUBLIC HEALTH. *Educ:* Johns Hopkins Univ, AB, 20; Univ Md, MD, 25. *Prof Exp:* Intern, Univ Hosp, Univ Md, 25-26; resident, Sydenham Hosp Contagious Dis, Baltimore, 26-27, St Vincent's Infant Home, 27-28 & health off, Dept Health, Baltimore, 28-41; from asst prof to assoc prof pediat, med sch, Univ Md, Baltimore, 48-71, emer prof, 71-90; RETIRED. *Concurrent Pos:* Practicing pediatrician, 28-; staff pediatrician, Baltimore City Dept Health, 65-, Johns Hopkins Hosp, Univ Md Hosp, Sinai Hosp. *Mem:* Fel Am Acad Pediatrics. *Res:* Nutritional diseases; study of child health services in Maryland; teaching medical students. *Mailing Add:* 3737 Clarks Lane Baltimore MD 21215

GLICK, STANLEY DENNIS, b New York, NY, Oct 10, 44; m 73. NEUROPHARMACOLOGY. *Educ:* NY Univ, BA, 65; Albert Einstein Col Med, PhD(pharmacol), 69, MD, 71. *Prof Exp:* From asst prof to prof pharmacol, Mt Sinai Sch Med, 71-84; PROF & CHMN PHARMACOL & TOXICOL, ALBANY MED COL, 84- *Mem:* AAAS; Am Soc Pharmacol & Exp Therapeut; Soc Neurosci; NY Acad Sci. *Res:* Changes in drug sensitivity following brain damage; cerebral asymmetry; drug addiction and dependence; basal ganglia function; function of dopamine in brain; function of histamine in brain. *Mailing Add:* Dept Pharmacol & Toxicol Albany Med Col Albany NY 12208

GLICKMAN, LAWRENCE THEODORE, b New York, NY, May 16, 42; m 67; c 2. EPIDEMIOLOGY, VETERINARY MEDICINE. *Educ:* State Univ NY, Binghamton, BA, 64, MA, 66; Univ Pa, VMD, 72; Univ Pittsburgh, MPH, 75, DrPH(epidemiol), 77. *Prof Exp:* Res assoc pharmacol, Endo Labs, 66-67; assoc veterinarian, Trooper Vet Hosp, 72-74; asst dir lab animal, Health Ctr, Univ Pittsburgh, 76-77; asst prof epidemiol & head div, NY State Col Vet Med, Cornell Univ, 77-80; assoc prof & chief epidemiol, Sch Vet Med, Univ Pa, 80-88; DEPT HEAD VET PATH, PURDUE UNIV, 88- *Concurrent Pos:* Fel cardiovasc dis, Grad Sch Pub Health, Univ Pittsburgh, 75-76. *Mem:* Am Pub Health Asn; Soc Epidemiol Res; Teachers Vet Prev Med & Pub Health; Am Vet Med Asn; Asn Vet Med Educrs. *Res:* Epidemiology; seroepidemiology; zoonotic diseases; parasitic diseases; nosocomial infections; preventive medicine; comparative medicine. *Mailing Add:* Dept Vet Path Purdue Univ West Lafayette IN 47907

GLICKMAN, SAMUEL ARTHUR, b New York, NY, June 21, 18; m 43; c 3. ORGANIC CHEMISTRY. *Educ:* City Univ NY, AB, 39; Columbia Univ, MA, 41, PhD(org chem), 44. *Prof Exp:* Asst chem, Nat Defense Res Comt Contract, Columbia Univ, 41-42, asst, Univ, 42-44; sr org res chemist, GAF Corp, 44-54, res fel, 54-60, asst prog mgr acetylene derivatives, 60-68; sr scientist, Reaction Motors Div, Thiokol Chem Corp, 68-71, sr scientist, Chem Div, 71-83, sr res scientist, Morton Chem Div, Morton Thicokol Inc, 83-84; RETIRED. *Concurrent Pos:* Consult, 84- *Mem:* Am Chem Soc. *Res:* Synthetic organic research; dyestuff organic research; acetylene reactions; vinyl and condensation polymerization; dyestuffs and intermediates; fluorochemistry; free radical reactions. *Mailing Add:* 796-A Sparta Rd Cranbury NJ 08512

GLICKMAN, WALTER A, b New York, NY, Sept 8, 38. PHYSICS. *Educ:* Alfred Univ, BA, 59; Pa State Univ, MS, 62, PhD(physics), 64. *Prof Exp:* Asst prof, 64-70, assoc prof, 70-77, PROF PHYSICS, LONG ISLAND UNIV, 77- *Res:* Molecular spectroscopy. *Mailing Add:* Dept Physics Long Island Univ Univ Plaza Brooklyn NY 11201

GLICKSMAN, ARVIN SIGMUND, b New York, NY, Mar 14, 24; m 56; c 5. MEDICINE, BIOLOGY. *Educ:* Chicago Med Sch, MB, 48, MD, 49. *Prof Exp:* Intern, Kings County Hosp, Brooklyn, 48-50; resident, Mem Ctr for Cancer & Allied Dis, 52-54; from asst to assoc, Radiobiol Sect, Sloan-Kettering Inst Cancer Res, 55-65; assoc dir dept radiation ther, Michael Reese Hosp & Med Ctr, Chicago, 65-67; assoc prof radiol, Mt Sinai Sch Med, 67-70, prof radiother & dir radiother res unit, 70-73; prof med sci & chmn dept radiation med, Brown Univ, 73-90; dir dept radiation oncol, RI Hosp, 73-84, dir dept radiation med & biol res, 84-89; DIR RADIATION ONCOL, ROGER WILLIAMS CANCER CTR, 89- *Concurrent Pos:* AEC fel med sci, Dept Biochem, Duke Univ, 50-52; res fel, Sloan-Kettering Inst Cancer Res, 54-55; Dillon fel, Royal Marsden Hosp, London, 61-62; NIH res career develop award, Royal Marsden Hosp & Inst Cancer Res, London, 62-64; guest physician, Brookhaven Nat Lab, 50-52; asst attend, physician & med consult, Mem Ctr Cancer & Allied Dis, 55-65; asst attend physician, Kings County Hosp, 55-60, James Ewing Hosp, 59-64, asst vis radiotherapist, 64-65; asst attend radiotherapist, Mem Hosp, 64-65; dep dir radiother ctr, Mt Sinai Hosp, 67-73; consult radiotherapist, Arden Hill Hosp & Elizabeth A Horton Mem Hosp, 68-; consult radiation ther, Univ Conn Health Ctr, 72-75; assoc, Div Oncol, Roger Williams Gen Hosp, 74-; physician-consult radiol, Vet Admin Hosp, Providence, 74-; mem cancer clin invest rev comt, Nat Cancer Inst, 75-79, chmn, community clin oncol prog eval comm, 84-86. *Mem:* AAAS; Radiation Res Soc; Am Asn Cancer Res; NY Acad Med; Am Soc Clin Oncol; Am Soc Therapeut Radiol Oncol. *Res:* Radiobiology; radiation therapy; oncology; internal medicine. *Mailing Add:* Dept Radiation Oncol Roger Williams Med Ctr Providence RI 02908

GLICKSMAN, MARTIN E, b New York, NY, Apr 4, 37; m 66. PHYSICAL METALLURGY. *Educ:* Rensselaer Polytech Inst, BMetE, 57, PhD(phys metall), 61. *Prof Exp:* Nat Acad Sci-Nat Res Coun fel, Naval Res Lab, 61-63, res metallurgist, 63-66, actg br head metal physics, 66-69, head transformation & kinetics br, 69-75; CHMN, DEPT MAT ENG, RENSSELAER POLYTECH INST, 75- *Concurrent Pos:* John Tod Horton distinguished prof, 86- *Honors & Awards:* Arthur S Flemming Award, 69; Grossman Award, Am Soc Metals, 71; Sci Res Soc Am Award, 67. *Mem:* AAAS; Am Inst Mining, Metall & Petrol Engrs; Am Soc Metals; Sigma Xi; Am Phys Soc. *Res:* Solidification and crystal growth of metals and alloys; surface energy studies; defects in metals; thermodynamics of stressed systems; interfacial phenomena in metals; metallurgy of superconductors; lattice vacancies. *Mailing Add:* Dept Mat Eng Mat Res Ctr Rensselaer Polytech Inst Troy NY 12180

GLICKSMAN, MAURICE, b Toronto, Ont, Oct 16, 28; nat US; m 49; c 3. EXPERIMENTAL SOLID STATE PHYSICS. *Educ:* Univ Chicago, SM, 52, PhD(physics), 54. *Prof Exp:* Researcher nuclear physics, Atomic Energy Proj, Chalk River, Can, 49-50; instr physics, Roosevelt Univ, 53-54; res assoc, Inst Nuclear Studies, Ill, 54; mem tech staff, RCA Labs, NJ, 54-62, head plasma physics, 61-63, dir res, Tokyo Lab, 63-67, head gen res, 67-69; dean grad sch, Brown Univ, 74-76, dean fac & acad affairs, 76-78, provost & dean fac, 78-86, provost, 86-90, UNIV PROF & PROF ENG, BROWN UNIV, 69-, PROF PHYSICS, 90- *Concurrent Pos:* Chmn comt mat for radiation detection devices, Nat Res Coun-Nat Mat Adv Bd, 71-74; chmn vis comt elec eng, Univ Pa, 77-84; chmn vis comt grad educ, Vanderbilt Univ, 77-81; mem, Solar Energy Res Inst Univ Adv Panel, 78-79; mem bd trustees, Northeast Radio Observ Corp, 78-; pres adv comt, Emory Univ, 81-; dir, Ctr Res Libr, 81-86 & chmn, 83-84; trustee, Unive Res Asn, 83-88, vchmn, 86-88, Fermilab Bd, 88-, vchmn 88-89, chmn 89-; mem higher educ policy adv comt, Online Comput Libr Ctr, Cleveland, Ohio, 84-90. *Mem:* Fel Am Phys Soc; fel Inst Elec & Electronics Engrs. *Res:* Nuclear properties; pimeson scattering; semiconductor alloys; band structure; semiconductor transport properties; galvanomagnetic effects; hot electrons; plasmas in solids; gaseous plasmas; semiconductor luminescence. *Mailing Add:* Box D Brown Univ Providence RI 02912

GLICKSTEIN, JOSEPH, b New York, NY, Aug 31, 17; m 41; c 1. ANALYTICAL CHEMISTRY. *Educ:* Mich State Col, BS, 38; NY Univ, PhD(chem), 51. *Prof Exp:* Asst gastric chem res, Mt Sinai Hosp, NY, 38-40 & City Dept Hosps, New York, 40-42; res engr, Bolsey Res & Develop Corp, 51; proj engr, Electro-Phys Labs, Conn, 52-55; prof chem, Adelphi Univ, 55-63; prof, 63-86, EMER PROF CHEM, BROOKLYN COL, CITY UNIV NEW YORK, 86- *Concurrent Pos:* Res collabr, Brookhaven Nat Lab, 58-62. *Mem:* AAAS; Am Chem Soc. *Res:* Instrumentation; electrochemical studies, including polarography; spectrophotometry. *Mailing Add:* Dept Chem Brooklyn Col City Univ NY Brooklyn NY 11210

GLICKSTEIN, MITCHELL, b Boston, Mass, July 13, 31; m 76; c 3. PHYSIOLOGICAL PSYCHOLOGY. *Educ:* Univ Chicago, BA, 51, PhD(psychol), 58. *Prof Exp:* USPHS fel biol, Calif Inst Technol, 58-60; res assoc, Sch Med, Stanford Univ, 60-61; assoc prof physiol & biophys, Sch Med, Univ Wash, 61-66, biol structure, 66-67; from assoc prof to prof psychol, 67-80, SR SCIENTIST MED RES COUN, BROWN UNIV, 80-; PROF NEUROSCI, DEPT ANAT, UNIV COL LONDON, 87- *Concurrent Pos:* Sr res assoc, Oxford Univ, 69-70; vis prof, Univ Parma, 83. *Mem:* Physiol Soc; Anat Asn Gt Brit & Ireland; Am Physiol Soc. *Res:* Brain mechanisms in vision and movement. *Mailing Add:* Dept Anat Univ Col London London England

GLICKSTEIN, STANLEY S, b Brooklyn, NY, Nov 12, 33; m 57; c 4. NUCLEAR PHYSICS, WELDING. *Educ:* Polytech Inst Brooklyn, BEE, 55; Pa State Univ, MS, 59, PhD(physics), 61. *Prof Exp:* Res assoc physics, Princeton Univ, 61-63, instr, 63-64; ADV SCIENTIST, BETTIS ATOMIC POWER LAB, WESTINGHOUSE ELEC CORP, 64- *Mem:* Am Welding Soc; Am Phys Soc; Laser Inst Am. *Res:* Welding arc physics; lasers for materials processing; neutron radiography. *Mailing Add:* Bettis Atomic Power Lab Westinghouse Elec Corp Pittsburgh PA 15122

GLIDDEN, RICHARD MILLS, b Irvington, NJ, Nov 3, 24; m 52; c 3. RUBBER CHEMISTRY. *Educ:* Rutgers Univ, BS, 50. *Prof Exp:* Tech dir, 50-75, vpres, 65-75, exec vpres opers, 75-78, pres, 78-85, CHMN BD DIRS, AMES RUBBER CORP, 85- *Mem:* Am Chem Soc. *Res:* Development and manufacture of rubber compounds. *Mailing Add:* 39 Summit Trail Sparta NJ 07871

GLIEDMAN, MARVIN L, b New York, NY, Aug 3, 29; m 54; c 2. SURGERY. *Educ:* Syracuse Univ, BA, 50; State Univ NY, MD, 54. *Prof Exp:* From asst instr to assoc prof surg, State Univ NY Downstate Med Ctr, 59-66; PROF SURG, ALBERT EINSTEIN COL MED, 67-, CHMN DEPT, 72-; SURGEON-IN-CHIEF COMBINED DEPTS SURG, MONTEFIORE HOSP & MED CTR, 76- *Concurrent Pos:* Markle scholar acad med, 64-69; asst attend Surgeon, Kings County Hosp, Brooklyn, 60-; chief dept surg, Montefiore Hosp & Med Ctr, 67-76. *Honors & Awards:* Dudley Mem Medal, 54; Linder Surg Prize, 54. *Mem:* Am Soc Nephrol; NY Acad Sci; Am Col Surg; Am Gastroenterol Asn; Am Soc Artificial Internal Organs; Am Med Asn; Am Surg Asn; Am Med Writers Asn; Asn Advan Med Instrumentation; Soc Surg Alimentary Tract. *Res:* Gastrointestinal surgery and physiology with emphasis on the liver and pancreas, particularly with regard to problems of cirrhosis, diabetes and transplantation. *Mailing Add:* 111 E 210th St Bronx NY 10467

GLIMCHER, MELVIN JACOB, b Brookline, Mass, June 2, 25; m; c 3. ORTHOPEDIC SURGERY, BIOPHYSICS. *Educ:* Purdue Univ, BS(gen sci) & BS(mech eng), 46; Harvard Univ, MD, 50. *Prof Exp:* From asst prof to assoc prof, 59-65, PROF ORTHOP SURG, HARVARD MED SCH, 65-; DIR, ORTHOP RES LABS, CHILDREN'S HOSP MED CTR, 80- *Concurrent Pos:* Res fel, Harvard Med Sch, 56; res fel biol, Mass Inst Technol & fel, Sch Advan Studies, 56-59; chief orthop serv, Mass Gen Hosp, 66-71; orthop surgeon-in-chief, Children's Hosp Med Ctr, 71-80. *Mem:* Electron Micros Soc Am; Orthop Res Soc; Biophys Soc; Am Acad Arts & Sci; Sigma Xi. *Res:* Biological mineralization. *Mailing Add:* 77 Florence St Apt N606 Chestnut Hill MA 02167

GLIME, JANICE MILDRED, b Cumberland, Md, Jan 27, 41. BRYOLOGY, ECOLOGY. *Educ:* Frostburg State Col, BS, 62; WVa Univ, MS, 64; Mich State Univ, PhD(bot), 68. *Prof Exp:* Asst instr botany, Mich State Univ, 68; from instr to asst prof, Plymouth State Col, 68-70, asst prof biol sci, 70-73; from asst prof to assoc, 73-84, PROF BIOL SCI, MICH TECHNOL UNIV, 84- *Concurrent Pos:* Assoc ed, Hikobia & J Hattori Bot Lab. *Mem:* Ecol Soc Am; Am Bryol & Lichenological Soc; Am Inst Biol Sci; Brit Bryol Soc; Int Asn Bryol. *Res:* Ecology of streams, especially bryophytes and insects; systems ecology and taxonomy of Fontinalis; experimental bryology; effects of acid rain and heavy metals on byrophytes; bryophyte niche width; geothermal plant communities. *Mailing Add:* Dept Biol Sci Mich Technol Univ Houghton MI 49931

GLIMM, JAMES GILBERT, b Peoria, Ill, Mar 24, 34; m 57; c 1. SCIENTIFIC COMPUTING. *Educ:* Columbia Univ, AB, 56, AM, 57, PhD(math), 59. *Prof Exp:* NSF fel, 59-60; from asst prof to prof math, Mass Inst Technol, 60-68; prof math, Rockefeller Univ, 74-82; vis prof math, Courant Inst Math, NY Univ, 80-82, prof, 82- 89; vis leading prof, DISTINGUISHED PROF & HEAD DEPT APPL MATH, STATE UNIV NY, STONY BROOK, 88- *Concurrent Pos:* Temp mem, Inst Advan Study, Princeton, NJ, 59-60; Guggenheim fel & vis assoc prof, Courant Inst Math, 64-65 & Aarhus Univ, Denmark, 65-66; vis mem, courant Inst Math NY Univ, 63-64, prof, 68-74. *Honors & Awards:* Phys & Math Sci Award, NY Acad Sci, 79; Dannie Heineman Prize, 80. *Mem:* Nat Acad Sci; Am Math Soc; Am Acad Arts & Sci; Soc Indust & Appl Math; Soc Petrol Engrs; Int Asn Math Physicists. *Res:* Non-linear differential equations; functional analysis; operators on Hilbert space; mathematical physics; quantum field theory; computational fluid dynamics. *Mailing Add:* Dept Appl Math & Statist State Univ NY Stony Brook NY 11794-3600

GLINSKI, RONALD P, b Windsor, Ont, Aug 4, 41. BIO-ORGANIC CHEMISTRY, PATHOLOGY. *Educ:* Univ Windsor, BS, 63; Wayne State Univ, PhD(org chem), 67; Univ Mich, MD, 75. *Prof Exp:* Sr res chemist org biochem, Ash Stevens Inc, 66, res supvr, 68-75; resident pathol, Univ Mich, 75-80; CONSULT 80- *Mem:* Am Chem Soc; Chem Inst Can; AAAS; AMA; NY Acad Sci; Sigma Xi. *Res:* Organic synthesis; carbohydrates; nucleotides; natural products; kinetics; enzymology; process development. *Mailing Add:* 720 Jefferson St Rte 2 Box 32H Whiteville NC 28472

GLINSKY, MARTIN JAY, animal nutrition, animal management, for more information see previous edition

GLINSMANN, WALTER H, m; c 3. NUTRITION. *Educ:* Columbia Univ, BA, 56, MD, 60. *Prof Exp:* Asst chief, Dept Metab, Walter Reed Army Inst & attend physician med & endocrinol, Walter Reed Army Hosp, 62-65; prog planner, Growth & Develop Prog, Nat Inst Child Health & Human Develop, NIH, 65-66, guest investr, Clin Endocrinol Br, Nat Inst Arthritis & Metab Dis, 66-67, sr res investr, Lab Biomed Sci, Nat Inst Child Health & Human Develop, 67-68, chief, Sect Physiol Controls, 68-78; chief, Exp Nutrit Br & dir res, 78-83, chief, Clin Nutrit Br, 83-87, ASSOC DIR CLIN NUTRIT, DIV NUTRIT, CTR FOOD SAFETY & APPL NUTRIT, FOOD & DRUG ADMIN, DEPT HEALTH & HUMAN SERV, 87- *Concurrent Pos:* Pvt consult res planning & prog develop, Growth & Develop Prog, Nat Inst Child Health & Human Develop, NIH, 65. *Mem:* Am Inst Nutrit; Am Soc Clin Nutrit; Am Physiol Soc; Am Diabetes Asn; AAAS. *Mailing Add:* Div Nutrit Food & Drug Admin 200 C St SW Washington DC 20204

GLISSON, ALLEN WILBURN, JR, b Meridian, Miss, June 26, 51; m 75; c 2. ELECTRICAL ENGINEERING, ELECTROMAGNETIC THEORY. *Educ:* Univ Miss, BS, 73, MS, 75, PhD(elec eng), 78. *Prof Exp:* Res asst physics, 72-73, res asst elec eng, 73-78, asst prof elec eng, 78-84, ASSOC PROF ELEC ENG, UNIV MISS, 84- *Concurrent Pos:* Consult, Northrop Corp, 82, Lincoln Lab, Mass Inst Technol, 83 & TRW, 83-85; Arthur J Schmitt Fac Grant to attend Nat Commun Forum, 83 & 85. *Mem:* Inst Elec & Electronics Engrs; Prof Soc Antennas & Propagation; Prof Soc Microwave Theory & Tech; Sigma Xi; Electromagnetics Soc; Int Union Radio Sci; Prof Soc Electromagnetic Compatibility. *Res:* Radiation and scattering; antennas; mathematical methods; numerical techniques; electromagnetic theory. *Mailing Add:* PO Box 1652 University MS 38677

GLISSON, SILAS NEASE, b Springfield, Ill, May 8, 41; m 68; c 2. CARDIOVASCULAR ANESTHESIA. *Educ:* Southern Ill Univ, BA, 64; Loyola Univ Chicago, Stritch Sch Med, PhD(pharmacol), 72. *Prof Exp:* Instr pharmacol, Univ Conn Med Ctr, 71-72; asst prof, 72-78, ASSOC PROF ANESTHESIOL & PHARMACOL, LOYOLA UNIV, STRITCH SCH MED, 78- *Concurrent Pos:* Consult, Hines Vet Admin Hosp, 77- *Mem:* Am Soc Anesthesiologists; Am Soc Pharmacol & Exp Therapeut; Int Anesthesia Res Soc; Soc Cardiovasc Anesthesiologists. *Res:* Pharmacology of anesthetic and related drugs; mechanisms of neuroendocrine responses associated with surgery and anesthesia; interaction of the sympathoadrenal, renin-angiotensin and adrenal cortical responses to perioperative stresses; anesthetic pharmacokinetics. *Mailing Add:* Dept Anesthesiol/Pharm Loyola Univ 2160 S First Ave Maywood IL 60153

GLITZ, DOHN GEORGE, b Buffalo, NY, Sept 28, 36; m 66; c 1. BIOCHEMISTRY. *Educ:* Univ Ill, BS, 58; Univ Wis, MS, 60; Univ Calif, PhD, 64. *Prof Exp:* Chemist, Swift & Co Res Labs, 58; from asst prof to assoc prof, 67-77, PROF BIOL CHEM, MED SCH, UNIV CALIF, LOS ANGELES, 77-, VICE CHAIR, 78- *Concurrent Pos:* Fel, Enzyme Inst, Univ Wis, 63-64; USPHS fel, Virus Res Unit, Cambridge Univ, 64-66; fel, Virus Lab, Univ Calif, Berkeley, 66-67; Guggenheim fel, Friedrich Miescher Inst, Basel, Switz, 74-75; vis scientist, Nat Inst Med Res, London, Eng, 81-82. *Mem:* Am Soc Biol Chem. *Res:* Nucleic acid biochemistry, including physical and chemical properties; nucleic acid enzymology; ribosome structure. *Mailing Add:* Dept Biol Chem Univ Calif Med Sch Los Angeles CA 90024

GLOBE, SAMUEL, applied physics; deceased, see previous edition for last biography

GLOBENSKY, YVON RAOUL, b Montreal, Que, Feb 6, 37; m 59; c 2. GEOLOGY. *Educ:* Univ Montreal, BSc, 59; Univ NB, MSc, 62, PhD(geol), 65. *Prof Exp:* Asst dir geol, Geol Explor Serv, 64-72, geologist in chg St Lawrence Lowland of Quebec, 77-84, geologist in chg southern div, 84-88, DIR MONTREAL OFF, SERV GEOL, QUEBEC MINISTRY ENERGY & RESOURCES, 88- *Mem:* Fel Geol Soc Am; fel Geol Asn Can; Paleont Soc; Soc Econ Paleont & Mineral; Geol Soc France. *Res:* Geology of the St Lawrence Lowlands; paleontology, stratigraphy and mineral deposits. *Mailing Add:* Que Min Energy & Resources 2100 rue Drummond Suite 240 Montreal PQ H3G 1X1 Can

GLOBUS, ALBERT, b New York, NY, Oct 13, 31; m 51; c 3. NEUROANATOMY, NEUROPHYSIOLOGY. *Educ:* Northwestern Univ, BA, 51, MD, 55. *Prof Exp:* Intern, Orange County Hosp, 55-56; pvt pract, Calif, 58-63; guest scientist, Max Planck Inst, 67-68; asst prof human morphol, psychobiol & psychiat, Univ Calif, Irvine-Calif Col Med, 68-74; assoc prof psychiat, Univ Calif, Davis, 74-; PVT PRACT PSYCHIAT. *Concurrent Pos:* USPHS fel neuroanat, Univ Calif, Los Angeles, 63-67. *Mem:* AMA; Am Asn Anat; Am Psychiat Asn. *Res:* Golgi analysis of cortical anatomy; intracellular physiology of spinal motoneurons; iontophoresis of radioactive amino acids in spinal motoneurons; maternal behavior in primates; psychiatry. *Mailing Add:* 1721 Second St Suite 103 Sacramento CA 95814

GLOCK, ROBERT DEAN, b York, Nebr, Nov 1, 36. VETERINARY PATHOLOGY. *Educ:* Iowa State Univ, DVM, 61, PhD(vet path), 71; Am Col Vet Pathologists, dipl, 74. *Prof Exp:* Vet virologist, US Army, 61-63; vet practr, Edgerton Vet Serv, 63-67; NIH fel, Iowa State Univ, 67-70, asst prof, 70-76, prof vet path, 76-81; DIR, CENT ARIZ VET LAB, 81- *Mem:* Am Vet Med Asn; Am Asn Swine Practrs (pres, 76-77); Sigma Xi; Comp Gastroenterol Soc. *Res:* Swine disease research with emphasis on pathogenesis of enteric diseases and Aujeszky's Disease. *Mailing Add:* Diag Lab Colo State Univ Ft Collins CO 80523

GLOCK, WALDO SUMNER, ecology, geology; deceased, see previous edition for last biography

GLOCKER, EDWIN MERRIAM, applied statistics; deceased, see previous edition for last biography

GLOCKLIN, VERA CHARLOTTE, b Edmonton, Alta, Oct 16, 26; US citizen. DRUG REGULATION, PHARMACOLOGY. *Educ:* Univ Toronto, BA, 48; McGill Univ, MSc, 50; Yale Univ, PhD, 59. *Prof Exp:* USPHS med res grant, Inst Neurol Dis & Blindness, 62-66; res technician pharmacol, Western Reserve Univ, 50-51, res asst ophthal, 51-53; res asst, Cancer Res Lab, Univ Fla, 53-55; res assoc, Eye Res Lab, Univ Chicago, 59-66; pharmacologist, Food & Drug Admin, 66-70, supvry pharmacologist, 70-80, asst dir, Off Drug Eval, 80-89; RETIRED. *Concurrent Pos:* Abstractor, Chem Abstr, 55-60. *Honors & Awards:* Distinguished Career Award, Drug Info Asn, 89. *Mem:* Soc Toxicol; Int Soc Study Xenobiotics; Drug Info Asn; Sigma Xi. *Res:* Neuropharmacology; comparative biochemistry; endocrinology. *Mailing Add:* 4409 Rosedale Ave Bethesda MD 20814

GLOCKNER, PETER G, b Moragy, Hungary, Jan 26, 29; nat US; m 55; c 1. APPLIED SOLID & CONTINUUM MECHANICS, STRUCTURAL ENGINNERRING. *Educ:* McGill Univ, BEng, 55; Mass Inst Technol, MSc, 56; Univ Mich, PhD(civil eng), 64. *Prof Exp:* Struct design engr, C C Parker, Whittaker & Co, Ltd, Alta, 56-58; from asst prof to assoc prof appl mech, Univ Alta, 58-68; prof civil eng, 68-76, prof mech eng & head dept, 76-87; PROF CIVIL & MECH ENG, UNIV CALGARY, 87- *Concurrent Pos:* Chmn, Res & Develop Div, Can Soc Mech Eng, 83-85. *Honors & Awards:* Gzowski Gold Medal, Eng Inst Can, 72; Moisseif Award, Am Soc Civil Engr, 83. *Mem:* Am Soc Civil Engrs; Eng Inst Can; Int Asn Shell Struct; Can Soc Mech Eng; Can Soc Civil Engr. *Res:* Fundamentals of continuum mechanics and solid mechanics and their application to linear and nonlinear plate and shell theory and the static and dynamic linear and nonlinear response of plates and shells. *Mailing Add:* Dept Civil & Mech Eng Univ Calgary Calgary AB T2N 1N4 Can

GLOD, EDWARD FRANCIS, b Youngstown, Ohio, June 28, 42; m 71; c 1. ENVIRONMENTAL & ANALYTICAL CHEMISTRY. *Educ:* Youngstown Univ, BS, 65; Ohio State Univ, MS, 69, PhD(org chem), 71. *Prof Exp:* Res asst, Ohio State Univ, 71-73; surv chemist lab cert, Ohio Dept Health Labs, 73-74, chief, Environ Labs, 74-78; hazardous waste coordr, Ohio Environ Protection Agency, 78-82; RES SCIENTIST, CHEM WASTE MGT INC, 82- *Concurrent Pos:* Mem, Int Joint Comn, 76- *Mem:* Am Chem Soc; Am Soc Testing & Mat. *Mailing Add:* Northrop Corp 600 Hicks Rd Rolling Meadows IL 60008-1098

GLODE, LEONARD MICHAEL, b Chadron, Nebr, July 19, 47; m 69; c 3. MEDICAL ONCOLOGY. *Educ:* Washington Univ, MD, 72. *Prof Exp:* Res assoc microbiol, Lab Microbiol & Immunol, NIH, 74-76; fel med oncol, Dana Farber Cancer Inst, Harvard Med Sch, 76-78; asst prof, 78-83, ASSOC PROF MED, DIV MED ONCOL, UNIV COLO HEALTH SCI CTR, 83- *Concurrent Pos:* Prin investr, Am Cancer Soc Inst Grant Comt, mem bd dirs, Colo Div; dir, Prostate Res Clin; mem, Clin Sci Depts Comt; clin fel, Am Cancer Soc Jr Fac, 80-82; Fogarty sr int fel, 86-87. *Mem:* Am Asn Cancer Res; Am Soc Clin Oncol. *Res:* Utility of ligand directed toxins on regulated expression of toxin genes in cancer therapy. *Mailing Add:* Dept Med Oncol B171 Univ Colo Health Sci Ctr 4200 E 9th Ave Denver CO 80220

GLOECKLER, GEORGE, b Odessa, Russia, Aug 10, 37; US citizen; m 58. PHYSICS. *Educ:* Univ Chicago, SB, 60, SM, 61, PhD(physics), 65. *Prof Exp:* Res assoc cosmic ray physics, Univ Chicago, 65-66, Enrico Fermi fel, 66-67; asst prof physics, 67-74, assoc prof, 74-80, PROF PHYSICS & ASTRON, UNIV MD, COLLEGE PARK, 80- *Concurrent Pos:* Res fel, NASA, 62-64, co-recipient res grant, 68-; Ennco Fermi fel, 64-65; Alfred P Sloan res fel, 69-71; Alexander von Humboldt Sr US Scientist Award, 77-78. *Honors & Awards:* Except Sci Achievement Medal & Group Achievement Award, NASA, 81. *Mem:* Fel Am Phys Soc; Am Astron Soc; Am Geophys Union; NY Acad Sci. *Res:* Cosmic plasma physics and heliospheric physics; cosmic rays; investigation of thermal and suprathermal plasma composition in the heliosphere, magnetospheres of Earth and the outer planets, and near the Giacobini-Zinner comet. *Mailing Add:* 1514-A Farlow Ave Crofton MD 21114

GLOEGE, GEORGE HERMAN, b Helena, Mont, Dec 7, 04; m 27; c 4. CHEMISTRY. *Educ:* Intermountain Union Col, BA, 26; Univ Wis, MS, 31; Mont State Col, DEd, 60. *Prof Exp:* Teacher pub sch, Mont, 27-28; prof chem, Intermountain Union Col, 28-36 & Billings Polytech Inst, 36-39; dean, Custer County Jr Col, 39-42, prin, 42-46; from asst prof to assoc prof, 46-48, chmn dept phys sci, 66-71, prof 48-74, EMER PROF CHEM, EASTERN MONT COL, 73- *Mem:* Fel AAAS. *Res:* Detection of carnuba wax in beeswax; science education, especially college general chemistry laboratory. *Mailing Add:* 2351 Solomon Ave Apt 282 Billings MT 59102-2887

GLOERSEN, PER, b Washington, Pa, Dec 19, 27; m 53; c 3. PHYSICS. *Educ:* Johns Hopkins Univ, MA, 52, PhD(physics), 56. *Prof Exp:* Physicist, Gen Elec Space Sci Lab, 56-70; PHYSICIST, GODDARD SPACE FLIGHT CTR, NASA, 70- *Mem:* Am Phys Soc; NY Acad Sci; Am Geophys Union; Int Glaciological Soc; Sigma Xi. *Res:* Plasma physics; physical properties of radiating shock waves in gases; high resolution molecular spectroscopy in the ultraviolet, visible and infrared; instrumentation; electric propulsion; remote sensing in the microwave regime. *Mailing Add:* 191 Topeg Dr Severna Park MD 21146

GLOGE, DETLEF CHRISTOPH, b Breslau, Ger, Feb 2, 36; m 69. OPTICAL COMMUNICATIONS. *Educ:* Brunswick Tech Univ, Dipl Ing, 61, Dr Ing(optical waveguides), 64. *Prof Exp:* Asst prof electronics, Brunswick Tech Univ, 64-65; mem tech staff optical commun res, 65-75, HEAD, LIGHTWAVE SYSTS DEVELOP, CRAWFORD HILL LAB, BELL LABS, 75- *Mem:* Fel Inst Elec & Electronics Engrs; fel Optical Soc Am. *Res:* Optical communication systems. *Mailing Add:* Bell Tel Labs Crawfords Corner Rd Holmdel NJ 07733

GLOGOVSKY, ROBERT L, b North Chicago, Ill, May 29, 36; m 61; c 4. PHYSICAL CHEMISTRY, BIOCHEMISTRY. *Educ:* Univ Northern Ill, BS, 59; Univ Colo, PhD(phys chem), 65. *Prof Exp:* Kettering Found intern chem, Kalamazoo Col, 64-65; from asst prof to assoc prof, Elmhurst Col, 65-74, chmn chem dept, 71-89, chmn Div Nat Sci, 75-78, PROF CHEM, ELMHURST COL, 74-; PROF, STATE UNIV NY, BUFFALO, MALAYSIA, 90- *Concurrent Pos:* Pres, Sea Educ Develop Group. *Mem:* AAAS; Am Chem Soc. *Res:* Energy transfer and storage; biophysical chemistry; trace metal uptake in plants. *Mailing Add:* Dept Chem Elmhurst Col Elmhurst IL 60126

GLOMB, WALTER L, b Glen Ridge, NJ, Feb 7, 25; m 48; c 6. RADIO COMMUNICATIONS. *Educ:* Columbia Univ, BS, 46, MS, 48. *Prof Exp:* Sr engr, Defense Commun Div, ITT, 50-70, vpres & dir, 70-; RES ENGR, UNITED TECHNOL RES. *Mem:* Fel Inst Elec & Electronics Engrs. *Res:* Radio communication at microwave frequencies via line of sight; tropospheric scatter; satellite communication systems including transmission, switching and system organization. *Mailing Add:* One Oakwood Circle Ellington CT 06029

GLOMSET, JOHN A, b Des Moines, Iowa, Nov 2, 28; m 54; c 2. BIOCHEMISTRY, PHYSIOLOGY. *Educ:* Univ Uppsala, med lic, MD, 60. *Prof Exp:* Docent med chem, Univ Uppsala, 60; from res asst prof to res prof, 60-74, PROF MED & ADJ PROF BIOCHEM, UNIV WASH, 74-, MEM STAFF, REGIONAL PRIMATE CTR, 64-, INVESTR, HOWARD HUGHES MED INST LAB, 76-, PROF BIOCHEM, 84- *Mem:* Nat Acad Sci; Am Soc Biochem & Molecular Biol; Am Soc Cell Biol. *Res:* Protein prenylation; phospholipid metabolism. *Mailing Add:* Howard Hughes Med Inst J611F Health Sci Bldg SL15 Univ Wash Seattle WA 98195

GLOMSKI, CHESTER ANTHONY, b Detroit, Mich, June 8, 28; m 66; c 2. ANATOMY, HEMATOLOGY. *Educ:* Detroit Inst Technol, BS, 51; Wayne State Univ, MS, 53; Univ Minn, PhD(anat), 61; Univ Miss, MD, 65. *Prof Exp:* Teaching asst anat, Sch Med, Univ Minn, 55-61; intern med, St Paul-Ramsey Hosp, St Paul, Minn, 65-66; from asst prof to assoc prof, 66-79, PROF ANAT, SCH MED & DENT, STATE UNIV NY, BUFFALO, 79- *Concurrent Pos:* Consult hemat, Health Care Plan, Inc. *Mem:* Am Asn Anat; Am Soc Hemat; Sigma Xi. *Res:* Experimental hypersplenism; Comparative hematology; erythrocyte kinetics by neutron activation analysis; blood dyscrasias; hemopoiesis and low dose X-irradiation. *Mailing Add:* 75 Windermere Buffalo NY 14226

GLONEK, THOMAS, b Chicago, Ill, Apr 28, 41. PHYSICAL CHEMISTRY, INORGANIC PHYSIOLOGY. *Educ:* Northern Ill Univ, BS, 63; Univ Ill, PhD(biochem), 69. *Prof Exp:* assoc prof biochem, 69-81, ASSOC PROF PHYSIOL, RES RESOURCES CTR, MED CTR, UNIV ILL, 81- *Mem:* Sigma Xi. *Res:* Nuclear magnetic resonance; second row element chemistry; biochemistry; physiology; medicine. *Mailing Add:* 803 Highland Ave Oak Park IL 60304

GLOOR, PIERRE, b Basel, Switz, Apr 5, 23; Can citizen; m 54; c 2. NEUROPHYSIOLOGY. *Educ:* Univ Basel, MD, 49; McGill Univ, PhD(neurophysiol), 57. *Prof Exp:* Lectr EEG, Mcgill Univ, 55-57, assoc electroencephalographer, 55-61, asst prof exp neurol, 57-62, assoc prof clin neurophysiol, 62-68; prof clin neurophysiol, Montreal Neurol Inst, McGill Univ, 68-88; Chief lab, 61-84, ELECTROENCEPHALOGRAPHER, MONTREAL NEUROL HOSP, 61- *Concurrent Pos:* Electroencephalographer, Jewish Gen Hosp, Montreal, 56-61; del, Int Fedn Soc EEG & Clin Neurophysiol, 61-69; physician in charge EEG, Royal Victoria Hosp, Montreal, 62-84. *Honors & Awards:* Robert Bing Prize, Swiss Acad Med Sci, 62; Michael Found Prize, 80; William G Lennox Award, Am Epilepsy Soc, 81; Herbert Jasper Award, Am EEG Soc, 88; Wilder Penfield Award, Can League Against Epilepsy, 90. *Mem:* Can Soc EEG (secy, 62-63, pres, 64-65); Can Physiol Soc; Can Neurol Soc; Am EEG Soc (pres, 78-79); Am Epilepsy Soc (pres, 76); Soc Neurosci; Int Brain Res Orgn. *Res:* Neurophysiology of limbic system and hypothalamus; experimental and clinical electroencephalographic studies on epilepsy; physiological basis and history of electroencephalography. *Mailing Add:* Montreal Neurol Inst 3801 University St Montreal PQ H3A 2B4 Can

GLOOR, WALTER ERVIN, b Cleveland, Ohio, Nov 9, 07; wid. CHEMISTRY. *Educ:* Case Inst Technol, BS, 29, ChE, 34. *Prof Exp:* Res chemist, Hercules Inc, NJ, 29-31, Exp Sta, Del, 31-36, NJ, 36-44, plastics supvr, 44-46, new prod develop, Del, 46-64, supvr prod develop, 64-70, mgr plastics technol, Hercules Inc, Del, 70-72; CONSULT, 72- *Mem:* Am Chem Soc; NY Acad Sci. *Res:* Plastics and new uses; new additives for plastics; better molding compounds from polyolefins; fire retardancy; polymer molecular weight distribution. *Mailing Add:* 4031 Kennett Pike No 168 Wilmington DE 19807-2036

GLOOR, WALTER THOMAS, JR, b Norfolk, Nebr, Sept 26, 24; m 48; c 6. PHARMACY. *Educ:* Univ Nebr, BS, 50; Univ Wash, PhD(pharmacog), 55. *Prof Exp:* Instr pharm, Col Pharm, Univ Nebr, 50-52; sr scientist pharmaceut chem, Smith Kline & French Labs, 55-56, unit head, Invest Prod Lab Mfg Unit, 56-68; assoc prof pharm, Col Pharm, Univ RI, 68-71 & Sch Pharm, WVA Univ, 71-72; asst dean, 72-80, dir allied health sci div, 79-82, PROF SCH PHARM, CREIGHTON UNIV, 72- *Res:* Alkaloid biosynthesis in fungi; problems in manufacture of parenterals, tablets and other pharmaceutical forms. *Mailing Add:* Pharm Sch Creighton Univ 2500 California St Omaha NE 68178

GLOOSCHENKO, WALTER ARTHUR, b Berkeley, Calif, Sept 9, 38; m 67; c 1. WETLANDS, AQUATIC ECOLOGY. *Educ:* Univ Calif, Berkeley, BS, 60; Ore State Univ, PhD(oceanog), 67. *Prof Exp:* Soil scientist, US Bur Reclamation, 61-62; teacher high sch, Calif, 62-64; asst prof oceanog, Fla State Univ, 67-70; res scientist, Fisheries Res Bd Can, 70-73, RES SCIENTIST, NAT WATER RES INST, ENVIRON CAN, 73- *Concurrent Pos:* Adj prof, dept land resource sci, Univ Guelph, Ont, 78-; vis assoc prof, fac environ studies, York Univ, Toronto, 85-; consult, Environ Impact Develop Projs, Mex, Brazil, Nicaragua. *Mem:* Soc Wetland Scientists (pres, 83-84); Int Peat Soc; Ecol Soc Am; Soc Conserv Biol. *Res:* Ecology and geochemistry of salt marshes; wetlands ecology; peat; toxic substances; tropical coastal ecology. *Mailing Add:* Nat Water Res Inst Box 5050 Burlington ON L7R 4A6 Can

GLORIG, ARAM, b Manchester, Eng, June 8, 06; nat US; m 44; c 2. OTOLOGY. *Educ:* Loma Linda Univ, MD, 38; Am Bd Otolaryngol, dipl, 47. *Prof Exp:* Intern, Mem Assoc Hosp, New London, Conn, 37-38; asst, Path Lab, Los Angeles County Hosp, 38-39; resident, Willard Parker Hosp, NY, 39-40; assoc otolaryngol, Emory Univ, 41-42, asst prof, 46-47; chief audiol & speech correction, Phys Med & Rehab Div, Dept Med & Surg, Vet Admin, Washington, DC, 47-53; assoc prof ear, nose & throat, Univ Southern Calif, 53-64; clin prof otolaryngol, Univ Tex Med Sch, Dallas, 64-77; dir, Callier Hearing & Speech Ctr, 64-77, dean sch human develop, 74-77; ASSOC DIR, EAR RES INST, 78- *Concurrent Pos:* Dir audiol & speech correction ctr, Walter Reed Army Hosp, 47-52; instr, Univ Md, 48-53; lectr, Episcopal Eye, Ear & Throat Hosp, 49-53; clin instr, George Washington Univ, 51-53; assoc prof, Loma Linda Univ, 54-64; asst prof, Univ Calif, Los Angeles, 55-64; consult to Surgeon Gen, US Army, Off Voc Rehab, Washington, DC, Civil Aeronaut Admin, Vet Admin. *Mem:* Fel Acoust Soc Am; Am Otol Soc; fel Am Speech & Hearing Asn; fel AMA; fel Am Acad Ophthal & Otolaryngol. *Res:* Audiology; psychoacoustics. *Mailing Add:* Otologic Med Group 2122 W Third St Los Angeles CA 90057

GLORIOSO, JOSEPH CHARLES, III, b Louisville, Ky, July 9, 45. MOLECULAR VIROLOGY, GENETICS THERAPY. *Educ:* Univ Southwestern La, BS, 68, MS 70,State Univ, PhD(microbiol), 74. *Prof Exp:* Post doc, La State Univ Med Ctr, 74-76; from asst prof to assoc prof, 76-84,

PROF MICROBIOL & IMMUNOL, UNIV MICH MED CTR 84-, ASST DEAN RES GRAD STUDIES, 88- *Concurrent Pos:* Spec reviewer Genetic Biol Study Sect, NSF, 81-, mem, Virol Study Sect, NIH, 86-; guest scientist, Ger Cancer Res Ctr, 82-83; mem, Virol study Sect, NIH, 86-; pres, Primagen, 88- *Mem:* Am Soc Microbiol; AAAS; Am Soc Virol; Am Asn Immunologists; Soc Gen Microbiol; Am Chem Soc. *Res:* Herpes simplex virus and neurotropic human pathogen; genetic engineering technologies employed to understand molecular basis of HSV latency; virus reactivation; neuropathogenesis; development of HSV as a gene transfer vector for treatment of autosomal neurologic diseases. *Mailing Add:* Molecular Genetics Biochem Univ Pittsburgh Sch Med Pittsburgh PA 15261

GLORIOSO, ROBERT M, b Danbury, Conn, Apr 18, 40; m 62; c 1. COMPUTER SCIENCE, ELECTRICAL ENGINEERING. *Educ:* Northeastern Univ, BSEE, 62; Univ Conn, MS, 64, PhD(comput sci), 67. *Prof Exp:* Instr elec eng, Univ Conn, Storrs, 64-67; proj eng comput tech team, US Army Electronics Command, Ft Monmouth, 67-69; assoc prof elec & comput eng, Univ Mass, Amherst, 69-76; mgr appl res & develop & tech dir comput syst develop group, 76-80, MGR CORP RES, DIGITAL EQUIP CORP, 80- *Concurrent Pos:* Consult, Conn Res Assocs, 63-72; lectr math, Stevens Inst Technol, Hoboken, 68-69. *Mem:* Sr mem Inst Elec & Electronics Engrs; Asn Comput Mach; Sigma Xi. *Res:* Computer system organization; theory of computer design; human factors in computer system design; intelligent systems. *Mailing Add:* 70 Birch Hill Rd Stow MA 01775

GLOSS, STEVEN PAUL, b Medford, Ore, May 1, 44. WATER RESOURCE MANAGEMENT, AQUATIC ECOLOGY. *Educ:* Mt Union Col, BS, 66; SDak State Univ, MS, 69; Univ NMex, PhD(biol), 77. *Prof Exp:* Proj leader, Utah Div Wildlife Res, 69-74; res biol fisheries, US Fish & Wildlife Serv, 78-87; DIR, WYO WATER RES CTR, 87- *Mem:* Am Fisheries Soc; Sigma Xi; AAAS; Am Water Resources Asn. *Res:* Water resources; water policy; environmental pollutants. *Mailing Add:* Wyo Water Res Ctr Univ Wyo 13th & Lewis PO Box 3067 Laramie WY 82071

GLOSSER, ROBERT, b Johnstown, Pa, Dec 14, 37; m 63; c 1. OPTICAL PROPERTIES OF SOLIDS, MODULATION SPECTROSCOPY. *Educ:* Mass Inst Technol, SB, 59; Univ Chicago, MS, 62, PhD(physics), 67. *Prof Exp:* Res scientist, Naval Weapons Ctr, 67-69; lectr physics, Univ Calif, Santa Barbara, 69-71; asst prof, Univ Md, 71-75; assoc prof, 75-90, PROF PHYSICS, UNIV TEX, DALLAS, 90- *Concurrent Pos:* Consult, Naval Weapons Ctr, 74, Varo, Inc, 84 & 87-90; fac fel, Naval Res Lab, 86, distinguished fac fel, 90. *Mem:* Sigma Xi; Am Phys Soc; Inst Elec & Electronics Engrs. *Res:* Application of optical techniques to study the electronic structure of solids and to characterize material properties; silicon and III-V compounds and alloys. *Mailing Add:* Univ Tex Dallas PO Box 830688 MS F023 Richardson TX 85083-0688

GLOSTEN, LAWRENCE R, ENGINEERING. *Prof Exp:* CHMN, GLOSTEN ASSOC, 58- *Mem:* Nat Acad Eng. *Mailing Add:* 7751 Hansen Rd Bainbridge Island WA 98110

GLOTH, RICHARD EDWARD, b Springfield, Mass, Aug 25, 41; m 66; c 3. ORGANIC & ORGANOMETALLIC CHEMISTRY. *Educ:* Univ Mass, Amherst, BA, 63, MS, 69, PhD(org chem), 70. *Prof Exp:* Sr res chemist, Hydrocarbon Sect, Goodyear Tire & Rubber Co, 69-76, sr res chemist, Rubber Chem Sect, 76-80, prod rep, Specialty Chem, 81-86, latex & resins, 86-89, PROD REP, NITRILE ELASTOMERS, GOODYEAR TIRE & RUBBER CO, 89- *Mem:* Am Chem Soc. *Res:* Synthesis and reactions of arene-chromium and arene-tungsten tricarbonyls; synthesis of organometallic catalysts; homogeneous and heterogeneous catalysis; accelerations; rubber antioxidants. *Mailing Add:* Chem Div Dept 743 Goodyear Tire & Rubber Co Akron OH 44316-0001

GLOVER, ALLEN DONALD, b Kirksville, Mo, Jan 6, 38; m 63; c 1. ORGANIC CHEMISTRY. *Educ:* Culver-Stockton Col, AB, 60; Carnegie Inst Technol, PhD(chem), 65. *Prof Exp:* Res fel chem, Calif Inst Technol, 65-66; asst prof, 66-71, ASSOC PROF CHEM, BRADLEY UNIV, 71-, COORDR MED TECHNOL & COOP COORDR SCI, 85- *Concurrent Pos:* Res assoc, dept org chem, Univ Adelaide, Australia, 73 & dept clin biochem, Univ Otago, Dunedin, NZ, 80 & 87. *Mem:* Am Chem Soc; Royal Soc Chem; Sigma Xi. *Res:* Structure elucidation of the products formed by photodimerizations; photochemical cycloaddition reactions of conjugated keto steroids. *Mailing Add:* Dept Chem Bradley Univ Peoria IL 61606

GLOVER, BENJAMIN HOWELL, b Chicago, Ill, Apr 29, 16; m 40; c 2. PSYCHIATRY. *Educ:* Northwestern Univ, BS, 37, MS, 39, MD, 43; Am Bd Psychiat, dipl, 52. *Prof Exp:* Instr biol, Allegheny Col, 40-42; asst prof, 47-58, Univ Wis-Madison asst dean, Med Sch, 72-76, assoc prof, 58-61; chief psychiat, Vet Admin Med Ctr, Tomah, Wis, 81-87; CONSULT, ROCK CO WIS MENT HEALTH CTR, 87-; EMER PROF PSYCHIAT, UNIV WIS-MADISON, 61- *Concurrent Pos:* Consult, Vet Hosp, Madison. *Mem:* Fel Am Psychiat Asn; Pan-Am Med Asn. *Res:* Psychotherapy; geriatrics; continuing medical education. *Mailing Add:* 3706 Spring Trail Madison WI 53711

GLOVER, CLAIBORNE V C, III, b Atlanta, Ga, July 8, 47; m 80; c 1. PROTEIN PHOSPHORYLATION. *Educ:* Duke Univ, BA, 69; Ga State Univ, MS, 74; Univ Rochester, PhD(biol), 79. *Prof Exp:* Post-doctoral fel biochem, Stanford Univ, 79-83; asst prof biochem, 83-88, asst prof genetics, 84-88, ASSOC PROF BIOCHEM & GENETICS, UNIV GA, 88- *Mem:* AAAS; Am Soc Biochem & Molecular Biol. *Res:* Biochemical, molecular and genetic analysis of protein kinases and their substrates in D melanogaster and S cevevisiae; regulation of nuclear protein function by phosphorylation/dephosphorylation. *Mailing Add:* Dept Biochem Univ Ga Athens GA 30602

GLOVER, DAVID VAL, b Farmington, Utah, May 18, 32; m 54; c 6. CYTOGENETICS. *Educ:* Utah State Univ, BS, 54, MS, 59; Univ Calif, PhD(genetics), 62. *Prof Exp:* Asst agron, Utah State Univ, 57-59; asst genetics & plant breeding, Univ Calif, 59-62; asst prof, 62-65, assoc prof, 65-72, PROF PLANT GENETICS & BREEDING, PURDUE UNIV, 72- *Concurrent Pos:* Vis Scholar, Univ MN, 84- *Mem:* AAAS; Am Genetic Asn; fel Am Soc Agron; Sigma Xi; fel Crop Sci Soc Am; fel AAAS; Coun Agr Sci & Technol. *Res:* Genetics and cytogenetics of crop plants; genetics and biochemistry of maize endosperm gene interactions; carbohydrate, protein and nutritional quality in plant genetics and breeding; autotetraploid inheritance in alfalfa; breeking and physiology of alfalfa; somoclonal genetics. *Mailing Add:* 220 Knox Dr West Lafayette IN 47906

GLOVER, ELSA MARGARET, b Rochester, NY, Nov 11, 39. PHILOSOPHY OF SCIENCE. *Educ:* Univ Rochester, BS, 61; Purdue Univ, MS, 63, PhD(physics), 67. *Prof Exp:* Asst prof, 67-71, assoc prof physics, 71-84, PROF PHYSICS, STILLMAN COL, 84- *Concurrent Pos:* Lectr, Int Hq Rosicrucian Fellowship, 74-89. *Mem:* Sigma Xi; Am Asn Physics Teachers. *Res:* Scientific method and other methods of searching for truth; the relation between physics and metaphysics; the nature of space, time and matter. *Mailing Add:* Dept Physics Stillman Col Tuscaloosa AL 35403

GLOVER, FRANCIS NICHOLAS, b Brooklyn, NY, Jan 11, 25. PHYSICS. *Educ:* Sacred Heart Col, Philippines, AB, 49, MA, 50; St Louis Univ, PhD, 55. *Prof Exp:* Instr physics, Ateneo de Manila Univ, 50-51; asst, St Louis Univ, 53-55; res assoc, Woodstock Col, Md, 55-59; res assoc ionosphere res sect, Manila Observ, 59-65; PROF PHYSICS, ATENEO DE MANILA UNIV, 65- *Mem:* Am Phys Soc; Inst Elec & Electronics Eng. *Res:* Radio propagation; physics of the ionosphere; computer applications. *Mailing Add:* Dept Physics Ateneo de Manila Univ Manila Philippines

GLOVER, FRED ARTHUR, wildlife management, for more information see previous edition

GLOVER, FRED WILLIAM, b Kansas City, Mo, Mar 8, 37; m 88; c 2. ARTIFICIAL INTELLIGENCE, MATHEMATICAL PROGRAMMING. *Educ:* Univ Mo, BBA, 60; Carnegie-Mellon Univ, PhD, 65. *Prof Exp:* Asst prof, Univ Calif, Berkeley, 65-66; assoc prof, Univ Tex, Austin, 66-69; prof, Univ Minn, Minneapolis, 69-70; John King prof, 70-87, US WEST CHAIR SYST SCI, UNIV COLO, BOULDER, 87- *Concurrent Pos:* Consult, over 70 US Corps & govt agencies, 65-, US Cong, 84 & Nat Bur Standards, 86; lectr, NATO, France, Italy, Ger, Denmark, 70-89, Inst Decision Sci, 84; chmn, Bd Analysis, Res & Computation, Austin, 71-83; bd dirs, Mgt Systs & Decision Analysis Inst, Boulder, 74-82; partic host vis exchange, Nat Acad Sci, 81; res dir, Artificial Intel Ctr, Boulder, 84-; head res assoc, Global Optimization Space Consternation Ctr, Boulder, 88-; vis res chair eng, Univ Tex, Austin, 89; invited distinguished lectr, Swiss Fed Inst Technol, Lausanne, 90-; res prin, US West Joint Res Initiative, Univ Colo, 90-; prin investr, Air Force Off Sci Res, Off Naval Res, 90-; mem, US Nat Adv Bd Univ Res Initiative on Combinatorial Optimization. *Honors & Awards:* Int Achievement Award, Inst Mgt Sci, 82; Outstanding Achievement Award, Am Inst Decision Sci, 84. *Mem:* Fel AAAS; Am Inst Decision Sci; Am Asn Collegiate Schs Bus. *Res:* Development of network and netform models for optimization, and of heuristic search methods linking artificial intelligence and mathematical programming. *Mailing Add:* Col Bus & Admin Univ Colo Box 419 Boulder CO 80309-0419

GLOVER, GEORGE IRVIN, b Oakland, Calif, Mar 8, 40; m 59; c 2. ENZYMOLOGY, PROTEIN CHEMISTRY. *Educ:* Univ Calif, Berkeley, BS, 62, PhD(org chem), 66. *Prof Exp:* Sr res chemist, Air Prod & Chem, Inc, 66-67; res collabr protein chem & enzyme, Brookhaven Nat Lab, 67-70; from asst prof to assoc prof, Tex A&M Univ, 70-80; res mgr biocatalysis, 80-83, res mgr protein biochem, bio sci dept, 84-89, SCI FEL, MONSANTO CO, 89-; ASSOC DIR PROTEIN BIOCHEM, SMITH KLINE BEECHAM, 89- *Concurrent Pos:* NIH fel, 67-69; asst vis scientist, Brookhaven Nat Lab, 69-70. *Mem:* Am Soc Biol Chemists; Am Chem Soc. *Res:* Irreversible enzyme inhibitors; purification, kinetics and active sites of enzymes; synthesis of potential pesticides; medicinal chemistry; protein biochemistry. *Mailing Add:* Res & Develop Div SmithKline Beecham PO Box 1539 King of Prussia PA 19406-0939

GLOVER, KENNETH MERLE, b Hamilton, Ohio, Oct 11, 28; m 76; c 4. ATMOSPHERIC PHYSICS, ELECTRONICS. *Educ:* Miami Univ, AB, 58, MS, 60. *Prof Exp:* Res engr, Corp Res Labs, P R Mallory & Co, 60-62; res physicist, Bedford, 62-70, supvry res physicist, Air Force Cambridge Res Labs, Off Aerospace Res, 70-74, CHIEF WEATHER RADAR BR, AIR FORCE GEOPHYS LAB, AIR FORCE SYST COMMAND, L G HANSCOM AFB, 74- *Mem:* Inst Elec & Electronics Engrs; Am Meteorol Soc; Sigma Xi. *Res:* Radar backscatter cross-sections of dielectric spheres; backscatter from the clear atmosphere; radar observations of the tropopause; radar observations of insects; remote sensing of clear air turbulence; Doppler radar data processing and display; use of Doppler radar in severe storms; atmospheric remote sensing. *Mailing Add:* 42 Jackson Dr Acton MA 01720

GLOVER, LEON CONRAD, JR, b Exeter, NH, July 5, 35; m 63; c 3. ORGANIC CHEMISTRY, POLYMER CHEMISTRY. *Educ:* Univ Notre Dame, BS, 57; Stanford Univ, PhD(org chem), 62. *Prof Exp:* Chemist, Shell Develop Co, Calif, 62-64; staff chemist, 64-73, mgr, 73-81, TECH DIR, RAYCHEM CORP, 81- *Mem:* Am Chem Soc; Sigma Xi; Soc Plastics Engrs. *Res:* Organic peroxides; vanadium catalyzed olefin polymerizations; lithium catalyzed living polymer systems; high temperature polymer synthesis and evaluation; polymer characterization; radiation chemistry; hot melt adhesives development; polymer additives evaluation. *Mailing Add:* 1131 Buckingham Dr Los Altos CA 94024

GLOVER, LYNN, III, b Washington, DC, Nov 29, 28; m 50. TECTONICS. *Educ:* Va Polytech Inst & State Univ, BS, 52, MS, 53; Princeton Univ, PhD(geol), 67. *Prof Exp:* Geologist, US Geol Surv, 53-70; PROF GEOL, VA POLYTECH INST & STATE UNIV, 70- *Mem:* Fel Geol Soc Am; Am Asn Petrol Geol; Soc Econ Paleont & Mineral; Am Geophys Union; Cosmos Club; Explorer's Club. *Res:* Appalachian geology and tectonics. *Mailing Add:* Dept Geol Sci Va Polytech Inst & State Univ Blacksburg VA 24061

GLOVER, ROBERT E(LLSWORTH), engineering; deceased, see previous edition for last biography

GLOVER, ROLAND LEIGH, b Peterborough, Ont, Dec 19, 11; nat US; m 37. ELECTROCHEMISTRY. *Educ:* Univ Toronto, BASc, 34, MASc, 35. *Prof Exp:* Asst, Univ Toronto, 34-35; lab technician, Nat Carbon Co Div Union Carbide Corp, Ont, 35-39, develop engr, Ohio, 39-44, head prod & process develop, 44-45, asst supt in chg eng & lab, 45-48, supt, Ont, 49-50, factory mgr, 51-54, asst dir res, Ohio, 54-56, asst to vpres & gen mgr NY, 56-59, gen mgr tech, Union Carbide Consumer Prod Div, 59-62, vpres, 62-73, technol consult, Battery Prod Div, Union Carbide Corp, 73-76; RETIRED. *Concurrent Pos:* Exec secy, Corp Technol Comt, 74-76. *Mem:* Electrochem Soc; fel Chem Inst Can. *Res:* Administration of chemical and electrochemical consumer products. *Mailing Add:* 2145 Lemon Ave Englewood FL 34223

GLOVER, ROLFE ELDRIDGE, III, b Wilmington, Del, Sept 6, 24; m 57; c 3. SOLID STATE PHYSICS. *Educ:* Bowdoin Col, BA, 48; Mass Inst Technol, BS, 48; Univ Gottingen, PhD(physics), 53. *Prof Exp:* NSF fel, Univ Calif, 55-57; asst prof, Univ NC, 57-61; assoc prof, 61-66, PROF PHYSICS, UNIV MD, COLLEGE PARK, 66- *Concurrent Pos:* Alfred P Sloan res fel, 58-62. *Mem:* Fel Am Phys Soc; Am Asn Physics Teachers. *Res:* Superconductivity; phase transitions; amorphous materials; thin films; chemical reactions at low temperatures. *Mailing Add:* Dept Physics & Astron Univ Md College Park MD 20742

GLOVER, ROY ANDREW, b Cleveland, Ohio, Apr 7, 41; m 63; c 3. NEUROANATOMY, MEDICAL EDUCATION. *Educ:* Calvin Col, BS, 63; Ohio State Univ, MSc, 65, PhD(anat), 68. *Prof Exp:* From instr to asst prof, 68-88, ASSOC PROF ANAT, UNIV MICH, ANN ARBOR, 88- *Honors & Awards:* Richard O'Connor Award for Res Excellence, Arthroscopy Asn NAm, 90. *Mem:* Am Asn Anatomists; Soc Neurosci; Am Asn Advan Sci; Soc Sigma Xi. *Res:* Development neurobiology of the neural tube in neurological mutations; biology of neuroglia; medical education; peripheral nerve degeneration and regeneration; role of substance P in the etiology of knee joint pain and the inflammatory changes accompanying degenerative joint disease; multiple substructure anatomical organ model patent. *Mailing Add:* Dept Anat Univ Mich Ann Arbor MI 48109

GLOVER, SANDRA JEAN, b Mansfield, La, Jan 19, 39. ENTOMOLOGY. *Educ:* Northwestern State Univ, BS, 61; Univ Ga, MEd, 65, PhD(entom), 68. *Prof Exp:* Teacher high sch, La, 61-64; asst prof biol, Miss State Col Women, 68; res fel entom, Univ Ga, 68-69; assoc prof, 69-80, PROF BIOL, APPALACHIAN STATE UNIV, 80- *Concurrent Pos:* Instr, Univ Ga, 64-65 & fel, 68-69. *Mem:* Entom Soc Am; Am Inst Biol Sci; AAAS. *Res:* Insect physiology and behavior. *Mailing Add:* Dept Biol Appalachian State Univ Boone NC 28608

GLOVSKY, M MICHAEL, b Boston, Mass, Aug 15, 36. ALLERGY, CLINICAL IMMUNOLOGY. *Educ:* Tufts Univ, BS, 57, MD, 62. *Prof Exp:* VIS ASSOC CHEM, CALIF INST TECHNOL, 77-; CLIN PROF MED, UNIV SOUTHERN CALIF, LOS ANGELES, 84- *Concurrent Pos:* Spec NIH res fel, Walter Reed Army Inst Res, 66-69; fel hemat & immunol, Univ Calif, San Francisco, 68-69; fel Kaiser Permanente Med Group, 69-71, chief dept allergy & clin immunol, 72-, prin investr, Allergy Res Lab, 70-84, co-dir residency prog allergy & clin immunol, 74-84; head, Asthmat & Allergy Ctr, Huntington Mem Hosp, Pasadena, Calif, 89- *Mem:* AAAS; Reticuloendothelial Soc; Am Asn Immunologists; fel Am Acad Allergy. *Res:* Interaction of immunoglobulins and complement; role of complement in human disease; mechanisms of anaphylaxis and allergy. *Mailing Add:* 1961 Oak St South Pasadena CA 91030

GLOWA, JOHN ROBERT, b Washington, DC, Sept 10, 46. NEUROENDOCRINOLOGY. *Educ:* Univ Md, BS, 72, MS, 76, PhD(biopsychol), 79. *Prof Exp:* Grad res asst, Dept Psychol, Univ Md, 75-79; res assoc, Dept Psychiat, Harvard Med Sch, 79-83; sr pharmacologist, Ayerst Labs Res, 83-84; sr staff fel, Clin Neurosci Br, 84-86, CHIEF, BIOPSYCHOL UNIT, CLIN NEUROENDOCRINOL BR, NIMH, 87-; ASSOC PROF PSYCHOL, UNIV COL, UNIV MD, 89- *Concurrent Pos:* AAAS environ sci fel, 86; vis asst prof, Dept Psychol, Univ Md, 86-87; coordr, Biopsychol Prog & assoc prof, Psychol Dept, Hood Col, Frederick, Md, 86-; res psychologist-in-residence, Dept Psychol, Am Univ, 87- *Mem:* Fel Am Psychol Asn; Am Soc Pharmacol & Exp Therapeut; Behav Pharmacol-Toxicol Soc; Soc Neurosci. *Mailing Add:* NIMH Bldg 14D Rm 311 Bethesda MD 20814

GLOWACKI, JULIE, b Boston, Mass, Sept 16, 44. ORTHOPEDIC. *Educ:* Boston Univ, BA, 66; Harvard Univ, PhD, 73. *Prof Exp:* Res fel med, Endocrine Unit, Mass Gen Hosp, Harvard Med Sch, 72-74, assoc med, Med Sch, 74-76, res fel surg, 75-76, assoc biol chem, 75-78, res assoc, Dept Surg, 77-82, asst prof, 82-88, ASSOC PROF, DEPT ORTHOP SURG, HARVARD MED SCH, 89-; SR INVESTR, DEPT ORTHOP SURG, BRIGHAM & WOMEN'S HOSP, BOSTON, 88- *Concurrent Pos:* Assoc med, Mass Gen Hosp, 74-76 & asst biochem, Endocrine Unit, 76-77; sr scientist & admin dir, Bay Biochem Res, 75-77; vis scientist, Univ Mex & Nat Inst Nutrit, 76-77; vis investr, Salk Inst, La Jolla, 82-83; vis res biologist, Vet Admin Med Ctr, La Jolla, 82-87; consult, US Army, 89-; mem, Sci-Technol Comt, Am Asn Tissue Banks, 89-91 & Dent Prod Panel, US Food & Drug Admin, 91- *Mem:* Am Asn Dent Res; Am Elasmobr Soc; Am Soc Bone & Mineral Res; Am Soc Cell Biol; Am Soc Zoologists; Endocrine Soc; Int Asn Dent Res; Orthop Res Soc; Plastic Surg Res Coun; Soc Exp Biol & Med. *Res:* Cell biology of osteoclasts, osteoblasts and chondroblasts; clinical use of demineralized bone in craniomaxillofacial orthopedic and periodontal surgery; regulation of bone formation; regulation and mechanisms of bone resorption; pathophysiology of metabolic bone disease; hemangiomas and vascular malformations; mast cell as a bone cell; comparative skeletal and mineral metabolism. *Mailing Add:* Dept Orthop Res Brigham & Women's Hosp 75 Francis St Boston MA 02115

GLOWER, DONALD D(UANE), b Shelby, Ohio, July 29, 26; m 53; c 4. MECHANICAL ENGINEERING, NUCLEAR ENGINEERING. *Educ:* US Merchant Marine Acad, BS, 46; Antioch Col, MS, 53; Iowa State Univ, MSc, 58, PhD, 60. *Prof Exp:* Asst engr officer, Grace Lines Inc, 47-49; res engr, Battelle Mem Inst, 53-54; asst prof, Col Eng, Iowa State Univ, 54-58, asst prof nuclear eng, 60-61; fel, NSF, 59-60; mem res staff, Sandia Corp, 61-63; head, Radiation Effects Dept, Gen Motors Corp, 63-64; prof & chmn nuclear eng, Ohio State Univ, 64-68, prof & chmn mech eng, 68-76, dean Col Eng, 76-90, VPRES COMMUN & DEVELOP, OHIO STATE UNIV, 90- *Concurrent Pos:* mem, Ohio Atomic Energy Bd, Ohio Indust Technol Adv Bd. *Mem:* Fel Am Nuclear Soc; Am Soc Eng Educ; fel Am Soc Mech Engrs; Sigma Xi. *Res:* Energy conversion. *Mailing Add:* 2338 Kensington Dr Columbus OH 43221

GLOWIENKA, JOHN CLEMENT, b Milwaukee, Wis, Apr 18, 48; m 74. PLASMA PHYSICS. *Educ:* Rensselaer Polytech Inst, BS, 71, ME, 73, PhD(plasma dynamics), 75. *Prof Exp:* Eng asst elec power, Cent Hudson Gas & Elec Corp, 67-71; res asst plasma dynamics, Rensselaer Polytech Inst, 71-75; electronics engr plasma diag, Technol Div, Simulation Br, Air Force Weapons Lab, 75-78; asst prof plasma physics, Univ Ill, Urbana-Champaign, 78-80; MEM RES STAFF, OAK RIDGE NAT LAB, 80- *Mem:* Am Phys Soc; Inst Elec & Electronics Engrs. *Res:* High energy density; magnetically confined plasma diagnostics by heavy ion beam probing; x-ray diagnostics. *Mailing Add:* 235 Briarcliff Ave Oak Ridge TN 37830

GLOYER, STEWART EDWARD, b Milwaukee, Wis, May 23, 42; m 67; c 2. INDUSTRIAL ORGANIC CHEMISTRY. *Educ:* Univ Rochester, BS, 64; Univ Mich, MS, 66, PhD(chem), 69; Univ Chicago, MBA, 77. *Prof Exp:* Asst org chem, Univ Mich, 64-69; res chemist, Kraft Res & Develop Div, 69-71, group leader, 71-76, com develop mgr, Humko Scheffield Chem Div, 77-78, tech dir, Humko Sheffield Chem, Div Kraft Inc, 78-80, tech dir, 80-83; VPRES RES & DEVELOP HUMKO CHEM, DIV WITCO CORP, 83-, VPRES BUS MGR, 89- *Mem:* Soc Automotive Engrs; Am Chem Soc; Am Oil Chemists Soc; Soc Plastics Engrs. *Res:* Organic nitrogen chemistry; fatty acid chemistry; synthetic lubricant base stocks; additives for plastics; homogeneous and heterogeneous catalysis of organic reactions. *Mailing Add:* Humko Chem Div Witco Corp PO Box 125 Memphis TN 38101

GLOYER, STEWART WAYNE, b Milwaukee, Wis, May 22, 10; m 39; c 3. CHEMISTRY. *Educ:* Beloit Col, BS, 32; Univ Wis, PhD(chem), 39. *Prof Exp:* Res chemist, Coatings & Resins Div, PPG Industs Inc, 39-44, group leader, 44-52, asst dir res, 52-58, assoc dir res & develop, 58-65, div dir coatings develop, 65-68, div tech dir indust prod, 68-75; mgt & tech consult, Coatings Indust, 75-83; RETIRED. *Mem:* Am Chem Soc; Forest Prod Res Soc; fel Am Inst Chemists; Fedn Socs Paint Technol. *Res:* Protective and industrial coatings; resins; automotive finishes; electrocoating; ultraviolet and radiation coatings; coil and container coatings. *Mailing Add:* 3145 Madison St Waukesha WI 53188-4131

GLOYNA, EARNEST F(REDERICK), b Vernon, Tex, June 30, 21; m 46; c 2. SANITARY & ENVIRONMENTAL ENGINEERING. *Educ:* Tex Tech Univ, BS, 46; Univ Tex, MS, 49; Johns Hopkins Univ, DrEng, 52. *Prof Exp:* Jr engr, Tex Hwy Dept, Wichita Falls, 46; off engr, Magnolia Petrol Co, Dallas, Tex, 46-47; from asst prof to assoc prof civil eng, Univ Tex, Austin, 47-59, prof & dir, Environ Health Eng Labs, 54-70 & Ctr Res Water Resources, 63-73, dean, Col Eng, 70-87, Bettie Margaret Smith chair environ health eng, 82-87, PROF CIVIL ENG, UNIV TEX, AUSTIN, 59- *Concurrent Pos:* Res asst, Johns Hopkins Univ, 50-52; guest engr, Brookhaven Nat Lab, 52; consult indust, USAF, USArmy, US Senate, Environ Protection Agency & UN; mem, Sci Adv Bd, US Environ Protection Agency, chmn, 82; dir, Environ Health Eng Labs, Univ Tex, Austin, 53-70, Ctr Res Water Resources, 63-76; gov coun, Nat Acad Eng; dir, Parker Drilling Co, Okla, 78-, bd dirs, US Nat Comt, World Energy Conf, 79-80, Nat Soc Prof Engrs, 81-88, Civil Eng Res Found, Am Soc Civil Engrs, 89-90, Res Found, Water Pollution Control Fedn, 89-90 & Epsey, Huston & Assocs, 90-; mem, Steinman Coun, Nat Soc Prof Engrs, 87- *Honors & Awards:* Environ Qual Award, Environ Protection Agency, 77; Harrison Prescott Eddy Medal, Water Pollution Control Fedn, 59 & Gordon Maskew Fair Medal, 79; Gordon Maskew Fair Award, Am Acad Environ Engrs, 82; J J King Prof Eng Achievement Award, 82; Order of Henri Pittier, Nat Conserv Medal, Repub Venezuela, 83; Award for Appln of Sci Knowledge in Pub Affairs, Nat Environ Develop Asn, 83; Simon W Freese Environ Eng Award, Am Soc Civil Engrs, 86; Nat Conserv Achievement Award in Sci, 86 & Nat Wildlife Fedn, 87. *Mem:* Nat Acad Sci; Nat Acad Eng; hon mem & fel Am Soc Civil Engrs; Asn Environ Eng Prof (past pres); Am Acad Environ Engrs (pres, 83); Am Inst Chem Engrs; hon mem Water Pollution Control Fedn (pres, 83-84); Nat Soc Prof Engrs; Am Soc Eng Educ; Am Water Works Asn; Asn Prof Environ Eng (past pres). *Res:* Water and wastewater treatment; author or editor of eight books, and author or coauthor of about 300 publications. *Mailing Add:* Dept Civil Eng Univ Tex Austin TX 78712

GLUCK, JEREMY V(ICTOR), metallurgical engineering, for more information see previous edition

GLUCK, LOUIS, b Newark, NJ, June 18, 24; m 47; c 4. BIOCHEMISTRY, PEDIATRICS. *Educ:* Rutgers Univ, BSc, 48; Univ Chicago, MD, 52; Am Bd Pediat, dipl, 63. *Prof Exp:* Intern med, Univ Clin, Univ Chicago, 53; resident pediat, Columbia-Presby Med Ctr, 56-59; instr, Sch Med, Stanford Univ, 59-60; from asst prof to assoc prof pediat, Sch Med, Yale Univ, 60-68; prof, Sch Med, Univ Miami, 68-69; prof reproductive med, 69-80, PROF PEDIAT, SCH MED, UNIV CALIF, SAN DIEGO, 69- *Concurrent Pos:* Ed-in-chief,

Current Probs Pediat; chmn, State of Calif Area XIII Ment Retardation Bd, 73-75; exec ed, Perinatology & Neonatology J. *Mem:* NY Acad Sci; Soc Study Reproduction; Perinatal Res Soc; Am Pediat Soc; Soc Pediat Res; Sigma Xi. *Res:* Developmental biochemistry; neonatal intensive care; fundamental studies on lung maturity of the fetus; L/S ratio test for lung maturity. *Mailing Add:* Med Ctr Pediat-Bldg 27 RT 81 Univ Calif Irvine 101 City Dr S Orange CA 92668

GLUCK, RONALD MONROE, b Brooklyn, NY, Jan 25, 37; m 58; c 3. SOLID STATE CHEMISTRY & ELECTRONICS. *Educ:* Polytech Inst Brooklyn, BS, 57; Univ Rochester, MS, 71. *Prof Exp:* Chemist, 57-68, res chemist, 68-75, sr res chemist, 76-81, RES ASSOC, EASTMAN KODAK CO, 81- *Mem:* Electrochem Soc. *Res:* Materials oriented research and process development for solid state microelectronic image sensors. *Mailing Add:* 55 Sandcastle Rochester NY 14622

GLUCKSTEIN, FRITZ PAUL, b Berlin, Ger, Jan 24, 27; US citizen; m 55; c 1. INFORMATION COLLECTION & RETRIEVAL. *Educ:* Univ Minn, BS, 53, DVM, 55; Am Bd Vet Pub Health, dipl, 66; Univ Md, MLS, 84. *Prof Exp:* Vet meat inspector, USDA, 55-56, asst vet pathologist, 58-59, vet analyst, 59-63; chief microbiol br, Sci Info Exchange, Smithsonian Inst, 63-66; COORDR VET AFFAIRS, NAT LIBR MED, US PUB HEALTH SERV, 66- *Mem:* Am Vet Med Asn; Am Pub Health Asn; US Animal Health Asn; Am Asn Lab Animal Sci; fel Royal Soc Health; Med Libr Asn. *Res:* Foreign animal diseases; laboratory animal science; veterinary public health; analysis and retrieval of scientific information. *Mailing Add:* 5603 Oak Pl Bethesda MD 20817

GLUCKSTEIN, MARTIN E(DWIN), b Brooklyn, NY, May 19, 28; m 51; c 5. CHEMICAL ENGINEERING, FUEL TECHNOLOGY. *Educ:* Univ Mich, BSE, 50, MSE, 51, PhD, 57. *Prof Exp:* Asst, Eng Res Inst, Univ Mich, 51, res assoc, 52-56, assoc res engr, 56; chem engr, Ethyl Corp, 56-63, res assoc, 63-64, res supvr, 64-86, res & develop mgr, 86-89, MGR PROCESS SAFETY, ETHYL CORP, 90- *Concurrent Pos:* Instr, Wayne State Univ, 57-62. *Mem:* AAAS; Am Chem Soc; Am Inst Chem Engrs; Combustion Inst. *Res:* Combustion; rate and transport processes; advanced structural ceramics; research management; fuels; lubricants; chemical process safety; catastrophe investigation. *Mailing Add:* 2625 W High Meadow Ct Baton Rouge LA 70816

GLUCKSTERN, ROBERT LEONARD, b Atlantic City, NJ, July 31, 24. PHYSICS. *Educ:* City Col New York, BEE, 44; Mass Inst Technol, PhD(physics), 48. *Prof Exp:* AEC fel physics, Univ Calif, Berkeley, 48-49 & Cornell Univ, 49-50; res assoc, Yale Univ, 50-51, from asst prof to assoc prof, 51-64; prof & head dept, Univ Mass, Amherst, 64-69, assoc provost, 69-71, vchancellor acad affairs & provost, 71-75; chancellor, 75-82, PROF, UNIV MD, COLLEGE PARK, 75- *Concurrent Pos:* Consult, Lawrence Radiation Lab, Livermore, 58-61; Yale Univ fac fel, European Orgn Nuclear Res, Switz, 61-62; consult, Brookhaven Nat Lab, 63-, Los Alamos Nat Lab, 64- & Fermi Nat Accelerator Lab, 68-70 & 84-87; vis prof, Univ Tokyo, 70. *Mem:* Am Phys Soc; Am Asn Physics Teachers. *Res:* Theoretical, nuclear and elementary particle physics; accelerator theory. *Mailing Add:* Univ Md College Park MD 20742

GLUDE, JOHN BRYCE, marine biology, for more information see previous edition

GLUECK, BERNARD CHARLES, b Baltimore, Md, Aug 26, 14; m 36; c 2. PSYCHIATRY. *Educ:* Columbia Univ, AB, 34, CPM, 51; Harvard Med Sch, MD, 38. *Prof Exp:* Instr psychiat, Col Physicians & Surgeons, Columbia Univ, 52-55, assoc psychoanalyst, Psychoanalysis Clin, 51-55; from assoc prof to prof, Univ Minn, 55-60; dir res psychiat, Inst Living, 60-79; RETIRED. *Concurrent Pos:* Lectr, Sch Med, Yale Univ, 66-79; prof, Dept Psychiat, Univ Conn, 78-82. *Mem:* Am Psychosom Soc; fel Am Psychiat Asn; Am Psychopath Asn (treas, 52-64, vpres, 64-65, pres, 66-67); Asn Psychoanal Med; fel Acad Psychoanal. *Res:* Psychoanalysis; computers; behavioral science. *Mailing Add:* 275 Steele Rd West Hartford CT 06117

GLUECK, HELEN IGLAUER, b Cincinnati, Ohio, Feb 4, 07; m 31; c 1. MEDICINE. *Educ:* Univ Wis, BA, 29; Univ Cincinnati, MD, 34. *Prof Exp:* From asst prof to assoc prof, 50-67, prof med, Col Med, 67-80, EMER PROF PATH & LAB MED, UNIV CINCINNATI, 80- *Concurrent Pos:* Dir coagulation lab, Cincinnati Gen Hosp, 56-, Student Health Ctr, 49-59. *Mem:* Am Soc Hemat; Am Fedn Clin Res; fel Am Col Physicians; Int Soc Hemat. *Res:* Blood coagulation. *Mailing Add:* Dept Path Univ Cincinnati Col Med Cincinnati OH 45267

GLUNTZ, MARTIN L, b Harrisburg, Pa, Feb 11, 31; m 77; c 5. ORGANIC CHEMISTRY, FOOD TECHNOLOGY. *Educ:* Lebanon Valley Col, BS, 53; Univ Del, MS, 57, PhD(org chem), 60. *Prof Exp:* Develop chemist, Am Cyanamid Co, NJ, 60-63; plant chemist, Tenneco Colors Div, Tenneco Chem, Inc, Pa, 63-66, mgr res, 66-68, plant mgr, SC, 68-69; chief prod develop technologist, Hershey Food Corp, 69-72, mgr planning & admin, 72-76, prod mgr, Hershey Plant, 76-77, mgr eng liaison, 77-81, dir int tech serv, 81-85, VPRES TECH SERV, HERSHEY INT, LTD, 85- *Mem:* Am Chem Soc. *Res:* Dyes and dye intermediates; food technology. *Mailing Add:* 1175 Draymore Ct Hummelstown PA 17036

GLUSHIEN, ARTHUR SAMUEL, b Brooklyn, NY, July 15, 11; wid; c 1. CARDIOLOGY, INTERNAL MEDICINE. *Educ:* NY Univ, BSc, 30, MD, 36; Am Bd Internal Med, dipl, 47. *Prof Exp:* Physician, US Vet Admin, 39-44, chief cardiol sect, Hosp, Aspinwall, 44-54, chief med serv, Hosp, Pittsburgh, 55-59; pvt pract, 59-64; chief cardiol sect, Vet Admin Hosp, East Orange, NJ, 64-74, asst chief med serv, 66-74; RETIRED. *Concurrent Pos:* Assoc prof med, NJ Col Med, 66-74. *Mem:* Am Heart Asn; fel Am Col Physicians; fel Am Col Cardiol. *Res:* Cardiovascular disease. *Mailing Add:* 6761 Caminito Del Greco San Diego CA 92120-2218

GLUSHKO, VICTOR, b Kiel, Ger, June 26, 46; m 78. BIOCHEMISTRY, ENDOCRINOLOGY. *Educ:* Earlham Col, AB, 68; Ind Univ, PhD(chem), 72. *Prof Exp:* Fel endocrinol, Sloan-Kettering Inst, 72-73, res assoc, 73-77; asst prof biochem, Sch Med, Temple Univ, 77-83, adj prof 83-88, vpres prod develop, 88-89; EXEC COORDINATOR, VITAPHORE WOUND HEALING JOINT VENTURE, VITAPHORE & SMITH & NEPHEW, 89-; EXEC COORDR, NEW TECHNOL APPLICATIONS, BIOMAT SYSTS, UNION CARBIDE, 91- *Concurrent Pos:* Instr biochem, Cornell Univ, 74-77. *Mem:* Biophys Soc; NY Acad Sci; Int Soc Magnetic Resonance. *Res:* Organization of plasma membranes and effect of drugs on membranes; structure-function relationships in polypeptide hormone action; physical biochemistry of fluorescent membrane probes. *Mailing Add:* VitaPhore Corp 887 Industrial Rd San Carlos CA 94070-3312

GLUSKER, JENNY PICKWORTH, b Birmingham, Eng, June 28, 31; nat US; m 55; c 3. PHYSICAL BIOCHEMISTRY, X-RAY CRYSTALLOGRAPHY. *Educ:* Oxford Univ, BA, 53, MA & DPhil(chem), 57. *Hon Degrees:* DSc, Col Wooster, Ohio, 85. *Prof Exp:* Res fel x-ray crystallog, Calif Inst Technol, 55-56; res assoc, 56-67, asst mem, 67, mem, 67-79, SR MEM, INST CANCER RES, 79- *Concurrent Pos:* Res assoc prof, Univ Pa, 69-79, adj prof, 80- *Honors & Awards:* Garvan Medal, Am Chem Soc, 79. *Mem:* AAAS; Am Crystallog Asn; Am Chem Soc; Am Soc Biol Chem; Biophys Soc. *Res:* Infrared spectroscopy; molecular structures in general; mechanisms of enzyme reactions from x-ray crystallographic studies of enzymes and their substrates and inhibitors; studies of polycyclic mutagens and carcinogens and the metabolic products of carcinogens. *Mailing Add:* Inst Cancer Res 7701 Burholme Ave Philadelphia PA 19111

GLUSKOTER, HAROLD JAY, b Chicago, Ill, May 8, 35; m 57; c 3. GEOLOGY. *Educ:* Univ Ill, BS, 56; Univ Iowa, MS, 58; Univ Calif, Berkeley, PhD(geol), 62. *Prof Exp:* From asst geologist to assoc geologist, Ill State Geol Surv, 62-70, geologist, Coal Sect, 70-78, head, 75-78; sr res assoc, Exxon Prod Res Co, Houston, 78-81, sr res supvr, 81-85; CHIEF, BR COAL GEOL, US GEOL SURV, 85- *Concurrent Pos:* Mem, Nat Comt Geochem, 75-80. *Mem:* Geol Soc Am; Am Inst Mining, Metall & Petrol Eng; Soc Econ Paleontologists & Mineralogists; Geochem Soc. *Res:* Geology of the Franciscan Formation, California; inorganic chemistry of coal; coal geology; mineral matter in coal. *Mailing Add:* 12201 Sunrise Valley Dr MS 956 Reston VA 22092

GLUSMAN, MURRAY, b New York, NY, Dec 31, 14; m 49; c 2. NEUROPSYCHIATRY. *Educ:* NY Univ, BS, 34, MD, 38. *Prof Exp:* Asst neurol, 49-50, from asst to assoc clin prof psychiat, 50-72, PROF CLIN PSYCHIAT, COL PHYSICIANS & SURGEONS, COLUMBIA UNIV, 72-, CHIEF PSYCHIAT RES BEHAV PHYSIOL, NY STATE PSYCHIAT INST, 66- *Concurrent Pos:* Dir postdoctoral res training prog, Col Physicians & Surgeons, Columbia Univ, 68- *Mem:* Am Acad Neurol; AAAS; Asn Res Nervous & Ment Dis; fel Am Psychiat Asn; Psychiat Res Soc. *Res:* Neural mechanisms in pain; brain mechanisms and emotional behavior; biogenic amines and aggression. *Mailing Add:* NY State Psychol Inst 722 W 168th St New York NY 10032

GLUYAS, RICHARD EDWIN, physical chemistry; deceased, see previous edition for last biography

GLYDE, HENRY RUSSELL, b Calgary, Alta, Oct 31, 37; c 2. THEORETICAL SOLID STATE PHYSICS. *Educ:* Univ Alta, BS, 60; Oxford Univ, PhD(physics), 64. *Prof Exp:* Ciba Found fel, Univ Brussels, 64-65; res fel physics, Univ Sussex, 65-69; physicist, Atomic Energy Can Ltd, 69-71 & 72-75; proj staff, Int Develop Res Ctr, Can, 71-72; from assoc prof to prof physics, Univ Ottawa, 78-82; prof physics & chmn, Univ Alta, 89-91; chmn, 82-89, PROF PHYSICS, UNIV DEL, 82-89 & 91- *Concurrent Pos:* Vis physicist, Nat Res Coun, Can, 68, Atomic Energy Can Ltd, 78, Inst Laue Langevin, France, 75, 76, 81 & 86 & Brookhaven Nat Lab, 77 & 83; chmn theoret physics div, Can Asn Physicists, 74-75, chmn, div physics & soc, 76-78; vis prof, Chulalongkorn Univ, Thailand, 79-84 & State Univ NY, 83. *Mem:* Fel Am Phys Soc; Can Asn Physicists. *Res:* Neutron scattering; theory of metals; solidified inert gases; lattice dynamics; liquid and solid helium. *Mailing Add:* Dept Physics Univ Del Newark DE 19716

GLYMPH, EAKIN MILTON, b Pomaria, SC, Sept 6, 15; m 44; c 3. RUBBER CHEMISTRY. *Educ:* Clemson Univ, BS, 36; Iowa State Univ, PhD(biophys chem), 41. *Prof Exp:* Res chemist, Firestone Plantations Co, WAfrica, 41-46 & Firestone Tire & Rubber Co, 46-48, head latex group, Res Lab, 48-66, mgr natural rubber res & develop, 66-70, mgr natural rubber res & develop, Firestone Natural Rubber & Latex Co, 70-80; RETIRED. *Mem:* Am Chem Soc. *Res:* Polarography; industrial fermentations; synthetic and hevea rubber latexes; latex preparations and applications. *Mailing Add:* 728 Mt Gilead Dr Murrells Inlet SC 29576

GLYNN, PETER W, b Coronado, Calif, Apr 20, 33; m 60; c 2. MARINE ECOLOGY. *Educ:* Univ SDak, BS, 55; Stanford Univ, MS, 60, PhD(biol), 63. *Prof Exp:* Asst prof marine ecol, Univ PR, Mayag ez, 60-67; ZOOLOGIST, SMITHSONIAN TROP RES INST, 67- *Concurrent Pos:* NSF res grant, 63-66. *Mem:* Sigma Xi. *Res:* Structure and interactions in coral reef communities; systematics of marine isopoda, especially sphaeromatids. *Mailing Add:* Opers Res Stanford Univ Stanford CA 94305-4022

GLYNN, WILLIAM ALLEN, b Nowata, Okla, Jan 21, 35; m 53; c 2. MATHEMATICS. *Educ:* Northeastern State Col, BS, 60; Okla State Univ, MS, 62, PhD(math), 65. *Prof Exp:* Asst prof, Western Ill Univ, 65-68; assoc prof math & chmn dept, 68-73, prof math sci & chmn dept, 73-75, PROF MATH SCI & ASSOC DEAN ARTS & SCI, VA COMMONWEALTH UNIV, 75-, ASSOC DEAN HUMANITIES & SCI, 81- *Mem:* Math Asn Am. *Res:* Topology. *Mailing Add:* Sch Arts & Sci Va Commonwealth Univ Richmond VA 23284

GNADE, BRUCE E, b St Charles, Mo, Nov 2, 55; m 82; c 1. SEMICONDUCTOR MATERIALS, SEMICONDUCTOR PROCESSING. *Educ:* St Louis Univ, BA, 76; Ga Inst Technol, PhD(nuclear chem), 82. *Prof Exp:* Mem tech staff mat characterization, 82-88, sr mem tech staff, 88-90, MGR ADVAN SCI MAT, CENT RES LABS, TEX INSTRUMENTS INC, 90- *Concurrent Pos:* Adj prof, Physics Dept, Univ NTex, 88- *Mem:* Am Phys Soc; Mat Res Soc. *Res:* Silicon heterostrute materials systems for quantum electronics; thin film ferroelectric materials; silicon surface chemistry; nuclear analytical techniques; low energy nuclear structure. *Mailing Add:* 13588 N Cent Expressway PO Box 655936 MS 147 Dallas TX 75265

GNAEDINGER, JOHN P(HILLIP), b Oak Park, Ill, Jan 11, 26; m 56; c 2. CIVIL ENGINEERING. *Educ:* Cornell Univ, BCE, 46; Northwestern Univ, MS, 47. *Prof Exp:* Structural designer, Shaw Metz & Dolio, Ill, 46-49; pres, Soil Testing Serv, Inc, 48-91, EMER CHMN STS CONSULT LTD, 91- *Concurrent Pos:* Pres, Ill Eng Coun, 62; mem, Spec Adv Comt Struct Response to Sonic Boom, comt Radioactive Waste Disposal in Enewetok, Nat Acad Sci-Nat Acad Eng-Nat Res Coun, 64; chmn, Fed Construct Coun, 68-69 & Bldg Res Adv Bd, 69-71; spec adv comts, Soil Compaction & Slab-on-Ground Const, Fed Housing Authority; hon mem, Bldg Res Adv Bd; Distinguished Scholar, Nat Acad Sci Exchange Prog, Peoples Repub China, 84. *Mem:* Am Soc Testing & Mat; fel Am Soc Civil Engrs; Nat Soc Prof Engrs; fel Asn Soil & Found Engrs (past pres). *Res:* Soil mechanics and foundation engineering, chemical stabilization of soils; design foundations and superstructures of buildings. *Mailing Add:* 111 Pfingsten Rd Northbrook IL 60062

GNAEDINGER, RICHARD H, b Pocahontas, Ill, Sept 15, 30; m 77. FOOD SCIENCE, ANIMAL HUSBANDRY. *Educ:* Southern Ill Univ, BS, 58; Mich State Univ, MS, 60, PhD(food sci), 62. *Prof Exp:* Food technologist, Bur Com Fisheries, US Fish & Wildlife Serv, 62-63, chemist, 63-68; asst dir pet foods nutrit res, Ralston Purina Co, 68-70; chief chemist, 73-88, ENVIRON SPECIALIST, SECT LAB SERV, MO DEPT HEALTH, 88- *Concurrent Pos:* Pest Control & Financial Serv. *Mem:* Sigma Xi. *Res:* Animal and pet nutrition; animal production; utilization of fish and fishery products; laboratory animal care; body composition; headspace blood alcohol; breath and blood alcohol; issued two patents. *Mailing Add:* 2709 Mohawk Dr Jefferson City MO 65101

GNANADESIKAN, RAMANATHAN, b Madras, India, Nov 2, 32; m 65; c 2. STATISTICAL ANALYSIS. *Educ:* Univ Madras, BSc, 52, MA, 53; Univ NC, PhD(math statist), 57. *Prof Exp:* Res asst math statist, Univ NC, 54-57; sr res statistician, Procter & Gamble Co, 57-59; mem tech staff, 59-68, head statist & data analysis res dept, Bell Tel Labs, 68-83, div mgr, info sciences res, 84-86, ASST VPRES MATH, COMMUN & COMPUT SCI RES, BELLCORE, 86- *Concurrent Pos:* Consult, Cincinnati Bd Educ, 57-59; lectr & consult, Sch Med, Univ Cincinnati, 58-59; adj assoc prof, Courant Inst Math Sci, NY Univ, 61-63; adv, US Bur Census, 67-69; vis prof, Imp Col Sci & Technol, Univ London, 69; vis prof, Princeton Univ, 71 & Inst Advan Studies, Vienna, Austria, 81; adv comt, NSF, Div Math Sci, 85-88; bd gov, Inst Math & Its Appln, Univ Minn, 86-89. *Mem:* Fel AAAS; Biomet Soc; fel Am Statist Asn; fel Inst Math Statist (pres, 88-89); Math Asn Am; fel Royal Statist Soc; Int Asn Statist Comput (pres, 81-83); Int Statist Inst. *Res:* Multivariate analysis, especially in the fields of the analysis and design of experiments and related areas of data analysis and theoretical statistics. *Mailing Add:* 425 Fairmount Ave Chatham NJ 07928-1369

GO, MATEO LIAN POA, b Amoy, China, Sept 17, 18; m 52; c 3. STRUCTURAL ENGINEERING. *Educ:* Cornell Univ, BCE, 42, PhD(struct & transp eng, soil mech), 46; Mass Inst Technol, SMCE, 43. *Prof Exp:* Construct engr, Mahony-Troast Construct Co, NJ, 42, 46-47; mgr Mateo L P Go Construct Co, Philippines, 47-53, tech consult, Go Occo & Co, 53-54, br mgr, 54-56; asst prof eng, Syracuse Univ, 57-59; assoc prof, 59-63, dir prof adv serv ctr, 67-72, chmn dept civil eng, 69-72, PROF ENG, UNIV HAWAII, 63-, CHMN DEPT CIVIL ENG, 81- *Concurrent Pos:* Spec lectr, Cebu Inst Technol Philippines, 47-50; pres, Eng Assocs, 71. *Mem:* Am Soc Civil Engrs; Am Soc Eng Educ; Am Concrete Inst. *Res:* Airport engineering; space technology; structural mechanics; reinforced concrete; transportation and construction engineering. *Mailing Add:* 2415 Ferdinand Ave Honolulu HI 96822

GO, VAY LIANG, b Ozamis City, Philippines, Aug 29, 38; US citizen; m 63; c 3. MEDICINE. *Educ:* Univ Santo Tomas, Manila, AA, 58, DR, 63. *Prof Exp:* Med internship, Northwestern Hosp, Minneapolis, 64; resident internal med, Mayo Clin & Mayo Found, Rochester, 65-66, fel gastroenterol, 67-68; res assoc, Banting & Best Dept Med Res, Univ Toronto, 69-71; from asst prof to prof med, Mayo Med Sch, 72-88; dir, Div Digestive Dis & Nutrit, Nat Inst Arthritis, Diabetes, Digestive & Kidney Dis, NIH, 85-88; EXEC CHAIR & PROF, DEPT MED, SCH MED, UNIV CALIF, LOS ANGELES, 88- *Concurrent Pos:* Mem, NIH Group Pancreas, 72-74, Immunol Comt, 73-75; mem comt res, Am Gastroenterol Asn, 77-80 & 83-86; consult, Mayo Clin & Mayo Found, 71-88; mem, Fed Drug Admin Adv Comt Gastrointestinal Drugs, 90-94. *Honors & Awards:* 189th Luis Guerrero Mem Lectr, Philippine Soc Gastroenterol, 79; Alimurong Lectr Award, Univ Santo Tomas Alumni Asn, 89. *Mem:* AAAS; Am Asn Cancer Res; Am Asn Clin Invest; Am Fedn Clin Res; Am Gastroenterol Asn; Am Pancreatic Asn (pres, 78-79 & 88-89); Am Soc Clin Nutrit; Endocrine Soc; Sigma Xi; fel Am Col Physicians. *Res:* Relationship between pancreatic blood flow and vasoactive intestinal polypeptide; effective gastric stimulus and distribution of active hypothalamic sites; role of glucagon in splanchnic hyperemia of chronic portal hypertension. *Mailing Add:* Dept Med 37-120 CHS UCLA 10833 Le Conte Ave Los Angeles CA 90024

GOAD, CLYDE CLARENTON, b Mount Airy, NC, Nov 20, 46; m 69; c 2. GEODESY. *Educ:* NC State Univ, BS, 69; Johns Hopkins Univ, MS, 71; Catholic Univ Am, PhD(aerospace eng), 77. *Prof Exp:* Sr analyst, Wolf Res & Develop Corp, Riverdale, Md, 69-75; geodesist, 75-81, CHIEF GRAVITY, ASTRON & SPACE GEOD DIV, NAT OCEAN SURV, NAT OCEANIC & ATMOSPHERIC ADMIN, 81- *Mem:* Am Geophys Union; Sigma Xi. *Res:* Satellite geodesy-effects of earth and ocean tides on the orbits of artificial satellites; numerical modeling of the geoid of the earth. *Mailing Add:* 744 Northbridge Lane Worthington OH 43085-1748

GOAD, WALTER BENSON, JR, b Marlowe, Ga, Sept 5, 25; m 52; c 3. BIOPHYSICS. *Educ:* Union Col, BS, 45; Duke Univ, PhD(physics), 54. *Prof Exp:* MEM STAFF, T-DIV, LOS ALAMOS NAT LAB, 50- *Concurrent Pos:* Sr fel biophys, Univ Colo, 64-65, vis prof, 68-; vis scientist, Med Res Coun Lab Molecular Biol, Cambridge Univ, Eng, 70-71; scholar, Eleanor Roosevelt Inst Cancer Res, 77-; fel, Los Alamos Nat Lab, 87- *Mem:* Fel AAAS; fel Am Phys Soc; Sigma Xi. *Res:* Informatics; algorithms for molecular genetics. *Mailing Add:* Rte 1 Los Alamos Nat Lab Espanola NM 87532

GOANS, RONALD EARL, b Knoxville, Tenn, Aug 12, 46; m 71; c 2. RADIATION PHYSICS, INSTRUMENTATION. *Educ:* Univ Tenn, BS, 68, MS, 69, PhD(physics), 74; George Washington Univ, MD, 83. *Prof Exp:* Instr electronics, Univ Tenn, 69-74; res assoc health physics, Oak Ridge Nat Lab, 74-79 & George Washington Med Sch, 79-83; res assoc, Dept Obstet & Gynec, Providence Hosp, Washington, DC, 83-87; LAB THEORET & PHYS BIOL, NAT INST CHILD HEALTH & DEVELOP, NIH, 87- *Concurrent Pos:* Consult, Brazilian Inst Atomic Energy, 76-79; mem, Defense Nuclear Agency Inspection Team for radiol clean-up of Enewetak Atoll, 77-79. *Honors & Awards:* IR-100 Award, Indust Res-Develop Mag, 78. *Mem:* Health Physics Soc; Sigma Xi; AMA; Am Burn Asn. *Res:* Radiation instrumentation; optimization techniques; medical ultrasonics; chemical physics; theoretical and experimental research in medical ultrasonics; tissue characterization relevent to burns and to obstetrics & gynecology; biomathematics. *Mailing Add:* 1422 Eagle Bend Rd Clinton TN 37716-4009

GOATES, JAMES REX, b Lehi, Utah, Aug 14, 20; m 48; c 3. PHYSICAL CHEMISTRY. *Educ:* Brigham Young Univ, BS, 42; Univ Wis, PhD(phys & soil chem), 47. *Prof Exp:* From asst prof to assoc prof, 47-54, chmn dept, 65-68, PROF CHEM, BRIGHAM YOUNG UNIV, 54-, DEAN, COL PHYS & MATH SCI, 81- *Mem:* Am Chem Soc; Sigma Xi. *Res:* Thermodynamic properties of metallic sulfides; thermodynamics and mechanisms of adsorption; thermodynamics of solutions. *Mailing Add:* Dept Chem Brigham Young Univ Provo UT 84601

GOATES, STEVEN REX, b Provo, Utah, Nov 8, 51; m 77; c 1. ANALYTICAL CHEMISTRY. *Educ:* Brigham Young Univ, BSc, 76; Univ Mich, MSc, 77, PhD(chem), 81. *Prof Exp:* Res assoc, Radiation Lab, Columbia Univ, 81-82; asst prof, 82-87, ASSOC PROF CHEM, BRIGHAM YOUNG UNIV, 87- *Mem:* Sigma Xi; Am Chem Soc; Am Phys Soc. *Res:* Chemical analysis by laser methods; investigation of supersonic jet spectroscopy and supercritical fluids for chemical analysis; laser spectroscopic interrogation of non-electrolyte solutions; laser photoframentation spectroscopy. *Mailing Add:* Dept Chem Brigham Young Univ Provo UT 84602

GOAZ, PAUL WILLIAM, b Lafayette, Ind, Apr 13, 22; m 48; c 2. DENTISTRY. *Educ:* Loyola Univ, Chicago, DDS, 50; Univ Chicago, SM, 53; Okla State Univ, BS, 59. *Prof Exp:* Intern, Zoller Mem Clin, Chicago, 50-51, asst dent surg, 51-53; clin asst dent surg, Sch Med, Univ Okla, 53-54, from instr to asst prof, 54-70; assoc prof oral diag & vchmn dept, Sch Dent, Loyola Univ Chicago, Maywood, 70-78; PROF ORAL DIAG & ROENTGENOL, BAYLOR COL DENT, DALLAS, 78- *Concurrent Pos:* Res fel, Okla Med Res Found & head dent sect, 53-62; pvt pract. *Mem:* Am Asn Phys Anthrop; Am Asn Dent Res; Sigma Xi. *Res:* Salivary, bacterial growth factors; clinical testing of potential anticariogenic agents; dental morphology and anthropology; oral radiology. *Mailing Add:* Diag Sci Dept Baylor Col Dent 3302 Gaston Ave Dallas TX 75246

GOBEL, FREDERICK L, b La Crosse, Wis, Jan 8, 35; c 3. CARDIOVASCULAR DISEASES. *Educ:* Univ Wis, BS, 57, MD, 59. *Prof Exp:* Gen pract med, Duluth, Minn, 60-62; chief resident, Med Serv, Vet Admin Hosp, 66; cardiac fel, Dept Med, Sch Med, Univ Minn, 66-69; dir cardiac cath lab med, Cardiovasc Sect, Vet Admin Hosp, 69-77, chief cardiovasc sect, 72-77; assoc prof med, 72-80, CLIN ASSOC PROF MED, SCH MED, UNIV MINN, 80- *Concurrent Pos:* From instr to asst prof med, Sch Med, Univ Minn, 67-72; pvt practr cardiol, 77- *Honors & Awards:* Mordse J Shapiro Mem Travel Award, 69. *Mem:* Am Heart Asn; fel Am Col Physicians; fel Am Col Cardiol; fel Coun Clin Cardiol. *Res:* Relationship between myocardial blood flow and myocardial oxygen consumption, and hemodynamic predictors of myocardial blood flow, and how each is altered by medical or surgical therapy. *Mailing Add:* Minn Soc IM 920 E 28th St Minneapolis MN 55407

GOBEL, STEPHEN, b New York, NY, Dec 27, 38; m 61; c 2. CYTOLOGY, NEUROBIOLOGY. *Educ:* NY Univ, DDS, 63. *Prof Exp:* Dent surgeon, Nat Inst Dent Res, 66-77, chief neurocytology & exp anat sect, 77-84; EXEC SECY, NEUROL SCI STUDY, NIH, 84- *Concurrent Pos:* Fel cytol, Col Dent, NY Univ, 63-66. *Honors & Awards:* Commendation Medal, US Pub Health Serv. *Mem:* Soc Neurosci; Int Asn Study Pain; Int Asn Dent Res; AAAS. *Res:* Electron microscopical studies of the neural circuitry of primary and second order neurons in the dorsal horns of the trigeminal system and spinal cord; basic mechanisms underlying pain and other somatic sensations; effects of nerve injury on the spinal cord and brain. *Mailing Add:* Nat Inst Health Westwood Bldg Rm 304 Bethesda MD 20892

GOBELI, GARTH WILLIAM, solid state physics, for more information see previous edition

GOBLE, ALFRED THEODORE, b River Falls, Wis, Jan 24, 09; m 35; c 3. OPTICS. *Educ:* Univ Wis, BA, 29, PhD(physics), 33. *Prof Exp:* Asst physics, Univ Wis, 29-33, Univ Alumni Res Found res assoc, 33-34; from instr to asst prof, Univ Tulsa, 34-37; from asst prof to assoc prof, Alfred Univ, 37-46; from lectr to prof, 45-74, chmn dept, 66-71, EMER PROF PHYSICS & RES PROF, UNION COL, NY, 74- *Concurrent Pos:* Vis asst prof, Princeton Univ, 42-43; res assoc, Harvard Univ, 44-45; consult, Standard Rolling Mills, Inc, 49-54, Revere Copper & Brass, Inc, 54-72 & Thompson-Ramo-Wooldridge Corp, 55-58; mem tech staff, Space Tech Lab, 58-59, consult, 59-60; consult, Aerospace Corp, 60-64; vis researcher, Clarendon Lab, Oxford, 64-65 & 71-72. *Mem:* Am Phys Soc; Optical Soc Am; Am Asn Physics Teachers. *Res:* Theoretical spectroscopy; intensities in platinum-like spectra; recombination spectra of potassium; cosmic rays; high resolution spectroscopy; isotope shifts. *Mailing Add:* Dept Physics Union Col Schenectady NY 12308

GOBLE, DAVID FRANKLIN, b Pincher Creek, Alta, July 31, 40; m 62; c 2. THEORETICAL PHYSICS. *Educ:* Univ Alta, BSc, 62, MSc, 63; Univ Toronto, PhD(theoret physics), 67. *Prof Exp:* NATO res fel, Inst Pure & Appl Phys Sci, Univ Calif, San Diego, 67-68; asst prof theoret physics, 68-77, ASSOC PROF THEORET PHYSICS, DALHOUSIE UNIV, 77- *Mem:* Am Asn Physics Teachers; Can Asn Physicists. *Res:* Development of a microscopic theory of liquid helium four using many-body techniques; explanation of finite size effects in liquid helium four. *Mailing Add:* Dept Physics Dalhousie Univ Halifax NS B3H 3J5 Can

GOBLE, FRANS CLEON, b Chicago, Ill, July 11, 13. ARCHEOZOOLOGY, COMPARATIVE OSTEROLOGY & ODONTOLOGY. *Educ:* Battle Creek Col, BS, 33; Univ Mich, MS, 34, ScD(zool), 39. *Prof Exp:* Asst, Univ Mich, 34, tech asst histol, 35-38; pathologist, Bur Game, State Conserv Dept, NY, 38-45; pathologist & parasitologist, Sterling-Winthrop Res Inst, 45-51; head lab path, Abbott Labs, 51-52; biol consult, St Thomas, Virgin Islands, 52-53; dir parasitol, Ciba Pharmaceut Co, 54-60, dir chemother, 61-68; dir biol, Smith, Miller & Path, Inc, 69-72; dir, Dept Infectious Dis, Res & Develop Div, Cooper Labs Inc, 72-78; ARCHEOLOGIST, DEPT ANTHROP, SCI MUS MINN, 83- *Concurrent Pos:* Mem, 4th, 6th, 7th & 8th Int Cong Trop Med & Malaria; mem, 1st Int Cong Chagas Dis, Adv Group Res, Pan-Am Health Orgn, 62, consult & chmn chemother res group, 63-; mem exp adv panel parasitic dis, WHO, 64-79; consult toxicol, 78-, osteol, Crown Canyon Archeol Ctr, 89- *Mem:* Am Anthrop Asn; Am Asn Pathologists; Soc Toxicol; Am Soc Mammalogists; Archaeol Inst Am; Soc Am Archaeol. *Res:* Pathology and chemotherapy of experimental infectious diseases; histology and hematology in subacute and chronic toxicity studies on synthetic organic compounds and antibiotics; identification of animal bones from archeological sites. *Mailing Add:* 1147 S Winthrop St Paul MN 55119

GOBLE, GEORGE G, b Eagle, Idaho, Sept 11, 29; m 53; c 2. STRUCTURAL ENGINEERING. *Educ:* Univ Idaho, BS, 51; Univ Wash, MS, 57, PhD(struct), 61. *Prof Exp:* Prof struct, Case Western Reserve Univ, 61-77; CHMN DEPT CIVIL & ARCHIT ENG, UNIV COLO, BOULDER, 77- *Concurrent Pos:* Vis prof, Univ Calif, Berkeley, 68- *Honors & Awards:* Collingwood Award, Am Soc Civil Engrs, 65; Lincoln Prof Struct Design Award, 66. *Mem:* Am Soc Civil Engrs; Am Concrete Inst; Soc Exp Stress Analysis. *Res:* Behavior of civil engineering structures; design of structures; optimum design of structures using mathematical programming; dynamic pile behavior. *Mailing Add:* Dept Civil & Archit Eng Univ Colo Boulder CO 80309

GOBLIRSCH, RICHARD PAUL, b Minneapolis, Minn, Jan 1, 30; m 52; c 3. MATHEMATICS. *Educ:* Col St Thomas, BA, 51; Univ Wis, MA, 52, PhD, 56. *Prof Exp:* Asst math, Univ Wis, 53-56; instr, Univ Va, 56-59; asst prof, Univ Rochester, 59-63; vis asst prof, Univ Colo, 63-64; assoc prof, 64-77, PROF MATH & ACTG DEAN, COL ST THOMAS, 77- *Mem:* Am Math Soc; Math Asn Am; Sigma Xi. *Res:* Geometric topology; real variables. *Mailing Add:* Mail No 4105 2115 Summit Ave St Paul MN 55105

GOBRAN, RAMSIS, b Sinbellawin, Egypt, Feb 20, 32; US citizen; m 59; c 3. ORGANIC CHEMISTRY. *Educ:* Univ Alexandria, BSc, 52; Univ Southern Calif, MS, 57, PhD(chem), 59. *Prof Exp:* Parke Davis Co fel, 57-59, Goodyear Rubber Co fel, 59; res polymer chemist, Minn Mining & Mfg Co, 59-67, res specialist, 67-72, sr res specialist, 72-74, prod develop mgr, Indust Tape Div, 74-80, researcher, Prod Develop & New Prod Mgr, Disposable Prod, 80-82, div scientist, Disposable Prod Div, 83-84, CORP SCIENTIST, DISPOSABLE PROD DIV, 3M CO, 85- *Mem:* Am Chem Soc; Sigma Xi. *Res:* Polymer, photographic and organic heterocyclic chemistry; adhesives. *Mailing Add:* 1026 Brenner Ave Roseville MN 55113

GOBUTY, ALLAN HARVEY, radiopharmacy, radiopharmacology, for more information see previous edition

GOCHENAUR, SALLY ELIZABETH, b Cleveland, Ohio, Mar 30, 32. MYCOLOGY. *Educ:* Univ Cincinnati, BS, 58, MS, 60; Univ Wis, PhD(bot), 64. *Prof Exp:* Asst prof microbiol, 64-70, assoc prof, 70-80, PROF BIOL, ADELPHI UNIV, 80- *Concurrent Pos:* Res assoc, Univ Wis, 64-; res collabr, Brookhaven Nat Lab, 68-69. *Mem:* Bot Soc Am; Mycol Soc Am; Brit Mycol Soc; Ecol Soc Am; Sigma Xi. *Res:* Fungal physiology; taxonomy and ecology of soil fungi. *Mailing Add:* 1580 Colonial Dr Cincinnati OH 45238

GOCHENOUR, WILLIAM SYLVA, b East St Louis, Ill, July 18, 16; m 40; c 1. VETERINARY MEDICINE. *Educ:* Univ Pa, VMD, 37; Am Bd Pub Vet Health, dipl; Am Bd Microbiol, dipl. *Prof Exp:* Lab vet, Bur Animal Indust, USDA, 37-38; res vet, Pitman Moore Co Div, Allied Labs, Inc, 39; chief vet bact, Med Serv Grad Sch, Walter Reed Army Med Ctr, US Army, 49-53, chief animal stand & liaison br, Off Surgeon Gen, 53-54, chief animal assessment br, Med Unit, Ft Detrick, 54-63, dep dir, Walter Reed Army Inst Res, Walter Reed Army Med Ctr, DC, 63-74; dir animal care serv, Med Ctr, Univ Ky, 74-77; RETIRED. *Mem:* Fel AAAS; Am Pub Health Asn; Am Vet Med Asn; Asn Mil Surg US. *Res:* Infectious diseases, especially respiratory diseases and the zoonoses. *Mailing Add:* 1286 Oaklawn Park Lexington KY 40502

GOCHMAN, NATHAN, b Brooklyn, NY, Nov 11, 33; m 55; c 2. BIOCHEMISTRY. *Educ:* Brooklyn Col, BS, 55; Northwestern Univ, PhD(chem), 58. *Prof Exp:* Supvr, Bioanal Lab, G D Searle & Co, 58-62; dir clin chem, Technicon Instruments Corp, 62-66, res lab dir, 66-68; clin chemist, Clin Ctr, NIH, 68-72; CLIN CHEMIST, LAB SERV, VET ADMIN HOSP, 72- *Concurrent Pos:* Assoc adj prof chem & pathol, Univ Calif, San Diego, 72-79, adj prof path, 79- *Mem:* AAAS; Am Asn Clin Chem (pres, 78); NY Acad Sci; Am Chem Soc; Sigma Xi. *Res:* Enzymology; analytical biochemistry; drug metabolism; development of methods and instruments for biochemical determinations with emphasis on automation. *Mailing Add:* 6409 E Lookout Lane Anaheim Hills CA 92807

GOCHNAUER, THOMAS ALEXANDER, b Appleton, Wis, July 29, 19; m 50; c 3. MICROBIOLOGY. *Educ:* Univ Wis, PhD(bact), 50. *Prof Exp:* Res assoc entom & econ zool, Univ Minn, 49-61; chief apicult sect, Entom Res Inst, 61-70, RES SCIENTIST, ENTOM SECT, OTTAWA RES STA, CAN DEPT AGR, 70- *Mem:* AAAS; Am Soc Microbiol; NY Acad Sci; Can Soc Microbiol; Soc Invert Path; Sigma Xi. *Res:* Honeybee diseases and control; antibiotics; phages; effects of gamma radiation; detection, distribution and infectivity of B larvae and Ascosphaera apis spores in the honeybee environment. *Mailing Add:* 14 Trinity Dr Nepean ON K2H 6H2 Can

GODAR, EDITH MARIE, b Chicago, Ill, June 19, 21. ORGANIC CHEMISTRY, ANALYTICAL CHEMISTRY. *Educ:* Rosary Col, BS, 42; Loyola Univ, Ill, MS, 56, PhD(org chem), 59. *Prof Exp:* Res chemist, Res Div, Am Can Co, 43-56; instr chem, Loyola Univ, Ill, 59-68; assoc prof chem, Mary Manse Col, 68-80, chmn dept sci & math, 71-80; RETIRED. *Concurrent Pos:* Instr, Rosary Col, 53-54; sr res chemist, Int Minerals & Chem Corp, 59-68; fel, Univ Detroit, 68. *Mem:* Am Chem Soc; Soc Appl Spectros; Sigma Xi. *Res:* Synthetic organic chemistry; instrumental methods of analysis. *Mailing Add:* 11704 103rd Ave Juanita WA 98034

GODARD, HUGH P(HILLIPS), b Montreal, Que, Oct 19, 14; m 38; c 1. CHEMICAL ENGINEERING, INDUSTRIAL CHEMISTRY. *Educ:* Univ BC, BASc, 36, MASc, 37; McGill Univ, PhD(indust & cellulose chem), 41. *Prof Exp:* Chief chemist, Chems & Explosives, Inspection Bd, UK & Can, 41-45; head, Chem Div, Aluminum Labs, Ltd, 45-66, dep dir, Kingston Lab, Alcan Int Ltd Res Ctr, Kingston, 66-73; RETIRED. *Concurrent Pos:* Bd dirs, Nat Asn Corrosion Engrs, 52-55; dir publ, Chem Inst Can, 55-58; ed, Mats Performance, 73-83; pres, Aluminum Corrosion Serv, 73-91; pres, Int Cong Metallic Corrosion, 84. *Honors & Awards:* Frank Newman Speller Award, Nat Asn Corrosion Engrs, 63; Plummer Medal, Eng Inst Can, 56. *Mem:* Nat Asn Corrosion Engrs (pres, 59-60). *Res:* Corrosion of aluminum and its prevention. *Mailing Add:* 309-5558 15B Ave Delta BC V4M 2H3 Can

GODBEE, H(ERSCHEL) W(ILLCOX), b Hinesville, Ga, Mar 4, 28; m 52; c 3. CHEMICAL ENGINEERING. *Educ:* Ga Inst Technol, BChE, 52, PhD(chem eng), 64. *Prof Exp:* RES ENGR, OAK RIDGE NAT LAB, 58- *Honors & Awards:* Sarge Ozker Award, Am Soc Mech Engrs. *Mem:* Sigma Xi. *Res:* Heat transfer, especially measurement of thermo-physical properties; safe economical treatment of radioactive wastes. *Mailing Add:* 104 Tidewater Lane Oak Ridge TN 37830

GODBEE, RICHARD GREENE, II, b Macon, Ga, May 6, 51; m 68; c 1. EQUINE NUTRITION. *Educ:* Univ Ga, BS, 73, MS, 75; Colo State Univ, PhD(equine nutrit), 78. *Prof Exp:* Aide nutrit, USDA, 71-73; res asst, Univ Ga, 73-75; teaching asst, Colo State Univ, 75-77; lectr, Calif State Univ, 77-78; ASST PROF ANIMAL SCI, CLEMSON UNIV, 78- *Mem:* Am Soc Animal Sci; Equine Nutrit & Physiol Soc; Coun Agr Sci & Technol; Sigma Xi. *Res:* Non-protein nitrogen utilization by the equine; supplemental amino acids for ruminant animals; broodmare nutrition; nutrition of the growing horse. *Mailing Add:* W R Grace & Co PO Box 1857 Hereford TX 79045

GODBEY, JOHN KIRBY, b Cisco, Tex, Nov 14, 21; m 43; c 3. ELECTRONICS ENGINEERING. *Educ:* Southern Methodist Univ, BS, 44; Univ Tex, MS, 47. *Prof Exp:* Res technologist, Field Res Labs, Magnolia Petrol Co, 47-55; sr res technologist, Mobil Oil Corp, 55-68, sr res engr, Mobil Res & Develop Corp, 68-; PRES, JOHN K GODBEY ENG CONSULT, INC. *Mem:* Inst Elec & Electronics Engrs; Am Inst Mining, Metal & Petrol Engrs; Nat Soc Prof Engrs. *Res:* Exploration and production well surveying; drilling methods and instrumentation; automatic control; borehole measurements; pumping well technology. *Mailing Add:* 4339 Hockaday Dr Dallas TX 75229

GODBEY, WILLIAM GIVENS, b Georgetown, Tex, Sept 24, 19; m 53; c 1. PHYSICAL CHEMISTRY, NUCLEAR CHEMISTRY. *Educ:* Southwestern Univ, BS(chem) & BS(educ), 40; Univ Mo, MS, 58, PhD(agr chem), 60. *Prof Exp:* Anal dir, Agr Consult Labs, 45-53 & Res Found Labs, Colo Sch Mines, 53-54; chief chemist, Am Smelting & Ref Co, 54-55; from instr to asst prof agr chem, Univ Mo, 55-62; prof chem, Southwest Mo State Col, 62-65; prof chem, Alaska Methodist Univ, 65-77; EMER PROF, ALASKA PAC UNIV, 85- *Concurrent Pos:* Agr consult, Tex Exp Sta, Dow Chem Co, Rio Grande Valley Cities & Texsun Citrus Growers, 45-51; NIH res grant molecular studies on casein, 60-62. *Mem:* AAAS; fel Am Inst Chem; Am Chem Soc; NY Acad Sci. *Res:* Molecular structure of biochemical macromolecules using electrophoresis, ultra centrifugation and wide and small angle x-ray scattering techniques. *Mailing Add:* PO Box 418 Skagway AK 99840

GODBOLE, SADASHIVA SHANKAR, b Indore, India, Sept 7, 39; m 62; c 2. INSTRUMENTATION & INTERACTIVE SIMULATION. *Educ:* Victoria Jubilee Tech Inst, India, BE, 58; Va Polytech Inst, MS, 69, PhD(elec eng), 71; George Washington Univ, MEA, 82. *Prof Exp:* Technician, Hindustan Elec Co, India, 58; sect engr, Heavy Elec Ltd, 59-67; sr res engr, 71-74, group supvr, 74-79, res specialist, 79-87, APPLN SPECIALIST, BABCOCK & WILCOX CO, 87- *Mem:* Sr mem Inst Elec & Electronics Engrs. *Res:* Development of math models and simulations for application in design, integration, familarization, demonstration, training, equipment checkout, initial start-up and operation of a new or retrofit system; design of intrumentation and control systems; practical application of electrical, nuclear, thermal, hydraulic and automatic control theory to analytical design and modeling of power plants; industrial processes. *Mailing Add:* PO Box 10935 Babcock & Wilcox Co Lynchburg VA 24506

GODCHAUX, WALTER, III, b New Orleans, La, Apr 15, 39; div; c 1. MEMBRANES PROTEIN SYNTHESIS. *Educ:* Mass Inst Technol, SB, 60, PhD(biol), 65. *Prof Exp:* NIH fel biochem, Univ Ore, 65-67; fel molecular biol, Sch Med, Yale Univ, 67-70; asst prof biol, Amherst Col, 70-78; ASSOC PROF, UNIV CONN, 78- *Mem:* Am Soc Biochem & Molecular Biol; Am Soc Microbiol; Am Sigma Xi. *Res:* Eukaryotes: mechanism of protein biosynthesis; biosynthesis of membrane proteins; photoreceptor function; lipids, membranes and motility of gliding bacteria. *Mailing Add:* Box U-131 Univ Conn Storrs CT 06269-2131

GODDARD, CHARLES K, b Detroit, Mich, Nov 29, 37. COMPUTER SCIENCE. *Prof Exp:* ASST ANALYST, COMPUT SCI ADVAN DIGITAL SYST, 79- *Mailing Add:* Advan Digital Syst Inc 10052 Mesa Ridge Ct San Diego CA 92121

GODDARD, EARL G(ASCOIGNE), b Mesilla Park, NMex, Nov 10, 17; m 47; c 2. ELECTRONIC SYSTEMS ENGINEERING. *Educ:* NMex State Univ, BS, 39; Stanford Univ, AM & EE, 47. *Prof Exp:* Instr elec eng, Rice Univ, 41-43; asst microwave res, Off Res & Inventions proj, Stanford Univ, 46-47; asst prof elec eng, Duke Univ, 47-48; from asst prof to assoc prof electronics, US Naval Post-Grad Sch, Calif, 48-55; sr res engr, Stanford Res Inst, 55-60, mgr field eng training, Microwave Tube Group, Varian Assocs, 61-63; mgr eng, Delcon Div, Hewlett-Packard, 64-65; electronic systs engr, Dalmo Victor Div, Textron, 65-67; engr, Appl Technol Div, Itek, 67-69; engr, Dalmo Victor Div, Textron, 69-72, sr scientist, Appl Technol Div, Itek, 72-78; CONSULT, 78- *Concurrent Pos:* Electronic systs consult, 78-84; Goddard adv bd, Col Eng, NMex State Univ. *Honors & Awards:* Centennial Bronze Medal, Inst Elec & Electronics Engrs, 84. *Mem:* Goddard Asn Am; Goddard Asn Europ; Nat Geog Soc; Nature Conservancy; US Naval Sailing Asn; Am Asn Retired Persons; Sierra Club. *Res:* Design and application of electronic systems. *Mailing Add:* 2522 Webster St Palo Alto CA 94301-4249

GODDARD, JOE DEAN, b Buncombe, Ill, July 13, 36; m 57; c 5. PHYSICS ON FLUIDS. *Educ:* Univ Ill, BS, 57; Univ Calif, PhD(chem eng), 61. *Prof Exp:* Res engr, French Inst Petrol, 62-63; from asst prof to prof chem eng, Univ Mich, Ann Arbor, 63-75; chmn dept, 76-86, FLUOR PROF CHEM ENG, UNIV SOUTHERN CALIF, 76- *Concurrent Pos:* NATO fel, 61-63; NSF sr fel, Univ Cambridge, 70-71; Fulbright fel, Belg, 84. *Mem:* Am Inst Chem Engrs; Soc Rheol (vpres, 89-91); Brit Soc Rheology; Am Math Soc; Am Phys Soc. *Res:* Applied mathematics; heat and mass transfer; rheology of non-Newtonian fluids; fluid mechanics. *Mailing Add:* Dept Chem Eng Univ Southern Calif Los Angeles CA 90089-1211

GODDARD, JOHN BURNHAM, b New Haven, Conn, Nov 7, 42; m 68; c 2. INORGANIC CHEMISTRY, ANALYTICAL CHEMISTRY. *Educ:* Mass Inst Technol, BS, 64; Northwestern Univ, PhD(inorg chem), 69. *Prof Exp:* Teaching asst chem, Northwestern Univ, 65-68; from res chemist to sr res chemist, 68-76, sr scientist, 76-81, proj mgr, metals div, 81-85, MGR ANALYTICAL SERV, LINDE DIV, UNION CARBIDE CORP, 85- *Mem:* Am Chem Soc; Metall Soc; Sigma Xi. *Res:* Atmospheric gas purification; gas analysis; aerosol particulate analysis. *Mailing Add:* Union Carbide Indust Gases Inc Union Carbide Corp PO Box 44 Tonawanda NY 14151

GODDARD, MURRAY COWDERY, b Cleveland, Ohio, May 14, 24; m 46; c 3. COLOR SCIENCE. *Educ:* Mass Inst Technol, SB, 48; Case Inst Technol, MS, 50. *Prof Exp:* Photog engr, Color Tech Div, 50-64, sr physicist, Res Labs, 64-65, res assoc image struct & comput appln, 65-68, res assoc, Res Labs, Computer-Based Learning Systs, 68-74, RES ASSOC, COLOR PRINTER EXPOSURE DETERMINATION, EASTMAN KODAK CO, 74- *Mem:* Am Phys Soc; Soc Photog Sci & Eng; Inst Elec & Electronics Eng; Optical Soc Am; Sigma Xi. *Res:* Image structure; computer applications; colorimetry; exposure determination; electronics; individualized learning systems for education and training; color science applied to photography. *Mailing Add:* 614 38th St Anacortes WA 98221

GODDARD, RAY EVERETT, b Lakeland, Fla, Sept 28, 22; m 43; c 4. FOREST GENETICS. *Educ:* Univ Fla, BSF, 47, MSF, 48; Agr & Mech Col, Tex, PhD(genetics), 60. *Prof Exp:* Silviculturalist, Tex Forest Serv, 48-59; from asst prof & asst geneticist to assoc prof & assoc geneticist, 59-76, PROF FOREST GENETICS, AGR EXP STA, UNIV FLA, 76- *Mem:* Soc Am Foresters; Am Genetic Asn; Sigma Xi. *Res:* Physiology and genetics of disease resistance in forest trees. *Mailing Add:* Sch Forest Resources & Conserv Univ Fla Gainesville FL 32601

GODDARD, STEPHEN, b Ogden, Utah, Jan 5, 37; m 66. ORNITHOLOGY, ECOLOGY. *Educ:* Utah State Univ, BS, 60, MS, 62, PhD(biol), 67. *Prof Exp:* Res asst zool, Okla State Univ, 63-66; asst prof, 66-70, ASSOC PROF BIOL, WIS STATE UNIV-RIVER FALLS, 70- *Mem:* Wildlife Soc; Wilson Ornith Soc; Am Ornith Union; Am Soc Mammal. *Res:* Ecology and management of game and nongame bird populations. *Mailing Add:* 1115 Oaklawn Base ID 83709

GODDARD, TERRENCE PATRICK, technical management, physical chemistry; deceased, see previous edition for last biography

GODDARD, WILLIAM ANDREW, III, b El Centro, Calif, Mar 29, 37; m 57; c 4. SUPERCONDUCTIVITY, REACTION MECHANISMS. *Educ:* Univ Calif, Los Angeles, BS, 60; Calif Inst Technol, PhD(eng sci), 64. *Prof Exp:* Arthur Amos Noyes res fel chem, Calif Inst Technol, 64-66, Noyes instr, 66, from asst prof to prof theoret chem, 67-78, prof chem & appl physics, 78-84, CHARLES & MARY FERKEL PROF CHEM & APPL PHYSICS, CALIF INST TECHNOL, 84- *Concurrent Pos:* Alfred P Sloan Found res fel, 67-69; mem numerous comts & bds, 72-; vis staff mem, Theoret Physics Div, Los Alamos Nat Lab, NMex, 73-; consult, Gen Motors Res Labs, 78-, Argonne Nat Lab, 78-82, Bell Labs, 79-83, Sandia Nat Lab, 79-84, Gen Elec Res & Develop Ctr, 82-, Shell Develop Co, 82-, Res & Develop Standard Oil Ohio, 84- & var other cos & insts; sci adv bd, Triton Biosci, 84-86. *Honors & Awards:* Buck Witney Medal, Am Chem Soc, 78, Comput in Chem Award, 88. *Mem:* Nat Acad Sci; Mat Res Soc; Am Vacuum Soc; Sigma Xi; fel AAAS; fel Am Phys Soc; Am Chem Soc; Int Acad Quantum Molecular Sci. *Res:* Electronic wave functions and properties of molecules and solids; chemical reactions; electronic states of surfaces; heterogeneous and homogeneous catalysis; superconductivity; polymers clanco. *Mailing Add:* Calif Inst Technol Mail Code 139-74 Pasadena CA 91125

GODDIN, C(LIFTON) S(YLVANUS), b Richmond, Va, July 23, 14. GENERAL ENGINEERING. *Educ:* Univ Mich, BS, 36, MS, 37, PhD, 65. *Prof Exp:* Chem engr, Res Dept, Standard Oil Co, Ind, 37-41; sr res assoc, Res Ctr, Amoco Prod Co, 46-84; CONSULT, GAS PROCESSING, 84- *Mem:* Am Chem Soc; Am Inst Chem Engrs. *Res:* Geothermal energy; petrochemicals; gas processing; thermodynamics. *Mailing Add:* Hyco Corp 10205 E 47 Pl Tulsa OK 74146

GODDU, ROBERT FENNO, b Winchester, Mass, Oct 5, 25; m 49; c 3. POLYMER CHEMISTRY. *Educ:* Harvard Univ, AB, 48; Mass Inst Technol, PhD(anal chem), 51. *Prof Exp:* From res chemist to sr res chemist, Hercules Res Ctr, 51-61, res supvr, 61-67, mgr, Fiber & Film Res Div, 67-69, Mat Sci Div, 69-74, New Enterprise Res Div, 74-75, dir res & develop, Coatings & Specialty Prod Dept, 75-78, dir, Plant Develop Labs, 78-80, admin dir, Develop Dept, Hercules Inc, 80-84; RETIRED. *Mem:* Am Chem Soc. *Res:* Near-infrared and infrared spectrophotometry; polymer chemistry processes and fabrication; polymer characterization; water soluble polymers and their applications. *Mailing Add:* 52 Ulverston Dr Kennett Sq PA 19348

GODEC, CIRIL J, b Ljubljana, Yugoslavia, Feb 16, 37; US citizen; c 3. MEDICINE. *Educ:* Univ Ljubljana, MD, 63. *Prof Exp:* Internship, Univ Ljubljana, 74-75, assoc prof urol, 75-76; resident, Hennepin County Med Ctr, Minn, 76-79, asst prof urol, 79-83; DIR UROL, LONG ISLAND COL HOSP, BROOKLYN, NY, 83- *Concurrent Pos:* Assoc prof, Dept Urol, State Univ NY Downstate Med Ctr, Brooklyn, 83- *Mem:* Am Urol Asn; Int Spina Bifida Soc; NY Acad Med. *Res:* Development of treatment for urinary incontinency and impotency. *Mailing Add:* Ramsey Med Ctr Dept U St Paul MN 55101

GODEFROI, ERIK FRED, organic chemistry, for more information see previous edition

GODENNE, GHISLAINE D, b Brussels, Belg, June 2, 24. ADOLESCENT YOUTH PSYCHIATRY. *Educ:* Univ Louvain, BS, 48, MD, 52. *Prof Exp:* Intern & resident pediat, Providence Hosp, Washington, DC, 52-54; from asst prof to assoc prof psychiat, Johns Hopkins Univ, 65-68, asst prof ment hyg, 62-68 & pediat, 63-68, assoc prof psychiat, ment hyg & pediat, 68-80, PROF PSYCHOL & DIR COUN & PSYCHIAT SERV, JOHNS HOPKINS UNIV, 73-, PROF PSYCHIAT, MENT HYG & PEDIAT, 80- *Concurrent Pos:* Fel, Mayo Clin, 54-57; Parke, Davis & Co res invest grant, 57-58; NIMH gen practitioner grant, 61-63, career teaching award, 63-65; resident psychiat, Johns Hopkins Hosp, 58-62, dir adolescent psychiat serv, 64-73, dir health serv, 78-88; consult, Baltimore City Hosps, 59-81 & Dept Ment Health, 64-72; assoc prof clin psychiat, Univ Md, 78-; fulbright grant, 51-52; consult, Loyola Col, 90- *Mem:* Fel Am Psychiat Asn; fel Am Col Physicians; fel Am Soc Adolescent Psychiat (pres, 81-82); fel Am Pub Health Asn; Am Psychoanalysis Asn; Am Asn Univ Professors; fel Am Orthopsychiat Asn; Int Soc Adolescent Psychiat. *Res:* Clinical papers on adolescence and youth. *Mailing Add:* Johns Hopkins Univ Baltimore MD 21218

GODER, HAROLD ARTHUR, b North Cape, Wis, July 3, 24; m 57; c 3. PLANT ECOLOGY. *Educ:* Univ Wis, BS, 50, MS, 51, PhD, 55. *Prof Exp:* Assoc prof biol, Wis State Col, 55-70; PROF BIOL, KEENE STATE COL, 70- *Mem:* Ecol Soc Am; Nat Asn Biol Teachers. *Res:* Phytosociological study of Tsuga Canadensis at the termination of its range in Wisconsin; taxonomic survey of the amphibians of New Hampshire. *Mailing Add:* Dept Biol Keene State Col Univ NH 229 Main St Keene NH 03431

GODFREY, ANDREW ELLIOTT, b Philadelphia, Pa, May 31, 40; m 68; c 2. GEOMORPHOLOGY. *Educ:* Franklin & Marshall Col, AB, 64; Johns Hopkins Univ, PhD(geog), 69. *Prof Exp:* Asst prof geol, Vanderbilt Univ, 68-72; mem staff, Ashley Nat Forest, 72-79; MEM STAFF, FISHLAKE NAT FOREST, 79- *Concurrent Pos:* Consult geologist, Nat Ecol Found, Nashville, 72-73, Brigham Young Univ, Mus Peoples Cult, Provo, 84-89. *Mem:* AAAS; Geol Soc Am; Am Quaternary Asn; Sigma Xi. *Res:* Climatic and geologic factors of pediment formation; geochemistry of small streams and its relationship to processes of erosion; environmental factors in planning (recreational) land use; rates of cave development in Uinta Mountains; magnitude and frequency of landslides in central Utah; quantitative erosion rates of shale in an arid environment. *Mailing Add:* 523 Valley View Dr Richfield UT 84701

GODFREY, BRENDAN BERRY, b Norfolk, Va, July 17, 45; m 72; c 3. PLASMA PHYSICS, NUMERICAL ANALYSIS. *Educ:* Univ Minn, BS, 67; Princeton Univ, MA, 69, PhD(physics), 70. *Prof Exp:* Off & staff scientist physics, Air Force Weapons Lab, 70-72; staff mem physics, Los Alamos Sci Lab, 72-77, group leader, T-15, 77-79; group leader, Mission Res Corp, 79-88, vpres, 88-89, chief scientist, Weapons Lab, 89-91; ADJ PROF, UNIV NMEX, 91- *Mem:* Fel Am Phys Soc; sr mem Inst Elec & Electronics Engrs. *Res:* Theoretical and computational study of intense relativistic electron beam phenomena with application to microwave sources, accelerator development, and inertial fusion. *Mailing Add:* PL/WS Kirtland AFB NM 87117-6008

GODFREY, CHARLES S, b San Francisco, Calif, Aug 12, 18; m 45, 63; c 5. PHYSICS. *Educ:* Mass Inst Technol, BS & MS, 41; Univ Calif, PhD(physics), 53. *Prof Exp:* From physicist to div leader physics, Lawrence Radiation Lab, Univ Calif, 53-62; VPRES & CHIEF SCIENTIST, PHYSICS INT CO, 62- *Res:* Continuum mechanics and explosive phenomena. *Mailing Add:* 1450 Euclid Ave Berkeley CA 94708

GODFREY, DONALD ALBERT, b Tarrytown, NY, Aug 26, 44; m 67; c 2. PHYSIOLOGY, BIOCHEMISTRY. *Educ:* Rensselaer Polytech Inst, BS, 66; Harvard Univ, PhD(physiol), 72. *Prof Exp:* Fel pharmacol, Sch Med, Wash Univ, 72-75, res asst prof, 75-78; res assoc physiol, Cent Inst Deaf, 75-78; from asst prof to assoc prof physiol, Oral Roberts Univ, 81-88; PROF OTOLARYNGOL, PHYSIOL & ANAT, MED COL OHIO, 88- *Concurrent Pos:* NIH fel, Sch Med, Wash Univ, 72-74. *Mem:* Sigma Xi; AAAS; Soc Neurosci; Int Brain Res Orgn; Acoust Soc Am; Asn Res Otolaryngol. *Res:* Biochemistry, physiology and anatomy of the central nervous system, especially those parts concerned with special senses, particularly audition vision and olfaction. *Mailing Add:* Dept Otolaryngol Med Col Ohio, PO Box 10008 Toledo OH 43699-0008

GODFREY, GARY LUNT, b San Francisco, Calif, Nov 4, 46. ELEMENTARY PARTICLE PHYSICS. *Educ:* Calif Inst Technol, BS, 68; Univ Calif, Berkeley, PhD(physics), 75. *Prof Exp:* Fel, Lawrence Berkeley Lab, 75; res assoc physics, 76-80, STAFF PHYSICIST, STANFORD LINEAR ACCELERATOR CTR, 80- *Res:* Experimental electron-positron storage ring physics using a highly segmented NaI detector known as the Crystal Ball. *Mailing Add:* 20 Tynan Way Menlo Park CA 94028

GODFREY, GEORGE LAWRENCE, b Shawnee, Okla, Apr 25, 43; m 64; c 6. SYSTEMATIC ENTOMOLOGY. *Educ:* Northern State Col, SDak, BS, 65; Cornell Univ, PhD(entom), 70. *Prof Exp:* Res asst insect ecol, Cornell Univ, 65-66, teaching asst ecol, 66-67, res asst systematics, 67-70; asst entomologist, Ill Natural Hist Surv, 70-76; res assoc agr entom, 70-71, asst prof, 71-76, int prog coord, Int Soybean Prog, 71-76, ASSOC PROF AGR ENTOM, UNIV ILL, 76-, AFFIL, DEPT ENTOM, 87-; ASSOC PROF SCIENTIST SYSTEMATICS, ILL NATURAL HIST SURV, 76- *Concurrent Pos:* Consult, Mo Bot Garden, 73-75; ed J of Lepidopterists' Soc, 75-77, Smithsonian Inst Libr, 87-88. *Mem:* Entom Soc Wash; Lepidopterists' Soc; Lepidoptera Res Found. *Res:* Systematics and life history studies of Lepidoptera, especially of the Noctuidae and Notodontidae. *Mailing Add:* Ill Natural Hist Surv Natural Res Bldg 607 E Peabody Champaign IL 61820

GODFREY, H(ENRY) P(HILIP), b Poughkeepsie, NY, Aug 7, 41; m 77; c 2. IMMUNOLOGY, EXPERIMENTAL PATHOLOGY. *Educ:* Harvard Univ, AB, 61, MD, 65; Birmingham Univ (UK) PhD, 80. *Prof Exp:* Intern med, Barnes Hosp, St Louis, Mo, 65-66; surgeon, Div Biol Stand, NIH, 66-70; hon res fel, Univ Birmingham, 70-72; lectr, Inst Exp Immunol, Univ Copenhagen , 72-75; asst prof path, Sch Basic Health Sci, State Univ NY Stony Brook, 75-82; assoc prof path, NY Med Col, Valhalla, NY, 82-88, PROF PATH, NY MED COL, 88- *Concurrent Pos:* Moseley traveling fel, Harvard Univ, 70-72. *Mem:* AAAS; Am Soc Microbiol; NY Acad Sci; Am Asn Immunol; Harvey Soc; Am Asn Pathologists. *Res:* Cellular and biochemical bases of the induction and elicitation of cellular immune response and other forms of chronic inflammation, especially as related to disease processes such as tuberculosis and Lyme disease. *Mailing Add:* Dept Path NY Med Col Basic Sci Bldg Valhalla NY 10595

GODFREY, JOHN CARL, b Cornelius, Ore, Mar 11, 29; m 54; c 3. ORGANIC CHEMISTRY. *Educ:* Pomona Col, BA, 51; Univ Rochester, PhD(org chem), 54. *Prof Exp:* Res chemist, Shell Develop Co, 54-55; Smith Kline & French fel org chem, Rutgers Univ, 55-57, instr chem, 57-59; res chemist & group leader pharmaceut chem, Bristol-Myers Co, 59-65, dir biochem res, 65-74, dir med chem process labs, 74-75, clin monitor, 75-77, mgr technol eval & new ventures, Bristol Labs Div, 77-79; asst dir clin res, Revlon Health Care Group, USV Pharmaceut Co, 79-85; PRES, GODFREY SCI & DESIGN, INC, 79-; ASSOC DIR CLIN RES, RORER PHARMACEUT CORP, 85- *Mem:* Am Chem Soc; Am Soc Microbiol; Asn Clin Pharmacol; AAAS; NY Acad Sci. *Res:* Design, organize, monitor, analyze and report clinical evaluations of new drugs in areas of antibiotics, cardiovascular agents, anti-asthmatics and other; synthesis of penicillins; organic synthesis; semisynthetic antibiotics; hypolipemic agents. *Mailing Add:* Rorer Pharmaceut Corp 800 Bus Ctr Dr Horsham PA 19044

GODFREY, JOHN JOSEPH, b New York, NY, Dec 19, 41; m 76; c 2. SPEECH SCIENCE. *Educ:* Fordham Univ, AB, 65, MA, 66; Georgetown Univ, PhD(ling), 71. *Prof Exp:* Res assoc speech recognition, Aerospace Med Res Lab, US Air Force, Dayton, Ohio, 70-72; res assoc, Univ Dayton Res Inst, 72-74; asst prof speech sci, Univ Tex, Dallas, 75-80, assoc prof, 80-; MEM STAFF, DEPT COMMUN DIS, UNIV TEX, RICHARDSON. *Concurrent Pos:* Res assoc, Nat Res Coun Nat Acad Sci, 70-72. *Mem:* Acoust Soc Am; Ling Soc Am; Am Asn Phonetic Scis. *Res:* Speech perception in normal and auditorily impaired populations. *Mailing Add:* 6557 Dykes Way Dallas TX 75230

GODFREY, PAUL JEFFREY, b New York, NY, Apr 2, 40; m 68. COASTAL PLANT ECOLOGY, BIOMETEOROLOGY. *Educ:* Univ Conn, BS, 62; Duke Univ, PhD(plant ecol), 69. *Prof Exp:* Biologist, Cape Lookout Nat Seashore, 68-69; biologist, US Nat Park Serv, 69-70; asst prof, 70-76, ASSOC PROF PLANT ECOL, UNIV MASS, 76- *Concurrent Pos:* Res biologist, US Nat Park Serv, 70-; leader, Nat Park Serv Coop Res Unit, Univ Mass, 73-77; appointee, Barrier Island Study Group, US Dept Interior, 77-79; vis prof, Marine Lab, Duke Univ, 78-81; appointee, UNESCO-Man & Biosphere Coastal Classification Comn, 79- *Mem:* Ecol Soc Am; Brit Ecol Soc; Int Soc Biometeorol; Sigma Xi. *Res:* Coastal ecology of eastern North America: geobotany, biogeography, biometeorology, autecology of Spartina patens and other coastal species, human impacts, and management of beaches and dunes; ecology of central New England forests. *Mailing Add:* 47 Harkness Rd Pelham MA 01002

GODFREY, PAUL RUSSELL, b Ventnor, NJ, Nov 26, 14; m 40; c 5. BIOCHEMISTRY. *Educ:* William Jewell Col, AB, 38; Purdue Univ, MS, 40, PhD(biochem), 52. *Prof Exp:* Asst prof chem, Ouachita Col, 41-42 & Furman Univ, 42-43; asst chemist, Purdue Univ, 43-46; prof chem, La Col, 46-70; prof chem, Tusculum Col. 70-80, dean fac, 81-85; CONSULT 85- *Concurrent Pos:* Res partic, Oak Ridge Inst Nuclear Studies, 60 & 61, res assoc, 63; res asst, Sch Med, Tulane Univ, 64, res partic, Case Inst Technol, 65; vis prof, Northwestern State Col La, 66; res chemist, Southern Forest & Range Exp Sta, US Forest Serv, 67, consult, Insul-Therm Co, 87- *Mem:* Am Chem Soc; Sigma Xi. *Res:* Analytical methods and nutritional studies of fluorine and iodine; lipid metabolism. *Mailing Add:* Rte 8 Box 385 Greeneville TN 37743

GODFREY, ROBERT ALLEN, b Sharon, Pa, Dec 28, 44; m 67. ANALYTICAL CHEMISTRY. *Educ:* Thiel Col, AB, 66. *Prof Exp:* Chemist prod characterization, 67-73, group leader, 73-76, SECT MGR POLYMER ANALYSIS & PHYSICS, MOBAY CORP, 76- *Mem:* Am Chem Soc; Soc Plastics Indust; Am Soc Testing & Mat. *Res:* Materials characterization; spectroscopy; chromatography; polymer characterization. *Mailing Add:* 306 Serendipity Dr Coraopolis PA 15108

GODFREY, SUSAN STURGIS, b Philadelphia, Pa, Nov 20, 44; m 71. MICROBIOLOGY. *Educ:* Wheaton Col, Norton, Ma, AB, 66; Univ Pa, PhD(microbiol), 73. *Prof Exp:* Postdoctoral fel, Dept Biochem, Univ Pittsburgh Sch Med, 73-76, res asst prof develop biol, Dept Biol Sci, 77-80, lab instr microbiol, 80-90, LECTR MICROBIOL, DEPT BIOL SCI, UNIV PITTSBURGH, 90- *Concurrent Pos:* Asst res fel, Inst Molecular Biol, Academia Sinica, Taipei, Taiwan, 90. *Mem:* AAAS; Am Soc Microbiol; Int Soc Develop Biol. *Res:* Comparative study aimed at understanding how proteins adapted to extreme environments, particularly such as habitats of the archaea, differ from their more familiar homologues and how this affects their function. *Mailing Add:* Dept Biol Sci Univ Pittsburgh Pittsburgh PA 15260

GODFREY, THOMAS NIGEL KING, b Madison, Wis, Dec 11, 27; m 50; c 1. PHYSICS, COMPUTER SCIENCE. *Educ:* Mass Inst Technol, BS, 50; Princeton Univ, PhD(physics), 54. *Prof Exp:* staff mem, Los Alamos Nat Lab, 54-89; RETIRED. *Mem:* AAAS. *Res:* Nuclear weapons; Monte Carlo neutron transport computer code. *Mailing Add:* 156 Tunyo Los Alamos NM 87544

GODFREY, W LYNN, b Washington, DC, Jan 24, 39; m 59; c 5. SPENT FUEL REPROCESSING, RADIOACTIVE WASTE MANAGEMENT. *Educ:* Brigham Young Univ, BES, 61. *Prof Exp:* Engr, Am Potash & Chem Co, 61-63; sr engr, Atlantic Richfield Hanford Co, 63-73; sr proj engr, Allied-Gen Nuclear Serv, 73-83; PRES, B E INC, 83- *Mem:* Am Nuclear Soc. *Res:* Generation, volume reduction, packaging, transportation, storage and disposal of solid, liquid and gaseous radioactive wastes. *Mailing Add:* 2405 Jackson St Barnwell SC 29812

GODFREY, WILLIAM EARL, b Wolfville, NS, Mar 18, 10; m 46; c 1. ORNITHOLOGY. *Educ:* Acadia Univ, BSc, 34. *Hon Degrees:* DSc, Acadia Univ, 69. *Prof Exp:* Res assoc ornith, Cleveland Mus Natural Hist, 40-57; cur ornith, 47-76, head vert zool sect, 68-76, chief, 76-77, EMER CUR ORNITH, NAT MUS NATURAL SCI, NAT MUS CAN, 77- *Honors & Awards:* Doris Heustic Speirs Award, Soc Can Ornithologists, 86. *Mem:* Fel Am Ornith Union; corresp mem Brit Ornithologists Union. *Res:* Distribution, taxonomy and ecology of North American birds. *Mailing Add:* Vert Zool Sect Nat Mus Natrl Sci Nat Mus Can Ottawa ON K1P 6P4 Can

GODHWANI, ARJUN, b Sind, India, Dec 29, 41; m 68; c 2. ELECTRICAL ENGINEERING, CONTROL SYSTEMS. *Educ:* Vikram Univ, India, BS, 63; Roorkee Univ, India, MS, 65; Univ Ark, PhD(elec eng), 71. *Prof Exp:* Lectr elec eng, Pilani Univ, India, 66-68; instr, Univ Ark, 71-72; asst prof, 72-73, ASSOC PROF ELEC ENG, SOUTHERN ILL UNIV, 73- *Concurrent Pos:* Staff specialist, McDonnell Douglas Astronaut, 77-; Instr Sci Equip Prog Proposal, NSF, 74. *Mem:* Inst Elec & Electronics Engrs; Am Soc Eng Educ. *Res:* Control systems; digital systems including microprocessors. *Mailing Add:* Dept Eng Southern Ill Univ Box 65E Edwardsville IL 62026

GODIN, CLAUDE, b Quebec City, Que, Mar 12, 26; m 53; c 5. BIOCHEMISTRY. *Educ:* Laval Univ, BA, 46, BScA, 50, DSc(chem), 53. *Prof Exp:* Res fel, Nat Res Coun, Can & Nat Inst Med Res, Eng, 53-55; from asst prof to assoc prof biol, Ottawa Univ, 56-70; assoc prof, 70-71, chmn dept, 80-87, PROF BIOCHEM, LAVAL UNIV, 71- *Concurrent Pos:* France-Can exchange fel, Inst Molecular Path, Paris, France, 67-68, Ctr Nat Res Sci, Marseille, France, 78-79. *Mem:* Can Biochem Soc (treas, 72-75); Chem Inst Can. *Res:* Chemistry and biochemistry of proteins and amino acids; membrane proteins; cell differentiation. *Mailing Add:* Dept Biochem Fac Sci Laval Univ Quebec PQ G1K 7P4 Can

GODINE, JOHN ELLIOTT, b Montreal, Que, Feb 14, 47; m 68; c 2. ENDOCRINOLOGY. *Educ:* Princeton Univ, AB, 67; Mass Inst Technol, PhD(physics), 72; Med Sch, Harvard Univ, MD, 76. *Prof Exp:* Resident, 76-78, fel, 78-81, INSTR, MASS GEN HOSP, MED SCH, HARVARD UNIV, 81- *Mem:* Am Col Physicians; AAAS. *Res:* Recombinant DNA studies of the biosynthesis of pituitary glycoprotein hormones; neuroendocrinology. *Mailing Add:* Mass Gen Hosp ACC 630 Fruit St Boston MA 02114

GODING, JAMES WATSON, b Melbourne, Australia, July 21, 46; m 80. CELL MEMBRANE RECEPTORS. *Educ:* Monash Univ, MB & BS, 71; Univ Melbourne, BMed Sci, 73, PhD(immunol), 77. *Prof Exp:* C J Martin fel immunol, Dept Genetics, Stanford Univ, 78-79; head, Immunochem Lab Immunol, Walter & Eliza Hall Inst, 80-84; PROF IMMUNOL & CHMN, DEPT PATH, MONASH UNIV, 85- *Concurrent Pos:* Mem, Sci Adv Comt, Anti-cancer Coun Victoria & Regional Grant Interviewing Comt & Assigner's Panel, Nat Health & Med Res Coun, 85-90, Biotechnol Prod Comt, Commonwealth Dept Health, Australia & Nat Biotechnol Comt, 86-90; assoc dean, Fac Med, Monash Univ, 88-90; vis prof, Ctr Blood Res, Harvard Univ, 91. *Mem:* Am Asn Immunologists; AAAS. *Res:* Structure and function of cell surface receptors, particularly those of the immune system. *Mailing Add:* Monash Med Sch Alfred Hosp Prahran Victoria 3181 Australia

GODINO, CHARLES F, b Brooklyn, NY, Mar 31, 34; m 56. MATHEMATICS. *Educ:* St Peter's Col, NJ, BA, 55; Univ Notre Dame, MS, 58, PhD(math), 62. *Prof Exp:* Instr math, Univ Notre Dame, 62; Joseph Fels Ritt instr, Columbia Univ, 62-65; asst prof, 65-74, ASSOC PROF MATH, BROOKLYN COL, 74- *Mem:* Am Math Soc. *Res:* Group theory; group representation theory; number theory. *Mailing Add:* 64 Uxbridge St Staten Island NY 10314

GODKE, ROBERT ALAN, b Kewanee, Ill, Sept 9, 44; m 65; c 3. REPRODUCTIVE PHYSIOLOGY. *Educ:* Southern Ill Univ, BS, 66, MS, 68; Univ Mo-Columbia, PhD(reprod physiol), 74. *Prof Exp:* Br chief preliminary toxicol, US Army Environ Hyg Agency, Edgewood Arsenal, Md, 69-70; asst prof reproduction physiol, 73-77, ASSOC PROF REPRODUCTION PHYSIOL, LA STATE UNIV, BATON ROUGE, 77- *Concurrent Pos:* Teaching fel, Nat Asn Col & Teachers Agr, 80. *Mem:* Am Soc Animal Sci; Int Soc Study Animal Reproduction; Int Embryo Transfer Soc; Soc Study Reproduction. *Res:* Farm and exotic animals; induced behavioral estrus studies; prostaglandins and progestins for estrus synchronization; embryo transfer, cloning and induced twinning in farm animals; ovarian-uterine relationships; luteal function during pregnancy; prostaglandins for inducing parturition. *Mailing Add:* Dept Animal Sci La State Univ Baton Rouge LA 70803

GODLEY, WILLIE CECIL, b Miley, SC, Oct 3, 22; m 44; c 3. GENETICS. *Educ:* Clemson Col, BS, 43; NC State Col, MS, 49, PhD, 55. *Prof Exp:* From instr to asst prof animal husb, 46-47, assoc prof, 52-57, assoc dir, SC Agr Exp Sta, 73-75, PROF ANIMAL SCIENCE, CLEMSON UNIV, 57-, ASSOC DEAN & DIR, SC AGR EXP STA, 75- *Concurrent Pos:* Assoc animal husbandman, SC Agr Exp Sta, 54-57, animal husbandman & geneticist, 57-64. *Mem:* Am Soc Animal Sci; Am Genetic Asn; Sigma Xi. *Res:* Animal breeding. *Mailing Add:* 103 Lewis Rd Clemson SC 29631

GODMAN, GABRIEL C, b Albany, NY, Jan 24, 21. CELL BIOLOGY, PATHOLOGY. *Educ:* NY Univ, AB, 41, MD, 44. *Prof Exp:* House officer med, Bellevue Hosp, NY, 44-45; asst path, Yale Univ, 48-50; res assoc surg path, from asst prof to assoc prof microbiol, 54-68, PROF MICROBIOL, COL PHYSICIANS & SURGEONS, COLUMBIA UNIV, 68-, PROF PATH, 69- *Concurrent Pos:* Fel path, Mt Sinai Hosp Sch Med, 50-52, NIH grant, 64-68; vis investr cell biol, Rockefeller Inst, 57-60; consult study sect, NIH, 64-68. *Mem:* Am Asn Path; Int Acad Path; Am Soc Cell Biol; Harvey Soc; Tissue Cult Asn. *Res:* Cellular pathology. *Mailing Add:* Col Physicians & Surgeons Columbia Univ New York NY 10032

GODOLPHIN, WILLIAM, b St Boniface, Man, June 25, 41; m 67; c 2. CLINICAL CHEMISTRY. *Educ:* Univ Man, BSc, 63; Univ Alta, PhD(clin biochem), 74. *Prof Exp:* Prin, Amaranth High Sch, Man, 62-64; sr sci master, Awe Boys High Sch, Nigeria, 64-66; prin & area adminr, Snowdrift Sch, 67-69; clin chemist, 74-78, DIR RES & DEVELOP, VANCOUVER GEN HOSP, 78-; PROF PATH, UNIV BC, 75-; VPRES RES & DEVELOP, ANDRONIC TECHNOL INC, 86- *Concurrent Pos:* Mem, Environ Appeal Bd, Govt BC, 83-90 & Assoc Comt Toxicol, Nat Res Coun, 88-91. *Honors & Awards:* MDS Health Group Award, Can Soc Clin Chemists, 86. *Mem:* Can Soc Clin Chemists (secy, 86-89); Can Acad Clin Biochem; Asn Clin Biochemists; Am Asn Clin Chem. *Res:* Investigation of prognostic factors associated with human breast cancer; robotic and other automation for preanalytic specimen handling and preparation in the clinical laboratory. *Mailing Add:* Clin Chem Div Vancouver Gen Hosp 855 W 12th Ave Vancouver BC V5Z 1M9 Can

GODRICK, JOSEPH ADAM, b Rutland, Vt, May 20, 42; m 68; c 2. MECHANICAL BEHAVIOR OF MATERIALS. *Educ:* Univ Vt, BS, 63; Mass Inst Technol, MS, 64, PhD(mech eng), 68. *Prof Exp:* Staff engr, Lodgemont Lab, Kennelst Co, 68-81; sr engr, Foster-Miller Assocs, 81-82; PRIN ENGR, RAYTHEON CO, 82- *Mem:* Am Inst Mech Engrs; Am Soc Metals. *Res:* Material systems for electronics, including thick film and thin film techniques, ceramic metalization, ceramics, metals and polymers in electronics packages, joining methods and materials in electronics; materials for solar energy applications. *Mailing Add:* 12 Parker Rd Wellesley MA 02181

GODSAY, MADHU, b Bombay, India, July 16, 32; US citizen; m 55; c 2. CELLULOSE CHEMISTRY. *Educ:* Univ Bombay, BSc, 55; Univ Manchester, DTC, 58; Queen's Univ, MSc, 59; Polytech Inst NY, PhD, 85. *Prof Exp:* Res chemist, Madhusudan Mills, 55-57; from res chemist to mgr tech serv, Courtauld's Can Ltd, 59-67; from sr res assoc to consult dissolving pulps, 67-74, mgr, primary process & cellulose res, 74-85, MGR, MAT RES & EXPLOR DEVELOP, INT PAPER CO, 85- *Mem:* Am Chem Soc; Tech Asn Pulp & Paper Indust. *Res:* Novel processes for dissolving cellulose and thereby producing cellulosic fibers and films; modification of cellulose via grafting or other chemical reactions to enhance its properties; R&D in pulping & bleaching of lignocellulosic materials. *Mailing Add:* Int Paper Co Long Meadow Rd Tuxedo NY 10987

GODSCHALK, DAVID ROBINSON, b Enid, Okla, May 14, 31; m 59; c 1. GROWTH MANAGEMENT, HAZARD MITIGATION. *Educ:* Dartmouth Col, BA, 53; Univ Fla, BArch, 59; Univ NC, MRP, 64, PhD(planning), 71. *Prof Exp:* Vpres, Milo Smith Assocs, 59-61; planning dir, City Gainesville, Fla, 64-65; asst prof planning, Fla State Univ, 65-67; assoc prof, 72-77, PROF PLANNING, UNIV NC, 77- *Concurrent Pos:* Ed, J Am Inst Planners, 68-71; lectr, Univ NC, 69-71; vis prof, Univ Hawaii, 73; scholar in residence, Southern Growth Policies Bd, 77-78; prin investr, NSF, 75-77, Off Ocean & Coastal Resource Mgt, 83-84, Nat Inst Dispute Resolution & US Dept Housing & Urban Develop, 85, Int Bus Mach Corp, 89-90. *Honors & Awards:* Serv Medal, Am Inst Planners, 71. *Mem:* Am Inst Cert Planners; Urban Land Inst; Am Planning Asn. *Res:* Analysis of public policy dealing with growth management, hazard mitigation, dispute resolution, land use planning, and geographic information systems. *Mailing Add:* Dept City & Regional Planning Univ NC Chapel Hill NC 27599-3140

GODSHALK, GORDON LAMAR, b Eugene, Ore, July 19, 48. AQUATIC ECOLOGY. *Educ:* Univ Calif, Davis, BS, 70; Mich State Univ, MS, 72, PhD(bot), 77. *Prof Exp:* Res assoc aquatic ecol, Dept Environ Eng Sci, Univ Fla, 77-78, adj asst prof bot, 78-79, vis asst prof zool, 79-80; asst prof biol, Univ Southern Miss, 80-85; res ecologist, Waterways Exp Sta, US Army Corps Engrs, 85-87; ASSOC PROF BIOL & DIR ENVIRON STUDIES, ALFRED UNIV, 87- *Mem:* Am Soc Limnol & Oceanog; Ecol Soc Am; Int Asn Theoret & Appl Limnol; Sigma Xi; Soc Wetlands Scientists; NAm Benthological Soc. *Res:* Heterotrophic metabolism of detritus in aquatic ecosystems; interactions of plants and animals in aquatic ecosystems; physiological ecology of aquatic plants. *Mailing Add:* Dept Biol Alfred Univ Box 456 Alfred NY 14802

GODSHALL, FREDRIC ALLEN, b Pottstown, Pa, Sept 4, 34; m 59; c 3. METEOROLOGY. *Educ:* Ursinus Col, BS, 56; NY Univ, MS, 58; Univ Md, PhD(civil eng), 73. *Prof Exp:* Meteorologist, Proj Mercury, 61-66 & Argonne Nat Lab, 66; proj scientist, Environ Data Serv, Environ Sci Serv Admin, 66-87; phys scientist, Computer Sci Corp, 88-91; PHYS SCIENTIST, SCI APPLN INT CORP, 91- *Res:* Application of satellite data in meteorology. *Mailing Add:* Sci Appln Int Corp 27 Tarzwell Dr Narragansett RI 02882

GODSON, GODFREY NIGEL, b London, Eng, June 20, 36; US citizen. BIOCHEMISTRY. *Educ:* Univ London, Eng, BSc, 57, MSc, 59, PhD(biochem), 62, DSc(biochem/molecular biol), 84. *Prof Exp:* Postdoctoral fel, Div Biol, Calif Inst Technol, Pasadena, 64-66; res assoc, Dept Radiol, Radiobiol Labs, Sch Med, Yale Univ, 66-67; mem sci staff, Dept Phys Chem, Royal Cancer Hosp, Inst Cancer Res, Chester Beatty Res Inst, London, Eng, 61-64; mem sci staff, Biochem Div, Nat Insts Med Res, London, Eng, 67-69; from asst prof to assoc prof radiobiol, Sch Med, Yale Univ, 69-80; PROF & CHMN, DEPT BIOCHEM, SCH MED, NY UNIV, 80- *Concurrent Pos:* Vis prof, Dept Physiol Chem, Univ Utrecht, Holland, 76; vis scientist, Med Res Coun, Cambridge, Eng, 76-78; ad hoc mem, Trop Med & Parasitol Study Sect, 84-85, mem, 85-90, chmn, 88-90; mem sci adv comt, Irma T Hirschl Found, 83-87; mem biochem sect, Nat Bds Med Educ Exam Comt, 85- *Mem:* Am Soc Biol Chemists; Harvey Soc; Am Soc Microbiol. *Res:* Control of DNA replication in Ecoli; transposition and gene expression in Ecoli; gene structure and control of gene expression in the malarial organism plasmodium; use of molecular biology to develop anti-parasite strategies. *Mailing Add:* Dept Biochem Med Ctr NY Univ 550 First Ave New York NY 10016

GODSON, WARREN LEHMAN, b Victoria, BC, May 4, 20; m 77; c 5. METEOROLOGY. *Educ:* Univ BC, BA, 39, MA, 41; Univ Toronto, MA, 44, PhD(phys meteorol), 48. *Prof Exp:* Asst, Meteorol Serv Can, 42-43, gen res & lectr, 43-51, supvr res unit, 51-54, supt atmospheric res sect, 54-71, dir atmospheric processes res br, Atmospheric Environ Serv, 71-73, dir-gen atmospheric res directorate, 73-84, SR SCI ADV, ATMOSPHERIC ENVIRON SERV, 84- *Concurrent Pos:* Spec lectr, Univ Toronto, 48-60, hon prof inst environ studies, 75-80; liaison officer, Int Union Geod & Geophys, World Meteorol Orgn, 60-75, pres, Comn Atmospheric Sci, 73-77. *Honors & Awards:* Darton Prize, Royal Meteorol Soc, 61, Buchan Prize, 64; Patterson Medal, 68; IMO Prize, World Meteorol Orgn, 75. *Mem:* Int Asn Meteorol & Atmospheric Physics (secy, 60-75, vpres, 75-79, pres, 79-83); fel Am Meteorol Soc; Royal Meteorol Soc (pres, Can Br, 57-59); fel Royal Soc Can. *Res:* Precipitation physics and induced precipitation; atmospheric infrared radiation; atmospheric ozone; numerical weather prediction; stratospheric fields of wind and temperature in the Arctic; atmospheric thermodynamics; long-period atmospheric oscillations; extreme value statistics. *Mailing Add:* 39 Dove Hawkway Willowdale ON M2R 3M8 Can

GODT, HENRY CHARLES, JR, b Ft Smith, Ark, July 26, 25; m 50; c 1. SYNTHETIC ORGANIC & NATURAL PRODUCTS CHEMISTRY. *Educ:* Univ Mich, BS, 50, MS, 51, PhD(pharmaceut chem), 54. *Prof Exp:* Res chemist, Monsanto Co, 53-58, res group leader, 58-70, technol appraisal mgr, 70-86; PVT CONSULT, 87- *Mem:* Am Chem Soc. *Res:* Organic chemical synthesis and process development; medicinal chemistry; agricultural chemicals; petroleum additives; technology evaluation and appraisal. *Mailing Add:* 12410 Ballas Meadows Dr Des Peres MO 63131-3041

GODT, ROBERT EUGENE, b Des Moines, Iowa, Nov 18, 42. PHYSIOLOGY, BIOPHYSICS. *Educ:* Carnegie Inst Technol, BS, 64; Univ Wash, PhD(physiol & biophys), 71. *Prof Exp:* Res asst physiol biochem, Univ Bern, Switz, 71-74; res asst physiol, Lunds Univ, Swed, 73-74; instr pharmacol, Mayo Found, Rochester, Minn, 74-75, asst prof physiol & biophys, 75-78, asst prof pharmacol, 77-78; asst prof, 78-81, ASSOC PROF PHYSIOL, MED COL GA, 81- *Concurrent Pos:* Res asst, Swiss Nat Fund, 71-74. *Mem:* Biophys Soc; Soc Gen Physiol; AAAS; Am Physiol Soc. *Res:* Muscle physiology, biophysics and biochemistry; physiology and biophysics of excitable membranes. *Mailing Add:* Dept Physiol Med Col Ga Augusta GA 30912

GODWIN, JAMES BASIL, JR, b Beaumont, Tex, Dec 11, 24; m 48; c 4. CAUSE & ORIGIN OF PETROLEUM EXPLOSIONS, CRASHWORTHNESS OF AIRCRAFT. *Educ:* Univ Tex, Austin, Aero Eng, 49. *Prof Exp:* Dist engr petrol, Mobil Oil Co, Magnolia Petrol Co, 49-60; air staff engr electrostatics in petrol, USAF, 60-85; OWNER & CHIEF EXEC OFFICER FORENSIC ENGR, GODWIN CONSULTS, INC, 80- *Concurrent Pos:* Adv & consult, Am Petrol Inst, 66-91; prin, Static Elec Comt, Nat Fire Protection Asn, 89-91. *Mem:* Fel Nat Acad Forensic Engrs; Am Soc Mech Engrs; Soc Automotive Engrs. *Res:* Electrostatic hazards in petroleum systems and cause/origin of fires and explosions caused by static electricity. *Mailing Add:* 4438 Centerview Dr Suite 300 San Antonio TX 78228

GODWIN, JOHN THOMAS, b Social Circle, Ga, Dec 2, 17; m 48; c 3. ANATOMIC PATHOLOGY, CLINICAL PATHOLOGY. *Educ:* Emory Univ, BS, 38, MD, 41; Am Bd Path, dipl, 49; Am Bd Clin Path, dipl, 61; Am Bd Nuclear Med, dipl, 76. *Prof Exp:* Res path, Touro Infirmary, New Orleans, La, 41-42, 47-48; pathologist & lectr, Ochsner Found Hosp & Sch Med,

Tulane Univ, 50-51; head, Div Path & Pathologist, Hosp, Brookhaven Nat Lab, 51-55; clin asst prof path, 55-58, assoc prof, 58-60, PROF DENT PATH, SCH DENT, EMORY UNIV, 60-; CLIN PROF, MOOREHOUSE SCH MED, 80- *Concurrent Pos:* Am Cancer Soc fel, Mem Hosp, 48-50; asst, Radioautographic Div, Physics Dept & preceptor, sr med students, Mem Hosp, Sloan Kettering Inst, New York, 49-50, asst attend pathologist, 51-55; head Biomed Reactor Res Proj & spec res scientist, Ga Inst Technol, 58-83; abstr ed, J Am Cancer Soc, 54-55; asst vis pathologist, James Ewing Hosp, 51-55; pathologist & dir labs, St Joseph's Infirmary, Atlanta, 55-78; res collarbr, Brookhaven Nat Lab, 55-58; mem vis staff, Grady Hosp, Atlanta, 58-; consult, Grady Clay Mem Eye Clin, 58-75 & Ministry Health, Saudi Arabia, 80-85; adj clin prof, Allied Health Sci Sch & med dir, Med Technol Prog, Ga State Univ, 71-78; mem, Gov Sci & Technol Comn & Radiation Control Comt, State of Ga; chmn, Dept Path & Lab Med, King Faisal Specialist Hosp & Res Ctr, Riyadh, Saudi Arabia, 80-85; founder, Riyadh Path Group, Saudi Arabia; pathologist, West Paces Ferry Hosp, Atlanta, 86- *Honors & Awards:* Silver Medal, Am Soc Clin Path, 52, Gold Medal, 53. *Mem:* Fel Col Am Pathologists; Soc Exp Biol & Med; fel Am Soc Clin Path; Am Thyroid Asn; AMA; fel Soc Nuclear Med; Am Asn Pathologists; Soc Head & Neck Surgeons; Soc Surg Oncol. *Res:* Pathology and natural history of human neoplasms; human and animal carcinogenesis, particularly by ionizing radiations; thyroid physiology; radiology; diagnosis and treatment of human cancer; original investigations in thermal neutron; capture therapy and use of short half radioactive isotopes. *Mailing Add:* 4691 Sentinel Post Rd NW Atlanta GA 30327

GODWIN, RICHARD P, mechanical engineering, for more information see previous edition

GODWIN, ROBERT O(WEN), b Dothan, Ala, July 31, 35; m 57; c 2. LASERS, ELECTRO-OPTICS. *Educ:* Ga Inst Technol, BEE, 58; Univ Southern Calif, MS, 60; Calif State Univ, Fullerton, MBA, 73. *Prof Exp:* Mem tech staff, Hughes Aircraft Co, 58-61; sr res engr, Autonetics Div, NAm Rockwell Corp, 62-64, proj engr, 64-66, group scientist, 66-69, mgr, 69-74; proj mgr, Lawrence Livermore Labs, 74-86, assoc dir, 86-90; ENG MGT CONSULT, 90- *Mem:* Inst Elec & Electronics Engrs; Laser Inst Am (treas, 75-77, pres, 79). *Res:* Laser research and development; range finders, illuminators and other applications; optical detection and signal processing; electronic design; design and development of laser fusion facilities. *Mailing Add:* 1057 Red Maple Way New Smyrna Beach FL 32168

GODWIN, ROBERT PAUL, b Harvey, Ill, Apr 4, 37; m 60; c 2. PHYSICS. *Educ:* DePauw Univ, AB, 59; Univ Ill, Urbana, MS, 61, PhD(physics), 66. *Prof Exp:* Volkswagen fel physics, Deutsches Elektronen Synchrotron, 66-68; STAFF MEM PHYSICS, LOS ALAMOS NAT LAB, 68-,. *Concurrent Pos:* Guest scientist & Fulbright grant, Laser Res Proj, Max Planck Inst, Ger, 76-77. *Mem:* Am Phys Soc; Am Asn Physics Teachers; Sigma Xi; Inst Elec & Electronics Engrs. *Res:* Laser-plasma interaction experiments and diagnostics with neodymium and carbon dioxide lasers basic to laser-fusion studies; vacuum ultraviolet and x-ray spectroscopy; surface physics; fluids. *Mailing Add:* Los Alamos Nat Lab MS-F663 PO Box 1663 Los Alamos NM 87545

GODYCKI, LUDWIG EDWARD, b Bethlehem, Pa, Dec 21, 21; m 46; c 3. PHYSICAL CHEMISTRY. *Educ:* Lehigh Univ, BS, 43, MS, 47; Iowa State Col, PhD(chem), 51. *Prof Exp:* Asst Lehigh Univ, 46-48 & Iowa State Col, 48-51; res chemist, Houdry Process Corp, 51-52; asst prof chem, St Louis Univ, 52-56; assoc chemist, Res Lab, Int Bus Mach Corp, NY, 56-64; mgr mat, Bourns, Inc, 64-67; CONSULT, SEEC SERV, 67- *Mem:* AAAS; Am Chem Soc; Am Phys Soc; Am Crystallog Asn; Sigma Xi. *Res:* X-ray crystallography; surface chemistry and physics; semiconductor chemistry; electrochemistry. *Mailing Add:* 1060 Granada Ave San Marino CA 91108

GODZESKI, CARL WILLIAM, b Champaign Ill, July 17, 26; m 49, 63; c 2. MICROBIOLOGY, BIOCHEMISTRY. *Educ:* Univ Ill, BS, 49; Pa State Col, PhD(bact), 53. *Prof Exp:* Pathogenic bacteriologist, Pitman-Moore Co, 53-54; from res microbiologist to sr res microbiologist, Eli Lilly & Co, 54-65, res assoc, 65-69, chief microbiol immunol serv, 69-74, res assoc bio res div, 74-85; RETIRED. *Mem:* Am Soc Microbiol; Sigma Xi. *Res:* Penicillin-cephalosporin antibiotics; bacterial L-forms and mycoplasma; infectious disease; infection in connective tissue diseases; virology. *Mailing Add:* 1914 Seclusion Dr Daytona Beach FL 32124-6826

GOE, GERALD LEE, b Kansas City, Mo, Aug 17, 42; m 66; c 1. ORGANIC CHEMISTRY. *Educ:* Univ Mo-Columbia, BS, 63; Mass Inst Technol, PhD(org chem), 67. *Prof Exp:* NIH fel, Iowa State Univ, 67-69; asst prof chem, Univ Notre Dame, 69-73; assoc dir org res, 77-80, RES CHEMIST, REILLY INDUST INC, 73-, DIR RES, 80- *Mem:* Am Chem Soc. *Res:* Heterogeneous catalysis; synthesis of heterocyclic compounds; organic photochemistry. *Mailing Add:* Reilly Indust Inc PO Box 41076 Indianapolis IN 46241

GOEBEL, CARL JEROME, b Milwaukee, Wis, Mar 24, 29; m 54; c 5. ENVIRONMENTAL SCIENCES, RANGE MANAGEMENT. *Educ:* Univ Idaho, BS, 55; Utah State Univ, PhD(range mgt), 60; Harvard Univ, MPA, 64. *Prof Exp:* Range conservationist, US Forest Serv, 58-59; asst prof forestry, Iowa State Univ, 59-63; range researcher, US Forest Serv, 63-68; chmn, dept environ sci, 71-73, PROF FORESTRY & RANGE MGT, WASH STATE UNIV, 68- *Concurrent Pos:* Fel, Australian Nat Univ, Canberra, 75; Morroco Range Exten, 79-81; Lesotho Res Proj, 83-85; fac, Univ Canterbury, 87-88. *Mem:* Soc Am Foresters; Soc Range Mgt. *Res:* Competition between the native bluebunch wheatgrass; introduced perennials by the annual wheatgrass and other less desirable annuals. *Mailing Add:* Forest & Range Mgt Dept Wash State Univ Pullman WA 99164

GOEBEL, CHARLES GALE, b Indianapolis, Ind, Nov 19, 17; m 45; c 3. ORGANIC CHEMISTRY. *Educ:* Ind Univ, AB, 39; Purdue Univ, MS, 41, PhD(org chem), 44. *Prof Exp:* Sr chemist, 43-52, mgr lab, 52-59, res dir, 59-66, corp tech dir, 66-69, vpres res & develop, 69-78, VPRES & GEN MGR, EMERY INDUSTS, INC, SUBSID NAT DISTILLERS & CHEM CORP, 78- *Mem:* Am Chem Soc; Am Oil Chemists' Soc. *Res:* Fatty acids and derivatives; plasticizers; synthetic lubricants, resins; plastics and detergents. *Mailing Add:* 543 Larmie Trail Cincinnati OH 45215

GOEBEL, CHARLES JAMES, b Chicago, Ill, Dec 16, 30; m 51; c 2. THEORETICAL PHYSICS. *Educ:* Univ Chicago, PhB, 49, PhD(physics), 56. *Prof Exp:* Res assoc, Lawrence Radiation Lab, Univ Calif, 54-56; asst prof, Univ Rochester, 56-59, assoc prof physics & astron, 59-61; assoc prof, 61-64, PROF PHYSICS, UNIV WIS-MADISON, 64- *Concurrent Pos:* Alfred E Sloan res fel quantum field theory, 59-61. *Mem:* Fel Am Phys Soc. *Res:* Quantum field theory, especially meson scattering; general relativity. *Mailing Add:* Dept Physics Univ Wis 14213 Chamberlin Hall Madison WI 53706

GOEBEL, EDWIN DEWAYNE, b Moline Ill, Dec 10, 23; m 50; c 2. ENVIRONMENTAL GEOLOGY. *Educ:* Augustana Col, AB, 49; Univ Iowa, MS, 51; Univ Kans, PhD(regional stratig, econ geol), 66. *Prof Exp:* Asst geol, Univ Iowa, 50-51; geologist, State Geol Surv, Kans, 51, div head oil & gas, 51-66, chief admin serv, 66-67, sr geologist, 67-71; PROF GEOL, UNIV MO-KANSAS CITY, 71-, CHMN GEOSCI DEPT, 77- *Concurrent Pos:* Tech dir, Nat Gas Surv, Fed Power Comn, 74-76; consult, 71- *Mem:* Geol Soc Am. *Res:* Petroleum geology; subsurface stratigraphy; structural geology; micropaleontology; gas reserves, proved and potential. *Mailing Add:* Dept Geosci Univ Mo 5100 Rockhill Rd Kansas City MO 64110

GOEBEL, JACK BRUCE, b Wallowa, Ore, Aug 22, 32; m 83; c 5. COMPUTER SCIENCE. *Educ:* Univ Ore, BA, 54, MA, 56; Ore State Univ, PhD(math), 62. *Prof Exp:* Instr math, Ore State Univ, 60-61; from mathematician to sr mathematician, Gen Elec Co, 61-65; res assoc appl math, Pac Northwest Lab, Battelle Mem Inst, 65-67; chmn grad sch, 69-73, prof math & head dept, 67-80, dean, Div Arts & Sci, 73-86, PROF MATH, MONT COL MINERAL SCI & TECHNOL, 86- *Mem:* Am Math Soc; Soc Indust & Appl Math; Math Asn Am; Asn Comput Mach; Sigma Xi. *Res:* Abstract algebra; applied mathematics. *Mailing Add:* Dept Math Mont Col Mineral Sci & Technol Butte MT 59701

GOEBEL, MARISTELLA, b Racine, WI, Sept 10, 15. ABNORMAL PSYCHOLOGY & TEACHING, HEALTH PSYCHOLOGY. *Educ:* Edgewood Col, Madison, BS, 44; Catholic Univ Am, MA, 47, PhD(clin psychol), 66. *Prof Exp:* Teacher Eng, Cathedral High Sch Sionx Falls, 46-47, Heart Mary High, Mobile, 47-49; assoc prof educ, 49, prof, 66-89, EMER PROF PSYCHOL, ROSARY COL, RIVER FOREST IL, 89-; CONSULT PSYCHOL, SINSINAWA DOMINICAN SISTERS, 66-; CLIN PSYCHOL, RENAL SECT, HINES VET ADMIN HOSP, ILL, 70- *Concurrent Pos:* Chmn & ed, Southeastern Curric Comt,Teachers guide, 48-51; prin investr, NIMH small res grant, 62-63, 65-66, Hines Vet Admin Hosp, 76-87, Nat Heart Lung & Blood Inst HIPP proj, 84-86; bd dirs, Am Asn Biofeedback Clinicians, 79-86; bd advocate, Biofeedback Soc Am, 81-83. *Mem:* Am Psychol Asn; Am Asn Biofeedback Clinicians; Biofeedback Soc Am; Soc Clin & Exp Hypnosis; AAAS; Soc Behav Med; Am Heart Asn; Int Soc Hypn. *Res:* Biobehavioral treatments in essential hypertension. *Mailing Add:* 7900 W Division St River Forest IL 60305

GOEBEL, RONALD WILLIAM, b Cincinnati, Ohio, Dec 23, 47; m 77. ASTROPHYSICS, ASTRONOMY. *Educ:* Ind Univ, BA, 70; State Univ NY, Albany, MS, 76, PhD(astrophys), 77. *Prof Exp:* Astronomer, US Naval Observ, 70-72; instr physics & astron, Williams Col, 77-78; RES ASSOC ASTRON, ALLEGHENY OBSERV, UNIV PITTSBURGH, 78- *Mem:* Am Astron Soc. *Res:* Spectroscopy; development of new astronomical equipment and computer interfacing to such equipment; astrometry. *Mailing Add:* Ten Mayflower Dr Albany NY 12205

GOEBEL, WALTHER FREDERICK, b Palo Alto, Calif, Dec 24, 1899; m 30; c 2. BIOCHEMISTRY. *Educ:* Univ Ill, AB, 20, AM, 21, PhD(org chem), 23. *Hon Degrees:* DSc, Middlebury Col, 59 & Rockefeller Univ, 78. *Prof Exp:* From asst to assoc chem immunologist, Rockefeller, Univ, 24-34, from assoc mem to mem, 34-74, prof, 57-74, EMER PROF BIOCHEM, ROCKEFELLER UNIV, 74- *Honors & Awards:* Avery-Landsteiner Award, Europ Soc Immunologists, 75. *Mem:* Nat Acad Sci; Am Chem Soc; Am Soc Biol Chemists; Am Soc Microbiol; Harvey Soc. *Res:* Bacterial antigens; bacterial polysaccharides; synthetic carbohydrate protein antigens; blood group specific substances; antigens of dysentery bacilli; glucuronic acid; structure of polysaccharides; heterophile antigens; bacteriophage and viral receptors; colicines. *Mailing Add:* Rockefeller Univ 66th & York Ave New York NY 10021-6399

GOECKNER, NORBERT ANTHONY, b Chicago, Ill, Sept 30, 30; m 54; c 7. ORGANIC CHEMISTRY. *Educ:* Univ Ill, BS, 54; Univ Iowa, MS, 57, PhD(chem), 59. *Prof Exp:* Chemist gas chromatography sect, Standard Oil Co Ind, 59-61; from asst prof to assoc prof, 61-73, PROF CHEM, WESTERN ILL UNIV, 73- *Mem:* Am Chem Soc. *Res:* Gas chromatography. *Mailing Add:* Dept Chem Western Ill Univ Macomb IL 61455

GOEDDEL, DAVID V, b Pasadena, Calif, May 3, 51. MOLECULAR BIOLOGY. *Educ:* Univ Calif, San Diego, BA, 72; Univ Colo, PhD(biochem), 77. *Prof Exp:* Teaching asst & res asst, Univ Colo, 73-77; postdoctoral fel, Stanford Res Inst, 77-78; sr scientist, 78-81, DIR, DEPT MOLECULAR BIOL, GENENTECH INC, 80-, STAFF SCIENTIST, 81- *Honors & Awards:* Eli Lilly Award Biol Chem, 84; Scheele Medal, Swedish Acad Pharmaceut Sci. *Mem:* Fel AAAS. *Res:* Enzymology; author of numerous scientific publications. *Mailing Add:* Dept Molecular Biol Genentech Inc 460 Pt San Bruno Blvd S San Francisco CA 94080

GOEDEN, RICHARD DEAN, b Neillsville, Wis, May 20, 35; m 62; c 3. ENTOMOLOGY, ECOLOGY. *Educ:* Univ Wis, BS, 62, MS, 63, PhD(entom), 65. *Prof Exp:* Asst entomologist & lectr, 65-69, asst prof & asst entomologist, 69-71, assoc prof & assoc entomologist, 71-77, PROF & ENTOMOLOGIST, UNIV CALIF, RIVERSIDE, 77- *Mem:* Entom Soc Am; Entom Soc Wash; Int Orgn Biol Control; Int Weed Sci Soc; Pac Coast Entom Soc. *Res:* Biological control and insect ecology of weeds; phytophagous insect and host-plant relationships; insect life histories; biosystematics of Tephritidae. *Mailing Add:* Dept Entom Univ Calif 900 Univ Ave Riverside CA 92521

GOEDERT, MICHEL G, analytical chemistry, for more information see previous edition

GOEDICKE, VICTOR ALFRED, b Holt, Wyo, Sept 25, 12; m 37. ASTRONOMY. *Educ:* Univ Mich, AB, 35, MS, 36, PhD(astron), 38. *Prof Exp:* Instr astron, Wesleyan Univ, 39-40; lectr astron, Brown Univ, 40; from instr to asst prof astron, Yale Univ, 40-46; from assoc prof to prof math & astron, 46-84, EMER PROF, OHIO UNIV, 84- *Concurrent Pos:* Prof, US Army, Italy, 45. *Mem:* Am Astron Soc. *Res:* Astronomical spectroscopy; astronomical photometry; spectrum of VV cephei; theory of statistics. *Mailing Add:* Five Boyd St Athens OH 45701

GOEDKEN, VIRGIL LINUS, b Independence, Iowa, Nov 29, 40; m 66; c 2. INORGANIC CHEMISTRY. *Educ:* Upper Iowa Col, BS, 63; Fla State Univ, PhD(chem), 68. *Prof Exp:* Fel chem, Ohio State Univ, 68-70; asst prof, Univ Chicago, 70-76; ASSOC PROF CHEM, FLA STATE UNIV, 76- *Mem:* Royal Soc Chem; AAAS; Am Chem Soc. *Res:* Inorganic transition metal chemistry; synthetic macrocyclic ligand complexes; reactions of coordinated ligands; correlation of stability and reaction rates with electronic structure; solid state studies via x-ray crystallography. *Mailing Add:* Dept Chem Fla State Univ Tallahassee FL 32306

GOEHLER, BRIGITTE HANNA, b Radebeul, Germany, July 24, 26; US citizen. MICROBIOLOGY. *Educ:* Wayne State Univ, MS, 60, PhD(bact), 64. *Prof Exp:* Biologist, Inst Tobacco Res, Germany, 51-52; biologist, German Acad Agr Sci, Berlin, 52-55; instr biol, Wayne State Univ, 60-62, asst prof, 64-65; NIH fel, 65-67; from asst prof to assoc prof, 67-75, PROF BIOL, CALIF STATE POLYTECH UNIV, 75- *Mem:* Am Soc Microbiol; Sigma Xi. *Res:* Bacterial physiology; molecular biology. *Mailing Add:* Dept Biol Sci Calif State Polytech Univ Pomona CA 91768

GOEHRING, JOHN BROWN, b Pittsburgh, Pa, Apr 24, 35; m 60; c 3. INORGANIC CHEMISTRY. *Educ:* Davidson Col, BS, 56; Univ NC, PhD(chem), 62. *Prof Exp:* E I du Pont Co asst chem, Univ NC, 58-59, instr, 60-63; from asst prof to assoc prof, 63-70, PROF CHEM, WASH & LEE UNIV, 70- *Concurrent Pos:* NSF sci fac fel, Univ Calif, Los Angeles, 69-70. *Mem:* AAAS; Am Chem Soc; Royal Soc Chem; Hist Sci Soc. *Res:* Properties of inorganic electrolytes; experimentation in educational media; historical development of chemistry. *Mailing Add:* Dept Chem Wash & Lee Univ Lexington VA 24450

GOEKE, GEORGE LEONARD, b Philadelphia, Pa, July 16, 42; m 69; c 1. ORGANIC & ORGANOMETALLIC CHEMISTRY. *Educ:* Pa State Univ, BS, 69; Princeton Univ, MA, 71, PhD(org chem), 73. *Prof Exp:* Indexer jour publ, Inst Sci Info, 67-70; RES CHEMIST, UNION CARBIDE CORP, 73- *Mem:* Am Chem Soc; Catalysis Soc; Sigma Xi. *Res:* Synthetic organometallic chemistry; polymerization catalysis; organic synthesis. *Mailing Add:* Hiland Dr Belle Meade NJ 08502

GOEKEN, NANCY ELLEN, INTERNAL MEDICINE. *Educ:* Univ Mo, BA, 68, PhD, 72. *Prof Exp:* NSF res fel, Univ Chicago, 65-67; teaching asst vert histol & gen biol, Univ Mo, 68-72; res assoc, Dept Med, Univ Iowa, 72-73, res investr, 73-75, asst res scientist, 75-76, res health sci specialist, Vet Admin Med Ctr, 77-78, tech supvr, Tissue Typing Lab, 78-80, dir, Histocompatibility Testing, 80-82, assoc res scientist, Dept Med, Univ Iowa, 80-83, asst prof, 83-86, DIR, TISSUE TYPING LAB, VET ADMIN MED CTR, 82-, ASSOC PROF, DEPT INTERNAL MED & PATH, COL MED, 86- *Concurrent Pos:* Mem, Nat Task Force Organ Transplantation, 85-86. *Mem:* AAAS; Sigma Xi; Am Soc Histocompatibility & Immunogenetics (vpres, 87-88, pres-elect, 88-89, pres, 89-90); Am Asn Immunologists; Transplantation Soc; Am Soc Transplant Physicians (pres, 86-87). *Res:* Immunobiology; immunoregulation; transplantation immunology; author of numerous scientific publications. *Mailing Add:* Dept Internal Med Vet Admin Med Ctr Hwy 6W Rm 10E-4 Iowa City IA 52246

GOEL, AMRIT LAL, b Meerut, India, Mar 4, 38; m 67. STATISTICS, OPERATIONS RESEARCH. *Educ:* Agra Univ, BSc, 57; Univ Roorkee, BEng, 61; Univ Wis, Madison, MS, 63, PhD, 68. *Prof Exp:* Asst mech engr, Dept Atomic Energy, India, 61-62; lectr eng, Univ Wis, Milwaukee, 67-68; from asst prof to assoc prof indust eng, 68-77, PROF INDUST ENG & OPERS RES, SCH COMPUT & INFO SCI, SYRACUSE UNIV, 77- *Concurrent Pos:* NSF res grants, 69-71 & 70-71; fel, Rome Air Develop Ctr, Griffiss AFB, 70-71; consult, res grants data mgt systs, USAF, 69-76, reliability & maintainability, 70-76; Air Force grants, 75-78. *Mem:* Am Statist Asn; Opers Res Soc Am; Asn Comput Mach; Am Soc Qual Control; fel Royal Statist Soc; Sigma Xi. *Res:* Reliability theory; software reliability and modelling; sequential analysis; cumulative sum chart procedures; time series analysis. *Mailing Add:* 5011 Woodside Rd Fayetteville NY 13066

GOEL, ANIL B, research management, for more information see previous edition

GOEL, KAILASH C(HANDRA), b Tasing, India, July 12, 37; m 62; c 2. ENVIRONMENTAL ENGINEERING, ENVIRONMENTAL SCIENCES. *Educ:* Banaras Hindu Univ, BS, 59; Univ Roorkee, MEng, 60; Okla State Univ, PhD(water pollution control), 68; Wilmington Col, MBA, 85. *Prof Exp:* Asst engr, Pub Health Eng, State Uttar Pradesh, India, 60-64; sanit engr, Water & Air Resources Comn, State Del, 68, dir labs, 68-69; dir labs, Edward H Richardson Assoc, Inc, 69-77, head, environ sci div & br mgr, 77-86; PRES, G&B ENVIRON, INC, 86- *Concurrent Pos:* Exten prof, Univ Del, 70-71. *Mem:* Water Pollution Control Fedn; Am Acad Environ Engrs. *Res:* Treatment of waste waters and sludges; wastewater sampling and analysis. *Mailing Add:* 15 Wingedfoot Rd Fox Hall Dover DE 19901

GOEL, KRISHAN NARAIN, chemistry, for more information see previous edition

GOEL, NARENDRA SWARUP, b Muzaffarnagar, India, June 12, 41; US citizen. THEORETICAL BIOCHEMISTRY, RELIABILITY. *Educ:* Agra Univ, India, BS, 57; Dehi Univ, India, MS, 59 Poona Univ, India, MS, 62; Univ Md, PhD(physics), 65. *Prof Exp:* Res asst, Indian Inst Technol, 59-62; res asst instr physics, Univ Md, 66; asst prof physics & assoc biophysics, Univ Rochester, 66-72; mgr & prin scientist eng & technol, Xerox Corp, 72-76; PROF SYST SCI, STATE UNIV NY, BINGHAMTON, 76- *Concurrent Pos:* Consult, Inst Defense Analyses, 64-67, Int Bus Mach Corp, 76-81, Gen Elec Corp, 77-78, Xerox Corp, 79- & NASA, 81-; ed, J Theoret Biol, 73-85; Jacobs Scholar, Marine Biol Lab, Univ Md, 66; assoc ed, Bulletin Math Biol, 77-87; pres, Aster Consulting Assoc Inc, 77-; exchange scholar, US-USSR Acad Sci Exchange Prog, 81; vis prof, San Luis Univ, Argentina, 83; regional ed, Remote Sensing Rev, 88- *Mem:* Bioeng Soc; Biophys Soc; Soc Indust & Appl Math; NY Acad Sci; Soc Origin Life; Agron Soc. *Res:* Remote sensing; modeling and analysis of biological and engineering systems; computer simulation of self-organization in biological systems; population biology; failures in digital systems; reliability of complex electro-mechanical machines; continuing education of scientists and engineers; computer graphics; remote sensing of vegetation. *Mailing Add:* Dept Systs Sci State Univ NY Binghamton NY 13902-6000

GOEL, OM PRAKASH, b Delhi, India, Sept 25, 43; nat US; m 68; c 3. ORGANIC CHEMISTRY, MEDICINAL CHEMISTRY. *Educ:* Univ Delhi, BSc, 62; Carnegie-Mellon Univ, MS, 65, PhD(org chem), 66. *Prof Exp:* Assoc res chemist, Parke-Davis Co, 66-68, res chemist, 68-72, res scientist, 73-76, res assoc, 77-81, sr res assoc, 82-89, SECT DIR, PARKE-DAVIS RES, DIV WARNER-LAMBERT, 90- *Concurrent Pos:* Adj assoc prof, Univ Mich, Dearborn, 87. *Mem:* Am Chem Soc. *Res:* Fischer indole synthesis; process research and scaled-up synthesis of new chemotherapeutic agents and chemical intermediates; catalytic hydrogenation and high pressure reactions; radiolabeled drugs. *Mailing Add:* Parke-Davis Div Warner-Lambert 2800 Plymouth Rd Ann Arbor MI 48106

GOEL, PREM KUMAR, b Bareilly, India, 43; US citizen; m 72; c 3. BAYESIAN DECISION ANALYSIS, STATISTICAL MODELING. *Educ:* Agra Univ, India, BSc, 62, MStat, 64; Carnegie-Mellon Univ, MS, 69, PhD(statist), 71. *Prof Exp:* Res fel statist, Inst Social Sci, Agra, 65-67, lectr, 67-68; res asst, Carnegie-Mellon Univ, 68-71; from asst prof to assoc prof statist, Purdue Univ, 71-83; prof statist & dir statist consult serv, 83-88, PROF & CHMN, DEPT STATIST, OHIO STATE UNIV, 88- *Concurrent Pos:* Res assoc, dept statist, Carnegie-Mellon Univ, 74 & 75, consult, 78, vis assoc prof,78-79; consult, Dept statist, Carnegie-Mellon Univ, 78, vis assoc prof, 78-79; assoc ed, J Am Statist Asn, 76-82; prin investr, NSF Res Award, 80-82, 82-84 & 89-91; prog dir, statist & Probability Prog, NSF, 82-83; prin investr, AFOSR Res Grant, 84-87, ARO Res Award, 85 & 87, NHISA-VTRC contracts, 84, 86, 87 & 89-91; elect mem, Int Statist Inst, Bharatiya Temple Soc, pres, 87-88; expert witness variety legal cases, 83-; co-prin investr educ grant, Howard Hughes Med Inst, 89-; mem comt Appl & Theoret Statist, Nat Res Coun, 90-, panel Combining Info, 90-91. *Mem:* Am Statist Asn; fel AAAS; Int Statist Inst; fel Inst Math Statist; fel Am Statist Assoc; fel Royal Statist Soc. *Res:* Bayesian decision analysis; probability modeling and statistical inference for applied problems in engineering and social sciences; linear models and time series analysis. *Mailing Add:* Dept Statist Ohio State Univ 1958 Neil Ave Columbus OH 43210

GOEL, RAM PARKASH, b India, Dec 15, 42; m 72; c 3. PLASTICITY, CONNECTOR MECHANICS. *Educ:* Punjabi Univ, India, BS, 64; Mich State Univ, MS, 66, PhD(eng), 69. *Prof Exp:* Design Engr, McDowell Wellman Eng Co, Cleveland, 65-67; asst prof teaching, Mich State Univ, 69-70; mem tech staff res, Bell Telephone Labs, Ohio, 70-79; mgr connector mech & res, 79-85, dir, Eng Assurance, 85-90, DIR, TECHNOL PLANNING, AMP, INC, 90- *Concurrent Pos:* Lectr, M C Polytech, India, 64-65; assoc prof, NC A&T State Univ, 75-76. *Mem:* Am Soc Metals. *Res:* Mechanics of electrical connectors; mathematical modeling; mechanical properties of materials; developing reliability testprograms on electrical connectors; technical management. *Mailing Add:* 515 Colony Rd Camp Hill PA 17011

GOELA, JITENDRA SINGH, b New Delhi, India, Apr 20, 51; m 79; c 2. CHEMICAL VAPOR DEPOSITION, LASER TECHNOLOGY. *Educ:* Indian Inst Technol, BTech, 72; Brown Univ, RI, MSc, 74, PhD(mech eng), 76; Northeastern Univ, MA & MBA, 91. *Prof Exp:* Prin scientist, Phys Sci Inc, Andover, Mass, 76-78; from lectr to asst prof mech eng, Indian Inst Technol, Kanpur, India, 78-84; PRIN SCIENTIST, MORTON INT-CVD INC, WOBURN, MASS, 84- *Concurrent Pos:* Vis scientist, Phys Sci Inc, 79 & 80, CVD Inc, 83; consult, Sanders Assocs, Nashua, NH & Aeromet Inc, Tulsa, Okla, 86-87; pres, Efficient Sys Inc, Andover, Mass, 87-90. *Honors & Awards:* Arthur L Williston Medal, Am Soc Mech Engrs, 78. *Mem:* Am Soc Mech Engrs; Soc Photo-Optical Instrumentation Engrs; Am Phys Soc; Optical Sco Am; Mat Res Soc; Am Ceramic Soc. *Res:* Development of advanced infrared optical materials, ceramic materials, SIC fiber, superconducting materials and CdZnTe substrates via chemical vapor deposition lasers and applications. *Mailing Add:* 12 Messina Dr Andover MA 01810

GOELL, JAMES E(MANUEL), b New York, NY, Oct 13, 39; m 60; c 2. FIBER OPTICS, ELECTRICAL ENGINEERING. *Educ:* Cornell Univ, BEE, 62, MS, 63, PhD(elec eng), 65. *Prof Exp:* Mem tech staff, Bell Tel Labs, 65-74; vpres & dir eng, Electro Optical Prod Div, Int Tel & Tel Corp, 74-81; pres, Lightwave Technol, Inc, 81-85; VPRES, PCO, 85- *Mem:* Fel Inst Elec & Electronics Engrs; Optical Soc Am; Am Phys Soc. *Res:* Ferrites, millimeter waves; periodic structures; communications; microwave tubes and circuits; microwaves; optics; integrated optical circuits; fiber optics. *Mailing Add:* 4052 Bon Homme Rd Woodland CA 91364

GOELLNER, KARL EUGENE, vertebrate zoology, for more information see previous edition

GOEPEL, CHARLES ALBERT, b Philadelphia, Pa, June 5, 53; m 73; c 2. FIBER OPTICS, FIBER OPTIC SENSORS. *Educ:* Univ Del, BEE, 76. *Prof Exp:* Sr engr, Circuit Prod Div, Gen Tel, 77-81; eng mgr, Optical Info Systs, Exxon Enterprises, 81-83; prin engr, Gould Ocean Systs, 83-88; SR STAFF ENGR, MARTIN MARIETTA, 88- *Concurrent Pos:* Prin investr fiber optics, Martin Marietta, 88- *Mem:* Inst Elec & Electronics Engrs; Soc Photo-Optical Instrumentation Engrs. *Res:* Fiber optic sensors; interferometric based optical hydrophones; optical communications; lasers; coherent multiplexing; acoustic modeling; loudspeaker design; high strength electro-optic cables; lithium niobate modulators. *Mailing Add:* 181 Berrywood Dr Severna Park MD 21146

GOEPFERT, JOHN MCDONNELL, microbiology, food bacteriology, for more information see previous edition

GOEPP, ROBERT AUGUST, b Chicago, Ill, Nov 3, 30; m 60; c 3. DENTISTRY, PATHOLOGY. *Educ:* Loyola Univ, Chicago, BS, 54, DDS, 57; Univ Chicago, MS, 61, PhD, 67; Am Bd Oral Path, dipl, 63; Am Bd Oral & Maxillofacial Radiol. *Prof Exp:* Instr dent & path, Univ Hosps & Clins, 61-64, asst prof oral path, Zoller Mem Dent Clin & Dept Path, 64-70, assoc prof, Pritzker Sch Med, 70-75, dir, Zoller Dent Clin, Univ Chicago, 78-87, PROF ORAL PATH, PRITZKER SCH MED, UNIV CHICAGO, 75- *Concurrent Pos:* Consult, Chicago Bd Health, 63-64 & Am Dent Asn, 72-; mem, Nat Coun Radiation Protection & Measurements, 77- & Coun Dent Res, Am Dent Asn, 78-; dir & vpres, Contracap Inc, 81-88. *Mem:* AAAS; Am Dent Asn; fel Am Acad Oral Path; fel Am Acad Dent Radiol (past pres); Am Asn Path & Bact; Sigma Xi; Rad Res Soc; fel Int Col Dentists; fel Am Col Dentists; Odontographic Soc. *Res:* Caries; oral pathology; dental radiology; radiation biology; barrier fertility control. *Mailing Add:* 5841 S Maryland Ave Chicago IL 60637

GOERING, CARROLL EUGENE, b Platte Center, Nebr, June 8, 34; m 60; c 3. AGRICULTURAL ENGINEERING. *Educ:* Univ Nebr, BS, 59; Iowa State Univ, MS, 62, PhD(agr eng, eng mech), 65. *Prof Exp:* Design engr, Int Harvester Co, 59-61; asst agr eng, Iowa State Univ, 61-62, res assoc, 62-65; from asst prof to prof, Univ Mo-Columbia, 65-77; PROF AGR ENG, UNIV ILL-URBANA, 77- *Mem:* Am Soc Agr Engrs; Am Soc Eng Educ; Am Soc Testing & Mat; Soc Automotive Engrs. *Res:* Biofuels; engine efficiency; author of one book. *Mailing Add:* Dept Agr Eng Univ Ill 1304 W Pennslyvania Urbana IL 61801

GOERING, HARLAN LOWELL, b McPherson, Kans, July 13, 21; m 44; c 1. ORGANIC CHEMISTRY. *Educ:* Bethel Col, AB, 43; Univ Colo, PhD(chem), 48. *Prof Exp:* Res assoc, Univ Calif, Los Angeles, 48-50; from instr to prof, 50-72, SAMUEL M MCELVAIN PROF CHEM, UNIV WIS-MADISON, 72- *Concurrent Pos:* Vis Prof, Harvard Univ, 63. *Mem:* Am Chem Soc; Royal Soc Chem. *Res:* Mechanisms of organic reactions. *Mailing Add:* Dept Chem Univ Wis Madison WI 53706

GOERING, JOHN JAMES, b Clifton, Kans, June 2, 34; m 56; c 3. MARINE ECOLOGY. *Educ:* Bethel Col, Kans, BS, 56; Univ Wis, MS, 60, PhD(zool), 62. *Prof Exp:* From asst prof to assoc prof, 62-68, PROF MARINE SCI, INST MARINE SCI, UNIV ALASKA, 68- *Concurrent Pos:* NSF res grant, 64-90; coun Agr Sci & Technol. *Mem:* AAAS; Am Soc Limnol & Oceanog; Artic Inst NAm. *Res:* Nitrogen cycle in lakes and the sea; silicon cycle in the sea. *Mailing Add:* Inst Marine Sci Univ Alaska Fairbanks AK 99775-1080

GOERING, KENNETH JUSTIN, b San Francisco, Calif, Dec 26, 13; m 36; c 3. CHEMISTRY. *Educ:* Mont State Col, BS, 36; Calif Inst Technol, MS, 39; Iowa State Col, PhD(biophys chem), 41. *Prof Exp:* Asst chemist, Exp Sta, Mont State Col, 36-37; asst org chem, Calif Inst Technol, 37-38 & Iowa State Col, 38-41; res chemist, Anheuser-Busch Inc, Mo, 41-42; instr chem, Iowa State Col, 42-43; chemist, Univ Nebr, 43-44; asst chemist, Farm Crops Processing Corp, Nebr, 44-45; vpres & gen mgr, Mold Bran Co & Enzymes, Inc, 45-49; from asst prof to prof chem, Mont State Univ, 49-75, asst dean Grad Div, 64-67, dean, Col Grad Studies, 67-75; CONSULT, 75- *Concurrent Pos:* Plant mgr, Agr Prods Corp, Iowa, 48-49; lectr, Tech Inst, Rio de Janero, Brazil; consult for several companies; assoc ed, Am Cereal Chemists. *Mem:* Am Chem Soc; Am Asn Cereal Chem; Sigma Xi. *Res:* Concentration and fractionation of enzymes and protein purification; isolation and study of properties of new starch sources. *Mailing Add:* 8383 Saddle Mountain Rd Bozeman MT 59715

GOERINGER, GERALD CONRAD, b Philadelphia, Pa, Jan 2, 33; m 57; c 2. CELL BIOLOGY. *Educ:* Univ Pa, AB, 55; Northwestern Univ, PhD(embryol), 59. *Prof Exp:* Asst prof biol, San Diego State Col, 63-66; asst prof, 66-70, actg chmn, 80-81, ASSOC PROF ANAT, SCH MED, GEORGETOWN UNIV, 70- *Concurrent Pos:* Life Ins Med Res Fund fel, 59-62; res fel immunol, Univ Stockholm, 62-63; life sci adv comt, NASA, 83-88. *Mem:* Teratol Soc; Soc Develop Biol; Am Soc Cell Biol; Am Inst Biol Sci; Sigma Xi; Am Soc Space Gravitational Biol. *Res:* Developmental biology; pancreas development; lung development; cell hybridization; antigens of subcellular fractions; metaplastic effects of ribonucleic acid; smoking and health; teratogenesis. *Mailing Add:* Dept Anat Sch Med Georgetown Univ Washington DC 20007

GOERKE, JON, b Newark, NJ, Jan 15, 29; m 54; c 3. BIOPHYSICS, SURFACE CHEMISTRY. *Educ:* Calif Inst Technol, BS, 51; Yale Univ, MD, 55. *Prof Exp:* Intern internal med, Univ Calif, San Francisco, 55-56; resident, Vet Admin Hosp, Boston, 59-60; res fel, Biophys Lab, Harvard Med Sch, 60-62; cardiol trainee, Univ Calif, San Francisco, 62-64; USPHS spec fel, Cardiovasc Res Inst, 64-66; asst res physician & asst clin prof med, 66-67, asst prof med, 67-72, asst prof physiol, 72-74, assoc prof physiol, 74-80, PROF PHYSIOL, UNIV CALIF, SAN FRANCISCO, 80-, SR STAFF, CARDIOVASC RES INST, 74- *Concurrent Pos:* Assoc staff, Cardiovasc Res Inst, 67-74; estab investr, Am Heart Asn, 67-72. *Mem:* AAAS; Biophys Soc; Am Physiol Soc. *Res:* Lung surfactant monolayer stability in model systems and excised lung; air-water interface as a model membrane; liposomes as model systems. *Mailing Add:* CVRI Univ Calif Sch Med San Francisco CA 94143-0130

GOERNER, JOSEPH KOFAHL, b Houston, Tex, Mar 20, 25; m 46; c 5. CHEMICAL ENGINEERING. *Educ:* Rice Univ, BS, 45. *Prof Exp:* Res chem engr, 46-52, supvr process develop div, 52-56, asst to gen mgr res & develop, 56-58, mkt mgr specialty prod, 58-61, mgr prod mkt, 61-66, asst gen mgr mkt, 66-67, gen mgr res & develop, 67-73, vpres res & planning, Jefferson Chem Co, Inc, 74-80, mgr res & develop div, Petrochem Dept, Texaco, Inc, 77-80, VPRES RES & DEVELOP, TEXACO CHEM CO, 80-; PROF CHEM DEPT, LA STATE UNIV. *Mem:* Am Chem Soc; Am Inst Chem Engrs. *Res:* Process development for production of derivatives of ethylene, propylene related olefins, and other petroleum and coal derived hydrocarbons. *Mailing Add:* Dept Chem La State Univ One University Pl Shreveport LA 71115

GOERTEMILLER, CLARENCE C, JR, b Big Stone Gap, Va, Feb 19, 28; wid. DEVELOPMENTAL BIOLOGY, CELL BIOLOGY. *Educ:* Univ Md, BS, 51; RI Col, EdB, 59; Brown Univ, ScM, 62, PhD(biol), 64. *Prof Exp:* Res fel biol, Harvard Med Sch, 64-65; asst prof, 65-70, assoc prof, 70-77, PROF ZOOL, UNIV RI, 77- *Mem:* AAAS; Am Soc Zoologists; Am Soc Cell Biol; Sigma Xi. *Res:* Development of secretory competence in salt-secreting epithelia. *Mailing Add:* Dept Zool Univ RI Kingston RI 02881

GOERTZ, CHRISTOPH KLAUS, b Danzig, Germany, June 1, 44; m 66; c 2. SPACE PHYSICS. *Educ:* Tech Univ Berlin, WGermany, Dipl, 67, Rhodes Univ, SAfrica, PhD(physiol), 72. *Prof Exp:* Lectr physics, Rhodes Univ, SAfrica, 69-73; res assoc, 73-75, from asst prof to assoc prof, 75-80, PROF PHYSICS, UNIV IOWA, 80- *Concurrent Pos:* Prin investr, NASA & NSF, 75-; sr res scientist, Max Planck Inst, 79-81; assoc editor, JGR & GRL. *Mem:* Am Geophys Union. *Res:* Physics of planetary magnetospheres; solar wind; basic plasmaphysics; plasma physics applied to astrophysics. *Mailing Add:* Dept Physics & Astron Univ Iowa Iowa City IA 52242

GOERTZ, GRAYCE EDITH, b Carnduff, Sask, July 9, 19; US citizen. FOOD SCIENCE. *Educ:* Kans State Univ, BS, 41, MS, 47, PhD(food sci), 52. *Prof Exp:* Instr pub sch, Kans, 41-44 & pub sch & jr col, 44-46; asst, Kans State Univ, 46-47, 49-52, instr foods, 47-48; instr, Univ Ill, 48-49; asst prof, Ore State Univ, 52-55; prof, Kans State Univ, 55-65; prof & head, Dept Food Sci & Syst Admin, 64-74, assoc dean, Col Home Econ, 74-80, EMER PROF, UNIV TENN, KNOXVILLE, 80- *Concurrent Pos:* Mem educ comn, Inst Food Technol, 74- *Mem:* Inst Food Technol; Am Dietetic Asn; Am Home Econ Asn. *Res:* Heat effects on proteins including pigments. *Mailing Add:* 1110 Turnberry Lane Maryville TN 37801

GOERTZ, JOHN WILLIAM, b Hackensack, NJ, Aug 25, 29; m 53; c 3. VERTEBRATE ZOOLOGY. *Educ:* Ore State Univ, BS, 57, MS, 59; Okla State Univ, PhD(zool), 62. *Prof Exp:* From asst prof to prof zool, La Tech Univ, 62-89; RETIRED. *Concurrent Pos:* Environ consult, 89- *Mem:* Am Soc Mammalogists. *Res:* Population dynamics; biology, habitats, parasites, distribution, reproductive; rates in rodent ecology; distribution and biology in mammalogy; nesting behavior in ornithology; wildlife and forest management. *Mailing Add:* Dept Zool 1420 Caddo St Ruston LA 71272

GOERTZEL, GERALD, b New York, NY, Aug 18, 19; m 41; c 3. IMAGE PROCESSING. *Educ:* Stevens Inst Technol, ME & MS, 40; NY Univ, PhD(physics), 47. *Prof Exp:* Physicist, Oak Ridge Nat Lab, 46-48; assoc prof physics, NY Univ, 48-53; tech dir, Nuclear Develop Corp Am, 53-60; vpres eng, Sage Instruments Inc, 60-64; RES STAFF, INT BUS MACH, 64- *Mem:* Am Phys Soc; Sigma Xi; AAAS; Soc Photog Scientists & Engrs. *Res:* Data compression; reactor physics; Fourier analysis; internal conversion; medical electronics; digital halftoning. *Mailing Add:* Seven Sparrow Circle White Plains NY 10605

GOERZ, DAVID JONATHAN, JR, b Los Angeles, Calif, Sept 25, 34; m 62; c 3. PHYSICS. *Educ:* Univ Calif, Los Angeles, AB, 56; Stanford Univ, MS, 57. *Prof Exp:* Res assoc microwave physics, W W Hansen Lab, Stanford Univ, 56 & 58, res assoc, 58-62; pres & tech dir, Vactite, Inc, 62-63; from sr engr to sr supv engr, Bechtel Corp, 63-68, res mgr, 68-71, mgr bus develop, 71-77, mgr coal prog & vpres, Bechtel Nat, Inc, 77-81, mgr, new proj develop, Bechtel Civil & Minerals Inc, 81-85, vpres, Bechtel China, Inc, 82, mgr Strategic Planning & Mkt & vpres, Bechtel Power Corp, 87-89, MGR DIV BUS DEVELOP & MKT, WESTERN PETROL & MINING & METALS DIVS, BECHTEL, INC, 85-; SR VPRES C ITOH & CO (AM) INC. *Concurrent Pos:* Consult, Hughes Aircraft Co, 59 & Phys Electronics Lab, 60; mem bd gov, Int Microwave Power Inst, 68-, chmn, 70-71; gen chmn, Microwave Power Conf, The Hague, 69; alt dir, China Am Int Eng Inc, 84; co-chmn, Comt Scholarly Commun People's Repub China, Nat Acad Sci/Nat Acad Eng; mem Steering Comt, Nat Res Coun. *Mem:* Inst Elec & Electronics Engrs; Sigma Xi. *Res:* Accelerator physics; microwave theory and applications; ultra high vacuum; cryogenics; industrial process development; environmental research; coal technology and regional planning; electric power generation; project financing; petrochemical and refining. *Mailing Add:* 11 Shasta Lane Menlo Park CA 94025

GOETHERT, B(ERNHARD) H(ERMANN), aerodynamics, space sciences; deceased, see previous edition for last biography

GOETINCK, PAUL FIRMIN, b Bruges, Belg, June 17, 33; US citizen; m 58; c 2. GENETICS. *Educ:* Univ Calif, Davis, BS, 56, PhD(genetics), 63. *Prof Exp:* Trainee biochem, Univ Calif, Los Angeles, 62-64; from asst prof to assoc prof, 64-72, PROF ANIMAL GENETICS, UNIV CONN, 72- *Concurrent Pos:* NIH res career develop award, 68-73. *Mem:* AAAS; Am Genetic Asn; Soc Develop Biol. *Res:* Developmental genetics; embryology; protein biochemistry. *Mailing Add:* La Jolla Cancer Res Found 10901 N Torrey Pines Rd La Jolla CA 92037

GOETSCH, DENNIS DONALD, b Brewster, Kans, Dec 6, 24; m 53; c 3. VETERINARY PHYSIOLOGY. *Educ:* Kans State Univ, BS & DVM, 52, MS, 55; Okla State Univ, PhD, 61. *Prof Exp:* From instr to asst prof physiol, Kans State Univ, 52-56; tech dir, Specified, Inc, Ind, 56-57; from asst prof to prof physiol, Okla State Univ, 57-69; PROF PHYSIOL & PHARMACOL, SCH VET MED, UNIV GA, 69- *Concurrent Pos:* NSF fel, 59-60; mem exec coun, Conf of Res Workers in Animal Dis, 72-77, pres, 77; mem coun educ, Am Vet Med Asn, 72-82. *Mem:* Am Soc Vet Physiol & Pharmacol (pres, 77); Am Physiol Soc; Am Vet Med Asn. *Res:* Carbohydrate and energy metabolism, especially of the ruminant; ketosis of the ruminant animal. *Mailing Add:* 25 Trace Way Sanford NC 27330

GOETSCH, GERALD D, b Colby, Kans, Apr 6, 23; m 47; c 3. ANIMAL PHYSIOLOGY. *Educ:* Kans State Univ, DVM, 45; Purdue Univ, MS, 55, PhD(animal physiol), 57. *Prof Exp:* Asst state veterinarian, Ill State Dept Agr, 45-46; pvt pract, Ill, 46-47; asst prof vet physiol, Univ Mo, 47 & Okla State Univ, 48-52; from instr to assoc prof, 52-59, PROF & HEAD DEPT VET PHYSIOL, SCH VET MED, PURDUE UNIV, 59- *Concurrent Pos:* Mem, NCent Regional Comt Ruminant Bloat, 54-62; mem coun biol & therapeut agents, Am Vet Med Asn, 62- *Mem:* Am Vet Med Asn; Am Soc Vet Physiol & Pharmacol (pres, 64); Conf Res Workers Animal Dis. *Res:* Metabolic diseases of animals; ruminant bloat; ketosis of sheep; gastric ulcers of swine. *Mailing Add:* 328 Fernleaf Dr West Lafayette IN 47906

GOETSCHEL, CHARLES THOMAS, b Chicago, Ill, Sept 21, 35; m 58; c 2. ORGANIC CHEMISTRY, INORGANIC CHEMISTRY. *Educ:* Northwestern Univ, BA, 57; Univ Mich, PhD(org chem), 62. *Prof Exp:* AEC res fel catalysis, Northwestern Univ, 62-64; chemist, Shell Develop Co, 64-70; chemist, Ctr Technol, Kaiser Aluminum & Chem Corp, Pleasanton, 70-72; INSTR CHEM, CHABOT COL, 72- *Concurrent Pos:* Mem, Int Cong Catalysis. *Mem:* Am Chem Soc. *Res:* Application of organic chemistry to catalysis; inorganic radiation chemistry; inorganic fluorine chemistry; inert gas chemistry and high energy inorganic oxidizers. *Mailing Add:* Dept Chem Chabot Col 25555 Hesperian Blvd Hayward CA 94545

GOETTLER, LLOYD ARNOLD, b Rockville Centre, NY, July 5, 39; m 61; c 5. ALLOYS BLENDS & COMPOSITES, RHEOLOGY. *Educ:* Cornell Univ, BChE, 62; Univ Del, PhD(chem eng), 67. *Prof Exp:* Sr res engr, 66-72, res specialist, 73-76, sr res specialist, 76-85, FEL, MONSANTO CO, 85- *Concurrent Pos:* Adj prof polymer eng, Univ Akron, 85- *Honors & Awards:* Pres Cup, Soc Plastics Engrs, 86. *Mem:* Soc Plastics Engrs; Am Chem Soc; Soc Rheology; Am Inst Chem Engrs; Polymer Process Soc. *Res:* Diffusion with simultaneous chemical reaction; diffusion and growth kinetics in crystallization from the vapor; processing of polymeric materials; short fiber composites; reinforced elastomers; thermoplastic elastomers; adhesion; polymer blends and alloys. *Mailing Add:* Monsanto Chem Co 730 Worcester St Springfield MA 01151

GOETTLICH RIEMANN, WILHELMINA MARIA ANNA, b Jaworow, Poland, June 25, 34; Polish & US citizen; m 65. PROTEIN CHEMISTRY, DAIRY & MEAT SCIENCE & TECHNOLOGY. *Educ:* Warsaw Agr Univ, BS, 54, MS, 57; Gdansk Polytech, PhD(chem), 77. *Hon Degrees:* LHD, World Univ, Tucson, Ariz, 82. *Prof Exp:* Res asst dairy sci, Warsaw Agr Univ, 54-56 & Warsaw Dairy Res Inst, 57-62; asst prof meat sci, Warsaw Meat Res Inst, 61-65; staff res assoc animal sci, Univ Calif, Davis, 66-68, res asst food sci, 68-70, staff res assoc nutrit, 70-82, researcher postgrad food sci, 86-87, EMER RESEARCHER FOOD SCI, UNIV CALIF, DAVIS, 89- *Concurrent Pos:* Consult, Polish Eng Asn, 57-65 & Polish Patent Comn, 63-65; coordr, Cameron Meat Res Inst, 62-64. *Mem:* Sigma Xi. *Res:* Protein chemistry; analytical chemistry; physico chemistry apparatus and instruments; author of 18 publications; dietetics; biochemistry; computer science. *Mailing Add:* 816 Miller Dr Davis CA 95616

GOETTSCH, ROBERT, b Atlantic, Iowa, Sept 27, 27; m 58; c 3. PHARMACEUTICS, BIOPHARMACEUTICS. *Educ:* Univ Colo, BS, 51; Univ Iowa, MS, 53. *Prof Exp:* Instr pharm, Univ Colo, 53-54; asst prof pharm & pharmaceut chem, Univ Kans, 57-60; asst prof pharmaceut chem, Northeast La State Col, 60-61; assoc prof pharmaceut, Univ Tenn, 61-65; assoc prof, 65-68, PROF PHARM, IDAHO STATE UNIV, 68- *Concurrent Pos:* Res grant, Burroughs Wellcome & Co, USA, Inc, 62-63; consult, Palmer Chem & Equip Co, Inc, 62-65. *Mem:* AAAS; Am Pharmaceut Asn; Am Chem Soc; Acad Pharmaceut Sci; NY Acad Sci; Sigma Xi. *Res:* Applications of physical-chemical principles to pharmaceutical systems; interaction of drugs with macromolecules; drug degradation and stabilization; analytical methods for drugs and drug mixtures; pharmacy. *Mailing Add:* Idaho State Univ Col Pharm Pocatello ID 83201

GOETZ, ABRAHAM, b Grybow, Poland, Apr 8, 26; nat; m 55; c 2. MATHEMATICS. *Educ:* Wroclaw Univ, MS, 49; Math Inst, Polish Acad Sci, PhD(math), 57. *Prof Exp:* Asst math, Wroclaw Univ, 48-53, adj, 53-58; lectr, 58-64; lectr math, Math Inst, Polish Acad Sci, 53-62; vis assoc prof, 62, ASSOC PROF MATH, UNIV NOTRE DAME, 65- *Mem:* Am Math Soc; Polish Math Soc; Math Asn Am. *Res:* Topological and Lie groups; differential geometry; universal algebras. *Mailing Add:* Dept Math Univ Notre Dame Notre Dame IN 46557

GOETZ, ALEXANDER FRANKLIN HERMANN, b Pasadena, Calif, Oct 14, 38; m 82; c 2. REMOTE SENSING, RESEARCH ADMINISTRATION. *Educ:* Calif Inst Technol, BS, 61, MS, 62, PhD(planetary sci), 67. *Prof Exp:* Mem tech staff lunar geol, Bell Tel Labs, 67-70; mem tech staff planetology, Calif Inst Technol, 70-73, group supvr earth resources, 73-75, sect mgr planetology & oceanog, 75-77, sr mem tech staff, 77-81, sr res scientist, Jet Propulsion Lab, 81-85; PROF GEOL SCI & DIR, CTR STUDY EARTH FROM SPACE, UNIV COLO, BOULDER, 85- *Concurrent Pos:* Assoc ed, Geophys Res Lett, 74-77, Remote Sensing Environ, 84-; pres, GeoImages Inc, Altadena, Calif, 75-78; vis prof, Univ Calif Los Angeles, 83; mem, Nat Res Coun Bd Earth Sci, 87-90; trustee, San Juan Capistrano Res Inst, 87-; distinguished vis scientist, Jet Propulsion Lab, 89-90; pres, Analytical Spectral Devices Inc, Boulder, Colo, 90- *Honors & Awards:* Charles E Ives Award, Soc Photog Scientists & Engrs, 74; Autometrics Award, Soc Photogram 82; William T Pecora Award, 82; NASA Space Act Award, 87; Exceptional Sci Achievement Medal, NASA, 82. *Mem:* AAAS; Am Geophys Union; Sigma Xi; Am Soc Photogram & Remote Sensing. *Res:* Space and airborne remote sensing in the ultraviolet to microwave spectrum applied to geologic mapping; global change manifestation studies from the standpoint of field observations, field instrument development and hyperspectral image data analysis techniques. *Mailing Add:* 3555 4th St Boulder CO 80304-1703

GOETZ, CHARLES ALBERT, analytical chemistry; deceased, see previous edition for last biography

GOETZ, FREDERICK CHARLES, b Fond du Lac, Wis; m 60; c 4. INTERNAL MEDICINE, DIABETES. *Educ:* Harvard Univ, BS, 43, MD, 46. *Prof Exp:* From instr to assoc prof, 55-68, PROF MED, MED SCH, UNIV MINN, MINNEAPOLIS, 68- *Mem:* Endocrine Soc; Am Diabetes Asn. *Res:* Diabetes mellitus in the human being, especially complications of diabetes and mechanisms of insulin secretion; endocrinology. *Mailing Add:* Univ Minn Hosp 412 Union St SE Minneapolis MN 55455

GOETZ, FREDERICK WILLIAM, JR, b Philadelphia, Pa, Jan 31, 50; m 72; c 3. COMPARATIVE ENDOCRINOLOGY, REPRODUCTIVE BIOLOGY. *Educ:* Colgate Univ, BA, 72; Univ Wyo, PhD(zool), 76. *Prof Exp:* Fel, Fisheries & Marine Serv, Nat Res Coun Can, 76-77; ASST PROF PHYSIOL & ENDOCRINOL, DEPT BIOL, UNIV NOTRE DAME, 77- *Mem:* AAAS; Am Soc Zoologists; Am Fisheries Soc; Soc Study Reproduction. *Res:* Hormonal control of oocyte final maturation and ovulation in piscine vertebrates and the evolution of this control in teleost fish. *Mailing Add:* Dept Biol Sci Univ Notre Dame Notre Dame IN 46556

GOETZ, HAROLD, b Halliday, NDak, Dec 20, 32; m 60; c 2. PLANT ECOLOGY, RANGE ECOLOGY. *Educ:* NDak State Univ, BS, 60, MS, 63; Utah State Univ, PhD(range sci), 68. *Prof Exp:* Range conservationist, Bur Indian Affairs, Nebr, 63; from asst prof to assoc prof bot, NDak State Univ, 63-75, dir, Tri-Col Ctr Environ Studies, 70-73, prof bot, 75-85; HEAD, RANGE SCI DEPT, COLO STATE UNIV, 85- *Concurrent Pos:* Mem, Great Plains Res Coun Comt, 66- & Great Plains Range & Livestock Adv Comt, 66-; US Govt consult, 67-; sr agr officer, fac renewable natural resources, Univ Tehran, 73-; Fulbright fel. *Mem:* Am Soc Range Mgt (secy-treas, 65-); Am Inst Biol Sci; Am Forestry Asn; Ecol Soc Am. *Res:* Rangeland ecology; effects of nitrogen fertilizer on the botanical composition of range sites as related to water use, phenological development, soil chemical and physical changes and other factors of the individual plant species. *Mailing Add:* Range Sci Dept Col Forestry-Natural Res Colorado State Univ Ft Collins CO 80523

GOETZ, KENNETH LEE, b Java, SDak, Jan 7, 32; m 62; c 2. CARDIOVASCULAR PHYSIOLOGY, BODY FLUID HOMEOSTASIS. *Educ:* Univ Wis, BS, 58, PhD(physiol), 63; Univ Kans, MD, 67. *Prof Exp:* From instr to asst prof physiol, Med Ctr, Univ Kans, 63-68; med intern, 69, HEAD EXP MED, ST LUKE'S HOSP, 70-, DIR RES, 81- *Concurrent Pos:* Adj prof, Dept Physiol, Med Ctr, Univ Kans. *Mem:* AAAS; fel Am Physiol Soc; fel Am Heart Asn; Soc Exp Biol & Med; Sigma Xi; Am Soc Hypertension. *Res:* Cardiovascular physiology; physiologic effects of changes in atrial pressures; regulation of body fluid volume; atrial natriuretic peptides; urodilatin. *Mailing Add:* Div Exp Med St Luke's Hosp 44th & Wornall Rd Kansas City MO 64111

GOETZ, RICHARD W, b Cincinnati, Ohio, July 29, 36; m 60; c 4. INDUSTRIAL ORGANIC CHEMISTRY. *Educ:* Univ Cincinnati, ChemE, 59, PhD(org chem), 63. *Prof Exp:* Res chemist, Monsanto Co, 63-64; res chemist, Nat Distillers & Chem Co, 64-73, from res assoc to sr res assoc, 73-81, mgr, US Indust Chem Co Div, Nat Distillers & Chem Co, 81-87; RES MGR, QUANTUM CHEM CORP, USI DIV, 87- *Mem:* Sigma Xi; Am Chem Soc; AAAS. *Res:* Homogeneous and heterogeneous catalysis of organic reactions by transition metal compounds. *Mailing Add:* Quantum Chem Corp USI Div 11530 Northlake Dr Cincinnati OH 45249

GOETZ, RUDOLPH W, b Worcester, Mass, Dec 24, 42; m 66; c 2. ORGANIC CHEMISTRY, ANALYTICAL CHEMISTRY. *Educ:* Mass Inst Technol, BS, 64; Boston Col, PhD(chem), 69. *Prof Exp:* Asst prof chem, Curry Col, 68-74; asst prof chem, Boston Col, 74-76; lab coordr, Boston Univ, 76-77; COORDR UNDERGRAD ORG LABS, MICH STATE UNIV, 77- *Concurrent Pos:* Consult, Tim Hennigan Eng, 70-72; lectr, Univ Mass, Boston, 74-75. *Mem:* Am Chem Soc. *Res:* Synthesis of bicyclic compounds; steric effects by spectroscopy. *Mailing Add:* 4505 Marlborough St Okemos MI 48864-1327

GOETZ, WILLIAM H(ARNER), b Ann Arbor, Mich, Sept 29, 14; m 38; c 3. HIGHWAY ENGINEERING, MATERIALS SCIENCE. *Educ:* Univ Mich, BS, 36; Purdue Univ, MS, 42. *Prof Exp:* Lab technician, State Hwy Dept, Mich, 36-38; from asst prof to prof hwy eng, 45-85, asst head Sch Civil Eng, 68-85, EMER PROF HWY ENG, PURDUE UNIV, 85- SCH CIVIL ENG, 68- *Concurrent Pos:* consult, Mc Connaughay Assoc, 48-, Corps Engrs, US

Army, 52- & Nat Co-op Hwy Res Prog, 62- *Honors & Awards:* Prevost Hubbard award, Am Soc Testing & Mat, 81. *Mem:* Am Soc Eng Educ; Nat Soc Prof Engrs; Asn Asphalt Paving Technol (pres, 58); Am Soc Testing & Mat. *Res:* Bituminous materials and pavements; bituminous mixture design; asphalt emulsions. *Mailing Add:* Sch Civil Eng Civil Eng Bldg Purdue Univ Lafayette IN 47907

GOETZEL, CLAUS G(UENTER), b Ger, July 14, 13; US citizen; m 38; c 2. MATERIALS SCIENCE, METALLURGICAL ENGINEERING. *Educ:* Tech Hochsch, Berlin, dipl ing, 35; Columbia Univ, PhD(metall), 39. *Prof Exp:* Head res lab, Charles Hardy, Inc, 37-39; tech dir & works mgr, Am Electro Metals Corp, 40-47; vpres & dir res, Sintercast Corp Am, 47-57; sr res scientist, NY Univ, 57-60; consult scientist, res lab, Lockheed Aircraft Corp, Palo Alto, 61-78; RETIRED. *Concurrent Pos:* Adj prof, NY Univ, 44-60; chmn prog comt, Int Powder Metall Conf, 60; lectr, Stanford Univ, 61-; US sr scientist, Alexander von Humboldt Found & Govt Fed Repub Ger, 77; vis prof, Univ Karlsruhe, Ger, 78-80. *Mem:* Fel Am Soc Mat; Am Inst Mining, Metall & Petrol Engrs; Am Powder Metall Inst; assoc fel Am Inst Aeronaut & Astronaut; Inst Mat UK. *Res:* Metals technology; high temperature materials; powder metallurgy; dispersion alloys; metal-nonmetal composites; nuclear reactor components; wear and corrosion-resistant parts; tools and dies; graphite composites; rocket propulsion components. *Mailing Add:* 250 Cervantes Rd Portola Valley CA 94028

GOETZINGER, CORNELIUS PETER, b Baltimore, Md, Feb 10, 11; m 42; c 2. AUDIOLOGY, PSYCHOLOGY. *Educ:* Washington Univ, BS, 41; Gallaudet Col, MA, 44; Univ Calif, Berkeley, MA, 50; Northwestern Univ, Evanston, PhD(audiol), 55. *Prof Exp:* Teacher deaf & audiologist, Cent Inst Deaf, 41-43; teacher deaf, audiologist & psychologist, Calif Sch for Deaf, Berkeley, 44-51; prof otolaryngol, 53-80, DIR AUDIOL-ELECTRONYSTAGMOGRAPHIC CLIN, SCH MED, UNIV KANS MED CTR, KANSAS CITY, 53-, EMER PROF, 80- *Concurrent Pos:* Consult psychol & audiol, Kans State Sch Deaf, 55-; consult audiol, Vet Admin Hosp, 59- & Menninger Found, Children's Hosp, 64- *Mem:* Fel Am Speech & Hearing Asn. *Res:* Hearing pathology with reference to diagnostic tests and rehabilitation in conjunction with research into the psychological impact of deafness upon intelligence and adjustment; author or coauthor of numerous publications and book reviews. *Mailing Add:* 14604 S Chalet D 312 Kansas City KS 66062

GOFF, CHARLES W, b Rowlesburg, WVa, July 25, 41; m 74; c 3. ELECTRON MICROSCOPY, CYTOCHEMISTRY. *Educ:* Fairmont State Col, BA, 64; Univ Tex, Austin, PhD(bot), 70. *Prof Exp:* Asst prof, 69-77, ASSOC PROF LIFE SCI, IND STATE UNIV, 78-, DIR, ELECTRON MICROS FAC, 81- *Mem:* AAAS; Am Soc Cell Biol; Am Soc Plant Physiologists; Bot Soc Am; Histochem Soc; Sigma Xi. *Res:* Activities and functions of the cytomembrane system, particularly during mitosis and cytokinesis, as studies by experimental treatment; ultrastructural and cytochemical techniques. *Mailing Add:* 211 Hulman-Salem Rd Terre Haute IN 47803

GOFF, CHRISTOPHER GODFREY, b Providence, RI, Apr 11, 45; m 68. MOLECULAR BIOLOGY, BIOCHEMICAL GENETICS. *Educ:* Amherst Col, BA, 67; Harvard Univ, MA, 68, PhD(biochem), 73. *Prof Exp:* Am Cancer Soc fel biochem, Med Res Coun Lab Molecular Biol, Cambridge, Eng, 73-75, Europ Molecular Biol Orgn fel, 75; from asst prof to assoc prof, 81-86, chmn dept, 83-86, PROF BIOL, HAVERFORD COL, 86- *Concurrent Pos:* Vis sr res scientist, Collab Genetics Inc, Waltham, Mass, 81-83; consult, Collab Genetics, Inc, 83-85 & Molecular Genetics, Res & Develop, Smithkline, Beckman, Inc, Philadelphia, Pa, 85; co-chmn, biol panel, NSF Col Sci Instrumentation Prog, 88. *Mem:* AAAS. *Res:* Regulation of dihydrofolate reductase gene expression in yeast; molecular mechanisms controlling transcription specificity after bacteriophage infection of E coli; expression of bovine prochymosin in yeast. *Mailing Add:* Dept Biol Haverford Col Haverford PA 19041-1392

GOFF, GERALD K, b Apache, Okla, June 26, 25; m 51; c 2. MATHEMATICS. *Educ:* Phillips Univ, BA, 50, MEd, 53; Okla State Univ, EdD(math), 62. *Prof Exp:* Instr pub schs, Okla, 50-53, prin, 53-56; from asst prof to assoc prof, Southwestern State Col, Okla, 57-65; assoc prof, 65-70, PROF MATH, OKLA STATE UNIV, 70- *Concurrent Pos:* Consult sch math study group, Stanford Univ, 63-; adj prof, Okla State Univ, 62-64; vis prof, Univ Carabobo, Valencia, Venezuela, 70. *Mem:* Math Asn Am. *Res:* Number theory; prime and composite numbers. *Mailing Add:* 2803 N Lincoln Stillwater OK 74074

GOFF, HAROLD MILTON, b St Louis, Mo, Sept 24, 47; m 71; c 2. BIOINORGANIC CHEMISTRY. *Educ:* Univ Mo-Columbia, BS, 69, MA, 71; Univ Tex, Austin, PhD(chem), 76. *Prof Exp:* Med lab instr, US Army, Ft Sam Houston, Tex, 70-73; fel chem, Univ Calif, Davis, 76; from asst prof to assoc prof, 76-85, PROF CHEM, UNIV IOWA, 85- *Concurrent Pos:* NIH Metallobiochem Study Sect, 85-89. *Mem:* Am Chem Soc. *Res:* Synthesis, ligand binding, redox properties and NMR of iron porphyrins as models for hemoproteins; chemistry of peroxidase enzymes. *Mailing Add:* Dept Chem Univ Iowa Iowa City IA 52242

GOFF, JAMES FRANKLIN, b Louisville, Ky, Aug 1, 28; m 59; c 2. SOLID STATE PHYSICS. *Educ:* Mass Inst Technol, BS, 50; Purdue Univ, MS, 53, PhD(physics), 62. *Prof Exp:* Group leader res, US Naval Surface Weapons Ctr/White Oak Lab, 61-77, ctr coordr nondestructive eval, 77-79, head mat applns, 80-90; RETIRED. *Concurrent Pos:* Tech asst, Naval Sea Systs Command, 75-76. *Mem:* Am Phys Soc; AAAS; Cosmos Club. *Res:* Thermoelectric energy conversion; failure of composite materials. *Mailing Add:* 3405 34th Pl NW Washington DC 20016

GOFF, JESSE PAUL, b Nov 12, 55. VETERINARY MEDICINE. *Educ:* Cornell Univ, BS, 77; Iowa State Univ, MS, 80, DVM, 84, PhD(nutrit physiol), 86. *Prof Exp:* Microbiologist, Poultry Sci Dept, Cornell Univ, 77-78; teaching asst, Dept Vet Physiol & Pharmacol, Iowa State Univ, 78-80; biol lab tech, Physiopath Res Unit, 80-84, res assoc, Dept Vet Physiol & Pharmacol, 84-86, VET MED OFFICER, MINERAL METAB & MASTITIS RES UNIT, NAT ANIMAL DIS CTR, USDA AGR RES SERV, 85- *Concurrent Pos:* Holco res excellence award, Animal Sci Dept, Iowa State Univ, 86, David R Griffith Res Award, 86; adj asst prof vet physiol & pharmacol, Iowa State Univ, 88. *Mem:* Am Dairy Sci Asn. *Res:* Metabolic diseases in domestic animals; parturient hypocalcemia in dairy cattle; hypomagnesemia of beef cattle; interaction of nutrition with immune competence during the periparturient period; author of numerous scientific publications. *Mailing Add:* Mineral Metab & Mastitis Res Unit Nat Animal Dis Ctr USDA/ARS PO Box 70 Ames IA 50010

GOFF, KENNETH W(ADE), b Salem, WVa, June 14, 28; m 50; c 3. ELECTRICAL ENGINEERING. *Educ:* Univ WVa, BS, 50; Mass Inst Technol, SM, 52, ScD(elec Eng), 54. *Prof Exp:* Consult engr, Bolt Beranek & Newman, Inc, 54-56; proj mgr, Gruen Precision Labs, 56-57; mgt, Leeds & Northrup Co, 57-85; VPRES ADVAN TECHNOL, PERFORMANCE CONTROLS INC, 85- *Mem:* Fel Instrument Soc Am; fel Inst Elec & Electronics Engrs. *Res:* Analysis of control systems; techniques for applying analog and digital computers to the control of industrial processes; hardware and software development for digital computer systems; design of motion control systems. *Mailing Add:* Performance Controls Inc 433 Caredean Dr Horsham PA 19044

GOFF, LYNDA JUNE, b Oakland, Calif, June 12, 49; m 77. PHYCOLOGY-SYMBIOSIS, HOST-PARASITE INTERACTIONS. *Educ:* Western Ore State Col, BSc, 71; Univ BC, PhD(bot), 75. *Prof Exp:* Teaching asst bot & phycol, Univ BC, 71-75, res assoc phycol, 75; asst prof, 75-81, ASSOC PROF BIOL, UNIV CALIF, SANTA CRUZ, 81- *Concurrent Pos:* Sr res award, Fulbright Found, 83; vis res prof, div bio & med, Brown Univ, 84-; assoc ed, J Phycol, Phycol Soc Am, 84- *Honors & Awards:* Harold Bold Award, Phycol Soc Am, 75. *Mem:* Phycol Soc Am; Soc Exp Mycol; Bot Soc Am; Int Phycol Soc; Electron Micros Soc Am; Am Soc Cell Biol. *Res:* Algae as model experimental systems in cellular and developmental biology; parasitic algae systems used to examine cell-cell interactions and control mechanisms in biotrophic parasitism. *Mailing Add:* Biol Dept & Ctr Coastal Marine Sci Univ Calif Santa Cruz CA 95064

GOFF, STEPHEN PAYNE, b Providence, RI, Oct 22, 51; m 77; c 2. VIROLOGY, ENZYMOLOGY. *Educ:* Amherst Col, AB, 73; Stanford Univ, PhD(biochem), 78. *Prof Exp:* Fel biol, Mass Inst Technol, 78-81; asst prof biochem, 81-85, assoc prof biochem & molecular biophys, 85-86, PROF BIOCHEM & MOLECULAR BIOPHYS, COL PHYSICIANS & SURGEONS, COLUMBIA UNIV, 86- *Honors & Awards:* Harold & Golden Lamport Res Award. *Mem:* Am Soc Microbiol; NY Acad Sci. *Res:* Genetics and molecular biology of two retroviruses (Moloney and Abelson murine leukemia viruses) using in vitro mutogenesis of cloned DNA copies of the viral genomes. *Mailing Add:* Dept Biochem Col Physicians & Surgeons Columbia Univ 630 W 168th St New York NY 10032

GOFFINET, EDWARD P(ETER), JR, b Louisville, Ky, Nov 15, 30; m 53; c 2. CHEMICAL REACTION ENGINEERING, CHEMICAL PROCESS MODELING. *Educ:* Univ Notre Dame, BS, 52. *Prof Exp:* Res engr, 52-62, E I Dupont de Nemours & Co, res div head, Elastomer Chem Dept Exp Sta, 63-72, sr consult, Eng Dept, 72-83, sr res assoc, 83-86, RES FEL, DUPONT POLYMERS, E I DUPONT DE NEMOURS & CO, 86- *Res:* Applied reaction kinetics; polyester polymerization processes; engineering computer applications; hydrocarbon elastomer processes; definition of new chemical processes. *Mailing Add:* 2403 Annwood Dr Wilmington DE 19803

GOFFMAN, MARTIN, b Philadelphia, Pa, June 22, 40; m 65; c 3. ENVIROMENTAL SCIENCES GENERAL, INTELLIGENT SYSTEMS. *Educ:* Temple Univ, AB, 61, MA, 63, PhD(chem), 65. *Prof Exp:* Sr res chemist phys chem & electrochem, Cent Res Lab, Asarco Inc, 65-86, proj leader, 81-85, sect head, 85-86; PRIN, MARTIN GOFFMAN ASSOC, 86-; PRES, DELLVIEW USA MED INC, 90- *Concurrent Pos:* Consult, proj dir, Small Buss Innovation Res & course dir, Lab Info Mgt Syst; fel, AEC, 63-65. *Mem:* AAAS; Am Chem Soc; Electrochem Soc; Am Soc Metals; Sigma Xi; Asn Consult Chemists & Chem Eng. *Res:* Information management systems and information retrieval; prototype development, biomedical engineering and fiber optics; detoxification of wastes; computer applications and training; electro-chemistry; corporate. *Mailing Add:* Three Dellview Dr Edison NJ 08820-2545

GOFMAN, JOHN WILLIAM, b Cleveland, Ohio, Sept 21, 18; m 40; c 1. CHROMOSOMAL DISEASES. *Educ:* Oberlin Col, AB, 39; Univ Calif, PhD(chem), 43, MD, 46. *Prof Exp:* Res assoc chem, 43-44, intern med, Dept Med, Univ Hosp, 46-47, from asst prof to prof, 47-74, assoc dir, Lawrence Radiation Lab, 63-69, dir biol & med, Lawrence Radiation Lab, 63-66, EMER PROF MED PHYSICS, DONNER LAB MED PHYSICS, UNIV CALIF, BERKELEY, 74- *Concurrent Pos:* Clin instr med, Med Sch, Univ San Francisco, 47-; chmn, Comt Nuclear Responsibility, 71- *Honors & Awards:* Modern Med Award, 54; Lyman Duff Lectureship Award, Am Heart Asn, 65; Stouffer Prize, 72. *Res:* Manhattan project plutonium chemistry; co-discovery of uranium-233 and its fissionability; separation and identification of lipoproteins and their causal relationships with atherosclerosis; chromosomal imbalances and cancer; radiation carcinogenesis; diagnostic medical x-rays. *Mailing Add:* PO Box 11207 San Francisco CA 94101

GOFORTH, DERETHA RAINEY, b San Antonio, Tex, Mar 14, 44; div; c 2. CEREAL CHEMISTRY, FOOD TECHNOLOGY. *Educ:* Incarnate Word Col, BA, 65; Okla State Univ, MS, 69; Kans State Univ, PhD(cereal chem), 76. *Prof Exp:* Instr chem, Univ Md Eastern Shore, 69-71; res chemist biochem, USDA Grain Mkt Res Ctr, 71-76; PROD DEVELOP SCIENTIST

FOOD CHEM, PILLSBURY CO, 76- *Concurrent Pos:* Lectr, Inver Hills Community Col, 84- *Mem:* Am Asn Cereal Chem; Inst Food Technol; AAAS. *Res:* Functional properties and biochemistry of wheat proteins and their interaction with each other and other biochemical constituents; semolina and pasta products; starch and its function in food systems; development of refrigerated dough products; development of miscellaneous dessert products in dry mix products. *Mailing Add:* 4909 W 96th St Minneapolis MN 55437

GOFORTH, RONALD R, modeling & computer simulations, for more information see previous edition

GOFORTH, THOMAS TUCKER, b Dallas, Tex, June 26, 37; m 72; c 3. GEOPHYSICS. *Educ:* Baylor Univ, BS, 59; Univ Tex, MA, 62; Southern Methodist Univ, PhD(geophys), 73. *Prof Exp:* Geophysicist, Teledyne-Geotech, 63-66, sr geophysicist, 66-69; res assoc geophysicist, Southern Methodist Univ, 69-72, sr res assoc geophysicist, 72-74, asst dir, 74-80, DIR DALLAS GEOPHYS LAB, SOUTHERN METHODIST UNIV, 80- *Concurrent Pos:* Consult, Teledyne-Geotech, 69- *Mem:* Soc Explor Geophysicists; Am Geophys Union. *Res:* Earthquake surface waves; environmental problems associated with exploiting geopressured water reservoirs. *Mailing Add:* Dept Geol Baylor Univ Main Campus Waco TX 76798

GOGAN, NIALL JOSEPH, b Dublin, Ireland, Feb 26, 41; m 64; c 3. INORGANIC CHEMISTRY, ACADEMIC ADMINISTRATION. *Educ:* Nat Univ Ireland, BSc, 62, PhD(inorg chem), 65. *Prof Exp:* Res assoc chem, Univ Chicago, 65-67; from asst prof to assoc prof, 67-76, PROF CHEM, MEM UNIV NFLD, 76- *Concurrent Pos:* Assoc vpres res, 83- *Mem:* Am Chem Soc; fel Chem Inst Can. *Res:* Transition element chemistry, especially organometallics derived from metal carbonyl and mostly group VI and VII; metal chelates, especially ligands containing a pyrrole nucleus; electron spin resonance spectroscopy; water quality and the chemistry of natural water systems. *Mailing Add:* Dept Chem Mem Univ Nfld St John's NF A1B 3X7 Can

GOGARTY, W(ILLIAM) B(ARNEY), b Provo, Utah, Apr 23, 30; m 51; c 5. CHEMICAL ENGINEERING, PHYSICAL CHEMISTRY. *Educ:* Univ Utah, BS, 53, PhD(chem eng), 60. *Prof Exp:* Jr engr, Shell Chem Corp, 52; sr res scientist, Marathon Oil Co, 59-67, mgr, Prod Technol Dept, 67-72, Oil Recovery Dept, 72-73, sr staff engr, 73-75, assoc res dir, Denver Res Ctr, 75-86; RETIRED. *Concurrent Pos:* Adj assoc prof, Dept Chem Eng & Metall, Univ Denver, 67-73; Distinguished lectr, Soc Petrol Engrs, 82-83; mem, Enhanced Oil Recovery Comt, Nat Petrol Coun Chem Task Group, 82-84; consult, Nat Inst Petrol & Energy Res, Enhanced Oil Recovery Inst, Univ Wyoming & Ciba-Geigy Corp; consult, Dept Tech Coop UN, India, 86, Rogaland Res Inst, Norsk Hydro & Statoil, Norway, 87, Petromer Trend Corp, Indonesia, 88 & Petrobas, Brazil, 89; dir, Soc Petrol Engrs, 88-90. *Honors & Awards:* Lester C Uren Award, Soc Petrol Engrs, 87, Enhanced Oil Recovery Pioneer Award, 90; Muskat Lectr, Univ Utah, 85. *Mem:* Nat Acad Eng; Am Inst Chem Engrs; Sigma Xi; Soc Petrol Engrs. *Res:* Micellar-polymer flooding and polymer flooding of oil as a means of secondary recovery; enhanced oil recovery; fluid flow; 57 United States patents, 80 foreign patents and published 36 technical papers. *Mailing Add:* 2754 E Long Pl Littleton CO 80122

GOGEL, GERMAINE E, b Jasper, Ind, Aug 21, 52; m. PROTEIN CHEMISTRY, PHOTOSYNTHESIS GROWTH FACTORS. *Educ:* Purdue Univ, BS, 74; Northwestern Univ, MS, 76, PhD(biochem), 79. *Prof Exp:* ASST PROF CHEM, COLGATE UNIV, 81- *Concurrent Pos:* Fel biophysics, Cornell Univ, 80-81; vis scientist, Dept Pure & Appl Biol, Imp Col, London, 85; vis prof, Dept Biochem, Univ Col, Cardiff, 88; fel biochem, Med Col Va, 89. *Mem:* NY Acad Sci; AAAS; Am Chem Soc. *Res:* Photosynthesis and proteins involved in photosynthesis; chemical modification of membrane proteins; characterization of growth factors. *Mailing Add:* Dept Chem Colgate Univ Hamilton NY 13346

GOGER, PAULINE ROHM, b Connellsville, Pa, Jan 22, 13; m 42. INTERNAL MEDICINE. *Educ:* Oberlin Col, AB, 34; Wellesley Col, MA, 36; Univ Pa, PhD(zool), 42; NY Univ, MD, 50. *Prof Exp:* Asst histol & embryol, Wellesley Col, 34-37; teacher, Punahoe Acad, Hawaii, 37-38; teacher biol & chem, Stevens Sch, Philadelphia, 39-41; instr biol, Simmons Col, 41-43; instr zool, NJ Col Women, Rutgers Univ, 43-46, asst prof, 46-50; from intern to resident med, Bellevue Hosp, 50-53; instr & asst clin vis physician, NY Univ-Bellevue Med Ctr, 53; from asst to assoc, 53-60, ASST DIR INTERNAL MED, HUNTERDON MED CTR, 60- *Concurrent Pos:* Assoc clin prof, Med Sch, Rutgers Univ & Sch Med, NY Univ. *Mailing Add:* PO Box 106 Stanton NJ 08885

GOGERTY, JOHN HARRY, b Laramie, Wyo, Mar 9, 33; m 56; c 5. PHARMACOLOGY. *Educ:* Univ Wyo, BS, 54; Univ Wash, MS, 57, PhD(pharmacol), 59. *Prof Exp:* Asst pharmacol, Sch Med, Univ Wash, 54-58; asst prof, Univ Okla, 58-63, dir poison info ctr, 60-61; neuropharmacologist, 63-68, head pharmacol sect, 68-71, dir pharmacol & biochem sect, 71-75, DIR CLIN RES, SANDOZ, INC, 75- *Mem:* AAAS; Am Soc Pharmacol & Exp Therapeut; Am Soc Pharmacol & Chemother; NY Acad Sci. *Res:* Neuropharmacology; neurophysiological and enzymological aspects of neuropharmacology; neuroendocrinology; behavioral pharmacology; neuropharmacology of sleep; clinical pharmacology of antiinflammatories; major and minor tranquilizers; sleep inducers and muscle relaxants. *Mailing Add:* Med-Lab Inc Box 1272 Morristown NJ 07962

GOGGANS, JAMES F, b Tifton, Ga, Oct 30, 20; m 50; c 2. FORESTRY, GENETICS. *Educ:* Univ Ga, BS, 42; Duke Univ, MF, 47; NC State Univ, PhD(forest genetics), 62. *Prof Exp:* From asst prof to assoc prof forestry, Auburn Univ, 47-63, prof forest genetics, 63-84; RETIRED. *Mem:* Soc Am Foresters. *Res:* Tree breeding and genetics of forest trees; variation in anatomy of wood cells. *Mailing Add:* 100 Kuderna Acres Auburn AL 36830

GOGGINS, JEAN A, b Bronx, NY, July 28, 47. BIOCHEMISTRY. *Educ:* Molloy Col, BS, 69; Fordham Univ, MS, 73; Case Western Reserve Univ, PhD(biomed eng), 85. *Prof Exp:* Instr, dept sci, Dominican Com High Sch, Jamaica, NY, 75-79; res asst, dept anat, mt Sinai Sch Med, NY, 79-80; sr res scientist, Vascular Implants Res & Develop, 85-86, MGR, RES & DEVELOP, MEADOX MED, INC, 86- *Mem:* Am Chem Soc; NY Acad Sci; Sigma Xi; Soc Biomat; Int Soc Artifical Organs. *Res:* Biomaterials; biocompatibility; surface interactions; cellular response; blood-material interactions; vascular grafts. *Mailing Add:* Meadox Med Inc 103 Bauer Dr Oakland NJ 07436

GOGGINS, JOHN FRANCIS, b Flint, Mich, Oct 26, 33; m 60; c 3. DENTAL RESEARCH. *Educ:* Marquette Univ, DDS, 58, MS, 65. *Prof Exp:* Pvt pract, 60-63; USPHS training grant, Marquette Univ, 63-65, instr oral path, Sch Dent, 64-65; USPHS fel, NIH, 65-66, investr, histochem sect, Exp Path Br, 66-73, chief, peridont & soft tissue dis prog, 73-74, assoc dir collab res, 74-81, dep dir, Nat Inst Dent Res, 81-84; DEAN, SCH DENT, MARQUETTE UNIV, 84- *Concurrent Pos:* Actg dir, Nat Inst Dent Res, NIH, 82-83, dir Extramural Progs, 83-84. *Mem:* AAAS; Am Dent Asn; Int Asn Dent Res; Am Asn Dent Res; Am Col Dentists; Pierra Fauchard Acad. *Res:* Chemistry of connective tissues, bones and teeth. *Mailing Add:* Sch Dent Marquette Univ 604 N 16th St Milwaukee WI 53233

GOGLIA, GENNARO LOUIS, b Hoboken, NJ, Jan 15, 21; m 42; c 2. THERMODYNAMICS. *Educ:* Univ Ill, BS, 42; Ohio State Univ, MS, 50; Univ Mich, PhD(mech eng), 59. *Prof Exp:* Jr engr, Rochester Ord Dist, 42-45; develop engr refrig, Gen Elec Co, 45-47; instr eng educ, Ohio State Univ, 47-51; asst prof mech eng, Univ Detroit, 51-59; assoc prof, NC State Col, 59-62; prof & head dept, Univ Maine, 62-64; prof eng, Old Dominion Univ, 64-65, chmn, Dept Thermal Eng, 65-70, prof thermal eng, 65-76, asst dean, Sch Eng, 70-76, prof & chmn dept, 76-79, prof mech eng & mech & chmn dept, 80-84, eminent prof mech eng, 79-86, EMER EMINENT PROF MECH ENG, OLD DOM UNIV, 86- *Concurrent Pos:* Tech writer, Cent Heating Dept, Detroit Edison Co, 51-54; consult, Overhead Heaters Inc, 52-55; co-dir summer fac progs, Langley Res Ctr, NASA, 69-81. *Mem:* Fel Am Soc Mech Engrs; Am Soc Eng Educ. *Res:* Heat transfer; supersaturation phenomena; fluids. *Mailing Add:* 7416 Gardner Rd Norfolk VA 23518

GOGLIA, M(ARIO) J(OSEPH), b Hoboken, NJ, Mar 30, 16; m 40; c 2. MECHANICAL ENGINEERING. *Educ:* Stevens Inst Technol, ME, 37, MS, 41; Purdue Univ, PhD(thermodyn), 48. *Prof Exp:* Instr mech eng, Stevens Inst Technol, 37-38; asst prof, Univ Ill, 38-46; appln engr, Repub Flowmeters, 46-47; assoc prof mech eng, Purdue Univ, 47-48; prof, Ga Inst Technol, 48-58; dean, Col Eng, Univ Notre Dame, 58-60; regents prof mech eng & assoc dean faculties, Ga Inst Technol, 60-66, dean grad div, 61-66; vchancellor res, Regents, Univ Syst, Ga, 66-81; ENG CONSULT, 81- *Concurrent Pos:* Asst, Munitions Develop Lab, Univ Ill, 45-46; mem adv panel sci & eng, NSF; consult, Nat Acad Sci-Nat Res Coun, NSF, AEC, US Off Educ & Southern Asn Cols & Schs. *Mem:* AAAS; Am Soc Mech Engrs. *Res:* Thermodynamics, heat transfer; fluid mechanics. *Mailing Add:* 3066 Arden Rd NW Atlanta GA 30305-1915

GOGOLEWSKI, RAYMOND PAUL, b Waterbury, Conn, Jan 25, 41; m 64; c 3. THEORETICAL AND APPLIED MECHANICS. *Educ:* Rensselaer Polytech Inst, BS, 62, MS, 64; Cornell Univ, PhD(theoret & appl mech), 70. *Prof Exp:* Staff engr, Tunnel Facil, Arnold Eng Dev Ctr, 64-65; res engr, systs div, United Technol Res Ctr, 65-67; fel, Cornell Univ, 67-70; asst prof eng mech, NC State Univ, 70- 73; proj mgr systems eval, Inst Defense Analysis, 73-77 & tactical warfare, Syst Planning Corp, 77-78; prog mgr tactical technol, Defense Adv Res Proj Agency, 78-83; sr vpres, Aeronaut Res Assocs Princeton, 83-85; pres, Cannonball Consults, 85-87; GROUP LEADER, PHYSICS DEPT, LAWRENCE LIVERMORE NAT LAB, 87- *Mem:* Am Acad Mech; Soc Ind & Appl Math; Opers Res Soc Am; Sigma Xi. *Res:* Basic research and exploratory development oriented toward application to military tactical technology; applied mathematics and optimal control theory in dynamical systems theory. *Mailing Add:* 228 Royal Saint Ct Danville CA 94526-5407

GOGOS, COSTAS G, b Athens, Greece, June 30, 38; m 65. CHEMICAL ENGINEERING. *Educ:* Princeton Univ, BSChE, 61, MS, 62, MA, 64, PhD(mech eng), 66. *Prof Exp:* From asst prof to assoc prof, 64-74, PROF CHEM ENG, STEVENS INST TECHNOL, 74- *Mem:* Am Chem Soc; Soc Plastics Engrs; Soc Rheology. *Res:* Physical properties of high polymers, in particular rheological properties; rheological aspects of polymer processing; kinetics and morphology of deformed polymer crystals. *Mailing Add:* 652 Cumberland Ave Teaneck NJ 07666-1814

GOGUEN, JOSEPH A, JR, b Pittsfield, Mass, June 28, 41; c 3. COMPUTER SCIENCE, INFORMATION SCIENCE. *Educ:* Harvard Univ, BA, 63; Univ Calif, Berkeley, MA, 66, PhD(math), 68. *Prof Exp:* From teaching asst to instr math, Univ Calif, Berkeley, 63-68; asst prof info sci, Univ Chicago, 68-73; asst prof, Univ Calif, Los Angeles, 72-74, assoc prof comput sci, 74-78, prof, 78-81; sr staff scientist, SRI WT, Palo Alto, Calif, 79-85, managing dir, structural semantics, 78-88, prog mgr, 85-88; prin mem, Ctr Study Lang & Info, Stanford Univ, 83-88; PROF COMPUT SCI, OXFORD UNIV, ENG, 88- *Concurrent Pos:* Int Bus Mach Corp fel, T J Watson Res Lab, NY, 71-72; fac mem, Naropa Inst, Boulder, Colo, 74-; sr vis fel, Univ Edinburgh, 76, 77; co-dir, Software Technol Seminars, Capri, Italy, 80, 81. *Mem:* Am Math Soc; Math Asn Am; Asn Comput Mach. *Res:* Multiprocessor computer architecture for ultra high level languages, software engineering, semantics of languages and natural languages; linguistics, especially discourse analysis; artificial intelligence; fuzzy sets; system theory; cybernetics. *Mailing Add:* Univ Oxford 8-11 Keble Rd Oxford England

GOH, EDWARD HUA SENG, b Sarawak, Malaysia, Jan 20, 42; US citizen; m 69; c 2. DRUG METABOLISM, TOXICOLOGY. *Educ:* Berea Col, BA, 68; Vanderbilt Univ, PhD(pharmacol), 74. *Prof Exp:* Fel pharmacol, Univ Mo, Columbia, 74-75, instr, 75-77; asst prof, 77-82, ASSOC PROF

PHARMACOL, IND UNIV, 82- *Mem:* Am Soc Pharmacol & Exp Therapeut. *Res:* Metabolism of blood cholesterol; drug metabolism and toxicology; testing of new drugs. *Mailing Add:* Pharmacol Sect Med Sci Prog Sch Med Ind Univ Myers Hall Bloomington IN 47405

GOHEEN, AUSTIN CLEMENT, b Bellingham, Wash, Apr 22, 17; m 42; c 5. PLANT PATHOLOGY. *Educ:* Univ Wash, BS, 47; Wash State Univ, PhD(plant path), 53. *Prof Exp:* Observer, US Weather Bur, 38-46; asst dept plant path, Wash State Univ, 47-50; agent, NJ Agr Exp sta, 50-53, plant pathologist, Plant Indust Sta, 53-55 & US Hort Field Sta, 55-56, assoc, Univ Calif Exp Sta, USDA, 56-, res plant pathologist, 57-, res leader crops path, agr res serv, 72-; RETIRED. *Concurrent Pos:* Asst res specialist, Agr Exp Sta, Rutgers Univ, 50-53. *Honors & Awards:* Ruth Allen Award, Am Phytopath Soc, 74. *Mem:* Am Phytopath Soc; Sigma Xi. *Res:* Virus diseases of grapes. *Mailing Add:* 2412 Pine Grove Rd Klamath Falls OR 97603

GOHEEN, DAVID WADE, b Bellingham, Wash, June 23, 20; m 43; c 2. BOTANY, PLANT BREEDING & GENETICS. *Educ:* Univ Wash, BS, 42, PhD(chem), 51. *Prof Exp:* Res chemist org chem, Eng Res Inst, Univ Mich, 51-52; chef chem res sect, Cent Res Dept, Crown Zellerbach Corp, 52-55, res specialist chem prod div, 55-67, sr res chemist, 67-78, proj leader, pioneering res, Cent Res Div, 78-81; VPRES, E M SEIDEL ASSOC, 81- *Concurrent Pos:* Vpres, Am Rhododendron Species Found, 80-81, pres, 82-84; mem vis comt, Solar Energy Res Inst, 81. *Honors & Awards:* Gold Medal, Am Rhododendron Soc, 88. *Mem:* Am Chem Soc; Tech Asn Pulp & Paper Indust. *Res:* Natural products; wood extractives; lignin; cellulose. *Mailing Add:* PO Box 826 Camas WA 98607

GOHEEN, HARRY EARL, mathematics; deceased, see previous edition for last biography

GOHEEN, LOLA COLEMAN, b Los Angeles, Calif; m 69; c 1. REQUIREMENTS ANALYSIS. *Educ:* Stanford Univ, BS, 70, MS, 72, PhD(opers res), 74. *Prof Exp:* Oper analyst, 74-85, SR RES ENGR & SR OPER ANALYST, SRI INT, 85- *Mem:* Soc Indust & Appl Math; Am Asn Artificial Intel; Opers Res Soc Am. *Res:* Modeling, analysis, and optimization of systems. *Mailing Add:* 876 Coleman Ave Menlo Park CA 94025

GOHEEN, STEVEN CHARLES, b Seattle, Wash, May 14, 51; m 81; c 3. ENVIRONMENTAL CHEMISTRY. *Educ:* Univ Wash, BS, 73, Northwestern Univ, PhD(mat sci), 78. *Prof Exp:* Res chemist biochem, 77-83, sr res chemist, 83-89, SR RES SCIENTIST, BIO-RAD LABS, MARTINEZ VET ADMIN MED CTR, BATTELLE NORTHWEST, 89- *Mem:* AAAS; Am Oil Chemists Soc; NY Acad Sci; Sigma Xi; Am Inst Nutrit; Am Soc Biochem & Molecular Biol. *Res:* Environmental organic chemistry; chromatography; high performance liquid chromotography. *Mailing Add:* 2016 Harris Ave Richland WA 99352

GOHR, FRANK AUGUST, b Boston, Mass, Sept 26, 22; m 49; c 4. ENVIRONMENTAL SCIENCES. *Educ:* Univ Calif, Los Angeles, BA, 48, BS, 49; Univ Calif, Berkeley, MPH, 50, DrPH(sanit sci), 62. *Prof Exp:* Supvr statewide sanit sci, Univ Calif, 50-60; environ health & safety officer, Univ Calif Med Ctr, San Francisco, 60-86; RETIRED. *Concurrent Pos:* Adv, Nat Adv Coun Health Facil, 67-70; consult health serv & ment health admin, USPHS, 67-73. *Honors & Awards:* Walter S Mangold Award, Nat Environ Health Asn, 63. *Mem:* Fel Am Pub Health Asn; Nat Environ Health Asn. *Mailing Add:* 15 Arroyo Dr Orinda CA 94563

GOIHMAN-YAHR, MAURICIO, b Caracas, Venezuela, Apr 8, 38; m 69; c 2. DERMATOLOGY, CELLULAR IMMUNOLOGY. *Educ:* Central Univ, Venezuela, MD, 60; Stanford Univ, PhD(med microbiol), 69. *Prof Exp:* Clin lectr dermat, Stanford Univ, 64-68; instr dermat, Central Univ, Venezuela, 69-71, asst prof, 71-72, adj prof, 72-73, assoc prof, 73-79, PROF DERMAT, CENT UNIV, VENEZUELA, 80- *Concurrent Pos:* Mem bd dirs, Venezuela Asn Advan Sci, 73-74, Nat Res Coun, 85-89; pres, Comt Clin Med & Res, Asst Res Coun, Venezuela, 74-89; alt mem Super Coun, Nat Res Coun, 90- *Honors & Awards:* Cesar Lilardo Prize, Venezuela Soc Dermat, 84; Order of Andres Bello, Venezuelan Govt, 85. *Mem:* Soc Investigative Dermat; Am Acad Dermat; Am Soc Microbiol; Soc Exp Biol & Med; Am Asn Immunologists; Int Leprosy Asn. *Res:* Cellular immunology and actions of phagocytes in granulomtory diseases due to live organisms. *Mailing Add:* Jet Int M-154 PO Box 020010 Miami FL 33102

GOINS, NEAL RODNEY, b Columbus, Ohio, Jan 4, 52. GEOPHYSICS, SEISMOLOGY. *Educ:* Princeton Univ, AB, 73; Mass Inst Technol, PhD(geophys), 78. *Prof Exp:* Res staff mem, Mobil Res & Develop Corp, 78-85, sr geophys adv, Tex & NMex, 85-87, mgr technol eval, 87-89, EXPLOR MGR, MOBIL E&P US INC, 89- *Concurrent Pos:* Mem, Viking Seismol Team, 77-, comt future comput needs & resources, Mass Inst Technol, 77-78. *Mem:* Am Geophys Union; Soc Explor Geophysicists; AAAS; Seismol Soc Am; Sigma Xi. *Res:* Solid earth geophysics, signal analysis and inversion; planetology, including internal structure, composition and seismicity; digital signal processing; wave propagation and attenuation; gravity field analysis and interpretation; plate tectonics. *Mailing Add:* Mobil E&P US Inc 10000 Ming Bakersfield CA 93311

GOINS, TRUMAN, b Peason, La, Nov 14, 20; m 43; c 1. ENGINEERING, NATURAL RESOURCES. *Educ:* La State Univ, BS, 50; Ohio State Univ, MS, 54. *Prof Exp:* Asst prof agr eng, Ohio Agr Exp Sta, 50-58; agr eng adv, US Oper Mission, Honduras, 58-61 & USAID, Ecuador, 61-63, regional dir, 62, agr adv, 63; res agr engr, Agr Res Serv, USDA, Va, 63-66; water resources engr, USAID, Laos, 66-67 & Washington, DC, 67-68; civil engr & prog officer, 68-77, WATER RESOURCES COORDR, DEPT HOUSING & URBAN DEVELOP, 77- *Concurrent Pos:* Secy alternate, Water Resources Coun; rep for Dept Housing & Urban Develop on various comts and task forces in govt, Nat Climate Prog Policy Bd & Interagency Geothermal Coord Coun. *Mem:* Am Soc Agr Engrs; Am Soc Civil Engrs; Nat Soc Prof Engrs. *Res:* Agricultural drainage and irrigation; soil and water conservation; crop tillage and rotation practices; farm building materials; land use and development; water resources planning and development; community planning and management. *Mailing Add:* 2111 Jefferson Davis Hwy 302N Arlington VA 22202

GOINS, WILLIAM C, JR, RESEARCH ADMINISTRATION. *Prof Exp:* SR VPRES, O'BRIEN-GOINS-SIMPSON & ASSOC INC, 77- *Mem:* Nat Acad Eng. *Mailing Add:* O'Brien-Goins-Simpson & Assoc Inc 6430 Hillcroft Suite 112 Houston TX 77081

GOISHI, WATARU, b Florin, Calif, Feb 6, 28; m 52; c 2. RADIOCHEMISTRY. *Educ:* Univ Calif, Berkeley, BS, 50; Univ Chicago, PhD(radiochem), 54. *Prof Exp:* Radiochemist, Univ Chicago, 53-54; radiochemist, 54-62, group leader, 63-77, ASST DIV LEADER, NUCLEAR WEAPONS DIAG, LAWRENCE LIVERMORE LAB, UNIV CALIF, 77- *Mem:* AAAS; Am Chem Soc. *Res:* Fast neutron reactions and measurements; kinetics and mechanisms of inorganic reactions. *Mailing Add:* 588 Emerald St Livermore CA 94550

GOITEIN, MICHAEL, b Broadway, Eng, Nov 14, 39; m 68; c 2. MEDICAL PHYSICS. *Educ:* Oxford Univ, BA, 61; Harvard Univ, PhD(physics), 68. *Prof Exp:* Res fel physics, Harvard Univ, 68-69; staff physicist, Lawrence Berkeley Lab, 69-72; from asst prof to assoc prof, 73-86, PROF MED PHYSICS, HARVARD MED SCH, 86- *Concurrent Pos:* Res career develop award, USPHS, 76- *Mem:* Am Phys Soc; Am Asn Physicists Med; Am Soc Therapeut Radiologists. *Res:* Radiation therapy, computerized tomography. *Mailing Add:* Dept Radiation Med Mass Gen Hosp Boston MA 02114

GOJMERAC, WALTER LOUIS, b Rib Lake, Wis, Apr 8, 25; m 55; c 5. ENTOMOLOGY. *Educ:* Univ Wis, BS, 49, PhD(entom), 55; Marquette Univ, MS, 53. *Prof Exp:* Teacher, pub sch, 49-51; asst prof entom & asst entomologist, NDak Agr Col, 55-57; res entomologist, Calif Spray Chem Corp, Standard Oil Co Calif, 57-65; assoc prof, 65-73, PROF ENTOM, UNIV WIS-MADISON, 73-, EXTEN ENTOMOLOGIST, 65- *Mem:* Fel AAAS; Entom Soc Am. *Res:* Control of insects affecting man and animals; bees and beekeeping; control of insects in food processing and handling establishments. *Mailing Add:* Dept Entom Univ Wis Madison WI 53706

GOJNY, FRANK, b San Francisco, Calif, Jan 24, 52; m 79; c 2. MECHANICAL ENGINEERING. *Educ:* San Francisco State, BS, 74. *Prof Exp:* Proj dir qual assurance, Nat Steel & Shipbldg, 75-78; qual & mfg engr, 78-80, supvr, 81-84, DEP PROJ ENGR, ADVAN METALLICS GROUP, ROHR INDUSTS, 78- *Concurrent Pos:* Metallurgist, alloy develop, advan processes & struct develop. *Mem:* Am Soc Metals Int. *Res:* Granted various patents; author of several publications. *Mailing Add:* 5388 Dressage Dr Bonita CA 91902

GOKCEN, NEV A(LTAN), b Turkey, Sept 21, 21; nat US; m 46; c 3. THERMODYNAMICS. *Educ:* Univ Pittsburgh, BS, 42, MS, 45; Mass Inst Technol, ScD(metall), 51. *Prof Exp:* Instr metall, Lafayette Col, 43-44 & Robert Col, 45-47; asst, Mass Inst Technol, 47-51, res assoc, 51; assoc prof, Mich Technol Univ, 51-56 & Univ Pa, 56-61; head chem thermodyn sect, Aerospace Corp, 61-77; RES SUPVR, THERMODYN LAB, BUR MINES, DEPT INTERIOR, ALBANY, 77- *Concurrent Pos:* Mem, Japan Soc Prom Sci. *Mem:* Am Chem Soc; Am Soc Metals; Am Inst Mining, Metall & Petrol Engrs. *Res:* Physical chemistry of metals and alloys; high temperature chemistry; thermodynamics of solutions; chemistry; materials science. *Mailing Add:* Bur Mines 1450 Queen Ave SW Albany OR 97321

GOKEL, GEORGE WILLIAM, b New York, NY, June 27, 46; m 78; c 1. ORGANIC CHEMISTRY. *Educ:* Tulane Univ, BS, 68; Univ Southern Calif, PhD(chem), 71. *Prof Exp:* Fel chem, Univ Southern Calif, 71-72 & Univ Calif, Los Angeles, 72-74; asst prof, Pa State Univ, University Park, 74-77; from assoc prof to prof chem, Univ Md, College Park, 78-84; PROF CHEM, UNIV MIAMI, 84- *Concurrent Pos:* Consult, W R Grace Co, 77- *Mem:* Am Chem Soc; Royal Soc Chem; Sigma Xi. *Res:* Reactions in mixed phase media; phase transfer chemistry mediated by quaternary salts and heteromacrocycles; novel heteromacrocycles; biologically active and chemically switchable macrocycles lariat ethers and cryptands. *Mailing Add:* Dept Chem Univ Miami Coral Gables FL 33124

GOKEN, GAROLD LEE, b Decatur, Ill, Aug 6, 37; m 58; c 3. POLYMER CHEMISTRY, ORGANIC CHEMISTRY. *Educ:* Millikin Univ, BA, 59; Univ Utah, PhD(chem, pharmaceut chem), 62. *Prof Exp:* Sr chemist, Tape Res Lab, Minn Mining & Mfg Co, 62-65, supvr, pioneering res group, 65-69, res supvr, Indust Spec Prod Dept, 69-71, res mgr, 71-72, new prod mgr, 72-74, tech mgr cushioning prod proj, 74-78, tech mgr cushioning & new prod, Indust Specialties Div, 78-79, mgr new bus develop, Indust Specialties Div, 79-80, tech mgr, Europ Tape Labs, 3M Europe, 81-84, tech dir Adhesives, Coatings & Sealers Div, 84-90, MGR LAB & MFG OPERS, I & E SECTOR NEW PROD DEPT, 3M, ST PAUL, MINN, 90- *Concurrent Pos:* Pres, Minn Acad Sci. *Mem:* Am Chem Soc; Soc Plastics Engrs; AAAS. *Res:* Commercialization of advanced new technologies, particularly materials and product systems for energy and environmental markets. *Mailing Add:* I & E New Prod 3M Co Bldg 218-35-04 St Paul MN 55144-1000

GOKHALE, DATTAPRABHAKAR V, b Poona, India, Mar 18, 36; m 61; c 2. STATISTICS. *Educ:* Univ Poona, BSc, 55, MSc, 57; Univ Calif, Berkeley, PhD(statist), 66. *Prof Exp:* Lectr statist, Univ Poona, 59-69, assoc prof, 69-70; from asst prof to assoc prof, 70-79, PROF STATIST, UNIV CALIF, RIVERSIDE, 79-, STATISTICIAN, 70- *Concurrent Pos:* Asst prof, Univ Calif, Berkeley, 66. *Mem:* Inst Math Statist, Int Statist Inst; Indian Statist Soc; Biomet Soc; Am Statist Asn. *Res:* Distribution theory; nonparametrics; analysis of frequency data; information theoretic approach to statistical problems; statistical ecology. *Mailing Add:* Dept Statist Univ Calif Riverside CA 92521

GOKHALE, VISHWAS VINAYAK, fluid mechanics, reaction engineering, for more information see previous edition

GOKSEL, MEHMET ADNAN, b Ankara, Turkey, May 27, 24; m 60; c 3. CHEMICAL ENGINEERING, EXTRACTIVE METALLURGY. *Educ:* Istanbul Univ, MSc, 49, DSc(chem eng), 51; Karlsruhe Tech Univ, Dr rer nat habil(chem eng), 56. *Prof Exp:* Res asst indust chem, Istanbul Univ, 49-52; sci mil adv, Turkish Gen Staff, 55-57; docent indust chem, Istanbul Univ, 57-60; Fulbright res fel extractive metall, Mich Technol Univ, 60-62; docent indust chem, Istanbul Univ, 62-63, prof, 63-65; assoc prof chem eng, 65-67, assoc prof extractive metall, 65-76, RES LEADER & ASSOC PROF MINERAL RES, MICH TECHNOL UNIV, 76- *Concurrent Pos:* Consult, Emayetas Co, Turkey, 63-65 & UN, 79 & 81. *Mem:* Sigma Xi; Am Inst Mining, Metall & Petrol Engrs; Inst Briquetting & Agglomeration. *Res:* Agglomeration; pollution; building materials; chemical processes; iron and steel making. *Mailing Add:* Inst Mineral Res Mich Technol Univ Houghton MI 49931

GOKTEPE, OMER FARUK, b Istanbul, Turkey, Aug 7, 39; US citizen; m 73; c 2. NUCLEAR ENGINEERING, ATOMIC PHYSICS. *Educ:* Univ Istanbul, BS, 62; Univ Mich, MS, 69; Univ Md, MS, 72, PhD(nuclear eng), 77. *Prof Exp:* Res physicist, Inst Nuclear Energy, 62-64; res assoc physics, Gen Technol Corp, 64-66; res asst nuclear, Univ Mich, 66-69; res asst plasma, Univ Md, 69-71, teaching asst kinetic theory, 71-72; res asst exp develop, NASA, Goddard Space Flight Ctr, 72-78; Nuclear Engr, Naval Surface Warfare Ctr, 79-91; ADJ ASSOC PROF, DEPT PHYSICS, GEORGETOWN UNIV, 86-; GEN ENGR, DIV HIGH ENERGY PHYSICS, DEPT HIGH ENERGY, 91- *Concurrent Pos:* Res physicist, Plasma Lab, 62-64; res assoc, Inst Phys Sci & Technol, 77-78. *Mem:* AAAS; Am Phys Soc; Am Nuclear Soc. *Res:* Physical phenomena produced by charged particle beams in a solid surface analysis, new material development techniques based on ion bombardment of solid surfaces; radiation transport and shielding; environmental science. *Mailing Add:* 7604 Bells Mill Rd Bethesda MD 20817

GOLAB, TOMASZ, b Dynow, Poland, Oct 18, 22; m 58; c 2. AGRICULTURAL BIOCHEMISTRY. *Educ:* Univ Basel, PhD(org chem), 60. *Prof Exp:* Res assoc, Univ Basel, 60; scientist, Worcester Found Exp Biol, Mass, 60-62; sr agr biochemist, 62-72, res scientist, Lilly Res Labs, Eli Lilly & Co, 72-88; RETIRED. *Mem:* AAAS; Am Chem Soc; NY Acad Sci; Swiss Chem Soc. *Res:* Pesticide metabolism. *Mailing Add:* 1430 Anemone Ct Indianapolis IN 46219

GOLAND, ALLEN NATHAN, b Chicago, Ill, Apr 26, 30; m 59; c 2. PARTICLE-SOLID INTERACTIONS, LATTICE DEFECTS. *Educ:* Roosevelt Univ, BS, 51; Northwestern Univ, PhD(physics), 56. *Prof Exp:* Res assoc physics, Brookhaven Nat Lab, 56-58, solid state physicist, ord mat res off, 58-63, assoc physicist, 63-67, physicist, 68-73, assoc chmn, Dept Appl Sci, 82-87, dep chmn, 87-89, SR PHYSICIST, BROOKHAVEN NAT LAB, 73- *Concurrent Pos:* Vis prof, State Univ NY Stony Brook, 70, adj prof, 74-; dir, NATO Advan Study Inst, 71; vis res consult, Univ Sao Paulo, 71; guest lectr, Enrico Fermi Int Sch Physics, Italy, 74; mem tech adv panel, Los Alamos Meson Physics Facil, Los Alamos Sci Lab, 75-78, mem solid state physics & mat sci comt, 76-78; secy & mem damage anal & fundamental studies task group, Off Fusion Energy, Dept Energy, 76-81; consult, Argonne Univ Asn, 76-77; mem ad hoc comt use fission reactors div mag fusion energy alloy develop prog, Div Mag Fusion Energy, ERDA, 76-77; mem Nat Acad Sci-Nat Res Coun panel thin film microstruct sci & technol, 78-; vis prof, State Univ NY Stony Brook, 78; mem panel, effects of radiation on struct & properties of mat, US Dept Energy, 81; mem rev panel, Chem & Mat Sci Div, Los Alamos Nat Lab, 80-85, chmn, 81-82; assoc ed, Phys Rev B, 82-; mem, Res & Rev Comt, NY State Sci & Technol Found, 83-, Nat Steering Comt Advan Neutron Source, 86-, Phys Sci Proj Adv Group, Gas Res Inst, 86-, Tech Rev Panel, Dept Energy-HFIR, 87-88 & Radiation Sci Panel, Nat Acad Sci, Nat Res Coun, 88; US rep, Eval Panel Int Fusion Mat Irradiative Facil, 89. *Mem:* AAAS; fel Am Phys Soc; Am Nuclear Soc; Mat Res Soc; Am Crystallog Asn. *Res:* Defects in solids; radiation effects; x-ray and neutron scattering; applications of particle accelerators to solid-state physics research; positron annihilation in solids. *Mailing Add:* Brookhaven Nat Lab Bldg 480 Upton NY 11973

GOLAND, MARTIN, b New York, NY, July 12, 19; m 48; c 3. APPLIED MECHANICS, AERONAUTICS. *Educ:* Cornell Univ, ME, 40. *Hon Degrees:* LLD, St Mary's Univ, San Antonio, 68. *Prof Exp:* Instr mech eng, Cornell Univ, 40-42; sect head appl mech, Res Lab, Curtiss Wright Co, NY, 42-46; chmn, Eng Mech Div, Midwest Res Inst, 46-50, dir eng sci, 50-55; chmn, Comn Eng & Tech Systs, Nat Res Coun, 83-86; PRES, SOUTHWEST INST, 59- *Concurrent Pos:* Mem adv coun, Sch Eng, George Washington Univ, 70; mem sci comt, Gen Motors, 71-81; pres, Southwest Found Biomed Res, 72-82; mem, Naval Res Adv Comt, 74-80, chmn, 77; mem, Comt Int Rels, Nat Acad Sci, 77-79; chmn, Bd Army Sci & Technol, Nat Acad Sci, 82-89; chmn, Mat & Struct Group, NASA, 79-83. *Honors & Awards:* Spirit of St Louis Jr Award, Am Soc Mech Engrs, 44, Jr Award, 46; Alfred Noble Prize, Am Soc Civil Engrs, 46; Nat Eng Award, Am Acad Eng Sci, 85; Herbert Hoover Medal, Am Soc Mech Engrs, 87. *Mem:* Nat Acad Eng; fel AAAS; hon men Am Soc Mech Engrs; fel Am Inst Aeronaut & Astronaut (pres, 71); Sigma Xi. *Res:* Applied mechanics; applied mathematics and engineering analysis; structures; aerodynamics; fluid flow; aircraft dynamics; vibration and impact problems; engineering analysis; operations research; industrial economics techniques. *Mailing Add:* Southwest Res Inst 6220 Culebra Rd Drawer 28510 San Antonio TX 78284

GOLARZ-DE BOURNE, MARIA NELLY, b Montevideo, Uruguay, Jan 21, 34; nat US; m 64. MICROSCOPIC ANATOMY, HISTOCHEMISTRY. *Educ:* Univ Repub, Uruguay, 54; Emory Univ, PhD(anat), 63. *Prof Exp:* Res assoc anat micros, Univ Repub, Uruguay, 55-58, asst to prof exp biol, 58; asst anat, Med Sch, 58-59, res assoc, 60-63, res assoc histochem, 63-66, ASST HISTOCHEMIST, YERKES PRIMATE RES CTR, EMORY UNIV, 66- *Concurrent Pos:* Res scientist, Colonial Res Inst, 64-66; co-prin investr, NASA grant, 67-; prof Cell Biol, Histol & Neuroscience, St George Univ, Sch Med, Gieuada , West Indies, 78- *Mem:* Histochem Soc; Am Asn Anat; Int Primatol Soc; Geront Soc; Sigma Xi; Am Asn Adv Sci. *Res:* Golgi apparatus; histochemistry of muscle diseases; space pathology. *Mailing Add:* 849 Lullwater Pkwy NE Atlanta GA 30307

GOLAY, MARCEL JULES EDOUARD, b Neuchatel, Switz, May 3, 02; nat US; c 4. PHYSICS. *Educ:* Fed Polytech Inst, Zurich, LicEE, 24; Univ Chicago, PhD(physics), 31. *Hon Degrees:* Dr, Fed Polytech Inst, Lausanne, 77. *Prof Exp:* Engr, Bell Tel Labs, 24-28 & Automatic Elec Co, 30-31; from assoc physicist to chief scientist, Signal Corps Labs, 31-55; SR SCIENTIST, PERKIN-ELMER CORP, 55- *Honors & Awards:* Diamond Award, Inst Radio Eng, 51; Sargent Award, Am Chem Soc, 60; Distinguished Accomplishment Award, Instrument Soc Am, 61; Jimmie Hamilton Award, Am Soc Naval Engrs, 72; Tswett medal, 76; Chromatography Award, Am Chem Soc, 81; Dal Nogare Award, 82. *Mem:* Inst Elec & Electronics Eng; emer mem Swiss Soc Nat Sci. *Res:* Acoustics; infrared; communications; nuclear magnetic resonance; gas chromatography. *Mailing Add:* Creux de Corsy La Conversion 1093 Switzerland

GOLAY, MICHAEL W, US citizen. NUCLEAR POWER PLANT ENGINEERING, FLUID MECHANICS. *Educ:* Univ Fla, BME, 64; Cornell Univ, PhD(nuclear eng), 69. *Prof Exp:* Res assoc nuclear eng, Rensselaer Polytech Inst, 69-71; asst prof, 71-75, assoc prof, 75-86, PROF NUCLEAR ENG, MASS INST TECHNOL, 86- *Concurrent Pos:* Consult, Stone & Webster Eng Co, 71-, Elec Power Res Inst, 83- & numerous others. *Mem:* Am Nuclear Soc; Am Soc Mech Engrs. *Res:* Pollutant transport; reactor engineering; nuclear energy policy; turbulent flow; reactor safety; cooling towers. *Mailing Add:* Dept Nuclear Eng Mass Inst Technol Cambridge MA 02139

GOLBEY, ROBERT (BRUCE), b New York, NY, July 15, 22; m 48; c 1. CHEMOTHERAPHY. *Educ:* Bethany Col, BS, 43; NY Univ, MD, 49. *Prof Exp:* Res assoc, 55-57, asst, 57-60, ASSOC MEM, DIV CLIN CHEMOTHER, SLOAN-KETTERING INST, 60-; ASSOC PROF CLIN MED, COL MED, CORNELL UNIV, 70- *Concurrent Pos:* Res fel, Div Clin Chemother, Sloan-Kettering Inst, 54; spec fel med, Mem Ctr & James Ewing Hosp, 54-57; from instr to asst prof clin med, Col Med, Cornell Univ, 57-70; clin asst, James Ewing Hosp, 57-59, asst vis physician, 59-62, assoc vis physician, 62-; clin asst physician, Dept Med, Mem Hosp, 57-60, from asst attend physician to assoc attend physician, 60-71, attending physician, 71-; consult, St Joseph's Hosp, 60-, Stamford Hosp, 65-; chief, Solid Tumor Serv, Mem Hosp, 79. *Mem:* AAAS; Am Asn Cancer Res; Am Soc Clin Oncol; Soc Surg Oncol; NY Acad Sci. *Res:* Cancer chemotherapy; biology of malignant tumors and its relationship to clinical course and effective therapy. *Mailing Add:* 34 Wiley Bottom Rd Savannah GA 31411

GOLBUS, MITCHELL S, b Chicago, Ill, Apr 6, 39; m 67; c 3. REPRODUCTIVE GENETICS. *Educ:* Ill Inst Technol, BS, 59; Univ Chicago, MD, 63; AM Bd Obstet Gynec, dipl. *Prof Exp:* Res fel med genetics & asst res geneticist, 71-73, clin instr, Dept Obstet & Gynec, 72-73, asst prof, 73-77, ASSOC PROF OBSTET, GYNEC & PEDIAT, UNIV CALIF, SAN FRANCISCO, 81- *Concurrent Pos:* Clin instr, Dept Obstet & Gynec, Univ Calif, San Francisco, 72-73; mem, Adv Comt Genetics, Calif State Dept Health, 73-, NCalif Tay-Sachs Dis Prevention Prog, 73-, Clin Res Grant Review Comt, March of Dimes, 78-, NIH Task Force Predictors Hereditary Dis & Congenital Defects, 78-79, Med Adv Comt, Nat Genetics Found, 81-; sci adv comt, Nat Tay-Sachs & Allied Dis Asn, 80- *Mem:* Fel Am Col Obstetricians & Gynecologists; Am Soc Human Genetics; Soc Gynec Invest; AAAS. *Res:* Relationship of chromosome heteroploidy to birth defects and reproductive inefficiency; developing techniques for prenatal diagnosis, specifically of defects not demonstrable in amniotic fluid cells. *Mailing Add:* Dept Obstet Gynec & Reproductive Sci Univ Calif San Francisco CA 94143

GOLCHERT, NORBERT WILLIAM, b Chicago, Ill, July 12, 35; m 58; c 4. ENVIRONMENTAL RADIOCHEMISTRY. *Educ:* Univ Ill, BS, 58; Ill Inst Technol, MS, 65, PhD(chem), 69. *Prof Exp:* Asst in chem, 58-66, asst chemist, 66-73, chemist, 73-88, ENVIRON SCIENTIST, ARGONNE NAT LAB, 88- *Mem:* Am Chem Soc; Sigma Xi; Am Soc Testing & Mat. *Res:* Measurement of radioactivity in the environment with primary interest in measurement of transuranium nuclides; radioanalytical chemistry of technetium, measurement of nuclear reaction cross-sections, and disposal of low-level radioactive waste. *Mailing Add:* Argonne Nat Lab 9700 S Cass Ave Argonne IL 60439

GOLD, ALBERT, b Philadelphia, Pa, July 2, 35; div; c 3. SCIENCE ADMINISTRATION, SOLID STATE PHYSICS. *Educ:* Lehigh Univ, BS, 56; Univ Rochester, PhD(physics), 60. *Prof Exp:* Res assoc physics, Univ Ill, 60-62; from asst prof to assoc prof optics, Univ Rochester, 62-69, assoc dean col eng & appl sci, 67-69; spec asst to pres, Rockefeller Univ, 69-72, dir postdoctoral affairs, 70-72, vpres, 72-78; provost, Polytech Inst New York, 78-81; vpres finance & admin, Desert Res Inst, 82-86. *Concurrent Pos:* Consult, Argonne Nat Lab, 60-69, United Aircraft Corp, 63-64 & Eastman Kodak Co, 64-69; dir, Univ Patents, Inc, 79-82. *Mem:* Am Phys Soc; AAAS. *Res:* Theory of optical properties of solids; theory of interaction of intense optical radiation with matter. *Mailing Add:* Div Appl Sci Harvard Univ 29 Oxford St Cambridge MA 02138

GOLD, ALLEN, b Montreal, Que, Dec 19, 18; m 42; c 5. MEDICINE, ENDOCRINOLOGY. *Educ:* McGill Univ, BSc, 40, MD, CM, 42, MSc, 48. *Prof Exp:* Lectr, 49-70, ASSOC PROF MED, MCGILL UNIV, 70- *Concurrent Pos:* Assoc physician, Montreal Gen Hosp, 51-65, sr physician, 65-; consult endocrinol, Reddy Mem Hosp & Lachine Gen Hosp. *Mem:* Endocrine Soc; fel Am Col Physicians; Am Thyroid Asn; Can Soc Clin Invest. *Res:* Clinical endocrinology, especially thyroid gland; development of new thyroid diagnostic tests. *Mailing Add:* 360 Victoria Ave Westmount PQ H3Z 2N4 Can

GOLD, ALLEN MORTON, b Chicago, Ill, May 25, 30; m 52; c 2. BIOCHEMISTRY, ORGANIC CHEMISTRY. *Educ:* Univ Chicago, BA, 50, MS, 52; Harvard Univ, PhD(org chem), 55. *Prof Exp:* Staff scientist org chem, Worcester Found Exp Biol, 56-58; asst prof biochem, 62-68, ASSOC PROF BIOCHEM, COLUMBIA UNIV, 68- *Concurrent Pos:* NIH fel, Swiss Fed Inst Technol, 55-56; spec fel neurochem, Columbia Univ, 58-62; NIH res career develop award, 62-72. *Mem:* AAAS; Am Soc Biol Chem; Am Chem Soc; Harvey Soc. *Res:* Structure and mechanism of enzymes; protein chemistry; biochemistry of parasites. *Mailing Add:* Dept Biochem & Molecular Biophys Columbia Univ New York NY 10032

GOLD, ALVIN HIRSH, b Tyler, Tex, May 26, 32; m 59; c 1. BIOCHEMISTRY, PHARMACOLOGY. *Educ:* Univ Tex, BA, 58, MA, 61; St Louis Univ, PhD(pharmacol), 65. *Prof Exp:* Res assoc biochem, State Univ NY Buffalo, 64-66; asst prof, Wake Forest Univ, 66-68; asst prof, 68-71, assoc prof, 71-81, PROF PHARMACOL, ST LOUIS UNIV, 81- *Concurrent Pos:* USPHS res grant, Bowman Gray Sch Med, 66-68; USPHS res grants, St Louis Univ, 68-71 & 75-84; Nat Acad Sci exchange scientist, Univ Zagreb, 71-72. *Mem:* AAAS; Am Soc Biol Chem & Molecular Biol; Am Soc Pharmacol & Exp Therapeut; Am Chem Soc; Brit Biochem Soc; Am Diabetes Asn. *Res:* Enzymology; metabolic control systems; endocrinology. *Mailing Add:* Dept Pharmacol St Louis Univ Med Sch St Louis MO 63104

GOLD, ANDREW VICK, b Inveresk, Scotland, Mar 20, 34; m 58; c 3. PHYSICS. *Educ:* Univ Edinburgh, BSc, 55; Cambridge Univ, PhD(physics), 58. *Prof Exp:* Fel low-temperature & solid state physics, Nat Res Coun Can, 58-60; res assoc, Iowa State Univ, 60-61, from asst prof to prof, 61-68; PROF PHYSICS, UNIV BC, 68- *Concurrent Pos:* Sloan res fel, 64-67. *Mem:* Fel Am Phys Soc. *Res:* Electronic and transport properties of metals; de Haas-van Alphen effect; ferromagnetism. *Mailing Add:* Dept Physics Univ BC 2075 Wesbrook Mall Vancouver BC V7S 1C9 Can

GOLD, ARMAND JOEL, b Baltimore, Md, June 7, 26; m 52; c 2. MEDICAL PHYSIOLOGY. *Educ:* Western Md Col, AB, 49; Univ Md, MS, 51, PhD(physiol), 55. *Prof Exp:* Asst zool, Univ Md, 50-52; physiologist, Chem Corps Med Labs, US Army, 52-55; asst prof physiol res, Sch Med, Univ Md, 56-59; sr develop engr, Goodyear Aircraft Corp, 59-60; physiologist, missile & space vehicle dept, Gen Elec Co, 60-63; staff scientist, res dept, Martin Co, 63-65; head dept environ physiol, Negev Inst Arid Zone Res, Beersheba, Isreal, 65-67; assoc prof, 68-77, PROF PHYSIOL, HOWARD UNIV COL MED, 77- *Mem:* AAAS; Am Physiol Soc; Aerospace Med Asn. *Res:* Heat stress; starvation; hypoxia; cellular metabolism. *Mailing Add:* Dept Physiol Howard Univ Col Med Washington DC 20059

GOLD, BARRY IRA, b Everett, Mass, Apr 30, 46; m 71; c 2. PHARMACOLOGY, NEUROCHEMISTRY. *Educ:* Univ Cincinnati, BS, 68; Boston Univ, PhD(pharmacol), 76. *Prof Exp:* Res assoc pharmacol, Squibb Inst Med Res, 68-71; fel, Sch Med, Yale Univ, 75-78; asst prof pharmacol, Sch Med, Uniformed Serv Univ, 78-84; RES PHARMACOLOGIST, ANAQUEST, DIV BOC GROUP, 84- *Honors & Awards:* Sandoz Award, Sandoz Pharmaceut, 75. *Mem:* AAAS; Soc Neurosci; Am Soc Pharmacol & Exp Therapeut. *Res:* Opiate and cholinergic receptors and the effects of centrally acting drugs and neuromuscular blocking drugs. *Mailing Add:* Anaquest Div BOC Group 100 Mountain Ave Murray Hill NJ 07974

GOLD, BARRY IRWIN, b Brooklyn, NY, Apr 16, 46; m 71; c 2. BIO-ORGANIC CHEMISTRY, CHEMICAL CARCINOGENESIS. *Educ:* Hunter Col, City Univ New York, BA, 66, Univ Nebr-Lincoln, PhD(org chem), 72. *Prof Exp:* Lectr org chem, Univ Toronto, 71-73; res assoc chem carcinogenics, 73-76, asst prof, 76-80, ASSOC PROF BIOMED CHEM, UNIV NEBR MED CTR, 80- *Concurrent Pos:* Prin investr grants, Nat Cancer Inst, NIH, 77-85. *Mem:* Am Chem Soc; Soc Toxicol; Am Asn Cancer Res. *Res:* Mechanisms by which N-nitroso and chlorinated hydrocarbon compounds initiate the induction of cancer in laboratory animals. *Mailing Add:* Dept Pharm Univ Nebr Med Ctr 42 & Dewey Omaha NE 68105

GOLD, BERNARD, b Brooklyn, NY, Mar 31, 23; m 45; c 3. SPEECH COMMUNICATION, COMPUTER SCIENCE. *Educ:* City Col New York, BEE, 44; Polytech Inst Brooklyn, PhD(elec eng), 49. *Prof Exp:* Mem staff radar, Avion Instrument Corp, 48-50; mem staff noise theory, Hughes Aircraft Co, 50-53; mem staff speech & comput, Lincoln Lab, Mass Inst Technol, 53-69, mem staff commun dept, 69-76, group leader, 76-80, LECTR, DIGITAL PROCESSORS, LINCOLN LAB, MASS INST TECHNOL, 80-, MEM SR STAFF, 80- *Concurrent Pos:* Fulbright fel to Italy, 54-55; vis prof, Mass Inst Technol, 65-66 & 88. *Honors & Awards:* Soc Award, Acoustic Speech & Signal Processing Inst Elec & Electronics Engrs, 86. *Mem:* Nat Acad Eng; fel Inst Elec & Electronics Engrs; Acoust Soc Am. *Res:* Noise theory; radar; pattern recognition; speech bandwidth compression; theory and design of computers. *Mailing Add:* Lincoln Lab Mass Inst Technol Lexington MA 02173

GOLD, DANIEL HOWARD, b New York, NY, Jan 8, 29; m 52; c 2. PHARMACEUTICAL MANUFACTURING. *Educ:* Brooklyn Col, BS, 50; Polytech Inst Brooklyn, PhD(chem), 57. *Prof Exp:* Chemist, Jewish Hosp, Brooklyn, 50-51 & Sylvania Elec Prod, Inc, NY, 51-53; sr chemist, Am Cyanamid Co, NJ, 56-62; sr process res specialist, Lummus Co, 62-67; gen supt rubber chem dept, Am Cyanamid Co, Bound Brook, NJ, 67-72, mat mgr, 72-73, mfg mgr, 73-75, res mgr, Stamford, Conn, 75-77, oper mgr pharmaceut, Bound Brook, NJ, 78-83, MGR VALIDATION, LEDERLE LABS, PEARL RIVER, NY, 84- *Concurrent Pos:* Dir, Parenteral Drug Asn. *Mem:* Am Chem Soc; NY Acad Sci; Parenteral Drug Asn. *Res:* Organic process development; statistical experimental design and analysis; chromatography; complexation; metal ion-polyelectrolyte interactions; pharmaceutical chemicals. *Mailing Add:* 441 Alpine Terr Ridgewood NJ 07450

GOLD, DANIEL P, b San Francisco, Calif, July 28, 54. EXPERIMENTAL BIOLOGY. *Educ:* Univ Calif, Los Angeles, BA, 76; Tufts Univ, PhD(immunol), 83. *Prof Exp:* Fel molecular immunol, Ctr Cancer Res, Mass Inst Technol, 83-84, Dana Farber Cancer Ctr, 84-86; asst mem, Div Immunol, Med Biol Inst, 86-90; ASST MEM, LA JOLLA INST EXP MED, 90- *Concurrent Pos:* Cancer Res Inst fel, 85-87. *Mem:* Am Asn Immunologists; Am Diabetes Asn. *Res:* Regulation of autoimmune T cell reactivity; identification of auto-antigens in Type I diabetes mellitus; regulation of pathogenic graft versus host disease; author of numerous scientific publications. *Mailing Add:* La Jolla Inst Exp Med 11099 N Torrey Pines Rd La Jolla CA 92037

GOLD, DAVID PERCY, b Durban, Natal, SAfrica, June 22, 33; Can citizen; m 59; c 4. GEOLOGY, PETROLOGY. *Educ:* Univ Natal, BSc, 53, Hons, 54, MSc, 58; McGill Univ, PhD(geochem), 63. *Prof Exp:* Geol asst mineral explor, Union Corp Ltd, SAfrica, 55-56; lectr geol, Loyola Col, Can, 62-64; res assoc geol & geochem, 64-68, assoc prof, 68-74, PROF GEOL, PA STATE UNIV, 74-, CHMN GEOL GRAD PROG, 77- *Concurrent Pos:* Nat Res Coun Can grant, 63-64; res assoc, Spec Crater Proj, NASA grants, 64-78, NASA earth resources & remote sensing res projs, 72-78; NSF res grants, 65-70; distinguished lectr, Am Geol Inst, 70; mem post-doctoral fel comt, Nat Acad Sci, 71; mem planetology comt, Int Geol Cong, 72; pres, Yellowstone Bighorn Res Asn, 73-75; mem, NASA & Dept Energy tech comt. *Honors & Awards:* Barlow Mem Medal, Can Inst Mining & Metal Eng, 67. *Mem:* Geol Soc Am; Mineral Asn Can; Geol Soc SAfrica; Geol Asn Can; Meteoritical Soc; Sigma Xi. *Res:* Geophysical exploration and mining geology in Southern Africa; petrology and geochemistry of carbonatites and alkaline rocks; geological mapping in Northern Quebec; structural analysis in Canadian Appalachians and around volcanic and impact craters; ore deposits; plate tectonics environments and remote sensing; fission track dating and analysis. *Mailing Add:* 117 Wildot Dr State College PA 16801

GOLD, EDWARD, b New York, NY, Nov 25, 41; m 63; c 2. PHYSICAL METALLURGY. *Educ:* Polytech Inst Brooklyn, BS, 63; Columbia Univ, MS, 65. *Hon Degrees:* ScD, Columbia Univ, 67. *Prof Exp:* Sr scientist phys metall, Appl Res Lab, Aeronutronic Div, Philco-Ford Corp, 67-68, prin scientist, Advan Develop Oper, 69-72; chief engr, Resource-Recovery Systs, Barbler Coleman Co, 73-74; vpres eng & mfg, Columbia Yacht Div, Whittaker Corp, 75-76; vpres eng, Chris Craft Corp, 77-79; PRES, CUSTOM MARBLE INC, 79- *Mem:* Am Soc Metall Int. *Res:* Field ion microscope study of short-range order in platinum base alloys; thermomechanical processing of aluminum alloys and superalloy development; physical metallurgy and mechanical properties of beryllium. *Mailing Add:* 20976 Cipres Way Boca Raton FL 33433

GOLD, ELI, b New Haven, Conn, July 20, 20; m 43; c 2. VIROLOGY, PEDIATRICS. *Educ:* Univ Conn, BS, 42; Western Reserve Univ, MD, 50; Am Bd Pediat, dipl. *Prof Exp:* Intern & resident pediat, Boston Children's Med Ctr, 50-53; resident, Cleveland City Hosp, 53-54; from instr to asst prof prev med & pediat, 54-65, from assoc prof to prof pediat, Sch Med, Case Western Reserve Univ, 65-74; chmn dept, 74-80, PROF PEDIAT, SCH MED, UNIV CALIF, DAVIS, 74- *Concurrent Pos:* USPHS spec res fel, Inst Virol, Scotland, 61-62, career develop award, 64-70. *Mem:* Soc Pediat Res. *Res:* Synthesis of herpes simplex particle; characteristics of meningococcus; virulence factors, especially immunity. *Mailing Add:* 7216 78th Ave SE Mercer Island WA 98040-5511

GOLD, ELIJAH HERMAN, b New York, NY, May 22, 36; m 62; c 3. ORGANIC CHEMISTRY. *Educ:* City Col New York, BS, 57; Yale Univ, MS, 58, PhD(org chem), 63. *Prof Exp:* Fel, Columbia Univ, 62-64; NIH fel, Israel Inst Tech, 64-66; sr res chemist, 66-69, prin chemist, 69-70, sect leader, Med Chem Dept, 70-73, mgr, 73-74, ASSOC DIR CHEM RES, MEDICINALS, SCHERING CORP, 74- *Concurrent Pos:* US patent agent, 85; med res & develop comt, US Army, 83. *Mem:* Am Chem Soc; NY Acad Sci; Am Inst Chemists. *Res:* Mechanistic and synthetic organic chemistry; organic photochemistry; small ring heterocycles and medicinal chemistry; 44 US patents. *Mailing Add:* Ten Roosevelt Ave West Orange NJ 07052

GOLD, GERALD, b Pittsburgh, Pa, Aug 8, 27; m 54; c 3. PHARMACY. *Educ:* Univ Pittsburgh, BSc, 51, MSc, 58, PhD(pharm), 61. *Prof Exp:* Res pharmacist, Consumer Health Care Div, Miles Home Prod, Miles Labs Inc, 61-80, head, Explor Res Sect, 80-; MEM STAFF, METAB RES UNIT, UNIV CALIF, SAN FRANCISCO. *Mem:* Am Pharmaceut Asn; fel Acad Pharmaceut Sci. *Res:* Emulsions; tableting technology; aerosols. *Mailing Add:* 2309 Kenilworth Dr Elkhart IN 46514

GOLD, HAROLD EUGENE, b St Louis, Mo, Jul 28, 28; m 60; c 2. METALLURGICAL RESEARCH, QUALITY ASSURANCE. *Educ:* Wash Univ, St Louis, BSME, 49. *Prof Exp:* Prod engr, 49-54, asst chief engr, Kennard Corp, 56-62; eng specialist, US Army, 54-56; eng prod mgr, Am Air Filter, 62-67, mgr eng & qual control, 67-80; eng & qual control mgr, Allis Chalmers, 80-91; MGR, ENERGY FILTRATION PROD GROUP, SNYDER GEN, 91- *Concurrent Pos:* Various engr comts, Air Conditioning Refrig Inst, 62- *Mem:* Am Soc Heating, Refrig & Air Conditioning Engrs. *Res:* Development of air conditioning equipment; heat transfer components; development of computerized research data. *Mailing Add:* PO Box 35690 Louisville KY 40232

GOLD, HARRIS, b New York, NY, Mar 20, 36; m 60; c 2. WATER PURIFICATION, SEPARATION SCIENCES. *Educ:* Polytech Inst Brooklyn, BME, 58; Columbia Univ, MS, 59; Calif Inst Technol, PhD(aeronaut), 63. *Prof Exp:* Asst aeronaut, Calif Inst Technol, 59-63; sr staff scientist, Avco Systs Div, 63-66, mgr, Aerophys Dept, 66-68, sr consult scientist, 68-70, prin staff scientist, Avco Systs Div, 70-74; partner, Water Purification Assocs, Cambridge, Mass, 74-82, pres, Water General, 82-83; MGR, FOSTER-MILLER, INC, WALTHAM, MASS, 83- *Concurrent Pos:* Consult, E H Plesset Assocs, Inc, Calif, 61-62. *Mem:* Am Inst Chem Engrs; Am Water Works Asn; Am Electroplaters Asn; AAAS; Am Chem Soc; Water Pollution Control Fedn. *Res:* Ion exchange; gas separation; membrane separation; protein purification; separation processes; water treatment; molecular sieves; metal complexation. *Mailing Add:* 18 Peachtree Rd Lexington MA 02173

GOLD, HARVEY JOSEPH, b Southampton, NY, Oct 20, 32; m 51; c 2. DECISION ANALYTIC MODELING. *Educ:* Univ Miami, BS, 54; Univ Wis, MS, 56, PhD(food technol, biochem), 58; NC State Univ, PhD(biomath), 65. *Prof Exp:* Asst, Univ Wis, 54-58; res biochemist, USDA, 58-63; fel biomath, 63-65, assoc prof, 65-74, dir biomath prog, 77-85, PROF STATIST & BIOMATH, NC STATE UNIV, 74- *Mem:* AAAS; Soc Gen Systs Res; Biomet Soc; Am Statist Asn; Opers Res Soc Am. *Res:* Mathematical modeling of biological systems; decision modeling; application to agricultural and forest pest control and to environmental science. *Mailing Add:* Biomath Prog Dept Statist NC State Univ Campus Box 8203 Raleigh NC 27695-8203

GOLD, HARVEY SAUL, b Neptune, NJ, Feb 23, 52. ANALYTICAL CHEMISTRY, PHYSICAL CHEMISTRY. *Educ:* Cornell Univ, BA, 74; Univ NC, Chapel Hill, PhD(chem), 78. *Prof Exp:* Res asst analytical chem, Univ NC, Chapel Hill, 74-78; asst prof chem, Univ Del, 78-85; RES CHEMIST, E I DU PONT DE NEMOURS & CO, 85- *Mem:* Am Chem Soc; Coblentz Soc; Soc Appl Spectros. *Res:* Analytical spectroscopy, vibrational and magnetic resonance; spectroscopy of catalysts and computer treatment of spectroscopic data; inelastic electron tunneling spectroscopy; polymer characterization. *Mailing Add:* 101 Lennie Ct Newark DE 19702-1909

GOLD, JAY JOSEPH, b New York, NY, Feb 17, 23; m 50; c 2. INTERNAL MEDICINE, ENDOCRINOLOGY. *Educ:* NY Univ, AB, 46; State Univ NY, MD, 50; Am Bd Internal Med, dipl, 58. *Prof Exp:* Intern, Brooklyn Jewish Hosp, 50-51, resident internal med, 51-52; resident, Bronx Vet Admin Hosp, 54-55; asst prof med, Chicago Med Sch, 56-65; assoc prof med, Med Sch, Northwestern Univ, Chicago, 65-72; CLIN PROF MED, ABRAHAM LINCOLN SCH MED, UNIV ILL MED CTR, 72-, ADJ PROF OBSTET & GYNECOL, 78- *Concurrent Pos:* Fel endocrinol, Col Physicians & Surgeons, Columbia Univ, 52-54; attend endocrinol, Westside Vet Admin Hosp, 59-, consult endocrinol, 72-; dir endocrinol & metab sect, St Francis Hosp, Evanston; staff consult, Lutheran Gen Hosp, Mercy Hosp; attend, St Francis Hosp, Rush North Shore Hosp; dir, Diab Treat Ctgr, St Francis Hosp. *Mem:* Endocrine Soc; Am Fertil Soc; Am Diabetes Asn; fel Am Col Physicians. *Res:* Chemistry and physiology of pituitary and adrenal glands, gonads, hirsutism and infertility. *Mailing Add:* 9669 Kenton Skokie IL 60076

GOLD, JOHN RUSH, b Brownwood, Tex, Dec 14, 46. GENETICS. *Educ:* Knox Col, AB, 68; Univ Calif, Davis, PhD(genetics), 73. *Prof Exp:* Asst res genetics, Dept Animal Sci, Univ Calif, Davis, 73-75; ASST PROF GENETICS, TEX A&M UNIV, 75- *Mem:* Genetics Soc Am; Am Soc Ichthyologists & Herpetologists; Soc Study Evolution; Soc Syst Zool. *Res:* Genetics, evolution and taxonomy of native North American fish. *Mailing Add:* Dept Wildlife & Fisheries Sci Tex A&M Univ College Station TX 77843

GOLD, JUDITH HAMMERLING, b New York, NY, June 24; 41; Can citizen; m 65. PSYCHIATRY. *Educ:* Dalhousie Univ, MD, 65; FRCP(C), 71. *Prof Exp:* Staff psychiatrist student health, Dalhousie Univ, 71-73; vis colleague psychiat, Welsh Nat Sch Med, Univ Wales, 73-75; asst prof, 75-77, assoc prof psychiat, Dalhousie Univ, 77-87; mem bd regents, 86-89, FIRST VPRES, AM COL PSYCHIATRISTS, 90- *Concurrent Pos:* Lectr psychiat, Dalhousie Univ, 71-75; MRC fel, Can, 73-75; consult psychiatrist, NS Hosp, Dartmouth, 75-77; health welfare res grant, Can, 77-80; mem review comt, Nat Health Res & Develop Prog, Can, 78-81; vchmn, bd gov, Mt St Vincent Univ, 84-86, chmn bd gov, 86-87; ed, Clin Practice Ser, APPI, 87- *Mem:* Can Psychiat Asn (pres, 81-82); Am Psychiat Asn; Can Med Asn; Royal Col Psychiatrists; NS Med Asn; Am Col Psychiatrists; fel Am Col Psychiatrists. *Res:* Crosscultural psychiatry. *Mailing Add:* 5991 Spring Garden Rd Suite 375 Halifax NS B3H 1Y6 Can

GOLD, LEWIS PETER, b Brockton, Mass, May 21, 35; m 65; c 3. PHYSICAL CHEMISTRY. *Educ:* Harvard Univ, AB, 57, AM, 59, PhD(chem), 62. *Prof Exp:* Res physicist, Columbia Univ, 61-62, res assoc physics, 62-65; asst prof, 65-70, ASSOC PROF CHEM, PA STATE UNIV, 70- *Mem:* AAAS; Am Chem Soc; Am Phys Soc. *Res:* Molecular spectroscopy and molecular structure; laser-induced fluorescence and optical-optical double resonance spectroscopy of small molecules. *Mailing Add:* Dept Chem Davey Lab Pa State Univ University Park PA 16802

GOLD, LORNE W, b Saskatoon, Sask, June 7, 28; m 51; c 4. ENGINEERING PHYSICS, GLACIOLOGY. *Educ:* Univ Sask, BSc, 50; McGill Univ, MSc, 52, PhD, 70. *Prof Exp:* Jr res officer snow & ice, 50-51, head sect, 52-69, head geotech sect, Div Bldg Res, 69-74, asst dir, 74-79, res mgt chmn & assoc comt geotech res, 76-83, ASSOC DIR, DIV BLDG RES, NAT RES COUN CAN, 79- *Concurrent Pos:* Can deleg, Int Union Testing & Res Labs Mat & Struct, Can, 82- *Mem:* Can Geotech Soc; Int Glaciol Soc (pres, 78-81); Can Asn Physicists; Eng Inst Can; fel Royal Soc Can. *Res:* Deformation and failure of ice; ice engineering; ground thermal regime; heat exchange at snow, ice, water and ground surfaces. *Mailing Add:* 1903 Illinois Ave Ottawa ON K1H 6W8 Can

GOLD, MARK STEPHEN, b New York, NY, May 6, 49; m 71; c 4. MEDICINE, PSYCHIATRY. *Educ:* Wash Univ, BA, 71; Univ Fla, MD, 75. *Prof Exp:* Fel neuropsychiat, Yale Univ Sch Med, 75-78; vpres basic res, Psychiat Inst Am, 78-79; lectr psychiat, Yale Univ Sch Med, 78-80; DIR RES FAIR OAKS HOSP, 78- *Concurrent Pos:* Biol sci training prog fel, Yale Univ Sch Med, 77-78; consult, Addiction Prev & Treatment, 76-78; founder & dir res, Psychiat Diag Labs, 79-84; founder, Nat Cocaine Hotline, 83- *Honors & Awards:* Mead Johnson & Roche, Nat Student Res Forum, 75; Seymour F Lustmann, Yale Univ Sch Med & Laughlin Nat Award, Psychiat Endowment Fund, 78; Found Fund Prize, Am Psychiat Asn, 81; Presidential Res Award, Nat Asn Pvt Psychiat Hosps, 82; Silver Anvil Award, 83. *Mem:* Soc Neurosci; Am Psychiat Asn; Int Soc Psychoneuroendocrinol; fel Am Col Clin Pharmacol; Soc Biol Psychiat; Am Acad Psychiatrists Alcoholism & Addiction. *Res:* Neuropsychopharmacology of psychiatric illness and treatment; neuropsychopharmacology and treatment of endogenous and drug-induced euphoric and withdrawal states; development and use of biological measures in psychiatric diagnosis and treatment. *Mailing Add:* Fair Oaks Hosp 5440 Linton Blvd Delray Beach FL 33484

GOLD, MARTIN, b Philadelphia, PA, 32. CLINICAL BIOCHEMISTRY, PHYSIOLOGY. *Educ:* Philadelphia Col Pharm & Sci, BSc, 54; Hahnemann Med Col, MS, 56, PhD(biochem), 59. *Prof Exp:* Res assoc physiol, Hahnemann Med Col, 59-64, res asst prof, 64-67; res assoc, Geront Res Inst, Philadelphia Geriat Ctr, 67-70; dir clin biochem, Dept Lab Med, Nazareth Hosp, 70-78; dir clin chem, 78-79, QUAL ASSURANCE COORDR, PATH DEPT, MERCER MED CTR, 89- *Concurrent Pos:* Nat Inst Arthritis & Metab Dis grant, 64-67; SCent Pa Heart Asn grant, 67-68. *Mem:* AAAS; Am Inst Chem; Am Chem Soc; Am Physiol Soc; Am Asn Clin Chemists. *Res:* Glycolytic enzymes; proteinases; lipid metabolism; specifically free fatty acid and glyceride metabolism. *Mailing Add:* 11000 Greiner Rd Philadelphia PA 19116

GOLD, MARTIN I, b Philadelphia, Pa, Nov 22, 28; m 51; c 3. MEDICINE, ANESTHESIOLOGY. *Educ:* Univ Pa, BA, 50; State Univ NY, MD, 54. *Prof Exp:* Resident, Grad Hosp, Univ Pa, 54-57; from instr to prof, Sch Med, Univ Md, 59-72, med dir inhalation ther & div acute respiratory care, 68-72; PROF ANESTHESIOL, SCH MED, UNIV MIAMI, 72-; CHIEF ANESTHESIOL SERV, VET ADMIN HOSP, MIAMI, 72- *Concurrent Pos:* NIH fel, 61-67 & spec res fel, 67-68; vis prof, Royal Postgrad Med Sch, London, Eng, 67-68. *Mem:* Am Soc Anesthesiol; Am Soc Pharmacol & Exp Therapeut; Asn Univ Anesthetists; Sigma Xi; Int Anesthesia Res Soc. *Res:* Pulmonary physiology as related to mechanics and gases during anesthesia; respiratory care. *Mailing Add:* Vet Admin Hosp 1201 NW 16th St Miami FL 33125

GOLD, MARVIN B, b New York, NY, June 23, 33; m 62; c 1. CHEMICAL EDUCATION. *Educ:* Ohio State Univ, BSc, 58; Univ Calif, Berkeley, PhD(chem), 62. *Prof Exp:* Res chemist, Univ Calif, Berkeley, 63; from asst prof to assoc prof, 63-71, chmn dept chem, 70-73, PROF INORG CHEM, CALIF STATE UNIV, CHICO, 71- *Mem:* Am Chem Soc; Nat Asn Sci Teachers. *Res:* Computer assisted learning. *Mailing Add:* Dept Chem Calif State Univ Chico CA 95926

GOLD, MARVIN H, b Toronto, Ont, Feb 2, 36; m 58; c 5. BIOCHEMISTRY. *Educ:* Univ Toronto, BA, 57, PhD(biophys), 62. *Prof Exp:* Asst prof develop biol, Albert Einstein Col Med, 65-67; assoc prof med biophys, 67-69, assoc prof med cell biol, 69-82, PROF MED GENETICS, UNIV TORONTO, 83- *Concurrent Pos:* Jane Coffin Childs fel, 62-63. *Res:* Nucleic acids of bacteria and bacteriophage; genetics; enzymology. *Mailing Add:* Dept Molecular & Med Genetics Med Ctr Bldg Rm 4180 One Kings Col Circle Toronto ON M5S 1A8 Can

GOLD, MARVIN H, b Buffalo, NY, June 23, 15; m 40; c 2. POLYMER CHEMISTRY. *Educ:* Univ Calif, Los Angeles, BA, 37; Univ Ill, PhD(org chem), 40. *Prof Exp:* Postdoctoral fel, Northwestern Univ, 40-42; res chemist, Visking Corp, 42-48; sr scientist, assoc dir chem & mgr, Propellant Chem Div, Aerojet Gen Corp, 48-72; CONSULT, 72- *Mem:* Am Chem Soc; Sigma Xi; Am Inst Aeronaut & Astronaut. *Res:* Polynuclear hydrocarbons; polynitro aliphatic chemistry; polymers; author of more than 30 publications; awarded more than 85 patents. *Mailing Add:* 2601 Latham Dr Sacramento CA 95864

GOLD, MICHAEL HOWARD, b Paterson, NJ, Jan 16, 41; m 64; c 2. LIGNIN BIODEGRADATION, FUNGAL GENETICS. *Educ:* Rutgers Univ, AB, 63; State Univ NY Buffalo, PhD(biochem), 70. *Prof Exp:* Asst prof biochem genetics, Rockefeller Univ, 75-76; from asst prof to assoc prof biochem sci, 76-82, PROF & CHAIR CHEM & BIOL SCI, ORE GRAD INST, 82- *Concurrent Pos:* Fel, Univ Calif, Davis, 70-73; res assoc, Rockefeller Univ, 74-75. *Mem:* Am Soc Microbiol; AAAS; Soc Indust Microbiol. *Res:* Biochemistry and genetics of fungi; lignin biodegradation; fungal peroxidases; molecular biology of fungi; oxygen metabolism; metabolic regulation; enzymology; biomass utilization. *Mailing Add:* Dept Chem & Biol Sci Ore Grad Inst Sci & Technol 19600 NW Von Neumann Dr Beaverton OR 97006-1999

GOLD, NORMAN IRVING, biochemistry; deceased, see previous edition for last biography

GOLD, PAUL ERNEST, b Detroit, Mich, Jan 7, 45; c 1. NEUROBIOLOGY & MEMORY, AGING. *Educ:* Univ Mich, BA, 66; Univ NC, MS, 68, PhD(psychol), 71. *Prof Exp:* Postdoctoral psychobiol, Univ Calif, Irvine, 70-72, lectr, 72-76; from asst prof to assoc prof, 76-81, PROF PSYCHOL, UNIV VA, CHARLOTTESVILLE, 81- *Concurrent Pos:* Postdoctoral fel, NIMH, 72; ed, Psychobiol J, Psychonomic Soc, 90-; mem comt, Am Psychol Soc, 90-91, mem prog comt, 91-; dir neurosci, Univ Va, 91- *Mem:* Fel Am Psychol Am; fel Am Psychol Soc; Soc Neurosci. *Res:* Brain mechanisms responsible for the formation of memory, and biological processes which regulate memory function, including meuroendocrine regulation of memory function, contribution of hormonal responses to age-related memory impairments and to memory pathologies. *Mailing Add:* Dept Psychol 102 Gilmer Hall Univ Va Charlottesville VA 22903-2477

GOLD, PHIL, b Montreal, Can, Sept 17, 36; m 60; c 3. CANCER IMMUNOLOGY. *Educ:* McGill Univ, BSc Hons, 57, MSc & MD, CM, 61, PhD(physiol), 65; FRCPS(C); Royal Col Physicians & Surgeons, cert internal med, 66. *Hon Degrees:* DSc McMaster Univ, 86. *Prof Exp:* Jr intern, Montreal Gen Hosp, 61-62, jr asst resident med, 62, sr resident, 65-66; fel, Med Res Coun, Can, 63-65, centennial fel, 67-68; from lectr to assoc prof physiol, McGill Univ, 64-74, from teaching fel to assoc prof med & clin med, 65-73, dir, Cancer Ctr, 78-80, chmn, Dept Med, 80-90, PROF MED & CLIN MED, MCGILL UNIV, 73-, PROF PHYSIOL, 74-, DOUGLAS G CAMERON PROF MED, 89-; PHYSICIAN-IN-CHIEF, DEPT MED, MONTREAL GEN HOSP, 80- *Concurrent Pos:* Vis scientist, Pub Health Res Inst, New York, 67-68; chmn, Grants Panel, Immunol & Chemother Sect, Am Cancer Soc, 72-73; assoc, Med Res Coun, Can, 68, chmn, Grants Panel Cancer Res, 72-77; sr investr, McGill Univ Med Clin & Montreal Gen Hosp Res Inst, 72-; sr physician, Montreal Gen Hosp, 73-; assoc ed, Cancer Res, 73-80; consult allergy & immunol, Mt Sinai Hosp, Que, 75-; chmn clin immunol, English Exam Bd, Royal Col Physicians & Surgeons, Can, 75-77;

mem const comt, Int Res Group Carcinoembryonic Proteins, 76-; dir, Div Clin Immunol & Allergy, Montreal Gen Hosp & McGill Univ Med Ctr, 77-80; pres, Sci Adv Bd, Israel Cancer Res Fund, 77-88; mem cancer grants panel B, Nat Cancer Inst, Can, 77-; mem bd dir, Mt Sinai Inst, Toronto, 78-; patron, Alzheimer Soc Montreal, 86-; hon dir, Can Magen David Adom Inst, 86; mem coun, Med Res Coun Can, 86-, exec comt, 87-, comt ethics & med res, 89-; mem bd dirs, Int Soc Prev Oncol, 89-; counr, Can Inst Acad Med, 90- *Honors & Awards:* Medal Med, Royal Col Physicians & Surgeons, Can, 65; E W R Steacie Sci Prize, Nat Res Coun, Can, 73; Can Silver Jubilee Medal, 77, Heath Mem Award, 80; Johann-Georg-Zimmerman Cancer Res Prize, Med Univ, 78; Gairdner Found Award, 78; Terry Fox Inaugural lectr, BC Med Asn, Penticton, 81; Ernest C Manning Found Award, 82; Med Achievement Award, Can Asn Mfrs Med Devices, 83; FNG Starr Award, Can Med Asn, 83; Isaak Walton Killam Award Med, Can Coun, 85; Tower of Hope Award, Israel Cancer Res Fund, 85; Sci Achievement Medal, Italian Govt, 90. *Mem:* Am Soc Clin Invest; Am Asn Cancer Res; Am Asn Immunologists; Am Acad Allergy; fel Royal Soc Can; fel Am Col Physicians; Can Fedn Biol Sci; AAAS; Am Fedn Clin Res; Asn Am Physicians; hon mem Can Asn Radiologists. *Res:* Immunologic studies dealing with the presence of specific materials present in cancer cells that are absent from corresponding normal cells, and the use of this information in diagnostic medicine; the study of AIDS in hemophiliacs - disease evolution and immuno-pathology. *Mailing Add:* Montreal Gen Hosp 1650 Cedar Ave Rm 648E Montreal PQ H3G 1A4 Can

GOLD, PHILIP WILLIAM, b Newport News, Va, Sept 23, 44; m 72; c 3. NEUROENDOCRINOLOGY. *Educ:* Duke Univ, BA, 66; Duke Univ Sch Med, MD, 70. *Prof Exp:* Intern med, Boston City Hosp, 70-71; resident psychiat, Harvard Med Sch, 71-74; clin assoc biol psychiat, 74-76, chief, Neuroendocrinol Unit, 76-81, Neuroendocrinol Sect, 81-88, CHIEF, CLIN NEUROENDOCRINOL BR, INTRAMURAL RES PROG, NIMH, 88- *Concurrent Pos:* Vis prof psychiat, Duke Univ Sch Med, 86-; coun scholars, Library Congress, 88- *Honors & Awards:* Curt Richter Prize, Int Soc Psychoneuroendocrinol, 85. *Mem:* Endocrine Soc; Am Col Neuropsychopharmacol; Int Soc Psychoneuroendocrinol; Am Soc Clin Res; AAAS; Am Psychiat Asn. *Res:* Physiological and molecular mechanisms of physical and emotional stress and their relevance to major psychiatric and endocrine disorders. *Mailing Add:* Nat Inst Health Clin Ctr 9000 Rockville Pike Bldg 10 Rm 45239 Bethesda MD 20814

GOLD, RAYMOND, b New York, NY, Oct 3, 27; m 51; c 4. EXPERIMENTAL NUCLEAR PHYSICS, REACTOR PHYSICS. *Educ:* NY Univ, BA, 51, MS, 54; Ill Inst Technol, PhD, 58. *Prof Exp:* Physicist, Columbia Univ, 52-54; group leader, Gen Elec Co, 54-55; res physicist, Armour Res Found, Ill Inst Technol, 55-58; prof & head nuclear sci & eng dept, Lowell Univ, 58-62; sect head, Argonne Nat Lab, 62-72; resident dir, Joint Ctr Grad Study, 72-74; pres, Radiation & Dosimetry Serv, Inc, 75-76; fel scientist, Westinghouse Hanford Co, 76-87; vpres, 88-90, PRES, METROL CONTROL CORP, 91- *Concurrent Pos:* Consult, US Naval Radiol Defense Lab, 59-60, Tech Opers Inc, 60-62, Avco Corp, 61-62, Martin Marrietta Co, 71-78 & Westinghouse Hanford Co, 73-75. *Mem:* Am Phys Soc; Am Nuclear Soc; Health Phys Soc. *Res:* Experimental nuclear physics; environmental radiation measurements; reactor physics. *Mailing Add:* 1982 Greenbrook Blvd Richland WA 99352

GOLD, RICHARD HORACE, b New York, NY, Nov 20, 35; m 65; c 2. BREAST CANCER, SKELETAL DISEASES. *Educ:* NY Univ, BA, 56; Univ Louisville, MD, 60. *Prof Exp:* Fel skeletal radiol, Univ Calif, San Francisco, 67-68, asst prof radiol, 68-72; from asst prof to assoc prof, 72-78, PROF RADIOL, UNIV CALIF, LOS ANGELES, 78-, CHIEF, GEN DIAGNOSTIC DIV, DEPT RADIOL, 86- *Concurrent Pos:* Mem, Working Group Rev Breast Cancer Demonstration Proj, Nat Cancer Inst, 77-78; mem, Nat Task Force Breast Cancer Control, Nat Cancer Soc, 77-82; consult, Dept Radiol, Sepulveda Vet Admin Hosp, 73-89, & Wadsworth Vet Admin Hosp, 79-89; mem bd dirs, Calif Div, Am Cancer Soc, 80-86; mem, comt breast imaging, Am Col Radiol, 73-88. *Mem:* Asn Univ Radiologists; Radiol Soc N Am; Am Roentgen Ray Soc; Am Col Radiol; Int Skeletal Soc; Soc Breast Imaging. *Res:* Radiologic diagnosis of diseases of the breast and musculoskeletal systems; breast imaging; breast cancer detection. *Mailing Add:* Dept Radiol Sci Univ Calif Sch Med Los Angeles CA 90024

GOLD, RICHARD MICHAEL, b New York, NY, Apr 20, 37; c 3. NEUROPSYCHOLOGY. *Educ:* Univ Chicago, PhD(bio-psychol), 66. *Prof Exp:* From asst prof to assoc prof psychol, State Univ NY Col Cortland, 66-74; PROF PSYCHOL, UNIV MASS, AMHERST, 74- *Mem:* Soc Neurosci. *Res:* Neuroanatomy of appetite regulation. *Mailing Add:* Dept Psychol Univ Mass Amherst MA 01003

GOLD, RICHARD ROBERT, b Chicago, Ill, Mar 27, 30; m 52; c 3. PHYSICS, APPLIED MATHEMATICS. *Educ:* Univ Ill, BS, 51, MS, 55, PhD(gas dynamics, math), 56. *Prof Exp:* Mem tech staff gas dynamics, McDonnell Aircraft Corp, 51-52; mem tech staff, Inst Math, US Armed Forces Inst, 52-54; mem tech staff aerodyn, Hughes Aircraft Co, 56-59, staff physicist, 59-61; head magnetogas dynamics sect, Lab Div, Aerospace Corp, 61-64; mgr aerospace physics res dept, Space Systs Div, Hughes Aircraft Co, 64-68; prof, 68-75, ADJ PROF MECH ENG, CALIF STATE COL, LONG BEACH, 75- *Concurrent Pos:* Lectr mech eng dept, Univ Southern Calif, 57-58 & phys sci exten, Univ Calif, Los Angeles, 57-62; adj prof mech & struct, Univ Calif, Los Angeles, 81-82. *Mem:* Assoc fel Am Inst Aeronaut & Astronaut; Am Phys Soc; AAAS; Sigma Xi. *Res:* Electromagnetic wave propagation through plasmas; magnetohydrodynamics; fluid dynamics. *Mailing Add:* 30239 Cheret Place Palos Verdes Peninsula CA 90274

GOLD, ROBERT, b Philadelphia, Pa, Mar 29, 42; m 68; c 2. NUMBER THEORY. *Educ:* Swarthmore Col, BA, 64; Mass Inst Technol, PhD(math), 68. *Prof Exp:* Chief mathematician, S Ross & Co, 68; from asst prof to assoc prof, 68-86, PROF MATH, OHIO STATE UNIV, 86- *Concurrent Pos:* Humboldt fel, 78-79. *Mem:* Am Math Soc. *Res:* Algebraic number theory; Iwasawa theory; structure of ideal class groups; imaginary quadratic fields; cyclotomic fields. *Mailing Add:* Dept Math Col Math & Phys Sci Ohio State Univ 231 W 18th Ave Columbus OH 43210

GOLD, ROGER EUGENE, b Salt Lake City, Utah, Sept 10, 44; m 66; c 3. ENTOMOLOGY. *Educ:* Univ Utah, BS, 68, MS, 70; Univ Calif, Berkeley, PhD(entom), 74. *Prof Exp:* Res assoc, Vectors Entom Res Group, Univ Calif, Berkeley, 73-74; exten pesticide specialist, asst prof entom & asst prof plant path, Univ Ariz, 74-76, prog dir, Agr Coop Exten Serv, 76-77; ASSOC PROF ENTOM & COORDR ENVIRON PROGS, INST AGR & NATURAL RESOURCES, UNIV NEBR-LINCOLN, 77- *Mem:* Entom Soc Am; Sigma Xi; Am Coun Sci & Health. *Res:* Insects as vectors of plant pathogens; insect control in the near environment; pesticide assessment in the environment. *Mailing Add:* 5840 LaSalle Lincoln NE 68516

GOLD, STEVEN HARVEY, b Philadelphia, Pa, Sept 3, 46; m 74; c 2. FREE ELECTRON LASERS & CYCLOTRON MASERS. *Educ:* Haverford Col, BA, 68; Univ Md, MS, 70, PhD(plasma physics), 78. *Prof Exp:* Nat Res Coun resident res assoc plasma physics, 78-80, res physicist, 80-87, SUPVRY RES PHYSICIST & SECT HEAD, NAVAL RES LAB, 87- *Concurrent Pos:* Assoc ed, Inst Elec & Electronics Engrs, Transactions on Plasma Sci, 88- *Mem:* Am Phys Soc; Inst Elec & Electronics Engrs. *Res:* High power microwave and millimeter-wave generation using the free-electron laser and cyclotron maser interactions. *Mailing Add:* Plasma Physics Div Code 4790 Naval Res Lab Washington DC 20375

GOLD, SYDELL PERLMUTTER, b New York, NY, Jan 3, 41; c 3. GEOMETRY, SYSTEM ANALYSIS. *Educ:* Columbia Univ, AB, 61; Univ NMex, MS, 62; Univ Calif, Berkeley, PhD(math), 73. *Prof Exp:* Staff mem math & tech systs analyst, Sandia Labs, 62-67; mem prof staff & systs analyst, Lawrence Livermore Labs, Univ Calif, 74-80; OFF OF SECY, AIR FORCE, 82- *Concurrent Pos:* Mem staff, Nat Security Coun, 80-82. *Honors & Awards:* David Rist Prize, Military Opers Res Soc, 80. *Mem:* Am Math Soc. *Res:* Analysis of United States Air Force weapons systems; scissor congruence of unbounded convex plane subsets. *Mailing Add:* 7301 Westerly Ln McLean VA 22101

GOLD, THOMAS, b Vienna, Austria, May 22, 20; nat US; m 47, 72; c 4. ASTRONOMY. *Educ:* Cambridge Univ, BA, 42, MA, 46, ScD, 69. *Hon Degrees:* MA, Harvard Univ, 57. *Prof Exp:* Exp officer radar res & develop, Brit Admiralty, 42-46; res grant, Cavendish Lab, Cambridge Univ, 46-47, fel, Trinity Col, 47-51, demonstr physics, 49-52, Med Res Coun res grant, zool lab, 47-49; sr prin sci officer, Royal Greenwich Observ, 52-56; prof astron, Harvard Univ, 57-58, Robert Wheeler Willson prof appl astron, 58-59; prof, 59-71, dir, Ctr Radiophysics & Space Res, 59-81, John L Wetherill prof astron, 71, EMER PROF ASTRON, CORNELL UNIV, 87- *Honors & Awards:* Gold Medal, Royal Astron Soc, 85. *Mem:* Nat Acad Sci; fel Royal Soc; fel Am Philos Soc; Am Astron Soc; fel Am Geophys Union. *Res:* Cosmology; geophysics; radio astronomy; magnetohydrodynamics. *Mailing Add:* Dept Astron Space Sci Bldg Cornell Univ Ctr Radio Phys & Space Res Ithaca NY 14853

GOLD, WARREN MAXWELL, b Chester, Pa, Jan 25, 34; c 4. PULMONARY PHYSIOLOGY, PULMONARY IMMUNOLOGY. *Educ:* Harvard Col, BA, 55; Harvard Med Sch, MD, 59. *Prof Exp:* Res fel pulmonary physiol, Cardiovasc Res Inst, Univ Calif, San Francisco, 62-65; dir pulmonary lab & res assoc cardiol, Children's Hosp Med Ctr, 65-69; asst prof med, Univ Calif, San Francisco, 69-72; assoc prof med, 72-77, PROF MED, UNIV CALIF, SAN FRANCISCO, 78-, PROF, CARDIOVASC RES INST; CHIEF, CLIN CHEST SECT, MOFFITT HOSP, 69- *Concurrent Pos:* Assoc pediat & tutor med sci, Harvard Med Sch, 65-69; from assoc mem to sr staff mem, Cardiovasc Res Inst, Univ Calif, San Francisco, 69-; vis scientist, Univ Sydney, Australia, 75; asst dir, Chronic Dis Airways, 74- *Mem:* Am Soc Clin Invest; Am Thoracic Soc; Am Fedn Clin Res; Am Physiol Soc; Am Acad Allergy. *Mailing Add:* Cardiovasc Res Inst Univ Calif 505 Parnassus Ave Box 0130 San Francisco CA 94143

GOLD, WILLIAM, b Buffalo, NY, Jan 2, 23; m 52; c 4. DENTISTRY. *Educ:* Cornell Univ, BS, 43; Univ Wis, MS, 48, PhD(bact), 50. *Prof Exp:* Instr bact & bact physiol, 50; fel, Rutgers Univ, 50-51; res assoc, E R Squibb & Sons, 51-57; microbiologist, Bzura, Inc, 57-63; supvr biol res, Beech Nut Lifesavers, Inc, 63-67; asst prof microbiol, Col Dent, NY Univ, 67-73; sr scientist oral biol, Cooper Labs, Inc, 73-81; SR MICROBIOLOGIST, BAXTER MICROSCAN, 82- *Concurrent Pos:* Prin investr, John Hartford Found Grant, 68-73. *Mem:* Am Chem Soc; Am Soc Microbiol; Sigma Xi. *Res:* Biochemistry of microbial processes in oral diseases; control of caries and periodontitis; development of microbiological diagnostics for laboratory use; developing media, equipment and processes to manufacture panels to identify bacteria and the antimicrobiols to which they are sensitive. *Mailing Add:* 3121 Hempstead Rd Sacramento CA 95864

GOLDAN, PAUL DAVID, b Reno, Nev, Nov 25, 33; m 59; c 3. ATMOSPHERIC PHYSICS, ATMOSPHERIC CHEMISTRY. *Educ:* Mass Inst Technol, BS, 55; Univ Ill, MS, 58, PhD(plasma physics), 64. *Prof Exp:* Physicist & Nat Res Coun res grant, Boulder Lab, Nat Bur Stand, 64-65; prof physics, Dartmouth Col, 65-66; PHYSICIST, AERONOMY LAB, NAT OCEANIC & ATMOSPHERIC ADMIN, 66- *Mem:* Am Phys Soc; Am Geophys Union. *Res:* Use of spectroscopic, laser and chromatographic techniques for the measurement of trace atmospheric constituents; characterization of undisturbed atmospheric composition and anthropogenic contributions thereto. *Mailing Add:* 325 Broadway Boulder CO 80303

GOLDBERG, A JON, b Baltimore, Md, Feb 18, 47; m 75. DENTAL MATERIALS. *Educ:* Drexel Univ, BS, 70; Univ Mich, MS, 71, PhD(dent mat & metall eng), 77. *Prof Exp:* Asst prof, 75-80, ASSOC PROF RESTORATIVE DENT, SCH DENT MED, UNIV CONN, 81- *Honors & Awards:* Edward H Hatton Award, Int Asn Dent Res, 74. *Mem:* Int Asn Dent Res; Am Soc Metals; Am Soc Testing & Mat. *Res:* Structure-property relationships in polymers and metals of interest in the fields of dentistry and medicine; clinical evaluations of dental restorative materials. *Mailing Add:* Dept Prosthodontics Univ Conn Health Ctr Farmington CT 06032

GOLDBERG, AARON, b Brooklyn, NY, Nov 4, 17. TAXONOMIC BOTANY. *Educ:* Brooklyn Col, AB, 39; DePaul Univ, MS, 54; George Washington Univ, PhD(bot), 62. *Prof Exp:* Res clerk, Census of Agr, US Bur Census, 40-41; vet parasitologist, Zool Div, Bur Animal Indust, USDA, Md, 47-51, parasitologist in charge, Zool Lab, Ill, 51-53, vet parasitologist, Vet Sci Res Div, Nat Animal Parasite Lab, Agr Res Serv, 53-71, zoologist, 71-72; COLLABR, DEPT BOT, SMITHSONIAN INST, 72- *Mem:* Soc Syst Zool; Am Soc Plant Taxon. *Res:* Phylogeny and taxonomy of angiosperms at family level and above. *Mailing Add:* Dept Bot Smithsonian Inst Washington DC 20560

GOLDBERG, AARON JOSEPH, b New Haven, Conn, Nov 4, 43; m 69; c 2. ELECTRICAL ENGINEERING. *Educ:* Mass Inst Technol, SB & SM, 67, PhD(elec eng), 70. *Prof Exp:* Electronic engr, Defense Commun Agency, 70-72; SR ENG SPECIALIST, GTE SYLVANIA ELECTRONIC SYSTS GROUP, 72- *Concurrent Pos:* Adj prof, Northeastern Univ, Boston, 75-77; lectr, Univ Lowell, Boston, 78-; adj prof eng, Boston Univ, 79- *Mem:* Inst Elec Electronics & Engrs; Acoust Soc Am. *Res:* Voice, speech processing, digital signal processing, realtime signal processing, microprocessing. *Mailing Add:* Mitre Corp Burlington Rd MSA 378 Bedford MA 01770

GOLDBERG, ALAN HERBERT, b Boston, Mass, Nov 29, 31; m 58, 81; c 3. ANESTHESIOLOGY. *Educ:* Brown Univ, AB, 53; Boston Univ, MD, 57; Georgetown Univ, PhD(physiol), 65; Am Bd Anesthesiol, dipl, 63. *Prof Exp:* Intern surg, Univ Hosp, Boston Univ, 57-58; resident anesthesia, Mass Gen Hosp, 58-60; clin instr anesthesia & lectr pharmacol, Georgetown Univ, 63-65; asst prof anesthesiol, Sch Med, Boston Univ & asst anesthesiologist, Univ Hosp, 65-69, assoc prof & assoc anesthesiologist, 69-70, lectr physiol, 70; assoc prof anesthesiol, Harvard Med Sch, 70-75; prof anesthesiol, Sch Med, Boston Univ, 75-81; clin dir, Dept Anesthesiol, Boston City Hosp, 75-81; chief, dept anesthesiol, Mt Sinai Med Ctr, Milwaukee, Wis, 81-84; dir, postgrad med educ, Dept Anesthesiol, 84-89, PROF ANESTHESIOL, MED COL WIS, 81- *Concurrent Pos:* NIH career develop award, 66-70; consult, Chelsea Naval Hosp, 69-73; supv physician, Resucitation Serv, Cardiovasc Unit, 70-75, dir anesthesia res & assoc vis physician, Boston City Hosp, 70-75, act dir dept anesthesiol, 74-75, dir, 75-81; mem, teaching staff, Cardiovasc Block, Harvard Med Sch, 71-72; sr vis anesthesiologist, Univ Hosp, 78-81, actg chmn, Dept Anesthesiol, 80, dir, Anesthesiol Residency Prog, 80-81; assoc ed, Soc for Educ in Anesthesia; sr res assoc, Nuffield Dept Anaesthetics, Radcliffe Infirmary, Univ Oxford, Eng, 89; mem, Educ Meeting Comt, Soc Educ Anesthesia, 87-90, assoc ed, Anesthesia Educ, 88-, prog chmn, 90, treas, 90- *Mem:* Cardiac Muscle Soc; Asn Univ Anesthetists; Am Soc Anesthesiol. *Res:* Myocardial contractility; effects of anesthetics on circulatory system; effects of ischemia and drugs, including anesthetics, on the myocardium; evaluation of myocardial contractility; myocardial preservation. *Mailing Add:* Milwaukee County Med Ctr 8700 W Wisconsin Ave Milwaukee WI 53226

GOLDBERG, ALAN MARVIN, b New York, NY, Nov 20, 39; m 60; c 2. PHARMACOLOGY, TOXICOLOGY. *Educ:* Long Island Univ, BS, 61; Univ Minn, PhD, 66. *Prof Exp:* Res asst, Univ Wis, 61-62; res asst, Univ Minn, 62-66; asst prof pharmacol, Inst Psychiat, Ind Univ, 67-69; from asst prof to prof environ health sci, 69-78, dir div toxicol, 80-82, DIR CTR FOR ALTERNATIVES TO ANIMAL TESTING, JOHNS HOPKINS UNIV, 81-, ASSOC DEAN RES, SCH PUB HEALTH, 84- *Concurrent Pos:* Res fel, Inst Psychiat, Ind Univ, 66-67; prin res scientist, Chesapeake Bay Instit, 80-83. *Mem:* AAAS; Am Soc Pharmacol & Exp Therapeut; Soc Neurosci; Am Soc Neurochem; Int Soc Neurochem; Soc Toxicol; Am Col Toxicol. *Res:* Acetylcholine metabolism; drug action on the central nervous system; neurotransmitters and environmental toxicology; in vitro toxicology. *Mailing Add:* Assoc Dean for Res Sch Hyg Johns Hopkins Univ Baltimore MD 21205

GOLDBERG, ALFRED, b Montreal, Que, Oct 2, 23; nat US; m 50; c 3. PHYSICAL METALLURGY, STEEL TECHNOLOGY. *Educ:* McGill Univ, BEng, 46; Carnegie Inst Technol, MS, 47; Univ Calif, Berkeley, PhD(metall eng), 55. *Prof Exp:* Res engr phys metall, Univ Calif, 47-53; from asst prof to assoc prof metall, US Naval Postgrad Sch, 53-64; SR SCIENTIST, LAWRENCE LIVERMORE NAT LAB, 64- *Honors & Awards:* First Metallog Award, Int Metallog Soc. *Mem:* Fel Am Soc Metals; Sigma Xi. *Res:* Calorimetry of alloys; deformation of metals; phase transformations; corrosion and erosion in energy systems; mechanical properties of elastomers; superplasticity in steels. *Mailing Add:* 1220 Glenwood St Livermore CA 94550

GOLDBERG, ALFRED L, b Providence, RI, Sept 3, 42; m 70; c 2. PHYSIOLOGY, BIOCHEMISTRY. *Educ:* Harvard Univ, AB, 63, PhD(physiol), 68. *Prof Exp:* Churchill scholar, Cambridge Univ, 63-64; From instr to assoc prof, 68-76, PROF PHYSIOL, HARVARD MED SCH, 77- *Concurrent Pos:* Nat Res Coun fel, 68-69; Med Found Inc fel, 69-71; vis prof, Univ Calif Berkeley, 76; mem, Res Allocations Comn, Am Heart Asn, 76-79; sci bd dirs, Biogen Inc, 82; co-chmn, Fedn Am Soc Exp Biol Conf Protein Degradation, 89; Res Career Dev Award, NIH, 72-74. *Honors & Awards:* Wise & Helen Burroughs lectr, Univ Iowa, 70. *Mem:* AAAS; Am Physiol Soc; Soc Biol Chem; Soc Gen Physiol; Biochem Soc. *Res:* Regulation of growth and protein metabolism in mammalian tissues; mechanisms of muscle hypertrophy and atrophy; biochemical mechanisms of protein degradation in animal and bacterial cells; metabolic regulation; amino acid metabolism; hormonal regulation. *Mailing Add:* Dept Cellular & Molecular Physiol Harvard Med Sch 25 Shattuck St Boston MA 02115

GOLDBERG, ALLAN ROY, b New York, NY, Dec 4, 41; m 65; c 2. VIROLOGY. *Educ:* Cornell Univ, BA, 63; Princeton Univ, PhD(biochem & biol), 67. *Prof Exp:* Fel, Albert Einstein Col Med, 67-71; res assoc, 71-72, asst prof, 72-78, ASSOC PROF, ROCKEFELLER UNIV, 78- *Concurrent Pos:* Asst ed, J Exp Med, 77- *Mem:* Fedn Am Soc Exp Biol; Am Soc Virol; Am Soc Microbiol; Am Soc Cell Biol; NY Acad Sci; Sigma Xi. *Res:* Mechanism of transformation by oncogenic viruses. *Mailing Add:* Rockefeller Univ 1230 York Ave New York NY 10021

GOLDBERG, ANDREW PAUL, b New York, NY, Mar 12, 45; m; c 2. GERIATRICS. *Educ:* Clark Univ, BA, 65, State Univ New York, MD, 69; Am Bd Internal Med, cert, 75. *Prof Exp:* Sr res fel, Metab Sect, Vet Admin Hosp, Seattle, 74-76; from instr to asst prof, Dept Med, Univ Wash, 76-83; assoc prof med & head res, Div Geriat Med, Johns Hopkins Univ, 83-90, Francis Scott Key Med Ctr, 83-90; adj assoc prof, Dept Food & Nutrit, 84-91, ADJ PROF, COL PHYS EDUC, HEALTH & RECREATION, UNIV MD, COLLEGE PARK, 84-; ASSOC CHIEF STAFF, GERIAT SERV, VET ADMIN MED CTR, BALTIMORE, 90- *Concurrent Pos:* Res fel nutrit lipids, Univ Hosp, Seattle, 76-77; asst physician, Barnes Hosp, St Louis, 77-83; chmn, Nutrit Comt, Francis Scott Key Med Ctr, 85-87, prog dir, Gen Clin Res Ctr, 85-90; mem, Nat Comt Sci & Med Progs, Am Diabetes Asn, 89-90, Res Comt, 89-90. *Res:* Effects of obesity on lipid metabolism; effects of exercise training on lipid and carbohydrate metabolism in hemodialysis patients; exercise training in type II diabetes mellitus; exercise, blood pressure and metabolism in the elderly; author of numerous scientific publications. *Mailing Add:* Dept Geriat Vet Admin Med Ctr 3900 Loch Raven Blvd Baltimore MD 21218

GOLDBERG, ARTHUR H, b New York, NY, Jan 21, 35; m 62; c 2. PHYSICAL PHARMACY, BIOPHARMACEUTICS. *Educ:* Columbia Univ, BS, 56, MS, 65; Univ Mich, PhD(pharm chem), 68. *Prof Exp:* Asst prof biopharmaceut, Col Pharmaceut Sci, Columbia Univ, 68-72; dir pharmaceut res & develop, Hoffman-La Roche, Inc, 72-85; vpres res & develop, Chelsea Labs, Lakeview, NY; VPRES SCI AFFAIRS, RUGBY-DARBY GROUP CO, ROCKVILLE CTR, NY. *Concurrent Pos:* Adj prof, Col Pharm, Univ Ky, 77- *Mem:* Am Pharmaceut Asn; Acad Pharmaceut Sci; Am Chem Soc; Sigma Xi. *Res:* The application of physical chemistry towards the absorption, distribution, metabolism and elimination of drugs. *Mailing Add:* Rugby-Darby Group Co 100 Banks Ave Rockville Centre NY 11570

GOLDBERG, BENJAMIN, b New York, NY, Apr 14, 15; m 38, 85; c 3. PHYSICS. *Educ:* City Col New York, BS, 36. *Prof Exp:* Physicist, eng res & develop labs, US Army, 41-46, chief, reflector res lab, 47-49, proj engr infrared, 50-51, chief, illum sect, 51-52, chief, far infrared sect, 52-53, night vision equip br, 54-58, mine detection br, 58-62, barrier & intrusion detection div, 63-65, warfare vision div, 65, dep dir, night vision lab, 65-68, dir, 68-73; CONSULT GOVT & INDUST, 73- *Honors & Awards:* Except Civilian Serv Awards, US Army, 45 & 72, Meritorious Civilian Serv Awards, 69 & 70. *Mem:* AAAS; Optical Soc Am; Sigma Xi. *Res:* Infrared; illumination; reflector optics; mine detection; intrusion detection; image intensification. *Mailing Add:* 11410 Strand Dr Rockville MD 20852

GOLDBERG, BURTON DAVID, b Milwaukee, Wis, Jan 6, 27. PATHOLOGY, CELL BIOLOGY. *Educ:* Northwestern Univ, BS, 48, MD, 50; Am Bd Path, dipl, 56. *Prof Exp:* Intern, Cincinnati Gen Hosp, 51-52; instr path, Sch Med, Boston Univ, 53-55; instr, Tufts Univ, 54-55; from asst prof to prof path, Sch Med, NY Univ, 57-85; PROF & CHMN PATH, UNIV WIS-MADISON, 85- *Concurrent Pos:* Res fel, Nat Found Infantile Paralysis, 55-57; teaching fel, Harvard Med Sch, 56-57; USPHS career develop award, 60-70; resident, Mallory Inst Path, Boston City Hosp, 52-55; res assoc, Mass Inst Technol, 56-57; assoc pathologist, Bellevue Hosp & Univ Hosp, 57-85. *Mem:* Am Soc Exp Path; Am Soc Biochem & Molecular Biol; Am Soc Cell Biol. *Res:* Electron microscopy; cell biology; collagen biosynthesis; biochemistry. *Mailing Add:* Dept Path Univ Wis-Madison 470 N Charter St Madison WI 53706

GOLDBERG, COLMAN, solid state electronics; deceased, see previous edition for last biography

GOLDBERG, CONRAD STEWART, b New York, NY, June 10, 43; m 67; c 2. PHYSICS, COMPUTER SCIENCE. *Educ:* City Col New York, BS, 64; Queens Col, MA, 67; City Univ New York, PhD(physics), 74. *Prof Exp:* Prof physics, Queens Col, 65-75; asst dir, Psychol Corp, 76-81; syst mgr, Harcourt, Brace, Jovanovich, 81-82; PROF PHYSICS, NY INST TECHNOL, 75-; SYST DIR, TAMBRANDS, INC, 82- *Concurrent Pos:* Guest physicist, Brookhaven Nat Labs, 65-69 & Lawrence Radiation Lab, 68-69; systs analyst, Control Data Corp, 69-73; coordr educ testing, Harcourt, Brace, Jovanovich, 73-74; sr analyst, Grumman Data Syst, 74-76; consult ed, Faim Inc & Westinghouse Learning Corp, 75. *Res:* Defects in crystals; computer operating systems; education; computer systems design. *Mailing Add:* 11 Peppermill Ct Commack NY 11725

GOLDBERG, DAVID C, b Dallas, Tex, June 27, 21; m 43; c 3. METALLURGY, MATERIALS SCIENCE. *Educ:* Antioch Col, BS, 43. *Prof Exp:* Engr, Battelle Mem Inst, 41-44 & Saunders Mach & Tool, 46-47; sr engr, Hamilton Watch Co, 47-51; supvr metall, Aviation Gas Turbine Div, 51-54, mgr, 54-60, adv engr, 60-61, dir space mat, 61-62, mgr struct mat & processes, Astronuclear Lab, 62-66, mgr mat dept, 66-74, mgr eng dept, 74-76, mgr advan prog, Advan Energy Syst Div, 76-82, PRES CONSULT SERV, WESTINGHOUSE ELEC CORP, 82- *Concurrent Pos:* Mem comt toxicity of beryllium, Mat Adv Bd-Nat Acad Sci, 53-54, mem refractory metal sheet rolling panel C, 64-66, chmn tubing comt, 66-67, mem comt tech aspects of strategic mat, 67-71; mem res adv comt, mat, NASA, 62-71, chmn, 67-70; mem, NATO Summer Support Eval Conf Refractory Metals, 67; tech ed, Air Force Struct Mat Handbk, 68-; chmn design panel, Nat Mat Adv Bd, 76-78. *Mem:* Am Inst Aeronaut & Astronaut; Am Soc Metals; Am Soc Testing & Mat; Sci Res Soc Am. *Res:* Development of alternate energy technology; solar, magnetohydrodynamics, wind; hydrogen technologies. *Mailing Add:* 17091 Edgewater Dr Port Charlotte FL 33948

GOLDBERG, DAVID ELLIOTT, b Scranton, Pa, July 26, 32; m 59; c 2. INORGANIC CHEMISTRY. *Educ:* George Washington Univ, BS, 54; Pa State Univ, PhD(inorg chem), 59. *Prof Exp:* Instr chem, Pa State Univ, 58-59; from instr to assoc prof, 59-70, dep chmn dept, 63-72, chmn dept, 72-78, PROF CHEM, BROOKLYN COL, 70- *Concurrent Pos:* F G Cottrell grant, 60-61; City Univ NY grants, 64-65 & 76-77; Nat Inst Gen Med spec fel, Univ Sussex, 66-67. *Mem:* Am Chem Soc; Royal Soc Chem; Sigma Xi. *Res:* Stability and bonding in coordination compounds; solvent extraction of heavy metal ions by coordination; x-ray diffraction of coordination compounds. *Mailing Add:* Dept Chem Brooklyn Col Brooklyn NY 11210

GOLDBERG, DAVID MYER, b Glasgow, UK, Aug 30, 33; m 64; c 2. BIOCHEMISTRY. *Educ:* Univ Glasgow, BSc, 58, PhD(biochem), 65, MD, 74. *Prof Exp:* House surgeon, Stobhill Hosp, Glasgow, 59-60; sr house officer biochem, Western Infirmary, Glasgow, 60-61, registrar, 61-63, sr registrar, 63-67; hon lectr, Univ Sheffield, 67-75; biochemist-in-chief, Hosp Sick Children, Toronto, 75-89; chmn, 77-89, PROF DEPT CLIN BIOCHEM, UNIV TORONTO, 77- *Concurrent Pos:* House physician, Southern Gen Hosp, Glasgow, 60; consult chem path, United Sheffield Hosps, 67-75; joint ed-in-chief, Clin Biochem. *Honors & Awards:* Van Slyke Award, Am Asn Clin Chem, 82; Roman Award, Australian Asn Clin Biochemists, 84; Lion of Venice Award for contribs to int med, Ital Ministry Health, 84; Nova Idea Int Prize Lab Med, Ital Asn Clin Pathologists, 85. *Mem:* Int Soc Clin Enzymol; Biochem Soc; Am Asn Clin Chem; Can Soc Clin Chemists; Can Biochem Soc. *Res:* Enzymes in diagnosis in liver, pancreatic disease, and heart disease; regulatory aspects of pancreatic enzyme secretion; enzymes in cancer with special emphasis upon regulation of glycolysis; microsomal enzyme induction; biosynthesis and peripheral catabolism of tracylglycerol. *Mailing Add:* Nine Harrison Rd Toronto ON M2L 1V3 Can

GOLDBERG, EDWARD B, b Bronx, NY, July 19, 35; wid; c 2. MOLECULAR GENETICS & TRANSPORT. *Educ:* Columbia Univ, BA, 56; Johns Hopkins Univ, PhD(biol), 61. *Prof Exp:* Jr instr biol, Johns Hopkins Univ, 56-57; guest investr bacteriophage genetics, Carnegie Inst, 61-63, fel, Genetics Res Unit, 63-65; asst prof microbiol, 65-69, assoc prof molecular biol & microbiol, 69-76, PROF MOLECULAR BIOL & MICROBIOL, SCH MED & DENT, 76-, HEAD OF SCI ADV BD APPL MICROBIOL, TUFTS UNIV, 89- *Concurrent Pos:* Fel, Nat Found Med Res, 61-63; NIH career develop awards, 65-75; vis prof, Hadassah Med Sch, Hebrew Univ, Israel, 69-70, Med Res Coun, Lab Molecular Biol, Cambridge, Eng, 76-77 & Dept Microbiol Ecol, Hebrew Univ, 83-84; Guggenheim fel, 83-84. *Mem:* AAAS; Am Soc Biol Chemists; Genetics Soc Am. *Res:* Mechanisms of control of phage attachment and DNA injection, restriction, recombination and transcription of bactgeriophage; ion transport in bacteria. *Mailing Add:* Dept Molecular Biol & Microbiol Tufts Univ 136 Harrison Ave Boston MA 02111

GOLDBERG, EDWARD D, b Sacramento, Calif, Aug 2, 21; m 45, 72; c 4. MARINE GEOCHEMISTRY. *Educ:* Univ Calif, BS, 42; Univ Chicago, PhD(chem), 49. *Prof Exp:* From asst prof to assoc prof, 49-60, PROF CHEM, SCRIPPS INST OCEANOG, UNIV CALIF, SAN DIEGO, 60- *Concurrent Pos:* Guggenheim fel, Univ Bern, 61; NATO fel, Univ Brussels, 70. *Honors & Awards:* Tyler Prize. *Mem:* Nat Acad Sci; Geochem Soc; Am Geophys Union; AAAS. *Res:* Geochemistry of marine waters; marine sedimentation; meteoritics; radiochemistry; atmospheric and marine pollution. *Mailing Add:* Scripps Inst Oceanog La Jolla CA 92093

GOLDBERG, EDWIN A(LLEN), b Dallas, Tex, Sept 13, 16; m 42; c 2. ELECTRONICS. *Educ:* Univ Tex, BS, 38, MS, 40. *Prof Exp:* Seismic engr, Magnolia Petrol Co, 38-39; asst elec eng, Univ Tex, 39-40; engr, Res Dept, 40-42, mem tech staff, Res Labs, 42-57, mgr missile & weapon systs, Defense Electronic Prods Div, Spec Systs Develop Dept, 57-58, eng & admin, Astro-Electronics Div, 58, design, 58-59, space vehicle systs, 59-62, spacecraft design & testing, 62-64, spacecraft tests, 64-65 & spec projs, 65-75, prof mgr hydrogen clock, RCA Astro Electronics, 75-77, MGR PROD & SYST SAFETY & MISSION ASSURANCE, DYNAMIC EXPLOR PROG, RCA ASTRO ELECTRONICS, RCA CORP, 78- *Mem:* Inst Elec & Electronics Engrs. *Res:* Electronic analog computer components and systems; color television systems; satellite and space electronics systems; hydrogen maser frequency standard. *Mailing Add:* RCA Corp Astro Electronics PO Box 800 Princeton NJ 08540

GOLDBERG, ERWIN, b Waterbury, Conn, Jan 14, 30; m 51, 85; c 3. BIOCHEMISTRY, REPRODUCTIVE BIOLOGY. *Educ:* State Univ NY, BA, 51; Univ Iowa, MS, 53, PhD(zool, biochem), 56. *Prof Exp:* Res assoc zool, Univ Iowa, 56-58; asst prof biol, WVa Univ, 58-61; asst prof zool, NDak State Univ, 61-63; from asst prof to prof biol, 63-74, PROF BIOCHEM, MOLECULAR BIOL & CELL BIOL, NORTHWESTERN UNIV, EVANSTON, 74- *Mem:* AAAS; Am Soc Biol Chemists; Am Soc Androl; Soc Study Reproduction; Int Soc Immunol Reproduction. *Res:* Biochemistry and immunology of mammalian male germ cells; reproductive immunology; developmental biology; immunocontraception. *Mailing Add:* Dept Biochem & Molecular & Cell Biol Northwestern Univ Evanston IL 60208

GOLDBERG, ESTELLE MAXINE, b Las Vegas, Nev, Apr 19, 34; m 54; c 2. MATHEMATICS. *Educ:* Univ Calif, Los Angeles, BA, 53; Columbia Univ, MA, 59, PhD(math), 65. *Prof Exp:* From asst prof to prof math, San Francisco State Univ, 62-79; RETIRED. *Res:* Algebra; homological algebra. *Mailing Add:* 110 Pine Needle Ln Altamonte Springs FL 32714

GOLDBERG, EUGENE, b Chicago, Ill, May 29, 27; c 2. NUCLEAR PHYSICS. *Educ:* Ill Inst Technol, BS, 48, MS, 50; Univ Wis, PhD(physics), 53. *Prof Exp:* Asst physics, Ill Inst Technol, 48-50; Alumni Res Found asst, Univ Wis, 50-51; sr physicist, Livermore Proj, 53-65, exp physics div leader, 65-73, assoc leader, B Div, 73-79, STAFF PHYSICIST, LAWRENCE LIVERMORE NAT LAB, UNIV CALIF, 79- *Mem:* Fel Am Phys Soc; Am Nuclear Soc. *Res:* Experimental nuclear physics; interaction of charged particles with light nuclei, using Van de Graaff generator; nuclear and thermonuclear weapon development; neutronic design of epithermal homogeneous reactors; experimental studies of heavy ion interactions; radiobiological studies involving fast neutrons; radiation dosimetry measurements; foreign energy technology assessment. *Mailing Add:* Lawrence Livermore Lab PO Box 808 L-22 Livermore CA 94550

GOLDBERG, EUGENE P, b Southampton, NY, Nov 11, 28; m 50; c 3. BIOMATERIALS, BIOPOLYMERS. *Educ:* Univ Miami, BS, 50; Ohio Univ, MS, 51; Brown Univ, PhD(chem), 53. *Prof Exp:* Res & develop chemist, Gen Elec Co, 53-60; assoc dir & head, Chem Res Dept, Res Ctr, Borg-Warner Corp, 60-65; dir, Chem Res Lab, Rochester Res Ctr, Xerox Corp, 66-75; PROF MAT SCI & PHARMACOL & DIR, BIOMED ENG CTR, UNIV FLA, 75- *Concurrent Pos:* Vis prof, Hebrew Univ, 72; mem solid state sci adv panel, Nat Acad Sci-Nat Res Coun, 72-80 & Mat Res Adv Comt, NSF, 80-86; US-USSR sci exchange prog, US Nat Acad Sci, 76; Japan Soc Promotion Sci fel, Kyoto Univ, 82. *Mem:* AAAS; Am Chem Soc; Sigma Xi; Am Acad Ophthal; Acad Surg Res; Mat Res Soc; Am Soc Cataract & Refractive Surg. *Res:* Polymer science; biopolymers; biomedical materials and devices; biomaterials; surgical research; biosurfaces; polymer surface modification; hydrophilc polymers; polymer biocompatibility; targeted drug delivery. *Mailing Add:* Dept Mat Sci & Eng Biomed Eng Ctr Univ Fla MAE 317 Gainesville FL 32611

GOLDBERG, GERSHON MORTON, b Paterson, NJ, Apr 5, 24; m 51; c 3. PHOTOGRAPHIC CHEMISTRY, MICROLITHOGRAPHY. *Educ:* Pa State Col, BS, 44, MS, 47, PhD(org chem), 49. *Prof Exp:* Fel, Northeastern Univ, 49-50; res chemist, petrol, Daugherty Refining Div, L Sonnenborn Sons, Inc, 50-54; res chemist, Tech Opers Inc, 54-75; prin scientist, Aerodyne Res, Inc, 75-78; sr scientist, GCA Corp, 81-85; prin scientist, Ionomet Res Corp, 78-87, proj mgr, 87-88; STAFF SCIENTIST, INNOVATIVE IMAGING SYSTS, INC, 88- *Honors & Awards:* Serv Award, Soc Imaging Sci & Technol, 75. *Mem:* AAAS; Am Chem Soc; Sigma Xi; Soc Imaging Sci & Technol; Royal Photog Soc Gt Brit. *Res:* Solubility of quaternary ammonium compounds; organosilicon compounds; petroleum sulfonates; sensitizing dyes; thin films; unconventional photographic systems; microencapsulation; microlithography; photoresists; infrared fiber optics. *Mailing Add:* 31 Grand View Rd Arlington MA 02174

GOLDBERG, HARRY, b Philadelphia, Pa, May 19, 18; m 46; c 2. CARDIOLOGY. *Educ:* Univ Pa, AB, 39; Long Island Col Med, MD, 44; Am Bd Internal Med, dipl, 53; Am Subspecialty Bd, dipl, 54. *Prof Exp:* Intern, Mt Sinai Hosp, Albert Einstein Med Ctr, 44-45; resident path, Metrop Hosp, NY, 45-46 & 48; resident internal med, Mt Sinai Hosp, Pa, 48-49; asst prof physiol, Hahnemann Med Col, 51-53, asst prof med, 53-60; dir, Cardiovasc Dept, 60-84, EMER CHMN CARDIOVASC DEPT, ALBERT EINSTEIN MED CTR, 84-; PROF MED, SCH MED, TEMPLE UNIV, 69- *Concurrent Pos:* Fel cardiovasc dis, Michael Reese Hosp, Ill, 50; dir cardiopulmonary lab, Hahnemann Med Col, 51-60; dir cardiol dept, Deborah Heart & Lung Ctr, 58-85, vpres-in-chg med affairs, 73-81; assoc prof, Temple Univ, 63-69. *Mem:* AAAS; Am Physiol Soc; Am Heart Asn (pres, 73); fel Am Col Physicians; fel Am Col Chest Physicians; fel Am Col Cardiol; Sigma Xi. *Res:* Clinical cardiovascular research; cardiology. *Mailing Add:* Albert Einstein Med Ctr N Div York and Tabor Rds Philadelphia PA 19141

GOLDBERG, HERBERT SAM, b New York, NY, July 23, 26; m 48; c 2. MICROBIOLOGY, MEDICAL EDUCATION. *Educ:* St Johns Univ, NY, BS, 48; Univ Mo, MA, 50; Ohio State Univ, PhD(bact), 53. *Prof Exp:* Asst bact, Ohio State Univ, 50-53; from asst prof to assoc prof microbiol, Sch Med, Univ Mo, Columbia, 53-61, chmn dept, 65-66, asst to dean, 66-67, asst dean, 67-71, dir sch health related prof, 78-83, assoc dean, 71-83, assoc vpres res, 85-86, PROF MICROBIOL, SCH MED, UNIV MO, COLUMBIA, 61-, ASSOC DEAN RES & ACAD AFFAIRS, 87- *Concurrent Pos:* Wellcome Trust traveling fel, 60 & 64; NIH grant electron micros, 62; vis scientist, Cambridge Univ, 60; vis prof, Southern Ill Univ, 61, Univ Calif, Los Angeles & Rutgers Univ, 84; consult, WHO, 62-72. *Honors & Awards:* Byler Admin Award, 76. *Mem:* AAAS; Am Soc Microbiol; NY Acad Sci; Soc Appl Bact; Brit Soc Gen Microbiol. *Res:* Antibiotic assay techniques; nontherapeutic uses of antibiotics; chemotherapy of Leptospiroses; Bacteroides identification; biomedical research and education administration. *Mailing Add:* Sch Med Univ Mo Columbia MO 65212

GOLDBERG, HOWARD S, b Chicago, Ill, Mar 26, 36; m 64; c 2. PARTICLE PHYSICS. *Educ:* Univ Mich, Ann Arbor, BSEE, 58; Univ Calif, Berkeley, PhD(physics), 64. *Prof Exp:* Asst prof, Tuskegee Inst, 64-66; asst prof, 66-72, assoc prof, 72-80, PROF PHYSICS, UNIV ILL, CHICAGO CIRCLE, 80- *Mem:* Am Phys Soc; Am Asn Physics Teachers. *Res:* Strong and weak interactions in elementary particle high energy physics. *Mailing Add:* Dept of Physics Univ Ill Box 4348 Chicago IL 60680

GOLDBERG, HYMAN, b Montreal, Que, May 21, 39; m 65. PHYSICS. *Educ:* McGill Univ, BSc, 59; Mass Inst Technol, PhD(physics), 63. *Prof Exp:* Instr physics, Brandeis Univ, 63-64, res assoc, 64; Nat Res Coun Can overseas fel, Theoret Physics Lab, Orsay, France, 64-65; instr & res assoc, Cornell Univ, 65-67; from asst prof to assoc prof, 67-75, PROF PHYSICS, NORTHEASTERN UNIV, 75- *Mem:* Am Phys Soc. *Res:* Elementary particle physics. *Mailing Add:* Dept Physics Northeastern Univ Boston MA 02115

GOLDBERG, IRA BARRY, b Brooklyn, NY, Apr 10, 43; m 66; c 3. PHYSICAL CHEMISTRY. *Educ:* Adelphi Univ, AB, 64; Univ Minn, PhD(phys chem), 69. *Prof Exp:* Fel chem, Univ Tex, Austin, 69-71; mem tech staff, 71-84, MGR PHYS CHEM, ROCKWELL INT SCI CTR, 84- *Mem:* Am Chem Soc; Am Phys Soc; Soc Magnetic Resonance; Inst Elec & Electronics Eng. *Res:* Electron spin resonance; magnetics and magnetic materials; ferromagnetic resonance; electrochemistry; chemical kinetics; radical ions; microwave measurement and instrumentation. *Mailing Add:* 54 Westbury Ct Thousand Oaks CA 91360

GOLDBERG, IRVIN H(YMAN), b Hartford, Conn, Sept 2, 26; m 56; c 2. MEDICINE, PHARMACOLOGY. *Educ:* Trinity Col, Conn, BS, 49; Yale Univ, MD, 53; Rockefeller Inst, PhD(biochem), 60. *Hon Degrees:* AM, Harvard Univ, 64. *Prof Exp:* Intern med, Columbia-Presby Med Ctr, NY, 53-54, asst resident, 54-56, chief resident, 56-57; from asst prof to assoc prof med & biochem, Univ Chicago, 60-64; assoc prof med, 64-68, chmn div med sci, 68-70, Gustavus Adolphus Pfeiffer prof pharmacol, 72-83, chmn dept pharmacol, 72-86 PROF MED, HARVARD MED SCH, 68-, OTTO KRAYER PROF, DEPT BIOL CHEM & MOLECULAR PHARMACOL, 83- *Concurrent Pos:* Am Cancer Soc fac res assoc award, 60-71; Guggenheim Mem Found fel, Oxford Univ, 70-71; instr, Col Physicians & Surgeons, Columbia Univ, 56-57; mem, BC Res Coun Can, 59; chief endocrinol metab unit, Beth Israel Hosp, Boston, 64-68, physician, 64-72, mem bd consult med, 72-; mem res comt, Med Found, Inc, Boston, 68-; clin consult, Harvard Med Serv, Boston City Hosp, 72-73; consult clin pharmacol, Children's Hosp Med Ctr, 72-91; sr fel, Bd Fels, Trinity Col, 73-76; mem exp therapeut study sect, NIH, 74-77; mem comt proposed legis restruct food & drug admin, Assembly Life Sci, Nat Acad Sci-Nat Res Coun & Inst Med, 76; physician, Div Med Oncol, Sidney Farber Cancer Inst, Boston, 80- *Mem:* AAAS; Am Soc Microbiol; Am Soc Pharmacol & Exp Therapeut; Brit Pharmacol Soc; Am Soc Biol Chem; Am Acad Arts & Sci; Am Chem Soc; Am Soc Clin Invest; Asn Am Physicians; Inst Med Nat Acad Sci. *Res:* Molecular mechanism of action of antibiotics affecting nucleic acid and protein synthesis and function; DNA damage and repair. *Mailing Add:* Dept Chem & Molecular Pharmacol Harvard Med Sch Boston MA 02115

GOLDBERG, IRVING DAVID, b Boston, Mass, July 28, 21; m 67. MENTAL HEALTH SERVICES. *Educ:* City Col New York, BS, 42; Univ Mich, MPH, 54. *Prof Exp:* Jr statistician, New York City Dept Health, 46-48; biostatistician, NY State Dept Health, 48-50, sr biostatistician, 50-57; asst chief, Biomet Br, Nat Inst Neurol Dis & Blindness, 57-66; chief eval studies sect, Biomet Br, NIMH, 66-73, chief, Appl Biomet Res Br, 73-82; RETIRED. *Mem:* Am Statist Asn; fel Am Pub Health Asn; Soc Epidemiol Res; fel Am Col Epidemiol. *Res:* Statistical and epidemiological research relating to occurrence and distribution of diseases, and evaluation of the delivery of health and mental health services. *Mailing Add:* 5101 River Rd Apt 402 Bethesda MD 20816

GOLDBERG, IVAN D, b Philadelphia, Pa, May 13, 34; m 79; c 5. MICROBIAL GENETICS, MOLECULAR GENETICS. *Educ:* Univ Pa, AB, 56; Univ Ill, PhD(microbiol), 61. *Prof Exp:* Fel microbial genetics, Inst Microbiol, Rutgers Univ, 61-62, Ore State Univ, 62-63 & Army Biol Labs, Ft Detrick, 63-64, microbial geneticist, 64-71; assoc prof, 71-77, PROF MICROBIOL, UNIV KANS MED CTR, KANSAS CITY, 77- *Mem:* Am Soc Microbiol; Sigma Xi. *Res:* Microbial genetics, including transformation and transduction; bacteriophages and lysogeny; plasmids; genetics of Neisseria gonorrhoeae; genetic bases of enzyme thermostability. *Mailing Add:* Dept Microbiol Univ Kans Med Ctr 39th & Rainbow Blvd Kansas City KS 66103

GOLDBERG, JACOB, b San Francisco, Calif, June 4, 26; m 48; c 4. COMPUTER SCIENCE, ELECTRICAL ENGINEERING. *Educ:* Univ Calif, BS, 50; Stanford Univ, MS, 54. *Prof Exp:* SR SCI CONSULT, SRI INT, 51- *Mem:* Fel Inst Elec & Electronics Engrs; Asn Comput Mach. *Res:* Computers; fault-tolerant computing, software engineering. *Mailing Add:* SRI Int Menlo Park CA 94025

GOLDBERG, JAY M, b Chicago, Ill, Nov 9, 35; c 4. NEUROPHYSIOLOGY. *Educ:* Univ Chicago, AB & SB, 56, PhD(psychol), 60. *Prof Exp:* From asst prof to assoc prof physiol, 63-72, PROF PHYSIOL & NEUROBIOL, UNIV CHICAGO, 72- *Concurrent Pos:* NSF fel, 60-62; NIH trainee, 62-63. *Honors & Awards:* Javits Neurosci Investr Award. *Mem:* Fel Acoustical Soc Am; Soc Neurosci; AAAS; Barany Soc; Asn Res Otolaryngol; Am Soc Gravitation Space Biol. *Res:* Vestibular neurophysiology. *Mailing Add:* Dept Pharmacol & Physiol Sci Univ Chicago Chicago IL 60637

GOLDBERG, JOSEPH, b New York, NY, Apr 30, 23; m 45; c 4. MECHANICAL & AERONAUTICAL ENGINEERING. *Educ:* NY Univ, BAE, 44; Princeton Univ, MSE, 59. *Prof Exp:* Aerodynamicist, Chance-Vought Aircraft, 44; flight test engr, Wright Field Air Develop Ctr, 44-46; aerodynamicist, Fairchild Aircraft, 46-47; sr aerodynamicist, Glenn L Martin Co, 47-49; aerodyn design specialist, Naval Air Develop Ctr, 49-52; res assoc flight dynamics, Princeton Univ, 52-60; asst prof aeronaut eng, Univ Ill, 60-61; assoc prof mech eng, Drexel Inst, 61-65; ASSOC PROF MECH ENG, VILLANOVA UNIV, 65- *Concurrent Pos:* Consult, Curtiss-Wright Corp, Boeing Vertol Div & Aeronaut Res Assocs, Princeton Univ, 52-60, Thiokol Chem Corp & Piasecki Aircraft, 61- *Mem:* AAAS; Am Inst Aeronaut & Astronaut; Sigma Xi; Am Soc Mech Engrs. *Res:* Flight dynamics of missiles and fixed-wing and rotary wing aircraft; automatic controls; systems dynamics. *Mailing Add:* 327 Vassar Ave Swarthmore PA 19081

GOLDBERG, JOSEPH LOUIS, b New York, NY, Sept 11, 09. ORGANIC CHEMISTRY. *Educ:* City Col New York, BS, 30; Columbia Univ, AM, 33, PhD(biochem), 47. *Prof Exp:* Microanalyst, Rockefeller Inst, 34-37; from instr to asst prof chem, City Col New York, 51-81; RETIRED. *Mem:* Am Chem Soc. *Res:* Quantitative organic microanalysis; amino acids and proteins; isolation of antibiotics. *Mailing Add:* PO Box L New York NY 10021

GOLDBERG, JOSHUA NORMAN, b Rochester, NY, May 30, 25; m 49; c 2. PHYSICS. *Educ:* Univ Rochester, AB, 47; Syracuse Univ, MS, 50, PhD(physics), 52. *Prof Exp:* Physicist, Armour Res Found, Ill Inst Technol, 52-56 & gen physics br, Aeronaut Res Lab, US Air Force, 56-63; chmn dept, 76-83, PROF PHYSICS, SYRACUSE UNIV, 63- *Concurrent Pos:* Instr, Grad Ctr, Ohio State Univ, 58-59; NSF sr fel, King's Col, Univ London, 60-61; adj assoc prof, Univ Cincinnati, 62-63; vis prof, Israel Inst Technol, Haifa, 71, Inst Henri Poincaré, Paris, 83-84; Int Comt on Gen Relativity & Gravitation, 83-92. *Mem:* AAAS; Am Phys Soc; Sigma Xi; Int Soc Gen Relativity & Gravitation. *Res:* General relativity; cosmology; theoretical physics in general. *Mailing Add:* Dept Physics Syracuse Univ Syracuse NY 13210

GOLDBERG, KENNETH PHILIP, b New York, NY, Dec 9, 45; m 77; c 3. MATHEMATICS, STATISTICS. *Educ:* NY Univ, BA, 67, MS, 69; Mich State Univ, PhD(math), 73. *Prof Exp:* Comput programmer, AEC Comput Ctr, NY Univ, 64; asst instr math, Mich State Univ, 69-72; assoc prof, 73-82, PROF MATH EDUC, NY UNIV, 82- *Concurrent Pos:* Consult math, Nat Lab Higher Educ, 73-75; instr math, Proj Upward Bound, VI, 74. *Mem:* Nat Coun Teachers Math; Math Asn Am; Am Math Soc. *Res:* Mathematics education making use of electronic hand-held calculators and microcomputers; technology. *Mailing Add:* Dept Math Educ NY Univ 239 Greene St New York NY 10003

GOLDBERG, LAWRENCE SPENCER, b St Louis, Mo, June 11, 40; m 69; c 3. QUANTUM OPTICS, LASERS. *Educ:* Wash Univ, BS, 61; Cornell Univ, PhD(physics), 66. *Prof Exp:* Res asst solid state physics, Phys Inst, Univ Frankfurt, 66-67; RES PHYSICIST, OPTICAL SCI DIV, OPTICAL PROBES, NAVAL RES LAB, 67- *Concurrent Pos:* Vis scientist, Dept Physics, Imperial Col, Univ London, 76-77; dir, prog quantum electronics, waves & beams, NSF, 85-86. *Mem:* Am Phys Soc; Optical Soc Am; Sigma Xi. *Res:* Picosecond spectroscopy; nonlinear optics and laser physics research. *Mailing Add:* Nat Sci Found Quantum Electron Waves & Beams Prog Washington DC 20550

GOLDBERG, LEE DRESDEN, b Pt Pleasant, NJ, July 29, 37; m 67; c 3. ENDOCRINOLOGY. *Educ:* Yale Univ, BS, 59; MD, 63. *Prof Exp:* Intern, Mt Sinai Hosp, New York, 63-64; resident internal med, Montefiore Hosp, New York, 64 & 66-68; USPHS fel endocrinol, Albert Einstein Col Med, 68-69; fel, Bellevue Hosp, NY Univ Med Ctr, 69-70; clin instr, Sch Med, Univ Miami, 70-71, clin asst prof, 71-80; chief internal med, S Shore Hosp, 75-79; CO-CHIEF ENDOCRINOL, MT SINAI HOSP, GREATER MIAMI, 74-; CLIN ASSOC PROF MED, SCH MED, UNIV MIAMI, 80- *Concurrent Pos:* Med officer, US Navy, 64-66; teaching asst med, Sch Med, NY Univ, 69-70; assoc chmn med serv, St Francis Hosp, 77-78; pract endocrinologist, 70- *Mem:* Am Fedn Clin Res; Am Physicians Fel Med Israel; Am Diabetes Asn; Endocrine Soc; Sigma Xi. *Res:* Diabetes mellitus; treatment of type II diabetes; calcium disorders; thyroid problems. *Mailing Add:* 1680 Meridian Ave Suite 204 Miami Beach FL 33139

GOLDBERG, LEON ISADORE, pharmacology, internal medicine; deceased, see previous edition for last biography

GOLDBERG, LEONARD H, b Pretoria, S Africa, Apr 27, 45. SKIN CANCER, MOHS SURGERY. *Educ:* Univ Pretoria, MBCB, 67; Royal Col Physicians, UK, MRCP, 72. *Prof Exp:* Res fel dermat, Stanford Univ, 76-79, surg, NY Univ 79-80; ASSOC PROF DERMAT, BAYLOR COL MED, 81- *Mem:* Am Acad Dermat; Soc Invest Dermat; Am Soc Dermat Surg; Int Soc Dermat Surg; Am Col Mohs Surg & Cutaneous Oncol. *Res:* Skin cancer, including basal cell cancer, squamores cell cancer and melanoma; effects of sunlight on the skin. *Mailing Add:* Dept Dermat Baylor Col Med, One Baylor Plaza Houston TX 77030

GOLDBERG, LOUIS J, b Middletown, NY, July 20, 36; m 63. NEUROSCIENCE, DENTISTRY. *Educ:* Brooklyn Col, BA, 56; NY Univ, DDS, 60; Univ Calif, Los Angeles, PhD(anat), 68. *Prof Exp:* From asst prof to assoc prof oral biol, 68-72, from asst prof to assoc prof anat, 69-72, dean res, sch dent, 72-79, PROF ORAL BIOL, SCH DENT, UNIV CALIF, LOS ANGELES, 77-, PROF ANAT, SCH MED, 77- *Concurrent Pos:* Nat Inst Dent Res fels, NY Univ, 61-62 & Sch Dent, Univ Calif, Los Angeles, 64-68; NIH career develop award, 68; USPHS res grant, 68- *Mem:* Sigma Xi. *Res:* Neurophysiological studies in oral pharyngeal reflexes. *Mailing Add:* Sch Dent Univ Calif Ctr for Health Sci Los Angeles CA 90024

GOLDBERG, MARK ARTHUR, b New York, NY, Sept 4, 34; m 69; c 1. PHARMACOLOGY, NEUROLOGY. *Educ:* Columbia Univ, SB, 55; Univ Chicago, PhD(pharmacol), 59, MD, 62. *Prof Exp:* Res assoc & instr pharmacol, Univ Chicago, 59-62; intern, Bronx Munic Hosp, NY, 62-63; resident neurol, Presby Hosp, 63-66; mem staff, US Army Med Res Lab, Md, 66-67; asst prof neurol, Col Physicians & Surgeons, Columbia Univ, 68-71; assoc prof neurol & pharmacol, 71-78, CHMN DEPT NEUROL, HARBOR GEN HOSP, 78-; PROF NEUROL & PHARMACOL, UNIV CALIF, LOS ANGELES, 78- *Concurrent Pos:* USPHS fel, 64-66. *Mem:* Am Acad Neurol; Am Soc Neurochem; Am Neurol Asn; Soc Neuroscience. *Res:* Neuropharmacology; epilepsy; stroke; aids; neuroscience. *Mailing Add:* Dept Neurol Harbor-UCLA Med Ctr 1000 W Carson St Torrance CA 90509

GOLDBERG, MARTIN, b Philadelphia, Pa, Sept 15, 30; m 51, 78; c 4. NEPHROLOGY. *Educ:* Temple Univ, BA, 51, MD, 55. *Hon Degrees:* MS, Univ Pa, 70. *Prof Exp:* Rotating intern, Philadelphia Gen Hosp, 55-56; asst resident internal med, Cleveland Clin Hosp, 56-57; resident, Philadelphia Gen Hosp, 57-58, sr resident, 58-59; from asst prof to assoc prof, Chem Sect, Univ Pa Sch Med, 59-67, assoc prof to prof med, Renal-Electrolyte Sect, 67-79, chief sect, 66-79; prof & chmn, Dept Internal Med, Univ Cincinnati Sch Med, 79-86; dean, 86-89, PROF MED & NEPHROLOGY, SCH MED, TEMPLE UNIV, 86- *Concurrent Pos:* NIH res career develop award, 62-70, res grant, 65-90 & training grant, 69-79; Hoechst Found res award, 66-72; Lederle Fund res award, 66-70; John Hartford Found res grant, 70-73; clin investr, Gen Clin Res Ctr, Hosp Univ Pa, 61-79, mem attend staff, 62-79; chmn coun on kidney in cardiovasc dis, Am Heart Asn; mem sci adv bd, Nat Kidney Found, 73-76; chmn nephrology subcomt, Am Bd Internal Med, 77-80; mem gov coun, Int Soc Nephrology, 79-85; prin investr, Gen Clin Res Ctr grant, NIH, Temple Univ Hosp, 86-90. *Honors & Awards:* Res Prize, Philadelphia Gen Hosp, 59. *Mem:* AAAS; Am Soc Clin Invest; fel Am Col Physicians; Am Soc Nephrology (secy-treas, 75-78); Int Soc Nephrology; Sigma Xi. *Res:* Renal physiology and pathophysiology; renal pharmacology; action of diuretics; regulation of sodium excretion; renal regulation of calcium and phosphate transport; action of parathyroid hormone on the kidney; computer-assisted diagnosis and teaching. *Mailing Add:* Parkinson Bldg Fifth Floor Temple Univ Hosp Philadelphia PA 19140

GOLDBERG, MARTIN A, b New York, NY, Sept 19, 29; m 51; c 2. APPLIED MECHANICS. *Educ:* NY Univ, BAE, 51; Univ Buffalo, MS, 55; Rensselaer Polytech Inst, PhD(appl mech), 58. *Prof Exp:* Struct engr, Bell Aircraft Corp, 53-55; instr appl mech, Rensselaer Polytech Inst, 55-58; res engr, Grumman Aircraft Corp, 58-62; specialist engr, Repub Aviation Corp, 62-64; assoc prof appl mech, 64-76, assoc prof mech & aerospace eng, Polytech Inst NY, 76-; PROF ENG, WEBB INST NAVAL ARCHIT. *Mem:* Assoc Am Soc Mech Eng; Soc Eng Sci; assoc Soc Naval Archit & Marine Eng. *Res:* Solid mechanics; applied elasticity, plates, shells and elastic stability; heat conduction. *Mailing Add:* 14 Chestnut Hill Roslyn NY 11576

GOLDBERG, MARVIN, b New York, NY, Sept 14, 39; m 61; c 2. HIGH ENERGY PHYSICS. *Educ:* City Col New York, BS, 60; Syracuse Univ, PhD(physics), 65. *Prof Exp:* From res asst to res assoc, 60-66, asst prof, 66-69, assoc prof, 69-74, PROF PHYSICS, SYRACUSE UNIV, 74- *Mem:* Am Phys Soc; Sigma Xi. *Res:* Experimental study of elementary particles and their interactions, especially classification of these particles and measurement of their quantum numbers. *Mailing Add:* Dept Physics Syracuse Univ Syracuse NY 13244

GOLDBERG, MELVIN LEONARD, b Chester, Pa, June 19, 32; m 54; c 2. BIOCHEMISTRY, PATHOLOGY. *Educ:* Calif Inst Technol, BA, 54; Univ Ill, Urbana, MA, 55; NY Univ, PhD(biochem), 62; Univ Calif, San Francisco, MD, 66; Am Bd Path, cert anat & clin path, 79, cert chem path, 82. *Prof Exp:* Intern, Sch Med, Univ Calif, San Francisco, 66-67, asst res pathologist, 67-69, asst clin prof path, 69-79, resident, 77-78; DIR CLIN CHEM, FLA HOSP, 79- *Concurrent Pos:* NIH & Univ Calif Cancer Res Coord Comt res grants, 68-78; NIH career develop award, 71-76. *Mem:* Am Asn Pathologists; Col Am Pathologists; Am Soc Clin Pathologists. *Res:* Role of cyclic nucleotides in disease; role of essential fatty acids in atherosclerosis. *Mailing Add:* Dept Path Fla Hosp 601 E Rollins St Orlando FL 32803

GOLDBERG, MERRILL B, b Minneapolis, Minn, June 14, 43; m 77; c 5. MATHEMATICS, COMPUTER SCIENCES. *Educ:* Univ Minn, Minneapolis, BA, 64; Univ Calif, San Diego, MA, 65, PhD(math), 69. *Prof Exp:* Specialist math, San Diego City Schs, 68-69; asst prof, Univ Colo, Boulder, 69-73; PROF MATH, ROCKHURST COL, MO, 73- *Concurrent Pos:* Consult, Proj Unified Sci & Math for Elementary Schs, 70-75. *Mem:* Math Asn Am; Asn Computer Math. *Res:* Abstracting Lp spaces; Orlicz spaces; math analysis; math education; math economics; micro-computers and software; co-author Pascal textbook. *Mailing Add:* Dept Math Computer Sci & Physics Rockhurst Col 1100 Rockhurst Rd Kansas City MO 64110-2508

GOLDBERG, MICHAEL ELLIS, b New York, NY, Aug 10, 41; m 66; c 2. NEUROLOGY, NEUROPHYSIOLOGY. *Educ:* Harvard Col, AB, 63, MD, 68. *Prof Exp:* House officer med, Peter Bent Brigham Hosp, Boston, 68-69; staff assoc neurol, NIMH, 69-72; resident, Children's Hosp Med Ctr, Boston, 72-74; res med officer, Armed Forces Radiobiol Res Inst, 74-78; from asst prof to assoc prof, 76-87, PROF NEUROL, GEORGETOWN UNIV, 87- *Concurrent Pos:* Fel neurol, Harvard Univ, 72-75; vis investr, Lab Neurobiol, NIMH, 74-78; res med officer neurol, 78-81, chief, Sect Neuro-ophthal Mechanisms, Lab Sensorimotor Res, Nat Eye Inst, NIH, 81- *Honors & Awards:* S Weir Mitchell Award, Am Acad of Neurol, 72. *Mem:* Soc Neurosci; Am Acad Neurol; Asn Res Vision & Ophthal; Am Neurol Asn. *Res:* Neurophysiology of vision, eye movements, and behavior in the primates; computer methods in neurophysiology. *Mailing Add:* 80 Post Rd Churchville PA 18966-1160

GOLDBERG, MICHAEL IAN, b New York, NY, Mar 18, 44; c 1. TECHNICAL MANAGEMENT, MOLECULAR BIOLOGY. *Educ:* Union Col, BS, 65; Yale Univ, PhD, 70. *Prof Exp:* Teaching fel biochem & biophys, Univ Calif, San Francisco, 70-75; health scientist adminr, Nat Inst Gen Med Sci, NIH, 75-78; sr health policy analyst, Off of Secy, Dept Health & Human Serv, 78-79; exec asst to comnr policy coord, Food & Drug Admin, 79-80; br chief, Div Legis Anal, NIH, 80-81, div dir, 81-84, assoc dir, Prog Planning & Eval, 82-84; EXEC DIR, AM SOC MICROBIOL, 84- *Concurrent Pos:* Damon Runyon fel; Gianini Found fel. *Honors & Awards:* Dir Award, NIH. *Mem:* Sigma Xi; Am Chem Soc; AAAS. *Mailing Add:* 7508 Hampden Lane Bethesda MD 20814-1332

GOLDBERG, MORTON EDWARD, b Philadelphia, Pa, July 11, 32; m 54; c 3. PHARMACOLOGY. *Educ:* Philadelphia Col Pharm, BS, 54, MS, 55, DSc, 58. *Prof Exp:* Sr res pharmacologist, Abbott Labs, 58-60; res fel, Mellon Inst, 60-63, sr fel, 64; sr pharmacologist, Union Carbide Corp, 65-67, from asst dir to dir pharmacol, 67-69; head, Gen Pharmacol Sect, Warner Lambert Res Inst, 69-71, dir pharmacodynamics, 71-73; dir pharmacol, Squibb Inst Med Res, 73-77; dir, 77-81, vpres biomed res, 81-84, VPRES RES, DEVELOP & REGULATORY AFFAIRS, STUART PHARMACEUT, ICI AMERICAS, INC, 84- *Concurrent Pos:* Adj assoc prof pharmacol, Med Col Pa, 78-; adj prof toxicol, Philadelphia Col Pharm, 82- *Mem:* Fel AAAS; fel NY Acad Sci; Am Pharmaceut Asn; Am Soc Pharmacol & Exp Therapeut; Soc Toxicol; Col Int Neuropsychopharmacol. *Res:* Central nervous system and behavioral pharmacology; neuropharmacology; neurochemistry; general drug development activities. *Mailing Add:* ICI Pharmaceut Group Div ICI Americas Inc Wilmington DE 19897

GOLDBERG, MORTON FALK, b Lawrence, Mass, June 8, 37; m 68. MEDICINE, OPHTHALMOLOGY. *Educ:* Harvard Univ, AB, 58, MD, 62; Am Bd Ophthal, dipl, 68. *Prof Exp:* Intern med, Peter Bent Brigham Hosp, 62-63; resident ophthal, Johns Hopkins Hosp, 63-67; asst prof, Med Sch, Yale Univ, 67-69; pres, Univ Ill Hosp Med Staff, 81-83, PROF OPHTHAL & HEAD DEPT, COL MED, UNIV ILL & OPHTHALMOLOGIST-IN-CHIEF, UNIV HOSP & EYE & EAR INFIRMARY, 70- *Concurrent Pos:* Fel ophthal, Johns Hopkins Hosp, 63-67, fel genetics, 69; prog dir, Nat Eye Inst training grants, 70; consult, Vet Admin Hosp, Chicago, 70-; Macy Found Fac Scholar Award, 79-80; vis prof, Pac Med Ctr, San Francisco & Mayo Clin, Rochester, Minn, 79; ed-in-chief, Arch Ophthal, 84- *Honors & Awards:* Sr Hon Award, Am Acad Ophthal, 85. *Mem:* Am Ophthal Soc; Asn Res Vision & Ophthal; Am Acad Ophthal & Otolaryngol; Retina Soc; Macula Soc; hon fel Royal Australian Col Ophthal. *Res:* Retinopathies; ocular genetics; diabetes mellitus; sickle cell retinopathy; ocular trauma; laser photocoagulation; retinal and vitreous surgery. *Mailing Add:* 111EE Infirm Dept Opthal Rm 250 1855 W Taylor Chicago IL 60612

GOLDBERG, MORTON HAROLD, oral surgery, maxillofacial surgery, for more information see previous edition

GOLDBERG, NELSON D, b Cleveland, Ohio, June 22, 31; m 58; c 2. BIOCHEMISTRY. *Educ:* Univ Toledo, BS, 53; Univ Wis, PhD(pharmacol), 63. *Hon Degrees:* DSc, Univ Toledo, 81. *Prof Exp:* From instr to prof pharmacol, 64-86, PROF PATH, 73-, PROF BIOCHEM, MED SCH, UNIV MINN, MINNEAPOLIS, 87- *Concurrent Pos:* NIH fel, Wash Univ, 62-64; Pfizer traveling fel, 78. *Mem:* Am Soc Neurochem; Am Soc Biochem & Molecular Biol; Am Chem Soc; Biophys Soc. *Res:* Regulation of cellular metabolism and function: excitation/contraction coupling, excitation/secretion coupling, phototransduction, adenine and guanine nucleotide metabolism. *Mailing Add:* 4-225 Millard Hall Dept Biochem Univ Minn Minneapolis MN 55455-0326

GOLDBERG, NORMAN, b Philadelphia, Pa, Aug 5, 21; m 63; c 3. PHYSICS. *Educ:* Pa State Teachers Col, West Chester, BS, 43; Univ Pa, MS, 48, PhD(physics), 54. *Prof Exp:* Res assoc, Wash Univ, 54-56; asst prof physics, Univ Pa, 56-59; PHYSICIST, UNIVAC DIV, SPERRY RAND CORP, 59 - *Mem:* Am Phys Soc; Inst Elec & Electronics Engrs; Optical Soc Am. *Res:* Nuclear physics; magnetism; thin films; magnetooptics; fiber optics. *Mailing Add:* 1580 Tralee Dr Dresher PA 19025

GOLDBERG, PAULA BURSZTYN, b Siedlce, Poland, Jan 9, 38. CARDIOVASCULAR RESEARCH. *Educ:* State Univ NY, PhD(pharmacol), 68. *Prof Exp:* From asst prof to assoc prof pharmacol, Med Col Pa, 72-85; ASSOC SR INVESTR CLIN RES & DEVELOP, SMITHKLINE FRENCH LABS, 85- *Concurrent Pos:* Vis assoc prof, Med Col Pa, 86- *Mem:* Am Soc Pharmacol & Exp Therapeut; Geront Soc Am; NY Acad Sci; Sigma Xi; Soc Exp Biol & Med. *Mailing Add:* Smith Kline & French Labs Clin R & D 709 Swedeland Rd L 223 Swedeland PA 19479

GOLDBERG, R J, b Lowell, Mass, Mar 20, 50; m 78; c 2. CARDIOVASCULAR EPIDEMIOLOGY, EPIDEMIOLOGIC METHODOLOGY. *Educ:* Univ Mass, BS, 72; Tufts Univ, MSPH, 73; Johns Hopkins Univ Sch Hyg & Pub Health, PhD(epidemiol), 78. *Prof Exp:* Fel epidemiol, Johns Hopkins Univ Sch Hyg & Pub Health, 78-79; asst prof pub health, Ohio State Univ Sch Med, 79-81; asst prof, 81-85, ASSOC PROF EPIDEMIOL, UNIV MASS MED SCH, 85- *Mem:* Soc Epidemiol Res; Am Pub Health Asn; Int Epidemiol Asn; Am Heart Asn; Europ Soc Prev Cardiol; Am Col Epidemiol. *Res:* Cardiovascular epidemiology; primary and secondary prevention of coronary heart disease; epidemiologic methodology. *Mailing Add:* Dept Med Univ Mass Med Sch Worcester MA 01605

GOLDBERG, RICHARD ARAN, b Boston, Mass, Jan 6, 36; m 65; c 1. ATMOSPHERIC PHYSICS. *Educ:* Rensselaer Polytech Inst, BS, 57; Pa State Univ, PhD(physics), 63. *Prof Exp:* Nat Acad Sci-Nat Res Coun resident res assoc, 63-64, PHYSICIST, GODDARD SPACE FLIGHT CTR, NASA, 64- *Concurrent Pos:* Mem comn G, US Nat Comt, Int Sci Radio Union. *Mem:* AAAS; Am Phys Soc; Am Geophys Union; Am Meteorol Soc; Sigma Xi. *Res:* Sounding rocket studies of the upper atmosphere with mass spectrometry, in-situ probes, radio propagation techniques, air-glow measurements; mesospheric and stratospheric aeronomy; solar-terrestrial relationships; middle atmospheric electrodynamics. *Mailing Add:* 1020 Kathryn Rd Silver Spring MD 20904

GOLDBERG, RICHARD ROBINSON, b Chicago, Ill, Sept 6, 31; m 53; c 4. MATHEMATICS. *Educ:* Northwestern Univ, BS, 51; Harvard Univ, AM, 52, PhD(math), 56. *Prof Exp:* From instr to prof, Northwestern Univ, 57-68; prof math, Univ Iowa, 68-76, chmn dept, 70-76; PROF MATH & CHMN DEPT, VANDERBILT UNIV, 76- *Concurrent Pos:* Managing ed, Proceedings Am Math Soc. *Mem:* Am Math Soc; Math Asn Am; Soc Indust & Appl Math. *Res:* Fourier analysis; integral transforms. *Mailing Add:* Dept Math Vanderbilt Univ Nashville TN 37240

GOLDBERG, ROBERT JACK, b Denver, Colo, May 22, 14; m 40; c 3. ZOOLOGY. *Educ:* Univ Ill, AB, 37; Ill Inst Technol, PhD(biol), 54. *Prof Exp:* Instr biol, Chicago City Jr Col, 41-54; assoc, Chicago Teachers Col, 54-59, prof, 59-64, dean grad studies, 62-64; prof biol, 64-77, vpres acad affairs & dean fac, 64-77, EMER PROF BIOL, NORTHEASTERN ILL UNIV, 77- *Mem:* Genetics Soc Am; Soc Study Evolution; Soc Syst Zool; Sigma Xi. *Res:* Taxonomy and speciation of Gastrotrichia. *Mailing Add:* 211 Elgin Apt 4C Forest Park IL 60130

GOLDBERG, ROBERT N, computer science, for more information see previous edition

GOLDBERG, ROBERT NATHAN, b Stamford, Conn, Dec 11, 43. PHYSICAL CHEMISTRY. *Educ:* Johns Hopkins Univ, BA, 65; Carnegie-Mellon Univ, PhD(chem), 68. *Prof Exp:* Asst, Carnegie-Mellon Univ, 65-68, fel electrolyte solutions, Mellon Inst, 68-69; CHEMIST, NAT INST STANDARDS TECHNOL, 69- *Concurrent Pos:* Guest mem staff, Univ Newcastle, Tyne, Eng, 80. *Mem:* Am Chem Soc; Am Inst Chemists; Calorimetry Conf. *Res:* Chemical thermodynamics and its applications to biochemistry and electrolyte solutions; microcalorimetry; solution calorimetry; thermodynamic data evaluation. *Mailing Add:* 21404 Davis Mill Rd Germantown MD 20874

GOLDBERG, SABINE RUTH, b Erlangen, Ger, Mar 25, 58; US citizen; m 80. SOIL CHEMISTRY, ENVIRONMENTAL CHEMISTRY. *Educ:* Univ Fla, BSA, 77; Univ Calif, Riverside, PhD(soil sci), 83. *Prof Exp:* Res asst, dept soil sci, Univ Calif, Riverside, 78-83, grad regents fel, 80-82; SOIL SCIENTIST, US SALINITY LAB, AGR RES SERV, USDA, RIVERSIDE, CALIF, 83- *Mem:* Soil Sci Soc Am; Am Chem Soc; Am Geophys Union; Clay Minerals Soc; Am Soc Agron. *Res:* Inorganic anion adsorption reactions on oxide minerals, clay minerals and soil materials; chemical modeling of anion adsorption reactions using chemical surface complexation models; chemical factors affecting soil structural stability. *Mailing Add:* US Salinity Lab Agr Res Serv USDA 4500 Glenwood Dr Riverside CA 92501

GOLDBERG, SAMUEL, b New York, NY, Mar 14, 25; m 53; c 1. MATHEMATICS. *Educ:* City Col New York, BS, 44; Cornell Univ, PhD(math), 50. *Prof Exp:* From instr to asst prof math, Lehigh Univ, 50-53; from asst prof to assoc prof, 53-60, PROF MATH, OBERLIN COL, 60- *Concurrent Pos:* Vis assoc prof, Grad Sch Bus Admin, Harvard Univ, 59-60. *Mem:* Math Asn Am; Opers Res Soc Am; Am Statist Asn. *Res:* Mathematical theory of probability; mathematical models in the social sciences. *Mailing Add:* Alfred P Sloan Fdn 630 Fifth Ave, Ste 2550 New York NY 10111

GOLDBERG, SAMUEL I, b Toronto, Ont, Aug 15, 23; nat US; m 51; c 3. MATHEMATICS. *Educ:* Univ Toronto, BA, 48, MA, 49, PhD, 51. *Prof Exp:* Sci officer, Defence Res Bd, Ottawa, 51-52; asst prof math, Lehigh Univ, 52-55; assoc prof, Wayne State Univ, 55-61; vis assoc prof, 60-61, assoc prof, 61-65, PROF MATH, UNIV ILL, URBANA, 65- *Concurrent Pos:* Res fel, Harvard Univ, 59-60; consult, Avco Corp, 63-66; NSF grant, 64-; on sabbatical, Univ Calif, Berkeley, 66-67 & Israel Inst Technol, 71-72, 79; vis prof, Univ Toronto, 68, Cambridge Univ & Col de France, 79, Queens Univ & Univ Lecce, 87; ed-at-large, Marcel Dekker, Inc, 69- & Tensor J, 74-; Sci Res Coun vis foreign scientist, Univ Liverpool, 73; Lady Davis fel, Israel Inst Technol, 79; adj prof, Queen's Univ, 81- & Queen's Quest Prof, 81; at Ctr Advan Study, Univ Ill, 83-84. *Mem:* Am Math Soc. *Res:* Differential geometry; topology; analysis. *Mailing Add:* Dept of Math Univ of Ill Urbana IL 61801

GOLDBERG, SEYMOUR, b Brooklyn, NY, Mar 24, 28; m 52; c 2. MATHEMATICS. *Educ:* Hunter Col, AB, 50; Ohio State Univ, MA, 52; Univ Calif, Los Angeles, PhD(math), 58. *Prof Exp:* Asst math, Ohio State Univ, 50-52; math analyst, Lockheed Aircraft Corp, 52-54; asst math, Univ Calif, Los Angeles, 54-58; Brown fel, Hebrew Univ, Israel, 58-59; asst prof math, NMex State Univ, 59-62; assoc prof, 62-67, PROF MATH, UNIV MD, 67- *Mem:* Am Math Soc; Math Asn Am. *Res:* Functional analysis, particularly theory of linear operators on a normed linear space. *Mailing Add:* Dept of Math Univ of Md College Park MD 20742

GOLDBERG, SEYMOUR, b Boston, Mass, July 12, 27; m 48; c 2. ELECTRICAL ENGINEERING, ELECTRONICS. *Educ:* Northeastern Univ, BS, 48; Mass Inst Technol, MS, 49. *Prof Exp:* Asst, Mass Inst Technol, 48-49; engr res & develop, Sylvania Elec Prod, Inc, 49-51; engr & dir tube res & develop, Edgerton, Germehausen & Grier, Inc, 51-64, dir component res & develop, 64-65, staff scientist, EG&G, Inc, 65-71, DIV STAFF ENGR, EG&G, INC, 71- *Mem:* Sr mem Inst Elec & Electronics Engrs; Sigma Xi. *Res:* Gas discharges and vacuum tube technology; hydrogen thyratrons; electron optics and wide band width transient instrumentation and electron devices; marine seismic exploration and semiconductor detectors; atomic frequency standards. *Mailing Add:* 1409 Sheffield Way Saugus MA 01906

GOLDBERG, STANLEY, b Cleveland, Ohio, Aug 4, 34; div; c 3. PHYSICS, HISTORY OF PHYSICS. *Educ:* Antioch Col, BS, 60; Harvard Univ, AMT, 61, PhD(educ), 69. *Prof Exp:* Asst prof hist sci, Antioch Col, 65-71; assoc prof, 71- 79, prof hist sci, Hampshire Col, 79-85; STAFF MEM, SMITHSONIAN INST, 85- *Concurrent Pos:* Sr lectr, Sci Educ Ctr, Univ Zambia, 68-70. *Mem:* AAAS; Hist Sci Asn. *Res:* History of physics of the 19th and 20th century, especially electrodynamics; science education for non-scientists in high school and college. *Mailing Add:* 504 Third St SE Washington DC 20003

GOLDBERG, STANLEY IRWIN, b New York, NY, Apr 12, 30; m 56; c 3. BIO-ORGANIC CHEMISTRY, ORGANIC CHEMISTRY. *Educ:* Univ Md, BS, 53; Ind Univ, PhD, 58. *Prof Exp:* Res assoc & instr chem, Univ Ill, 59-60; from asst prof to assoc prof chem, Univ SC, 60-71; PROF CHEM, UNIV NEW ORLEANS, 71- *Concurrent Pos:* Vis prof, Univ Bristol, 69-70. *Honors & Awards:* Matheson, Coleman & Bell Award, 71. *Mem:* Am Chem Soc; Royal Soc Chem. *Res:* Origins of the configurationally one-sidedness of life. *Mailing Add:* Dept of Chem Univ of New Orleans New Orleans LA 70148

GOLDBERG, STEPHEN, b Brooklyn, NY, Dec 20, 42; m 67; c 4. NEUROANATOMY, NEUROEMBRYOLOGY. *Educ:* Yeshiva Col, BA, 63; Albert Einstein Col Med, MD, 67. *Prof Exp:* Intern, Montefiore Hosp, Bronx, NY, 67-68; surgeon neurol, Staten Island USPHS Hosp, 68-70; resident ophthal, New York Med Col, 71-75, asst, 73-75; asst prof anat & family med, 75-79, ASSOC PROF ANAT, SCH MED, UNIV MIAMI, 80- *Concurrent Pos:* Guest researcher, Nat Eye Inst, NIH, 70-71; dir, Ambulatory Care Unit, Jackson Mem Hosp, 78; prin investr grants, NIH, 79-82. *Mem:* Int Soc Develop Neurosci; Am Asn Anatomists; Asn Res Vision & Ophthal; AAAS; Soc Neurosci. *Res:* Mechanisms of development, regeneration, and degeneration of the nervous system with special emphasis on the visual system. *Mailing Add:* M D Albert Einstein Col Med 2900 Bridgeport Ave Coconut Grove FL 33133

GOLDBERG, STEPHEN ROBERT, b New York, NY, Mar 4, 41; m; c 1. ZOOLOGY. *Educ:* Boston Univ, BA, 62; Univ Ariz, MS, 65, PhD(zool), 70. *Prof Exp:* From asst prof to assoc prof, 70-83, PROF BIOL, WHITTIER COL, 83- *Concurrent Pos:* Res assoc herpet, Los Angeles County Natural Hist Mus, 76- *Mem:* Am Soc Parasitol; Helm Soc Wash; Am Inst Biol Sci; Am Soc Ichthyologists & Herpetologists; Soc Study Amphibians & Reptiles; Wildlife Dis Asn; Herpetologist's League; Entom Soc Am. *Res:* Herpetology; vertebrate reproductive cycles; comparative pathology; parasitology; entomology. *Mailing Add:* Dept Biol Whittier Col Whittier CA 90608

GOLDBERG, STEPHEN ZALMUND, b Brooklyn, NY, May 26, 47. INORGANIC CHEMISTRY, STRUCTURAL CHEMISTRY. *Educ:* Cornell Univ, AB, 68; Univ Calif, Berkeley, PhD(chem), 73. *Prof Exp:* Instr & fel, Univ Rochester, 73-75; asst prof, 75-80, assoc prof chem, Adelphi Univ, 80-86, vis scientist, Weizmann Inst, 82-83; PROF, ADELPHI UNIV, 86- *Mem:* NY Acad Sci; Am Crystallog Asn; Am Chem Soc; AAAS. *Res:* Computer programming synthetic and structural studies of transition metal compounds and complexes; structural studies of drug compounds. *Mailing Add:* Dept of Chem Adelphi Univ Garden City NY 11530

GOLDBERG, STEVEN R, b Boston, Mass, July 5, 41. PHARMACOLOGY. *Educ:* Northeastern Univ, BA, 64; Univ Mich, DPhil, 69. *Prof Exp:* NIH fel, Dept Psychol & Lab Organismic & Quant & Biol, Univ Miami, Coral Gables, 69; res fel pharmacol & psychobiol, Harvard Med Sch, Boston, 70-73, from instr to asst prof, Dept Psychiat, 73-80; pharmacologist, Lexington, Ky, 79-81, PHARMACOLOGIST, NAT INST DRUG ABUSE, ADDICTION RES CTR, BALTIMORE, MD, 81-, CHIEF, BEHAV PHARMACOL & GENETICS LAB & PRECLIN PHARMACOL BR, 84-; ASSOC PROF, DEPT PHARMACOL & EXP THERAPEUT, SCH MED, UNIV MD, 81- *Concurrent Pos:* Assoc scientist, New Eng Regional Primate Res Ctr, 73-80; actg chief, Neuropsychopharmacol Lab, Nat Inst Drug Abuse, Addiction Res Ctr, 86-87. *Mem:* Am Soc Pharmacol & Exp Therapeut; AAAS; Behav Pharmacol Soc; NY Acad Sci. *Res:* Pharmacology; author of numerous scientific publications. *Mailing Add:* Pharmacol Br Nat Inst Drug Abuse Addiction Res Ctr PO Box 5180 Baltimore MD 21224

GOLDBERG, VLADISLAV V, b Moscow, USSR, Jan 4, 36; US citizen; m 58; c 2. GEOMETRY. *Educ:* Moscow State Univ, BS, 56, MS, 58, PhD(math), 61. *Prof Exp:* Assoc prof math, Yaroslavl' State Pedagogical Inst, USSR, 61-64; prof, Inst Steel & Alloys, Moscow, 64-78; vis prof, Lehigh Univ, 79-81; prof, 81-85, DISTINGUISHED PROF MATH, NJ INST TECHNOL, 85- *Concurrent Pos:* Sr sci ed, Pub House MIR, Moscow, 61; sr sci consult, Metall Lab, Inst Steel & Alloys, Moscow, 68-78; prin invest geom grant, NSF, 80-82; vis prof, Univ Waterloo, 80, Univ Stuttgart, WGer, 82, Univ Messina, Italy, 86; corresp mem, Academia Peloritana, Messina, Italy, 83-; co-chmn, web geom session, Math Res Inst, Oberwolfach, WGer, 84. *Honors & Awards:* Harlan J Perlis Res Award, 85. *Mem:* Am Math Soc; Tensor Soc Japan. *Res:* Projective differential geometry and web geometry; linear algebra, tensor calculus and engineering mathematics; mathematical education; author of 7 books and more than 60 science papers. *Mailing Add:* Dept Math NJ Inst Technol Newark NJ 07102

GOLDBERG, WALTER M, b Mass, Jan 9, 46; m 71; c 2. CORAL BIOLOGY, ENVIRONMENTAL IMPACTS. *Educ:* Am Univ, BS, 68; Fla Atlantic Univ, MS, 70; Univ Miami, PhD(marine biol), 73. *Prof Exp:* Postdoctoral biochem, Papanicolaou Cancer Res Inst, 73; from asst prof to assoc prof, 73-88, PROF ZOOL, FLA INT UNIV, 88-, CHAIR BIOL SCI, 88- *Mem:* Western Soc Naturalists; AAAS. *Res:* Coelenterate biology; ecology, biochemistry, histology and ultrastructure of corals; community structure, effects of nearshore construction on, effects of beach restoration and sedimentation on reefs. *Mailing Add:* Fla Int Univ Tamiami Trail Miami FL 33199

GOLDBERGER, AMY, b Memphis, Tenn, Aug 15, 57. CHROMATIN BIOCHEMISTRY, HEMATOPOIETIC DIFFERENTIATION. *Educ:* Univ Tenn, BA, 79; Vanderbilt Univ, PhD(biochem), 83. *Prof Exp:* Fel, dept biochem & molecular biol, Mayo Clin, 84-87; INSTR CELL BIOL, UNIV MASS SCH MED, 87- *Mem:* Sigma Xi; Am Soc Cell Biol. *Res:* The role of chromatin proteins in gene expression, using hematopoietic differentiation and steroid hormone action as model systems. *Mailing Add:* Investr Blood Ctr Southeastern Wis 1701 W Wisconsin Ave Milwaukee WI 53233

GOLDBERGER, MARVIN LEONARD, b Chicago, Ill, Oct 22, 22; m 45; c 2. THEORETICAL HIGH ENERGY PHYSICS. *Educ:* Carnegie Inst Technol, BS, 43; Univ Chicago, PhD(physics), 48. *Hon Degrees:* DSc, Carnegie-Mellon Univ & Univ Notre Dame, 79; DHL, Hebrew Union Col, 80, & Univ Judaism, 82; LLD, Occidental Col, 80. *Prof Exp:* Physicist, Radiation Lab, Univ Calif, 48-49; res assoc physics, Mass Inst Technol, 49-50; from asst prof to prof physics, Univ Chicago, 50-57; Higgins prof math physics, Princeton Univ, 57-77, chmn dept, 70-76, Joseph Henry prof physics, 77-78; pres, Calif Inst Tech, 78-87; dir, Inst Adv Studies, Princeton, NJ, 87-91; PROF PHYSICS, UNIV CALIF, LOS ANGELES, 91. *Concurrent Pos:* Higgins vis assoc prof, 53-54; mem, President's Sci Adv Comt, 65-69; chmn, Nat Acad Sci Comt Int Security & Arms Control, 87-; Super Collider Site Eval Comt, Nat Acads Sci & Eng, 87-; mem, Adv Coun, Plasma Physics Lab, Princeton Univ, 87- *Honors & Awards:* Dannie Heineman Prize, 61. *Mem:* Nat Acad Sci; Am Phys Soc; Am Acad Arts & Sci; Fedn Am Scientists; Inst Adv Study (bd mem, 86). *Mailing Add:* Dept Physics Univ Calif Los Angeles CA 90024

GOLDBERGER, MICHAEL ERIC, b New York, NY, Oct 21, 35; div. EXPERIMENTAL NEUROLOGY. *Educ:* Brooklyn Col, BA, 58; Emory Univ, MS, 62; Univ Pa, PhD(anat), 64. *Prof Exp:* USPHS fel, 66-67; asst prof neuroanat, Univ Chicago, 67-73; assoc prof, 73-80, PROF ANAT, MED COL PA, 80- *Concurrent Pos:* USPHS gen res support fel, 67-69, career develop award, 68-73; Pa Plan for Develop Scientists in Med Res fel, 67-68; NSF res grant psychobiol, 68-70. *Mem:* AAAS; Am Asn Anatomists. *Res:* Restitution of function and sprouting following lesions of the motor systems; methods of increasing restitution through training and secondary lesions; anatomical and behavioral analysis of motor function. *Mailing Add:* Dept Anat Med Col Pa 3300 Henry Ave Philadelphia PA 19129

GOLDBERGER, ROBERT FRANK, b New York, NY, June 2, 33; m 58; c 3. BIOCHEMISTRY. *Educ:* Harvard Univ, AB, 54; NY Univ, MD, 58. *Prof Exp:* Intern, Mt Sinai Hosp, New York, 58-59; trainee, Inst Enzyme Res, Univ Wis, 59-61; vis scientist, Weizmann Inst, 63; med officer & biochemist, Lab Chem Biol, Nat Inst Arthritis, Metab & Digestive & Kidney Dis, 63-66, chief biosynthesis & control sect, 66-73, actg chief, 69-70, chief, Lab Biochem, Nat Cancer Inst, 73-79; dep dir sci, NIH, 79-81; vpres health sci, 81-83, PROF BIOCHEM & MOLECULAR BIOPHYS, COLUMBIA UNIV, 81-, PROVOST, 81- *Mem:* Am Fedn Clin; Am Soc Biol Chemists; Am Soc Cell Biol; Am Soc Microbiol; Biophys Soc. *Res:* Protein chemistry; biochemical genetics and evolution; biological regulatory mechanisms; enzyme induction and repression; mechanism of action of steroid hormones; structure of the eukaryotic genome. *Mailing Add:* Columbia Univ 205 Low Library New York NY 10027

GOLDBERGER, W(ILLIAM) M(ORGAN), b Perth Amboy, NJ, Dec 26, 28; m 56; c 4. CHEMICAL ENGINEERING, METALLURGY. *Educ:* Ga Inst Technol, BChE, 50; Polytech Inst Brooklyn, MS, 56, PhD(chem eng), 61. *Prof Exp:* Res engr, Appl Res Labs, US Steel Corp, 56-57; instr chem eng, Stevens Inst Technol, 59-60; sr chem engr, 60-65, chief, Div Minerals & Metall Processing, 65-77, SR RES LEADER, COLUMBUS DIV, BATTELLE MEM INST, 77-, MEM RES COUN, 78-; dir res & develop, 79-89, VPRES RES, SUPER GRAPHITE CO, 89- *Mem:* Am Inst Chem Engrs; Am Inst Mining, Metall & Petrol Engrs; Am Ceramic Soc; Am Carbon Soc. *Res:* High and ultra-high temperature chemical process development; development of new processing for separating and purifying chemical compounds; mineral processing and assessment of mineral resources and methods for recovery; carbon/graphite production technology. *Mailing Add:* 2175 E Broad St Columbus OH 43209

GOLDBLATT, IRWIN LEONARD, b Bronx, NY, Jan 22, 40; m 62; c 3. TRIBOLOGY, CHEMICAL ENGINEERING. *Educ:* City Col New York, BChE, 62; Brandeis Univ, MA, 67, PhD(chem), 73. *Prof Exp:* Chemist, Polaroid Corp, 62-64; chemist tribology & combustion sci, Exxon Corp, 68-86; CHEMIST TRIBOLOGY & COMBUSTION SCI, CASTROL INC, 87- *Concurrent Pos:* Consult, Sanders Assocs, 64-68. *Honors & Awards:* W D Hudson Mem Award, Am Soc Lubrication Eng, 74. *Mem:* Am Chem Soc; Am Soc Lubrication Eng; Am Soc Mech Engrs; Am Soc Testing & Mat. *Res:* Study of the interactions between lubricant base stock, additives and metal in order to derive an understanding of how to control wear; combustion research in internal combustion engines; polymer grafting; analytical chemistry; mechanical engineering. *Mailing Add:* Eight Celler Rd Edison NJ 08817

GOLDBLATT, LEO ARTHUR, organic chemistry, for more information see previous edition

GOLDBLATT, PETER, b Johannesburg, SAfrica Oct 8, 43; US citizen. SYSTEMATIC BOTANY. *Educ:* Univ Witwatersrand, BSc, 65, BSc Hons, 66; Univ Cape Town, PhD(bot), 70. *Prof Exp:* Jr lectr bot, Univ Cape Town, 68-72; res botanist, 72-74, B A KRUKOFF CUR AFRICAN BOT, MO BOT GARDEN, 74- *Mem:* SAfrican Asn Botanists; Am Soc Plant Taxonomists; Int Asn Plant Taxonomists. *Res:* Systematics and evolution of Iridaceae; cytology of angiosperms in relation to phylogeny; African flora, especially floristics of southern Africa and the Cape region. *Mailing Add:* Mo Bot Garden PO Box 299 St Louis MO 63166

GOLDBLATT, PETER JEROME, PATHOLOGY. *Educ:* Case Western Reserve Univ, AB, 55, MD, 59; Am Bd Path, dipl anat. *Prof Exp:* From instr to asst prof, Dept Path, Sch Med, Univ Pittsburgh, 64-69; assoc prof, Univ Conn, 69-77; prof, Sch Med, Univ Md, 77-79; PROF & CHMN, DEPT PATH, MED COL OHIO, TOLEDO, 79- *Concurrent Pos:* Mem staff, Univ-McCook Hosp, Hartford, Conn, 70-74; mem, Conn Cancer Epidemiol Coun, 73-75; asst med examr, State Conn, 76-77; assoc chmn educ, Dept Path, Sch Med, Univ Md, 77-79; pres, Northwest Ohio Med Mgt, Inc, 84-; consult. *Mem:* Am Soc Clin Pathologists; AAAS; Electron Micros Soc Am; Col Am Pathologists; Am Asn Pathologists; Am Asn Cancer Res; AMA; Am Soc Cytol; Asn Clin Scientists; Soc Toxicol Pathologists. *Res:* Pathology; author of numerous scientific publications. *Mailing Add:* Dept Path Med Col Ohio PO Box 10008 Toledo OH 43614

GOLDBLITH, SAMUEL ABRAHAM, b Lawrence, Mass, May 5, 19; m 41; c 2. FOOD SCIENCE. *Educ:* Mass Inst Technol, SB, 40, SM, 47, PhD, 49. *Prof Exp:* Mem staff, Arthur D Little, Inc, 40-41; res assoc food sci, Mass Inst Technol, 49-52, asst prof, 52-55, assoc prof & exec officer dept, 55-59, from actg head to assoc head dept nutrit & food sci, 59-74, prof, 59-72, dir indust liaison, 74-78, Underwood-Prescott prof food sci, 72-78, prof food sci & vpres resource develop, 78-86, prof & sr adv to pres, 86-91, EMER PROF, MASS INST TECHNOL, 91- *Concurrent Pos:* Tech aide comt radiation preservation food, Nat Acad Sci-Nat Res Coun, 53-56, mem comt radiation preservation food, 59-63 & 71-73, chmn, 63-71, mem ad hoc subcomt high level dosimetry, 56-59, ad hoc subcomt radionuclides in foods, food protection comt, 60-61, comt nutrit, 60-62, gen comt foods, 63-71, US adv comt foot-and-mouth dis, 62-71 & gen comt, Dept Defense Food Prog, 70-, chmn task group feeding study protocol, 70-; mem Am Inst Biol Sci adv comt radiation pasteurization of foods to AEC, 60-62 & 65-70; mem Nat Pub Health Serv comt, Surgeon Gen, US, 61; officer US AEC team to Japan, Conf Radioisotopes, 61, 64, 66 & 67; mem sci adv coun, Refrig Res Found, 70-73; fel, Inst Food Sci & Technol, UK, 74. *Honors & Awards:* Monsanto Award, Inst Food Technol, 53; Babcock-Hart Award, 69; Nicholas Appert Award, 70; Distinguished Food Scientist Award, 69 & 76. *Mem:* Am Chem Soc; fel Inst Food Technol; fel AAAS; foreign mem, Royal Swedish Acad Eng Sci; Foreign Mem Swiss Acad Eng Sci. *Res:* Radiation preservation of foods; freeze dehydration of foods; microwaves and their application in food processing; food technol. *Mailing Add:* Resource Develop Mass Inst Technol Cambridge MA 02139

GOLDBLOOM, DAVID ELLIS, b London, Eng, 1933; US citizen. NUTRITION, BIOCHEMISTRY. *Educ:* Cambridge Univ, BA, 55, MA, 61; Univ Calif, Berkeley, PhD(nutrit), 68. *Prof Exp:* Res asst biochem, Unilever Food Res Labs, Eng, 58-60; asst mkt res, F W Berk Chem Co, London, 61; chemist qual control, Watney's Brewery, 61-62; asst prof nutrit, 68-74, ASSOC PROF HOME ECON, CALIF STATE UNIV, FRESNO, 74- *Res:* Comparative biochemistry of heme proteins. *Mailing Add:* Dept of Home Economics Calif State Univ Fresno CA 93740

GOLDBLOOM, RICHARD B, b Montreal, Que, Dec 16, 24; m 46; c 3. MEDICINE. *Educ:* McGill Univ, BSc, 45, MD, CM, 49; FRCP(C), 55. *Prof Exp:* Asst prof pediat, Fac Med, McGill Univ, 62-67; PROF PEDIAT & HEAD DEPT, DALHOUSIE UNIV, 67-; PHYSICIAN-IN-CHIEF, IZAAK WALTON KILLAM HOSP FOR CHILDREN, 67- *Concurrent Pos:* Lederle med fac award, 62; assoc physician, Montreal Children's Hosp, 62-67; chmn med adv bd, Can Cystic Fibrosis Found, 69-71; mem, Med Res Coun Can, 70-73; dir, Atlantic Res Ctr Ment Retardation, 67-75. *Mem:* Soc Pediat Res; Am Acad Pediat; Can Pediat Soc; Am Pediat Soc; Can Soc Clin Invest. *Mailing Add:* Dept Pediat IWK Hosp/Children 5070 Stn Halifax S Halifax NS B3J 3G9 Can

GOLDBURG, ARNOLD, b New York, NY, Aug 11, 27; m 58; c 3. AEROSPACE ENGINEERING, PETROLEUM ENGINEERING. *Educ:* Princeton Univ, BSE, 48, PhD(aerospace eng), 60; Mass Inst Technol, MS, 51, Eng, 52. *Prof Exp:* Asst dir Proj SQUID, Princeton Univ-Off Naval Res, 58-60; prin res scientist, Avco-Everett Res Lab, 60-66; head flight sci lab, Boeing Sci Res Labs, 66-70, chief scientist, Supersonic Transport Prog, 70-71, dir sci & technol, Off Corp Bus Develop, Boeing Co, 71-74, dir int technol, Boeing Aerospace Co, 74-75; mgr, Enhanced Oil Recovery, Gary Oil Co, 75-79, vpres, 80-81; Naval Air Systs Command Res Prof, US Naval Acad, 84-85; CONSULT, 82- *Concurrent Pos:* Mem res & tech adv subcomt, fluid mech, NASA, 69-70; mem, Proj Independence Panels, 74; mem res & develop subcomt, Interstate Oil Compact Comn, 79-82. *Mem:* Am Phys Soc; Am Inst Aeronaut & Astronaut; Am Geophys Union; Soc Petrol Engrs. *Res:* Aerothermochemistry of combustion; boundary layer fluid mechanics; reentry heat transfer; flow fields of hypersonic bodies; clear air turbulence; sonic boom; magnetohydrodynamics; stratospheric fluid physics and environmental impact; petroleum reservoir fluid physics; aircraft design. *Mailing Add:* 4 Carriage Lane Littleton CO 80121

GOLDBURG, WALTER ISAAC, b New York, NY, Sept 27, 27; m 56; c 2. SOLID STATE PHYSICS. *Educ:* Cornell Univ, BA, 51; Duke Univ, PhD, 54. *Prof Exp:* Res physicist, Carnegie Inst Technol, 54, 56, instr physics, 56-59; from asst prof to assoc prof, Pa State Univ, 59-63; assoc prof, 63-66, PROF PHYSICS, UNIV PITTSBURGH, 66- *Mem:* Fel Am Phys Soc; AAAS. *Res:* Critical phenomena, light scattering, turbulence. *Mailing Add:* Dept Physics Univ Pittsburgh Pittsburgh PA 15260

GOLDE, DAVID WILLIAM, b New York, NY, Oct 23, 40; m; c 2. HEMATOLOGY, ONCOLOGY. *Educ:* Fairleigh Dickinson Univ, BS, 62; McGill Univ, MD & CM, 66. *Prof Exp:* Dir, Aids Ctr, 86- 90, PROF MED, SCH MED, UNIV CALIF, LOS ANGELES, 79-, CHIEF, DIV HEMATOL & ONCOL, 81-, DIR, CLIN RES CTR, 87- *Mem:* Am Col Physicians; Am Soc Clin Invest; Am Soc Hematol; Am Asn Cancer Res; Asn Am Phys. *Res:* Hormonal modulation of normal and neoplastic hematopoiesis and the mechanism of normal and leukemic blood cell function. *Mailing Add:* Dept Hemat & Oncol Univ Calif Sch Med Los Angeles CA 90024-1678

GOLDE, HELLMUT, b Berlin, Ger, Feb 6, 30; nat US; m 57; c 3. COMPUTER SCIENCE, ELECTRICAL ENGINEERING. *Educ:* Munich Tech Univ, Dipl Ing, 53; Stanford Univ, MS, 55, PhD(elec eng), 59. *Prof Exp:* Asst microwave studies, Stanford Univ, 53-55, 56-59, res assoc, 59; from asst prof to assoc prof elec eng, 60-69, PROF ELEC ENG & COMPUT SCI, UNIV WASH, 69-, VICE PROVOST COMPUT, 84- *Mem:* Asn Comput Mach. *Res:* Computer languages and language processors. *Mailing Add:* Dept of Comput Sci/Elec Eng Univ of Wash Seattle WA 98195

GOLDEMBERG, ROBERT LEWIS, b Passaic, NJ, Sept 18, 25; c 3. COSMETIC CHEMISTRY & FORMULATING, DERMATOLOGICAL FORMULATING. *Educ:* Princeton Univ, BSE, 48. *Prof Exp:* Textile chemist, United Piece Dye Works, 48-50; sr chemist, Coty Inc, 50-58; res dir, Shulton Inc, 58-64 & Lanvin-Charles Ritz, 64-67; dir tech serv, Van Dyk & Co, 67-73; PRES & OWNER, RAKUMA LABS, INC, 73- *Concurrent Pos:* Chmn bd, Soc Cosmetic Chemists, 74; dir, Continuing Educ Ctr Inc, 75-84; pres, Trion Chem Corp, 81-85. *Mem:* Fel Soc Cosmetic Chemists (pres, 73); Am Chem Soc; Dermal Clin Eval Soc; NY Acad Sci. *Res:* Anti-irritants in cosmetic formulating; emulsion technology; sunscreen formulating. *Mailing Add:* Rakuma Labs Inc PO Box 2083 South Hackensack NJ 07606

GOLDEN, ABNER, b New York, NY, July 22, 18; m 43; c 3. PATHOLOGY. *Educ:* Columbia Univ, AB, 39; Harvard Med Sch, MD, 42. *Prof Exp:* From instr to prof path, Emory Univ, 48-61; prof, Georgetown Univ, 61-76, chmn dept, 62-76; prof path & chmn dept, Univ Ky, 76-88; RETIRED. *Mem:* Am Soc Clin Path; Am Asn Path; Col Am Path; Am Fedn Clin Res. *Res:* Endocrine pathology; renal disease. *Mailing Add:* Dept Path Med Col Univ Ky Lexington KY 40506

GOLDEN, ALFRED, b New York, NY, Aug 4, 08. PATHOLOGY. *Educ:* Univ Wis, BS, 30, MS, 34; Wash Univ, MD, 37. *Prof Exp:* Res assoc bact & immunol, Univ Wis, 31-35; res assoc, Sch Med, Wash Univ, 36-38; resident path, Western Reserve & Allied Serv, 38-40; pathologist & exec officer, Armed Forces Inst Path & lectr, Div Trop Med, Army Med, Sch, 41-45, med officer in chg trop dis, Trop Dis Invest Unit, Off Inter-Am Affairs, 45-46; assoc prof path, Col Med, Univ Tenn, 46-50; assoc prof, Sch Med, Univ Buffalo, 50-55; assoc prof, Col Med, Wayne State Univ, 55-56; dir labs, Jennings Mem Hosp & Alexander Blain Mem Hosp, Detroit, 55-74; chief of staff, Dept Path, St Vincent Hosp & Med Ctr, Toledo, 74-80; VPRES, ARIZ DIV, AM CANCER SOC, 80- *Concurrent Pos:* Consult pathologist, Washington Sanitarium, DC, 41-45, Buffalo Eye Bank & Res Soc, 50-55 & Mem Hosp, Niagara Falls, NY, 51-54; chief lab serv, Vet Admin Hosp, 49-55, consult pathologist, Detroit, 55-56. *Honors & Awards:* ARM Commendation Medal; Alexander Berg Prize Microbiol, Wash Univ Sch Med. *Mem:* Fel Am Soc Clin Path; Am Soc Exp Path; Am Asn Path & Bact; fel Am Col Physicians; fel Col Am Path. *Res:* Pathology of diseases of the lungs; infectious and tropical diseases; experimental tumor pathology. *Mailing Add:* 7527 N Del Norte Dr Scottsdale AZ 85258

GOLDEN, ALVA MORGAN, b Milledgeville, Ga, July 13, 20; m 57; c 1. NEMATOLOGY. *Educ:* Univ Ga, BSA, 50, MSA, 51; Univ Md, PhD(plant path & nematol), 56. *Prof Exp:* Asst plant path, Univ Ga, 50-51; plant pathologist fungicide develop, Vanderbilt Co, NY, 51; from asst nematologist to assoc nematologist, 52-56, nematologist, Calif, 56-59, NEMATOLOGIST, NEMATOL LAB, AGR RES SERV, BELTSVILLE AGR RES STA, USDA, 59- *Honors & Awards:* Int Honor Award, USDA, 82. *Mem:* AAAS;

fel Soc Nematologists, (treas, 63-66, vpres, 66-67, pres, 67-68); fel Am Phytopath Soc; Soc Syst Zool; Am Soc Sugar Beet Technol; Europ Soc Nematologists. *Res:* Plant nematology, with emphasis on taxonomy of plant parasitic nematodes; taxonomy, morphology, life cycle and identification of plant nematodes, hosts, occurence and distribution; curate and expand USDA nematode collection, one of largest in existence; over 165 publications in subject areas. *Mailing Add:* Nemat Lab Agr Res Serv USDA Beltsville Agr Res Ctr W Bldg 011A Rm 159 Beltsville MD 20705

GOLDEN, ARCHIE SIDNEY, b Danbury, Conn, Feb 9, 31; m 63; c 3. PEDIATRICS. *Educ:* Univ Conn, BA, 53; Univ Vt, MD, 57; Johns Hopkins Univ, MPH, 66. *Prof Exp:* Dir community health progs, Proj Hope, 62-70; assoc prof maternal & child health, 70-76, assoc prof & dir health assoc prog, Sch Health Serv, 73-79, ASSOC PROF INT HEALTH & PEDIAT, SCH HYG & PUB HEALTH & SCH MED, JOHNS HOPKINS UNIV, 79-; MED DIR, CHESAPEAKE HEALTH PLAN, 81-; CHIEF PEDIAT, FRANCIS SCOTT KEY MED CTR, 84- *Concurrent Pos:* Vis prof pediat & pub health, Med Sch, Univ Trujillo, Peru, 62-65 & Sch Med, Univ Cartagena, Colombia, 66-69; chmn interpersonal skills comt, Nat Bd Med Examrs, 72-76; consult, Latin Am Comt Pediat Residencies, 72-79; chmn, Comt Int Child Health, Am Acad Pediat & Comt New Health Practr, Am Pub Health Asn, 76-80; hon prof, Sch Med, Univ Cartagena, Colombia, 69. *Mem:* Am Acad Pediat; Asn Teachers Prev Med; Ambulatory Pediat Asn; Am Pub Health Asn; Asn Physicians Assts Progs (pres, 78-79). *Res:* Utilization of job and task analysis technique in the development of curricula for health professionals; translation of public health needs into curricula for health professionals; content of primary health care practice; education of primary care physicians; doctor-patient relationship. *Mailing Add:* Francis Scott Key Med Ctr 4940 Eastern Ave Baltimore MD 21224

GOLDEN, BEN ROY, b Bainbridge, Ga, Nov 19, 37; m 64; c 4. GENETICS. *Educ:* Mid Tenn State Univ, BS, 58; George Peabody Col, MA, 60; Brown Univ, PhD(genetics), 71. *Prof Exp:* Teacher, High Sch, Ill, 58-66; partic, NSF Acad Year Inst, Brown Univ, 66-67; asst prof genetics develop, Skidmore Col, 70-76; asst prof genetics develop, 76-80, ASSOC PROF BIOL, KENNESAW COL, 80- *Mem:* AAAS. *Res:* Analysis of the meiotic behavior of x-ray induced chromosome aberrations in an attempt to determine the mechanisms of pairing and disjunction in the spermatocytes of the Drosophila male. *Mailing Add:* Dept Biol Kennesaw Col PO Box 444 Marietta GA 30061

GOLDEN, CAROLE ANN, b Los Angeles, Calif, Sept 23, 42. IMMUNOASSAY DEVELOPMENT, IN VITRO DIAGNOSTICS. *Educ:* Okla Col Lib Arts, AB, 63; Miami Univ, MS, 69, PhD(microbiol), 73. *Prof Exp:* Instr microbiol, Miami Univ, 72-73; res asst prof, Med Sch, Univ Utah, 73-79; sci dir, Microbiol Res Corp, Bountiful, Utah, 78-87; VPRES, RES & DEVELOP, ENVIRON DIAGNOSTICS, INC, BURLINGTON, NC, 87- *Concurrent Pos:* Microbial geneticist, Nat Inst Occup Safety & Health, 76; sr consult scientist, Univ Utah Res Inst, 76-82; chmn, Res & Develop Mgt Comt, Asn Biotechnology Co. *Honors & Awards:* Distinguished Tech Commun Award, Soc Tech Commun. *Mem:* AAAS; Sigma Xi; Am Soc Microbiol; Reticuloendothelial Soc; NY Acad Sci; Asn Off Anal Chemists; Asn Biotechnology Co. *Res:* Development of immunological-based on-site tests for commercial application in the diagnosis of viral infections and the detection of toxic substances and drugs of abuse in animals and humans. *Mailing Add:* Environmental Diagnostics Inc 1238 Anthony Rd Burlington NC 27215

GOLDEN, DAVID E, b New York, NY, May 27, 32; m 62; c 2. ATOMIC PHYSICS, SPECTROSCOPY. *Educ:* NY Univ, BA, 54, PhD(physics), 61. *Prof Exp:* Asst prof physics, NY Univ, 60-61; asst prof physics, Adelphi Univ, 61-62; eng specialist, Gen Tel & Electronics Labs, Inc, Calif, 62-63; from res scientist to staff scientist phys electronics, Lockheed Palo Alto Res Labs, 63-67; prof physics, Univ Bari, 67-70; from assoc prof to prof physics, Univ Nebr, Lincoln, 70-75; George Lynn Cross prof physics & chmn dept physics & astron, Univ Okla, 75-85; provost & vpres acad affairs, 85-88, PROF PHYSICS, UNIV NTEX, 88- *Concurrent Pos:* Instr physics, City Col New York, 56-58 & NY Univ, 58-59, res assoc, 60-61; consult, Autometric Corp, NY, 61-62, Advan Res Instrument Systs, Inc, Tex, 70-72, Tracor, Austin, Tex, 72-74, Lawrence Radiation Lab, Livermore, Calif, 75-77 & Minn Mining & Mfg, 84-85; sr scientist, Sylvania Elec Prod, Mass, 69-70; hon lectr, Mid-Am State Univ Asn, 82-83; mem, Comt Atomic & Molecular Sci, Nat Acad Sci, 82-85; bd mem & treas, Say It Straight Found, 84- & Aranta-The Satir Network, 88- *Mem:* Fel Am Phys Soc; Sigma Xi. *Res:* Lasers; atomic collisions; electron scattering from atoms and molecules; electron spectroscopy; resonances in electron scattering; ion-molecule reactions; gaseous electronics; plasma physics; residual gas analyzers. *Mailing Add:* Dept Physics Univ NTex Denton TX 76203-5371

GOLDEN, DAVID MARK, b New York, NY, July 18, 35; m 63; c 3. CHEMICAL KINETICS. *Educ:* Cornell Univ, AB, 56; Univ Minn, PhD(phys chem), 61. *Prof Exp:* Phys chemist, 63-70, sr phys chemist, 70-76, DIR, DEPT CHEM KINETICS, SRI INT, 76- *Concurrent Pos:* Consult prof, Stanford Univ, 74-; ed, Int J Chem Kinetics, 83- *Mem:* Am Chem Soc; fel Am Phys Soc; fel AAAS. *Res:* Chemical kinetics and thermochemistry; molecular spectroscopy. *Mailing Add:* SRI Int Menlo Park CA 94025-3346

GOLDEN, GERALD SEYMOUR, b Hartford, Conn, June 18, 33; m 55; c 3. ANALYTICAL CHEMISTRY, INDUSTRIAL WASTE RECYCLING. *Educ:* Mass Inst Technol, SB, 54; Rensselaer Polytech Inst, PhD(chem), 57. *Prof Exp:* Anal res chemist radiochem, Nuclear Div, Combustion Eng, Inc, 57-61, supvr chem sect, 61-62; head, Anal Chem Lab, 62-65, chief, Anal Radioisotope Labs, 65-80, MGR, MAT CHARACTERIZATION & PROCESSING LAB, UNITED TECHNOL RES CTR, 81- *Mem:* Sigma Xi; Am Soc Testing & Mat; Mats Res Soc; Anal Lab Mgrs Asn. *Res:* Analysis of trace elements in materials; general analytical and radiochemistry; resource recovery; industrial chemical processing. *Mailing Add:* United Technol Res Ctr East Hartford CT 06108

GOLDEN, JOHN O(RVILLE), b Nashville, Tenn, Jan 8, 37; m 58; c 4. ACADEMIC ADMINISTRATION, CHEMICAL ENGINEERING. *Educ:* Vanderbilt Univ, BE, 59, MS, 60; Iowa State Univ, PhD(chem eng), 64. *Prof Exp:* Res asst chem eng, Vanderbilt Univ, 59-60 & Iowa State Univ, 60-64; res scientist assoc thermodyn, Huntsville Res & Eng Ctr, Lockheed Missiles & Space Co, 64-67; from asst prof to assoc prof, Colo Sch Mines, 67-70, dir res develop, 75-79, dean grad studies & res, 79-83, vpres acad affairs & dean fac, 83-90, PROF CHEM & PETROL REFINING ENG, COLO SCH MINES, 75- *Concurrent Pos:* Consult, Samsonite Corp, 68-70, Martin Marietta, 71 & EG&G, 90-; fel, Am Coun Educ Acad Admin, 78-79. *Mem:* Am Chem Soc; Am Inst Chem Engrs; Am Soc Eng Educ. *Res:* Coal liquefaction; coal processing. *Mailing Add:* 2009 Goldenvue Dr Golden CO 80401

GOLDEN, JOHN TERENCE, b Ft Wayne, Ind, Jan 31, 32. COLLOID CHEMISTRY, SURFACE CHEMISTRY. *Educ:* Univ Mich, BSCh, 53; Ohio State Univ, MSc, 58, PhD(phys org chem), 63. *Prof Exp:* Teaching asst, res asst & asst instr chem, Ohio State Univ, 56-62; res chemist, BF Goodrich Res Ctr, 62-63; indexer, Chem Abstracts Serv, 63-64; chemist, 64-75, tech & mfg mgr chem, 75-90, VPRES TECH, RENITE CO, 90- *Concurrent Pos:* NSF grant, Ohio State Univ, 62. *Mem:* Am Chem Soc; Soc Tribologists & Lubrication Engrs; Am Ceramic Soc. *Res:* Engaged in development and manufacture of high temperature lubricants, coatings, and release agents for the glass and metalworking industries. *Mailing Add:* Renite Co PO Box 30830 Columbus OH 43230

GOLDEN, KELLY PAUL, b Detroit, Mich, Oct 12, 43; m 64; c 2. ELECTRONICS ENGINEERING, COMPUTER SCIENCE. *Educ:* Mich State Univ, BS, 65, & MS, 67, PhD(elec eng), 71. *Prof Exp:* Comput operator, Mich State Univ, 65-66, teaching asst elec eng, 66-67; consult, Okemos Res Lab, Owens-Ill Inc, 67-73; SR RES ASSOC ELECTRONIC IMAGING, E I DU PONT DE NEMOURS & CO, INC, 73- *Mem:* Inst Elec & Electronics Engrs; Sigma Xi; Instrument Soc Am. *Res:* Electronic circuit design; instrumentation and product development; electric properties of materials and devices; hardware and software; laser and other electronic imaging systems. *Mailing Add:* 204 Caravel Dr Bear DE 19701-1629

GOLDEN, KENNETH IVAN, b Chicago, Ill, Oct 24, 32; m 65; c 2. PLASMA PHYSICS. *Educ:* Northwestern Univ, BS, 55; Mass Inst Technol, SM, 56, ME, 57; Inst Henri Poincare, Univ Paris, Dr, 64. *Prof Exp:* Res assoc physics, Brandeis Univ, 65-67, vis asst prof, 67-69; from assoc prof to prof, Northeastern Univ, 69-80, William L Smith prof elec eng, 80-83, George A Snell prof elec eng 83-86; PROF & CHMN, COMPUT SCI & ELEC ENG, UNIV VT, 86- *Concurrent Pos:* Consult, Lincoln Lab, Mass Inst Technol, 69-72; prin investr, NSF grants, 81-83, Air Force Off Sci Res grants, 76-81 & Air Force Geophysics Lab contract, 73-76; sci collabr, Physics & Plasma Res Ctr, Switz, 76; vis prof, Lab Physics Theoret & High Energy, Univ Paris, 75-76; invited lectr, NATO Advan Inst Strongly Coupled Plasmas, 77; res leader, Int Ctr Theoret Physics, 81; invited lectr, Nanjing & Hoefi Inst Plasma Physics, China, 82; vis prof, Univ New S Wales, 82, vis prof theoret physics, 84-85; Fulbright sr scholar Australia, 84-85. *Mem:* NY Acad Sci; Sigma Xi; Am Phys Soc. *Res:* Plasma many body theory; statistical mechanics and kinetic theory; strongly coupled plasmas; shock waves in plasmas. *Mailing Add:* Two Barstow Court Shelburne VT 05482

GOLDEN, LARON E, b Arkadelphia, Ark, Apr 20, 20; m 41; c 4. SOIL FERTILITY, PLANT NUTRITION. *Educ:* Univ Ark, BS, 49; La State Univ, MS, 51, PhD(agron), 59. *Prof Exp:* Instr agron & hort, Southern State Col, 49-53; asst prof agron, Univ Houston, 53-58; from asst prof to prof agron, La State Univ, Baton Rouge, 58-84; RETIRED. *Concurrent Pos:* Checchi & Co-Agency Int Develop consult, Repub Guinea, WAfrica, 64; Bahamas Sugar Proj consult, Owens-Ill Inc, 65-67. *Mem:* Am Soc Agron; Am Soc Sugar Cane Technol; Int Soc Sugar Cane Technol. *Res:* Sugar cane nutrition; tropical crops. *Mailing Add:* 12145 Pecan Grove Ct Baton Rouge LA 70810

GOLDEN, MICHAEL STANLEY, b West Point, Ga, Aug 23, 42; m 72; c 1. PLANT ECOLOGY, FOREST ECOLOGY. *Educ:* Trevecca Col, AB, 64; Auburn Univ, MS, 68; Univ Tenn, PhD(plant ecol), 74. *Prof Exp:* Asst & assoc prof forest ecol, Ala A&M Univ, 72-74; ASST PROF FOREST ECOL & SILVICULT, AUBURN UNIV, 75- *Mem:* Soc Am Foresters; Ecol Soc Am; Am Inst Biol Sci. *Res:* Descriptive ecology of forested ecosystems; forest site evaluation and classification; factors affecting height growth of loblolly pine and yellow-poplar; silviculture. *Mailing Add:* Dept Forestry Auburn Univ Auburn AL 36849

GOLDEN, ROBERT K, b Brooklyn, NY, Feb 16, 25; c 1. VENTURE CAPITAL, CORPORATE FINANCE. *Educ:* Pa State Univ, BS, 47; Univ Pa, MS, 49, PhD(physics), 55. *Prof Exp:* Instr physics, Univ Pa, 49-50; physicist, Eckert-Mauchly Div, Remington Rand Corp, 51-53; res assoc physics, Univ Pa, 53-55; consult physicist, NY, 55-57; staff engr, Int Bus Mach Corp, 57-62; res dir, Computronics Inc, 62-64; dir, NY Off, Data Systs Analysts, Inc, NJ, 64-65; pres, Compunet, Inc, 65-69; vpres & dir, Comput Profile Group Ltd, 69-70; pres & dir, Dwygofax, Inc, 70-72; vpres, Shearson Hayden Stone Inc, 73-77 & Rooney Pace Inc, 78-79; sr vpres, John Muir & Co, 80-81; PRES, R K GOLDEN ASSOCS INC, 81- *Concurrent Pos:* Mem adj fac, grad dept elec eng, Polytech Inst Brooklyn, 61-68; adj fac comput sci, City Col New York, 66-67. *Mem:* Am Phys Soc; Inst Elec & Electronics Engrs; Asn Comput Mach; Financial Analysts Fedn. *Res:* Computer and information systems; physics of communications; nuclear isomers. *Mailing Add:* 15 W 72nd St New York NY 10023-3402

GOLDEN, SIDNEY, b Boston, Mass, June 23, 17; m 41; c 2. PHYSICAL CHEMISTRY, THEORETICAL CHEMISTRY. *Educ:* City Col New York, BS, 38; Purdue Univ, 40-42; Harvard Univ, PhD(phys chem), 48. *Prof Exp:* Asst chem, Purdue Univ, 41-42; res assoc, Nat Defense Res Comt, George Washington Univ, 42-46; phys chemist, Hydrocarbon Res, Inc, 48-51; from asst prof to prof, 51-81, EMER PROF CHEM, BRANDEIS UNIV, 81- *Concurrent Pos:* Fulbright sr scholar, 59-60; Guggenheim fel, 59-60; lectr, polytech Inst Brooklyn, 49-50; res assoc, Mass Inst Technol, 52; consult, Nat Bur Standards, 57-64; vis prof, Univ Calif, Berkeley, 63, Hebrew Univ Jerusalem, 67-68 & 74-75; sr Weizmann fel, Weizmann Inst Sci, 74-75; exchange prof, Univ Paris-Sud, 75. *Mem:* Am Chem Soc; fel Am Phys Soc; fel Am Acad Arts & Sci. *Res:* Theoretical chemical kinetics; quantum mechanics; free radicals; ionic solvation; metal-ammonia solution; quantum statistical inequalities. *Mailing Add:* 8614 N 84th St Scottsdale AZ 85258

GOLDENBAUM, GEORGE CHARLES, b New York, NY, Aug 11, 36; m 61; c 3. PLASMA PHYSICS. *Educ:* Muhlenberg Col, BS, 57; Univ Md, PhD(physics), 66. *Prof Exp:* Physicist, Nat Bur Standards, 58-61; res assoc plasma physics, Univ Md, 66-69; res physicist, US Naval Res Lab, 69-74; assoc prof, 74-80, PROF ASTRON & PHYSICS, UNIV MD, COLLEGE PARK, 80- *Concurrent Pos:* Vis scientist, Culham Lab, Eng, 66-67, Lawrence Livermore Nat Lab, Calif, 80-81. *Mem:* Fel Am Phys Soc. *Res:* Dynamics of low density plasmas; collision free shock waves in plasmas; plasma turbulence; interaction of radiation with plasmas; magnetic confinement of plasmas. *Mailing Add:* Dept Physics/Astron Univ Md College Park MD 20742

GOLDENBAUM, PAUL ERNEST, b Newport News, Va, Aug 25, 43; m 69; c 2. CLINICAL MICROBIOLOGY, MICROBIAL PHYSIOLOGY. *Educ:* Lynchburg Col, BS, 66; Western Mich Univ, MA, 68; NC State Univ, PhD(microbiol), 72. *Prof Exp:* Asst prof biol, Southern Ill Univ, Edwardsville, 74-78, assoc prof, 78-81; sr microbiologist, 81-83, mgr, clin microbiol proj, 83-85, MGR CLIN MICROBIOL, BECTON DICKINSON DIAG INSTRUMENT SYSTS, MD, 85- *Concurrent Pos:* Fel res assoc, AB Chandler Med Sch, Dept Biochem, Univ Ky, 72-73; res fel, NIH, Nat Cancer Inst, Pub Health Serv, 73-74; res scholar, Southern Ill Univ, 77-78 & prin investr res grants, 74-; vis assoc prof microbiol & immunol, Med Sch, Wash Univ, St Louis, 78-79. *Honors & Awards:* Mary Posten Award, NC Br of Am Soc Microbiol, 68. *Mem:* Am Soc Microbiol; Sigma Xi; AAAS; fel Am Acad Microbiol; NY Acad Sci. *Res:* Microbial physiology; regulation of gene activity in bacteria including catabolite repression; morphogenesis; membrane physiology; automated, rapid detection of bacteria; clinical microbiology: detection and recovery of pathogens. *Mailing Add:* 732 Silver Creek Rd Baltimore MD 21208

GOLDENBERG, ANDREW AVI, b Romania, Aug, 45; Can citizen; m 70; c 2. CONTROL SYSTEMS, ROBOTIC BASED AUTOMATION. *Educ:* Israel Inst Technol, BSc, 69, MSc, 72; Univ Toronto, PhD(elec eng), 76. *Prof Exp:* Staff eng, control systs, Spar Aerospace, Ltd, 75-81; asst prof elec eng, 81-82, assoc prof mech eng & elec eng, 82-87, PROF MECH ENG & ELEC ENG & BIOMED ENG, UNIV TORONTO, 87- *Concurrent Pos:* Consult, Eng Serv, Inc, 80- *Mem:* Inst Elec & Electronics Engrs; Am Soc Mech Engrs; Soc Mech Eng. *Res:* Robots; design; kinematics; control; manufacturing; real-time; sensors. *Mailing Add:* Dept Mech Eng Univ Toronto 5 Kings Col Rd Toronto ON M5S 1A4 Can

GOLDENBERG, BARBARA LOU, b Trenton, NJ, Feb 22, 52; m 86; c 2. SOLID STATE SENSOR DEVELOPMENT. *Educ:* Occidental Col, AB, 73; Brandeis Univ, PhD(chem physics), 82. *Prof Exp:* Fel chem dept, Univ Minn, 82-84; prin res scientist, Sensors & Signal Processing Lab, 84-91, SR PRIN RES SCIENTIST, SENSOR & SYST DEVELOP CTR, HONEYWELL INC, 91. *Mem:* Am Phys Soc; Am Chem Soc; Mat Res Soc. *Res:* Solid state sensor development based on novel semiconducting materials and sensor designs; metallorganic chemical vapor deposition growth and characterization of III-V semiconductors. *Mailing Add:* 10701 Lyndale Ave S MN09 A200 Bloomington MN 55420-5601

GOLDENBERG, DAVID MILTON, b Brooklyn, NY, Aug 2, 38; m 61; c 5. ONCOLOGY, NUCLEAR MEDICINE. *Educ:* Univ Chicago, SB, 58; Univ Erlangen-Nuremburg, ScD(natural sci), 65; Univ Heidelberg, MD, 66. *Prof Exp:* Intern-resident med & path, Univ Hosp, Erlangen, Germany, 66-67, head clin exp oncol, Surg Clin, 67-68; assoc prof path, Sch Med, Univ Pittsburgh, 68-70 & Temple Univ, 70-72; from assoc prof to prof, Univ Ky, 72-73; PRES, CTR MOLECULAR MED & IMMUNOL & ADJ PROF MED & SURG, UNIV MED & DENT NJ, 83- *Concurrent Pos:* Staff pathologist, Vet Admin Hosp, Pittsburgh, 68-70; Coun Tobacco res grant, 68-71; Nat Cancer Inst grants, 69-; Damon Runyon Fund cancer res grant, 71-75; Am Cancer Soc grant, 73-76; mem & chmn, Vet Admin Merit Rev Bd Oncol, 74-77; consult, Vet Admin Hosp, Lexington, Ky, 74-83 & Brookhaven Nat Labs, 83-87; exec dir, Ephraim McDowell Community Cancer Network, 75-80; mem bd trustees, Assoc Community Cancer Ctr, 77-79; pres, Ephraim McDowell Cancer Res Found, Inc, 78-80; Univ Ky Found res award, 78; ed, J Cancer Res & Clin Oncol, 79-87, Diag Path, 80- & Tumor Diag, 80-; mem exp immunol study sect, NIH, 80-83; mem sci adv bd, German Fund Cancer Res, 80-; assoc ed, Cancer Res, 82- *Honors & Awards:* Outstanding Investr Award, Nat Cancer Inst, NIH, 85. *Mem:* Am Soc Exp Path; Am Asn Path & Bact; Tissue Cult Asn; Am Soc Cell Biol; Am Soc Human Genetics; Am Asn Cancer Res; Am Asn Pathologists; Soc Exp Biol & Med; Soc Nuclear Med. *Res:* Cancer, improved diagnosis and treatment; immunology; treatment of human tumors with radiolabeled monoclonal antibodies. *Mailing Add:* Ctr Molecular Med & Immunol One Bruce St Newark NJ 07002

GOLDENBERG, GERALD J, b Brandon, Man, Nov 27, 33; m 59; c 4. INTERNAL MEDICINE, ONCOLOGY. *Educ:* Univ Man, MD, 57; Univ Minn, PhD(morphol, hemat), 65. *Prof Exp:* Lectr internal med, Univ Man, 64-66, from asst prof to prof internal med, 75-90; PROF MED PHARMACOL, UNIV TORONTO, 90-, DIR ONCOL, 90- *Concurrent Pos:* Res asst, Man Cancer Found, 64-73; clin res assoc, Nat Cancer Inst Can, 67-75; consult, Winnipeg Children's Hosp, 67-90, Princess Margaret Hosp, Toronto, 90-; dir, Man Inst Cell Biol, 73-88. *Mem:* Can Soc Clin Invest; Am Asn Cancer Res; Am Soc Exp Path; fel Am Col Physicians; fel Royal Col Physicians Can. *Res:* Cancer chemotherapy; mechanism of action of alkylating agents; molecular pharmacology of drug resistance; membrane transport. *Mailing Add:* Dir Oncol Univ Toronto 92 College St Toronto ON M5G 1L4 Can

GOLDENBERG, MARTIN IRWIN, b Jersey City, NJ, May 10, 33; m 54; c 2. MEDICAL BACTERIOLOGY, LABORATORY MEDICINE. *Educ:* Rutgers Univ, BS, 54; Yale Univ, MS, 57, PhD(med microbiol), 60; Univ Md, BA, 82. *Prof Exp:* Microbiologist, State Dept Health, Hawaii, 60-61; scientist, Commun Dis Ctr, USPHS, San Francisco, 61-68 & Ft Collins, Colo, 68-70; lab dir, US Med Ctr, 70-75; mem hq staff, Div Hosp & Clin, USPHS, 75-81, mem staff grants admin, 81-85, asst dir spec prog clin ctr, 85-88, CHIEF, OFF MED BD SERV CLIN CTR, NIH, USPHS, 88- *Concurrent Pos:* Assoc mem grad fac, Colo State Univ, 68-70; exec secy, med bd, NIH. *Honors & Awards:* Commendation Medal, USPHS, 80. *Mem:* Fel AAAS; fel Am Pub Health Asn. *Mailing Add:* Bldg 10 Rm IC121 Warren Grant Magnuson Clin Ctr NIH Bethesda MD 20892

GOLDENBERG, MARVIN M, b New York, NY, July 7, 35; m 57; c 2. PHARMACOLOGY, PHYSIOLOGY. *Educ:* Long Island Univ, BS, 57; Temple Univ, MS, 59; Woman's Med Col, Pa, PhD(pharmacol), 65. *Prof Exp:* From sr res scientist I to sr res scientist II, 65-73, res assoc, Eaton Labs, Norwich-Eaton Pharmaceut, 73-, Controlled Substances Officer, 75-; dir immunopharmacol, 80-84, ASST DIR CLIN PHARMACOL, MERCK SHARP & DOHME RES LABS, 85- *Concurrent Pos:* Ad hoc reviewer, Div Res Grants, NIH; instr pharmacol, State Univ NY Binghamton. *Mem:* Am Soc Pharmacol & Exp Therapeut; Europ Biol Res Asn; Int Inflammation Res Soc; NY Acad Sci; Sigma Xi; Am Gastoenterol Asn; Am Soc Clin Pharmacol; Inflammation Res Asn. *Res:* Autonomic nervous system pharmacology in relation to the gastrointestinal area; toxicology with regard to the antidotal effectiveness of agents against cyanide; anti-inflammatory drug research; clinical protocol monitoring and analysis; immunosuppression and analgesic areas of basic and applied research; clinical monitor for phase I and phase II studies in man in the field of inflammation (rheumatoid arthritis, analgesia), psoriasis, wound healing, ulcerative colitis, hypertension, Alzheimers disease and Parkinsons disease; studies in man on pharmacokinetics and pharmodynamics; bioequivalence in man. *Mailing Add:* Device Dept Am Cyanamid Middletown Rd Pearl River NY 10965

GOLDENBERG, NEAL, b Brooklyn, NY, Apr 29, 35; m 57; c 3. PHYSICAL CHEMISTRY. *Educ:* City Col New York, BS, 56; Univ Ark, MS, 58, PhD(phys chem), 61. *Prof Exp:* Instr chem, Univ Ark, 61; sr res chemist, Mound Lab, Monsanto Res Corp, 61-65, in chg plutonium isotope fuels group, Snap Prog, 65-67; asst prof chem, Fairleigh Dickinson Univ, 67-69; tech coordr, Space Nuclear Systs Div, 69-72, chief, Isotope Technol Br, 72-76, asst dir advan isotope separation, US Energy Res & Develop Admin, 76-79, dir Advan Nuclear Systs & Projs Div & dir Off Plans & Resource Mgt, 80-82, DIR DIV SAFETY, QUAL ASSURANCE & SAFEGUARDS, US DEPT ENERGY, 82- *Mem:* Am Chem Soc. *Res:* Uranium enrichment; advanced and laser isotope separations; photochemistry; plutonium chemistry; reactor fuel reprocessing; nuclear waste management; thermodynamics; plasmas; development and production of nuclear power systems for use in the exploration of space; high temperature gas reactor program; evaluation of advanced nuclear energy system concepts; development of the national fission energy plan for reactor development; strategic analyses; integration and development of planning and budget for reactor programs; providing oversight to international activities involving government-sponsored research and development. *Mailing Add:* 9328 Garden Ct Potomac MD 20854

GOLDENSOHN, ELI SAMUEL, b New York, NY, June 25, 15; m 40; c 3. NEUROLOGY. *Educ:* George Washington Univ, AB, 37, MD, 40. *Prof Exp:* Instr, Med Sch, Univ Colo, 45-48, asst prof physiol, 49-53, attend staff, 51-53, dir epilepsy serv, 53; from asst prof to assoc prof neurol, Col Physicians & Surgeons, Columbia Univ, 53-63; prof, Sch Med, Univ Pa, 63-67; PROF NEUROL, COL PHYSICIANS & SURGEONS, COLUMBIA UNIV, 67-; PROF NEUROL, ALBERT EINSTEIN COL MED, 83- *Concurrent Pos:* Consult, Vet Admin Hosp, Bronx, 59-61. *Honors & Awards:* W G Lennox Award, Am Epilepsy Soc, 89. *Mem:* Am Electroencephalog Soc (pres, 71); Am Neurol Asn; Am Epilepsy Soc (pres, 68); Am Acad Neurol; Epilepsy Found Am (pres, 81). *Res:* Brain physiology; electrical activity of the brain; convulsive disorders. *Mailing Add:* Dept Neurol Albert Einstein Col Med 1300 Morris Park Ave Bronx NY 10461

GOLDENSON, JEROME, b Greensburg, Pa, July 23, 12. CHEMISTRY. *Educ:* Carnegie Inst Technol, BS, 34. *Prof Exp:* Res chemist, Nat Alloy Steel Co, Pa, 35-37, Am Gas & Elec Corp, WVa, 37-38 & Acme Protection Equip Co, Pa, 38-39; design engr, E W Voss Mach Co, 39-40; phys chemist, Chem Ctr, US Army, 40-55, chief colloid br, Physicochem Res Div, Chem Res & Develop Labs, 55-71; RETIRED. *Concurrent Pos:* Mem mining adv bd, Carnegie Inst Technol; chem consult, 71- *Mem:* Am Chem Soc; Coblentz Soc; Sigma Xi; Am Ord Asn. *Res:* Infrared spectroscopy; organic phosphorus compounds; aluminum soaps; aerosols; colloid chemistry; alloy steels; chemical analytical methods of traces of gases in air. *Mailing Add:* 5 Cobbler Court Pikesville MD 21208-1321

GOLDENTHAL, EDWIN IRA, b Plainfield, NJ, Feb 2, 30; m 70; c 5. PHARMACOLOGY. *Educ:* George Washington Univ, BS, 52, MS, 53, PhD(pharmacol), 56. *Prof Exp:* Chemist, Geol Surv, US Dept Interior, 51-53; pharmacologist, Food & Drug Admin, HEW, 56-63, chief drug rev br, Div Toxicol Eval, 63-66, dep dir, Off New Drugs, 66-70; dir safety eval, 70-72, VPRES & DIR RES, INT RES & DEVELOP CORP, MATTAWAN, MICH, 71- *Mem:* Soc Toxicol; Am Soc Pharmacol & Exp Therapeut; Europ Soc Study Drug Toxicity; Am Col Toxicol. *Res:* Toxicology; drug metabolism. *Mailing Add:* 4330 Old Field Trail Kalamazoo MI 49008

GOLDER, RICHARD HARRY, b Philadelphia, Pa, July 6, 22; m 57. BIOCHEMISTRY. *Educ:* Univ Wis, BS, 44; Temple Univ, PhD(biochem), 57. *Prof Exp:* Chemist, Barrett Div, Allied Chem & Dye Corp, 44-45; chemist, Philadelphia Rust-Proof Co, 45-46; tech asst, Gen Chem Co, 47-49; asst Inst Cancer Res, 49-56; assoc prof, Sch Dent, Temple Univ, 64-74, assoc prof biochem, Sch Pharm, 74-87; RETIRED. *Concurrent Pos:* Fel, Johnson Found, Univ Pa, 56-59, Physiol-Chem Inst, Univ Marburg, 59-60; fel chem,

Univ Pa, 60-64. *Mem:* AAAS; Am Chem Soc; Sigma Xi. *Res:* Controlling factors in glucose catabolism; nucleic acid changes following drug administration. *Mailing Add:* Dept Med Chem Temple Univ Sch Pharm Philadelphia PA 19140

GOLDER, THOMAS KEITH, b San Francisco, Calif, Nov 3, 42; c 1. CELL BIOLOGY. *Educ:* Calif State Univ, San Francisco, AB, 66; Univ Calif, Davis, PhD(zool), 75. *Prof Exp:* Instr biol sci, Calif State Univ, San Francisco, 68-69; lectr zool, Univ Calif, Davis, 74-75, 78, res physiologist, 75-78; RES SCIENTIST, INT CTR INSECT PHYSIOL & ECOL, 78- *Concurrent Pos:* Consult, NIH grant, 75. *Mem:* AAAS; Am Soc Cell Biol; Sigma Xi; EAfrica Soc Parasitol. *Res:* Localization of acetylcholinesterase in normal and trypanosome infected tsetses; development and cultivation of infective salivarian trypanosomes and relationship to tsetse salivary gland physiology; effect of trypanosome development on tsetse physiology. *Mailing Add:* 789 Green Valley Rd No 36 Watsonville CA 95076

GOLDEY, JAMES MEARNS, b Wilmington, Del, July 3, 26; m 51; c 2. PHYSICS. *Educ:* Univ Del, BS, 50; Mass Inst Technol, PhD(physics), 55. *Prof Exp:* Mem tech staff, Bell Tel Labs, 54-59, engr transistor develop, 59-61, head silicon transistor & integrated circuit dept, 61-66, head device technol dept, 66-69, dir mat & process technol lab, 69-74, dir, Solid State Device & Mat Lab, 74-78, dir, Electronic Compound Subsyst Lab, 78-81, dir, Integrated Circuit Customer Serv Lab, 81-84, DIR, LINEAR & HIGH VOLTAGE INTEGRATED CIRCUIT LAB, AT&T BELL LABS, 84- *Mem:* Fel Inst Elec & Electronics Engrs. *Res:* Semiconductor physics; semiconductor device physics. *Mailing Add:* RD 7 Allentown PA 18103

GOLDFARB, DAVID S, b Mar 14, 56. EXPERIMENTAL BIOLOGY. *Educ:* Yale Col, BA, 77, MD, 81; Am Bd Internal Med, dipl, 85, dipl nephrology, 88. *Prof Exp:* Intern & resident, Dept Internal Med, Dept Vet Affairs Med Ctr & NY Univ Med Ctr, 81-84, res fel, Nephrology Sect, 85-86, clin fel, 86-87, instr, 87-91; STAFF PHYSICIAN, DEPT INTERNAL MED, DEPT VET AFFAIRS MED CTR & DIR, RENAL/HYPERTENSION CLIN, 87-; ASST PROF MED, SCH MED, NY UNIV, 91- *Concurrent Pos:* Attend physician, Adult Emergency Serv, Bellevue Hosp, 85-87, clin asst attend, Dept Med, 87-90, assoc attend, 90-; mem, End Stage Renal Dis Comt, Dept Vet Affairs Med Ctr, 87- *Mem:* Am Col Physicians; Am Fedn Clin Res; assoc mem Am Physiol Soc; Int Soc Nephrology; Am Soc Nephrology. *Res:* Effect of systemic acid-base balance on ileal secretion; effect of acid-base variables on airway epithelia; author of numerous scientific publications. *Mailing Add:* Nephrology Sect/111 Dept Vet Affairs Med Ctr 423 23rd St New York NY 10010

GOLDFARB, DONALD, b New York, NY, Aug 14, 41; m 68; c 2. MATHEMATICAL PROGRAMMING. *Educ:* Cornell Univ, BChE, 63; Princeton Univ, MA, 65, PhD(chem eng), 66. *Prof Exp:* Asst res scientist, Courant Inst Math Sci, NY Univ, 66-68; from asst prof to prof comput sci, City Col N Y, 71-83, chmn, 78-79; PROF INDUST ENG & OPER RES, COLUMBIA UNIV, 82-, CHMN, 84- *Concurrent Pos:* NSF fel, 63-66; res assoc, NY Sci Ctr, IBM, 70, T J Watson Res Ctr, 72, 76, 91 & Atomic Energy Res Estab, Harwell, UK, 74-75; assoc ed, Math of Computation, Soc Indust & Appl Math, 69-, Opers Res, 83-, Math Prog, 83-; NSF grantee, 73-75 & 80-, Army Res Off, 77-80 & 82-, Off Naval Res, 87-; ed, J Numerical Anal, Soc Indust & Appl Math, 82-84, J on Optimization, 89-; mem coun, Mem Am Math Soc, 85-87, Math Prog Soc, 82-85; mem comt, Recommend US Army Basic Sci Res, Nat Res Coun, 83-86. *Mem:* Am Math Soc; Math Prog Soc; Soc Indust & Appl Math; Oper Res Soc Am. *Res:* Mathematical programming/optimization; numerical analysis. *Mailing Add:* Dept Indust Eng Opers Res Columbia Univ 316 SW Mudd Bldg New York NY 10027

GOLDFARB, JOSEPH, b 1943. NEUROPHARMACOLOGY, NEUROPHYSIOLOGY. *Educ:* Mass Inst Technol, BS, 63; Yeshiva Univ, PhD(biomed sci), 69. *Prof Exp:* From asst prof to assoc prof, 72-89, PROF PHARMACOL, MT SINAI SCH MED, 89- *Mem:* Soc Neurosci; Am Soc Pharmacol & Exp Therapeut. *Mailing Add:* Dept Pharmacol Box 1215 Mt Sinai Sch Med One Gustave L Levy Pl New York NY 10029

GOLDFARB, NATHAN, b New York, NY, Apr 28, 13; m 35; c 3. STATISTICS. *Educ:* NY Univ, MA, 37, PhD(statist), 55. *Prof Exp:* Statistician econ statist, US Bur Old Age & Survivors Ins, 36-51; dir med care, Spec Res Proj Med Care, Commonwealth Found, 51-54; dir mkt res, Forbes, 54-56; prof statist & admin, Hofstra Univ, 56-78, emer prof, 78-88; dir eval & res, Rockland Children's Psychial Ctr, 78-83; RETIRED. *Concurrent Pos:* Dir psychiat res, Henry Ittleson Ctr Res, 57-; Ford Found fel, Univ Chicago, 62-67; IBM Systs Res Inst fel, 66. *Res:* Growth of schizophrenic children; business administration. *Mailing Add:* 101 Gedney St No 2T Nyack NY 10960

GOLDFARB, RONALD B, MAGNETICS, SUPERCONDUCTIVITY. *Educ:* Rice Univ, BA, 73, MS, 75; Colo State Univ, MS, 76, PhD (physics), 79. *Prof Exp:* SR RES PHYSICIST, NAT INST STANDARDS & TECHNOL. *Concurrent Pos:* Lectr, physics dept, Univ Colo. *Mem:* Inst Elec & Electronics Engrs; Am Phys Soc; Am Soc Testing Mat. *Res:* Magnetism, superconductivity; instrument design. *Mailing Add:* Nat Inst Standards & Technol 325 Broadway Boulder CO 80303-3328

GOLDFARB, ROY DAVID, b New York, NY, Sept 23, 47; m 73; c 2. CARDIOVASCULAR PHYSIOLOGY. *Educ:* Colgate Univ, AB, 68; Hahnemann Med Col, MSc, 71, PhD(physiol), 73. *Prof Exp:* NIH fel pharmacol, Col Med, Univ S Ala, 73-74, instr, 74-75; asst prof, 75-80, assoc prof, 87-91, PROF PHYSIOL, ALBANY MED COL, 91-, PROF PHYSIOL & MED, RUSH MED COL. *Concurrent Pos:* Vis scientist, Univ Calif-San Diego, La Jolla, Calif, Heart Res Inst, Sydney, Australia. *Mem:* AAAS; Am Heart Asn; Am Physiol Soc. *Res:* Mechanisms of cardiac injury induced by regional ischemia and septemic sepsis. *Mailing Add:* Dept Med Rush Med Col Sect Cardiol Chicago IL 60612

GOLDFARB, STANLEY, b Mar 20, 31; wid. CHEMICAL CARCINOGENESIS, HEPATOLOGY. *Educ:* Univ NY, Buffalo, MD, 55. *Prof Exp:* PROF PATH, SCH MED, UNIV WIS-MADISON, 69- *Mem:* Am Asn Cancer Res; Am Asn Study Liver Dis; Am Soc Exp Path. *Res:* Histogenesis, growth kinetics and pathobiology of experimental rodent hepatocellular carcinomas with emphasis on factors involved in tumor promotion and progression; histogenesis and pathobiology of estrogen induced and estrogen dependent renal carcinoma in the hamster.. *Mailing Add:* Dept Path Sch Med Univ Wis 1300 University Ave Madison WI 53706

GOLDFARB, THEODORE D, b New York, NY, May 6, 35; m 85; c 4. ENVIRONMENTAL PHYSICAL CHEMISTRY. *Educ:* Cornell Univ, AB, 56; Univ Calif, PhD(chem), 59. *Prof Exp:* Asst, Univ Calif, 56-57; asst prof, 59-67, ASSOC PROF CHEM, STATE UNIV NY, STONY BROOK, 67-, assoc vice provost, 84-88. *Mem:* AAAS; Am Chem Soc; Sci for the People. *Res:* Environmental effects of energy production and use, agricultural chemicals and waste disposal technologies; science policy and scientific decision making. *Mailing Add:* Dept Chem State Univ NY Stony Brook NY 11794

GOLDFEDER, ANNA, b Poland, July 25, 97; nat US. EXPERIMENTAL PATHOLOGY. *Educ:* Univ Prague, DSc, 23. *Prof Exp:* Asst path, Masaryk Univ, 23-25, res assoc, Dept Physiol, 25-28, in chg lab cancer res, 28-31; res worker biol chem, Harvard Med Col, 33; res worker bact & immunol, Col Physicians & Surgeons, Columbia Univ, 34; in chg cancer div, Dept Hosps, New York, 34-50, sr biologist, 50-60, prin res scientist & dir cancer & radiobiol res lab, Cancer Div, 60-77; ADJ PROF BIOL, GRAD SCH ARTS & SCI & DIR CANCER & RADIOBIOL RES LAB, NY UNIV, 77- *Concurrent Pos:* Fel, Univ Vienna, 28-29; res fel cancer, Lenox Hill Hosp, New York, 31-33; res fel, Cancer Div, Dept Hosps, New York, 34; from res assoc to prof, NY Univ, 34- *Honors & Awards:* Award, Radiol Soc NAm, 48; Presidential Gold Medal, NY Acad Sci, 78. *Mem:* AAAS; Soc Exp Biol & Med; Radiol Soc NAm; Radiation Res Soc; Am Asn Cancer Res. *Res:* Relationship between radiation, viral and chemical carcinogenesis and inherent or genetic biological properties of a specific inbred strain of mice. *Mailing Add:* Dept Biol Cancer & Radiobiol Res Lab NY Univ Two Washington Square Village F Apt 91 New York NY 10012

GOLDFELD, DORIAN, b Marburg, Germany, Jan 21, 47; US Citizen; m 85; c 2. MATHEMATICS. *Educ:* Columbia Univ, BS, 67, PhD, 69. *Prof Exp:* Lectr, Tel-Aviv Univ, 72-73; asst prof, MIT, 76-82; assoc prof, Univ Texas, 83-85; PROF MATH, COLUMBIA UNIV, 85- *Concurrent Pos:* Sloan Fel, NSF, 77-79, prin investr, 83- *Honors & Awards:* Vaughn Prize, 85; Cole Prize in Number Theory, 87. *Mem:* Am Math Soc; Math Assoc Am. *Res:* Analytic number theory; automorphic forms and ecliptic curves. *Mailing Add:* Columbia Univ Math Dept New York NY 10027

GOLDFIELD, EDWIN DAVID, b New York, NY, Oct 26, 18. DEMOGRAPHY, ECONOMICS. *Educ:* City Univ New York, BS, 39; Columbia Univ, MA, 40. *Prof Exp:* Statist adv, City Court Spec Session, New York, 39; chief, stat reports div, US Bur Census, 55-68, asst dir, 68-71, chief, Int Stat Prog Ctr, 71-75; study dir, 75-78, exec dir, 78-87, SR ASSOC, COMT NAT STATIST, NAT ACAD SCI, 87- *Concurrent Pos:* Proj mgr, manpower res, US Econ Coop Admin, 51-52; tech adv, Social Sci Res Coun, 53; staff dir, subcomt Census & Statist, US House Rep, 59-60 & 67; mem bd dirs, Am Statist Asn, 85-89. *Mem:* Fel Am Statist Asn; Pop Asn Am; Int Asn Surv Statisticians; InterAm Statist Inst; Am Econ Asn; Int Statist Inst. *Res:* Federal statistical organization, policy and programs; demography; labor force economics; international development; survey and census procedures; privacy and confidentiality. *Mailing Add:* Nat Acad Sci 2101 Constitution Ave NW Washington DC 20418

GOLDFIEN, ALAN, b Brooklyn, NY, Apr 16, 23; m 50; c 6. MEDICINE. *Educ:* Univ Calif, AB, 46, MD, 50. *Prof Exp:* Asst meteorol, NY Univ, 43-44; res fel med, Harvard Univ, 53-56; from asst prof to assoc prof, 58-67, PROF MED, UNIV CALIF, SAN FRANCISCO, 67-, RES ASSOC PHYSIOL, 56-, MEM CARDIOVASC RES INST, 58-, ASSOC DEAN, SCH MED, 75- *Concurrent Pos:* Nat Cancer Inst res fel, 53-55, Nat Inst Arthritis & Metab Dis res fel, 56-57; Gianini fel med sci, Univ Calif, 57-58; asst, Peter Bent Brigham Hosp, Mass, 53-56. *Honors & Awards:* Bennett Prize, 58. *Mem:* AAAS; Am Physiol Soc; Endocrine Soc; Am Fedn Clin Res; Am Soc Clin Invest; Sigma Xi; Soc Gynec Invest. *Res:* Adrenal medullary secretion; nervous control of intermediary metabolism; physiology of reproduction; medical complications of pregnancy; signal transmission in smooth muscle. *Mailing Add:* 377 Vista Linda Mill Valley CA 94941

GOLDFINE, HOWARD, b Brooklyn, NY, May 29, 32; m 63; c 2. BIOCHEMISTRY, MICROBIOLOGY. *Educ:* City Col New York, BS, 53; Univ Chicago, PhD(biochem), 57. *Prof Exp:* Res fel, dept chem, Harvard Univ, 59-62; instr bact & immunol, Harvard Med Sch, 62-63, assoc, 63-66, asst prof, 66-68; assoc prof, 68-76, PROF MICROBIOL, SCH MED, UNIV PA, 76- *Concurrent Pos:* USPHS res fel, 58-60; Am Cancer Soc scholar, 60-63; tutor biochem sci, Harvard Univ, 60-67; res career develop award, Nat Inst Allergy & Infectious Dis, 65-68; mem physiol chem study sect, Div Res Grants, NIH, 69-73; chmn, Lipid Metabolism, Gordon Res Conf, 75; Josiah Macy Jr Found fac scholar, 76-77; assoc ed, J Lipid Res, 83-86; Fogarty sr int fel, 85; exec ed, Anal Biochem. *Honors & Awards:* Fel, Am Acad of Microbiol. *Mem:* AAAS; Am Soc Biochem & Molecular Biol; Am Soc Microbiol; Soc Gen Microbiol. *Res:* Microbial metabolism; lipid chemistry and metabolism; cell membrane structure and function. *Mailing Add:* Dept Microbiol Sch Med Univ Pa Philadelphia PA 19104

GOLDFINE, IRA D, b Chicago, Ill, Aug 10, 43. CELL BIOLOGY. *Educ:* Univ Ill, MD, 67; Am Bd Internal Med, cert, 72, cert endocrinol & metab, 72. *Prof Exp:* Intern, Michael Reese Hosp, Chicago, 67-68; resident med, Univ Chicago, 68-70; NIH fel, Nat Inst Arthritis & Metab Dis, Bethesda, Md, 70-71, staff assoc, 71-73; from asst prof to assoc prof, Dept Med, Univ Calif, San Francisco, 73-85, dir, Cell Biol Res Lab, Mt Zion Med Ctr, 79-87, PROF,

DEPT MED & PHYSIOL, UNIV CALIF, SAN FRANCISCO, 85-, ASSOC CHIEF MED, MT ZION MED CTR, 85-, DIR, DIV DIABETES & ENDOCRINE RES, 87- *Concurrent Pos:* Vis scientist, Dept Endocrinol, Karolinska Hosp, Stockholm, Sweden, 70; clin investr, Vet Admin Hosp, San Francisco, 77-78; mem sci rev comt, Juvenile Diabetes Found, 85-89; mem, Metab Study Sect, NIH, 86-87. *Honors & Awards:* Mary Jane Kugel Award, Juvenile Diabetes Asn, 88; Rosenthal Award, Am Col Physicians, 89. *Mem:* Fel Am Col Physicians; Am Fedn Clin Res; AAAS; Endocrine Soc; Am Thyroid Asn; Am Diabetes Soc; Am Soc Clin Invest; Am Soc Cell Biol; Am Asn Physicians. *Res:* Diabetes research; author of numerous scientific publications. *Mailing Add:* Div Diabetes & Endocrine Res Mt Zion Med Ctr Univ Calif 1600 Divisadero St PO Box 7921 San Francisco CA 94120

GOLDFINE, LEWIS JOHN, b London, Eng. PHYSICAL MEDICINE & REHABILITATION. *Educ:* Univ London, MB, BS, 61; Royal Col Physicians & Surgeons, dipl phys med, 67. *Prof Exp:* Asst prof, 69-73, ASSOC PROF REHAB MED, SCH MED, UNIV MD, 73- *Concurrent Pos:* Consult, Vet Admin Hosp, Fort Howard, Md, 85- *Mem:* Fel Am Acad Phys Med & Rehab; Am Asn Acad Physiatrists. *Res:* Arthritis rehabilitation; undergraduate and postgraduate education in rehabilitation medicine. *Mailing Add:* Univ Md Hosp Baltimore MD 21201

GOLDFINGER, ANDREW DAVID, b New York, NY, Mar 12, 45. PHYSICS, SYSTEMS ENGINEERING. *Educ:* Rensselaer Polytech Inst, BS, 65; Brandeis Univ, PhD(physics), 72; Johns Hopkins Univ, MS, 80. *Prof Exp:* PHYSICIST RES & DEVELOP, APPL PHYSICS LAB, JOHNS HOPKINS UNIV, 72- *Mem:* Am Geophys Union; Inst Elec & Electronics Engrs. *Res:* Remote sensing; atmosphere and ocean physics; orbital mechanics; navigation; image processing; radar; microwave systems; space technology. *Mailing Add:* Appl Physics Lab Johns Hopkins Rd Laurel MD 20723-6099

GOLDFISCHER, SIDNEY L, b New York, NY, Dec 28, 26; div; c 4. PATHOLOGY, CYTOCHEMISTRY. *Educ:* Columbia Univ, BS, 58; NY Univ, MD, 61. *Prof Exp:* From asst prof to assoc prof, 65-74, PROF PATH, ALBERT EINSTEIN COL MED, 74-, ACTG CHMN PATH, 84- *Concurrent Pos:* attend pathologist, Bronx Munic Hosp Ctr, 71-; dir, Office of Indust Liason, Albert Einstein Col Med, 83-; Burroughs Wellcome Prof, 88; assoc dean, Albert Einstein Col Med, 90- *Mem:* NY Acad Sci; Histochem Soc; Am Soc Cell Biol; Int Acad Path; Am Asn Path. *Res:* Development of cytochemical staining procedures for light and electron microscopy; application of these procedures which permit a synthesis of functional and morphological studies to problems in physiology and pathology at the subcellular level; detection and elucidation of peroxisomal diseases. *Mailing Add:* Dept Path Albert Einstein Col Med Bronx NY 10461

GOLDFRANK, MAX, b Glen Ridge, NJ, Jan 16, 11; m 34; c 2. CHEMICAL ENGINEERING. *Educ:* Columbia Univ, AB, 31, BS, 32, ChE, 33. *Prof Exp:* Chemist, Stein, Hall & Co, Celanese Corp, 34-44, asst mgr lab, NY, 44-45, chief chemist, Starch & Allied Res Lab, 45-46, chief chemist, Adhesive Lab, 46-47, exec asst to tech dir, 47-50, mgr labs, NY, 50-54, mgr develop, Starch & Allied Prod, 54-70, mgr patents & regulatory compliance, 70-76; RETIRED. *Concurrent Pos:* Consult. *Mem:* Am Chem Soc; Am Asn Cereal Chemists; Inst Food Technologists. *Res:* Starches; dextrines; adhesives; carbohydrate gums; food additives; hydrocolloids; food regulations. *Mailing Add:* 108 Castillion Terr Santa Cruz CA 95060-3254

GOLDFRIED, MARVIN R, b Brooklyn, NY, Jan 24, 36; m 67; c 2. PSYCHOTHERAPY, CLINICAL PSYCHOLOGY. *Educ:* Brooklyn Col, BA, 57; State Univ NY-Buffalo, PhD(clin psychol), 61; Am Bd Prof Psychol, dipl, 69. *Prof Exp:* Instr psychol, State Univ NY, Buffalo, 60-61; asst prof psychol, Univ Rochester, 61-64; PROF PSYCHOL, STATE UNIV NY, STONY BROOK, 64- *Concurrent Pos:* Vis assoc prof, Bar-Ilan Univ, Ramat Gan, Israel, 70-71; vis scholar, Univ Calif, Berkeley, 77-78; assoc ed, Cognitive Ther & Res, 77-82; prin investr, NIMH grant, 74, 66-68, 67-71, 73-84 & 84-88; mem adv bd, Soc Explor Psychother Integration, 83- *Mem:* Am Psychol Asn; Asn Advan Behav Ther; Soc Psychother Res; Soc Explor Psychother Integration. *Res:* Delineation of common therapeutic principles of change that can cut across various therapeutic orientations; overall objective is to get beyond the constraints of those language problems that prevent communication across the different orientations; focus on those various activities that effective clinicians use that may be functionally linked to ultimate successful outcome. *Mailing Add:* Dept Psychol State Univ NY-Stony Brook Stony Brook NY 11794

GOLDHABER, ALFRED SCHARFF, b Urbana, Ill, July 4, 40; m 69; c 2. THEORETICAL PHYSICS. *Educ:* Harvard Univ, AB, 61; Princeton Univ, PhD(physics), 64. *Prof Exp:* Miller fel pysics, Univ Calif, Berkeley, 64-66, lectr, 66-67; asst prof, 67-72, assoc prof, 72-77, PROF PHYSICS, STATE UNIV NY STONY BROOK, 77- *Concurrent Pos:* Vis staff mem, Los Alamos Sci Lab, 64-; US-USSR Acad Sci exchange vis, 70; NSF sr fel, Europ Orgn Nuclear Res, 71; sr vis, Lawrence Berkeley Lab, 77; vis prof, Univ Sussex, 78, Cambridge Univ, 86-87; Coorganizer, Int Sem High Energy Collisions Nuclei, Hakone, Japan, 80. *Mem:* Fel Am Phys Soc. *Res:* High energy theory; weak interactions; optical models; electrodynamics; nuclear physics; classical limits; magnetic monopoles. *Mailing Add:* Inst Theoret Physics State Univ NY at Stony Brook Stony Brook NY 11794-3840

GOLDHABER, GERSON, b Chemnitz, Ger, Feb 20, 24; nat US; m 69; c 3. ELEMENTARY PARTICLE PHYSICS. *Educ:* Hebrew Univ, MSc, 47; Univ Wisc, PhD, 50. *Hon Degrees:* PhD, Univ Stockholm, 86. *Prof Exp:* Asst physics, Hebrew Univ, 47 & Univ Wis, 48-50; instr, Columbia Univ, 50-53; from actg asst prof to assoc prof, 53-64, PROF PHYSICS, UNIV CALIF, BERKELEY, 64- *Concurrent Pos:* Asst res prof, Miller Inst Basic Res, 57-58; Ford Found fel, Europ Orgn Nuclear Res, 60-61; group leader, Lawrence Berkeley Lab, Berkeley, 62-; Guggenheim fel, Europ Orgn Nuclear Res, 72-73; prof, Miller Inst Basic Res, Univ Calif, 75-76 & 84-85; Morris Loeb lectr, Harvard Univ, 76-77; vis fel, CERN, 86. *Honors & Awards:* Pandofsky Prize, Am Phys Soc, 91. *Mem:* Nat Acad Sci; Fel Am Phys Soc; Swed Royal Acad Sci. *Res:* Experimental meson and antiproton interactions; charmed meson studies in electron-positron annihilation; high energy and elementary particle physics; astro-particle physics. *Mailing Add:* Dept Physics Lawrence Berkeley Lab Univ Calif Berkeley CA 94720

GOLDHABER, GERTRUDE SCHARFF, b Mannheim, Ger, July 14, 11; nat US; m 39; c 2. PHYSICS. *Educ:* Univ Munich, PhD(physics), 35. *Prof Exp:* Res assoc physics, Imp Col, Univ London, 35-39; res physicist, Univ Ill, 39-48, spec res asst prof physics, 48-50; from assoc physicist to physicist, Brookhaven Nat Lab, 50-62, sr physicist, 62-79; adj prof, Cornell Univ, 80-82; ADJ PROF, JOHNS HOPKINS UNIV, 82- *Concurrent Pos:* Consult, Argonne Nat Lab, 48-50 & Los Alamos Nat Lab, 53-; chmn, panel on eval of nuclear data compilations, Nat Acad Sci-Nat Res Coun, 69-71; mem bd trustees, Fermi Nat Accelerator Lab, 72-77; mem res adv comt, NSF, 72-74; mem report rev comt, Nat Acad Sci, 73-81, mem forum comt, 74-81; sci consult, Arms Control Disarmament Agency, 74-77; mem nom comt, Presidential Medal of Sci, 77-79; NAm rep, Europhysics J, Inst Physics, Bristol, London, 78-80; mem comt educ & employ women in sci & eng, Nat Res Coun Comn Human Resources, 78-83; mem Educ Adv Comt, NY Acad Sci, 82-; mem Nat Acad Sci Comt on Human Rights, 84-87; vis scholar, Nat Adv Comt Sci, Technol & Soc, 83-88. *Mem:* Nat Acad Sci; fel Am Phys Soc; fel AAAS; Sigma Xi. *Res:* Spontaneous fission neutrons; identity of beta-rays with atomic electrons; K-forbiddenness; odd-odd nuclei; long-lived isomers; variable moment of inertia law; parity violation in electromagnetic transitions; relation of moment of inertia to quadrupole moment; heavy ion physics; study of band structure in relation to nuclear dynamics, neutron proton interaction in pseudo-magic nuclei; relation of variable moment of inertia law to interacting boson approximation; discovery of heaviest pseudomajor nucleus at major number. *Mailing Add:* Dept Physics Brookhaven Nat Lab Upton NY 11973

GOLDHABER, JACOB KOPEL, b Brooklyn, NY, Apr 12, 24; m 51; c 3. ALGEBRA. *Educ:* Brooklyn Col, BA, 44; Harvard Univ, MA, 45; Univ Wis, PhD(math), 50. *Prof Exp:* Instr math, Univ Conn, 50-53 & Cornell Univ, 53-55; from asst prof to assoc prof math, Wash Univ, 55-60; res assoc prof, 60-61, head dept, 68-77, PROF MATH, UNIV MD, COLLEGE PARK, 61- *Concurrent Pos:* NSF sci fac fel, 66-67; exec secy, off math sci, Nat Res Coun, 75-82, actg dean grad studies & res, 84-85, 87- *Mem:* AAAS; Am Math Soc; Math Asn Am. *Mailing Add:* Dept Math Univ Md College Park MD 20740

GOLDHABER, MAURICE, b Lemberg, Austria, Apr 18, 11; nat US; m 39; c 2. NUCLEAR PHYSICS, FUNDAMENTAL PARTICLES. *Educ:* Cambridge Univ, PhD(physics), 36. *Hon Degrees:* PhD, Tel-Aviv Univ, 74; Dr, Univ Louvain-La-Neuve, 82, State Univ NY, Stony Brook, 83. *Prof Exp:* Charles Kingley Bye fel, Magdalene Col, Cambridge, 36-38; from asst prof to prof physics, Univ Ill, 38-50; sr scientist, Brookhaven Nat Lab, Assoc Univs, Inc, 50-60, chmn, Dept Physics, 60-61, dir, 61-73, distinguished scientist, 73-85, EMER DISTINGUISHED SCIENTIST, BROOKHAVEN NAT LAB, ASSOC UNIV, INC, 85- *Concurrent Pos:* Assoc ed, Phys Rev, 51-53; consult labs, AEC; mem nuclear sci comt, Nat Res Coun; mem bd gov, Weizmann Inst & Tel-Aviv Univ; adj prof physics, State Univ NY, Stony Brook, 61-; vis fel, Clare Hall, Cambridge, Eng, 67; chmn, Nuclear Phys Div, Am Phys Soc, 68 & Sect B Physics, AAAS, 80; consult, Nat Labs, Nat Res Coun, Adv Panel Physics, NSF & NY State Adv Coun Advan Indust Res & Develop; mem adv bd, Fla State Univ, Univ Ga & Univ Mich; mem sci coun, Ctr Theoret Studies, Univ Miami. *Honors & Awards:* Morris Loeb Lectr, Harvard Univ, 55; Tom W Bonner Prize in Nuclear Physics, Am Phys Soc, 71; Pauli Mem Lectr, Zurich, 80; J Robert Oppenheimer Mem Prize, 82; Nat Medal of Sci, 83; Leonard Schiff Mem Lectr, Stanford Univ, 84; Henry Newson Mem Lectr, Duke Univ, 84; Brickwedde Lectr, Johns Hopkins Univ, 85; Am Acad Achievement Award, 85; Rutherford Mem Lectr, Royal Soc, Can, 87; Peter Axel Mem Lectr, Univ Ill, 89; Samuel Goudsmit Mem Lectr, Univ Nev, 90; Boris Jacobsen Mem Lectr, Univ Wash, 90; Wolf Prize in Physics, Jerusalem, 91. *Mem:* Nat Acad Sci; fel Am Phys Soc (pres, 83); Am Acad Arts & Sci; AAAS; Am Philos Soc. *Res:* Nuclear physics; radioactivity; nuclear isomers, photoelectric effect and models; fundamental particles; electromagnetic transitions in nuclei and their role in elucidating nuclear structure; slow neutrons; nuclear theory; weak interactions; astrophysics. *Mailing Add:* Brookhaven Nat Lab Assoc Univs Inc Upton NY 11973

GOLDHABER, PAUL, b New York, NY, Mar 16, 24; m 49; c 2. DENTISTRY, PERIODONTAL DISEASES. *Educ:* New York Univ, DDS, 48; City Col New York, BS, 54; Am Bd Periodont, dipl, 54. *Hon Degrees:* MA, Harvard Univ, 62. *Prof Exp:* Asst ophthal res, Harvard Med Sch, 48-50; asst dent, Sch Dent & Oral Surg, Columbia Univ, 50; res assoc oral path, Harvard Sch Dent Med, 55-56, assoc, 56-59, from asst prof to assoc prof, 59-66, dir postdoctoral studies, 62-68, dean, 68-90, PROF PERIODONT, HARVARD SCH DENT MED, 66-, EMER DEAN, 90- *Concurrent Pos:* Res fel dent med, Harvard Sch Dent Med, 54-55; vis res fel, Sloan-Kettering Inst Cancer Res, 54-55, res fel, 55-56; USPHS sr res fel, 56-61 & res career develop award, 61-66; vol tissue-cult technician, Sloan-Kettering Inst Cancer Res, 46-48; asst ophthal res, Mass Eye & Ear Infirmary, 48-50; dent consult, Fulbright Med Sci Comt, 65; mem, Periodont Dis Adv Comt, Nat Inst Dent Res, 75-; mem, Nat Adv Dent Res Coun, NIH, 88-; vis prof, Univ Southern Calif, 90-91, Univ Calif, Los Angeles, 90-91, Fac Dent, Louis Pasteur Univ, France, 91. *Mem:* Inst Med-Nat Acad Sci; fel AAAS; Tissue Cult Asn; Int Asn Dental Res (pres, 85-86); Am Soc Cell Biol; fel NY Acad Sci; Am Dent Asn; Am Asn Cancer Res; Am Acad Periodont; Am Asn Dent Res. *Res:* Periodontal disease; oral carcinogenesis; bone transplantation; bone resorption and formation in tissue culture. *Mailing Add:* Harvard Sch Dent Med 188 Longwood Ave Boston MA 02115

GOLDHAMMER, PAUL, b Portland, Ore, Nov 10, 29. NUCLEAR PHYSICS. *Educ:* Reed Col, BA, 52; Wash Univ, PhD(physics), 56. *Prof Exp:* Asst prof physics, Univ Del, 56-57; from asst prof to prof physics, Univ Nebr, 57-64; PROF PHYSICS, UNIV KANS, 64- *Mem:* Fel Am Phys Soc. *Res:* Theoretical nuclear physics. *Mailing Add:* Dept Physics Univ Kans Lawrence KS 66045

GOLDHEIM, SAMUEL LEWIS, chemistry; deceased, see previous edition for last biography

GOLDHIRSH, JULIUS, b Philadelphia, Pa, Oct 2, 35; m 71; c 4. ELECTRICAL ENGINEERING. *Educ:* Drexel Inst, BS, 58; Rutgers Univ, MS, 60; Univ Pa, PhD(elec eng), 64. *Prof Exp:* Asst instr elec eng, Rutgers Univ, 58-59, res asst plasmas, 59-60; instr elec eng, Univ Pa, 60-65, asst prof, 65-71; assoc prof, Holon Univ Technol, Israel, 71-72; SR ENGR, APPL PHYSICS LAB, JOHNS HOPKINS UNIV, 72-, PRIN STAFF, 79-, GROUP SUPVR, SPACE GEOPHYS GROUP, SPACE DEPT, 79- *Concurrent Pos:* Consult, Raytheon Co, Mass, 63-68; mem Int Radio Consult Comt; instr, Johns Hopkins Univ, 85-; vchmn, Int Union Radio Sci, 91- *Mem:* Fel Inst Elec & Electronics Engrs; Int Union Radio Sci; Sigma Xi; Consultative Comt Int Radio. *Res:* Radio wave propagation; remote sensing from ground and satellites; radar meteorology, rain attenuation; mobile satellite system propagation. *Mailing Add:* Appl Physics Lab Johns Hopkins Rd Laurel MD 20707-6099

GOLDHOR, SUSAN, b Brooklyn, NY, Mar 24, 39. SCIENCE EDUCATION. *Educ:* Barnard Col, AB, 60; Yale Univ, MS, 62, PhD(biol), 67. *Hon Degrees:* DHL, Wyndham Col, 74. *Prof Exp:* Am Cancer Soc res fel, Dept Biol Sci, Stanford Univ, 67-68, NIH res fel, 68-69; asst prof biol, Hacettepe Univ, Turkey, 69-71; vis fel biol, Yale Univ, 71-73; assoc prof biol & dean natural sci, 73-77, dir New Eng Farm Ctr, Hampshire Col, 78-81; PRES, CTR FOR APPL REGIONAL STUDIES, 81- *Concurrent Pos:* Mem, Adv Comt Sci Educ, NSF, 75-78; mem, Biomass Conversion, Advan Study Inst, NATO, 82; leader, Northeastern Regional Aquaculture Ctr Proj, USDA, 90-93. *Mem:* Inst Food Technologists; New Eng Fisheries Develop Found; World Aquacult Soc. *Res:* Economic development of natural resources based industries; technology transfer; waste utilization; developing environmentally benign and economically acceptable technologies for utilising high protein processing waste and developing products from those waters. *Mailing Add:* 45B Museum St Cambridge MA 02138-1921

GOLDICH, SAMUEL STEPHEN, b Grand Forks, NDak, Jan 17, 09. GEOLOGY. *Educ:* Univ Minn, AB, 29, PhD(geol), 36; Syracuse Univ, AM, 30. *Prof Exp:* Asst geol, Mo Sch Mines, 30-32; fel, Wash Univ, 32-34; asst & chemist, Rock Anal Lab, Univ Minn, 34-36; from instr to assoc prof geol, Agr & Mech Col, Tex, 36-41; geologist US Geol Surv, 42-45, 47-48 & 59-60, chief, Br Isotope Geol, 60-64; from assoc prof to prof geol, Univ Minn, 48-50; prof geol & geochem & dir mineral constitution labs, Pa State Univ, 64-65; prof geol, State Univ NY Stony Brook, 65-68; prof, 68-77, EMER PROF GEOL, NORTHERN ILL UNIV, 77- *Concurrent Pos:* Assoc prof & actg prof, Agr & Mech Col, Tex, 45-46; geologist, Bur Econ Geol, Tex, 46-48; adj prof, Colo Sch Mines, 77- *Honors & Awards:* Goldschmidt Award, Geochem Soc, 83. *Mem:* Fel Am Geophys Union; fel Geol Soc Am; fel Mineral Soc Am; Am Chem Soc; Geochem Soc; Am Asn Petrol Geologists. *Res:* Areal geology in Trans-Pecos Texas; origin of laterite and bauxite; petrology; geochemistry; geochronology. *Mailing Add:* US Geol Surv Denver Fed Ctr Box 25046 MS 963 Denver CO 80225

GOLDIE, JAMES HUGH, b Windsor, Ont, Jan 16, 37; m 70; c 1. MEDICAL ONCOLOGY. *Educ:* Univ Toronto, MD, 61; FRCPS(C), 66. *Prof Exp:* Asst prof med, Univ Toronto, 70-76; from asst prof to assoc prof med, 76-84, CLIN PROF MED, UNIV BC, 84-; HEAD ONCOL, DIV ADVAN THERAPEUT, CANCER CONTROL AGENCY OF BC, 76- *Concurrent Pos:* R S McLaughlin Found fel, 69-70; Med Res Coun Can grant, 70-73; Ont Cancer Treatment & Res Found grant, 73-76; chmn invest drugs comt, Nat Cancer Inst Can, 78-82, grant, 80-; mem bd sci counrs, Nat Cancer Inst, 82-85; assoc ed, Cancer Res, 83-86; mem adv coun res, Nat Cancer Inst Can, 86-89; mem, ad hoc drug screening comt, Nat Cancer Inst US, 86-89. *Honors & Awards:* Terry Fox Medal, 82; Camo Lectr, 90. *Mem:* Am Soc Clin Oncol; Am Asn Cancer Res; Can Oncol Soc; Royal Col Physicians Can. *Res:* Clinical and cellular pharmacology of antineoplastic drugs. *Mailing Add:* Cancer Control Agency BC 600 Tenth Ave Vancouver BC V5Z 4E6 Can

GOLDIE, MARK, developmental biology; deceased, see previous edition for last biography

GOLDIN, ABRAHAM SAMUEL, b Brooklyn, NY, Apr 22, 17; m 45; c 3. CHEMISTRY. *Educ:* Columbia Univ, AB, 37, AM, 41; Univ Tenn, PhD, 51. *Prof Exp:* Control chemist, Eastern Wine Corp, NY, 41; prod chemist, Mutual Chem Co, NJ, 42; res asst, sam labs, Columbia Univ, 42-45; tech engr & chemist, Carbide & Carbon Chem Corp, 45-50; radiochemist, USPHS, 51-60; chem dir, Nat Lead Co, Inc, 60-61; assoc prof, NY Univ, 61-62; dep officer in charge, Northeastern Radiol Health Lab, 62-68; assoc prof, Sch Pub Health, Harvard Univ, 68-74; dir tech opers, Radiation Mgt Corp, 74-76; ENVIRON SCIENTIST, US ENVIRON PROTECTION AGENCY, 77- *Mem:* AAAS; Am Chem Soc; Health Physics Soc; Am Inst Chemists; Am Pub Health Asn. *Res:* Assessment of envionmental radioactivity and radiation; radio analytical quality assurance; radiation protection standards; measurement of radioactivity and radiation; radioactive waste management and disposal. *Mailing Add:* 1505 Columbia Ave Rockville MD 20850

GOLDIN, EDWIN, b Philadelphia, Pa, Oct 12, 38; m 75; c 5. QUANTUM OPTICS, COMPUTER EDUCATION. *Educ:* Temple Univ, BA, 59; Polytech Inst Brooklyn, MA, 61; Polytech Inst NY, PhD(physics), 73. *Prof Exp:* Asst prof physics, Maritime Col, State Univ NY, 66-68; dir, Univ Air, Queens Col, NY, 68-70; asst prof, Fordham Univ NY, 70-76; asst prof math & physics, Ramapo Col NJ, 76-81; proj dir, NSF Cause Comput grant, Bethany Col, WVa, 81-84, dir Acad Comput Ctr & Dual Degree Eng Prog, 81-88, prof physics & head dept, 81-88; prof officer, Fund Improv Post-Sec Educ, US Dept Educ, 88-90; ASST ED, DIV MGR & ASSOC DIR SOCIAL PHYSICS STUDENTS, AM INST PHYSICS, 90- *Concurrent Pos:* Cong scientist fel, Am Phys Soc, 87-88; sci adv & legis asst to US Rep, Edward Markey, 87-88; consult solar design, Hasco, Inc, 80; chnm, Comt Physics Higher Educ, Am Asn Physics Teachers, 84-86; bd mem, comp in educ, Carnegie-Mellon Univ, 86-87; dir, hon Sci Appln Prog, 84-87. *Mem:* Am Asn Physics Teachers; AAAS; Am Phys Soc. *Res:* Coherent states of quantized fields and correlation in quantum optics; computer graphics and animation for instructional quantum theory films; solar design and engineering; computer education. *Mailing Add:* Am Inst Physics 1825 Connecticut Ave NW Suite 213 Washington DC 20009

GOLDIN, GERALD ALAN, b Brooklyn, NY, Oct 16, 43; m 68; c 2. MATHEMATICAL PHYSICS, MATHEMATICS EDUCATION. *Educ:* Harvard Univ, BA, 64; Princeton Univ, MA, 66, PhD(physics), 69. *Prof Exp:* Res assoc physics, Univ Pa, 68-70, asst prof educ, 71-77; from asst prof to assoc prof math sci, Northern Ill Univ, 77-85; assoc prof, 84-85, PROF MATH, PHYSICS & COMPUTER EDUC, RUTGERS UNIV, 85- *Concurrent Pos:* US Off Educ fel, Univ Pa, 70-71; vis lectr, Cabrini Col, 74-78 & Beaver Col, 75-78; vis staff mem, Los Alamos Nat Lab, 77-; vis fel, Princeton Univ, 82-83; vis assoc prof, Teachers Col, Columbia Univ, 82-83; guest prof, Tech Univ Clausthal, 86-; vis distinguished prof, Northern Ill Univ, 90-91. *Mem:* Nat Coun Teachers Math; Math Asn Am; Am Educ Res Asn; Am Math Soc; Int Asn Math Physics. *Res:* Group representations in mathematical physics; science and mathematics education; the psychology of mathematical problem solving. *Mailing Add:* Ctr Math Sci & Computer Educ Rutgers Univ Piscataway NJ 08855

GOLDIN, MILTON, b New York, NY, July 26, 17; m 51; c 3. BACTERIOLOGY. *Educ:* Brooklyn Col, BS, 38; Univ Ill, MS, 48; Chicago Med Sch, PhD, 70. *Prof Exp:* Bacteriologist, USPHS, 41-42; chief bacteriologist, Hektoen Inst, Cook County Hosp, 46-48; DIR MICROBIOL DEPT, MT SINAI HOSP MED CTR, 48- *Concurrent Pos:* Assoc prof, Rush Med Sch, 75- *Mem:* Am Soc Microbiol; Sigma Xi. *Res:* Factors determining virulence of infectious organisms; immunological studies in collagen diseases; mycotic infections in humans. *Mailing Add:* 1500 Sheridan Rd No 9D Wilmette IL 60091

GOLDIN, STANLEY MICHAEL, b New York, NY, 48; m 75; c 2. BIOCHEMISTRY, NEUROBIOLOGY. *Educ:* Mass Inst Technol, SB, SM, 70; Harvard Univ, PhD(biochem), 77. *Prof Exp:* Proj leader membrane develop, Millipore Corp, 70-72; Harvard Soc Fel biochem, 76-79, from asst prof to assoc prof pharmacol, 79-89, LECTR BIOCHEM, HARVARD MED SCH, 90-; DIR PHARMACOL, CAMBRIDGE NEUROSCI INC, 90- *Concurrent Pos:* Consult, Millipore Corp, 71-75; McKnight scholar, 81 & Rita Allen Scholar, 85. *Honors & Awards:* Chem Commendation, Am Chem Soc, 65; Searle Award, 81. *Mem:* Sigma Xi; AAAS. *Res:* Membrane biochemistry and physiology; biochemistry and organization of nervous system; molecular basis of neuronal electrical activity. *Mailing Add:* Cambridge Neurosci Inc One Kendall Sq Bldg 700 Cambridge MA 02139

GOLDING, BRAGE, b Chicago, Ill, Apr 28, 20; m 41; c 3. POLYMER CHEMISTRY, EDUCATION ADMINISTRATION. *Educ:* Purdue Univ, BS, 41, PhD(chem eng), 48. *Hon Degrees:* LLD, Wright State Univ. *Prof Exp:* Asst dir res, Lilly Varnish Co, 48-57, dir, 57-59; prof chem eng & head sch, Purdue Univ, 59-66, res assoc, 48-57, vis prof, 57-59; vpres, Wright State Campus, Miami Univ-Ohio State Univ, 66, pres, Wright State Univ, 67-72; pres, San Diego State Univ, 72-77; pres, Kent State Univ, 77-82; pres, Metrop State Col, 84-85; ACTG PRES, WESTERN STATE COL, 85- *Mem:* Fel AAAS; Am Chem Soc; Soc Plastics Engrs; Am Soc Testing & Mat; Am Inst Chem Engrs. *Res:* Surface coatings technology; polymer technology and research; organic technology. *Mailing Add:* 17 Dorset Lane Bedminster NJ 07921

GOLDING, BRAGE, JR, b Ft Bragg, NC, Sept 24, 42; m 64; c 2. EXPERIMENTAL SOLID STATE & ACOUSTICS PHYSICS. *Educ:* Purdue Univ, BMetE, 63; Mass Inst Technol, PhD(mat sci), 66. *Prof Exp:* Res asst mat sci, Mass Inst Technol, 63-66, res assoc, 66-67; mem tech staff, 67-81, head, Condensed State Phys Res Dept, 81-84, HEAD, NONEQUILIBRIUM PHYSICS RES DEPT, BELL LABS, 85- *Honors & Awards:* Humboldt Prize. *Mem:* AAAS; fel Am Phys Soc. *Res:* Experimental research on the behavior of high frequency sound waves propagating in solids undergoing phase transitions; the nature of excitations in amorphous solids at ultra low temperatures; optical properties of glasses; novel superconductors; low temperature physics; resonance; noise in mesoscopic systems. *Mailing Add:* Bell Lab Rm 1A-161 600 Mountain Ave Murray Hill NJ 07974

GOLDING, DOUGLAS LAWRENCE, b St John, Can, May 5, 31; m 56; c 3. FOREST HYDROLOGY. *Educ:* Univ NB, Fredericton, BSc, 53; Purdue Univ, MS, 61; Univ BC, PhD(forest hydrol), 68. *Prof Exp:* From forester mgt to forester inventory, Sask Dept Natural Resources, 54-64; res scientist forest hydrol, Can Forestry Serv, Can Dept Environ, 67-78; PROF FOREST HYDROL, UNIV BC, 78- *Mem:* Can Inst Forestry; NZ Hydrol Soc; Can Meteorol Soc. *Res:* Snowpack management for water yield and regime through forest cover manipulation; snow ablation during chinook conditions; energy balances in forest and forest openings; erosion and slope stability; water yield and regime changes related to forest management. *Mailing Add:* Fac Forestry MacMillan Bldg Univ BC 2075 Westbrook Mall Vancouver BC V6T 1W5 Can

GOLDING, HANA, b Jerusalem, Israel, Dec 26, 49; US citizen; m; c 2. VIROLOGY. *Educ:* Hebrew Univ, Jerusalem, BSc, 72; Ore Health Sci Univ, PhD(immunol), 81. *Prof Exp:* Res asst, Cell & Clin Immunol Lab, SAfrican Inst Med Res, 73-75; sr staff fel, 87-89, SR INVESTR, DIV VIROL, CBER, FOOD & DRUG ADMIN, 89- *Concurrent Pos:* Tartar Found res award, 78-80; NIH AIDS res grants, 87-88 & 91. *Mem:* Am Asn Immunologists. *Res:* AIDS; author of numerous scientific publications. *Mailing Add:* Dept Virol CBER Food & Drug Admin Bldg 29A Rm 2C-15 8800 Rockville Pike Bethesda MD 20892

GOLDING, LEONARD S, b New York City, NY, June 8, 35; US citizen. SATELLITE COMMUNICATIONS, TELEVISION TRANSMISSION. *Educ:* Columbia Univ, BA, 57, BS, 58; Yale Univ, MS, 60, DEngr, 66. *Prof Exp:* Res asst, Bell Tel Labs, 60-62; sr tech staff, Raytheon Corp, 62-67; dir, Transmission Systs Lab, 67-75; founder, Digital Commun, 71-75, vpres & dir, Macon Res Ctr, 82-85, VPRES, RES & DEVELOP, HUGHES AIRCRAFT, 75-, VPRES, ENG NETWORK SYSTS, 87- *Concurrent Pos:* Chmn, US Deleg Working Group, Inter Telecommun Union, 70-75; mem, Fed Commun Comn Advan Tel Inquiry, 88. *Mem:* Sigma Xi; fel Inst Elec & Electronics Engrs. *Mailing Add:* Hughes Network Syst Inc 11717 Exploration Lane Germantown MD 20874

GOLDINGS, HERBERT JEREMY, b Boston, Mass, May 28, 29; m 55; c 3. PSYCHIATRY. *Educ:* Harvard Univ, AB, 50, MD, 54; Boston Psychoanal Soc Inst, grad, 65; Am Bd Psychiat & Neurol, dipl, 61, cert child psychiat, 64. *Prof Exp:* Intern med, Boston City Hosp, 54-55; asst psychiat, Harvard Med Sch, 60-65, instr psychiat, 65-69, res psychiat, 55-58, assoc dir child psychiat, 65-77; SR PHYSICIAN, CHILD PSYCHIAT SERV, MASS MENT HEALTH CTR, 61- *Concurrent Pos:* Teaching fel, Harvard Med Sch, 55-57, res fel psychiat, 57-58 & 60-61; consult, Div Legal Med, Mass, 57-58, 5040th US Air Force Hosp, Alaska, 58-60, & Parents Asn Retarded Children, Alaska, 58-60; asst examr, Am Bd Psychiat & Neurol, 64-; mem fac, Boston Psychoanal Soc & Inst, 68-, training & supv analyst, 73-; asst clin prof psychiat, Harvard Med Sch, 69-77; training & supv analyst, Psychoanal Inst New Eng, 75- *Mem:* Fel Am Psychiat Asn; AMA; fel Am Acad Child Psychiat; Am Psychoanal Asn; Int Psychoanal Asn. *Res:* Child psychiatry; psychoanalysis; medical education. *Mailing Add:* 65 Sterling St West Newton MA 02165-2614

GOLDISH, DOROTHY MAY (BOWMAN), b NJ, May 6, 34; m 58; c 2. ORGANIC CHEMISTRY. *Educ:* Stanford Univ, BS, 55; Univ Calif, PhD(chem), 58. *Prof Exp:* From asst prof to assoc prof, 58-73, PROF CHEM, CALIF STATE UNIV, LONG BEACH, 73- *Mem:* Am Chem Soc. *Res:* Heterocyclic chemistry; natural products. *Mailing Add:* Dept Chem Calif State Univ Long Beach CA 90840

GOLDISH, ELIHU, b Marietta, Ohio, Oct 18, 28; m 58; c 2. PHYSICAL CHEMISTRY. *Educ:* Marietta Col, BSc, 49; Calif Inst Technol, PhD(chem), 56. *Prof Exp:* Fel, Ohio State Univ, 56-57; res fel chem, Univ Southern Calif, 58-60; res chemist, Res Ctr, Union Oil Co Calif, 60-86; ADJ PROF, DEPT GEOL SCI, CALIF STATE UNIV, LONG BEACH, 86- *Concurrent Pos:* Fel, Univ Calif, Los Angeles, 66; fel, Calif Inst Technol, 74. *Mem:* Mineral Soc Am; Royal Soc Chem; Clay Minerals Soc; Am Crystallog Asn. *Res:* Materials characterization by x-ray diffraction and x-ray spectrometry. *Mailing Add:* Dept Geol Sci Calif State Univ Long Beach Long Beach CA 90840

GOLDKNOPF, IRA LEONARD, b Mar 13, 46; m; c 4. CHEMISTRY. *Educ:* Hunter Col, BA, 67; Kans State Univ, PhD(biochem), 71. *Prof Exp:* Postdoctoral fel develop therapeut, M D Anderson Hosp, Houston, 71-72; fel pharmacol, Col Med, Baylor Univ, 72-74, Nat Cancer Inst instr, 74-75, asst prof, 75-85; qual assurance supvr, 85-88, MGR QUAL ASSURANCE, GAF CHEM PLANT, CALVERT CITY, KY, 88- *Res:* Structure and function of proteins of the cell nucleus and their roles in the control of gene expression; management development. *Mailing Add:* GAF Chem Corp PO Box 37 Calvert City KY 42029

GOLDMACHER, VICTOR S, b Chernowitz, USSR, Dec 25, 52; US citizen; m 80; c 2. PHARMACOLOGY, CELL BIOLOGY. *Educ:* Moscow Univ, MS, 74, PhD(enzym), 77. *Prof Exp:* Assoc genetic toxicol, Mass Inst Technol, 81-83; assoc path, 83-85, asst prof path, 85-88, LECTR PATH, HARVARD MED SCH, 88-; HEAD CELL BIOL DEPT & LECTR PATH, IMMUNOGEN, INC, 88- *Mem:* Am Soc Cell Biol. *Res:* Cell biology of protein toxins; mechanisms of action of immunotoxins; mechanisms of endocytosis; interaction of monoclonal antibodies and their toxin conjugates with cultured cells. *Mailing Add:* Immunogen Inc 148 Sidney St Cambridge MA 02139

GOLDMAN, AARON SAMPSON, b Red Lion, Pa, Feb 8, 32. STATISTICS. *Educ:* Okla State Univ, PhD, 61. *Prof Exp:* Math statistician, Los Alamos Nat Lab, 60-65; assoc prof, Gonzaga Univ, 65-69; prof math, Univ Nev, Las Vegas, 69-79; mem nuclear safeguards, Los Alamos Nat Labs, 79-88; mem, Int Atomic Energy Agency, 88-91; LAB ASSOC, LOS ALAMOS NAT LABS, 91- *Mem:* Am Statist Asn; Sigma Xi. *Res:* Nuclear safeguards. *Mailing Add:* 4723 Sandia Los Alamos NM 87544

GOLDMAN, ALLAN LARRY, b Minneapolis, Minn, June 3, 43; m 69; c 4. PULMONARY DISEASES, CRITICAL CARE MEDICINE. *Educ:* Univ Minn, Minneapolis, BA & BS, 64, MD, 68. *Prof Exp:* PROF MED & DIR, DIV PULMONARY, CRITICAL CARE & OCCUP MED, COL MED, UNIV S FLA, TAMPA, 74- *Mem:* Am Thoracic Soc; Am Col Physicians; Am Col Chest Physicians. *Res:* Respiratory surveillance; carbon monoxide and smoking; cyclic adenosine monophosphates in lung tissue and drug effects. *Mailing Add:* Tampa Vet Admin Hosp 13000 Bruce B Downs Blvd Tampa FL 33612

GOLDMAN, ALLEN MARSHALL, b New York, NY, Oct 18, 37; m 60; c 3. CONDENSED MATTER PHYSICS. *Educ:* Harvard Univ, AB, 58; Stanford Univ, PhD(physics), 65. *Prof Exp:* From asst prof to assoc prof, 65-75, PROF PHYSICS, UNIV MINN, MINNEAPOLIS, 75- *Concurrent Pos:* Alfred P Sloan Found fel, 66-70; co-chmn, Adv Study Inst, NATO, Les Arcs, France, 83. *Mem:* Fel Am Phys Soc; fel AAAS; Sigma Xi. *Res:* Experimental condensed matter physics; superconductivity, electron tunneling in superconductors; time dependent effects, fluctuation phenomena, superconducting devices and materials; properties of disordered and dimensionally constrained materials; high temperature superconductors. *Mailing Add:* Sch Physics & Astron Univ Minn Minneapolis MN 55455

GOLDMAN, ALLEN S, b Providence, RI, Oct 25, 29; m 81; c 3. TERATOLOGY, PHARMACOLOGY. *Educ:* Brown Univ, AB, 51, ScM, 53; State Univ NY, MD, 58. *Hon Degrees:* ScM, Univ Pa, 71. *Prof Exp:* Instr pediat, Univ Pa, 61-64, assoc, 64-66, from asst prof to assoc prof pediat, 66-78, res prof pediat & pediat pharmacol, 78-85; PROF PEDIAT & GENETICS, UNIV ILL, CHICAGO, 86- *Concurrent Pos:* Res career develop award, NIH, 66; reviewer, Nat Inst Dent Res, NADRC, 77 & 84; mem, Res Comt Ment Retardation, Nat Inst Child Health & Human Develop, 79, consult, five year plan for congenital defects, 80; reviewer, Toxicol Study Sect, NIH, 85; mem adv comt, Nat Inst Dent Res, 89- *Mem:* Sigma Xi. *Res:* Causes and prevention of birth defects: discovered that some birth defects results from a deficiency of phosphatidylinositol turnover and arachidonic acid leading to prostagladins, the defects can be prevented by supplemental myoinsitol, arachidonic acid or prostaglandins. *Mailing Add:* Ctr Craniofacial Anomalies Univ Ill MC 588 808 S Wood St Chicago IL 60612

GOLDMAN, ANNE IPSEN, b Copenhagen, Denmark, Apr 13, 35; US citizen; m 67; c 3. DESIGN DATABASES. *Educ:* Radcliffe Col, BA, 56; Harvard Univ, PhD(statist), 71. *Prof Exp:* Asst prof, 70-78, assoc prof biomet, 78-84, PROF BIOSTAT, SCH PUB HEALTH, UNIV MINN, MINNEAPOLIS, 84- *Concurrent Pos:* Biostatistician, Nat Inst Allergy & Infectious Dis, Comm Proj Clin Res in AIDS. *Mem:* Am Statist Asn; Biomet Soc; Soc Clin Trials. *Res:* Design, management and analysis of research databases, especially for clinical research in AIDS; design, management and analysis of multiclinic clinical trials; multivariate methods of prediction of survival based on risk factors. *Mailing Add:* Sch Public Health Univ Minn Box 197 Minneapolis MN 55455

GOLDMAN, ARMOND SAMUEL, b San Angelo, Tex, May 26, 30; m 50; c 5. PEDIATRICS, IMMUNOLOGY. *Educ:* Univ Tex, MD, 53; Am Bd Pediat, dipl, 60; Am Bd Allergy & Immunol, dipl, 74. *Prof Exp:* From instr to assoc prof pediat, 58-72, PROF PEDIAT & HUMAN BIOL CHEM & GENETICS, UNIV TEX MED BR, GALVESTON, 72-, DIR DIV IMMUNOL-ALLERGY, 59- *Concurrent Pos:* Prin investr, Nat Inst Child Health & Human Development, 64-67, 68-71, & 78-81; vis lectr, Univ Olu, Finland, 75; vis prof, Univ Tromso, Norway & Univ Goteborg, Sweden, 75; invited speaker, Nat Inst Child & Human Develop, 76; mem Nat comt, Nat Inst Child Health & Human Develop, 76-77; co-organizer, Int Conf, Maternal-Environ Effects Human Lactation, Oaxaca, Mex, 86, prin organizer, Effects Human Milk on Recipient Infant, Konstanz, Ger, 86; inv speaker, Columbia Univ, Nutrit Inst, 86, hon speaker, Int Congress Neonatal Hemat & Immunol, Siena, Italy, 90, inv speaker, Int Congress Mucosal Immunol, Tokyo, Japan, 90, inv partic & speaker, Immunol Milk & Neonate Conf, Miami, Fla, 90. *Mem:* Am Asn Immunol; Reticuloendothelial Soc; Soc Pediat Res; Am Pediat Soc; Am Acad Pediat. *Res:* Immunobiology of human milk; the ontogeny of the immune systems; molecular biology of genetic immunologic defects. *Mailing Add:* Dept Pediat Child Health Ctr-230 Univ Tex Med Br Galveston TX 77550

GOLDMAN, ARNOLD I, cell biology, biophysics, for more information see previous edition

GOLDMAN, ARTHUR JOSEPH, b New York, NY, Aug 14, 34; m 57; c 2. NUCLEAR & CHEMICAL ENGINEERING. *Educ:* City Col NY, BChE, 57; New York Univ, MChE, 61, PhD(chem eng), 66; Univ Chicago, MBA, 80. *Prof Exp:* Consult nuclear eng, United Nuclear Corp, 57-66; proj mgr chem eng, Exxon Corp, 66; vpres & tech dir nuclear eng, Nuclear Technol Corp, 67-70; dir fuel mgt servs, S M Stoller Corp, 71; vpres appl sci, Transfer Systs Inc, 72; ASSOC DIV DIR & SR NUCLEAR ENG, ARGONNE NAT LAB, 73- *Concurrent Pos:* AEC fel nuclear eng, 64-65, adj assoc prof chem eng, NY Univ, 66-71. *Mem:* Am Nuclear Soc. *Res:* Nuclear reactor safety; nuclear fuel design and performance; engineering of energy production systems. *Mailing Add:* Argonne Nat Lab 9700 South Cass Ave Argonne IL 60439

GOLDMAN, BRUCE DALE, b Gary, Ind, Dec 11, 40; c 3. PINEAL GLAND, PHOTOPERIODISM. *Educ:* Univ Mich, BS, 62; Univ Wis, MS, 66, Med Col Ga, PhD(endocrinol), 68. *Prof Exp:* Postdoctoral fel, Univ Tex Southwestern Med Sch, 68-69, Univ Calif, Los Angeles, 69-70; prof endocrinol, Worcester Found Exp Biol, 81-87; asst prof biobehav sci, 70-80, PROF PHYSIOL & NEUROBIOL, UNIV CONN, 87- *Mem:* Soc Study Biol Rhythms; Soc Study Reproduction. *Res:* Neuroendocrine basis of photoperiodism in vertebrates, pineal gland, regulation of pituitary gonadotropins and prolactin. *Mailing Add:* Dept Physiol & Neurobiol Univ Conn Storrs CT 06269

GOLDMAN, CHARLES REMINGTON, b Urbana, Ill, Nov 9, 30; m 74; c 3. LIMNOLOGY. *Educ:* Univ Ill, Urbana, BA, 52, MS, 55; Univ Mich, PhD(fisheries & limnol), 58. *Prof Exp:* Asst aquatic biol, State Natural Hist Surv, Ill, 54-55; asst fisheries, Univ Mich, 55-57; biologist fisheries res, US Fish & Wildlife Serv, Alaska, 57-58; from instr to prof zool, 58-71, dir Inst Ecol, 66-69, PROF DIV ENVIRON STUDIES, UNIV CALIF, DAVIS, 71-, CHMN, 88- *Concurrent Pos:* Consult biologist, Eng Res Inst, Univ Mich, 56-57; consult, Fisheries Res Inst, Univ Wash, 61 & Eng-Sci, Inc, Calif, 62-63; NSF sr fel, 64; Guggenheim fel, 65; consult, Atty Gen, State of Minn, 70 & Environ Protection Agency, 71-; mem sci & technol adv coun, Calif State Assembly, 70-75; mem sci exchange comn to Russia, Food & Agr Orgn & UN Develop Prog, Environ Protection Agency, 73; mem, Calif State Solid Waste Mgt Bd, 73-77; consult to design environ improvement, UN Develop Prog, Papua New Guinea, 74-75, Parana River Flood Control, Argentina, 79, El Cajon Dam & Reservoir studies, Honduras, 79-84, Paute Mazar Dams Ecuador, 81; mem subcomt, US Nat Comt for Man & the Biosphere Prog, UNESCO, 75-, chmn, 88-; Fulbright distinguished prof, Yugoslavia, 85; adv comt, Oak Ridge Nat Lab, 85-87. *Mem:* Am Inst Biol Sci; Int Asn Trop Ecol; Int Soc Theoret & Appl Limnol; Am Soc Limnol & Oceanog (pres); Int Asn Astacology; Orgn Trop Studies. *Res:* Aquatic biology-limnology; biological productivity, eutrophiation and nutrient limiting factors; field work on African, Alaskan, Antarctic, Argentine, Brazil (Amazonia), Ecuadorian, Central American, Philippine, Oregonian, Californian, Swedish, Lapland, Italian, New Guinea, USSR, Lake Baikal and New Zealand; documentary films on Lake Tahoe and Tropics. *Mailing Add:* Div Environ Studies Univ Calif Davis CA 95616

GOLDMAN, DANIEL WARE, b Mar 28, 52; m 76; c 2. PHARMACOLOGY, CELL BIOLOGY. *Educ:* Harvard Univ, PhD(biochem), 79. *Prof Exp:* ASST PROF IMMUNOL, JOHNS HOPKINS UNIV, 86- *Mem:* Am Asn Immunol; Am Asn Pathol; Am Soc Cell Biol. *Res:* Elucidate the signal transduction mechanisms which mediate the activation of leukocytes by lipid mediators of inflammation; to define the biochemical events which regulate leukocyte activation. *Mailing Add:* Div Clin Immunol Good Samaritan Hosp Johns Hopkins Univ 301 Bayview Blvd Baltimore MD 21224

GOLDMAN, DAVID, b Galveston, Tex, July 7, 52; m 72; c 3. BEHAVIORAL GENETICS, POPULATION GENETICS. *Educ:* Yale Univ, BS, 74; Univ Tex, MD, 78. *Prof Exp:* INSTR HUMAN GENETICS, GEORGE UNIV, 84-; CHIEF, SECTION GENETIC STUDIES, LAB CLIN STUDIES, NAT INST ALCOHOL ABUSE & ALCOHOLISM, BETHESDA, MD, 88- *Concurrent Pos:* Clin assoc, Lab Clin Sci, Nat Inst Ment Health, Bethesda, Md, 80-81, staff physician, 81-84; chief, Unit on Genetic Studies, Lab Clin Studies, Nat Inst Alcohol Abuse & Alcoholism, Bethesda, Md, 84-87; adj prof, Dept Biol Sci, George Washington Univ, 87- *Mem:* Am Soc Human Genetics; Soc Alcoholism. *Res:* Genetic linkage studies for alcoholism and neurobehavioral differences and molecular cloning of alcohol dehydrogenases and genes involved in serotonergic function. *Mailing Add:* Clinic Studies Lab Ten NIAAA Bldg Bethesda MD 20205

GOLDMAN, DAVID ELIOT, b Boston, Mass, Aug 11, 10; m 38; c 1. BIOPHYSICS, PHYSIOLOGY. *Educ:* Harvard Univ, AB, 31; Columbia Univ, PhD(physiol), 43. *Prof Exp:* Lab asst physiol, Mass Dept Ment Health Comn, Boston, 31-36; instr physiol, Col Physicians & Surgeons, Columbia Univ, 40-41, asst surg, 42-43, res assoc physics, Nat Defense Res Comt, 43-44; mem staff, Naval Med Res Inst, 44-47, head biophys div, 47-55, sci liaison officer, Off Naval Res, London, 55-57, head biophys div, Naval Med Res Inst, 57-67; prof, 67-76, dir grad prog, 70-76, EMER PROF PHYSIOL & BIOPHYS, MED COL PA, 76- *Concurrent Pos:* Mem comt on hearing & bioacoust, Nat Res Coun, 53-74, mem comt biol & agr, 68-71, mem coun, 70-74; mem biophys study sect, NIH, 58-64, mem study comt biophys, President's Off Sci & Technol, 64, mem adv coun comt head injury, Nat Inst Neurol Dis & Stroke, 66-70, mem physiol training grant comt, 70-74; mem sect comt bioacoust & vibration, Am Nat Standards Inst, 61-65, mem sect comt microwave hazards, 68-73; mem comt handbook environ biol, Fedn Am Socs Exp Biol, 63-66; adj prof biol, State Univ NY Binghamton, 78-82. *Honors & Awards:* K S Cole Award, Biophys Soc, 73. *Mem:* Am Phys Soc; Acoust Soc Am; Biophys Soc; Am Physiol Soc; Soc Gen Physiol; fel AAAS. *Res:* Electrical and mechanical phenomena in cells and tissues; theoretical biology; shock-vibration-bioacoustics; physical factors in health and safety. *Mailing Add:* 63 Loop Rd Falmouth MA 02540

GOLDMAN, DAVID TOBIAS, b Brooklyn, NY, Jan 25, 33; m 57; c 5. SCIENCE ADMINISTRATION, NUCLEAR PHYSICS. *Educ:* Brooklyn Col, AB, 52; Vanderbilt Univ, MS, 54; Univ Md, PhD(physics), 58. *Prof Exp:* Physicist, Evans Signal Lab, 52-53; asst physics, Univ Md, 54-58; res assoc, Univ Pa, 58-59; theoret physicist, Knolls Atomic Power Lab, Atomic Energy Comn, 59-63, supv physicist, 63-65; chief theoret physics, Reactor Radiation Div, Dept Com, 65-69, prog leader, standard nuclear ref data, 65-72, prog analyst, sci & technol prog, 70-72, actg dir, 73-74, dep dir, Inst Basic Standards, 72-77, assoc dir planning, Nat Measurement Lab & metric coordr, Nat Bur Standards, 77-82, exec assoc sci & technol, 82-85; mgr, Argonne Area Off, 85-87, ASST MGR LABS, CHICAGO OPERS OFF, DEPT ENERGY, 87 - *Concurrent Pos:* Physicist, Naval Res Lab, 56 & Oak Ridge Nat Lab, 57-58; adj assoc prof, Rensselaer Polytech Inst, 60-65; adj prof nuclear eng, Univ Md, 65 -; sci policy analyst, Bur Budget, 69-70; sci fel, Dept Commerce, 69-70. *Mem:* Fel Am Phys Soc; Am Nuclear Soc; Am Nat Metric Coun. *Res:* Theoretical nuclear physics; nuclear reactions and structure; reactor physics; effect of chemical binding; technology assessment. *Mailing Add:* Dept Energy Chicago Opers Off 9800 South Cass Ave Argonne IL 60439

GOLDMAN, DEXTER STANLEY, b Boston, Mass, June 17, 25; m 64; c 7. BIOLOGICAL CHEMISTRY. *Educ:* Univ Calif, Los Angeles, BS, 48; Univ Calif, PhD(comp biochem), 51. *Prof Exp:* Asst, Univ Calif, 48-51; trainee, Nat Heart Inst, 51-53, asst prof, Enzyme Inst, 54-59, clin assoc prof, Univ Wis, 59-71; chief biochemist, Vet Admin Hosp, 53-71; prof & dir inst biol, Haifa Univ, 71-73; dir res & develop, Makor Chem Ltd, Israel, 73-80; toxicologist, Prog Operations br, Nat Toxicol Prog, Nat Inst Environ Health, NIH, Bethesda, MD, 80-; sr proj officer, Nat Toxicol Prog/NIEHS NIH, 80-84; dir, Lab Data Integrity Assurance Div, Off Compliance Monitoring, US Environ Protection Agency, Washington, DC, 84-88; PRES, GOLDMAN ASSOCS INT INC, ROCKVILLE, MD, 88- *Concurrent Pos:* Vis res prof, Div Biochem, Scripps Clin & Res Found, Univ Calif, La Jolla, 63; vis prof, Hebrew Univ, Israel, 68-69. *Mem:* Am Soc Biol Chem; Am Acad Microbiol; Am Soc Microbiol. *Res:* Physiology and enzyme systems of the tubercle bacillus; enzymatic basis of disease and chemotherapy. *Mailing Add:* Goldman Assocs Int Inc PO Box 1853 Rockville MD 20849-1853

GOLDMAN, EMANUEL, b New York, NY, Feb 19, 45. PROTEIN SYNTHESIS, REGULATION OF GENE EXPRESSION. *Educ:* Brandeis Univ, BA, 66; Mass Inst Technol, PhD(biochem), 72. *Prof Exp:* Fel viral oncol, Public Health Res Inst, New York, 72-73; res fel path, Harvard Med Sch, 73-75; assoc microbiol, Col Med, Univ Calif, Irvine, 75-77, asst res prof, 77-79; asst prof, 79-83, ASSOC PROF MICROBIOL, UNIV MED & DENT NJ, 83 - *Concurrent Pos:* contrib ed, New Univ, 75-79; res assoc, Dept Chem, Mass Inst Technol, 76; consult, Dept Path, Harvard Med Sch, 76; res career develop award, NIH, 83-88. *Res:* Regulation of protein synthesis in bacteria, particularly transfer RNA decoding properties, mistranslation, efficiency of codon recognition, transfer RNA isoacceptor utilization, and ribosomal mutations using in vitro systems, and also phage MS2 infected cells. *Mailing Add:* Dept Microbiol NJ Med Sch 185 South Orange Ave Newark NJ 07103

GOLDMAN, ERNEST HAROLD, b Lynn, Mass, Oct 4, 22; m 48; c 2. COMPUTER COMMUNICATIONS, INFORMATION SYSTEMS. *Educ:* US Coast Guard Acad, BS, 43; Harvard Univ, MS, 49,. *Hon Degrees:* DSc, Harvard UNiv, 52. *Prof Exp:* Instr physics, Univ Bridgeport, 47-48; develop engr, 51-52, sr engr, 52-55, mgr comput test planning & testing, 55-57, res mgr exp systs, IBM T J Watson Res Ctr, 57-67, res dir, 61-64, mgr E Fishkill Lab, 64-70, mgr component indust eval, 70-81, TECH CONSULT, IBM CORP HQ, 82-; BANNOW-WAHLSTROM PROF COMPUT ENG, UNIV BRIDGEPORT. *Concurrent Pos:* Adj prof info sci, Pace Univ, 73-82. *Mem:* AAAS; sr mem Inst Elec & Electronics Engrs; Asn Comput Mach; Comput Soc. *Res:* Computer communications; computer technology; computer applications; electronic systems and circuits; business modelling; information retrieval and management systems. *Mailing Add:* Col Sci & Eng Univ Bridgeport Bridgeport CT 06602

GOLDMAN, HAROLD, b Chicago, Ill, May 29, 27; m 53; c 2. NEUROPHARMACOLOGY. *Educ:* Univ Chicago, MS, 53; Univ Ill, PhD(physiol), 57. *Prof Exp:* From instr to asst prof psychiat, Col Med, Ohio State Univ, 57-66, asst prof pharmacol, 64-66, from assoc prof to prof psychiat & pharmacol, 66-74; PROF PHARMACOL, SCH MED, WAYNE STATE UNIV, 74- *Concurrent Pos:* Res neuroendocrinologist, Res Div, Columbus Psychiat Inst, 57-62. *Mem:* Biophys Soc; Am Physiol Soc; Endocrine Soc; Am Soc Pharmaceut Exp Therapeuts. *Res:* Interregulation of nervous and endocrine systems; metabolism of nervous tissue; small organ blood flow; neurohumors; regional cerebral blood flow; head injury. *Mailing Add:* Dept Pharmacol Wayne State Univ Sch Med Detroit MI 48201

GOLDMAN, HARVEY, b Philadelphia, Pa, May 25, 32; m 67; c 3. GASTROINTESTINAL PATHOLOGY. *Educ:* Temple Univ, BA, 53, MD, 57. *Prof Exp:* Intern med, Philadelphia Gen Hosp, 57-58; resident & fel path, Beth Israel Hosp, Boston, 58-62; pathologist, US Naval Base, Great Lakes, Ill, 62-64; asst pathologist, Beth Israel Hosp, Boston, 64-68, assoc pathologist, 68-75, pathologist, 75-84; instr & assoc, 64-67, from asst prof to assoc prof, 67-75, PROF PATH, MED SCH, HARVARD UNIV & DIV HEALTH SCI & TECHNOL, MASS INST TECHNOL, 76-; FAC DEAN, MED EDUC, HARVARD, 88- *Concurrent Pos:* Consult path, Children's Hosp & Brigham & Women's Hosp, Boston, Mass, 74-; sr pathologist, Beth Israel Hosp, Boston, Mass, 85- *Mem:* Gastrointestinal Path Club (pres, 83-84); Am Gastroenterol Asn; Am Asn Pathologists; Int Acad Path. *Res:* General pathology and gastrointestinal pathology. *Mailing Add:* Beth Israel Hosp 330 Brookline Ave Boston MA 02215

GOLDMAN, HENRY M(AURICE), oral pathology, periodontology; deceased, see previous edition for last biography

GOLDMAN, ISRAEL DAVID, b Jersey City, NJ, Nov 17, 36; m 64; c 3. BIOPHYSICS. *Educ:* NY Univ, BA, 58; Univ Chicago, MD, 62. *Prof Exp:* From asst prof to assoc prof med & pharmacol, Sch Med, Univ NC, Chapel Hill, 69-74; assoc prof med, 74-75, vchmn dept, 74-80, PROF MED & PHARMACOL, MED COL VA, VA COMMONWEALTH UNIV, 75-, CHMN DEPT, DIV MED ONCOL & BIOCHEM PHARMACOL, DEPT MED & PHARMACOL, 80- *Concurrent Pos:* Fel, Biophys Lab, Harvard Med Sch, 65-66; Pharmaceut Mfrs Asn Found fac develop award, 71; Nat Cancer Inst res career develop award, 73; prin investr, Nat Cancer Inst res grants, 70-86; prog dir, Membrane Biol Training Grant, 75-85. *Mem:* AAAS; Am Soc Pharmacol & Exp Therapeut; Am Physiol Soc; Am Asn Cancer; fel Am Col Physicians. *Res:* Am Soc Clin Invests. Res: Transport and biochemistry of the folate compounds; mechanisms of membrane transport. *Mailing Add:* Dept Med Med Col Va MCV Box 230 Richmond VA 23298

GOLDMAN, JACK LESLIE, b Chicago, Ill, Nov 20, 35. HEALTH EDUCATION. *Educ:* Univ Chicago, BA & BS, 58; Loyola Univ, Ill, MS, 61, PhD(chem), 66. *Prof Exp:* Instr chem, Mundelein Col, 63-65; res phys chemist, Velsicol Chem Corp, 66-67; mem fac natural sci & chem, Shimer Col, 67-71; lectr chem, 72-82, lectr math, 82-87, DIR, POST-BACCALAUREATE HEALTH SCI PROG, LOYOLA UNIV CHICAGO, 83-, DIR PRE-HEALTH PROF PROG, 87- *Concurrent Pos:* Chmn natural sci, Shimer Col, 67-71. *Mem:* fel Am Inst Chem; Am Asn Physics Teachers; NY Acad Sci; Am Phys Soc; Nat Asn Adv Health Prof. *Res:* Conceptual development of quantum mechanics; methodology of science, especially the interrelations between biological and physical sciences; integrative principles in the sciences and humanities; health ethics and education. *Mailing Add:* Health Prof Prog Loyola Univ 6525 North Sheridan Rd Chicago IL 60626

GOLDMAN, JACOB E, b New York, NY, July 18, 21; m 43; c 3. PHYSICS. *Educ:* Yeshiva Univ, AB, 40; Univ Pa, MA & PhD(physics), 43. *Hon Degrees:* LLD, Yeshiva Univ, 61. *Prof Exp:* Asst physics, Univ Pa, 41-42, asst instr, 43, res assoc, Nat Defense Res Comt proj, 43; res physicist, Res Labs, Westinghouse Elec Corp, 43-50; lectr, Carnegie Inst Technol, 50, asst prof, 51-55; mgr physics dept, Sci Lab, Ford Motor Co, 55-60, from assoc dir to dir, 60-68; group vpres res & develop, Xerox Corp, 68-74, group vpres, 74-80, chief scientist, 74-83, sr vpres, 80-83; RETIRED. *Concurrent Pos:* Westinghouse lectr, Grad Sch, Univ Pittsburgh, 45-50; group leader, Westinghouse Elec Corp, 46-50; head, lab magnetics res, Carnegie Inst Technol, 53-55; chmn sci adv comt, Detroit Inst Cancer Res; gen chmn, Nat Conf Magnetism & Magnetic Mat, 56-58; chmn panel mat res & mem solid state adv panel, Nat Acad Sci; mem adv panel select comt govt res, US House Rep; adv comt solid state, Off Sci Res, US Air Force; Edwin Webster prof, Mass Inst Technol, 59; consult, Oak Ridge Nat Lab, Brookhaven Nat Lab, Naval Ord Lab, US Steel Corp & Allis-Chalmers Mfg Co; chmn, Softstrip Lab, 84- *Mem:* Fel Am Phys Soc; Am Asn Physics Teachers; Inst Elec & Electronics Engrs. *Res:* Solid state physics; semiconductors; physics of metals; ferromagnetism; magnetic materials; magnetostriction phenomena; low temperature properties of metals; administration of research. *Mailing Add:* Eight Landmark Sq Norwalk CT 06850

GOLDMAN, JAMES ALLAN, b Chicago, Ill, Nov 20, 35; m 78. PHILOSOPHY OF SCIENCE, SCIENCE POLICY. *Educ:* Univ Chicago, BS & BA, 58; Northwestern Univ, PhD(chem), 63. *Prof Exp:* Instr chem, Northwestern Univ, 61-62; asst prof, Polytech Inst Brooklyn, 62-71; coordr tech & indust progs, Div Continuing Educ, New York City Community Col, 71-81, coordr & higher educ assoc, 81-86; dir admin, Res & Develop & Higher Educ Officer, 86-90, DEP DEAN, DIV CONTINUING EDUC & DIR, PROG PLANNING & ECON DEVELOP, NEW YORK CITY TECH COL, 90- *Concurrent Pos:* Mem exec bd, ChemTech, 73-77; sci & technol ed, USA Today, 75-86; abstractor, J Col Sci Teaching; Interfaith relation comts & conferences, 88- *Mem:* Sigma Xi; AAAS; Am Chem Soc; Am Asn Physics Teachers; NY Acad Sci. *Res:* Science education; philosophy and contemporary sociology of science; science, technology and society interactions between science and the humanities; theory of titration curves, especially redox titrations; science/religion relationships; Christian/Jewish relations. *Mailing Add:* Div Continuing Educ 300 Jay St New York City Tech Col Brooklyn NY 11201-2983

GOLDMAN, JAMES ELIOT, b Washington, DC, Oct 26, 46; m 71; c 2. NEUROBIOLOGY, NEUROPATHOLOGY. *Educ:* Amherst Col, BA, 68; NY Univ, MD & PhD(med & neurobiol), 76. *Prof Exp:* Teaching fel neurochem, Albert Einstein Col Med, Bronx, NY, 76-77, resident path, 77-78, resident neuropath, 78-80, asst prof, 80-85, assoc prof neuropath, 85-87; ASSOC PROF PATH & PSYCHIAT, COLUMBIA UNIV, COL PHYSICIANS & SURGEONS, 87- *Concurrent Pos:* Attend neuropathologist, Albert Einstein Col Hosp, Bronx Munic Hosp Ctr & Bronx Lebanon Hosp, 80-87; assoc attend neuropathologist, Presby Hosp, New York; NIH teacher-investr award, 80-85, NIH Jarvits neurosci award, 88-95. *Mem:* Sigma Xi; Am Soc Neurochem; Int Soc Neurochem; Soc Neurosci; Am Asn Neuropathologists; AAAS. *Res:* Molecular mechanisms and pathogenesis of Alzheimer's disease and other neurodegenerative disorders; histogenesis of the central nervous system, especially glial cell development; molecular mechanisms of glial scar formation in neurological diseases. *Mailing Add:* Dept Path Columbia Univ Col Physician & Surgeons 630 West 168th St New York NY 10032

GOLDMAN, JAY, b Norfolk, Va, Apr 15, 30; m 59. INDUSTRIAL ENGINEERING. *Educ:* Duke Univ, BSME, 50; Mich State Univ, MSME, 51; Wash State Univ, DSc(indust eng), 55. *Prof Exp:* Lectr indust mgt, Wash Univ, 52-55, lectr indust eng, 55-56, asst prof, 56-63, actg chmn human & org factor area, 63, dir indust engr, Jewish Hosp St Louis, 60-64; prof indust eng, NC State Univ, 64-68; prof indust eng & chmn dept, Univ Mo-Columbia, 68-84; DEAN & PROF ENG, UNIV ALA BIRMINGHAM, 84- *Concurrent Pos:* Consult, Off Naval Res, Ill, Artcraft Venetian Blind Mfg Co, Mo, Goffre Carbon Co, Arg, Boeing Airplane Co, Wash, Dr LeGear Med Co, Mo, Shampaine Industs, McDonnell Aircraft Corp, Banner Hardware Co, Beltx Corp, Mead Johnson & Co, Ind, Med Ctr, Duke Univ, DC, Nat Heart, Lung & Blood Inst, Nat Can Inst, Nat Libr Med; mem health care systs study sect, Nat Ctr Health Serv Res & Develop, Dept Health, Educ & Welfare; mem, adv panels grad progs, NC State Univ & Rensselaer Polytech Inst, Safety & Occup Health Study Sect, Ctr Dis Control & Accreditation Bd Eng & Technol. *Honors & Awards:* Award, InstIndust Engrs, 81. *Mem:* Am Soc Eng Educ; Fel Inst Indust Engrs; Coun Indust Eng Acad; Health Systs Soc; Oper Res Soc Am; Ala Soc Prof Engrs. *Res:* Administrative organization and system design for health care delivery; design of man-machine work systems; measurement of human performance; human factors engineering. *Mailing Add:* Dean Eng Univ Ala Birmingham UAB Sta Birmingham AL 35294

GOLDMAN, JOEL HARVEY, b Brooklyn, NY, May 26, 43. ELEMENTARY PARTICLE PHYSICS. *Educ:* Mass Inst Technol, BS, 66; Univ Minn, MS, 67; Univ Md, PhD(physics), 72. *Prof Exp:* Res assoc, New York Univ, 72-75; assoc physicist, Brookhaven Nat Lab, 75-78; asst prof physics, 79-83, ASSOC RES SCIENTIST, FLA STATE UNIV, 83- *Mailing Add:* 410 Victory Garden Dr Apt 70 Tallahassee FL 32301

GOLDMAN, JOHN ABNER, b Cincinnati, Ohio, Jan 9, 40; m 81; c 2. IMMUNOLOGIC & RHEUMATIC DISEASES. *Educ:* Univ Wis-Madison, BS, 62; Cincinnati Med Ctr, MD, 66. *Prof Exp:* Intern, Med Ctr, Univ Ore, 66-67; resident, Med Ctr, Univ Cincinnati, 67-69, fel immunol & rheumatol, 69-71; mem staff, Med Corps, US Army, 71-73; asst prof, Sch Med, Emory Univ, 73-77, assoc prof med, 77-82; PRIVATE PRACTICE RHEUMATOLOGY & IMMUNOL, 82- *Concurrent Pos:* Instr med, Med Col Ga, 72-73; mem, Med Adv Bd, Systemic Lupus Erythematosus Found, Inc, 76-; Dir, Med Sects, Am Soc Laser Med & Surg, 81-83. *Mem:* Fel Am Rheumatism Asn; fel Am Soc Laser Med & Surg; fel Am Col Physicians; Am Acad Allergy; AAAS; Am Med Asn. *Res:* Understanding immune mechanisms of immunologic and connective tissue diseases; role of immune complexes and pathogenesis; pathogenesis of infectious arthritis and the immunologic aspects of laser on immune response and in the treatment of connective tissue diseases. *Mailing Add:* Med Quarters 5555 Peach Tree Dunwoody Rd Suite 293 Atlanta GA 30342-1711

GOLDMAN, JOSEPH L, b San Francisco, Calif, Aug 25, 32; div; c 3. MESO-METEOROLOGY, AGRO-METEOROLOGY. *Educ:* Tex A&M Univ, BS, 58, MS, 60; Univ Okla, PhD, 71. *Prof Exp:* Mathematician, WCoast Res Co, Calif, 56; assoc meteorologist, Gulf Consult, Tex, 56-58; res scientist, Tex A&M Res Found, 58-60; res asst meteorol, Univ Chicago, 60-62, meteorologist, 62-65; tech staff prin investr, Nat Eng Sci Co, Tex, 65; assoc dir, Inst Storm Res, 66-75; assoc prof physics & meteorol, Univ St Thomas, Tex, 68-75; TECH DIR, INT CTR FOR SOLUTION OF ENVIRON PROB, HOUSTON, 76- *Concurrent Pos:* Gen chmn, Houston Sci & Eng Fair, 75; chmn, Disaster Serv, Greater Houston Chapter Red Cross, 82-84. *Mem:* AAAS; Am Meteorol Soc; Am Geophys Union; NY Acad Sci; Sigma Xi; Soc Risk Anal. *Res:* Severe weather phenomena; effective forecasting and control through physical-hydrodynamic principles, effects on ocean environment and ecological, sociological and induced behavioristic aspects; urban pollution control; meteorological engineering; urbanization and climate change; mathematical modeling of environmental processes; drift of agricultural herbicide sprays; groundwater contamination by surface released chemicals, through surface pounding and soil seepage; mathematics of the human genetic code; determination of soil temperature and moisture from atmospheric temperature, rainfall and soil type; morphology of soil boone pathogeous due to temperature and moisture. *Mailing Add:* Int Ctr Solution Environ Probs 535 Lovett Blvd Houston TX 77006

GOLDMAN, KENNETH M(ARVIN), b Pittsburgh, Pa, Dec 8, 22; m 51; c 3. METALLURGICAL ENGINEERING. *Educ:* Carnegie Inst Technol, BS, 43, DSc(metall), 52. *Prof Exp:* adv scientist, Bettis Lab, Westinghouse Elec Corp, 80-89; RETIRED. *Concurrent Pos:* Adj proj, Univ Pittsburgh. *Honors & Awards:* Hunt Award, Am Inst Mining, Metall & Petrol Engrs, 52; Eng Mat Achievement Award, Am Soc Metals, 72. *Mem:* Fel Am Soc Metals; Am Inst Mining, Metall & Petrol Engrs. *Res:* Physical chemistry of steelmaking; radioactive tracers applied to metallurgy; fissionable materials; metallurgy of zirconium and uranium. *Mailing Add:* 2223 Shady Ave Pittsburgh PA 15217

GOLDMAN, LAWRENCE, b Boston, Mass, May 6, 36; c 1. BIOPHYSICS. *Educ:* Tufts Univ, BS, 58; Univ Calif, Los Angeles, PhD(membrane biophys), physiol), 64. *Prof Exp:* NIH fel, Col Physicians & Surgeons, Columbia Univ, 64-65; asst prof zool, 65-67, from asst prof to assoc prof physiol, Sch Med, 67-77, PROF PHYSIOL & BIOPHYS, SCH MED, UNIV MD, BALTIMORE, 77- *Concurrent Pos:* Vis scientist, Lab Biophys, NIH, 66-68; NATO sr fel, Queen Mary Col, London, 70; NIH spec res fel, Univ Cambridge, 73; Fulbright sr prof, I Physiol Inst, Univ Saarlandas, 87-88. *Mem:* Fel AAAS; Soc Gen Physiol; Am Physiol Soc; Soc Neurosci; NY Acad Sci; Am Inst Biol Sci; Biophys Soc. *Res:* Membrane biophysics; theory of excitable membranes; impulse initiation and propagation; membrane transport and selectivity; mathematical modeling. *Mailing Add:* Dept Physiol Univ Md Sch Med Baltimore MD 21201

GOLDMAN, LEONARD MANUEL, b New York, NY, Mar 22, 25; m 52; c 3. PLASMA PHYSICS. *Educ:* Cornell Univ, AB, 45; McGill Univ, MSc, 48; Univ Rochester, PhD, 52. *Prof Exp:* Asst physics, McGill Univ, 47-48 & Univ Rochester, 48-52; res assoc, Princeton Univ, 52-56; res assoc, Res & Develop Ctr, Gen Elec Co, 56-75; prof mech & aerospace sci, Univ Rochester, 75-88; PRIN SCIENTIST, BECHTEL NAT INC, SAN FRANCISCO, 88- *Concurrent Pos:* Vis fel, Culham Lab, UK Atomic Energy Authority, 65-66; actg mgr appl physics, res & develop div, Bechtel Inc, San Francisco, 85-86. *Mem:* Fel Am Phys Soc; AAAS. *Res:* High temperature plasma physics; low energy nuclear physics; laser fusion. *Mailing Add:* Bechtel Nat Inc 50 Beale St PO Box 3965 San Francisco CA 94119

GOLDMAN, MALCOLM, b Brooklyn, NY, July 18, 29; m 66. MATHEMATICS. *Educ:* Univ Minn, BA, 49, MA, 51; Univ Chicago, PhD(math), 55. *Prof Exp:* Mem tech staff, Bell Tel Labs, Inc, 54-55; instr math, Univ Mich, 55-60; vis asst prof math, Reed Col, 60-62; assoc prof, 62-75, admin officer, Dept Econ, 75-76, ASSOC PROF MATH, NY UNIV, 76- *Mem:* Am Math Soc. *Res:* Probability theory; functional analysis. *Mailing Add:* Dept Math New York Univ Wash Sq New York NY 10003

GOLDMAN, MANUEL, b Montreal, Que, May 19, 32; US citizen; m 69; c 2. MICROBIAL PHYSIOLOGY, METABOLISM. *Educ:* Roosevelt Univ, BS, 53; Univ Chicago, SM, 55; Univ Mich, PhD(microbiol), 62. *Prof Exp:* Fel, Inst Sci Technol, Univ Mich, 61-62; asst prof biol & microbiol, Wayne State Univ, 62-64; microbiologist, Stanford Res Inst, 64-67; ASST PROF BIOL SCI, UNIV ILL, CHICAGO, 67- *Concurrent Pos:* Vis prof microbiol, La State Univ, 85; asst dean, Col Liberal Arts & Sci, Univ Ill, Chicago, 87- *Mem:* Soc Indust Microbiol; Am Soc Microbiol; Am Chem Soc; Mycological Soc Am. *Res:* Metabolism and physiology of sporeforming bacteria; environmental modification of cellular metabolism and development; mode of action of psychopharmacological drugs; regulation of carbohydrate metabolism; metabolism of smut fungi. *Mailing Add:* Dept Biol Sci Univ Ill Box 4348 Chicago IL 60680

GOLDMAN, MARVIN, b New York, NY, May 2, 28; m 53; c 3. RADIOBIOLOGY, HEALTH RISK ASSESSMENT. *Educ:* Adelphi Col, AB, 49; Univ Md, MS, 51; Univ Rochester, PhD(radiation biol), 57. *Prof Exp:* Asst, Univ Md, 50-51; biologist, Phys Biol Lab, NIH, 51-52; res assoc, Univ Rochester, 53-57, jr scientist, Atomic Energy Proj, Sch Med & Dent, 57-58; assoc res radiation biologist, Sch Med, Univ Calif, Davis, 58-64, res radiation biologist & lectr physiol sci, Sch Vet Med, 64-72, adj prof radiobiol, Sch Med, 69-73, dir, lab energy-related health res, 73-85, PROF RADIOL SCI, SCH VET MED & PROF RADIOL, SCH MED, UNIV CALIF, DAVIS, 73-, CO-DIR, WESTERN REGIONAL CTR, NAT INST GLOBAL ENVIRON CHANGE, UNIV CALIF, DAVIS, 90- *Concurrent Pos:* Consociate mem, Nat Coun Radiation Protection; chmn, Interagency Nuclear Safety, Rev Panel, Biomed and Environ Panel, 73; co-chmn, Nuclear Emergency Procedures, Calif State Senate Task Force, 87-88; chmn, Comt Assessment of Environ & Health Consequences in Exposed Chernobyl Pop, 86; prin investr, Pharm Kenetics New Investigational Drugs, US Army Med Res & Develop Command. *Honors & Awards:* E O Lawrence Award, US Atomic Energy Comn, 72; Distinguished Sci Award, Health Physics Soc, 88. *Mem:* AAAS; Soc Risk Anal; Radiation Res Soc; Health Physics Soc. *Res:* Biologic effects of radiation; toxicologic risk assessment; environmental and occupational health; radionuclide toxicity and metabolism; radiation dosimetry; bone and lung physiology and pathology; trace element metabolism; chemical toxicology; health physics; tumor biology; energy related health effects; pharmacokinetics; genotoxicology; cell toxicology. *Mailing Add:* Nat Inst Global Environ Change Univ Calif Davis CA 95616-8756

GOLDMAN, MAX, b Newtown, Conn, Nov 28, 20; m 58; c 2. ENDOCRINOLOGY, BIOLOGY. *Educ:* NY Univ, BA, 48; Univ Mo-Columbia, MA, 53; Univ Calif, Berkeley, PhD(endocrinol), 64. *Prof Exp:* Asst prof biol, Long Island Univ, 64-68; assoc prof, 68-75, PROF BIOL, UNIV S DAK, 75- *Mem:* AAAS; Am Inst Biol Sci; NY Acad Sci. *Res:* Toxicological and behavioral effects of chemical substances on endocrine function, such as dimethyl sulfoxide, oxalic acid, povidone-iodine, heavy metals, (lead, mercury) sodium acetate, DDT, and acidic amino acids, monosodium glutamate, aspartic acid on endocrine function and thyroid function; lewisite, its chemistry, toxicology and biological effects. *Mailing Add:* Dept Biol Univ SDak Vermillion SD 57069

GOLDMAN, NORMAN L, b New York, NY, Aug 11, 33. CHEMISTRY. *Educ:* City Col New York, BS, 54; Harvard Univ, AM, 56; Columbia Univ, PhD(org chem), 59. *Prof Exp:* NSF fel, Imp Col, Univ London, 59-60; NIH fel, Columbia Univ, 61; from asst prof to prof, 61-76, chmn dept, 72-77, actg assoc dean & actg dean fac, Div Math & Natural Sci, 77-79, PROF CHEM, QUEENS COL, NY, 76-, DEAN FAC, DIV MATH & NATURAL SCI, 79- *Mem:* Am Chem Soc; Royal Soc Chem. *Res:* Synthetic Organic Chemistry. *Mailing Add:* Div Math & Natural Sci Remsen Hall 125 Queens Col Flushing NY 11367

GOLDMAN, OSCAR, mathematics; deceased, see previous edition for last biography

GOLDMAN, PETER, b New York, NY, May 23, 29; m 59; c 2. BIOCHEMISTRY, PHARMACOLOGY. *Educ:* Cornell Univ, BEngPhys, 52; Harvard Univ, MA, 53; Johns Hopkins Univ, MD, 57. *Prof Exp:* Res assoc biochem, Nat Heart Inst, 59-63; sr investr biochem pharmacol, Nat Inst Arthritis, Metab & Digestive Dis, 63-72; prof clin pharmacol, 72-, PROF HEALTH SCI, DEPT NUTRIT, HARVARD SCHOOL PUB HEALTH, 82-, ACTG CHMN, 84- *Concurrent Pos:* NIH fel, 59-60; mem adv comt biotechnol resources, NIH, 77- *Mem:* Am Soc Biol Chemists; Am Chem Soc; Am Soc Pharmacol & Exp Therapeut; Am Soc Clin InvestS. *Res:* Nutrition; drug metabolism in the intestinal microflora; anaerobic bacteria, nitroimidazoles, laboratory quality control, clinical pharmacology and toxicology. *Mailing Add:* Dept Pharmacol & Nutrit Harvard Univ 665 Huntington Ave Boston MA 02115-6021

GOLDMAN, RALPH, internal medicine; deceased, see previous edition for last biography

GOLDMAN, RALPH FREDERICK, b Boston, Mass, Mar 3, 28; m 56; c 2. PHYSIOLOGY. *Educ:* Univ Denver, AB, 50; Boston Univ, AM, 51, PhD(physiol), 55; Northeastern Univ, SM, 62. *Prof Exp:* USPHS res fel, 55; physiologist, USArmy Res & Eng Command, 56-62; dir, Mil Ergonomics Lab, US Army Res Inst Environ Med, 62-82; sr vpres, Multi-Tech Corp, Natick, Ma, 82-89; CHIEF SCIENTIST, COMFORT TECHNOL, 89- *Concurrent Pos:* Adj prof, Boston Univ, 69-; lectr, Mass Inst Technol, 73-; adj prof, NC State Univ, 89- *Mem:* Biophys Soc; Am Physiol Soc; Inst Elec & Electronics Engrs; Am Soc Heating, Refrig & Air Conditioning Eng; Brit Ergonomics Res Soc; Am Soc Testing & Mat; Am Col Sports Med. *Res:* Environmental physiology; human heat, cold and work stresses; clothing; heat transfer; body composition. *Mailing Add:* Comfort Technol PO Box 847 Framingham MA 01701

GOLDMAN, ROBERT BARNETT, b Philadelphia, Pa, June 21, 27; m 47; c 3. APPLIED PHYSICS. *Educ:* Temple Univ, BA, 48, MA, 50, PhD, 52. *Prof Exp:* Asst, Villanova Col, 49-50 & Temple Univ, 50-51; physicist, Philco Corp, 51-62, mem corp eng & res staff, 62-68, dir advan technol, Electronics Group, Philco-Ford Corp, 68-69; tech dir govt & indust div, Urbana Opers, Magnavox Co, 69-71; dir res & develop, Govt Electronics Div, Magnavox Co, 71-72; assoc dir advan eng resources, Leeds & Northrup Co, 72-74; chief scientist, Systs & Opers Anal Div, 75-79, DIR SCI ANAL DIV, ANALYTICS INC, 79- *Concurrent Pos:* Lectr, Temple Univ, 54-65. *Mem:* Acoust Soc Am; Inst Elec & Electronics Engrs. *Res:* Acoustics-propagation; transducers; noise control; communications; information theory; control systems; transmission and reception of electromagnetic energy for detection, surveillance and guidance; systems analysis; operations research; applied physics; electronics; electrooptics; military systems. *Mailing Add:* Analytics 2500 Maryland Rd Willow Grove PA 19090

GOLDMAN, ROBERT DAVID, b Port Chester, NY, July 23, 39; m 65; c 2. CELL BIOLOGY. *Educ:* Univ Vt, BA, 61, MS, 63; Princeton Univ, PhD(biol), 67. *Prof Exp:* Am Cancer Soc fel histochem, Royal Postgrad Med Sch, London & Med Res Coun Gt Brit Exp Virus Res Unit, Glasgow, 67-69; asst prof biol, Case Western Reserve Univ, 69-74; prof biol, Mellon Inst Sci, Carnegie-Mellon Univ, 74-81; STEPHEN WALTER RANSON, PROF & CHMN, DEPT CELL BIOL & ANAT, MED SCH, NORTHWESTERN UNIV, 81- *Concurrent Pos:* Assoc ed, Cell Motion & Cytoskeletal; dir physiol course, Marine Biol Lab. *Mem:* Biophys Soc; Soc Develop Biol; Am Soc Cell Biol; NY Acad Sci. *Res:* Cytoskeletal systems of mammalian cells. *Mailing Add:* Dept Cell Biol/Anat Sch Med Northwestern Univ 303 E Chicago Ave Chicago IL 60611

GOLDMAN, RONALD, b Brooklyn, NY, Sept 22, 33; m 54; c 3. SPEECH PATHOLOGY, AUDIOLOGY. *Educ:* Birmingham Southern Col, BA, 55; Univ Pittsburgh, MS, 57, PhD(speech path, audiol), 60. *Prof Exp:* Asst prof speech path & audiol, Tulane Univ, 60-64; prof, Sch Med, Vanderbilt Univ, 64-74; PROF BIOCOMMUN & STAFF PATHOLOGIST SPEECH, CTR DEVELOP & LEARNING DISORDERS, UNIV ALA, BIRMINGHAM, 74-, TRAINING DIR, 81-, PROF EDUC. *Concurrent Pos:* Assoc prof spec educ, George Peabody Col, 66-71; prof, Ignacio Barraquer Film Award, 67; Am Speech & Hearing Asn fel, 69; res consult, US Off Educ, 70-; consult, NIH, 71; vpres planning Am Speech & Hearing Asn, 78-80. *Mem:* Am Speech & Hearing Asn; Coun Exceptional Children; Am Asn Ment Deficiency. *Res:* Development of new procedures for studying auditory perceptual and processing disorders; development of new therapeutic procedures for articulatory deficits and auditory perceptual disorders. *Mailing Add:* Dept Biocommunication Univ Ala Sch Med Univ Sta Birmingham AL 35294

GOLDMAN, S ROBERT, b New York, NY, Nov 16, 37; m 73; c 2. FLUID PHYSICS, ATMOSPHERIC SCIENCES. *Educ:* Harvard Univ, AB, 59; Yale Univ, MS, 61, PhD(physics), 63. *Prof Exp:* Fel plasma physics, Princeton Plasma Lab, 63-64; res assoc fluid physics, Univ Md, 64-66, res asst prof, 66-69; staff scientist field testing, Los Alamos Sci Lab, 69-71; asst prof physics, Hunter Col, 71-74; scientist plasma physics, Sci Applns, Inc, 74-76; SR SCIENTIST ENERGY, JAYCOR, 76- *Mem:* Am Phys Soc; Am Geophys Soc. *Res:* Ionospheric physics; plasma physics; fluid physics relating to fossil energy processes. *Mailing Add:* Los Alamos Lab MS E-531 P O Box 1663 Los Alamos NM 87545

GOLDMAN, STANFORD, b Cincinnati, Ohio, Nov 14, 07; m 35; c 3. THEORETICAL PHYSICS, COMMUNICATION THEORY. *Educ:* Univ Cincinnati, AB, 26, AM, 28; Harvard Univ, PhD(physics), 33. *Prof Exp:* Sound engr, Photophone, Inc, Radio Corp Am, NY, 30-31; sr engr, Gen Elec Co, Conn, 35-46; res assoc elec eng, Mass Inst Technol, 46-49; prof, 49-75, EMER PROF ELEC ENG, SYRACUSE UNIV, 75-; RES ASSOC, UNIV CALIF, SAN DIEGO, 77- *Concurrent Pos:* Consult physicist, Electronics Res Lab, US Air Force, 46-54, Gen Elec Co 54- & Eng & Res Develop Lab, US Army, 58-; vis Makay prof, Univ Calif, Berkeley, 62; vis prof, Univ Calif, San Diego, 70, res assoc, 77- *Mem:* Fel AAAS; Am Phys Soc; Biophys Soc; fel Inst Elec & Electronics Engrs. *Res:* Mathematical biology; theoretical physics and biology; noise and information theory. *Mailing Add:* Dept Elec Eng & Comput Sci Univ Calif San Diego La Jolla CA 92093

GOLDMAN, STEPHEN ALLEN, b New York, NY, Apr 2, 46. PHYSICAL CHEMISTRY, MAGNETIC RESONANCE. *Educ:* Brooklyn Col, BS, 67; Cornell Univ, MS, 69, PhD(phys chem), 73. *Prof Exp:* STAFF SCIENTIST, PROCTER & GAMBLE CO, 72- *Mem:* Am Chem Soc; AAAS. *Res:* Magnetic resonance spectroscopy; molecular orientation and motion in liquid crystalline systems; colloid chemistry; physical chemistry of polymer gels, polymer physical chemistry. *Mailing Add:* Procter & Gamble Co Miami Valley Labs PO Box 39175 Cincinnati OH 45247

GOLDMAN, STEPHEN L, b New York, NY, Sept 18, 42. GENETICS. *Educ:* Brooklyn Col, BS, 63; Univ Mo-Columbia, MS, 67, PhD(genetics), 68. *Prof Exp:* Res asst genetics, Univ Mo-Columbia, 64-68; NIH fel, Univ Tex, Dallas, 69-71; asst prof genetics, Univ Toledo, 71-75, assoc prof, 75-80; PROF BIOL & ADJ PROF MICROBIOL, MED COL OHIO, 80- *Mem:* AAAS; Genetics Soc Am; Am Inst Biol Sci; Am Soc Microbiol. *Res:* Genetic recombination in maize and gene conversion in Schizosaccharromyces pombe. *Mailing Add:* Dept Biol Univ Toledo 2801 West Bancroft Toledo OH 43606

GOLDMAN, STEPHEN SHEPARD, b Brockton, Mass, Nov 24, 41. NEUROSURGERY. *Educ:* Northeastern Univ, BA, 64; Univ Ill, MS, 66, PhD(physiol), 70. *Prof Exp:* Fel neurochem, NIH, 70-72, staff fel, 72-77; ASST PROF EXP NEUROSURG, MED CTR, NY UNIV, 77- *Mem:* Am Physiol Soc; Am Soc Neurochem; NY Acad Sci; AAAS; Sigma Xi. *Res:* Catoin and lipid metabolism in brain with special reference to cerebral ischemia and thyroid hormone. *Mailing Add:* 89 Morningside Dr Croton-on-Hudson NY 10520

GOLDMAN, STEVEN, CARDIOLOGY. *Educ:* Univ Med Ctr, Tucson, MD, 86. *Prof Exp:* CHIEF CARDIOL, UNIV MED CTR, TUCSON, 86- *Mailing Add:* Dept Internal Med Univ Ariz Sch Med 111C Vet Admin Med Ctr Tucson AZ 85723

GOLDMAN, TERRENCE JACK, b Winnipeg, Man, Can, Feb 20, 47; m 68; c 4. STRONG INTERACTIONS, GRAVITY. *Educ:* Univ Man, BSc, 68; Harvard Univ, AM, 69, PhD(theoret physics), 73. *Prof Exp:* Fel, Nat Res Coun Can, Stanford Linear Accelerator Ctr, 73-75; staff fel, 75-78, STAFF MEM, LOS ALAMOS NAT LAB, 78- *Concurrent Pos:* Sr res fel, Calif Inst Technol, 78-80; panel mem, Solar Neutrino Proj Rev Panel, US Dept Energy, 81; vis assoc prof, Univ Calif, Santa Cruz, 83; distinguished vis scholar, Univ Adelaide, 87. *Honors & Awards:* Delphasus Lectr, Univ Calif, Santa Cruz, 88. *Mem:* Am Phys Soc. *Res:* Weak radiative corrections to leptonic pseudoscalar meson decay; rare decays of mesons and leptons; neutrino mixing properties; precise proton decay lifetime prediction; grand unified field theories; observable fractional charge hadrons; gravitational properties of anti-matter; quark structure of nuclei. *Mailing Add:* T-5 MS B283 Los Alamos Nat Lab PO Box 1663 Los Alamos NM 87545

GOLDMAN, THEODORE DANIEL, b Washington, DC, Oct 31, 46; m 72. ORGANIC POLYMER CHEMISTRY. *Educ:* Wash Col, BS, 69; NY Univ, MS, 72, PhD(org chem), 74. *Prof Exp:* NIH fel, 74-75; SCIENTIST POLYMER CHEM, ROHM AND HAAS CO, 75- *Mem:* Am Chem Soc; Sigma Xi. *Res:* Design and synthesis of polymeric modifiers for plastics; organic photochemistry. *Mailing Add:* 1290 Gen Derfermoy Rd Washington Crossing PA 18977

GOLDMAN, VLADIMIR J, b Moscow, USSR; US citizen. FRACTIONAL QUANTUM HALL EFFECT, RESONANT TUNNELING. *Educ:* Moscow State Pedag Inst, dipl physics, 77; Univ Md, PhD(physics), 85. *Prof Exp:* Res assoc physics, Princeton Univ, 85-87, staff mem, 88; asst prof, 88-91, ASSOC PROF PHYSICS, STATE UNIV NY, STONY BROOK, 91- *Concurrent Pos:* NSF presidential young investr, 89; Alfred P Sloan Found fel, 89. *Mem:* Am Phys Soc; AAAS. *Res:* Physics of semiconductors; heterostructure devices; fractional quantum Hall effect; Wigner crystallization; single-electron tunneling. *Mailing Add:* Dept Physics State Univ NY Stony Brook NY 11794-3800

GOLDMAN, YALE E, MUSCLE PHYSIOLOGY, BIOPHYSICS. *Educ:* Univ Pa, MD, 75, PhD(physiol), 76. *Prof Exp:* ASSOC PROF PHYSIOL, SCH MED, UNIV PA, 85- *Res:* Optics. *Mailing Add:* Dept Physiol Univ Pa Philadelphia PA 19104

GOLDMANN, KURT, b Eschwege, Ger, Sept 8, 21; nat US; m 55; c 3. MECHANICAL ENGINEERING. *Educ:* Pa State Col, BS, 42; Mass Inst Technol, MS, 46. *Prof Exp:* Chief analyst free piston mach, Lima-Hamilton Corp, 46-49; sr engr nuclear energy aircraft, Fairchild Eng & Aircraft Co, 49-50; res engr, Carrier Corp, 51; mgr, MCR Primary Systs Dept Nuclear Reactors, United Nuclear Corp, 51-71, mgr liquid metal systs dept, Gulf United Nuclear Fuels Co, 71-74; chief engr, Transnuclear, Inc, 74-87; CONSULT, 87- *Mem:* Am Soc Mech Engrs; Am Nuclear Soc; Sigma Xi. *Res:* Thermodynamics; fluid flow and heat transfer; liquid metal technology; free piston; gas turbine and nuclear power plants. *Mailing Add:* 62 Holbrooke Rd White Plains NY 10605

GOLDMAN-RAKIC, PATRICIA S(HOER), b Salem, Mass. DEVELOPMENTAL NEUROBIOLOGY, PSYCHOBIOLOGY. *Educ:* Vassar Col, AB, 59; Univ Calif, Los Angeles, PhD(psychol), 63. *Hon Degrees:* AM, Yale Univ, 79. *Prof Exp:* Staff fel, NIMH, 65-68, res physiologist, 68-78, chief, Sect Develop Neurobiol, Lab Neuropsychol, 78-79; dir grad studies, Sect Neuroanat, 81-86, actg chmn, 86-87, PROF NEUROSCI, YALE UNIV, SCH MED, 79- *Concurrent Pos:* Mem, Nat Adv Res Task Force NIMH, 72-74; mem, Biol Sci Fel & Training Grants Initial Rev Group, 74-77; pres, NIMH, Nat Inst Neurol & Commun Disorders & Stroke Assembly Scientists, 74-75; mem, Int Res Fel Rev Group, Fogarty Ctr, 76-78; mem, Long-Range Strategy Res Develop Disorders Panel, Adv Panel, Nat Inst Neurol & Commun Disorders & Stroke, 78-79, Develop Neurobiol Adv Panel, 81-84; sr scientist award, NIMH, 80 & 85; numerous lectrs, Univs & Res Ctrs, 72-87; mem, Third World Cong Comt, Int Brain Res Orgn, 90-91. *Honors & Awards:*

Herbert Birch Mem lectr, Int Neuropsychobiol Soc, 81; Kendon Smith Mem lectr, Univ NC, 85; Bernard Sachs Mem lectr, Child Neurol Soc, 85; Fyssen Found Prize, Neurosci, Paris, 90; Sally Harrington Goldwater lectr, Barrow Inst, 90; Rushton lectr, Fla State Univ, 90. *Mem:* Nat Acad Sci; Soc Neurosci (pres, 89-90); fel AAAS; Am Anat Asn; Int Brain Res Orgn; fel Am Psychol Asn. *Res:* Organization, development and plasticity of primate frontal lobe. *Mailing Add:* Sect Neurobiol Yale Univ Sch Med 333 Cedar St New Haven CT 06510

GOLDNER, ADREAS M, b Zurich, Switz, June 14, 34; US citizen; m 88; c 1. PHYSIOLOGY. *Educ:* Oberlin Col, BA, 56; Stanford Univ, MA, 57; George Washington Univ, PhD(physiol), 66. *Prof Exp:* Instr physiol sci, Menlo Col, Calif, 59-61; lectr biol, Montgomery Jr Col, Md, 65-66; res assoc, Yale Univ, 69-70; asst prof human physiol, Sch Med, Univ Calif, Davis, 70-75; ASSOC PROF PHYSIOL & ASSOC DEAN STUDENT AFFAIRS, UNIV ARIZ COL MED, 75- *Concurrent Pos:* USPHS fel biophys, Harvard Med Sch, 66-67 & physiol, Sch Med, Yale Univ, 67-69; consult, Biol Commun Proj, George Washington Univ, 66-69; lectr, Southern Conn State Col, 68-69 & SCent Community Col, 69-70; vis scientist, Max Planck Inst Biophys, Frankfurt, 72. *Mem:* AAAS; Am Physiol Soc; Biophys Soc; Biomed Eng Soc. *Res:* Membrane phenomena; intestinal permeability-relationship between nutrients and electrolyte transport; effects of hormones and divalent cations on epithelial permeability. *Mailing Add:* Off Student Affairs Univ Ariz Col Med Tucson AZ 85724

GOLDNER, HERMAN, b Wilkes-Barre, Pa, Aug 20, 28; m 55. MEDICAL MICROBIOLOGY, BIOHAZARDS CONTROL. *Educ:* King's Col, BS, 50; Cath Univ Am, MS, 52; Univ Pa, PhD(med microbiol), 57. *Prof Exp:* Lectr bact, King's Col, 52; asst cancer immunol, Wistar Inst, 55-57; res immunologist, Biochem Res Found, 57-59; res assoc, Merck Inst Therapeut Res, Div Merck & Co, Inc, 59-63; asst mem, S Jersey Med Res Found, 63-65; assoc mem, Res Labs, Northern Div, Albert Einstein Med Ctr, 65-70, admin officer res, 71-76; BIOHAZARDS OFFICER, HEALTH SCI CAMPUS, TEMPLE UNIV, 73-, DIR SPONSORED PROG ADMIN 76- *Concurrent Pos:* Leukemia scholar Leukemia Soc, 67-70; adj assoc prof, Dept Microbiol & Immunol, Temple Univ, 74. *Mem:* Nat Asn Univ Res Adminr; Sigma Xi. *Res:* Virology; immunology. *Mailing Add:* Old Med Sch Rm 100 3400 N Broad St Philadelphia PA 19140

GOLDNER, JOSEPH LEONARD, b Omaha, Nebr, Nov 18, 18; m; c 2. ORTHOPEDIC SURGERY. *Educ:* Univ Minn, AB, 39; Univ Nebr, BScMed & MD, 43; Am Bd Orthop Surg, dipl, 52. *Prof Exp:* Resident orthop, Med Ctr, 46-50, assoc, 50, from asst prof to assoc prof, 51-56, chmn div orthop surg, 68-77, PROF ORTHOP SURG, MED CTR, DUKE UNIV, 56-, CHIEF ORTHOPAMPUTEE CLINS, 55- *Concurrent Pos:* Fel, Amputee Training Sch, Univ Calif, Los Angeles, 55; exchange fel, Am & Brit Orthop Asn, Eng, 55; res, Ga Warm Springs Found, 47-48; mem appl physiol study sect, NIH, 63-67; attend orthop surgeon, Vet Admin Hosp, Durham, NC, Watts Hosp & NC Cerebral Palsy Hosp, consult, McCain Tuberc Sanatorium, NC & Army Hosp, Ft Bragg; mem comt clean air in oper rm, Nat Res Coun; consult, Surgeon Gen, US Navy; vis prof, Children's Hosp, Australia, 66, Univ Calif, Los Angeles, 67, Harvard Univ, 68, Univ NMex, 71 & Crippled Children's Div, Vt State Bd Health, 71. *Honors & Awards:* Gibson Mem lectr, Winnipeg, Can, 69; Hike-Kite lectr, 71. *Mem:* Am Soc Surg of Hand (pres, 69-70); Am Orthop Asn; Can Orthop Asn; Am Acad Orthop Surg; Am Acad Cerebral Palsy; Sigma Xi. *Res:* Tendon healing; hand reconstruction; amputee problems; fat embolism; ultraviolet light in operating room. *Mailing Add:* Dept Surgery Box 3706 Med Ctr Duke Univ Med Ctr Durham NC 27710

GOLDNER, RONALD B, b New York, NY, Mar 24, 35; m 59; c 1. ELECTRICAL ENGINEERING, SOLID STATE PHYSICS. *Educ:* Mass Inst Technol, SB & SM, 57, EE, 59; Purdue Univ, PhD(elec eng), 62. *Prof Exp:* Asst, Mass Inst Technol, 57-59, res assoc, Lab Insulation Res, 59; instr elec eng, Purdue Univ, 59-62; asst prof, Mass Inst Technol, 62-64; from asst prof to assoc prof, 64-77, PROF ELEC ENG, TUFTS UNIV, 77- *Concurrent Pos:* Ford Found fel eng, 62-64; consult various private industs, 63- *Mem:* Am Phys Soc; Inst Elec & Electronics Engrs; Solar Energy Soc; Electrochem Soc; Optical Soc Am; Sigma Xi. *Res:* Optoelectronic properties of solids; optoelectronic applications. *Mailing Add:* Dept Elec Eng Tufts Univ Medford MA 02155

GOLDREICH, PETER, b New York, NY, July 14, 39; m 60; c 2. ASTROPHYSICS. *Educ:* Cornell Univ, BS, 60, PhD(physics), 63. *Prof Exp:* Instr astron, Cornell, 61-63; Nat Acad Sci-Nat Res Coun fel, Cambridge Univ, 63-64; asst prof astron & geophys, Univ Calif, Los Angeles, 64-66; assoc prof, 66-69, PROF ASTRON & PLANETARY SCI, CALIF INST TECHNOL, 69-, LEE A DUBRIDGE PROF ASTROPHYS & PLANETARY PHYS, 81- *Concurrent Pos:* Sloan Found fel, 68-70; T Gold lectr, Cornell Univ, 87-88; regents fel, Smithsonian Inst Wash, 88-89 & 89-90; Miller prof, Univ Calif, Berkeley, 90. *Honors & Awards:* Henry Norris Russell Lectr, Am Astron Soc, 79; Chapman Medal, Royal Astron Soc, 85; Dirk Brouwer Award, 86; Amos de Shalit Lectr, Weltzmann Inst, Rehovot, Israel, 86. *Mem:* Nat Acad Sci; Am Acad Arts & Sci. *Res:* Theoretical study of solar system; galactic structure and cosmology. *Mailing Add:* Div Geol & Planetary Sci Calif Inst Technol 1201 E California Blvd Pasadena CA 91125

GOLDRICH, STANLEY GILBERT, b New York, NY, Sept 22, 37. PHYSIOLOGICAL PSYCHOLOGY, NEUROSCIENCE. *Educ:* Queens Col, NY, BA, 59, MA, 65; City Univ New York, PhD(psychol), 66; Mass Col Optom, OD, 74. *Prof Exp:* Lectr psychol, Queens Col, NY, 64-65; res assoc, Primate Ctr, Univ Wis, 65-67; asst prof, Ohio State Univ, 67-72; ASSOC CLIN PROF OPTOM, STATE COL OPTOM, STATE UNIV NY, 74-, DIR BIOFEEDBACK LAB, 84- *Mem:* Fel Am Acad Optom; Am Psychol Assoc; Am Optom Assoc. *Res:* Biofeedback control of eye movement disorders including nystagmus, strabismus and reading dysfunction; inventor of Orthotone Biofeedback Eye Training System, and Goldrich Contour Rotator for treatment of oculomotor disorders; control of primate visually guided behavior. *Mailing Add:* State Col Optometry State Univ New York, 100 E 24th St New York NY 10016

GOLDRING, LIONEL SOLOMON, b Los Angeles, Calif, Oct 24, 22; m 49; c 1. PHYSICAL CHEMISTRY, RESEARCH ADMINISTRATION. *Educ:* Univ Calif, Los Angeles, BA, 43; Mass Inst Technol, PhD(phys chem), 50; NY Univ, MBA, 73. *Prof Exp:* Jr chemist, Oak Ridge Nat Lab, 43-46; res assoc, Mass Inst Technol, 46, asst, 46-49; res chemist, Brookhaven Nat Lab, 50-52; head radiochem group, United Nuclear Corp, 53-57, mem adv proj group, 58-60; sr res chemist, AMF, Inc, 60-64, sr tech specialist, 64-66, sr res specialist, Res Dept, 66-67, lab mgr, 67-70, mgr chem phys res, 70-74, dir chem lab, 74-78; dir appl res, Orion Res, Inc, 78-81; dir Transducer develop, Novametrix Med Systs, Inc, 81-88; SR CHEMIST, SPACELABS, 89- *Concurrent Pos:* Consult, Brookhaven Nat Lab, 47-49; adj assoc prof, NY Univ Grad Sch Business Admin, 74-78. *Mem:* Am Chem Soc; Soc Plastics Engrs. *Res:* Electrochemical sensors; polymer and plastics science and engineering; inorganic chemistry; nuclear science; air and water pollution; polymers and their application; data acquisition and analysis. *Mailing Add:* 11240 109th Ave NE Kirkland WA 98033-4503

GOLDRING, ROBERTA M, b June 18, 29; wid; c 3. MEDICINE. *Educ:* Vassar Col, AB, 49; Columbia Univ, MD, 53. *Prof Exp:* Asst med, Col Physicians & Surgeons, 62-66; from asst prof to assoc prof, 66-84, PROF MED, SCH MED, NY UNIV, 84- *Concurrent Pos:* Asst vis physician, Presby Hosp, 56-66, Bellevue Hosp, Univ Hosp, 66-73; dir, Pulmonary Function Lab, Bellevue Hosp, 75-, co-dir, Med Inst Care Unit, 87-; assoc attend physician, Bellevue Hosp, Univ Hosp, 73-84, attend physician, 84- *Mem:* Am Physiol Soc; Am Fedn Clin Res; AAAS; Am Heart Asn; Am Thoracic Soc. *Res:* Medicine. *Mailing Add:* Dept Med NY Univ Sch Med 550 First Ave New York NY 10016

GOLDRING, SIDNEY, b Kremnitz, Poland, Apr 2, 23; nat US; m 45; c 2. NEUROSURGERY. *Educ:* Wash Univ, MD, 47. *Prof Exp:* Instr, Med Unit to Thailand, Univ Wash, 52-53, from instr to assoc prof neurosurg, 56-64; prof & head dept, Univ Pittsburgh, 64-66; head dept, 74-90, PROF NEUROL SURG, SCH MED, WASH UNIV, 66- *Concurrent Pos:* USPHS div res grants; neurosurgeon-in-chief, Barnes & Allied Hosps, St Louis, 74-90. *Mem:* Am Physiol Soc. *Res:* Electrophysiological studies in animal and human cerebral cortex, employing direct current amplifiers and non-polarizable recording electrodes for correlation of slow electrical changes with various physiological and pathological states; microelectrode and computer techniques in study of experimental epilepsy. *Mailing Add:* Dept Neurol Surg Wash Univ Sch Med 660 S Euclid Ave St Louis MO 63110

GOLDSACK, DOUGLAS EUGENE, b London, Ont, July 17, 39; m 64; c 2. PHYSICAL CHEMISTRY. *Educ:* Univ Western Ont, BSc, 61; Univ Wis, PhD(chem), 66. *Prof Exp:* Res assoc biophys chem, Univ Wis, 66-67; res scientist, Vancouver Lab, Fisheries Res Bd Can, 67-68; asst prof biophys chem, 68-74, chmn dept chem, 75-80, assoc prof, 74-83, PROF CHEM, LAURENTIAN UNIV, 83-, DEAN SCI & ENG, 81- *Res:* Physical chemistry of proteins; physical chemistry of solutions. *Mailing Add:* 13 Aspenwood Ct Sudbury ON P3E 5T6 Can

GOLDSBOROUGH, JOHN PAUL, b Newark, NJ, May 19, 34; m 56; c 2. LASERS. *Educ:* Lehigh Univ, BS, 56; Stanford Univ, PhD(physics), 61. *Prof Exp:* Res staff mem, Int Bus Mach Corp, 60-66; dir res, Spectra-Physics Inc, 66-68, sr proj engr, 68-74, eng dept mgr, 74-89; ENG MGR, LICONIX, 89- *Mem:* Am Phys Soc. *Res:* Magnetic resonance; photoconductivity; low temperature physics. *Mailing Add:* Liconix 3281 Scott Blvd Santa Clara CA 95054

GOLDSBY, ARTHUR RAYMOND, b Flora, Ill, Nov 20, 04; m 34; c 1. ORGANIC CHEMISTRY. *Educ:* Ill Col, AB, 28; Northwestern Univ, MS, 30, PhD(org chem), 32. *Prof Exp:* Asst chem, Northwestern Univ, 28-32; chemist, Nat Aluminate Corp, Ill, 32-33; res chemist, Nat Aniline & Chem Co, NY, 33-36; dir explor res gasoline, Tex Co, 36-40, tech adv, Texaco Develop Corp, 40-55; dir res, Stratford Eng Corp, Mo, 55-57; tech adv, Texaco Inc, 57-69; consult & dir, Stratford-Graham Eng Corp, 69-79. *Mem:* Am Chem Soc; Am Inst Chemists; Am Inst Chem Engrs. *Res:* Reactions of furan and furfural; pyrolysis of olefins and paraffins; synthetic organic detergents; aviation gasoline; alkylation; isomerization; catalytic cracking; fuel composition in relation to engine performance; recovery of sulfuric acid licensing. *Mailing Add:* 76 Marcourt Dr Chappaqua NY 10514

GOLDSBY, RICHARD ALLEN, b Kansas City, Mo, Dec 19, 34. BIOCHEMISTRY, IMMUNOLOGY. *Educ:* Univ Kans, BS, 57; Univ Calif, Berkeley, PhD(chem), 61. *Prof Exp:* Jr org chemist, Monsanto, Inc, 57-58; biochemist-virologist, E I du Pont de Nemours & Co, Inc, 61-66; from asst prof to assoc prof biol, Yale Univ, 66-72; prof chem, Univ Md, College Park, 72-; AT DEPT BIOL, AMHERST COL. *Concurrent Pos:* Morse fel, Yale Univ, 70-71, master, Pierson Col, 71-72; bd dirs, Carver Found, Tuskegee Inst, 73-; Nat Res Coun Sr fel, Ames Res Ctr, 75- *Mem:* Am Soc Microbiol. *Res:* Somatic cell genetics of thymidine kinase; senescence in cultured somatic cells; immunoglobin synthesis. *Mailing Add:* Dept Biol Amherst Col Amherst MA 01003

GOLDSCHMID, OTTO, b Vienna, Austria, May 1, 10; nat US; m 42; c 2. WOOD CHEMISTRY. *Educ:* Univ Stuttgart, Dipl Ing, 34; Univ Calif, PhD(chem), 40. *Prof Exp:* Res supvr, Olympic Res Div, ITT Rayonier, Inc, 39-75; RETIRED. *Mem:* Am Chem Soc. *Res:* Viscosity of suspensions and solutions; absorption spectra of organic compounds; composition and structure of cellulose fibers; lignin chemistry. *Mailing Add:* 726 S Ninth St Shelton WA 98584

GOLDSCHMIDT, BERNARD MORTON, b New York, NY, Feb 7, 36; m 65; c 2. CANCER. *Educ:* City Univ New York, BS, 57; Univ Wis-Madison, PhD(org chem), 62. *Prof Exp:* NIH res fel, State Univ NY Stony Brook, 62-63; NIH res fel, 63-65, res asst, 65-68, asst prof, 68-74, ASSOC PROF ENVIRON MED, MED SCH, NY UNIV, 74- *Concurrent Pos:* Adj assoc prof, Adelphi Univ, 68-72; adj asst prof, Manhattan Community Col, 72-73. *Mem:* Am Asn Cancer Res; Am Chem Soc; Royal Soc Chem. *Res:* Chemical carcinogenesis, isolation and identification of chemical carcinogens; mechanism of cancer induction. *Mailing Add:* Dept Environ Med NYU Sch Med 550 First Ave New York NY 10016

GOLDSCHMIDT, ERIC NATHAN, b Berlin, Ger, Nov 25, 27; nat US; m 54; c 3. INFORMATION SCIENCE, PHARMACEUTICAL INDUSTRY. *Educ:* Brooklyn Col, BS, 50; Harvard Univ, AM, 51, PhD(org chem), 55. *Prof Exp:* Lectr chem, Brooklyn Col, 50; scientist, Warner-Chilcott Labs, 56-57; sr scientist, Warner Lambert Res Inst, 57-62, sr literature chemist, 62-67; res specialist, Off Eng Res & lectr, dept chem, Univ Pa, 67-68; asst to dir res, Endo Labs, 68-70, asst to vpres res, 70-72, dir res serv, 72-78, dir sci info, 78-80; Mgr prod planning, Mead Data Central, 80-82; sr literature scientist, 82-84, SR INFO ASSOC, WARNER LAMBERT CO, 84- *Concurrent Pos:* Lectr, Brooklyn Col, 50 & 69, Univ Pa, 67-68 & Hofstra Univ, 69-79; consult, Ergo Assoc, 82- *Mem:* AAAS; fel Am Inst Chemists; Am Chem Soc; NY Acad Sci; Royal Soc Chem. *Res:* Antimetabolites; biosynthesis inhibitors; nitrogen heterocycles; chemotherapeutic agents; new synthetic methods; information storage and retrieval; research administration. *Mailing Add:* 147-29 68th Rd Kew Gardens Hills NY 11367

GOLDSCHMIDT, MILLICENT, b Erie, Pa, June 11, 26; m 49; c 2. ORAL MEDICAL, CLINICAL MICROBIOLOGY, IMMUNOLOGY. *Educ:* Case Western Reserve Univ, BA, 47; Purdue Univ, MS, 50, PhD(microbiol), 52. *Prof Exp:* Asst zool, Purdue Univ, 50-51; Purdue Res Found fel, 51-52; res assoc, Dept Biochem, Sch Med, Case Western Reserve Univ, 53; sr res assoc, Biophys Res Labs, George Washington Univ, 56-59; instr chem, Hood Col, 59-60; asst prof med in bact, Sch Med, Univ Md & US Army Med Unit, Ft Detrick, 60-61; NIH fel microbiol, Univ Tex, 61-63; res instr microbiol, Col Med, Baylor Univ, 63-67; asst prof microbiol & asst microbiologist, Dept Path, Univ Tex M D Anderson Hosp & Tumor Inst, 67-71, assoc microbiologist & actg chief clin microbiol, 70-71; assoc prof, Univ Tex Grad Sch Biomed Sci, 71-73; assoc prof clin microbiol, Prog Infectious Dis & Clin Microbiol, Med Sch, Univ Tex, Houston, 73-76, Nat Inst Dent Res fel microbiol, 77-78, dir grad prog microbiol, 83-90, ASSOC PROF MICROBIOL, DENT SCI INST & DEPT PERIOD, DENT BR, UNIV TEX HEALTH SCI CTR HOUSTON, 79-, GRAD SCH BIOMED SCI, 70- *Concurrent Pos:* Assoc microbiologist, Dept Med & Dept Lab Med, Univ Tex M D Anderson Cancer Ctr, 70-; dir, Tex Outstanding Biol Teachers Awards Comt, Nat Asn Biol Teachers, 71-80; pres, Tex Br, Am Soc Microbiol, 80-81, Houston Asn Med Microbiologists, 82-84, Rice Univ/Tex Med Ctr Chap Sigma Xi, 85-86 & Gulf Coast Chap, Asn Women in Sci, 87-88; chmn, Prog Comt, Am Acad Microbiol, 81-83 & mem bd gov, 85-88; treas, Southeast Tex Sigma Xi Coun, 84-89; found lectr, Am Soc Microbiol, 87-88; mem eval panel, Biomed Sci Grad Fel Prog, NSF, 90- *Mem:* Sigma Xi; Am Soc Microbiol; fel Am Acad Microbiol; Asn Women Sci; Int Asn Dent Res; Am Asn Dent Res; Asn Clin Immunol; Asn Clin Microbiologists; Am Soc Microbiol. *Res:* Instrumentation and immunologic methods for rapid detection and characterization of microorganisms; oral, medical, biochemical and biophysical aspects of microbiology, immunology, microbial physiology and nutrition; biochemistry. *Mailing Add:* Dental Sci Inst Dent Br Univ Tex Health Sci Ctr Houston PO Box 22068 Houston TX 77225

GOLDSCHMIDT, PETER GRAHAM, b Cardiff, Wales, Feb 18, 45; US citizen. QUALITY MANAGEMENT, HEALTH POLICY. *Educ:* Polytech Cent London, DMS, 68; Univ London, MB & BS, 70; Johns Hopkins Univ, MPH, 71, DrPH, 80. *Prof Exp:* Res assoc, Sch Hyg & Pub Health, Johns Hopkins Univ, 72-75; vpres, Policy Res Inc, 74-81; dir health serv res & develop, US Dept Vet Affairs, 81-86; vpres res & develop, Qual Standards Med, 86-90; PRES, WORLD DEVELOP GROUP, 86- *Concurrent Pos:* Pres, Health Improv Inst, 90- *Mem:* AMA; Am Pub Health Asn; Opers Res Soc Am. *Res:* Health care quality management and effectiveness; technology assessment, quality measurement, assurance, and improvement. *Mailing Add:* 5101 River Rd No 1913 Bethesda MD 20816

GOLDSCHMIDT, RAUL MAX, b Santiago, Chile, Sept 23, 41. MICROBIAL GENETICS. *Educ:* Univ Chile, Licenciado, 64; Mass Inst Technol, MSc, 67; Columbia Univ, PhD(biol), 70. *Prof Exp:* Assoc prof microbiol, Univ Ala, Birmingham, res assoc, 74, vis prof microbiol, 74-82; fac res assoc, Dept Biol, Wash Univ, St Louis, 82-90; RES SCIENTIST, THE BLOOD CTR, 90- *Mem:* Am Soc Microbiol. *Res:* Elucidation of the molecular mechanisms responsible for bacterial contact, gene transfer and recombination of genetic material during bacterial conjugation and in the in-vivo expression of genes contained on bacterial plasmids; genetical and structural characterization of plasmids responsible for drug-resistance among clinical and animal gram-negative bacterial isolates. *Mailing Add:* Blood Ctr 310 E 67th New York NY 10021

GOLDSCHMIDT, VICTOR W, b Montevideo, Uruguay, Apr 20, 36; US citizen; m 58; c 3. FLUID MECHANICS, THERMOSCIENCES. *Educ:* Syracuse Univ, BS, 57, PhD(eng), 65; Univ Pa, MS, 60. *Prof Exp:* Appl engr, Control Valve Div, Honeywell, Inc, 57-59, develop engr, 59-60; instr mech eng, Syracuse Univ, 60-64; asst prof civil eng, 64-68, assoc prof civil eng, 68, assoc prof, 68-71, PROF MECH ENG, PURDUE UNIV, 71- *Concurrent Pos:* Dir Purdue Fels, Latin Am, 67; UN Indust Develop Orgn consult, Argentina, 79 & 81; ambassador, Am Soc Heating, Refrig & Air-Conditioning Engrs, NZ & Australia, 79; Fulbright Sr Scholar, 79. *Honors & Awards:* Freeman Fund Award, Am Soc Civil Engrs, 71. *Mem:* Am Soc Mech Engrs; Am Soc Civil Engrs; Am Soc Eng Educ; fel Am Soc Heating, Refrig & Air-Conditioning Engrs; Am Phys Soc. *Res:* Fundamental studies in turbulence, particularly related to transport, interaction with acoustics and measurements of properties characterizing turbulence; energy utilization in heating and air conditioning; performance of heating and air conditioning systems. *Mailing Add:* Sch Mech Eng Purdue Univ Lafayette IN 47907

GOLDSCHMIDT, YADIN YEHUDA, b Kfar-Saba, Israel, Aug 25, 49; m 75; c 3. STATISTICAL MECHANICS, PHASE TRANSITIONS. *Educ:* Hebrew Univ, Jerusalem, BSc, 73, PhD(physics), 78; Weizmann Inst Sci, Rehovot, Israel, MSc, 75. *Prof Exp:* Vis scientist physics, Ctr Nuclear Studies, Saclay, France, 78-80; asst prof res, Brown Univ, 80-82; asst prof, 82-86, ASSOC PROF PHYSICS, UNIV PITTSBURGH, 86- *Concurrent Pos:* Vis scientist, CEA-Saclay, France, 89-90. *Mem:* Am Phys Soc. *Res:* Phase transitions and critical phenomena; random magnetic systems; random fields; random anisotropies; spin glasses; applications of field theory in statistical mechanics; investigation of two dimensional systems. *Mailing Add:* Dept Physics & Astron Univ Pittsburgh Pittsburgh PA 15260

GOLDSCHNEIDER, IRVING, b Philadelphia, Pa, Feb 26, 37; m; c 4. PATHOLOGY. *Educ:* Univ Pa, BA, 58, MD, 62. *Prof Exp:* Res asst, Wistar Inst Anat & Biol, Philadelphia, 58-59; fel, Med Student Res Training Prog, Sch Med, Univ Pa, 59-62; fel, exp path & immunol, Western Reserve Univ, NIH, 63-66; sr investr, Dept Bact, Walter Reed Army Inst Res, Washington, DC, 66-68; spec res fel develop biol, Nat Cancer Inst, Univ Chicago, 68-69; from asst prof to assoc prof, 69-79, PROF, DEPT PATH, SCH MED, UNIV CONN, 79-, DIR, CELL SORTING LAB, 79-, CHMN, DEPT PATH, 82- *Concurrent Pos:* Assoc ed, J Immunol, 74-82. *Mem:* Am Cancer Soc. *Res:* Pathology; numerous publications. *Mailing Add:* Dept Path Univ Conn Health Ctr 263 Farmington Ave Farmington CT 06032

GOLDSMITH, CARL, internal medicine, nephrology, for more information see previous edition

GOLDSMITH, CHARLES HARRY, b Flin Flon, Man, Aug 27, 39; m 66; c 2. DATA ANALYSIS, APPLIED STATISTICS. *Educ:* Univ Man, BSc, 61, MSc, 63; NC State Univ, PhD(exp statist), 69. *Prof Exp:* Lectr statist, Carleton Univ, Ont, 63-65; res assoc biostatist, Univ NC, Chapel Hill, 65-69; asst prof biostatist & appl math, 69-73, assoc prof, 73-79, PROF BIOSTATIST, DEPT CLIN EPIDEMIOL & BIOSTATIST, MCMASTER UNIV, 79- *Concurrent Pos:* Consult, Dept Hwys & Educ, Man Prov Govt, 63; abstractor, Exec Sci Inst, 67-71; adv, Med Commun Comt, Mohawk Col, Ont, 70-81; Ont Ministry Health fel & External Affairs Can travel fel, 75; chmn demonstration model grants rev comt, Ont Ministry Health, 79-82; ed consult, J Can Physiother Asn, 80-87; chmn, rev comt 53- Epidemiol Chronic Dis & Occup Health, Nat Health Res & Develop Prog, Health & Welfare Can, 83-; chair, Biopharmaceut Sect, Am Statist Asn, 87; mem study sect, Found Chiropractic Educ Res, 87-; Can ed, J Biopharm Statist, 90- *Honors & Awards:* Shewell Award, Am Soc Qual Control, 68, Frank J Wilcoxon Award, 70. *Mem:* Fel Am Statist Asn; Biomet Soc; Royal Statist Soc; Statist Soc Can (secy, 77-81); Int Statist Inst; Soc Clin Trials; Can Rheumatism Asn. *Res:* Design and analysis of health surveys and clinical trials; experimental designs; regression analysis; statistical data analysis; statistical problems in therapeutic compliance; statistical epidemiology and pharmacolepidemiology. *Mailing Add:* Dept Clin Epidemiol & Biostatist McMaster Univ Hamilton ON L8N 3Z5 Can

GOLDSMITH, DALE PRESTON JOEL, b Catasauqua, Pa, Oct 30, 16. BIOCHEMISTRY. *Educ:* Lehigh Univ, BS, 38; Harvard Univ, MA, 40; Pa State Col, PhD(org chem), 42; Univ Rochester, MS, 57. *Prof Exp:* Chemist, Nat Aniline & Chem Co, NY, 37; chemist, Res Labs, Socony-Vacuum Oil Co, NJ, 38-39; instr org chem, Pa State Col, 40-42, Parke, Davis & Co fel, 42-43; sr res chemist, Merck & Co, Inc, 43-55; from instr to asst prof physiol, Sch Med, Univ Rochester, 57-64; ASSOC PROF BIOCHEM, COL MED, UNIV NEBR, OMAHA, 64- *Mem:* AAAS; Am Chem Soc; Am Physiol Soc; NY Acad Sci; fel Am Inst Chemists; Sigma Xi. *Res:* Mucosal regeneration; atherosclerosis; medicinal chemicals; gastrointestinal biochemistry; lipid metabolism. *Mailing Add:* 11716 Mayberry St Omaha NE 68154

GOLDSMITH, DAVID JONATHAN, b Flushing, NY, Aug 25, 31; m 59; c 3. SYNTHETIC ORGANIC CHEMISTRY. *Educ:* Univ Mich, BS, 52, MS, 53; Columbia Univ, PhD(chem), 58. *Prof Exp:* Res fel, Harvard Univ, 57-59; asst prof chem, Wayne State Univ, 59-63; from asst prof to assoc prof chem, 63-75, chmn, 83-90, PROF CHEM, EMORY UNIV, 75- *Concurrent Pos:* Vis prof, Univ Strasbourg, 71-72. *Mem:* Am Chem Soc. *Res:* Synthesis of natural products and synthetic methods. *Mailing Add:* Dept Chem Col Arts & Sci Emory Univ Atlanta GA 30332

GOLDSMITH, DONALD LEON, b Philadelphia, Pa, Aug 12, 37; m 63; c 2. SOFTWARE DEVELOPMENT, MATHEMATICS. *Educ:* Univ Pa, BA, 59, MA, 60, PhD(math), 64. *Prof Exp:* Asst prof, Fordham Univ, 64-68; assoc prof, 68-75, PROF MATH, WESTERN MICH UNIV, 75- *Concurrent Pos:* NSF res grant, 66-68; vis mem, Courant Inst Math Sci, NY Univ, 68-69. *Mem:* Am Math Soc; Math Asn Am; Asn Comput Mach. *Res:* Number theory, combinatorics, graph theory; computer science. *Mailing Add:* Dept Math Western Mich Univ Kalamazoo MI 49008

GOLDSMITH, EDWARD, b Baltimore, Md, Sept 9, 23; m 50; c 2. FIRE RETARDANT COATINGS. *Educ:* Univ Md, BS, 48. *Prof Exp:* Sr chemist, Baltimore Paint & Chem Corp, 58-76; TECH DIR, OCEAN CHEM, INC, 76- *Concurrent Pos:* Consult plastics, chemicals & fire retardants, 70- *Mem:* Soc Coatings Technol; Am Soc Testing & Mat. *Res:* Intumescent fire retardant coatings. *Mailing Add:* 117 Todd St Savannah GA 31402

GOLDSMITH, ELI DAVID, b New York, NY, Apr 10, 07; m 40; c 1. DEVELOPMENTAL BIOLOGY, RESEARCH ADMINISTRATION. *Educ:* City Col NY, BS, 26; Harvard Univ, AM, 28, PhD(biol), 34; NY Univ, MS, 31. *Prof Exp:* Asst zool, Harvard Univ, 28-29; lab instr comp anat, NY Univ, 29-31; asst zool, Harvard Univ, 31-33, Austin teaching fel, 33-34; instr biol & chem, City Col NY, 34-45; asst prof anat, 45-48, from assoc prof to prof histol, Col Dent, 48-73, from asst prof to prof, Grad Sch Arts & Sci, 45-73, res coordr, 48-69, EMER PROF HISTOL, COL DENT & EMER PROF, GRAD SCH ARTS & SCI, NY UNIV, 73-, SPEC CONSULT RES & TRAINING TO DEAN, COL DENT, 78-, DIR RES TRAINING & PROG DEVELOP, 78- *Concurrent Pos:* Harvard grant, Marine Biol Lab, Woods Hole, 32 & Bermuda Biol Sta, 34; Commonwealth Fund grant, 44; Thompson Fund grant, 44, 46; USPHS grant, 47-70; Am Cancer Soc grant, 51-55; biomed res coordr, Off Naval Res, US Dept Navy, NY, 51-53; actg dir, Off Res Serv, NY Univ, 63-64; consult to surgeon gen, USPHS; consult to dir, Nat Inst Dent Res; sci consult, Osborn Labs, Marine Sci; mem corp, Bermuda Biol Sta; mem selection panel sr res fels, NIH; mem dent prog proj comt, Nat Inst Dent Res. *Mem:* AAAS; fel Am Col Dent; fel, NY Acad Med; Int Asn Dent Res; hon mem Acad Gen Dent; Am Chem Soc. *Res:* Regeneration effect of chemicals and endocrines; carcinogenesis; anti-vitamin; vitamin-endocrinology relationship; goitrogenic drugs; radioactive iodine; blood depressors; dental histology; antimetabolites and nucleic acid metabolism in Drosophila; chemosterilants and control of insect populations; teratology. *Mailing Add:* 418 Beach 133rd St Far Rockaway NY 11694-1416

GOLDSMITH, GEORGE JASON, b Newburyport, Mass, Mar 29, 23; m 45; c 4. PHYSICS. *Educ:* Univ Vt, BS, 44; Purdue Univ, MS, 48, PhD(physics), 55. *Prof Exp:* Instr physics, Purdue Univ, 48-55; physicist, Labs, RCA Corp, 55-68; ASSOC PROF PHYSICS, BOSTON COL, 68- *Concurrent Pos:* Dir, Boston Col Environ Ctr, 75- *Mem:* AAAS; Am Phys Soc; Sigma Xi. *Res:* Photoconductivity in insulators; solid state and nuclear physics; science and society; energy and environment. *Mailing Add:* 45 Eliot St Sherborn MA 01770

GOLDSMITH, HARRY L, b Nurnberg, Ger, May 11, 28; Can citizen; m 58; c 2. BIOPHYSICS, PHYSICAL CHEMISTRY. *Educ:* Oxford Univ, BA, 50, BSc, 51; McGill Univ, PhD(chem), 61. *Prof Exp:* Tech officer phys chem, Dyestuffs Div, Imp Chem Industs, Ltd, Eng, 50-56; fel suspension rheology, McGill Univ, 61-64, Med Res Coun Can scholar blood rheology, 64-67, from asst prof to assoc prof exp med, 66-73, PROF EXP MED, MCGILL UNIV, 73-, MED RES COUN CAN CAREER INVESTR BLOOD RHEOLOGY, 67- *Honors & Awards:* Landis Award, Microcirculatory Soc, Inc, 84. *Mem:* Am Physiol Soc; Can Soc Clin Invest; Int Soc Biorheology; European Soc Microcirculation; Microcirculatory Soc; Soc Rheology; NAm Soc Rheology. *Res:* Solution kinetics; lyophopic colloids; flow of suspensions of model rigid and deformable particles through tubes; flow behavior and interactions of human blood cells; measurement of force of adhesion between cells and between cells and the vessel wall. *Mailing Add:* Dept Med McGill Univ Montreal Gen Hosp 1650 Cedar Ave Montreal PQ H3G 1A4 Can

GOLDSMITH, HARRY SAWYER, b Newton, Mass, Sept 30, 29; m 61; c 3. NEUROSURGERY. *Educ:* Dartmouth Col, AB, 52; Boston Univ, MD, 56. *Prof Exp:* Prof & chmn, Dept Surg, Jefferson Med Col, 70-77; prof surg, Dartmouth Med Sch, 78-83; PROF SURG, BOSTON UNIV SCH MED, 83- *Concurrent Pos:* Adj prof neurosurg, Boston Univ Sch Med, 83- *Mem:* Am Col Surgeons; Soc Vascular Surg; Int Soc Surg; Soc Surg Oncol. *Res:* Use of the omentum and its effects on the central nervous system. *Mailing Add:* Univ Hosp 75 E Newton St Boston MA 02118

GOLDSMITH, HENRY ARNOLD, b Berlin, Ger, Dec 17, 10; nat US; m 41; c 2. PHYSICAL CHEMISTRY. *Educ:* Univ Berlin, Dipl, 32; Univ Genoa, PhD(chem), 34; Univ Pavia, dipl, 35. *Prof Exp:* From chemist to chief chemist, Glyco Prod Co, Inc, NY, 35-44; coatings chemist, Standard Varnish Works, Inc, 44-46; chief chemist cleaning compounds, Phipps Prod, Inc, Mass, 47-48 & org chem, Colgate-Palmolive-Peet 48-50; chief chemist detergents, Theobald Industs, 50-54 & metal cleaners, Solvent or Chem Prod Co, Mich, 54-58; from group leader to mgr surface prep, Turco Prod CDiv, Purex Corp, Calif, 58-64; sr scientist, US Borax Res Corp, Anaheim, 65-67; res chemist, Purex Corp, 68-72, mgr res & develop, 72-81; RETIRED. *Mem:* Am Chem Soc; AAAS. *Res:* Organic chemistry and applications in the fields of fat derivatives, detergents, surfactants, emulsifiers and resins. *Mailing Add:* 421 Via Colusa Torrance CA 90505

GOLDSMITH, JEWETT, b Baltimore, Md, Apr 1, 19; m 56; c 3. PSYCHIATRY. *Educ:* Johns Hopkins Univ, AB, 38; Univ Md, MD, 42; Am Bd Psychiat & Neurol, dipl, 48. *Prof Exp:* Intern, Kings County Hosp, Brooklyn, 42-43; Vet Admin resident psychiat, Duke Univ Hosp, 46-48; instr, Med Sch, Duke Univ, 48-49, assoc, 50-54, asst prof, 54-59; asst prof, 59-64, ASSOC PROF CLIN PSYCHIAT, MED SCH, NORTHWESTERN UNIV, 64-; SERV CHIEF UNIT 7E, ILL STATE PSYCHIAT INST, 84- *Concurrent Pos:* Staff, Ment Hyg Clin, Vet Admin, 49-50; dir, Psychiat Outpatient Clin, Duke Univ Hosp, 50-56; psychiat attend, Vet Admin Hosp, Durham, 54-56, ward physician, 56-59; consult, Family Serv Agency, 54-59; dir northwestern serv, Ill State Psychiat Inst, 59-70, dir inpatient serv, 70-72; clin dir, Forensic Psychiat Prog, Ill State Psychiat Inst, 73-79 & 5E, 80-82, serv chief, Educ-Serv Unit, 82-84. *Mem:* AAAS; Am Psychiat Asn; Am Acad Psychiat & Law. *Res:* Psychiatric treatment and education. *Mailing Add:* Ill State Psychiat Inst 1601 W Taylor St Chicago IL 60612

GOLDSMITH, JOHN ROTHCHILD, b Portland, Ore, Jan 5, 22; m 47; c 4. MEDICAL & SCIENCE EDUCATION. *Educ:* Reed Col, BA, 42; Harvard Univ, MD, 45, MPH, 57. *Prof Exp:* Res fel prev med, Harvard Med Sch, 55-57; head air pollution med studies, Calif State Dept Pub Health, 57-65; epidemiologist, WHO, Switz, 64-66; Epidemiol Studies Lab, Calif State Dept Health, 66-78; vis prof, 78-80, PROF, EPIDEMIOL UNIT, FAC HEALTH SCI, BEN GURION UNIV NEGEV, 80- *Concurrent Pos:* Consult, WHO, 63; spec consult, Adv Comt Smoking & Health, Div Air Pollution & Surgeon Gen, USPHS; regents lectr, Univ Calif, Irvine; Mem, Int Occup Health; distinguished vis scientist, US Environ Protection Agency, 85-89; chmn adv panel, Israel Ministry Environ. *Mem:* AMA; fel Am Pub Health Asn; Int Epidemiol Asn; Int Soc Environ Epidemiol; Int Comt Occup Health. *Res:* Occupational and environmental epidemiology; air quality standard setting; environmental cancer; internal medicine; health effects of energy production; evaluation of health services; epidemiological methods; planning occupational health services. *Mailing Add:* Fac Health Sci Ben Gurion Univ Negev PO Box 653 Beer Sheva 84120 Israel

GOLDSMITH, JULIAN ROYCE, b Chicago, Ill, Feb 26, 18; m 40; c 3. GEOLOGY, GEOCHEMISTRY. *Educ:* Univ Chicago, SB, 40, PhD(geol), 47. *Prof Exp:* Asst petrol, Univ Chicago, 41-42; res chemist, Corning Glass Works, 42-46; asst petrol, 46-47, res assoc geochem, 47-51, from asst prof to prof, 51-69, Charles E Merriam distinguished serv prof, 69-88, EMER PROF GEOCHEM, UNIV CHICAGO, 88- *Concurrent Pos:* Assoc dean phys sci div, Univ Chicago, 60-72, chmn dept geophys sci, 63-71, assoc chmn, 61-62, actg dean phys sci div, 62; co-ed, J Geol, 57-62; mem, Earth Sci Panel, NSF, 58-61, chmn, 60-61, mem, Nat Sci Bd, 64-70; consult ed, Encycl Britannica & McGraw-Hill Encycl Sci & Technol; Gov sci adv comt, 89- *Honors & Awards:* Mineral Soc Award, Mineral Soc Am, 55, Roebling Medal, 88; Harry H Hess Medal, Am Geophys Union, 87. *Mem:* Fel Am Geophys Union; fel Geol Soc Am (vpres, 73-74, pres, 74-75); Mineral Soc Am (vpres, 68-69, pres, 70-71); Geochem Soc (vpres, 55, pres, 65-66); Am Chem Soc; Sigma Xi. *Res:* Phase equilibria and crystal chemistry of silicates and carbonates. *Mailing Add:* Dept Geophys Sci 5734 Ellis Ave Chicago IL 60637

GOLDSMITH, LAURA TOBI, b New York, NY, June 2, 50. REPRODUCTIVE ENDOCRINOLOGY, CELL BIOLOGY. *Educ:* Drexel Univ, BS, 71; NY Univ, MS, 78, PhD(biol), 81. *Prof Exp:* Postdoctoral fel physiol, Med Sch, Univ Pittsburgh, 80-82; res asst prof obstet & gynec, Med Sch, NY Univ, 82-86; ASST PROF OBSTET & GYNEC, NJ MED SCH, UNIV MED & DENT NJ, 86-, ASST PROF BIOCHEM & MOLECULAR BIOL, 89- *Concurrent Pos:* Vis lectr, Med Sch, NY Univ, 86-88. *Mem:* AAAS; Soc Study Reproduction; Endocrine Soc; Soc Gynec Invest. *Res:* Mechanisms by which ovarian cells recognize extracellular signals enabling proper hormone secretion and the actions of the ovarian hormone relaxin. *Mailing Add:* NJ Med Sch Univ Med & Dent NJ 185 S Orange Ave Newark NJ 07103

GOLDSMITH, LOWELL ALAN, b Brooklyn, NY, Mar 29, 38; m 60; c 2. DERMATOLOGY. *Educ:* Columbia Col, AB, 59; State Univ NY Downstate Med Ctr, MD, 63. *Prof Exp:* Resident, Harvard-Mass Gen Hosp, 67-69, asst prof, 70-73; from assoc prof to prof, Duke Med Ctr, 73-81; PROF & CHAIR DERMATOL, UNIV ROCHESTER SCH MED, 81- *Concurrent Pos:* Assoc ed, Dialogues in Dermatol, 76-; Res Career & Develop award, NIH, 76 & 81; co-ed, Progress Dermatol, 76-82; Macy Found fel, 78; consult, Roswell Park Mem Inst, 82- *Honors & Awards:* Res Career Develop Award, NIH. *Mem:* Am Soc Biol Chem; Am Dermatol Asn; Am Soc Clin Invest; Am Asn Physicians. *Res:* Biochemistry of epidermis and epidermal derivatives; mechanism of genetic disorders of human skin; monoclonal antibody probes to human skin; changes in epidermis in neoplasia. *Mailing Add:* Sch Med & Dent Box 697 Univ Rochester Rochester NY 14642

GOLDSMITH, MARY HELEN MARTIN, b Boston, Mass, May 2, 33; m 55; c 2. BIOLOGY, PLANT PHYSIOLOGY. *Educ:* Cornell Univ, BA, 55; Radcliffe Col, AM, 56, PhD(biol), 60. *Prof Exp:* Nat Cancer Inst fel, Harvard Univ, 59-60; fel, King's Col, Univ London, 60-61; res assoc, 61-73, lectr, 63-64 & 66-73, assoc prof, 74-84, PROF BIOL, YALE UNIV, 84-, DIR MARSH BOT GARDEN, 86- *Concurrent Pos:* Carnegie fel, Carnegie Inst Wash, Dept Plant Biol, Stanford, 78; Guggenheim fel, John Simon Guggenheim Found, 87; master, Silliman Col, Yale Univ, 87-; Brenda Ryman vis fel, Girton Col, Univ Cambridge, 87. *Mem:* Am Soc Plant Physiol; Soc Gen Physiol. *Res:* Plant physiology; polar transport and action of auxin; gravitropic and phototropic responses; role of proton pumps and ion channels in cell growth and plant development; their contribution to electrical properties of plasma membrane and tonoplast. *Mailing Add:* Dept Biol Yale Univ Kline Biol Tower New Haven CT 06511

GOLDSMITH, MERRILL E, b Chicago, Ill, Aug 8, 46. MOLECULAR BIOLOGY. *Educ:* Stanford Univ, BA, 68; Yale Univ, MPhil, 71, PhD(biol), 77. *Prof Exp:* USPHS fel, Lab Molecular Biol, Nat Inst Arthritis, Metab & Digestive Dis, NIH, 76-79; staff fel, Clin Hemat Br, Nat Heart Lung & Blood Inst, NIH, 79-81; staff fel pharmacol, Nat Cancer Inst, NIH, 82-86, MICROBIOLOGIST, CLIN PHARMACOL BR, NAT CANCER INST, NIH, 87- *Mem:* Am Soc Cell Biol; Am Soc Microbiol. *Res:* Isolation, regulation and expression of genes involved in resistance of tumor cells to chemotherapeutic drugs (dihydrofolate reductase, mdr (P-glycoprotein) glutathione-S- transferases and topoisomerase I and II. *Mailing Add:* Clin Pharmacol Br Bldg 10 Rm 6N119 Nat Cancer Inst NIH Bethesda MD 20892

GOLDSMITH, MICHAEL ALLEN, b Bronx, NY, Jan 28, 46; m 71; c 4. MEDICAL ONCOLOGY, PHASE I CLINICAL TRIALS. *Educ:* Yeshiva Univ, BA, 67; Albert Einstein Col Med, MD, 71. *Prof Exp:* Staff assoc, Nat Cancer Inst, NIH, 72-74; ASST CLIN PROF MED & NEOPLASTIC DIS, MT SINAI SCH MED, 77-; ATTEND PHYSICIAN, ONCOL CONSULT, PC, 77- *Mem:* Fel Am Col Physicians; Am Soc Clin Oncol; Am Asn Cancer Res; AAAS; NY Acad Sci. *Res:* Early clinical trials of novel anticancer drugs in evaluation of clinical pharmacology and toxicology in man; animal models as quantitative predictors of human toxicology. *Mailing Add:* Dept Oncol Oncol Consult PC 1045 5th Ave New York NY 10028

GOLDSMITH, PAUL FELIX, b Washington, DC, Nov 5, 48; m 88. ASTROPHYSICS, ELECTRICAL ENGINEERING. *Educ:* Univ Calif, Berkeley, AB, 69, PhD(physics), 75. *Prof Exp:* Mem tech staff, AT&T Bell Labs, 75-77; from asst prof to assoc prof, 77-87, PROF PHYSICS & ASTRON, UNIV MASS, AMHERST, 87- *Concurrent Pos:* Consult, Lincoln Lab, Mass Inst Technol, 77-80; vpres res & develop, Millitoch Corp, South Deerfield, Mass, 82- *Mem:* Am Astron Soc; fel Inst Elec & Electronics Engrs; Union Radio Sci Int. *Res:* Structure and composition of interstellar molecular clouds, particularly concerning process of star formation; development of instrumentation for millimeter and submillimeter radio astronomy; developed quasioptical technology for millimeter wavelengths. *Mailing Add:* Five Leadmine Hill Rd RFD 3 Amherst MA 01002

GOLDSMITH, PAUL KENNETH, b Washington, DC, Nov 28, 42; m 73. COMPARATIVE MARINE BIOCHEMISTRY. *Educ:* Univ Md, BS, 65, PhD(cytol), 77. *Prof Exp:* Biologist, Nat Cancer Inst, 65-71, BIOLOGIST, LAB BIOL MED, NAT INST ARTHRITIS, DIGESTIVE DIS & KIDNEYS, NIH, 71- *Concurrent Pos:* Investr, Marine Biol Lab, 77-80. *Mem:* AAAS; NY Acad Sci. *Res:* Enzyme activites in hepatocyte endoplasmic reticulum; carbohydrate metabolism in hepatopancreus of marine invertebrates; characterization of immunoglobulin E responses in rheumatic diseases; food allergy and recurrant infections. *Mailing Add:* Metab Dis Br Bldg 10 Rm 8C206 Nat Inst of Arthritis Diabetics Digestive & Kidney Dis NIH Bethesda MD 20892

GOLDSMITH, RALPH SAMUEL, b Baltimore, Md, Feb 23, 31; m 53; c 3. MEDICINE, ENDOCRINOLOGY. *Educ:* Franklin & Marshall Col, BS, 50; Univ Md, MD, 54. *Prof Exp:* Intern, Walter Reed Army Hosp, 54-55, resident physician internal med, 56-58, student investr metab, Inst Res, 58-59; chief metab br, US Army Surg Res Unit, 59-63; instr med, Harvard Med Sch, 63-64, assoc, 64-68, asst prof, 68, res assoc, Thorndike Mem Lab, 63-68; from asst prof to prof med, Mayo Grad Sch Med, Univ Minn, 68-74, dir clin study unit & consult div endocrinol, Mayo Clin & Found, 68-74, from assoc prof to prof med, Mayo Med Sch, 72-74; prof med, Sch Med, Univ Tex, San Antonio, 74-79, dep chmn dept, 74-78; chief med, Audie Murphy Mem Vet Hosp, 74-78; PROF MED & ASSOC DEAN, SCH MED, UNIV CALIF, SAN FRANCISCO, 79- *Concurrent Pos:* Civilian scientist, US Army, 62-63; consult, US Army Res Inst Environ Med, 63-68; asst physician, Second & Fourth Med Serv, Boston City Hosp, 64-68; ed, Bone, 77-86; mem, B Study Sect Gen Med, NIH, 78-82; chief staff, Vet Admin Med Ctr, San Francisco, 79-89. *Mem:* AAAS; Endocrine Soc; Am Fedn Clin Res; Am Soc Clin Invest; Am Soc Bone & Mineral Res; Am Physiol Soc. *Res:* Calcium and phosphorus metabolism; parathyroid hormone; calcitonin; bone diseases; vitamin D. *Mailing Add:* 2995 Woodside Rd No 400 Woodside CA 94062

GOLDSMITH, RICHARD, b Salem, Mass, Sept 30, 18; m 55; c 3. GEOLOGY. *Educ:* Univ Maine, AB, 40; Univ Wash, PhD(geol), 52. *Prof Exp:* geologist, US Geol Surv, 52-86; RETIRED. *Mem:* AAAS; Geol Soc Am; Am Quaternary Asn. *Res:* Igneous and metamorphic petrology; areal geology; geomorphology and glacial ecology; economic geology. *Mailing Add:* 2208 Burgee Ct Reston VA 22091

GOLDSMITH, TIMOTHY HENSHAW, b New York, NY, May 1, 32; m 55; c 2. VISUAL PHYSIOLOGY. *Educ:* Cornell Univ, BA, 54; Harvard Univ, PhD(biol), 58. *Prof Exp:* Jr fel, Soc Fels, Harvard Univ, 58-61; instr, Marine Biol Lab, Woods Hole, 59-61; from asst prof to assoc prof, 61-70, chmn dept, 71-77, PROF BIOL, YALE UNIV, 70- *Concurrent Pos:* Guggenheim fel, 67-68; trustee, Marine Biol Lab, Woods Hole, 72- *Mem:* Fel AAAS; Soc Gen Physiol; Biophys Soc; fel Am Acad Arts & Sci; Asn Res Vision & Ophthal. *Res:* Neurophysiology and biochemistry of light sensitive systems, particularly vision of arthropods and hummingbirds. *Mailing Add:* Dept Biol 736 Kbt Yale Univ PO Box 6666 New Haven CT 06520

GOLDSMITH, VICTOR, geological oceanography, for more information see previous edition

GOLDSMITH, WERNER, b Düsseldorf, Ger, May 23, 24; US citizen; m 73; c 3. APPLIED MECHANICS, BIOMECHANICS. *Educ:* Univ Tex, Austin, BSME, 44, MSME, 45; Univ Calif, Berkeley, PhD(mech eng), 49. *Prof Exp:* Engr turbines, Westinghouse Elec Corp, 45-47; from instr to prof, 47-87, PROF MECH ENG, UNIV CALIF, BERKELEY, 88- *Concurrent Pos:* Mech engr, Detonation Physics Div, US Naval Weapons Ctr, China Lake, Calif, 51-; Guggenheim Mem Found fel, 53-54; consult, numerous gov agencies and priv industs, 53-; prin investr res grants, NIH, NSF, US Army & US Navy, 55-; chmn head injury model construct comt, Nat Inst Neurol Dis & Stroke, NIH, 66-70; Fulbright res fel, US Educ Found, Tech Univ Athens, 74-75, Univ Patras, Greece, 81-82. *Honors & Awards:* Nat Acad Eng. *Mem:* Fel Am Soc Mech Engrs; fel Am Acad Mech. *Res:* Collisions; wave propagation; dynamic properties of materials; biomechanics; rock mechanics; head and neck injury; experimental mechanics; protective devices. *Mailing Add:* 450 Gravatt Dr Berkeley CA 94705

GOLDSPIEL, SOLOMON, b Brody, Poland, June 1, 13; US citizen; m 38, 79; c 2. ENGINEERING. *Educ:* City Col New York, BS, 34, Chem Eng, 36; Polytech Inst Brooklyn, MS, 54. *Prof Exp:* From eng aide metall to jr chemist, Indust Test Lab, Philadelphia Naval Shipyard, US Navy, 37-40, from sr phys sci aide metall to metallurgist, Mat Lab, NY Naval Shipyard, 40-43, from physicist to sr physicist & head, Emission Spectros & X-ray Diffraction Sect, 43-54, supv metallurgist & head Casting Develop & Foundry Control Sect, 54-70, sr nondestructive test specialist & coordr, Mat Sci Div, Naval Appl Sci Lab, Brooklyn, NY, 63-70; MECH ENGR, BD WATER SUPPLY NEW YORK, 70- *Concurrent Pos:* Consult, Naval Ship Res & Develop Ctr, Annapolis, 70-73, Cadcom, Inc, Annapolis, Md, 73-75 & DEP, Bd Water Supply, 82-; instr, Pratt Inst Sch Eng, Brooklyn, NY; adj prof, Polytech Inst Brooklyn, NY; lectr, Brooklyn Col, NY; chmn, EO7.02 Ref Radiographs, Am Soc Testing & Mat, mem, E07 Coun, A01 Steel, G01 Corrosion & F16 Fasteners; mem, Nat Radiation Comt, Am Soc Nondestructive Testing & Res Coun; mem, Comt PH2, Am Nat Standards Inst & chmn, PH2-34; mem, Brass & Bronze Div Papers & Publ & Recommended Pract Comts, Am Phys Soc. *Mem:* Soc Appl Spectros. *Res:* Physical & foundry metallurgy; nondestructive stability; spectroscopy and x-ray diffraction; reference radiographs; granted two patents; author of several publications. *Mailing Add:* 732 Gerald Ct Brooklyn NY 11235

GOLDSTEIN, ABRAHAM M B, b Iasi, Romania, Aug 10, 14; US citizen; m 47; c 1. UROLOGY, HISTOLOGY. *Educ:* Univ Iasi, Dr Med & Surg, 36. *Prof Exp:* Instr histol, Univ Iasi, 32-36; prof, Onesco Col, Bucharest, 41-44; sect physician, Caritas Hosp, 44-48; asst prof, Sch Med, Univ Bucharest, 49-54; primary physician, Oncol Inst, 50-60; asst, Allgemeines Hosp, Hamburg, Germany, 60-61; instr anat, histol & path, Univ Southern Calif, 61-62; assoc prof histol, Calif Col Med, 62-68; assoc prof, 69-70, ADJ PROF SURG & HISTOL, SCH MED, UNIV SOUTHERN CALIF, 70- *Concurrent Pos:* German Govt fel, 64; NIH res grant, 64-; Nat Acad Sci travel awards, Europe, 68 & 70; consult, Los Angeles County Gen Hosp, 63-; urologist, Ross-Loos Med Group. *Mem:* Am Asn Anat; fel Royal Soc Health; NY Acad Sci; AMA; Am Urol Asn. *Res:* Microscopic three dimensional architecture of tissues and organs; regeneration mechanism of urinary bladder; changes in the urinary tract following extensive subtotal cystectomy; tumors of the urogenital systems; mechanism of regeneration of the urothelium; function of the straight segments of the nephron under normal and increased intratubular pressure; seeding mechanism of Urothelium; urothelial dynamics under the influence of a surgical stimulus; microscopic architecture and intrinsic physiology of the corpora cavernosa of the penis. *Mailing Add:* Dept Urol Univ Southern Calif Sch Med Los Angeles CA 90033

GOLDSTEIN, ALBERT, b New York, NY, May 26, 38; m 68; c 2. MEDICAL PHYSICS, DIAGNOSTIC ULTRASOUND. *Educ:* City Col New York, BS, 60; Mass Inst Technol, PhD(physics), 65. *Prof Exp:* Staff mem physics, Nat Magnet Lab, Mass Inst Technol, 65; res assoc, Ecole Normale Superieure, Paris, 65-67; staff mem, T J Watson Res Ctr, IBM Corp, 67-71; asst prof, Dept Radiol, Univ Kans Med Ctr, 72-76, adj asst prof, Dept Elec Eng, Univ Kans, 73-76; chief, div med physics, Henry Ford Hosp, 76-84; ASSOC PROF, SCH MED, WAYNE STATE UNIV, 85- *Concurrent Pos:* Assoc scientist, Mid-Am Cancer Ctr, Kans, 75-76; adj assoc prof radiol, Med Sch, Wayne State Univ, 80-84; adj asst res scientist radiol, Med Sch, Univ Mich, Detroit, 81-82. *Mem:* Am Asn Physicists Med; Inst Elec & Electronics Engrs; fel Am Inst Ultrasound Med; Am Col Radiol; Radiol Soc NAm. *Res:* Physics and quality assurance of diagnostic ultrasound and computed tomography, the information content of medical images; diagnostic ultrasound equipment standardization and performance calibration. *Mailing Add:* Radiology Detroit Receiving Hosp 4201 St Antoine Detroit MI 48201

GOLDSTEIN, ALBERT, b Brooklyn, NY, Feb 20, 28; m 52; c 4. CHEMICAL CONSULTING, HYDROPHILIC RESINS. *Educ:* Rutgers Univ, BSc, 51; Cornell Univ, PhD(org chem), 54. *Prof Exp:* Res chemist, Cent Res Labs, Gen Foods Corp, 54-55 & Catalin Corp Am, 55-59; chief chemist, Chemirad Corp, 59-65; sr scientist, Devro, Johnson & Johnson, 65-69; dir res, Hydron Labs, Nat Patent Develop Corp, NB, 69-71; CHEM CONSULT, GOLDSTEIN ASSOCS, 71-,; PRES, VENTURE CHEM CO, 72- *Concurrent Pos:* Dir, Vanguard Res, Inc, 87- *Mem:* Am Chem Soc; Inst Food Technol; Am Asn Textile Chemists & Colorists; Asn Consult Chemists & Chem Engrs. *Res:* Process and product development, reactive and hydrophilic resins, specialty organics, flame retardants, organic and inorganic coatings, acrylic resins, ink formulation and product formulation. *Mailing Add:* Goldstein Assocs PO Box 88 Adelphia NJ 07710

GOLDSTEIN, ALLAN L, b Bronx, NY, Nov 8, 37; m 75; c 3. BIOCHEMISTRY. *Educ:* Wagner Col, BS, 59; Rutgers Univ, MS, 61, PhD(physiol, biochem), 64. *Prof Exp:* Teaching asst zool, Rutgers Univ, 59-61, from asst instr to instr, 61-64; USPHS res fel biochem, Albert Einstein Col Med, 64-66, from instr to assoc prof biochem, 66-72; prof biochem & dir div, Univ Tex Med Br, Galveston, 72-78; PROF BIOCHEM & CHMN DEPT, SCH MED & HEALTH SCI, GEORGE WASHINGTON UNIV, 78- *Concurrent Pos:* Consult, Hoffmann-La Roche Inc, 74-82, Genentech Inc, 79-80 & Battelle Mem Inst, 80; mem, Med Res Serv Merit Rev Bd Oncol, Vet Admin, 77-80; adv ed, Cancer Biol Rev, 78; mem, Biol Response Modifier Prog, Div Cancer Treatment, Nat Cancer Inst; mem sci adv comt, Found Integrative Biomed Res, 80- & Papanicolaou Cancer Res Inst, Miami, 81-84; mem sci adv bd, Gerimmun Found, Fribourg, Switz, 81-; assoc ed, Human Lymphocyte Differentiation, 81- *Honors & Awards:* Gordon Wilson Medal, Am Clin & Climat Soc, 76; Van Dyke Award, Burroughs Wellcome, 84- *Mem:* AAAS; Am Asn Immunol; Endocrine Soc; Am Soc Biol Chem; Transplantation Soc. *Res:* Isolation, purification, mechanism of action and biological role of thymic humoral factors; lymphoid tissue biochemistry; hormonal control of metabolic processes; co-discoverer of the thymosins, the family of hormones of the thymus gland. *Mailing Add:* Dept Biochem Ross 531A George Washington Univ 2300 Eye St NW Washington DC 20037

GOLDSTEIN, ALLEN A, b Baltimore, Md, Jan 7, 25; m 49; c 4. APPLIED MATHEMATICS. *Educ:* St John's Col, Md, BA, 47; Georgetown Univ, MA, 52, PhD, 54. *Prof Exp:* Engr & physicist, Kellex Corp, 48-51; physicist math comput, Appl Math Lab, Nat Bur Stand, 51-54; res assoc astron, Observ, Georgetown Univ, 54-55; design specialist, Convair Div, Gen Dynamics Corp, 55-60; res assoc elec eng, Mass Inst Technol, 60-63; assoc prof math, Univ Tex, 63-64; assoc prof, 64-65, PROF MATH, UNIV WASH, 65- *Mem:* Am Math Soc. *Res:* Application of functional analysis to numerical analysis and applied mathematics. *Mailing Add:* Dept Math Univ Wash Seattle WA 98195

GOLDSTEIN, ANN L, CARDIOVASCULAR PHARMACOLOGY. *Educ:* Univ Pa, BA, 65; Vassar Col, MS, 68; NY Univ, PhD(biol), 79. *Prof Exp:* Teaching asst, Vassar Col, 67-68; instr biol & bact, Rockland Community Col, 68-69; res assoc, Vassar Col, 76-79, NIH postdoctoral fel, 79-82; sr scientist, Dept Pharmacol, 82-88, RES INVESTR, DEPT PHARMACOL & CHEMOTHER, HOFFMANN-LA ROCHE INC, 88- *Concurrent Pos:* NIH res grants, 76-79 & 79-; vis asst prof, Vassar Col, 79-82. *Mem:* Sigma Xi; AAAS; NY Acad Sci; Tissue Cult Asn; Am Soc Cell Biol. *Res:* Regulation of cell metabolism including cell proliferation, lipid and protein turnover, and cell-cell interactions; in vitro systems applicable to pharmacological approaches to cardiovascular disease; author of numerous technical publications. *Mailing Add:* Hoffmann-La Roche Inc 340 Kingsland St Nutley NJ 07110-1199

GOLDSTEIN, ARTHUR MURRAY, b New York, NY, Mar 13, 22; m 50; c 3. CARBOHYDRATE CHEMISTRY, POLYMER CHEMISTRY. *Educ:* City Col New York, BS, 43; Ohio State Univ, MS, 47. *Prof Exp:* Chemist natural gum, Stein, Hall & Co, 47-51, chief chemist paper, 51-53, lab mgr gums, paper & resins, 53-57, tech dir, 57-72; tech mgr water soluable polymers, 72-81, mgr Ventures & New Prod Develop, 81-83, MGR MKT OIL FIELD CHEM, CELANESE POLYMER SPECIALITIES CO, 83- *Mem:* Am Chem Soc; Tech Asn Pulp & Paper Indust; Inst Food Technologists; Am Asn Textile Chemists & Colorists; Am Inst Mech Engrs. *Res:* Development of chemically modified guar gum polymers for industrial uses; synthetic water soluble polymers; application of polymers to the oil industry. *Mailing Add:* 9261 Vista Del Lago 19D Boca Raton FL 33428

GOLDSTEIN, AUGUST, JR, b Shreveport, La, Dec 3, 20; m 45; c 3. PETROLEUM GEOLOGY. *Educ:* La State Univ, BS, 40, MS, 42; Univ Colo, PhD(geol), 48. *Prof Exp:* Geologist, Remvo Superior Vermiculite Co, Colo, 45-46 & Winter-Weiss Co, 46-47; asst geol, Univ Colo, 47-48; res geologist, Stanolind Oil & Gas Co, 48-57; chief geologist, Bell Oil & Gas Co, 57-63, mgr explor, 63-65; gen mgr, Lubell Oil Co, 65-85. *Concurrent Pos:* Instr, Denver exten, Univ Colo, 48; Consult, 85- *Mem:* Fel Geol Soc Am (treas, 73-76); Soc Econ Paleont & Mineral; hon mem Am Asn Petrol Geologists (vpres, 73-74). *Res:* Sedimentary petrography; petroleum geology; geology of Ouachita Mountains, Oklahoma and Arkansas. *Mailing Add:* 1033 Mayo Bldg Tulsa OK 74103-3902

GOLDSTEIN, AVRAM, b New York, NY, July 3, 19; m 47; c 4. NEUROSCIENCES. *Educ:* Harvard Univ, AB, 40, MD, 43. *Prof Exp:* Intern, Mt Sinai Hosp, NY, 44; instr pharmacol, Harvard Univ, 47-49, assoc, 49-51, asst prof pharmacol & tutor biochem sci, 52-55; chmn dept, 55-70, prof, 55-89, EMER PROF PHARMACOL, STANFORD UNIV, 89- *Concurrent Pos:* Moseley traveling fel, Harvard Univ, 49-50; asst, Univ Bern, 51-52; founder & ed, Molecular Pharmacol, 65-68; dir, Addiction Res Found, 74-87. *Honors & Awards:* Franklin Medal, 80; Sollmann Award, Am Soc Pharmacol & Exp Therapeut, 81. *Mem:* Nat Acad Sci; Inst Med-Nat Acad Sci; Am Soc Pharmacol & Exp Therapeut; Am Soc Biol Chem. *Res:* Mechanism of drug action; opiate receptors and opioid peptides; narcotics; narcotic addiction; methadone treatment. *Mailing Add:* 735 Dolores Stanford CA 94305

GOLDSTEIN, BERNARD, surface physics, semiconductor lasers, for more information see previous edition

GOLDSTEIN, BERNARD, b San Francisco, Calif, Oct 21, 35; m 61; c 1. HUMAN SEXUALITY. *Educ:* San Francisco State Col, BA, 62, MA, 64; Univ Calif, Davis, PhD(functional morphol), 68. *Prof Exp:* From asst prof to assoc prof, 68-75, chmn, Dept Physiol & Behav Biol, 73-75, PROF BIOL, SAN FRANCISCO STATE UNIV, 75-, ACTG DIR RES & PROF DEVELOP, 89- *Concurrent Pos:* NSF instnl grant, 69, NIH summer grant, 69; chmn, Statewide Acad Senate, Calif State Univ, 84-87. *Mem:* AAAS; Am Soc Zool; Am Asn Sex Educr, Counr & Therapists; Soc Sci Study Sex; Sigma Xi. *Res:* Burrowing mechanisms of fossorial mammals; population dynamics; ecological physiology of vertebrates; biology of man; human sexuality; reproductive physiology. *Mailing Add:* 111 Park Ave San Carlos CA 94070

GOLDSTEIN, BERNARD DAVID, b Bronx, NY, Feb 28, 39; m 63; c 2. ENVIRONMENTAL MEDICINE, HEMATOLOGY. *Educ:* Univ Wis-Madison, BS, 58; NY Univ, MD, 62. *Prof Exp:* Intern & resident, Bellevue Hosp, New York, 62-65; teaching asst, Sch Med, NY Univ, 63-66; instr med, Univ Southern Calif, 66-68; from instr to assoc prof, Sch Med, NY Univ, 68-80; PROF & CHMN ENVIRON MED & PROF MED, 80-, DIR, ENVIRON & OCCUP HEALTH SCI INST, ROBERT WOOD JOHNSON MED SCH, UNIV MED & DENT NJ. *Concurrent Pos:* NIH fel, Med Ctr, NY Univ, 65-66; attend physician, Bellevue & NY Univ Hosps, NY, 68-; res collabr, Med Dept, Brookhaven Nat Lab, 73-83; chmn, Health Res Coun, Environ Pollution Working Group, New York, NY, 73-75; consult, Nat Acad Sci, 74-76, comt mem, 76-80; NIH Fogarty Sr Int fel biochem, Brunel Univ, Eng, 77-78; asst adminr res & develop, US Environ Protection Agency, 83-85; chmn, NAS/NAC Comt Biomarkers Environ Health Res, 86-, NAS/IOM Comt Role Primary Care Practr Environ & Occup Med, 86-89; mem, MH Nat Adv Environ Health Sci Coun, 88-, NAS/NAC bd, Environ Sci & Toxicol; mem, Comn Health & Environ, WHO, chmn, Panel Industrialization; chmn, Panel Risk Assessment Methodology, NAS. *Mem:* Inst Med-Nat Acad Sci; Am Fedn Clin Res; Am Soc Hemat; Soc Toxicol; Am Pub Health Asn; Am Soc Clin Invest. *Res:* Effects of free radicals, peroxides and active states of oxygen on cellular membranes; biochemical toxicity of environmental agents; inhalation toxicology of air pollutants; mechanisms of hemolysis. *Mailing Add:* Robert Wood Johnson Med Sch Univ Med & Dent NJ 675 Hoes Lane Piscataway NJ 08854-5635

GOLDSTEIN, BURTON JACK, b Baltimore, Md, Sept 23, 30; div; c 3. PSYCHIATRY, PSYCHOPHARMACOLOGY. *Educ:* Univ Md, BS, 53, MD, 60. *Prof Exp:* Instr psychiat, 64-68, from asst prof to assoc prof, 68-73, assoc prof pharmacol, 71-74, dir div addiction sci, 75-80, actg chmn dept psychiat, 83-86, PROF PSYCHIAT, UNIV MIAMI SCH MED, 73-, PROF PHARMACOL, 74- *Concurrent Pos:* NIMH training fel psychiat, Univ Miami-Jackson Mem Hosp, 61-64; consult psychiat res, Fla State Hosp, 66-, Indust Security Prog, Dept Defense, 67- & Vet Admin Psychiat Serv, Fla, 70-; examr, Am Bd Psychiat Neurol, 72-; vchmn dept psychiat, Univ Miami Sch Med, 71- *Mem:* Fel Am Col Psychiat; fel Int Col Psychopharmacol; AMA; Fel Am Psychiat Asn; Am Col Neuropsychopharmacol. *Res:* Clinical psychopharmacology, especially epidemiology, usage-nature, extent and character of psychotropic drug prescribing; clinical measurement of anxiety and depression and response to pharmacotherapy. *Mailing Add:* Dept Psychiat D4-12 Univ Miami Sch Med 1500 NW 12 Ave Suite 1103 Miami FL 33136

GOLDSTEIN, BYRON BERNARD, b New York, NY, Nov 24, 39. BIOPHYSICS. *Educ:* City Col New York, BS, 61; NY Univ, PhD(plasma physics), 67. *Prof Exp:* Instr physics, NY Univ, 67 & 68; lectr, Queens Col, NY, 68; from assoc prof to prof physics, Farleigh Dickinson Univ, 68-77; STAFF MEM, THEORET BIOL & BIOPHYSICS GROUP, LOS ALAMOS NAT LAB, 77- *Concurrent Pos:* NIH fel chem, Univ Calif, San Diego, 69-70; vis staff mem, Los Alamos Sci Lab, 75-77. *Mem:* AAAS; Biophys Soc; Am Phys Soc; Am Asn Immunologists. *Res:* Ligand receptor interactions; biophysics of the immune response. *Mailing Add:* M5 K710 Group T10 Los Alamos Nat Lab PO Box 1663 Los Alamos NM 87545

GOLDSTEIN, CHARLES IRWIN, b New York, NY, Nov 21, 40; m 75; c 2. APPLIED MATHEMATICS. *Educ:* City Col New York, BS, 62; NY Univ, MS, 64, PhD(math), 67. *Prof Exp:* SR RES SCIENTIST MATH, BROOKHAVEN NAT LAB, 67- *Concurrent Pos:* Vis res mathematician, Math Res Ctr, Univ Wis, 72, Inst Comput Appl Sci & Eng, NASA Langley Res Ctr, Hampton, Va, 78, Math Dept, Univ Calif, Berkeley, 79, Naval Res Lab, Washington, DC, 83. *Mem:* Am Math Soc; Soc Indust Appl Math; Sigma Xi. *Res:* Scattering and perturbation theory; microwave physics; the numerical solution of differential equations and mathematical analysis of the finite element method; acoustics and electromagnetic wave propagation. *Mailing Add:* Brookhaven Nat Lab 61 Brookhaven Ave Upton NY 11973

GOLDSTEIN, CHARLES M, b Chicago, Ill, Sept 27, 29; m 61; c 3. INFORMATION SCIENCE, COMPUTER TECHNOLOGY. *Educ:* Purdue Univ, BS, 53, MS, 54. *Prof Exp:* Plasma electron physics numerical math, NASA, 55-68, br chief off of computerized info systs info sci & technol, 68-73; dir data mgt tech info technol, Informatics, 73-74; CHIEF, INFO TECHNOL BR, TECHNOL RES & DEVELOP, INFO SCI, LISTER HILL NAT CTR BIOMED COMMUN, NAT LIBR MED, 74- *Concurrent Pos:* Fulbright award, Ger, 54; chief sci comput numerical anal, NASA, 72-73; bd dir, Health Educ Network Inc, 76-78; subcomt technol, Nat Fedn Abstracting & Indexing Socs, 78- *Honors & Awards:* Brad Rogers Award, Med Libr Asn, 88. *Mem:* Am Soc Info Sci; AAAS; Med Libr Asn. *Res:* Information technology. *Mailing Add:* Nat Libr Med Lister Hill Nat Ctr Biomed Bldg 38A Rm 8N803 Bethesda MD 20894

GOLDSTEIN, DAVID, b New York, NY, Apr 20, 29; m 53; c 4. CHEMICAL ENGINEERING, INORGANIC CHEMISTRY. *Educ:* City Col New York, BChEng, 51; Iowa State Col, MS, 53. *Prof Exp:* Design engr filters, Micro-Metallic Corp, 51-52; res asst thermal res, Eng Exp Sta, Iowa State Col, 52-53; res proj supvr extractive metall, Chem Construct Corp, 53-56; process design engr, Nichols Eng & Res Corp, 56; assoc res chem engr stainless & alloy steels, Crucible Steel Co Am, 56-61; mgr eng develop indust chem, FMC Corp, 61-83, process eng consult, 83-86; RETIRED. *Mem:* Am Inst Chem Engrs. *Res:* Extractive metallurgy; industrial chemicals including soda ash and phosphates; fertilizers; process development; detergents; waste disposal; solvent extraction; alloy steel surface conditioning; corrosion; refractories; magnesia; calcination and heat transfer. *Mailing Add:* 13 Perry Rd East Brunswick NJ 08816

GOLDSTEIN, DAVID JOEL, b New York, NY, June 25, 47; div; c 2. PEDIATRICS, BIOCHEMISTRY. *Educ:* Franklin & Marshall Col, BA, 69; Univ Tenn, MD, 73, PhD(biochem), 75. *Prof Exp:* Residency pediat, Mayo Grad Sch Med, 75-78; fel, Univ Pa, 78-81; asst prof med genetics, Ind Univ, 81-87; CLIN RES PHYSICIAN, ELI LILLY & CO, 87- *Concurrent Pos:* NIH fel. *Honors & Awards:* Quigley Award in Physiol. *Mem:* Am Soc Human Genetics; AAAS; Soc Neurosci; Am Soc Clin Pharmacol Ther; Sigma Xi. *Res:* Clinical pharmacology; medical genetics; analgesia; obesity. *Mailing Add:* Lilly Res Labs Corp Ctr Indianapolis IN 46285-0001

GOLDSTEIN, DAVID LOUIS, b Providence, RI, Aug 22, 57; m 83; c 3. COMPARATIVE OSMOREGULATORY PHYSIOLOGY, HUMAN ANATOMY & PHYSIOLOGY. *Educ:* Univ Pa, BA, 79; Univ Calif, Los Angeles, PhD(biol), 83. *Prof Exp:* Postdoctoral, Dept Physiol, Univ Ariz, 83-86; ASST PROF ANAT & PHYSIOL, DEPT BIOL SCI, WRIGHT STATE UNIV, 86- *Mem:* Am Soc Zool; Am Ornithologists Union; AAAS. *Res:* Research focuses on adaptive and ecological physiology: physiological systems vary in animals from different habitats and how they respond to environmental change; osmoregulation particularly in birds. *Mailing Add:* Dept Biol Sci Wright State Univ Dayton OH 45435

GOLDSTEIN, DAVID STANLEY, b Newark, NJ, June 23, 48; m 72; c 5. HYPERTENSION, DYSAUTONOMIAS. *Educ:* Yale Col, BA, 70; Johns Hopkins Sch Med, MD, PhD(behav sci), 76. *Prof Exp:* Med resident internal med, Univ Wash Affil Hosp, Seattle, 76-78; clin assoc, Nat Heart, Lung & Blood Inst, NIH, 78-82, med staff fel, 82-83, sr investr, 83-90; SR INVESTR, NAT INST NEUROL DIS & STROKE, NIH, 90- *Mem:* Fel Am Col Physicians; Am Fedn Clin Res; Am Soc Clin Invest. *Res:* Sympathetic nervous system activity in cardiovascular diseases; metabolism of catecholamines; pathophysiology of hypertension; neuroendocrinology of stress. *Mailing Add:* 6929 Race Horse Lane Rockville MD 20852

GOLDSTEIN, DORA BENEDICT, b Milton, Mass, Apr 25, 22; m 47; c 4. PHARMACOLOGY. *Educ:* Harvard Med Sch, MD, 49. *Prof Exp:* Fel bact, Harvard Med Sch, 53-55; res assoc, 55-70, sr scientist, 70-74, adj prof, 74-78, PROF PHARMACOL, SCH MED, STANFORD UNIV, 78- *Honors & Awards:* Award Sci Excellence, Res Soc Alcoholism, 81. *Mem:* Am Soc Biol Chem; Am Soc Pharmacol & Exp Therapeut; Soc Neurosci; Res Soc Alcoholism (pres, 79-81). *Res:* Induction of cholinesterase synthesis in bacteria; biochemical effects of barbiturates; alcohol; drug dependence; anticonvulsants; effects of drugs on biomembranes. *Mailing Add:* Dept Pharmacol Stanford Univ Sch Med Stanford CA 94305

GOLDSTEIN, E BRUCE, b Washington, DC, Mar 31, 41; div; c 2. PSYCHOPHYSICS. *Educ:* Tufts Univ, BS, 63; Brown Univ, ScM, 65, PhD(psychol), 68. *Prof Exp:* Res asst psychol, Brown Univ, 63-65; USPHS res fel, Harvard Univ, 67-69; asst prof psychol & pharmacol, 69-75, ASSOC PROF PSYCHOL, UNIV PITTSBURGH, 75- *Mem:* Psychonomic Soc; Asn Res Vision & Ophthal. *Res:* Perception of depth in pictures; visual perception; mechanisms of form perception; sensation and perception; introductory psychology; psychology of gender. *Mailing Add:* Dept Psychol Univ Pittsburgh 7455 Langley Hall Pittsburgh PA 15260

GOLDSTEIN, EDWARD, b Stanislawow, Poland, Jan 28, 23. TELECOMMUNICATIONS, SYSTEMS ENGINEERING. *Educ:* Univ Minn, BS, 49. *Prof Exp:* Dir, Military Commun Systs & Eng, Bell Labs, 49-66,; eng dir, AT&T, 60-70, vpres mkt, NY Tel Co, 70-72, mem staff, 72-83, corp vpres strategy develop, 83-85; VPRES, MAC GROUP, 85- *Mem:* Fel Inst Elec & Electronics Engrs. *Mailing Add:* Mt Auburn St University Pl Cambridge MA 02138

GOLDSTEIN, ELISHEVA, b Bucharest, Rumania, July 12, 47; US citizen; m 68; c 2. MOLECULAR ORBITAL CALCULATIONS, SPECTROSCOPY. *Educ:* City Col, City Univ New York, BS, 71; Calif Polytech Univ, MS, 76; Univ Southern Calif, PhD(chem), 79. *Prof Exp:* Lectr chem, Calif Polytech Univ, 79-80; asst prof, Claremont Col, 80-81; ASSOC PROF, CALIF POLYTECH UNIV, 81- *Mem:* Sigma Xi; Am Chem Soc. *Res:* Molecular arbital calculations, involving fluoro and chloro carbenes and their reactions with substituted olefins; molecular orbital calculations of elctron scattering of small molecules; ab initio calculations on the Diels-Alder reaction of x-pyrone and acetylene. *Mailing Add:* 19844 E Navilla Pl Covina CA 91724

GOLDSTEIN, ELLIOT, b New York, NY, Dec 29, 34; m 58; c 5. INFECTIOUS DISEASES. *Educ:* Cornell Univ, BA, 56, MD, 60. *Prof Exp:* Res fel infectious dis, Channing Lab, Harvard Med Sch, 64-66, instr internal med, 68-69; from asst prof to assoc prof, 69-77, PROF INTERNAL MED, SCH MED, UNIV CALIF, DAVIS, 77- *Concurrent Pos:* Consult, Sacramento Med Ctr, 71-, San Joaquin County Hosp, 74- & Nat Res Coun-Nat Acad Sci, 74-76; Fulbright-Hays fel, Comt Int Exchange Persons, 75-76 & 81-82. *Mem:* Infectious Dis Soc Am; Am Soc Microbiol; Am Thoracic Soc. *Res:* Pathogenesis of AIDS infections. *Mailing Add:* Sect Infectious & Immunol Dis Univ Calif Davis CA 95616

GOLDSTEIN, ELLIOTT STUART, b Brooklyn, NY, July 7, 42; m 63; c 2. DEVELOPMENTAL GENETICS, MOLECULAR BIOLOGY. *Educ:* Univ Hartford, BS, 67; Univ Minn, MS, 70, PhD(genetics), 72. *Prof Exp:* Fel molecular biol, Mass Inst Technol, 72-74; asst prof, 74-78, ASSOC PROF ZOOL, ARIZ STATE UNIV, 78- *Concurrent Pos:* NIH grant, 76-81; vis assoc prof, Calif Inst Technol, 81. *Mem:* Sigma Xi; AAAS; Genetics Soc Am; Soc Develop Biol. *Res:* Use of cloned regulated genes to understand the mechanism by which genes control early embryogenesis and by which the genes themselves are regulated; mechanism by which the cancer oncogeneium transforms a normal cell into a tumor cell. *Mailing Add:* Dept Zool Ariz State Univ Tempe AZ 85287-1501

GOLDSTEIN, FRANZ, b Ger, Feb 23, 22; US citizen; m 49; c 3. GASTROENTEROLOGY, INTERNAL MEDICINE. *Educ:* Jefferson Med Col, MD, 53. *Prof Exp:* From intern to chief med resident & resident gastroenterol, Grad Hosp, Univ Pa, 53-57; from instr to assoc prof, 57-70, PROF MED, JEFFERSON MED COL, 70- *Concurrent Pos:* Am Trudeau Soc fel, Grad Hosp, Univ Pa, 55-56; mem bd trustees & sci adv bd, Nat Found Ileitis & Colitis, 74-; chief gastroenterol, Lankenau Hosp, 70-87. *Honors & Awards:* Rorer Award, Am Col Gastroenterol, 74 & 81. *Mem:* Am Gastroenterol Asn; fel Am Col Physicians; fel Am Col Gastroenterol (pres, 81-82); Bockus Int Soc Gastroenterol; AMA (pres, 85-87). *Res:* Bacterial flora of gastrointestinal tract in relation to malabsorption; gall bladder kinetics; inflammatory bowel disease. *Mailing Add:* 707 Arlington Rd Narberth PA 19072

GOLDSTEIN, FRED BERNARD, b Philadelphia, Pa, Dec 11, 24; m 45; c 3. BIOCHEMISTRY. *Educ:* Philadelphia Col Pharm, BSc, 51; McGill Univ, PhD(biochem), 55. *Prof Exp:* assoc res scientist, NY State Dept Ment Hyg, 55-76; ASSOC RES SCIENTIST, LAB DIAG & TREATMENT INHERITED METAB DIS, NY UNIV MED CTR, 76- *Concurrent Pos:* Lectr, Letchworth Village, 57-71. *Mem:* AAAS; Am Asn Ment Deficiency; Am Acad Ment Retardation; Am Chem Soc; Brit Biochem Soc. *Res:* Biochemical aspects of mental retardation with reference to intermediary metabolism and neurochemistry. *Mailing Add:* PO Box 482 Thiells NY 10984

GOLDSTEIN, FREDERICK J, b Philadelphia, Pa, Jan 12, 42; c 2. NEUROPHARMACOLOGY. *Educ:* Philadelphia Col Pharm & Sci, BSc, 63, MSc, 65, PhD(pharmacol), 68. *Prof Exp:* Asst, Philadelphia Col Pharmacol & Sci, 64-67, from instr to asst prof, 65-72, chmn, Continuing Educ Prog for Pharmacists, 69-76, assoc prof, 72-80, PROF PHARMACOL, PHILADELPHIA COL PHARMACOL & SCI, 80-; ASSOC STAFF MEM, DEPT MED, FOX CHASE CANCER CTR, 84- *Concurrent Pos:* Reviewer, J Pharm Sci, 71-; lectr, continuing educ, Pa Col Optometry, 72-78; consult, drug abuse legis, Pa State Senate Comt Pub Health & Welfare, 73-76; chmn prog comt, Pharmacol & Toxicol Sect, Acad Pharmaceut Sci, 75; lectr pharmacol, Univ Pa Sch Dent Med, 75-; consult, Franklin Inst Res Labs, 77-79; clin pharmacologist, Walter Reed Army Med Ctr, 83; vis assoc prof pharmacol, Med Col Pa, 84-; chmn prog comt, Pharmacol & Toxicol Sect, Acad Pharmaceut Sci, 75; lectr pharmacol, Univ Pa Sch Dent Med, 75-77; consult, Franklin Inst Res Labs, 77- *Mem:* AAAS; Am Soc Pharmacol & Exp Therapeut; fel Am Col Clin Pharmacol. *Res:* Effects of acute and chronic tricyclic antidepressants on opioid analgesia; mechanisms of narcotic analgesia. *Mailing Add:* 853 Beverly Rd Rydal PA 19046

GOLDSTEIN, GERALD, b New York, NY, Apr 23, 30; m 51; c 4. ANALYTICAL CHEMISTRY. *Educ:* Brooklyn Col, BA, 51; Univ Tenn, MS, 59, PhD(chem), 65. *Prof Exp:* Anal chemist, Ledoux & Co, 52-53 & Lucius Pitkin, Inc, 53-56; chemist, Oak Ridge Nat Lab, 56-66, group leader anal chem div, 66-72, res staff mem, 72-76; PHYS SCIENTIST, OFF HEALTH & ENVIRON RES, DEPT ENERGY, 76 - *Mem:* AAAS; Am Chem Soc; Sigma Xi; fel Am Inst Chemists. *Res:* Analytical and environmental chemistry; instrumentation. *Mailing Add:* Off Health & Environ Res Dept Energy Washington DC 20545

GOLDSTEIN, GERALD, b New York, NY, Oct 20, 22; m 53; c 3. MICROBIOLOGY, MEDICINE. *Educ:* Moravian Col, BS, 48; Univ Pa, MD, 52. *Prof Exp:* Intern, Univ Va Hosp, 52-53; resident internal med, Vet Admin Hosp, Dallas, Tex, 53-54 & Martinsburg, WVa, 54-56; instr microbiol, 57-59, asst prof microbiol & instr med, 59-63, asst prof med, 63-69, assoc prof microbiol, 63-73, assoc prof internal med, 69-80, PROF MICROBIOL, SCH MED, UNIV VA, 73-, PROF INTERNAL MED, 80- *Concurrent Pos:* Damon Runyon cancer res fel, Univ Va, 56-58, USPHS sr res fel, 59- *Res:* Immunology with respect to infectious diseases and tumors. *Mailing Add:* Biol Sci Dept Ohio Wesleyan Univ Delaware OH 43015

GOLDSTEIN, GIDEON, b Kaunas, Lithuania, Jan 21, 37; m 65; c 2. IMMUNOLOGY, INTERNAL MEDICINE. *Educ:* Univ Melbourne, MB, BS, 59, MD, 63, PhD(immunopath), 67. *Prof Exp:* Resident med & path, Royal Melbourne Hosp, 61-64; pathologist, Walter & Eliza Hall Inst, 64-66; vis scientist, Lab Immunol, Nat Inst Allergy & Infectious Dis, 67-68; res assoc prof path, Sch Med & Postgrad Med Sch, NY Univ, 68-74; mem, Sloan-Kettering Cancer Res Inst, 74-77; DIR IMMUNOSCI, ORTHO PHARMACEUT CORP, 77- *Concurrent Pos:* Mem med adv bd, Myasthenia Gravis Found, 71. *Mem:* Am Asn Immunologists; Am Soc Exp Path. *Res:* Thymic hormones; immunoregulatory therapy; immunodiagnosis. *Mailing Add:* Immunobiol Res Inst Rte 22 E PO Box 999 Annandale NJ 08801-0999

GOLDSTEIN, GILBERT, b Brooklyn, NY, Apr 23, 14; m 52; c 1. BIOCHEMISTRY. *Educ:* Brooklyn Col, BS, 34; Polytech Inst Brooklyn, MS, 41, PhD(chem), 51. *Prof Exp:* Chemist, Dextran Corp, 50-52; asst clin biochemist, Mem Hosp & asst, Sloan-Kettering Inst, 52-57; assoc chemist, Mt Sinai Hosp, 57-66, RES BIOCHEMIST, BETH ISRAEL MED CTR, 66- *Mem:* AAAS; Sigma Xi. *Res:* Enzymes, proteins; vasoactive and smooth-muscle acting polypeptides. *Mailing Add:* 1245 Park Ave Apt Mailbox 40 New York NY 10128

GOLDSTEIN, GORDON D(AVID), b Rochester, NY, Apr 7, 17; m 42; c 2. ELECTRONICS ENGINEERING. *Educ:* Clarkson Col Technol, BEE, 39. *Prof Exp:* Inspector equip, Radio Navig Systs, US Sig Corps, 41-44; develop engr instrumentation, Wash Inst Technol, Inc, 44-50; chief engr computer, US Bur Census, 50-51; electronic scientist, US Naval Ord Lab, Dept Navy, 51, head res & develop br, Appl Math Lab, David Taylor Model Basin, 52-56, electronic engr & ed digital comput newslett, Info Systs, Off Naval Res, 56-80; CONSULT, 80-; COMMUN ENGR, MONTGOMERY COUNTY GOVT, MD, 84- *Concurrent Pos:* Treas, Spring Joint Comput Conf, Am Fedn Info Processing Socs, 71; mem subcomt travel grants comput sci, Div Math Sci, Nat Acad Sci; mem comt to develop talent bank of sr scientists & engrs, Am Asn Retired Persons, 82-84. *Mem:* AAAS; Asn Comput Mach; Inst Elec & Electronics Engrs. *Res:* Research administration information systems; computers and their applications. *Mailing Add:* 9520 Saybrook Ave Silver Spring MD 20901-3034

GOLDSTEIN, HAROLD WILLIAM, b Jersey City, NJ, Aug 23, 31; m 59; c 2. AIR POLLUTION. *Educ:* Univ Ala, BS, 53, MS, 55; Ohio State Univ, PhD(phys chem), 60. *Prof Exp:* Staff mem, Union Carbide Corp Res Inst, 60-63; sr res scientist, Martin Co, Martin-Marietta Corp, Fla, 63-65; mgr space exp & appln, Space Sci Lab, Gen Elec Co, 65-91; RETIRED. *Mem:* Am Chem Soc; Am Inst Aeronaut & Astronaut. *Res:* High temperature thermodynamics; mass spectrometry; ablation kinetics; material properties; space experiments; spacecraft contamination; design and development of instrumentation for the measurement of trace atmospheric species. *Mailing Add:* 1012 N Ocean Blvd Pompano Beach FL 33062

GOLDSTEIN, HARRIS SIDNEY, b Chicago, Ill, Mar 6, 34; m 61; c 4. CHILD PSYCHIATRY, PSYCHOSOMATIC DISEASES. *Educ:* Univ Ill, Urbana, BS, 55; Univ Ill, Chicago, MD, 59; Johns Hopkins Univ, MLA, 66; State Univ NY, DMSc, 71. *Prof Exp:* From asst prof to assoc prof psychiat, Downstate Med Ctr, State Univ NY, 68-75; ASSOC PROF PSYCHIAT, ROBERT WOOD JOHNSON MED SCH, 75- *Mem:* Am Psychiat Asn; Am Psychosomatic Soc; Soc Res Child Develop. *Res:* Epidemiological research in child psychopathology; psychophysiological studies of cardiovascular disorders. *Mailing Add:* Robert Wood Johnson Med Sch Univ Med & Dent NJ 675 Hoes Lane Piscataway NJ 08854

GOLDSTEIN, HERBERT, b New York, NY, June 26, 22. PHYSICS. *Educ:* City Col New York, BS, 40; Mass Inst Technol, PhD(physics), 43. *Prof Exp:* Tutor physics, City Col New York, 40-41; mem staff, Radiation Lab, Mass Inst Technol, 42-46, fel, AEC, 49-50; instr physics, Harvard Univ, 46-49; sr physicist, Nuclear Develop Corp Am, 50-61; PROF NUCLEAR SCI & ENG, COLUMBIA UNIV, 61- *Concurrent Pos:* Vis assoc prof, Brandeis Univ, 52-53. *Mem:* Am Phys Soc; Sigma Xi. *Res:* Electromagnetic theory; radio wave propagation; theory of nuclear forces; neutron cross sections; radiation shielding; nuclear reactors. *Mailing Add:* Columbia Univ 520 W 120th St New York NY 10027

GOLDSTEIN, HERBERT JAY, b New York, NY, Sept 20, 23; m 47; c 4. ORGANIC CHEMISTRY, POLYMER CHEMISTRY. *Educ:* NY Univ, AB, 46, PhD(chem), 52. *Prof Exp:* Res chemist org, E I du Pont de Nemours Co, Jackson Lab, 52-56; sr res chemist polymer, Synpol Inc, 56-57, sr res scientist, 58-64, mgr org res, 64-69, mgr polymerization res, 69-72, mgr polymer develop, 72-78, mgr res & develop, Tex Res Ctr, 78-82, mem indust res lab, 82-87; RETIRED. *Mem:* Am Chem Soc; Sigma Xi. *Res:* Synthetic elastomers; polymers; polymerization chemistry; olefin chemistry; butadiene chemistry. *Mailing Add:* 17 Dowlen Pl Beaumont TX 77706

GOLDSTEIN, HERMAN BERNARD, b Providence, RI, June 19, 17; m 42; c 2. TEXTILE CHEMISTRY, PAPER CHEMISTRY. *Educ:* Brown Univ, AB, 40. *Prof Exp:* Dir res, Providence Textile Chem Co, 40-44; tech mgr, Warwick Chem Div, Sun Chem Corp, 45-52, mgr prod & res, Textile Chem Dept, 53-63, res mgr, Chem Prod Div, 63-67, gen mgr, Chem Div, 68-72, vpres, Chem Div, 72-84; pres, HBG Export Corp, 85-90; RETIRED. *Concurrent Pos:* Adv, USDA, 65-68. *Honors & Awards:* Olney Award, Am Asn Textile Chemists & Colorists, 73 & Millson Award, 85. *Mem:* Am Asn Textile Chemists & Colorists; Tech Asn Pulp & Paper Industs; Am Chem Soc; Am Soc Testing & Mat; Am Inst Chemists; Sigma Xi. *Res:* Chemical specialties, auxiliaries, and processing aids for related industries including water and oil repellants, soil release agents, permanent press resins, softeners, dyehouse chemicals, paper sizes and coating insolubilizers for United States, Europe and Far East. *Mailing Add:* 138 Park Dr Chester SC 29706

GOLDSTEIN, INGE F, b Plauen, Ger, Nov 25, 30; US citizen; m 54; c 3. EPIDEMIOLOGY, BIOPHYSICS. *Educ:* Wellesley Col, BA, 51; Univ Pittsburgh, MA, 56; Columbia Univ, MS, 68, DrPH, 76. *Prof Exp:* Res asst biophys, Univ Mich, 61-64; res assoc, 68-78, SR RES ASSOC EPIDEMIOL & ASSOC PROF CLIN PUB HEALTH, SCH PUB HEALTH, COLUMBIA UNIV, 78- *Mem:* Soc Environ Geochem & Health; fel Am Col Epidemiol; Soc Epidemiol Res; Sigma Xi; Air Pollution Control Asn. *Res:* Acute and chronic health effects of environmental pollutants; methodological approaches in assessing the physical environment in their effects on human health; asthma. *Mailing Add:* Div Epidemiol Columbia Univ Sch Pub Health New York NY 10032

GOLDSTEIN, IRVING SOLOMON, b Bronx, NY, Aug 20, 21; m 45; c 3. WOOD CHEMISTRY, PAPER CHEMISTRY. *Educ:* Rensselaer Polytech Inst, BS, 41; Ill Inst Technol, MS, 44; Harvard Univ, PhD(org chem), 48. *Prof Exp:* Asst chem, Ill Inst Technol, 41-42; res chemist, NAm Rayon Corp, 48-51; sr chemist, Res Dept, Koppers Co, Inc, 51-53; leader, Wood Preserving Group, 53-55, mgr wood chem, 55-63; sr res chemist, Nalco Chem Co, 63-66; mgr paper res, Corp Res & Develop Dept, Continental Can Co, 66-68; prof forest sci, Tex A&M Univ, 68-71; head dept, 71-78, PROF WOOD & PAPER SCI, NC STATE UNIV, 71- *Mem:* AAAS; Am Chem Soc; Forest Prod Res Soc; Soc Wood Sci & Technol; Tech Asn Pulp & Paper Indust; fel, Int Acad Wood Sci. *Res:* Wood, cellulose and lignin chemistry; pulp chemistry; chemical utilization of wood. *Mailing Add:* Dept Wood & Paper Sci Col Forest Res NC State Univ Raleigh NC 27695-8005

GOLDSTEIN, JACK, b Philadelphia, Pa, June 24, 30; m 67; c 3. BIOCHEMISTRY. *Educ:* Brooklyn Col Pharm, BS, 52; Cornell Univ, MNS, 57, PhD(biochem), 59. *Prof Exp:* Asst biochem, Cornell Univ, 55-57; res chemist, US Plant, Soil & Nutrit Lab, 57-59; vis investr biochem, Rockefeller Inst, 59-61, res assoc, 61-64, asst prof, 64-65; ADJ ASSOC PROF BIOCHEM, MED COL, CORNELL UNIV, 65-; INVESTR, NEW YORK BLOOD CTR, 66- *Mem:* Am Soc Biol Chem; Am Chem Soc; Am Soc Cell Biol. *Res:* Ribonucleic acids; protein structure and biosynthesis; Cell surface antigens; structure and function. *Mailing Add:* New York Blood Ctr 310 E 67th St New York NY 10021

GOLDSTEIN, JACK STANLEY, b New York, NY, May 10, 25; m 48; c 3. PHYSICS. *Educ:* City Col New York, BS, 47; Univ Okla, MS, 48; Cornell Univ, PhD(physics), 53. *Prof Exp:* Res physicist, Cornell Aeronaut Lab, 48-50; asst, Cornell Univ, 50-52; mem, Inst Advan Study, 52-53; res assoc, Mass Inst Technol, 53-54; sr physicist, Baird-Atomic, Inc, Mass, 54-57; vis asst prof, 56-57, from asst prof to assoc prof, 57-66, chmn, Dept Physics, 67-69, dir, Astrophys Inst, 63-73, sci dir, African Primary Sci Prog, 65-72, dean, Grad Sch, 72-74, dean fac, 74-81, PROF PHYSICS, BRANDEIS UNIV, 66- *Concurrent Pos:* Fulbright scholar, Israel, 60-61; Fulbright scholar & Guggenheim fel, Univ Rome, 66-67; consult div sci educ, UNESCO, 67-74; vis scholar, Kyoto Univ, 82. *Mem:* Am Phys Soc; Am Astron Soc; NY Acad Sci. *Res:* Quantum mechanics; field theory; astrophysics; energy; history of science; science policy. *Mailing Add:* Physics Dept Brandeis Univ Waltham MA 02254

GOLDSTEIN, JACOB HERMAN, b Atlanta, Ga, Dec 18, 15; m 53. PHYSICAL CHEMISTRY. *Educ:* Emory Univ, AB, 43, MS, 45; Harvard Univ, MS, 47, PhD(chem), 49. *Prof Exp:* From asst prof to prof, 49-60, CANDLER PROF CHEM, EMORY UNIV, 60- *Honors & Awards:* Charles Herty Medal, Am Chem Soc, 81, Charles H Stone Award, 84. *Mem:* Am Chem Soc; Am Phys Soc; Biophys Soc. *Res:* Nuclear magnetic resonance spectroscopy, including biological applications; liquid crystals; computer applications in chemistry. *Mailing Add:* 820 Oakdale Rd NE Emory Univ 1364 Clifton Rd NE Atlanta GA 30307-1210

GOLDSTEIN, JEFFREY JAY, b New York, NY, Dec 3, 57. PLANETARY ATMOSPHERES, HETERODYNE SPECTROSCOPY. *Educ:* City Univ NY, BA, 80; Univ Pa, MA, 87, PhD(astrophys), 89. *Prof Exp:* Instr astron & astrophys, Univ Pa, 81-85; grad student researcher, Lab Extraterrestrial Physics, NASA/Goddard Space Flight Ctr, 85-88; ASTROPHYSICIST, LAB ASTROPHYS, NAT AIR & SPACE MUS, SMITHSONIAN INST, 88- *Concurrent Pos:* Adj lectr, Physics Dept, Queensborough Community Col, City Univ NY, 80, 81 & Queens Col, 82; lectr, Resident Associateship Prog, Smithsonian Inst, 89. *Mem:* Am Astron Soc; AAAS; Am Phys Soc; Am Geophys Union. *Res:* Observations of planetary atmospheric circulation using infrared heterodyne and microwave spectroscopy; science education lecturer visiting elementary and secondary classrooms nationwide to foster interest in science through space sciences. *Mailing Add:* Lab Astrophys Nat Air & Space Mus Smithsonian Inst Washington DC 20560

GOLDSTEIN, JEFFREY MARC, b Bronx, NY, May 9, 47; c 2. PSYCHOPHARMACOLOGY, NEUROPHARMACOLOGY. *Educ:* Colo State Univ, BS, 70; Seton Hall Univ, MS, 73; Univ Del, PhD(neurosci), 80. *Prof Exp:* Assoc scientist, Schering Corp, 70-76; PRIN PHARMACOLOGIST, DEPT PHARMACOL, ICI PHARMACEUT GROUP, 76- *Mem:* AAAS; NY Acad Sci; Am Soc Pharmacol & Exp Therapeut; Soc Neurosci; Sigma Xi. *Res:* Psychiatric and neurologic drugs, discovery and development. *Mailing Add:* Dept Pharmacol ICI Pharmaceut Group ICI Am Inc Wilmington DE 19897

GOLDSTEIN, JEROME ARTHUR, b Pittsburgh, Pa, Aug 5, 41. DIFFERENTIAL EQUATIONS, QUANTUM THEORY. *Educ:* Carnegie-Mellon Univ, BS, 63, MS, 64, PhD(math), 67. *Hon Degrees:* SMD, Loyola Univ, 73. *Prof Exp:* Mem math, Inst Advan Study, 67-68; from asst prof to prof math, Tulane Univ, 68-91; res prof math, Math Sci Res Inst, 90-91; PROF MATH, LA STATE UNIV, 91- *Concurrent Pos:* Prin investr, NSF grants, 68-92; vis prof, Fed Univ Rio de Janiero & Univ Brasilia, 75, Univ Padora, 81, Univ London, 80-81, Univ Graz, 88 & Univ Tübingen, 89; ed, three journals, 82-92. *Mem:* Am Math Soc; Math Asn Am; Soc Indust & Appl Math; Asn Women math. *Res:* Nonlinear partial differential equations; mathematical quantum theory; operator semi groups; functional analysis; stolnastic processes. *Mailing Add:* Dept Math La State Univ Baton Rouge LA 70803

GOLDSTEIN, JEROME CHARLES, b Glens Falls, NY, Nov 4, 35; m 65; c 3. OTOLARYNGOLOGY. *Educ:* Univ Rochester, AB, 57; State Univ NY Upstate Med Ctr, MD, 63. *Prof Exp:* Intern, Philadelphia Gen Hosp, 63-64; resident surg, Bronx Munic Hosp, 64-65; resident otolaryngol, State Univ NY, Syracuse, 65-68; asst prof otolaryngol & maxillofacial surg, Northwestern Univ, 68-71; instr otolaryngol, 71-74, PROF OTOLARYNGOL & HEAD DIV, ALBANY MED COL, 74- *Mem:* Fel Am Acad Ophthal & Otolaryngol; fel Am Col Surg; fel Am Soc Head & Neck Surg; fel Am Acad Facial Plastic & Reconstructive Surg; Am Soc Head & Neck Surg (secy, 76-). *Mailing Add:* One Prince St Alexandria VA 22314

GOLDSTEIN, JOHN CECIL, b New York, NY, May 28, 44. LASERS. *Educ:* Univ Ill, BS, 65; Mass Inst Technol, PhD(physics), 71. *Prof Exp:* Fel physics, Mass Inst Technol, 71; fel, 71-73, STAFF MEM PHYSICS, LOS ALAMOS SCI LAB, 73- *Mem:* Am Phys Soc. *Res:* High-power short pulse lasers; pulse propagation in amplifiers and attenuators; nonlinear optics; theory of the free-electron laser; theory and numerical simulation of free-electron oscillators and amplifiers, nonlinear phenumena such as sideband generation and refractive index effects; short pulse effects and theory of soft-x-ray free electron lasers. *Mailing Add:* Los Alamos Sci Lab Group X-1 MS-F645 Los Alamos NM 87545

GOLDSTEIN, JORGE ALBERTO, b Buenos Aires, Arg, July 17, 49; m 77; c 1. ORGANIC CHEMISTRY, ENZYMOLOGY. *Educ:* Rensselaer Polytech Inst, BS, 71; Harvard Univ, MS, 73, PhD(chem), 76. *Prof Exp:* Res asst chem, Harvard Univ, 71-76; res assoc, Mass Inst Technol, 76-77; PATENT CHEMIST & CONSULT, OBLON, FISHER, SPIVAK, MCCLELLAND & MAIER, 77- *Mem:* Am Chem Soc; AAAS. *Res:* Enzymology; phosphorus chemistry; organic chemical mechanisms; patent law. *Mailing Add:* 7723 Heatherton Lane Potomac MD 20854

GOLDSTEIN, JOSEPH I, b Syracuse, NY, Jan, 6, 39; m 63; c 2. METALLURGY, MATERIALS SCIENCE. *Educ:* Mass Inst Technol, BS, 60, MS, 62, ScD(metall), 64. *Prof Exp:* Metallurgist, Smithsonian Astrophys Observ, 63-64; aerospace technologist, Goddard Space Flight Ctr, NASA, 64-68; from asst prof to assoc prof, 68-75, vpres Grad Studies & Res, Whitaker Lab, 83-90, THEODORE L DIAMOND PROF METALL & MAT SCI, LEHIGH UNIV, 76-, ASST VPRES RES, 79-, RD STOUT PROF MAT SCI, 90- *Concurrent Pos:* Lectr, Univ Md, 66-67. *Mem:* Am Inst Mining, Metall & Petrol Engrs; fel Am Soc Metals; Am Geophys Union; Meteoritical Soc; Microbeam Anal Soc (pres, 77, past pres, 78); Sigma Xi. *Res:* Solid state kinetics; meteorites; phase equilibria; analytical electron microscopy, scanning electron microscopy. *Mailing Add:* Whitaker Labs #5 Lehigh Univ Bethlehem PA 18015

GOLDSTEIN, JOSEPH LEONARD, b Sumter, SC, Apr 18, 40. GENETICS & GENETIC ENGINEERING, MEDICINE. *Educ:* Washington & Lee Univ, BS, 62; Univ Tex, Dallas, MD, 66; Am Bd Internal Med, dipl. *Hon Degrees:* DSc, Univ Chicago, 82, Rensselaer Polytechnic Inst, 82, Washington & Lee Univ, 86 & Univ Paris-Sud, 88. *Prof Exp:* Intern & resident internal med, Mass Gen Hosp, 66-68; clin assoc biochem, NIH, 68-70; spec NIH fel med genetics, Univ Wash, 70-72; from asst prof to assoc prof, 72-76, PROF MED, UNIV TEX HEALTH SCI CTR, DALLAS, 76-, PAUL J THOMAS PROF MED & CHMN, DEPT MOLECULAR GENETICS, 77- *Concurrent Pos:* Mem, Mammalian Cell Lines Nat Adv Comt, NIH, 75-78, sci rev bd, Howard Hughes Med Inst, 78-84 & adv bd, 85-; lectr, Harvey Soc, 77; non-resident fel, Salk Inst, 83-; distinguished res award, Asn Am Med Cols, 84; mem, sci adv bd, Welch Found & Jane Coffin Child Mem Fund Med Res, Prog Adv Comt Human Genome, NIH. *Honors & Awards:* Nobel Prize Physiol Med, 85; Heinrich Wieland Prize, WGer, 74; Pfizer Award, Am Chem Soc, 76; Passano Award, Passano Found, 78; Lounsbery Award, Nat Acad Sci, 79; Biol & Med Sci Award, NY Acad Sci, 81; Int Award, Gairdner Found, 81; Lita Annenberg Hazen Award, 82; Res Achievement Award, Am Heart Asn, 84; Louisa Gross Horwitz Award, 84; 3M Life Sci Award, Fedn Am Socs Exp Biol, 85; William Allan Award & Albert D Lasker Award, Am Soc Human Genetics, 85; Am Col Physicians Award, 86; Nat Medal of Sci, 88. *Mem:* Nat Acad Sci; Inst Med-Nat Acad Sci; Am Acad Arts & Sci; Am Soc Clin Invest (pres, 85-86); Asn Am Physicians; Am Soc Biol Chemists; Am Fedn Clin Res; Am Col Physicians; Am Soc Human Genetics; Am Philosophical Soc. *Res:* Human biochemical genetics; regulation of cholesterol and lipoprotein metabolism; membrane receptors. *Mailing Add:* Univ Tex Southwestern Med Ctr 5323 Harry Hines Blvd Dallas TX 75235

GOLDSTEIN, JOYCE ALLENE, b Whittier, Calif, Mar 8, 41; m 63. DRUG-METABOLIZING ENZYMES, MECHANISMS OF TOXICITY. *Educ:* Southwest Mo State Univ, BS, 62; Univ Tex Southwestern Med Sch Dallas, PhD(pharmacol), 68. *Prof Exp:* Res pharmacologist, Commun Dis Ctr & Food & Drug Admin, 68-72 & Environ Protection Agency, 72-77; PHARMACOLOGIST, NAT INST ENVIRON HEALTH SCI, 77- *Concurrent Pos:* Fel, Emory Univ Med Sch, 67-68. *Mem:* Am Soc Pharmacol & Exp Therapeut; Soc Toxicol. *Res:* Effects of pesticides and environmental chemicals on mammalian hepatic drug-metabolizing enzymes; purification and characterization of subspecies of cytochrome P-450; drug-induced porphyria; structure activity relationships. *Mailing Add:* Nat Inst Environ Health Sci Bldg 17 Research Triangle Park NC 27709

GOLDSTEIN, JULIUS L, b Brooklyn, NY, July 9, 35; m 62; c 4. ELECTRICAL ENGINEERING, SENSORY PERCEPTION. *Educ:* Cooper Union, BEE, 57; Polytech Inst Brooklyn, MEE, 60; Univ Rochester, PhD(elec eng), 66. *Prof Exp:* Staff engr, Polytech Res & Develop, NY, 57-60; lectr, Univ Rochester, 60-61; fel hearing res, Lab Psychophys, Harvard Univ, 66-68; from asst prof to prof, elec eng, Mass Inst Technol, 68-73; chmn bioeng prog, 73-76, chmn dept electronics, 76-78, from assoc prof to PROF, FAC ENG, TEL AVIV UNIV, 73-; SR RES SCIENTIST, CENT INST DEAF, 88- *Concurrent Pos:* Lectr, Polytech Inst Brooklyn, 59-60; NSF fel, Inst Perception Res, Eindhoven, Netherlands, 65-66; mem comt hearing bioacoust & biomed, Nat Res Coun, 77- *Mem:* Sr mem Inst Elec & Electronics Engrs; fel Acoust Soc Am; Israeli Soc Med & Biol Eng (pres, 75-77); Asn Res Ophthal. *Res:* Auditory psychology and physiology; bioengineering; human sensory communication. *Mailing Add:* Cent Inst Deaf 818 S Euclid Ave St Louis MO 63110

GOLDSTEIN, LARRY JOEL, b Philadelphia, Pa, Dec 4, 44; m 67; c 2. NUMBER THEORY. *Educ:* Univ Pa, BA & MA, 65; Princeton Univ, MA & PhD(math), 67. *Prof Exp:* Josiah Willard Gibbs instr math, Yale Univ, 67-69; assoc prof, 69-72, prof math, 72-84, DIST SCHOLAR & TEACHER, UNIV MD, COLLEGE PARK, 79- *Concurrent Pos:* NSF res grants, 69-82; adj prof, Goldstein Software Inc, 84- *Honors & Awards:* Allan C Davis Medal. *Mem:* Am Math Soc; Math Asn Am; Asn Comput Mach. *Res:* Algebraic number theory; theory of automorphic functions. *Mailing Add:* 21231 Georgia Ave Brookeville MD 20833

GOLDSTEIN, LAWRENCE HOWARD, b New York, NY, Jan 7, 52. ELECTRICAL ENGINEERING, COMPUTER-AIDED DESIGN. *Educ:* Cooper Union, BE, 73; Princeton Univ, MSE, 74, PhD(elec eng), 76. *Prof Exp:* mem tech staff comput-aided design, Sandia Labs, 76-80; MGR SCI COMPUT, INMOS CORP, 80-; AT SIERRA SEMICONDUCTOR CORP. *Honors & Awards:* Browder J Thompson Award, 81. *Mem:* Inst Elec & Electronics Engrs; Asn Comput Mach; Soc Indust & Appl Math. *Res:* Computer-aided design of large scale integrated circuits. *Mailing Add:* Standard Microsyst 35 Marcus Blvd Hauppauge NY 11788

GOLDSTEIN, LAWRENCE S B, b Buffalo, NY, Feb 20, 56. CELL & DEVELOPMENTAL BIOLOGY. *Educ:* Univ Calif, BA, 76; Univ Wash, PhD(genetics), 80. *Prof Exp:* Postdoctoral fel genetics & biochem, Univ Colo, 80-83; res assoc molecular biol, Harvard Univ & Mass Inst Technol, 83-84; asst prof cell & develop biol, 84-88, John L Loeb assoc prof natural sci, 88-90, PROF CELL & DEVELOP BIOL, HARVARD UNIV, 90- *Concurrent Pos:* Ad hoc mem, Nucleic Acids & Protein Synthesis Adv Comt, Am Cancer Soc, 85 & 88, Cell Biol Study Sect, NIH, 90 & Molecular Cytol Study Sect, 91; NIH grants, 86-91, 88-91 & 88-93; Am Cancer Soc fac res award, 90-95. *Mem:* Genetics Soc Am; Am Soc Cell Biol. *Res:* Molecular, biochemical, and genetic analysis of protein motors, mitotic mechanisms, and cytoskeletal function in Drosophila; author of more than 30 technical publications. *Mailing Add:* Dept Cellular & Develop Biol Harvard Univ 16 Divinity Ave Cambridge MA 02138

GOLDSTEIN, LEON, b Malden, Mass, Feb 6, 33; m 58; c 3. PHYSIOLOGY. *Educ:* Northeastern Univ, BS, 54; Boston Univ, PhD(pharmacol), 58. *Hon Degrees:* MA, Univ Oxford, 83. *Prof Exp:* Res fel zool, Dartmouth Col, 58-59, instr, 59-60; res assoc, Cancer Res Inst, New Eng Deaconess Hosp, 60-62; assoc physiol, Harvard Med Sch, 62-64, asst prof, 64-68; assoc prof med sci, 68-72, PROF MED SCI, BROWN UNIV, 73-, CHMN PHYSIOL & BIOPHYS, 78- *Concurrent Pos:* lectr, Harvard Med Sch, 68-69; ed, Physiol & Biochem, J Exp Zool, 78-; dir, Mount Desert Island Biol Lab, 79-83, vpres, 86-; Res Career Develop Award, NIH, 63; Fogarty Sen Int Fel, 83. *Mem:* Am Chem Soc; Am Soc Zool; Soc Gen Physiol; Am Physiol Soc; fel AAAS. *Res:* Kidney; comparative physiology; membrane transport; metabolic regulation. *Mailing Add:* Div Biol & Med Sci Brown Univ Providence RI 02912

GOLDSTEIN, LEONIDE, biological psychiatry; deceased, see previous edition for last biography

GOLDSTEIN, LESTER, b Brooklyn, NY, June 28, 24; div; c 2. CELL BIOLOGY. *Educ:* Brooklyn Col, BA, 48; Univ Pa, PhD, 53. *Prof Exp:* Lectr biol, Queens Col (NY), 50; asst instr zool, Univ Pa, 50-51; USPHS fel, Univ Calif, Berkeley, 53-55, asst res physiologist, Med Ctr, Univ Calif, San Francisco, 55-59; assoc prof zool, Univ Pa, 59-64, prof biol, 64-67; prof, Inst Develop Biol, Univ Colo, Boulder, 67-68, & Dept Molecular, Cellular & Develop Biol, 68-82; PROF & DIR, T H MORGAN SCH BIOL SCI, UNIV KENTUCKY, 82- *Concurrent Pos:* Damon Runyon Mem Fund fel, 55-56; Am Cancer Soc scholar, Univ Wash, Seattle, 77-78. *Mem:* Am Soc Cell Biol; AAAS; Genetics Soc Am. *Res:* Interrelationships of nucleus and cytoplasm; nuclear RNA's and proteins and cell cycle. *Mailing Add:* T H Morgan Sch Biol Sci Univ Ky Lexington KY 40536

GOLDSTEIN, LEWIS CHARLES, b Paterson, NJ, Dec 31, 17. CYTOLOGY. *Educ:* Univ Richmond, BS, 38, MS, 40; Univ Va, PhD(biol), 47. *Prof Exp:* Instr zool, Univ Va, 46-47; asst prof biol, Sampson Col, 47-49, Champlain Col, 49-53, Univ Mass, 53-54 & Smith Col, 54-55; head dept, 55-71, asst dean arts & sci, 69-76, PROF BIOL, VA COMMONWEALTH UNIV, 55- *Mem:* AAAS; Am Soc Zoologists; Am Inst Biol Sci; NY Acad Sci. *Res:* Histology of the oyster; nematocysts of hydra; oogenesis of hamster. *Mailing Add:* 1243 Gaskins Rd B Richmond VA 23233

GOLDSTEIN, LOUIS, b Dombrad, Hungary, Mar 25, 04; nat US; m 33; c 1. THEORETICAL PHYSICS, LOW TEMPERATURE PHYSICS. *Educ:* Col City Nagyvarad, Testimonium Maturitatis, 22; Univ Paris, lic es S, 26, DSc(theoret physics), 32. *Prof Exp:* Researcher, Inst Henri Poincare, Univ Paris, 28-32, res assoc, 32-39; independent researcher, NY Univ, 39-41; instr physics, City Col, New York, 41-44; mem staff, Wave Propagation Group, Off Sci Res & Develop contract, Columbia Univ, 44-46; physicist, Fed Telecommunications Lab, Int Tel & Tel, NY, 46; mem staff, 46-71, CONSULT MEM STAFF, LOS ALAMOS NAT LAB, 71- *Mem:* Fel Am Phys Soc. *Res:* Theory of atoms; atomic nuclei, statistical mechanics; properties of matter at very low temperatures; theory of the helium isotopes. *Mailing Add:* LANL MS-B210 Los Alamos NM 87545

GOLDSTEIN, LOUIS ARNOLD, b Spring Valley, NY, Apr 15, 07; m 35; c 2. SURGERY. *Educ:* Alfred Univ, BS, 28; Univ Rochester, MD, 32. *Prof Exp:* From instr to assoc prof, Sch Med & Dent, 37-67, clin prof, Med Ctr, 67-69, chmn div, 69-74, prof, 69-85, EMER PROF ORTHOP SURG, MED CTR, UNIV ROCHESTER, 85- *Concurrent Pos:* Consult, Genesee Hosp. *Mem:* AMA; Am Orthop Asn; Am Col Surg; Am Acad Orthop Surg; Scoliosis Res Soc. *Res:* Scoliosis. *Mailing Add:* Strong Mem Hosp 601 Elmwood Ave Rochester NY 14642

GOLDSTEIN, MARCUS SOLOMON, b Philadelphia, Pa, Aug 22, 06; m 45; c 1. PHYSICAL ANTHROPOLOGY. *Educ:* George Washington Univ, AB, 30, AM, 32; Columbia Univ, PhD(anthrop), 37. *Prof Exp:* Sci aide phys anthrop, Mus Natural Hist, Washington, DC, 28-30; res assoc, Dept Orthodontia, Col Dent, NY Univ, 33-39 & Dept Anthrop, Univ Tex, Austin, 39-41; analyst, Off Strategic Serv, 44-46 & USPHS, 46-61; anthropologist, Nat Inst Ment Health, 61-67; analyst med care, Social Security Admin, 67-71; ASSOC PROF & CONSULT ANAT & ANTHROP, MED SCH, TEL AVIV UNIV, 71- *Concurrent Pos:* Grant, Inst Latin-Am Studies, Univ Tex, Austin, 41-43; chmn, Israel Anthrop Asn, 73-75, ed, newsletter, 73-77; field work fel, Isle d'Orleans, Que. *Mem:* Am Anthrop Asn; Israel Anthrop Asn. *Res:* Anthropological research in modern societies and effective public dissemination of contributions of diverse ethnic groups to enrichment of culture; practical means to full medical care services to total population. *Mailing Add:* Dept Anat & Anthrop Tel Aviv Univ Sackler Med Sch Tel Aviv Israel

GOLDSTEIN, MARGARET ANN, b Sinton, Tex, Mar 13, 39; m 59; c 1. CELL BIOLOGY, MUSCULAR PHYSIOLOGY. *Educ:* Rice Univ, BA, 65, PhD(cell biol), 69. *Prof Exp:* Res asst psychol, Baylor Col Med, 60-61; cardiovasc technician, Vet Admin Hosp, 61-62; head technician pulmonary physiol,Tex Inst Rehab & Res, 63-64; res asst biol & physiol, Rice Univ, 64; instr biol, Univ Tex M D Anderson Hosp & Tumor Inst Houston, 69-70; instr cell biol, Baylor Col Med, 70-73, asst prof cell biophys & exp med, 72-77, from asst prof to assoc prof, 77-89, PROF MED & CELL BIOL, BAYLOR COL MED, 89- *Concurrent Pos:* Vis lectr, Rice Univ, 69-74; Am Heart Asn grant, Baylor Col Med, 71-73, Nat Heart, Lung & Blood Inst grants, 71-82 & 74-90; mem, Coun Basic Sci, Am Heart Asn, estab investr; NIH res & career develop award; rev comt A, Nat Heart Lung & Blood Inst; panelist, Nat Res Coun; biol dir, Electron Micros Soc Am. *Mem:* AAAS; Am Soc Cell Biol; Electron Micros Soc Am; Am Heart Asn; Biophys Soc; Int Soc Heart Res. *Res:* Cytochemistry of striated muscle; ultrastructure of muscle cells in developing hearts, normal and diseased hearts; optical diffraction analysis of Z lattice in muscle and role of Z substance. *Mailing Add:* Dept Med Baylor Col Med Houston TX 77030-3498

GOLDSTEIN, MARK KANE, b Allentown, Pa, May 29, 38; m 61; c 3. BEHAVIORAL MEDICINE, SENSORY PSYCHOPHYSICS. *Educ:* Muhlenberg Col, AB, 61; Columbia Univ, MA, 62; Cornell Univ PhD, 71. *Prof Exp:* Staff psychologist, exp psychol, US Vet Admn Med Ctr, 71-82; DIR MED SCIENTIST BIOMED RES, CTR AMBULATORY STUDIES, UNIV FLA, 82- *Res:* Tactile thresholds in breast cancer detection; improved methods for tactile detection of lesions; ambulatory monitoring in chronic physical disease; neurlogical monitoring and neuro protheses in chronic patents. *Mailing Add:* 1512 NW Seventh Pl Gainesville FL 32603

GOLDSTEIN, MARTIN, b New York, NY, Nov 18, 19; m 54; c 3. PHYSICAL CHEMISTRY. *Educ:* City Col New York, BS, 40; Columbia Univ, PhD(chem), 50. *Prof Exp:* USPHS fel, Polytech Inst Brooklyn, 50-51; fel, Harvard Univ, 51-53 & Mellon Inst, 53-58; vis scientist, Nat Phys Lab Israel, 58-60; staff scientist chem dept, Sci Lab, Ford Motor Co, 60-65; vis assoc prof ceramics, Mass Inst Technol, 64-65; PROF CHEM, DIV NATURAL SCI & MATH, BELFER GRAD SCH SCI, YESHIVA UNIV, 65- *Concurrent Pos:* Chmn, Gordon Res Conf on Glass, 65; vis prof, Univ Ill, 67-68, Israel Inst Technol, 72 & Univ Bristol, 72 & 73; co-chmn workshop conf glassy state, NY Acad Sci, 75, 81, 85. *Honors & Awards:* Forrest & Meyer Award, Am Ceramic Soc, 65. *Mem:* AAAS; Am Chem Soc; fel Am Phys Soc. *Res:* Glassy state, thermodynamics and statistical mechanics of glass transition, secondary relaxations in glasses. *Mailing Add:* 1222 Kensington Rd Teaneck NJ 07666

GOLDSTEIN, MARVIN E, b Cambridge, Mass, Oct 11, 38; m 65; c 2. FLUID MECHANICS, APPLIED MATHEMATICS. *Educ:* Northeastern Univ, BS, 61; Mass Inst Technol, MS, 62; Univ Mich, PhD(mech eng), 65. *Prof Exp:* Res assoc fluid mech, Mass Inst Technol, 65-67; aerospace engr, 67-79, CHIEF SCIENTIST, NASA-LEWIS RES CTR, 80- *Concurrent Pos:* Assoc ed, J Am Inst Aeronaut & Astronaut, 77- *Honors & Awards:* Except Sci Achievement Award, NASA, 79; Aeroacoustics Award, Am Inst Aeronaut & Astronaut, 83, Pendray Award, 83. *Mem:* Nat Acad Eng; fel Am Inst Aeronaut & Astronaut; fel Am Phys Soc. *Res:* Fluid mechanics, heat transfer, aeroacoustics. *Mailing Add:* NASA-Lewis Res Ctr 21000 Brookpark Rd MS 3-7 Cleveland OH 44135

GOLDSTEIN, MARVIN SHERWOOD, b Newark, NJ, Feb 13, 34; m 66; c 4. PETROLEUM CHEMISTRY, CATALYSIS. *Educ:* NY State Col Teachers, Albany, BS, 56; Rensselaer Polytech Inst, PhD(phys chem), 60. *Prof Exp:* Res engr, Allison Div, Gen Motors Corp, 59-60; res chemist, org chem div, 60-69, group leader, Refining Catalyst Res, Stamford Res Labs, 69-72, proj leader, chem res div, 72-83, dir res, Catalyst Recovery Inc, 83-84, PRIN RES CHEMIST, STAMFORD RES LABS, AM CYANAMID CO, 85- *Concurrent Pos:* Lectr, Univ Conn, 62-63. *Mem:* Am Chem Soc; Catalysis Soc; Sigma Xi. *Res:* Catalysis; surface chemistry; kinetics and thermodynamics of oxidation and combustion; high energy fuels; boron hydrides. *Mailing Add:* Stamford Res Labs Am Cyanamid Co Stamford CT 06904

GOLDSTEIN, MAX, b New York, NY, Jan 4, 20; m 46. MATHEMATICS. *Educ:* Brooklyn Col, BA, 40; NY Univ, MS, 62. *Prof Exp:* Asst head numerical anal group, Theoret Physics Div, Nat Res Coun Can, 43-46; alternate group leader, Los Alamos Sci Lab, 46-57; sr res scientist, 57-66, from adj assoc prof to adj prof math, 63-67, assoc dir, AEC Comput & Appl Math Ctr, 66-77, PROF COMPUT SCI, COURANT INST MATH SCI, NY UNIV, 67-, ASSOC CHMN DEPT, 70- *Mem:* Am Math Soc; Asn Comput Mach. *Res:* Numerical mathematical analysis; general purpose automatic digital calculating machines. *Mailing Add:* Dept Comput Sci NY Univ Washington Square 251 Mercer St New York NY 10012

GOLDSTEIN, MELVIN E, b Glen Ridge, NJ, July 27, 36. PHYCOLOGY. *Educ:* Ind Univ, BA, 58, PhD(bot), 63. *Prof Exp:* Asst prof bot, 63-74, ASSOC PROF BIOL, MCGILL UNIV, 74- *Mem:* Bot Soc Am; Am Phycol Soc; Am Soc Protozool; Int Phycol Soc. *Res:* Speciation, sexuality, cytology and genetics of the colonial green flagellate Eudorina; taxonomy, morphology and ecology of the marine algae of Barbados. *Mailing Add:* Dept Biol McGill Univ 1205 Dr Penfield Ave Montreal PQ H3A 1B1 Can

GOLDSTEIN, MELVIN JOSEPH, b New York, NY, Dec 28, 33; div; c 3. ORGANIC CHEMISTRY. *Educ:* Columbia Univ, BA, 54; Yale Univ, PhD(chem), 58. *Prof Exp:* NSF fel, Harvard Univ, 58-59; from instr to prof chem, Cornell Univ, 59-85; AT BROMINE COMPOUNDS LTD, 84- *Concurrent Pos:* NATO sr sci fel, 68; Guggenheim Mem Found fel, 74; Japan Soc Prom Sci Prof, 77. *Mem:* Am Chem Soc; Brit Chem Soc; Swiss Chem Soc; Israel Chem Soc. *Res:* Organic reaction mechanisms; applications of kinetic isotope effects and isotope tracer studies; aromatic synthesis, process development and optimization. *Mailing Add:* Bromine Compounds Ltd PO Box 180 Beer Sheva 84101 Israel

GOLDSTEIN, MELVYN L, b New York, NY, Jan 18, 43; c 1. SPACE PLASMA PHYSICS, ASTROPHYSICS. *Educ:* Columbia Col, AB, 64; Univ Md, PhD(physics), 70. *Prof Exp:* Vis lectr space physics, Tel Aviv Univ, 70-72; res assoc, Nat Res Coun, 72-74; SPACE SCIENTIST SPACE PHYSICS, NASA, GODDARD SPACE FLIGHT CTR, 74- *Concurrent Pos:* Vis scientist astron, Univ Fla, 80. *Mem:* Am Geophys Union; Am Phys Soc. *Res:* Jovian radio science; theory of cosmic ray propagation; magnetohydrodynamic turbulence. *Mailing Add:* NASA Goddard Space Flight Ctr Code 692 Greenbelt MD 20771

GOLDSTEIN, MENEK, b Kolmya, Poland, Apr 8, 24. BIOCHEMISTRY. *Educ:* Univ Bern, PhD, 55. *Hon Degrees:* MD, Karolinska Inst, Stockholm, Sweden. *Prof Exp:* Res instr biochem, Univ Bern, 54-56; mem res staff, Worcester Found Exp Biol, 56-57; from instr to assoc prof, 58-69, PROF NEUROCHEM, MED CTR, NY UNIV, 69- *Mem:* Am Chem Soc; Am Soc Biol Chem. *Res:* Neuroch- emistry; metabolism and biosynthesis of biogenic amines; role of hormones in the central nervous system; mode of action of hormones and enzymes in vivo; steroid hormones; neurotransmitters and their interaction with specific receptors; purification of neurotransmitter synthesizing enzymes; immunohistochemical mapping of neuronal systems. *Mailing Add:* Dept Psychiat NY Univ Med Ctr 560 First Ave New York NY 10016

GOLDSTEIN, MILTON NORMAN, b Cleveland, Ohio, Jan 15, 25; m 52; c 2. CELL BIOLOGY. *Educ:* Western Reserve Univ, BS, 46, MS, 47, PhD(zool), 52. *Prof Exp:* USPHS fel path, Western Reserve Univ, 52-54; investr cancer res, Roswell Park Mem Inst, 54-55, sr cancer res scientist, 55-62; sr cancer res scientist, Cell Biol Labs, St Jude Children's Res Hosp, Memphis, Tenn, 62-66; assoc prof anat, Sch Med, Wash Univ, 66-88; RETIRED. *Concurrent Pos:* Grant reviewer oncol, US Vet Admin, 74-76. *Mem:* Tissue Cult Asn; Am Asn Anat; Am Asn Cancer Res; Am Soc Cell Biol. *Res:* Cytology; differentiation of cells in vitro; histochemistry; electron microscopy applied to cell growing in tissue culture. *Mailing Add:* 8145 Amherst University City MO 63130

GOLDSTEIN, MINDY SUE, PROTEIN PURIFICATION. *Educ:* NY Univ, PhD(med sci), 83. *Prof Exp:* SR STAFF SCIENTIST, APPLIED GENETICS INC, 85- *Mailing Add:* Biochem Dept Life Sci Bldg State Univ NY Stony Brook NY 11794

GOLDSTEIN, MOISE HERBERT, JR, b New Orleans, La, Dec 26, 26; m 70; c 4. ELECTRICAL ENGINEERING. *Educ:* Tulane Univ, BS, 49; Mass Inst Technol, SM, 51, ScD, 57. *Prof Exp:* Asst elec eng, Mass Inst Technol, 49-51, mem defense staff, 52-55; mem staff, Tex Co, 51-52; instr elec eng, Mass Inst Technol, 55-56, from asst prof to assoc prof, 56-63; assoc prof biomed eng, 63-75, prof elec eng, 67-75, PROF BIOMED ENG & EDWARD J SCHAEFER PROF ELEC ENG, JOHNS HOPKINS UNIV, 75- *Concurrent Pos:* NSF fel, Univ Pisa, 59-60; vis lectr, Mass Inst Technol, 63-64; Guggenheim Found fel, Jerusalem Israel, 70-71; consult, FDA, NIH, NSF; NIH sr fel, 81-82 & 89-90, Univ Col London, 82, UCSD, 90. *Mem:* Fel Acoustical Soc Am; Inst Elec & Electronics Engrs; Am Auditory Soc; Am Speech Hearing & Lang Soc. *Res:* Speech signal processing; prostheses for the deaf. *Mailing Add:* ECE Dept Barton Hall Johns Hopkins Univ Baltimore MD 21218

GOLDSTEIN, MURRAY, b New York, NY, Oct 13, 25; m 57; c 2. MEDICAL ADMINISTRATION. *Educ:* NY Univ, BA, 47; Still Col Osteop & Surg, DO, 50; Univ Calif, MPH, 59. *Hon Degrees:* Various from US & foreign univs, 67-90. *Prof Exp:* Intern, Still Osteop Hosp, Des Moines, Iowa, 50-51, resident, 51-53; asst chief grants & training br, Nat Heart Inst, 53-58, asst chief res grants rev br, 59-60, dir extramural prog, Nat Inst Neurol Dis & Stroke, 61-76, dir stroke & trauma prog, 76-78, dep dir, 78-81, actg dir, 81-82, DIR, NAT INST NEUROL & COMMUN DIS & STROKE, 82- *Concurrent Pos:* Vis scientist, Mayo Grad Sch Med, Univ Minn, 67-68. *Honors & Awards:* Cert of Merit, Am Neurol Asn; Meritorious Serv Medal, USPHS, Distinguished Serv Medal. *Mem:* AAAS; Am Osteop Asn; Am Pub Health Asn; Am Acad Neurol; Am Heart Asn; Soc Neurosci; Am Neurol Asn (vpres, 82-83); Am Osteop Col Prev Med (pres, 89-90). *Res:* Nervous system trauma; cerebrovascular diseases; public health; epidemiology. *Mailing Add:* Nat Inst Neurol Dis & Stroke NIH Bldg 31 Rm 8A52 Bethesda MD 20892

GOLDSTEIN, MYRON, b Los Angeles, Calif, July 5, 35. MATHEMATICS. *Educ:* Univ Calif, Los Angeles, BS, 57, MA, 60, PhD(math), 63. *Prof Exp:* Instr math, Univ Calif, Los Angeles, 63; assoc prof, 63-77, PROF MATH, ARIZ STATE UNIV, 77- *Mem:* Math Asn Am; Am Math Soc. *Res:* Riemann surfaces. *Mailing Add:* 5914 Newberry Tempe AZ 85283

GOLDSTEIN, NORMA ORNSTEIN, b Bronx, NY, Dec 13, 21. CONNECTIVE TISSUE GENES, HUMAN BOVINE & SHEEP ELASTIN. *Educ:* NY Univ, AB, 42; Columbia Univ, AM, 49; Univ Pa, PhD(molecular biol), 71. *Prof Exp:* Chemist mfg control, Nat Starch Prod, Inc, NJ & NY, 42-44; chemist anal biochem, Food & Drug Res Labs, NY, 44-47; res asst biochem, Montefiore Hosp, Bronx, NY, 49-50, chem embryol, dept zool, Columbia Univ, 50-53 & cytochem, dept zool, Univ Calif, Berkeley, 54-59; res investr exp chemother & biochem, dept chem, Univ Pa, 60-65, trainee molecular biol, Nat Inst Gen Med Sci, USPHS, Grad Sch Arts & Sci, 65-70, res assoc, Cancer Res Unit, dept animal biol, 71; E H Bobst Penn plan scholar, 72-74; res assoc virol, Sloan-Kettering Inst Cancer Res, Walker Lab, Rye, NY, 75-76; res asst prof biochem, dept path, Mt Sinai Sch Med, City Univ New York, 77; res asst prof biochem, dept animal biol, Sch Vet Med, Univ Pa, 78-82, molecular & cell res biologist, Dept Histol & Embryol, Sch Dent Med, 82-89; RETIRED. *Concurrent Pos:* Res fel cancer res, Inst Cancer Res, Columbia Univ, 70; adj res assoc, Sch Dent Med, Univ Pa, 89- *Mem:* Am Soc Cell Biol; Am Soc Microbiol; AAAS; Asn Women Sci. *Res:* Characterization of human elastin genes; i.e., normal vs those from patients with connective tissue pathologies. *Mailing Add:* Dept Histol & Embryol Sch Dent Med Univ Pa 4010 Locust St Philadelphia PA 19104

GOLDSTEIN, NORMAN N, BIOLOGICAL SCIENCE. *Educ:* Univ Calif, Berkeley, AB, 51, MA, 52, PhD(sci educ), 64. *Prof Exp:* Dir, Proj Instrumentation Methods Physiol Studies, Dept Phsyiol & Anat, Univ Calif, Berkeley, 61-66; mem staff, 64-71, prof, 71-86, EMER PROF, DEPT BIOL SCI, CALIF STATE UNIV, HAYWARD, 86- *Concurrent Pos:* Indust consult, Biocom Inc & Tektronix Inc, 63-; lectr biomed instrumentation, US, Can & Europe, 63-; mem, Conf Electronics Scientists, NSF, 66, Conf Electronics Med & Cell Membrane Symp, Am Physiol Soc, 70; vis prof, Dept Physiol & Anat, Univ Calif, Berkeley, 87. *Mem:* Am Phsyiol Soc. *Res:* Human Anatomy and physiology; computer applications for patient care; author of numerous technical publications. *Mailing Add:* Dept Biol Sci Calif State Univ Hayward CA 94542

GOLDSTEIN, NORMAN PHILIP, b Brooklyn, NY, Mar 31, 21; m 43; c 3. NEUROLOGY, NEUROCHEMISTRY. *Educ:* NY Univ, BA, 41; George Washington Univ, MA, 42, MD, 46. *Prof Exp:* Intern, Mt Sinai Hosp, NY, 46-47; instr biochem, George Washington Univ, 47-49; fel med, Mayo Clin, 49-50, fel psychiat, 50-53, asst to staff neurol, 53; chief sect clin biochem, NIMH, 53-54, chief lab clin sci, 54-55; from instr to prof, Mayo Clin & Mayo Grad Sch Med, Univ Minn, 55-73, head sect, 66-76, prof neurol, 73-83, vchmn dept, 76-82, consult, 55-83; RETIRED. *Concurrent Pos:* Assoc, Sch Med, George Washington Univ, 53-55, prof lectr, 67-83. *Mem:* Sigma Xi; Am Neurol Asn; AMA; Asn Res Nerv & Ment Dis; Am Acad Neurol. *Res:* Metal neuropathy; Wilson's disease; cerebral edema; multiple sclerosis. *Mailing Add:* Route One Box 72 Red Wing MN 55066

GOLDSTEIN, NORMAN PHILLIP, b Montreal, Que, Aug 11, 40. NUCLEAR PHYSICS. *Educ:* McGill Univ, BSc, 61, MSc, 63, PhD(nuclear physics), 67. *Prof Exp:* Sr fel physicist, Westinghouse Elec Corp, 67-73, fel physicist, 73-81, adv physicist, Westinghouse Res Labs, 81-88, ADV PHYSICIST, ELECTRONICS SYSTS GROUP, WESTINGHOUSE ELEC CORP, 89- *Mem:* Am Nuclear Soc; Health Phys Soc. *Res:* Short lifetime measurements with a fast-slow coincidence system; nuclear scattering from low atomic weight nuclei at 100 million electron volts; radiation damage studies; neutron detectors for fast breeder reactors; design of radiation monitors, and new self-powered incore detectors; new core designs for improved fuel utilization in reactors; development of a monitor to measure the burnip of spent fuel. *Mailing Add:* Westinghouse Elec Corp Electronics Systs Group PO Box 746 MS 235 Baltimore MD 21203

GOLDSTEIN, PHILIP, b Bronx, NY, Jan 18, 30. APPLIED PHYSICS. *Educ:* City Col New York, BS, 51; Carnegie-Mellon Univ, MS, 56, PhD(physics), 58. *Prof Exp:* Adminr electronics, Dept Physics, Princeton Univ, 58-60; tech staff mem comput, RCA Lab, 60-68; prof physics, 68-81, PROF COMPUT SCI, JERSEY CITY STATE COL, 81- *Concurrent Pos:* Mem, Phase III Master Plan Comt Comput & Info Sci, Dept Higher Educ NJ, 72-73. *Mem:* Inst Elec & Electronics Engrs; Int Elec & Electronics Engrs Comput Soc. *Res:* Computer applications in the sciences; development of computer-based educational materials; computer applications in medicine; ada language technology consultant. *Mailing Add:* Comput Sci Dept Jersey City State Col 2039 Kennedy Blvd Jersey City NJ 07305

GOLDSTEIN, R(ICHARD) J(AY), b New York, NY, Mar 27, 28; m 63; c 4. MECHANICAL ENGINEERING. *Educ:* Cornell Univ, BME, 48; Univ Minn, MS, 50, MS, 51, PhD, 59. *Prof Exp:* Instr, Univ Minn, 48-51; develop engr, Oak Ridge Nat Lab, 51-54; sr res engr, Lockheed Aircraft Corp, 56; instr, Univ Minn, 56-58; asst prof eng, Brown Univ, 59-61; assoc prof mech eng, 61-65, PROF MECH ENG, UNIV MINN, MINNEAPOLIS, 65-, HEAD DEPT, 77-, JAMES J RYAN PROF, 89-, REGENTS PROF, 90- *Concurrent Pos:* NATO fel, Nat Ctr Sci Res, France, 60-61; NSF sr fel, Univ Cambridge, Eng, 71-72; vis prof, Imperial Col, London, Eng, 84; chmn, Coun Eng, Am Soc Mech Eng, Coun Energy, Engineering Res, Dept of Energy. *Honors & Awards:* Award for Serv, Am Soc Mech Eng, 77, Centennial Medallion, 78; Heat Transfer Mem Award, Am Soc Mech Engrs, 78; Technical Innovation Award, NASA, 77. *Mem:* Nat Acad Eng; fel Am Soc Mech Engrs (sr vpres, 89-); Am Phys Soc; Am Soc Eng Educ; fel AAAS. *Res:* Heat transfer; thermodynamics; fluid mechanics. *Mailing Add:* Dept Mech Eng 125 Mech Engr Univ Minn 111 Church St SE Minneapolis MN 55455

GOLDSTEIN, RAYMOND, b Dec 15, 31. SPACE PLASMA, PHYSICS. *Educ:* City Col New York, BS, 53; Lehigh Univ, MS, 57, PhD(physics), 62. *Prof Exp:* Staff scientist physics, Boeing Sci Res Labs, 62-67; sr scientist, 67-74, MEM TECH STAFF, JET PROPULSION LAB, CALIF INST TECHNOL, 74- *Concurrent Pos:* Instr phys sci, Art Ctr Col Design, 76-; fel, Japan Soc Prom Sci, Gunma Univ, Japan, 79; detailee, NASA Hq, 80-81; vis prof, Univ Bern, Switz, 83-85. *Mem:* Am Phys Soc; Am Asn Physics Teachers; Sigma Xi. *Res:* Shock waves; condensation; magnetohydrodynamics; gaseous conduction; photoionization; fiber optics; electric propulsion; space plasma physics; mass spectrometry. *Mailing Add:* 169-506 Jet Propulsion Lab 4800 Oak Grove Dr Pasadena CA 91109

GOLDSTEIN, RICHARD, b Philadelphia, Pa, Mar 18, 39. COMPUTER SCIENCE FUNCTION. *Educ:* Univ Pa, PhD(math), 66; Tampa Univ, BA, 69. *Prof Exp:* PROF, STATE UNIV NY, ALBANY, 70- *Mem:* Am Math Soc; Am Math Asn. *Mailing Add:* State Univ NY Albany Albany NY 12203

GOLDSTEIN, RICHARD NEAL, b Philadelphia, Pa, Apr 22, 43; div; c 1. GENE EXPRESSION, BACTERIAL VIRUS. *Educ:* Franklin & Marshall Col, BS, 65; Univ Pa, PhD(biol), 71. *Prof Exp:* Am Cancer Soc res fel virol, Dept Molecular Biol, Virus Labs, Univ Calif, Berkeley, 71-73; Silas Arnold Houghton asst prof, 73-78, SILAS ARNOLD HOUGHTON ASSOC PROF, DEPT MICROBIOL & MOLECULAR GENETICS, MED SCH, HARVARD UNIV, 78- *Concurrent Pos:* Prin investr res grants, NIH, NSF & Am Chem Soc, 73-82; Mellon Fund fac award, Med Sch, Harvard Univ, 74, tutor, Dept Biochem & Molecular Biol, 74-; NIH Career Develop Award, US Pub Health Serv, 75-80; mem, Planning Comt, Acad Forum on Recombinant DNA, Nat Acad Sci, 76-77, & Ethical Values in Sci & Technol Prog, 76-78; mem, Steering Comt, Risk Assessment Panel for Recombinant

DNA Studies, NIH, 77-78, Recombinant DNA Adv Comt, 79-82. *Mem:* Am Soc Microbiol; Electron Micros Soc Am; AAAS; Am Asn Univ Professors. *Res:* Regulation of gene expression by bacteria, bacterial viruses, and plasmids; helper-satellite virus P2:P4 interactions; viral morphogenesis; genetics of protein breakdown and protease gene expression in bacteria; regulation of pathogenic virulence factors by E coli and S pyogenes; high resolution electron microscopy. *Mailing Add:* Dept Microbiol & Molecular Genetics Bldg D-1 Med Sch Harvard Univ 25 Shattuck St Boston MA 02115

GOLDSTEIN, ROBERT, b Plymouth, Pa, Sept 7, 24; m 48; c 2. AUDIOLOGY, PSYCHOPHYSIOLOGY. *Educ:* Pa State Univ, BS, 48; Wash Univ, PhD(audiol), 52. *Prof Exp:* Res assoc, Cent Inst Deaf, 53-57; dir div audiol & speech path, Jewish Hosp, St Louis, 58-67; PROF COMMUN DIS, UNIV WIS-MADISON, 67- *Concurrent Pos:* From asst prof to assoc prof, Wash Univ, 53-67; mem, int elec response audiometry study group, Sigma Xi. *Mem:* AAAS; Acoust Soc Am; fel Am Speech-Lang-Hearing Asn (pres, 71-72); Alexander Graham Bell Asn for Deaf; Am EEG Soc; Am Acad Audiol. *Res:* Electrophysiologic tests of hearing; neonatal hearing evaluation; central auditory function; hearing disorders in prison populations; genetic hearing loss. *Mailing Add:* 1975 Willow Dr Madison WI 53706

GOLDSTEIN, ROBERT, b New York, NY, Aug 6, 12; m 41; c 2. MEDICINE. *Educ:* Princeton Univ, AB, 33; Harvard Univ, MD, 37; Am Bd Internal Med, dipl, 49, dipl hemat, 72. *Prof Exp:* Asst med, Harvard Med Sch, 46-51, instr, 51-54, clin assoc, 54-58; asst prof, Med Sch, Tufts Univ, 57-61; assoc prof, 61-65, chmn dept, 71-77, PROF MED, NY MED COL, 65-, ASSOC DEAN STUDENT AFFAIRS, 76- *Concurrent Pos:* Asst, Med Sch, Tufts Univ, 46-50, instr, 50-51; estab investr, Am Heart Asn, 58- *Mem:* AAAS; Am Soc Hemat; Am Heart Asn; Am Physiol Soc; Int Soc Hemat. *Res:* Internal medicine, especially hematology; blood coagulation. *Mailing Add:* Dept Med New York Med Col Elmwood Hall Valhalla NY 10595

GOLDSTEIN, ROBERT ARNOLD, b Brooklyn, NY, Dec 1, 41; m 81; c 4. ALLERGY & CLINICAL IMMUNOLOGY, INTERNAL MEDICINE. *Educ:* Brandeis Univ, AB, 62; Jefferson Med Col, MD, 66; George Washington Univ, PhD(micro immunol), 76. *Prof Exp:* Rotating internship med, Philadelphia Gen Hosp, 66-67; asst chief, Tripler Army Med Ctr, Honolulu, 70-71, chief, 71-72; resident, internal med, 67-69, fel pulmonary dis, 69-70, mem staff, 72-76, assoc chief, 77-78, MEM STAFF, PULMONARY DIS, VET ADMIN MED CTR, WASHINGTON, DC, 78- *Concurrent Pos:* Asst clin prof, Sch Med, Univ Hawaii, 70-72; consult internal med, Queens Hosp, 70-72 & Washington Hosp Ctr, 73-; staff physician, Pulmonary Dis Div, George Washington Univ Med Ctr, 72-74, from asst prof to assoc prof, 72-78, actg dir, 74-76, mem staff, 77-, assoc clin prof, internal medicine, 78-; med care consult, Nat Heart, Lung & Blood Inst, 72-75, extramural prog chief, Nat Inst Allergy & Infectious Dis, NIH, 78-; consult, Children's Nat Med Ctr, 77- *Mem:* Am Thoracic Soc; Am Col Chest Physicians; Am Asn Immunologists; fel Am Col Physicians; Am Acad Allergy & Immunol; Am Fedn Univ Prof. *Res:* Mechanisms of cellular immune responses in infections and non-infectious systemic granulomatous diseases, including sarcoidosis, tuberculosis, aspergillosis and nontuberculous mycobacterial infections. *Mailing Add:* Allergy Immunol & Transplant NIAID NIH Bldg 31 Rm 7A52 Bethesda MD 20205

GOLDSTEIN, ROBERT LAWRENCE, b New York, NY, July 3, 31; m 55; c 2. SOLID STATE LASERS, CRYSTAL GROWTH. *Educ:* Newark Col Eng, BSEE, 63; Fla State Univ, PhD, 79. *Prof Exp:* Proj engr stellar invertial navigation, Kearfot Div, Gen Precision, 58-63 & proj engr appolo navigation syst, Barnes Eng, 63-64; dept mgr electronics, Isomet Corp, 64-65; PRES LASERS & ELECTRO-OPTICS, LASERMETRICS, INC, 65- *Concurrent Pos:* Dir & chmn educ comt, Laser Inst Am, 81- *Mem:* Laser Inst Am; Am Asn Crystal Growth; Am Inst Physics; Inst Elec & Electronics Engrs; Optical Soc Am. *Res:* Development of techniques for growth of improved crystals from aqueous solutions; electro-optic crystals and components; development of picosecond optical switching devices. *Mailing Add:* 196 Coolidge Ave Englewood NJ 07631

GOLDSTEIN, ROBIN SHERYL, b Philadelphia, Pa, Oct 18, 53; m. INVESTIGATIVE TOXICOLOGY. *Educ:* Mich State Univ, BS, 75, MS, 78, PhD(human nutrit), 81. *Prof Exp:* Res & teaching asst, Dept Food Sci & Human Nutrit, Mich State Univ, 76-78 & 79-82; postdoctoral fel, Dept Gen & Biochem Toxicol, Chem Indust Inst Toxicol, 82-83; assoc sr investr & supvr, Nephrotoxicity Lab, Dept Invest Toxicol, Smith Kline & French Labs, 83-88; sr investr & supvr, Nephrotoxicity Lab, 88-90, ASST DIR INVEST TOXICOL, SMITH KLINE BEECHAM PHARMACEUT, 90- *Concurrent Pos:* Lectr, Med Col Pa & Thomas Jefferson Univ, 87-88, Temple Univ, 88 & 89, Tex A&M Univ, 90; mem, Animals Res Comt, Soc Toxicol, 87-90 & In Vitro Toxicol Task Force, Pharmaceut Mfg Asn, 88- *Mem:* Soc Toxicol; Am Inst Nutrit; Am Soc Pharmacol & Exp Therapeut; Pharmaceut Mfg Asn; Am Soc Nephrology. *Res:* Mechanisms of nephrotoxicity; age-dependent nephrotoxicity; biochemical mechanisms of cell injury and death; design of chronic toxicity studies; author of more than 70 technical publications. *Mailing Add:* SmithKline Beecham Pharmaceut 709 Swedeland Rd PO Box 1539 L66 King of Prussia PA 19406-0939

GOLDSTEIN, RUBIN, b New York, NY, Mar 29, 33; m 56; c 3. MATHEMATICAL PHYSICS, APPLIED STATISTICS. *Educ:* Princeton Univ, AB, 55; Harvard Univ, AM, 56, PhD(physics), 60. *Prof Exp:* Res assoc appl physics & instr appl math, Harvard Univ, 60; asst prof nuclear eng, Univ Calif, Berkeley, 60-65; physicist, Dept Appl Sci, Brookhaven Nat Lab, 65-71; consult staff physicist, dept physics, Combustion Eng, Inc, 71-73, sr consult physicist, 73-76, sr consult physicist & area mgr, dept performance anal, 77-79, prin consult scientist, nuclear power systs & mgr statist margin develop, 80-82, sr prin consult scientist, Nuclear Power Systs, 83-86; CONSULT SCIENTIST, 87- *Concurrent Pos:* Consult, Gen Elec Co, 62-65; adj prof, Columbia Univ, 70; consult, Div Reactor Safety Res, US Nuclear Regulatory Comn, Washington, DC, 76-, Combustion Eng, Inc, 87-89, Asea Brown Boveri, 90- & Northeast Utilities, 90-; adj prof eng & comput info sci divs, Hartford Grad Ctr, 77- *Mem:* Am Phys Soc; Am Nuclear Soc. *Res:* Theoretical reactor physics; resonance absorption theory; analytic approximation techniques; transport theory; variational and iterative methods; statistical design of experiments; statistical analysis of scientific and engineering problems; market forecasting, strategic planning; quality control methods and procedures; nuclear material safeguards and accountability; statistical process control. *Mailing Add:* Eight E Normandy Dr West Hartford CT 06107-1405

GOLDSTEIN, SAMUEL JOSEPH, (JR), b Indianapolis, Ind, June 23, 25; m 56; c 3. RADIO ASTRONOMY. *Educ:* Purdue Univ, BS, 48; Stanford Univ, MS, 54, PhD(elec eng), 58. *Prof Exp:* Res engr, Jet Propulsion Lab, Calif Inst Technol, 51-53; actg instr elec eng, Stanford Univ, 55; res asst astron, Harvard Univ, 55-58, res assoc, 58-61, lectr, 61-65; ASSOC PROF ASTRON, UNIV VA, 65- *Concurrent Pos:* Consult, Jet Propulsion Lab, Calif Inst Technol, 60-62; vis res off, Astrophys Br, Nat Res Coun Can, 70-71. *Mem:* Inst Elec & Electronics Engrs; Am Astron Soc; Int Astron Union. *Res:* Galactic structure. *Mailing Add:* Dept Astron Rm 314 Math Astron Bldg Univ Va Charlottesville VA 22903

GOLDSTEIN, SANDU M, b Iasi, Romania, May 22, 36; m 66; c 1. PHARMACEUTICAL CHEMISTRY. *Educ:* Polytech Inst Iasi, MSc, 58, PhD(anal chem), 75. *Prof Exp:* Chem engr, process eng res & develop, Antibiotics & Vitamines Lab, Iasi, Romania, 58-75; chemist res & develop anal chem, ICN Pharmaceut, Can, 76-77, assoc dir quality control, 77-79, quality control & regulatory affairs, 79-80, dir regulatory affairs, 80-81, dir sci & regulatory affairs, 81-86; DIR RES, ADMIN & PLANNING, MERCK FROSST CTR THERAPEUT RES, CAN, 86- *Concurrent Pos:* Nominee, Comt Revision, US Pharmacopeia, 80-85 & 85-90. *Mem:* Chem Inst Can; Am Chem Soc; NY Acad Sci; Can Soc Chem Eng; Royal Soc Chem. *Res:* Applied analytical chemistry in biosynthesis; isolation and purification of cobalt corrinoids (vitamin B12 group substances); biotechnology; biosynthesis: manufacturing of pharmaceutical products (antibiotics, vitamins). *Mailing Add:* Merck Frosst Ctr Therapeut Res PO Box 1005 Pointe Claire-Dorval PQ H9R 4P8 Can

GOLDSTEIN, SELMA, b New York, NY, Feb 28, 49. HYDRODYNAMICS, SOLID STATE PHYSICS. *Educ:* NY Univ, AB, 69; Yale Univ, MPhil, 71, PhD(appl physics), 74. *Prof Exp:* Battelle Inst fel solid state physics, Battelle Columbus Labs, 74-76; MEM STAFF DETONATION PHYSICS, LOS ALAMOS SCI LAB, 77- *Mem:* Am Phys Soc. *Res:* Properties of materials at high pressure; physics of detonations; properties of explosives. *Mailing Add:* 22600 Savi Ranch Pkwy Yorba Linda CA 92686

GOLDSTEIN, SETH RICHARD, b New York, NY Dec 10, 39; m 65; c 1. BIOMEDICAL INSTRUMENTATION, DYNAMIC SYSTEMS & CONTROL. *Educ:* Mass Inst Technol, BS, 61, MS, 62, MS, 63, ScD(mech eng), 65. *Prof Exp:* Eng scientist, Aerospace Systs Div, RCA, 65-67; proj engr, Foster Miller Assocs, Consult Engrs, 67-71; mech engr, 71-80, CHIEF, MECH ENG SECT, BIOMED ENG & INSTRUMENTATION BR, DIV RES SERV, NIH, 80- *Concurrent Pos:* Assoc ed, Transaction J Dynamic Systs Measurement & Control, Am Soc Mech Engrs, 74-77 & J Biomed Eng, 77-83. *Honors & Awards:* IR 100 Award, 84. *Mem:* Fel Am Soc Mech Engrs. *Res:* Conception, analysis, design, development and usage of novel, complex biomedical instrumentation and clinical systems; miniaturized and minimally invasive techniques and sensors. *Mailing Add:* Biomed Eng & Instrumentation Br NIH Bldg 13 Rm 3W13 Bethesda MD 20892

GOLDSTEIN, SHELDON, b Augusta, Ga, Oct 24, 47. MATHEMATICS, PHYSICS. *Educ:* Yeshiva Univ, BA, 69, BS, 71, PhD(math),73. *Prof Exp:* PROF MATH, RUTGERS UNIV, 77- *Mem:* Am Math Asn. *Mailing Add:* Rutgers Univ New Brunswick NJ 08903

GOLDSTEIN, SIDNEY, b Utica, NY, Oct 18, 30; m 56; c 2. CARDIOLOGY. *Educ:* Cornell Univ, BA, 52, MD, 56. *Prof Exp:* Intern, NY Hosp-Cornell Med Ctr, 56-57, resident, 59-61, NY Heart Asn res fel, 61-62; chief cardiol, Rochester Gen Hosp, 62-74; from asst prof to assoc prof med, Univ Rochester, 66-74; HEAD CARDIOVASCULAR UNIT, HENRY FORD HOSP, DETROIT, 74-; PROF CLIN MED, UNIV MICH, 75- *Concurrent Pos:* Mem coun clin cardiol, Am Heart Asn. *Mem:* Am Fedn Clin Res; fel Am Col Physicians; fel Am Col Cardiol. *Res:* Cardiac physiology. *Mailing Add:* Henry Ford Hosp 2799 W Grand Blvd Detroit MI 48202

GOLDSTEIN, SIDNEY, b Philadelphia, Pa, Mar 27, 32; m 55; c 3. PHARMACOLOGY. *Educ:* Philadelphia Col Pharm, BSc, 54, MSc, 55, DSc, 58. *Prof Exp:* Sr res pharmacologist, Eaton Labs, Norwich Pharmacal Co, 58-59 & Lederle Labs, Am Cyanamid Co, 59-61; sr res pharmacologist, Nat Drug Co, 61-64, dir pharmacol, 64-67, dir prod develop, 67-70; sect head anti-inflammatory & respiratory sects & clin monitor, Med Res Dept, William S Merrell Co, 70-72, assoc group dir, Dept Clin Pharmacol, 72-73, exec asst to vpres res, Merrell-Nat Labs, 74-76, DIR PHARMACEUT SCI, MERRELL DOW PHARMACEUTICALS INC, 77- *Concurrent Pos:* Lectr, Philadelphia Col Pharm, 67-70; adj assoc prof pharmaceut, Univ Cincinnati, 84- *Mem:* Acad Pharmaceut Sci; Am Soc Clin Pharmacol & Therapeut; Am Pharmaceut Asn; Am Soc Pharmacol & Exp Therapeut; Soc Exp Biol & Med. *Res:* Respiratory research; topical steroids; anti-inflammatory, central nervous system, cardiovascular and biopharmaceutics drugs; pharmaceutics. *Mailing Add:* Merrell Dow Pharmaceuticals Inc 2110 E Galbraith Rd Cincinnati OH 45215

GOLDSTEIN, SOLOMON, b New York, NY, May 2, 29; m 54. MYCOLOGY. *Educ:* City Col New York, BA, 52; Univ Mich, MS, 53, PhD(bot), 59. *Prof Exp:* Instr bot, Univ Mich, 58-59; asst prof biol, State Univ NY Stony Brook, 59-60; res assoc microbiol, Yale Univ, 60-64; dep chmn, Grad Div, 64-69, actg dean, Sch Sci, 76-77, PROF BIOL, BROOKLYN COL, 69-; DEAN OF RES, GRAD CTR, CITY UNIV NY, 79- *Mem:* Bot Soc Am; Mycol Soc Am; Am Soc Microbiol; Am Soc Protozool; Brit Soc Gen Microbiol. *Res:* Physiology and morphology of marine phycomycetes. *Mailing Add:* Dean Grad Ctr Brooklyn Col 33 W 42nd St New York NY 10036

GOLDSTEIN, STANLEY P(HILIP), b Brooklyn, NY, Feb 3, 23; m 49; c 2. ASTRONAUTICAL ENGINEERING. *Educ:* Univ Okla, BS, 49; NY Univ, MS, 56; Polytech Inst Brooklyn, PhD(astronaut), 69. *Prof Exp:* Develop engr, Vapor Recovery Systs, Co, 50-52; proj engr, Aldison Res Labs, 52-54; prof eng, Hofstra Univ, 54-84, chmn dept, 57-69, 71-73 & 80-83, dir acad comput ctr, 71-74, assoc provost, 74-76, assoc dean, 77-79, dir, C Step Prog, 87-89, Emer Prof Eng, Hofstra Univ, 84-; PRES, TECHMARK INDUSTS INC, 89- *Concurrent Pos:* Dir domestic opers, Hofstra Int Trade & Develop Corp, 85-87. *Mem:* Sigma Xi. *Res:* Compressible flow; free convection heat transfer; computer science. *Mailing Add:* 21 Harvard St Westbury NY 11590

GOLDSTEIN, STUART FREDERICK, b Beloit, Wis, Sept 11, 39; m 78; c 2. CELL PHYSIOLOGY. *Educ:* Univ Minn, BA, 62; Calif Inst Technol, PhD(cell biol), 68. *Prof Exp:* Med Res Coun Gt Brit grant, Queen Elizabeth Col, London, 68-69; USPHS grant, Univ Calif, Berkeley, 69-71; asst prof zool, Univ Minn, Minneapolis, 71-77; ASSOC PROF GENETICS & CELL BIOL, UNIV MINN, ST PAUL, 77- *Concurrent Pos:* NSF grant, Univ Poitiers, France, 77-78. *Mem:* Am Soc Cell Biol; AAAS; Sigma Xi. *Res:* Flagellar and bacterial motility. *Mailing Add:* Dept Genetics & Cell Biol Univ Minn St Paul MN 55108

GOLDSTEIN, THEODORE PHILIP, b Baltimore, Md, Feb 23, 28; m 68. ORGANIC CHEMISTRY, GEOCHEMISTRY. *Educ:* Johns Hopkins Univ, BA, 51, MA, 58, PhD(chem), 61. *Prof Exp:* Chemist, US Army Chem Corps, 52-55; chemist, Sinai Hosp, Baltimore, 55-56; res chemist, Cent Res Lab, Mobil Oil Co, Inc, 61-63, sr res chemist, Mobil Res & Develop Corp, 63-75, ASSOC, MOBIL RES & DEVELOP CORP, 75- *Concurrent Pos:* Assoc ed, Organic Geochem, 77-82. *Mem:* AAAS; Sigma Xi; Am Chem Soc; NY Acad Sci; Catalysis Soc; Am Inst Chem. *Res:* Histochemistry; catalysis; organic geochemistry; origin of oil; low temperature catalytic reactions; enzyme models. *Mailing Add:* Mobil Res & Develop Corp Box 1025 Princeton NJ 08540

GOLDSTEIN, WALTER ELLIOTT, b Chicago, Ill, Nov 28, 40; m 62; c 2. PLANT CELL CULTURE, PHARMACEUTICALS. *Educ:* Ill Inst Technol, BS, 61; Mich State Univ, MBA, 68; Univ Notre Dame, MS, 71, PhD(chem eng), 73. *Prof Exp:* Develop engr cryogenics, Linde Div, Union Carbide Corp, 61-64; assoc proj engr pharmaceut & chem, Corp Eng Div, Miles Labs Inc, 64-67, assoc res scientist clin diag tests, Ames Div, 67-72, res scientist, 72-74, res supvr, 74-76, mgr chem eng res & pilot serv, Marschall Div, 76-78, dir chem eng res & pilot serv, biotechnol group res & develop, 78-82, vpres res & develop, biotechnol group, 82-87; VPRES & DIR RES, ESCA GENETICS CORP, 87- *Concurrent Pos:* Adj asst prof chem eng, Univ Notre Dame, 74-75; advisor qual assurance, Nat Acad Sci, 86-87, mkt & com develop, Nat Inst Standards & Technol, 87-91; indust eng consult, 87. *Mem:* Sigma Xi; AAAS; Am Inst Chem Engrs; NY Acad Sci; Inst Food Technologists. *Res:* Membrane separation; industrial applications of immobilized enzymes; optimization of fermentation and processing for micro-organisms and plant cells; computer control; instrumented biotesting; plant breeding and genetics; immunological-based tests; food ingredients and pharmaceuticals from plant cell culture. *Mailing Add:* Four E Court Lane Foster City CA 94404

GOLDSTICK, THOMAS KARL, b Toronto, Ont, Aug 21, 34; m 82. PHYSIOLOGICAL OXYGEN TRANSPORT. *Educ:* Mass Inst Technol, BS, 57, MS, 59; Univ Calif, Berkeley, PhD(eng sci), 66. *Prof Exp:* Res engr iron ore, Jones & Laughlin Steel Corp, 59-61; asst res engr mech eng & surg, Univ Calif, Berkeley & San Francisco, 66-67; from asst prof to assoc prof chem eng & biol sci, 67-81, PROF CHEM ENG, BIOMED ENG, NEUROBIOL & PHYSIOL, NORTHWESTERN UNIV, 81- *Concurrent Pos:* NIH spec res fel, Univ Calif, San Diego, 71-73; assoc staff mem, Evanston Hosp, 71-; mem bd dir, Biomed Eng Soc, 83-86; chmn Publ Bd, 85-87; mem exec comt, Int Soc Oxygen Transport Tissue, 86-89 & 90- *Mem:* Biomed Eng Soc; Int Soc Oxygen Transport Tissue (secy, 81-86); Am Inst Chem Engrs; Microcirc Soc; Asn Res Vision & Opthal. *Res:* Mass transport in biological and medical systems, applying traditional engineering techniques to living systems; oxygen transport in the cornea, retina and choroid of the eye; diabetes mellitus; oxygen transport tissue, retinal vascular disease and impaired oxygen transport; perfluorocarbon artificial blood and AIDS correlated with retinal microvascular disease; whole blood coagulation parameters by thrombelastography in diabetes. *Mailing Add:* Dept Chem Eng Northwestern Univ Evanston IL 60208-3120

GOLDSTINE, HERMAN HEINE, b Chicago, Ill, Sept 13, 13; m 41, 66; c 2. NUMERICAL ANALYSIS. *Educ:* Univ Chicago, BS, 33, MS, 34, PhD(math), 36. *Hon Degrees:* PhD, Lund Univ, Sweden, 74; DSc Adelphi Univ, 78 & Amherst Col, 78. *Prof Exp:* Assoc & instr math, Univ Chicago, 36-39; instr, Univ Mich, 39-45, asst prof, 45-50; asst proj dir, Electronic Comput Proj, 46-58, MEM, INST ADVAN STUDY, 50-; IBM FEL, THOMAS J WATSON RES CTR, IBM CORP, 69- *Concurrent Pos:* Dir sci develop, Math Sci, IBM Corp, 58-65 & Data Processing Div, 65-67, consult to dir res, 67-69; mem adv coun, Phys Sci Div, Univ Chicago; exec officer, Am Philos Soc, 84- *Honors & Awards:* Harry Goode Award, 79; Charter Pioneer Award, 82; Nat Medal Sci, 83. *Mem:* Nat Acad Sci; Am Math Soc; Math Asn Am; Am Philos Soc; AAAS. *Res:* Applied mathmatics; electronic computers; history of computers and mathematics. *Mailing Add:* Am Philos Soc 104 S 5th St Philadelphia PA 19106-3387

GOLDSTONE, ALFRED D, MECHANISM OF HORMONE ACTION, BIOCHEMISTRY. *Educ:* St Louis Univ, PhD(pharmacol), 64. *Prof Exp:* ASSOC PROF NEUROL & BIOCHEM, NORTHWESTERN MED SCH, 80- *Mailing Add:* Dept Neurol Mole Biol NW Univ Med Sch 333 E Huron St Chicago IL 60611

GOLDSTONE, JEFFREY, b Manchester, Eng, Sept 3, 33; m 80; c 1. THEORETICAL PHYSICS, ELEMENTARY PARTICLE PHYSICS. *Educ:* Cambridge Univ, BA, 54, PhD(theoret physics), 58. *Prof Exp:* Res fel, Trinity Col, Cambridge, 56-60; lectr appl math & theoret physics, Cambridge Univ, 61-76, reader math physics, 76; PROF PHYSICS, MASS INST TECHNOL, 77- *Concurrent Pos:* Staff fel, Trinity Col, Cambridge, 62-81; dir, Ctr Theoret Physics, Mass Inst Technol, 83-89. *Honors & Awards:* Heineman Prize, Am Phys Soc, 81; Guthrie Medal, Inst Physics, London, 83. *Mem:* Am Acad Arts & Sci; Am Phys Soc; Royal Soc London. *Res:* Theoretical particle physics and quantum field theory. *Mailing Add:* 6-313 Mass Inst Technol Cambridge MA 02139

GOLDSTONE, PHILIP DAVID, b Brooklyn, NY, Mar 5, 50; m 72, 90. PHYSICS. *Educ:* Polytech Inst Brooklyn, BS, 71, MS, 72; State Univ NY, Stony Brook, PhD(physics), 75. *Prof Exp:* Appointee nuclear physics, Los Alamos Nat Lab, Univ Calif, 76-77, mem staff laser & plasma physics, 77-81, group leader, laser matter interaction exp, 81-89, prog mgr inertial confinement fusion exp, 87-89, MEM STAFF, WEAPONS TECHNOL PROG, LOS ALAMOS NAT LAB, UNIV CALIF, 89- *Mem:* Am Phys Soc; Sigma Xi; Optical Soc Am. *Res:* Study of interactions of matter with intense laser beams; physics of inertial confinement fusion including physics of dense plasmas. *Mailing Add:* 103 Mateo Circle N Sante Fe NM 87501-4029

GOLDTHWAIT, DAVID ATWATER, b Providence, RI, Nov 7, 21; m 49; c 4. BIOCHEMISTRY, MOLECULAR BIOLOGY. *Educ:* Columbia Univ, MD, 45. *Prof Exp:* From asst prof to assoc prof biochem & med, 51-68, PROF BIOCHEM, SCH MED, CASE WESTERN RESERVE UNIV, 68-, PROF MED, 78. *Mem:* Am Soc Biol Chemists; Am Soc Clin Invest; Sigma Xi. *Res:* Nucleic acids; metabolism; cancer. *Mailing Add:* Dept Biochem Sch Med Case Western Reserve Univ Cleveland OH 44106

GOLDTHWAIT, R(ICHARD) G(RAHAM), b Duluth, Minn, Feb 9, 19; m 49; c 2. CHEMICAL ENGINEERING. *Educ:* Pa State Univ, BS, 40. *Prof Exp:* Asst, Pa State Univ, 40-41; buyer, Westinghouse Elec Corp, 41-43; proj leader & chem engr, Gulf Res & Develop Co, 43-49, asst group leader, 49-53, group leader, 53-56, asst sect head, 56-60, sect supvr, 60-65, staff engr, 65-66, tech consult, 66-72, Gulf Res Lab, Netherlands, 69-72, dir planning & coord, refinery dept, Gulf Oil-Eastern Hemisphere, London, 72-75, safety adv, 75-76, sr staff engr, Gulf Sci & Technol Co, 76-82; RETIRED. *Mem:* Am Chem Soc; Am Inst Chem Engrs. *Res:* Pilot plant work in petroleum refining and related fields; catalytic reforming; thermal and catalytic cracking; isomerization; distillation; hydrogenative processes; correlation development; process control; coal liquefaction. *Mailing Add:* 114 Kinvara Ct Pittsburgh PA 15237

GOLDTHWAIT, RICHARD PARKER, b Hanover, NH, June 6, 11; m 37; c 4. GEOLOGY. *Educ:* Dartmouth Col, AB, 33; Harvard Univ, MA, 37, PhD(geol), 39. *Prof Exp:* Instr geol, Dartmouth Col, 34-35; asst, Harvard Univ, 36-38; instr, Brown Univ, 39-40, asst prof, 41-42; tech consult, Air Transp Command, US Air Force, 42-43, mat engr, Air Materiel Command, 43-46; from assoc prof to prof, 46-77, EMER PROF GEOL, OHIO STATE UNIV, 77- *Concurrent Pos:* Dir, Inst Polar Studies, Ohio State Univ, 60-65, chmn dept geol, 65-69, actg dean, Col Math & Phys Sci, 71-72; Fulbright res scholar, NZ, 57; NATO sr fel, 69; chmn, Alaska Earthquake Comn, Nat Acad Sci, 64-74, Comt Int Quaternary Assoc, 74-77; auditor, Int Quaternary Assoc, 73; mem lectr, Antarctican Soc, 84; consult, Pollution Control Bd, Manatee County, 85. *Honors & Awards:* Antarctic Medal, US Cong, 68. *Mem:* AAAS; fel Geol Soc Am; fel Arctic Inst NAm; hon mem Glaciol Soc. *Res:* Glacial geology of western Ohio and central New Hampshire; glacier studies in Alaska, the Alps, Baffin Island, Greenland and Antarctica; river floodplain studies in New Hampshire and Ohio; reconnaissance geomorphology of northeast Tibet, central Baffin Island, New Zealand, Antarctica; international geophysical year glaciology data reduction. *Mailing Add:* PO Box 656 Anna Maria FL 34216

GOLDTHWAITE, DUNCAN, b New York, NY, Mar 31, 27; m 56; c 4. PETROLEUM GEOLOGY. *Educ:* Oberlin Col, BA, 50; Harvard Univ, MA, 52. *Prof Exp:* Geologist, Chevron Oil Co, NDak & SDak, 52-54, geophysicist, Miss & NLa, 54-57, dist geologist, NLa, 57-58 & Mich & Appalachians, 58-64, sr geologist, SLa & Atlantic Coast Phosphates & Appalachians, 64-77, sr geologist, Tex Coast, 78-85; CONSULT GEOLOGIST, 85- *Mem:* Fel Geol Soc Am; Am Asn Petrol Geol. *Res:* Structure and stratigraphy as related to petroleum in Michigan, Appalachians, and Gulf Coast; phosphate investigations. *Mailing Add:* 4608 James Dr Metairie LA 70003

GOLDWASSER, EDWIN LEO, b New York, NY, Mar 9, 19; m 40; c 5. PHYSICS. *Educ:* Harvard Univ, BA, 40; Univ Calif, Berkeley, PhD(physics), 50. *Prof Exp:* Physicist, Degaussing Ships, Bur Ord, Dept Navy, 41-43, sr physicist, Navy Yard, Mare Island, 43-45; asst physics, Univ Calif, Berkeley, 46-49, res assoc, 50-51; dep dir, Fermi Lab, 67-78; assoc dir, Cent Design Group, Superconducting Super-Collider, 86-88; res assoc, Univ, Ill, Urbana, 51-53, from asst prof to prof, 53-67, vchancellor res & dean, Grad Col, 78-79, vchancellor acad affairs, 79-86, ACTG DIR, COMPUTER BASED EDUC RES LAB, UNIV ILL, URBANA, 90- *Concurrent Pos:* Consult, Phys Sci Study Comt, 56-61; Fulbright & Guggenheim fels, Univ Rome, 57-58; vchmn div phys sci, Nat Res Coun, 61-66, chmn, 66-68; mem panel high energy accelerator physics, Gen Adv Comt, AEC & President's Sci Adv Comt, 62-63; mem physics surv comt, Nat Acad Sci, 63-65, site selection comt, 65-66; mem sci adv comn, State of Ill, 67-70; mem comn on particles & fields, Int Union Pure & Appl Physics, 72-81, secy, 75-78, chmn, 78-81; mem, US-USSR Joint Coord Comt Fundamental Properties of Matter, 75-78 & Int Comn Future Accelerators, 78-81; chmn, Sci & Educ Adv comt, Univ Calif, 86- *Mem:* Fel Am Phys Soc; Fedn Am Sci; Sigma Xi. *Res:* Elementary particle physics; cosmic rays; electron and photon interactions; ion physics. *Mailing Add:* 612 W Delaware Urbana IL 61801

GOLDWASSER, EUGENE, b New York, NY, Oct 14, 22; m 49; c 3. BIOCHEMISTRY. *Educ:* Univ Chicago, SB, 43, PhD(biochem), 50. *Prof Exp:* Res assoc, 52-56, from asst prof to assoc prof, 56-63, PROF BIOCHEM, UNIV CHICAGO, 63-, CHMN COMT DEVELOP BIOL, 76- *Concurrent Pos:* Am Cancer Soc fel, Copenhagen Univ, 50-52; Guggenheim fel, Dept Biochem, Oxford Univ, 66-67. *Honors & Awards:* Ann Langer Award, Cancer Res, 87; Annual Prize of Int Soc Blood Purification, 88. *Mem:* AAAS;

Am Soc Biol Chem; Biochem Soc; Endocrine Soc; Int Soc Exp Hemat; Am Soc Hematology. *Res:* Biochemistry of differentiation; biochemistry of erythropoiesis. *Mailing Add:* Dept Biochem & Molecular Biol Univ Chicago 920 E 58 St Chicago IL 60637

GOLDWASSER, SAMUEL M, b Philadelphia, Pa. IMAGE PROCESSING & DISPLAY, COMPUTER ARCHITECTURE. *Educ:* Drexel Univ, BS, 74; Mass Inst Technol, MS, 76, PhD(elec eng), 79. *Prof Exp:* ASST PROF, COMPUT SCI & ELEC ENG, UNIV PA, 80-, CO-DIR, GEN ROBOTICS & ACTIVE SENSORY PROCESSING LAB & DIR, DIGITAL SYSTS LAB, 85- *Concurrent Pos:* IBM fel, 78; consult, Assoc Press, Providence Gravure Inc, Psychol Dept, Mass Inst Technol, OCR Syst, E I du Pont, Glassboro State Col & RCA. *Mem:* Sigma Xi; Inst Elec & Electronics Engrs. *Res:* Special purpose computer architecture and digital hardware design for image processing, graphics, machine perception and robotics; real-time interactive display and manipulation of 3-D medical objects. *Mailing Add:* Dynamic Digital Displays Inc 3508 Market St Philadelphia PA 19104

GOLDWATER, LEONARD JOHN, b New York, NY, Jan 15, 03; m 78. OCCUPATIONAL MEDICINE. *Educ:* Univ Mich, AB, 24; NY Univ, MD, 28, ScD(internal med), 37; Columbia Univ, MS, 41; Am Bd Internal Med, dipl; Am Bd Prev Med, dipl. *Prof Exp:* Intern & resident physician, Bellevue Hosp, NY, 29-32; instr med, NY Univ Col Med, 32-36, instr & asst prof prev med, 38-41; prof indust hyg, Sch Pub Health, Columbia Univ, 46-52, prof occup med, 52-69; vis prof environ sci, Univ NC, Chapel Hill, 69-73; prof community health sci, Med Ctr, Duke Univ, 69-73, chief occup med, 73-79, consult, 79-83; EMER PROF, SCH PUB HEALTH, COLUMBIA UNIV, 69-; EMER PROF, MED CTR, DUKE UNIV, 83- *Concurrent Pos:* Sr indust hyg physician, NY State Dept Labor, 36-38; consult, Div Indust Hyg, Off Voc Rehab, US Dept Health, Educ & Welfare, 52-60 & AEC, 48-49; Harben Lectr, Royal Inst Pub Health & Hyg, 64; praelector, Univ St Andrews, 66; mem, Exp Adv Panel Social & Occup Health, WHO, 51-73; vis scholar, Duke Univ, 68-69; mem, Nat Air Pollution Res & Develop Adv Coun, 70-72; adj prof environ sci, Univ NC, Chapel Hill, 73-88. *Honors & Awards:* William S Knudsen Award, Am Occup Med Asn, 80; Robert Kehoe Award, Am Acad Occup Med, 83. *Mem:* Fel AMA; fel Am Pub Health Asn; Am Indust Hyg Asn; Am Occup Med Asn; fel Am Acad Occup Med (pres, 59). *Res:* Industrial toxicology; mercury in the environment. *Mailing Add:* Duke Univ Med Ctr Durham NC 27710

GOLDWATER, WILLIAM HENRY, b Plattsburgh, NY, Apr 4, 21; m 48; c 2. BIOCHEMISTRY. *Educ:* Columbia Univ, AB, 41, PhD(biochem), 47. *Prof Exp:* Res technician, Mt Sinai Hosp, NY, 41-42; asst, Columbia Univ, 42-43, div war res, 43-45, med sch, 46-47; res chemist, Mt Sinai Hosp, 47-48; res assoc, Sch Med, Tulane Univ, 48-49, assoc, 49-51, asst prof biochem & med, 51; radiol biochemist, US Naval Radiol Defense Lab, 52-59; exec secy, Metab Study Sect, Div Res Grants, NIH, 59-62, chief, Spec Res Projs, Nat Heart Inst, 62-69, assoc dir extramural Progs, Nat Inst Environ Health Sci, 69-70, SPEC ASST TO ASSOC DIR EXTRAMURAL RES & TRAINING, & TO DIR, EXTRAMURAL PROGS MGT OFF, OFF DIR, NIH, 70- *Concurrent Pos:* Mem res & develop & grants study groups, Comn Govt Procurement, 71-72; mem, Coun Epidemiol, Am Heart Asn. *Honors & Awards:* Super Serv Award, Pub Health Serv, 89. *Mem:* Fel AAAS; Am Chem Soc; fel Am Col Cardiol; fel Am Heart Asn; NY Acad Sci; Soc Res Adminr; Sigma Xi; Nat Assistance Mgt Asn. *Res:* Biomedical research administration and policy; peer review; statutory, regulation and legal issues in research and development contract and grant policies and procedures. *Mailing Add:* NIH Bldg 1 Rm 328 Bethesda MD 20892

GOLDWEIN, MANFRED ISAAC, b Bochum, Ger, Apr 19, 24; US citizen; m 49; c 3. INTERNAL MEDICINE, HEMATOLOGY. *Educ:* Univ Del, BS, 50; Univ Vt, MD, 54. *Prof Exp:* Asst instr, 55-58, instr, 58-60, assoc, 60-62, asst prof, 62-75, ASSOC PROF MED, UNIV PA, 75- *Mem:* Am Col Physicians; Am Soc Hemat. *Res:* Clinical hematology; vitamin B12 metabolism; ineffective erythropoiesis. *Mailing Add:* Hosp Univ Pa 3400 Spruce St Philadelphia PA 19104

GOLDWHITE, HAROLD, b London, Eng, Dec 25, 31; m 58; c 4. INORGANIC CHEMISTRY. *Educ:* Univ Cambridge, BA, 53, PhD(chem), 56. *Prof Exp:* Res assoc chem, Cornell Univ, 56-58; lectr, Univ Manchester, 58-62; from asst prof to assoc prof, 62-67, PROF CHEM, CALIF STATE UNIV, LOS ANGELES, 67- *Mem:* Am Chem Soc; Royal Chem Soc; Sigma Xi; AAAS. *Res:* Chemistry of compounds containing phosphorus; fluorine chemistry; organometallic chemistry. *Mailing Add:* Dept Chem Calif State Univ Los Angeles CA 90032

GOLDWYN, ROGER M(ARTIN), b Tulsa, Okla, Dec 18, 36; m 64; c 2. COMPUTER SCIENCE, APPLIED MATHEMATICS. *Educ:* Rice Univ, BA, 58, BS, 59, MS, 60; Harvard Univ, AM, 61, PhD(appl math), 64. *Prof Exp:* Asst prof eng, Rice Univ, 64-66; mem res staff, Thomas J Watson Res Ctr, Int Bus Mach Corp, 67-70, mgr, Comput Sci Dept, 71-77, sr mgr prototype systs, Thomas J Watson Res Ctr, IBM Corp, 77-84, mgr adv systs prod, IBM, 84-87; PROG MGR, IBM ENTRY SYSTS DIV, BOCA RATON, FL. *Concurrent Pos:* NSF grant, 64-66; consult, Guid & Control Div, Manned Spacecraft Ctr, NASA, Vet Admin Hosp, Houston & Esso Prod Res Lab, 64-66; asst clin prof, Albert Einstein Col Med & lectr, Dept Elec Eng, Univ Conn, 69-71. *Mem:* Sigma Xi; Inst Elec & Electronics Engrs. *Res:* Stability problems in control theory; wave propagation; multivariate data analysis; biomedical data analysis; computer sciences; communications. *Mailing Add:* 5428 NW 20th Ave Boca Raton FL 33496

GOLDYNE, MARC E, b San Francisco, Calif, Oct 15, 44. DERMATOLOGY, PROSTTAGLANDINS. *Educ:* Univ Calif, San Francisco, MD, 70; Univ Minn, PhD(dermat), 80. *Prof Exp:* ASSOC PROF MED, UNIV CALIF, SAN FRANCISCO, 84-; ASST CHIEF DERMAT, SAN FRANCISCO VET ADMIN MED CTR, 85- *Mailing Add:* Dept Med & Dermat Univ Calif 4150 Clement St San Francisco CA 94121

GOLE, JAMES LESLIE, b Chicago, Ill, Sept 17, 45; m 70. HIGH TEMPERATURE CHEMICAL PHYSICS. *Educ:* Univ Calif, Santa Barbara, BS, 67; Rice Univ, PhD(chem), 71. *Prof Exp:* Asst prof chem, Mass Inst Technol, 73-77; from assoc prof to prof chem, 77-83, PROF PHYSICS, GA INST TECHNOL, 83- *Concurrent Pos:* NSF fel, 71-73; mem, Nat Res Coun, 85-; consult, Sanders Assocs, 85- *Honors & Awards:* Sustained Res Award, Sigma Xi. *Mem:* Am Chem Soc; Am Physical Soc; AAAS; Sigma Xi. *Res:* Laser induced fluorenscence applied to the study of small metal clusters and the products of their oxidation; chemiluminescent techniques applied as a probe of metal cluster and surface oxidation, molecular electronic structure, ultrafast energy transfer and molecular dynamics; chemical lasers; quantum chemistry. *Mailing Add:* Dept Physics Ga Inst Technol Atlanta GA 30332

GOLEMAN, DENZIL LYLE, b Coles Co, Ill, Jan 3, 24; m 52; c 3. ECONOMIC ENTOMOLOGY. *Educ:* Eastern Ill State Univ, BS, 49; Iowa State Univ, MS, 51, PhD(entom), 54. *Prof Exp:* Assoc, Agr Exp Sta & Agr Exten Serv, Iowa State Univ, 51-54; asst exten entomologist, Ohio State Univ, 54-56, exten entomologist, 56-59; entomologist, Am Cyanamid Co, 59-65; exten specialist agr chem, 65-68, chmn dept entom, 68-83, PROF DEPT ENTOM, OHIO STATE UNIV, 83- *Mem:* AAAS; Entom Soc Am. *Res:* Management of insect pests; use of insecticides including benefits and risks. *Mailing Add:* Dept Entom Col Biol Sci Ohio State Univ Columbus OH 43210

GOLEMBA, FRANK JOHN, b Hamilton, Ont, Dec 3, 43; m 67; c 2. POLYMER CHEMISTRY, REINFORCED PLASTICS. *Educ:* Univ Toronto, BSc, 65, MSc, 67, PhD(polymer chem), 70. *Prof Exp:* Res scientist & mgr anal, polyester chem & res & develop, 70-89, GEN MGR MIA CHEM & MGR TEXTILES & REINFORCEMENTS MKT, 89- *Mem:* Soc Plastics Engrs; Am Chem Soc; fel Chem Inst Can. *Res:* Polymer synthesis, structure and properties; reinforced polymers; polymer photochemistry; chemistry of the matrix reinforcement interface; rheology and cure behavior of thermosetting resins; unsaturated polyesters; analytical research characterizing glass bonding thermoset resin behavior; cure and high temperature behavior; industrial hygiene. *Mailing Add:* Fiberglas Can Inc 4100 Yonge St Suite 600 Willowdale ON M2P 2B6 Can

GOLES, GORDON GEORGE, b Chicago, Ill, Mar 6, 34; m 63; c 2. GEOTHERMAL ENERGY, ARCHAEOLOGY. *Educ:* Harvard Univ, AB, 56; Univ Chicago, PhD(chem), 61. *Prof Exp:* Asst res chemist, Univ Calif, San Diego, 61-62, asst prof chem, 62-67; assoc prof, 67-72, dir ctr volcanology, 71-74, PROF CHEM & GEOL, UNIV ORE, 72- *Mem:* AAAS; Sigma Xi. *Res:* Origin and history of the solar system; trace-element geochemistry; activation analysis; archaeology; origin of ores. *Mailing Add:* Dept Geol Univ Ore Eugene OR 97403

GOLESTANEH, AHMAD ALI, b Isfahan, Iran, June 2, 20; nat US; m 51; c 3. THEORETICAL PHYSICS. *Educ:* Univ Tehran, MSc, 42; Univ Manchester, PhD(physics of metals), 52. *Prof Exp:* Chief tech officer, Iranian State Rwys, 42-47, chief engr, 52-54; sr res officer, Brit Welding Res Asn, 55-57; res assoc, Gen Elec Co, Pa, 57-59; asst prof physics, Utica Col, 59-60; asst prof, Polytech Inst Brooklyn, 61-64; res worker, Columbia Univ, 64-66; from assoc prof to prof physics, Mt Union Col, 66-75; resident scientist, Argonne Nat Lab, 75-81; RETIRED. *Concurrent Pos:* Instr, Univ Tehran, 42-46 & 53-54, Behrend Ctr, Pa State Univ, 59 & Fordham Univ, 63; consult, Gen Eng Lab, Gen Elec Co, NY, 60; Nat Found grants; vis scholar, Northwestern Univ, 81- *Mem:* Am Phys Soc. *Res:* Theory of elementary particles; field theory; nuclear physics; materials science. *Mailing Add:* 1010 E Hyde Park Blvd No 2 Chicago IL 60615

GOLET, FRANCIS CHARLES, b Middletown, Conn, Nov 20, 45; m 72; c 1. WILDLIFE ECOLOGY, WETLAND ECOLOGY. *Educ:* Brown Univ, BA, 67; Cornell Univ, MS, 69; Univ Mass, PhD(wildlife biol), 73. *Prof Exp:* ASSOC PROF NATURAL RESOURCES SCI, UNIV RI, 72- *Mem:* Wildlife Soc; Wilson Ornithol Soc; Ecol Soc Am. *Res:* Wetland classification and evaluation; wetland plant ecology and wetland wildlife ecology; wetland habitat management. *Mailing Add:* 313 Woodward Hall Univ RI Kingston RI 02881

GOLIBER, EDWARD WILLIAM, b Buffalo, NY, Nov 21, 20; m 52; c 4. CHEMISTRY. *Educ:* Canisius Col, BS, 43; Mass Inst Technol, PhD(chem), 48. *Prof Exp:* Lab technician, Nat Aniline Div, Allied Chem & Dye Corp, NY, 41-42; res chemist, Linde Air Prod Co, NY, 43-46; sr res chemist, Carboloy Dept, Gen Elec Co, 48-58, consult engr, Carboloy Systs Dept, 58-60, mgr mat res, 60-83; CONSULT, 83- *Concurrent Pos:* Consult, 83- *Mem:* Am Chem Soc; Am Ceramic Soc. *Res:* Cemented carbides; powder metallurgy; oxides; refractory materials; inorganic chemistry. *Mailing Add:* 1722 Green Leaf Dr Royal Oak MI 48067

GOLIBERSUCH, DAVID CLARENCE, b Buffalo, NY, Jan 20, 42; m 68; c 2. KNOWLEDGE BASED SYSTEMS, TECHNICAL MANAGEMENT. *Educ:* Rensselaer Polytech Inst, BS, 63; Univ Pa, MS, 65, PhD(physics), 69. *Prof Exp:* NSF fel physics, Imp Col Sci & Technol, Univ London, 69-70; physicist, Gen Elec Res & Develop Ctr, 70-74, mgr tech relations, 74-75, Energy Sci Br, 75-77, Signal Lab, 77-83, consult, strategic anal & planning, 83-84, INFO SCIENTIST, GEN ELEC CORP RES & DEVELOP, 85- *Mem:* Inst Elec & Electronics Engrs; Am Asn Artificial Intel; Asn Comput Mach. *Res:* Information systems; knowledge based systems; artificial intelligence; expert systems; physics. *Mailing Add:* Corp Res & Develop Gen Elec Co PO Box 8 Schenectady NY 12301

GOLIKE, RALPH CROSBY, physical chemistry, for more information see previous edition

GOLIN, STUART, b Chicago, Ill, Jan 15, 38; m 70. IMAGE COMPRESSION & IMAGE PROCESSING, SOFTWARE SYSTEMS. *Educ:* Univ Chicago, MSc, 59, PhD(physics), 63. *Prof Exp:* Res assoc physics, Univ Chicago, 63-64; NSF fel, Nuclear Res Ctr, Saclay, France, 64-65; asst prof metall eng, Univ Ill, 65-66; asst prof physics, Univ Pittsburgh, 66-69; mem tech staff, 70-88, INTEL DAVID SARNOFF RES CTR, 88- *Mem:* Asn Comput Mach; Inst Elec & Electronics Engrs Computer Soc. *Res:* Image processing; computer systems. *Mailing Add:* 626 Dutch Neck Rd Hightstown NJ 08520

GOLINKIN, HERBERT SHELDON, b Chicago, Ill, May 24, 40; m 61; c 2. PHYSICAL ORGANIC CHEMISTRY, POLYMER CHEMISTRY. *Educ:* Johns Hopkins Univ, AB, 61; Univ Alta at Calgary, PhD(phys org chem), 66. *Prof Exp:* Fel phys org chem, Div Pure Chem, Nat Res Coun Can, 66-68; asst prof chem, Univ Minn, 68-70; res chemist, Res & Develop Dept, Amoco Chem Corp, 70-83, STAFF RES CHEMIST, RES & DEVELOP DEPT, AMOCO PETROL ADDITIVES CO, 83- *Concurrent Pos:* Petrol Res Fund grant, 68-70; counr, Am Chem Soc, 82- *Mem:* Am Chem Soc; Chem Inst Can; Soc Petrol Engrs; Can Soc Chem. *Res:* Aqueous solutions; reaction kinetics; reaction mechanisms; piezochemistry; effects of external fields on chemical reactions; condensation polymers; oil production chemicals; petroleum additives; addition polymers; lubricants. *Mailing Add:* Amoco Petrol Additives Co PO Box 3011 Naperville IL 60566-7011

GOLL, DARREL EUGENE, b Garner, Iowa, Apr 19, 36; m 58; c 3. MOLECULAR BIOLOGY, BIOCHEMISTRY. *Educ:* Iowa State Univ, BS, 57, MS, 59; Univ Wis, PhD(biochem), 62. *Prof Exp:* From asst prof to prof biochem, Iowa State Univ, 62-76; PROF BIOCHEM, UNIV ARIZ, 76- *Concurrent Pos:* USPHS res grant, 64-67 & 69-; NIH spec fel, Univ Calif, Los Angeles, 66-67 & Univ Oxford, Eng, 72. *Honors & Awards:* Samuel Cate Prescott Res Award, Inst Food Technol, 71; Distinguished Res Award, Am Meat Sci Asn, 72; Meats Res Award, Am Soc Animal Sci, 73. *Mem:* AAAS; Am Chem Soc; Biophys Soc; Am Soc Biochem & Molecular Biol. *Res:* Chemistry of muscle proteins; muscle contraction; protein turnover and intracellular proteases; cell motility; connective tissue protein; protein structure; nucleic acid and protein biosynthesis. *Mailing Add:* Muscle Biol Group Univ Ariz Tucson AZ 85721

GOLL, ROBERT JOHN, b Milwaukee, Wis, Jan 14, 26. PHYSICAL CHEMISTRY. *Educ:* Beloit Col, BS, 47; Northwestern Univ, BS, 48; Univ Wis, PhD(chem), 60. *Prof Exp:* Res chemist cent res dept, E I du Pont de Nemours & Co, Inc, Del, 59-69; chmn dept, 70-75, PROF PHYSICS, MIAMI-DADE JR COL, 69- *Mem:* AAAS. *Res:* Molecular structure; science education. *Mailing Add:* Dept Physics Miami-Dade Jr Col 11380 NW 27th Ave Miami FL 33167

GOLLAHALLI, SUBRAMANYAM RAMAPPA, b Sadali, India, Nov 26, 42; m 67; c 2. MECHANICAL ENGINEERING, PROPULSION ENGINEERING. *Educ:* Univ Mysore, BE, 63; Indian Inst Sci, ME, 65; Univ Waterloo, MASc, 70, PhD(mech eng), 73. *Prof Exp:* Lectr internal combustion eng, Indian Inst Sci, 65-68; res & develop engr mech eng, John Fowlers Ltd, India, 68; res asst prof, Univ Waterloo, 73-75, asst prof, 75-76; from asst prof to assoc prof, 76-84, PROF MECH ENG, UNIV OKLA, 84- *Concurrent Pos:* Fel Univ Waterloo, 73; Nat Res Coun Can res grant, 75-76; Nat Sci Found grant, 84-87; Dept Energy grant, 87-89. *Honors & Awards:* Robert Angus Medal, Can Soc Mech Eng, 78; Ralph Teetor Award, Soc Automotive Engrs, 78; Haliburton Distinguished Lectr, 84. *Mem:* Am Soc Mech Engrs; Soc Automotive Engrs; Int Combustion Inst; Soc Eng Educrs; Am Soc Aeronaut Engrs. *Res:* Combustion; fires; internal combustion engines; gas turbines. *Mailing Add:* Sch Aerospace Mech & Nuclear Eng Univ Okla Norman OK 73019

GOLLAMUDI, RAMACHANDER, b India, June 27, 34; m 60; c 2. ANTITHROMBOTIC DRUGS, DRUG-ENZYME INTERACTIONS. *Educ:* Osmania Univ, India, PhD(chem), 60; Fairleigh Dickinson Univ, MBA, 76. *Prof Exp:* Group leader drug metab, USV Pharmaceut Corp, 65-71; sr scientist drug disposition, Warner-Lambert Co, 71-75; sect head biotransformation & anal chem, 75-76, ASSOC PROF MED CHEM, CTR HEALTH SCI, UNIV TENN, 76- *Concurrent Pos:* Res assoc biochem & Fulbright scholar, Northwestern Univ Sch Med, 62-65. *Mem:* Am Soc Pharmacol & Exp Therapeut; Am Chem Soc; Sigma Xi. *Res:* Drug metabolism; mechanism of drug action; chemical constitution and biological activity. *Mailing Add:* Dept Med Chem Univ Tenn Ctr Health Sci Memphis TN 38163

GOLLER, EDWIN JOHN, b Lawrence, Mass, Jan 28, 40; m 84; c 1. ORGANIC CHEMISTRY. *Educ:* Merrimack Col, BS, 61; Northeastern Univ, MS, 64; Univ NH, PhD(org chem), 69. *Prof Exp:* Instr chem, Merrimack Col, 63-66; from asst prof to assoc prof, 69-78, PROF CHEM, VA MIL INST, 78-, HEAD, CHEM DEPT, 89- *Mem:* Sigma Xi; Am Chem Soc. *Res:* Mechanistic organometallic chemistry; stereochemistry; organic synthesis. *Mailing Add:* RFDS Box 217 Lexington VA 24450

GOLLEY, FRANK BENJAMIN, b Chicago, Ill, Sept 24, 30; m 53; c 3. ECOLOGY. *Educ:* Purdue Univ, BS, 52; Wash State Univ, MS, 54; Mich State Univ, PhD(zool), 58. *Prof Exp:* Asst prof zool, Univ NC, 59; prof biol, NC Col, 59; instr zool, 59-62, asst prof, 62-64, assoc prof zool & dir Savannah River Ecol Lab, 64-68, prof zool & exec dir Inst Ecol, 68-81, dir, Inst Ecol 84-87, RES PROF ECOL & PROF ZOOL & ENVIRON DESIGN, UNIV GA, 81- *Concurrent Pos:* Mem, Orgn Trop Studies; pres, Intecol. *Mem:* AAAS; Ecol Soc Am; Int Soc Trop Ecol; Int Asn Ecol; Int Asn Landscape Ecol. *Res:* Ecosystem and tropical ecology; energy flow; mineral cycling; ecological succession; landscape ecology. *Mailing Add:* Inst Ecol Univ Ga Athens GA 30602

GOLLIN, SUSANNE MERLE, CYTOGENETICS, CELL BIOLOGY. *Educ:* Northwestern Univ, PhD(biol sci), 80. *Prof Exp:* ASST PROF PATH & PEDIAT & DIR, CYTOGENETICS LAB, ARK CHILDREN'S MED CTR, UNIV ARK MED SCI, LITTLE ROCK, 84- *Mailing Add:* Human Genetics Prog Grad Sch Univ Pittsburgh 130 De Soto St Pittsburgh PA 15261

GOLLMAR, DOROTHY MAY, b Big Bend, Wis, May 4, 27; m 62. MATHEMATICS. *Educ:* Univ Wis, BS, 49, MA, 51. *Prof Exp:* Teacher pub sch, Wis, 49-50 & Ill, 50-51; instr math, Monticello Col, 51-54; from asst prof to assoc prof, State Univ NY Col Cortland, 54-62; ASSOC PROF MATH, WIS STATE UNIV-WHITEWATER, 62- *Mem:* AAAS; Am Math Soc; Math Asn Am; Nat Coun Teachers Math; Am Asn Univ Profs. *Mailing Add:* Box 337 Rochester WI 53167

GOLLNICK, PHILIP D, exercise physiology; deceased, see previous edition for last biography

GOLLOB, FRED, b New York, NY, Oct 7, 27; m 51; c 3. PHYSICAL CHEMISTRY. *Educ:* Columbia Univ, BA, 50, MA, 51, PhD, 54. *Prof Exp:* Res chemist, Marshall & Newburgh Labs, E I du Pont de Nemours & Co, 53-58; group leader cent res lab, Air Reduction Co, Inc, 58-62; dir, 62-74, pres, 74-76, DIR, GOLLOB ANAL SERV, INC, 76- *Mem:* AAAS; Instrument Soc; Am Asn Consult Chemists & Chem Engrs; Am Chem Soc; Soc Appl Spectros; Sigma Xi. *Res:* Gas analysis. *Mailing Add:* 261 Highwood Rd Mountainside NJ 07092

GOLLOB, LAWRENCE, b Yonkers, NY. POLYMER CHARACTERIZATION, PRODUCTION EFFICIENCY ANALYSIS. *Educ:* State Univ NY, Stony Brook, BS, 74; Duke Univ, MS, 76; Ore State Univ, PhD(wood technol & polymer chem), 82. *Prof Exp:* Sr develop chemist, 82-86, RES & DEVELOP GROUP LEADER, GA-PAC RESINS, INC, 86- *Concurrent Pos:* Partic, bus, prod efficiency & qual teams. *Mem:* Am Chem Soc; Soc Plastics Engrs. *Res:* The interaction of chemical structure with performance of wood adhesive systems; technical management. *Mailing Add:* L F Bornstein Lab Ga-Pac Corp 2883 Miller Rd Decatur GA 30035

GOLLON, PETER J, b New York, NY. PARTICLE PHYSICS, HEALTH PHYSICS. *Educ:* Columbia Univ, BA, 63, MA, 65, PhD(physics), 69. *Prof Exp:* Physicist, 68-74, group leader and sr radiation safety officer, 74-77, liason physicist, Proton Dept, Fermi Nat Accelerator Lab, 78-79; physicist, Isabelle Proj, 79-82, PHYSICIST, SAFETY & ENVIRON PROTECTION DIV, BROOKHAVEN NAT LAB, 82- *Concurrent Pos:* Lectr, Sch High Energy Radiation Dosimetry & Protection, Erice, Italy, 75. *Mem:* Am Phys Soc; Sigma Xi. *Res:* Accelerator shielding, personnel dosimetry, neutron spectroscopy; accelerator radiation dosimetry. *Mailing Add:* 15 Eleanor Pl Huntington NY 11743

GOLLUB, JERRY PAUL, b St Louis, Mo, Sept 9, 44. NONLINEAR DYNAMICS & TURBULENCE. *Educ:* Oberlin Col, AB, 66; Harvard Univ, AM, 67, PhD(physics), 71. *Prof Exp:* From asst prof to prof, 70-87, KENAN PROF PHYSICS, HAVERFORD COL, 87- *Concurrent Pos:* Adj prof physics, Univ Pa, 81-; secy-treas, Div Fluid Dynamics, Am Phys Soc, 85-88; Sigma Xi nat lectr, 83-85; Guggenheim fel, 84-85; vis prof, Ecole Normale, Paris, 85; mem mat res adv comt, NSF, 86-87; provost, Haverford Col, 88-90; dir, Mid Atlantic Pew Sci Prog, 88-91. *Honors & Awards:* Am Phys Soc Award Res Undergrad Inst, 85. *Mem:* Fel Am Phys Soc; Fedn Am Scientists; AAAS; Sigma Xi. *Res:* Non-equilibrium condensed matter physics, especially pattern formation, hydrodynamic instabilities, turbulence and growth of solids; nonlinear dynamics and the origins of noise in physical systems; computer-based instrumentation for research and education. *Mailing Add:* Physics Dept Haverford Col Haverford PA 19041

GOLLUB, SEYMOUR, b Philadelphia, Pa, Feb 21, 25; m 50; c 2. NUCLEAR MEDICINE. *Educ:* Univ Pa, BA, 47; Jefferson Med Col, PhD(physiol), 54; Hahnemann Med Col, MD, 58; Am Bd Nuclear Med, cert, 73. *Prof Exp:* Res asst path, Mt Sinai Hosp, Philadelphia, 49-50; res assoc, Albert Einstein Med Ctr, 50-53, assoc dir res surg, Southern Div, 55-63, dir, Mandell Lab, 56-63; dir hemat, St Barnabas Hosp, 63-73, dir nuclear med, 73-; PVT PRACT. *Concurrent Pos:* Pvt & govt res grants physiol & hemat, 55-; dir coagulation lab, Hahnemann Hosp, 56-63, assoc res surg, Med Col, 58-60; consult, Roswell Park Mem Hosp, Buffalo, NY, 57-59; dir res lab, Hillman Med Ctr, 57-63; mem, Comn Plasma, Nat Res Coun, 68-71. *Mem:* Int Soc Hemat; Am Soc Hemat; Microcirculatory Soc; Am Soc Exp Path. *Res:* Hematology, coagulation, surgery circulation; enzymology. *Mailing Add:* 170 Berrian Dr New Rochelle NY 10804

GOLOD, WILLIAM HERSH, b New York, NY, June 28, 33; m 61; c 3. PHARMACY. *Educ:* Fordham Univ, BS, 54; St Louis Col Pharm, MS, 55; Purdue Univ, PhD(phys pharm), 58. *Prof Exp:* From asst prof to assoc prof, 58-66, dir pharm serv, 59-63, asst dean, 63-65, DEAN COL PHARM, MED UNIV SC, 65-, PROF PHARM, 66- *Mailing Add:* Col Pharm Med Univ SC 171 Ashley Ave Charleston SC 29425-2301

GOLOMB, FREDERICK M, b New York, NY, Dec 18, 24; m 54; c 2. SURGERY. *Educ:* Yale Univ, BS, 45; Rochester Univ, MD, 49; Am Bd Surg, dipl. *Prof Exp:* From asst prof to assoc prof, 56-77, dir tumor serv, Univ Hosp, 70-86, DIR SURG CHEMOIMMUNOTHER DIV, TUMOR SERV, MED CTR, NY UNIV, 67-, PROF CLIN SURG, SCH MED, 77- *Mem:* AMA; Am Asn Cancer Res; fel Am Col Surg; Soc Head & Neck Surg; Soc Surg Alimentary Tract; Sigma Xi. *Res:* Cancer surgery, immunotherapy and chemotherapy; regional techniques for administration of cancer chemotherapeutic agents; techniques of reconstructive surgery following radical exenterations; tumor immunology. *Mailing Add:* 910 Fifth Ave New York NY 10021

GOLOMB, HARVEY MORRIS, b Pittsburgh, Pa, Feb 13, 43; m 65; c 2. HEMATOLOGY, ONCOLOGY. *Educ:* Univ Chicago, BA, 64; Univ Pittsburgh, MD, 68; Am Bd Internal Med, dipl, 75. *Prof Exp:* Intern internal med, Boston City Hosp, 68-69; resident internal med, Baltimore City Hosp, 71-72; fel med genetics, Johns Hopkins Univ Hosp, 72-73; fel hemat & oncol, 73-75, asst prof, 75-79, ASSOC PROF, UNIV CHICAGO HOSP, 79-, CHIEF HEMAT & ONCOL, 81- *Mem:* Am Soc Human Genetics; AAAS; Electron Micros Soc Am; Am Soc Hemat; Am Fedn Clin Res. *Res:* Hairy cell leukemia; acute leukemia and malignant lymphomas; surface characteristics of malignant cells; cells; staging and chemotherapy of lung cancer. *Mailing Add:* Univ Chicago Hosps Box 420 5841 S Maryland Ave Chicago IL 60637

GOLOMB, MICHAEL, b Munich, Germany, May 3, 09; nat US; m 39; c 2. MATHEMATICS. *Educ:* Univ Berlin, PhD(math), 33. *Prof Exp:* Res assoc elec eng, Cornell Univ, 39-42, instr math, 40-42; instr, Purdue Univ, 42-44; res engr & chief anal sect, Franklin Inst, 44-46; assoc prof, 46-50, prof, 50-76, EMER PROF MATH, PURDUE UNIV, 76- *Concurrent Pos:* Vis prof math res ctr, Univ Wis, 56-57, 58, 67, 70 & 78, Brown Univ, 75; distinguished vis prof, San Diego State Univ, 84-85; mem, Coun Am Math Soc (73-75); consult, appl math div, Argonne Nat Lab, 61-70. *Mem:* Am Math Soc; Math Asn Am; Soc Indust & Appl Math; fel AAAS; Sigma Xi. *Res:* Approximation theory; differential and integral equations; applied mathematics. *Mailing Add:* Dept Math Purdue Univ West Lafayette IN 47907

GOLOMB, SOLOMON WOLF, b Baltimore, Md, May 31, 32; m 56; c 2. DISCRETE MATHEMATICS, INFORMATION THEORY & CODING. *Educ:* Johns Hopkins Univ, AB, 51; Harvard Univ, AM, 53, PhD(math), 57. *Prof Exp:* Fulbright fel, Univ Oslo, 55-56; sr res engr, Jet Propulsion Lab, 56-58, leader, Info Processing Group, 58-60, asst chief, Telecommun Res Sect, 60-63; vprovost res, 86-89, PROF ELEC ENG & MATH, UNIV SOUTHERN CALIF, 63- *Concurrent Pos:* Consult, aerospace & electronics indust, 54-; res fel inst commun sci, Mass Inst Technol, 59; mem Comn C, Int Sci Radio Union, 60-, US deleg gen assembly; vis prof, Calif Inst Technol, 71-72; fel, Sackler Inst Advan Study, Tel Aviv Univ, 90. *Honors & Awards:* Shannon Lectr, Inst Elec & Electronics Engrs & Int Symp Brighton, Eng, 85. *Mem:* Nat Acad Eng; Am Math Soc; Math Asn Am; Soc Indust & Appl Math; fel Inst Elec & Electronics Engrs; Int Union Radio Sci; fel AAAS; Am Asn Univ Prof. *Res:* Applications of mathematical techniques to problems involving the storage, transfer, communication of information and computer technology; social implications of developments in science and technology; technology forecating, particularly with reference to communications and computer technology; probability; prime number theory; combinatorial analysis, mathematical games and modelling; author of five books; author of over 250 technical articles. *Mailing Add:* Powell Hall 506 Univ Southern Calif Los Angeles CA 90089-0272

GOLOMSKI, WILLIAM ARTHUR, APPLIED MATHEMATICS, STATISTICS. *Educ:* Milwaukee Sch Eng, MSEM, 69; Univ Chicago, MBA, 72; Columbia Col, BA, 87. *Prof Exp:* PRES, W A GOLOMSKI & ASSOC, 70- *Concurrent Pos:* Mgt consult, 49-70; lectr, Univ Wis, 58, mem adv coun, Sch Educ, 58-64; consult, US Bur Census, 60; mem, Comn on Prod Qual, 67-70; sr lectr bus policy, Univ Chi, 89- *Honors & Awards:* Testimonial Award, Am Soc Qual Control, 70; Edwards Medal, 76; W A Golomski Medal, Am Soc Qual Control, 86. *Mem:* Fel AAAS; fel Am Soc Qual Control (vpres, 64-66, pres, 66-67); fel NY Acad Sci; Opers Res Soc Am; fel Am Statist Asn; Int Acad for Quality (vpres, 91-); fel IIE, 89. *Res:* Industrial operations research; marketing; management science; mathematical programming; research & development management; applied mathematics and statistics in quality, nutrition, financial growth and sociology. *Mailing Add:* W A Golomski & Assoc 59 E Van Buren St Chicago IL 60605

GOLOSKIE, RAYMOND, b Providence, RI, Oct 30, 30; m 54; c 2. NUCLEAR PHYSICS. *Educ:* Brown Univ, ScB, 53; Harvard Univ, MA, 55, PhD(physics), 61. *Prof Exp:* Instr physics, Colgate Univ, 59-62; asst prof, 62-74, ASSOC PROF PHYSICS, WORCESTER POLYTECH INST, 74- *Mem:* Am Phys Soc. *Res:* Experiments in electron and photon interactions with nuclei at intermediate energies; ion micro probes. *Mailing Add:* Dept Physics Worcester Polytech Inst 100 Institute Rd Worcester MA 01609

GOLOVIN, MICHAEL N, b Harbin, China, June 30, 23; US citizen; m 56; c 3. ENGINEERING PHYSICS. *Educ:* Univ Calif, Berkeley, BS, 52. *Prof Exp:* Physicist rocket propulsion, Naval Weapons Ctr, China Lake, Calif, 52-53; thermodynamicyst, McDonnell Aircraft Corp, St Louis, 54-55; prin physicist aeronaut sci, 53-54, eng physicist aerospace sci, 58-78, RES SCIENTIST, BATTELLE COLUMBUS DIV, OHIO, 78- *Res:* Pulsed power supplies for advanced weapons systems; advanced armor and anti-armor technologies; ball lightning phenomena; directed energy technologies, aerosol science, electromagnetic launchers and power supplies; geophysical electromagnetic phenomena. *Mailing Add:* Battelle Columbus Div 505 King Ave Columbus OH 43201-2693

GOLOWICH, EUGENE, b Mt Vernon, NY, July 21, 39; m 61; c 2. PHYSICS. *Educ:* Rensselaer Polytech Inst, BS, 61; Cornell Univ, PhD(physics), 65. *Prof Exp:* Instr physics, Cornell Univ, 65; res assoc, Carnegie-Mellon Univ, 65-67; asst prof, 67-70, assoc prof, 70-76, PROF PHYSICS, UNIV MASS, AMHERST, 76- *Concurrent Pos:* Vis assoc prof, Cornell Univ, 73-74; vis res scientist, Stanford Linear Accelerator Ctr, 74; vis res scientist, Inst Theoret Physics, Univ Calif, Santa Barbara, 81; vis prof, Univ Hawaii, 87, Univ Paris-sud, 88. *Mem:* AAAS; Am Phys Soc. *Res:* Theoretical physics; high energy physics. *Mailing Add:* Dept Physics LGRT-C Univ Mass Amherst MA 01003

GOLTON, WILLIAM CHARLES, b Chicago, Ill, Mar 27, 36; m 64; c 3. ANALYTICAL CHEMISTRY. *Educ:* Univ Ky, BS; Univ Iowa, MS, 59, PhD(anal chem), 62. *Prof Exp:* Res chemist, 62-68, staff chemist, 68-73, res supvr, 73-80, qual mgr, fabrics & finishes dept, 81-85, SR RES ASSOC, MARSHALL LAB, E I DU PONT DE NEMOURS & CO, INC, 85- *Mem:* Am Chem Soc; Sigma Xi; Am Soc Testing & Mat; Am Soc Qual Control. *Res:* General analytical chemistry; physical measurements of paint and related materials; quality assurance; organic finishes, polymers, consumer products. *Mailing Add:* Automotive Prod Dept Marshall Lab PO Box 3886 Philadelphia PA 19146

GOLTZ, ROBERT W, b St Paul, Minn, Sept 21, 23; m 45; c 2. DERMATOLOGY, HISTOPATHOLOGY. *Educ:* Univ Minn, Minneapolis, BS, 43, MB, 44, MD, 45. *Prof Exp:* From clin instr to clin assoc prof dermat, Univ Minn, Minneapolis, 51-65; prof & dir div, Med Ctr, Univ Colo, Denver, 65-71; PROF DERMAT & CHMN DEPT, MED SCH, UNIV MINN, MINNEAPOLIS, 71- *Concurrent Pos:* Am Cancer Soc res grant, 58-59; USPHS res grant, 59-66; res grant, Grad Sch, Univ Minn, Minneapolis, 62-64. *Mem:* Am Dermat Asn; Am Acad Dermat; Soc Invest Dermat; Histochem Soc; Am Soc Dermatopath. *Res:* Venereology; dermatopathology; histochemistry. *Mailing Add:* H-8115 Univ Calif San Diego Med Ctr 225 Dickinson St San Diego CA 92103

GOLTZMAN, DAVID, b Montreal, Que, Sept 22, 44; m 68; c 3. ENDOCRINOLOGY & METABOLISM. *Educ:* McGill Univ, BSc, 66, MD, 68. *Prof Exp:* Intern, Royal Victoria Hosp, Montreal, 68-69, asst physician, 75-78; resident, Columbia-Presby Med Ctr, NY, 69-71; clin & res fel endocrinol, Mass Gen Hosp, 71-75; instr med, Med Sch, Harvard Univ, 74-75; from asst prof to assoc prof, 75-83, PROF MED, MCGILL UNIV, 83-, PROF PHYSIOL, 88-, CHMN, DEPT PHYSIOL, 88-; SR PHYSICIAN, ROYAL VICTORIA HOSP, MONTREAL, 87- *Concurrent Pos:* Chmn, comt for exp med, Med Res Coun Can, 84-; scientist Award, Med Res Coun Can, 83-88; assoc physician, Royal Victoria Hosp, Montreal, 80-87; mem, gen med B study sect, NIH, 88-; assoc ed, Bone, 89-; chmn, prog comt, Am Soc Bone & Mineral Res, 90- *Honors & Awards:* Andre Lichtwitz Prize for Res in Calcium, Nat Inst Health & Med Res, 87. *Mem:* Am Soc Clin Invest; Am Fedn Clin Res; Am Soc Bone & Mineral Res; Endocrine Soc; Can Soc Endocrinol & Metab (secy-treas, 81-84, pres, 90-); Am Physiol Soc. *Res:* Examination of the humoral control of calcium and skeletal homeostasis; exploration of the chemical structural, biosynthesis, metabolism and mechanism of action of hormones and of other humoral factors involved in calcium homeostasis. *Mailing Add:* Calcium Res Lab Royal Victoria Hosp Rm H4-67 687 Pine Montreal PQ H3A 1A1 Can

GOLUB, ABRAHAM, b Brooklyn, NY, May 16, 21; m 50; c 2. OPERATIONS RESEARCH, SCIENCE ADMINISTRATION. *Educ:* Brooklyn Col, BA, 41; Univ Del, MA, 49. *Prof Exp:* Sect chief, Surveillance Br, Ballistic Res Labs, Aberdeen Proving Ground, 46-50, asst chief, 50-55, chief, Artillery Weapon Systs Br, 55-62, assoc tech dir, Ballistic Res Labs, 62-64; dep spec asst opers res, Off Asst Secy Army, 64-67, asst dep undersecy army, 67-69, sci adv to asst chief of staff for force develop, Dept Army, Washington, DC, 69-73, tech adv to dep chief of staff for opers & plans, Hq, 73-76; PRES, ABRAHAM GOLUB INC, CONSULT SERV, 76- Instr, Univ Del, 50-51. *Honors & Awards:* Army Res & Develop Tech Achievement Award, 62. *Mem:* Opers Res Soc Am; Am Math Soc; Inst Math Statist; Am Ord Asn. *Res:* Weapons systems analysis; ordnance; statistical estimation; acceptance sampling; experimental design; analysis of variance; force analysis and design; cost and operational effectiveness analysis. *Mailing Add:* Apt 607 203 Yoakum Parkway Alexandria VA 22304

GOLUB, EDWARD S, b Chicago, Ill, Oct 6, 34; m 59; c 2. IMMUNOLOGY, CELL BIOLOGY. *Educ:* Roosevelt Univ, BS, 56; Miami Univ, MS, 59; Univ NC, PhD(bact), 64. *Prof Exp:* Res fel immunol, Duke Univ, 64-65 & Scripps Clin & Res Found, 65-68; from asst prof to prof biol sci, Purdue Univ, 68-88; dir res, Ortho Pharmaceut, 88-90; DIR, J&J LABS, SCRIPPS, 90- *Concurrent Pos:* USPHS career develop award, 68; Merck Found fac develop award, 69; assoc ed, J Immunol Methods & Modern Methods in Immunol; mem study sect, USPHS, 77-81; mem rev group, NSF, 82-86; mem study sect, Nat Cancer Inst, 87- *Mem:* AAAS; Am Asn Immunol; Brit Soc Immunol; Hist Sci Soc; Soc Develop Biol. *Res:* Induction of immunological tolerance; cell interactions in the immune response; mechanisms of tumor immunity; history of science. *Mailing Add:* Dept Molecular Biol Scripps Clin & Res Found La Jolla CA 92037

GOLUB, EFIM I, b Minsk, USSR, July 15, 34; m 68; c 2. MOLECULAR BIOLOGY, VIROLOGY. *Educ:* State Univ, USSR, BS, 57; Inst Gen Genetics, USSR, PhD(biophys), 65, SciD, 75. *Prof Exp:* Res microbiologist, Inst Gen Genetics, Moscow, 63-70; assoc prof genetics, Pedagogical Inst, USSR, 70-71; sr res microbiologist, All-Union Res Inst Antiobiotics, USSR, 71-79; res asst prof, Col Med, Northeastern Ohio Univ, 80-81; res scientist microbiol, Sch Med, Yale Univ, 81-88; res scientist, Columbia Univ, 88-90; RES SCIENTIST, YALE UNIV, 90- *Res:* Physical-chemical properties of nucleic acids; genetics of temperate phages of E coli; R-factors and other plasmid of E coli; transposable elements of bacteria; mechanism of genetical recombination. *Mailing Add:* 100 York St 16R New Haven CT 06511

GOLUB, ELLIS ECKSTEIN, b New York, NY, Nov 1, 42; m 64; c 3. BIOCHEMISTRY. *Educ:* Brandeis Univ, BA, 63; Tufts Univ, PhD(biochem), 69. *Prof Exp:* NIH res fel biol, Calif Inst Technol, 69-71, res fel, Jet Propulsion Lab, 71-72; mem fac oral biol, Health Ctr, Univ Conn, 72-77; from asst prof to assoc prof, 77-90, PROF BIOCHEM, SCH DENT MED, UNIV PA, 90- *Concurrent Pos:* Consult, Merck, Sharp & Dohme Res Labs, 87- *Mem:* AAAS; Am Chem Soc; NY Acad Sci; Am Soc Bone & Mineral Res; Protein Soc; Am Soc Biochem & Molecular Biol. *Res:* Calcification of bone and cartilage; computer analysis of protein structure. *Mailing Add:* Dept Biochem Univ Pa Sch Dental Med 4001 Spruce St A1 Philadelphia PA 19104-6003

GOLUB, GENE H, b Chicago, Ill, Feb 29, 32. MATHEMATICS. *Educ:* Univ Ill, BS, 53, MA, 54, PhD(math), 59. *Hon Degrees:* DrTech, Linköping Univ Sweden, 84; Dr, Technol & Med Univ Grenoble, 86; LLD, Univ Dundee, 87. *Prof Exp:* NSF fel, Math Lab, Univ Cambridge, 59-60; mem staff, Lawrence Radiation Lab, Univ Calif, 60-61; mem tech staff, Space Tech Labs, Inc, TRW, Inc, 61-62; vis asst prof, 62-64, from asst prof to assoc prof, 64-70, chmn dept, 81-84, PROF COMPUT SCI, STANFORD UNIV, 70- *Concurrent Pos:* Adj asst prof, Courant Inst Math Sci, NY Univ, 65-66; assoc ed, Linear Algebra & Its Appln, 72-, Appl Math & Optimization, 74-85, Soc Indust & Appl Math Rev, 74-80, J Am Statist Asn, 76-79; mem coun, Soc Indust & Appl Math, 75-78, gov bd, Ger Soc Appl Math & Mech & Adv Comt Comput Sci, NSF, 82-85, US Nat Comt Math, 84-86 & Adv Comt Eng & Math Div, Oak Ridge Nat Lab, 84-87; Guggenheim fel, 87-88. *Honors & Awards:* Forsythe lectr, 78; A R Mitchell Lectr, 91. *Mem:* Nat Acad Eng; fel AAAS; hon mem Inst Elec & Electronics Engrs; Soc Indust & Appl Math (pres, 85-87); Math Asn Am. *Res:* Numerical analysis; derivation of numerical algorithms for solving problems arising in linear algebra; statistical application and the solution of sparse linear systems arising in approximations to elliptic partial differential equations. *Mailing Add:* Dept Computer Sci Bldg 460 Rm 306 Stanford Univ Stanford CA 94305

GOLUB, LORNE MALCOLM, b Winnipeg, Man, Jan 13, 41; m 64; c 2. PERIODONTOLOGY, ORAL BIOLOGY. *Educ:* Univ Man, DMD, 63, MSc, 65; Harvard Univ, Cert Periodont, 68. *Prof Exp:* Res fel periodont, Harvard Sch Dent Med, 65-68; assoc prof, Fac Dent, Univ Man, 68-73; assoc prof, 73-77, PROF ORAL BIOL & PATH, SCH DENT MED, STATE UNIV NY, STONY BROOK, 77- *Concurrent Pos:* Prin investr, Nat Inst Dent Res, NIH, 76-, Kroc Found, New York Diabetes Asn, 77-79 & 81-83; consult, Periodont Dis Prog, Nat Inst Dent Res, NIH, 77, 78 & 80. *Honors & Awards:* Merit Award Res, NIH, 87. *Mem:* Int Asn Dent Res; Am Acad Periodont; AAAS; Int Group Peridont Res; Am Asn Oral Biologists. *Res:* Periodontitis; diabetes and collagen metabolism; anti-collagenases. *Mailing Add:* Dept Oral Biol & Path Sch Dent Med State Univ NY Stony Brook NY 11794

GOLUB, MORTON ALLAN, b Montreal, Que, June 11, 25; nat US; m 54; c 2. POLYMER CHEMISTRY, PHYSICAL CHEMISTRY. *Educ:* McGill Univ, BSc, 44; Univ NB, MSc, 47; Univ Mo, PhD(chem), 51. *Prof Exp:* Chemist, Defense Indust, Ltd, Que, 44-45; teacher pvt sch, Montreal, 45-46; demonstr physics, Univ NB, 46-47; asst chem, Univ Mo, 47-48, instr math, 49-51; res chemist, B F Goodrich Chem Co, 51-54 & Res Ctr, 54-60; sr polymer chemist, Stanford Res Inst, 60-68; Nat Res Coun-NASA sr resident res assoc, 68-70, RES SCIENTIST, AMES RES CTR, NASA, 70- *Concurrent Pos:* Instr, Foothill Col, 63-68; adv bd, J Appl Polymer Sci, 79-89. *Mem:* AAAS; Am Chem Soc. *Res:* Reactions and characterization of high polymers; photochemistry; radiation chemistry; elastomers; plasma chemistry. *Mailing Add:* Ames Res Ctr Mail Stop 239-4 NASA Moffett Field CA 94035-1000

GOLUB, SAMUEL J(OSEPH), b Middleboro, Mass, July 25, 15; m 40; c 2. TEXTILE TECHNOLOGY, BIOLOGY. *Educ:* Univ Mass, BS, 38, MS, 40; Harvard Univ, PhD(biol), 45. *Prof Exp:* Inspector biol mat, Dept Ord, US Navy, 42; asst prof bot, Univ Mass, 46-47, assoc prof, 47; bot ed, G & C Merriam Co, 47-49; lectr biol, Brandeis Univ, 49-50, asst prof, 50-56; sr res assoc, Fabric Res Labs, Inc, 57-59; assoc dir, ACH Fiber Serv, Inc, 59-63; asst dir biol sci, FRL-AN Albany Int Co, 63-85; TEXTILE CONSULT, 85- *Concurrent Pos:* Chmn Comt Textiles D13, Am Soc Testing & Mat, 72-78. *Mem:* Hon mem Am Soc Testing & Mat; Am Asn Textile Chemists & Colorists; Soc Indust Microbiol; Fiber Soc; fel Am Inst Chem. *Res:* Plant morphology and pathology; textile fiber analysis; textile and cordage defect analysis, textile flammability; wool, flax, jute, specialty fibers. *Mailing Add:* 38 Myerson Lane Newton Ctr Dedham MA 02159

GOLUB, SHARON BRAMSON, b New York, NY, Mar 25, 37; m 58; c 2. HEALTH PSYCHOLOGY. *Educ:* Columbia Univ, BS, 59, MA, 66; Fordham Univ, PhD(psychol), 74. *Prof Exp:* Head nurse psychiat, Mt Sinai Hosp, 57-59; ed, RN Mag, 67-74; from asst prof to assoc prof, 74-86, PROF PYSCHOL, COL NEW ROCHELLE, 86- *Concurrent Pos:* Dir women's studies, Col New Rochelle, 78-79, chmn, 79-82; pvt pract, 76-; adj prof psychiat, NY Med Col, 80-; dir, Soc Menstrual Cycle Res, 81- *Honors & Awards:* Distinguished Publ Award, Asn Women Psychol, 84. *Mem:* Fel Am Psychol Asn; Soc Menstrual Cycle Res (pres 81-83); Asn Women in Psychol. *Res:* The menstrual cycle, health psychology and the psychology of women. *Mailing Add:* Dept Psychol Col New Rochelle New Rochelle NY 10805

GOLUB, SIDNEY HARRIS, b Hartford, Conn, July 28, 43; m 66. IMMUNOLOGY. *Educ:* Brandeis Univ, BA, 65; Temple Univ, PhD(microbiol), 70. *Prof Exp:* Damon Runyon Mem Fund fel cancer res, Karolinska Inst, Sweden, 69-71; from asst prof to assoc prof surg & microbiol & immunol, 71-83, PROF SURG, DIV ONCOL & PROF MICROBIOL & IMMUNOL, SCH MED, UNIV CALIF, LOS ANGELES, 83- *Concurrent Pos:* NIH career develop award, 74; mem, Nat Breast Cancer Task Force, 75-; vis investr, Sloan-Kettering Inst, 78-79; assoc dean career develop, Sch Med, Univ Calif, Los Angeles, 86- *Mem:* Am Soc Microbiol; Am Asn Cancer Res; Am Asn Immunol; Reticuloendothelial Soc; Sigma Xi. *Res:* Tumor immunology; cellular immunology; tumor biology; viral oncology; functioning of the reticuloendothelial system. *Mailing Add:* 54 140 CHS Univ Calif Sch Med Los Angeles CA 90024

GOLUBA, RAYMOND WILLIAM, b Streator, Ill, Oct 5, 39; m 61; c 3. MECHANICAL ENGINEERING. *Educ:* Univ Ill, BS, 61, MS, 62; Univ Wis, PhD(mech eng), 68. *Prof Exp:* Engr, ARO, Inc, Arnold Air Force Sta, Tenn, 62-63; instr thermodyn, Univ Wis, Madison, 63-64; engr, 68-85, dep div leader, 85-90, PROJECT MANAGER, DEPT MECH ENG, LAWRENCE LIVERMORE NAT LAB, 90- *Mem:* Am Soc Mech Engrs. *Res:* Heat transfer in oscillating flows; heat transfer from capillary wetted surfaces with applications toward heat pipes; computerized data bases. *Mailing Add:* Lawrence Livermore Nat Lab PO Box 808 L-30 Livermore CA 94550

GOLUBIC, STEPHAN, b Zagreb, Yugoslavia, May 4, 34. ECOLOGY, PHYCOLOGY. *Educ:* Univ Zagreb, dipl biol, 56, PhD(biol sci), 63. *Prof Exp:* Res asst inst exp biol, Yugoslavia Acad Sci, Univ Zagreb, 56-59; res asst, Inst Marine Biol, Rovinj, 59-63; Alexander von Humboldt Found fel, Inst Limnol, Univ Freiburg, 63-65; res assoc geol, Princeton Univ, 65-66; from instr to asst prof biol, Yale Univ, 66-69; assoc prof, Peterson State Col, 69-70; assoc prof, 70-80, PROF BIOL, BOSTON UNIV, 80- *Concurrent Pos:* NSF res grant, 67-69 & 70-71, 71-75, 74-76, 76-79; res grant, NASA, 79 & 81. *Mem:* Brit Phycol Soc; Phycol Soc Am; Int Phycol Soc; Int Asn Theoret & Appl Limnol; Sigma Xi. *Res:* Phycology, ecology and taxonomy of Cyanophyta; limnology; marine biology; ecology of carbonate deposition and dissolution; precambrian microfossils and stromatolites; water pollution. *Mailing Add:* Dept Biol Boston Univ Two Cummington St Boston MA 02215

GOLUBITSKY, MARTIN AARON, b Philadelphia, Pa, May 4, 45; m 76; c 2. BIFURCATION THEORY, SINGULARITY THEORY. *Educ:* Univ Pa, AB & AM, 66; Mass Inst Technol, PhD(math), 70. *Prof Exp:* From asst prof to assoc prof math, Queens Col, NY, 74-79; prof math, Ariz State Univ, 79-83; PROF MATH, UNIV HOUSTON, 83- *Concurrent Pos:* Vis mem, Courant Inst, New York Univ, 77-78, Inst Advan Study, 78; assoc prof, Univ Nice, 80; vis prof, Duke Univ, 81; vis sr scientist, Univ Calif, Berkeley, 82. *Mem:* Fel AAAS; Am Math Soc; Soc Indust & Appl Math; Soc Natural Philos. *Res:* Application of singularity theory & group theory techniques to bifurcation theory & their application to physical problems, such as pattern formation in fluid systems. *Mailing Add:* Dept Math Univ Houston Houston TX 77204

GOLUBJATNIKOV, RJURIK, b Kuressaare, Estonia, June 19, 31; US citizen. MICROBIOLOGY, EPIDEMIOLOGY. *Educ:* Millikin Univ, BA, 54; Univ Mich, MPH, 58, PhD(epidemiol sci), 64. *Prof Exp:* Asst prof prev med & dir, Vaccine Eval Prog Lab, State Univ NY Buffalo, 64-67; ASST PROF PREV MED & CHIEF, IMMUNOL SECT, STATE LAB HYG, UNIV WIS-MADISON, 67- *Concurrent Pos:* Ecol Studies Mycoplasma grant, 68-69; assoc investr, USPHS res grant, 64-68; prin investr, Huixquilucan Serum Surv grant, 67-68; adv, Inst Nutrit Cent Am & Panama, 68; Alpha-Fetoprotein Maternal Serum Screening Proj grant, 77-; partic, Int Schoolchildren Study of Serum Cholesterol Levels, 78; AIDS Studies, 84-; effects of aldicarb contaminated well water on immunity, 86-; specialist microbiologist, Pub Health & Med Lab Microbiol, Am Acad Microbiol. *Mem:* AAAS; Am Pub Health Asn; Am Soc Microbiol; Asn Teachers Prev Med; Am Venereal Dis Asn; Soc Epidemiol Res. *Res:* Immunology, including primary and secondary antibody responses and immunologic memory; infectious disease epidemiology; serological epidemiology; vaccine studies; perinatal health; childhood antecedents of coronary heart disease, epidemiology of AIDS; effects of aldicarb contaminated well water on immunity; clinical trials of immunodiagnostic devices. *Mailing Add:* State Lab Hygiene 465 Henry Mall Madison WI 53706

GOLUBOVIC, ALEKSANDAR, b Maribor, Yugoslavia, Feb 19, 20; US citizen; m 60; c 1. CHEMISTRY. *Educ:* Univ Zagreb, dipl, 43, PhD(chem), 60. *Prof Exp:* Res chemist, Inst Indust Res, Univ Zagreb, 45-56; res assoc chem, Mass Inst Technol, 56-62; sr scientist, Tyco Labs, Mass, 62-64; RES SCIENTIST CHEM, ROME LAB, HANSCOM AFB, BEDFORD, MASS, 64- *Mem:* Am Chem Soc; Am Vacuum Soc. *Res:* Semiconductors and photoconductors; thin film technology and phenomena. *Mailing Add:* 31 Heath St Brookline MA 02146

GOLUBOW, JULIUS, b New York, NY, June 2, 29; m 58; c 3. BIOCHEMISTRY. *Educ:* City Col New York, BS, 52; Purdue Univ, MS, 55; Univ Pittsburgh, PhD(biochem), 60. *Prof Exp:* Asst, Purdue Univ, 53-55 & Univ Pittsburgh, 55-59; res assoc biochem, Med Col, Cornell Univ, 59-61, from instr to asst prof, 61-68; ASSOC PROF BIOL SCI, LEHMAN COL, 68- *Mem:* AAAS; Am Chem Soc; Harvey Soc; Brit Biochem Soc; NY Acad Sci. *Res:* Enzymology; mineral metabolism. *Mailing Add:* Dept Biol Sci Herbert H Lehman Col Bronx NY 10468

GOLUEKE, CLARENCE GEORGE, b Menominee, Mich, Sept 28, 11; m 48; c 1. ENVIRONMENTAL SCIENCES. *Educ:* St Louis Univ, BA, 39; Univ Ill, MA, 41; Univ Calif, PhD(mycol), 53. *Prof Exp:* Instr gen sci, Nazareth Col, 45-46; asst bot, Univ Ill, 48; asst bot, Sanit Eng Res Lab, Univ Calif, Berkeley, 48-51, lectr & res biologist, 51-78; DIR RES & DEVELOP, CAL RECOVERY SYSTS, INC, 78- *Concurrent Pos:* Consult, Erco, Inc, Cambridge, Mass, Calif State Depts Water Resources & Pub Health, US Environ Protection Agency & UN Indust Develop Orgn, Vienna, Austria; sr ed, Biocycle, Emmaus, Pa. *Honors & Awards:* Arthur M Wellington Prize, Am Soc Civil Engrs, 66. *Mem:* Sigma Xi; Am Pub Health Asn; AAAS. *Res:* Environmental health sciences; waste treatment and resource recovery; closed environmental systems; on-site waste disposal. *Mailing Add:* 2691 McMorrow Rd San Pablo CA 94806

GOLUMBIC, CALVIN, b Lock Haven, Pa, Nov 8, 12; m 46; c 1. BIOCHEMISTRY. *Educ:* Pa State Col, BS, 34; Rutgers Univ, MS, 35, PhD(org chem), 37. *Prof Exp:* Res assoc antioxygens, Univ Iowa, 37-42; chemist, Rockefeller Inst, 42-45; asst res prof chem, Univ Pittsburgh, 45-47; chemist, Bur Mines, 47-53; chemist, Naval Stores Res Sect, Agr Res Serv, USDA, 53-54, head, Qual Eval Sect, Biol Sci Br, Mkt Res Div, 54-60, chief field crops & animal prod br, 60-67, asst dir, Mkt Qual Div, Md, 67-68, asst dep adminr, 68-72, staff officer, 72-82; RETIRED. *Concurrent Pos:* Consult, 82- *Honors & Awards:* Cert Merit, USDA, 63 & 69. *Mem:* Am Chem Soc; NY Acad Sci; Inst Food Technol; Wash Acad Sci; Am Asn Cereal Chemists. *Res:* Antioxidants and autooxidation of fats; vitamin E; nitrogen mustards; countercurrent distribution; coal, food and agricultural chemistry; synthetic liquid fuels; terpenes. *Mailing Add:* 6000 Highboro Dr Bethesda MD 20817-6008

GOLUMBIC, MARTIN CHARLES, b Erie Pa, Sept 30, 48; c 4. COMPUTER SCIENCE & MATHEMATICS. *Educ:* Pa State Univ, BS, 70, MA (Math), 70; Columbia Univ, Ma, 71; PhD, 75. *Prof Exp:* Jr mathematician, U S Army Computer Systems Command, 70; vis scientist, IBM Watson Res Ctr, Yorktown Heights, Ny, 74 & 89-90; mem tech staff, Bell Tel Lab, Murray Hill, NJ, 80-82; RES STAFF MEM, IBM ISRAEL SCI CTR, HAIFA, ISRAEL, 83- *Concurrent Pos:* Asst prof Computer Sci, New York Univ, Courant Instit of Math, 75-80; vis prof, Universite de Paris VI, Instutut de Programmation, 76-77; res grants, NSF, 78-81; adj prof, Bar-Ilan Univ, Ramat Gan, Israel, 85-; mem adv bd, Int J Expert Syst: Res & Appln, 86-90; ed-in-chief, J Annals Math & Artificial Intel, 88-; Found fel, Inst Combinatorics & Appln, 91. *Honors & Awards:* Rensselaer Medal for Excellence in Mathematics, 66; Hon Mention, NSF, 70. *Res:* Computer science and mathematics; patents; author of book and numerous publications. *Mailing Add:* IBM Watson Research Center PO Box 704 Yorktown Heights NY 10598

GOLUMBIC, NORMA, b Brooklyn, NY, Apr 30, 16; m 46; c 1. CANCER. *Educ:* Brooklyn Col, BA, 36; Univ Iowa, MS, 38. *Prof Exp:* Chemist, Bur Mines, 42-50; sci writer, Agr Res Serv, USDA, 55; sci writer, Nat Cancer Inst, 56-58, head, Res & Prog Reports Sec, 58-69, asst chief, Res Info Br, 65-69; info officer, Div Nursing, NIH, 69-74, sr sci ed, Nat Cancer Inst, 74-79; FREE LANCE MEDICAL WRITER, 79- *Mem:* AAAS; Am Chem Soc; fel Am Inst Chemists; Nat Asn Sci Writers. *Res:* Synthetic liquid fuels; coal chemistry; catalysis; cancer research literature; medical writing. *Mailing Add:* 6000 Highboro Dr Bethesda MD 20817-6008

GOLWAY, PETER L, ANIMAL CARE RESEARCH. *Educ:* Fla State Univ, BA, 65; Univ Pa, VMD, 69. *Prof Exp:* Med & surg pract, Berliner Animal Hosp, Washington, DC, 69-73; staff vet, Lab Animal Sci Div, Litton Bionetics, Inc, Md, 74-80; CHIEF, ANIMAL CARE BR, NAT INST ALLERGY & INFECTIOUS DIS, NIH, BETHESDA, 80- *Mem:* Asn Gnotobiotics; Am Vet Med Asn; Am Soc Lab Med Practr; Am Asn Lab Animal Sci; Nat Asn Biomed Res; Am Heartworm Soc. *Mailing Add:* Nat Inst Allergy & Infectious Dis NIH Animal Care Br Bldg 14B S Rm 228 Bethesda MD 20892

GOMATOS, PETER JOHN, b Cambridge, Mass, Feb 13, 29. BIOCHEMISTRY, VIROLOGY. *Educ:* Mass Inst Technol, SB, 50; Johns Hopkins Univ, MD, 54; Rockefeller Inst, PhD(virol), 63. *Prof Exp:* Med asst resident, Boston City Hosp, 57-58; sr med resident, Mass Gen Hosp, 58-59; guest investr, Rockefeller Inst, 63-64; mem, Sloan-Kettering Inst, 64-; AT LETTERMN ARMY MED CTR. *Concurrent Pos:* Vis investr, Max Planck Inst Virus Res, 64 & Inst Virol, Univ Glasgow, 64-65. *Mem:* Am Chem Soc; Harvey Soc; Am Asn Immunologists; Am Col Physicians. *Res:* Study of physical-chemical characteristics of viruses; effects of viruses on cells; neoplastic potential of various viruses. *Mailing Add:* Assoc Dir Health & Human Serv Treatment Res Prog Div AIDS NIH 6003 Exec Bldg Bethesda MD 20892

GOMBAR, CHARLES T, b New York, NY, Dec 13, 52; m. EXPERIMENTAL BIOLOGY. *Educ:* Pace Univ, NY, BS, 74; Albany Med Col Union Univ, PhD(pharmacol), 79. *Prof Exp:* Fel, Fels Res Inst, Sch Med, Temple Univ, 79-82; assoc sr investr, Dept Drug Metab, Smith Kline & French Labs, 82-85, sr investr, 85-87, asst dir, 87-89, ASSOC DIR PROJ MGT, SMITHKLINE BEECHAM PHARMACEUT, 89- *Concurrent Pos:* Consult, Cancer Info Dissemination & Anal Ctr, Franklin Inst, Philadelphia, 80-82; adj asst prof, Dept Chem, Temple Univ, 86-; assoc ed, Cancer Res, 89-90. *Mem:* Am Asn Cancer Res; Am Asn Pharmaceut Scientists; Am Chem Soc; Am Soc Pharmacol & Exp Therapeut; Int Soc Study Xenobiotics; NY Acad Sci; Sigma Xi. *Res:* Worldwide development of carvedilol, a cardiovascular drug; pharmacokinetos and metabolsim of carcinogenic N-nitroso compounds; numerous publications. *Mailing Add:* Smith Kline & French Labs PO Box 1539 L-311 King of Prussia PA 19406-0939

GOMBERG, HENRY J(ACOB), b New York, NY, Apr 16, 18; m 67; c 5. NUCLEAR ENGINEERING. *Educ:* Univ Mich, BSE, 41, MSE, 43, PhD(elec eng), 51. *Hon Degrees:* DSc, Albion Col, 68. *Prof Exp:* Asst dir, Mem Phoenix Proj, Univ Mich, 52-59, dir, 59-61, prof nuclear eng & chmn dept, 58-61; prof physics & dep dir, PR Nuclear Ctr, Univ PR, 61-66, dir, 66-71; pres, KMS Fusion Inc, 71-81; PRES, ANN ARBOR NUCLEAR, INC, 81-; MANAGING DIR, ANN ARBOR NUCLEAR ISRAEL LTD, 85-; PRES, PENETRON INC, 87- *Concurrent Pos:* Res proj dir new high resolution methods of radiation detection, AEC, 50-58; lectr, Oak Ridge Inst Nuclear Studies, 51-53; Carnegie prof, Univ Hawaii, 61; consult, AEC, Lockheed Aircraft Corp & Argonne Nat Lab; chmn comt res reactors, Nat Acad Sci-Nat Res Coun; mem comt radiation effects, Div Phys Sci, Nat Res Coun. *Mem:* Fel Am Nuclear Soc; Am Phys Soc; Am Soc Eng Educ; Inst Elec & Electronics Engrs; fel AAAS. *Res:* High resolution detection of radiation; effect of radiation on matter; biological effects of radiation; reactor safety and accident analysis; reactor site selection; nuclear energy; development of nuclear fusion for power and other peaceful purposes; synthetic fuels through nuclear energy; neutron scatter for search and analysis. *Mailing Add:* Ann Arbor Nuclear Inc PO Box 8618 Ann Arbor MI 48107

GOMBERG, JOAN SUSAN, b Chicago, Ill, Nov 1, 57; m 89. REGIONAL NETWORK SEISMOLOGY, SEISMIC HAZARD ASSESSMENT. *Educ:* Mass Inst Technol, BS, 79; Univ Calif, San Diego, PhD(geophys), 86. *Prof Exp:* Postdoctoral, Univ Nev, Reno, 87-88; GEOPHYSICIST, US GEOL SURV, 88- *Mem:* Am Geophys Union; Seismol Soc Am. *Res:* Seismic network operations; analysis of seismicity data; assessing regional seismic hazard. *Mailing Add:* US Geol Surv MS 966 Denver Fed Ctr Box 25046 Denver CO 80225

GOMBOS, ANDREW MICHAEL, JR, b Beaver Falls, Pa, June 10, 48. MICROPALEONTOLOGY, MARINE GEOLOGY. *Educ:* Washington & Lee Univ, BS, 70; Univ Ill, MS, 73; Fla State Univ, PhD(geol), 76. *Prof Exp:* Develop geologist, Chevron USA, Inc, 76-78; RES SPECIALIST, EXXON PROD RES CO, 78- *Concurrent Pos:* Penrose grant, Geol Soc Am, 74; res assoc, Carnegie Museum Natural Hist, 81- *Mem:* Soc Economic Paleontologists & Mineralogists; Sigma Xi; Geol Soc Am; Gulf Coast Asn Geol Sci; Am Asn Petrol Geologists. *Res:* Stratigraphical distribution and evolution of diatoms; antarctic marine geology and paleohistory; carbonate facies interpretation; seismic stratigraphy. *Mailing Add:* Exxon Prod Res Co PO Box 2189 Houston TX 77252

GOMER, RICHARD HANS, b Chicago, Ill, Oct 11, 56; m; c 1. CELL BIOLOGY, DICTYOSTELIUM. *Educ:* Pomona Col, BA, 77; Calif Inst Technol, PhD(biol), 83. *Prof Exp:* FEL, DEPT BIOL, UNIV CALIF, SAN DIEGO, 83- *Concurrent Pos:* Vis scientist, Mt Wilson & Las Campanas Observ, Carnegie Inst Wash, 83. *Honors & Awards:* Tileston Physics Prize. *Mem:* Am Soc Cell Biol. *Res:* Using combined immunological and molecular biology techniques to study differential developmental gene regulation and oncogenes in Dictyostelium discoideum. *Mailing Add:* Dept Biol Rice Univ PO Box 1892 Houston TX 77251

GOMER, ROBERT, b Vienna, Austria, Mar 24, 24; nat US; m 55; c 2. PHYSICAL CHEMISTRY. *Educ:* Pomona Col, BA, 44; Univ Rochester, PhD(chem), 49. *Prof Exp:* AEC fel chem, Harvard Univ, 49-50; from instr to prof chem, 50-77, dir, James Franck Inst, 77-83, C W EISENDRATH DISTINGUISHED PROF, JAMES FRANCK INST & UNIV CHICAGO, 84- *Concurrent Pos:* Sloan fel, 58-61; Guggenheim fel, 69-70; Fulbright prof, Austria, 81. *Honors & Awards:* Bourke lectr, Eng, 59; Kendall Award, Am Chem Soc, 75; Humboldt Sr US Scientist Award, 78; Davisson-Germer Prize, Am Phys Soc, 81; M W Welch Award, Am Vacuum Soc, 89. *Mem:* Nat Acad Sci; Am Chem Soc; Am Acad Arts & Sci; Lepoldina Akademie. *Res:* Surface chemistry and physics; field and ion emission; surface diffusion; electron stimulated desorption; metal overlayers. *Mailing Add:* James Franck Inst 5640 Ellis Ave Chicago IL 60637

GOMES, WAYNE REGINALD, b Modesto, Calif, Nov 15, 38; m 64; c 2. REPRODUCTIVE BIOLOGY, ANIMAL SCIENCES. *Educ:* Calif Polytech State Univ, BS, 60; Wash State Univ, MS, 62; Purdue Univ, PhD(endocrine physiol), 65. *Prof Exp:* From asst prof to prof dairy sci, Animal Reproduction Ctr, Ohio State Univ, 65-81; prof & head dairy sci, 81-85, prof & head animal sci, 85-89, DEAN, COL AGR, UNIV ILL, 89- *Concurrent Pos:* Res award, Ohio Sigma Xi, 70; Fulbright-Hays traveling fel, 74. *Mem:* AAAS; Am Dairy Sci Asn; Am Soc Animal Sci; Am Physiol Soc; Soc Study Reproduction; fel Japan Soc Prom Sci. *Res:* Biochemistry and endocrinology of testis function. *Mailing Add:* Univ Ill 1301 W Gregory Dr Urbana IL 61801

GOMEZ, ARTURO, b Monterrey, NL, Sept 27, 51; m 82; c 3. CARDIOPULMONARY INTERACTIONS IN SECONDARY PULMONARY HYPERTENSION. *Educ:* Univ Nuevo León, Bachelor, 69, MD, 76; Univ Man, PhD(physiol), 86. *Prof Exp:* ASSOC PROF MED, PROF MED CARDIOL & SR SCIENTIST, NAT HEART INST, MEX, 89-, ASSOC PROF MED, NURSING SCH, 90- *Concurrent Pos:* Sr clin fel crit care med, ABC Hosp, Mex, 80-82, chest med, Univ Man, 86-87; sr res fel cardiopulmonary serv, Nat Heart Inst, 83-84, Univ Man, 87-89; sr scientist, Nat Res Syst, Mex, 89-, Nat Health Care Syst, Mex, 90- *Honors & Awards:* St Boniface Res Found Inc Award, 88. *Mem:* Am Thoracic Soc; Fedn Am Socs Exp Biol. *Res:* Heart and lung interactions in patients; experimental designs in animals with pulmonary hypertension, especially the right ventricle; teaching cardiopulmonary interactions. *Mailing Add:* Dept Cardioneumolog Inst Nac Cardiol Ignacio Chaves Juan Baduano I Tlalpan Mexico DF 14080 Mexico

GOMEZ, ILDEFONSO LUIS, b Cardenas, Cuba, Jan 23, 28; m 54; c 6. ORGANIC CHEMISTRY, POLYMER CHEMISTRY. *Educ:* Univ Havana, DSc(chem), 52. *Prof Exp:* Anal chemist, Cia Rayonera Cuba, 51-52, res chemist, 52-56; tech supvr, Plinex SA, 56-58, tech supt, 58-61; sr res chemist, Monsanto Co, 61-67, res specialist, 67-68, specialist, Tech Dept, Plastic Prod & Resins Div, 68-70, specialist, Packaging Div, 70-74, group leader, New Prod Develop, Bloomfield, 74-79, sr technol specialist, 79-82, SCI FEL, INDUST RES DIV, MONSANTO CO, MASS, 82- *Concurrent Pos:* Consult, Ministry Com, Cuba, 58-60; vis lectr, Univ Havana, 59-60; ed-at-large, Marcel Dekker, Inc, 84- *Mem:* Am Chem Soc; Soc Plastics Engrs; NY Acad Sci. *Res:* Rheology and melt processing of high nitrile and polyester polymers for carbonated beverage and food containers fabrication; coloring and recycle of polymers; prototype equipment development for processing polymers into containers; phenolic resins for glass fiber bonding applications; polyvinylbutyral interlayers for architectural applications. *Mailing Add:* 223 Franklin Rd Longmeadow MA 01106

GOMEZ-CAMBRONERO, JULIAN, b Manzanares, Spain, Sept 29, 59; m 82; c 1. SIGNAL TRANSDUCTION, CELL BIOLOGY. *Educ:* Univ Complutense, Madrid, BS, 82, MS, 83, PhD(biochem), 86. *Prof Exp:* Postdoctoral fel cell biol, 86-90, INSTR PHYSIOL, UNIV CONN HEALTH SCI CTR, 91- *Concurrent Pos:* Postdoctoral fel, Am Heart Asn, 87-88. *Mem:* Am Soc Cell Biol; Soc Leukocyte Biol. *Res:* Molecular mechanisms of cell activation and signal transduction in leukocytes; intracellular changes by cytokines and colony-stimulating factors; tyrosine phosphorylation induced by several cell agonists; inflammatory mediators of lipidic nature and mechanisms of actuation; enzyme purification and characterization; tyrosine phosphorylation. *Mailing Add:* Dept Physiol Univ Conn Health Ctr Farmington CT 06030

GOMEZPLATA, ALBERT, b Colombia, SAm, July 2, 30; nat US; c 3. CHEMICAL ENGINEERING. *Educ:* Polytech Inst Brooklyn, BSc, 52; Rensselaer Polytech Inst, MSc, 54, PhD(chem eng), 59. *Prof Exp:* Process engr, Gen Chem Div, Del Works, 52-53; after 58-68, prof chem eng, 68-82, assoc dean, 82-85, EMER PROF, UNIV MD BALTIMORE COUNTY, 85- *Concurrent Pos:* Year in indust prof, E I du Pont de Nemours & Co, Inc, 65-66; vis prof, Univ PR, 70-71; consult, Naval Ship Res Develop Ctr, Anapolis Md. *Mem:* Fel Am Inst Chem; Am Inst Chem Engrs; Am Soc Eng Educ; NY Acad Sci; Sigma Xi. *Res:* Kinetics; heterogeneous flow systems; fluidized beds. *Mailing Add:* Dept Chem Eng Univ Md Baltimore County Baltimore MD 21228

GOMEZ-RODRIGUEZ, MANUEL, b Ponce, PR, Oct 15, 40; US citizen; m 65; c 2. THEORETICAL PHYSICS, SOLID STATE PHYSICS. *Educ:* Univ Puerto Rico, BSc, 62; Cornell Univ, PhD(theoret physics), 68. *Prof Exp:* Fel solid state physics, Naval Res Lab, Washington, DC, 67-69; scientist, PR Nuclear Ctr, 70-74; asst prof physics, 69-71, chmn dept, 71-75, dean, Col Nat Sci, 75-86, PROF PHYSICS, UNIV PR, 75-, DIR RESOURCE CTR SCI & ENG, 86- *Concurrent Pos:* Fel Nat Res Coun-Nat Acad Sci, 67-69; NSF grant, Off Naval Res, 74 & Army Res Off, 81-; proj dir comput in teaching sci, NSF, 76-, minority ctr grad educ sci, 77-79 & dir, Resource Ctr Sci & Eng, PR, 80-, dir, Exp Prog Stimulate Competitive Res, 86. *Mem:* Am Phys Soc; Sigma Xi; Sci Teachers Asn Puerto Rico (pres, 74). *Res:* Theoretical solid state physics; optical properties of solids; optical properties of CERMET materials utilized in thermal solar collectors; phase transitions of ferroelectrics materials. *Mailing Add:* Fac Natural Sci Univ Puerto Rico Rio Piedras PR 00931

GOMMEL, WILLIAM RAYMOND, b Indianapolis, Ind, Aug 16, 24; m 43; c 3. METEOROLOGY. *Educ:* Univ Calif, Los Angeles, AB, 51, MA, 52; Purdue Univ, PhD(atmospheric sci), 73. *Prof Exp:* Prod draftsman, Atkins Saw Div, Borg-Warner Corp, Ind, 40-42; asst sta weather officer, US Air Force, 44-50, weather cent anal supvr, Ger, 52-53, chief, Tech Info & Ed, Hqs 2nd Weather Wing, 53-55, Tech Pubs Br, Sci Servs, Hqs Air Weather Serv, 55-56, asst to dir, Sci Serv, 56-58, staff meteorologist, Ballistic Missile Div, 60-61, Space Systs Div, 62-63, staff scientist, Directorate Sci & Tech, Hqs, 64-65; asst prof math, 65-69, assoc prof math & earth sci, 69-75, PROF MATH & EARTH SCI, IND CENT UNIV, 76-, CHMN DEPT EARTH SCI, 66- *Concurrent Pos:* Meteorologist, Channel 13, Video Ind, Inc, Indianapolis, 75; cooperator, NOAA Weather Radio Broadcast Facil, Indianapolis, 77- *Mem:* AAAS; Am Geophys Union; Am Meteorol Soc; Sigma Xi. *Res:* Interaction between sea and atmosphere; climatic change; synoptic meteorology; cloud physics; upper atmosphere; satellites; instrumentation; tropical storms; hurricanes; eclipse meteorology. *Mailing Add:* 2301 Lawrence Ave Indianapolis IN 46227

GOMOLL, ALLEN W, b Chicago, Ill, July 10, 33; m 55; c 2. PHARMACOLOGY. *Educ:* Univ Ill, BS, 55, MS, 58, PhD(pharmacol), 61. *Prof Exp:* From instr to asst prof pharmacol, Col Med, Univ Ill, 61-66; sr scientist, Mead Johnson Pharmaceut, 66-67, group leader cardiovasc pharmacol, 67-69, sect leader, 69-70, prin investr cardiovasc pharmacol, 70-76, prin res assoc, 76-80, sect mgr, 80- 81, prin res scientist, Biol Res, 81-84; res fel, 84-90, SR RES FEL, BRISTOL-MYERS SQUIBB PRI, 90- *Concurrent Pos:* Mem, Basic Sci Coun & fel Coun Circulation, Am Heart Asn. *Honors & Awards:* Grad Res Award, Sigma Xi, 61. *Mem:* AAAS; NY Acad Sci; Am Heart Asn; Am Soc Pharmacol & Exp Therapeut; Soc Exp Biol & Med; Sigma Xi; fel Am Col Cardiol; Int Soc Heart Res. *Res:* Cardiovascular pharmacology; renal pharmacology; effects of drugs on adrenal cortical hormone biosynthesis. *Mailing Add:* Bristol-Myers Squibb PRI PO Box 5100 Five Research Pkwy #307 Wallingford CT 06492

GOMORY, RALPH E, b New York, NY, May 7, 29; div; c 3. APPLIED MATHEMATICS. *Educ:* Williams Col, BA, 50; Princeton Univ, PhD(math), 54. *Hon Degrees:* DSc, Williams Col, 73, Polytechnic Univ, 86; LHD, Pace Univ, 86. *Prof Exp:* Asst prof math & Higgins lectr, Princeton Univ, 57-59; mgr mgt sci & combinatorial studies, T J Watson Res Ctr, IBM Corp, 59-64, dir, Math Sci Dept, 65-70, mem, Corp Tech Comt, 70, dir res, 70-86, vpres, 73-84, sr vpres, 85-86, sr vpres sci & technol, 86-89; PRES, ALFRED P SLOAN FOUND, 89- *Concurrent Pos:* Mem adv panel appl math div, Nat Bur Standards, 64-68; mem, Mayor's Opers Res Coun New York, 66-70, vchmn, 69-70; consult, NSF, 67-69; mem comt appln math, Div Math Sci, Nat Res Coun, 67-70, chmn, 69-70, mem-at-large, 69-72; adj prof, Courant Inst, NY Univ, 70-71; Andrew D White prof-at-large, Cornell Univ, 70-76; mem vis comt, Sloan Sch Mgt, Mass Inst Technol, 71-77; mem adv coun, Dept Math, Princeton Univ, 82-85, chmn, 84-85; mem coun grad sch bus, Univ Chicago, 71-80; bd trustees, Conf Bd Math Sci, 72-74; bd trustees, T A Edison Found, 73-77 & Hampshire Col, 77-86; chmn, Nat Acad Sci Adv Comt for Int Inst Appl Systs Res, 73-75; mem comt soc sci, Yale Univ Coun, 74-77; mem adv coun, Sch Eng, Stanford Univ, 78-85; dir, Bank NY, 81-, Wash Post Co, Ashland Oil, Inc & Lexmark Int, Inc; mem, President's Coun Adv Sci & Technol, 90- *Honors & Awards:* Lanchester Prize, Opers Res Soc Am, 63; John von Neumann Theory Prize, Opers Res Soc & Inst Mgt Sci, 84; Nat Medal of Sci, 88. *Mem:* Nat Acad Sci; Nat Acad Eng; Am Acad Arts & Sci; fel Economet Soc; Am Math Soc; Am Philos Soc. *Res:* Linear and integer programming; network flow theory; nonlinear differential equations; computers; nature of technology and product development; research in industry; industrial competitiveness. *Mailing Add:* Alfred P Sloan Found 630 Fifth Ave Suite 2550 New York NY 10111

GOMPERTZ, MICHAEL L, b New Haven, Conn, Dec 10, 12; m 41; c 2. INTERNAL MEDICINE, GASTROENTEROLOGY. *Educ:* Yale Univ, BA, 33; Columbia Univ, MD, 37. *Prof Exp:* Asst chief gastroenterol, Vet Admin Hosp, Memphis, 53-55, chief, 55-83; from instr to assoc prof, 55-68, prof, 68-87, EMER PROF MED GASTROENTEROL, COL MED, UNIV TENN, MEMPHIS, 87- *Mem:* Fel Am Col Physicians; Am Gastroenterol Asn. *Res:* Clinical research in liver disease and peptic ulcer disease. *Mailing Add:* Univ Tenn 951 Court Ave Memphis TN 38163

GONA, AMOS G, b Nandyal, India, July 16, 33; US citizen; m 62; c 2. ENDOCRINOLOGY, DEVELOPMENTAL ANATOMY. *Educ:* Andhra Univ, India, BSc, 54; City Col New York, MA, 65; Albert Einstein Col Med, PhD(anat), 67. *Prof Exp:* Lectr demonstr zool, Andhra Christian Col, India, 54-55 & Univ Rangoon, 55-57; sci master, Anglo-Chinese Sch, Malaysia, 57-60; sr biol master, Opoku Ware Sch, Ghana, 60-63; NIH fel anat, Albert Einstein Col Med, 68; from asst prof to assoc prof, 68-77, PROF ANAT, COL MED & DENT NJ, 77- *Mem:* Am Asn Anat; Am Soc Zool; Soc Develop Biol; NY Acad Sci. *Res:* Electron microscopy; endocrines and neuroendocrines in amphibian metamorphosis; normal and thyroid-induced maturation of cerebellum. *Mailing Add:* Dept Anat UMDJ-NJMS 185 S Orange Ave Newark NJ 07103-2757

GONA, OPHELIA DELAINE, b Columbia, SC; m 62; c 2. OPHTHALMIC RESEARCH, ENDOCRINOLOGY. *Educ:* Johnson C Smith Univ, BS, 57; Yeshiva Univ, MS, 64; City Col New York, MA, 69; City Univ New York, PhD(biol), 71. *Prof Exp:* Lab technician, New York Univ, 57-58; high sch instr biol, New York City Public Schs, 58-61; vol, US Peace Corps, 61-63; instr, City Col New York, 69-71; asst prof, Montclair State Col, 71-77; ASST PROF ANAT, NJ MED SCH, COL MED & DENT NJ, 77- *Mem:* Am Asn Anatomists; AAAS; Asn Res Vision & Ophthal; Am Inst Biol Sci; NY Acad Sci. *Res:* Structure and function of the eye, with particular reference to the lens and cornea after injury or disease. *Mailing Add:* Dept Anat UMDNJ NJ Med Sch 185 S Orange Ave Newark NJ 07103

GONANO, JOHN ROLAND, b Winchester, Va, Jan 21, 39; m 59; c 3. PHYSICS. *Educ:* Univ WVa, BS, 60; Duke Univ, PhD(physics), 67. *Prof Exp:* Fel, Univ Fla, 66-68; res physicist, Nat Bur Standards, 68-71; res physicist, US Army Mobility Equip Res & Develop Command, 71-82; res & develop mgr, US Army Mat Command, 82-86, MGR & CHAIR, US ARMY ADV CONCEPTS & TECHNOL COMT, 86- *Mem:* Am Phys Soc; Sigma Xi. *Res:* Investigation of magnetic interactions by means of specific heat, nuclear magnetic resonance and thermal expansion measurements; low temperature thermometry; application of physical principles to detection of concealed explosives; program management; development of techniques for detection of land minefields; camouflage; counter surveillance and deception. *Mailing Add:* 10401 Regina Ct Clarksburg MD 20871

GONCZ, JOHN HENRY, geophysics, seismology, for more information see previous edition

GONDA, MATTHEW ALLEN, b Cheverly, Md, Aug 13, 49; m 83; c 2. AIDS VIROLOGY, AIDS ANIMAL MODELS. *Educ:* Univ Va, BS, 71; George Mason Univ, MS, 76; Johns Hopkins Univ, PhD(virol), 82. *Prof Exp:* Res asst electron micros, Meloy Labs, 71-73; head electron micros, Litton Bionetics, Frederick Cancer Res Facil, 73-87; HEAD LAB CELL & MOLECULAR STRUCT, NCI-FREDERICK CANCER RES FACIL, 88- *Concurrent Pos:* Lectr grad prog, Hood Col, 79; consult, Kirkegard & Perry Labs, Inc & Bethesda Res Labs, 81-82, Johns Hopkins Univ, 85-86, Oncor, 84-85, Serano Labs, 87, Centocor, 88 & NY State Inst Basic Res, 90; lectr, Sch Hyg & Pub Health, Johns Hopkins Univ, 88-91. *Mem:* AAAS; Sigma Xi; Am Soc Microbiol. *Res:* To understand the molecular mechanisms of pathogenesis and persistence of lentiviruses by uning animal retroviruses (lentiviruses) related to AIDS; develop animal models for AIDS to test vaccines and therapeutic agents; retrovirus biology; gene expression; gene regulation. *Mailing Add:* Lab Cell & Molecular Struct Prog Resources Inc Dyn Corp NCI Frederick Cancer Res & Develop Ctr Frederick MD 21702

GONDER, ERIC CHARLES, b Jefferson, Iowa, July 7, 50; m 87. POULTRY, POULTRY PRODUCTION. *Educ:* Iowa State Univ, DVM, 74; Univ Minn, MS, 76; NC State Univ, PhD(vet path), 91. *Prof Exp:* Vet med asst I&II, Dept Path, Col Vet Med, Univ Minn, 74-76; vet virol, USAMRIID, US Army, 76-78; Net poultry, ASL/Schering Plough, 78-83; vet med asst I&II, FAE, Col Vet Med, NC State Univ, 83-86; VET POULTRY, GOLDSBORO MILLING CO, 86- *Mem:* Am Asn Avian Pathologists; Poultry Sci Asn; US Animal Health Asn; Asn Vet Turkey Prod. *Res:* Focal ulcerative dermatitis of turkeys; enteric disease of turkeys. *Mailing Add:* Goldsboro Milling Co PO Drawer 10009 Goldsboro NC 27532

GONDOS, BERNARD, b New York, NY, Jan 6, 35. EXPERIMENTAL BIOLOGY. *Educ:* Yale Univ, BS, 56; George Washington Univ, MD, 59. *Prof Exp:* Asst prof path, Sch Med, Univ Calif-Los Angeles, 67-73; assoc clin prof path, Sch Med, Univ Calif, San Francisco, 73-75, assoc prof path, 75-78; dir anat path, Health Ctr Univ Conn, 78-86; SR SCIENTIST, SANSUM MED RES FOUND, SANTA BARBARA, CALIF, 86-, DIR PATH, 86- *Concurrent Pos:* Chief cytol & gynec path, Harbor Gen Hosp, Torrance, Calif, 67-73; fac mem, prog reproductive biol, Harbor Gen Hosp & Sch Med, Univ Calif, Los Angeles, 68-70; dir, Sch Cytotechnol, Harbor Gen Hosp, 69-73; chief cytol serv & gynec path, San Francisco Med Ctr, Univ Calif, 73-78; fac mem & dir, ultrastruct studies fac, Reproductive Endocrinol Ctr, Univ Calif, San Francisco, 74-78. *Res:* Experimental biology; numerous publications. *Mailing Add:* Sansum Med Res Found 2219 Bath St Santa Barbara CA 93105

GONG, WILLIAM C, b San Francisco, Calif, Aug 22, 48; m; c 3. CLINICAL PHARMACY. *Educ:* San Jose State Univ, BA, 70; Univ Southern Calif, PharmD(pharmacy), 74. *Prof Exp:* Lectr, 74-75, asst prof, 75-82, ASSOC PROF CLIN PHARMACY, UNIV SOUTHERN CALIF, 82- *Concurrent Pos:* Residency in clin pharmacy, Los Angeles County, Univ Southern Calif Med Ctr, Los Angeles, 74-75, coordr residency prog, Sch Pharmacy, 78-90, sect leader ambulatory care progs, 78-85; course coordr ambulatory care & skill nursing, 85-; dir, residency & fel training progs. *Honors & Awards:* Outstanding Serv Award, Am Diabetes Asn, 89. *Mem:* Am Col Clin Pharmacol; Am Soc Hosp Pharmacists; Am Asn Col Pharmacy; Am Pharmaceut Asn. *Res:* Investigation of the utilization of pharmacists as an active member of the health care team in providing direct patient care services, especially in the ambulatory care settings; developing residency training programs for pharmacists so that they may become proficient educators, researchers, and health care practitioners in the area of pharmacy practice; training programs for family practice residents; identifying predictors for control of diabetes. *Mailing Add:* Sch Pharmacy Univ Southern Calif 1985 Zonal Ave Los Angeles CA 90033

GONICK, ELY, b Detroit, Mich, May 24, 25; m 52; c 2. INORGANIC CHEMISTRY. *Educ:* Drew Univ, AB, 48; Pa State Col, PhD(inorg chem), 51. *Prof Exp:* Res chemist, E I du Pont de Nemours & Co, Inc, 51-55, tech supvr, 55-58, asst lab dir, 58-62, tech sect mgr, 62-64, asst dir res, 65-70, mgr, Inorg Fibers Div, Pigments Dept, 70-72, dir int develop, 73-74, dir res, Pigments Dept & asst dir, Finishes Div, Fabrics & Finishes Dept, 78-79, dir res, Finishes Dept, 80-81; vpres technol, Int Paper Co, NY, 81-89; RETIRED. *Mem:* Am Chem Soc; Sigma Xi. *Res:* Coordination compounds; polymers; pigments; surface chemistry; fiber reinforcement of plastics; light absorption and scattering; high temperature reactions. *Mailing Add:* 611 Millersville Lancaster PA 17603-6025

GONICK, HARVEY C, b Apr 10, 30; US citizen; m 67; c 2. NEPHROLOGY. *Educ:* Univ Calif, Los Angeles, BS, 51; Univ Calif, San Francisco, MD, 55. *Prof Exp:* Intern internal med, Peter Bent Brigham Hosp, Boston, Mass, 55-56; asst med, Sch Med, Boston Univ, 56-57; resident internal med, Wadsworth Vet Admin Hosp, 59-61, clin investr, 61-64, sect chief, Metab Balance Ward, 64-67; from instr to assoc prof, 64-76, ADJ PROF MED, SCH MED, UNIV CALIF, LOS ANGELES, 76-; DIR, TRACE ELEMENT LAB, CEDARS-SINAI MED CTR, 79- *Concurrent Pos:* Nat Heart Inst fel nephrol, Mass Mem Hosp, Boston, 56-57; fel nephrol, Wadsworth Vet Admin Hosp, Los Angeles, 59-61; Nat Kidney Found traveling fel, 63; mem, Nat Kidney Found, 63-, chmn res comt, 67, actg secy nat med adv coun, 69-70, regional rep, 70-73; chmn sci adv coun, SCalif Kidney Found, 68-70, mem bd dirs, 74-83; mem coun circulation, Am Heart Asn, 64-; mem renal dialysis & transplantation adv comt, Calif State Dept Pub Health, 69-71; mem ad hoc steering comt, Kidney Dis Planning Proj, Calif Regional Med Prog, 70, area rep, 71; Clin Chief nephrol, Cedars-Sinai Med Ctr, 83-85. *Mem:* AAAS; fel Am Col Physicians; Am Fedn Clin Res; AMA; Am Soc Nephrology. *Res:* Renal pathophysiology; trace element metabolism; effects of transport inhibitors on renal tubular function. *Mailing Add:* 2080 Century Pk East 707 Los Angeles CA 90067

GONNELLA, PATRICIA ANNE, b Philadelphia, Pa, Nov 25, 49. CELL BIOLOGY. *Educ:* Temple Univ, AB, 71, MA, 76, PhD(biol), 79. *Prof Exp:* Fel, Sch Med, Univ Pa, 78-81; fel, 82-85, instr, 86, ASST PROF, HARVARD MED SCH, HARVARD UNIV, 87- *Mem:* AAAS; Am Soc Cell Biologists. *Res:* Barrier function in the developing intestine and immune function in response to trauma and sepsis. *Mailing Add:* Brigham & Women's Hospital 75 Francis St Thor-14 Boston MA 02115

GONOR, JEFFERSON JOHN, b Lafayette, La, Nov 8, 32; m 69; c 2. INVERTEBRATE ZOOLOGY, MARINE BIOLOGY. *Educ:* Univ Southwestern La, BS, 53; Univ Wash, PhD(invert zool), 64. *Prof Exp:* Actg instr zool, Univ Wash, 59 & 60; vis asst prof biol, Univ of the Pac, 60 & 62; asst prof marine sci, Univ Alaska, 62-64; asst prof biol oceanog, 65-74, assoc prof oceanog & zool, 74-82, PROF BIOL OCEANOG, ORE STATE UNIV, 82-, HEAD ADV, COL OCEANOG, 87-, DIR, MARINE RESOURCE MGT GRAD PROG, 87- *Concurrent Pos:* Res partic, Orgn Trop Studies, 64; resident staff, OSU Marine Sci Ctr, Newprot, Ore, 65-; consult mitigation methods estuarine land use planning & develop, 77-; regional ed, Marine Biol, Int J Life Oceans & Coastal Waters, 76-77. *Mem:* AAAS; Soc Syst Zool; Ecol Soc Am. *Res:* Systematics of opisthobranch gastropods; molluscan functional anatomy; neurosecretion in barnacles; marine invertebrate reproduction and larval biology; biological oceanography; population biology of introduced marine wood boring isopods. *Mailing Add:* 33267 White Oak Rd Corvallis OR 97333

GONSALVES, DENNIS, b Kohala, Hawaii, Apr 2, 43; m 65; c 2. PLANT VIROLOGY. *Educ:* Univ Hawaii, BS, 65, MS, 68; Univ Calif, Davis, PhD(plant path), 72. *Prof Exp:* Asst plant path, Univ Hawaii, 65-66; asst prof, Univ Fla, 72-77; assoc prof, 77-86, PROF PLANT PATH, CORNELL UNIV, 86- *Mem:* Am Phytopath Soc; Soc Gen Microbiol. *Res:* Characterization of fruit tree viruses; genetics of plant viruses; cross protection. *Mailing Add:* Dept Plant Path Agr Exp Sta Cornell Univ Geneva NY 14456

GONSALVES, LENINE M, b New Bedford, Mass, Nov 23, 27; m 52; c 4. ELECTRICAL ENGINEERING. *Educ:* US Naval Acad, BS, 52; Northeastern Univ, MSEE, 60. *Prof Exp:* Proj engr, Res Dept, Aerovox Corp, 52-53; instr elec eng, 53-56, assoc prof, 56-57, chmn elec eng dept, 57-69 & 77-78, PROF ELEC ENG, SOUTHEASTERN MASS UNIV, 57-, CHMN ELEC ENG TECHNOL DEPT, 87- *Concurrent Pos:* Consult, Eng & Res Sect, Aerovox Corp, 57, 61-62; prof & chmn dept elec eng, Col Petrol & Mineral, Univ Dharan, Saudi Arabia, 69-71; acad adv eng technol, Nat Inst Elec & Electronics Engrs, 78-79. *Mem:* Nat Soc Prof Engrs; Am Soc Eng Educ; Inst Elec & Electronics Engrs. *Res:* Circuit theory; electrical power systems. *Mailing Add:* Dept Elec Eng Technol Southeastern Mass Univ North Dartmouth MA 02747

GONSALVES, NEIL IGNATIUS, b Georgetown, Brit Guiana, Feb 1, 38; US citizen; m 61; c 2. GENETICS. *Educ:* Georgetown Univ, BS, 59; Brown Univ, PhD(biol), 69. *Prof Exp:* Asst prof genetics, Lowell State Col, 68-69; asst prof genetics, 69-73, assoc prof biol, 73-78, PROF BIOL, RI COL, 78-, CHMN DEPT, 81- *Concurrent Pos:* Vis prof, Brown Univ, 76- *Mem:* Am Soc Zool; Genetics Soc Am; Am Soc Cell Biol; Am Genetics Asn; Soc Invest Dermat; Sigma Xi. *Res:* Radiation biology of mammalian skin; mechanisms of wound healing in mammalian skin; developmental biology and genetics of anophthalmia in the mouse. *Mailing Add:* 281 River Ave Providence RI 02908

GONSER, BRUCE WINFRED, b Hudson, Ind, Sept 9, 99; m 25; c 3. METALLURGY. *Educ:* Purdue Univ, BS, 23; Univ Utah, MS, 24; Harvard Univ, SD(metall), 33. *Hon Degrees:* DrEng, Purdue Univ, 67. *Prof Exp:* Res engr, Am Smelting & Refining Co, 24-31; metall engr, Nat Radiator Corp, 33-34; res chief, Battelle Mem Inst, 34-50, asst dir metall, 50-53, tech dir, 53-64; consult, 64-84; RETIRED. *Concurrent Pos:* Secy-treas bd dir, Tin Res Inst Inc, 50-84. *Honors & Awards:* James Douglas Gold Medal, Am Inst Mining, Metall & Petrol Engrs, 71; Award Merit, Am Soc Testing & Mat. *Mem:* Am Inst Mining, Metall & Petrol Engrs; Am Soc Metals; Am Soc Testing & Mat; Wire Asn; Electrochem Soc. *Res:* Nonferrous metallurgy; uncommon metals; extractive metallurgy. *Mailing Add:* 1812 Riverside Dr Apt 22 Columbus OH 43212

GONSHOR, HARRY, b Montreal, Que, Sept 26, 28; US citizen; m 62. PURE MATHEMATICS. *Educ:* McGill Univ, BSc, 48, MSc, 49; Harvard Univ, PhD(math), 53. *Prof Exp:* Vis instr math, Univ Southern Calif, 53-54; asst prof, Univ Miami, 54-55 & Pa State, 55-57; from asst prof to assoc prof, 57-86, PROF MATH, RUTGERS UNIV, 86- *Mem:* Am Math Soc; Can Math Cong. *Res:* Functional analysis; non-standard analysis; algebra. *Mailing Add:* Dept Math Rutgers State Univ New Brunswick NJ 08903

GONTIER, JEAN ROGER, b Lens Pas de Calais, France, Mar 8, 27. PHYSIOLOGY OF RESPIRATION, RESPIRATION & DIVING. *Educ:* Col d'Etampes France, AB, 45; Col Sci Paris, MS, 47; Sch Med Paris, MD, 65. *Prof Exp:* Prof physiol, UGSEL, Paris, 57-62; instr med, Sch Med Paris, 60-65; dir physiol, Sch Med Reims, 66-68; prof physiol, Sch Med, Univ Montreal, 70-78; CONSULT INTERNAL MED, 79- *Concurrent Pos:* Consult & ed, var publ, New York, NY, 75-79. *Mem:* Am Physiol Soc; Can Physiol Soc; NY Acad Sci; AAAS. *Res:* physiology of respiration in man; respiratory adaptions to diving in man; mechanical factors in breathing; author of textbooks on human physiology and respiratory physiology. *Mailing Add:* 133 rue Michel-Ange Paris F-75016 France

GONYEA, WILLIAM JOSEPH, b Springfield, Mass, Aug 18, 42; m 66; c 2. ANATOMY, EXERCISE BIOLOGY. *Educ:* Univ Miami, BEd, 68, MA, 69; Univ Chicago, PhD(anat), 73. *Prof Exp:* From asst prof to assoc prof, 72-80, PROF ANAT, UNIV TEX HEALTH SCI CTR, 80-, CHMN DIV ANAT, 85- *Mem:* Am Asn Anatomists; Am Col Sports Med; Am Soc Zoologists; Nat Strength & Conditioning Asn. *Res:* Structure and function of skeletal muscle and the cardiovascular system; evolutionary biology of vertebrates especially mammaliam carnivores. *Mailing Add:* Dept Cell Biol Univ Tex Health Sci Ctr Dallas TX 75235

GONZALES, CIRIACO Q, b Socorro, NMex, Oct 22, 33; m 58; c 4. COMPARATIVE PHYSIOLOGY. *Educ:* NMex State Univ, BS, 54; Univ Ariz, MS, 58; Univ Calif, Berkeley, PhD(entomol & physiol), 62. *Prof Exp:* Asst res entomologist, Univ Calif, Berkeley, 61-63; from asst prof to assoc prof biol, Col Santa Fe, NMex, 63-72; health scientist adminr, NIH, 72-75; DIR, MINORITY BIOMED RES SUPPORT PROG, NIH, 75- *Concurrent Pos:* Mem, Nat Acad Sci Panel, Peru, 75. *Mem:* AAAS; Am Soc Microbiol. *Res:* Pesticide toxicology; physiological effects on humans and animals and effects on the environment; transmission of viruses by insect vectors to fruit trees. *Mailing Add:* 1508 Aintree Dr Rockville MD 20850

GONZALES, ELWOOD JOHN, b New Orleans, La, Oct 19, 27; m 63; c 4. CELLULOSE CHEMISTRY. *Educ:* Loyola Univ, La, BS, 53; Tulane Univ, MS, 55, PhD(chem), 58. *Prof Exp:* Res chemist cellulose chem, USDA, New Orleans, 57-63, res chemist chem invests, 63-72, res chemist, Polymer Finishes Res Unit, Southern Reg Res Ctr, 72-78, res chemist, Cellulose Polymer Res Div, Agr Res Serv, 78-84, res chem ist textile finishing chem, 85-87; RETIRED. *Mem:* AAAS; Am Chem Soc; Sigma Xi; Am Asn Textile Chemists & Colorists; Am Inst Chemists. *Res:* Kinetics and mechanisms of cellulose recactions; modification of cotton cellulose; flame retardant and cellulose chemistry; application of statistics for cellulose reactions; dyeing modified cotton celluloses. *Mailing Add:* 2137 Graham Dr Gretna LA 70056

GONZALES, FEDERICO, b San Antonio, Tex, July 16, 21; m 45; c 7. CELL BIOLOGY. *Educ:* St Louis Univ, BS, 48, MS, 51, PhD(biophys), 54. *Prof Exp:* Sr res fel biol, Univ Tex M D Anderson Hosp & Tumor Inst, 53-54, res assoc, 54-55; from instr to asst prof anat, Dent Br, Univ Tex, 55-61; asst prof anat, Col Med, Baylor Univ, 57-63, asst prof exp biol, 61-63; ASSOC PROF ANAT, SCHS DENT & MED, NORTHWESTERN UNIV, CHICAGO, 63- *Concurrent Pos:* USPHS sr res fel, 58-62 & career develop award, 62-63. *Mem:* Am Soc Cell Biol; Am Physiol Soc; Am Asn Anat; Tissue Cult Asn; Soc Develop Biol; Sigma Xi. *Res:* Differentiation in tissue culture; structure and function of the nucleolus; survival of frozen tissues; ultrastructure of bone and other hard tissues; ultrastructure of heart. *Mailing Add:* Northwestern Univ 303 E Chicago Ave Chicago IL 60611

GONZALES, SERGE, b Boston, Mass, Feb 23, 36; m 68; c 1. STATIGRAPHY-SEDIMENTATION. *Educ:* Duke Univ, AB, 58; Miami Univ, MS, 60; Cornell Univ, PhD(stratig paleont), 63. *Prof Exp:* Assoc, Miami Univ, 59-60; asst, Cornell Univ, 60-63; petrol geologist, Humble Oil & Refining Co, La, 63-66; asst prof, Miami Univ, 66-71; asst prof geol & staff geologist, Inst Area & Community Develop, Univ Ga, 71-78, assoc prof & assoc dir, Inst Nat Resources, 78-81; CONSULT, 81- *Concurrent Pos:* Staff consult, Earth Sci Labs, Ohio, 67-69; consult, Oak Ridge Nat Lab, 74-, US Sen Ga, 74-79 & Battelle Mem Inst, 78-81; pres, Earth Resource Assocs, Ga & Okla, 74-; mem, Argonne Nat Lab, 81-; Am Inst Profs Geol (nat secy 88-89). *Mem:* Am Inst Profs Geol; Am Asn Petrol Geologists; Sigma Xi; AAAS; Am Inst Mining Metall & Petrol Engrs. *Res:* Geological sciences; energy and mineral resources; environmental and engineering geology; geologic energy resources-exploration for and development of; radioactive and hazardous waste disposal in geologic environment; industrial-minerals geology; underground hydrocarbon-storage caverns-siting of. *Mailing Add:* Earth Resource Assocs Inc 125 Cedar Creek Dr Athens GA 30605-3356

GONZALEZ, EDGAR, b San Juan, PR, Nov 29, 24; US citizen; m 53; c 1. MECHANICAL ENGINEERING. *Educ:* Univ Wis-Madison, BSEE, 47. *Prof Exp:* Elec engr, US Steel Corp, 47-50; proj engr, Skidmore, Owings & Merrill A/E, 50-53; proj engr, Fay, Spofford & Thorndike Engrs, 54-56; proj engr, Guy B Panero Engrs, 56-68; contract engr, E Gonzalez Consult, 58-64; sr proj engr, Am Can Co, 64-80; CONSULT ENGR, E GONZALEZ ENGRS, PROF CONSULTS, 80- *Concurrent Pos:* Consult engr, E Gonzalez Prof Engrs-Consult Engrs, 87- *Mem:* Inst Elec & Electronics Engrs; Am Soc Mech Eng; Soc Am Mil Engrs. *Mailing Add:* Consult Engr 112-18 86th Ave Richmond Hill NY 11418

GONZALEZ, ELMA, b June 6, 42; US citizen. CELL BIOLOGY. *Educ:* Tex Woman's Univ, BS, 65; Rutgers Univ, PhD(cell biol), 72. *Prof Exp:* NIH fel plant physiol, Univ Calif, Santa Cruz, 72-74; asst prof cell biol, 74-81, ASSOC PROF BIOL, UNIV CALIF, LOS ANGELES, 81- *Concurrent Pos:* Fel, Nat Chicano Coun Higher Educ Ford Found, 78. *Mem:* Am Soc Plant Physiologists; AAAS; NY Acad Sci; Soc Advan Chicanos & Native Am in Sci. *Res:* Mechanisms and regulation of the formation of microbodies and other metabolic compartments; intracellular protein traffic and sorting; calcification and golgi apparatus in marine micro algae. *Mailing Add:* Dept Biol Univ Calif Los Angeles CA 90024-1606

GONZALEZ, EULOGIO RAPHAEL, b Arecibo, PR, April 26, 48; US citizen; m 79; c 3. NEUROENDOCRINOLOGY. *Educ:* St Francis Col, BS, 70; NY Univ, MS, 72, Albert Einstein Col Med, MD, 86, PhD(biol), 79. *Prof Exp:* Teaching asst physiol, dept biol, NY Univ, 72-78 & hematol technol, Med Ctr, 75-, NY Univ Med Ctr, 75-78; REASTS PHYSIOLOGIST, VET ADMIN CTR, E ORANGE, NJ, 79- *Concurrent Pos:* Adj asst prof, dept natural sci, City Univ NY, 79-; assoc surg, dept neurosurg, Col Med & Dent NJ, 80- *Mem:* NY Acad Sci; AAAS; Soc Neurosci; Am Acad Opthamol; AMA. *Res:* Central mechanisms subserving cardiovascular and respiratory control; hypertension; electrophysiological recording of autonomic reflexes in normal and hypertensive animals; techniques incorporated include stimulation and/or recording of aortic, carotidsinus, splanchnic, renal or phrenic nerves in decerebrate preparations. *Mailing Add:* 3224 Grand Concourse 3C Bronx NY 10458

GONZALEZ, FRANCISCO MANUEL, b San Juan, PR, Sept 28, 33; c 5. NEPHROLOGY. *Educ:* Va Mil Inst, BA, 53; Med Col Va, MD, 57. *Prof Exp:* Intern, Georgetown Univ Hosp, 57-58, resident internal med, 58-61; dir, Renal Div, US Naval Hosp, Oakland, Calif, 63-65; from asst prof to assoc prof med, 66-74, PROF MED & PHARMACOL, MED CTR, LA STATE UNIV, NEW ORLEANS, 74-; DIR DEPT NEPHROLOGY, CHARITY HOSP, 66- *Concurrent Pos:* Clin & res fel nephrology, Georgetown Univ Hosp, 61-63; vis physician, Hotel Dieu Hosp, New Orleans, 66-, E Jefferson Hosp, 66- & Charity Hosp, 66-; consult nephrology, Earl K Long Hosp, Baton Rouge, 66-, Lake Charles Charity Hosp, 66- & Lafayette Charity Hosp, 66-; mem, Nat Kidney Found, 66- *Mem:* Am Soc Nephrology; Am Fedn Clin Res; Am Soc Artificial Internal Organs; Am Heart Asn; fel Am Col Physicians. *Res:* Renal disease; physiology and pharmacology; artificial kidney and transplantation. *Mailing Add:* 2000 Tulane Ave New Orleans LA 70112

GONZALEZ, FRANK J, b Tampa, Fla, Nov 30, 53. MOLECULAR CARINOGENETICS. *Educ:* Univ SFla, Tampa, BA, 75, MA, 77; Univ Wis-Madison, PhD(oncol), 81. *Prof Exp:* Fel, McArdle Lab Cancer Res, Univ Wis-Madison, 81-82; staf fel, Lab Develop Pharmacol, Nat Inst Child Health & Human Develop, NIH, Bethesda, 82-84, sr staff fel, Lab Molecular Carcinogenesis, Nat Cancer Inst, 84-88, actg chief, Nucleic Acids Sect, 86-89, SUPVRY RES CHEMIST, LAB MOLECULAR CARCINOGENESIS, NAT CANCER INST, NIH, BETHESDA, 88-, CHIEF, NUCLEIC ACIDS SECT, 89- *Concurrent Pos:* Ad hoc mem, Phys Biochem Study Sect, NIH, 86 & 88, Nat Inst Drug Abuse, 89, Prom Rev Comt, Nat Inst Environ Health Serv, 89, ad hoc reviewer, Chem Path Study Sect, 89, adv comt, Pharmacol Res Assoc Training Prog, 90-94; ad hoc mem, Carcinogenesis & Nutrit Study Sect, Am Cancer Soc, 91. *Honors & Awards:* Rawls-Palmer Progress Med Award, Am Soc Clin Pharmacol & Therapeut, 91. *Mem:* Am Soc Biochem & Molecular Biol; Int Soc Study Xenobiotics; Am Fedn Clin Res. *Res:* Molecular carinogenetics; numerous publications. *Mailing Add:* Nucleic Acids Sec Lab Nat Cancer Inst NIH Bldg 37 Rm 3E24 Bethesda MD 20892

GONZALEZ, GLADYS, b San Germán, PR, Oct 11, 55; US citizen; m 77; c 2. LAND RESOURCE ECONOMICS, AGRICULTURAL PRODUCTION ECONOMICS. *Educ:* Univ PR, BA, 76, MS, 81; Univ MO, PhD(agr econ), 84. *Prof Exp:* Res asst agr econ, Univ MO, Columbia, 79-83; res asst, 77-79, asst prof, 84-87, ASSOC PROF AGR ECON, UNIV PR, MAYAQUEZ, 87- *Concurrent Pos:* Dept chairperson, Dept Agr Econ, Univ PR, Mayaquez, 84-, actg assoc dean for resident instr, Fac Agr, 91- *Mem:* Am Agr Econ Asn; Asn Economist PR; Col Agron PR. *Res:* Tropical agriculture and food consumption in Puerto Rico; three years of research experience in US agricultural land use problems. *Mailing Add:* Dept Agr Econ PO Box 5000 Mayaguez PR 00708

GONZALEZ, GUILLERMO, b Havana, Cuba, June 25, 44; US citizen; c 2. ELECTRICAL ENGINEERING. *Educ:* Univ Miami, BSEE, 65, MSEE, 66; Univ Ariz, PhD(elec eng), 69. *Prof Exp:* Sr engr elec eng, Bell Aerosyst Co, 67-69; from asst prof to assoc prof, 69-79, PROF ELEC ENG, UNIV MIAMI, 79- *Concurrent Pos:* Prin investr, Univ Miami Inst grant, 69-70, NSF grant, 71-72, 73-75, 76-77 & Ryder Prog Transp grant, 78. *Mem:* Sigma Xi. *Res:* Microwave electronics and electromagnetics. *Mailing Add:* Dept Elec Eng Univ Miami Coral Gables FL 33124

GONZALEZ, LUIS L, b Edinburg, Tex, June 7, 28; m; c 4. SURGERY. *Educ:* Univ Tex, BA, 49, MD, 53; Univ Cincinnati, DS(surg), 54. *Prof Exp:* Chief surg serv, Vet Admin Hosp, Cincinnati, 61-65; from instr to asst prof, 59-69, ASSOC PROF SURG, COL MED, UNIV CINCINNATI, 69-; ASST CHIEF SURG SERV, VET ADMIN HOSP, 65- *Concurrent Pos:* Asst attend surgeon, Cincinnati Gen Hosp, 61-; attend surgeon, Christian R Holmes Hosp, 64-; assoc attend surgeon, Christ Hosp, 64- *Mem:* Am Col Surg; Int Cardiovasc Soc. *Res:* Cardiovascular research; thoracic surgery. *Mailing Add:* Christian Holmes Hosp Cincinnati OH 45219

GONZALEZ, MARIO J, b Laredo, Tex, Nov 11, 41; m 66; c 1. HARDWARE SYSTEMS, SOFTWARE SYSTEMS. *Educ:* Univ Tex, Austin, BS, 64, MS, 69, PhD(elec eng), 71. *Prof Exp:* Mem tech staff comput systs, Tex Instruments, 72-73; asst prof comput sci, Northwestern Univ, 73-77; assoc prof, 77-81, div dir comput eng, 82-86, ASSOC DEAN ENG, UNIV TEX, SAN ANTONIO, 86- *Concurrent Pos:* Mem, Equal Opportunity Sci & Technol Comt, NSF, 84- *Mem:* Inst Elec & Electronics Engrs; Am Soc Eng Educ; Asn Comput Mach. *Res:* Design and implementation of distributed processor systems. *Mailing Add:* Col Eng Eng 236 Univ Tex Austin TX 78712

GONZALEZ, NORBERTO CARLOS, b La Plata, Arg, Aug 1, 37; m 71; c 2. MEDICAL PHYSIOLOGY. *Educ:* Nat Univ La Plata, MD, 62. *Prof Exp:* Instr physiol, Nat Univ La Plata, 62-65; fel, Univ Kans Med Ctr, 65-67; asst prof, Nat Univ La Plata, 67-70; from asst prof to assoc prof, 70-78, PROF PHYSIOL, COL HEALTH SCI & HOSP, UNIV KANS MED CTR, 78- *Concurrent Pos:* Fel, Nat Coun Invests, Arg, 65-66; USPHS fel, 66-67; fel, Alexander von Humboldt Found, Bonn, WGer, 77-78 & 80. *Mem:* Sigma Xi; Am Physiol Soc. *Res:* Intra- and extracellular acid-base regulation; tissue and lung gas exchange; acid-base effects on cardiovascular system; myocardiol hypertrophy; adaptation to hypoxig. *Mailing Add:* Dept Physiol Col Health Sci Univ Kans Med Ctr Kansas City KS 66103

GONZALEZ, PAULA, b Albuquerque, NMex, Oct 25, 32. ENVIRONMENTAL SCIENCE, BIOETHICS. *Educ:* Col Mt St Joseph, AB, 52; Cath Univ Am, MS, 62, PhD(cell physiol), 66. *Prof Exp:* Instr anat, physiol & chem, Regina Sch Nursing, NMex, 52-54; biol teacher, Seton High Sch, Ohio, 55-60; from asst prof to prof, 65-82, chmn dept biol, 68-73, adj prof biol, 82-86, EDUC BIOLOGIST, COL MT ST JOSEPH, 86- *Concurrent Pos:* Freelance Futurist, Col Mt St Joseph, 86- *Mem:* World Future Soc; Inst Soc, Ethics & Life Sci; Sigma Xi. *Res:* Nucleolar changes during the cell cycle; course development; energy and environment; biomedical advances and their human implications; science, technology and human values; global ecology; development of personalized learning system for human anatomy and physiology, (including use of audio-visual components); planetary consciousness, hope through alternatives, passive solar design. *Mailing Add:* 5820 Bender Rd Cincinnati OH 45233

GONZALEZ, RAFAEL C, b Havana, Cuba, Aug 26, 42; US citizen; m 65; c 2. DIGITAL IMAGE PROCESSING, PATTERN RECOGNITION. *Educ:* Univ Miami, Coral Gables, Fla, BSEE, 65; Univ Fla, Gainesville, ME, 67, PhD(elec eng), 70. *Prof Exp:* Design engr, Gen Tel & Electronics, 66-67; grad asst, Univ Fla, Gainesville, 67-70; from asst prof to prof, 70-81, IBM PROF, UNIV TENN, KNOXVILLE, 81-; PRES, PERCEPTICS CORP, KNOXVILLE, TENN, 80- *Concurrent Pos:* Consult, Oak Ridge Nat Lab, Tenn, 72-82, Lockheed Corp, Sunnyvale, Calif, 81-84 & Tex Instruments, Inc, Dallas, 84-87; chmn bd, Perceptics Corp, Knoxville, Tenn, 80-; prof eng, Magnavox, 80; M E Brooks distinguished prof, Univ Tenn, Knoxville, 81 & distinguished serv prof, 84-; prof, IBM, 81. *Mem:* Fel Inst Elec & Electronics Engrs; Pattern Recognition Soc. *Res:* Image processing; pattern recognition; machine learning. *Mailing Add:* Perceptics Corp PO Box 22991 Knoxville TN 37933

GONZALEZ, RAMON RAFAEL, JR, b Los Angeles, Calif, May 25, 40; m 64; c 1. BIOENGINEERING & BIOMEDICAL ENGINEERING. *Educ:* Walla Walla Col, BS, 62, MA, 65; Wake Forest Univ, PhD(physiol), 73. *Prof Exp:* Teaching asst, Dept Biol, Walla Walla Col, 62-64; physiologist, Virginia Mason Res Ctr, Seattle, Wash, 66-68; asst prof, 73-85, ASSOC PROF PHYSIOL, SCH MED, LOMA LINDA UNIV, 85- *Concurrent Pos:* Res fel, Dept Neurosurg, Univ Basel, 73-75; consult cardiovasc pharmacol & biomed electronics, Delalande Int, Paris, France, 74-75; assoc res, Univ Calif Irvine, 88-89. *Mem:* Am Physiol Soc; Am Heart Asn; Sigma Xi; Am Inst Ultrasound Med. *Res:* Autonomic control of the circulation of blood in mammals, especially the neurovascular smooth muscle interface in skeletal muscle; human brain blood flow as studied with ultrasound. *Mailing Add:* Dept Physiol Loma Linda Univ Sch Med Loma Linda CA 92350

GONZALEZ, RAUL A(LBERTO), b Talca, Chile, Feb 8, 33; US citizen; m 58; c 3. CHEMICAL ENGINEERING. *Educ:* Santa Maria Univ Chile, BS, 57; Univ Va, MSc, 60, DSc(chem eng), 62. *Prof Exp:* Res engr, Santa Maria Univ Chile, 57-58; prof chem eng, Cath Univ Chile, 62; ENG FEL, JACKSON LAB, E I DU PONT DE NEMOURS & CO, INC, NJ, 62- *Res:* Reaction kinetics; reverse osmosis; process evaluation and design; high temperature reactions. *Mailing Add:* 206 Dallas Ave Newark DE 19711

GONZALEZ, RICHARD DONALD, b New York, NY, Apr 11, 32; m 62; c 1. PHYSICAL CHEMISTRY. *Educ:* Rensselaer Polytech Inst, BChE, 61; Johns Hopkins Univ, MA, 63, PhD(chem), 65. *Prof Exp:* From asst prof to assoc prof chem, 65-77, prof chem, Univ RI, 77-; AT DEPT CHEM ENG, UNIV ILL. *Mem:* Am Chem Soc. *Res:* Surface chemistry; heterogeneous catalysis. *Mailing Add:* Dept Chem Eng Univ Ill Box 4348 Chicago IL 60680-4348

GONZALEZ, RICHARD RAFAEL, b El Paso, Tex, Sept 15, 42; m 65; c 4. PHYSIOLOGY. *Educ:* Univ Tex, BS, 62; Univ San Francisco, MS, 66; Univ Calif, Davis, PhD(physiol, biophys), 70. *Prof Exp:* Vis asst fel physiol, 70-72, asst fel, 72-75, assoc fel bioeng, John B Pierce Found, 75-; asst prof, 72-76, assoc prof environ physiol, Sch Med, Yale Univ, 76-; AT DEPT ERGONOMICS, ENVIRON MED, US ARMY RES INST. *Concurrent Pos:* USPHS fel physiol, Sch Med, Yale Univ, 70-72; fel, Branford Col, Yale Univ, 72-77; assoc prof, Dept Kinesiology, Univ Calif, Los Angeles, 73-75. *Mem:* Am Physiol Soc. *Res:* Thermal physiology; peripheral circulation; environmental physiology; temperature regulation. *Mailing Add:* 129 Woodland St Sherborn MA 01770

GONZALEZ, ROBERT ANTHONY, b Victoria de las Tunas, Cuba, Jan 17, 45; US citizen; m 66; c 3. NUCLEAR PHYSICS, COMPUTER SCIENCE. *Educ:* Fla Southern Univ, BS, 71; State Univ NY, MS, 79. *Prof Exp:* Nuclear analyst radiation, 71-75, nuclear engr, 75-80, LEAD ENGR RADIATION ANAL & MGR, SHIELDING METHODS DEVELOP, KNOLLS ATOMIC POWER LAB, 80-, MGR SERV ENG, 85- *Res:* Radiation analysis, specifically the shielding of all radiation, ie gamma, neutron, emanating from fission reactors; large scientific computers. *Mailing Add:* Knolls Atomic Power Lab PO Box 1072 Schenectady NY 12301

GONZALEZ-ANGULO, AMADOR, b Tijuana, Mex, May 29, 33; m 58; c 2. NEUROPATHOLOGY. *Educ:* Univ Sonora, BS, 51; Nat Univ Mex, MD, 58; Am Bd Path, dipl, 63. *Prof Exp:* Instr path, Col Med, Baylor Univ, 61-63; ASST PROF PATH, MED SCH, NAT UNIV MEX, 66-; CHIEF ELECTRONIC MICROS SECT, DIV PATH, DEPT SCI RES, NAT MED CTR, 66- *Concurrent Pos:* Consult, Methodist Hosp, Houston, Tex, 62-63, Ben Taub Gen Hosp, 62-63 & Biomed Inst, Nat Univ Mex, 67-68; vis prof path, Pittsburgh, Pa, 73. *Mem:* Electron Micros Soc Am; Am Asn Anat; Mex Asn Path (pres, 69). *Res:* Ultrastructure of neoplasia with special emphasis on gynecological lesions; mechanisms of implantation and blocking agents in the experimental animal; tissue reaction to contraceptive medication in women (scanning and transmission electronmicroscopic studies). *Mailing Add:* Dept Sci Res Nat Univ Mexico Med Sch Unidad de Investigacion Biomed Nat Med Ctr Apartado 73-032 Mexico City DF 73 Mexico

GONZALEZ DE ALVAREZ, GENOVEVA, b Mayaguez, PR, Oct 21, 26; m 51; c 2. ORGANIC CHEMISTRY, PHARMACEUTICAL CHEMISTRY. *Educ:* Univ PR, Rio Piedras, BS, 47; Radcliffe Col, MA, 50; Univ Madrid, MS & PhD(chem), 66. *Prof Exp:* Asst instr chem, Univ PR, Rio Piedras, 46-47, instr, 47-51; from instr to prof chem, Univ PR, Mayaguez, 51-84; RETIRED. *Mem:* Am Chem Soc. *Res:* Alkaloids of Retama sphaerocarpa Boiss; synthesis of potential hypotensive agents. *Mailing Add:* 79 Oviedo St Belmonte Mayaguez PR 00708

GONZALEZ-FERNANDEZ, JOSE MARIA, b Buenos Aires, Arg, Nov 10, 22; US citizen; m 57; c 2. MATHEMATICS, PHYSIOLOGY. *Educ:* Univ Buenos Aires, MD, 47; Northwestern Univ, MSc, 54, PhD(math), 58. *Prof Exp:* Instr math, Univ Wash, 57-58; res assoc physiol, Mayo Clin, 58-59; temp mem staff math, Courant Inst Math Sci, NY Univ, 59-60; mem staff math res ctr, Univ Wis, 60-62; RES MATHEMATICIAN, NAT INSTS HEALTH, 62- *Mem:* Am Math Soc. *Res:* Integrability theorems in trigonometric series; tauberian theorems; integral equations; transport and consumption of oxygen in blood capillary-tissue systems. *Mailing Add:* Math Res Br Bldg 31 Rm 413-54 NIH 9000 Rockville Pike Bethesda MD 20892

GONZALEZ-LIMA, FRANCISCO, b Havana, Cuba, Dec 7, 55; US citizen; m 81. FUNCTIONAL NEUROANATOMY. *Educ:* Tulane Univ, BS, 76, BA, 77; Univ PR, PhD(anat), 80. *Prof Exp:* Res fel neurophysiol, Sch Med, Univ PR, 80-81; from asst prof to assoc prof neuroanat, Ponce Sch Med, 80-85; asst prof med neuroanat, Col Med, Tex A&M Univ, 85-90; ASSOC PROF PSYCHOL & NEUROSCI, UNIV TEX AUSTIN, 91- *Concurrent Pos:* Prin investr, NSF, 81-83, Resource Ctr Sci & Eng, PR, 82-83, NIH, 82-86, NATO, 87-88 & NIMH, 88-92; consult, dept radiol, PR Med Ctr, 81-85; Humboldt res fel, Alexander Von Humboldt Found, WGermany, 82-83; vis prof, German Sci Found, Inst Zool, Tech Univ Darmstadt, WGermany, 84-85; mem, Grant Rev Comt, NIMH, 90-93; Presidential Award lectr, Tex

A&M Chap, Soc Neurosci, 90. *Honors & Awards:* Int Soc Neuroethology Prize, 87. *Mem:* Soc Neurosci; Europ Neurosci Asn; Int Soc Neuroethology. *Res:* Neuroanatomical bases of integrative processes, such as arousal, motivation, learning and memory; application of 2-deoxyglucose method and histochemistry for functional brain mapping of behavioral activities. *Mailing Add:* Dept Psychol & Inst Neurosci Univ Tex Austin Austin TX 78712

GOO, EDWARD KWOCK WAI, b Honolulu, Hawaii, Nov 25, 56; m 86; c 2. CERAMICS ENGINEERING, METALLURGY & PHYSICAL METALLURGICAL ENGINEERING. *Educ:* Cornell Univ, BS, 78; Univ Calif, Berkeley, MS, 80; Stanford Univ, PhD(mat sci), 85. *Prof Exp:* staff scientist, Lawrence Berkeley Lab, 85; asst prof, 85-90, ASSOC PROF MAT SCI, UNIV SOUTHERN CALIF, 90- *Concurrent Pos:* NSF presidential young investr, 88; Off Naval Res young investr, 90. *Honors & Awards:* Hardy Gold Medal, Metall Soc, 86. *Mem:* Am Phys Soc; Am Ceramic Soc; Am Soc Metals; Metall Soc; Electron Micros Soc Am; AAAS. *Res:* Transmission electron micrscopy of materials; structure-property relationship. *Mailing Add:* Dept Mat Sci & Eng Univ Southern Calif Los Angeles CA 90089-0241

GOOCH, VAN DOUGLAS, b Oakland, Calif, July 23, 45; m 73. CELL PHYSIOLOGY. *Educ:* Calif State Univ, Hayward, BS, 67; Univ Calif, Berkeley, PhD(biophysics), 73. *Prof Exp:* Fel circadian rhythm, Harvard Univ, 73-76; asst prof biol, Univ Minn, 76-77; asst prof, Hiram Col, 78; ASST PROF BIOL, UNIV MINN, MORRIS, 78- *Concurrent Pos:* Fel, Nat Inst Health, 74-76. *Mem:* Biophys Soc; Int Soc Chronobiol. *Res:* Circadian rhythms; biological oscillators; bioluminescence; membrane biology. *Mailing Add:* Dept Math & Sci Univ Minn Morris MN 56267

GOOCHEE, HERMAN FRANCIS, b Emporium, Pa, Oct 13, 21; m 49; c 3. INORGANIC CHEMISTRY. *Educ:* St Vincent Col, BS, 43; Univ Tenn, MS, 57. *Prof Exp:* Lab engr, Stupakoff Ceramic & Mfg Co, 43-45; engr, Speer Carbon Co, 46-52; supvr process eng, Union Carbide Nuclear Co, Tenn, 52-57; consult & supvr cent eng, Speer Carbon Co Div, Air Reduction Co, Inc, 57-60, dir process eng, 60-61, dir develop & tech serv, 61-67, dir mfg carbon & graphite, 67-70, vpres mfg, 70-76, vpres tech opers, Carbon-Graphite, Airco Carbon Div, 76-88; RETIRED. *Mem:* AAAS; Am Ceramic Soc; Am Chem Soc; Am Inst Mech Engr. *Res:* Technical management; ceramic research; carbon and graphite development; utilities operations; nuclear engineering and operations; combustion engineering and furnace design; technical supervision and planning. *Mailing Add:* 301 Russ Lane St Mary's PA 15857

GOOD, A L, b Marysville, Pa, Jan 8, 21; m 43; c 4. VETERINARY PHYSIOLOGY. *Educ:* Univ Pa, VMD, 43; Kans State Univ, MS, 50; Univ Minn, PhD(vet physiol), 56. *Prof Exp:* Asst prof physiol, Col Vet Med, Kans State Univ, 46-51; instr, 51-56, assoc prof, 56-61, PROF PHYSIOL, COL VET MED, UNIV MINN, ST PAUL, 61- *Mem:* AAAS; Am Vet Med Asn; Am Physiol Soc. *Res:* Cardiovascular physiology of domestic animals; temperature regulation in animals; effects of cold environmental temperatures on new-born calves. *Mailing Add:* Col Vet Med Univ Minn St Paul MN 55108

GOOD, ANNE HAINES, b Everett, Wash, June 25, 31; m 64; c 3. IMMUNOCHEMISTRY, EDUCATION. *Educ:* Wellesley Col, BA, 52; Yale Univ, MD, 57; Western Reserve Univ, PhD(immunol & exp path), 63. *Prof Exp:* Fel immunol, dept biol, San Diego, 63-66, lectr, dept microbiol & immunol, Berkeley, 66-79, SR LECTR IMMUNOL, DEPT MICROBIOL & IMMUNOL, UNIV CALIF, BERKELEY, 79- *Mem:* NY Acad Sci; Am Asn Immunologists; Reticuloendothelial Soc; AAAS. *Res:* Immunochemistry; immunoregulatory properties of bacterial and parasite products. *Mailing Add:* Dept Microbiol & Cell Biol Rm 142 LSA Box 4 Univ Calif Berkeley CA 94720

GOOD, CARL M, III, b Akron, Ohio, Mar 7, 44; m 69. BIOCHEMISTRY, GENETICS. *Educ:* Ohio Wesleyan Univ, BS, 66; Iowa State Univ, PhD(biochem genetics), 70. *Prof Exp:* Fel bact genetics, Univ Wis Sch Med, 70-71; fel biochem, Univ Mass Sch Med, 71-72; technol mgr, 72-74, res group leader, 74-76, tech dir, 76-80, ASSOC TECH DIR, MILLIPORE CORP, 76- *Mem:* Am Soc Microbiol; Am Asn Clin Chem; Sigma Xi. *Res:* Molecular genetics; clinical biochemistry. *Mailing Add:* 560 Longley Rd Groton MA 01450

GOOD, DON L, b Van Wert, Ohio, Oct 8, 21; m 47; c 3. ANIMAL SCIENCE. *Educ:* Ohio State Univ, BS, 47; Kans State Univ, MS, 50; Univ Minn, PhD(animal sci), 56. *Prof Exp:* From instr to assoc prof, 47-61, PROF ANIMAL SCI, KANS STATE UNIV, 61-, HEAD DEPT ANIMAL SCI & INDUST, 66- *Concurrent Pos:* Mem res comt, Am Hereford Asn, 63-; USAID consult, Ahmadu Bello Univ, Zaria, Nigeria, 68 & 69; consult, Govt Turkey, 71. *Mem:* Am Soc Animal Sci. *Res:* Production and relationship of live animal evaluation to carcass characteristics and economical production traits in beef cattle. *Mailing Add:* Animal Sci-Ind Dept Kansas State Univ Manhattan KS 66502

GOOD, ERNEST EUGENE, b Van Wert, Ohio, Jan 7, 13; m 40; c 3. WILDLIFE MANAGEMENT. *Educ:* Ohio State Univ, BSc, 40, MSc, 47, PhD(zool), 52. *Prof Exp:* Proj biologist, Soil Conserv Serv, USDA, 35-40, asst agr aide, 40-41; proj leader, Ind Dept Conserv, 41-42; farmer, 42-48; instr, Ohio State Univ, 48-53, from asst prof to assoc prof, 53-70, chmn, Div Fisheries & Wildlife Mgt, 69-80, prof zool, 70- 80, emer prof natural resources, 80-; RETIRED. *Concurrent Pos:* Collabr, US Fish & Wildlife Serv, Dept Interior, 49- *Mem:* Wildlife Soc; Am Soc Mammal; Wilson Ornith Soc; Am Ornith Union; Sigma Xi. *Res:* Animal ecology and behavior; ecology. *Mailing Add:* 7085 Linworth Rd Worthington OH 43235

GOOD, HAROLD MARQUIS, b Brantford, Ont, Jan 16, 20; m 45; c 4. PLANT PATHOLOGY. *Educ:* Univ Toronto, BA, 43, PhD, 47. *Prof Exp:* Lectr bot, Univ Toronto, 47-49; from asst prof to prof biol, Queens Univ, Ont, 49-83, assoc head dept, 70-73; RETIRED. *Concurrent Pos:* Dir, Ont Univ Prog Instr Develop, 73- *Mailing Add:* Centreville ON K0K 1N0 Can

GOOD, IRVING JOHN, b London, Eng, Dec 9, 16. MATHEMATICS, STATISTICS. *Educ:* Cambridge Univ, BA, 38, PhD(math), 41, MA, 43, ScD(math, statist), 63; Oxford Univ, DSc, 64. *Prof Exp:* Civil servant, Foreign Off, 41-45; lectr math & electronic comput, Univ Manchester, 45-48; civil servant, Govt Commun HQ, 48-59; spec merit dep chief sci officer, Admiralty Res Lab, 59-62; consult math, statist & electronic comput, Commun Res Div, Inst Defense Anal, 62-64; sr res fel, Trinity Col, Oxford & Atlas Comput Lab, Sci Res Coun, 64-67; res prof, 67-69, DISTINGUISHED PROF STATIST, VA POLYTECH INST & STATE UNIV, 69- *Concurrent Pos:* Mem commun theory comt, Ministry Supply, Gt Brit, 53-59; vis res assoc prof, Princeton Univ, 55; distinguished mem, Crypto-Math Inst, 67; mem adv bd, Ctr Study Pub Choice, Va. *Mem:* Am Math Soc; Math Asn Am; hon mem Int Statist Inst; fel Inst Math Statist; fel Am Statist Asn; Am Acad Arts & Sci. *Res:* Mathematical analysis; cryptology; computers; foundations of statistical and scientific inference; machine intelligence; scientific speculation; information theory. *Mailing Add:* Dept Statist Va Polytech Inst & State Univ Blacksburg VA 24061

GOOD, MARY LOWE, b Grapevine, Tex, June 20, 31; m 52; c 2. INORGANIC CHEMISTRY, RADIOCHEMISTRY. *Educ:* Ark State Teachers Col, BS, 50; Univ Ark, MS, 53, PhD(inorg chem, radiochem), 55. *Hon Degrees:* Dr, Univ Ark, 79, Univ Ill, Chicago, 83 & Clarkson Univ, 84, E Mich Univ, 86, Duke Univ, 87, St Marys Col, 87 & Kenyon Col, 88, Lehigh Univ, 89, NJ Inst Technol, 89, Northeastern Ill Univ, 89, Stevens Inst Technol, 89, Univ SC, 89. *Prof Exp:* From instr to asst prof chem, La State Univ, Baton Rouge, 54-58; from assoc prof to prof, Univ New Orleans, 58-74, Boyd prof, 74-80; vpres & dir res, UOP, Inc, 80-85; dir res, Signal Res Ctr, 85-86, pres, Eng Mat Res, Allied Signal Inc, 86-88, SR VPRES TECHNOL, ALLIED SIGNAL RES & TECHNOL LAB, 88- *Concurrent Pos:* Mem bd dirs, Oak Ridge Assoc Univ, 71-76, Am Chem Soc, 71-80, Indust Res Inst, 82-87, Nat Inst Petrol & Energy Res; mem, NSF Chem Adv Panel, 72-76; mem & chmn med chem B study sect, NIH, 72-76; chmn bd dirs, Am Chem Soc, 78 & 80; pres, Inorganic Div, Int Union Pure & Appl Chem; mem, res coord coun, Gas Res Inst; Nat Acad Sci Panel Sci Communs & Nat Security, 83, NRC/Nat Acad Sci Panel Impact Nat Security Controls Int Technol Transfer, 83-86; vchmn, Nat Sci Bd, 84-86, chmn, 88-92; mem, Joint High Level Adv Panel, US-Japan agreement Res & Develop in Sci & Technol. *Honors & Awards:* Garvan Medal, Am Chem Soc, 73; Gold Medal, Am Inst Chemists, 83; Delmer S Fahrney Medal, Franklin Inst, 88; Charles Lathrop Parsons Award, Am Chem Soc, 91. *Mem:* Nat Acad Eng; Int Union Pure & Appl Chem; Am Chem soc (pres, 87); fel AAAS. *Res:* Spectroscopy and structure determination of inorganic complexes; coordination chemistry; radiochemistry and isotope applications, particularly Mossbauer spectroscopy; supported metal catalysts; reactions of penicillin with metal ions; organometallic antifouling coatings; low dimensional magnetic materials. *Mailing Add:* Allied Signal Inc Res & Technol Lab 101 Columbia Rd Morristown NJ 07962-1021

GOOD, MYRON LINDSAY, b Buffalo, NY, Oct 25, 23; m 50; c 3. PHYSICS. *Educ:* Univ Buffalo, BA, 43; Duke Univ, PhD(physics), 51. *Prof Exp:* Physicist radiation lab, Univ Calif, 51-59; from assoc prof to prof physics, Univ Wis, 60-67; PROF PHYSICS, STATE UNIV NY STONY BROOK, 67- *Mem:* Am Phys Soc. *Res:* Beta decay; high energy physics. *Mailing Add:* Dept Physics Grad Physics Bldg/Rm D137 State Univ NY Stony Brook NY 11794

GOOD, NORMA FRAUENDORF, b Elizabeth, NJ, Aug 24, 39; m 62; c 1. PLANT ECOLOGY. *Educ:* Rutgers Univ, BA, 61, PhD(plant ecol), 65. *Prof Exp:* Teaching asst biol, Rutgers Univ, 61-65; asst prof, Wagner Col, 65-67; ED, BIOL ABSTRACTS, 68- *Mem:* Ecol Soc Am; Torrey Bot Club; Sigma Xi. *Res:* Forest ecology; population dynamics; productivity. *Mailing Add:* c/o Judy Personnel Dept Bio Sci Info Serv 2100 Arch St Philadelphia PA 19103

GOOD, NORMAN EVERETT, b Brantford, Ont, May 20, 17; m 52; c 6. PLANT PHYSIOLOGY. *Educ:* Univ Toronto, BA, 48; Calif Inst Technol, PhD(biochem), 51. *Prof Exp:* Res fel photosynthesis, Univ Minn, 51-52; Agr Res Coun Gt Brit res fel, Univ Cambridge, 52-54; plant physiologist sci serv, Can Dept Agr, 54-57; res fel photosynthesis, Univ Minn, 57-58; plant physiologist sci serv, Can Dept Agr, 58-62; from assoc prof bot & plant path to prof bot & plant path, 62-86, EMER PROF, MICH STATE UNIV, 86- *Concurrent Pos:* Guggenheim fel, Bristol Univ, 71-72. *Mem:* Sigma Xi. *Res:* Photosynthesis, especially chloroplast reactions; metabolism of growth substances and aromatic acids in general; amino acid metabolism. *Mailing Add:* Dept Bot/Plant Path Mich State Univ East Lansing MI 48823

GOOD, RALPH EDWARD, b Chicago, Ill, Feb 24, 37; m 62. PLANT ECOLOGY. *Educ:* Univ Ill, BS, 60, MS, 61; Rutgers Univ, PhD(ecol), 65. *Prof Exp:* Res assoc forestry, Univ Ill, 61; teaching asst biol, Rutgers Univ, 64; lectr, Queens Col, NY, 65-66, asst prof, 66-67; from asst prof to assoc prof, 71-77, PROF BIOL, RUTGERS UNIV, 77-, DISTINGUISHED PROF, 83-, GRAD PROG DIR, 88- *Concurrent Pos:* Bus mgr, Ecol Soc Am, 73-79; chmn biol dept, Rutgers Univ, 78-82, dir, div pinelands res, CCES, 81. *Mem:* AAAS; Ecol Soc Am (vpres, 81-82); Am Inst Biol Sci. *Res:* Ecology of wetlands, including coastal and fresh-water tidal marshes; the role of wetland plants in ecosystem function; also the ecology of the New Jersey Pine Barrens; overall objective of an understanding of the coastal plain ecosystem complex of southern New Jersey. *Mailing Add:* Dept Biol Rutgers Univ Camden NJ 08102

GOOD, RICHARD ALBERT, b Ashland, Ohio, Sept 24, 17; m 46; c 3. ALGEBRA. *Educ:* Ashland Col, AB, 39; Univ Wis, MA, 40, PhD(algebra), 45. *Prof Exp:* Asst math, Univ Wis, 41-43, actg instr, 43-45; from asst prof to prof, 45-88, EMER PROF MATH, UNIV MD, 89- *Mem:* AAAS; Soc Indust & Appl Math; Am Math Soc; Math Asn Am. *Res:* Semigroups; matrices; games; linear programing. *Mailing Add:* Dept Math Univ Md College Park MD 20742-4015

GOOD, ROBERT ALAN, b Crosby, Minn, May 21, 22; m 46, 86; c 5. PEDIATRICS. *Educ:* Univ Minn, BA, 44, MB, 46, MD & PhD(anat), 47. *Hon Degrees:* Dr Med, Univ Uppsala, 66; DSc, NY Med Col, 73, Med Col Ohio, 73, Col Med & Dent NJ, Hahnemann Med Col & Univ Chicago, 74, St John's Univ NY, 77, Univ Health Sci Chicago Med Sch, 78 & Univ Minn, 89. *Prof Exp:* Asst anat, Med Sch, Univ Minn, Minneapolis, 44-45, from instr to assoc prof pediat, 50-54, Am Legion Mem res prof pediat, 54-73, prof microbiol, 62-72, regents prof pediat & microbiol, 69-73, prof path & chmn dept, 70-72; pres & dir, Sloan-Kettering Inst Cancer Res, 73-80, vpres, 80-82; dir & prof path, Sloan-Kettering Div, Grad Sch Med Sci, Cornell Univ, 73-82, prof med & pediat, Med Col, 73-82; res prof med, prof microbiol & head clin immunol, dept pediat, Univ Okla, 82-85; PROF PEDIAT, PROF MICROBIOL & CHMN PEDIAT DEPT, UNIV SFLA, ST PETERSBURG, 85-, GRAD RES PROF, 89-; CHIEF PHYSICIAN, ALL CHILDREN'S HOSP, ST PETERSBURG, 85- *Concurrent Pos:* Res fel poliomyelitis, Med Sch Univ Minn, Minneapolis, 47-48, Whitney fel rheumatic fever, 48-49; Markle Found scholar med sci, 50-55; vis investr, Rockefeller Inst, 49-50, asst physician, Hosp, 49-50, adj prof & vis physician, Rockefeller Univ, 73-; mem, President's Cancer Panel, 72-73; dir res & attend physician, Mem Hosp Cancer & Allied Dis, 73- attend pediat, NY Hosp, 73-; emer prof, Catholic Med Col, Seoul, SKorea, 77-; chmn, Int Bone Marrow Registry, 77-79, mem, 80-, chmn, Gov Task Force Acquired Immune Deficiency Syndrome, 85-87, mem, 88-; first foreign, Chinese Acad Med Sci, People's Repub China, 80-; head, Cancer Res Prog, Okla Med Res Found, Oklahoma City, 82-85; mem, WHO Sci Group Immunodeficiency, 82-; dir, Children's Res Inst, All Children's Hosp, St Petersburg, 85-, chief physician, All Children's Hosp; hon mem med staff, Miami Children's Hosp, 86; Wellcome vis prof, Fedn Am Scientists Exp Biol, 86; Theodore & Venette Askounes-Ashford distinguished scholar award, Univ S Fla, 90. *Honors & Awards:* E Mead Johnson First Award, 55; Theobald Smith Award, AAAS, 55; Parke-Davis Award, Soc Exp Path, 62; Cooke Medal, Am Acad Allergy, 68; Squibb Award, Infectious Dis Soc Am, 68; Gairdner Found Award, 70; Albert Lasker Med Res Award, 70; Am Col Physicians Award, 72; Swed Med Soc Silver Medal, 72; David A Karnofsky Medal, Am Soc Clin Oncol, 73; Col Medalist, Am Col Chest Physicians, 74; Lila Gruber Mem Award Cancer Res, Am Acad Dermat, 74. *Mem:* Nat Acad Sci; Inst Med-Nat Acad Sci; AAAS; Am Soc Exp Path (pres, 74-75); Am Asn Immunol (pres, 75-76); Am Soc Clin Invest (pres, 66); Cent Soc Clin Res (pres 66); Reticuloendothelial Soc (pres, 75); hon mem Am Soc Transplant Surgeons; Am Pediat Soc; Soc Exp Biol & Med; Soc Pediat Res; Int Soc Blood Transfusion; Infectious Dis Soc Am; Asn Am Physicians. *Res:* Natural and acquired resistance to gram-negative endotoxins, primary and secondary immunodeficiencies, agammaglobulinemias and hypergammaglobulinemias; rheumatic, rheumatoid and autoimmune diseases; acute-phase reactions, immunology and hypersensitivity reactions; cellular basis of immunity; development of lymphoid systems; bone marrow transplantation; thymus function; immunologic tolerance; biology of retroviruses. *Mailing Add:* 801 S Sixth St St Petersburg FL 33701

GOOD, ROBERT CAMPBELL, b Erwin, Tenn, Aug 23, 26; m 51; c 2. MEDICAL MICROBIOLOGY, MYCOBACTERIOLOGY. *Educ:* Univ Tenn, BA, 49, MS, 50; Northwestern Univ, PhD(med microbiol), 54. *Prof Exp:* Res bacteriologist, Bauer & Black Div, Kendall Co, 50-51; sr res microbiologist, Int Minerals & Chem Corp, 54-59; sr res asst exp tuberc, Christ Hosp Inst Med Res, 59-60, res assoc, 60-63; assoc res bacteriologist, Nat Ctr Primate Biol, Univ Calif, Davis, 63-69; mem staff microbiol, Life Sci Ctr, 69-71, sr proj mgr, Hazleton Labs Am, 71-76, dir microbiol-virol dept, 72-76; chief, mycobact br, 76-82, chief respiratory spec pathogens lab br, 82-85, CHIEF RESPIRATORY DIS BR, CTR INFECTIOUS DIS, CTR DIS CONTROL, 85- *Concurrent Pos:* Chmn, Sci Assembly Microbiol & Immunol, 71-72; rep, Comt on Guidance of Tuberc/Respiratory Dis Prog, Am Lung Asn, 72-73; chmn, Mycobateriol Div, Am Soc Microbiol, 83-84; adj assoc prof, dept parasitol lab pract, Sch Pub Health, Univ NC, 83-; adj fac, Dept Pathol Lab Med, Emory Univ Sch Med, 85- *Mem:* AAAS; Am Soc Microbiol; Am Thoracic Soc; fel Am Acad Microbiol. *Res:* Mycobacteria; chemotherapy; virulence; bacterial respiratory diseases, epidemology. *Mailing Add:* 8095 Habersham Waters Rd Dunwoody GA 30350

GOOD, ROBERT HOWARD, b Ann Arbor, Mich, Aug 26, 31; m 64; c 3. HIGH ENERGY PHYSICS. *Educ:* Univ Mich, AB, 53; Univ Calif, Berkeley, PhD(physics), 61. *Prof Exp:* Physicist, Saclay Nuclear Res Ctr, France, 61-62 & Univ Calif, San Diego, 62-66; PROF PHYSICS, CALIF STATE COL, HAYWARD, 66- *Res:* Experimental high energy nuclear physics. *Mailing Add:* Dept Physics Calif State Univ Hayward CA 94542

GOOD, ROBERT JAMES, b Lincoln, Nebr, Aug 13, 20; m 64; c 3. SURFACE CHEMISTRY, CHEMICAL ENGINEERING. *Educ:* Amherst Col, BA, 42; Univ Calif, MS, 43; Univ Mich, PhD(chem), 50. *Prof Exp:* Asst chem, Univ Calif, 42-43; teaching fel, Univ Mich, 46-48; chemist, Dow Chem Co, Calif, 43-44 & Am Cyanamid & Chem Co, Calif, 44-46; res chemist, Monsanto Chem Co, Ala, 49-53; asst prof appl sci, Univ Cincinnati, 53-56; sr staff scientist, Convair Sci Res Lab, Gen Dynamics Corp, 57-64; PROF CHEM ENG, STATE UNIV NY BUFFALO, 64- *Concurrent Pos:* Vis prof, Univ Bristol, Eng, 70-71, City Univ, London, 71 & Univ Col London, 80; vis scientist, NIH, Bethesda, Md, 79-80. *Honors & Awards:* Kendall Award, Am Chem Soc, 76; Schoellkopf Award, 79. *Mem:* Am Chem Soc; The Chem Soc; Nat Asn Corrosion Engrs. *Res:* Surface chemistry; interfacial tension; adhesion; wetting; porosity and pore penetration; thermodynamics; intermolecular forces; solubility of nonelectolytes; polymer solvent-swelling, biophysics; electron diffraction of solids; fracture; emulsions; corrosion; materials science, coal. *Mailing Add:* Dept Chem Eng State Univ NY Buffalo NY 14260

GOOD, ROLAND HAMILTON, JR, b Toronto, Ont, Can, Oct 22, 23; US citizen; m 44; c 3. THEORETICAL PHYSICS. *Educ:* Lawrence Inst Technol, BME, 44; Chrysler Inst Eng, MAE, 46; Univ Mich, MS, 48, PhD(physics), 51. *Prof Exp:* Engr, Chrysler Corp, 42-47; instr physics, Univ Calif, 51-53; asst prof, Pa State Univ, 53- 56; from assoc prof to prof, Iowa State Univ, 56-70, distinguished prof, 70-72; head, Physics Dept, Pa State Univ, 72-81, prof physics, 72-88; RETIRED. *Concurrent Pos:* Vis lectr, Univ Colo, 58; mem, Inst Advan Study, 60-61; NSF sr fel, 60-61; vis prof, Inst Math Sci, Madras, India, 68; guest, Stanford Linear Accelerator Ctr, 68-69; vis prof, Seoul Nat Univ, Korea, 79. *Mem:* Fel Am Phys Soc. *Res:* Elementary particles; solid state. *Mailing Add:* 653D Avenida Sevilla Laguna Hills CA 92653

GOOD, WILFRED MANLY, b Hope, Kans, Oct 13, 13; m 40; c 3. RESEARCH ADMINISTRATION. *Educ:* Univ Kans, AB, 36 & MS, 38; Mass Inst Tech, PhD(physics), 44. *Prof Exp:* Res asst nuclear physics, Oak Ridge Nat Lab, 46-48, dir, High Voltage Acceleration Lab, 48-66; head, Nuclear Data Sect, Int Atomic Energy Agency, 66-70; sr physicist, Oak Ridge Nat Lab, 70-78; RETIRED. *Mem:* Sigma Xi; AAAS; NY Acad Sci. *Res:* Properties of beta-decay including angular correlations of emitted radiation; particle reaction with light nuclei; neutron reactions; nuclear structure. *Mailing Add:* 13 Essex Circle Sherwood Forest Brevard NC 28712

GOOD, WILLIAM BRENEMAN, b Middletown, Pa, Nov 19, 20; m 51; c 2. PHYSICS. *Educ:* Univ NC, BS, 48, MS, 50; Univ Ala, PhD(physics), 57. *Prof Exp:* Physicist, Rohm & Haas Co, Redstone Arsenal, Ala, 50-54; assoc prof physics, Baylor Univ, 56-57; from asst prof to assoc prof, 57-66, PROF PHYSICS, NMEX STATE UNIV, 66-, HEAD DEPT, 71- *Mem:* Fel AAAS; Am Phys Soc; Am Asn Physics Teachers; Am Meteorol Soc; Am Geophys Union; Sigma Xi. *Res:* Upper atmosphere; condensation and nucleation phenomena; properties of supercooled liquids. *Mailing Add:* 2515 S Solano Dr Las Cruces NM 88001

GOOD, WILLIAM E, b Hillsdale, Mich, Apr 25, 16; m 41; c 6. COLOR TELEVISION PROJECTION SYSTEMS. *Educ:* Kalamazoo Col, AB, 37; Univ Ill, MS, 39; Univ Pittsburgh, PhD(physics), 46. *Prof Exp:* Engr, Westinghouse Res Labs, 41-50; engr, Gen Elec Co, 50-65; mgr, Advan Develop, TV Receiver Dept, 65-77; CONSULT, GEN ELEC CO, 77- *Mem:* Fel Inst Elec & Electronics Engrs; Soc Info Display; Am Radio Relay League. *Res:* Color television large screen projection using diffraction grating written on a fluid control layer to determine the hue and brightness of each picture element in real time; radio controlled model airplanes. *Mailing Add:* Good Electronics Inc 700 Tulip St Liverpool NY 13088

GOODACRE, ROBERT LESLIE, b Grantham, Eng, Feb 3, 47; m 72; c 3. GASTROENTEROLOGY. *Educ:* London Univ, MB, BS, 70; FRCP(C), 75. *Prof Exp:* Res fel gastroenterol, 74-77, assoc prof, 78-84, ASSOC PROF MED, McMASTER UNIV, 84- *Mem:* Am Gastroenterol Asn; Can Asn Gastroenterol; Royal Col Physicians. *Res:* Gastroenterology; inflammatory bowel disease; mucosal immunology. *Mailing Add:* 14 Ravine Dr Dundas ON L8N 3Z4 Can

GOODALE, FAIRFIELD, b Framingham, Mass, May 4, 23; m 45; c 5. ANATOMIC PATHOLOGY, CLINICAL PATHOLOGY. *Educ:* Western Reserve Univ, MD, 50. *Prof Exp:* Asst in path, Harvard Med Sch, 57-58; asst prof, Dartmouth Med Sch, 58-60; assoc prof, Albany Med Col, 60-63; prof path & chmn dept, Med Col Va, Commonwealth Univ, 63-76, asst dean curriculum, 72-76; DEAN & MED DIR, MED COL GA, 76-, PROF PATH, 76- *Concurrent Pos:* Teaching fel, Harvard Med Sch, 54-55; Am Cancer Soc res fel, Mass Gen Hosp, Boston, 54-55; USPHS res fels, St Mary's Hosp, London, 55-56, Oxford Univ, 56-57 & Mass Gen Hosp, Boston, 57-58; assoc dir labs, Mary Hitchcock Mem Hosp & Clin, Hanover, NH, 58-60; Nat Bd Med Examr; chmn path test comt, Int Path Coun, 78-81, vpres, 74-77, pres, 77-79. *Mem:* Int Acad Path; Am Soc Clin Path; Am Asn Path & Bact; fel Col Am Path; Am Soc Exp Path. *Res:* Pathogenesis of fever. *Mailing Add:* Sch Med Med Col Ga Augusta GA 30912

GOODALL, MARCUS CAMPBELL, biophysics, systems theory, for more information see previous edition

GOODARZI, FARIBORZ, b Tehran, Iran, Feb 20, 40; Can citizen; m 74; c 3. ORGANIC PETROLOGY, SCIENCE EDUCATION. *Educ:* Univ Tehran, BS, 63; Univ Newcastle-Upon-Tyne, MS, 71, PhD(geol), 75. *Prof Exp:* Geologist, ASK Coal Co, 63-64; sect head, geol, Ministry Water & Power; asst prof geol, Univ Technol, Can, 75-80; dir, coal & petrol, FG consultant, 80-82; SR RES SCIENTIST, COAL & PETROL, INST SEDIMENTARY & PETROL GEOL, 82-, CO-COORD, GLOBAL CHANGE, 88- *Concurrent Pos:* Adj prof geol, Univ Regina, Sask, 85-, Univ Waterloo, 87-, Univ Western Ont, 89-; adv, Geol Surv Greece, 86- *Mem:* Royal Microscopical Soc; Int Comt Coal Petrology; Am Soc Organic Petrol; Can Mineralogical Assoc. *Res:* Investigation of inorganic and organic compositional characteristics of carbonaceous material, for example coal, bitumen, and kerogen; results of research relevant to understanding utilization behavior of coal,; geological interpretation of burial and thermal histories with reference to oil exploration; global change and environmental impact of elements. *Mailing Add:* 3303 33rd St NW Calgary AB T2L 2A7 Can

GOODDING, JOHN ALAN, b Lincoln, Nebr, Mar 11, 22; m 45; c 2. AGRONOMY. *Educ:* Univ Nebr, BS, 47; Kans State Univ, MS, 50; Wash State Univ, PhD, 57. *Prof Exp:* Instr, Kans State Univ, 48-51 & Wash State Univ, 53-55; asst prof agron & asst agronomist, Univ Nebr, 55-61; asst dir resident instr, Univ Minn, St Paul, 61-70, from assoc prof to prof agron, 61-89, asst dean, Col Agr, 70-73, actg dean, 73-76, EMER PROF AGRON, DEPT AGRON & PLANT GENETICS, UNIV MINN, ST PAUL, 89- *Mem:* Am Soc Agron; Nat Asn Col & Teachers Agr. *Res:* Range management; seed technology; miscellaneous legumes. *Mailing Add:* Dept Agron & Plant Genetics Univ Minn St Paul MN 55108

GOODE, HARRY DONALD, b Newark, NJ, May 31, 12; m 46. GEOLOGY. *Educ:* Univ Ariz, BS, 51; Univ Colo, PhD(geol), 59. *Prof Exp:* Field asst geol, US Geol Surv, 49-52, geologist, 52-59, asst dist geologist, 59-62; from assoc prof to prof geol, Univ Utah, 62-77; consult, 77-85; RETIRED. *Mem:* Fel AAAS; fel Geol Soc Am; Am Inst Prof Geologists. *Res:* Ground water geology, especially springs in bedrock; geomorphology; Quaternary geology, especially deposits of Lake Bonneville; study of thermal waters of Utah. *Mailing Add:* 2275 South 2200 East Salt Lake City UT 84109-1134

GOODE, JOHN WOLFORD, b Paxton, Ill, Apr 19, 39; m 67; c 4. TOXICOLOGY. *Educ:* Univ Ill, Urbana, BS, 62; Loyola Univ, MS, 66, PhD(pharmacol), 67; Duquense Univ, MBA, 82. *Prof Exp:* Dir pharmacol, Bio-Test, Nalco Chem, 67-71, dir, Decatur Res Labs, 71-77; dir, Gulf Toxicol Fel Lab, Carnegie-Mellon Inst, 78-80; dir, Gulf Life Sci Lab, Gulf Sci & Technol Co, 80-85; dir, Sri Int, 85-87; DIR, AT KEARNEY, 87- *Mem:* Am Indust Hygiene Asn; Am Chem Soc; Am Col Toxicol; Soc Environ Toxicol & Chem. *Res:* Toxicity of industrial compounds. *Mailing Add:* 1903 Somersworth Dr San Jose CA 95124

GOODE, JULIA PRATT, b Carnesville, Ga, Feb 26, 29. ANALYTICAL CHEMISTRY. *Educ:* Agnes Scott Col, BA, 50; Emory Univ, MS, 52. *Prof Exp:* Chemist, Ga Dept Pub Health, 52-53; asst to dir agr res, Tenn Corp, 53-56; chmn dept sci, Fulton County Bd Ed, 56-66; assoc prof chem, Baptist Col, Charleston, 66-69; asst prof, 69-73, assoc prof med technol & assoc dean, Col Allied Health Sci, 73-83, REGISTR & DIR ADMIS, MED UNIV SC, 83- *Mem:* Am Chem Soc. *Res:* Quantitative techniques in paper chromatography. *Mailing Add:* 5490 Storoview Dr John's Island SC 29455

GOODE, LEMUEL, b Saulsville, WVa, Jan 2, 21; m 47; c 2. PHYSIOLOGY, ANIMAL PHYSIOLOGY. *Educ:* WVa Univ, BS, 42, MS, 46; Univ Fla, PhD, 61. *Prof Exp:* Asst county agent, NC State Univ, 46-47, from instr to assoc prof, 47-65, prof animal sci, 65-86; RETIRED. *Concurrent Pos:* Southern Fels Fund fac fel. *Mem:* Fel Am Soc Animal Sci. *Res:* Physiology of reproduction; effects of nutrition and environment on ovarian activity, reproductive rate and milk production; roughage and pasture utilization. *Mailing Add:* Dept Animal Sci NC State Univ Raleigh NC 27650

GOODE, MELVYN DENNIS, b New Albany, Ind, Feb 18, 40; m 68. CELL BIOLOGY, BIOPHYSICS. *Educ:* Univ Kans, BS, 63; Iowa State Univ, PhD(cell biol), 67. *Prof Exp:* Asst biomed res, Argonne Nat Labs, 63; res assoc anat, Univ Pa, 67-68; asst prof, 68-73, ASSOC PROF ZOOL, UNIV MD, COLLEGE PARK, 73- *Concurrent Pos:* Vis investr, Univ Tex MD Anderson Hosp & Tumor Inst, 69, Kings Col, London, 75, Sta Zool, Villefranche-sur-mer, 83; res grants, Univ Md Gen Res Bd, 69-70, biomed sci comt, 70-71 & 84-85, Am Cancer Soc, 70-73 & 87-88, Am Heart Asn, 74-75, NIH, 76-78, NSF, 78-83, Juv Diabetes Found, 89-91. *Honors & Awards:* Nikon Award, 79. *Mem:* Am Soc Cell Biol; Electron Micros Soc Am; Int Soc Evolution Protistology (pres, 85-87). *Res:* Mechanism of kinesin-based movement of secretory granules; Mechanism and evolution of mitosis; relationship between differentiation and DNA synthesis in cultured cells. *Mailing Add:* Dept Zool Univ Md College Park MD 20742

GOODE, MONROE JACK, b Whitney, Ala, Feb 15, 28; m 50; c 3. PHYTOPATHOLOGY. *Educ:* Miss State Univ, BS, 52, MS, 54; NC State Univ, PhD(plant path), 57. *Prof Exp:* From asst prof to assoc prof, 57-66, PROF PLANT PATH, UNIV ARK, FAYETTEVILLE, 66- *Concurrent Pos:* Mem, Southern Task Force Veg Res, 74- *Mem:* Am Phytopath Soc. *Res:* Diseases of vegetable crops; quantitative disease resistance breeding; resistance ultrastructure. *Mailing Add:* Dept Plant Path Univ Ark Main Campus Fayetteville AR 72701

GOODE, PHILIP RANSON, b San Francisco, Calif, Jan 4, 43. SYSTEMS ENGINEERING. *Educ:* Univ Calif, Berkeley, AB, 64; Rutgers Univ, PhD(physics), 69. *Prof Exp:* Res assoc physics, Rutgers Univ, 69; res assoc, Univ Rochester, 69-71; asst prof physics, Rutgers Univ, 71-77, mem tech staff, Bell Tel Labs, 77-80; assoc prof, Univ Ariz, 80-84; PROF & CHMN PHYSICS, NJ INST TECHNOL, 84- *Mem:* Am Phys Soc; Sigma Xi; Am Astron Soc; AAAS; Inst Am Univ. *Res:* Astrophysics. *Mailing Add:* 116 Jefferson Ave Westfield NJ 07090

GOODE, ROBERT J, b McDonald, Pa, Dec 21, 32. FRACTURE MECHANICS. *Educ:* Carnegie Inst Technol, BS, 56. *Prof Exp:* Assoc supt mat sci technol, Naval Res Lab, Washington, DC, 80-87; RETIRED. *Honors & Awards:* Award of Merit, Am Soc Testing & Mat, 83. *Mem:* Fel Am Soc Metals; fel Am Soc Testing & Mat. *Mailing Add:* 2402 Kegwood Lane Bowie MD 20715

GOODE, ROBERT P, b New York, NY, Sept 12, 36. BIOLOGY, EMBRYOLOGY. *Educ:* NY Univ, BS, 57; Columbia Univ, MA, 60, PhD(exp embryol), 64. *Prof Exp:* Asst prof embryol & histol, Otterbein Col, 63-64; from asst prof to assoc prof, 64-82, PROF EMBRYOL, CITY COL NEW YORK, 82-, DIR, PROG IN PREMED STUDIES, 80- *Mem:* Am Soc Zool; Soc Develop Biol. *Res:* Regeneration of limbs in amphibia, particularly of adults in several species of Anurans under natural conditions; induction of regeneration in amphibians and mammals. *Mailing Add:* Dept Biol City Col NY Convent Ave at 138th St New York NY 10031

GOODE, SCOTT ROY, b Chicago, Ill, July 9, 48; m 75. ANALYTICAL CHEMISTRY. *Educ:* Univ Ill, Urbana, BS, 69; Mich State Univ, PhD(chem), 74. *Prof Exp:* Res asst chem, Mich State Univ, 69-74; asst prof, 74-80, ASSOC PROF CHEM, UNIV SC, 80- *Mem:* Am Chem Soc; Soc Appl Spectros; Optical Soc Am. *Res:* Atomic flame and furnace spectrometry; analytical instrumentation and interactive computer control of experimentation. *Mailing Add:* Dept Chem Univ SC Columbia SC 29208

GOODEARL, KENNETH RALPH, b Weymouth, Mass, Sept 4, 45; m 86. ALGEBRA. *Educ:* Amherst Col, BA, 67; Univ Wash, MS, 69, PhD(math), 71. *Prof Exp:* Instr math, Univ Chicago, 71-72; instr, 72-75, from asst prof to assoc prof, 75-84, PROF MATH, UNIV UTAH, 84- *Concurrent Pos:* Sr vis fel, Univ Leeds, Eng, 77; prin investr res grants, Nat Sci Found, 74-; Humboldt res fel, Univ Passau, Germany, 84-85. *Mem:* Am Math Soc. *Res:* Noncommutative ring theory; nonsingular rings; von Neumann regular rings; differential operator rings; C*-algebras; ordered K-theory; noetherian rings. *Mailing Add:* Dept Math Univ Utah Salt Lake City UT 84112

GOODELL, HORACE GRANT, b Decatur, Ill, Oct 12, 25; m 56; c 3. GEOLOGY. *Educ:* Southern Methodist Univ, BS, 54; Northwestern Univ, PhD(geol), 58. *Prof Exp:* From asst prof to prof geol, Fla State Univ, 57-70; chmn dept, 71-79, PROF ENVIRON SCI, UNIV VA, CHARLOTTSVILLE, 70, CHMN MARINE AFFAIRS, 80- *Mem:* Geol Soc; Soc Econ Paleont & Mineral; Geochem Soc; Am Asn Petrol Geologists; fel Geol Soc Am. *Res:* Sedimentary petrology; lithostratigraphy; geochemistry; marine geology; hydrogeology. *Mailing Add:* Dept Environ Sci Clark Hall Univ Va Charlottesville VA 22903

GOODELL, RAE SIMPSON, b Cambridge, Mass, May 16, 44; m 65; c 3. SCIENCE & THE MEDIA. *Educ:* Pomona Col, AB, 67; Stanford Univ, MA, 71, PhD(commun), 75. *Prof Exp:* Res assoc oral hist prog, Mass Inst Technol, 75-77, asst prof, 71-81, assoc prof sci writing, 81-86; PARENT EDUCR, 86- *Mem:* Fel AAAS; Nat Asn Sci Writers; Soc Social Studies Sci; Coun Advan Sci Writing. *Res:* Relationship between science and the media; public understanding of science; science policy issues. *Mailing Add:* Mass Inst Technol Rm 4-144 77 Massachusetts Ave Cambridge MA 02139

GOODENOUGH, DANIEL ADINO, b New Haven, Conn, July 6, 44; m 68; c 2. CELL BIOLOGY. *Educ:* Harvard Univ, BA, 66, PhD(anat), 70. *Prof Exp:* NSF grant, Cardiovasc Res Inst, Univ Calif, San Francisco, 70-71; from instr to assoc prof, 71-81, PROF ANAT, HARVARD MED SCH, 81- *Concurrent Pos:* Takeda Prof, 89. *Mem:* Am Soc Cell Biol; Biophys Soc. *Res:* Structure and function of intercellular junctions, particularly the gap junction; electron microscopy; protein and lipid chemistry; x-ray diffraction; subcellular isolation; molecular biology. *Mailing Add:* Dept Anat & Cell Biol Harvard Med Sch 220 Longwood Ave Boston MA 02115

GOODENOUGH, DAVID GEORGE, b Victoria, BC, Nov 12, 42; m 65, 88; c 2. REMOTE SENSING. *Educ:* Univ BC, BSc, 64; Univ Toronto, MSc, 67, PhD(astron), 69. *Prof Exp:* Fel, Univ Victoria, BC, 69-70; asst prof astron, Wheaton Col, Mass, 70-73; res scientist, Can Ctr Remote Sensing, Ottawa, Can, 73-74, res scientist head, Methodol Sect, 74-78, sr res scientist & head, 78-85, chief res scientist & head, 85-90, CHIEF RES SCIENTIST & HEAD, KNOWLEDGE-BASED METHODS & SYSTS SECT, CAN CTR REMOTE SENSING, OTTAWA, CAN, 90-; HEAD METHODOLOGY SECT, CAN CTR REMOTE SENSING, 74- *Concurrent Pos:* Consult, 75-; assoc ed, J Can Aeronaut & Space Inst, 75-78; adj prof, Physics Dept, York Univ, 76-86; adj prof, Dept Elec Eng, Univ Ottawa, 79-; mem adv comt, Inst Elec & Electronics Engrs Geosci & Remote Sensing Soc, 87- *Mem:* Am Astron Soc; Royal Astron Soc Can; Can Aeronaut & Space Inst; Pattern Recognition Soc; sr mem Inst Elec & Electronics Engrs; Geosci & Remote Sensing Soc (vpres, 87-); Asn Artificial Intel; AAAS. *Res:* Multiple expert systems for integrating remote sensing data, geographic information systems, domain knowledge for resource management and environmental monitoring; physical and mathematical bases for microwave, visible and infrared remote sensing systems. *Mailing Add:* Can Ctr Remote Sensing 1547 Merivale Rd Ottawa ON K1A 0Y7 Can

GOODENOUGH, DAVID JOHN, b Reading, Eng, Oct 3, 44; m 63; c 2. MEDICAL PHYSICS, RADIOLOGICAL PHYSICS. *Educ:* Univ Chicago, BS, 67, PhD(med physics), 72. *Prof Exp:* Instr radiol, Ctr Radiologic Image Res, Univ Chicago, 72-73; vis assoc, Bur Radiol Health, 73; asst prof radiol, Sch Med, Johns Hopkins Univ, 74-75; assoc prof & dir div radiol physics, dept radiol, 75-84, PROF RADIOL, GEORGE WASHINGTON UNIV, 85-; CO-DIR, INDUST MED IMAGING & IMAGE ANAL, 90- *Concurrent Pos:* Adj asst prof, Hood Col, Md, 73-75; pres, Inst Radiol Image Sci, 75-; mem, Nat Adv Comt Diag Radiol, nat Cancer Inst, 77-81; IAEA expert, 89- *Honors & Awards:* Serv Citation, Soc Photo-Optical Instrument Eng, 77; Medal Award, Int Cong Radiol, 89. *Mem:* NY Acad Sci; Asn Univ Radiologists; Soc Photo-Optical Instrumentation Engrs; Soc Magnetic Resonance Med. *Res:* Image analysis and signal detection theory applied to diagnostic procedures; application of information and communication theory to analysis of radiologic system performance; computed tomography; magnetic resonance imaging. *Mailing Add:* Div Radiol Physics Dept Radiol George Washington Univ Med Ctr Washington DC 20037

GOODENOUGH, EUGENE ROSS, b Neptune, NJ, Jan 8, 46; m 66; c 1. FOOD SCIENCE. *Educ:* Rutgers Univ, BS, 69, MS, 71, PhD(food sci), 75. *Prof Exp:* Res food technologist, Merck Chem Corp, 66; res asst, Rutgers Univ, 70-75; food scientist res & develop, 75-82, dir prod develop, 82-88, DIR OPERS SUPPORT, BEST FOODS RES CTR, UNION, NJ, 89- *Mem:* Sigma Xi; Inst Food Technologists; Am Dairy Sci Asn. *Res:* Shelf stable food products; packaging technology and quality assurance management. *Mailing Add:* Best Foods Div CPC Int PO Box 8000 Int Plaza Englewood Cliffs NJ 07632

GOODENOUGH, JOHN BANNISTER, b Jena, Ger, July 25, 22; US citizen; m 51. PHYSICS. *Educ:* Yale Univ, AB, 43; Univ Chicago, PhD(physics), 52. *Hon Degrees:* Dr, Univ Bordeaux, 67. *Prof Exp:* Res engr, Westinghouse Elec Corp, 51-52; res physicist, Lincoln Lab, Mass Inst Technol, 52-76; prof inorg chem, Univ Oxford, 76-86; PROF, UNIV TEX, 86- *Concurrent Pos:* Fel, Neurosci Res Prog, 61-72, mem exec comt, Div Solid State Physics, 65-68; mem bd, Univ Without Walls, Roxbury, Mass, 71-73; mem, Solid State Sci Panel & Comt, 74-76 & Nat Mat Adv Bd, 75-77; assoc ed, Mat Res Bull, J Solid State Chem, Struct & Bonding, Nouveau J de Chimie, J Solid State Ionics, Superconductor Sci & Technol; gen ed, Int Ser Monogrs Chem, 78-86; res award, Am Soc Eng Educ, 90. *Honors & Awards:* Solid State Chem Prize, Royal Soc Chem, 80; Centenary Lectr, Royal Soc Chem, 76; Von Hippel Award, Mat Res Soc, 89. *Mem:* Nat Acad Eng; AAAS; Phys Soc Japan; fel Am Phys Soc; Am Chem Soc; Royal Soc Chem; hon mem Res Soc India. *Res:* Theory of solids; transition metal compounds; magnetism; superconductivity; materials for energy conversion and storage. *Mailing Add:* Ctr Mat Sch & Eng ETC 5-160 Univ Tex Austin TX 78712-1084

GOODENOUGH, JUDITH ELIZABETH, b Geneva, NY, 1948; m 71; c 2. BIOLOGY. *Educ:* Wagner Col, NY, BS, 70; NY Univ, PhD(biol), 77. *Prof Exp:* Instr biol, NY Univ, 74; lectr-staff asst, 75-83, LECT-STAFF ASSOC, UNIV MASS, 83- *Concurrent Pos:* Fac res grant, 80 & 82; consult, Saunders Col Publ, Benjamin Cummings/Addison Wesley Publ, Prentice Hall Publ, W C Brown Publ, Times Mirror & Harper Collins Worth, McGraw Hill/Random House. *Mem:* Sigma Xi; AAAS; Animal Behav Soc. *Res:* Circadian and monthly biological rhythms. *Mailing Add:* Dept Zool Univ Mass Amherst MA 01003

GOODENOUGH, URSULA WILTSHIRE, b Queens Village, NY, Mar 16, 43; m 80; c 5. CELL BIOLOGY, GENETICS. *Educ:* Barnard Col, BA, 63, Columbia Univ, MA, 65; Harvard Univ, PhD(biol), 69. *Prof Exp:* NIH fel, Biol Lab, 69-71, from asst prof to assoc prof biol, Harvard Univ, 71-78; assoc prof, 78-82, PROF BIOL, WASHINGTON UNIV, 82- *Concurrent Pos:* Ed, J Cell Biol. *Mem:* Am Soc Cell Biol. *Res:* Cell biology and genetics of gametic differentiation in Chlamydomonas reinhardi; structure and function of cilia. *Mailing Add:* Dept Biol Washington Univ St Louis MO 63130

GOODER, HARRY, b Castleford, Eng, May 7, 28; nat US; m 54, 79; c 2. MICROBIAL STRUCTURE, MICROBIAL PHYSIOLOGY. *Educ:* Univ Leeds, BSc, 49, PhD(biochem), 52; FRSC. *Prof Exp:* Res fel bact, Harvard Med Sch, 52-54, teaching fel, 54-55; mem sci staff, Cent Pub Health Lab, Med Res Coun, Eng, 55-61; from asst prof to assoc prof, 61-68, PROF MICROBIOL & IMMUNOL, SCH MED, UNIV NC, CHAPEL HILL, 68-, ACTG CHAIR, 79-81 & 89- *Concurrent Pos:* Mem coun policy comt, Am Soc Microbiol, 78-88, chmn bd meetings, 78-88; chmn fac, Univ NC, 88-91. *Mem:* Am Soc Microbiol; Brit Soc Gen Microbiol; Brit Soc Immunol; fel Am Acad Microbiol. *Res:* Bacterial structure and chemistry, specifically cell walls and L forms; bacterial physiology in relation to pathogenicity; bacterial genetics; medical bacteriology. *Mailing Add:* Dept Microbiol & Immunol CB 7290 622 FLOB 231H Univ NC Chapel Hill NC 27514

GOODEVE, ALLAN MCCOY, b Toronto, Ont, Aug 12, 23; wid. PHARMACOGNOSY. *Educ:* Univ Toronto, PhB, 48, MSc, 56; Univ Sask, BSP, 50; Purdue Univ, PhD(pharmacog), 60. *Prof Exp:* Asst prof pharmacog, Univ NC, 59-60; asst prof pharmacog, Univ BC, 60-86, emer prof, 86; RETIRED. *Mem:* AAAS; Am Soc Pharmacog; Am Chem Soc; Can Pharmaceut Asn. *Res:* Alkaloid biosynthesis; solanaceous alkaloids. *Mailing Add:* 4055 W 29th Ave Vancouver BC V6S 1V4 Can

GOODEY, PAUL RONALD, b Hull, Eng, Oct 16, 46; m 70; c 2. CONVEX SETS THEORY. *Educ:* London Univ, BSc, 68, PhD(math), 70. *Prof Exp:* Lectr math, Royal Holloway Col, London Univ, 70-83; ASSOC PROF MATH, OKLA UNIV, 83- *Concurrent Pos:* Vis prof, Okla Univ, 79-80, Siegen Univ, WGer, 82-83. *Mem:* Am Math Soc. *Res:* Geometric analysis; convexity; geometric probability and stochastic geometry. *Mailing Add:* Phys Sci 423 Univ Okla 601 Elm Ave Norman OK 73019-0315

GOODFRIEND, LAWRENCE, PROTEIN CHEMISTRY, ALLERGIES. *Educ:* McGill Univ, PhD(biochem), 61. *Prof Exp:* ASSOC PROF, DEPT IMMUNOL, MCGILL UNIV, 63- *Res:* Human immuno-genetics. *Mailing Add:* 687 Pine Ave W Montreal PQ H3A 1A1 Can

GOODFRIEND, LEWIS S(TONE), b New York, NY, May 21, 23; m 50, 61, 85; c 4. ACOUSTICS. *Educ:* Stevens Inst Technol, ME, 47; Polytech Inst Brooklyn, MEE, 52. *Prof Exp:* Staff engr, Stevens Inst Technol, 47-49; chief elec engr acoust instruments, Audio Instruments Co, 50-53; vpres noise control eng, Zurn Environ Engrs, 69-72; PRES, L S GOODFRIEND & ASSOCS, 72- *Concurrent Pos:* Ed, Noise Control, 55-58 & Sound & Vibration, 67-71; mem bd dir, Inst Noise Control Eng, 71-76; mem bd of trustees, Stevens Inst Technol, 78- *Mem:* Fel Acoust Soc Am; Am Indust Hyg Asn; Inst Elec & Electronics Engrs; Inst Noise Control Eng, (pres, 83). *Res:* Noise control by design of products and systems; environmental noise impact on people and societal goals; architectural acoustics. *Mailing Add:* L S Goodfriend & Assocs 301 E Hanover Ave Morristown NJ 07962-2453

GOODFRIEND, PAUL LOUIS, b Dallas, Tex, Aug 10, 30; m 53; c 2. PHYSICAL CHEMISTRY. *Educ:* Univ Va, BS, 52; Ga Inst Technol, PhD(chem), 57. *Prof Exp:* Fel chem, Univ Rochester, 56-58; asst prof, Col William & Mary, 58-61; res chemist, Texaco Exp, Inc, 61-66; assoc prof, 66-70, PROF CHEM, UNIV MAINE, ORONO, 70- *Mem:* Am Chem Soc. *Res:* Molecular spectroscopy and quantum mechanics; excited states of free radicals; kinetics. *Mailing Add:* Dept Chem Univ Maine Orono ME 04473

GOODFRIEND, THEODORE L, b Philadelphia, Pa, Sept 30, 31; div; c 2. INTERNAL MEDICINE, BIOCHEMISTRY. *Educ:* Swarthmore Col, AB, 53; Univ Pa, MD, 57. *Prof Exp:* Intern, Univ Hosps Cleveland, 57-58; asst resident med, Barnes Hosp, Wash Univ, 60-62; from asst prof to assoc prof med & pharmacol, 65-74, PROF MED & PHARMACOL, UNIV WIS-MADISON, 74-, HEAD SECT CLIN PHARMACOL, 73-; ASSOC CHIEF STAFF RES, VET ADMIN HOSP, MADISON, 75- *Concurrent Pos:* Helen Hay Whitney res fel biochem, Brandeis Univ, 62-65; NIH res career develop award, 68-73; mem med adv bd coun high blood pressure res, Am Heart Asn, 69-, mem res study comt, Cardiovascular A, 73-; mem endocrinol & metab adv comt, Food & Drug Admin, 72- *Mem:* Am Soc Pharmacol & Exp Therapeut; Am Soc Clin Pharmacol & Therapeut; NY Acad Sci; Soc Exp Biol & Med; Sigma Xi. *Res:* Receptors for polypeptide hormones; angiotensin; pharmacology; antibodies to small polypeptides; kinins. *Mailing Add:* Vet Admin Hosp 2500 Overlook Terr Madison WI 53705

GOODGAL, SOL HOWARD, b Baltimore, Md, July 25, 21; div; c 2. MOLECULAR BIOLOGY, GENETICS. *Educ:* Univ Md, BS, 42; Johns Hopkins Univ, PhD(biol), 50. *Prof Exp:* Res assoc, Inst Coop Res, Johns Hopkins Univ, 50-52, res assoc biochem, 52-58, asst prof sch hyg & pub health, 58-61; assoc prof, 61-66, PROF MICROBIOL, SCH MED, UNIV PA, 66- *Concurrent Pos:* Fel, Pasteur Inst, 59-60; vis prof, L'Institut de Biolgie Physico-Chemique, 68-69; mem, Johns Hopkins Soc Scholars, 70. *Mem:* AAAS; Genetics Soc Am; Am Soc Microbiol; Am Soc Nat; Am Soc Biol Chem. *Res:* Transformation of bacteria; bacteriophage; role of nucleic acid in genetic determination; radiation effects on genetics of microorganisms; biochemistry. *Mailing Add:* Dept Microbiol Med Univ PaSch Med 346 Johnson/G2 Philadelphia PA 19104

GOODGAME, THOMAS H, b Camden, Ark, May 23, 21; m 46; c 1. ENVIRONMENTAL ENGINEERING, CHEMICAL ENGINEERING. *Educ:* La Tech Univ, BS, 42; La State Univ, Baton Rouge, MS, 47; Mass Inst Technol, ScD(chem eng), 53; Am Acad Environ Engrs, dipl, 70. *Prof Exp:* Engr, Jones Mills Works, Aluminum Co Am, 42-43; asst prof mech & chem eng, La Tech Univ, 47-49; group leader res & develop & sr chem engr, Cabot Corp, 53-55; head res & develop, Tex Butadiene & Chem Corp, 55-57; assoc prof chem eng, Ga Inst Technol, 57-59; sr proj engr, Cabot Corp, 59-62, gen mgr chem plant, Cabot Titania Corp, 63-64; sr res engr, Whirlpool Corp, 64-67, mgr eng res, 67-70, dir environ control, 70-84; PRES, ENVIRON & CHEM CONSULT ENGRS, INC, 84- *Concurrent Pos:* Consult engr, 47-49 & 57-59; prof, Mich State Univ, 65- *Honors & Awards:* Environ Award, Am Inst Chem Engrs, 78; President's Award, Porcelain Enamel Inst, 84. *Mem:* Sigma Xi; fel Am Inst Chem Engrs; Am Acad Environ Engrs; AAAS; Am Chem Soc; Rog Prof Engr NMex. *Res:* Air and water pollution control; solid waste management; resource recovery, membrane separation processes; process and equipment design and evaluation; engineering economy; transport phenomena; hydrocarbon emission control; leaching of solid waste. *Mailing Add:* Environ & Chem Consult Engrs Inc PO Box 914 Alamogordo NM 88310

GOODGE, WILLIAM RUSSELL, b Seattle, Wash, Apr 26, 28; m 57; c 2. HUMAN ANATOMY, VERTEBRATE MORPHOLOGY. *Educ:* Univ Wash, BS, 49, PhD(zool), 57; Univ Mich, MA, 50. *Prof Exp:* Asst zool, Univ Wash, 53-55, actg instr, 55-57; instr gross & neurol anat, Sch Med, WVa Univ, 57-61, asst prof, 61-64; asst prof, 64-70, interim dept chmn, 70-80, ASSOC PROF ANAT, SCH MED, UNIV MO-COLUMBIA, 70- *Mem:* AAAS; Am Asn Anat; Cooper Ornith Soc; Am Ornith Union. *Res:* Vertebrate functional anatomy, musculature, sense organs and glands; histology and histochemistry. *Mailing Add:* Univ Mo Columbia Med Sch Univ Mo Sch Med Columbia MO 65212

GOODGOLD, JOSEPH, b New York, NY, Mar 21, 20; m 42; c 2. MEDICINE. *Educ:* Brooklyn Col, BA, 41; Middlesex Univ, MD, 45. *Prof Exp:* Chief dept phys med & rehab, Ft Campbell Hosp, Ky, 52-53; lectr rehab med, Sch Educ, 56, assoc prof phys med & rehab, Sch Med, 58-63, prof rehab med, 63-75, DIR ELECTRODIAG DEPT, INST REHAB MED, MED CTR, NY UNIV, 56-, HOWARD A RUSK PROF REHAB MED, 75-, CHMN DEPT, 81- *Concurrent Pos:* Consult, Div Voc Rehab, State Educ Dept, 59, Brookdale Hosp Ctr, Brooklyn, 64- & Manhattan Vet Admin Hosp, 69; attend, Bellevue Hosp, 65; dir rehab med serv, Sch Med, NY Univ, 81- *Honors & Awards:* Krusen Award, 85. *Mem:* AMA; Am Asn Electromyog & Electrodiag; Am Rheumatism Asn; Am Cong Rehab Med; NY Acad Sci. *Res:* Electrophysiology. *Mailing Add:* Inst Rehab Med NY Univ Med Ctr 400 E 34th St New York NY 10016

GOODHART, ROBERT STANLEY, b Altoona, Pa, July 19, 09; m 35; c 2. MEDICINE. *Educ:* Lafayette Col, BS, 30; NY Univ, MD, 34, DMedSc, 40. *Prof Exp:* Intern, Brooklyn Hosp, NY, 34-36; res physician, Psychiat Div, Bellevue Hosp, 36-39; instr med, Col Med, NY Univ, 39-42, asst prof, 42-47, asst dean, 39-40; sci dir, Nat Vitamin Found, Inc, 46-67, pres, 62-67; exec secy comt med educ, NY Acad Med, 67-78; RETIRED. *Concurrent Pos:* Milbank Mem Fund res fel, 41-42; mem comt nutrit, Indust, Food & Nutrit Bd, Nat Res Coun, 41-49, vchmn, 42-46, chmn, 46-49; chief indust feeding prog div, War Food Admin, 43-46; physician in chg, Washington Heights Nutrit Clin, City Health Dept, New York, 46-68; lectr, Columbia Univ, 50-68; chmn adv comt, Emergency Food Authority, Off Civil Defense, NY, 50-; mem comt nutrit, Adv Bd Qm Res & Develop, Nat Acad Sci-Nat Res Coun, 60-62; adj prof, Mt Sinai Med Sch, City Univ New York, 69-72. *Mem:* Am Soc Clin Invest; Harvey Soc; Am Inst Nutrit; fel NY Acad Med. *Res:* Nutritional disturbances and related disorders in man. *Mailing Add:* 67 Forest Rd Tenafly NJ 07670

GOODHEART, CLARENCE F(RANCIS), b Porterville, Calif, Jan 24, 16; m 41; c 2. ELECTRICAL ENGINEERING. *Educ:* Calif Inst Technol, BS, 36; Ohio State Univ, MS, 38. *Prof Exp:* Asst elec eng, Ohio State Univ, 36-38; instr, Agr & Mech Col, Tex, 38-42; proj engr & sect head, Naval Ord Lab, 42-47; chmn dept elec eng, 47-69, PROF ELEC ENG, UNION COL, NY, 47- *Mem:* Am Soc Eng Educ; Inst Elec & Electronics Engrs. *Res:* Electrical engineering education; underwater ordnance, particularly in magnetic devices. *Mailing Add:* Dept Elec Eng & Comput Sci Union Col Schenectady NY 12308

GOODHEART, CLYDE RAYMOND, b Erie, Pa, June 9, 31; m 53; c 3. BIOLOGY, MEDICINE. *Educ:* Northwestern Univ, BS, 53, MS & MD, 57, MM, 87. *Prof Exp:* Res fel animal virol & cell biol, Calif Inst Technol, 58-61; asst prof pediat, Sch Med, Univ Southern Calif, 61-64, assoc prof, 65; assoc mem staff, Inst Biomed Res, AMA-Educ & Res Found, Ill, 65-69, mem staff, 69-70; sr microbiologist & vis prof, Rush-Presby-St Luke's Med Ctr, Chicago, 70-80; PRES, BIOLABS, INC, 70- *Mem:* AAAS; Am Soc Microbiol; Tissue Cult Asn. *Res:* Experimental embryology; respiratory physiology; tissue culture; autoradiography; animal virology and viral oncology. *Mailing Add:* BioLabs Inc 15 Sheffield Ct Lincolnshire IL 60069-3161

GOODHUE, CHARLES THOMAS, b Ames, Iowa, Apr 30, 32; m 77; c 4. BIOCONVERSIONS, ENZYMOLOGY. *Educ:* Univ Ill, BS, 54; Univ Calif, PhD(biochem), 61. *Prof Exp:* Res biochemist, Distillation Prod Indust, 61-65; res microbiologist, Eastman Kodak Co, 65-74, res assoc, 74- 80, mem sci bd, 88-90, SR RES ASSOC, EASTMAN KODAK CO, 80-; MEM, GENENCOR INT, 90- *Mem:* Am Chem Soc; Am Soc Microbiol; Sigma Xi. *Res:* Biochemistry of microorganisms; bioconversions, enzymology; microbiol sources for enzymes useful for analytical systems; bioconversion processes useful for combined organic and biological syntheses of natural products and photographic chemicals; fermentation processes for production of enzymes and bioconversion products; chemistry and biochemistry of vitamin E. *Mailing Add:* 89 Hidden Spring Circle Rochester NY 14616

GOODIN, JOE RAY, b Claude, Tex, July 28, 34. PLANT PHYSIOLOGY. *Educ:* Tex Tech Univ, BS, 55; Mich State Univ, MS, 58; Univ Calif, Los Angeles, PhD(plant physiol), 63. *Prof Exp:* Asst prof plant physiol, Univ Calif, Riverside, 63-70; ASSOC PROF PLANT PHYSIOL, TEX TECH UNIV, 70- *Mem:* Sigma Xi (secy-treas, 66-68); Soc Develop Biol; Bot Soc Am; Am Soc Plant Physiol; Am Soc Range Mgt (secy-treas, 65-67). *Res:* Salinity and drought tolerance; environmental stress; physiological aging and juvenility in plants. *Mailing Add:* Dept Biol Sci Tex Tech Univ Lubbock TX 79409

GOODING, GRETCHEN ANN WAGNER, b Columbus, Ohio, July 2, 35; m 61; c 3. ULTRASONOGRAPHY. *Educ:* Col St Mary of Springs, Ohio, BA, 57; Col Med, Ohio State Univ, MD, 61. *Prof Exp:* Asst chief, 78-87, CHIEF ULTRASONOGRAPHY, VET ADMIN MED CTR, SAN FRANCISCO, 75-, CHIEF DEPT RADIOL, 87-; PROF RADIOL, UNIV CALIF, SAN FRANCISCO, 86-, VCHMN DEPT, 87- *Mem:* Fel Am Col Radiol; fel Am Inst Ultrasound in Med; Am Asn Wome Radiologists (pres, 84-85); World Fedn Ultrasound in Med; Asn Univ Radiologists; Radiol Soc NAm. *Res:* Ultrasonography; diagnostic radiology. *Mailing Add:* Vet Admin Med Ctr San Francisco Eight Overhill Rd Mill Valley CA 94941

GOODING, GUY V, JR, b Kinston, NC, Mar 16, 31; m 51; c 5. PHYTOPATHOLOGY. *Educ:* NC State Col, BS, 54, MS, 58; Univ Calif, Davis, PhD(plant path), 62. *Prof Exp:* Plant pathologist, US Forest Serv, 62-65; from asst prof to assoc prof plant path, 65-73, PROF PLANT PATH, NC STATE UNIV, 73- *Mem:* Am Phytopath Soc; Sigma Xi. *Res:* Plant virology; general phytopathology. *Mailing Add:* Dept Plant Path NC State Univ Raleigh NC 27650

GOODING, JAMES LESLIE, b Tehachapi, Calif, Sept 13, 50; div; c 3. GEOCHEMISTRY. *Educ:* Calif Polytech State Univ, San Luis Obispo, BS, 72; Univ Hawaii, Manoa, MS, 75; Univ NMex, PhD(geol), 79. *Prof Exp:* Teaching asst geol, Univ NMex, 75-76, res asst geol & meteoritics, 76-79; sr scientist planetary geochem & res grant prin; investr, NASA/Caltech Jet Propulsion Lab, 79-81, SPACE SCIENTIST, NASA/JOHNSON SPACE CTR, 81- *Mem:* NAm Thermal Anal Soc; Meteoritical Soc. *Res:* Chemical-mineralogical analysis of extraterrestrial materials; origin and geochemical evolution of the solar system. *Mailing Add:* NASA Johnson Space Ctr SN21 Houston TX 77058

GOODING, LINDA R, b Portland, Maine, Nov 23, 45. IMMUNOLOGY, VIROLOGY. *Educ:* Bowling Green State Univ, BS, 67; Cornell Univ, PhD(biochem), 72. *Prof Exp:* Fel immunol, Johns Hopkins Univ, 72-74; assoc, Duke Univ Med Ctr, 74-75, asst prof immunol, 76-80; assoc prof, 81-84, PROF MICROBIOL, SCH MED, EMORY UNIV, 84- *Mem:* Am Asn Immunol; Am Soc Microbiol; AAAS. *Res:* Immune recognition and response to DNA tumor viruses. *Mailing Add:* Dept Microbiol & Immunol Emory Univ 1510 Clifton Rd NE Atlanta GA 30322

GOODING, ROBERT C, b New Orleans, La, June 27, 18. NAVAL ARCHITECTURE, ENGINEERING ADMINISTRATION. *Educ:* US Naval Acad, BSc, 41; Mass Inst Technol, MSc, 46. *Prof Exp:* VAdm, USN, 40-76; CONSULT, 76-; CHMN, COLUMBIA RES CORP, 80- *Concurrent Pos:* Instr calculus, Pratt Inst, Brooklyn, NY, 58-62. *Mem:* Nat Acad Eng; Sigma Xi. *Mailing Add:* 1305 Trinity Dr Alexandria VA 22314

GOODING, RONALD HARRY, b Edmonton, Alta, Oct 18, 36; m 59; c 1. ENTOMOLOGY, BIOCHEMISTRY. *Educ:* Univ Alta, BSc, 57; Rice Inst, MA, 60; Johns Hopkins Univ, ScD(pathobiol), 64. *Prof Exp:* Res assoc biochem, Vanderbilt Univ, 64-66; from asst prof to assoc prof, 66-76, PROF ENTOM, UNIV ALTA, 76-, CHAIR, 89- *Concurrent Pos:* Oper grants, Univ Alta Gen Res Fund, 67-71, Nat Res Coun Can, 67-, Can Dept Agr 67-71 & 76-79, Alta Agr Res Trust, 67-71 & 74-76, Defense Res Bd Can, 69-74 & WHO, 75-77. *Mem:* Entom Soc Can; Can Soc Zool; Genetics Soc Can; Entom Soc Am; Soc Study Evolution. *Res:* Insect biochemistry; medical entomology; tsetse fly genetics. *Mailing Add:* Dept Entom Univ Alta Edmonton AB T6G 2E2 Can

GOODINGS, DAVID AMBERY, b Toronto, Ont, July 8, 35; m 60, 80; c 3. SOLID STATE PHYSICS. *Educ:* Univ Toronto, BA, 57; Cambridge Univ, PhD(physics), 61. *Prof Exp:* Res assoc theoret physics, Atomic Energy Res Establish, Harwell, 60-61; res assoc physics, Univ Pittsburgh, 61-64; lectr, Univ Sussex, 64-68; asst prof, Am Univ Beirut, 68-69; assoc prof, 69-74, PROF PHYSICS, MCMASTER UNIV, 74- *Concurrent Pos:* Vis lectr, Univ Ife, Nigeria, 66; vis prof, KFA Julich, Fed Repub Ger, 76. *Mem:* Can Asn Physicists. *Res:* Nonlinear dynamics and chaos in physical systems; condensed matter theory. *Mailing Add:* Dept Physics McMaster Univ Hamilton ON L8S 4M1 Can

GOODINGS, JOHN MARTIN, b Toronto, Ont, Feb 18, 37; Can citizen; m 61; c 2. PHYSICAL CHEMISTRY, FLAME CHEMISTRY. *Educ:* Univ Toronto, BA, 58; Univ Cambridge, PhD(phys chem), 61. *Prof Exp:* Asst prof chem, McGill Univ, 62-65; assoc prof, 65-72, chmn dept, 79-82, PROF CHEM, YORK UNIV, ONT, 72- *Concurrent Pos:* Thornton res fel, Thornton Res Ctr, Eng, 71-72; vis prof, Dept Chem Eng & Fuel Technol, Univ Sheffield, Eng, 78-79; vis fel, dept chem engr, Univ Cambridge, Eng, 87. *Mem:* Fel Chem Inst Can; Combustion Inst. *Res:* Flame ionization; combustion; flame-ion mass spectrometry; ion chemistry in hydrocarbon flames and hydrogen flames with additives; involving nitrogen and sulphur pollutants, soot formation, metallic ions, and negative ions formed by electron scavenging. *Mailing Add:* Dept Chem York Univ 4700 Keele St North York ON M3J 1P3 Can

GOODISMAN, JERRY, b Brooklyn, NY, Mar 22, 39; m 63; c 2. CHEMICAL PHYSICS. *Educ:* Columbia Col AB, 59; Harvard Univ, AM, 60, PhD(chem), 63. *Prof Exp:* Instr phys chem, Univ Ill, Urbana, 63-65, asst prof chem, 65-69; assoc prof, 69-75, PROF CHEM, SYRACUSE UNIV, 75- *Mem:* Am Phys Soc. *Res:* Theoretical description of the ideally polarisable interface; use of quantum statistical theories for electronic structure problems; statistical mechanics of electrode interfaces; scattering from noncrystalline systems. *Mailing Add:* Dept Chem Syracuse Univ Syracuse NY 13210

GOODJOHN, ALBERT J, b Calgary, Alta, Feb 18, 28; m 51; c 6. REACTOR PHYSICS. *Educ:* Univ Alta, BSc, 50, MSc, 51; Queen's Univ, Ont, PhD(physics), 56. *Prof Exp:* Design specialist, Canadqir Div, Gen Dynamics Corp, 56-59; mem staff theoret physics, Gen Atomics, 59-61, div head, 61-63, proj mgr advan concepts, 63-66, asst dept chmn reactor physics, 64-66, assoc div mgr, 66-70, gen mgr mkt, 70-75, dir reactor progs, 76-86; RETIRED. *Concurrent Pos:* Energy consult, 86- *Mem:* Fel Am Nuclear Soc. *Res:* Experimental and theoretical reactor and nuclear physics. *Mailing Add:* PO Box 1767 Rancho Santa Fe CA 92067

GOODKIN, JEROME, b New York, NY, Mar 22, 29; m 52; c 2. PHYSICAL CHEMISTRY, ELECTROCHEMISTRY. *Educ:* NY Univ, BA, 52, MS, 54; Rensselaer Polytech Inst, PhD(phys chem), 59. *Prof Exp:* Res asst phys chem, NY Univ, 52-54; asst, Rensselaer Polytech Inst, 54, 56-58; sect coordr, Catalyst Res Corp, 58-61; res chem res div, FMC Corp, 62-63; sr res scientist, Yardney Elec Corp, 63-68; PROF CHEM, TRENTON STATE COL, 68- *Concurrent Pos:* Adj prof, Mercer County Community Col, 63-68. *Mem:* Am Chem Soc; Electrochem Soc. *Res:* Molten salt chemistry; constitution of solution and electrochemical phenomena; corrosion of steel; electrode processes; batteries. *Mailing Add:* 22 Camelia Ct Lawrenceville NJ 08648

GOODKIND, JOHN M, b New York, NY, Aug 27, 34; m 63; c 2. PHYSICS. *Educ:* Amherst Col, AB, 56; Duke Univ, PhD(physics), 60. *Prof Exp:* Res asst physics, Stanford Univ, 60-62; from asst prof to assoc prof, 62-74, PROF PHYSICS, UNIV CALIF, SAN DIEGO, 74- *Concurrent Pos:* Sloan Found fel, 62-66; NSF res grant, 62- *Mem:* Am Phys Soc; Am Geophys Union; Sigma Xi. *Res:* Low temperature physics; liquid and solid helium three; nuclear adiabatic demagnetization; superconductivity; geophysics; gravimetry. *Mailing Add:* 1535 Forest Way Del Mar CA 92014

GOODKIND, MORTON JAY, b New York, NY, Apr 29, 28; div; c 2. INTERNAL MEDICINE. *Educ:* Princeton Univ, AB, 49; Columbia Univ, MD, 53. *Prof Exp:* Instr internal med, Sch Med, Yale Univ, 58-62, asst prof, 62-64; from asst prof to assoc prof, 64-74, CLIN ASSOC PROF MED, SCH MED, UNIV PA, 74- *Concurrent Pos:* Am Heart Asn advan fel, 59-61, grants-in-aid, 59-61 & 61-67; USPHS grant-in-aid, Nat Heart Inst, 61-67; consult, Vet Admin Hosp, Philadelphia & Walson Army Hosp, Ft Dix, NJ; dir coronary care unit, Philadelphia Gen Hosp, 71-74; attending assoc, Mercer Med Ctr, Trenton, NJ. *Mem:* Am Fedn Clin Res; Am Physiol Soc; Am Col Physicians. *Res:* Effects of thyroid hormone on myocardial metabolism of norepinephrine and on myocardial contractility and hemodynamics; cardiovascular hemodynamics; renal function in congestive heart failure; cardiology; effects of alcohol on heart. *Mailing Add:* 416 Bellevue Ave Mercer Med Ctr Trenton NJ 08618

GOODKIND, RICHARD JERRY, b Brooklyn, NY, Oct 5, 37; m 60; c 2. PROSTHODONTICS. *Educ:* Columbia Univ, 55-58; Tufts Univ, DMD, 62; Univ Mich, MS, 64; Am Bd Prosthodont, dipl, 67. *Prof Exp:* Dir grad restorative dent, 66-77, PROF DENT, SCH DENT, UNIV MINN, MINNEAPOLIS, 77- *Concurrent Pos:* Consult & mem prosthodontic staff, Univ Minn Hosp, 68- *Mem:* Midwest Prosthodont Soc; Am Col Prosthodontists. *Res:* Fixed and removable prosthodontics. *Mailing Add:* Div Removable Prosthodontics Univ Minn Sch Dent Minneapolis MN 55455

GOODLAND, ROBERT JAMES A, b Sept 26, 45; Can citizen; m 86; c 1. ECOLOGY, ENVIRONMENTAL MANAGEMENT. *Educ:* McGill Univ, BSc, 64, MSc, 65, PhD(trop ecol), 69. *Prof Exp:* Prof ecol, Univ Brasilia, 69-71; prof environ studies, McGill Univ, 71-72; ecologist, Cary Arboretum, NY Bot Garden, 72-75, chmn & asst dir environ assessment dept, 75-78; ecologist Off Environ & Health Affairs, 78-87, chief, Environ Div, 86-90, HEAD, ENVIRON ASSESSMENT, WORLD BANK, 90- *Concurrent Pos:* Ecologist, Amazonian ecosystems course, Nat Inst for Amazonian Res, Manaus, 73, World Bank Environ Missions, Indonesia & Malaysia, 74 & 75 & Environ Mission, El Salvador & Guatemala, Pan Am Health Orgn,75, Bangladesh, 76. *Honors & Awards:* Logan Prize. *Mem:* Ecol Soc NAm(pres, 88); Am Inst Biol Sci. *Res:* Environmental assessment of development projects, especially tropical; biodiversity and ecosystem conservation. *Mailing Add:* Environ Dept The World Bank Washington DC 20433

GOODLOE, PAUL MILLER, II, b Berea, Ky, May 21, 11; m 35; c 3. CHEMISTRY. *Educ:* East Ky Univ, BS, 32; Johns Hopkins Univ, PhD(chem), 37. *Prof Exp:* Asst org chem, Johns Hopkins Univ, 34-35; res chemist, Socony-Vacuum Oil Co, Inc, NJ, 36-43, supvr, NY, 43-46; supvr tech sales, Brown Co, NH, 46-53, asst dir res, 53-55, dir control, 55-57, asst tech dir, 57-58, dir res & develop, 58-60, gen mgr, Chem Prod Div, 60-62; pres, Kenrich Petrochem, Inc, NJ, 62-65; consult & eastern vpres, Vensearch Corp, Tex, 65-68; mgr spec projs, White Chem Corp, NJ, 68-72; vpres, Nolan Co, Md, 68-72, pres, 72-77; pres, 68-84, CHMN & CHIEF EXEC OFFICER, ASSOC CHEM INDUSTS, INC, 84- *Concurrent Pos:* Mem, Assoc Estate Planning, Upsala Col, NJ, 77- *Mem:* Am Chem Soc; Am Inst Chemists; Am Inst Chem Engrs; Tech Asn Pulp & Paper Indust. *Res:* Rubber and plastics compounding; plasticizers and fillers; cellulose products; emulsions; detergents; production of specialty chemicals; research administration; market development; mergers and acquisitions. *Mailing Add:* 175 Prospect St East Orange NJ 07017

GOODMAN, A(LVIN) M(ALCOLM), b Philadelphia, Pa, July 8, 30; m 66; c 2. ELECTRICAL ENGINEERING, SOLID STATE PHYSICS. *Educ:* Drexel Inst Technol, BS, 52; Princeton Univ, MA, 55, PhD(elec eng), 58. *Prof Exp:* Asst elec eng, Princeton Univ, 56-57; asst prof, Case Inst Technol, 57-59; mem tech staff, Solid State Devices Res Lab, David Sarnoff Res Ctr, RCA Corp, 59-82, sr mem tech staff, 82-87; SCI OFFICER, ELECTRONICS DIV, OFF NAVAL RES, 87- *Concurrent Pos:* lectr, Ariz State Univ, 84. *Honors & Awards:* Charles Ira Young Medal, 58. *Mem:* Am Phys Soc; Inst Elec & Electronics Engrs; Sigma Xi; Mat Res Soc. *Res:* Metal-semiconductor physics and devices; insulator physics and chemistry; physical measurements and instrumentation; semiconductor device processing and fabrication; author of various scientific articles; 57 publications; 23 patents. *Mailing Add:* Off Naval Res-Code 1114 800 N Quincy St Arlington VA 22217-5000

GOODMAN, ABRAHAM H(ARRISON), chemical engineering, food technology; deceased, see previous edition for last biography

GOODMAN, ADOLPH WINKLER, b San Antonio, Tex, July 20, 15; m 47; c 3. MATHEMATICS, GEOMETRIC FUNCTION THEORY. *Educ:* Univ Cincinnati, BSc, 39, MA, 41; Columbia Univ, PhD(math), 47. *Prof Exp:* Prin eng draftsman, US Navy Yard, Pa, 41-43; instr math, Syracuse Univ, 43-44; sr stress engr, Repub Aviation Corp, NY, 44-46; instr math, Rutgers Univ, 47-49; from assoc prof to prof, Univ Ky, 49-64; PROF MATH, UNIV S FLA, 64- *Concurrent Pos:* Mem, Inst Advan Study, 56-57. *Mem:* Am Math Soc; Math Asn Am. *Res:* Theory of functions of a complex variable; conformal mapping; polynomials; graph theory; number theory. *Mailing Add:* Dept Math Univ SFla Tampa FL 33620

GOODMAN, ALAN LAWRENCE, b Miami Beach, Fla, July 27, 38; m 64; c 2. POLYMER CHEMISTRY, SOFTWARE SYSTEMS. *Educ:* Univ Del, BS, 59; Stanford Univ PhD(org chem), 64. *Prof Exp:* Res fel, Brandeis Univ, 63-64; sr res chemist, Plastics Div, Allied Chem Corp, 64-66; res chemist, Elastomer Chem Dept, 66-80, SR RES CHEMIST, POLYMER PROD DEPT, E I DU PONT DE NEMOURS & CO, INC, 80- *Mem:* Am Chem Soc; Am Soc Qual Control. *Res:* Polymer properties and synthesis; statistical quality control. *Mailing Add:* One Eagles Circle Raven Crest Chadds Ford PA 19317-9193

GOODMAN, ALAN LEONARD, b Newark, NJ, July 28, 41; m 73. MEAN FIELD THEORIES, PHASE TRANSITIONS. *Educ:* Cornell Univ, BS, 64; Univ Calif, Berkeley, MS, 66, PhD(nuclear sci), 69. *Prof Exp:* Fel physics, Argonne Nat Lab, 69-71 & Carnegie-Mellon Univ, 73-76; from asst prof to assoc prof, 76-81, PROF PHYSICS, TULANE UNIV, 81- *Concurrent Pos:* Prin investr, NSF grants theoret nuclear physics, 77-; consult, Oak Ridge Nat Lab, 77-; ed, Proceedings Int Conf Band Structure & Nuclear Dynamics, 80; vis prof, Mich State Univ, 77, Niels Bohr Inst, Copenhagen, 79. *Mem:* Am Phys Soc; Sigma Xi. *Res:* Quantum-mechanical many-body theories are applied to the nucleus; phase transitions in excited nuclear energy levels. *Mailing Add:* Physics Dept Tulane Univ New Orleans LA 70118

GOODMAN, ALBERT, b New York, NY, Jan 22, 27; m 48; c 3. POLYMER CHEMISTRY. *Educ:* City Col, BS, 46; NY Univ, MS, 51; Polytech Inst Brooklyn, PhD(chem), 56. *Prof Exp:* Instr biol, City Col, 47-51; res chemist, Dextran Corp, 51-55; res chemist, E I du Pont de Nemours & Co Inc, 55-66, res supvr, 66-69, supvr plant technol, Nylon Div, 69-73, supvr textile res, 73-85, sr res assoc, 85-90; RETIRED. *Concurrent Pos:* Consult, 91- *Mem:* AAAS; Am Chem Soc; Sigma Xi. *Res:* Structure-property relationships in fibers; polymer physics and rheology; polymer chemistry; enzymatic polymerizations; ionic and stereoregular polymerizations; kinetics of polymerizations; fiber-fabric structural relationships; electrical properties of fibers and fabrics; textile processability of fibers; dye stability; UV stability of polymers. *Mailing Add:* 2806 Bodine Dr Wilmington DE 19810

GOODMAN, BARBARA EASON, b Hanover, NH, Nov 17, 49; m 72; c 2. RESPIRATORY PHYSIOLOGY IN MEMBRANE TRANSPORT. *Educ:* Duke Univ, BA, 72; Univ Minn, PhD(physiol), 81. *Prof Exp:* Asst res prof med, Univ Calif, Los Angeles, 84-86; ASST PROF PHYSIOL, SCH MED, UNIV SDAK, 86- *Concurrent Pos:* Prin investr, 88 & 89; vis lectr, 90-; mem, Cardiopulmonary Coun, Am Heart Asn. *Mem:* Am Physiol Soc; Am Thoracic Soc; Sigma Xi; Am Heart Asn. *Res:* Characteristics and regulation of active transport across the pulmonary alveolar epithelium which may be important for physiological and pathological lung fluid balance. *Mailing Add:* Dept Physiol Sch Med Univ SDak Vermillion SD 57069-2390

GOODMAN, BERNARD, b Philadelphia, Pa, June 14, 23; m 51; c 3. THEORETICAL SOLID STATE PHYSICS. *Educ:* Univ Pa, AB, 43, PhD(physics), 55. *Prof Exp:* Res stress analyst, Int Harvester Co, 47-52; res assoc physics, Univ Mo, 52-54, from asst prof to prof, 54-64; PROF PHYSICS, UNIV CINCINNATI, 65- *Concurrent Pos:* Consult, Argonne Nat Lab, 61-70; Guggenheim fel, 62; guest prof, Inst Theoret Physics, Univ Uppsala, 62 & Gothenberg Univ, 71 & 85; Fulbright scholar & guest prof, Inst Theoret Physics, Trieste, 79; Gordon Godfrey fel, Univ New South Wales, 90. *Mem:* Fel Am Phys Soc. *Res:* Lattice dynamics; quantum liquids; electron correlations and many-body theory; disordered systems; semi-conductors. *Mailing Add:* Dept Physics Univ Cincinnati Cincinnati OH 45221

GOODMAN, BILLY LEE, b Pauls Valley, Okla, Jan 21, 30; m 51; c 5. POULTRY SCIENCE. *Educ:* Okla State Univ, BS, 51, MS, 53; Ohio State Univ, PhD, 59. *Prof Exp:* Asst poultry, Okla State Univ, 51-52, instr, 52-55; instr, Ohio State Univ, 55-58; PROF ANIMAL INDUST, SOUTHERN ILL UNIV, CARBONDALE, 58- *Mem:* Poultry Sci Asn. *Res:* Poultry genetics and breeding. *Mailing Add:* Dept Animal Sci Southern Ill Univ Carbondale IL 62901

GOODMAN, CHARLES DAVID, b New York, NY, May 9, 28; m 52; c 2. EXPERIMENTAL NUCLEAR PHYSICS. *Educ:* Clark Univ, AB, 49; Univ Rochester, PhD, 55. *Prof Exp:* From physicist to sr physicist, Oak Ridge Nat Lab, 55-80; PROF, IND UNIV, 80- *Concurrent Pos:* Vis scientist, Weizmann Inst Sci, 66; vis prof, Univ Colo, Boulder, 72-73. *Honors & Awards:* Tom W Bonner Prize, Am Phys Soc, 83; Humboldt Res Award, 91. *Mem:* Fel Am Phys Soc; AAAS; Inst Elec & Electronics Engrs. *Res:* Experimental nuclear physics; mainly using nuclear charge exchange reactions to study nuclear structure models; with emphasis on Gamow-Teller transitions; the missing strength problem, and applications to solar neutrino detection; neutron spectroscopy in the 50-100 MeV range by the time of flight; neutron polarimetry. *Mailing Add:* Ind Univ 2401 Milo Sampson Lane Bloomington IN 47408-0768

GOODMAN, COREY SCOTT, b Chicago, Ill, June 29, 51; m 84. DEVELOPMENTAL NEUROBIOLOGY, DEVELOPMENTAL BIOLOGY. *Educ:* Stanford Univ, BS, 72; Univ Calif, Berkeley, PhD(neurobiology), 77. *Prof Exp:* From asst prof to assoc develop neuobiology, Dept Biol Sci, Stanford Univ, 79-87; prof, dept biochem, 87-89, PROF DEPT MOLECULAR & CELL BIOL, UNIV CALIF, BERKELEY, 89-; INVESTR, HOWARD HUGHES MED INST, 88- *Concurrent Pos:* Fel, H H Whitney Found, 77-79; McKnight Found Scholar; Sloan Found Fel; McKnight scholars award rev comt, 84-, chmn, 89-; ed bd reviewing, Science, 86-, ed, Develop Neurobiol sect, J Neurosci, 89-; Searle scholars adv comt, 88- *Honors & Awards:* Charles Judson Herrick Award, 82; Alan T Waterman Award, Nat Sci Bd, 83; Javits Neurosci Investr Award, 85; Merit Award, NIH, 86. *Mem:* Neurosciences Soc; Soc Develop Biol. *Res:* Developmental neurobiology; cell recognition during neuronal development; growth cone guidance; molecular genetics of cell adhesion molecules during morphogenesis. *Mailing Add:* 140 Panoramic Way Berkeley CA 94704

GOODMAN, DAVID BARRY POLIAKOFF, b Lynn, Mass, June 1, 42; m 65; c 2. ENDOCRINOLOGY. *Educ:* Harvard Univ, AB, 64; Univ Pa, MD, 68, PhD(biochem), 72. *Prof Exp:* Fel, Am Heart Asn, 70-71; assoc biochem, Univ Pa, 72-73, assoc pediat, 72-73, res asst prof, 73-76; asst prof med, Sch Med, Yale Univ, 76-79, assoc prof, 79-80; assoc prof & dir, div lab med, 80-83, PROF, DEPT PATH & LAB MED, SCH MED, UNIV PA, 84- *Concurrent Pos:* Consult neurobiol & biol, NSF, 73- & consult endocrinol, Vet Admin, 79-; ed, Bone & Related Res. *Honors & Awards:* Lamport Award, NY Acad Sci, 81. *Mem:* AAAS; Soc Neurosci; Soc Develop Biol; Am Fedn Clin Res; Am Asn Path; Am Physiol Soc. *Res:* Biochemistry of hormone action, ion transport, cell activation and oxygen toxicity; laboratory medicine; clinical pathology. *Mailing Add:* Dept Path & Lab Med Hosp Univ Pa 3400 Spruce St Philadelphia PA 19104

GOODMAN, DAVID JOEL, b Brooklyn, NY, June 9, 39; m 80; c 2. ELECTRICAL & COMPUTER ENGINEERING. *Educ:* Rensselaer Polytech Inst, BEE, 60; NY Univ, MEE, 62; Univ London, PhD(elec eng), 67. *Prof Exp:* Dept head, AT&T Bell Lab, 67-88; PROF & CHAIR ELEC & COMPUTER ENG, RUTGERS UNIV, 88- *Concurrent Pos:* Vis prof, Imp Col, Univ London, 83-88 & Southampton, 87-89; dir, Rutgers Wireless Info Network Lab, 89. *Mem:* Fel Inst Elec & Electronics Engrs; fel Inst Elec Engrs; Radio Club Am. *Res:* Wireless access to information networks; communication network architecture; digital signal processing; speech communication. *Mailing Add:* Dept Elec & Computer Eng Rutgers Univ PO Box 909 Piscataway NJ 08855-0909

GOODMAN, DAVID WAYNE, b Glen Allen, Miss, Dec 14, 45; m 67; c 1. SURFACE CHEMISTRY, MOLECULAR SPECTROSCOPY. *Educ:* Miss Col, BS, 68; Univ Tex-Austin, PhD(phys chem), 74. *Prof Exp:* NATO fel chem, Tech Univ, Parmstadt, Ger, 74-76; NRC fel, Nat Bur Standards, Washington, DC, 76-78, res scientist, 78-80; mem tech staff, Sandia Nat Labs, Albuquerque, NMex, 80-85 head surface sci div, 85-88; PROF CHEM, TEX A&M UNIV, COL STA, TX, 88- *Concurrent Pos:* Distinguished lectr, 82, adj prof, Univ Tex, Austin, 85-86; frontiers chem lectr, Tex A&M Univ, 87. *Honors & Awards:* Ipatieff Prize, Am Chem Soc, 83. *Mem:* Am Chem Soc; Am Phys Soc; Am Vacuum Soc; Mat Res Soc; AAAS. *Res:* Studies of chemisorption & catalytic reactions on atomically clean & chemically modified metal single crystal surfaces using modern surface science techniques. *Mailing Add:* Dept Chem Tex A&M Univ Col Sta TX 77843

GOODMAN, DEWITT STETTEN, b New York, NY, July 18, 30; m 57; c 2. NUTRITION, INTERNAL MEDICINE. *Educ:* Harvard Univ, AB, 51, MD, 55. *Hon Degrees:* MDr, Univ Oslo, Norway, 86. *Prof Exp:* Intern med, Presby Hosp, New York, 55-56; investr, Nat Heart Inst, 56-58; asst resident med, Presby Hosp, 58-59; investr, Nat Heart Inst, 60-62; from asst prof to prof, 62-71, TILDEN-WEGER-BEILER PROF MED & DIR, DIV METAB & NUTRIT, COLUMBIA UNIV MED CTR, 71-, DIR, INST HUMAN NUTRIT, COLUMBIA UNIV, 88- *Concurrent Pos:* Helen Hay Whitney Found fel, Hammersmith Hosp, London, Eng, 59-60; career scientist, City of New York Health Res Coun, 64-74; mem metab study sect, NIH, 66-70; assoc ed, J Clin Invest, 67-70, ed-in-chief, 70-72; mem exec comt, Coun on Arteriosclerosis, Am Heart Asn, 69-72 & 77-85; Guggenheim Found fel & vis fel, Clare Hall, Cambridge Univ, 72-73; adj prof, Rockefeller Univ, 74-77; dir, Arteriosclerosis Res Ctr, 76-; mem, acteriosclerosis, hypertension & lipid metab adv comt, Nat Heart, Lung & Blood Inst, NIH, 78-87, chmn, 85-87; chmn, Coun on Arteriosclerosis, Am Heart Asn, 79-81; chmn, Gordon Conf on Lipids, 67 & Arteriosclerosis, 81; vis prof, Hebrew Univ, 81-82; chmn, Nat Cholesterol Educ Prog Expert Panel, 86-87; vchmn, comt diet & health, Nat Res Coun, 86-89; mem, comt A, Nat Res Agenda Aging, Inst Med, 88-, Food & Nutrit Bd, 88- *Honors & Awards:* Samuel Meltzer Award, Soc Exp Biol & Med, 64; Osborne & Mendel Award, Am Inst Nutrit, 74; Award of Merit, Am Heart Asn, 90. *Mem:* Inst Med-Nat Acad Sci; Am Soc Clin Invest; Asn Am Physicians; Am Soc Biol Chemists; Am Inst Nutrit; Harvey Soc (vpres, 81-82 & pres, 82-83); Am Soc Clin Nutrit; Endocrine Soc; Am Oil Chemists Soc; fel AAAS. *Res:* Lipid metabolism; cholesterol turnover and metabolism; lipid transport; vitamin A and retinoid metabolism and transport; retinoid binding proteins; fat-soluble vitamins; arteriosclerosis. *Mailing Add:* Dept Med Col Physicians & Surgeons Columbia Univ 630 W 168th St New York NY 10032

GOODMAN, DONALD, b Boston, Mass, Apr 7, 33; m 61; c 2. POLYMER CHEMISTRY. *Educ:* Harvard Univ, AB, 54. *Prof Exp:* Res assoc res & develop polymer chem, W R Grace & Co, Dewey & Almy Chem Div, 56-65; group leader res & develop polymer chem, Continental Oil Co, Thompson Apex Div, 65-68; mgr res res & develop polymer chem, Olin Corp, Olin Plastics Div, 68-72; mgr polymers res & develop, Tenneco Inc, Tenneco Chem Div, 72-86; MGR TECH PROJS, OCCIDENTAL CHEM CORP, 86- *Mem:* Am Chem Soc; AAAS; Soc Plastics Engrs. *Res:* Polymer science and technology, especially polymer structure versus properties relationships; polymer kinetics and mechanisms of polymerization; emulsion polymerization; polymer characterization; environmental science; industry and regulatory affairs; issues management. *Mailing Add:* Rd One 63-D Valley Hill Rd Malvern PA 19355

GOODMAN, DONALD CHARLES, b Chicago, Ill, Nov 24, 27; m 68; c 6. NEUROANATOMY, COMPARATIVE NEUROLOGY. *Educ:* Univ Ill, BS, 49, MS, 50, PhD(zool), 54. *Prof Exp:* Instr anat, Sch Med, Univ Pa, 54-56; instr, Univ Mich, 56; asst prof, Univ Fla, 56-59, assoc prof, 59-63, prof, 63-68, chmn dept, 65-68; dean med sci & vpres acad affairs, 75-76, prof & chmn dept, 68-82, dean col grad studies, 73-82, vpres res & acad affairs, 76-82, DEAN & CHMN DEPT, STATE UNIV NY UPSTATE MED CTR, 83-, PROVOST, 85- *Concurrent Pos:* Fel, Inst Neurol Sci, Univ Pa, 54-55; Nat Inst Neurol Dis & Stroke res grant, 58-; mem study sect neurol B, NIH, 68-71, mem anat training prog, 71-75; adj prof neurosci, State Univ NY, Binghamton. *Honors & Awards:* Res Award, Sigma Xi, 62. *Mem:* Nat Coun Univ Res Admin; AAAS; Am Asn Anat; Am Soc Allied Health Prof. *Res:* Neuromorphological plasticity of central nervous system and its relationship to recovery of brain function; comparative neurology of cerebellum and basal ganglia and of brain structure of reptiles. *Mailing Add:* Provost State Univ NY Upstate Med Ctr Syracuse NY 13210

GOODMAN, ELI I, b Bronx, NY, June 26, 29; m 51; c 3. NUCLEAR ENGINEERING, CHEMICAL ENGINEERING. *Educ:* Mass Inst Technol, BS, 50, MS, 51. *Prof Exp:* Jr chem engr, Brookhaven Nat Lab, 51-53, assoc chem engr, 53-55; sr chem engr, Nuclear Sci & Eng Corp, Pa, 55-59; sr engr, Westinghouse Elec Corp, 59-65; opers analyst, USAEC, 65-67, chief plans & forecast br, 67-70, asst sci rep, Tokyo, 70-72, consult, off planning & anal, 72-74; tech adv, Int Atomic Energy Agency, Vienna, 74-78; consult, Off Prog Planning & Anal, 78-82, SR TECH ADV, DIV NUCLEAR ENERGY, US DEPT ENERGY, 82- *Mem:* Fel Am Nuclear Soc. *Res:* Nuclear power reactors; nuclear fuel cycle; international aspects of nuclear energy; applications of radioactivity. *Mailing Add:* 3304 Rittenhouse St NW Washington DC 20015

GOODMAN, EUGENE MARVIN, b Buffalo, NY, June 26, 37; m 57; c 4. CELL BIOLOGY. *Educ:* State Univ NY Buffalo, BA, 63, PhD(biol), 67. *Prof Exp:* Fel cell biol, McArdle Lab Cancer Res, 66-69, from asst prof to assoc prof, 69-79, PROF CELL BIOL, UNIV WIS-PARKSIDE, 80, DIR, BIOMED RES INST, 81- *Mem:* AAAS; Bioelectromagnetics Soc. *Res:* Molecular effects of extremely low frequency electromagnetic fields on growth; cell cycle controls. *Mailing Add:* Biomed Res Inst Univ Wis-Parkside PO Box 2000 Kenosha WI 53141

GOODMAN, FRANK R, b Norfolk, Va, Dec 26, 43; m 67; c 2. PHARMACOLOGY. *Educ:* Univ Richmond, BA, 66; Med Col Va, PhD(pharmacol), 70. *Prof Exp:* Instr pharmacol, Univ Tex Health Sci Ctr, Dallas, 72-73, asst prof, 73-77; pharmacologist, Dow Chem Co, 77-83; mgr, Dept Cardiopulmonary Res, 83-86, DIR, STRATEGY & BUS DEVELOP, CIBA-GEIGY CORP, 89- *Concurrent Pos:* Adj assoc prof, Med Sch, Ind Univ, Indianapolis, 78- *Mem:* Physiol Soc; Sigma Xi; Am Soc Pharmacol & Exp Therapeut; Biophys Soc; Am Heart Asn. *Res:* Cellular pharmacology; mechanisms of drug interactions in smooth and lung muscle; action of drugs in asthma and diabetes; lipid metabolism. *Mailing Add:* Strategy Bus & Develop Dept Ciba-Geigy Corp 556 Morris Ave Summit NJ 07901

GOODMAN, FRED, b New York, NY, Nov 13, 28; m 53; c 2. MOLECULAR GENETICS. *Educ:* Brooklyn Col, AB, 49, MA, 52; Columbia Univ, PhD(zool), 58. *Prof Exp:* Lectr asst biol, Brooklyn Col, 49-50, substitute instr, 50-52; res asst biochem, Columbia Univ, 52-54, lectr zool, 54-56, res worker biochem, 58-60; from asst prof to assoc prof, Yeshiva Univ, 60-67, prof biol, Stern Col Women, 67-; AT BEIT CAN, ISRAEL. *Concurrent Pos:* NIH grants, 61-67 & 67-70; vis assoc prof, Univ Calif, Los Angeles, 66; vis prof, Bar-Ilan Univ, Israel, 68-69. *Mem:* AAAS; Am Soc Microbiol; Sigma Xi. *Res:* Nucleic acid and protein synthesis; radiation biology; metabolic events in bacteriophage-infected cells. *Mailing Add:* Nine Rozin St Harnof Jerusalem 93870 Israel

GOODMAN, GERALD JOSEPH, b Winthrop, Mass, Oct 24, 42. INFORMATION SCIENCE. *Educ:* Rutgers Univ, BA, 64; Johns Hopkins Univ, PhD, (physics), 69. *Prof Exp:* Res assoc physics, Johns Hopkins Univ, 69-70; MGR, AM INST PHYSICS, 70- *Mem:* Am Phys Soc. *Mailing Add:* Am Inst Physics 335 E 45th St New York NY 10017

GOODMAN, GORDON LOUIS, b Chicago, Ill, Aug 19, 33; m 62; c 2. ELECTRONIC STRUCTURE OF ATOMS, MOLECULES & SOLID-STATE SYSTEMS. *Educ:* Harvard Univ, AB, 55, PhD, 59. *Prof Exp:* Chemist, Argonne Nat Lab, 59-71 & 83-90; HEAD, COMMUNAISSANCE, 70- *Concurrent Pos:* Vis prof, Cornell Univ, 64, Univ Leuven, Belgium, 87; prof, Southern Ill Univ, 70; sci assoc, CEA, Saclay, France, 74. *Mem:* AAAS; Asn Comput Mach; fel Am Phys Soc; Sigma Xi. *Res:* Modeling electronic structure of solid-state systems, such as actinide compounds and ceramic copper oxides; interpretation of photoionization and high resolution spectra of open-shell atoms and molecules. *Mailing Add:* 5834 Middaugh St Downers Grove IL 60516

GOODMAN, HAROLD ORBECK, b Minneapolis, Minn, Sept 8, 24; m 43; c 4. GENETICS. *Educ:* Univ Minn, BA, 48, MA, 50, PhD, 53. *Prof Exp:* Fel med genetics, Bowman Gray Sch Med, 53-54; from instr to asst prof zool, Mich State Col, 54-58; from asst prof to assoc prof, 58-70, PROF & HEAD SECT MED GENETICS & ASSOC DEAN BIOMED GRAD STUDIES, BOWMAN GRAY SCH MED, 70- *Mem:* Am Soc Human Genetics; Am Genetic Asn; Brit Eugenics Soc; Sigma Xi. *Res:* Genetics of dental caries; salivary components; atherosclerosis; mongolism. *Mailing Add:* Dept Preventive Med Gen Bowman Gray Sch Med Winston-Salem NC 27103

GOODMAN, HENRY MAURICE, b Glen Cove, NY, May 4, 34; m 61; c 3. PHYSIOLOGY, ENDOCRINOLOGY. *Educ:* Brandeis Univ, AB, 56; Harvard Univ, AM, 57, PhD(physiol), 60. *Prof Exp:* From instr to assoc prof physiol, Harvard Med Sch, 62-70; PROF PHYSIOL & CHMN DEPT, MED SCH, UNIV MASS, 70- *Concurrent Pos:* Jr fel, Harvard Univ, 60-62; teaching fel med, Tufts Univ, 61-62; USPHS career develop award, 66-70; assoc dean sci affairs, Med Sch, Univ Mass, 79- 81, assoc dean, 81- *Mem:* Am Physiol Soc; Endocrine Soc. *Res:* Physiological actions of growth hormone; endocrine regulation of metabolism; physiology of adipose tissue. *Mailing Add:* Dept Physiol Univ Mass Med Sch Worcester MA 01605

GOODMAN, HOWARD CHARLES, b Rochester, NY, July 18, 20; m 42; c 3. IMMUNOLOGY. *Educ:* Harvard Univ, AB, 41; Johns Hopkins Univ, MD, 44. *Prof Exp:* Intern & asst resident med serv, Peter Bent Brigham Hosp, Boston, 44-45, 47-48; resident med serv, Sawtelle Vet Admin Hosp, Los Angeles, 48-49; res assoc, Inst Med Res, Cedars of Lebanon Hosp, 49-53; res assoc, Nat Heart Inst, 53-60, head clin immunol sect, Nat Inst Allergy & Infectious Dis, 60-63; chief immunol unit, WHO, 63-75; prof path, Fac Med, Univ Geneva, 72-75; dir trop dis res & training & res prom & develop, 75-77, dir, Trop Med Ctr, 77-86, EMER PROF, DEPT IMMUNOL & INFECTIOUS DIS, JOHNS HOPKINS UNIV, 86- *Concurrent Pos:* Howard Hughes med res fel, 51-53; asst med unit, St Mary's Hosp, London, 51-52; clin assoc prof, Med Sch, Univ Southern Calif, 52-53. *Mem:* AAAS; Am Fedn Clin Res; Am Asn Immunol. *Res:* Internal medicine; basic and clinical immunology. *Mailing Add:* PO Box 69 Key Colony Beach FL 33051

GOODMAN, HOWARD MICHAEL, b Brooklyn, NY, Nov 29, 38; m 69; c 1. BIOCHEMISTRY. *Educ:* Williams Col, BA, 60; Mass Inst Technol, PhD(biophys), 64. *Prof Exp:* Lab teaching asst genetics, Mass Inst Technol, 62; Helen Hay Whitney fel, 64-67, Am Cancer Soc fel, 67-69; asst prof, Univ Geneva, 69-70; from asst prof to assoc prof, 70-76, PROF BIOCHEM, UNIV CALIF, SAN FRANCISCO, 76- *Concurrent Pos:* Investr, Howard Hughes Med Inst, 78- *Mem:* Am Soc Biol Chemists. *Res:* Mechanisms of specific recognition of DNA by restriction endonucleases; study of eucaryotic gene organization using molecular cloning techniques; regulation of transcription and control. *Mailing Add:* 61 Frances St Needham MA 02192-1532

GOODMAN, IRVING, b Denver, Colo, May 22, 17; m 43; c 4. BIOCHEMISTRY. *Educ:* Univ Colo, BA, 39, MA, 41, PhD(chem), 44; MA, Univ Cambridge, England, 50. *Prof Exp:* Asst chem, 39-43, from instr to asst prof, Univ Colo, 43-49; Guggenheim fel, Cambridge Univ & Free Univ Brussels, 49-51; sr res chemist, Wellcome Res Labs Div, Burroughs, Wellcome & Co, Inc, 52-59; asst prof biochem, Col Physicians & Surgeons, Columbia Univ, 59-84; RETIRED. *Concurrent Pos:* Res assoc, Paris, 51-52; vpres, Coherin Res Found, Inc, 80-; pres, Intesco Labs, Inc, 84- *Mem:* AAAS; Am Chem Soc; Am Soc Biol Chemists; The Chem Soc; NY Acad Sci. *Res:* Nucleic acid derivatives and homologues; amino acids and peptides; carbohydrates; mechanisms of drug action; bivalent sulfur compounds; metabolic inhibitors; pituitary hormones. *Mailing Add:* Intesco Labs PO Box 676 Waldoboro ME 04572

GOODMAN, JACOB ELI, b Lynn, Mass, Nov 15, 33; m 73; c 2. DISCRETE GEOMETRY, CONFIGURATIONS. *Educ:* New York Univ, BA, 53; Columbia Univ, MA, 55, PhD(math), 67. *Prof Exp:* from inst to asst prof math, New York Univ, 60-67; from asst prof to assoc prof, 67-80, PROF MATH, CITY COL, CITY UNIV NEW YORK, 81- *Concurrent Pos:* Prin investr, NSF res grants, 67-; co-ed-in-chief, Discrete & Computational Geom, 86- *Honors & Awards:* Lester R Ford Award, Math Asn Am, 90. *Mem:* Am Math Soc. *Res:* Classification of configurations of points in the plane and higher-dimensional space, and of topological generalizations; combinatorial and computational questions relating to configurations. *Mailing Add:* Dept Math City Col CUNY New York NY 10031

GOODMAN, JAMES R, b Riverton, Wyo, May 23, 33; m 53; c 2. STRUCTURAL & CIVIL ENGINEERING. *Educ:* Univ Wyo, BS, 55; Colo State Univ, MS, 61; Univ Calif, Berkeley, 63, PhD(struct eng), 67. *Prof Exp:* Bridge designer, Wyo State Hwy Dept, 55-57; from instr to prof civil eng, Colo State Univ, 57-87; PVT CONSULT, 87- *Concurrent Pos:* Field tech, US geol Surv, 51 & 52. *Honors & Awards:* L J Markwardt Award, Am Soc Testing & Mat, 74. *Mem:* Am Soc Civil Engrs; Am Soc Eng Educ; Forest Prod Res Soc. *Res:* Properties of wood and wood engineering; application of computers to engineered structures. *Mailing Add:* 3404 Terry Point Dr Ft Collins CO 80521

GOODMAN, JAY IRWIN, b Brooklyn, NY, Apr 4, 43; m 65; c 2. PHARMACOLOGY, TOXICOLOGY & CHEMICAL CARCINOGENESIS. *Educ:* Brooklyn Col Pharm, BS, 65; Univ Mich, PhD(pharmacol), 69; Am Bd Toxicol, dipl. *Prof Exp:* Nat Cancer Inst res fel, McArdle Lab Cancer Res, Univ Wis, 69-71; from asst prof to assoc prof, 71-82, PROF PHARMACOL & TOXICOL, MICH STATE UNIV, 82- *Concurrent Pos:* fel, Nat Cancer Inst, NIH, 69-71. *Mem:* AAAS; Am Soc Pharmacol & Exp Therapeut; Am Asn Cancer Res; Soc Toxicol. *Res:* Chemical carcinogenesis; genetic toxicology. *Mailing Add:* Dept Pharmacol & Toxicol Mich State Univ East Lansing MI 48824

GOODMAN, JEROME, b Brooklyn, NY, Apr 2, 26; m 67; c 2. RESEARCH ADMINISTRATION, WRITING. *Educ:* Polytech Inst Brooklyn, BS, 49, PhD(chem), 61. *Prof Exp:* Dir res, Radiation Res Corp, 55-57, 59; sr scientist, Radiation Appln Inc, 59-60; dir chem res, Nuclear Res Assocs, Inc, 60-71; sr analyst, Edwards & Hanly, 71-74; GRANTS ADMINR, AM HEALTH FOUND, 74- *Concurrent Pos:* Fulbright grant, Lab de Chim Phys, Paris, 58-59; sr instr, Polytech Inst Brooklyn, 64-65, chmn, human subjects, Inst Rev Bd. *Mem:* AAAS; Am Chem Soc; Sigma Xi. *Res:* Research administration in preventive medicine; grant writer. *Mailing Add:* 35 Pond Rd Great Neck NY 11024-1018

GOODMAN, JOAN WRIGHT (MRS CHARLES D), b El Paso, Tex, May 14, 25; m 52; c 2. PHYSIOLOGY. *Educ:* Barnard Col, Columbia Univ, BA, 45; Univ Rochester, PhD(physiol), 52. *Prof Exp:* Chemist, Tidewater Assoc Oil Co, 45; asst chem, Manhattan proj, Univ Rochester, 45-46, technician electromyog, Dept Orthop, 46-47, jr scientist cell physiol, Div Radiation Biol, Univ Atomic Energy Proj, 53-54; res assoc biol, Mass Inst Technol, 56; sr res biologist, Div Biol, Oak Ridge Nat Lab, 57-78; prof, Grad Sch Biomed Sci, Univ Tenn, 69-78; staff scientist, 78-80, MEM GROUP IMMUNOL, LAWRENCE BERKELEY LAB, UNIV CALIF, BERKELEY, 80- *Mem:* Am Physiol Soc; Radiation Res Soc; Soc Exp Biol & Med; Fedn Am Socs Exp Biol; Int Soc Exp Hemat. *Res:* Transplantation; immunogenetics; hemopoiesis; cellular immunology. *Mailing Add:* 1885 Grand View Dr Oakland CA 94618

GOODMAN, JOEL MITCHELL, b New York, NY, Oct 13, 48; m 77; c 1. PROTEIN SORTING, ORGANELLE ASSEMBLY. *Educ:* Univ Calif San Diego, BA, 71; Univ Southern Calif, PhD(pharmacol), 80. *Prof Exp:* fel molecular biol, Univ Calif Los Angeles Molecular Biol Inst, 79-82; asst prof, 82-88, ASSOC PROF PHARMACOL, UNIV TEX SOUTHWESTERN MED SCH, 88- *Concurrent Pos:* Prin investr, Nat Inst Health, 84-; fac researcher, Am Cancer Soc, 87- *Mem:* Sigma Xi; Am Soc Biochemists & Molec Biologists; Am Soc Cell Biologists. *Res:* Assembly of mitracellular organelles, particularly the mechanism of translocation of proteins across organellar membranes. *Mailing Add:* 1507 Jeanette Way Carrollton TX 75006-2977

GOODMAN, JOEL WARREN, b New York, NY, Feb 2, 33; m 64; c 2. IMMUNOLOGY. *Educ:* Brooklyn Col, BA, 53; Columbia Univ, PhD(microbiol), 59. *Prof Exp:* Fel, Nat Inst Med Res, London, 59-60; from asst prof to assoc prof, 60-70, PROF MICROBIOL, SCH MED, UNIV CALIF, SAN FRANCISCO, 70- *Concurrent Pos:* Mem, Study Sect, NIH, 72-76; fel Rev Panel, Am Cancer Soc, 78-81; consult, Space Prog, NASA, 74-; NIH study sect, 90- *Mem:* AAAS; Am Asn Immunol; The Planetary Soc. *Res:* Chemical substructure of antigens; mechanisms of lymphocyte activation; mechanism of immune induction. *Mailing Add:* Dept Microbiol & Immunol Univ Calif Box 0414 San Francisco CA 94143

GOODMAN, JOSEPH WILFRED, b Boston, Mass, Feb 8, 36; m 62; c 1. OPTICS, ELECTRICAL ENGINEERING. *Educ:* Harvard Univ, AB, 58; Stanford Univ, MS, 60, PhD(elec eng), 63. *Prof Exp:* Res asst elec eng, Stanford Univ, 58-62; fel, Norweg Defense Res Estab, 62-63; res assoc, 63-67, from asst prof to assoc prof, 67-88, dir, Info Systs Lab, 81-82, WILLIAM E AYER PROF ELEC ENG, STANFORD UNIV, 88-, CHMN DEPT, 88- *Concurrent Pos:* Ed adv, Optics Commun, 72-; vis prof, Univ Paris XI, 73-74; assoc ed, Optical Eng, 76-79; ed, J Optical Soc Am, 78-83; vpres, Int Comn Optics, 85-87, pres, 87-90. *Honors & Awards:* F E Terman Award, Am Asn Eng Educ, 71; Max Born Award, Optical Soc Am, 83, Frederik Ives Medal, 90; Dennis Gabor Award, Int Optical Eng Soc, 87; Education Medal, Inst Elec & Electronics Engrs, 87. *Mem:* Nat Acad Eng; fel Soc Photo-Optical Engrs; fel Optical Soc Am (pres elect, 91); fel Inst Elec & Electronics Engrs. *Res:* Optical computing; fourier optics; holography; statistical optics. *Mailing Add:* Durand 127 Stanford Univ Stanford CA 94305

GOODMAN, JULIUS, risk analysis, reliability & statistics; deceased, see previous edition for last biography

GOODMAN, L(AWRENCE) E(UGENE), b New York, NY, Mar 12, 20; m 51; c 3. APPLIED MECHANICS, STRUCTURAL ENGINEERING. *Educ:* Columbia Univ, AB, 39, BS, 40, PhD(appl mech), 49; Univ Ill, MS, 42, MA, Cambridge Univ, 62. *Prof Exp:* Electro-mech engr, Appl Physics Lab, Johns Hopkins Univ, 43-44, 46; instr civil eng, Sch Eng, Columbia Univ, 47-48; from res asst prof to res assoc prof, Univ Ill, 49-53; prof mech, Univ Minn, Minneapolis, 54-65, head, Dept Civil Eng, 65-72, prof civil eng, 65-80, James L Record prof civil eng, 80-88; RETIRED. *Concurrent Pos:* NSF sr fel, 62-63; mem sci comt, Conf Mech of Contact, Int Union Theoret & Appl Mech, 73-75. *Mem:* Fel Am Soc Civil Engrs; fel Am Soc Mech Engrs. *Res:* Stress analysis; vibration of structures; material damping; structural analysis; contact stresses. *Mailing Add:* 1589 Vincent St St Paul MN 55108

GOODMAN, LEON, b Livingston, Mont, Dec 16, 20; m 56; c 2. ORGANIC CHEMISTRY. *Educ:* Univ Calif, Berkeley, BS, 41; Univ Calif, Los Angeles, PhD(chem), 50. *Prof Exp:* Res asst explosive chem, Off Sci Univ & Develop, 42-45; res chemist, Los Alamos Sci Lab, 50-55; chemist, Stanford Res Inst, 55-61, chmn, Dept Bioorg Chem, 61-70; prof chem & chmn dept, Univ RI, 70-76, prof chem, 76-89; RETIRED. *Concurrent Pos:* Res assoc, Univ Southern Calif, 53-54; mem, Med Chem Study Sect, NIH, 70-73; Eleanor Roosevelt Int Cancer Fel, 77-78; mem bd sci counr, Div Cancer Treat, Nat Cancer Inst, 82-86. *Mem:* Am Chem Soc; Royal Soc Chem; NY Acad Sci. *Res:* Cancer chemotherapy; nucleoside chemistry; carbohydrate chemistry; nitrogen heterocyclic chemistry; silicon chemistry; organic chemistry of high explosives; organic sulfur chemistry. *Mailing Add:* Dept Chem Univ RI Kingston RI 02881

GOODMAN, LEON JUDIAS, b New York, NY, Sept 5, 30; m 51; c 2. RADIOLOGICAL PHYSICS, DOSIMETRY. *Educ:* Cooper Union, BS, 59. *Prof Exp:* Detail draftsman, Northern Elec Co Ltd, Que, Can, 50, jr toolmaker, 50-51; draftsman, W Green Elec Co, 51-52; engr's asst, Wright Aeronaut Corp, 52-55; design draftsman, Walter Motor Truck Co, 55-56; indust hygienist health & safety lab, AEC, 56-58; res scientist radiol res lab, Columbia Univ, 58-67; proj mgr, Brookhaven Nat Lab, 67-80; PHYS SCIENTIST, CTR RADIATION RES, NAT BUR STANDARDS, 80- *Concurrent Pos:* Assoc radiol physics, Columbia Univ, 67-80. *Mem:* Inst Elec & Electronics Engrs; Health Physics Soc; Radiation Res Soc; Am Asn Physicists Med. *Res:* Design and development of instrumentation, systems and experiments for radiation dosimetry in connection with radiobiological research into the effects of radiation on living systems. *Mailing Add:* 2515 Boston St No 908 Baltimore MD 21224

GOODMAN, LEONARD SEYMOUR, nuclear physics; deceased, see previous edition for last biography

GOODMAN, LIONEL, b Brooklyn, NY, Apr 23, 27; m 50; c 2. CHEMICAL PHYSICS, PHOTOBIOLOGY. *Educ:* Iowa State Univ, PhD, 54. *Prof Exp:* Fel, Fla State Univ, 54-55; from asst prof to assoc prof chem, Pa State Univ, 56-66; PROF CHEM, RUTGERS UNIV, 66- *Concurrent Pos:* Vis prof, Sorbonne, 64; vis prof molecular biophysics, Stockholm Univ, 68; NSF Sr Fel, 61-62; Guggenheim fel, 65-66. *Honors & Awards:* Frontiers Chem Lectr, Wayne State Univ, 88. *Mem:* Am Chem Soc. *Res:* Laser spectroscopy, especially involving multiphoton excitation in supersonic jets; coupling of vibrational and electronic motions in molecules; molecular potential surfaces; theoretical simulation of molecular spectra. *Mailing Add:* Dept Chem Rutgers Univ New Brunswick NJ 08903

GOODMAN, LOUIS E, surgery, oncology; deceased, see previous edition for last biography

GOODMAN, LOUIS SANFORD, b Portland, Ore, Aug 27, 06; m 33; c 2. PHARMACOLOGY. *Educ:* Reed Col, BA, 28; Univ Ore, MD & MA, 32. *Hon Degrees:* DSc, Univ Man, 65, Univ Utah, 69 & Med Col Wis, 73. *Prof Exp:* Asst psychol, Reed Col, 27-29; asst neurol & pharmacol, Sch Med, Univ Ore, 29-32; house officer med, Johns Hopkins Hosp, 32-33; Nat Res Coun fel, Sch Med, Yale Univ, 34, instr pharmacol & toxicol, 35-37, asst prof, 37-43; prof pharmacol & physiol & chmn dept, Col Med, Univ Vt, 43-44; prof pharmacol & chmn dept, 44-72, DISTINGUISHED PROF PHARMACOL, COL MED, UNIV UTAH, 72- *Concurrent Pos:* Mem pharmacol study sect, USPHS, 48-52, pharmacol & exp therapeut study sect, 54; ed-in-chief, Pharmacol Rev, Am Soc Pharmacol & Exp Therapeut, 49-53; mem res comt sci coun, Am Heart Asn, 53-55; mem med bd, Myasthenia Gravis Found, 53-58; mem, Nat Adv Neurol Dis & Blindness Coun, 54-58; mem pharmacol test comt, Nat Bd Med Exam, 55-59; mem adv coun, Life Inst Med Res Fund, 56-59; mem sci bd, Nat Neurol Res Found, 57-64; chmn pharmacol training comt, NIH, 58-61, mem comt res career awards, Div Gen Med Sci, mem nat adv ment health coun, 62-66 & mem nat adv coun, Health Res Facil, 66-70; mem adv comt psychopharmacol serv ctr, NIMH, 58-62 & panel neuropharmacol, Int Brain Res Orgn, 60; mem, Depts State & Health, Educ & Welfare Neurol Sci Mission to USSR, 58; US rep, Int Union Physiol Sci, 63-66. *Mem:* Nat Acad Sci; AAAS; Am Soc Pharmacol & Exp Therapeut (pres, 59-60); fel Am Col Neuropsychopharmacol; fel NY Acad Sci. *Res:* Anticonvulsant and autonomic drugs; pharmacodynamics. *Mailing Add:* Dept Pharmacol Univ Utah Col Med Salt Lake City UT 84132

GOODMAN, MADELENE JOYCE, b New York, NY, Sept 11, 45; m 65; c 2. HUMAN BIOLOGY, BIOPHILOSOPHY-BIOETHICS. *Educ:* Barnard Col, BA, 67; Oxford Univ, Eng, MSc, 68; Univ Hawaii, PhD(genetics), 73. *Prof Exp:* From asst prof to assoc prof, 74-85, PROF GEN SCI & WOMEN'S STUDIES, UNIV HAWAII, 85-; ASST VPRES ACAD AFFAIRS. *Concurrent Pos:* Res assoc, Pac Health Res Inst, Honolulu, 75-80, pres, 80-85; prin investr, NSF, PHS & Am Can Soc grants, 77-91; mem, Human Biol Coun. *Honors & Awards:* Nat Lectr, Sigma Xi. *Mem:* NY Acad Sci; AAAS; Sigma Xi; Soc Study Human Biol; Asn Women in Sci. *Res:* Cross-cultural variation in breast cancer epidemiology, menopause, menarche and women's reproductive patterns; biophilosophy, bioethics and feminist biology. *Mailing Add:* Acad Affairs Univ Hawaii Honolulu HI 96822

GOODMAN, MAJOR M, b Des Moines, Iowa, Sept 13, 38; m 70; c 2. GENETICS, EVOLUTIONARY BIOLOGY. *Educ:* Iowa State Univ, BS, 60; NC State Univ, MS, 63, PhD(genetics), 65. *Prof Exp:* NSF fel, Inst Genetics, Advan Sch Agr, Univ Sao Paulo, 65-67; vis asst prof statist, 67-68, from asst prof to prof, 68-88, WILLIAM NEAL REYNOLDS DISTINGUISHED UNIV PROF BOT, STATIST & GENETICS, NC STATE UNIV, 88- *Concurrent Pos:* Del, Int Genetics Cong, Tokyo, 67; mem, Maize Germ Plasm Resources Comt, Rockefeller Found, 69-72; chmn, Maize Coop Adv Comt, 81-86; panel dir, Competitive Grants Panel Plant Genetics & Molecular Biol, USDA, 87-88. *Mem:* Nat Acad Sci; Genetics Soc Am; Soc Study Evolution; Soc Econ Bot; Soc Syst Zool; Crop Sci Soc Am. *Res:* Evolution of cultivated plants; numerical taxonomy; history and evolution of maize; applied multivariate statistics. *Mailing Add:* PO Box 7620 Raleigh NC 27695

GOODMAN, MARK WILLIAM, b Columbia, Mo, Jan 6, 60; m 84; c 1. SUPERSTRING THEORY, FIELD THEORY. *Educ:* Brown Univ, BA, 81; Princeton Univ, MA, 83, PhD(physics), 86. *Prof Exp:* Postdoctoral mem, Inst Theoret Physics, Univ Calif, Santa Barbara, 86-88; POSTDOCTORAL FEL, RUTGERS UNIV, 88- *Mem:* Am Phys Soc; AAAS; Fedn Am Scientists. *Res:* Theory of elementing particles and superstring theory, particularly symmetries in four dimensional string models. *Mailing Add:* Dept Physics & Astronomy Rutgers Univ PO Box 849 Piscataway NJ 08854-0849

GOODMAN, MICHAEL GORDON, b Denver, Colo, July 4, 46. IMMUNOLOGY, BIOLOGY GENERAL. *Educ:* Yale Univ, BA, 68; Univ Calif, San Francisco, MD, 72; dipl, Am Bd Internal Med, 75. *Prof Exp:* Intern, Univ Miami Affil Hosps, 73; med resident internal med, Jefferson Univ Hosp, 74; med res II internal med, Univ Calif, Los Angeles & Wadsworth Hosp, 75; res fel, 75-78, asst mem, 78-83, ASSOC MEM IMMUNOL, SCRIPPS CLIN & RES FOUND, 83-; CONSULT, ORTHO PHARMACEUT CORP, 82-

Concurrent Pos: US Pub Health Serv Nat Res Serv Awardee, Scripps Clin & Res Found, 75-78, prin investr, 78-, arthritis found fel, 78-80, US Pub Health Serv Career Develop Awardee, 80-85; vis physician, Univ Calif, San Diego, 80-83; ad hoc mem, Med Biochem Study Sect, NIH, 87-88. *Mem:* NY Acad Sci; fel Am Col Physicians; Am Asn Immunologists; Am Asn Pathologist. *Res:* Biochemistry of B lymphocyte activation to proliferation and immunoglobulin production; post-transduction events and consequences of intracellular receptor-ligand interactions. *Mailing Add:* Dept Immunol Imm-9 Scripps Clin & Res Found 10666 N Torrey Pines Rd La Jolla CA 92037

GOODMAN, MORRIS, b Milwaukee, Wis, Jan 12, 25; m 46; c 3. IMMUNOLOGY. *Educ:* Univ Wis, BS, 48, MS, 49, PhD(zool), 51. *Prof Exp:* USPHS res fel, Calif Inst Technol, 51-52; res assoc, Med Col, Univ Ill, 52-54; res assoc, Detroit Inst Cancer Res, 54-58; res assoc & assoc, Lafayette Clin, 58-60, res assoc prof, 60-66, PROF ANAT, COL MED, WAYNE STATE UNIV, 66- *Concurrent Pos:* Dir res, Plymouth State Home & Training Sch, 66-72; mem adv panel syst biol, Div Biol & Med Sci, NSF, 69-72; co-ed, J Human Evolution, 72-77. *Honors & Awards:* Lady Margaret lectr, Christ's Col, Cambridge Univ, Eng, 84; von Hofsten Mem lectr, Univ Uppsala, 87. *Mem:* Am Asn Immunol; Am Soc Zool; Am Asn Phys Anthrop; Am Soc Naturalists; Soc Study Evolution; Sigma Xi. *Res:* Molecular evolution and systematics of primates, other mammals and vertebrates; analysis of the evolution of genes for globins and other protein families; analysis of the evolution of elements for globins and other protein families. *Mailing Add:* 24211 Oneida St Oak Park MI 48237

GOODMAN, MURRAY, b New York, NY, July 6, 28; m 51; c 3. ORGANIC CHEMISTRY. *Educ:* Brooklyn Col, BS, 49; Univ Calif, PhD(org chem), 52. *Prof Exp:* Res assoc org chem, Mass Inst Technol, 52-55; fel, Am Cancer Soc Nat Res Coun, Cambridge Univ, 55-56; from asst prof to prof chem, Polytech Inst Brooklyn, 56-71, dir, Polymer Res Inst, 67-71; chmn, Chem Dept, 76-81, PROF CHEM, UNIV CALIF, SAN DIEGO, 71- *Concurrent Pos:* Ed, Biopolymers, 63-; mem, steering comt human reproduction, WHO, 74-79 & US Nat Comt Int Union Pure & Appl Chem, 80-86; Goldberg Chair, Bio-Med Eng, Technion, Israel Inst Technol, 82; Humboldt Award, 86-87. *Honors & Awards:* Scoffone Medal, Biopolymer Res Ctr, Univ Padova, 80; Alberta Heritage Vis Prof Award, Med Res, Univ Alta, 81; William H Rauscher Lectr, Rensselaer Polytech Inst, 82. *Mem:* AAAS; Am Chem Soc; Am Soc Biol Chemists; The Chem Soc; Sigma Xi; Biophys Soc. *Res:* Peptide synthesis; biopolymers; conformational analyses; carrier drug conjugates; biospectroscopy; biologically-active peptides and their analogs; conformational calculations. *Mailing Add:* 9760 Blackgold Rd La Jolla CA 92037

GOODMAN, MYRON F, b New York, NY, Dec 31, 39. BIOCHEMISTRY. *Educ:* Johns Hopkins Univ, PhD(elec eng), 68. *Prof Exp:* PROF MOLECULAR BIOL, UNIV SOUTHERN CALIF, 80- *Mailing Add:* 4719 Alminar Ave La Canada CA 91011

GOODMAN, NELSON, b Brooklyn, NY, Aug 16, 32; m 56; c 3. INDUSTRIAL MICROBIOLOGY. *Educ:* Brooklyn Col, BS, 56; Brandeis Univ, PhD(biol), 62. *Prof Exp:* NIH fel, 62-64; microbiologist, Bioferm, Int Minerals & Chem Co, 64-67; microbiologist, Midwest Res Inst, 67; microbiologist, Accent Int & Supvr Indust Microbiol, San Jose Res Lab, 68-79, SUPVR, FOOD & BIOTECHNOL, WESTERN RES LABS, STAUFFER CHEM CO, 79- *Mem:* Sigma Xi; Am Soc Microbiol; Soc Indust Microbiol; Soc Invert Path. *Res:* Monosodium glutamate fermentation; microbial insecticides; functionalized food fermentations. *Mailing Add:* Stauffer Chem Co 1200 S 47th St Richmond CA 94804

GOODMAN, NICOLAS DANIELS, b Berlin, Germany, June 23, 40; US citizen; m 62; c 2. PROOF THEORY, INTUITIONISM. *Educ:* Harvard Univ, AB, 61; Stanford Univ, MS, 63, PhD(math), 68. *Prof Exp:* Instr math, Univ Ill at Chicago Circle, 65-66; asst prof, Univ Santa Clara, 68-69; from asst prof to assoc prof, 69-89, PROF MATH, STATE UNIV NY BUFFALO, 89- *Mem:* Am Math Soc; Asn Symbolic Logic; Math Asn Am. *Res:* Constructive and numerical analysis, with an emphasis on their metamathematical and philosophical aspects; recursion-theoretic and proof-theoretic techniques; modal logic. *Mailing Add:* Dept Math State Univ NY Buffalo NY 14214

GOODMAN, NORMAN L, b Milburn, Okla, Sept 29, 31; m 57; c 2. MICROBIOLOGY, BIOCHEMISTRY. *Educ:* Southeast State Col, BS, 54; Univ Okla, MS, 60, PhD(microbiol), 65; Am Bd Med Microbiol, dipl. *Prof Exp:* Teacher pub sch, Okla, 56-57; microbiologist, Oper Res, Tuberc Br, Commun Dis Ctr, USPHS, 63-68; asst prof microbiol & dir diag bact labs, Med Univ SC, 68-70; assoc prof, 70-78, prof community med, med mycol prog, 78-80, DIR, CLINICAL MICROBIOL LABS, 79-, PROF PATH & MED MICROBIOL & IMMUNOL, COL MED, UNIV KY, 80- *Concurrent Pos:* Chmn comt continuing educ, Am Soc Microbiol, 73-79, mem bd educ & training, 76-79, chmn, Div Med Mycol, 76-77 & Comt Undergrad & Grad Educ, 81-84; mem standards & exam comt, Am Bd Med Microbiol, 78-84; mem comt postdoctoral educ progs, Am Acad Microbiol, 79-82, bd gov; mem US nat comt, Int Union Microbiol Socs, 82-86; mem exec comt & chmn, Mycol Div, Int Union Microbiol Soc, 71-90 & 82-86; assoc ed, Diag Microbiol & Infectious Dis, 82-88; chmn, Abbott Lab Award Comt, Am Soc Microbiol, 89-90; mem, Standards & Exam Comt, Am Bd Med Microbiol; counr, Am Soc Microbiol, 90-92. *Honors & Awards:* Meridian Award, Med Mycrological Soc Am, 89. *Mem:* AAAS; Am Soc Microbiol; Mycological Soc Am (pres, 82); fel Am Acad Microbiol; Int Soc for Human & Animal Mycoses; Soc for Experimental Biol & Med. *Res:* Pathogenesis of systemic mycoses and epidemiology of histoplasmosis. *Mailing Add:* Dept Path Univ Ky Col Med Lexington KY 40536-0840

GOODMAN, PHILIP, physical chemistry, chemical instrumentation; deceased, see previous edition for last biography

GOODMAN, RICHARD E, b New York, NY, Dec 25, 35; m 57; c 3. CIVIL & GEOLOGICAL ENGINEERING. *Educ:* Cornell Univ, BA, 55, MS, 58; Univ Calif, PhD(geol eng), 63. *Prof Exp:* from asst prof to assoc prof 62-77, PROF GEOL ENG, UNIV CALIF, BERKELEY, 77- *Concurrent Pos:* Grants, NSF & others; Gugenheim fel; pres, Geol Eng Found; consult, Defense Nuclear Agency, Corps Engrs, Columbian govt. *Honors & Awards:* Rock Mech Award, Assoc Inst Mining & Metal Engrs, 76; Burwell Award, Geol Soc Am, 77; Basic Res Award, US Nat Comn Rock Mech, 84. *Mem:* Nat Acad Eng; Geol Soc Am; Am Inst Mining, Metall & Petrol Engrs; Asn Eng Geologists; Am Soc Civil Engrs; AAAS. *Res:* Applications of geological data in civil engineering design, dams and underground works; block theory; physical and mathematical model studies of jointed rock masses; author of four books and 180 publications. *Mailing Add:* Dept Civil Eng 434B Davis Hall Univ Calif Berkeley CA 94720

GOODMAN, RICHARD E, b Los Angeles, Calif, Jan 28, 38; div; c 1. MICROBIAL PHYSIOLOGY, GENERAL BIOLOGY. *Educ:* Univ Calif, Los Angeles, BA, 60, PhD(microbiol), 65. *Prof Exp:* USPHS fel microbiol, Sch Med, Univ Wash, 65-67; from asst prof to assoc prof biol, Calif State Col, San Bernardino, 67-77; DIR LABS, DEPT BIOL SCI, UNIV SOUTHERN CALIF, 78- *Res:* Delayed lactose fermentation by Enterobacteriaceae; physiology of streptomycin-dependent and thermophilic bacteria. *Mailing Add:* Dept Biol Sci Univ Southern Calif Univ Park Los Angeles CA 90089

GOODMAN, RICHARD HENRY, b Brooklyn, NY. NUMERICAL ANALYSIS, COMPUTER SCIENCE. *Educ:* Harvard Univ, AB, 63, AM, 64, PhD(math), 71. *Prof Exp:* Asst prof, 70-78, ASSOC PROF MATH, UNIV MIAMI, 78- *Mem:* Sigma Xi; Soc Indust & Appl Math; Asn Comput Mach. *Res:* Numerical solution of differential equations; theory of computer arithmetic, with applications to computer design. *Mailing Add:* Dept Math Univ Miami PO Box 249085 Coral Gables FL 33124

GOODMAN, RICHARD S, b Brooklyn, NY, Aug 4, 34; m 86; c 5. MEDICINE, LAW. *Educ:* Alfred Univ, BA, 55; NY Univ, MD, 60; Am Bd Orthop Surg, cert, 69; Touro Col, JD, 87. *Prof Exp:* Intern surg, Ind Univ Med Ctr, 60-61; asst resident, Bronx Munic Hosp Ctr, 61-62; resident orthop, NY Univ Med Ctr & Bellevue Hosp, 64-67; asst prof, Dept Anat, State Univ NY, Stony Brook, 71-86; ATTEND PHYSICIAN, ST JOHN'S HOSP, SMITHTOWN, NY, 67-, COMMUNITY HOSP WESTERN SUFFOLK, 67- *Concurrent Pos:* Solo practr lectr; mem bd dirs, United Physicians Ins & Univ Club. *Mem:* Fel Am Acad Orthop Surgeons; Am Col Sports Med; Am Col Legal Med; Am Rheumatism Asn; Am Soc Law & Med; Int Col Surgeons. *Res:* Author of numerous technical articles & publications, lecturers and writes on medical-legal issues such as risk management, malpractice and insurance. *Mailing Add:* 285 E Main St Smithtown NY 11787

GOODMAN, ROBERT M(ENDEL), b Philadelphia, Pa, Nov 21, 20; m 49; c 3. ELECTRICAL ENGINEERING. *Educ:* Univ Pa, BS, 43. *Prof Exp:* Design engr, Hazeltine Electronics Corp, 43-47; res assoc, group supvr & instr, Univ Pa, 47-51; co-founder, dir, vpres, treas & admin dir, Am Electronic Labs, Inc, 51-59; lab mgr biodynamics, 59-74, PRIN SCIENTIST, LABS RES & DEVELOP, FRANKLIN INST, 74- *Concurrent Pos:* Am Deleg Workshop Common Health Care, Dubrornik, Yugoslavia, 72. *Mem:* AAAS; Sigma Xi; Inst Elec & Electronics Engrs; Biophys Soc; Am Asn Med Instrs. *Res:* Instrumentation for the life sciences; implementation of effort, administration and technical prosecution. *Mailing Add:* 7811 Mill Rd Elkins Park PA 19117

GOODMAN, ROBERT MERWIN, b Ithaca, NY, Dec 30, 45. VIRUSES, PLANT DISEASES. *Educ:* Cornell Univ, BSc, 67, PhD(plant virol), 73. *Prof Exp:* From asst prof to prof plant pathol, Univ Ill, Urbana-Champaign, 74-84; exec vpres res & develop, Calgene Inc, 82-90; scholar in residence, Nat Res Coun, 90-91; VIS PROF, UNIV WIS-MADISON, 91- *Concurrent Pos:* Mem, Nat Acad Sci-Nat Res Coun Bd Agr, 86-; adj prof, Univ Calif, Berkeley, 86-89; panel mem, Nat Res Coun, Biosci Res Agr, 84-85, Alt Farming Methods in Prod Agr, 86-; chmn, Comt Exam Plant Sci Res Prog US, 90- *Mem:* AAAS; NY Acad Sci; Am Chem Soc; Am Soc Virol; Int Soc Plant Molecular Biol; Am Phytopath Soc. *Res:* Discoverey of the plant viruses which contain as their genetic material single-stranded DNA; epidemiology and yield impact of crop viruses, virus structure, and plant diseases caused by plant pathogenic spiroplasmas; incorporation of soybean virus resistance into high-yielding soybean varieties; replication of DNA plant viruses; tropical legume virology; plant diseases caused by viruses and spiroplasmas; plant molecular biology & genetics; genetics of disease resistance in plants; sustainable agriculture. *Mailing Add:* Russell Labs Rm 284 Univ Wis 1630 Linden Dr Madison WI 53706

GOODMAN, ROBERT NORMAN, b Yonkers, NY, Dec 15, 21; m 49; c 3. PLANT PATHOLOGY. *Educ:* Univ NH, BS, 48, MS, 50; Univ Mo, PhD(plant path), 52. *Prof Exp:* Asst hort, Univ NH, 48-50; from asst to assoc prof, 50-61, chmn dept, 68-79, PROF PLANT PATH, UNIV MO-COLUMBIA, 68- *Concurrent Pos:* Guggenheim fel, Swiss Fed Inst Technol, 58-59; NIH spec fel, Astbury Dept Biophys, Univ Leeds, 65-66; vis scientist, Volcanic Res Ctr, Israel Ministry Agr, 72-73, Weizmann Inst Sci, Israel, 79-80. *Honors & Awards:* Lalor Found Award, 59. *Mem:* Fel Am Phytopath Soc; Soc Am Microbiol. *Res:* Control of bacterial pathogens with antibiotics and antibacterial substances; adsorption and translocation of organic antimicrobial substances by foliage; bacterial toxins; ultrastructural changes in plants caused by bacteria; induced resistance in plants to bacterial pathogens. *Mailing Add:* Dept Plant Path 108 Waters Hall Univ Mo Columbia MO 65211

GOODMAN, ROE WILLIAM, b Pasadena, Calif, Jan 9, 38; m 61; c 1. MATHEMATICAL ANALYSIS. *Educ:* Fla Southern Col, BS, 58; Mass Inst Technol, PhD(math), 63. *Prof Exp:* Instr math, Mass Inst Technol, 62-63; Nat Acad Sci-Nat Res Coun fel, Harvard Univ, 63-64; lectr, Mass Inst Technol, 64-66, asst prof, 66-71; assoc prof, 71-75, PROF MATH, RUTGERS UNIV, 75- *Concurrent Pos:* Vis mem, Inst Advan Study, 68-69 & Inst Advan Study Sci, 75-76. *Mem:* Am Math Soc; Math Asn Am; Math Soc France. *Res:* Functional analysis and operator theory; representations of Lie groups and harmonic analysis; quantum field theory. *Mailing Add:* Dept Math Rutgers Univ New Brunswick NJ 08903

GOODMAN, RONALD KEITH, b Yuma, Ariz, Oct 31, 29; m 59; c 4. PLASMA PHYSICS, THOMSON SCATTERING. *Educ:* Wheaton Col, BS, 53; San Jose State Col, MS, 67. *Prof Exp:* Engr, Fargo Co, Calif, 57-58, chief engr, 58-61; res operator, 61-66, PHYSICIST, LAWRENCE LIVERMORE NAT LAB, UNIV CALIF, 66- *Mem:* Am Phys Soc; Am Vacuum Soc. *Res:* Ultra-high vacuum and related surface physics; plasma physics in controlled thermonuclear research. *Mailing Add:* 3923 Fordham Way Livermore CA 94550

GOODMAN, SARAH ANNE, b Anderson, SC, June 9, 45; m 88. FACTORS AFFECTING VACCINE, HEALTH EFFECTS OF FOOD. *Educ:* Agnes Scott Col, BA, 66; Univ Ga, MS, 69; Univ Tenn, PhD (biomed sci), 73. *Prof Exp:* Asst instr microbiol & immunol, Univ Tex, Southwestern Med Sch, Dallas, 73-75; asst prof microbiol & immunol, 75-80, res assoc, Emory Univ Sch Med, 81-85; assoc libr info sci, Nat Libr Med, 80-81; subcontractor immunol & microbiol, USAID, 85-88; consult toxicol, Coca Cola Co, Int Life Sci Inst, 85-88; SR ASSOC TOXICOL, CLEMENT ASSOC INC, 88-; REVIEWER CLIN PROTOCOL, FDA, 88- *Res:* Immunology, microbiology, reproductive physiology laboratory research; toxicology of food additives and environmental chemicals; possible factors affecting malaria sporozoite vaccine efficacy; clinical protocol for drug biologic development. *Mailing Add:* 968 Yachtsman Way Annapolis MD 21403

GOODMAN, SEYMOUR, b New York, NY, Nov 12, 33. COMPUTER SCIENCE, PHYSICAL CHEMISTRY. *Educ:* City Col New York, BS, 54; Univ Chicago, MS, 54; Columbia Univ, PhD(chem), 62. *Prof Exp:* From instr to assoc prof, 62-70, chmn, Dept Comput Sci, 70-76, PROF COMPUT SCI, QUEENS COL NY, 70-, DIR, ACAD COMPUT CTR, 65- *Concurrent Pos:* NSF grant, 65-67. *Mem:* AAAS; Am Chem Soc; Am Phys Soc; Asn Comput Mach; Pattern Recognition Soc; Sigma Xi. *Res:* Data acquisition and signal processing; computer architecture. *Mailing Add:* 67-44 171st St Hillcrest NY 11365

GOODMAN, STEVEN RICHARD, b New York, NY, Dec 29, 49; c 1. BIOCHEMISTRY. *Educ:* State Univ NY, Stony Brook, BS, 71; St Louis Univ, PhD(biochem), 76. *Prof Exp:* Res fel biochem, Sidney Farber Cancer Inst, Harvard Univ, 76-77, res fel cell biol, 77-; AT DEPT PHYSIOL, PA STATE UNIV MED SCH. *Concurrent Pos:* Nat Heart, Lung & Blood Inst fel, 77- *Res:* Membrane biochemistry, especially the association of peripheral cytoskeletal proteins with the membrane surface and the role of peripheral proteins in regulating the mobility of integral membrane proteins and hence cell surface topography. *Mailing Add:* Dept Struct & Cell Biol Rm 2042 Univ SAla Col Med Mobile AL 36688

GOODMAN, SUE ELLEN, b East St Louis, Ill, Nov 26, 46. MATHEMATICS. *Educ:* St Louis Univ, AB, 68, MA, 71, PhD(math), 74. *Prof Exp:* Pvt br tel exchange engr, Southwestern Bell, 68-69; res asst math, Wash Univ, 73-74; ASST PROF MATH, UNIV NC, 74- *Concurrent Pos:* NSF res grant, 74-; vis mem, Inst Advan Study, 76-77. *Mem:* Am Math Soc. *Res:* Foliations and dynamical systems. *Mailing Add:* Dept Math Univ NC Chapel Hill NC 27514

GOODMAN, THEODORE R(OBERT), b Brooklyn, Ny, Mar 21, 25. FLUID DYNAMICS, APPLIED MATHEMATICS. *Educ:* Rensselaer Polytech Inst, BAE, 45; Calif Inst Technol, MAE, 46; Cornell Univ, PhD(aeronaut eng), 54. *Prof Exp:* From jr aerodynamicist to assoc aerodynamicist, Cornell Aeronaut Lab, 46-51, prin engr, 53-55; prin engr, Allied Res Assocs, Inc, 56-61; vpres, Oceanics, Inc, 62-76; sr res scientist, 76-80, assoc res prof, 78-82, chief, Fluid Mech Div, Davidson Lab, Stevens Inst Technol, 80-82; RETIRED. *Mem:* Assoc fel Am Inst Aeronaut & Astronaut; Am Soc Mech Engrs; Sigma Xi. *Res:* Hydrodynamics; theoretical aerodynamics; fluid mechanics; boundary layer and wing theory; heat transfer; aeroelasticity; ballistics; system identification. *Mailing Add:* 21 Chapel Pl Great Neck NY 11021

GOODMAN, VICTOR HERKE, b Kansas City, Mo, Nov 11, 18; m 43. PLANT ANATOMY, PHYCOLOGY. *Educ:* Univ Mo, BA, 47; Cornell Univ, PhD(plant physiol), 51. *Prof Exp:* Asst bot, Cornell Univ, 47-51; assoc plant physiologist, Agr Exp Sta, Miss State Col, 51-53; from asst prof to prof bot, 54-80, EMER PROF BOT, UNIV CALIF, RIVERSIDE, 80- *Res:* Physiological anatomy of plants; terrestrial algae. *Mailing Add:* 764 S University Dr Riverside CA 92507

GOODMAN, VICTOR WAYNE, b Wellington, Kans, Oct 1, 43. PURE MATHEMATICS. *Educ:* Univ Kans, BA, 65; Cornell Univ, PhD(math), 70. *Prof Exp:* Fel math, Univ NMex, 70-72; asst prof, 72-80, ASSOC PROF MATH, INDIANA UNIV, 80- *Concurrent Pos:* NSF res grant, 72. *Mem:* Am Math Soc; Sigma Xi. *Res:* Investigation of mathematical models of brownian motion and deriving such models as asymptotic limits of certain sampling situations in the theory of statistics. *Mailing Add:* Dept Math Ind Univ Bloomington IN 47405

GOODNER, CHARLES JOSEPH, b Seattle, Wash, Aug 19, 29; m 51; c 3. ENDOCRINOLOGY. *Educ:* Reed Col, BA, 51; Univ Utah, MD, 55. *Prof Exp:* From asst prof to assoc prof, 62-73, PROF MED, SCH MED, UNIV WASH, 73- *Mem:* Am Physiol Soc; Endocrine Soc; Am Diabetes Asn; Am Fedn Clin Res; Am Soc Clin Invest. *Mailing Add:* Harborview Med Ctr 325 Ninth Ave Seattle WA 98104

GOODNER, DWIGHT BENJAMIN, b What Cheer, Iowa, Aug 15, 13; m 36. PURE MATHEMATICS. *Educ:* William Penn Col, BA, 34; Haverford Col, MA, 35; Univ Ill, PhD(math), 49. *Prof Exp:* Instr math, SDak State Col, 37-42, asst prof, 42-46; mem consult bur comt undergrad prog math, 62-78, from asst prof to prof, 49-54, EMER PROF MATH, FLA STATE UNIV, 78- *Concurrent Pos:* Asst dean, Grad Sch, Fla State Univ, 53-55, assoc dean, 55-58; mem, Comt Exam, Educ Testing Serv, NJ, 65-70; mem, Comt Undergrad Prog Math, 67-71; consult mem, Bd Gov, Math Asn Am, 67-71, consult, 71- *Mem:* Am Math Soc; Math Asn Am. *Res:* Normed Linear spaces; non-Euclidean geometry. *Mailing Add:* Dept Math Fla State Univ Tallahassee FL 32306

GOODNEY, DAVID EDGAR, b Wilmington, Del, Dec 2, 49; m 71; c 2. INSTRUMENTAL ANALYSIS. *Educ:* Austin Col, BA, 71; Univ Hawaii, PhD(chem), 77. *Prof Exp:* Res asst chem, Univ Hawaii, 77; from asst prof to assoc prof, 77-88, PROF CHEM, WILLAMETTE UNIV, 88 - *Concurrent Pos:* Vis prof chem, Ore Grad Ctr, 86. *Mem:* Am Chem Soc; AAAS. *Res:* Solving analytical problems related to environmental concerns, such as the composition of carbonaceous aerosols, trace metals in natural waters and trace chemicals in foods. *Mailing Add:* Willamette Univ 900 State St D140 Salem OR 97301

GOODNO, BARRY JOHN, b Madison, Wis, July 19, 47. SOLID MECHANICS, CIVIL ENGINEERING. *Educ:* Univ Wis, Madison, BS, 70; Stanford Univ, MS, 71, PhD(civil eng), 75. *Prof Exp:* asst prof, 74-79, ASSOC PROF CIVIL ENG, GA INST TECHNOL, 79- *Concurrent Pos:* Res grant, NSF, 75; consult, US Army Corps Engrs, 76; mem Comt Electronic Comput, Struct Div & Dynamics Comt, Eng Mech Div, Am Soc Civil Engrs, 76- *Mem:* Am Soc Civil Engrs; Earthquake Eng Res Inst, Seismol Soc Am; Sigma Xi. *Res:* Analysis of structural dynamics of buildings with applications to wind and earthquake engineering; finite element methods in structural engineering; matrix methods of structural analysis; analysis of cladding on buildings. *Mailing Add:* 2893 Northbrook Dr Ga Inst Technol 225 North Ave NW Atlanta GA 30340

GOODPASTURE, JESSIE CARROL, b Oak Park, Ill, Nov 27, 52; m 89. REPRODUCTION, ENDOCRINOLOGY. *Educ:* Southern Ill Univ, BA, 73, MS, 77; Univ Ill, PhD(physiol & biophys), 80. *Prof Exp:* Teaching asst physiol, Southern Ill Univ, 74-75; teaching asst, Med Ctr, Univ Ill, 75-76, res asst, 76-80; asst prof, dept obstet-gynec, Sch Med, Univ Essen, 80-81; staff researcher I, 81-83, staff researcher II, 83-89, CLIN PROG DIR, SYNTEX RES, SYNTEX CORP, 89- *Concurrent Pos:* Consult, VLI Corp, 84-85, HDC Corp, 85-; mem sci & math adv bd, Mills Col, Oakland, CA,; mem fel panel, Am Asn, Univ Women. *Mem:* Am Soc Andrology; Am Fertil Soc; Soc Advan Contraception; Int Soc Andrology; Soc Study Reproduction. *Res:* Clinical trials of new drugs for treatment of endocrine-related disorders. *Mailing Add:* Syntex Res 3401 Hillview Ave Palo Alto CA 94304

GOODRICH, CECILIE ANN, b Denver, Colo, Apr 27, 41. GENERAL PHYSIOLOGY. *Educ:* Univ Mich, BS, 62; Harvard Univ, PhD(physiol), 67. *Prof Exp:* Fel physiol, Lab Neuropharmacol, Nat Inst Med Res, Eng, 67-68; res fel anat, Harvard Med Sch, 68-71; from asst prof to assoc prof biol, 71-84, PROF BIOL, CLEVELAND STATE UNIV, 84- *Concurrent Pos:* NIH fels, 67-71, res grant, 69-72. *Mem:* Am Physiol Soc; Am Soc Zoologists. *Res:* Maturation of serotonin system; temperature regulation; cerebrospinal fluid. *Mailing Add:* Dept Biol Cleveland State Univ Cleveland OH 44115

GOODRICH, JUDSON EARL, b Seneca, Kans, Aug 14, 22; m 51; c 3. ORGANIC CHEMISTRY. *Educ:* Univ Kans, AB, 47, MA, 48; Univ Calif, PhD(chem), 51. *Prof Exp:* Asst instr, Univ Kans, 47-48 & Univ Calif, 48-51; assoc res chemist, Calif Res Corp, Standard Oil Co Calif, 51-54, res chemist, 54-60, sr res chemist, 60-64, sr res assoc, Chevron Res Co, 64-85; RETIRED. *Mem:* Am Chem Soc. *Res:* Positive halogen salts; desulfurization of thioamides; cyclic ketals of diketones; fuel oil stability; lubricating grease gelling agents; cationic polymerization; stabilization of polyolefins; polyurethanes; asphalt emulsions; industrial asphalt chemistry. *Mailing Add:* 6396 Stone Bridge Rd Santa Rosa CA 95409-5824

GOODRICH, MAX, b Calhoun, Mo, Dec 11, 05; m 29; c 2. EXPERIMENTAL PHYSICS. *Educ:* Westminster Col, Mo, BA, 27; Univ Minn, PhD(physics), 36. *Prof Exp:* Instr math, Salt Lake Collegiate Inst, 27-29; asst instr physics, Univ Minn, 29-36; from instr to prof & dean grad sch, 36-73, EMER PROF PHYSICS & EMER DEAN GRAD SCH, LA STATE UNIV, 73- *Concurrent Pos:* Physicist, War Res Lab, Univ Tex, 45; sr physicist, Oak Ridge Nat Lab, 49-50; chief, Grad Acad Prog Br, US Off Educ, 68-69; mem comput comt, Southern Regional Educ Bd, 68-72; mem coun, Oak Ridge Assoc Univs, 69-72 & mem bd dir, 72-78. *Mem:* Fel AAAS; fel Am Phys Soc; Am Asn Physics Teachers. *Res:* Electron impact phenomena; electron diffraction; electron scattering in helium; spectrometry of beta- and gamma-rays. *Mailing Add:* 2700 Burcham Dr No 406 East Lansing MI 48823

GOODRICH, MICHAEL ALAN, b New York, NY, Apr 24, 33; m 55; c 3. SYSTEMATICS. *Educ:* Bucknell Univ, BS, 55; Pa State Univ, MEd, 61, PhD(entom), 64. *Prof Exp:* Asst zool & entom, Pa State Univ, 61-64, instr entom, 63; assoc prof, 64-71, chmn dept, 82-88, PROF ZOOL, EASTERN ILL UNIV, 71- *Mem:* Entom Soc Am; Entom Soc Can; Soc Syst Zool; Animal Behav Soc; Coleopterists Soc; Am Entom Soc; Sigma Xi. *Res:* Taxonomy of Coleoptera; utilization of sense organs and feeding behavior in snakes; evolution of behavior; systematics of Coleoptera, especially Byturidae, Biphyllidae, Erotylidae and Scarabaeidae of the world; behavior of reptiles, especially utilization of sense organs and feeding behavior of snakes. *Mailing Add:* Dept Zool Eastern Ill Univ Charleston IL 61920

GOODRICH, RICHARD DOUGLAS, b New Richmond, Wis, July 2, 36; m 56; c 2. ANIMAL NUTRITION. *Educ:* Wis State Univ, River Falls, BS, 58; SDak State Univ, MS, 62; Okla State Univ, PhD(animal nutrit), 65. *Prof Exp:* PROF ANIMAL SCI, UNIV MINN, ST PAUL, 65- *Mem:* Am Soc Animal Sci; Am Inst Nutrit. *Res:* Ruminant and mineral nutrition; forage evaluation; characterization of silage fermentation; adaptation to and utilization of nonprotein nitrogen by ruminants; feedlot nutrition and management; beef cow nutrition; feedlot housing systems. *Mailing Add:* Dept Animal Sci 120 Peters Hall Univ Minn St Paul MN 55108

GOODRICH, ROBERT KENT, b Lapoint, Utah, July 5, 41; m 65; c 1. MATHEMATICS. *Educ:* Univ Utah, BA, 63, PhD(math), 66. *Prof Exp:* ASSOC PROF MATH, UNIV COLO, BOULDER, 66- *Mem:* Am Math Soc. *Res:* Functional analysis. *Mailing Add:* Dept Math Univ Colo Box 426 Boulder CO 80309

GOODRICH, ROY GORDON, b Dallas, Tex, Sept 17, 38; m 59; c 3. SOLID STATE PHYSICS. *Educ:* La Polytech Inst, BS, 60; Univ Calif, Riverside, MA, 63, PhD(physics), 65. *Prof Exp:* Instr physics, Univ Southern Miss, 60-61; from asst prof to assoc prof, 65-72, chmn dept physics & astron, 73-76, PROF PHYSICS, LA STATE UNIV, BATON ROUGE, 72- *Mem:* Am Phys Soc. *Res:* Thermal and magnetic properties of solids at low temperatures. *Mailing Add:* Dept Physics & Astron La State Univ Baton Rouge LA 70803-4001

GOODRICK, RICHARD EDWARD, b Verdun, Que, Feb 19, 41; US citizen; m 66. MATHEMATICS. *Educ:* Univ Wash, BS, 62; Univ Wis, MS, 64, PhD(math), 66. *Prof Exp:* Asst prof math, Univ Utah, 66-69; asst prof, 69-71, assoc prof, 71-78, PROF MATH, CALIF STATE UNIV, HAYWARD, 80- *Concurrent Pos:* Vis fel, Univ Warwick, 68-69. *Mem:* Am Math Soc. *Res:* Piece wise linear topology. *Mailing Add:* 9348 32nd SW Seattle WA 98126

GOODRIDGE, ALAN G, b Peabody, Mass, Apr 2, 37; m 60; c 2. METABOLIC REGULATION, GENE EXPRESSION. *Educ:* Tufts Univ, BS, 58; Univ Mich, MS, 63, PhD(zool), 64. *Prof Exp:* Asst prof physiol, Med Ctr, Univ Kans, 66-68; assoc prof med sci, Univ Toronto, 68-76, prof, 76-77; prof pharmacol, Case Western Reserve Univ, 77-87, prof biochem, 80-87; PROF & HEAD DEPT BIOCHEM, UNIV IOWA, 87- *Concurrent Pos:* Nat Inst Arthritis & Metab Dis res fel, Harvard Med Sch, 64-66; Josiah Macy Jr fac scholar, 75-76. *Mem:* AAAS; Sigma Xi; Am Soc Biochem & Molecular Biol; Am Thyroid Asn. *Res:* Regulation of metabolism; nutritional and hormonal regulation of gene expression; hormone action. *Mailing Add:* Dept Biochem Univ Iowa Iowa City IA 52242

GOODSON, ALAN LESLIE, b London, Eng, Apr 11, 33; US citizen; m 76. INFORMATION SCIENCE, ORGANIC CHEMISTRY. *Educ:* Univ London, BSc, 57, PhD(chem), 61; ARIC, 58; Woolwich Polytech, AWP, 61. *Prof Exp:* Control chemist, Prince Regent Tar Co, 55-57; res chemist, Shell Chem Co, 60-63, info officer, 63-64; ed, 64-69, mem ed develop, 69-83, NOMENCLATOR, CHEM ABSTRACTS SERV, 83- *Concurrent Pos:* Rep Am Chem Soc, Am Nat Standards Inst, 84- *Mem:* Am Chem Soc; fel Royal Soc Chem; AAAS; Soc Chem Indust. *Res:* Chemical nomenclature; chemical literature; allenes; acetylenes; aluminum alkyls; polyalkylene glycols. *Mailing Add:* Chem Abstracts Serv Dept 64 PO Box 3012 Columbus OH 43210-0012

GOODSON, JAMES BROWN, JR, b Sandersville, Ga, July 10, 15; m 37; c 1. SANITARY ENGINEERING. *Educ:* Univ Fla, BS, 37, PhD(chem), 50. *Prof Exp:* Chemist, Int Paper Co, SC, 37-41; proj engr, Sheppard T Powell & Assocs, Md, 50-63; proj engr, 63-68, vpres, 68-76, CONSULT CHEM ENGR, BLACK, CROW & EIDSNESS, INC, 76- *Mem:* Am Inst Chem Engrs; Am Chem Soc; Am Water Works Asn; Water Pollution Control Fedn. *Res:* Water and waste water treatment; pulp and paper manufacturing process. *Mailing Add:* 327 SW 40th St Gainesville FL 32607

GOODSON, LOUIE AUBREY, JR, b Providence, NC, Dec 20, 22; m 45; c 4. TEXTILE CHEMISTRY. *Educ:* NC State Col, BS, 43; Georgetown Univ, JD, 51. *Prof Exp:* Res chemist, Dan River Mills, Inc, 46-48, asst dir res, 51-53; patent searcher, Fisher & Christen, 48-51, patent lawyer & partner, Fisher, Christen & Goodson, 53-59; vpres & dir res, 59-80, SR VPRES DEVELOP & GOVT REGULATIONS & PRES CHEM PROD DIV, DAN RIVER, INC, 80- *Concurrent Pos:* Mem nat adv comt, Flammable Fabrics Act, Consumer Prod Safety Comn, 75-77; hon fel, Textile Res Inst, 80. *Mem:* Am Chem Soc; Am Asn Textile Chem & Colorists; Am Asn Textile Technol. *Mailing Add:* 6005 Old Jonesboro Rd Bristol TN 37620-3017

GOODSON, RAYMOND EUGENE, b Canton, NC, Apr 22, 35; m 57; c 2. SYSTEMS & CONTROL ENGINEERING. *Educ:* Duke Univ, AB, 57, BSME, 59; Purdue Univ, MSME, 61, PhD(fluids, automatic control), 63. *Prof Exp:* From asst prof to assoc prof, Purdue Univ, 63-70, prof mech eng, 70-81, dir, Energy Policy Res & Info Ctr & Energy Policy Prog, 77-81; GROUP VPRES, AUTOMOTIVE PROD GROUP, HOOVER UNIV, INC, SALINE, MICH, 81- *Concurrent Pos:* Consult various industs & govt, 63-; res grants, NSF, NASA & Off Naval Res; vis prof, Weizmann Inst Sci, Rehovot, Israel, 72; Japanese Soc Promotion Sci traveling scholar, 72; chief scientist, US Dept Transp, 73-75, mem Fed Interagency Task Force Motor Vehicle Goals Beyond 1980, 75-76; mem NASA Adv Comt Guid & Control, 73-74; chmn transp subcomt, Energy Storage Comt, Nat Acad Sci, 75; dir, Inst Interdisciplinary Eng Studies, Purdue Univ, 76-, dir, Opportunity Risk Anal Proj, Energy Res & Develop Admin, 77- & assoc dir, Eng & Exp Sta, 78-; rep US/USSR Environ Agreement, Transp Source Air Pollution Control Technol, USSR, 76; mem Comt Advan Energy Storage Systs, Nat Res Coun, Nat Acad Eng, 76; tech rep, UN Motor Vehicle conf, Int Metalworkers Fedn, Paris, 76; mem, Comt on Nuclear & Alternative Energy Systs Study, Nat Acad Sci/Nat Acad Eng & chmn, Transp Resource Group, Demand-Conservation Panel, 76-77; mem energy eng bd, Nat Res Coun Assembly Engrs; chmn, Off Tech Assessment, Automobile Adv Panel. *Mem:* Am Soc Mech Engrs; Instrument Soc Am; Inst Elec & Electronics Engrs; Sigma Xi. *Res:* Energy systems and energy conservation; transportation systems; particularly motor vehicles and urban transportation, glass industry research, and distributed systems. *Mailing Add:* 1545 Arboretum Dr Oshkosh WI 54901

GOODSPEED, FREDERICK MAYNARD (COGSWELL), b St John, NB, Sept 14, 14; m 46; c 2. MATHEMATICS. *Educ:* Univ Man, BSc, 35, MA, 36; Univ Cambridge, PhD(math), 42. *Prof Exp:* Exp officer, Projectile Develop Estab, 40-46 & Can Armament Res Estab, 46-47; asst prof math, Queen's Univ, Ont, 47-50; assoc prof, Univ BC, 50-59; from assoc prof to prof math, Laval Univ, 59-83; RETIRED. *Mem:* Can Math Cong. *Res:* Analysis; armaments. *Mailing Add:* 1307 Lauigerie Ste-Foy Quebec PQ G1W 3X4 Can

GOODSPEED, ROBERT MARSHALL, b Somerville, Mass, Feb 11, 38; div; c 2. ENVIRONMENTAL SCIENCES. *Educ:* Tufts Univ, BS, 60; Univ Maine, MS, 62; Rutgers Univ, PhD(geol & geochem), 68. *Prof Exp:* from asst prof to assoc prof geol, Susquehanna Univ, 66-83, chmn, Dept Geol Sci, 69-81 & 85-88, chmn, Sci Div, 70-74, PROF GEOL, SUSQUEHANNA UNIV, 83- *Concurrent Pos:* Air pollution researcher & consult, 90- *Mem:* AAAS; Mineral Soc Am; Geol Soc Am; Nat Asn Geol Teachers. *Res:* Concentrations of radon in air and water associated with lithologic units; concentrations of outdoor-indoor air pollutants in industrial and rural areas. *Mailing Add:* Dept Geol & Environ Sci Susquehanna Univ Selinsgrove PA 17870

GOODSTEIN, DAVID LOUIS, b New York, NY, Apr 5, 39; m 60; c 2. CONDENSED MATTER PHYSICS. *Educ:* Brooklyn Col, BS, 60; Univ Wash, Seattle, PhD(physics), 65. *Prof Exp:* Asst physics, Univ Wash, Seattle, 60-61, res instr, 65-66; res fel, Calif Inst Technol, 66-67; NSF fel, Univ Rome, 67-68; from asst prof to assoc prof, 68-75, PROF PHYSICS, CALIF INST TECHNOL, 75-, VPROVOST, 87- *Concurrent Pos:* Sloan fel, Calif Inst Technol, 69-71; NATO vis prof, Frascati Nat Lab, Italy, 72 & res collabr, 73-; creator & host, The Mech Universe & Beyond, PBS Educ Ser, 82-87; bd dir, Calif Coun Sci & Technol, 88- *Mem:* Am Acad Arts & Sci; Sigma Xi. *Res:* Thermal properties of helium and other gases in thin films; two dimensional matter; surfaces; interfaces; phase transtitions. *Mailing Add:* Calif Inst Technol Pasadena CA 91125

GOODSTEIN, MADELINE P, b New York, NY, Oct 23, 20; m 47; c 3. SCIENCE EDUCATION, CHEMISTRY. *Educ:* Brooklyn Col, BA, 41; Polytech Inst Brooklyn, MS, 48; Columbia Univ, EdD(chem), 68. *Prof Exp:* Chemist, Trubek Labs, 41-47, consult, 48; asst prof, 68-71, assoc prof, 71-77, prof chem, 78-83; dir, US Dept Educ, Nat Diffusion Network, Sci-Math Proj, 83-87; CONSULT, 87- *Concurrent Pos:* Dir, NSF Teacher Inst, 72-73 & NSF Curric Proj, 77-79 & 79-82; ed, Conn J Sci Educ, 76-79. *Mem:* Am Chem Soc; Nat Sci Teachers Asn. *Res:* Oxidation-reduction concept; science education for non-scientist; art chemistry. *Mailing Add:* 6 Woodland Dr Woodbridge CT 06525

GOODSTEIN, ROBERT, b Brooklyn, NY, June 16, 26; m 51; c 3. ENGINEERING MECHANICS. *Educ:* Mass Inst Technol, BS, 46; Ohio State Univ, MS, 49, PhD(eng mech), 57. *Prof Exp:* Instr theoret & appl mech, Iowa State Col, 49-53; res engr, Boeing Co, 53-55; from instr to asst prof eng mech, Ohio State Univ, 55-58; from res specialist to sr group engr, 58-69, GUID & CONTROL TECH MGR, BOEING CO, 69- *Concurrent Pos:* Lectr, Univ Wash, 63-71. *Mem:* Aerospace Indust Asn Am; Sigma Xi. *Res:* Inertial navigation components and systems; dynamics and vibrations. *Mailing Add:* 4247 87th Ave SE Mercer Island WA 98040

GOODWILL, ROBERT, b Ridgeway, Pa, Dec 13, 36; m 65; c 2. QUANTITATIVE GENETICS, POPULATION GENETICS. *Educ:* State Univ NY Col Fredonia, BS, 63; Univ Minn, PhD(genetics), 69. *Prof Exp:* Asst prof biol, Bemidji State Col, 68-69; asst prof animal sci, 69-75, ASSOC PROF ANIMAL SCI, UNIV KY, 75- *Mem:* AAAS; Genetics Soc Am; Am Genetic Asn; Am Inst Biol Scientists; Am Soc Animal Sci. *Res:* Experimental evaluation of population and quantitative genetic theory using Tribolium castaneum and computer simulation. *Mailing Add:* 1001 Elmendorf Dr Agr Sci Univ Ky Lexington KY 40502

GOODWIN, ARTHUR VANKLEEK, b Hartford, Conn, Feb 1, 40; m 64; c 2. SOLENOID DESIGN, VALVE DESIGN. *Educ:* Bates Col, BS, 63. *Prof Exp:* Engr, Chandler-Evans, Div of Colt Indust, 63-68, Hamilton Standard, Div of United Technol, 68-72; engr, Wright Components Inc, 72-74, chief engr, 74-77, chief engr & dir, 77-83, chief engr, EG&G Wright Components, 83-86, vpres eng, 86-90; CONSULT, 90- *Mem:* Am Inst Aeronaut & Astronaut. *Res:* Magnetic circuit design and solenoid equations. *Mailing Add:* 28 Charter Oaks Dr Pittsford NY 14534

GOODWIN, BRUCE K, b Providence, RI, Oct 14, 31; m 56; c 3. GEOLOGY. *Educ:* Univ Pa, AB, 53; Lehigh Univ, MS, 57, PhD(geol), 59. *Prof Exp:* Instr geol, Univ Pa, 59-63; from asst prof to assoc prof, 63-71, chmn dept, 70-76 & 83-88, PROF GEOL, COL WILLIAM & MARY, 71- *Concurrent Pos:* Coun, undergraduate res, Acad of Sci, Va. *Mem:* AAAS; fel Geol Soc Am; Nat Asn Geol Teachers; Sigma Xi; Am Inst Prof Geologists. *Res:* Geology of the eastern piedmont of North America; Triassic Richmond basin and other Triassic-Jurassic basins; structural geology. *Mailing Add:* Dept Geol Col William & Mary Williamsburg VA 23185

GOODWIN, CHARLES ARTHUR, b Oneida, NY, Sept 6, 47; m 69; c 2. SOLID STATE ELECTRONICS, CERAMICS. *Educ:* Alfred Univ, BS, 69; Mass Inst Technol, PhD(ceramics), 73. *Prof Exp:* mem tech staff, 73-80, SUPVR ELECTRONICS, BELL LABS, 80- *Mem:* Am Ceramic Soc; Electrochem Soc. *Res:* Electronic devices including metal-oxide-silicon integrated circuits. *Mailing Add:* 819 Carman Dr Wyomissing PA 19610

GOODWIN, FRANCIS E, b Hastings, Nebr, Sept 7, 27; m 49; c 3. LASER APPLICATIONS, MICROWAVE RADAR & RADIOMETRY. *Educ:* Univ Calif, Los Angeles, AB, 56, MS, 57. *Prof Exp:* Mem tech staff, Hughes Aircraft Co. 57-65, sr staff engr, Hughes Res Lab, 65-74, dept mgr, Hughes Space & Commun Group, 74-82, chief scientist, Hughes Electrooptical Develop Div, 82-; VPRES PROD DEV, DIGITAL SIGNAL CORP, 82- *Honors & Awards:* Century Award, Int Elec Electronic Eng. *Mem:* Sr mem Inst Elec & Electronics Engrs; Am Phys Soc; Optical Soc Am. *Res:* Microwave antennas and devices; microwave masers; lasers and laser communications; millimeter wave radiometers for meteorological applications; coherent laser imaging radar. *Mailing Add:* Digital Signal Corp 5554 Port Royal Rd Springfield VA 22151

GOODWIN, FRANK ERIK, b Bethlehem, Pa, Jan 6, 54; m 87; c 1. CASTING TECHNOLOGY, COATINGS TECHNOLOGY. *Educ:* Cornell Univ, BS, 75; Mass Inst Tech, SM, 76, PhD(mat eng), 79. *Prof Exp:* Metallurgist, Chambersburg Eng Co, 79-80; asst dir prod & process develop, Chromalloy Am Corp, 80-82; mgr develop, Int Lead Zinc Res Orgn, 82-84; mgr metall, 84-91; VPRES, INT LEAD ZINC RES ORGN, 86- *Mem:* Am Soc Mats Int; Can Inst Mining & Metall; Am Inst Metall Engrs; Am Foundrymen's Soc; Am Soc Testing & Mat; Am Mgt Asn. *Res:* Development of new uses and improvement of existing uses of lead and zinc metals. Areas of specialization are die casting, steel castings, nuclear waste cantainment, composite structures and resistance welding. *Mailing Add:* PO Box 12036 2525 Meridian Pkwy Research Triangle Park NC 27709-2036

GOODWIN, FREDERICK KING, b Cincinnati, Ohio, Apr 21, 36; m 63; c 3. NEUROLOGY. *Educ:* Georgetown Univ, BS, 58; St Louis Univ, MD, 63. *Prof Exp:* Fel philos, St Louis Univ, 58-59, Wash Sch Psychiat, 69-71; res asst, Lab Clin Biochem, Nat Heart Inst, NIH, 60, 61 & 62, spec res fel, 67-68; chief, Clin Res Unit, Sect Psychiat, Lab Clin Sci, NIMH, 68-73, chief, Sect Psychiat, 73-77, chief, Clin Psychol Br, 77-81, DIR, INTRAMURAL RES PROG, NIMH, 82- *Concurrent Pos:* Res psychiatrist, Univ NC, Chapel Hill, 64-65; clin assoc, Adult Psychiat Br, NIMH, 65-67; vis prof, Univ Wis, 76, Boston Univ, 76, Univ Calif, 77, Univ Southern Calif, 79, Duke Univ, 79-, Univ Tenn, 86. *Honors & Awards:* A E Bennett Award for Clin Res, Soc Biol Psychiat, 70; Psychopharmacol Res Prize, Am Psychol Asn, 70; Hofheimer Prize, Am Psychiat Asn, 71; Taylor Manor Award, 76; Int Anna-Monika Prize, 71; Edward A Strecker Award, 83; Centennial Lectr, Univ NC, Sch Med; George C Ham Mem Lectr, Univ NC, Chapel Hill, 84; David C Wilson Lectr, Univ Va, Charlottesville, 85; Theodore L Dehne MD Mem Lect, Friends Hosp Clin Conf, 87. *Mem:* Inst Med-Nat Acad Sci; AAAS; Am Philos Asn; fel Am Psychiat Soc; Am Psychosomatic Soc; Soc Biol Psychiat; Am Acad Psychoanal; Soc Neurosci; Am Psychopath Asn. *Res:* Author of seven books and over 368 publications. *Mailing Add:* Intramural Res Prog NIMH NIH Bldg 10 9000 Rockville Pike Bethesda MD 20892

GOODWIN, GENE M, b York, Maine, Dec 2, 41; m 63; c 1. JOINING. *Educ:* Rensselaer Polytech Inst, BMetE, 63, PhD, 68. *Prof Exp:* MEM, RES STAFF, OAK RIDGE NAT LAB, 68- *Honors & Awards:* Spraragen Award, Am Welding Soc, 75, McKay-Helm Award, 79 & 80. *Mem:* Fel Am Soc Metals; Am Welding Soc. *Res:* Joining research and development in the areas of stainless steels, nickle-base alloys and other metals and alloys. *Mailing Add:* Oak Ridge Nat Lab PO Box X Oak Ridge TN 37830

GOODWIN, JAMES CRAWFORD, b Camden, Ark, Mar 21, 26; m 52; c 3. ORGANIC & CARBOHYDRATE CHEMISTRY. *Educ:* Philander Smith Col, Ark, BS, 49; Ore State Univ, MS, 59. *Prof Exp:* Teacher chem, Hempstead County Sch Dist, Ark, 49-54, Little Rock Sch Dist, 57-58 & Grambling State Univ, La, 59-60; RES CHEMIST ORG CHEM, USDA NORTHERN REGIONAL RES CTR, 63- *Concurrent Pos:* Fel, NSF, 58-59; teacher chem, Ill Cent Col, East Peoria, 80-81. *Mem:* Nat Orgn Prof Advan Black Chemists & Chem Engrs; Am Chem Soc; Nat Inst Sci. *Res:* Organic-carbohydrate chemistry; correlation of stereochemical structure of simple sugars and sugar analogs with their sweet and bitter tastes. *Mailing Add:* Chem Dept Mabee Kresge Sci Hall 812 W 13th St Little Rock AR 72202-3799

GOODWIN, JAMES GORDON, JR, b Walterboro, SC, Dec 14, 45; m 81. HETEROGENEOUS CATALYSIS, KINETICS. *Educ:* Clemson Univ, BS, 67; Ga Inst Technol, MS, 69; Univ Mich, PhD(chem eng), 76. *Prof Exp:* Instr math, Nasson Col, 68-69, Middle East Tech Univ, 69-70 & Univ Liberia, 70-71; res asst catalysis, Univ Mich, 73-76; exchange scientist, French Inst Catalysis, 77-78; asst prof chem eng, Univ SC, 78-79; from asst prof to assoc prof, 79-87, PROF CHEM ENG, UNIV PITTSBURGH, 87- *Concurrent Pos:* guest prof, Tech Univ, Vienna, 87- *Mem:* Am Inst Chem Engrs; Am Chem Soc; NAm Catalysis Soc. *Res:* Synthesis of hydrocarbons from carbon monoxide and hydrogen; analysis of catalytically active surfaces. *Mailing Add:* Dept Chem & Petrol Eng 1249 Benedum Hall Univ Pittsburgh Pittsburgh PA 15261-2212

GOODWIN, JAMES SIMEON, b Cleveland, Ohio, May 19, 45; m; c 4. GERIATRICS. *Educ:* Amherst Col, BA, 67; Harvard Univ, MD, 71; Am Bd Internal Med, cert, 76, cert rheumatology, 78. *Prof Exp:* Intern, Harbor Gen Hosp, 71-72; staff assoc, Nat Heart & Lung Inst, NIH, 72-74; resident internal med, Sch Med, Univ NMex, 74-76, fel rheumatology, 76-78, from asst prof to res prof, Dept Med, 78-87, chief, Div Geront, 80-85; prof & vchmn, Dept Med, Med Col Wis, 85-88; PROF MED & CHIEF GERIAT, UNIV WIS-MILWAUKEE, 88-; MED DIR, SINAI SAMARITAN GERIAT INST, 88- *Concurrent Pos:* Nat Inst Allergy & Infectious Dis young investr award, 78-81; Nat Inst Aging res grants, 79-96; mem, Ad Hoc Study Sect Immunol Sci, NIH, 79, Subcomt Nutrit & Rheumatic Dis, Nat Inst Aging Res Planning Panel, 82, Fel Selection Comt, Nat Arthritis Found, 85-89, Arthritis Ctr Rev, NIH, 85, Nat Prog Comt, Am Rheumatologic Asn, 87, Clin Liaison Team, Nat Res Agenda Aging, Nat Acad Sci, Inst Med, 88-89 & Prog Comt, Am Asn Immunologists, 88-91; assoc ed, J Immunol, 85-89; chmn, Inflammation-Immunopharmacol Block, Am Asn Immunologists, 88-91. *Mem:* Am soc Clin Invest; Am Rheumatologic Asn; Am Fedn Clin Res; Am Asn Immunologists; Am Geront Asn; Am Geriat Soc. *Res:* Immunobiology of aging; role of arachidonic acid metabolites in modulation of the immune response; effects of subclinical malnutrition in the elderly; patterns of cancer care in the elderly; author of more than 180 technical publications. *Mailing Add:* Geriat Inst Sinai Samaritan Med Ctr 950 N 12th St Milwaukee WI 53233

GOODWIN, JAMES THOMAS, aquatic entomology, for more information see previous edition

GOODWIN, JESSE FRANCIS, b Greenville, SC, Feb 7, 29; m 59; c 3. CLINICAL BIOCHEMISTRY, ANALYTICAL CHEMISTRY. *Educ:* Xavier Univ, La, BS, 51; Wayne State Univ, MS, 53, PhD(chem), 57. *Prof Exp:* Res assoc, Col Med & spec instr, Col Pharm, Wayne State Univ, 58-59; clin biochemist, Wayne County Gen Hosp, Eloise, Mich, 58-62; instr pediat, Col Med, Wayne State Univ, 64-66, asst prof biochem, 66-70, dir, Core Lab, Gen Clin Res Ctr Childrens Hosp & Sch Med, 63-73, asst prof pediat, Col Med, 70-73; DIR LABS, DETROIT HEALTH DEPT, 73- *Concurrent Pos:* Biochemist, Dept Labs, Children's Hosp, Detroit; mem, Toxic Substance Control Comn, State of Mich, 83, Bd Trustees, Horizon Health Systs, Oak Park, Mich, 82-, Marygrove Col, Detroit Mich, 77-83,85-; bd dirs, Am Asn Clin Chem, 87-89, Nat Acad Clin Biochem, 86. *Mem:* AAAS; Am Chem Soc; Am Pub Health Asn; fel Am Inst Chemists; Am Asn Clin Chem; fel Nat Acad Clin Biochem. *Res:* Methodology in clinical chemistry procedures; metabolism of muco-polysaccharides, chelates and gold; clinical chemistry methodology, trace metal estimation, amino acid estimation and carbohydrate interaction with various amines; methodology in carbohydrate, protein, drug and trace metal estimation; seroprevalence studies; AIDS. *Mailing Add:* 19214 Appoline Detroit MI 48235

GOODWIN, JOHN THOMAS, JR, b San Diego, Calif, May 28, 14; m 38; c 5. ORGANIC CHEMISTRY. *Educ:* Okla Agr & Mech Col, BS, 36; Univ Pittsburgh, PhD(org chem), 48. *Prof Exp:* Chemist, Mid-Continent Petrol Corp, 36-39; asst org chem, Univ Pittsburgh, 39-40; res chemist, Gulf Res & Develop Co, Pa, 40-41; fel, Mellon Inst, 41-43, indust fel, 46, sr fel, 46-49; group leader, Dow Corning Corp, 49-51; chemist, Gen Elec Co, 51-54; mgr chem res, Midwest Res Inst, 54-57; vpres, Corn Industs Res Found, Inc, DC, 57-65; vpres chem & chem eng, Southwest Res Inst, 65-79; CONSULT, 79- *Mem:* AAAS; Am Chem Soc; fel NY Acad Sci. *Res:* Microencapsulation; membrane technology; radiation polymerization; organosilicon chemistry; carbohydrates. *Mailing Add:* 815 Serenade Dr San Antonio TX 78216

GOODWIN, KENNETH, b New York, NY, Sept 30, 20; m 47; c 1. GENETICS. *Educ:* Cornell Univ, BS, 48, MS, 50, PhD(animal genetics), 52. *Prof Exp:* Asst animal genetics, Cornell Univ, 48-52; geneticist, Kimber Farms, Inc, Calif, 52-64 & Heisdorf & Nelson Farms, Inc, 64-66; prof poultry sci & head dept, 66-84, EMER PROF, PA STATE UNIV, 84- *Mem:* AAAS; Genetics Soc Am; Am Genetic Asn; Poultry Sci Asn. *Res:* Genetics of poultry; improvement of economic characters; genetics and disease resistance. *Mailing Add:* Dept Poultry Sci Pa State Univ 222 Henning Bldg University Park PA 16802

GOODWIN, LESTER KEPNER, b Casper, Wyo, July 2, 28; m 58; c 1. NUCLEAR SCIENCE. *Educ:* Calif Inst Technol, BS, 50; Univ Calif, Berkeley, MA, 57, PhD(physics), 60. *Prof Exp:* Physicist, Los Alamos Sci Lab, 51-53 & 56 & Lawrence Radiation Lab, Univ Calif, 60; physicist, Aeronutronic Div, Ford Motor Co, 60-63, sect supvr, Aeronutronic Div, Philco Corp, 63-69; sr scientist, KMS Technol Ctr, 69-73; PHYSICIST, AERONUTRONIC-FORD, 73- *Mem:* AAAS; Am Phys Soc; Sigma Xi. *Res:* Laser effects; nuclear weapons effects; neutron, gamma and meson cross-section measurements; aerosol research; nuclear weapon testing; rfion-source development; application of gamma rays to reentry erosion measurements. *Mailing Add:* 303 Esperanza Newport Beach CA 92660

GOODWIN, MELVIN HARRIS, JR, b Thomasville, Ga, Jan 9, 17; m 42; c 2. EPIDEMIOLOGY, PARASITOLOGY. *Educ:* Univ Ga, BS, 41, MS, 51; Emory Univ, PhD(parasitol), 55. *Prof Exp:* Asst, Rabies Lab, State Dept Health, Ga, 38-39, malaria control biologist, 42-43; mem staff, Entom Div, Commun Dis Ctr, USPHS, 44-46, chief lib & reports div, 47-48, chief malaria invests sect, 49-53, asst chief invests, Tech Br, 53-57, chief, Phoenix Field Sta, 57-66; dir, Div Prev Med Serv, Ariz State Health Dept, 66-69; dir, Ariz Health Planning Authority, 69-74; consult, health & environ sci, 74-75; dir health studies, Copley Int Corp, 76-77; prof, family & community med & sr epidemiologist, 77-87, SR CLIN LECTR, UNIV ARIZ, COL MED, 88- *Concurrent Pos:* Biologist-in-chg malaria res sta, Emory Univ, 39-42, dir, 44-57, assoc, Sch Med, 48-57. *Mem:* AAAS; Am Pub Health Asn; Sigma Xi; Am Soc Trop Med & Hyg. *Res:* Epidemiology; ecology of enteric diseases, malaria and other infectious diseases. *Mailing Add:* 327 W Orchid Lane Phoenix AZ 85021

GOODWIN, PAUL NEWCOMB, b Evanston, Ill, Nov 3, 26; m 51; c 2. RADIOLOGICAL PHYSICS. *Educ:* Harvard Univ, BS, 48; Johns Hopkins Univ, MA, 56; Univ London, PhD(med physics), 59; Am Bd Radiol & Am Bd Health Physics, dipl, 61. *Prof Exp:* Physicist, USPHS Hosp, Baltimore, 49-56; USPHS fel, Royal Cancer Hosp, London, Eng, 56-59, radiol dept, Johns Hopkins Hosp, 59-66; asst prof radiol, Sch Med & Sch Hyg & Pub Health, Johns Hopkins Univ, 62-66; asst prof, Col Physicians & Surgeons, Columbia-Presby Med Ctr, Columbia Univ, 66-71; ASSOC PROF RADIOL, ALBERT EINSTEIN COL MED, YESHIVA UNIV, 71- *Mem:* Health Physics Soc; Am Asn Physicists in Med; Soc Nuclear Med; Radiol Soc NAm; Am Col Radiol. *Res:* Radiological physics; medical application of radiation and radioactive isotopes; radiation dosimetry and protection; calorimetric measurement of x-rays; computed tomography; bone mineral measurement. *Mailing Add:* Dept Radiol Albert Einstein Col Med Bronx NY 10461

GOODWIN, PETER WARREN, b Wilmington, Del, Apr 12, 36; m 84; c 2. STRATIGRAPHY. *Educ:* Dartmouth Col, AB, 58; Univ Iowa, MS, 61, PhD(geol), 64. *Prof Exp:* From instr to assoc prof, 63-83, PROF GEOL, TEMPLE UNIV, 83- *Concurrent Pos:* Chmn geol dept, Temple Univ, 76-87. *Mem:* Geol Soc Am; Soc Econ Palentologists & Mineralogists; Int Asn Sedimentologists. *Res:* Paleozoic stratigraphy of Appalachian Basin; paleoenvironments and paleogeography; punctuated aggradational cycles (PACs); general model of episodic stratigraphic accumulation. *Mailing Add:* Dept Geol Temple Univ Philadelphia PA 19122

GOODWIN, RICHARD HALE, b Brookline, Mass, Dec 14, 10; m 36; c 2. BOTANY, MANAGEMENT OF NATURAL AREAS. *Educ:* Harvard Univ, AB, 33, MA, 34, PhD(biol), 37. *Prof Exp:* Am-Scand Found fel, Copenhagen, 37-38; instr bot, Univ Rochester, 38-41, asst prof, 41-44; dir arboretum, 44-65, prof, 44-76, EMER PROF BOT, CONN COL, 76- *Concurrent Pos:* Comnr, Conn Geol & Natural Hist Surv, 45-72; pres, Conserv & Res Found, 53- & Nature Conservancy, 56-58 & 64-66; mem biol coun, Nat Res Coun, 53-57; bd dirs, Am Inst Biol Sciences, 62-71. *Honors & Awards:* Environ Merit Award, Environ Protection Agency, 90. *Mem:* AAAS; Bot Soc Am; Soc Study Develop & Growth (secy, 49-53); fel Am Acad Arts & Sci; Inst Ecol (treas, 75-77). *Res:* Effect of light on growth of plants; physiology of growth; plant hormones; fluorescent substances in plants; morphogenesis of roots; bioecology of natural areas. *Mailing Add:* Dept Bot Conn Col New London CT 06320

GOODWIN, ROBERT ARCHER, JR, b Kuling, China, Aug 3, 14; US citizen; m 44; c 3. INTERNAL MEDICINE. *Educ:* Univ Va, BS, 36; Johns Hopkins Univ, MD, 40. *Prof Exp:* Asst res physician, Thorndike Mem Lab, Boston City Hosp, 41-42; from asst resident to resident physician, Hosp, 46-47, from instr to asst prof, Univ, 46-56, assoc prof clin med, 56-74, PROF, VANDERBILT UNIV, 74-, EMER PROF MED, 81- *Concurrent Pos:* Chief pulmonary dis serv, Vet Admin Hosp, 47-81. *Mem:* Am Thoracic Soc; fel Am Col Physicians. *Res:* Infectious diseases; pulmonary diseases. *Mailing Add:* 3720 Benham Ave PO Box 158543 Nashville TN 37215-8543

GOODWIN, ROBERT EARL, b Rensselaer, NY, May 16, 26; m 54; c 2. VERTEBRATE ECOLOGY & SYSTEMATICS. *Educ:* Hartwick Col, BS, 51; Cornell Univ, MS, 53, PhD(vert zool), 60. *Prof Exp:* Instr comp anat, 59-67, assoc prof, 67-70, chmn dept biol, 81-84, PROF ZOOL, COLGATE UNIV, 70- *Mem:* Am Soc Mammal; Am Ornith Union. *Res:* Ecology, behavior and systematics of vertebrates, particularly bats. *Mailing Add:* Dept Biol Colgate Univ Hamilton NY 13346

GOODWIN, RONALD HAYSE, b Los Angeles, Calif, Oct 15, 33; m 66; c 2. INSECT PATHOLOGY, CELL BIOLOGY. *Educ:* Univ Calif, Berkeley, BS, 56, MS, 61, PhD(insect path), 66. *Prof Exp:* Res scientist insect path, Commonwealth Sci & Indust Res Org, Div Entom, Australia, 66-69; RES ENTOMOLOGIST INSECT VIROL & TISSUE CULT, INSECT PATH LAB, AGR RES SERV, USDA, 69- *Concurrent Pos:* Res fel entomol, Nat Res Coun, Agr Res Serv, 69-71; adv, Subcomt Invert Viruses, Int Comt Nomenclature Viruses, 68- *Mem:* Soc Invert Path; Tissue Cult Asn; Entom Soc Am. *Res:* Insect cell biology and cell culture for the study of insect pathogens. primarily arthropod specific viruses of the baculovirus and entomopoxvirus groups. *Mailing Add:* Montana State Univ Rangeland Insect Lab USDA/ARS Bozeman MT 59717-0001

GOODWIN, SIDNEY S, b Corbin, Ky, May 3, 06. MINING GEOLOGY. *Educ:* Univ Ky, BS, 28. *Prof Exp:* From mining geologist to vpres explor & mining, NJ Zinc Co, 30-71; consult, 71-73; RETIRED. *Mem:* Fel Geol Soc Am; Soc Econ Geologists; Geochem Soc; Am Inst Mining & Metall Engrs; Mining & Metall Soc Am. *Mailing Add:* 1902 Huntington Chase Dunwoody GA 30350

GOODWIN, THOMAS ELTON, b Nashville, Ark, Aug 8, 47; m 70; c 2. ORGANIC CHEMISTRY. *Educ:* Ouachita Baptist Univ, BS, 69; Univ Ark, PhD(chem), 74. *Prof Exp:* Robert A Welch fel org chem, Rice Univ, 74-75; res scientist, Continental Oil Co, 75-76; lectr & res assoc, Tex A&M Univ, 76-78; ASST PROF ORG CHEM, HENDRIX COL, 78- *Mem:* Am Chem Soc. *Res:* Organic synthesis; natural products synthesis; the synthesis of therapeutically active organic chemicals. *Mailing Add:* Dept Chem Hendrix Col Conway AR 72032-3099

GOODWIN, TOMMY LEE, b Little Rock, Ark, Apr 6, 36; m 59; c 2. FOOD SCIENCE. *Educ:* Univ Ark, BSA, 58, MS, 60; Purdue Univ, PhD(food technol), 62. *Prof Exp:* From asst prof to prof poultry prod technol, Univ Ark, Fayetteville, 64- *Honors & Awards:* Res Award, Poultry & Egg Inst Am, 75. *Mem:* Inst Food Technol; Poultry Sci Asn; World's Poultry Sci. *Res:* Factors affecting nutrition, quality and tenderness of poultry meat and improving quality of eggs; meat yields. *Mailing Add:* Pilgrims Pride 110 S Texas St Pittsburg TX 75686

GOODWIN, WILLIAM JENNINGS, b Bybee, Va, Apr 30, 25; m 47; c 2. MEDICAL ENTOMOLOGY. *Educ:* Okla State Univ, BS, 50; Cornell Univ, MS, 51, PhD, 53. *Prof Exp:* Assoc prof entom, Clemson Col, 53-57; vector control adv, Int Coop Admin, Tripoli, Libya, 57-58, malaria adv, 59-61; malaria adv, Agency Int Develop, Port-Au-Prince, Haiti, 61-63; scientist adminr, NIH, 64-65; scientist adminr ord, Bur State Serv, USPHS, 65-67; chief, Regional Primate Res Ctr, NIH, Bethesda, Md, 67-75; res coordr, Med Res Found, Ore, 75-78; assoc dir, 78-80, DIR, DEPT LAB ANIMAL MED, SOUTHWEST FOUND BIOMED RES, 80- *Concurrent Pos:* Assoc Sci Dir, Southwest Found Biomed Res, 86. *Mem:* Entom Soc Am; Am Soc Primatology; Am Asn Lab Animal Sci. *Res:* Medical and veterinary entomology; malaria eradication; primatology; science administration; public health. *Mailing Add:* Southwest Found Biomed Res PO Box 21847 San Antonio TX 78280

GOODWINE, JAMES K, JR, b Evanston, Ill, Mar 9, 30; m 59; c 2. MECHANICAL & AUTOMOTIVE ENGINEERING, TECHNICAL MANAGEMENT. *Educ:* Purdue Univ, BS, 52, MS, 56, PhD(mech eng), 60. *Prof Exp:* Res asst mech eng, Purdue Univ, 56-57, instr, 57-59; res engr petrol prod, Chevron Res Co, 59-67; staff engr, Eng Dept, United Airlines, 67-70, mgr, power plant eng, 70-79, new aircraft & oper eng, 79-82, dir, engine tech serv, 82-87, mgr new technol eng, 87-89; CONSULT, AVIATION MGT SYSTS INC & MICHAEL GOLDFARB ASSOCS, 89- *Mem:* AAAS; Soc Automotive Engrs; Sigma Xi. *Res:* Petroleum products, especially jet fuels and gasolines; fundamentals of combustion, including computer simulation of engine cycles; turbine engine maintenance; aircraft performance; aircraft maintenance systems and facilities. *Mailing Add:* 1423 Enchanted Way San Mateo CA 94402

GOODWYN, JACK RAY, b Center, Tex, June 28, 34; m 55; c 3. POLYMER CHEMISTRY. *Educ:* Baylor Univ, BA, 56, PhD(phys chem), 60. *Prof Exp:* Chemist, Tex Eastman Co, 60-65, sr chemist, Plastics Lab, 65-73, asst to mgr, 73-76, mem staff prod dept, 76-77, coordr clean environ prog, 77-81. *Mem:* Am Chem Soc; Soc Plastics Engrs; Air Pollution Control Fedn. *Res:* Molecular structure of high polymers. *Mailing Add:* 2308 Kentucky Dr Longview TX 75601

GOODY, RICHARD (MEAD), b Eng, June 19, 21; nat US; m; c 1. ATMOSPHERIC PHYSICS. *Educ:* Univ Cambridge, BA, 42, PhD, 49. *Hon Degrees:* AM, Harvard Univ, 58. *Prof Exp:* Sci officer, Ministry Aircraft Prod, Eng, 42-46; fel, St John's Col, Univ Cambridge, 50-53; reader, Imp Col, Univ London, 53-58; Abbott Lawrence Rotch prof dynamic meteorol & dir, Blue Hill Meteorol Observ, Harvard Univ, 58-68, Mallinkrodt prof planetary physics, 70-91, dir, Ctr Earth & Planetary Physics, 71-74, Gordon McKay prof appl physics, 80-91; RETIRED. *Honors & Awards:* Buchan Prize, Royal Meteorol Soc, 58; 50th Anniversary Medal, Am Meteorol Soc, 70 & Cleveland Abbe Award, 77; Pub Serv Medal, NASA, 80. *Mem:* Nat Acad Sci; Am Meteorol Soc; Am Acad Arts & Sci; Royal Meteorol Soc. *Res:* Physics and dynamics of the atmospheres of the earth and other planets; infrared spectroscopy. *Mailing Add:* Box 430 Falmouth MA 02541

GOODYEAR, WILLIAM FREDERICK, JR, b Camden, NJ, Aug 13, 29; m 53; c 2. WOOD AND TECHNOLOGY. *Educ:* Rutgers Univ, BS, 57, PhD(org chem), 59. *Prof Exp:* Res chemist, 59-63, res supvr, 63-69, res mgr, 69-73, res gen mgr, 73-82, RES ASSOC, ARMSTRONG CORK CO, 82- *Mem:* Soc Plastics Eng; Forest Prods Res Soc; Sigma Xi. *Res:* Reaction kinetics; organophosphorus-organic peroxide reaction mechanisms; filled plastics; wood technology; lumber processing; furniture manufacturing methods and material. *Mailing Add:* 1707 Marietta Ave Apt 1E Lancaster PA 17603

GOODYER, ALLAN VICTOR, b New York, NY, Nov 21, 18; m 46; c 3. CARDIOVASCULAR DISEASES. *Educ:* Yale Univ, BS, 39, MD, 42. *Prof Exp:* Intern, New Haven Hosp, 42-43, asst resident med, 46-47; fel, 47-48, instr internal med, 48-50, from asst prof to assoc prof, 50-66, PROF MED, YALE UNIV, 66- *Concurrent Pos:* Markle Found scholar, 49-54; estab investr, Am Heart Asn, 54-59. *Mem:* Am Soc Clin Invest; Am Fedn Clin Res; Asn Univ Cardiol; Am Col Cardiol; Asn Am Physicians. *Res:* Cardiovascular physiology and disease. *Mailing Add:* Dept Med Yale Univ Hosp 333 Cedar St New Haven CT 06510

GOODZEIT, CARL LEONARD, b Newark, NJ, Mar 19, 28; m 64; c 4. MECHANICAL ENGINEERING. *Educ:* Rutgers Univ, BSc, 50; Brown Univ, ScM, 59. *Prof Exp:* Res engr, Gen Motors Res Labs, 50-59; MECH ENGR, BROOKHAVEN NAT LAB, 59- *Res:* Structural mechanics; computer methods in mechanical design; high energy physics. *Mailing Add:* 1409 Yardley Pl DeSoto TX 75115

GOOGIN, JOHN M, b Lewiston, Maine, May 2, 22; m 49; c 4. MATERIALS SCIENCE ENGINEERING. *Educ:* Bates Col, BS, 44; Univ Tenn, PhD(phys chem), 53. *Hon Degrees:* DSc, Bates Col, 68. *Prof Exp:* Jr chemist Manhattan Proj, Oak Ridge Y-12 Plant, Tenn Eastman Div, Eastman Kodak Co, 43-47, Union Carbide Corp, 47-83; SR STAFF CONSULT, DEVELOP DIV, OAK RIDGE Y-12 PLANT, MARTIN MARIETTA ENERGY SYSTS, INC, 83, SR CORP FEL, 87- *Concurrent Pos:* Resource person, Dept Energy; res fel, Union Carbide Corp, 76. *Honors & Awards:* Ernest Orlando Lawrence Mem Award, AEC, 67; William J Kroll Zirconium Medal, 87; Gold Medal Award, Am Soc Metals, 89. *Mem:* Nat Acad Eng; Am Chem Soc; Am Soc Advan Sci; fel Am Soc Metals; Sigma Xi; Soc Advan Mat & Process Eng. *Res:* Separation of the isotopes of hydrogen; refractory ceramics and metals; holds over a dozen patents. *Mailing Add:* Martin Marietta Energy Systems Inc Bldg 9202 MS 8097 PO Box 2009 Oak Ridge TN 37831-8097

GOON, DAVID JAMES WONG, b LaCrosse, Wis, Jan 15, 42. ORGANIC CHEMISTRY. *Educ:* Univ Minn, Minneapolis, BCh, 64; Univ Ill, Urbana, MS, 66, PhD(org chem), 68. *Prof Exp:* Fel chem, Univ Ill, Chicago Circle, 68-69, instr, 68-70; teaching fel, Univ Guelph, 70-71; sr res chemist, Gen Mills, Inc, Minneapolis, 72; asst scientist, Gen Clin Res Ctr, 72-73, res fel, Dept Med Chem, Univ Minn, Minneapolis, 74-76; instr, Normandale Community Col, Bloomington, Minn, 76-84; asst prof, Univ Wisc-Eau Claire, 85-87; CONSULT CHEMIST, VET ADMIN HOSP, MINNEAPOLIS, MN, 87- *Concurrent Pos:* Post-doc assoc, Dept Med Chem, Univ Minn, Minneapolis, 89- *Mem:* AAAS; Am Chem Soc. *Res:* Medicinal chemistry; ethanol metabolism; kinetics and mechanism. *Mailing Add:* 8624 Columbus Ave S Bloomington MN 55420

GOONEWARDENE, HILARY FELIX, b Colombo, Sri Lanka, Apr 9, 25; US citizen; m 65; c 3. PLANT RESISTANCE, PLANT BREEDING. *Educ:* Univ Sydney, BSc, 53; Univ NZ, BAgrSc, 57; Rutgers Univ, MS, 60, PhD(entom, plant path), 61. *Prof Exp:* Res officer, Comn Sci & Indust Res Org, Australia, 52-53; crop protection officer, Coconut Res Inst, Sri Lanka, 56-58; res asst entom, Rutgers Univ, 58-61, agr chem, 61; tech dir pesticides, Corona Chem Div, Pittsburgh Plate Glass Co, 61-63; head pesticides res & develop, Smith Kline & French Labs, Inc, 63-64; res entomologist, Agr Res Serv, USDA, 65-88, collabr, 88-90; EMER PROF ENTOMOLOGY, DEPT ENTOMOLOGY, PURDUE UNIV, 90- *Concurrent Pos:* Agr res grants adminstr for several cols, univs & industs, 61-64; adj assoc prof, Col Biol Sci, Ohio State Univ & dept entom, Ohio Agr Res & Develop Ctr, 71-74; adj assoc prof, Purdue Univ, 74-; mem Crop Adv Comt, Malus, 87-88; consult, Marriott Corp, Fac Mgt Div, 88- *Mem:* Emer mem AAAS; Am Pomol Soc; emer mem Entom Soc Am; emer mem Sigma Xi; emer mem Am Soc Hort Sci. *Res:* Fruit tree breeding; insect and/or mite resistance to high quality disease-resistant cultivars and development of new cultivars with these characteristics; pest management of resistant germ plasm. *Mailing Add:* Dept Entom Entom Hall Purdue Univ Rm 106 West Lafayette IN 47907

GOOR, CHARLES G, applied statistics; deceased, see previous edition for last biography

GOOR, RONALD STEPHEN, b Washington, DC, May 31, 40; m 67; c 2. BIOCHEMISTRY, NUTRITION. *Educ:* Swarthmore Col, BA, 62; Harvard Univ, PhD(biochem), 67, MPH, 76. *Prof Exp:* NIH fel, 67-69; staff fel biol viruses, Nat Inst Allergy & Infectious Dis, 69-70; spec asst to dir, Nat Mus Natural Hist, Smithsonian Inst, 70-72; prog mgr, Environ Aspects Trace Contaminants, Res Appl to Nat Needs, NSF, 72-76; int coordr, coronary primary prev trial, Nat Heart, Lung & Blood Inst, NIH, 76-83, coordr, Nat Cholesterol Educ Prod, 83-85; DIR, HEALTH PROSPECT ASSOCS, 86- *Mem:* AAAS; Am Chem Soc; Am Soc Biol Chemists; NY Acad Sci; Am Dietetic Asn. *Res:* Environmental contamination; public health; nutrition. *Mailing Add:* 9301 Cedarcrest Dr Bethesda MD 20814

GOOREY, NANCY REYNOLDS, b Davenport, Iowa; m; c 4. DENTISTRY. *Educ:* Wooster Col, BS, 40; Ohio State Univ, DDS, 55. *Prof Exp:* From instr to prof dent, 55-86, dir & chmn div dent hyg, 69- 86, asst dean, 76-86, EMER PROF DENT & ASST DEAN AUXILIARY PROGS, COL DENT, OHIO STATE UNIV, 86- *Concurrent Pos:* Pvt pract, Gen Dent & Gen Anesthesiol, 56-74; consult. *Mem:* Am Dent Asn; Int Asn Dent Res; Am Asn Dent Schs. *Res:* Dental hygiene. *Mailing Add:* 2201 Castle Crest Dr Worthington OH 43085

GOORVITCH, DAVID, b San Pedro, Calif, July 16, 41; m 66. ASTROPHYSICS. *Educ:* Univ Calif, Berkeley, BA, 63, PhD(physics), 67. *Prof Exp:* PHYSICIST, AMES RES CTR, 67- *Mem:* Am Phys Soc; Optical Soc Am; Sigma Xi; Am Astron Soc. *Res:* Isotope shift and hyperfine structure of radioactive elements; spectroscopy of highly stripped elements; accurate wavelength determinations of thorium transitions; airborne infrared astronomy; astrophysics; infrared Fourier spectroscopy; infrared astronomy of the planets, cool stars, H-II regions and our galactic center from high altitudes. *Mailing Add:* N245-6 Ames Res Ctr Moffett Field CA 94035

GOOS, ROGER DELMON, b Beaman, Iowa, Oct 29, 24; m 46; c 2. MICROBIOLOGY. *Educ:* Univ Iowa, BA, 50, MS, 55, PhD, 58. *Prof Exp:* Teacher pub sch, Iowa, 51-53; asst bot, Univ Iowa, 53-58; mycologist, Cent Res Labs, United Fruit Co, Mass, 58-62; scientist, NIH, 62-64; cur fungi, Am Type Cult Collection, 64-68; assoc researcher bot, Univ Hawaii, 68-69, vis assoc prof, 69-70; assoc prof, 70-72, chmn dept, 71-86, PROF BOT, UNIV RI, 72- *Concurrent Pos:* Vis res assoc, Dept Bot, Univ BC, 77 & Dept Bot, Univ Hawaii, 77, Univ Exeter, Eng, 84; vis fel, Univ Madras, 81; vis res, Bishop Mus, Honolulu, 90. *Mem:* Mycol Soc Am (secy-treas, 80-83, vpres, 83-84, pres-elect, 84-85, pres, 85-86); Am Phytopath Soc; Brit Mycol Soc; Am Soc Microbiol; Mycol Soc Japan; AAAS. *Res:* Classification and life histories of the fungi imperfecti and ascomycetes; tropical fungi; soil mycology. *Mailing Add:* Dept Bot Univ RI Kingston RI 02881-0812

GOOSMAN, DAVID R, b Portland, Ore, Feb 6, 41. RADIOGRAPHY, ELECTRO-OPTICS. *Educ:* Reed Col, BA, 62; Calif Inst Technol, PhD(physics), 67. *Prof Exp:* Physicist, 74-80, GROUP LEADER, LAWRENCE LIVERMORE LABS, 80- *Mem:* Am Phys Soc. *Mailing Add:* Lawrence Livermore Nat Lab L-368 PO Box 808 Livermore CA 94550

GOOSSENS, JOHN CHARLES, b Chicago, Ill, Aug 19, 28; m 56; c 3. ORGANIC POLYMER CHEMISTRY. *Educ:* Univ Notre Dame, BS, 50; Univ Md, PhD(org chem), 57. *Prof Exp:* Asst, Wayne State Univ, 50-51; chemist, Standard Oil Co, Ind, 55-60; res chemist, Silicone Prod Dept, Gen Elec Co, Pittsfield, 60-71, res chemist, Plastics Dept, 71-77; COATINGS RES CHEMIST, STRUCTURED PROD DEPT, GEN ELEC PLASTICS GROUP, MT VERNON, IND, 77- *Mem:* Am Chem Soc. *Res:* Polymers; lubrication; organometallics. *Mailing Add:* 1612 Magnolia Ct Mt Vernon IN 47620-9304

GOOTENBERG, JOSEPH ERIC, b Boston, Mass, Aug 9, 49; m. PEDIATRIC HEMATOLOGY & ONCOLOGY. *Educ:* Harvard Col, AB, 71; Albert Einstein Col Med, MD, 75. *Prof Exp:* Pediat resident, Children's Hosp Med Ctr, Boston, 75-78; clin assoc, Pediat Oncol Br, Nat Cancer Inst, NIH, 78-80 & Lab Tumor Cell Biol, 80-82, cancer expert, 82-83; asst prof pediat, 83-89, MEM, VINCENT T LOMBARDI CANCER RES CTR, SCH MED, GEORGETOWN UNIV, 83-, CHIEF, DIV PEDIAT HEMAT-ONCOL, 88-, ASSOC PROF PEDIAT, 90- *Concurrent Pos:* Guest worker, Metab Br, Nat Cancer Inst, NIH, 75, prin investr, 83-85 & 84-88; prin investr, Div Res Resources, NIH, 83-84, Naval Med Res Inst, 86-87, Nat Inst Dent Res, 85-89, Lombardi Cancer Ctr Develop Funds, 88-89 & Hoechst-Roussel Pharmaceut, Inc, 88-90; Leukemia Soc Am spec fel, 84-86; Am Cancer Soc jr fac clin fel, 84-86; med adv, Asn Retarded Citizens, 85-; vis res assoc, Naval Med Res Inst, 86-; mem, Med Adv Bd, Ronald McDonald House, 88- *Mem:* Am Soc Hemat; Am Soc Clin Oncol; Int Asn Comp Res Leukemia & Related Dis; Am Asn Cancer Res; Am Asn Immunologists; Am Soc Pediat Hemat-Oncol; Am Acad Pediat. *Res:* Pediatric hematology-oncology; author of numerous technical publications. *Mailing Add:* Div Pediat Hemat-Oncol Vincent T Lombardi Cancer Res Ctr Georgetown Univ Hosp 3800 Reservoir Rd NW Washington DC 20007

GOOTMAN, NORMAN LERNER, b Philadelphia, Pa, Feb 6, 33; m 58; c 2. CARDIOVASCULAR DISEASES & PHYSIOLOGY. *Educ:* Univ Vt, BS, 54; Univ Vt Col Med, MD, 58. *Prof Exp:* Clin assoc prof pediat, Downstate Med Col, 65-72; assoc prof 72-75, PROF PEDIAT, SCH MED, HEALTH SCI CTR, STATE UNIV NY STONY BROOK, 75-; ASSOC DIR PEDIAT, SCHNEIDER CHILDRENS HOSP, LONG ISLAND JEWISH MED CTR, 75- *Concurrent Pos:* Chief pediat cardiol, Long Island Jewish-Hillside Med Ctr, 65- *Mem:* Am Heart Asn; Soc Pediat Res; Am Col Cardiol; fel Am Acad Pediat; Am Pediat Soc. *Res:* Developmental aspects of neural control circulation; postnatal development of cardiovascular system. *Mailing Add:* Dept Pediat Schneider Childrens Hosp Long Island Jewish Med Ctr New Hyde Park NY 11042

GOOTMAN, PHYLLIS MYRNA ADLER, b New York, NY, June 8, 38; m 58; c 2. NEUROPHYSIOLOGY, PHYSIOLOGY. *Educ:* Columbia Univ, BA, 59; Albert Einstein Col Med, PhD(physiol), 67. *Prof Exp:* Res assoc physiol, Sch Med, Univ Wash, 63; from instr to asst prof, Albert Einstein Col Med, 68-73; from asst prof to assoc prof, 73-81, PROF PHYSIOL, STATE UNIV NY HEALTH SCI CTR AT BROOKLYN, 81- *Concurrent Pos:* NIH fels, 68-69, 70-71, prin investr, 74-77, 78-81, 82-91, mem, Study Sect Cardiovasc Renal B, 81-85; vis asst prof physiol, Albert Einstein Col Med, 73-76, vis prof, 89-; staff app pediat & consult pediat cardiol, Schneider Children's Hosp, Long Island Jewish Med Ctr, 76- *Mem:* AAAS; Am Physiol Soc; Biophys Soc; Soc Neurosci; Am Assoc Lab Animal Sci; Am Inst Biol Sci; Soc Exp Biol & Med; Int Soc Develop Neurosci; Sigma Xi. *Res:* Central nervous system regulation of cardiovascular function; interrelations between sympathetic vasomotor and central respiratory rhythm generators; postnatal maturation of central nervous system regulation of cardiovascular function; postnatal development of the respiratory pattern generator; electrophysiological and neuroanatomical studies of the autonomic nervous system. *Mailing Add:* Dept Physiol State Univ NY Health Sci Ctr Brooklyn Brooklyn NY 11203

GOPAL, RAJ, b Bangalore, India, Feb 1, 42; US citizen; m 67; c 2. ENERGY SYSTEMS, DEMAND SIDE MANAGEMENT-TECHNICAL. *Educ:* Univ Mysore, India, BS, 63; Indian Inst Tech, Madras, India, MTech, 65; Univ Akron, PhD(mech eng), 73. *Prof Exp:* Prod mgr, Hin Dustan Ferobo Ltd, Bombay, India, 65-69; res asst, Dept Mech Eng, Univ Akron, 69-73; res scientist, Johnson Contruls Inc, 73-84; pres, Eng Consult Serv, 84-87; RES & DEVELOP MGR, ANCO CONSULT GROUP, 87- *Concurrent Pos:* Lectr, Milwaukee Sch Eng, 85-86, Univ Wis, 86-87. *Mem:* Am Soc Mech Eng; Am Soc Heating, Refrig & Air Conditioning Engrs; Sigma Xi; Asn Energy Eng. *Res:* Technology-analysis; assessment of promotion farming; demand side management to electric utilities; thermal energy storage; energy management control system; HVAC and knowledge based technology evaluation and technology impact evaluation for electric utilities. *Mailing Add:* 6267 W Silverbrook Lane Brown Deer WI 53223

GOPALAKRISHNA, K V, b Mysore, India, Jan 11, 44; m 71; c 4. INFECTIOUS DISEASES, INTERNAL MEDICINE. *Educ:* Mysore Univ, MBBS, 65. *Prof Exp:* Chief gen med sect, Vet Admin Hosp, 72-74, chief infectious dis sect, 73-76; CHIEF INFECTIOUS DIS SECT, FAIRVIEW GEN HOSP, 76-; ASSOC CLIN PROF MED, CASE WESTERN UNIV, 77- *Concurrent Pos:* Consult, Lutheran Med Ctr, 72-, St John's Hosp, 72-, Lakewood Hosp, 72-, Parma Community Hosp, 72- & Southwest Gen Hosp, 75- *Mem:* Am Soc Microbiol; Asn Practicioners Infection Control; Infectious Dis Soc; Fel, Am Col Physicians. *Res:* Pathogenesis of endocarditis; antimicrobial agents and mode of action. *Mailing Add:* 1139 Richmar Dr Cleveland OH 44145

GOPALAKRISHNAN, KAKKALA, b Kundoor, India, Mar, 42; m 71; c 1. BIOLOGICAL OCEANOGRAPHY, INVERTEBRATE ZOOLOGY. *Educ:* Univ Kerala, India, BSc, 63, MSc, 65; Scripps Inst Oceanog, Univ Calif, PhD(biol oceanog), 73. *Prof Exp:* Jr sci asst, Nat Inst Marine Biol, Indian Ocean Biol Ctr, India, 65-67; fel Inst Marine Biol, Univ Hawaii, 73-75; INSTR OCEANOG, HONOLULU COMMUNITY COL, UNIV HAWAII, 75- *Mem:* Ecol Soc Am. *Res:* Krill (euphausid) biology, ecology and zoogeography; aquaculture, particularly crustaceans. *Mailing Add:* Dept Nat Sci Honolulu Community Col 874 Dillingham Blvd Honolulu HI 96817

GOPIKANTH, M L, b Mysore, India, Aug 9, 54; US citizen. ELECTROCHEMICAL ENGINEERING, ELECTROCHEMISTRY. *Educ:* Univ Mysore, India, BS, 72; Birla Inst Technol & Sci, Pilani, India, MS, 74; Indian Inst Sci, PhD(electrochem), 78; Northeastern Univ, Boston, Mass, MBA, 85. *Prof Exp:* Postdoctoral res asst, Indian Inst Sci, 78 & I Dept Chem Eng, Ill Inst Technol, Chicago, 78-80; sr res scientist, Cardiac Pacemakers, Inc, Minn, 80-82; prin scientist, Duracell Inc, Needham, Mass, 82-85; prin engr, Life Systs, Inc, Cleveland, Ohio, 85-86; VPRES, CSI, 86- *Mem:* Sigma Xi; Am Chem Soc; Electrochem Soc; NY Acad Sci; AAAS. *Res:* Batteries and chemical sensors. *Mailing Add:* CSI PO Box 1067 Burlington MA 01803

GOPLEN, BERNARD PETER, b Griffin, Sask, March 6, 30; m 56; c 4. GENETICS, PLANT BREEDING. *Educ:* Univ Sask, BSA, 52, MSc, 55; Univ Calif, PhD(genetics), 58. *Prof Exp:* Asst agron, Univ Calif, 55-57; PRIN RES SCIENTIST & HEAD FORAGE SECT, RES STA, CAN DEPT AGR, 58- *Concurrent Pos:* Adj prof crop sci, Univ Sask, 79-; Pres, NAm Alfalfa Improv Conf, 84-86. *Honors & Awards:* Merit Cert, Am Forage & Grassland Coun, 86. *Mem:* Fel Agr Inst Can (vpres, 83-84); Genetics Soc Can; Am Soc Agron; Crop Sci Soc Am; Can Soc Agron; hon life, mem, Can Seed Growers Assoc, 82. *Res:* Bloat research in legumes; developing bloat-safe alfalfa; developing low-coumarin sweetclover. *Mailing Add:* Res Sta Agr Can 107 Sci Crescent Saskatoon SK S7N 0X2 Can

GOPLERUD, CLIFFORD P, b Osage, Iowa, Dec 6, 24; m 47; c 4. OBSTETRICS & GYNECOLOGY. *Educ:* Carlton Col, 42-43; Washington & Lee Univ, 43-44; Univ Iowa, MD, 48. *Prof Exp:* Intern med, Cincinnati Gen Hosp, Ohio, 48-49; asst resident obstet & gynec, Grady Mem Hosp, 49-50, 51-52, resident, 52-53; from asst prof to assoc prof, 58-66, PROF OBSTET & GYNEC, UNIV IOWA HOSPS, 66- *Mem:* AMA; Am Col Obstet & Gynec; Am Gynec Soc; Am Asn Obstet & Gynec; Asn Profs Gynec & Obstet. *Res:* Erythroblastosis; amniocentesis; diabetes in pregnancy; infectious problems in obstetric patients. *Mailing Add:* State Univ Iowa Hosps Obstet & Gynec Univ Iowa Hosp Univ Iowa Col Med Iowa City IA 52240

GORA, EDWIN KARL, b Bielsko, Poland, Oct 22, 11; nat US; m 45; c 5. THEORETICAL PHYSICS. *Educ:* Jagellonian Univ, Poland, PhM, 34; Univ Leipzig, Ger, ScD, 43. *Prof Exp:* Lectr physics, St Xavier's Col, India, 35-37; asst theoret physics, Univ Lwow, Poland, 37-38 & Univ Warsaw, 38-39; asst physics, Univ Munich, 44-45, lectr, 45-47, asst theoret physics, 47-48; asst prof physics, Col Steubenville, 48-49; from asst prof to prof, 49-82, chmn physics dept, 68-72, EMER PROF PHYSICS, PROVIDENCE COL, 82- *Concurrent Pos:* Vis prof, Univ RI, 62-63; consult, US Army Missile Command, 63-70; vis scientist, Max Planck Inst for Physics & Astrophys, 72. *Mem:* Am Phys Soc; Europ Phys Soc. *Res:* Classical and quantum theory of radiation; molecular spectroscopy; cosmology. *Mailing Add:* Physics Dept Providence Col Providence RI 02918

GORA, THADDEUS F, JR, b Elizabeth, NJ, Nov 16, 41; m 66; c 1. THEORETICAL SOLID STATE PHYSICS. *Educ:* Rensselaer Polytech Inst, BS, 63; Univ Del, PhD(solid state physics), 68. *Prof Exp:* Physicist, US Army, 68-88, chief, Elec Armaments Div, 89-90, SR RES SCIENTIST, ELECTROMECH, HYPERVELOCITY, & PULSE POWER, US ARMY RES & DEVELOP CTR, PICTANNY ARSENAL, 91- *Concurrent Pos:* Mem, Nat Adv Panel Electromagnetic Launch; mem tech comt, Pulse Power Conf, Inst Elec & Electronics Engrs. *Honors & Awards:* Army Res & Develop Award, US Army, 70. *Mem:* Am Phys Soc. *Res:* Band structure of psuedostable solids; x-ray photoelectron spectroscopy; graded band structure in heterogeneous materials; field effects in reactive solids; physics of perception; physics of electromagnetic and electrothermal launchers; pulse power physics. *Mailing Add:* 41 Crane Rd Mountain Lakes NJ 07046

GORADIA, CHANDRA P, b Rajula, India, Aug 6, 39; m 69. ELECTRICAL ENGINEERING, PHYSICS. *Educ:* Univ Bombay, BSc, 60, MSc, 62; Univ Okla, MEE, 64, PhD(elec eng), 68. *Prof Exp:* Asst res scientist, Biophys Res Lab, Tulsa Div, Avco Corp, 64-65; asst prof elec eng, 67-71, ASSOC PROF ELEC ENG, CLEVELAND STATE UNIV, 71- *Mem:* AAAS; Inst Elec & Electronics Engrs; Sigma Xi. *Res:* Semiconductor surfaces; surface barrier nuclear radiation detectors. *Mailing Add:* Dept Elec Eng Cleveland State Univ Cleveland OH 44115

GORAN, MICHAEL I, b Glasgow, Scotland, June 19, 61. MEDICINE. *Educ:* Univ Manchester, UK, BSc Hons, 82, PhD, 86. *Prof Exp:* Res asst, Dept Pharmacol, Hebrew Univ, Israel, 82-83; res fel, Dept Biochem, Univ Bath, UK, 86-87 & Metab Unit, Shriners Burn Inst, Tex, 87-89; RES ASST PROF, DEPT MED, DIV ENDOCRINOL, METAB & NUTRIT, UNIV VT, BURLINGTON, 89- *Concurrent Pos:* Prod biochemist, Makor Chemicals, Jerusalem, Israel, 82-83; reserve instr chem, Galveston Community Col, Tex, 88-89; instr, Div Human Nutrit, Med Br, Univ Tex, Galveston, 88-89; clin nutrit fel award, Am Soc Clin Nutrit, 90. *Mem:* Am Soc Clin Nutrit; Am Inst Nutrit. *Mailing Add:* Div Endocrinol Metab & Nutrit Univ Vt Dept Med Burlington VT 05405

GORAN, MORRIS, b Chicago, Ill, Sept 4, 16; m 51; c 2. CHEMISTRY. *Educ:* Univ Chicago, BS, 36, MS, 39, PhD(sci educ), 57. *Prof Exp:* Chemist, Dearborn Chem Co, Ill, 41-43; instr physics, Ind Univ, 43-44; assoc prof chem, 46-58, PROF PHYS SCI, ROOSEVELT UNIV, 58-, CHMN DEPT, 46- *Concurrent Pos:* Lectr, George Williams Col, 51-53; res chemist, Manhattan Proj, Oak Ridge, Tenn; sr consult, Ctr Sci Anomaly Res. *Mem:* AAAS; Am Chem Soc; Am Phys Soc; Hist Sci Soc; Am Inst Chemists; Am Asn Physics Teachers. *Res:* Natural science history and methods; science education. *Mailing Add:* 7330 N Kilbourne Ave Chicago IL 60646

GORAN, ROBERT CHARLES, b St Louis, Mo, Feb 7, 17; m 47; c 2. STRUCTURAL ENGINEERING, MATERIALS SCIENCE. *Educ:* Wash Univ, BS, 37. *Prof Exp:* Chem engr, Titanium Div, Nat Lead Co, 37-40 & Fraser-Brace Eng Co, 41-43; sr supvr, Trojan Powder Co, 43-44; stress anal engr, Consol-Vultee Aircraft Corp, 44-45; stress engr, McDonnell Douglas Corp, 45-51, sr engr & supvr struct design & anal, 51-55, engr & supvr proj strength, 55-60, chief strength engr & dept head, 60-70, dir struct technol, 70-77, dir eng technol, struct & mat, 77-85; RETIRED. *Concurrent Pos:* Mem res adv comt, Space Vehicle Struct, NASA, 64-67, aircraft struct, 69-71; mem indust adv group, US Air Force Aircraft Struct Integrity Prog, 68-; mem spec staff to chief engr, Assess Shuttle Flight Cert Prog, NASA, 79; consult aerospace struct & mat, 85- *Honors & Awards:* Structures Design Award, Am Inst Aeronaut & Astronaut, 76. *Mem:* Assoc fel Am Inst Aeronaut & Astronaut. *Res:* Structural design, analysis and development of aerospace vehicles. *Mailing Add:* Four Georgian Acres St Louis MO 63131

GORANSON, EDWIN ALEXANDER, b New Westminister, BC, Nov 12, 04; m 35; c 3. EXPLORATION GEOLOGY, SOLID MINERALS. *Educ:* Univ BC, BASc, 28; Harvard Univ, AM, 30, PhD(geol), 33. *Prof Exp:* From asst prof to prof, E S Larsen, Harvard UNIV, 29, instr mining geol, 30-33; geologist, Labrador Explor Co, 33; geologist & engr, Base Metals Mining Corp, 33-37; geologist, E Malartic Mines, Ltd, Can, 37-43 & J P Norrie, Ltd, 43-45; consult geologist, 45-49; geologist, NJ Zinc Explor, Ltd, 49-50 & NJ Zinc Explor Co Can Ltd, 50-70, consult, 70-80; RETIRED. *Mem:* Fel Soc Econ Geologists; Am Inst Mining, Metall & Petrol Engrs; fel Geol Soc Can; Can Inst Mining & Metall. *Mailing Add:* 66 Billings Ave Ottawa ON K1H 5K7 Can

GORANSON, H T, b Annapolis, Md, Apr 3, 47; m 70; c 1. UNIFIED MODELS, OPEN SYSTEMS. *Educ:* Mass Inst Technol, BSEE, 69, BS, 70, BSAD, 71. *Prof Exp:* Sr scientist, Arcturus, Inc, 69-81 & Sirius Inc, 81-88; SR SCIENTIST, SCI APPLICATIONS INC, 88- *Concurrent Pos:* Govt tech adv, Sematech; engr, Enterprise Integration; US agent, Esprit Collab, Dept Defense. *Mem:* Am Asn Artificial Intel; Asn Comput Mach; Computer Professionals for Social Responsibility; Inst Elec & Electronics Engrs Computer Soc; Soc Logistics Engrs. *Res:* National government and industry program in models and integrated model services; morphology of perception. *Mailing Add:* 1976 Munden Pt Virginia Beach VA 23457-1227

GORBACH, SHERWOOD LESLIE, b Hartford, Conn, Oct 25, 34; m; c 3. INFECTIOUS DISEASES, INTERNAL MEDICINE. *Educ:* Brandeis Univ, BA, 55; Tufts Univ, MD, 62; London Sch Hyg & Trop Med, dipl appl parasitol & entom, 67; Am Bd Internal Med, dipl, 72. *Prof Exp:* Intern, Cornell Univ 2nd Med Div, Bellevue Hosp, 63, asst resident med, 64; vis scientist bact & gastroenterol, Postgrad Med Sch, Hammersmith Hosp, London, 66-67; instr med & pathobiol, Sch Med, Johns Hopkins Univ, 67-69; assoc prof med & microbiol, Univ Ill Col Med & dir dept parasitol, Univ Ill Hosp, 69-72; prof med, Sch Med, Univ Calif, Los Angeles, 72-75, prof microbiol & immunol, 73-75; CHIEF, INFECTIOUS DIS SECT, NEW ENG MED CTR HOSP & PROF MED, SCH MED, TUFTS UNIV, 75- *Concurrent Pos:* Vis staff mem, Infectious Dis Hosp & Sch Trop Med, Calcutta, India, 66-67; chief dept infectious dis, Cook County Hosp, Chicago, 70-72; chief infectious dis sect, Vet Admin Hosp, Sepulveda, Calif, 72-75. *Mem:* Am Soc Microbiol; Am Gastroenterol Asn; Am Soc Clin Invest; Am Fedn Clin Res; Infectious Dis Soc Am. *Res:* Anaerobes in intraabdominal sepsis; enterotoxin of Escherichia coli in diarrhea; cholera-like toxigenic coliform diarrhea in children; intestinal bacteria in acute diarrhea; Escherichia coli isolated from food; mucosal bacteria in colonic cancer. *Mailing Add:* New England Med Ctr Hosps Boston MA 02111

GORBATKIN, STEVEN M, b Chicago, Ill, July 13, 60. MATERIALS SCIENCE ENGINEERING, METALLURGY. *Educ:* Univ Ill, BS, 81, PhD(mat sci), 87. *Prof Exp:* Res asst, Coord Sci Lab, Univ Ill, 81-87, instr thermodyn, 84-85; Eugene Wigner fel, 87-89, RES STAFF, OAK RIDGE NAT LAB, 89- *Concurrent Pos:* Prin investr, Sematech, Joint Advan Etch Prog, Oak Ridge Nat Lab, 89-91. *Mem:* Am Vacuum Soc; Mat Res Soc. *Res:* Thin film plasma processing, including deposition and etching using electron cyclotron resonance microwave plasmas. *Mailing Add:* Solid State Div Oak Ridge Nat Lab PO Box 2008 MS-6057 Oak Ridge TN 37831-6057

GORBATSEVICH, SERGE N, b Poland, Mar 15, 22; US citizen; div. PULP & PAPER TECHNOLOGY. *Educ:* State Univ NY Col Forestry, Syracuse, BS, 54, MS, 55. *Prof Exp:* Res chemist, Consol Water Power & Paper Co, Wis, 55-56; from instr to asst prof pulp & paper technol, 56-65, ASSOC PROF PAPER SCI & ENG, STATE UNIV NY COL ENVIRON SCI & FORESTRY, 65- *Mem:* Tech Asn Pulp & Paper Indust. *Res:* Pulping; bleaching; paper properties. *Mailing Add:* 209 Arnold Ave Syracuse NY 13210

GORBATY, MARTIN LEO, b Brooklyn, NY, Nov 17, 42; m 68; c 3. COAL SCIENCE, PETROLEUM CHEMISTRY. *Educ:* City Col NY, BS, 64; Purdue Univ, PhD(org chem), 69. *Prof Exp:* Res chemist org synthesis, 69-72, sr res chemist, polymer sci, 72-75, group head, coal sci, 75-78, lab dir heavy hydrocarbons, 78-84, SCI COORDR, EXXON, 84- *Concurrent Pos:* Chmn petrol chem div, Am Chem Soc, 83-84; coun petrol chem div, Am Chem Soc, 88-; chmn, Gordon Conf Fuel Sci, 88. *Mem:* AAAS; Sigma Xi; NY Acad Sci; Am Chem Soc. *Res:* Fundamental chemical and physical structures of refractory hydrocarbons including heavy oils, shale and coal; the relationships between structure and reactivity, and their conversion to clean usable fuels and chemicals. *Mailing Add:* Exxon Res & Eng Co PO Box 998 Annadale NJ 08801-0998

GORBET, DANIEL WAYNE, b Corpus Christi, Tex, Oct 16, 42; m 62; c 2. AGRONOMY, PLANT BREEDING. *Educ:* Tex A&I Univ, BS, 65; Okla State Univ, MS, 68, PhD(crop sci), 71. *Prof Exp:* From asst prof to assoc prof, 70-84, PROF AGRON AGR RES & EDUC CTR, UNIV FLA, 84- *Concurrent Pos:* NSF fel. *Mem:* Am Soc Agron; Crop Sci Soc Am; Am Peanut Res & Educ Soc (pres, 87-88). *Res:* Sorghum breeding and genetics; breeding, genetics, management and variety development of peanuts; sorghum and soybean variety testing. *Mailing Add:* N Fla Res & Educ Ctr 3925 Hwy 71 Marianna FL 32446

GORBMAN, AUBREY, b Detroit, Mich, Dec 13, 14; m 38; c 4. ZOOLOGY. *Educ:* Wayne Univ, AB, 35, MS, 36; Univ Calif, PhD(zool), 40. *Prof Exp:* Asst zool, Univ Calif, 36-40, res assoc Inst Exp Biol, 40-41; instr biol, Wayne Univ, 41-44; Childs fel, Yale Univ, 44-46; from asst prof to prof zool, Barnard Col, Columbia Univ, 46-63, exec officer dept univ, 52-55, col, 57-60; prof & chmn, Dept Zool, Univ Wash, 63-66; RETIRED. *Concurrent Pos:* Fulbright scholar, Col of France, 51-52; res collabr, Brookhaven Nat Lab, 52-60; Guggenheim fel, Univ Hawaii, 55-56; vis prof, Nagoya Univ, 56; USPHS fel & vis prof, Univ Tokyo, 60; ed-in-chief, Gen & Comp Endocrinol, 60-; mem, Endocrinol Study Sect, NIH, 68-72 & Regulatory Biol Panel, NSF, 74-77; vis prof biol, Tokyo Univ & Toho Univ, Japan, 84; James chmn, Biol Dept, St Francis Xavier Univ, Nova Scotia, Can, 85; vis prof zool, Univ Alberta, 86. *Mem:* Am Soc Naturalists; Soc Exp Biol & Med; Am Soc Ichthyologists & Herpetologists; Am Soc Zoologists (pres, 77); fel NY Acad Sci. *Res:* Comparative endocrinology; actions of hormones on nervous system. *Mailing Add:* 4218 55th Ave NE Seattle WA 98105

GORBSKY, GARY JAMES, b Jan 2, 55; m; c 2. CELL BIOLOGY, MYTOSIS. *Educ:* Princeton Univ, PhD(biol), 82. *Prof Exp:* RES ASSOC MOLECULAR BIOL, UNIV WIS, 83- *Mem:* Am Soc Cell Biol. *Res:* Cell junctions, mitosis, cell cycle. *Mailing Add:* Box 439 Health Sci Ctr Univ Va Charlottesville VA 22908

GORBUNOFF, MARINA J, b Moscow, Russia, May 25, 27; US citizen; m 53; c 1. ORGANIC CHEMISTRY. *Educ:* Syracuse Univ, BS, 51; Yale Univ, MS, 54, PhD(chem), 56. *Prof Exp:* Res fel org chem, Bryn Mawr Col, 56-57 & Univ Pa, 57-59; res chemist, Eastern Regional Res Lab, Agr Res Serv, USDA, 61-67; SR RES ASSOC BIOCHEM, BRANDEIS UNIV, 67- *Mem:* Am Chem Soc; Am Soc Biol Chemists; Sigma Xi. *Res:* Structure of proteins; chemical modification of active groups on proteins; synthetic organic chemistry. *Mailing Add:* Grad Dept Biochem Brandeis Univ Waltham MA 02254

GORCHOV, DAVID LOUIS, b Philadelphia, Pa, July 12, 58. PLANT ECOLOGY, SEED DISPERSAL. *Educ:* Princeton Univ, AB, 80; Univ Mich, MS, 81, PhD(biol), 87. *Prof Exp:* Vis asst prof biol, Rutgers Univ, 87-88; res assoc, Princeton Univ, 88-89; ASST PROF BOT, MIAMI UNIV, 90- *Mem:* Ecol Soc Am; Bot Soc Am; Asn Trop Biol; Am Soc Naturalists. *Res:* Natural regeneration of tropical rain forest; seed dispersal; ecological interactions between plants and their vertebrate seed dispersal agents; population biology of economic plants. *Mailing Add:* Dept Bot Miami Univ Oxford OH 45056

GORCZYNSKI, REGINALD MEICZYSLAW, b Plymouth, Eng, Oct 23, 47; m 70; c 2. IMMUNOLOGY, CELL BIOLOGY. *Educ:* Oxford Univ, BA, 69; Toronto Univ, PhD(immunol), 72. *Prof Exp:* Fel immunol & virol, Imperial Cancer Res Fund, London, 72-74; MEM RES STAFF IMMUNOL & CELL BIOL, ONT CANCER INST, 74- *Concurrent Pos:* Can Med Res Coun res grant, 74-80, Can Nat Cancer Inst, 74-79; consult, Cedarlove Labs, London, Ont, 76- *Mem:* Europ Asn Cancer Res; Can Soc Immunol; Am Soc Immunol. *Res:* Self-non-self discrimination or tolerace as studied in neonates; role of response to fetal antigens in neoplasia; regulation of immune responses by antigen presenting cells. *Mailing Add:* Dept Immunol & Surg Clin Sci Div Rm 7352 Med Sci Bldg Univ Toronto Toronto ON M5S 1A8 Can

GORDAN, GILBERT SAUL, b San Francisco, Calif, July 8, 16; m 78. MEDICINE. *Educ:* Univ Calif, AB, 37, MD, 41, PhD(endocrinol), 47. *Prof Exp:* Intern, Univ Hosp, 40-41, asst resident med, Med Sch, 41-42, clin instr, 45-47, from instr to asst prof med & pharmacol, 48-54, from assoc prof to prof, 54-85, EMER PROF MED, MED SCH, UNIV CALIF, SAN FRANCISCO, 85-; RETIRED. *Concurrent Pos:* Commonwealth fel, Harvard Med Sch, 47-48; ed, Yr Bk Endocrinol, 51-63; consult, Cancer Chemother Nat Serv Ctr, USPHS, 57-59, chmn coop breast cancer group, 56-59, mem & spec consult hormone evaluation joint comt clin & endocrinol panels, 57-69; Commonwealth Fund fel, Col Physicians & Surgeons, Columbia Univ, 62-63 & Strangeways Lab, Cambridge Univ; consult, Manned Orbiting Lab Proj, US Air Force, Aerospace Corp, 64-; Guggenheim fel & vis prof, Makerere Univ, Univ Athens & Hebrew Univ Jerusalem, 67-68; William Beaumont vis prof, Sch Med, Wash Univ, 70; Lady Davis vis prof, Hebrew Univ, 78; assoc chief staff educ, VAMC, Martinez, Calif, 85-88. *Honors & Awards:* Arnold O Beckman lectr, 79. *Mem:* Fel AAAS; Am Soc Clin Invest; Endocrine Soc; hon mem Royal Soc Med; fel Am Col Physicians; Sigma Xi. *Res:* Clinical and experimental endocrinology; calcium metabolism; bone physiology; parathyroid function. *Mailing Add:* 1100 West K St Benicia CA 94510

GORDEE, ROBERT STOUFFER, b Chicago, Ill, June 12, 32; m 61; c 2. MICROBIOLOGY, CHEMOTHERAPY. *Educ:* Mich State Univ, BS, 54; Purdue Univ, MS, 59, PhD(mycol, biochem), 61. *Prof Exp:* Fisheries biologist, Great Lakes Fishery Invest, US Fish & Wildlife Serv, 54; res scientist, Biophys-Res Div, Lockheed-Calif Co, 61-64; sr microbiologist, Biol-Pharamacol Res Div, 64-68, res scientist, Biol Res Div, 69-71, head exp chemother, Res Labs, 71-76, head, Microbiol & Fermentation Prods Res, 76-79, RES ASSOC, ELI LILLY & CO, 79- *Concurrent Pos:* Vis scholar, Univ Mich, 62. *Mem:* AAAS; Am Soc Microbiol; NY Acad Sci. *Res:* Antimicrobial agents; chemotherapy of infectious diseases. *Mailing Add:* 5166 E 74 Pl Indianapolis IN 46240

GORDEN, BERNER J, b Detroit, Mich, Jan 22, 39; m 62; c 2. ORGANIC CHEMISTRY. *Educ:* Luther Col, Iowa, BA, 61; Wayne State Univ, MS, 64, PhD(org chem), 65. *Prof Exp:* Asst prof chem, Muskingum Col, 65-67; chmn dept chem, 67-74, from asst prof to assoc prof, 67-75, PROF CHEM, SAGINAW VALLEY STATE UNIV, 75- *Concurrent Pos:* Instr, Dow Forman's Acad, 74-76; Univ Hawaii res grant, 75; sabbaticals, Mich State Univ, 80 & Univ Ga, 81. *Mem:* Am Chem Soc; Am Inst Chemists; Sigma Xi. *Res:* Conformational analysis; organic synthesis; mechanisms and reagents in organic reactions; heats of hydrogenation. *Mailing Add:* Dept Chem Saginaw Valley State Univ University Center MI 48710

GORDEN, PHILLIP, b Baldwyn, Miss, Dec 22, 34; m 59; c 2. ENDOCRINOLOGY & METABOLISM, INTERNAL MEDICINE. *Educ:* Vanderbilt Univ, BA, 57, MD, 61; Am Bd Internal Med, dipl; Am Bd Endocrinol & Metab, dipl. *Hon Degrees:* DMed, Univ Geneva, 86. *Prof Exp:* Intern, Yale Univ, 61-62, asst resident, 62-64, USPHS clin fel metab, 64-65, USPHS res fel metab, 65-66; sr investr, Clin Endocrinol Br, NIAMDD, NIH, 66-74, Diabetes Br, NIADDK, 74-78, clin dir, Nat Inst Arthritis, Metab & Digestive Dis, 74-76; vis prof, Sch Med, Univ Geneva, 76-78; clin dir, 80-89, chief, Diabetes Br, 83-89, CHIEF, SECT CLIN & CELLULAR BIOL, DIABETES BR, NAT INST ARTHRITIS, DIABETES & DIGESTIVE & KIDNEY DIS, NIH, 78-, DIR, 86- *Concurrent Pos:* Vis prof, Univ Geneva, 78-, Nat Univ La Plata, Arg, 80, Univ Mich Sch Med, 87, Sch Med, Univ Calif-Los Angeles, 89 & Univ Tenn Col Med, 90; attend physician, Howard Univ Sch Med, Wash, DC, 70-76, clin assoc prof, 78-; clin prof med, Uniformed Serv Med Sch, Bethesda, Md, 78-; plenary lectr, 78-89; Thomas Amatruda vis prof, Yale affil Waterbury Hosp, 90; chmn, Med Bd Clin Ctr, NIH, 75-76, dir, Endocrine-Metab Student Elective Prog, 72-76; attend physician, Nat Naval Med Ctr, 79-; mem, Nat Comt Res, Am Diabetes Asn, 80-83, Nat Comn Orphan Dis, 87-89. *Honors & Awards:* Distinguished Serv Medal, USPHS, 86 & Commendation Medal, 88; Nesbitt Lectr, Univ Minn Sch Med, Minneapolis, 83; James Grant Thompson Mem Lectr, Am Diabetes Asn, Chicago, 88; Alexander Marble Lectr, Joslin Diabetes Ctr, Boston, 90. *Mem:* Fel Am Col Physicians; Am Soc Clin Invest; Endocrine Soc; Asn Am Physicians; Am Soc Biochem & Molecular Biol; Am Diabetes Asn; Am Fedn Clin Res; Am Soc Cell Biol. *Res:* Clinical studies of diabetes and related endocrine disorders; biochemical and cell biological approaches to the study of the insulin receptor and insulin action in normal and pathologic states. *Mailing Add:* Nat Inst Health Bldg 31 Rm 9A-52 Bethesda MD 20892

GORDEN, ROBERT WAYNE, b LaPorte Co, Ind, Aug 22, 32; m 54, 84; c 4. MICROBIOLOGY, ECOLOGY. *Educ:* Manchester Col, BS, 57; Univ Ga, MS, 62, PhD(microbiol), 67. *Prof Exp:* Teacher-coach high sch, Ind, 57-61; instr biol, Manchester Col, 62-64; asst prof microbiol, Tex Tech Univ, 67-70; prof biol, Univ Southern Colo, 70-86; INHS, CHAMPAIGN, ILL, 86- *Mem:* Ecol Soc Am; Am Soc Microbiol. *Res:* Ecology of microorganisms aquatic; ecosystems. *Mailing Add:* State Natural Hist Survey 607 E Peabody Champaign IL 61820

GORDH, GEORGE RUDOLPH, JR, b Macon, Ga, May 12, 44; m 79. GENERAL MATHEMATICS. *Educ:* Guilford Col, BA, 66; Univ Calif, Riverside, PhD(topology), 71. *Prof Exp:* Asst, Univ Calif, Riverside, 71; fel math, Univ Ky, 71-72, vis asst prof, 72-73; from asst prof to assoc prof, 74-86, PROF MATH, GUILFORD COL, 86- *Concurrent Pos:* Exchange scientist, Nat Acad Sci, Yugoslavia, 73-74. *Mem:* Math Asn Am; Am Math Soc; Sigma Xi. *Res:* Continua; ordered spaces; upper semicontinuous decompositions; inverse limits; shape theory. *Mailing Add:* Dept Math Guilford Col Greensboro NC 27410

GORDH, GORDON, b Augusta, Ga, Jan 7, 45. ENTOMOLOGY, SYSTEMATICS. *Educ:* Univ Colo, BA, 67; Univ Kans, MA, 72; Univ Calif, Riverside, PhD(entom), 74. *Prof Exp:* res entomologist, syst entom lab, US Nat Mus, 74-77; from asst prof to assoc prof, 77-85, PROF ENTOM, UNIV CALIF, RIVERSIDE, 85- *Concurrent Pos:* Asst adj prof entom, NC State Univ, 75-76; res assoc, Georgetown Univ, 75-76; adj asst prof, Univ Md, 74-77. *Mem:* Royal Entom Soc London; Entom Soc Can; Sigma Xi; Entom Soc SAfrica; Soc Study Evolution. *Res:* Systematics, biology, behavior and evolution of the parasitic Hymenoptera belonging to the super-family Chalcidoidea. *Mailing Add:* Dept Entom Univ Calif Riverside CA 92521

GORDIS, ENOCH, b Feb 21, 31. ALCOHOL ABUSE. *Educ:* Columbia Univ, BA, 50, MD, 54. *Prof Exp:* From asst prof to assoc prof med, Rockefeller Univ, 63-71; assoc prof clin med, 71-85, ASST PHYSICIAN & DIR ALCOHOLISM PROG, MT SINAI HOSP, 86-; DIR, NAT INST ALCOHOL ABUSE & ALCOHOLISM, 86- *Concurrent Pos:* Adj prof & vis physician, Rockefeller Univ, NY, 71-; prog chmn, Nat Meeting Alcoholism, Am Med Soc, New Orleans, 81, Washington, DC, 82, Houston, Tex, 83; mem, treatment & prev study sect, Nat Inst Alcohol Abuse & Alcoholism, 85-86; corresp, comt human rights, Inst Med-Nat Acad Sci, 88-89. *Mem:* Inst Med-Nat Acad Sci; Am Col Neuropsychopharmacol; Am Fedn Clin Res; Am Gastroenterol Asn; Am Soc Addiction Med; Am Physiol Soc; fel Am Col Physicians; Int Soc Biomed Res Alcoholism; Sigma Xi. *Res:* Author or co-author of over 40 publications. *Mailing Add:* Nat Inst Alcohol Abuse & Alcoholism 5600 Fishers Lane Rockville MD 20857

GORDIS, LEON, b New York, NY, July 19, 34; m 55; c 3. PEDIATRICS, EPIDEMIOLOGY. *Educ:* Columbia Col, BA, 54; State Univ NY Downstate Med Ctr, MD, 58; Johns Hopkins Univ, MPH, 66, DrPH, 68. *Prof Exp:* Chief dept community med, Sinai Hosp, Baltimore, 68-69; vis prof med ecol, Hebrew Univ Jerusalem, 69-71; ASSOC PROF PEDIAT, SCH MED, JOHNS HOPKINS UNIV, 72-, PROF EPIDEMIOL & CHMN DEPT, SCH HYG & PUB HEALTH, 73- *Honors & Awards:* T Dukette Jones Lectr, Am Heart Asn, 84. *Mem:* Fel Am Acad Pediat; Am Heart Asn; Soc Epidemiol Res (pres, 79-80); Am Epidemiol Soc; Soc Pediat Res. *Res:* Epidemiology of chronic diseases; evaluation of health care. *Mailing Add:* Dept Epidemiol Sch Hyg & Pub Health Johns Hopkins Univ Baltimore MD 21205

GORDON, ADRIENNE SUE, NEUROBIOLOGY, MEMBRANE BIOCHEMISTRY. *Educ:* Mass Inst Technol, PhD(chem), 66. *Prof Exp:* ADJ PROF PHARMACOL & NEUROL, MED SCH, UNIV CALIF, SAN FRANCISCO, 74- *Mailing Add:* Dept Pharm/Neurol Bldg One Rm 101 Univ Calif Sch Med San Francisco Gen Hosp San Francisco CA 94110

GORDON, ALBERT MCCAGUE, b Tanta, Egypt, Sept 23, 34; US citizen; m 78; c 2. MUSCULAR PHYSIOLOGY, BIOPHYSICS. *Educ:* Univ Rochester, BS, 56; Cornell Univ, PhD(solid state physics), 61. *Prof Exp:* Nat Found fel, Univ Col, Univ London, 60-62; fel biophys, 62-64, from instr to assoc prof physiol & biophys, 64-75, PROF PHYSIOL & BIOPHYS, UNIV WASH, 75- *Concurrent Pos:* Jacob Javits Neuroscience Investr Award.. *Mem:* AAAS; Biophys Soc; Am Physiol Soc. *Res:* Mechanical properties of muscle and the contractile mechanism; muscle pathophysiology; control of contraction in muscle. *Mailing Add:* Dept Physiol & Biophys Univ Wash SJ-40 Seattle WA 98195

GORDON, ALBERT RAYE, b McKeesport, Pa, Sept 22, 39; m 65; c 2. CELL BIOLOGY, ELECTRON MICROSCOPY. *Educ:* Western Reserve Univ, BA, 61, MA, 67, Case Western Reserve Univ, PhD(biol), 68. *Prof Exp:* Fel biol, Case Western Reserve Univ, 67-68; asst prof, Knox Col, Ill, 68-69; asst prof, 69-74, assoc prof, 74-78, PROF LIFE SCI, SOUTHWEST MO STATE UNIV, 78- *Mem:* AAAS; Am Soc Cell Biol; Electron Micros Soc Am; Am Soc Plant Physiol; Am Inst Biol Sci; Sigma Xi. *Res:* Significance and functions of multiple enzyme systems; cellular and ultrastructural localization of peroxidase isoenzymes and their relationship to plant tissue differentiation and function. *Mailing Add:* Biomed Sci Dept Southwest Mo State Univ Springfield MO 65804

GORDON, ALBERT SAUL, b Brooklyn, NY, Aug 8, 10; m 35; c 1. PHYSIOLOGY. *Educ:* City Col New York, BS, 30; NY Univ, MS, 31, PhD(physiol), 34. *Hon Degrees:* DSc, Adelphi Univ, 81. *Prof Exp:* From asst instr to prof, 31-75, RESEARCHER & EMER PROF BIOL, GRAND SCH ARTS & SCI, NY UNIV, 75- *Concurrent Pos:* Instr, Newark Inst Arts & Sci, 34-35; fel, Dazian Found Med Res, 49-51; mem study sect hemat, NIH, 58-62 & 63-65, mem erythropoietin comt, Nat Heart Inst, 64-71, NY State Health Res Coun, 76-; sci adv, New Eng Inst Med Res, 59-70; grants, Am Cancer Soc, Damon Runyon Mem Fund Cancer Res, NSF, Commonwealth Fund, US Air Force, NY State Health Res Coun, NIH; ed-in-chief, J Reticuloendothelial Soc, 64-66. *Honors & Awards:* Morrison Prize, NY Acad Sci, 48 & 66; Dameshek Award, Am Soc Hemat; Stratton Lectr. *Mem:* Am Soc Zoologists; Am Physiol Soc; Soc Exp Biol & Med; fel NY Acad Sci; fel Royal Soc Med. *Res:* Red and white blood cell formation release and destruction; physiology of reticuloendothelial system; endocrine interrelations; humoral regulation of hematopoiesis. *Mailing Add:* A S Gordon Lab Exp Hemat NY Univ 100 Washington Sq E New York NY 10003

GORDON, ALVIN S, b New York, NY, Oct 25, 14; m 42. PHYSICAL CHEMISTRY. *Educ:* Polytech Inst Brooklyn, BS, 37; NY Univ, PhD(phys chem), 41. *Prof Exp:* Asst phys chemist, Boyce Thompson Inst, 41-42; res assoc, Nat Defense Res Comt, Carnegie Inst Technol, 42-45; phys chemist, US Bur Mines, 45-51; phys chemist, Naval Weapons Ctr, 51-75, sci liaison officer, Off Naval Res, London, Eng, 57-59; adj prof eng chem, Univ Calif San Diego, 76-82; RETIRED. *Concurrent Pos:* Vis sr res scientist, Princeton Univ, 64-65; mem Calif Air Resources Bd, 76-82. *Mem:* Am Chem Soc; fel Am Phys Soc. *Res:* Kinetics and mechanism of reactions. *Mailing Add:* Dept Appl Mech & Eng Sci Univ Calif San Diego AMES Mail Stop B010 La Jolla CA 92093

GORDON, ANNETTE WATERS, b Sylvania, Ga, Aug 9, 37; m 62; c 1. ORGANIC CHEMISTRY, BIOCHEMISTRY. *Educ:* Duke Univ, BS, 59; Vanderbilt Univ, PhD(org chem), 68. *Prof Exp:* Asst prof chem, 62-68, assoc prof, 68-78, PROF CHEM, MURRAY STATE UNIV, 78- *Mem:* Am Chem Soc. *Res:* Stereochemistry. *Mailing Add:* 1515 South Fairway Springfield MO 65804

GORDON, ARNOLD J, b Boston, Mass, Dec 13, 37; m 64; c 2. PUBLIC HEALTH & EPIDEMIOLOGY. *Educ:* Northeastern Univ, BSc, 60; NY Univ, MS, 62, PhD(org chem), 64. *Prof Exp:* Instr gen & org chem, NY Univ, 64; res chemist, US Army Edgewood Arsenal, 65-66; asst prof chem, Cath

Univ Am, 66-71; group assoc dir, Sci Affairs, Pfizer Pharmaceut, 71-75, dir sci affairs, 75-86, EXEC DIR SCI AFFAIRS & SAFETY ASSURANCE, PFIZER TNT, 86- *Concurrent Pos:* Res grants, Cath Univ, 67 & Res Corp, Am Chem Soc Petrol Res Fund & NSF, 67-71; manuscript referee, J Org Chem, 67- & Chem Rev, 68; Am Chem Soc-Du Pont small grants award, 68. *Mem:* Am Chem Soc; AAAS; NY Acad Sci; Am Inst Chemists; Sigma Xi (pres, 70-71); Drug Info Asn; Asn Res Dirs (vpres, 86-87, pres, 87-88). *Res:* Chemistry and biochemistry of drugs; stereochemistry of di-coordinated oxygen; synthesis and properties of imides and other nitrogen compounds; chemical education; new drug development and research; practical data, techniques and references for chemists; clinical research on new drugs; preclinical pharmacology and toxicology research; clinical data storage and analysis. *Mailing Add:* Pfizer Int Inc 235 E 42nd St New York NY 10017

GORDON, ARNOLD L, b New York, NY, Feb 4, 40; m 70; c 4. PHYSICAL OCEANOGRAPHY. *Educ:* Hunter Col, BA, 61; Columbia Univ, PhD(oceanog), 65. *Prof Exp:* Res asst, 61-65, assoc prof, 71-77, RES ASSOC OCEANOG, LAMONT-DOHERTY GEOL OBSERV, COLUMBIA UNIV, 65-, PROF, 77- *Concurrent Pos:* Ford Found fel, 61-65, NSF res grant, 65-; asst prof oceanog, Columbia Univ, 66-71. *Honors & Awards:* Henry Bigelow Medal, Woods Hole Oceanog Inst, 84. *Mem:* AAAS; Am Geophys Union; Sigma Xi. *Res:* Circulation of the Caribbean Sea; physical oceanographic investigations in the Southern Hemisphere oceans. *Mailing Add:* Lamont-Doherty Geol Observ Columbia Univ Palisades NY 10964

GORDON, BARRY MAXWELL, b Chicago, Ill, Jan 26, 30; m 58; c 2. PHYSICAL CHEMISTRY, NUCLEAR CHEMISTRY. *Educ:* Univ Calif, Los Angeles, BS, 51; Wash Univ, PhD(chem), 55. *Prof Exp:* Res assoc chem, Brookhaven Nat Lab, 55-57; asst prof, State Univ NY Stony Brook, 57-63; assoc chemist, 63-82, CHEMIST, BROOKHAVEN NAT LAB, 82- *Mem:* Am Chem Soc; Sigma Xi. *Res:* Kinetic studies of oxidation-reduction reactions; nuclear reactions; radiochemistry; trace-elemental analysis in medicine; applications of synchrotron radiation to chemical analysis. *Mailing Add:* Brookhaven Nat Lab Upton NY 11973

GORDON, BARRY MONROE, b New York, NY, July 15, 47; m 81; c 1. PHYSICAL METALLURGY, CORROSION. *Educ:* Carnegie-Mellon Univ, BS, 69, MS, 71; Nat Asn Corrosion Engrs, cert. *Prof Exp:* Assoc engr, Westinghouse Elec Corp, 69-70; teaching asst thermodynamics, Carnegie-Mellon Univ, 70-71; engr corrosion, Westinghouse Elec Corp, 71-75; prog mgr, Mat Corrosion Prog, 75-88, LEAD ENGR, CORROSION PERFORMANCE, GEN NUCLEAR ENERGY, 88- *Honors & Awards:* Waldemar Cancer Res Award, Waldemar Cancer Res Inst, 65. *Mem:* Nat Asn Corrosion Engrs; Am Soc Metals. *Res:* Mitigation of intergranular stress corrosion cracking; crevice corrosion and pitting of nickel-chromium-iron alloys stainless steels and ferritic alloys in high temperature agueous environments; corrosion control in power systems. *Mailing Add:* GE Nuclear Energy MC785 175 Curtner Ave San Jose CA 95125

GORDON, BENJAMIN EDWARD, b New York, NY, May 8, 16; m 40; c 3. ANALYTICAL CHEMISTRY, RADIOCHEMISTRY. *Educ:* Univ Ill, BS, 40, MS, 42. *Prof Exp:* Res chemist, Shell Oil Co, 42-54, sr technologist radiochem, 54-57, group leader radiochem & radiation chem, 57-61, supvr radiochem, Shell Develop Co, 61-71 & Koninklijke Shell Labs, 71-74; SUPVR RADIOCHEM, LAWRENCE RADIATION LAB, UNIV CALIF, BERKELEY, 74- *Concurrent Pos:* Mgr, Nat Tritium Labeling Facil. *Mem:* Am Chem Soc. *Res:* novel methods of labeling with tritium; gas chromatography; product and process research; reaction mechanism; kinetics; activation analysis; instrumentation; applying nuclear techniques to problems in environmental conservation; hot atom chemistry of Carbon-14; excitation labeling with tritium. *Mailing Add:* 1330 Brewster Dr El Cerrito CA 94530-2506

GORDON, BERNARD LUDWIG, b Westerly, RI, Nov 6, 31; m 59; c 2. MARINE BIOLOGY. *Educ:* Univ RI, BSc, 55, MSc, 58. *Prof Exp:* Instr biol, RI Col, 56-60; teaching fel, Boston Univ, 60-61; instr natural sci, 61-65, asst prof earth sci, 65-68, ASSOC PROF EARTH SCI, NORTHEASTERN UNIV, 69- *Concurrent Pos:* Consult, Mass State Dept Educ, 63 & Quincy Marine Sci Prog, Mass, 65-; mem bd dirs, Southern New Eng Marine Sci Asn, 65- Littauer Found grant, 66; consult, Wally Sea Prod Co, 67- & Oceanog Educ Ctr, Woods Hole Oceanog Inst, Bur Com Fisheries & Marine Biol Lab, 68-; pres of fels, Mystic Seaport Libr, 90-91. *Mem:* AAAS; Nat Asn Biol Teachers; Am Soc Limnol & Oceanog; Marine Technol Soc; Am Fisheries Soc; Explorers Club. *Res:* Marine ichthyology; parasitology; oceanography; marine sciences; fish surveys; life history studies; history of sciences; author of 16 science books. *Mailing Add:* Dept Earth Sci Northeastern Univ Boston MA 02115

GORDON, BERNARD M, b 1927. ANALOG-DIGITAL INTERFACE. *Prof Exp:* PRES, CHIEF EXEC OFFICER & CHMN BD, ANALOGIC CORP, 67- *Honors & Awards:* Nat Medal of Technol, 86. *Mem:* Nat Acad Eng; fel Inst Elec & Electronics Engrs. *Mailing Add:* Analogic Corp Eight Centennial Dr B-1 Peabody MA 01961

GORDON, BURTON LEROY, b Asotin, Wash, Feb 13, 20; m 41; c 4. BIOGEOGRAPHY. *Educ:* San Francisco State Col, AB, 42; Univ Calif, PhD(geog), 54. *Prof Exp:* Asst prof geog, Univ NMex, 55-60, assoc prof & chmn dept, 60-65; prof geog, Calif State Univ, San Francisco, 65-87; RETIRED. *Concurrent Pos:* Collabr, US Forest Serv, 49-54; consult, Jicarilla Apache Indian Tribe, 59-65; exchange lectr geog, Queen Mary Col, Univ London, 62-63. *Mem:* AAAS; Am Asn Univ Prof; NY Acad Sci. *Res:* Biogeography of central California coast and Western Europe. *Mailing Add:* 113 David Way San Cruz CA 95060

GORDON, CAROLYN SUE, b Charleston, WVa, Dec 26, 50. RIEMANNIAN GEOMETRY, LIE GROUPS. *Educ:* Purdue Univ, BS, 71, MS, 72; Wash Univ, PhD(math), 79. *Prof Exp:* Lady Davis fel, Technion-Israel Inst Technol, 79-80; ASST PROF MATH, LEHIGH UNIV, 80- *Mem:* Am Math Soc; Math Asn Am; Asn Women Math. *Mailing Add:* Dartmouth Col Hanover NH 03755

GORDON, CHESTER DUNCAN, b Ankerton, Alta, Oct 13, 20; nat US; m 50; c 3. ORGANIC CHEMISTRY. *Educ:* Univ Alta, BSc, 49; Univ Notre Dame, PhD(org chem), 52. *Prof Exp:* Res chemist, Calif Res Corp Div, Standard Oil Co Calif, 52-60, sr res chemist, 60-66, sr res assoc, Chevron Res Co, 66-84; RETIRED. *Concurrent Pos:* Consult. *Mem:* Am Chem Soc; Sigma Xi. *Res:* Organic synthesis; vinyl polymerizations; nitrogen chemistry; process development. *Mailing Add:* 2579 Patra Dr El Sobrante CA 94803-3627

GORDON, CHESTER MURRAY, b Brooklyn, NY, Mar 31, 18; m 42; c 2. MATHEMATICS, STATISTICS. *Educ:* City Col New York, BS, 38, MSEd, 43. *Prof Exp:* Indust specialist, AEC, 48-51; mobilization specialist, Off Defense Mobilization, Exec Off of Pres, 51-53; defense coordr & procurement asst, AEC, 53-71, chief property mgt br & relocation officer, 71-72, chief logistics staff, 73; consult, Dept Energy, 74; RETIRED. *Concurrent Pos:* Asst opers chief, Bur Census, 40-42; statist analyst, War Prod Bd, 42-48; mgt analyst, Dept Navy, 48. *Res:* Industrial mobilization; priorities; civil defense. *Mailing Add:* 3330 N Leisure World Blvd No 609 Silver Spring MD 20766

GORDON, CHRISTOPHER JOHN, b Ventura, Calif, Mar 24, 53; m 77; c 2. THERMAL PHYSIOLOGY, TOXICOLOGY. *Educ:* Univ Idaho, BS, 75, MS, 77; Univ Ill, Urbana, PhD(physiol), 80. *Prof Exp:* RES PHYSIOLOGIST, US ENVIRON PROTECTION AGENCY, 80- *Concurrent Pos:* Trainee physiol, NIH, 78; consult, Gen Elec. *Mem:* AAAS; Am Physiol Soc. *Res:* Mechanisms for the central neural control of body temperature; physiological effects of exposure to microwave radiation; pharmacology and toxicology of thermoregulation. *Mailing Add:* Neurotoxicol Div US Environ Protection Agency Mail Drop 74B Research Triangle Park NC 27711

GORDON, CLAIRE CATHERINE, b New Orleans, La, June 12, 54; m 79; c 1. ANTHROPOMETRY, HUMAN ENGINEERING. *Educ:* Univ Notre Dame, BS, 76; Northwestern Univ, Ba, 77, PhD(bioanthrop), 82; Harvard Univ, MS, 90. *Prof Exp:* Res assoc phys anthrop & archaeol, Pac Studies Inst, 80; res coordr human genetics, Northwestern Univ Med Sch, 82-83; res anthropologist, 83-87, group leader, 88-90, SR ANTHROPOLOGIST, US ARMY NATICK RES, DEVELOP & ENG CTR, 90- *Concurrent Pos:* Rep, Air Standardization Coord Comt, US Army, 85- *Mem:* Am Anthrop Asn; Am Asn Phys Anthropologists; Sigma Xi; Biomet Soc; Human Biol Coun. *Res:* Anthropometric instrumentation and methodology; age-sex-racial components of human physical variation; dimensional relationships between body parts; statistics of human engineering design, sizing and tariffing; human factors. *Mailing Add:* Attn STRNC-YB US Army Natick Res Develop & Eng Ctr Natick MA 01760-5020

GORDON, COURTNEY PARKS, b Bridgeport, Conn, Dec 3, 39; m 64; c 3. ASTRONOMY. *Educ:* Vassar Col, AB, 61; Univ Mich, MA, 63, PhD(astron), 67. *Prof Exp:* Res assoc radio astron, Nat Radio Astron Observ, Va, 67-69, asst scientist, 69-70; from asst prof to assoc prof astron, Hampshire Col, 70-85, asst dean, 73-74, assoc dean, 77-80; AT DEPT COMPUTER SCI, UNIV MASS, 85- *Mem:* Am Astron Soc; Int Astron Union; Danforth Found. *Res:* Spectroscopy of carbon stars; galactic neutral hydrogen; pulsars. *Mailing Add:* 480 Middle St Amherst MA 01002

GORDON, DANIEL ISRAEL, b Norwich, Conn, Aug 12, 20; m 43; c 3. ELECTROMAGNETISM, COMPUTER SCIENCES. *Educ:* Yale Univ, BE, 42, MEng, 47. *Prof Exp:* Elec engr, 42-64, res physicist, Naval Surface Weapons Ctr, 64-80; prog mgr, Naval Sea Systs Command, 80-84; CONSULT, VELA ASSOC, 84- *Concurrent Pos:* Guest scientist, Israel Inst Technol, 68-69 & Weizmann Inst Sci, 68-69. *Honors & Awards:* Centennial Medal, Inst Elec & Electrinonics Engrs, 84- *Mem:* Am Phys Soc; sr mem Inst Elec & Electronics Engrs; Am Soc Testing & Mat; Magnetics Soc (secy-treas, 75-76, vpres, 77-78, pres, 79-80); Sigma Xi. *Res:* Magnetism and magnetic materials; measurements; thin films; magnetic device research including magnetometers, amplifiers, transducers, and memory elements; electron devices and hardware systems; very high speed integrated circuits for computers. *Mailing Add:* 2711 Colston Dr Chevy Chase MD 20815

GORDON, DAVID BUDDY, b Chicago, Ill, Dec 17, 18; m 48; c 3. PHYSIOLOGY. *Educ:* Univ Chicago, BS, 45; Univ Southern Calif, MS, 49, PhD(physiol), 51. *Prof Exp:* Instr physiol, Univ Southern Calif, 51-55; asst prof, Univ Miami, 57-72; from asst chief to chief physiol res lab, Vet Admin Hosp, San Francisco, 64-71; res physiologist, 71-72, Vet Admin Hosp, Livermore, 72-82, dir res lab, 72-82; RETIRED. *Concurrent Pos:* Am Heart Asn fel, 55-56, adv fel, 62-64; lectr physiol, Univ Calif, Berkeley, 75-76. *Mem:* AAAS; Am Physiol Soc; Am Heart Asn; Int Soc Hypertension; Am Soc Hypertension. *Res:* Cardiovascular physiology; hypertension; endocrine function of kidneys. *Mailing Add:* Res Lab Vet Admin Hosp Livermore CA 94550

GORDON, DAVID STEWART, BONE MARROW TRANSPLANTATION, HEMATOLOGY. *Educ:* Univ Colo, MD, 67. *Prof Exp:* PROF MED HEMAT & ONCOL, SCH MED, EMORY UNIV, 82- *Mem:* Am Soc Hemat; Am Asn Immunol. *Res:* Immunology. *Mailing Add:* Dept Med/Hemat 308 Woodruff Mem Bldg Emory Univ Sch Med Atlanta GA 30322

GORDON, DENNIS T, b Chicago, Ill, Nov 13, 41. FOOD SCIENCE AND NUTRITION. *Educ:* Univ Ill, Champaign, BS, 63; Univ Conn, MS, 69, PhD(nutrit biochem), 74. *Prof Exp:* Lt, US Army, Quartermaster Corps, Ger, 63-66, Capt, Vietnam, 66-67; grad asst, Dept Animal Indust, Univ Conn, 67-69, res asst, Dept Nutrit Sci, 70-73; res assoc, Dept Food Sci & Technol, 74-77, asst prof, 77-79; asst prof, 79-82, ASSOC PROF, DEPT FOOD SCI & NUTRIT, UNIV MO-COLUMBIA, 82- *Concurrent Pos:* Adj prof, Food Sci & Human Nutrit Dept, Univ Fla, 86. *Mailing Add:* Dept Food Sci & Nutrit Univ Mo Col Agr 122 Eckles Hall Columbia MO 65211

GORDON, DONALD, b Indianapolis, Ind, June 5, 39; m 62; c 1. PLANT TAXONOMY. *Educ:* Hanover Col, BA, 61; Ind Univ, MA, 64, PhD(bot), 66. *Prof Exp:* Asst prof biol, NMex State Univ, 66-68; from asst prof to assoc prof, 58-74, PROF BIOL, MANKATO STATE COL, 74- *Concurrent Pos:* NSF res grant, 68-70. *Mem:* AAAS; Bot Soc Am; Am Soc Plant Taxon; Int Soc Plant Taxon. *Res:* A revision of the genus Gleditsia; biosystematics of Maurandya and related genera. *Mailing Add:* Dept Biol Scis Mankato State Univ Mankato MN 56001

GORDON, DONALD THEILE, b Cincinnati, Ohio, June 10, 35; m 61; c 3. PLANT PATHOLOGY, VIROLOGY. *Educ:* Univ Cincinnati, BS, 60; Univ Wis, PhD(plant path), 66. *Prof Exp:* assoc prof, 66-80, PROF PLANT PATH, OHIO STATE UNIV & OHIO AGR RES & DEVELOP CTR, WOOSTER, 80- *Mem:* AAAS; Am Phytopath Soc. *Res:* Isolation, characterization and identification of plant viruses, especially those of corn. *Mailing Add:* Dept Plant Path 201 Kottman Hall Ohio State Univ Main Campus Columbus OH 43210

GORDON, DONOVAN, b Beloit, Wis, Aug 15, 34; m 63; c 2. VETERINARY MEDICINE, VETERINARY PATHOLOGY. *Educ:* Iowa State Univ, DVM, 59; Univ Wis, MS & PhD(vet path), 67; Am Col Vet Path, dipl. *Prof Exp:* Vet, Pines Meadow Vet Clin, 61-63; pathologist, Abbott Labs, 68-71, head path sect, Indust Bio-Test Labs, 71-84; CONSULT VET PATH, 85- *Concurrent Pos:* Mem, Ill State Bd Vet Examr, 74-78; consult, vet path, 85- *Mem:* Am Vet Med Asn; Soc Toxicol Pathologists; Am Col Vet Pathologists. *Res:* Viral oncology; chemical carcinogens; diagnostic and toxicologic pathology. *Mailing Add:* 833 Kimball Rd Highland Park IL 60035

GORDON, DOUGLAS LITTLETON, b Baton Rouge, La, Mar 15, 24; m 47, 61; c 4. MEDICINE, ENDOCRINOLOGY. *Educ:* Tulane Univ, MD, 46; Am Bd Internal Med, dipl, 53. *Prof Exp:* Intern, Charity Hosp, New Orleans, La, 46-47; fel internal med, Ochsner Clin, 47-49; instr med, Sch Med, Tulane Univ, 49-51, 53-57; from clin asst prof to clin assoc prof, 57-67, CLIN PROF MED, SCH MED, LA STATE UNIV MED CTR, NEW ORLEANS, 67- *Concurrent Pos:* Clin assoc, Endocrinol Res Lab, mem staff, Sect Endocrinol, clin & mem vis staff, Hosp, Alton Oschner Med Found, 49-51, 53-57; asst vis physician, Tulane Univ, Charity Hosp, 51, vis physician, 53-57; mem vis staff, Baton Rouge Gen & Our Lady of the Lake Hosps, 54-; pvt pract, 54-; dep coroner, East Baton Rouge Parish, 55-71; vis physician, La State Univ, 57-61, sr vis prof, Charity Hosp, 61-; mem staff, Sect Med, Baton Rouge Clin, 59- *Mem:* AAAS; AMA; fel Am Col Physicians; Am Diabetes Asn; Endocrine Soc. *Res:* Internal medicine and endocrinology. *Mailing Add:* La State Univ Med Sch Baton Rouge Gen Hosp La 8415 Goodwood Blvd Baton Rouge LA 70806

GORDON, ELLIS DAVIS, b Johnson Co, Kans, Aug 6, 13; m 37; c 3. HYDROLOGY, HYDROGEOLOGY. *Educ:* Univ Kans, AB, 43; Univ Nebr, MSc, 49. *Prof Exp:* Chief party, test drill unit, Kans Geol Surv, 39-42; geologist, Kans State Hwy Comn, 42-44; geologist & ground water hydrologist, Nebr Geol Surv, 44-49; geologist, Ground Water, Br, US Geol Surv, NDak, 49-51, Mineral Deposits Br, Colo, 51-52, Ground Water Br, Washington DC, 52-54 & NMex, 54-58, dist geologist, Water Resources Div, Wyo, 58-67, staff hydrologist, regional staff Hq, Water Resources Div, US Geol Surv, Cent Region, 67-79; RETIRED. *Concurrent Pos:* Consult, USAID, Tanzania, 72. *Mem:* Am Geophys Union; Asn Prof Geol Scientists; fel Geol Soc Am. *Res:* Ground water hydrology; subsurface research in ground water; stratigraphy. *Mailing Add:* 333 N Second St Lindsborg KS 67456

GORDON, ERIC MICHAEL, b New York, NY, Feb 10, 46; m 71; c 2. ENZYME INHIBITORS, AMINO ACIDS & PEPTIDES. *Educ:* Beloit Col, BS, 67; Univ Wis, MS, 69, PhD(med chem), 73. *Prof Exp:* Res chemist org chem, Hoffmann-LaRoche Inc, 69-70; assoc nat prod chem, Yale Univ, 73-74; assoc, Squibb Inst Med Res, 74-75, res investr, 75-79, sr res investr, 79-80, res group leader, cardiovasc chem, 80-86, sect head mem chem, 86-87, DIR MED CHEM, SQUIBB INST MED RES. *Concurrent Pos:* NIH fel, 67-69; Henry S Wellcome Mem Fel, Am Found Pharmaceut Educ, 72.. *Mem:* Am Chem Soc; Sigma Xi; AAAS. *Res:* Rational design of enzyme inhibitors of medicinal importance; amino acids and peptide chemistry; chemistry and biology of naturally occurring materials. *Mailing Add:* Squibb Inst Med Res PO Box 4000 Princeton NJ 08540

GORDON, EUGENE IRVING, b New York, NY, Sept 14, 30; m 56; c 2. PHYSICS. *Educ:* City Col New York, BS, 52; Mass Inst Technol, PhD(physics), 57. *Prof Exp:* Mem staff, Mass Inst Technol, 57; tech staff, Bell Tel Labs, Inc, 57-64, head optical device dept, 64-68, dir electrooptical device lab, 68-73, dir pattern generation technol lab, 73-80, dir optical devices lab, 80-83; chmn & pres, Lytel Inc, 84-87; vpres & dir res labs, Hughes Aircraft Co, 87-88; founder, chief exec officer & pres, Photon Imaging Corp, 88-89; DISTINGUISHED PROF ELEC ENG & DEP EXEC DIR, CTR MFG SYSTS, NJ INST TECHNOL, 89- *Concurrent Pos:* Chmn, Adv Group Electron Devices, Dir Defense, 71-84; consult, 83-84. *Honors & Awards:* Vladimir K Zworykin Award, Inst Elec & Electronics Engrs, 75, Edison Medal & Centennial Medal, 84. *Mem:* Nat Acad Eng; Electron Devices Soc; fel Inst Elec & Electronics Engrs. *Res:* Properties of ionized media; microwave tubes; optical masers; coherent light techniques; image and display devices; pattern generation tools and techniques for the manufacture of semiconductors devices and packaging of silicon integrated circuits; manufacturing systems. *Mailing Add:* 50-B New England Ave Summit NJ 07901

GORDON, FLORENCE S, b Montreal, Que; m 65; c 2. MATHEMATICAL STATISTICS. *Educ:* McGill Univ, BSc, 63, MSc, 64, PhD(math), 68. *Prof Exp:* Asst prof, 68-72, assoc prof math, C W Post Col, Long Island Univ, 75-77; adj prof math, Adelphi Univ, 77-82, asst prof statist & electronic data processing, 82-83; ASSOC PROF MATH, NY INST TECHNOL, 83- *Mem:* Math Asn Am. *Res:* Characterization problems for both univariate and multivariate statistical distributions using regression properties of one statistic on another; computer usage in statistical education; author of "Statistics through the Eye of the Computer". *Mailing Add:* 61 Cedar Rd East Northport NY 11731

GORDON, GARY DONALD, b Elkins, WVa, May 28, 28; m 56; c 5. SPACECRAFT TECHNOLOGY, SPACE PHYSICS. *Educ:* Wesleyan Univ, BA, 50, Harvard Univ, MA, 51, PhD(physics), 54. *Prof Exp:* Instr physics, Harvard Univ, 54; res physicist, Biol Warfare Labs, US Army, 55-56; mem staff, Opers Res, Inc, 57-58; sr engr astro-electronics div, RCA Corp, NJ, 59-64, adminstr course develop, Current Concepts Sci & Eng Prog, Prod Eng, 65-68; sr staff scientist, Comsat Clarksburg, 69-83; AEROSPACE CONSULT, 84- *Concurrent Pos:* mem, Tech Comt Space Systs, Am Inst Aeronaut & Astronaut, 74-75. *Mem:* Am Inst Aeronaut & Astronaut. *Res:* Spacecraft technology; geostationary orbit; spacecraft thermal design; satellite reliability; spacecraft attitude; spacecraft bearings; in-orbit servicing; space and plasma physics; Fortran and Speakeasy programming; operations research; cosmic rays; piezoelectricity. *Mailing Add:* 400 Center St Box 125 Washington Grove MD 20880-0125

GORDON, GEOFFREY ARTHUR, b St Louis, Mo, Feb 5, 48; m 70. CLIMATOLOGY, COMPUTER APPLICATIONS. *Educ:* Univ Mo, BS, 71, MS, 73, PhD(atmospheric sci), 79. *Prof Exp:* Res assoc, Lab Tree-Ring Res, Univ Ariz, 77-81; fac assoc, Inst Quaternary Studies, Univ Maine, 81-89. *Concurrent Pos:* Consult, Lab Tree-Ring Res, Univ Ariz, 81; fel, Nat Oceanic & Atmospheric Admin, 81-82. *Mem:* Am Meteorol Soc; Am Quaternary Asn; AAAS; Asn Am Geographers; Tree-Ring Soc. *Res:* Exploitation of proxy records of climate variation including tree rings and historical documents and the development of statistical methodology for interpreting those records. *Mailing Add:* 91 Mill St Orono ME 04473

GORDON, GEORGE SELBIE, b Pittsfield, Mass, Apr 28, 19; m 46; c 2. PHYSICAL CHEMISTRY, ENVIRONMENTAL ENGINEERING. *Educ:* Princeton Univ, AB, 41; Northwestern Univ, PhD(chem), 49. *Prof Exp:* Chemist, Tenn Eastman Corp, 41-42; develop engr, Gen Elec Co, 48-51; vpres & dir res & develop, Titanium Zirconium Co, Inc, 51-55; dir res, US Potash Co, 55-56; assoc dir chem res, US Borax Res Corp, 56-58; dir chem res & vpres, IIT Res Inst, 59-64; chief, Textile & Apparel Tech Ctr, Nat Bur Stand, 64-66, chief officer eng standards & anal, 66-67, chief officer indust servs, 67-70; Off Environ Affairs, US Dept Com, 70-81; RETIRED. *Mem:* AAAS; Am Chem Soc. *Res:* Catalysis; mechanism of catalytic alkylation reactions; semiconductors; chemistry of zirconium compounds. *Mailing Add:* PO Box 1527 Vineyard Haven MA 02568

GORDON, GERALD ARTHUR, b Chicago, Ill, Mar 5, 34; m 67; c 2. POLYMER SCIENCE. *Educ:* Purdue Univ, BS, 56; Mass Inst Technol. MS, 57, ScD(chem eng), 61. *Prof Exp:* Res chemist, Munising Div, Kimberly-Clark Corp, 61-63; sr res chemist, Continental Can Co, 63-70, sr res scientist, 70-74, adv scientist, 74-80; RES CONSULT, SONOCO FIBRE DRUM, 80- *Concurrent Pos:* Lectr dept chem, Roosevelt Univ, Ill Inst Technol & Univ Ill, Chicago Circle, 71-81. *Mem:* AAAS; Am Chem Soc; Soc Plastic Engrs; Fedn Am Scientists; Tech Asn Pulp & Paper Indust. *Res:* Permeability of polymers; structure of highly cross-linked polymers; polymer morphology; product development in regard to fibre drums. *Mailing Add:* Sonoco Fibre Drum Inc 245 Eisenhower Lane Lombard IL 60148-5412

GORDON, GERALD BERNARD, b Harrisburg, Pa, June 4, 34; m 56; c 4. PATHOLOGY. *Educ:* Franklin & Marshall Col, BS, 56; Yale Univ, MD, 59. *Prof Exp:* Rotating intern, Univ Hosps, Cleveland, Ohio, 59-60; resident path, Yale-New Haven Med Ctr, 60-63; instr, Yale Univ, 64; asst pathologist, Armed Forces Inst Pathol, Washington, DC, 64-66; from asst prof to assoc prof, Rutgers Med Sch, 66-70; assoc prof, 70-77, PROF PATH, STATE UNIV NY UPSTATE MED CTR, 77- *Concurrent Pos:* USPHS trainee & fel, 61-64, grant, 67-; res fel, Sch Med, Yale Univ, 63-64; consult, Vet Admin Hosp, Lyons, NJ & Muhlenberg Hosp, Plainfield, 66-70; assoc attending pathologist, State Univ Hosp, Syracuse, NY, 70-77; attend pathologist, State Univ Hosp, Syracuse, NY, 77- & Vet Admin Hosp, 71- *Mem:* Int Acad Path; Am Asn Path & Bact; Am Soc Exp Path. *Res:* Experimental cellular pathology utilizing electron microscopy, tissue culture, histochemistry and biochemistry; studies in phagocytosis, intracellular digestion, lipid metabolism and accumulation; muscular diseases. *Mailing Add:* Dept Path State Univ NY Upstate Med Ctr 155 Elizabeth Blackwell St Syracuse NY 13210

GORDON, GERALD M, b Detroit, Mich, May 24, 31; m 79; c 5. METALLURGICAL ENGINEERING, CORROSION. *Educ:* Wayne State Univ, BS, 56; Ohio State Univ, PhD(metall), 59. *Prof Exp:* Sr metallurgist, Stanford Res Inst, 59-63; sr metallurgist, Vallecitos Nuclear Ctr, 63-69, mgr metall develop, 69-73, mgr zircaloy performance, Nuclear Energy Div, 73-75, mgr, Plant Component Behav Anal, mgr mats technol, 79-85, MGR, PLANT MAT ENG, NUCLEAR ENERGY DIV GEN ELEC CO, 76-, MGR FUEL & PLANT MATS TECHNOL, 86- *Mem:* Am Inst Mining, Metall & Petrol Engrs; fel Am Soc Metals; Nat Asn Corrosion Engrs. *Res:* Physical metallurgy of nuclear reactor materials; stress corrosion and electrochemistry of stainless steels and zirconium alloys; oxidation mechanism in high temperature alloys; materials engineering management. *Mailing Add:* Gen Elec Co Nuclear Energy Div 175 Curtner Ave San Jose CA 95125

GORDON, GILBERT, b Ill, Nov 11, 33; m 57; c 2. INORGANIC CHEMISTRY. *Educ:* Bradley Univ, BS, 55; Mich State Univ, PhD(inorg chem), 59. *Prof Exp:* Asst, Mich State Univ, 55-58; res assoc, Univ Chicago, 59-60; from asst prof to prof, Univ Md, 60-67; prof Univ Iowa, 67-73; prof chem, 73-84, VOLWILER DISTINGUISHED RES PROF, MIAMI UNIV, 84- *Concurrent Pos:* Chem abstr sect ed, Catalysis & Reaction Kinetics Sect, Am Chem Soc, 71-; consult, Olin Corp, 72- & Scopa Technol, 83- *Mem:* Am Chem Soc; Chem Soc London; Int Union Pure Appl Chem. *Res:* Electron transfer processes; coordination chemistry; equilibria in aqueous solutions; mechanisms of inorganic reactions; water purification processes; reactions and interactions of chlorine-containing oxidizing agents. *Mailing Add:* Dept Chem Miami Univ Oxford OH 45056

GORDON, GLEN EVERETT, b Keokuk, Iowa, Oct 13, 35; m 58; c 2. TRACE ELEMENT ANALYSIS, RECEPTOR MODELING. *Educ:* Univ Ill, BS, 56; Univ Calif, PhD(chem), 60. *Prof Exp:* From instr to assoc prof chem, Mass Inst Technol, 60-69; assoc prof, 69-73, PROF CHEM, UNIV MD, 73- *Concurrent Pos:* Prog mgr, NSF, 75-76; fel natural sci, Resources Future, 82-83. *Honors & Awards:* Award Nuclear Appl, Am Chem Soc, 77. *Mem:* Am Chem Soc; AAAS; Sigma Xi. *Res:* Development of nuclear methods of analysis to determine trace element concentrations on airborne particles; atmospheric receptor models; stable isotope tracers for study of trace element nutrition in humans. *Mailing Add:* Dept Chem & Biochem Univ Md College Park MD 20742

GORDON, HAROLD THOMAS, b Holyoke, Mass, Nov 21, 18; m 55; c 3. BIOCHEMISTRY. *Educ:* Mass State Col, BS, 39; Harvard Univ, MA, 40, PhD(biol), 47. *Prof Exp:* Asst, Off Sci Res & Develop Proj, Harvard Med Sch, 42-45, res assoc insecticides univ, 45-46; asst insect toxicologist, 47-50, assoc insect toxicologist, 50-58, insect toxicologist, 58-63, insect biochemist, 63-77, LECTR ENTOM, UNIV CALIF, BERKELEY, 77- *Mem:* Entom Soc Am; NY Acad Sci; Brit Biochem Soc. *Res:* Insect growth, nutrition, reproduction and toxicology; ultramicroanalysis. *Mailing Add:* Dept Entom Univ Calif Berkeley CA 94720

GORDON, HARRY HASKIN, pediatrics; deceased, see previous edition for last biography

GORDON, HARRY WILLIAM, b New York, NY, Mar 31, 24; m 50; c 1. BIOCHEMISTRY. *Educ:* Long Island Univ, BS, 48; Georgetown Univ, MS, 51, PhD(biochem), 52. *Prof Exp:* Instr indust med, NY Univ-Bellevue Med Ctr, 52-55; res assoc enzym, Colgate-Palmolive Co, 55-57; asst dir res & develop, Block Drug Co, 57-59; dir exp res lab, St Barnabas Med Ctr, 59-62; vpres res & develop, Julius Schmid Labs, Inc, 62-76; VPRES RES & DEVELOP COSMETIC & DRUGS, DEL LABS INC, 76- *Concurrent Pos:* Mem Clin Res Comt, St Barnabas Med Ctr, 62-65; res consult, Interboro Gen Hosp, 65-70. *Mem:* Fel AAAS; Am Chem Soc; Electron Micros Soc Am; NY Acad Sci; Am Pharmaceut Asn. *Res:* Pharmacology-toxicology; clinical studies; international as well as domestic regulatory affairs; transplantation and teratology. *Mailing Add:* Del Labs Inc 565 Broad Hollow Rd Farmingdale NY 11735

GORDON, HAYDEN S(AMUEL), b Whitebird, Idaho, June 18, 10; div. MECHANICAL ENGINEERING. *Educ:* Univ Calif, BS, 37, MS, 38, PhD, 50. *Prof Exp:* Res engr, Shell Oil Co, Calif, 37; asst mech eng, Univ Calif, 37-38; asst prof physics, San Bernardino Jr Col, 38-39; instr & jr engr, Univ Calif, 39-42; sr engr & asst chief develop, Ryan Aeronaut Co, Calif, 42-45; sr engr, Lawrence Radiation Lab, Univ Calif, 45-55, nuclear propulsion div leader, 55-57, chief engr, 57-64; MEM STAFF, W M BROBECK & ASSOC, 65- *Concurrent Pos:* Consult, 65-; dir Eng Develop Corp, Calif; dir & chmn, Berkeley Sci Capital Corp. *Mem:* AAAS; Am Soc Mech Engrs. *Res:* Evaporation of water from quiet surfaces; heat transfer by free convection from small diameter wires. *Mailing Add:* 17 Culver Ct Orinda CA 94563

GORDON, HELMUT ALBERT, b Malinska, Austria, May 5, 08; nat US; m 42; c 2. PHYSIOLOGY, PHARMACOLOGY. *Educ:* Pazmany Peter Univ, MD, 32, Dr Habil, 44. *Prof Exp:* Asst prof physiol, Col Med, Univ Budapest, 32-37, adjunctus, 38-44, privatdocent, 44; res assoc biol, NY Univ, 37-38; asst res prof, Lobound Labs, Univ Notre Dame, 46-52, assoc res prof, 52-62; prof, 62-80, EMER PROF PHARMACOL, COL MED, UNIV KY, 80- *Honors & Awards:* A Cressy Morrison Award, NY Acad Sci, 66. *Mem:* Am Physiol Soc; Asn Gnotobiotics (past pres); fel Geront Soc; fel NY Acad Sci. *Res:* Effects of microbial flora on cardiovascular, digestive and defensive systems; aging phenomena; germ-free life studies. *Mailing Add:* Richmond Pl 3051 Rio Dosa Dr Apt 201 Lexington KY 40509

GORDON, HOWARD ALLAN, b Chicago, Ill, Oct 1, 43; m 68; c 2. ELEMENTARY PARTICLE PHYSICS. *Educ:* Univ Ill-Urbana, BS, 64, MS, 66, PhD(physics), 70. *Prof Exp:* Res assoc, 70-71, from assoc physicist to physicist, 71-82, SR PHYSICIST, BROOKHAVEN NAT LAB, 82- *Concurrent Pos:* Mem, high energy adv comt, Brookhaven Nat Lab, 82-84, elem physics panel, Nat Res Coun, 83-85, Lawrence Berkeley Lab, 84-87, Fermilab Prog, 84-88 & high energy physics adv panel, Dept Energy, 85-90. *Mem:* Fel Am Physical Soc. *Res:* Experimental study of elementary particle physics; large transverse momentum phenomena; rare K decays; search for objects beyond the standard model. *Mailing Add:* Physics Dept 510 A Brookhaven Nat Lab Upton NY 11973

GORDON, HOWARD R, b Plattsburg, NY, May 21, 40; m 76. OPTICS, OCEANOGRAPHY. *Educ:* Clarkson Col Technol, BS, 61; Pa State Univ, MS, 63, PhD(physics), 65. *Prof Exp:* Asst prof physics, Col William & Mary, 65-67; asst prof physics & marine sci, 67-70, assoc prof, 70-76, PROF PHYSICS & MARINE SCI, UNIV MIAMI, 76- *Mem:* Fel Optical Soc Am; Am Geophys Union. *Res:* Oceanography; light scattering; underwater optics; radiation transfer; ocean remote sensing; atmospheric optics; digital simulations. *Mailing Add:* Dept Physics Univ Miami Box 248046 Coral Gables FL 33124

GORDON, HUGH, b New York, NY, Nov 23, 30. MATHEMATICAL ANALYSIS. *Educ:* Columbia Univ, AB, 51, AM, 52, PhD(math), 58. *Prof Exp:* Asst math, Columbia Univ, 52-53, lectr, 53-58: Peirce instr, Harvard Univ, 58-61; asst prof, Univ Pa, 61-66; assoc prof, 66-69, PROF MATH, STATE UNIV NY ALBANY, 69- *Mem:* Am Math Soc. *Res:* Abstract analysis. *Mailing Add:* Dept Math State Univ NY Albany NY 12222

GORDON, HYMIE, b Cape Town, SAfrica, Sept 20, 26. MEDICAL GENETICS, INTERNAL MEDICINE. *Educ:* Univ Cape Town, BSc, 46, MB, ChB, 50, MD, 58; FRCP(E), 66, FRCP(L), 76. *Prof Exp:* Intern med, Groote Schuur Hosp, Cape Town, 51-52; registr, Addington Hosp, Durban, 52-55 & Groote Schuur Hosp, 55-58; Frank Forman res fel, Postgrad Med Sch, London, 58-59; asst physician, Johns Hopkins Hosp, Baltimore, 59-61; sr lectr & consult med, Univ Cape Town, 61-69; consult genetics & med, Mayo Clin, 69-89, assoc prof 69-74, prof med genetics & chmn dept, Mayo Med Sch & Mayo Clin, 74-89, CONSULT HIST MED, MAYO CLIN, 90- *Concurrent Pos:* Eli Lilly fel, 59-60. *Mem:* AAAS; Am Soc Human Genetics; Brit Genetical Soc; fel Brit Interplanetary Soc; fel Royal Soc Med; Sigma Xi. *Res:* Clinical cardiology; nutrition and heart disease; clinical biochemical and population genetics; genetics counseling. *Mailing Add:* Dept Hist Med Mayo Clin Rochester MN 55905

GORDON, IRVING, b St Paul, Minn, Jan 29, 22; m 52; c 2. ORGANIC CHEMISTRY, INFORMATION SCIENCE. *Educ:* Univ Minn, BChem, 43, PhD(org chem), 50. *Prof Exp:* Fel amino acids & peptides, Univ Pittsburgh, 50-51; res chemist, org & inorg phosphorus chem, Oldbury Electro-Chem Co, 51-56; res chemist, Hooker Electro-Chem Co, 56-64, sr chemist, Hooker Chem Corp Div, Occidental Petrol Corp, 64-69, SUPVR TECH INFO CTR, CORP RES, HOOKER CHEM, 69- *Mem:* Am Chem Soc; Am Soc Info Sci; Mfg Chemists Asn. *Res:* Electrochemistry; polymer chemistry; chemical patents; statistics; information science; computer science and programming; chemical information center design. *Mailing Add:* 1334 Garrett Ave Niagara Falls NY 14305-2547

GORDON, IRVING, b Cleveland, Ohio, June 20, 14; m 39; c 3. MEDICAL MICROBIOLOGY. *Educ:* Univ Mich, MD, 37; Am Bd Path, Am Bd Microbiol & Nat Bd Med Examrs, dipl. *Prof Exp:* Intern & resident, State Univ NY Down State, 37-39; jr med bacteriologist, NY State Dept Health, 39-40; fel, Rockefeller Found, 41; from instr to assoc path & bact, Albany Med Col, 42-43 & 46-48, assoc prof med & bact, 48-55; PROF MICROBIOL, SCH MED, UNIV SOUTHERN CALIF, 55-; SR ATTEND PHYSICIAN, LOS ANGELES COUNTY GEN HOSP, 56- *Concurrent Pos:* Vol asst, Trudeau Sanitorium, NY, 42; from sr med bacteriologist to prin med bacteriologist, Div Lab & Res, NY State Dept Health, 46-53, asst dir, 53-55; bacteriologist & immunologist comn acute respiratory dis, Epidemiol Bd, US Army, 43-46, assoc mem, comn liver dis, 50-55, comn enteric infections, 53-59, mem comn viral infections, 53-71; consult, USPHS, 51-55, 56-57 & 66-71, Chem Corps, US Army, 56-62 & Virus Lab, Calif State Dept Pub Health, 56-64; sr lectr, Univ Sheffield, Eng & Holme lectr, Univ Col Hosp Sch Med, Univ London, 54; mem bact test comt, Nat Bd Med Exam, 56-59; mem allergy & infectious dis training grant comt, NIH, 62-66, Spec Study Sect, 76-, mem, Comt Etiology of Cancer, Am Cancer Soc, 60-63 & Personnel in Res, 64-69; Coun Res & Invest Awards, 70-76; mem spec adv comt prophylaxis of poliomyelitis, Calif State Dept Pub Health, 60-63; assoc dean, Univ Southern Calif, 63-66. *Mem:* AAAS; Am Soc Clin Invest; Am Soc Microbiol; Soc Exp Biol & Med; Am Asn Immunol. *Res:* Virology; immunology; host-parasite relationships. *Mailing Add:* Dept Microbiol HMR 401 Univ Southern Calif 2025 Zonal Ave Los Angeles CA 90033-1054

GORDON, JAMES POWER, b New York, NY, Mar 20, 28. LASERS, OPTICS. *Educ:* Mass Inst Technol, BS, 49; Columbia Univ, MA, 51, PhD(physics), 55. *Prof Exp:* Asst physics, Columbia Univ, 53-55; mem tech staff electronics res, 55-59, head, Quantum Electronics Res Dept, 59-80, SR TECH STAFF CONSULT, AT&T BELL LABS, 80- *Honors & Awards:* Max Born Award, Optical Soc Am. *Mem:* Nat Acad Eng; fel Am Phys Soc; sr mem Inst Elec & Electronics Engrs. *Res:* Quantum electronics; interaction of electromagnetic waves with matter; communication theory. *Mailing Add:* AT&T Bell Labs Holmdel NJ 07733

GORDON, JAMES SAMUEL, b New York, NY, Oct 12, 41. PSYCHOSOMATIC MEDICINE, HOLISTIC MEDICINE. *Educ:* Harvard Univ, AB, 62, MD, 67. *Prof Exp:* Resident & chief resident psychiat, Albert Einstein Col Med, 68-71; res psychiatrist, Nat Inst Mental Health, 71-82; CLIN PROF, DEPTS PSYCHIAT, COMMUNITY FAMILY MED, GEORGETOWN UNIV SCH MED, 80- *Concurrent Pos:* Teaching fel gen educ, Harvard Col, 63-67; vol physician, Haight-Ashbury Free Clin, 67-68; vis lectr, Grad Sch Psychol, Catholic Univ Am, 74-75, Community Ther Training Ctr, 75-76; dir spec study & alternative servs, Presidents Comn Mental Health, 78-79; Blanche Ittleson consult, Group Advan Psychiat, 79-80; chief adolescent servs br, St Elizabeths Hosp, 80-82; author of Health for the Whole Person: the Complete Guide to Holistic Medicine. *Mem:* Am Psychiat Asn; Am Holistic Med Asn; Physicians Social Responsibility. *Mailing Add:* 5225 Conn Ave NW Washington DC 20015

GORDON, JAMES WYLIE, b Topeka, Kans, Jan 21, 34; m 63; c 3. PHYSICS. *Educ:* Univ Kans, BS, 61, PhD(physics), 68. *Prof Exp:* Res scientist physics, Kaman Sci Corp, 68-74; mem physics staff, 74-85, group leader, Thermonuclear Applications, 85-88, LAB FEL, LOS ALAMOS NAT LAB, 88- *Honors & Awards:* Ernest Orlando Lawrence Award, Dept Energy, 87. *Mem:* Am Phys Soc; Sigma Xi. *Res:* Nuclear weapon development. *Mailing Add:* 149 Piedra Loop Los Alamos NM 87544

GORDON, JEFFREY I, MOLECULAR BIOLOGY, INTESTINAL DEVELOPMENT. *Educ:* Univ Chicago, MD, 73. *Prof Exp:* ASSOC PROF MED & BIOL CHEM, SCH MED, WASH UNIV, 81- *Mailing Add:* 12 Stacy Dr Olivette MO 63132

GORDON, JEFFREY MILES, b Brooklyn, NY, June 10, 49; m 77; c 3. SOLAR ENERGY, APPLIED OPTICS. *Educ:* Columbia Col, BA & MA, 70; Brown Univ, PhD(chem), 76. *Prof Exp:* Fel polymer sci, Weizmann Inst Sci, 76-78; lectr, Inst Desert Res, Ben-Gurion Univ, 78-80, res scientist solar energy, 78-82, sr lectr, 82-87, assoc prof, 87-90, PROF, INST DESERT RES, BEN-GURION UNIV, ISRAEL, 90- *Concurrent Pos:* Mem, solar energy comt, Israel Standards Inst, 82-85; vis res fel, Solar Energy Res Inst, Golden, Colo, 85; vis assoc prof, Univ Colo, Boulder, 85-86. *Honors & Awards:* Ben-Gurion Prize, Israel, 80. *Mem:* Int Solar Energy Soc. *Res:* Solar thermal energy systems; thermodynamics and irreversible thermodynamics; heat engines and heat pumps; nonimaging optics; solar power generation; photovoltaic systems. *Mailing Add:* Inst Desert Res Ben-Gurion Univ Sede-Boqer 84993 Israel

GORDON, JOAN, b Pine Island, Minn, Feb 8, 23. FOOD SCIENCE, NUTRITION. *Educ:* Univ Minn, BS, 45, MS, 47, PhD(home econ), 53. *Prof Exp:* Asst home econ, Univ Minn, 45-47, instr, 47-53; asst prof, Iowa State Col, 53-54; from asst prof to assoc prof, Univ Minn, 54-60; prof foods & nutrit, Pa State Univ, 60-67; chmn dept food sci & nutrit, 72-74, PROF FOOD SCI & NUTRIT, UNIV MINN, ST PAUL, 67- *Mem:* AAAS; Am Home Econ Asn; Am Dietetic Asn; Inst Food Technol; Am Chem Soc; Sigma Xi. *Res:* Vitamins; food quality. *Mailing Add:* Food Sci & Nutrit Bldg Univ Minn St Paul MN 55108

GORDON, JOEL ETHAN, b Denver, Colo, May 9, 30; m 56; c 2. PHYSICS. *Educ:* Harvard Univ, AB, 52; Univ Calif, Berkeley, PhD(physics), 58. *Prof Exp:* From asst prof to assoc prof physics, 57-68, prof physics, 68-81, William R Kenan Jr prof, 83-87, STONE PROF NATURAL SCI, AMHERST COL, 81- *Concurrent Pos:* Attached staff, Atomic Energy Res Estab, Harwell, Eng, 64-65; spec staff mem, Rockefeller Found, 68-69; vis prof, Univ Valle, Colombia, 68-69; book review ed, Am J Physics, 74-; vis Scientist, Ceng Grenoble, France, 82; vis Scholar, Univ Calif, Berkeley & Lawrence Berkeley Lab, 87. *Mem:* Am Phys Soc; Am Asn Physics Teachers. *Res:* Low temperature and solid state physics. *Mailing Add:* Dept Physics Amherst Col Amherst MA 01002

GORDON, JOHN C, b Nampa, Idaho, June 10, 39; m 64; c 1. PLANT PHYSIOLOGY, SILVICULTURE. *Educ:* Iowa State Univ, BS, 61, PhD(plant physiol, silviculture), 66. *Prof Exp:* Instr forestry, Iowa State Univ, 65-66; res plant physiologist pioneering res proj, NCent Forest Exp Sta, US Forest Serv, 66-70; assoc prof forestry, Iowa State Univ, 70-73, prof forestry, 73-77; prof forest sci & head dept, Ore State Univ, 77-83; SCH FORESTRY, YALE UNIV, 83- *Mem:* AAAS; Soc Am Foresters; Sigma Xi. *Res:* Photosynthesis and translocation in trees; enzymes in woody plants; nitrogen fixation. *Mailing Add:* Sch Forestry Yale Univ New Haven CT 06511

GORDON, JOHN EDWARD, b Columbus, Ohio, Aug 5, 31; m 56; c 2. ORGANIC CHEMISTRY, PHYSICAL CHEMISTRY. *Educ:* Ohio State Univ, BSc, 53; Univ Calif, PhD(chem), 56. *Prof Exp:* Instr chem, Brown Univ, 56-58; fel, Mellon Inst, 58-65; from asst to assoc scientist, Woods Hole Oceanog Inst, 65-68; assoc prof, 68-70, chmn dept, 78-81, PROF CHEM, KENT STATE UNIV, 70- *Mem:* Am Chem Soc. *Res:* Organic electrolytes; chemical information science. *Mailing Add:* Dept Chem Kent State Univ Kent OH 44242-0002

GORDON, JOHN P(ETERSEN), b Port Washington, NY, Aug 29, 28; m 56; c 4. ELECTRICAL ENGINEERING. *Educ:* Va Polytech Inst, BS, 53, MS, 59; Univ Ill, Urbana, PhD(elec eng), 69. *Prof Exp:* Test engr, Appalachian Elec Power Co, 53-55; assoc prof elec eng, Va Polytech Inst, 55-65; from instr to asst prof, Univ Ill, Urbana, 65-71; prof area coordr elec eng & technol, 71-80, ASSOC PROF ENG, NORTHERN ARIZ UNIV, 71- *Mem:* Am Soc Eng Educ; Inst Elec & Electronics Engrs. *Mailing Add:* Dept Elec Eng Va Mil Inst Lexington VA 24450

GORDON, JOHN S(TEVENS), b Mt Kisco, NY, Sept 7, 31; m 67; c 2. CHEMICAL ENGINEERING, INDUSTRIAL ENGINEERING. *Educ:* Cornell Univ, BChE, 53; Ohio State Univ, MS, 56. *Prof Exp:* Process develop technologist, Socony-Mobil Labs, 53-54; proj engr liquid rocket fuels, Wright-Patterson AFB, 54-57; unit supvr propellant liaison & thermochem, Reaction Motors Div, Thiokol Chem Corp, NJ, 57-62; mgr phys chem technol, Astrosyst Int, Inc, 62-67; prin res scientist, Atlantic Res Corp, 67-70; prog mgr, Pope, Evans & Robbins, 70-72; staff engr, Res Cottrell Inc, 72-73; prin investr, TRW Inc, 73-74; group leader, Mitre Corp, 74-77; chem engr, TRW Inc, 77-81; chem engr, Engr Soc Comn Energy, 81-85; chem engr, Nat Syst Mgt Corp, 86-89, Versar Inc, 89-90; SUPPORT COORDR, CLEAN COAL TECHNOL, EG&G INC, 90- *Mem:* Am Chem Soc; Am Inst Aeronaut & Astronaut; Am Inst Chem Engrs; Combustion Inst. *Res:* Thermochemical calculations and propellant properties; high temperature thermodynamic, radiative and electromagnetic properties of gases; synthetic fuels; integrated combustion/environmental control systems; ordnance technology; hazardous waste technology. *Mailing Add:* 2364 Gallant Fox Ct Reston VA 22091-2611

GORDON, JON W, b Pasadena, Calif, Jan 12, 49; m 78. DEVELOPMENTAL BIOLOGY. *Educ:* Columbia Univ, BA, 71; Yale Univ, PhD(biol), 78, MD, 80. *Prof Exp:* Hudson Brown fel obstet & gynec, Sch Med, 79-80, FEL, DEPT BIOL, YALE UNIV, 80-; MEM STAFF, DEPT OBSTET & GYNEC, MT SINAI SCH MED, NY, 82-, MATHERS PROF GERIATRIC & ADULT DEVELOP, 87- *Concurrent Pos:* Fel reviewer, NSF. *Mem:* AAAS; Soc Develop Biol. *Res:* Production of allophenic mice; study of mammalian sex determination and gainetogenesis; transfer of cloned genes into developing mice. *Mailing Add:* Dept Obstet & Gynec Mt Sinai Sch Med One Gustave Levy Plaza New York NY 10029

GORDON, JOSEPH GROVER, II, b Nashville, Tenn, Dec 25, 45; m 72; c 1. ELECTROCHEMISTRY, SPECTROSCOPY. *Educ:* Harvard Col, AB, 66; Mass Inst Technol, PhD(inorg chem), 70. *Prof Exp:* Asst prof chem, Calif Inst Technol, 70-75; mgr interfacial electrochem, 78-84, mgr interfacial sci, 84-87, tech asst to dir res, 86-87, MEM RES STAFF, IBM, ALMADEN RES CTR, SAN JOSE, CALIF, 75-, MGR MAT SCI & ANAL, 90- *Concurrent Pos:* Chmn grad fel evaluation panels, NSF, 79-81; chmn, Gordon Conf Electrochem, 87. *Mem:* Am Chem Soc; Chem Soc; Nat Orgn Black Chemists & Chem Engrs; AAAS; Electrochem Soc. *Res:* Interfacial electrochemistry; surface plasmon and vibrational spectroscopy of the electrode: solution interface; structure of adsorbed layers using synchnotron radiation. *Mailing Add:* IBM Res Div Dept K34 San Jose CA 95120-6099

GORDON, JOSEPH R, b Boston, Mass, May 17, 24; m 47; c 3. ORGANIC CHEMISTRY, POLYMER CHEMISTRY. *Educ:* Amherst Col, BA, 44; Univ Ill, PhD(chem), 49. *Prof Exp:* Res chemist, Johnson & Johnson, NJ, 49-51 & Kolker Chem, 51-52; res chemist, Gen Chem Div, Allied Chem Corp, 52-59, res supvr, 59-63, mgr polymer res, 63-64; dir develop res, 64-70, admin asst res & develop, 70-77, DIR ENVIRON AFFAIRS, LUBRIZOL CORP, 77- *Mem:* Am Chem Soc. *Mailing Add:* 3170 Bayou Sound Longboat Key FL 34228-4096

GORDON, JULIUS, b Budapest, Hungary, Nov 22, 32; Can citizen; m 54; c 2. IMMUNOLOGY. *Educ:* Sir George Williams Col, BS, 54; McGill Univ, PhD(biochem), 58. *Prof Exp:* Fel immunol, Chester Beatty Res Inst, London, 58-60; res assoc, Montreal Cancer Inst, 61-64; from asst prof to assoc prof, 65-80, PROF, DEPT SURG, MCGILL UNIV, 81- *Concurrent Pos:* Res assoc, Med Res Coun Can, 66. *Mem:* Transplantation Soc; Am Asn Immunologists; Can Soc Immunol. *Res:* Cell-mediated immunity as applied to transplantation and tumor immunology. *Mailing Add:* Dept Surg Donner Bldg McGill Univ 740 Docteur Penfield Ave Montreal PQ H3A 1A4 Can

GORDON, KATHERINE, b Las Cruces, NMex, Sept 20, 54. MOLECULAR BIOLOGY, DEVELOPMENTAL BIOLOGY. *Educ:* Simon Fraser Univ, BS, 77; Wesleyan Univ, PhD(molecular biol), 82. *Prof Exp:* Res fel, Yale Univ, 82-84; SR SCIENTIST, INTEGRATED GENETICS, 84- *Mem:* AAAS; Soc Develop Biol; Am Soc Cell Biol. *Mailing Add:* 393 Broadway St 31 New York Ave Cambridge MA 01701

GORDON, KENNETH RICHARD, b Oakland, Calif, Feb 15, 45. VERTEBRATE FUNCTIONAL MORPHOLOGY. *Educ:* San Jose State Univ, BS, 72, MA, 74; Univ Calif, Davis, PhD(zool), 79. *Prof Exp:* Fel & res assoc, Univ Ill Med Ctr, 79-82; asst prof, 82-87, ASSOC PROF, DEPT BIOL SCI, FL INT UNIV, MIAMI, 87- *Mem:* Am Soc Zoologists; Am Soc Mammalogists; Sigma Xi. *Res:* Analysis of the development and regeneration of bone in response to loading; mechanics of locomotion and feeding of mammals; evolution of feeding and locomotion, especially in marine mammals. *Mailing Add:* Dept Biol Sci Fla Int Univ Tamiami Campus Miami FL 33199

GORDON, KURTISS JAY, b New York, NY, July 20, 40; m 64; c 3. COMPUTER SCIENCE, ASTRONOMY. *Educ:* Antioch Col, BS, 64; Univ Mich, MA, 66, PhD(astron), 69; Univ Mass, MSECE, 85. *Prof Exp:* Jr res assoc, Nat Radio Astron Observ, Va, 67-69, res assoc, 69-70; from asst prof to assoc prof astron, Hampshire Col, 70-85; sr res assoc comput sci, 85-87, SCI IMAGE APPLN COORDR, UNIV COMPUT SERV, UNIV MASS, 87- *Concurrent Pos:* Danforth Assoc, 81-86. *Mem:* Am Astron Soc; Asn Comput Mach; Inst Elec & Electronics Engrs; Sigma Xi. *Res:* Computer graphics and image processing; modeling and simulation; systems design. *Mailing Add:* Univ Comput Serv Univ Mass Amherst MA 01003

GORDON, LANCE KENNETH, b Chicago, Ill, Dec 11, 47; m 69; c 2. HUMAN VACCINE, CLINICAL TRIALS MANAGEMENT. *Educ:* Univ Calif, Humboldt, BS, 73; Univ Conn, Farmington, PhD(biomed sci), 77. *Prof Exp:* Postdoctoral immunol, Wash Univ Med Sch, 77-79; sect head immunol, Connaught Labs, Inc, Pa, 80-83; res dir vaccines, Connaught Labs Ltd, Toronto, Can, 83-87; assoc med dir infectious dis, E R Squibb & Sons, 87-88; chief exec officer admin, Selcore Labs, Inc, 88-89; sr vpres res & develop vaccines, NAm Vaccines, 89-90; PRES & CHIEF EXEC OFFICER MUCOSAL IMMUNITY, ORAVAX, INC, 90- *Mem:* Am Asn Immunologists; Am Soc Microbiol. *Res:* First successful carrier-hapten vaccine; haemophilus influenzae b vaccine licensed; new pertussis vaccine through phase II clinical trials; awarded three US patents. *Mailing Add:* 15416 Merrifields Lane Silver Spring MD 20906

GORDON, LOUIS, b Philadelphia, Pa, Dec 13, 46. MATHEMATICAL STATISTICS. *Educ:* Mich State Univ, BS, 68, MS, 68; Stanford Univ, MS, 69, PhD(statist), 71. *Prof Exp:* Asst prof statist, Stanford Univ, 71-73; statistician, Alza Corp, 73-78 & US Dept Energy, 78-83; assoc prof, 83-88, PROF MATH, UNIV SOUTHERN CALIF, 88- *Concurrent Pos:* Mem, Comt Nat Statist, 87-; fel, John S Guggenheim Found, 89-90; scholar, Fulbright-Hays Scholar, 89-90. *Mem:* Am Math Soc; fel Inst Math Statist; Am Statist Asn; Soc Indust & Appl Math. *Res:* Nonparametric inference in high dimensional parameter spaces. *Mailing Add:* Math Dept DRB-155 Univ Southern Calif Los Angeles CA 90089-1113

GORDON, LOUIS IRWIN, b Los Angeles, Calif, Aug 17, 28; m 51; c 3. CHEMICAL OCEANOGRAPHY. *Educ:* Univ Calif, Los Angeles, BS, 51, San Diego, MS, 53; Ore State Univ, PhD(chem oceanog), 73. *Prof Exp:* Asst geochem, Scripps Inst Oceanog, Univ Calif, 58, res chemist, 58-62, asst specialist marine chem, San Diego, 62-64, assoc specialist, 64-66; instr oceanog, 69-72, asst prof, 72-77, ASSOC PROF OCEANOG, ORE STATE UNIV, 77- *Mem:* Am Geophys Union; Am Soc Limnol & Oceanog; Sigma Xi. *Res:* Chemical oceanography; dissolved gases; nutrients; carbon; analytical chemistry; stable isotopes in oceanography; atmospheric and biological exchange processes, coastal processes. *Mailing Add:* 2941 Ashwood Dr Corvallis OR 97330

GORDON, LYLE J, b Rupert, Idaho, Aug 19, 26; m 46; c 3. CHEMICAL ENGINEERING, REASEARCH MANAGEMENT. *Educ:* Univ Wash, Seattle, BS, 48, MS, 50, PhD(chem eng), 62. *Prof Exp:* Proj engr, Scott Paper Co, 53-55, develop engr, 55-57, develop mgr, 57-59, mgr pulp res & develop, West Coast, 59-69, corp mgr, 70-75, DIR PULPING RES & DEVELOP, SCOTT PAPER CO, 75- *Concurrent Pos:* Chmn, Comt P3, Am Nat Stand Inst, 77- *Honors & Awards:* Shibley Award, Tech Asn Pulp & Paper Indust, 56. *Mem:* Am Inst Chem Engrs; fel Tech Asn Pulp & Paper Indust (pres, 71-72). *Res:* Liquid-solid fluidization; sulfite and high yield pulping; wood (cellulose) pulping research. *Mailing Add:* Scott Paper Co Scott Plaza Three Philadelphia PA 19113

GORDON, MALCOLM STEPHEN, b Brooklyn, NY, Nov 13, 33;; wid; c 1. COMPARATIVE ECOLOGICAL PHYSIOLOGY, MARINE BIOLOGY. *Educ:* Cornell Univ, BA, 54; Yale Univ, PhD(zool), 58. *Prof Exp:* Asst oceanog, Cornell Univ, 51-53; from instr to assoc prof, 58-68, dir, Inst Evolutionary Environ Biol, 70-76, PROF BIOL, UNIV CALIF, LOS ANGELES, 68- *Concurrent Pos:* Guggenheim fel, 61-62; Nat Acad Sci exchange vis, USSR, 65; asst dir res, Nat Fisheries Ctr & Aquarium Dept of Interior, DC, 68-69; vis prof, Chinese Univ, Hong Kong, 71-72; sr Queen's fel marine sci, Dept Sci Govt Australia, 76; chair, div comp physiol biochem, Am Soc Zoologists, 88-89. *Mem:* AAAS; Am Soc Ichthyol & Herpetol; Am

Soc Zool; Soc Exp Biol; Am Physiol Soc. *Res:* Comparative vertebrate ecological physiology with emphasis on fishes and amphibians; activity metabolism; amphibious fishes; mechanisms of osmoregulation and salinity adaptation; adaptation to low temperatures and high hydrostatic pressures; ecological physiology; nutrient recycling from wastewater; fish aquaculture. *Mailing Add:* Dept Biol Univ Calif Los Angeles CA 90024-1606

GORDON, MALCOLM WOFSY, b Stamford, Conn, Dec 2, 17; m 46; c 3. BIOCHEMISTRY. *Educ:* City Col New York, BS, 40; Univ Tex, MA, 47, PhD(biochem), 48. *Prof Exp:* Asst biochem, Univ Tex, 46-47, res assoc, 48; Merck fel, Calif Inst Technol, 48-50 & Dept Physiol Chem, Univ Wis, 50-53; res assoc, Inst Living, 53-56; dir, Biochem Res Labs, 56-58; mem grad fac, 58-70, prof biobehav sci, Univ Conn, 70-; dir res, Abraham Ribicoff Res Labs, Norwich Hosp, Conn, 68-; RETIRED. *Concurrent Pos:* Vis prof, Trinity Col, Conn, 57-58; adj prof, Univ Hartford, 63- *Mem:* AAAS; Am Chem Soc; Am Biochem Soc; Am Physiol Soc; Am Soc Biol Chem. *Res:* Neurochemistry. *Mailing Add:* Box 508 Norwich CT 06360

GORDON, MANUEL JOE, b Cleveland, Ohio, July 3, 22; m 43; c 2. GENETICS, CELL BIOLOGY. *Educ:* Ohio State Univ, BSc, 49; Univ Calif, MA, 52, PhD(zool), 55. *Prof Exp:* Res fel zool, Univ Calif, 55-57; asst prof res, Dairy Dept, Mich State Univ, 57-61; MGR APPLN, RES DEPT, SPINCO DIV, BECKMAN INSTRUMENTS, INC, 61- *Mem:* NY Acad Sci; Am Soc Human Genetics; AAAS; Am Inst Biol Sci; Soc Sci Study Sex; Sigma Xi. *Res:* Molecular structure of proteins, methods for characterization of human serum lipoproteins; physico-chemical biology; reproductive physiology, genetics. *Mailing Add:* 3640 Evergreen Dr Palo Alto CA 94303

GORDON, MARK A, b Springfield, Mass, Oct 13, 37; m 80; c 2. RADIO ASTRONOMY. *Educ:* Yale Univ, BA, 59; Univ Colo, PhD(astron-geophys), 66. *Prof Exp:* Scientist, Arctic Inst NAm, 59-61; staff, Lincoln Lab, Mass Inst Technol, 66-69; asst scientist, 69-72, assoc scientist, 72-77, asst dir, Ariz Opers, 73-84, SCIENTIST, NAT RADIO ASTRON OBSERV, 77- *Mem:* Am Astron Soc; Int Union Radio Sci; Int Astron Union. *Res:* Interstellar medium, galactic structure. *Mailing Add:* Nat Radio Astron Observ 949 N Cherry Ave Campus Bld 65 Tucson AZ 85721

GORDON, MARK STEPHEN, b New York, NY, Jan 18, 42; m 65; c 1. QUANTUM CHEMISTRY. *Educ:* Rensselaer Polytech Inst, BS, 63; Carnegie-Mellon Univ, PhD(chem), 68. *Prof Exp:* Assoc, Iowa State Univ, 67-70; asst prof, 70-73, assoc prof, 73-78, PROF CHEM, NDAK STATE UNIV, 78-, CHMN DEPT, 81- *Mem:* Am Chem Soc; Sigma Xi; Am Phys Soc. *Res:* Molecular structure and chemical bonding in ground and excited states; calculations on excited states, intramolecular hydrogen bonding; internal rotation; localized orbital calculations. *Mailing Add:* Dept Chem LADD 104J NDak State Univ Fargo ND 58102

GORDON, MAXWELL, b USSR, Feb 13, 21; nat US; wid; c 2. MEDICINAL CHEMISTRY. *Educ:* Philadelphia Col Pharm, AB & BS, 41; Univ Pa, MS, 46, PhD(org chem), 48; Imp Col, London, dipl, 51. *Prof Exp:* Anal chemist, US Naval Med Supply Depot, NY, 41-42; instr org chem, Philadelphia Col Pharm, 46-47; res fel, Radiation Lab, Univ Calif, 49-50, Swiss Tech Inst, Zurich, 48-49 & Imp Col, London, 50-51; res assoc, Radio-Isotope Lab, Squibb Inst Med Res, 51-55; sr scientist, Smith Kline & French Labs, 55-57, head phys sci sect, 57-68, assoc dir res activities, 68-70; dir res planning, 70-74, vpres, 74-82, sr vpres, Sci & Technol Group, Bristol Labs, 82-86; RETIRED. *Mem:* Am Chem Soc; Royal Soc Chem; Swiss Chem Soc; Soc Ger Chem; Austrian Chem Soc. *Res:* Chemical kinetics; azulenes; heterocycles; cancer; radioactive tracers; antibiotic biosynthesis; medicinal chemistry; scientific documentation; research planning; bio-electronics; analgetics and antagonists; central nervous system drugs; recombinant DNA, interferon, antiviral drugs. *Mailing Add:* 345 Park Ave 6-71 New York NY 10154

GORDON, MICHAEL ANDREW, MEMBRANE BIOPHYSICS, MOLECULAR GRAPHICS. *Educ:* Univ CAlif, PhD(pharmacol), 76. *Prof Exp:* ASSOC PROF PHARMACOL, MED CTR, KANS UNIV, 82- *Mailing Add:* Dept Pharm Kans Univ Med Ctr Kansas City KS 66103

GORDON, MILDRED KOBRIN, b New York, NY. CELL BIOLOGY. *Educ:* Brooklyn Col, BA, 41; Tulane Univ, MS, 46; Yale Univ, PhD(anat), 66. *Prof Exp:* Asst prof biol, Univ Hartford, 56-62; asst prof zool, Conn Col, 65-68; asst prof obstet, gynec & anat, Sch Med, Yale Univ, 68-75; assoc prof anat, State Univ NY Buffalo, 75-78; MED PROF, CTR BIOMED EDUC, NY, 78- *Mem:* Am Soc Cell Biol; Electron Micros Soc Am; Am Asn Anatomists; Soc Study Reproduction; AAAS. *Res:* Properties of the plasma membrane of mammalian spermatozoa and modifications during maturation, capacitation, fertilization; membrane of unfertilized and fertilized mammalian ovum. *Mailing Add:* 938 St Nicholas Ave New York NY 10032

GORDON, MILLARD F(REEMAN), b Eastvale, Pa, Sept 3, 21; m 43; c 2. ELECTRICAL ENGINEERING. *Educ:* Carnegie Inst Technol, BS, 42, MS, 47, DSc, 48. *Prof Exp:* Assoc prof eng, Brown Univ, 50-54; head penetrations systs dept, Thompson-Ramo-Wooldridge, Inc, 54-60; sr staff adv, Bissett-Berman Corp, 60-67; chief scientist radar, Hughes Aircraft Co, Culver City, 67-82; RETIRED. *Mem:* Inst Elec & Electronics Engrs. *Res:* Research, development and management. *Mailing Add:* PO Box 637 Cherokee Village AR 72525

GORDON, MILTON, b New York, NY, May 28, 29; m 52; c 3. MICROBIOLOGY. *Educ:* Univ Calif, Los Angeles, AB, 50; George Washington Univ, MS, 55, PhD, 62. *Prof Exp:* Microbiologist, US Army Biol Labs, 51-70; health scientist adminr, Div Res Grants, NIH, 70-86; RETIRED. *Mem:* AAAS; Am Soc Microbiol; Sigma Xi. *Res:* Immunology; medical bacteriology. *Mailing Add:* 1409 Pinewood Dr Frederick MD 20701-4263

GORDON, MILTON ANDREW, b Chicago, Ill, May 25, 35; m 62; c 2. MATHEMATICAL STATISTICS. *Educ:* Xavier Univ La, BS, 57; Univ Detroit, MA, 60; Ill Inst Technol, PhD(math), 68. *Prof Exp:* Mathematician, Labs Appl Sci, Univ Chicago, 59-62; from instr to asst prof math, Loyola Univ Chicago 66-71, dir afro-am studies prog, 71-78; PROF MATH, CHICAGO STATE UNIV & DEAN, COL ARTS & SCI, 78- *Mem:* Math Asn Am; Am Math Soc. *Res:* General algebraic systems and statistics, especially as they relate to the social and behavioral sciences; curriculum for Afro-American Studies Program. *Mailing Add:* 800 N State College Blvd Fullerton CA 92634

GORDON, MILTON PAUL, b St Paul, Minn, Feb 8, 30; m 55; c 4. BIOCHEMISTRY. *Educ:* Univ Minn, BA, 50; Univ Ill, PhD(biochem), 53. *Prof Exp:* Asst, Univ Ill, 52-53; res fel, Nat Cancer Inst, Sloan-Kettering Inst, 53-55, asst, 55-57; asst res biochemist virus lab, Univ Calif, 57-59; from asst prof to assoc prof, 59-66, PROF BIOCHEM, UNIV WASH, 66- *Concurrent Pos:* Ed, Biochemistry, Am Chem Soc, 61-; vis res chemist, Princeton Univ, 66-67. *Mem:* AAAS; Am Chem Soc; Am Soc Biol Chemists. *Res:* Molecular basis of crown gall tumorigenesis of plants and biological engineering of plants; exobiology. *Mailing Add:* Dept Biochem Univ Wash Seattle WA 98195

GORDON, MORRIS AARON, b Waterbury, Conn, Apr 3, 20; div; c 3. MEDICAL MYCOLOGY. *Educ:* City Col New York, BS, 40; Univ Chicago, MS, 42; Duke Univ, PhD(bot, mycol), 49. *Prof Exp:* Mycologist, Commun Dis Ctr, USPHS, 49-53, head, Airborne Pathogens Lab, 53-54; res specialist, Chem Corps Training Command, Ft McClellan, Ala, 54-55; asst prof microbiol & mycol, Med Col, SC, 55-57, assoc prof, 57-59; sr res scientist, NY State Health Dept, 59-61, assoc res scientist, 61-68, prin res scientist, Wadsworth Ctr Labs & Res, 68-87, chief, lab clin microbiol, 84-87; RES PROF MICROBIOL, ALBANY MED COL, 64-; PROF, GRAD SCH PUB HEALTH SCI, STATE UNIV NY, ALBANY, 85- *Concurrent Pos:* Instr med sch, Emory Univ, 49-54; La State Univ Inter-Am fel, Cent Am, 59; consult, Grant Comts, NIH, 59-; mem bact & mycol study sect, NIH, 71-75; mem adv comt, Brown-Hazen Prog of Res Corp, 75-; Fulbright prof, Uruguay, 78. *Mem:* Fel Am Acad Microbiol; Am Soc Microbiol; Sigma Xi; Int Soc Human & Animal Mycol (vpres, 82-85); Med Mycol Soc Americas (pres, 78-79). *Res:* Fungal serology; cryptococcosis; nocardia; dermatophilus; antifungal antibiotics; fluorescent antibody; lipophilic yeasts. *Mailing Add:* 9 Leaf Rd Delmar NY 12054-2607

GORDON, MORTON MAURICE, b Atlantic City, NJ, Nov 8, 24; m 50; c 2. ACCELERATOR THEORY, CYCLOTRON DESIGN. *Educ:* Univ Chicago, BS, 46; Wash Univ, PhD(physics), 50. *Prof Exp:* From instr to prof physics, Univ Fla, 50-59; assoc prof, 59-62, PROF PHYSICS, MICH STATE UNIV, 62- *Concurrent Pos:* Physicist, Oak Ridge Nat Lab, 57-58; vis prof & consult, Ind Univ, 72-73 & TRUME Cyclotron, 83; consult, TRIUMF Cyclotron, 83. *Mem:* Am Phys Soc; Am Asn Physics Teachers. *Res:* Atomic and nuclear scattering theory; accelerator theory; migma fusion theory; super conducting cyclotron design. *Mailing Add:* Cyclotron Labs Mich State Univ East Lansing MI 48824

GORDON, MYRA, b Mt Vernon, NY, Dec 12, 39. ORGANIC CHEMISTRY. *Educ:* Mt Holyoke Col, AB, 60; Univ Pittsburgh, PhD(chem), 65. *Prof Exp:* Asst prof chem, Tulane Univ, 66-70 & Univ Western Ont, 70-74; Isotopes Div, Merck, Sharp & Dohme Can Ltd, 74-87; Isotec Inc, 87-90; PRIN, MG ASSOC, 90- *Mem:* Am Chem Soc; Chem Inst Can. *Res:* Intermolecular interactions; molecular spectroscopy, particularly nuclear magnetic resonance spectroscopy; applications of stable isotopes in chemistry and medicine. *Mailing Add:* MG Assoc 612-1209 Richmond St London ON N6A 3L7 Can

GORDON, NATHAN, b New York, NY, Sept 12, 17; m 44; c 2. BIOCHEMISTRY. *Educ:* Brooklyn Col, BA, 38; Georgetown Univ, MS, 48, PhD(biochem), 52. *Prof Exp:* Chemist, US Food & Drug Admin, 46-48 & USDA, 48-52; intel res specialist, US Army Chem Corps, 52-55, dir tech opers, Intel Agency, 55-62, chief atomic, biol & chem div, US Army Foreign Sci & Tech Ctr, DC, 62-67, chief chem br, US Army Sci & Liaison Adv Group, 67-72; MGR, MIDWEST RES INST, 72- *Concurrent Pos:* Lectr, Montgomery Jr Col, 64-69; independent consult, 83- *Mem:* AAAS; Am Inst Chemists; Am Chem Soc. *Res:* Coal tar colors; pesticides; chemical warfare agents; drugs; foodstuffs; natural products; animal behavior. *Mailing Add:* 4990 Sentinel Dr Apt 301 Bethesda MD 20816

GORDON, P(AUL), b Hartford, Conn, Jan 1, 18; m 41; c 2. PHYSICAL METALLURGY. *Educ:* Mass Inst Technol, SB & SM, 40, ScD(metall), 49. *Prof Exp:* Asst steel metall, Mass Inst Technol, 40-43, group leader, Manhattan Proj & Atomic Energy Comn, 43-48, res assoc metall of uranium, 48-49; asst prof metall, Ill Inst Technol, 49-50; asst prof, Inst Study of Metals, Chicago, 50-54; chmn, Dept Metall, 66-76, PROF METALL, ILL INST TECHNOL, 54- *Mem:* AAAS; Am Soc Eng Educ; Am Soc Metals; Am Inst Mining, Metall & Petrol Engrs; Sigma Xi. *Res:* Transformations in metals; recrystallization; grain growth; order-disorder phenomena; precipitation; metal failures; surface reactions. *Mailing Add:* Metall Eng Dept Ill Inst Technol Chicago IL 60616

GORDON, PAUL, applied mathematics, for more information see previous edition

GORDON, PHILIP N, b Boston, Mass, April 21, 19; m 44; c 4. FORESTRY, ECOLOGY. *Educ:* Polytech Inst Brooklyn, BS, 42; Univ Minn, MS, 50. *Prof Exp:* Jr anal chemist, Tenn Valley Authority, 42-43; res org chemist, Pfizer, Inc, 50-67, residue chemist, 67-73, sr res scientist, 73-82; res assoc, Dept Biol, Yale Univ, 82-90; RES ASSOC, INST ECON BOT, NY BOT GARDEN, 88-, CONN FOREST & PARK ASN, 90- *Mem:* AAAS; Am Chem Soc; Soc Econ Bot; Int Asn Plant Tissue Cult; Am Inst Biol Sci; Am Soc Plant Physiol; Sigma Xi. *Res:* Forest ecology; plant genetics; antibiotics; trace analysis; plant tissue culture. *Mailing Add:* Conn Forest & Park Asn Middlefield 16 Meridian Rd Rockfall CT 06481-2961

GORDON, PHILIP RAY, b Sacramento, Calif, Mar 4, 55. CUTANEOUS GERONTOLOGY, NUTRITION. *Educ:* Univ Calif, Davis, BS, 77; Univ Mo-Columbia, MS, 80, PhD(biochem nutrit), 82. *Prof Exp:* Res fel, Ctr Biochem & Biophys Sci & Med, Harvard Med Sch, 81-84; scientist III, USDA Human Nutrit Res Ctr Aging, Tufts Univ, 84-87, instr, Dept Biochem & Pharmacol, Sch Med, 85-88, ASST PROF, SCH NUTRIT, TUFTS UNIV, BOSTON, MASS, 86-, SCIENTIST II, USDA HUMAN NUTRIT RES CTR AGING, 87-, ASST PROF, DEPT BIOCHEM, SCH MED, 88- *Concurrent Pos:* Reviewer, J Nutrit, 86- & J Invest Dermat, 87-; mem, Spec Rev Comt, Nat Inst Arthritis & Musculoskeletal & Skin Dis, 88, Nat Cancer Inst, 89; reviewer, Vet Health Serv & Res Admin, 90. *Honors & Awards:* Ruth L Pike Frontiers Nutrit Lect, Pa State Univ, 89. *Mem:* Soc Invest Dermat; Fedn Am Socs Exp Biol; Am Inst Nutrit; AAAS. *Mailing Add:* Cutaneous Geront Lab USDA Human Nutrit Res Ctr Aging Tufts Univ 711 Washington St Boston MA 02111

GORDON, PORTIA BEVERLY, b Brooklyn, NY, May 7, 52. BIOCHEMISTRY. *Educ:* Smith Col, BA, 74; Albert Einstein Col Med, MSc, 77, PhD(molecular pharmacol), 80. *Prof Exp:* Postdoctoral fel, Albert Einstein Col Med, 79-83, res assoc biochem, 83-85, ASST PROF MED, MONTEFIORE MED CTR, ALBERT EINSTEIN COL MED, 85- *Mem:* AAAS; Am Women Sci; NY Acad Sci; Am Soc Cell Biol; Fedn Am Soc Cell Biol. *Res:* Author of numerous publications. *Mailing Add:* Dept Biochem Montefiore Med Ctr Albert Einstein Col Med 111 E 210th St Bronx NY 10467

GORDON, R(OBERT), b New York, NY, June 23, 17; m 41; c 2. NUCLEAR ENGINEERING. *Educ:* Cooper Union, BSME, 40; Univ Calif, Berkeley, MS, 57, PhD(nuclear eng), 62. *Prof Exp:* Engr, Power Plant Lab, Air Mat Command, Wright Field, Ohio, 40-42 & 45; engr, Consol Vultee Aircraft Corp, 43; asst chief engr, Adv Liquid Propulsion Dept, Aerojet-Gen Corp, Gen Tire & Rubber Co, 45-51, Liquid Engine Div, 51-56; supv rep, Aerojet-Gen Nucleonics, 57-58, tech staff mem, 59-61, from sci adv to vpres & gen mgr, 61, vpres & tech dir, 61-65, mgr, Snap 8 Div, Aerojet-Gen Corp, Von Karmen Ctr, 65-68, Mech Systs Opers, 68-70; PRES, STRUCT COMPOSITES INDUSTS, INC, 71- *Concurrent Pos:* Mem tech adv comt space power, NASA/Off Advan Res & Technol, chmn subcomt dynamic power systs. *Mem:* Am Nuclear Soc; Am Inst Aeronaut & Astronaut; Soc Aerospace Mat & Process Engrs. *Res:* Nuclear reactor technology; propulsion; rocketry; power conversion; systems analysis; research and research managment; advanced composite materials and structures. *Mailing Add:* Wallace Labs Div Carter-Wallace Inc Half Acre Rd Cranbury NJ 08512

GORDON, RICHARD, b New York, NY, Nov 6, 43; m 74; c 6. MORPHOGENESIS, MEDICAL IMAGING. *Educ:* Univ Chicago, BSc, 63; Univ Ore, PhD(chem physics), 67. *Prof Exp:* Res assoc theoret biol, Ctr Theoret Biol, State Univ NY, Buffalo, 69-72; fel, Math Res Br, NIH, 72-75, expert, Image Processing Unit, 75-78; PROF RADIOL, UNIV MAN, 78-, PROF BOT, 84- *Concurrent Pos:* Assoc res prof, Dept Radiol, George Washington Univ, 74-78; adj prof, dept elec eng, Univ Manitoba, 82-; vis prof, solid mech div, Univ Waterloo, 86-88; adj prof physics & elec & computer eng, Univ Man. *Mem:* Can Asn Theoret Biologists (pres); Int Soc Diatomic Res; Micros Soc Can; Can Fedn Biol Socs; Soc Math Biol. *Res:* Morphogenesis, developmental control of genetics and differentration; neural tube defects; molecular basis of diatom shell patterns, diatom gliding motility; medical imaging; theory of computed tomography, position tomography, teleradiology; detection of early breast cancer; AIDS and condoms. *Mailing Add:* Dept Bot Univ Manitoba Winnepeg MB R3T 2N2 Can

GORDON, RICHARD LEE, b Lewiston, Idaho, Nov 6, 35; m 59; c 3. PHYSICS. *Educ:* Wash State Univ, BS, 58, PhD(physics), 66. *Prof Exp:* Physicist, Lawrence Radiation Lab, Univ Calif, 58-61; SR RES SCIENTIST SOLID STATE PHYSICS, PAC NORTHWEST LABS, BATTELLE MEM INST, 65- *Concurrent Pos:* Lectr physics, Joint Ctr Grad Study, 68- *Res:* Shock hydrodynamics; surface physics; stimulated Brillouin scattering; solid state physics. *Mailing Add:* Dept Mineral Econ Penn St Univ University Park PA 16802

GORDON, RICHARD SEYMOUR, b New York, NY, June 28, 25; m 51; c 5. AQUACULTURE, DEVELOPMENT OF NEW FOODS. *Educ:* Univ Rochester, BS, 47; Harvard Univ, Med Sci, 54; Mass Inst Technol, PhD(biochem & biophys), 54. *Prof Exp:* Chief scientist, Monsanto Co, 51-71; sr adv, Food & Drug Admin, 71-73; pres, Inst Urban Develop, 72-76; proj dir & vis prof, Harvard Univ, 77-79; dir, Newcast Ctr, 81-90, PROF AGRIBUS, SCH AGRIBUS & ENVIRON RESOURCES, ARIZ STATE UNIV, 80-; PRES, GORDON GROUP, 77- *Concurrent Pos:* Vchmn, Bd Agr, Nat Acad Sci & Nat Res Coun, 63-69; vis prof, Wash Univ, Princeton Univ & Calif Inst Technol; panel chair, White House Conf Food, Health & Nutrit, 67-70. *Mem:* Am Chem Soc; Am Soc Microbiol; AAAS; Soc Chem Indust; Poultry Sci Asn. *Res:* Mulitple use of natural resource; physiological and environmental control of growth and development in plants and animals; novel approaches to organizing and managing technically driven ventures particularly in agribusiness and technology; economic incentives for environmental restoration. *Mailing Add:* Sch Agribus & Environ Resources Ariz State Univ Tempe AZ 85287-3306

GORDON, ROBERT BOYD, b East Orange, NJ, Dec 25, 29; div; c 2. GEOPHYSICS, ENGINEERING. *Educ:* Yale Univ, BS, 52, DEng(metall), 55. *Prof Exp:* Asst prof metall, Columbia Univ, 55-57, asst prof metall, 57-60, assoc prof eng & appl sci, 60-68, prof appl sci & geol, 68-69, prof geophys & appl sci, 69-80; prof geophys & chmn dept geol, 80-83, PROF GEOPHYS & APPLIED MECH, YALE UNIV, 83- *Concurrent Pos:* Consult, Nat Acad Sci, 57-59; Regents fel, Smithsonian Inst, 91. *Mem:* Am Inst Mining, Metall & Petrol Eng; Am Geophys Union; Sigma Xi. *Res:* Applied Science; solid state science of earth materials; archaeometallurgy. *Mailing Add:* Kline Geol Lab Box 6666 Yale Univ New Haven CT 06511

GORDON, ROBERT DIXON, b Toronto, Ont, Dec 15, 36; m 60; c 2. PHYSICAL CHEMISTRY. *Educ:* McMaster Univ, BSc, 59, MSc, 61; Univ London, PhD(phys chem), 64. *Prof Exp:* Lectr chem, Univ Ibadan, 64-66; asst prof, 66-71, ASSOC PROF CHEM, QUEEN'S UNIV, ONT, 71- *Res:* Ultraviolet spectroscopy of gaseous molecules and of molecular crystals. *Mailing Add:* Dept Chem Queen's Univ Kingston ON K7L 3N6 Can

GORDON, ROBERT EDWARD, b New York, NY, June 20, 25; m 48; c 2. ZOOLOGY. *Educ:* Emory Univ, AB, 49; Univ Ga, MS, 50; Tulane Univ, PhD, 56. *Hon Degrees:* LLD, Univ Notre Dame, 89. *Prof Exp:* Cur mus, Highlands Biol Sta, 49-50; asst prof biol, Northeastern La State Univ, 54-58; from asst prof to prof biol, Univ Notre Dame, 58-90, head dept biol, 64-67, assoc dean sci, 67-71, vpres advan study, 71-89; RETIRED. *Concurrent Pos:* Ed, Am Midland Naturalist, 58-64; sect ed, Biol Abstr, 63, pres bd trustees, 78; mem gov bd, Am Inst Biol Sci, 69-77, exec comt, 71, vpres, 75, pres, 76; mem US nat comt, Int Fedn Doc, 69-71; mem US nat comt, Int Union Biol Sci, 70-75; trustee, Argonne Univ Asn, 73-82 & Univs Res Asn, 83-90; mem bd dir, Coun Grad Schs, 81-86, chmn bd, 85; mem bd dir, Grad Rec Exam Bd, 81-85. *Mem:* Am Soc Naturalists; Soc Study Amphibians & Reptiles (pres, 71); AAAS; Am Soc Ichthyol & Herpet; Ecol Soc Am; Am Inst Biol Sci (pres, 76). *Res:* Scientific communication; ecology, systematics and behavior of cold-blooded vertebrates. *Mailing Add:* 100 Chinaberry Lane LaGrange GA 30240-9569

GORDON, ROBERT JAY, b Brooklyn, NY, Feb 29, 44; m 69; c 3. PHYSICAL CHEMISTRY, CHEMICAL PHYSICS. *Educ:* Harvard Univ, AB, 65, AM, 66, PhD(chem physics), 70. *Prof Exp:* Res assoc chem physics, Calif Inst Technol, 70-72; res assoc chem, Naval Res Lab, 72-73; asst prof, 73-78, PROF CHEM, UNIV ILL, CHICAGO, 78- *Concurrent Pos:* NSF grant, 75-; Dept Energy grant, 77-; vis scholar, Stanford Univ, 80-81; Petroleum Res Fund grant, 82- *Mem:* Am Chem Soc; Am Phys Soc. *Res:* Experimental studies in chemical kinetics and laser-induced chemical reactions and theoretical studies in chemical kinetics and scattering. *Mailing Add:* Dept Chem Univ Ill Chicago Chicago IL 60680

GORDON, ROBERT JULIAN, b Seattle, Wash, July 31, 23; m 48; c 2. ENVIRONMENTAL CHEMISTRY. *Educ:* Univ Calif, Los Angeles, BS, 47, PhD(chem), 52. *Prof Exp:* Technologist petrol res, Shell Oil Co, 52-58, sr technologist, 58-59, group leader, 59-61, res supvr, Shell Develop Co, 61-64; supvry physicist, Vehicle Pollution Lab, Calif State Dept Health, 64-68; asst prof path, Sch Med, Univ Southern Caiif, 68-77, assoc clin prof, 77-78; lab mgr, Pac Environ Serv, 78-80; div mgr, Global Geochem Corp, 80-88; INDEPENDENT CONSULT, 88- *Mem:* AAAS; Am Chem Soc. *Res:* Reaction mechanisms in Grignard alkylations, aromatic sulfonation, preflame combustion reactions and photochemical air pollution; mass spectrometry; infrared spectrophotometry; chromatography; environmental carcinogens; toxic pollutants. *Mailing Add:* 335 Montecito Ave Pismo Beach CA 93449-1924

GORDON, RONALD E, b Brooklyn, NY, May 6, 49; m; c 2. LUNG PATHOLOGY. *Educ:* State Univ NY, Stony Brook, PhD(exp path), 76. *Prof Exp:* Asst prof, 79-84, DIR ELECTRON MICROSCOPY PROG, MT SINAI SCH MED, 79-, ASSOC PROF PATH, 84- *Mem:* Am Asn Pathologists; Am Soc Cell Biol; Electron Micros Soc Am; Am Acad Forensic Sci. *Mailing Add:* Dept Path Mt Sinai Med Ctr One Gustave L Levy Pl New York NY 10029

GORDON, RONALD STANTON, b Oakland, Calif, Sept 29, 37; m 63. MATERIALS SCIENCE, ENGINEERING. *Educ:* Univ Calif, Berkeley, BS, 59, MS, 61; Mass Inst Technol, ScD(ceramics), 64. *Prof Exp:* Res asst cement chem, Univ Calif, Berkeley, 56-61; res asst ceramics, Mass Inst Technol, 61-64; asst prof ceramic eng, Univ Utah, 64-69, assoc prof mat sci & eng, 69-74; pres, Ceramatec, Inc, 77-84; PROF MAT SCI & ENG, UNIV UTAH, 74-; CHMN, CERAMATEC, INC, 84- *Concurrent Pos:* Res initiation grant, NSF, 65-67; res grant, Dept Energy, 66-81; mem, staff lighting res lab, Gen Elec Co, Ohio, 71-73; NSF, Dept Energy contracts, 73-85. *Mem:* Fel Am Ceramic Soc; Nat Inst Ceramic Engrs; Electrochem Soc; Mat Res Soc. *Res:* Properties of ceramic materials; thermodynamics of solids; mechanical behavior of ceramics at elevated temperatures; solid state electrochemistry; processing and characterization of solid electrolytes and advanced ceramics. *Mailing Add:* Mat Eng Va Polytech Inst State Univ Blackburg VA 24061

GORDON, ROY GERALD, b Akron, Ohio, Jan 11, 40; m 61; c 3. PHYSICAL CHEMISTRY, SOLID STATE PHYSICS. *Educ:* Harvard Univ, BA, 61, MA, 62, PhD(chem physics), 64. *Prof Exp:* Fel, Harvard Univ, 64-66, asst prof, 66-69, prof, 69-82, THOMAS D CABOT PROF CHEM, HARVARD UNIV, 82- *Concurrent Pos:* Assoc ed, J Chem Physics & Chem Physics Letters; distinguished vis prof, Berkeley, 69, Wis, 70, Yale, 73, Oxford, 77 & Paris, 80; Einstein fel, Israel, 84; chmn, Div Chem Physics, Am Phys Soc. *Honors & Awards:* Pure Chem Award, Am Chem Soc, Baekland Award; Bourke Award, Faraday Soc. *Mem:* Nat Acad Sci; fel Am Phys Soc; fel Am Chem Soc; Am Acad Arts & Sci; Union Concerned Scientists; Europ Acad Sci. *Res:* Theory of intermolecular forces; bonding in molecules and solids; crystal structures and phase transitions; dynamics of chemical reactions; mechanisms of chemical vapor deposition; materials for solar energy and energy conservation. *Mailing Add:* Dept Chem Harvard Univ 12 Oxford St Cambridge MA 02138

GORDON, RUTH EVELYN, b Richmondville, NY. BACTERIOLOGY. *Educ:* Cornell Univ, AB, 32, MS, 33, PhD(bact), 34. *Prof Exp:* Instr, NY Vet Col, Cornell Univ, 34-37; asst bacteriologist, Div Soil Microbiol, USDA, 38-42, bacteriologist, 50-51; asst bacteriologist, Army Med Ctr, 43-44; bacteriologist, Am Type Cult Collection, 44-47, cur, 47-50; assoc res specialist, NJ Agr Exp Sta, 51-54; assoc prof, Waksman Inst Microbiol, Rutgers Univ, 54-71, prof microbiol, 71-81; VIS INVESTR, AM TYPE CULT COLLECTION, 81- *Honors & Awards:* J Roger Porter Award, US Fedn Cult Collections, 83. *Mem:* AAAS; Am Soc Microbiol; Am Acad

Microbiol; Brit Soc Gen Microbiol; Can Soc Microbiol; US Fedn Cult Collections. *Res:* Taxonomy of aerobic, sporeforming bacteria; mycobacteria; nocardiae; streptomycetes. *Mailing Add:* Am Type Cult Collection 12301 Parklawn Dr Rockville MD 20852

GORDON, RUTH VIDA, b Seattle, Wash, Sept 19, 26; m 49; c 3. STRUCTURAL ENGINEERING. *Educ:* Stanford Univ, BS, 48, MS, 49. *Prof Exp:* Struct designer, Isadore Thompson, Consult Engr, San Francisco, Calif, 50-51, K P Norrie, Consult Engr, Spokane, Wash, 51 & Bechtel Corp, San Francisco, Calif, 51-53; civil engr, Caltrans, San Francisco, Calif, 53-54; struct designer, Russell Fuller, Consult Engr, San Francisco, Calif, 54 & Western Knapp Eng Corp, San Francisco, Calif, 54-55; struct eng assoc, State Calif, Off State Architect, Struct Safety Sect, San Francisco, Calif, 56-57, sr struct designer, 57-59, sr struct engr, 59-76 & dist struct engr, 76-86; PRES, PEGASUS ENG, INC, 84- *Concurrent Pos:* Mem, Adv Comt Master Plan Use Educ Facil, San Francisco Unified Sch Dist, 71-72; vpres, Golden Gate Sect, Soc Women Engrs, 76-77, pres, 78-79, vchair & actg chair, Nat Conv, 79, chair, Affirmative Action, 79-80, 84-; dir, San Francisco Bay Area Eng Coun, 77-79, treas, 79-80, secy, 80-81, vpres, 81-82, pres, 82-83, co-chair prof, 77-87; chair, Legis Comt, Struct Engrs Asn Northern Calif, 78-79, deleg Calif Legis Coun, Prof Engrs, 78-79, chair, Prof Policies Comt, 82-83, dir, 84-86, deleg, San Francisco Bay Area Eng Coun, 85-87; affirmative action coordr, Nat Soc Women Engrs, 79-80; mem, Adv Panel Calif Bd Archit Examr Exam Rev Proj, 79-81; deleg, San Francisco Bay Area Eng Coun, Am Soc Civil Engrs, 81-85. *Mem:* Am Soc Civil Engrs; Earthquake Eng Res Inst; Asn Women Sci. *Res:* Author of one publication. *Mailing Add:* PO Box 210425 San Francisco CA 94121-0425

GORDON, SAMUEL MORRIS, b Boston, Mass, Sept 27, 98; m 23; c 2. CHEMISTRY. *Educ:* Tufts Col, BS, 22; Univ Iowa, MS, 23; Univ Wis, PhD(biochem), 26. *Prof Exp:* Asst chem, Univ Iowa, 22-23; chemist, Portland Cement Asn, Ill, 23-24; Nat Res Coun fel biol, Univ Wis, 26-28; chief bur chem & secy coun dent therapeut, Am Dent Asn, Ill, 28-38; dir res & vpres, Endo Labs, Inc, NY, 38-59; assoc, exp sta, Univ Calif, Berkeley, 59-70; RETIRED. *Concurrent Pos:* Consult drug admin, NY Health Dept, 30; ed, Dent Sci & Dent Art; Accepted Dent Remedies. *Mem:* Fel AAAS; fel NY Acad Sci; fel Am Inst Chemists; fel Am Col Angiol; fel Am Med Writers Asn; AMA; Am Chem Soc. *Res:* Medicinal chemistry. *Mailing Add:* 2270 Sloat Blvd San Francisco CA 94116

GORDON, SAMUEL ROBERT, b Alton, Ill, Feb 19, 43. MATHEMATICS. *Educ:* Calif Inst Technol, BS, 64; Yale Univ, MA, 66, PhD(math), 69. *Prof Exp:* Actg instr math, Yale Univ, 68-69; asst prof in residence, Univ Calif, Los Angeles, 69-70; asst prof, Univ Calif, Riverside, 70-77; MEM TECH STAFF, AEROSPACE CORP, 77- *Concurrent Pos:* Vis asst prof, Univ Va, 73. *Mem:* Math Asn Am; Am Math Soc. *Res:* Jordan algebras and Lie algebras and the connections between them, especially the use of techniques from algebraic groups to study the automorphism groups and Lie algebras of Jordan algebras. *Mailing Add:* Aerospace Corp PO Box 92957 MI-043 Los Angeles CA 90009-2957

GORDON, SAUL, b Bronx, NY, Nov 29, 25; m 46, 74, 82; c 7. SCIENCE EDUCATION. *Educ:* Ohio State Univ, BA, 46; Univ Ky, MS, 49, PhD(chem), 51. *Prof Exp:* Chemist, Picatinny Arsenal, 51-52, group leader, 52-55, chief basic chem res unit, 55-59, staff chemist specialist, Pyrotech Lab, 59-61; from asst prof to prof chem, Florham-Madison Campus, Fairleigh Dickinson Univ, 60-67, chmn dept, 63-67, dir annual thermoanal insts, 62-67; PRES, CTR PROF ADVAN, DIV TECHNOL ADVAN CTRS, INC, 67- *Concurrent Pos:* Govt & indust consult; chmn bd trustees, Inst Advan thru Educ. *Mem:* AAAS; Am Chem Soc; fel Am Inst Chem; Am Soc Eng Educ; Sigma Xi. *Res:* Continuing professional education for scientists, engineers and technical managers. *Mailing Add:* Ctr Prop Adv 19920 Milan Terr Boca Raton FL 33434

GORDON, SHEFFIELD, b Chicago, Ill, Feb 10, 16; m 64. RADIATION CHEMISTRY. *Educ:* Univ Chicago, SB, 37; Univ Notre Dame, PhD(phys chem), 53. *Prof Exp:* Chemist, Alton RR Co, 38-41; asst, Univ Chicago, 41-42; res assoc, Metall Lab, 42-46 & Univ Notre Dame, 46-49; chemist, Argonne Nat Lab, 50-84; RETIRED. *Concurrent Pos:* Foreign collabr, Saclay Nuclear Res Ctr, France, 65. *Mem:* Am Phys Soc; Am Chem Soc; Radiation Res Soc. *Res:* Photochemistry; chemical dynamics using fast reaction techniques. *Mailing Add:* 5000 S East End Ave Chicago IL 60615

GORDON, SHELDON P, b New York, NY, July 11, 42; m 65. APPLIED MATHEMATICS, STATISTICS. *Educ:* Polytech Inst Brooklyn, BS, 63; McGill Univ, MSc, 65, PhD(math), 69. *Prof Exp:* Lectr math, McGill Univ, 63-68; asst prof, Queens Col, NY, 68-74; PROF MATH, SUFFOLK COUNTY COMMUNITY COL, 74- *Concurrent Pos:* State Univ NY fac res fel & improv undergrad instr res award, 75, 78, 81, 85, 87 & 89; proj dir, NSF, Sci Educ proj & Course Improvement proj, Suffolk County Community Col, 77-80; pres, Mathegraphics Software, 86-; Harvard Univ Calculus Reform Proj, 89- *Mem:* Soc Indust Appl Math; Math Asn Am. *Res:* Stability theory of both differential equations and difference equations including partial difference equations; characterizations of multivariate statistical distributions using regression properties; applications of discrete mathematics to undergraduate calculus; uses of computers in mathematics and science education; computers in statistics education. *Mailing Add:* 61 Cedar Rd East Northport NY 11731

GORDON, SHELDON ROBERT, b Detroit, Mich, Sept 13, 49; m 78; c 4. HISTOCHEMISTRY, CYTOCHEMISTRY. *Educ:* Oakland Univ, Mich, BA, 72; Univ Vermont, PhD(zool), 80. *Prof Exp:* Res asst, Kresge Eye Inst, Wayne State Univ, 78-80, res assoc, 80-83; asst prof, 83-89, ASSOC PROF, OAKLAND UNIV, 89- *Mem:* Am Soc Cell Biol; Asn Res Vision & Ophthal; Sigma Xi; Histochem Soc. *Res:* Cell biology of corneal endothelial cell regeneration. *Mailing Add:* Dept Biol Scis Oakland Univ Rochester MI 48063

GORDON, STANLEY H(OWARD), electrooptics; deceased, see previous edition for last biography

GORDON, STEPHEN L, b Philadelphia, Pa, July 25, 44; m 68; c 2. MUSCULOSKELETAL DISEASES, ORTHOPEDICS. *Educ:* Drexel Univ, BS, 68, MS, 70, PhD(biomed eng), 73. *Prof Exp:* Investr, Naval Air Develop Ctr, 68-73; group head, Nat Hwy Traffic Safety Admin, 73-77; grants assoc, 78-79, PROG DIR, NIH, 79- *Concurrent Pos:* Lectr, Univ Md, 74-77. *Mem:* Am Soc Bone & Mineral Res; Orthop Res Soc. *Res:* Approve and monitor the scientific and fiscal aspects of research projects related to the musculoskeletal system and associated injuries and diseases. *Mailing Add:* NIAMS/NIH Westwood Bldg 407 Bethesda MD 20892

GORDON, SYDNEY MICHAEL, b Pretoria, SAfrica, Apr 18, 39; m 64, 84; c 4. MASS SPECTROMETRY, GAS CHROMATOGRAPHY. *Educ:* Univ Pretoria, BS, 59, MS, 62, PhD(phys chem), 65. *Prof Exp:* Chief scientist & head kinetics subdivision, Atomic Energy Bd, SAfrica, 63-77; head, Mass Spectrometry, IIT Res Inst, Chicago, 77-91, sci adv, 82-91; RES LEADER, BATTELLE COLUMBUS, 91- *Concurrent Pos:* Mem adv bd, Inst Chromatography, Pretoria, SAfrica, 75-77; mem sr tech comt, Nat Inst Petrol Energy Res, Bartlesville, OK, 85-86; chmn peer rev panel, US Environ Protection Agency, 85 & 87. *Mem:* Am Soc Mass Spectrometry; Am Chem Soc. *Res:* Analytical mass spectrometry; combined gas chromatography-mass spectrometry and applications; computer applications; analysis of environmental pollutants; analysis of biological fluids; computer applications. *Mailing Add:* Battele Columbus Labs 505 King Ave Columbus OH 43201

GORDON, TAVIA, b Chicago, Ill, Dec 14, 17; m 48; c 4. ANALYTICAL STATISTICS. *Educ:* Univ Calif, Berkeley, BA, 38. *Prof Exp:* Statistician, USPHS, 52-77; sr scientist, Gen Elec Corp, 80-82; RES PROF, GEORGE WASHINGTON UNIV, 81- *Concurrent Pos:* Statistician, Nat Heart, Lung & Blood Inst. *Mem:* Fel Am Statist Asn; fel Am Heart Asn. *Res:* Biomedical statistics; health and vital statistics. *Mailing Add:* 12901 Bluehill Rd Silver Spring MD 20906

GORDON, WAYNE ALAN, b New York, NY, Feb 2, 46; m 79; c 1. PSYCHOLOGY, REHABILITATION MEDICINE. *Educ:* NY Univ, BA, 66; Yeshiva Uinv, PhD(educ psychol), 72; Am Bd Prof Psychol, dipl, 85; Am Bd Clin Neuropsychol, dipl, 85. *Prof Exp:* Sr psychologist, 72-77, ASST PROF CLIN REHAB MED, SCH MED, NY UNIV, 77-; ASSOC PROF REHAB MED & PSYCHIAT & ASSOC DIR REHAB MED, MT SINAI SCH MED, 86- *Concurrent Pos:* Consult, United Cerebral Palsy, NY State, 79- & Rehab Sect, Pa Cancer Control Prog, 82-; prin investr, NIH, NIDRR & RSA grants. *Honors & Awards:* Licht Award, Am Cong Rehab Med, 86. *Mem:* Am Psychol Asn; Am Cong Rehab Med; Acad Behav Med Res; Soc Behav Med; Int Neuropsychol Soc. *Res:* Neuropsychological rehabilitation following traumatic brain damage; diagnosis and treatment of post-stroke depression; rehabilitation of persons with disabilities. *Mailing Add:* 420 West End Ave Apt 9B New York NY 10024

GORDON, WAYNE LECKY, b Stamford, Conn, Jan 15, 52; m 76; c 2. OVARIAN POLYPEPTIDES. *Educ:* Westminister Col, BS, 73; Univ Ill, MS, 77, PhD(physiol), 82. *Prof Exp:* Fel, Syst Cancer Ctr, Univ Tex, 81-83, res assoc, M D Anderson Hosp & Tumor Inst, 83-87; SR SCIENTIST, TANOX BIOSYSTS, 87. *Mem:* Soc Study Reproduction; Endocrine Soc; AAAS; NY Acad Sci; Sigma Xi; Am Chem Soc. *Res:* Purification and biochemical characterization of the ovarian polypeptide inhibitors inhibin-F and luteinizing hormone-receptor binding inhibitor; antibody purification and chemical modification. *Mailing Add:* 2843 E Pebble Beach Dr 10301 Stella Link Suite 110 Missouri City TX 77459-2528

GORDON, WILLIAM ANTHONY, paleontology, for more information see previous edition

GORDON, WILLIAM BERNARD, b Washington, DC, Nov 16, 35; m 61; c 2. GLOBAL ANALYSIS, RADAR SYSTEMS. *Educ:* George Washington Univ, BS, 59, MS, 60; Johns Hopkins Univ, PhD(math), 68. *Prof Exp:* Mathematician, Naval Res Lab, 60-62; asst prof math, Towson State Col, 66-68; instr, Johns Hopkins Univ, 68-69; res mathematician, Math Res Ctr, 69-72, RES MATHEMATICIAN, RADAR DIV, NAVAL RES LAB, 72- *Mem:* Am Math Soc; Inst Elec & Electrical Engs. *Res:* Global analysis; analysis on manifolds; Riemannian geometry; dynamical systems; optics. *Mailing Add:* Naval Res Lab Code 5311 Washington DC 20375

GORDON, WILLIAM E(DWIN), b Paterson, NJ, Jan 8, 18; m 41; c 2. RADIO SCATTERING. *Educ:* Montclair State Col, BA, 39, MA, 42; NY Univ, MS, 46; Cornell Univ, PhD(elec eng), 53. *Hon Degrees:* PhD, Austin Col. *Prof Exp:* Assoc dir elec eng, Res Lab, Univ Tex, 46-48; res assoc, Cornell Univ, 48-53, from assoc prof to prof elec eng, 53-65; dir, Arecibo Ionospheric Observatory, PR, 60-65; Walter R Read prof eng, 65, prof elec eng, space physics & astron, 66-86, dean sci & eng, 66-75, dean, Sch Natural Sci, 75- 80, provost & vpres, 80-86, DISTINGUISHED EMER PROF SPACE SCI, RICE UNIV, 86- *Concurrent Pos:* Vpres to pres, Int Sci Radio Union, 75-84; chmn bd trustees, Upper Atmosphere Res Corp, 71-78; mem bd trustees, Univ Corp Atmospheric Res, 77-86, 91-; trustee, Cornell Univ, 76-80; mem Arecibo Observ Adv Bd, 77-80, 90-; vpres, Int Coun Sci Unions, 88- *Honors & Awards:* Balth van der Pol Award, 66; Artowski Medal, Nat Acad Sci, 84; Medal Geophys, USSR, 85. *Mem:* Nat Acad Sci (foreign secy, 86-90); Nat Acad Eng; fel Inst Elec & Electronics Engrs; Am Meteorol Soc; fel Am Geophys Union; Am Acad Arts & Sci. *Res:* Radio physics, meteorology and waves; concept design and construction of world's largest antenna reflector. *Mailing Add:* Space Sci Rice Univ PO Box 1892 Houston TX 77251-1892

GORDON, WILLIAM EDWIN, b Lumsden, Sask, Mar 31, 19; nat US; m 43; c 3. PHYSICAL CHEMISTRY. *Educ:* Univ Sask, BA, 37, Hons, 38, MA, 40; Harvard Univ, PhD(phys chem), 43. *Prof Exp:* Res assoc & group leader, Off Sci Res & Develop Contract, Oceanog Inst Woods Hole, 43-46; asst prof chem, Univ Mo, 46-50; group & sect leader, Res & Develop Div, Arthur D Little, Inc, 50-61; mem staff, Inst Defense Anal, 61-62 & Combustion & Explosives Res, Inc, 62-65; assoc prof chem, Pa State Univ, McKeesport, 66-83; VIS RES ASSOC PROF, UNIV PITTSBURGH, 83- *Mem:* AAAS;

Am Chem Soc; Combustion Inst. *Res:* Combustion and explosion phenomena; chemical process technology; physical and numerical analysis; computer analysis of acid-base titration. *Mailing Add:* 1351 Terrace Dr Pittsburgh PA 15228-1636

GORDON, WILLIAM JOHN, b East McKeesport, Pa, Dec 4, 39; m 65. APPLIED MATHEMATICS. *Educ:* Univ Pittsburgh, BS, 61; Brown Univ, PhD(appl math), 65. *Prof Exp:* Reactor analyst, Westinghouse Elec Corp, 60-61; res mathematician, Res Labs, Boeing Co, 62; res asst appl math, Brown Univ, 62-65; sr res mathematician, Res Labs, Gen Motors Corp, 65-71, asst head dept math, 71-76; MEM STAFF, OFF NAVAL RES, LONDON, 76- *Mem:* Am Math Soc; Soc Indust & Appl Math; Asn Comput Mach. *Res:* Numerical analysis; optimization techniques; approximation theory. *Mailing Add:* PO Box 52 Palmerton PA 18071

GORDON, WILLIAM LIVINGSTON, b Tanta, Egypt, Jan 17, 27; US citizen; m 49; c 3. SOLID STATE PHYSICS. *Educ:* Muskingum Col, BSc, 48; Ohio State Univ, MSc, 50, PhD, 54. *Prof Exp:* Instr physics, Ohio State Univ, 54-55; from instr to assoc prof, 55-67, PROF PHYSICS, CASE WESTERN RESERVE UNIV, 67-, CHMN DEPT, 79- *Mem:* Am Phys Soc. *Res:* Cryogenics; structure of liquid helium; Fermi surfaces in metals; transport properties in metals; electrical properties of polymers. *Mailing Add:* Dept Physics Case Western Reserve Univ Cleveland OH 44106

GORDON, WILLIAM RANSOME, b Richmond, Va, June 1, 43; m 67; c 1. PLANT PHYSIOLOGY, DEVELOPMENTAL BOTANY. *Educ:* Tuskegee Univ, BS, 65, MS, 71; Univ Minn, PhD(plant physiolphysiol), 77. *Prof Exp:* Res assoc biol, Brookhaven Nat Lab, 77-78; ASST PROF BOT & PLANT PHYSIOL, HOWARD UNIV, 78- *Concurrent Pos:* Mem ecol comt, Sci Adv Bd, Environ Protection Agency, 79-81. *Mem:* Am Soc Plant Physiologists; Bot Soc Am; Int Soc Chronobiol; Scandinavian Soc Plant Physiol; Sigma Xi. *Res:* Physiology of plant growth and development; influence of heavy metals on nitrogen metabolism; secondary metabolism and allelopathy; phytochrome physiology; circadian rhythmicity. *Mailing Add:* Dept Bot Howard Univ Washington DC 20059

GORDUS, ADON ALDEN, b Chicago, Ill, Mar 23, 32; m 58. ANALYTICAL CHEMISTRY. *Educ:* Ill Inst Technol, BS, 52; Univ Wis, PhD, 56. *Prof Exp:* Asst, 56-57, from instr to assoc prof, 57-70, PROF CHEM, UNIV MICH, ANN ARBOR, 70-, ASSOC DIR HONORS PROG, 64- *Concurrent Pos:* John Simon Guggenheim Mem Found fel, 74. *Mem:* AAAS; Am Chem Soc; Sigma Xi. *Res:* Neutron activation analysis of archaeological artifacts, ancient and medieval coins; environmental and forensic chemistry; trace element analysis of environmental, clinical, archaeological, historical and forensic samples; synthetic fuel production by nuclear radiation. *Mailing Add:* Dept Chem Univ Mich Ann Arbor MI 48109

GORDY, EDWIN, b Philadelphia, Pa, May 17, 25; m; c 3. BIOMEDICAL ENGINEERING. *Educ:* Jefferson Med Col, MD, 48; Univ Pa, DSc(physiol chem), 52. *Prof Exp:* Intern, Jewish Hosp, Philadelphia, 48-49; head instrument design & develop dept, Roswell Park Mem Inst, 54-65; sr scientist, instrument design & develop lab, Worcester Found Exp Biol, 65-67; dir res, Lexington Instruments Corp, 67-73; vpres res & develop, Kalvex Inc, 73-80; CONSULT, BIOMED INSTRUMENTATION, 71- *Concurrent Pos:* Consult, Biomed Instrumentation, 71- *Mem:* Biophys Soc; assoc Inst Elec & Electronics Eng. *Res:* Development of new research tools for use in biological research. *Mailing Add:* 17 Otis St Boston MA 02160

GORDY, THOMAS D(ANIEL), b High Point, NC, Oct 8, 15; wid; c 4. ELECTRICAL ENGINEERING. *Educ:* Univ NC, BSEE, 36; Rensselaer Polytech Inst, MEE, 48, PhD(elec eng), 55. *Prof Exp:* Test engr, Gen Elec Co, 36-37, elec engr, 37-41, develop engr magnetics, 41-53, statist analyst, 53-56, specialist opers res & synthesis, 56-60, adminr bus res, 60-61, consult engr logistics mgt, 61-77; RETIRED. *Concurrent Pos:* Lectr, Williams Col, Mass, 56-57; adj prof, Rensselaer Polytech Inst, 57-65; adj prof, Union Col, 71-77. *Mem:* Fel Inst Elec Engrs. *Res:* Research, development and application of oriented magnetic core steels; business operations, their integration and inter-relations; design of logistics systems. *Mailing Add:* 239 Cascade Dr High Point NC 27265-8612

GORDZIEL, STEVEN A, b Santa Monica, Calif, Nov 19, 46; m 70; c 3. ANALYTICAL CHEMISTRY, PHARMACEUTICAL PRODUCT DEVELOPMENT. *Educ:* Philadelphia Col Pharm & Sci, BS, 70; Univ Conn, PhD(pharmaceut chem), 76. *Prof Exp:* Sr res scientist pharmaceut prod develop, Wyeth Labs Inc, 74-77; scientist, Ortho Pharmaceut, Inc, 77-79; mgr pharmaceut develop, 79-81, dept head, 81-84, DIR DEVELOP RES, PHARMACEUT PROD DEVELOP, CARTER-WALLACE, INC, 84- *Mem:* Am Asn Pharmaceut Scientists; Acad Pharmaceut Sci; Am Pharmaceut Asn; Parenteral Drug Asn; Controlled Release Soc. *Res:* Direct pharmaceutical scientists and analytical chemists in the physical-chemical characterization of drug substances; the development of pharmaceutical products and analytical methods and specifications. *Mailing Add:* 32 Cheston Ct Belle Mead NJ 08502

GORE, BRYAN FRANK, b Berwyn, Ill, Dec 3, 38; m 63; c 3. REACTOR OPERATIONS AND SAFETY ANALYSIS. *Educ:* Cornell Univ, BEngPhys, 61; Univ Mich, MS, 64, PhD(physics), 67. *Prof Exp:* Asst prof physics, Univ Idaho, 67-68; asst prof, Cent Wash State Col, 68-72; STAFF SCIENTIST, PAC NORTHWEST LABS, BATTELLE MEM INST, 72- *Concurrent Pos:* Consult, Bunker Hill Co, Kellogg, Idaho, 67-68; vis scientist, Ctr Theoret Physics, Univ Md, College Park, 69-70; fac researcher, Northwest Col & Univ Asn Sci (NORCUS), 72-73; mem fac, Tri Cities Br, Wash State Univ, 75-; reactor operator licensing examr, Nuclear Regulatory Comn, 81-; power plant inspections proj mgr, 84-; lead instr, Probabilistic Risk Assessment Methods & Applns Courses, US Dept Energy & US Nuclear Regulatory Comn, 88- *Mem:* Am Nuclear Soc; Am Asn Univ Prof; Am Phys Soc; Am Asn Physics Teachers. *Res:* Nuclear reactor safety analysis; nuclear reactor operations, staffing, training and qualifications; PRA applications to power plant inspections; generic PRA applications; nuclear waste management; environmental analysis; nuclear criticality; neutron interactions and transport; blanket design and environmental effects of controlled nuclear fusion and hybrid fusion-fission reactors. *Mailing Add:* Pac Northwest Labs Battelle Mem Inst Richland WA 99352

GORE, DOROTHY J, b Oklahoma City, Okla, May 1, 26. GEOLOGY, GEOGRAPHY. *Educ:* Principia Col, BS, 48; Univ Ill, MS, 52; Univ Wis, PhD(geol), 63. *Prof Exp:* Asst geologist, Ill State Geol Surv, 48-52; asst prof geol, Principia Col, 52-65; from asst prof to assoc prof earth sci, Southern Ill Univ, 65-81; PROF GEOL, HARDIN-SIMMONS UNIV, ABILENE, 81- *Mem:* Mineral Soc Am; Geol Soc Am; Nat Asn Geol Teachers; Am Asn Petrol Geologists. *Res:* Petrology and mineralogy of granite and related rocks. *Mailing Add:* Box 1304 Brackettville TX 78832

GORE, ERNEST STANLEY, b Toronto, Can, May 8, 42; m 71. INORGANIC CHEMISTRY, PHYSICAL CHEMISTRY. *Educ:* Univ Toronto, BS, 64; Univ Ill, Urbana, MS, 66, PhD(phys chem), 68. *Prof Exp:* Res chemist phys chem, E I du Pont de Nemours & Co, Inc, 73-77; res chemist inorg chem, Matthey Bishop, Inc, 77-81; GROUP LEADER INORG CHEM, JOHNSON MATTHEY, INC, 81- *Concurrent Pos:* Lectr chem, Widener Col, 77- *Mem:* Am Chem Soc. *Res:* Transition metal chemistry; organometallic compounds; homogeneous and heterogeneous catalysis. *Mailing Add:* Johnson Matthey Inc 2001 Nolte Dr West Deptford NJ 08066-1727

GORE, IRA, b New York, NY, Sept 10, 13; m 54; c 2. PATHOLOGY. *Educ:* Cornell Univ, AB, 34, MD, 37. *Prof Exp:* From intern to resident, 37-41; pathologist, Armed Forces Inst Path, 46-50; pathologist, Col Med, Baylor Univ, 50-51; pathologist, Henry Ford Hosp, 51-52; pathologist, Col Med, Univ Utah, 52-53; pathologist, Mt Sinai Hosp, Ill, 53-54; from asst clin prof to assoc clin prof path, Sch Med & assoc nutrit, Sch Pub Health, Harvard Univ, 54-62; prof path, Sch Med, Boston Univ, 62-68; prof path, Sch Med, Univ Ala, Birmingham, 68-74; dir path, Park Ridge Hosp, 74-79; clin prof path, obstet & gynec, Med Sch, Univ Ala, 82-84; RETIRED. *Concurrent Pos:* Pathologist, Vet Admin Hosp, 54-62; lectr, Harvard Med Sch, 62-68; clin prof path, Sch Med, Rochester Univ, 74-80. *Mem:* Am Soc Clin Path; AMA; Am Asn Path & Bact; Col Am Path. *Res:* Hemopoietic and cardiovascular diseases. *Mailing Add:* 5101 Kirkwall Lane Birmingham AL 35242

GORE, PAMELA J(EANNE) W(HEELESS), b Washington, DC, Oct 10, 55; m 77. PALEOENVIRONMENTAL ANALYSIS, NON-MARINE PALEOECOLOGY. *Educ:* Univ Md, Col Park, BS, 77; George Washington Univ, MS, 81, MPhil, 83, PhD(geol), 83. *Prof Exp:* Res asst geol, Univ Md, 76-77; res asst, George Washington Univ, 77-78, teaching asst, 77-79 & 80-83; asst prof geol, 83-89, ADJ PROF GEOSCI EMORY UNIV, 89-; ASST PROF GEOL, DEKALB COL, 89- *Concurrent Pos:* consult, Texaco, 85-; chmn, US working group, Int Geol Correlation Prog, Working Group 219, 85-90; prin investr, Nat Geog Soc grant, Triassic Deep River Basin, NC, 85-86, Texaco grant, Sedimentary Structures & Textures of Petrol Source Rocks, 86-88. *Mem:* Geol Soc Am; Soc Sedimentary Geol; AAAS; Int Asn Sedimentologists; Sigma Xi. *Res:* Sedimentology, stratigraphy and invertebrate paleontology of the Triassic and Jurassic Newark Supergroup in eastern North America; lacustrine sequences, fluvial sequences, paleoenvironmental and paleoecological interpretations; diagenesis of lacustrine black shales, and non-marine carbonate rocks; field trip coordinator for International Geological Congress Trip T351. *Mailing Add:* Dept Geol DeKalb Col 555 N Indian Creek Dr Clarkston GA 30021

GORE, ROBERT CUMMINS, physical chemistry; deceased, see previous edition for last biography

GORE, WILBERT LEE, b Meridian, Idaho, Jan 25, 12; m 35; c 5. PHYSICAL CHEMISTRY. *Educ:* Univ Utah, BS, 33, MS, 35. *Hon Degrees:* HHD, Westminster Col, Salt Lake City, 71. *Prof Exp:* Analyst & chem engr, Am Smelting & Refining Co, 35-41; eng supvr, Remington Arms Co, 41-45; res supvr, E I du Pont de Nemours & Co, Inc, Del, 45-57; pres, 59-76, CHMN, W L GORE & ASSOC, INC, 76- *Concurrent Pos:* Tech consult, UN Tech Mission to India, 62. *Mem:* Am Chem Soc; Inst Elec & Electronics Engrs. *Res:* Polytetrafluoroethylene structure, processing and application; rheology of thermoplastics; applications of statistical methods; sociology of enterprise; medical prosthetics; filtration membranes; transmission of electronic signals. *Mailing Add:* 487 Paper Mill Rd Newark DE 19711

GORE, WILLIAM EARL, b Kirbyville, Tex, Mar 7, 46; m 68; c 1. ORGANIC CHEMISTRY, ANALYTICAL CHEMISTRY. *Educ:* La State Univ. Baton Rouge, BS, 68, MS, 72; State Univ NY, Syracuse, PhD(org chem), 75. *Prof Exp:* Res chemist & group leader, Lederle Lab, Am Cyanamid Co, 75-81, mgr anal & res serv, Chem Res Div, 81-86; MGR ANAL CHEM, POLAROID CORP, 86- *Mem:* Am Chem Soc. *Res:* Isolation and identification of natural products; medicinal chemistry; insect chemistry; spectroscopy; organic analysis; analytical chemistry. *Mailing Add:* 35 Gray St Boston MA 02116

GORE, WILLIS C(ARROLL), b Baltimore, Md, May 20, 26; m 59; c 3. ELECTRICAL ENGINEERING. *Educ:* Johns Hopkins Univ, BE, 48, DrEng, 52. *Prof Exp:* From jr instr to assoc prof, 47-73, chmn dept, 74-80 & 86-87, PROF ELEC ENG, JOHNS HOPKINS UNIV, 73- *Mem:* Sr mem Inst Elec & Electronics Engrs; Sigma Xi. *Res:* Digital computers; information theory; algebraic coding theory. *Mailing Add:* Dept Elec Eng Johns Hopkins Baltimore MD 21218

GOREE, JAMES GLEASON, b Birmingham, Ala, June 21, 35; m 61; c 2. COMPOSITE MATERIALS, STRESS ANALYSIS. *Educ:* Univ Fla, BSME, 60; Univ Wash, Seattle, MSAE, 62; Univ Ala, PhD(eng mech), 66. *Prof Exp:* Res engr, Redstone Arsenal Res Div, Rohm & Haas Co, 61-62; asst prof eng mech, Univ Ky, 62-63; from asst prof to assoc prof, 66-75, PROF ENG MECH, CLEMSON UNIV, 75- *Mem:* Am Soc Mech Engrs; Am Soc Composites; Am Soc Testing & Mat; Am Acad Mech. *Res:* Elasticity; mathematical analysis of composite materials; fracture mechanics. *Mailing Add:* Dept Mech Eng Clemson Univ Clemson SC 29634-0921

GORELIC, LESTER SYLVAN, b Chicago, Ill, Oct 28, 40; m 64; c 3. BIOCHEMISTRY, MOLECULAR BIOLOGY. *Educ:* Ill Inst Technol, BS, 62; Univ Chicago, PhD(chem), 69. *Prof Exp:* NIH fel microbiol, Sch Med, Wash Univ, 69-71; asst prof biol, Southern Ill Univ, Edwardsville, 71-72; asst prof chem, Wayne State Univ, 72-78; ASSOC FOUND SCIENTIST, DEPT CELLULAR & MOLECULAR BIOL, SOUTHWEST FOUND RES & EDUC, 78- *Concurrent Pos:* NIH Career Develop Award, 74. *Mem:* Am Soc Biol Chemists; Sigma Xi; Am Chem Soc. *Res:* Photochemical probes of nucleic acid-protein interactions in intact nucleoprotein complexes; biochemical studies of androgen regulations of cell functions in normal and neoplastic tissue. *Mailing Add:* 11530 Old Manse San Antonio TX 78230

GORELICK, JERRY LEE, b Los Angeles, Calif, Oct 21, 46. RADIATION EFFECTS, RADIATION HARDNESS. *Educ:* Univ Calif Los Angeles, BS, 68; Pa State Univ, PhD(physics), 76. *Prof Exp:* Sr staff scientist, 85-88, SCIENTIST, HUGHES AIRCRAFT CO, 88- *Mem:* Am Phys Soc; Inst Elec & Electronics Engrs. *Mailing Add:* Hughes Aircraft Co S & CG MS 370 Bldg S33 PO Box 92919 Los Angeles CA 90009

GORELICK, KENNETH J, b New York, NY, Dec 14, 52; m 84; c 1. CRITICAL CARE MEDICINE, PULMONARY DISEASE. *Educ:* State Univ NY, Buffalo, BS, 73; Cornell Univ, MD, 78. *Prof Exp:* Asst dir, pulmonary div, Palo Alto Vet Admin Med Ctr, 83-84; med dir, Intensive Care Unit, Community Hosp, Sacramento, 84-85; assoc med dir, Fisons Corp, 85-87; DIR, CRITICAL CARE & INFECTIOUS DIS, XOMA CORP, 87- *Concurrent Pos:* Clin instr, Stanford Univ, 84-86. *Mem:* Fel Am Col Chest Physicians; Soc Critical Care Med; Am Thoracic Soc. *Res:* Development of monoclonal antibodies for the prevention and treatment of gram negative sepsis and septic shock. *Mailing Add:* Xoma Corp 2910 Seventh St Berkeley CA 94710

GORELL, THOMAS ANDREW, b Chicago, Ill, Oct 9, 40; m 65; c 2. ENDOCRINOLOGY, DEVELOPMENTAL BIOLOGY. *Educ:* Quincy Col, BS, 63; Univ Ark, MS, 66; Northwestern Univ, PhD(biol sci), 70. *Prof Exp:* Teaching asst biol sci, Northwestern Univ, 66-67, res assoc, 70; res fel biochem, Ben May Lab Cancer Res, Univ Chicago, 70-72, res assoc & asst prof, 72-75; from asst prof to assoc prof, zool & entomol, 75-81, asst dean, Col Nat Sci, 84-86, PROF BIOL & ASSOC DEAN, COL NAT SCI, COLO STATE UNIV, 86- *Mem:* AAAS; Am Soc Zoologists; Sigma Xi; Soc Study Reproduction; Endocrine Soc. *Res:* Relationship of steroid hormones and their receptor proteins during the growth, development and functioning of hormonal target tissues; enzyme induction and regulation by steroids. *Mailing Add:* Dept Biol Colo State Univ Ft Collins CO 80523

GOREN, ALAN CHARLES, b Brookline, Mass, Dec 8, 46. PHYSICAL CHEMISTRY, ENVIRONMENTAL CHEMISTRY. *Educ:* Univ Mass, BA, 68; Univ Del, PhD(chem), 75. *Prof Exp:* Res engr polyester fibers, Fiber Industs, Inc, 74-78; asst prof chem, Hollins Col, 78-79; vis prof chem, Va Polytech Inst & State Univ, 79-81; assoc prof environ chem, New England Col, 81-85; ASSOC PROF CHEM, TRANSYLVANIA UNIV, 85- *Concurrent Pos:* Assoc prof chem, Johnson C Smith Univ, 76-77. *Mem:* Am Chem Soc; Am Soc Agron. *Res:* Synthetic polymer research, flame retardant polyester fibers; surface chemistry of polyester tire cord adhesion to rubber; thermochemistry of ion-molecule reactions; soil chemistry associated with surface mine reclamation; enviromental chemistry. *Mailing Add:* Dept Chem Transylvania Univ Lexington KY 40508-1797

GOREN, HOWARD JOSEPH, b Bialocerkwe, Ukraine, Apr 9, 41; Can citizen; m 65; c 2. BIOCHEMISTRY, ENDOCRINOLOGY. *Educ:* Univ Toronto, BS, 64; State Univ NY Buffalo, PhD(biochem, pharmacol), 69. *Prof Exp:* from asst prof to assoc prof, 70-82, PROF MED BIOCHEM, UNIV CALGARY, 82- *Concurrent Pos:* Nat Res Coun Can fel, Weizmann Inst Sci, 68-70; Med Res Coun Can operating grants, 71-92, Can Diabetes Asn grant, 79-84; asst ed, Molecular Pharmacol, 74-77; vis scientist, NIH, 77-78, Med Res Coun Can, 84-85; vis prof, Harvard Med Sch, 84-85. *Mem:* Am Diabetes Asn; Am Soc Biochem & Molecular Biol; Can Soc Clin Invest; Can Biochem Soc; Am Soc Pharmacol & Exp Therapeut. *Res:* Insulin signaling mechanisms; structure-function insulin receptor and tyrosine kinase substrates. *Mailing Add:* Fac Med Univ Calgary Calgary AB T2N 4N1 Can

GOREN, MAYER BEAR, b Tomaszow, Poland, Mar 19, 21; US citizen; m 43; c 2. MICROBIOLOGY, ORGANIC CHEMISTRY. *Educ:* Rice Univ, AB, 42, AM, 43; Harvard Univ, PhD(org chem), 49. *Prof Exp:* Jr chemist, Shell Develop Co, 43-44; asst prof chem, Northeastern Univ, 50-52; chief res chemist, Kerr-McGee Corp, 52-60, chief microbiol div, 60-63; sr res scientist, Nat Jewish Ctr Immunol & Respiratory Med, 63-75 & Margaret Regan investr chem path, 75-90; from asst prof to prof microbiol & immunol, Sch Med, Univ Colo, Denver, 78-90; RETIRED. *Concurrent Pos:* Mem, Bact & Mycol Study Sect, NIH, 77-81, US Tuberculosis Panel, US-Japan Coop Med Sci Prog, 77-86, chmn, 80-86. *Mem:* AAAS; Am Chem Soc; Am Soc Microbiol. *Res:* Chemistry of lipids from tubercle bacilli-mechanisms in virulence and biological activities; anti-tumor activities; macrophage function. *Mailing Add:* 125 Locust St Denver CO 80220

GOREN, SIMON L, b Baltimore, Md, Aug 31, 36; m 62; c 1. CHEMICAL ENGINEERING. *Educ:* Johns Hopkins Univ, BES, 58, DEng(chem eng), 62. *Prof Exp:* Engr, Esso Res & Eng Co, Standard Oil Co NJ, 61-62; from asst prof to assoc prof chem eng, 62-71, PROF CHEM ENG, UNIV CALIF, BERKELEY, 71- *Concurrent Pos:* Prog dir, Particulate & Multiphase Processes Prog, NSF, 77-78. *Res:* Formation, hydrodynamic behavior and separation of particulates systems. *Mailing Add:* Dept Chem Eng Univ Calif Berkeley CA 94720

GORENSTEIN, DANIEL, MATHEMATICS. *Prof Exp:* DIR, DIMACS CTR, RUTGERS UNIV, 88- *Mem:* Nat Acad Sci. *Mailing Add:* Dimacs Ctr Rutgers Univ PO Box 1179 Piscataway NJ 08855

GORENSTEIN, DAVID GEORGE, b Chicago, Ill, Oct 6, 45; m 67. BIO-ORGANIC CHEMISTRY. *Educ:* Mass Inst Technol, SB, 66, AM, 67, Harvard Univ, PhD(chem), 69. *Prof Exp:* From asst prof to prof chem, Univ Ill, Chicago, 69-73; PROF CHEM & DIR BIOL & MED RES LAB, PURDUE UNIV, 85- *Concurrent Pos:* Nat Inst Gen Med Sci grant, 76-; NSF grant, 76-; Fulbright sr res fel, Oxford Univ, 77- *Honors & Awards:* AP Sloan Found Fel, 75; Guggenheim Fel, 86. *Mem:* Soc Biol Chemists; Int Soc Magnetic Resonance; Am Chem Soc; Sigma Xi. *Res:* Application of nuclear magnetic resonance spectroscopy to enzymology and molecular biology; physical organic and bioorganic studies of biologically important phosphate esters. *Mailing Add:* Dept Chem Purdue Univ W Lafayette IN 47907

GORENSTEIN, MARC VICTOR, b Boston, Mass, Sept 21, 50. ASTROPHYSICS. *Educ:* Mass Inst Technol, BS, 72; Univ Calif, Berkeley, PhD(physics), 78. *Prof Exp:* ASSOC ASTROPHYSICS, MASS INST TECHNOL, 78- *Mem:* Am Phys Soc; Am Astron Soc. *Res:* Cosmology; anisotropy of cosmic microwave background radiation; radio astronomy; quasars; conduct observations of compact radio sources using Very Long Baseline Interferometry technique; observations and understanding of gravitationally produced quasar images. *Mailing Add:* Chromatography Div RDE Millipore Corp 34 Maple St Milford MA 01757

GORENSTEIN, PAUL, b New York, NY, Aug 15, 34. ASTROPHYSICS, NUCLEAR PHYSICS. *Educ:* Cornell Univ, BEng Phys, 57; Mass Inst Technol, PhD(physics), 62. *Prof Exp:* Instr physics, Mass Inst Technol, 62-63; Fulbright fel & spec consult, Nat Comt Nuclear Energy, Italy, 63-65; sr scientist, Am Sci & Eng, Inc, 65-70, sr staff scientist, 70-73; ASTROPHYSICIST, CTR ASTROPHYS, 73- *Concurrent Pos:* Lectr astron, Harvard Univ, 73- *Honors & Awards:* Medal for Except Sci Achievement, NASA, 73. *Mem:* Am Phys Soc; Am Astron Soc. *Res:* X-ray astronomy and planetology, especially using nuclear techniques; nuclear instrumentation and high energy nuclear physics. *Mailing Add:* 100 Memorial Dr Cambridge MA 02142

GORENSTEIN, SHIRLEY SLOTKIN, b NY, Mar 4, 28; m 48; c 2. MATERIAL CULTURE, ANTHROPOLOGY. *Educ:* Queens Col, BA, 49; Columbia Univ, MA, 53, PhD(anthrop), 63. *Prof Exp:* Lectr anthrop, Columbia Univ, 63-71, from asst prof to assoc prof, 71-75; assoc prof anthrop, 75-78, PROF SCI & TECHNOL STUDIES, RENSSELAER POLYTECH INST, 78-, CHAIR, DEPT SCI & TECHNOL STUDIES, 82- *Concurrent Pos:* Prin investr, NSF, 65-67 & 76-79 & Grad Ctr, New Sch Social Res, 75-76; bd mem, NY State Bd Hist Preserv, 76-84; mem standards bd, Soc Prof Archaeol, 82-85. *Mem:* Am Anthrop Asn; Soc Prof Archaeologists; Asn Field Archaeol; Hist Sci Soc; Soc Hist Technol. *Res:* Science and technology studies; cultural and social contexts of science and technology; material culture research evaluates these contexts for scientific instrumentation and for technology. *Mailing Add:* Rensselaer Polytech Inst 5508 Sage Lab Troy NY 12180-3590

GORES, GREGORY J, b Devils Lake, NDak, May 22, 55; m 80; c 2. GASTROENTROLOGY. *Educ:* Univ NDak, Williston Br, AA, 74; Univ NDak, Grand Forks, BS, 76, BS, 78, MD, 80; Am Bd Internal Med, dipl, 83. *Prof Exp:* Resident, Dept Med, 80-83, fel, Div Gastroenterol, 83-86, ASST PROF MED, MAYO MED SCH, MAYO CLIN, 86-, SR ASSOC CONSULT, DIV GASTROENTEROL, DEPT INTERNAL MED, 88- *Mem:* Fel Am Col Physicians; Am Gastroenterol Asn; Am Asn Study Liver Dis; Am Soc Cell Biol; Gastroenterol Res Group; Am Fedn Clin Res; Sigma Xi; AMA; AAAS; Am Physiol Soc. *Res:* Gastroenterology; numerous publications. *Mailing Add:* Div Gastroenterol Mayo Clin Rochester MN 55905

GORESKY, CARL A, b Mundare, Alta, Aug 25, 32; m 55; c 7. INTERNAL MEDICINE. *Educ:* McGill Univ, BSc, 53, MD & CM, 55, PhD(physiol), 65; FRCP(C), 60. *Prof Exp:* Sessional lectr physiol, 63-64, from asst prof to assoc prof, 64-72, PROF MED, McGILL UNIV, 72-; DEP DIR, UNIV MED CLIN, MONTREAL GEN HOSP, 70-, HEAD, DIV GASTROENTEROL, 88- *Honors & Awards:* Royal Col Physicians & Surgeons Can Award, 63; Landis Award, Microcirculatory Soc, 82; Gold Medal, Can Liver Found, 88. *Mem:* Fel Am Col Physicians; Am Physiol Soc; Am Asn Physicians; fel AAAS; Am Soc Clin Invest. *Res:* Physiology of transcapillary exchange; multiple indicator dilution techniques to investigate exchange of materials across the sinusoids of the liver and the capillaries of the heart; analysis of dilution curves. *Mailing Add:* Univ Med Clin Montreal Gen Hosp Rm 1068 Montreal PQ H3G 1A4 Can

GORETTA, LOUIS ALEXANDER, b Portland, Ore, Aug 2, 22; m 51; c 3. ORGANIC CHEMISTRY. *Educ:* Univ Portland, BS, 43; Univ Notre Dame, MS, 44, PhD(chem), 51. *Prof Exp:* Res chemist fats & oils, Armour & Co, 50-53; res chemist petrol, Standard Oil Co, 53-59; sr chemist, 59-66, GROUP LEADER, NALCO CHEM CO, 66- *Mem:* Am Chem Soc; Sigma Xi; fel Am Inst Chemists. *Res:* Hydrocarbons; industrial chemicals; polymers. *Mailing Add:* 321 Osage Lane Naperville IL 60540-7820

GORFIEN, HAROLD, b New York, NY, Aug 29, 24; m 51; c 2. PLANT PHYSIOLOGY, FOOD PROCESSING & PRESERVATION. *Educ:* City Col New York, BS, 48; Univ Mass, MS, 51. *Prof Exp:* Processed food inspector, Prod & Mkt Admin, USDA, 50-51; res assoc, Univ Mass, 51-53; food spec supvr, USN Res & Develop Facil, 53-56; food technologist, DCA Food Industs, Inc, 56-62 & US Navy Res & Develop Facil, 62-66; FOOD TECHNOLOGIST, US ARMY NATICK RES & DEVELOP CTR, 66- *Concurrent Pos:* Teacher, NY Inst Dietetics, 63-66, Essex Agr & Tech Inst, 71-73. *Mem:* Am Chem Soc; Inst Food Technologists; fel AAAS. *Res:* Post harvest physiology and biochemistry; enzyme and controlled atmosphere studies; development of cereal, hydrocolloid, nutritional and dairy products; development of processes for adhesion and coating. *Mailing Add:* Food Eng Lab US Army Natick Res & Develop Ctr Natick MA 01760

GORFIEN, STEPHEN FRANK, b Staten Island, NY, Feb 4, 58; m 85; c 2. CELL CULTURE. *Educ:* State Univ NY, BS, 80, PhD(exp path), 85. *Prof Exp:* Res assoc, Connective Tissue Res Inst, Philadelphia, 85-88, Am Heart Asn spec investr, 88-89; res scientist, 89-90, SR SCIENTIST, GIBCO LAB, CELL CULT RES & DEVELOP, GRAND ISLAND, 90- *Mem:* Am Soc Cell Biol; NY Acad Sci; Am Col Nutrit; AAAS; Tissue Cult Asn. *Res:* Cell culture; numerous publications. *Mailing Add:* Media Develop Dept Cell Cult Res & Develop Gibco Labs 2086 Grand Island Blvd Grand Island NY 14072

GORGES, HEINZ A(UGUST), b Stettin, Ger, July 22, 13; US citizen; m 57. MECHANICAL ENGINEERING. *Educ:* Dresden Tech Univ, ME, 38; Hanover Tech Univ, PhD(mech eng), 46. *Prof Exp:* Group leader supersonics, Aero Res Estab, Ger, 40-45; scientist, Royal Aircraft Estab, Eng, 46-49; prin sci officer, Weapons Res Estab, SAustralia, 49-59; sci asst aeroballistics, Marshall Space Flight Ctr, NASA, Ala, 59-61; dir adv projs, Cook Technol Ctr, Ill, 61-62; sci adv, IIT Res Inst, 62-66; asst vpres, Environ & Phys Sic Div, Tracor, Inc, 66-72, vpres & chief engr, Tracor-Jitco, Inc, 72-75; PRES, VINETA INC, 75- *Concurrent Pos:* Prof, Redstone Exten, Univ Ala, 60-61. *Mem:* Assoc fel Am Inst Aeronaut & Astronaut; Am Soc Mech Engrs; Acoust Soc Am. *Res:* Supersonics and hypersonics; aerodynamics of propulsion; energy management; energy systems; life cycle economics. *Mailing Add:* 3705 Sleepy Hollow Rd Falls Church VA 22041

GORGONE, JOHN, b Johnstown, Pa, Dec 23, 41. COMPUTER INFORMATION SYSTEMS. *Educ:* Univ Southern Colo, BA, 66; Univ Northern Colo, MA, 66; Southern Ill Univ, PhD(computer info systs), 74. *Prof Exp:* Chmn dept, 77-85, PROF COMPUTER INFO, BENTLEY COL, 77- *Concurrent Pos:* Actg dean, Grad Sch, Bentley Col, 89-90. *Mem:* Asn Comput Mach; Inst Elec & Electronics Engrs; Soc Info Mgt. *Mailing Add:* Computer Info Systs Dept Bentley Col 175 Forest St Waltham MA 02154-4705

GORHAM, ELAINE DEBORAH, b Brattleboro, Vt, July 19, 45. GEOPHYSICS. *Educ:* Brown Univ, ScB, 67; Northeastern Univ, PhD(physics), 74. *Prof Exp:* Fel, 74-76, staff physicist, 76-79, mem tech staff, 81-87, TECH SUPVR, SANDIA LABS, 87- *Concurrent Pos:* Foreign sci attache, Nuclear Regulatory Comn, Caderache, France. *Mem:* Am Phys Soc. *Res:* Blasting technology for oil shale mining; hydrogen diffusion in metals; electrical conductivity in semiconductors and low temperature metals; reactor safety core debris coolability; equations of state and other materials properties of reactor fuel. *Mailing Add:* Div 6344 Sandia Labs Albuquerque NM 87185

GORHAM, EVILLE, b Halifax, NS, Oct 15, 25; m 48; c 4. LIMNOLOGY, BIOGEOCHEMISTRY. *Educ:* Dalhousie Univ, BSc, 45, MSc, 47; Univ London, PhD(plant ecol), 51. *Prof Exp:* Lectr bot, Univ Col, Univ London, 50-54; ecologist, Brit Freshwater Biol Asn, 54-58; lectr bot, Univ Toronto, 58-59, asst prof, 59-62; assoc prof, Univ Minn, 62-65; prof biol & head dept, Univ Alta, Calgary, 65-66; head dept bot, 67-71, prof bot, 66-75, prof ecol, 75-84, REGENTS' PROF ECOL & BOT, UNIV MINN, ST PAUL, 84- *Concurrent Pos:* Mem vis comt to rev progs in toxicol, Nat Acad Sci-Nat Res Coun, 74-75; mem coord comt sci & tech assessment environ pollutants, Environ Studies Bd, Nat Acad Sci, Nat Res Coun, 75-78, mem comt med & biologic effects environ pollutants, Assembly Life Sci, Nat Acad Sci-Nat Res Coun, 76-77; mem NATO, Adv Res Inst Acid Precipitation, Can, 78 & Adv Res Wkshp Acid Deposition, Can, 85; mem comt atmosphere & biosphere, Bd Agr & Renewable Resources, Nat Acad Sci, Nat Res Coun, 79-81; mem, diesel impact study comt, Nat Res Coun-Nat Acad Eng, 80-81; mem, US, Can, Mex Tri-Acad Comt, Acid Precipitation, Environ Studies Bd, Nat Acad Sci, Nat Res Coun, Royal Soc Can, Mex Acad Sci, 81-84; mem panel to rev final reports for US/Can Memorandum Intent Transboundary Air Pollution, Royal Soc Can, 82-83 & rev panel Can fed govt progs on long-range transport air pollution, 84; environ consult for var US & Can govt agencies; mem workshops on global change, Royal Soc Can, 85, 90. *Honors & Awards:* George C Wheeler Distinguished Lectr, Univ NDak, 70; CP Snow Lectr, Ithaca Col, 80 & 89; Moore Lectr, Univ Va, 81; Oosting Mem Lectr, Duke Univ, 83; G Evelyn Hutchinson Medal, Am Soc Limnol & Oceanog; Sigurd Olson Award, Sierra Club, 86. *Mem:* Fel Sci Inst Pub Info; Ecol Soc Am; Am Soc Limnol & Oceanog; Int Asn Limnol; Int Asn Ecol; Soc Conserv Biol; fel AAAS; Acid Rain Found (secy-treas, 82-84); hon mem Swed Phytogeographical Soc; British Ecol Soc; fel Royal Soc Can. *Res:* Biogeochemical aspects of ecology, limnology and soil science; wetland peatland ecology; chemistry of atmospheric precipitation with special attention to acid deposition; history of ecology; biochemistry. *Mailing Add:* Dept Ecol Evolution & Behavior Univ Minn 318 Church St Minneapolis MN 55455

GORHAM, JOHN FRANCIS, b Medford, Mass, Sept 24, 21; m 48; c 2. CHEMICAL ENGINEERING. *Educ:* Univ Maine, BS, 50, MS, 52. *Prof Exp:* Jr engr, Stamford Lab, Am Cyanamid Co, 52-53; from instr to assoc prof chem eng, 62-86, chmn dept, 81-86, EMER PROF CHEM ENG, UNIV MAINE, 86- *Mem:* Am Inst Chem Engrs; Tech Asn Pulp & Paper Indust. *Res:* Process dynamics and control; pulp and paper technology; electronic computer specialties. *Mailing Add:* Five Maplewood Ave Orono ME 04473

GORHAM, JOHN RICHARD, b Puyallup, Wash, Dec 19, 22; m 44; c 2. VETERINARY MEDICINE. *Educ:* State Col Wash, BS & DVM, 46, MS, 47; Univ Wis, PhD, 54. *Prof Exp:* Asst prof path, State Col Wash, 47-51; asst virol, Univ Wis, 51-53; vet-in-chg, Fur Animal Dis Sta, 53-66, DIR, ANIMAL DIS UNIT, AGR RES SERV, WASH STATE UNIV, USDA, 66- *Concurrent Pos:* Assoc prof path, Wash State Univ, 53-57, prof, 57-79. *Mem:* Am Soc Exp Path; Am Vet Med Asn. *Res:* Virology and epizootiology; virus and rickettsial diseases of carnivores; slow virus diseases. *Mailing Add:* FDA HFF 237 200 C St SW Washington DC 20204

GORHAM, JOHN RICHARD, b Montgomery Co, Ohio, July 27, 31; m 52; c 7. MEDICAL ENTOMOLOGY. *Educ:* Miami Univ, AB, 53, MS, 56; Ohio State Univ, PhD(entom), 60; Malaria Eradication Training Ctr, Jamaica, dipl, 61. *Prof Exp:* Res asst invert zool, Univ NMex, 53; prev med technician, Army Health Nursing Serv, Walter Reed Army Med Ctr, 54-55; grad asst zool, Miami Univ, 55-56, instr, 56-57; grad asst, Ohio State Univ, 57, asst instr, 58-60; consult entomologist malaria eradication, Pan Am Health Orgn, Paraguay, 61-63; res assoc, Inst Int Med, Univ Md, 63-64; res assoc, Pakistan Med Res Ctr, 64-65; sr scientist, Vector Borne Dis Training Sect, Ctr Dis Control, Ga, 66-69, res entomologist, Arctic Health Res Ctr, Alaska, 69-73; RES ENTOMOLOGIST, FOOD & DRUG ADMIN, 73- *Concurrent Pos:* USPHS trainee, Sch Trop Med, Univ PR, 60-61; res assoc, Conserv Found, NY, 58-59; aquatic entomologist, Maine Forest Serv, 58-59; vis lectr, Atlanta Baptist Col, 68-69; consult entomologist, Alaska Air Command, 70-73; lectr, Univ Alaska, 71-73; Food & Drug Admin liaison rep, Armed Forces Pest Mgt Bd, 74-; scientist dir, USPHS, 76-; adj assoc prof prev med, Uniformed Serv Univ Health Sci, 90- *Mem:* Entom Soc Am; Int Asn Ecol; Am Registry Prof Entomologists; Int Asn Milk, Food & Environ Sanitarians; Soc Vector Ecol; Nat Environ Health Asn. *Res:* Ecology and systematics of mosquitoes; ecology of Paraguay; arctic ecology; pest management; terrestrial and aquatic ecology; zoonoses; ecology of vector-borne diseases; malariology; arbovirology; ecology of pesticides; myrmecology; venomous arthropods; food pests. *Mailing Add:* Food & Drug Admin 200 C St SW Washington DC 20204-0001

GORHAM, PAUL RAYMOND, b Fredericton, NB, Apr 16, 18; m 43; c 3. PLANT PHYSIOLOGY, PHYCOLOGY. *Educ:* Univ NB, BA, 38; Univ Maine, MS, 40; Calif Inst Technol, PhD(plant physiol), 43. *Hon Degrees:* DSc, Univ NB, 73. *Prof Exp:* Agr asst plant physiol div bot & plant path, Dom Dept Agr, Ottawa, 43-45; jr res officer div biosci, 45-46, asst res officer plant sci invests, 46-51, from assoc res officer plant physiol sect to prin res officer plant physiol sect, Nat Res Coun Can; prof bot, 69-83, chmn dept, 71-79, assoc dir, bot garden, 76-83, EMER PROF BOT, UNIV ALTA, 83- *Concurrent Pos:* Head plant physiol sect, Nat Res Coun Can, 52-69; mem Can comt, Int Biol Prog, 67-74, secy, 68-69; mem adv comt, Alta Oil Sands Environ Res Prog, 75-76. *Honors & Awards:* Centennial Medal, 67; Silver Jubilee Medal, 78; Mary E Elliot Award, Can Bot Soc, 79, George Lawson Medal, 88; Gold Medal, Can Soc Plant Physiol, 87. *Mem:* Phycol Soc Am; Can Soc Plant Physiol (pres, 58-59); Can Biochem Soc; Am Soc Plant Physiologists; Can Bot Asn (pres, 77-78); Int Soc Limnol; fel Royal Soc Can; fel Rawson Acad Aquatic Sci. *Res:* Phloem translocation; physiology of submerged aquatic macrophytes; toxic blue-green algae; hydrobiology. *Mailing Add:* 12408 49th Ave Edmonton AB T6H 0H2 Can

GORHAM, R C, ELECTRICAL ENGINEERING. *Prof Exp:* FAC MEM, UNIV PITTSBURGH. *Mem:* Inst Elec & Electronics Engrs. *Mailing Add:* Univ Pittsburgh Benedum Hall Rm 437 Beh Pittsburgh PA 15261

GORHAM, WILLIAM FRANKLIN, b Brandon, Vt, Aug 13, 26; m 53; c 4. ORGANIC CHEMISTRY, POLYMER CHEMISTRY. *Educ:* Univ Miami, Ohio, AB, 48; Mass Inst Technol, PhD(chem), 51. *Prof Exp:* Sr res chemist, Bakelite Corp, 51-56; group leader, Plastics Div, 56-65, asst dir res, 65-72, ASSOC DIR RES & DEVELOP, CHEM & PLASTICS, UNION CARBIDE CORP, 72-; MGR RES & DEVELOP, ENG POLYMER, 81- *Mem:* Am Chem Soc. *Res:* Synthetic organic chemistry; management of research and development on thermoplastic polymers, new product invention and market development, parylene vapor deposition technology, applications for parylene, analytical chemistry and the combustibility properties of plastics; product safety. *Mailing Add:* Union Carbide Plastics 39 Old Ridgebury Rd Danbury CT 06817-0001

GORI, GIO BATTA, b Tarcento, Italy, Feb 23, 31; US citizen; m 58; c 2. TOXICOLOGY, EPIDEMIOLOGY. *Educ:* Univ Camerino, ScD, 56; Harvard Univ, SMG cert, 76; Johns Hopkins Univ, MPH, 78; Acad Toxicol Sci, dipl, 82. *Prof Exp:* Assoc microbiol, Ist Superiore Sanita, Italy, 56-58; assoc virol, Univ Pittsburgh, 58-59; assoc dir, Ist Sclavo, Italy, 59-60; assoc virol, Wistar Inst, Univ Pa, 60-62; dir qual control & asst to pres, Microbiol Assocs Inc, 62-63, dir prod, 63-65; head virol, toxicol & immunol dept, Melpar Inc, 65-67; dir biol res lab, Litton Systs Inc, 67-68; assoc sci dir prog etiology, Nat Cancer Inst, 68-73, dep dir, Div Cancer Causes & Prev, 73-80; vpres, 80-88, DIR, POLICY ANALYSIS CTR, FRANKLIN INST, 88-, DIR, HEALTH POLICY CTR, 88- *Honors & Awards:* Super Serv Award, USPHS, 76. *Mem:* AAAS; Am Pub Health Asn; Soc Toxicol; Int Soc Regulatory Toxicol & Pharmacol. *Res:* Algae; halophilic bacteria; virus epidemiology; production and control of polio vaccines; cell physiology and transformation; continuous cultivation of cells and viruses; environmental analysis and toxicology; chemical carcinogenesis; tobacco smoking and health; nutrition and cancer epidemiology; health economics management; policy analysis of health and safety regulation. *Mailing Add:* 6503 Pyle Rd Bethesda MD 20817

GORIN, GEORGE, b Como, Italy, Aug 19, 25; m 52; c 2. PHYSICAL BIOCHEMISTRY, CHEMICAL INFORMATION. *Educ:* Brooklyn Col, AB, 44; Princeton Univ, MA, 47, PhD(chem), 49. *Prof Exp:* Asst, Princeton Univ, 44; chemist, Heyden Chem Corp, 45; res assoc, Rutgers Univ, 49-50; fel, Purdue Univ, 51; asst prof chem, Univ Ore, 52-55; from asst prof to assoc prof, 55-61, PROF CHEM, OKLA STATE UNIV, 62- *Concurrent Pos:* NIH career develop award, 63-73. *Mem:* AAAS; Am Chem Soc; Am Soc Biol Chemists. *Res:* Enzymes; kinetics and thermodynamics of biochemical reactions; radiation damage; sulfur compounds; chemical information. *Mailing Add:* Dept Chem Okla State Univ Stillwater OK 74078

GORIN, PHILIP ALBERT JAMES, b Bristol, Eng, Dec 26, 31; m 56; c 4. CHEMISTRY. *Educ:* Bristol Univ, BSc, 52, PhD(carbohydrate chem), 56. *Prof Exp:* Sr res officer. Prairie Regional Lab, Nat Res Coun Can, 55-; ADJ PROF, DEPT BIOCHEM, FED UNIV PARANA. *Mem:* Fel Chem Inst Can. *Res:* Carbohydrates, especially the structure of microbiological products and their properties; proton and C nmr of mono-, oligo- and polysaccharides yeast lichen and protozoal polysaccharides. *Mailing Add:* Dept Biochem CP19046 UFPR 80 000 Curitiba Brazil

GORING, DAVID ARTHUR INGHAM, b Toronto, Ont, Nov 26, 20; m 48; c 3. WOOD CHEMISTRY. *Educ:* Univ Col, London, BSc, 42; McGill Univ, PhD(phys chem), 49; Cambridge Univ, PhD(colloid chem), chem), 53. *Prof Exp:* Asst res officer phys chem, Atlantic Regional Lab, Nat Res Coun Can, 51-55; scientist II, Pulp & Paper Res Inst, 55-60, res group leader, 60-71, dir res, 71-76, vpr es sci 76-83, prin scientist, 69-85, vpres acad, 83-85; PROF CHEM ENG, UNIV TORONTO, 86- *Concurrent Pos:* Res assoc, McGill Univ, 56-68, sr res assoc, 68-86. *Honors & Awards:* Anselme Payen Award, Cellulose, Wood & Fiber Chem Div, Am Chem Soc, 73; Gunnar Nicholson Tappi Gold Medal, Tech Asn Pulp & Paper Indust, 86. *Mem:* Fel Chem Inst Can; Can Pulp & Paper Asn; fel Royal Soc Can; fel Int Acad Wood Sci; fel Tech Asn Pulp & Paper Indust. *Res:* Chemistry of wood and wood pulp. *Mailing Add:* Dept Chem Eng & Appl Chem Univ Toronto Toronto ON M5S 1A4 Can

GORING, GEOFFREY E(DWARD), b New York, NY, Nov 8, 20; m 55; c 3. CHEMICAL ENGINEERING, PHYSICS. *Educ:* Yale Univ, BE, 42; Mass Inst Technol, ScD(chem eng), 49; Fairleigh Dickinson Univ, MS, 62. *Prof Exp:* Proj leader, Pittsburgh Consol Coal Co, Pa, 49-52; proj scientist, Standard Oil Co Ind, 52-54; proj dir, Am Messer Corp, NY, 54-56; tech liaison, Argonne Nat Lab, Ill, Union Carbide Corp, 56-69, sr scientist, NY, 59-62; asst div dir, Atlantic Res Corp, Va, 62-63; assoc prof eng sci, 63-67, PROF ENG SCI, TRINITY UNIV, 67- *Mem:* Am Phys Soc. *Res:* Thermodynamics; applied mathematics; environmental science; materials science. *Mailing Add:* 224 Gardenview San Antonio TX 78213

GORLA, RAMA S R, b India; US citizen. HEAT TRANSFER, FLUID MECHANICS. *Educ:* S V Univ, BS, 63; Indian Inst Technol, MS, 65; Univ Toledo, PhD(mech eng), 72. *Prof Exp:* Lectr mech eng, S V Univ, 65-66; asst prof space eng & rocketry, Birla Inst Technol, 66-68; res asst mech eng, Univ Toledo, 68-72; res engr, Teledyne CAE, 72-73 & Chrysler Corp, 73-74; asst prof mech eng, Gannon Univ, 74-77; PROF MECH ENG, CLEVELAND STATE UNIV, 77- *Concurrent Pos:* Prin investr, Lewis Res Ctr, NASA, 79- *Mem:* Am Soc Mech Engrs; Am Soc Eng Educ. *Res:* Computational and experimental aspects of heat transfer, fluid flow, rheology, finite elements and numerical analysis. *Mailing Add:* 8041 Oxford Dr Strongsville OH 44136

GORLAND, SOL H, b Brooklyn, NY, Sept 14, 41; m 66; c 2. POWER SYSTEMS, PROPULSION SYSTEMS. *Educ:* Cooper Union, BS, 62; Univ Toledo, MS, 70. *Prof Exp:* Aerospace engr, Power Syst Div, 62-75, prog mgr, Off Energy Prog, Hq, 75-76, head, Power Syst Sect, 77-80, mgr syst anal off, Lewis Res Ctr,81-84, CHIEF, LAUNCH VEHICLE TECHNOL BR, NASA, 84- *Concurrent Pos:* Mem, Energy Res & Develop Admin, Dept Energy Geothermal Adv Panel, 75-78 & Space Syst Technol Adv Comt, NASA, 78-80 & Aerospace Power Systs Tech Comt, Am Inst Aeronaut & Astronaut, 78-81; assoc ed, J Energy, 80-84. *Mem:* Assoc Fel Am Inst Aeronaut & Astronaut. *Res:* Systems analysis for space power and propulsion; technology for primary propulsion systems for space. *Mailing Add:* Lewis Res Ctr NASA 21000 Brookpark Rd MS 500-219 Cleveland OH 44135

GORLICK, DENNIS, b Detroit, Mich, Dec 17, 44. ANIMAL BEHAVIOR, NEUROBIOLOGY. *Educ:* Wayne State Univ, BSc, 67, MS, 71; Univ Hawaii, PhD(zool), 78. *Prof Exp:* Actg asst prof biol, Univ Hawaii, 77-78, vis asst prof zool, 78-79; NIMH fel, Princeton Univ, 80-81; NIMH fel, 82-84, ASSOC RES SCIENTIST, COLUMBIA UNIV, 84- *Mem:* Animal Behav Soc; AAAS; Soc Neurosci. *Res:* Neural basis of aggressive behavior; ontogeny of behavior; evolution and function of interspecific communication systems. *Mailing Add:* 601 W 113th St New York NY 10027

GORLIN, RICHARD, b Jersey City, NJ, June 30, 26. MEDICINE. *Educ:* Harvard Med Sch, MD, 48; Am Bd Internal Med, dipl. *Prof Exp:* Intern & med house officer, Peter Bent Brigham Hosp, 48-49, asst, 49-51, sr asst res, 51-52, chief resident physician, 53-54, from assoc to sr assoc, 57-66, physician, 66-69, chief cardiol, 69-74; ROSENBERG PROF MED & CHMN DEPT, MT SINAI SCH MED, 74-, PHYSICIAN-IN-CHIEF, MT SINAI HOSP, 74- *Concurrent Pos:* Res fel med, Harvard Med Sch, 49-51, teaching fel, 51-52; Moseley traveling fel, St Thomas' Hosp, London, 52-53; Brower traveling scholar, Am Col Physicians, 60; instr med, Harvard Med Sch, 56-58, assoc, 58-61, asst prof, 61-68, assoc prof, 68-74; attend physician, US Vet Hosp, Rutland, 58-; lectr, US Naval Hosp, Chelsea, 59-; clin asst, St Thomas' Hosp, London, 60; consult, Nat Heart & Lung Inst. *Mem:* Am Physiol Soc; Am Soc Clin Invest; Am Heart Asn; Am Fedn Clin Res; fel Am Col Cardiol. *Res:* Academic medicine; cardiac physiology. *Mailing Add:* Mt Sinai Sch Med Mt Sinai Med Ctr One Gustave Levy Pl New York NY 10029

GORLIN, ROBERT JAMES, b Hudson, NY, Jan 11, 23; m 52; c 2. ORAL PATHOLOGY, CLINICAL GENETICS. *Educ:* Columbia Univ, AB, 43; Wash Univ, DDS, 47; Univ Iowa, MS, 56; Am Bd Oral Path, dipl, 56; Am Bd Med Genetic, dipl, 84. *Hon Degrees:* DSc, Univ Athens, 82. *Prof Exp:* Instr dent, Columbia Univ, 50-51; from assoc prof to prof Oral Path, chmn div, 58-79, chmn dept oral path, 79-91, REGENTS PROF, UNIV MINN, 78- *Concurrent Pos:* Fulbright exchange prof & Guggenheim fel, Royal Dent Sch, Copenhagen, 61; consult, Univ Minn Hosps, 60-, dir human genetics clin, 71-; consult, US Vet Admin Hosp, Minneapolis, Nat Fedn Birth Defects, Glenwood Hills Med Ctr, Mt Sinai Hosp, Hennepin County Gen Hosp, Ramsey County Gen Hosp, Minneapolis Children's Hosp, St Paul Children's Hosp. WHO, NIH & coun dent educ, Am Dent Asn & Armed Forces Inst Path; assoc ed, Am J Human Genetics, 70-73, J Oral Path, 72- & J Maxillo-Facial Surg, 73-, Cleft Palate J, 76-, Embryol & Pathogenesis & Prenatal Diag, 77, Am J Med Genetics, 77-, J Craniofacial Genetic Develop Biol, 80-, J Clin Dysmorphol, 82-85, J Gerodontics, 84-89, Dysmorph Clin Genet, 86-, Birth Defects Encycl, 86-; mem, Minn Human Genetics League; consult dent study sect, NIH; dir, Am Bd Oral Path, 70-75; mem adv comt, Nat Found Clin Res, 74-90; mem bd dirs, Group Health, Inc; prof path, prof derm & prof pediat, Sch Med, Univ Minn, 71-, prof obstet & gynec & prof otolaryngol, 73-; vis prof, Sch Med, Tel-Aviv Univ, 81, Sch Dent, Jerusalem, 81, Univ Mo, 84, Univ Ill, 84, Univ Pittsburgh, 87; Robert J Gorlin vis prof dysmorphol, 89, Burroughs Wellcome vis prof, Royal Soc Med, London, 90-91. *Honors & Awards:* Howard Fox Lectr, NY Acad Med, 68; Carl M Lingamfelter Lectr, Univ Va, 71; Frank Hoopes Lectr, Wilmington Del, 77; Wincompleck Lectr, Univ Okla, 82; Chase Mem Lectr, Univ Minn, 83; Samuel Charles Miller Award, Am Acad Oral Med, 85; Frederick Birnberg Res Award, Columbia Univ, 87; Col Harland Sanders Award, March of Dimes, 89; Second Edward Sheridan Lectr, Dublin, Ireland, 89; Windermere Lectr, British Pediat Asn, 90. *Mem:* Am Dent Asn; fel Am Acad Oral Path (secy-treas, 58-64, vpres, 64-65, pres, 66-67); Am Acad Dermat; Int Asn Dent Res; Int Soc Craniofacial Biol (pres, 69-70); Am Soc Human Genetics; Birth Defects & Clin Genetics Soc; fel Royal Col Surgeons Eng. *Res:* Pediatrics; relationships between oral and systemic disease; oral syndromes; human genetics. *Mailing Add:* Depts Oral Path & Genetics Univ Minn Health Sci Ctr Minneapolis MN 55455

GORMAN, ARTHUR DANIEL, b Chicago, Ill, Oct 31, 46; m 68, 85; c 2. ASYMPTOTICS, CAUSTICS. *Educ:* Univ Ill, BS, 68; Pa State Univ, PhD(physics), 80. *Prof Exp:* Syst engr develop, Univac, 68-69; prog engr, Gen Elec, 69-71; res asst, Wash Univ, 71-74; res asst develop, Appl Res Lab, Pa State Univ, 74-82; ASST PROF, LAFAYETTE COL, 82-, CHMN DEPT, 88- *Mem:* Am Math Soc; Sigma Xi. *Res:* Asymptotic solution of differential equations near caustics. *Mailing Add:* 312 March St Apt 5 Easton PA 18042

GORMAN, COLUM A, b Mayobridge, North Ireland, June 27, 36; m 61; c 4. ENDOCRINOLOGY, INTERNAL MEDICINE. *Educ:* Queens Univ, MB & BCh, 59; Univ Minn, PhD(med, biochem), 68. *Prof Exp:* NIH fel, 64-65; consult internal med, Mayo Clin, 66, from instr to asst prof med, Mayo Grad Sch, 66-73, from asst prof to assoc prof, 73-81, chmn, div endocrinol, 85, PROF MED, MAYO MED SCH, 81- *Mem:* Am Fedn Clin Res; Am Thyroid Asn (secy, 84-); Endocrine Soc; fel Am Col Physicians. *Res:* Thyroxine and triiodothyronine; ophthalmopathy of Graves' disease; thyroid carcinoma; thyroid radiation dosimetry. *Mailing Add:* Mayo Clin W18A Rochester MN 55905

GORMAN, CORNELIA M, b El Paso, Tex, Dec 16, 51. BIOCHEMICAL GENETICS. *Educ:* Univ Tex, El Paso, BS, 74; Wash State Univ, MS, 76, PhD(genetics), 81. *Prof Exp:* Teaching asst biol & zool, Wash State Univ, 74-77, res asst biochem, 78-80; AM CANCER SOC FEL, NAT CANCER INST, 81- *Mem:* AAAS; Asn Women Sci; Am Soc Cell Biol. *Res:* Construction of eucaryotic vectors which will allow for increased efficiency in modifying mammalian cells, involving development of new cloning vehicles. *Mailing Add:* Bldg 37 - Rm 2E10 NIH Bethesda MD 20814

GORMAN, EUGENE FRANCIS, b Brooklyn, NY, Nov 15, 26; m 53; c 4. WELDING TECHNOLOGY, ROLL FORMED METAL SHAPES. *Educ:* Columbia Univ, BS, 52. *Prof Exp:* Develop engr, Welding Lab, Linde Div, Union Carbide Corp, 52-65, proj engr, 65-71; owner, Weld Accessories Co, 71-73; sr engr, Cent Res & Eng Lab, Airco, Inc, 73-79; mfg engr, Teledyne, Inc, 79-81, mgr eng, Metal Forming, 81-91; CONSULT, MFG ENG CO, 91- *Concurrent Pos:* Lectr, welding technol, Fairleigh Dickenson Univ & welding metall, Polytech Inst NY, 77. *Honors & Awards:* Lincoln Gold Medal, Am Welding Soc, 62. *Mem:* Am Soc Mech Engrs; Nat Soc Prof Engrs; Am Welding Soc; Am Soc Metals; Soc Mfg Engrs. *Res:* Development and application of arc welding processes and equipment; manufacturing of roll formed shapes and welded tubing. *Mailing Add:* 54722 Glenwood Park PO Box 1374 Elkhart IN 46515-1374

GORMAN, GEORGE CHARLES, b Brooklyn, NY, July 9, 41; m 69; c 1. HERPETOLOGY. *Educ:* Cornell Univ, BA, 62; Harvard Univ, PhD(biol), 68. *Prof Exp:* Miller fel, Univ Calif, Berkeley, 68-70; res assoc herpet, Mus Comp Zool, Harvard Univ, 71-72; assoc prof biol, 73-80, PROF BIOL, UNIV CALIF, LOS ANGELES, 80- *Concurrent Pos:* NATO fel, Hebrew Univ Jerusalem, Israel, 70. *Mem:* Am Soc Ichthyol & Herpet; Herpet League; Soc Study Amphibians & Reptiles; Soc Syst Zool; Ecol Soc Am. *Res:* Evolutionary genetics; ecology of lizards; cytotaxonomy. *Mailing Add:* 952 The Alameda Berkeley CA 94707

GORMAN, JOHN RICHARD, b Fairbank, Iowa, Nov 11, 13; m 46; c 2. MATHEMATICS. *Educ:* Los Angeles City Col, AA, 34; Univ Calif, Los Angeles, BA, 36, MA, 38. *Prof Exp:* Instr, Compton City Col, 39-41 & 46-47; engr, US Navy, 42-46, chief engr, 45-46; from instr to assoc prof math, US Naval Acad, 47-79; RETIRED. *Mem:* Math Asn Am. *Res:* Analysis. *Mailing Add:* 217 Norwood Rd Wardour Annapolis MD 21401-1203

GORMAN, MARVIN, b Detroit, Mich, Sept 24, 28; m 76; c 3. BIO-ORGANIC CHEMISTRY. *Educ:* Univ Mich, BS, 50; Wayne State Univ, PhD(org chem), 55. *Prof Exp:* Res fel, Israel Inst Technol, 54-55 & Wayne State Univ, 56; org chemist, Eli Lilly & Co, 56-64, res assoc, 64-69, res adv res labs, 69-82; EXEC VPRES, ONCOGEN DIV, BRISTOL-MYERS CO, 82- *Mem:* Am Soc Microbiol; AAAS; Am Chem Soc. *Res:* Natural products; isolation, characterization and structure determination, primarily of alkaloids and antibiotics; chemistry of beta-lactam antibiotics; ionophoretic substances; antiviral agents immunoregulators; pharmaceutical development of antibiotics and proteins. *Mailing Add:* 722 38th Ave Seattle WA 98121

GORMAN, MELVILLE, b San Francisco, Calif, Nov 18, 10; m 37; c 2. INORGANIC CHEMISTRY, HISTORY OF SCIENCE. *Educ:* Univ San Francisco, BS, 31; Univ Calif, MS, 39; Stanford Univ, PhD(chem), 46. *Prof Exp:* From instr to assoc prof chem, 31-52, PROF CHEM, UNIV SAN FRANCISCO, 52- *Concurrent Pos:* Fac fel, Ford Found, 54-55; NSF fel, 59-60. *Mem:* Am Chem Soc; Hist Sci Soc; Soc Hist Technol. *Res:* Transmission of scientific ideas from Europe to other cultures in the nineteenth century; properties of thiocyanic acid and its salts. *Mailing Add:* Dept Chem Univ San Francisco San Francisco CA 94117-1080

GORMAN, ROBERT ROLAND, b Mojave, Calif, Aug 19, 44; m 66; c 2. BIOCHEMISTRY. *Educ:* Univ Idaho, BS, 66, MS, 68; Colo State Univ, PhD(biochem), 70. *Prof Exp:* Fel pharmacol, State Univ NY Upstate Med Ctr, 70-72; SR RES SCIENTIST BIOCHEM, UPJOHN CO, 72- *Mem:* AAAS; Am Soc Biol Chemists. *Res:* Mechanism of action of prostaglandins and the role of cyclic nucleotides in this area. *Mailing Add:* Dept Cell Biol Upjohn Co 300 Henrietta St Kalamazoo MI 49001

GORMAN, SUSAN B, phosgene chemistry, for more information see previous edition

GORMAN, WILLIAM ALAN, b Montreal, Que, Oct 18, 25; m 50; c 3. QUATERNARY GEOLOGY. *Educ:* McGill Univ, BSc, 49, MSc, 52, PhD(geol), 56. *Prof Exp:* Field geologist, Que Dept Mines, 51-56; from asst prof to assoc prof geol sci, 55-71, PROF GEOL SCI, QUEEN'S UNIV, ONT, 71- *Concurrent Pos:* Geologist, Steep Rock Iron Mines, 57-59; spec lectr, Royal Mil Col, Can, 59-64 & 68-77; photogeologist, Neilson Assoc, 79-80; geologist, Mobil Oil, 80-81. *Mem:* Fel Geol Asn Can. *Res:* Holocene geochronology in Eastern Ontario; geochemistry of Holocene lake sediments. *Mailing Add:* Dept Geol Queen's Univ Kingston ON K7L 3N6 Can

GORMICAN, ANNETTE, b Fond du Lac, Wis, Apr 26, 24. NUTRITION. *Educ:* Col St Catherine, BS, 46; Univ Iowa, MS, 47, PhD(nutrit), 65. *Prof Exp:* Chief dietitian, Mercy Hosp, Jackson, Mich, 48-50; instr sch home econ, Univ Minn, St Paul, 52-54, asst prof nutrit, 55-60; educ dir dietetics, Sch Med, Univ Wis, Madison, 61-62; therapeut dietitian, Univ Iowa, 62-64; asst prof nutrit, Col Med, Univ Nebr, 65-66; assoc prof dept nutrit sci & dir dietetic internship, 67-68; dir dietetic internship & residency prog, Univ Wis, Madison, 68-, assoc prof nutrit sci, 80-; AT DEPT FOOD SCI & NUTRIT, UNIV MINNEAPOLIS. *Mem:* Am Soc Clin Nutrit; Am Inst Nutrit; Am Inst Chemists; NY Acad Sci; AAAS. *Res:* Trace elements in foods, mineral absorption, computer applications in nutrition, clinical dietetics. *Mailing Add:* Dept Food Sci & Nutrit Univ Minn 1334 Eckles Ave St Paul MN 55108

GORMLEY, MICHAEL FRANCIS, b Somerville, Mass, Apr 9, 39. ELEMENTARY PARTICLE PHYSICS. *Educ:* Boston Col, BS, 60; Trinity Col, MS, 65; Columbia Univ, PhD(physics), 70. *Prof Exp:* Asst prof physics, Univ Ill, 73-76; STAFF PHYSICIST, FERMI NAT ACCELERATOR LAB, 76- *Mem:* Am Phys Soc. *Res:* Experimental particle physics; electromagnetic interactions; computer-oriented control systems. *Mailing Add:* PO Box 695 Batavia IL 60510

GORMLEY, WILLIAM THOMAS, b Versailles, Ky, May 24, 15; m 47; c 5. ORGANIC CHEMISTRY. *Educ:* Univ Ky, BS, 37, MS, 45; NY Univ, PhD(chem), 52. *Prof Exp:* Chemist, State Testing Labs, Ky, 37-42; Cincinnati Ord Dist, 42-43; instr chem, Army Student Training Prog, Univ Ky, 43-44; chemist, Wm S Merrell Co, 44-46; instr chem, Hunter Col, 46-52; res chemist, E I du Pont de Nemours & Co, 52-53; fel, Mellon Inst, 53-60; RES CHEMIST, KOPPERS CO, INC, MONROEVILLE, 61- *Mem:* Am Chem Soc. *Res:* Mannich reaction; organic peroxides; synthetic lubricants; high polymers. *Mailing Add:* 406 W Swissvale Ave Pittsburgh PA 15218-1637

GORMUS, BOBBY JOE, b Buckingham Co, Va, Nov 7, 41; m 60; c 1. INFECTIOUS DISEASES, BIOLOGY OF CELL MEMBRANES. *Educ:* Univ Richmond, BS, 64; Duke Univ, PhD(biochem), 71. *Prof Exp:* Fel, Univ Fla, Gainesville, 71-74; res chemist, Vet Admin Hosp, Minneapolis, 74-79; RES SCIENTIST, DELTA REGIONAL PRIMATE RES CTR, TULANE UNIV, COVINGTON, LA, 79- *Concurrent Pos:* Prin investr, NIH grant, 82- *Mem:* Am Asn Immunologists; Am Leprosy Asn; Am Soc Trop Med & Hyg. *Res:* Immunology and pathology of experimental leprosy in monkeys; immununology of AIDS in monkeys (Simian AIDS or SAIDS). *Mailing Add:* Dept Microbiol Delta Primate Res Ctr Tulane Univ 18703 Three Rivers Rd Covington LA 70433

GORN, SAUL, b Boston, Mass, Nov 10, 12; m 43. COMPUTER SCIENCES, INFORMATION SCIENCES. *Educ:* Columbia Univ, BA, 31, PhD(math), 42; Univ Bordeaux, dipl d'etudes sup, 32. *Prof Exp:* Asst, Columbia Univ, 37; instr math, Brooklyn Col, 38-42; staff mathematician, Aircraft Radiation Lab, Air Material Command, 46-51; math adv, Comput Lab, Aberdeen Proving Ground, 51-55; from assoc prof to prof elec eng, 55-70, prof comput & info sci, 70-83, EMER PROF COMPUT & INFO SCI, UNIV PA, 83- *Honors & Awards:* Van Amringe Prize, Columbia Univ; Distinguished Serv Award, Asn Comput Mach, 74. *Mem:* AAAS; Am Math Soc; Soc Indust & Appl Math; Asn Comput Mach. *Res:* Foundations of geometry; lattice theory; mathematics of high speed computing; learning models; mechanical languages. *Mailing Add:* 1901 Walnut St No 2E Philadelphia PA 19103-4664

GORNALL, ALLAN GODFREY, b River Hebert, NS, Aug 28, 14; m 41; c 4. BIOCHEMISTRY. *Educ:* Mt Allison Univ, BA, 36; Univ Toronto, PhD(path chem), 41. *Hon Degrees:* DSc, Mt Allison Univ, 78. *Prof Exp:* Clin chemist, Can Naval Med Serv, 42-46; lectr path chem, Univ Toronto, 46, from asst prof to prof, 46-63, chemn dept, 66-76, prof, 63-80, EMER PROF CLIN BIOCHEM, UNIV TORONTO, 80- *Honors & Awards:* Reeve Prize, 41; Nuffield Fel, 49; Ames award, Can Soc Clin Chemists, 77. *Mem:* Can Soc Clin Chemists; Can Physiol Soc (treas, 54-57); Can Soc Endocrinol & Metab; Can Fedn Biol Soc (hon treas, 57-62); fel Royal Soc Can; fel Can Acad Clin Biochem. *Res:* Urea synthesis; liver function; protein and steroid methodology; metabolic effects of hormones; aldosterone; electrolytes; hypertension; pregnancy toxemia; active site of carbonic anhydrase; histamine receptor proteins. *Mailing Add:* Dept Clin Biochem Univ Toronto 100 Col St Toronto ON M5G 1L5 Can

GORNEY, RODERIC, b East Grand Rapids, Mich, Aug 13, 24; m 86. INTERPERSONAL CONFLICTS, PSYCHOANALYSIS. *Educ:* Stanford Univ, BS, 48, MD, 49; Southern Calif Psychoanal Inst, PhD(psychoanal), 77. *Prof Exp:* From asst prof to assoc prof psychiat, 71-73, assoc dir, Outpatient Dept, 71-76, DIR, PROG PSYCHOSOCIAL ADAPTATION & THE FUTURE, DEPT PSYCHIAT, UNIV CALIF, LOS ANGELES, 71-, PROF PSYCHIAT, NEUROPSYCHIAT INST, 80- *Concurrent Pos:* Lectr psychiat, Sch Social Welfare, Univ Calif, Los Angeles, 72-73, mem, Div Biobehav Res, 75-; mem, Comt Social Issues, Group Advan Psychiat. *Mem:* Fel Am Acad Psychoanal; fel AAAS; fel Am Psychiat Asn. *Res:* Effects of television and motion picture drama on adults; cultural determinants of achievement, aggression and psychological distress; past and future evolution of values and behavior; marital-relationship problems; neurotic character disorders. *Mailing Add:* Neuropsychiat Inst Univ Calif 760 Westwood Plaza Rm C8-150 Los Angeles CA 90024-1759

GORNIAK, GERARD CHARLES, b Erie, Pa, Apr 23, 49; m 71; c 2. ANATOMICAL SCIENCES, FUNTIONAL MORPHOLOGY. *Educ:* State Univ NY, Buffalo, BS, 71, PhD(anat), 76. *Prof Exp:* Asst clin res, Rehab Med Eng Lab, E J Meyer Mem Hosp, Buffalo, 71-72; chief phys ther & treat, Columbus Hosp, Buffalo, 72-73; NIH fel, Univ Mich, 76-80; asst prof, Dept Biol Sci, 80-86, ASSOC DIR, PROG MED SCI, FLA STATE UNIV, 86- *Concurrent Pos:* CPNIH fel, Inst Arthritis, Metab & Digestive Dis, 76-79; assc ed, J Morphol, 78- *Mem:* Sigma Xi; Am Asn Anatomists; Am Soc Zoologists. *Res:* The functional morphology and neural control of mammalian feeding mechanisms; the morphology, physiology and function of growing masticatory muscle; comparative muscle morphology and physiology. *Mailing Add:* Dept Biol Sci Fla State Univ Tallahassee FL 32306

GORNICK, FRED, b New York, NY, Mar 12, 29; m 50; c 2. PHYSICAL CHEMISTRY. *Educ:* City Col New York, BS, 51; Univ Pa, PhD(phys chem), 59. *Prof Exp:* Chemist, Allied Chem & Dye Corp, 51-52 & Rohm & Haas Co, 52-53; asst instr chem, Univ Pa, 53-54, asst, 56-59; phys chemist, Nat Bur Standards, 59-65; assoc prof mat sci, Univ Va, 65-67; assoc prof chem, Univ Md, Baltimore County, 67-69, prof & chmn dept, 69-75; CONSULT, NAT BUR STANDARDS, 65- *Concurrent Pos:* Nat Res Coun fel, 59-60; lectr, Georgetown Univ, 64-65; prog mgr, Chem & Hydrogen Energy Systs, Dept Energy, 80-81; vis scientist, Appl Physics Lab, Johns Hopkins Univ, 89-90. *Mem:* AAAS; Am Chem Soc; Am Phys Soc. *Res:* Physical chemistry of macromolecules; thermodynamics and kinetics of crystallization in high polymers; investigations of synthetic polypeptides; sol-gel chemistry. *Mailing Add:* Dept Chem Univ Md Baltimore Co 5401 Wilkens Ave Baltimore MD 21228

GORODETZKY, CHARLES W, b Boston, Mass, May 31, 37; m 61; c 4. CLINICAL PHARMACOLOGY, DRUG DEVELOPMENT. *Educ:* Mass Inst Technol, BS, 58; Boston Univ, MD, 62; Univ Ky, PhD, 75. *Prof Exp:* Intern med, Boston City Hosp, Mass, 62-63; med officer pharmacol, Addiction Res Ctr, Nat Inst Drug Abuse, 63-65 & outside-serv pharmacol, 65-68, chief sect drug metab & kinetics, 68-81, dep dir, 77-81, assoc dir, 81-84; head, Neuropharmacol Sect, Clin Res Div, Burroughs Wellcome Co, 84-90; EXEC DIR, CLIN RES, DEVELOP (CNS), CIBA-GEIGY CORP, 90- *Concurrent Pos:* adj prof pharmacol, Univ Ky, 66-, pharmacol & toxicol, Univ Louisville, 77- & psychiat, Univ NC, 85-; mem, Comt Probs of Drug Dependence. *Honors & Awards:* Outstanding Serv Medal, USPHS Commissioned Corps, 83. *Mem:* AAAS; Am Soc Pharmacol & Exp Therapeut; Soc Neurosci; Am Soc Clin Pharmacol & Therapeut; NY Acad Sci. *Res:* Clinical research and development of neuropharmacologic drugs. *Mailing Add:* Ciba-Geigy Corp 556 Morris Ave Summit NJ 07901

GORODY, ANTHONY WAGNER, b Zurich, Switz, Oct 13, 49; US citizen; m 81. EXPLORATION, GEOCHEMISTRY. *Educ:* Rutgers State Univ, BA, 71; Rice Univ, MS, PhD, 80. *Prof Exp:* Res asst, Lamont/Doherty Observ, 72-75; curatorial rep, Scripps Inst Oceanog, 75-76; sr res adv, Tex Eastern, 80-86; PRIN SCIENTIST, GAS RES INST, 86- *Concurrent Pos:* Consult, Neosho Oil & Gas Co, 79, Hanna Mining Co, 77-80 & Amax Coal Co, 87-90; explor geologist, Hanna Mining Co, 77-80. *Mem:* Am Chem Soc; AAAS; Soc Petrol Eng; Am Geophys Union. *Res:* Technological development for incremental reserve growth in natural gas. *Mailing Add:* 7833 Cortland Pkwy Elmwood Park IL 60635

GOROFF, DIANA K, BIOTECHNOLOGY. *Prof Exp:* RES ASSOC DIAG PROD RES, ORGANON TECKNIKA CORP. *Mem:* Am Asn Immunologists; Am Soc Microbiologists. *Res:* Mechanism of IgA blockage of immune lysis and its role in disease caused by Neisseria meningitides; immunochemical studies of the Group B Streptococcal antigen. *Mailing Add:* Diag Prod Res Organon Tecknika Corp Biotech Res Inst 133-A Piccard Dr Rockville MD 20850-4373

GOROG, ISTVAN, b Budapest, Hungary, Mar 13, 38; US citizen. ELECTRICAL ENGINEERING, PHYSICS. *Educ:* Univ Calif, Berkeley, BSc, 61, MSc, 62, PhD(elec eng), 64. *Prof Exp:* Asst, Univ Calif, Berkeley, 61-64; mem tech staff, RCA Labs, 64-70, head optical electronics, 70-78, head, Mfg Res Group, 78-82, dir, Mfg Technol Res Lab, 82-87; AT DAVID SCRONOFF RES CTR, SRI INT, 87- *Concurrent Pos:* NSF fel, Italy, 68- *Mem:* Am Phys Soc. *Res:* Lasers and quantum electronics; electro-optics; plasmas; video systems; psycho-physics and visual perception; video discs; manufacturing technology. *Mailing Add:* David Scronoff Res Ctr SRI Int CNN5300 Princeton NJ 08543-5300

GOROVSKY, MARTIN A, b Chicago, Ill, Apr 26, 41; m 67; c 2. BIOLOGY, CELL BIOLOGY. *Educ:* Univ Chicago, AB, 63, PhD(cell biol), 68. *Prof Exp:* NSF fel, 68-70; from asst prof to assoc prof, 70-80, PROF BIOL, UNIV ROCHESTER, 80-, CHMN DEPT, 81-, RUSH RHEES PROF, 90- *Mem:* AAAS; Am Soc Cell Biol. *Res:* Molecular biology; study of structure and function of eukaryotic nuclei; regulation of gene expression and organelle biogenesis in tetrahymena. *Mailing Add:* Dept Biol Univ Rochester Rochester NY 14627

GOROZDOS, RICHARD E(MMERICH), b Chicago, Ill, May 20, 28; m 54; c 6. ELECTRICAL ENGINEERING. *Educ:* Univ Ill, BS, 50, MS, 51; Univ Md, PhD(elec eng), 60. *Prof Exp:* Elec engr, Naval Res Lab, 51-54 & Ballistic Res Lab, 54-56; ELEC ENGR, APPL PHYSICS LAB, JOHNS HOPKINS UNIV, 56- *Res:* Tactical missile guidance and control systems; nonlinear control systems; precision tracking; network synthesis. *Mailing Add:* Dept Applied Physics Lab Johns Hopkins Univ, John Hopkins Rd Laurel MD 20723

GORRAFA, ADLY ABDEL-MONIEM, b Alexandria, Egypt, Sept 5, 35; m 63. TEXTILE SCIENCE & TECHNOLOGY. *Educ:* Alexandria Univ, BSc Hons, 56; Manchester Univ, Eng, PhD(textile technol), 61. *Prof Exp:* Instr eng design, Alexandria Univ, Egypt, 56-58; res engr, 62-68, sr res engr, 68-72, res assoc, 72-86, SR RES ASSOC, TEXTILE TECHNOL, EI DU PONT DE NEMOURS & CO INC, 87- *Concurrent Pos:* Mgr, worldwide licensing Taslan, Air Jet Bulking Technol, 86- *Mem:* Fel Textile Inst, Eng. *Res:* Textile technology development leading to diversified uses of synthetic fibers. *Mailing Add:* Lancaster Pike & Loveville Rd Wilmington DE 19899

GORRELL, THOMAS EARL, b Ft Wayne, Ind, Aug 7, 50; m 74. BIOCHEMICAL PARASITOLOGY & CYTOLOGY. *Educ:* Purdue Univ, BS, 72; Mich State Univ, PhD(microbiol), 78. *Prof Exp:* Fel, Dept Microbiol, Univ Iowa, 77-79; res assoc, Rockefeller Univ, 79-80, fel, Dept Biochem & Cytol, 80-85; DIR RES, EPCO INC, 85- *Mem:* Soc Protozool; Am Soc Microbiol. *Res:* Physiology and biochemistry of anaerobic microorganisms; hydrogenosomes; pyruvate catabolism; activation of nitroheterocyclic drugs. *Mailing Add:* c/o NTV 101 Middle Quarter Woodbury CT 06798

GORRILL, WILLIAM R, b Holbrook, Mass, Oct 1, 21; m 43; c 4. CIVIL ENGINEERING. *Educ:* Northeastern Univ, BS, 48; Univ Maine, MS, 56. *Prof Exp:* Engr soils eng, Maine State Highway Comn, 51-57; assoc prof civil eng, 57-64, prof, 64-75, DIR, SCH ENG TECHNOL, UNIV MAINE, 75- *Concurrent Pos:* Owner, W R Gorrill & Assoc, 57-68; pres, Jordan Gorrill Assoc, 68-74, consult, 74-76. *Mem:* Am Soc Civil Engrs; Nat Soc Prof Engrs; Am Soc Eng Educ. *Res:* Use of peat as a construction material. *Mailing Add:* 15 Grove Orono ME 04469

GORRY, G ANTHONY, MEDICAL INFORMATICS. *Educ:* Yale Univ, BSE, 62; Univ Calif, Berkeley, MS, 62; Mass Inst Technol, PhD(computer sci), 67. *Prof Exp:* Asst prof, Sloan Sch Mgt, Mass Inst Technol, 67-70, assoc prof, 70-73, assoc prof computer sci, 73-75; assoc prof community med, Baylor Col Med, 75-78, dir, Eval Res Group, Nat Heart & Blood Vessel Res & Demonstration Ctr, 75-82, dir, Health Mgt Res, 78-80, prof health mgt, 78-85, assoc dean, 79-83, vpres, 80-83, vpres instnl develop, 83-86, PROF MED INFORMATICS, DEPT COMMUNITY MED, BAYLOR COL MED, 78-, PROF, DIV NEUROSCI & VPRES INFO TECHNOL, 86- *Concurrent Pos:* Assoc fac mem, Opers Res Ctr, Mass Inst Technol, 71-75; lectr, Dept Med, Tufts Univ Sch Med, 71-75; adj assoc prof math sci, Rice Univ, 75-78, adj prof, 78-85, adj prof, Dept Computer Sci, 85-; adj prof bus & econ, Tex Woman's Univ, 78-79; mem, Biomed Libr Rev Comt, Nat Libr Med, 84-88; prin fac mem, Prog Info Technol, Asn Am Med Cols; dir, W M Keck Ctr Comput Biol, Baylor Col Med & Rice Univ. *Mem:* Inst Med-Nat Acad Sci; fel Am Col Med Informatics. *Res:* Information technology; computational biology; community medicine. *Mailing Add:* Baylor Col Med Tex Med Ctr One Baylor Plaza Houston TX 77030

GORSE, JOSEPH, b Warren, Ohio, Sept 13, 45; m 82; c 3. CAPILLARY ELECTROPHORESIS, HIGH PERFORMANCE LIQUID CHROMATOGRAPHY. *Educ:* Ohio State Univ, BS, 67; Cleveland State Univ, MS, 70; Univ Ariz, PhD(chem), 85. *Prof Exp:* Teacher math & chem, J F Kennedy High Sch & Walsh High Sch, 67-71; chemist, Morgan Adhesives, 72-73 & Firestone Tire & Rubber, 73-76; asst prof chem, Knox Col, 85-88; ASSOC PROF CHEM, BALDWIN-WALLACE COL, 88- *Concurrent Pos:* Res assoc, Dept Chem, Univ Tenn, 86, 87 & 90 & NASA Lewis, 91; co-prin, Dept Chem, Knox Col, 87 & 88. *Mem:* Am Chem Soc; Soc Appl Spectros. *Res:* Solid-liquid interface responsible for retention and selectivity in bonded-phase liquid chromatography; solid-liquid interface responsible for electroosmotic flow in capillary electrophoresis. *Mailing Add:* Dept Chem Baldwin-Wallace Col Berea OH 44017

GORSE, ROBERT AUGUST, JR, b San Diego, Calif, Apr 7, 42; m 77; c 2. ATMOSPHERIC CHEMISTRY & PROCESSES, VEHICLE EMISSIONS. *Educ:* San Diego State Univ, BS, 65, MS, 67; Univ Calif, Davis, PhD(chem), 72. *Prof Exp:* Teaching & res asst chem, San Diego State Univ, 66-67 & Univ Calif, Davis, 67-72; res assoc, Univ Tex, 72-75 & Argonne Nat Lab, Univ Chicago, 75-78; sr res scientist, 78-80, prin res scientist chem, Sci Res Lab, 81-90, PRIN RES ENGR, AUTOMOTIVE EMISSIONS OFF, FORD MOTOR CO, 91- *Concurrent Pos:* Fel, Off Sci Res, US Air Force, 72-73 & Robert A Welch Foun, 73-75; chmn, Coord Res Coun Comt Acid Rain, 80-; mem comt pyrenes, Nat Acad Sci, 81-83; chmn comt Heterogeneous Drobiet, 84-; chmn comt Cloud Chem, 83-84; chmn comt Mobile Emissions, 86- *Mem:* Am Chem Soc; AAAS; Sigma Xi. *Res:* Characterization and environmental importance of mobile source emissions; formation and reactions of atmospheric particulates; atmospheric chemistry and gas phase kinetics; reactions of free radicals; modeling of atmospheric process; analysis of field measurements; auto/oil air quality improvement; research program analysis and reporting. *Mailing Add:* 5714 Blue Grass Lane Saline MI 48176

GORSIC, JOSEPH, b Ponova vas, Slovenia, Jan 6, 24; nat US; m 66; c 6. PLANT GENETICS. *Educ:* Univ Ill, BS, 52; Univ Chicago, PhD, 57; State Univ Agr & Forestry, Austria, Dipl Ing, 65. *Prof Exp:* Instr biol, Marquette Univ, 57-59; assoc prof biol, 59-78, PROF BIOL, ELMHURST COL, 78- *Mem:* AAAS; Am Genetic Asn. *Res:* Comparative genetics in plants. *Mailing Add:* 442 Ida Lane Elmhurst IL 60126

GORSICA, HENRY JAN, b Brno, Czech, June 24, 07; nat US; m 31; c 3. BIOCHEMISTRY. *Educ:* Univ Wis, BS, 29, MS, 31, PhD(biochem), 34. *Prof Exp:* Asst, Univ Wis, 31-35; res chemist, Pabst Brewing Co, Wis, 35-43; dir labs, Northwestern Yeast Co, Ill, 43-46; chief chemist & plant supt, Fearn Labs, Inc, 46-51, vpres in charge res & prod, Fearn Foods, Inc, 51-54; dir labs, B Heller & Co, Ill, 55-66; asst prof chem, Wis State Univ, White Water, 66-76; RETIRED. *Mem:* Fel AAAS; Am Chem Soc. *Res:* Biochemistry of processed meat products; spices and spice extractives; flavorings; malting, brewing, fermentations and production of active dry yeast; production of soup bases and seasonings; chemistry of foods and nutrition. *Mailing Add:* 301 Halcysn Pl Ft Atkinson WI 53538-2430

GORSKI, ANDRZEJ, b Wroclaw, Poland, Aug 11, 46; m 88; c 2. TRANSPLANTATION IMMUNOLOGY, IMMUNOMODULATION. *Educ:* Warsaw Med Sch, MD, 70, PhD(immunol), 73, bd cert, 80; Educ Coun For Med Graduates, ECFMG cert, 75. *Prof Exp:* Assoc scientist, Sloan-Kettering Inst Cancer Res, New York, NY, 76; HEAD, DEPT IMMUNOL, WARSAW MED SCH, POLAND, 84-, PROF MED & IMMUNOL, 88- *Concurrent Pos:* Vis prof, Weizmann Inst Sci, Rehovot, Israel, 88, Guy's Hosp, Univ London, UK, 90; head, Senate Comn Sci, Warsaw Med Sch, Poland, 90-; head, Nat Comt Coop, Int Union Immunol Socs, 91-; pres, immunol comt, Polish Acad Sci, 91-; prin investr res projs. *Honors & Awards:* Meller Award for Excellence in Cancer Res, Sloan-Kettering Inst Cancer Res, 76; J Sniadecki Mem Award, Polish Acad Sci, 88. *Mem:* Polish Acad Sci; Int Union Immunol Socs; Am Asn Immunologists; Transplantation Soc; Int Soc Exp Hemat; NY Acad Sci. *Res:* Immunobiology of graft rejection and the development of novel modalities to control this process; immunomodulating action of heparin and the perspectives for its application in clinical immunosuppressive therapy; role of cell adhesion molecules in the immune response. *Mailing Add:* Nowogrodzka 59 Warsaw 02006 Poland

GORSKI, JACK, b Green Bay, Wis, Mar 14, 31; m 55; c 2. BIOCHEMISTRY, ENDOCRINOLOGY. *Educ:* Univ Wis, BS, 53; Wash State Univ, MS, 56, PhD(animal sci), 58. *Prof Exp:* USPHS fel, Univ Wis, 58-61; from asst prof to assoc prof physiol & biophys, 61-69, prof physiol & biochem, Univ Ill, Urbana, 69-73; PROF BIOCHEM, UNIV WIS, MADISON, 73- *Concurrent Pos:* NSF sr fel, Princeton Univ, 66-67; mem endocrinol study sect, NIH, 67-71; mem biochem & chem carcinogenesis, Adv Comt, Am Cancer Soc, 73-76, Adv Comt, Dept Exp Therapeut, Roswell Park Inst, 73-75 & Molecular Biol Study Sect, NIH, 77-81; adv comt, Am Cancer Soc, 84- *Honors & Awards:* Oppenheimer Award, Endocrine Soc, 71; Wellcome Lectr, 81. *Mem:* Am Soc Biol Chemists; Endocrine Soc. *Res:* Molecular mechanisms of hormone action; estrogen control of the uterus; regulation of protein synthesis; mechanisms of estrogenic hormone action, biosynthesis of pituitary hormones. *Mailing Add:* S83 W24555 Artesian Ave Mukwonago WI 53149

GORSKI, JEFFREY PAUL, b May 17, 47; m; c 2. EXTRACELLULAR MATRIX, CELL BIOLOGY. *Educ:* Univ Wis, Madison, PhD(biochem), 75. *Prof Exp:* Asst prof biochem, Mayo Clin & Med Sch, 79-87; ASSOC PROF, UNIV Mo, KANS CITY, 87- *Honors & Awards:* Charnley Award, 86. *Mem:* Am Soc Biochem & Molecular Biol; Orthodol Res Soc; Am Chem Soc; Am Soc Bone Mineral Res. *Res:* Bone matrix biochemistry and bone cell biology. *Mailing Add:* Div Biochem & Molecular Biol Sch Basic Life Sci Univ Mo-KC 5100 Rockhill Rd Kansas City MO 64110

GORSKI, LEON JOHN, b New Britain, Conn, Sept 29, 38; m 62; c 2. ECOLOGY, BIOLOGY. *Educ:* Cent Conn State Col, BS, 61; Univ Conn, MS, 63, PhD, 69. *Prof Exp:* From asst prof to assoc prof ecol & biol, 65-77, asst to dean arts & sci, 74-77, PROF BIOL & COORDR GRANTS, CENT CONN STATE COL, 77- *Mem:* Am Ornith Union; Am Inst Biol Sci; Soc Syst Zool; Sigma Xi. *Res:* Systematics and ecology of sibling species of the Traill's flycatchers; significance of vocalizations as reproductive isolating mechanisms; tropical overwintering behavior of the two song forms of the Traill's flycatcher in Panama and Peru; multidisciplinary environmental education programs at college and state levels. *Mailing Add:* Dept Biol Sci Cent Conn State Col New Britain CT 06050

GORSKI, ROBERT ALEXANDER, b Passaic, NJ, Nov 24, 22; m 44; c 5. PHYSICAL ORGANIC CHEMISTRY. *Educ:* La Salle Col, BA, 47; Univ Pa, MS, 49, PhD(phys chem), 51. *Prof Exp:* Instr algebra, trigonom & calculus, La Salle Col, 46-48, instr phys chem, 48-49; res asst thermodynam res lab, Univ Pa, 49-51; res chemist, Org Phys Chem, Jackson Labs, E I du Pont de Nemours & Co, Inc, 51-54, res chemist, Freon Prods Lab, 54-90, tech assoc, 67-71, sr res chemist, 71-85, res assoc, 85-87, consult, Freon Prods Lab, 87-90; RETIRED. *Concurrent Pos:* Mem, Am Standards & Testing Mat Comt on Vapor Degreasing & Solvent Anal, & reviewer, Fire Res, Nat Acad Sci-Nat Res Coun, 58-78, mem, 71- *Mem:* Am Chem Soc; Sigma Xi. *Res:* Interaction coefficients of binary gas mixtures; organic physical chemistry; development of technical and material compatibility data for fluorinated compounds for sales promotion. *Mailing Add:* 735 Harvard Lane Newark DE 19711

GORSKI, ROGER ANTHONY, b Chicago, Ill, Dec 30, 35; m 59; c 3. NEUROENDOCRINOLOGY. *Educ:* Univ Ill, BS, 57, MS, 59; Univ Calif, Los Angeles, PhD(anat), 62. *Prof Exp:* Vchmn grad affairs, 67-74, from asst prof to assoc prof, 62-70, PROF ANAT, SCH MED, UNIV CALIF, LOS ANGELES, 70-, CHMN DEPT, 80-; DIR, LAB NUEROENDOCRINOL BRAIN RES INST, 81. *Honors & Awards:* Ernst Oppenheimer Mem Award, Endocrine Soc, 76; Res Award, Soc Study Reproduction. *Mem:* Int Brain Res Orgn; Int Soc Neuroendocrinol; Soc Exp Biol Med; AAAS; Am Asn Anatomists; Sigma Xi; Am Acad Arts Sci. *Res:* Sexual differentiation of hypothalamic control of reproduction; hypothalamic regulation of ovulation; effect of steroids on brain; electrical activity of hypothalamus; regulation of sexual behavior. *Mailing Add:* Dept Anat & Cell Biol Univ Calif Sch Med Los Angeles CA 90024-1763

GORSLINE, DONN SHERRIN, b Los Angeles, Calif, Dec 15, 26. MARINE GEOLOGY. *Educ:* Mont Sch Mines, BS, 50; Univ Southern Calif, MS, 54, PhD(geol, oceanog), 58. *Prof Exp:* Asst oceanog, Allan Hancock Found, Univ Southern Calif, 54-56, res assoc geol, 53-57; asst prof marine geol, Oceanog Inst, Fla State Univ, 58-61, assoc prof & actg dir, 61-62; assoc prof geol, 62-66, PROF GEOL, UNIV SOUTHERN CALIF, 66- *Concurrent Pos:* Mem adv panel coastal geog, Nat Acad Sci-Nat Res Coun, 61 & chmn adv panel on ocean waste disposal, 74-76; gen chmn & ed, Nat Shallow Water & Coastal Res Inst, 61-62 & 71; mem adv panel geol & geophys continental margins, 77-78, adv comn ocean resources, Calif, 65-69; ed, Marine Geol, 67-70; mem adv panel on oceanog, NSF, 72-74. *Mem:* Fel AAAS; fel Geol Soc Am; Am Soc Limnol & Oceanog; Soc Econ Paleont & Mineral (ed, J Sedimentary Petrol, 70-76, pres, 77-78); Coastal Res Soc; Sigma Xi. *Res:* Sedimentology; marine geology; coastal studies; shallow water oceanography; continental margin. *Mailing Add:* Dept Geol Sci Univ Southern Calif Los Angeles CA 90089

GORSON, ROBERT O, b Philadelphia, Pa, July 16, 23; m 46. RADIOLOGY, MEDICAL PHYSICS. *Educ:* Univ Pa, BA, 49, MS, 52; Am Bd Radiol, dipl, 54, Am Bd Health Physics, dipl, 60. *Prof Exp:* Univ health physicist, Univ Pa, 49-51, instr radiol physics, 52-53, assoc, 53-59, assoc grad sch med, 55-59; assoc prof radiol, 59-65, PROF MED PHYSICS, THOMAS JEFFERSON UNIV, 65-, CHIEF DIV MED PHYSICS, COL HOSP, 59- *Concurrent Pos:* Vis lectr, Grad Sch Med, Univ Pa, 59-71; consult, Philadelphia Dept Pub Health, 57- & div radiation health, USPHS, 63-; chmn subcomt 3, Nat Comt

Radiation Protection, 61-77, mem, 77-, mem subcomt 16 & sci comt 9, 63- & sci comt 44, 73-; mem comn units, standards and protection, Am Col Radiol, 59-; mem, Nat Coun Radiation Protection & Measurements, 64-; mem adv comt med x-ray protection, USPHS, 64-66; mem, Am Bd Health Physics, 64-65, chmn, 65-67; deleg, coun Alliance for Eng in Med & Biol, 70-; mem diag res adv group, Nat Cancer Inst, 75-; asst ed, Med Physics, 77- *Mem:* Radiol Soc N Am; Am Asn Physicists in Med; Radiation Res Soc; Health Physics Soc; fel Am Col Radiol; Sigma Xi. *Res:* Radiation and health physics. *Mailing Add:* 2113 Pine St Philadelphia PA 19103-6513

GORTATOWSKI, MELVIN JEROME, b Chicago, Ill, Oct 30, 25. ORGANIC BIOCHEMISTRY, HEALTH SCIENCES. *Educ:* Univ Ill, BS, 50, PhD(chem), 55; Wash State Col, MS, 52. *Prof Exp:* Asst chem res, Div Fluorine Chem, Ill State Geol Surv, 52-53; res instr & fel, Lab for Study Hereditary & Metab Disorders, Col Med, Univ Utah, 56-58, res assoc dept psychiat, 58-59, res instr biochem, 59-65; asst prof pediat, Univ Southern Calif, 65-71; chief clin chem sect, Utah State Health Dept Lab, 71; RETIRED. *Concurrent Pos:* Biochemist, Vet Admin Hosp, Salt Lake City, 59-65; assoc investr, Clin Res Ctr, Childrens Hosp, Los Angeles, 65-71. *Mem:* Am Chem Soc; Sigma Xi. *Res:* Metabolic disease associated with mental retardation including inherited disorders of amino acid, carbohydrate and fat metabolism; environmental chemistry, especially health hazards from drugs, pesticides, pollutants in air, water and food; screening for metabolic diseases in newborn population. *Mailing Add:* 4045 Foubert Ave Salt Lake City UT 84124

GORTHY, WILLIS CHARLES, b Buffalo, NY, Dec 4, 34; m 59; c 3. CYTOPATHOLOGY, HISTOLOGY. *Educ:* Columbia Univ, AB, 56; NY Univ, MS, 61; Princeton Univ, MA, 63, PhD(biol), 65. *Prof Exp:* Res aide, Sloan-Kettering Inst Cancer Res, 57-60; Nat Acad Sci-Nat Res Coun res assoc, Human Nutrit Res Div, USDA, 64-66; asst prof 66-72, ASSOC PROF ANAT, COLO STATE UNIV, 72- *Concurrent Pos:* Res grants, Nat Inst Neurol Dis & Blindness, 68-70 & Nat Eye Inst, 71-74, 74-77, 77-81 & 87-90, US Dept Health, Educ & Welfare. *Mem:* Electron Microscope Soc Am; Asn Res Vision & Ophthal. *Res:* Light and electron microscopical studies of mammalian eye; morphological and histochemical studies of the lens; studies involve morphological and histochemical changes in mammalian lenses during development of hereditary and senescent cataracts; histochemical experiments are aimed at exploration of transport and digestive properties of the lens. *Mailing Add:* Dept Anat & Neurobiol Colo State Univ Ft Collins CO 80523

GORTLER, LEON BERNARD, b Des Moines, Iowa, Jan 30, 35; m 60; c 3. ORGANIC CHEMISTRY, HISTORY OF SCIENCE. *Educ:* Univ Chicago, AB, BS & MS, 57; Harvard Univ, PhD(chem), 62. *Prof Exp:* Fel, Univ Calif, Berkeley, 61-62; from instr to assoc prof chem, 62-74, chmn, 84-87, PROF CHEM, BROOKLYN COL, 74- *Mem:* Am Chem Soc (chmn, Div Hist Chem, 82-83); Sigma Xi; Hist Sci Soc. *Res:* Synthesis and decomposition of peresters; study of reaction mechanisms; history of chemistry. *Mailing Add:* Dept Chem Brooklyn Col Brooklyn NY 11210

GORTNER, ROSS AIKEN, JR, biochemistry; deceased, see previous edition for last biography

GORTNER, SUSAN REICHERT, b San Francisco, Calif, Dec 23, 32; m 60; c 2. CARDIAC RECOVERY. *Educ:* Stanford Univ, Palo Alto, AB, 53; Western Reserve Univ, MN, 57; Univ Calif Berkeley, PhD(higher educ), 64. *Prof Exp:* Asst supvr & instr surg, Johns Hopkins Hosp & Sch Nursing, 57-58; instr nursing, Col Nursing, Univ Hawaii, 58-63, asst prof nursing & chmn, Unit Med Surg Nursing, 63-64; spec consult, Nursing Educ & Training Br, Div Nursing, Pub Health Serv, Dept Health, Educ & Welfare, 66-67, exec secy res nursing patient care review, Nursing Res Br, 67-72, chief, Res Grants Sect, Nursing Res Br & actg br chief, 71-73, chief, 73-78 & chief, Nursing Pract Br, 75-76; PROF FAMILY HEALTH CARE, SCH NURSING, UNIV CALIF SAN FRANCISCO, 78-, ADJ PROF, INTERNAL MED, 89- *Concurrent Pos:* Mem cabinet nursing res, Am Nurses' Asn, 84-86; assoc dean res, Sch Nursing, Univ Calif San Francisco, 78-86; clin assoc cardiovasc surg, Dept Nursing Serv, Moffitt-Long Hosp, San Francisco, 81-; mem Coun Cardiovasc Nursing, Am Heart Asn, 87-91. *Honors & Awards:* Helen Nahm Res lectr, Sch Nursing, Univ Calif San Francisco, 87; Fulbright lectr, Inst Nursing Sci, Univ Oslo, 88. *Mem:* Fel Am Acad Nursing; Am Nurses Asn; Am Heart Asn; Soc Behav Med. *Res:* Heart surgery patients and families treatment beliefs and outcomes and recovery processes; protocols for assisting recovery; study of elders undergoing heart surgery; history of nursing science; philosophy of health science. *Mailing Add:* Univ Calif San Francisco N411Y San Francisco CA 94143-0606

GORTNER, WILLIS ALWAY, b Cold Spring Harbor, NY, Dec 20, 13; m 60; c 4. SCIENCE COMMUNICATIONS. *Educ:* Univ Minn, AB, 34; Univ Rochester, PhD(biochem), 40. *Prof Exp:* Res chemist, Gen Mills, Inc, 34-37 & 40-42; asst biochem, Univ Rochester, 37-40; asst prof biochem & chem eng, Cornell Univ, 43-45, assoc prof biochem, 45-48; head chem dept, Pineapple Res Inst, Univ Hawaii, 48-64; dir human nutrit res div, 64-72, staff scientist human nutrit, nat prog staff, AGR RES SERV, USDA, 72-76; exec officer, Am Inst Nutrit, 76-78; consult, 78-82; RETIRED. *Concurrent Pos:* With USDA, 44, Bjorksten Res Found, 53, Nat Acad Sci-Nat Res Coun, 57, Nat Canners Asn, 61 & Univ Calif, 63. *Honors & Awards:* Hoffman-LaRoche Lectr, Nutrit Soc Can, 71. *Mem:* Fel AAAS; Am Chem Soc; Am Soc Biochem & Molecular Biol; fel Inst Food Technol; Am Soc Clin Nutrit; Am Inst Nutrit; Am Rock Art Res Asn; Soc Am Archaeol. *Res:* Food biochemistry; plant enzymes and growth regulators; fruit development and composition; human nutrition. *Mailing Add:* 470 Cervantes Rd Portola Valley CA 94028

GORTON, ROBERT LESTER, b Houston, Tex, Oct 19, 31; m 60; c 4. MECHANICAL ENGINEERING. *Educ:* La Polytech Inst, BS, 53; La State Univ, MS, 60; Kans State Univ, PhD(mech eng), 66. *Prof Exp:* Engr, Schlumberger Well Surv Corp, 55-58; assoc mech eng, La State Univ, 59-60; from instr to assoc prof, 60-74, asst head dept, 66-69, PROF MECH ENG, KANS STATE UNIV, 74- *Concurrent Pos:* NASA faculty fel, 70; tech consult, Black & Veach Consult Engrs, 74-, Oak Ridge Nat Lab, 80-82, Wilson Co Consult Engrs & Meritek Eng Corp; Halliburton prof mech eng, 82-85. *Honors & Awards:* Centennial Medallion, Am Soc Mech Engrs; Distinguished Serv Award, Am Soc Heating, Refrig & Air Conditioning Engrs. *Mem:* Fel Am Soc Mech Engrs; fel Am Soc Heating, Refrig & Air Conditioning Engrs. *Res:* Environmental engineering; energy systems, heat transfer, fluid mechanics; industrial air conditioning. *Mailing Add:* Dept Mech Eng Col Eng Durland Hall Kans State Univ Manhattan KS 66506

GORTSEMA, FRANK PETER, b Grand Rapids, Mich, Dec 25, 33; m 59; c 2. INORGANIC CHEMISTRY. *Educ:* Calvin Col, AB, 55; Purdue Univ, PhD(phys chem), 60. *Prof Exp:* Sr res chemist, Parma Res Ctr, Union Carbide Corp, Ohio, 59-63, sr res chemist, Res Inst, 63-79, develop assoc, Linde Div, Molcular Sieve, NY, 79-89; CHEM FEL, TECH OPERS, MERCK & CO, RAMWAY, NJ, 90- *Mem:* Am Chem Soc; Mat Res Soc. *Res:* Chemistry of ruthenium in solution; thermoelectric properties of rare earth nitrides and monosulfides; preparation and properties of reduced valency transition metal fluorides; platinum hexafluoride chemistry; solar energy conversion; preparation of fine powders, fibrous ceramics; catalysis, molecular sieves; hydrocracking; catalytic preparation of drug intermediates. *Mailing Add:* Seven Brairwood Lane Pleasantville NY 10570

GORZ, HERMAN JACOB, b Eagle River, Wis, Nov 22, 20; m 51; c 2. PLANT GENETICS, PLANT BREEDING. *Educ:* Univ Wis, BS, 42, MS, 48, PhD(agron, genetics), 51. *Prof Exp:* Assoc prof agron & assoc agronomist, NDak State Univ, 51-54; SUPVRY RES GENETICIST & RES LEADER, WHEAT & SORGHUM RES UNIT, AGR RES SERV, USDA & PROF AGRON, UNIV NEBR, 54- *Mem:* Fel Am Soc Agron; Genetics Soc; Soc Agron. *Res:* Genetic, breeding and biochemical studies in Sorghum, Melilotus, Trifolium and other forage species; insect resistance and quality factors in forages. *Mailing Add:* 6126 Leighton Ave Lincoln NE 68507

GORZYNSKI, EUGENE ARTHUR, b Buffalo, NY, Oct 8, 19; m 46; c 3. IMMUNOLOGY. *Educ:* State Univ NY Buffalo, BA, 49, MA, 51, PhD(microbiol), 68. *Prof Exp:* Assoc res bacteriologist, Children's Hosp, 47-68; sr cancer res scientist, Springville Labs, Roswell Park Mem Inst, 68-69; asst dir, Pub Health Div, Erie County Lab, 69-75; PROF MICROBIOL, SCH MED, STATE UNIV NY, BUFFALO, 69-, PROF PATH, 89- *Concurrent Pos:* mem, Ctr Immunol, State Univ NY, Buffalo, 74-; Microbiologist & clin lab coordr, Vet Admin Med Ctr, 75-80; chief, microbiol sect, Vet Admin Med Ctr, Buffalo, 80- *Mem:* Fel AAAS; Am Soc Microbiol; Asn Mil Surg US; Am Asn Immunologists; fel Am Acad Microbiol, 77; Sigma Xi; fel Infectious Dis Soc Am. *Res:* Enterobacterial antigens and immunity; host defense mechanisms; infectious diseases. *Mailing Add:* Clin Lab Serv Vet Admin Med Ctr 3495 Bailey Ave Buffalo NY 14215

GORZYNSKI, JANUSZ GREGORY, b Warsaw, Poland, Aug 6, 40; US citizen. PSYCHIATRY. *Educ:* Acad Med, Warsaw, MD, 63. *Prof Exp:* Resident psychiat, Montefiore Hosp Med Ctr, Bronx, 71-74; instr, Albert Einstein Col Med, 74-77; asst clin prof psychiat, Cornell Univ Med Col, 77-84; CLIN ASST PROF, COL PHYSICIANS & SURGEONS, COLUMBIA UNIV, 84- *Concurrent Pos:* Sci dir, Barbara P Johnson Found, New York, 76-; J J Johnson & B P Johnson Found res grant, 76-78; Asst Attend, St Luke's Roosevelt Med Ctr, 84-; coordr res, Mem-Sloan Kettering Caner Ctr, 78-84. *Mem:* Am Psychiat Asn; Am Psychosom Soc; AAAS; NY Acad Sci. *Res:* Psychological adjustment in cancer patients; endocrinology of anorexia nervosa; psychosexual findings in polycystic ovary syndrome. *Mailing Add:* 115 E 87th New York NY 10128

GORZYNSKI, TIMOTHY JAMES, b March 19, 50; m; c 3. IMMUNOLOGY. *Educ:* State Univ NY, BA, 76, MA, 78, PhD(microbiol), 80. *Prof Exp:* Lab technologist, Vet Admin Med Ctr, 73-81; res fel, Mayo Clin, 81-84, res assoc, 85; res immunologist, 85-89, SR RES IMMUNOLOGIST, E I DU PONT DE NEMOURS, GLASGOW, 89- *Concurrent Pos:* Fel, State Univ NY, 76-80; res grants, Minn Arthritis Found, 85-86, DuPont Seed Grant, 86-87. *Honors & Awards:* McGrath Biol Award. *Res:* Immunology; numerous publications; enchancement of antibody response in mice; T-cells-B-cell epitope peptides as immunogens; identification of high affinity monoclonal antibodies during primary screeing; enchancing ascites yield in mice; mouse model for testing anti-arthritic drugs. *Mailing Add:* Med Prod Dept E I du Pont de Nemours Glasgow Site 714 PO Box 6101 Newark DE 19714-6101

GOSE, EARL E(UGENE), b Aberdeen, Wash, Mar 14, 34; m 62. ENGINEERING. *Educ:* Carnegie Inst Technol, BS, 56; Univ Calif, PhD(chem eng), 60. *Prof Exp:* NSF fel, Sorbonne, 60-61; fel, Mass Inst Technol, 61-62; asst prof eng, Case Inst Technol, 62-65; assoc prof info eng, 65-77, PROF INFO ENG, UNIV ILL, CHICAGO CIRCLE, 77- *Concurrent Pos:* Assoc prof physiol, Univ Ill Med Ctr; staff mem, Presby St Luke's Hosp, Chicago; prof, Rush Med Col. *Mem:* AAAS; Am Soc Cybernet; Asn Comput Mach; Inst Elec & Electronics Engrs; Instrument Soc Am. *Res:* Pattern recognition; natural and artificial intelligence; neurophysiology; biomedical engineering. *Mailing Add:* Dept Bioeng MC/063 Univ Ill Box 4348 Chicago IL 60680

GOSFIELD, EDWARD, JR, b New York, NY, July 17, 18; m 41, 85; c 3. INTERNAL MEDICINE, CARDIOLOGY. *Educ:* Univ Pa, AB, 39, MS, 40, MD, 44; Am Bd Internal Med, dipl, 52, recert, 77. *Prof Exp:* From asst instr to asst prof internal med, 50-63, chmn med bd, Grad Hosp, 75-77, actg chmn, Dept Med, 79-80, ASSOC PROF CLIN MED, SCH MED, UNIV PA, 63-, ASSOC CHMN, 80- *Concurrent Pos:* Consult, Wills Eye Hosp, 50-77; chief hypertension clin, Grad Hosp, Univ Pa, 57-79, assoc physician, Hosp, 59-; assoc physician, Presby-Univ Pa Med Ctr, 71-87. *Mem:* Am Heart Asn; fel Am Col Physicians; fel Am Col Cardiol; NY Acad Med. *Res:* Cardiovascular disease. *Mailing Add:* Univ Pa Pa Grad Hosp 2113 Spruce St Philadelphia PA 19103

GOSHAW, ALFRED THOMAS, b West Bend, Wis, Aug 26, 37; m 66; c 2. EXPERIMENTAL HIGH ENERGY PHYSICS. *Educ:* Univ Wis-Madison, BS, 59, MS, 61, PhD(physics), 66. *Prof Exp:* Instr physics, Princeton Univ, 66-69; vis scientist, Nuclear Physics Apparatus Div, Europ Orgn Nuclear Res, 69-70, mem staff, 70-73; from asst prof to assoc prof, 73-84, PROF PHYSICS, DUKE UNIV, 84- *Concurrent Pos:* Alfred P Sloan Found fel, 57-58; NSF fel, 59-61; Swiss Am Fedn Sci Exchange fel, 69. *Mem:* Am Phys Soc; Am Asn Univ Prof. *Res:* Hyperon resonance studies; inclusive pion and photon production in pion proton and pion nucleus collisions; high energy electromagnetic processes; heavy quark photo and hadro production. *Mailing Add:* Dept Physics Duke Univ Durham NC 27706

GOSINK, JOAN P, b Jamaica, NY, Mar 26, 41; m 61; c 4. ARCTIC HYDROLOGY. *Educ:* Mass Inst Technol, BS, 62; Old Dominion Univ, MS, 73; Univ Calif, Berkeley, PhD(mech eng), 79. *Prof Exp:* Teaching asst thermodynamics & fluid dynamics, Univ Calif, Berkeley, 75-76, res asst, 76-78; fel, 79-81, ASST PROF GEOPHYSICS, GEOPHYS INST, UNIV ALASKA, 81- *Concurrent Pos:* Res fel, NASA, 72-73 & Fulbright-Hayes Found, 74-75; mem, Glaciol Comt, Nat Res Coun, 81- & Polar Res Bd, 81- *Mem:* Am Soc Civil Engrs; Am Soc Mech Engrs; Am Meteorol Soc; Int Glaciol Soc. *Res:* Mass, heat and momentum transport in ice and water systems. *Mailing Add:* Geophys Inst Univ Alaska Fairbanks AK 99775

GOSLIN, ROY NELSON, b Lincoln, Nebr, Nov 21, 04; m 34; c 2. PHYSICS, MATHEMATICS. *Educ:* Nebr Wesleyan Univ, BA, 28; Univ Wyo, MA, 30. *Hon Degrees:* DSc, Oglethorpe Univ, 71. *Prof Exp:* Instr physics, Ala Polytech Inst, 30-39, asst prof, 40-44; res physicist, Clinton Eng Works & Tenn Eastman Corp, 44-46; chmn div sci, 46-77, PROF PHYSICS, OGLETHORPE UNIV, 46-, PROF MATH, 65- *Concurrent Pos:* Consult, Carbide & Carbon Chem Corp, 49-53, Southern Res Inst, Ala, 55-56 & Oak Ridge Nat Lab, 59-; dir admis, Oglethorpe Univ, 54-58. *Mem:* AAAS; Am Phys Soc; Am Asn Physics Teachers. *Res:* Mass spectroscopy; electrical discharges; high voltage insulation; plasma physics. *Mailing Add:* 4040 Navajo Trail NE Atlanta GA 30319

GOSLINE, JOHN M, b Oakland, Calif, Dec 14, 43; m 69; c 2. BIOMECHANICS. *Educ:* Univ Calif, Berkeley, BA, 66; Duke Univ, PhD(zool), 70. *Prof Exp:* Res fel zool, Cambridge Univ, 70-73; from asst prof to assoc prof 73-86, PROF ZOOL, UNIV BC, 86- *Concurrent Pos:* Sr Killam scholar, Univ BC Killam Found, 83. *Mem:* Fel AAAS; Am Soc Biomech; Am Soc Zoologists; Biophys Soc Can. *Res:* Molecular biomechanics of structural biomaterials, including elastin and other rubber-like proteins, invertebrate connective tissues, silks and related fibrous proteins; mechanics of jet-propelled swimming in animals. *Mailing Add:* Dept Zool Univ BC Vancouver BC V6T 2A9 Can

GOSLING, JOHN RODERICK GWYNNE, obstetrics & gynecology, for more information see previous edition

GOSLING, JOHN THOMAS, b Akron, Ohio, July 10, 38; wid; c 2. SPACE PLASMA PHYSICS. *Educ:* Ohio Univ, BS, 60; Univ Calif, Berkeley, PhD(physics), 65. *Prof Exp:* Staff mem, Los Alamos Sci Lab, 65-67; staff mem, High Altitude Observ, Nat Ctr Atmospheric Res, 67-75; MEM STAFF, LOS ALAMOS NAT LAB, 75- *Mem:* AAAS; Am Geophys Union; Int Astron Union. *Res:* Solar wind and solar physics; space plasma physics; collisionless shocks, reconnection. *Mailing Add:* MS D438 Los Alamos Nat Lab Los Alamos NM 87545

GOSLOW, GEORGE E, JR, b Tacoma, Wash, May 16, 39; m 60; c 3. VERTEBRATE MORPHOLOGY. *Educ:* Univ Calif, Los Angeles, AB, 62, PhD(zool), 67; Humboldt State Col, MA, 65. *Prof Exp:* Assoc zool, Univ Calif, Davis, 65-66; asst prof, 67-74, assoc prof, 74-80, PROF BIOL, NORTHERN ARIZ UNIV, 80- *Mem:* Am Soc Zoologists; Soc Neurosci; Sigma Xi. *Res:* Functional vertebrate morphology, especially bone-muscle systems as they relate to the control of locomotion; motor units and muscle proprioceptors. *Mailing Add:* Box 5640 Flagstaff AZ 86011

GOSMAN, ALBERT LOUIS, b Detroit, Mich, May 27, 23; m 46; c 2. THERMODYNAMICS, MECHANICAL ENGINEERING. *Educ:* Univ Mich, BS, 50; Univ Colo, Boulder, MS, 55; Univ Iowa, PhD(thermodyn), 65. *Prof Exp:* Instr mech eng, Colo Sch Mines, 50-53, asst prof, 53-55; res engr, Northrop Aircraft, Inc, Calif, 56-58; asst prof mech eng, Colo Sch Mines, 58-62; instr, Univ Iowa, 62-65; assoc prof, Wayne State Univ, 65-67; head, dept mech eng, 67-71, assoc dean col eng & assoc dir eng res, 71-80, PROF MECH ENG, WICHITA STATE UNIV, 67-,. *Concurrent Pos:* Res engr, Cryogenics Div, Nat Bur Standards, Colo, 59-69; panel deleg, Nat Res Coun & Nat Bur Standards Conf Thermodyn, 69. *Mem:* Am Soc Mech Engrs; Am Soc Eng Educ; AAAS. *Res:* Determination of thermodynamic properties of liquids and gases; equation of state for liquid and gaseous argon from triple point to 1000 atmospheres and 1000 degrees Kelvin. *Mailing Add:* Col Eng Wichita State Univ Wichita KS 67208

GOSNELL, AUBREY BREWER, b Provo, Ark, Sept 1, 29; m 48; c 3. ANALYTICAL CHEMISTRY, POLYMER CHEMISTRY. *Educ:* Henderson State Univ, BS, 51; Univ Ark, MS, 62; NC State Univ, PhD(org chem), 67. *Prof Exp:* Teacher high sch, Mansfield, Ark, 49-51; tester refinery lab, El Dorado Refinery, Am Oil Co, Ark, 51-58, refinery operator, 58-60; res chemist, Camile Dreyfus Lab, Res Triangle Inst, 62-67; assoc prof chem, 67-72, PROF CHEM, HENDERSON STATE UNIV, 72- *Concurrent Pos:* Pres, Go Chem, Inc, 73- *Mem:* Am Chem Soc; Sigma Xi. *Res:* Gas chromatographic analysis of chlorinated fatty acids; synthesis and characterization of branched polystyrenes; solution properties of polymers; synthesis and properties of well defined graft copolymers; anionic polymerization; drug analysis; pesticide toxicology; metabolism of antihistamines; analysis of doxylamine and metabolites in rats. *Mailing Add:* 2075 Elaine Circle Arkadelphia AR 71923

GOSS, CHARLES RAPP, JR, b Bridgeton, NJ, Apr 15, 37; m 62; c 1. ELECTRICAL ENGINEERING. *Educ:* NMex State Univ, BS, 73, MS, 74. *Prof Exp:* Develop engr electronic warfare, 75-76, actg dep chief electronic warfare div, 76-77, actg dept chief syst avionics div, 77-78, DEP CHIEF ELECTRONIC TECHNOL DIV, AIR FORCE AVIONICS LAB, 78-; AT ARNOLD ENG DEVELOP CTR. *Mem:* Inst Elec & Electronics Engrs. *Res:* Microwave and microelectronics technology. *Mailing Add:* Bldg 260 Teas Group Sverdrup Technol PO Box 1935 Eglin AFB FL 32542

GOSS, DAVID, b July 20, 37; m 70. THEORETICAL PHYSICS, GEOPHYSICS. *Educ:* Southwestern Univ, Tex, BS, 59; Mich State Univ, MS, 61; Univ Tex, PhD(nuclear physics), 64; Univ Nebr-Lincoln, BS, 82. *Prof Exp:* from asst prof to assoc prof, 64-75, PROF PHYSICS, NEBR WESLEYAN UNIV, 75- *Concurrent Pos:* Consult & vis scientist, SSC/CDG (URA), 85-89, consult, SSC Lab, 89-90. *Mem:* AAAS; Am Phys Soc; Am Asn Physics Teachers; Fedn Am Sci; Am Geophys Union. *Res:* Accelerator geophysics; radon as a probe of geol structure; general geophysics and geology related to SSC construction and operation. *Mailing Add:* Dept Physics Nebr Wesleyan Univ 5000 St Paul St Lincoln NE 68504-2796

GOSS, DAVID A, b Joliet, Ill, July 22, 48. PHYSIOLOGICAL OPTICS. *Educ:* Ill Wesleyan Univ, BA, 70; Pac Univ, BS, 72, OD, 74; Ind Univ, PhD(physiol optics), 80. *Prof Exp:* Optometrist, Storm Lake, Iowa, 74-75; assoc instr optom, Ind Univ, 76-78, res assoc, 78-80; from asst prof to assoc prof, 80-89, PROF PHYSIOL OPTICS & OPTOM, NORTHEASTERN STATE UNIV, 89- *Concurrent Pos:* mem Nat Acad Sci, working group on myopia, 84-87; mem j rev bd, J Am Optom Asn, 88- *Mem:* Am Acad Optom; Asn Res Vision & Ophthal; Optical Soc Am; Am Optom Asn; Optometric Hist Soc; Asn Optometric Educators; AAAS. *Res:* Ocular refractive errors; myopia; ocular optics; clinical accommodative convergence data. *Mailing Add:* Col Optom Northeastern State Univ Tahlequah OK 74464

GOSS, GARY JACK, b South Bend, Ind, Apr 8, 46. BIOLOGY, ENTOMOLOGY. *Educ:* Fla Atlantic Univ, BS, 69, MS, 73; Univ Miami, PhD(biol), 77. *Prof Exp:* from asst prof to assoc prof, 76-84, PROF BIOL, PALM BEACH ATLANTIC COL, 84-, CHAIR, DIV NATURAL SCI, MATH, 85- *Concurrent Pos:* Adj asst prof biol, Fla Atlantic Univ, 80-; consult, Palm Beach County Sch Bd, 84-87. *Mem:* Am Inst Biol Sci. *Res:* Plant and animal interaction including pollination biology and association between Lepidoptera and pyrrolizidine plants; spermatophore nutrient transfer in insects and parental investment; Marine Biology; Botany. *Mailing Add:* Palm Beach Atlantic Col PO Box 3353 West Palm Beach FL 33402-3353

GOSS, GEORGE ROBERT, b Kewanee, Ill, Dec 16, 52; m 75; c 3. CHEMISTRY. *Educ:* Western Ill Univ, BS, 74, MS(chem), 78, MS(bot), 82. *Prof Exp:* Technician, Ill Environ Protection Agency, 77-78; res assoc, Kalo Inc, 78-81; res mgr, Kalo Agr Chems Inc, 81-84; chief chemist, Oil Dri Corp, 84-87, TECH DIR, 87- *Mem:* Am Chem Soc; Am Oil & Chemists Soc; Am Soc Testing & Meas. *Res:* Clay minerals for industrial use; pet box absorbents; agricultural chemical carriers; vegetable oil purification; fertilizer production. *Mailing Add:* 22149 N Pet Lane Prairie View IL 60069

GOSS, JAMES ARTHUR, feed analysis; deceased, see previous edition for last biography

GOSS, JOHN DOUGLAS, b Columbus, Ohio, June 14, 42. EXPERIMENTAL NUCLEAR PHYSICS. *Educ:* Ohio State Univ, BSc. 65, MS, 67, PhD(physics), 70. *Prof Exp:* Fel physics, Univ Notre Dame, 70-73, staff fac fel, 73-74, asst fac fel, 74-75; DEVELOP PHYSICIST SENSOR GROUP, MEASUREX CORP, 75- *Mem:* Am Phys Soc. *Res:* Measurements of reaction Q-values, excitation energies, angular distributions and study of heavy-ion reactions using broad-range magnetic spectrography; application of these techniques in nuclear gauging development. *Mailing Add:* 6151 Hancock Ave San Jose CA 95123

GOSS, JOHN R(AY), b Winona, Minn, May 30, 23; m 47; c 4. MACHINERY ENGINEERING, GASIFICATION OF BIOMASS. *Educ:* Univ Calif, Los Angeles, BS, 52; Univ Calif, Davis, MS, 55. *Prof Exp:* Asst specialist agr eng, Univ Calif, Davis, 53-57, assoc specialist, 57-58, asst prof eng & asst agr engr, 58-62, assoc prof eng & assoc agr engr, 62-67, acad asst to chancellor, 63-66, chmn dept agr eng, 68-74, PROF ENG & AGR ENGR, UNIV CALIF, DAVIS, 67- *Concurrent Pos:* Consult, US Dept Energy, USAID, Calif Energy Comm & pvt firms on gasification biomass. *Mem:* Fel Am Soc Agr Engrs; Am Soc Eng Educ; fel AAAS; Soc Am Foresters; NY Acad Sci; Nat Soc Prof Engrs. *Res:* Combine harvester performance in legume, cereal and dry bean crops for seed and food; pneumatic conveying and forage harvesting machinery research; microclimatic factors; gasification of agricultural and forest residues and utilization of producer gas in boilers, heated air burners; gasoline, natural gas and dual fuel diesel engines. *Mailing Add:* 754 Plum Lane Davis CA 95616

GOSS, RICHARD JOHNSON, b Marblehead, Mass, July 19, 25; m 51; c 2. DEVELOPMENTAL ANATOMY. *Educ:* Harvard Univ, AB, 48, AM, 51, PhD(biol), 52. *Prof Exp:* From instr to assoc prof, 52-64, chmn sect develop biol, div biol & med sci, 72-77, dean biol sci, 77-84, PROF BIOL, BROWN UNIV, 64-, ROBERT P BROWN PROF BIOL, 85- *Concurrent Pos:* Fel, Carnegie Inst, 60; trustee, Mt Desert Island Biol Lab, 60-64; pres, RI Zool Soc, 71-74; dir, Roger Williams Park Zoo, Providence, RI, 75. *Honors & Awards:* Marcus Singer Award. *Mem:* AAAS (secy sect biol sci, 70-74); Soc Develop Biol; Am Soc Zool; Am Asn Anat; Int Soc Cell Biol. *Res:* Regeneration; organ growth regulation; deer antler development. *Mailing Add:* Div Biol Brown Univ Providence RI 02912

GOSS, ROBERT CHARLES, b Huntington, Ind, Mar 31, 29; m 50; c 3. MICROBIOLOGY. *Educ:* Huntington Col, BS, 51; Purdue Univ, MS, 53, PhD(plant sci), 57. *Prof Exp:* Asst biol & plant path, Purdue Univ, 51-55, instr zool & chem exten, 55-56; plant pathologist, United Fruit Co, Costa Rica, 56;

instr gen sci & math, Woodworth Sch, Mich, 57-58; asst prof microbiol, Loyola Univ, La, 58-64; assoc prof, 64-71, prof microbiol, 71-77, PROF BIOL, UNIV NORTHERN IOWA, 77- *Mem:* Soc Indust Microbiol; Am Phytopath Soc; Nat Asn Biol Teachers. *Res:* Soil microbiology; plant pathogens, control and cancer. *Mailing Add:* Dept Sci Iowa State Col Cedar Falls IA 50613

GOSS, ROBERT NICHOLS, b Des Moines, Iowa, Jan 7, 21; m 42; c 3. MATHEMATICS. *Educ:* Drake Univ, AB, 42; Iowa State Univ, MS, 47, PhD(appl math), 50. *Prof Exp:* Instr math, Iowa State Univ, 46-50 & Univ Tulsa, 50-51; mathematician, US Navy Electronics Lab Ctr, 51-77; RETIRED. *Concurrent Pos:* Lectr, Univ Calif, Los Angeles, 51-54 & 56-60. *Mem:* Am Math Soc; Soc Indust & Appl Math; Math Asn Am; Asn Comput Mach. *Res:* Digital computer programming; partial differential equations; mathematical linguistics. *Mailing Add:* 650 W Harrison Ave Claremont CA 91711

GOSS, WILBUR HUMMON, b Tacoma, Wash, June 16, 11; c 3. PHYSICS. *Educ:* Univ Puget Sound, BS, 32; Univ Wash, PhD(physics), 39. *Prof Exp:* Lectr physics, Univ BC, 39-40; asst prof, NMex State Univ, 40-42; res physicist, Appl Physics Lab, Johns Hopkins Univ, 42-61, asst dir, 61-67; CONSULT, 67- *Concurrent Pos:* Vpres, Seatek Corp, 76-87. *Honors & Awards:* Presidential Cert Merit; Potts Medal, Franklin Inst; Distinguished Pub Serv Award, USN. *Mem:* Fel Am Phys Soc. *Res:* Electron physics; combustion; ordnance development. *Mailing Add:* 1050 Monte Dr Santa Barbara CA 93110

GOSS, WILLIAM PAUL, b Milford, Conn, May 23, 38; c 1. HEAT TRANSFER, THERMAL INSULATION. *Educ:* Univ Conn, BSE, 61, MS, 62, PhD(thermal sci), 67. *Prof Exp:* Res asst heat transfer, Univ Conn, 61-62; sr anal engr, Pratt & Whitney Div, United Aircraft Corp, 62-64; instr mech eng, Univ Conn, 64-67; asst prof, Va Polytech Inst, 67-70; assoc prof, 70-77, PROF MECH ENG, UNIV MASS, AMHERST, 78- *Concurrent Pos:* Consult, Mitre Corp, 67-71, Pratt & Whitney Commerical Div, United Technol Corp & Alfa-Thermal Systs Div, Alfa-Laval Thermal Inc, 78-81, Owens-Corning Fiberglass, 81-86, Cold Regions Res & Eng Lab, 82-87, Nat Bur Standards, 83-88, Jim Walter Res, 85- & US Dept Energy, 88- *Honors & Awards:* Centenial Medal, Am Soc Mech Engr. *Mem:* Am Soc Testing & Mat; Am Soc Eng Educ; Am Soc Mech Engrs; Am Soc Heat, Refrig & Air-Conditioning Engrs. *Res:* Window thermal performance; thermal insulation; thermal measurements and modeling. *Mailing Add:* Dept Mech Eng Univ Mass Amherst MA 01003

GOSSAIN, VED VYAS, b Jhang, India, Mar 25, 41; c 3. INTERNAL MEDICINE, ENDOCRINOLOGY. *Educ:* DAV Col, India, BSc, 58; Med Col Amritsar, MBBS, 63; All India Inst Med Sci, New Delhi, MD, 67. *Prof Exp:* Asst prof, St Louis Univ Sch Med, 73-75; from asst prof to assoc prof, 73-82, PROF MED, COL HUMAN MED, MICH STATE UNIV, 82-, ASSOC CHMN, DEPT MED, 88- *Concurrent Pos:* Asst res officer, Indian Coun Med Res, 65-67; clin fel endocrinol, Univ Cincinnati Med Ctr, res fel, 71-72; res fel Med Col Wis, 72-73; fel, All India Inst Diabetes. *Mem:* Fel Am Col Physicians; fel Royal Col Physicians Can; Endocrine Soc; Am Fedn Clin Res; Am Diabetes Asn; Int Diabetes Fedn; Asn Acad Minority Physicians. *Res:* Diabetes mellitus; insulin; glucagon; renin; angiotensin; aldosterone system. *Mailing Add:* Dept Med CHM Mich State Univ East Lansing MI 48824

GOSSARD, ARTHUR CHARLES, b Ottawa, Ill, June 18, 35. PHYSICS. *Educ:* Harvard Univ, BA, 56; Univ Calif, Berkeley, PhD(physics), 60. *Prof Exp:* Mem tech staff, AT&T Bell Labs, 60-87; PROF, MATS, ELEC & COMPUT ENG, UNIV CALIF, SANTA BARBARA, 87- *Concurrent Pos:* NSF fel, Saclay Nuclear Res Ctr, France, 62-63. *Honors & Awards:* Oliver Buckley Condensed Matter Physics Prize, Am Physics Soc, 84; Page Prize lectr, Physics Dept, Yale Univ, 86. *Mem:* Fel Am Phys Soc; Nat Acad Eng. *Res:* Solid state physics, nuclear magnetic resonance; ferromagnetism; transition metals; superconductivity; semiconductor films; heterostructures; interfaces; superlattices; molecular beam epitaxy. *Mailing Add:* Mat Dept Col Eng Univ Calif Santa Barbara Santa Barbara CA 93106

GOSSARD, EARL EVERETT, b Eureka, Calif, Jan 8, 23; m 48; c 3. PHYSICAL OCEANOGRAPHY, RADIO METEOROLOGY. *Educ:* Univ Calif, Los Angeles, AB, 48; Univ Calif, San Diego, MS, 51, PhD, 56. *Prof Exp:* From physicist to head radio physics div, US Navy Electronics Lab, San Diego, 48-71; chief geoacoust & chief radar meteorol, Wave Propagation Lab, Environ Res Lab, Nat Oceanic & Atmospheric Admin, Boulder, Colo, 71-83; RES ASSOC, UNIV COLO, 83- *Concurrent Pos:* Chmn, US Nat Comn F Int Radio Sci Union; consult, Interunion Comt Radio Meteorol; assoc ed, J Appl Meteorol & Radio Sci. *Honors & Awards:* Dept Com Silver Medal; Excellence in Refereeing Award, Am Geophys Union. *Mem:* Nat Acad Eng; Am Geophys Union; Int Radio Sci Union; fel Am Meteorol Soc. *Res:* Micrometeorology; lower ionosphere physics; internal gravity waves; radar meteorology. *Mailing Add:* CIRES Campus Box 449 Univ Colo Boulder CO 80309

GOSSEL, THOMAS ALVIN, b Lancaster, Ohio, Oct 20, 41; m 62; c 3. PHARMACOLOGY. *Educ:* Ohio Northern Univ, BS, 63; Purdue Univ, MS, 70, PhD(pharmacol), 72. *Prof Exp:* PROF PHARMACOL & TOXICOL, OHIO NORTHERN UNIV, 72-, CHMN DEPT PHARMACOL & BIOMED SCI, 77- *Concurrent Pos:* Consult, Nat Asn Bd Pharm, 73- *Mem:* AAAS; Sigma Xi; Soc Exp Biol & Med; Am Pharmaceut Asn. *Res:* Adult continuing education. *Mailing Add:* Ohio Northern Univ Col Pharm Ada OH 45810

GOSSELIN, ARTHUR JOSEPH, ELECTROPHYSIOLOGY, INTERVENTIONAL MEDICINE. *Educ:* Univ Ottawa, MD, 54. *Prof Exp:* DIR CARDIOVASC LAB, MIAMI HEART INST, 66- *Mailing Add:* Cardiovasc Lab Miami Heart Inst 4701 N Meridian Ave Miami Beach FL 33140

GOSSELIN, EDWARD ALBERIC, b Rutland, Vt, Feb 12, 43; m 70; c 2. HISTORY OF SCIENCE, SCIENCE EDUCATION. *Educ:* Yale Univ, BA, 65; Columbia Univ, MA, 66, PhD(early mod hist), 73. *Prof Exp:* Asst prof, 69-74, assoc prof, 74-79, PROF HIST, CALIF STATE UNIV, LONG BEACH, 79-, CHAIR, DEPT HIST, 86- *Concurrent Pos:* Co-dir hist & sci proj, Calif State Univ, Long Beach, 73-74, & dir, Modernization Global Perspective Proj, 80-81; ed, Hist Teacher, 85- *Mem:* Hist Sci Soc; Renaissance Soc Am. *Res:* Giordano Bruno; Bruno's and Galileo's trials; Hermetism and scientific revolution; French scientific and mathematical reform, 1495-1540. *Mailing Add:* Dept Hist Calif State Univ Long Beach CA 90840

GOSSELIN, RICHARD PETTENGILL, b Springfield, Mass, June 29, 21; m 44; c 2. MATHEMATICAL ANALYSIS. *Educ:* Univ Chicago, BS, 44, PhD, 51; Univ Rochester, MA, 48. *Prof Exp:* Sr mathematician, Inst Air Weapons Res, Univ Chicago, 51-52; prof math & head dept, Youngstown Col, 52-55; from assoc prof to prof, 55-88, EMER PROF MATH, UNIV CONN, 88- *Concurrent Pos:* Res contract, US Air Force, 59-60; consult, Inst Air Weapons Res, Univ Chicago. *Mem:* Am Math Soc; Math Asn Am. *Res:* Trigonometric series; trigonometric interpolating polynomials; localization theory and LP Fourier series. *Mailing Add:* 80 Oakwood Dr Windham CT 06280

GOSSELIN, ROBERT EDMOND, b Springfield, Mass, Sept 2, 19; m 48, 81; c 2. PHARMACOLOGY, TOXICOLOGY. *Educ:* Brown Univ, AB, 41; Univ Rochester, PhD(mammalian physiol), 45, MD, 47. *Prof Exp:* Asst, Off Sci Res & Develop contract, Univ Rochester, 42-44; med intern, Grace-New Haven Community Hosp, Conn, 47-48; from instr to asst prof pharmacol, Sch Med, Univ Rochester, 48-56; chmn dept, 56-75, prof, 56-89, EMER PROF PHARMACOL, DARTMOUTH MED SCH, 89- *Concurrent Pos:* Researcher, Atomic Energy Proj, 48-52 & 54-56; mem toxicol study sect, USPHS, 64-68; consult, Food & Drug Admin, 66-69, Consumer Prod Safety Comn, 79-85. *Mem:* Am Physiol Soc; Soc Pharmacol & Exp Therapeut; Soc Toxicol. *Res:* Atropine metabolism; ultraphagocytosis of radiocolloids; physiology and pharmacology of cilia; drug receptor kinetics; microcirculatory function; clinical toxicology. *Mailing Add:* Dept Pharmacol & Toxicol Dartmouth Med Sch Hanover NH 03756

GOSSELINK, EUGENE PAUL, b Pella, Iowa, Jan 13, 37; m 64; c 3. POLYMER CHEMISTRY. *Educ:* Cent Col, Iowa, BA, 59; Univ Wis, PhD(org chem), 64. *Prof Exp:* Fel org chem, Univ Toronto, 64-65; fel, Yale Univ, 65-66; RES STAFF CHEMIST ORG CHEM, PROCTER & GAMBLE CO, 66- *Mem:* Am Chem Soc. *Res:* Synthesis of new organic compounds; study of their effects at interfaces. *Mailing Add:* Miami Valley Lab PO Box 398707 Cincinnati OH 45061

GOSSELINK, JAMES G, b Kodaikanal, India, Sept 14, 31; US citizen; m 55; c 3. PLANT ECOLOGY, MARINE SCIENCES. *Educ:* Oberlin Col, AB, 53; Rutgers Univ, MS, 58, PhD(hort), 59. *Prof Exp:* Res horticulturist, New Crops Res Br, USDA, 59-61; asst prof biol, State Univ NY, Binghamton, 61-64; asst prof plant physiol, La State Univ, 64-66, asst prof bot & dir, Sci Training Progs, 66-68, assoc prof bot & marine sci, 68-73, chmn, dept marine sci, 74-79 & 79-85, dir, Coastal Ecol Lab, Ctr Wetland Resources, 77-79, prof Marine Sci & Coastal Ecol Lab, 73-90, EMER PROF, MARINE SCI, LA STATE UNIV, BATON ROUGE, 90- *Mem:* Fel AAAS; Am Inst Biol Sci; Ecol Soc Am; Soc Wetland Scientists; Estuarine Res Fedn. *Res:* Physiological systems and landscape ecology, especially of wetlands. *Mailing Add:* 7133 Sevenoaks Ave Baton Rouge LA 70806

GOSSER, LAWRENCE WAYNE, b Seattle, Wash, Aug 21, 38. PHYSICAL ORGANIC & ORGANOMETALLIC CHEMISTRY. *Educ:* Univ Wash, BS, 60; Univ Calif, Los Angeles, PhD(chem), 64. *Prof Exp:* Instr chem, Cent Ore Col, 64-65; CHEMIST, CENT RES DEPT, 65-, DIR CORP AFFIL, E I DU PONT DE NEMOURS & CO, INC, 80- *Mem:* Am Chem Soc. *Res:* Organic reactions involving carbanions; homogeneous catalysis; chemistry of organo-transition metal complexes. *Mailing Add:* 22 Brandywine Blvd Wilmington DE 19809

GOSSER, LEO ANTHONY, analytical & pharmaceutical chemistry, for more information see previous edition

GOSSETT, BILLY JOE, b Holladay, Tenn, Dec 13, 35; m 63; c 2. WEED SCIENCE. *Educ:* Univ Tenn, BS, 57; Univ Ill, MS, 59, PhD(agron), 62. *Prof Exp:* From asst prof to assoc prof, 62-72, PROF AGRON, AGRON & SOILS DEPT, CLEMSON UNIV, 72- *Mem:* Weed Sci Soc Am; Am Soc Agron. *Res:* Weed biology and control in agronomic cropping systems. *Mailing Add:* Dept Agron Clemson Univ Main Campus Clemson SC 29634

GOSSETT, CHARLES ROBERT, b Manila, PI, Sept 29, 29; m 52; c 4. NUCLEAR PHYSICS. *Educ:* Duke Univ, BS, 51; Rice Univ, MA, 53, PhD(physics), 55. *Prof Exp:* Physicist nucleonics div, Naval Res Lab, 55-58, supvry physicist radiation div, 58-66, res physicist, condensed mat & radiation sci div, 66-89; RES PHYSICIST, SFA, INC, 89- *Mem:* Am Phys Soc; Sigma Xi. *Res:* Application of nuclear techniques to surface and near-surface materials analysis. *Mailing Add:* CM&RS Div Code 4673 Naval Res Lab Washington DC 20375-5000

GOSSETT, DORSEY MCPEAKE, b Benton Co, Tenn, Dec 10, 31; m 57; c 1. AGRONOMY. *Educ:* Univ Tenn, Martin, BS, 55; Univ Ill, MS, 57; NC State Univ, PhD(crop physiol), 61. *Prof Exp:* Ext asst prof agron, NC State Univ, 61-65; asst prof agron, 65-70, supt, W Tenn Exp Sta, 70-72, asst dean, 72-76, DEAN, AGR EXP STA, UNIV TENN, 76- *Res:* Administering the Tennessee Agricultural Experiment Station program; weed science. *Mailing Add:* 8843 Cove Point Lane Knoxville TN 37922

GOSSETT, JAMES MICHAEL, b San Rafael, Calif, May 3, 50; m 81; c 2. WASTEWATER TREATMENT. *Educ:* Stanford Univ, BS, 73, MS, 73, PhD(civil eng), 77. *Prof Exp:* asst prof, 76-82, ASSOC PROF ENVIRON ENG, CORNELL UNIV, 82- *Concurrent Pos:* Fel, US Air Force Fac Res Prog, 80; vis prof, Univ Resident Res Prog, USAF Systs Command, 84-85. *Mem:* Am Soc Microbiol; Water Pollution Control Fedn; Int Asn Water Pollution Res; Asn Environ Eng Professors. *Res:* Anaerobic biological processes; biological wastewater treatment; methane production from biomass; biomass energy conversion; treatment and restoration of contaminated groundwaters. *Mailing Add:* Sch Civil/Environ Engr Cornell Univ Ithaca NY 14853

GOSSLEE, DAVID GILBERT, b Fargo, NDak, June 22, 22; m 47; c 3. BIOMETRICS. *Educ:* Moorhead State Col, BS, 47; Iowa State Univ, MS, 50; NC State Col, PhD, 56. *Prof Exp:* Instr math, NDak State Univ, 50-54, statistician, 54-58; assoc prof biomet, Univ Conn, 58-61; statist sect head, Comput Sci Div, Union Carbide Corp, 61-87; RETIRED. *Concurrent Pos:* Instr, Univ Tenn, Oak Ridge Biomed Grad Sch, 68- *Mem:* Fel Am Statist Asn; Sigma Xi; Biomet Soc; fel AAAS. *Res:* Consulting and collaboration; statistical analysis of toxicological assays. *Mailing Add:* 106 Indian Lane Oak Ridge TN 37831

GOSSLING, JENNIFER, b Eng, July 25, 34; m 56. INDIGENOUS BACTERIA, PATHOGENIC BACTERIA. *Educ:* Cambridge Univ, BA, 55; WVa Univ, PhD(microbiol), 73. *Prof Exp:* Clin microbiologist, Ottawa Civic Hosp, 57-59; vet microbiologist, Ont Vet Col, 59-61; asst vet microbiol, Univ Ill, 62-66; bacteriologist, Univ Manchester, 66-69; asst instr med microbiol, WVa Univ, 69-73; biochemist, Med Col Ohio, 75; fel, Dental Res Inst, Univ Mich, 78-79; clin microbiologist, Ind Hosp, 79-80; CLIN MICROBIOLOGIST, JEWISH HOSP, ST LOUIS, 80- *Concurrent Pos:* Asst prof dental microbiol, Sch Dent Med, Wash Univ, 81-91. *Mem:* Am Soc Microbiol; AAAS; NY Acad Sci; Am Soc Clinical Path. *Res:* Investigation of oral and intestinal bacteria of man and other animals; the predominant anaerobes and some potential pathogens, analyzed by serology, culture and serial studies. *Mailing Add:* Jewish Hosp St Louis 216 S Kings Hwy St Louis MO 63110

GOSSMANN, HANS JOACHIM, b Schwandorf, Fed Rep Ger, Sept 15, 55; m 87. SURFACE SCIENCE, ION SCATTERING. *Educ:* Univ Wurzburg, Ger, Vordipl, 77, dipl, 81; State Univ NY, Albany, MS, 79, PhD, 84. *Prof Exp:* Fel, 84-85, MEM TECH STAFF, AT&T BELL LABS, 85- *Honors & Awards:* Morton H Traum Award, Am Vacuum Soc, 84. *Mem:* Am Vacuum Soc; Mat Res Soc. *Res:* Fundamental aspects of epitaxial growth and strained layer hetero structures; investigation of phenomena and processes occurring during initial stage of interface formation; silicon molecular beam epitaxy. *Mailing Add:* AT&T Bell Labs 600 Mountain Ave Murray Hill NJ 07974

GOSWAMI, AMIT, b Faridpur, India, Nov 4, 36; m 60; c 2. THEORETICAL PHYSICS. *Educ:* Presidency Col, Univ Calcutta, India, BSc, 55, MSc, 60, DPhil(physics), 64. *Prof Exp:* Instr physics, Case Western Reserve Univ, 63-65, asst prof, 65-67; assoc prof, 68-74, PROF PHYSICS, UNIV ORE, 74- *Concurrent Pos:* Res assoc, Inst Theoret Sci, Univ Ore, 68- *Mem:* Am Phys Soc. *Res:* Generalized pairing in light nuclei; bootstrap theory of vibrations; use of surface-delta-interaction in nuclear structure calculations; theory of fine structure of giant dipole resonance; gapless superconductivity in nuclei; pion condensation in nuclear matter; quantum theory of consciousness. *Mailing Add:* Dept Physics Univ Ore Eugene OR 97403

GOSWAMI, BHUVENESH C, b Bannu, Pakistan, Oct 13, 37; US citizen. TEXTILE PHYSICS, TEXTILE TECHNOLOGY. *Educ:* Delhi Univ, BS, 59; Univ Bombay, MSc, 63; Univ Manchester, PhD(textile physics), 66. *Hon Degrees:* FTI, Textile Inst, Eng, 72. *Prof Exp:* Spinning supvr spinning mills, Arvind Mills, India, 59-61; res engr prod develop, Calico Chem & Plastics, Bombay, 66-67; vis lectr textile sci, Clemson Univ, 67-68; staff scientist textile physics, Textile Res Inst, Princeton, 69-75; assoc prof, 75-80, PROF TEXTILE PHYSICS, UNIV TENN, KNOXVILLE, 80- *Concurrent Pos:* Fel Textile Res Inst, 69; res fel, Univ Manchester, Inst Sci & Technol, 72-73. *Mem:* Textile Inst; Fiber Soc; Am Asn Textile Chemists & Colorists; Am Soc Mech Eng. *Res:* Structural mechanics of fiber, yarns and fabrics; thermal behavior of textile structures. *Mailing Add:* Dept Textiles Technol Clemson Univ Main Campus Clemson SC 29634

GOSWITZ, FRANCIS ANDREW, b St Paul, Minn, Sept 8, 31; m 56; c 3. NUCLEAR MEDICINE, HEMATOLOGY. *Educ:* Marquette Univ, BS, 53, MD, 56. *Prof Exp:* Intern, Univ Iowa Hosps, 56-57, resident internal med, 57-58, 60 & 61-63; clinician internal med & coordr med radioisotope courses, Spec Training Div, 65-68 & Med Div, 68-75, SR CLINICIAN INTERNAL MED, MED DIV, OAK RIDGE MED CLIN, 75- *Concurrent Pos:* Fel nuclear med, Oak Ridge Inst Nuclear Studies, 60-61; fel hemat, Med Ctr, Univ Utah, 63-65. *Mem:* Soc Nuclear Med; Am Med Asn; Am Soc Hemat; NY Acad Sci. *Res:* Internal medicine; effects of irradiation on hematopiesis; radioprotective cytologic agents; diagnostic applications of radioisotopes in medicine; biochemical and physiological characteristics of the erythrocyte and lymphocyte; cancer chemotherapy. *Mailing Add:* 170 W Tennessee Ave Oak Ridge TN 37830

GOSWITZ, HELEN VODOPICK, b Milwaukee, Wis, Apr 29, 31; m 56; c 3. HEMATOLOGY, NUCLEAR MEDICINE. *Educ:* Marquette Univ, BS, 53, MD, 56. *Prof Exp:* Intern, Univ Iowa Hosp, 56-57, resident internal med, 57-60, instr internal med, Hosp, 61-63; clinician internal med, Oak Ridge Assoc Univs, 65-74; INTERNIST, OAK RIDGE MED CLIN, 74- *Concurrent Pos:* Fel nuclear med, Oak Ridge Inst Nuclear Studies, 60-61; fel hemat, Univ Utah, 63-65. *Mem:* Fel Am Col Physicians; Soc Nuclear Med; AMA; Am Soc Hemat. *Res:* Internal medicine; granulocytokinetics; histocompatibility testing; oncology. *Mailing Add:* Oak Ridge Med Clin 170 W Tenn Ave Oak Ridge TN 37830

GOSZ, JAMES ROMAN, b Menasha, Wis, May 4, 40. ECOLOGY. *Educ:* Mich Technol Univ, BS, 63; Univ Idaho, PhD(forest sci), 68. *Prof Exp:* Res assoc ecol, Dartmouth Col, 68-69 & Cornell Univ, 69-70; asst prof, 70-74, assoc prof, 75-78, prof ecol, Univ NMex, 78-; dir, Ecosystems Studies Prog, 84-86. *Concurrent Pos:* Univ Founder Rep, Inst Ecol, 73- *Mem:* Ecol Soc Am; Int Asn Ecol. *Res:* Ecosystem analysis through mineral cycling and energy flow dynamics; physiological plant ecology. *Mailing Add:* Dept Biol Univ NMex Albuquerque NM 87131

GOTCHER, JACK EVERETT, b Wichita Falls, Tex, May 11, 49; m 72; c 2. ORAL & MAXILLOFACIAL SURGERY, BONE HISTOMORPHOMETRY. *Educ:* Midwestern Univ, BS, 71; Harvard Sch Dent Med, DMD, 75; Univ Utah, PhD(anat), 79. *Prof Exp:* Teaching asst anat, Col Med, Univ Utah, 75-77, lectr histol, 77; asst prof oral-maxillofacial surg, Sch Dent, Emory Univ, 82-83; clin instr, 78-82, ASSOC PROF ORAL-MAXILLOFACIAL SURG, CTR HEALTH SCI, UNIV TENN, 83- *Concurrent Pos:* Prin investr, Nat Inst Dent Res, Univ Utah, 75-78 & Upjohn Co Clin Trial Alveolar Atrophy, 84-; vis prof oral surg, Sch Dent Loyola Univ, 85; guest lectr, III Int Sem Maxillofacial Surg, Guadalajaha, 86; chmn, Accredited Dent Educ, Tenn Dent Asn, treas, 2nd Dist, Tenn Dent Soc, 88. *Honors & Awards:* Res Travel Award, Johnson & Johnson Co, 75. *Mem:* Int Asn Dent Res; Am Bd Oral & Maxillofacial Surg; Am Asn Oral & Maxillofacial Surgeons; Am Col Oral & Maxillofacial Surgeons; Am Dent Asn; Am Asn Dent Schs. *Res:* Hard and soft tissue histomorphometry in areas of interest to oral and maxillofacial surgeons; temporomandibular joint pathology; alveolar atrophy. *Mailing Add:* 1928 Alcoa Hwy Suite 305 Knoxville TN 37920

GOTELLI, DAVID M, b Stockton, Calif, Jan 1, 43; m 65; c 1. BOTANY, MYCOLOGY. *Educ:* Univ Calif, Berkeley, BA, 64; Univ Wash, PhD(bot), 69. *Prof Exp:* Asst prof biol, 70-72 assoc prof biol, 72-80, PROF BOTANY, CALIF STATE COL, STANISLAUS, 72- *Mem:* Mycol Soc Am; Bot Soc Am. *Res:* Ultrastructure and development of biflagellate fungi; culture and development of higher basidiomycetes. *Mailing Add:* 800 W Monte Vista Rd Turlock CA 95380

GOTH, ANDRES, pharmacology; deceased, see previous edition for last biography

GOTH, JOHN W, b Ree Heights, SDak, Apr 22, 27. METALLURGY & PHYSICAL METALLURGICAL ENGINEERING. *Educ:* SDak Sch Mines, BS, 50, PhD(bus admin), 78; McGill Univ, MS, 51. *Prof Exp:* Sr exec vpres, Amax Corp, 85; EXEC DIR, MINERAL INFO INST, 86- & DENVER GOLD GROUP, 91- *Concurrent Pos:* Dir, var mining co; mem, Nat Strategic Mat & Minerals Prog Adv Comt. *Mem:* Fel Am Soc Metals. *Res:* Molybdenum market development. *Mailing Add:* Mineral Info Inst 1536 Cole Blvd Suite 320 Golden CO 80401

GOTH, ROBERT W, b Phillips, Wis, May 10, 27; m 54; c 3. PLANT PATHOLOGY. *Educ:* Wis State Univ, Superior, BS, 54; Univ Minn, MS, 57, PhD(plant path), 61. *Prof Exp:* Plant pathologist, 61-68, res plant pathologist potato invest, 68-72, RES PLANT PATHOLOGIST VEG LAB, USDA, 72- *Concurrent Pos:* Vpres, Potmac Div, Am Phytopath Soc, 84-85, pres, 85-86. *Mem:* Am Phytopath Soc (sect-treas, 81-83); Am Potato; Mycol Soc. *Res:* Nature of resistance, mode of infection, physical and chemical nature of the causal organisms of virus, fungal and bacterial diseases of potatoes, beans, peas and related legumes. *Mailing Add:* Headhouse 13 Bldg 011 Veg Lab Agr Res Serv USDA Beltsville MD 20705

GOTHELF, BERNARD, b Chicago, Ill, May 8, 28; m 58; c 5. PHARMACOLOGY, TOXICOLOGY. *Educ:* Univ Ill, BS, 50; Univ Iowa, MS, 52; Loyola Univ Chicago, PhD(pharmacol), 65. *Prof Exp:* Asst prof pharmacol, Loyola Univ, La, 63-64; instr, Sch Med, Marquette Univ, 64-66; asst prof, Dent Br, Univ Tex, Houston, 66-69; res assoc, Univ Tex Med Br, Galveston, 70-73 & assoc prof pharmacol, Tex Col Osteopath Med, 73; asst prof pharmacol, Baylor Col Dent, 73-77; vis prof, Univ de Caribe, Escuela de Medicina de Cayey, Inc, Cayey, PR, 77-78; CONSULT & DIR, 5-STAR TOXICOL ANALYSIS & CONSULT, DALLAS, 78- *Concurrent Pos:* Post-doctoral fel, Marquette Univ. *Mem:* Am Col Toxicol; Am Indust Hyg Asn. *Res:* Neuropharmacology; neurochemistry; analysis and distribution of chlorpromazine in tissues. *Mailing Add:* 6406 Dykes Way Dallas TX 75230

GOTLIEB, AVRUM I, b Montreal, Que, Jan 17, 46. PATHOLOGY. *Educ:* McGill Univ, BSc, 67, MD, 71; Am Bd Path, cert, 76. *Prof Exp:* Staff pathologists, St Michael's Hosp, 78-89; from asst prof to assoc prof, 78-88, PROF PATH, UNIV TORONTO, 88- *Concurrent Pos:* Assoc pathologist, Toronto Hosp, 80-82, consult staff pathologist, 82-88; dir, Vascular Res Lab, Dept Path & Banting & Best Diabetes Ctr, 83- *Mem:* Soc Cardiovascular Path (secy, 86); Am Asm Pathologists; fel Am Heart Asn; Am Soc Cell Biol; Sigma Xi; NY Acad Sci. *Res:* Pathology. *Mailing Add:* Dept Path Toronto Gen Hosp Res Ctr CCRW 1-857 Univ Toronto Fac Med 200 Elizabeth St Toronto ON M5G 2C4 Can

GOTLIEB, C(ALVIN) C(ARL), b Toronto, Ont, Mar 27, 21; m 49; c 3. COMPUTER SCIENCE, SOFTWARE SYSTEMS. *Educ:* Univ Toronto, BA, 42, MA, 46, PhD(physics), 47. *Hon Degrees:* DrMath, Univ Waterloo, 68; DEng, NS Tech Univ, 85. *Prof Exp:* Asst, Nat Res Coun Can, Toronto, 42-47; from asst prof to prof physics, Univ Toronto, 48-64, dir, Inst Comput Sci, 62-70, head, Dept Comput Sci, 64-67, prof, 64-88, EMER PROF COMPUT SCI, UNIV TORONTO, 88- *Concurrent Pos:* Civilian sci officer, Nat Res Coun Can, 42-47; mem, Admiralty Sig Estab & Ministry Supply, Eng, 43-44; consult, Defense Res Med Labs, Royal Can Air Force, 46-56; Can deleg, Int Fedn Info Process Socs, 59-65, vchmn prog comt, Int Fedn Info Processing Cong, 71; ed-in-chief, J Asn Comput Mach, 66-68, ed, 69-; chmn tech comt, Int Fedn Info Processing, 76-85; consult, UN & Can. *Mem:* Fel Royal Soc Can; Asn Comput Mach; hon mem Can Info Processing Soc (vpres, 58-59, pres, 59-60); fel Brit Comput Soc. *Res:* Computer systems and applications; business data processing; social implications of computers; economics of computers; author of over one hundred publications-research papers, books and articles. *Mailing Add:* Dept Comput Sci Sandford Fleming Bldg Univ Toronto Toronto ON M5S 1A4 Can

GOTOFF, SAMUEL P, b New York, NY, Mar 22, 33; m 56; c 3. IMMUNOLOGY. *Educ:* Amherst Col, BA, 54; Univ Rochester, MD, 58. *Prof Exp:* Instr, Sch Med, Yale Univ, 60-61; clin instr pediat, Abraham Lincoln, Sch Med, Univ Ill Col, 61-63, from asst to assoc prof, 65-71, assoc prof microbiol, 68-71, prof pediat & microbiol, Abraham Lincoln Sch Med, Univ Ill Col Med, 71-73; prof pediat, Univ Chicago & chmn dept pediat, Michael Reese Med Ctr, 73-86; PROF PEDIAT, RUSH MED COL, CHMN, DEPT PEDIAT, RUSH-PRESBY-ST LUKE'S MED CTR, 86- *Concurrent Pos:* Attend physician, Munic Contagious Dis Hosp, 67-73. *Mem:* AAAS; Am Acad Pediat; Soc Pediat Res; Am Asn Immunol; Am Soc Microbiol. *Res:* Neonatal infections; neonatal immunity; group b streptococci. *Mailing Add:* Rush Presby St Luke's Med Ctr 1753 W Congress Pkwy Chicago IL 60612

GOTOLSKI, WILLIAM H(ENRY), b Newark, NJ, Sept 6, 26; m 51; c 2. CIVIL ENGINEERING. *Educ:* Columbia Univ, BS, 46, MS, 47; Pa State Univ, PhD(civil eng), 59. *Prof Exp:* Instr civil eng, Ohio Univ, 47-52; instr, 52-60, assoc prof, 60-65, asst dean & res instr, 74-84, spec asst to dean, 84-85, assoc dean & undergrad instr, 87-88, PROF CIVIL ENG, PA STATE UNIV, UNIVERSITY PARK, 65- DIR, ENG ADVAN CTR, 85- *Concurrent Pos:* Mem Transp Res Bd, Nat Acad Sci-Nat Res Coun; consult civil engr, 50- *Mem:* Am Soc Civil Engrs; Am Soc Testing & Mat. *Res:* Soil mechanics, foundations and earth structures; bituminous concrete. *Mailing Add:* 623 W Hillcrest Ave State College PA 16803

GOTS, JOSEPH SIMON, b Phila, Pa, Oct 12, 17; m 42; c 2. MICROBIOLOGY, BIOCHEMICAL GENETICS. *Educ:* Temple Univ, AB, 39; Univ Pa, MS, 41, PhD(med bact), 48; Am Bd Microbiol, dipl. *Prof Exp:* Bact lab aide, Univ Pa, 39-41; spec field agent Bang's dis res, USDA, 41-42; from instr to assoc bact, 48-51, from asst prof to assoc prof microbiol, 51-63, PROF MICROBIOL, SCH MED, UNIV PA, 63- *Concurrent Pos:* Vis investr, Inst Radium, Paris, 65-66; consult, Parke Davis & Co, Mich; vis prof, Peking Union Med Col, Beijing, China, 85, Univ Hawaii, 90. *Mem:* AAAS; Am Soc Microbiol; Soc Exp Biol & Med; Genetics Soc Am; Am Soc Biol Chem; Am Asn Cancer Res; Sigma Xi; Am Acad Microbiol; NY Acad Sci. *Res:* Mechanism of action and resistance to chemotherapeutics; bacterial metabolism; biochemical mutations in bacteria; regulation of gene expression; biosynthesis of purines and amino acids. *Mailing Add:* Dept Microbiol Univ Pa Sch Med Philadelphia PA 19104-6076

GOTSCHLICH, EMIL C, b Bangkok, Thailand, Jan 17, 35. IMMUNOLOGY. *Educ:* NY Univ, AB, 55; NY Univ Sch Med, MD, 59. *Prof Exp:* From asst prof to assoc prof, 68-78, PROF & SR PHYSICIAN, THE ROCKEFELLER UNIV, 78-, CHMN, LAB BACTERIOL/IMMUNOL, 81- *Concurrent Pos:* Staff, Med Corp, US Army, Walter Reed Army Inst Res, Washington, DC, 66-68. *Honors & Awards:* Albert Lasker Award, 78. *Mem:* Inst Med, Nat Acad Sci; Sigma Xi; Am Asn Immunol. *Res:* Published numerous articles inv arious journals. *Mailing Add:* Dept Immunol Rockefeller Univ 1230 York Ave New York NY 10021-6399

GOTSHALL, DANIEL WARREN, b Springfield, Ill, Dec 20, 29; m 52. MARINE ECOLOGY. *Educ:* Humboldt State Col, BS, 57, MS, 70. *Prof Exp:* Fisheries biologist, Marine Resources Opers, 57-59, from marine biologist to sr marine biologist, Marine Resources Region, Long Beach, 60-72, sr marine biologist, Marine Res Br, Calif Dept Fish & Game, Avila Beach, 72-83, Monterey, 84-87; RETIRED. *Concurrent Pos:* Mem adv bd underwater parks & reserves, Calif Dept Parks & Recreation, 71, vice chmn, 74-; assoc invert zool, Calif Acad Sci, San Francisco, 75-, chmn, 87-88. *Honors & Awards:* Fel, Explorers Club, 80. *Mem:* Am Fisheries Soc; fel Am Inst Fishery Res Biologists; Western Soc Naturalists. *Res:* Life history of blue rockfish; marine sport fishing; population dynamics of the ocean shrimp, Pandalus jordani and the Dungeness crab, Cancer magister; benthic ecology using scuba and underwater photography; conducting baseline marine ecological surveys around nuclear power plants to determine effects of construction and operation of plants. *Mailing Add:* 4 Somerset Rise Skyline Forest Monterey CA 93940-4112

GOTSHALL, ROBERT WILLIAM, b Carrollton, Ohio, Apr 28, 45. PHYSIOLOGY. *Educ:* Mt Union Col, BS, 67; Ohio State Univ, MS, 69, PhD(physiol), 71. *Prof Exp:* Fel physiol, Sch Med, Univ Mo, 71-73; asst prof, Med Col Wis, 73-75; asst prof physiol, 75-77, ASSOC PROF PHYSIOL & BIOPHYSICS, SCH MED, WRIGHT STATE UNIV, 78- *Mem:* Am Physiol Soc; Am Heart Asn; Soc Exp Biol & Med. *Res:* Cardiac function and control in adults and newborns. *Mailing Add:* Dept Physiol & Biophys Wright State Univ PO Box 927 Dayton OH 45401

GOTT, EUYEN, b Kweilin, China, Jan 2, 15; US citizen; m 48; c 2. ELECTRICAL ENGINEERING, SOFTWARE DESIGN. *Educ:* Nat Kwangsi Univ, China, BS, 38; Stanford Univ, AM, 45; Johns Hopkins Univ, DrEng, 59. *Prof Exp:* Instr elec eng, Nat Kwangsi Univ, China, 38-41; head design engr, Yee-Chong Mfg Co, 41-44; develop engr, Radio Corp Am, 45-49; proj engr, Sierra Electronics Corp, 50-53; res assoc radiation lab, Johns Hopkins Univ, 53-59; assoc prof elec eng, Univ Hawaii, 59-63, prof, 63-65; mem tech staff, Hughes Aircraft Co, 65-66, staff engr, 66-67; sr tech specialist, Autonetics Div, NAm Rockwell Corp, 67-69; sr res scientist, Norden Div, United Aircraft Corp, 71-72; staff scientist, Lockheed Missiles & Space Co, 72-86; RETIRED. *Concurrent Pos:* Vis prof, Calif State Col, Fullerton, 70-71. *Mem:* Inst Elec & Electronics Engrs; Am Soc Eng Educ. *Res:* Detection of signal in noise; information processing systems; circuit theory; computer-aided active network analysis and synthesis; software design and development; emulation, automatic fault isolation, automatic flowcharting. *Mailing Add:* 1221 Eichler Ct Mountain View CA 94040

GOTT, J RICHARD, III, b Louisville, Ky, Feb 8, 47; m 78; c 1. COSMOLOGY, GALAXY FORMATION. *Educ:* Harvard Univ, AB, 69; Princeton Univ, PhD(astrophys), 72. *Prof Exp:* Fel, Calif Inst Technol, 73-74; vis fel, Cambridge Univ, 75; from asst prof to assoc prof, 76-87, PROF, PRINCETON UNIV, 87- *Concurrent Pos:* Alfred P Sloan Found res fel, 77; chmn judges, Westinghouse Sci Talent Search, 86- *Honors & Awards:* Trumpler Award, Astron Soc Pac, 75. *Mem:* Int Astron Union; Am Astron Soc. *Res:* Cosmology, galaxy formation and galaxy clustering; theory that the universe is open and will continue expanding forever. *Mailing Add:* Dept Astrophys Sci Princeton Univ Princeton NJ 08544

GOTT, PRESTON FRAZIER, b Waxahachie, Tex, Nov 21, 19; m 42, 90; c 2. ASTRONOMY, DIRECTOR OBSERVATORY. *Educ:* Univ Tex, BS, 44, MA, 47. *Prof Exp:* Tutor, Univ Tex, 44-46; from instr to asst prof, Hardin Col, 47-49; from asst prof to assoc prof physics, 49-89, EMER ASSOC PROF PHYSICS & DIR, TEX TECH OBSERV, TEX TECH UNIV, 89- *Concurrent Pos:* Engr II, Arnold Eng Develop Ctr, Tullahoma, Tenn, 57; prin investr res contract, Tex Tech Univ, 59-60; physicist, Land-Air Div, Dynalectron Corp, White Sands Missile Range & Holloman AFB, NMex, 61-64 & consult, 63 & 65; sr scientist space instruments sect, Jet Propulsion Lab, Calif Inst Technol, 65, 66, 67, 70; co-ed proc, Southwest Regional Conf Astron & Astrophys, 75-84, mem bd, 76-78. *Mem:* Fel AAAS; Soc Photo-Optical Instrument Eng; Optical Soc Am; Int Amateur Prof Photom; Am Astron Soc; Sigma Xi. *Res:* Photography; atmospheric optics; space physics; experimental optical design. *Mailing Add:* Dept Physics Tex Tech Univ Lubbock TX 79406

GOTT, VINCENT LYNN, b Wichita, Kans, Apr 14, 27; m 54; c 3. CARDIOVASCULAR SURGERY. *Educ:* Univ Wichita, BS, 51; Yale Univ, MD, 53. *Prof Exp:* Instr surg, Med Sch, Univ Minn, 59-60; from asst prof to assoc prof, Sch Med, Univ Wis, 60-65; chief cardiac surg, 65-82, assoc prof, 65-68, PROF SURG, SCH MED, JOHNS HOPKINS UNIV, 68- *Concurrent Pos:* Markle scholar acad med, 62-67. *Honors & Awards:* Holstoen Gold Medal, Am Med Asn, 57. *Mem:* Soc Univ Surg; fel Am Col Surg; Am Soc Artificial Internal Organs; Am Asn Thoracic Surg. *Res:* Cardiovascular research, including myocardial metabolism, artificial heart valves; artificial pacemakers; coronary artery disease; intravascular clotting problems; clinical problems; clinical surgery of the heart and blood vessels. *Mailing Add:* Johns Hopkins Hosp 614 Blalock Bldg Baltimore MD 21205

GOTTERER, GERALD S, b New York, NY, Oct 17, 33; m 78; c 4. MEDICAL EDUCATION, BIOCHEMISTRY. *Educ:* Harvard Univ, AB, 55; Univ Chicago, MD, 58; Johns Hopkins Univ, PhD(biochem), 64. *Prof Exp:* Intern internal med, Grace-New Haven Community Hosp, 58-59; asst dean student affairs, Sch Med, Johns Hopkins Univ, 70-72, from instr to asst prof physiol chem, 65-78, dean predoctoral progs, 72-78; assoc prof biochem & assoc dean med student progs, Rush Med Col, 78-86; ASSOC DEAN & PROF MED ADMIN, VANDERBILT SCH MED, 86- *Concurrent Pos:* USPHS fels, 59-64. *Mem:* Am Chem Soc; Am Soc Biol Chem; Sigma Xi. *Res:* medical education. *Mailing Add:* Vanderbilt Univ Sch Med 201 Light Hall Nashville TN 37232-0685

GOTTERER, MALCOLM HAROLD, b New York, NY, Mar 11, 24; m 57; c 1. INFORMATION SCIENCE, COMPUTER SCIENCE. *Educ:* Suffolk Univ, BS, 55, MS, 56; Harvard Univ, DBA, 60. *Prof Exp:* Instr bus, Grad Sch Bus, Harvard Univ, 56-57; asst prof, Univ Calif, Berkeley, 59-62; assoc prof indust mgt, Ga Inst Technol, 62-64; prof bus admin & comput sci, Pa State Univ, University Park, 65-74; dir info systs, Fla Int Univ, 77-78, prof comput sci, 73-86, dir Grad Studies in Comput Sci, 79-86, COMPUT CONSULT, FLA INT UNIV, 65-, EMER PROF, 91- *Concurrent Pos:* Soc Sci Res Coun res grant, Info Processing Div, Syst Develop Corp, Calif, 65; Int Bus Mach Systs Res Inst fel, NY, 64; consult, Ctr Comput Sci & Technol, Nat Bur Standards, 66-68 & Human Resources Res Orgn, George Washington Univ, 67; prin sect comt comput & info processing & comt credit card stand, US Am Standards Inst, 68-70; mem US comt admin data processing group, Int Fedn Info Processing; fel comput sci, Johns Hopkins Univ, 71-72; mem comt professionalism, Am Fedn Info Processing Socs; mem UN tech asst mission to Jamaica, 69; Nat lectr, Asn Comput Mach, 70-73. *Mem:* Asn Comput Mach; Inst Elec & Electronics Engrs. *Res:* Computer system evaluation; performance improvement, and configuration administration; data structures. *Mailing Add:* 6400 SW 112th St Miami FL 33156

GOTTESFELD, ZEHAVA, b Tel Aviv, Israel. NEUROBIOLOGY, NEUROCHEMISTRY. *Educ:* Hebrew Univ, Jerusalem, MSc, 60; Univ NC, Chapel Hill, PhD(physiol), 67. *Prof Exp:* Fel neurochem, McGill Univ, 68-70; investr neurobiol, Weizmann Inst Sci, 71-74, sr scientist, 75-76; vis scientist histopharmacol, NIH, Bethesda, Md, 76-78; ASSOC PROF NEUROBIOL, UNIV TEX MED SCH, HOUSTON, 78- *Concurrent Pos:* Assoc prof biomed res grant, Univ Tex, 78-79; vis scientist, NIH. *Mem:* Int Soc Neurochem; Soc Neurosci; Am Soc Neurochem. *Res:* Neurocytology; neuroanatomy; histochemistry; neurochemistry; neuroendocrinology. *Mailing Add:* Dept Neurobiol & Anat Univ Tex Med Sch Houston TX 77025

GOTTESMAN, ELIHU, b Brooklyn, NY, June 6, 19; m 49; c 3. PHARMACEUTICAL CHEMISTRY, SAFETY. *Educ:* Johns Hopkins Univ, BA, 39; Fordham Univ, MS, 40. *Prof Exp:* Chemist, Food & Drug Res Labs, Inc, 41-42; supvr, Vitamin Lab, Endo Labs, Inc, 46-52, mgr, parenteral prod div, 52-73; safety supvr, E I du Pont de Nemours & Co, Garden City, 73-85; RETIRED. *Concurrent Pos:* Ed, Int Blue Book, 47-52. *Mem:* AAAS; Am Chem Soc; Am Pharmaceut Asn; Am Soc Safety Engrs; Fel Am Inst Chemists. *Res:* Development of pharmaceutical equipment and procedures and of pharmaceutical dosage forms; equipment and special building construction for pharmaceutical industry. *Mailing Add:* 148 Mead Ct Wantagh NY 11793

GOTTESMAN, MICHAEL, b Jersey City, NJ, Oct 7, 46; m 66; c 2. CELL BIOLOGY. *Educ:* Harvard Univ, MD, 70. *Prof Exp:* CHIEF LAB CELL BIOL, NAT CANCER INST, 80- *Honors & Awards:* Soma Weiss, 70; James Tolbert Shipley Prize, 70; Milken Award, 90. *Mem:* Am Soc Cell Biologist; AAAS; Genetics Soc Am; Am Asn Cancer Res; Am Soc Biochem & Molecular Biol. *Res:* Molecular basis of drug resistance, growth regulation, invasiveness and metastasis in cancer. *Mailing Add:* NIH Bldg 37 Rm 1B22 Bethesda MD 20892

GOTTESMAN, ROY TULLY, b Bayonne, NJ, March 6, 28; m 54; c 2. ORGANIC & POLYMER CHEMISTRY. *Educ:* Rutgers Univ, BS, 47, MS, 50, PhD(chem), 51. *Prof Exp:* Asst chem, Univ Minn, 47-48; asst, Rutgers Univ, 48-51; chemist res & develop, Heyden Chem Corp, 51-56, group leader, Res & Develop Dept, Heyden Div, Heyden Newport Chem Corp, 56-61, supvr org res, 61-65; mgr, Tenneco Chem, Inc, 65-70, mgr, tech develop intermediates div, 70-71, mgr res & develop, org & polymers div, 71-76, dir, chem develop, 76-77, dir, 77-80, vpres, environ & regulatory affairs, 80-83; EXEC DIR, VINYL INST, 83- *Mem:* Am Chem Soc; NY Acad Sci; Sigma Xi. *Res:* Synthetic organic chemistry; preparation and reactions of polyols; chemistry of salicylic acid; process development and improvement; agricultural chemicals; research administration; biocides, paint and plastics additives; polymers; environmental regulations; occupational health and safety; product safety; hazardous waste disposal; vinyl polymers. *Mailing Add:* Vinyl Inst Wayne Interchange Plaza II 155 Rt 46 W Wayne NJ 07470

GOTTESMAN, STEPHEN T, b New York, NY, Feb 23, 39; m 68, 90; c 5. RADIO ASTRONOMY. *Educ:* Colgate Univ, BA, 60; Manchester Univ, PhD(radio astron), 67. *Prof Exp:* Lectr physics, Univ Keele, 68-69; res assoc radio astron, Nat Radio Astron Observ, 69-71; res fel, Calif Inst Technol, 71; from asst prof to assoc prof, 72-81, PROF ASTRON, UNIV FLA, 81-, CHMN DEPT ASTRON, 88- *Concurrent Pos:* Chmn, Int Astron Union Working Group Internal Motion Galaxies, 79-85; Fulbright Scholar, 60-61; Leverhulme fel, 61-64. *Mem:* Am Astron Soc; Royal Astron Soc; Int Astron Union; Comn V, Int Union Radio Sci. *Res:* Neutral hydrogen and other radio spectral line observations of the interstellar medium and of the structure, kinematics and dynamics of the galaxy and of extragalactic nebulae; the mass of galaxies. *Mailing Add:* Dept Astron Univ Fla-211 SSRB Gainesville FL 32611

GOTTFRIED, BRADLEY M, b Philadephia, Pa, Mar 1, 50; c 1. BEHAVIORAL ECOLOGY, ORNITHOLOGY. *Educ:* W Chester State Col, BA, 71; Western Ill Univ, MS, 73; Miami Univ, PhD(zool), 76. *Prof Exp:* Instr biol, Mount Mercy Col, 76-78; assoc prof biol, Col St Catherine, 78-81, prof & chmn dept, 81-; AT DEPT BIOL, ARMSTRONG STATE COL, GA. *Concurrent Pos:* Student res zool, Miami Univ, 76. *Mem:* AAAS; Am Ornithologists' Union; Am Soc Zool; Ecol Soc Am; Sigma Xi. *Res:* Ecological aspects of nest predation; factors influencing the size of avian territories; winter ecology of birds; biogeography of small mammals; feeding ecology of birds. *Mailing Add:* Univ Wis Ctr FDL Campus Dr Fond Du Lac WI 54935

GOTTFRIED, BYRON S(TUART), b Detroit, Mich, May 24, 34; m 59; c 3. COMPUTER APPLICATIONS, COAL TECHNOLOGY. *Educ:* Purdue Univ, BS, 56; Univ Mich, MS, 58; Case Inst Technol, PhD(chem eng), 62. *Prof Exp:* Assoc engr, Lewis Res Ctr, NASA, 59-62; assoc engr, Gulf Res & Develop Co, 62-65; asst prof mech eng, Carnegie-Mellon Univ, 65-68; sect supvr, Econ & Comput Sci Div, Gulf Res & Develop Co, 68-70; assoc prof, 70-75, dir energy resources prog, 75-76, PROF INDUST ENG, ENG MGT OPERS RES, UNIV PITTSBURGH, 75- *Concurrent Pos:* Consult, Lord Corp, 65-68, Westinghouse Elec Corp, 66-, US Dept Energy, 75- & US Civil Serv Comn, 75-, Syst Modeling Corp, 89- *Mem:* Am Inst Indust Engrs; Soc Comput Simulation; Soc Mfg Engrs. *Res:* Applied operations research; computer simulation; optimization; applications to manufacturing and coal utilization. *Mailing Add:* Dept Indust Eng Univ Pittsburgh Pittsburgh PA 15261

GOTTFRIED, EUGENE LESLIE, b Passaic, NJ, Feb 26, 29; m 57. HEMATOLOGY. *Educ:* Columbia Univ, AB, 50, MD, 54; Am Bd Internal Med, dipl & cert hemat. *Prof Exp:* Intern, Presby Hosp, NY, 54-55, asst resident, 57-58; resident, Bronx Munic Hosp Ctr, 58-59; from asst instr to instr med, Albert Einstein Col Med, 59-61, assoc, 61-65, asst prof, 65-69; assoc prof med, Med Col, Cornell Univ, 69-81, assoc prof path, 75-81; CLIN PROF LAB MED, UNIV CALIF, SAN FRANCISCO, 81- *Concurrent Pos:* Fel med, Bronx Munic Hosp Ctr, 59-60; asst vis physician, Bronx Munic Hosp Ctr, 60-66, assoc attend physician, 66-69; asst vis physician, Lincoln Hosp, 63-69; Health Res Coun NY career scientist, 64-72; assoc attend physician & dir lab clin hemat, NY Hosp, 69-81, assoc attending pathologist, 75-81; attending physician, Burke Rehab Ctr, White Plains, NY, 75-81; dir clin labs, San Francisco Gen Hosp Med Ctr, 81- *Mem:* Am Soc Hemat; Acad Clin Lab Physicians & Scientists; fel Am Col Physicians; fel Int Soc Hemat. *Res:* Internal medicine; laboratory medicine; plasmalogens; lipids of blood cells; hemolytic phosphatides. *Mailing Add:* San Francisco Gen Hosp Med Ctr 1001 Potrero Ave San Francisco CA 94110

GOTTFRIED, KURT, b Vienna, Austria, May 17, 29; m 55; c 2. THEORETICAL PHYSICS. *Educ:* McGill Univ, BEng, 51, MSc, 53; Mass Inst Technol, PhD(theoret physics), 55. *Prof Exp:* Asst physics, Mass Inst Technol, 52-55; jr fel, Soc of Fels, Harvard Univ, 55-58; res fel, Inst Theoret Physics, Copenhagen, 58-59; res fel physics, Harvard Univ, 59-60, asst prof, 60-64; from assoc prof to PROF PHYSICS, CORNELL UNIV, 64- *Concurrent Pos:* Guggenheim fel, European Orgn Nuclear Res, Geneva, 63-64, mem staff, 70-73; mem bd dirs, Union Concerned Scientists; chmn, div particles & fields, Am Phys Soc, 81, coun-at-large, 90- *Mem:* Fel Am Phys Soc; fel Am Acad Arts & Sci; AAAS. *Res:* Theory of nuclear structure; quantum-mechanical many body problems; elementary particles; arms control & international security. *Mailing Add:* Lab Nuclear Studies Newman Lab Cornell Univ Ithaca NY 14853

GOTTFRIED, PAUL, b Vienna, Austria, Nov 7, 27; US citizen; m 51; c 2. RELIABILITY, SYSTEM SAFETY. *Educ:* Rose-Hulman Inst Technol, BS, 49. *Prof Exp:* Design engr, Charles N Debes & Assocs, 49-50; specialty engr, Chambers Corp, 50-53; sr engr, Inland Testing Labs, Cook Elec Co, 53-54, proj dir, 56-59; components engr, Chicago Aerial Industs, Inc, 54-55; partner, Reliability Eng Assocs, 59-61; prin scientist, Booz, Allen & Hamilton, Inc, 61-71; SELF-EMPLOYED, 71- *Concurrent Pos:* Newslett ed, Inst Elec & Electronics Engrs Reliability Soc, 68-74; trans assoc ed, 86. *Honors & Awards:* Centennial Medal, Inst Elec & Electronics Engrs, 84. *Mem:* Fel AAAS; Am Statist Asn; sr mem Inst Elec & Electronics Engrs (treas, 75-78, vpres, 89-91); Opers Res Soc Am. *Res:* Methodology for prediction of reliability and safety in systems involving new technology. *Mailing Add:* 9251 Three Oaks Dr Silver Spring MD 20901-3366

GOTTHEIL, EDWARD, b Montreal, Que, Aug 6, 24; US citizen; m 51; c 2. PSYCHIATRY, PSYCHOLOGY. *Educ:* Queen's Univ, Ont, BA, 46; McGill Univ, MA, 48; Univ Tex, PhD(psychol), 51, MD, 55. *Prof Exp:* Chief psychologist, Austin State Hosp, Tex, 49-51; intern, Roanoke Mem Hosp, Va, 55-56; resident psychiat, Letterman Gen Hosp, San Francisco, Calif, 56-59; post psychiatrist, Ft Riley, Kans, 60-62; chief med res proj, West Point, NY, 62-64; assoc prof, 64-69, PROF PSYCHIAT, JEFFERSON MED COL, 69- *Concurrent Pos:* Consult, Friend's Hosp, Philadelphia, 64-68, East Pa Psychiat Inst, 66-69, Del State Hosp, 65-, Coatesville Vet Admin Hosp, 67- *Mem:* AAAS; Am Psychol Asn; Am Psychiat Asn; Sigma Xi. *Res:* Small group interactions; psychiatric decision making; diagnosis; preventive social psychiatry; behavioral recording and data reduction; alcoholism; schizophrenia; medical education; emotional communication; sexual beliefs and behavior. *Mailing Add:* 1201 Chestnut St 15th Floor Philadelphia PA 19107-4192

GOTTLIEB, A ARTHUR, US citizen; m 58; c 2. IMMUNOLOGY, MOLECULAR BIOLOGY. *Educ:* Columbia Univ, AB, 57; NY Univ, MD, 61. *Prof Exp:* Asst prof med, Harvard Med Sch, 69; from assoc prof to prof microbiol, Inst Microbiol, Rutgers Univ, New Brunswick, 69-75; PROF MICROBIOL & IMMUNOL & CHMN DEPT, SCH MED, TULANE UNIV, 75-, PROF MED, 76-; PRES & CHIEF EXEC OFFICER, IMREG, INC, 85- *Concurrent Pos:* Nat Inst Gen Med Sci spec fel, Harvard Univ, 65-67 & career develop award, Harvard Med Sch, 67-69; consult, Squibb Inst Med Res, 69- & Vet Admin Hosp, East Orange, NJ, 73-75; mem ed adv bd, Immunol Commun; ed-in-chief, Uniscience Series Immunol, CRC Press, Cleveland; mem breast cancer task force, Nat Cancer Inst, 76-80; consult, Nat Ctr Toxicol Res, Jefferson, Ark, 76-79; vis prof, Walker & Elizabeth Hall Inst, Melbourne, 79. *Honors & Awards:* F S Burns Award, Mass Div, Am Cancer Soc, 68. *Mem:* Sigma Xi; fel Am Col Physicians; fel Am Acad Microbiol; Am Chem Soc; Int Asn Comp Res Leukemia & Related Dis; AAAS. *Res:* Biology and biochemistry of lymphoid cells; DNA polymerases of myeloma cells; leukemia markers; immunomodulators macrophage function; catabolism of antigens; binding of antigens to ribonucleoproteins; structure and function of immunoregulatory molecules in man; development of new approaches to immunotherapy. *Mailing Add:* Dept Microbiol & Immunol Tulane Univ Sch Med 1430 Tulane Ave New Orleans LA 70112

GOTTLIEB, ABRAHAM MITCHELL, b Chicago, Ill, Feb 22, 09; m 34; c 2. MEDICINE, CARDIOLOGY. *Educ:* Univ Ill, BS, 30, MD, 33; Am Bd Internal Med, dipl, 48. *Prof Exp:* Ward physician, Med Serv, Vet Admin Hosp, Tuscaloosa, Ala, 38-39, Milwaukee, Wis, 39-40, sect chief cardiol, 40-42; sect chief cardiol, Vet Admin Hosp, Dearborn, 46; asst prof med, Col Med, Wayne State Univ, 48-60; from assoc prof to prof, Univ Wis, Madison, 60-68; prof, 68-75, EMER PROF CLIN MED, STANFORD UNIV, 75-; CONSULT, VET ADMIN HOSP, PALO ALTO, 75- *Concurrent Pos:* Chief prof serv, Vet Admin Hosp, Dearborn, 46-60, attend consult, 48-60, hosp dir, Madison, Wis, 60-68; dir, Vet Admin Hosp, Palo Alto, 68-75. *Honors & Awards:* Dist Serv Medal, Vet Admin, 75. *Mem:* AMA; Am Heart Asn; fel Am Col Physicians. *Res:* Cardiovascular; pulmonary; sarcoidosis. *Mailing Add:* 101 Alma St, Apt 103 Palo Alto CA 94301

GOTTLIEB, ALLAN, b Queen's NY, Aug 2, 45; m 72; c 2. HIGHLY PARALLEL COMPUTERS, SHARED MEMORY COMPUTERS. *Educ:* Mass Inst Technol, BS, 67; Branders Univ, MA, 68, PhD(math), 73. *Prof Exp:* Instr math, State Col Mass, North Adams, 72-73; from asst prof to assoc prof math, York Col, City Univ NY, 73-81; assoc res prof, Courant Inst, 81-85, assoc prof computer sci, 85-90, assoc dir, Ultracomputer Res Lab, 86-89, DIR, ULTRA COMPUTER RES LAB, COURANT INST, 89-, PROF COMPUTER SCI, NY UNIV, 90- *Concurrent Pos:* Ed, Technol Rev, 67; vis mem, Courant Inst Math Sci, NY Univ, 79-81; sci adv bd mem, NCR, 88-, Parallel Processing Adv Comt, 90- *Mem:* Am Math Soc; Asn Comput Mach; Inst Elec & Electronics Engrs Computer Soc; NY Acad Sci. *Res:* Large-scale parallel processing emphasizing shared-memory designs; fetch-and-add and combining memory references to avoid hot-spot slowdowns; prototype hardware; parallel operating systems; full-custom very large scale integration switches. *Mailing Add:* Ultracomputer Res Lab 10th Floor NY Univ 715 Broadway New York NY 10003

GOTTLIEB, ARLAN J, b New York, NY, July 22, 33; m 64; c 3. HEMATOLOGY, ONCOLOGY. *Educ:* Columbia Univ, AB, 54, MD, 58; Am Bd Internal Med, dipl. *Prof Exp:* Intern med, Mt Sinai Hosp, NY, 58-59, from asst resident to chief resident, 59-62; asst prof med, Sch Med, Univ Pa, 67-71; assoc prof, 71-75, PROF MED, STATE UNIV NY UPSTATE MED CTR, 75-, CHIEF SECT HEMAT, 71- *Concurrent Pos:* Res fel molecular biol, Sect Chem Genetics, Nat Inst Arthritis & Metab Dis, 62-66, fel biophys, Lab Chem, Nat Inst Neurol Dis & Blindness, 66-67; Nat Inst Arthritis & Metab Dis res grant, 67-70; res career develop award, 68-71; asst clin prof, Sch Med, George Washington Univ, 66-67; attend, DC Gen Hosp George Washington Serv, 66-67 & Hosp Univ Pa, 67-71; consult & attend, Vet Admin Hosp, Pa, 67-71, hematologist & attend, NY, 71-; mem Clin Cancer Investigative Review Comt, Nat Cancer Inst, 82-86, chmn, 84-86. *Mem:* Am Soc Clin Oncol; Am Fedn Clin Res; fel Am Col Physicians; NY Acad Sci. *Res:* Relationship between structure and function of biologically active genetically determined proteins in health and diseases; design and execution of clinical trials in the etiology, diagnosis and therapy of malignancies. *Mailing Add:* Dept Med State Univ NY Health Sci Col Med 750 E Adams St Syracuse NY 13210

GOTTLIEB, CHARLES F, b Paris, Tex, Jan 17, 44; m 64; c 1. CANCER RESEARCH, RADIOBIOLOGY. *Educ:* Univ Tenn, Knoxville, PhD(radiobiol), 70. *Prof Exp:* Instr, Univ Tenn, 70-76, mem Shadd Fac, 73-76; from asst prof to assoc prof radiol, 76-86, ASSOC PROF RADIATION ONCOL, UNIV MIAMI, 86-, DIR ENVIRON HEALTH & SAFETY, 89- *Mem:* Radiation Res Soc; Am Soc Therapeut Radiol & Oncol; Bioelectromagnetics Soc; Am Soc Safety Engrs. *Mailing Add:* 8385 SW 90 St Miami FL 33156

GOTTLIEB, DANIEL HENRY, b Hollywood, Calif, Dec 7, 37; div; c 2. MATHEMATICS. *Educ:* Univ Calif, Los Angeles, BA, 59, MA, 61, PhD(math), 62. *Prof Exp:* Instr math, Univ Ill, 62-64; res mathematician, Inst Defense Anal, NJ, 64-67; PROF MATH, PURDUE UNIV, 67- *Mem:* Am Math Soc; Math Asn Am. *Res:* Algebraic topology; study of fibre bundles. *Mailing Add:* Dept Math Purdue Univ West Lafayette IN 47906

GOTTLIEB, DAVID, plant pathology, for more information see previous edition

GOTTLIEB, FREDERICK JAY, b New York, NY, June 17, 35; m 58; c 3. GENETICS, DEVELOPMENTAL BIOLOGY. *Educ:* Hofstra Col, BA, 56; Wesleyan Univ, MA, 58; Univ Calif, Berkeley, PhD(genetics), 62; Univ Pittsburgh, MS. *Prof Exp:* NIH trainee entom, Univ Calif, Berkeley, 60 & NIH trainee genetics, 60-62; instr zool, 62-63, asst prof biol, 64-67, ASSOC PROF BIOL, UNIV PITTSBURGH, 67-, ASSOC PROF BIOL & BIOSTATIST, GRAD SCH PUB HEALTH, 71- *Concurrent Pos:* Consult, Proj Solo; Am Cancer Soc instnl res grants, 63-66; USPHS res grant, 63-67; mem exec comt, Comput Ctr, Univ Pittsburgh, 69-73; Alexander von Humboldt Found sr scientist award & sr res scholar, Fulbright-Hays Prog, W Germany, 73-74. *Mem:* AAAS; Soc Develop Biol; Genetics Soc Am. *Res:* Genetic control of behavior in Drosophila; genetic control of host-symbiont recognition in Drosophila paulistorum and Ephestia kuehniella; genetic control of development of eye structure and pigmentation mutants in Ephestia kuehniella and Drosophila melanogaster; genetic and developmental effects of environmental pollutants. *Mailing Add:* Dept Biol A-234 Langley Hall Univ Pittsburgh Main Campus Pittsburgh PA 15260

GOTTLIEB, GERALD LANE, b New York, NY, Feb 6, 41; m 68; c 2. ELECTRICAL ENGINEERING, PHYSIOLOGY. *Educ:* Mass Inst Technol, BS, 62, MS, 64; Univ Ill Med Ctr, PhD(physiol), 70. *Prof Exp:* Res & develop engr, Aeronutronics Div, Philco-Ford Corp, 64-66; instr bioeng, Univ Ill, Chicago Circle, 67-73; asst prof biomed eng, 70-73, assoc prof physiol & biomed eng, 73-80, PROF PHYSIOL, RUSH MED COL, 80- *Concurrent Pos:* Adj assoc prof biomed eng, Univ Ill, Chicago Circle, 73- *Mem:* AAAS; Soc Neurosci; fel Inst Elec & Electronics Eng; Sigma Xi. *Res:* Nervous control of human motor system; computer applications for research and clinical testing in a medical scientific environment. *Mailing Add:* Dept Physiol Rush Univ 600 S Paulina St Chicago IL 60612

GOTTLIEB, GILBERT, b Brooklyn, NY, Oct 22, 29; m 61; c 4. PSYCHOBIOLOGY. *Educ:* Univ Miami, AB, 55, MS, 56; Duke Univ, PhD(psychol), 60. *Prof Exp:* Res assoc med psychol, Med Ctr, Duke Univ, 58-59; mem staff clin psychol, Dorothea Dix Hosp, Raleigh, 59-61, res scientist biopsychol, Psychol Lab, Res Sect, NC Div Ment Health, 61-82; HEAD & EXCELLENCE PROF PSYCHOL DEPT, UNIV NC, GREENSBORO, 82- *Concurrent Pos:* Res award, Nat Inst Child Health & Human Develop, 85, NSF, 85-88, NIMH, 89-92; res prof psychol, Univ NC, Chapel Hill, 73-82; assoc ed, J Comp & Physiol Psychol, 74-80; contrib ed, NC J Ment Health, 65-72 & 77-80; US del, Int Ethological Cong Comt, 77-83. *Mem:* AAAS; Int Soc Develop Psychobiol (pres, 87-88); fel Animal Behav Soc; Int Ethological Cong; fel Am Psychol Asn; Sigma Xi. *Res:* Developmental comparative psychology; behavioral embryology. *Mailing Add:* Psychol Dept Eberhart Bldg Univ NC 1000 Spring Garden St Greensboro NC 27412-5001

GOTTLIEB, IRVIN M, b Philadelphia, Pa, July 15, 21. INORGANIC CHEMISTRY, ANALYTICAL CHEMISTRY. *Educ:* Univ Pa, BS, 43, MS, 47, PhD, 53. *Prof Exp:* Dep chief textile finishes lab, Qm Res & Develop Lab, Pa, 51-54; group leader, Textile Res Inst, NJ, 55-58; prof chem, Trenton State Col, 58-61; head dept, 67-70, PROF CHEM, WIDENER COL, 61- *Mem:* Fiber Soc; The Chem Soc; fel Am Inst Chem; fel Royal Inst Chem; Sigma Xi. *Res:* Flame and thermal protection; functional finishes for textiles. *Mailing Add:* Dept Chem Widener Univ Chester PA 19013

GOTTLIEB, KAREN ANN, b Milwaukee, Wis, Jan 10, 51. HUMAN POPULATION STRUCTURE, ECOGENETICS. *Educ:* Univ Wis-Madison, BS, 72; Univ Colo, Boulder, MA, 76, PhD(phys anthrop), 78. *Prof Exp:* Res asst, Pop Genetics Lab, Sch Med, Health Sci Ctr, Univ Colo, 76-78, res fel pediat pharmacol, 79-81; ASST PROF PHYS ANTHROP, PA STATE UNIV, 81- *Concurrent Pos:* Instr, Univ Colo, Denver, 77-78. *Mem:* Am Asn Phys Anthropologists; Am Soc Human Genetics; Soc Med Anthrop; Human Biol Coun. *Res:* Human population variability; population structure; genetic demography of English Channel Islands; sources of variability in placental enzymes. *Mailing Add:* Dept Anthrop Carpenter Bldg Pa State Univ University Park PA 16802

GOTTLIEB, LEONARD SOLOMON, b Boston, Mass, May 26, 27; c 3. PATHOLOGY. *Educ:* Bowdoin Col, AB, 46; Tufts Univ, MD, 50; Harvard Sch Pub Health, MPH, 69. *Prof Exp:* From instr to prof path, Tufts Univ, 53-71; assoc dir, 66-72, DIR, MALLORY INST PATH, BOSTON CITY HOSP, 72-; PROF PATH, SCH MED, BOSTON UNIV, 71-, CHIEF PATH, UNIV HOSP, BOSTON UNIV MED CTR, 73-, CHMN DEPT, 80- *Concurrent Pos:* Instr, Med Sch, Boston Univ, 53-61, lectr path, 67-71; consult & lectr, US Naval Hosp, Chelsea, Mass, 57-58; from asst pathologist to assoc pathologist, Mallory Inst Path, Boston City Hosp, 57-66; lectr, Harvard Med Sch, 63-; hon mem, Fac Med, Hebrew Univ, Jerusalem, 87. *Mem:* Am Soc Clin Path; Am Soc Cell Biol; Int Acad Path; Am Soc Exp Path; Am Asn Study Liver Dis. *Res:* Gastrointestinal and liver diseases; nutritional pathology. *Mailing Add:* Dept Path & Lab Med Boston Univ Sch Med 80 E Concord St Boston MA 02118

GOTTLIEB, MARISE SUSS, b New York, NY, July 16, 38; m 58; c 2. PREVENTIVE MEDICINE, EPIDEMIOLOGY. *Educ:* Barnard Col, AB, 58; NY Univ, MD, 62; Harvard Univ, MPH, 66; Am Bd Prev Med, dipl. *Prof Exp:* Intern med, Univ Hosp, Boston, Mass, 62-63; med officer, Nat Heart Inst, NIH, 63-65; resident prev med epidemiol, Sch Pub Health, Harvard Univ, 65-68; instr med, Harvard Med Sch, 69-70; dir chronic dis serv, NJ State Dept Health, 70-72; asst prof community med, Col Med & Dent NJ, Rutgers Med Sch, 72-75; chief, Chronic Dis Control, 75-85, La Div Health & Environ Serv, Health & Human Resources; assoc prof epidemiol, Sch Pub Health & Trop Med, 75-80, ASSOC PROF MED, TULANE UNIV SCH MED, 75- *Concurrent Pos:* Spec res fel, Joslin Res Lab, Harvard Med Sch, 66-68; mem heart coun, NJ Regional Med Prog, 71-73; mem pulmonary dis adv coun, Nat Heart & Lung Inst, NIH, 75-77; mem neurosci rev panel, Health Res, Environ Protection Agency, 81-82; mem epidemiol & Dis Control Study Sect, NIH, 82-85; vpres med affairs, Imreg, Inc, New Orleans. *Mem:* Am Fedn Clin Res; Soc Epidemiol Res; Am Diabetes Asn; Am Pub Health Asn; fel Am Col Prev Med; fel Am Col Epidemiol. *Res:* Chronic disease epidemiology, field trials, coronary heart disease, hypertension, diabetes mellitus, experimental drug development clinical trials, environment and cancer. *Mailing Add:* Imreg Inc 144 Elk Pl Suite 1400 New Orleans LA 70112

GOTTLIEB, MELVIN BURT, b Chicago, Ill, May 25, 17; m 48; c 2. PHYSICS. *Educ:* Univ Chicago, BS, 40, PhD, 50. *Prof Exp:* Res assoc, Harvard Univ, 43-45; instr phys sci, Univ Chicago, 45-46, asst physics, 46-50; asst prof, Univ Iowa, 50-54; assoc dir, Proj Matterhorn, 54-61, assoc chmn astrophys sci, 74-80, DIR, PLASMA PHYSICS LAB & PROF ASTROPHYS SCI, PRINCETON UNIV, 61- *Mem:* Sigma Xi; Am Phys Soc. *Res:* Cloud chamber studies of penetrating showers; rocket and balloon-borne studies of primary cosmic ray intensities; low momentum cut off in intensity of heavy nuclei in cosmic rays; plasma physics; controlled thermonuclear reactors. *Mailing Add:* Plasma Physics Lab Princeton Univ PO Box 451 Princeton NJ 08540

GOTTLIEB, MELVIN HARVEY, b New York, NY, May 2, 29; m 60; c 3. BIOPHYSICS. *Educ:* Brooklyn Col, BS, 49; Polytech Inst Brooklyn, PhD(chem), 54. *Prof Exp:* Chemist textile finishes, United Merchants & Mfrs, 49-50; scientist, Nat Heart Inst, NIH, 53-56; chemist, Interchem Corp, 56-59; mem tech staff, Bell Tel Labs, 59-65; res scientist, 65-87, HEALTH SCI ADMIN, NAT INST ARTHRITIS & METAB DIS, NIH, 87- *Mem:* Am Chem Soc. *Res:* Ion exchange; ion exchange membranes; surface chemistry; electrochemistry; lipid physical chemistry; biological membranes. *Mailing Add:* Nat Inst Arthritis & Musculo Dis NIH Bethesda MD 20892

GOTTLIEB, MICHAEL STUART, b New Brunswick, NJ, Dec 26, 47. IMMUNE DEFICIENCY, HUMAN IMMUNOLOGY. *Educ:* Rutgers Univ, AB, 69; Univ Rochester, MD, 73. *Prof Exp:* Intern med & surg, Strong Mem Hosp, 73-74, resident internal med, 74-77; fel immunol, Sch Med, Stanford Univ, 77-79 & Howard Hughes Med Inst, 79-80; ASST PROF MED, SCH MED, UNIV CALIF, LOS ANGELES, 80- *Mem:* AAAS; Am Fedn Clin Res. *Res:* Human immunology and immune deficiency; experimental transplantation; advanced immune deficiency syndrome. *Mailing Add:* Univ Calif Los Angeles Santa Monica Hosp 4955 Van Nuys Blvd No 715 Sherman Oaks CA 91403

GOTTLIEB, MILTON, b New York, NY, July 2, 33; m 57; c 2. PHYSICS, OPTICS. *Educ:* City Col New York, BS, 54; Univ Pa, MS, 56, PhD(physics), 59. *Prof Exp:* Res assoc physics, Gen Atomic Div, Gen Dynamics Corp, 56-59; CONSULT SCIENTIST, RES LABS, WESTINGHOUSE ELEC CORP, 59- *Mem:* Optical Soc Am; Am Phys Soc. *Res:* Solid state physics superconductivity; interactions of light with acoustic waves; materials and techniques for optical signal processing, fiber optic sensors; acoustics. *Mailing Add:* Westinghouse Res & Develop Ctr 1310 Beulah Rd Pittsburgh PA 15235

GOTTLIEB, OTTO RICHARD, b Brno, Czech, Aug 31, 20; Brazilian citizen; m 47; c 3. PHYTOCHEMISTRY, BIOCHEMICAL EVOLUTION. *Educ:* Univ Brazil Indust Chem, Rio de Janeiro, 45; Univ Fed Rural Rio Jan, Liore Docente, 66. *Hon Degrees:* Dr, Univ Fed Pariba & Univ Hamburg, 88. *Prof Exp:* Prod chemist, Ornstein & Cia, Brazil, 46-54; from res assoc to res head natural prod chem inst agr chem, Ministry Agr & Nat Res Coun, 55-63; prof org chem & coordr cent inst chem, Univ Brasilia, 64-65; prof org chem, Fed Rural Univ, Rio de Janeiro, 66-73; PROF, UNIV SAO PAULO, 74- *Concurrent Pos:* Res fel org chem, Weizmann Inst Sci, 60; prof & adv grad org chem, Univ Minas Gerais, 62-75; vis prof, Univ Sheffield, 64; prof & coordr grad org chem, Fed Univ Pernambuco, 66-67; vis prof, Univ Sao Paulo, 68-73; coordr phytochem, Nat Res Inst Amazonia, 67-73; hon prof, Univ Fed Alagoas, 81. *Honors & Awards:* Feigl Prize Chem, 77; Sci Medal Amazonia, 78; Golden Retort Chem Syndicate, 80; Alvaro Alberto Prize Sci Technol, 90. *Mem:* Brazil Acad Sci; Latin Am Acad Sci; Int Acad Wood Sci. *Res:* Analytical chemistry; gasometric titrations; natural products chemistry; chemistry of Brazilian Lauraceae, Leguminosae, Guttiferae, Myristicaceae and Gnetaceae; micromolecular evolution; systematics and ecology; chemical variability of plants with environmental pressure; medicinal plants; basic principles and methodology for the application of secondary plant metabolites as markers; chemical nomenclature; basic principles and methodology for the application of secondary plant metabolites as systematic markers; replacement-nodal-subtractive nomenclature and codes of chemical compounds; author or co-author of over 570 articles and three books. *Mailing Add:* Rua Cinco de Julho 323 Apt 1001 Rio de Janeiro 22051 Brazil

GOTTLIEB, PAUL DAVID, b New Brunswick, NJ, Dec 4, 43; m 69; c 1. IMMUNOLOGY, IMMUNOGENETICS. *Educ:* Princeton Univ, BA, 65; Rockefeller Univ, PhD(life sci), 71. *Prof Exp:* Fel immunol, Rockefeller Univ, 71; spec fel biochem, Stanford Univ Sch Med, 71-73; asst prof immunol, Mass Inst Technol, 73-77, assoc prof immunol, 77-80; PROF MICROBIOL, UNIV TEX, AUSTIN, 80- *Concurrent Pos:* Mem adv panel for human cell biol, NSF, 76-79; mem exp immunol study sect, NIH, 78-; assoc ed, J Immunol, 78- & Molecular Immunol, 78-; mem, immunol & immunother panel, Am Cancer Soc, 81- *Mem:* Sigma Xi; Am Asn Immunol; AAAS; NY Acad Sci. *Res:* Cell surface immunogenetics of the mouse; genetics and structure of mouse immunoglobulin light chain polymorphisms and Lyt alloantigens; lymphocyte ontogeny and fetal mouse thymus development. *Mailing Add:* 104 W 32nd St Austin TX 78705

GOTTLIEB, PETER, b Cleveland, Ohio, Nov 29, 35; m 57; c 3. STATISTICS, PHYSICS. *Educ:* Calif Inst Technol, BS, 56; Mass Inst Technol, PhD(physics), 59. *Prof Exp:* Tech staff physics, Hughes Res Labs, 59-64 & Librascope, 64-66; res specialist syst anal, Jet Propulsion Lab, 66-73; assoc statist, Dames & Moore, 73-80, dir comput serv, 80-82; dir productivity, 83-86, dir computer integrated mfg, 86-87, SR STAFF SCIENTIST, TRW, 88- *Concurrent Pos:* Res assoc, Univ Calif, San Diego, 62-63; lectr, Univ Calif, Los Angeles, 60-67; vis asst prof, Calif State Univ, Los Angeles, 67; dir syst eng & environ sci, West Coast Univ, 66-80; lectr, Univ Calif, Los Angeles, 81-84. *Mem:* Inst Elec & Electronics Engrs. *Res:* Oil spill risk; lunar gravity; bacterial growth statistics; statistical mechanics; ferromagnetism. *Mailing Add:* 246 S Anita Ave Los Angeles CA 90049

GOTTLIEB, SHELDON F, b New York, NY, Dec 22, 32; m 56; c 4. PHYSIOLOGY, MICROBIOLOGY. *Educ:* Brooklyn Col, BA, 53; Univ Mass, MS, 56; Univ Tex, PhD(physiol), 59. *Prof Exp:* Res physiologist, Res Labs, Linde Div, Union Carbide Corp, 59-64; asst prof physiol & anesthesiol, Jefferson Med Col, 64-68; assoc prof to prof dept biol sci, Purdue Univ, 68-80; dean grad sch & dir res, 80-85, PROF BIOL SCI, UNIV SALA, MOBILE, 85- *Concurrent Pos:* Consult, Comt High Oxygen Pressure Equip, Am Soc Anesthesiol, 64 & US Air Force Sch Aviation Med, 75. *Mem:* AAAS; Am Soc Microbiol; Am Physiol Soc; Soc Gen Physiol; Aerospace Med Asn. *Res:* Physiological and biochemical effects of gaseous environments on living systems. *Mailing Add:* 2532 Tahoe Dr Mobile AL 36695

GOTTLIEB, STEVEN ARTHUR, b Brooklyn, NY, Jan 28, 52; m 82; c 1. LATTICE GAUGE THEORY, QCD. *Educ:* Cornell Univ, AB, 73; Princeton Univ, MA, 75, PhD(physics), 78. *Prof Exp:* Appointee, Argonne Nat Lab, 78-80; res assoc, Fermi Nat Lab, 80-82; res assoc, Univ Calif, San Diego, 82-85; asst prof, 85-88, ASSOC PROF, IND UNIV, 88- *Mem:* Am Phys Soc; Sigma Xi. *Res:* Field theory models of the strong interactions; lattice gauge theory. *Mailing Add:* Dept Physics Ind Univ Bloomington IN 47405

GOTTLING, JAMES GOE, b Baltimore, Md, Dec 11, 32; m 56; c 3. SOLID STATE ELECTRONICS. *Educ:* Lehigh Univ, BS, 53 & 54; Mass Inst Technol, SM, 56, ScD(elec eng), 60. *Prof Exp:* Res asst elec eng, Mass Inst Tech, 54-56 & 57-60, div sponsored res, 60, asst prof, 60-65; assoc prof, 65-70, PROF ELEC ENG, OHIO STATE UNIV, 70- *Concurrent Pos:* Ford Found fel, 61-63. *Mem:* Inst Elec & Electronics Eng. *Res:* Computer analysis of electronic circuits; modeling and analysis of electronic devices. *Mailing Add:* Dept Elec Eng 205 Electronics Hall Ohio State Univ 375 Caldwell Ohio State Univ Main Campus OH 43210

GOTTO, ANTONIO MARION, JR, b Nashville, Tenn, Oct 10, 35; m 59; c 3. BIOCHEMISTRY, METABOLISM. *Educ:* Vanderbilt Univ, BA, 57, MD, 65; Oxford Univ, DPhil(biochem), 61. *Prof Exp:* Res assoc biochem, Vanderbilt Univ, 61-64, res assoc molecular biol, 64-65; intern, Mass Gen Hosp, 65-66, med resident, 66-67; staff assoc, Nat Heart Inst, 67-69, head sect molecular struct, Molecular Dis Br, Nat Heart & Lung Inst, 69-71; PROF MED & BIOCHEM & CHIEF DIV ARTERIOSCLEROSIS & LIPOPROTEIN RES, BAYLOR COL MED & THE METHODIST HOSP, HOUSTON, 71-, CHMN DEPT INTERNAL MED, 77- *Concurrent Pos:* Fel, Coun Arteriosclerosis, Am Heart Asn; dir specialized ctr res arteriosclerosis & lipid res clin, NIH, Houston; mem metab study sect, NIH, 72-; mem, Nat Diabetes Adv Bd, 77-80, 81-; vpres, Int Soc Atherosclerosis, 79-; ed, Atherosclerosis Reviews; CO-PRIN INVESTR & SCI DIR, NAT HEART & BLOOD VESSEL RES & DEMONSTRATION CTR, 74-,. *Mem:* Inst Med-Nat Acad Sci; Am Col Cardiol; Am Soc Clin Invest; Asn Am Physicians; Asn Prof Med. *Res:* Lipid metabolism; structure and function of the plasma lipoproteins. *Mailing Add:* Dept Med Baylor Col Med Methodist Hosp 6565 Fannin MS A601 Houston TX 77030

GOTTSCHALK, ALEXANDER, b Chicago, Ill, Mar 23, 32; m 60; c 3. RADIOLOGY, NUCLEAR MEDICINE. *Educ:* Harvard Univ, BA, 54; Wash Univ, MD, 58. *Prof Exp:* Res assoc radiol, Donner Lab, Lawrence Radiation Lab, Univ Calif, Berkeley, 62-64; from asst prof to prof radiol, Univ Chicago Hosps, 64-74, chief nuclear med sect, 64-74; dir nuclear med, 74-77, PROF DIAG RADIOL, SCH MED, YALE UNIV, 74-, VCHMN DIAG RADIOL, 77- *Concurrent Pos:* Fac res assoc, Am Cancer Soc, 65-; dir, Argonne Cancer Res Hosp, 67-74. *Mem:* Fel Am Col Radiol; Radiol Soc NAm; Asn Univ Radiol (secy-treas, 68-69, pres elect, 69-70, pres, 70-71); Soc Nuclear Med (pres, 74-75); Am Roentgen Ray Soc. *Res:* Isotope development and clinical applications; instrumentation in nuclear medicine; image processing. *Mailing Add:* Dept Diag Radiol Yale Univ Sch Med Yale-New Haven Hosp Box 3333 333 Cedar St New Haven CT 06510

GOTTSCHALK, BERNARD, b Frankfurt, Ger, Jan 6, 35; US citizen; m 62; c 3. MEDICAL PHYSICS, ACCELERATOR PHYSICS. *Educ:* Rensselaer Polytech Inst, BS, 55; Harvard Univ, AM, 57, PhD, 62. *Prof Exp:* Res fel physics, Harvard Univ, 62-65; from asst prof to assoc prof physics, Northeastern Univ, 65-73, prof, 73-81; SR RES FEL, HARVARD UNIV, 81- *Concurrent Pos:* Vis scientist, Stanford Linear Accelerator Ctr, 71-72, Max Planck Inst, Munich, 74-75 & Stanford Linear Accelerator Ctr, 79-81. *Mem:* Am Phys Soc. *Res:* Medium-energy nuclear physics; experimental high energy physics; particle accelerator design. *Mailing Add:* Harvard Cyclotron Lab 44 Oxford St Cambridge MA 02138

GOTTSCHALK, CARL WILLIAM, b Salem, Va, Apr 28, 22; m 47; c 3. PHYSIOLOGY, NEPHROLOGY. *Educ:* Roanoke Col, BS, 42; Univ Va, MD, 45. *Hon Degrees:* DSc, Roanoke Col, 66. *Prof Exp:* Intern med, Mass Gen Hosp, 45-46, asst resident & resident med, 50-52; from instr to prof med, 53-61, from assoc prof to prof physiol, 59-61, KENAN PROF MED & PHYSIOL, UNIV NC, CHAPEL HILL, 69-, ACTG CHAIR, DEPT PHYSIOL, 90- *Concurrent Pos:* Res fel physiol, Harvard Med Sch, 48-50; fel cardiol, NC Mem Hosp, Univ NC, Chapel Hill, 52-53; consult to Surgeon Gen, 51; mem physiol study sect, NIH, 61-65; estab investr, Am Heart Asn, 57-61, career investr, 61-; Harvey lectr, 62; mem, Res Career Awards Comt, 65-69, Physiol Training Comt, 69-73, Nat Adv Coun, Nat Inst Gen Med Sci, 77-81 & Nat Inst Arthritis, Diabetes, Digestive & Kidney Dis, 82-86, Physiol Test Comt, Nat Bd Med Exam, 66-70; biol & med sci adv comt, NSF, 67-69, vchmn, 68 & chmn, 69; Burroughs-Wellcome Fund Adv Comt Clin Pharmacol, 80- *Honors & Awards:* Horsley Mem Prize, Univ Va, 56; NC Award, 67; Homer W Smith Award, NY Heart Asn, 70; David M Hume Award, Nat Kidney Found, 76; O Max Gardner Award, Univ NC, 78; Hon Mem, Hungarian Physiol Soc, 88; A N Richards Award, Int Soc Nephrology, 90. *Mem:* Nat Acad Sci; Inst Med; Am Soc Nephrology (pres, 75-76); Am Soc Clin Invest; Am Clin & Climat Asn; Am Acad Arts & Sci; Asn Am Physicians. *Res:* Renal physiology, utilizing micropuncture techniques. *Mailing Add:* Dept Med CB No 7155 Univ NC Sch Med Chapel Hill NC 27599-7155

GOTTSCHALK, JOHN SIMISON, b Berne, Ind, Sept 27, 12; m 37; c 2. CONSERVATION. *Educ:* Earlham Col, AB, 34; Ind Univ, MA, 43; Earlham Col, LLD, 66. *Prof Exp:* Naturalist, Ind State Dept Conserv, 34-38, supt fisheries, 38-41; sr bacteriologist, Schenley Labs, 44-45; aquatic biologist, US Fish & Wildlife Serv, 45-51, asst chief br fed aid, 51-57, chief div sport fisheries, 58-59, regional dir, 59-64, dir, Bur Sport Fisheries & Wildlife, 64-70, asst to dir, Nat Marine Fisheries Serv, 70-73; exec vpres, Int Asn Game, Fish & Conserv Comnrs, 73-78, counsel, 78-81; RETIRED. *Honors & Awards:* Seth Gordon Award, Int Asn Game Fish & Conserv Comnrs, 75; Leopold Award, Wildlife Soc, 76. *Mem:* Wildlife Soc (vpres, 55); Am Fisheries Soc (vpres, 41, 63, pres, 64). *Res:* Limnology; fish population dynamics; wildlife management. *Mailing Add:* 4664 34th St N Arlington VA 22207

GOTTSCHALK, LOUIS AUGUST, b St Louis, Mo, Aug 26, 16; m 44; c 4. PSYCHIATRY, PSYCHOANALYSIS. *Educ:* Wash Univ, AB, 40, MD, 43; S Calif Psychoanalytic Inst, PhD, 77. *Prof Exp:* Asst neuropsychiat, Sch Med, Wash Univ, 44-46; clin psychiatrist, chmn & dir EEG lab, US Pub Health Serv Hosp, Fort Worth, Tex, 46-48; res assoc & instr EEG & clin neurophysiol, Inst Psychosom & Psychiat Res, Michael Reese Hosp, Chicago, Ill, 48-51; res psychiatrist, NIMH, 51-53; from assoc prof to res prof psychiat & coordr res, Med Col, Univ Cincinnati, 53-67; chmn Dept Psychiat & Human Behav, Col Med, 67-77, dir clin psychiat, Irvine Med Ctr, 71-77, dir, Psychiat Consult & Liaison Div, Irvine Med Ctr, Orange, 77-86, PROF PSYCHIAT & HUMAN BEHAV, COL MED, UNIV CALIF, IRVINE, 67- *Concurrent Pos:* Asst chief child psychiat clin, Michael Reese Hosp, 50-51; attend physician clin psychiat, Cent Psychiat Clin, Cincinnati Gen Hosp, 53-67; supv & training analyst, Chicago Inst Psychoanal, 57-67 & Southern Calif Psychoanal Inst & Soc, 74-; NIMH US Pub Health Serv res career award, 61-67; mem clin psychopharmacol res rev comt, NIMH, 68-71; mem res adv comt, Calif State Dept Ment Hyg, 68-71 & 84-; physician consult, Fairview State Hosp, Costa Mesa, Calif, 68-76; consult psychiat, Vet Admin Hosp, Long Beach, Calif, 68-; consult res, Metrop State Hosp, Norwalk, Calif, 70-77; prin investr, Nat Inst Drug Abuse, 72-77, chmn psychosocial panel & mem extramural res rev comt, 73-74; sci co-dir, Alcohol Res Ctr, Nat Inst Alcohol Abuse & Alcoholism, 78-84; res scholar, Rockefeller Study & Conf Ctr, Bellagio, Italy, 85. *Honors & Awards:* Hofheimer Award, Am Psychiat Asn, 55; Franz Alexander Essay Prize, Southern Calif Psychoanal Inst, 73; Found Fund Prize Res Psychiat, Am Psychiat Asn, 78; Ripley lectr, Univ Wash, Seattle, 81. *Mem:* AAAS; Am Psychosom Soc; fel Am Col Neuropsychopharmacol; Am Psychoanal Asn; life fel Am Psychiat Asn; Nat Acad Pract Med. *Res:* Improving the quality of measurement in psychiatry and the medical sciences; from the form and content of verbal behavior, developing a method of objective assessment of psychological states and traits, validated independently by clinical, physiological and biochemical data, including the effects of psychopharmacological agents; development of artificial intelligence software to computerize the content analysis of natural language. *Mailing Add:* Dept Psychiat & Human Behav Univ Calif Irvine CA 92717

GOTTSCHALK, ROBERT NEAL, b Milwaukee, Wis, Apr 24, 28; m 52; c 4. PROJECT MANAGEMENT, MANUFACTURING MANAGEMENT. *Educ:* Univ Wis, BS, 52. *Prof Exp:* Pilot plant engr, Chem Prod Div, Ansul Co, 53-60, sr proj engr, 60-65, eng mgr, 65-69, mgr chem mfg eng, 69-70, mgr chem environ eng, 70-74, int proj mgr, 74-75, mfg mgr, 76-77, sr proj mgr, 78-79; gen mgr, Wormald US Inc, 79-82, facil mgr, 83-87; CERT ASBESTOS INSPECTOR & MGMT PLANNER, AHERA INSPECTIONS, 88- *Concurrent Pos:* Mgr design, construct & start-up of herbicide plants, Malaysia, 68-70, 74-75, 91. *Res:* Design, development and production; process for nitrogen heterocyclics; glycol ethers; organic arsenic compounds. *Mailing Add:* 628 Carney Blvd Marinette WI 54143

GOTTSCHALL, ROBERT JAMES, b Mt Vernon, NY, Apr 28, 35. RESEARCH & SCIENCE ADMINISTRATION. *Educ:* Rensselaer Polytech Inst, BS, 57; Polytech Inst Brooklyn, MS, 62; Columbia Univ, PhD(mat sci), 75. *Prof Exp:* Metallurgist, Lycoming Div, Avco Corp, 57-59, Nuclear Develop Corp, 59-60; staff scientist, T J Watson Res Ctr, IBM, 60-61, Philips Labs, 62-65; res assoc, Univ Ill, 75-77; ceramist, Div Mat Sci, 77-88, BRANCH CHIEF, METALL & CERAMICS BRANCH, US DEPT ENERGY, 88- *Mem:* Am Ceramic Soc; Nat Inst Ceramic Engrs; Am Inst Mining, Metall & Petrol Engrs Metall Soc; Am Phys Soc; Mat Res Soc; AAAS. *Res:* Superconductors; electron beam microanalysis; surface and interface behavior; corrosion; sintering; powder metallurgy and ceramics; coating processes; thin films; metallic glasses; semiconductors; physical and mechanical behavior; creep, fatigue; fracture mechanics; internal friction; non-destructive evaluation. *Mailing Add:* Div Mat Sci Off Basic Energy Sci Mail Stop G-236 US Dept Energy Washington DC 20545

GOTTSCHALL, W CARL, b Pittsburgh, Pa, Nov 15, 38; m 62; c 3. RADIATION CHEMISTRY, HEALTH & SAFETY. *Educ:* Calif Inst Technol, BS, 60; Univ Colo, PhD(inorg chem), 64. *Prof Exp:* Res assoc radiation chem, Argonne Nat Lab, 64-66; from asst prof to prof chem, Univ Denver, 66-86; TRAINING COORDR PUB SERV, COLO, 86- *Concurrent Pos:* Vis scholar, Linus Pauling Inst Sci & Med, 74; fel, Princeton Univ, Nuclear Regulatory Comn, 83-85; chmn, Nat Am Chem Soc Comt Chem Safety. *Mem:* Am Chem Soc; Radiation Res Soc; Sigma Xi. *Res:* Radiation chemistry of chelates and salts; radiation damage and protection in biochemical compounds and in vivo; chemical health safety. *Mailing Add:* 17 Bradbury Lane Littleton CO 80120

GOTTSCHANG, JACK LOUIS, b Woodland, Calif, Feb 16, 23; m 42; c 4. MAMMALOGY, HERPETOLOGY. *Educ:* San Jose State Col, BA, 47; Cornell Univ, PhD(zool), 50. *Prof Exp:* Asst zool, Cornell Univ, 47-50; from asst prof to assoc prof, 50-70, head, dept biol sci, 75-83, PROF ZOOL, UNIV CINCINNATI, 70- *Mem:* AAAS; Am Soc Mammal; Am Soc Zool. *Res:* Taxonomy; ecology; life history studies; small mammals. *Mailing Add:* Dept Biol Univ Cincinnati Cincinnati OH 45221

GOTTSCHLICH, CHAD F, b Laura, Ohio, Feb 26, 29; m 50; c 4. CHEMICAL ENGINEERING. *Educ:* Univ Cincinnati, ChE, 51, Phd(chem eng), 61; Princeton Univ, MSE, 53. *Prof Exp:* Engr petrol refining, Standard Oil Co, Ohio, 52-54; instr chem eng, Univ Cincinnati, 54-60; res assoc mech eng, Northwestern Univ, 60-61, asst prof, 61-63; asst prof, Univ Pa, 63-66; sect mgr res, 66-81, DIR RES, SELAS CORP AM, 81- *Mem:* Am Inst Chem Engrs; Am Chem Soc. *Res:* Measurements of the properties of high temperature gases; industrial heating. *Mailing Add:* 4616 Hazel Ave Philadelphia PA 19143-2104

GOTTSCHO, ALFRED M(ORTON), b Brooklyn, NY, Apr 29, 19; m 41; c 3. TOBACCO SCIENCE, SHEET MANUFACTURING. *Educ:* City Col New York, BChE, 40; Franklin & Marshall Col, MS, 54. *Prof Exp:* Tech serv engr rubber & plastic printing plates, Mosstype Corp, NJ, 40-41; engr fuel consumption study, US Corps Engrs, 41-42; prod supvr explosives, Plumbrook Ord Works, Ohio, 42-44; chem engr tobacco processing, Res Lab, 46-57, asst dir develop, 57-68, dir res & develop & asst vpres, 68-75, vpres, Gen Cigar Co, Div of Culbro Corp, 75-86; vpres, Helme Tobacco Co, Div Am Maize-Prods Co, 86-89; RETIRED. *Honors & Awards:* Res Award Cigar Res Coun, 73. *Mem:* AAAS; Am Chem Soc; Am Inst Chem Engrs; Soc Chem Indust. *Res:* Alkaloids of tobacco and their changes during industrial processing; changes in nitrogenous constituents in cigar tobaccos; development of tobacco sheets and smokeless tobacco products; tobacco fermentation. *Mailing Add:* 223 Suncrest Rd Lancaster PA 17601-4800

GOTTSCHO, RICHARD ALAN, b Lancaster, Pa, May 19, 52. PLASMA CHEMISTRY, LASER SPECTROSCOPY. *Educ:* Pa State Univ, BS, 74; Mass Inst Technol, PhD(phys chem), 79. *Prof Exp:* Chaim S Weizmann fel, Dept Physics, Mass Inst Technol, 79-80; MEM TECH STAFF, BELL LABS, 80- *Honors & Awards:* Peter Mark Mem Award, Am Vacuum Soc; Tegal Thinker Award. *Mem:* Am Phys Soc; Electrochem Soc; Am Vacuum Soc; Sigma Xi; AAAS. *Res:* Plasma chemistry; laser diagnostics for glow discharges; mechanisms for plasma etching; deposition in microelectronic fabrication; photochemical microelectronic processing techniques. *Mailing Add:* 118 Midland Blvd Maplewood NJ 07040

GOTTSEGEN, ROBERT, b New York, NY, June 21, 19; m 52; c 3. DENTISTRY. *Educ:* Univ Mich, AB, 39; Columbia Univ, DDS, 43, cert, 47. *Prof Exp:* From asst clin prof to assoc clin prof dent, Columbia Univ 49-69, prof dent, chmn dept periodont & postdoctoral periodont, 69-89, EMER PROF DENT, COLUMBIA UNIV, 89- *Concurrent Pos:* USPHS res fel, NIH, 48-50; attend dent, Presby Hosp. *Honors & Awards:* Isadore Hirschfeld Medal, NE Soc Periodontists, 84; Gold Medal, Am Acad Periodont, 88. *Mem:* AAAS; hon fel Am Acad Periodont (pres, 70-71); Int Asn Dent Res; fel Am Col Dent. *Res:* Metabolic influences on periodontal diseases; diabetes and periodontal diseases. *Mailing Add:* Sch Dent & Oral Surg Columbia Univ 630 W 168th St New York NY 10032

GOTTSTEIN, WILLIAM J, b Syracuse, NY, July 15, 29; m 55; c 3. PHARMACEUTICAL CHEMISTRY. *Educ:* Le Moyne Col, NY, BS, 51. *Prof Exp:* sr res scientist, Bristol-Myers Inc, Syracuse, 52-86; SR RES SCIENTIST, LE MOYNE COL, SYRACUSE, NY, 86- *Mem:* Am Chem Soc. *Res:* Chemical alteration of natural products. *Mailing Add:* 116 Woodmancy Lane Fayetteville NY 13066

GOTTWALD, JIMMY THORNE, b San Antonio, Tex, May 29, 38; m 60; c 3. ACOUSTICS. *Educ:* Baylor Univ, BS, 61, MS, 62. *Prof Exp:* Asst prof physics, Southwestern Univ, 62-65; engr scientist, 65-72, DEPT DIR OCEAN SCI, TRACOR INC, 73- *Mem:* Am Inst Physics; Acoust Soc Am. *Res:* Underwater acoustic propagation, background noise and signal coherence; underwater sound generation by optical means. *Mailing Add:* 1109 Claget St 7210 Pindell School Rd Laurel MD 20707

GOTTWALD, TIMOTHY R, b Lynnwood, Calif, Feb 14, 53; m 79. MYCOLOGY, EPIPHYTOLOGY. *Educ:* Long Beach State Univ, BS, 75; Ore State Univ, PhD(plant path), 80. *Prof Exp:* Res asst plant path, Dept Bot & Plant Path, Ore State Univ, 75-79, res assoc dis diag, 79; PLANT PATHOLOGIST, AGR RES SERV, US DEPT AGR, 79- *Mem:* Am Phytopath Soc; Mycological Soc Am. *Res:* Diseases of pecan especially those caused by fungi; basic biology of host-parasite interaction; epiphytology of disease, especially spore release and dispersal; mycology and physiology. *Mailing Add:* 3913 Bibb La Orlando FL 32817

GOTWALD, WILLIAM HARRISON, JR, b Trenton, NJ, May 6, 39; div; c 2. SYSTEMATIC ENTOMOLOGY. *Educ:* Millersville State Col, BS, 61; Pa State Univ, MS, 64; Cornell Univ, PhD(entom), 68. *Prof Exp:* Instr zool, Pa State Univ, Altoona, 63-65; from asst prof to assoc prof, 68-78, PROF BIOL, UTICA COL, SYRACUSE UNIV, 78- *Concurrent Pos:* NSF grants, 70-73, 73-76, 76-79, 79-81, 82-84 & 84-86; consult, sex & aging. *Mem:* Int Union Study Social Insects; Am Asn Univ Professors. *Res:* Phylogeny and systematics of the ants, especially the morphology and behavior of tropical ants belonging to the subfamilies Ponerinae and Dorylinae. *Mailing Add:* Dept Biol Utica Col Burrstone Rd Utica NY 13502

GOUBAU, WOLFGANG M, b Jena, EGer, Oct 17, 44; US citizen. PLASMA PROCESSING, PROCESS CONTROL & STATISTICS. *Educ:* Rutgers Univ, AB, 66. *Prof Exp:* Teaching asst physics, Cornell Univ, 66-68, res asst solid state physics, 68-74; IBM fel physics, Univ Calif, Berkeley, 74-76; staff physicist geophysics, Earth Sci Div, Lawrence Berkeley Lab, 76-83; res staff mem packaging technol, Yorktown, NY, 83-86,; PROCESS ENGR MAGNETIC RECORDING, IBM SAN JOSE, 86- *Mem:* Am Phys Soc. *Res:* Thermaland acoustic properties of solids with pulsed phonons; design of the first thin film direct current superconducting quantum interferance device (SQUID) with Nb-oxide-PB tunnel junctions; superconducting quantum interference device magnetometers and gradiometers for geophysics; magnetotelurics removal of bias and error analysis; packaging technology for computer circuits. *Mailing Add:* IBM E-12 015 5600 Cottle Rd San Jose CA 95193

GOUD, PAUL A, b The Hague, Netherlands, Sept 2, 37; Can citizen; m 61; c 3. COMMUNICATIONS. *Educ:* Univ Alta, BSc, 59; Univ BC, MASc, 61, PhD(elec eng), 64. *Prof Exp:* Mem sci staff, Res & Develop Labs, Bell Northern Res, Ont, 65-66; from asst prof to assoc prof elec eng, 66-72, PROF ELEC ENG, UNIV ALTA, 72- *Concurrent Pos:* Natural Sci & Eng Res Coun Can res grants, 66-; Dept Commun Can res grant, 71- *Mem:* Sr mem Inst Elec & Electronics Engrs. *Res:* Microwave electronics; radio and optical fiber telecommunications; microwave and optical industrial measurments. *Mailing Add:* Dept Elec Eng Univ Alta Edmonton AB T6G 2G7 Can

GOUDARZI, GUS (HOSSEIN), b Iran, Mar 27, 18; nat US; m 44; c 4. MINING & GEOLOGICAL ENGINEERING. *Educ:* Mont Sch Mines, BS, 39, MS, 41. *Hon Degrees:* Dr, Mont Tech, 80. *Prof Exp:* Miner hard rock mining, Anaconda Copper Mining Co, 37-41, sampler, 41-42, mining engr, 42-46, geol engr, 46-48, supvr underground mining & develop, 48-50; econ geologist mineral explor, AID, US Geol Surv, Ghana, 50-52, Arabia, 52-54 & Libya, 54-57, chief party & tech adv minerals & ground water invest, 57-62, actg dir dept mines & geol, Libyan Govt, 61-62 & regional geol, Ky, 64-68; econ explor geologist, US Geol Surv-USAID, Brazil, 68-70, econ geologist, Off Resource Anal, 71-78, dep chief, Off Mineral Resources, US Geol Surv, Reston, Va, 78-85; RETIRED. *Honors & Awards:* Herbert Hoover Award, Am Inst Mech Engrs, 85. *Mem:* Fel Geol Soc Am; Am Inst Mining, Metall & Petrol Engrs; Am Inst Prof Geologists; Am Inst Mining Engrs. *Res:* Mining methods; mapping and evaluation of mineral occurrences; mine examination, evaluation and development; regional geology; economic geology; exploration; aerial photography. *Mailing Add:* 9336 Ashley Dr Brooksville FL 34613

GOUDSMIT, ESTHER MARIANNE, b Ann Arbor, Mich, July 29, 33. MOLLUSCAN NEUROENDOCRINOLOGY, BIOCHEMISTRY. *Educ:* Univ Mich, Ann Arbor, BA, 55, MS, 59, PhD(zool), 64; USPHS trainee, Univ London, 60-61. *Prof Exp:* Res asst marine ecol, Woods Hole Oceanog Inst, 59; USPHS fel biochem, Nat Inst Arthritis & Metab Dis, Bethesda, MD, 64-66; res assoc pathobiol, Sch Pub Health, Johns Hopkins Univ, 67-69; asst prof biol, Brooklyn Col, 69-72; from asst to assoc prof, 72-87, PROF BIOL, OAKLAND UNIV, ROCHESTER, MICH, 88- *Concurrent Pos:* NIH Res Grant, 75-80, NSF Res Grant, 87- *Mem:* Soc Develop Biol; Soc Cell Biologists; Soc Neurosci. *Res:* Carbohydrate metabolism in snails; neurohormonal regulation of galactogen and glycogen synthesis. *Mailing Add:* Dept Biol Oakland Univ Rochester MI 48309-4401

GOUGE, EDWARD MAX, b Marion, NC, Aug 30, 47; m 70; c 1. INORGANIC CHEMISTRY. *Educ:* Western Carolina Univ, 69; Clemson Univ, PhD(inorg chem), 76. *Prof Exp:* from asst prof to assoc prof, 76-88, PROF CHEM, PRESBY COL, 89- *Mem:* Am Chem Soc. *Res:* Coordination chemistry-preparation, characterization and properties of polydentate ligands and their coordination compounds. *Mailing Add:* Dept Chem Presby Col Clinton SC 29325

GOUGÉ, SUSAN CORNELIA JONES, b Chicago, Ill, Apr 18, 24; m 43; c 3. MEDICAL MICROBIOLOGY, PUBLIC HEALTH. *Educ:* George Washington Univ, BS, 48, Norwich Univ, MA, 84. *Prof Exp:* Med technician, Children's Hosp, Washington, DC, 48-49; bacteriologist, George Washington Univ Res Lab, DC Gen Hosp, 50-53; med microbiologist, Walter Reed Army Inst Res, 53-61, res asst, 61-62; microbiologist, Antibiotics Div, HEW, Food & Drug Admin, 62-63; supvr qual control, John D Copanos Co, Baltimore, Md, 63-64; res teaching asst, Howard Univ Med Sch, 64-65; res assoc & chief technologist, Georgetown Univ Lab Infectious Dis, DC Gen Hosp, 66-69; mycologist, Georgetown Univ Hosp, 69-70; microbiologist, Res Found, Wash Hosp, Ctr, 71-73, dir qual control, Bio-Medium Corp, 73-76; staff microbiologist, Alcolac Inc, Baltimore, Md, 76-77; microbiologist, div labs, Dept Human Resources, Community Health & Hosps Admin, Washington, DC, 78-79; MICROBIOLOGIST, DIV OPHTHAL DEVICES, OFF DEVICE EVAL, FOOD & DRUG ADMIN, 79- *Concurrent Pos:* Tech consult, Volunteers Int Tech Asst Proj, 70-71 & Wesley-Jessen Inc, 74; Zacchaeus Free Clinic, 79-84. *Mem:* Am Soc Microbiol; AAAS; Am Chem Soc; Am Pub Health Asn; NY Acad Sci. *Res:* Antibiotic sensitivity; reversal of resistance; methodology for antibiotic assays; rapid diagnostic methods; microbial ecology; sterilization and disinfection of contact lenses. *Mailing Add:* Div Ophthal Devices Off Device Eval Food & Drug Admin 1390 Piccard Dr Rockville MD 20850

GOUGH, DAVID ARTHUR, b Salt Lake City, Utah, Aug 16, 46; m 70; c 4. BIOMEDICAL ENGINEERING, PHYSIOLOGY. *Educ:* Univ Utah, BS, 70, BA, 71, PhD(mat sci, eng), 74. *Prof Exp:* Res asst mat sci & eng, Univ Utah, 71-74; res assoc med, Harvard Med Sch, 74-76; PROF BIOENG, UNIV CALIF, SAN DIEGO, 76-, VCHAIR, DEPT APPL MECH & ENG SCI, 89- *Concurrent Pos:* Res assoc, E P Joslin Res Lab, Joslin Diabetes Found, Inc, Boston; assoc, Peter Bent Brigham Hosp, Boston; NIH fel, Harvard Med Sch, 74-76. *Mem:* AAAS; Am Soc Artificial Internal Organs; Am Inst Chem Engrs. *Res:* Biochemical specific sensors; mass transfer in physiologic systems; correction of metabolic disturbances in diabetes; artificial internal organs; biomaterials; immobilized catalysts. *Mailing Add:* Dept Appl Mech & Eng Sci Univ Calif San Diego La Jolla CA 92093

GOUGH, DENIS IAN, b Port Elizabeth, SAfrica, June 20, 22; m 45; c 2. GEOPHYSICS. *Educ:* Rhodes Univ Col, SAfrica, BSc, 43, MSc, 47; Univ Witwatersrand, PhD(geophys), 53. *Prof Exp:* Res officer geophys, SAfrican Nat Phys Res Lab, 47-55, sr res officer, 55-58; lectr physics, Univ Col

Rhodesia & Nyasaland, 58-60, sr lectr, 61-63; assoc prof geophys, Southwest Ctr for Advan Studies, Dallas, 64-66; dir, Inst Earth & Planetary Physics, 75-80, PROF PHYSICS, UNIV ALTA, 66- *Concurrent Pos:* Mem, Nat Comt Geod & Geophys, Cent African Fedn, 60-63; mem subcomt gravity, adv comt geophys & geod, Nat Res Coun Can, 67-, mem subcomt geomag & aeronomy, 67-; mem, Can Nat Comt Develop & Evolution Lithosphere. *Mem:* Am Geophys Union; fel Royal Soc Can; fel Royal Astron Soc; fel Geol Asn Can; chmn Int Asn Geomagnetism. *Res:* Studies of crustal structure through gravity and magnetic anomalies; studies of paleomagnetism, especially in relation to continental drift; study of crustal structure and upper mantle temperature distribution through electromagnetic induction; study of induced earthquakes and of stress in the lithosphere. *Mailing Add:* Dept Physics Univ Alta Edmonton AB T6G 2E2 Can

GOUGH, FRANCIS JACOB, b Grafton, WVa, Apr 9, 28; m 50; c 2. PLANT PATHOLOGY. *Educ:* Univ WVa, BS, 52, MS, 54, PhD(plant path), 57. *Prof Exp:* Asst plant path, Univ WVa, 52-57; plant pathologist cereal rusts, USDA, NDak State Univ, 57-67; res plant pathologist, USDA, Tex A&M Univ, 67-74; RES PLANT PATHOLOGIST, USDA, OKLA STATE UNIV, 74- *Concurrent Pos:* Dir, Plant Sci & Water Conserv Lab, USDA-Agr Res Serv. *Mem:* Am Phytopath Soc; Sigma Xi; Int Phytopath Soc; NY Acad Sci. *Res:* Genetics of host-parasite relationships. *Mailing Add:* Agr Res Serv Plant Res Lab USDA 1301 N Western Rd Stillwater OK 74075-2714

GOUGH, LARRY PHILLIPS, b New Castle, Ind, Dec 28, 44; c 2. PLANT ECOLOGY, BIOGEOCHEMISTRY. *Educ:* Carroll Col, BS, 67; Univ Louisville, MS, 68; Univ Colo, PhD(bot), 73. *Prof Exp:* Asst prof biol, Oakland City Col, 73-74; botanist, 74-87, SUPVRY BOTANIST, US GEOL SURV, 87- *Concurrent Pos:* Assoc ed, Reclamation & Revegetation Res. *Mem:* Am Bryological & Lichenological Soc; Am Soc Agron; Sigma Xi. *Res:* Geochemical survey of the Western coal regions; impact of industrial emissions on the element content of plants and soils; availability measures and soil-plant element transfer processes. *Mailing Add:* US Geol Surv Denver Fed Ctr Mail Stop 973 Denver CO 80225

GOUGH, LILLIAN, b Detroit, Mich, May 15, 18. MATHEMATICS. *Educ:* Univ Buffalo, BA, 39, MA, 49, PhD(math), 53. *Prof Exp:* Instr math, Univ Buffalo, 46-53; asst prof, Oswego State Teachers Col, 53-55; from assoc prof to prof, 55-83, chmn dept, 56-80, EMER PROF MATH, UNIV WIS-RIVER FALLS, 83- *Mem:* Math Asn Am; Nat Coun Teachers Math. *Res:* Topology. *Mailing Add:* 1105 Wasson Circle River Falls WI 54022

GOUGH, MICHAEL, b Springfield, Mo, Feb 4, 39; m 64; c 2. SCIENCE POLICY. *Educ:* Grinnell Col, BA, 61; Brown Univ, PhD(biol), 66. *Prof Exp:* Fel microbial genetics, Dept Human Genetics, Univ Mich, 65-68; asst prof microbiol, Baylor Col Med, 68-72; from asst to assoc prof microbiol, State Univ NY, Stony Brook, 72-76; health scientist adminr, NIH, Bethesda, 76-78; dir, Risk Sci Inst, Washington, DC, 85; sr fel, Ctr Risk Mgt, Resources for the Future, 87-90, dir, 90; sr analyst, 78-85, MGR, BIOL APPLN PROG, OFF TECHNOL ASSESSMENT, US CONG, WASHINGTON, DC, 91- *Concurrent Pos:* Fulbright lectr, Peru, 71 & India, 75; mem adv panel, Health Prog, Off Technol Assessment, 85-90, Assessment Agt Orange, 85-90; chmn, Adv Comt Health Effects Herbicides, Dept Vet Affairs, 87-90 & Comt Oper Ranch Hand Studies, Dept Health & Human Serv. *Mem:* Soc Risk Anal (secy, 85-86). *Res:* Interrelationships among science, technology and public policy. *Mailing Add:* Off Technol Assessment US Cong Washington DC 20510-8025

GOUGH, ROBERT EDWARD, b Wakefield, RI, Jan 31, 49; m 73; c 2. POMOLOGY, DEVELOPMENTAL ANATOMY. *Educ:* Univ RI, BA, 70, MS, 73, PhD(bot), 77. *Prof Exp:* Co agt, Va Polytech Inst & State Univ, 73-74, instr, 76-77, asst prof, 77-81, ASSOC PROF PLANT & SOIL SCI, UNIV RI, 81- *Concurrent Pos:* Fel, Lilly Endowment, 78-79. *Mem:* Am Soc Hort Sci; Sigma Xi; Bot Soc Am. *Res:* Small fruit production; developmental anatomy of fruit crops; tissue culture; somatic embryogenesis. *Mailing Add:* PO Box 95 Charleston RI 02813

GOUGH, ROBERT GEORGE, b Kitchener, Ont, May 20, 39; US citizen; m 67. ORGANIC CHEMISTRY. *Educ:* Univ Waterloo, BSc, 62; Pa State Univ, PhD(chem), 67. *Prof Exp:* Res assoc chem, Cincinnati Milacron Inc, 67-77, mgr regulatory affairs, 77-; AT ERVIN, VARN, JACOBS, ODOM & KITCHEN. *Mem:* Am Chem Soc. *Res:* Polymer rheology; polymer extrusion lubricants and processing aids; heat stabilizers; light stabilizers and fire retardants. *Mailing Add:* 6456 Needles Trail Tallahassee FL 32308

GOUGH, SIDNEY ROGER, b Staffordshire, Eng, Oct 30, 38; m 62; c 3. PHYSICAL CHEMISTRY, SCIENTIFIC NUMERIC DATA BASES. *Educ:* Univ Wales, BSc, 61, PhD(phys chem), 64. *Prof Exp:* Fel, 64-66, from asst res officer to sr res officer, 66-85, SERV COORDR, SCI NUMERIC DATABASE SERV, NAT RES COUN CAN, 85- *Res:* Dielectric properties of solids and liquids; molecular motions and structure; clathrate hydrates; scientific numeric databases. *Mailing Add:* CISTI Nat Res Coun Can Montreal Rd Ottawa ON K1A 0R6 Can

GOUGH, STEPHEN BRADFORD, b New Castle, Ind, Sept 13, 50; m 71; c 3. PHYCOLOGY, SOFTWARE AND SYSTEMS ENGINEERING. *Educ:* Carroll Col, Wis, BS, 72; Univ Wis, Madison, PhD(bot), 76. *Prof Exp:* Res asst bot, Univ Wis, Madison, 72-75, teaching asst, 75-76; comput systs specialist, Syst Develop Corp, Fredericksburg, Va, 82-86; RES ASSOC ENVIRON RES, ENVIRON SCI DIV, OAK RIDGE NAT LAB, 76-; sr software engr, Int Comput Equip, Inc, Fredericksburg, Va, 86-87; PROD DEVELOP MGR, TRW INC, 87- *Concurrent Pos:* Adj fac mem, Dept Biol Sci, Mary Washington Col, Fredericksburg, Va, 90- *Mem:* Phycol Soc Am; Sigma Xi; Am Soc Limnol & Oceanog; AAAS; Asn Comput Mach. *Res:* Biology of legionnaires disease bacterium; restoration and enhancement of natural and artificial ecosystems; bacterial ecology and physiology; ecological toxicology; algal ecology, systematics and physiology; software and systems engineering; diagnostics, system optimization and risk analysis. *Mailing Add:* Eight Norman Ct Fredericksburg VA 22407

GOUIN, FRANCIS R, b Laconia, NH, June 3, 38; m 62; c 2. PLANT PHYSIOLOGY. *Educ:* Thompson Sch Agr, cert, 58; Univ NH, BS, 62; Univ Md, College Park, MS, 65, PhD(hort, bot), 69. *Prof Exp:* Exten specialist, 65-69, from asst prof to assoc prof, 69-79, PROF HORT, UNIV MD, COLLEGE PARK, 79- *Honors & Awards:* Nursery Exten Award, Am Acad Nutrit & Am Soc Hort Sci, 82. *Mem:* Fel Am Soc Hort Sci; Int Plant Propagators. *Res:* Mineral nutrition of woody ornamental plants; winter hardiness of roots of container grown woody ornamental plants; chemical weed control in and around woody ornamental plants; composted sewage sludge on non-food crops. *Mailing Add:* 420 E Bay Front Rd Deal MD 20751

GOULARD, BERNARD, b Paris, France, May 9, 33; m 59; c 1. THEORETICAL PHYSICS, NUCLEAR PHYSICS. *Educ:* Univ Nancy, lic, 57; Univ Grenoble, Dr(nuclear physics), 60; Univ Pa, PhD(physics), 64. *Prof Exp:* Fel theoret physics, Bartol Res Found, 64-65; from asst prof to assoc prof, 65-73; assoc prof, 73-75, PROF PHYSICS, UNIV MONTREAL, 75- *Mem:* Am Phys Soc. *Res:* Weak interaction of muons and neutrinos with nuclei; electron scattering and isobaric properties of nuclei. *Mailing Add:* Lab Nuclear Physics Univ Montreal Montreal PQ H3C 3J7 Can

GOULD, A LAWRENCE, b Feb 6, 41; US citizen; m 73; c 2. BIOMETRICS. *Educ:* Case Western Reserve Univ, AB, 62, PhD(biomet, statist), 67. *Prof Exp:* Statistician, Res Triangle Inst, 67-69; sect head, Merck Sharp & Dohme Res Labs, 69-73, mgr clin biostatist, Div Med Affairs, 73-77, dir invest res, Clin Biostatist & Res Data Syst, 76-83, SR SCIENTIST, MERCK SHARPE & DOHME RES LABS, 83- *Concurrent Pos:* Assoc bus admin, Wharton Sch, Univ of Pa, 75; adj assoc prof, Biomet Dept, Temple Univ, 75- *Mem:* Fel Am Statist Asn (secy, Biopharmaceut subsect, 75-76); Biomet Soc. *Res:* Statistical methodology pertinent to problems arising in reality, especially in the biological and medical sciences. *Mailing Add:* Biostatist & Res Data Syst Merck Sharpe & Dohme Res Labs West Point PA 19486

GOULD, ADAIR BRASTED, b New York, NY, Feb 18, 16; m 40; c 3. GENETICS. *Educ:* Barnard Col, AB, 36; Univ Rochester, AM, 38, PhD(genetics), 40. *Prof Exp:* Res assoc genetics, Univ Del, 64-69, lab tech biol, 70-72, from instr to assoc prof, 72-81; RETIRED. *Mem:* AAAS; Genetics Soc Am; Sigma Xi. *Res:* Developmental genetics, especially as related to the genetic control of longevity. *Mailing Add:* 106 Hoiland Dr Wilmington DE 19803-3228

GOULD, ANNE BRAMLEE, b Bangor, Maine, Jan 30, 28. BIOCHEMISTRY, BIOLOGY. *Educ:* Univ Maine, BS, 50; Rutgers Univ, MS, 54; Brown Univ, PhD(biol), 61. *Prof Exp:* Res biochemist, VA Hosp, Cleveland, 62-66, Mt Sinai Hosp, Cleveland, 66-73; asst prof med, 73-81, RES ASSOC PROF MED, HAHNEMANN UNIV, 80- *Concurrent Pos:* Fel, Coun High Blood Pressure, Am Heart Asn. *Mem:* Am Heart Asn; Am Soc Nephrology; Int Soc Hypertension. *Res:* Biochemistry and physiology of high blood pressure; the question of whether or not high blood pressure is the result of a blood vessel adaptation to a renal energy deficiency caused by disease, poor nutrition or heredity. *Mailing Add:* Renal Res Lab Rm 6216 Hahnemann Univ Philadelphia PA 19102

GOULD, CHARLES JAY, b Eaton, Ohio, Feb 28, 12; m 40; c 2. PLANT PATHOLOGY. *Educ:* Marshall Col, AB, 34; Iowa State Col, MS, 37, PhD(plant path), 42. *Prof Exp:* Instr bot, Iowa State Col, 37-41; asst prof, Okla Agr & Mech Col, 40; from asst plant pathologist to plant pathologist, 41-77, EMER PLANT PATHOLOGIST, WESTERN WASH RES & EXTEN CTR, WASH STATE UNIV, 77- *Concurrent Pos:* Collabr, USDA, 47-77; Fulbright scholar, Netherlands, 51; sabbatical leave, Europe, 67. *Honors & Awards:* Res Award, Soc Am Florists, 50. *Mem:* Fel Am Phytopath Soc, 78; Sigma Xi. *Res:* Diseases of turf grasses, iris, tulips, narcissus, pyracantha and syringa. *Mailing Add:* 12316 123 St Ct E Puyallup WA 98374

GOULD, CHRISTOPHER M, b Oakland, Calif, Nov 24, 51. ULTRALOW TEMPERATURE, SUPERFLUID HELIUM THREE. *Educ:* Stanford Univ, BS, 73; Cornell Univ, MS & PhD(physics), 79. *Prof Exp:* res asst prof, 78-81, asst prof, 81-86, ASSOC PROF PHYSICS, UNIV SOUTHERN CALIF, 86- *Concurrent Pos:* Res fel, Alfred P Sloan Found, 83. *Mem:* Am Phys Soc. *Res:* Ultralow temperature condensed matter physics, especially superfluid helium 3; electronic transport at ultralow temperatures, especially thermalization problems. *Mailing Add:* Dept Physics Seaver Sci Ctr Univ Southern Calif Los Angeles CA 90089-0484

GOULD, CHRISTOPHER ROBERT, b Newbury, Eng, Mar 11, 44. POLARIZED NEUTRON PHYSICS, POLARIZED TARGET PHYSICS. *Educ:* Imp Col, London, BS, 65; Univ Pa, MS, 66, PhD(physics), 69. *Prof Exp:* Res assoc, dept physics, Duke Univ, Durham, NC, 69-71; from asst prof to assoc prof, 71-83, PROF PHYSICS, NC STATE UNIV, 83- *Concurrent Pos:* Alexander von Humboldt Award, Univ Frankfurt, 76-77; guest lectr, Inst Atomic Energy, Beijing, People's Repub China, 84, Univ Petrol & Minerals, Dhahran, Saudi Arabia, 86; mem, Panel Basic Nuclear Data Needs, Nat Res Coun, 87-90, Tech Comt Computer Applications in Nuclear & Plasma Sci, Inst Elec & Electronics Engrs, 87- & Instrumentation Subcomt, Nuclear Safety Anal Ctr, 88-89; Assoc Western Univs sabbatical fel, Los Alamos Meson Physics Facil, Los Alamos Nat Lab, 90-91. *Mem:* Am Phys Soc; Am Asn Univ Professors; Am Asn Physics Teachers; Sigma Xi; Inst Elec & Electronics Engrs. *Res:* Experimental nuclear physics; accelerators; scattering of polarized and unpolarized fast neutron beams; cryogenic polarized target development; computer based data acquisition systems for nuclear physics. *Mailing Add:* Dept Physics NC State Univ Box 8202 Raleigh NC 27695-8202

GOULD, DAVID HUNTINGTON, b New York, NY, Nov 23, 21; m 45, 82; c 3. TOXICOLOGY, STRUCTURE-ACTIVITY RELATIONSHIPS. *Educ:* Yale Univ, BS, 42, MS, 44, PhD(org chem), 45. *Prof Exp:* Lab asst, Yale Univ, 42-44; res chemist, Nopco Chem Co, 44-48; res assoc, Hickrill Chem Res Found, 48-49; sr res chemist, Schering Corp, 49-57, adminr extramural sci res, 57-59; grant res, Colgate-Palmolive Co, 59-68, head tech info serv, 68-70; clin res assoc, Merck Sharpe & Dohme Res Labs, 70-72; sr scientist, Off Toxic

Substances, US Environ Protection Agency, 77-88; RETIRED. *Concurrent Pos:* Dir, NValley Consumers Coop, 54-58; exec secy, Schering Found, 58-59; mem, Chem Select Working Group, Nat Cancer Inst-Nat Tox Prog, 77-88. *Honors & Awards:* Bronze Medal, US Environ Protection Agency. *Mem:* Soc Toxicol; Am Chem Soc; fel AAAS; Am Col Toxicol; Soc Risk Anal; Asn Govt Toxicologists; Int Soc Study Xenobiotics. *Res:* Natural products; cortical and sex hormones; chemotherapeutics; chemical structure retrieval; mechanisms of toxicity; metabolism; chemical carcinogenicity; risk assessment; xenobiotic toxicity; structure activity relationships; toxicology. *Mailing Add:* Swan's Way Rte 2 Box 67 Scottsville VA 24590

GOULD, DOUGLAS JAY, b San Francisco, Calif, May 29, 23; m 47; c 3. MEDICAL ENTOMOLOGY. *Educ:* Univ Calif, BA, 44, PhD(parasitol), 53. *Prof Exp:* Asst parasitol, Univ Calif, 47-48; parasitologist, State Dept Pub Health, 48-50; asst parasitol, Univ Calif, 50-51; entomologist, Walter Reed Army Inst Res, 51-64; chief, Dept Entom, Southeast Asia Treaty Orgn Med Res Lab, 64-77; chief, dept entom, Walter Reed Army Inst Res, 77-79; RETIRED. *Concurrent Pos:* Collabr, US Nat Mus, 53-55. *Mem:* Am Soc Trop Med & Hyg; Am Mosquito Control Asn; Sigma Xi; Soc Vector Ecologists. *Res:* Arthropod transmission of infectious diseases; microbiology of medically important arthropods; ecology of arthropod borne diseases. *Mailing Add:* PO Box 519 Plymouth CA 95669

GOULD, EDWIN, b Newark, NJ, Nov 21, 33; m 68; c 1. ECOLOGY, ANIMAL BEHAVIOR. *Educ:* Cornell Univ, BS, 55; Tulane Univ, PhD(zool), 62. *Prof Exp:* From asst prof to prof ment hyg, Lab Comp Behav, Johns Hopkins Univ, 62-80; CUR MAMMALS, DEPT MAMMAL, NAT ZOO, SMITHSONIAN INST, 80- *Concurrent Pos:* NIH fel, Yale Univ, 62-63; career develop award, 71-76; mem expeds to Madagascar; Fulbright scholar, Univ Malaya, 74-75. *Mem:* Am Soc Mammal; Am Soc Zoologists; Animal Behav Soc; Sigma Xi. *Res:* Echolocation and communication in the Insectivora and bats; homing orientation in turtles; rodent population dynamics; feeding efficiency in bats; comparative behavior. *Mailing Add:* Nat Zool Park Smithsonian Inst Mammals Iron House Washington DC 20008

GOULD, EDWIN SHELDON, b Los Angeles, Calif, Aug 19, 26; m 52; c 2. INORGANIC CHEMISTRY, ORGANIC CHEMISTRY. *Educ:* Calif Inst Technol, BS, 46; Univ Calif, Los Angeles, PhD(chem), 50. *Prof Exp:* Instr chem, Polytech Inst Brooklyn, 50, from asst prof to assoc prof, 52-59; sr inorg chemist, Stanford Res Inst, 59-66; prof chem, San Francisco State Univ, 66-67; PROF CHEM, KENT STATE UNIV, 67- *Concurrent Pos:* NSF sci fac fel, 57-58, sr fel, Stanford Univ, 62-63; NIH spec fel, Univ Calif, 63-64. *Mem:* Am Chem Soc; Royal Soc Chem. *Res:* Chelating agents; electron transfer reactions in solution; metalloaromatics; catalysis of oxidation by transition metals. *Mailing Add:* Chem Dept Kent State Univ Kent OH 44242-0002

GOULD, ERNEST MORTON, JR, forest economics, resource management; deceased, see previous edition for last biography

GOULD, G(ERALD) G(EZA), b Budapest, Hungary, Nov 17, 13; m 41; c 3. ELECTRICAL ENGINEERING. *Educ:* City Col, BS, 35; Purdue Univ, MS, 36. *Prof Exp:* Elec engr develop high voltage equip, Porcelain Prods, Inc, 36-40 & signal syst, New York Subways, 40-42; proj engr frequency control, Acoustics, US Naval Ord Lab, 46-51; chief engr torpedo res & develop, US Navy Bur Ord, 51-55, tech dir, US Naval Underwater Weapons Res & Engr Sta, 55-70, dep tech dir, US Naval Underwater Systs Ctr, 70-72, tech dir, Naval Coastal Systs Ctr, 72-81; ENGR CONSULT, GERALD GOULD ASSOC, INC, 81- *Concurrent Pos:* Chmn, Undersea Warfare Res & Develop Planning Coun, 65-66 & US Navy Lab Dirs Coun, 67-69. *Mem:* Fel, Inst Elec & Electronics Engrs; Nat Soc Prof Engrs; Acoust Soc Am. *Res:* Management of research and development; underwater weapons systems; electrical generation; high voltage engineering. *Mailing Add:* Gerald Gould Assocs, Inc 2329 Magnolia Dr Panama City FL 32408

GOULD, GEORGE EDWIN, b Concordia, Kans, Apr 22, 05; m 29, 73. ENTOMOLOGY. *Educ:* Univ Kans, AB, 27, MA, 29; Purdue Univ, PhD(entom), 42. *Prof Exp:* Asst entom mus, Univ Kans, 27-28; asst entomologist, Va Truck Exp Sta, 28-31; from asst entomologist to assoc entomologist, exp sta, 31-44, from asst prof to prof, 44-71, EMER PROF ENTOM, PURDUE UNIV, 71- *Concurrent Pos:* Consult, W B McCloud & Co, Ill, 45 & 46. *Mem:* AAAS; Entom Soc Am. *Res:* Economic entomology; soil and vegetable insects. *Mailing Add:* 2741 N Salisbury St West Lafayette IN 47906-1431

GOULD, HARRY J, III, b Columbus, Ohio, Mar 1, 47; m 71; c 2. COMPARATIVE NEUROLOGY. *Educ:* State Univ NY Stony Brook, BS, 69; Brown Univ, PhD(biol), 74; La State Univ Med Sch, MD, 90. *Prof Exp:* Instr anat, Col Med, Univ Cincinnati, 74-76, asst prof, 76-80; from asst to assoc prof anat, 80-90, RESIDENT NEUROL, LA STATE MED CTR, 90- *Concurrent Pos:* Vis scientist, Vanderbilt Univ 82 & 85. *Mem:* AAAS; AMA; Cajal Club; Sigma Xi; Am Asn Anatomists. *Res:* Neuroanatomical and physiological analysis of the intrinsic circuitry of the motor and somatosensory cerebral cortex. *Mailing Add:* Dept Neurol Med Ctr La State Univ 1542 Tulane Ave New Orleans LA 70112

GOULD, HARVEY A, b Oakland, Calif, Sept 4, 38; m 74; c 3. STATISTICAL MECHANICS. *Educ:* Univ Calif, Berkeley, PhD(physics), 66. *Prof Exp:* Nat Res Coun-Nat Acad Sci res fel, Nat Bur Standards, DC, 66-67; asst prof physics, Univ Mich, Ann Arbor, 67-71; assoc prof, 71-81, PROF PHYSICS, CLARK UNIV, 81- *Concurrent Pos:* Vis assoc prof physics, Bar-Ilan Univ, Israel, 74-75; vis scientist, Univ Chicago, 78-79; vis res prof, Boston Univ, 80-86, vis prof, 86-87. *Mem:* Am Phys Soc; Am Asn Physics Teachers. *Res:* Computer simulation and dynamics of first-order phase transitions. *Mailing Add:* Dept Physics Clark Univ Worcester MA 01610-1477

GOULD, HARVEY ALLEN, b Brooklyn, NY, Feb 28, 45. FEW-ELECTRON VERY-HEAVY IONS, RELATIVISTIC ATOMIC COLLISIONS. *Educ:* Stevens Inst Technol, BS, 65; Brandeis Univ, MS, 67, PhD(physics), 70. *Prof Exp:* Teaching fel physics, Brandeis Univ, 65 & 66; STAFF SCIENTIST PHYSICS, MAT & MOLECULAR RES DIV, LAWRENCE BERKELEY LAB, UNIV CALIF, BERKELEY, 71- *Concurrent Pos:* Sr vis fel physics, Oxford Univ, 80. *Mem:* Am Phys Soc; AAAS. *Res:* Experimental atomic physics; quantum electrodynamics in few-electron very-heavy ions; electron, electric-dipole-moment as a test of CP violation. *Mailing Add:* Lawrence Berkeley Lab Bldg 71 Berkeley CA 94720

GOULD, HENRY WADSWORTH, b Portsmouth, Va, Aug 26, 28; m 69. MATHEMATICS, COMBINATORICS. *Educ:* Univ Va, BA, 54, MA, 56. *Prof Exp:* From instr to assoc prof, 58-69, PROF MATH, WVA UNIV, 69- *Concurrent Pos:* Consult Sci Serv, DC, 59-60; NSF res grant, 60-67; assoc ed, Fibonacci Quart, 62-; lectr, Math Asn Am, 67-70 & Soc Indust & Appl Math, 74-76; ed-in-chief, Proc WVa Acad Sci, 74-79; dep ed-in-chief, J Math Res & Expos, Hefei, People's Repub China. *Mem:* Fel AAAS; Am Math Soc; Math Asn Am; Soc Indust & Appl Math; Fibonacci Asn; Sigma Xi; fel Inst Combinatrics & Appln. *Res:* Theory of combinatorial identities; combinatorial analysis; special functions; number theory; theory of binomial coefficient summations; history of mathematics. *Mailing Add:* Dept Math WVa Univ Morgantown WV 26506

GOULD, HOWARD ROSS, b Adrian, WVa, Nov 10, 21; m 48; c 2. GEOLOGY. *Educ:* Univ Minn, BA, 43; Univ Southern Calif, PhD(geol), 53. *Prof Exp:* Training assoc, Div War Res, Univ Calif, 43-45, assoc marine geologist, 46; asst geol, Scripps Inst, Calif, 46-47, geologist, US Geol Surv, DC, 47-54; asst prof oceanog, Univ Wash, 53-56, sr geologist, Geol Res Sect, Humble Oil & Ref Co, 56-63, staff geologist, 63-64, chief, 64; mgr stratig & struct geol div, Exxon Prod Res Co, 64-66, mgr stratig geol div, 66-67, res scientist, 67-86; INDEPENDENT GEOLOGIST, 86- *Concurrent Pos:* Spec consult, US Navy. *Mem:* Fel AAAS; fel, Geol Soc Am (vpres, 80 & pres, 81); hon mem Soc Econ Paleontologists & Mineralogists (vpres, 68); hon mem Am Asn Petrol Geol; Am Geophys Union; Am Geol Inst (vpres, 83, pres, 84). *Res:* Sedimentation, especially in marine and lake environments; general geology of the sea floor; petroleum geology. *Mailing Add:* 5231 Piping Rock Lane Houston TX 77056

GOULD, JACK RICHARD, b Brooklyn, NY, Feb 28, 22; m 45; c 2. ENVIRONMENTAL SCIENCE & TECHNOLOGY. *Educ:* Brooklyn Col, BA, 43; Pa State Univ, MS, 47, PhD(org chem), 49. *Prof Exp:* Chemist res & develop, Montrose Chem Co, 43-45; asst chem, Pa State Univ, 45-48; res chemist petrochems, Houdry Process Corp, 49-51; proj leader, Reaction Motors, Inc, Olin Mathieson Chem Corp, 51-53, sect head, Chem Dept, 54-56; head org chems sect, Eastern Res Ctr, Stauffer Chem Co, 56-63; dir res, Metalsalts Corp, 63-64; sr staff contract res, Eng Res & Develop Dept, M W Kellogg Co, 64-69; sr environ coordr, 69-80, sr environ scientist, environ affairs dept, 80-85, SR ENVIRON SCIENTIST, HEALTH & ENVIRON SCI DEPT, AM PETROL INST, 85- *Concurrent Pos:* Pres greater Washington chapt, Nat Retinitis Pigmentosa Found Fighting Blindness, 71-85, mem bd trustees, 79-81; managing ed, Div Petrol Chem, Am Chem Soc, 77-79; mem adv bd, Vol Visually Handicapped, 78-83. *Mem:* Am Chem Soc; NY Acad Sci; Sigma Xi; AAAS; Marine Tech Soc. *Res:* Organic and organometallic chemistry and industrial applications; high energy liquid rocket propellants; water pollution control technology and regulations related to the petroleum industry; prevention, control and effects of oil spills; alternate fuels; fate and environmental effects of petroleum and related substances in marine environment and human health implications; natural resource damage assessment. *Mailing Add:* Am Petrol Inst 1220 L St NW Washington DC 20005

GOULD, JAMES L, b Tulsa, Okla, July 31, 45; m 70. ETHOLOGY. *Educ:* Calif Inst Technol, BS, 70; Rockefeller Univ, PhD(ethology), 75. *Prof Exp:* Asst prof animal behav, 75-80, assoc prof, 80-83, PROF BIOL, PRINCETON UNIV, 83- *Concurrent Pos:* NSF fel; Guggenheim Found fel. *Mem:* Fel AAAS. *Res:* Communication, learning and orientation behavior. *Mailing Add:* 11 Herrontown Cir Princeton NJ 08540

GOULD, JAMES P, ENGINEERING ADMINISTRATION. *Educ:* Univ Wash, BSCE, 44; Mass Inst Technol, MSCE, 46; Harvard Univ, MS, 48, ScD, 49. *Prof Exp:* US Army Corps Engrs, 44-45; eng asst, Harvard Univ, 48-49; US Bur Reclamation, Earth Dams Sect, 50-53; staff, 53-55, assoc, 55-73, PARTNER, MUESER RUTLEDGE CONSULT ENGRS, 73- *Concurrent Pos:* Transp Res Bd, 62-66; mem, Underground Technol Res Coun; adj prof, Purdue Univ, 69; fel, Am Consult Engrs Coun; consult various subway projs. *Honors & Awards:* Kapp Mem Lectr, Am Soc Civil Engrs, 74 & 82; Crom Lectr, Univ Fla, 78; Haley Mem Lectr, 85. *Mem:* Nat Acad Eng; Am Soc Civil Engrs. *Res:* Author of about three dozen technical articles or publications. *Mailing Add:* Mueser Rutledge Consult Engrs 708 Third Ave 5th Floor New York NY 10017

GOULD, JOHN MICHAEL, b Dayton, Ohio, Dec 15, 49; m 74; c 2. MICROBIAL BIOTECHNOLOGY, PLANT BIOCHEMISTRY. *Educ:* Univ Cincinnati, BS, 71, MS, 72; Mich State Univ, PhD(plant biochem), 74. *Prof Exp:* Res assoc, Dept Biochem, Cell & Molecular Biol, Cornell Univ, 74-75; NIH fel, Dept Biol Sci, Purdue Univ, 75-77; asst prof biochem, Dept Chem, Univ Notre Dame, 77-81; res biochemist, 81-83, res leader, Northern Regional Res Ctr, 83-90, DIR RES, BIOTECHNOL RES & DEVELOP CORP, USDA, 90- *Concurrent Pos:* Consult, US Fed Trade Comn (biodegradation), 90- *Honors & Awards:* Fed Lab Consortium Award, 87; R & D 100 Award, 89. *Mem:* Biophys Soc; Am Soc Microbiol. *Res:* Biodegradation of plastics and polymer composites; properties of plant cell walls; modification of starch and cellulose to produce novel products; photoacoustic spectrometry; polymer biochemistry; microbial biotechnology. *Mailing Add:* Biotechnol Res & Develop Corp 1815 N University St Peoria IL 61604

GOULD, K LANCE, b Wilsonville, Ala, Oct 28, 38. CARDIOVASCULAR MEDICINE. *Educ:* Western Reserve Med Sch, MD, 64. *Prof Exp:* PROF & DIR, CTR CARDIOVASC & IMAGING RES, UNIV TEX HEALTH SCI CTR, HOUSTON, 80- *Concurrent Pos:* Mem bd trustees, Am Col Cardiol. *Mem:* Asn Am Physicians; Am Asn Univ Cardiologists; Am Soc Clin Invest. *Mailing Add:* Div Cardiol Univ Tex Med Sch Box 20708 Houston TX 77225

GOULD, KENNETH ALAN, b Brooklyn, NY, May 18, 49; m 75. FUEL SCIENCE, HYDROCARBON CHEMISTRY. *Educ:* Cornell Univ, BA, 71; Ohio State Univ, PhD(org chem), 75. *Prof Exp:* RES CHEMIST FUEL SCI, EXXON RES & ENG CO, 75- *Mem:* Am Chem Soc. *Res:* Coal conversion to clean fuels and improved processing of heavy crude oils. *Mailing Add:* 320 River Bend Rd Berkeley Heights NJ 07922

GOULD, KENNETH G, b Woodfort, Eng, Aug 18, 43. REPRODUCTIVE BIOLOGY, CONTRACEPTIVE DEVELOPMENT. *Educ:* London Univ, Eng, DVM, 66, PhD, 73. *Prof Exp:* CHIEF, DIV REPRODUCTIVE BIOL, YERKES CTR, EMORY UNIV. *Mem:* Am Fertil Soc; Am Physiol Soc; Brit Vet Asn; Int Primate Soc. *Mailing Add:* Dept Reproduction Biol Yerkes Reg Primate Res Ctr Emory Univ Atlanta GA 30322

GOULD, LAWRENCE A, b Brooklyn, NY, Dec 15, 30; m 57; c 2. INTERNAL MEDICINE, CARDIOLOGY. *Educ:* Brooklyn Col, BA, 52; NY Univ, MD, 56; Am Bd Internal Med, dipl & Am Bd Cardiovasc Dis, dipl. *Prof Exp:* Clin instr med, NY Med Col, 62-65; from asst chief to chief cardiol serv, Bronx Vet Admin Hosp, NY, 65-69; chief cardiac catheterization lab, Misericordia-Fordham Hosp, 69-74; assoc prof med, NY Med Col, 72-75; clin assoc prof, 75-81, CLIN PROF MED, DOWNSTATE MED CTR, STATE UNIV NY, 81-; CHIEF CARDIOL SERV, METHODIST HOSP, 74- *Mem:* Am Heart Asn; fel Am Col Physicians; fel Am Col Cardiol; Am Fedn Clin Res. *Res:* Clinical research of the cardiovascular system; cardiac hemodynamic studies. *Mailing Add:* Methodist Hosp 506 6th St Brooklyn NY 11215

GOULD, LEONARD A(BRAHAM), b Brooklyn, NY, Nov 20, 27; m 59. ELECTRICAL ENGINEERING. *Educ:* Mass Inst Technol, SB, 48, ScD(elec eng), 53. *Prof Exp:* Asst, 48-50, from instr to assoc prof, 50-68, PROF ELEC ENG, MASS INST TECHNOL, 68- *Concurrent Pos:* Fulbright res fel, Tech Univ Denmark, 57; vis prof, Dept Appl Physics, Michelson Inst, Bergen, Norway, 68-69. *Mem:* Fel Inst Elec & Electronics Engrs. *Res:* Control and estimation. *Mailing Add:* 47 Shaw Dr Wayland MA 01778

GOULD, MARK D, b Washington, DC, Feb 1, 46; m 72; c 2. BENTHIC ECOLOGY, TROPICAL MARINE BIOLOGY. *Educ:* Univ RI, BS, 67, MS, 70, PhD(biol sci), 73. *Prof Exp:* Lectr aquatic ecol, Univ RI, 70-72, asst prof, 74-78; biol area, 76-79, PROF MARINE BIOL & ZOOL, COORDR DIV NATURAL SCI & DEAN SCI & MATH, ROGER WILLIAMS COL, 72- *Concurrent Pos:* Sr biologist, Grumman Ecosysts Corp, 74-75; res biologist, Environ Protection Agency, 81-; lectr, Mystic Aquarium, 85. *Honors & Awards:* Chautauqua Award, NSF, Univ Hartford, 79 & Hampshire Col, 81. *Mem:* Sigma Xi; AAAS; Ecol Soc Am; Soc Nematologists. *Res:* Interrelationships between the micro, meio and macrofauna and the effect of various pollutants; taxonomy of tropical marine organisms; indicator organisms in fresh water. *Mailing Add:* Sch Sci & Math Roger Williams Col Old Ferry Rd Bristol RI 02809

GOULD, MICHAEL NATHAN, b New York, NY, Sept 9, 47. RADIATION BIOLOGY, EXPERIMENTAL ONCOLOGY. *Educ:* Univ Wis, Madison, BS, 69, MS, 73, PhD(radiation biol), 77. *Prof Exp:* Fel cell biol, Argonne Nat Lab, 77-78; INSTR HUMAN ONCOL RADIOL, UNIV WIS, MADISON, 78- *Concurrent Pos:* Fel Am Cancer Soc, 77-78. *Mem:* Radiation Res Soc; Tissue Cult Asn. *Res:* Normal tissue response to radiation; chemical and radiation carcinogenesis; biology of the mammary gland. *Mailing Add:* 13 S Blackhawk Ave Madison WI 53705-3317

GOULD, PHILLIP, b New York, NY, Feb 19, 40; m 64, 80; c 2. MECHANICAL ENGINEERING. *Educ:* City Col New York, BME, 61; Mass Inst Technol, SM, 63, ScD(mech eng), 65. *Prof Exp:* Asst prof mech eng, Mass Inst Technol, 65-67; ASST DIR, SYSTS EVAL DIV, INST FOR DEFENSE ANAL, 85-, MEM, 67- *Concurrent Pos:* Fel eng, Mass Inst Tech, 65-67. *Mem:* NY Acad Sci; Am Inst Aeronaut & Astronaut; Am Soc Mech Engrs, Operations Res Soc Am; fel AAAS. *Res:* Continuum mechanics; operations analysis. *Mailing Add:* 4590 Indian Rock Terrace Washington DC 20007-2567

GOULD, PHILLIP L, b Chicago, Ill, May 24, 37; m 61; c 4. STRUCTURAL ENGINEERING, SOLID MECHANICS. *Educ:* Univ Ill, Urbana, BS, 59, MS, 60; Northwestern Univ, PhD(civil eng), 66. *Prof Exp:* Struct designer, Skidmore, Owings & Merrill, Ill, 60-63; prin struct engr, Westenhoff & Novick, 63-64; from asst prof to assoc prof, 66-74, chmn dept, 74-81, PROF CIVIL ENG, WASH UNIV, ST LOUIS, 75-, HAROLD D JOLLEY CHAIR CIVIL ENG, 81- *Concurrent Pos:* Consult, design of shell roof struct, 66-; vis Prof, Univ Sydney, Australia, 81; adv prof, dept civil eng, Shaghai Inst Technol, China, 86. *Honors & Awards:* Sr Scientist Award, Alexander von Humboldt Found, 74-75. *Mem:* Am Soc Civil Engrs; Am Soc Eng Educ; Am Acad Mech; Int Asn Shell Struct. *Res:* Analysis and design of multistory building frames, highway bridges, shell roof structures, and pressure vessels; development of computer-based numerical techniques for the analysis of thin shells and human hearts; earthquake engineering. *Mailing Add:* Sch Eng & Appl Sci Wash Univ St Louis MO 63130

GOULD, ROBERT GEORGE, b Plattsburg, NY, June 12, 47; m 77. MEDICAL RADIOLOGIC PHYSICS. *Educ:* Col Wooster, BA, 69; Univ Pa, MS, 71; Harvard Univ, ScD, 78. *Prof Exp:* Hosp physicist, Beth Israel Hosp, Boston, 72-77; ASST PROF RADIOL, UNIV CALIF, SAN FRANCISCO, 77- *Concurrent Pos:* Prin investr, NIH grants, 80-81 & 82-84. *Mem:* Am Asn Physicists Med; Soc Photo-Optical Instrumentation Engrs. *Res:* Medical x-ray imaging instrumentation; digital radiography; fluoroscopy; computed tomography. *Mailing Add:* Dept Radiol Univ Calif San Francisco Med Sch 513 Parnassus Ave San Francisco CA 94143

GOULD, ROBERT HENDERSON, physics; deceased, see previous edition for last biography

GOULD, ROBERT JAMES, b Hong Kong, Dec 9, 54; Can citizen; m 78; c 1. ANTITHROMBOTICS, PLATELETS. *Educ:* Spring Arbor Col, BA, 76; Univ Iowa, PhD (biochem), 81. *Prof Exp:* Res fel, dept neurosci, Johns Hopkins Univ, 81-84; sr res pharmacologist, 84-87, res fel, 87-90, ASSOC DIR, MERCK, SHARP & DOHME RES LABS, 90- *Mem:* Am Soc Biochem & Molecular Biol; Am Soc Pharmacol & Exp Therapeut; Soc Neurosci; NY Acad Sci. *Res:* Membrane receptors, platelets and antithrombotics. *Mailing Add:* Dept Biol Chem Merck Sharp & Dohme Res Labs W26-265 West Point PA 19486

GOULD, ROBERT JOSEPH, b Providence, RI, May 31, 35. THEORETICAL PHYSICS, ASTROPHYSICS. *Educ:* Providence Col, BS, 57; Cornell Univ, PhD(theoret physics), 63. *Prof Exp:* From res physicist to asst res physicist, 63-65, from asst prof to assoc prof, 65-76, PROF PHYSICS, UNIV CALIF, SAN DIEGO, 76- *Concurrent Pos:* Sr lectr, Univ Sydney, 68-69; vis scholar, Stanford Univ, 75-76. *Mem:* Am Phys Soc; Am Astron Soc. *Res:* Theoretical astrophysics; atomic processes in low-density plasmas; high-energy phenomena in astrophysics. *Mailing Add:* Dept Physics B-019 Univ Calif La Jolla CA 92093

GOULD, ROBERT K, b Gloucester, Mass, Jan 4, 29; div; c 3. ACOUSTICS. *Educ:* Univ Maine, BS, 51; Brown Univ, MS, 56, PhD(physics), 61. *Prof Exp:* Asst prof physics, Muskingum Col, 57-59; from asst prof to assoc prof, Lafayette Col, 60-68; chmn dept, 68-80, PROF PHYSICS, MIDDLEBURY COL, 80- *Mem:* Acoust Soc Am; Am Asn Physics Teachers. *Res:* High intensity sound. *Mailing Add:* Dept Physics Middlebury Col Middlebury VT 05753

GOULD, ROBERT KINKADE, b Lawrenceville, Ill, Sept 6, 40; m 65; c 2. CHEMICAL PHYSICS, FLUID MECHANICS. *Educ:* Wash Univ, AB, 62; Univ Wis, PhD(physics), 69. *Prof Exp:* Staff scientist, Aerochem Res Labs, 69-78, head, Appl Physics Group, 78-81; res dir, Universal Silicon Inc, 81-86; SR RES SCIENTIST, STANDARD OIL ENG MATS CO, 86- *Mem:* Am Phys Soc; Mat Res Soc; Fire Particle Soc; AAAS. *Res:* Turbulence; two-phase flows; combustion and high temperature chemistry; aerosol science. *Mailing Add:* 9835 Hollingson Rd Clarence NY 14031

GOULD, ROBERT WILLIAM, b Shanghai, China, Feb 5, 34; US citizen; m 55; c 4. MATERIALS FAILURE. *Educ:* Univ Fla, BS, 55, MS, 62, PhD(metall), 64. *Prof Exp:* Field engr, Am Cyanamid Co, 57; res engr, Kaiser Aluminum & Chem Corp, 57-59; from instr to prof metall, 59-80, prof mat sci & eng, 80-89, EMER PROF MAT SCI & ENG, UNIV FLA, 89- *Concurrent Pos:* Pres PLC Inc & Aeroforensics Inc. *Mem:* Am Soc Testing & Mat; Am Chem Soc; Soc Automotive Engrs; Soc Plastics Engrs. *Res:* Characterization of materials, especially x-ray diffraction and spectroscopy; failure analysis of materials; product liability and engineering. *Mailing Add:* Gould Lewis & Proctor 6712 NW 18th Dr Gainesville FL 32606

GOULD, ROY W(ALTER), b Los Angeles, Calif, Apr 25, 27; m 52; c 2. PHYSICS, ELECTRICAL ENGINEERING. *Educ:* Calif Inst Technol, BS, 49, PhD(physics), 56; Stanford Univ, MS, 50. *Prof Exp:* Res engr missile guid, Jet Propulsion Lab, Calif Inst Technol, 51-52; res engr electron tubes, Hughes Aircraft Co, 53-55; asst prof elec eng, 55-58, from assoc prof to prof elec eng & physics, 58-73, prof appl physics, 74-79, exec officer appl physics, 73-79, chmn, Div Eng & Appl Sci, 79-85, SIMON RAMO PROF ENG, CALIF INST TECHNOL, 79- *Concurrent Pos:* NSF sr fel, 63-64; asst dir res, US AEC, DC, 70-72. *Mem:* Nat Acad Sci; Nat Acad Eng; fel Am Phys Soc; fel Inst Elec & Electronics Engrs; fel Am Acad Arts & Sci. *Res:* Electron and ion dynamics; plasma oscillation and wave phenomena; physics of ionized gases; electromagnetism; microwaves; plasma physics; fusion energy. *Mailing Add:* Mail Sta 128-95 Calif Inst Technol Pasadena CA 91125

GOULD, STEPHEN JAY, b New York, NY, Sept 10, 41; m 65; c 2. PALEONTOLOGY, EVOLUTIONARY BIOLOGY. *Educ:* Antioch Col, AB, 63; Columbia Univ, PhD(paleont), 67. *Hon Degrees:* From US Cols & Univs, 82-90. *Prof Exp:* Asst prof geol, Antioch Col, 66; from asst prof to assoc prof, 67-73, PROF GEOL, HARVARD UNIV, 73-, ALEXANDER AGASSIZ PROF ZOOL, 82- *Concurrent Pos:* McArthur Prize Fel, 81-86; assoc cur invert paleont, Mus Comp Zool, Harvard Univ, 71-73, cur invert paleont, 73-, adj mem, dept hist sci, 73-; assoc ed, Evolution, 70-72; mem, Smithsonian Coun, 76-, adv bd, Children's TV Workshop, 78-81 & adv bd, NOVA, 80-; prin investr, var NSF grants, 69-; counr, Paleont Soc, 73-75; mem coun, AAAS, 74-76 Nat Portrait Gallery, 89- & Space Explor Coun, NASA, 89-; bd dirs, Biol Sci Curric Study, 76-79; Int bd dirs, Gallery of Evolution, Mus Natural Hist, Paris, 89. *Honors & Awards:* Schuchert Award, Paleont Soc, 75; F V Haydn Medal, Philadelphia Acad Natural Sci, 82; J Priestly Award & Medal, Dickinson Col, 83; Distinguished Serv Award, Am Humanists' Asn, 84 & Am Geol Inst, 86; Meritorious Serv Award, Am Asn Systs Collections; Silver Medal, Zool Soc London, 84; Bradford Washburn Award & Gold Medal, Mus Sci, Boston, 84; John & Samuel Bard Award Med & Sci, Bard Col, 84; Tanner lectrs, Cambridge Univ, Eng, 84; Terry lectrs, Yale Univ, 86; Glenn T Seaborg Award, 86; Anthrop Media Award, Am Anthrop Asn, 87; Hist Geol Award, Geol Soc Am, 88; Pub Serv Award, Am Inst Prof Geologists, 89; Brittanica Award & Gold Medal, 90. *Mem:* Nat Acad Sci; Paleant Soc (pres, 85-86); Soc Study Evolution (vpres, 75-76, pres, 90); Soc Syst Zool; Am Soc Naturalists (pres, 79-80); fel Am Acad Arts & Sci; hon foreign fel, Europ Union Geosci; foreign, Linnean Soc London; Sigma Xi; Hist Sci Soc; fel AAAS. *Res:* Evolutionary paleontology; quantitative studies of form, function and ontogeny in relation to phylogeny; evolution and speciation in land snails; systematic zoology; history and philosophy of geology and evolutionary biology; wrote six books. *Mailing Add:* Mus Comp Zool Harvard Univ Cambridge MA 02138

GOULD, STEVEN JAMES, b New York, NY, Feb 18, 46; m 88. BIO-ORGANIC CHEMISTRY. *Educ:* Univ Calif, Los Angeles, BS, 66; Mass Inst Technol, PhD(org chem), 70. *Prof Exp:* Fel, Swiss Fed Inst Technol, 70-72; sr res chemist org chem, Syva Res Inst, Palo Alto, 72-74; asst prof, Sch Pharm, Univ Conn, 74-80, assoc prof pharmacog, 80-82; assoc prof chem, 82-86, PROF CHEM, ORE STATE UNIV, 86- *Concurrent Pos:* Res career develop award, Nat Cancer Inst, NIH, 79-84; Fulbright fel, 89-90; Am Cancer Soc, 89-90. *Mem:* Am Chem Soc; Am Soc Pharmacog; AAAS; Sigma Xi. *Res:* Biosynthesis of natural products; enzyme reaction mechanisms as studied with isotopically-labeled substrates; synthesis of natural products; steptomyces molecular genetics. *Mailing Add:* 7707 NW McDonald Circle Corvallis OR 97330

GOULD, WALTER LEONARD, b Burlington, Wyo, Apr 5, 23; m 47; c 3. WEED SCIENCE, NATURAL RESOURCES. *Educ:* Univ Wyo, BS, 48, MS, 57; Oregon State Univ, PhD(field crops), 64. *Prof Exp:* Conservationist, Soil Conserv Serv, USDA, 48-56, agronomist, Agr Res Serv, 64-66; instr, 57-60, asst prof, 66-69, ASSOC PROF AGRON, N MEX STATE UNIV, 69- *Mem:* Weed Sci Soc Am; Soc Range Mgt. *Res:* Weed control in field and vegetable crops and on rangeland; control of brush on rangeland; reclamation of surface-mine spoils. *Mailing Add:* 1812 Imperial Ridge St Las Cruces NM 88003

GOULD, WALTER PHILLIP, b North Adams, Mass, July 14, 25; m 54; c 2. FORESTRY, WILDLIFE MANAGEMENT. *Educ:* Univ Mass, BS, 50; Yale Univ, MF, 51; Syracuse Univ, PhD(forest shrub ecol), 66. *Prof Exp:* Forester, Brown Co, 51-54; assoc prof forest & wildlife mgt & chmn dept, 54-87, EMER PROF, UNIV RI, 87- *Concurrent Pos:* Fel, Mem Univ Nfld, 67-68. *Mem:* Soc Am Foresters; Wildlife Soc. *Res:* Effects of silvicultural practices on the ecology of forest wildlife species. *Mailing Add:* Johnny Cake Trail Indian Lake Shores RI 02881

GOULD, WILBUR ALPHONSO, b Colebrook, NH, Aug 7, 20; m 44; c 3. FOOD SCIENCE, HORTICULTURE. *Educ:* Univ NH, BS, 42; Ohio State Univ, MS, 47, PhD(hort, food technol), 49. *Prof Exp:* Plant breeder, Ferry-Morse Seed Co, 42; food inspector, USDA, 43-44; from instr to prof hort, Ohio State Univ, 47-85, head div food processing, 56-85; RETIRED. *Concurrent Pos:* Food consult annual fine food expos, Ger, USDA, 59. *Mem:* Fel AAAS; Am Soc Hort Sci; fel Inst Food Technologists. *Res:* Quality evaluation and control in food processing, especially plant efficiencies; new product development; food regulations and standards. *Mailing Add:* PO Box 312 Worthington OH 43085

GOULD, WILLIAM ALLEN, b Clearfield, Pa, Dec 3, 41; m 65; c 5. MATHEMATICS, COMPUTER SCIENCE. *Educ:* Elizabethtown Col, BS, 63; Pa State Univ, MA, 66, EdD(math), 71. *Prof Exp:* From instr to assoc prof math, 66-70, prof & asst dir comput ctr, 70-82, DIR COMPUT CTR, SHIPPENSBURG UNIV, 82- *Mem:* Math Asn Am. *Res:* Solution preserving operators for three-dimensional second order partial differential equations; computer simulation of the academic course loadings. *Mailing Add:* Comput Ctr Shippensburg Univ Shippensburg PA 17257

GOULD, WILLIAM DOUGLAS, Orillia, Ont, Apr 13, 44. MICROBIOLOGY. *Educ:* Univ Manitoba, BSc, 65; Univ Alta, MSc, 70, PhD(soil microbiol), 76. *Prof Exp:* Chemist, Kalium Chem, Ltd, 66-67; fel, Colo State Univ, 76-79; environ microbiologist, Agr Can, 79-82; res microbiologist, Allied Corp, 82-86; RES MICROBIOLOGIST, ENERGY MINES & RESOURCES, CAN, 86- *Concurrent Pos:* Adj asst prof, State Univ NY, Syracuse, 83- *Mem:* Sigma Xi; Am Soc Microbiol; Can Soc Microbiol; Am Soc Agron; Int Soc Soil Sci. *Res:* Metabolism of thiobacilli; treatment of wastes produced by the mining industry. *Mailing Add:* Energy Mines & Resource Co 555 Booth St Ottowa ON K1G 0G1 Can

GOULD, WILLIAM E, b Orange, NJ, May 7, 34. MATHEMATICS. *Educ:* Rutgers Univ, BA, 56, MS, 58; Princeton Univ, MA, 64, PhD(math), 66. *Prof Exp:* Instr math, Rutgers Univ, 61-62 & Wash Col, 62-66; assoc prof, Bradley Univ, 66-69; assoc prof, 69-74, PROF MATH, CALIF STATE UNIV, DOMINGUEZ HILLS, 74- *Mem:* Math Asn Am. *Res:* Logic. *Mailing Add:* 12529 Sleepyhollow Lane Cerritos CA 90701

GOULD, WILLIAM RICHARD, b Provo, Utah, Oct 31, 19; c 4. MECHANICAL ENGINEERING, ELECTRICAL ENGINEERING. *Educ:* Univ Utah, BS, 42; USN, dipl naval archit, 47. *Prof Exp:* Mech engr, 48-49, asst supt, 49-54, supt, 54-58, gen supt, 58-59, asst mgr eng dept, 59-62, mgr eng dept, 62-63, vpres, 63-67, sr vpres, 67-73, exec vpres, 73-78, pres, 78-80, chief exec officer & chmn bd, 80-84, EMER CHMN BD, SOUTHERN CALIF EDISON CO, 84- *Concurrent Pos:* Fel Inst Advan Eng, 71-, chmn bd, 76-; US comt pres, Int Conf on Large High Voltage Elec Systs, 73; gen adv comt, US Dept Energy, 75; trustee, Calif Inst Technol, 78-; dir, Mono Power Co, Environ & Energy Serv Co, Joy Technol Inc, dir & consult, Southern Calif Edison Co. *Honors & Awards:* Power-Life Award, Inst Elec & Electronics Engrs, 78, Centennial Medal, 84; George Westinghouse Gold Medal, Am Soc Mech Engrs, 79, Centennial Award for Serv, 79; Achievement Award, Nat Energy Found, 81; Distinguished Contrib Award, Inst Advan Eng, 82. *Mem:* Nat Acad Eng; fel Am Soc Mech Engrs; Elec Power Res Inst. *Res:* Electric power generation, transmission and distribution. *Mailing Add:* Southern Calif Edison Co 2244 Walnut Grove Ave Rosemead CA 91770

GOULD, WILLIAM ROBERT, III, b Cincinnati, Ohio, Nov 16, 31; m 62; c 2. ZOOLOGY. *Educ:* Colo State Univ, BS, 54; Okla State Univ, MS, 60, PhD(zool), 62. *Prof Exp:* NIH fel, 62-63; prog leader, Bur Com Fisheries, US Fish & Wildlife Serv, 63; asst leader, Mont Coop Fishery Res Unit, Mont State Univ, 64-91; RETIRED. *Mem:* Am Fisheries Soc; Am Soc Icthyol & Herpet. *Res:* Biology and ecology of fishes. *Mailing Add:* 323 N 20th Ave Bozeman MT 59715

GOULDEN, CLYDE EDWARD, b Kansas City, Kans, Nov 30, 36; m 57; c 2. ECOLOGY, LIMNOLOGY. *Educ:* Kans State Teachers Col, BS, 58, MS, 59; Ind Univ, PhD(zool), 62. *Prof Exp:* Res assoc limnol, Yale Univ, 62-64, lectr biol, 63-64, asst prof, 64-66; assoc cur, Dept Limnol, 66-72, vpres res, 72-73, dir, Div Environ Res, 73-77, CUR, ACAD NAT SCI, PHILADELPHIA, 81-; PROF BIOL, UNIV PA, 81- *Concurrent Pos:* Res grants, Am Philos Soc Penrose Fund & Soc Sigma Xi, 64; exchange scientist, Cult Exchange Prog, Acad Sci USSR-Nat Acad Sci, 66 & NSF, 67-; adj assoc prof biol, Univ Pa, 70-71, ed, Aquatics & Ecol & Ecol Soc dir, Div Limnol & Ecol, Acad Natural Sci Philadelphia, 73-77; NSF grant, 76-78, 79 & 84-85; EPA grant, 81-82. *Mem:* Fel AAAS; Am Soc Naturalists; Soc Study Evolution; Ecol Soc Am; Int Asn Theoretical & Appl Limnol. *Res:* Ecology and systematics of the Cladocera; paleolimnology; biology of the zooplankton; evolution of the aquatic community. *Mailing Add:* Acad Natural Sci 19th & Pkwy Philadelphia PA 19103

GOULDEN, PETER DERRICK, analytical chemistry, chemical engineering; deceased, see previous edition for last biography

GOULDING, CHARLES EDWIN, JR, b Tampa, Fla, Nov 23, 16; m 50, 72; c 1. BIOELECTRONICS, BIOMAGNETICS. *Educ:* Univ Tampa, BS, 39; Univ Fla, Gainsville, MS, 44, PhD(chem & chem eng), 46. *Hon Degrees:* DSc, Chem Eng, Marquis Giuseppe Sciclura Univ, 87; PhD, Albert Einstein Int Inst, 90. *Prof Exp:* Fel Univ Pa, 66-68; pres, T&M Machining, 78-85; CHIEF ENGR, TWINCITIES ENG ASSOC, 77-; PRES, BRISTOL RES CORP, 85- *Mem:* Am Chem Soc; Sigma Xi; Nat Asn Prof Engrs; Inst Elec & Electronics Engrs; Am Inst Chem Engrs; AAAS. *Res:* Use of cowave energy a natural viricide to stop herpes, Acquired Immune Deficiency Syndrome, flue, virus cancer, Alzheimer's Syndrome, virus pneumonia, common colds and Multiple Sclerosis; biomedicine. *Mailing Add:* 2569 Volunteer Pkwy Bristol TN 37620

GOULDING, ROBERT LEE, JR, entomology; deceased, see previous edition for last biography

GOULD-SOMERO, MEREDITH, b Billings, Mont, Feb 2, 40; m 68. DEVELOPMENTAL BIOLOGY. *Educ:* Mt Holyoke Col, BA, 61; Stanford Univ, PhD(biol), 67. *Prof Exp:* USPHS fel biochem, Univ Wash, 67-68; fel zool, Univ BC, 68-70; res biologist, Univ Calif, San Diego, 70-, lectr, 73-; AT BIOQUIMICA, MEXICO CITY, MEX. *Mem:* Am Soc Cell Biologists; Am Soc Zool; Soc Develop Biol. *Res:* Gametogenesis, fertilization and early development. *Mailing Add:* Universidad Autonoma de Baja Calif Ensenada BC AP 2921 Mexico

GOULET, JACQUES, b Quebec, Can, June 22, 46; m 70; c 4. FOOD SCIENCE, FOOD MICROBIOLOGY. *Educ:* Laval Univ, BA, 66, BSA, 70; McGill Univ, PhD(microbiol), 74. *Prof Exp:* Res coordr, Vermette & Fils Dairy, 73-74; from asst prof to assoc prof, 74-86, head dept, 80-82, PROF FOOD SCI, LAVAL UNIV, 86-; RES DIR, LALLEMAND, INC, 82- *Concurrent Pos:* Consult, dairy & fermentation industs. *Mem:* Am Soc Microbiol; Can Inst Food Sci & Technol; Inst Food Technol. *Res:* Fermented dairy products; biological treatment of industrial effluents for biomass and fermentation products recovery; probiotics. *Mailing Add:* Dept Food Sci Laval Univ Quebec PQ G1K 7P4 Can

GOULIAN, DICRAN, b Weehawken, NJ, Mar 31, 27; m 54; c 4. PLASTIC SURGERY. *Educ:* Columbia Col, AB, 48; Columbia Univ, DDS, 51; Yale Univ, MD, 55; Am Bd Plastic Surg, dipl, 62. *Prof Exp:* Intern gen surg, Grace-New Haven Community Hosp, 55-56, resident, 56-58; resident plastic surg, NY Hosp, 58-60; instr plastic surg, 60-61, from clin asst prof to clin assoc prof surg, 61-71, assoc prof, 71-74, PROF SURG, MED COL, CORNELL UNIV, 74- *Concurrent Pos:* NSF fel surg res, Yale Univ, 56-57; USPHS fel, Cornell Univ, 60-61; attend surgeon-in-chg plastic surg, NY Hosp-Cornell Univ Med Ctr; adj surgeon, Lenox Hill Hosp, NY; consult plastic surgeon, Bronx Vet Admin Hosp; clin abstr ed, Transplantation Bull, 61 & Transplantation, 62-66; pvt pract, 64-; treas, NY Regional Soc Plastic Surgeons, 69-72; pres sect plastic surg, NY State Med Soc, 71-72. *Honors & Awards:* Ella Marie Ewell Medal Outstanding Proficiency Dent, Sch Dent & Oral Surg, Columbia Univ. *Mem:* AAAS; Am Asn Cleft Palate Rehab; Am Soc Plastic & Reconstruct Surg; Am Soc Maxillofacial Surg (secy-treas, 67-71); fel NY Acad Med (secy sect plastic surg, 71-). *Mailing Add:* Cornell Univ NY Hosp Cornell 525 E 68th St New York NY 10021

GOULIAN, MEHRAN, b Weehawken, NJ, Dec 31, 29; m 61; c 3. HEMATOLOGY. *Educ:* Columbia Univ, AB, 50, MD, 54. *Prof Exp:* Intern, Barnes Hosp, St Louis, Mo, 54-55; resident med, Mass Gen Hosp, Boston, 58-59 & 61; instr, Harvard Univ, 63-65; assoc prof med & biochem, Univ Chicago, 67-70; PROF MED, UNIV CALIF, SAN DIEGO, 70- *Concurrent Pos:* Fel hemat, Sch Med, Yale Univ, 59-60; res fel, Harvard Univ & Mass Gen Hosp, 60 & 62-63; fel biochem, Sch Med, Stanford Univ, 65-67. *Mem:* Am Soc Biol Chem; Am Soc Clin Invest; Am Soc Hemat; Asn Am Physicians. *Res:* Biochemistry of nucleic acids. *Mailing Add:* Dept Med Univ Calif San Diego 0613G La Jolla CA 92093-0613

GOULIANOS, KONSTANTIN, b Salonica, Greece, Nov 9, 35; m. HIGH ENERGY PHYSICS. *Educ:* Columbia Univ, MA, 60, PhD(physics), 63. *Prof Exp:* Res assoc physics, Columbia Univ, 63-64; instr, Princeton Univ, 64-67, asst prof, 67-71; assoc prof, 71-81, PROF PHYSICS, ROCKEFELLER UNIV, 81- *Concurrent Pos:* Fulbright scholar, 58-59. *Mem:* Fel Am Phys Soc. *Res:* High energy neutrino interactions; time reversal invariance; elastic scattering and diffraction dissociation; hadron collider experiments. *Mailing Add:* Rockefeller Univ New York NY 10021

GOULSON, HILTON THOMAS, b Montevideo, Minn, May 4, 30; m 54; c 2. PARASITOLOGY. *Educ:* Luther Col, AB, 52; Univ NC, MSPH, 53, PhD(parasitol), 57. *Prof Exp:* Instr, 54-57, res assoc, 57-59, from asst prof to assoc prof, 59-69, PROF PARASITOL & LAB PRACT, SCH PUB HEALTH, UNIV NC, CHAPEL HILL, 69- *Concurrent Pos:* Prog dir lab pract training prog, Univ NC, Chapel Hill. *Mem:* Am Soc Trop Med & Hyg;

Am Soc Parasitol; fel Am Pub Health Asn. *Res:* Medical parasitology; immunity studies and host-parasite relationships of nematode parasites. *Mailing Add:* Dept Parasitol & Lab Practice B#7400 Univ NC Chapel Hill NC 27599-7400

GOUNARIS, ANNE DEMETRA, b Boston, Mass, Oct 27, 24. BIOCHEMISTRY. *Educ:* Boston Univ, AB, 55; Radcliffe Col, PhD(chem), 60. *Prof Exp:* NIH res fels, Brookhaven Nat Lab, 60-62; NIH fel, Carlsberg Lab, Denmark, 62-63; Rast-Osted Found fel, 63-64; res assoc enzyme action & protein structure, Rockefeller Inst, 64-66; from asst prof to prof, 66-90, EMER PROF BIOCHEM, VASSAR COL, 90- *Concurrent Pos:* NIH grant, 68-71& A&M Inst, 72-76. *Mem:* AAAS; Am Chem Soc; NY Acad Sci; Am Soc Biol Chemists; Sigma Xi. *Res:* Thiamine pyrophosphate requiring enzymes; alpha-keto decarboxylase; cathepsin B; cysteine proteinase inhibitor; protein sequence analysis. *Mailing Add:* 144 Marble St No 501 Stoneham MA 02180

GOURARY, BARRY SHOLOM, b Rostov, Russia, Feb 10, 23; nat US; m 53; c 2. MICROCOMPUTER APPLICATIONS SOFTWARE. *Educ:* Columbia Univ, AM, 50. *Prof Exp:* Asst radiation lab, Columbia Univ, 47-50, lectr math, 50-51; physicist, Sound Sect, Nat Bur Standards, 51; mem sr staff, Appl Physics Lab, Johns Hopkins Univ, 51-59, prin staff, 59-60; fel physicist, Res Labs, Westinghouse Elec Corp, 60, supvry physicist, 61, mgr luminescence sect, 61-62; sr staff scientist, Res Labs, United Aircraft Corp, 62-68; dir electronic & optical technol, Defense-Space Group, Int Tel & Tel Corp, 68-70, tech dir, electron tube div, 70, dir technol, Defense-Space Group, 70-71; mgt consult, 71-74; PRES, GOURARY ASSOC INC, 73- *Concurrent Pos:* Consult, Inst Defense Anal, 72-87. *Mem:* Fel Am Phys Soc; Inst Elec & Electronics Eng; Inst Mgt Sci; Am Defense Prep Asn. *Res:* Operations research; microcomputer applications software; technology assessments; planning studies; technical management; solid state physics; quantum theory; cost studies. *Mailing Add:* Gourary Assocs Inc 187 Gates Ave Montclair NJ 07042

GOURAS, PETER, b Brooklyn, NY, Apr 15, 30; m 59; c 3. PHYSIOLOGY. *Educ:* Johns Hopkins Univ, AB, 51, MD, 55. *Prof Exp:* Intern surg, Johns Hopkins Univ, 55-56; res assoc physiol, NIH, 56-57; assoc physiol, Med Sch, Univ Pa, 59-60; res assoc ophthal & physiol, NIH, 60-68, chief sect ophthal physiol, Nat Inst Neurol Dis & Stroke, 68-71, head sect physiol, Lab Vision Res, Nat Eye Inst, 71-76; HEAD, LAB NEUROBIOL, EDWARD S HARKNESS EYE INST, 76-; PROF OPHTHAL, COLUMBIA UNIV COL PHYSICIANS & SURGEONS, 81- *Concurrent Pos:* Nat Found fel, Cambridge Univ, 58-59; mem comt vision res & training, Nat Eye Inst, 70-74; comt sensory prosthesis, Nat Inst Neurol Dis & Stroke, 70-74; vis prof, Neurol Inst, Univ Freiburg, 74-75; mem, Am Comt Optics & Visual Physiol, Sect Ophthal, AMA, 75-78; mem sci coun, Nat Retinitis Pigmentosa Soc, 72-78; mem NIH visual sci study sect B, 78-82. *Honors & Awards:* Award, Res to Prevent Blindness Inc; Alcon Res Award, 83, 91. *Mem:* AAAS; fel Optical Soc Am; Am Physiol Soc; Int Soc Clin Electroretinography; Asn Res Vision & Ophthal. *Res:* Neurophysiology; retina; visual pathways; vision. *Mailing Add:* Ophthal Res Columbia Univ New York NY 10032

GOURDINE, MEREDITH CHARLES, b Newark, NJ, Sept 26, 29. DIRECT ENERGY CONVERSION, MAGNETOHYDRODYNAMICS. *Educ:* Cornell Univ, BS, 53; Calif Inst Technol, PhD(eng sci), 60. *Prof Exp:* Res scientist aerospace, Jet Propulsion Lab, 59-60; lab dir aerospace, Plasmadyne Corp, 60-62; chief scientist aeronaut, Curtiss Wright Corp, 62-64; pres energy conversion res & develop, Gourdine Systs, Inc, 64-73; PRES ENERGY CONVERSION RES & DEVELOP, ENERGY INNOVATIONS, INC, 74- *Mem:* Nat Acad Eng; NY Acad Eng; Am Inst Aeronaut & Astronaut; Sigma Xi. *Res:* Direct energy conversion; electrogasdynamic which is related to interaction between an electric field and charged particles in gases; applications include: airport fog clearing, air pollution control systems, coating systems etc; thermovoltaic batteries and vortadynamic energy conversion systems; 70 US/foreign patents; author and co-author of over 20 publications. *Mailing Add:* 8709 Knight Rd Houston TX 77054

GOURISHANKAR, V (GOURI), b Simla, Punjab, India, Jan 22, 29; m 66; c 2. ELECTRICAL ENGINEERING, CONTROL SYSTEMS. *Educ:* Univ Madras, BE, 50; Univ Ill, MS, 58, PhD, 61. *Prof Exp:* Asst prof elec eng, Univ Ill, Urbana-Champaign, 61-65; assoc prof, 65-70, PROF ELEC ENG, UNIV ALTA, 70- *Mem:* Inst Elec & Electronics Engrs. *Res:* Process control systems; identification; adaptive control; robotics. *Mailing Add:* Dept Elec Eng Univ Alta Edmonton AB T6G 2G7 Can

GOURLEY, DESMOND ROBERT HUGH, b Thunder Bay, Ont, Nov 2, 22; US citizen; m 46; c 5. PHARMACOLOGY. *Educ:* Univ Toronto, BA, 45, PhD(cellular physiol), 49. *Prof Exp:* Lab instr comp anat, Univ Toronto, 45-46, asst cellular physiol, 46-47, demonstr, 47-49; res asst, Sch Med, Univ Va, 49-51, from asst prof to prof pharmacol, 51-73, chmn dept, 67-68; prof, 73-86, EMER PROF PHARMACOL, EASTERN VA MED SCH, 86- *Concurrent Pos:* Fedn Am Soc Exp Biol travel grant, Brussels, 56 & Int Union Physiol Soc travel grant, Buenos Aires, 59; mem, US Pharmacopeial Conv, 60-65 & 80-86; Humboldt Found fel, Univ Freiburg, 68-69; adj instr, Sch Continuing Educ, Univ Va, 70-; adj prof chem, Old Dominion Univ, 75-86. *Honors & Awards:* Horsley Prize for Res, 52. *Mem:* Am Soc Pharmacol & Exp Therapeut; Am Physiol Soc; Soc Exp Biol & Med; Pharmacol Soc Can. *Res:* Effects of insulin on isolated muscle tissue; active transport and phosphate metabolism in erythrocytes; effects of drugs at cell membranes; molecular mechanism of drug tolerance and dependence. *Mailing Add:* 255 College Cross Suite 45 Norfolk VA 23510-1136

GOURLEY, EUGENE VINCENT, b Detroit, Mich, Nov 15, 40; m 62; c 2. ZOOLOGY. *Educ:* Eastern Mich Univ, BA, 62, MS, 64; Univ Fla, PhD(turtle behav), 69. *Prof Exp:* Interim instr zool, Univ Fla, 68-69; assoc prof, 69-79, PROF BIOL, RADFORD COL, 79- *Mem:* AAAS; Am Soc Zoologists; Animal Behav Soc; Sigma Xi. *Mailing Add:* 1009 9th St Radford VA 24141

GOURLEY, LLOYD EUGENE, JR, b Bergheim, Tex, Apr 22, 23; m 48, 76. PHYSICS. *Educ:* Univ Tex, BS, 46, MA, 48, PhD, 59. *Prof Exp:* Asst physicist, Atlantic Refining Co, 48-51; res physicist, NMex Inst Mining & Technol, 51-56; from assoc prof to prof, 59-88, assoc dean & chmn sci area, 74-77, dean sci, 77-80, EMER PROF PHYSICS, AUSTIN COL, 88- *Res:* Properties of solids under high pressures. *Mailing Add:* 3731 Jefferson Dr Denison TX 75020

GOURLEY, PAUL LEE, b Fargo, ND, Mar 15, 52; m 81; c 2. PHYSICS OF SEMICONDUCTORS, OPTOELECTRONIC MATERIAL & ARTIFICALLY STRUCTURED MATERIALS. *Educ:* Univ NDak, BS, 74; Univ Ill, MS, 76, PhD(physics), 80. *Prof Exp:* Res asst, Univ NDak, 72-73, US Bur Mines, 73-74, Univ Ill, 76-79; mem tech staff, 80-87, DIV SUPVR, SANDIA NAT LAB, 87- *Mem:* Am Phys Soc; Optical Soc Am; Mat Res Soc; Sigma Xi. *Res:* Artifically structured materials, semiconductors, surface-emitting and strained-layer lasers; optical switching phenomenon and devices; semiconductor optical interference structure devices; lattice mismatched semiconductor epitaxy, excitonic phenomenon, electronic transport and recombination; defects in semiconductors. *Mailing Add:* Sandia Nat Labs Albuquerque NM 87185

GOURSE, JEROME ALLEN, b Bristol, Va, June 15, 29; m; c 2. ORGANIC CHEMISTRY. *Educ:* Va Polytech Inst, BS, 50; Univ Del, MS, 51; Univ Ill, PhD(chem), 59. *Prof Exp:* res chemist & group leader, Velsicol Chem Corp, 58-82. *Mem:* Am Chem Soc; fel Am Inst Chem; Sigma Xi. *Res:* Agricultural chemistry. *Mailing Add:* 16606 Calico Pl Tampa FL 33618

GOURSE, RICHARD LAWRENCE, b Fall River, Mass, Apr 5, 49; m 76. MOLECULAR BIOLOGY, GENETICS. *Educ:* Brown Univ, AB, 71, MAT, 73, PhD(molecular biol), 80. *Prof Exp:* Fel, Brown Univ, 80-82; FEL, UNIV WIS, 82- *Res:* Methods for introducing and analyzing site-directed mutations in ribosomal RNA in order to examine ribosome structure and function; control of gene expression; ribosome structure and function. *Mailing Add:* 149 Hardin Dr Athens GA 30605-1519

GOURZIS, JAMES THEOPHILE, b Boston, Mass, Mar 30, 28; m 62; c 2. CLINICAL PHARMACOLOGY, RESEARCH ADMINISTRATION. *Educ:* Harvard Univ, AB, 49; Boston Univ, AM, 51; Univ Man, PhD, 62, MD, 63. *Prof Exp:* Jr pharmacologist, Ciba Pharm Prod Inc, 51-52; sr pharmacologist, Riker Lab, Inc, 52-57; res assoc pharmacol & therapeut, Univ Man, 57-62, lectr, 62-63; lectr, Col Med, Univ Cincinnati, 63-66; asst dir clin invest, McNeil Labs Inc, 66-68, dir clin pharmacol, 68-70; dir clin pharmacol, Schering Corp, 70-73, sr dir clin res, 73-77; dir clin pharmacol, Merrell Res Ctr, 77-80, exec dir med serv, 80-81; vpres med affairs, Gensia Pharmaceut, 88-89; PRES, MEDRAND, INC, 81- *Concurrent Pos:* Assoc dir, Bethesda Med Res Ctr, 64-66; mem bd dirs, Sperti Drug Prod Inc, 81-83, Itesco, Inc 85-87; gen mgr & dir, Hill Top Pharmatest, 86-88. *Mem:* Am Soc Pharmacol & Exp Therapeut; Drug Info Asn; Can Pharmacol Soc; Am Soc Clin Pharmacol & Therapeut; Asn Clin Pharmacol. *Res:* Cardiovascular physiology and pharmacology; drug development; consultation. *Mailing Add:* Medrand Inc 9744 Claiborne Square La Jolla CA 92037

GOUSE, S WILLIAM, JR, b Utica, NY, Dec 15, 31; m 55; c 2. MECHANICAL ENGINEERING. *Educ:* Mass Inst Technol, SB & SM, 54, ScD(mech eng), 58. *Prof Exp:* From instr to assoc prof mech eng, Mass Inst Technol, 56-67; prof, Carnegie-Mellon Univ, 67-69; tech asst for civilian technol to dir, Off Sci & Technol, Off of President, 69-70; assoc dean Carnegie Inst Technol & Sch of Urban & Pub Affairs, Carnegie-Mellon Univ, 71-73; dir, Off of Res & Develop & Sci Adv to Secy, US Dept of Interior, 73-75, actg dir, Off of Coal Res, 74-75; dep asst adminr Fossil Energy, Energy Res & Develop Admin, 75-77; chief sci, 77-79, vpres Metrek Div, 79-80, vpres & gen mgr, Metrek Div, 80-84, SR VPRES & GEN MGR, CIVIL SYSTS DIV, MITRE CORP, VA, 84- *Concurrent Pos:* Adj prof eng & pub policy, Carnegie-Mellon Univ, 80- *Mem:* AAAS; Am Soc Mech Engrs; Soc Automotive Engrs; Am Soc Heating, Refrig & Air-Conditioning Engrs; Am Soc Eng Educ; Sigma Xi; NY Acad Sci; Int Asn Energy Economists. *Res:* Energy; transportation; housing; environment; heat transfer; engineering education; automotive propulsion; information systems. *Mailing Add:* PO Box 9085 McLean VA 22102-0085

GOUST, JEAN MICHEL, b Paris, France, Jan 21, 41; m 63; c 2. NEUROIMMUNOLOGY, CLINICAL IMMUNOLOGY. *Educ:* Facul Med, Univ Paris, BA, 58, Dr, 71, Am Bd Med Lab Immunol, dipl, 80. *Prof Exp:* Resident med, Univ Paris VI, 68-72, instr exp path, 69-72, asst prof med, 72-74, asst prof immunol, 72-74; asst prof immunol, 75-80, med, 76-80, PROF IMMUNOL, MED UNIV SC, 85-, ASSOC PROF NEUROL, 81- *Concurrent Pos:* Prin investr, Nat Inst Neurol Commun Disorders & Stroke, 80-83 & Nat Multiple Sclerosis Soc,78-85. *Mem:* Am Asn Immunologists; Am Soc Microbiol; Soc Exp Biol & Med; Am Fedn Clin Res. *Res:* Clinical immunology: investigation of abnormal immunoregulation in various neurological disorders, primarily multiple sclerosis. *Mailing Add:* Dept Immunol Med Univ SC 171 Ashley Ave Charleston SC 29425

GOUSTIN, ANTON SCOTT, b Minneapolis, Minn, July 8, 53; m 83; c 2. DNA CLONING, SITE-DIRECTED MUTAGENESIS. *Educ:* Univ Minn, BA, 74; Univ Calif, Berkeley, PhD(zool), 79. *Prof Exp:* Res Asst biochem, Univ Minn, St Paul, 81-82; asst appl cell & molecular biol, Univ Umea, Sweden, 82-85; res assoc biochem & molecular biol, Mayo Clin & Found, Rochester, Minn, 85-86; INSTR PEDIAT, HARVARD MED SCH, 87-; INVESTR, CTR BLOOD RES, BOSTON, 86- *Concurrent Pos:* Prin investr award, Soc Develop Biol, 79. *Mem:* Am Soc Cell Biol; Soc Develop Biol; AAAS. *Res:* Peptide growth factors, their mechanism of action and role in embryonic development and in neoplasia; transforming growth factors alpha and beta, and their platelet derived growth factor; structural-functional analysis of platelet-derived growth factor-like proteins. *Mailing Add:* Wayne State Univ 2727 Second Ave Rm 4114 Detroit MI 48201

GOUTERMAN, MARTIN (PAUL), b Philadelphia, Pa, Dec 26, 31. CHEMICAL PHYSICS. *Educ:* Univ Chicago, BA, 51, MS, 55, PhD(physics), 58. *Prof Exp:* Res fel chem, Harvard Univ, 58-59, instr, 59-61, asst prof, 61-66; assoc prof, 66-68, PROF CHEM, UNIV WASH, 68- *Mem:* Fel Am Phys Soc; Am Chem Soc. *Res:* Electronic spectra of porphyrins; electronic phenomena porphyrin solids; porphyrin luminesence sensors. *Mailing Add:* Dept Chem Univ Wash BG-10 Seattle WA 98195

GOUTMANN, MICHEL MARCEL, b Lyon, France, Jan 2, 39; US citizen; m 68; c 2. COMMUNICATION SYSTEMS, INFORMATION THEORY. *Educ:* Mass Inst Technol, BS & MS, 61; Cornell Univ, PhD (elec eng), 65. *Prof Exp:* Ford fel & asst prof elec eng, Mass Inst Technol, 65-67; mem res & develop staff, Magnavox/Gen Atronics Corp, 67-73, dept mgr, 73-81, tech dir, 81-85, DIR ENG, MAGNAVOX/GEN ATRONICS CORP, 85- *Concurrent Pos:* Adj pro, Univ Pa, 80-86. *Mem:* Fel Inst Elec & Electronics Engrs. *Res:* Design and implementation of digital communication systems and networks operating in the presence of distortion, noise and electronic counter measures. *Mailing Add:* 710 Pennsylvania Ave Ft Washington PA 19034

GOUW, T(AN) H(OK), b Jakarta, Indonesia, Jan 17, 33; US citizen; m 60; c 2. ANALYTICAL CHEMISTRY, CHEMICAL ENGINEERING. *Educ:* Delft Technol Univ, ChemIng, 58, DSc, 62. *Prof Exp:* Asst chem eng, Delft Technol Univ, 58-62; assoc res chemist, Calif Res Corp, 62-63; res chemist, 63-68, sr res engr, 68-76, SR ENGR ASSOC, CHEVRON RES CO, 76- *Concurrent Pos:* Mem staff, letters & sci exten, Univ Calif, Berkeley, 64-72. *Mem:* Am Chem Soc; Netherlands Royal Inst Eng; Am Soc Testing & Mat. *Res:* Chromatography; analytical physico-chemical separation methods; instrumental methods of analysis; design small pilot plants. *Mailing Add:* 7783 Duke Ct El Cerrito CA 94530-2544

GOVAN, DUNCAN EBEN, b Winnipeg, Man, Aug 23, 23; m 50; c 6. UROLOGY. *Educ:* Univ Man, MD, 48; FRCP(C), 56; Univ Chicago, PhD(surg), 57. *Prof Exp:* Instr urol, Sch Med, Univ Chicago, 53-54; pvt pract, Can, 54-61; from asst prof to assoc prof, 61-70, PROF SURG, SCH MED, STANFORD UNIV, 70- *Concurrent Pos:* Surg consult, Vet Admin Hosp, Palo Alto, Calif, 61-67 & Santa Clara County Hosp, San Jose, 65-67. *Mem:* Fel Am Col Surg. *Res:* Neurophysiology of the urinary tract; pediatric urology; hydronephrosis and urinary infection. *Mailing Add:* Dept Urol S287 Stanford Univ Med Ctr Stanford CA 94305

GOVE, HARRY EDMUND, b Niagara Falls, Ont, Can, May 22, 22; nat US; m 45; c 2. PHYSICS. *Educ:* Queen's Univ, BSc, 44; Mass Inst Technol, PhD(physics), 50. *Prof Exp:* Res assoc, Mass Inst Technol, 50-52; assoc res officer, Atomic Energy Can, Ltd, 52-58, br head nuclear physics, 56-63, sr res officer, 58-63; chmn dept physics, 77-80, PROF PHYSICS & DIR, NUCLEAR STRUCT RES LAB, UNIV ROCHESTER, 63- *Mem:* Fel Am Phys Soc; Can Asn Physicists. *Res:* Structure of nuclei and mechanisms of nuclear reactions principally employing particle accelerators; ultrasensitive atom spectrometry with electrostatic accelerators. *Mailing Add:* Univ Rochester River Campus Sta Rochester NY 14627

GOVE, NORWOOD BABCOCK, b New York, NY, Oct 23, 32; m 59; c 3. NUCLEAR PHYSICS. *Educ:* Harvard Univ, BA, 53; Univ Ill, PhD(physics), 58. *Prof Exp:* COMPUT ANALYST, OAK RIDGE NAT LAB, 58- *Mem:* Am Phys Soc; Am Soc Info Sci; Sigma Xi. *Res:* Decay schemes of radioactive nuclei; information science. *Mailing Add:* 120 Dana Dr Oak Ridge TN 37830

GOVER, JAMES E, b Bronston, Ky, Dec 30, 40; m 64; c 2. RADIATION EFFECTS IN MICROELECTRONICS. *Educ:* Univ Ky, BS, 63; Univ NMex, MS, 65, PhD(nuclear eng), 71. *Prof Exp:* Staff mem, 63-73, DIV SUPVR, SANDIA NAT LABS, 73- *Concurrent Pos:* Lectr, Inst Elec & Electronics Engrs, 80, 81, 84, 86 & 87; gen chem, Inst Elec & Electronics Engrs, 83, steering comt, 81-84, Nuclear & Spec Radiation Effects Conf, cong fel, 88 & 91, vchmn, 91; gen chmn, Heart Conf, 83, tech prog chmn, 86, invited speaker 89; US House Reps, 88. *Mem:* Fel Inst Elec & Electronics Engrs; Heart Soc; AAAS. *Res:* Radiation effects and radiation effects phenomena; pulsed power technology; thermomechanical shocks; energy systems technology; electromagnetic pulse phenomena; competitiveness of American technology and industry. *Mailing Add:* Sandia Nat Lab Box 5800 Albuquerque NM 87185

GOVETT, GERALD JAMES, b Barry, South Wales, July 30, 32; m 68. GEOCHEMISTRY. *Educ:* Univ Wales, BSc, 55; Univ London, DIC & PhD(geol), 58; Univ Wales, DSc, 73. *Prof Exp:* Assoc res off & geochemist, Res Coun Alta, 58-65; UN tech expert & vis prof, Univ Philippines, 65-66; vis prof, Univ NB, 66-67; UN consult, Cyprus, 67-68; from assoc prof to prof geol, Univ NB, 68-77; head, Sch Appl Geol, 79-84, PROF GEOL, UNIV NEW SOUTH WALES, 78-, DEAN, FAC APPL SCI, 84- *Concurrent Pos:* Dir, Delta Gold NL, 83- *Mem:* Brit Inst Mining & Metall; Asn Explor Geochemists; Can Inst Mining & Metall; Australian Inst Mining & Metall; Geol Soc Australia. *Res:* Exploration geochemistry, primary and secondary dispersion; geochemistry and genesis of mineral deposits, especially sulfides; availability and supply of world mineral resources. *Mailing Add:* Dept Appl Geol PO Box 1 Univ New South Wales Kensington NSW 2033 Australia

GOVIER, G(EORGE) W(HEELER), b Nanton, Alta, June 15, 17; m 40; c 3. CHEMICAL ENGINEERING. *Educ:* Univ BC, BASc, 39; Univ Alta, MSc, 45; Univ Mich, ScD(chem eng), 49. *Hon Degrees:* LLD, Univ Calgary. *Prof Exp:* Plant operator, Standard Oil Co, BC, 39-40; from lectr to asst prof chem eng, Univ Alta, 40-48; prof chem eng & head dept chem & petrol eng, Univ Alta, 48-59, dean, Fac Eng, 59-63; chmn, Energy Resources Conserv Bd, 62-78; PRES, GOVIER CONSULT SERV LTD, 78- *Concurrent Pos:* Consult, 40-62; mem permanent coun, World Petrol Cong, 60-; part-time prof, Univ Alta & Univ Calgary, 63-; vpres & mem bd dirs, Petrol Recovery Res Inst, 66-; chmn sci prog comt, World Petrol Cong, 75-83; dir, Canadian Foremost Ltd & Texaco, Inc, Can, 79-, Canadian-Mont Gas Co Ltd, Canadian-Mont Pipe Line Co, Roan Resources Ltd, C-E Combustion Eng Ltd & Stoned Webster Ltd, Can, 80-, Bow Valley Resources Ltd, 81-, Co-operative Energy Develop Corp, 82- *Honors & Awards:* R S Jane Mem Award, Chem Inst Can, 66, Selwyn G Blaylock Medal, 71; Anthony F Lucas Gold Medal, Soc Petrol Engrs, Am Inst Mech Engrs, 89. *Mem:* Nat Acad Eng; Am Inst Chem Engrs; fel Chem Inst Can. *Res:* Pipeline flow of complex mixtures, especially non-Newtonians, gas-liquid, liquid-liquid and solid-liquid mixtures. *Mailing Add:* Govier Consult Serv Ltd 1507 Cavanaugh NW Calgary AB T2L 0M8 Can

GOVIER, WILLIAM CHARLES, b Nashville, Tenn, Apr 6, 36; m 59; c 2. RESEARCH ADMINISTRATION. *Educ:* Kalamazoo Col, BA, 57; McGill Univ, MD, 61; Univ Miss, PhD(pharmacol), 65. *Prof Exp:* From instr to asst prof pharmacol, Univ Miss, 63-66; asst prof, Univ Tex Southwestern Med Sch, 66-68; sr surgeon, Exp Therapeut Br, Nat Heart Inst, 68-70; dir res systs develop, Worley & Ringe, Inc, 70-71; mem staff, Clin Pharmacol Div, Ciba-Geigy Corp, 71-73, exec dir biol res, Pharmaceut Div, 73-75; dir, pharmaceut res, Lederle Labs,75-77, dir, med res, 77-80; mem staff, med res, Am Cyanamid Co, 80-; DIR PHARMACEUT R & D, E I DU PONT DE NEMOURS & CO. *Concurrent Pos:* Fel pharmacol, Med Ctr, Univ Miss, 62-63; fel, Oxford Univ, 65-66. *Mem:* AAAS; NY Acad Sci; Am Soc Pharmacol & Exp Therapeut; Soc Exp Biol & Med; Am Soc Clin Pharmacol & Therapeut. *Res:* Cardiovascular; autonomic; peptides; tissue regeneration. *Mailing Add:* 16575 Parklane Dr Los Angeles CA 90049

GOVIER, WILLIAM MILLER, pharmacology; deceased, see previous edition for last biography

GOVIL, NARENDRA KUMAR, b Aligarh, UP India, Jan 05, 40; m 64; c 2. COMPLEX ANALYSIS, APPROXIMATION THEORY. *Educ:* Agra Univ, India, BSc, 57; Aligarh Univ, India, MSc, 59; Univ Montreal, PhD(math), 68. *Prof Exp:* Lect math, 67-68, asst prof math, Concordia Univ Montreal, 68-70; from asst prof to prof math, Indian Inst Technol, New Delhi, India, 70-85; assoc prof math, 85-86, PROF MATH, AUBURN UNIV, 86- *Concurrent Pos:* fel, Univ Montreal, Can, 72-73; vis scientist, Dalhousie Univ, Halifax, Can, 80; vis prof, Univ Alta, Edmonton, Can, 81 & Auburn Univ, 83-85. *Mem:* Fel Nat Acad Sci; Indian Math Soc; Am Math Soc. *Res:* Published papers in extremal problems for polynomials and related entire functions of exponential type; location of the zeros of a polynomial; functions of a complex variable, Fourier series. *Mailing Add:* Dept Algebra Combinatorics & Analysis Auburn Univ Auburn AL 36849-5307

GOVIL, SANJAY, b New Delhi, India, Oct 4, 62; m 89. GEOTECHNICAL ENGINEERING, STRUCTURAL ENGINEERING. *Educ:* Indian Inst Technol, Kanpur, BTech, 85; Asian Inst Technol, Bangkok Thailand, MEng, 87. *Prof Exp:* Res assoc, Asian Inst Technol, 87; teaching assoc eng graphics & statist, 87-89, res assoc & computer consult, 89, RES ASSOC SOIL MECH, ARIZ STATE UNIV, 89- *Mem:* Am Soc Civil Engrs. *Res:* Soil dynamics; earthquake engineering; soil structure interaction; structural dynamics; shear strength of soils and seismic slope stability including state-of-art lab testing of soils; numerical modelling; software development. *Mailing Add:* Dept Civil Eng Ariz State Univ Tempe AZ 85287-5306

GOVIN, CHARLES THOMAS, JR, b Evanston, Ill, May 6, 46; m 68; c 4. GEOLOGY, HYDROLOGY. *Educ:* Univ Wis, Madison, BS, 68; Univ Tex, Austin, MA, 73. *Prof Exp:* Asst geologist, Dames & Moore, 73-74, staff geologist, 74-75, proj geologist, 75-77; sr geologist, Wis Elec & Power Co, 77-89; DIR ADMIN SERV DEPT, WIS NATURAL GAS, 89- *Concurrent Pos:* Rep utility solid waste act group, Edison Elec Inst, 77-78. *Mem:* Am Water Resources Asn; Geol Soc Am. *Res:* Ground water flow in the surficial saturated zone as relates to land disposal of industrial waste; behavior of industrial waste materials in land disposal situations; methodology for siting of industrial waste disposal facilities. *Mailing Add:* 601 Cheyenne Dr Waukesha WI 53186

GOVIND, CHOONILAL KESHAV, b Durban, SAfrica, Oct 14, 38; Can citizen; m 67; c 2. NEUROBIOLOGY DEVELOPMENT. *Educ:* Rhodes Univ, BSc, 61; Natal Univ, MSc, 65; Univ Man, PhD(zool), 71. *Prof Exp:* Asst prof biol, Med Sch, Natal Univ, 63-67; from asst prof to assoc prof, 72-81, PROF ZOOL, SCARBOROUGH COL, UNIV TORONTO, 82- *Concurrent Pos:* Vis res assoc, Marine Prog, Boston Univ, 73-78, adj assoc prof, 79; mem & chmn, grant selection comt, NSERC, 83-85. *Mem:* Soc Neurosci; Am Soc Zoologists; Can Soc Zoologists. *Res:* Electrophysiological and ultrastructural analysis of the mechanisms governing the development, growth and aging of specific neurons, muscle fibers and neuromuscular synapses in the lobster where they are easily identified and manipulated. *Mailing Add:* Scarborough Col 1265 Military Trail Scarborough ON M1C 1A4 Can

GOVINDARAJULU, ZAKKULA, b Atmakur, India, May 15, 33; m 61; c 3. MATHEMATICS, STATISTICS. *Educ:* Univ Madras, BA, 52, MA, 53; Univ Minn, PhD(math, statist), 61. *Prof Exp:* Lectr asst math, Madras Christian Col, Univ Madras, 52-54, res asst statist, Univ, 53-56; asst biostatist, Univ Minn, 56-57, asst math, 57-60, instr, 60-61; from asst prof to assoc prof, Case Inst Technol, 61-68; PROF STATIST, UNIV KY, 68- *Concurrent Pos:* Lectr asst math & statist, Govt Arts Col, Madras, 54-56; NSF grants, 62-64, 68-69; US Air Force grant & vis assoc prof, Univ Calif, Berkeley, 64-65; USPHS Nat Ctr Health Statist res grant, 70-71; vis prof, Univ Mich, Ann Arbor, 72; res contracts, Off Naval Res, 73-75, 80-81, Water Resources Inst, 74-76; Univ Ky Found res award, 75; NSF grant, Northwestern Univ, 76; vis prof, Indian Statist Inst, 76; Ger Res Asn grant award, 77. Brit fel Royal Statist Soc; Bernoulli Soc Math Statist & Off Naval Res, 83-84; vis prof, Stanford Univ, 84; distinguished summer fac fel, Naval Auxiliary Aircraft Carrier Develop Ctr, Warminster, Pa, 88, 89, 90; vis prof, Columbia Univ, 88-89; res contract, Naval Air Develop Ctr, 89-90. *Mem:* Fel AAAS; fel Inst Math Statist; Int Statist Inst; fel Am Statist Asn; Nat Acad Sci; Allahabad Math Soc; Indian Soc Probability & Statist. *Res:* Large sample theory; nonparametric inference; sequential procedures. *Mailing Add:* Dept Statist Univ Ky Lexington KY 40506

GOVINDJEE, M, b Allahabad, India, Oct 24, 33; nat US; m 57; c 2. BIOPHYSICS, PLANT PHYSIOLOGY. *Educ:* Univ Allahabad, BSc, 52, Univ Ill, PhD(biophys), 60. *Prof Exp:* Lectr bot, Univ Allahabad, 54-56; fel physico-chem biol, 56-59, res asst bot, 59-60, USPHS fel biophys, 60-61, asst prof bot, 61-65, assoc prof bot & biophys, 65-69, assoc head dept, 73-74, actg head dept bot, 74-75, chmn biophys div, 77, PROF BOT & BIOPHYS, UNIV ILL, URBANA, 69- *Concurrent Pos:* distinguished lectr, Sch Life & Sci, Univ Ill, Urbana, 78. *Mem:* Fel AAAS; Am Soc Plant Physiol; Biophys Soc; Am Soc Photobiol (pres, 81); fel Nat Acad Sci India; Sigma Xi. *Res:* Photosynthetic mechanisms of green plants and the relation of chlorophyll fluorescence to photosynthesis; mechanisms of excitation energy transfer and electron transfer in photosynthesis, particularly, the role of anions (bicarbonate and chloride) in the oxygen evolving system; use of chlorophyll and fluorescence in monitoring the effect of stress during photosynthesis. *Mailing Add:* Dept Plant Biol 289 Morrill Hall Univ Ill 505 S Goodwin Ave Urbana IL 61801-3793

GOW, K V, b Ottawa, Ont, Aug 29, 19; m 61; c 3. PHYSICAL METALLURGY, CERAMICS. *Educ:* Univ Toronto, BASc, 43, PhD(phys metall), 51; Rutgers Univ, MSc, 49; Univ Birmingham, MSc, 53. *Prof Exp:* Sr sci officer phys metall, Can Dept Energy & Mines, 55-62; from assoc prof to prof, 64-84, EMER PROF PHYS METALL, TECH UNIV NOVA SCOTIA, 85- *Concurrent Pos:* Consult, metall failure Anal. *Mem:* Brit Ceramic Soc; Can Inst Mining & Metall. *Res:* Fracture mechanics, transformations in alloy steels. *Mailing Add:* Dept Metall Tech Univ NS Halifax NS B3H 3J5 Can

GOW, WILLIAM ALEXANDER, b Toronto, Ont, Nov 11, 20; m 45; c 2. METALLURGICAL ENGINEERING. *Educ:* Univ Toronto, BASc, 43. *Prof Exp:* Res engr, Radioactivity Div, Can Ctr Mineral & Energy Technol, Dept Energy, Mines & Resources, 46-53, sect head, Hydrometall Sect, Extraction Metall Div, 53-70, asst chief, 70-74, dir, Mineral Sci Labs, 74-85; RETIRED. *Concurrent Pos:* Nipissing fel, 46. *Honors & Awards:* H T Airey Award, Can Inst Mining & Metall, 80. *Mem:* Can Inst Mining & Metall. *Res:* Hydrometallurgical treatment of ores and concentrates; development of uranium ore treatment processes; application of leaching, ion exchange and solvent extraction methods to various metallic ores; planning and managing mineral processing and mineral utilization research and development. *Mailing Add:* 30 Cramer Dr Nepean ON K2H 5X5 Can

GOWANS, CHARLES SHIELDS, b Salt Lake City, Utah, Sept 17, 23; m 50; c 3. GENETICS, PHYCOLOGY. *Educ:* Univ Utah, AB, 49; Stanford Univ, PhD(biol), 57; Indiana Univ, NIH fel, 56-57. *Prof Exp:* Asst biol, Stanford Univ, 52-55; from asst prof to prof, 68-90, EMER PROF BOT, UNIV MO-COLUMBIA, 90- *Concurrent Pos:* Vis prof, Dept Bot, Nat Taiwan Univ, 65-66; spec chair, Inst Bot, Acad Sinica, Nankang, Taiwan, 65-66; Fulbright-Hays Res fel, Inst Microbiol, Göttingen Univ, 72-73. *Honors & Awards:* Lalor Found Award, Harvard Univ, 58. *Mem:* Genetics Soc Am; Phycol Soc Am; Am Soc Microbiol. *Res:* Genetics of algae, specifically biochemical genetics of Chlamydomonas. *Mailing Add:* 701 Redbud Lane Columbia MO 65203

GOWANS, JAMES L, b Sheffield, Eng, May 7, 24; m 56; c 3. CELLULAR IMMUNOLOGY, LYMPHOCYTES. *Educ:* London Univ, MB & BS, 47; Oxford Univ, BA, 48, DPhil(path), 53. *Hon Degrees:* ScD, Yale Univ, 66; DSc, Univ Chicago, 71, Univ Birmingham, UK, 78, Univ Rochester, 87; MD, Univ Edinburgh, 79; DM, Univ Southampton, UK, 87; LLD, Univ Glasgow, 88. *Prof Exp:* Dir, Cellular Immunol Univ, Med Res Coun, Sch Path, Oxford Univ, 63-77, secy & chief exec, 77-87; SECY-GEN, HUMAN FRONTIER SCI PROG, STRASBOURG, FRANCE, 89- *Concurrent Pos:* Res prof, Royal Soc, 62-77; Mem, Gov Coun, Int Agency Res Cancer, 80-87, Res Progs Adv Comt, Nat Mult Sclerosis Soc, NY, 88-90 & Awards Comt, Gen Motors Cancer Res Found, 88-; chmn, Europ Med Res Coun, 85-87; consult, WHO Global Prog AIDS, Geneva, 87-88. *Honors & Awards:* Gairdner Found Int Award, Can, 68; Paul Ehrlich Prize, Ger, 74; Feldberg Found Award, 79; Wolf Prize in Med, Israel, 80; Medansar Prize, Int Transplantation Soc, 90. *Mem:* Foreign assoc Nat Acad Sci; hon mem Am Asn Immunologists; hon mem Am Asn Anatomists. *Res:* Recirculation of small lymphocytes from blood to lymph by way of the lymph nodes and their interaction with antigens to initiate immune response. *Mailing Add:* 75 Cumnor Hill Oxford 0X2 9HX England

GOWDA, NETKAL M MADE, b Mysore, India, Apr 10, 47; c 3. PHYSICAL CHEMISTRY, ANALYTICAL & ENVIRONMENTAL CHEMISTRY. *Educ:* Univ Mysore, India, BSc, 69, MSc, 71, PhD(phys chem), 78. *Prof Exp:* Lectr chem, Yuvaraja's Col, India, 71-72; lectr & res chemist, Univ Mysore, India, 72-79; James W McLaughlin fel, Div Environ Toxicol, Dept Prev Med & Community Health, Univ Tex Med Br, 80-82, res assoc, Div Biochem, Dept Human Biol Chem & Genetics, 82-86; asst prof chem, Purdue Univ Sch Sci, Ind Univ-Purdue Univ, Indianapolis, 86-89; mem fac, Dept Sci, Univ VI; ASST PROF CHEM, WEST ILL UNIV, 89- *Concurrent Pos:* Robert A Welch fel,Univ Tex Med Br, 79; reserve instr chem, Galveston Col, Galveston, Tex, 81-84; res asst prof, Dept Chem, Old Dominion Univ, Norfolk, VA, 84. *Mem:* Am Chem Soc; Sigma Xi; fel Am Inst Chem; Int Union Physicists & Chemists. *Res:* Analysis and reactions of sterol hydroperoxides; mutagenicity and microbiological transformations of sterols; spectrophotometric determinations of transition and platinum group metals using phenothiazines; analytical applications of chloramine-T and other halomines; chemical kinetics of redox reactions; chloramine-T as alternative disinfectant to chlorine in drinking water treatment; analysis of THMs and other organics in chlorinated water samples; chemical, biological and analytical abstracts. *Mailing Add:* Dept Chem 214 Currens Hall West Ill Univ Macomb IL 61455

GOWDEY, CHARLES WILLIS, b St Thomas, Ont, Sept 3, 20; m 46; c 4. PHARMACOLOGY. *Educ:* Univ Western Ont, BA, 44, MSc, 46; Oxford Univ, DPhil(pharmacol), 48. *Prof Exp:* Demonstr pharmacol, Oxford Univ, 46-48; lectr, 48-49, from asst prof to assoc prof, 50-60, prof & head dept, 60-81, EMER PROF PHARMACOL, MED SCH, UNIV WESTERN ONT, 81- *Mem:* Am Soc Pharmacol & Exp Therapeut; Can Pharmacol Soc; Brit Pharmacol Soc. *Res:* Decompression sickness; narcotic dependence; cardiovascular pharmacology; adverse drug reactions. *Mailing Add:* Health Sci Ctr Univ Western Ont London ON N6A 5C1 Can

GOWDY, JOHN NORMAN, b Elk City, Okla, Feb 7, 45; m 70; c 3. DIGITAL SIGNAL PROCESSING, SPEECH SIGNAL PROCESSING. *Educ:* Mass Inst Technol, BS, 67; Univ Mo-Columbia, MS, 68, PhD(elec eng), 70. *Prof Exp:* PROF ELEC ENG, CLEMSON UNIV, 71- *Concurrent Pos:* Interim dept head, Elec Eng, Computer Sci Dept, Clemson Univ, 88-90. *Mem:* Inst Elec & Electronics Engrs; Inst Elec & Electronics Engrs Signal Processing Soc; Inst Elec & Electronics Engrs Computer Soc; Inst Elec & Electronics Engrs Commun Soc. *Res:* Digital signal processing; speech recognition; speaker recognition; speech synthesis. *Mailing Add:* Elec Eng & Computer Sci Dept Clemson Univ Clemson SC 29634

GOWDY, KENNETH KING, b Memphis, Tenn, June 25, 32; m 52; c 4. MECHANICAL ENGINEERING. *Educ:* Kans State Univ, BS, 55, MS, 61; Okla State Univ, PhD(mech eng), 65. *Prof Exp:* Engr trainee, Continental Pipeline, 55; instr & asst to dean, Kans State Univ, 57-62, asst prof mech eng, 62-70, assoc prof mech eng, 70-79, asst dean eng, 65-75, head Dept Eng Technol, 75-79; prof & head Dept Eng Technol, Tex A&M Univ, Col Station, Tex, 79-84; PROF & ASSOC DEAN ENG, KANS STATE UNIV, MANHATTAN, 84- *Honors & Awards:* Outstanding Serv Award, Am Soc Eng Educ. *Mem:* Am Soc Eng Educ; Am Soc Mech Engrs; Sigma Xi. *Res:* Automatic controls; systems analysis. *Mailing Add:* Col Eng Kans State Univ Manhattan KS 66506

GOWDY, ROBERT HENRY, b Putnam, Conn, Apr 7, 41; m 66; c 3. GENERAL RELATIVITY, COSMOLOGY. *Educ:* Worcester Polytech Inst, BS, 63; Yale Univ, MS, 64, PhD(physics), 68. *Prof Exp:* AEC fel physics res, Yale Univ, 68-70; sr fel, Dept Physics & Astron, Univ Md, 70-72, asst prof, 72-78; asst prof, 78-82, ASSOC PROF PHYSICS, VA COMMONWEALTH UNIV, 82-, CHMN DEPT, 82- *Mem:* Am Inst Physics; Am Phys Soc; Am Astron Soc; AAAS; Sigma Xi. *Res:* General relativity and cosmology with a particular emphasis on the application of geometrical methods. *Mailing Add:* Three Clarke Rd Richmond VA 23226

GOWDY, SPENSER O, b Philadelphia, Pa, July 6, 41. MATHEMATICS. *Educ:* West Chester State Col, BS, 63; Villanova Univ, MA, 65; Temple Univ, PhD(math), 71. *Prof Exp:* Asst math, Villanova Univ, 63-65; from instr to asst prof, 65-74, ASSOC PROF MATH, ST JOSEPH'S COL, PA, 74- *Mem:* Math Asn Am. *Res:* Group theory; small cancellation groups. *Mailing Add:* 216 E Manoa Rd Havertown PA 19083

GOWE, ROBB SHELTON, b St Boniface, Man, Oct 9, 21; c 3. POULTRY BREEDING, ANIMAL BREEDING. *Educ:* Univ Toronto, BSA, 45; Cornell Univ, MS, 47, PhD(genetics & physiol), 49. *Prof Exp:* Asst genetics, Dept Poultry, Cornell Univ, 45-49; head poultry breeding, Poultry Div, Exp Farms Serv, 49-59, chief genetics sect, 59-65, dir Animal Res Ctr, Can Dept Agr, 65-86; dir res, Shaver Poultry Breeding Farm Ltd, 87-90; ADJ PROF, UNIV GUELPH, 90- *Concurrent Pos:* Chmn Ottawa Regional Hosp Planning Bd, 74, Queensway Carleton Hosp Bd, 78- *Honors & Awards:* Tom Newman Mem Int Award; Sir John Hammond Mem Lect Award, Sir John Hammond Mem Fund, 74; Merit Award, Can Soc Animal Sci, 84. *Mem:* Fel Poultry Sci Asn; Am Genetic Asn; Genetics Soc Am; World Poultry Sci Asn; Genetics Soc Can; Agr Inst Can; Can Soc Animal Sci. *Res:* Poultry genetics. *Mailing Add:* 14 Fernbank Pl Guelph ON N1C 1A7 Can

GOWEN, RICHARD J, b New Brunswick, NJ, July 6, 35; m 55; c 5. ELECTRICAL & BIOMEDICAL ENGINEERING. *Educ:* Rutgers Univ, BS, 57; Iowa State Univ, MS, 61, PhD(elec eng), 62. *Prof Exp:* USAF, 57-77, ground electronics officer, 57-59, asst prof elec eng, USAF Acad, 62-63, res assoc bioeng, 63-64, from asst prof to prof elec eng, 64-77, dep head dept, 69-77; pres Dak State Univ, 77-84; vpres & dean eng, 77-84, prof elec eng, 80-84, PRES, SDAK SCH MINES & TECHNOL, 87- *Concurrent Pos:* Consult numerous cos, 63-; dir, Joint Air Force-NASA Med Instrumentation Lab, 65-77, Mgt Eval, Dept Defense Mil Health Care Syst Spec Study Group, Off Asst Secy Defense, 75-76; mem numerous comts, Inst Elec & Electronics Engrs, 69-; dir & prin investr, SDak Sch Mines & Technol Res Proj Support Space Cardiovasc Studies, 77-80; prin investr, Effects Electromagnetic Energy Biol Tissue, SDak Sch Mines & Technol, 78-; co-chmn, Joint Nuclear Regulatory Comt & Inst Elec & Electronics Conf Identify Applns Adv Electrotechnol Nuclear Power Plants, 79-80, Joint Indust Probabilistic Risk Assessments Guidelines Nuclear Power Plants Proj, 80-83; co-chmn, Nat Eng Leadership Forum, Nat Acad Eng, 88; pres, Tech Assistance Prog, 88- *Honors & Awards:* Centennial Medal, Inst Elec & Electronics Engrs, 84. *Mem:* Fel Inst Elec & Electronics Engrs; Nat Soc Prof Engrs; Inst Elec & Electronics Engrs Eng Med & Biol Soc (vpres, 74, pres-elect, 75, pres, 76); Am Asn Eng Soc (secy-treas, 86). *Res:* Physiological monitoring; biological simulation; computer design; semiconductor circuits; health care systems; author of numerous publications. *Mailing Add:* 1609 Palo Verde Rapid City SD 57701

GOWGIEL, JOSEPH MICHAEL, b Summit, Ill, May 19, 26; c 1. DENTISTRY, ANATOMY. *Educ:* Loyola Univ Chicago, DDS, 50; Univ Chicago, PhD, 58. *Prof Exp:* Intern, Zoller Dent Clin, Univ Chicago, 50-51, from instr to asst prof, 58-63; ASSOC PROF ANAT, SCH DENT, LOYOLA UNIV CHICAGO, 64-, CHMN DEPT, 69- *Mem:* Fel AAAS; Am Asn Anat; Int Asn Dent Res; Sigma Xi; Am Asn Dent Sch. *Res:* Normal histophysiology of the oral tissues, especially of tooth development and the adult periodontal ligament. *Mailing Add:* Loyola Univ 2160 S First Ave Maywood IL 60153

GOY, ROBERT WILLIAM, b Detroit, Mich, Jan 25, 24; m 48; c 3. PSYCHOPHYSIOLOGY. *Educ:* Univ Mich, BS, 47; Univ Chicago, PhD, 53. *Prof Exp:* USPHS fel anat & psychol, Univ Kans, 54-56, from instr to assoc prof anat, 56-63; assoc scientist, Ore Regional Primate Res Ctr, 63-65, chmn dept reproductive physiol & behav, 65-71, sr scientist, 67-71; DIR, WIS

REGIONAL PRIMATE RES CTR, 71- *Concurrent Pos:* Vis scientist, Wis Regional Primate Res Ctr, 61-63; from assoc prof to prof, Med Sch, Univ Ore, 65-71; prof psychol, Univ Wis-Madison, 71- *Mem:* Am Asn Anatomists; Endocrine Soc; Animal Behav Soc; Brit Soc Study Fertil; Int Soc Psychoneuroendocrinol. *Res:* Comparative endocrinology of reproduction and reproductive behavior. *Mailing Add:* 1845 Summit Ave Madison WI 53705

GOYAL, MEGH R, b Sangrur, India, Aug 1, 49; m 70; c 3. IRRIGATION, DRIP-TRICKLE IRRIGATION. *Educ:* Punjab Agr Univ, India, BSc, 71; Ohio State Univ, MSc, 77, PhD(agr eng), 79. *Prof Exp:* Lectr irrig, Haryana Agr Univ, India, 72-75; res asst, soil & water, Ohio State Univ, 76-79; asst agr eng, 79-83, assoc agr engr irrig, 83-88, PROF UNIV PR, MAYAGUEZ, 88- *Concurrent Pos:* Consult, Col Engrs & Survrs, PR, 82-; mem, US Comt Irrig & Drainage, Pan Am Fedn Eng Socs (UPADI), 84-, Comt Agr Eng, 84- *Honors & Awards:* Blue Ribbon Award, Am Soc Agr Engrs, 83 & 86, Eng Achievement Award, 87. *Mem:* Am Soc Agr Engrs; Int Soc Soil Sci; Am Soc Agron; Coun Agr Sci & Technol. *Res:* Drip irrigation in vegetables; small farm mechanization; soil crusting. *Mailing Add:* Box 5984 Mayaguez PR 00709-5984

GOYAL, RAJ K, DISEASES OF THE ESOPHAGUS. *Educ:* Maulana Azad Med Col, New Delhi, India, MD, 65. *Prof Exp:* CHIEF GASTROENTEROL, BETH ISRAEL HOSP, BOSTON, 81- *Concurrent Pos:* Ed-in-chief, Gastroenterol. *Mailing Add:* Beth Israel Hosp 330 Brookline Ave Boston MA 02215

GOYAL, SURESH, b India, Dec 6, 60; m. MATHEMATICAL MODELING & APPLIED MECHANICS, SOFTWARE DEVELOPMENT. *Educ:* Indian Inst Technol, Kharagpur, B Tech, 82; Univ Iowa, Iowa City, MS, 84; Cornell Univ, Ithaca, NY, PhD(mech eng), 89. *Prof Exp:* Trainee engr, Int Tractor, Bombay, 81; res & training asst mech, Univ Ky, Lexington, 82-83; res asst bio-mech, Univ Iowa, Iowa City, 83-84; res & teaching asst mech & mech eng, Cornell Univ, Ithaca, NY, 84-88, postdoctoral res assoc computer sci, 88-89; POSTDOCTORAL MEM TECH STAFF, INTERACTIVE SYSTS RES, AT&T BELL LABS, MURRAY HILL, 89- *Mem:* Inst Elec & Electronics Engrs. *Res:* Software development for simulation of multiple body dynamics with friction for robotics, automation and computer reality applications; characterization and modeling of friction; mathematical modeling, analysis and geometry in applied mechanics; algorithms and complexity theory; control theory; experimental spinal biomechanics. *Mailing Add:* AT&T Bell Labs Rm 2D-404 600 Mountain Ave Murray Hill NJ 07974-0636

GOYAN, JERE EDWIN, b Oakland, Calif, Aug 3, 30; div; c 3. PHARMACEUTICAL CHEMISTRY. *Educ:* Univ Calif, San Francisco, BS, 52; Univ Calif, Berkeley, PhD(pharmaceut chem), 57. *Hon Degrees:* DSc, Albany Col Pharm, Union Univ, 79, Philadelphia Col Pharm & Sci, 80; LLD, Mass Col Pharm & Allied Health Prof, 80. *Prof Exp:* From asst prof to assoc prof pharm, Univ Mich, 56-63; assoc prof pharm, 63-65, chmn dept, 65-67, assoc dean sch pharm, 66-67, PROF PHARM, PHARMACEUT CHEM, UNIV CALIF, SAN FRANCISCO, 65-, DEAN SCH PHARM, 67- *Concurrent Pos:* Mem, Div Physician Manpower, Bur Health Manpower, Dept Health, Educ & Welfare, 67-69, mem pharm rev comt, Bur Health Prof Educ & Manpower Training, 69-71; Am Pharmaceut Asn rep on US Adopted Names Rev Bd, 70-79; consult Alaska Health Manpower Corp, Univ Alaska, 74-75; commnr, Food & Drug Admin, Dept Health, Educ & Welfare, Pub Health Serv, 79-81; mem Pew Health Professions Comn, 90. *Honors & Awards:* T Edward Hicks lectr, State Univ NY, 72; Hugo H Schaefer Med, Am Pharmaceut Asn, 80; Herman Lubin lectr, Univ Tenn, 81; Kremer's Mem lectr, Univ Wis, 81; Samuel W Melendy mem lectr, Univ Minn, 82; Albertson's mem lectr, Idaho State Univ, 88. *Mem:* Inst Med-Nat Acad Sci; NY Acad Sci; fel AAAS; fel Acad Pharmaceut Sci; Am Asn Col Pharm (pres, 78-79); Am Asn Pharmaceut Soc. *Res:* Kinetics of drug degradation, properties of drugs in solution. *Mailing Add:* Sch Pharm Univ Calif San Francisco CA 94143-0446

GOYER, GUY GASTON, meteorology, for more information see previous edition

GOYER, ROBERT ANDREW, b Hartford, Conn, June 2, 27; m 55; c 4. PATHOLOGY. *Educ:* Col of Holy Cross, BS, 50; St Louis Univ, MD, 55; Am Bd Path, dipl, 64. *Prof Exp:* Instr path, St Louis Univ, 60-62, asst prof, 62-65; from asst prof to prof path, Sch Med, Univ NC, Chapel Hill, 71-74; dep dir, Nat Inst Environ Health Sci, 79-87; PROF PATH & CHMN DEPT, HEALTH SCI CTR, UNIV WESTERN ONT, 74-79, 87- *Concurrent Pos:* Nat Found fel, Sch Med, St Louis Univ, 59-60; res fel, Metab Unit, Univ Col Hosp, Med Sch, Univ London, 61-62; clin pathologist, Cardinal Glennon Mem Hosp Children, St Louis, 62-63, dir labs, 63-65; chmn, World Health Orgn Task Group on Environ Health Criteria for Lead, 75, Task Group on Health Criteria for Cadmium, 84 & 89; mem, Int Regist Potentially Toxic Chemicals, UN Environ Prog Expert Group Meeting, Geneva, 86 & 91, Metals Subcomt Sci Adv Bd, US Environ Protection Agency, 85-, Res Grants Comt, Med Res Coun Can, 87-; pres, Metals Specialty Sect, Soc Toxicol, 90-91. *Mem:* fel Col Am Path; Soc Exp Biol & Med; Soc Toxicol. *Res:* Environmental pathology; metal toxicology; nephrotoxicity; co-author of one document and author of one report. *Mailing Add:* Dept Path Univ Western Ontario Health Scis Ctr London ON N6A 5C1 Can

GOYER, ROBERT G, b Montreal, Que, May 13, 38; m 60; c 2. PHARMACOLOGY. *Educ:* Univ Montreal, BA, 58, BSc, 62; Univ Sorbonne, PhD(pharmacol), 65. *Prof Exp:* PROF FAC PHARM, UNIV MONTREAL, 65- *Concurrent Pos:* Grants, Med Res Coun Can, Med Res Coun Que & Can Found Advan Pharm, 66-; consult, Desbergers & Nadeau Pharmaceut Labs, sci dir, 71-77; pres, Clinipharm Inc, 77-87; vpres, Can Med Rev Bd, 87- *Mem:* AAAS; NY Acad Sci; Fr-Can Asn Advan Sci; Fr Soc Therapeut & Pharmacodyn; Pharm Soc Can. *Res:* Clinical evaluation of new drugs. *Mailing Add:* Fac Pharm Univ Montreal PO Box 6128 Sta A Montreal PQ H3C 3J7 Can

GOYERT, SANNA MATHER, MOLECULAR BIOLOGY PROTEIN STRUCTURE. *Educ:* NY Univ, PhD, 83. *Prof Exp:* SR SCIENTIST, DEPT RHEUMATOL, DIV CELLULAR & MOLECULAR BIOL, HOSP FOR JOINT DIS, 85- *Mailing Add:* Dept Med Northshore Univ Hosp Cornell Univ Med Col 350 Community Dr Manhasset NY 11030

GOYINGS, LLOYD SAMUEL, b White Cloud, Mich, Dec 26, 33; m 54; c 5. PATHOLOGY, BIOCHEMISTRY. *Educ:* Mich State Univ, BS, 58, DVM, 60, MS, 61, PhD(path), 65. *Prof Exp:* Instr path, Mich State Univ, 61-65, asst prof path, surg & med, 65-66; res assoc,Upjohn Co, 66-70, res head plant & animal path , 70-; RETIRED. *Concurrent Pos:* Consult. *Mem:* Am Vet Med Asn; Conf Res Workers Animal Dis; Am Asn Lab Animal Sci. *Res:* Canine dermatology; atypical mycobacteniosis in cattle, dogs, cats and chickens; canine hypothyroidism; causation of canine leukemia; canine cancer registry. *Mailing Add:* 9491 W Milo Rd Plainwell MI 49080

GOZ, BARRY, b Brooklyn, NY, Oct 21, 37; m 61; c 2. PHARMACOLOGY, BIOCHEMISTRY. *Educ:* Columbia Univ, BA, 58; State Univ NY, PhD(pharmacol), 65. *Prof Exp:* Asst prof pharmacol, Sch Med, Yale Univ, 67-74; ASSOC PROF PHARMACOL, SCH MED, UNIV NC, CHAPEL HILL, 74- *Concurrent Pos:* USPHS fel pharmacol, Yale Univ, 65-67. *Mem:* Am Soc Pharmacol & Exp Therapeut; Am Soc Biol Chem; Am Asn Cancer Res. *Res:* Cellular and viral replication and control mechanisms; pyrimidine analogues and nucleic acid metabolism; cancer chemotherapy; antiviral agents. *Mailing Add:* Dept Pharmacol 1122 Fac Lab Off Bldg CB 7365 Sch Med Univ NC Chapel Hill NC 27514

GOZZO, JAMES J, b Hartford, Conn, May 30, 43; m; c 3. MEDICAL SCIENCE. *Educ:* Univ Conn, BS, 65; Boston Col, PhD, 69. *Prof Exp:* Teaching asst, Boston Col, 65-66, teaching fel, 66-67, instr, 67-68, NSF fel, 67-69; lectr, Northeastern Univ, 70-73, from asst prof to assoc prof health sci, 72-80, dir, Med Lab Sci Grad Prog, 76-88, assoc dean, 85-88, actg dean, 88-89, ELEANOR W BLACK PROF HEALTH SCI, NORTHEASTERN UNIV, 80-, DEAN, COL PHARM & ALLIED HEALTH PROFESSIONS, 89- *Concurrent Pos:* Sr res assoc, Cancer Res Inst, New Eng Deaconess Hosp, 75-82, vis asst lab med, 75-80, assoc mem, 83- *Mem:* Transplantation Soc; Am Asn Immunologists; Am Soc Microbiol; Am Diabetes Asn; AAAS; Fedn Am Soc Exp Biologists. *Res:* Immunodiagnostic assay of bladder carcinoma. *Mailing Add:* Dept Med Lab Sci Col Pharm & Allied Health Prof Northeastern Univ 360 Huntington Ave Boston MA 02115

GRAAE, JOHAN E A, b Copenhagen, Denmark, July 1, 09; US citizen; m 41; c 3. MECHANICAL & CHEMICAL ENGINEERING. *Educ:* Tech Univ Denmark, MSc, 35. *Prof Exp:* Mech engr, Burmeister & Wain, Denmark, 35-38; field engr, Vacuum Oil Co, Denmark, 38-40; develop engr, F L Smith & Co, NY, 40-43; proj engr, Lummus Co, 43-56; sr mech engr, 56-75, CONSULT, ARGONNE NAT LAB, 75- *Mem:* Am Inst Chem Engrs; Am Nuclear Soc; Sigma Xi. *Res:* Design and development of processes and equipment for production of fissionable material; equipment and facilities for remote handling, reprocessing and refabrication of nuclear, radioactive fuel; development of electric batteries and cars. *Mailing Add:* 315 N Myrtle Ave Elmhurst IL 60126

GRAB, EUGENE GRANVILLE, JR, b Nashville, Tenn, Apr 17, 14; m 45; c 2. FOOD SCIENCE. *Educ:* Univ Pa, BS, 34, MS, 35. *Prof Exp:* Chemist, Adams Apple Prod Corp, Pa, 35-37, prod mgr, 37-41; assoc mkt specialist, USDA, Ill, 41-45; food processing specialist, Tenn Valley Auth, 45-48; in charge custromer res, South Dist, Diamond Int, Inc, 48-55, from assoc dir res to dir res, Heekin Can Div, 55-78, vpres, 64-78; RETIRED. *Concurrent Pos:* Mem Nat Canners Asn-Can Mfrs Inst, Comt Lead in Canned Foods & chmn, Processing Sub-comt, 74-; consult food & container indust, 78- *Mem:* Am Chem Soc; Am Soc Brewing Chem; Master Brewers Asn Am; Inst Food Technol; Am Inst Chem. *Res:* Development of new or improvement of existing food products; determination of cause of food spoilage in tin, tin free steel and aluminum containers; evaluation quality of foodstuffs; calculation of heat processes; development of new lacquers, fluxes and plate types in the manufacture of tin cans; development of drawn and ironed cans for beer and beverage, fruits, vegetables and aerosols. *Mailing Add:* 6251 Berkinshaw Dr Cincinnati OH 45230-3663

GRABEL, ARVIN, b New York, NY, Mar 8, 35; m 69. ELECTRONICS. *Educ:* NY Univ, BEE, 56, MEE, 57, ScD(elec eng), 64. *Prof Exp:* Instr elec eng, NY Univ, 57-63; from asst prof to assoc prof, 64-78, PROF ELEC ENG, NORTHEASTERN UNIV, 78- *Concurrent Pos:* Ed, Circuit Theory Newslett, Inst Elec & Electronics Engrs, 59-61. *Mem:* AAAS; Inst Elec & Electronics Engrs; Am Soc Eng Educ; Sigma Xi. *Res:* Solid state circuits; active network synthesis. *Mailing Add:* Dept Elec Eng Northeastern Univ 360 Huntingten Ave Boston MA 02115

GRABEN, HENRY WILLINGHAM, b Talladega, Ala, Nov 9, 34; m 61; c 2. THEORETICAL PHYSICS. *Educ:* Birmingham-Southern Col, BS, 57; Univ Tenn, MS, 61, PhD(physics), 62. *Prof Exp:* Physicist plasma physics, Oak Ridge Nat Lab, 62-63; from asst prof to assoc prof, 63-74, PROF PHYSICS, CLEMSON UNIV, 74- *Mem:* Am Phys Soc; Am Asn Physics Teachers. *Res:* Theoretical study of intermolecular forces; statistical mechanics. *Mailing Add:* Dept Physics Clemson Univ Clemson SC 29631

GRABENSTETTER, JAMES EMMETT, b Cincinnati, Ohio, Sept 18, 46; Can citizen. OIL RESERVOIR SIMULATION. *Educ:* Reed Col, BA, 68; McGill Univ, PhD(chem), 74; Univ Calgary, MSc, 88. *Prof Exp:* Postdoctoral fel quantum chem, Univ NB, 73-75; res assoc theoret chem, Univ Waterloo, 75-79; res chemist corrosion chem, 79-82, atmospheric chem, Ont Hydro, 82-84; RES SCIENTIST, COMPUTER MODELLING GROUP, 89- *Mem:* Soc Indust & Appl Math; Soc Petrol Engrs. *Res:* Numerical methods for oil reservoir simulation. *Mailing Add:* Computer Modeling Group 3512-33 St NW Calgary AB T2L 2A6

GRABER, CHARLES DAVID, b Pomeroy, Ohio, Dec 19, 17; m 44; c 3. BACTERIOLOGY, IMMUNOLOGY. *Educ:* Ohio State Univ, BS, 39, PhD(bact, immunol), 57; Univ Colo, MS, 54. *Prof Exp:* Bacteriologist & immunologist, Surg Res Unit, Brooke Army Med Ctr, 57-62; asst prof microbiol, Col Med, Baylor Univ, 62-66; from assoc prof to prof, 66-84, EMER PROF MICROBIOL, MED UNIV SC, 85- *Concurrent Pos:* Fel, Am Bd Microbiol. *Mem:* Am Acad Microbiol; Fr Soc Immunol; Soc Exp Biol & Med; Am Asn Immunol. *Res:* Burn infection immunology; effect of intestinal microflora on bile acid metabolism in atherosclerosis; the role of ureaplasma and chlamydia in infertility and nongonococcal urethritis. *Mailing Add:* 436 Trapier Dr Charleston SC 29412

GRABER, GEORGE, b New York, NY, Nov 24, 40; m 65. ANIMAL NUTRITION. *Educ:* Rutgers Univ, BS, 63, MS, 67; Univ Ill, Urbana, PhD(animal sci), 71. *Prof Exp:* ANIMAL HUSBANDMAN, DIV DRUGS SWINE & MINOR SPECIES, METAB PROD BR, BUR VET MED, FOOD & DRUG ADMIN, 71- *Mem:* Am Inst Nutrit. *Res:* Amino acid metabolism in chick and pig; vitamin-amino acid interrelation; swine metabolism studies; grain source-antimicrobial agent interrelationship in chick. *Mailing Add:* Div Animals Feeds Food & Drug Admin Ctr Vet Med 5600 Fisher Lane Rockville MD 20857

GRABER, HARLAN DUANE, b Newton, Kans, July 19, 35; m 58; c 3. NUCLEAR PHYSICS. *Educ:* Bethel Col, BS, 57; Univ Kans, PhD(physics), 64. *Prof Exp:* From asst prof to assoc prof, 62-76, PROF PHYSICS, CORNELL COL, 76- *Mem:* Am Phys Soc; Am Asn Physics Teachers. *Res:* Low energy nuclear physics; gamma ray spectroscopy. *Mailing Add:* Dept Physics Cornell College Mt Vernon IA 52314

GRABER, LELAND D, b Marion, SDak, Nov 5, 24; m 55; c 3. MATHEMATICS. *Educ:* Wheaton Col, Ill, BS, 48; Univ Minn, MA, 50; Iowa State Univ, PhD(math), 64. *Prof Exp:* Res engr, Aeronaut Res Dept, Minneapolis-Honeywell Regulator Co, 51-53; instr math, Am Univ Beirut, 53-59 & Iowa State Univ, 59-64; asst prof, Fla Presby Col, 64-70; ASSOC PROF MATH, CENT COL, IOWA, 70- *Mem:* Math Asn Am; Am Math Soc. *Res:* Functional analysis, linear operators in banach spaces. *Mailing Add:* Dept Math Cent Col Pella IA 50219

GRABER, RICHARD REX, b Kingman Co, Kans, Aug 25, 24; m 45. ZOOLOGY, WILDLIFE. *Educ:* Washburn Univ, BS, 48; Univ Mich, MA, 49; Univ Okla, PhD, 55. *Prof Exp:* Asst zool, Univ Mich, 48-50; mus field collector, La State Univ, 51-52; asst, Biol Surv, Univ Okla, 52-55; instr biol, Southwestern Tex State Teachers Col, 55-56; wildlife specialist, Ill Natural Hist Surv, 56-83, emer prin scientist, 84-; RETIRED. *Mem:* Cooper Ornith Soc; Wilson Ornith Soc; Am Ornith Union. *Res:* Evolution, migration and ecology of birds. *Mailing Add:* RR 1 Box 216 A Golconda IL 62938

GRABER, ROBERT PHILIP, b Glen Ullin, NDak, Oct 4, 18; m 59; c 2. STEROIDS, FERMENTATIONS. *Educ:* Univ Minn, BChem, 41; Univ Wis, PhD(chem), 49. *Prof Exp:* Jr chemist, Merck & Co Inc, 41-46, sr chemist, 49-56; proj leader, Res Labs, Gen Mills Inc, 57-58, sect leader, 58-63; mem res staff, Searle Chem Inc, Ill, 63, sci dir, Searle Mex, SA CV, 64-77, G D Searle & Co, Ltd, 77-80, SCI DIR, COMPANIA ESPANOLA SINTESIS QUIMICA, SA, 81- *Mem:* Am Chem Soc; AAAS. *Res:* Antibiotics; synthetic estrogens; steroids; natural products; fermentations; enzymes. *Mailing Add:* Velarde No 17-2 Apartado Postal 190 Segovia 40003 Spain

GRABER, T(OURO) M(OR), b St Louis, Mo, May 27, 17; m 41; c 5. ORTHODONTICS. *Educ:* Wash Univ, DDS, 40; Northwestern Univ, MSD, 46, PhD, 50; Am Bd Orthod, dipl, 54. *Hon Degrees:* Dr(odontol), Univ Gothenburg, Sweden, 89; DSc, Wash Univ, 91. *Prof Exp:* From instr to assoc prof orthod, Northwestern Univ, 45-58, dir res cleft lip & palate inst, 49-58; assoc attend orthodontist, Children's Mem Hosp, 58-62; spec lectr, Univ Mich, 58-67; prof orthod, Univ Chicago, 69-80, prof biol sci, pediat & anthrop, 74-80, chmn dept orthod, 69-80, res assoc prof plastic & reconstruction surg, 80-83; adj prof, Univ Mich, 84-88; RETIRED. *Concurrent Pos:* Grieve Mem Lectr, 64-84; dir, Kenilworth Res Found, 68-; Australian Res Found lectr, 71; vis scientist, Am Dent Asn, 79-90; ed, Am J Orthod, 85- *Honors & Awards:* Ketcham Award, 75; Mem lectr, Univ Witwatersrand, 75; Benno Lischer Award, 88; Strang Mem Lectr, 88; Mershon Award, 89; Northcroft Lectr, 89; Wylie Mem Award, 89. *Mem:* AAAS; Am Asn Orthod; Am Dent Asn; Europ Orthod Soc. *Res:* Maxillo-facial deformities; diagnosis of orthodontic therapy; craniofacial growth and development; use of orthopedic appliances in orthodontic therapy; temporomandibular joint dysfunction. *Mailing Add:* 2895 Sheridan Pl Evanston IL 60201

GRABIEL, CHARLES EDWARD, b Alliance, Ohio, July 31, 27; m 49; c 4. ORGANIC CHEMISTRY. *Educ:* Col of Wooster, AB, 50; Brown Univ, PhD(chem), 54. *Prof Exp:* Asst, Brown Univ, 50-53; chemist, 54-65, MGR, CHEM MKT RES, DOW CHEM CO, 65- *Mem:* Chem Mkt Res Asn; Am Chem Soc. *Res:* Vapor-liquid equilibrium; polynitroparaffins; polyelectrolytes. *Mailing Add:* Dow Chem Co Midland MI 48674

GRABINER, JUDITH VICTOR, b Los Angeles, Calif, Oct 12, 38; m 64; c 2. HISTORY OF SCIENCE. *Educ:* Univ Chicago, BS, 60; Harvard Univ, MA, 62, PhD(hist sci), 66. *Prof Exp:* Instr hist sci, Harvard Univ, 66-69; lectr hist, Univ Calif, Santa Barbara, 69-70; lectr hist & math, Calif State Univ, Los Angeles, 70-71; asst prof, 72-75, assoc prof, 75-80, PROF HIST SCI, SMALL COL, CALIF STATE UNIV, DOMINGUEZ HILLS, 80- *Concurrent Pos:* Fel, Am Coun Learned Soc, 71-72; mem comn hist math, Div Hist Sci, Int Union Hist & Philos Sci, 74- *Mem:* Sigma Xi; Hist Sci Soc; Math Asn Am. *Res:* History of mathematics, especially analysis in 18th and 19th centuries; Scopes trial; mathematics in America. *Mailing Add:* Dept Math Pitzer Col Claremont CA 91711

GRABINER, SANDY, b New York, NY, Dec 15, 39; m 64; c 2. BANACH ALGEBRAS, OPERATOR THEORY. *Educ:* Rice Univ, BA, 60; Harvard Univ, AM, 61, PhD(math), 67. *Prof Exp:* Instr math, Mass Inst Technol, 67-69; asst prof, Claremont Grad Sch, 69-74; assoc prof, 74-82, PROF MATH, POMONA COL, 82- *Concurrent Pos:* Vis assoc prof math, Ind Univ, 81-82; chmn prog comt, Math Asn Am, 84-85, mem, comt vis lectrs. *Mem:* AAAS; Am Math Soc; Math Asn Am; London Math Soc. *Res:* Banach algebras and operators on Banach spaces. *Mailing Add:* Dept Math Pomona Col Claremont CA 91711

GRABLE, ALBERT E, b San Bernardino, Calif, Mar 10, 39; m 62; c 3. SYSTEMATIC BOTANY, INSECT ECOLOGY. *Educ:* La Sierra Col, BA, 59; Univ Minn, MS, 62, PhD(entom), 65. *Prof Exp:* From instr to asst prof, 63-70, ASSOC PROF BIOL, WALLA WALLA COL, 70- *Concurrent Pos:* Consult, US Bur Reclamation, 74, Tropical Insects, Dominican Repub, 87. *Mem:* Entom Soc Am; Bot Soc Am; Am Soc Plant Taxonomists; Sigma Xi; Am Registry Prof Entomologists. *Res:* Biology of social hymenoptera; taxonomy of flowering plants. *Mailing Add:* Dept Biol Walla Walla Col College Place WA 99324

GRABOIS, NEIL, b New York, NY, Dec 11, 35; m 56; c 2. ALGEBRA, NUMBER THEORY. *Educ:* Swarthmore Col, BA, 57; Univ Pa, MA, 59, PhD(math), 63. *Prof Exp:* Asst instr math, Univ Pa, 57-61; instr, Lafayette Col, 61-63; from asst prof to assoc prof, William Col, 63-73, dean fac, 75-77, provost, 77-80, prof math, 73-88, chmn dept math sci, 81-88; PRES, COLGATE COL, 88- *Mem:* NY Acad Sci; Am Math Soc; Math Asn Am. *Res:* Algebraic number theory; graph theory and combinatorial mathematics with a particular interest in matchings and matroids. *Mailing Add:* Pres Off Colgate Col Colgate Hall 13 Oak Dr Hamilton NY 13346-1398

GRABOWSKI, CASIMER THADDEUS, b Cleveland, Ohio, Aug 16, 27; m 80; c 1. EMBRYOLOGY, ZOOLOGY. *Educ:* Case Western Reserve Univ, BS, 50; Johns Hopkins Univ, PhD(zool), 54. *Prof Exp:* Jr instr biol, Johns Hopkins Univ, 50-52; from instr to asst prof anat, Univ Pittsburgh, 54-60; assoc prof zool, 60-67, chmn, 87-90, PROF BIOL, UNIV MIAMI, 67- *Mem:* Am Soc Zoologists; Teratol Soc (pres, 87-88); Soc Develop Biol; Int Soc Develop Biol. *Res:* Experimental embryology; teratology and embryonic physiology, as related to normal and abnormal morphogenesis; effects of teratogenic agents on cardiovascular physiology of mammalian embryos; teratogenic effects of atmospheric pollutants and pesticides. *Mailing Add:* Dept Biol Univ Miami Coral Gables FL 33124

GRABOWSKI, EDWARD JOSEPH JOHN, b Jamaica, NY, Apr, 23, 40; m 61; c 1. PROCESS RESEARCH. *Educ:* Mass Inst Technol, SB, 61; Univ Rochester, PhD(chem), 65. *Prof Exp:* Sr chemist, Merck, Sharp & Dohme Res Labs, Merck & Co Inc, 65-72, res fel, 72-75, sr res fel, 75-79, sr investr, 79-86, dir, 86-88, SR DIR, PROCESS RES, MERCK, SHARP & DOHME RES LABS, MERCK & CO INC, 88- *Mem:* Am Chem Soc; Royal Soc Chem; Eur Photochem; Int Soc Heterocyclic Chem. *Res:* Design and development of synthesis for new drug candidates; development of processes for drug manufacture with emphasis on beta-lactam, heterocyclic, amino acid, aromatic and natural product chemistry; reaction mechanisms; physical organic chemistry. *Mailing Add:* 741 Marcellus Dr Westfield NJ 07090

GRABOWSKI, JOSEPH J, b Elkridge, Md, Nov 4, 56; m 77. GAS-PHASE ION CHEMISTRY, PHOTOACOUSTIC CALORIMETRY. *Educ:* Univ Md, Baltimore County, BA, 78; Univ Colo, PhD(org chem), 83. *Prof Exp:* Teaching asst chem, Univ Colo, 78-79, res asst, 79-83; postdoctoral fel, 83-84, asst prof, 84-88, ASSOC PROF ORG CHEM, HARVARD UNIV, 88- *Concurrent Pos:* NSF presidential young investr award, 86-91; chmn, Dept Teaching Fel Training Prog, 88-91. *Honors & Awards:* VG Instruments Res Award, Am Soc Mass Spectrometry, 86. *Mem:* Fel AAAS; Am Chem Soc; Am Soc Mass Spectrometry. *Res:* Investigations to elucidate the intrinsic factors that control organic reactions; utilization of the flowing afterglow technique to analyze ionic organic reactions; development and exploitation of photoacoustic calorimetry for bioorganic problems. *Mailing Add:* Dept Chem Harvard Univ 12 Oxford St Cambridge MA 02138

GRABOWSKI, KENNETH S, b Detroit, Mich, Aug 15, 52; m 76; c 2. SURFACE MODIFICATION. *Educ:* Univ Mich, BSE, 74, MSE, 75, PhD(nuclear eng), 80. *Prof Exp:* Mem lab staff, Argonne Nat Lab, 76-79, res assoc, 79-80; RES PHYSICIST, NAVAL RES LAB, 81- *Mem:* Am Vacuum Soc; AAAS; Am Inst Mining, Metall & Petrol Engrs; Mat Res Soc. *Res:* Thin film growth using ion and laser beams, including their influence on chemical composition, microstructure, and materials properties; ion-beam and x-ray characterization of materials. *Mailing Add:* Naval Res Lab Code 4671 Washington DC 20375-5000

GRABOWSKI, SANDRA REYNOLDS, b Chicago, Ill, Mar 3, 43; m 65. NEUROBIOLOGY. *Educ:* Purdue Univ, BS, 65, PhD(biol), 73. *Prof Exp:* NIH fel & res fel neurobiol, 73-77, INSTR BIOL SCI, PURDUE UNIV, WEST LAFAYETTE, IND, 77- *Mem:* AAAS; Asn Women Sci. *Res:* Visual electrophysiology. *Mailing Add:* Dept Biol Purdue Univ Main Campus West Lafayette IN 47907

GRABOWSKI, THOMAS J, b New Brunswick, NJ, Aug 21, 50; m 70; c 3. SAMPLE PREP TECHNIQUES. *Educ:* Southwestern Univ, BSME, 82. *Prof Exp:* PRES, COMPUTER WIZARD INC, 89- *Concurrent Pos:* Sr tech, NL Industs Corp Labs, 69-79; engr, Revere Co, Revere Res inc, 79-82; mgr metall eng, Exide Corp, 82-89. *Mem:* Am Soc Metals. *Res:* Metallographic laboratory techniques. *Mailing Add:* 400 S Ninth St Quakertown PA 18951

GRABOWSKI, WALTER JOHN, b Philadelphia, Pa, May 19, 49; m 70; c 1. FLUID DYNAMICS, PHYSICAL OCEANOGRAPHY. *Educ:* Univ Pa, BS, 71; Univ Calif, Berkeley, MS, 72, PhD(mech eng), 74. *Prof Exp:* Scientist fluid dynamics, Sci Appln Inc, 74-75; scientist, Flow Res Inc, 75-76; physicist, David Taylor Naval Ship Res & Develop Ctr, 76-77; scientist phys oceanog, Sci Applns Inc, 77-81, proj mgr, 81-89; PROJ MGR, AT&T BELL LABS, 89- *Mem:* Am Geophys Union. *Res:* Theoretical analysis and numerical modeling of fluid dynamic phenomena; emphasis on oceanic phenomena such as internal waves. *Mailing Add:* 9718 Laurel St Fairfax VA 22032

GRABOWSKI, ZBIGNIEW WOJCIECH, b Plock, Poland, June 22, 31; US citizen; m 65. NUCLEAR PHYSICS. *Educ:* Jagiellonian Univ, MS, 54; Univ Uppsala, Fil lic, 61, PhD(physics), 62. *Prof Exp:* Teaching asst, Tech Univ Gliwice, Poland, 54-55; res asst nuclear physics, Nuclear Res Inst, Poland, 55-58; Univ Uppsala, 58-62; proj mgr physics of Aurora, Kiruna Geophys Observ, Sweden, 62-63; res assoc, 63-65, from asst prof to assoc prof, 65-75, PROF PHYSICS, PURDUE UNIV, 75- *Mem:* Am Phys Soc; Sigma Xi; Am Asn Physics Teachers. *Res:* Nuclear spectroscopy; angular correlation of nuclear radiation; internal conversion; magnetic moments of excited nuclear states; hyperfine fields; nuclei at high angular momentum. *Mailing Add:* Dept Physics Purdue Univ W Lafayette IN 47907

GRACE, DONALD J, b Oklahoma City, Okla, Feb 21, 26; m 49; c 2. RESEARCH ADMINISTRATION. *Educ:* Ohio State Univ, BSEE, 48, MSEE, 49; Stanford Univ, PhD, 62. *Prof Exp:* Lectr, Ohio State Univ, 48-49; res engr, Airborne Instruments Lab, 59-61; res assoc, Stanford Univ, 62-63, sr res assoc, Systs & Techniques Lab, 63-66, assoc prof elec eng, 63-67, dir, 66-67, assoc dean eng, 67-69; dir res, Kentron Hawaii, Ltd, 69-73; dir, Ctr Eng Res, Univ Hawaii, 73-76; DIR, ENG EXP STA, GA INST TECHNOL, 76- *Concurrent Pos:* Mem, FORECAST Panel, USAF, 63, mem reconnaissance adv bd, 64-65; tech adv, US Army Security Agency, 65-66; dir, Stanford Univ Instructional TV Network, 67-69; asst secy, Ga Tech Res Inst & rep, Pub Serv Sattelite Consortium, 77-; univ adv panel, Nat Solar Energy Res Inst, 78-79; mem forum, Nat Security Affairs, Pentagon, 80. *Mem:* Inst Elec & Electronics Engrs; Sigma Xi; AAAS. *Res:* Microwave components and subsystems; defense electronics; alternate energy; instructional television; research administration, management and planning. *Mailing Add:* Off Dir Ga Tech Res Inst Cenntennial Res Bldg Atlanta GA 30332

GRACE, EDWARD EVERETT, b Gulfport, Miss, Apr 15, 27; m 54; c 3. POINT SET TOPOLOGY. *Educ:* Univ NC, BS, 51, PhD(math), 56. *Prof Exp:* Instr math, Univ NC, 55-56; asst prof, Emory Univ, 56-62; vis assoc prof, Univ Ga, 62-63; PROF MATH, ARIZ STATE UNIV, 63- *Concurrent Pos:* NSF fac fel, Univ Wis, 61-62, res assoc, 62; vis prof, Univ Wis, 69-70 & Univ Houston, 77. *Mem:* Am Math Soc; Math Asn Am; AAAS; Sigma Xi; Am Asn Univ Prof. *Res:* Point set topology, particularly continuum theory. *Mailing Add:* Dept Math Ariz State Univ Tempe AZ 85287-1804

GRACE, HAROLD P(ADGET), b Parsons, Kans, Apr 19, 19; m 42; c 2. CHEMICAL ENGINEERING, COLLON CHEMISTRY. *Educ:* Univ Pa, BS, 41. *Prof Exp:* Field engr, Eng Dept, E I du Pont de Nemours & Co, Inc, 41-44, res engr, Eng Res Lab, 46-52, res proj engr, 52-56, res assoc, 56-71, res fel, Eng Res Lab, 71-79; CONSULT, 80- *Honors & Awards:* Colburn Award, Franklin Inst, 56. *Mem:* Am Chem Soc; Am Inst Chem Engrs. *Res:* Chemical engineering research and development involving particulate solids, particle mechanics, mechanical separations, particle size analysis, size reduction and flocculation; dispersion, mixing in polymer melts. *Mailing Add:* 108 N Concord Ave Havertown PA 19083

GRACE, JOHN NELSON, nuclear engineering, for more information see previous edition

GRACE, JOHN ROSS, b London, Ont, June 8, 43; m 64; c 2. CHEMICAL ENGINEERING. *Educ:* Univ Western Ont, BESc, 65; Cambridge Univ, PhD(chem eng), 68. *Prof Exp:* From asst prof to prof chem eng, McGill Univ, 68-79; head dept, 79-87, PROF CHEM ENG DEPT, UNIV BC, 79-, DEAN GRAD STUDIES, 90- *Concurrent Pos:* Sr proj engr, Surveyer, Nenniger & Chenevert Inc, 74-75; Nat Res Coun Can sr indust fel, 74-75; ed, Chem Eng Sci. *Honors & Awards:* Erco Award; L S Lauchland Medal; Killam Res Prize. *Mem:* Chem Inst Can; Brit Inst Chem Engrs; Can Soc Chem Eng; Asn Prof Engrs BC. *Res:* Fluidization; circulating beds; dynamics of bubbles, drops and particles; reactor design; energy technology; spouted beds. *Mailing Add:* Dept Chem Eng Univ BC 2216 Main Mall Vancouver BC V6T 1W5 Can

GRACE, MARCELLUS, b Selma, Ala, Oct 17, 47; m 73; c 3. PHARMACY ADMINISTRATION. *Educ:* Xavier Univ, La, BS, 71; Univ Minn, MS, 75, PhD(pharm admin), 76. *Prof Exp:* Hosp pharm resident, USPHS Hosp, Md, 71-72, staff pharmacist, Mass, 72-73; staff pharmacist, Thrifty Drug Stores, Los Angeles, 73 & Methodist Hosp, St Louis Park, Minn, 74; asst dir pharm, Bethesda Hosps, Cincinnati, 75; dir pharm serv, Tulane Univ Med Ctr, Nola, 76-77; asst dean & assoc prof pharm, Howard Univ, Wash, DC, 79-82; asst prof clin pharm, 76-78, DEAN & PROF ADMIN, XAVIER UNIV, LA, 83- *Concurrent Pos:* Chmn bd, Minority Health Prof Found, 89-91; mem, Adv Coun, Nat Heart, Lung & Blood Inst, NIH, Inst Rev Bd, Clin Res Ctr, New Orleans & bd dirs, St Thomas Clin, 90-; pharm consult, United Teachers New Orleans Health & Welfare, 90- *Mem:* Am Asn Cols Pharm (secy, 83-86); Asn Minority Health Prof Schs (secy, 85-87, pres, 87-89). *Res:* Management with emphasis on motivation; patient education; hypertension with emphasis on patient compliance; pharmacist communication skills; author of 35 technical publications. *Mailing Add:* Xavier Univ La Col Pharm 7325 Palmetto St New Orleans LA 70125

GRACE, NORMAN DAVID, b Worcester, Mass, Sept 23, 36; m 60; c 2. GASTROENTEROLOGY, INTERNAL MEDICINE. *Educ:* Brown Univ, AB, 58; Tufts Univ, MD, 62. *Prof Exp:* Asst clin instr med, Albany Med Col, 63-65; from instr to asst prof, 65-75, asst dean hosp affairs, 71- 85, assoc prof, 75-88, PROF MED, SCH MED, TUFTS UNIV, 88- *Concurrent Pos:* Clin fel gastroenterol, Albany Med Col, 64-65; Nat Inst Arthritis & Metab Dis res fel, Sch Med, Tufts Univ, 65-67; chief gastroenterology, Faulkner Hosp, 71- & Lemuel Shattuck Hosp, 72- *Mem:* AAAS; Am Gastroenterol Asn; Am Asn Study Liver Dis; Am Fedn Clin Res; Am Soc Gastrointestinal Endoscopy; Int Asn Study of Liver Dis; fel Am Col Physicians; fel Am Col Gastroenterology. *Res:* Liver disease; portal hypertension; iron metabolism; hemochromatosis. *Mailing Add:* 48 Rosalie Rd Needham MA 02194

GRACE, O DONN, b Kansas City, Mo, Feb 10, 36. SYSTEM SCIENCE, ACOUSTICS. *Educ:* Portland State Col, BS, 63; Colo State Univ, MS, 66; Univ Calif, San Diego, PhD(appl physics), 75. *Prof Exp:* Assoc res scientist acoust, Appl Res Lab, Univ Tex, Austin, 65-69; physicist acoust, Naval Ocean Syst Ctr, San Diego, 69-80; sr prin engr, Orincon Corp, 80-83; staff scientist, Phys Dynamics, La Jolla, Calif, 83-87; INDEPENDENT CONSULT, 87- *Concurrent Pos:* Lectr, Univ Calif, San Diego & San Diego State Univ. *Mem:* Acoust Soc Am; Inst Elec & Electronics Engrs. *Res:* System analysis, data processing signal physics, random processes. *Mailing Add:* 20503 Fortuna Del Sur Escondido CA 92029

GRACE, OLIVER DAVIES, b Washington, DC, Dec 21, 14; m 48; c 2. VETERINARY PATHOLOGY. *Educ:* Colo Agr & Mech Col, DVM, 40; Univ Ill, MS, 51; Am Col Vet Microbiol, dipl. *Prof Exp:* Field veterinarian, Bur Animal Indust, USDA, 40-42, veterinarian in charge serol lab, NH, 42-46; food & drug admin, Dept Health, Educ & Welfare, Ill, 46-53; head dept vet med, Baxter Labs, Inc, 53-55; from assoc prof to prof vet sci, Univ Nebr-Lincoln, 55-82, interim head, 76-77; RETIRED. *Mem:* Am Vet Med Asn; Am Asn Avian Path; Conf Res Workers Animal Dis. *Res:* Etiology, pathogenesis and pathology of the infectious diseases of large animals and poultry. *Mailing Add:* 1720 Donald Circle Lincoln NE 68505

GRACE, RICHARD E(DWARD), b Chicago, Ill, June 26, 30; m 55; c 2. ENGINEERING EDUCATION. *Educ:* Purdue Univ, BSMetE, 51; Carnegie Inst Technol, PhD, 54. *Prof Exp:* From asst prof to prof metall eng, Purdue Univ, 54-62, head sch, 64-72, head div interdisciplinary eng studies, 70-82, head dept freshman eng & asst dean eng, 81-87, PROF METALL ENG, PURDUE UNIV, 62-, VPRES STUDENT SERV, 87- *Concurrent Pos:* Consult, Midwest Industs; officer eng educ & accreditation comn, Engrs Coun Prof Develop. *Honors & Awards:* Bradley Stoughton Award, Am Soc Metals, 62; Grinte Award, Accreditation Bd Eng & Technol, 89. *Mem:* Fel Am Soc Metals; fel Am Soc Eng Educ; Am Inst Mining, Metall & Petrol Engrs. *Res:* diffusion and mass transport; corrosion and oxidation. *Mailing Add:* Schleman Hall Purdue Univ West Lafayette IN 47907

GRACE, ROBERT ARCHIBALD, b London, Ont, Jan 1, 38; m 82; c 4. OCEAN ENGINEERING. *Educ:* Univ Western Ont, BESc, 60; Mass Inst Technol, SM, 62, PhD(hydrodyn & water resources), 66. *Prof Exp:* Assoc res scientist, Hydronautics, Inc, Md, 61-63; engr, Electricite de France, Chatou, 63, 68; asst prof civil eng, 66-69, assoc prof, 70-76, PROF CIVIL ENG, UNIV HAWAII, 76- *Concurrent Pos:* Vis prof, Ore State Univ, 77, 79-81; vis lectr, Univ Cape Town, 83. *Res:* Coastal engineering; ocean wave forces on structures; marine disposal of wastes; artificial reefs. *Mailing Add:* Dept Civil Eng Univ Hawaii Manoa Honolulu HI 96822

GRACE, THOM P, b Evergreen Park, Ill, Jan 30, 55; m 90; c 1. COMPUTER GRAPHICS, PROGRAMMING LANGUAGES. *Educ:* Univ Ill, BS, 76, MS, 79, PhD(math), 82. *Prof Exp:* Res & teaching asst math, Math Dept, Univ Ill, Chicago, 76-82; asst prof, 82-86, ASSOC PROF COMPUTER SCI, COMPUTER SCI DEPT, ILL INST TECHNOL, 86- *Concurrent Pos:* Consult, Dewar Info Systs Corp, 90-91; expert witness, 91. *Mem:* Asn Comput Mach; Math Asn Am; Am Math Soc; Inst Elec & Electronics Engrs Computer Soc. *Res:* Realistic computer-generated imagery using approximations of physically correct behavior; programming languages, programming environments and novel programming paradigms; numerical and nonnumerical computation; discrete mathematics: graph theory, algebra, error-correcting codes. *Mailing Add:* Dept Computer Sci Ill Inst Technol Chicago IL 60616

GRACE, THOMAS MICHAEL, b Beaver Dam, Wis, Oct 3, 38; m 63, 80; c 3. CHEMICAL ENGINEERING. *Educ:* Univ Wis-Madison, BS, 60; Univ Minn, Minneapolis, PhD, 63. *Prof Exp:* Aerospace technologist, Lewis Res Ctr, NASA, 63-65; assoc prof chem eng & res assoc, Inst Paper Chem, 66-76, chmn dept, 70-76, prof chem eng & sr res assoc, 76-89; PRES, T M GRACE CO, INC, 89- *Honors & Awards:* Forest Prod Div Award, Am Inst Chem Eng, 84. *Mem:* Fel Tech Asn Pulp & Paper Ind; Am Inst Chem Engrs; Tech Asn Pulp & Paper Indust. *Res:* Chemical recovery technology; heat and mass transfer. *Mailing Add:* 2517 S Harmon St Appleton WI 54915

GRACIE, G(ORDON), b Toronto, Ont, Sept 11, 30; m 58; c 2. PHOTOGRAMMETRY. *Educ:* Univ Toronto, BASc, 52; Univ Ill, PhD(civil eng), 63. *Prof Exp:* Proj engr aerial surv, Photog Surv Corp, 52-55; from instr to asst prof civil eng, Univ Ill, 57-65; sr scientist photogram autometric opers, Raytheon Co, 65-71; PROF SURV SCI, UNIV TORONTO, 71- *Concurrent Pos:* Dir, Ctr Surv Sci, Univ Toronto, 88- *Mem:* Am Soc Civil Engrs; Am Soc Photogram; Am Cong Surv & Mapping; Can Inst Surv & Mapping. *Res:* Aerial surveying; survey analysis; analytical photogrammetry; non-topographic photogrammetry; statistics. *Mailing Add:* Erindale Campus Univ Toronto Mississauga ON L5L 1C6 Can

GRACY, ROBERT WAYNE, b McKinney, Tex, Dec 30, 41; m 63. ENZYMOLOGY. *Educ:* Calif State Polytech Col, BS, 64; Univ Calif, Riverside, PhD(biochem), 68. *Prof Exp:* Chemist, Space Gen, El Monte, Calif, 63-64; Damon Runyon fel molecular biol, Albert Einstein Col Med, 68-70; asst prof, 70-73, assoc prof, 73-75, PROF BIOCHEM & CHMN DEPT, NTEX STATE UNIV, 75-; PROF BIOCHEM & CHMN DEPT, TEX COL OSTEOP MED, 75- *Concurrent Pos:* Lectr, Calif State Polytech Col, 63-64; res career develop award, NIH. *Mem:* AAAS; Am Chem Soc; Am Asn Biol Chem. *Res:* Protein chemistry and enzymology; structure and function of enzymes and relation to catalytic mechanisms; physical, chemical and catalytic properties of enzymes, especially metalloproteins and metabolic diseases; molecular basis of altered enzymes during aging; aging and the failure of the immune system. *Mailing Add:* Dept Biochem Tex Col Osteop Med 3500 Camp Bowie Blvd Ft Worth TX 76107-2690

GRAD, ARTHUR, b Stryj, Austria, Jan 31, 18; nat US; m 46; c 2. MATHEMATICS. *Educ:* City Col New York, BS, 38; Columbia Univ, AM, 39; Stanford Univ, PhD(math), 48. *Prof Exp:* Mathematician, US Coast & Geod Surv, 41-46; res assoc & actg instr math, Stanford Univ, 46-48; mathematician, Off Naval Res, 48-53 & Comput Facility, AEC, NY Univ, 53-54; head math br, Off Naval Res, 54-59; prog dir math sci & head math sci sect, NSF, 59-63; assoc dean grad div, Stanford Univ, 63-64; prof math & dean grad sch, Ill Inst Technol, 64-71; pres, Polytech Inst Brooklyn, 71-74; PROF, CITY COL NEW YORK, 74- *Concurrent Pos:* Lectr, Univ Md, 49-53; mem comt expert exam, US Civil Serv Comn, 51-53; liaison rep, Div Math, Nat Acad Sci-Nat Res Coun, 59-63, mem comt on travel grants, 60-63, liaison rep, Comt on Use of Electronic Comput, 62-63, mem comt on sources & forms of support, Div Math, 69-, chmn, 70- *Mem:* AAAS; Am Math Soc; Math Asn Am. *Res:* Functions of a complex variable, conformal mapping; schlicht functions. *Mailing Add:* 19 Mountain View Ave Ardsley NY 10502

GRAD, BERNARD RAYMOND, b Montreal, Que, Feb 4, 20; m 48; c 2. EXPERIMENTAL BIOLOGY. *Educ:* McGill Univ, BSc, 44, PhD(exp morphol), 49. *Prof Exp:* Asst endocrinol, geront & cancer, 49-55, lectr, 55-61, asst prof, 61-65, assoc prof biol, McGill Univ, 65-85; ASSOC PROF, INST ARMAND FRAPPIER, UNIV QUEBEC, MONTREAL, 85- *Concurrent Pos:* Ed, Excerpta Medica, 58-; Ciba Found Award, 55; assoc scientist, Royal Victoria Hosp, 72-91; Consult, 86- *Mem:* Am Asn Cancer Res. *Res:* Variables effecting quantitative changes in coronal images; stress and aging; placebo effect; bioenergetics; biomagnetics; anti-cancer compounds; biopoiesis. *Mailing Add:* 5317 Snowdon Montreal PQ H3X 1Y3 Can

GRADIE, JONATHAN CAREY, b Putnam, Conn, June 20, 51; m 77; c 2. PLANETARY SCIENCES, ASTRONOMY. *Educ:* Univ Ariz, BS, 73, PhD(planetary sci), 78. *Prof Exp:* res assoc planetary sci, Lab Planetary Studies, Cornell Univ, 78-; AT PLANETARY GEOSCI, UNIV HAWAII, HONOLULU. *Honors & Awards:* Asteroid named in hon, #3253 Gradie. *Mem:* AAAS; Am Astron Soc; Int Astron Union; Am Geophys Union. *Res:* Planetary surfaces; asteroids, comets and natural satellites; optical design visible and infrared spectrometers; geophysics. *Mailing Add:* Planetary Geosci Univ Hawaii 2525 Correa Rd Honolulu HI 46822

GRADSTEIN, FELIX MARCEL, b Nov 8, 41; Canadian citizen; m 68; c 2. MICROPALEONTOLOGY, MARINE GEOLOGY. *Educ:* Utrecht State Univ, PhD(micropaleont, biostratig), 72. *Prof Exp:* Asst paleont, Utrecht State Univ, 64-72; res scientist, Imperial Oil Ltd, Calgary, 72-74; RES SCIENTIST BIOSTRATIG, GEOL SURV CAN, 74- *Concurrent Pos:* Consult, Arps Consult Ltd, 70-72; vis fel, Woods Hole Oceanog Inst, 78; chief scientist, ocean drilling expeds, 81 & 88. *Mem:* Koninklijk Nederlands Geologisch Mijinbouwkundig; Can Soc Petrol Geologists; Comt Quantitative Stratigraphy. *Res:* Foraminiferal biostratigraphy; paleo-oceanography; agglutinated benthonic foraminifera, particularly as applied in stratigraphic/paleoecologic evolutions of petroleum basins such as North Sea and Labrador Sea; quantitative biostratigraphy; development of new stratigraphic correlation methods; Canadian quantitative stratigraphy. *Mailing Add:* Atlantic Geosci Ctr Geol Surv Can Bedford Inst Dartmouth NS B2Y 4A2 Can

GRADY, CECIL PAUL LESLIE, JR, b Des Arc, Ark, June 25, 38; m 61; c 2. BIOLOGICAL WASTEWATER TREATMENT, BIODEGRADATION. *Educ:* Rice Inst, BA, 60; Rice Univ, BSCE, 61, MS, 63; Okla State Univ, PhD(environ eng), 69. *Prof Exp:* Res asst environ eng, Rice Univ, 60-61, res fel, 61-63; design engr, Charles R Haile Assoc Consult Engrs, 63; environ engr, US Army Environ Hygiene Agency, 63-65; res fel environ eng, Okla State Univ, 65-68; from asst prof to prof, Purdue Univ, 68-81; prof environ eng, 81-83, R A BOWEN PROF ENVIRON ENG, CLEMSON UNIV, 83- *Concurrent Pos:* Career develop award, Merck Found, 71; vis scholar, Dept Chem Eng, Univ Tex, 75-76. *Honors & Awards:* Simon W Freese Award, Am Soc Chem Eng, 89. *Mem:* Water Pollution Control Fedn; Am Inst Chem Eng; Am Soc Microbiol; Am Chem Soc; Int Asn Water Pollution Res & Control. *Res:* Microbial kinetics; application of biochemical engineering principles to wastewater treatment; microbial ecology; biodegradation of toxic organic pollutants. *Mailing Add:* Environ Systs Eng Clemson Univ 501 Rhodes Clemson SC 29634-0919

GRADY, HAROLD JAMES, b Excelsior Springs, Mo, May 3, 20; m 44; c 5. CLINICAL BIOCHEMISTRY. *Educ:* St Louis Univ, PhD(biochem), 51. *Prof Exp:* Asst prof biochem & med, Univ Kans Med Ctr, Kansas City, 50-55, from assoc prof to prof biochem in path, 55-70; CLIN BIOCHEMIST, BAPTIST MEM HOSP, KANSAS CITY, 70- *Concurrent Pos:* Consult, Vet Admin & Independence Hosps, Kansas City. *Mem:* AAAS; Am Soc Clin Path; Am Chem Soc; Am Asn Clin Chemists. *Res:* Steroid metabolism and clinical chemistry. *Mailing Add:* Baptist Mem Hosp 6601 Rockhill Rd Kansas City MO 64131-1197

GRADY, HAROLD ROY, b Wooster, Ohio, Aug 8, 22; m 45; c 3. PHYSICAL CHEMISTRY. *Educ:* Wooster Col, AB, 43; Brown Univ, PhD(chem), 49. *Prof Exp:* Res chemist, Manhattan Proj, Oak Ridge, Tenn, 44-46; prof chem & head dept, Muskingum Col, 49-53, 54-55; cryog engr, H L Johnston, Inc, 53-54; dir chem res, Vanadium Corp Am, Ohio, 55-67; mgr vanadium chem res, Foote Mineral Co, 67-71, vpres & gen mgr, Lithium Battery Dept, 71-88, VPRES RES, CYPRUS FOOTE MINERAL CO, 88- *Mem:* Am Chem Soc; Electrochem Soc. *Res:* Infrared spectroscopy; vanadium compounds; analytical chemistry; metallurgy. *Mailing Add:* 725 Fox Lane Chester Springs PA 19425

GRADY, JOSEPH EDWARD, b Plains, Pa, Apr 1, 27; m 54; c 4. MICROBIOLOGY, BIOCHEMISTRY. *Educ:* Univ Scranton, BS, 48; Purdue Univ, MS, 51, PhD, 58. *Prof Exp:* Microbiologist, Lederle Labs Div, Am Cyanamid Co, 51-54; microbiologist, Upjohn Co, 58-78, mgr infectious dis res, 78-88; RETIRED. *Mem:* AAAS; Am Soc Microbiol; NY Acad Sci. *Res:* Antibiotics, screening, in vitro and in vivo evaluation. *Mailing Add:* 3825 Greenleaf Circle Kalamazoo MI 49008

GRADY, LEE TIMOTHY, b Chicago, Ill, Mar 21, 37; m 64; c 2. ANALYTICAL CHEMISTRY, PHARMACEUTICAL CHEMISTRY. *Educ:* Univ Ill, BS, 59, PhD(org chem), 63. *Prof Exp:* Sr res pharmacologist, Merck Inst Therapeut Res, 65-68; sr supvr drug stand lab, Am Pharmaceut Asn Found, 68-71, dir, 71-74; dir, Drug Res & Testing Lab, 75-78, DIR, DRUG STANDARDS DIV, US PHARMACOPEIA, 79- *Concurrent Pos:* Mem adj fac, Sch Pharm, Univ Md, Baltimore, 72-; expert adv, WHO, 79-87 & Pan-Am Health Orgn, 84-85. *Honors & Awards:* Res Award, Am Soc Hosp Pharmacists, 82; Justin Powers Award, 90. *Mem:* Fel AAAS; Am Chem Soc; Acad Pharmaceut Sci; Fed Int Pharm; fel Am Asn Pharm Sci. *Res:* Physical organic chemistry; drug metabolism; chromatography; analytical chemistry; drug assay; chemical purity and methods; analysis of packaging materials; drug bioequivalence; standards setting. *Mailing Add:* USP Drug Standards Div 12601 Twinbrook Pkwy Rockville MD 20852

GRADY, PERRY LINWOOD, b Mt Olive, NC, Sept 10, 40; m 75; c 2. POLYMER SCIENCE, ELECTRICAL ENGINEERING. *Educ:* NC State Univ, BS, 62, MS, 67, PhD(fiber & polymer sci), 73. *Prof Exp:* Res asst, 62-67, res instr, 67-73, from asst prof to assoc prof, 73-82, PROF & ASSOC DEAN, COL TEXTILES, NC STATE UNIV, 82- *Honors & Awards:* Fiber Soc Award, Inst Soc Am Award, Exten Award. *Mem:* Instrument Soc Am; Fiber Soc; Inst Elec & Electronics Engrs; Am Phys Soc; Am Asn Textile Chemists & Colorists. *Res:* Instrument design and development; physical properties of fibers; electrostatic and charge transport properties of fibers; open end spinning; computer applications in textiles; ballistic testing of textile materials; energy utilization and conservation. *Mailing Add:* Col Textiles NC State Univ PO Box 8301 Raleigh NC 27650-8301

GRAEBEL, WILLIAM P(AUL), b Manitowoc, Wis, July 15, 32; m 54; c 2. ENGINEERING MECHANICS, APPLIED MATHEMATICS. *Educ:* Univ Wis, BS, 54, MS, 55; Univ Mich, PhD(eng mech), 59. *Prof Exp:* Mem tech staff, Bell Labs, Inc, 55-56; from instr to assoc prof eng mech, 56-67, PROF APPL MECH, UNIV MICH, 67- *Concurrent Pos:* Mem, Col Eng Exec Comt, 84-, chmn, thermal fluids div, 87- *Mem:* Am Phys Soc; Am Soc Mech Eng; assoc fel Am Inst Aeronaut & Astronaut; Sigma Xi. *Res:* Fluid mechanics. *Mailing Add:* Dept Mech Eng & Appl Mech Univ Mich Ann Arbor MI 48109-2125

GRAEBER, EDWARD JOHN, b Denver, Colo, May 14, 34; m 56; c 4. X-RAY CRYSTALLOGRAPHY. *Educ:* Colo Sch Mines, BS, 56; Univ NMex, PhD(geol), 70. *Prof Exp:* STAFF MEM CRYSTALLOG, SANDIA LABS, 56- *Mem:* Am Inst Physics; Am Crystallog Asn. *Res:* X-ray diffraction; organic, inorganic and mineral structures; solid state. *Mailing Add:* Org 5214 Sandia Nat Labs Albuquerque NM 87185

GRAEBERT, ERIC W, b Laren, Holland, Jan 20, 24; US citizen; m 46; c 3. TOXICOLOGY MANAGEMENT. *Educ:* Univ Pittsburgh, BSCheE, 49. *Prof Exp:* Mgt positions in detergent & food prod & paper fields, Procter & Gamble Co, Eng & US, 49-78; vpres opers, Bio/Dynamics, Inc, 78-80, sr vpres, 80-82; PRES, GRAEBERT ASSOCS, 82 - *Res:* Industrial food product development; development of raw material sources in Europe; reorganization and formalization programs for carrying out safety (toxicology) evaluations of product materials prior to marketing; contract research in the life sciences. *Mailing Add:* Graebert Assocs Rte 1 Box 457-5 St Michaels MD 21663

GRAEDEL, THOMAS ELDON, b Portland, Ore, Aug 23, 38; m 66; c 2. ATMOSPHERIC CHEMISTRY, CORROSION OF METALS. *Educ:* Wash State Univ, BS, 60; Kent State Univ, MS, 64; Univ Mich, MA, 67, PhD(astrophys), 69. *Prof Exp:* Res assoc space physics, Radio Astron Observ, Univ Mich, 68-69; mem, Tech Staff Atmospheric Chem, 69-83, DISTINGUISHED MEM TECH STAFF, BELL LABS, 84- *Concurrent Pos:* Atmospheric Sci Rev Comt, Langley Res Ctr, NASA, 79; oxidant panel, Nat Comn on Air Quality, 79-80; Air Quality Div Rev Panel, Nat Ctr for Atmospheric Res, 80 & 86; publ comt, Am Geophys Union, 80-82, chmn, 82-84, budget & finance comt, 86-88, assoc ed, J Atmospheric Environ, 79-82; invitee & rapporteur, Dahlem Conf Atmospheric Chem, Berlin, 82 & Chemrawn 4, Keystone, Colo, 87; mem, tech adv group, Nat Acid Deposition Modeling Proj, 84-87, Gov Panel Acid Deposition NJ, 84-85; corrosion consult, Statue Liberty Restoration Proj, 84-86; mem, comt conserv hist & artistic works, Nat Asn Corrosion Engrs, 86-; bd atmospheric sci & climate, Nat Res Coun, 87- *Mem:* AAAS; Am Chem Soc; Am Geophys Union; Am Meteorol Soc. *Res:* Chemistry and physics of atmospheric gases and aerosols; effects of atmospheric contaminants on materials and electrical and mechanical equipment. *Mailing Add:* AT&T Bell Labs 1D 349 Murray Hill NJ 07974-2070

GRAEF, PHILIP EDWIN, b New York, NY, Oct 29, 23; m 49; c 1. BIOLOGY. *Educ:* Emory & Henry Col, AB, 47; George Peabody Col, MA, 48; Univ Va, PhD(biol), 55. *Prof Exp:* Asst prof chem, Bridgewater Col, 48-50; asst prof biol, ECarolina Col, 55-57; PROF BIOL, COLUMBIA COL, SC, 57-, CHMN DEPT BIOL, 71- *Mem:* Nat Asn Biol Teachers; Am Inst Biol Sci; Nat Sci Teachers Asn. *Res:* Megasporogenesis, ovule development and megagametogenesis of flowering plants. *Mailing Add:* Dept Biol Columbia Col Columbia SC 29203

GRAEF, WALTER L, b Staten Island, NY, Jan 17, 38; m 65; c 1. ORGANIC CHEMISTRY, PHARMACEUTICAL CHEMISTRY. *Educ:* Albany Col Pharm, NY, BS, 59; Univ Wis, MS, 61, PhD(pharmaceut chem), 63. *Prof Exp:* Res chemist org chem, Dyes Div, E I du Pont de Nemours & Co, 63-64; asst prof pharmaceut chem, Fordham Univ, 64-66; asst ed, 66-67, dept head, Org Index Ed Dept, 67-69, mgr, 69-71, mgr, Org Index Dept, 71-72, mgr, Phys Inorg Anal Chem Dept, 72-73, mgr, Chem Substance Handling Dept, 73-77, MGR, CHEM TECHNOL DEPT, CHEM ABSTR SERV, 77- *Mem:* Am Chem Soc. *Res:* Stereochemical requirements for ganglionic blockade; synthesis of short chain mono- and bis- quaternary ammonium compounds. *Mailing Add:* 164 S Stanwood Rd Columbus OH 43209-1859

GRAESSLEY, WILLIAM W(ALTER), b Muskegon, Mich, Sept 10, 33; m 53; c 3. POLYMER SCIENCE, RHEOLOGY. *Educ:* Univ Mich, BS & BSE, 56, MSE, 57, PhD(chem eng), 60. *Prof Exp:* Sr chemist, Air Reduction Co, NJ, 59-63; from asst prof to prof, 63-81, Walter Murphy prof chem eng & mat sci, Northwestern Univ, 81-83; sr sci adv, Corp Res Lab, Exxon Res & Eng Co, 82-88; PROF CHEM ENG, PRINCETON UNIV, 87- *Concurrent Pos:* Sr vis fel, Cambridge Univ, 79-80. *Honors & Awards:* Bingham Medal, Soc Rheol; Whitley Lectr, Akron, 86; Physics Prize, Am Phys Soc, 90. *Mem:* Am Chem Soc; Am Inst Chem Engrs; fel Am Phys Soc; Soc Rheol. *Res:* Relationship between molecular structure and properties of polymer solutions, melts and networks. *Mailing Add:* Dept Chem Eng Princeton Univ Princeton NJ 08544

GRAETZ, DONALD ALVIN, b Pound, Wis, Nov 19, 42. SOIL CHEMISTRY, WATER CHEMISTRY. *Educ:* Univ Wis, BS, 64, MS, 67, PhD(soils), 70. *Prof Exp:* Proj assoc soils, Univ Wis, 70-71; asst prof soils, 71-77, ASSOC PROF SOIL SCI, UNIV FLA, 77- *Mem:* Am Soc Agron; Soil Sci Soc Am; Sigma Xi. *Res:* Soil and water pollution; nitrogen transformations and transport in soils; nutrient cycles in aquatic systems. *Mailing Add:* Dept Soil Sci Univ Fla Gainesville FL 32611

GRAETZER, HANS GUNTHER, b Ger, Feb 13, 30; US citizen; m 57; c 4. PHYSICS. *Educ:* Oberlin Col, BA, 52; Yale Univ, MS, 53, PhD(physics), 56. *Prof Exp:* From asst prof to assoc prof, 56-67, chmn dept, 76-80, PROF PHYSICS, SDAK STATE UNIV, 67- *Concurrent Pos:* NSF fel, Univ Colo, 63-64; Fulbright teacher exchange, Switzerland, 87. *Mem:* Am Asn Physics Teachers; Fedn Am Sci; Am Phys Soc; Sigma Xi. *Res:* Nuclear reactions; gamma ray spectroscopy; history of science and technology; neutron activation analysis of biological and soil samples; author of articles on science topics for the general public. *Mailing Add:* Dept Physics SDak State Univ Brookings SD 57007

GRAETZER, REINHARD, b Ger, Sept 28, 33; US citizen; m 62; c 2. DNA REPAIR MECHANISMS. *Educ:* Oberlin Col, AB, 55; Univ Wis, MA, 57, PhD(physics), 62. *Prof Exp:* Ford Found fel, Niels Bohr Inst, Denmark, 62-64; res assoc & instr physics, Univ Wis, 64-65; asst prof, 65-73, ASSOC PROF PHYSICS, PA STATE UNIV, 73- *Concurrent Pos:* Sabbatical, Univ Colo, Boulder, 71-72; fac res partic, Oak Ridge Nat Lab, 73, 85, Argonne Nat Lab, 82, & USAF Sch Aerospace Med, 90; sabbatical, Nat Inst Environ Health Sci, Res, Triangle Park, NC, 87. *Mem:* Am Phys Soc. *Res:* Radiation-and-chemically induced DNA damage and repair in microorganisms. *Mailing Add:* Dept Physics 104 Davey Lab Pa State Univ University Park PA 16802

GRAF, DONALD LEE, b Howard Co, Iowa, Jan 24, 25; m. GEOCHEMISTRY. *Educ:* Colo Sch Mines, Geol E, 45; Columbia Univ, AM, 47, PhD(mineral), 50. *Prof Exp:* Geologist, Indust Minerals Div, Ill State Geol Surv, 48-63, geochemist, Chem Group, 63-65; prof geol & geophys, Univ Minn, Minneapolis, 65-69; prof, 69-82, EMER PROF GEOL, UNIV ILL, URBANA, 82- *Concurrent Pos:* Res fel geophys, Harvard Univ, 61-63. *Honors & Awards:* Mineral Soc Am Award, 60. *Mem:* Fel Am Geophys Union; fel Mineral Soc Am. *Res:* Sedimentary mineralogy and geochemistry. *Mailing Add:* 2836 E First St Tucson AZ 85716

GRAF, E(DWARD) R(AYMOND), b Cullman, Ala, Sept 26, 31; m 61; c 1. ELECTRICAL ENGINEERING. *Educ:* Auburn Univ, BSc, 57, MSc, 58; Stuttgart Tech Univ, PhD(elec eng), 63. *Prof Exp:* Asst prof elec eng, Auburn Univ, 58-59; asst, Ger Res Found, 60-63; assoc prof, 63-69, PROF ELEC ENG, AUBURN UNIV, 69- *Concurrent Pos:* Consult, US Army Missile Command, Redstone Arsenal & George C Marshall Space Flight Ctr, 63- *Mem:* Asn Ger Engrs. *Res:* Electromagnetic field theory; antennas and propagation; laser. *Mailing Add:* Dept Elec Eng Auburn Univ Auburn AL 36830

GRAF, ERLEND HAAKON, b Oslo, Norway, Oct 21, 39; US citizen; m 68. LOW TEMPERATURE PHYSICS. *Educ:* Mass Inst Technol, BS, 61; Cornell Univ, PhD(physics), 68. *Prof Exp:* Asst prof, 67-74, ASSOC PROF PHYSICS, STATE UNIV NY STONY BROOK, 74- *Mem:* Am Phys Soc. *Res:* Liquid and solid helium; solid hydrogen; nuclear magnetic resonance. *Mailing Add:* Dept Physics State Univ NY Stony Brook Main Campus Stony Brook NY 11794

GRAF, GEORGE, PROTEIN DYNAMICS, NUCLEAR SPECTROSCOPY. *Educ:* Univ Budapest, PhD(biochem), 44. *Prof Exp:* PROF BIOCHEM, NDAK STATE UNIV, 68- *Mailing Add:* Dept Biochem NDak State Univ Fargo ND 58102

GRAF, HANS PETER, b Rebstein, Switz, Jan 28, 52; m 80; c 3. ARTIFICIAL NEURAL NETWORKS, PATTERN RECOGNITION. *Educ:* Swiss Fed Inst Technol, dipl physics, 76, Dr ScNat, 81. *Prof Exp:* Res asst, Swiss Fed Inst Technol, 76-81; mem tech staff res, Spreches & Schuh, Aaraw, Switz, 81-83; MEM TECH STAFF RES, AT&T BELL LABS, 83- *Concurrent Pos:* Mem, Darpa Panel on Neural Networks, 87-88; lectr, Course Analog, MOS Integrated Circuits, 90 & 91 & Summer Sch on Neural Networks, NATO, 89. *Mem:* Sr mem Inst Elec & Electronics Engrs; Am Phys Soc. *Res:* Design of microelectronic circuits implementing artificial neural networks; applications of artificial neural networks to pattern recognition, in particular to machine vision. *Mailing Add:* AT&T Bell Labs Rm 4G320 Holmdel NJ 07733-1988

GRAF, LLOYD HERBERT, b Wis, Sept 14, 19; m 43; c 6. BIOCHEMISTRY. *Educ:* Univ Wis, BS, 41, MS, 43, PhD(biochem), 48. *Prof Exp:* Biochemist, 48-53, SUPVRY BIOCHEMIST, PHYS DEFENSE DIV, US ARMY BIOL LABS, 53- *Mem:* Sigma Xi. *Res:* Microbial nutrition and physiology; analytical biochemistry. *Mailing Add:* 618 Wilson St Frederick MD 21701

GRAF, PETER EMIL, b Baltimore, Md, Mar 31, 30; m 58; c 3. PHYSICAL CHEMISTRY. *Educ:* Univ Rochester, BS, 51; Univ Wis, PhD(phys chem), 56; Golden Gate Univ, JD, 75. *Prof Exp:* Res chemist, Calif Res Corp, 56-66; SR RES CHEMIST, CHEVRON RES CORP, 66- *Mem:* Am Chem Soc. *Res:* Chemical kinetics; photochemistry; non-electrolyte solutions; colloid chemistry; emulsion theory and technology. *Mailing Add:* 253 Manzanita Dr Orinda CA 94563-1723

GRAF, WILLIAM L, b Zanesville, Ohio, Feb 7, 47. GEOLOGY. *Educ:* Univ Wis, BA, 69, MS, 71, PhD(geog & water resources), 74. *Prof Exp:* Intel officer, US Air Force, 71-74; from asst prof to assoc prof geog, Univ Iowa, 74-78; PROF GEOG, ARIZ STATE UNIV, TEMPE, 78- *Concurrent Pos:* Consult geomorphologist, 79- *Honors & Awards:* G K Gilbert Award, Asn Am Geographers, 84; Cole Mem Award, Geol Soc Am, 85. *Mem:* Geol Soc Am; Asn Am Geographers; Brit Geomorphol Group; AAAS. *Res:* River mechanics; flooding; river channel changes; heavy metals and radionuclides in river systems. *Mailing Add:* Dept Geog Ariz State Univ Tempe AZ 85287

GRAFF, DARRELL JAY, b Cedar City, Utah, Sept 8, 36; m 62; c 4. PHYSIOLOGY, NUTRITION. *Educ:* Utah State Univ, BS, 58, MS, 60; Univ Calif, Los Angeles, PhD(parasitol), 63. *Prof Exp:* Asst, Zool, Physiol & Parasitol Labs, Utah State Univ, 57-60 & Univ Calif, Los Angeles, 60-62; NSF fel, Rice Univ, 63-64, NIH fel, 64-65; from asst prof to assoc prof, 65-72, PROF PHYSIOL, WEBER STATE COL, 72- *Concurrent Pos:* Res asst, Communicable Dis Ctr, USPHS, 57-58; partic, Int Cong Parasitol, Washington, DC, 70 & Conf on Aging, NIH, Univ Calif, Riverside, 70; tech writer, Allied Health Prof Projs, Univ Calif, Los Angeles, 70; co-dir, NSF Undergrad Res Proj Grant to Weber State Col, 70, partic, NSF Col Sci Improvement Proj Grant, 71-; consult, Albion Labs, Clearfield, Utah. *Mem:* Am Soc Parasitologists. *Res:* Carbohydrate and protein metabolism of parasites; active absorption of amino acids and minerals by animals and plants (foliar sprays); behavioral effects of abnormal trace metals mercury and lead. *Mailing Add:* Dept Zool Weber State Col Ogden UT 84408

GRAFF, GEORGE STEPHEN, b New York, NY, Mar 16, 17; m 42; c 5. AERODYNAMICS. *Educ:* DeSales Col, BA, 39; Univ Detroit, BS, 42. *Prof Exp:* Draftsman, Continental Aviation & Eng Corp, 40-42; aerodynamicist, McDonnell Aircraft Corp, 42-45, proj aerodynamicist, 45-50, chief aerodynamicist, 50-54, chief aerodyn engr, 54-57, mgr aeromech, 57-60, asst chief engr, 60-61, dir syst tech, 61-64, vpres eng technol, 64-68, vpres eng, 68-70, exec vpres, 70-71, pres, 71-82, vpres & dir, McDonnell Douglas Corp, 71-86; RETIRED. *Concurrent Pos:* Mem subcomt stability & control, Nat Adv Comt Aeronaut, 51-56, mem subcomt aerodyn stability & control, NASA, 56-58 & missile & spacecraft aerodyn, 59-61, chmn res & technol adv subcomt aircraft aerodyn, 67-68, mem res & technol adv comt aeronaut, 67-; vpres, McDonnell Douglas Corp, 71, dir, 73, mem exec comt, 74, mem subcomt, 74- *Mem:* Nat Acad Eng; Am Inst Aeronaut & Astronaut. *Res:* Stability and control. *Mailing Add:* 761 Kent Rd St Louis MO 63124

GRAFF, GUSTAV, ENZYMOLOGY, PROSTAGLADINS CHEMISTRY. *Educ:* Univ Ill, PhD(biochem), 74. *Prof Exp:* SR RESEARCHER, A H ROBINS CO, 83- *Mailing Add:* Dept Molecular Biol Res & Develop Div A H Robins Co 1211 Sherwood Ave Richmond VA 23261

GRAFF, MORRIS MORSE, b Lafayette, Ind, Mar 3, 10; m 39; c 3. ENDOCRINOLOGY. *Educ:* NY Univ, BA, 39; Tulane Univ, MSc, 46. *Prof Exp:* Asst biochem, Columbia Univ, 29-42; from asst chemist to assoc chemist, USDA, La, 42-47; chemist, US Army Chem Ctr, Md, 47-49; resident, Med Br, Endocrinol Serv, NIH, 49-56, resident, Endocrinol Sect Cancer Chemother, Nat Serv Ctr, Nat Cancer Inst, 56-59, Exec Secy, Endocrinol Study Sect, Div Res Grants, NIH, 59- 85; RETIRED. *Mem:* Am Chem Soc; Endocrine Soc. *Res:* Fats and sterols; food chemistry; rosin acids; naval stores; chromatography; rubber chemistry; cancer chemotherapy; biochemistry. *Mailing Add:* 10023 Raynor Rd Silver Springs MD 20014

GRAFF, ROBERT A, b 1933; US citizen; m 56; c 3. COAL CONVERSION TECHNOLOGY, SYNTHETIC FUELS. *Educ:* NY Univ, BE, 55; Columbia Univ, MS, 57, DeSc, 63. *Prof Exp:* From lectr to prof chem eng, 56-80, MICHAEL POPE PROF ENERGY RES, CITY UNIV NEW YORK, 80-, DIR, CLEAN FUELS INST, 76-, CHAIR, DEPT CHEM ENG, 89- *Concurrent Pos:* Vis prof, Imperial Col Sci & Technol, London, 70-71; dir, NY State Consortium Advan Coal Res, 79-81; assoc ed, Fuel Processing Technol, 81- *Mem:* Sigma Xi; Am Inst Chem Engrs; Am Chem Soc; Am Soc Eng Educ. *Res:* Coal conversion technology; pyrolysis; liquification; hot gas cleaning. *Mailing Add:* Dept Chem Eng City Col NY New York NY 10031

GRAFF, SAMUEL M, b New York, NY, May 22, 45. APPLIED MATHEMATICS. *Educ:* Rensselaer Polytech Inst, BS, 66; NY Univ, MS, 68, PhD(math), 71. *Prof Exp:* From asst to assoc prof math, 71-80, chmn, Math Dept, 84-90, PROF MATH, JOHN JAY COL CRIMINAL JUSTICE, CITY UNIV NEW YORK, 81- *Mem:* Am Math Soc; Math Asn Am; Soc Indust & Appl Math; Sigma Xi. *Res:* The application of stochastic processes to the modelling of criminal justice systems and the computer simulation of rail transportation networks; a text on differential equalieres and stability being is in progress. *Mailing Add:* John Jay Col Criminal Justice Dept Math 445 W 59 St New York NY 10019

GRAFF, THOMAS D, b Paoli, Pa, Feb 10, 26; m 49; c 2. ANESTHESIOLOGY. *Educ:* Haverford Col, AB, 49; Temple Univ, MD, 53. *Prof Exp:* Intern, USPHS Hosp, 53-54; resident anesthesia, Detroit Receiving Hosp, 54-56; instr, Sch Med, Univ Md, 56; anesthesiologist, Hosp Women of Md, Baltimore, 57; jr asst resident pediat, Baltimore City Hosps, 61-62; from asst prof to assoc prof anesthesia, Sch Med, Johns Hopkins Univ, 62-86; RETIRED. *Mem:* AMA; Am Soc Anesthesiol. *Mailing Add:* 917 Jamieson Rd Lutherville MD 21091-4801

GRAFF, WILLIAM (ARTHUR), b Highland, Ill, Dec 25, 23; m 45; c 2. CERAMICS. *Educ:* Univ Ill, BS, 46, MS, 47, PhD(ceramics), 49. *Prof Exp:* Asst ceramics, Univ Ill, 46-47; res glass technologist, Glass Tech Lab, Gen Elec Co, 49-53, res supvr, 53-66, mgr glass res lab, Lamp Glass Dept, 66-74, mgr mat lab, Lamp Glass Prod Dept, 74-80, mgr eng support, Lamp Glass & Components Dept, 80-85, mgr glass sci projs, Glass & Metall Prods Dept, 85-86; RETIRED. *Concurrent Pos:* Glass consult, 86- *Mem:* Am Chem Soc; Am Ceramic Soc; Brit Soc Glass Technol. *Res:* Formation, properties, structure, reactions and uses of glasses and semi-vitreous materials. *Mailing Add:* Seven Lakes Box 1089 West End NC 27376

GRAFF, WILLIAM J(OHN), JR, b Marshall, Tex, May 10, 23; m 44, 77; c 3. STRUCTURAL & MECHANICAL ENGINEERING. *Educ:* Tex A&M Univ, BS, 47, MS, 48; Purdue Univ, PhD(mech eng), 51. *Prof Exp:* Instr mech, Tex A&M Univ, 46-48; instr mech, design, Purdue Univ, 48-51; sr propulsion engr, Gen Dynamics-Convair, 51-54, nuclear group engr, 54-56; prof mech eng & chmn dept, Southern Methodist Univ, 56-61; dean instr, Tex A&M Univ, 61-65, dean acad admin, 65-66; PROF CIVIL ENG, UNIV HOUSTON, 66- *Concurrent Pos:* Conviar, grant, Oak Ridge Sch Reactor Technol, 52-53; consult, Gen Dynamics-Conair, 56-57 & Tex Instruments, Inc, 58-61; partic, Scand & Russian comp educ field study, 63; vis prof, Aalborg Univ Ctr, Denmark, 77-78; Fulbright scholar, Comn Educ Exchange, Denmark & US, 77-78; invited lectr, offshore structs, Tokyo Japan, 81. *Mem:* Am Soc Mech Engrs; Am Soc Eng Educ; Am Soc Civil Engrs. *Res:* Structures; vibrations; properties of materials; offshore structures; welded tubular joints; dynamics; heat transfer. *Mailing Add:* 16635 Cairnway Dr Houston TX 77084

GRAFFEO, ANTHONY PHILIP, b Boston, Mass, Oct 26, 47; m 73. TOXICOLOGY. *Educ:* Northeastern Univ, BA, 70, PhD(anal chem), 75. *Prof Exp:* Res chemist anal & marine chem, Battelle New England Marine Lab, 75-86; MGR, CHEM & LIFE SCI, ARTHUR D LITTLE INC, 87- *Concurrent Pos:* Dir, Mass Ctr Excellence, Marine Sci Bd. *Mem:* Am Chem Soc; Sigma Xi; AAAS. *Res:* Isolating and separating organic and biochemical compounds from complex mixtures using high performance separation techniques; environmental and marine science. *Mailing Add:* PO Box 2220 Duxbury MA 02332

GRAFFIS, DON WARREN, b Royal Center, Ind, Feb 17, 28; m 56; c 3. ANIMAL SCIENCE & NUTRITION. *Educ:* Purdue Univ, BS, 50, MA, 56; Univ Ill, PhD(agron), 60. *Prof Exp:* Voc agr instr, Van Buren Twp Schs, Ind, 50; field rep, Plant Food Div, Swift & Co, 52-53; voc agr instr, Noble Twp Schs, Ind, 53-55; res assoc forage crops, Univ Ill, 58-60; exten agronomist, Rutgers Univ, 60-63; exten agronomist, Ohio State Univ, 63-66; EXTEN & RES AGRONOMIST, DEPT AGRON, UNIV ILL, URBANA, 66- *Concurrent Pos:* Pvt consult, 88- *Honors & Awards:* Fel, Am Soc Agron, 84; Fel, Crop Sci Soc Am, 85. *Mem:* Am Soc Agron; Crop Sci Soc Am; Am Forage & Grassland Coun. *Res:* Production management of forage crops in humid regions. *Mailing Add:* Dept Agron Univ Ill 1102 S Goodwin Ave Urbana IL 61801

GRAFFIUS, JAMES HERBERT, b Pitcairn, Pa, June 8, 28; m 67. BOTANY. *Educ:* Univ Pittsburgh, BS, 54, MS, 58; Mich State Univ, PhD(bot), 63. *Prof Exp:* Asst biol, Univ Pittsburgh, 54-58; asst bot, Mich State Univ, 58-62; asst prof, 62-68, ASSOC PROF BOT, OHIO UNIV, 68- *Mem:* Phycol Soc Am; Int Phycol Soc; Am Micros Soc; Bot Soc Am; Sigma Xi. *Res:* Phycology; taxonomy and ecology of fresh-water algae; bog algae; aquatic biology. *Mailing Add:* Dept Bot Ohio Univ Athens OH 45701

GRAFIUS, EDWARD JOHN, b Brookings, SDak, Oct 16, 48; m 70; c 2. ENTOMOLOGY, ECOLOGY. *Educ:* Mich State Univ, BS, 70; Ore State Univ, MS, 73, PhD(entom), 77. *Prof Exp:* Grad res asst, Dept Entom, Ore State Univ, 70-77, res assoc, Agr Res Sta, 77; asst prof, Dept Entom, 77-82, assoc prof, 82-90, ENTOM EXTEN PROJ LEADER, MICH STATE UNIV, 89-, PROF, 90- *Mem:* Entom Soc Am; Ecol Soc Am. *Res:* Insect ecology; population dynamics; economic thresholds; agricultural entomology; pest management. *Mailing Add:* Dept Entom Mich State Univ East Lansing MI 48824

GRAFSTEIN, BERNICE, b Toronto, Ont, Sept 17, 29; m 63; c 1. REGENERATION, DEVELOPMENT. *Educ:* Univ Toronto, BA, 51; McGill Univ, PhD(physiol), 54. *Prof Exp:* Asst physiol, McGill Univ, 51-54, lectr, 54-55; hon res asst anat, Univ Col London, 55-57; asst prof physiol, McGill Univ, 57-62; asst prof, Rockefeller Univ, 62-69; assoc prof, 69-73, PROF PHYSIOL, MED COL, CORNELL UNIV, 73-, VINCENT & BROOKE ASTOR DISTINGUISHED PROF NEUROSCI, 84- *Concurrent Pos:* Nat Res Coun Can fel, Univ Col London, 55-56; Grass Found fel, Woods Hole Marine Biol Lab, 61; trustee, Grass Found, 65-; adj prof, Rockefeller Univ, 69-; chmn comt brain sci, Nat Res Coun-Nat Acad Sci, 75-78; nat adv coun, Nat Inst Neurol & Commun Dis & Stroke, 83-87; sci adv bd, Pew Found, 87-90. *Mem:* AAAS; Am Physiol Soc; Int Brain Res Orgn; Soc Neurosci (treas, 77-80, pres, 85-86). *Res:* Development of the nervous system; growth and regeneration of nervous tissue; electrical activity of the central nervous system. *Mailing Add:* Dept Physiol Cornell Univ Med Col 1300 York Ave New York NY 10021

GRAFSTEIN, DANIEL, b New York, NY, Dec 24, 27; m 47; c 1. CHEMISTRY. *Educ:* City Col NY, BS, 48; Purdue Univ, MS, 49, PhD(chem), 51. *Prof Exp:* Res chemist, Westinghouse Elec Corp, 51-56; res chemist, Reaction Motors Div, Thiokol Chem Corp, 56-57, group leader, 57-58, unit supvr, Adv Res Group, 59-61; sr staff scientist, Aerospace Res Ctr, Gen Precision, Inc, 62-63, prin staff scientist & mgr chem dept, 63-66, mgr, Mat Dept, 66-69; mgr, Chem Res Prog, Govt Res Lab, Exxon Res & Eng Co, 69-75, mgr laser fusion, 75-77, mgr biomass fuels proj, 77-79, sr res assoc, Corp Res Sci Labs, 79-82; H H Seedoffs Straede, Denmark; mgr dir, London Corp, Senetek DLC, 82-85; RETIRED. *Mem:* AAAS; Am Chem Soc; Am Phys Soc; NY Acad Sci. *Res:* Fluorine, boron and structural chemistry; organometallics; carboranes; photochemistry; lasers; fluorescence; laser fusion. *Mailing Add:* 4100 N Ocean Dr Singer Island FL 33404-2855

GRAFTON, ROBERT BRUCE, b Rochester, NY, May 15, 35; m 67; c 2. APPLIED MATHEMATICS, NUCLEAR ENGINEERING. *Educ:* Brown Univ, ScB, 58, PhD(appl math), 67. *Prof Exp:* Engr, Bur Ships, Atomic Energy Comn, 58-62; asst prof math, Univ Mo, Columbia, 67-71, asst prof math, Trinity Col, 71-75; mathematician, Off Naval Res, 75-78, computer scientist, 78-86; PROG DIR, NSF, 86- *Mem:* Math Asn Am; Inst Elec & Electronics Engrs; Sigma Xi. *Res:* Differential equations; periodic solutions. *Mailing Add:* MIRS Div Rm 411 1800 G St NW Washington DC 20550

GRAFTON, THURMAN STANFORD, b Chicago, Ill, Dec 20, 23; m 46; c 4. LABORATORY ANIMAL MEDICINE. *Educ:* Mich State Univ, DVM, 47; Am Col Lab Animal Med, dipl, 66. *Prof Exp:* Instr surg & med, Vet Col, Mich State Univ, 47-48; vet lab officer, Walter Reed Army Inst Res, 48-50; asst chief, Dept Virus & Rickettsial Dis, 406th Med Gen Lab, Japan, 50-52; assoc prof virol, Univ Md, 52-53; course supvr vet serv, USAF Sch Aviation Med, 53-56; exec officer clin lab, 7520th Med Clin Lab, Eng, 56-60; chief vivarium br, 6571st Aeromed Res Lab, Holloman AFB, NMex, 60-64; asst chief nutrit br, Food Div, US Army Natick Labs, Mass, 64-66; prof lab animal sci & chmn dept, Sch Health Rel Professions & res assoc prof microbiol, Sch Med & dir lab animal facil, State Univ NY, Buffalo, 66-76; exec dir, Nat Soc Med Res, 76-80; CONSULT, MED RES MGT, 81- *Concurrent Pos:* USAF Rep, Int Conf Vet Educ, London, 59; consult, univs, hosps & med res insts. *Honors & Awards:* distinguished vis prof, State Univ NY, Buffalo, 76- *Mem:* Am Vet Med Asn; Asn Gnotobiotics; Conf Pub Health Vets; Am Soc Lab Animal Practitioners (pres, 71-72); Am Asn Zoo Vets. *Res:* Improved support of the use of animals in biomedical research; diseases and treatment of aquatic animals; medical research management; education and utilization of animal technicians. *Mailing Add:* 4939 Floramar Terr Apt 907 New Port Richey FL 34652-3314

GRAGE, THEODOR B, b Munster, Ger, Mar 24, 27; US citizen; m 53; c 8. SURGERY. *Educ:* Creighton Univ, MD, 55; Univ Minn, MS & PhD(surg), 63. *Prof Exp:* Intern med, Creighton Mem St Joseph's Hosp, 55-56; from instr to asst prof, 61-69, ASSOC PROF SURG, UNIV MINN HOSPS, 69- *Concurrent Pos:* Fel surg, Univ Minn Hosps, 56-61; Am Cancer Soc adv clin fel, 62-65. *Mem:* Soc Head & Neck Surg; Am Col Surg; Am Asn Clin Oncol; Am Asn Cancer Res. *Res:* Surgical treatment of malignant diseases and chemotherapy of solid tumors. *Mailing Add:* Univ Surg Oncol Pa 5712 Clinten Ave S Minneapolis MN 55419

GRAGG, WILLIAM BRYANT, JR, b Bakersfield, Calif, Nov 2, 36; m 58, 78; c 3. MATHEMATICS. *Educ:* Univ Denver, BS, 57; Stanford Univ, MS, 59; Univ Calif, Los Angeles, PhD(math), 64. *Prof Exp:* Mathematician numerical anal, Bell-Comm, Inc, 63-64; mathematician, Oak Ridge Nat Lab, 64-67; from asst prof to assoc prof math, Univ Calif, San Diego, 67-75, prof, 75-80; PROF MATH, UNIV KY, 80- *Concurrent Pos:* Vis asst prof, Univ Tenn, 65-67; vis prof math, Univ Colo, 74-75; vis scientist, Inst Comput Appln Sci & Eng, Langley Res Ctr, NASA, 75-76. *Mem:* Soc Indust & Appl Math; Am Math Soc. *Res:* Numerical analysis, especially ordinary value problems, Pade table, continued fractions and matrices. *Mailing Add:* Code 53GR Naval Postgrad Sch Monterey CA 93943-5100

GRAHAM, A RICHARD, b Wichita, Kans, Aug 5, 34; m 62; c 2. ADVANCED MANUFACTURING SYSTEMS, INSTRUMENTATION. *Educ:* Kans State Univ, BSME, 57, MS, 60; Univ Iowa, PhD(mech eng), 66. *Prof Exp:* Struct engr, Cessna Aircraft Co, 57-58; instr mech eng, Kans State Univ, 58-60, Univ Mo, Rolla, 60-63, & Univ Iowa, 63-64; from asst prof to prof mech eng, 65-79, chmn dept, 78-84, assoc dir, 84-86, PROF, WICHITA STATE UNIV, 84-, DIR, CTR FOR PRODUCTIVITY ENHANCEMENT, 86- *Concurrent Pos:* Prin investr, City Wichita Energy Off, 76-80 & Energy Planning Proj, US Off Educ, 77-80; dir, Am Soc Eng Educ, 87-89. *Mem:* Am Soc Mech Engrs; Soc Mfg Engrs; Am Soc Eng Educ; Am Soc Qual Control. *Res:* Application of advanced manufacturing systems in small and medium size companies. *Mailing Add:* Ctr Productivity Enhancement Wichita State Univ 1845 Fairmont Box 146 Wichita KS 67208-1595

GRAHAM, A(LBERT) RONALD, b Red Deer, Can, Feb 15, 17; m 42; c 1. MINERALOGY, MINING ENGINEERING. *Educ:* Univ Alberta, BA, 37; Queen's Univ, Ont, BSc, 40, MSc, 47; Univ Toronto, PhD(mineral), 50. *Prof Exp:* Asst engr, MacKenzie Red Lake Gold Mines Ltd, 40, shift boss, 41-42; mines scientist, Mines Br, Govt of Can, 49-52; res mineralogist, geol explor, Dominion Gulf Co, 52-55; sect head, Metall Labs, Falconbridge Nickel Mines Ltd, 55-70, asst mgr, 70-77, consult, 77-82; RETIRED. *Concurrent Pos:* Consult, 77-82. *Honors & Awards:* Berry Medal, Mineral Asn of Can, 87. *Mem:* Mineral Asn Can (pres, 70-72); Can Geosci Coun (secy-treas, 72-75); Can Inst Mining & Metall; fel Mineral Soc Am; fel Geol Asn Can. *Res:* Instrumentation for mineral exploration; applications for use in deposit evaluations of X-ray diffraction and fluorescence, neutron activation analysis, microscopy (optical and electron). *Mailing Add:* 515 St George St E Fergus ON N1M 1L1 Can

GRAHAM, ALAN KEITH, b Houston, Tex, May 5, 34; m 60. PLANT MORPHOLOGY, PALEOBOTANY. *Educ:* Univ Tex, BA, 56, MA, 58; Univ Mich, PhD(bot), 62. *Prof Exp:* Instr bot, Univ Mich, 62-63; fel, Evolutionary Biol Prog, Harvard Univ, 63-64; from asst prof to assoc prof, 64-72, PROF BOT, KENT STATE UNIV, 72- *Concurrent Pos:* NSF res grants, 66-68, 69-71, 79-81, 82-84, 85-87 & 87-89; vis res scientist, Univ Amsterdam, 69-70, Univ Tex, 79-80. *Mem:* AAAS; Bot Soc Am; Int Asn Plant Taxon. *Res:* Tertiary history of Latin American vegetation; palynology and taxonomy. *Mailing Add:* Dept Biol Sci Kent State Univ Main Campus Kent OH 44242

GRAHAM, ALOYSIUS, organic chemistry, for more information see previous edition

GRAHAM, ANGUS FREDERICK, b Toronto, Ont, Mar 28, 16; nat US; m 54; c 3. MOLECULAR BIOLOGY, GENTICS. *Educ:* Univ Toronto, BASc, 38, MASc, 39; Univ Edinburgh, PhD(biochem), 42, DSc(microbiol), 52. *Prof Exp:* Lectr biochem, Univ Edinburgh, 40-47; res assoc microbiol, Connaught

Med Res Labs, Univ Toronto, 47-58, assoc prof, 53-58; prof, Univ Pa, 58-70; mem, Wistar Inst, 58-70; chmn dept, 70-80, Gilman Cheyney prof, Dept Biochem, 70-86, EMER PROF BIOCHEM, MED SCH, MCGILL UNIV, 86- *Concurrent Pos:* Eleanor Roosevelt Int Cancer fel, 64; mem study sect on virol & rickettsiology, NIH, 65-69; ed-in-chief, J Cellular Physiol, 65-70; mem comt recombinant DNA, Med Res Coun Can, 75; Josiah Macy, Jr Found fac scholar award, 77-78; mem bd dir, W Alton Jones Cell Sci Ctr, 80- *Mem:* Am Soc Microbiol; Royal Soc Can. *Res:* Replication of mammalian viruses; bacterial bioluminescence. *Mailing Add:* 4300 De Maisonneuve O Westmount PQ H3G 3C7 Can

GRAHAM, ARTHUR H(UGHES), b Philadelphia, Pa, Nov 18, 33; m 60; c 1. METALLURGY, ELECTROCHEMISTRY. *Educ:* Pa State Univ, BS, 58, MS, 60, PhD(metall), 63. *Prof Exp:* Res metallurgist, 63-66, sr res metallurgist, 66-70, sr res specialist, 70-74, res assoc, 74-75, sr res assoc, 75-83, res fel, Eng Technol Lab, 83-90, SR RES FEL, E I DU PONT DE NEMOURS & CO, INC, 90- *Mem:* Am Soc Metals; Electrochem Soc; Am Electroplaters Soc. *Mailing Add:* E I du Pont de Nemours & Co PO Box 80304 Wilmington DE 19880-0304

GRAHAM, ARTHUR RENFREE, b London, Ont, Nov 2, 19; m 50; c 4. ANIMAL PHYSIOLOGY. *Educ:* Univ Western Ont, BSc, 48, MSc, 51, PhD(physiol), 54. *Prof Exp:* Assoc prof zool, Mem Univ, 54-57; from assoc prof to prof physiol sci, Ont Vet Col, Univ Guelph, 71-84; RETIRED. *Mem:* AAAS; NY Acad Sci; Can Physiol Soc. *Res:* Physiology of central nervous system; ruminant digestive system. *Mailing Add:* 15 Lockyer Rd Guelph ON N1G 1J9 Can

GRAHAM, BEARDSLEY, b Berkeley, Calif, Apr 24, 14; m 76; c 2. SCIENCE COMMUNICATIONS. *Educ:* Univ Calif, Berkeley, BS, 35. *Prof Exp:* Asst dir, Stanford Res Inst, 53-57; staff mem, Lockheed Missile & Space Co, 57-61; pres, Spindletop Res Inc, Lexington, Ky, 61-67; CONSULT, MGT, 67- *Concurrent Pos:* Vchmn, Cent Ore Community Col Eng Technol Adv Comt; incorporator & mem bd, app by Pres Kennedy, Commun Satellite Corp; vchmn, Eng Technol Adv Comt, Cent Ore Community Col; mem, Energy Adv Comt, League Ore Cities; mem, Cent Ore Coun Higher Educ. *Mem:* Fel Inst Elec & Electronics Eng; assoc fel Am Inst Aeronaut & Astronaut. *Res:* Instrumental in the creation of Communication Satellite Corp; broad-band communication systems and television; video programming; satellite systems and applications; nuclear weapons and power; hydro-electric policy and its implementation; solar energy, applications and implementation; electronic devices. *Mailing Add:* 214 Hillcrest Pl Baker City OR 97814

GRAHAM, BENJAMIN FRANKLIN, b East Milton, Mass, Sept 24, 20; m 47; c 4. BOTANY, PLANT ECOLOGY. *Educ:* Univ Maine, BS, 43, MS, 48; Duke Univ, PhD(bot), 59. *Prof Exp:* Instr bot, Univ Maine, 48-51; instr, Miami Univ, 54-56; res fel, Dartmouth Col, 57-59; from assoc prof to prof, 68-91, EMER PROF BIOL, GRINNELL COL, 91- *Concurrent Pos:* Vis biologist, Brookhaven Nat Lab, 65-66; consult, land restoration, 65-; dir, Conrad Environ Res Area, 70-86. *Mem:* Fel AAAS; Am Inst Biol Sci; Ecol Soc Am. *Res:* Forest ecology; natural root grafts; palynology; radiation ecology. *Mailing Add:* 1514 West St Grinnell IA 50112-0806

GRAHAM, BETTIE JEAN, b Beaumont, TX, July 11, 41. TROPICAL MEDICINE, BIOMEDICAL RESEARCH. *Educ:* Texas Southern Univ, BS, 62; Baylor Col Med, PhD(virology), 71. *Prof Exp:* Fel, Albert Einstein Col of Med, 71-74; STAFF FEL, NIH, 74- *Mem:* Am Soc Microbiol; Am Soc Tropical Medicine Hygiene; AAAS. *Res:* Defining gaps in biomedical research areas; genomic research. *Mailing Add:* Rm 610 Bldg 38A Nat Ctr Human Genome Res NIH Bethesda MD 20892

GRAHAM, BRUCE ALLAN, b Menzies, Western Australia, Aug 27, 38. ORGANIC CHEMISTRY. *Educ:* Univ Western Australia, BSc, 59; Univ Alta, PhD(chem), 68. *Prof Exp:* Develop chemist, Assoc Pulp & Paper Mfrs, 60-61; from res & develop chemist agr chem to sect mgr pollution control, 67-76, sect mgr org & anal chem, 76-81, DIR RES & DEVELOP, UNIROYAL CHEM LTD, RES LAB, GUELPH, ONT, 81- *Mem:* Chem Inst Can. *Res:* Synthesis of organic chemicals for evaluation as agricultural and rubber chemicals. *Mailing Add:* Uniroyal Res Lab 120 Huron St Guelph ON N1E 5L7 Can

GRAHAM, BRUCE DOUGLAS, b Roberts, Wis, Dec 15, 15; m 46; c 3. PEDIATRICS. *Educ:* Univ Ala, AB, 39; Vanderbilt Univ, MD, 42. *Prof Exp:* From asst prof to prof pediat, Univ Mich Hosp, 51-59, dir pediat labs, 54-59; prof & head dept, Fac Med, Univ BC, 59-64; chief staff, Children's Hosp, 64-70, med dir, 70-74, dir ambulatory serv, 74-82; prof pediat, Col Med, Ohio State Univ, 64-82; RETIRED. *Concurrent Pos:* Pediatrician-in-chief, Health Ctr Children, Vancouver, BC, 59-64; chief pediat, Children's Hosp, 61-64; head dept pediat, Ohio State Univ, 64-76. *Mem:* AAAS; Soc Pediat Res; Am Acad Pediat(exec bd, 70-78, vpres, 78-79, pres, 79-80); Am Pediat Soc; AMA. *Res:* Acid base metabolism of premature infants, full term and sick children; oxygen tension study of newborn infants and sick children. *Mailing Add:* 4915 Brand Rd Dublin OH 43017

GRAHAM, C BENJAMIN, b Hannibal, Mo, Jan 15, 31; m 56; c 1. RADIOLOGY, PEDIATRIC RADIOLOGY. *Educ:* Univ Ill, Urbana, BA, 54; Univ Wash, MD, 58; Am Bd Radiol, dipl, 66. *Prof Exp:* From asst to assoc radiol, Univ Wash, 59-63, from instr to assoc prof radiol & pediat, 63-74, dir, Children's Hosp & Med Ctr Radiol, 78-90, PROF RADIOL & PEDIAT, SCH MED, UNIV WASH, 74- *Concurrent Pos:* Am Cancer Soc clin fel, 60; James Picker Found adv fel acad radiol, 62-64 & scholar radiol res, 64-66. *Mem:* Am Roentgen Ray Soc; Soc Pediat Radiol; fel Am Col Radiol; Radiol Soc NAm; fel Am Acad Pediat. *Res:* Bone development; neonatal diagnostic imaging. *Mailing Add:* Dept Radiol Children's Hosp & Med Ctr PO Box C5371 Seattle WA 98105-0371

GRAHAM, CHARLES, b Atlantic City, NJ, Nov 21, 37; div; c 3. COGNITIVE SCIENCES, PSYCHOPHYSIOLOGY. *Educ:* Univ Md, BS, 66; Pa State Univ, MS, 68, PhD(psychol), 70. *Prof Exp:* Res assoc psychol & fac mem, Inst Pa Hosp, 70-74; instr psychiat, Univ Pa Sch Med, 70-74; sr exp psychologist, 74-79, PRIN EXP PSYCHOLOGIST, MIDWEST RES INST, 79- *Concurrent Pos:* Lectr psychol, Univ Pa, 70-74; prin investr, Nat Inst Drug Abuse, 74-76, Nat Cancer Inst, 77-79, Nat Inst Gen Med Sci, 79-82, US Army Med Res & Develop Comn, 79-82, NY Dept Health, 82-84 & Dept Energy, 84-92; reviewer, 4 sci journals. *Honors & Awards:* Henry Guze Award, Soc Clin & Exp Hypnosis, 78. *Mem:* Sigma Xi; Am Psychol Asn; NY Acad Sci; Soc Psychophysiol Res. *Res:* Neurobehavioral studies of human health, behavior and productivity under stress. *Mailing Add:* Midwest Res Inst 425 Volker Blvd Kansas City MO 64110

GRAHAM, CHARLES D(ANNE), JR, b Philadelphia, Pa, Oct 15, 29; m 52; c 4. PHYSICAL METALLURGY, MAGNETIC MATERIALS. *Educ:* Cornell Univ, BMetE, 52; Univ Birmingham, Eng, PhD(metall), 54. *Prof Exp:* Res metallurgist, res & develop ctr, Gen Elec Co, NY, 54-69; chmn, 79-84, PROF MAT SCI & ENG, UNIV PA, 69- *Concurrent Pos:* Guggenheim fel, inst solid state physics, Univ Tokyo, 61-62; UK Sci Res Coun sr fel, Wolfson Ctr, Univ Col, Cardiff, 78; coop res, Res Develop Corp, Japan, 85. *Mem:* Am Phys Soc; Metall Soc; Phys Soc Japan; Inst Elec & Electronics Engrs; Am Soc Metal Int. *Res:* Magnetic materials; magnetic measurements. *Mailing Add:* Dept Mat Sci & Eng Univ Pa 3231 Walnut St Philadelphia PA 19104-6272

GRAHAM, CHARLES EDWARD, b St Paul, Minn, Nov 22, 19; m 43; c 3. GEOLOGY. *Educ:* State Col Wash, BS, 47, MS, 49; Univ Iowa, PhD(geol), 54. *Prof Exp:* Asst geologist, State Bur Mines & Geol, Mont, 49-50; asst instr geol, Univ Iowa, 51-53; from instr to prof geol, 53-81, EMER PROF GEOL, DENISON UNIV, 81- *Mem:* Geol Soc Am; Nat Asn Geol Teachers; Am Geophys Union. *Res:* Structural and urban geology. *Mailing Add:* 166 Hermosa Dr Durango CO 81301

GRAHAM, CHARLES LEE, b White Lake, SDak, Aug 12, 31; m 58; c 3. ENTOMOLOGY, PARASITOLOGY. *Educ:* Northern State Col, BS, 59; Utah State Univ, MS, 62, PhD(entom, parasitol), 64. *Prof Exp:* RES ENTOMOLOGIST, PLANT DIS RES LAB, SCI & EDUC ADMIN, USDA, 63- *Concurrent Pos:* Instr, Frederick Community Col. *Mem:* Entom Soc Am; Am Mosquito Control Asn; Sigma Xi. *Res:* Insect flight and aging as related to the epidemiology of insect borne diseases; current authority on the biology and taxonomy of the bot fly genus Cuterebra. *Mailing Add:* Bldg 1301 Fort Detrick Frederick MD 21701

GRAHAM, CHARLES RAYMOND, JR, b Baltimore, Md, June 17, 40; m 62; c 4. ICHTHYOLOGY, PHYSIOLOGY. *Educ:* Loyola Col, BS, 62; Univ Del, MS, 64, PhD(ichthyol, cytol), 67. *Prof Exp:* Lab dir, Med Eye Bank Md, 74-84; sci consult, Tissue Banks Int, 84-87; from instr to assoc prof, 66-80, chmn dept, 75-79, PROF BIOL, LOYOLA COL, MD, 80- *Concurrent Pos:* Res assoc, Sch Med, Marquette Univ, 69 & 70. *Mem:* Am Soc Zool; Am Asn Tissue Banks (secy, 81-86); Nat Asn Adv Health Professions; Asn Res Vision & Ophthalmol. *Res:* Morphology and physiology of chondrichthian reproductive systems; morphology and physiology of cellular membranes; corneal physiology; preservation of corneal tissue. *Mailing Add:* Two Deer Pass Ct Cockeysville MD 21030-2605

GRAHAM, COLIN C, b Jan 16, 42; US citizen. MATHEMATICS. *Educ:* Harvard Univ, BS, 64; Mass Inst Technol, PhD(math), 68. *Prof Exp:* Asst prof, 68-73, assoc prof, 73-81, PROF MATH, NORTHWESTERN UNIV, 81- *Mem:* Am Math Soc; Math Asn Am; AAAS. *Res:* Abstract harmonic analysis. *Mailing Add:* Dept Math Sci Lakehead Univ Thunder Bay ON P7B 5E1 Can

GRAHAM, DALE ELLIOTT, b Newark, NJ, Aug 15, 44; m 80. MOLECULAR BIOLOGY. *Educ:* Univ Tenn, Knoxville, BS, 66; Univ Tenn, Oak Ridge, PhD(biomed sci), 71. *Prof Exp:* Res fel molecular biol, Calif Inst Technol, 71-74; asst prof molecular biol, Purdue Univ, W Lafayette, 74-80; mem staff, Cancer Expert Lab Molecular Biol, Nat Cancer Inst, 80-84, mem staff, Spec Expert, Molecular, Cellular & Nutrit Endocrinol Br, Nat Inst Diabetes & Digestive & Kidney Dis, 84-91, MEM STAFF & BIOLOGIST, PERSONAL COMPUT BR, DIV COMPUTER RES & TECHNOL, NIH, 91- *Concurrent Pos:* Damon Runyon Mem Found Cancer Res fel, 71-73; Carnegie Inst Wash fel, 73-74. *Res:* Mechanisms of molecular evolution; evolving DNA sequences as related to genome structure and function; nucleic acid reassociation; gene expression; nuclei acid technology. *Mailing Add:* DCRT NIH Bldg 12A Rm 3039 Bethesda MD 20892

GRAHAM, DAVID LEE, b New York, NY, Dec 5, 39; m 73; c 2. VETERINARY PATHOLOGY, WILDLIFE DISEASES. *Educ:* Pa State Univ, BSc, 61; Cornell Univ, 65; Iowa State Univ, PhD(vet Path), 73. *Prof Exp:* Intern, Animal Med Ctr, 65-66, resident, 66-67; from instr to prof vet path, Iowa State Univ, 67-81; prof vet path, Cornell Univ, 81-87; DIR, SCHUBOT EXOTIC BIRD HEALTH CTR & PROF VET PATHOBIOL, COL VET MED, TEX A&M UNIV, 87- *Mem:* Am Vet Med Asn; Wildlife Dis Asn; Am Col Vet Pathologists; Am Asn Zoo Vets. *Res:* Diseases of wildlife, avian herpesvirus infections, pseudorabies; diseases of zoo animals; diseases of birds of prey and cage and aviary birds. *Mailing Add:* Schbot Exotic Bird Health Ctr Col Vet Med Texas A&M Univ College Station TX 77843-4467

GRAHAM, DAVID TREDWAY, b Mason City, Iowa, June 20, 17; m 41; c 3. INTERNAL MEDICINE. *Educ:* Princeton Univ, BA, 38; Yale Univ, MA, 41; Wash Univ, MD, 43. *Prof Exp:* Asst prof med, Wash Univ, 51-57; assoc prof, 57-63, chmn dept, 71-80, prof med, Sch Med, Univ Wis-Madison, 63-85; ADJ PROF PSYCHOL, UNIV DEL, 86- *Concurrent Pos:* Commonwealth fel med, Med Col, Cornell Univ, 48-51; ed, Proc, Am Fedn Clin Res, 54-59. *Mem:* Am Psychosom Soc (pres, 78-79); AMA; Am Fedn Clin Res; Soc Psychophysiol Res (pres, 69-70); Am Soc Internal Med. *Res:* Psychosomatic medicine. *Mailing Add:* Dept Psychol Univ Del, Wolf Hall Newark DE 19716

GRAHAM, DEE MCDONALD, b Dixon, Miss, Oct 11, 27; m 48; c 7. FOOD SCIENCE, FOOD TECHNOLOGY. *Educ:* Miss State Col, BSc, 50; Iowa State Col, MSc, 51, PhD(dairy bact), 54. *Prof Exp:* Res assoc dairy bact, Iowa Agr Exp Sta, 51-52; from asst prof to assoc prof dairy mfg, Clemson Col, 53-58; div chief evaporated milk res, Pet Milk Co, 58-60, mgr milk prod develop, 61-65, assoc dir res, 65-68, tech dir, Grocery Prod Div, 68-69; chmn dept food sci & nutrit, Univ Mo, Columbia, 69-75; asst dir sci res, Del Monte Corp, 75-80, dir cent res, 80-84, dir technol develop, 84-86, dir technol servs & int prod develop, 86-90, SR CONSULT, DEL MONTE CORP, 91- *Concurrent Pos:* Mem food protect comt & comt on rev of food additives, Nat Acad Sci-Nat Res Coun; dir, R&D Assoc; dir, Coun Agr Sci & Technol. *Mem:* Am Dairy Sci Asn; Inst Food Technologists; fel Inst of Food Technologists; Soc Nutrit Educ. *Res:* Product and process development; flavor chemistry; milk manufacturing technology; human nutrition; fruits and vegetable technology. *Mailing Add:* Del Monte Corp Res Ctr PO Box 9004 Walnut Creek CA 94598

GRAHAM, DONALD C W, b Thomasville, Ga, Jan 11, 32; m 53; c 4. FOOD SCIENCE, MICROBIOLOGY. *Educ:* Ft Valley State Col, BS, 54; Tuskegee Univ, MS, 58; Cornell Univ, PhD(food sci), 71. *Prof Exp:* Clerk, Off Registrar, Ft Valley State Col, 54-56; res asst physiol, Tuskegee Univ, 58-61, instr nutrit, 59-61; instr nutrit & gen sci, Ala State Univ, 61-65; res technician nutrit, 65-68, asst prof, 71-77, ASSOC PROF FOOD SCI, CORNELL UNIV, 77-, MEM FAC, 71- *Concurrent Pos:* Mem, Grad Fac Food Sci & Technol & Fac Inst Food Sci, Cornell Univ, 71-; asst dir off instr, Col Agr & Life Sci, Cornell Univ, 76-82. *Mem:* AAAS; Inst Food Technol; NY Acad Sci; Sigma Xi; Am Soc Microbiol. *Res:* Characterization of Plasmid DNA and bacteriophages from Pediococci; use of Pediococci in the manufacture of fermented dairy products; unusual food fermentations, their characterization and nutritional evaluation. *Mailing Add:* Dept Food Sci Cornell Univ Ithaca NY 14853-7201

GRAHAM, DOYLE GENE, b Coeur D'Alene, Idaho, Aug 18, 42; m 84; c 3. NEUROTOXICOLOGY, ANATOMIC PATHOLOGY. *Educ:* Duke Univ, MD, 66, PhD(exp path), 71. *Prof Exp:* Lectr, 70-71, asst prof, 71-78, assoc clin prof, 78-81, assoc prof, 81-86, PROF PATH, DUKE UNIV, 86- *Concurrent Pos:* Vis asst prof path, Sch Med, Temple Univ, 71-72; asst prof, Sch Med, Emory Univ, 72-73; vis prof path, Univ Edinburgh, 80-81; mem, NIH Toxicol Study Sect, 84-88, EPA Sci Rev Panel Health Res, 85-87; dir, Integrated Toxicol Prog, 86-; chmn, Comt Neurotoxicol & Risk Assessment, Comn Life Sci, Bd Environ Studies & Toxicol, Nat Res Coun, 88- *Mem:* Am Asn Pathologists; AAAS; Soc Toxicol; Am Asn Neuropathologists; Sigma Xi. *Res:* Development of anti-cancer agents; Parkinson's disease; neurotoxicity of hexane, carbon disulfide and other toxicants; mechanisms of metabolic and degenerative nervous system disease. *Mailing Add:* Dept Path Duke Univ Med Ctr Durham NC 27710

GRAHAM, (FRANK) DUNSTAN, b Princeton, NJ, Aug 17, 22; m 44; c 2. AERONAUTICAL ENGINEERING. *Educ:* Princeton Univ, BSE, 43, MSE, 47. *Prof Exp:* Aerodynamicist, Boeing Airplane Co, 47-48; flight res engr, Cornell Aeronaut Lab, 48-50; supvry gen engr, Wright Air Develop Ctr, 50-55; chief engr flight controls, Lear, Inc, 55-59; assoc prof aeronaut eng, 59-66, PROF AERONAUT ENG, PRINCETON UNIV, 66-; TECH DIR, SYSTS TECHNOL, INC, 59- *Mem:* Am Inst Aeronaut & Astronaut; Inst Elec & Electronics Engrs. *Res:* Aeronautical instrumentation and automatic guidance and control; system engineering. *Mailing Add:* Princeton Univ Princeton NJ 08540

GRAHAM, EDMUND F, b Mesilla Park, NMex, July 25, 24; m 48; c 6. PHYSIOLOGY, CRYOBIOLOGY. *Educ:* Utah State Univ, BS, 49; SDak State Col, MS, 52; Univ Minn, PhD(physiol), 55. *Prof Exp:* Asst dairy biochem, SDak State Col, 50-52; asst, 52-53, res fel, 53-55, from instr to assoc prof, 55-61, PROF ANIMAL PHYSIOL, UNIV MINN, ST PAUL, 61- *Mem:* Am Soc Animal Sci; Am Dairy Sci Asn; Soc Study Reproduction; Soc Cryobiol; Brit Soc Study Fertil; Sigma Xi. *Res:* Physiology and biochemistry of reproductive processes in animal industries; cryobiology and preservation of cells and tissue at low temperature. *Mailing Add:* Dept Animal Sci Univ Minn 130 Haecker St Paul MN 55108

GRAHAM, EDWARD UNDERWOOD, b Washington, DC, Sept 27, 43; div. ENVIRONMENTAL SYSTEMS, INFORMATION SCIENCE. *Educ:* Mass Inst Technol, BS, 64; Carnegie-Mellon Univ, MS, 65, PhD(systs & commun sci), 69. *Prof Exp:* Consult systs eng & info processing, Auerbach Corp, Arlington, Va, 68-72; environ planner, Off Environ Planning, Montgomery County, Md, 72-76, asst dir, 76-79; head, Sewer Syst Planning Sect, 79-83, head, Water Resources Planning Sect, 83-84, ASST TO GEN MGR, WASH SUBURBAN SANITARY COMN, HYATTSVILLE, MD, 84- *Concurrent Pos:* Dir, Off Technol Resource Mgt, Washington Suburban Sanitary Comn, Laurel, Md, 86- *Mem:* Simulation Coun; Inst Elec & Electronics Engrs; Water Pollution Control Fedn; Am Water Works Asn. *Res:* Environmental systems engineering and modeling; information systems; optimization theory. *Mailing Add:* Wash Suburban Sanitary Comn 8103 Sandy Spring Rd Laurel MD 20707

GRAHAM, FRANCES KEESLER, b USA, Aug 1, 18; m 41; c 3. PSYCHOPHYSIOLOGY. *Educ:* Pa State Univ, BA, 38; Yale Univ, PhD(psychol), 42. *Prof Exp:* Asst & instr med psychol, Wash Univ, 42-48, instr & res assoc, 53-57; instr psychol, Barnard Col, Columbia Univ, 48-51; res assoc, Univ Wisconsin, Madison, 57-64, from assoc prof to prof pediat, 64-80, prof psychol, 69-80, Hillsdale res prof, 80-86; RES PROF PSYCHOL, UNIV DEL, 86- *Concurrent Pos:* Psychologist & actg dir, St Louis Psychiat Clin, 42-44; res scientist award, Nat Inst Ment Health, 64-89, mem, Exp Psychol Study Sect, 70-74, mem, Bd Sci Counr, 77-81, chair, 79-81; Am Psychol Asn rep, Nat Res Coun, 71-74; spec consult, Nat Inst Neurol Dis & Blindness; mem, Study Ethics in Med & Biomed & Behav Res, Pres' Comn, 80-82; William James fel, Am Psychol Soc, 90. *Honors & Awards:* Distinguished Contrib Award, Soc Psychophysiol Res, 81; G Stanley Hall Medal, Am Psychol Asn, 82; Award Distinguished Sci Contrib, Am Psychol Asn, 90; Distinguished Sci Contrib Award, Soc Res Child Develop, 91. *Mem:* Nat Acad Sci; AAAS; Soc Psychophysiol Res (pres, 73-74); Soc Res Child Develop (pres, 75-77); Am Psychol Asn; Soc Exp Psychol; Soc Neuroscience; Am Phychol Soc; Psychonomic Soc. *Res:* Psychophysiology and developmental psychophysiology with special interest in attentional processes, reflex modifiability and brain-behavior relations. *Mailing Add:* Dept Physiol Univ Del Newark DE 19716

GRAHAM, FRANK LAWSON, b Preeceville, Sask, Mar 23, 42; m 68. VIROLOGY, MOLECULAR BIOLOGY. *Educ:* Univ Man, BS, 64; Univ Toronto, MA, 65, PhD(biophys), 70. *Prof Exp:* Res assoc virol, State Univ Leiden, 70-74; from asst prof to assoc prof, 75-83, PROF VIROL, MCMASTER UNIV, 83- *Concurrent Pos:* Fel, Nat Cancer Inst Can, 70-72, res scholar, 75-81, res assoc, 81- *Mem:* NY Acad Sci; Am Asn Microbiol; Cancer Soc Cell Biol. *Res:* Transformation of mammalian cells by oncogenic DNA viruses; adenovirus DNA structure and replication; adenovirus based expression vectors and recombinant vaccines. *Mailing Add:* Dept Biol Sci McMaster Univ Hamilton ON L8S 4L8 Can

GRAHAM, GEORGE ALFRED CECIL, b Ireland, Feb 25, 39; m 68. APPLIED MATHEMATICS, MECHANICS. *Educ:* Univ Dublin, BA, 61; Brown Univ, MS, 64; Glasgow Univ, PhD(math), 66. *Prof Exp:* Asst prof math, NC State Univ, 66-67; from asst prof to assoc prof, 67-80, dept chmn, 81-86, PROF MATH, SIMON FRASER UNIV, 80- *Res:* Mathematical theories of elasticity and viscoelasticity. *Mailing Add:* Dept Math Simon Fraser Univ Burnaby BC V5A 1S6 Can

GRAHAM, GEORGE G, b Hackensack, NJ, Oct 4, 23; m 49; c 6. PEDIATRICS, NUTRITION. *Educ:* Univ Pa, AB, 41, MD, 45. *Prof Exp:* Mem staff pediat, Brit Am Hosp, Lima, Peru, 47-50, chief, 52-55; res resident, Hosp Univ Pa, 51; resident, Baltimore City Hosps, Md, 55-57; mem staff, Cleveland Clin, 57-59; dir res, Brit Am Hosp, 60-71; dir Nutrit Prog, 76-85, ASSOC PROF PEDIAT, SCH MED, JOHNS HOPKINS UNIV, 65-, PROF INT HEALTH, SCH HYG & PUB HEALTH, 68- *Concurrent Pos:* Lectr, Mass Inst Technol, 62-65; vis prof, Agrarian Univ, Peru, 62-65; assoc chief pediat, Baltimore City Hosps, 65-68; mem comt amino acids, Food & Nutrit Bd, Nat Res Coun, 66-71, mem, US Nat Comt, 77-84, mem, Comt Int Nutrit Progs, 78-80, mem, Food & Nutrit Bd, 81-84; consult nutrit, NIH, 66-71, mem nutrit study sect, 71-75, chmn, 73-75; consult nutrit, US AID, 69-; dir res, Inst Invest Nutrit, Lima, Peru, 71- *Honors & Awards:* Goldberger Award, AMA, 72; Borden Award, Am Acad Pediat, 77. *Mem:* Am Inst Nutrit; Am Soc Clin Nutrit; Soc Pediat Res; Am Pediat Soc. *Mailing Add:* Dept Int Health Johns Hopkins Univ Sch Hyg & Pub Health 615 Wolfe St Baltimore MD 21205

GRAHAM, HAROLD L(AVERNE), b Cleburne, Tex, July 16, 30; m 52; c 2. CHEMICAL ENGINEERING. *Educ:* Vanderbilt Univ, BE, 55, MS, 58; Iowa State Univ, PhD, 60. *Prof Exp:* Gas field reservoir engr, Phillips Petrol Co, 55-56; from instr to assoc prof chem eng, Vanderbilt Univ, 56-62; res engr, Jersey Prod Res Co Div, Standard Oil Co, NJ, 62-64, sr res engr, Esso Prod Res Co Div, Esso Middle East, 64-66, sr res specialist, 66-67, res supvr, 67-71, tech adv drilling, Humble Oil & Refining Co, 71-72, eng mgr, Baytown Dist, Exxon Corp, 72-73, supply planning adv, Esso Explor, 73-75, drilling planning & anal mgr, 75-77, proj mgr, Esso Middle East, 77-90; PRES, GLOBAL ADVANTAGE, 90- *Mem:* Am Inst Chem Engrs; Am Inst Mining, Metall & Petrol Engrs. *Mailing Add:* PO Box 434 Seabrook TX 77586

GRAHAM, HAROLD NATHANIEL, b New York, NY, July 27, 21; m 44; c 5. ORGANIC CHEMISTRY. *Educ:* Cornell Univ, BA, 41; Univ Chicago, PhD(org chem), 46. *Prof Exp:* Res chemist, 52-60, mgr food res, 60-62, asst dir res, 62-67, dir food res & nutrit, 67-77, VPRES RES & PROD DEVELOP, THOMAS J LIPTON, INC, 77- *Concurrent Pos:* Dir Ceytea Ltd, Colombo, Ceylon, 66. *Mem:* AAAS; Am Chem Soc; fel Am Inst Chem; Inst Food Tech. *Res:* Free radical chemistry; chemistry of natural products; biochemistry of plants and microorganisms; petroleum and synthetic rubber research; flavor chemistry; gas-liquid chromatography; food chemistry; food product development; plant proteins. *Mailing Add:* 256 Broad Ave Englewood NJ 07631

GRAHAM, HARRY MORGAN, b Whittier, Calif, June 18, 29. ENTOMOLOGY. *Educ:* Univ Calif, BS, 51, MS, 53, PhD(entom), 59. *Prof Exp:* Entomologist, Agr Res Serv, USDA, 58-71, location-res leader, Cotton Insects Lab, Brownsville, Tex, 71-77, res leader, Biol Control Insects Lab, 77-87; RETIRED. *Concurrent Pos:* Vis prof entom, Univ Ariz, 75-76, adj prof, 77-87. *Mem:* AAAS; Entom Soc Am; Int Orgn Biol Control; Sigma Xi. *Res:* Insect ecology and economic entomology; ecology of insect pests of cotton. *Mailing Add:* 5655 E Towner Tucson AZ 85712

GRAHAM, HENRY ALEXANDER, JR, immunohematology, for more information see previous edition

GRAHAM, HENRY COLLINS, b Pittsburgh, Pa, Aug 16, 34; m 57; c 3. CERAMIC ENGINEERING. *Educ:* Alfred Univ, BS, 56, MS, 58; Ohio State Univ, PhD(ceramic eng), 65. *Prof Exp:* Mem staff crystalizable glasses, Res Lab, Pittsburgh Plate Glass, Pa, 57-60; res physicist, Aerospace Res Labs, 60-77, MEM STAFF, AIR FORCE MAT LAB, WRIGHT PATTERSON AFB, 77- *Mem:* Am Ceramic Soc. *Res:* Determining the defect structure of both metals and metal oxides by making measurements such as electrical conductivity and continuous weight change under various ambient conditions; behavior of materials in reactive atmospheres including the determination of the reaction mechanisms and the influence of the defect structure and the microstructural features on these mechanisms. *Mailing Add:* 1148 Ludlow Rd Xenia OH 45385

GRAHAM, JACK BENNETT, b Superior, Nebr, Oct 16, 13; m 41; c 6. PHYSICAL GEOLOGY. *Educ:* York Col, AB, 35; Univ Nebr, BS, 37; Univ Iowa, MS, 40, PhD(geol), 42. *Prof Exp:* Asst geol, Univ Iowa, 38-42; jr geologist, Ground Water Br, US Geol Surv, 42-43, from asst geologist to assoc geologist, 43-48, dist geologist, 48-52, chief water utilization sect, Tech Coord Br, 52-54; ground water geologist, Leggette & Brashears, 54-55, partner, 55-76, pres, Leggette, Brashears & Graham, Inc, 76-79; RETIRED. *Concurrent Pos:* Mem adv bd, Am Inst Hydrol. *Mem:* AAAS; Am Inst Prof Geol (vpres, 67); Geol Soc Am; Am Inst Mining, Metall & Petrol Eng; Am Inst Hydrol. *Res:* Ground water geology and hydrology; hydrogeology. *Mailing Add:* One Detmer Rd Setauket NY 11733-1912

GRAHAM, JACK RAYMOND, b High Point, NC, Sept 15, 25; m 48; c 3. ORGANIC CHEMISTRY. *Educ:* Univ NC, AB, 50, MA, 52. *Prof Exp:* Chemist, Westvaco Chem Div, 52-54, Niagara Chem Div, 54-55, proj leader, 55-58, group leader, 58-62, supvr, 62-63, MGR, AGR CHEM DIV, FMC CORP, 63- *Mem:* Am Chem Soc. *Res:* Synthesis, residue analysis, metabolism, environmental impact, process development and formulation of organic chemicals for use as agricultural pesticides. *Mailing Add:* PO Box 13528 2505 Meriden Pkwy Research Triangle Park NC 27709-3528

GRAHAM, JAMES CARL, b Clarendon, Tex, May 20, 41; m 64; c 2. AGRONOMY, BOTANY. *Educ:* Tex Tech Col, BS, 63; Univ Wis, MS, 64, PhD(agron), 66. *Prof Exp:* Res asst, Univ Wis, 63-66; sr res chemist, Monsanto Co, 66-69, res specialist, 69-75, develop assoc, 75-77, regional mgr prod develop, 77-79, dir, 79-89, dir herbicide technol, 89-91, DIR PLANT PROTECTION, IMPROV & LICENSING, MONSANTO CO, 91- *Mem:* Am Soc Agron; Crop Sci Soc Am; Weed Sci Soc. *Res:* Crop physiology, physiological aspects of yield of major field crops; practical aspects of stress physiology and plant growth regulation. *Mailing Add:* Monsanto Co 800 N Lindbergh Blvd V2B St Louis MO 63167

GRAHAM, JAMES W, b Copper Cliff, Ont, Jan 17, 32; m 55; c 3. MATHEMATICS, COMPUTER SCIENCE. *Educ:* Univ Toronto, BA, 54, MA, 55. *Prof Exp:* Appl sci rep, Int Bus Mach Corp, 55-59; from asst prof to assoc prof, 59-66, PROF MATH, UNIV WATERLOO, 66-, DIR COMPUT CTR, 61- *Mem:* Asn Comput Mach; Am Math Soc; Math Asn Am; Comput Soc Can (pres, 65-66). *Res:* Non-numeric computing including sorting; specialized computer languages; information retrieval. *Mailing Add:* Dept Comput Sci Univ Waterloo Waterloo ON N2L 3G1 Can

GRAHAM, JOHN, b Wishaw, Scotland, June 10, 40; m 60; c 6. CATALYSIS, LOW TEMPERATURE REACTIONS. *Educ:* Detroit Inst Technol, BS, 67; Wayne State Univ, PhD(phys & organ chem), 71. *Prof Exp:* Sr res chemist, Occidental Petrol (Hooker), 70-76; tech mgr, de Soto Inc, 76-78; res dir, Seibert Oxidermo, 78-80; coordr polymers & coating, 80-86, RES DIR, COATINGS RES INST, EASTERN MICH UNIV & PAINT RES ASSOCS, 86- *Mem:* Am Chem Soc; Fedn Soc Coatings Technol. *Res:* Polymer synthesis, low temperature cross-linking reactions & electrophillic aromatic substitution reactions. *Mailing Add:* 430 W Forest Ypsilanti MI 48093

GRAHAM, JOHN, b Windsor, Eng, Jan 25, 33; Brit & US citizen; m 58; c 2. GENERAL PHYSIOLOGY. *Educ:* Univ Wales, BSc, 54. *Prof Exp:* Sr sci officer, UK Atomic Energy Authority, 58-68; fel scientist, Westinghouse Elec Corp, 68-71, mgr, 71-84; mgr, Rockwell Hanford Opers, 84-87; mgr, Westinghouse Hanford Opers, 87-88; DIR, ATOMIC ENERGY CAN LTD, 88- *Concurrent Pos:* Consult, Nat Acad Sci, 82; assoc prof nuclear safety, Carnegie-Mellon Univ, Pittsburgh, 70-74; mem bd, Am Nuclear Soc, 77-80 & 86-92. *Mem:* Fel Am Nuclear Soc; Am Col Sports Med. *Res:* Nuclear energy applied to raise the standard of our life and our environment; innovative and safe licensing of large and small nuclear plants. *Mailing Add:* 344 Slater St Atomic Energy Can Ltd Ottawa ON K0J 1J0 Can

GRAHAM, JOHN BORDEN, b Goldsboro, NC, Jan 26, 18; m 43; c 3. PATHOLOGY. *Educ:* Davidson Col, BS, 38; Cornell Univ, MD, 42; Am Bd Path, dipl, 61. *Hon Degrees:* DSc, Davidson Col, 84. *Prof Exp:* Intern path, NY Hosp, 42-43; asst, Cornell Univ, 43-44; from instr to prof, 46-66, chmn bd pop ctr, 64-67, mem, 72-76 & 76-80, assoc dean med sch, 68-70, dir genetics training prog, 61-85, chmn curric in genetics, 63-85, coodr interdepartmental grad progs biol, 68-85, ALUMNI DISTINGUISHED PROF PATH UNIV NC, CHAPEL HILL, 66- *Concurrent Pos:* Markle scholar, 49-54; mem res career award comt, USPHS, 59-62, mem genetics training comt, 62-66, chmn, 67-71, mem genetic basis disease comt, 77-80; mem path test comt, Nat Bd Med Exam, 63-67; Int Comt Hemostasis & Thrombosis, 63-67; mem med & sci adv comt, Nat Hemophilia Found, 75-78. *Honors & Awards:* O Max Gardner Award, Univ NC, 68. *Mem:* AAAS; Am Soc Exp Path; Soc Exp Biol & Med; Am Soc Human Genetics (secy, 64-67, pres, 72); AMA. *Res:* Physiology of blood coagulation; hemorrhagic diseases; human genetics; human population dynamics. *Mailing Add:* Box 607 Chapel Hill NC 27514

GRAHAM, JOHN ELWOOD, b Kingston, Ont, Jan 30, 33; div; c 1. STATISTICS, MATHEMATICS. *Educ:* Carleton Univ, BSc, 55; Queen's Univ, Ont, MA, 57; Iowa State Univ, MS, 60, PhD(statist), 63. *Prof Exp:* Statistician, Dominion Bur Statist, 57-58; asst, Statist Lab, Iowa State Univ, 58-63; statistician, Dominion Bur Statist, 63-65; from asst prof to assoc prof math, 65-82, PROF STATIST, CARLTON UNIV, ONT, 83- *Concurrent Pos:* Vis assoc prof, NC State Univ, 78-79. *Mem:* Am Statist Asn. *Res:* Sample survey methods and theory. *Mailing Add:* Dept Math & Statist Carleton Univ Ottawa ON K1S 5B6 Can

GRAHAM, JOHN W, METALLIC & CERAMIC PROCESSING. *Educ:* Univ Ill, BS, 46. *Prof Exp:* Res Staff, Univ Ill, 46-49; res engr, Kennametal, Inc, 49-55; res engr, mgr & proj dir, GE Aircraft Gas Turbine Div, 55-61; FOUNDER & PRES, ASTRO MET ASSOCS, INC, 61- *Mem:* Am Ceramic Soc; fel Am Soc Metals; Sigma Xi; Soc Adv Mat & Processing Engrs; Soc Biomat; Orthop Res Soc. *Res:* Sintered ceramics and cermets for jet engine and rocket applications; titanium carbide cermets for jet engine and abrasion applications; coating for refractory metal, intermetallic compounds, oxidation and synthesis at elevated temperature plasma and arc spray of refractory materials for elevated temperature applications; unusual porous metal and ceramic systems; high density wear resistant ceramics and wear resistant cermet for high temperatures; granted seven patents. *Mailing Add:* Astro Met Assoc Inc 9974 Springfield Pike Cincinnati OH 45215

GRAHAM, JOSEPH H, b Richmond, Va, Sept 23, 33; m 58; c 1. ORGANIC CHEMISTRY. *Educ:* Va Union Univ, BS, 53; Howard Univ, MS, 56, PhD(org chem), 59. *Prof Exp:* Res chemist, Div Drug Chem, 59-78, referee drugs, steroids & terpinoids, Asn Off Anal Chemists, 60-81, dir, Nat Ctr Antibiotics Anal, 78-82, CHIEF, ANTIMICROBIAL DRUGS BR, FOOD & DRUG ADMIN, 82- *Mem:* Am Chem Soc; Sigma Xi. *Res:* Preparation and properties of benzyl-o-nitrophenylglyoxals; development of analytical procedures for equine conjugated estrogens, synthetic estrogens and iodoamino acids in pharmaceuticals; gel filtration. *Mailing Add:* 7516 Greer Dr Ft Washington MD 20744

GRAHAM, JOSEPH HARRY, b Anderson, SC, Sept 11, 21; m 46; c 2. PHYTOPATHOLOGY. *Educ:* Clemson Univ, BS, 42; NC State Col, PhD(plant path), 50. *Prof Exp:* Plant pathologist, US Regional Pasture Res Lab, 50-66; asst br chief, Veg & Ornamentals Res Br, Agr Res Ctr, USDA, 66-72, Plant Stress Lab, Beltsville Md, 72-77; plant pathologist, Waterman-Loomis Co, 77-80 & Waterman-Loomis Res, Inc, 80-88; RETIRED. *Mem:* Am Phytopath Soc. *Res:* Diseases of forage grasses, legumes and field crops; research administration; production of vegetables and ornamentals; plant stress-pathogen relationships. *Mailing Add:* 900 Balmoral Dr Silver Spring MD 20903

GRAHAM, KENNETH JUDSON, b Modesto, Calif, Jan 17, 47; m 71; c 2. ENERGETIC MATERIALS CHEMISTRY, DETONATION PHYSICS. *Educ:* Univ Calif, Berkeley, AB, 68; Naval Postgrad Sch, MS, 79. *Prof Exp:* Chemist, Naval Postgrad Sch, 70-78; res chemist, Naval Weapons Ctr, China Lake, 78-88; PRIN SCIENTIST, ATLANTIC RES CORP, 89- *Concurrent Pos:* Instr chem, Monterey Peninsula Col, 74-78; adj res prof aeronaut eng, Naval Postgrad Sch, 83-84; lectr & course coordr, Computational Mech Assocs, 86-; mem, var subcomts, Joint-Army-Navy-NASA-Air Force Interagency Propulsion Comt. *Mem:* Sigma Xi. *Res:* Propellants and explosives performance and sensitivity tailoring; insensitive munitions; detonation physics & chemistry; internal & external blast; explosives testing; detonation modeling; co-author of book on explosive shock in air and one patent. *Mailing Add:* 551 Appomattox Dr Warrenton VA 22186

GRAHAM, LE ROY CULLEN, b Meeker, Colo, Dec 14, 26; m 49; c 4. SYNTHETIC APERTURE RADAR APPLICATIONS, APPLIED MATHEMATICS. *Educ:* Univ Colo, BS, 50; Stevens Inst Technol, MS, 56. *Prof Exp:* Mem tech staff, Bell Tel Labs, 50-56; sr develop engr, Loral Defense Systs-Ariz, 56-57, eng specialist, 57-60, sr eng specialist, 60-76, dir recon systs, 76-89; CONSULT, 89- *Mem:* Inst Elec & Electronics Engrs. *Res:* Coherent radar system; synthesis of new systems and image analysis; radar elevation measurement; interferometry and stereo. *Mailing Add:* 2038 W Solano Dr Phoenix AZ 85015

GRAHAM, LINDA KAY EDWARDS, b Springfield, Mo, Mar 4, 46; m 69; c 1. BOTANY, CTYOLOGY. *Educ:* Wash Univ, BA, 67; Univ Tex, Austin, MA, 69; Univ Mich, Ann Arbor, PhD(bot), 75. *Prof Exp:* Asst prof, 76-82, ASSOC PROF BOT, UNIV WIS-MADISON, 82- *Concurrent Pos:* Lectr electronmicros, Eastern Mich Univ, 76. *Mem:* Am Inst Biol Sci; Phycol Soc Am; Bot Soc Am. *Res:* Origin of lang plants (embryophytes) from ancestral green algae; elctronmicroscopy, cell biology, chemical systematics of advanced green algae as compared to that of lower plants; evolutionary origins of plant reproduction. *Mailing Add:* Dept Bot 132 Birge Hall Univ Wis 430 Lincoln Dr Madison WI 53706

GRAHAM, LOREN R, b Hymera, Ind, June 29, 33; m 55; c 1. HISTORY OF SCIENCE, SCIENCE POLICY. *Educ:* Purdue Univ, BS, 55; Columbia Univ, MA, 60, PhD(hist), 64. *Hon Degrees:* DHL, Purdue Univ, 86. *Prof Exp:* Asst prof hist sci, Ind Univ, 63-66; from assoc prof to prof hist, Columbia Univ, 66-78; PROF HIST SCI, MASS INST TECHNOL, 78-; PROF HIST SCI, HARVARD UNIV, 85- *Concurrent Pos:* Guggenheim Found fel, 69-70; mem Inst Advan Study, Princeton Univ, 69-70; Rockefeller Found fel, 76-77; Prog on Sci Int Affairs res fel, Harvard Univ, 76-77; consult, Nat Acad Sci, 76-78, Nat Endowment Humanities, 76-77 & NSF, 76. *Mem:* AAAS; Hist Sci Soc; Am Asn Advan Slavic Studies (treas, 66-67); Am Hist Asn; Soc Hist Technol. *Res:* History of science especially in Russia; science and values. *Mailing Add:* 7 Francis Ave Cambridge MA 02138

GRAHAM, LOUIS ATKINS, b Kenbridge, Va, Mar 27, 25; m 55; c 1. TEXTILES, COLOR SCIENCE. *Educ:* Univ Va, BCh Eng, 49; Univ Louisville, MChEng, 50. *Prof Exp:* Chem engr, fibers, Am Viscose Div, FMC Corp, 50-52, mgr qual control, 52-54, color specialist, 53-56, corp color specialist, 56-63, div color specialist, 63-66, sect leader fibers res & develop, 66-67; res & develop mgr textiles, Burlington Industs, 67-87; PRES, LOU GRAHAM & ASSOCS, INC, 87- *Concurrent Pos:* Assoc adj prof textile chem, NC State Univ, Raleigh, 70-79; co-founder, Color Mkt Group. *Honors & Awards:* Forrest L Dimicic Award, Color Mkt Group, 81. *Mem:* Am Asn Textile Chemists & Colorists; Optical Soc Am; Color Mkt Group (pres, 62-65); Inter-Soc Color Coun (pres, 82-84); Sigma Xi. *Res:* Textiles and color science (dyeing, color technology). *Mailing Add:* 1207 Colonial Ave Greensboro NC 27408

GRAHAM, MALCOLM, b Pa, Nov 26, 23; m 52; c 2. MATHEMATICS. *Educ:* NJ State Col, Trenton, BS, 46; Univ Mass, MS, 48; Columbia Univ, EdD(math, educ), 54. *Prof Exp:* Instr math, Marion Inst, 48-49; assoc prof, Longwood Col, 51-55; instr, E Carolina Col, 55-56; from asst prof to assoc prof math, 56-64, chmn div sci, math & appl sci, 59-64, chmn dept math, 65-66, PROF MATH, UNIV NEV, LAS VEGAS, 64- *Mem:* Math Asn Am; Am Math Soc. *Res:* Applied science. *Mailing Add:* Dept Math Univ Nev Las Vegas NV 89154

GRAHAM, MARGARET HELEN, b Dallas, Tex; m 50; c 3. INFORMATION SCIENCE. *Educ:* Univ Va, BS, 47. *Prof Exp:* Info chemist, Sharples Chem Co, 47-50; supvr lit searching, Ethyl Corp, 50-56; consult tech lit, 56-63; info scientist, Gen Motors Res Labs, 63-66; mgr info serv, 66-85, SR STAFF ADV, EXXON RES & ENG CO, 85- *Concurrent Pos:* Mem bd dirs & secy, Eng Info, 82- *Mem:* Am Chem Soc; Am Soc Info Sci; fel Am Inst Chemists; Am Petrol Inst. *Res:* Computer technology in the handling of documented information for building of knowledge information data bases for online access. *Mailing Add:* 92 Pine Way New Providence NJ 07974-1821

GRAHAM, MARTIN H(AROLD), b Jamaica, NY, July 12, 26; m 49; c 2. INSTRUMENTATION, COMPUTER SCIENCE. *Educ:* Polytech Inst Brooklyn, BEE, 47, DEE(electronics), 52; Harvard Univ, MSc, 48. *Prof Exp:* Instr, Polytech Inst Brooklyn, 47-50; res assoc, Brookhaven Nat Lab, 50-52, electronic engr, 52-57; from assoc prof to prof elec eng, Rice Univ, 57-66; assoc dir comput ctr, 66-69, PROF COMPUT SCI, UNIV CALIF, BERKELEY, 66- *Concurrent Pos:* Consult, Brookhaven Nat Lab, 57-61, US Atomic Energy Comn, 60-63, Manned Spacecraft Ctr, NASA, 63-64 & Lawrence Berkeley Lab, 74-77; vis prof, Univ Calif, Berkeley, 64-65. *Mem:* Fel Inst Elec & Electronics Engrs. *Res:* Electronic instrumentation; medical electronics; digital computers. *Mailing Add:* Elec Eng Dept Univ Calif Cory Hall Berkeley CA 94720

GRAHAM, MICHAEL JOHN, b Carlisle, Eng, Mar 21, 40; Can citizen; c 3. PHYSICAL CHEMISTRY. *Educ:* Liverpool Univ, BSc, 61, Hons, 62, PhD(surface chem), 65. *Prof Exp:* Fel chem, Nat Res Coun Can 65-67; res officer mat, Cent Elec Generating Bd, UK, 67-69; res officer chem, 69-77, head metallic corrosion & oxidation, 77-90, HEAD CHEMICAL CHARACTERIZATION, NAT RES COUN CAN, 91- *Honors & Awards:* T P Hoar Prize, Inst Corrosion Sci & Technol, 83. *Mem:* Nat Asn Corrosion Engrs; Royal Soc Chem; Royal Inst Chem; Inst Corrosion Sci & Technol; Am Soc Metals; Chem Inst Can; Electrochem Soc. *Res:* Kinetics and mechanism of metallic corrosion and oxidation; application of surface-analytical techniques to oxidation and corrosion research and microelectronics degradation. *Mailing Add:* Inst Microstruct Sci Nat Res Coun Can Ottawa ON K1A 0R9 Can

GRAHAM, PAUL WHITENER LINK, b Elmira, NY, Sep 19, 39; m 60; c 2. CERAMICS ENGINEERING. *Educ:* Pa State Univ, BS, 61, MS, 62, PhD(ceramic technol), 65. *Prof Exp:* Sr technologist, Corp Res, 65-69, sr scientist, Glass Container Div, 69-75, SECT CHIEF PROD & PROCESS PERFORMANCE, GLASS CONTAINER DIV, OWENS ILL, INC, 75- *Mem:* Am Ceramic Soc; AAAS; Soc Glass Technol; Am Soc Testing & Mat. *Res:* New product development; glass container design; surface treatments; chemical properties; quality control; glass strength; computer aided design; computer simulation. *Mailing Add:* 3105 Pelham Rd Toledo OH 43606

GRAHAM, RAY LOGAN, b Tex, July 1, 34. MATHEMATICS. *Educ:* WTex State Univ, BS, 56; NMex State Univ, MAT, 60, PhD(math), 68. *Prof Exp:* Mathematician, White Sands Missile Range, 56-58; teacher math, Las Cruces High Sch, 59; asst prof, Hardin-Simmons Univ, 59-62 & Appalachian State Univ, 63-65; res assoc, Ctr Res Sci & Math, 67-68; prof, Appalachian State Univ, 69-70, chmn math dept, 70-73; dir instrnl syst, Drexel Univ, 73-74; PROF MATH, APPALACHIAN STATE UNIV, 74- *Mem:* Math Asn Am; Nat Coun Teachers Math; Nat Educ Asn. *Res:* Effects of computer usage upon elementary analysis courses. *Mailing Add:* Dept Math Appalachian State Univ Boone NC 28608

GRAHAM, RAYMOND, b Clinton, Ind, July 1, 35. PHYSICAL CHEMISTRY, MOLECULAR SPECTROSCOPY. *Educ:* Ind State Univ, Terre Haute, BS, 62; Mont State Univ, PhD(chem), 70. *Prof Exp:* ASST PROF CHEM, ROCKY MOUNTAIN COL, 70- *Mem:* Am Chem Soc. *Res:* Absorption spectroscopy of forbidden transitions in organic molecules; luminescence spectra of transition metal complexes; polarized luminescence of aromatic molecules. *Mailing Add:* 1511 Poly Dr Billings MT 59102-1739

GRAHAM, RICHARD CHARLES BURWELL, b Miami Beach, Fla, Apr 17, 26; Can citizen; m 80; c 4. PHARMACOLOGY. *Educ:* Univ Western Ont, BSc, 49, MSc, 51, PhD(endocrinol, pharmacol), 55. *Prof Exp:* Pharmacologist, Food & Drug Labs, Dept Nat Health & Welfare, Can, 55-60; investr med sci, Inst Exp Med, Caracas, Venezuela, 60-62; pharmacologist, Food & Drug Labs, Dept Nat Health & Welfare, Can, 62-66, asst chief div med & pharmacol, 66-70, asst dir Bur Drugs, 70-87, actg dir Bur Human Prescription Drugs, 87-88, special adv, Bur Human Prescription Drugs, Health Protection Br, 88-91; PVT CONSULT, PHARMACEUT, 91- *Concurrent Pos:* Cairncross & Lawrence Prize, Ont, 55. *Res:* Toxicology of pesticides and drugs; biological activity of plant products; endocrinology. *Mailing Add:* 2117 Kingsley Rd Ottawa ON K2C 2X6 Can

GRAHAM, ROBERT (KLARK), b Harbor Springs, Mich, June 9, 06; m 49; c 8. OPTICS. *Educ:* Mich State Univ, AB, 31; Ohio State Univ, BSc, 37. *Hon Degrees:* DSc, Pasteur Found, S Calif Col Optom, OD, Ohio State Univ. *Prof Exp:* Spec rep ophthal instruments, Bausch & Lomb Optical Co, 37-40; from western mgr to sales mgr, Univis Lens Co, 40-46; dir res, Plastic Optics Co, 46-47; DIR RES, ARMORLITE LENS CO, INC, 47-; PRES, GRAHAM INT, INC, 79- *Concurrent Pos:* Assoc prof, Southern Calif Col Optom, 48-68; spec lectr, Sch Med, Loma Linda Univ, 50-; dir, Advan Concepts Tech, Inc, 64-; comt mem, Am Nat Standards Inst, 70-; secy, Intrasci Res Found, 74-, vpres, 81; CO-FOUNDER SPERM BANKING, REPOSITORY FOR GERMINAL CHOICE, 79. *Honors & Awards:* Friedrich Wilhelm Herschel Gold Medal, Int Soc Contact Lens Specialists, 58; Nat Eye Res Found Award, 66; Wiiliam Feinbloom Award, Am Acad Optom. *Mem:* AAAS; Optical Soc Am; Am Ord Asn; Am Asn Physics Teachers; NY Acad Sci; Sigma Xi. *Res:* Variable focus lenses; hard resin optics; contact lenses; prismatic effects of ophthalmic lenses; lens manufacturing; future of man. *Mailing Add:* 2141 Palomar Airport Rd Suite 300 Carlsbad CA 92008

GRAHAM, ROBERT, b Pueblo, Colo, Feb 15, 43; m. MEDICINE. *Educ:* Earlham Col, AB, 65; Univ Kans, MD, 70. *Prof Exp:* Asst adminr agency goals, Health Serv & Ment Health Admin, Dept Health Educ & Welfare, Washington, DC, 70-73; asst dir, Div Educ, Am Acad Family Physicians, Kansas City, Mo, 73-76; dep dir, Bur Health Manpower, Health Resources Admin, Dept Health Educ & Welfare, 76-78, dep adminr, 78-79; prof staff mem, Subcomt Health & Sci Res, Comt Labor & Human Resources, US Senate, 79-80; actg adminr, Health Resources Admin, Dept Health & Human Serv, 81-82, adminr, 82-85; EXEC VPRES, AM ACAD FAMILY PHYSICIANS, 85- *Concurrent Pos:* Consult, Off of Dir, Nat Health Serv Corps, 73-74, Social Security Studies, Inst Med-Nat Acad Sci, 74-76; resident family pract, Baptist Mem Hosp, 74-75; staff, Prog Health Mgt, Baylor Col Med, 76; exec secy, Grad Med Educ Nat Adv Comt, 78-79; mem, Am Asn Med Soc Execs Adv Comt to Exec Vpres of AMA, 86-; mem, Rural Health Adv Comt, Off Technol Assessment, Washington, DC, 88-; bd dirs, Sun Valley Forum Nat Health, 89- *Mem:* Inst Med-Nat Acad Sci; Asn Am Med Cols; AMA; Am Acad Family Physicians; Am Acad Med Dirs; Am Asn Med Soc Execs; Am Soc Asn Execs. *Res:* Family medicine. *Mailing Add:* Am Acad Family Physicians 8880 Ward Pkwy Kansas City MO 64114

GRAHAM, ROBERT ALBERT, b Dallas, Tex, Feb 11, 31; m 51; c 3. PHYSICS. *Educ:* Univ Tex, BS, 54, MS, 58; Tokyo Inst Technol, DSc, 90. *Prof Exp:* Res engr, Southwest Res Inst, 56-57; mem res staff, 58-83, DISTINGUISHED MEM TECH STAFF, SANDIA LABS, 83- *Honors & Awards:* Excellence Award, Dept Energy, 83; C B Sawyer Award, 84. *Mem:* Fel AAAS; fel Am Phys Soc; Inst Elec & Electronics Engrs. *Res:* Mechanical and physical properties of solids under shock wave loading conditions; shock compression science; high pressure physics. *Mailing Add:* 383 La Entrada Rd Tome Los Lunas NM 87031

GRAHAM, ROBERT LESLIE, b Saratoga, Iowa, Jan 9, 26; m 57; c 1. INORGANIC CHEMISTRY, PHYSICAL CHEMISTRY. *Educ:* Mankato State Col, BA, 52; Univ Minn, MS, 54; Univ Va, PhD(inorg chem), 58. *Prof Exp:* Res chemist, Orlon Res Dept, E I du Pont de Nemours & Co, 58-59; assoc prof chem, Va Polytech Inst, 59-63; assoc prof, 63-70, chmn dept, 64-80, PROF CHEM, MANKATO STATE COL, 70-, MEM FAC, DEPT CHEM & GEOL, 77- *Mem:* Am Chem Soc; Sigma Xi. *Res:* Thermochemistry of inorganic compounds by solution calorimetry. *Mailing Add:* Dept Chem PO Box 40 Mankato State Col Mankato MN 56001

GRAHAM, ROBERT LOCKHART, b Peterborough, Ont, Feb 17, 21; m 44; c 3. PHYSICS. *Educ:* McMaster Univ, BA, 43, MA, 45; Univ London, PhD(physics), 49; Imp Col, Univ London, Dipl, 49. *Prof Exp:* Res physicist, Nat Res Coun Can, Atomic Energy Can, Ltd, 43-46, asst res officer, 49-54, assoc res officer, 54-61, sr res officer, Nuclear Labs, 62-85; adj prof, Univ Guelph, 85-90; RETIRED. *Concurrent Pos:* Vis physicist, Lawrence Radiation Lab, Univ Calif, 62-63. *Mem:* Fel Am Phys Soc; Can Asn Physicists. *Res:* Low energy nuclear physics, chiefly beta and gamma ray spectroscopy and associated techniques; positron annihilation in liquids and solids; computer control system for superconducting cyclotron. *Mailing Add:* Four Tweedsmuir Rd Deep River ON K0J 1P0 Can

GRAHAM, ROBERT M, b Sydney, Australia, Apr 2, 48. HYPERTENSION. *Educ:* Univ NSW, Sydney, BS & MB, 72, MD, 88; FRCP(A), 79. *Prof Exp:* Intern med, St Vincent's Hosp, Sydney, Australia, 72-73, resident, 73-74, registrar, 74-75; postdoctoral fel, Dept Pharmacol, Health Sci Ctr, Univ Tex, Dallas, 77-78, asst prof pharmacol & internal med, 78-82, fac mem, Grad Sch Biomed Sci, 81-82; assoc prof med, Harvard Med Sch, Boston, 82-89; PROF, DEPT PHYSIOL & BIOPHYS, CASE WESTERN RESERVE UNIV, 89-; ROBERT C TARAZI CHMN, DEPT HEART & HYPERTENSION RES, RES INST, CLEVELAND CLIN FOUND, 89- *Concurrent Pos:* Physician, Parkland Mem Hosp, Dallas, 78-82; asst med, Mass Gen Hosp, Boston, 82-86, dir, Cardiac Immunodiagnostic Lab, 84-87, consult, 87-89; staff physician, Dept Hypertension & Nephrology, Cleveland Clin Found, 89-; consult. *Honors & Awards:* Pfizer Lectr. *Mem:* Brit Med Asn; Int Soc Nephrology; Am Fedn Clin Res; Am Soc Pharmacol & Exp Therapeut; AAAS; Am Soc Clin Invest. *Res:* Antihypertensive drug pharmacology; role of the sympathetic nervous system in circulatory homeostasis; molecular characterization of adrenergic receptors; adrenergic mechanisms and lipoprotein metabolism; molecular and cellular biology of atrial natriuretic factor; author of numerous scientific publications. *Mailing Add:* Dept Heart & Hypertension Res Res Inst Cleveland Clin Found One Clinic Ctr 9500 Euclid Ave Cleveland OH 44195-5071

GRAHAM, ROBERT MONTROSE, b St Johns, Mich, Sept 26, 29. SOFTWARE DEVELOPMENT METHODS. *Educ:* Univ Mich, BA, 56, MA, 57. *Prof Exp:* Mem staff, Univ Mich, 57-61, res assoc, 61-63; prog coordr, Proj Mac, Mass Inst Technol, 63-66, staff mem, Elec Eng Dept, 65-67, assoc prof comput sci, 67-72; assoc prof, City Col New York, 72-75; PROF COMPUT SCI, UNIV MASS, 75- *Concurrent Pos:* Vis assoc prof, Univ Calif, Berkeley, 70-72; chmn, Dept Comput & Info Sci, Univ Mass, 75-81; mem, Comput Sci Bd, 80-83; consult, Gen Motors Tech Ctr, 61-63 & 68, Honeywell Info Syst, 72-74 & 77 & Bell Telephone Labs, 74-75; vis scholar, Stanford Univ, 81. *Mem:* Asn Comput Mach. *Res:* Programming languages and compiler building systems; operating systems; software development methods and environments; computer science education. *Mailing Add:* Dept Comput Sci Univ Mass Amherst Campus Amherst MA 01003

GRAHAM, ROBERT REAVIS, b Hiawatha, Kans, Jan 8, 25; m 50; c 3. ELECTRICAL ENGINEERING. *Educ:* Univ Kans, BS, 49. *Prof Exp:* Dept head microwave eng, McDonnell Douglas Electronics Co, 61-64, dir foliage progs, 64-65, vpres foliage progs, 65-66, vpres & group exec, Mil Radar Div, Mich, 66-68, vpres & gen mgr, Ann Arbor Div, 68-71, div vpres, Mo, 71-75, prog mgr, F-15 Avionics Automatic Test Equip Div, McDonnell Douglas Corp, 75-85, dir aircraft support eng, McDonnell Aircraft Co, 85-89; RETIRED. *Mem:* Sigma Xi. *Mailing Add:* 25 Arrowhead Chesterfield MO 63017

GRAHAM, ROBERT WILLIAM, b Cleveland, Ohio, Oct 10, 22; m 51; c 3. HEAT TRANSFER. *Educ:* Case Inst Technol, BS, 48; Purdue Univ, MS, 50, PhD(mech eng), 52. *Prof Exp:* Instr mech eng, Purdue Univ, 48-52; aero res scientist, Lewis Res Ctr, NASA, 53-57, res supvr, 57-80, sr staff scientist, 81-84, chief, Off Technol Aassessment, 84-90; RETIRED. *Concurrent Pos:* Adj prof, NC State Univ, 65-72; consult, 90- *Honors & Awards:* Centennial Medal, Am Soc Mech Engrs, 80; Except Eng Achievement Medal, NASA, 82. *Mem:* Fel Am Soc Mech Engrs; Sigma Xi; assoc fel Am Inst Aeronaut & Astronaut. *Res:* Rotating stall of axial flow compressors; liquid rocket heat transfer; cryogenic fluid heat transfer; boiling and two phase flow; turbulent boundary layers; curvature effects on heat transfer; alternate fuels and gas; turbine heat transfer. *Mailing Add:* 22895 Haber Dr Fairview Park OH 44126

GRAHAM, ROGER KENNETH, b New York, NY, May 24, 29; m 50; c 2. POLYMER CHEMISTRY. *Educ:* Mass Inst Technol, BS, 50; Univ Chicago, PhD, 53. *Prof Exp:* Res chemist, 53-60, head lab, 60-73, sect mgr, 73-84, PATENT LIAISON MGR, ROHM & HAAS CO, 84- *Mem:* Am Chem Soc. *Res:* Graft and block copolymers; polymer chemistry; monomer and polymer synthesis; modifiers for polyvinyl chloride and other plastics. *Mailing Add:* Rohm & Haas Co PO Box 219 Bristol PA 19077

GRAHAM, ROGER NEILL, b Augusta, Ga, Feb 21, 41. COMPUTER SCIENCE. *Educ:* Valdosta State Col, BS, 63; Univ NC, PhD(physics), 71. *Prof Exp:* Asst prof physics, Concord Col, 68-76, assoc prof, 76-78; WRITER, 78- *Mem:* Inst Elec & Electronics Engrs Comput Soc; Asn Comput Mach; Sigma Xi. *Res:* Computer science. *Mailing Add:* 913 Papaya Street Augusta GA 30904

GRAHAM, RONALD A(RTHUR), b College Point, NY, Jan 6, 24; m 50; c 3. CHEMICAL ENGINEERING. *Educ:* Columbia Univ, BS, 44. *Prof Exp:* Glass technol engr, Corning Glass Works, 44-47; res proj engr, Wyandotte Chem Corp, 47-50, sect head eng, 50-53, asst to dir contract res, 53, mgr dept, 54-57, dir, 57-61, asst to vpres res, 61-64 & mfg, 64-66; mgr lab facil, Celanese Res Co, 66-68, dir admin & tech support, 68-73; PRES, GRAHAM ASSOCS INC, 73- *Mem:* Am Chem Soc; Am Inst Chem Engrs; fel Am Inst Chemists; Soc Chem Indust. *Res:* Rocket propellant technology; glass technology; silicates; high temperature reactions; heat transfer; chemical process development; management of research and industrial chemicals manufacturing. *Mailing Add:* 92 Pine Way New Providence RI 07974-1821

GRAHAM, RONALD LEWIS, b Taft, Calif, Oct 31, 35; div; c 2. MATHEMATICS. *Educ:* Univ Alaska, BS, 58; Univ Calif, Berkeley, MA & PhD(math), 62. *Hon Degrees:* LLD, Western Mich Univ, 84; DSc, St Olaf Col, 85, Univ Alaska, 88. *Prof Exp:* dir, Math Sci Res Ctr, 62-88, ADJ DIR, INFO SCI DIV, AT&T BELL LABS, 88-; UNIV PROF, RUTGERS UNIV, 87- *Concurrent Pos:* Regents prof math, Univ Calif, Los Angeles, 74; vis prof comput sci, Stanford Univ, 79 & 81; Fairchild dist scholar, Calif Inst Technol, 82; vis prof comput sci, Princeton Univ, 87. *Honors & Awards:* Polya Prize in Combinatorics, 72. *Mem:* Nat Acad Sci; Math Asn Am; Sigma Xi; Soc Indust & Appl Math; Asn Comp Mach; Am Math Soc; Inst Elec & Electronics Engrs; Oper Res Soc Am; fel Am Acad Arts & Sci. *Res:* Combinatorics; graph theory; algorithms; number theory; combinatorial geometry. *Mailing Add:* 2T-102 Bell Tel Labs Murray Hill NJ 07974

GRAHAM, RONALD POWELL, b Ottawa, Ont, Apr 25, 15. ANALYTICAL CHEMISTRY. *Educ:* Queen's Univ, Ont, BA, 37, MA, 38; Columbia Univ, AM, 40, PhD(chem), 42. *Prof Exp:* Asst chem, Columbia Univ, 39-42; lectr, 42-44, from asst prof to assoc prof, 44-53, chmn dept, 52-58, dean sci, 62-68, dean sci studies, 67-77, PROF CHEM, MCMASTER UNIV, 53- *Honors & Awards:* Chem Educ Award, Chem Inst Can, 61, Fisher Sci Lectr Award, 70. *Mem:* Am Chem Soc; fel Chem Inst Can; fel Royal Soc Chem. *Res:* Less-common metals. *Mailing Add:* Dept Chem McMaster Univ Hamilton ON L8S 4K1 Can

GRAHAM, SHIRLEY ANN, b Flint, Mich, Mar 20, 35; m 60; c 3. BOTANY, TAXONOMY. *Educ:* Mich State Univ, BS, 57; Univ Mich, PhD(bot), 63. *Prof Exp:* Botanist, Harvard Univ, 63-64; asst prof bot, Univ Akron, 65-66; res assoc, 67-74, ADJ PROF BIOL SCI, KENT STATE UNIV, 84- *Mem:* Bot Soc Am; Am Soc Plant Taxon (treas, 82-85); Asn Trop Biol; Int Asn Plant Taxon. *Res:* Taxonomic studies in family Lythraceae, especially genus Cuphea. *Mailing Add:* Dept Biol Sci Kent State Univ Main Campus Kent OH 44242

GRAHAM, STEPHAN ALAN, b Evansville, Ind, Apr 25, 50; div; c 2. SEDIMENTARY TECTONICS, PETROLEUM GEOLOGY. *Educ:* Ind Univ, AB, 72; Stanford Univ, MS, 74, PhD(geol), 76. *Prof Exp:* Instr field geol, Ind Univ Geol Field Sta, 72 & Stanford Univ Geol Surv, 74; res geologist plate tectonics, Exxon Prod Res Co, 76; explor geologist petrol, Chevron USA, Inc, 76-80; assoc prof, 80-88, PROF APPL EARTH SCI & GEOL, STANFORD UNIV, CALIF, 88- *Concurrent Pos:* Petrol geologist consult, 80- *Honors & Awards:* Sproule Award, Am Asn Petrol Geologists. *Mem:* Fel Geol Soc Am; Soc Econ Paleontologists & Mineralogists; Sigma Xi; Am Asn Petrol Geologist; Am Geophys Union. *Res:* Interplay of sedimentation and tectonics, particularly sedimentary response to plate tectonic processes; application of sedimentary tectonics to exploration for energy resources. *Mailing Add:* Dept Appl Earth Sci Stanford Univ Stanford CA 94305

GRAHAM, SUSAN LOIS, b Cleveland, Ohio, Nov 16, 42; m 71. COMPUTER SCIENCE. *Educ:* Harvard Univ, AB, 64; Stanford Univ, MS, 66, PhD(comput sci), 71. *Prof Exp:* Assoc res scientist & adj asst prof comput sci, Courant Inst Math Sci, New York Univ, 69-71; from asst prof to assoc prof, 71-81, PROF COMPUT SCI, UNIV CALIF, BERKELEY, 81- *Concurrent Pos:* NSF res grant, 74-; co-ed, Communications, Asn Comput Mach, 75-79, ed-in-chief, Transactions on Prog Lang & Systs, 78- *Mem:* Asn Comput Mach; Inst Elec & Electronics Engrs. *Res:* Programming language design and implementation; syntax error recovery, parsing, code generation and optimization. *Mailing Add:* Comput Sci Div Eecs Univ Calif 2120 Oxford St Berkeley CA 94720

GRAHAM, TERRY EDWARD, b Grand Rapids, Mich, Sept 19, 40; m 64; c 2. PHYSIOLOGICAL ECOLOGY. *Educ:* Suffolk Univ, AB, 65; Univ NH, MS, 67; Univ RI, PhD(zool), 72. *Prof Exp:* Instr, 70-73, from asst prof to assoc prof, 73-83, PROF BIOL, WORCESTER STATE COL, 83- *Concurrent Pos:* Prin investr, US Fish & Wildlife Serv, 79-81, Mass Div Fisheries & Wildlife, 81-88 & NY Zool Soc Conserv grant, 81-82. *Mem:* Am Soc Ichthyologists and Herpetologists; Soc Study Reptiles & Amphibians; Herpetologists' League. *Res:* Turtle ecology; determination of the effects of temperature and photoperiod on acclimatization of temperature preference and locomotor activity patterns in reptiles and amphibians; analysis of growth, morphometric variation and population structure in freshwater turtles; zoogeography of New England amphibians and reptiles. *Mailing Add:* Human Biol Univ Guelph Guelph ON N1G 2W1 Can

GRAHAM, TOM MANESS, b Paducah, Ky, Mar 16, 37; m 76. BIOLOGY. *Educ:* Florence State Univ, BS, 59; Univ Ala, MS, 66, PhD(zool, physiol), 70. *Prof Exp:* Asst prof physiol, Tex Col Osteop Med, 70-71 & Chicago Col Osteop Med, 71-72; from asst prof to assoc prof, 72-83, PROF BIOL, UNIV ALA, 83- *Res:* Physiology and morphogenesis of Hydra. *Mailing Add:* Box 870344 Tuscaloosa AL 35487-0344

GRAHAM, W(ALTER) DONALD, b Ottawa, Ont, June 2, 19; m 42; c 2. BIOCHEMISTRY. *Educ:* Ont Agr Col, BSA, 40; McGill Univ, MSc, 42; Univ Toronto, PhD(biochem), 45. *Prof Exp:* Ont Res Found res fel pharmaceut, Toronto, 45-47; biochemist nutrit res, Wash State Col, 48-49; chemist pharmacol res, Food & Drug Lab, Can Dept Nat Health & Welfare, 49-58; dir res, Midwest Med Res Found, 58-64; dir prod & process res, Farmland Indust, Inc, 64-78, exec dir res & develop, 78-82; RETIRED. *Concurrent Pos:* Mem shock & plasma expander panel, Defence Res Bd Can, 56-58. *Mem:* Am Chem Soc; Am Soc Pharmacol & Exp Therapeut. *Res:* Feeds; fertilizers; petroleum; chemical engineering; agricultural engineering; pesticides; plant sciences; pollution control; waste utilization; small scale fuel alcohol production. *Mailing Add:* 2565 S Gleneagles Dr Deland FL 32724-8458

GRAHAM, WALTER WAVERLY, III, b Nashville, Tenn, Apr 30, 33; m 55; c 2. NUCLEAR ENGINEERING, TECHNICAL MANAGEMENT. *Educ:* US Naval Acad, BS, 55; Vanderbilt Univ, MS, 60; Ga Inst Technol, PhD(nuclear eng), 65. *Prof Exp:* Res & develop staff assoc, Gen Atomic Div, Gen Dynamics Corp, 60-62; asst prof nuclear eng, Ga Inst Technol, 65-68, assoc prof, 68-70, res scientist, Nuclear Res Ctr, 65-70; pres, Tech Anal Corp, 70-83; exec vpres, Digital Commun Asn, 83-85. *Concurrent Pos:* Consult comput systs. *Mem:* Sigma Xi. *Res:* Computer-aided experimentation; minicomputer, microcomputer and time-sharing project management. *Mailing Add:* 13505 Freemanville Rd Alpharetta GA 30201

GRAHAM, WILLIAM ARTHUR GROVER, b Rosetown, Sask, Aug 23, 30; m 51; c 5. METAL CARBONYLS, HYDROCARBON ACTIVATION. *Educ:* Univ Sask, BA, 52, MA, 53; Harvard Univ, PhD(chem), 56. *Prof Exp:* Lectr & res assoc chem, Univ Southern Calif, 56-57; res chemist & consult, Arthur D Little, Inc, 57-62; assoc prof, 62-67, PROF CHEM, UNIV ALTA, 67- *Honors & Awards:* Noranda Award, Chem Inst Can, 70; Centenary Lectr, Royal Soc Chem, 87-88. *Mem:* Am Chem Soc; Chem Inst Can; Royal Soc Chem. *Res:* Synthesis, reactions and structure of organometallic and coordination compounds; transition metal hydrides and carbonyls; activation of carbon-hydrogen bonds under mild conditions. *Mailing Add:* Dept Chem Univ Alta Edmonton AB T6G 2G2 Can

GRAHAM, WILLIAM DOYCE, JR, b Clarendon, Tex, Feb 22, 39; m 62; c 2. PLANT GENETICS, PLANT BREEDING. *Educ:* Purdue Univ, MS, 65, PhD(genetics), 67. *Prof Exp:* From asst prof to prof, plant genetics & breeding, Clemson Univ, 66-88; CONSULT, 88- *Mem:* Am Soc Agron; Crop Sci Soc Am. *Res:* Small grain genetics, including disease resistance; genetics of yield and variety development. *Mailing Add:* Dept Agron & Soils Clemson Univ Clemson SC 29634-0359

GRAHAM, WILLIAM JOSEPH, b Cle Elum, Wash, Jan 6, 32; m 75; c 1. BEHAVIORAL ECOLOGY. *Educ:* Whitman Col, BA, 54; Univ Mich, MS, 62, PhD(zool), 68. *Prof Exp:* Teacher pub schs, 56-58 & 59-60; res asst, Univ Mich, 65-67; from lectr to asst prof biol, City Col New York, 67-71; asst prof biol, State Univ NY Col Geneseo, 71-75; asst prof, 75-85, ASSOC PROF BIOL & CHMN, MONROE COMMUNITY COL, ROCHESTER, NY, 85- *Mem:* Am Soc Zool; Ecol Soc Am; Am Soc Mammal. *Res:* Daily activity patterns and social interactions in small mammals; effects of behavior on numbers and densities of animals. *Mailing Add:* Dept Biol Monroe Community Col Rochester NY 14623

GRAHAM, WILLIAM MUIR, b Wainwright, Alta, June 30, 29; m 55; c 3. ECOLOGY, INSECT BEHAVIOR. *Educ:* Univ Alta, BSc, 53; Imp Col, Univ London, Dipl, 56; St Andrews Univ, PhD(insect behav), 64. *Prof Exp:* Res asst, Forest Biol Lab, Can Dept Agr, summers 49-55; stored prod entomologist, Maize & Produce Bd, Govt Kenya, 57-61; storage specialist, Trop Stored Prod Centre, Ministry of Overseas Develop, UK Govt, 64-67, head ecol sect, 67-68; ASSOC PROF BIOL, LAKEHEAD UNIV, 68- *Mem:* Entom Soc Can. *Res:* Forest and stored products entomology; general ecological problems. *Mailing Add:* Ten Wishart Crescent Thunder Bay ON P7A 6G3 Can

GRAHAM, WILLIAM RENDALL, b Melbourne, Australia, Nov 22, 38; m 84; c 2. PHYSICS, MATERIALS SCIENCE. *Educ:* Univ Melbourne, BSc, 59, MSc, 61; Oxford Unit, PhD(exp nuclear physics), 65. *Prof Exp:* Res staff molecular biophysicist, Yale Univ, 65-68, asst prof molecular biophys, 68-72; vis asst prof metall, Univ Ill, Urbana, 72-74; assoc prof, 74-85, PROF MAT SCI & ENG, UNIV PA, PHILADELPHIA, 86- *Mem:* Am Vacuum Soc; Sigma Xi. *Res:* Surfaces and interfaces; materials science; ion scattering, electron spectroscopics, auger, leed, ups, field ion microscopy. *Mailing Add:* Dept Metall Mat Sci Univ Pa 3231 Walnut St Philadelphia PA 19104

GRAHAM, WILLIAM RICHARD MONTGOMERY, b New Westminster, BC, Jan 22, 44. MOLECULAR SPECTROSCOPY. *Educ:* Univ Western Ont, BSc, 66, MSc, 68; York Univ, PhD(physics), 71. *Prof Exp:* Nat Res Coun Can fel, dept chem, Univ Fla, 71-75; res assoc molecular spectros, Herzberg

Inst Astrophysics, Nat Res Coun Can, 75-77; asst prof, 77-81, assoc prof, 81-87, chmn, Dept Physics, 84-86, PROF PHYSICS, TEX CHRISTIAN UNIV, 87- *Mem:* Am Phys Soc; Am Chem Soc; Am Astron Soc; Sigma Xi. *Res:* Molecular spectroscopy; spectroscopic studies of molecules of astrophysical interest; solid state electron paramagnetic resonance studies; characterization and utilization of fossil fuels. *Mailing Add:* Dept Physics Box 32915 Tex Christian Univ Ft Worth TX 76129

GRAHAM-ELLIS, AVIS, minority aged suicide & depression, for more information see previous edition

GRAHN, DOUGLAS, b Newark, NJ, Apr 25, 23; m 46, 73; c 3. GENETICS, RADIOBIOLOGY. *Educ:* Rutgers Univ, BS, 48; Iowa State Univ, MS, 50, PhD(genetics), 52. *Prof Exp:* Asst zool, Rutgers Univ, 48; asst genetics, Iowa State Col, 48-51; assoc scientist, Argonne Nat Lab, 53-58; geneticist, US AEC, 58-61; assoc biologist, Argonne Nat Lab, 61-62, assoc dir, 62-66, sr biologist, 66-78, div dir, 78-81, sr biologist, Div Biol & Med Res, 66-87, scientist, spec term appointee, 87; RETIRED. *Concurrent Pos:* Mem radiation control adv bd, Md State Dept Health, 60-61; mem radiobiol adv panel, Space Sci Bd, Nat Acad Sci-Nat Res Coun, 62-71, chmn, 71-74; consult, McDonnell Aircraft Corp, Mo, 63-65 & Off Manned Space Flight, NASA, 64-70; sci prog coordr foreign exhibit prog, US AEC, 66-69, sr analyst, Div Biomed & Environ Res, US AEC, 73-74; mem comt biol sci, Ill Bd Higher Educ, 69-70; mem sci comt biol aspects radiation protection criteria, Nat Coun Radiation Protection, 72-90, coun mem, 78-84; chmn working group life sci, Comt Space Res, Int Coun Sci Unions, 74-78; mem, comt Biol Effects Ionizing Rad, Nat Acad Sci & Nat Res Coun, 87-89. *Mem:* Radiation Res Soc; Environ Mutagen Soc. *Res:* Mammalian radiation genetics; genetic effects of transuranic elements; external radiation toxicology; neutron radiobiology. *Mailing Add:* Biol & Med Res Div Argonne Nat Lab Argonne IL 60439

GRAHN, EDGAR HOWARD, b Bremerton, Wash, Nov 19, 19; m 50. INORGANIC CHEMISTRY. *Educ:* Col Puget Sound, BS, 41; Univ Idaho, MS, 48; Univ Ill, PhD(chem), 55. *Prof Exp:* From instr to assoc prof chem, 46-62, exec secy res coun, 61-71, from asst dean to assoc dean grad sch, 65-77, PROF CHEM, UNIV IDAHO, 62-, EMER DEAN GRAD SCH, 77- *Mem:* Am Chem Soc. *Res:* Coordination compounds and optical isomerism. *Mailing Add:* 1414 Borah Ave Moscow ID 83843-2404

GRAIFF, LEONARD B(ALDINE), b Litchfield, Ill, Dec 16, 33; m 63; c 3. MECHANICAL ENGINEERING. *Educ:* Univ Ill, BSc, 55, MSc, 56; Purdue Univ, PhD(combustion eng), 59. *Prof Exp:* Res engr fuels res, Shell Oil Co, Ill, 59-62, res engr, Thornton Res Ctr, Shell Res Ltd, Eng, 62-64, res group leader, Shell Oil Co, Ill, 64-69, res group leader, Tex, 69-70, staff engr fuels res, 70-79, sr staff engr, 79-84, sr prin scientist, Thornton Res Centre, Shell Res Ltd, Eng, 84-87, RES ADV, SHELL DEVELOP CO, 87- *Honors & Awards:* Henry Ford Mem Award. *Mem:* Soc Automotive Engrs; Combustion Inst. *Res:* Combustion phenomena; petroleum fuels; gasoline additives; automotive exhaust emissions. *Mailing Add:* Fuels Dept Shell Develop Co PO Box 1380 Houston TX 77251-1380

GRAIKOSKI, JOHN T, b Wakefield, Mich, June 7, 24; m 58; c 4. ENVIRONMENTAL MICROBIOLOGY. *Educ:* Univ Mich, BS, 49, MS, 52, PhD(bact), 61. *Prof Exp:* Res asst botulium studies, Univ Mich, Ann Arbor, 52-55, res assoc, 55-64; supvry res microbiol, Bur Com Fisheries, Ann Arbor, 64-70; SUPVRY RES MICROBIOL ECOL, NAT MARINE FISHERIES SERV, MILFORD, CONN, 70- *Concurrent Pos:* Lectr radiation biol, Univ Mich, 56-58 & fisheries, 65-67; consult, Bur Com Fisheries, 63-64 & Governor's Comt Botulism Control, 63-66; mem, Toxic Microorganism Panel, US-Japan, 64-75, comt compendium food microbiol methods, 71- *Mem:* Am Soc Microbiol; Wildlife Dis Asn; Int Asn Great Lakes Res. *Res:* Botulism; radiation sterilization; radiation biology; microbial ecology, especially marine; anaerobic bacteria; heat sterilization of foods. *Mailing Add:* 379 W River Rd Orange CT 06477

GRAINGER, CEDRIC ANTHONY, b Juneau, Alaska, Mar 1, 44; m 65; c 4. CLOUD PHYSICS, WEATHER MODIFICATION. *Educ:* Mont State Univ, BS, 66, MS, 68; State Univ NY, Albany, PhD(atmospheric sci), 73. *Prof Exp:* Res assoc, Mont State Univ, 72-73; asst prof meteorol, State Univ NY, Oswego, 73-76; res scientist meteorol, Environ Res & Technol, 77-79; DIR ATMOSPHERIC RES, CTR AEROSPACE SCI, UNIV NDAK, 79- *Mem:* Am Meteorol Soc; Royal Meteorol Soc; Am Geophys Union; Weather Modification Asn. *Res:* Atmospheric sciences; cloud physics, employing research aircraft and radar; formation of precipitation and the potential of clouds for precipitation augmentation; wind shear and turbulence measurements in cloud and aircraft icing. *Mailing Add:* Dept Meteorol Univ NDak Main Campus Grand Forks ND 58202

GRAINGER, JOHN JOSEPH, b Dublin, Ireland, Jan 29, 34; US citizen; m 63; c 4. ENERGY ENGINEERING, ELECTRICAL ENGINEERING. *Educ:* Nat Univ Ireland, BE, 61; Univ Wis, Madison, MSEE, 64, PhD(elec eng), 68. *Prof Exp:* Asst prof elec eng, Ill Inst Technol, 68-70; prin engr, Wis Elec Power Co, Milwaukee, 70-77; assoc prof, 77-81, PROF ELEC ENG, NC STATE UNIV, 81- *Concurrent Pos:* Consult, Ill Res Inst Technol, 68, Commonwealth Edison Co, Chicago, 69, Res Triangle Inst, Raleigh, 77-, NC Power & Light Co, 78- *Mem:* Inst Elec & Electronics Engrs; Nat Soc Prof Engrs; Am Soc Eng Educ; Sigma Xi. *Res:* Electric power engineering; electrical machinery; transmission and distribution engineering; solar energy systems; switching overvoltages. *Mailing Add:* Dept Elec-Comput Eng NC State Univ PO Box 7911 Raleigh NC 27695-7911

GRAINGER, ROBERT BALL, b Centerview, Mo, Mar 3, 23; m 47; c 2. NUTRITION, BIOCHEMISTRY. *Educ:* Cent Mo State Col, BS, 47; Univ Mo, MA, 49, PhD(physiol chem), 54. *Prof Exp:* Asst, Univ Mo, 47-52, instr, 52-53; from asst prof to assoc prof animal nutrit, Univ Ky, 53-60; res chemist, Monsanto Co, St Louis, 60-61, prod develop, 61-63, res group leader, 63-68; dir res, Diamond A Cattle Industs, 68-70; vpres, Agr Technol, Inc, 70-80; researcher, Great Plains Consult Inc, 80-87; RETIRED. *Mem:* Am Soc Animal Sci; Poultry Sci Asn Am; Am Dairy Sci Asn; Sigma Xi; Fedn Am Socs Exp Biol. *Res:* Monogastric and ruminant intermediary metabolism. *Mailing Add:* 2850 E Serendipity Circle Colorado Springs CO 80917

GRAINGER, ROBERT MICHAEL, b San Francisco, Calif, Sept 10, 48. DEVELOPMENTAL BIOLOGY. *Educ:* Stanford Univ, AB, 70; Univ Calif, Berkeley, PhD(zool), 74. *Prof Exp:* Fel biol, Yale Univ, 74-76; from asst prof to assoc prof, 76-89, PROF BIOL, UNIV VA, 89- *Concurrent Pos:* Fel USPHS, 70-74; fel J C Childs Mem Fund for Med Res, 74-76; NIH grant, 78-94. *Mem:* Am Soc Cell Biol; Am Soc Zoologists; Soc Develop Biol; AAAS. *Res:* Molecular basis of early development; determination of the eye, lens and other anterior structures. *Mailing Add:* Dept Biol Gilmer Hall Univ Va Rm 264 Charlottesville VA 22901

GRAINGER, THOMAS HUTCHESON, JR, b Bethlehem, Pa, Dec 14, 13; m 41; c 1. MEDICAL BACTERIOLOGY. *Educ:* Lehigh Univ, BA, 36, MS, 38, PhD(bact), 46. *Prof Exp:* Asst, Sharp & Dohme, Pa, 36-38; asst instr, Med Sch, Univ Pa, 39-41; from instr to assoc prof bact, Lehigh Univ, 46-59; asst dir biol labs, Nat Drug Co, Swiftwater, Pa, 59-70, mgr biol serv, 70-71; BACTERIOLOGIST, MONROE PATH ASN, 71- *Concurrent Pos:* Lectr, Inst Microbiol, Univ Colo, 64. *Mem:* Fel Am Pub Health Asn; Hist Sci Soc; NY Acad Sci; Am Med Writers' Asn; Sigma Xi. *Res:* General bacteriology; history of bacteriology. *Mailing Add:* PO Box 26 Swiftwater PA 18370

GRALLA, EDWARD JOSEPH, b Wyoming Co, Pa, Mar 14, 32; m 59; c 3. EXPERIMENTAL TOXICOLOGY. *Educ:* Univ Pa, VMD, 61. *Prof Exp:* Vet practitioner, 61-62; toxicologist, Med Res Lab, Chas Pfizer & Co, Inc, 62-69; assoc prof comp med & pharmacol, Sch Med, Yale Univ, 69-75; dir pharmacol & toxicol, Mason Res Inst, 75-76; chief toxicol & mem staff, Chem Indust Inst Toxicol, 77-82; PRES, TOXICOCONSULTANTS, INC, 82- *Mem:* AAAS; Am Col Vet Comp Toxicol; Am Vet Med Asn; Am Asn Lab Animal Sci; Soc Toxicol. *Res:* Experimental toxicology; improving animal models for predicting adverse chemical effects in man; toxicology of anti-cancer agents. *Mailing Add:* 3025 Sylvania Dr Raleigh NC 27607

GRALLA, JAY DOUGLAS, b Brooklyn, NY, Jan 28, 48. BIOCHEMISTRY. *Educ:* Clarkson Col, BS, 69; Yale Univ, PhD(chem), 73. *Prof Exp:* Jane Coffin Childs fel, Harvard Univ, 73-75; from asst prof to assoc prof, 75-85, PROF CHEM & MOLECULAR BIOL, UNIV CALIF, LOS ANGELES, 85- *Res:* Regulation of gene expression; protein-nucleic acid interactions. *Mailing Add:* Dept Chem Univ Calif Los Angeles 405 Hilgard Ave Los Angeles CA 90024

GRALNICK, SAMUEL LOUIS, b New York, NY, Feb 27, 44; m 66; c 2. PLASMA PHYSICS, FUSION ENGINEERING. *Educ:* City Col New York, BE, 66, ME, 68; Columbia Univ, PhD(plasma physics), 72. *Prof Exp:* Res assoc fusion reactor design, Princeton Plasma Physics Lab, 72-74, mem tech staff, 74-77, res engr advan syst design, 77-78; PRIN SCIENTIST PLASMA PHYSICS & ENG, GRUMMAN AEROSPACE CORP, 78- *Concurrent Pos:* Adj assoc prof appl physics & nuclear eng, Columbia Univ. *Mem:* Am Phys Soc; Am Nuclear Soc; Soc Indust & Appl Math. *Res:* Plasma physics of fusion reactor devices; plasma engineering and fusion reactor and research equipment design. *Mailing Add:* 12 Pinewood Dr Commack NY 11725

GRAM, THEODORE EDWARD, b Minneapolis, Minn, Sept 26, 34; m 59; c 3. PHARMACOLOGY, BIOCHEMISTRY. *Educ:* Univ Minn, MS, 62, PhD(pharmacol, biochem), 64. *Prof Exp:* Asst pharmacol, Univ Minn, 58-64; res assoc chem pharmacol, Nat Inst Gen Med Sci, Nat Heart Inst, Md, 67-70; supvry pharmacologist & head sect enzyme-chem interaction, Pharmacol-Toxicol Br, Nat Inst Environ Health Sci, 70-72; supvry pharmacologist & head drug interactions sect, Lab Med Chem & Biol, 72-84, DEP CHIEF, TOXICOL BR, NAT CANCER INST, 84- *Concurrent Pos:* USPHS fel, Univ Iowa, 65-67; NIH fel, Lab Chem Pharmacol, Nat Heart & Lung Inst, 67-70; counr, Am Soc Pharmacol & Exp Therapeut, 84- *Mem:* AAAS; Soc Exp Biol & Med; Am Soc Biol Chemists; Sigma Xi; Soc Toxicol; Am Soc Pharmacol & Exp Therapeut; Biochem Soc; NY Acad Sci; Int Soc Study Xenobiotics. *Res:* Hepatic microsomal structure and function; biochemistry and enzymology of microsomal drug metabolism and electron transport; cell biology; biochemistry; toxicology. *Mailing Add:* Toxicol Br-DTP-DCT-NCT Landow Bldg Rm 5806 Bethesda MD 20205

GRAMBLING, JEFFREY A, b Milwaukee, Wis, Apr 1, 53; m 75; c 2. METAMORPHIC PETROLOGY, PRECAMBRIAN GEOLOGY. *Educ:* Colgate Univ, BA, 75; Princeton Univ, PhD(geol), 79. *Prof Exp:* Asst prof, Univ Okla, 79-80; asst prof, 80-85, ASSOC PROF GEOL, UNIV NMEX, 85- *Concurrent Pos:* NSF fel, 76-79; mem, exec comt, NMex Geol Soc, 82-85. *Mem:* Geol Soc Am; Mineral Soc Am; Mineral Asn Can. *Res:* Metamorphic petrology, especially phase relations and volatile behavior in regionally metamorphosed terranes; Precambrian geology; silicate mineralogy. *Mailing Add:* Dept Geol Univ NMex Albuquerque NM 87131

GRAMERA, ROBERT EUGENE, b Joliet, Ill, Feb 6, 36; m 57; c 2. BIOCHEMISTRY, ORGANIC CHEMISTRY. *Educ:* Lewis Col, BS, 57; Purdue Univ, MS, 61, PhD(biochem), 63. *Prof Exp:* State anal control chemist, Purdue Univ, 57-63; res scientist, Moffett Tech Ctr, Corn Prod Co, 63-65, res sect head polymer res & develop, 65-69; tech supvr appl packaging, Crown Zellerbach Corp, Calif, 69-70, supvr plant & customer trials, packaging res & develop lab, 70-71; dir tech develop & serv, Great Western Sugar Co, 71-75, exec dir mfg res & develop, 75-76; tech consult & pres, World Future, Inc, 76-80; PRES, BARLEY PROD INT, INC, 80- *Concurrent Pos:* Tech consult & vpres, Agribus Blackstone Group Ltd, Consults, Denver, Colo, 82- *Mem:* Am Chem Soc. *Res:* Isolation, purification and structural investigation of polysaccharides; synthesis of polymerizable monosaccharide monomers; organic reaction mechanisms of carbohydrates; chemistry of urethane foams from carbohydrate based polyols. *Mailing Add:* 704 County Rd 220 Durango CO 81301

GRAMIAK, RAYMOND, b Philadelphia, Pa, Mar 23, 24; m 49; c 4. MEDICINE, RADIOLOGY. *Educ:* Univ Rochester, MD, 49. *Prof Exp:* Resident radiol, Univ Rochester, 53-56, instr, 56-57; pvt pract, NY, 57-65; assoc prof, 65-67, PROF RADIOL, UNIV ROCHESTER, 67- *Concurrent Pos:* Fel cinefluorography, Univ Rochester, 52-53. *Mem:* Am Col Radiol; Radiol Soc NAm. *Res:* Diagnostic radiology; cardiac ultrasonics. *Mailing Add:* Dept Radiol Univ Rochester Strong Mem Hosp 601 Elmwood Ave Box 648 Rochester NY 14642

GRAMICK, JEANNINE, b Philadelphia, Pa, Aug 8, 42. MATHEMATICS. *Educ:* Col Notre Dame, Md, BA, 65; Univ Notre Dame, MS, 69; Univ Pa, PhD(educ), 75. *Prof Exp:* Teacher sr high sch math, Notre Dame Prep Sch, 65-68; asst prof math, Col Notre Dame, Md, 72-76; co-dir, New Ways Ministry, Mt Rainier, Md, 76-84; CONSULT, 84- *Res:* Whether arithmetic algorithms should be taught to conform to observed behaviors. *Mailing Add:* 4012-29th St Mt Rainier MD 20712

GRAMINSKI, EDMOND LEONARD, b Dunkirk, NY, Oct 14, 29; m 56; c 8. PHYSICAL ORGANIC CHEMISTRY, RHEOLOGY. *Educ:* Univ Buffalo, BS, 52, PhD(chem), 56. *Prof Exp:* Sr res chemist, Olin Mathieson Chem Co, 55-60 & Harris Res Labs Div, Gillette Razor Co, 60-64; proj leader rheology of paper, Nat Bur Standards, 64-80; CHIEF, OFF RES, BUR ENGRAVING & PRINTING, 80- *Mem:* Tech Asn Pulp & Paper Indust; Tech Asn Graphic Arts; Am Chem Soc. *Res:* Counterfeit deterrence materials, inks, graphic arts. *Mailing Add:* Bur Engraving & Printing 14th & C Sts SW Washington DC 20228

GRAMLICH, JAMES VANDLE, b Charleston, Ark, July 25, 39; m 58; c 2. PLANT PHYSIOLOGY. *Educ:* Univ Ark, BS, 61, MS, 62; Auburn Univ, PhD(bot), 65. *Prof Exp:* Instr bot, Auburn Univ, 64-65; sr plant physiologist, Eli Lilly & Co, 65-69, res scientist, 69, mgr plant sci res, Lilly Res Ctr, Ltd, 69-75, mgr plant sci res, Lilly Res Labs, 75-77, dir plant sci res, & agr prod develop, 77-80; dir discovery, 80-83, DIR AGR RES DIV, AM CYANAMID, 83- *Mem:* Weed Sci Soc Am. *Res:* Absorption, translocation, metabolism and degradation of herbicides; commercial development of herbicides, fungicides, insecticides, plant growth regulators and animal health products. *Mailing Add:* PO Box 400 Princeton NJ 08543

GRAMLING, LEA GENE, pharmaceutical chemistry; deceased, see previous edition for last biography

GRAMS, ANNE P, b Albany, Ga, May 28, 47; m 68. MATHEMATICS. *Educ:* Fla State Univ, BA, 68, PhD(math), 72; Cath Univ Am, MA, 70. *Prof Exp:* From asst prof to assoc prof math, Univ Tenn, Nashville, 77-80, coordr pub serv activ, Div Arts & Sci, 75-77; assoc prof math, Embry-Riddle Aero Univ, Daytona Beach, Fla, 80-82. *Mem:* Am Math Soc; Math Asn Am; Sigma Xi. *Res:* Commutative rings, problems relating to factorization, Dedekind domains, power series rings. *Mailing Add:* 3909 Cardinal Blvd Daytona Beach FL 32127

GRAMS, GARY WALLACE, b Moline, Ill, June 13, 42; m 64; c 2. ORGANIC CHEMISTRY, INORGANIC CHEMISTRY. *Educ:* Valparaiso Univ, BS, 64; Northwestern Univ, PhD(org chem), 68. *Prof Exp:* Chemist, Cereal Properties Lab, Northern Mkt & Nutrit Res Div, Agr Res Serv, USDA, 68-72; dir res & develop, Benz Oil Inc, 72-76; mgr res & develop, Pearsall Chem Corp, 76-79, vpres res & develop, 79-80, VPRES, ARGUS DIV, PEARSALL PRODUCTS, WITCO CORP, 80- *Mem:* Am Chem Soc; Am Inst Chemists; Am Soc Testing & Mat; AAAS. *Res:* Organochemical approaches to the synthesis of ribonucleotides; processing effects on cereal nutrients; photooxidation of vitamin E; lubricant formulation and synthesis; flame retardent and plastic additives; aluminum chloride catalysis. *Mailing Add:* Witco Argus Div Eight Wright Way Oakland NJ 07436-3175

GRAMS, GERALD WILLIAM, b Mankato, Minn, Dec 7, 38; m 88; c 3. METEOROLOGY, LASERS. *Educ:* Mankato State Col, BS, 60; Mass Inst Technol, PhD(meteorol), 66. *Prof Exp:* Teacher high sch, Minn, 60-61; res assoc meteorol & laser applns, Mass Inst Technol, 66-67; aerospace technologist, Electronics Res Ctr, NASA, Cambridge, 67-70; scientist meteorol & laser applns, Nat Ctr Atmospheric Res, 70-77; prof geophys sci, Ga Inst Technol, 77-90; PRES, GRAMS ENVIRON LABS, 90- *Concurrent Pos:* Lectr, Northeastern Univ, 67-70; res affil, Res Lab Electronics, Mass Inst Technol, 67-70; mem various comts, Nat Acad Sci & NASA. *Mem:* Am Meteorol Soc; Am Geophys Union; fel Optical Soc Am; AAAS; Soc Photo-Optical Instrumentation Engrs. *Res:* Meteorology and physics of the upper atmosphere; laser atmospheric probing techniques; stratospheric aerosols; noctilucent clouds; optical properties of atmospheric particulates; satellite observations of atmospheric constituents; atmospheric optics and radiation transfer. *Mailing Add:* Grams Environ Labs PO Box 88340 Atlanta GA 30356-8340

GRAMZA, ANTHONY FRANCIS, b Milwaukee, Wis, June 26, 36; m 77; c 1. ZOOLOGY, ETHOLOGY. *Educ:* Marquette Univ, BS, 62; Univ Wis, MS, 65, PhD(zool), 69. *Prof Exp:* Asst prof psychobiol, Children's Res Ctr, Univ Ill, 68-74; fac psychobiol, Nat Col Educ, 74-76; asst prof, 76-80, ASSOC PROF BIOL, MUNDELEIN COL, 80-, CHMN DEPT, 82- *Mem:* AAAS; Am Inst Biol Sci; Animal Behavior Soc; Sigma Xi. *Res:* Evolution and adaptive significance of behavioral systems; human ethology; social behavior; communication; ontogeny of behavior. *Mailing Add:* Mundelein Col Dept Biol 6363 Sheridan Rd Chicago IL 60660

GRAN, RICHARD J, b Brooklyn, NY, Apr 16, 40; m 84; c 1. CONTROL THEORY, IMAGE & SIGNAL PROCESSING. *Educ:* Polytech Inst Brooklyn, BS, 61, MS, 65, PhD(systs sci), 69. *Prof Exp:* Design engr, Grumman Aerospace Corp, 62-70, sr res scientist, Corp Res Ctr, 70-75, lab head, 75-88, DIR ADVAN CONCEPTS, GRUMMAN CORP, 88- *Concurrent Pos:* Adj prof, Polytech Univ, 65-85, State Univ NY, Stony Brook, 75-; consult, Math Anal Co, Legal Consult & Pension & Annuity Anal, 85-; mem, Maglev Technol Adv Comt, US Senate, 88-89; vpres & chief tech officer, Maglev USA, 89- *Mem:* Inst Elec & Electronics Engrs. *Res:* Control systems design methods; image processing; multidimensional signal processing; tracking filters. *Mailing Add:* 218 Willard Ave Farmingdale NY 11735

GRANATA, WALTER HAROLD, JR, petrology, stratigraphy, for more information see previous edition

GRANATEK, ALPHONSE PETER, b Hartford, Conn, Feb 13, 20; m 43; c 2. PHARMACEUTICAL CHEMISTRY. *Educ:* Trinity Col, Conn, BS, 42; Syracuse Univ, MS, 53. *Prof Exp:* Chemist, US Indust Chem, Md, 42-44; chemist, Ernest Bischoff Pharmaceut Co, Conn, 44-46; dir prod develop labs, Bristol-Myers Co, 46-67, dir prod develop res, 67-80; vpres, IMS Ltd, 80-85; RETIRED. *Concurrent Pos:* Consult, 85- *Mem:* Am Chem Soc; Am Pharmaceut Asn. *Res:* Insecticides; resins; antacid resins; antibiotics; pharmacy; development of pharmaceuticals; medical biochemistry; physics; electronics. *Mailing Add:* 13449 N 82nd St Scottsdale AZ 85260

GRANATH, JAMES WILTON, b Chicago, Ill, Oct 3, 49; m 81. STRUCTURAL GEOLOGY, REMOTE SENSING. *Educ:* Univ Ill, Urbana, BS, 71, MS, 73; Monash Univ, Australia, PhD(structural geol), 77. *Prof Exp:* Asst prof geol, State Univ NY, Stony Brook, 76-81; sr res geologist, 81-84, sr res scientist, Explor Res Div, 84-87, GEOL ADV, INT EXPLOR, CONOCO, INC, 87- *Concurrent Pos:* Vis prof, Field Camp Wyo, Univ Mo, Columbia, 80 & Univ Ill, 81. *Mem:* Geol Soc Am; Am Geophys Union; AAAS; Sigma Xi. *Res:* evolution of structural geometries; geotectonics. *Mailing Add:* Conoco Inc Int Explor PO Box 1267 Ponca City OK 74603

GRANATO, ANDREW VINCENT, b Cleveland, Ohio, May 9, 26; m 56; c 4. PHYSICS. *Educ:* Rensselaer Polytech Inst, BS, 48, MS, 50; Brown Univ, PhD(appl math), 55. *Prof Exp:* Res assoc appl math, Brown Univ, 53-57; res asst prof, 57-59, assoc prof, 61-64, PROF PHYSICS, UNIV ILL, URBANA, 64- *Concurrent Pos:* Guggenheim Found fel, Ger, 59-60; vis prof, Aachen Tech Univ, 60-61. *Honors & Awards:* Alexander von Humboldt US Sr Scientist Award, 76. *Mem:* Fel Am Phys Soc; Acoust Soc Am. *Res:* Ultrasonic wave propagation; dislocations in crystals; radiation damage. *Mailing Add:* Physics Dept Univ Ill 1110 W Green St Urbana IL 61801

GRANATSTEIN, VICTOR LAWRENCE, b Toronto, Ont, Feb 8, 35; m 55; c 3. PLASMA PHYSICS, MICROWAVE PHYSICS. *Educ:* Columbia Univ, BS, 60, MS, 61, PhD(eng, plasma physics), 63. *Prof Exp:* Res assoc plasma waves, Columbia Univ, 63-64; mem tech staff, Bell Tel Labs, Inc, 64-72; head high power electromagnetic radiation br, Plasma Physics Div, Naval Res Lab, 72-83; PROF ELEC ENG, 83-, DIR, LAB FOR PLASMA RES, UNIV MD, 88- *Concurrent Pos:* Vis sr lect, Dept Physics, Hebrew Univ, Jerusalem, 69-70. *Honors & Awards:* E O Hulburt Award, 80. *Mem:* Sr mem Inst Elec & Electronics Engr; fel Am Phys Soc. *Res:* Microwave generation with intense relativistic electron beams; millimeter and submillimeter wave generation; radiofrequency plasma heating; turbulence of fluids and plasmas; electromagnetic propagation through random media; plasma waves; free electron lasers; gyrotrons; radiofrequency sources for TeV linear colliders. *Mailing Add:* Lab Plasma Res Univ Md College Park MD 20742

GRANBERG, CHARLES BOYD, b Wessington, SDak, May 6, 21; m 55; c 3. PHARMACY, PHARMACOLOGY. *Educ:* SDak State Col, BS, 42; Univ Ill, MS, 47, PhD(pharmacol), 50. *Prof Exp:* Instr pharm, Univ Ill, 47-49; prof pharm, Drake Univ, 50-84, dean col pharm, 77-84; RETIRED. *Concurrent Pos:* Ed, Am J Pharmaceut Educ, 61-74. *Res:* Gastrointestinal x-ray contrast media; pharmacodynamics. *Mailing Add:* 1390 S State Farm Rd Des Moines IA 50265

GRANBERRY, DARBIE MERWIN, b Dothan, Ala, Jan 23, 43; m 62; c 2. PLANT BREEDING. *Educ:* Auburn Univ, BS, 68, ME, 69, PhD(plant breeding), 75. *Prof Exp:* Teaching asst, Dept Voc & Adult Educ, Auburn Univ, 68-69, res asst, Dept Hort, 69-72; res assoc hort, Dept Plant & Soil Sci, Tuskegee Inst, 72-75; asst prof hort, Edisto Br Res Sta, Clemson Univ, 75-76; ASST PROF EXTEN HORT, ATTAPULGUS EXTEN-RES CTR, UNIV GA, 76- *Mem:* Am Soc Hort Sci; Am Phytopath Soc; AAAS. *Res:* Increase efficiency of existing vegetable production and determine the potential of other vegetable crops not presently commercially produced in Georgia. *Mailing Add:* Rural Dev Ctr Tifton GA 31794

GRANBORG, BERTIL SVANTE MIKAEL, b Stockholm, Sweden, Aug 9, 23; US citizen; m 63; c 1. ELECTRICAL ENGINEERING. *Educ:* Royal Inst Technol, Sweden, MS, 53; Univ Wis, PhD(elec eng, math), 61. *Prof Exp:* Invest engr, Swed State Power Bd, 53-56; instr elec eng, Univ Wis, 56-60; res engr, Gen Elec Co, 61-63; ASSOC PROF ELEC ENG, UNIV HAWAII, 64- *Concurrent Pos:* Mem staff, high tension lab, Royal Inst Technol, Sweden, 51-52 & high tension & short circuit labs, Electricite de France, 52; vis teacher, Trade Sch Math, Sweden, 54-55; Nat Res Coun sr resident res assoc & vis scientist, Manned Spacecraft Ctr, NASA, 70-71. *Mem:* Inst Elec & Electronics Engrs; Swed Asn Eng & Architects. *Res:* Automatic feedback control; industrial process control; steel mill and steam station automation; power production, distribution and control; magnetic materials. *Mailing Add:* Dept Elec Eng Univ Hawaii Honolulu HI 96822

GRANCHELLI, FELIX EDWARD, b Cambridge, Mass, Oct 22, 23; m 48; c 4. MEDICINAL CHEMISTRY. *Educ:* Northeastern Univ, BS, 48, PhD(med chem), 72; Boston Univ, MA, 50. *Prof Exp:* Sr staff res & develop chem, Arthur D Little Inc, 51-75; sr scientist med chem, Col Pharm, Northeastern Univ, 75-77; SUPVR & SR STAFF MEM, NEW ENG NUCLEAR, 77- *Concurrent Pos:* Consult, Arthur D Little Inc, 75- *Mem:* Am Chem Soc; Sigma Xi. *Res:* Synthesis of heterocyclics, chiefly alkaloids, as potential antitumor agents, central nervous system drugs and antifertility compounds; dopaminergic, anti-Parkinsonian and analgesic activities; radiochemical synthesis utilizing 14C and 3H isotopes. *Mailing Add:* 120 Spring St Arlington MA 02174-7949

GRANCHI, MICHAEL PATRICK, b Uniontown, Pa, June 6, 46; m 72; c 1. MICROCHEMICAL ANALYSIS, CHROMATOGRAPHY. *Educ:* Waynesburg Col, BS, 68; Univ Pittsburgh, PhD(inorg chem), 77. *Prof Exp:* ASSOC CHEMIST, MOBIL RES & DEVELOP CORP, 77- *Mem:* Am Chem Soc. *Res:* Chromatography, microchemical analysis, laboratory automation. *Mailing Add:* Mobil Res & Develop Corp Billingsport Rd Paulsboro NJ 08066

GRAND, RICHARD JOSEPH, b New York, NY, Feb 1, 37; c 3. PEDIATRICS. *Educ:* Harvard Univ, BA, 58; NY Univ, MD, 62. *Prof Exp:* Intern, Bellevue Hosp, New York, 62-63; jr asst resident, Children's Hosp Med Ctr, Boston, 63-64, teaching fel pediat, 66-67, fel med, 67-70; res fel med, 69-70, asst prof, 70-76, ASSOC PROF PEDIAT, HARVARD MED SCH, 76-; CHIEF, DIV GASTROENTEROL, CHILDREN'S HOSP MED CTR, BOSTON, 75-, SR ASSOC MED, 76- *Concurrent Pos:* Med Found Res Award, NIH, 70, Nat Res Serv Award, 70-71; Acad Career Develop Award, Nat Inst Arthritis, Metab & Digestive Dis, 72-76; assoc med, Children's Hosp Med Ctr, Boston, 70-76; asst pediat, Beth Israel Hosp, Boston, 72-; asst pediatrician, 72, consult pediatrician, Boston Hosp Women, 76- *Mem:* Am Acad Pediat; Am Fedn Clin Res; NY Acad Sci; Soc Pediat Res; Am Gastroenterol Asn. *Mailing Add:* Div Ped GI/Nutr New England Med Ctr 171 Harrison Ave Box 213 Boston MA 02111

GRAND, STANLEY, b Paterson, NJ, Jan 5, 27; m 47; c 2. BIOCHEMISTRY. *Educ:* NY Univ, BS, 48; Polytech Inst Brooklyn, PhD(phys chem), 51. *Prof Exp:* Sect leader anal chem, Mineral Benefits Lab, Columbia Univ, 54-57; dir res & develop, Radiation Res Corp, 57-60; head dept chem & physics, W Orange Lab, Vitro Labs, 60-66; vpres, Nuclear Res Assocs, Inc, 66-71; dir labs, Brooklyn-Cumberland Med Ctr, 71-73; CONSULT CLIN CHEM, STANLEY GRAND INC, 74- *Mem:* Am Chem Soc; Am Asn Clin Chem; Clin Ligand Assay Soc. *Res:* High temperature materials and coatings; nuclear weapons effects, arcs and plasmas; clinical chemistry methods. *Mailing Add:* 2040 Wellington Ct Westbury NY 11590

GRAND, THEODORE I, b Newark, NJ, Feb 10, 38. GROSS ANATOMY. *Educ:* Brown Univ, BA, 59; Univ Calif, Berkeley, PhD(anthrop), 64. *Prof Exp:* Asst scientist phys anthrop, Ore Regional Primate Res Ctr, 63-66, assoc scientist, 67-78, scientist, 78-83; RES ASSOC, DEPT ZOOL RES, NAT ZOO, SMITHSONIAN INST, 83- *Concurrent Pos:* Vis asst prof, Univ Ore, 65, vis lectr, 74; vis lectr, Univ Western Australia, 69; res assoc, Smithsonian Inst, 76 & 80-81; vis fel, Cornell Univ, 78; vis prof, Univ Colo, 79, NECOM, 84-; adj assoc prof anat, Uniformed Serv Univ Health Sci. *Mem:* Am Asn Phys Anthrop; Am Soc Mammal. *Res:* Comparative anatomy of musculo-skeletal system; primate anatomy and evolution; history of anatomy; growth and development of mammals. *Mailing Add:* Dept Zool Res Nat Zoo Washington DC 20009

GRANDA, ALLEN MANUEL, b Belvedere, Calif, Feb 20, 29; div; c 3. VISION, SENSORY PROCESSES. *Educ:* Univ Calif, Los Angeles, AB, 50; Brown Univ, PhD, 59. *Prof Exp:* Res physiologist, Walter Reed Army Inst Res, 58-65; assoc prof, 65-71, dir, Inst Neurosci, 75-84, PROF NEUROSCI, UNIV DEL, 71- *Concurrent Pos:* Vis prof, Sch Med, Keio Univ, Japan, 71-72; vis prof, Physiol Inst, Free Univ Berlin, 78-79; lectr, Taniguchi Found Int; vis prof, Nat Res Coun, Pisa, Italy. *Honors & Awards:* Humboldt Prize. *Mem:* Sigma Xi; Asn Res Vision & Ophthal; Soc Neurosci. *Res:* Vision; sensory processes. *Mailing Add:* Sch Life Sci Univ Del Newark DE 19716

GRANDAGE, ARNOLD HERBERT EDWARD, b Springfield, Mass, Nov 12, 18; m 45; c 2. EXPERIMENTAL STATISTICS. *Educ:* Lehigh Univ, BA, 42; NC State Col, PhD(statist), 54. *Prof Exp:* From assoc prof to prof statist, NC State Univ, 55-81; RETIRED. *Concurrent Pos:* Consult var industs & govt agencies. *Mem:* Biomet Soc; Sigma Xi. *Res:* Industrial, physical science, behavioral science and highway construction applications of statistics; use of computers in statistics. *Mailing Add:* 223 Port Noble Dr Bloomsburg PA 17815-2503

GRANDEL, EUGENE ROBERT, b Chicago, Ill, Jan 16, 33; m 62; c 2. DENTISTRY. *Educ:* Univ Ill, BS, 56, DDS, 58, MS, 62. *Prof Exp:* Instr oral path, Univ Ill, 61-63; asst prof, 63-68, assoc prof oral path, 68-80, assoc prof pedodont, 70-80, PROF PEDODONT, SCH DENT, LOYOLA UNIV CHICAGO, 80- *Mem:* Am Acad Oral Path; Int Asn Dent Res; Sigma Xi. *Res:* Cytology; histochemistry; enzymatic cytology. *Mailing Add:* Dent Sch Loyola Univ 2160 S First Ave Maywood IL 60153

GRANDHI, RAJA RATNAM, b Kullur, India, July 1, 38; m 61; c 2. ANIMAL NUTRITION. *Educ:* Venkateswara Univ, India, BVSc, 60; Univ Guelph, Ont, MSc, 70, PhD(animal nutrit), 73. *Prof Exp:* Vet asst surg & animal husb exten officer, Dept Animal Husb, Ministry Agr & Food, India, 61-68; dir animal nutrit, Agron Consult Ltd, Ont, 73-74; res assoc, Dept Animal Sci, Univ Guelph, 74-77; RES SCIENTIST SWINE NUTRIT, AGR CAN RES STA, BRANDON, MAN, 77- *Mem:* Can Soc Animal Sci; Agr Inst Can; Am Soc Animal Sci; Am Registry Prof Animal Scientists. *Res:* Nutrition and physiology; study of nutritional and metabolic factors and interrelationships affecting the nutrition and reproduction of domestic animals (swine). *Mailing Add:* Agr Can Res Sta PO Box 610 Brandon MB R7A 5Z7 Can

GRANDI, STEVEN ALDRIDGE, b Santa Monica, Calif, Nov 9, 50. ASTRONOMY. *Educ:* Calif Inst Technol, BS, 72; Univ Ariz, PhD(astron), 75. *Prof Exp:* Res astronr, Lick Observ, Univ Calif, Santa Cruz, 76-78; asst prof astron, Univ Calif, Los Angeles, 78-; AT NAT OPTICAL ASTRON OBSERV, ARIZ. *Mem:* Am Astron Asn. *Res:* Observational and theoretical studies of QSOs, Seyfert galaxies, H II regions, planetary nebulae and other emission line objects. *Mailing Add:* KPNO PO Box 26732 Tucson AZ 85726

GRANDJEAN, CARTER JULES, b Socorro, NMex, Sept 10, 41; m 70; c 2. BIOCHEMISTRY, PHARMACOLOGY. *Educ:* NMex Inst Mining & Technol, BS, 65; Tex Tech Univ, MS, 68, PhD(biochem), 70. *Prof Exp:* Trainee, Eppley Cancer Inst & Col Pharm, Univ Nebr Med Ctr, from asst prof to assoc prof cancer res, 71-79; mem staff, Midwest Res Inst, 79-; ASSOC PROF PEDIAT & BIOCHEM, UNIV NEBR MED CTR, OMAHA. *Concurrent Pos:* NIH res grant, 73. *Mem:* Am Chem Soc; Am Soc Pharmacol & Exp Therapeut; Am Inst Chemists; AAAS; Am Asn Cancer Res. *Res:* Basic biochemical studies concerning the mechanisms of biological activation of probable environmental chemical carcinogens such as polycyclic aromatic hydrocarbons and N-nitrosamines; pharmacokinetics of xenobiotics and carcinogens and their metabolism in rodents. *Mailing Add:* 1201 Skyline Dr Elkhorn NE 68022

GRANDOLFO, MARIAN CARMELA, b Chicago, Ill, May 27, 43; m 78; c 1. PETROLEUM ENGINEERING. *Educ:* Marquette Univ, BS, 65; Univ Notre Dame, MS, 72, PhD(inorg chem), 74. *Prof Exp:* Technician chem, Mt Sinai Hosp, 65; sci asst, Argonne Nat Lab, 65-69; sr chemist anal chem, anal & info div, Exxon Res & Eng Co, 74-82; COMPUT CONSULT & TEACHER, DATA DETECHTIVE, 84- *Mem:* Soc Appl Spectros; Am Chem Soc. *Res:* Vibrational analysis of carbonates and transition metal and lanthanide trifluorides; chromatography of naturally occurring substances and investigation of their structure; analysis of petroleum, its fractions; inorganic, organic and polymeric materials; method development to aid in functional group analysis. *Mailing Add:* 15 King St Fanwood NJ 07023-1517

GRANDTNER, MIROSLAV MARIAN, b Liptovska-Teplicka, Slovak, Aug 23, 28; Can citizen; div; c 1. FOREST ECOLOGY. *Educ:* Univ Louvain, Ing E & F, 55, Dr Sc(agron), 62; Laval Univ, MSc, 59. *Prof Exp:* Chief cartographer, Ctr Phytosociol Mapping, Belg, 56-57; asst forest ecol, 57-58, from asst prof to assoc prof bot, 58-67, PROF ECOL, LAVAL UNIV, 67- *Concurrent Pos:* Consult, Eastern Quebec Mgt Off, 63-66; asst ed, Le Naturaliste Can, 68-79; lectr, Univ Montreal & Univ Que Rimuski, 71-84; chmn, Port-au-Saumon Ecol Ctr, 73-87; ed, Phytocoenologia, 74-, Etudes Ecol, Dossiers Foresterie Int, 90- *Mem:* Int Asn Veg Sci; Ecol Soc Am; Int Asn Ecol. *Res:* Forest ecology and phyto-sociology; vegetation mapping; cyto-ecology and bio-indicators. *Mailing Add:* Forest Sci Laval Univ Quebec PQ G1K 7P4 Can

GRANDY, CHARLES CREED, b Alamosa, Colo, Dec 6, 28; m 51, 83; c 6. MATHEMATICS, PHYSICS. *Educ:* Colo State Univ, BS, 52; Northeastern Univ, MS, 62; Am Univ, MS, 84. *Prof Exp:* Mem tech staff digital comput applns, Lincoln Lab, Mass Inst Technol, 52-58, assoc group leader air defense systs, 58; head, Dept Systs Testing & Eval, 58-60, head, Dept Command & Control Systs, 60-62, assoc tech dir systs eng, 62-65, systs planning, 65-67, tech dir, Nat Command & Control Systs Div, 67-71, asst vpres Washington opers, 71-72, VPRES, CIVIL SYSTS DIV, MITRE CORP, 72- *Mem:* AAAS; Math Asn Am; Soc Indust & Appl Math; Am Inst Aeronaut & Astronaut. *Res:* Applications of digital computers to real time control problems; systems engineering in military command and control systems; operations research; statistical inference and experimental design. *Mailing Add:* Mitre Corp 7525 Colshire Dr McLean VA 22102

GRANDY, WALTER THOMAS, JR, b Philadelphia, Pa, June 1, 33; m 55; c 4. THEORETICAL PHYSICS. *Educ:* Univ Colo, BS, 60, PhD(physics), 64. *Prof Exp:* Phys sci aide, Nat Bur Standards, Colo, 58-60, mathematician, 60-61, physicist, 61-63; from asst prof to assoc prof, 63-69, chmn dept, 71-78, PROF PHYSICS, UNIV WYO, 69- *Concurrent Pos:* Fulbright scholar, Brazil, 66-67; vis prof, Univ Sao Paulo, Brazil, 82, Univ Tubingen, Ger, 78-79 & Univ Sydney, Australia, 88. *Mem:* Am Asn Physics Teachers; Am Phys Soc; AAAS. *Res:* Electrodynamics; nonequilibrium statistical mechanics. *Mailing Add:* Dept Physics & Astron Univ Wyo Laramie WY 82071

GRANEAU, PETER, b Mar 13, 21; Brit citizen; m 55; c 1. ELECTRICAL ENGINEERING, PHYSICS. *Educ:* Univ Nottingham, BScEE, 55, PhD(elec eng), 62. *Prof Exp:* Dept head instr & control, BICC Res Lab, Eng, 55-62, asst res mgr elec eng & physics, 62-67; tech dir cable res, Simplex Wire & Cable Co, Mass, 67-70; PRES RES & DEVELOP, UNDERGROUND POWER CORP, 70- *Concurrent Pos:* Consult, A D Little Inc, 71-76, Lincol Lab, Mass Inst Technol, 73-75; vis scientist, Nat Magnetic Lab, Mass Inst Technol, 72-; vis prof, Northeastern Univ, 84- *Mem:* Fel Brit Inst Physics; Inst Elec & Electronics Engrs. *Res:* Underground power transmission; conductor metal; dielectrics for high voltage cables; electromagnetism; superconductors. *Mailing Add:* 205 Holden Wood Rd Concord MA 01742

GRANET, IRVING, b New York, NY, July 2, 24; m 51; c 3. MECHANICAL & NUCLEAR ENGINEERING. *Educ:* Cooper Union, BME, 44; Polytech Inst Brooklyn, MME, 48. *Prof Exp:* Proj engr, Res Dept, Foster Wheeler Corp, 47-53, sr engr, Nuclear Energy Dept, 53-55, dir staff eng, 55-59; assoc scientist plasma propulsion proj, Repub Aviation Corp, 59-64 & Power Conversion Div, 64-70, proj mgr, Repub Aviation Div, Fairchild-Hiller Corp, 70-73; vpres & chief engr, EMS Develop Corp, NY, 73-76; assoc prof mech eng, NY Inst Technol, 76-77; from asst prof to assoc prof, 77-82, PROF MECH ENG TECHNOL, QUEENSBOROUGH COMMUNITY COL, CITY UNIV NEW YORK, 82- *Concurrent Pos:* Instr, Polytech Inst Brooklyn, 51-52; adj asst prof, C W Post Col, Long Island, 59-; adj assoc prof, NY Inst Technol, 68- *Mem:* Am Soc Mech Engrs; Nat Soc Prof Engrs; NY Acad Sci; Am Soc Naval Engrs; Am Soc Eng Educ. *Res:* Physics; heat transfer and fluid mechanics; applied magnetohydrodynamics and space propulsion; thermodynamics. *Mailing Add:* Queensborough Community Col Bayside NY 11364

GRANEY, DANIEL O, b Los Angeles, Calif, Oct 29, 36; m 60; c 2. ANATOMY. *Educ:* Univ Calif, Berkeley, AB, 58, Univ Calif, San Francisco, MA, 62, PhD(anat), 65. *Prof Exp:* From instr to asst prof, 66-74, ASSOC PROF BIOL STRUCT, SCH MED, UNIV WASH, 74- *Mem:* Am Asn Anat; Am Soc Cell Biol. *Res:* Fine structure of gastrointestinal absorption and secretion; transport by capillary endothelia; clinical anatomy. *Mailing Add:* Dept Biol Structure Univ Wash Sch Med Seattle WA 98195

GRANGE, RAYMOND A, b Darlington, Wis, Aug 14, 10. THERMOMECHANICAL TREATMENT. *Educ:* Univ Wis-Madison, BS, 35. *Prof Exp:* Scientist, US Steel, 38-74, sr scientist, 69-74; RETIRED. *Mem:* Am Soc Metals Int. *Mailing Add:* 4733 Flournoy Valley Roseburg OR 97470

GRANGER, CARL V, b Brooklyn, NY, Nov 26, 28; m 51, 83; c 2. PHYSICAL MEDICINE & REHABILITATION. *Educ:* Dartmouth Col, AB, 48; NY Univ, MD, 52. *Prof Exp:* Asst chief phys med, Letterman Gen Hosp, 58-61; from instr to asst prof, Sch Med, Yale Univ, 61-67, assoc clin prof, 67-68, assoc dir dept phys med, New Haven Hosp, 61-66, actg chief, 66-67; prof phys & rehab med & chmn dept, Sch Med, Tufts Univ, 68-77; dir phys med & rehab, Mem Hosp, Pawtucket, 78-83; prof community health, phys med & rehab, Brown Univ, 78-83; HEAD, REHAB MED, BUFFALO GEN HOSP, 83-; PROF REHAB, STATE UNIV NY, BUFFALO, 83- *Concurrent Pos:* Survey consult, Comn Accreditation of Rehab Facil, 68; mem expert med comt, Am Rehab Found, 68; physiatrist in chief, New Eng Med Ctr Hosp, 68-77; mem adv bd, Nat Multiple Sclerosis Soc, 69-75. *Mem:* Am Asn Electromyog & Electrodiag (pres, 68-69); Am Cong Rehab Med; Am Acad Phys Med & Rehab (pres, 75-76). *Res:* Electrodiagnosis; peripheral nerve disorders; patient care outcome measures; braces and splints. *Mailing Add:* Rehab Ctr Buffalo Gen Hosp 100 High St Buffalo NY 14203

GRANGER, D NIEL, OXYGEN RADICALS, MICROCIRCULATION. *Educ:* Univ Miss, PhD(physiol), 77. *Prof Exp:* PROF PHYSIOL, UNIV S ALA, 78- *Mailing Add:* Dept Physiol LSU Med Ctr 15011 Kings Hwy Box 33932 Shreveport LA 71130-3932

GRANGER, DONALD LEE, b Salt Lake City, Utah, Apr 12, 43; m 68; c 4. INFECTIOUS DISEASES. *Educ:* Univ Utah, BA, 66, MD, 72; Univ Rochester, MS, 68. *Prof Exp:* Intern med, Univ Rochester, 72-73; res assoc, Rocky Mountain Lab, Nat Inst Allergy & Infectious Dis, NIH, 73-75; resident med, Univ Rochester, 75-77; fel med, Univ Utah, 77-80; fel biochem, Johns Hopkins Univ, 80-82; ASST PROF MED, DUKE UNIV, 82- *Concurrent Pos:* Prin investr, Nat Cancer Inst, NIH, 83-86, Nat Inst Allergy & Infectious Dis, 88-91. *Mem:* Infectious Dis Soc Am. *Res:* Elucidating the biochemical mechanisms used by macrophages to stop proliferation of microbes. *Mailing Add:* Dept Physiol/Chem Johns Hopkins Univ Sch Med Baltimore MD 21205

GRANGER, GALE A, b San Pedro, Calif, June 18, 37; m 60; c 2. MICROBIOLOGY. *Educ:* Univ Wash, BS, 62, MS, 63, PhD(microbiol), 65. *Prof Exp:* From asst prof to assoc prof, 67-73, PROF IMMUNOL & MICROBIOL, UNIV CALIF, IRVINE, 73- *Concurrent Pos:* NIH res grants & career develop award, 67- *Mem:* Transplantation Soc; Am Asn Immunol; Reticuloendothelial Soc; NY Acad Sci. *Res:* Immunology; cell biology especially in vitro studies of immune cell induced graft cell destruction; cellular and sub cellular mechanisms. *Mailing Add:* Dept Molecular Biol & Biochem Univ Calif Irvine CA 92717

GRANGER, JOEY PAUL, CARDIOVASCULAR & RENAL PHYSIOLOGY. *Educ:* Univ Miss, PhD(physiol), 83. *Prof Exp:* Asst prof physiol, Mayo Med Sch, 83-88; ASSOC PROF PHYSIOL, EASTERN VA MED SCH, 88- *Mailing Add:* Dept Physiol Biophys 2500 N State St Jackson MS 39216

GRANGER, JOHN VAN NUYS, b Cedar Rapids, Iowa, Sept 14, 18; m 45, 83; c 3. ELECTRONICS ENGINEERING. *Educ:* Cornell Col, AB, 41; Harvard Univ, MS, 42, PhD(physics), 48. *Prof Exp:* Instr math, Cornell Col, 41; res assoc, Harvard Univ, 42-47; head radio systs lab, Stanford Res Inst, 49-53, asst dir eng div, 53-56; pres, Granger Assocs, 56-70; dep dir, Bur Int Sci & Technol Affairs, US Dept State, 71-74; dep asst dir, NSF, 74-77; counr sci & technol, US Embassy, London, 77-81; sci attache, US Permanent Deleg to UNESCO, 81-83; RETIRED. *Mem:* Nat Acad Eng; Sigma Xi; fel Inst Elec & Electronics Eng (pres, 71). *Res:* Antennas and propagation; communications systems; science policy; technical management. *Mailing Add:* The Well House Ampney St Mary Cirencester GL7-5SN England

GRANGER, ROBERT A, II, b Evanston, Ill, Aug 7, 28; m 51; c 2. MATHEMATICAL PHYSICS, FLUID DYNAMICS. *Educ:* Pomona Col, BA, 55; Drexel Inst Technol, MS, 59; Univ Md, PhD(physics), 70. *Prof Exp:* Res scientist aeroelasticity, Martin Co, Md, 55-60; PROF ENG, US NAVAL ACAD, 60- *Concurrent Pos:* Consult, US Naval Reserve Officer Candidate Sch, Md, 64-, Trident Eng Assoc, Md, 65-, Bay-Tech Eng, Md, 69-70 & Cadcom Inc, Md, 70-; guest lectr, Royal Aircraft Estab, Eng, & Munich Tech Univ, 68-69 & Eugenides Found, 70-71; prin engr & consult, Boeing Com Airplane Co, Wash, 75; vis prof, Univ Petrol & Minerals, Dhahran, Saudi Arabia, 77-79; organizer & dir, lect ser vortex dynamics, Von Karman Fluid Dynamics Inst, 86; vis prof mech eng, Yale Univ, 90. *Mem:* Hon mem Inst Mod Physics Greece. *Res:* Applied and theoretical mathematical physics; boundary layer research; vortex dynamics; solutions of Navier-Stokes equations; experimental fluid dynamics; turbulence; author of numerous books and over 400 technical papers. *Mailing Add:* Dept Mech Eng US Naval Acad Annapolis MD 21204

GRANICK, STEVE, b Hanover, NH, July 10, 53; m 78; c 2. POLYMER SURFACES, TRIBOLOGY. *Educ:* Princeton Univ, BA, 78; Univ Wis-Madison, PhD(chem), 82. *Prof Exp:* Researcher, Col France, 82-83 & Univ Minn, 83-84; asst prof, 85-90, ASSOC PROF, DEPT MAT SCI, UNIV ILL, URBANA-CHAMPAIGN, 90- *Mem:* Am Chem Soc; Am Phys Soc. *Res:* Physical chemistry of macromolecules, especially at interfaces; surface and colloid chemistry; diffusion, surface forces and tribology. *Mailing Add:* Dept Mat Sci Univ Ill Urbana IL 61801

GRANIRER, EDMOND, b Constantsa, Romania, Feb 19, 35. MATHEMATICS. *Educ:* Hebrew Univ, MSc, 59, PhD(math), 62. *Prof Exp:* prof math, Univ Ill, 62-63 & Cornell Univ, 64-65; PROF MATH, UNIV BC, 65- *Mem:* Fel Royal Soc Can. *Res:* Analysis and functional analysis. *Mailing Add:* Univ Brit Columbia Vancouver BC V6T 1W5 Can

GRANITO, CHARLES EDWARD, b Brooklyn, NY, Nov 14, 37; m 59; c 3. CHEMISTRY, INFORMATION SCIENCE. *Educ:* Univ Miami, BS, 60, MS, 62. *Prof Exp:* Chemist, Indust Liaison Off, US Army Res Labs, Edgewood Arsenal, Md, 62-66; from chemist to sr res chemist, T R Evans Res Ctr, Diamond Shamrock Corp, 66-69; mgr info serv, Inst Sci Info, 69-70, dir chem info serv, 70-72; consult, C G Assocs, 73-75; pres, Chem Info Mgt, Inc, 75-80; PRES & CHIEF RES EXEC, INT PROD CORP, 81- *Mem:* AAAS; Am Chem Soc; Chem Notation Asn (vpres, 68, pres, 69); Sigma Xi; Soap & Detergent Asn; President's Asn. *Res:* Organic chemistry; physical chemistry. *Mailing Add:* Inst Prod Corp PO Box 70 Burlington NJ 08016-0070

GRANNEMANN, GLENN NIEL, mining engineering; deceased, see previous edition for last biography

GRANNEMANN, W(AYNE) W(ILLIS), b New Haven, Mo, Oct 11, 23; m 47; c 3. SOLID STATE PHYSICS, ELECTRICAL ENGINEERING. *Educ:* Univ Tex, BS, 47, MA, 49, PhD(physics), 53. *Prof Exp:* Asst prof physics, Ark Polytech Col, 49-51; res physicist, Defense Res Lab, Tex, 52-53; res physicist, Calif Res Corp, Standard Oil Co, Calif, 53-56; assoc prof elec eng, 56-63, dir eng exp sta, 59-63, dir bur eng res, 60-71, PROF ELEC ENG, UNIV NMEX, 63- *Concurrent Pos:* Consult, McAllister & Assocs, Inc & Sandia Corp. *Mem:* Inst Elec & Electronics Engrs. *Res:* Semiconductors; solid state devices; waves in solids; physical electronics; seismic instrumentation; radiation effects on electronics. *Mailing Add:* Star Rte Box 762A Sandia Park NM 87111

GRANNER, DARYL KITLEY, b Algona, Iowa, Dec 12, 36; m 58; c 2. ENDOCRINOLOGY. *Educ:* Univ Iowa, BA, 58, MS & MD, 62. *Prof Exp:* Asst prof med, Univ Wis, 69-70; asst prof, Univ Iowa, 70-72, assoc prof, 72-75, prof med & biochem & dir, Div Endocrinol & Metab, 75-, dir, Diabetes & Endocrinol Res Ctr, 79-; PROF PHYSIOL, VANDERBILT UNIV. *Mem:* Endocrine Soc; Am Soc Biol Chemists; Am Fedn Clin Res; Am Soc Clin Invest. *Res:* Action of hormones such as insulin, glucocorticoids and cyclic nucleotides at the molecular level. *Mailing Add:* Dept Molecular Physiol & Biophys Vanderbilt Univ Med Sch Nashville TN 37064

GRANNIS, PAUL DUTTON, b Dayton, Ohio, June 26, 38; m 72; c 4. HADRONIC INTERACTIONS. *Educ:* Cornell Univ, BEng, 61; Univ Calif, Berkeley, PhD(physics), 65. *Prof Exp:* Res assoc, Lawrence Radiation Lab, Univ Calif, Berkeley, 65-66; from asst prof to assoc prof, 69-74, PROF PHYSICS, STATE UNIV NY, STONY BROOK, 74. *Concurrent Pos:* Vis scientist, Europ Ctr Nuclear Res, Geneva, 73; mem, High Energy Adv Comt, Brookhaven Nat Lab, 74-76; sr res fel, Sci Res Coun, UK, 76. *Mem:* Am Phys Soc. *Res:* Experimental high energy physics. *Mailing Add:* Dept Physics 3800 State Univ NY Stony Brook NY 11794

GRANOFF, ALLAN, b New Haven, Conn, June 26, 23; c 4. VIROLOGY. *Educ:* Univ Conn, BS, 48; Univ Pa, MS, 49, PhD(virol), 52; Am Bd Med Microbiol, dipl. *Prof Exp:* From asst to assoc, Div Infectious Dis, Pub Health Res Inst NY, 52-62; chmn virol lab, 62-69, chmn biol, 65-69, chmn virol & immunol labs, 69-75, CHMN DEPT VIROL & MOLECULAR BIOL, 75-, DEP DIR, ST JUDE CHILDREN'S RES HOSP, 84- *Concurrent Pos:* Mem, Virol Study Sect, NIH, 69-73, chmn, Small Bus Grants Study Sect, 85-; assoc ed, Virol, 70 & Intervirol, 73-; mem, Virol & Cell Biol Study Sect, Am Cancer Soc, 76-79, Inst Grant Study Sect, 86-; assoc dir, Basic Sci, St Jude Children's Res Hosp, 79-83; from assoc prof to prof microbiol, Univ Tenn, Memphis, 62-85. *Honors & Awards:* Stuart Mudd Lectr, 87. *Mem:* AAAS; Am Soc Microbiol; Soc Exp Biol & Med; Am Asn Cancer Res; Am Asn Immunol. *Res:* Molecular biology of virus replication. *Mailing Add:* Box 318 Memphis TN 38101

GRANOFF, BARRY, b Jersey City, NJ, June 30, 38; m 89; c 2. APPLIED MATHEMATICS. *Educ:* Fairleigh Dickinson Univ, BS, 60; NY Univ, MS, 62, PhD(math), 65. *Prof Exp:* Assoc res scientist, Courant Inst Math Sci, NY Univ, 65-66; asst prof, 66-70, ASSOC PROF MATH, BOSTON UNIV, 70- *Concurrent Pos:* Assoc prof math, Harvard Univ Summer Sch, 84-91. *Mem:* Soc Indust & Appl Math; Math Asn Am. *Res:* Applied mathematics and analysis, partial and ordinary differential equations, asymptotic methods, methods of mathematical physics; singular perturbation techniques. *Mailing Add:* Dept Math Boston Univ 111 Cummington St Boston MA 02215

GRANOFF, BARRY, b Brooklyn, NY, Sept 7, 40; m 65; c 2. FUEL SCIENCE, CHEMICAL KINETICS. *Educ:* City Col New York, BS, 62; Princeton Univ, MA, 64, PhD(x-ray crystallog), 66. *Prof Exp:* Res chemist, E I du Pont de Nemours & Co, Inc, 66-69; MEM TECH STAFF, SANDIA LABS, 69- *Mem:* Am Chem Soc; Inst Elec & Electronics Engrs; Am Crystallog Asn; Sigma Xi. *Res:* Structure of carbon and graphite; oil shale conversion; pyrolysis kinetics; coal liquefaction; hydrodesulfurization; structure and reactivity of coal; hydrogenation. *Mailing Add:* 2219 Camino de los Artesanos NW Albuquerque NM 87107

GRANOFF, DAN MARTIN, b New York, NY, Jan 22, 44. IMMUNOGENETICS, HAEMOPHILUS INFLUENZAE. *Educ:* Johns Hopkins Univ, AB, 65, MD, 68. *Prof Exp:* Intern, pediat, Children's Hosp, Philadelphia, 68-69; res, pediat, Johns Hopkins Hosp, 69-71; postdoctoral fel, infectious dis, Case Western Reserve Univ, 73-75; from instr to asst clin prof, pediat, Univ Calif, San Francisco, 76-79; assoc prof, 80-85, PROF, PEDIAT & ASSOC PROF MICROBIOL, WASH UNIV MED SCH, ST LOUIS, 85- *Concurrent Pos:* Dir, infectious dis, St Louis Children's Hosp, 80-; consult, Nat Acad Sci, 84-; assoc ed, Pediat Res, 89-; prin investr, NIH, 79- *Mem:* Am Soc Clin Invest; Infectious Dis Soc Am; Soc Pediat Res; Am Pediat Soc; Am Soc Microbiol; Am Acad Pediat. *Res:* Haemophilius influenzae, bacterial meningitis in children, the molecular epidemiology of disease. *Mailing Add:* 400 S Kingshighway Blvd St Louis MO 63110

GRANSTROM, MARVIN L(E ROY), b Anaconda, Mont, Sept 25, 20; m 44; c 3. SANITARY ENGINEERING, CIVIL ENGINEERING. *Educ:* Morningside Col, BS, 42; Iowa State Col, BS, 43; Harvard Univ, MS, 47, PhD(sanit eng), 55. *Prof Exp:* Instr sanit eng, Case Inst Technol, 47-49; assoc prof, Univ NC, 49-58; prof civil & environ eng, Rutgers Univ, 58-83; RETIRED. *Concurrent Pos:* Consult, Nat Univ Eng, Peru, 55-57 & WHO, Peru, Chile, Arg & Brazil, 66-; mem, sanit eng subcomt, Nat Acad Sci-Nat Res Coun, 63- *Mem:* Am Soc Civil Engrs Environ Engrs; Sigma Xi. *Res:* Ecology; water resources planning and design. *Mailing Add:* 931 Oakwood Pl Plainfield NJ 07060

GRANT, ALAN LESLIE, b Watertown, NY, May 8, 62; m 85. GROWTH BIOLOGY. *Educ:* Cornell Univ, BS, 84; Mich State Univ, MS, 87, PhD(animal sci), 90. *Prof Exp:* Grad res asst dairy nutrit, Mich State Univ, 84-87, growth biol, 87-90; ASST PROF GROWTH BIOL, PURDUE UNIV, 90- *Mem:* Am Soc Animal Sci; AAAS; Am Meat Sci Asn; Sigma Xi. *Res:* Identification and characterization of cellular and molecular mechanisms that regulate skeletal muscle growth and development in food-producing animals. *Mailing Add:* Dept Animal Sci Purdue Univ West Lafayette IN 47907

GRANT, ARTHUR E, b Freeport, Ill, Oct 27, 23; m 47; c 3. PHYSICAL MEDICINE & REHABILITATION. *Educ:* Case Western Reserve Univ, MD, 50. *Prof Exp:* Intern, Fitzsimons Gen Hosp, 50-51, resident phys med & rehab, Letterman Gen Hosp, 51-54, staff physician, Phys Med Serv, 54-55; staff physician, Brooke Gen Hosp, 55-57; chief, Outpatient Serv, 5th Gen Hosp, Ger, 58-59 & div surgeon, 24th Inf Div, Ger, 59-60; chief, Phys Med Serv, Brooke Gen Hosp, 60-66 & Letterman Gen Hosp, 66-67; PROF PHYS MED & REHAB & CHMN DEPT, UNIV TEX MED SCH, SAN ANTONIO, 67- *Mem:* Am Cong Rehab Med; Am Acad Phys Med & Rehab; Am Pain Soc; Int Rehab Med Asn; Int Asn Study Pain. *Res:* Clinical and research electromyography and electrodiagnosis; rehabilitation of neurologic disorders. *Mailing Add:* Dept Phys Med & Rehab Univ Tex Health Sci Ctr 13147 N Hunters Circle San Antonio TX 78230

GRANT, BARBARA DIANNE, b Phoenix, Ariz. SYNTHETIC ORGANIC CHEMISTRY. *Educ:* Ariz State Univ, BS, 71; Stanford Univ, PhD(chem), 74. *Prof Exp:* Res chemist, 75-77, MGR SYNTHETIC ORG CHEM, IBM CORP, 77- *Concurrent Pos:* Fel, Syntex Res, Inst Molecular Biol, 74-75. *Res:* Synthesis of organic compounds in the solid state; synthesis of electrochromic and photochromic materials. *Mailing Add:* 1034 Belder Dr San Jose CA 95120

GRANT, BRUCE S, b New York, NY, Apr 17, 42; m 64; c 2. GENETICS. *Educ:* Bloomsburg State Col, BS, 64; NC State Univ, MS, 66, PhD(genetics), 68. *Prof Exp:* From asst prof to assoc prof, 68-82, PROF BIOL, COL WILLIAM & MARY, 82- *Mem:* Am Genetics Asn; Soc Study Evolution. *Res:* Behavioral genetics as it pertains to questions of population genetics and evolution. *Mailing Add:* Dept Biol Col William & Mary Williamsburg VA 23185

GRANT, CLARENCE LEWIS, b Dover, NH, July 8, 30; m 52; c 3. ANALYTICAL CHEMISTRY, ENVIRONMENTAL SCIENCES. *Educ:* Univ NH, BS, 51, MS, 56; Rutgers Univ, PhD(soils, chem), 60. *Prof Exp:* Instr, Pub Sch, NJ, 51-52; res asst chem, Univ NH, 52-53, instr, 53-55, res assoc, 55-57, asst prof, 57-58; asst instr soils & chem, Rutgers Univ, 58-60, assoc prof soils, 60-61; res assoc prof, Chem Eng Exp Sta, Univ NH, 61-64, res prof, Ctr Indust & Instnl Develop, 64-76, assoc dir, 73-81, chmn dept, 76-79, prof chem, 76-89, EMER PROF CHEM, UNIV NH, 89- *Concurrent Pos:* Consult, US Army & industs on environ & qual control problems. *Honors & Awards:* Distinguished Serv Award, Soc Appl Spectros, 87. *Mem:* Am Chem Soc; Soc Appl Spectros (pres, 69); Am Soc Testing & Mat. *Res:* Development of analytical methods for trace pollutants; distribution of trace metals and organics in aquatic systems; spectrochemistry; radio tracers; statistics in chemistry. *Mailing Add:* 5634 Sam Snead Dr Harlingen TX 78552

GRANT, CONRAD JOSEPH, b Corpus Christi, Tex, Apr 2, 56; m 78; c 3. APPLIED PHYSICS. *Educ:* Univ Md, College Park, BS, 78; Johns Hopkins Univ, MS, 82 & 84. *Prof Exp:* SR PHYSICIST, APPL PHYSICS LAB, JOHNS HOPKINS UNIV, 78- *Mem:* Inst Elec & Electronics Engrs; Am Soc Naval Engrs. *Res:* Combat system design; display system engineering; command, control and communications applications. *Mailing Add:* Appl Physics Lab Johns Hopkins Rd Laurel MD 20723-6099

GRANT, CYNTHIA ANN, b Minnedosa, Man, Mar 24, 58; m 80; c 1. SOIL CHEMISTRY & FERTILITY, SOIL MANAGEMENT. *Educ:* Univ Man, BSA, 80, MSc, 82, PhD(soil chem), 86. *Prof Exp:* Info officer, 82-83, RES SCIENTIST, BRANDON RES STA, AGR CAN, 86- *Mem:* Agr Inst Can; Can Soc Soil Sci. *Res:* Effects of reduced tillage on crop yield and soil physical and chemical properties; effect of tillage and soil properties on nutrient availability and on the effects of fertilizer placement on nutrient uptake and crop yield. *Mailing Add:* Brandon Res Sta Agr Can Box 610 Brandon MB R7A 5Z7 Can

GRANT, DALE WALTER, b Woodland, Maine, Dec 22, 23; m 48; c 3. MICROBIOLOGY. *Educ:* Colo State Univ, BS, 52, MS, 53; Purdue Univ, PhD, 65. *Prof Exp:* Res microbiologist, Bioferm Corp, Calif, 53-57, head microbiol res unit, 57-61; res asst microbiol, Purdue Univ, 61-65; asst prof, 65-70, ASSOC PROF MICROBIOL, COLO STATE UNIV, 70- *Mem:* AAAS; Am Soc Microbiol; Sigma Xi. *Res:* Polypeptide biosynthesis; antibiotic biosynthesis; mycotoxins; microbial decomposition of agricultural wastes. *Mailing Add:* 608 Birky Pl Ft Collins CO 80521

GRANT, DAVID EVANS, b Milwaukee, Wis, Jan 15, 24; m 45; c 1. POLYMER SCIENCE, PHYSICAL CHEMISTRY. *Educ:* Haverford Col, BA, 45; Univ Wis, PhD(phys chem), 51. *Prof Exp:* Res chemist, Hercules Inc, 51-66, res supvr, 66-71, res assoc polymer sci, 71-86; CONSULT, 86- *Mem:* Soc Plastics Engrs. *Res:* Structure-property relations of polymers; Ziegler polymerization. *Mailing Add:* 108 Banbury Dr Wilmington DE 19803

GRANT, DAVID GRAHAM, b Fall River, Mass, May 14, 37; m 59; c 5. RADIATION PHYSICS. *Educ:* Southeastern Mass Univ, BS, 59; Univ Md, MA, 66. *Prof Exp:* Prin prof staff, Appl Physics Lab, 59-75, assoc prof radiation oncol, 75-76, DIR PHYSICS, SCH MED, JOHNS HOPKINS UNIV, 75-, ASST PROF BIOMED ENG, ONCOL & RADIOL, 76-, DIGITAL IMAGE ANALYST, 80- *Mem:* Sigma Xi; Am Asn Physicists Med; AAAS. *Res:* 3-D radiotherapy treatment planning; 3-D radiographic imaging techniques; optical image and signal processing techniques. *Mailing Add:* 14865 Triadelphia Rd Glenleg MD 21737

GRANT, DAVID JAMES WILLIAM, b Walsall, Eng, Mar 26, 37. PHARMACEUTICS, PHYSICAL CHEMISTRY OF SOLIDS & SOLUTIONS. *Educ:* Oxford Univ, Eng, BA, 61, MA & DPhil(phys chem), 63, DSc, 90. *Prof Exp:* Lectr chem, Univ Col Sierra Leone, 63-65; lectr pharmaceut chem, Univ Nottingham, Eng, 65-74, sr lectr, 74-81; prof phys pharm, Univ Toronto 81-88; PROF & HEAD PHARMACEUT, COL PHARM, UNIV MINN, 88- *Concurrent Pos:* Vis prof pharmaceut chem, Univ Kans, 80, Pharmaceut Mfr Asn Can, 84 & Med Res Coun Can, 87; mem grants comt pharmaceut sci, Med Res Coun Can, 83-87; mem, Ont Univ Comt Health Res, 85-87. *Honors & Awards:* Res Award, Leverhulme Found, 69. *Mem:* Fel Royal Soc Chem; Am Pharmaceut Asn; Am Asn Col Pharm; fel Am Asn Pharmaceut Scientists; Am Chem Soc. *Res:* Improving drug delivery; the fundamental science of the preparation of solid dosage forms; the interactions of drug molecules in the solid state and in solution. *Mailing Add:* Dept Pharmaceut Col Phar Univ Minn Health Sci Unit F 308 Harvard St S E Minneapolis MN 55455

GRANT, DAVID MILLER, MOLECULAR GENETICS. *Educ:* State Univ NY, BS, 71; Univ Chicago, PhD(genetics), 77. *Prof Exp:* Anna Fuller Fund fel, Duke Univ, 77-78, NIH fel, 78-80; NIH trainee, St Louis Univ, 80-81, asst res prof, 81-83; PROJ LEADER MOLECULAR GENETICS, PIONEER HI-BRED INT, 83- *Res:* Genetic analysis of chloroplast DNA function in Chlamydomonas; mitochordrial ribosomal RNA genes; author of numerous scientific publications. *Mailing Add:* Pioneer Hi-Bred Int 7250 NW 62nd Ave Johnston IA 50131

GRANT, DAVID MORRIS, b Salt Lake City, Utah, Mar 24, 31; m 53; c 5. PHYSICAL CHEMISTRY. *Educ:* Univ Utah, BS, 54, PhD(chem), 57. *Prof Exp:* Instr chem, Univ Ill, 57-58; from asst prof to prof, 58-85, chmn dept, 62-73, dean sci, 76-85, DISTINGUISHED PROF CHEM, UTAH STATE UNIV, 85-, ASSOC VPRES, ACAD & RES COMPUT, 88- *Concurrent Pos:* Sherman Fairchild distinguished scholar, Calif Inst Technol, 73-74; adj prof fuels eng, Univ Utah, 85-, chmn computer task force, 85- *Honors & Awards:* Am Chem Soc award in Petrol Chem, 91. *Mem:* AAAS; Am Chem Soc; Am Phys Soc; Sigma Xi. *Res:* High resolution nuclear magnetic resonance; carbon-13 magnetic resonance; molecular and electronic structure. *Mailing Add:* Dept Chem Univ Utah Salt Lake City UT 84112

GRANT, DONALD LLOYD, b Pictou, NS, Dec 31, 38; m 59; c 2. TOXICOLOGY. *Educ:* McGill Univ, BSc, 60, MSc, 62, PhD(biochem), 66. *Prof Exp:* Res scientist, 65-74, TOXICOLOGIST, HEALTH PROTECTION BR, BUR CHEM SAFETY, CAN, 74- *Mem:* Am Chem Soc; Soc Toxicol; Soc Toxicol Can. *Res:* Pesticides, particularly toxicity, metabolism and analysis; polychlorinated biphenyls, dioxin. *Mailing Add:* Toxicol Eval Div Bur Chem Safety Ottawa ON K1A 0L2 Can

GRANT, DONALD R, b Cut Knife, Sask, Aug 4, 32; m 55; c 6. BIOCHEMISTRY, ORGANIC CHEMISTRY. *Educ:* Univ Sask, BSA, 55; Wash State Univ, PhD, 64. *Prof Exp:* Chemist, Maple Leaf Milling Co, Toronto, 55-58; jr chemist, Wash State Univ, 58-63; from asst prof to assoc prof, 63-74, head dept chem eng, 81-82, head dept chem, 81-84, PROF ORG & CEREAL CHEM, UNIV SASK, 74- *Mem:* Fel Chem Inst Can; Am Asn Cereal Chemists; Can Inst Food Sci & Technol. *Res:* Separation and properties of seed proteins of agricultural crops, particularly wheat and peas; functions of additives in bread flour; enzymes in cereal crops. *Mailing Add:* Dept Chem Univ Sask Saskatoon SK S7N 0W0 Can

GRANT, DOUGLAS HOPE, b Glasgow, Scotland, Apr 29, 34; m 70; c 4. CHEMISTRY. *Educ:* Glasgow Univ, BSc, 56, PhD(polymer chem), 62. *Prof Exp:* Res fel polymer chem, Appl Chem Div, Nat Res Coun Can, 59-61; NATO-Dept Sci & Indust Res fel & res asst phys chem, Bristol Univ, 61-62, Imp Chem Indust res fel, 62-64; from asst prof to assoc prof, 64-83, PROF CHEM, MT ALLISON UNIV, 83- *Concurrent Pos:* Vis lectr, Loughborough Univ Technol, 71-72. *Mem:* Chem Inst Can; Can Asn Univ Teachers. *Res:* Kinetics and mechanism of thermal decomposition reactions of synthetic polymers in solid state and in solution, by appropriate techniques; analytical chemistry. *Mailing Add:* Dept Chem Mt Allison Univ Sackville NB E0A 3C0 Can

GRANT, DOUGLAS RODERICK, b Toronto, Ont, Mar 4, 39; m 64; c 2. ENVIRONMENTAL GEOLOGY. *Educ:* Dalhousie Univ, BSc, 60, MSc, 63; Cornell Univ, PhD(sea level changes), 70. *Prof Exp:* Sci officer, 68-69, RES SCIENTIST, GEOL SURV CAN, 70- *Concurrent Pos:* Secy, Can Quaternary Asn, 79- & Can Nat Comt for Int Union Quaternary Res, 82-; assoc ed, J Coastal Res. *Mem:* Fel Geol Asn Can. *Res:* Quaternary, surficial and applied geology; glacial history; environmental change; crustal, eustatic, tidal and geodetic movements. *Mailing Add:* Five Birchview Ct Nepean ON K2G 3M7 Can

GRANT, DOUGLASS LLOYD, b Glace Bay, NS, Nov 6, 46; m 72; c 2. TOPOLOGICAL GROUPS & SEMIGROUPS. *Educ:* Acadia Univ, BSc, 66; Syracuse Univ, MSc, 68; McMaster Univ, PhD(math), 71. *Prof Exp:* Asst prof math, Col Cape Breton, 71-77; assoc prof, 77-87, PROF MATH, UNIV COL CAPE BRETON, 87- *Concurrent Pos:* Vis assoc prof, Wesleyan Univ, 79-80. *Mem:* Can Math Soc; Am Math Soc; Math Asn Am. *Res:* Miniality, weak forms of compactness, open mapping and closed graph theorems; applications of ultrafilters in topological groups and semigroups. *Mailing Add:* Box 2649 RR 3 Sydney NS B1P 6L2 Can

GRANT, EDWARD R, b Tacoma, Wash, Sept 23, 47; m 80; c 2. CHEMISTRY. *Educ:* Occidental Col, BS, 69; Univ Calif, Davis, PhD, 74. *Prof Exp:* Fel, Univ Calif, Irvine, 74-76 & Berkeley, 76-77; from asst prof to assoc prof, Cornell Univ, 77-86; PROF CHEM, PURDUE UNIV, 86- *Concurrent Pos:* Vis prof physics, Dept Physics & Res Ctr, Univ Crete, 88. *Honors & Awards:* Nobel Laureate Signature Award, 86. *Mem:* Am Chem Soc; Am Phys Soc. *Res:* Experimental and theoretical studies of the dynamics of elementary chemical rate processes especially the unimolecular relaxation and fragmentation of highly excited states; multiresonant multiphoton spectroscopy; vibronic coupling; gas-phase kinetics of organometallic reactions. *Mailing Add:* Dept Chem Purdue Univ West Lafayette IN 47907

GRANT, ERNEST WALTER, b Brockton, Mass, July 17, 18; m 45; c 2. PHARMACEUTICAL CHEMISTRY. *Educ:* Mass Col Pharm, BS, 39, MS, 48; Purdue Univ, PhD(pharmaceut chem), 51. *Prof Exp:* chemist, Eli Lilly & Co, 51-83; RETIRED. *Mem:* Am Chem Soc. *Res:* Analytical chemistry; alkaloids. *Mailing Add:* 2265 Rippling Way N Apt G Indianapolis IN 46260-6542

GRANT, EUGENE F(REDRICK), b Baker, Ore, June 15, 17; wid; c 2. ELECTRICAL ENGINEERING. *Educ:* Ore State Col, BS, 41, MS, 42. *Prof Exp:* Res engr res lab, Westinghouse Elec Corp, Pa, 42-45; proj engr, Sperry Gyroscope Co, NY, 45-46; chief appl math br, electronic res lab, Air Force, US Army, 46-51; sect mgr electronics, W L Maxson Corp, NY, 51-54; vpres eng, Nat Co Inc, Mass, 54-62; chief scientist, Space Systs Div, Hughes Aircraft Co, El Segundo, 62-78, Space & Commun Group, Systs Labs, 71-88; Consult, 86-; RETIRED. *Concurrent Pos:* Consult, Hughes Aircraft Co. *Mem:* Sr mem Inst Elec & Electronics Engrs; NY Acad Sci. *Res:* Space systems and technology; communication systems; atomic frequency standards. *Mailing Add:* 1304 Marinette Rd Pac Palisades CA 90272-2625

GRANT, EUGENE LODEWICK, b Chicago, Ill, Feb 15, 1897; m 23; c 1. ENGINEERING ECONOMY, STATISTICAL QUALITY CONTROL. *Educ:* Univ Wis-Madison, BS, 17, CE, 28; Columbia Univ, MAEcon, 28. *Hon Degrees:* Dr Eng, Mont State Univ, 73. *Prof Exp:* From instr civil eng to prof indust eng, Mont State Col, 20-30; from assoc prof to prof econ eng, Stanford Univ, 30-62; RETIRED. *Concurrent Pos:* Chmn, Comt Eng Econ, Am Soc Eng Educ, 36-42; dir, Eng Econ Div, Am Soc Civil Engrs, 39-40; consult, US CEngrs, Seattle Dist, 44-45, US Bur Reclamation, 52, Calif Div Water Resources, 58 & EBay Munic Utilities Dist, 61; exec head, Dept Civil Eng, Stanford Univ, 47-56. *Honors & Awards:* Shewhart Medal, Am Soc Qual Control, 52. *Mem:* Nat Acad Eng; hon mem Am Soc Qual Control; Am Soc Eng Educ; Am Soc Civil Engrs; Am Statist Asn. *Res:* Engineering economy; statistical quality control, depreciation, accounting and cost accounting; industrial engineering; author and co-author of numerous publications and books. *Mailing Add:* 850 Webster St Apt 809 Palo Alto CA 94301

GRANT, FREDERICK CYRIL, b Boston, Mass, July 18, 25; m 53, 76. PHYSICS. *Educ:* Mass Inst Technol, BS, 47; Col William & Mary, MA, 62; Va Polytech Inst, PhD(physics), 67. *Prof Exp:* Scientist, Nat Adv Comt Aeronaut, 48-58; PHYSICIST, NASA, 58- *Concurrent Pos:* Adj prof physics, Christofer Newport Col, 80- *Mem:* Am Phys Soc; Sigma Xi. *Res:* Theoretical and experimental supersonic aerodynamics; flight mechanics of entry into planetary atmospheres; plasma physics, especially Vlasov equations; tornado structure, dynamics. *Mailing Add:* 399 Stanton Rd Newport News VA 23606

GRANT, GEORGE C, b Medford, Mass, Aug 13, 29; m 52; c 2. BIOLOGICAL OCEANOGRAPHY. *Educ:* Univ Mass, BS, 56; Col William & Mary, MA, 62; Univ RI, PhD(biol oceanog), 67. *Prof Exp:* Fishery res biologist, Bur Com Fisheries, US Fish & Wildlife Serv, NC, 56-60; asst prof marine sci, Univ Va, 67-77; from asst prof to assoc prof, 68-81, PROF MARINE SCI, COL WILLIAM & MARY, 81- *Concurrent Pos:* From assoc marine scientist to sr marine scientist, Va Inst Marine Sci, 67-81, actg asst dir, 81-82, asst dir, 82-86. *Mem:* Am Soc Limnol & Oceanog; Atlantic Estuarine Res Soc; fel Am Inst Fishery Biologists; Crustacean Soc. *Res:* Distribution, ecology and morphometrics of North Atlantic Chaetognatha; marine zooplankton; population dynamics and ecology of commercially important fisheries; early life history and population dynamics of marine fishes. *Mailing Add:* Va Inst Marine Sci Gloucester Point VA 23062

GRANT, GREGORY ALAN, b Decorah, Iowa, Oct, 18, 49; m 71. PROTEIN CHEMISTRY. *Educ:* Iowa State Univ, BS, 71; Univ Wis-Madison, PhD(biochem), 75. *Prof Exp:* Res assoc, Sch Med, Wash Univ, 75-78, res asst prof, 78-82, asst prof, 82-88, ASSOC PROF BIOCHEM & DIR PROTEIN CHEM LAB, SCH MED, WASH UNIV, 89-, PRIN INVESTR, 81- *Concurrent Pos:* Ad hoc mem, Biochem Study Sect, NIH, 90-91; chmn, Protein Sequence Comt, Asn Biomolecular Resource Facil, 90-91. *Mem:* AAAS; Am Chem Soc; Am Soc Biochemists & Molecular Biologists; Protein Soc; Am Peptide Soc; Asn Biomolecular Resource Facil. *Res:* Structure-function relationships in proteins; collagenases, dehydrogenases, neurotoxins; protein sequencing, site-directed mutagenesis. *Mailing Add:* Dept Biochem Sch Med Wash Univ Box 8231 St Louis MO 63110

GRANT, IAN S, b Auckland, NZ, 1940; US citizen; m 62, 74; c 5. TRANSMISSION DESIGN, LIGHTNING & TRANSIENT ANALYSIS. *Educ:* Univ Nz, Auckland, BE, 62; Univ New S Wales, ME, 67. *Prof Exp:* Elec Comn NS Wales, 62-67, head, Transmission Line Elec Design Group, 67-69; staff mem, Lightning & Transient Group & Proj UHV, Elec High Voltage Lab, 69-72; staff mem, Power Technol Inc, 72-74, sr engr, 74-84, mgr, 84-86, DEPT MGR SOFTWARE PROD, POWER TECHNOL INC, 86-, COMPANY VPRES, 89- *Concurrent Pos:* US mem & convener, Lightning Work Group, Int Conf Large High Voltage Elec Systs & Co; vchmn, T & D Comt, Inst Elec & Electronics Engrs. *Mem:* Fel Inst Elec & Electronics Engrs. *Res:* Applications of compact lines; electrical and structural uprating; advanced insulator; structure applications; high phase order; line optimization; development of advanced computer hardware application; author of over 40 technical publications; computer and software applications. *Mailing Add:* Dept Software Products Power Technol Inc PO Box 1058 Schenectady NY 12301

GRANT, JAMES ALEXANDER, b Inverness, Scotland, Oct 3, 35; m 64; c 2. PETROLOGY. *Educ:* Aberdeen Univ, BSc, 57; Queen's Univ, Ont, MSc, 59; Calif Inst Technol, PhD(geol), 64. *Prof Exp:* Asst prof geol, Univ Minn, Minneapolis, 64-69; assoc prof, 69-74, PROF GEOL, UNIV MINN, DULUTH, 74- *Concurrent Pos:* Geologist, Ont Dept Mines, 56-62 & Minn Geol Surv, 64-71; NSF grants, 66-68, 76-78, 82-86 & 88-90; dir, Wasatch-Uinta Geol Field Camp, 72-79 & Study in Eng Prog, Univ Minn, Duluth, 80-81. *Mem:* Geol Soc Am; Mineral Soc Am; Sigma Xi. *Res:* Metamorphic petrology; field, petrologic and isotopic studies of the Grenville front and the Precambrian of southwestern Minnesota; phase equilibria in high-grade metamorphism and partial melting; petrology of the aureole of the Laramie anorthosite complex. *Mailing Add:* Dept Geol Univ Minn Duluth MN 55812

GRANT, JAMES J, JR, b Teaneck, NJ, May 24, 35; m 57; c 4. THEORETICAL PHYSICS. *Educ:* Manhattan Col, BS, 56; Rensselaer Polytech Inst, MS, 58; Fordham Univ, PhD(physics), 64. *Prof Exp:* From instr to assoc prof, 58-74, chmn dept, 63-81, acad dean, 81-85, PROF PHYSICS, ST PETER'S COL, NJ, 74-, ACAD VPRES, 85- *Mem:* NY Acad Sci. *Res:* Atomic scattering theory; theoretical astrophysics, optics. *Mailing Add:* Acad Vpres St Peter's Col 2641 Kennedy Blvd Jersey City NJ 07306

GRANT, JAMES WILLIAM ANGUS, b London, Ont, Jan 23, 55; m 81; c 1. TERRITORIALITY, FISH BEHAVIOR. *Educ:* Univ Western Ont, BSc, 77; Queen's Univ, MSc, 80; Univ Guelph, PhD(zool), 87. *Prof Exp:* Postdoctoral fel, McGill Univ, 87-89; ASST PROF ECOL, CONCORDIA UNIV, 89- *Mem:* Animal Behav Soc; Can Soc Zoologists; Int Soc Behav Ecol. *Res:* Behavioral ecology of agression in animals; effect of resource distribution on defence behavior; consequences of aggression on population regulation. *Mailing Add:* Biol Dept Concordia Univ 1455 de Maisonneuve W Montreal PQ H3G 1M8 Can

GRANT, JOHN ANDREW, JR, b Tallahassee, Fla, Jan 2, 41; m 62; c 3. IMMUNOLOGY, INTERNAL MEDICINE. *Educ:* Harvard Univ, AB, 62; Duke Univ, MD, 66; Am Bd Internal Med, dipl, 72; Am Bd Allergy & Immunol, dipl, 74; Diag Lab Immunol, dipl, 90. *Prof Exp:* Intern & resident, NY Hosp & Cornell Med Ctr, 66-68; clin assoc immunol, NIH, 68-71; fel, Johns Hopkins Univ, 71-73; asst prof internal med, 73-77, assoc prof internal med & genetics, 77-83, PROF INTERNAL MED & MICROBIOL, UNIV TEX MED BR GALVESTON, 83-, DIR ALLERGY & IMMUNOL SECT, 73- *Concurrent Pos:* NIH fel, 71-73; career develop award, Nat Inst Allergy & Infectious Dis, 75-80, prin investr, 75-93; prin investr, Merrell Ddow, Janssen, Fisons, Merck, Smith-Kline; consult immunol, Vet Admin, 77-80 & Food & Drug Admin, 81-84. *Mem:* Fel Am Acad Allergy; Am Asn Immunol, fel Am Col Physicians; Sigma Xi. *Res:* Mechanisms of hypersensitivity reactions, especially the role of the basophil; activation of basophils by complement fragments and by lymphokines; diagnosis and treatment of rhinitis, asthma and urticaria. *Mailing Add:* Univ Tex Med Br Adult Allergy Sect Clin Sci Bldg Rm 409 Galveston TX 77550

GRANT, JOHN WALLACE, b Weston, WVa, May 12, 46; m 69; c 3. BIOMECHANICS, VESTIBULAR MECHANICS. *Educ:* WVa Inst Technol, BS, 65; Tulane Univ, MS, 70, PhD(mech eng), 74. *Prof Exp:* Proj engr, Chem Plastics Div, Union Carbide Corp, 65-69; sr res engr, instrument prod, DuPont Co, 74-80; ASSOC PROF, MECH & BIOMED ENG, ENG SCI & MECH DEPT, VA POLYTECH INST & STATE UNIV, 80- *Concurrent Pos:* Fel, Naval Acceleration Physiol Res Lab, 85 & Naval Aerospace Med Res Lab, 90; mem bd dirs, Biomed Eng Soc, 85-87. *Mem:* Biomed Eng Soc; Am Soc Mech Eng; Am Soc Eng Educ; Am Aerospace Med Soc; Asn Res Otolaryngol. *Res:* Development of mathematic descriptions which quatify how the inner ear measure linear and rotational motion of the skull. *Mailing Add:* 4469 Pearman Rd Blacksburg VA 24060

GRANT, LELAND F(AUNTLEROY), b Etowah, Tenn, Oct 30, 13; m 39; c 2. GEOLOGICAL ENGINEERING. *Educ:* Univ Tenn, BS, 47. *Prof Exp:* Geologic aide, Tenn Valley Authority, 36-39, geologist, 39-57; secy & geologist, Schmidt Eng Co, Inc, 57-63, chief geologist, Hensley-Schmidt, Inc, 63-83; RETIRED. *Concurrent Pos:* Indust consult; mem US Comt Large Dams. *Mem:* Fel Geol Soc Am; Soc Mining Engrs; Am Soc Civil Engrs. *Res:* Engineering geology; foundation investigation and treatment, including computer analysis systems, of grouting operations. *Mailing Add:* 216 West Eighth St Suite 400 Chattanooga TN 37402-2505

GRANT, LOUIS RUSSELL, JR, b Macon, Mo, July 9, 28; m 55; c 2. SYNTHETIC INORGANIC & ORGANOMETALLIC CHEMISTRY. *Educ:* Lincoln Univ, Mo, BS, 49; Howard Univ, MS, 55; Univ Southern Calif, PhD(chem), 61. *Prof Exp:* Sr res engr, NAm Aviation, Inc, 60-62, res specialist, 62, prin scientist, 62-63, res specialist, 63-68, mgr explor, propellant & anal chem, Advan Prog, 73-78, prog mgr, Chem Progs, 78-89, MEM TECH STAFF, ROCKETDYNE DIV, ROCKWELL INT CORP, 68- *Concurrent Pos:* Consult, 89- *Mem:* Am Chem Soc; Royal Soc Chem; Sigma Xi. *Res:* Synthesis and characterization of simple and complex metal hydrides, Lewis acid base complexes, organo-metallic compounds of groups III and V elements and thermostable compounds; chemistry of explosives; energetic polymers; pilot plant production of energetic chemicals. *Mailing Add:* 2278 Ronda Vista Dr Los Angeles CA 90027

GRANT, MARTIN, b St John, NB, Nov 30, 56; m 83. NONEQUILIBRIUM STATISTICAL PHYSICS, COMPUTATIONAL PHYSICS. *Educ:* Univ PEI, BSc, 78; Univ Toronto, MSc, 80, PhD(physics), 82. *Prof Exp:* Res assoc, Physics Dept, Temple Univ, 83-86; asst prof, 86-90, ASSOC PROF, PHYSICS DEPT, MCGILL UNIV, 90- *Mem:* Can Asn Physicists; Am Phys Soc. *Res:* Kinetics of ordering systems: domain growth during first-order transitions, interface dynamics and crystal growth; analytic theory and large-scale numerical modelling. *Mailing Add:* Physics Dept Rutherford Bldg McGill Univ Montreal PQ H3A 2T8 Can

GRANT, MICHAEL CLARENCE, b Louisville, Ky, Oct 20, 42; m 61; c 1. POPULATION BIOLOGY, BOTANY. *Educ:* Tex Tech Univ, BA, 69, MS, 70; Duke Univ, PhD(bot), 74. *Prof Exp:* Dir, Mountain Res Sta, 74-76, from asst prof to assoc prof, 74-84, chmn, 82-85, PROF, DEPT ENVIRON POP & ORGANIC BIOL, UNIV COLORADO, BOULDER, 84- *Concurrent Pos:* Statist consult. *Mem:* Phycol Soc Am; Bot Soc Am; Ecol Soc Am; Soc Study Evolution; Soc Syst Zool. *Res:* Genetic and ecological structure of natural plant populations, life history adaptations, evolutionary mechanisms. *Mailing Add:* Dept Environ Pop Organismic Biol Univ Colo Boulder CO 80309-0334

GRANT, MICHAEL P(ETER), b Oshkosh, Wis, Feb 26, 36; m 61; c 3. ELECTRICAL ENGINEERING. *Educ:* Purdue Univ, BS, 57, MS, 58, PhD(elec eng), 64. *Prof Exp:* Engr, Westinghouse Res Labs, 53-57; instr elec eng, Purdue Univ, 58-64; mem tech staff, Aerospace Corp, 61; sr engr, 64-66, mgr systs & control, 66-74, asst gen mgr, 74-76, mgr systs design, Accuray

Corp, 76-87; VPRES, SYNGENICS CORP, 87-; DIR, COMPUTER INTEGRATED OPERS, NAT CTR MFG SCI, ANN ARBOR, MICH, 88- *Mem:* Inst Elec & Electronics Engrs; Sigma Xi. *Res:* Statistical communication theory and application to radar systems; control of industrial processes in real time employing digital computer systems. *Mailing Add:* 4461 Sussex Dr Columbus OH 43220

GRANT, NEIL GEORGE, b Chicago, Ill, Dec 7, 37. PLANT PHYSIOLOGY. *Educ:* Univ Ill, Urbana, BS, 60; Univ NC, Chapel Hill, PhD(bot), 72. *Prof Exp:* Asst prof biol, City Col New York, 71-77; ASSOC PROF BIOL SCI, WILLIAM PATERSON COL NJ, 77- *Concurrent Pos:* Plant quarantine inspector, USDA, 60-64. *Mem:* Am Soc Plant Physiologists; NY Acad Sci; Biophys Soc; AAAS. *Res:* Investigation of the roles of respiration and photosynthesis in the energy economy of algal cells. *Mailing Add:* Dept Biol William Paterson Col Wayne NJ 07470

GRANT, NICHOLAS J(OHN), b South River, NJ, Oct 21, 15; m 63; c 5. METALLURGY, MATERIALS SCIENCE. *Educ:* Carnegie Inst Technol, BS, 38; Mass Inst Technol, ScD(metall), 44. *Prof Exp:* Metall engr, Bethlehem Steel Co, 38-40; instr steel making, Mass Inst Technol, 42-43, from asst prof to prof metall, 44-57, dir, Ctr Mat Sci & Eng, 68-77; chmn US Side of US-USSR Sci Technol Agreement, Electrometall & Mat, State Dept, 77-82; Abex prof, 75-85, EMER PROF & SR LECTR, MASS INST TECHNOL, 85- *Concurrent Pos:* Chmn mat adv comt, Nat Adv Comt Aeronaut & NASA, 48-67; tech dir, Invest Casting Inst, 56; tech dir, Titanium Adv Bd; pres, New Eng Mat Lab, Inc, 56-67; lectr, Am Powder Metall Inst & Metal Powder Indust Fedn, 60; lectr, Japanese Advan Mat & Processes Comt, 86; Nicholas J Grant Ann Grad Res Fel, Mass Inst Technol, 87. *Honors & Awards:* J Wallenberg Award, Royal Swed Acad Engr Sci; Krumb Lectr, Am Inst Mining, Metall & Petrol Engrs, 78. *Mem:* Nat Acad Eng; fel Am Inst Mining, Metall & Petrol Engrs; Metal Powder Indust Fedn; Am Soc Testing & Mat; fel Am Soc Metals; Am Acad Arts & Sci; hon mem Japan Inst Metals; Mat Res Soc. *Res:* Heat resistant materials; powder metallurgy; deformation and fracture; structure-property control; rapid soldification; author of numerous technical publications. *Mailing Add:* Mass Inst Technol Rm 8-407 Cambridge MA 02139

GRANT, NORMAN HOWARD, b Chicago, Ill, July 21, 27; m 50; c 2. BIOCHEMISTRY. *Educ:* Univ Chicago, SB, 47; Univ Ill, MS, 48, PhD(chem), 50. *Prof Exp:* Sr res biochemist, Labs, Armour & Co, 50-56; sr res scientist biochem, 56-57, asst mgr, 68-74, mgr biochem, 74-78 assoc dir res, 78-82, DIR RES & DEVELOP ADMIN, WYETH LABS, 83- *Mem:* Soc Tech Commun; Am Soc Biol Chemists. *Res:* Chemotherapy; low temperature biochemistry; hormonal peptides; anti-infectious agents; inflammation; connective tissue proteins. *Mailing Add:* Res & Develop Admin Wyeth-Ayerst Res Box 8299 Philadelphia PA 19101

GRANT, PATRICK MICHAEL, b Oakland, Calif, Sept 20, 44; m 69; c 2. RADIOCHEMISTRY, ANALYTICAL CHEMISTRY. *Educ:* Univ Calif, Santa Barbara, BS, 67; Univ Calif, Irvine, PhD(chem), 73. *Prof Exp:* Teaching & res asst chem, Univ Calif, Irvine, 67-73; res radiochemist nuclear med, Sch Med, Univ NMex, 73-74; staff mem nuclear & radiochem, Los Alamos Sci Lab, 74-80, assoc group leader med radioisotopes, 80-81; res chemist, Anal Res & Serv, Chevron Res Co, 81-83; RADIOCHEMIST, NUCLEAR CHEM DIV, LIVERMORE NAT LAB, 83- *Concurrent Pos:* Adj asst prof chem, Univ NMex, 77-81; consult, Los Alamos Nat Lab, 81-83. *Mem:* Am Chem Soc; Am Nuclear Soc; Am Phys Soc; AAAS. *Res:* Radiochemistry; nuclear chemistry; nuclear medicine; analytical chemistry; inorganic chemistry; lanthanide-actinide thermodynamics; hot atom chemistry. *Mailing Add:* Nuclear Chemistry Div L-234 Lawrence Livermore Nat Lab Livermore CA 94550

GRANT, PAUL MICHAEL, b Poughkeepsie, NY, May 9, 35; m 58; c 4. SOLID STATE PHYSICS. *Educ:* Clarkson Col Technol, BS, 60; Harvard Univ, AM, 61, PhD(physics), 65. *Prof Exp:* STAFF MEM SOLID STATE PHYSICS, RES LAB, IBM CORP, 65- *Mem:* Am Phys Soc. *Res:* Solid state physics of high temperature superconductors. *Mailing Add:* IBM Almaden Res Ctr 650 Harry Rd San Jose CA 95120

GRANT, PETER MALCOLM, b Cleethorpes, Eng, Sept 30, 33; m 58. SCIENCE POLICY, POLYMER CHEMISTRY. *Educ:* Univ Birmingham, BSc, 54, PhD(org chem), 57. *Prof Exp:* UK Atomic Energy Res Estab fel, Univ Birmingham, 57-58; plant mgr, Gen Chem Div, Imp Chem Industs, Ltd, 58-59; res chemist, Textile Fibers Div, Du Pont Can, Ltd, 59-61; res chemist, Distillation Prod Industs Div, Eastman Kodak Co, 61-62, head, Chem Use Lab, 62-69, sr res chemist, Eastman Chem, 69-73, res assoc, 73-81, tech info assoc, 81-88, SR TECH INFO ASSOC, EASTMAN CHEM, 88- *Mem:* Am Chem Soc. *Res:* Opportunity analysis; business analysis; strategy analysis; international technical liaison; polymers; coatings. *Mailing Add:* 3632 Hemlock Park Kingsport TN 37663

GRANT, PETER RAYMOND, b London, Eng, Oct 26, 36; m 62; c 2. ZOOLOGY. *Educ:* Cambridge Univ, BA, 60; Univ BC, PhD(zool), 64. *Prof Exp:* Fel zool, Yale Univ, 64-65; from asst prof to prof, McGill Univ, 65-77; prof zool, Univ Mich, Ann Arbor, 77-85; PROF, PRINCETON UNIV, 85- *Honors & Awards:* Brewster Award, Am Ornith Union, 83. *Mem:* AAAS; Soc Study Evolution; Am Ornith Union; Ecol Soc Am; Soc Am Naturalists; fel Royal Soc. *Res:* Evolutionary significance of interactions among animal species, approached through ecology, behavior, systematics and genetics. *Mailing Add:* Dept Biol Princeton Univ Princeton NJ 08544

GRANT, PHILIP, b New York, NY, Sept 22, 24; m 46; c 4. DEVELOPMENTAL BIOLOGY, NEUROEMBRYOLOGY. *Educ:* Columbia Univ, AM, 49, PhD(zool), 52. *Prof Exp:* Tutor & instr biol, City Col New York, 47-54; res assoc embryol, Inst Cancer Res, 54-57; asst prof pathobiol, Sch Hyg & Pub Health, Johns Hopkins Univ, 57-62; prog dir develop biol, NSF, 62-66; prof, 66-91, EMER PROF BIOL, UNIV ORE, 91- *Concurrent Pos:* USPHS fel, Columbia Univ, 52-54, sr res fel, 58; ed, J Exp Zool, 78-; trustee, Marine Biol Lab, 76-79 & 80-84 & Biosis, 80-86; spec expert, NIH, 88-91. *Mem:* AAAS; Am Soc Zool; Soc Develop Biol; Marine Biol Asn; Am Soc Cell Biol; Soc Neurosci. *Res:* Ontogeny of fiber patterns between eye and brain in amphibia; growth and guidance of optic fibers in vitro and in vivo. *Mailing Add:* Dept Biol Univ Ore Eugene OR 97403

GRANT, PHILIP R, JR, GEOLOGY. *Educ:* Univ NMex, BSc, 51. *Prof Exp:* From jr petrol geologist to regional explor geologist, Sinclair Oil & Gas Co, 51-65; CONSULT GEOLOGIST & PRES, ENERGY RESOURCES EXPLOR INC, 65-; INSTR GEOSCI, SANDIA LABS, 79- *Concurrent Pos:* Mem bd dirs, Albuquerque Bd Realtors, 69-71; Realtor's Asn NMex, 70-81 & vpres, 70, acad adv comt, Am Asn Petrol Geologists, 72-, sci & technol comt, Nat Conf States Legislatures, 75-78, Albuquerque Indust Develop Serv Inc, 75-81. *Mem:* Fel Geol Soc Am; Am Asn Petrol Geologists. *Res:* Author of over 30 publications. *Mailing Add:* 1101 Rocky Point NE Albuquerque NM 87123

GRANT, RHODA, Hopewell, NS, Jan 12, 02. GASTRIC CANCER. *Educ:* McGill Univ, BA, 24, MA, 30, PhD(exp med), 32. *Prof Exp:* Res biochem, Tech Royal Victoria Hosp, Montreal, 25-29; Dem Physiol Banting & Best Res Inst Univ Toronto, 34-35; asst prof physiol, Med Col, Dalhousie Univ Halipax, 37-38; instr physiol, Med Col McGill Univ, 39-47; res assoc clin sci, 48-61, res assoc path, Univ Ill, Chicago, 61-66; RETIRED. *Concurrent Pos:* Partic invite NSF, Int Exp Gastric Cancer, Hakone, Japan, 69. *Mem:* Sigma Xi (secy, 40-43); Can Physiol Soc; Am Physiol Soc. *Res:* Biochemistry of gastric secretions-regeneration of gastric surf epithelium injury and repair of gastric lesions-gastric carcinogenesis; biochem res on gastric secretions led to the incidental discovery of the rapid regeneration of gastric surface epithelium and to researdh on injury and repair of mucosal; lesions of clinical size and finally to research on experimental gastric cancer, mast cells in gastric mucosa. *Mailing Add:* 525 University Dr East Lansing MI 48823

GRANT, RICHARD EVANS, b St Paul, Minn, June 18, 27; m 58; c 3. INVERTEBRATE PALEONTOLOGY. *Educ:* Univ Minn, MS, 53; Univ Tex, PhD(geol), 58. *Prof Exp:* Instr geol, Univ Tex, 53-54; res asst invert paleont, Smithsonian Inst, 57-61; geologist, US Geol Surv, 61-72; chmn dept paleobiol, 72-77, cur paleobiol, 72-83, SR GEOLOGIST, NAT MUS NATURAL HIST, SMITHSONIAN INST, 83- *Honors & Awards:* Daniel Giraud Elliot Medal, Nat Acad Sci, 79. *Mem:* AAAS; Geol Soc Am; Paleont Soc (pres, 79); Int Paleont Asn (treas, 80-89). *Res:* Permian Brachiopods; their biostratraphy, functional morphology and biogeography. *Mailing Add:* Dept Paleobiol E-206 Natural Hist Bldg Washington DC 20560

GRANT, RICHARD J, b Boness, Scotland, Oct 6, 18; US citizen; m 55; c 2. PHYSICAL CHEMISTRY. *Educ:* Waynesburg Col, BS, 50; Carnegie-Mellon Univ, MS, 55. *Prof Exp:* Res chemist, Pittsburgh Coke & Chem Co, 50-58, Pittsburgh Chem Co, 58-64 & Pittsburgh Activated Carbon Co, 64; fel chem, Mellon Inst, 64-70; res assoc chem, Calgon Corp, 70-76, sr res assoc, 76-84; RETIRED. *Mem:* Am Chem Soc. *Res:* Adsorption, pollution, pore size, hydrocarbon oxidation catalysis, activated carbon. *Mailing Add:* 492 Salem Dr Pittsburgh PA 15243-2076

GRANT, RODERICK M, JR, b Chicago, Ill, July 9, 35; m 57; c 3. SOLID STATE PHYSICS, MEDICAL PHYSICS. *Educ:* Denison Univ, BS, 57; Univ Wis, MS, 59, PhD(physics), 65. *Prof Exp:* Teaching asst physics, Univ Wis, 57-59, instr, Marathon County Ctr, 59-61, res asst, Univ, 61-65; from asst prof to assoc prof, 65-77, chmn dept, 70-74 & 76, PROF PHYSICS & HENRY CHISHOLM CHAIR PHYSICS, DENISON UNIV, 77- *Mem:* AAAS; Am Asn Physics Teachers (secy, 77-83); Am Asn Physicists Med; Am Phys Soc; Am Inst Physics (secy, 82-). *Res:* Problems of medical physics; problems of computer-assisted instruction in physics and astronomy. *Mailing Add:* Dept Physics Denison Univ Granville OH 43023

GRANT, RONALD, neurophysiology; deceased, see previous edition for last biography

GRANT, RONALD W(ARREN), b Cleveland, Ohio, Nov 4, 37; m 59; c 3. SEMICONDUCTOR INTERFACE, EPITAXIAL GROWTH OF III-V SEMICONDUCTORS. *Educ:* Case Inst Technol, BS, 59; Univ Calif, Berkeley, PhD(nuclear chem), 63. *Prof Exp:* Mem tech staff, 63-76, group leader solid state physics, Sci Ctr, 76-79, mgr, mat properties sect, Microelectronics Res & Develop Ctr, 79-87, MGR, EPITAXIAL MAT DEPT, SCI CTR, ROCKWELL INT, 87- *Concurrent Pos:* Mem ad hoc comt, Nat Res Coun, 70-74; chmm, 18th Ann Physics & Chem Semiconductor Interface Conf, 91. *Mem:* Am Phys Soc; Sigma Xi; Am Vacuum Soc. *Res:* Semiconductor surface and interface characterization by using X-ray photoelectron spectroscopy; epitaxial growth of III-V semiconductor materials. *Mailing Add:* Sci Ctr Rockwell Int 1049 Camino Dos Rios Thousand Oaks CA 91360

GRANT, SHELDON KERRY, b Cedar City, Utah, Apr 29, 39; m 60; c 3. MINERALOGY, PETROLOGY. *Educ:* Univ Utah, BS, 61, PhD(geol eng), 66. *Prof Exp:* Asst prof geol eng, 65-70, assoc prof, 70-83, PROF GEOL, UNIV MO, ROLLA, 83- *Mem:* Am Inst Mining, Metall & Petrol Eng; Mineral Soc Am; Geol Soc Am; Sigma Xi. *Res:* Stratigraphy, mineralogy, alteration, relation to ore deposits and tectonic evolution of ash-flow tuffs; petrology and chemistry of igneous rocks; field tectonics in the Cordillera. *Mailing Add:* Dept Geol & Geophys Univ Mo Rolla MO 65401

GRANT, STANLEY CAMERON, b Cedar Rapids, Iowa, Apr 21, 31; m 54, 84; c 3. GEO-TECHNICAL MANAGEMENT, ENVIRONMENTAL ADMINISTRATION. *Educ:* Coe Col, BA, 53; Univ Wyo, MA, 55; Univ Idaho, PhD(geol), 71. *Prof Exp:* Petrol geologist, Calif Co, Standard Oil of Calif, 55; chief geologist, Gas Hills Uranium Co, Am Nuclear, 55-56; instr, Howard County Jr Col, Big Spring, Tex, 61-62; asst prof aerospace study, Univ Idaho, 66-69; assoc prof geol, Univ Northern Iowa, 70-75; state geologist & dir, Iowa Geol Surv, 75-80; vpres operations, Bishop Oil & Refining Co, 80-81; partner, Grant Geol Serv, 81-87; CABINET SECY,

KANS DEPT HEALTH & ENVIRON, 87- *Concurrent Pos:* Sci adv to Gov of State of Iowa, Iowa Interagency Resources Coun, 75-80, Kans Water Authority, 88-; Environ Protection Agency/State Opers coun, 90-; Kans Geog Info Syst Policy bd, 89- *Mem:* Geol Soc Am; Am Inst Mining, Metall & Petrol Engrs; Asn Eng Geologists; Am Inst Prof Geologists; Sigma Xi. *Res:* Thirtyfive millimeter color oblique aerial photographic techniques in geologic reconnaissance; photographic remote sensing of the physical environment; remote sensing applications to petroleum geology; environmental remediation in the Midwest and Western US. *Mailing Add:* 8251 SW 61st St Topeka KS 66610-9652

GRANT, VERNE (EDWIN), b San Francisco, Calif, Oct 17, 17; m 46, 60; c 3. BOTANY, EVOLUTION. *Educ:* Univ Calif, AB, 40, PhD(bot), 49. *Prof Exp:* Asst bot, Univ Calif, 46-49; vis investr, Carnegie Inst Wash, Stanford, Calif, 49-50; geneticist & exp taxonomist, Rancho Santa Ana Bot Garden, Claremont, Calif, 50-67; prof biol, Inst Life Sci, Tex A&M Univ, 67-68; prof biol sci & dir Boyce Thompson Southwest Arboretum, Univ Ariz, Superior, 68-70; prof, 70-87, EMER PROF BOT, UNIV TEX, AUSTIN, 87- *Concurrent Pos:* Nat Res Coun fel, 49-50; from asst prof to prof bot, Claremont Col, 51-67. *Honors & Awards:* Cert Merit, Bot Soc Am, 71. *Mem:* Nat Acad Sci; fel Am Acad Arts & Sci; Genetics Soc Am; Soc Study Evolution (vpres, 66, pres, 68); Bot Soc Am; Am Soc Naturalists; Int Soc Plant Taxonomists; Am Soc Plant Taxonomists. *Res:* Cytotaxonomy and phylogeny of Polemoniaceae; fertility relationships in annual Gilias; effects of pollinating animals on flower evolution; evolutionary theory; speciation in higher plants; population biology of cacti (opuntia); systematics of the Ipomopsis aggregata group and the alpine Polemoniums (Polemoniaceae). *Mailing Add:* Dept Bot Univ Tex Austin TX 78713

GRANT, WALTER MORTON, b Lawrence, Mass, July 12, 15; m 36; c 3. OPHTHALMOLOGY. *Educ:* Harvard Univ, SB, 36, MD, 40. *Prof Exp:* Asst ophthal res, 41-42, instr, 42-46, assoc, 46-51, from asst prof to prof, 51-74, David Glendenning Cogan prof ophthal, DAVID GLENDENNING COGAN EMER PROF OPHTHAL, HARVARD MED SCH, 82- *Concurrent Pos:* Dir Glaucoma Consult Serv, Mass Eye & Ear Infirmary, 60-81, Ophthal Res Adminr, 71; mem study sect, NIH, 60-63; chmn adv panel, Hazardous Substances Labeling Act, Fed Drug Admin, US Dept HEW, 63-64; mem ophthal panel, Drug Efficacy Study, Nat Res Coun, 69; consult, Med Letter, 59-, Nat Coun to Combat Blindness, 59-62, US Pharmacopeia, 60-88, Boston Childrens Med Ctr, 70-81 & Nat Eye Inst, 71-73; mem, Nat Adv Eye Coun, NIH, 73-77; mem, Ophthal Drugs Adv Comt, Food & Drug Admin, 73-76. *Honors & Awards:* NE Ophthal Soc Award, 50; Proctor Medal, 56; Knapp Medal, 61; Howe Medal, Am Ophthal Soc, 68; Res to Prevent Blindness Trustees' Award, 69; Eye Found Am Award, 76. *Mem:* Am Chem Soc; Am Ophthal Soc; AMA; Am Acad Ophthal. *Res:* Physiology of aqueous humor; chemical injuries of eyes; toxicology of the eyes; investigation and development of new drugs for treatment of glaucoma; development of colorimetric microchemical analytical methods; investigation of factors controlling outflow of aqueous humor from normal and glaucomatous eyes; clinical study of glaucoma. *Mailing Add:* 243 Charles St Boston MA 02114

GRANT, WARREN HERBERT, b New Orleans, La, Oct 16, 33; m 59; c 3. PHYSICAL CHEMISTRY, POLYMER CHEMISTRY. *Educ:* Talladega Col, BS, 55; Howard Univ, MS, 62, PhD(chem), 68. *Prof Exp:* RES CHEMIST, NAT BUR STANDARDS, 62-, PROJ LEADER, 68- *Concurrent Pos:* Sci adv, Nat Acad Sci-Nat Res Coun, 73- *Mem:* Am Chem Soc; Soc Biomat. *Res:* Polymers; polymer adsorption, flocculation and fractionation; surface properties; interfaces; biomaterials; conformation of polymers. *Mailing Add:* 4917 Morning Glory Ct Rockville MD 20853

GRANT, WILLARD H, b Springfield, Mass, Jan 19, 23; m 50; c 4. GEOLOGY. *Educ:* Emory Univ, AB, 48, MS, 49; Johns Hopkins Univ, PhD(geol), 55. *Prof Exp:* Assoc prof, 52-69, PROF GEOL, EMORY UNIV, 69- *Mem:* AAAS; Geochem Soc; Clay Minerals Soc. *Res:* Chemistry and mineralogy of the weathering environment; structural petrology. *Mailing Add:* 3971 Ashleywood Ct Tucker GA 30084

GRANT, WILLIAM B, b Fresno, Calif, Feb 4, 42; m 81. LIDAR SYSTEMS. *Educ:* Univ Calif, Berkeley, BA, 64, PhD(physics), 71. *Prof Exp:* Postdoctoral res assoc, Inst Physics, Free Univ Berlin, 71- 73; sr physicist, SRI Int, 73-79; mem tech staff, Jet Propulsion Lab, Calif Inst Technol, 79-89; SR RES SCIENTIST, ATMOSPHERIC SCI DIV, NASA LANGLEY RES CTR, 89- *Mem:* AAAS; Optical Soc Am; Am Meteorol Soc; Am Geophys Union. *Res:* Development and demonstration of laser systems for the remote measurement of atmospheric gases and aerosols. *Mailing Add:* Atmospheric Sci Div NASA Langley Res Ctr Hampton VA 23665-5225

GRANT, WILLIAM CHASE, JR, b Baltimore, Md, Aug 24, 24; m 47; c 2. ZOOLOGY. *Educ:* Dartmouth Col, AB, 49; Yale Univ, PhD(zool), 53. *Prof Exp:* Asst zool, Yale Univ, 49-53; asst prof, Gettysburg Col, 53-54, Col William & Mary, 54-55 & Dartmouth Col, 55-56; from asst prof to prof, 56-68, chmn dept, 62-77 & 87-91, S F CLARKE PROF BIOL, WILLIAMS COL, 68- *Concurrent Pos:* NIH spec fel, Oxford Univ, 65-66; corp mem, Mt Desert Island Biol Lab, secy; mem bd dirs, Chase Forests & Sanctuary; mem, Consults Bur, Off Biol Educ, 68-71; vis prof, Edinburgh Univ, 72-73. *Mem:* Fel AAAS; Ecol Soc Am; Am Soc Zoologists. *Res:* Physiological ecology; endocrinology of lower vertebrates; invertebrate behavior. *Mailing Add:* Box 647 Williams Col Williamstown MA 01267

GRANT, WILLIAM FREDERICK, b Hamilton, Ont, Oct 20, 24; m 49; c 1. CYTOGENETICS. *Educ:* McMaster Univ, BA, 47, MA, 49; Univ Va, PhD(biol), 53. *Prof Exp:* Botanist, Colombo Plan, Dept Agr, Malaya, 53-55; from asst prof to prof, 55-90, EMER PROF GENETICS, MCGILL UNIV, MACDONALD CAMPUS, 90- *Concurrent Pos:* Ed, Lotus Newslett, 70-84 & Can J Genetics & Cytol, 74-82; treas, Biol Coun Can, 74-78; mem, environ contaminants adv comt, Ministers Environ & Nat Health & Welfare, Ottawa, 78-; adv WHO int prog chem safety collab study on short-term tests for genotoxicity & carcinogenicity, 84-; archivist, Gen Soc Can; sr fel lectr, Japan Soc Prom Sci, 89. *Honors & Awards:* Andrew Fleming Award, 53; Gov Gen Silver Medal, 77; George Lawson Medal, Can Bot Asn, 89. *Mem:* Fel AAAS; fel Linnean Soc; Genetics Soc Can (pres, 74-75); Int Orgn Plant Biosyst (pres, 81-87); Soc Study Evolution (vpres, 72); Sigma Xi; Can Bot Asn; Am Bot Asn; Am Soc Plant Taxonomists. *Res:* Lotus cytogenetics; cytogenetic effects of environmental agents; biosystematics. *Mailing Add:* Dept Plant Sci Box 4000 McGill Univ Macdonald Campus Ste-Anne-de-Bellevue PQ H9X 1C0 Can

GRANTHAM, JARED JAMES, b Dodge City, Kans, May 19, 36; m 58; c 4. MEDICINE, NEPHROLOGY. *Educ:* Baker Univ, AB, 58; Univ Kans, MD, 62. *Prof Exp:* Assoc prof, 69-76, PROF MED, SCH MED, UNIV KANS, 76-, HEAD, NEPHROL SECT, 70- *Concurrent Pos:* Fel nephrology, NIH, 64-66; Kaw Valley Heart Asn grant, 69-70; Nat Inst Arthritis & Metab Dis grant, 69-83. *Mem:* Am Soc Nephrology; Am Soc Clin Invest; Am Physiol Soc; Am Fedn Clin Res; Asn Am Phys. *Res:* Fluid and electrolyte metabolism; electrolyte transport; mechanism of action of antidiuretic hormone; polycystic kidney disease. *Mailing Add:* Med Dept Med Ctr Univ Kans 39th & Rainbow Blvd Kansas City KS 66103

GRANTHAM, LEROY FRANCIS, b Chadron, Nebr, Nov 23, 29; m 52; c 5. INORGANIC CHEMISTRY, PHYSICAL CHEMISTRY. *Educ:* Chadron State Col, BS, 51; Iowa State Univ, MS, 54; Kans State Univ, PhD(chem), 59. *Prof Exp:* Jr chemist, Ames Lab, Iowa State Univ, 54; chemist, Phosphate Develop Works, 54-56; sr chemist, 59-63, res specialist, 63-67, SR SCIENTIST, ATOMICS INT, ROCKWELL INT CORP, 67- *Mem:* Am Chem Soc; Sigma Xi; Am Nuclear Soc. *Res:* Isotopic and electron exchange; corrosion; fused salts; metal-salt melts; high temperature chemistry; mass spectrometry; radioactive waste handling, volume reduction and insolubilization; pollution control; hazardous waste management; nuclear fuel cycle. *Mailing Add:* 26117 Hatmor Calabasas CA 91302

GRANTHAM, PRESTON HUBERT, toxicology, pharmacology; deceased, see previous edition for last biography

GRANTMYRE, EDWARD BARTLETT, b Sydney, NS, July 30, 31; m 56; c 3. RADIOLOGY. *Educ:* Dalhousie Univ, MD, CM, 56; FRCP(C). *Prof Exp:* Resident radiol, St John Gen Hosp, 56-58; chief resident pediat radiol, Childrens Hosp Med Ctr, Boston, Mass, 58-59; radiologist, Childrens Hosp Halifax, 59-66; lectr, 60-62, from asst prof to assoc prof, 62-66, PROF RADIOL, DALHOUSIE UNIV, 66-; RADIOLOGIST-IN-CHIEF, IWK HOSP FOR CHILDREN & GRACE MATERNITY HOSP, HALIFAX, 66- *Mem:* Can Med Soc; Can Asn Radiol; Soc Pediat Radiol; Am Inst Ultrasonics Med. *Res:* Diseases of children; ultrasound. *Mailing Add:* Dept Radiol Victoria Gen Hosp 1278 Tower Rd Halifax NS B3H 2Y9 Can

GRANTZ, ARTHUR, b New York, NY, Nov 9, 27; m 51; c 4. GEOLOGY. *Educ:* Cornell Univ, AB, 49; Stanford Univ, MS, 61, PhD(geol), 66. *Prof Exp:* Asst, Cornell Univ, 49; geologist, Br Alaska Geol, 49-66, chief, Br Pac Environ Geol, 66-71, GEOLOGIST, BR ALASKAN GEOL & BR PAC-ARCTIC MARINE GEOL, US GEOL SURV, 71- *Concurrent Pos:* Mem & vchmn, Calif State Mining & Geol Bd, 76-79. *Mem:* Geol Soc Am; Am Asn Petrol Geol; Am Geophys Union. *Res:* Structural geology and stratigraphy of Alaska; geologic structure and petroleum resources of northern Alaska's continental shelves and Arctic Ocean. *Mailing Add:* US Geol Surv 345 Middlefield Rd Menlo Park CA 94025

GRANTZ, DAVID ARTHUR, b Nashville, Tenn, Dec 29, 51; m 80; c 2. STOMATAL PHYSIOLOGY, AIR POLLUTION. *Educ:* Univ Calif, Santa Cruz, AB, 73, Riverside, MS, 79; Univ Ill, Urbana, PhD(plant physiol), 83. *Prof Exp:* Res assoc, dept biol sci, Stanford Univ, 83-85; plant physiologist, Agr Res Serv, USDA, 85-90; ASSOC RES PLANT PHYSIOLOGIST, STATEWIDE AIR POLLUTION RES CTR & ASSOC EXTEN AGR, DEPT BOT & PLANT SCI, UNIV CALIF RIVERSIDE, 90- *Concurrent Pos:* Res fel, Hebrew Univ Jerusalem, 86. *Mem:* Am Soc Plant Physiologists; Am Soc Agron; AAAS; Crop Sci Soc Am; Sigma Xi; Plant Growth Soc Am. *Res:* Plant responses to environmental stress including water deficit and air pollution; regulation of leaf gas exchange and stomatal physiology. *Mailing Add:* Kearney Agr Ctr Univ Calif Riverside 9240 S Riverbend Ave Parlier CA 93648

GRANZOW, KENNETH DONALD, b Oak Park, Ill, Mar 26, 33; m 52; c 4. PHYSICS, ELECTRICAL ENGINEERING. *Educ:* Univ Ill, BS, 58, MS, 59. *Prof Exp:* Staff mem, Sandia Corp, 59-63; from res physicist to sr res physicist, The Dikewood Corp, 63-68, leading scientist, 68-80, prin scientist & dir, 80-85; sci asst div mgr, Mission Res Corp, 85-87; SCI CONSULT, 87- *Mem:* AAAS; Am Phys Soc. *Res:* Nuclear weapon effects; nuclear safeguards; electromagnetism; quantum field theory; atmospheric physics; gas discharges. *Mailing Add:* 1079 Haverhill Pl Colorado Springs CO 80919

GRAPER, EDWARD BOWEN, b Los Angeles, Calif, June 1, 47; m 68. THIN FILMS, HIGH VACUUM FOR SCIENCE & TECHNOLOGY. *Educ:* Univ Calif, Berkeley, BSME, 65. *Prof Exp:* Engr, Unified Sci Inc, Pasadena, Calif, 65-67; proj engr, Sloan Technol, Santa Barbara, Calif, 67-73; PRES, LEBOW CORP, GOLETA, CALIF, 73. *Concurrent Pos:* Expert thin films, UN Indust Develop Orgn, Vienna, 86- *Honors & Awards:* I-R 100 Award, 77 & 79. *Mem:* Am Vacuum Soc; Am Optical Soc; Soc Photo-Optical Instrumentation Engrs. *Res:* Deposition of thin films by evaporating and sputtering; vacuum technology; application of thin films and vacuum to industrial and laboratory products and processes. *Mailing Add:* RR 1 Box 230A Goleta CA 93117

GRASDALEN, GARY, b Albert Lea, Minn, Oct 7, 45. ASTRONOMY. *Educ:* Harvard Col, AB, 67; Univ Calif, Berkeley, MA, 70, PhD(astron), 72. *Prof Exp:* Asst astronr, Kitt Peak Nat Observ, 72-75, assoc astronr, 75-77; asst prof, 77-80, ASSOC PROF ASTRON, UNIV WYO, 80- *Mem:* Am Astron Soc. *Res:* Infrared astronomy; premain sequence evolution; composite stellar systems; observational cosmology. *Mailing Add:* Dept Physics & Astron Univ Wyo Laramie WY 82071

GRASS, ALBERT M(ELVIN), b Quincy, Mass, Sept 3, 10; m 36; c 2. ELECTRONICS. *Educ:* Mass Inst Technol, BS, 34. *Hon Degrees:* DSc, Drexel Inst, 63. *Prof Exp:* PRES, GRASS INSTRUMENT CO, 35- *Concurrent Pos:* Res engr, Harvard Med Sch, 35-43; staff mem, Radiation Lab, Mass Inst Technol, 41-45. *Mem:* AAAS; Biophys Soc; Soc Photo-Optical Instrument Eng; Am Electroencephalog Soc; Inst Elec & Electronics Engrs; Sigma Xi. *Res:* Electro-medical instruments; radar range circuits; electroencephalograph devices; servomechanisms. *Mailing Add:* 77 Reservoir Rd Quincy MA 02170

GRASSE, PETER BRUNNER, b Chicago, Ill, Feb 3, 56. CHEMISTRY. *Educ:* Carleton Col, BA, 78; Univ Ill, PhD(org chem), 83. *Prof Exp:* Sr prod develop chemist, indust tape div lab, 83-89, PROD ENG SUPVR, AUTOMOTIVE DESIGN SYSTS, 3M CO, 89- *Mem:* Sigma Xi; Am Chem Soc. *Res:* Pressure sensitive adhesive development; products for electronics market. *Mailing Add:* Automotive Design Systs Lab 3M Co Bldg 209-2N-02 St Paul MN 55144-1000

GRASSELLI, JEANETTE GECSY, b Cleveland, Ohio, Aug 4, 28; m 87. SPECTROSCOPY. *Educ:* Ohio Univ, BS, 50; Western Reserve Univ, MS, 58. *Hon Degrees:* DSc, Ohio Univ, 78 & Clarkson Univ, 86; DEng, Mich Technol Univ, 89. *Prof Exp:* Chemist infrared spectros, BP Am Res & Develop, 50-56, proj leader, Absorption Spectros Group, 56-70, supvr molecular spectros, 70-81, dir, Anal Sci Lab, 81-83, dir, Tech Support Dept, 83-85, dir, Corp, Res & Anal Sci, 85-89; DIR, RES ENHANCEMENT & DISTINGUISHED VIS PROF, OHIO UNIV, 90- *Concurrent Pos:* adv bd, Univ Southern Miss, 85-88; adv bd, US Dept Energy, 87-88; vis comt, NIST, 88-; mem bd, Chem Sci & Technol, NRE, 86- & Edison Biotech Ctr, 88-; adv comt chem, NSF, 88-; exhibition adv bd, Smithsonian, 90-; bd dirs, Nic Instr Co, 82-; chmn bd trustees, Ohio Univ, 85-; vpres, NE Ohio Sci & Eng Fair, 77-; chmn, Anal Div, Am Chem Soc, 90-91. *Honors & Awards:* Anachem Award, 78; Williams-Wright Award, 80; Harry Hallan Mem lectr, Swansea, UK, 80; Lucy Pickett Mem lectr, Mt Holyoke Col, 80; Garvan Medal, Am Chem Soc, 86. *Mem:* Am Chem Soc; Soc Appl Spectros (pres, 70); Coblentz Soc; Fedn Anal Chem & Spectros Socs (secy, 76-80); AAAS. *Res:* Molecular spectroscopy; infrared; raman. *Mailing Add:* 150 Greentree Rd Chagrin Falls OH 44022

GRASSELLI, ROBERT KARL, b Celje, Yugoslavia, June 7, 30; nat US; m 57. PHYSICAL CHEMISTRY & CATALYSIS. *Educ:* Harvard Univ, AB, 52; Case Western Reserve Univ, MS, 55, PhD(phys chem), 59. *Prof Exp:* Chemist, Standard Oil Co, Ohio, 52-55, sr chemist & proj dir, 55-59, sr res chemist & group leader, 60-65, res assoc, 65-69, sr res assoc, 69-74, res supvr explor & fundamental catalysis, 74-78, sr scientist, 78-80, sci fel & dir, Catalysis & Solid State Sci Lab, 81-85; dir, Chem Div, Off Naval Res, 86-89; RES SCIENTIST, MOBIL RES & DEVELOP CORP, PRINCETON, 89- *Concurrent Pos:* Adj prof chem, Case Western Reserve Univ, 83-89; guest prof indust chem, Univ Bologna, Italy, 87-; assoc ed, Catalysis Today, 87-; adj res prof chem, Georgetown Univ, 88-; dir, Phil Catalysts Soc, 90- *Honors & Awards:* Charles D Hurd Lectr, Northwestern Univ, 82; E V Murphree Award, Am Chem Soc, 84 & Petrol Chem Award, 90. *Mem:* Am Chem Soc; Catalysis Soc; fel AAAS; Sigma Xi. *Res:* Heterogeneous catalysis; kinetics of solid state and surface dynamics; reaction mechanisms; petrochemicals; surface reactions; controlled sequencing of catalytic sites; synthesis; alternate fuels. *Mailing Add:* Mobil Cent Res & Develop Corp PO Box 1025 Princeton NJ 08543-1025

GRASSER, BRUCE HOWARD, b Cleveland, Ohio, Mar 8, 41; m 61; c 3. CHEMICAL ENGINEERING. *Educ:* Iowa State Univ, BSChE, 64; Case Western Reserve Univ, MBA, 71. *Prof Exp:* Process engr chem process develop, Goodrich-Gulf Chem, Inc, 64-68; group leader chem processing res & develop, 70-77, group leader maintenance & eng, 72-77, asst dept head, 77-78, DEPT HEAD PILOT PLANT CHEM PROCESS ENG RES & DEVELOP, LUBRIZOL CORP, 78- *Mem:* Am Chem Soc; Am Inst Chem Engrs. *Res:* Chemical process research and development relative to lubricant and fuel additive chemistry. *Mailing Add:* 10055 Woodview Dr Chardon OH 44024-9105

GRASSETTI, DAVIDE RICCARDO, b Padua, Italy, Aug 24, 20; US citizen; m 52; c 4. BIOPHARMACEUTICS, PHARMACOLOGY. *Educ:* Univ Lausanne, BS, 41, PhD(chem), 45, ChE, 45. *Prof Exp:* Instr pharmaceut chem, Univ Padua, 48-50; res assoc org chem, Mass Inst Technol, 50-52; res assoc, Yale Univ, 52-54; res chemist, Nopco Chem Co, 54-56; assoc res chemist, Sch Med, Univ Calif, San Francisco, 57-62; proj dir cancer chemother, Inst Chem Biol, Univ San Francisco, 62-66; dir res, Arequipa Found, San Francisco, 67-74; pres, Newcell Biochem, Berkeley, 75-80; prof chem, Univ Pac, Stockton, Calif, 84-86; RETIRED. *Concurrent Pos:* Lectr, Med Ctr, Univ Calif, San Francisco, 60-64; clin assoc prof biochem, Sch Dent, Univ of the Pac, 73-; vis prof pharmacol, Univ Padua Med Sch, Italy, 79-82. *Mem:* Am Chem Soc. *Res:* Sulfhydryl groups on cell surfaces; modification of cell surfaces by heterocyclic disulfides; prevention of cancer metastases; cancer therapy. *Mailing Add:* 26 Northgate Ave Berkeley CA 94708

GRASSHOFF, JURGEN MICHAEL, b Berlin, WGer, May 7, 36; US citizen; m 63; c 3. POLYMER CHEMISTRY, ORGANIC CHEMISTRY. *Educ:* Hannover Tech Univ, MS, 61; Swiss Fed Inst Technol, PhD(org chem), 65. *Prof Exp:* Head, Prod Develop Lab, Steding & Co, WGer, 61-62; sci co-worker, Swiss Fed Inst Technol, 65; res chemist, Pennsalt Chem Corp, Pa, 65-67; scientist, 67-77, SR SCIENTIST, POLAROID CORP, CAMBRIDGE, 79- *Concurrent Pos:* Abstractor, Chem Abstr Serv, 68- *Mem:* Am Chem Soc. *Res:* Aromatic diepoxides and diglycols; radical copolymerization of highly fluorinated monomers; swelling of water-soluble polymers; grafting of cellulose derivatives; release of photographic reagents; synthesis of polymeric mercaptides; emulsion polymers. *Mailing Add:* Seven John Robinson Rd Hudson MA 01749-2821

GRASSI, RAYMOND CHARLES, b Highland Park, Mich, Nov 27, 18; m 42. MECHANICAL ENGINEERING. *Educ:* Univ Calif, BS, 40, MS, 44. *Prof Exp:* Inspector, Nordstrom Valve Co, 40-41, tool design engr, 41-42; instr mech eng, 42-44, 45-46, asst prof, 46-52, assoc prof indust eng, 52-60, PROF INDUST ENG, UNIV CALIF, BERKELEY, 60-, PROF OPER RES, 70- *Concurrent Pos:* Ed supvr, Bur Ships Contract, 43-44, 46-, res engr, 48-49; engr, Boeing Aircraft Co, Wash, 46; res engr, Off Sci Res & Develop contract, 44-46, Off Naval Res contract, 49-50, Atomic Energy Comn, Oak Ridge, 50-52 & Off Ord Res, US Army, 54- *Mem:* Am Soc Mech Engrs; Am Soc Metals; Am Soc Eng Educ; Sigma Xi. *Res:* Welding; residual stresses; structure; copper brazing of steel; effect of combined stresses on materials; production processes involving metal working or forming; stress rupture tests of various metals subjected to liquid-metal environment. *Mailing Add:* 4129 Etcheverry Hall Univ Calif Berkeley CA 94720

GRASSI, VINCENT G, b Geneva, NY, Feb 21, 56. CHEMICAL PROCESS CONTROL, CHEMICAL PROCESS DEVELOPMENT & DESIGN. *Educ:* Univ Rochester, BS, 78; Lehigh Univ, MS, 84, PhD(chem), 91. *Prof Exp:* Comp control analyst, MIS, Air Prod & Chem Inc, 78-80, process engr, chem group engr, 80-83, sr process engr, process systems, 83-86, prin process engr, process technol, 86-89, GROUP LEADER, PROCESS TECHNOL MODELING & PROCESS DYNAMICS, AIR PROD & CHEM INC. *Concurrent Pos:* Lectr, distillation dynamics & control short course, Lehigh Univ, 86-, sem lectr, Chem Process Modeling & Control Ctr. *Mem:* Am Inst Chem Engrs. *Res:* Development of computer models and analysis methods of chemical process dynamics and control; distillation processes including the process design and control of extraction distillation. *Mailing Add:* Air Prod & Chemicals Inc 7201 Hamilton Blvd Allentown PA 18195-1501

GRASSINO, ALEJANDRO E, RESPIRATORY MEDICINE. *Educ:* Univ Rosario, Arg, MD, 64. *Prof Exp:* PROF MED, UNIV MONTREAL & MCGILL UNIV, 85- *Res:* Respiratory muscle physiology; chest wall mechanics. *Mailing Add:* Dept Med Univ Montreal & McGill Univ 3775 University St Montreal PQ H3A 2B4 Can

GRASSL, STEVEN MILLER, b East Lansing, Mich, Mar 3, 52; m 75; c 3. PHARMACOLOGY, PHYSIOLOGY. *Educ:* Dickinson Col, BS, 74; Rutgers Univ, MS, 79; Cornell Univ, PhD(physiol), 83. *Prof Exp:* Fel physiol, Yale Sch Med, 80-83; INSTR PHARMACOL, HEALTH SCI CTR, STATE UNIV NY, SYRACUSE, 83- *Mem:* Am Physiol Soc. *Res:* Membrane transport physiology. *Mailing Add:* Dept Pharmacol State Univ NY Health Sci Ctr 766 Irving Ave Syracuse NY 13214

GRASSLE, JOHN FREDERICK, b Cleveland, Ohio, July 14, 39; m 64; c 1. ECOLOGY, BIOLOGICAL OCEANOGRAPHY. *Educ:* Yale Univ, BS, 61; Duke Univ, PhD(zool), 67. *Prof Exp:* Fulbright-Hays grant, Univ Queensland, 67-69; from asst scientist to assoc scientist, 69-83, SR SCIENTIST, WOODS HOLE OCEANOG INST, 83-; DIR, INST MARINE & COASTAL SCI, 89- *Mem:* Am Soc Limnol & Oceanog; Soc Study Evolution; Ecol Soc Am. *Res:* Population biology of marine benthic organisms; coastal, deep-sea and coral reef communities. *Mailing Add:* Inst Marine & Coastal Sci Rutgers Univ PO Box 231 New Brunswick NJ 08903

GRASSLE, JUDITH PAYNE, b Brisbane, Australia, Dec 4, 36; m 64; c 1. POPULATION GENETICS. *Educ:* Univ Queensland, BSc, 58, Hons, 60; Duke Univ, PhD(zool), 68. *Prof Exp:* Res asst physiol larval crustacea, Marine Lab, Duke Univ, 60-61, teaching asst zool, 61-62, res asst hemocyanin, 62-67; res assoc coelenterate toxins, Univ Queensland, 68-69; res assoc marine invert, 70, independent investr pop genetics, 72-86, sr scientist, Marine Biol Lab, Woods Hole, 86-89; PROF, RUTGERS UNIV, 89- *Mem:* AAAS; Sigma Xi; Am Soc Zoologists. *Res:* Population genetics, dispersal, and recruitment of marine invertebrates. *Mailing Add:* Inst Marine & Coastal Sci Rutgers Univ PO Box 231 New Brunswick NJ 08903-0231

GRASSMICK, ROBERT ALAN, b Gering, Nebr, Aug 25, 36; m 62; c 2. PROTOZOOLOGY, PARASITOLOGY. *Educ:* Univ Nebr, BS, 58; Univ Mich, Ann Arbor, MA, 64; Iowa State Univ, PhD(zool), 71. *Prof Exp:* Instr life sci, Sutherland High Sch, Sutherland, Nebr, 58-59; instr, Deuel County High Sch, Chappell, Nebr, 61-63; instr life sci, Dept Biol, Mankato State Col, 64-67; from grad teaching asst zool to instr, Iowa State Univ, 67-71; asst prof, 71-75, ASSOC PROF ZOOL, MIAMI UNIV, 75- *Concurrent Pos:* NSF fel, 75. *Mem:* Soc Protozoologists; Am Inst Biol Sci; AAAS; Soc Invert Pathologists; Sigma Xi. *Res:* Potential of ciliated protozoa as possible biological control agents for medically important mosquitoes; isolation of new strains of parasitic ciliates, host specificity and mode of infection. *Mailing Add:* Seven Mckee Ave Oxford OH 45056

GRASSO, JOSEPH ANTHONY, b Cambridge, Mass, Sept 17, 35; m 63; c 1. ANATOMY, CYTOLOGY. *Educ:* Tufts Univ, BS, 57; Ohio State Univ, PhD(anat), 61. *Prof Exp:* From instr to asst prof anat, Ohio State Univ, 59-61; from sr instr to asst prof, Sch Med, Case Western Reserve Univ, 63-70; assoc prof anat, Sch Med, Boston Univ, 70-75; assoc prof, 75-78, PROF ANAT, UNIV CONN HEALTH CTR, 78- *Concurrent Pos:* USPHS fel, Univ Chicago, 61-64; USPHS Career Develop Award, 65-73. *Mem:* NY Acad Sci; Am Soc Hematol; Soc Develop Biol; Am Soc Cell Biol; AAAS. *Res:* Biochemistry and ultrastructure of red blood cell development; iron metabolism; receptor mediated iron delivery. *Mailing Add:* 31 Walnut Farm Rd Farmington CT 06032

GRATCH, SERGE, b Monte San Pietro, Italy, May 2, 21; nat US; m 51; c 10. THERMODYNAMICS. *Educ:* Univ Pa, BS, 43, MS, 45, PhD(mech eng), 50. *Prof Exp:* Asst instr mech eng, Univ Pa, 43-44, instr, 44-47, assoc, 47-49, asst prof, 49-51; sr scientist res, Rohm & Haas Co, 51-59; assoc prof mech eng, tech inst, Northwestern Univ, 59-61; supvr appl sci, Ford Motor Co, 61-62, mgr, chem Process & Develop Appl Res Off, 62-69, asst dir eng sci, sci res staff, 69-72, dir, Chem Sci Lab, 72-83 & Mat Sci Lab, 84-85, dir vehicle, power train & compressor res, 85-86; PROF MECH ENG, GMI ENG & MGT INST, 86- *Concurrent Pos:* Regional ed, Int S Fracture, 65-; mem, Air

Pollution Res Adv Comt Coord Res Coun, Inc, 67-86 & chmn, 83-85; mem, Bd Lubricant Rev, Inst Soc Automotive Engrs, 78-85 & chmn, 83-84; presidential app, Nat Alcohol Fuels Comn, 79-81; fel, Soc Automotive Engrs. *Mem:* Nat Acad Eng; AAAS; Am Chem Soc; hon mem Am Soc Mech Engrs (pres-elect, 81-82, pres, 82-83); Soc Automotive Engrs; Am Soc Eng Educ. *Res:* Thermodynamic properties of moist air; intermolecular forces in gas mixtures; zero-pressure thermodynamic properties of gases from spectroscopic data; chemical kinetics; polymerization kinetics; viscoelasticity; pollution control. *Mailing Add:* 32475 Bingham Rd Birmingham MI 48010

GRATIAN, J(OSEPH) WARREN, b Hartford, Conn, Sept 5, 18; m 43; c 3. ACOUSTICS, MAGNETICS. *Educ:* Univ Ill, BS, 41; Univ Rochester, MS, 63. *Prof Exp:* From jr engr to asst elec engr, Naval Ord Lab, DC, 41-43; from res asst to res assoc physics, Univ Mich, 43-45; elec engr, Res Div, Stromberg Carlson Co, 45-50, sr engr, 50-56, asst sect head, Electroacoust Lab, Res & Develop Dept, Stromberg Carlson Div, Gen Dynamics Corp, 56-58, sect head info storage sect, Appl Physics Lab, 58-61, prin engr, Electronics Div, 62, eng staff specialist, 63-71; consult, 71-80; RETIRED. *Mem:* Acoust Soc Am; Audio Eng Soc; Inst Elec & Electronics Engrs; Sigma Xi. *Res:* Analog and digital magnetic recording; electronic circuits; electroacoustics; signal analysis; solid-state devices and digital information storage techniques. *Mailing Add:* 156 Willowbend Rd Rochester NY 14618

GRATTAN, JAMES ALEX, b New Brunswick, NJ, Feb 17, 48; m 72; c 2. IMMUNODIAGNOSTICS, BIO-ORGANIC CHEMISTRY. *Educ:* Col of the Holy Cross, AB, 70; Mass Inst Technol, PhD(org chem), 74. *Prof Exp:* Res scientist bio-org chem, Union Carbide Corp, 74-76, proj leader chem/immunochem, 76-78, group leader immunochem, 78-79; sr res scientist diagnostics, Bristol Labs, 79-82, dir, diagnostics res & develop, 82-83, ASSOC DIR, IMMUNOL, BRISTOL-MYERS PHARMACEUT RES & DEVELOP DIV, 83- *Mem:* Am Chem Soc; Sigma Xi; AAAS. *Res:* Cancer diagnosis; tumor markers; immunoassay; immunochemistry; isotopic and non-isotopic immunologic methods; immunocytochemistry; tissue culture; immunology; protein purification; diagnostics clinical research; product development; protein conjugation; chemo-immuno therapy; antibody targeted therapy; hybridomas; monoclonal antibodies; tumor biology. *Mailing Add:* 100 Grist Mill Rd Wethersfield CT 06109-3649

GRATTAN, JEROME FRANCIS, b Southold, LI, NY, Aug 25, 12; m 44; c 6. BIOCHEMISTRY. *Educ:* Holy Cross Col, BS, 35, MS, 36. *Prof Exp:* Instr chem, Holy Cross Col, 35-36; res chemist, Boston Dispensary, 36-38; asst biochem, Squibb Inst Med Res, 38-41; res chemist, Int Vitamin Corp, 41-43; res biochemist, Carroll Dunham Smith Pharmacal Co, 46-50, dir res, 51-58, vpres, 58-59; vpres res, Smith, Miller & Patch, Inc, 59-67, adminr, Instrument Div, 67-72; MKT TECH MGR, COOPER CO, INC, 72- *Mem:* Am Chem Soc; Soc Toxicol; NY Acad Sci; AAAS; fel Am Inst Chem. *Res:* Polythionates; depressants; relaxants; vitamins; hormones. *Mailing Add:* 12 Kerschner Lane E Brunswick NJ 08816-2424

GRATTON, ENRICO, b Merate, Italy, May 23, 46; m 70; c 2. BIOPHYSICS, BIOCHEMISTRY. *Educ:* Univ Rome, DPhys, 69. *Prof Exp:* Fel biophys, Inst Advan Health, Rome, 69-70; researcher, Snam Prog, Rome, 70-77; res assoc biochem, 77-78, ASST PROF PHYSICS, UNIV ILL, URBANA, 78- *Res:* Enzyme mechanism; fluorescence life time; enzyme dynamics. *Mailing Add:* Dept Physics 407 S Goodwin Ave Univ Ill Urbana IL 61801

GRATZ, NORMAN G, b Minneapolis, Minn, May 16, 25; m 58; c 1. MEDICAL ENTOMOLOGY. *Educ:* Univ Calif, Berkeley, BSc, 48, MSc, 50; Univ Geneva, DSc(zool), 66. *Prof Exp:* Dir vector control, Ministry Health, Israel, 53-58; proj leader, Res Unit, WHO, Liberia, 58-61 & Nigeria, 61-62, chief ecol & control vectors, 62-81, dir, Div Vector Biol & Control, Switz, 81-86; CONSULT, USAID & INDUST, 87- *Concurrent Pos:* Guest lectr, Hebrew Univ, Israel, 55-58, vis prof dept med ecol, Med Sch, 70; consult, WHO, 87-90. *Honors & Awards:* Medal of Honor, Am Mosquito Control Asn, 85. *Mem:* Am Mosquito Control Asn. *Res:* Biology and control of arthropod vectors and rodent reservoirs of human disease. *Mailing Add:* Four Chemin du Ruisseau Commugny 1291 Switzerland

GRATZ, RONALD KARL, b Northfield, NJ, June 1, 46. VERTEBRATE PHYSIOLOGY, HERPETOLOGY. *Educ:* Univ Notre Dame, BS, 68, MS, 72; Univ Okla, PhD(zool), 76. *Prof Exp:* Sci asst physiol div, Max Planck Inst Exp Med, 76-78; asst prof, 78-83, ASSOC PROF BIOL SCI, MICH TECHNOL UNIV, 83- *Mem:* Am Soc Zoologists; Am Physiol Soc; Sigma Xi; Nat Asn Adv Health Professions. *Res:* Comparative vertebrate physiology/physiological ecology. *Mailing Add:* Dept Biol Sci Mich Technol Univ Houghton MI 49931

GRATZ, ROY FRED, b Pittsburgh, Pa, Nov 6, 42; m 69; c 1. ORGANIC CHEMISTRY, POLYMER CHEMISTRY. *Educ:* Univ Pittsburgh, BS, 64; Duke Univ, MA, 68, PhD(org chem), 70. *Prof Exp:* Assoc pharmaceut chem, Med Univ SC, 70-71; asst prof chem, Salem Col, 71-75; from asst prof to assoc prof, 75-86, PROF CHEM, MARY WASHINGTON COL, 86- *Concurrent Pos:* Fac fel, NASA Lewis Res Ctr, Cleveland, 83, 84; fac res prog, Naval Res Lab, Washington, DC, 85, 86. *Mem:* Am Chem Soc. *Mailing Add:* Dept Chem & Geol Mary Washington Col Fredericksburg VA 22401-5358

GRATZEK, JOHN B, b St Paul, Minn, Jan 23, 31; m 57; c 4. VIROLOGY. *Educ:* St Mary's Col, Minn, BS, 52; Univ Minn, DVM, 56; Univ Wis, MS, 59, PhD(virol), 61. *Prof Exp:* Instr vet sci, Univ Wis, 56-61; assoc prof virol, Iowa State Univ, 61-66; PROF MICROBIOL & PREV MED & HEAD DEPT, UNIV GA, 66-, MEM FAC MICROBIOL, 76- *Mem:* Am Vet Med Asn; Sigma Xi. *Res:* Bovine virus diarrhea; infectious bovine rhinotracheitis; infectious enteritis of turkeys; fish diseases. *Mailing Add:* Dept Microbiol & Prev Med Univ Ga Sch Med Athens GA 30601

GRATZER, GEORGE, b Budapest, Hungary, Aug 2, 36; m 61; c 1. MATHEMATICS. *Educ:* Eotvos Lorand Univ, Budapest, PhD, 60. *Prof Exp:* Researcher algebra, Math Inst, Hungarian Acad Sci, 59-63; vis asst prof math, Pa State Univ, 63-64, from assoc prof to prof, 64-67; PROF MATH, UNIV MAN, 67- *Concurrent Pos:* Can Res Coun Can fels, 61, grant, 67-; NSF grant, 65-67; vis prof math, Univ Man, 66-67; ed-in-chief, Algebra Universalis; mem math grant comt, Nat Res Coun. *Honors & Awards:* Grunwald Mem Prize, 60; Steacie Prize, Nat Res Coun Can, 71; Zubec Res Award, Univ Man, 74. *Mem:* Am Math Soc; Can Math Cong; fel Royal Soc Can. *Res:* Lattice theory; universal algebra; applications of logic to lattices and algebras. *Mailing Add:* Dept Math & Astron Univ Man Winnipeg MB R3T 2N2 Can

GRATZNER, HOWARD G, b Philadelphia, Pa, July 21, 34; m 60. GENETICS, GENETIC TOXICOLOGY. *Educ:* Pa State Univ, BS, 56; Temple Univ, AM, 60; Fla State Univ, PhD(genetics), 64. *Prof Exp:* Res asst biochem, Albert Einstein Med Ctr, 56-58; asst prof zool, Univ SFla, 64-67; asst prof, Univ Miami, 68-71; RES SCIENTIST, PAPANICOLAOU CANCER RES INST, 71-; ASST PROF PATH, SCH MED, UNIV MIAMI, 75-, ASST PROF SURG, 80- *Concurrent Pos:* NIH spec res fel, Calif Inst Technol, 67-68. *Mem:* AAAS; Genetics Soc Am; Tissue Cult Asn; Soc Anal Cytol. *Res:* Genetic control of DNA and protein synthesis; mutagenesis/carcinogenesis; chromosome structure; monoclonal antibodies; flow cytometry. *Mailing Add:* 2601 Bellefontaine Apt C201 Houston TX 77025

GRAU, ALBERT A, b Zurich, Switz, Oct 25, 18; US citizen; m 65; c 1. MATHEMATICS, COMPUTER SCIENCE. *Educ:* Univ Mich, BS, 40, MS, 41, PhD(math), 44. *Prof Exp:* Instr math, Univ Mich, 44, Rackham fel, Inst Adv Study, 44-45; instr math, Drake Univ, 45-46; asst prof, Univ Ky, 46-47; assoc prof, Univ Ala, 47-48; from assoc prof to prof, Univ Okla, 48-57; sr mathematician, Oak Ridge Nat Lab, 56-63; PROF MATH & ENG SCI, NORTHWESTERN UNIV, 63-, PROF COMPUT SCI, 70- *Concurrent Pos:* Lectr, George Washington Univ, 47; lectr, Univ Tenn, 57-60; consult, Oak Ridge Nat Lab, 63-66; consult, Argonne Nat Lab, 67-72. *Mem:* Am Math Soc; Math Asn Am; Soc Indust & Appl Math; Asn Comput Mach. *Res:* Boolean algebra; numerical analysis; programming languages and compilers. *Mailing Add:* 109 Glengary Dr Flat Rock NC 28731

GRAU, CHARLES RICHARD, b National City, Calif, Nov 5, 20; m 41; c 4. POULTRY NUTRITION. *Educ:* Univ Calif, BS, 42, PhD(animal nutrit), 46. *Prof Exp:* From instr to assoc prof, Univ Calif, Davis, 46-58, chmn dept, 69-76, prof avian sci, 58-89, nutritionist, Exp Sta, 58-89, EMER PROF AVIAN SCI, UNIV CALIF, DAVIS, 89- *Mem:* Soc Develop Biol; Teratol Soc; Am Soc Cell Biol; Soc Exp Biol & Med; Poultry Sci Asn. *Res:* Amino acid requirements and metabolism in the chick; protein concentrates as amino acid sources; metabolism of phenylalanine and tyrosine in the chick and mouse; metabolism of gossypol in laying hens; energy needs and food intake; nutrition of chick embryos; egg formation; seabird reproductive ecology; silt pollution and seabird reproduction. *Mailing Add:* Dept Avian Sci Univ Calif Davis CA 95616-0532

GRAU, CRAIG ROBERT, b Manning, Iowa, Dec 5, 46; m 69; c 2. PHYTOPATHOLOGY. *Educ:* Iowa State Univ, BS, 69, MS, 71; Univ Minn, PhD(plant path), 75. *Prof Exp:* Res assoc plant path, NC State Univ, 75-76; ASST PROF PLANT PATH, UNIV WIS, 76- *Concurrent Pos:* NDEA Title IV fel, Univ Minn. *Mem:* Am Phytopath Soc; Am Inst Biol Sci; Sigma Xi. *Res:* Biology and control of soil-borne plant pathogens and the mechanism and use of host resistance to this group of plant pathogens. *Mailing Add:* Dept Plant Path Univ Wis Madison WI 53706

GRAU, EDWARD GORDON, b Baltimore, Md, May 3, 46; m 69; c 1. COMPARATIVE ENDOCRINOLOGY, NEUROENDOCRINOLOGY. *Educ:* Loyola Col, BS, 68; Morgan State Univ, MS, 73; Univ Del, PhD(biol sci), 78. *Prof Exp:* NIH fel, Univ Calif, Berkeley, 79-82; from asst prof to assoc prof, 82-87, PROF ZOOL, UNIV HAWAII, 87-, DIR, HAWAII INST MARINE BIOL, 85- *Concurrent Pos:* NIH fel, 78; mem, US-Japan Collaborative Study Salmon Biol, NSF, 80-81. *Mem:* Am Soc Zoologists; AAAS. *Res:* Comparative endocrinology of reproduction, growth and development: special interest in pituitary control of thyroid function and in the mechanisms by which hypothalamic neurosecretory factors regulate the secretion of thyrotropin, prolactin, growth hormone. *Mailing Add:* Dept Zool Univ Hawaii Honolulu HI 96822

GRAU, FRED V, turfgrass, turf cultivation; deceased, see previous edition for last biography

GRAUBARD, MARK AARON, b Plock, Poland, Jan 5, 04; nat US; c 2. HISTORY OF SCIENCE. *Educ:* City Col New York, BS, 26; Columbia Univ, MA, 27, PhD(genetics, zool), 30. *Prof Exp:* Asst zool, Columbia Univ, 28-31; Nat Res Coun fel, Manchester & London, Eng, 31-33; asst genetics, Columbia Univ, 33-34, res assoc biol chem, 34-38; res assoc physiol, Clark Univ, 38-41; nutrit ed, Food Distribution Admin, US Dept Agr, 42-46; asst prof natural sci, Univ Chicago, 46-47; assoc prof, Univ Minn, 47-58, prof natural sci & chmn natural sci prog, 59-72; RETIRED. *Honors & Awards:* George Washington Hon Medal, 55. *Mem:* AAAS; Am Physiol Soc; Genetics Soc Am; Am Soc Zool; Hist Sci Soc; NY Acad Sci. *Res:* Chemistry and physiology of the pigment reaction; oxidases and hormone metabolism; history of science. *Mailing Add:* 2928 Dean Pkwy Minneapolis MN 55416

GRAUE, DENNIS JEROME, b Minot, NDak, Sept 12, 39; m 59; c 2. CHEMICAL ENGINEERING, PETROLEUM. *Educ:* Univ Colo, BS, 61; Calif Inst Technol, MS, 62, PhD(chem eng), 65. *Prof Exp:* Res engr, Chevron Oil Field Res Co, 65-69, engr, Chevron Standard Ltd, Alta, Can, 69-72, sr res engr, Chevron Oil Field Res Co, 72-76, sr staff engr, Chevron USA, Inc, 76-78; mgr, Sci Software Co, 78-81,DIV VPRES, SCI SOFTWARE INTERCOMP, 81- *Concurrent Pos:* NSF fel, 61-65; distinguished lectr, Enhanced Oil Recovery, Soc Petrol Engrs, 83-84. *Mem:* Am Inst Chem Engrs; Soc Petrol Engrs. *Res:* Fluid flow through porous media; oil reservoir behavior; diffusion; transport phenomena. *Mailing Add:* Sci Software Intercomp 1801 California St Denver CO 80202

GRAUE, LOUIS CHARLES, b Louisiana, Mo, Dec 23, 23; m 49; c 2. MATHEMATICS. *Educ:* Univ Chicago, BS, 47, MS, 48; Ind Univ, PhD(math), 50. *Prof Exp:* Asst prof math, Sacramento State Col, 50-56; assoc prof, Coe Col, 56-59; assoc prof, 59-70, chmn dept, 65-74, PROF MATH, BOWLING GREEN STATE UNIV, 70- *Mem:* Math Asn Am. *Res:* Mathematics, algebra and differential geometry and computers. *Mailing Add:* Dept Math Bowling Green State Univ Main Campus Bowling Green OH 43403

GRAUER, ALBERT D, b Chicago, Ill, May 21, 42; c 2. ASTRONOMY. *Educ:* Concordia Col, Seward, Nebr, BS, 64; NC State Univ, PhD(physics), 71. *Prof Exp:* Woodrow Wilson fel, 68, assoc prof physics, North Ga Col, 71-77; ASSOC PROF ASTRON, UNIV ARK, LITTLE ROCK, 77- *Concurrent Pos:* Assoc vis prof astron, La State Univ, 77- *Mem:* Am Astron Soc; Royal Astron Soc; Am Asn Variable Star Observers. *Res:* Astronomical photoelectric photometry of variable stars; development of instrumentation for astronomical measurements. *Mailing Add:* PO Box 26732 Tucson AZ 85726-6732

GRAUER, AMELIE L, b Vienna, Austria, May 7, 99; US citizen; m 44. BIOCHEMISTRY. *Educ:* Univ Vienna, PhD(chem), 23. *Prof Exp:* Res chemist, Polytech Inst Brooklyn, 50-52; res chemist, NY Med Col, 52-57; res chemist, Albert Einstein Col Med, 57; res chemist ophthal, Col Physicians & Surgeons, Columbia Univ, 58-66; res chemist path, NY Univ, 66-70; CONSULT BIOCHEMIST, 70- *Mem:* Am Chem Soc. *Res:* Enzymes; steroids; general chemistry and organic preparation. *Mailing Add:* 720 West End Ave Apt 618 New York NY 10025-0299

GRAUL, WALTER DALE, b Wichita, Kans, Jan 5, 44; m 63; c 1. BEHAVIORAL BIOLOGY, VERTEBRATE ECOLOGY. *Educ:* Emporia Kans State Col, BS, 66, MS, 68; Univ Minn, PhD(ecol), 73. *Prof Exp:* High sch teacher biol, Bear Creek High Sch, 67-68 & Hastings High Sch, 70-71; asst prof, Univ NC, Charlotte, 73-75; nongame bird specialist, 75-79, nongame res leader, 79-81, NE REGIONAL MGR, WILDLIFE MGT, COLO DIV WILDLIFE, 84-; ASSOC FAC MEM, COLO STATE UNIV, 75- *Mem:* Sigma Xi; Am Ornithologists Union; Wildlife Soc; Wilson Ornith Soc; Cooper Ornith Soc. *Res:* Behavioral adaptations in birds with an emphasis on social behavior of shorebirds; applied research regarding endangered avian species and ecosystem management. *Mailing Add:* 317 W Prospect Ft Collins CO 80526

GRAULTY, ROBERT THOMAS, b Troy, NY, July 22, 28; m 50; c 7. QUALITY ASSURANCE, DEVELOPMENT OF EMPLOYEE INVOLVEMENT MANAGEMENT METHODS. *Educ:* US Merchant Marine Acad, BS, 49. *Prof Exp:* Proj engr, Am Locomotive Co, 49-55; engr, Westinghouse Bettis Atomic Power Lab, 55-69, mgr, Submarine Reactor Design, 69-77, mgr mfg, 77-82, plant mgr mfg, Fuel Div, 82-86; pvt consult, 86-91; CONSULT, WESTINGHOUSE SAVANNAH RIVER CO, 91- *Mem:* Nat Soc Prof Engrs; Soc Mfg Engrs. *Res:* Design and development of nuclear reactor cores, vessels, components and systems; development of unique manufacturing processes; design of highly automated manufacturing plants; quality assurance in design and manufacturing. *Mailing Add:* 109 Miles Rd Columbia SC 29223

GRAUMAN, JOSEPH URI, b Hadera, Israel, Oct 1, 41; US citizen; m 69; c 2. TOTAL QUALITY MANAGEMENT. *Educ:* Stevens Inst Technol, BS, 63, MS, 65, PhD(physics), 69; Rutgers Univ, MBA, 80. *Prof Exp:* Vis asst prof physics, Stevens Inst Technol, 69-70; asst prof physics, Jersey City State Col, 70-77; intermediate engr, Singer-Kearfott, 77-78; sr scientist, Xybion Corp, 78-80; DISTINGUISHED MEM TECH STAFF, AT&T BELL LABS, 80- *Concurrent Pos:* Res assoc, Stevens Inst Technol, 70-77. *Mem:* Am Phys Soc; NY Acad Sci; Sigma Xi; Am Inst Aeronaut & Astronaut; Am Soc Qual Control. *Res:* Implementation of total quality management in research and development environment; development of metrics for measurement of performance on research and development projects; development of business and financial reports, proposals and cost estimates on research and development projects; digital signal processing in connection with passive surveillance ASW program; communications systems engineering. *Mailing Add:* 2244 Copper Hill Dr Union NJ 07083

GRAUMANN, HUGO OSWALT, b Granite, Okla, May 31, 13; m 35; c 1. AGRONOMY. *Educ:* Okla Agr & Mech Col, BSc, 38, MSc, 40; Univ Nebr, PhD(agron), 50. *Prof Exp:* Asst agron, Okla Agr & Mech Col, 38-40, instr, 40-41, asst prof & asst agronomist, 41-46, assoc prof, 46-47; agronomist, USDA, 47-53, res agronomist, Forage & Range Res Br, Agr Res Serv, 53-58, chief, 58-64, assoc dir, Crops Res Div, 64-70, dir plant sci res div, 70-72, asst adminr, 72-78, actg asst adminr, Sci & Educ Admin, 78-80; RETIRED. *Concurrent Pos:* Agronomist, Univ Nebr, 47-53. *Mem:* AAAS; Am Soc Agron; Am Genetic Asn; Crops Sci Soc Am. *Res:* Forage crops. *Mailing Add:* 1714 Edgewater Pkwy Silver Spring MD 20903

GRAUPE, DANIEL, b Jerusalem, Israel, July 31, 34; m 68; c 3. CONTROL SYSTEMS, SYSTEMS & BIOMEDICAL ENGINEERING. *Educ:* Israel Inst Technol, BSME, 58, BSEE, 59, Dipl Ing, 60; Univ Liverpool, PhD(elec eng), 63. *Prof Exp:* Engr automatic control, Israel Govt Industs, Tel Aviv, 59-60; lectr elec eng, Univ Liverpool, 63-67; sr lectr mech eng, Israel Inst Technol, 67-70; from assoc prof to prof elec eng, Colo State Univ, 70-78; Bodine chair & distinguished prof elec & comput eng, Ill Inst Technol, 78-85; PROF ELECT ENG, COMPUTER SCI & BIOMED, UNIV ILL, CHICAGO, 85-, PROF PHYS MED & REHAB, 90-; VPRES, SIGMEDICS INC, NORTHBROOK. *Concurrent Pos:* Vis prof elec eng, Univ Notre Dame, 76; Russell Springer vis chair & prof mech eng, Univ Calif, Berkeley, 77; mem med staff res, Michael Reese Hosp & Med Ctr, Chicago. *Mem:* Fel Inst Elec & Electronics Engrs; NY Acad Sci. *Res:* Problems of automatic control, specifically adaptive control and artificial intelligence problems; methods of identification and estimation of processes and signals; applications of same to industrial and medical systems; systems theory; biomedical engineering; time series analysis; artificial limbs and powered braces for high-level amputees and hemiplegics; self-adaptive filter of noise from speech; EMG control of electrical stimulation to enable paraplegics to ambulate independently. *Mailing Add:* Dept Elec Eng & Comput Sci Univ Ill at Chicago Chicago IL 60680

GRAVA, JANIS (JOHN), b Vecgulbene, Latvia, Jan 24, 20; US citizen; m 42, 68; c 2. SOIL SCIENCE. *Educ:* Univ Gottingen, MS, 48, PhD(agron), 50. *Prof Exp:* Res asst soil sci, Kans State Univ, 52-53, instr, 53-54; res fel, 54-57, from asst prof to prof, 57-85, EMER PROF SOIL SCI, UNIV MINN, ST PAUL, 85- *Concurrent Pos:* Mem, Regional Soil Test Comt, 55-, chmn, 58. *Mem:* Am Soc Agron; Soil Sci Soc Am; Int Soc Soil Sci. *Res:* Crop production; soil fertility, chemical analyses and testing; grass seed and wild rice production. *Mailing Add:* Dept Soil Sci Univ Minn St Paul MN 55108

GRAVANI, ROBERT BERNARD, b Jersey City, NJ, Aug 11, 45; m 80; c 2. FOOD SCIENCE. *Educ:* Rutgers Univ, BS, 67; Cornell Univ, MS, 69, PhD(food sci), 75. *Prof Exp:* Asst dir, Inst Food Sci & Mkt, Cornell Univ, 73-75; sci dir Cereal Inst, Inc, 75-78; dir, Empire State Food & Agr Leadership Inst, 87-89; from asst prof to assoc prof, 78-90, PROF FOOD SCI, CORNELL UNIV, 91- *Concurrent Pos:* Continuing educ comt, Inst Food Technologists, 82-86, vchmn, 83-84, chmn, 84-85 & past chmn, 85-86; comt, Inter Am Conf Food Protection, NAS & Food & Nutrit Bd, 84-86; tuna processing & standards review comt, Can Dept Fisheries & Oceans, 86; vis prof, dept food sci, Univ Minn, 86; vis scholar, dept food sci, Univ Nebr, 87; chmn, Educ Award Subcomt, Int Asn Milk, Food & Environ Sanitarians, 84-88, prog chmn, 88; mem bd trustees, Food Processors Inst, 88-; leadership develop fel, Nat Ctr Food & Agr Policy Resources for the Future, Wash, DC, 90; mem, Nat Adv Comt Microbiol Criteria for Foods, 90-; mem, Nat Food Safety & Qual Implementation Team, USDA, 91- *Honors & Awards:* William V Hickey Award, Asn Milk & Food Sanitarians, 84; Norbert F Sherman Award, Nat Restaurant Asn Educ Found, 89. *Mem:* NY Acad Sci; Inst Food Technologists; Int Asn Milk, Food & Environ Sanitarians (secy, 84-86, vpres, 87, pres elect, 88, pres, 89); Am Soc Microbiol; Nat Restaurant Asn; Asn Food & Drug Officials. *Res:* Food safety and sanitation; food microbiology; food processing and service; food regulations, industry and regulatory agency training; consumer food safety; retail food industries. *Mailing Add:* Dept Food Sci Cornell Univ Ithaca NY 14853

GRAVANIS, MICHAEL BASIL, b Krokylion Doridos, Greece, Jan 27, 29; US citizen; m 77; c 2. PATHOLOGY. *Educ:* Univ Thessaloniki, MD, 52. *Prof Exp:* From asst to assoc prof, 65-70, prof & chmn dept, 70-85, PROF PATH, EMORY UNIV, 85- *Mem:* Col Am Path; Int Acad Path; Am Asn Path & Bact; AMA. *Res:* Cardiovascular pathology. *Mailing Add:* Dept Anatomic Path 1364 Cliften Rd NE Rm F-167A Atlanta GA 30322

GRAVATT, CLAUDE CARRINGTON, JR, b Washington, DC, Dec 12, 39; m 64; c 2. PHYSICAL CHEMISTRY. *Educ:* Univ Richmond, BS, 62; Duke Univ, PhD(phys chem), 66. *Prof Exp:* Res grant chem physics, Cornell Univ, 65-67; mem tech staff, Bell Tel Labs, NJ, 67-69; res chem physicist, Nat Bur Standards, 69-73, asst to dir, 73-75, dep chief, Ctr Anal Chem, 75-77, dir, Off Environ Measures, 77-79, dep dir, Nat Measurement Lab, Nat Inst Standards & Technol, Dept Com, 80-83, dir, Landsat Transition Group, 83-86, dep dir, Off Space Com, 86-88, assoc dir planning, 89-90, DIR, CRIT TECH GROUP, DEPT COM, NAT INST STANDARDS & TECHNOL, 90- *Mem:* Am Phys Soc; Am Chem Soc; Optical Soc Am. *Res:* Electromagnetic scattering investigations of critical phenomena, liquid crystals, liquids and solutions; air pollution analysis by light scattering; remote sensing and image analysis; earth sciences; critical and emerging technologies. *Mailing Add:* 7064 Wolftree Lane Rockville MD 20852

GRAVE, GILMAN DREW, b Rhineback, NY, Jan 3, 41; m; c 1. ENDOCRINOLOGY. *Educ:* Harvard Univ, AB, 62, MD, 66; Am Bd Internal Med, dipl, 72. *Prof Exp:* Hemat technician, Mass Gen Hosp, Boston, 61-66, intern internal med, 66-67, resident, 67-68; instr med, Harvard Univ, 67-68; res assoc, Lab Cerebral Metab, Div Biol & Biochem Res, Intramural Res Prog, NIMH, NIH, 68-70, staff fel, Lab, 70-72, med officer, Growth & Develop Br, Nat Inst Child Health & Human Develop, 72-76, Develop Biol & Nutrit Br, 76-77, actg chief, 77-79, chief, Sect Nutrit & Endocrinol, Clin Nutrit & Early Develop Br, 79-85, CHIEF ENDOCRINOL, NUTRIT & GROWTH BR, NAT INST CHILD HEALTH & HUMAN DEVELOP, NIH, BETHESDA, 85- *Concurrent Pos:* Res assoc, Huntington Mem Labs, 63; med staff, Alexandria Gen Hosp, Va, 68-75, Doctor's Hosp, Washington, DC, 75-79 & Shady Grove Adventist Hosp, Md, 79-82; inst dir rep, Nat Diabetes Adv Bd, 78-; asst ed, Am J Clin Nutrit, 80-81; chmn, Subcomt Med Appln Diabetes Res, 81 & Clin Rev Subpanel, Nat Inst Child Health & Human Develop, 88-; mem, Sci Adv Comt, Nat Diabetes Res Interchange, 82- *Mem:* Am Diabetes Asn; Am Fedn Clin Res; Am Soc Neurochem. *Mailing Add:* NIH Nat Inst Child Health & Human Develop Endocrinol Nutrit & Growth Br Exec Plaza N Rm 637 6130 Executive Blvd Bethesda MD 20892

GRAVEEL, JOHN GERARD, b Mishawaka, Ind, Dec 8, 53; m 79; c 3. SOIL MICROBIOLOGY, SOIL BIOCHEMISTRY. *Educ:* Purdue Univ, BS, 77, MS, 79, PhD(soil microbiol), 84. *Prof Exp:* Soil scientist, Ind Dept Natural Resources, 76; lab technician bionucleonics, Purdue Univ, 76, res asst soil microbiol, 77-84, teaching asst soil sci, 79-84; ASST PROF SOIL MGT, UNIV TENN, 83- *Concurrent Pos:* Instr, Ind Univ & Purdue Univ, 81. *Mem:* Am Soc Agron; Am Chem Soc; Soil Conserv Soc Am; Nat Asn Col Teachers Agr; Sigma Xi. *Res:* Decomposition of anthropogenic substances in the environment; effect of no-tillage farming practices on the physical, chemical and biological properties of soil. *Mailing Add:* Dept Plant & Soil Sci Univ Tenn Knoxville TN 37901-1071

GRAVEL, DENIS FERNAND, b St Lambert, Que, Nov 24, 35; m 66; c 2. PHOTOCHEMISTRY. *Educ:* Univ Montreal, BSc, 58, MS, 59, PhD(org chem), 62. *Prof Exp:* Nat Res Coun Can fel org chem, Swiss Fed Inst Technol, 62-64; from asst prof to assoc prof, 64-73, PROF ORG CHEM, UNIV MONTREAL, 73- *Mem:* Am Chem Soc; Chem Inst Can; Brit Chem Soc. *Res:* Synthetic organic chemistry; organic photochemistry. *Mailing Add:* Dept Chem Univ Montreal PO Box 6128 Montreal PQ H3C 3J7 Can

GRAVELL, MANETH, b Tamagua, Pa, Aug 15, 32; m 63; c 3. VIROLOGY. *Educ:* Muhlenberg Col, BS, 59; Lehigh Univ, MS, 65, PhD(biol), 66. *Prof Exp:* Res assoc virus & tissue cult, Merck Inst Ther Res, 59-61; spec technologist, St Jude's Children's Res Hosp, 65-66; instr microbiol, Hahnemann Med Col, 66-67; res assoc, St Jude Children's Res Hosp, 67-68, asst mem virol, 68-72; virologist, Microbiol Assocs Inc, Bethesda, Md, 72-73; dir, biol develop & exp oncol & prin investr, Nat Cancer Inst contract, Litton Bionetics, Inc, Kensington, Md, 73-74; res microbiologist, Viral Biol Br, Div Cancer Cause & Prev, Nat Cancer Inst, 74-75, CHIEF SECT NEUROVIROL, INFECTIOUS DIS BR, NAT INST NEUROL & COMMUN DIS & STROKE, NIH, BETHESDA, MD, 76- *Concurrent Pos:* From instr to asst prof, Col Med, Univ Tenn, 67-72. *Mem:* AAAS; Am Soc Microbiol; Sigma Xi. *Res:* Cell and virus interactions. *Mailing Add:* 14901 Plainfield Lane Germantown MD 20874

GRAVELY, SALLY MARGARET, b Thorpe, WVa, Oct 13, 47; m 76; c 1. ELECTRON MICROSCOPY. *Educ:* Bluefield State Col, BS, 69; Ohio State Univ, MS, 73, PhD(microbiol), 76. *Prof Exp:* Rockefeller Found fel, 76-78; ASST PROF MICROBIOL, HOWARD UNIV, 78- *Mem:* Am Soc Microbiol; AAAS. *Res:* Malaria-host parasite interaction including the host's immune response to Plasmodium (malaria) antigens. *Mailing Add:* Dept Microbiol Col Med Howard Univ 520 West St NW Washington DC 20059

GRAVEN, STANLEY N, b Greene, Iowa, May 20, 32; m 54; c 4. PEDIATRICS, BIOCHEMISTRY. *Educ:* Wartburg Col, BS, 55; Univ Iowa, MD, 56; Am Bd Pediat, dipl, 61. *Prof Exp:* Intern med, Madigan Army Hosp, Tacoma, Wash, 56-57; resident pediat, Cincinnati Children's Hosp, Ohio, 57-58 & Univ Iowa Hosps, 58-60; chief pediat serv, USAF Hosp, Fairchild AFB, Wash, 60-62, dir newborn & premature serv, Wilford Hall, Lackland AFB, Tex, 62-64; from asst prof to prof pediat, 66-76; co-dir, Wis Perinatal Ctr, 66-76; prof, Dept Pediat & Obstet/Gynec, Univ SDak, 76-80; prof, dept child health, Univ Mo, 80-; AT COL PUB HEALTH, UNIV SFLA. *Concurrent Pos:* USPHS fel biochem & pediat, Univ Wis-Madison, 64-66; consult, Lakeland Village, Wash, 60-62; sr prog consult, Robert Wood Johnson Found, 79-, consult, Mo Div Health, 80-; med dir, Off Maternal-Child Health, SDak Dept Health, 76-79. *Mem:* Am Acad Pediat; Soc Pediat Res; Am Pediat Soc; Am Pub Health Asn; Sigma Xi. *Res:* Neonatology; newborn and premature physiology and biochemistry. *Mailing Add:* Col Pub Health Univ SFla 13301 N 30th St Tampa FL 33612

GRAVENSTEIN, JOACHIM STEFAN, b Berlin, Ger, Jan 25, 25; nat US; m 49; c 8. MEDICINE, ANESTHESIOLOGY. *Educ:* Univ Bonn, Dr med, 51; Harvard Univ, MD, 58; Am Bd Anesthesiol, dipl. *Hon Degrees:* Dr, Univ Graz, 88. *Prof Exp:* Intern, Surg Univ Hosp, Switz, 51; assoc anesthetist, Mass Gen Hosp, 54-58; res assoc anesthesia, Harvard Univ, 55-56; prof surg & chief anesthesia, Col Med, Univ Fla, 58-65, prof anesthesiol & chmn dept, 65-69; prof anesthesiol & chmn dept, Sch Med, Case Western Reserve Univ, 69-79; GRAD RES PROF, COL MED, UNIV FLA, GAINESVILLE, 79- *Concurrent Pos:* Clin fel anesthesia, Mass Gen Hosp, 52-54; fel, Mass Gen Hosp, 54-58. *Mem:* Am Soc Anesthesiol; Am Soc Pharmacol & Exp Therapeut. *Res:* Anesthetics and cardiovascular pharmacology; monitoring. *Mailing Add:* Dept Anesthesiol Col Med Univ Fla Gainesville FL 32610

GRAVER, JACK EDWARD, b Cincinnati, Ohio, Apr 13, 35; m 61; c 3. MATHEMATICS. *Educ:* Miami Univ, BA, 58; Ind Univ, MA, 61, PhD(math), 64. *Prof Exp:* Lectr math, Ind Univ, 64; res instr, Dartmouth Col, 64-66; from asst prof to assoc prof, 66-75, chmn dept, 79-82, PROF MATH, SYRACUSE UNIV, 75- *Concurrent Pos:* Vis prof math, Univ Nottingham, Eng, 71-72. *Mem:* AAAS; Am Math Soc; Math Asn Am; Soc Indust & Appl Math; Am Asn Univ Professors. *Res:* Combinatorics; graph theory; systems of subsets of a finite set; integer programming. *Mailing Add:* Dept Math Syracuse Univ Syracuse NY 13244

GRAVER, RICHARD BYRD, b Cambridge City, Ind, Apr 5, 32; m 52; c 5. ORGANIC POLYMER CHEMISTRY. *Educ:* Purdue Univ, BSChE, 54; Univ Mich, MSE, 55, PhD(chem eng), 58. *Prof Exp:* Res chemist resins, Archer Daniels Midland Co, 57-59, proj leader, 59-61, group leader, 61-67; mgr polymer res, 67-73, tech mgr powder coatings, 73-76, tech mgr res, 76-78, TECH MGR SPECIALTY RESINS, CELANESE POLYMER SPECIALTIES CO, 78- *Concurrent Pos:* Prod safety specialist, Rhone-Paulenc. *Mem:* Am Chem Soc; Fedn Paint Soc; Sigma Xi; Am Soc Testing & Mat. *Res:* Polymers for coatings; water soluble polymers. *Mailing Add:* 9004 Fawn Ct Louisville KY 40242

GRAVER, WILLIAM ROBERT, b Allentown, Pa, Apr 24, 47. ELECTRO-OPTICS, LASERS. *Educ:* Muhlenberg Col, BS, 49; Am Univ, MS, 74; Georgetown Univ, PhD(physics), 88. *Prof Exp:* Res physicist, Riverside Res Inst, 78-83; sr engr, W J Schafer Assoc, 83-85; SR SCIENTIST, BALL CORP, 85- *Mem:* Am Phys Soc; Optical Soc Am; Acoust Soc Am. *Mailing Add:* 6137 Ninth Rd N Arlington VA 22205

GRAVES, ANNE CAROL FINGER, b Ludlow, Miss, Nov 29, 33; m 56; c 3. PLANT CELL TRANSFORMATION. *Educ:* Millsaps Col, BS, 55; Northwestern Univ, MS, 56; Bowling Green State Univ, PhD(biol sci), 82. *Prof Exp:* From instr to assoc prof biol, Flint Community Col, 57-65; from instr to asst prof, Bowling Green State Univ, 67-83; asst prof, Univ Toledo, 83-89; ASST PROF, BOWLING GREEN STATE UNIV, 89- *Mem:* Am Soc Microbiol; AAAS; Sigma Xi; Electron Micros Soc Am. *Res:* Ecology of fungus-inhabiting insects; transformation of monocotyledonous plants by Agrobacterium tumefaciens; bacterial attachement to plant cells; bacteria-plant cell interactions, plant cell culture. *Mailing Add:* 627 Crestview Dr Bowling Green OH 43402

GRAVES, BRUCE BANNISTER, b Lafayette, Ind, Dec 7, 28; m 53; c 3. ELECTROCHEMISTRY, PHYSICS. *Educ:* Swarthmore Col, BA, 51; Univ Louisville, MS, 64, PhD(chem), 67. *Prof Exp:* Sci glassworker, Purdue Univ, 52-59; chemist, Radiochem, Inc, Ky, 60-62, chief chemist, 62-65, lab mgr, 64-65; from asst prof to assoc prof chem, Ky Southern Col, 66-68, actg head dept, 67-68; from asst prof to assoc prof, 68-74, PROF CHEM, EASTERN MICH UNIV, 74- *Concurrent Pos:* Prin investr, NSF Grant, 69-73. *Mem:* AAAS; Electrochem Soc; Sigma Xi. *Res:* Differential thermal analysis; differential scanning calorimetry; catalysis and transient processes in electrochemistry; glass surface chemistry; history of glassworking and science; carbon 14 and tritium dating; isotope substitutions in spectrophotometry; photosensitivity; thermodyanmics. *Mailing Add:* 1209 Roosevelt Blvd Ypsilanti MI 48197

GRAVES, CARL N, PERIODONTAL DISEASE, ROOT CARIES. *Educ:* Univ Ala, Birmingham, PhD(physiol), 78. *Prof Exp:* ASST PROF RESPIRATORY PHYSIOL, UNIV ALA, BIRMINGHAM, 78- *Mailing Add:* Dept Physiol & Biophysics Univ Ala University Sta Birmingham AL 35294

GRAVES, CHARLES NORMAN, b Fitchburg, Mass, Feb 6, 30; m 63; c 4. PHYSIOLOGY, BIOCHEMISTRY. *Educ:* Okla State Univ, BS, 58; Univ Ill, Urbana, MS, 59, PhD(dairy sci), 62. *Prof Exp:* Res assoc, 61-64, asst prof, 64-68, ASSOC PROF DAIRY SCI, UNIV ILL, URBANA, 68- *Concurrent Pos:* Res fel, Johns Hopkins Univ, 68-69. *Mem:* AAAS; Am Dairy Sci Asn; Am Soc Animal Sci; Soc Study Reprod. *Res:* Oogenesis; fertilization; early embryonic development; alterations in spermatozoa during storage. *Mailing Add:* Dept Animal Sci 110 Animal Genetics Univ Ill 1301 Taft Dr Urbana IL 61801

GRAVES, CLINTON HANNIBAL, JR, b Ackerman, Miss, July 22, 27; m 64; c 1. PLANT PATHOLOGY. *Educ:* Miss State Univ, BS, 50; Univ Wis, PhD(phytopath), 54. *Prof Exp:* Asst plant pathologist, Miss Agr Exp Sta, 53-60, assoc plant pathologist, 60-66; PROF PLANT PATH, MISS STATE UNIV, 66- *Concurrent Pos:* Gen Educ Bd scholar; bd dir, Assoc Southern Agr Scientists, 71-72, exec comt, 72-; counr, Am Phytopath Soc, 72-77. *Mem:* Am Phytopath Soc, Southern Div; Am Phytopath Soc. *Res:* Diseases of fruits and nuts; chemical control; breeding for disease resistance; interspecific crosses; molecular and cellular genetics; etiology. *Mailing Add:* Drawer PG Miss State Univ Mississippi State MS 39762

GRAVES, DAVID E, b Cullman, Ala, Dec 12, 50; m 78; c 2. BIOCHEMISTRY, BIOPHYSICAL CHEMISTRY. *Educ:* Univ Ala, Birmingham, BS, 74, PhD(biochem), 79. *Prof Exp:* NIH fel biophys, Univ Rochester, 80-84; ASST PROF CHEM, UNIV MISS, 85- *Mem:* Am Chem Soc; AAAS; Biophys Soc; Sigma Xi. *Res:* Structural and functional properties of nucleic acids; interactions of drugs and carcinogens with DNA; characterization of ligand-DNA interactions on high resolution nuclear magnetic resonance techniques. *Mailing Add:* Dept Chem Univ Miss University MS 38677

GRAVES, DAVID J(AMES), b Niagara Falls, NY, Feb 25, 41; m 66; c 2. CHEMICAL & BIOMEDICAL ENGINEERING. *Educ:* Carnegie-Mellon Univ, BS, 63; Mass Inst Technol, MS, 65, DSc(chem eng), 67. *Hon Degrees:* MA, Univ Pa, 74. *Prof Exp:* Asst prof, 69-74, ASSOC PROF CHEM ENG, UNIV PA, 74- *Concurrent Pos:* Consult legal & indust concerns, 70-; NSF, NIH & Dept Energy Grants; Humboldt Found fel, Ger, 76-77; Fulbright-Hays award, Coun Int Exchange Scholars, Sweden, 76-77. *Mem:* AAAS; Am Chem Soc; Am Inst Chem Engrs. *Res:* Applied chemistry, particularly polymer, surface and enzyme chemistry; biomedical applications of chemical engineering; enzyme technology; ultrasound; gas diffusion through the skin. *Mailing Add:* Univ Pa 311 A Towne Bldg Philadelphia PA 19104

GRAVES, DONALD C, b Detroit Lakes, Minn, Jan 4, 42; m 64; c 2. MEDICAL VIROLOGY, MOLECULAR PARASITOLOGY. *Educ:* St John's Univ, BS, 64; NDak State Univ, MS, 68; Mich State Univ, PhD(microbiol), 73. *Prof Exp:* Res assoc virol, Univ Pa, 74-76 & Southern Ill Univ, 76-77; asst prof, 77-83, ASSOC PROF VIROL, UNIV OKLA HEALTH SCI CTR, 83- *Concurrent Pos:* Prin investr, Nat Inst Allergy & Infectious Dis, 78-; mem, Bd Educ & Training, Nat Am Soc Microbiol, 80-85. *Mem:* Am Soc Microbiol; Sigma Xi; AAAS; Soc Gen Microbiol. *Res:* Studies on the basic biology of Pneumocystis carinii; immunological studies using antibodies and molecular biology studies using Pc/DNA probes are being conducted to demonstrate strain variation in this organism. *Mailing Add:* Dept Microbiol & Immunol Univ Okla Health Sci Ctr PO Box 26901 Oklahoma City OK 73190

GRAVES, DONALD J, b Evanston, Ill, Oct 15, 33; m 58; c 5. BIOCHEMISTRY. *Educ:* Univ Ill, BS, 55; Univ Wash, PhD(biochem), 59. *Prof Exp:* NIH fel enzymol, Univ Minn, 59-61; from asst prof to assoc prof, 61-68, PROF BIOCHEM, IOWA STATE UNIV, 68- *Concurrent Pos:* NIH career develop award, 65-69 & 70- *Mem:* Am Soc Biol Chemists; Am Chem Soc. *Res:* Mechanism of enzyme action; protein chemistry. *Mailing Add:* Dept Biochem & Biophys Iowa State Univ Ames IA 50011-0061

GRAVES, GLEN ATKINS, b Monroe Co, Ind, Nov 11, 27; m 51; c 3. NUCLEAR PHYSICS. *Educ:* Ind Univ, AB, 48, MS, 50, PhD(physics), 53. *Prof Exp:* Res assoc, Los Alamos Sci Lab, 49, mem staff, 52-67, asst group leader, 67-74; at off energy res & develop policy, NSF, 74-77; MEM STAFF, LOS ALAMOS SCI LAB, 77- *Concurrent Pos:* Prof, Univ NMex, 54-74; head physics sect, Int Atomic Energy Agency, Austria, 69-70. *Mem:* Am Phys Soc; Am Inst Aeronaut & Astronaut; Am Nuclear Soc. *Res:* Nuclear spectroscopy; beta and gamma emission; nuclear reactor physics; radiation problems of propulsion reactors. *Mailing Add:* 80 Barranca Rd Los Alamos NM 87544

GRAVES, GLENN WILLIAM, b Detroit, Mich, June 22, 29; m 55. MATHEMATICS. *Educ:* Western Mich Col Educ, BA, 51; Mich State Col, MA, 52; Univ Mich, PhD(math), 63. *Prof Exp:* Asst, Willow Run Res Ctr, Univ Mich, 52-54, asst, Statist Res Lab, 54-62; mem staff, Aerospace Corp, Calif, 62-64 & Rand Corp, 64-65; assoc prof bus admin, 65-71, assoc prof, 71-74, PROF QUANT METHODS, UNIV CALIF, LOS ANGELES, 74- *Mem:* Am Math Soc; Asn Comput Mach. *Res:* Linear programming; numerical analysis; large scale digital computers; statistics. *Mailing Add:* 3642 Seahorn Dr Malibu CA 90265

GRAVES, HANNON B, b Independence, Va, Mar 7, 43; m 63; c 2. ETHOLOGY. *Educ:* Va Polytech Inst, BS, 65, PhD(genetics), 68. *Prof Exp:* From asst prof to assoc prof, Dept Poultry Sci, Pa State Univ, University Park, 68-85; CHIEF SCIENTIST, CROWELL & MORING, WASHINGTON, DC, 85- *Mem:* AAAS; Genetics Soc Am; Animal Behav Soc; Poultry Sci Asn. *Res:* Behavior genetics; ecology; psychology; evolution and ecology of social sytems. *Mailing Add:* 730 Tussey Lane State College PA 16801

GRAVES, HAROLD E(DWARD), b Beardsley, Minn, Feb 20, 09; m 35; c 2. CHEMICAL ENGINEERING. *Educ:* Univ Minn, BS & MS, 32, PhD(chem eng), 35. *Prof Exp:* Res chemist, Calco Chem Div, Am Cyanamid Co, 35-36; assoc prof chem eng, Miss State Col, 36-38; instr, Yale Univ, 38-40; asst prof, Worcester Polytech Inst, 40-41, prof, 41-48; prof & head dept, RI State Col, 48-52; chief chem engr, Jackson & Church Co, 52-56; supvr process eng sect, Dow Chem Co, 56-64, chief process engr, 64-67, staff asst, Midland Div, 67-74; RETIRED. *Mem:* Am Chem Soc; Am Inst Chem Engrs. *Res:* Process design. *Mailing Add:* 4305 Berkshire Ct Midland MI 48640

GRAVES, HARVEY W(ILBUR), JR, b Rochester, NY, June 18, 27; m 60; c 3. NUCLEAR ENGINEERING. *Educ:* Dartmouth Col, BA, 50, MS, 51; Univ Mich, PhD, 73. *Prof Exp:* Engr, Westinghouse Elec Corp, 51-53, nuclear engr, 53-55, supvry engr, 55-56, mgr nuclear eng, 56-66, mgr advan reactor develop, 66-68; consult engr, 68-79; PRES, ENERGY ANALYSIS SOFTWARE SERV, INC, 79- *Concurrent Pos:* Lectr, Dept Nuclear Eng, Univ Mich, 68-73; adj prof, Dept Chem & Nuclear Eng, Univ Md, 81-84, 88- *Mem:* Fel Am Nuclear Soc. *Res:* Development of interactive software and expert systems software for engineering applications; nuclear reactor physics and engineering; nuclear fuel management. *Mailing Add:* 7723 Curtis St Chevy Chase MD 20815

GRAVES, JERRY BROOK, b Tylertown, Miss, Feb 28, 35; m 60; c 1. ENTOMOLOGY. *Educ:* Miss State Univ, BS, 55, MS, 58; La State Univ, PhD(entom), 62. *Prof Exp:* Res assoc, 61-63, from asst prof to assoc prof, 63-71, head dept, 77-85, PROF ENTOM, LA STATE UNIV, BATON ROUGE, 77- *Mem:* Entom Soc Am. *Res:* Control of cotton insects; insecticide resistance; insect toxicology; insecticide-wildlife relations. *Mailing Add:* Dept Entom La State Univ Baton Rouge LA 70803

GRAVES, LEROY D, b Kokomo, Ind, Mar 12, 12; m 34; c 2. CIVIL ENGINEERING. *Educ:* Purdue Univ, BS, 33, MS, 41. *Prof Exp:* Field engr, State Dept Conserv, Ind, 33-37; res engr, joint hwy res proj, Purdue Univ, 37-41; soils engr, Ohio River Div Labs, Corps Engrs, 41-46; from asst prof to assoc prof civil eng, 46-77, acting head dept, 60-61, asst chmn dept, 68-77, EMER PROF CIVIL ENG, UNIV NOTRE DAME, 77- *Concurrent Pos:* Assoc, Hwy Res Bd, Nat Acad Sci-Nat Res Coun, 38-; soil mech consult, 46-; mem, Int Coun Soil Mech & Found Eng, 48-; secy-treas, Shilts, Graves & Assocs, Inc, 77- *Mem:* Fel Am Soc Civil Engrs; Nat Soc Prof Engrs; Am Soc Eng Educ. *Res:* Soil mechanics and foundation engineering; highways and airports. *Mailing Add:* 19681 Brick Rd South Bend IN 46637

GRAVES, ROBERT CHARLES, b Evanston, Ill, Oct 24, 30; m 56; c 3. ENTOMOLOGY. *Educ:* Northwestern Univ, BS, 52, MS, 53, PhD(biol sci), 56. *Prof Exp:* Asst biol, Northwestern Univ, 52-56; instr, Lake Forest Col, 56-57; prof, Flint Community Col, 57-66; assoc prof, 66-68, PROF BIOL, BOWLING GREEN STATE UNIV, 68- *Concurrent Pos:* Ed, Cicindela, 69- *Mem:* Am Entom Soc; Entom Soc Can; Coleopterists Soc; Sigma Xi. *Res:* Ecology, systematics and distribution of Cicindelidae and Carabidae; fungus-inhabiting insects; Coleoptera. *Mailing Add:* Dept Biol Sci Bowling Green State Univ Bowling Green OH 43403-0212

GRAVES, ROBERT EARL, b Haynesville, La, Oct 25, 38. PHYSICAL CHEMISTRY, ORGANIC CHEMISTRY. *Educ:* Ouachita Baptist Univ, BS, 60; Baylor Univ, PhD(chem), 72. *Prof Exp:* Asst prof, Moody Col, Tex A&M Univ, 71-77; PROF CHEM, E TEX BAPTIST COL, 77- *Mem:* Am Chem Soc. *Res:* Environmental quality; nitrogen fixation. *Mailing Add:* Dept Chem E Tex Baptist Col 1209 N Grove St Marshall TX 75670

GRAVES, ROBERT GAGE, b Gilroy, Calif, May 16, 42; m 62; c 3. EXPERIMENTAL NUCLEAR PHYSICS. *Educ:* Bucknell Univ, BS, 65; State Univ NY, Stony Brook, MA, 69, PhD(physics), 71. *Prof Exp:* Jr res assoc nuclear physics, Brookhaven Nat Lab, 69-71; res scientist nuclear physics, Cyclotron Inst, Tex A&M Univ, 71-80; mem fac, Univ Rochester Cancer Ctr, 80-88; PHYSICIST, RADIO THER UPSTATE NY, 88- *Mem:* Am Phys Soc. *Res:* Measurements of polarization transfer, analyzing power and neutron cross section in three nucleon systems and in proton-neutron reactions using light nuclei; production of polarized and unpolarized monoenergetic neutron beams. *Mailing Add:* Radio Ther Upstate NY 815 James St Syracuse NY 13203

GRAVES, ROBERT JOHN, b Buffalo, NY, Sept 25, 45; m 68; c 3. INDUSTRIAL ENGINEERING, OPERATIONS RESEARCH. *Educ:* Syracuse Univ, BS, 67; State Univ NY, Buffalo, MS, 69, PhD(opers res), 74. *Prof Exp:* Dir Scheduling & Inventory Div, State Univ NY, Buffalo, 68-71, asst to vpres facil planning, 71-73; instr indust eng, Sch Indust & Syst Eng, Ga Inst Technol, 73-74, asst prof, 74-79; assoc prof, 79-88, PROF INDUST ENG, UNIV MASS, 88- *Concurrent Pos:* Consult engr, Clorox Corp, Bath Iron Works Shipyard, Draper Labs & Peterson Builders Shipyard, Digital Equip Corp, 77- *Honors & Awards:* Outstanding Res Contribution Award, Sigma Xi, 78; Spec Citation Award, Inst Indust Engrs, 85. *Mem:* Am Inst Indust Engrs; Soc Naval Architects & Marine Engrs; Inst Mgt Sci; Sigma Xi. *Res:* Location theory and facilities planning, production planning and control, computerized layout, project planning and control; flexible assembly systems. *Mailing Add:* Dept Indust Eng & Opers Res Univ Mass Amherst MA 01003

GRAVES, ROBERT JOSEPH, b Hammond, Ind, Aug 2, 52. ANALYTICAL CHEMISTRY. *Educ:* Univ Ala Birmingham, BS, 78, MS, 81. *Prof Exp:* Mass spectrometrist & lectr chem, Univ Ala Birmingham, 80-82; prod develop scientist, Craven Labs, Austin, Tex, 82-83; DIR, GC/MS LAB, UNIV ALA, 83- *Concurrent Pos:* Consult, IBM Corp, Dionex Corp, Sunnyvale, Calif, P E LaMoreaux & Assocs, Tuscaloosa, Ala, Tuscaloosa Testing Lab, Jim Walter Resources Corp, Merichem Chem Co, Tex, Koppers Corp Pittsburgh & South Eastern Anal Serv, Huntsville, Ala. *Mem:* Am Chem Soc; AAAS; Asn Off Anal Chemists. *Res:* Development of new and superior analytical instrumentation and methodologies in the fields of biomass energy conversion; alternate fuel and chemical feedstock sources and environmental investigations. *Mailing Add:* 3011 Singletary Baton Rouge LA 70809-1825

GRAVES, ROBERT LAWRENCE, b Chicago, Ill, Sept 1, 26; m 51; c 4. MATHEMATICS. *Educ:* Oberlin Col, BA, 47; Harvard Univ, MA, 48, PhD(math), 52. *Prof Exp:* Sr proj supvr, Standard Oil Co, Ind, 51-58; from asst prof to assoc prof, 58-65, assoc dean, Grad Sch Bus, 72-73 & 75-81, PROF APPL MATH, UNIV CHICAGO, 65-, DEP DEAN, 81- *Mem:* Am Math Soc; Opers Res Soc Am; Math Asn Am; Asn Comput Mach; Inst Mgt Sci. *Res:* Operations research, especially linear programming; digital computers. *Mailing Add:* Grad Sch Bus Univ Chicago Chicago IL 60637

GRAVES, ROY WILLIAM, JR, b Ada, Okla, Dec 29, 15; m 42; c 3. GEOLOGY. *Educ:* Agr & Mech Col, Tex, BS, 39; Mo Sch Mines, MS, 41; Univ Tex, PhD(geol), 49. *Prof Exp:* Lab asst mineral, Mo Sch Mines, 39-41; instr geol, Univ Tex, 46-48; res geologist, Calif Res Corp, Standard Oil Co, Calif, 49-55; div stratigrapher, Calif Co, 55-57, area geologist, 57-61; sr geologist, Monsanto Chem Co, 61-63; adj assoc prof geol & explor ed, Info Serv Div, Univ Tulsa, 63-83, adj prof & actg dir, Info Serv, 80-82; RETIRED. *Mem:* Fel Geol Soc Am; Am Asn Petrol Geol; Geosci Info Soc; Sigma Xi; Asn Earth Sci Ed. *Res:* Sedimentary petrology and paleoecology. *Mailing Add:* 9760 Capilano Rd Desert Hot Springs CA 92240-1101

GRAVES, SCOTT STOLL, b Oxnard, Calif, Sept 11, 52. CELLULAR IMMUNOLOGY, CELL BIOLOGY. *Educ:* Univ Calif, Davis, BS, 74; San Diego State Univ, MS, 79; Univ Ga, PhD(microbiol & immunol), 84. *Prof Exp:* Res asst marine biol, Scripps Inst Oceanog, La Jolla, 75-76 & immunol, Salk Inst, 79-81; lab technician & teaching asst, Dept Med Microbiol, Univ Ga, 81-84; postdoctoral, Dept Microbiol & Immunol, Univ Calif, 84-87; SR SCIENTIST, NEORX CORP, 87- *Mem:* Am Asn Immunologists. *Res:* Monoclonal antibody based cancer therapy; novel cytokine; bacterial toxins as cancer therapeutics; new technologies for cancer therapy. *Mailing Add:* 23302 171st Ave SE Monroe WA 98272

GRAVES, TOBY ROBERT, b Stillwater, Okla, Jan 25, 46; m 66; c 2. POLYMER CHEMISTRY. *Educ:* Okla State Univ, BS, 67, MS, 70, PhD(chem eng), 72. *Prof Exp:* Develop engr, Petrolite Corp, 72-74, develop mgr, 75-80, tech mgr, 80-88, VPRES & GEN MGR, PETROLITE SPECIALTY POLYMERS GROUP, PETROLITE CORP, 88- *Mem:* Am Inst Chem Engrs; Nat Soc Prof Eng. *Res:* Development of new polymers, derivations and applications for such products. *Mailing Add:* 8717 S 69 E Ave Tulsa OK 74133

GRAVES, VICTORIA, b Houston, Tex, Feb 14, 41. ORGANOMETALLIC CHEMISTRY, INORGANIC CHEMISTRY. *Educ:* Incarnate Word Col, BA, 67; Univ Tex, Austin, MA, 69, PhD(inorg chem), 75. *Prof Exp:* Instr chem, Incarnate Word Col, 69-71; teaching asst, Univ Tex, Austin, 71-74; asst prof, Incarnate Word Col, 74-75; res assoc, Mass Inst Technol, 75-76; teaching fel, Latin Am Found, Technol Inst Monterrey, Mex, 76-77; res chemist plastics res & develop, Gulf Oil Chem Co, 78-79; res chemist, Plastics Technol Div, Exxon Chem Co, 79-89; SR RESIN DEVELOP ENGR, PLASTICS TECH CTR, PHILLIPS 66 CO, 89- *Mem:* Am Chem Soc; Soc Plastics Engrs. *Res:* Synthetic reactions of transition metal atoms; electronic structure and chemical reactions of transition metal pi-complexes; use of organometallic compounds in homogeneous and heterogeneous catalysis. *Mailing Add:* 30000 Monticello Bartlesville OK 74006

GRAVES, WAYNE H(AIGH), b Des Moines, Iowa, Dec 6, 25. PHYSICS, ELECTRICAL ENGINEERING. *Educ:* Iowa State Univ, BS, 50; Univ Iowa, MS, 58, PhD(elec eng), 61. *Prof Exp:* Physicist, Eng Res Assoc, Minn, 50-51; staff engr commun, Collins Radio Co, Iowa, 52-63; staff engr electronics, Viron Div, Geophys Corp Am, 64-66; electronics res engr, N Star Res & Develop Inst, Minneapolis, 66-71; consult, 72-74; tech rep, 74-80, CONSULT, AM MED SYSTS, INC, 80- *Mem:* Inst Elec & Electronics Engrs; NY Acad Sci; Sigma Xi. *Res:* Nonlinear circuit theory and electromagnetic propagation; neurology and urology instrumentation. *Mailing Add:* 11511 Lakeview Lane Minnetonka MN 55343

GRAVES, WILLIAM EARL, b Conway, Mass, June 1, 41; m 60; c 3. ENDOCRINOLOGY, PHYSIOLOGY. *Educ:* Univ Mass, Amherst, BS, 63; Univ Wis-Madison, MS, 65, PhD(endocrinol, reprod physiol), 67. *Prof Exp:* From asst prof to assoc prof vert physiol, State Univ NY Col Environ Sci & Forestry, 67-76; asst vpres, 73-76, PRES, GRAVES FARMS INC, 76- *Res:* Agriculture; dairy management. *Mailing Add:* Graves Farms Inc Bardswell Ferry Rd Conway MA 01341

GRAVES, WILLIAM EWING, b Louisville, Ky, Mar 17, 30. REACTOR PHYSICS. *Educ:* Univ Louisville, AB, 51; Ind Univ, MS, 53, PhD(physics), 55. *Prof Exp:* Physicist, Eng Div, Savannah River Lab, E I du Pont de Nemours & Co, Inc, 55-61, res supvr, 61-80, res assoc, 80-85, sr res assoc, 85-88; RETIRED. *Mem:* Am Nuclear Soc. *Res:* Experimental and theoretical physics of nuclear reactors. *Mailing Add:* 908 West Ave North Augusta SC 29841

GRAVES, WILLIAM HOWARD, b Lebanon, Ky, July 4, 40; m 62. VECTOR-VALUED MEASURES. *Educ:* Ind Univ, AB, 62, MS, 65, PhD(math), 66. *Prof Exp:* Res assoc programming, Jet Propulsion Lab, Calif Inst Technol, 62-63; from asst prof to assoc prof, 67-79, assoc dean, 81-87, PROF MATH, UNIV NC CHAPEL HILL, 79-, SPEC ASST TO PROVOST, 87- *Concurrent Pos:* NSF res grant, 69-70; consult scholar, IBM, 87-88. *Mem:* Am Math Soc; Math Asn Am; Asn Develop Comput-Based Instrs Systs. *Res:* Theory of vector-valued measures. *Mailing Add:* Univ NC Chapel Hill NC 27599-3250

GRAVETT, HOWARD L, b Normal, Ill, Sept 21, 11; m 37. GENETICS. *Educ:* James Millikin Univ, AB, 33; Univ Ill, MA, 34, PhD(genetics), 39. *Prof Exp:* Asst biol, Univ Ill, 33-35, asst genetics, 35-37; from asst prof to prof biol, Elon Col, 37-46, head dept, 46; from asst prof to prof, 46-76, EMER PROF BIOL, TEX A&M UNIV, 76- *Mem:* AAAS; Am Inst Biol Sci; Am Soc Zool; Nat Asn Biol Teachers. *Res:* Genetics of Drosophila; embryology. *Mailing Add:* Box 387 Danvers IL 61732

GRAVITZ, SIDNEY I, b Baltimore, Md, June 28, 32; m 64; c 2. SYSTEMS ENGINEER, PROGRAM DEVELOPMENT. *Educ:* Mass Inst Technol, BS, 53, MS, 54. *Prof Exp:* Dynamics group engr, NAm Aviation, 57-60; syst & prog eng mgr, Boeing Co, 60-90; CONSULT ENGR, 90- *Concurrent Pos:* Res engr aeroelasticity, Aerospace Eng Dept, Mass Inst Technol, 52-57; mem, NASA-Indust Space Shuttle Design Criteria Working Group, 70, Radio Tech Comn for Aeronaut Panel, 80-81 & Naval Res Adv Comt Panel, 90; adj lectr, Cogswell Col, 90. *Mem:* AAAS; Am Soc Mech Eng; Am Inst Aeronaut & Astronaut; Sigma Xi. *Res:* Author, supervisor and consultant of papers and studies concerned with concept formulation and program development. *Mailing Add:* 8428 SE 62nd St Mercer Island WA 98040

GRAY, A(UGUSTINE) H(EARD), JR, b Long Beach, Calif, Aug 18, 36; m 59; c 1. ENGINEERING, MATHEMATICS. *Educ:* Mass Inst Technol, SM & SB, 59; Calif Inst Technol, PhD(eng sci), 64. *Prof Exp:* Instr physics, San Diego State Col, 59-60; instr eng sci, Calif Inst Technol, 64; from asst prof to assoc prof elec eng, Univ Calif, Santa Barbara, 64-76, prof, 76-80; SR SCIENTIST, SIGNAL TECHNOL INC, 80- *Concurrent Pos:* Consult, Delco Electronics, Gen Motors Corp & Culler Harrison Labs. *Mem:* Inst Elec & Electronics Engrs; Soc Indust & Appl Math; Acoust Soc Am; Am Soc Mech Engrs. *Res:* Applied mathematics; stochastic processes; applied mechanics; numerical analysis; signal processing. *Mailing Add:* Smartstar Corp 120 Cremona Dr Goleta CA 93116

GRAY, ALAN, b Brooklyn, NY, Oct 11, 26; m 51; c 2. VIROLOGY. *Educ:* Pa State Univ, BA, 48, MS, 50; Univ Pa, PhD(pub health, prev med), 53; Am Bd Microbiol, dipl. *Prof Exp:* Asst, Pa State Univ, 49-50; USPHS fel, Univ Pa, 53-55; chief virus dept, Microbiol Assocs, Washington, DC, 55-60; mgr biol develop, 60-63, mgr bact & viral vaccine prod, 63-70, dir biol prod, 70-74, sr dir biologics, Merck Sharp & Dohme, 74-88; PVT CONSULT, 88- *Concurrent Pos:* Res assoc, Children's Hosp, Philadelphia, 53-55; instr, Univ Md, 58; mem rev comt, US Pharmacopeia, 70-80. *Mem:* AAAS; Am Soc Microbiol; Tissue Cult Asn; NY Acad Sci; Int Asn Biol Standardization; Sigma Xi. *Res:* Viral diagnosis; veterinary and human vaccines; viral tumors; blood products. *Mailing Add:* 652 Mulford Rd Wyncote PA 19095

GRAY, ALFRED, b Dallas, Tex, Oct 22, 39; m 64. GEOMETRY. *Educ:* Univ Kans, BA, 60, MA, 61; Univ Calif, Los Angeles, PhD(math), 64. *Prof Exp:* From instr to asst prof math, Univ Calif, Berkeley, 64-68; assoc prof, 68-70, PROF MATH, UNIV MD, COLLEGE PARK, 70- *Concurrent Pos:* NSF fel, Univ Calif, Berkeley, 65-66; Univ Md fac develop fel, 69; partic, Int Cong Mathematicians, Moscow, 66 & Nice, 70; partic, Oberwolfach Conf, 67 & 69. *Mem:* AAAS; Am Math Soc; Math Asn Am. *Res:* Differential geometry; complex analysis. *Mailing Add:* Dept Math Univ Md College Park MD 20742

GRAY, ALLAN, JR, b San Angelo, Tex, Aug 27, 30; m 54; c 2. MATHEMATICS. *Educ:* NMex State Univ, BS, 52, MS, 55, PhD(math), 60. *Prof Exp:* Instr math, NMex State Univ, 55-58; mathematician, White Sands Missile Range, 58-61; assoc prof, 61-80, PROF MATH, NORTHERN ARIZ UNIV, 80- *Mem:* Am Math Soc; Math Asn Am. *Res:* Monomial and permutation groups. *Mailing Add:* Dept Math Northern Ariz Univ Flagstaff AZ 86011

GRAY, ALLAN P, b New York, NY, May 14, 22; m 47; c 1. ENVIRONMENTAL SCIENCES, TOXICOLOGY. *Educ:* Cornell Univ, AB, 43; Columbia Univ, AM, 47, PhD(chem), 50. *Prof Exp:* Jr chemist, Cent Labs, Gen Foods Corp, 43-44; asst, Columbia Univ, 47-50; fel, Univ Chicago, 50-51; res chemist, Neisler Labs, Inc, 51-56, dir chem res, 56-66, sect head, Neisler Labs, Subsid Union Carbide, 66-69; assoc prof pharmacol, Col Med, Univ Vt, 69-73; proj mgr & sci adv, ITT Res Inst, 73-79; dept mgr, Environ Sci, Dynamac Corp, 79-84, dir environ & health sci, 84-90; ENVIRON & HEALTH CONSULT, A P GRAY ASSOCS, 90- *Concurrent Pos:* Reviewer, NIH, NIDA, NIEHS contracts & grant appls; chmn, Decatur-Springfield Subsect, Am Chem Soc, 55, vchmn, Div Med Chem, 74 & 75, chmn div, 76; mem, Develop Therapeut Contract Rev Comt, Nat Cancer Inst, 85-88; Adj Prof, Dept Pharmacol, Sch Med, Georgetown Univ, 87- *Mem:* Fel AAAS; NY Acad Sci; Am Chem Soc. *Res:* Synthesis and properties of nitrogen heterocycles; hypotensive agents; muscle relaxants; analgesics; central agents; narcotic antagonists; toxic environmental contaminants; cancer chemotherapy; environmental chemistry; health and environmental effects of industrial chemicals; risk assessment; design, synthesis and evaluation of potential medicinal agents. *Mailing Add:* A P Gray Assoc Environ Consults 11905 Renwood Lane Rockville MD 20852

GRAY, ALLEN G(IBBS), b Birmingham, Ala, July 28, 15; m 48; c 2. PHYSICAL CHEMISTRY, METALLURGY & CRITICAL & STRATEGIC MATERIALS. *Educ:* Vanderbilt Univ, BS, 37, MS, 38; Univ Wis, PhD(phys chem), 40. *Prof Exp:* Lab instr chem, Vanderbilt Univ, 37-38; instr, Univ Wis, 38-40; res chemist, E I du Pont de Nemours & Co, 40-52; tech ed, Steel, 52-57; tech dir, 76-83, ED, METAL PROGRESS, AM SOC METALS, 57-, DIR PUBL, 61- & CONSULT, 83- *Concurrent Pos:* Consult & prog dir, Manhattan Dist, 42-46; ed-in-chief, Mod Electroplating, 53; chmn comt tech aspects of critical & strategic mat, Nat Mat Adv Bd, Nat Acad Sci, 68-69, chmn, 70; ed ser, Monogr on Metall in Nuclear Technol, Atomic Energy Comn, mem adv comt indust info; adj prof mech & mat eng, Vanderbilt Univ, 80- *Honors & Awards:* William Hunt Eisenman Medal, Am Soc Metals, 67; Nat Mat Advan Award, Fedn Mat Socs. *Mem:* Am Chem Soc; Electrochem Soc; Am Soc Testing & Mat; Am Electroplaters Soc; Am Soc Metals. *Res:* Organic coatings and dispersions; electropolishing; heat treatment; surface active agents; materials and corrosion; chemicals in metal working; ferrous and nonferrous metallurgy; coatings for uranium; chemistry of steelmaking; high temperature materials; vacuum melting; materials selection and process engineering for manufacturing; recognized as an authority on critical and strategic materials. *Mailing Add:* 4301 Esteswood Dr Nashville TN 37215

GRAY, ANDREW P, b Bonner Springs, Kans, July 20, 16; m 54; c 1. VETERINARY MEDICINE, PATHOLOGY. *Educ:* Kans State Univ, BS & DVM, 53, MS, 63, PhD(path), 66. *Prof Exp:* Pvt pract vet med, Ind, 53-61; assoc prof vet path & vet pathologist, Kans State Univ, 66-86; RETIRED. *Mem:* Am Vet Med Asn; US Animal Health Asn. *Res:* Lungs, upper respiratory tract, eye and adnexa of domesticated animals. *Mailing Add:* 3011 Wayne Dr Manhattan KS 66506

GRAY, BRAYTON, b Chicago, Ill, Dec 19, 40; div. MATHEMATICS. *Educ:* Univ Chicago, PhD(math), 65. *Prof Exp:* Lectr math, Manchester Univ, 64-66; asst prof, 66-69, ASSOC PROF MATH, UNIV ILL, CHICAGO CIRCLE, 71- *Concurrent Pos:* Lectr, Univ Aarhus, Denmark, 69-70; vis prof, Univ Heidelberg, 77-78. *Mem:* Am Math Soc. *Res:* Algebraic topology; homotopy groups of spheres; homotopy theory; cobordism theory. *Mailing Add:* Dept Math Univ Ill Chicago Circle Chicago IL 60680

GRAY, BRUCE WILLIAM, b Ithaca, NY, May 27, 37; m 58; c 3. CYTOLOGY. *Educ:* Cornell Univ, DVM, 61, PhD(vet anat), 70. *Prof Exp:* Asst prof, Okla State Univ, 69-72; from asst prof to assoc prof, 72-89, PROF VET HIST, AUBURN UNIV, 89- *Mem:* Am Asn Vet Anatomists; World Asn Vet Anatomists; Am Asn Anatomists; Int Embryo Transfer Soc. *Res:* Adenosine triphosphatase activity in bovine rumen; clinical sensory innervation of the equine metacarpophalangeal joint; fine structure of the excurrent duct system of the male goat. *Mailing Add:* Dept Anat Col Vet Med Auburn Univ 109 Greene Hall Auburn AL 36849

GRAY, CHARLES A(UGUSTUS), b Washington, DC, Oct 15, 38; m 65; c 3. CHEMICAL ENGINEERING. *Educ:* Cornell Univ, BChE, 61; Mass Inst Technol, PhD(chem eng), 66. *Prof Exp:* Instr indust chem, Mass Inst Technol, 63; res engr, Cent Res Ctr, 65-69, eng supvr, Inorg Chem Div, 69-71, mgr eng res, 70-71, supt tech dept, SCharleston Plant, 71-74, asst dir process eng, FMC Indust Chem Div, NJ, 74-76, Dir Process Res & Eng, FMC Agr Chem Res & Develop, 76-80, pyrethroid projs mgr, FMC Indust Chem Group, DIR TECHNOL, FMC CORP, 82- *Mem:* Am Inst Chem Engrs. *Res:* Process development; reaction engineering; process modeling. *Mailing Add:* 196B Springdale Rd Princeton NJ 08540

GRAY, CLARKE THOMAS, b Norwood, Ohio, May 7, 19; m 42; c 2. MICROBIOLOGY, BIOCHEMISTRY. *Educ:* Eastern Ky Univ, BS, 41; Ohio State Univ, PhD(bact), 49. *Hon Degrees:* MA, Dartmouth Col, 64; DSc, Eastern Ky State Univ, 89. *Prof Exp:* Instr bact, Ohio State Univ, 48-49; res assoc, Harvard Med Sch, 49-59; assoc prof, 60-62, chmn dept, 66-80, PROF MICROBIOL, DARTMOUTH MED SCH, 62- *Concurrent Pos:* Guggenheim fel, Oxford Univ, 59-60; biochemist, Leonard Wood Mem Found, Harvard Med Sch, 49-59; adj prof biol, Dartmouth Col, 71- *Mem:* AAAS; Am Soc Microbiol; NY Acad Sci; Am Soc Biol Chem. *Res:* Physiology of mycobacteria; oxidative-phosphorylation; biological formation of hydrogen; nitrate reduction; cytochromes; aerobic and anaerobic electron transport. *Mailing Add:* Dept Microbiol Dartmouth Med Sch Hanover NH 03755

GRAY, CLIFFTON HERSCHEL, JR, b Riverside, Calif, Dec 27, 25; m 55; c 1. GEOLOGY. *Educ:* Univ Calif, Los Angeles, BA, 49; Claremont Grad Sch, MA, 53. *Prof Exp:* Asst geol sci, Pomona Col, 50-51; geologist, Mineral Deposits Br, US Geol Surv, 51-54; jr mining geologist, State Div Mines & Geol, Calif, 54-55, asst mining geologist, 55-58, assoc mining geologist, 58-70, sr geologist, 70-74, supv dist geologist, 70-88; RETIRED. *Mem:* Fel AAAS; fel Geol Soc Am; Mineral Soc Am; Am Asn Petrol Geologists; Soc Econ Geologists; Am Inst Pro Geologists; Am Inst Mining, Metall, Petrol Engrs. *Res:* Economic geology; nonmetallic industrial minerals, especially limestone and dolomite in California; engineering geology. *Mailing Add:* 4464 Edgewood Pl Riverside CA 92506

GRAY, CONSTANCE HELEN, b Medway, Mass, Feb 27, 26; c 2. ANATOMY. *Educ:* Univ Mass, BS, 47; Univ Hawaii, MS, 51; Univ Calif, Berkeley, PhD(anat), 74. *Prof Exp:* Lectr anat, Univ Calif, Berkeley, 74-76; lectr, Calif Polytech State Univ, 76-78, from asst prof to prof biol, 85-88; RETIRED. *Mem:* Soc Neurosci; AAAS; Am Pub Health Asn. *Res:* Neuropathology; cancer education. *Mailing Add:* 2630 Bonifacio Dr Tracy CA 95376

GRAY, D ANTHONY, b Los Angeles, Calif, May 17, 50; m 74; c 3. ENVIRONMENTAL CHEMISTRY & MODELING. *Educ:* Drexel Univ, BS, 73; Syracuse Univ, PhD(org chem), 81. *Prof Exp:* Res assoc, 79-84, fel, 84-88, CTR DIR, SYRACUSE RES CORP, 88- *Mem:* Am Chem Soc; Soc Environ Toxicol & Chem. *Res:* Development of documentation to support the Environmental Protection Agency's regulatory actions; determination of environmental testing needs for commercial chemicals, modeling the fate and transport of chemicals in the environment; development of information systems. *Mailing Add:* Syracuse Res Corp Merrill Lane Syracuse NY 13210-4080

GRAY, DAVID BERTSCH, b Grand Rapids, Mich, Feb 7, 44; m 67; c 3. EXPERIMENTAL ANALYSIS OF BEHAVIOR. *Educ:* Lawrence Univ, BA, 66; Western Mich Univ, MA, 70; Univ Minn, PhD(psychol), 74. *Prof Exp:* Instr psychol, Seton Hill Col, 68-70; prog dir admin, Rochester State Hosp, 75-79, res dir, 80-81; health scientist adminr, 82-86, DIR, NAT INST REHAB RES, HSA, NIH, 87- *Mem:* Am Psychol Asn; AAAS; Nat Spinal Cord Injury Asn; Behav Genetics Asn. *Res:* Applied behavioral fields; mental retardation and developmental disabilities language development, behavior genetics, developmental behavioral pharmacology and learning disabilities. *Mailing Add:* Mental Retardation & Develop Disabilities Br 6130 Executive Blvd Exec Plaza N Rm 631 Bethesda MD 20892

GRAY, DAVID ROBERT, b Victoria, BC, Oct 4, 45; m 78; c 4. MAMMALOGY, ARCTIC RESEARCH. *Educ:* Univ Victoria, BSc, 67; Univ Alta, PhD(zool), 73. *Prof Exp:* From asst cur to assoc cur, Nat Mus Natural Sci, Can, 73-88; CUR VERT ETHOL, CAN MUS NATURE, 88- *Concurrent Pos:* Comt mem, Tundra Panel, Can Comt Int Biol Prog, 73-75; prin investr, Nat Mus Natural Sci, High Arctic Res Sta, Can, 73-79, dir, 88-; mem, comt Polar Bear Pass, Nat Wildlife Area. *Mem:* Arctic Inst NAm. *Res:* Long-term ethological research on arctic mammals, particularly social behavior of muskoxen, arctic hare; collection of visual and sound documentation of avian and mammalian behavior patterns. *Mailing Add:* Ethol Sect Can Mus Nature Ottawa ON K1P 6P4 Can

GRAY, DENNIS JOHN, b Juneau, Alaska, Nov 17, 53; m 83. CELL CULTURE, PLANT DEVELOPMENTAL BIOLOGY. *Educ:* Calif State Col, Stanislaus, BA, 76; Auburn Univ, MS, 79; NC State Univ, PhD(bot), 82. *Prof Exp:* Res asst mycol, Auburn Univ, 77-79; res asst bot & path, NC State Univ, 79-82; res assoc cell cult, Univ Tenn, 82-84; asst prof, 84-89, ASSOC PROF DEVELOP BIOL, UNIV FLA, 89- *Concurrent Pos:* Chmn, Tissue Cult Working Group, Am Soc Hort Sci, 88; consult, Inst Paper Sci & Technol, 88-, Goodyear Tire & Rubber Co, 88-89; assoc ed, Plant Cell Tissue & Organ Cult, 88-; prin investr, USDA & Int Bd Plant Genetic Resources, 89- & USAID, 90- *Mem:* AAAS; Am Inst Biol Sci; Am Soc Hort Sci; Bot Soc Am; Tissue Cult Asn; Coun Agr Sci & Technol. *Res:* Nonconventional methods of crop germ plasm improvement; conservation and propagation using cell, tissue and organ culture; synthetic seed technology; genetic transformation; grape, watermelon, canteloupe, squash, corn, mango and coniferous trees. *Mailing Add:* Cent Fla Res & Educ Ctr Univ Fla 5336 University Ave Leesburg FL 34748

GRAY, DEREK GEOFFREY, b Belfast, Ireland, Feb 9, 41; Can citizen; m 67; c 2. PHYSICAL CHEMISTRY, POLYMER SCIENCE. *Educ:* Queen's Univ, Belfast, BSc, 63; Univ Man, MSc, 65, PhD(chem), 68. *Prof Exp:* Fel electrochem, Univ Newcastle, Eng, 68-69; fel polymer chem, Univ Toronto, 69-71; res scientist, 72-84, dir, Appl Chem Div, 84-89, PRIN SCIENTIST, PULP & PAPER RES INST CAN, 89- *Concurrent Pos:* Paprican adj prof, Dept Chem, McGill Univ, 75-; mem bd dirs, Can Soc Chem, 85-88. *Mem:* Am Chem Soc; Can Pulp & Paper Asn; Chem Inst Can. *Res:* Surface chemistry of paper and cellulose; properties of wood fibres; liquid crystalline derivatives of cellulose. *Mailing Add:* Pulp & Paper Ctr McGill Univ 3420 University St Montreal PQ H3A 2A7 Can

GRAY, DON NORMAN, b Carlyle, Ill, July 28, 31; m 59; c 3. ORGANIC POLYMER CHEMISTRY. *Educ:* Colo State Univ, BS, 53; Univ Colo, PhD(org chem), 56. *Prof Exp:* Res org chemist, Dow Chem Co, 56-57; res chemist, Denver Res Inst, Univ Denver, 57-64, asst prof chem, 60-64; mem staff, Martin Marietta Corp, Md, 64-66; sr res scientist, Owens-Ill Res Lab, Owens-Ill Tech Ctr, Mich, 66-68, sect head, 68-70, mgr biotechnol & toxicol, 70-85; PRES, ANATRACE INC, 85- *Concurrent Pos:* Instr, Univ Colo Exten, 57-64. *Mem:* AAAS; Am Chem Soc; Brit Chem Soc. *Res:* Fluorine and polymer chemistry; membrane chemistry; biopolymers; high temperature polymers; clinical instrumentation. *Mailing Add:* Anatrace Inc 1280 Dussel Dr Maumee OH 43537-1640

GRAY, DONALD HARFORD, b San Salvador, El Salvador, Dec 14, 36; US citizen; m 60; c 2. SOIL MECHANICS, GEOLOGICAL ENGINEERING. *Educ:* Univ Calif, Berkeley, BS, 59, MS, 61, PhD(civil eng), 66. *Prof Exp:* Res engr petrol, Chevron Res Corp, 61-63; from asst prof to assoc prof, 66-75, PROF CIVIL ENG, UNIV MICH, ANN ARBOR, 75- *Concurrent Pos:* Proj dir res grants, NSF, 68-79, Alcoa Found, 73-75 & Rockefeller Found, 75-76; actg dir Inst Environ Qual, Univ Mich, 71-72; consult, State Mich Atty Gen Off, 73-74. *Mem:* Am Soc Civil Engrs. *Res:* Soil reinforcement and slope stabilization by woody vegetation; engineering properties and utilization of residuals; creep behavior and rheological properties of soils. *Mailing Add:* Dept Civil Eng Univ Mich Main Campus Ann Arbor MI 48109-2125

GRAY, DONALD JAMES, anatomy; deceased, see previous edition for last biography

GRAY, DONALD M, physics, for more information see previous edition

GRAY, DONALD MELVIN, b Milton, Pa, Apr 4, 38; m 70. BIOPHYSICAL CHEMISTRY. *Educ:* Susquehanna Univ, BA, 60; Yale Univ, MS, 63, PhD(molecular biophys), 67. *Prof Exp:* Nat Inst Gen Med Serv fel, Univ Calif, Berkeley, 67-69; from asst prof to assoc prof, 70-83, PROF MOLECULAR BIOL, UNIV TEX, DALLAS, 83-, HEAD MOLECULAR & CELL BIOL, 89- *Concurrent Pos:* Fogarty Sr Int fel, Europ Molecular Biol Lab, 77-78. *Mem:* Fel AAAS; Biophys Soc; Fedn Am Scientists; Am Chem Soc; Am Asn Univ Profs. *Res:* The circular dichroism of polynucleotides; the relationship of polynucleotide sequence to secondary and tertiary conformations; the structure of DNA-binding proteins. *Mailing Add:* Dept Molecular & Cell Biol Univ Tex-Dallas Mail Sta F031 Box 830688 Richardson TX 75083-0688

GRAY, DOUGLAS CARMON, b Poplar Bluff, Mo, Dec 19, 38; m 60; c 2. INDUSTRIAL TOXICOLOGY, GOVERNMENTAL RELATIONS. *Educ:* Univ Calif, Davis, BS, 60; Univ Cincinnati, MS, 70, PhD(environ health), 74. *Prof Exp:* Comdr, 33rd Chem Detachment, US Army, 61-62; indust hyg, Aerojet Gen Corp, 63-67; assoc prof community health, Northwestern Univ Med Sch, 72-74; indust hygienist, Los Alamos Sci Lab, 74-81; mgr environ hyg, Olin Corp, 81-82; dir occup safety & health, Univ New Haven, 82-83; MGR, CORP INDUSTR HYG, CABOT CORP, 84- *Concurrent Pos:* Chmn, Aeresol Technol Comt, Am Indust Hyg Asn, 85, Comt E34.18, Am Soc Testing & Mat, 85- & Synthetic Amorphous Silicas & Silicates Indust Asn, 86-88. *Mem:* Air Pollution Control Asn; Am Asn Aerosol Res; Am Indust Hyg Asn; Am Soc Testing & Mat; Am Conf Govt Indust Hygenists; Synthetic Amorphous Silica & Silicate Indust Asn. *Res:* Collection characteristics of aerosol inertial size analyzers (cascade impactors and cyclones), prediction of fluid evaporation and diffusion, and respiratory protection devices. *Mailing Add:* 267 Pine Hill Rd Chelmsford MA 01824

GRAY, EARL E, b Milliken, Colo, Jan 30, 29; m 49; c 2. ELECTRICAL ENGINEERING. *Educ:* Colo State Univ, BSEE, 55, MEE, 60. *Prof Exp:* Specialist elec eng, Gen Elec Co, 55-56; instr, Mich Col Mining & Technol, 56-57; asst prof, Colo State Univ, 57-62; asst prof, 62-65, NSF sci fac fel, 69-70, assoc prof, 66-80, PROF ELEC ENG, UNIV IDAHO, 80- *Concurrent Pos:* Elec engr, Nat Bur Standards, 61. *Mem:* Inst Elec & Electronics Engrs; Am Soc Eng Educ. *Res:* Electronics; logic design; information theory. *Mailing Add:* PO Box 8342 Moscow ID 83843

GRAY, EDWARD RAY, b Ava, Ill, Nov 20, 38; m 60; c 2. EXPERIMENTAL NUCLEAR PHYSICS, PARTICLE ACCELERATOR PHYSICS. *Educ:* Univ Ill, BS, 60, MS, 62, PhD(physics), 67. *Prof Exp:* Physicist, Aeronaut Div, Philco-Ford Corp, 66-67; physicist, Fermi Nat Accelerator Lab, Energy Res Develop Admin, 68-81; physicist, Technicare Corp, 81-84; PHYSICIST, LOS ALAMOS NAT LABS, 84- *Mem:* Am Phys Soc; Sigma Xi. *Res:* Design and development of particle accelerator components. *Mailing Add:* Los Alamos Nat Lab MS H817 Los Alamos NM 87545

GRAY, EDWARD THEODORE, JR, b Little Falls, NY, June 25, 50; m 73; c 1. INORGANIC CHEMISTRY. *Educ:* Union Col, Schenectady, NY, BS, 72; Purdue Univ, PhD(anal chem), 77. *Prof Exp:* ASSOC PROF CHEM, UNIV HARTFORD, 77-, CHAIR, DEPT CHEM, 90- *Mem:* Am Chem Soc; Sigma Xi. *Res:* Bio-inorganic analytical chemistry; analytical measurement and/or separation of chloramines and chloramino acids to nanomolar concentrations; redox properties of metal-peptide complexes. *Mailing Add:* Dept Chem Univ Hartford 200 Bloomfield Ave West Hartford CT 06117

GRAY, EDWIN R, b New Britain, Conn, May 30, 31; m 59; c 2. HUMAN ANATOMY, ELECTROMYOGRAPHY. *Educ:* Univ Conn, BS, 55; Univ Vt, MSc, 65; Queens Univ, Ont, PhD(anat, electromyography), 67. *Prof Exp:* Asst prof anat, Univ Conn, 67-69; assoc prof biochem, Pa State Univ, 69-72; DIR PHYS THER PROG, UNIV WIS-LA CROSSE, 72- *Concurrent Pos:* Speaker, Second Int Cong Electromyographic Kinesiology. *Mem:* Am Asn Anat; Can Asn Anat; Am Phys Ther Asn; Int Soc Electromyog & Kinesiology. *Res:* Electromyographic kinesiology; single motor unit research. *Mailing Add:* Los Alamos Nat Lab MS H817 Los Alamos NM 87545

GRAY, ELMER, b Gray Hawk, Ky, Mar 29, 34; m 57; c 1. PLANT BREEDING. *Educ:* Berea Col, BS, 56; Univ Ky, MS, 58; Cornell Univ, PhD(plant breeding), 62. *Prof Exp:* From asst prof to assoc prof agron, Univ Tenn, 62-68; assoc prof, 68-71, PROF AGRON & DEAN GRAD COL, WESTERN KY UNIV, 71- *Mem:* Am Soc Agron; Crop Sci Soc Am; Am Genetic Asn; World Pop Soc. *Res:* Plant science; genetics; statistics. *Mailing Add:* Dept Agr Western Ky Univ Bowling Green KY 42101

GRAY, EOIN WEDDERBURN, plasma physics, high temperature chemistry; deceased, see previous edition for last biography

GRAY, ERNEST DAVID, b Winnipeg, Man, Oct 3, 30; m 57; c 2. BIOCHEMISTRY, IMMUNOLOGY. *Educ:* Univ Man, BSc, 52; Univ Minn, PhD(physiol chem), 58. *Prof Exp:* From asst to instr physiol chem, Univ Minn, 52-58; asst lectr biochem, Glasgow Univ, 58-59; res assoc, Col Physicians & Surgeons, Columbia Univ, 60-62; asst prof, 62-68, ASSOC PROF PEDIAT & BIOCHEM, UNIV MINN, MINNEAPOLIS, 68- *Concurrent Pos:* Res fel, Imp Chem Industs, 59-60. *Mem:* AAAS; Brit Biochem Soc; Am Soc Biol Chem; Am Chem Soc; Am Soc Microbiol. *Res:* Nucleic acid; metabolism, nuclease action; streptococcal products; purification and characterization of biological and chemical properties; analysis of cellular immune response to these products and its relationship to the pathogenesis of rheumatic heart disease. *Mailing Add:* Univ Minn Box 296 Mayo Minneapolis MN 55455

GRAY, ERNEST PAUL, b Vienna, Austria, Mar 12, 26; nat US; m 54; c 2. ELECTROMAGNETIC SCATTERING, FLUID SURFACE & INTERNAL WAVES. *Educ:* Cornell Univ, AB, 47, PhD(theoret physics), 52. *Prof Exp:* Asst physics, Cornell Univ, 49-52; sr staff mem & physicist, Johns Hopkins Univ, 51-58, chief theoret plasma physics staff, 66-73, mem staff, Res Ctr, 73-83, mem staff space dept, 84-90, PRIN STAFF MEM, APPL PHYSICS LAB, JOHNS HOPKINS UNIV, 58-, ADJ PROF, 59- *Concurrent Pos:* Parsons fel, 57-58; Parsons vis prof, Johns Hopkins Univ, 68-69 & 84-85; consult, Argonne Nat Lab, 75-77; mem staff, Submarine Technol Dept, Johns Hopkins Univ, 91- *Mem:* AAAS; Am Phys Soc. *Res:* Problems in the propagation and scattering of microwaves; atomic excitation, ionization and recombination; highly ionized plasmas; charged particle trajectories; fluid dynamics; internal waves and surface waves in fluids. *Mailing Add:* Johns Hopkins Univ Appl Phys Lab Johns Hopkins Rd Laurel MD 20723

GRAY, F(ESTUS) GAIL, b Moundsville, WVa, Aug 16, 43; m 68; c 3. COMPUTER DESIGN. *Educ:* WVa Univ, BSEE, 65, MSEE, 67; Univ Mich, PhD(comput info control eng), 71. *Prof Exp:* Instr, elec eng, WVa Univ, 66-67; teaching fel, comput eng, Univ Mich, 67-70; from asst prof to assoc prof, 71-84, PROF ELEC ENG, VA POLYTECH INST & STATE UNIV, 84- *Concurrent Pos:* Trainee, US Aviation Mat Lab, 62, 63 & 64; res asst, Argonne Nat Lab, 67 & Univ Mich, 69-70; fac fel, NASA-Am Soc Elec Eng, Langley Res Ctr, 75; prin investr, NSF, 72-79, Rome Air Develop Ctr, 80-81, NASA Langley Res Ctr, 81-82, Naval Surface Weapons Ctr, 73-74, Dahlgren, 82-84 & Off Naval Res, 80-83, Army Res Off, 82-85; vis scientist, Res Triangle Inst, 84-85, consult, 86- *Mem:* Inst Elec & Electronics Engrs; Sigma Xi; Asn Comput Mach. *Res:* Fault tolerant computer architectures reconfigurable structures; parallel computer architectures; built in self test; simulation of fault tolerant systems; algebraic coding theory for data storage and retrieval. *Mailing Add:* Dept Elec Eng Va Polytech Inst & State Univ Blacksburg VA 24061-0111

GRAY, FAITH HARRIET, b Mt Vernon, NY, Jan 15, 40. ZOOLOGY. *Educ:* Chatham Col, BS, 62; Mt Holyoke Col, MA, 64; Ohio State Univ, PhD(entom), 68. *Prof Exp:* Fel physiol, Univ Miami, 68-70; fel entom, Ohio State Univ, 70-71; asst prof biol, Washington & Lee Univ, 71-73; asst prof,

Wellesley Col, 73-74; ASSOC PROF BIOL, HOLLINS COL, 74- *Mem:* AAAS; Am Inst Biol Sci; Am Entom Soc; Am Soc Zool; Tissue Cult Asn; Sigma Xi. *Res:* Insect physiology; comparative endocrinology; mechanisms of cellular aging. *Mailing Add:* Dept Biol Hollins Col PO Box 9616 Hollins College VA 24026

GRAY, FENTON, b Santa Clara, Utah, Aug 12, 16; m 38; c 5. SOIL SCIENCE. *Educ:* Univ Utah, BS, 38; Ohio State Univ, PhD(soils), 51. *Prof Exp:* Soil surveyor, Utah Agr Exp Sta, 39-41 & Soil Conserv Serv, USDA, 41-48; asst soils, Ohio State Univ, 48-51; from asst prof to prof soils, Okla State Univ, 59-82; RETIRED. *Concurrent Pos:* Sr soil scientist & proj leader, Food & Agr Admin, UN, Brazil, 61-62; travel grant to NZ, Australia & Orient, 68. *Honors & Awards:* Ed Award, Soil Conserv Soc Am, 69. *Mem:* Fel AAAS; Am Soc Agron; Soil Sci Soc Am; Soil Conserv Soc Am; Clay Minerals Soc. *Res:* Basic chemical, physical and mineralogical properties of Oklahoma soils with relationships to morphology, genesis, classification, soil and water conservation, productivity and good land use. *Mailing Add:* 1017 S Orchard Lane Stillwater OK 74074

GRAY, FRANK DAVIS, JR, b Marshall, Minn, Aug 24, 16; m 41. INTERNAL MEDICINE. *Educ:* Northwestern Univ, BS, 38; Columbia Univ, MD, 43. *Prof Exp:* Intern & asst resident med, Bellevue Hosp, 43-44; asst surg, Johns Hopkins Univ, 46-47; asst resident med, Yale Univ, 47-48, res fel, 48-49, from instr to assoc prof, 49-68; dir, Div Med, Lankenau Hosp, 68-76; prof med, Jefferson Med Col, 68-76, Magee prof med & chmn dept, 76-81, interim dean, 81-82; RETIRED. *Concurrent Pos:* Fel med, Yale Univ, 48-49; physician-in-chief, Thomas Jefferson Univ Hosp, 76-81. *Mem:* Am Soc Clin Invest; AMA; fel Am Col Physicians; fel Am Col Chest Physicians; Am Fedn Clin Res; Sigma Xi. *Res:* Clinical physiology of heart and lungs. *Mailing Add:* 67 Llanfair Circle Ardmore PA 19003

GRAY, FREDERICK WILLIAM, b Wakefield, Mass, Oct 4, 18; m 49; c 2. ORGANIC CHEMISTRY. *Educ:* Tufts Col, BS, 41, MS, 43; Pa State Col, PhD(chem), 49. *Prof Exp:* Asst, Tufts Col, 41-43; res chemist, Cent Res Lab, Gen Aniline & Film Co, 43-46; asst, Pa State Univ, 46-49; res chemist, Colgate-Palmolive Co, Piscataway, 49-63, res assoc, 63-68, sr res assoc, 68-78, sr scientist, 78-83; RETIRED. *Mem:* Am Chem Soc; fel Am Inst Chem; Am Oil Chemists Soc. *Res:* Organic synthesis; dye intermediates; vinyl polymers; synthetic detergents; optical brighteners; chlorine and oxygen bleaches; development of household specialty product. *Mailing Add:* 14 Stockton Rd Summit NJ 07901

GRAY, FRIEDA GERSH, b Hartford, Conn, Nov 30, 17; m 41. INTERNAL MEDICINE. *Educ:* Hunter Col, AB, 39; NY Med Col, MD, 44. *Prof Exp:* Asst bact, Col Physicians & Surgeons, Columbia Univ, 39-41; intern & asst resident med, Bellevue Hosp, 44-46; asst path, Johns Hopkins Univ & Hosp, 46-47; instr med, Sch Med, Yale Univ, 49-53, from asst clin prof to assoc clin prof, 53-60, from asst prof to assoc prof, 60-68; assoc prof, Jefferson Med Col, 68-73; assoc prof med, Sch Med, Univ Pa, 73-77; prof community med & environ health, 77-79, PROF MED, HAHNEMANN UNIV SCH MED, 77- *Concurrent Pos:* Res fel med, Sch Med, Yale Univ, 47-49; supt & chief med, Woodruff Hosp, 53-60; attend physician & dir, Med Clins, Yale-New Haven Hosp, 60-68; dir div ambulatory care, Lankenau Hosp, 68-73; chief ambulatory serv & sr attend physician, Philadelphia Gen Hosp, 73-77; assoc vpres health affairs, Hahnemann Univ Hosp, 77-82, hon staff, 88- *Mem:* AAAS; AMA; Am Heart Asn; Am Rheumatism Asn; Am Fedn Clin Res; Sigma Xi. *Res:* Quality assurance and utilization review. *Mailing Add:* 67 Llanfair Circle Ardmore PA 19003

GRAY, GARY D, b St Louis, Mo, Aug 6, 36; m 59; c 2. BIOCHEMISTRY, IMMUNOLOGY. *Educ:* Cent Methodist Col, AB, 59; Univ Mo, MS, 63, PhD(biochem), 64. *Prof Exp:* Res scientist, 64-67, SR RES SCIENTIST, UPJOHN CO, KALAMAZOO, 67- *Res:* Enzymology; immunosuppression; transplantation; infectious diseases; immunopotentiation; hypersensitivity diseases. *Mailing Add:* Dept Infectious Dis Upjohn Co Kalamazoo MI 49001

GRAY, GARY M, b Seattle, Wash, June 4, 33; m 57; c 5. GASTROENTEROLOGY, BIOCHEMISTRY. *Educ:* Seattle Univ, BS, 55; Univ Wash, MD, 59. *Prof Exp:* Fel gastroenterol, Univ Hosp, Boston, 62-64; gastroenterologist, US Army Trop Res Med Lab, 64-65, chief, Med Div, 65-66; from asst prof to assoc prof, 66-78, actg dir, Gen Clin Res Ctr, 68-71, head div gastroenterol, 71-88, PROF MED, STANFORD UNIV, SCH MED, 78-, DIR, DIGESTIVE DIS CTR, 87- *Concurrent Pos:* NIH grants, 67-, 72- & 84-86; prog dir, NIH training grant gastroenterol, 72-85, 86-88, Digestive Dis Ctr, 87-; mem, NIH Gen Med A Study Sect, 72-76, Vet Admin Merit Rev Bd, Gastroenterol, 72-76, Cellular Molecular Dis Rev Group, Nat Inst Gen Med Sci, 81-85, res career develop award, 70-75; counr, Am Gastroenterol Asn, 80-84; vis prof & lectr, Ohio State Univ, 69, Univ Wash, 70, Univ Pittsburgh, 72, Univ Pa, 74, Univ Fla, 74-87, Wayne State Univ, 77, Univ Tex, San Antonio, 77, Duke Univ, 77, Harvard Univ, 77, Univ Calif, Davis, 79, Univ Kans, 80, Mex Asn Gastroenterol, 81, Univ Calif, San Diego, 82, Univ Southern Calif, 82 & 86, Cent Am Cong Gastroenterol, 82, Univ Mich, 85, Univ Chicago, 87, Univ Hawaii, 87 & Yale Univ, 89; chmn, Digestive Dis Ctr, Rev Comt, NIH, NIDDK, 90; mem, NIH Panel Technol Assessment Conf Bovine Somatotropin, 90. *Mem:* Am Fedn Clin Res; Am Soc Clin Invest; Am Chem Soc; Am Soc Biol Chem & Molecular Biol; Am Gastroenterol Asn; Asn Am Physicians; Am Clin & Climatol Soc; Am Inst Nutrit. *Res:* Regulation of expression of intestinal membrane glycoproteins and definition of their role in nutrition. *Mailing Add:* Div Gastroenterol Sch Med Stanford Univ Stanford CA 94305

GRAY, GARY RONALD, b Coushatta, La, Dec 4, 42; m 80; c 3. CARBOHYDRATE CHEMISTRY. *Educ:* Ouachita Baptist Univ, BS, 64; Univ Iowa, MS, 67, PhD(biochem), 69. *Prof Exp:* Fel biochem, Univ Iowa, 69; NIH fel, Univ Calif, Berkeley, 69-71; from asst prof to assoc prof, 72-83, PROF CHEM, UNIV MINN, MINNEAPOLIS, 83- *Mem:* Am Chem Soc. *Res:* Structural studies of polysaccharides and lipopolysaccharides; carbohydrate chemistry and reaction mechanisms; chemical immunology of bacterial infections. *Mailing Add:* Dept Chem Univ Minn 207 Pleasant St SE Minneapolis MN 55455

GRAY, GEORGE A(LEXANDER), b Armagh, Ireland, Apr 16, 21; nat US; m 44; c 2. CIVIL ENGINEERING. *Educ:* Tech Univ, BCE, 43; Yale Univ, MEng, 48, DEng, 59. *Prof Exp:* Instr math, Clarkson Tech Univ, 46-47, asst prof civil eng, 48-51; asst prof, Yale Univ, 53-59; prof civil eng, Va Polytech Inst & State Univ, 59-84, assoc dean eng acad affairs, 77-84; RETIRED. *Mem:* Am Soc Civil Engrs; Am Soc Eng Educ; Nat Soc Prof Engrs. *Res:* Structural engineering; structural mechanics. *Mailing Add:* 304 Franklin Dr NE Blacksburg VA 24060-7204

GRAY, GRACE WARNER, b Chicago, Ill, Nov 20, 24. PHARMACOLOGY. *Educ:* Mt Holyoke Col, BA, 45; Univ Mich, PhD(pharmacol), 51. *Prof Exp:* Sr res pharmacologist, Bristol Labs, 51-54; from instr to asst prof pharmacol, Sch Med, Marquette Univ, 54-63; USPHS trainee lipid metab, Col Med, Univ Tenn, 63-64; asst prof pharmacol, Woman's Med Col Pa, 64-67; assoc prof pharmacol, Col Vet Med, Univ Minn, St Paul, 67-83; RETIRED. *Mem:* AAAS; Am Chem Soc; Am Soc Pharmacol & Exp Therapeut. *Res:* Autonomic and gastrointestinal pharmacology. *Mailing Add:* 5719 Donegal Dr Shoreview MN 55126

GRAY, GREGORY EDWARD, b Los Angeles, Calif, Sept 27, 54; m 77; c 1. BIOLOGICAL PSYCHIATRY, PSYCHIATRIC EDUCATION. *Educ:* Univ Calif, Davis, BS, 75, MS, 76; Univ Southern Calif, PhD(biomet), 80, MD, 83. *Prof Exp:* Teaching asst biol, 76-77, nutritionist, dept family & prev med, 77-79 & dept prev med, 83, physician, dept psychiat, 83-87, ASST PROF, DEPT PSYCHIAT & DEPT PREV MED, UNIV SOUTHERN CALIF, 86- *Concurrent Pos:* Instr, Azusa Pac Col, 80; lectr, Med Sch, Univ Southern Calif, 84-86; mem, Nat Coun Against Health Fraud; dir, Pac Geriat Educ Ctr, 88-; asst dir, Los Angeles County, Univ Southern Calif Psychiat Hosp, 89- *Mem:* Acad Psychosom Med; Am Pub Health Asn; Asn Acad Psychiat; Am Geriat Soc. *Res:* Nutritional assessment; diet and cancer; diet and behavior; clinical epidemiology; geriatric psychiatry and nutrition; psychopharmacology. *Mailing Add:* Los Angeles County-USC Psychiat Hosp 1934 Hosp Pl Los Angeles CA 90033

GRAY, H(ARRY) J(OSHUA), b St Louis, Mo, June 24, 24; m 49; c 4. ELECTRICAL ENGINEERING, ELECTRONICS ENGINEERING. *Educ:* Univ Pa, BS, 44, MS, 47, PhD, 53. *Prof Exp:* Staff mem elec eng, Moore Sch, Univ Pa, 47-53, asst prof, 53-54; proj engr, Remington Rand Corp & Univac, 54-55, staff engr, 55-57; assoc prof, 57-64, PROF ELEC ENG, MOORE SCH ELEC ENG, UNIV PA, 64- *Concurrent Pos:* Consult, Philco Corp, 58-63, Int Tel & Tel Corp, 59, Curtiss-Wright Electronics Div, 60, Xerox Corp, 66-67 & Burroughs Corp, 67-70. *Mem:* Inst Elec & Electronics Engrs; Am Soc Eng Educ. *Res:* Digital, analog, and interface circuits and their associated electromagnetic circuit problems. *Mailing Add:* 412 Colonial Park Dr Springfield PA 19064

GRAY, HARRY B, b Woodburn, Ky, Nov 14, 35; m 57; c 3. INORGANIC CHEMISTRY. *Educ:* Western Ky Univ, BS, 57; Northwestern Univ, PhD(chem), 60. *Hon Degrees:* DSc, Northwestern Univ, 84, Univ Rochester, 87, Univ Chicago, 87. *Prof Exp:* NSF fel chem, Copenhagen Univ, 60-61; from asst prof to prof, Columbia Univ, 61-66; ARNOLD O BECKMAN PROF CHEM & DIR, BECKMAN INST, 66- *Concurrent Pos:* Mem, Adv Bd, Oak Ridge Nat Lab, 84-; vis comt, Los Alamos Nat Lab, 84-; mem coun, Nat Acad Sci, 86. *Honors & Awards:* E C Franklin Mem Award, 67; Fresenius Award & Shoemaker Award, 70; Am Chem Soc Award Pure Chem, 70 & Inorganic Chem, 78; Harrison Howe Award, 72; Remsen Award, 79; Bailar Medal, Univ Ill, 84; Nat Medal Sci, 86; Priestley Medal, 91. *Mem:* Nat Acad Sci; Royal Danish Acad Sci & Lett; Am Chem Soc; Am Acad Arts & Sci. *Res:* Electronic structure of metal complexes; inorganic reaction mechanisms. *Mailing Add:* Beckman Inst Calif Inst Technol Pasadena CA 91125

GRAY, HARRY EDWARD, b Nashville, Tenn, Mar 24, 43. POLYMERS USED AS CELL CULTURE SUBSTRATES. *Educ:* Univ Pa, BA, 65; Univ Fla, PhD(neurosci), 82. *Prof Exp:* Res technician, Dept Biophys, Univ Pa, 67-68, Dept Psychol, Brandeis Univ, 71 & Univ Rochester, NY, 71-73; res fel cell biol, Dept Pharmacol, Harvard Med Sch, 83-85; vis scientist cell biol & med genetics, Univ Uppsala, Sweden, 85, French Inst Health, 85-86; SCIENTIST CELL BIOL, LABWARE DIV, BECTON-DICKINSON & CO, 86- *Mem:* Am Chem Soc; Am Soc Cell Biol; Tissue Cult Asn; Soc Neurosci; AAAS. *Res:* Improvement of polymers to be used as substrates for cell culture; effects of substrate surface chemistry on cell morphology and physiology; effects on oncogene expression on cell proliferation. *Mailing Add:* Labware Div Becton Dickinson & Co Two Bridgewater Lane Lincoln Park NJ 07035

GRAY, HENRY HAMILTON, b Terre Haute, Ind, Mar 18, 22; m 44; c 2. GEOLOGY. *Educ:* Haverford Col, BS, 43; Univ Mich, MS, 46, Ohio State Univ, PhD(geol), 54. *Prof Exp:* Geologist, US Geol Surv, 43-45; from instr to asst prof geol, Kent State Univ, 48-53; coal geologist, 54-60, map ed, 55-61, head stratigrapher, 60-87, EMER HEAD STRATIGRAPHER, IND GEOL SURV, 87- *Mem:* Soc Econ Paleontologists & Mineralogists; Am Asn Petrol Geologists. *Res:* Stratigraphy; sedimentary petrology; geomorphology; geologic mapping. *Mailing Add:* 5431 E King Rd Bloomington IN 47408

GRAY, HORACE BENTON, JR, b DeLand, Fla, Oct 29, 41; m 72; c 2. PHYSICAL BIOCHEMISTRY. *Educ:* Fla State Univ, BS, 63; Univ Calif, Berkeley, PhD(chem), 67. *Prof Exp:* Nat Cancer Inst fel, Calif Inst Technol, 67-69; asst prof, 69-72, assoc prof biophys, 72-81, PROF, DEPT BIOCHEM & BIOPHYS SCI, UNIV HOUSTON, 81- *Mem:* Sigma Xi; Am Soc Biol Chemists; Biophys Soc; AAAS. *Res:* Hydrodynamics and thermodynamic properties of circular DNA and other forms of DNA; interaction of nucleic acids with nucleases and other enzymes of nucleic acid metabolism. *Mailing Add:* Dept Biochem & Biophys Sci Univ Houston Houston TX 77204-5500

GRAY, IAN, b Cookstown, NIreland, Dec 26, 44. FOOD CHEMISTRY. *Educ:* Queens Univ, Valfast, BS, 68, PhD(food sci), 71. *Prof Exp:* Res assoc agr, Univ Guelph, Ont, 74-78; res assoc, 72-74, ASSOC DIR, AGR EXP STA, MICH UNIV, 78- *Mem:* Am Oil Chemists Soc; Inst Food Technol. *Res:* Liquid oxidation; food safety; food packaging. *Mailing Add:* Agr Exp Sta Mich Univ 109 Agr Hall East Lansing MI 48823

GRAY, IRVING, b Boston, Mass, Apr 27, 20; m 47; c 2. BIOCHEMISTRY, BIOPHYSICS. *Educ:* Va Polytech Inst, BS, 41; Mass Inst Technol, PhD(biochem), 48. *Prof Exp:* Develop electrochemist, Westinghouse Elec & Mfg Co, Pa, 41; res biochemist, Med Serv, US Army, 48-52, head dept biochem, Walter Reed Army Inst Res, 52-56, sci dir res & develop, Qm Res & Eng Ctr Labs, 56-59, chief phys sci div, Med Unit, 59-64; PROF BIOL, GEORGETOWN UNIV, 64- *Mem:* AAAS; Am Physiol Soc; Biophys Soc; Soc Exp Biol & Med; NY Acad Sci. *Res:* Biochemical response to stress; enzyme kinetics in temperature adaptation. *Mailing Add:* Dept Biol Georgetown Univ Washington DC 20057

GRAY, JAMES CLARKE, developmental anatomy; deceased, see previous edition for last biography

GRAY, JAMES EDWARD, b Santa Barbara, Calif, June 3, 32; m 76. ELECTRONICS ENGINEERING, PHYSICS. *Educ:* Univ Calif, BA, 54, MA, 60. *Prof Exp:* Physicist dielec, Nat Bur Standards, 60-66, physicist atomic time, 66-73,; ENGR INSTRUMENTATION, NAT INST STANDARDS & TECHNOL, 73- *Mem:* Sigma Xi. *Res:* Time and frequency metrology. *Mailing Add:* 942 Seventh St Boulder CO 80302

GRAY, JAMES P, b Los Angeles, Calif. DATA COMMUNICATIONS, SYSTEMS NETWORK ARCHITECTURE. *Educ:* Yale Univ, BE, 65, PhD(elec eng), 70. *Prof Exp:* Res staff mem, IBM Res Div, 70-72, adv engr, Systs Commun Div, 72-79, sr engr, Commun Prods Div, 79-82, sr tech staff, 82-84, IBM FEL, IBM COMMUN PRODS DIV, 84- *Mem:* Fel Inst Elec & Electronics Engrs; Asn Comput Mach. *Res:* Research and advanced technology on distributed operating systems, especially as applied to SNA, SAA and OSI; software development tools, defect reduction and productivity. *Mailing Add:* IBM Corp Cpd-F92/673 PO Box 12195 Research Triangle Park NC 27709

GRAY, JAMES S, US citizen. ELECTRICAL ENGINEERING. *Educ:* Southwestern Univ, Memphis, BS, 62; Ga Inst Technol, MS, 65, PhD(elec eng), 67. *Prof Exp:* From assoc prin engr to prin engr, Electron Systs Div, Harris Corp, 67-77; PRIN ENGR, SATELLITE COMMUN DIV, SCIENTIFIC-ATLANTA, INC, 77- *Honors & Awards:* Harris Corp Eng Award, 76. *Mem:* Inst Elec & Electronics Engrs; Sigma Xi. *Res:* Statistical communications; circuit theory; digital processing; filtering. *Mailing Add:* 4718 Pine Acres Ct Dunwoody GA 30338

GRAY, JANE, b Omaha, Nebr, Apr 19, 31. PALEOECOLOGY, EVOLUTION PALEOBIOLOGY. *Educ:* Harvard Univ, BA, 51; Univ Calif, PhD(paleont), 58. *Prof Exp:* Ed asst, J Sedimentary Petrol, Univ Ill, 51-52; instr geol, Univ Tex, 56-58; res assoc palynology, Univ Ariz, 58-61; PROF BIOL, UNIV ORE, 66- *Concurrent Pos:* New world secy, Proj Ecostratigraphy, IUGS & IGCP, UNESCO, 74- *Mem:* Bot Soc Am; Soc Study Evolution; Int Soc Taxon. *Res:* Ecosystem evolution; evolution of early land plants; nonmarine paleoecology; Silurian biostratigraphy; late Cenozoic pollen analysis. *Mailing Add:* Dept Biol Univ Ore Eugene OR 97403

GRAY, JOE WILLIAM, b Hobbs, NMex, Apr 26, 46; m 67; c 1. BIOMATHEMATICS. *Educ:* Kans State Univ, PhD(physics), 72. *Prof Exp:* SR SCIENTIST, LAWRENCE LIVERMORE LAB, 72-, SECT LEADER, 82- *Concurrent Pos:* Adj asst prof radiol, Univ Calif, Davis, 76-; adj prof lab med, Univ Calif, San Francisco, 84-; NIH Genome coun, 90; prin investr, NIH & DOE grants. *Honors & Awards:* Res Award, Radiation Res Soc, 85; E O Lawrence Award, Dept Energy, 87. *Mem:* Cell Kinetics Soc (pres, 83); Soc Anal Cytol; AAAS; Radiation Res Soc; Am Soc Human Genetics. *Res:* Molecular cytogenics; analytical cytology; genetic disease diagnosis; prenatal diagnosis; tumor cytogenetics; biological dosimetry. *Mailing Add:* Biomed Div Lawrence Livermore Lab PO Box 5507 Livermore CA 94550

GRAY, JOEL EDWARD, b Carlisle, Pa, Aug 14, 43; m 82; c 2. RADIOLOGY. *Educ:* Rochester Inst Technol, BS, 70; Univ Ariz, MS, 72; Univ Toronto, PhD(radiol sci), 77. *Prof Exp:* Res asst, Optical Sci Ctr, Univ Ariz, 71-73, res fel, diag radiol & optical sci, Dept Radiol, 73-74; res fel radiol sci, Inst Med Sci, Univ Toronto, 74-77; from asst prof to assoc prof, 77-87, PROF RADIOL PHYSICS, MAYO MED SCH, 87-; ADJ ASST PROF, DEPT MED PHYSICS, UNIV WIS, MADISON, 82- *Concurrent Pos:* Ed, J Appl Photog Eng, 75-79; review ed, Med Physics, Am Asn Physicists Med, 76-81; consult, diag radiol, Mayo Clinc, 77-; mem, bd dir, Am Asn Physicists Med, 83-85; prog comt, Radiol Soc NAm, 85-89, chmn, physics prog subcomt, 89-; mem, Int Comn Radiol Protection, 85-, Nat Coun Radiation Protection & Measurements, 87- *Mem:* Radiol Soc NAm; fel Am Asn Physicists Med; Am Col Radiol; fel Soc Photog Scientists & Engrs; Asn Univ Radiologists; Am Roentgen Ray Soc; Sigma Xi; Soc Motion Picture & TV Engrs. *Res:* Medical imaging; application of optical image evaluation techniques in diagnostic radiology; application of psychophysical techniques in the evaluation of radiological imaging systems; magnetic resonance imaging. *Mailing Add:* 726 19th Ave SW Rochester MN 55902

GRAY, JOHN AUGUSTUS, III, b Waterbury, Conn, Aug 13, 24; m 52; c 3. PHYSICAL CHEMISTRY. *Educ:* Yale Univ, BS, 45, PhD(chem), 49. *Prof Exp:* Res chemist, Miami Valley Labs, Procter & Gamble Co, 48-80; EXEC DIR, INT ASSOC DENT RES, 80- *Concurrent Pos:* Trustee, Children's Hosp Med Ctr, 74-; mem bd sci counrs, Nat Inst Dental Res, 74-78; trustee, Children's Dent Care Found. *Mem:* Fel AAAS; Am Chem Soc; Int Asn Dent Res (vpres, 77-78); Europ Orgn Res Fluorine & Dent Caries Prev; Am Asn Dent Res (secy-treas, 73-76); Am Dent Asn; hon fel Am Col Dentists. *Res:* Physical chemical studies of detergents, solutions, dentifrices, tooth structure and caries mechanism. *Mailing Add:* 9300 Meadow Ridge Dr Cincinnati OH 45241

GRAY, JOHN EDWARD, b Syracuse, NY, May 24, 49. MOLECULAR BIOLOGY, MICROBIOLOGY. *Educ:* Ursinus Col, BS, 71; Univ Del, PhD(biol), 79. *Prof Exp:* Fel genetics, Microbiol Dept, Mt Sinai Sch Med, 78-81; staff scientist, Genetics Div, Bethesda Res Labs, Gaithersburg, Md, 81-89; SR RES SCIENTIST, E I DU PONT DE NEMOURS & CO, 89- *Mem:* Sigma Xi; AAAS; Am Soc Microbiol. *Res:* Modification of bacteria for industrial production of amino acids; genetic regulation of metabolism in bacteria; translational processes in cyanobacteria. *Mailing Add:* E I du Pont de Nemours & Co Stine-Haskell Res Ctr PO Box 30 Elkton Rd Newark DE 19714

GRAY, JOHN LEWIS, forest economics, forest management, for more information see previous edition

GRAY, JOHN MALCOLM, b Thurnscoe, UK, Mar 28, 40; m 63; c 3. WELDING METALLURGY, CORROSION OF PIPELINE MATERIALS. *Educ:* Univ Sheffield, Eng, AMet, 61, MMet, 62, PhD(metall), 65. *Prof Exp:* Metallurgist, Brit Steel, Rotherham, UK, 56-65; sr researcher, US Steel Res, Monroeville, Pa, 66-70; vpres, Molycorp Div, Union Oil Calif, 71-76; PRES, MICROALLOYING INT, 76- *Mem:* Fel Am Soc Metals; fel Metals Soc; Nat Asn Corrosion Engrs; Metall Soc; Am Welding Soc. *Res:* Physical metallurgy of low carbon microalloyed steels; corrosion in wet H2S environment. *Mailing Add:* 9025 Briar Forest Dr Houston TX 77024

GRAY, JOHN PATRICK, electron microscopy, for more information see previous edition

GRAY, JOHN STEPHENS, b Chicago, Ill, Aug 11, 10; m 35; c 2. PHYSIOLOGY. *Educ:* Knox Col, BS, 32; Northwestern Univ, MS, 34, PhD(physiol), 36, MD, 46. *Prof Exp:* From instr to prof, 36-74, chmn dept, 46-70, EMER PROF PHYSIOL, MED SCH, NORTHWESTERN UNIV, 74- *Concurrent Pos:* Guggenheim fel, 62; res physiologist, US Air Force, Randolph Field, 42-46. *Mem:* AAAS; Am Physiol Soc; Soc Exp Biol & Med. *Res:* Gastrointestinal and respiratory physiology. *Mailing Add:* 1408 W Lake St Ft Collins CO 80521

GRAY, JOHN WALKER, b St Paul, Minn, Oct 3, 31; m 57; c 2. MATHEMATICS. *Educ:* Swarthmore Col, BA, 53; Stanford Univ, PhD(math), 57. *Prof Exp:* Mem, Inst Advan Study, 57-59; Ritt instr math, Columbia Univ, 59-62; from asst prof to assoc prof, 61-66, PROF MATH, UNIV ILL, URBANA, 66- *Concurrent Pos:* NSF sr fel, 66-67; Fulbright-Hays sr scholar, Australian-Am Educ Found, 75-76. *Mem:* AAAS; Am Math Soc; ACM. *Res:* Category theory; sheaf theory and its geometrical applications; categorical semantics of programming languages. *Mailing Add:* Dept Math Univ Ill Urbana IL 61801

GRAY, JOSEPH B(URNHAM), b Annapolis, Md, Aug 8, 15; m 41; c 3. CHEMICAL ENGINEERING. *Educ:* St John's Col, Md, BA, 36; Johns Hopkins Univ, BE, 38, PhD(chem eng), 41. *Prof Exp:* Chem engr, process design div, res lab, Standard Oil Co, Ind, 41-43, group leader, 43-47; from asst prof to assoc prof chem eng, Syracuse Univ, 47-51; res engr, E I Du Pont de Nemours & Co, Inc, 51-60, consult, 61-81; mixing consult, Beechwood Consults, Inc, 81-88; RETIRED. *Mem:* Am Chem Soc; fel Am Inst Chem Engrs. *Res:* Fluid mechanics; agitation and mixing. *Mailing Add:* 1909 Beechwood Dr Westwood Manor Wilmington DE 19810

GRAY, KENNETH EUGENE, b Herrin, Ill, Jan 11, 30; m 55; c 3. PETROLEUM ENGINEERING. *Educ:* Univ Tulsa, BS, 56, MS, 57; Univ Tex, PhD(petrol eng), 63. *Prof Exp:* Drilling engr, Calif Co, 57-59; reservoir engr, Sohio Petrol Co, 59-60; from asst prof to assoc prof petrol eng, 62-68, chmn dept, 66-74, HALLIBURTON PROF PETROL ENG, UNIV TEX, AUSTIN, 68-, DIR CTR EARTH SCI, 68- *Concurrent Pos:* Res grants, Petrol Res Fund, Am Chem Soc, 63-, Tex Petrol Res Comt, 63-, Am Petrol Inst, 64- & Gulf Res & Develop Co, 64-; consult, Res Dept, Continental Oil Co, 63-; mem, US Nat Comt Rock Mech. *Mem:* Fel Am Inst Chemists; NY Acad Sci; Am Acad Mech. *Res:* Rock mechanics; properties and behavior of rocks under conditions of elevated temperature and pressure; reservoir transients; unsteady state reservoir pressure analysis. *Mailing Add:* Mat Sci Div Bldg 223 Rm A125 9700 S Cass Ave Argonne IL 60439

GRAY, KENNETH STEWART, b Teaneck, NJ, June 28, 45; m 67; c 2. SPECIAL PURPOSE MEMORY DESIGN, LOGIC DESIGN. *Educ:* Norwich Univ, BS, 67; Univ Vt, MS, 78. *Prof Exp:* ADV ENGR ELECTRONICS ENG, IBM CORP, 67- *Mem:* Nat Soc Prof Engrs; Inst Elec & Electronic Engrs. *Res:* Two patents and forty publications in the areas of NMOS, CMOS, BICMOS circuit, memory and chip design. *Mailing Add:* 288 Griswold St Jericho VT 05465

GRAY, KENNETH W, b DeRidder, La, Apr 2, 44; m 65; c 2. CORROSION RESISTANT COATINGS, COATING APPLICATION EQUIPMENT. *Educ:* Ouachita Baptist Univ, BS, 66. *Prof Exp:* Officer, Chem, Biol, Radiol Warfare, US Army Chem Corps, 66-67; res scientist inks/coatings, Continental Can Co, 67-70; sr chemist coatings, Polymer Corp, 70-78, product mgr, 78-80; vpres & gen mgr pipecoatings, ICO-Spincote Plastic Coatings, 80-91; CONSULT COATINGS, CLOUD-COTE, INC, 91- *Concurrent Pos:* Pres, IWM Inc, 89-91. *Mem:* Nat Asn Corrosion Engrs. *Res:* Computer color matching; coatings formulation; testing and application; coating application equipment; oilfield equipment; issued 8 patents. *Mailing Add:* 2403 Bobwhite Dr Odessa TX 79761

GRAY, LAWRENCE FIRMAN, b Santa Monica, Calif, May 25, 49; m 72; c 4. PROBABILITY. *Educ:* Univ Calif, San Diego, BA, 73; Cornell Univ, PhD(math), 77. *Prof Exp:* ASST PROF MATH, UNIV MINN, 77- *Mem:* Inst Math Statist. *Res:* Analysis of Markov processes which model systems of large numbers of interacting components; percolation theory; statistical mechanics. *Mailing Add:* Sch Math Univ Minn Minneapolis MN 55455

GRAY, LEWIS RICHARD, b Madison, Wis, Feb 14, 36; m 58; c 3. PLANT MORPHOLOGY. *Educ:* Univ Miami, BS, 59; Univ Cincinnati, MS, 61; Univ Ill, PhD(bot), 65. *Prof Exp:* From instr to assoc prof biol, Mayfair Col, 65-76, actg chmn dept, 71, PROF BIOL, CITY COLS CHICAGO, TRUMAN COL, 76- *Mem:* Am Asn Stratigraphic Palynologists. *Res:* Paleozoic palynology. *Mailing Add:* 811 W Hintz Rd Arlington Heights IL 60004

GRAY, LINSLEY SHEPARD, JR, b Sandwich, Ill, Oct 11, 29; m 51; c 4. SPECTROSCOPY, PHYSICAL PROPERTIES. *Educ:* Beloit Col, BS, 51; Iowa State Univ, PhD(chem), 58. *Prof Exp:* Jr chemist, Ames Lab, Iowa State Univ, 51-52, asst, 52-58; chemist, E I du Pont de Nemours & Co, 58-59; chemist, Armour & Co, 59-60, sr res chemist, Akzo Chem Inc, 60-66, sect head phys chem res, 66-85, sect head phys chem res, 74-91; CONSULT. *Mem:* Am Chem Soc; Coblentz Soc; Soc Appl Spectros; Am Oil Chemists' Soc. *Res:* Spectroscopy; surface chemistry; material characterization. *Mailing Add:* 3948 Forest Ave Downers Grove IL 60515

GRAY, MARY JANE, b Columbus, Ohio, June 13, 24; div; c 4. MEDICINE. *Educ:* Swarthmore Col, BA, 49; Wash Univ, MD, 49; Am Bd Obstet & Gynec, dipl, 59. *Hon Degrees:* ScD, Columbia Univ, 53. *Prof Exp:* Intern, Barnes Hosp, 49-50; asst resident obstet & gynec, Presby Hosp, 50-53, chief resident, 56; instr, Col Physicians & Surgeons, Columbia Univ, 56-59, assoc, 59-60; from asst prof to prof obstet & gynec, Col Med, Univ Vt, 60-77; prof obstet & gynec, 77-90, asst dean, 86-90, EMER PROF OBSTET & GYNEC, UNIV NC, 90- *Concurrent Pos:* Barnes Foster res fel, Columbia Univ, 53-54 & Karolinska Inst, Sweden, 55; attend physician, Med Ctr Hosp Vt; trainee, Marriage Coun Philadelphia, dept psychiat, Univ Pa, 70-71. *Mem:* AAAS; Am Med Asn; Soc Gynec Invest; Am Col Obstet & Gynec; Am Obstet & Gynec Soc. *Res:* Toxemia of pregnancy; electrolyte changes of menstrual cycle; gynecologic cancer; sex education; marriage counseling. *Mailing Add:* Dept Obstet & Gynec Univ NC Chapel Hill NC 27514

GRAY, MARY WHEAT, b Hastings, Nebr, Apr 8, 39; m 64. MATHEMATICS. *Educ:* Hastings Col, AB, 59; Univ Kans, MA, 62, PhD(math), 64. *Prof Exp:* From asst prof to assoc prof math, Calif State Col, Hayward, 65-68; assoc prof, 68-71, PROF MATH, AM UNIV, 71-, CHAIR, 77- *Concurrent Pos:* Translator, Am Math Soc, 64-66; consult statist, computer sci, law, 79- *Mem:* AAAS; Am Math Soc; Math Asn Am; NY Acad Sci; Asn Women Math; Sigma Xi; Am Statist Asn; Asn Comput Mach; Am Asn Univ Prof. *Res:* Category theory; homological algebra; ring theory; applied statistics; computers and the law. *Mailing Add:* Dept Math Statist & Comput Sci Am Univ Washington DC 20016

GRAY, MICHAEL WILLIAM, b Medicine Hat, Alta, July 18, 43; m 68; c 2. ORGANELLE MOLECULAR BIOLOGY, EVOLUTION. *Educ:* Univ Alta, BSc, 64, PhD(biochem), 68. *Prof Exp:* From asst prof to assoc prof, 70-83, PROF BIOCHEM, DALHOUSIE UNIV, 83- *Concurrent Pos:* Nat Res Coun Can fel, Stanford Univ, 68-70; Med Res Coun Can grant, Dalhousie Univ, 70- & Med Res Coun Can scholar, 73-78; Natural Sci & Eng Res Coun Can grant, 82-91; vis scientist, Med Res Coun Can, Stanford Univ, 84-85; mem ed bd, Plant Sci, 83-87. *Honors & Awards:* Fraser Medal, Atlantic Prov Coun Sci, 82; Max Forman Res Prize, Dalhousie Mes Res Found, 86; Boehringer Mannheim Can Prize, Canadian Biochem Soc, 87. *Mem:* Can Biochem Soc; Am Soc Biochem & Molecular Biol; Int Soc Plant Molecular Biol; Can Soc Plant Molecular Biol (vpres, 86 & pres 87); fel Can Inst Advan Res; AAAS. *Res:* Structure, function and evolution of mitochondrial DNA, especially in plants and algae; evolution of ribosomal RNA structure; phylogeny of organisms and organelles based on ribosomal RNA. *Mailing Add:* Dept Biochem Dalhousie Univ Halifax NS B3H 4H7 Can

GRAY, NANCY M, b Trenton, NJ, Nov 7, 54; m 78; c 2. MEDICINAL CHEMISTRY, PHARMACOLOGY. *Educ:* Bucknell Univ, BS, 76; Univ Ill, PhD(med chem), 82. *Prof Exp:* Post doctoral chem, Purdue Univ, 81-82, vis instr med chem, 82-84; sr res chemist, Monsanto Chem Co, 84-85; RES SCIENTIST, G S SEARLE & CO, 85- *Mem:* Am Chem Soc; Sigma Xi; Am Assoc Adv Sci. *Res:* Design and synthesis of novel antiphsychotic and anti-ischemic agents with non-classical modes of action. *Mailing Add:* 1182 Sandhurst Buffalo Grove IL 60089

GRAY, PAUL, b Vienna, Austria, Dec 8, 30; US citizen; m 52; c 1. INFORMATION SYSTEMS, MANAGEMENT SCIENCE. *Educ:* NY Univ, BA, 50; Univ Mich, MA, 54; Purdue Univ, MS, 62; Stanford Univ, PhD(opers res), 68. *Prof Exp:* Asst systs eng, Willow Run Labs, Univ Mich, 51-52, asst tech ed, 52-55; nuclear engr, Convair/Gen Dynamics, 55-58; proposal engr, Solar Aircraft Co, 58-60; instr elec eng, Purdue Univ, 60-62; res engr, Systs Eng, SRI Inst, 62-64, sr res engr, 65-69, prog dir transp, 69-70; prof, Ga Inst Technol, 71-72; prof, Univ Southern Calif, 72-79; prof & chmn, mgt sci & comput, Southern Methodist Univ, 79-83; PROF & CHMN, INFO SCI, CLAREMONT GRAD SCH, 83- *Concurrent Pos:* Lectr, Stanford Univ, 68-69, consult assoc prof, 69-70, assoc prof, 70-71; vis scholar, Mass Inst Technol, 86; mem coun, Opers Res Soc Am, 88- *Honors & Awards:* Syst Sci Prize, NATO, 78. *Mem:* Inst Mgt Sci (secy, 75-79, vpres, 83-86); Opers Res Soc Am; Asn Comput Mach; Soc Info Mgt. *Res:* Decision support systems; technology assessment; modeling and simulation; urban systems; transportation; public safety; air traffic control; mixed integer programming; site location; information and decision processes; numerical analysis and digital computation. *Mailing Add:* Claremont Grad Sch Claremont CA 91711

GRAY, PAUL EDWARD, b Newark, NJ, Feb 7, 32; m 55; c 4. ELECTRICAL ENGINEERING. *Educ:* Mass Inst Technol, SB, 54, SM, 55, ScD(elec eng), 60. *Hon Degrees:* LHD, Wheaton Col, 80; DEng, Northeastern Univ, 81 & Tech Univ NS, 84; PhD, Cairo Univ, 85; ScD, Rensselaer Polytech Inst, 88. *Prof Exp:* From instr to prof elec eng, Mass Inst Technol, 57-71, assoc dean student affairs, 65-67, asst provost, 67-69, Class 1922 prof, 68-71, assoc provost, 69-70, dean, Sch Eng, 70-71, chancellor, 71-80, pres, 80-90, CHMN CORP, MASS INST TECHNOL, 90- *Concurrent Pos:* Ford Found fel, 61-63; mem, White House Sci Coun, 82-86. *Mem:* Nat Acad Eng; fel Am Acad Arts & Sci; AAAS; corresp mem, Mex Nat Acad Eng; fel Inst Elec & Electronics Engrs. *Res:* Physical electronics and circuit characterization of semiconductor devices; electric and magnetic properties of high-field superconductors; solid-state energy conversion; author of numerous technical publications. *Mailing Add:* Mass Inst Technol Rm 5-205 77 Massachusetts Ave Cambridge MA 02139

GRAY, PAUL EUGENE, b Sturgis, Ky, June 6, 38; m 63; c 2. ELECTRICAL ENGINEERING. *Educ:* Va Polytech Inst, BS, 61, MS, 64; Kans State Univ, PhD(elec eng), 69. *Prof Exp:* Instr elec eng, Va Polytech Inst, 61-65 & Kans State Univ, 65-69; assoc prof, NC A&T Univ, 69-74, prof & head elec eng, 74-76; prof & head elec eng, Portland State Univ, 76-80; PROF, DEPT ELEC ENG, UNIV WIS, 81- *Concurrent Pos:* Engr, Gen Elec Co, 65-66; adj prof, NC State Univ, 70-71; engr, Gilbarco, 71-72, Western Elec Co, 72-73. *Res:* Networks and group theory; digital firmware architecture. *Mailing Add:* Dept Elec Eng Univ Wis Platteville WI 53818

GRAY, PAUL R, INTEGRATED CIRCUITS. *Prof Exp:* ASSOC PROF, DEPT ELEC & ELECTRONICS ENG, UNIV CALIF, BERKELEY. *Mem:* Nat Acad Eng; fel Inst Elec & Electronics Engrs. *Mailing Add:* 108 Scenic Dr Orinda CA 94563

GRAY, PAULETTE S, b Chattanooga, Tenn, Feb 21, 43; m 64; c 2. CELL & DEVELOPMENTAL BIOLOGY, CYTOGENETICS. *Educ:* Tuskegee Inst, BS, 66; Atlanta Univ, MS, 76, PhD(cell & develop biol), 78. *Prof Exp:* Asst prof & dir, Electron Micros Lab, Atlanta Univ, 78-79; postdoctoral res assoc, Univ Kaiserslautern, Fed Repub Ger, 79-81; instr biol, Europ Div, Univ Md, 80-82; supvr, Clin Microbiol Sect, Landstuhl Army Med Ctr, Fed Repub Ger, 81-82; exec secy, 83-84, spec rev officer, 84-88, CHIEF, REV LOGISTICS BR, NAT CANCER INST, NIH, 88- *Concurrent Pos:* Mem, Adv Comt Women, NIH, 85-, chmn, Organizing Subcomt, 86-87; mem, Sci Educ Comt, Am Asn Cancer Res, 91-92. *Honors & Awards:* H E Finley Mem Award, 78. *Mem:* Am Asn Cancer Res; Women Cancer Res; Fed Exec Inst. *Mailing Add:* Westwood Bldg Rm 850 5333 Westbard Ave Bethesda MD 20892

GRAY, PETER NORMAN, b Trenton, NJ, Nov 16, 40; m 65; c 2. BIOMEDICAL SCIENCES, HUMAN BIOCHEMICAL GENETICS. *Educ:* Univ Del, BA & BS, 62; Northwestern Univ, MS, 65; Univ Tex, Houston, PhD(biomed), 70. *Prof Exp:* Asst prof biochem, Col Med, Col Dent & Grad Col, Univ Okla, 73-75, assoc prof molecular biol, Col Med & assoc prof biochem, Col Dent & Grad Col, Health Sci Ctr, 76-82, actg head, dept biochem & molecular biol, 78, interim head, dept biochem, 81-82; dir lif sci, Int Minerals & Chem, 82-87; dir res, 87-88, VPRES, PITMAN-MOORE, INC, 88- *Concurrent Pos:* Europ Molecular Biol Orgn longterm fel, Nat Ctr Sci Res, Marseille, France, 70-71; Damon Runyon Mem Fund Cancer Res fel, Calif Inst Technol, 71-72; USPHS fel, 73. *Honors & Awards:* O B Williams Award, Southwest Br, Am Soc Microbiol, 69; Sigma Xi Res Award, 73; Mary Jennifer Selznick Award, 78 & 80. *Mem:* AAAS; Am Soc Microbiol; Am Soc Cell Biol; Neurosci Soc. *Res:* Molecular oncology; Huntington's disease; regulation of human alkaline phosphatase induction; veterinary health. *Mailing Add:* Pitman-Moore Inc PO Box 207 Terre Haute IN 47808

GRAY, PETER VANCE, b Oak Park, Ill, July 17, 28; m 56; c 3. SOLID STATE PHYSICS, ELECTRONICS ENGINEERING. *Educ:* Union Col, BS, 58; Univ Ill, MS, 59, PhD(physics), 62. *Prof Exp:* Res assoc elec eng, Univ Ill, 62-63; PHYSICIST, GEN ELEC CO, 63- *Mem:* Inst Elec & Electronics Engrs; Am Phys Soc. *Res:* Semiconductor surface physics. *Mailing Add:* Gen Elec Res Lab Schenectady NY 12301

GRAY, RALPH DONALD, JR, b Akron, Ohio, May 15, 38; m 68; c 2. CHEMICAL ENGINEERING, THERMODYNAMICS. *Educ:* Case Inst Technol, BSChE, 60; Univ Del, MChE, 63, PhD(chem eng), 65. *Prof Exp:* Engr, 64-67, res engr, 67-71, sr res engr, 71-77, staff engr, 77-80, SR STAFF ENGR, HYDROCARBON THERMODYNAMICS, EXXON RES & ENG CO, 80- *Concurrent Pos:* Adj assoc prof, Manhattan Col, 81- *Mem:* Am Inst Chem Engrs; Am Chem Soc. *Res:* Chemical engineering thermodynamics with emphasis on hydrocarbon thermodynamics, equation of state methods and computer calculation techniques. *Mailing Add:* 20 Colonial Rd Morristown NJ 07960

GRAY, RALPH J, b Wheeling, WVa, Oct 28, 23; m 49; c 4. COAL & COKE PETROGRAPHY, GEOLOGY. *Educ:* WVa Univ, BS, 51, MS, 52. *Prof Exp:* Geol asst, WVa Geol & Econ Surv, 50-51; cartog photogram aide, Photogram Div, Army Map Serv, 51-52; geologist, Fuels Br, US Geol Surv, 52-57; assoc res consult, Appl Res Lab, US Steel Corp, 57-83; INDEPENDENT CONSULT, RALPH GRAY SERVS, 83- *Concurrent Pos:* Tech adv to Dept of Energy on Coal Characterization Evaluation Int Coal Classification, 83- *Honors & Awards:* Joseph Becker Award, Iron & Steel Soc, 84; Gilbert H Cady Award, Geol Soc Am, 88. *Mem:* Geol Soc Am; Am Asn Petrol Geologists; Mineral Soc Am; Am Asn Stratig Palynologists; Asn Inst Mining Engrs; Am Soc Testing & Mat. *Res:* Use of palynology in coal correlation; application of coal petrographic techniques to the prediction of coking properties of coal; use of microscopic techniques in coke evalutation; raw materials utilization; development of automatic microscope techniques for coal and coke analysis. *Mailing Add:* Ralph Gray Serv 303 Drexel Dr Monroeville PA 15416-1511

GRAY, RAYMOND FRANCIS, b Morrillton, Ark, Mar 3, 26; m 46; c 3. MICROBIOLOGY. *Educ:* Univ Mo, BS, 52, MA, 54, PhD(microbiol), 60. *Prof Exp:* Asst instr microbiol, Sch Med, Univ Mo, 56-60; head dept, Clin Labs, 60-69; vpres, Lab Exp Biol, Int Bio-Res, Inc, 69-72; dir, R F Gray Lab, 72-84; CONSULT, 84- *Concurrent Pos:* Consult, Lab Exp Biol, Mo, 60-, Pleasant View Tuberc Sanitorium, Ill, 65-, St Mary's Hosp, East St Louis, Ill & De Paul Hosp, St Louis, 72- *Mem:* Am Soc Microbiol; Am Pub Health Asn. *Res:* Studies on various hemolysins of Staphylococci and their inter-reactions with hemolysins from other organisms; classification of atypical Mycobacterium. *Mailing Add:* PO Box 544 La Rose LA 70373

GRAY, REED ALDEN, b Santa Clara, Utah, Jan 12, 21; m 51, 77; c 2. BIOCHEMISTRY, WEED SCIENCE. *Educ:* Univ Utah, BS, 43, MS, 44; Calif Inst Technol, PhD(biochem & plant physiol), 48. *Prof Exp:* Assoc biochemist, Pineapple Res Inst, Hawaii, 48-53; plant physiologist, Merck & Co, NJ, 53-60; plant physiologist & mgr herbicide res & biochem, 60-74, SR SCIENTIST, AGR RES CTR, STAUFFER CHEM CO, 74- *Honors & Awards:* Glycerine Res Award, 55. *Mem:* Fel AAAS; Am Chem Soc; Plant

Growth Regulator Soc Am; Am Soc Plant Physiol; Weed Sci Soc Am. *Res:* Plant biochemistry; study of plant growth regulators; antibiotics and plant diseases; metabolism and mode of action of herbicides and herbicide antidotes; behavior and persistence of herbicides in soils; herbicide soil extenders. *Mailing Add:* 19327 Portos Ct Saratoga CA 95070-5119

GRAY, RICHARD C, animal breeding, animal physiology, for more information see previous edition

GRAY, ROBERT DEE, b Evansville, Ind, May 7, 41; m; c 2. BIOCHEMISTRY. *Educ:* DePauw Univ, BA, 63; Fla State Univ, PhD(chem), 68. *Prof Exp:* Res assoc biochem, Cornell Univ, 68-70; from asst prof to assoc prof, 71-84, PROF BIOCHEM, SCH MED, UNIV LOUISVILLE, 84- *Mem:* Am Soc Biol Chemists; AAAS. *Res:* Reaction kinetics of hemoproteins; biochemistry of collagenases. *Mailing Add:* Dept Biochem Univ Louisville Sch Med Louisville KY 40292

GRAY, ROBERT H, b Beacon, NY, Feb 15, 40; m 65; c 2. ECOLOGY, FISH & WILDLIFE SCIENCE. *Educ:* Winona State Col, BA, 64; Univ Ore, MA, 67; Ill State Univ, PhD(zool, biol), 71. *Prof Exp:* Med technician gastroenterol, Mayo Clinic, 64-65; teaching asst biol, Univ Ore, 65-67 & Ill State Univ, 67-71; NSF traineeship, 68-71, asst prof, lake Erie Col, 71-73; prog mgr, environ health & safety res, 79-87, PROG MGR, OFF AANFORD ENVIRON, PAC MORTHWEST LABS, BATTLELE MEN INST, 87- *Concurrent Pos:* Lectr biol, Joint Ctr Grad Study & Wash State Univ, 78-; fel, Am Inst Fisheries Res Biologists, 83 & mem, People to People Fisheries Res Deleg, Citizen Ambassador Prog to East Asia, 85; chmn, Richland Ecol Comn. *Mem:* Am Fisheries Soc; Am Inst Fisheries Res Biologist; Ecol Soc Am; Nat Wildlife Fedn. *Res:* Environmental effects of nuclear and nonnuclear power production and hazardous materials release; environmental, health and engineering aspects of energy technology development; environmental monitoring and assesement; toxicol behavior-ethology. *Mailing Add:* PO Box 999 Mail Stop K1-33 Richland WA 99352

GRAY, ROBERT HOWARD, b Meadville, Pa, Sept 14, 37; m 61; c 3. SUB-CELLULAR TOXICOLOGY, CELL BIOLOGY. *Educ:* Ohio Univ, BSEd, 60, MS, 62; Univ Ill, PhD(bot), 67. *Prof Exp:* NIH fel, Univ Wis, 67-69; asst prof, 69-75, actg asst dean curric, 78-79, assoc prof, 75-79, RES ASSOC, INST ENVIRON & INDUST HEALTH, UNIV MICH, ANN ARBOR, 69-, PROF ENVIRON & INDUST HEALTH, 79-, RES SCIENTIST, 79- *Concurrent Pos:* Vis sr scientist, Lovelace Inhalation Toxicol Res Inst, 79-80; consult, Lovelace Inhalation Toxicol Res Inst, 80 & Warner-Lambert/Parke-Davis, 81. *Mem:* AAAS; Microbeam Anal Soc; Am Soc Cell Biol; Electron Micros Soc Am. *Res:* Biochemical and ultrastructural studies on cellular accommodation to environmental stress; sterological studies on hepatic subcellular responses to hypolipidemic compounds; development of an animal model for Reye's Syndrome. *Mailing Add:* Dept Environ & Indust Health Univ Mich Ann Arbor MI 48109-2029

GRAY, ROBERT J, b Sterling, Kans, Nov 15, 18. METALLURGY. *Educ:* Sterling Col, BS, 41. *Prof Exp:* Sr staff mem, Metallog Div, Oak Ridge Nat Lab, 48-85; CONSULT & MEM STAFF, AM SOC METALS INT, 85- *Concurrent Pos:* Mem bd dirs, Int Metallog Soc, 73-89. *Mem:* Am Soc Metals; Int Metallog Soc (pres, 77-79). *Mailing Add:* Am Soc Metals Int 137 Orchard Lane Oak Ridge TN 37830

GRAY, ROBERT MOLTEN, b San Diego, Calif, Nov 1, 43; m 73; c 2. DATA COMPRESSION, INFORMATION THEORY. *Educ:* Mass Inst Technol, BS, 66, MS, 66; Univ Southern Calif, PhD(elec eng), 69. *Prof Exp:* From asst prof to assoc prof, 69-80, PROF ELEC ENG, STANFORD UNIV, 80- *Concurrent Pos:* Assoc ed & ed-in-chief, Trans Info Theory, Inst Elec & Electronic Engrs, 77-80, ed, 80-; Guggenheim fel, 81; dir, Info Systs Lab, Stanford Univ, 84-87; assoc ed, Math Control & Systs Sci. *Honors & Awards:* Sr Award, Inst Elec & Electronic Engrs ASSP, 83, Centennial Medal, 84. *Mem:* Fel Inst Elec & Electronic Engrs; Inst Math Statist; Soc Indust & Appl Math; AAAS; Soc Sci Engrs France; Sigma Xi. *Res:* Image compression and enhancement; classification trees; information theory, the mathematical theory of communication including coding theorems for probabilistic sources and channels and applications of ergodic theory to information theory; computer aided design of data compression systems; quantization and analog-to-digital conversion theory; author of four books. *Mailing Add:* Elec Eng Dept 133 Durand Bldg Stanford Univ Stanford CA 94305-4055

GRAY, ROBIN B(RYANT), b Statesville, NC, Dec 4, 25; m 49; c 2. AEROSPACE ENGINEERING. *Educ:* Rensselaer Polytech Inst, BAE, 46; Ga Inst Technol, MS, 47; Princeton Univ, PhD(aeronaut), 57. *Prof Exp:* Res engr, Ga Inst Technol, 47-49; from asst to res assoc, Princeton Univ, 49-56; assoc prof aeronaut eng, 56-61, prof aerospace eng, 61-76, REGENT'S PROF AEROSPACE ENG, GA INST TECHNOL, 76-, ASSOC DIR, 67- *Mem:* Am Inst Aeronaut & Astronaut; Sigma Xi. *Res:* Rotary and propeller aerodynamics. *Mailing Add:* Dept Aerospace Eng Ga Inst Techno Atlanta GA 30322

GRAY, RUSSELL HOUSTON, b Akron, Ohio, May 30, 18; m 42; c 4. COLOR-SPECTROPHOTOMETRIC SCIENCE. *Educ:* Univ Akron, BSc, 40. *Prof Exp:* Chemist, Quaker Oats Co, 40-41; photo chemist, Wright Field Aerial Photo Lab, 41-42; officer, US Army Air Corps, 42-46; res assoc, Photo Prod Dept, E I Du Pont de Nemours, 46-80; RETIRED. *Concurrent Pos:* Consult, Appl Color Systs, 80-81, Mead Paper Co, 83- & various other indust cos, 80-89. *Honors & Awards:* Serv Award, Soc Photog Scientists & Engrs, 65. *Mem:* Sr mem Soc Photog Scientists & Engrs; Sigma Xi; Inter-Soc Color Coun; Tech Asn Graphic Arts; Am Chem Soc. *Res:* Colorants and color reproduction, particularly for graphic arts applications; color proofing systems; photographic and color sciences research and development; color imaging. *Mailing Add:* Nine Circle Dr Rumson NJ 07760

GRAY, SAMUEL HUTCHISON, b Avalon, Pa, Oct 22, 48. MATHEMATICS. *Educ:* Ga State Univ, BS, 70; Univ Del, PhD(math), 78. *Prof Exp:* RESEARCHER, FUEL & PETROL, AMOCO PROD CO, 82- *Mailing Add:* Amoco Prod Co PO Box 591 Tulsa OK 74102

GRAY, SARAH DELCENIA, b Newark, NJ, Aug 1, 34. CARDIOVASCULAR PHYSIOLOGY. *Educ:* Barnard Col, BA, 56; NY Univ, MS, 60; Univ Calif, San Francisco, PhD(physiol), 66. *Prof Exp:* Res assoc physiol, Univ Calif Med Ctr, San Francisco, 66; assoc, Mt Sinai Sch Med, 68-71, asst prof, 71-72; from asst prof to assoc prof, 72-83, PROF PHYSIOL, SCH, MED, UNIV CALIF, DAVIS, 83-, LECTR, LAW SCH, 77- *Concurrent Pos:* NIH fel, 66-67, consult, minority biomed support prog, 76-, ischemic heart dis, Spec Ctr Res, 78-79 & study sect, 82-86; sr investr, NY Heart Asn, 69-72; assoc ed, Microvascular Res, 77-84 & Blood Vessels, 85- *Honors & Awards:* Nat Res Coun travel award, Int Union Physiol Sci, India, 74. *Mem:* Sigma Xi; Microcirculatory Soc; Europ Soc Microcirculation; Am Physiol Soc; Am Heart Asn; Int Soc Hypertension. *Res:* Mechanisms governing microvascular behavior, especially reactivity to neural and hormonal stimuli which influence patterns of pressure and flow in skeletal muscle during neonatal development, exercise, and the development of genetic hypertension. *Mailing Add:* Dept Human Physiol Univ Calif Sch Med Davis CA 95616

GRAY, STEPHEN WOOD, b Oakland, Calif, Apr 27, 15; m 38, 73. ANATOMY, EMBRYOLOGY. *Educ:* Lake Forest Col, AB, 36; Univ Ill, AM, 37, PhD(zool), 39. *Hon Degrees:* DSc, Lake Forest Col, 90. *Prof Exp:* Instr physiol, Univ Ill, 39-42; prof anat, Sch Med, Emory Univ, 46-83; ASSOC DIR, THALIA & MICHAEL CARLOS CTR SURG ANAT & TECHNIQUE, 83- *Concurrent Pos:* Mem working group comt space res, Int Coun Sci Unions. *Mem:* Am Soc Zool; Am Physiol Soc; Am Asn Anat; Am Asn Clin Anat. *Res:* Biological effects of gravity; dynamics of growth; embryological defects; surgical anatomy. *Mailing Add:* Emory Univ Sch Med 1462 Clifton Rd NE, Ste 303 Atlanta GA 30322

GRAY, THEODORE FLINT, JR, b Anniston, Ala, Feb 25, 39; m 60; c 3. POLYMER CHEMISTRY. *Educ:* Centre Col, BA, 60; Univ Fla, PhD(org & polymer chem), 64. *Prof Exp:* Res chemist, Eastman Kodak Co, 65-66, sr res chemist, 66-72, res assoc, 72-79, sr res assoc, 79-84, staff asst to vpres res & develop, 84-85, mgr tech serv & develop animal nutrit suppl, 85-86, DIR, POLYMER RES DIV, EASTMAN CHEM DIV, EASTMAN KODAK CO, 86- *Mem:* Am Chem Soc; Soc Plastics Engrs; AAAS. *Res:* Kinetics and stereochemistry of intra-intermolecular polymerization; polymer structure in relation to glass transitions, melt rheology and mechanical properties; structure influence on polymer stability; relation of polymer and reinforcement variables to composite's properties. *Mailing Add:* 2153 Westwind Dr Kingsport TN 37660

GRAY, THOMAS JAMES, b Atherstone, Eng, July 28, 17; nat US; m 41; c 3. PHYSICAL CHEMISTRY, CERAMIC ENGINEERING. *Educ:* Bristol Univ, BSc & PhD(chem), 38. *Prof Exp:* Lectr chem, Bristol Univ, 46-53; from assoc prof to prof phys chem, State Univ NY Col Ceramics, Alfred Univ, 53-68, adminr, Off Res, 64-68; dir, Atlantic Indust Res Inst, NS Tech Col, 68-76; chief phys chemist, Metals Res Lab, Olin Corp, 76-80; CONSULT, 81- *Concurrent Pos:* Consult scientist, 48- *Mem:* Am Chem Soc; fel Am Ceramic Soc; fel Am Soc Metals. *Res:* Catalysis; semiconductors; dielectric and magnetic properties of materials; high temperature materials; fuel cells; strontium containing ceramics; tidal power; unconventional energy systems. *Mailing Add:* 971 Hoop Pole Rd Guilford CT 06437

GRAY, THOMAS MERRILL, b Washington, DC, Oct 9, 29; m 59; c 3. ENTOMOLOGY, ECOLOGY. *Educ:* Kans State Col, BS, 56; Kans State Univ, MS, 57; Univ Okla, PhD(zool), 72. *Prof Exp:* Prof, 58-87, EMER PROF BIOL SCI, SOUTHWESTERN OKLA STATE UNIV, 87- *Mem:* Entom Soc Am; Am Inst Biol Sci. *Res:* Biology and taxonomy of tachinid flies. *Mailing Add:* Dept Biol Sci Southwestern Okla State Univ 100 Campus Dr Weatherford OK 73096

GRAY, TIMOTHY KENNEY, b Baltimore, Md, Oct 4, 39; m 63; c 3. INTERNAL MEDICINE, ENDOCRINOLOGY. *Educ:* Loyola Col, BS, 61; Univ Md, MD, 65. *Prof Exp:* Resident, Sch Med, Univ Md, 65-68, chief resident & instr, 70-71; from asst prof to assoc prof med, 71-79, dir, Clin Res Ctr, 76-79, vchmn, Dept Med, 77-80, PROF MED, SCH MED, UNIV NC, 79- *Concurrent Pos:* Spec USPHS res fel, Dept Pharmacol, 68-70. *Honors & Awards:* fel, Jefferson Pilot award, Sch Med, Univ NC, 73; investr NIAMDD res grant award, 74-80, NSF Clin res award, 75-79, & NSF basic res award, 78-80. *Mem:* Am Fedn Clin Res; Endocrine Soc; AAAS; fel Am Col Physicians. *Res:* Mineral metabolism and its hormonal regulation; mineral metabolism of pregnancy and fetal development; vitamin D metabolism and fetal skeletal development. *Mailing Add:* NC Meml Hosp Box 501 Chapel Hill NC 27514

GRAY, TOM J, b Newton, Kans, Dec 2, 37; div; c 2. PHYSICS. *Educ:* NTex State Univ, BS, 60, MS, 62; Fla State Univ, PhD(physics), 67. *Prof Exp:* From asst prof to prof, NTex State Univ, 67-75; vis prof, 75-77, PROF PHYSICS, KANS STATE UNIV, 77- *Concurrent Pos:* Consult, Tex Instruments Inc, 73-75. *Mem:* Fel Am Phys Soc. *Res:* Cryogenics, superconducting linear accelerator design and fabrication; experimental ion/atom and ion molecule collisions studies; molecular ion production and cluster ion production studies. *Mailing Add:* Dept Physics Cardwell Hall Kans State Univ Manhattan KS 66506-2601

GRAY, TRUMAN S(TRETCHER), b Spencer, Ind, May 3, 06; m 31. ELECTRICAL ENGINEERING, ELECTRONICS. *Educ:* Univ Tex, BS, 26, BA, 27; Mass Inst Technol, MS, 29, ScD(elec eng), 30. *Prof Exp:* Asst physics, Univ Tex, 24-27; asst elec eng, 27-28, from instr to prof, 30-71, EMER PROF ELEC ENG, MASS INST TECHNOL, 71- *Concurrent Pos:* Consult to govt labs & indust. *Honors & Awards:* Centennial Medal, Inst Elec & Electronics Engrs, 84. *Mem:* Am Soc Eng Educ; fel Inst Elec & Electronic Engrs; Sigma Xi. *Res:* Photoelectric integraph; electronic devices, circuits and applications; nuclear reactor instrumentation and control; photocomposing machines. *Mailing Add:* 1010 Waltham St No A411 Lexington MA 02173-8002

GRAY, WALTER C(LARKE), b Roanoke, Va, Oct 4, 19; m 54; c 6. CHEMICAL ENGINEERING. *Educ:* Va Polytech Inst, BS, 41, MS, 49, PhD(chem eng), 52. *Prof Exp:* Supvr tech dept, Radford Arsenal, Hercules Powder Co, 51-52; res engr, Carothers Res Lab, Exp Sta, E I du Pont de Nemours & Co, Del, 52-54, res engr, Dacron Res Lab, 54-62, sr res engr res & develop, Kinston, NC, 62-85; RETIRED. *Mem:* Nat Soc Prof Engrs; Sigma Xi; Am Inst Chem Engrs. *Res:* Ultrasonics; polymers such as polyamides, polycaproamides and polyethylene terephthalate by continuous polymerization; propellants for guided missiles; polymer melt spinning and quenching processes. *Mailing Add:* 202 Greenbriar Dr Greenville NC 27834-6336

GRAY, WALTER STEVEN, b Ponca City, Okla, Feb 21, 38. NUCLEAR PHYSICS. *Educ:* Okla State Univ, BS, 58, MS, 60; Univ Colo, PhD(physics), 64. *Prof Exp:* Res assoc nuclear physics, 64-66, asst prof physics, 66-74, ASSOC PROF PHYSICS, UNIV MICH, 74- *Mem:* Am Phys Soc. *Res:* Nuclear spectroscopy and accelerator-induced nuclear reactions. *Mailing Add:* Dept Physics Univ Mich Ann Arbor MI 48109

GRAY, WILLIAM DAVID, b Gilford, Northern Ireland, May 12, 16; nat US; m 44; c 2. PHARMACOLOGY. *Educ:* Clark Univ, BA, 38; Univ Toronto, MA, 40; Yale Univ, PhD(pharm), 50. *Prof Exp:* Res assoc, Ciba Pharmaceut Prods, Inc, NJ, 40-52; res assoc, Merck Inst Therapeut Res, 46-47; pharmacologist, Lederle Labs Div, Am Cyanamid Co, 50-54, group leader pharmacol, 54-56, head dept exp pharmacol, 56-71, dir cent nerv syst dis ther res, 71-73; mgr toxicol, Schering Corp, 73-75, assoc dir, 75-77, assoc dir lab compliance, 77-80; dir res, Biosphere Res Ctr, Inc, 80-81, consult, 81-82; RETIRED. *Mem:* Am Soc Pharmacol & Exp Therapeut; Soc Neurosci; Soc Toxicol; Sigma Xi. *Res:* Neuropharmacology; general pharmacodynamic and behavioral actions of CNS agents; biochemical pharmacology of CNS agents; biogenic amines and anticonvulsant action of carbonic anhydrase inhibitors. *Mailing Add:* 31 Diane Dr New City NY 10956

GRAY, WILLIAM DUDLEY, mycology; deceased, see previous edition for last biography

GRAY, WILLIAM GUERIN, b San Francisco, Calif, Jan 9, 48; m 70; c 5. CIVIL ENGINEERING, HYDROLOGY. *Educ:* Univ Calif, Davis, BS, 69; Princeton Univ, PhD(chem eng), 74. *Prof Exp:* Res hydrol, US Geol Surv, 73-74; lectr, 74-75, from asst prof to assoc prof civil eng, Princeton Univ, 75-84; PROF & CHMN, CIVIL ENG, UNIV NOTRE DAME, 84- *Concurrent Pos:* Hydrologist, US Geol Surv, 74-75. *Mem:* Inst Assoc Hydraul Res; Sigma Xi; Am Inst Chem Engrs; Am Geophys Union. *Res:* Numerical solution of differential equations; finite element method; surface flow modeling; physics of porous media flow. *Mailing Add:* Civil Eng Dept Univ Notre Dame Notre Dame IN 46556-0767

GRAY, WILLIAM HARVEY, b Nashville, Tenn, Dec 15, 48; m 70; c 2. SUPERCONDUCTING MAGNETS, COMPUTER GRAPHICS. *Educ:* Vanderbilt Univ, BE, 70, MS, 73, PhD(mech eng), 75. *Prof Exp:* Grad res asst, Los Alamos Nat Lab, 73-74; DEVELOP ENGR, OAK RIDGE NAT LAB, 74- *Mem:* Inst Elec & Electronics Engrs. *Res:* Structural analysis problems associated with toroidal and polidal superconducting and normal magnets for fusion and physics research. *Mailing Add:* Martin Marietta Energy Systs PO Box 2009 Mail Stop 8160 Bldg 9103 Oak Ridge TN 37831

GRAY, WILLIAM MASON, b Detroit, Mich, Oct 9, 29; m 54; c 3. METEOROLOGY. *Educ:* George Washington Univ, BA, 52; Univ Chicago, MS, 59, PhD(geophys sci), 64. *Prof Exp:* Res asst meteorol, Univ Chicago, 57-61; asst meteorologist, 61-64, from asst prof to assoc prof, 64-74, PROF ATMOSPHERIC SCI, COLO STATE UNIV, 74- *Concurrent Pos:* NSF res grant, Japan, 65-66, England 70-71. *Mem:* Am Meteorol Soc. *Res:* Atmospheric science; tropical meteorology and hurricanes; cumulus convection; meteorological observations; tornadoes. *Mailing Add:* Dept Atmospheric Sci Colo State Univ Ft Collins CO 80523

GRAYBEAL, JACK DANIEL, b Detroit, Mich, May 16, 30; m 54; c 3. MICROWAVE SPECTROSCOPY. *Educ:* WVa Univ, BS, 51; Univ Wis, MS, 53, PhD(phys chem), 55. *Prof Exp:* Mem tech staff, Bell Tel Labs, 55-57; from asst prof to assoc prof chem, WVa Univ, 57-68; assoc prof, 68-69, asst head dept, 74-75, PROF CHEM, VA POLYTECH INST & STATE UNIV, 69-, ASSOC HEAD DEPT, 75- *Mem:* Am Chem Soc; Am Phys Soc; Sigma Xi. *Res:* Microwave spectroscopy; nuclear quadrupole resonance spectroscopy; molecular structure; dielectric behavior. *Mailing Add:* Dept Chem Va Polytech Inst & State Univ Blacksburg VA 24061-0212

GRAYBEAL, WALTER THOMAS, mathematics; deceased, see previous edition for last biography

GRAYBIEL, ANN M, NEUROANATOMY. *Prof Exp:* res assoc, 71-73, from asst prof to assoc prof psychol, 73-80, PROF NEUROANAT, DEPT BRAIN & COGNITIVE SCI, MASS INST TECHNOL, 83- *Concurrent Pos:* Javits neurosci investr, 88. *Honors & Awards:* Williams & Wilkins Award, Am Asn Anatomists, 70, Charles Judson Herrick Award, 78; McKnight Award, 85. *Mem:* Nat Acad Sci; assoc Neurosci Res Prog; hon mem Royal Acad Med. *Res:* Co-author of numerous publications. *Mailing Add:* Dept Brain & Cognitive Sci Mass Inst Technol 45 Carlton St Cambridge MA 02139

GRAYBIEL, ASHTON, b Port Huron, Mich, July 24, 02; wid; c 2. MEDICINE. *Educ:* Univ Southern Calif, AB, 24, AM, 25; Harvard Med Sch, MD, 30. *Prof Exp:* Instr cardiol, Mass Gen Hosp, 34-42; dir res, 45-70, spec asst progs & head, Dept Biol Sci, 70-80, RESEARCHER, US NAVAL SCH AVIATION MED, NAVAL AEROSPACE MED INST, 42-, CHIEF, SCI ADV NAVAL AEROSPACE MED RES LAB, 81- *Concurrent Pos:* Moseley traveling fel, Univ Col Hosp, London, 32-33; Dalton fel, Mass Gen Hosp, 33-34; instr, Harvard Med Sch, 40-42; res assoc, Fatigue Lab, Harvard Univ, 36-42; at Nat Res Coun, 44; lectr, Med Col, Univ Ala, 58-68; adj prof psychol, Brandeis Univ, 81- *Mem:* AMA; Am Heart Asn; Aerospace Med Asn (pres, 57); AAAS; Int Acad Astronaut. *Res:* Electrocardiography in practice; clinical electrocardiography; effect of acceleration on semicircular canals and otolith organs. *Mailing Add:* Naval Aerospace Med Res Lab Naval Air Sta Pensacola FL 32508

GRAYBILL, BRUCE MYRON, b Council Bluffs, Iowa, Oct 2, 31; m 52; c 3. PHYSICAL ORGANIC CHEMISTRY. *Educ:* Iowa State Univ, BS, 55; Fla State Univ, PhD(chem), 59. *Prof Exp:* Res chemist, Rohm and Haas Co, 59-61; from asst prof to assoc prof chem, 61-70, chmn div sci & math, 66-74, actg dean fac, 74-75, prof chem, 70-77, DISTINGUISHED PROF CHEM, GRACELAND COL, 77-, CHMN, DIV SCI & MATH, 81- *Concurrent Pos:* AEC comn fac fel, PR Nuclear Ctr, 70. *Mem:* Am Chem Soc; Am Inst Chem; Sigma Xi. *Res:* Organic reaction mechanisms. *Mailing Add:* Dept Chem Graceland Col Lamoni IA 50140

GRAYBILL, DONALD LEE, b Harrisburg, Pa, Feb 25, 43; m 65; c 3. RESOURCE MANAGEMENT, INSTITUTIONAL STRENGTHENING. *Educ:* Bucknell Univ, BA, 65; Univ Pittsburgh, ME, 66, PhD(ecol), 70. *Prof Exp:* Asst prof biol, Alliance Col, Cambridge Springs, Pa, 70-73; res assoc ecol, Univ Mass, 73-74; ecologist, sr mgr consult & supvr, Environ Sci Sect, Gilbert/Commonwealth, 74-85; sr mgt consult & mgr int opers, MGT Resources Int, Inc, 85-88; VPRES, INT AFFAIRS, 88- *Concurrent Pos:* Res assoc, dept biol, Univ Pittsburgh, 71-73; fel, E-W Ctr Environ & Policy Inst, Honolulu, Hawaii, 85-86; US comt large dams, Environ Effects Subcomt, 86- *Mem:* Ecol Soc Am; Nat Asn Environ Professionals; Int Asn Impact Assessment. *Res:* Ecology. *Mailing Add:* PO Box 201 Mgt Resources Int Inc Reading PA 19603-0201

GRAYBILL, FRANKLIN A, b Carson, Iowa, Sept 23, 21; m 47; c 2. MATHEMATICAL STATISTICS. *Educ:* William Penn Col, BS, 47; Okla State Univ, MS, 49; Iowa State Univ, PhD(math statist), 52. *Prof Exp:* Prof math & sta statistician, Agr Exp Sta, Okla State Univ, 52-60; dir statist lab, 60-88, PROF MATH STATIST, COLO STATE UNIV, 88- *Concurrent Pos:* Consult, Standard Oil Co, NJ, 54-; mem adv panel, Am Inst Biol Sci, NASA, 71-74; ed, Biometrics, J Biometric Soc, 71-75. *Mem:* Fel Am Statist Asn (pres, 76-); fel Inst Math Statist; fel AAAS; Int Statist Inst; Biomet Soc. *Res:* Variance component analysis. *Mailing Add:* Dept Statist Colo State Univ Ft Collins CO 80523

GRAYBILL, HOWARD W, b Bearville, Pa, Aug 2, 15. HIGH-POWER ELECTRICAL EQUIPMENT. *Educ:* Drexel Univ, BS, 37. *Prof Exp:* Design eng, Westinghouse Elec Co, 37-45, New Holland Machine Co, 47-49; various positions, ITE Circuit Breaker Co, 50-68, Mgr, gas Insulated Prods Eng, 68-78; RETIRED 80. *Honors & Awards:* Fel Inst Elec & Electronic Engrs. *Res:* Design & marketing high-voltage & high-power electrical equipment; design of high-voltage, gas-insulated substations. *Mailing Add:* 5158 Don Mata Dr Carlsbad CA 92008

GRAYDON, WILLIAM FREDERICK, b Toronto, Ont, June 27, 19; m 45; c 5. PHYSICAL CHEMISTRY. *Educ:* Univ Toronto, BASc, 42, MASc, 45; Univ Minn, PhD(chem), 49. *Prof Exp:* Asst prof, 49-59, assoc dean fac appl sci, 66-77, chmn dept chem eng & appl chem, 70-77, PROF CHEM ENG, UNIV TORONTO, 59- *Mem:* Am Chem Soc; Royal Soc Chem; Chem Inst Can. *Res:* Ion exchange; corrosion; process dynamics; applied physical chemistry. *Mailing Add:* Dept Chem Eng Univ Toronto 35 St George St Toronto ON M5S 1A4 Can

GRAYHACK, JOHN THOMAS, b Kankakee, Ill, Aug 21, 23; m 50; c 5. UROLOGY. *Educ:* Univ Chicago, BS, 45, MD, 47. *Prof Exp:* Instr urol, Sch Med, Johns Hopkins Univ, 52-54; from asst prof to assoc prof, 56-63, PROF UROL & CHMN DEPT, MED SCH, NORTHWESTERN UNIV, CHICAGO, 63- *Concurrent Pos:* Am Cancer Soc fel, 50-51; Runyon fel, 52-54. *Mem:* AAAS; AMA; Am Asn Genito-Urinary Surg; Am Surg Asn; Endocrine Soc. *Res:* Factors in normal and abnormal prostatic growth; evaluation and treatment of diseases of genito-urinary tract. *Mailing Add:* Dept Urol Northwestern Mem Hosp 303 E Chicago Ave Chicago IL 60611

GRAYSON, HERBERT G, b New York, NY, Jan 5, 26; m 47; c 3. CHEMICAL ENGINEERING. *Educ:* City Col New York, BS, 46; Polytech Inst Brooklyn, MS, 52. *Prof Exp:* Sr develop engr, Eng Dept, Socony Mobil Oil Co. Inc, 47-64, prod planner, Prod Dept, 64-74, environ assoc, 74-79, MGR INDUST/ENERGY PROG, MOBIL OIL CORP, 79- *Mem:* Am Inst Chem Engrs; Sigma Xi; Air Pollution Control Asn; Am Petrol Inst. *Res:* Properties of petroleum; vapor-liquid equilibria and enthalpy; unit operations; fluid flow; heat transfer; tray efficiency; properties of granular solids; air pollution; energy and technological forecasting; automobile emissions reduction. *Mailing Add:* 325 Laurel Ln Laurel Hollow NY 11791

GRAYSON, JOHN, b Huddersfield, Eng, Jan 4, 19; m 61. PHYSIOLOGY. *Educ:* Victoria Univ Manchester, BSc, 40, MSc, 41, MB & ChB, 43, MD, 49, DSc(physiol), 66. *Prof Exp:* Res asst surg, Royal Victoria Infirmary, Eng, 47-49; lectr physiol, Univ Bristol, 49-55; prof & chmn dept, Univ Ibadan, 55-67; vis prof, Univ Alta, 67-68; PROF PHYSIOL, UNIV TORONTO, 68- *Mem:* Can Physiol Soc; Brit Physiol Soc; Europ Soc Microcirc; Can Soc Microcirc (vpres, 71). *Res:* Human skin blood vessels, environmental change, cold and frostbite; gastrointestinal and liver blood vessels in general and glucose homeostasis; autoregulation and brain blood flow; heart blood vessels, nutritional role of anastomotic network, effect of acute occlusion and atheroma. *Mailing Add:* Dept Physiol & Med Sci Univ Toronto Toronto ON M5S 2R8 Can

GRAYSON, JOHN FRANCIS, b Bay City, Mich, Mar 23, 28; m 51; c 4. PALYNOLOGY, GEOCHEMISTRY. *Educ:* Univ Mich, BS, 51, MS, 52, PhD(bot), 56. *Prof Exp:* Sr res palynologist, Res Labs, Socony Mobil Oil Co, Tex, 54-62; res group supvr, Res Ctr, Pan Am Petrol Corp, 61-68, res assoc, 68-71, res assoc, Res Ctr, Amoco Prod Co, 71-83; CONSULT, 83- *Honors & Awards:* John F Grayson Palynology Libr, Nat Museums Can, Ottawa, Can. *Mem:* AAAS; Am Asn Petrol Geol; Am Geol Inst; Am Asn Stratig

Palynologists; Geol Soc Am. *Res:* Plant geography and ecology; stratigraphic geology; paleontology; paleobotany; microspectrofluorescence; organic matter present in rocks as it pertains to petroleum exploration. *Mailing Add:* 4149 S Sandusky Tulsa OK 74135

GRAYSON, LAWRENCE P(ETER), b Brooklyn, NY, May 16, 37; m 64; c 6. ELECTRICAL ENGINEERING. *Educ:* Polytech Inst Brooklyn, BEE, 58, MEE, 59, PhD(elec eng), 62. *Hon Degrees:* DEng, Milwaukee Sch Eng, 88. *Prof Exp:* Asst prof elec eng, Johns Hopkins Univ, 62-67; Ford Found researcher eng practice, Thomas J Watson Res Ctr, IBM Corp, 67-68; assoc prof elec eng, Manhattan Col, 68-69; specialist comput in educ, Bur Res, US Off Educ, 69-70, dep dir div educ technol, Bur Libr & Educ Technol, 70-72, dir div technol, 72-73; mem, task force productivity & technol, Nat Inst Educ, 73-75, dir div technol develop, 75-78, head info technol, media & pub commun, 78-81, asst dir regional progs, 81-82, inst advan math, 82-86; Inst Elec & Electronic Engrs Cong fel, Off Congressman Jack Kemp, 86-87; vis prof, Elec & Eng, Cath Univ Am, 88-89; dep dir, 87-88, DIR POSTSEC RELS STAFF, US DEPT EDUC, 90- *Concurrent Pos:* Consult, Math Br, Frankford Arsenal, 62-66, Johns Hopkins Hosp, 67, UNESCO, 73-82, Gen Elec Co & Lawrence Livermore Nat Labs; mem educ systs comt, Nat Acad Eng Comn on Educ, 68-73; mem, Nat Comt Full Develop Instructional TV Fixed Serv, 69-73; mem educ comt, World Fedn Eng Orgn, 76-81; adj prof libr & info sci, Catholic Univ Am, 79-87, mem vis bd, 80- *Honors & Awards:* William Elgin Wickenden Award, Am Soc Eng Educ, 79; Centennial Medal, Inst Elec & Electronic Engrs, 84, Roland J Schmidt Award, 85; Distinguished Serv Citation, Am Soc Eng Educ, 81; Herbert Hoover Medal, 87; George Washington Medal, Freedom Found, 89. *Mem:* Fel Inst Elec & Electronic Engrs; fel Am Soc Eng Educ (vpres, 72-74, pres, 88-); fel AAAS; Sigma Xi; Am Asn Eng Soc; Nat Coalition Eng Soc. *Res:* Nonlinear oscillations; nonlinear and adaptive control systems; engineering education; educational technology; computers, satellites; societal implications of technology. *Mailing Add:* Off Postsecondary Educ US Dept Educ Washington DC 20202

GRAYSON, MARTIN, b New York, NY; m 70; c 3. CHEMISTRY. *Educ:* NY Univ, BS; Purdue Univ, PhD(chem). *Prof Exp:* Chemist, Nitrogen Div, Allied Corp, 52-56; sr res chemist, Am Cyanamid Co, 56-60, group leader, 60-71, prin res scientist, 72-75; publ, John Wiley & Sons, 75-76, PRES, VCH PUBLS, 87- *Concurrent Pos:* Adj prof, Univ Bridgeport, 65-72; ed, Topics Phosphorus Chem, 60-; ed-in-chief, Phosphorus & Sulfur, 70-87, ed emer, 87- *Mem:* Fel NY Acad Sci; fel AAAS; Am Chem Soc. *Res:* Organophosphorus chemistry; chemical technology. *Mailing Add:* 82 Valleywood Rd Cos Cob CT 06807

GRAYSON, MERRILL, b New York, NY, Apr 19, 19. OPHTHALMOLOGY, CORNEAL SURGERY. *Educ:* NY Univ, BA, 38, MD, 41; Am Bd Ophthal, dipl, 55. *Prof Exp:* Asst prof ophthal, Med Ctr, Ind Univ, 57-60; assoc prof & chmn dept, Sch Med, Univ Ark, 60-61; assoc prof, 61-68, prof ophthal, Med Ctr, Ind Univ Sch Med, 68-88; surg dir Lions Eye Bank of Ind, 61-88; DISTINGUISHED PROF IND UNIV SCH MED, 81-, ACTG CHMN DEPT OPHTHAL, 84- *Mem:* Am Acad Ophthal & Otolaryngol; AMA; Sigma Xi. *Res:* Diseases and surgery of the cornea. *Mailing Add:* Dept Ophthal RO Second Floor Ind Univ Sch Med 702 Rotary Circle Indianapolis IN 46223

GRAYSON, MICHAEL A, b Wichita, Kans, July 18, 41; m 64; c 4. ENVIRONMENTAL AGING OF POLYMERIC MATERIALS, MASS SPECTROMETRY INSTRUMENTS. *Educ:* St Louis Univ, BA, 63; Univ Mo, Rolla, MA, 65. *Prof Exp:* Res scientist, 66-70, scientist, 70-85, SR SCIENTIST, MCDONNELL DOUGLAS RES LABS, 85- *Concurrent Pos:* Lectr, Wash Univ Sch Continuing Educ, 69-71. *Mem:* Am Soc Mass Spectrometry; NY Acad Sci; Sigma Xi; Am Chem Soc; AAAS. *Res:* Development and application of mass spectrometric techniques to the characterization and study of environmental aging in polymeric and composite material systems. *Mailing Add:* 3433 Bluff View Dr St Charles MO 63303

GRAYSON, WILLIAM CURTIS, JR, b Decatur, Miss, Nov 17, 29; m 51, 76; c 5. NUCLEAR WEAPONS PHYSICS. *Educ:* Univ Chicago, SB, 50; Duke Univ, PhD(physics), 55. *Prof Exp:* Physicist, Lawrence Livermore Nat Lab, Univ Calif, 54-56, group leader, 66-68, div, leader, 68-71, dept div leader, 71-76, dept assoc dir, 76-78, tech adv, 78-81; sr scientist, R & D Assoc, 81-90; SR SCIENTIST, MERIDIAN CORP, 90- *Mem:* Am Phys Soc; Am Nuclear Soc. *Res:* Nuclear physics; physics of nuclear explosives; computational physics; technical management; arms control. *Mailing Add:* 5644 Bent Branch Rd Bethesda MD 20186

GRAYSTON, J THOMAS, b Wichita, Kans, Sept 6, 24; m 47, 80; c 3. EPIDEMIOLOGY. *Educ:* Univ Chicago, BS, 47, MD, 48, MS, 52; Am Bd Internal Med, dipl, 57; Am Bd Prev Med, dipl, 68. *Prof Exp:* Epidemiologist, Commun Dis Ctr, USPHS, 51-53; from instr to assoc prof med, Univ Chicago, 53-60; prof prev med & chmn dept, Sch Med, Univ Wash, 60-70, founding dean, Sch Pub Health & Community Med, 70-71, vpres health sci, 71-83, PROF EPIDEMIOL, UNIV WASH, 70- *Concurrent Pos:* Chief div microbiol & epidemiol, Naval Med Res Unit, Taiwan, 57-60; assoc mem, Comn Acute Respiratory Dis, 61-65, mem, 65-74; mem panel on biol & med, Dept Defense, 62-66, virol & rickettsiology study sect, NIH, 63-67, int ctrs comt, 67-71, mem nat adv coun health professions educ, 72-75; mem expert comt trachoma, WHO, 71- *Honors & Awards:* Joseph E Smadel Award & Medal, Infectious Dis Soc Am, 89. *Mem:* Inst Med-Nat Acad Sci; fel Am Pub Health Asn; Am Epidemiol Soc; Am Asn Physicians; fel Infectious Dis Soc Am. *Res:* Epidemiology and prevention of infectious diseases; international health. *Mailing Add:* SC-36 Univ Wash Seattle WA 98195

GRAYZEL, ARTHUR I, b New York, NY, Mar 8, 32; m 57; c 3. MEDICINE. *Educ:* Harvard Col, AB, 53; Harvard Med Sch, MD, 57. *Prof Exp:* PROF MED, ALBERT EINSTEIN COL MED, 77- *Concurrent Pos:* Head, Div Rheumatol, Montefiore Med Ctr, 69-89. *Mem:* Am Asn Immunologists; fel Am Rheumatism Asn; fel Am Col Physicians; Brit Soc Rheumatologists. *Res:* Pathogenesis and immunologic basic and treatment of rheumatic diseases including rheumatoid arthritis and systemic lupus erythematosus. *Mailing Add:* Arthritis Found 1314 Spring St NW Atlanta GA 30309

GRAZIANO, FRANK M, MEDICINE, ALLERGY. *Educ:* Univ Va, MD & PhD(microbiol), 73. *Prof Exp:* ASSOC PROF MED & IMMUNOL, DEPT MED, UNIV WIS HOSP, 78- *Res:* Rheumatology. *Mailing Add:* 3309 Heatherdell Lane Madison WI 53713

GRDINA, DAVID JOHN, b Hammond, Ind, Oct 26, 44; m 67; c 4. RADIATION BIOLOGY, TUMOR BIOLOGY. *Educ:* St Mary's Col, Winona, Minn, BA, 66; Univ Kans, Lawrence, MS, 69, PhD(radiation biophys), 71; Univ Houston, Tex, MBA, 80. *Prof Exp:* Res fel DNA repair, M D Anderson Hosp & Tumor Inst, Houston, Tex, 71-72, res assoc tumor radiobiol, 72-75, from asst prof to assoc prof exp radiother, 75-83; scientist, 83-87, SR SCIENTIST RADIATION CARCINOGENESIS, ARGONNE NAT LAB, 87-; PROF RADIATION ONCOL, UNIV CHICAGO, ILL, 87- *Concurrent Pos:* Vis lectr nuclear med, Baylor Col Med, Houston, Tex, 76-83; vis assoc prof therapeut radiol, Rush Med Col, Rush-Presbyterian-St Lukes Med Ctr, Chicago, Ill, 84-85; assoc prof radiation oncol, Univ Chicago, 85-87; mem radiation study sect, NIH, 90-94; mem Ill Div study sect, Am Cancer Soc, 91-95. *Honors & Awards:* Japanese Govt Res Awards, Foreign Specialists, 88. *Mem:* Am Asn Cancer Res; Radiation Res Soc; Cell Kinetics Soc. *Res:* Chemical radiation modifying agents to study mechanisms involved in radiation carcinogenesis; tumor heterogeneity and the associated factors involved in the treatment and control of tumors in experimental systems. *Mailing Add:* Div Biol & Med Res Argonne Nat Lab 9700 S Cass Ave Argonne IL 60439

GRDIS, ENOCH, b Feb 21, 31; c 2. ALCOHOLISM. *Educ:* Columbia Univ, BA, 50, MD, 54. *Prof Exp:* Clin fel, Mt Sinai Hosp, 59-60, chief resident, 60-65; assoc prof, Rockefeller Univ, 65-71; assoc prof, 71-79, PROF CLIN MED, MT SINAI SCH MED, DEPT MED, 79-; DIR, NAT INST ALCOHOL ABUSE & ALCOHOLISM, 86- *Concurrent Pos:* vis physician, Rockefeller Hosp & adj prof, Rockefeller Univ, 71-; Grand Rounds lectr, Elmhurst Hosp, Manhattan VA Hosp, Rockefeller Univ, Syracuse Univ, Univ Wash, St Luke's Hosp, Cornell Univ Sch Med; dir, Alcoholism Treat Prog, Elmhurst Hosp, NY, 71-86. *Mem:* Inst Med-Nat Acad Sci; Am Fedn Clin Res; Am Col Neuropsychopharmacol; Am Physiol Soc; fel Am Col Physicians; Am Gastroenterol Asn; Sigma Xi. *Mailing Add:* Nat Inst Alcohol Abuse & Alcoholism 5600 Fishers Lane Rm 16-105 Rockville MD 20857

GREAGER, OSWALD HERMAN, b Hyattsville, Md, Aug 23, 05; m 26; c 1. PHYSICAL CHEMISTRY. *Educ:* Univ Md, BS, 25; Univ Mich, MS, 27, PhD(chem), 29. *Prof Exp:* Chemist, EI du Pont de Nemours & Co, 29-42; tech officer, Manhattan Eng Dist, 42-46; head separations div, Tech Dept, Hanford Atomic Prod Oper, Gen Elec Co, 47-48, asst mgr tech sect, 48-51, mgr tech eng dept, 51-54 & pile tech eng dept, 54-56, mgr res & eng, Irradiation Processing Dept, 56-65, consult engr, Atomic Prod Div, 65-68, mgr div planning opers, 68-70; RETIRED. *Concurrent Pos:* Chmn, Wash State Thermal Power Plant Site Eval Coun, 70-74; consult nuclear, 70-89. *Mem:* Fel Am Nuclear Soc; emermem Am Chem Soc. *Res:* Liquid-solid interfacial relationships; pigments and paints; cement and related building materials; nuclear technology. *Mailing Add:* PO Box 202 Richland WA 99352

GREANEY, WILLIAM A, b Memphis, Tenn, Jan 10, 31; m 54; c 2. OPERATIONS RESEARCH. *Educ:* Univ Tenn, BSChE, 52; Cornell Univ, MS, 59; Case Inst Technol, PhD(opers res), 65. *Prof Exp:* Proj engr, Wright Air Develop Ctr, 53-55; nuclear engr, Internuclear Co, 57-61; asst prof indust eng, State Univ NY, Buffalo, 64-67; assoc prof, Univ Pittsburgh, 67-70; tech assoc econ & comput sci div, Gulf Res & Develop Co, 70; mgr anal serv, Computation & Commun Serv Dept, Gulf Oil Corp, 71-75, mgr, Gulf Mgt Sci Group, Gulf Sci & Technol Co, 75-83, MGR, DECISION SUPPORT SERV, GULF OIL CORP, 83- *Mem:* Opers Res Soc Am; Inst Mgt Sci. *Res:* Nuclear engineering. *Mailing Add:* 2021 Spenwick Dr No 217 Houston TX 77055

GREASER, MARION LEWIS, b Vinton, Iowa, Feb 10, 42; m 65; c 2. MUSCLE BIOLOGY, MEAT SCIENCE. *Educ:* Iowa State Univ, BS, 64; Univ Wis, MS, 67, PhD(biochem-muscle biol), 69. *Prof Exp:* Fel, Boston Biomed Res Found, 68-71; from asst prof to assoc prof, 71-77, PROF, UNIV WIS, MADISON, 77- *Concurrent Pos:* NIH fel, Boston Biomed Res Found, 68-70. *Honors & Awards:* Distinguished Res Award, Am Meat Sci Asn, 81; Outstanding Researcher Award, Am Heart Asn Wis, 85. *Mem:* Am Soc Biochem Molecular Biol; Biophys Soc; Am Soc Cell Biol; Am Meat Sci Asn; Inst Food Technologists. *Res:* Mechanism of muscle contraction and its control; mechanisms of muscle growth and assembly. *Mailing Add:* Muscle Biol Lab Univ Wis 1805 Linden Dr Madison WI 53706

GREASHAM, RANDOLPH LOUIS, b Evansville, Ind, Apr 7, 42; m 65. MICROBIAL BIOCHEMISTRY. *Educ:* Univ Evansville, BA, 64; Villanova Univ, MS, 67; Hahnemann Med Col, PhD(microbiol), 70. *Prof Exp:* Res assoc indust microbiol, Mass Inst Technol, 70-72; RES MICROBIOLOGIST INDUST MICROBIOL, INT MINERALS & CHEM CORP, 72-; AT MERCK & CO. *Mem:* Am Chem Soc; Am Soc Microbiol. *Res:* Fermentation biochemistry; microbial genetics. *Mailing Add:* Merck & Co PO Box 2000 Rahway NJ 07065

GREATBATCH, WILSON, b Buffalo, NY, Sept 6, 19; m 45; c 5. BIOMEDICAL ENGINEERING. *Educ:* Cornell Univ, BEE, 50; Univ Buffalo, MSEE, 57. *Hon Degrees:* ScD, Houghton Col, 71; State Univ N Y, Buffalo, 84; Clarkson Univ, 86 & Roberts Wesleyan Col, 87. *Prof Exp:* Proj engr, Cornell Aeronaut Lab, Inc, 50-52; asst prof elec eng, Univ Buffalo, 52-57; mgr, Electronics Div, Taber Instrument Corp, 57-60; vpres, Mennen Greatbatch Electronics, Inc, 62-78, pres, Wilson Greatbatch, Ltd, 70-86; PRES, GREATBATCH GEN-AID LTD, 86- *Concurrent Pos:* Adj prof phys sci, Houghton Col, 70-; adj prof elec eng, State Univ NY, Buffalo, 81-; adj prof eng, Cornell Univ, 87- *Honors & Awards:* William Morlock Award, Inst Elec & Electronic Engrs, 68, Centennial Medal, 84; Laufman Award, Asn Advan Med Instrumentation, 85; Holley Award, Am Soc Mech Engr, 87; Nat Technol Medal, Pres Bush, 90. *Mem:* Nat Acad Engrs; fel Am Col Cardiol; fel Royal Soc Health; Asn Advan Med Instrumentation; NY Acad Sci; Sigma Xi; fel AAAS; fel Inst Elec & Electronics Engrs. *Res:* Invention of implantable cardiac pacemaker; implantable power supplied for medical uses; biomass energy. *Mailing Add:* Greatbatch Gen-Aid LTD 10871 Main St Clarence NY 14031

GREATHOUSE, TERRENCE RAY, b Hindsboro, Ill, Nov 7, 32; m 61; c 2. ANIMAL NUTRITION. *Educ:* Univ Ill, Urbana, BS, 55, MS, 58; Univ Ky, PhD(animal nutrit), 64. *Prof Exp:* Asst prof beef cattle res & animal scientist, Univ Ill, Urbana, 58-67; from asst prof to prof beef cattle res, Mich State Univ, 67-73; head dept animal sci, Colo State Univ, 73-77; assoc dean agr, Col Agr, Tex A&M Univ, 77-78, assoc vpres agr & renewable resources & coordr int affairs, 78-79, vpres int affairs, 79-82, assoc dean admin, 82-89, PROF ANIMAL SCI, COL AGR, TEX A&M UNIV. *Mem:* Fel Am Soc Animal Sci; Am Dairy Sci Asn; Am Poultry Sci Asn; AAAS; Sigma Xi. *Res:* Beef cattle feeding, breeding and management. *Mailing Add:* Dept Animal Sci Col Agr Tex A&M Univ College Station TX 77843-2471

GREAVES, BETTINA BIEN, b Washington, DC, July 21, 17; m 71. HISTORY OF WORLD WAR II, HISTORY OF ECONOMIC THOUGHT. *Educ:* Wheaton Col, Norton Mass, BA, 38; Columbia Univ, MLS, 67. *Hon Degrees:* Dr Soc Sci, Univ Francisco Marroquin, Guatemala, 83. *Prof Exp:* Secretarial, Bd Econ Warfare, Foreign Econ Admin, 43-46; ed asst, Found Freedom, 46-47; SR STAFF MEM, FOUND ECON EDUC, 51-, PROF ECON, NY INST CREDIT, 86- *Res:* Economic theory, Austrian School of Economics, background and revelation concerning the Japanese attack on Pearl Harbor, Dec 7, 1941. *Mailing Add:* Found Econ Educ 30 S Broadway Irvington NY 10533

GREAVES, WALTER STALKER, b New York, NY, Feb 7, 37; m 67; c 1. VERTEBRATE PALEONTOLOGY. *Educ:* State Univ NY, Oswego, BS, 66; Ohio Univ, MS, 67; Univ Chicago, PhD(evolutionary biol), 71. *Prof Exp:* Asst prof biol, Indiana Univ, Pa, 71-75, assoc prof, 75; asst prof, 75-80, ASSOC PROF ORAL ANAT, UNIV ILL, CHICAGO, 80- *Concurrent Pos:* Res assoc, Sect Vert Fossils, Carnegie Mus Nat Hist, 74- *Mem:* Sigma Xi; Soc Study Evolution; Soc Vert Paleont; AAAS; Am Soc Zoologists; Am Soc Phys Anthropologists. *Res:* Functional morphology of fossil and recent mammalian jaw mechanisms. *Mailing Add:* Dept Oral Anat Univ Ill at Chicago Chicago IL 60612

GREBE, JANICE DURR, b Alton, Ill, Nov 24, 40; m 79. GENETICS, ANATOMY. *Educ:* Univ Tex, Austin, BA, 62, MA, 63, PhD(zool), 65. *Prof Exp:* Vis asst prof biol, Univ Kans, 67-68; asst prof, Avila Col, 69-71; instr & chmn biol, Pa Valley Commun Col, 72-79; asst prof, Med Ctr, Kans Univ, 79-81; asst prof, Univ Health Sci, Kans City, Mo, 81-85; CONSULT, 85- *Concurrent Pos:* NSF fel, Univ Wash, 65-67; NSF sci fac prof develop grant, Univ Kans Med Ctr, Kansas City, 78-79. *Mem:* Am Asn Anat; Sigma Xi; Am Soc Law & Med. *Res:* Developmental genetics; cytogenetics. *Mailing Add:* 4820 W 57th St Shawnee Mission KS 66205

GREBENAU, MARK DAVID, b Newport News, Va, Mar 26, 51; m; c 2. EXPERIMENTAL BIOLOGY. *Educ:* Yeshiva Col, BA, 72, NY Univ, MS, 76, MD, 78, PhD(immunol), 79. *Prof Exp:* Residency internal med, King's County Hosp Ctr/Downstate Univ Med Ctr/Brooklyn Vet Admin Hosp Ctr, 78-81; Lita Annenberg Hazen fel clin res & assoc physician, Rockefeller Univ Hosp, 81-83; asst dir, 83-86, assoc dir, 86-90, ASSOC MED DIR, SANDOZ MED OPERS, 90- *Concurrent Pos:* Assoc med staff, Dept Pediat, United Hosps Med Ctr, Newark, NJ, 85-87, courtesy staff, 87- *Mem:* Am Col Physicians; NY Acad Sci; Am Asn Immunologists; Am Asn Clin Immunol & Allergy; Am Soc Clin Pharmacol & Therapeut; Am Col Allergists. *Res:* Immunoglobulin therapy; development of immunological products. *Mailing Add:* Sandoz Res Inst 59 Rte 10 East Hanover NJ 07936

GREBENE, ALAN B, b Istanbul, Turkey, Mar 13, 39; m 67; c 1. ELECTRONICS. *Educ:* Robert Col, Istanbul, BSc, 61; Univ Calif, Berkeley, MSc, 63; Rensselaer Polytech Inst, PhD(elec eng), 68. *Prof Exp:* Mem tech staff, Fairchild Semiconductor, 63-64 & Sprague Elec Co, 64-65; lectr elec eng, Rensselaer Polytech Inst, 65-68; mgr circuit res, Signetics Corp, 68-71; SR VPRES & FOUNDER, EXAR INTEGRATED SYSTS, INC, 71-; PRES, MICRO-LINEAR CORP. *Concurrent Pos:* Lectr, Univ Santa Clara, 68-71. *Mem:* AAAS; Inst Elec & Electronics Engrs; Am Soc Eng Educ. *Res:* Integrated circuits research and development. *Mailing Add:* ABG Assoc 15479 Belnap Dr Saratoga CA 95070

GREBER, ISAAC, b Poland, Sept 20, 28; US citizen; m 53; c 2. MECHANICAL ENGINEERING, BIOENGINEERING & BIOMEDICAL ENGINEERING. *Educ:* City Col New York, BME, 50; Univ Mich, MSAE, 52; Mass Inst Technol, PhD(aerodyn), 59. *Prof Exp:* Group leader theoret aerodyn, United Aircraft Res Labs, 52-54; sr aerodyn engr, Chance Vought Aircraft Corp, 54-55; assoc prof eng, 59-70,chmn,dept mech & aerospace eng, 79-84, PROF ENG, CASE WESTERN RESERVE UNIV, 70- *Concurrent Pos:* Fulbright prof, appl math dept, Univ Tel-Aviv, 66-67; vis prof, dept aeronaut & astronaut, Mass Inst Technol, 84-85. *Mem:* AAAS; Am Inst Aeronaut & Astronaut; Am Soc Eng Educ; Am Phys Soc; Sigma Xi. *Res:* Viscous flows; viscid-inviscid interactions; molecular dynamics. *Mailing Add:* 17937 Sherrington Rd Shaker Heights OH 44122

GREBNER, EUGENE ERNEST, b Pittsburgh, Pa, Feb 6, 31; m 55; c 2. BIOCHEMISTRY, BIOCHEMICAL GENETICS. *Educ:* Hiram Col, AB, 52; Univ Pittsburgh, MS, 60, PhD(biochem), 64. *Prof Exp:* Res biochemist, Albert Einstein Med Ctr, 67-74; RES ASSOC PROF, THOMAS JEFFERSON UNIV, 74- *Concurrent Pos:* USPHS fel biochem, NIH, Md, 64-67. *Mem:* Soc Complex Carbohydrates; Sigma Xi; AAAS. *Res:* Lysosomal storage diseases. *Mailing Add:* 1100 Walnut St 4th Floor Thomas Jefferson Univ Philadelphia PA 19107

GREBOGI, CELSO, b Curitiba, Parana, Brazil, July 29, 47; nat US; m 80. CHAOTIC DYNAMICS, SYMPLECTIC GEOMETRY. *Educ:* Fed Univ Parana, BS, 70; Univ Md, MS, 75, PhD(physics), 78. *Prof Exp:* Instr physics, Pontificia Catolica Univ, 71-74; staff scientist physics, Lawrence Berkeley Lab, 78-81; res assoc physics, 81-85, res scientist physics & math, 85-90, ASSOC PROF MATH, UNIV MD, 90- *Concurrent Pos:* Consult, Naval Surface Warfare Ctr, 83-, Lawrence Livermore Lab, 84-85, Science Applications Intern, Corp, 86-89; vis scientist, Univ Calif, Santa Barbara, 85; comt mem, US Dept Energy, 84-85; panel, Office Naval Technol, 87-90; prin invest, US Dept Energy, 84- *Mem:* Am Math Soc; Soc Indust & Appl Math; AAAS. *Mailing Add:* Dept Math Univ Maryland College Park MD 20742

GREBOW, PETER ERIC, b New York, NY, Nov 25, 46; m 69; c 2. BIOPHARMACEUTICS. *Educ:* Cornell Univ, AB, 67; Rutgers Univ, MS, 69; Univ Calif, Santa Barbara, PhD(chem), 73. *Prof Exp:* Res fel pharmacol, Col Physicians & Surgeons, Columbia Univ, 73-75; group leader drug metab biopharmaceut, USV Pharmaceut Corp, 75-81, from assoc dept dir to dept dir drug disposition, Revlon Health Care Group, 81-88; vpres, New Drug Develop, Rorer, 88-90; DIR, DRUG DEVELOP, CEPHALON, INC, 91- *Mem:* Am Chem Soc; AAAS. *Res:* drug and product development. *Mailing Add:* 704 Buckley Dr Penllyn PA 19422

GRECCO, WILLIAM L, b Brockway, Pa, Aug 28, 24; m 47; c 9. CIVIL ENGINEERING, URBAN PLANNING. *Educ:* Univ Pittsburgh, BS, 47, MS, 51; Mich State Univ, PhD(transp), 62. *Prof Exp:* From instr to assoc prof civil eng, Univ Pittsburgh, 47-61; res asst, Mich State Univ, 61-62; from assoc prof to prof urban planning & eng, Purdue Univ, 62-72; prof civil eng & head dept, 72-86, ASSOC DEAN, UNIV TENN, KNOXVILLE, 86- *Concurrent Pos:* Consult, Dines & Grecco, 50-52, Found Assocs, 52 & Donald M McNeil, 53-58; mem info syst storage & retrieval comt, Hwy Res Bd, Nat Acad Sci-Nat Res Coun, 64; mem bd dirs, Purdue Calument Found, 67-71; ed, Am Soc Civil Engrs J, 68-71; eng accreditation comt, Accreditation Bd Eng & Technol, 80-85. *Mem:* Am Soc Civil Engrs; Am Soc Eng Educ; Inst Traffic Engrs; Am Planning Asn; Am Inst Cert Planners. *Res:* Urban traffic forecasting by system engineering; recreational travel; synthetic travel patterns. *Mailing Add:* 102 Heskins Lib Univ Tenn Knoxville TN 37996-4007

GRECO, CLAUDE VINCENT, b Bronx, NY, Sept 13, 30. ORGANIC CHEMISTRY. *Educ:* Manhattan Col, BS, 52; NMex Highlands Univ, MS, 55; Fordham Univ, PhD(chem), 60. *Prof Exp:* Asst chem, Wellcome Res Labs, 52-53 & Fordham Univ, 56-57; assoc chemist, Midwest Res Inst, 59-61; fel, State Univ NY, 61-62; from asst prof to assoc prof, 62-73, PROF ORG CHEM, ST JOHN'S UNIV, NY, 73- *Mem:* Am Chem Soc; Sigma Xi. *Res:* Synthetic organic chemistry; substitution reactions; mesoionic compounds; condensed heterocyclic systems; chemotherapy. *Mailing Add:* Dept Chem St John's Univ Jamaica NY 11432

GRECO, EDWARD CARL, b Marsala, Italy, Nov 2, 11; US citizen; m 38; c 2. CORROSION CHEMISTRY, CORROSION ENGINEERING. *Educ:* Northwestern State Univ, La, BS, 34. *Hon Degrees:* DSc, Centenary Col, La, 63. *Prof Exp:* State dir, Fed Surplus Commodities Corp, 35-38; anal chemist, United Gas Pipeline Co, 38-43; develop chemist, E I Dupont de Nemours, 43-45; res chemist, United Gas Corp, 45-55, sr res assoc, 55-68; dir chem res, Northwestern State Univ, 68-74; VPRES & SR CORROSION ENGR, FW CORROSION CONTROL, INC, 74- *Concurrent Pos:* Dir, La State Sci Fairs, 56-65; pres, 2nd Int Cong, Metallic Corrosion, 62; head team, Scientists & Engrs Reciprocal Exchange, USSR, 62; chmn, Permanent Int Corrosion Coun, 63-68; ed, J, Mat Protection, 63-74. *Honors & Awards:* Prof Engrs Award, Eng & Sci Coun, 68. *Mem:* Emer mem Am Chem Soc; emer mem Am Inst Chemists; Nat Assn Corrosion Engrs (pres, 62-63). *Res:* Effect of hydrogen sulfide in aqueous solutions and various concentrations on high strength steels. *Mailing Add:* 1406 Captain Shreve Dr Shreveport LA 71105

GRECO, SALVATORE JOSEPH, b Richmond, Calif, Jan 25, 21; m 46; c 6. PHARMACY. *Educ:* Duquesne Univ, BS, 42; Univ Md, PhD(pharm), 48. *Prof Exp:* Asst prof chem, Temple Univ, 48-49; from asst prof to assoc prof pharm, George Washington Univ, 49-56; assoc prof & asst to dean, Sch Pharm, 56-58, dean, 58-71, PROF PHARM, CREIGHTON UNIV, 58- *Mem:* Am Pharmaceut Asn; Am Asn Cols Pharm. *Res:* Product development of pharmaceuticals. *Mailing Add:* Sch Pharm Creighton Univ Omaha NE 68178

GRECO, WILLIAM ROBERT, b Detroit, Mich, Oct 14, 51; m 71; c 2. PHARMACO-MATHEMATICS, DATA ANALYSIS. *Educ:* Rensselaer Polytech Inst, BS, 73; State Univ NY Buffalo, PhD(pharmacol), 79. *Prof Exp:* Fel pharmacol, State Univ NY Buffalo, 79; RES SCIENTIST CANCER DATA ANAL, DEPT BIOMATH & DEPT EXP THERAPEUT, ROSWELL PARK MEM INST, 80-; PROF BIOMET, NIAGRA UNIV, 81- *Concurrent Pos:* Assoc prof biomet, State Univ NY at Buffalo, 80- *Mem:* Asn Comput Mach; Am Statist Asn; AAAS; Am Asn Cancer Res; Am Soc Biol Chemists. *Res:* Biomathematics in pharmacology; biostatistics; modelling of cancer growth; assessment of synergism for drug interactions; physiological pharmacokinetic modelling. *Mailing Add:* Dept Biomath Roswell Mem Inst 666 Elm St Buffalo NY 14263

GREDEN, JOHN F, b Winona, Minn, July 24, 42. PSYCHIATRICS. *Educ:* Univ Minn, BS, 65; Univ Minn Med Sch, 67. *Prof Exp:* Asst chief, Psychiat Outpatient Clin, Fort Lee, US Army, 69-70; assoc dir, Walter Reed Army Med Ctr, 72-73, dir psychiat res, 73-74; asst prof psychiat, 74-77, dir clin studies Unit Affective disorders Inpatient prog, 77-80, assoc prof psychiat, 77-81, dir Clin Studies Unit Affective Disorders, 80-85, PROF PSYCHIAT, UNIV MICH MED SCH, 81-, CHMN, 85- *Concurrent Pos:* Mem, Am Psychiat Assoc Task Force to Eval Melancholia in Diag & Stat Manual, 83-86; mem, Samuel G Hibbs Bd, Am Psychiat Assoc; mem, Nat Inst Mental Health Psychopharmacol, Biol, & Physician Treatments Subcomt, 86-89; prin invest, NIAAA Ctr Grant, 88-, NIMH Grant, 85-88. *Honors & Awards:* A E Bennett Award, A E Bennet Cent Neuropsychiat Found, 74; Ralph Patterson Mem Award, Ohio State Univ, 80; Nolan D C Lewis vis Scholar Award, Carrier Found, Belle Meade, 82. *Mem:* Am Psychiat Assoc fel; Biol Psychiat; Am Col Neuropsychopharmacol; Am Asn Adv Sci; Asn Med Educ & Res Substance Abuse; Psychiat Res Soc (counr, 88-89, pres elect, 89-70). *Res:* Neuroendocrine regulation in patients with affective disorders; cholinergic function in depression; pharmacological manifestations of catteinism; interaction between aging and alcoholism in the production of central nervous system abnormalities. *Mailing Add:* Dept Psychiat B4950 Univ Mich Hosps 1500 Medical Center Dr Ann Arbor MI 48109-0704

GREDING, EDWARD J, JR, b Mar 30, 40; US citizen; m 61. HERPETOLOGY. *Educ:* Univ Tex, Arlington, BS, 61; ETex State Univ, MS, 62; Univ Tex, Austin, PhD(zool), 68. *Prof Exp:* Instr biol, Tarleton State Col, 62-65; asst prof, Pan Am Col, 68-71; prof, Univ El Salvador, 71-73; PROF BIOL, DEL MAR COL, 73- *Mem:* Herpetologists League; Soc Study Amphibians & Reptiles; Am Soc Ichthyologists & Herpetologists. *Res:* Ecology and evolution of frogs of the American tropics. *Mailing Add:* Dept Biol Del Mar Col Corpus Christi TX 78404

GREEAR, PHILIP FRENCH-CARSON, b Troutdale, Va, Aug 25, 18; m 43; c 5. ECOLOGY, GEOENVIRONMENTAL SCIENCE. *Educ:* Univ Ga, BSA, 49, MS, 59, PhD(bot-ecol), 67. *Prof Exp:* From asst prof to assoc prof biol, 61-63, from actg head to head dept biol & earth sci, 64-86, prof biol, 71-86, EMER PROF BIOL, SHORTER COL, 86- *Concurrent Pos:* NSF res partic, 64; mem, Ga Natural Areas Coun & Ga Environ Educ Coun; mem bd & exec comt, Ga Conservancy; chmn, Coosa Basin Water & Resources Group; trustee, Ga Chap, Nature Conservancy, Ossahaw Island Found. *Honors & Awards:* Nat Conservationist, Am Motors Corp, 73. *Mem:* AAAS. *Res:* Effect of Hydroperiod on plant zonation. *Mailing Add:* 330 Mt Alto Rd Rome GA 30161

GREECHIE, RICHARD JOSEPH, b Boston, Mass, Apr 12, 41. MATHEMATICS. *Educ:* Boston Col, BA, 62; Univ Fla, PhD(math), 66. *Prof Exp:* Asst prof math, Univ Mass, Boston, 65-67; from asst prof to assoc prof math, 67-77, dir, ASN MATH FOUND EMPIRICAL STUDIES, 70-, PROF MATH, KANS STATE UNIV, 77- *Concurrent Pos:* NSF grant, 70-73; vis prof, Inst Phys Theory, Univ Geneva, 73-74, Tech Hochschule Darmstadt, WGer, 80, McMaster Univ, 80, Univ Denver, 81, Univ Genova, Italy, 81. *Mem:* AAAS; Am Math Soc; Math Asn Am; Soc Indust & Appl Math. *Res:* Orthomodular lattice theory; empirical logics ranging from the classical logics to the non-classical quantum-mechanical logics, especially the way in which the classical sub-logics interrelate in the non-classical logic. *Mailing Add:* Dept Math Kans State Univ Manhattan KS 66506

GREEDAN, JOHN EDWARD, b Beaver, Pa, June 4, 42. SOLID STATE CHEMISTRY. *Educ:* Bucknell Univ, BA, 64; Tufts Univ, PhD(chem), 69. *Prof Exp:* Fel chem, Univ Pittsburgh, 69-71, res asst prof chem, 71-72; asst prof, Dalhousie Univ, 72-74; assoc prof, 74-83, PROF CHEM, MCMASTER UNIV, 83- *Mem:* Am Chem Soc; Can Inst Chem; Can Asn Physics. *Res:* Synthesis, crystal growth and solid state properties of inorganic materials; emphasis on rare earth and transition metal oxides and intermetallics; x-ray and neutron diffraction; magnetic, transport and optical properties. *Mailing Add:* Dept Chem McMaster Univ Hamilton ON L8S 4M1 Can

GREELEY, FREDERICK, b Winnetka, Ill, Aug 26, 19; m 44; c 4. BIOLOGY, WILDLIFE MANAGEMENT. *Educ:* Kenyon Col, BA, 41; Univ Wis, MS, 49, PhD(zool, wildlife mgt), 54. *Prof Exp:* Res assoc endocrine studies of pheasants, Univ Wis, 54-55; res assoc deer nutrit, Univ NH, 55-56; proj leader pheasant range anal & nutrit, Ill Natural Hist Surv, 56-60; assoc prof, 60-81, EMER PROF, WILDLIFE BIOL, UNIV MASS, AMHERST, 81- *Concurrent Pos:* Mem adv comt, Mass Natural Heritage & Endangered Species Prog, 82- *Mem:* Wildlife Soc; Wilson Ornith Soc; Am Ornithologists Union; Am Soc Mammalogists. *Res:* Avian ecology; forest wildlife ecology and management. *Mailing Add:* Teawaddle Hill Rd Amherst MA 01002

GREELEY, GEORGE H, JR, b Detroit, Mich, Oct 15, 47. GASTROINTESTINAL ENDOCRINOLOGY. *Educ:* Med Col Ga, PhD(endocrinol), 74. *Prof Exp:* ASSOC PROF SURG, UNIV TEX MED BR, GALVESTON, 83- *Mailing Add:* Dept Surg & Pharmacol & Toxicol Univ Tex Med Br 6 118 McCullough Bldg G25 Galveston TX 77550

GREELEY, RICHARD STILES, b Framingham, Mass, Dec 25, 27; m 51; c 2. PHYSICAL CHEMISTRY. *Educ:* Harvard Univ, BS, 49; Northwestern Univ, MS, 51; Univ Tenn, PhD, 59. *Prof Exp:* Develop engr, Oak Ridge Nat Lab, 54-60; engr advan design, Mitre Corp, Bedford, Mass, 60-63, assoc dept head strategic systs, 63-68, dept head spec projs, 68-70, assoc tech dir systs develop, 70-75; tech dir energy, resources & environ, Metrek Div, Mitre Corp, McLean, Va, 76-80; vpres, Econenvironics Div, Roy F Weston, Inc, West Chester, Pa, 80-82; PRES, THE GREELEY-POLHEMUS GROUP, INC, ST DAVIDS, PA, 82- *Concurrent Pos:* Prof lectr chem, West Chester Univ, Pa, 83-84. *Mem:* Am Chem Soc. *Res:* Energy and environmental systems engineering; high temperature aqueous electrochemistry. *Mailing Add:* 418 Roundhill St Davids PA 19087

GREELEY, RONALD, b Columbus, Ohio, Aug 25, 39; m 60; c 2. GEOLOGY, PLANETARY SCIENCE. *Educ:* Miss State Univ, BS, 62, MS, 63; Univ Mo-Rolla, PhD(geol), 66. *Prof Exp:* Instr geol, Univ Mo-Rolla, 65-66; geologist, Standard Oil Co Calif, 66-67; US Army res scientist, Ames Res Ctr, NASA, 67-69, Nat Res Coun geologist, 69-71, Univ Santa Clara res prof, 71-77; chmn geol, 86-90, PROF GEOL, ARIZ STATE UNIV, 77- *Concurrent Pos:* Prof geol, Foothill Col, 70-77; consult, Jet Propulsion Lab, Pasadena, Calif & Lunar Planetary Inst, Houston, Tex, 72-75; hon res fel, Univ London, 75; vis assoc prof planetary sci, Calif Inst Technol, 77; adj prof geol, State Univ NY, Buffalo, 78; consult, Kuwait Inst Sci Res, 85-88. *Honors & Awards:* Pub Serv Medal, NASA, 77. *Mem:* Fel Geol Soc Am; Am Geophys Union; AAAS; Meteoritical Soc; Sigma Xi. *Res:* Fundamental research on the geology of the planets through the analysis of spacecraft data coupled with laboratory simulation and geological field work of terrestrial analogs to planetary surface features; emphasis on volcanology and windblown sand. *Mailing Add:* Dept Geol Ariz State Univ Tempe AZ 85287-1404

GREEN, ALBERT WISE, b Jackson, Miss, Dec 15, 38; m 67; c 2. PHYSICAL OCEANOGRAPHY. *Educ:* Vanderbilt Univ, BA, 60; Mass Inst Technol, PhD(phys oceanog), 69. *Prof Exp:* Mem staff, Inst Geophys, Univ Oslo, Norway, 68-70; asst prof oceanog & meteorol, Dept Atmospheric & Oceanic Sci, Univ Mich, 70-77; br head phys oceanog, 77-84, HEAD OCEANOG DIV, NAVAL OCEAN & ATMOSPHERIC RES LAB, 84- *Concurrent Pos:* Prin Investr, Circulation Modelling Proj, Sea Grant, Univ Mich, 70-76; consult, Aeromatrix, Inc & Environ Eng Co, 75-76 & Environ Studio & Argonne Nat Lab, 76; mem steering comt, Nat Marine Bd, 80-81. *Mem:* Am Geophys Union; Am Meteorol Soc; Sigma Xi. *Res:* Turbulent oceanic boundary layer processes; experimental study of non-linear wave interactions; experimental and theoretical research in mesoscale ocean hydrodynamics; design, testing and modelling of physical oceanographic measuring instruments. *Mailing Add:* Oceanog Div Naval Ocean & Atmospheric Res Lab Stennis Space Center MS 39529

GREEN, ALEX EDWARD SAMUEL, b New York, NY, June 2, 19; m 46; c 5. PHYSICS. *Educ:* City Col New York, BS, 40; Calif Inst Technol, MS, 41; Univ Cincinnati, PhD(physics), 48. *Prof Exp:* Physicist & instr, Calif Inst Technol, 40-44; instr physics & math, Newark Col Eng, 45-46; asst prof physics, Univ Cincinnati, 46-53; from assoc prof to prof & actg head dept, Fla State Univ, 53-59, sci dir, Tandem Van de Graaff Lab, 58-59; chief physics, Convair Div, Gen Dynamics & mgr space sci lab, Astronaut Div, 59-63; GRAD RES PROF PHYSICS, NUCLEAR ENG SCI, ELEC ENG & ENG SCI, UNIV FLA, 63-, DIR, CTR AERONOMY & OTHER ATMOSPHERIC SCI, 74- *Concurrent Pos:* Consult, Jet Propulsion Lab, Marshall Space Flight Ctr & Inst Defense Anal; consult & group leader theoret div, Los Alamos Sci Lab, 57-58; lectr exten div, Univ Calif, 60-63. *Honors & Awards:* Medal of Freedom Award, 47. *Mem:* Fel Am Phys Soc; Am Asn Physics Teachers; fel Optical Soc Am; Am Geophys Union; Am Soc Photobiol. *Res:* Theoretical nuclear, atomic, radiation and atmospheric physics and ultraviolet photoclimatology. *Mailing Add:* Dept Mech/Nuclear Eng 311 SSRB Univ Fla Gainesville FL 32611

GREEN, ALLEN T, b Chicago, Ill, Mar 30, 34; m 56; c 3. NONDESTRUCTIVE TESTING, ACOUST EMISSION. *Educ:* Univ Ill, Urbana, BS, 56. *Prof Exp:* Test eng Convair, 56-61; prog engr, Aerojet Gen Corp, 61-70; vpres, Dunegan Res Corp, 70-72; pres, Acoust Emission Tech Corps, 72-88; PRES, HSB INSPECTION TECHNOL, 88- *Concurrent Pos:* Founder, Acoust Emission Working Group, 67- *Mem:* Instrument Soc Am; Am Soc Testing & Mat; fel Am Soc Nondestructive Testing; Soc Exp Mech; Am Soc Mech Engrs. *Res:* Development of acoustic emission for uses in material science; structural testing; nondestructive testing. *Mailing Add:* 1600 Tribute Rd Sacramento CA 95815

GREEN, ALWIN CLARK, b Meadville, Pa, Dec 19, 30; m; c 3. MATHEMATICS. *Educ:* Hiram Col, BA, 56; Syracuse Univ, MA, 68, PhD(math), 72. *Prof Exp:* Design engr mercury vapor arcs, Large Lamp Div, Gen Elec Co, 56-64; asst prof math, State Univ NY, Buffalo, 72-78, prof math & chmn math dept, 78-81. *Concurrent Pos:* Panel vis lectrs, Math Asn Am, 78-90. *Mem:* Math Asn Am. *Res:* Theoretical research in connectedness of networks; applications, networks and finite mathematics in the social sciences, tiling and geometry. *Mailing Add:* Dept Math State Univ Col Buffalo 1300 Elmwood Ave Buffalo NY 14222

GREEN, ARTHUR R, b Loma Linda, Calif, May 21, 34; div. GEOLOGY. *Educ:* Wash State Univ, BS, 57; Univ Ore, MS, 62. *Prof Exp:* Explor & prod, Humble Co, 62-69; sr scientist oceanog, 69-72, global geol studies mgr, 72-82, RES SCIENTIST, EXXON PROD RES CO, 82- *Concurrent Pos:* Sr Adv Comt, Inc Res Inst Semisol; adv comt, US Geol Surv & French Petrol Inst; chmn, Nat Sci Found Adv Comt, 88-; mem adv bd, Int Union Geol Sci; mem, Ocean Drilling Prog Safety & Pollution Prev Comt, Deep Observ & Sampling of the Earth's Continental Crust, NSF; mem, Marine Geol Comt, Am Asn Petrol Geologists. *Mem:* AAAS; Am Geophys Union; Sigma Xi; Fel Geol Soc Am. *Res:* Author of several articles. *Mailing Add:* Exxon Prod Res Co PO Box 2189 Houston TX 77252

GREEN, BARRY A, b Los Angeles, Calif, Mar 17, 40; m 70; c 3. THEORETICAL SOLID STATE PHYSICS. *Educ:* Pomona Col, BA, 62; Northwestern Univ, MS, 64; Univ Ariz, PhD(nuclear eng), 72. *Prof Exp:* Res engr compact reactor res & develop, Atomics Int, 64-65; sr physicist, Gulf Radiation Technol, Div Gen Atomic, 72-73; sr physicist theoret solid state physics, IRT Corp, 73-76; sr physicist, Air Force Mat Lab, Wright-Patterson AFB, 77-78; TECH STAFF, ELECTRONIC SYSTS GROUP, ROCKWELL INT, 78- *Concurrent Pos:* vis theorist, Air Force Mat Lab, Wright-Patterson AFB, 78- *Mem:* Am Phys Soc. *Res:* Solid state studies of metal oxide insulators and ternary alloy semiconductors, including radiation effects, electrical transport, photon absorption, lifetime and defect studies; hole mobility and Hall resistance factor in silicon; infrared detector development. *Mailing Add:* 525 S 30th St Mesa AZ 85204

GREEN, BARRY GEORGE, b Gloucester, Mass, Mar 17, 49. PSYCHOPHYSICS. *Educ:* Univ Calif, Riverside, AB, 71; Ind Univ, PhD(psychol), 75. *Prof Exp:* NIH fel psychol, John B Pierce Found Lab, 75-78; res assoc psychol, Princeton Univ, 78-80; res assoc psychol, Ind Univ, 80-83; MEM STAFF, MONELL CHEM SENSES CTR, PHILADELPHIA, 83- *Concurrent Pos:* Lectr, South Conn State Col, 77; Vis asst prof epidemiology & public health, Yale Univ, 75-78. *Mem:* AAAS; Sigma Xi; Psychonomic Soc. *Res:* Psychophysics of the skin senses, including the interactions among vibrotactile, tactile and thermal stimuli; cutaneous communication of speech. *Mailing Add:* 2212 Fairhill Ave Glenside PA 19038

GREEN, BEVERLEY R, b Vancouver, BC, Apr 17, 38. BIOCHEMISTRY, PLANT PHYSIOLOGY. *Educ:* Univ BC, BSc, 60; Univ Wash, PhD(biochem), 65. *Prof Exp:* NATO fel, Free Univ Brussels, 66-67; from asst prof to assoc prof, 67-84, PROF BOT, UNIV BC, 84- *Concurrent Pos:* Res fel, Biol Labs, Harvard Univ, 75; vis res scientist, Lawrence Berkeley Labs, 82-83; Killam Fac Res Fel, 87-88. *Mem:* AAAS; Can Soc Plant Physiologists; Am Soc Plant Physiologists; Int Soc Plant Molecular Biol. *Res:* Structure of photosynthetic membranes; chlorophyll-protein complexes; synthesis and assembly of photosynthetic membranes. *Mailing Add:* Dept Bot Univ BC Vancouver BC V6T 2B1 Can

GREEN, BRIAN, b Liverpool, Eng, Feb 21, 35; m 62; c 2. ORGANIC CHEMISTRY. *Educ:* Univ Liverpool, BSc, 56, PhD(org chem), 59. *Prof Exp:* Res asst org chem, Univ Maine, 59-61, res assoc, 61-64; Alexander von Humboldt fel, Max Planck Inst Biochem, Munich, Ger, 64-65; from asst prof to assoc prof, 65-75, PROF CHEM, UNIV MAINE, 75- *Concurrent Pos:* Alexander von Humboldt fel org chem, Univ Bonn, 71-72; coop prof oceanog, Univ Maine, 77- *Mem:* Am Chem Soc; The Chem Soc. *Res:* Chemistry of natural products, especially terpenoids and steroids with emphasis on both synthetic and degradative aspects of the field. *Mailing Add:* 387 Aubert Hall Univ Maine Orono ME 04469

GREEN, BYRON DAVID, b Philadelphia, Pa, June 25, 50. CHEMICAL PHYSICS. *Educ:* Univ Pa, BA, 71; Mass Inst Technol, PhD(phys chem), 76. *Prof Exp:* PRIN SCIENTIST RES, PHYS SCI INC, 76- *Honors & Awards:* Marcus O'Day Award, Air Force Geophys Lab. *Mem:* Optical Soc Am; Sigma Xi; Am Geophys Union. *Res:* Molecular spectroscopy; energy transfer; chemiluminescence, kinetics; radiative transfer; atmospheric modeling; optical detection of trace species; laser development; electron irradiation of gases. *Mailing Add:* New Eng Bus Ctr Andover MA 01810

GREEN, CHARLES E, b San Diego, Calif, Mar 27, 12; m 37; c 2. UNDERWATER ACOUSTICS, ELECTROACOUSTICS. *Educ:* Univ Calif, Los Angeles, AB, 36, EB, 37, BS, 42, PhD(eng), 81. *Prof Exp:* Instr pub schs, San Diego, Calif, 37-46; res physicist, Navy Electronics Lab, 46-59, supvry physicist, 59-67, head sonar test & eval, Naval Undersea Ctr, 67-75; INDUST & GOVT CONSULT, 75- *Concurrent Pos:* Counr pub schs, San Diego, Calif, 44-46; consult sonar, Navy & Acoust Soc Publ, 46-75, Channel Industs Inc, 75- & Helle Eng Inc, 83- *Honors & Awards:* Presidential Citation, 64. *Mem:* Acoust Soc Am; AAAS. *Res:* Underwater transducers; design measurement and evaluation of sonar components and systems; automated electronic systems and eleven patents. *Mailing Add:* 3427 Florida St San Diego CA 92104

GREEN, CHARLES RAYMOND, b Fredericksburg, Va, Aug 15, 42; m 64; c 2. ANALYTICAL CHEMISTRY, TOBACCO CHEMISTRY. *Educ:* Univ Va, BS, 64, PhD(chem), 68. *Prof Exp:* sr res chemist, 68-80, MASTER SCIENTIST, R J REYNOLDS TOBACCO CO, 80- *Mem:* Am Chem Soc; Sigma Xi. *Res:* Tobacco smoke chemistry; gas chromatography; gas chromatography-mass spectrometry; glass capillary gas chromatography. *Mailing Add:* 430 Burkeridge Ct Winston-Salem NC 27104-2602

GREEN, CLAUDE CORDELL, b Ft Worth, Tex, Dec 26, 41. COMPUTER SCIENCE. *Educ:* Rice Univ, BA & BS, 64; Stanford Univ, MS, 65, PhD(elec eng), 69. *Prof Exp:* Res mathematician artificial intel, Stanford Res Inst, 66-69; res & develop prog mgr & info processing techniques officer, Advan Res Projs Agency, 70-71; asst prof comput sci, Stanford Univ, 72-78; chief scientist comput sci, systs control inc, 78-81; DIR, KESTREL INST, 81- *Concurrent Pos:* Consult, Stanford Res Inst, 71-73, Xerox Corp, 71-78 & Systs Control Inc, 75-76; ed artificial intel area, J Asn Comput Mach, 72- *Mem:* Asn Comput Mach. *Res:* Artificial intelligence, especially automatic programming, problem solving and answering. *Mailing Add:* Kestrel Inst 1801 Page Mill Rd Palo Alto CA 94304

GREEN, DANIEL G, b New York, NY, Sept 3, 37; m 57; c 2. PHYSIOLOGICAL OPTICS. *Educ:* Univ Ill, BSEE, 59; Northwestern Univ, MS & PhD(elec eng), 64. *Prof Exp:* Asst elec eng & bioeng, Northwestern Univ, 59-64; NSF fel, Physiol Lab, Cambridge Univ, 64-65; USPHS fel, Nobel Insts Neurophysiol, Stockholm, Sweden, 65-66; asst prof physiol optics, 66-70, assoc prof physiol optics, psychol & elec eng, 71-76, PROF PHYSIOL OPTICS, PSYCHOL & ELEC ENG, UNIV MICH, ANN ARBOR, 76- *Concurrent Pos:* Vis scientist, Biol Labs, Harvard Univ, 72-73; vis prof ophthal, Univ Calif, San Francisco, 80-81; vis scientist, Physiol Lab, Cambridge Univ, 81. *Mem:* AAAS; Am Physiol Soc; Asn Res Vision & Ophthal; fel Optical Soc Am; Soc Neurosci; fel Inst Elec & Electronic Engrs. *Res:* Biomedical engineering and physiology of vision; synaptic mechanisms and functional interactions used by the vertebrate retina to process visual information; neural basis for the desensitizing effects of light adaptation. *Mailing Add:* Univ Mich Neurosci Bldg 1103 E Huron Ann Arbor MI 48109

GREEN, DAVID, b Philadelphia, Pa, Oct 1, 34; m 58; c 3. HEMATOLOGY. *Educ:* Univ Pa, AB, 56; Jefferson Med Col, MD, 60; Am Bd Internal Med, dipl, 67, cert hemat, 72; Northwestern Univ, PhD(biochem), 74. *Prof Exp:* ATTEND PHYSICIAN, REHAB INST CHICAGO & NORTHWESTERN UNIV, CHICAGO, 67-, PROF INTERNAL MED, MED SCH, 67- *Res:* Hemostasis with particular emphasis on relation to atherosclerosis. *Mailing Add:* Northwestern Univ Rm 1407 Northwestern Mem Hosp 345 E Superior St Chicago IL 60611

GREEN, DAVID CLAUDE, b Ft Wayne, Ind, June 7, 45; m 67; c 3. ORGANIC CHEMISTRY. *Educ:* DePauw Univ, BA, 67; Univ Wash, MS, 69, PhD(org chem), 74. *Prof Exp:* Res chemist, 74-81, SR RES ASSOC, PROCESS RES DEPT, CHEVRON RES CO, CALIF, 81- *Mem:* Am Chem Soc. *Res:* Petroleum hydroprocessing; residuum processing; fluid catalytic cracking technology. *Mailing Add:* Chevron Res Co PO Box 1627 Richmond CA 94802-0627

GREEN, DAVID FRANCIS, physiological chemistry, for more information see previous edition

GREEN, DAVID J, b Reading, Eng, Jan 7, 40. BIOTECHNOLOGY. *Prof Exp:* PRES & SCI DIR, BIOTECHNOL MGT ASN INC, 88- *Mem:* Am Asn Clin Chem; Am Biol Asn. *Mailing Add:* Biotechnol Mgt Asn Inc 25 S Olympia Ave Woburn MA 01801

GREEN, DAVID M(ARTIN), b Vancouver, BC, Jan 20, 53. EVOLUTIONARY BIOLOGY, HERPETOLOGY. *Educ:* Univ BC, BSc, 76; Univ Guelph, MSc, 79, PhD(zool), 82. *Prof Exp:* Postdoctoral fel zool, Mus Vert Zool, Univ Calif, Berkeley, 81-83; lectr biol, McMaster Univ, 83-84, asst prof, 84-85; asst prof biol, Univ Windsor, 85-86; ASST PROF BIOL, REDPATH MUS, MCGILL UNIV, 86- *Concurrent Pos:* Res assoc, Herpet Sect, Can Mus Nature, 84-, mem, Collections Comt, 90-; affil prof, Dept Biol, McGill Univ, 86-; ed, Can Asn Herpetologists Bull, 86- *Mem:* Am Soc Ichthyologists & Herpetologists; Herpetologists League; Soc Study Amphibians & Reptiles; Soc Study Evolution; Sigma Xi. *Res:* Biosystematics and evolution; interspecific hybridization, geographic variation and chromosome evolution in frogs; evolution of sex determination in tetrapods; evolution of supernumerary chromosomes; dynamical systems and fractals as models of evolution; functional morphology of adhesive organs in vertebrates. *Mailing Add:* Redpath Mus McGill Univ 859 Sherbrooke St W Montreal PQ H3A 2K6 Can

GREEN, DAVID WILLIAM, b Hudson, Mich, Nov 19, 42; m 67; c 5. PHYSICAL CHEMISTRY. *Educ:* Albion Col, BA, 64; Univ Calif, Berkeley, PhD(chem), 68; Univ Chicago, MBA, 85. *Prof Exp:* Lectr chem, Univ Calif, Berkeley, 68; res assoc physics, Lab Molecular Struct & Spectra, Univ Chicago, 68-71; asst prof chem, Albion Col, 71-74; chemist, Chem Eng Div, 82, MGR, ANALYTICAL CHEMLAB, ARGONNE NAT LAB, 82-, ASSOC DIR, CHEM TECH DIV, 88- *Concurrent Pos:* Fel, Lawrence Radiation Lab, Univ Calif, Berkeley, 68. *Mem:* AAAS; Am Chem Soc; Anal Lab Mgrs Asn (treas). *Res:* Molecular electronic spectroscopy and structure; high resolution optical spectroscopy; high temperature chemistry; molecular hyperfine structure; matrix isolation spectroscopy. *Mailing Add:* Chem Technol Div Argonne Nat Lab Argonne IL 60439

GREEN, DETROY EDWARD, b Zalma, Mo, Mar 26, 30; m 51; c 5. AGRONOMY. *Educ:* Univ Mo, BS, 54, MS, 61, PhD(field crops), 65. *Prof Exp:* High sch instr voc agr, Mo, 54-59; instr field crops, Univ Mo, 61-64; from asst prof to assoc prof, 64-70, PROF AGRON, IOWA STATE UNIV, 70- *Honors & Awards:* Agron Resident Educ Award, Am Soc Agron, 80; Fel Award, Am Soc Agron, 81; Fel Award, Crop Sci Soc Am, 85. *Mem:* Am Soc Agron; Crop Sci Soc Am; Nat Asn Col & Teachers Agr; Am Soybean Asn. *Res:* Genetics, breeding, and physiology related to soybean improvement. *Mailing Add:* Dept Agron Iowa State Univ Ames IA 50011

GREEN, DON WESLEY, b Tulsa, Okla, July 8, 32; m 54; c 3. CHEMICAL & PETROLEUM ENGINEERING. *Educ:* Univ Tulsa, BS, 55; Univ Okla, MS, 59, PhD(chem eng), 63. *Prof Exp:* Res reservoir engr, Continental Oil Co, 62-64; from asst prof to assoc prof chem & petrol eng, Univ Kans, 64-71, actg chmn dept, 67-68, chmn dept, 70-74, PROF CHEM & PETROL ENG, UNIV KANS, 71-, CONGER-GABEL PROF CHEM & PETROL ENG, 82- *Concurrent Pos:* Co-dir, Tertiary Oil Recovery Proj, 74- *Honors & Awards:* Distinguished Fac Achievement Award, Soc Petrol Engrs. *Mem:* Am Soc Eng Educ; fel Am Inst Chem Engrs; distinguished mem Soc Petrol Engrs. *Res:* Fluid flow through porous media; dispersion of heat and mass in porous media; mathematical modeling of natural resource systems; enhanced oil recovery. *Mailing Add:* 1020 Sunset Dr Lawrence KS 66044

GREEN, DONALD EUGENE, b Napa, Calif, Nov 25, 26; m 51; c 5. BIOCHEMICAL PHARMACOLOGY, ANALYTICAL CHEMISTRY. *Educ:* Univ Calif, Berkeley, BS, 48; Univ Calif, San Francisco, MS, 52, BS, 55; Wash State Univ, PhD(med chem), 62. *Prof Exp:* Instr pharmaceut chem, Idaho State Univ, 55-57; instr, Wash State Univ, 57-58; instr, Idaho State Univ, 58-60; res chemist, Varian Assocs, 62-64, mgr dept biophys, 64-66; sr res chemist, Syva Res Inst, 66-67; sr res chemist, Anal Instrument Res, Varian Assocs, 67-70; res biochemist, Drug Res Lab, Vet Admin Hosp, 70-74, res scientist, Biochem Res Lab, 71-80 & Drug Metab Res Lab, 80-85; SR SCIENTIST, BIOORG CHEM DEPT, APPL IMMUNE SCI, 85- *Concurrent Pos:* Res fel, Sch Pharm, Univ Calif, San Francisco, 62; biochemist, Biochem Res Lab, Vet Admin Hosp, 62-64; res assoc, Sch Med, Stanford Univ, 71-74; consult, Universal Monitor Corp, Pasadena, 71-76; sr res scientist, Inst Chem Biol, Univ San Francisco, 74-82. *Mem:* Am Chem Soc; Am Pharmaceut Asn; Western Pharmacol Soc; Am Soc Pharmacol & Exp Therapeut; Int Asn Forensic Toxicol. *Res:* Chemical structure-pharmacological activity relationships of central nervous system drugs; mechanisms of drug actions; metabolism of cannabinoids and phenothiazine tranquilizers; development of gas chromatography/mass spectrometry interfaces and automated mass fragmentography instrumentation; immobilization of DNA or proteins on polymer surfaces for use in medical therapy devices. *Mailing Add:* Dept Bioorg Chem Appl Immune Sci 200 Constitution Dr Menlo Park CA 94025

GREEN, DONALD MACDONALD, b Poughkeepsie, NY, Apr 6, 30; m 57; c 3. GENETICS. *Educ:* Oberlin Col, AB, 54; Univ Rochester, PhD, 58. *Prof Exp:* Res assoc, Biol Div, Oak Ridge Nat Lab, 58-60; res fel chem & tutor biochem sci, Harvard Univ, 60-64; assoc prof biol, Univ Pittsburgh, 64-67; PROF BIOCHEM & GENETICS, UNIV NH, 67-, CHMN BIOCHEM, 85- *Concurrent Pos:* Prog dir biochem, NSF, 79-80; chmn genetics, Univ NH, 70-72, 85-88. *Res:* Microbial genetics; genetic structure of bacteria and viruses; bacterial transformation and transfection. *Mailing Add:* Dept Biochem Univ NH Durham NH 03824

GREEN, DONALD WAYNE, b Coldwater, Mich, June 19, 24; c 2. PHYSICS. *Educ:* Kalamazoo Col, BA, 49; Ohio State Univ, PhD(physics), 54. *Prof Exp:* Assoc prof, 54-70, PROF PHYSICS, KNOX COL, ILL, 70- *Mem:* Am Asn Physics Teachers. *Res:* Low energy particle accelerators; stopping power of various materials for protons. *Mailing Add:* Dept Physics Knox Col Galesburg IL 61401

GREEN, DOUGLAS R, b New York, NY, Feb 15, 55. IMMUNOLOGY. *Educ:* Yale Univ, BS, 77, PhD(biol), 81. *Prof Exp:* Fel, Dept Path, Yale Univ, 81-83, res assoc, 83-84, assoc res biologist, 84-85; asst prof, 85-87, ASSOC PROF, DEPT IMMUNOL, UNIV ALTA, 87-; MEM & HEAD, DIV CELLULAR IMMUNOL, LA JOLLA INST ALLERGY & IMMUNOL, 90- *Concurrent Pos:* Assoc ed, Cytokines & J Exp Zool, 89- *Mem:* Am Asn Immunologists; Sigma Xi. *Res:* Thermal trauma and immunity; bacterial sepis and immunity; inhibitory lymphokines in thermal trauma; antigen specific T cell factors; antigen-specific regulatory factors; author of numerous scientific publications. *Mailing Add:* La Jolla Inst Allergy & Immunol 11149 N Torrey Pines Rd La Jolla CA 92037

GREEN, EDWARD JEWETT, b Bakersfield, Calif, Oct 31, 37. CHEMICAL OCEANOGRAPHY. *Educ:* Univ Calif, Santa Barbara, AB, 58; Mass Inst Technol, PhD(geochem), 65. *Prof Exp:* Asst prof geochem, Carnegie Inst Technol, 65-71; assoc prof oceanog & geol sci, Univ Maine, 71-78; dir chem oceanog prog, 76-81, group leader oceanog chem & biol group, 81-83, PROJ MGR OCEANIC CHEM, OFF NAVAL RES, 84- *Concurrent Pos:* Staff specialist, Off Undersecretary Defense Res & Eng, 83-84. *Mem:* Am Soc Limnol & Oceanog; AAAS; Am Geophys Union. *Res:* Marine chemistry, especially sources and sinks of dissolved gases in natural waters; scale lengths of variability of trace elements in the oceans. *Mailing Add:* Off Naval Res Code 1122C Oceanic Chem Prog Arlington VA 22217-5000

GREEN, EDWARD LEWIS, b Brooklyn, NY, Apr 4, 46; m 66; c 2. ALGEBRA. *Educ:* Cornell Univ, AB, 67; Brandeis Univ, MA, 68, PhD(math), 73. *Prof Exp:* Instr math, Univ Pa, 73-75; vis lectr, Univ Ill, Urbana, 75-77; asst prof, 77-79, assoc prof, 80-82, PROF MATH, VA POLYTECH INST & STATE UNIV, 83- *Mem:* Am Math Soc. *Res:* Representation theory of rings and algebras, particularly the study of what internal structural properties of a module make it direct sum indecomposable. *Mailing Add:* Dept Math Va Polytech Inst & State Univ Blacksburg VA 24060

GREEN, EDWIN ALFRED, b Atlanta, Ga, Feb 24, 31; m 54; c 3. EXPERIMENTAL ZOOLOGY. *Educ:* Morehouse Col, BS, 54; Atlanta Univ, MS, 60; Univ Okla, PhD(zool), 72. *Prof Exp:* Instr biol, Ft Valley State Col, 60-65; asst prof, Albany State Col, 65-67 & 69-70; asst, Univ Okla, 70-71; from asst prof to assoc prof, 71-78, PROF BIOL, ALBANY STATE COL, 78- *Mem:* AAAS; Sigma Xi; Am Inst Biol Sci. *Mailing Add:* Biol Dept Albany State Col Albany GA 31705

GREEN, EDWIN JAMES, b Oceanside, NY, Nov 11, 54; m 76; c 2. FOREST BIOMETRICS. *Educ:* State Univ NY, Col Environ Sci & Forestry, 76, MS, 78; Va Polytech Inst & State Univ, PhD(forest biometrics), 81. *Prof Exp:* ASST PROF FORESTRY, RUTGERS UNIV, 81- *Mem:* Soc Am Foresters; Am Statist Asn; Biometric Soc. *Res:* Developing growth and yield systems; applying modern quantitative methods to forestry research. *Mailing Add:* Dept Forestry Rutgers Univ New Brunswick NJ 08903

GREEN, ERIC DOUGLAS, b St Louis, 59; m 84. HUMAN GENETICS, CLINICAL PATHOLOGY. *Educ:* Univ Wis, Madison, BS, 81; Wash Univ, MD, 87, PhD(cell biol), 87. *Prof Exp:* RESIDENT LAB MED, WASH UNIV SCH MED, 87-, FEL DEPT GENETICS, 88- *Res:* Structure, synthesis and function of protein-bound oligosaccharides; development and application of recombinant DNA technologies for studying human diseases. *Mailing Add:* Dept Path Box 8118 Wash Univ Med Sch 660 S Euclid Ave St Louis MO 63110

GREEN, ERIKA ANA, b Lucenec, Czech, May 27, 28; US citizen; m 61; c 2. MEDICAL MICROBIOLOGY, SCIENCE WRITING. *Educ:* Cent Univ Ecuador, BS, 55; Hunter Col, MA, 59; Rutgers Univ, PhD(microbiol), 71. *Prof Exp:* Res technician org chem, Sloan Kettering Inst Cancer & Allied Dis, 55-59; res assoc biochem, Rutgers Univ, 59-62; lit scientist, Coun Tobacco Res, 71-72 & Carter-Wallace, Inc, 72-73; lit scientist, 73-81, clin res scientist, Hoffmann-La Roche Inc, 81-85; MED WRITER, THERADEX SYSTS INC, DU PONT MED PROD, SANDOS PHARMACEUT, 85- *Mem:* Am Soc Microbiol; NY Acad Sci; Sigma Xi. *Res:* Literature research in health sciences, especially immunology, medical microbiology and biomedical pharmacology; clinical development of new drugs, especially antibiotics. *Mailing Add:* 409 Grant Ave Highland Park NJ 08904

GREEN, EUGENE L, b Minneapolis, Minn, Oct 15, 27; m 51, 69; c 5. PHYSICS, OPTICS. *Educ:* Carnegie Inst Technol, BS, 47, MS, 49; Temple Univ, PhD(physics), 65. *Prof Exp:* Physicist, US Army, Frankford Arsenal, Pa, 49-65; res physicist, Naval Underwater Systs Ctr, New London Lab, 65-87; CONSULT, SARASOTA, FLA, 90- *Mem:* AAAS; Optical Soc Am. *Res:* Optical information processing, properties of alloys, image analysis and instrumentation; laser development; holography; acoustic array signal processing; optical hydrophones; fiber optics; granted seven patents. *Mailing Add:* 96 Glenwood Ave New London CT 06320

GREEN, FLOYD J, b Sharonville, Ohio, Dec 11, 17; m 41; c 3. ORGANIC & ANALYTICAL CHEMISTRY. *Educ:* Maryville Col, Tenn, AB, 41; Univ Cincinnati, MS, 50; St Thomas Inst, PhD, 69. *Prof Exp:* Res chemist bldg mat, Phillip Carey Mfg, 41-46; chemist reagent chem, Matheson, Coleman & Bell Div, Matheson Co Inc, 46-50, chief chemist, 50-54; tech dir, MC&B Mfg Chemists, Will Ross Co, Inc, 54-73, vpres, 68-73; partner, vpres & treas, Aristo Custom Chem Inc, 73-77; vpres, Aldrich Chem Co, 77-86; RETIRED. *Concurrent Pos:* Adj asst prof, St Thomas Inst. *Mem:* AAAS; fel Am Inst Chemists; Am Chem Soc; Am Soc Testing & Mat. *Res:* Dyes, particularly oxazones; high purity organic reagents. *Mailing Add:* 340 Colony Ct Poinciana Kissimee FL 32758

GREEN, FRANK ORVILLE, b Toledo, Iowa, Nov 2, 08; m 35; c 3. CHEMISTRY. *Educ:* Greenville Col, BS, 31; Northwestern Univ, MS, 37, PhD(org chem), 39. *Prof Exp:* Pub sch teacher, Ohio, 31-35; asst instr, Northwestern Univ, 37-39; prof & dir dept, Greenville Col, 39-41; res chemist, Bauer & Black Co, Ill, 41-42 & Swift & Co, 42-45; from asst prof to prof, 45-82, EMER PROF CHEM, WHEATON COL, 82- *Concurrent Pos:* Fulbright vis lectr, Univ Cairo & Ibrahim Univ, Egypt, 52-53; res assoc, Radiobiol Unit, Mt Vernon Hosp, Eng, 57-58; chem consult, Daubert Chem Co, 60-66; reader, Advan Placement Exam Chem, Educ Testing Serv, NJ, 61, 63, 64, 66, 67 & 68; instr, Sch Nursing, W Suburban Hosp, 75-82; Childs Mem Fund award, 57. *Mem:* Am Chem Soc. *Res:* Acylals; phenoxthins; biochemistry; organic chemistry. *Mailing Add:* Dept Chem Wheaton Col Wheaton IL 60187-5593

GREEN, G W, b Philadelphia, Pa, June 15, 27; m 53; c 2. MICROSCOPY. *Educ:* Drexel Inst, BS, 62. *Prof Exp:* PROD DEVELOP ENGR, GEORGIA PACIFIC CORP, 62- *Mailing Add:* Decatur Gypsum Lab Ga Pac Corp, 2861 Miller Rd Decatur GA 30033

GREEN, GARY MILLER, b Bakersfield, Calif, Dec 28, 40; m 67; c 2. NUTRITION, PHYSIOLOGY. *Educ:* Univ Calif, Berkeley, BS, 66, PhD(nutrit), 71. *Prof Exp:* Fel nutrit & digestion res, Univ Calif, Berkeley, 71-72; asst res physiol, B Lyon Mem Res Lab, Children's Hosp Med Ctr, Oakland, Calif, 72-79; asst prof med, 79-84, asst prof, 84-88, ASSOC PROF PHYSIOL, UNIV TEX HEALTH SCI CTR, SAN ANTONIO, 88- *Concurrent Pos:* NIH fel, dept nutrit sci, Univ Calif, Berkeley, 71-72; NIH grant, B Lyon Mem Res Lab, Children's Hosp Med Ctr, 74-80, cystic fibrosis found grant, 78-80. *Mem:* Am Inst Nutrit; Am Physiol Soc. *Res:* Physiology of digestion and absorption in non-ruminants; regulation of exocrine pancreatic secretion; bile acid metabolism; nutrition and gastrointestinal function in digestive diseases, particularly cystic fibrosis. *Mailing Add:* Dept Physiol Univ Tex Health Sci Ctr 7703 Floyd Curl Dr San Antonio TX 78284

GREEN, GEORGE G, b Sayre, Okla, Mar 21, 22; m 46; c 1. ANIMAL SCIENCE, ANIMAL NUTRITION. *Educ:* Okla State Univ, BS, 50, MS, 56; Tex A&M Univ, PhD(nutrit), 58. *Prof Exp:* Asst exten agt agr, Okla State Univ, 50-51, exten agt, 51-52, farm mgr, 52-54; assoc prof animal sci, Va Polytech Inst, 58-63; exten specialist livestock, Auburn Univ, 63-64; ASSOC PROF ANIMAL SCI, VA POLYTECH INST & STATE UNIV, 64- *Mem:* Am Soc Animal Sci. *Res:* Appetite and growth among ruminants. *Mailing Add:* 2504 Meadowbrook Dr Blacksburg VA 24060

GREEN, GERALD, b New York, NY, June 29, 41; m 76; c 2. IMMUNOCHEMISTRY, CLINICAL CHEMISTRY. *Educ:* Brooklyn Col, BSc, 64; Ind Univ, Bloomington, PhD(biol chem), 70; New Eng Col Optom, OD, 89. *Prof Exp:* Res assoc chem, Univ Calif, Los Angeles, 70-73; sr scientist, Union Carbide Corp, Tarrytown, NY, 73-77; sr scientist, Wampole Labs, Princeton, NJ, 77-79; sr res investr, Squibb Inst Med Res, Princeton, NJ, 79-84; proj mgr, Ventrex Labs, Portland, Maine, 84-87; PUT PRAC OPTOMETRIST, NJ, 89- *Concurrent Pos:* Adj assoc prof health sci, C W Post Col, Greenvale, NY, 74-86. *Mem:* Am Chem Soc; Am Asn Clin Chemists; Clin Radioassay Soc; AAAS; Am Optom Asn. *Res:* Physico-chemical studies on antibody-antigen interactions; development and production of antibodies to molecules of clinical interest; characterization of antibodies with regard to titer, specificity, binding constant and suitability in radioimmunoassays; manual and automated radioimmunoassays. *Mailing Add:* 16 Wallingford Dr Princeton NJ 08540

GREEN, HAROLD D, MECHANISM OF ACTION & CONTROL OF HEART CIRCULATION. *Educ:* Case Western Reserve Univ, MD, 31. *Prof Exp:* Chmn dept physiol & pharmacol, Bowman Gray Sch Med, 45-80; Retired. *Mailing Add:* 3619 Dewsbury Rd Winston-Salem NC 27104

GREEN, HAROLD RUGBY, b Hallettsville, Tex, Feb 19, 26; m 53. NUMBER THEORY. *Educ:* Tex Wesleyan Col, BA, 46; Tex Christian Univ, MA, 48; NTex State Univ, MS, 61. *Prof Exp:* Pub sch teacher, Tex, 46-55; asst prof math, Arlington State Col, 55-59 & McNeese State Col, 59-60; asst prof, 60-63, ASSOC PROF MATH, UNIV TEX, ARLINGTON, 63-; DIV CHMN, INDUST & TECHNOL DIV, ORANGEBURG CALHOUN TECH COL. *Mailing Add:* Orangeburg Calhoun Tech Col 3250 St Matthews Rd SE Orangeburg SC 29115-8299

GREEN, HARRY, b Philadelphia, Pa, Sept 7, 17; m 45; c 2. BIOCHEMISTRY. *Educ:* Univ Pa, AB, 38, MS, 39, PhD(org chem), 42. *Prof Exp:* Sr res chemist, Lion Oil Refining Co, Ark, 41-44 & Whitemarsh Res Labs, Pennsalt Mfg Co, 44-47; res assoc physiol chem, Univ Pa, 47-52; chief biochem res, Wills Eye Hosp, 52-58; sr res scientist, Smithkline Corp, 58-64, head neurobiochem, 64-67, dir biochem, 67-75, dir sci liaison, 75-80, vpres sci liaison, 80-83; RETIRED. *Concurrent Pos:* Harrison fel chem, 40-41; asst prof biochem, Grad Sch Med, Univ Pa, 54-68. *Mem:* AAAS; Am Chem Soc; Am Soc Biol Chem; Asn Res Nerv & Ment Dis; Am Soc Pharmacol & Exp Therapeut. *Res:* Intermediary metabolism; ocular biochemistry and physiology; enzymology; corticosteroids; neurobiochemistry; drug metabolism; biochemistry of respiratory glycoproteins; diabetes. *Mailing Add:* 5771 Fairway Park Ct Boynton Beach FL 33437

GREEN, HARRY J(AMES), JR, b St Louis, Mo, Dec 7, 11; m 39; c 2. CHEMICAL ENGINEERING. *Educ:* Ohio State Univ, BChE, 32, PhD(chem eng), 43; Mass Inst Technol, MS, 38. *Prof Exp:* Instr chem, Agr & Tech Col, NC, 34-37, asst prof, 38-41, prof, 43-44; sr engr, Mat Eng Dept, Stromberg-Carlson Co, 44-59, supvr mfg res & develop, Prod Eng Dept, 59-67, prin engr microelectronics, Electronics Div, Gen Dynamics, 67-70; SCIENTIST RES DEPT, XEROX CORP, 70- *Mem:* Am Chem Soc; Am Soc Metals. *Res:* Materials engineering; metals; polymer applications; wire and insulation; telephone transmitter materials; electrical properties of plastics; microelectronic packaging of thick and thin film hybrid circuits; xerographic materials development. *Mailing Add:* 307 Greeley St Rochester NY 14609

GREEN, HARRY WESTERN, II, b Orange, NJ, Mar 13, 40; m 75; c 7. GEOPHYSICS, STRUCTURAL GEOLOGY. *Educ:* Univ Calif, Los Angeles, AB, 63, MS, 67, PhD(geol), 68. *Prof Exp:* Res assoc geol & metall, Case Western Reserve Univ, 68-70; from asst prof to assoc prof, 70-80, chmn dept, 84-88, PROF GEOL, UNIV CALIF, DAVIS, 80- *Concurrent Pos:* NSF res grants, 69-; US-France exchange scientist, 73; vis prof, Univ Nantes, France, 78-79 & Monash Univ, Australia, 84; spec adv, China Univ Geosci, 88; DOE res grants, 88-; adj sr researcher, Lamont-Doherty Geol Observ, Columbia Univ, 89- *Mem:* Am Geophys Union; fel Mineral Soc Am. *Res:* Experimental rock deformation at high temperature and pressure; elucidation of deformation mechanisms utilizing high voltage transmission electron microscopy; theoretical studies of deformation in the earth's interior; solid earth geophysics; role of volatiles in the dynamics of the earths mantle; mechanism of deep focus earthquakes; effect of stress on phase transformations. *Mailing Add:* Dept Geol Univ Calif Davis CA 95616

GREEN, HOWARD, b Toronto, Ont, Sept 10, 25; nat US; m 54. CELLULAR PHYSIOLOGY, MOLECULAR BIOLOGY. *Educ:* Univ Toronto, MD, 47; Northwestern Univ, MS, 50. *Hon Degrees:* DSc, Univ Conn, 85; MD, Univ Göteborg, 89. *Prof Exp:* Instr biochem, Univ Chicago, 51-53; instr, Dept Pharmacol, Sch Med, NY Univ, 54, asst prof chem path, 56-59, from assoc prof to prof path, 59-68, prof & chmn, Dept Cell Biol, 68-70; capt, Immunol Div, Walter Reed Army Inst Res, USAR, 55-56; prof cell biol, Dept Biol, Mass Inst Technol, 70-80; GEORGE HIGGINSON PROF CELLULAR PHYSIOL & CHMN, DEPT CELLULAR & MOLECULAR PHYSIOL, HARVARD MED SCH, 80- *Concurrent Pos:* NIH vis scientist, Cornell Univ, 54. *Honors & Awards:* J Howard Mueller Mem Lectr, Harvard Univ, 69; J N Taub Int Mem Award, 77; Selman A Waksman Award, 78; Lewis S Rosenstiel Award, 80; Lila Gruber Res Award, Am Acad Dermat, 80; Passano Award, 85; Howard Fox Mem Lectr, New York Univ, 78; Harvey Lectr, 79; Marchon Lectr, Newcastle, Eng, 82; Unilever Lectr, Eng, 84; Ravdin Lectr, Am Col Surgeons, 85; H M Evans Mem Lectr, Univ Calif, San Francisco, 88. *Mem:* Nat Acad Sci; Am Soc Cellular Biol; Am Acad Arts & Sci. *Res:* Differentiation; genetics; cancer; epidermal drafts in burn treatments. *Mailing Add:* Dept Cellular & Molecular Biol Harvard Med Sch 25 Shattuck St Boston MA 02115

GREEN, HUBERT GORDON, b Dallas, Tex, Oct 31, 38; m 69; c 4. PUBLIC HEALTH, PEDIATRICS. *Educ:* Rice Univ, BA, 62; Univ Tex Southwestern Med Sch, MD, 68; Univ Calif, MPH, 72. *Prof Exp:* Assoc prof pediat, Col Med, Univ Ark, Little Rock, 72-77, assoc prof biomet, 75-77; med dir, Ark Children's Hosp, 73-77; dir Handicapped Children's Ctr, Ark Dept Health, 75-76; dep dir, Div Health Serv, US Pub Health Serv, Dallas, Tex, 77-83; DIR, DALLAS COUNTY HEALTH DEPT, 83- *Mem:* Am Pub Health Asn; Am Col Prevent Med; Am Acad Pediat; AMA. *Res:* Teratogenic effects, drugs and chemicals; organization of health services; handicapped children. *Mailing Add:* Dallas County Health Dept 1936 Amelia Ct Dallas TX 75235

GREEN, IRA, TUMOR IMMUNOLOGY, CLINICAL IMMUNOLOGY. *Educ:* State Univ NY, Brooklyn, MD, 53. *Prof Exp:* SR INVESTR IMMUNOL, NAT INST ALLERGY & INFECTIOUS DIS, NIH, 68- *Mailing Add:* Pub Health Serv Parklawn Build 5600 Fishers Lane Rm 18-40 Rockville MD 20857

GREEN, JACK PETER, b New York, NY, Oct 4, 25; m 58. PHARMACOLOGY. *Educ:* Pa State Univ, BS, 47, MS, 49; Yale Univ, PhD(pharmacol), 51, MD, 57. *Prof Exp:* From asst prof to assoc prof pharmacol, Yale Univ, 57-66; assoc prof, Med Sch, Cornell Univ, 66-68; PROF PHARMACOL & CHMN DEPT, MT SINAI SCH MED, 68- *Concurrent Pos:* Res fel, Polytech Inst, Denmark, 53-55; USPHS res career prog award, 58-66; Eleanor Roosevelt fel, 64-66; vis scientist, Inst Phys Chem Biol, Univ Paris, 64-66; consult, chem/biol info handling rev comt, Div Res Facil & Resources, NIH, 66-68, mem, 68-70, mem preclin psychopharmacol res rev comt, NIMH, 69-73; mem steering comt, exec comt, Biochem Pharmacol Discussion Group, 66-70; mem prog comt, Gordon Res Conf on Med Chem, 67-68, vchmn, 73, chmn, 74; mem res coun, Pub Health Res Inst New York, 72-75; consult, Health Res Coun New York, 72-75, chmn, 75; mem med adv bd, Dysautonomia Found, 74-77; mem sci adv comt, Irma T Hirschl Trust, 77-; sci counr, Nat Inst Environ Health Serv, 81-84. *Honors & Awards:* Claude Bernard Award, 66. *Mem:* NY Acad Sci; Soc Drug Res; Int Soc Quantum Biol; fel Am Col Neuropsychopharmacol; Asn Med Sch Pharmacol (treas, 73-75); Am Soc Neurochem. *Res:* Mechanism of action of drugs. *Mailing Add:* Dept Pharmacol Mt Sinai Sch Med New York NY 10029-6574

GREEN, JAMES WESTON, b Elkins, WVa, May 16, 13; m 61; c 2. PHYSIOLOGY. *Educ:* Davis-Elkins Col, BS, 35; Princeton Univ, MA, 47, PhD(biol), 48. *Prof Exp:* Instr chem, Potomac State Col, 41-42; asst biol, Princeton Univ, 48-49; from asst prof to assoc prof physiol, 48-61, prof physiol, 61-80, chmn dept, 66-72, dir grad physiol, 68-73, actg dean grad sch, 74-75, EMER PROF PHYSIOL, RUTGERS UNIV, NEW BRUNSWICK, 80- *Concurrent Pos:* Ford fel, 53-54. *Mem:* Soc Gen Physiol (secy, 59-61); Am Physiol Soc; NY Acad Sci. *Res:* Cell permeability; cation regulation in cells and tissues; tissue cell membranes; membrane biochemistry; photo-oxidation in cells; red cell aging. *Mailing Add:* 409 Grant Ave Highland Park NJ 08904

GREEN, JEFFREY DAVID, b Worcester, Mass, Feb 22, 47. CELL BIOLOGY, FERTILIZATION & EARLY DEVELOPMENT. *Educ:* Clark Univ, Worcester, Mass, AB, 69; Univ NC, Chapel Hill, MA, 76; State Univ NY, Buffalo, PhD(anat), 81. *Prof Exp:* NIH fel, Univ Wash, Seattle, 81-83; asst prof, 83-88, ASSOC PROF ANAT, MED CTR, LA STATE UNIV, NEW ORLEANS, 88- *Mem:* Am Asn Anatomists; Am Soc Cell Biol; AAAS; Am Soc Zoologists. *Res:* Cell biology of fertilization; role of proteolytic enzymes during sperm-egg interaction and early development. *Mailing Add:* Dept Anat La State Univ Med Ctr 1901 Perdido St New Orleans LA 70112-1393

GREEN, JEFFREY SCOTT, b Colorado Springs, Colo, Oct 22, 47; m 75; c 5. PREDATOR RESEARCH, WILDLIFE RESEARCH. *Educ:* Brigham Young Univ, BS, 71, MS, 76, PhD(wildlife & range resources), 78. *Prof Exp:* Res assoc, Brigham Young Univ, 78-79; res wildlife biologist, Agr Res Serv, 79-88, WILDLIFE BIOLOGIST, ANIMAL & PLANT HEALTH INSPECTION SERV, ANIMAL DAMAGE CONTROL, US DEPT AGR, 88- *Mem:* Am Soc Mammalogists; Soc Range Mgt; Wildlife Soc. *Res:* Methodology for reducing predation of sheep by coyotes: livestock guarding dogs, coyote attractants and antifertility agents. *Mailing Add:* US Dept Agr Animal Damage Control Div 12345 W Alamedo Park Suite 313 Lakewood CO 80228

GREEN, JEROME, b Cameron, Mo, June 3, 53; m 76; c 4. PLANT PHYSIOLOGY. *Educ:* Univ Iowa, BA, 75, MS, 77, PhD(bot), 79. *Prof Exp:* Res biologist, 79-91, SR RES BIOLOGIST, E I DU PONT DE NEMOURS & CO, INC, 91- *Mem:* Am Soc Plant Physiol; Plant Growth Regulator Soc Am; Weed Sci Soc Am. *Res:* Plant growth regulators; herbicides. *Mailing Add:* Stine-Haskell Res Ctr E I du Pont de Nemours & Co Inc PO Box 30 Elkton Rd Newark DE 19714

GREEN, JEROME GEORGE, b Brooklyn, NY, June 20, 29; m 52; c 2. MEDICINE. *Educ:* Brooklyn Col, BS, 50; Albany Med Col, MD, 54. *Prof Exp:* Extramural Scientist, NIH, 55-57; resident med, PHS Hosp, San Francisco, 57-59; res fel cardiol, Univ Calif, San Francisco, 59-60; clin invest cardiol, Cleveland Clin Res Div, 60-65; assoc dir, Nat Heart & Lung Inst, 66-72, div dir, 72-86, DIR, DIV RES GRANTS, NIH, 86- *Honors & Awards:* Distinguished Serv Medal, PHS; Distinguished Serv Award, Am Col Cardiology. *Mem:* Am Col Cardiol; fel Am Heart Assoc. *Res:* Cardiopulmonary. *Mailing Add:* Div Res Grants NIH Westwood Bldg 450 Westbard Ave Bethesda MD 20892

GREEN, JEROME JOSEPH, b Chicago, Ill, Oct 10, 32; m 58; c 4. MAGNETISM. *Educ:* Northwestern Univ, BS, 54; Harvard Univ, AM, 55, PhD(appl physics), 59. *Prof Exp:* Asst, Harvard Univ, 57-59; mem staff, 59-62, mgr magnetic group, 70-84, PRIN RES SCIENTIST, RES DIV, RAYTHEON, INC, 62- *Mem:* Sr mem Inst Elec & Electronics Engrs. *Res:* Microwave physics; ferromagnetic resonance; magnetic circuits; phase shifters; magnetic memories. *Mailing Add:* 28 Winchester Dr Lexington MA 02173

GREEN, JOHN ARTHUR SAVAGE, b Ballycastle, N Ireland, Dec 29, 39; US citizen; m 68; c 2. ELECTROCHEMISTRY, MATERIALS SCIENCE. *Educ:* Queen's Univ, Belfast, BSc, 62, PhD(chem), 65. *Prof Exp:* Res fel corrosion, Mellon Inst, 65-66; res assoc, Univ Newcastle upon Tyne, 66-67; scientist, Res Inst Adv Studies, 67-72; sr scientist, 72-74, assoc dir & mgr, Aluminum Res, 74-84, DIR, ADV MAT, MARTIN MARIETTA LABS, 84- *Honors & Awards:* AB Campbell Award, Nat Asn Corrosion Eng, 69. *Mem:* Nat Asn Corrosion Eng; Electrochem Soc; Am Inst Metall Eng. *Res:* Hydrogen palladium system; corrosion of aluminum and titanium alloys; stress corrosion cracking; hydrogen embrittlement; bauxite refining; aluminum smelting and fabrication; metal matrix composites; particle technology and signiture reduction. *Mailing Add:* Martin Marietta Labs 1450 S Rolling Rd Baltimore MD 21227

GREEN, JOHN CHANDLER, b West Hartford, Conn, Feb 7, 32; m 58; c 2. VOLCANOLOGY, GEOCHEMISTRY. *Educ:* Dartmouth Col, AB, 53; Harvard Univ, MA, 56, PhD, 60. *Prof Exp:* From asst prof to assoc prof, 58-68, head dept, 74-77, PROF GEOL, UNIV MINN, DULUTH, 68- *Concurrent Pos:* Geologist, Minn Geol Survey, 62-78 & 85 & Minn Dept Natural Resources, 72 & 84. *Mem:* AAAS; Geol Soc Am; Mineral Soc Am; Norweg Geol Soc. *Res:* Petrology of igneous rocks; origin of large, higb-T hynolites; Pre-Cambrian geology of Minnesota; Keweenawan lavas and dikes; evolution of midcontinent rift system. *Mailing Add:* 1754 Old North Shore Rd Duluth MN 55804

GREEN, JOHN H, b Pittsburgh, Pa, Jan 23, 29. FOOD MICROBIOLOGY, FERMENTATION BIOCHEMISTRY. *Educ:* Univ Rochester, BA, 51; State Univ NY Col Teachers, Albany, 55; Mich State Univ, PhD(microbiol), 63; Univ Md, MSc, 76. *Prof Exp:* Instr chem & bact, State Univ NY Agr & Tech Inst, Alfred Univ, 56-58; fel, Dept Environ Sci, Univ Mass, 63-64, NIH res assoc, 64-66; res microbiologists, US Nat Marine Fisheries Serv, Nat Oceanic & Atmospheric Agency, 66-76; sr res assoc, Dept Food Sci, Cornell Univ, 76-77; auth & consult food waste mgt, 77-79; environ & energy specialist, Animal & Plant Health Inspection Serv, 79-81, MICOROBIOLOGIST, FOOD MICROBIOL BR, FOOD SAFETY & INSPECTION SERV, USDA, 81- *Concurrent Pos:* Instr food microbiol & food processing waste mgt, Univ Md, College Park. *Mem:* Am Soc Microbiol; Inst Food Technol; Soc Indust Microbiol; Brit Soc Appl Microbiol; Sigma Xi. *Res:* Microbial physiology; food science; industrial (food) waste management; author of over 40 publications; author of one book. *Mailing Add:* 255 Snapfinger Dr Athens GA 30605-4432

GREEN, JOHN IRVING, b Cedarhurst, NY, Mar 25, 24; m 60; c 2. ECOLOGY, BOTANY. *Educ:* State Univ NY, BS, 49; Syracuse Univ, MS, 51; Cornell Univ, PhD(sci educ), 61. *Prof Exp:* High sch teacher, NY, 51-53; asst prof biol, bot & zool, State Univ NY Col Brockport, 55-56; exten biologist wildlife educ, Wildlife Div, Mines & Resources Dept, Prov of Nfld, 58-60; asst prof, NJ State Sch Conserv, 61-62; asst prof conserv, Exten Div, Cornell Univ, 62-65; from asst prof to assoc prof ecol & bot, 65-77, chmn dept, 77-80, PROF BIOL, ST LAWRENCE UNIV, 77- *Concurrent Pos:* Guest lectr, Mem Univ Nfld, 59-60; mem, Temp State Comn Youth Educ in Conserv, 70-73. *Mem:* Am Wildlife Soc; Ecol Soc Am; Am Nature Study Soc (pres, 73). *Res:* Wildlife and sociological impact; development of techniques and methods for teaching conservation; wildlife ecology; waterfowl; wildlife-plant ecology. *Mailing Add:* Dept Biol St Lawrence Univ Canton NY 13617

GREEN, JOHN M, b Carlisle, Pa, Jan 10, 40; m 66; c 3. MARINE SCIENCES GENERAL. *Educ:* Univ Mich, BSc, 61; Univ Miami, MSc, 64; Univ BC, PhD(ichthyol), 68. *Prof Exp:* Asst prof marine biol, 68-73, assoc prof biol, 74-78, PROF BIOL, MEM UNIV, 79- *Concurrent Pos:* Consult, marine biol; instr, Bamfield Marine Sta; vis prof, Simon Fraser Univ, 80-, Univ BC, 81 & Univ Mauritius, 89. *Mem:* Am Soc Ichthyol & Herpet; Can Soc Zool; Sigma Xi. *Res:* Behavioral ecology of fishes, particularly with respect to reproductive behavior, social behavior and mechanisms of orientation and homing; ecology of arctic fishes. *Mailing Add:* Dept Biol Mem Univ Nfld St John's NF A1B 3X9 Can

GREEN, JOHN ROOT, b Alameda, Calif, Sept 19, 20; m 51; c 3. PHYSICS. *Educ:* Univ Calif, BS, 41, PhD(physics), 50. *Prof Exp:* From asst prof to prof, 50-81, EMER PROF PHYSICS, UNIV NMEX, 81- *Concurrent Pos:* Fulbright fel, Univ Aleppo, 66-67; Fulbright lectr physics, Univ Jordan, 73-74. *Mem:* Am Phys Soc. *Res:* Cosmic radiation; design of cloud chambers; plastic crystals; phase transformations; dielectric properties. *Mailing Add:* Dept Physics Univ NMex Albuquerque NM 87106

GREEN, JOHN WILLIAM, b Garrett, Ind, Dec 21, 35; c 2. INFORMATION SCIENCE & SYSTEMS. *Educ:* Miami Univ, BA, 57; Univ Wis, PhD(phys chem), 62. *Prof Exp:* Res chemist, Calif Res Corp, 62-70, sr res chemist, 70-76, sr res assoc, Chevron Res Co, 76-88, RES SCIENTIST, CHEVRON RES & TECHNOL CO, CHEVRON CORP, 88- *Mem:* Soc Appl Spectros. *Res:* High temperature thermodynamic studies; mass spectrometry; neutron activation analysis; radiotracer methods; optical and electron microscopy; x-ray diffraction; analytical instrumentation; laboratory data management systems. *Mailing Add:* Chevron Res & Technol Co 100 Chevron Way Richmond CA 94802

GREEN, JOHN WILLIE, b Hearne, Tex, Mar 8, 14; m 38; c 2. MATHEMATICS. *Educ:* Rice Inst, BA, 35, MA, 36; Univ Calif, PhD(math), 38. *Prof Exp:* Asst math, Rice Inst, 35-36; asst, Pierce Inst, Harvard Univ, 38-39; from instr to asst prof, Univ Rochester, 39-43; mathematician, Ballistic Res Lab, Aberdeen Proving Grounds, 43-45; from asst prof to prof, 45-84, EMER PROF MATH, UNIV CALIF, LOS ANGELES, 84- *Concurrent Pos:* Mem, Inst Adv Study, 51; chmn, Conf Bd Math Sci, 71-72. *Mem:* Am Math Soc (assoc secy, 47-55, secy, 57-67); Math Asn Am. *Res:* Potential theory; theory of functions; convex bodies; harmonic functions. *Mailing Add:* 6363 Math Sci Bldg Univ Calif Los Angeles Los Angeles CA 90024

GREEN, JONATHAN P, b New York, NY, June 16, 35; m 59; c 2. ZOOLOGY, PHYSIOLOGY. *Educ:* Pa State Univ, BS, 57; Univ Minn, PhD(zool), 63. *Prof Exp:* NIH training grant pathobiol, Johns Hopkins Univ, 63, fel, 63-64; asst prof biol, Brown Univ, 64-72; lectr physiol, Lab Comp Physiol, Dept Zool, Univ Malaya, 71-78; assoc prof biol, Reed Col, 78-79; CHMN & PROF BIOL, ROOSEVELT UNIV, 79- *Concurrent Pos:* Mem, Marine Biol Lab, Woods Hole, Mass, instr invertebrate zool, 66-70; fel, Thailand Nat Acad Sci, 84-86 & 88-90. *Mem:* AAAS; Am Soc Zool; Malaysian Nature Soc; Am Inst Biol Sci. *Res:* Physiology and ultrastructure of the crustacean epidermis; physiological and morphological color changes in Crustacea. *Mailing Add:* Dept Biol Roosevelt Univ Chicago IL 60605

GREEN, JOSEPH, b Brooklyn, NY, Oct 5, 28; m 51; c 3. FLAME RETARDANTS, PLASTICS. *Educ:* City Col New York, BS, 50; Univ Kans, MS, 52. *Prof Exp:* Chemist, US Rubber Reclaiming Co, Inc, 51-55 & Food Mach & Mfg Corp, Inc, 55-56; supvr appl chem res sect, Reaction Motors Div, Thiokol Chem Corp, 56-66, mgr synthesis & polymer chem dept, 66-67; mgr polymerization res, Petrochem Res Group, Cities Serv Res Ctr, Cranbury, 67-78; vpres mkt & technol, Saytech, Inc, Cranbury, 78-80; mgr polymer additives res & develop, 80-88, PRIN SCIENTIST, FMC CORP, PRINCETON, NJ, 88- *Concurrent Pos:* Sem fac, Soc Plastics Eng; mem bd dirs, Fire Retardant Chem Asn. *Mem:* Fel Am Inst Chemists; Am Chem Soc; Soc Plastics Eng; Fire Retardant Chem Asn (pres, 89-). *Res:* Polymer chemistry, synthesis and evaluation and organic chemistry; flame retardancy; thermoplastics, thermosets, polymer additives, rubber chemistry and formulation; thermal stability and chemical resistance; propellant chemistry; boron and nitroso polymers; carboranes; fluorocarbons; polyurethanes; polyesters; polypropylene; polyethylene; acrylonitrile butadiene styrene; polystyrene, PVC, engineering plastics and polymer alloys; organobromine chemistry; organophosphorus chemistry and polymer additives; over 175 technical publications. *Mailing Add:* 3 New Dover Rd East Brunswick NJ 08816

GREEN, JOSEPH MATTHEW, b New York, NY, Nov 29, 26; m; m 64; c 4. THEORETICAL PHYSICS. *Educ:* Calif Inst Technol, BS, 49, PhD(physics), 57. *Prof Exp:* Asst physics, Univ Chicago, 49-50; asst math, Calif Inst Technol, 52-54, asst hydrodyn, Hydrodyn Lab, 50-57; physicist, Rand Corp, 57-71; physicist, res & develop assocs, 71-90; CONSULT PHYSICS, LAWRENCE LIVERMORE NAT LABS, 90- *Mem:* Am Phys Soc; Sigma Xi. *Res:* Interaction of radiation with matter; equation of state; plasma physics. *Mailing Add:* 24617 Eliat St Woodland Hills CA 91364

GREEN, JUDY, b Brooklyn, NY, Sept 6, 43; m 64; c 2. HISTORY OF MATHEMATICS. *Educ:* Cornell Univ, BA, 64; Yale Univ, MA, 66; Univ Md, PhD(math), 72. *Prof Exp:* Instr math, Howard Univ, 66-67; from asst prof to assoc prof math, Rutgers Univ, Camden, 72-90, chair, Dept Math Sci, 83-86; PROF MATH, MARYMOUNT UNIV, 90- *Concurrent Pos:* Vis asst prof, Howard Univ, 75; hon res fel, div math, Nat Mus Am Hist, Smithsonian Inst, 79- *Mem:* Am Math Soc; Asn Women Math (secy, 71-72, vpres, 77-78); Am Asn Univ Profs (1st vpres, 88-90); Asn Symbolic Logic; Math Asn Am. *Res:* History of women in American mathematics; history of the algebra of logic. *Mailing Add:* 10106 Leder Rd Silver Spring MD 20902

GREEN, KEITH, b Nuneaton, Eng, Aug 16, 40; m 64; c 2. PHYSIOLOGY, BIOPHYSICS. *Educ:* Univ Leicester, BSc, 61; Univ St Andrews, PhD(pharmacol), 64, DSc, 84. *Prof Exp:* From instr to assoc prof ophthal, Sch Med, Johns Hopkins Univ, 66-74; from assoc prof to prof, 74-78, REGENTS PROF OPHTHAL, SCH MED, MED COL GA, AUGUSTA, 78-, DIR OPHTHAL RES, 74-, PROF PHYSIOL, 80- *Concurrent Pos:* NIH res fel ophthal, Sch Med, Johns Hopkins Univ, 64-66; Nat Eye Inst res career develop award, 70-74; lectr physiol, Med Col Ga, 74-80; vis A study sect, NIH, 78-82. *Mem:* Am Acad Ophthal; Am Physiol Soc; Biophys Soc; Asn Res Vision & Ophthal; Sigma Xi. *Res:* Solute and solvent transfer across biological membranes; mechanisms underlying control of corneal thickness; mechanism of aqueous humor formation; hormonal effects on membrane transport; mechanisms of fluid dynamics in the eye; ocular toxicology. *Mailing Add:* Dept Ophthal Med Col Ga Augusta GA 30912-3400

GREEN, LARRY J, b Memphis, Tenn, Jan 1, 31; m 58; c 3. ORTHODONTICS. *Educ:* Univ Pittsburgh, BS, 53, DDS, 56, MS, 60; Univ Iowa, PhD(anat growth), 65; Am Bd Orthod, dipl. *Prof Exp:* Asst prof orthod, Univ Pittsburgh, 60-62; from asst prof to assoc prof, 65-71, PROF ORTHOD, SCH DENT, STATE UNIV NY BUFFALO, 71- *Concurrent Pos:* Clin pract, 60-62, 65-; consult, NY State Dent Rehab Prog, 84-; pres, Erie County Dent Soc, 85. *Mem:* Am Dent Asn; Am Asn Orthod; Int Asn Dent Res; fel Int Col Dent; fel Am Col Dent; Nat Adv Dent Res Coun; Nat Inst Dent Res. *Res:* Anatomic growth, especially facial. *Mailing Add:* Dept Orthodont Sch Dent Med 140 Squire Hall State Univ NY Buffalo NY 14214

GREEN, LAWRENCE, b Gelenes, Hungary, May 3, 37; US citizen; m 63; c 2. EXPERIMENTAL NUCLEAR PHYSICS. *Educ:* City Col New York, BS, 59; Pa State Univ, PhD(nuclear physics), 63. *Prof Exp:* Sr scientist, Bettis Atomic Power Lab, 64-73, fel scientist, 73-78, fel scientist, Fusion Power Systs Div, 78-84, FEL SCIENTIST, NUCLEAR FUELS DIV, WESTINGHOUSE ELEC CORP, 84- *Concurrent Pos:* Vis prof, Ben Gurion Univ, Israel, 77-78; vis scientist, Swiss Fed Inst Technol, Lausanne, 83-84. *Mem:* AAAS; Am Nuclear Soc; Am Phys Soc. *Res:* Low energy nuclear physics; reactors. *Mailing Add:* 151 Dutch Lane Pittsburgh PA 15236

GREEN, LAWRENCE WINTER, b Bell, Calif, Sept 16, 40; m 62; c 2. PUBLIC HEALTH EDUCATION. *Educ:* Univ Calif, Berkeley, BS, 62, MPH, 66, DrPH, 68. *Prof Exp:* USPHS trainee pub health educ, Calif State Dept Pub Health, 62-63; training assoc family planning, Ford Found, Dacca & Karachi, Pakistan, 63-65; lectr pub health educ, Sch Pub Health, Univ Calif, Berkeley, 68-70; from asst prof to prof pub health admin, pop dynamics & behavioral sci, Sch Hyg & Pub Health, Johns Hopkins Univ, 70-81, asst dean continuing educ, 72-76; prof & head, Health Educ Div, Health Serv Res & Develop Ctr, Johns Hopkins Med Insts, 76-81, dir health educ studies, 75-81; dir, US Off Health Info & Health Prom, Dept Health & Human Serv, 79-81; prof community med & dir, Ctr Health Prom Res & Develop, & co-dir, Southwest Center for Prev Res, Univ Tex Health Sci Ctr, Houston, 81-88; vpres, Henry J Kaiser Family Found, Menlo Park, Calif, 88-91; POLICY SCHOLAR, INST HEALTH POLICY STUDIES, UNIV CALIF, SAN FRANCISCO, 91- *Concurrent Pos:* Res specialist, Family Res Ctr, Langley Porter Neuropsychiat Inst, 68-70; ed, Soc Pub Health Educ, 72-75; consult, WHO & AID, 73-74, 81-84, Vet Admin, Am Heart Asn, Kellogg Found & Dept Health, Educ & Welfare, 74-76 & Nat Ctr Health Serv Res, 75-; mem nat policy comt, Nat Ctr Health Educ, Nat Health Coun, 74-75; counr, Coun Educ for Pub Health, 74-78; consult & task force mem, Nat Heart & Lung Inst, 75-76; mem expert panel on consumer health educ, Am Col Prev Med & NIH, 76; vis lectr, Div Health Policy Res & Educ, Harvard Univ, 81-82; mem, Nat Adv Comt Vital & Health Statist, 85-87; chmn, Comt on Drug Abuse Prev Res, Nat Res Coun, Nat Acad Sci, 89-91; Alliance Scholar, Am Alliance Health, Phys Educ & Recreation. *Honors & Awards:* Award for Advan Fertil Control, Excerpta Medica Found & Syntex Labs, 70; Beryl J Roberts Mem Award for Res, Soc Pub Health Educ, 72; Distinguished Career Award, Am Public Health Asn, 78. *Mem:* Fel Am Pub Health Asn; distinguished fel Soc Pub Health Educ (pres, 83-84); Acad Behav Med Res; Soc Behav Med; Int Union Health Educ; assoc fel Acad Phys Ed; hon fel Am Sch Health Asn. *Res:* Cardiovascular disease prevention; secondary prevention of stroke and cancer; diffusion of health and family planning innovations in the public; health attitudes and behavior; economic and administrative analyses of health education; health promotion. *Mailing Add:* 236 W Portal Ave Suite No 339 San Francisco CA 94127

GREEN, LEON WILLIAM, b Passaic, NJ, Dec 12, 25; m 56, 83; c 2. MATHEMATICS. *Educ:* Harvard Univ, AB, 48; Yale Univ, MA, 49, PhD(math), 52. *Prof Exp:* Instr math, Princeton Univ, 52-53; from instr to asst prof, 53-63, PROF MATH, UNIV MINN, MINNEAPOLIS, 63- *Mem:* Am Math Soc; Math Asn Am. *Res:* Differential geometry; topological dynamics. *Mailing Add:* Dept Math/127 Vincent Hall Univ Minn Inst Technol Minneapolis MN 55455

GREEN, LISLE ROYAL, b Ogden, Utah, Nov 18, 18; m 46; c 4. RANGE MANAGEMENT. *Educ:* Utah State Univ, BS, 41, MS, 48. *Prof Exp:* Ranch planner, Soil Conserv Serv, USDA, 46-47, range conservationist, US Forest Serv, 48-54; asst prof range mgt, Calif State Polytech Col, 55-60; fuel-break proj leader & supvry range scientist, US Forest Serv, 60-77, range scientist, forest fie res, 77-82; CONSULT, 83- *Concurrent Pos:* Assoc, Agr Exp Sta, Univ Calif, Riverside, 64- *Mem:* Soil Conserv Soc Am; Soc Range Mgt; Soc Am Foresters. *Res:* Ecology and management of annual plant and chaparral covered land, including revegetation, use of prescribed fire, goats and herbicides, irrigation of chaparral with sewage effluent, and effects of these practices. *Mailing Add:* 22586 Maine St Grand Terrace CA 92324

GREEN, LOUIS CRAIG, b Macon, Ga, Feb 2, 11; m 40. ASTROPHYSICS. *Educ:* Princeton Univ, AB, 32, MA, 33, PhD(astron), 37. *Hon Degrees:* DSc, Haverford Col, 83. *Prof Exp:* From instr to asst prof math & astron, Allegheny Col, 37-41; from instr to prof astron, 41-76, chmn dept astron, 42-76, chmn dept physics, 63-65, provost, 65-68, prof physics, 79-80, prof astron, 80-85, EMER PROF ASTRON, HAVERFORD COL, 76- *Concurrent Pos:* Vis asst prof, Swarthmore Col, 44; lectr, Bryn Mawr Col, 44-46; Guggenheim fel, 55-56; vis prof, Max Planck Inst Munich, 59; mem, Inst Advan Study, 62-63 & 68-69. *Mem:* Fel Am Phys Soc; Am Astron Soc; Sigma Xi. *Res:* Far ultraviolet spectroscopy; rapidly rotating single stars and close binaries; atomic wave functions; solar and stellar pulsations. *Mailing Add:* Dept Astron Haverford Col Haverford PA 19041-1392

GREEN, LOUIS DOUGLAS, b Birmingham, Ala, July 4, 16; m 39; c 2. PATHOLOGY. *Educ:* Tuskegee Inst, BS, 37; Fisk Univ, MA, 40; Meharry Med Col, MD, 48. *Prof Exp:* Teacher gen sci, Bd Educ, Birmingham, 37-38; instr chem, Fisk Univ, 38-40; instr org chem, Iowa State Col, 40-42; prof chem, Tenn State Univ, 42-44; intern med, Homer G Phillips Hosp, St Louis, Mo, 48-49; pvt pract, Birmingham, Ala & Cleveland, Ohio, 49-61; resident path, Marymount Hosp, Cleveland, 61-64; resident, Cleveland Clin & Hosp, 64-67; dir path, Hillcrest Hosp, 67-73; prof path & chmn dept, Meharry Med Col, 73-81; pathologist, George W Hubbard Hosp, 73-81; RETIRED. *Concurrent Pos:* Consult lab serv, Riverside Hosp, Nashville, 73-81; pathologist, Cloverbottom Develop Ctr, 73-81. *Mem:* Fel Col Am Path; fel Am Soc Clin Path; Int Acad Path. *Mailing Add:* 5467 Granny White Pike Brentwood TN 37027

GREEN, MARGARET, b Shamrock, La, Mar 27, 17. MICROBIOLOGY. *Educ:* La State Univ, BS, 37; Emory Univ, MS, 39; Univ Wis, PhD, 52. *Prof Exp:* Med bacteriologist, Charity Hosp, Shreveport, La, 40-42; instr bact, La State Univ, 42-47; asst, Univ Wis, 47-52; actg head dept microbiol, Univ Ala, 68-70, chmn dept microbiol, 70-80, from asst prof to emer prof bact, 52-80;

RETIRED. *Mem:* AAAS; Am Chem Soc; Am Soc Microbiol; Sigma Xi. *Res:* Physiology of bacteria; nitrogen-fixation; arthrobacteriology; morphogenesis; iron bacteria; herpes simplex virus. *Mailing Add:* Seven Rhonda Dr Tuscaloosa AL 35401

GREEN, MARIE RODER, b Vienna, Austria, March 19, 29; US citizen; m 52; c 2. HISTOCHEMISTRY. *Educ:* Brooklyn Col, NY, BS, 50; Univ Wis, Madison, MS, 52; Albany Med Col, NY, PhD(exp path), 56. *Prof Exp:* Res biochemist, NY State Dept Health, Albany, 52-55, Nat Inst Health, 56-57, Mt Zion Hosp, San Francisco, 57-58 & Western Res Univ, Cleveland, 62-66; res assoc, Armed Forces Inst Path, 66-67; res chemist, Molecular Carcinogenesis Lab, Nat Cancer Inst, NIH, Bethesda, MD, 70-88, health scientist adminr, 88. *Concurrent Pos:* NIH fel, Nat Inst Arthritis, Metab & Digestive Dis, 67-69. *Mem:* Soc Develop Biol; Histochem Soc; Am Dairy Sci Asn. *Res:* Differentiation and development of mammary gland; hormonal, chemical and viral factors in mammary gland carcinogenesis; altered expression of macromolecules in neoplasia; histochemistry of metachromatic dyes. *Mailing Add:* NHLBI NIH Fed Bldg Rm 4A08 7550 Wisconsin Ave Bethesda MD 20892

GREEN, MARK ALAN, b Sidney, Ohio, Sept 10, 56; m 90. RADIOPHARMACEUTICAL CHEMISTRY, NUCLEAR PHARMACY. *Educ:* Rose-Hulman Inst Technol, BS, 78; Ind Univ, PhD(inorg chem), 82. *Prof Exp:* Postdoctoral assoc radiol, Sch Med, Wash Univ, 82-85; asst prof radiol, Univ Minn, 85-87; asst prof, 87-90, ASSOC PROF MED CHEM, SCH PHARM, PURDUE UNIV, 90- *Concurrent Pos:* Prin investr, Nat Heart, Lung & Blood Inst Res Career Develop Award, 86-91, grant, Nat Cancer Inst, 87-, res grant, US Dept Energy, 89-; adj prof, Sch Med, Ind Univ, 90- *Mem:* Am Chem Soc; Soc Nuclear Med; Int Soc Cerebral Blood Flow & Metab. *Res:* Radiopharmaceuticals labeled with metal radionuclides; diagnostic imaging by positron emission tomography; design and synthesis of radiotracers for use in diagnostic nuclear medicine and biomedical research. *Mailing Add:* Dept Med Chem Purdue Univ West Lafayette IN 47907

GREEN, MARK LEE, b Minneapolis, Minn, Oct 1, 47; m; c 2. MATHEMATICS. *Educ:* Mass Inst Technol, BS, 68; Princeton Univ, MA, 70, PhD(math), 72. *Prof Exp:* Lectr, Univ Calif, Berkeley, 72-74, Mass Inst Technol, 74-75; from asst prof to assoc prof, 75-82, PROF, UNIV CALIF, LOS ANGELES, 82- *Concurrent Pos:* Fel, NSF, 68-72, Woodrow Wilson Found, 68-72, Procter fel, Princeton Univ, 70-71 & Alfred P Sloan Found, 76-80. *Mem:* Am Math Soc. *Res:* Several complex variables; differential geometry; algebraic geometry. *Mailing Add:* Univ Calif Los Angeles CA 90024

GREEN, MARK M, b New York, NY, Apr 6, 37; m 85; c 3. STEREOCHEMISTRY. *Educ:* City Univ New York, BSc, 58; Princeton Univ, PhD(chem), 66. *Prof Exp:* Asst prof chem, Univ Mich, 67-74 & Mich State Univ, 74-76; assoc prof, Clarkson Col, 76-79; assoc prof, 80-86, PROF CHEM, POLYTECH INST NEW YORK, 86- *Concurrent Pos:* NIH fel, 66-67; vis prof, Technion Haifa, Israel, 71-72, Chem Inst, Barcelona, 71 & Indian Inst Exp Med, Calcutta, 78; Fulbright fel, India, 78; US-Japan NSF fel, 89-90. *Mem:* Am Chem Soc; AAAS. *Res:* Stereochemistry of macromolecules; physical organic chemistry. *Mailing Add:* Dept Chem Polytech Univ Brooklyn NY 11201

GREEN, MARTIN DAVID, b Detroit, Mich, Aug 29, 47; m 75; c 2. EXPERIMENTAL DESIGN, COMPUTER MODELING. *Educ:* Univ Calif, Los Angeles, BA, 70, PhD(pharmacol), 75. *Prof Exp:* Prin investr, 80-81, chief, appl pharmacol, 81-84, chief, drug assessment div, 84-86, PROD MGR, USA MED MAT DEVELOP ACTIVITY, 86- *Mem:* Am Soc Pharmacol & Exp Therapeut; Am Chem Soc; Sigma Xi; NY Acad Sci. *Res:* Drug extrapolates; utilization of pharmacological data. *Mailing Add:* 5699 Glenrock Dr Frederick MD 21701

GREEN, MARTIN LAURENCE, b New York, NY, Jan 24, 49; m 70; c 2. ELECTRONIC MATERIALS, MATERIALS SCIENCE. *Educ:* Polytech Inst, Brooklyn, BS, 70, MS, 72; Mass Inst Technol, PhD(mat sci), 78. *Prof Exp:* Assoc mem tech staff, 70-74, MEM TECH STAFF, BELL LABS, 77- *Mem:* Am Inst Mining, Metall & Petrol Eng; Electrochem Soc; Mat Res Soc. *Res:* Metallurgy; thin films; electronic materials; phase transformations; materials science. *Mailing Add:* Bell Labs Rm 1A-122 Murray Hill NJ 07974

GREEN, MARY ELOISE, b East Liberty, Ohio, June 10, 03. NUTRITION. *Educ:* Ohio State Univ, BS, 28, MS, 33; Iowa State Col, PhD(foods, nutrit), 49. *Prof Exp:* Pub sch teacher, Ohio, 23-26 & 28-37; instr educ, Ohio Wesleyan Univ, 37-39; from instr to prof, 39-72, EMER PROF FOODS & NUTRIT, OHIO STATE UNIV, 72- *Mem:* AAAS; Am Home Econ Asn; Am Dietetic Asn; Inst Food Technol. *Res:* Physical properties of meats, flour mixtures and starchy foods; food preservation. *Mailing Add:* 116 W Como Ave Columbus OH 43202-1028

GREEN, MAURICE, b New York, NY, May 5, 26; m 50; c 3. BIOCHEMISTRY. *Educ:* Univ Mich, BS, 49; Univ Wis, MS, 52, PhD(biochem), 54. *Prof Exp:* Instr biochem, Univ Pa, 55-56; from asst prof to assoc prof, 56-63, PROF MICROBIOL, SCH MED, ST LOUIS UNIV, 63-, CHMN, INST MOLECULAR VIROL, 64- *Concurrent Pos:* Nat Found Infantile Paralysis res fel, Univ Pa, 54-55, Lalor fel, 55-56; sr fel, USPHS, 58-62; lifetime res career award, Nat Inst Allergy & Infectious Dis, 63-; Burroughs-Wellcome fel, 87. *Honors & Awards:* Dyer Award, 72; Howard Taylor Ricketts Award, 76. *Mem:* AAAS; Am Soc Biol Chem; Am Chem Soc; Am Soc Microbiol. *Res:* Molecular biology of eucaryotic cells and tumor viruses; virus-cell transformation; oncology; cell regulation. *Mailing Add:* Inst Molecular Virol St Louis Univ Sch Med 3681 Park Ave St Louis MO 63110

GREEN, MELVIN HOWARD, b Pittsburgh, Pa, Feb 21, 37; m 65; c 4. BIOCHEMISTRY, VIROLOGY. *Educ:* Univ Pittsburgh, BS, 58; Univ Ill, PhD(biochem), 62. *Prof Exp:* NIH fel virol, Calif Inst Technol, 62-63; asst prof, 63-67, assoc prof, 67-77, PROF BIOL, UNIV CALIF, SAN DIEGO, 77- *Concurrent Pos:* Am Cancer Soc res scholar, Imp Cancer Res Fund, London, Eng, 70-71. *Mem:* Am Soc Cell Biol. *Res:* Regulation of genetic expression of lytic and temperate bacteriophages; development of DNA tumor viruses; chromatin structure and function; cancer cell biology; wound repair. *Mailing Add:* Univ Calif San Diego Dept Biol PO BOX 109 La Jolla CA 92093

GREEN, MELVIN MARTIN, b Minneapolis, Minn, Aug 24, 16; m 46. GENETICS. *Educ:* Univ Minn, BA, 38, MA, 40, PhD(zool), 42. *Hon Degrees:* Dr, Univ Umeá, Sweden, 72. *Prof Exp:* Asst prof zool, Univ Mo, 46-50; from asst prof to prof, 50-82, geneticist, Exp Sta, 69-82, EMER PROF GENETICS & EMER GENETICIST, UNIV CALIF, DAVIS, 82- *Mem:* Nat Acad Sci; Genetics Soc Am (pres, 73); Am Soc Naturalists. *Res:* Drosophila genetics; mutation; pseudoallelism. *Mailing Add:* Dept Genetics Univ Calif Davis CA 95616

GREEN, MICHAEL, c 2. MEMBRANE BIO-GENESIS, PROTEIN SORTING. *Educ:* Univ Wis, PhD(molecular biol), 72. *Prof Exp:* PROF MICROBIOL, SCH MED, ST LOUIS UNIV,81- *Mem:* Am Soc Biochem & Molecular Biol. *Res:* Synthesis and assembly of Biological membranes and protein sorting. *Mailing Add:* Dept Microbiol St Louis Univ Med Col 1402 S Grand Blvd St Louis MO 63104

GREEN, MICHAEL ENOCH, b New York, NY, Nov 5, 38; m 74; c 1. PHYSICAL CHEMISTRY, LABORATORY SAFETY. *Educ:* Cornell Univ, BA, 59; Yale Univ, MS, 61, PhD(phys chem), 64. *Prof Exp:* Res assoc exciton transport, Calif Inst Technol, 63-64; Peace Corps vis lectr chem, Mid East Tech Univ, Ankara, 64-66; from asst prof to assoc prof, 66-83, PROF PHYS CHEM, CITY COL NEW YORK, 84- *Concurrent Pos:* Vchmn, Biophys Sect, NY Acad Sci, 85-86, chmn, 87-88. *Mem:* AAAS; Am Chem Soc; Am Phys Soc; NY Acad Sci; Biophys Soc. *Res:* Electrical noise generated during ion transport across membranes; membrane transport; mechanisms of ion channel gating. *Mailing Add:* Dept Chem City Col City Univ New York NY 10031

GREEN, MICHAEL H(ENRY), b Fresno, Calif, Feb 9, 44; m; c 1. NUTRITION, BIOLOGY. *Educ:* Univ Calif, Berkeley, BA, 67, PhD(nutrit sci), 73. *Prof Exp:* Lab asst, Lawrence Radiation Lab, Berkeley, 65-67, postdoctoral fel, Dept Nutrit Sci, Univ Calif, 73; postdoctoral assoc, Div Nutrit Sci, Cornell Univ, 73-75, asst prof nutrit Sci, 75-84; prof in charge, 77-79, ASSOC PROF NUTRIT SCI, PA STATE UNIV, NUTRIT DEPT, 85- *Concurrent Pos:* Physiol fac, Pa State Univ, 76-; vis scientist, NIH, Lab Math Biol, 82; Fulbright res scholar, Eon Found Norway & Ctr Int Exchange Scholars, 85; vis researcher, Inst Nutrit Res, Univ Oslo, Norway, 85-86; vis scientist, Univ Pa, Sch Vet Med, 86; Mead Johnson Award Comt, 87-90. *Mem:* Am Inst Nutrit; Am Soc Clin Nutrit; Soc Math Biol; Am Heart Asn; Soc Exp Biol & Med. *Res:* The application of model-based compartmental analysis to biological systems, for example, metabolism of retinol, triglyceride-rich lipoproteins, copper, receptor mediated endocytosis; lipid absorption. *Mailing Add:* S-126 Henderson Bldg S Pa State Univ University Park PA 16802

GREEN, MICHAEL JOHN, b Slough, Eng, Nov 30, 42; m 68; c 2. MEDICINAL CHEMISTRY. *Educ:* Sheffield Univ, BS, 64, PhD(org chem), 67. *Prof Exp:* Res assoc org chem, Ben May Lab Cancer Res, Univ Chicago, 67-69 & Brookhaven Nat Lab, 69-71; DIR ALLERGY & INFLAMMATION, SCHERING-PLOUGH RES, 71- *Mem:* Am Chem Soc; The Chem Soc. *Res:* Steroid chemistry; synthesis and structure-activity relationships of biologically active organic compounds. *Mailing Add:* Schering Res Div Schering-Plough Corp 60 Orange St Bloomfield NJ 07003-4799

GREEN, MICHAEL PHILIP, b London, Eng, Feb 1, 61. ELECTROCHEMISTRY, ATOMIC SURFACE STRUCTURE. *Educ:* Yale Univ, BS, 83; Stanford Univ, MS, 87, PhD(appl physics), 90. *Prof Exp:* POSTDOCTORAL MEM TECH STAFF, AT&T BELL LABS, 90- *Mem:* Am Phys Soc. *Res:* Application of scanning tunneling microscopy to electrochemical reactions; atomic resolution investigation of metal plating and stripping. *Mailing Add:* AT&T Bell Labs Rm 7A-326 600 Mountain Ave Murray Hill NJ 07974-2070

GREEN, MILTON, b Pueblo, Colo, Jan 13, 12; m 44; c 2. APPLIED PHYSICS. *Educ:* Univ Wyo, BS, 35; Univ Calif, MA, 37, PhD(physics), 41. *Prof Exp:* Asst physics, Univ Calif, 36-39; jr physicist, Nat Bur Standards, 42-43, asst physicist, 43-45; physicist, Res & Develop Labs, US Army Signal Corps, Ft Monmouth, 45-54, phys scientist & group leader, 54-60; sr staff scientist, Res Labs, Burroughs Corp, Pa, 60-64; res physicist, US Navy Underwater Sound Lab, 64-70; RES ASSOC, NAVAL UNDERWATER SYSTS CTR, 70- *Mem:* Am Phys Soc; Optical Soc Am. *Res:* Electrical conduction in solids, mainly semiconductors; transport phenomena; thermoelectricity; photoelectricity; photographic photometry; underwater optics; instrumentation; signal processing. *Mailing Add:* 201 Gardner Ave New London CT 06320

GREEN, MILTON, b Boston, Mass, Apr 30, 20; m 54; c 4. PHOTOGRAPHIC CHEMISTRY. *Educ:* Mass Inst Technol, BS, 40; Columbia Univ, PhD(chem), 51. *Prof Exp:* Chemist, Atlantic Gelatin Co, 40-41; res chemist biochem & proteins, Burroughs-Wellcome Co, 41-42; res chemist sulfa drugs, Hoffmann-La Roche, Inc, 42-43; res chemist pharmaceut & amino acids, Wyeth Inst, 43-44; res chemist textile finishing, US Finishing Corp, 46-47; sr chemist & group leader photog & org chem, Polaroid Corp, 51-59, asst mgr, Org Res Div, 59-65, mgr photog chem res, 65-68, asst dir chem res & develop, 68-82; RETIRED. *Concurrent Pos:* Consult, toxicol, 82- *Honors & Awards:* fel Soc Photog Scientists & Engrs, 83. *Mem:* Am Chem Soc; fel Soc Photog Scientists & Engrs. *Res:* Chemistry of the photographic process; pharmaceuticals; proteins; amino acids; dyes; chemistry of diffusion transfer photography. *Mailing Add:* 38 Winston Rd Newton Center MA 02159

GREEN, MORRIS, b Brooklyn, NY, Apr 27, 31. RADIATION BIOLOGY. *Educ:* Brooklyn Col, BA, 52; Univ Rochester, PhD(radiation biol), 58. *Prof Exp:* Coffin res fel radiation biol, Univ Rochester, 58-64; from asst prof to assoc prof biol, 64-70, PERRY W LASH PROF PEDIAT, HUNTER COL, 70-, PHYSICIAN-IN-CHIEF & CHMN, DEPT PEDIAT. *Concurrent Pos:* Vis radiation biologist, Am Inst Biol Sci-AEC, 73- *Mem:* Am Chem Soc. *Res:* Biological effects of radiation; blood and liver proteins and their functions; tracer chemistry; diabetes, action of insulin; amino acid and protein metabolism in normal, aging and cataractous rat lenses. *Mailing Add:* Dept Pediat Ind Univ-Purdue Univ Indianapolis IN 46223

GREEN, MORTON, b Brooklyn, NY, Oct 25, 17; m 46; c 3. PALEONTOLOGY. *Educ:* Univ Kans, AB, 40, MA, 42; Univ Calif, PhD(paleont), 54. *Prof Exp:* Res assoc, SDak Sch Mines & Technol, 50-61, chmn dept biol, 50-80, assoc dir, Mus Geol, 68-75, cur vert paleont, 62-80; RES ASSOC, MUS NATURAL HIST, UNIV KANS, 80- *Mem:* Soc Vert Paleont; Sigma Xi; Paleont Soc. *Res:* Mammals. *Mailing Add:* 1933 Hillview Rd Lawrence KS 66046-2653

GREEN, NANCY R, NUTRITION AND FOOD SCIENCE. *Educ:* Univ Tenn, BS, 71, DPhil(nutrit), 74. *Prof Exp:* From asst prof to assoc prof, 74-87, asst dean, 89-90, PROF, DEPT NUTRIT & FOOD SCI, FLA STATE UNIV, 88-, ASSOC DEAN GRAD STUDIES & RES, COL HUMAN SCI, 90- *Concurrent Pos:* mem, Educ comt, Inst Food Technologist, 80-83, secy-treas, Nutrit Div, 91-93; actg dept head, Dept Nutrit & Food Sci, Fla State Univ, 88, grad prog adminr, 88-89. *Mem:* Inst Food Technologist; Am Inst Nutrit; Am Coun Sci & Health; Environ Mutagen Soc; AAAS; Sigma Xi. *Res:* Sodium reduction in hypertensive elderly; influence of dietary fats on mutagenicity; author of numerous scientific publications. *Mailing Add:* Col Human Sci Fla State Univ Tallahassee FL 32306-2033

GREEN, NORMAN EDWARD, b Electric City, Wash, Aug 22, 38; m 85; c 2. PLANT PATHOLOGY. *Educ:* Wash State Univ, BS, 67, PhD(plant path), 74; Colo State Univ, MS, 69. *Prof Exp:* Res asst range sci, Colo State Univ, 67-69; res technician plant path, Wash State Univ, 69-74; ASSOC PROF SOIL MICROBIOL & RANGE MGT, HUMBOLDT STATE UNIV, 76- *Concurrent Pos:* Fel libr, Dept Environ Sci, Wash State Univ, 74-75. *Mem:* Soc Range Mgt. *Res:* The influence of endomycorrhizal fungi on growth and development of selected range grasses and subclover. *Mailing Add:* Dept Range Mgt Humboldt State Univ Arcata CA 95521

GREEN, ORVILLE, b Oak Park, Ill, Jan 14, 26; m 54; c 3. PEDIATRICS, ENDOCRINOLOGY. *Educ:* Harvard Univ, AB, 49; Northwestern Univ, MD, 54. *Prof Exp:* From asst prof to assoc prof pediat, Col Med, Ohio State Univ, 60-63, asst prof med, 60-63; assoc prof, 63-68, PROF PEDIAT, MED SCH, NORTHWESTERN UNIV, CHICAGO, 68-; CONSULT, 89- *Concurrent Pos:* Clin res fel pediat med, Johns Hopkins Hosp, 57-60; NIH res fel, 58-60; dir div endocrinol, Children-s Mem Hosp, 63-89. *Mem:* Am Endocrine Soc; Am Fedn Clin Res; Am Pediat Soc. *Res:* Pediatric endocrinology; growth, sexual development and disorders of children and adolescents; sex hormones metabolism and secretion. *Mailing Add:* Dept Pediat Sch Med Northwestern Univ 100 E Bellevue Pl 11E Chicago IL 60611

GREEN, PAUL BARNETT, b Philadelphia, Pa, Feb 15, 31; m 57. BOTANY. *Educ:* Univ Pa, BA, 52; Princeton Univ, PhD, 57. *Prof Exp:* From asst prof to prof bot, Univ Pa, 58-70; PROF BIOL, STANFORD UNIV, 70- *Honors & Awards:* Darbaker Prize, Botanical Soc of Am, Pelton Award. *Mem:* Soc Develop Biol(pres, 82); Bot Soc Am; Am Soc Plant Physiologists; foreign mem Royal Belgian Acad of Arts & Sci. *Res:* Plant development; growth of cell walls. *Mailing Add:* Dept Biol Stanford Univ Stanford CA 94305

GREEN, PAUL E(LIOT), JR, b Durham, NC, Jan 14, 24; m 48; c 5. ELECTRONICS. *Educ:* Univ NC, AB, 44; NC State Col, MS, 48; Mass Inst Technol, ScD, 53. *Prof Exp:* Staff mem, Lincoln Lab, 53-58, group leader, 58-69; sr mgr, Comput Sci Dept, 69-74 & 77-81, mem corp tech comt, 75-76 & 81-83, STAFF MEM, IBM RES, 83- *Honors & Awards:* Aerospace Electronics Pioneer Award, Inst Elec & Electronics Engrs, 80. *Mem:* Nat Acad Eng; fel Am Inst Elec & Electronics Engrs. *Res:* Communication theory applied to radar astronomy, seismology, computer network architecture, optica fiber networking. *Mailing Add:* Roseholm Pl RFD 3 Mt Kisco NY 10549

GREEN, PHILIP S, b Youngstown, Ohio, Aug 4, 36; m 60; c 3. ULTRASOUND, ELECTRONICS. *Educ:* Johns Hopkins Univ, BESc, 58; Stanford Univ, MS, 67. *Prof Exp:* Assoc engr, Johns Hopkins Appl Physics Lab, 58-60; sr res engr, Lockheed Res Labs, 60-68; mgr, Ultrasonics Prog, 68-78, DIR, BIOENG RES LAB, SRI INT, 78- *Concurrent Pos:* Consult & prin investr grants, NIH, 72-; lectr, Med Schs, US & Europe, 75-; chmn, 5th Int Symposium Acoustical Holography & Imaging, 73; mem, Task Group Med Ultrasound, Nat Sci Found, 75. *Mem:* Am Inst Ultrasound Med. *Res:* Ultrasonic imaging methods for medical diagnosis; holography and optical imaging. *Mailing Add:* 585 California Way Redwood City CA 94062

GREEN, PHILLIP JOSEPH, II, b July 12, 41; US citizen; m 60; c 2. COSMIC RAY PHYSICS. *Educ:* Southwestern at Memphis, BS, 63; La State Univ, PhD(physics), 67. *Prof Exp:* ASSOC PROF PHYSICS, TEX A&M UNIV, 67- *Mem:* Am Phys Soc; Am Geophys Union; Am Inst Physics. *Res:* High energy cosmic ray research; muon intensity measurements and their implications; high frequency wave form digitizing systems. *Mailing Add:* 206 Brookside Dr Bryan TX 77843

GREEN, R(ALPH) V(ERNON), b Litchfield, Ill, Feb 11, 13; m 39; c 2. CHEMICAL ENGINEERING. *Educ:* Univ Ill, BS, 35, MS, 36. *Prof Exp:* Tech asst, Exp Sta, Univ Ill, 33-34, instr chem, 35-36; chem engr, Indust & Biol Chem Dept, E I du Pont de Nemours & Co, Inc, Del, 36-39, & WVa, 40-43, group leader, 43-45, asst tech supt, 45-52, tech supt, 52-54, tech specialist, 54-58, staff engr, 58-77; CONSULT, 77- *Mem:* Am Chem Soc; Am Inst Chem Engrs; fel Am Inst Chem. *Res:* Synthesis gas generation and processing. *Mailing Add:* 210 Sheller Dr Charleston WV 25314-1060

GREEN, RALPH, b Johannesburg, SAfrica, Sept 18, 40; m 64; c 5. HEMATOLOGY, NUTRITION. *Educ:* Univ Witwatersrand, MB, BCh, 63, MD, 76. *Prof Exp:* Inter med & surg, Johannesburg Gen Hosp, 64; NIH res fel, Univ Witwatersrand, 65-66; Wellcome res fel lab hemat, St Bartholomew's Hosp, London, Eng, 67-69; resident hemat, SAfrican Inst Med Res, 70-71; lectr & sr lectr, Sch Path, Univ Witwatersrand, 72-74; assoc hemat, Scripps Clin & Res Found, 75-80, asst mem, 80-; CHMN, DEPT LAB HEMAT, CLEVELAND CLIN FOUND. *Concurrent Pos:* Adj assoc prof path, Univ Cal, San Diego, 79-; mem nat comt clin lab standards, iron sect, Nat Inst Arthritis & Metabolic Dis, NIH. *Mem:* Am Soc Hemat; Am Fedn Clin Res; Am Inst Nutrit; Soc Exp Biol & Med. *Res:* Nutritional anemias; iron cobalamin and folate metabolism; iron ferritin and vitamin assays; drug targeting with erythrocyte ghosts; iron and metal storage diseases; neurobiology of cobalamin, other vitamins and trace minerals; fruit bat animal model for cobalamin deficiency; drug-nutrient interactions. *Mailing Add:* Dept Lab Hematol Cleveland Clin Found One Clinic Ctr Cleveland OH 44195-5139

GREEN, RALPH J, JR, b Naylor, Mo, Aug 17, 23; m 44; c 3. PLANT PATHOLOGY. *Educ:* Ind State Univ, BS, 48; Purdue Univ, MS, 50, PhD, 54. *Prof Exp:* Instr plant path, Purdue Univ, 51-53; instr, Univ Chicago, 53-55; from asst prof to prof, 55-88, EMER PROF PLANT PATH, PURDUE UNIV, 88- *Mem:* Fel Am Phytopath Soc (pres, 74-75); Coun Agr Sci Technol. *Res:* Soilborne plant pathogens and soil microbiology. *Mailing Add:* Dept Bot & Plant Path Purdue Univ Lafayette IN 47907

GREEN, RAY CHARLES, b Ecorse, Mich, Jan 22, 30; m 52; c 2. CHEMICAL ENGINEERING. *Educ:* Univ Mich, BS, 52. *Prof Exp:* Tech supt, Kerr McGee Chem Corp, 66-68, mill supt, 68-71, mining supt, 71-73, plant mgr, 73-78, asst to vpres, 78-79, mgr opers, 79-80, MGR RES & DEVELOP, KERR MCGEE CORP, 80- *Mem:* Am Inst Chem Engrs. *Res:* Applied research in chemical operations with special focus on surface chemistry, pigment, fertilizers, borox and soda ash. *Mailing Add:* Kerr McGee Corp 3301 NW 130th St PO Box 25861 Oklahoma City OK 73125

GREEN, RAYNA DIANE, b Dallas, Tex, July 18, 42. ETHNOSCIENCE, THIRD WORLD SCIENTIFIC DEVELOPMENT. *Educ:* Southern Methodist Univ, BA, 63, MA, 66; Ind Univ, PhD(Am studies), 73. *Prof Exp:* Prof folklore, Univ Ark, 71-72 & Univ Mass, 72-75; dir, Proj Native Am Sci, AAAS, 75-79; prof native Am studies & dir, Native Am Sci Resource Ctr, Dartmouth Col, 79-83, prof, 80-84; ASSOC, AM HIST, DIV SCI & TECHNOL, SMITHSONIAN INST, 83-; DIR, AM INDIA PROG, NAT MUS AM HIST, 86- *Concurrent Pos:* Res fel, Smithsonian Inst, 70-71; planner, US Dept Energy, 76-80 & NIH & NSF, 77-79; mem bd, Fund Improv Post-Sec Educ, 77-81; res fel, Ford Found-NRC, 83-84; mem bd, Indian Law Resource Ctr, 84- *Mem:* Am Indian Sci & Eng Soc; Soc Advan Chicano & Native Am Scientists; AAAS; Soc Ethnobot; Am Folklore Soc (pres, 86); Am Anthrop Asn. *Res:* Native American scientific and technical traditions, including ethnoscience, ethnobotany, ethnomedicine; native American scientific and technical development, energy, agriculture, medicine; philosophy of non-Western scientific traditions. *Mailing Add:* 814 G St SE Washington DC 20003

GREEN, RICHARD, b Brooklyn, NY, June 6, 36. PSYCHIATRY. *Educ:* Syracuse Univ, AB, 57; Johns Hopkins Univ, MD, 61. *Prof Exp:* Resident, Univ Calif, Los Angeles, 62-64; clin assoc, NIMH, 64-66; fel, Maudsley Hosp, 66-67; from asst prof to assoc prof, Univ Calif, Los Angeles, 68-74; PROF PSYCHIAT, STATE UNIV NY, STONY BROOK, 74- *Concurrent Pos:* Prin investr, NIMH res grants, 68-81, mem, study sect appl res, 74-78; res scientist develop award, NIMH, 68-73; psychiat sabbatical fel, Cambridge Univ, 80-81. *Mem:* Am Psychiat Asn; Soc Sci Study Sex (pres, 76-78); Int Acad Sex Res (pres, 75); Psychiat Res Soc; World Asn Sexology (secy, 78). *Res:* Psychosexual development in children; environmental and biological influences. *Mailing Add:* Dept Psychiat Behav Sci Univ Calif Los Angeles Los Angeles CA 90024

GREEN, RICHARD D, b Trenton, NY, Mar 1, 40. PHARMACOLOGY. *Educ:* Philadelphia Col, BS, 61; Univ Minn, PhD(pharmacol), 65. *Prof Exp:* PROF PHARMACOL, DEPT PHARMACOL, SCH MED, UNIV ILL, 86- *Mem:* Pharmacol Soc. *Mailing Add:* Dept Pharmacol Sch Med Univ Ill PO Box 6998 Chicago IL 60680

GREEN, RICHARD E, b Seward, Nebr, Mar 23, 31; m 55; c 4. SOIL PHYSICS. *Educ:* Colo State Univ, BS, 53; Univ Nebr, MS, 57; Iowa State Univ, PhD(soil mgt), 62. *Prof Exp:* Instr soil sci, Univ Nebr, 57-58; asst agronomist, Maui Br, 62-65, assoc prof, 65-73, PROF SOIL SCI, MANOA CAMPUS, UNIV HAWAII, 73- *Mem:* Am Soc Agron; Soil Sci Soc Am. *Res:* Behavior of pesticides in soils and water; water quality. *Mailing Add:* Dept Agron Univ Hawaii Manoa Honolulu HI 96822

GREEN, RICHARD H, b Baker, Mont, July 22, 36; m 58; c 2. MICROBIOLOGY. *Educ:* Whitman Col, BA, 58; Wash State Univ, MS, 61, PhD(microbiol & environ sci), 65. *Prof Exp:* Asst sanit chemist, Inst Environ Sci, Wash State Univ, 61-63; sr engr, Boeing Co, Wash, 65-67; vpres Inst advan, Corwin D Denney prof energy & dir, Energy Inst, Univ La Verne, 83-85; various positions, Offices Civil Systs, Energy & Technol Applications, Planning, Eng & Review Assessment, 67-83, 85-89, MGR, PROG ENG & REV OFF, OFF TECHNOL & APPLICATIONS, JET PROPULSION LAB, CALIF INST TECHNOL, 89- *Concurrent Pos:* Mem, Appollo 604 Accident Invest Team, NASA Planning Team; prin investr, Apollo 12 mission & moon landing; chmn, Intersoc Conf Environ Systs, 78; mem, Life Sci Comt, Am Inst Aeronaut & Astronaut, 75-78, Spec Proj Comt, Int Energy Info Forum & Workshop Educators, 82, Educ Adv Coun, Southern Calif Edison Co, 85- & adv comt, Higher Educ Doctoral Prog, Univ Calif, Berkeley; bd mem, One Touch Eng & SFE Technologies; co-chmn, Energy for Africa Conf, 88, pres, Africa-1000, 88- & bd mem, 89-; bd dirs, Camfel Prod, 87-; tech adv/sci consult, several motion pictures; partner, Film Effects Int, 83-85; pres, Global Energy Soc, vpres, Int Cogeneration Soc, 83-85. *Mem:* Am Soc

Microbiol; Am Asn Contamination Control. *Res:* Environmental health, especially water pollution; space microbiology, including life detection, environmental simulation, planetary quarantine and spacecraft sterilization. *Mailing Add:* 4715 Hillard Ave La Canada CA 91011

GREEN, RICHARD JAMES, b Newark, NJ, Apr 15, 28; m 57; c 4. RESEARCH & DEVELOPMENT MANAGEMENT, SCIENCE POLICY. *Educ:* Col Holy Cross, BS, 49; Fordham Univ, MS, 55; Harvard Bus Sch, AMP, 77. *Prof Exp:* Eng positions, Pratt & Whitney Aircraft & Socony Mobil Oil Co, 55-61; tech asst to assoc adminr & mgr, Apollo Lunar Surface Exp Prog, NASA, 61-70; exec asst, NSF, 70-72, dep asst dir appl res, 72-79; assoc dir res, Presidential Appointee, Fed Emergency Mgt Agency, 79-81; asst dir sci technol & int affairs, 82-89, DIR, RES FACILITIES PROG, NSF, 89- *Honors & Awards:* Exceptional Sci Achievement Award, NASA, 69; Commendation award, AEC, 70; Meritorious Serv award, NSF, 77; President's Meritorious Serv Achievement Award, US Govt, 87. *Mem:* AAAS. *Res:* Aerospace engineering; general management; international science and technology; solar and geothermal energy; advanced high temperature lubricants; lunar and planetary exploration; nuclear systems. *Mailing Add:* 3304 Carpenter St SE Washington DC 20020

GREEN, RICHARD STEDMAN, b Somerville, Mass, Mar 2, 14; m 43; c 3. SANITARY ENGINEERING. *Educ:* Harvard Univ, SB, 36, SM, 37; Am Acad Environ Engrs, dipl. *Prof Exp:* Sr sanit eng aide, Mass Dept Pub Health, 37-38; supt water purification, Panama Canal, 38-40; res assoc, Sch Med, Univ Pa, 40-41; sanit engr, USPHS, Md , Wash & Maine, 41-42, dir div pub health eng, Dept Health, Alaska, 42-46, sanit engr dir, hq, Washington, DC, 46-67, dir, Off Environ Health, Indian Health Serv, Md, 67-71, asst surg gen & chief eng officer, Washington, DC, 71-73; CONSULT ENGR, 73- *Honors & Awards:* William B Hatfield Award, Water Pollution Control Fedn, 70. *Mem:* Fel Am Soc Civil Engrs; fel Am Pub Health Asn; Am Water Works Asn; Am Acad Environ Engrs (pres, 76, treas, 81-83); Water Pollution Control Fedn. *Res:* Water quality control. *Mailing Add:* 9209 E Parkhill Dr Bethesda MD 20814

GREEN, ROBERT A, b Brooklyn, NY, May 13, 25; m 51; c 5. INTERNAL MEDICINE. *Educ:* Univ Ill, BS, 46, MD, 48. *Prof Exp:* Chief tuberc sect, Talihina Med Ctr, Okla, 52-54; asst chief pulmonary dis sect, Vet Admin Hosp, Bronx, NY, 54-58; assoc prof, 63-70, assoc dean student affairs, 68-74, assoc dean, 74-77, PROF INTERNAL MED, MED SCH, UNIV MICH, ANN ARBOR, 70- *Concurrent Pos:* Chief pulmonary dis sect, Vet Admin Hosp, Ann Arbor, 58-72, staff physician, 78- *Mem:* Am Thoracic Soc. *Res:* Pulmonary diseases; tuberculosis; unclassified mycobacterial disease; lung cancer. *Mailing Add:* Vet Admin Med Ctr 100 Martin Pl Ann Arbor MI 48104

GREEN, ROBERT BENNETT, b Coffeyville, Kans, Dec 22, 43; m 76. LASER SPECTROSCOPY. *Educ:* Okla State Univ, BS, 66; Ohio Univ, PhD(anal chem), 74. *Prof Exp:* Process chemist, Monsanto Co, 66-68; analytical chemist, Jefferson Chem Co, Inc, 68-70; res assoc, Nat Bur Standards, 74-76; asst prof anal chem, Dept Chem, WVa Univ, 76-79; asst prof, 79-81, ASSOC PROF ANALYTICAL CHEM, DEPT CHEM, UNIV ARK, 81- *Concurrent Pos:* Prin investr, Laser Intra-Cavity Detection for Gas Chromatography, Dept Interior, 80-82 & Laser Enhanced Ionization Spectrometry, NSF, 81-84. *Mem:* Sigma Xi; Soc Appl Spectros; Am Chem Soc. *Res:* Applications of lasers to analytical chemistry: laser enchanced ionization, optogalvanic, intra-cavity absorption, atomic and molecular fluorescence. *Mailing Add:* Code 385 Naval Weapons Ctr China Lake CA 93555

GREEN, ROBERT E(DWARD), JR, b Clifton Forge, Va, Jan 17, 32; m 62; c 2. MATERIALS SCIENCE, SOLID STATE PHYSICS. *Educ:* Col William & Mary, BS, 53; Brown Univ, ScM, 56, PhD(metal physics), 59. *Prof Exp:* Physicist, underwater explosions res div, Norfolk Naval Shipyard, 59; Fulbright grant metal physics, Aachen Tech Univ, 59-60; from asst prof to assoc prof, Mech Dept, 60-70, prof mech & mat sci & chmn dept, 70-73, chmn dept, 80-85, PROF MAT SCI & ENG, JOHNS HOPKINS UNIV, 70-, DIR CTR NONDESTRUCTIVE EVAL, 85- *Concurrent Pos:* Ford Found residency as sr engr, Radio Corp Am, Pa, 66-67; consult, US Army Ballistic Res Labs, 73-74 & Johns Hopkins Appl Physics Lab, 77-79; physicist, Nat Bur Standards, 74-80. *Honors & Awards:* Mehl Honor Lectr Award, Am Soc Nondestructive Testing, Lester Honor Lect Award. *Mem:* Acoust Soc Am; Am Phys Soc; fel Am Soc Metals Int; fel Am Soc Nondestructive Testing; Nat Mat Adv Bd; Am Inst Mining Metall Petrol Engrs. *Res:* Elasticity; plasticity; crystal growth and orientation; x-ray diffraction; electrooptical systems; linear and non-linear elastic waves; light-sound interactions; ultrasonic attenuation; fatigue; acoustic emission; nondestructive testing; residual stress; high-power ultrasonics composites; polymers; biomaterials. *Mailing Add:* Dept Mat Sci & Eng Johns Hopkins Univ Baltimore MD 21218

GREEN, ROBERT I, b New York, NY, May 7, 29; m 57; c 2. UNIT OPERATIONS IN PARTICULATE SOLIDS PROCESSING & HANDLING. *Educ:* Clarkson Univ, BChE, 50; Pa State Univ, MS, 52. *Prof Exp:* Process/proj engr, Singmaster & Breyer, 55-59; tech eng salesman, Vibro Dynamics Co, 59-73; pres, E-V Systs, Inc, 73-83; PRES MKT, ROBERT I GREEN ASSOCS, 83- *Concurrent Pos:* Lectr, NJ Ctr Prof Advan, 82-85. *Mem:* Com Develop Asn; NY Acad Sci; Asn Consult Chemists & Chem Engrs; Am Chem Soc; Am Inst Chem Engrs. *Res:* International marketing of chemical processes and process equipment; project management. *Mailing Add:* 168 E 74th St New York NY 10021

GREEN, ROBERT LEE, JR, b Fairfield, Ala, Aug 17, 21; div; c 2. PSYCHIATRY. *Educ:* Univ Ala, BS, 43; Hahnemann Med Col, MD, 46. *Prof Exp:* Intern, Jeff-Hillman Hosp, Birmingham, Ala, 46-47; resident obstet & gynec, Carraway Methodist, De Paul & St Vincent's Hosp, 49-52; pvt pract, Ala, 52-53; staff psychiatrist, Kennedy Vet Admin Hosp, Memphis, Tenn, 53, staff psychiatrist, Phys Med & Rehab Serv, 54-55, actg chief serv, 55-56; chief, Dept Psychiat, Vet Admin Hosp, Durham, 61-74, chief of staff, 72-80; resident psychiatrist, 56-59, from instr to assoc prof, 59-74, PROF PSYCHIAT, MED CTR, DUKE UNIV, 72- *Concurrent Pos:* Clin investr, Vet Admin, 59-60, mem cent off res comt, 63-65; lectr serv training prog, Cherry State Hosp, Goldsboro, 64-66; from resident psychiatrist to staff psychiatrist, Vet Admin Hosp, Durham, 56-61; med dir, Holly Hills Hosp, 80-90. *Mem:* AMA; Am Psychiat Asn; Acad Psychosom Med; Am Electroencephalographic Soc; Am Col Psychiat. *Res:* Electroencephalography; psychophysiology; neurophysiology. *Mailing Add:* 4100 Chapel Hill Rd I Innisfree Durham NC 27707-5058

GREEN, ROBERT PATRICK, b New York, NY, Mar 17, 25; m 47; c 4. PULP & PAPER TECHNOLOGY. *Educ:* Syracuse Univ, BS, 49; Miami Univ, MBA, 59. *Prof Exp:* Res chemist pulp res & develop, Champion Papers, Inc, 49-54, pilot plant engr, 54-58, group leader, 58-59, mgr, 59-61; res chemist, Kimberly-Clark Corp, 61-62, mgr pulp process controls, 62-71, sr develop engr, 71-73; process sales engr, 73-78, mkt mgr, 78-83, SR ASSOC, MEAD CHEM SYSTS, 83- *Mem:* Fel Tech Asn Pulp & Paper Indust; Can Pulp & Paper Asn. *Res:* Pulping and bleaching, including wood and agricultural fibers; end use of pulp fibers; research management. *Mailing Add:* Mead Cent Res PO Box 2500 Chillicothe OH 45601

GREEN, ROBERT S(MITH), b Lafayette, Ind, Dec 17, 14; m 38; c 1. CIVIL ENGINEERING. *Educ:* Purdue Univ, BS, 36, MS, 42. *Prof Exp:* Detailer, Am Bridge Co, 36; draftsman, Sinclair Refining Co, 36-37; technician, Carnegie-Ill Steel Corp, 37-38; instr gen eng, Purdue Univ, 38-40; asst struct engr, Spec Eng Div, Panama Canal, 40-41; engr, Blaw Knox Corp, Pa, 41-42 & Weirton Steel Co, 42-47; asst prof indust eng, 47, prof welding eng & chmn dept, 48-54, prof civil eng, 70-78, exec dir eng exp sta, 54-74, assoc dean col eng, 58-74, prof archit, 74-78, EMER PROF, OHIO STATE UNIV, 78- *Concurrent Pos:* Consult, Nat Cert Pipe Welding Bur, NY, 48- *Mem:* Nat Soc Prof Engrs; Am Soc Civil Engrs; Am Soc Mech Engrs; Am Soc Eng Educ. *Res:* Indeterminate structures; metal joining; power piping and pressure vessels; materials for high temperature service. *Mailing Add:* Tenn Valley Auth - W8-D180 C-K 400 W Summit Hill Dr Knoxville TN 37902

GREEN, ROBERT WOOD, b La Grange, Ill, Aug 7, 22; m 43; c 2. CRYOGENICS, OPTICS. *Educ:* Morningside Col, BS, 43; Univ Iowa, MS, 49; Iowa State Univ, PhD(physics), 60. *Prof Exp:* Assoc prof physics, Iowa Wesleyan Col, 49-50; from instr to asst prof, Morningside Col, 50-55; instr, Iowa State Univ, 55-56, res assoc, 56-58; head dept, 60-70, PROF PHYSICS, MORNINGSIDE COL, 60- *Mem:* Am Phys Soc; Am Asn Physics Teachers; Optical Soc Am; Sigma Xi. *Res:* Ultrasonic properties of single crystals of silver and copper; electric and magnetic properties of rare earth metal crystals; superconductivity transition temperatures. *Mailing Add:* 3801 Sixth Ave Sioux City IA 51105

GREEN, ROGER HARRISON, b New York, NY, June 22, 39; c 3. ECOLOGY. *Educ:* Col William & Mary, BS, 61; Cornell Univ, PhD(zool), 65. *Prof Exp:* Fulbright fel, Dept Zool, Univ Queensland, 65-66; resident ecologist, Syst-Ecol Prog, Marine Biol Lab, Woods Hole, 66-68; asst prof zool, Univ Man, 68-71, assoc prof, 71-76; adj prof biol, City Col New York, & vis prof biol Upsala Col, 76-77; assoc prof zool, 77-85, PROF ZOOL, UNIV WESTERN ONT, 85- *Concurrent Pos:* Consult, Can Dept Environ, Can Dept Indian & Northern Affairs, Acad Natural Sci Philadelphia, Ont Govt Ministries, Sask Res Coun, US Environ Protection Serv, US Minerals Mgt Serv, Battelle Marine Res, Du Pont de Nemours, Smithsonian Tropical Res Inst, Seakem Oceanog, Booth Aquatic Res Inc & Gartner-Lee Ltd; vis prof, City Col, New York, 76, Upsala Col, NJ, 76-77 & Nat Univ Singapore, 82-83; comt mem, US Nat Res Coun, US Nat Oceanic & Atmospheric Admin; workshops organizer, UNESCO, Singapore, 85 & Philippines, 87 & Univ Auckland, NZ, 85 & 87; ed, Can J Fisheries & Aquatic Sci, 86- *Mem:* Can Soc Zool; Ecol Soc Am; Biometric Soc. *Res:* Ecology of marine and freshwater populations and communities; statistical methods in ecological research. *Mailing Add:* 92 Windsor Ave London ON N6C 1Z8 Can

GREEN, RONALD W, b Mexia, Tex, Nov 15, 48; m 70; c 2. VETERINARY MEDICINE, RADIOLOGY. *Educ:* Abilene Christian Univ, BS, 71; Tex A&M Univ, DVM, 74, MS, 77; Am Col Vet Radiol, dipl, 78. *Prof Exp:* Res assoc radiol, Tex A&M Univ, 74-75, vet clin assoc, 75-77; asst prof biol, Ohio State Univ, 77-80, staff radiologist, Vet Teaching Hosp, 77-80; asst prof, 80-86, ASSOC PROF RADIOL, TEX A&M UNIV, 86- *Concurrent Pos:* Consult radiologist, Acres North Animal Hosp, San Antonio, Tex, 75-, Med Col, Ohio State Univ, 77-, Oak Hills Vet Hosp, San Antonio, Tex & Animal Med Clin, Columbus, Ohio, 78- *Mem:* Am Vet Med Asn; Am Col Vet Radiol; Am Animal Hosp Asn; Am Asn Vet Clinicians. *Res:* Radiation therapy and oncology; nuclear medicine, tumor scanning and clinical imaging; diagnostic radiology. *Mailing Add:* Col Vet Med Tex A&M Univ Col Sta College Station TX 77843

GREEN, SAUL, b New York, NY, Jan 8, 25. BIOCHEMISTRY, IMMUNOLOGY. *Educ:* City Col New York, BS, 48; Univ Iowa, MS, 50, PhD(biochem), 52. *Prof Exp:* Instr biochem, Univ Va, 52-54; res assoc, Dept Med, Med Col, Cornell Univ, 54-59; from asst mem to assoc mem, Sloan-Kettering Inst, 59-61, assoc, 61-67; asst prof biochem, Sloan-Kettering Div Grad Sch Med Sci, Cornell Univ, 62-67, assoc prof, 67-; assoc mem, Sloan-Kettering Inst, 67-; DIR RES & DEVELOP, US BIO PROD, INC, 82-; PRES, ZOL CONSULT, SCI DIR, EMPRISE INC, 88- *Concurrent Pos:* NIH res career develop award, 63; dir clin chem lab, Univ Va Hosp, 52-54; lectr, City Col NY, 56-59; vis prof, dept surg, Univ Va Hosp, 84-86. *Mem:* Am Soc Biol Chem; Am Asn Cancer Res. *Res:* Isolation, characterization and mechanism of action of naturally occurring antitumor agents (TNF) (NHG) in serum and tissues of bacillus Calmette-Guerin vaccine or c parvum primed, endotoxin treated mice; science education. *Mailing Add:* ZOL Consults Inc 340 W 57th St New York NY 10019

GREEN, SHERRY MERRILL, algebra, for more information see previous edition

GREEN, SIDNEY, b New Orleans, La, Dec 12, 39; m 63; c 3. TOXICOLOGY. *Educ:* Dillard Univ, BA, 61; Howard Univ, PhD(pharmacol), 72. *Prof Exp:* Lab technician cancer chemother, Microbiol Assocs, Inc, 61-65; pharmacologist, Food & Drug Admin, 65-78; assoc prof pharmacol, Med Sch, Howard Univ, 78-79; pharmacologist, Environ Protection Agency, 79-80; PHARMACOLOGIST, FOOD & DRUG ADMIN, 80- *Concurrent Pos:* Adj assoc prof pharmacol, Howard Univ, 79- *Mem:* Soc Toxicol; Tissue Cult Asn; Environ Mutagen Soc; Soc Risk Anal. *Res:* Mutagenic assay systems; toxicology of food additives; cancer chemotherapy; genetic effects of chemicals in mammals and cells in culture. *Mailing Add:* 5753 Desert View Dr La Jolla CA 92037

GREEN, SIDNEY J, b Rolla, Mo, Nov 17, 37; m 62; c 4. MECHANICAL ENGINEERING, MATERIALS SCIENCE. *Educ:* Univ Mo, Rolla, BS, 59; Univ Pittsburgh, MS, 60; Stanford Univ, Engr, 64. *Prof Exp:* Instr, Univ Mo, 58-59; engr, Westinghouse, Pa, 59-62 & Gen Motors Defense Res Lab, Calif, 64-67; head dept mfg develop staff, Gen Motors Tech Ctr, Mich, 67-70; PRES & CHIEF EXEC OFFICER, TERRA TEK, INC, SALT LAKE CITY, 71- *Concurrent Pos:* Mem, Geol Mt Property Comt, 72-74; adj prof mech eng, Univ Utah, 74-; chmn, Int Comt Rock Mech, Nat Acad Sci, 75-; consult, Drilling Technol Comt, Nat Acad Eng, 75-76; chmn, NSF Thermal-mech Fragmentation Rock Prog Rev Comt & Rock Mech Subcomt, Petrol Div, Am Soc Mech Engrs, 74-75; vchmn, Paper Comt, Exp Mech, 74-76; chmn, Panel Limitations Imposed by Rock Mech on Energy Resource Recovery, Nat Acad Sci, 76 & Gordon Res Conf, 76-78; US alt & deleg, Int Soc Rock Mech, Stockholm, Rio de Janeiro & Montreux, Switz, 77-79; guest lectr, Oxford Univ & Imperial Col, London, 78; dir, Native Plants, Inc, Utah, 77- & Plant Resources Inst, Inc, Utah, 78-; mem, Drilling, Offshore & Arctic Workshop, Dept Energy, 81. *Mem:* Am Soc Mech Engrs; Soc Exp Stress Anal; Geothermal Resources Coun; Am Underground Space Asn. *Res:* Author or coauthor of over 40 papers and numerous reports. *Mailing Add:* 400 Wakara Way Salt Lake City UT 84108

GREEN, STANLEY J(OSEPH), b New York, NY, Mar 11, 20; m 51; c 3. CHEMICAL ENGINEERING. *Educ:* City Col New York, BChE, 40; Drexel Inst, MS, 53; Univ Pittsburgh, PhD(chem eng), 68. *Prof Exp:* Inspector, US Corps Engr, NY, 41; concrete engr, Sci Concrete Serv Corp, 41-42; from jr chem engr to assoc chem engr, US Bur Mines, Md, 42-45; chem engr, Fercleve Corp, Oak Ridge, 45 & Acme Coppersmithing & Machine Co, 45-54; mgr thermal & hydraul eng sect, Bettis Atomic Power Lab, Westinghouse Elec Corp, 55-74, mgr, reactor develop & anal, 74-77; sr prog mgr, 77-79, DIR, STEAM GENERATOR PROJ OFF, ELEC POWER RES INST, 79- *Honors & Awards:* Centennial Medallion, Am Soc Mech Engrs, 80; Donald Q Kern Award, Am Inst Chem Engrs, 85. *Mem:* Fel Am Soc Mech Engrs; Am Inst Chem Engrs. *Res:* Phase rule chemistry applied to separation of compounds from solution by evaporation and crystallization; heterogeneous kinetics; production of alcohol from potatoes; batch solvent extraction of oil seeds; design chemical process equipment; liquid-vapor equilibrium; heat transfer and fluid flow research and design; single phase and two phase heat transfer and pressure drop; thermal, hydraulics, chemistry and materials of pressurized water nuclear steam generators. *Mailing Add:* 3348 Middlefield Rd Palo Alto CA 94306

GREEN, TERRY C, b Abilene, Tex, Nov 4, 35; m 63; c 2. ELECTRICAL ENGINEERING, APPLIED PHYSICS. *Educ:* Univ Tex, BSEE, 58; Trinity Univ, Tex, MS, 69. *Prof Exp:* Design engr, Chance Vought Aircraft Co, Tex, 58; capt, electronic warfare, USAF, 58-62; res engr, Dept Electronics & Elec Eng, Southwest Res Inst, 62-65, sr res engr, Dept Appl Electromagnetics, 65-71, mgr intercept & direction finding res, 71-74, asst dir, Electromagnetics Div, 74-75, DIR, DEPT ELECTROMAGNETIC ENG, ELECTROMAGNETICS DIV, SOUTHWEST RES INST, 75- *Mem:* Inst Elec & Electronic Engrs; Sigma Xi. *Res:* Design and development of low frequency through ultra high frequency direction finding antennas and associated electronics subsystems, including ferrite core and air core quadrupole and dipole mode antennas, interferometer antenna arrays, ring goniometers, low-noise preamplifiers and antenna control circuits; electromagnetic polarization measurements and analysis in the high frequency and very high frequency range including measurements on the non-thermal high frequency radiation from the planet Jupiter. *Mailing Add:* 14306 Clear Creek San Antonio TX 78232

GREEN, THEODORE, III, b Buffalo, NY, Mar 7, 38; m 65; c 2. PHYSICAL OCEANOGRAPHY, FLUID MECHANICS. *Educ:* Amherst Col, AB, 59; Stanford Univ, MS, 61, PhD(eng mech), 65. *Prof Exp:* Asst prof oceanog, Naval Postgrad Sch, 65-69; assoc prof meteorol & civil eng, 69-72, PROF METEOROL & CIVIL & ENVIRON ENG, UNIV WIS-MADISON, 72- *Mem:* AAAS; Am Geophys Union; Am Soc Civil Eng. *Res:* Convection; coastal engineering; small-scale air-sea interaction; lake circulations; remote sensing; water waves. *Mailing Add:* Depts Civil & Environ Eng Univ Wis 1415 W Johnson Madison WI 53706

GREEN, THEODORE JAMES, b North Adams, Mass, Oct 29, 35; m. MICROBIOLOGY. *Educ:* Cornell Univ, BS, 57; Ohio State Univ, MS, 75, PhD(microbiol), 77. *Prof Exp:* Res asst pharmacol, Sterling Winthrop Res Inst, 57-66, res assoc, 66-68; head animal care & res pharm, Warren-Teed Pharmaceut, Div Rohm & Haas Chem, 68-74; sr res scientist immunol, Parke-Davis Co, Div Warner-Lambert, 77-80; ASSOC PROF VET MICROBIOL, UNIV MO, COLUMBIA, 80- *Concurrent Pos:* Prin investr, Malaria Proj, Agency Int Develop, 78-87. *Mem:* Am Soc Trop Med & Hyg; Am Soc Parasitologists; NY Acad Sci; AAAS; Am Soc Microbiol. *Res:* Immunoparasitology of protozoa and helminthis including isolation and characterization of antigens, immune mechanisms, serology and vaccination studies; diagnosis. *Mailing Add:* Dept Vet Microbiol Col Vet Med Univ Mo Columbia MO 65211

GREEN, THOM HENNING, b Steamboat Springs, Colo, Sept 1, 15; m 42, 60; c 3. PETROLEUM GEOLOGY. *Educ:* Univ Colo, BA, 39; Univ Calif, Los Angeles, cert, 43. *Prof Exp:* Geophysicist, Shell Oil Co, 39-43, geologist, 46-48; from dist geologist to div geologist, Sunray Oil Corp, 48-52; chief geologist, Fargo Oils, Ltd, 52-56; consult geologist, 56-60; explor mgr, Wilcox Explor Co, 60-87; CONSULT & PETROL EXPLOR RES, 87- *Mem:* Am Asn Petrol Geol. *Res:* Structural geology; petroleum reservoir fluid mechanics as related to migration and accumulation. *Mailing Add:* Box 52801 Tulsa OK 74152

GREEN, THOMAS ALLEN, b Cleveland, Ohio, Mar 21, 25; m 49, 73; c 4. ALKALI HALIDES, AUGER THEORY. *Educ:* Case Inst Technol, BS, 47; Univ Geneva, ScD(physics), 51. *Prof Exp:* Asst, Univ Geneva, 49-50; mem theoret staff, Radiation Lab, Univ Calif, 51; assoc physics, Columbia Univ, 51-53; from asst prof to assoc prof, Wesleyan Univ, 53-64; mem staff, Lab, 63-69, supvr div 5261 atomic & molecular processes theory, 69-76, MEM STAFF LAB, SANDIA CORP, 76- *Mem:* Fel Am Phys Soc; Sigma Xi. *Res:* Solid state theory. *Mailing Add:* 8108 Northridge NE Albuquerque NM 87109

GREEN, THOMAS L, b Burlington, Iowa, Jan 7, 47; m 69; c 2. FORESTRY. *Educ:* Western Ill Univ, BS, 69, MS, 71; Iowa State Univ, PhD(plant path), 79. *Prof Exp:* Teacher sci, Iowa Sch Dist, Burlington, 71-72; teaching grad asst, plant path, Iowa State Univ, Ames, 76-78; arborist, City Burlington, Iowa, 79-80; RES PLANT PATHOLOGIST, MORTON ARBORETUM, LISLE, ILL, 80- *Concurrent Pos:* Second lieutenant, chem officer, Chem, Biol & Radiol Defense, 49th ADA Unit, Ft Lawton, Wash, 72-74; exten grad asst, plant path coop exten, Iowa State Univ, Ames, 74-76; consult, tree assessment & inventory, 81- *Mem:* Int Soc Arboricult; Int Ornamental Crabapple Soc. *Res:* National crabapple evaluation program; studying the effects of mulch on the growth of woody plants. *Mailing Add:* Morton Arboretum Rt 53 Lisle IL 60532

GREEN, VERNON ALBERT, b McClain County, Okla, Apr 13, 21; m 49; c 2. BIOCHEMICAL PHARMACOLOGY. *Educ:* Univ Okla, BS, 49, MS, 50; Univ Tex, PhD(pharmacol), 60. *Prof Exp:* Instr pharm, Univ Okla, 49-50; from instr to asst prof pharmacol, Univ Tex, 50-62; PROF PHARMACOL, UNIV MO, KANSAS CITY, 62-; TOXICOLOGIST, CHILDREN'S MERCY HOSP, 68- *Mem:* AAAS; Am Pharmaceut Asn; Am Soc Pharmacol & Therapeut; Am Acad Clin Toxicol; Soc Toxicol. *Res:* Pharmacodynamics; mechanisms of drug absorption and activity; cholinesterase activity in cellular permeability. *Mailing Add:* 24th at Gillham Rd Kansas City MO 64108

GREEN, VICTOR EUGENE, JR, b De Ridder, La, Sept 3, 22; m 45; c 2. FIELD CROPS. *Educ:* La State Univ, BS, 47, MS, 48; Purdue Univ, PhD(soil sci), 51. *Prof Exp:* Asst agronomist, Agr Exp Sta, La, 47-49; from asst agronomist to assoc agronomist, Everglades Exp Sta, 51-65, agronomist, Univ Fla, 65-; emer prof, Agron, 87; RETIRED. *Concurrent Pos:* Adv, AID-Univ Fla Contract, Costa Rican Govt, 65-68; curric adv, Sch Agr, Jamaica, WI, 70; chief agr officer, Transp Feasibility Surv for Panama, Int Bank for Reconstruction & Develop, 74, Cape Verde, 79 & Bahamas, 82. *Mem:* Am Soc Agron; Soil Sci Soc Am; Int Soc Soil Sci; Latin Am Railways Asn; Am Soc Hort Sci. *Res:* Organic soils; tropical crops including rice, aloe, dioscorea, sugarcane and coffee; corn; sorghum; millets; sunflower; rural development and crop campaigns in tropical areas. *Mailing Add:* 3915 SW Third Ave Gainesville FL 32607-2709

GREEN, WALTER L(UTHER), b Roanoke Rapids, NC, Mar 13, 34; m 56; c 2. ELECTRICAL & SYSTEMS ENGINEERING. *Educ:* Auburn Univ, BS, 57, MS, 60; Tex A&M Univ, PhD(elec eng), 65. *Prof Exp:* Instr elec eng, Auburn Univ, 58-61; staff engr, Missile Systs Proj Eng, Sandia Corp, NMex, 61-62; instr elec eng, Tex A&M Univ, 62-65; assoc prof, Miss State Univ, 65-66; adv engr, Fed Systs Div, IBM Corp, Ala, 66-68; prof, 68-83, HEAD DEPT ELEC ENG, UNIV TENN, KNOXVILLE, 83- *Concurrent Pos:* Consult, Nuclear Div, Union Carbide Corp, Oak Ridge, Tenn, 69- & IBM, 74- *Mem:* sr mem Inst Elec & Electronics Engrs; Am Soc Eng Educ; Instrument Soc Am. *Res:* Control system theory and application, including guidance and control of space vehicles, multivariable and machine-tooling control systems; process control; system simulation. *Mailing Add:* Dept Elec Eng Univ Tenn Knoxville TN 37946

GREEN, WALTER VERNEY, metal physics, for more information see previous edition

GREEN, WILLARD WYNN, animal breeding; deceased, see previous edition for last biography

GREEN, WILLIAM LOHR, b Harrisonburg, Va, May 1, 45; m 84. MATHEMATICAL ANALYSIS. *Educ:* Yale Univ, BA, 67; Univ Pa, MA, 70, PhD(math), 73. *Prof Exp:* Vis lectr math, Inst Math, Univ Oslo, 72-74; asst prof, Ga Inst Technol, 74-75; asst prof, Williams Col, 75-77; asst prof, 77-82, coordr Undergrad Math Progs, 86-88, ASSOC PROF MATH, GA INST TECHNOL, 82- *Concurrent Pos:* Vis assoc prof, Tulane Univ, 81. *Mem:* Am Mat Soc; Mat Asn Am. *Res:* Automorphisms of operator algebras and their relations to topological dynamics and to ergodic theory; applications of Banach algebras to problems in Banach spaces and in systems theory and circuits. *Mailing Add:* Sch Math Ga Inst Technol Atlanta GA 30332

GREEN, WILLIAM ROBERT, b Toledo, Ohio, Jan 25, 50; m 71; c 2. IMMUNOLOGY, MICROBIOLOGY. *Educ:* Univ Mich, BS, 72; Case Western Reserve Univ, PhD(microbiol), 77. *Prof Exp:* Fel, Med Sch, Johns Hopkins Univ, 77-78; fel immunol, Fred Hutchinson Cancer Res Ctr, 78-79, assoc basic immunol, 79-80, asst mem basic immunol, 80-83; res asst prof microbiol, Univ Wash, 81-83; asst prof, 83-85, assoc prof microbiol, 85-90, PROF, DARTMOUTH MED SCH, 90- *Concurrent Pos:* Fel, Nat Cancer Inst, 77-79. *Mem:* Am Asn Immunologists; Am Soc Microbiol. *Res:* Characterization of the generation, regulation, and specificity of cytolytic murine T lymphocytes, especially those that arise in response to virally induced tumors; mechanism of enhancement of histocompatibility antigen expression by interferon. *Mailing Add:* Dept Microbiol Dartmouth Med Sch Hanover NH 03756

GREEN, WILLIAM WARDEN, b Akron, Ohio, Mar 12, 39; m 63; c 3. AUDIOLOGY. *Educ:* Kent State Univ, BS, 62, MA, 64; Case Western Reserve Univ, PhD(audiol), 70. *Prof Exp:* From instr to asst prof audiol, Kent State Univ, 64-70; assoc prof pediat, 73, assoc prof spec educ, 73-79, DIR CLIN COMMUN DISORDERS, UNIV KY, 73-, PROF NEUROL, PEDIAT & SPEC EDUC, 79- *Concurrent Pos:* Consult, Vet Admin Hosp, 70- & Ky Dept Labor & Educ, 72-; consult, Cardinal Hill Hosp, 81- *Mem:* Am Speech & Hearing Asn; fel Am Speech Lang & Hearing Asn; Am Auditory Soc. *Res:* Clinical audiology; pediatric audiology; noise induced hearing loss. *Mailing Add:* Neurosensory & Commun Disorders Univ Ky Med Ctr Lexington KY 40536-0084

GREENAWAY, FREDERICK THOMAS, b Rakaia, NZ, Aug 18, 47. BIOINORGANIC CHEMISTRY, MAGNETIC RESONANCE. *Educ:* Univ Canterbury, NZ, BSc, 69, PhD(chem), 73. *Prof Exp:* Res assoc chem, Mich State Univ, 73-74; res assoc, Syracuse Univ, 74-77, vis asst prof chem, 77-80; asst prof, 80-86, ASSOC PROF CHEM, CLARK UNIV, 86-, CHAIR, 90- *Concurrent Pos:* Vis scientist, Boston Univ Sch Med, 87-88. *Mem:* Am Chem Soc; NY Acad Sci; Int Soc Magnetic Resonance. *Res:* Bioinorganic chemistry with special emphasis on metal-drug interactions; structural studies of metalloenzymes; magnetic resonance spectroscopy of transition metal compounds. *Mailing Add:* Dept Chem Clark Univ Worcester MA 01610-1477

GREENAWAY, KEITH R(OGERS), b Woodville, Ont, Apr 8, 16; m 44; c 2. ELECTRONICS. *Hon Degrees:* DMilS, Royal Rds Mil Col, Victoria, BC, 78. *Prof Exp:* Instr electronics & navig, Royal Can Air Force, 40-45, researcher, Navig Projs, with US Navy, 45-46, with US Air Force, 46-48, researcher arctic res, Defence Res Bd, 48-54, lectr navig, Strategic Air Command, US Air Force, 54-56, adminstr, hqs, 56-59, commanding officer, Cent Navig Sch, Royal Can Air Force, 59-63, Royal Can Air Force Sta Clinton, 63-67, air adv to chief of air staff, Royal Malaysian Air Force, 67-70; consult, 70-75, sr sci adv, 75-78, CONSULT, DEPT INDIAN AFFAIRS & NORTHERN DEVELOP, 78- *Concurrent Pos:* Chmn, Comt Navig Res, Defense Res Bd, 52; chmn bd, Land-Sea Resources Can, 80-84; fel, Lady Eaton Col, Trent Univ, 83. *Honors & Awards:* Pres Prize, Royal Meteorol Soc, 50; Thurlow Award, Inst Navig, 51; Can Trophy, McKee Trans-Can, 52; Massey Medal, Royal Can Geog Soc, 60. *Mem:* Inst Navig; fel Arctic Inst; fel Can Aeronaut & Space Inst, (pres, 77-78); assoc fel Can Meteorol & Oceanog Soc; fel Brit Inst Navig. *Res:* Aerial navigation, especially polar navigation; arctic research; seven publications. *Mailing Add:* 472 Wellesley Ave Ottawa ON K2A 1B4 Can

GREENBAUM, ELIAS, b Brooklyn, NY, May 12, 44; m 73; c 3. BIOLOGICAL ENERGY PRODUCTION, BIOELECTRONIC COMPOSITE MATERIALS. *Educ:* Brooklyn Col, BS, 65; Columbia Univ, MS, 67, PhD(physics), 70. *Prof Exp:* Res assoc, Univ Ill, Urbana-Champaign, 70-72; asst prof quantum mech, Rockefeller Univ, 72-77; staff scientist, Corp Res Lab, Union Carbide Corp, 77-79; staff scientist, 79-81, GROUP LEADER, OAK RIDGE NAT LAB, 81- *Concurrent Pos:* Adj assoc prof, Rockefeller Univ, 77-79; prin investr, Dept Energy, Washington, DC, 79-, Gas Res Inst, Chicago, 82- & Solar Energy Res Inst, Golden, Colo, 82-84; deleg, Int Solar Energy Soc, UN Conf on New & Renewable Sources Energy, Nairobi, Kenya, 81; Watkins vis prof, Wichita State Univ, 91. *Mem:* Fel Am Phys Soc; Am Chem Soc; Biophys Soc; fel AAAS; Sigma Xi. *Res:* Production of fuels and chemical feedstocks from renewable inorganic resources; fundamental studies of the physics and chemistry of photosynthesis; development of composite photocatalytic bioelectronic materials. *Mailing Add:* Chem Technol Div Oak Ridge Nat Lab PO Box 2008 Oak Ridge TN 37831-6194

GREENBAUM, IRA FRED, b Brooklyn, NY, Feb 27, 51; m 71; c 1. EVOLUTIONARY ZOOLOGY, CYTOGENETICS. *Educ:* Hofstra Univ, BA, 73; Tex Tech Univ, MS, 75, PhD(zool), 78. *Prof Exp:* From asst prof to assoc prof biol, 78-88, PROF BIOL & GENETICS, TEX A&M UNIV, 88- *Concurrent Pos:* NSF grant, 76-78, 82-85 & 87-90; NIH grant, 80-85 & 87-90. *Mem:* Am Soc Mammalogists; Sigma Xi; Soc Systematic Zoologists; AAAS; Soc Study Evolution. *Res:* Biochemical genetics; cytogenetics; evolutionary and systematic genetics; vertebrate speciation. *Mailing Add:* Dept Biol Sci Tex A&M Univ College Station TX 77843

GREENBAUM, LEON J, JR, b Baltimore, Md, Sept 24, 23; m 62; c 2. PHYSIOLOGY. *Educ:* Loyola Col, Md, BS, 47; Univ Md, MS, 55, PhD(physiol), 63. *Prof Exp:* Res asst neurophysiol, Johns Hopkins Hosp, 47-49; physiologist, Off Naval Res, 50-53; res assoc pharmacol, Inst Study Analgesic-Sedative Drugs, 53-55; physiologist, US Naval Med Res Inst, 57-60; res assoc, Med Col Va, 60-64; physiologist, US Naval Med Res Inst, 64-70; PHYSIOLOGIST, NEUROL DISORDERS PROG PROJ, NAT INST NEUROL DIS & STROKE, 70- *Concurrent Pos:* NIMH fel, Sch Med, Univ Md, 62-64; guest scientist, Naval Med Res Inst, 70- *Mem:* AAAS; Aerospace Med Asn; Undersea Med Soc. *Res:* Neurophysiology of inert gas narcosis; cortical excitability; respiratory physiology of divers; hyperbaric pharmacology; pathophysiology of stroke. *Mailing Add:* Nat Inst Neurol & Commun Dis & Strokes Div Extramural Activities Bldg Fed Rm 9C14 Bethesda MD 20892

GREENBAUM, LOWELL MARVIN, b Brooklyn, NY, June 13, 28; m 50; c 3. PHARMACOLOGY, BIOCHEMISTRY. *Educ:* City Col New York, BS, 49; Tufts Col, PhD(physiol), 53. *Prof Exp:* From instr to asst prof pharmacol, Col Med, State Univ NY Downstate Med Ctr, 56-64; from asst prof to prof pharmacol, Col Physicians & Surgeons, Columbia Univ, 64-79; prof & chmn pharmacol, 79-85, VPRES RES & DEAN SCH GRAD STUDIES, MED COL GA, 85- *Concurrent Pos:* Am Cancer Soc res fel biochem, Sch Med, Yale Univ, 54-56; vis prof, Osaka Univ, 70; chmn pub affairs comt, Am Soc Pharmacol & Exp Therapeut, 71-78. *Mem:* AAAS; Am Soc Biol Chem; Am Soc Pharmacol & Exp Therapeut; Am Chem Soc; Am Col Clinical Pharmacol; Sigma Xi; Asn Med Sch Pharmacologists (pres, 84-86). *Res:* Intracellular proteinases; pharmacologically active polypeptides; protein precursors of pharmacologically active polypeptides; inflammation and injury; ascites fluid; T-kinin and T-kininogen. *Mailing Add:* Sch Grad Studies Med Col Ga Augusta GA 30912

GREENBAUM, SHELDON BORIS, b Brooklyn, NY, Apr 15, 23; m 48; c 3. ORGANIC CHEMISTRY. *Educ:* City Col New York, BS, 44; Univ Tenn, MS, 46; Univ Md, PhD(chem), 52. *Prof Exp:* Res chemist, Gordon-Lacey Chem Prod Co, 43-44; res chemist, Pyridium Corp, 44; asst, Univ Md, 46-50; res assoc & instr med chem, Dept Pharmacol, Sch Med, Western Reserve Univ, 52-53; asst, Yale Univ, 53-56; res chemist, Hooker Chem Corp, 56-66; group mgr org & fermentation labs, Diamond Shamrock Chem Co, 66-76; mgr biosci technol, New Technol & Licensing Div, DIamond Shamrock Corp, 76-83; vpres spec serv, Howsafe Corp, 84-88; LECTR CHEM, BALDWIN-WALLACE COL, CUYAHOGA COMMUNITY COL, 88- *Concurrent Pos:* Vis assoc prof, Univ Tenn. *Mem:* Am Chem Soc; Licensing Execs Soc; Pac Indust Properties Asn. *Res:* Chemistry of vitamin D; vitamin E; riboflavin; calcium pantothenate; animal health products; antibiotics; pesticides; pyrimidines; chemical patents; technology transfer; licensing. *Mailing Add:* 24139 Shelburne Rd Shaker Heights OH 44122

GREENBAUM, STEVEN GARRY, b New York, NY, Oct 17, 54; m 83; c 2. MAGNETIC RESONANCE, STRUCTURAL PROBES OF DISORDERED SOLIDS. *Educ:* Clark Univ, BA, 76; Brown Univ, SM, 78, PhD(physics), 81. *Prof Exp:* Grad teaching asst physics, Brown Univ, 76-77, grad res asst, 77-81; Naval Res Ctr postdoctoral fel, Naval Res Lab, 81-83; from asst prof to assoc prof, 83-91, PROF PHYSICS, GRAD CTR, HUNTER COL, CITY UNIV NEW YORK, 92-, DEPT CHMN PHYSICS, 88- *Concurrent Pos:* Fulbright scholar vis scientist, Dept Chem Physics, Weitmann Inst Sci, 90-91. *Mem:* Am Phys Soc; Electrochem Soc; Mat Res Soc. *Res:* Structural studies of disordered solids (inorganic glasses, polymers); ion transport in solids for lectrochemical applications; magnetic resonance (nuclear magnetic resonance, nuclear quadrupole resonance, electron spin resonance) spectroscopy of solids. *Mailing Add:* Physics Dept Hunter Col City Univ New York 695 Park Ave New York NY 10021

GREENBERG, ALLAN S, b Brooklyn, NY, Oct 14, 43. LOW TEMPERATURE PHYSICS. *Educ:* Rensselaer Polytech Inst, BS, 65; Cornell Univ, MS, 68, PhD(physics), 71. *Prof Exp:* Teaching asst physics, Cornell Univ, 65-66, NDEA fel, 66-69; res assoc & instr, Univ Fla, 71-72, vis asst prof, 72-74; asst prof physics, Colo State Univ, 74-78; scientist, Nat Ctr Sci Res, France, 78-85; CONSULT, 85- *Concurrent Pos:* Vis assoc prof, Northwestern Univ, 85; consult, 86- *Mem:* Am Phys Soc; Sigma Xi. *Res:* Low temperature research with specific interests in the nuclear and thermodynamic properties of solid 3He, solid isotopic mixtures, polarized liquid 3He and spin polarized atomic hydrogen. *Mailing Add:* 11923 Darlington Ave No 2 Los Angeles CA 90049

GREENBERG, ARNOLD HARVEY, b Winnipeg, Man, Sept 29, 41. CANCER, IMMUNOLOGY. *Educ:* Univ Man, BS & MD, 65; Univ London, PhD(immunol), 74. *Prof Exp:* Asst prof pediat, Univ Manitoba, 74-80, asst prof immunol, 74-80, assoc prof pediat & immunol, 80-85, PROF PEDIAT, UNIV MAN, 86-; DIR, MAN INST CELL BIOLOGY, 88- *Concurrent Pos:* Terry Fox cancer res scholar. *Mem:* Can Soc Immunol; Am Asn Immunologists; Am Asn Cancer Res. *Res:* Molecular mechanisms of cytotoxicity; TGF-B in inflammation; genetic basis of metastasis formation. *Mailing Add:* Dept Pediat & Immunol Univ Manitoba Manitoba Inst Cell Biol 100 Olivia St Winnipeg MB R3E 0V9 Can

GREENBERG, ARTHUR, b Brooklyn, NY, Sept 27, 46; m 68; c 2. ORGANIC CHEMISTRY, ENVIRONMENTAL CHEMISTRY. *Educ:* Fairleigh Dickinson Univ, BS, 67; Princeton Univ, MA, 70, PhD (chem), 71. *Prof Exp:* Vis asst prof, Fairleigh Dickinson Univ, 71-72; chmn sci dept, Englewood Cliffs Col, 71-72; from asst prof to assoc prof, Frostburg State Col, 72-77; from asst prof to prof chem, NJ Inst Technol, 77-89; PROF ENVIRON SCI, RUTGERS UNIV, 89- *Concurrent Pos:* Ed, J Struct Chem; dep dir, Environ Occup Health Sci Inst, Rutgers-Univ Med & Dent NJ; ed book ser, Molecular Structure & Energetics. *Honors & Awards:* Harlan J Perlis Res Award, 86; Joseph Bittyman Award, 90. *Mem:* Am Chem Soc. *Res:* Stereochemistry; strained organic molecules; theoretical chemistry; thermochemistry; organic substituent effects; organic pollutants on airborne particulates. *Mailing Add:* Dept Environ Sci Rutgers Univ Cook Col New Brunswick NJ 08903-0231

GREENBERG, ARTHUR, physics, for more information see previous edition

GREENBERG, ARTHUR BERNARD, b Brooklyn, NY, Mar 28, 29; div; c 3. ENERGY SYSTEMS, SYSTEMS ENGINEERING. *Educ:* Purdue Univ, BSME & BSAeroE, 50, MS, 52, PhD(aeronaut eng), 55. *Prof Exp:* Sr engr, Aerojet-Gen Corp, 55-57; instr mech eng, Univ Dayton, 57-58; mem tech staff, Space Tech Labs, 58-60; head performance anal dept, Aerospace Corp, 60-62, dir propulsive vehicle systs, 62-63, asst dir group II study progs, 63-64, group dir tech anal & planning, 64, asst gen mgr, syst planning div, 64-69, assoc to vpres corp planning, 69-73, gen mgr, energy & transp div, 73-78, vpres & gen mgr, Govt Support Opers, 78-85; PRES SR STAFF, RES & DEVELOP ASN, 85- *Concurrent Pos:* Adv, Apollo Large Launch Vehicle Planning Group, 61, NASA, Dept Defense Manned Spaceflight Exp Bd & Aeronaut & Astronaut Coord Bd, subcomtreusable launch vehicles & technol, 62-63. *Mem:* Am Inst Aeronaut & Astronaut. *Res:* Rocket propulsion; advanced aerospace launch vehicles; satellite systems; energy systems and technology; energy resource development; air transportation. *Mailing Add:* 125 N Lee St 304 Alexandria VA 22314-3213

GREENBERG, BARRY H, b Brooklyn, NY, June 24, 44; m. MEDICINE. *Educ:* Brooklyn Col, BA, 66; Upstate Med Ctr, MD, 70; Am Bd Internal Med, 75 & 77. *Prof Exp:* Staff assoc, Nat Inst Heart, Lung & Blood, Lipid Metab Br, NIH, 71-73; resident med, Yale-New Haven Hosp, 73-75; fel cardiol, Univ Calif, 75-77; asst res physician I, Cardiovasc Res Inst, 76-77; from asst prof to assoc prof, 77-86, PROF MED, DIV CARDIOL, ORE HEALTH SCI UNIV, 87-, COLLAB SCIENTIST, 88- *Concurrent Pos:* Dir, Coronary Care Unit, Ore Health Sci Univ, 77-; vis colleague, Respiratory Unit, Royal Postgrad Med Sch, Hammersmith Hosp, 84-85. *Mem:* Fel Am Col Cardiol;

Am Fedn Clin Res; Am Heart Asn; Int Soc Heart Res; Am Soc Pharmacol & Exp Therapeut. *Res:* Long term vasodilator therapy of mitral regurgitation; long term vasodilator therapy inaortic insufficiency; left ventricular dysfunction; expression of vasoactive genes in heart failure; author of numerous scientific publications. *Mailing Add:* Div Cardiol L-462 Ore Health Sci Univ 3181 SW Sam Jackson Park Rd Portland OR 97201

GREENBERG, BERNARD, b Springfield, Mass, Oct 5, 24; m 49; c 2. SYSTEMS ANALYSIS. *Educ:* Univ Mass, AB, 48; George Washington Univ, MA, 54. *Prof Exp:* Chief sci intel sect, US Army Corp Engrs, DC, 48-57; eval specialist, Martin Co, Md, 57-58; weapons systs anal engr, Missile Systs Div, Repub Aviation Corp, NY, 58-59; opers analyst, Weapons Systs Lab, Stanford Res Inst, Menlo Park, Calif, 59-60, sr opers analyst, 60-62, prog mgr sea based systs, 62-65, sr systs analyst, Systs Res Div, 65-67, prog mgr, Opers Res Dept, 67-70, asst dir, Regional Security Studies Ctr, 68-72, sr opers analyst, Systs Eng Div, 72-77; mgr, security consult serv, Burns Int Security Serv Inc, 78-82; sr indust engr, Blue Cross Calif, Oakland, 82-87; DIR RES & DEVELOP, BAY PACIFIC HEALTH CORP, 87- *Concurrent Pos:* Consult & lectr on crime control & systs & procedures anal, 77-78. *Mem:* Opers Res Soc Am; Sigma Xi. *Res:* Scientific intelligence; weapons systems; systems requirements and evaluation analysis; criminal justice system studies; residential crimes analysis; innovative investigation procedure development; security systems research; police operations analysis; information systems requirements and operations (methods) analysis; business systems analysis. *Mailing Add:* 896 Southampton Dr Palo Alto CA 94303

GREENBERG, BERNARD, b New York, NY, Apr 24, 22; m 49; c 4. ENTOMOLOGY, ENVIRONMENTAL BIOLOGY. *Educ:* Brooklyn Col, BA, 44; Univ Kans, MA, 51, PhD, 54. *Prof Exp:* From instr to assoc prof, 54-66, PROF BIOL SCI, UNIV ILL, CHICAGO, 66- *Concurrent Pos:* Vis scientist, Istituto Superiore Sanitá, Rome, 60-61, Fulbright res fel, 67-68; Inst Pub Health & Trop Disease, Mex, 62-63; consult, Ill Inst Technol & US Navy, 71-76, Metrop Sanit Dist of Gtr Chicago, 73-75, Elec Power Res Inst, 75-83, forensic entomol, consult & expert witness, 76-, forensic entomol, Sci Gov, Chicago Acad Sci, 81-; NSF grant, 59-60, 79-83; NIH grant, 60-63, 63-67; USA Med Res Develop Command grant, 66-68, 69-72, 84, 85; ONR grant, 77-78. *Mem:* Fel AAAS; Entom Soc Am; Sigma Xi. *Res:* Fly biology; insect transmission of disease; biological impact of electromagnetic fields; forensic entomology. *Mailing Add:* Dept Biol Sci M/C 066 Box 4348 Univ Ill Chicago IL 60680

GREENBERG, CHARLES BERNARD, b Elizabeth, NJ, Dec 20, 39; m 67; c 3. THIN FILM TECHNOLOGY, ENVIRONMENTAL GLASSES. *Educ:* Rutgers Univ, BS, 61; Univ Ill, MS, 62, PhD(ceramics), 65. *Prof Exp:* Sr res ceramist, 65-69, res assoc, 69-71, sr res assoc, 71-77, STAFF SCIENTIST, GLASS RES CTR, PPG INDUSTS, 77- *Mem:* Electrochem Soc; Am Ceramic Soc; AAAS. *Res:* Thin transparent films on flat glass deposited by chemical and physical means; passive solar control in buildings and for variable transmission; product and process development and life projections in use. *Mailing Add:* PPG Industs Inc Glass Res Ctr Pittsburgh PA 15238

GREENBERG, DANIEL, b New York, NY, Dec 31, 27; m 53; c 3. PHYSICAL CHEMISTRY. *Educ:* Yale Univ, BS, 48; Univ Chicago, MS, 49; Columbia Univ, PhD(chem), 54. *Prof Exp:* Res assoc radiochem, Columbia Univ, 54-55; res scientist nuclear physics, Armour Res Found, 55-57; proj leader radiation chem res, Indust Reactor Labs, 58-68, sr res assoc, Polymer Serv Lab, 68-71, group leader, Pilot Plant, 71-73, asst mgr, Polymer Compounds Res Dept, 73-74, mgr, 74-75, ASST DIR RES, US INDUST CHEMS CO, 75- *Mem:* AAAS; Am Nuclear Soc; Am Chem Soc; Am Phys Soc. *Res:* Nuclear and radiation chemistry; polymer chemistry; microbiology; technical management. *Mailing Add:* 6244 Caribou Ct Cincinnati OH 45243-2947

GREENBERG, DANIELLE, b Paris, France, Dec 18, 48; m 80; c 2. INGESTIVE BEHAVIOR, BEHAVIORAL NEUROSCIENCE. *Educ:* Columbia Univ, BS, 75; City Univ New York, MPhil, 83, PhD(physiol psychol), 84. *Prof Exp:* Adj prof psychol, Hunter Col, City Univ New York, 81-82; res asst psychiat, Montefiore Med Ctr, 81-83; postdoctoral fel, 84-86, instr, 86-89, ASST PROF PSYCHIAT, MED COL, CORNELL UNIV, 89- *Concurrent Pos:* Nat res serv award, NIH, 85-86. *Mem:* AAAS; Am Physiol Soc; Am Psychol Soc; Soc Neurosci; NAm Soc Study Obesity. *Res:* Physiological controls of ingestion of nutritional fats; mechanism for the action of peptides such as cholecystokinin or bombesin on feeding behavior; ingestive behavior of genetically obese rodents; developmental nutrition; adequate stimulus for fat-induced satiety. *Mailing Add:* Bourne Lab NY Hosp-Cornell Med Ctr 21 Bloomingdale Rd White Plains NY 10605

GREENBERG, DAVID B(ERNARD), b Norfolk, Va, Nov 2, 28; div; c 3. BIOTECHNOLOGY. *Educ:* Carnegie Inst Technol, BS, 52; Johns Hopkins Univ, MS, 59; La State Univ, PhD, 64. *Prof Exp:* Res engr, Victor Div, Radio Corp Am, 52-53, Nat Distillers Prods Corp, 53-55 & Food Mach & Chem Corp, 55-56; asst, Johns Hopkins Univ, 56-58; asst prof chem, US Naval Acad, 58-61; from instr to prof chem eng, La State Univ, 61-74, res fel, 64-65; prof chem & nuclear eng & head dept, Univ Cincinnati, 74-81, prof nuclear eng & dir grad studies, 81-86; RETIRED. *Concurrent Pos:* Prog dir eng div, NSF, 72-73; sr scientist, US Army Chem Res, Develop & Eng Ctr, Aberdeen PG, MD, 81-86; prog dir, Eng Div, NSF, 89-90. *Mem:* Am Chem Soc; Am Inst Chem Engrs; Am Soc Eng Educ; fel Am Soc Laser Med & Surg. *Res:* Heat, mass, momentum transfer; reaction kinetics; laser research; water treatment; biotechnology; biomedical engineering. *Mailing Add:* Dept Chem Eng Univ Cincinnati Cincinnati OH 45221-0171

GREENBERG, DAVID MORRIS, biochemistry; deceased, see previous edition for last biography

GREENBERG, DONALD P, COMPUTER GRAPHICS. *Educ:* Cornell Univ, BCS, 58, PhD(struct eng), 68. *Prof Exp:* Consult engr, Severud Assocs, 60-65; JACOB GOULD SCHURMAN PROF COMPUTER GRAPHICS, CORNELL UNIV, 68-, DIR, PROG COMPUTER GRAPHICS, 68- *Concurrent Pos:* Guest prof, ETH, Zurich, Switz, 70-71; vis prof, Yale Univ; dir, Nat Sci & Technol Ctr Computer Graphics & Sci Visualization. *Honors & Awards:* Steven A Coons Award, Asn Comput Mach, 87. *Mem:* Nat Acad Eng; Asn Comput Mach; Inst Elec & Electronics Engrs. *Res:* Advancement of state-of-the-art computer graphics; hidden surface algorithms; geometric modeling; color science; realistic image generation. *Mailing Add:* 580 ETC Bldg Cornell Univ Ithaca NY 14853-0258

GREENBERG, ELLIOTT, b New York, NY, Mar 14, 27; m 51; c 3. PHYSICAL CHEMISTRY, INORGANIC CHEMISTRY. *Educ:* City Col New York, BS, 47; Univ Mich, MS, 48, PhD(chem), 55. *Prof Exp:* Instr, Univ Mich, 51-54; assoc chemist, Argonne Nat Lab, Ill, 54-69; prof chem, Prairie State Col, 69-88; RETIRED. *Concurrent Pos:* Chmn, Indust Sponsors, Two-Yr Col Chem Conf, 78-87. *Mem:* Am Chem Soc. *Res:* Fluorine bomb calorimetry; general chemistry. *Mailing Add:* 203 Berry St Park Forest IL 60466

GREENBERG, EVERETT PETER, b Hempsted, NY, Nov 7, 48. MICROBIOLOGY. *Educ:* Western Wash Univ, BA, 70; Univ Iowa, MS, 72; Univ Mass, PhD(microbiol), 77. *Prof Exp:* Fel biol, Biol Labs, Harvard Univ, 77-78; from asst prof microbiol to assoc prof microbiol, Cornell Univ, 78-88; resident instr microbiol ecol, 79-83, CO-DIR, MARINE BIOL LABS, WOODS HOLE, 85-; PROF MICROBIOL, UNIV IOWA, 90- *Concurrent Pos:* Prin investr, NSF, 80-, NIH, 86-90 & ONR, 88-; assoc prof microbiol, Cornell Univ, 84-88, assoc prof, 88-90. *Mem:* Am Soc Microbiol; fel AAAS; NY Acad Sci; Marine Biol Labs; Am Acad Microbiol. *Res:* Mechanisms of bacterial behavior, particularly studies of chemosensory transduction in bacteria and studies of chemical communication between bacterial cells. *Mailing Add:* Dept Microbiol Univ Iowa Iowa City IA 52242

GREENBERG, FRANK, b Perth Amboy, NJ, Aug 24, 48. MEDICAL GENETICS, DYSMORPHOLOGY & TERATOLOGY. *Educ:* Univ Mich, BA, 70, Rutgers Med Sch, MMS, 72, Univ Pa, MD, 74. *Prof Exp:* Pediat Resident, Childrens Hosp Pittsburgh, 74-76; pediat resident, St Christophers Children Hosp, 76-77, genetics fel, 77-79; EIS off, Ctr Dis Control, 79-81; ASSOC PROF, BAYLOR COL MED, 87-; DIR BIRTH DEFECTS, TEX CHILDRENS HOSP, 81- *Concurrent Pos:* Adj asst prof, Univ Tex Sch Pub Health, 85-; mem, Social Issues Comt, Am Soc Human Genetics, 85-88; co-chair, Tex Genetics Network Adv Comt, 86-; secy Planning Coun, Regional Networks, Genetics Servs, 89- *Mem:* AAAS; Am Fedr Clin Res; Am Acad Pediat; Am Soc Human Genetics; Teratol Soc. *Res:* Birth defects with a particular interest in Williams syndrome, Prader-Willi Syndrome, Digeorge anomaly and lissencephaly. *Mailing Add:* Baylor Col Med Birth Defects Ctr 6621 Fannin St Houston TX 77030

GREENBERG, GOODWIN ROBERT, b Danube, Minn, June 23, 18; m 65; c 4. BIOCHEMISTRY. *Educ:* Univ Minn, BA, 41, MS, 42, PhD, 44. *Prof Exp:* Asst physiol chem, Univ Minn, 42-44; res fel med, Univ Utah, 44-46; sr instr biochem, Western Reserve Univ, Cleveland, 46-48, from asst prof to assoc prof, 48-57; PROF BIOL CHEM, UNIV MICH, ANN ARBOR, 57- *Concurrent Pos:* NSF sr fel, Univ Durham, Eng, 56-57; vis scientist, Univ Geneva, Switz, 66; Josiah Macy Jr found scholar & vis prof, Mass Inst Technol, 74-75. *Mem:* Am Soc Microbiol; Am Soc Biol Chemists. *Res:* Nucleic acid and bacteriophage metabolism; microbial genetics. *Mailing Add:* Dept Biol Chem Box 034 Univ Mich Ann Arbor MI 48109

GREENBERG, HERBERT JULIUS, b Chicago, Ill, Nov 28, 21; m 46; c 3. APPLIED MATHEMATICS. *Educ:* Northwestern Univ, BS, 40, MS, 41; Brown Univ, PhD(appl math), 46. *Prof Exp:* Asst, Brown Univ, 43-45, res assoc, 45-47, asst prof math, 47-49; from asst prof to assoc prof, Carnegie Inst Technol, 49-56; assoc dir, AEC Comput & Appl Math Ctr & assoc prof, Inst Math Sci, NY Univ, 56-58; mem staff, Res Div, Int Bus Mach Corp, 58-65, asst dir math sci, 60-65, dir, 65; chmn dept math, 65-74, PROF MATH, UNIV DENVER, 65- *Concurrent Pos:* Mem appl math group, Off Sci Res & Develop, Brown Univ, 43-45; assoc prof, Inst Math Sci, NY Univ, 56-68; head math sci div, Denver Res Inst, 66-74. *Mem:* AAAS; Soc Indust & Appl Math; Math Asn Am. *Res:* Applied mechanics; numeric analysis; computing; mathematical education. *Mailing Add:* Math/Comput Sci Dept Univ Denver Denver CO 80208

GREENBERG, HERMAN SAMUEL, b Philadelphia, Pa, Jan 13, 39; m 70. SCIENCE EDUCATION, ORGANIC CHEMISTRY. *Educ:* Temple Univ, AB, 60, AM, 64; Univ Pa, PhD(org chem), 69. *Prof Exp:* Jr medicinal chemist, Smith Kline & French Labs, 60-63; chemist, Polysci Inc, 69-70; fel pharmacol, Sch Med, Univ Pa, 70-72; res assoc, Inst Cancer Res, 72-73; TEACHER SCI, SCH DIST PHILADELPHIA, 73- *Mem:* Am Chem Soc; Royal Soc Chem; Nat Sci Teachers Asn; Am Asn Physics Teachers. *Res:* General chemistry; chemical education; organic chemistry. *Mailing Add:* 1701 Fox Chase Rd Philadelphia PA 19152

GREENBERG, HOWARD, b New York, NY, Jan 16, 28; m 57. THEORETICAL PHYSICS. *Educ:* City Col New York, BS, 49; NY Univ, MS, 51, PhD(physics), 57. *Prof Exp:* Asst physics, NY Univ, 51-55; from instr to asst prof, 55-74, ASSOC PROF PHYSICS, CITY COL NEW YORK, 74- *Mem:* Am Phys Soc; Am Asn Physics Teachers; AAAS; NY Acad Sci. *Res:* Atomic physics; scattering of electrons by atoms. *Mailing Add:* Dept Physics City Col NY Covent at 138th St New York NY 10031

GREENBERG, IRWIN, b New York, NY, Sept, 18, 35; m 56; c 2. OPERATIONS RESEARCH, STATISTICS. *Educ:* NY Univ, BIE, 56, EngScD, 64; Northeastern Univ, MS, 60. *Prof Exp:* Opers analyst, Avco Res & Adv Develop Div, 56-60; from instr to assoc prof indust eng, NY Univ, 61-73; adj prof math, Pace Univ, 73-77; dir, Statistics Res Ctr, Mathtech Div Mathematica, Inc, 77-79; assoc prof, 79-85, PROF OPERS RES, GEORGE

MASON UNIV, 85- *Concurrent Pos:* Res analyst, Port NY Authority, 60-66; Fulbright travel grant & vis res scientist, Delft Univ, 66-67; assoc prof mgt sci, Univ New Haven, 76-77. *Mem:* Opers Res Soc Am; Inst Mgt Sci; Sigma Xi. *Res:* Stochastic processes; queuing; applied probability; weapon systems analysis; applied statistics. *Mailing Add:* George Mason Univ 4400 Univ Drive Fairfax VA 22030-4444

GREENBERG, JACK SAM, b Warsaw, Poland, May 23, 27; m 52; c 2. NUCLEAR SPECTROSCOPY. *Educ:* McGill Univ, BEng, 50, MSc, 51; Mass Inst Technol, PhD(physics), 55. *Prof Exp:* Rutherford mem fel, 55-56; from instr to asst prof, 56-64, dir grad studies, 67-69, assoc prof, 64-75, PROF PHYSICS, YALE UNIV, 75- *Concurrent Pos:* NSF grant, 61-64; vis scientist, Weizmann Inst Sci, 69-70; vis prof, Gesellschaft f r Schwerionenforschung, Darmstadt, 76-77. *Honors & Awards:* Alexander von Humbolt US Sr Scientist Award, 76-77. *Mem:* Fel Am Phys Soc; Sigma Xi; AAAS. *Res:* Weak interactions; quantum electrodynamics; nuclear structure physics; kaon and hyperon physics; inner-shell vacancy formation; electronic quasimolecules; electrodynamics of strong fields; vacuum decay in over critical fields. *Mailing Add:* Dept Physics/519 JWG Yale Univ New Haven CT 06511

GREENBERG, JACOB, b Mar 10, 29; US citizen; m 57; c 2. SCIENCE ADMINISTRATION REGULATORY AFFAIRS. *Educ:* Hebrew Univ, MSc, 58, PhD(biochem), 60; Adelphi Univ, Paralegal, 79. *Prof Exp:* Jr asst, Hebrew Univ, Israel, 58-60; res assoc biochem, NY Univ, 62-67; res assoc, Mt Sinai Sch Med, 67-68; res asst prof, NY Univ, 68-69, assoc res scientist path, 70-71; asst prof med, New York Med Col, 72-76; dir, Res & Develop Qual Assurance, ABC, New York, 71-83; PRES, PROTUS CO, 76- *Concurrent Pos:* Pilzer travel award, 61; NIH fel biol, Univ Utah, 61-62; NIH fel biochem, NY Univ, 62-63; Am Chem Soc travel award, 67; NIH grant hemat, 72; consult, Col Physicians & Surgeons, Columbia Univ, 65-67; mem, White House Inner Circle, Washington, DC, 84-, Autistic Orgn, Am Asn Univ Profs. *Mem:* AAAS; Am Chem Soc; Am Heart Asn; NY Acad Sci; Am Asn Univ Profs. *Res:* Connective tissue; protein, enzymes-purification, structure, kinetics, cations; diseases of collagen and mucopolysaccharides, collagenase; development of therapy: burns, inflammation and enzyme injection in intervertebral disc; blood clotting; secretory IgA; tissue culture; autism. *Mailing Add:* 130-16 Francis Lewis Blvd Laurelton NY 11413

GREENBERG, JAMES M, b Chicago, Ill, July 30, 40; m 90. NONLINEAR CONSERVATION LAWS. *Educ:* Cornell Univ, BCE, 63; Brown Univ, PhD, 66. *Prof Exp:* Vis mem, Courant Inst Math Sci, 73-74; assoc chmn math, State Univ NY, Buffalo, 75-76, prof, 76-82; prof, Ohio State Univ, 82-85; consult, Dept Energy, 85-86; PROF & CHMN, MATH DEPT, UNIV MD, BALTIMORE COUNTY, 86- *Concurrent Pos:* Prog dir, NSF, 78-79 & 80-82; mem, Adv Comt Super Computer Computations Res Inst, Fla State Univ & Comt Appln Math, Nat Res Coun, 86-88. *Mem:* Am Math Soc; Soc Indust & Appl Math. *Res:* Nonlinear conservation laws; continuum mechanics; transport theory; discrete particle systems. *Mailing Add:* Math & Statist Dept Univ Md Baltimore MD 21228

GREENBERG, JAY R, b Davenport, Iowa, Aug 23, 43. CELL BIOLOGY, MOLECULAR BIOLOGY. *Educ:* Univ Chicago, BS, 64, PhD(cell biol), 68. *Prof Exp:* Res assoc molecular biol, Inst Cancer Res, 70-73; staff scientist, 73-83, SR SCIENTIST, WORCESTER FOUND EXP BIOL, 83- *Concurrent Pos:* NIH fel molecular biol, Inst Cancer Res, 68-70; cancer res scholar, Am Cancer Soc, 74-77; res grants, Nat Cancer Inst, 74-77, NSF, 78 & 81, Nat Inst Gen Med Sci, NIH, 81-; mem biomedical sci study sect, NIH, 81- *Mem:* Am Soc Cell Biol; Am Soc Microbiol; AAAS; Am Soc Biol Chemists. *Res:* Messenger RNA production and function in eukaryotes; protein-nucleic acid interactions. *Mailing Add:* Dept Biol Yale Univ PO Box 6666 New Haven CT 06511

GREENBERG, JEROME HERBERT, b Trenton, NJ, Sept 1, 23; m 45; c 3. PREVENTIVE MEDICINE. *Educ:* Georgetown Univ, MD, 49; Am Bd Prev Med, dipl, 58. *Prof Exp:* Chief dept epidemiol, Walter Reed Army Inst Res, 57-58, asst chief, Commun Dis Br, Prev Med Div, Off Surgeon Gen, 58-60, prev med adv, Korean Mil Adv Group, 60-62, chief prev med br, Hq Fourth US Army, Tex, 64-65, dir dept prev med, Med Field Serv Sch, Brooke Army Med Ctr, 65-69, assoc commandant, Walter Reed Army Inst Res, Med Ctr, Washington, DC, 69-70, chief, Prev Med Div, Off Surgeon Gen, 70-71, dir health environ, Off Surgeon Gen, Dept Army, 71-73; dir pub health, Indianapolis-Marion County Health Dept, 73-75; chief, Bur Tuberc Servs, Tex Dept Health Resources, 76, assoc comnr preventable dis, 76-86; RETIRED. *Concurrent Pos:* Fel trop med, Sch Trop Med, Univ Calcutta, 59; partic interam trop med prog, La State Univ, 64. *Mem:* AMA; Am Pub Health Asn; Am Col Prev Med; Soc Med Consult Armed Forces. *Res:* Epidemiology; immunology. *Mailing Add:* 1900 Glendobbin Rd Winchester VA 22601

GREENBERG, JEROME MAYO, b Baltimore, Md, Jan 14, 22; m 47; c 4. ASTROPHYSICS. *Educ:* Johns Hopkins Univ, PhD(physics), 48. *Prof Exp:* Physicist, Nat Adv Comt Aeronaut, 44-46; instr physics, Exten Div, Univ Va, 45-46; asst prof, Univ Del, 48-51; res assoc, Inst Fluid Dynamics & Appl Math, Univ Md, 51-52; from asst prof to prof physics, Rensselaer Polytech Inst, 52-70; prof astrophys, State Univ NY Albany, 70-75, chmn dept astron & space sci, 71-74, res prof astrophysics, Res Div, 75; PROF LAB ASTROPHYSICS, UNIV LEIDEN, 75- *Concurrent Pos:* Orgn Europ Econ Coop sr vis fel, Univ Leiden, 61; mem Sch Math, Inst Advan Study, Princeton Univ, 65-66; dir astron, Rensselaer Polytech Inst, 67-70; prof lab astrophys, Univ Leiden, 68-69; mem comn interstellar matter & planetary nebulae, Int Astron Union; mem comn six, Int Radio & Sci Union; sr res assoc, Dudley Observatory, 70-75; adj prof, Rensselaer Polytech Inst, 81-; distinguished res prof, Univ Fla, 81- *Mem:* AAAS; Am Astron Soc; fel Am Phys Soc; Int Astron Union. *Res:* Theory of particle scattering and wave scattering; photochemistry of low temperature solids; photochemistry; interstellar matter; comets; origin of life. *Mailing Add:* Huygens Lab Univ Leiden Wassenaarseweg 78 Leiden 2300 RA Netherlands

GREENBERG, JERROLD, b New York, NY, Feb 20, 47; m 77; c 3. BIOCHEMISTRY, MOLECULAR BIOLOGY. *Educ:* Brooklyn Col, BS, 67; Columbia Univ, PhD(biochem), 75. *Prof Exp:* Res assoc molecular biol, Univ Wis-Madison, 75-79; vis asst prof, Southern Ill Univ-Carbondale, 79-80; CHMN, SCI DIV, YESHIVA OF ATLANTA, 80- *Concurrent Pos:* NIH fel genetics, Univ Wis-Madison, 75-76. *Mem:* NY Acad Sci; Am Soc Microbiol. *Res:* Bacterial antibiotic resistance plasmids, transposons, miniplasmids; enzymatic methylation of bacterial and phage DNA. *Mailing Add:* 3044 Stantondale Dr Atlanta GA 30341

GREENBERG, JUDITH HOROVITZ, b Philadelphia, Pa, Apr 2, 47; m 69; c 1. DEVELOPMENTAL GENETICS & BIOLOGY. *Educ:* Univ Pittsburgh, BS, 67; Boston Univ, MA, 70; Bryn Mawr Col, PhD(biol), 72. *Prof Exp:* DIR GENETICS PROG, NAT INST GEN MED SCI, NIH, 88- *Mem:* Am Soc Human Genetics. *Mailing Add:* Westwood Bldg Rm 910 Nat Inst Gen Med Sci Bethesda MD 20892

GREENBERG, LEONARD JASON, b Roxbury, Mass, July 8, 26; m 72. BIOCHEMISTRY, IMMUNOLOGY. *Educ:* Northeastern Univ, BS, 52; Univ Rochester, MS, 54; Univ Minn, PhD(biochem), 58. *Prof Exp:* Asst prof biochem, Univ Minn, 60-61; asst prof path, Stanford Univ, 61-62; sr investr exp path, Naval Radiol Defense Lab, San Francisco, 62-65; asst prof path, NY Univ Med Ctr, 65-67; group leader biochem, Union Carbide Res Inst, New York, 67-69, dir virus res, 69-72; assoc prof, 72-77, PROF PATH, UNIV MINN, MINNEAPOLIS, 77-, PROF LAB MED, 80- *Concurrent Pos:* Assoc prof path, NY Univ Med Ctr, 67-72. *Mem:* AAAS; Am Chem Soc; Am Soc Microbiol; Histochem Soc; Sigma Xi. *Res:* Immunogenetic aspects of susceptibility or resistance to disease. *Mailing Add:* Dept Lab Med Med Sch Univ Minn Mayo Univ Hosp Minneapolis MN 55455-0100

GREENBERG, LES PAUL, b Brooklyn, NY, July 13, 46. SOCIOBIOLOGY. *Educ:* Brooklyn Col, City Univ New York, BS, 67; City Col, City Univ NY, MA, 75; Univ Kans, PhD(entom), 81. *Prof Exp:* FEL, DEPT ENTOM, TEX A&M UNIV, 81- *Mem:* Entom Soc Am; Animal Behavior Soc; AAAS; Int Union Study Social Insects; Sigma Xi. *Res:* Social behavior of insects, especially nestmate and kin recognition; contributions of the environment and insect's genetic history to recognition odors. *Mailing Add:* 3905 B Olive Bryan TX 77801

GREENBERG, MARK SHIEL, b Hamilton, Ohio, May 28, 48; m 71. ANALYTICAL CHEMISTRY. *Educ:* Univ Cincinnati, BSc, 70; Mich State Univ, PhD(anal chem), 74. *Prof Exp:* Asst chem, Mich State Univ, 70-74; Killam fel chem, Univ Alta, 74-75; from asst prof to assoc prof anal chem, Kent State Univ, 75-81; RES SCIENTIST, PROCTER & GAMBLE CO, 80- *Mem:* Am Chem Soc; Sigma Xi. *Res:* Analytical chemistry of food products and raw materials. *Mailing Add:* 9486 Tramwood Ct Cincinnati OH 45242

GREENBERG, MARVIN JAY, b New York, NY, Dec 22, 35; div; c 1. MATHEMATICS. *Educ:* Columbia Univ, AB, 55; Princeton Univ, PhD, 59. *Prof Exp:* Asst math, Princeton Univ, 55-57 & Univ Chicago, 58; instr, Rutgers Univ, 58-59; asst prof, Univ Calif, Berkeley, 59-64; NSF res fel, Harvard Univ, 64-65; assoc prof, Northeastern Univ, 65-67; assoc prof, 67-69, PROF MATH, UNIV CALIF, SANTA CRUZ, 69- *Concurrent Pos:* NSF res fel, Harvard Univ, 61 & Univ Paris, 62; res asst, Erhard Sem Training, 75-76, cert in hypn, 88. *Res:* Algebraic geometry; algebraic number theory; algebraic topology; non-Euclidean geometry; mathematical logic. *Mailing Add:* Dept Math Univ Calif Santa Cruz CA 95064

GREENBERG, MICHAEL D(AVID), b Brooklyn, NY, Nov 15, 35; m 57; c 3. FLUID MECHANICS. *Educ:* Cornell Univ, BME, 58, MS, 60, PhD(theoret mech), 64. *Prof Exp:* Staff scientist, Therm Advan Res, Inc, 63-69; from asst prof to assoc prof, 69-79, PROF MECH & AEROSPACE ENG, UNIV DEL, 79- *Concurrent Pos:* Vis asst prof, Cornell Univ, 64-66, asst prof, 68-69. *Mem:* Sigma Xi. *Res:* Nonlinear mechanics; water wave mechanics; singular perturbation methods. *Mailing Add:* Dept Mech Eng Univ Del Newark DE 19716

GREENBERG, MICHAEL JOHN, b Brooklyn, NY, Sept 28, 31; m 54; c 3. COMPARATIVE PHYSIOLOGY & PHARMACOLOGY. *Educ:* Cornell Univ, AB, 53; Fla State Univ, MA, 55; Harvard Univ, PhD(biol), 58. *Prof Exp:* From instr to asst prof invert zool, Univ Ill, 58-64; from assoc prof to prof biol, Fla State Univ, 64-81, sci dir, Marine Lab, 78-81; DIR, WHITNEY LAB & PROF PHARMACOL & THERAPEUT, COL MED, UNIV FLA, 81- *Concurrent Pos:* NSF sr fel, Melbourne & Misaki Marine Labs, Japan, 64-65; dir exp invert zool course, Marine Biol Lab, Woods Hole, 75-77; vis prof, Dept Physiol, Med Sch, Hiroshima Univ, 78, Dept Zool, Univ Hong Kong, 82; mem, Regulatory Biol Panel, NSF, 83-85; ed-in-chief, Biol Bull, 89- *Honors & Awards:* Merit Award, Nat Heart Lung Blood Inst, NIH, 87. *Mem:* Fel AAAS; Soc Neuroscience; Soc Gen Physiol; Am Soc Pharmacol & Exp Therapeut; Am Soc Zool; Tallahassee, Sopchoppy & Gulf Coast Marine Biol Asn (pres, 69-). *Res:* Comparative physiology and pharmacology of invertebrate muscle; invertebrate neuropeptides; intracellular volume regulation. *Mailing Add:* Whitney Lab Univ Fla 9505 Ocean Shore Blvd St Augustine FL 32086-8623

GREENBERG, MILTON, b Carteret, NJ, April 21, 18; m 48; c 3. ATMOSPHERIC SCIENCES. *Educ:* City Col New York & NY Univ, BA, 43; Harvard Univ, MPA, 54. *Hon Degrees:* ScD, Canaan Col, New Hampshire, 61 & Merrimack Col, Mass, 81, DHL, Univ Lowell, Mass, 85. *Prof Exp:* Dep dir, Geophysics Res Directorate, 50-54, dir, 54-58; pres & chmn, GCA Corp, 58-86; RETIRED. *Concurrent Pos:* Consult, Nat Acad Sci Spec Study Group Geophysics, 57-58; mem & chmn, Physics Atmosphere & Space Comt, Am Rocket Soc, 58-59; ed in chief, Planetary & Space Sci, 57-62; mem, State Indust Adv Coun, Mass Inst Technol Sea Grant Prog, 73-74; Govt Affairs Comt, Sci Apparatus Makers Asn, 65-75. *Mem:* Sigma Xi; fel AAAS; assoc fel Am Inst Aeronaut & Astronaut. *Mailing Add:* 4941 Mission Hill Pl Tuscon AZ 85718-1818

GREENBERG, NEIL, b Newark, NJ, Oct 30, 41; c 1. ETHOLOGY. *Educ:* Drew Univ, BA, 63; Rutgers Univ, MS, 67, PhD(zool), 73. *Prof Exp:* Res scientist ethology, Lab Brain Evolution & Behav, NIMH, 73-78; asst prof, 78-84, ASSOC PROF ETHOLOGY, DEPT ZOOL, UNIV TENN, KNOXVILLE, 84- *Concurrent Pos:* NIMH Grant Found fel, 73-75; res assoc, Mus Comp Zool, Harvard Univ, 77-83; Danforth assoc, 81-86. *Honors & Awards:* Stokely Inst lectr. *Mem:* Animal Behav Soc; Am Soc Zoologists; Soc Neurosci; Int Soc Neuroethology. *Res:* Neural and endocrine aspects of social behavior, behavioral and physiological ecology and the evolution of behavior; interactions of physiological stress, aggression, and reproductive behavior. *Mailing Add:* Dept Zool Univ Tenn Knoxville TN 37996-0810

GREENBERG, NEWTON ISAAC, b Brooklyn, NY, Feb 26, 36; m 60; c 2. THEORETICAL PHYSICS. *Educ:* Brooklyn Col, BS, 57; Univ Md, PhD(physics), 61. *Prof Exp:* Asst physics, Univ Md, 57-61; fel, Bartol Res Found, 61-63; from asst prof to assoc prof, 63-75, PROF PHYSICS, STATE UNIV NY BINGHAMTON, 75- *Mem:* Am Phys Soc; Am Asn Physics Teachers. *Res:* Many body problem; nuclear structure theory. *Mailing Add:* Dept Physics & Astron State Univ NY Binghamton Binghamton NY 13901

GREENBERG, OSCAR WALLACE, b New York, NY, Feb 18, 32; m 69; c 3. ELEMENTARY PARTICLE PHYSICS, HIGH ENERGY PHYSICS. *Educ:* Rutgers Univ, BS, 52; Princeton Univ, PhD(physics), 57. *Prof Exp:* Instr physics, Brandeis Univ, 56-57; NSF fel, Mass Inst Technol, 59-61; from asst prof to assoc prof, 61-67, PROF PHYSICS, UNIV MD, COLLEGE PARK, 67- *Concurrent Pos:* Mem, Inst Advan Study, 64; Sloan Found fel, 64-66; vis assoc prof, Rockefeller Univ, 65-66; vis prof, Tel-Aviv Univ, 68-69 & Johns Hopkins Univ, 77; Guggenheim fel, 68-69; assoc ed, Phys Rev Lett, 76-78; vis scientist, NASA/Goddard Space Flight Ctr, 78 & Fermi Nat Accelerator Lab, 84-85; vis scholar, Univ Chicago, 84-85. *Honors & Awards:* Phys Sci Award, Wash Acad Sci, 71. *Mem:* Am Phys Soc; Am Math Soc; Int Asn Math Phys. *Res:* Quantum field theory; elementary particle and high energy physics; introduction of color degree of freedom in elementary particle physics and the symmetric quark model of baryons; study of subquark models of quarks and leptons; study of small violations of Pauli exclusion principle. *Mailing Add:* Dept Physics Univ Md College Park MD 20742-4111

GREENBERG, PHILIP D, b Brooklyn, NY, Nov 26, 46. MICROBIOLOGY, IMMUNOLOGY. *Educ:* Wash Univ, St Louis, BA, 67; State Univ NY, MD, 71; Am Bd Internal Med, cert, 74. *Prof Exp:* Intern & resident med, Univ Calif, San Diego, 71-74, res fel immunol, 74-76; sr fel, Div Oncol, Univ Wash & Fred Hutchinson Cancer Res Ctr, Seattle, 76-78, from asst prof to prof, 78-88, PROF MED, DIV ONCOL, SCH MED, UNIV WASH, 88- & PROF IMMUNOL, 89- *Concurrent Pos:* Asst mem, Fred Hutchinson Cancer Res Ctr, 78-82, assoc mem, 82-88, mem, 88-; assoc ed, J Immunol, 82-85; mem, US-Japan Cancer Res Coop Prog Comt Tumor Immunol, 85-89; NIH merit award, 91-01. *Mem:* Am Asn Cancer Res; Am Asn Immunologists; Am Soc Clin Invest. *Res:* Bone marrow transplantation. *Mailing Add:* Div Immunol Univ Wash Mail Stop Rm 17 1959 NE Pacific St Seattle WA 98195

GREENBERG, RALPH, b Chester, Pa, Sept 2, 44. MATHEMATICS. *Educ:* Univ Pa, BA, 66; Princeton Univ, PhD(math), 71. *Prof Exp:* Asst prof, Univ Md, 70-74, Brandeis Univ, 74-78; asst prof, 78-80, ASSOC PROF MATH, UNIV WASH, 80- *Mem:* Am Math Soc. *Res:* Algebraic number theory; theory of cyclotomic fields, p-adic L-functions. *Mailing Add:* Dept Math Univ Wash Seattle WA 98195

GREENBERG, RICHARD AARON, b Chicago, Ill, Aug 21, 28; m 54; c 3. BACTERIOLOGY. *Educ:* Univ Ill, AB, 48, MS, 50, PhD(bact), 54. *Prof Exp:* Res bacteriologist, Swift & Co, 56-58, head bact res div, 58-66, chief microbiologist, 62-66, assoc dir res, 66-70, dir, 70-72, vpres res, 72-80; CONSULT, FOOD & ALLIED INDUST, 81- *Mem:* Am Soc Microbiol; Am Acad Microbiol; Inst Food Technol. *Res:* Bacterial spores; food microbiology, especially thermally processed meat products; food poisoning; anti-microbial substances. *Mailing Add:* 809 Highland Dr Louisville KY 40206

GREENBERG, RICHARD ALVIN, b Hartford, Conn, Oct 28, 27; m 50; c 2. BIOSTATISTICS, EPIDEMIOLOGY. *Educ:* Univ Conn, BA, 50; Yale Univ, MPH, 59, PhD(biometry), 63. *Prof Exp:* From instr to assoc prof pub health, Yale Univ, 69-75; prof epidemiol & biostatist, Univ Louisville, 75-89; RETIRED. *Concurrent Pos:* Vis prof epidemiol, Tel Aviv Univ, 71-72. *Mem:* Am Pub Health Asn; Am Statist Asn. *Res:* Biometric aspects of epidemiological research; cancer epidemiology; chronic disease epidemiology. *Mailing Add:* 7360 S Oriole Blvd Apt 402 Delray Beach FL 33446

GREENBERG, RICHARD JOSEPH, b New York, NY, June 19, 47; m 68; c 3. PLANETARY SCIENCES, CELESTIAL MECHANICS. *Educ:* Mass Inst Technol, BS, 68, PhD(planetary sci), 72. *Prof Exp:* Res asst geophys, Mass Inst Technol, 68-70, teaching asst planetary sci, 70-72; res assoc, 72-75, asst prof planetary sci, Univ Ariz, 75-76; res scientist, Planetary Sci Inst, Sci Applns Inc, 76-86; sr res scientist, Lunar & Planetary Lab, 86-88, assoc dean sci, 88-89, PROF PLANETARY SCI & TEACHING & TEACHER EDUC & CHAIR INTER-COL SCI EDUC COMM, UNIV ARIZ, 89- *Mem:* Am Astron Soc; Royal Astron Soc; Sigma Xi. *Res:* Celestial mechanics and planetary dynamics with emphasis on processes of long-term evolution of the solar systems, resonant coupling between motions of planetary bodies and dynamics of the formation of the solar system. *Mailing Add:* Lunar & Planetary Lab Univ Ariz Space Sci Bldg Tucson AZ 85721

GREENBERG, ROLAND, b Winnipeg, Man, June 15, 35; m 65; c 2. PHARMACOLOGY. *Educ:* Univ Man, BSc, 60, MSc, 64, PhD(pharmacol), 68. *Prof Exp:* Res fel, Univ Aberdeen, 68-70; sr pharmacologist, Ayerst Res Labs, 70-71, res assoc, 71-77; sr res investr, Squibb Inst Med Res, 77-81, res fel pharmacol, 81-83, sci dir licensing, 83-88, dir pre clin reg affairs, 88-89, DIR CLIN PHARMACOL, BRISTOL-MYERS SQUIBB RES INST, 90- *Concurrent Pos:* Res fel, Can Med Res Coun, 68-70. *Mem:* Am Soc Pharmacol & Exp Therapeut; Pharmacol Soc Can; NY Acad Sci. *Res:* Respiratory and autonomic pharmacology; prostaglandins. *Mailing Add:* Bristol-Myers Res Inst PO Box 4000 Princeton NJ 08540

GREENBERG, RUVEN, b Columbus, Ohio, Mar 15, 18; m 48; c 3. PHYSIOLOGY. *Educ:* Ohio State Univ, BS, 38, PhD(physiol), 48; Northwestern Univ, MSc, 40. *Prof Exp:* Instr physiol, Med Sch, Ohio State Univ, 48-49; asst prof, Univ Tex Med Br, 49-53; asst prof, 53-59, assoc prof, 59-69, PROF PHYSIOL, UNIV ILL COL MED, 69- *Concurrent Pos:* Fulbright lectr, Univ Madrid, 63. *Mem:* Am Physiol Soc; Am Soc Pharmacol & Exp Therapeut; Soc Neurosci; Soc Exp Biol & Med; Am Soc Cell Biol. *Res:* Neuro-mediators; histo-physiology; blood-brain barrier; psychopharmacology. *Mailing Add:* Dept Physiol M/C 901 Univ Ill Col Med PO Box 6998 Chicago IL 60680

GREENBERG, SEYMOUR SAMUEL, b Brooklyn, NY, Feb 20, 30; m 64. GEOLOGY. *Educ:* Brooklyn Col, BS; Ind Univ, AM & PhD, 59. *Prof Exp:* Petrographer, Ind Geol Surv, 52-62; geologist, Va Div Mineral Resources, 62-64; assoc prof sci, 64-69, prof geol, 69-81, PROF EARTH SCI, WEST CHESTER STATE COL, 80- *Mem:* Geol Soc Am; Am Mineral Asn; Soc Econ Paleontologists & Mineralogists; Geochem Soc; Mineral Asn Can; Sigma Xi. *Res:* Petrology and mineralogy. *Mailing Add:* 317 W Virginia Ave West Chester PA 19380

GREENBERG, SIDNEY ABRAHAM, b New York, NY, Nov 26, 18; m 46; c 2. PHYSICAL CHEMISTRY. *Educ:* Wash Univ, BA, 39; Polytech Inst Brooklyn, MS, 47, PhD(phys chem), 50. *Prof Exp:* Chemist indust res, 41-43; res assoc, Polytech Inst Brooklyn, 46-50; res chemist, Sylvania Elec Co, 50-51; res chemist, Johns Manville Res Ctr, 52-55; res assoc, Univ Leiden, 55-56; assoc prof chem, Seton Hall Univ, 56-58; sr res chemist, Portland Cement Asn, 58-62; mgr appl res & adv develop, Ampex Corp, 62-64; res mgr, Mechrolab Div, Hewlett-Packard, 64-66; assoc prof phys sci, Dominican Col, Calif, 66-72; vis res prof, Zagreb Univ, Yugoslavia, 72-73; CONSULT, MACROMEASURE, FAIRFAX, CALIF, 73- *Mem:* Fel AAAS; Am Chem Soc. *Res:* Kinetics of chemical reactions and crystallization; properties of luminescent solids; structure and properties of inorganic materials; colloid chemistry of silicates; magnetic materials; chemical research and analytical instruments. *Mailing Add:* 187 Bothin Rd Fairfax CA 94930

GREENBERG, STANLEY, b Brooklyn, NY, Sept 14, 45; c 1. CARDIOVASCULAR DISEASES. *Educ:* Brooklyn Col Pharm, BS, 68; Univ Iowa, MS, 70, PhD(pharmacol), 72. *Prof Exp:* Orientation instr pharm, Long Island Univ, 67-68; NIH fel pharmacol, Univ Iowa, 68-72, fel internal med, 72-73; scholar internal med, Univ Mich, 73-74, Mich & Am Heart Asn fel physiol, 74; instr cell biophys, Baylor Col Med, 74-75; asst prof pharmacol, Ohio State Univ, 75-77; assoc prof pharmacol, Univ S Ala, 77-83; AT DEPT PHARMACOL, BERLEX LABS, 83-; PROF PHYSIOL, UNIV MED & DENT, NJ, 83- *Concurrent Pos:* Fel med adv bd, High Blood Pressure Coun, Am Heart Asn; Res Career Develop awards hypertension, Hypertension Br, NIH, Nat Heart, Lung & Blood Inst, 76-77 & 78-82. *Mem:* Am Heart Asn; Am Soc Pharmacol & Exp Therapeut; Am Physiol Soc; Int Study Group Res Cardiac Metab; Soc Exp Biol & Med; Am Soc Hypertension. *Res:* Study of the mechanism of blood vessel function in normal and disease states and the role of ions and prostaglandins in these processes. *Mailing Add:* 23 Lakeview Dr Morris Plains NJ 07950

GREENBERG, STANLY DONALD, b Beaumont, Tex, July 27, 30; m 53; c 3. CYTOPATHOLOGY, RESPIRATORY PATHOLOGY. *Educ:* Univ Tex, BA, 54; Baylor Univ, MD, 54; Univ Iowa, MS, 58. *Prof Exp:* Surg pathologist, Jefferson Davis Hosp, Houston, Tex, 62, dir path labs, Tuberc Div, 62-63; instr path, 62-63, from asst prof to assoc prof path & otolaryngol, 63-74, PROF PATH, MED & OTOLARYNGOL, BAYLOR COL MED, 75- *Concurrent Pos:* Attend, Ben Taub Gen Hosp, 63-; attend, Vet Admin Hosp, 64-; attend, Methodist Hosp, 65-; dir cytopath, Harris County Hosp Dist, 73-78. *Mem:* AMA; Am Soc Clin Path; Col Am Path; Am Soc Exp Path; Am Thoracic Soc. *Mailing Add:* Dept Path Baylor Col Med One Baylor Plaza Houston TX 77030

GREENBERG, STEPHEN B, MEDICINE. *Prof Exp:* INTERNAL MED, SUB-INFECTIOUS DIS, BAYLOR COL MED, 72- *Mailing Add:* Dept Med Baylor Col Med One Baylor Plaza Houston TX 77030

GREENBERG, STEPHEN ROBERT, b Omaha, Nebr, May 5, 27; m 52; c 2. PATHOLOGY, MEDICAL EDUCATION. *Educ:* St Louis Univ, BS, 51, MS, 52, PhD(path), 54. *Prof Exp:* Asst path, Clarkson Hosp, Omaha, Nebr, 54-55; instr, 55-57, assoc, 57-62, asst prof, 62-69, ASSOC PROF PATH, CHICAGO MED SCH, 69- *Concurrent Pos:* Grant, Forensic Sci Found. *Mem:* Fel Asn Clin Scientists; Int Acad Path; Am Geriat Soc; fel AAAS; Sigma Xi; Am Soc Clin Path; Am Acad Forensic Sci; Am Asn Pathologists. *Res:* Anatomic and clinical pathology; environmental pathology; experimental pulmonary disease; pathology of retained metals; tissue effects of metals implanted experimentally in animals and retained for long periods, simulating retained bullets in man. *Mailing Add:* Dept Path Chicago Med Sch North Chicago IL 60064

GREENBERG, WILLIAM, b Lakewood, NJ, Aug 4, 41; m 68; c 3. MATHEMATICAL PHYSICS. *Educ:* Princeton Univ, BA, 63; Harvard Univ, MA, 65, PhD(physics), 70. *Prof Exp:* Vaclav Hlavaty asst prof math, Ind Univ, 70-73; from asst prof to assoc prof, 73-80, PROF MATH, VA POLYTECH INST & STATE UNIV, 80- *Concurrent Pos:* Fel, Nat Sci Res Ctr, Marseille, France, 70; vis res assoc, Physics Theory Lab, Fed Polytech Sch, Lausanne, Switz, 72; vis prof, Univ Firenze, Italy, 82-83. *Mem:* Am Math Soc; Am Phys Soc. *Res:* Mathematical foundations of physics; rigorous statistical mechanics and solution of transport equations. *Mailing Add:* Dept Math Va Polytech Inst & State Univ Blacksburg VA 24061

GREENBERG, WILLIAM MICHAEL, b Toledo, Ohio, Nov 6, 38; m 74. APPLIED PHYSICS. *Educ:* Univ Toledo, BS, 60; Univ Mich, MS, 61, PhD(physics), 67. *Prof Exp:* RES PHYSICIST THIN FILMS, LIBBEY-OWENS-FORD CO, 67- *Mem:* Am Vacuum Soc. *Res:* Applied research on the physics of large-scale production of thin film optical coatings. *Mailing Add:* 1701 E Broadway Toledo OH 43605

GREENBERGER, DANIEL MORDECAI, b Bronx, NY, Sept 29, 33; m 87. THEORETICAL PHYSICS, QUANTUM THEORY. *Educ:* Mass Inst Technol, BS, 54; Univ Ill, Urbana, MS, 56, PhD(physics), 58. *Prof Exp:* Asst prof physics, Ohio State Univ, 60-61; NSF vis fel, Univ Calif, Berkeley, 61-62, asst prof, 62-63; from asst prof to assoc prof, 63-78, PROF PHYSICS, CITY COL NY, 79- *Concurrent Pos:* Vis scientist, Oxford Univ, 71; vis prof, Mass Inst Technol, 79-80; Fulbright guest prof, Atom Inst Austrian Univ, 86, Humboldt prof, Max Planck Inst Quantum Optics, 88. *Honors & Awards:* Fulbright fel, 86; A V Humboldt Prize, 88. *Mem:* AAAS; Am Phys Soc; Am Asn Physics Teachers; Sigma Xi. *Res:* Foundations of quantum theory; relativity and gravitation; neutron diffraction. *Mailing Add:* Dept Physics City Col NY New York NY 10031

GREENBERGER, JOEL S, b Pittsburgh, Pa, Dec 10, 46. RADIATION ONCOLOGY. *Educ:* Columbia Univ, BA, 67; Harvard Med Sch, MD, 71. *Prof Exp:* Staff assoc, Viral Carcinogenesis Br, Nat Cancer Inst, 72-74; resident, Joint Ctr Radiation Ther, Harvard Med Sch, 74-76; instr radiation ther, Sidney Farber Cancer Inst, 76-79; from asst prof to assoc prof radiation ther, Harvard Med Sch, 79-84, assoc dir res, Joint Ctr Radiation Ther, Dept Radiation Therapy, 82-84; PROF & CHMN, DEPT RADIATION ONCOL, MED CTR, UNIV MASS, 84- *Concurrent Pos:* Chmn, Comt Selection Fuller Jr Res Fel, Am Cancer Soc, 79-82, res grant sci adv comt, 83-88. *Mem:* AMA; AAAS; Soc Microbiol; Am Fedn Clin Res; Tissue Cult Asn; Am Soc Hemat; Int Asn Comp Res Leukemia & Related Dis; Radiation Res Soc; Int Asn Exp Hemat; Am Soc Clin Oncol. *Res:* Radiation oncology; numerous publications. *Mailing Add:* Dept Radiation Oncol Univ Mass Med Ctr 55 Lake Ave N Worcester MA 01605

GREENBERGER, LEE M, b Brooklyn, NY, June 5, 55; m 77. BIOCHEMISTRY. *Educ:* Univ Rochester, BA, 77; Emory Univ, PhD(anat), 84. *Prof Exp:* Res asst neurosci, Rochelle Univ, 77-79; postdoctorate neurosci, Columbia Univ, 84-86; postdoctorate molecular pharmacol, Albert Einstein Col Med, NY, 86-88, instr, 88-90; sr scientist, 90-91, GROUP LEADER CANCER PHARMACOL, LEDERLE LABS, NY, 91- *Concurrent Pos:* Vis asst prof med pharmacol, Albert Einstein Col Med, NY, 90- *Mem:* AAAS; Am Asn Cancer Res; Am Soc Biochem & Molecular Biol. *Res:* Identify and understand mechanisms and drug resistance in tumor cells; develop novel chemotherapeutic agents for the treatment of cancer. *Mailing Add:* Lederle Labs Bldg 60B Rm 209 401 N Middletown Rd Pearl River NY 10965

GREENBERGER, MARTIN, b Elizabeth, NJ, Nov 30, 31; m 82; c 4. POLICY ANALYSIS. *Educ:* Harvard Univ, AB, 55, AM, 56, PhD(appl math), 58. *Prof Exp:* Mgr appl sci, Int Bus Mach Corp, 57-58; from asst prof to assoc prof indust mgt, Mass Inst Technol, 58-67; prof comput sci, chmn & dir info processing, Johns Hopkins Univ, 67-72, prof math sci, 72-82; AT GRAD SCH MGT, UNIV CALIF, LOS ANGELES, 82-; IBM PROF COMPUT & INFO SYSTS, PROF PUB POLICY & ANALYSIS. *Concurrent Pos:* Guggenheim fel, Univ Calif, Berkeley, 65-66; mem bd trustees, EDUCOM, 68-73, chmn coun, 68-69; mem comput sci & eng bd, Nat Acad Sci, 70-72; adv, Off of Technol Assessment, US Cong, 74-75 & consult, Gen Acct Off, 75; chmn sect T on info, comput & commun, AAAS, 74-76; mem, Harvard Bd Overseers Vis Comt on Off for Info Technol, 75-81; mgr systs prog, Elec Power Res Inst, 76-77; chmn adv comt, Nat Ctr Anal Energy Systs, Brookhaven Nat Lab, 78; Isaac Taylor vis prof energy, Technion-Israel Inst Technol, 78-79 & Stanford, 80; mem rev comt, Appl Sci Div, Lawrence Berkeley Lab, 84-87; pres, Coun Technol & Individual, 84- *Mem:* Fel AAAS. *Res:* Multimedia; future technology; computer systems. *Mailing Add:* Anderson Grad Sch Mgt Univ Calif 405 Hilgard Ave Los Angeles CA 90024

GREENBERGER, NORTON JERALD, b Cleveland, Ohio, Sept 13, 33; c 3. INTERNAL MEDICINE, PHYSIOLOGY. *Educ:* Yale Univ, AB, 55; Western Reserve Univ, MD, 59. *Prof Exp:* From asst prof to prof med, Col Med, Ohio State Univ, 65-72, dir div gastroenterol, 67-72; PROF MED & CHMN DEPT, UNIV KANS MED CTR, KANSAS CITY, 72- *Concurrent Pos:* Attend physician, Ohio State Univ Hosp, 65-; consult, Vet Admin Hosp & Wright Patterson AFB Hosp, Dayton, Ohio. *Mem:* Am Fedn Clin Res; Am Gastroenterol Asn (pres, 84-85); Asn Am Physicians; Am Soc Clin Invest; Am Col Physicians (pres, 90-91). *Res:* Intestinal absorption of lipids and iron; intestinal transport of digitalis glycosides. *Mailing Add:* Dept Med Univ Kans Med Ctr 39th St & Rainbow Blvd Kansas City KS 66103

GREENBLATT, CHARLES LEONARD, b Youngstown, Ohio, Jan 17, 31; m 53; c 4. MICROBIOLOGY, PARASITOLOGY. *Educ:* Harvard Univ, AB, 52; Univ Pa, MD, 56. *Prof Exp:* Intern, Mary Imogene Bassett Hosp, 56-57; surgeon, Sect Photobiol, Lab Phys Biol, Nat Inst Arthritis & Metab Dis, 57-67, chief, Sect Cell Biol & Immunol, Nat Inst Allergy & Infectious Dis, 67-68; vis assoc prof, 68-69, assoc prof parasitol, 69-78, PROF PARASITOL, HADASSAH MED SCH, HEBREW UNIV, 78- *Concurrent Pos:* Vis prof, Fac Med, Univ El Salvador, 61 & 62; chmn, Kuvin Cent Study Infectious & Trop Dis, 77-, dir, Sch Pub Health & Community Med, 86-89. *Res:* Chlorophyll synthesis and destruction; chloroplast structure and function; hemoflagellate physiology; immune responses to parasites; genetic variation of parasites. *Mailing Add:* Microbiol Inst Hadassah Med Sch Hebrew Univ PO Box 1172 Ein Karem Jerusalem 91010 Israel

GREENBLATT, DAVID J, b Boston, Mass, Apr 8, 45. PHARMACOLOGY. *Educ:* Amherst Col, BA, 66; Harvard Med Sch, MD, 70. *Prof Exp:* Asst prof med, Harvard Med Sch, 74-79; chief, clin pharmacol unit, Mass Gen Hosp, 75-79; CHIEF, DIV CLIN PHARMACOL, NEW ENG MED CTR & PROF PSYCHIAT, MED & PHARMACOL, TUFTS UNIV SCH MED, 79- *Honors & Awards:* Rawls-Palmer Award, Am Soc Clin Pharmacol Therapeut; McKeen-Cattell Award, Am Col Clin Pharmacol. *Mem:* Am Soc Clin Invest; Am Col Neuopsychopharmacol; Am Soc Pharmacol & Exp Therapeut; Am Soc Clin Pharmacol Therapeut; Am Col Clin Pharmacol. *Res:* Investigating the mechanisms of altered drug sensitivity in the elderly. *Mailing Add:* 171 Harrison Ave Boston MA 02111

GREENBLATT, EUGENE NEWTON, pharmacology, toxicology; deceased, see previous edition for last biography

GREENBLATT, GERALD A, b Los Angeles, Calif, May 19, 32; m 61; c 1. PLANT PHYSIOLOGY. *Educ:* Los Angeles State Col, BA, 55; Univ Calif, Davis, PhD(plant physiol), 65. *Prof Exp:* Instr biol, Princeton Univ, 65-66; asst prof, Tex Tech Univ, 66-70; sr res fel, Inst Life Sci, 70-74, asst prof plant sci, 74-82, ASST PROF VET PHYSIOL & PHARMACOL, TEX A&M UNIV, 82- *Mem:* AAAS; Am Soc Plant Physiologists. *Res:* Abscission and senescence; growth hormones; secondary material products of plants; byssenosis; physiology host and pothogen interactions. *Mailing Add:* Western Wheat Lab Wash State Univ 7 Wilson Hall Pullman WA 99163

GREENBLATT, HELLEN CHAYA, US citizen. RESOURCE MANAGEMENT. *Educ:* City Univ New York, BA, 68; Univ Okla, MS, 71; State Univ Ny, Brooklyn, PhD(microbiol & immunol), 77. *Prof Exp:* Resident res assoc, Walter Reed Army Inst Res, Washington, DC, 78-80; sr res immunoparisitologist, Merck, Sharp & Dohme, 80-81; assoc, Dept Med, Albert Einstein Col Med, 81-84; dir res develop, Clin Sci, Inc, 84-86, dir sci & new bus develop, 86-87; sr develop virol, E I du Pont, 88-90, MANAGING DIR M-CAP TECHNOLOGIES, DUPONT CHEM, 90- *Res:* Expansion of technology of microencapsulation of materials. *Mailing Add:* 2109 Ferguson Dr Wilmington DE 19808

GREENBLATT, IRWIN M, b Brooklyn, NY, June 4, 30; m 55, 83; c 3. GENETICS, BOTANY. *Educ:* Ohio State Univ, BS, 53; Univ Wis, MS, 55, PhD(genetics), 59. *Prof Exp:* Res assoc genetics, Univ Wis, 59-60; asst prof biol, Marquette Univ, 60-66 & Northwestern Univ, 66-68; assoc prof, 68-84, PROF GENETICS, UNIV CONN, 84- *Concurrent Pos:* NSF & NIH grants, 61-; consult, Teweles Seed Co, Wis, 64-68. *Mem:* Fel AAAS; Genetics Soc Am; Am Genetic Asn; Bot Soc Am. *Res:* Genetics of regulatory mechanisms as seen in higher plant material; cytogenetics of maize and alfalfa; root tissue culture of maize callus; alfalfa; tissue culture of maize and tobacco; transposons in maize. *Mailing Add:* Dept Molecular Univ Conn Main Campus U-125 75 N Eaglevil Storrs CT 06268

GREENBLATT, JACK FRED, b Montreal, Que, June 6, 46; m 69; c 2. TRANSCRIPTION, REGULATION OF GENE EXPRESSION. *Educ:* McGill Univ, BSc, 67; Harvard Univ, PhD(biophys), 73. *Prof Exp:* Fel molecular biol, dept molecular biol, Univ Geneva, 72-74, sr researcher, 74-76; from asst prof to assoc prof, 77-84, PROF MOLECULAR BIOL, BANTING & BEST DEPT MED RES, UNIV TORONTO, 84-, PROF, DEPT MED BIOPHYS & DEPT MED GENETICS, 84- *Concurrent Pos:* Embo fel, Pasteur Inst, 77; prin investr, Med Res Coun Can, 77- & Nat Cancer Inst Can, 81- *Honors & Awards:* Ayerst Award, Can Biochem Soc, 83. *Res:* Regulation of the termination of transcription in eubacteria; regulation of transcriptional initiation by human RNA polymerase II. *Mailing Add:* Banting & Best Dept Med Res Univ Toronto 112 Col St Toronto ON M5G 1L6 Can

GREENBLATT, JAYSON HERSCHEL, b Montreal, Que, Mar 5, 22; m 44; c 2. EARTH SCIENCES. *Educ:* Dalhousie Univ, BSc, 42, MSc, 43; McGill Univ, PhD, 48. *Prof Exp:* From res chemist to head chem sect, Naval Res Estab, 48-62, head, Dockyard Lab, 63-64, supt eng physics wing, 64-68; dir physics div, Defence Res Estab Ottowa, 68-70, dir earth sci div, 70-75; counr defence res & develop, Can Defence Liaison Staff, Can Embassy, Washington, DC, 75-81; sr scientist to chief, Res & Develop Br, Dept Nat Defense, Can, 81-82; RETIRED. *Concurrent Pos:* Adj prof, Tech Univ NS, 82-91. *Mem:* Fel Chem Inst Can. *Res:* Physical properties of explosives and propellants; homogeneous gas phase reactions; physical electrochemistry as applied to the protection of metals in a marine environment; oceanography; arctic geophysics, especially water-air-ice interactions, properties of arctic terrain and remote sensing; studies of the combustion of cool liquid slurries, wood and other non oil fuels. *Mailing Add:* 6095 Coburg Rd Apt 101 Halifax NS B3H 4K1 Can

GREENBLATT, MARSHAL, b Philadelphia, Pa, Dec 11, 39; m 62; c 3. RADIATION DEVICES. *Educ:* Columbia Univ, BA, 61; Princeton Univ, PhD(aerodyn eng), 71. *Prof Exp:* Vpres, Fusion Systs, 74-78; vpres mkt, Isomet Corp, Springfield, Va, 78-79; MGR CONSULT, GREENBLATT ASSOC, 79-; PRES, MAT ENG ASSOCS, INC, 81- *Res:* Radiation and photochemistry; materials testing. *Mailing Add:* Greenblatt Assoc 10830 Spring Knoll Dr Potomac MD 20854

GREENBLATT, MARTHA, b Hungary, Jan 1, 41; US citizen; m 59; c 2. SOLID STATE CHEMISTRY, CRYSTALLOGRAPHY. *Educ:* Brooklyn Col, BSc, 62; Polytech Inst Brooklyn, PhD(inorg chem), 67. *Prof Exp:* Asst prof chem, Polytech Inst Brooklyn, 67-70, res assoc, 70-72; vis scientist, Weizmann Inst Sci, 72-73; vis asst prof, Polytech Inst Brooklyn, 73-74; from asst prof to assoc prof, 74-85, PROF CHEM, RUTGERS UNIV, 85- *Mem:* Am Chem Soc; Am Crystallog Asn. *Res:* Preparation of solid materials; crystal growth, study relationship between structural and physical properties by electrical resistivity, magnetic susceptibility; optical spectroscopy; x-ray diffraction. *Mailing Add:* Chem Dept Rutgers Univ PO Box 939 Piscataway NJ 08854

GREENBLATT, MILTON, b Boston, Mass, June 29, 14; wid; c 2. PSYCHIATRY. *Educ:* Tufts Col, AB, 35, MD, 39. *Prof Exp:* Instr physiol, Sch Med, Tufts Univ, 39-40, prof psychiat, 63-67; intern med, Beth Israel Hosp, 40-41; resident psychiat, Mass Ment Health Ctr, 41-42, dir EEG lab, 42-63; comnr, Mass Dept Ment Health, 67-73; asst dean, 79-84, dir, Neuropsychiat Inst Hosp & Clins, 79-84, PROF PSYCHIAT, SCH MED, UNIV CALIF, 73-, VCHMN, DEPT PSYCHIAT & BIOBEHAV SCI, 84- *Concurrent Pos:* Charleton res fel, Tufts Univ, 39-40; asst, Harvard Med Sch, 43-48, clin assoc, 50-53, asst clin prof, 53-58, assoc clin prof, 58-63, lectr, 63-73, lectr, Dept Soc Rels, Univ, 57-63; lectr, Sch Med, Boston Univ, 63, sr physician, Mass Ment Health Ctr, 43-45, dir labs & res, 46-63, clin psychiat, 53-57, asst supt, 57-63; supt, Boston State Hosp, 63-67; ed-in-chief, Seminars in Psychiat, 68-74, series ed, 74-; dir, Am Bd Psychiat & Neurol, 68-76, vpres, 75, pres, 76; dir psychiat, Sepulveda Vet Admin Hosp, 73-77 & 84-86; chief staff, Brentwood Vet Admin Hosp, 74-79; assoc ed, Am J Social Psychiat, 81-87; chief, Univ Calif, Los Angeles-San Fernando Valley Prog in Psychiat,

84; chief Psychiat, Olive View Med Ctr, Los Angeles, 86; dir, Prev Psychiat Ctr,Univ Calif, Los Angeles, 86-88. *Honors & Awards:* Hofheimer Prize, Am Psychiat Asn, 52 & Gold Medal Award, 64; Eugene Barrera mem lectr, Albany Med Sch, NY, 55; Israel Strauss mem lectr, Hillside Hosp, NY, 62; Bergendahl mem lectr, Tufts Univ Sch Med, Boston, 63; Lowell Lectr, Lowell Inst & Tufts Univ Sch Med, 68; Samuel Hamilton Award, Am Psychopath Asn, 71; Lemuel Shattuck Award, Mass Pub Health Asn, 73; Arnold L van Ameringen Award in Rehab Psychiat, Am Psychiat Asn, 86; Psychiat Inst Am Found Award, Hosp Psychiat Res, Am Psychiat Asn, 86. *Mem:* Life fel Am Psychiat Asn (vpres, 72-73); Am Asn Social Psychiat (pres, 74-76); Am Col Neuropsychopharmacol (asst secy-treas, 62, pres, 64); Am Psychopath Asn (pres, 70-71); fel Am Col Psychiatrists. *Res:* Administrative psychiatry, homeless adolescents. *Mailing Add:* 3326 Longridge Terr Sherman Oaks CA 91423

GREENBLATT, SAMUEL HAROLD, b Potsdam, NY, May 16, 39; m 63; c 3. ALEXIA, NEUROHISTORY. *Educ:* Cornell Univ, BA, 61, MD, 66; Johns Hopkins Univ, MA, 64. *Prof Exp:* Intern surg, Boston City Hosp, 66-67; resident neurol, Boston Vet Admin Hosp, 67-68; neurologist, Walson Army Hosp, Ft Dix, NJ, 68-70; resident neurol surg, Dartmouth Affil Hosps, 70-74; instr, Albert Einstein Med Col, 74-77; from asst prof to assoc prof neurosurg, Med Col Ohio, Toledo, 80-89; ASSOC PROF NEUROSURG, BROWN UNIV, 89-, CHIEF NEUROSURG, MEM HOSP RI, PAWTUCKET, 89- *Concurrent Pos:* Hon sr registr, Neurosurg Unit Guy's, Maudsley & Kings Col Hosps, London, 72; Tiffany Blake fel, Hitchcock Found, 72-73. *Mem:* Int Neuropsychol Soc; Cong Neurol Surgeons; Am Asn Neurol Surgeons; fel Am Col Surgeons; Acad Aphasia. *Res:* Anatomical basis of reading disorders; history of the neurosciences; medical education; degenerative disease of the spine. *Mailing Add:* Brown Univ Prog Neurosurg Mem Hosp Pawtucket RI 02860

GREENBLATT, SETH ALAN, b Bath, Maine, Mar 26, 60; m 81; c 1. NONLINEAR PARAMETER ESTIMATION, NONLINEAR MODEL SOLUTION. *Educ:* Barry Univ, BPS, 85; George Washington Univ, MPhil, 89, PhD(econ), 91. *Prof Exp:* Consult, 74-81; proj mgr, Martin Marietta Corp, 81-82; mgr, Realtron Corp, 82-83; proj mgr, Viewdata Corp Am, Knight-Ridder, 83-86; mgr, Entre Computer Centers, Inc, 86-87; sr consult, Systemhouse, Ltd, 87-89; VPRES ECONOMET, SORITES GROUP, INC, 89- *Concurrent Pos:* Lectr, var prof conferences, 84-; res consult, George Washington Univ, 88-89; vpres, Wash-Baltimore Sect, Soc Indust & Appl Math, 89-91, pres, 90-; vis scholar, Univ Indonesia, 91. *Mem:* Soc Indust & Appl Math; Am Math Soc; Am Econ Asn; Sigma Xi. *Res:* Development of numerical methods for unconstrained and constrained optimization; application of mathematical and statistical techniques to solve problems in economics and econometrics; mathematical modeling of behavioral systems; articles in various professional journals. *Mailing Add:* Sorites Group Inc PO Box 2939 Springfield VA 22152

GREENE, ALAN CAMPBELL, b Boston, Mass, Mar 31, 35; m 59; c 3. SOLID STATE SCIENCE. *Educ:* Northeastern Univ, BS, 58; Brown Univ, PhD(physics), 64. *Prof Exp:* Asst prof physics, Rensselaer Polytech Inst, 64-70; staff physicist, Comn Col Physics, 70-71; assoc prof, 71-74, PROF PHYSICS, CALIF STATE COL, 74- *Mem:* Am Asn Physics Teachers. *Res:* Defect interactions in ionic solids; production of defects by radiation; magnetic properties of crystalline defects. *Mailing Add:* Dept Physics & Earth Sci Calif State Univ 9001 Stockdale Hwy Bakersfield CA 93311

GREENE, ARTHUR E, b Philadelphia, Pa, Aug 8, 23; m 53; c 2. CELL BIOLOGY. *Educ:* Univ Pa, AB, 47; Phila Col Pharm, BSc, 49, MSc, 50, DSc(bact), 52; Am Bd Microbiol, dipl, 65. *Prof Exp:* Res fel virol, Children's Hosp of Philadelphia, 52-53, res assoc, Polio Res Lab, 53-56, res assoc, S Jersey Med Res Found, 56-57; res virologist, Nat Drug Co of Philadelphia, 57-59; virologist, Smith, Kline & French Labs, 59-60; assoc dir, Nat Inst Gen Med Sci, 72-83, dir, human genetic mutant cell repository, 83-88; head, dept cell biol, Nat Inst Aging, 61-88, dir aging cell culture repository, 80-88; VPRES, RES & SCI AFFAIRS, CORNELL INST MED RES, 85- *Concurrent Pos:* Thomas H Powers scholar, Philadelphia Col Pharm & Sci, 49, lectr virol, 56-85, vis asst res prof vet virol, 62-64; res assoc surg, Pa Med Sch, 64-; vis assoc prof microbiol, Jefferson Med Col, 68-73; assoc virol, Philadelphia Col Osteopath Med, 61-76. *Mem:* AAAS; Am Soc Microbiol; Tissue Cult Asn; Soc Cryobiol; NY Acad Sci. *Res:* Preservation and characterization of cell cult; cryobiology of cells and organs; techniques for the species identification of insect and animal cell lines; studies on genetic and biochemical cell cultures; virus chemotherapy and vaccines; aging syndrome cell cultures. *Mailing Add:* Dept Cell Biol Coriell Inst Med Res 401 Haddon Ave Camden NJ 08103

GREENE, ARTHUR EDWARD, b Chicago, Ill, Dec 10, 45; m 70; c 1. COMPUTATIONAL PHYSICS, LASER PHYSICS. *Educ:* Ohio State Univ, BS, 67, PhD(astron), 71. *Prof Exp:* Officer, USAF, 71-75; staff mem, theoret chem & molecular physics, 75-81, staff mem, thermonuclear applications, 81-86, TECH STAFF MEM, PULSED ENERGY APPLICATIONS, LOS ALAMOS NAT LAB, 86- *Concurrent Pos:* Lectr, Univ Colo, Colorado Springs, 72-73. *Honors & Awards:* Award Excellence, US Dept Energy, 86. *Mem:* Am Phys Soc. *Res:* Theoretical and computational support for the development of an intense, soft, x-ray source from a magnetically driven imploding liner. *Mailing Add:* Los Alamos Nat Lab X-10 B259 Los Alamos NM 87545

GREENE, ARTHUR FRANKLIN, b Hartford, Conn, Dec 18, 39; m 63; c 2. ELEMENTARY PARTICLE PHYSICS. *Educ:* Worcester Polytech Inst, BS, 61, MS, 63; Tufts Univ, PhD(exp elem particle physics), 67. *Prof Exp:* Asst, Worcester Polytech Inst, 61-63 & Tufts Univ, 63-67; physicist, div res, AEC, 67-69; res assoc high energy physics, Argonne Nat Lab, Ill, 69-72; physicist, Fermi Nat Accelerator Lab, 72-79, asst dir prog planning, 76-79; PHYSICIST, BROOKHAVEN NAT LAB, 79- *Mem:* Am Phys Soc. *Res:* Experimental high energy physics; elementary particle research using bubble chambers and electronic techniques; administration, with emphasis on planning for experiments related to high energy physics research; development of superconducting magnets for a high energy particle accelerator. *Mailing Add:* Accelerator Develop Dept Bldg 902A Brookhaven Nat Lab Upton NY 11973

GREENE, ARTHUR FREDERICK, JR, b Cleveland, Ohio, Sept 28, 27; m 57. ANALYTICAL CHEMISTRY, INORGANIC CHEMISTRY. *Educ:* Ohio State Univ, BSc, 51; Case Western Reserve Univ, MSc, 61. *Prof Exp:* Tech serv, R O Hull & Co, Inc, 55-61; analyst, 62-64, RES ANALYTICAL CHEMIST, ENGELHARD CORP, 64- *Honors & Awards:* George B Hogaboom Mem Award, 69. *Mem:* Am Chem Soc; Royal Soc Chem; NAm Thermal Anal Soc; Int Confederation Thermal Anal; Int Union Pure & Appl Chem. *Res:* Development of new analytical methods; thermal analysis techniques for synthesis and characterization of inorganic materials. *Mailing Add:* 1056 Lakeland Ave Lakewood OH 44107-1228

GREENE, BARBARA E, b Joliet, Ill, May 9, 35. FOOD SCIENCE, NUTRITION. *Educ:* Fla Southern Col, BS, 57; Fla State Univ, MS, 62, PhD(food, nutrit), 66. *Prof Exp:* Instr food sci & nutrit, Colo State Univ, 66-68; asst prof food & nutrit & dairy sci & indust, Iowa State Univ, 69; ASST PROF FOOD & NUTRIT, UNIV GA, 69- *Mem:* Inst Food Technologists; Am Oil Chemists Soc; Am Meat Sci Asn; Am Dietetics Asn; Sigma Xi. *Res:* Lipid oxidation in relation to color and flavor in fresh. *Mailing Add:* 1112 Hastings Ave E Cour D'Alene ID 83814-4453

GREENE, BETTYE WASHINGTON, b Palestine, Tex, Mar 20, 35; m 55; c 3. PHYSICAL CHEMISTRY, COLLOID CHEMISTRY. *Educ:* Tuskegee Inst, BS, 55; Wayne State Univ, PhD(phys chem), 65. *Prof Exp:* Res chemist, Dow Chem Co, 65-70, sr res chemist, 70-75, res assoc, 75-81, assoc scientist, 81-86; RETIRED. *Mem:* AAAS; Am Chem Soc; Sigma Xi. *Res:* Light scattering method for determining size distributions in colloid systems; determination of electrokinetic properties of pigment dispersions used for paper coatings; rheology of paper coating dispersions; design of composites for building products. characterization of polymers and latexes; repeptization in hydrophobic colloids; latex fiber interactions; heterofloeculation studies. *Mailing Add:* 5213 Cortland St Midland MI 48640

GREENE, BRUCE EDGAR, b Detroit, Mich, June 24, 33; m 56; c 4. PLASTICS, AUTOMOTIVE ENGINEERING. *Educ:* Wayne State Univ, BSME, 55; Chrysler Inst Eng, MAE, 57; Univ Detroit, MBA, 62. *Prof Exp:* Test & develop engr, Chrysler Corp, 57-60; cost analyst, 60-64, prod planning mgr, 64-69, design mgr, 69-75, exec engr, 75-77, tractor prod planning mgr, 77-80, CHIEF PLASTICS ENG, FORD MOTOR CO, 80- *Mem:* Soc Plastics Engrs; Soc Automotive Engrs. *Res:* Plastics development and applications. *Mailing Add:* 24945 Fairmount Dearborn MI 48124

GREENE, CHARLES EDWIN, b Aurora, Ill, May 17, 19; m 43; c 6. ORGANIC CHEMISTRY. *Educ:* Univ Notre Dame, BS, 41, MS, 47, PhD(chem), 49. *Prof Exp:* Chemist, Reilly Tar & Chem Co, 41-46; sr res chemist, Gen Tire & Rubber Co, 49-69; sr engr, Brunswick Corp, 69-75; sr res chemist, Freeman Chem Co, 78-82; RETIRED. *Mem:* Am Chem Soc. *Res:* Monomers and polymers; radomes and defense products. *Mailing Add:* 229 W Washington St, No 304 Port Washington WI 53074-1837

GREENE, CHARLES RICHARD, b Chicago, Ill, Oct 3, 23; m 47; c 2. PHYSICAL CHEMISTRY, CHEMICAL ENGINEERING. *Educ:* Univ Chicago, BS, 48, SM, 49, PhD(phys & org chem), 52. *Prof Exp:* Res chemist dermat cancers, Chicago Med Sch, 51; prof chem, George Williams Col, 51-52; res chemist, Standard Oil Co Ind, 52-54; res chemist, Shell Develop Co, 54-64, res supvr, 64-67, mgr prod, Shell Chem Co, NY, 67-68, mgr exp opers, Shell Develop Co, 68-69, mgr chem eng, 69-72, mgr surface projs, 72-80; DIR, ENERGY INDUST PROG, SRI INT, 80- *Concurrent Pos:* Consult, res appl nat needs, NSF, 73- *Honors & Awards:* Norton Prize. *Mem:* Am Chem Soc; fel Am Inst Chem Engrs; Sigma Xi; AAAS; fel Am Inst Chemists. *Res:* Petrochemicals; research in coal as source of chemicals; project management in development of new processes for manufacture of chemical intermediates; energy consulting. *Mailing Add:* 4512 Via Huerto Santa Barbara CA 93110

GREENE, CHARLOTTE HELEN, b Philadelphia, Pa, Mar 10, 43; m 71. PHYSIOLOGY, HISTOLOGY. *Educ:* West Chester State Col, BS, 65; Thomas Jefferson Univ, PhD(physiol), 74. *Prof Exp:* Instr, 75-77, from asst prof to assoc prof, 81-88, PROF PHYSIOL, PHILADELPHIA COL OSTEOP MED, 88- *Concurrent Pos:* Investr, Am Heart Asn, 77-79, dir electron micros, 88- *Mem:* Am Zool Soc; Am Physiol Soc; Sigma Xi. *Res:* Cardiovascular, endocrine physiology; lasers in medicine. *Mailing Add:* Dept Physiol & Pharmacol Philadelphia Col Osteop Med4150 City Ave Philadelphia PA 19131

GREENE, CHRISTOPHER HENRY, b Lincoln, Nebr, Aug 1, 54; m 77. THEORETICAL ATOMIC PHYSICS. *Educ:* Univ Nebr, Lincoln, BS, 76; Univ Chicago, MS, 77, PHD(physics), 80. *Prof Exp:* Asst prof, 81-84, ASSOC PROF PHYSICS, LA STATE UNIV, 84- *Concurrent Pos:* Alfred P Sloan Found Fel. *Mem:* Am Phys Soc; Sigma Xi. *Res:* Electron correlations in atoms and small molecules; threshold laws in photoabsorption and electron scattering. *Mailing Add:* Univ Colo La State Univ Boulder CO 80309

GREENE, CHRISTOPHER STORM, b Aurora, Ill; c 2. AERONAUTICAL ENGINEERING. *Educ:* Univ Colo, Boulder, BS(elec eng) & BS(bus), 73; Mass Inst Technol, PhD(elec eng), 78. *Prof Exp:* Res engr, 78-81, sect chief, Honeywell Syst & Res Ctr, 81-83, SR PROG MGR, HONEYWELL MIL AVIONICS, 83- *Mem:* Am Inst Aeronaut & Astronaut; Inst Elec & Electronics Engrs; Sigma Xi. *Res:* Guidance, navigation and control. *Mailing Add:* 2701 13th Terr NW New Brighton MN 55112

GREENE, CURTIS, b Philadelphia, Pa, Nov 10, 44; m 72. MATHEMATICS. *Educ:* Harvard Col, AB, 66; Calif Inst Technol, PhD(math), 69. *Prof Exp:* C L E Moore instr, Mass Inst Technol, 69-71, asst prof math, 71-76; assoc prof math, State Univ NY Buffalo, 75-78; assoc prof, 78-81, PROF MATH & CHMN DEPT, HAVERFORD COL, 81- *Mem:* Am Math Soc; Math Asn Am. *Res:* Combinatorial analysis; discrete applied mathematics. *Mailing Add:* 715 College Ave Haverford PA 19041

GREENE, DARYLE E, b Garfield, Ark, June 27, 32; m 54; c 4. NUTRITION, BIOCHEMISTRY. *Educ:* Univ Ark, BS, 54, MS, 55; Univ Ill, PhD(animal sci), 60. *Prof Exp:* Asst mgr turkey res div, Ralston Purina Co, 60-62, mgr, 62-64; assoc prof poultry nutrit, Univ Ark, 64-65; dir poultry res, 66-75, dir chow res & develop, 75-83, vpres & dir res, Purina Mills Inc, 83-88; RETIRED. *Mem:* AAAS; Poultry Sci Asn; Animal Nutrit Res Coun; Am Soc Animal Sci; Am Registry Cert Prof Animal Scientists. *Res:* Amino acid nutrition; unidentified growth factors; mineral metabolism; biological availability of nutrients in feed ingredients. *Mailing Add:* 514 Webster Forest Dr St Louis MO 63119

GREENE, DAVID C, b Elyria, Ohio, Nov 24, 22; m 49; c 4. PHYSICS, ACOUSTICS. *Educ:* Oberlin Col, BA, 49; Pa State Univ, MS, 59. *Prof Exp:* Physicist, US Naval Ord Lab, Md, 50-53; sr engr, Cook Res Labs, Ill, 53-55; physicist & res assoc, Ord Res Lab, Pa State Univ, 55-62; physicist & head acoust & electronics sect, Torpedo Supporting Res Br, Antisubmarine Warfare Off, US Bur Naval Weapons, 62-66, actg head acoust br, Acoust & Electromagnetics Div, Res & Technol Directorate, Naval Ord Systs, Command Hq, Washington, DC, 66-67; sr res scientist, Pac Northwest Labs, Battelle Mem Inst, 67-69; sr staff engr, Sensor Systs Dept, Electronic Systs Lab, Trw Inc, Washington, 69-87; systs anal, Fairfax County, Va, 87; RETIRED. *Mem:* Acoust Soc Am. *Mailing Add:* 382 Elm St Oberlin OH 44074

GREENE, DAVID GORHAM, b Buffalo, NY, Feb 5, 15; m 46; c 4. CARDIOLOGY. *Educ:* Princeton Univ, AB, 36; Harvard Univ, MD, 40. *Prof Exp:* Intern med, Presby Hosp, NY, 41-42, asst resident, 45-46; assoc, 50-51, asst prof, 51-55, asst physiol, 50-57, asst clin prof, 57-70, Harry M Dent Prof clin res in cardiovasc dis, 56-70, PROF MED, STATE UNIV NY BUFFALO, 70-, ASSOC PROF PHYSIOL, 70- *Concurrent Pos:* Fel path, Banting Inst, Toronto, Can, 40-41; res fel, Col Physicians & Surgeons, Columbia Univ, 46-48; res fel, State Univ NY, Buffalo, 48-50. *Mem:* Am Heart Asn; fel Am Col Physicians; fel Am Col Cardiol; Am Physiol Soc; Am Fedn Clin Res; Sigma Xi; Soc Cardiac Angiography & Interventions. *Res:* Cardiovascular diseases. *Mailing Add:* 100 High St Buffalo NY 14203

GREENE, DAVID LEE, b Louisville, Ky, Dec 19, 44; m 68; c 1. INORGANIC CHEMISTRY. *Educ:* Univ Notre Dame, 67; Univ Ky, PhD(chem), 71. *Prof Exp:* Fel inorg chem, WVa Univ, 71-72; chmn dept phys sci, 76-80, dean arts & sci, 80-89, FAC MEM CHEM, RI COL, 72- *Concurrent Pos:* Postdoctoral fel, WVa Univ, 71-72; Student Sci Training Prog grantee, NSF, 77; vis scientist, Oxford Univ, 89-90. *Mem:* Am Chem Soc; Sigma Xi. *Res:* Synthesis and characterization of coordination compounds; utilization of microwave heating in synthesis. *Mailing Add:* Dept Phys Sci RI Col Providence RI 02908

GREENE, DAVID LEE, b Denver, Colo, Aug 23, 38; m 62; c 1. PHYSICAL ANTHROPOLOGY, DENTAL ANTHROPOLOGY. *Educ:* Univ Colo, BA, 60, MA, 62, PhD(anthrop), 65. *Prof Exp:* Asst prof anthrop & orthod, State Univ NY Buffalo, 64-65; asst prof anthrop & head dept, Univ Wyo, 65-67; assoc prof, 67-71, PROF ANTHROP, UNIV COLO, BOULDER, 71- *Concurrent Pos:* Chmn, dept anthrop, Univ Colo, Boulder, 75-77, 81-83, 90- *Mem:* Am Asn Phys Anthrop. *Res:* Human evolution; primatology; population genetics; nubian prehistory. *Mailing Add:* Dept Anthrop Univ Colo Boulder CO 80303

GREENE, DONALD MILLER, b Dallas, Tex, Jan 9, 49; m 72; c 3. AGRICULTURAL CLIMATOLOGY, AGRICULTURAL STATISTICS. *Educ:* ETex State Univ, BS, 73; Univ Okla, MA, 77, PhD(geog), 79. *Prof Exp:* Lectr, 79-81, asst prof, 81-88, ASSOC PROF EARTH SCI, BAYLOR UNIV, 88- *Concurrent Pos:* Asst ed, Baylor Geol Studies Bull, 80- *Mem:* Asn Am Geographers. *Res:* Relationship between agricultural productivity and weather; statistical models that predict yield based upon the climate experienced during the growing season. *Mailing Add:* Baylor Univ Main Campus Baylor Univ PO Box 367 Waco TX 76798

GREENE, EDWARD FORBES, b New York, NY, Dec 29, 22; m 49; c 4. PHYSICAL CHEMISTRY. *Educ:* Harvard Univ, AB, 43, AM, 47, PhD(chem), 49. *Prof Exp:* Chemist, Shell Oil Co, 43-44; mem staff, Los Alamos Sci Lab, 49; res assoc, 49-51, from instr to assoc prof, 51-63, PROF CHEM, BROWN UNIV, 63-, JESSE H & LOUISA D SHARPE METCALF PROF, 85- *Concurrent Pos:* NSF sr fel, Univ Bonn, 59-60 & Calif Inst Technol, 66-67; vis prof, Tougaloo Col, 65; res vis, Bell Labs, Murray Hill, 76-77; Chmn, Chem Dept, Brown Univ, 80-84. *Mem:* Am Chem Soc; Am Phys Soc. *Res:* Chemical kinetics; scattering in molecular beams; reactions on surfaces. *Mailing Add:* Dept Chem Box H Brown Univ Providence RI 02912

GREENE, ELIAS LOUIS, b New York, NY, Jan 7, 32; m 54; c 2. IMMUNOLOGY, VIROLOGY. *Educ:* Brooklyn Col, BS, 53; Cornell Univ, PhD(virol immunol), 64. *Prof Exp:* Res asst virol, Sloan Kettering Inst Cancer Res, 55-64; virologist-immunologist, Res Inst, Henry Putnam Mem Hosp, Bennington, Vt, 64-65; res immunologist pediat, Long Island Jewish Hosp, NY, 65-67; from instr to asst prof pediat, Sch Med, Univ Miami, 67-73; asst dir diag eval, Behring Diag, Am Hoechst Corp, 73-77; asst dir med oper, 77-80, MGR REGULATORY AFFAIRS INT, CUTTER LABS, INC, BERKELEY, 80-, DIR, 85- *Concurrent Pos:* Nat Cystic Fibrosis Res Found fel, 68; lab technologist, Mercy Hosp, Rockville Centre, NY, 60-67; lectr, New York Community Col, 67; lab technologist, clin lab, Miami Heart Inst, 67-68, consult immunologist, Organ Transplant Prog, 68-73; dir qual control & responsible head biol prog, NAm Biol, Inc, Fla, 70-73; consult immunologist, United Labs, NY, 74- *Mem:* Am Acad Cosmetic Surg; Am Soc Microbiol; Int Asn Biol Standardization; Am Asn Immunol. *Res:* Immunologic and histochemical identification and enumeration of tissue antigens in normal and diseased states, especially in cystic fibrosis and liver diseases; properties of arthropod-borne and H-group viruses. *Mailing Add:* Regulatory Affairs Int Cutter Biol Miles Inc Fourth & Parker St Berkeley CA 94701

GREENE, FRANK CLEMSON, b Memphis, Tenn, Sept 2, 39; div; c 2. BIOCHEMISTRY. *Educ:* Morehouse Col, BS, 58; Atlanta Univ, MS, 65; Univ Calif, Davis, PhD(biochem), 69. *Prof Exp:* Chemist, Western Regional Res Lab, USDA, Albany, 61-65; chemist, USDA & Univ Calif, Davis, 65-69; RES CHEMIST, WESTERN REGIONAL RES CTR, USDA, BERKELEY, 69- *Mem:* Am Chem Soc; AAAS; Sigma Xi; Am Soc Plant Physiol. *Res:* Physical chemistry of proteins; fluorescent probe mechanisms; biosynthesis of proteins; gene expression. *Mailing Add:* Western Regional Res Ctr USDA Albany CA 94710

GREENE, FRANK EUGENE, DEVELOPMENTAL PHARMACOLOGY, DRUG METABOLISM. *Educ:* Univ Fla, PhD(pharmacol), 62. *Prof Exp:* ASSOC PROF PHARMACOL, MILTON S HERSHEY MED CTR OF PA STATE, 71- *Mailing Add:* Dept Pharmacol Pa State Univ Hershey PA 17033

GREENE, FRANK T, b Saginaw, Mich, Nov 26, 32; m 59; c 3. CHEMICAL PHYSICS. *Educ:* Univ Mich, BS, 55; Univ Calif, MS, 57; Univ Wis, PhD(phys chem), 61. *Prof Exp:* Res assoc chem, Univ Kans, 61-62; assoc physicist, 62-63, sr physicist, 63-68, PRIN CHEMIST, MIDWEST RES INST, 68- *Mem:* Am Chem Soc; Combustion Inst; Am Soc Mass Spectrometry. *Res:* High temperature chemistry and thermodynamics; mass and optical spectroscopy; boron chemistry; molecular beams; combustion; radiation-material interaction. *Mailing Add:* 169 Terrace Trail West Lake Quivira Kansas City KS 66106-9504

GREENE, FREDERICK DAVIS, II, b Glen Ridge, NJ, July 9, 27; m 53; c 4. ORGANIC CHEMISTRY. *Educ:* Amherst Col, AB, 49; Harvard Univ, AM, 51, PhD(chem), 53. *Hon Degrees:* ScD, Amherst Col, 69. *Prof Exp:* Res assoc chem, Univ Calif, Los Angeles, 52-53; from instr to assoc prof, 53-62, PROF ORG CHEM, MASS INST TECHNOL, 62- *Concurrent Pos:* Ed, J Org Chem, 62-88. *Mem:* Am Chem Soc; The Chem Soc; Am Acad Arts & Sci. *Res:* Mechanisms of organic reactions. *Mailing Add:* Dept Chem Mass Inst Technol Cambridge MA 02139

GREENE, FREDERICK LESLIE, b Norfolk, Va, Dec 18, 44; m 70; c 2. SURGICAL ONCOLOGY, GENERAL SURGERY. *Educ:* Univ Va, BA, 66, MD, 70. *Prof Exp:* Instr surg, Sch Med, Yale Univ, 75-76; attend surgeon, Portsmouth Naval Hosp, 77-78; clin instr, Univ Md, 78-80; chief surg, William J Bryan Dorn Vet Admin Hosp, 80-85; assoc prof, 80-85, PROF SURG, UNIV SC SCH MED, 85-, CHIEF SURG, RICHLAND MEM HOSP, COLUMBIA, SC, 88- *Concurrent Pos:* Vis prof, Portsmouth Naval Hosp, 86. *Mem:* Am Col Surgeons; Asn Acad Surg; Am Soc Clin Oncol; Am Asn Cancer Educ; Soc Am Gastrointestinal Endoscopic Surgeons. *Res:* Effects of nutrition and trace metals on production of colon cancer in animal models; methods of breast-sparing for breast cancer. *Mailing Add:* Two Medical Park Rd Suite 402 Columbia SC 29203

GREENE, GEORGE C, III, ENVIRONMENTAL CONSULTING. *Educ:* Univ Fla, BS, 67; Columbia Univ, MS, 69; Tulane Univ, PhD(chem eng), 73. *Prof Exp:* Engr, Lurgi Mineraloetechnik, Frankfurt, WGer, 73-77; sr engr, Exxon Res & Eng, Florham Park, NJ, 77-81; VPRES ENVIRON CONSULT, GEN ENG LABS, CHARLESTON, SC, 81- *Mem:* Am Inst Chem Engrs; Am Soc Testing & Mats; Am Chem Soc; Am Waterwell Asn. *Res:* Environmental consulting: waste characterization and site evaluation for impact to soils, surface water and groundwater; groundwater studies, metals, inorganic and organic analyses at trace levels; development and design of contaminant removal processes. *Mailing Add:* Gen Eng Labs PO Box 30712 Charleston SC 29417-0172

GREENE, GEORGE W, JR, b Brooklyn, NY, Aug 5, 19; c 2. ORAL PATHOLOGY. *Educ:* Univ Notre Dame, BS, 41; Columbia Univ, DDS, 44; Am Bd Oral Path, dipl. *Prof Exp:* Assoc prof oral path, Sch Dent, Georgetown Univ, 50-62; prof oral path & chmn dept, Sch Dent, State Univ NY, Buffalo, 63-85, dir, Oral Diag Serv, 73-85; RETIRED. *Concurrent Pos:* Sr oral pathologist, Cent Lab Path Anal & Res, Armed Forces Inst Path, 53-60, chief, Environ Oral Path Br, 54-60; secy res prog comt oral dis, Vet Admin, 58-62, asst dir dent prof serv, 60-62; chmn bd examr, Am Bd Oral Path, 63-65, pres, 65-, secy treas, 71-73; USPHS cancer coordr, dir continuing educ & dir diag serv, State Univ NY Buffalo, 63-68, dir grad & postgrad educ, 66-67; consult oral tumors, AMA, oral cytol reproducibility study, Cancer Control Prog, Nat Ctr Chronic Dis Control, 67-, path dept, Roswell Park Mem Inst & res & develop div, Off Surgeon Gen, US Army. *Honors & Awards:* Am Cancer Soc Nat Award, 70. *Mem:* Am Dent Asn; fel Am Col Dent; fel Am Acad Oral Path (pres, 64). *Res:* Bone physiology; cryogenics; oral diseases; oral cancer; enosseous implantology. *Mailing Add:* 544 Flotilla Rd North Palm Beach FL 33408

GREENE, GERALD L, b Jewell, Kans, July 7, 37; m 60; c 2. ENTOMOLOGY. *Educ:* Kans State Univ, BS, 59, MS, 61; Ore State Univ, PhD(entom, bot), 66. *Prof Exp:* Res asst entom, Kans State Univ, 59-61 & Ore State Univ, 61-64; asst entomologist, Univ Ky, 64-66 & Cent Fla Exp Sta, 66-70; asst prof entom, agr res & educ ctr, Univ Fla, 70-76; head Garden City Exp Sta, 76-82, PROF LIVESTOCK ENTOM, KANS STATE UNIV, 82- *Mem:* Entom Soc Am. *Res:* Insect ecology and life tables; field crop insect, biological control; university administration; livestock entomology. *Mailing Add:* SWKS Res-Ext Ctr Garden City KS 67846

GREENE, GORDON WILLIAM, b San Francisco, Calif, Feb 8, 21; m 42; c 8. GEOPHYSICS. *Educ:* Fresno State Col, AB, 54; Stanford Univ, MS, 57. *Prof Exp:* Technician, Pac Cement & Aggregates, Inc, 46-47, dist engr, 47-55, geologist, US Geol Surv, 56-57, geophysicist, 57-70; geophysicist, Off Earthquake Studies, 70-81; RETIRED. *Mem:* AAAS; Am Geophys Union; Geol Soc Am; Seismol Soc Am. *Res:* Seismology; tectonics; heat flow; permafrost; remote sensing. *Mailing Add:* 303 W 220 S Orem UT 84058-5475

GREENE, HARRY LEE, GENETIC LIVER DISEASES, INTESTINAL TRANSPORT OF NUTRIENTS. *Educ:* Emory Univ, MD, 64. *Prof Exp:* PROF PEDIAT & BIOCHEM, MED CTR, VANDERBILT UNIV, 79- *Mailing Add:* Dept Pediat Vanderbilt Univ Med Ctr Nashville TN 37232

GREENE, HOKE SMITH, b Gray, Ga, Aug 20, 06; m 36. PHYSICAL CHEMISTRY, ORGANIC CHEMISTRY. *Educ:* Mercer Univ, AB, 27; Univ Cincinnati, MS, 28, PhD(chem), 30. *Hon Degrees:* DSc, Mercer Univ, 63. *Prof Exp:* Res chemist, Roessler & Hasslacher Chem Dept, E I du Pont de Nemours & Co, 30-31; Am-Ger exchange fel, Inst Int Educ, Tech Hochschule, Ger, 31-32; res & develop chemist, E I du Pont de Nemours & Co, 32-34; from asst prof to assoc prof chem eng, 34-45, chmn grad studies, Dept Chem, 40-45, prof chem & head dept, 45-56, dean grad sch arts & sci, 47-59, dean acad admin, 56-59, vpres & dean fac, 59-67, vpres for res, 67-71, EMER PROF CHEM & VPRES, UNIV CINCINNATI, 71- *Honors & Awards:* Blalock Medal; William Howard Taft Medal, 71. *Mem:* AAAS; Am Chem Soc; fel Am Inst Chemists. *Res:* Synthetic organic chemistry; electric moments of organic compounds; electrolytic oxidation and reductions; industrial microbiology; synthesis of heterocyclic compounds and medicinals. *Mailing Add:* 2509 Evergreen Ridge Dr Cincinnati OH 45215

GREENE, HOWARD LYMAN, b Hackensack, NJ, Apr 3, 35; m 58; c 3. CHEMICAL ENGINEERING. *Educ:* Cornell Univ, BChemE, 59, MChemE, 63, PhD(chem eng), 66. *Prof Exp:* Asst prof chem tech, Broome Tech Community Col, 59-63; consult, Int Bus Mach Corp, NY, 63-64; from asst prof to assoc prof chem eng, 65-74, PROF CHEM ENG, UNIV AKRON, 74-, HEAD DEPT, 77- *Concurrent Pos:* HEW grant biomed polymers, 69-75; consult, Chemstress Consult Co, 72- *Mem:* Am Inst Chem Engrs; Sigma Xi. *Res:* Reaction engineering; mixing; chemical kinetics; direction of graduate study in these areas; biomedical engineering. *Mailing Add:* 2238 Randolph Rd Mogadore OH 44260

GREENE, JACK BRUCE, b Bloomington, Ind, June 9, 15; m 40; c 2. PHYSICS. *Educ:* Ind Univ, AB, 37; Univ Pittsburgh, MS, 40, PhD(physics), 42. *Prof Exp:* Asst physics, Univ Pittsburgh, 37-41; instr, Univ Ill, 41-43; from instr to assoc prof physics, Marquette Univ, 45-80, assoc chmn dept, 70-76; RETIRED. *Mem:* Am Phys Soc; Am Asn Physics Teachers. *Res:* Electronic structure of metals; cyclotron assembly and operation; electronic energy bands in the face-centered iron. *Mailing Add:* 933 N 34th St Milwaukee WI 53208

GREENE, JAMES H, b Elmwood, Nebr, Mar 12, 15; m 43; c 2. INDUSTRIAL & MANUFACTURING ENGINEERING. *Educ:* Univ Iowa, BS, 47, MS, 48, PhD, 57. *Prof Exp:* PROF INDUST ENG, PURDUE UNIV, 48- *Concurrent Pos:* Vis prof & Fulbright scholar, Finland Sch Technol. *Mem:* Fel Am Prod & Inventory Control Soc; Soc Mfg Engrs; Am Soc Eng Educ. *Res:* Industrial engineering; author four industrial engineering books, some translated in Japanese, Spanish and Italian. *Mailing Add:* Sch Indust Eng Purdue Univ West Lafayette IN 47905

GREENE, JANICE L, b Adams, NY, Mar 12, 30. ORGANIC CHEMISTRY. *Educ:* Alfred Univ, BA, 51; Pa State Univ, PhD(org chem), 58. *Prof Exp:* From asst prof to assoc prof chem, Univ NC, Greensboro, 55-60; sr res chemist, Standard Oil Co, Ohio, 60-66, res assoc, 66-79, sr res assoc, 79-81, sr scientist, 81-88; RETIRED. *Mem:* AAAS; Am Chem Soc; Soc Plastics Engrs. *Res:* Organic reactions; mechanisms and processes; small ring compounds; nitrile chemistry. *Mailing Add:* 28649 Jackson Rd Chagrin Falls OH 44022

GREENE, JOHN CLIFFORD, b Ashland, Ky, July 19, 26; c 3. DENTISTRY. *Educ:* Univ Louisville, DMD, 52; Univ Calif, Berkeley, MPH, 61. *Hon Degrees:* DSc, Univ Ky, 72, Boston Univ, 75 & Univ Louisville, 80. *Prof Exp:* Intern, USPHS, Chicago, 52-53, staff, Pub Health Serv Hosp, 53-54, asst regional dent consult, Region 9, San Francisco, 54-56; epidemic intel, Ctr Dis Control, 56-57; res staff, Epidemiol & Biomet Br, Nat Inst Dent Res, NIH, 57-58; asst to chief dent officer, Pub Health Serv, 58-60; chief, Epidemiol Prog, Dent Health Ctr, San Francisco, 61-66; dep dir, Div Dent Health Care, Bethesda, 66-73; actg dir, Bur Health Resources Develop & chief dent officer, Pub Health Serv, 73-75, spec asst dent affairs, Off Asst Secy Health & chief dent officer, 75-78, dep surgeon gen, 78-81; PROF, DEPT DENT PUB HEALTH & HYG & DEAN, SCH DENT, UNIV CALIF, SAN FRANCISCO, 81- *Concurrent Pos:* Spec consult, WHO, India, 57-; nutrit surv team mem, Ecuador, 59; vis lectr, Sch Dent, Univ Calif, 65-72; lectr, Sch Dent, Univ Pa & Sch Pub Health, Univ Mich, 72-75; chmn, Comn Dent Pub Health, Fed Dent Int; panel experts dent health, World Health Orgn, 76-; mem comt, Inst Med, Nat Acad Sci, 82-, sect leader, 83-87, Gov Coun, 85-; dean, search comt, Sch Med, Univ Calif, San Francisco, 82-83, bd adv, Off Res, Sch Nursing, 86- & dean, search comt, 87-; nat affairs comt, Am Asn Dent Res, 83-, prog comt, 84-; chmn, syst health sci comt, Univ Calif, 82-86 & 83-85, acad planning & prog rev bd, 83-84; vpres, Western Conf Dent Examr & Dent Sch Deans, 84-85, pres, 85-86; mem, US Prev Serv Task Force, 85- *Mem:* Inst Med-Nat Acad Sci; Am Asn Dent Res (vpres, 84-85, pres-elect, 85-86 & pres, 86-87); Inst Asn Dent Res (pres-elect, 91); fel Am Col Dentists; Am Asn Pub Health Dentists; Inst Health Policy Studies; Am Pub Health Asn; Nat Acad Pract; fel Int Col Dentists; fel AAAS. *Res:* Author of over 70 publications. *Mailing Add:* Sch Dent Univ Calif San Francisco CA 94143

GREENE, JOHN M, b Pittsburgh, Pa, Sept 22, 28; m 57; c 1. PLASMA PHYSICS. *Educ:* Calif Inst Technol, BS, 50; Univ Rochester, PhD(physics), 56. *Prof Exp:* Res physicist, Plasma Physics Lab, Princeton Univ, 56-82; SR TECH ADV, GEN ATOMIC CO, 82- *Mem:* Fel Am Phys Soc. *Res:* Magnetohydrodynamics; nonlinear equations. *Mailing Add:* Gen Atomic Co PO Box 85608 San Diego CA 92186-9784

GREENE, JOHN PHILIP, b Chicago, Ill, Oct 13, 55; m 85; c 2. THIN FILM PHYSICS, METALLURGY. *Educ:* Univ Ill, Chicago, BA, 78; DePaul Univ, MS, 82. *Prof Exp:* Proj chemist, DeSoto Inc, 78-82; SCI ASSOC, ARGONNE NAT LAB, 82- *Mem:* Am Chem Soc; Am Phys Soc; Soc Appl Spectros; Sigma Xi; Int Nuclear Target Develop Soc. *Res:* Nuclear target development; heavy-ion research. *Mailing Add:* 1200 Andover Circle Aurora IL 60504

GREENE, JOHN W, JR, b East Orange, NJ, July 25, 26; m 54; c 3. OBSTETRICS & GYNECOLOGY. *Educ:* Univ Pittsburgh, BS, 48; Univ Pa, MD, 52. *Prof Exp:* From intern to resident obstet & gynec, Hosp Univ Pa, 52-56; res assoc, Sch Med, Univ Pa, 57-59, asst prof, 59-63; PROF OBSTET & GYNEC & CHMN DEPT, MED CTR, UNIV KY, 63- *Concurrent Pos:* Res fel, Sch Med, Univ Pa, 56-57. *Mailing Add:* Dept Obstet & Gynec Univ Ky Med Ctr Lexington KY 40536

GREENE, JOSEPH LEE, JR, synthetic organic chemistry, for more information see previous edition

GREENE, KENNETH TITSWORTH, b Alfred, NY, Feb 7, 14; m 58; c 2. PHYSICAL CHEMISTRY. *Educ:* Alfred Univ, BS, 35; Rutgers Univ, MS, 37, PhD(ceramics), 40. *Prof Exp:* Instr ceramics, Rutgers Univ, 39-40; phys chemist & Portland Cement Asn fel, Nat Bur Standards, 40-45; chemist-petrogr, US Bur Reclamation, 45-52, chemist, 52-55; res petrogr, res dept, Ideal Cement Co, 55-82; RETIRED. *Mem:* Am Chem Soc; fel Mineral Soc Am. *Res:* Physical chemistry of silicate glasses; phase equilibria of Portland cement clinker; petrography and chemistry of cement, concrete, rocks and other engineering materials; hydration reactions of cement. *Mailing Add:* 1037 Glenmoor Dr Ft Collins CO 80521-4333

GREENE, KINGSLEY L, b New York, NY, Nov 26, 26; m 48; c 3. ECOLOGY. *Educ:* Cornell Univ, BS, 60, MS, 61. *Prof Exp:* High sch teacher, NY, 49-51 & 56-58; prof biol, Eastern Baptist Col, 61-66; dir outdoor educ, Rose Tree Media Sch Dist, Media, Pa, 66-67; prof natural resources, State Univ NY Agr & Tech Col Morrisville, 67-85; RETIRED. *Mem:* AAAS; Am Nature Study Soc. *Res:* Prey-predator relationships in aquatic communities. *Mailing Add:* PO Box 579 Dexter NY 13634

GREENE, LAURA H, b Cleveland, Ohio, June 12, 52; m 89; c 1. CONDENSED MATTER PHYSICS, MATERIALS PHYSICS. *Educ:* Ohio State Univ, BS, 74, MS, 77; Cornell Univ, MS, 80, PhD(physics), 84. *Prof Exp:* Mem tech staff, Hughes Aircraft Co, 74-75; teaching assoc, Ohio State Univ, 75-76, res assoc, 76-77; teaching assoc, Cornell Univ, 77-79, res assoc, 79-83; postdoctoral fel, 83-85, MEM TECH STAFF, BELLCORE, 85- *Mem:* Am Phys Soc; Mat Res Soc. *Res:* Physics of novel superconducting materials. *Mailing Add:* Rm 3X-281 Bellcore 331 Newman Springs Rd Red Bank NJ 07701-7040

GREENE, LEWIS JOEL, b New York, NY, Aug 10, 34; m 58; c 2. PROTEIN CHEMISTRY, ENZYMOLOGY. *Educ:* Amherst Col, BA, 55; Rockefeller Inst, PhD(biochem), 62. *Prof Exp:* From asst to assoc biochemist, Brookhaven Nat Lab, 62-68, biochemist, 68-76; prof biochem, 74-85, PROF PHARMACOL, FAC MED, UNIV SAO PAULO, RIBEIRAO, 85- *Concurrent Pos:* Affil, Rockefeller Univ, 68-72; adj staff, Cleveland Clin, 87- *Mem:* Am Chem Soc; Am Soc Cell Biol; Am Soc Biol Chemists; Nat Acad Sci Brazil; Brazilian Soc Biol Chemists; Brazilian Soc Pharm & Exp Theory. *Res:* Protein chemistry; amino acids sequence determination of biologically active peptides and proteins; metabolism of peptide hormones; characterization of protein and peptide drugs; prepared by conventional on modern method of biotechnology. *Mailing Add:* Dept Pharmacol Univ Sao Paulo 14049 Ribeirao Preto Sao Paulo Brazil

GREENE, LLOYD A, b Chicago, Ill, Aug 28, 44; m 74; c 1. NEUROBIOLOGY, NEUROCHEMISTRY. *Educ:* Univ Chicago, ScB, 65; Univ Calif, San Diego, PhD(chem), 70. *Prof Exp:* Asst prof neuropath, Harvard Med Sch, 74-77, assoc prof, 77-79; assoc prof, Sch Med, NY Univ, prof pharmacol, 82-87; PROF PATH, COLUMBIA UNIV COL PHYSICIANS & SURGEONS. *Res:* Neurochemical and cell biological studies on development of the nervous system. *Mailing Add:* Dept Path Columbia Univ Col Physicians & Surgeons 630 W 168th St New York NY 10032

GREENE, MARK IRWIN, b Winnipeg, Man, Aug 3, 48; m; c 3. IMMUNOLOGY. *Educ:* Univ Man, MD, 72, PhD(immunol), 77; FRCP(C), 76. *Prof Exp:* Intern internal med, Health Sci Ctr, Winnipeg, 72-73, resident, 73-76; res fel path, Harvard Med Sch, Boston, 76-77, from instr to assoc prof, 77-85, assoc prof immunol, Dept Cancer Biol, 82-85; prof med & chief div rheumatol/immunol, Sch Med, Tufts Univ, 84-85; DIR, DIV IMMUNOL, DEPT PATH & PROF, SCH MED, UNIV PA, 86-, ASSOC HEAD FUNDAMENTAL RES, CANCER CTR, 87-, FAC SENATE, 88- *Concurrent Pos:* Med Res Coun fel, Can, 73-75, Boston, 76-78; consult med, Dana Farber Cancer Ctr, 80-86, Hosp Univ Pa, 85-; chmn rev comt, Struct & Molecular Biol, Univ Pa, 88-89, mem, Howard Hughes Adv Comt, 89; ed, DNA & Cell Biol & Pathobiol, 90. *Honors & Awards:* Math Asn Prize, 66; Lotte Strauss Award, 86. *Mem:* Am Soc Clin Invest; Brit Soc Immunol; Am Asn Immunologists; Am Asn Pathologists. *Mailing Add:* Div Immunol Dept Path & Lab Med Univ Pa 252 Med Labs 36th & Hamilton Walk Philadelphia PA 19104-6082

GREENE, MICHAEL P, b New York, NY, June 20, 38; m 63; c 2. SOLID STATE PHYSICS, SCIENCE POLICY. *Educ:* Cornell Univ, BEngPhys, 60; Univ Calif, San Diego, MS, 62, PhD(physics), 65. *Prof Exp:* Asst prof physics, Univ Calif, Davis, 65; res assoc, Brown Univ, 65-67; asst prof, Univ Md, College Park, 67-74, assoc chmn dept physics & astron, 71-74; tech dir, Vol In Tech Assistance, 74-76; dep dir sci & technol, Org Am States, 76-80; ASSOC DIR, BD SCI & TECHNOL IN INT DEVELOP, NAT ACAD SCI, 81- *Concurrent Pos:* Fulbright-Orgn Am States lectr, Nat Univ Eng, Peru, 70-71; resident dir, Indonesia Prog, Nat Acad Sci, 90-93. *Mem:* Fel AAAS; Am Phys Soc; Soc Int Develop. *Res:* Science and technology policy for developing countries; technology transfer; project management. *Mailing Add:* Nat Res Coun HA476 Washington DC 20418

GREENE, NATHAN DOYLE, b Steele, Ala, Mar 2, 38; m 60; c 2. IMMUNOLOGY, IMMUNOTOXICOLOGY. *Educ:* Berea Col, BA, 60; NC State Univ, MS, 62; Emory Univ, PhD(biol), 65. *Prof Exp:* Med microbiologist, Nat Commun Dis Ctr, USPHS, Ga, 65-68; ASSOC FOUND SCIENTIST, SOUTHWEST FOUND RES & EDUC, 68- *Concurrent Pos:* Adj asst prof microbiol, Univ Tex Health Sci Ctr, San Antonio; adj assoc prof, Univ Tex, San Antonio. *Mem:* AAAS; Am Inst Biol Sci; Am Soc Parasitol; fel Royal Soc Trop Med & Hyg; Am Asn Immunologists. *Res:* Immunology of non-human primates; immunological aspects of pulmonary defense systems; inhalation toxicology; immunotoxicology related to environmental agents; mechanisms if immunomodulation by toxic agents. *Mailing Add:* Dept Biol Res Animal Health Div Mobay Corp PO Box 390 Shawnee KS 66201

GREENE, NEIL E(DWARD), b Toledo, Ohio, July 17, 36; m 59; c 4. HEAT TRANSFER, THERMODYNAMICS. *Educ:* Dartmouth Col, AB, 59, MS, 60; Univ Mich, PhD(phase equilibrium), 66. *Prof Exp:* Sr res scientist, 66-77, dept mgr, Tech Ctr, 77-80, RES FEL, OWENS-CORNING FIBERGLAS CORP, 80- *Mem:* Am Soc Mech Engrs; Sigma Xi. *Res:* Heat transfer and fluid mechanics associated with production of attenuating jets of molten glass; seven patents. *Mailing Add:* 288 Granview Granville OH 43023

GREENE, NICHOLAS MISPLEE, b Milford, Conn, July 11, 22; m 46; c 3. ANESTHESIOLOGY. *Educ:* Yale Univ, BS, 44; Columbia Univ, MD, 46; Am Bd Anesthesiol, dipl. *Hon Degrees:* MA, Yale Univ, 55;. *Prof Exp:* Surg intern, Presby Hosp, 46-47; instr anesthesia, Harvard Med Sch, 51-53; from assoc prof to asst prof pharmacol, Sch Med, Univ Rochester, 53-55; prof, 55-87, EMER PROF ANESTHESIOL, YALE UNIV, 87- *Concurrent Pos:* Resident, Mass Gen Hosp, 49-51, asst, 51-53; vis fel, Royal Infirmary of Edinburgh, 51; anesthetist in chief, Strong Mem Hosp, Rochester, 53-55; dir anesthesiol, Yale-New Haven Hosp, 55-73; ed, Anesthesiology, 65-73, ed-in-chief, 73-76; ed-in-chief, Anesthesia & Analgesia, 77- *Mem:* Asn Univ Anesthetists; Am Soc Anesthesiol; Int Anesthesia Res Soc; Sigma Xi. *Res:* Academic anesthesiology. *Mailing Add:* 333 Cedar St New Haven CT 06510

GREENE, PAUL E, chemistry, for more information see previous edition

GREENE, PETER RICHARD, b Huntington, NY, Mar 23, 51. MECHANICAL ENGINEERING, APPLIED PHYSICS. *Educ:* Cornell Univ, BS, 73; Harvard Univ, MS, 75, PhD(eng), 78. *Prof Exp:* Res engr fluid mech, Grumman Aerospace, 69-76; res scientist biomech, Harvard Univ, 73-80; asst prof biomech eng, Johns Hopkins Univ, 80-83; DIR RES ENGR, BGKT CONSULT, LTD, 83- *Mem:* Am Soc Mech Engrs; Am Inst Physics; Asn Res Vision & Ophthal; Inst Elec & Electronics Engrs. *Res:* Applied mechanics and bioengineering; VTOL; fluid mechanics; ocular mechanics; track and shoe design; muscle mechanics. *Mailing Add:* BGKT Consult Ltd 153 Main St Huntington NY 11743

GREENE, REGINALD, b Mar 27, 34. RADIOLOGY. *Educ:* Harvard Col, AB, 56; NY Med Col, MD, 60; Am Bd Radiol, dipl nuclear med, 68. *Prof Exp:* Intern, US Naval Hosp, Philadelphia, 60-61, resident, Oakland, 61-62, Mass Gen Hosp, Boston, 65-68; advan fel acad radiol, James Picker Found, Nat Acad Sci-Nat Res Coun, 68-71; from instr to assoc prof, 71-85, PROF RADIOL, HARVARD MED SCH, 85- *Concurrent Pos:* Clin asst radiol, Mass Gen Hosp, 70-71, asst radiologist, 71-75, dir thoracic radiol, 71-82, assoc radiologist, 75-77, radiologist, Mass Gen Hosp, 78-, dir, Wang Ambulatory Care Ctr & assoc radiologist-in-chief, 82- *Honors & Awards:* Silver Medal, Am Roentgen Ray Soc; Fleischner Medal Lectr, Interlaken, Switz. *Mem:* Radiol Soc NAm; Fleischner Soc (pres, 87); Am Physiol Soc; Am Thoracic Soc; Royal Col Radiologists; Am Roentgen Ray Soc; Am Col Radiol; Am Col Chest Physicians; AMA; Soc Computer Appln Radiol. *Res:* Diffuse lung damage, functional and anatomic consequences; radiologic screening methodology for the pneumoconioses; digital radiography; author of numerous scientific publications. *Mailing Add:* Dept Radiol Mass Gen Hosp Fruit St Boston MA 02114

GREENE, RICHARD L, b Bridgeport, Conn, Aug 26, 38. SOLID STATE PHYSICS. *Educ:* Mass Inst Technol, BS, 60; Stanford Univ, PhD(physics), 67. *Prof Exp:* Res assoc physics, Stanford Univ, 67-70; res scientist & mgr, IBM Res Lab, 70-89; PROF & DIR, DEPT PHYSICS, SUPERCONDUCTIVITY CTR, UNIV MD, 89- *Mem:* Fel Am Phys Soc. *Res:* Physical properties of one-dimensional organic and inorganic conductors; superconductivity; thin films. *Mailing Add:* Dept Physics Univ Md College Park MD 20742

GREENE, RICHARD LORENTZ, b Bridgeport, Conn. SOLID STATE PHYSICS. *Educ:* Mass Inst Technol, BS, 60; Stanford Univ, PhD(physics), 67. *Prof Exp:* Res staff mem, IBM, 70-74, mgr, 74-89; PROF & DIR, CTR SUPERCONDUCTIVITY, UNIV MD, 89- *Mem:* Am Phys Soc. *Res:* Experimental study of superconductors, low dimension metals and other novel states of matter. *Mailing Add:* Dept Physics Univ Md College Park MD 20742

GREENE, RICHARD WALLACE, b San Francisco, Calif, Oct 29, 41; m 65; c 2. INVERTEBRATE BIOLOGY, ALGAL PHYSIOLOGY. *Educ:* Univ Calif, Berkeley, BA, 64; Univ Calif, Los Angeles, MA, 66, PhD(zool), 69. *Prof Exp:* USPHS training grant, Univ Calif, Los Angeles, 69-70; asst prof, 70-76, ASSOC PROF BIOL, UNIV NOTRE DAME, 76- ASSOC DIR, ENVIRON RES CTR, 75- *Concurrent Pos:* Mem exec comt, Am Midland Naturalist, 72- *Mem:* AAAS; Am Soc Zool; NAm Benthological Soc; Sigma Xi. *Res:* Symbiosis between algae and chloroplasts with invertebrates; biology of heavy metals; environmental biology. *Mailing Add:* 619 Highlander Ave Placentia CA 92670-3229

GREENE, ROBERT CARL, b Bridgeport, Conn, Aug 14, 32; m 55; c 5. GEOLOGY. *Educ:* Cornell Univ, AB, 55; Univ Tenn, MS, 59; Harvard Univ, PhD(geol), 64. *Prof Exp:* Explor geologist, NJ Zinc Co, 55-57; GEOLOGIST, US GEOL SURV, 61- *Mem:* Geol Soc Am. *Res:* Precambrian rocks in Saudi Arabia. *Mailing Add:* US Geol Surv 345 Middlefield Rd Menlo Park CA 94025

GREENE, ROBERT EVERIST, b Knoxville, Tenn, May 9, 43; m 69. GEOMETRY. *Educ:* Mich State Univ, BS, 64; Univ Calif, Berkeley, PhD(math), 69. *Prof Exp:* Instr math, Courant Inst, NY Univ, 69-71; From asst prof to assoc prof, 71-76, PROF MATH, UNIV CALIF, LOS ANGELES, 76- *Concurrent Pos:* Res fel, Alfred P Sloan Found, 75-79; vis mem, Inst Advan Study, Princeton Univ, 76 & 79; vis prof, Univ Bonn, WGer, 80. *Mem:* Am Math Soc. *Res:* Geometry and function theory of noncompact Riemannian and Kahler manifolds; application of geometric methods to several complex variables. *Mailing Add:* Univ Calif 405 Hilgard Ave Los Angeles CA 90024

GREENE, ROBERT MORRIS, b Dorpen, Ger, Dec 15, 45; US citizen; m 67; c 2. DEVELOPMENTAL BIOLOGY, ANATOMY. *Educ:* Utica Col Syracuse Univ, BA, 67; Univ Va, PhD(anat), 74. *Prof Exp:* Staff fel biochem, Nat Inst Dent Res, NIH, 76-78; ASST PROF ANAT, JEFFERSON MED COL, 78- *Concurrent Pos:* Fel, Nat Inst Dent Res, NIH, 74-76. *Mem:* Soc Develop Biol; Am Soc Cell Biol; Teratology Soc; Int Asn Dent Res; Sigma Xi. *Res:* Developmental biology; craniofacial development; teratology; cleft palate. *Mailing Add:* Dept Anat Jefferson Med Col 1020 Locust St Philadelphia PA 19107

GREENE, RONALD C, b Los Angeles, Calif, June 7, 28; m 50; c 2. BIOCHEMISTRY. *Educ:* Calif Inst Technol, BS, 49, PhD(biochem), 54. *Prof Exp:* Asst scientist biochem, NIH, 54-55, sr asst scientist, 55-57; instr biochem, Sch Med, Duke Univ, 57-58, assoc, 58-62, asst prof, 62-71; biochemist radioisotope serv, 57-69, CHIEF BASIC SCI LAB, VET ADMIN HOSP, 69-; ASSOC PROF BIOCHEM, SCH MED, DUKE UNIV, 71- *Mem:* AAAS; Am Chem Soc; Am Soc Biol Chem; Am Soc Microbiol; Sigma Xi. *Res:* Biochemistry of sulfonium compounds; regulation of methionine biosynthesis; metabolism of methionine and related compounds. *Mailing Add:* Dept Biochem Sch Med Duke Univ Durham NC 27710

GREENE, THOMAS FREDERICK, b Chicago, Ill, Nov 5, 38; m 62; c 4. ASTROPHYSICS. *Educ:* Univ Wash, BS, 60, MS, 61, PhD(astron), 68. *Prof Exp:* Proj engr, Standard Pressed Steel Co, 63-64; engr, Boeing Co, 64-68, res astronr, Boeing Sci Res Labs, 68-72; SR SPECIALIST ENGR, BOEING AEROSPACE CO, 72- *Concurrent Pos:* Affil asst prof, Univ Wash, 71-; consult, Univ Utah, 73-74. *Mem:* Am Inst Aeronaut & Astronaut. *Res:* Aerosols; Jovian atmospheric aerosol content; ballistic missile defense, reentry vehicle penetration aid design. radiometry; radiative transfer; molecular spectroscopy; stellar abundances; x-ray telescopes. *Mailing Add:* Boeing Aerospace Co PO Box 3999 MS 3C-CF Seattle WA 98124

GREENE, VELVL WILLIAM, b Winnipeg, Man, July 5, 28; nat US; m 56; c 5. NOSCOMIAL INFECTIONS, MEDICAL ETHICS. *Educ:* Univ Man, BSA, 49; Univ Minn, MS, 51, PhD(bact), 56, MPH, 83. *Prof Exp:* Instr bact, Univ Sask, 51-52; asst animal indust, NC State Col, 52-53; asst prof bact, Univ Southwestern La, 56-59; asst prof pub health, Univ Minn, 59-61; mgr life sci res, Litton Industs, 61-64; assoc prof pub health, 65-70, prof pub health & microbiol, Univ Minn, Minneapolis, 70- 86; prof epidemiol, 86-90, CARLIN PROF PUB HEALTH & EPIDEMIOL, BEN GURION UNIV, 90- *Concurrent Pos:* Consult epidemiol, WHO, 78 & 82; Bush fel, 83; vis prof epidemiol, Ben Gurion Univ, Israel, 83-84. *Honors & Awards:* Fulbright Sr Lectr, 83. *Mem:* Am Soc Microbiol; fel Am Pub Health Asn; fel Am Acad Microbiol; fel Am Acad Sanitarians; Soc Hosp Epidemiol; Soc Epidemiol Res; Hosp Info Soc. *Res:* Psychrophiles; environmental microbiology; aerobiology; exobiology; hospital infections; institutional sanitation; disinfection and sterilization; environmental epidemiology. *Mailing Add:* Med Sch Ben Gurion Univ Beer-Sheva Israel

GREENE, VIRGINIA CARVEL, b Warrenton, Va, Jan 8, 34; m 77. TOXICOLOGY. *Educ:* Sweet Briar Col, AB, 55; Tulane Univ, MS, 57; Univ Va, PhD(chem), 63. *Prof Exp:* Instr chem, Tulane Univ, 57-59; res assoc mat sci, Univ Va, 63-64, res assoc clin chem, univ hosp, 65-67; assoc prof chem, Longwood Col, 67-69; res chemist, Fed Bur Invest, 69-77; res asst, Univ Va, 77-78; CHEMIST, US ARMY FOREIGN SCI & TECHNOL CTR, 79- *Mem:* Am Chem Soc. *Res:* Analytical spectrophotometry and chromatography; toxicology and drug analysis; gas chromatography/mass spectrometry; clinical chemistry; geomagnetism and aeronomy. *Mailing Add:* 540 E Rio Rd Charlottesville VA 22901

GREENE, WILLIAM ALLAN, b Worcester, Mass, June 15, 15; m 45; c 2. PSYCHOSOMATIC MEDICINE. *Educ:* Harvard Col, BA, 36; Harvard Med Sch, MD, 40; Am Bd Internal Med, cert, 49. *Prof Exp:* Rotating intern, Mary Imogene Bassett Hosp, Cooperstown, NY, 40-42; asst med res, Strong Mem Hosp, Rochester, NY, 46-48; from instr to prof med & psychiat, Sch Med, Univ Rochester, 50-90; physician & psychiatrist, Strong-Mem Hosp, 76-90; RETIRED. *Concurrent Pos:* Commonwealth Fund fel med & psychiat, Sch Med, Univ Rochester, 48-50. *Mem:* Fel Am Col Physicians; Am Psychosom Soc (secy-treas, 64-68, pres, 68). *Res:* Psychological factors in the development and course of organic disease; neoplasias, coronary artery disease, hemodialysis and renal transplantation. *Mailing Add:* Seven Roby Dr Rochester NY 14618

GREENEBAUM, BEN, b Chicago, Ill, Nov 30, 37; m 63; c 3. BIOLOGICAL EFFECTS OF ELECTROMAGNETIC FIELDS. *Educ:* Oberlin Col, AB, 59; Harvard Univ, AM, 61, PhD(physics), 65. *Prof Exp:* Res assoc & instr physics, Harvard Univ, 65-66 & Princeton Univ, 66-70; from asst prof to assoc prof, 70-80, actg vchancellor, 82-83 & 84-85, assoc dean fac, 79-89, head, Sci Div, 87-89, PROF PHYSICS, UNIV WIS-PARKSIDE, 80-, DEAN, SCH SCI & TECHNOL, 89- *Concurrent Pos:* Fac res prog mem, Argonne Nat Lab, 70-76; assoc ed, Bioelectromagnetics, 90- *Mem:* Am Phys Soc; Am Asn Physics Teachers; Bioelectromagnetics Soc. *Res:* Biological effects of weak, low frequency electromagnetic radiation. *Mailing Add:* Univ Wis-Parkside PO Box 2000 Kenosha WI 53141

GREENEBAUM, MICHAEL, b Brooklyn, NY, Aug 10, 40; m 68; c 5. OPTICAL PHYSICS, ATMOSPHERIC PHYSICS. *Educ:* Yeshiva Univ, NY, BA, 61; Mass Inst Technol, Cambridge, PhD(physics), 67. *Prof Exp:* Gen physicist, Naval Appl Sci Lab, Brooklyn, NY, 67-69; PRIN MEM RES STAFF, RIVERSIDE RES INST, NY, 69- *Concurrent Pos:* Lectr physics, Sch Gen Studies, Brooklyn Col, City Univ New York, 70-71, adj asst prof physics, 71-75; consult, Columbus Labs, Battelle Mem Inst, 77-78 & 80. *Mem:* Optical Soc Am; Am Phys Soc; Sigma Xi. *Res:* Effects of environmental (eg, atmospheric) attenuation, scattering, and dispersion on various aspects of the design of both coherent and non-coherent optical systems in the region between the ultraviolet and microwave wavelengths; thermal effects on ultrasonic light modulators; analysis of medical ultrasonic spectral analysis systems. *Mailing Add:* 1177 E 19th St Brooklyn NY 11230

GREENER, EVAN H, b Brooklyn, NY, Sept 8, 34; m 57; c 2. DENTAL MATERIALS. *Educ:* Polytech Inst Brooklyn, BMetE, 55; Northwestern Univ, MS, 57, PhD(mat sci), 60. *Prof Exp:* Lab instr, Polytech Inst Brooklyn, 55; Aitcheson fel metall, Northwestern Univ, 55-58, res fel, 58-59, prin investr, 59-60; from asst prof to assoc prof mat sci, Marquette Univ, 60-64; assoc prof, 64-69, PROF BIOL MAT, NORTHWESTERN UNIV, 69-, CHMN DEPT, 64- *Concurrent Pos:* Sr Fogarty Int fel, Turner Dent Sch, Univ Manchester, 77-78. *Mem:* Fel AAAS; NY Acad Sci; Am Soc Metals; Am Inst Mining, Metall & Petrol Engrs; Int Asn Dent Res. *Res:* Materials science. *Mailing Add:* Dept Biol Mat Northwestern Univ 303 E Chicago Ave Chicago IL 60611

GREENEWALT, CRAWFORD HALLOCK, b Cummington, Mass, Aug 16, 02; m 26; c 3. CHEMICAL ENGINEERING. *Educ:* Mass Inst Technol, BS, 22. *Hon Degrees:* Twenty from US cols, univs & insts, 40-70. *Prof Exp:* Res chemist, E I du Pont de Nemours & Co, Inc, 22-33 res supvr, 33-39, asst dir exp sta, chem dept, 39-42, mem bd dirs, 42, dir chem div, Grasselli Chem Dept, 42-43, mgr tech div, explosives dept, 43-45, asst dir, develop dept, 45, asst gen mgr, pigments dept, 45, vpres, 46-47, pres, 48-62, chmn bd, 62-67, chmn finance comt, 67-74, mem, Finance Comt, 74-88; RETIRED. *Concurrent Pos:* Emer trustee, Carnegie Inst, 52 & Nat Geog Soc, 59; dir, Morgan Guaranty Trust Co New York City, 62-72 & Boeing Co, 64-74; emer mem bd regents, Smithsonian Inst, 56-64. *Honors & Awards:* William Proctor Prize, Sci Res Soc Am, 57; Advan Res Medal, Am Soc Metals, 58; Chem Indust Medal, Soc Indust Chem, 52, Soc Medal, 63; Gold Medal, Am Inst Chemists, 59; John Fritz Medal, Am Inst Chem Engrs, 61. *Mem:* Nat Acad Sci; Am Chem Soc; Soc Indust Chem; Am Inst Chemists; Am Inst Chem Engrs. *Res:* High pressure reactions; catalysis; the separation of gaseous hydrocarbons; the partial pressure of water out of aqueous solutions of sulfuric acid; absorption of water vapor by sulfuric acid solutions. *Mailing Add:* E I du Pont de Nemours & Co Inc 9058 Du Pont Bldg Wilmington DE 19898

GREENEWALT, DAVID, b Wilmington, Del, Mar 26, 31; m 60; c 6. GEOPHYSICS. *Educ:* Williams Col, BA, 53; Mass Inst Technol, PhD(geophys), 60. *Prof Exp:* Instr geophys, Mass Inst Technol, 60-64, lectr, 64-66; GEOPHYSICIST, US NAVAL RES LAB, 66- *Mem:* Am Geophys Union. *Res:* Rock magnetism; electrical prospecting methods; gravity interpretation; sea floor magnetic field; physical oceanography. *Mailing Add:* 2509 Foxhall Rd NW Washington DC 20007

GREENFELD, SIDNEY H(OWARD), b Baltimore, Md, Apr 25, 23; m 48; c 4. SCIENCE ADMINISTRATION, CHEMICAL ENGINEERING. *Educ:* Univ Del, BChE, 44; Mass Inst Technol, SM, 45. *Prof Exp:* Res assoc colloid chem, Mass Inst Technol, 45-48; res chemist soap chem, Fels & Co, 48; res assoc phys & org chem, Asphalt Roofing Indust Bur, Nat Bur Standards, 49-57; res engr, Calif Res Corp, 57-59; res assoc, Asphalt Roofing Indust Bur, Nat Bur Standards, 59-68, mat engr, bur, 68-69, spec asst to chief, Off Flammable Fabrics, 69-73; tech asst to dir, Off Standards Coord & Appraisal, 73-76, tech asst to assoc exec dir eng & sci, Consumer Prod Safety Comn, 76-85; RETIRED. *Mem:* Am Chem Soc. *Res:* Surface chemistry of limestone, chalk, lime and related materials; colloid chemistry of Portland cement, soaps and detergents; asphalts and related substances; fabric flammability; product safety; residential building construction. *Mailing Add:* 23200 Wilderness Walk Ct Gaithersburg MD 20879-9042

GREENFIELD, ARTHUR JUDAH, b Oil City, Pa, Oct 3, 34; m 58; c 3. SOLID STATE PHYSICS. *Educ:* Wayne State Univ, BS, 56; Univ Chicago, MS, 57, PhD(physics), 63. *Prof Exp:* Res assoc physics, Univ Chicago, 63-64; mem tech staff, Bell Tel Labs, Inc, NJ, 64-67; assoc prof, 67-74, PROF PHYSICS, BAR-ILAN UNIV, ISRAEL, 74- *Mem:* Am Phys Soc. *Res:* Electromagnetic and thermal properties of liquid and solid metals. *Mailing Add:* Dept Physics Bar-Ilan Univ Ramat-Gan Israel

GREENFIELD, DAVID WAYNE, b Carmel, Calif, Apr 21, 40; m 71; c 1. ICHTHYOLOGY. *Educ:* Humboldt State Univ, AB, 62; Univ Wash, PhD(fisheries), 66. *Prof Exp:* Asst prof zool, Calif State Univ, Fullerton, 66-70; assoc prof, Northern Ill Univ, 70-77, prof biol sci, 77-84; assoc vchancellor, Univ Colo, Denver, 84-87; DEAN, GRAD DIV, UNIV HAWAII-MANOA, 87- *Concurrent Pos:* Res assoc, Los Angeles County Mus Natural Hist; res assoc, Field Mus Nat Hist, Chicago; prin aquatic ecologist, ENCAP, Inc. *Mem:* Am Soc Ichthyologists & Herpetologists; Soc Study Evolution; Soc Syst Zool. *Res:* Zoogeography; systematics of coral-reef fishes; zoogeography of marine and freshwater fishes; systematics of fishes of Belize; systematics of mosquitofishes (Gambusia). *Mailing Add:* Dean-Grad Div Univ Hawaii-Manoa 2540 Maile Way Honolulu HI 96822

GREENFIELD, EUGENE W(ILLIS), b Baltimore, Md, Nov 27, 07; m 29; c 2. PHYSICS, ELECTRICAL ENGINEERING. *Educ:* Johns Hopkins Univ, BE, 29, DE, 34. *Prof Exp:* Asst, Nat Elec Light Asn, Johns Hopkins Univ, 29-33, res assoc, 33-34; transmission engr, Pa RR, 34-35; res engr, Anaconda Wire & Cable Co, 35-41, supvr elec lab, 41-50; asst tech dir, Kaiser Aluminum & Chem Corp, 50-51, asst plant mgr, 51-52, head elec eng res, 52-57, res supvr, 57-58; prof elec eng & dir div indust res, Wash State Univ, 58-73, chmn elec power res & develop ctr, 64-73, asst dean col eng, 70-73; RETIRED. *Concurrent Pos:* Mem comt elec insulation, Nat Res Coun, 32-, exec comt, 48-, vchmn, 48-49; consult power transmission, 73- *Honors & Awards:* Distinguished Engr, Inst Elec & Electronics Engrs, 65, Centennial Medal, 84. *Mem:* Am Phys Soc; Am Soc Eng Educ; Am Soc Testing & Mat; Acoust Soc Am; fel Inst Elec & Electronics Engrs (vpres, 62-64). *Res:* Electrical insulation; electrical measuring instruments; application of insulation in the fields of power and communication; research and development of underground cables and conductors; research administration; electrical accident analyses. *Mailing Add:* 4747 Oakcrest Rd Fallbrook CA 02028

GREENFIELD, GEORGE B, b Brooklyn, NY, May 4, 28. RADIOLOGY, MEDICINE. *Educ:* NY Univ, BA, 49; State Univ Utrecht, MedDs, 56. *Prof Exp:* Attend radiologist, Cook County Hosp, Ill, 61-69, asst dir diag radiol, 66-69; prof radiol & chmn dept, Chicago Med Sch & Mt Sinai Hosp, 69-; PROF RADIOL, COOK COUNTY GARD SCH MED, 66-; AT DEPT RADIOL, RUSH UNIV. *Concurrent Pos:* Consult radiologist, Vet Admin Hosp, Dwight, Ill, 63-65; attend radiologist, Res & Educ Hosp, Chicago, 63-69; clin assoc prof, Col Med, Univ Ill, 67-69. *Mem:* Am Col Radiol; Asn Univ Radiol. *Mailing Add:* 12438 First St W Treasure Island FL 33706-5044

GREENFIELD, HAROLD, b New York, NY, May 6, 23; div; c 3. ORGANIC CHEMISTRY. *Educ:* City Col New York, BS, 43; Polytech Inst Brooklyn, MS, 48; Univ Pittsburgh, PhD(chem), 55. *Prof Exp:* Org chemist, Biochem Res Corp, NY, 48; phys chemist, Explosives Res Sect, US Bur Mines, Pa, 48-50, org chemist, Org Chem Sect, 50-55 & Coal Hydrogenation Sect, 55-58; sr res scientist, Uniroyal Inc, 58-80, res assoc Chem Div, 80-86; SR RES CHEMIST, FIRST CHEM CORP, 87- *Honors & Awards:* Chamberland Award, Am Chem Soc, 91. *Mem:* AAAS; Am Chem Soc; Catalysis Soc. *Res:* Heterogeneous and homogeneous catalysis; hydrogenation and reduction; oxo reaction; metal carbonyl chemistry; acetylene chemistry; coal chemistry; organic synthesis; high pressure reactions; rubber chemistry. *Mailing Add:* 4514 Concord St Pascagoula MS 39581

GREENFIELD, HARVEY STANLEY, b Pass-a-Grille, Fla, Dec 7, 24; m 53; c 3. MEDICAL BIOPHYSICS. *Educ:* Utah State Univ, BS, 50; Brigham Young Univ, MS, 52; Glasgow Univ, PhD(aeronaut & fluid mech), 65. *Prof Exp:* Asst physics, Utah State Univ, 52-54, res physicist, Dugway Proving Ground, 54-55, chief math anal br, 55-56, physics div, 56-58; sr rocket engr, Thiokol Chem Corp, 58-59, sr physicist, 59-60, staff specialist, 60-62, scientist, 62-63, consult, 63-64; lectr, Glasgow Univ, 64-66; sr numerical analyst, 68-69, assoc res prof comput sci & asst res prof surg, 69-70, ADJ PROF COMPUT SCI & ASSOC RES PROF SURG, UNIV UTAH, 70-, ADJ PROF BIOENG, 75- *Concurrent Pos:* Mem adv bd, Nat Asn Professions; adj prof mech eng, Univ Utah, 70- *Mem:* Assoc fel Am Inst Aeronaut & Astronaut; NY Acad Sci; assoc fel Royal Aeronaut Soc. *Res:* Turbulence; hemodynamics; computer graphics. *Mailing Add:* 3691 S 2235 E Univ Utah Salt Lake City UT 84109

GREENFIELD, IRWIN G, b Philadelphia, Pa, Nov 30, 29; m 51; c 3. SURFACE PROPERTIES, METAL MATRIX COMPOSITES. *Educ:* Temple Univ, BA, 51; Univ Pa, MS, 54, PhD(metall eng), 62. *Prof Exp:* Metallurgist, Naval Air Exp Sta, Philadelphia, 51-53; sr scientist, Franklin Inst Labs Res & Develop, Philadelphia, 53-63; dean, col eng, 75-84, asst to pres, 84-86; from asst prof to assoc prof, 63-68, PROF METALL, MECH & AEROSPACE ENG, UNIV DEL, 68-, PROF MAT SCI, 84- *Concurrent Pos:* Grants, Univ Del, 63-64, NSF, 64-69, 70-74, 78-79, 77-83, NASA, 69-70, Air Force Off Sci Res, 70-74, Argonne Nat Lab, 81-83; vis lectr, Univs Sendai, Tokyo, Osaka, Nagoya & Kyoto, Japan, 65; NSF travel grant, 65-73; vis prof, Stanford Univ, 69-, Oxford Univ, 70, Eindhoven Tech Univ, 78; mem, Sen Roth's Energy Adv Comt, 75-79, Del Energy Resources Comn, 75-76; chmn exec comt, Del Prog Minority Engrs, 74-77. *Mem:* Am Inst Mining, Metall & Petrol Engrs; Electron Micros Soc Am; Sigma Xi; Am Soc Metals; Am Soc Eng Educ. *Res:* Surfaces; electron microscopy; mechanical properties; fatigue; erosion; processing and properties of metal matrix composites; interfaces; extrusion. *Mailing Add:* Mech Eng Dept Univ Del Newark DE 19711

GREENFIELD, JOSEPH C, JR, b Atlanta, Ga, July 20, 31; m 55; c 3. CARDIOVASCULAR PHYSIOLOGY. *Educ:* Emory Univ, BA, 54, MD, 56; Am Bd Internal Med, dipl; Am Bd Cardiovasc Dis, dipl. *Prof Exp:* From intern to sr asst resident med, Duke Univ Hosp, 56-59; clin assoc, Sect Clin Biophys, Cardiol Br, Nat Heart Inst, 59-62; assoc med, Med Ctr, Duke Univ, 62-63, from asst prof to prof, 63-75; clin investr, 62-63, asst chief med serv, Vet Admin Hosp, 63-75; chief, Cardiol Sect, 81-88, DIR HEART STA, MED CTR, DUKE UNIV, 75-, JAMES B DUKE PROF MED, 81- *Concurrent Pos:* Career develop award, Nat Heart Inst, 66-75; mem cardiovasc & pulmonary study sect, NIH, 74-78, chmn, 75-78; consult, IBM, 71-84, Hewlett Packard Corp, 85-89, Spacelabs Corp, 88-; mem, Coun Clin Cardiol, Am Heart Asn, 72-74, Coun Basic Sci, 74-80 & Coun Circulation, 77-80; mem bd gov, Am Bd Internal Med, 88-; res prog specialist cardiol, Vet Admin, 89- *Honors & Awards:* Distinguished Scientist Award, Am Col Cardiol, 85. *Mem:* Inst Med-Nat Acad Sci; Am Physiol Soc; fel Am Heart Asn; fel Am Col Physicians; Am Soc Clin Invest; AAAS; Am Fedn Clin Res; fel Am Col Cardiol; Asn Am Physicians; Asn Professors Med. *Res:* Coronary blood flow. *Mailing Add:* Box 3246 Duke Univ Med Ctr Durham NC 27710

GREENFIELD, LAZAR JOHN, b Houston, Tex, Dec 14, 34; m 56; c 3. VASCULAR SURGERY. *Educ:* Baylor Univ, MD, 58; Am Bd Surg, dipl, 67; Am Bd Thoracic Surg, dipl, 67; Am Bd Gen Vascular Surg, dipl, 82. *Prof Exp:* Intern surg, Johns Hopkins Hosp, 58-59, from asst resident to resident, 61-66;

sr asst surgeon, Nat Heart Inst, Md, 59-61; chief surg serv, Vet Admin Hosp, Oklahoma City, 66-74; PROF SURG & CHMN DEPT, MED COL VA, VA COMMONWEALTH UNIV, 74- *Concurrent Pos:* Prof surg, Med Ctr, Univ Okla, 66-74; John & Mary R Markle scholar, 68- *Mem:* Soc Univ Surg; Am Physiol Soc; Am Thoracic Soc. *Res:* Pulmonary embolism; shock; cardiac function. *Mailing Add:* Prof & Chmn Dept Surg Univ Mich 2101 Taubman Ctr Ann Arbor MI 48109

GREENFIELD, LEONARD JULIAN, b New York, NY, May 18, 26. BIOCHEMISTRY. *Educ:* City Col New York, BS, 49; Univ Miami, MS, 51; Stanford Univ, PhD, 59. *Prof Exp:* Asst biochem marine wood borers, Marine Lab, Univ Miami, 51-54; asst productivity of Gulf Stream, 54-55; asst biochem marine invert, Hopkins Marine Sta, Stanford Univ, 55-59; from res asst prof to assoc prof, 59-74, assoc dean grad sch, 66-76, PROF BIOCHEM & ECOL PHYSIOL, MARINE LAB, UNIV MIAMI, 74-, CHMN DEPT BIOL, UNIV, 70-; PRES, MEGALINE SCI CORP, 82- *Concurrent Pos:* Consult, 77-; mem sci adv bd, US Environ Protection Agency, 79-82. *Mem:* AAAS. *Res:* Ecological physiology and ecological biochemistry of marine organisms; biological and chemical ecology of south Florida wetlands. *Mailing Add:* 6721 SW 69th Terr South Miami FL 33143-3134

GREENFIELD, RICHARD SHERMAN, b New York, NY, May 29, 33; m 66; c 3. METEOROLOGY. *Educ:* New York Univ, BS, 55, MS, 62, PhD(meteorol), 66. *Prof Exp:* Asst res scientist meteorol, Res Div, Col Eng, NY Univ, 55-57; chief forecaster, Kirtland AFB, USAF, 57-58, asst staff meteorologist, Air Force Ballistic Missiles Div, 58-59; asst res scientist, Geophys Sci Lab, NY Univ, 59-66; sr res scientist, Travelers Res Ctr, 66-69, Ctr Environ & Man, 70-74; prog dir global atmospheric res prog, 74-80, HEAD GRANT PROG SECT, DIV ATMOSPHERIC SCI, NSF, 80- *Concurrent Pos:* Consult, Res Found, State Univ NY Albany; lectr, Hartford Grad Ctr, Rensselaer Polytech Inst, 72-74; sr exec fel, Harvard Univ, 82. *Mem:* AAAS; fel Am Meteorol Soc. *Res:* Numerical modeling of atmospheric phenomena; dynamic meteorology; computer programming techniques; scientific administration; science policy. *Mailing Add:* NSF 1800 G St NW Washington DC 20550

GREENFIELD, ROY JAY, b New York, NY, Apr 8, 36; m 61; c 3. GEOPHYSICS. *Educ:* Mass Inst Technol, BS, 58, MS, 62, PhD(geophys), 65. *Prof Exp:* Assoc engr, IBM Corp, 60-61, staff engr, 61-62; staff mem seismol, Lincoln Lab, Mass Inst Technol, 65-68; from asst prof to assoc prof, 68-78, PROF GEOPHYS, PA STATE UNIV, 78- *Mem:* Am Geophys Union; Soc Explor Geophys; Seismol Soc Am. *Res:* Seismology; atmospheric waves; geoelectricity. *Mailing Add:* Pa State Univ University Park PA 16802

GREENFIELD, SEYMOUR, microbiology, biochemistry, for more information see previous edition

GREENFIELD, SEYMOUR, b Brooklyn, NY, Aug 25, 42; m 69; c 1. NEUROCHEMISTRY. *Educ:* Earlham Col, AB, 64; Fordham Univ, PhD(biochem), 71. *Prof Exp:* USPHS fel neurochem, Albert Einstein Col Med, 70-72; asst prof biochem, Med Sch, Aarhus Univ, 72-74; asst prof neurochem, 74-78, ASSOC PROF NEUROL, MED UNIV SC, 78- *Res:* Metabolism of neural membranes. *Mailing Add:* Dept Neurol & Biochem Med Univ SC Charleston SC 29425

GREENFIELD, STANLEY MARSHALL, b New York, NY, Apr 16, 27; m 52; c 2. ENVIRONMENTAL SCIENCE, ENVIRONMENTAL MANAGEMENT. *Educ:* NY Univ, BS, 50; Univ Calif, Los Angeles, PhD(meteorol), 67. *Prof Exp:* Meteorol aid, US Weather Bur, 44-47; asst, Eng Res Div, NY Univ, 48-50; phys scientist, Rand Corp, 50-71, head dept geophys & astron, 64-71; asst adminr res & monitoring, Environ Protection Agency, Washington, DC, 71-74; pres, Greenfield, Attaway & Tyler, Inc, 74-78; PRIN SCIENTIST, TEKNEKRON INC, 78- *Concurrent Pos:* Sci adv to dir res & develop, USAF, 59-61; mem, USAF Sci Adv Bd, 65- *Mem:* Am Meteorol Soc; Am Geophys Union; NY Acad Sci; Pan Am Med Asn. *Res:* Physics of the atmosphere; radioactive fallout; planetary sciences; radio wave propagation; cloud physics; environmental analysis of potential impacts; energy and environmental mangement and planning. *Mailing Add:* 2118 Milvia St Berkeley CA 94704

GREENFIELD, SYDNEY STANLEY, b Brooklyn, NY, Nov 28, 15. BOTANY, PLANT PHYSIOLOGY. *Educ:* Brooklyn Col, BA, 36; Columbia Univ, MA, 37, PhD(bot), 41. *Prof Exp:* Res assoc plant physiol, Columbia Univ, 41-45, instr bot, 43; from asst prof to assoc prof biol, 46-59, prof bot, 59-84, chmn, bot dept, 61-72, EMER PROF BOT, RUTGERS UNIV, NEWARD, 84- *Concurrent Pos:* Pub sch teacher, NY, 37-45; chmn, sci dept, Harlem Eve High Schs, 43-45; ed, Plant Sci Bull, 59-62. *Mem:* AAAS; Bot Soc Am. *Res:* Photosynthesis in Chlorella; mineral toxicity in plants; plant-growth substances; selenium poisoning and effects on plants; physiology of plant cells; economic botany. *Mailing Add:* Ten Huron Ave 11A Jersey City NJ 07306-3625

GREENFIELD, WILBERT, b Seven Springs, NC, July 18, 33; m 59; c 2. PHYSIOLOGY. *Educ:* A&T State Univ NC, BS, 56; Univ Iowa, MS, 58, PhD(physiol), 60. *Prof Exp:* Res asst physiol, Univ Iowa, 57-60; head dept biol, 60-67, assoc dean sch liberal studies, 67, prof biol, Jackson State Univ, 60-73, dean acad affairs, 70-73; PRES, JOHNSON C SMITH UNIV, 73-; PRES, VA STATE UNIV. *Concurrent Pos:* Consult, Paper Co Found grant, High Sch, Miss, 63-67; acad dean's inst, Am Coun Educ, Univ Chicago, 67; mentor, Acad Admin Internship Prog, 68-69. *Res:* Cardiovascular and respiratory physiology dealing with cross circulation and peripheral blood flow; sickle cell anemia. *Mailing Add:* Box T Va State Univ Petersburg VA 23803

GREENGARD, OLGA, b Arad, Hungary, Jan 13, 26; US citizen; m 54; c 2. BIOCHEMISTRY. *Prof Exp:* Univ London, BSc, 51, PhD(biochem), 55. Prof Exp: Res asst biochem, Inst Psychiat, Maudsley Hosp, Univ London, 55-56; biochemist, Courtauld Inst Biochem, Middlesex Hosp, London, 56-58; res assoc, Col Physicians & Surgeons, Columbia Univ, 62-65; res assoc, 65-69, prin assoc biochem, sr res assoc, Harvard Med Sch, 69-78; sr res assoc, mem cancer res inst, New Eng Deaconess Hosp, Boston, 65-78; RES PROF PEDIAT & PROF PHARMACOL, MT SINAI SCH MED, 78- *Mem:* Am Asn Cancer Res; Am Soc Biol Chemists; Am Soc Cell Biol; Brit Biochem Soc; Harvey Soc. *Res:* Regulation of enzyme synthesis in mammalian liver, in neoplastic tissues and in normal, developing organs during various stages of differentiation. *Mailing Add:* Dept Pediat Mt Sinai Sch Med One Gustave Levy Pl New York NY 10029

GREENGARD, PAUL, b New York, NY, Dec 11, 25; m 54; c 2. BIOCHEMISTRY. *Educ:* Hamilton Col, AB, 48; Johns Hopkins Univ, PhD(neuro physiol), 53. *Prof Exp:* Dir dept biochem, Geigy Res Labs, NY, 58-67; dir dept neuropharmacol, Inst Basic Res Ment Retardation, NY State Dept Ment Hyg, 67-68; prof pharmacol, Sch Med, Yale Univ, 68-83; Andrew D White prof-at-large, Cornell Univ, 81-87; PROF & HEAD, LAB MOLECULAR & CELLULAR NEUROSCI, ROCKEFELLER UNIV, 83- *Concurrent Pos:* NSF fel neurochem, Inst Psychiat, Univ London, 53-54; Nat Found Infantile Paralysis fel enzymol, Molteno Inst, Cambridge Univ, 54-55; Paraplegia Found fel neurochem, Nat Inst Med Res, Eng, 55-56, Nat Inst Neurol Dis & Blindness fel, 56-58; vis scientist, Nat Heart Inst, 58-59; vis assoc prof, Albert Einstein Col Med, 61-68, vis prof, 68-; vis prof, Vanderbilt Univ, 67-68; founder & ser ed, Advances Biochem Psychopharmacol, 68- & Advances Cyclic Nucleotide & Protein Phosphorylation Res, 71-; assoc ed, J Cyclic Nucleotide Res, 74-84; first distinguished lectr, Soc Gen Physiologists, Int Cong Physiol Sci, Paris, 77; assoc, Neurosci Res Prog; lectr, Harvey Soc, 80; mem, Med Adv Bd, Am Parkinson Dis Asn, 84-, Extramural Sci Adv Bd, NIMH, 86- & Bd Sci Counselors, Nat Inst Alcohol Abuse & Alcoholism, 86-88. *Honors & Awards:* Mayor's Gold Medallion, City of Milan; Lamson Mem Lectr, Vanderbilt Univ, 74; Louis B Flexner Lectr, Univ Pa, 76; Dickson Prize & Medal Med, 77; Ciba-Geigy Drew Award, 79; Biol & Med Sci Award, NY Acad Sci, 80; Oscar Bodansky Basic Sci Award & Lectr, Mem Sloan-Kettering Cancer Ctr, 82; Schueler Lectr, Tulane Univ, 86; 3M Life Sci Award, Fedn Am Socs Exp Biol, 87; Oliver H Lowry Lectr, Wash Univ, 87; William R McAlpin, Jr Ment Health Res Achievement Award, Nat Ment Health Asn, 87; Hughlings Jackson Mem Lectr, Montreal Neurol Inst, 87; Nat Acad Sci Award in Neurosci, 91. *Mem:* Nat Acad Sci; Am Acad Arts & Sci; fel Am Col Neuropsychopharmacol; Soc Neurosci; Soc Gen Physiologists; Int Brain Res Orgn; Am Soc Biol Chemists; Am Soc Pharmacol & Exp Therapeut. *Res:* Neurochemistry; cyclic nucleotides; protein phosphorylation; chemical neurophysiology; microchemistry; enzymology; biochemical pharmacology. *Mailing Add:* Lab Molecular & Cellular Neurosci Rockefeller Univ 1230 York Ave New York NY 10021

GREENHALGH, ROY, b Stockport, Eng, May 25, 26; Can citizen; m 54; c 1. PESTICIDE, MYCOTOXIN & ANALYTICAL CHEMISTRY. *Educ:* Univ Manchester, Eng, BSc(hons), 51; Univ Queensland, Australia, PhD(chem), 54. *Prof Exp:* Fel natural prod, Nat Res Coun Can, 55-57; res chemist, Defense Res Bd Can, 57-67; vis prof organophosphorus chem, Univ Kent, Eng, 67-68; res chemist pesticide chem, Agr Can Chem & Biol Res Inst, 69-80; res chemist, mycotoxin prog, Plant Res Ctr, 81-91; CONSULT, XENOS LABS, 91- *Concurrent Pos:* Mem, Can Adv Comt Common Names for Pesticides, 71; chmn tech comt, Int Standards Orgn, 73-; secy comn terminal pesticide residues, Int Union Pure & Appl Chem, 74-; Int Cong Pesticide Chem, Ottawa, 86; FAO expert, JMPR, 86, 87, 88, 89 & 90; adj prof, Ottawa Univ, 83-, Carleton Univ, 84-; secy, Int Union Pure & Appl Chem Appl Chem Div, exec comt, 87; consult, IARC, 90. *Honors & Awards:* Caledon Award, 83. *Mem:* Royal Soc Chem; Chem Inst Can; Am Chem Soc. *Res:* Development of new analytical procedures for pesticides and herbicide residues in animal, plant tissues and soil; metabolism and chemistry of pesticides; biosynthesis of mycotoxins; new analytical procedures for the toxins and metabolites; NMR of trichothecenes; characterization of fungal secondary metabolites. *Mailing Add:* 321 Cloverdale Ottawa ON K1M 0Y3 Can

GREENHALL, ARTHUR MERWIN, b New York, NY, Aug 6, 11; m 42; c 2. MAMMALOGY, CHIROPTERA. *Educ:* Univ Mich, BA, 34, MS, 35. *Prof Exp:* Dir, Portland Zool Park, Ore, 42-47; gen cur animal div, Detroit Zool Park, Mich, 47-53; cur mus, Royal Victoria Inst, Trinidad, WI, 53-63; res assoc Am Mus Nat Hist, 57-67; chief mammal sect, Bird & Mammal Labs, US Sport Fisheries & Wildlife, 63-68; bat ecologist, Food & Agr Orgn UN, 68-72; staff mem, Div Wildlife Res, US Sport Fisheries & Wildlife, US Fish & Wildlife Serv, 63-88, zoologist, Nat Fish & Wildlife Lab & Off Sci Auth, 85-88; res assoc, US Nat Mus, Smithsonian Inst, 67-88; RES ASSOC, AM MUS NAT HIST, 88- *Concurrent Pos:* Dir, Emperor Valley Zoo, Trinidad, 54-56; zoologist, Ministry Agr, Govt Trinidad & Tobago, 54-63; expert rabies investr, Govt Grenada, 55; lectr, Univ Col West Indies, 55-56; Rockefeller Found travel grant, 56; mammalogist, Trinidad Regional Virus Lab, 56-63; WHO fel, 57; expert rabies investr, Govt British Guiana, 59; mem Food & Agr Orgn-WHO Mission assess vampire bat rabies prob in Arg, Brazil, Venezuela, Trinidad & Mex, 66; staff specialist bat res, 77-81; staff specialist zoologist, Conv Int Trade Endangered Species Wild Fauna & Flora, 81-; consult, Pan Am Health Orgn/WHO, 89- *Mem:* AAAS; Am Soc Icthyol & Herpet; Am Soc Mammal; Wildlife Dis Asn; fel Linnean Soc London. *Res:* Life history and ecology of bats, including bat associated diseases affecting man and animals; control and management of bats which enter buildings, damage fruit crops, and vampire bats which attack livestock and humans, bat rabies, bat conservation, flying fox fruit bats. *Mailing Add:* 171 W 12th St New York NY 10011

GREENHALL, CHARLES AUGUST, b New York, NY, May 5, 39. MATHEMATICAL STATISTICS. *Educ:* Pomona Col, BA, 61; Calif Inst Technol, PhD(fourier series), 66. *Prof Exp:* Physicist, US Naval Ord Testing Sta, Calif, 62; Nat Res Coun resident res assoc, Jet Propulsion Lab, 66-68; asst prof math, Univ Southern Calif, 68-73; consult, 73-77, MEM TECH STAFF, JET PROPULSION LAB, 77- *Mem:* Am Math Soc; Soc Indust & Appl Math; Math Asn Am. *Res:* Statistics of precise time and frequency; NASA award for patent (frequency stability measurement). *Mailing Add:* Jet Propulsion Lab 298 4800 Oak Grove Dr Pasadena CA 91109

GREENHILL, STANLEY E, medicine; deceased, see previous edition for last biography

GREENHOUSE, GERALD ALAN, b New York, NY, Oct 18, 42; m 68; c 2. CYTOLOGY. *Educ:* Queens Col, City Univ New York, BA, 64; City Univ NY, PhD(biol), 69. *Prof Exp:* Fel molecular biol, Mass Inst Technol, 69-71; fel, Univ Geneva, Switz, 71-72; asst prof anat, cell & develop biol, Univ Calif, Irvine, 72-76; grants assoc, NIH, 76-77; EXEC SECY, CELL BIOL STUDY SECT, NIH, 77-, REFERRAL OFFICER, DIV RES GRANTS, 83- *Mem:* Teratology Soc; Am Soc Cell Biol. *Mailing Add:* NIH Westwood Bldg 1A06 Bethesda MD 20892

GREENHOUSE, HAROLD MITCHELL, b Chicago, Ill, Nov 20, 24; m 48; c 3. PHYSICAL INORGANIC CHEMISTRY, HYBRID MICROELECTRONICS. *Educ:* Ohio State Univ, BS, 48, MS, 51. *Prof Exp:* Res assoc, Res Found, Ohio State Univ, 50-53; res chemist, Ferroxcube Corp, 53-55; dir semiconductor res & magnetics, Aladdin Electronics Co, 55-59; dir microelectronics lab, 59-71, SR STAFF ENGR, BENDIX COMMUN DIV, ALLIED BENDIX AEROSPACE CO, 71- *Mem:* Int Soc Hybrid Microelectronics; Inst Elec & Electronics Eng; Sigma Xi; Am Crystallog Asn. *Res:* Materials research; solid state physical chemistry; x-ray diffraction; emission and adsorption spectroscopy; high temperature materials; ferrites; magnetic materials and devices; thin films; microelectronics; thick film hybrid microcircuits; precision crystal oscillators. *Mailing Add:* One Mica Ct Baltimore MD 21209

GREENHOUSE, NATHANIEL ANTHONY, b Washington, DC, June 20, 40. HEALTH PHYSICS, ENVIRONMENTAL SCIENCE. *Educ:* Catholic Univ Am, BA, 61; Univ Rochester, MS, 66. *Prof Exp:* Health physicist, Lawrence Livermore Lab, Univ Calif, 66-71; SR HEALTH PHYSICIST, BROOKHAVEN NAT LAB, 71- *Concurrent Pos:* Consult, Dames & Moore, 75-; mem, Am Bd Health Physics, 78- *Mem:* Health Physics Soc; Sigma Xi. *Res:* Transport of man-made environmental radioactive materials to man; modeling of radionuclide exposure pathways; internal and external radiation dosimetry. *Mailing Add:* Univ Calif LBL Bldg 75-112 Berkeley CA 94720

GREENHOUSE, SAMUEL WILLIAM, b New York, NY, Jan 13, 18; m 44; c 4. STATISTICS. *Educ:* City Col New York, BS, 38; George Washington Univ, MA, 54, PhD(math statist), 59. *Prof Exp:* Jr statistician, US Census Bur, 40-42; statistical analyst, UNRRA, 45-48; math statistician, Nat Cancer Inst, NIH, 48-54, chief, Statist & appl math sect, NIMH, 54-66, assoc dir epidemiol & biomet, Nat Inst Child Health & Human Develop, 66-74, actg assoc dir planning & eval, 69-74; chmn, 76-79, prof, 74-88, EMER PROF STATIST, GEORGE WASHINGTON UNIV, 88-; CONSULT, 88- *Concurrent Pos:* Mem psychopharmacol adv comt, NIH, 57-59, mem accident prev study sect, 58-62, mem statist & math res review panel, 63-70; med res adv coun, Fed Aviation Asn, 59-64; vis prof, Stanford Univ, 60-61; NSF vis lectr, 66-68; mem biomet & epidemiol methodol adv comt, Food & Drug Admin, 67-73, chmn, 71-73; appointed mem study sect, Comput Sci, Biomath & Statist, NIH, 75-77; vis prof, Dept Biostatist, Sch Pub Health, Harvard Univ, 81-82; mem Clin Trials Rev Comt, Nat Heart, Lung & Blood Inst, NIH, 83-86. *Mem:* Fel AAAS; Am Math Asn; fel Am Statist Asn; fel Inst Math Statist; Royal Statist Sec; Int Statist Inst; Biomet Soc (pres, 69). *Res:* Statistical methods and theory; application of statistical methods and mathematical models to the medical, biological and behavioral sciences. *Mailing Add:* 1724 Ladd St Silver Spring MD 20902

GREENHOUSE, STEVEN HOWARD, b Rockville Center, NY, Oct 10, 47. MASS SPECTROMETRY, SURFACE ANALYSIS. *Educ:* Adelphi Univ, Garden City, NY, BA, 69. *Prof Exp:* Chemist, Grumman Aerospace, 69-74; chemist, 74-77, res chemist, 77-88, SR RES CHEMIST, AM CYANAMID CO, 88- *Mem:* Am Soc Mass Spectrometry; Am Vacuum Soc. *Res:* Spectroscopic characterization of inorganic, organic and polymeric substances including combinations with chromatographic separation techniques. *Mailing Add:* Am Cyanamid Co 1937 W Main St Stamford CT 06904

GREENIDGE, KENNETH NORMAN HAYNES, botany; deceased, see previous edition for last biography

GREENKORN, ROBERT A(LBERT), b Oshkosh, Wis, Oct 12, 28; m 52; c 4. CHEMICAL ENGINEERING. *Educ:* Univ Wis, BS, 54, MS, 55, PhD(chem eng), 57. *Prof Exp:* Asst chem eng & math, Univ Wis, 56-57; NSF fel, Norway's Tech Univ, Trondheim, 57-58; res engr sec recovery, Jersey Prod Res Co, 58-63; assoc prof eng, Marquette Univ, 63-65; assoc prof chem eng, 65-67, head dept, 67-72, asst dean, Schs Eng, 72-76, assoc dean, Schs Eng & Dir, Eng Exp Sta, 76-80, vpres & assoc provost, 80-86, PROF CHEM ENG, PURDUE UNIV, 67-, VPRES RES, 86-, VPRES, PURDUE RES FOUND, 80- *Concurrent Pos:* Lectr, Univ Tulsa, 59-63; consult, Jersey Prod Res Co, 63-65, Esso Prod Res Co, 65-73, Occidental Petrol Corp, 83-84 & World Bank, 84; vpres & mem bd, Inventure, Inc, 85-; mem bd, Bemis Co, Inc, 84- *Mem:* Fel Am Inst Chem Engrs; Soc Petrol Engrs; Am Soc Eng Educ; Am Geophys Union; Am Chem Soc; AAAS. *Res:* Flow in porous media; thermodynamic properties. *Mailing Add:* Sch Chem Eng Purdue Univ West Lafayette IN 47907

GREENLAW, JON STANLEY, b Masardis, Maine, Aug 6, 39; m 62; c 1. ORNITHOLOGY, POPULATION ECOLOGY. *Educ:* Univ Maine, BA, 63; Rutgers Univ, PhD(ecol), 69. *Prof Exp:* Instr ornith, Douglass Col, Rutgers Univ, 67; asst prof, 69-74, ASSOC PROF BIOL, C W POST COL, LONG ISLAND UNIV, 74- *Mem:* Am Inst Biol Sci; Ecol Soc Am; Am Ornith Union; Brit Ornith Union; Cooper Ornith Soc. *Res:* Organization and evolution of avian social systems; communication in birds; avian population and community structure and function; the systematics of Emberizinae. *Mailing Add:* Five Etna Lane Dix Hills NY 11746

GREENLEAF, FREDERICK P, b Allentown, Pa, Jan 8, 38; m 64. MATHEMATICAL ANALYSIS. *Educ:* Pa State Univ, BS, 59; Yale Univ, MA, 62, PhD(math), 64. *Prof Exp:* Instr math, Yale Univ, 63-64; from instr to asst prof, Univ Calif, Berkeley, 64-68; asst prof, 68-69, assoc prof, 69-79, PROF MATH, NY UNIV, 79- *Concurrent Pos:* Prin investr, NSF res contract math anal, 71-84, 85-; vis prof math, Univ Calif, Los Angeles, 79-80, 81-82 & Univ Calif, Berkeley, 85. *Mem:* AAAS; Am Math Soc. *Res:* Geometry of groups; analysis and integration theory, with special emphasis on noncommutative harmonic analysis; especially, convolution of measures, Fourier analysis on nilpotent groups, solvability of differential operators on nilpotent groups, structure of conjugacy classes in lie groups. *Mailing Add:* Dept Math NY Univ 251 Mercer St New York NY 10012

GREENLEAF, JAMES FOWLER, BIOMEDICAL IMAGING, ULTRASONIC IMAGING. *Educ:* Purdue Univ, PhD(eng sci), 70. *Prof Exp:* STAFF CONSULT, MAYO FOUND, 76- *Res:* Ultrasound tissue characterization. *Mailing Add:* Dept Physiol & Biophysiol Med Sci Bldg Mayo Clin Rochester MN 55901

GREENLEAF, JOHN EDWARD, b Joliet, Ill, Sept 18, 32; m 60. EXERCISE PHYSIOLOGY, ENVIRONMENTAL & SPACE PHYSIOLOGY. *Educ:* Univ Ill, Urbana, BS, 55, MS, 62, PhD(physiol), 63; NMex Highlands Univ, MA, 56. *Prof Exp:* RES PHYSIOLOGIST, LAB HUMAN ENVIRON PHYSIOL, LIFE SCI DIV, AMES RES CTR, NASA, 63-, PRIN INVESTR. *Concurrent Pos:* Swed Med Res Coun fel, Stockholm, 66-67; Nat Acad Sci exchange fel, Warsaw, Poland, 73, 74, 77, 89; NIH fel, Warsaw, Poland, 80; ed bd, J Appl Physiol, Aviat Space Environ Med; vis prof, Univ Occup Environ Health, Kitakyushu, Japan, 89. *Honors & Awards:* Ellingson Award, Aerospace Med Asn, 83, 84, Eric Liljencrantz Award, 90. *Mem:* Fel Am Col Sports Med; Am Physiol Soc; fel Aerospace Med Asn; Am Inst Nutrit. *Res:* Thirst and drinking; heat acclimatization; exercise temperature regulation; body water metabolism; dehydration; space physiology, deconditioning. *Mailing Add:* Lab Human Environ Physiol NASA Ames Res Ctr Moffett Field CA 94035-1000

GREENLEAF, ROBERT DALE, b Salt Lake City, Utah, April 10, 48; m 78; c 1. CELL CULTURE. *Educ:* Utah State Univ, BS, 71; Univ Utah, PhD(anat), 75. *Prof Exp:* Fel, Cardiovasc Res Inst, Univ Calif, San Francisco, 76-78; ASST PROF ANAT, TUFTS UNIV, 78- *Concurrent Pos:* Guest ed, Am Rev Respiratory Dis, 81; guest reviewer, Am Thoracic Soc, 82. *Mem:* Am Thoracic Soc; Tissue Cult Asn; AAAS. *Res:* Regulation of metabolism of the lung. *Mailing Add:* 6019 Bear Creek Ct Elk Grove CA 94524

GREENLEAF, WALTER HELMUTH, b Stuttgart, Ger, Apr 3, 12; nat US; m 39; c 3. PLANT BREEDING. *Educ:* Univ Calif, BS, 36, PhD(genetics), 40. *Prof Exp:* Plant breeder, Vaughan Seed Store, Ill, 41-42; assoc horticulturist, Exp Sta, Univ Ga, 44-47; prof, Exp Sta, 47-82, EMER PROF HORT, AUBURN UNIV, 82- *Mem:* Fel AAAS; Am Soc Hort Sci. *Res:* Plant breeding; virology; cytology; horticulture; Nicotiana polyploids from auxin-induced callus; inheritance of resistance to tobacco etch virus in Capsicum; breeding multiple disease resistant quality tomatoes and peppers; Atkinson tomato, Bighart pimento, Greenleaf Tabasco and Auburn 76 FMN tomato. *Mailing Add:* Dept Hort Auburn Univ Auburn AL 36849-5408

GREENLEE, HERBERT BRECKENRIDGE, b Rockford, Ill, Sept 6, 27; m 55; c 4. SURGERY. *Educ:* Beloit Col, AB, 51; Univ Chicago, MD, 55. *Prof Exp:* From intern to resident, Univ Chicago Hosps, 56-62; pvt pract, Chicago, Hosps, 56-62; pvt pract, Chicago, Ill, 64-66; asst prof surg, Univ Wis-Madison, 66-67; assoc prof, 67-72, PROF SURG, STRITCH SCH MED, LOYOLA UNIV CHICAGO, 72-; CHIEF SURG SERV, VET ADMIN HOSP, HINES, 72- *Concurrent Pos:* Am Cancer Soc fel, Univ Chicago clins, 60-61; staff surgeon, Vet Admin Hosp, Madison, Wis, 66-67; asst chief surg serv, Vet Admin Hosp, Hines, 67-72; consult, Henrotin Hosp, Chicago, 73- *Mem:* Fel Am Col Surgeons; Am Gastroenterol Asn; Soc Surg Alimentary Tract; Asn Acad Surg; Int Col Digestive Surg. *Res:* Gastro-intestinal physiology; bacterial flora of gastrointestinal tract. *Mailing Add:* Dept Surg Loyola Univ Sch Med Maywood IL 60153

GREENLEE, JOHN EDWARD, b Mercedes, Tex, Sept 24, 40; m 65; c 3. NEUROLOGY, NEUROVIROLOGY. *Educ:* Hamilton Col, BA, 62; Univ Rochester, MD, 69. *Prof Exp:* Fel path, Sch Med, Univ Rochester, 67-68; intern med, Univ Va Hosp, 67-70, asst resident med, 70-71, resident neurol, 71-74; fel neurovirol, Johns Hopkins Univ Hosp, 74-76; ASST PROF NEUROL, UNIV VA HOSP, 76- *Concurrent Pos:* Attend neurologist, Baltimore City Hosp, 75-76; consult neurologist, Vet Admin Hosp, Salem, Va, 77- *Honors & Awards:* USPHS Teacher-Investr Award, NIH-NINCDS, 77. *Mem:* Am Acad Neurol; Am Soc Microbiol. *Res:* Central nervous system infections; papovaviruses and progressive multifocal leukoencephalopathy. *Mailing Add:* Dept Neurol Univ Hosp Charlottesville VA 22901

GREENLEE, KENNETH WILLIAM, b Leon WVa, Jan 23, 16; m 47; c 3. LIQUID AMMONIA REACTIONS, PHEROMONES SYNTHESIS. *Educ:* Antioch Col, BS, 38; Ohio State Univ, PhD(chem), 42. *Prof Exp:* Res assoc chem, Ohio State Univ, 42-59, lectr, 59-63; pres, Chem Samples Co, 63-78, Chemsampco, 78-80; vpres, Albany Int Corp, 80-83; RETIRED. *Concurrent Pos:* Assoc dir, Hydrocarbon Res Lab, Ohio State Univ Res Found, 50-59, dir, 59-63; consult, Goodyear Tire & Rubber Co, 57-82; founder, Am Chem Soc, Div Small Chem Bus, 78. *Mem:* Am Chem Soc; AAAS; Sigma Xi. *Res:* Synthesis of gasoline-range hydrocarbons; monomers; acetylenes; pheromones; drug intermediates; reactions in liquid ammonia; mechanism of combustion; organic peroxides; safety in lab and plant. *Mailing Add:* 98 Blenheim Rd Columbus OH 43214-3230

GREENLEE, THEODORE K, b Rockford, Ill, May 14, 34. ORTHOPEDIC ONCOLOGY. *Educ:* Northwestern Univ, MD, 59. *Prof Exp:* ASSOC PROF ORTHOP SURG, UNIV WASH, SEATTLE, 71- *Mailing Add:* Dept Orthop RK1D Univ Wash Seattle WA 98195

GREENLEE, THOMAS RUSSELL, b Lakewood, Ohio, Oct 25, 48. ATOMIC PHYSICS. *Educ:* Mich Technol Univ, BS, 70; Calif Inst Technol, MS, 73, PhD(physics), 78. *Prof Exp:* Teaching & res asst physics dept, Calif Inst Technol, 71-78; lectr physics, Univ Wis, 78-80; MEM FAC, BETHEL COL, 80- *Concurrent Pos:* NSF fel, physics dept, Calif Inst Technol, 71-74. *Mem:* Am Phys Soc; Am Asn Physics Teachers; Am Sci Affil. *Res:* Beam-foil spectroscopy; solar abundances of elements; oscillator strengths. *Mailing Add:* Dept Physics Bethel Col 3900 Bethel Dr St Paul MN 55112

GREENLER, ROBERT GEORGE, b Kenton, Ohio, Oct 24, 29; m 54; c 3. SURFACE PHYSICS, METEOROLOGICAL OPTICS. *Educ:* Univ Rochester, BS, 51; Johns Hopkins Univ, PhD(physics), 57. *Prof Exp:* Res assoc, Radiation Lab, Johns Hopkins Univ, 57; res physicist, Allis-Chalmers Mfg Co, 57-62; assoc prof, 62-67, chmn lab surface studies, 66-71, PROF PHYSICS, UNIV WIS-MILWAUKEE, 67- *Concurrent Pos:* Sci Res Coun sr vis fel, Sch Chem Sci, Univ EAnglia, 71-72; traveling lectr, Optical Soc Am, 73-74; sr Fulbright scholar, Fritz Haber Inst Max Planck Soc, WBerlin, 83. *Honors & Awards:* Glover Award, 75; Millikan Lectr Award, Am Asn Physics Teachers, 88. *Mem:* Am Asn Physics Teachers; fel AAAS; fel Optical Soc Am (vpres, 85, pres, 87); Coun Sci Soc (pres, 86-88). *Res:* Infrared interferometry; gas adsorption by infrared spectroscopy; physics and chemistry of solid surfaces; optical phenomena of the sky; structural colors in nature. *Mailing Add:* Dept Physics Univ Wis-Milwaukee Milwaukee WI 53201

GREENLEY, ROBERT Z, b Chicago, Ill, Jan 25, 34; m 56; c 3. BIOLOGICAL APPLICATIONS OF POLYMERS. *Educ:* John Carroll Univ, BS, 56; Univ Ill, PhD(org chem), 60. *Prof Exp:* Chemist, Lewis Labs, Nat Adv Comt Aeronaut, 56-57; sr res chemist, 60-70, sr res specialist & group leader, 70- 79, FEL, MONSANTO CO, 80- *Mem:* Am Chem Soc. *Res:* Vinyl copolymers; ring opening polymerizations; thermally stable polymers and monomer synthesis; bioactive polymers. *Mailing Add:* Monsanto Corp Res Monsanto Co 800 N Lindbergh Blvd St Louis MO 63167

GREENLICK, MERWYN RONALD, b Detroit, Mich, Mar 12, 35; m 56; c 3. PUBLIC HEALTH. *Educ:* Wayne State Univ, BS, 57, MS, 61; Univ Mich, PhD(med care orgn), 67. *Prof Exp:* From spec instr to instr pharm admin, Col Pharm, Wayne State Univ, 58-62; DIR, CTR HEALTH RES, KAISER PERMANENTE, NORTHWEST REGION, 64-, VPRES RES, KAISER FOUND HOSPS, 81- *Concurrent Pos:* Adj prof, Portland State Univ, 65-; assoc clin prof prev med & pub health, Ore Health Sci Univ, 71-84, clin prof, 84-90, prof & actg chair, 90-; consult, Israel Ministry Health, Jerusalem, 72, WHO, Colombia, 76 & 78; vis fac, Belgrade Fac Med, Yugoslavia, 72, Ben Gurion Univ, 73; mem, Inter-Soc Comn Heart Dis Res, 80-; mem, Comt Opportunities Res Prev & Treatment Alcohol Related Probs, Inst Med-Nat Acad Sci, 87-89, Steering Comt Study Treatment Alcohol Probs, 87-89, Nat Res Agenda Aging, 88- *Mem:* Inst Med-Nat Acad Sci; AAAS; Am Pub Health Asn; Am Heart Asn; Am Sociol Asn; Am Statist Asn; Am Soc Aging; Asn Health Serv Res. *Res:* Medical care organization, especially investigation of the relationship between the organization of the medical care system and individuals' utilization of medical care services; health behavior, cardiovascular and pidemology. *Mailing Add:* Ctr Health Res 4610 SE Belmont Portland OR 97215

GREENLIEF, CHARLES M, b Perkins, WVa, July 27, 37; m 60; c 3. SURFACE CHEMISTRY. *Educ:* Univ Calif, Berkeley, BS, 65; Univ Wash, PhD(phys chem), 70. *Prof Exp:* Welch fel, Univ Tex, Austin, 70-72; asst prof, 72-80, ASSOC PROF CHEM, EMPORIA STATE UNIV, 80-, CHAIR DEPT, 85- *Concurrent Pos:* Consult reviewer, Oceana Publ, Inc, 71-; vis prof, Univ Wash, 82-83 & Univ Colo, 85. *Mem:* AAAS; Am Chem Soc; Am Phys Soc; Sigma Xi. *Res:* Thermodynamics and statistical mechanics of physical adsorption; thermodynamics and transport properties of electrolytes in pure and mixed solvents. *Mailing Add:* Dept Chem Emporia State Univ Emporia KS 66801

GREENLIEF, CHARLES MICHAEL, b Oakland, Calif, June 8, 61; m 82; c 2. SURFACE CHEMISTRY, MATERIALS CHEMISTRY. *Educ:* Emporia State Univ, Kans, BS, 83; Univ Tex, Austin, PhD(chem), 87. *Prof Exp:* Res asst, Univ Tex Austin, 83-87; res assoc, IBM TJ Watson Res Ctr, 87-89; ASST PROF, UNIV MO, COLUMBIA, 89- *Honors & Awards:* Grad Student Award, Mat Res Soc, 86. *Mem:* Am Chem Soc; Am Vacuum Soc; Mat Res Soc. *Res:* Surface chemistry; electronic materials; deposition of thin films; electronic structure and morphology of surfaces; catalysis. *Mailing Add:* Dept Chem Univ Mo Columbia MO 65211

GREENMAN, DAVID LEWIS, b Williamston, Mich, Jan 19, 34; m 56; c 2. CANCER, ENDOCRINOLOGY. *Educ:* Asbury Col, AB, 56; Purdue Univ, MS, 59, PhD(endocrinol), 62. *Prof Exp:* Res assoc enzym, Oak Ridge Nat Lab, 62-63, USPHS fel, 63-64; asst prof & res assoc biol, Johns Hopkins Univ, 64-70; pharmacologist, Food & Drug Admin, 70; pharmacologist, Environ Protection Agency, 70-71, actg chief pharmacologist, Pesticides Regulation Div, 71-72; pharmacologist, 72-76, actg chief chronic studies, 74-75, RES PHYSIOLOGIST, NAT CTR TOXICOL RES, 76- *Mem:* AAAS; Am Soc Zool; Sigma Xi; Soc Study Reproduction; Am Inst Biol Sci. *Res:* Hormonal influences on nucleic acid metabolism of estrogen target tissues; effect of age, genetic, nutritional and environmental factors on chemical carcinogenesis; mechanism and nutritional modulation of estrogen carcinogenesis, risk assessment. *Mailing Add:* Nat Ctr Toxicol Res Jefferson AR 72079

GREENMAN, NORMAN, b Chicago, Ill, Nov 5, 20; m 49; c 4. GEOLOGY. *Educ:* Univ Chicago, BA, 41, MS, 48, PhD(geol), 51. *Prof Exp:* Geologist, Shell Oil Co, 51-61 & McDonnell Douglas Astronaut Co, 61-73; consult geologist, 73-76; geologist, Santa Fe Energy Co, 76-85; RETIRED. *Mem:* AAAS; Am Geophys Union; Geol Soc Am; assoc Soc Econ Paleont & Mineral; Sigma Xi. *Res:* Lunar and planetary geology; cosmic dust; mineral and Apollo lunar sample luminescence; sedimentary petrology. *Mailing Add:* 1437 Ninth St Manhattan Beach CA 90266

GREENOUGH, RALPH CLIVE, b Medford, Mass, June 1, 32; m 61; c 1. ANALYTICAL CHEMISTRY. *Educ:* Mass Inst Technol, SB, 53; Calif Inst Technol, PhD(chem), 62. *Prof Exp:* Sr res engr, Rocketdyne Div, NAm Aviation, Inc, 61-66; sr scientist, Warner-Lambert Co, 66-77; res scientist, Uniroyal, Inc, 77-85; TECH DIR, CRYODYNE TECHNOLOGIES, 85- *Mem:* Am Chem Soc; Am Soc Mass Spectrometry; Sigma Xi. *Res:* Nuclear magnetic resonance of fluorine compounds; infrared, ultraviolet, nuclear magnetic resonance and mass spectrometry of drugs and metabolites; infrared and chromatographic analysis of rubbers, plastics, gases and solvents. *Mailing Add:* 29 Lynwood Dr Milldale CT 06467-0019

GREENOUGH, WILLIAM TALLANT, b Seattle, Wash, Oct 11, 44; div; c 1. PHYSIOLOGICAL PSYCHOLOGY, NEUROANATOMY. *Educ:* Univ Ore, BA, 64; Univ Calif, Los Angeles, MA, 66, PhD(psychol), 69. *Prof Exp:* From instr to assoc prof, 68-78, PROF PSYCHOL & CELL STRUCT BIOL, UNIV ILL, URBANA-CHAMPAIGN, 78- *Concurrent Pos:* NIMH grants, Univ Ill, 83-, ONR grant, 85-; vis asst prof psychobiol, Univ Calif, Irvine, 72; NSF grant, 74-82; vis assoc prof, Univ Wash, 75-76; J McKeen Cattell Found fel, Ctr for Adv Study, Univ Ill, 75-76; chair, Univ Ill Neurol & Behav Biol Prog, 76-87, assoc dir, Beckman Inst Advan Sci & Technol, 86-90; prog chair, Winter Conf Brain Res, 84-85; ed, Behav & Neurol Biol J; NSF Integrative Neurol Systs Panel, 86-90; Nat Inst Mental Health prog eval panel, 83-84; forum res mgt, Fedn Behav, Psychol & Cognitive Sci, 88-90; vpres & exec coun, Fedn Behav & Cognitive Sci, 91-93. *Honors & Awards:* Merit Award, NIMH, 90. *Mem:* Fel AAAS; Soc Neurosci; Soc Develop Psychobiol; Soc Develop Neurosci; Soc Exp Psychologists; Am Psychol Soc. *Res:* Role of experience and other extra-nervous influences in the development of brain and behavior, with particular interest in changes in neuronal organization brought about by long-term behavioral treatment; quantitative methods in biol; neural bases of learning and memory. *Mailing Add:* Beckman Inst Univ Ill 405 N Mathews Urbana IL 61801-2325

GREENSHIELDS, JOHN BRYCE, b Bridgeport, Ill, June 19, 26. PHYSICAL CHEMISTRY. *Educ:* Carnegie-Mellon Univ, BS, 50, MS, 53, PhD(chem), 56. *Prof Exp:* Asst prof, 56-61, ASSOC PROF CHEM, DUQUESNE UNIV, 61- *Concurrent Pos:* Res fel, Univ Chicago, 62-64. *Mem:* AAAS; Am Chem Soc; Am Phys Soc; Sigma Xi. *Res:* Molecular quantum mechanics; statistical mechanics. *Mailing Add:* 5620 Fifth Ave Apt A6 Pittsburgh PA 15232

GREENSLADE, FORREST C, b Endicott, NY, Sept 23, 39; m 63; c 1. PHARMACOLOGY. *Educ:* State Univ NY, BA, 63; Tulane Univ, MS, 65, PhD(develop biol), 66. *Prof Exp:* Harpur Col Alumni asst biol & NSF asst, Tulane Univ, 63-65; Argonne Nat Lab fel, 66-67; sr scientist & group leader reprod biol, Ortho Res Found, 67-73, dir pharmacol, 73-75, prin scientist biochem pharmacol, asst dir med res, Ortho Pharmaceut Corp, 75-78; sr assoc dir, Pfizer Pharmaceut, 78-82; sr assoc, Int Prog, Pop Coun, 82-84; pres, Hunterdon Biomed Res Ctr Inc, 84-90; PRES, INT PROJS ASSISTANCE SERV, 90- *Mem:* Am Soc Zool; Soc Study Reproduction; Develop Biol Soc; NY Acad Sci; AAAS; Am Soc Pharmacol & Exp Therapeut. *Res:* Biochemical mechanisms of drug action. *Mailing Add:* 102 Barton Hollow Rd Flemington NJ 08822

GREENSLADE, THOMAS BOARDMAN, JR, b Staten Island, NY, Dec 23, 37; m 59; c 2. PHYSICS, HISTORY OF PHYSICS. *Educ:* Amherst Col, BA, 59; Rutgers Univ, MS, 61, PhD(physics), 65. *Prof Exp:* From instr to asst prof, 64-69, assoc prof physics, 69-88, PROF PHYSICS, KENYON COL, 88- *Concurrent Pos:* Lectr, Univ WI, 72-73; vis assoc prof, Kans State Univ, 85-86. *Honors & Awards:* Distinguished Serv Citation, Am Asn Physics Teachers. *Mem:* Am Asn Physics Teachers. *Res:* Nineteenth century physics teaching and apparatus. *Mailing Add:* Dept Physics Kenyon Col Gambier OH 43022

GREENSPAN, BERNARD, b New York, NY, Dec 17, 14; m 39; c 2. MATHEMATICS EDUCATION. *Educ:* Brooklyn Col, BS, 35, MA, 36; Rutgers Univ, PhD(math), 58. *Prof Exp:* Instr math, Brooklyn Col, 35-44, Polytech Inst Brooklyn, 43-44; instr math, Drew Univ, 44-47, asst prof math & physics, 47-58, chmn dept math, 59-75, from assoc prof to prof, 58-81, EMER PROF MATH, DREW UNIV, 81- *Concurrent Pos:* Consult & lectr, Bell Tel Labs, Whippany, NJ, 53-58; vis prof, Univ Santa Clara, 61, Rutgers Univ, 71; dir & prof, NSF Math Inst, Drew Univ, 61-75, NSF Math Summer Inst, 62-74; reader math advan placement exams, EONL Testing Serv, Princeton, NJ, 66-71, table leader, 72. *Mem:* Am Math Soc; Math Asn Am. *Res:* Differential algebra; theory of numbers and mathematics education for secondary school teachers; review mathematics texts. *Mailing Add:* 9164 Tangerine St San Ramon CA 94583-3921

GREENSPAN, DANIEL S, b Jersey City, NJ, Aug 31, 51. MOLECULAR BIOLOGY, BIOCHEMISTRY. *Educ:* NY Univ, BA, 74; NY Univ Sch Med, MS, 78, PhD(med sci), 81. *Prof Exp:* Postdoctoral, dept human genetics, Yale Univ Sch Med, 81-84, assoc res scientist, 84-86; ASST PROF, DEPT PATH & LAB MED, UNIV WIS SCH MED, 86- *Concurrent Pos:* Postdoctoral trainee, USPHS, 81-84; fel, Arthritis Found, 84-87; prin investr, Nat Inst Health, 87- *Mem:* Sigma Xi; Am Soc Biochem & Molecular Biol; Am Soc Microbiol. *Res:* Cloning and mutagenesis of genes encoding extra cellular matrix proteins; subsequent expression analyses in tissue culture systems and transgenic mice for characterization of domains involved in secretion; macromolecular interactions; pathogenesis of heritable diseases. *Mailing Add:* Dept Path 505 Serv Mem Inst 1300 Univ Ave Madison WI 53706

GREENSPAN, DONALD, b New York, NY, Jan 24, 28; m 57; c 3. MATHEMATICS. *Educ:* NY Univ, BS, 48; Univ Wis, MS, 49; Univ Md, PhD, 56. *Prof Exp:* Instr math, Univ Md, 49-53 & 54-56; res engr, Hughes Aircraft Co, 56-57; from asst prof to assoc prof math, Purdue Univ, 57-62; permanent mem, Math Res Ctr, Univ Wis-Madison, 62-80, prof comput sci, 65-80, mem, Comput Ctr, 69-80; PROF MATH, UNIV TEX, ARLINGTON, 80- *Concurrent Pos:* Consult, Radio Corp Am, 59. *Mem:* Am Math Soc; Soc Indust & Appl Math; Math Asn Am. *Res:* Numerical analysis; computer physics. *Mailing Add:* Box 19408 Arlington TX 76019-0408

GREENSPAN, FRANK PHILIP, b New York, NY, May 7, 17; m 43; c 1. ORGANIC CHEMISTRY. *Educ:* City Col New York, BS, 38; Polytech Inst Brooklyn, MS, 41; Univ Buffalo, PhD(org chem), 51. *Prof Exp:* Chief chemist, Lane Bryant, Inc, 37-41; chemist, Hart Prod Corp, 41 & Chem Warfare Serv, 42-45; res chemist, Buffalo Electro-Chem Co, 45, group leader, 47-52, mgr org res & develop, Becco Chem Div, FMC Corp, 53-56, dir develop, Chem & Plastics Div, 56-58, tech dir, Epoxy Dept, 58-61, dir res & develop, Plastics Dept, Org Chem Div, 61-64, mgr new prod develop, Chem Div, 65-66; vpres res & develop, Dexter Chem Corp, 66-67; VPRES RES & DEVELOP,

REEVES BROS, INC, 67- *Concurrent Pos:* Instr, Univ Buffalo, 52-56. *Mem:* AAAS; Am Chem Soc; Am Inst Chemists; Am Oil Chemists Soc; Com Develop Asn. *Res:* Chemistry, reactions and applications of hydrogen peroxides and peracids; organic peroxides; organic oxidative reactions; epoxidation-hydroxylation reactions; fats, oils and derivatives; plasticizers; epoxy, vinyl and allylic resins; polymers and polymerization; polyurethane foam; textiles; coated fabrics. *Mailing Add:* 2525 Lynbridge Dr Charlotte NC 28226

GREENSPAN, HARVEY PHILIP, b New York, NY, Feb 22, 33; m 53; c 2. APPLIED MATHEMATICS. *Educ:* City Col, BS, 53; Harvard Univ, MS, 54, PhD(appl math), 56. *Prof Exp:* Asst prof appl math, Harvard Univ, 57-60; assoc prof, 60-64, PROF APPL MATH, MASS INST TECHNOL, 64- *Mem:* Am Acad Arts & Sci. *Res:* Fluid dynamics. *Mailing Add:* Dept Math Mass Inst Technol Cambridge MA 02139

GREENSPAN, JOHN SIMON, b London, Eng, Jan 7, 38; m 62; c 2. ORALPATHOLOGY. *Educ:* Univ London, BSc, 59, BDS, 62, PhD(exp path), 67; MRCPath, 71; FRCPath, 83. *Hon Degrees:* ScD, Georgetown Univ, 90. *Prof Exp:* Asst lectr oral path, Royal Dent Hosp, London, Sch Dent Surg, Univ London, 63-65, lectr, 65-68, sr lectr, 68-76; dir, Grad Prog Oral Biol, 76-90, chmn div, 81-89, PROF ORAL BIOL & PATH, SCH DENT, UNIV CALIF, SAN FRANCISCO, 76-, PROF, DEPT PATH, SCH MED, 76-, CHMN, DEPT STOMATOL, 89- *Concurrent Pos:* Fel path, Royal Postgrad Med Sch, 63-64, Med Sch, St George's Hosp, 68-71; consult & dent surgeon, St George's Hosp, London, 72-76; consult oral path, St John's Hosp & Inst Dermat, London, 73-76; chmn, Oral Soft Tissue Dis Panel, Nat Inst Dent Res, 80-; assoc ed, J Oral Path, 80-; counr, Am Asn Dent Res, 81-83, prog chair, 87-88; vis scholar, Univ Wash, Seattle, 83 & Univ Iowa, Iowa City, 84; prin investr, numerous insts, 64-; consult, Coun Dent Therapeut, Am Dent Asn, 88-; vis assoc prof oral path, Sch Dent & Sch Med, Univ Calif, San Francisco, 72-73, assoc dir, Dent Clin Epidemiol Prog, 87- *Honors & Awards:* Gurley lectr, Univ Calif, San Francisco, 80; Pindborg lectr, Scand Soc Oral Path & Oral Med, 85; Ship Mem Lectr, World Dent Cong, Jerusalem, 86; Seymour J Krehover Lect Award, Nat Inst Dent Res, NIH, 89. *Mem:* Am Asn Dent Res (vpres, 86-87, pres-elect, 87-88, pres, 88-); Int Asn Dent Res; Am Acad Oral Path; Histochem Soc; Path Soc UK; fel Am Col Dentists; fel AAAS; Int Acad Oral Path; Am Dent Asn; Am Asn Pathologists. *Res:* Pathogenesis of oral soft tissue diseases notably AIDS, oral cancer, aphthous ulceration, periodontal disease and Sjogren's syndrome; immunological mechanisms as indicated by the mononuclear cells within the lesions; in site evidence of viral etiology; author of numerous publications. *Mailing Add:* Dept Stomatol Univ Calif San Francisco CA 94143-0512

GREENSPAN, JOSEPH, b Brooklyn, NY, July 8, 09; m 41. INSTRUMENTATION, PHYSICAL CHEMISTRY. *Educ:* City Col, BS, 29; Columbia Univ, MA, 30, PhD(phys chem), 33. *Prof Exp:* Asst chem, Columbia Univ, 33-36; assoc dir, Biochem Lab, Jewish Hosp, Brooklyn, 36-37; lectr, Grad Div Chem, Brooklyn Col, 37-47; DIR RES, PROCESS & INSTRUMENTS CORP, 45- *Concurrent Pos:* Instr, Undergrad Div, Brooklyn Col, 39-42; spec investr, US Dept Navy, Rockefeller Inst, 42-43; physicist & proj engr, Manhattan Proj, Kellex Corp, 43-45. *Mem:* Am Chem Soc; Am Phys Soc; Instrument Soc Am; Soc Appl Spectros; Inst Elec & Electronics Engrs. *Res:* Instrument design; mass spectrometers; DC amplifiers; cathode ray equipment; automatic analytical instruments; high vacuum techniques; optical and electronic instruments. *Mailing Add:* One Lincoln Plaza New York NY 10023

GREENSPAN, KALMAN, b Bariez, Poland, Apr 27, 25; US citizen; m 50; c 3. ELECTROPHYSIOLOGY, PHARMACODYNAMICS. *Educ:* Long Island Univ, BS, 48; Boston Univ, MA, 50; Columbia Univ, MA, 53; State Univ NY Brooklyn, PhD(physiol), 60. *Prof Exp:* Teaching asst biol, Boston Univ, 49-50; instr, Long Island Univ, 50; teaching asst physiol, State Univ NY Downstate Med Ctr, 57-60, instr, 60-62; from asst prof med to prof med & physiol, Sch Med, Ind Univ, Indianapolis, 62-74, PROF PHYSIOL & MED, SCH MED, IND UNIV, TERRE HAUTE CTR MED EDUC, 74- *Concurrent Pos:* Mem coun basic sci, Am Heart Asn; sr res assoc, Krannert Inst Cardiol, 62-65, head sect physiol, 65-74, dir instrumentation lab, 68-74. *Mem:* Fel Am Col Clin Pharmacol; fel Am Col Cardiol; Am Soc Pharmacol & Exp Therapeut; Am Physiol Soc; Am Heart Asn. *Res:* Cardiac research in the determination of the basic mechanisms for cardiac dysrhythmias and possible modes of therapy. *Mailing Add:* Terre Haute Ctr Med Educ Ind Univ Sch Med Terre Haute IN 47809

GREENSPAN, RICHARD H, b New York, NY, Apr 25, 25; m 52; c 3. RADIOLOGY. *Educ:* Columbia Univ, AB, 45; Syracuse Univ, MD, 48. *Prof Exp:* From instr to asst prof radiol, Univ Minn, 57-60; asst prof, Sch Med, Yale Univ, 60, attend radiologist, 60-61, from assoc prof to prof radiol & actg chmn dept, 61-68; prof, Univ Calif, San Francisco, 68-73; prof diag radiol & chmn dept, ASSOC DEAN CLIN AFFAIRS, SCH MED, YALE UNIV, 86- *Mem:* Am Roentgen Ray Soc; Asn Univ Radiol; Fed Am Socs Exp Biol; Sigma Xi. *Res:* Vascular radiologic studies and the application of isotopes to cardiovascular disease. *Mailing Add:* Clin Affairs Yale Univ Sch Med 333 Cedar St New Haven CT 06510

GREENSPAN, STANLEY IRA, b New York, NY, June 1, 41; m 75; c 3. PSYCHIATRY, PSYCHOANALYSIS. *Educ:* Harvard Univ, BA, 62; Yale Univ, MD, 66. *Prof Exp:* Resident psychiat, Psychiat Inst, Columbia Presby Med Ctr, New York, 67-69; res psychiatrist, Lab Psychol, 70-72, Ment Health Study Ctr, 72-74, asst chief, 74, acting chief, 74-75, chief, ment health study ctr, NIMH, 75-84, chief, Clin Infant-Child Develop Ctr, Health Resources & Serv Admin & NIMH, 84-86; PSYCHIATRIST, SUPV CHILD PSYCHOANALYST & CLIN PROF CHILD HEALTH DEVELOP PSYCHIAT, GEORGE WASHINGTON UNIV MED SCH, 86- *Concurrent Pos:* Fel, Hillcrest Children's Ctr, Children's Hosp Nat Med, 69-71; clin assoc prof psychiat, behav sci & child health & develop, Med Sch, George Washington Univ, 71-; acad fac, Children's Hosp Nat Med Ctr, 71-; pres & mem bd dir, Nat Ctr Clin Infant Progs, 77- *Honors & Awards:* Lucie Jessner Prize in Child Psychoanal, 74; Blanche F Ittleson Award, Am Psychiat Asn, 81- *Mem:* Am Psychiat Asn; Am Psychoanal Asn; Am Acad Child Psychiat. *Res:* Personality development, particularly its roots in the first three years of life and the development of clinical approaches for infants and toddlers; integrating different theories of personality development, especially learning theory and psychoanalytic theory; change in the therapeutic process. *Mailing Add:* 7201 Glenbrook Rd Bethesda MD 20814

GREENSPON, JOSHUA EARL, b Baltimore, Md, Sept 3, 28; m 54; c 3. ACOUSTICS, APPLIED MECHANICS. *Educ:* Johns Hopkins Univ, BE, 49, MS, 51, DEng, 56. *Prof Exp:* From jr instr to instr mech eng, Johns Hopkins Univ, 49-53; physicist vibrations, David Taylor Model Basin, 53-56; res engr aeroelasticity, Martin Co, 56-58; OWNER J G ENG RES ASSOC, 58- *Concurrent Pos:* Assoc Ed, J Acoust Soc. *Honors & Awards:* Silver Medal in Eng Acoust, Acoust Soc Am, 89. *Mem:* Inst Aerospace Sci; Soc Naval Archit & Marine Engrs; Am Soc Mech Engrs; fel Acoust Soc Am. *Res:* Underwater acoustics, statics and dynamics of structures. *Mailing Add:* 3831 Menlo Dr Baltimore MD 21215

GREENSTADT, MELVIN, b New York, NY, Jan 18, 18; m 41; c 3. INORGANIC CHEMISTRY, SCIENCE EDUCATION. *Educ:* City Col New York, BS, 38; Univ Southern Calif, BA, 48, MA, 49, PhD(sci ed), 56. *Prof Exp:* Chemist, Littauer Fund, Sch Med, NY Univ, 38-40 & War Dept, 41-42; teacher high schs, Calif, 50-66; assoc prof chem, Calif State Col Long Beach, 66-69; teacher high schs, Calif, 69-80; RETIRED. *Honors & Awards:* Western Regional Award, Am Chem Soc, 72, James Bryant Conant Award, 73; Western Regional Award, Chem Mfg Asn, 78. *Res:* Sulfa drugs; photolytic reactions; chemical education. *Mailing Add:* 6531 W Fifth St Los Angeles CA 90048

GREENSTEIN, BENJAMIN JOEL, b Rochester, NY, Feb 28, 59; m 86; c 2. EVOLUTIONARY PALEOBIOLOGY, TROPICAL MARINE ECOSYSTEMS. *Educ:* Univ Rochester, BA, 83; Univ Cincinnati, MS, 86, PhD(geol), 90. *Prof Exp:* ASST PROF GEOL, SMITH COL, 90- *Mem:* Sigma Xi; Paleont Soc; Soc Econ Paleontologists & Mineralogists; Geol Soc Am. *Res:* Processing affecting the preservation potential of marine invertebrates, primarily echinoids; quantitative assessment of these processes and their effects on the evolutionary histories of several echinoid groups. *Mailing Add:* Dept Geol Smith Col Northampton MA 01063

GREENSTEIN, DAVID SNELLENBURG, b Wilmington, Del, Mar 26, 28; m 52; c 6. MATHEMATICS. *Educ:* Univ Del, BS, 49; Univ Pa, AM, 53, PhD(math), 57. *Prof Exp:* Res student biophys, Univ Pa, 50-52; eng physicist, Philco Corp, 52-53; from asst instr to instr elec eng, Univ Pa, 53-56; prof engr, Radio Corp Am, 56-57; from instr to asst prof math, Univ Mich, 57-60; asst prof, Northwestern Univ, 60-63, assoc prof, 63-68, PROF MATH, NORTHEASTERN ILL UNIV, 68- *Mem:* Am Math Soc; Math Asn Am. *Res:* Approximation theory; moment problems; complex variables; real variables. *Mailing Add:* Northeastern Ill Univ Chicago IL 60625

GREENSTEIN, EDWARD THEODORE, laboratory animal medicine, toxicology, for more information see previous edition

GREENSTEIN, GEORGE, b Williams Bay, Wis, Sept 28, 40. ASTRONOMY, PHYSICS. *Educ:* Stanford Univ, BS, 62; Yale Univ, PhD(physics), 68. *Prof Exp:* Res assoc, Yeshiva Univ, 68-70 & Princeton Univ, 70-71; from asst prof to prof, 71-83, PROF ASTRON, AMHERST COL, 83- *Mem:* Am Astron Soc; Int Astron Union. *Res:* Cosmology; neutron star structure and pulsar physics; x-ray astronomy. *Mailing Add:* Amherst Col Amherst MA 01002

GREENSTEIN, JEFFREY IAN, b Durban, SAfrica, July 27, 47; US citizen; m 74. NEUROIMMUNOLOGY. *Educ:* Univ Cape Town, MBChB. *Prof Exp:* Resident neurol, Cleveland Metrop Gen Hosp, Case Western Reserve Univ, 76-79; staff fel neuroimmunol, Neuroimmunol Br, Nat Inst Neurol & Commun Dis & Stroke, NIH, 79-82, sr staff fel, 82-83; from asst prof to assoc prof neurol, 83-89, actg chmn dept, 87-89, DIR MULTIPLE SCLEROSIS CTR, TEMPLE UNIV SCH MED, 86-, PROF & CHMN, DEPT NEUROL, 89- *Mem:* Am Acad Neurol; Am Asn Immunologists; Am Soc Microbiol; NY Acad Sci; Am Soc Advan Sci. *Res:* Study of immune regulation in multiple sclerosis; involved in experimental immunotherapy for multiple sclerosis. *Mailing Add:* Dept Neurol Temple Univ Sch Med 3401 N Broad St Philadelphia PA 19140

GREENSTEIN, JESSE LEONARD, b New York, NY, Oct 15, 09; m 34; c 2. ASTROPHYSICS. *Educ:* Harvard Univ, AB, 29, AM, 30, PhD(astron), 37. *Hon Degrees:* DSc, Univ Ariz, 87. *Prof Exp:* Nat Res Coun fel, Yerkes Observ, Univ Chicago, 37-39, from instr to assoc prof astrophys, 39-47; res assoc, McDonald Observ, Chicago & Tex, 39-48; staff mem, Hale Observ, 48-80, prof astrophys, 48-70, Lee A Dubridge prof, 70-79, LEE A DUBRIDGE EMER PROF ASTROPHYS, CALIF INST TECHNOL, 80- *Concurrent Pos:* Chmn panel astron, NSF, 52-57, mem div comt math, phys & eng sci, 57-60; pres comn stellar spectra, Int Astron Union, 52-58; foreign ed, Annales d' Astrophysiques, 53-59; mem bd overseers, Harvard Univ, 65-71; mem coun, Nat Acad Sci, 66-69, chmn astron surv, 69-71; vis prof, Princeton Univ, 54, Inst Advan Studies, 68-69, Nordita, 72, Bohr Inst, Copenhagen, 79 & Univ Hawaii, 79, Univ Del, 81; Russell Lectr, Am Astron Soc; ed, Stellar Atmospheres; corres ed, Astrophys Lett, Comments on Astrophys & Space Sci; mem sci adv bd, Itek Corp, 72-76; chmn bd, Assoc Univ Res Astron, 74-77. *Honors & Awards:* Bruce Medal, Astron Soc Pac, 71; Gold Medal, Royal Astron Soc, 75; Medal, Univ Liege, 75. *Mem:* Nat Acad Sci; Am Astron Soc (vpres, 55-57); Am Philos Soc; assoc Royal Astron Soc; fel Royal Belgian Acad Sci, Liege. *Res:* Nature of interstellar matter; quantitative analysis of stellar atmospheres; effect of nuclear reactions on the origin and abundances of elements; spectra, colors, temperatures, radii, compositions, redshifts, magnetic fields and rotations of white dwarfs; the redshift and nature of the quasi-stellar objects; age and evolution of stars; final stages in the life of stars; the luminosities and compositions of faint main sequence and subluminous stars; ultraviolet spectra of stars; brown dwarfs; white dwarfs in binary systems. *Mailing Add:* Calif Inst Technol Pasadena CA 91125

GREENSTEIN, JULIA L, b Nov 6, 56; m 77; c 1. IMMUNOREGULATION, DIFFERENTIATION. *Educ:* Univ Rochester, PhD(microbiol & immunol), 81. *Prof Exp:* Asst prof, Dana-Farber Cancer Inst, Sch Med, Harvard Univ, 84-87; SR SCIENTIST, IMMUNOLOGIC PHARMACEUT CORP 87- *Mem:* Am Asn Immunol. *Res:* Cellular immunology. *Mailing Add:* Immulogic Pharmaceut Corp Kendall Sq Bldg 600 Cambridge MA 02139

GREENSTEIN, JULIUS S, b Boston, Mass, July 13, 27; m 54; c 5. ZOOLOGY, ACADEMIC ADMINISTRATION. *Educ:* Clark Univ, AB, 48; Univ Ill, MS, 51, PhD(zool), 55. *Prof Exp:* Asst zool, Univ Ill, 50-54; from asst res prof to assoc res prof animal sci, Univ Mass, 54-59; from assoc prof to prof biol, Duquesne Univ, 59-70, chmn dept, 61-70; prof & chmn dept, State Univ NY Col, Fredonia, 70-74, actg dean arts & sci, 73-74; dean, Sch Math & Nat Sci, Shippensburg State Col, 74-80, dir, Ctr Sci & Citizen, 75-80; PRES, CENT OHIO TECH COL & DEAN & DIR, OHIO STATE UNIV, NEWARK, 80- *Concurrent Pos:* NIH grant, 61-64; NSF grant, 64-65; fel, Sch Med, Harvard Univ, 66; Am Inst Biol Sci vis lectr, 66-; ed, Int J Fertil & Proc Pa Acad Sci; consult, Human Life Found; grant evaluator, Basic Sci, NSF & NATO. *Mem:* Am Pub Health Asn. *Res:* Embryology, histology and histochemistry of female reproductive organs and related endocrine glands; semen biochemistry and male accessory gland physiology; placental physiology; endocrinology; information science; mammalian reproduction particularly fertility and sterility. *Mailing Add:* Cent Ohio Tech Col Ohio State Univ Newark OH 43055

GREENSTEIN, TEDDY, b Czech, Mar 16, 37; US citizen; m 82; c 2. BIOCHEMICAL ENGINEERING. *Educ:* City Col New York, BChE, 60; NY Univ, MChE, 62, PhD(chem eng), 67. *Prof Exp:* Rating engr, Davis Eng, 60; high sch teacher, 63-64; res asst chem eng, NY Univ, 64-67; from asst prof to assoc prof, 67-84, PROF CHEM ENG, NJ INST TECHNOL, 84- *Concurrent Pos:* Found Advan Grad Study Eng res grant, 67-69. *Mem:* Am Soc Microbiol; Am Soc Eng Educ; Am Asn Univ Professors; Am Inst Chem Engrs; Sigma Xi. *Res:* Low Reynolds number hydrodynamics; biochemical engineering; fermentation; enzymes; antibiotics; bioelectrochemical fuel cells. *Mailing Add:* Dept Chem Eng Chem & Environ Sci NJ Inst Technol 323 M L King Blvd Newark NJ 07102

GREENSTOCK, CLIVE LEWIS, b High Wycombe, UK, Aug 14, 39; m 65; c 2. MEDICAL BIOPHYSICS, RADIATION BIOCHEMISTRY. *Educ:* Univ Leeds, BSc, 60; Univ London, MSc, 63; Univ Toronto, PhD(med biophys), 68. *Prof Exp:* Med physicist, Cardiff Radiother Ctr, UK, 60-61, Ont Cancer Inst, 63; radiol physics, Nat Phys Lab, UK, 63-64; res officer, Med Biophys Br, Pinawa, 70-88, SR RES OFFICER, RADIATION BIOL BR, AECL RES, CHALK RIVER, 88- *Concurrent Pos:* Fel Nat Cancer Inst Can, 68-70; lectr, Atomic Energy Can Ltd, 70-; consult, Radiation Chem Data Ctr, Notre Dame, 75-, Nat Cancer Inst, 77-, Med Res Coun, 80- & Can Cancer Soc, 83-; ed, Radiation Res, 77-80 & Adv Oxygen Radicals & Radioprotectors, 84; vis scientist, Nat Res Coun Can, 82; appointee radiosensitize-protector, Radiation Oncol Coord Comt, Nat Cancer Inst, 82-; fel Christie Hosp of Holt Radium Inst, Manchester, 83-84; deleg People to People Inst Cancer Res & Treat, SAfrica, 85; Heineman Found Fel; vis prof, Wallac Oy, Finland, 87; exchange prof, USSR, 87. *Honors & Awards:* Res Award, Royal Soc. *Mem:* Brit Asn Cancer Res; Radiation Res Soc; Brit Inst Radiol; Biophys Soc; Am Asn Cancer Res; Sigma Xi; fel Inst Physics. *Res:* Radiation damage in DNA, lipids and proteins; radiosensitization and protection; free radical mechanisms of carcinogenesis and its prevention; redox processes in metabolism and drug toxicity; fluorescence studies of cancer diagnosis and treatment; pulse radiolysis; chemical kinetic probes of structure, conformation, drug interactions; molecular radiobiology, activated oxygen and antioxidant defense; biological dosimetry. *Mailing Add:* Radiation Biol Br AECL Res Chalk River ON K0J 1J0 Can

GREENSTONE, REYNOLD, b New York, NY, Sept 30, 24; m 52; c 5. METEOROLOGY, PHYSICS. *Educ:* NY Univ, BS, 47; Univ Md, MS, 58. *Prof Exp:* Atmospheric scientist, Nat Bur Standards, 49-55; res scientist opers res, Tech Opers Inc, 55-59, Booz Allen Appl Res Inc, 59-60; sr scientist, 60-72, PRIN SCIENTIST ATMOSPHERIC SCI, ORI, INC, 72- *Mem:* Am Meteorol Soc; Am Phys Soc; Am Geophys Union; Am Inst Aeronaut & Astronaut. *Res:* Upper atmospheric physics; air pollution analysis. *Mailing Add:* ORI Inc 1375 Piccard Dr Rockville MD 20850

GREENSTREET, WILLIAM (B) LAVON, b Cora, Mo, Feb 22, 25; m 50; c 4. MECHANICAL ENGINEERING, SOLID MECHANICS. *Educ:* Colo State Univ, BS, 50; Univ Tenn, MS, 58; Yale Univ, PhD(solid mech), 68. *Prof Exp:* Design engr, Boeing Airplane Co, Wash, 50-51; engr, Oak Ridge Nat Lab, 51-58, head appl mech sect, Reactor Div, 58-74, dir res progs, Eng Technol Div, 74-90; RETIRED. *Concurrent Pos:* Mem pressure vessel res coun, 63-90; policy comt, Southeastern Conf on Theoret & Appl Mech, 68-; Am Soc Mech Engrs comt on Oper & Maintenance of Nuclear Power Plants, 85-90. *Mem:* Fel Am Soc Mech Engrs; Sigma Xi. *Res:* Applied solid mechanics; plasticity and creep. *Mailing Add:* 106 Westlook Circle Oak Ridge TN 37830

GREENWALD, GILBERT SAUL, b New York, NY, June 24, 27; m 50; c 3. PHYSIOLOGY, OBSTETRICS & GYNECOLOGY. *Educ:* Univ Calif, AB, 49, MA, 51, PhD, 54. *Prof Exp:* Assoc zool, Univ Calif, 54; from instr to asst prof anat, Univ Wash, 56-61; assoc prof anat, obstet & gynec, 61-64, distinguished prof, 73, prof anat, obstet & gynec, 64-77, res prof human reproduction, 61-77, prof gynec & obstet, 77-90, DISTINGUISHED PROF & CHMN DEPT PHYSIOL, UNIV KANS MED CTR, KANSAS CITY, 77- *Concurrent Pos:* Fel, Carnegie Inst, 54-56; mem reproductive biol study sect, NIH, 66-70, population res adv comt, 67-71, consult, Ctr Population Res, 68-72; assoc ed, Anat Rec, 68-74; referee ed, Soc Exp Biol Med, 73-76; ed-in-chief, Biol of Reproduction, 74-76; hon lectr, Mid-Am State Univ Asn, 78-79; mem, Regulatory Biol Panel, NSF, 83-86. *Honors & Awards:* Higuchi Res Award, Univ Kans, 84; Distinguished Serv Award, Soc Study Reproduction, 88. *Mem:* AAAS; Am Physiol Soc; Brit Soc Study Fertil; Soc Study Reproduction (vpres, 70-71, pres, 71-72); Endocrine Soc. *Res:* Reproductive physiology; endocrinology; embryology; ovarian function; fertility control; pituitary-ovarian relationships. *Mailing Add:* Dept Physiol Univ Kans Med Ctr Kansas City KS 66103

GREENWALD, HAROLD LEOPOLD, b Monticello, NY, June 10, 17; m 53; c 3. HYDROLOGY & WATER RESOURCES, TECHNICAL MANAGEMENT. *Educ:* Pa State Col, BS, 39; Columbia Univ, PhD(phys chem), 52. *Prof Exp:* Res asst & assoc rockets, Div 8, Nat Defense Res Comt, 41-45; chemist polymers & surfactants, Rohm & Haas Co, 50-75, patent agent polymers & miscellaneous, 75-82; patent agent, Pvt Pract & Consult, 82-91; CHEMIST, VOL, FOX CHASE CANCER CTR, 86- *Mem:* Am Chem Soc; AAAS; Sigma Xi. *Res:* Physical and chemical properties of polymers, especially polymer emulsions; surface active agents, especially nonionic agents, and the utilization of these materials in commerce. *Mailing Add:* Kendal at Hanover 80 Lyme Rd Apt 324 Hanover NH 03755-1218

GREENWALD, LEWIS, b Bronx, NY, Aug 23, 43; m 67. COMPARATIVE PHYSIOLOGY. *Educ:* Syracuse Univ, BA, 64, MS, 67; Duke Univ, PhD(zool), 70. *Prof Exp:* Fel zool, Wash State Univ, 70-72; asst prof, 72-77, ASSOC PROF ZOOL, OHIO STATE UNIV, 77- *Concurrent Pos:* Vis asst prof biol, Harvard Univ, 75. *Mem:* Am Soc Zoologists; Sigma Xi; AAAS; Am Physiol Soc. *Res:* Salt and water balance of aquatic animals; functions of fish gills; ion transport. *Mailing Add:* 2323 Johnston Rd Columbus OH 43220-4742

GREENWALD, PETER, b Newburgh, NY, Nov 7, 36; m 68; c 3. CANCER PREVENTION & CONTROL, CANCER EPIDEMIOLOGY. *Educ:* Colgate Univ, BA, 57; State Univ NY, Syracuse, MD, 61; Harvard Sch Pub Health, MPH, 67, PhD(cancer epidemiol), 74. *Prof Exp:* Teaching fel, Harvard Sch Pub lHealth, 67-68; asst med, Peter Bent Brigham Hosp, 67-68; dir, Cancer Control Bur, NY State Dept Health, 68-78, Div Epidemiol, 77-81; DIR, DIV CANCER PREV & CONTROL, NAT CANCER INST, NIH, 81- *Concurrent Pos:* Ed-in-chief, Jour Nat Cancer Inst, 81-87; bd dirs, Am Col Epidemiol, 81-83. *Honors & Awards:* Redway Medal, NY State J Med, 78. *Mem:* Fel Am Col Epidemiol; fel Am Col Physicians; fel Am Col Prev Med; Am Asn Cancer Res; Am Epidemiol Soc. *Res:* Human cancer prevention trials; cancer epidemiology; cancer control. *Mailing Add:* Div Cancer Prev & Control NCI NIH 9000 Rockville Pike Bldg 31 Rm 10A52 Bethesda MD 20892-3100

GREENWALD, STEPHEN MARK, b Louisville, Ky, May 14, 47. PLANT PHYSIOLOGY, IMMUNOLOGY. *Educ:* Bellarmine Col, Ky, BA, 69; Duke Univ, PhD(bot), 72. *Prof Exp:* Teacher, Jones Co Sr High Sch, 72-73; asst prof, Southside Va Community Col, 73-77; PROF BIOL, VALLEY CITY STATE COL, 77- *Res:* Exotoxins of Toxoplasma gondii. *Mailing Add:* Dept Biol Valley City State Col Valley City ND 58072

GREENWALT, TIBOR JACK, b Budapest, Hungary, Jan 23, 14; nat US; m 71; c 1. TRANSFUSION MEDICINE, HEMATOLOGY. *Educ:* NY Univ, BA, 34, MD, 37; Am Bd Internal Med, dipl, 46. *Prof Exp:* Intern path & bact, Mt Sinai Hosp, NY, 37-38; intern med, Kings County Hosp, 38-40; med resident, Montefiore Hosp, 40-41; hemat fel, New England Med Ctr & instr med, Med Col, Tufts Univ, 41-42; med dir, Milwaukee Blood Ctr, 47-66; clin instr med, Sch Med, Marquette Univ, 48-51, from asst clin prof to assoc clin prof, 51-60, from assoc prof to prof, 60-66; nat med dir, Am Red Cross Blood Serv, 66-68, sr sci adv, 78-79; dir, 79-87, DIR RES, HOXWORTH BLOOD CTR, 87- *Concurrent Pos:* Res asst, New Eng Med Ctr, 41-42; consult, Vet Admin Hosp, Wood, Wis, 46-66 & Milwaukee County Gen Hosp, 46-66; ed, Vox Sanguinis, 56-77; mem, hemat study sect, NIH, 60-63, chmn, 70-72, consult clin ctr, 67-, mem adv coun, Nat Heart, Lung & Blood Inst, 86-91; founding ed, Transfusion, 60-; chmn, Comt Blood & Transfusion Probs, Nat Acad Sci-Nat Res Coun, 63-66; clin prof med, Sch Med, George Washington Univ, 67-; ed, Gen Principles of Blood Transfusion; emer prof med & path, Univ Cincinnati, Col Med; historian, Int Soc Blood Transfusion, 75- *Honors & Awards:* John Elliot & Morten Grove-Rasmussen Awards, Am Asn Blood Banks, 66. *Mem:* Inst Med-Nat Acad Sci; fel Am Col Physicians; Int Soc Blood Transfusion (pres, 66-72); Am Soc Hemat (treas, 63-67); fel AAAS; fel NY Acad Sci. *Res:* Immunohematology; blood banking; blood group genetics; seventeen books, one hundred forty-three papers and sixty three articles related to blood transfusion. *Mailing Add:* Hoxworth Blood Ctr 3231 Burnett Ave Cincinnati OH 45267-0055

GREENWAY, CLIVE VICTOR, b Gloucester, Eng, Mar 6, 37; m 69; c 3. CARDIOVASCULAR, LIVER. *Educ:* Cambridge Univ, BA, 58, PhD(pharmacol), 61. *Prof Exp:* Demonstr pharmacol, Cambridge Univ, 60-63; lectr physiol, Aberdeen Univ, 63-67; asst prof, Univ Alta, 67-68; assoc prof, 68-79, PROF PHARMACOL, UNIV MAN, 79- *Mem:* Am Physiol Soc; Am Soc Pharmacol & Exp Therapeut; Can Physiol Soc; Pharmacol Soc Can; Am Asn Study Liver Dis. *Res:* Vascular beds of the liver, intestine and spleen; control of blood flow, volume and fluid exchange; hepatic drug metabolism; control of cardiac output and its modification by drugs. *Mailing Add:* Dept Pharmacol Fac Med Univ Man 770 Bannatyne Ave Winnipeg MB R3E 0W3 Can

GREENWOOD, ALLAN NUNNS, b Leeds, Eng; m 44; c 3. ELECTRICAL ENGINEERING. *Educ:* Cambridge Univ, BA, 43, MA, 48; Univ Leeds, PhD(elec eng), 52. *Prof Exp:* Develop engr high voltage, Steatite & Porcelain Prod, Stourport, UK, 46-48; lectr elec engr, Leeds Univ, 48-54; vis prof, Univ Toronto, 54-55; sr res engr & consult, Gen Elec Co, 55-72; prof elec eng, 72-80, dir, Ctr Elec Power Eng, 72-87, PHILIP SPORN PROF ENG, RENSSELAER POLYTECH INST, 80- *Concurrent Pos:* Consult elec power res inst, state & fed govt labs & numerous pvt co, 72-; vis fel, Churchill Col, Cambridge, 88; vpres, US Nat Comt, Int Conf Large High Voltage Elec Systs. *Honors & Awards:* Attwood Award, Int Conf Large High Voltage Elec Systs. *Mem:* Fel Inst Elec & Electronics Engrs; Sigma Xi. *Res:* Power switching technology. *Mailing Add:* 2102 Massachusetts Ave Troy NY 12180

GREENWOOD, BRIAN, b Berkeley, Gloucestershire, UK, Jan 1, 43; m 67. GEOMORPHOLOGY, SEDIMENTOLOGY. *Educ:* Univ Bristol, BSc, 64, PhD(geog/geol), 70. *Prof Exp:* From lectr to asst prof, 67-72, assoc prof geog, 72-79, PROF GEOG, SCARBOROUGH COL, UNIV TORONTO, 79-, PROF GEOL, 81- *Concurrent Pos:* Vis fel, Univ Bristol, 75-76; vis prof, Univ Uppsala, 82 & Univ Sydney, 83; vis scientist, US Geol Surv, 83; mem, Assoc Comt for Res on Shoreline Erosion & Sedimentation, Nat Res Cound Can. *Mem:* Int Asn Great Lakes Res; Can Asn Geogrs; Soc Econ Paleontologists & Mineralogists; Int Asn Sedimentology; Brit Geomorphol Res Group. *Res:* Coastal geomorphology and sedimentology; nearshore hydrodynamics and process-response modelling in modern coastal environments; use of modern analogues in the reconstruction of paleoenvironments; assessment of environmental impact in the coastal zone. *Mailing Add:* Scarborough Campus Univ Toronto 1265 Military Trail Scarborough ON M1C 1A4 Can

GREENWOOD, D(ONALD) T(HEODORE), b Clarkdale, Ariz, Dec 8, 23; m 51; c 2. ENGINEERING. *Educ:* Calif Inst Technol, BS, 44, MS, 48, PhD(elec eng), 51. *Prof Exp:* Lectr elec eng, Univ Southern Calif, 54-55; from asst prof to assoc prof, 56-63, PROF AERONAUT ENG, UNIV MICH, ANN ARBOR, 63- *Concurrent Pos:* Group engr, Lockheed Aircraft Corp, 51-56. *Mem:* Am Soc Mech Engrs; Am Inst Aeronaut & Astronaut; AAAS; Sixma Xi. *Res:* Flight mechanics; dynamics and the dynamical simulation of vehicles. *Mailing Add:* Dept Aerospace Eng Univ Mich Ann Arbor MI 48109

GREENWOOD, DONALD DEAN, b Milwaukee, Wis, Apr 22, 31; m 60; c 2. PSYCHOACOUSTICS. *Educ:* Univ Wis, BA, 51; Harvard Univ, PhD(exp psychol), 60. *Prof Exp:* Fel neurophysiol, Univ Wis, 60-63, proj assoc, 63-64; asst prof physiol, Duke Univ, 64-66; assoc prof psychol, 66-71, res assoc prof, 71-86, PROF AUDIOL & SPEECH SCI, UNIV BC, 87- *Mem:* AAAS; Soc Neurosci; Acoust Soc Am. *Res:* Hearing; sensory systems; research in audition. *Mailing Add:* Dept Speech Path & Audiol Univ BC 2075 Westbrook Pl Vancouver BC V6T 1W5 Can

GREENWOOD, FREDERICK C, b Portsmouth, Eng, July 5, 27; m 50; c 3. BIOCHEMISTRY, ENDOCRINOLOGY. *Educ:* Univ London, BSc, 50, MSc, 51, PhD(biochem), 53, DSc(endocrinol), 67. *Prof Exp:* Mem acad staff steroid biochem, Imp Cancer Res Fund, 53-58, head protein chem sect, 58-68; PROF BIOCHEM, UNIV HAWAII, 73-, DIR, PAC BIOMED RES CTR. *Concurrent Pos:* Rockefeller fel protein chem, Univ Wis, 58-59; consult, Med Sch, St Mary's Hosp, 66-68; mem, Coord Comt Human Tumor Invests, 65-67; secy sub-comt polypeptide hormones, Med Res Coun, 67-68. *Mem:* AAAS; Brit Soc Endocrinol; Brit Biochem Soc. *Res:* Steroid biochemistry and its application to clinical research; peptide hormones; chemistry and measurements in biological fluids. *Mailing Add:* Dept Biochem & Biophys Univ Hawaii Burns Med Sch 1960 East-West Rd Honolulu HI 96822

GREENWOOD, GEORGE W(ATKINS), b North New Portland, Maine, Nov 28, 29; m 51; c 3. CIVIL ENGINEERING. *Educ:* Univ Maine, BS, 51; Univ Ill, MS, 60, PhD(traffic eng), 63. *Prof Exp:* Eng aide, Bridge Div, Maine State Hwy Comn, 51, 53-54; instr eng graphics, Univ Maine, 54-56; instr eng, Univ Ill, 56-58, res assoc civil eng, 58-62; ASSOC PROF CIVIL ENG, UNIV MAINE, ORONO, 63- *Res:* Highway and traffic engineering; intercommunity traffic estimation models. *Mailing Add:* Dept Civil Eng Boardman Hall Univ Maine Orono Orono ME 04469

GREENWOOD, GIL JAY, b Lincoln, Nebr, Jan 10, 49; m 71; c 3. ANALYTICAL & ORGANIC CHEMISTRY. *Educ:* Okla State Univ, BS, 71, PhD(chem), 77. *Prof Exp:* SR RES CHEMIST MASS SPECTROMETRY, PHILLIPS PETROL CO, 77- *Mem:* Am Chem Soc; Am Soc Mass Spectrometry. *Res:* Mass spectrometry as applied to the analysis of alternate fuels, petroleum and petrochemicals; emphasis on FI/MS, FD/MS, LV/MS and GC/MS; chemistry of HDM/HDS; HDM/HDS catalysis. *Mailing Add:* Phillips Petrol Co 231 PLB Bartlesville OK 74004

GREENWOOD, HUGH J, b Vancouver, BC, Mar 17, 31; m 55; c 3. GEOLOGY, PETROLOGY. *Educ:* Univ BC, BASc, 54, MASc, 56; Princeton Univ, PhD, 60. *Prof Exp:* Staff scientist, Carnegie Inst, 60-63; assoc prof geol, Princeton Univ, 63-67; assoc prof, 67-69, head dept, 78-85, PROF GEOL SCI, UNIV BC, 69- *Honors & Awards:* Steacie Prize, Nat Res Coun Can, 70. *Mem:* Fel Royal Soc Can; fel Geol Soc Am; fel Mineral Soc Am; Geochem Soc (pres, 81-82). *Res:* Thermodynamics; phase equilibria; chemical kinetics; petrology. *Mailing Add:* Dept Geol Univ BC 2075 Westbrook Pl Vancouver BC V6T 1W5 Can

GREENWOOD, IVAN ANDERSON, b Cleveland, Ohio, Jan 31, 21; wid; c 1. ATOMIC PHYSICS, INSTRUMENTATION. *Educ:* Case Inst Technol, BS, 42. *Prof Exp:* From staff mem to assoc group leader, Radiation Lab, Mass Inst Technol, 42-46; proj engr, Singer Co, 46-60, mgr physics res, 60-80 & advan res projs, Kearfott Div, 80-85; CONSULT, 85- *Concurrent Pos:* Tech & vol ed, Radiation Lab Series, Mass Inst Technol & Gen Precision Lab, Inc, 45-48; consult, NY Univ, 58-61 & Yeshiva Univ, 61-64; founding dir, Bio-Instrumentation Inst, Inc, 62-; dir, Syntha Corp, 69-84. *Mem:* AAAS; Am Phys Soc; Inst Elec & Electronics Engrs; Am Inst Navig; Fedn Am Sci. *Res:* Electronic instrumentation; medical ultrasonics; magnetic resonance; optical pumping; lasers; magnetic resonance gyroscopes; fiber optics gyroscopes. *Mailing Add:* Six Weed Circle Stamford CT 06902

GREENWOOD, JAMES ROBERT, b Jersey City, NJ, Mar 27, 43. PLASMA PHYSICS, COMPUTER SCIENCE. *Educ:* Rutgers Univ, New Brunswick, BA, 69; Univ Wis-Madison, PhD(plasma physics), 75. *Prof Exp:* Systs programmer, Bell Tel Labs, Murray Hill, 70; PHYSICIST FUSION, LAWRENCE LIVERMORE NAT LAB, 75- *Mem:* Am Phys Soc. *Res:* Laser fusion; systems programming; control systems; programming language development. *Mailing Add:* 27481 Silver Creek Dr San Juan Capistrano CA 92675-1528

GREENWOOD, JOSEPH ALBERT, b Breckenridge, Mo, Sept 18, 06; m 30; c 3. MATHEMATICS. *Educ:* Univ Mo, AB, 27, AM, 29, PhD(math), 31. *Prof Exp:* From instr to asst prof math, Duke Univ, 30-41, math consult parapsychol lab, 38-41; statistician, Bur Aeronaut Navy Dept, 46-57, mathematician, Off Naval Collab Air Intel, 57-62; chmn anal dept, Armed Forces Radiobiol Res Inst, 62-66; head, Biomet Br, Food & Drug Admin, 66-68; math statistician, Bur Narcotics & Dangerous Drugs, 68-76; RETIRED. *Mem:* Fel AAAS; fel Am Soc Qual Control; Inst Math Statist; fel Am Statist Asn. *Res:* Problems in mathematical statistics; probability. *Mailing Add:* Rte 1 Box 73 Broad Run VA 22014-9501

GREENWOOD, MARY RITA COOKE, b Gainesville, Fla, Apr 11, 43; div; c 1. PHYSIOLOGY, NUTRITION. *Educ:* Vassar Col, AB, 68; Rockefeller Univ, PhD(physiol develop biol & neurosci, 73). *Prof Exp:* Fel, Rockefeller Univ, 68-73; fel, Columbia Univ, 73-74, res assoc, 74-75, asst prof genetics & develop, Inst Human Nutrit, 75-78; assoc prof, 78-81, chmn dept, 85, PROF BIOL, VASSAR COL, 81- *Concurrent Pos:* Res asst, Rutgers Univ, 62-65 & Vassar Col, 65-68; adj instr, Pratt Inst, 74; adj res assoc, Rockefeller Univ, 74-; adj assoc prof, Inst Human Nutrit, Columbia Univ; career develop award, NIH, 78-83; dir, Animal Model Core Lab, Obesity Res Ctr; mem, Nutrit Study Sect, NIH; mem, Food & Nutrit Bd, Nat Acad Sci; Pres, NAm Asn Study Obesity; prin investr, NIH grants, Am Heart Asn grant & prof dir, NIH CORE grant. *Honors & Awards:* Award in Exp Nutrit, Am Inst Nutrit, 82; John Guy Vassar Chair, 86. *Mem:* Am Physiol Soc; Am Inst Nutrit; AAAS; Am Asn Diabetes; Soc Exp Biol Med; NAm Asn Study Obesity. *Res:* Determination of the effect of environmental, nutritional and genetic influences on the development of obesity and of adipose tissue; study of behavioral correlates; pregnancy, lactation and sexual dimorphism in obesity. *Mailing Add:* Dept Nutrit & Internal Med Univ Calif 252 Mrak Hall Davis CA 95616

GREENWOOD, PAUL GENE, b Peoria, Ill, Aug 1, 58; m 88. CELL PHYSIOLOGY. *Educ:* Knox Col, BA, 80; Fla State Univ, MS, 83, PhD(biol sci), 87. *Prof Exp:* ASST PROF CELL BIOL, COLBY COL, 87- *Mem:* Am Soc Cell Biol; Am Soc Zoologists; AAAS; Sigma Xi; NY Acad Sci; Western Soc Naturalists. *Res:* Physiology and biochemistry of nematocysts; stinging structures characteristic of the Cnidaria; organisms that feed on Cnidarians incorporate nematocysts into their own cells for use later. *Mailing Add:* Dept Biol Colby Col Waterville ME 04901

GREENWOOD, REGINALD CHARLES, b Manchester, Eng, Nov 27, 35; US citizen; m 58. NUCLEAR PHYSICS. *Educ:* Univ Manchester, BSc, 56; Western Reserve Univ, MA, 58; Univ Western Ont, PhD(nuclear physics), 60. *Prof Exp:* Assoc physicist, Armour Res Found, 60-62, res physicist & group leader, IIT Res Inst, 62-65, sr physicist & group leader, 65-66; sr res physicist, Nat Reactor Testing Sta, Idaho Nuclear Corp/Aerojet Nuclear Co, 66-72, sci assoc, 72-82, PRIN SCIENTIST, IDAHO NAT ENG LAB, EG&G IDAHO, 82- *Concurrent Pos:* Tech expert, Int Atomic Energy Agency, Greece, 72 & 74; res collabr, Brookhaven Nat Lab, 75-76. *Mem:* Fel Am Phys Soc; Am Nuclear Soc; Sigma Xi. *Res:* Nuclear structure from neutron capture gamma ray and radioactive decay schemes; mass spectroscopy on line; nuclear metrology; dosimetry in fast reactors; decay heat in reactors. *Mailing Add:* Idaho Nat Eng Lab PO Box 1625 Idaho Falls ID 83415

GREENWOOD, ROBERT EWING, b Navasota, Tex, June 21, 11; m 51; c 1. MATHEMATICS. *Educ:* Univ Tex, AB, 33, Princeton Univ, AM, 38, PhD(math), 39. *Prof Exp:* From instr to prof, 34-81, EMER PROF MATH, UNIV TEX, AUSTIN, 81- *Mem:* Am Math Soc; Math Asn Am; Math Asn Gt Brit. *Res:* Numerical methods; combinatory analysis; probability. *Mailing Add:* Dept Math Univ Tex Austin TX 78712

GREENWOOD, WILLIAM R, b St Louis, Mo, Sept 15, 38; m 61; c 2. GEOLOGY. *Educ:* Univ Idaho, BS, 61, MS, 66, PhD(geol), 68. *Prof Exp:* Proj chief, Idaho Bur Mines & Geol, 66-68; geologist, Manned Spacecraft Ctr, NASA, 68-70; geologist, US Geol Surv, 70-81, dep chief, Off Mineral Resources, 81-84, assoc chief, Off Int Geol, 84-88, ASSOC CHIEF GEOLOGIST, US GEOL SURV, 88- *Mem:* Geol Soc Am; Sigma Xi; Am Geog Union; Nat Asn Biol Teachers; Nat Earth Sci Teachers Asn; Nat Sci Teachers Asn. *Res:* Metamorphic petrology and structural geology; lunar geology; Saudi Arabian geology; wilderness resource assessment. *Mailing Add:* US Geol Surv Nat Ctr MS 913 Reston VA 22090

GREEP, ROY ORVAL, b Longford, Kans, Oct 8, 05; m 31; c 3. ANATOMY. *Educ:* Kans State Col, BS, 30; Univ Wis, MS, 32, PhD(zool), 34. *Hon Degrees:* MA, Harvard Univ, 46; DSc, Univ Buffalo, 60; ScD, Kans State Univ, 68, Univ Sheffield, 71. *Prof Exp:* Asst zool, Univ Wis, 30-35; asst, Harvard Univ, 35-37, instr, 38; res assoc, Squibb Inst Med Res, 38-44; asst prof dent sci, Sch Dent Med, Harvard Univ, 44-46, lectr endocrinol, Dept Physiol, 46-47, from assoc prof to prof dent sci, 46-55, prof anat, 55-67, 72-74, dean, 52-67, actg chmn dept anat, 56-59, John Rock Prof Pop Studies, Sch Pub Health, 67-72, dir, Human & Reproductive Biol, 67-74; RETIRED. *Concurrent Pos:* Teaching fel anat, Harvard Med Sch, 44-46; Schering scholar, 57; consult, Opers Res Off, with Off Sci Res & Develop, 44; ed, Endocrinol, 52-62, chmn comt dent, Nat Res Coun, 56-61; mem endocrinol study sect, NIH, 56-59, 63-67; mem med adv bd, Nat Pituitary Agency, 63-67; expert adv panel biol of human reprod, WHO, 65-82; mem basic sci comt, Int Planned Parenthood Fedn, 71-; hon pres, Int Cong Endocrinol, 72-76, pres, Laurentian Hormone Conf, 72-85; proj specialist, Ford Found, 74-76. *Honors & Awards:* Henry Dale Medal, 67; Fred Conrad Koch Award, 71; Marshall Medal, Fel Inst Health Serv Admin, 86. *Mem:* Soc Exp Biol & Med (pres, 69-71); Endocrine Soc (pres, 65-66); fel Am Acad Arts & Sci; NY Acad Sci; Int Soc Endocrinol (pres, 68-72). *Res:* Endocrinology; adrenal cortex; gonadotropins; parathyroids; growth hormone; mineral metabolism; hermaphroditism; genetics; dental education. *Mailing Add:* 135 Oak St Foxboro MA 02035

GREER, ALBERT H, b New York, NY, Jan 30, 20; m 46; c 1. ORGANIC CHEMISTRY. *Educ:* City Col, BS, 41; Polytech Inst Brooklyn, MS, 45, PhD(chem), 49. *Prof Exp:* Chemist biol chem, State Hosp, Brooklyn, NY, 42-43; res chemist fungicides, Centro Res Co, 43-44; res chemist pharmaceuts, Day Chem Co, 44-45; res chemist textile chems, Commonwealth Color & Chem, 45-49; res group leader ion exchange & org res, 49-59, sr res chemist, 59-65, mgr adv res, 65-69, chief res chemist, 69-75, RES ASSOC, IONAC CHEM CO, 49- *Mem:* Am Chem Soc; Sigma Xi. *Res:* Synthetic ion exchangers; organic monomers and polymers; pyridines; guanidines; detergents; fungicides. *Mailing Add:* 228 Warwick Rd Haddonfield NJ 08033

GREER, DAVID STEVEN, b Brooklyn, NY, Oct 12, 25; m 50; c 2. INTERNAL MEDICINE. *Educ:* Univ Notre Dame, BS, 48; Univ Chicago Sch Med, MD, 53. *Hon Degrees:* Brown Univ, MS, 75; Southeastern Mass Univ, DHL, 81. *Prof Exp:* Fel med, Univ Chicago, USPHS, 55-56, instr endocrinol & med, 57; prac internal med, Truesdale Clin, 57-74; assoc dean med, 74-81, dir, Family Pract Residency Prog, 75-78, chmn, Sect Commun Health, 78-81, PROF COMMUN HEALTH, PROG MED, BROWN UNIV, 75-, DEAN MED, 81- *Concurrent Pos:* Sr clin instr med, Tufts Univ Col Med, 69-71, asst clin prof med, 71-78; chief staff, Dept Med, Fall River Gen Hosp, 59-62; med dir, Earle E Hussey Hosp, 62-75; chief of staff, Dept Med, Truesdale Clin & Truesdale Hosp, 71-74; chmn, bd trustees, Southeastern Mass Univ, 73-74, mem, bd trustees, 70-81; clin assoc prof commun health, Prog Med, Brown Univ, 73-75; delegate, White House Conf Aging, 71 & 81. *Mem:* Inst Med-Nat Acad Sci; Am Col Physicians; Am Cong Rehab Med; Int Soc Rehab Med. *Res:* Study of long-term care, including the impact of housing and community based services; rehabilitation medicine, including functional assessment; health care evaluation. *Mailing Add:* Brown Univ Box G Providence RI 02912

GREER, DONALD LEE, b Silver City, NMex, June 14, 36. MEDICAL MYCOLOGY. *Educ:* Col Idaho, BS, 58; Univ Wash, MS, 61; Tulane Univ, PhD(microbiol), 65. *Prof Exp:* Sr asst scientist mycol, Ctr Dis Control, USPHS, 65-68; prof microbiol, Univ Valle, 68-71; assoc prof mycol, Univ Valle, 71-77; actg asst dir, Int Ctr Med Res & Training, 73-74, scientist mycol, 68-81; prof mycol, Univ Valle, 77-81; assoc prof mycol, 81-88, PROF MYCOL, LA STATE UNIV, 89-; DIR, TUBERC & MYCOL DIAG LAB, CHND, 81- *Concurrent Pos:* Consult, Ochsner Hosp, ND & Vet Admin Hosp, ND. *Mem:* Am Soc Microbiol; Int Soc Human & Animal Mycol; Med Mycol Soc of the Americas; Sigma Xi; fel Am Acad Microbiol. *Res:* Epidemiology of paracoccidioidomycosis and dermatophytosis in tropical America. *Mailing Add:* Dept Dermat La State Univ Med Ctr 1542 Tulane Ave New Orleans LA 70112

GREER, EARL VINCENT, mathematics; deceased, see previous edition for last biography

GREER, GALEN LEROY, biology, for more information see previous edition

GREER, GEORGE GORDON, b Toronto, Ont, Apr 16, 46. MICROBIOLOGY. *Educ:* Univ Guelph, BSc, 70, MSc, 72; Queen's Univ, Ont, PhD(microbiol), 75. *Prof Exp:* RES SCIENTIST MICROBIOL, AGR CAN, 78- *Mem:* Int Soc Milk, Food & Environ Sanitarians. *Res:* Current interests concern the bacterial spoilage and safety of meats with emphasis on the contribution of meat processing procedures to the microbial load and shelf life of retail meat cuts. *Mailing Add:* Res Sta Agr Can Lacombe AB T0C 1S0 Can

GREER, HOWARD A L, b Warrensville, NC, Feb 7, 36; m 58; c 2. WEED SCIENCE. *Educ:* Berea Col, BS, 58; Univ Ky, MS, 61; PhD(agron, plant physiol), Iowa State Univ, 64. *Prof Exp:* Instr agron, Iowa State Univ, 60-63; instr biol, Appalachian State Teachers Col, 64-65; asst prof exten agron, 65-74, PROF AGRON, OKLA STATE UNIV, 74- *Honors & Awards:* Geigy Award Agron, 70. *Mem:* Am Soc Agron; Crop Sci Soc Am; Weed Sci Soc Am. *Res:* Effects of herbicides on plant growth and development. *Mailing Add:* Dept Agron Okla State Univ Stillwater OK 74078

GREER, MONTE ARNOLD, b Portland, Ore, Oct 26, 22; m 43; c 2. ENDOCRINOLOGY. *Educ:* Stanford Univ, AB, 44, MD, 47. *Prof Exp:* Asst med, Tufts Univ, 47-49, instr, 50-51, sr asst surgeon, NIH, 51-55; asst clin prof med, Univ Calif, Los Angeles, 55-56; assoc prof, Univ Ore, 56-62, head div endocrinol, 56-80, head div endocrinal metab & clin nutrit, 80-84, head sect endocrinol, 84-90, PROF MED, ORE HEALTH SCI UNIV, 62- *Concurrent Pos:* NIH res career award, 62-81; asst, Boston Univ, 49-50; assoc, George Washington Univ, 51-55; dir radioisotope serv, Long Beach Vet Hosp, 55-56; mem pharmacol & endocrinol fel comt, NIH, 68-72; dir, Am Thyroid Asn, 73-77, mem, Endocrinol Study Sect, NIH, & Thyroid Task Force, NIH Comt Eval Endocrinol & Metab Dis, 77-80, dir. *Honors & Awards:* Oppenheimer Award, Endocrine Soc, 58; Discovery Award, Med Res Found Ore, 85. *Mem:* Endocrine Soc (vpres, 76-77); Am Soc Clin Invest; Am Thyroid Asn (2nd vpres, 68-69, 1st vpres, dir, 73-77, pres, 80); Am Fedn Clin Res; AAAS; hon mem Endocrine Soc Japan; hon mem Czech Endocrine Soc; Europ Thyroid Asn; Int Brain Res Orgn; Asn Am Physicians; Soc Exp Biol & Med. *Res:* Normal and abnormal thyroid physiology; neural control of pituitary function. *Mailing Add:* Dept Med Ore Health Sci Univ Portland OR 97201

GREER, RAYMOND T(HOMAS), b East Orange, NJ, Apr 26, 40; m 65; c 1. SOLID STATE & MATERIALS SCIENCE. *Educ:* Rensselaer Polytech Inst, BS, 63; Pa State Univ, PhD(solid state sci), 68. *Prof Exp:* Res asst atomic develop, NY State Off Atomic & Space Develop, 62-63; res asst mat sci, Mat Res Lab, Pa State Univ, 63-68, res assoc, 68-69; resident res assoc, Jet Propulsion Lab, Calif Inst Technol, 69-70; from asst prof to assoc prof eng, 70-73, PROF ENG & SCI, IOWA STATE UNIV, 77- *Concurrent Pos:* Nat Res Coun fel, 69. *Res:* Biomaterials. *Mailing Add:* 3611 Woodland Ames IA 50010

GREER, SANDRA CHARLENE, b Greenville, SC, Jan 7, 45; m 68; c 2. PHYSICAL CHEMISTRY. *Educ:* Furman Univ, BS, 66; Univ Chicago, MS, 68, PhD(chem), 69. *Prof Exp:* Res chemist, Nat Bur Standards, 69-78; assoc prof, 78-83, PROF CHEM, UNIV MD, COLLEGE PARK, 83-, DEPT CHAIR, 90- *Mem:* Am Chem Soc; fel Am Phys Soc; AAAS; Asn Women Sci. *Res:* Experimental thermodynamics of phase transitions and critical phenomena and of fluid mixtures. *Mailing Add:* Dept Chem Univ Md College Park MD 20742

GREER, SHELDON, b Brooklyn, NY, July 11, 28; m 57; c 2. MOLECULAR GENETICS. *Educ:* Brooklyn Col, BA, 50; Columbia Univ, MA, 52, PhD(zool), 57. *Prof Exp:* Asst, Columbia Univ, 52-56; res assoc biochem, Col Physicians & Surgeons, Columbia Univ, 58-61; assoc prof microbiol, 61-74, PROF MICROBIOL & BIOCHEM, SCH MED, UNIV MIAMI, 74- *Mem:* Sigma Xi. *Res:* Ultra-violet light effects on unnatural deoxyribonucleic acid; regulatory mechanisms; sensitization of tumors to x-ray; pyrimidine metabolism-catabolism-transformation in Bacillus subtilis; lysogenic conversion in Streptococcus mutans; curing cells of their proviruses; mutagenesis and repair; selective chemotherapy of tumors and of herpes viruses by pyrimidine analogs. *Mailing Add:* 302 Royal Palm Way Apt 17 Boca Raton FL 33432-7944

GREER, WILLIAM LOUIS, b Bardstown, Ky, Apr 7, 43; m 68; c 2. CHEMICAL PHYSICS, DEFENSE ANALYSIS. *Educ:* Vanderbilt Univ, AB, 65; Univ Chicago, PhD(chem), 69. *Prof Exp:* Nat Res Coun-Nat Bur Standards res assoc & res chemist, Nat Bur Standards, 69-71; lectr chem, Georgetown Univ, 71; res assoc, Inst Molecular Physics, Univ Md, 72; asst prof, Dept Chem, George Mason Univ, 72-76, assoc prof, 76-78; defense analyst, Ctr Naval Anal, 78-84; defense analyst, Off Secy Defense, 84-86; DEFENSE ANALYST, INST DEFENSE ANALYSIS, 86- *Concurrent Pos:* Chem lectr, George Mason Univ, 80-85. *Mem:* Am Phys Soc; Am Chem Soc. *Res:* Theoretical research on the nature of electronic states in organic solids and fluids; naval defense issues; strategic defense analysis, command and control analysis. *Mailing Add:* 19 Mercy Ct Potomac MD 20854-4540

GREESON, PHILLIP EDWARD, b Lexington, Ky, Aug 11, 40; m 64; c 2. LIMNOLOGY. *Educ:* Univ Ky, BS, 62, MS, 63; Univ Louisville, PhD(limnol), 67. *Prof Exp:* LIMNOLOGIST, WATER RESOURCES DIV, US GEOL SURV, 67- *Mem:* Int Asn Theoret & Appl Limnol; Am Water Resources Asn. *Res:* Aquatic ecology; lake productivity and eutrophication; algal ecology. *Mailing Add:* US Geol Surv WRD/SR Spalding Woods Off Park Suite 160 3850 Holcomb Bridge Rd Norcross GA 30092-2202

GREEVER, JOE CARROLL, b Kansas City, Mo, Nov 5, 44; m 65; c 1. ORGANIC CHEMISTRY, ENVIRONMENTAL CHEMISTRY. *Educ:* Northeastern State Univ, Okla, BS, 65; Univ Ark, PhD(chem), 70. *Prof Exp:* From asst prof to assoc prof, Drury Col, Mo, 69-74; environ chemist, City Springfield, Mo, 74, asst supvr wastewater treatment, 74-75; assoc prof, 75-80, PROF CHEM, DELTA STATE UNIV, 80- *Mem:* Sigma Xi; Am Chem Soc. *Res:* Acid rain-alkalinity; improving the quality of the undergraduate laboratory experience in organic chemistry. *Mailing Add:* Dept Phys Sci Delta State Univ Cleveland MS 38733

GREEVER, JOHN, b Pulaski, Va, Jan 30, 34; m 53; c 3. MATHEMATICAL MODELLING, GENERALIZATIONS OF COMPACTNESS PROPERTIES. *Educ:* Univ Richmond, BS, 53; Univ Va, MA, 56, PhD(math), 58. *Prof Exp:* Asst prof math, Fla State Univ, 58-61; from asst prof to assoc prof, 61-70, chmn dept, 72-75, dir math clinic, 73-75, dir, Freshman Div, 81-83, PROF MATH, HARVEY MUDD COL, 70- *Concurrent Pos:* Instr, Univ Va, 53-58; vis prof, Res Inst Math Sci, Kyoto Univ, Japan, 67-68 & Inst Animal Resource Ecol, Univ BC, 84-85; res assoc dept biol, Univ Calif, Riverside, 75-78, vis res mathematician dept entom, 78; coun undergrad res, AAAS. *Mem:* Am Math Soc; Math Asn Am; AAAS. *Res:* mathematical modelling of insect populations, with consideration of control strategies; modelling of the effect of migration on the size of ungulate populations. *Mailing Add:* Dept Math Harvey Mudd Col Claremont CA 91711-5990

GREGER, JANET L, b Joliet, Ill, Feb 18, 48. HUMAN NUTRITION, MINERAL METABOLISM. *Educ:* Univ Ill, Urbana, BS, 70; Cornell Univ, MS, 71, PhD(human nutrit), 73. *Prof Exp:* Asst prof nutrit, Purdue Univ, 73-78; from asst prof to assoc prof, 78-83, PROF NUTRIT, UNIV WIS-MADISON, 83- *Concurrent Pos:* Prin investr grants, NIH, USDA & W K Kellogg Found, 75-91; cong sci fel, AAAS, 84-85; mem, Nutrit Study sect, NIH, 85-89; mem, Nat Nutrit Adv Comt, Pew Found, 86- *Mem:* Am Inst Nutrit; Am Soc Clin Nutrit; Inst Food Technologists; AAAS; Am Dietetics Asn. *Res:* Mineral metabolism from a nutritional and toxicological point of view; minerals studied include aluminum, calcium, zinc, tin and copper. *Mailing Add:* Dept Nutrit Sci Univ Wis 1415 Lin Dr Madison WI 53706

GREGERMAN, ROBERT ISAAC, b Boston, Mass, Apr 18, 30; m 57; c 2. ENDOCRINOLOGY, GERONTOLOGY. *Educ:* Harvard Univ, AB, 51; Tufts Univ, MD, 55. *Prof Exp:* Intern med, New Eng Med Ctr, Tufts Univ, 55-56; clin assoc, Geront Res Ctr, Nat Insts Health, 56-58; resident med, Vet Admin Hosp, Washington, DC, 58-59; resident, Univ Mich Hosp, 59-60, instr, 60-61; asst med, 63-64, from instr to assoc prof, 64-88, PROF MED, SCH MED, JOHNS HOPKINS UNIV, 88- *Concurrent Pos:* Fel endocrinol, Univ Mich Hosp, 60-61; chief endocrinol sect, Geront Res Ctr, Nat Inst Aging, 63-88; head, div endocrinol, Francis Scott Key Med Ctr, 65-70 & 75- *Honors & Awards:* Clin Res Award, Am Aging Asn, 90. *Mem:* Am Soc Biochemists; Am Soc Clin Invest; Am Thyroid Asn; Endocrine Soc; Geront Soc. *Res:* Clinical endocrinology; gerontology; geriatrics. *Mailing Add:* Dept Med AAC-Pulm 4B37 Johns Hopkins Univ Baltimore MD 21224-2780

GREGERSEN, HANS MILLER, b Pasadena, Calif, Dec 9, 38; m 61; c 2. FOREST ECONOMICS, RESOURCE MANAGEMENT. *Educ:* Pa State Univ, BS, 61; Univ Wash, MS, 63; Univ Mich, MA, 69, PhD(forest resource econ), 69. *Prof Exp:* Res forester, Forest Serv, USDA, 61-63, forestry officer planning, Food & Agr Orgn UN, 65-67; res assoc forestry, Univ Mich, 67-69; asst prof, 70-72, assoc prof forestry, 72-77, PROF FOREST SERVS, UNIV MINN, 78- *Concurrent Pos:* Consult, World Bank, 69- & Food & Agr Orgn UN, 72-, Inter Am Bank, 74- *Mem:* Soc Am Foresters; Am Agr Econ Asn; Soc Int Develop; Sigma Xi. *Res:* Forest economics, benefit-cost analysis and project analysis, economic development; economic policy; research evaluation. *Mailing Add:* Sch Forestry Univ Minn St Paul MN 55108

GREGERS-HANSEN, VILHELM, b Odense, Denmark, Sept 30, 34; US citizen; m 59; c 2. RADAR SYSTEM ENGINEERING. *Educ:* Tech Univ Denmark, MSc, 59. *Prof Exp:* Scientist, SHAPE Tech Ctr, Hague, Neth, 60-66; asst prof common theory, Tech Univ Denmark, 66-69; CONSULT SCIENTIST, EQUIP DIV, RAYTHEON CO, 68- *Concurrent Pos:* Lectr, Northeastern Univ, Mass, 71-80 & George Washington Univ, 78-86. *Mem:* Fel Inst Elec & Electronics Engrs. *Res:* Analysis & design of modern radar systems for military & civilian applications. *Mailing Add:* Raytheon Co Equip Div Box A-1 Boston Post Rd Wayland MA 01778

GREGG, CHARLES THORNTON, b Billings, Mont, July 27, 27; m 47; c 4. BIOCHEMISTRY. *Educ:* Ore State Univ, BS, 52, MS, 55, PhD(biochem), 59. *Prof Exp:* Instr, Ore State Univ, 55-59; USPHS res fel physiol chem, Sch Med, Johns Hopkins Univ, 59-63; staff mem biochem, Los Alamos Sci Lab, 63-85, vpres res, Los Alamos Diag, 86-90; PRES, BETHCO INC, 72- *Concurrent Pos:* Vis prof, Free Univ Berlin, 74-75. *Mem:* Fel AAAS; Am Soc Microbiol; Am Soc Biol Chemists. *Res:* Energy metabolism of mammalian cells; biological applications of stable isotopes; biosynthesis of stable isotope labeled compounds; mammalian toxicology; biological luminescence; author of three books. *Mailing Add:* PO Box 148 Los Alamos NM 87544

GREGG, CHRISTINE M, b Rochester, NY, Nov 4, 38; m 70. ENDOCRINE PHYSIOLOGY, RENAL PHYSIOLOGY. *Educ:* Case Western Reserve Univ, BS, 62; Univ Mich, MS, 68, PhD(physiol), 73. *Prof Exp:* Fel, Univ Rochester, NY, 73-77, asst prof, 77-80; asst prof, 80-87, SR RES ASSOC, PA STATE UNIV, UNIV PARK, 87- *Mem:* Am Physiol Soc; AAAS; Sigma Xi; NY Acad Sci. *Res:* Regulation of vasopressin release using in vivo and in vitro techniques. *Mailing Add:* 433 Glenn Rd State College PA 16803

GREGG, DAVID HENRY, b Minneapolis, Minn, Oct 19, 26; m 50; c 1. PHARMACEUTICAL CHEMISTRY, ORGANIC CHEMISTRY. *Educ:* Univ Minn, BS, 49, PhD(pharmaceut chem), 53. *Prof Exp:* Asst pharm, Univ Minn, 49-51; res scientist pharmaceut chem & develop, Upjohn Co, 53-57, head fine chems dept, 57-61, head chem sales, 61-63, mgr Europ agency dist, Upjohn Int, 63-67, mgr eng, distrib & prod, 67-72, group mgr, 72-74, dir, 74-78; dir pharmaceut mfg technol & dir prod mfg opers admin, Upjohn Co, 78-85, exec dir pharmaceut mfg technol, 85-87; RETIRED. *Mem:* AAAS; Am Hemat Soc; Am Pharmaceut Asn. *Res:* Cardiac glycosides and phytochemistry; steroid synthesis. *Mailing Add:* 1917 Argyle Ave Kalamazoo MI 49008

GREGG, JAMES HENDERSON, b Mobile, Ala, Mar 17, 20; m 45; c 2. DEVELOPMENTAL BIOLOGY. *Educ:* Univ Ala, BS, 43; Princeton Univ, MS, 48, PhD(biol), 49. *Prof Exp:* Res assoc zool, Univ Chicago, 49-50; interim asst prof biol, Vanderbilt Univ, 50-51; from asst prof to assoc prof, 51-63, prof biol & zool, 64-76, PROF MICROBIOL & CELL SCI, UNIV FLA, 76- *Concurrent Pos:* Mem corp, Marine Biol Lab, 54; NIH career develop award, 62-72. *Mem:* Soc Develop Biol. *Res:* Developmental biology of cellular slime molds. *Mailing Add:* 1502 NW 36th Way Gainesville FL 32605

GREGG, JAMES R, b Napoleon, Ohio, Oct 26, 14; m 38; c 2. OPTOMETRY. *Educ:* Ohio State Univ, BS, 37 & 42; Los Angeles Col Optom, DO, 48, D Ocular Sci, 53. *Prof Exp:* From assoc prof to prof optom, Los Angeles Col Optom, 47-72; prof optom & dir tech prog, 75-76, GRANTS ADMINR, SOUTHERN CALIF COL OPTOM, 72-, INTERIM DEAN, 76- *Mem:* Fel Am Acad Optom. *Res:* Physiological optics; practice management; author of over 500 articles. *Mailing Add:* Dept Optom Southern Calif Col Optom 2001 Assoc Rd Fullerton CA 92631

GREGG, JOHN BAILEY, b Sioux Falls, SDak, June 5, 22; m 46; c 4. OTOLARYNGOLOGY, PHYSICAL ANTHROPOLOGY. *Educ:* Univ Iowa, BA, 43, MD, 46, Am Bd Otolaryngol & Maxillo-Facial Surg, dipl, 67. *Hon Degrees:* DSc, Univ SDak, 89. *Prof Exp:* Extern anesthesia, Univ Iowa Hosps, 45-46; intern, Univ Md Hosp, 46-47; resident gen surg, Univ Iowa Hosps, 49-51, resident otolaryngol, 51-53, instr, 53-54; assoc prof, Univ SDak, 55-59; asst prof, Med Sch, Univ Iowa, 59-60; assoc prof, 60-62, dir anesthesia, internship & residency proj, McKennan Hosp, 61-63, prof otolaryngol, 62-78, chmn dept, 71-77, PROF SURG, SCH MED, UNIV SDAK, 78-; PRES MED STAFF & MEM SURG SECT, MCKENNAN HOSP, 63- *Concurrent Pos:* Chief otolaryngol, Vet Admin Hosp, Iowa City, 53-54; designated examr com pilots, Fed Aviation Agency, 54-78 & airline transport pilots, 61-78; consult, 54-59 & 60-77, mem staff, Vet Admin Hosp, Sioux Falls, SDak, 80-, Crippled Children's Field Clins, SDak, 54-56 & 60-72, Crippled Children's Hosp & Pub Sch Syst, Sioux Falls, 54-59 & 60, Indian Div, USPHS, 56-59 & 60-78 & Speech & Hearing Clin, Univ SDak, 61-; mem intern comt, McKennan Hosp, 57-58; mem intern & extern comt, Sioux Valley Hosp, 58-59; mem preceptorship comt, Col Med, Univ Iowa, 59-60; dir broncho-esophagol clin, Univ Iowa Hosps, 59-60; chmn Deafness Res Found & Temporal Bone Bank, N & SDak, 63-69; prof anthrop, Univ Tenn, Knoxville, 72-; assoc prof otolaryngol, Sch Med, Univ Nebr, 71-81; coordr, Surg Spec, Sch Med, Univ SDak, 80; vchmn Surg Dept, Univ SDak, 85; fel, otolaryngic path, Armed Forces Inst Path, 67. *Mem:* AMA; Am Acad Otolaryngol-Head & Neck Surg; Am Laryngol, Rhinol & Otol Soc; fel Am Col Surgeons; Am Acad Facial Plastic & Reconstruct Surg; Int Col Surgeons; Deafness Res Found; Soc Univ Otolaryngologists; AAAS; Am Asn Phys Anthropologists. *Res:* Bronchoesophagology; facial plastic surgery; otorhinolaryngology; ancient osteopathology, paleopathology; epidemiology; biomedical research. *Mailing Add:* 2807 S Phillips Ave Sioux Falls SD 57105

GREGG, JOHN RICHARD, b Mobile, Ala, Dec 23, 16; m 42. ZOOLOGY. *Educ:* Univ Ala, BS, 42; Princeton Univ, PhD(biol), 45. *Prof Exp:* Lab asst biol, Univ Ala, 40-42 & histol, 41-42; asst, Off Sci Res & Develop Contract, Princeton Univ, 42-45 & Columbia Univ, 45-46; instr biol, Johns Hopkins Univ, 46-47; from asst prof to assoc prof zool, Columbia Univ, 47-57; assoc prof, 57-60, prof, 60-, EMER PROF ZOOL, DUKE UNIV. *Concurrent Pos:* Mem corp, Marine Biol Lab, Woods Hole Oceanog Inst, 47-72; Killough fel, 49; Rockefeller fel, Carlsberg Labs, Denmark, 49 & Middlesex Hosp, London, 53-54. *Mem:* Harvey Soc; Am Soc Nat; Asn Symbolic Logic; Int Soc Develop Biol. *Res:* Chemical embryology of amphibia; mathematical and philosophical biology. *Mailing Add:* 3702 Randolph Rd Durham NC 27705

GREGG, MICHAEL CHARLES, b Knoxville, Tenn, May 13, 39; m 63; c 3. PHYSICAL OCEANOGRAPHY. *Educ:* Yale Univ, BS, 61; Univ Calif, San Diego, PhD(oceanog), 71. *Prof Exp:* Asst res oceanogr, Scripps Inst Oceanog, Univ Calif, San Diego, 71-74; asst res prof, dept oceanog & sr oceanog, 74-77, assoc res prof, 77-80, RES PROF & PRIN, DEPT OCEANOG, APPL PHYSICS LAB, UNIV WASH, 80-, PROF, 89- *Mem:* Fel Am Geophys Union; Am Meteorol Soc. *Res:* Small scale mixing processes in the ocean; instrumentation; signal processing; internal waves; equatorial dynamics. *Mailing Add:* Dept Oceanog Univ Wash Seattle WA 98195

GREGG, ROBERT EDMOND, b Chicago, Ill, Apr 19, 12; m 39. SYSTEMATIC ENTOMOLOGY. *Educ:* Univ Chicago, SB, 35, PhD(zool), 41. *Prof Exp:* Instr zool, Minn State Teachers Col, Duluth, 40-44; from instr to prof, 44-77, EMER PROF BIOL, UNIV COLO, BOULDER, 77- *Mem:* Sigma Xi. *Res:* Geographic distribution, ecology and taxonomy of ants; origin of castes in ants. *Mailing Add:* 762 Eighth St Boulder CO 80302

GREGG, ROBERT VINCENT, b Long Beach, Calif, Feb 16, 28; m 48; c 3. ANATOMY. *Educ:* Univ Calif, Los Angeles, AB, 52; Univ Southern Calif, PhD(anat), 62. *Prof Exp:* Mus aid, Vet Admin Hosp, Long Beach, Calif, 52-54; from instr to assoc prof anat, Sch Dent, Univ Southern Calif, 54-70, chmn dept, 62-70; PROF ANAT, HEALTH SCI CTR, SCH MED, UNIV LOUISVILLE, 70- *Mem:* AAAS; Am Asn Anat. *Res:* Gastrointestinal physiology and pathology; peptic ulcers and enterochromaffin cellular responses; biological specimen preservation by plastic embedding. *Mailing Add:* Dept Oral Biol Univ Louisville PO Box 35260 Louisville KY 40292

GREGG, ROGER ALLEN, b Lenoir, NC, July 19, 38; m 61; c 3. PHYSICAL METALLURGY. *Educ:* NC State Univ, BS, 60, MS, 62; Univ Fla, PhD(metall eng), 68. *Prof Exp:* Res asst alloy develop, Advan Mat Res & Develop Lab, Pratt & Whitney Aircraft Co, 62-63; res engr, E I du Pont de Nemours & Co, Inc, 68-69, sr engr, 69-70, sr res supvr reactor fuel & target elements, 70-75, chief supvr, 75-77, dept supt raw mat technol, 77-78, dept supt fuel reprocessing, Savannah River Plant, 78-79, sect dir solid waste technol, 79, supt employee rels, 80, mgr DMT Plant, Old Hickory, Tenn, 81-83, tech mgr polymer intermediates, 83-86, dir res & develop, Petrochem Dept, 86-89, DIR MFG COMPUTER SYSTS, DUPONT INFO SYSTS, E I DU PONT DE NEMOURS & CO, INC, WILMINGTON, DEL ,89- *Mem:* Am Inst Mining, Metall & Petrol Engrs; Sigma Xi. *Res:* Alloy development of nickel and cobalt base superalloys; sintering phenomena and surface tension of metals; irradiation stability of actinide oxide-aluminum systems; powder metallurgy; reprocessing of irradiated reactor fuel; polymer intermediates. *Mailing Add:* 435 Fox Meadow Lane West Chester PA 19380

GREGG, THOMAS G, US citizen, Dec 3, 31; m 56; c 4. GENETICS. *Educ:* Col Wooster, BA, 54; Univ Tex, MA, 56, PhD(genetics), 58. *Prof Exp:* NIH fel genetics, Univ Wis, 58-60; from asst prof to assoc prof, 60-71, PROF ZOOL, MIAMI UNIV, 71- *Concurrent Pos:* NSF grant, 60-62; NIH grants, 63-65 & 71-73, spec fel, Univ Tex, 67-68. *Mem:* Genetics Soc Am; Sigma Xi. *Res:* Drosophila genetics. *Mailing Add:* Dept Zool Miami Univ Oxford OH 45056

GREGO, NICHOLAS JOHN, b New York, NY, Feb 10, 45. PHYSIOLOGY, PHARMACOLOGY. *Educ:* Fairfield Univ, BS, 66; Adelphi Univ, MS, 68; Thomas Jefferson Univ, PhD(physiol), 74. *Prof Exp:* CLIN ASST PROF PHYSIOL & PHARMACOL, PHILADELPHIA COL OSTEOP MED, 72- *Mem:* AAAS; Sigma Xi. *Res:* Gastric secretion in cats; peripheral blood flow and osteopathic manipulative therapy. *Mailing Add:* 593 Wigard Ave Philadelphia PA 19128

GREGOR, CLUNIE BRYAN, b Edinburgh, Scotland, Mar 5, 29. GEOLOGY. *Educ:* Cambridge Univ, BA, 51, MA, 54; State Univ Utrecht, DSc(earth sci), 67. *Prof Exp:* From instr to assoc prof geol, 59-68, chmn dept, Am Univ Beirut, 67-68; res asst crystallog 64-65, head dept, Technol Univ Delft, 65-67; vis prof, Case Western Reserve Univ, 68-69; from assoc prof to prof, WGa Col, 69-72; PROF GEOL, WRIGHT STATE UNIV, 72- *Concurrent Pos:* Mem, Comt Natural Resources, Nat Coun Sci Lebanon, 67-68; Neth deleg, Comn Non-Metallic Minerals, Orgn Econ Coop & Develop, 68-69; consult, Food & Agr Orgn, UN, 69-; chmn, USA Work Group Geochemical Cycles, 72-; ed, Geochem News, 80-; vchmn panel Geochem Cycles, Nat Res Coun, Nat Acad Sci, 87- 90. *Mem:* Geol Soc Am; Am Geophys Union; Geochem Soc (secy, 83-89). *Res:* Geochem cycles of the elements; models of sedimentary cycle. *Mailing Add:* 136 W North College St Yellow Springs OH 45387

GREGOR, HARRY PAUL, b Minneapolis, Minn, Dec 16, 16; m 41; c 5. APPLIED CHEMISTRY. *Educ:* Univ Minn, BA, 39, PhD(phys chem), 45. *Prof Exp:* From asst prof to prof chem, Polytech Inst Brooklyn, 46-67; PROF CHEM ENG, COLUMBIA UNIV, 67- *Concurrent Pos:* Prof appl chem, 68-76. *Res:* Electrochemistry; colloids, including ion exchange; semipermeable membranes; ion binding by polyelectrolytes; multilayer electrodes; colloid and polymer chemistry; pure and applied studies on synthetic membranes, ultrafiltration and enzyme coupled membranes. *Mailing Add:* Dept Chem Eng & Appl Chem Columbia Univ Broadway & W 116 New York NY 10027

GREGORICH, DAVID TONY, b Crawford, Nebr, Feb 11, 37; div; c 4. INFRARED ASTRONOMY. *Educ:* Univ Calif, Los Angeles, BA, 62; Los Angeles State, MS, 64; Univ Calif, Riverside, PhD(physics), 68. *Prof Exp:* Jr engr, Consolidated Systs Corp, 60-62; assoc res physicist, Bell & Howell Res Labs, 62-64; res asst physics, Univ Calif, Riverside, 64-68; assoc prof, 68-77, PROF PHYSICS & ASTRON, CALIF STATE UNIV, LOS ANGELES, 77- *Concurrent Pos:* Consult, Bell & Howell Res Labs, 68-72; IRAS Sci Opers Team, Jet Propulsion Labs, 82-85; sci support staff, IPAC, Calif Inst Technol, 85- *Mem:* Am Phys Soc; Am Astron Soc; Sigma Xi. *Res:* Far-infrared astronomy. *Mailing Add:* 945 Swiss Trails Rd Duarte CA 91010-2181

GREGORY, A(LVIN) R(AY), b Gainesville, Tex, Nov 26, 15; div; c 5. RESERVOIR ENGINEERING, ROCK PHYSICS. *Educ:* Univ Tex, BS, 38. *Prof Exp:* Res engr prod eng & geophys, Phillips Petrol Co, 45-51; sr res engr petrophys, Gulf Res & Develop Co, 51-68, res assoc, Explor Dept, 68-73; res scientist assoc, 73-80, RES SCIENTIST & ENGR, UNIV TEX, AUSTIN, 80- *Concurrent Pos:* Radar proj officer, Radiation Lab, Mass Inst Technol, 44-45. *Mem:* Am Inst Mining, Metall & Petrol Engrs; Soc Explor Geophys. *Res:* Sonic and electrical well logging, propagation of acoustic energy in rocks; physical properties of rocks; fluid flow in porous media; geothermal energy from geopressured reservoirs on the Texas Gulf Coast. *Mailing Add:* 11515 Fast Horse Dr Austin TX 78759

GREGORY, ARTHUR ROBERT, b Binghamton, NY, Oct 19, 25; m 75; c 15. INHALATION TOXICOLOGY, ONCOLOGY. *Educ:* Cornell Univ, AB, 45; Univ NC, MSPH, 63, PhD(pub health), 65; Am Bd Tocixol, dipl, 84. *Prof Exp:* Asst toxicol, Med Ctr, Univ Calif, 65-66; physicist, Cutter Labs, 66-67; res prof physiol, Univ Calif, 67-69; lab mgr toxicol, NASA Labs, Houston, Tex, 69-72; sr scientist, Roswell Park Mem Inst, 72-73; dir biostatist, Erie County Health Dept, 73-76; pharmacologist, Nat Inst Occup Health, 77-80; pharmacologist, Consumer Prod Safety Comn, 80-87; PRES, TECHTO ENTERPRISES, 87- *Concurrent Pos:* Indust epidemiol, Buffalo Med Sch, 72-77; adv, chem selection working group, Nat Cancer Inst, NIH, 77-86; toxicologist, interagency testing comt, Toxic Substances Control Act, 80-84; state adv, US Cong Adv Bd, 82-86. *Mem:* Fel Am Inst Chemists; fel NY Acad Sci; Soc Toxicol; Asn Govt Toxicologists; Am Col Toxicologists; Int Soc Philos Inquiry; Am Indust Hyg Asn; Am Asn Cancer Res; Int Soc Regulatory Toxicol; Sigma Xi. *Res:* Toxicology of dyes; oncogenicity of benzidine and its dyes; lung edema-producing toxicant gases; effects of neonatal thymectomy on blood parameters; aerospace toxicology research on food additives and food colorants and maintenance of toxicology data base for the Food and Drug Administration. *Mailing Add:* PO Box 193 Sterling VA 22170

GREGORY, BOB LEE, b Allen, Okla, Sept 30, 38; m 61; c 3. ELECTRICAL ENGINEERING. *Educ:* Carnegie Inst Technol, BS, 60, MS, 61, PhD(elec eng), 64. *Prof Exp:* Supvr, Radiation Effects Div, Am Tel & Tel Co, 63-79, dir, microelectronics, Sandia Lab, 79-88; DIR MICROELECTRONICS, POLAROID, 88- *Concurrent Pos:* Mem sci adv bd, USAF; mem Jet Propulsion Lab microelectronics SAB. *Honors & Awards:* Centennial Award, Inst Elec & Electronics Engrs. *Mem:* Fel Inst Elec & Electronics Engrs. *Res:* Properties of radiation defects in materials and semiconductor devices; semiconductor devices; semiconductor device physics, integrated circuit technology; high power semi lasers; electronic imaging. *Mailing Add:* Microelectronics Org 2100 Polaroid Corp 21 Osborn St Cambridge MA 02139

GREGORY, BRIAN CHARLES, b Toronto, Ont, July 9, 38; m 84; c 2. PLASMA PHYSICS, ELECTRONIC PHYSICS. *Educ:* Univ Toronto, BASc, 60; Cambridge Univ, PhD(elec eng), 63. *Prof Exp:* Res engr, Compagnie Generale de Telegraphie sans Fil, France, 63-66; from asst prof to assoc prof physics, Trent Univ, 66-70; dir, 70-78, scientist, Ctr Nuclear Studies, Grenoble, 78-79, PROF, INST NAT RES SCI, UNIV QUE, 79-; RES DIR, CENTRE CANADIEN DE FUSION MAGNÉTIQUE, 88- *Mem:* Fel Am Phys Soc; assoc Can Asn Physicists; Inst Elec & Electronics Engrs; Nuclear & Plasma Sci Soc. *Res:* Electron beams and guns; microwave interaction with plasmas; plasmas and electron beams; physics of ionized gases; vacuum ultraviolet spectroscopy of tokamak plasmas; engineering physics; tokamak and fusion physics. *Mailing Add:* Inst Nat Res Sci Energy Univ Que CP 1020 Varennes PQ J0L 2P0 Can

GREGORY, BROOKE, b Boston, Mass, Apr 27, 41. LOW TEMPERATURE PHYSICS, ASTRONOMICAL INSTRUMENTATION. *Educ:* Amherst Col, BA, 63; Brown Univ, PhD(physics), 73. *Prof Exp:* Asst prof physics, Trinity Col, Conn, 71-80; ASSOC SUPPORT SCIENTIST, CERRO TOLOLO INT-AM OBSERV, 80- *Concurrent Pos:* Fulbright scholar & vis researcher, Bariloche Atomic Ctr, Nat Comn Atomic Energy, San Carlos de Bariloche, Arg, 75. *Mem:* Am Phys Soc. *Res:* Superconducting semiconductors; superconducting metrology with Josephson devices. *Mailing Add:* 950 N Clerry Ave Cerro Tololo Int-Am Observ Tucson AZ 85719

GREGORY, CLARENCE LESLIE, JR, b Stamford, Conn, Mar 14, 30. CHEMICAL ENGINEERING. *Educ:* Mass Inst Technol, BS, 51, ScD(chem eng), 57. *Prof Exp:* Res asst, Mass Inst Technol, 54-55; engr, Knolls Atomic Power Lab, Gen Elec Co, 57-61, engr supv eng, 61-65, proj engr, 65-80. *Mem:* Am Chem Soc. *Res:* Mass transfer; nucleate and film boiling, two-phase pressure drop, vapor fractions, and flow oscillations; thermal and hydraulic design of nuclear reactors; chemistry of high pressure, high temperature water. *Mailing Add:* 2112 Baker Ave E Schenectady NY 12309-2336

GREGORY, CONSTANTINE J, b Brockton, Mass, June 17, 39; m 62; c 3. ENVIRONMENTAL SCIENCES, AIR POLLUTION. *Educ:* Northeastern Univ, BA, 62; Rutgers Univ, MS, 64, PhD(environ sci), 68. *Prof Exp:* From asst prof to assoc prof environ sci, 67-76, assoc prof, 76-80, PROF CIVIL ENG, NORTHEASTERN UNIV, 80- *Concurrent Pos:* Mem bd dirs, New Eng Consortium on Air Pollution, 70- *Mem:* AAAS; Am Chem Soc; Air Pollution Control Asn. *Res:* Fate and effects of atmospheric pollutants. *Mailing Add:* Dept Civil Eng Northeastern Univ 360 Huntington Av Boston MA 02115

GREGORY, DALE R(OGERS), b Lake Village, Ark, Aug 1, 34; m 56; c 3. CHEMICAL ENGINEERING. *Educ:* Va Polytech Inst, BS, 56, PhD(chem eng), 66. *Prof Exp:* Prod supvr, Carbide & Carbon Chem Co, WVa, 56-57, 59-60; develop engr, E I du Pont de Nemours & Co, Inc, Va, 60-62; chem engr, Tenn Eastman Co, 65-66, sr chem engr, 66-67, sr res chem engr, 67-74, res assoc, 74-87, sr res assoc, 87-88, SR TECH ASSOC, TENN EASTMAN CO, 88- *Concurrent Pos:* Instr, Dept Chem & Metall Eng, Univ Tenn, 71- *Mem:* Am Inst Chem Engrs; Am Chem Soc; Am Soc Qual Control; Am Asn Textile Technol. *Res:* Fibers research and development; organic chemicals production; polymers engineering; melt spinning research. *Mailing Add:* Textile Fibers Div B-226 Tenn Eastman Co Kingsport TN 37660

GREGORY, DANIEL HAYES, b Watertown, NY, Dec 18, 33; m 60; c 4. INTERNAL MEDICINE, GASTROENTEROLOGY. *Educ:* Hamilton Col, AB, 57; Univ Va, MD, 62. *Prof Exp:* From instr to asst prof med, Univ Minn, 67-69; chief gastroenterol, Vet Admin Hosp, Albuquerque, NMex, 69-72; asst prof, Med Col Va, 72-73, assoc chmn, Dept Med, 76-78, assoc prof med, 73-78; assoc chief staff & dir res, McGuire Vet Admin Hosp, Richmond, chief med, 77-78; clin prof med, Univ WVa, Morgantown & Univ Pittsburgh, 78-91; SR ATTEND PHYSICIAN, DEPT MED, MARY IMOGENNE BASSETT HOSP, 91- *Concurrent Pos:* Fel med, Sch Med, Univ Minn, 63-66; fel gastroenterol, Minneapolis Vet Hosp, 66-67; assoc chief radioisotopes, Minneapolis Vet Admin Hosp, 67-69; asst prof med, Sch Med, Univ NMex, 69-72; asst chief gastroenterol, McGuire Vet Admin Hosp, Richmond, 72-78; sr attend physician & prog dir gastroenterol, Allegheny Gen Hosp, Pittsburgh, Pa, 78-; attend physician, North Hills Pesserant Hosp, Pittsburgh, Pa; bd dirs, Physicians Health Plan Pa. *Mem:* Am Gastroenterol Asn; Am Asn Study Liver Dis; Am Soc Gastrointestinal Endoscopy; Am Fedn Clin Res; Am Med Asn. *Res:* Biliary lipid metabolism. *Mailing Add:* Pt Vivian Alexandria Bay NY 13607

GREGORY, DONALD CLIFFORD, b Tyler, Tex, Sept 12, 49; m 71; c 2. ATOMIC PHYSICS. *Educ:* Univ Tex, Austin, BS, 71, MA, 73, PhD(physics), 76. *Prof Exp:* Exchange scientist chem, Ctr Study Molecular Struct, NSF & Hungarian Acad Sci, Budapest, 74-75; researcher physics & chem, Univ Tex, Austin, 75-76; res assoc physics, Joint Inst Lab Astrophys, Nat Bur Standards & Univ Colo, Boulder, 76-78; asst scientist physics, Brookhaven Nat Lab, 78-80; res assoc, 80-83, STAFF SCIENTIST PHYSICS, OAK RIDGE NAT LAB, 83- *Concurrent Pos:* Fel Welch Found, Ctr Struct Studies & Physics Dept, Univ Tex, Austin, 75-76; vis worker, Oak Ridge Nat Lab, 76-78. *Mem:* Am Phys Soc; Health Physics Soc. *Res:* Electron scattering from atoms and molecules (elastic), and from ions (inelastic); high energy ion beams for beam-foil-spectroscopy and ion-atom and ion-electron charge changing collisions. *Mailing Add:* Oak Ridge Nat Lab Bldg 6003 PO Box 2008 Oak Ridge TN 37831-6372

GREGORY, ERIC, b Golborne, Eng, Jan 5, 28; US citizen; m 56; c 1. METALLURGY, RESEARCH ADMINISTRATION. *Educ:* Cambridge Univ, BA, 48, MA, 52, PhD(metall), 54. *Prof Exp:* Res engr, Manganese Bronze Co, Eng, 54-56; tech dir, Sintercast Corp Am, 56-61, phys res, 61-68, dir metals res, Cent Res Labs, 68-72, dir corp res & develop, Cent Res & Develop Labs, 72-79, GEN MGR, AIRCO, INC, 79- *Mem:* Am Soc Metals; Am Inst Mining, Metall & Petrol Engrs; Am Ceramic Soc; Am Iron & Steel Inst; Brit Inst Metals. *Res:* High temperature alloys; powder metallurgy; high pressure, superconducting alloys; corrosion of stainless steels; materials produced by high vacuum electron beam processing. *Mailing Add:* 336 Sterling Rd Jefferson MA 01522

GREGORY, EUGENE MICHAEL, b Thirsk, Gt Brit, Mar 1, 45; US citizen. ENZYMOLOGY. *Educ:* Appalachian State Univ, BA, 67; Univ NC, Chapel Hill, PhD(biochem), 71. *Prof Exp:* Res assoc biochem, Med Ctr, Duke Univ, 72-74; asst prof, Ohio Univ, 74-75; asst prof, 75-80, ASSOC PROF BIOCHEM, VA POLYTECH INST & STATE UNIV, 80- *Mem:* Am Chem Soc; Am Soc Biol Chemists. *Res:* Mechanism of action of antioxidant enzymes and their physiological roles. *Mailing Add:* Dept Biochem Va State Univ Blacksburg VA 24060

GREGORY, FRANCIS JOSEPH, b Brooklyn, NY, June 21, 21; c 2. MICROBIOLOGY, BIOCHEMISTRY. *Educ:* Brooklyn Col, BA, 42; Rutgers Univ, PhD(microbiol), 54. *Prof Exp:* Res scientist, 55-81, mgr, Microbiol Dept, Wyeth Labs, Inc, 81-87; RETIRED. *Concurrent Pos:* Fel microbiol, Rutgers Univ, 54-55; dir res, Mushroom Growers Asn of Pa, 87- *Mem:* Am Asn Cancer Res; Tissue Cult Asn; Am Soc Parasitol; Am Soc Microbiol. *Res:* Cancer research; screening for antineoplastic agents; biological response modifiers; anti-metastatic spread; drugs that restore immune competence; cancer and immunity; in vivo assays for anaerobic infections; athymic mice; effect of drugs on experimental systemic lupus erythematosus; monobeta-lactams; quinolone anti-infectives; reverse transcriptase; cephalosporin antibiotics; enzyme inhibitors; basidiomiycete strain development; mushroom supplements; FIV research (feline AIDS). *Mailing Add:* 11 Cypress Lane Berwyn PA 19312

GREGORY, GAROLD FAY, b Arkansas City, Kans, Aug 15, 26; m 53; c 2. PHYTOPATHOLOGY. *Educ:* Kans State Univ, BS, 51; Iowa State Univ, MS, 56; Cornell Univ, PhD(plant path), 62. *Prof Exp:* Plant pathologist, 62-69, prin plant pathologist, 69-81, SUPVRY PLANT PATHOLOGIST, NORTHEASTERN FOREST EXP STA, US FOREST SERV, 81- *Mem:* Am Phytopath Soc; Int Soc Arboricult. *Res:* Research on chemical and biological control of tree diseases, particularly Dutch Elm disease; insect and disease problems of urban trees; microbiology and genetic manipulation to produce enhanced amounts of antibiotics by antigenistic organisms. *Mailing Add:* 11604 Mackel Dr Oklahoma City OK 73170

GREGORY, GARRY ALLEN, b London, Ont, July 27, 41; m 63; c 4. CHEMICAL ENGINEERING. *Educ:* Univ Waterloo, BASc, 64, MASc, 65, PhD(chem eng), 69. *Prof Exp:* From asst prof to prof chem eng, Univ Calgary, 68-83; vpres, 76-82, PRES, NEOTECHNOL CONSULTS LTD, 82- *Concurrent Pos:* Pres, Gregory Eng Ltd, 73- *Mem:* Can Soc Chem Eng; mem

Am Inst Chem Engrs; Soc Petrol Engrs; Petrol Soc Can; fel Chem Inst Can. *Res:* Multiphase gas-liquid flow in pipelines and producing wells. *Mailing Add:* Neotechnol Consult Ltd 510 1701 Centre St Nw Calgary AB T2E 8A4 Can

GREGORY, IAN (WALTER DE GRAVE), b London, Eng, July 14, 26; US citizen; m 50; c 4. PSYCHIATRY. *Educ:* Cambridge Univ, BA, 46, MBCh, 48, MA, 51; Univ Toronto, DPsych, 54; Cambridge Univ, MD, 56; Univ Mich, MPH, 59; FRCP(C). *Prof Exp:* Intern & res, Verdun Protestant Hosp, Montreal, 48-50; asst supt, Hollywood Sanitarium, BC, 50-52; instr psychiat, Univ Western Ont, 55-58; from asst prof to assoc prof, Univ Minn, 59-65; PROF PSYCHIAT & CHMN DEPT, COL MED, OHIO STATE UNIV, 65- *Concurrent Pos:* Psychiatrist, Ont Hosp, London, 54-58; consult psychiatrist, Carleton Col, 59-65; consult, Div Res Grants, NIH, 63-67, Vet Admin, 64- *Mem:* Fel Am Psychiat Asn; Can Psychiat Asn; fel Royal Col Psychiat. *Res:* Etiological research in psychiatry and abnormal psychology. *Mailing Add:* Dept Psychiat Ohio State Univ Hosps 473 W 12th Ave Columbus OH 43210

GREGORY, JESSE FORREST, III, b Columbus, Ohio, Aug 12, 50; m 76; c 2. FOOD CHEMISTRY, NUTRITION. *Educ:* Cornell Univ, BS, 72, MS, 75; Mich State Univ, PhD(food sci & human nutrit), 78. *Prof Exp:* Asst, Cornell Univ, 72-74, Mich State Univ, 74-77; from asst prof to assoc prof, 77-86, PROF FOOD CHEM, UNIV FLA, 86- *Honors & Awards:* Prescott Award Res, Inst Food Technologists, 83. *Mem:* Inst Food Technologists; Am Chem Soc; AAAS; Am Inst Nutrit; Soc Exp Biol & Med. *Res:* Nutritional quality and safety of foods; food chemistry and analysis; isotopic methods of vitamin research; stability, bioavailability, analysis of nutrients. *Mailing Add:* Dept Food Sci & Human Nutrit Univ Fla Gainesville FL 32611

GREGORY, JOHN, chemical engineering, for more information see previous edition

GREGORY, JOHN DELAFIELD, b New York, NY, May 18, 23; m 58; c 2. BIOCHEMISTRY. *Educ:* Yale Univ, BS, 44, PhD(org chem), 47. *Prof Exp:* Asst, Rockefeller Inst, 47-49; from asst biochemist to assoc biochemist, Mass Gen Hosp, 49-57; assoc prof biochem, Rockefeller Univ, 57-91; RETIRED. *Mem:* AAAS; Am Soc Biol Chem; Soc Complex Carbohydrates (pres, 76); Harvey Soc; Biochem Soc. *Res:* Sulfate metabolism; enzyme chemistry; glycosaminoglycans; connective tissues; structure of proteoglycans. *Mailing Add:* Rockefeller Univ New York NY 10021-6399

GREGORY, JOSEPH TRACY, b Eureka, Calif, July 28, 14; m 49; c 2. VERTEBRATE PALEONTOLOGY. *Educ:* Univ Calif, AB, 35, PhD(vert paleont), 38. *Prof Exp:* Lectr zool, Columbia Univ, 39; technician, Paleont Lab, Bur Econ Geol, Univ Tex, 39-41; instr geol, Univ Mich, 41-46; asst prof geol, Yale Univ, 46-52, assoc prof vert paleont, 52-60, cur vert paleont, Peabody Mus Nat Hist, 46-60; cur amphibians & reptiles Mus Paleont, 60-79, dir, 71-75, prof paleont, 60-79, EMER PROF PALEONT, UNIV CALIF, BERKELEY, 79- *Concurrent Pos:* Cur, Mus Paleont, Univ Mich, 41-46; asst ed, Am J Sci, 54-60; mem, US Nat Comt Hist Geol, Nat Asn Geol Teachers, 77-83; vis prof, Paleont Inst, Johannes-Gutenberg Univ Mainz, FRG, 67-68; ed, Bibliog Fossil Vert, 69-89. *Mem:* Fel Paleont Soc; Soc Vert Paleont (pres, 58); fel Geol Soc Am; Am Soc Mammal; Soc Study Evolution; Am Soc Zoologists; Hist Earth Sci Soc; Nat Asn Geol Teachers. *Res:* Fossil reptiles and amphibians; history of paleontology. *Mailing Add:* Mus Paleont Univ Calif Berkeley CA 94720

GREGORY, KEITH EDWARD, b Franklin, NC, Oct 27, 24; m 51; c 2. ANIMAL BREEDING. *Educ:* NC State Univ, BS, 47; Univ Nebr, MS, 49; Univ Mo, PhD(animal breeding), 51. *Prof Exp:* Res asst animal husb, Univ Nebr, 47-49; res asst, Univ Mo, 49-51; assoc prof, Auburn Univ, 51-55; animal geneticist & regional coordr, Beef Cattle Breeding Res, Agr Res Serv, 55-66, dir, US Meat Animal Res Ctr, US Dept Agr, 66-77, dir, Kans-Nebr area, Agr Res Serv, 72-77; res geneticist, US meat Animal Res Ctr, 77-84, supvr res geneticist & res leader prod systs, 84-87; RES GENETICIST, PROF ANIMAL SCI, UNIV NEBR, 87- *Concurrent Pos:* Vis prof, Tex A&M Univ, 77-78. *Honors & Awards:* Animal Breeding & Genetics Award, Am Soc Animal Sci, 67; Res Award, Polled Hereford Asn, 71; Res Award, Nat Cattlemens Asn, 85. *Mem:* Fel AAAS; fel Am Soc Animal Sci; Am Genetic Asn; Sigma Xi. *Res:* Beef cattle breeding research--heterosis, selection, heterosis retention, maternal effects, breed characterization, twinning. *Mailing Add:* US Meat Animal Res Ctr Clay Center NE 68933

GREGORY, KENNETH FOWLER, b Calgary, Alta, May 12, 26; m 53; c 3. MICROBIOLOGY. *Educ:* Univ BC, BSA, 47; Univ Wis, MSc, 49, PhD(bact), 51. *Prof Exp:* Asst bact, Univ Wis, 47-51, res assoc, 51-52; asst prof, Dalhousie Univ, 52-54; asst prof, Ont Agr Col, 54-56, assoc prof microbiol, 56-67; prof microbiol, Univ Guelph, 67-86, chmn, 81-86; RETIRED. *Mem:* Can Soc Microbiol. *Res:* Bacterial genetics; the genetic improvement of microorganisms for use as single-cell protein and the use of thermotolerant fungi as food and feed. *Mailing Add:* 156 Maple St Guelph ON N1G 2G7 Can

GREGORY, M DUANE, b Weatherford, Okla, June 24, 42; m 63; c 2. PHYSICAL CHEMISTRY. *Educ:* Southwestern State Col, Okla, BS, 64; Univ Okla, PhD(phys chem), 68. *Prof Exp:* SCIENTIST, RES & DEVELOP DEPT, CONTINENTAL OIL CO, 68- *Mem:* Am Chem Soc; Soc Petrol Eng. *Res:* Hydration of organic amines in nonaqueous solvents; surfactant waterflooding; resolution of oil in water dispersions. *Mailing Add:* 115 Orchard Lane Ponca City OK 74604

GREGORY, MAX EDWIN, b Yorkville, Tenn, Jan 14, 31; m 54; c 3. FOOD SCIENCE. *Educ:* Univ Tenn, BS, 53; NC State Col, MS, 56, PhD(dairy mfg), 59. *Prof Exp:* Dairy prod specialist, Ohio State Univ, 59-62; dairy prod specialist, 62-80, EXTEN PROF FOOD SCI, NC STATE UNIV, 80- *Res:* Dairy products. *Mailing Add:* 4133 White Pine Dr Raleigh NC 27650

GREGORY, MICHAEL BAIRD, b Sacramento, Calif, Apr 15, 44; div; c 2. MATHEMATICAL ANALYSIS. *Educ:* Univ Conn, BA, 66, MS, 67, PhD(math), 71. *Prof Exp:* PROF MATH, UNIV NDAK, 71- *Concurrent Pos:* Vis prof math, Mich State Univ, 87-88. *Mem:* Am Math Soc; Math Asn Am. *Res:* Classical analysis, including real functions, measure and integration; complex analysis. *Mailing Add:* Univ NDak Grand Forks ND 58202

GREGORY, MICHAEL VLADIMIR, b Shanghai, China, Jan 31, 45; US citizen; m 68; c 2. REACTOR PHYSICS, THERMAL-HYDRAULIC METHODS. *Educ:* Columbia Univ, AB, 66, BS, 67; Mass Inst Technol, PhD(nuclear eng), 73. *Prof Exp:* Res asst nuclear eng, Brookhaven Nat Lab, 67; nuclear engr, US Army Engrs Reactor Group, 69-71; res asst reactor physics, Mass Inst Technol, 72-73; staff analyst reactor physics, 73-90, RES ASSOC NUCLEAR ENG, SAVANNAH RIVER LAB, E I DU PONT DE NEMOURS & CO, 90-; SR ADV ENGR, WESTINGHOUSE SAVANNAH RIVER CO, 90- *Concurrent Pos:* Vis scientist, Risoe Nat Lab, Denmark, 79; chmn math & comp div, Am Nuclear Soc, 84-85. *Mem:* Am Nuclear Soc; Soc Computer Simulation. *Res:* Development of digital computation methods for reactor physics, thermal-hydraulic and safety analyses; applications of neutron diffusion theory; neutron resonance theory; large scale data handling techniques; reactor simulator software. *Mailing Add:* Savannah River Lab Westinghouse Savannah River Co Aiken SC 29808-0001

GREGORY, NORMAN WAYNE, b Albany, Ore, June 23, 20; m 43; c 3. PHYSICAL CHEMISTRY, THERMODYNAMICS. *Educ:* Univ Wash, Seattle, BS, 40, MS, 41; Ohio State Univ, PhD(chem), 43. *Prof Exp:* Chemist, Radiation Lab, Univ Calif, 44-46; from instr to prof chem, 46-89, chmn, Dept Chem, 70-75, EMER PROF CHEM, UNIV WASH, 89- *Mem:* Am Chem Soc; Sigma Xi. *Res:* Molecular composition of vapors; thermodynamics of metal halide systems; crystal structure; vaporization reactions. *Mailing Add:* Dept Chem 8610 Univ Wash Seattle WA 98195

GREGORY, PETER, b London, UK, Apr 11, 47; m 72; c 2. PLANT BIOCHEMISTRY, AGRICULTURE. *Educ:* Univ London, UK, BSc, 69, PhD(plant biochem), 72; London Col Music, LLCM, 72. *Prof Exp:* Assoc, Dept Bot, Imp Col, Univ London, 72; assoc, Dept Biochem Molecular & Cell Biol, Cornell Univ, 72-74, res assoc, 74-75, from asst prof to assoc prof biochem, 75-85, asst dir, Agr Exp Sta, 80-82; DIR RES, INT POTATO CTR, LIMA, PERU, 85- *Concurrent Pos:* Adj assoc prof, Cornell Univ, 86-89, adj prof, 89- *Mem:* Europ Potato Asn; Potato Asn Am. *Res:* Research management in international agriculture; specifically studies on pest and disease resistance in crops and on natural toxicants in potatoes; development of sustainable agriculture for the developing world; application of plant biochemistry to problems in agriculture. *Mailing Add:* Int Potato Ctr (CIP) Apartado 5969 Lima Peru

GREGORY, R(OBERT) LEE, b Mishawaka, Ind, Oct 22, 36. CONTROL SYSTEMS. *Educ:* Purdue Univ, BSME, 59, MSME, 61; Univ Notre Dame, PhD(mech eng), 70. *Prof Exp:* Test engr, Studebaker-Packard Corp, 60-61; eng analyst, Energy Controls Div, 61-69, staff engr, Bendix Res Labs, 70-80, MEM TECH STAFF, BENDIX AEROSPACE TECHNOL CTR, ALLIED-SIGNAL CORP, 80- *Mem:* Am Soc Mech Engrs; Nat Soc Prof Engrs. *Res:* Design and analysis of multivariable control systems. *Mailing Add:* Ohio Mach Co 3993 E Royalton Rd Broadview Heights OH 44147

GREGORY, RAYMOND (LESLIE), internal medicine; deceased, see previous edition for last biography

GREGORY, RICHARD ALAN, JR, b Ayer, Mass, June 2, 43; m 68; c 3. SEPARATION SCIENCE. *Educ:* Drexel Univ, BS, 66; Princeton Univ, MA, 68, PhD(chem eng), 70. *Prof Exp:* Sr engr, Union Carbide Corp, 80-74, proj scientists, 74-76, group leader, 76-79, sr group leader, 79-81, ASSOC DIR, UNION CARBIDE CORP, 81- *Mem:* Am Inst Chem Engrs; Sigmi Xi. *Res:* Separation systems including cyclicial absorption processes and gas-liquid and liquid-liquid absorption with emphasis on acid gas removal. *Mailing Add:* Union Carbide Corp PO Box 670 Bound Brook NJ 08805

GREGORY, RICHARD WALLACE, b Chicago, Ill, Sept 28, 36; m 57; c 2. FISHERIES. *Educ:* Colo State Univ, BS, 58, PhD(fish mgt), 69; Univ Wash, MS, 62. *Prof Exp:* Fish biologist, Colo Game, Fish & Parks Dept, 63, wildlife researcher cand, 63-64, asst wildlife researcher, 64-66, wildlife researcher, 66-69; asst prof zool & asst leader Coop Fisheries Unit, Univ Maine, Orono, 69-74; assoc prof biol & leader Coop Fish Res Unit, Mont State Univ, 74-79; prof wildlife ecol & leader Coop Fish & Wildlife Res Unit, 79-84, chief Info Transfer Section, 84-87, deputy chief, Off Info Transfer, 87-88, CHIEF, OFF INFO TRANSFER, UNIV FLA, 88- *Concurrent Pos:* Prin investr, Decker Coal Co, US Forest Serv, US Fish & Wildlife Serv. *Mem:* Am Fisheries Soc (pres elect, 90-91, pres, 91-92); Am Inst Fishery Res Biol. *Res:* Biological and economic value of wetlands; limnology of Florida lakes; wetland habitat requirements of gallinules and river otters; impacts of rock mining on fish and wildlife; impacts of exotic fishes on native fishes; cryo preservation of fish sperm; habitat use by bass in heavily vegetated Florida lakes. *Mailing Add:* Off Info Tranfer USFWS 1025 Pennock Pl Ste 212 Ft Collins CO 80524

GREGORY, ROBERT AARON, b Hudson Falls, NY. PLANT PHYSIOLOGY, BOTANY. *Educ:* Cornell Univ, BS, 52; Yale Univ, MF, 54; Ore State Univ, PhD(bot), 68. *Prof Exp:* Res forester, Pac Northwest Forest & Range Exp Sta, Southeast Alaska, 54-58, proj leader, 58-68, res plant physiologist, Northeastern Forest Exp Sta, Forest Physiol Lab, Beltsville, Md, 68-74, res plant physiologist, Northeastern Forest Exp Sta, USDA, Burlington, Vt, 74-88, proj leader, 82-88; RETIRED. *Concurrent Pos:* Adj assoc prof, George D Aiken Sch Natural Resources, Univ Vt, 81- *Mem:* AAAS; Int Asn Wood Anatomists. *Res:* Xylem translocation; circulation, storage and mobilization of assimilates; effects of stress on growth, development and physiological processes in trees. *Mailing Add:* Ferry Rd PO Box 173 Charlotte VT 05445

GREGORY, STEPHEN ALBERT, b Paris, Ill, Nov 2, 48; div; c 2. ASTROPHYSICS. *Educ:* Univ Ill, BS, 70; Univ Ariz, PhD(astron), 74. *Prof Exp:* Asst prof astron, State Univ NY Col Oswego, 73-77; asst prof physics, Bowling Green State Univ, 77-; ASSOC PROF DEPT PHYSICS & ASTRON, UNIV NMEX, ALBUQUERQUE, 89- *Mem:* Am Astron Soc; Am Inst Physics. *Res:* Dynamical and static properties of galaxy clusters; present emphasis is on the structure and evolution of galaxy superclusters and their environments. *Mailing Add:* Dept Physics & Astron Univ Nmex Albuquerque MN 87131

GREGORY, THOMAS BRADFORD, b Traverse City, Mich, Dec 13, 44. LIE ALGEBRAS. *Educ:* Oberlin Col, AB, 67; Yale Univ, MA, 69, MPhil, 75, PhD(math), 77. *Prof Exp:* Head, Systs Design Br, Mgt Info Ctr, US Naval Commun Command, 69-72; lectr math & comput sci, 77-78, asst prof math, 78-84, ASSOC PROF MATH, OHIO STATE UNIV, MANSFIELD, 84- *Concurrent Pos:* Translator, Am Math Soc, 74-82; exec officer, commanding officer & eng duty officer, US Naval Reserve, 79- *Mem:* Am Math Soc; Am Soc Naval Engrs. *Res:* Classification of the simple finite-dimensional Lie algebras over algebraically closed fields of prime characteristic, in extending from the restricted case to the general case. *Mailing Add:* 930 Maumee Ave Mansfield OH 44906-2909

GREGORY, WESLEY WRIGHT, JR, b Camden, SC, Sept 9, 42; m 65; c 2. ECONOMIC ENTOMOLOGY. *Educ:* Wofford Col, BS, 64; Clemson Univ, MS, 66, PhD(entom, environ health & bot), 69. *Prof Exp:* Res asst invert zool, Clemson Univ, 65-66; asst prof entom, Univ Ky, 69-72, assoc prof, 72-80. *Mem:* AAAS; Entom Soc Am. *Res:* Bioaccumulation and transferal of pesticide residues in biological systems; development of insect management systems for soil and foliar pests of corn and vegetable crops, arthropod ecology. *Mailing Add:* 1632 Linstead Drive Lexington KY 40504

GREIBACH, SHEILA ADELE, b New York, NY, Oct 6, 39; c 1. COMPUTER SCIENCE. *Educ:* Radcliffe Col, AB, 60, AM, 62; Harvard Univ, PhD(appl math), 63. *Prof Exp:* Lectr appl math, Harvard Univ, 63-65, asst prof, 65-69; from assoc prof to prof comput sci, 69-85, VCHMN COMPUT SCI, UNIV CALIF, LOS ANGELES, 85- *Concurrent Pos:* Consult, Systs Develop Corp, 64-70. *Mem:* Am Math Soc; Asn Comput Math; Inst Elec & Electronics Engrs Computer Soc. *Res:* Theoretical computer science in general with emphasis on formal languages and computational complexity. *Mailing Add:* Dept Comput 3731 Boelter Hall Univ Calif Los Angeles CA 90024

GREICHUS, YVONNE A, pesticide chemistry, toxicology, for more information see previous edition

GREIDER, KENNETH RANDOLPH, clifford algebras & particle physics; deceased, see previous edition for last biography

GREIDER, MARIE HELEN, b Newark, Ohio, Jan 15, 22. CYTOLOGY. *Educ:* Ohio State Univ, BSc, 49, MSc, 55, PhD(zool), 60. *Prof Exp:* Asst, Ohio State Univ, 51-60, res assoc, 60-64, asst prof, 64-68; asst prof, 68-70, ASSOC PROF PATH, WASH UNIV, 70- *Mem:* AAAS; Am Soc Cell Biol; Am Asn Anat; NY Acad Sci; Electron Micros Soc Am. *Res:* Electron microscopy; cytochemistry; endocrinology. *Mailing Add:* 5516 Lancaster Rd Hebron OH 43025

GREIF, MORTIMER, b New York, NY, Aug 13, 26; m 48; c 2. SURFACE CHEMISTRY, POLYMER CHEMISTRY. *Educ:* City Col New York, BS, 47; Univ Ky, MS, 49; Polytech Inst Brooklyn, PhD(phys chem), 53. *Prof Exp:* Staff chemist phys res, Fabrics & Finishes Div, E I du Pont de Nemours & Co, 52-63; tech dir, 63-78, ASST GEN MGR, STAHL FINISH CO, 78- *Mem:* Am Chem Soc; AAAS; Am Leather Chemists Asn. *Res:* Alternating current polarography; film formation; emulsion paints; pigment dispersions; leather finishes. *Mailing Add:* 20 Ocean Ave Unit 4 Marblehead MA 01945-3614

GREIF, RALPH, b New York, NY, Nov 28, 35; m 58; c 3. MECHANICAL ENGINEERING, HEAT & MASS TRANSFER. *Educ:* NY Univ, BS, 56; Univ Calif, Los Angeles, MS, 58; Harvard Univ, MA & PhD(eng), 62. *Prof Exp:* Mem tech staff, Hughes Res & Develop Labs, 56-58, Raytheon Res Labs, 58; res fel gas dynamics, Harvard Univ, 62-63; from asst prof to assoc prof mech eng, 63-73, vchmn dept, 74-76, PROF MECH ENG, UNIV CALIF, BERKELEY, 73- *Concurrent Pos:* Consult var industs, 61-; Guggenheim fel, 69-70; vis scholar, Imp Col Sci & Technol, London, 69-70; vis prof & Lady Davis fel, Technion, Israel Inst Technol, 77; chmn Heat Transfer Div, Am Soc Mech Engrs, 70-73, computer technol policy comt, 75-79, assoc tech ed, J Heat Transfer, 83- 89, tech comt aerospace heat transfer, 67- *Honors & Awards:* Heat Transfer Mem Award, Am Soc Mech Engrs, 85. *Mem:* Am Inst Aeronaut & Astronaut; fel Am Soc Mech Engrs. *Res:* Thermal radiation; natural convection; combustion; unsteady heat transfer; two phase flow; solar collection; rotating flows; chemical vapor deposition. *Mailing Add:* Dept Mech Eng Etcheverry Hall Univ Calif Berkeley Berkeley CA 94720

GREIF, ROBERT, b New York, NY, Jan 17, 38; m 63; c 2. APPLIED MECHANICS, MECHANICAL ENGINEERING. *Educ:* NY Univ, BME, 58; Harvard Univ, SM, 59, PhD(appl mech), 63. *Prof Exp:* Staff scientist, Missile Systs Div, Avco Corp, 63-65, sr staff scientist, 65-67; from asst prof to assoc prof, 67-78, chmn dept, 81-89, PROF MECH ENG, TUFTS UNIV, 78- *Concurrent Pos:* Consult, Beverly Res Lab, USM Corp, 67-68, Stone & Webster Eng Corp, 71-78 & Transp Systs Ctr, US Dept Transp, 77-; vis res fel, Univ Sussex, 74; vis scholar, Harvard Univ, 81; NRC sr res assoc, NASA, Langley, 88. *Mem:* Am Soc Mech Engrs; assoc fel Am Inst Aeronaut & Astronaut; Am Soc Eng Educ; Am Asn Univ Professors. *Res:* Vibration and dynamics; stress analysis; applied mechanics; fracture mechanics; rail mechanics. *Mailing Add:* Dept Mech Eng Tufts Univ Medford MA 02155

GREIF, ROGER LOUIS, b Baltimore, Md, Aug 23, 16; m 50, 73; c 2. ANIMAL PHYSIOLOGY. *Educ:* Haverford Col, BS, 37; Johns Hopkins Univ, MD, 41. *Prof Exp:* Intern med, Johns Hopkins Hosp, 41-42; asst resident, Lakeside Hosp, 42-43; asst physician, Rockefeller Inst Hosp, 47-53; from asst prof to assoc prof physiol, 53-65; prof, 65-82, EMER PROF PHYSIOL, MED COL, CORNELL UNIV, 82- *Concurrent Pos:* Fel, Johns Hopkins Hosp, 46-47; asst physician outpatients, NY Hosp, 53-60; consult, Metab Sect, Health Res Coun & NY, 84; vis investr, Univ Aix-Marseille, France, 65-66 & 76; vis prof, Nat Defense Med Ctr & Med Sch, Nat Taiwan Univ, 82; mem bd dirs, Am Mem Hosp, Reims, France, 84. *Mem:* Soc Exp Biol & Med; Am Physiol Soc; Endocrine Soc; Am Thyroid Asn; Soc Gen Physiol. *Res:* Thyroid and endocrine physiology; hormone-enzyme relationships. *Mailing Add:* Dept Physiol Cornell Univ Med Col New York NY 10021-4885

GREIFER, AARON PHILIP, b Passaic, NJ, Sept 29, 19; m 43; c 2. PHYSICAL INORGANIC CHEMISTRY, MAGNETISM. *Educ:* Ohio State Univ, BA, 42; Columbia Univ, MA, 48; Am Inst Chem, cert, 71. *Prof Exp:* Chemist, Elwood Ord Plant, Ill, 42-44; res assoc cryogenics, Res Found, Ohio State Univ, 44-45; chemist, Kellex Corp, NY, 48-49 & Fed Tel & Radio Corp, 51; scientist, Gen Elec Co, 51-56; sr scientist, Radio Corp Am, 56-61; proj chemist, Clevite Corp, 61-64; staff chemist, Univac Div, Sperry Rand Corp, 64-67, mgr ferrite core res, 67-71, staff scientist, 71-81; RETIRED. *Concurrent Pos:* Jr scientist, Los Alamos Sci Lab, Calif; lectr, Villanova Univ, 84- *Honors & Awards:* Cert Appreciation, Corps Engrs, Manhattan Dist, War Dept, 45. *Mem:* Am Chem Soc; Am Phys Soc; fel Am Inst Chem. *Res:* Magnetic materials; crystal growth; piezomagnetic ferrites; low loss ferrites; memory cores; material and process development for bubble memory; LPE growth of garnet films; thin film circuits. *Mailing Add:* 51 E Golf View Rd Ardmore PA 19003

GREIFF, DONALD, b Toronto, Ont, Aug 20, 15; nat US; m 42; c 2. CRYOBIOLOGY. *Educ:* Marquette Univ, BS, 38; Johns Hopkins Univ, ScD(genetics), 42. *Prof Exp:* From instr to prof biol, St Louis Univ, 42-57; PROF PATH, MED COL WIS, 57-, ASSOC DEAN GRAD AFFAIRS, 68- *Concurrent Pos:* Mem Comn C1, Int Inst Refrig. *Mem:* Asn Am Med Cols; affil Royal Soc Med; Soc Cryobiol (pres, 67). *Res:* Effects of enzyme inhibitors and activators on the multiplication of viruses and Rickettsiae; freezing and freeze-drying of biologic materials; studies on freeze-dried biologic materials. *Mailing Add:* Dept Path Med Col Wis 8701 Watertown Plank Rd Milwaukee WI 53226

GREIFINGER, CARL, b Poland, Apr 19, 26; nat US; m 47; c 3. THEORETICAL PHYSICS. *Educ:* Cornell Univ, AB, 48, PhD(physics), 54. *Prof Exp:* Instr physics, Univ Pa, 53-55; asst prof, Univ Southern Calif, 55-58; physicist, Rand Corp, 58-71; physicist, R&D Assocs, 71-91; RETIRED. *Concurrent Pos:* Consult, Hughes Aircraft Co, 55-58. *Mem:* Am Geophys Union. *Res:* Theoretical nuclear physics; magneto-hydrodynamics; ionospheric physics; low frequency electromagnetics; e-m propagation ocean electromagnetics and magnetic surveillance. *Mailing Add:* 16948 Dulce Ynez Lane Pacific Palisades CA 90272

GREIG, DOUGLAS RICHARD, b Oakland, Calif, Oct 26, 50; m 77. ANALYTICAL CHEMISTRY, COMPUTER SCIENCE. *Educ:* Univ Calif, Berkeley, BS, 72; Northwestern Univ, PhD(chem), 76. *Prof Exp:* asst prof chem, Lake Forest Col, 77-80; mem staff, Varian Assoc, 80-; AT NELSON ANALYTICAL. *Concurrent Pos:* Consult, Northwestern Univ, 77- *Mem:* Am Chem Soc; Sigma Xi. *Res:* Computer control of chemical instrumentation; Raman spectroscopy. *Mailing Add:* PE Nelson 10040 Bubb Rd Cupertino CA 95014-4132

GREIG, J ROBERT, b Maidenhead, Eng, Apr 12, 38; m 64; c 3. PLASMA PHYSICS. *Educ:* Univ London, BSc & ARCS, 59, PhD(physics) & DIC, 65. *Prof Exp:* Res officer, Cent Electricity Res Lab, Eng, 62-65; asst prof physics, Univ Md, 65-70; res physicist, Imp Chem Industs, Ltd, Eng, 70-73; res physicist, Naval Res Lab, Washington, DC, 73-86; SR SCIENTIST, GT-DEVICES, ALEXANDRIA, VA, 86- *Mem:* Am Phys Soc; Inst Physics London. *Res:* Plasma spectroscopy and diagnostic methods used in plasma physics; laser produced plasmas and intense relativistic electron beams; electrothermal guns and interior ballistics. *Mailing Add:* 103 Northway Rd Greenbelt MD 20770

GREIG, JAMES KIBLER, JR, b Van Buren, Ark, Apr 9, 23; m 47; c 3. HORTICULTURE. *Educ:* Univ Ark, MS, 50; Kans State Univ, PhD(agron), 60. *Prof Exp:* Asst, Univ Ark, 49-50, instr & jr horticulturist, 50-52; from asst prof & asst olericulturist to assoc prof & assoc olericulturist, 52-70, PROF HORT & OLERICULTURIST, KANS STATE UNIV, 70- *Mem:* Fel Am Soc Hort Sci; Weed Sci Soc Am; Sigma Xi. *Res:* Vegetable culture, nutrition and physiology. *Mailing Add:* 1728 Little Kitten Ave Manhattan KS 66502

GREIM, BARBARA ANN, b Philadelphia, Pa. ALGEBRA, COMPUTER SCIENCE EDUCATION. *Educ:* Ursinus Col, BS, 64; Univ NC, Chapel Hill, PhD(math), 70. *Prof Exp:* ASSOC PROF MATH SCI, UNIV NC, WILMINGTON, 69- *Mem:* Math Asn Am; Asn Comput Mach. *Res:* Semigroup rings; use of computers in scientific education. *Mailing Add:* Dept Math Univ NC PO 601 S College Rd Wilmington NC 28403

GREIN, FRIEDRICH, b Freudenberg am Main, Ger, Dec 22, 29; m 59; c 3. QUANTUM CHEMISTRY. *Educ:* Univ Gottingen, 54, MSc, 58; Univ Frankfurt, PhD(chem physics), 60. *Prof Exp:* Fel quantum chem, Univ NB, 60-61 & Nat Res Coun Can, 61-62; asst prof physics, 62-63, from asst prof to assoc prof chem, 63-72, PROF CHEM, UNIV NB, 72- *Concurrent Pos:* Vis prof, Uppsala Univ, 68-69, Univ Fla, 76 & Univ Bonn, 83; fel, Chem Inst Can, 81. *Mem:* Can Asn Physicists; Chem Inst Can. *Res:* Quantum theoretical studies on electronic structure and spectra of diatomic and polyatomic molecules; development of multi-configuration methods for ground and excited states of molecular systems. *Mailing Add:* Dept Chem Univ NB Bag Serv No 45222 Fredericton NB E3B 6E2 Can

GREINER, DALE L, CELLULAR IMMUNOLOGY, AUTOIMMUNITY. *Educ:* Univ Iowa, PhD(microbiol), 78. *Prof Exp:* ASST PROF IMMUNOL, UNIV CONN HEALTH CTR, 79- *Mailing Add:* Dept Path Univ Conn Health Ctr Farmington CT 06032

GREINER, DOUGLAS EARL, b Los Angeles, Calif, Jan 29, 39; m 61; c 3. EXPERIMENTAL NUCLEAR PHYSICS, COSMIC RAY PHYSICS. *Educ:* Univ Calif, BA, 61, PhD(physics), 64. *Prof Exp:* Physicist space physics, Space Sci Lab, 64-74, PHYSICIST EXP NUCLEAR PHYSICS, LAWRENCE BERKELEY LAB, UNIV CALIF, 74- *Mem:* Am Phys Soc; Am Geophys Union; Sigma Xi. *Res:* Study of relativistic heavy ion reactions; fragmentation cross sections; search for abnormal matter, multiparticle inclusive reactions; isotopic abundances of the cosmic rays. *Mailing Add:* 483 Western Dr Richmond CA 94801

GREINER, NORMAN ROY, b Muskegon, Mich, Aug 30, 38; m 64; c 2. CHEMICAL KINETICS, LASERS. *Educ:* St Mary's Univ, Tex, BS, 60; Univ Tex, Austin, PhD(phys chem), 64. *Prof Exp:* Sect leader chem, 64-74, alt group leader laser chem, 76-77, ASST GROUP LEADER CHEM LASERS, LOS ALAMOS SCI LAB, 74- *Mem:* AAAS; Am Chem Soc. *Res:* Gas kinetics; photochemistry; air pollution control; reactions of internally excited molecules; reactions initiated by flash photolysis; reactions of excited noble gas atoms; chemistry of planetary atmospheres; chemical lasers; laser isotope separation; environmental chemistry of natural radioisotopes. *Mailing Add:* Nine Loma Vista Los Alamos NM 87544-3066

GREINER, PETER CHARLES, b Budapest, Hungary, Nov 1, 38; Can citizen; m 65, 89; c 3. MATHEMATICS. *Educ:* Univ BC, BSc, 60; Yale Univ, MA, 62, PhD(math), 64. *Prof Exp:* Instr math, Princeton Univ, 64-65; asst prof, 65-70, assoc prof, 70-77, PROF MATH, UNIV TORONTO, 77- *Concurrent Pos:* Mem, Inst Advan Study, 73-74; vis prof, Univ Paris, 80-81; mem, Max-Planck Inst, Bonn Ger, 87-88. *Honors & Awards:* Steacie Prize in Natural Sci, Can, 77. *Mem:* Am Math Soc; Can Math Soc; Fel Royal Soc Can. *Res:* Partial differential equations; functional analysis. *Mailing Add:* Dept Math Univ Toronto Toronto ON M5S 1A1 Can

GREINER, RICHARD A(NTON), b Milwaukee, Wis, Feb 13, 31. ELECTRICAL ENGINEERING. *Educ:* Univ Wis, BS, 54, MS, 55, PhD(elec eng), 57. *Prof Exp:* From asst prof to assoc prof, 57-63, PROF ELEC ENG, UNIV WIS-MADISON, 63- *Mem:* Audio Eng Soc; Inst Elec & Electronics Engrs. *Res:* Solid state devices, circuits and electroacoustics. *Mailing Add:* Dept Elec Eng Univ Wis Madison WI 53706

GREINER, RICHARD WILLIAM, b New York, NY, Feb 24, 32; m 49; c 2. PHYSICAL CHEMISTRY, POLYMER CHEMISTRY. *Educ:* Bucknell Univ, BS, 53; Univ Wis, PhD(org chem), 57. *Prof Exp:* Res chemist, Hercules Powder Co, 57-69, sr res chemist, 69-76, res scientist, 76-83, res assoc, res ctr, Hercules Inc, 83-90; ADJ PROF, UNIV MINN, 90- *Mem:* Am Chem Soc; Sigma Xi. *Res:* Colloid and surface chemistry; organic polymer chemistry; chemistry of free radicals; physical organic chemistry. *Mailing Add:* 48 Clifton Dr Kennett Sq PA 19348

GREINKE, EVERETT D, b Elmhurst, Ill, Oct 31, 29; m 51; c 3. SPACE ELECTRONICS. *Educ:* Northern Ill Univ, BS, 51, MS, 56. *Prof Exp:* Asst br head res & develop, Photo Div Bur Naval Weapons, Navy Dept, 56-61, tech adv data proc, Tech Anal & Adv Group, Dep Chief Naval Opers, 56-65, asst dir command control, 65-67, staff specialist, 67-75, asst dir combat support, 75-77, dir combat support, 77-80, dir NATO/Europ affairs, defense res & eng, 80-82, actg dep undersecy defense, Int Progs, 82, sci adv Sageur Shape/NATO, 82-86, DEP UNDERSECY DEFENSE, INT PROGS, 86- *Mem:* Am Inst Chem. *Res:* Development, testing, evaluation and engineering programs in the fields of command and control, navigation, electronic warfare, reconnaissance and intelligence; armaments cooperation, international scientific cooperation. *Mailing Add:* 8315 Toll House Rd Annandale VA 22003

GREINKE, RONALD ALFRED, b Mt Prospect, Ill, Aug 20, 35; m 67; c 2. ANALYTICAL CHEMISTRY. *Educ:* Univ Ill, BS, 63; Univ Mich, MS, 65, PhD(anal chem), 67. *Prof Exp:* Anal chemist, Chem & Plastics Div, 67-69, res scientist, Carbon Prod Div, 69-70, head anal res & develop, 71-75, RES ASSOC, CARBON PROD DIV, UNION CARBIDE CORP, 75- *Mem:* Am Chem Soc. *Res:* Kinetics in analytical chemistry; gas chromatography; thermal analysis; carbon chemistry; environmental polycyclic hydrocarbon analysis; mesophase chemistry; carbonization kinetics and mechanisms; graphite intercalation chemistry. *Mailing Add:* Union Carbide Corp PO Box 6116 Cleveland OH 44101

GREISEN, ERIC WINSLOW, b Los Alamos, NMex, Apr 14, 44; m 67. RADIO ASTRONOMY. *Educ:* Cornell Univ, AB, 66; Calif Inst Technol, PhD(astron), 73. *Prof Exp:* Res assoc radio astron, 72-74, asst scientist, 74-76, ASSOC STAFF SCIENTIST RADIO ASTRON, NAT RADIO ASTRON OBSERV, ASSOC UNIVS, INC, 76- *Mem:* Am Astron Soc. *Res:* Radio spectral-line interferometry of interstellar neutral hydrogen including galactic structure, interstellar medium, HII region, and high velocity cloud questions. *Mailing Add:* 208 Georgetown Rd Charlottesville VA 22901

GREISEN, KENNETH I, b Perth Amboy, NJ, Jan 24, 18; m 41, 76; c 2. COSMIC RAYS. *Educ:* Franklin & Marshall Col, BS, 38; Cornell Univ, PhD(physics), 42. *Prof Exp:* Asst physics, Cornell Univ, 38-42, instr, 42-43; mem staff, Manhattan Proj, AEC, Los Alamos Sci Lab, 43-45, group leader, 45-46; from asst prof to assoc prof, 46-50, prof astron, 75-79, chmn astron dept, 76-79, prof physics, 50-84, dean fac, 78-83, EMER PROF PHYSICS, CORNELL UNIV, 84- *Concurrent Pos:* Adj prof physics, Univ Utah, 75- *Mem:* Nat Acad Sci; Am Phys Soc; Am Astron Soc; Int Astron Union. *Res:* Cosmic rays; nuclear physics; high energy astrophysics. *Mailing Add:* 336 Forest Home Dr Ithaca NY 14850

GREISMAN, SHELDON EDWARD, b New York, NY, Jan 24, 28; m 57; c 5. PHYSIOLOGY, MEDICINE. *Educ:* NY Univ, MD, 49. *Prof Exp:* From instr to assoc prof, 54-72, PROF MED, SCH MED, UNIV MD, BALTIMORE CITY, 72-, PROF PHYSIOL, 76- *Concurrent Pos:* Assoc mem comn epidemiol surv, Armed Forces Epidemiol Bd, 64- *Mem:* Asn Am Physicians; Am Clin & Climat Asn; Am Soc Clin Invest; Soc Exp Biol & Med. *Res:* Mechanisms of tolerance to bacterial endotoxins in man; role of bacterial endotoxins in human gram-negative bacterial infections. *Mailing Add:* Univ Md Sch Med Lombard & Greene Sts Baltimore MD 21201

GREISS, FRANK C, JR, b Philadelphia, Pa, July 13, 28; m 53; c 4. OBSTETRICS & GYNECOLOGY. *Educ:* Univ Pa, BA, 49, MD, 53. *Prof Exp:* From instr to assoc prof, 60-70, PROF OBSTET & GYNEC, BOWMAN GRAY SCH MED, WAKE FOREST UNIV, 70- *Honors & Awards:* Found Prize Thesis, Am Asn Obstet & Gynec, 68. *Mem:* Fel Am Col Obstet & Gynec; AMA; Am Gynec Soc; Am Asn Obstetricians & Gynecologists; Soc Gynec Invest; Sigma Xi. *Res:* Obstetric and fetal physiology; uterine blood flow during pregnancy; infertility. *Mailing Add:* Rte 6 Box 586 Mooresville NC 28115

GREIST, JOHN HOWARD, b Shoals, Ind, Feb 1, 06; m 36; c 3. PSYCHIATRY. *Educ:* DePauw Univ, AB, 26; Ind Univ, MD, 29. *Prof Exp:* From instr to asst clin prof, 30-69, assoc prof, 69-75, clin prof psychiat, Sch Med, Ind Univ, Indianapolis, 75-; AT DEPT PSYCHIAT, UNIV WIS. *Concurrent Pos:* Fel, Sch Med, Johns Hopkins Univ; mem, Asn of Consult to Surgeons Gen, Dept Defense; pres, Ind Ment Health Found. *Mem:* AMA; fel Am Psychiat Asn. *Res:* Psychotherapy; electroencephalography. *Mailing Add:* Ind Univ Methodist Hosp 4343 Washington Blvd Indianapolis IN 46205

GREITZER, EDWARD MARC, b New York, NY, May 8, 41; m 66; c 2. AERONAUTICAL ENGINEERING, MECHANICAL ENGINEERING. *Educ:* Harvard Col, BA, 62; Harvard Univ, MS, 64, PhD(eng), 70. *Prof Exp:* Res asst, Div Eng & Appl Physics, Harvard Univ, 66-69; res engr, Pratt & Whitney Aircraft, 69-76; indust fel commoner, Cambridge Univ, 75; sr res engr, United Technol Res Ctr, 76-77; from asst prof to assoc prof, 77-84, PROF & DIR, GAS TURBINE LAB, MASS INST TECHNOL, 84-, H N SLATER PROF AERONAUT/ASTRONAUT, 88- *Concurrent Pos:* Consult, United Technol Res Ctr, 77-, Cummins Engine Co, 78- & Sundstrand Corp, 81-; Freeman scholar award fluids eng, Am Soc Mech Engrs, 80; guest fel, Royal Soc; overseas fel, Churchill Col, Cambridge Univ, 83-84; vis fel, Peterhouse, Cambridge Univ, 90-91 & Japan Soc Promotion Sci, 87-; mem, Aeronaut Adv Comt, NASA, 90- *Honors & Awards:* Gas Turbine Power Award, Am Soc Mech Engrs, 77 & 79; T Bernard Hall Prize, Inst Mech Engrs, 78. *Mem:* Fel Am Soc Mech Engrs; Sigma Xi; Am Inst Aeronaut & Astronaut. *Res:* Internal fluid dynamics; fluid mechanics and thermodynamics of gas turbine engines; propulsion; flow control. *Mailing Add:* Dept Aeronaut & Astronaut Mass Inst Technol 77 Massachusetts Ave Cambridge MA 02139

GREIZERSTEIN, HEBE BEATRIZ, b Buenos Aires, Arg; US citizen. POLYMER COATINGS, BIOCHEMICAL PHARMACOLOGY. *Educ:* Univ Buenos Aires, Lic, 58, PhD(org chem), 61. *Prof Exp:* Res assoc pharmacol, State Univ NY Buffalo, 69-70; sr res scientist, 70-73, assoc res scientist pharmacol, Res Inst Alcoholism, NY State Dept Ment Hyg, 73-75, res scientist V, 75-79, actg dir, 79-80, res scientist VII, 80-83; PRES, E B ASSOC LABS, INC, 83-; DIR ANALYTICAL LAB, TOXICOL RES CTR, STATE UNIV NY, BUFFALO, 88- *Concurrent Pos:* Res instr pharmacol, State Univ NY Buffalo, 73-75, res asst prof, 75-; consult, Inst Child Health & Human Develop, 74-75; res grant, Nat Inst Alcohol Abuse & Alcoholism, 74-76 & Am Heart Asn, 81-84. *Mem:* AAAS; Am Chem Soc; NY Acad Sci; Am Soc Pharmacol & Exp Therapeut. *Res:* Research administration; adaptions of state-of-the-art instrumentation to the analysis of industrial products; management information systems, data bases and computer communications. *Mailing Add:* 72 Brandywine Dr Williamsville NY 14221

GREIZERSTEIN, WALTER, b Buenos Aires, Arg, Sept 30, 35; US citizen; m 60; c 3. ORGANIC CHEMISTRY. *Educ:* Univ Buenos Aires, PhD(org chem), 60; State Univ NY, MBA, 75. *Prof Exp:* Res fel org chem, Univ Buenos Aires, 60-61; Petrol Res Fund fel, Brown Univ, 61-62; res chemist, Rohm and Haas Co, 62-68, lab mgr indust finishes, 68-75, dir int develop, 75-78, TECH DIR, PIERCE & STEVENS CHEM CORP, 78-; AT PRATT & LAMBERT INC. *Mem:* Am Chem Soc. *Res:* Electronic and steric effects in nucleophilic substitutions; London Forces; correlations of structure and properties in polymers; reactions of ligands in organometallic complexes. *Mailing Add:* Pratt & Lambert Specialty Pro Box 1505 Buffalo NY 14240-0022

GREK, BORIS, b Vinperk, Czechoslovakia, Feb 2, 46; US citizen; m 69; c 3. LASER DEVELOPMENT, VERY LOW LIGHT LEVEL DETECTION. *Educ:* Royal Mil Col Can, BSc, 69; Princeton Univ, PhD(physics). *Prof Exp:* Res physicist laser development, Defense Res Estab, 72-75; prof plasma physics, Inst Nat Res Sci Can, 75-80; RES PHYSICIST, PRINCETON PLASMA PHYSICS, PRINCETON UNIV, 80- *Concurrent Pos:* Vis prof, Princeton Plasma Physics Lab, PRinceton Univ, 78-80; invited prof, Univ Quebec, Montreal, 82- *Mem:* Am Phys Soc. *Res:* Laser development; plasma physics and tokamak transport; non-linear plasma physics; Thomson scattering. *Mailing Add:* Princeton Univ PO Box 451 Princeton NJ 08544

GRELECKI, CHESTER, b Newton Twp, Pa, June 22, 27; m 50; c 7. PHYSICAL CHEMISTRY, CHEMICAL ENGINEERING. *Educ:* King's Col, Pa, BS, 50; Duquesne Univ, MS, 52; Catholic Univ Am, PhD(chem), 57. *Prof Exp:* Res chem, Reaction Motors Div, Olin Matheson Corp, 56-59; sr supvr, Reaction Motors Div, Thiokol Chem Corp, 59-67, res dir, 67-69; PRES & CHIEF SCIENTIST CHEM, HAZARDS RES CORP, 69- *Mem:* Am Chem Soc; Am Soc Testing & Mat. *Res:* Combustion and explosion phenomenon; fire and explosion hazards evaluation; shock waves and blast effects. *Mailing Add:* 141 Halsey Ave Rockaway NJ 07866-3008

GRELEN, HAROLD EUGENE, b Bryan, Tex, Nov 13, 29; m 53; c 3. RANGE SCIENCE. *Educ:* Agr & Mech Col Tex, BS, 52, MS, 56. *Prof Exp:* Range conservationist, US Soil Conserv Serv, 54-55; range scientist, US Forest Serv, 56-85; RETIRED. *Mem:* Soc Range Mgt. *Res:* Range ecology; plant taxonomy; effects of prescribed burning and grazing on southern pine range. *Mailing Add:* 607 Edgewood Dr Pineville LA 71360

GRELLER, ANDREW M, b New York City, NY, Mar 18, 41; m 64; c 2. BIOCLIMATOLOGY, CONSERVATION. *Educ:* City Col, New York, 62; Columbia Univ, MA, 64, PhD(bot), 67. *Prof Exp:* From asst prof to assoc prof, 67-86, PROF BOT, QUEENS COL, NY, 86- *Concurrent Pos:* Collabr, Rocky Mt Nat Park, US Nat Park Serv, 70-72; vis asst prof, Univ Colo, 71, Univ Calif, Davis, 74; vis scientist, Am Mus Natural Hist, 76-80; Fulbright sr lectr, Dept Bot, Univ Peradeniya, Sri Lanka, 80-81; vis assoc prof, Univ Peradiniya, 85. *Mem:* Torrey Bot Club (pres, 88); Bot Soc Am; Int Soc Plant Morphologists; Sigma Xi. *Res:* Cape ecology, classification, climatology and conservation of tropical Asian forests and eastern North American forests. *Mailing Add:* Dept Biol Queens Col Flushing NY 11367

GREMINGER, GEORGE KING, JR, b Syracuse, NY, Feb 4, 16; m 46, 85; c 8. PULP CHEMISTRY, PAPER CHEMISTRY. *Educ:* State Univ NY, BS, 38. *Prof Exp:* With Mead Corp, Ohio, 37-38; Dow chem group leader, Dow Chem Co, 38-57, sect head tech serv & develop, 57-65, chem develop & serv specialist, 65-66 & Switz, 66-68, develop assoc, Designed Prod Dept, 68-72, assoc scientist, 72-86; RETIRED. *Concurrent Pos:* sr consult scientist, Omni Tech Int Ltd, Midland, Mich. *Mem:* Am Chem Soc; Tech Asn Pulp & Paper Indust; AAAS. *Res:* Water soluble gums, especially cellulose ethers. *Mailing Add:* 802 W Larkin Midland MI 48640

GRENANDER, ULF, b Vastervik, Sweden, July 23, 23; US citizen; m 46; c 3. APPLIED MATHEMATICS, PATTERN THEORY. *Educ:* Univ Stockholm, Fil Dr, 50. *Prof Exp:* Docent math statist, Univ Stockholm, 50-51 & 54-57, prof, 59-66; asst prof statist, Univ Chicago, 51-52; assoc prof, Univ Calif, 53; prof, 57-58 & 66-69, L HERBERT BALLOU UNIV PROF APPL MATH, BROWN UNIV, 69- *Mem:* Fel Inst Math Statist; Royal Swedish Acad Sci; Int Statist Inst; hon fel Royal Statist Soc (London). *Res:* Probability; theoretical statistics; operations research and insurance mathematics; pattern theory. *Mailing Add:* 26 Barberry Hill Providence RI 02912

GRENCHIK, RAYMOND THOMAS, b Whiting, Ind, Aug 24, 22; m 57; c 4. ASTROPHYSICS. *Educ:* St Procopius Col, BA, 43; Univ NMex, MS, 49; Ind Univ, PhD(astrophys), 56. *Prof Exp:* Lab instr elec & electronics, Signal Corps Sch, Chicago, 43, jr physicist, Instrument Sect, Metall Lab, 43-46; instr physics, Univ NMex, 46-50, Sault Br, Mich Tech Univ, 50-52 & Vanderbilt Univ, 55-57; asst prof, La State Univ, Baton Rouge, 57-61, assoc prof physics & astron, 61-88; RETIRED. *Concurrent Pos:* Mem comt educ astron, 66-72; consult, La Arts & Sci Ctr, 73-75. *Mem:* Am Astron Soc; AAAS. *Res:* Stellar atmospheres; radiative and convective transport; solar system physics. *Mailing Add:* 5960 Menlo Dr Baton Rouge LA 70808-5049

GRENDA, STANLEY C, b Chicago, Ill, Aug 12, 34. INORGANIC CHEMISTRY. *Educ:* DePaul Univ, BS, 58; Univ Ariz, MS, 62; Lehigh Univ, PhD(chem), 64. *Prof Exp:* Asst prof chem, Wis State Univ, Superior, 64-65, Whitewater, 65-67; chmn dept chem, 72-73, asst prof, 67-70, ASSOC PROF CHEM, UNIV NEV, LAS VEGAS, 70 - *Concurrent Pos:* Treas, Boulder Dam Sect, Am Chem Soc, 84. *Mem:* Am Chem Soc. *Res:* Educational research in qualitative inorganic chemistry; organometallic chemistry closely related to Grignard reagents; molybdenum chemistry. *Mailing Add:* Dept Chem Univ Nev Las Vegas NV 89109

GRENDA, VICTOR J, b Boston, Mass, May 18, 33. ORGANIC CHEMISTRY. *Educ:* Northeastern Univ, BS, 55; Mass Inst Technol, PhD(org chem), 60. *Prof Exp:* Sr res chemist, Merck Sharp & Dohme Res Labs, 60-66, sect leader, 66-72, sr proj coordr, proj planning & mgt, 72-77, dir process res, 77-84, sr dir, 84-90; RETIRED. *Mem:* Am Chem Soc; Chem Soc London. *Res:* Research and development of chemical processes; project management and planning. *Mailing Add:* 15 Loriann Rd Warren NJ 07059-5444

GRENDER, GORDON CONRAD, geology, for more information see previous edition

GRENFELL, RAYMOND FREDERIC, b West Bridgewater, Pa, Nov 23, 17; m 44; c 4. INTERNAL MEDICINE. *Educ:* Univ Pittsburgh, BS, 39, MD, 41. *Prof Exp:* Intern, Western Pa Hosp, 41-42; clin instr, 55-59, head Hypertension Clin, 56-79, CLIN ASST PROF MED, SCH MED, UNIV MISS, 59-, VIS PROF, 77- *Concurrent Pos:* Mem staff, Hinds Gen Hosp, Riverside Hosp, St Dominic-Jackson Mem Hosp, Miss Baptist Hosp & Doctor's Hosp; pvt pract, 46- *Honors & Awards:* Bronze Medal, Am Heart Asn, 63, Silver Medal, 65. *Mem:* Fel Am Col Angiol; fel Int Col Angiol; fel Am Col Chest Physicians; Am Soc Clin Pharmacol & Therapeut (vpres, 76); Am Fedn Clin Res. *Res:* Investigation of antihypertensive drugs. *Mailing Add:* 514 H-E Woodrow Wilson Jackson MS 39216

GRENGA, HELEN E(VA), b Newnan, Ga, Apr 11, 38. PHYSICAL CHEMISTRY & METALLURGY. *Educ:* Shorter Col, Ga, BA, 60; Univ Va, PhD(phys chem), 67. *Prof Exp:* Fel, 67, from asst prof to assoc prof, 68-77, PROF METALL, SCH CHEM ENG, GA INST TECHNOL, 77-, PROF CHEM ENG & ACTG ASSOC DEAN, GRAD DIV, 80- *Concurrent Pos:* NSF res grants, 69-70 & 71- *Mem:* Am Soc Metals; Am Inst Mining, Metall & Petrol Engrs; Sigma Xi. *Res:* Chemistry and physics of solid surfaces; structure-property correlations of gas-solid and solid-solid interfaces; oxidation; corrosion; catalysis; extractive metallurgy; field-ion and emission microscopy. *Mailing Add:* Grad Studies Ga Inst Tech Atlanta GA 30332

GRENIER, CLAUDE GEORGES, b Les Rousses, France, Feb 24, 23; m 53; c 2. SOLID STATE PHYSICS. *Educ:* Sorbonne, DSc, 56. *Prof Exp:* Res assoc, 56-57, from asst prof physics to assoc prof physics, 57-65, PROF PHYSICS & ASTRON, LA STATE UNIV, BATON ROUGE, 65- *Res:* Low temperature physics; electron transport phenomena. *Mailing Add:* Dept Physics La State Univ Baton Rouge LA 70803

GRENNEY, WILLIAM JAMES, b Saginaw, Mich, Aug 10, 37; m 59; c 3. CIVIL ENGINEERING. *Educ:* Mich Technol Univ, BS, 60; Ore State Univ, MS, 69, PhD(civil eng), 72. *Prof Exp:* Proj engr civil eng, Bethlehem Steel Corp, 60-62; consult engr, Consult Engrs, 67-68; asst prof, Calif State Polytech Univ, 71-72; res engr water res, Utah Water Res Lab, 72-74; assoc prof civil eng, 74-76, div head, environ eng, 76-77, PROF & DEPT HEAD CIVIL & ENVIRON ENG, UTAH STATE UNIV, 77-; WATER QUALITY SYSTS DIR, US FISH & WILDLIFE SERV, 78- *Concurrent Pos:* Pres, Intermountain Consult & Planners, 74-79; water quality systs dir, US Fish & Wildlife Serv, 78-79. *Honors & Awards:* Outstanding Res Award, Col Eng, Utah State Univ, 76. *Mem:* AAAS; Am Soc Civil Engrs; Am Water Res Asn; Water Pollution Control Fedn; Asn Environ Eng Profs; Sigma Xi. *Res:* Mathematical simulation and stochastic modeling of the chemical and biological responses of aquatic systems. *Mailing Add:* 1716 E 1600 N Logan UT 84321

GRENS, EDWARD A(NTHONY), II, chemical engineering, mass & heat transfer, for more information see previous edition

GRENVIK, AKE N A, b Sunne, Sweden, July 10, 29; m 52; c 4. CRITICAL CARE MEDICINE. *Educ:* Karolinska Inst, Sweden, MD, 56; Univ Uppsala, Sweden, PhD, 66. *Prof Exp:* Asst prof surg, Univ Uppsala, Sweden, 66-68; vis assoc prof, 68-70, assoc prof, 70-74, PROF ANESTHESIOL, SCH MED, UNIV PITTSBURGH, 74-, PROF SURG, 81-, PROF MED, 90- *Concurrent Pos:* Prin investr grants, Swedish Asn Against Heart & Lung Dis, 64-66 & NIH; med dir, Intensive Care Unit, Presby-Univ Hosp & Critical Care Med Training Prog, Univ Pittsburgh Med Ctr, 73-; ed, Clins Critical Care Med, 77-89, co-ed Textbk Critical Care, 85-; co-ed Contemp Mgt Critical Care, 89- *Mem:* Soc Critical Care Med (secy, 75-76, pres, 77-78); World Fedn Soc Intensive Critical Care Med (treas, 81-85); Am Col Chest Physicians; Am Soc Anesthesiologists; AMA. *Res:* Pathophysiology and management of cardiac, respiratory and brain failure; intensive care unit design and organization; critical care medicine education and certification; brain death, foregoing life sustaining therapy, ethical problems in organ donation and transplantation; author or coauthor of over 300 publications. *Mailing Add:* Dept Anesthesiol Univ Pittsburgh Sch Med Pittsburgh PA 15260

GRESHAM, ROBERT MARION, b Akron, Ohio, Aug 7, 43; m 66; c 2. MATERIALS SCIENCE ENGINEERING. *Educ:* Emory Univ, BS, 65, PhD(chem), 69. *Prof Exp:* Sr res chemist res & develop, DuPont Co, 80-81; lab mgr res & develop, 81-82, tech dir, 82-84, VPRES TECHNOL RES & DEVELOP, E M CORP, 84- *Concurrent Pos:* Indust coun coordr, Soc Tribologists & Lubrication Engrs, 88-, chmn solid lubricants tech comt, 84-88, chmn aerospace indust coun, 84-88; mem Asn Finishing Processes, Soc Mfg Engrs. *Mem:* Am Soc Testing & Mat; Armed Forces Commun & Electronics Asn; Soc Advan Mat & Process Engrs; Nat Lubricating Grease Inst; Soc Mfg Engrs; Soc Tribologists & Lubrication Engrs. *Res:* Design and development of solid lubrication materials primarily used in the aerospace industry; materials for shielding electromagnetic radiation. *Mailing Add:* E M Corp 2801 Kent Ave Box 2400 West Lafayette IN 47906

GRESIK, EDWARD WILLIAM, b Chicago, Ill, Oct 14, 39. ANATOMY. *Educ:* Xavier Univ, Ohio, BA, 61; Univ Ill, MS, 66, PhD(anat), 68. *Prof Exp:* Instr anat, State Univ NY Downstate Med Ctr, 67-74; assoc, Mt Sinai Sch Med, 74-76, from asst prof to assoc prof anat, 76-90; PROF CELL BIOL & ANAT, MED SCH, CITY UNIV NY, 90- *Mem:* Histochem Soc; Am Asn Anatomists; Sigma Xi; AAAS. *Res:* Cell fine structure; histochemistry; developmental biology; influence of drugs and hormones on the fine structure of the developing and mature salivary glands; growth factors. *Mailing Add:* Dept Cell Biol & Anat Sci City Univ NY Med Sch New York NY 10031

GRESKOVICH, CHARLES DAVID, b Fredericktown, Pa, June 13, 42; m 66; c 3. CERAMICS. *Educ:* Pa State Univ, BS, 64, MS, 66, PhD(ceramic sci), 68. *Prof Exp:* STAFF SCIENTIST CERAMICS, CORP RES & DEVELOP, GEN ELEC CO, 69- *Concurrent Pos:* NSF fel, 68. *Honors & Awards:* Ross Coffin Purdy Award Ceramic Lit, Am Ceramic Soc, 78; Richard M Fulrath Award, 83; Dushman Award, 86. *Mem:* Am Ceramic Soc. *Res:* Preparation and processing of ionic and covalent polycrystalline ceramics, and the effect of these on microstructural development and optical, magnetic and mechanical properties; ceramic optical elements, microwave ferrites, ceramics for turbine engines and cores/molds for casting alloys; x-ray scintillators for modern x-ray detectors are of major interest. *Mailing Add:* 1229 Viewmont Dr Schenectady NY 12301

GRESS, MARY EDITH, b Pensacola, Fla, July 11, 46. RESEARCH ADMINISTRATION, PHYSICAL CHEMISTRY. *Educ:* Sweet Briar Col, AB, 68; Iowa State Univ, PhD(phys chem), 73. *Prof Exp:* Res chemist, E I du Pont de Nemours & Co, 73-74; res assoc, Brookhaven Nat Lab, 74-77; vis asst prof phys chem, Univ Vt, 77-79; asst chemist, Ames Lab, Iowa State Univ, 79-82; STAFF MEM, DIV CHEM SCI, OFF BASIC ENERGY SCI, US DEPT ENERGY, 80- *Mem:* Am Chem Soc; Am Crystallographic Asn; AAAS. *Res:* Crystal structure determination by X-ray and neutron diffraction. *Mailing Add:* Div Chem Sci ER 141 US Dept Energy Washington DC 20585

GRESS, RONALD E, b Genoa, Nebr, Dec 5, 49. CELLULAR IMMUNOLOGY. *Educ:* Baylor Univ, MD, 75. *Prof Exp:* SR INVESTR, IMMUNOL BR, NIH, 83- *Mem:* Am Asn Immunologists. *Mailing Add:* Exp Immunol Br Bldg 10 Rm 41317 NCI NIH Bethesda MD 20892

GRESSEL, JONATHAN BEN, b Cleveland, Ohio, Oct 30, 36; m 58; c 4. DEVELOPMENTAL BIOLOGY, BIOCHEMISTRY. *Educ:* Ohio State Univ, BSc, 57; Univ Wis, MSc, 59, PhD(bot), 63. *Prof Exp:* Jr scientist, 62-67, res assoc, 67-68 & 69-70, sr scientist, 70-79, assoc prof, 79-85, PROF BIOL & GILBERT DE BOTTON PROF CHAIR PLANT SCI, WEIZMANN INST SCI, 85- *Concurrent Pos:* Consult, biotechnol chem co; Oxford Surveys, Plant Molecular & Cell biol; vis res fel, Res Sch Biol Sci, Australian Nat Univ, 74-75; vis scientist, Metab & Radiation Res Lab, USDA, ND, 81; co-vchmn,

Gordon Conf Agr Sci, 91- *Mem:* Am Soc Plant Physiologists; hon fel Weed Sci Soc Am; Int Plant Tissue Cult Asn; Sigma Xi; Am Chem Soc; Int Soc Plant Molecular Biol. *Res:* Herbicide resistance and metabolism; secondary metabolite biosyntheses metabolically synergizing chemical and mycoherbicides. *Mailing Add:* Dept Plant Genetics Weizmann Inst Sci Rehovot Israel

GRESSER, ION, b New York, NY, Oct 25, 28; m 68; c 2. VIROLOGY, EXPERIMENTAL PATHOLOGY. *Educ:* Harvard Univ, BA, 48; Yale Univ, MD, 55; Am Bd Microbiol, dipl. *Prof Exp:* Res assoc virol, Childrens Cancer Res Found, Boston, Mass, 59-65; CHIEF LAB VIRAL ONCOL, INST SCI RES CANCER, FRANCE, 65- *Concurrent Pos:* USPHS spec fel, 61-62 & career develop award, Childrens Cancer Res Found, 62-65. *Honors & Awards:* Leon Etancelin Prize, French Acad Sci, 70; Jean-Louis Camus Prize, 72; Avery-Landsteiner Award, German Immunol Soc, 83; Silver Medal, Nat Sci Res Ctr, 85; Delahautemaison Prize, Med Res Found, France, 88; Antoine Lacassagne Prize, French Nat League Against Cancer, 88. *Mem:* Soc Exp Biol & Med; Am Soc Microbiol; Am Asn Immunologists; Am Asn Immunol; French Soc Microbiol; Soc Interferon Res. *Res:* Inhibitory effect of interferon on growth of viral induced and transplantable tumors in experimental animals; effect of interferon on cellular physiology; ecology of Japanese encephalitis virus; pathogenesis of myxoviral infection. *Mailing Add:* Viral Oncol Inst Sci Res Cancer Boite Postale eight Villejuif 94801 France

GRESSER, MICHAEL JOSEPH, b Booneville, Mo, May 6, 45; m 68. BIOENERGETICS, PHYSICAL ORGANIC CHEMISTRY. *Educ:* Univ Kans, BA, 67; Brandeis Univ, PhD(biochem), 76. *Prof Exp:* Fel, Univ Calif, Los Angeles, 76-80; asst prof, 80-84, assoc prof biochem, Simon Fraser Univ, 84-88, DIR BIOCHEM, MERCK FROSST CTR THERAPEUT RES, 88- *Mem:* Am Chem Soc; Am Soc Biochemists; Can Biochem Soc; Can Soc Chem. *Res:* Energy transducing mechanisms in biological systems; enzymic and nonenzymic acyl and phosphoryl transfer reaction mechanisms; regulatory mechanisms; biochemistry of vanadium. *Mailing Add:* Dept Biochem Merck Frosst Canada Inc PO Box 1005 Pointe Claire-Dorval PQ V5A 1S6 Can

GREST, GARY STEPHEN, b New Orleans, La, Nov 22, 49; m 70; c 4. CONDENSED MATTER PHYSICS. *Educ:* La State Univ, BS, 71, MS, 73, PhD(physics), 74. *Prof Exp:* Res asst, Rutgers Univ, 74-77; Chaim Weizmann fel & res asst, James Franck Inst, Univ Chicago, 77-78; asst prof physics, Purdue Univ, 79-81; staff physicist, 81-84 SR STAFF PHYSICIST, EXXON, 84- *Concurrent Pos:* Alfred P Sloan Found fel, 81. *Mem:* Am Phys Soc. *Res:* Problems in condensed matter and many-body physics including spin-glasses, complex fluids and polymers, structural and dynamical properities of glasses; molecular dynamics and Monte Carlo simulation techniques. *Mailing Add:* Exxon Res & Eng Co Rte 22 E Annadale NJ 08801

GRETHE, GUENTER, b Hannover, Ger, Oct 13, 33; US citizen; m 60; c 2. COMPUTER APPLICATIONS IN SYNTHESIS, DATA BASE MANAGEMENT SYSTEMS IN CHEMISTRY. *Educ:* Braunschweig Tech Univ, Dipl chem, 60, PhD(org chem), 61. *Prof Exp:* Fel chem, Univ Wis, 62-63; sr res chemist, Chem Res Div, Hoffman-La Roche, Inc, 63-71, res fel, 72-80, res group chief, 80-85; DIR, SCI APPLNS, MOLECULAR DESIGN LTD, 85- *Concurrent Pos:* Adv, NY Univ, 74-77. *Mem:* Am Chem Soc; Ger Chem Soc. *Res:* Synthetic work in the fields of tetracyclines, alkaloids and heterocyclic compounds; drug design, SAR; computer assisted synthesis; computers in chemistry; reaction indexing. *Mailing Add:* 352 Channing Way Alameda CA 94501

GRETSKY, NEIL E, b Boston, Mass, Mar 17, 41; div. MATHEMATICS. *Educ:* Calif Inst Technol, BS, 62; Carnegie Inst Technol, MS, 64, PhD(math), 67. *Prof Exp:* Asst prof, 67-74, ASSOC PROF MATH, UNIV CALIF, RIVERSIDE, 74- *Concurrent Pos:* Vis, Dept Math, Univ Calif, Los Angeles, 70-71, Univ Calif, Berkeley, 75-76, Univ Ill, Urbana, 78-79; consult, Mathematica, Inc, 74-75; consult, 75- *Mem:* Am Math Soc; Math Asn Am; Soc Indust & Appl Math; Asn Comput Mach. *Res:* Functional analysis; operator representation theorems on Banach function spaces; vector measures; mathematical economics. *Mailing Add:* Univ Calif Riverside CA 92521

GRETTIE, DONALD POMEROY, b Salem, Ore, June 23, 00; m 30; c 2. ORGANIC CHEMISTRY. *Educ:* Willamette Univ, BA, 24; Univ Ore, MA, 27; Univ Pittsburgh, PhD(chem), 30. *Prof Exp:* Teacher pub sch, 24-25; asst, Univ Ore, 25-27 & Univ Pittsburgh, 27-29; res chemist, Swift & Co, 29-65; clin instr, Univ Ore Med Sch, 65-70; sr chemist, United Med Labs, Portland, 70-73; res assoc, Med Sch, Univ Ore Health Sci Ctr, 73-80; RETIRED. *Mem:* Am Chem Soc. *Res:* Adsorption; vitamin C; chemistry of fats and oils; chemistry of collagenous proteins; endocrinology; insulin and glucose tolerance tests; radioimmunoassay of hormones and drugs; relating polyamine levels in body fluids and tissues with physiological diseases, particularly cancer; a cancer detection method by radioimmunoassay for polyamines. *Mailing Add:* 16465 S W King Charles Tigard OR 97224

GREUB, LOUIS JOHN, b Humbird, Wis, Feb 18, 33; m 65; c 2. AGRONOMY. *Educ:* Wis State Univ, River Falls, BS, 63; Iowa State Univ, MS, 66, PhD(crop prod, plant physiol), 68. *Prof Exp:* Res asst agron, Iowa State Univ, 63-66, assoc, 66-68; assoc prof plant sci, 68-77, PROF AGRON, UNIV WIS-RIVER FALLS, 77- *Mem:* Am Soc Agron; Coun Agr Sci & Technol. *Res:* Crop physiology, especially forage physiology and management; leaf area, yield, carbohydrate reserves and photosynthesis of forage legumes; forage quality. *Mailing Add:* Dept Plant & Earth Sci Univ Wis-River Falls River Falls WI 54022

GREUER, RUDOLF E A, b Guetzlaffshagen, Germany, Apr 6, 27; m 63; c 2. MINING ENGINEERING. *Educ:* Clausthal Tech Univ, Dipl Ing, 53, Dr Ing, 55. *Prof Exp:* Res student mine ventilation, Univ Witwatersrand, 56; lectr mining eng, Tech Univ Istanbul, 57; res engr, Mining Res Estab, WGermany, 57-67; PROF MINING ENG, MICH TECHNOL UNIV, 67-, CHMN DEPT, 79- *Concurrent Pos:* Mem subcomts fires in mineshafts & mine ventilation, Europ Community Coal & Steel, 60-67. *Mem:* Am Inst Mining, Metall & Petrol Engrs. *Res:* Thermodynamics of mine ventilation; computer simulation of ventilation systems. *Mailing Add:* Dept Mining Eng & Petrol Mich Technol Univ Houghton MI 49931

GREULICH, RICHARD CURTICE, b Denver, Colo, Mar 22, 28; m 58; c 4. ANATOMY, GERONTOLOGY. *Educ:* Stanford Univ, AB, 49; McGill Univ, PhD(anat), 53. *Prof Exp:* From instr to prof anat, Sch Med, Univ Calif, Los Angeles, 53-66, from assoc prof to prof oral biol, Sch Dent, 61-66; sci dir intramural res, Nat Inst Dent Res, 66-74; staff dir, President's Biomed Res Panel, Nat Inst Aging, NIH, 74-75, actg dir, 75-76, dir, Geront Res Ctr, 76-89; RETIRED. *Concurrent Pos:* Bank Am-Giannini Found fel, 55-57; USPHS spec fel, 62-63; res assoc, Karolinska Inst, Sweden, 55-57; vis investr, Univ London, 62 & McGill Univ, 63; consult, Nat Inst Dent Res, 64-66 & Procter & Gamble Co, 64-66. *Honors & Awards:* Res Award, Int Asn Dent Res, 63; Super Serv Award, Dept Health, Educ & Welfare, 71. *Mem:* AAAS; Am Soc Cell Biol; Am Inst Biol Sci; Am Asn Anat; Geron Soc. *Res:* Physiology of growth, differentiation and aging; autoradiography; microradiography; histochemistry; physiology of mineralization. *Mailing Add:* 137 St Andrews Rd Severn Park MD 21146

GREVE, JOHN HENRY, b Pittsburgh, Pa, Aug 11, 34; m 56; c 3. VETERINARY PARASITOLOGY. *Educ:* Mich State Univ, BS, 56, DVM, 58, MS, 59; Purdue Univ, PhD(vet parasitol), 63. *Prof Exp:* Res assoc vet path, Mich State Univ, 58-59; instr vet parasitol, Purdue Univ, 59-62; from asst prof to assoc prof, 63-68, PROF VET PARASITOL, IOWA STATE UNIV, 68- *Concurrent Pos:* Secy-treas, Ann Midwestern Conf Parasitologists, 67-75, pres, 75-76. *Mem:* Am Vet Med Asn; Am Asn Vet Parasitol (pres, 68-70); Am Soc Parasitol; World Asn Advan Vet Parasitol; Asn Am Vet Med Cols. *Res:* Pathological response and host-parasite relationships in arthropods and nematodes of veterinary medical importance. *Mailing Add:* Dept Vet Path Iowa State Univ Ames IA 50011-1250

GREVILLE, THOMAS NALL EDEN, b New York, NY, Dec 27, 10; m 34, 51; c 2. MATHEMATICS. *Educ:* Univ of the South, BA, 30; Univ Mich, AM, 32, PhD(math), 33. *Prof Exp:* Actuarial asst, Acacia Mutual Life Ins Co, DC, 33-37; instr math, Univ Mich, 37-40; actuarial mathematician, US Bur Census, 40-46; USPHS, 46-52; statist consult, Int Coop Admin, 52-54; asst chief actuary, US Social Security Admin, 54-58; dep chief mathematician, US Army Qm Corps, 58-60, chief mathematician, 60-61; vpres, S A Miller Co, DC, 61-62; prof, Math Res Ctr, 63-81, prof, Sch Bus, 64-81, EMER PROF, UNIV WIS-MADISON, 81- *Concurrent Pos:* Statist ed, J Parapsychol, 44-67; vis prof, Univ Mich, 62-63; ed, J Appl Math, Soc Indust Appl Math, 68-77 & J Math Anal, 77-81; actuarial adv, Nat Ctr Health Statist, 73-76. *Mem:* Fel Soc Actuaries; Am Math Soc; fel Am Statist Asn; Opers Res Soc Am. *Res:* Matrices, approximation and interpolation; actuarial mathematics. *Mailing Add:* 228 Turkey Ridge Rd Charlottesville VA 22901-9726

GREW, EDWARD STURGIS, b Boston, Mass, May 29, 44; m 75. GRANULITE FACIES METAMORPHISM. *Educ:* Dartmouth Col, BA, 65; Harvard Univ, PhD(geol), 71. *Prof Exp:* Field asst geol, US Geol Surv, Denver, 65, Boston, 66, res assoc, Washington, DC, 71-72; proj assoc, Geophys & Polar Res Ctr, Univ Wis, 72-75; asst res geologist, Univ Calif, Los Angeles, 75-83; Humboldt fel, Ruhr-Univ Bochum, Fed Repub Ger, 83-84; RES ASSOC PROF, UNIV MAINE, ORONO, 84- *Concurrent Pos:* US exchange scientist, Soviet Antarctic Exped, Molodezhnaya Sta, 72-74; Fulbright scholar, Australia, 78; Indo-Am Fel Prog, 80-81; US exch scientist, Japan Antarctic Res Expedition, 87-88; mem, interacad exchange, USSR, 87 & 90; res scientist, Nat Inst Polar Res, Tokyo, 88-89. *Honors & Awards:* Bellingshausen Medal; US Antarctic Serv Medal. *Mem:* Geol Soc Am; Am Geophys Union; Mineral Soc Am; Sigma Xi; AAAS; Explorer's Club. *Res:* Petrology and mineralogy of granulite-facies rocks in Antarctica, US and USSR; regional geology of the Precambrian shield of East Antarctica; petrologic significance of the light elements (lithium, beryllium, boron and fluorine) in rock-forming minerals. *Mailing Add:* Dept Geol Univ Maine Orono ME 04469

GREW, PRISCILLA CROSWELL PERKINS, b Glens Falls, NY, Oct 26, 40; m 75. GEOLOGY, EARTH SCIENCES. *Educ:* Bryn Mawr Col, BA, 62; Univ Calif, Berkeley, PhD(geol), 67. *Prof Exp:* Instr geol, Boston Col, 67-68, asst prof, 68-72; asst res geologist, Inst Geophys, Univ Calif, Los Angeles, 72-77; dir, Calif Dept Conserv, 77-81; comnr, Calif Pub Utilities Comn, 81-86; DIR, MINN GEOL SURV, 86-; PROF GEOL, UNIV MINN, 86- *Concurrent Pos:* Secy, Geosci Adv Panel, Los Alamos Sci Lab, Univ Calif, 72-75, vis staff mem, 72-77; mem, Comt Minority Partic Earth Sci & Mineral Eng, US Dept Interior, 72-75; exec secy & co-ed, Bull Lake Powell Res Proj, 73-77; vis asst prof geol, Univ Calif, Davis, 73-74; adj asst prof environ sci & eng, Univ Calif Los Angeles, 75-76; chmn, Calif State Mining & Geol Bd, 76-77, Calif Geothermal Resources Task Force, 77 & Calif Resources Bd, 77-81; chmn, Calif Oil Shale Task Force, Geol Soc Am, 77, comt pub policy, 81-84, counr, 87-90; mem, comt eng, Nat Asn Regulatory Utility Comn, 81-82, comt gas, 82-86, exec comt, 84-86, comt energy conserv, 83-84, comt Mineral Resource Eval, Nat Res Coun, 81-82 & US Geol Surv Earthquake Studies Adv Panel, 79-83; trustee, Nat Parks & Conserv Asn; mem, adv coun, Gas Res Inst, 82-86, Bd on Mineral & Energy Resources, Nat Res Coun, 82-88, Comt Adv to US Geol Surv, 82-86 & Subcomt on Earthquake Res, 84-88, Comt on Equal Opportunities Sci & Technol, NSF, 85-86 & US Nat Comt on Geol, 85-; mem, bd Earth Sci & Resources, Nat Res Coun, 88-90, comt global change, 91-; chmn, comt pub affairs, Am Geophys Union, 84-89; mem, res coord coun, Gas Res Inst, 86-; mem, Minn Minerals Coord comt, 86-; mem, adv comt earth sci, NSF, 87-90, sci & technol ctrs develop, 87-90; mem, adv bd earth sci, Stanford Univ, 89- *Mem:* Fel AAAS; fel Geol Soc Am; Nat Asn Regulatory Utility Comnrs; Am Geophys Union; fel Mineral Soc Am; fel Geol Asn Can; Wilderness Soc. *Res:* Policy analysis for energy resource management; oil, gas, geothermal and coal resource development; federal land management issues; geothermal and other alternative energy development; public utility regulation. *Mailing Add:* Minn Geol Survey 2642 University Ave St Paul MN 55114

GREWAR, DAVID, b Dundee, Scotland, July 4, 21; Can citizen; m 46; c 8. MEDICINE, PEDIATRICS. *Educ:* Univ St Andrews, MB, ChB, 45; FRCP(E)(C), 71. *Prof Exp:* House physician med, Royal Infirmary Dundee, 45-46, registr pediat, 48-52; lectr child health, Univ St Andrews, 52-53; chief resident pediat, Childrens Hosp Winnipeg, 53-55; lectr, 55-59, from asst prof to assoc prof, 59-73, PROF PEDIAT, UNIV MAN, 73-; HEAD DEPT PEDIAT, ST BONIFACE HOSP, 77- *Concurrent Pos:* Attend staff, Childrens Hosp, Winnipeg, 56- *Mem:* Can Med Asn; Can Pediat Soc. *Res:* Prognois of prematurity; neonatology, especially erythroblastosis foetalis; nutrition, especially scurvy and vitamin deficiency in infancy. *Mailing Add:* St Boniface Hosp 409 Tache Ave Winnipeg MB R2H 2A6 Can

GREWE, ALFRED H, JR, b St Cloud, Minn, Mar 21, 26. ZOOLOGY, BOTANY. *Educ:* St Cloud State Col, BA, 50; Univ Minn, MA, 54; Univ SDak, PhD(zool), 66. *Prof Exp:* Mus scientist & teaching asst zool, Univ Minn, 51-56; instr biol, St Cloud State Col, 58-59; instr Itasca Jr Col, 59-62; asst prof, 65-71, PROF BIOL, ST CLOUD STATE COL, 71- *Mem:* AAAS; Am Ornith Union; Sigma Xi. *Res:* Natural history of the bald eagle. *Mailing Add:* 35948 Co Rd one St Cloud MN 56301

GREWE, JOHN MITCHELL, b Eau Claire, Wis, Feb 6, 38; m 59; c 4. ORTHODONTICS, ANATOMY. *Educ:* Univ Minn,Minneapolis,BS, 60, DDS, 62, MSD, 64, PhD(anat), 66. *Prof Exp:* Asst prof pediat & dent, acting chmn dept, Sch Dent & Med Sch, Univ Minn, 66-67; asst prof orthod & anat, Cols Dent & Med, Univ Iowa, 67-69; assoc prof, 69-75, chmn dept, 75-79, PROF ORTHOD, SCH DENT, UNIV MD, BALTIMORE, 75-; STAFF ORTHODONTIST, JOHNS HOPKINS HOSP, 79- *Concurrent Pos:* USPHS nonserv fel dent & anat, Sch Dent & Med Sch, Univ Minn, 62-66; consult, Univ Minn, 67; pres, Md Soc Dent Children, 73-74 & Md Soc Orthod, 75-76; consult, Fedn Dentaire Int, 73- & WHO, 75; consult, Nat Inst Dent Res, NIH, 75- *Mem:* AAAS; Am Dent Asn; Am Soc Dent for Children; Am Asn Orthod; Am Acad Oral Path. *Res:* Genetic and environmental influences on the growth of the craniofacial complex; dental development; malocclusion indices; bone development; orthogathic surgery. *Mailing Add:* 2601 Merrymans Mill Rd Phoenix MD 21131

GREY, ALAN HOPWOOD, b Auckland, NZ, Oct 8, 32; US citizen; m 57; c 2. PHYSICAL GEOGRAPHY. *Educ:* Brigham Young Univ, BA, 59; Univ Wis, MA, 60, PhD(geog), 63. *Prof Exp:* Asst prof geog & geol, Western Ill Univ, 63-64; from asst prof to assoc prof, 64-80, PROF GEOG, BRIGHAM YOUNG UNIV, 80- *Concurrent Pos:* Vis lectr geog, Univ Canterbury, 71. *Mem:* NZ Geog Soc; Asn Am Geographers. *Res:* Climatology; historical geography of the Western United States; regional and historical geography of Australasia. *Mailing Add:* Dept Geog Brigham Young Univ 690 swkt Provo UT 84602

GREY, GOTHARD C, b Sheridan, Wyo, Oct 9, 57; m 80; c 2. SCIENCE EDUCATION. *Educ:* Univ Utah, BS(math), BS(chem) & BS(physics), 80; Calif Inst Technol, MS, 82; Univ Wis-Madison, PhD(chem), 87. *Prof Exp:* Fac intern phys chem, Dept Chem, Univ Utah, 87-88; teacher chem, Alta High Sch, 88-89; ASST PROF PHYSICS, WESTMINSTER COL, SALT LAKE CITY, 89- *Mem:* Am Phys Soc; Am Chem Soc; Am Asn Physics Teachers; Nat Sci Teachers Asn. *Res:* Science education and curriculum development at all levels, kindergarten through undergraduate; new laboratory exercises; strong experimentally motivated curricula; coupled relaxation in nuclear magnetic resonance. *Mailing Add:* Westminster Col 1840 S 1300 E Salt Lake City UT 84105

GREY, HOWARD M, b New York, NY, Aug 16, 32. EXPERIMENTAL BIOLOGY. *Educ:* Univ Pa, BA, 53, NY Univ, MD, 57. *Prof Exp:* Med intern, Johns Hopkins Hosp, Baltimore, 57-58; res fel, Dept Med & Path, Univ Pittsburgh, 58-61, Dept Exp Path, Scripps Clin & Res Found, La Jolla, 61-63; guest investr, Rockefeller Univ, New York, 63-64, asst prof, 64-65; assoc prof, Dept Exp Path, Scripps Clin & Res Found, 65-67, assoc mem, 67-70; from assoc prof to prof path, Med Ctr, Univ Colo, Denver, 70-88; VPRES RES & DEVELOP, CYTEL, LA JOLLA, 88- *Concurrent Pos:* Mem, Dept Med, Nat Jewish Ctr Immunol & Respiratory Med, Denver, 70-88, actg dir, 77-78, head, Basic Immunol Div, 78-88; mem, Bd Sci Counr, NIH, 88-91. *Honors & Awards:* William B Coley Award, Cancer Res Inst, 89. *Mem:* Am Soc Exp Path; Am Asn Immunologists; Am Soc Clin Invest. *Res:* Molecular and cellular mechanisms of antigen recognition by T cells; structure-function relationships of immunoglobulins; author of numerous scientific publications. *Mailing Add:* Res & Develop Cytel Corp 3525 John Hopkins Ct San Diego CA 92121

GREY, JAMES TRACY, JR, b Newstead Twp, NY, May 27, 14; m 41; c 3. INORGANIC CHEMISTRY, RESEARCH MANAGEMENT. *Educ:* Univ Buffalo, BA, 36, PhD(chem), 40. *Prof Exp:* Asst chem, Univ Buffalo, 36-39; res chemist, Durez Plastics & Chem, Inc, NY, 39-43; plastics res engr, Res Lab, Aeroplane Div, Curtiss-Wright Corp, 42-46; sr res chemist, Cornell Aeronaut Lab, Inc, 46-47, head, Chem & Fuels Sects, 47-57; sci adv to dir res & develop, Hqs, US Air Force, Wash, DC, 57-59; dir res planning staff, Rocket Div, Thiokol Chem Corp, 57-64, dir res opers, 64-65, asst to pres res & develop, 65-81, corp dir res & develop, 81-82; RETIRED. *Mem:* Am Chem Soc; Am Inst Aeronaut & Astronaut; Combustion Inst. *Res:* Resins and plastics; catalytic oxidation and reduction; chlorination of hydrocarbons; fundamentals of combustion; catalytic combustion; heat transfer at extreme conditions; fuels; magnetic susceptibility measurements. *Mailing Add:* 107 Dolington Pd Yardley PA 19067

GREY, JERRY, b New York, NY, Oct 25, 26; m 74; c 2. AEROSPACE SCIENCE & ENGINEERING. *Educ:* Cornell Univ, BME, 47, MS, 49; Calif Inst Technol, PhD(aeronaut eng), 52. *Prof Exp:* Mem tech staff, Engine Div, Fairchild Engine & Airplane Corp, 49-50; hypersonic aerodynamicist, Calif Inst Technol, 50-52; res assoc aerospace eng, Princeton Univ, 52-56, from asst prof to assoc prof, 56-67; pres, Greyrad Corp, 67-71; pres, Calprobe Corp, 72-80, chmn, 80-83; adminr pub policy, 71-82, publ, Aerospace Am, 82-87, DIR, SCI & TECHNOL POLICY, AM INST AERONAUT & ASTRONAUT, 87- *Concurrent Pos:* Sr engr, Marquardt Aircraft Co, 51-52; consult various univs, pvt industs & govt agencies; vpres publ, Am Inst Aeronaut & Astronaut, 66-71; chmn solar adv panel, Off Technol Assessment, US Cong, 74-79; adj prof, Long Island Univ, Southampton, 76-81; dir, Scientists Inst Pub Info, 77- & Appl Solar Energy Corp, 79-; chmn coord comt energy, Asn Coop Eng, 78; vis prof, Princeton Univ, 90- *Mem:* AAAS; Int Solar Energy Soc; Int Astron Fedn (vpres, 78-84, pres, 84-86); Int Acad Astronaut (vpres, 82-85); Am Astronaut Soc (dir, 78-82). *Res:* Aerospace propulsion systems; heat transfer; combustion, nuclear power generation, instrumentation; solar energy; conservation technology; plasma dynamics. *Mailing Add:* One Lincoln Plaza 25-0 New York NY 10023

GREY, PETER, b Karachi, Pakistan, May 1, 38; m 61; c 3. ANALYTICAL CHEMISTRY. *Educ:* McGill Univ, BS, 61, PhD(anal chem), 67. *Prof Exp:* Res chemist, Mobil Res & Develop Corp, 66-68, sr res chemist, 68-70, supvry chemist, Tech Serv Div, 70-81, res assoc, 81-89, SR RES ASSOC, MOBIL RES & DEVELOP CORP, PAULSBOROUGH, 89-, SR SUPV CHEMIST, ANAL CHEM, RES SERV DIV, 81- *Concurrent Pos:* Adj prof, Temple Univ. *Mem:* Am Chem Soc; Am Soc Qual Control. *Res:* Chemistry of metal chelates in nonaqueous solvents; total quality management. *Mailing Add:* 11 Hollybrook Way West Berlin NJ 08091

GREY, ROBERT DEAN, b Liberal, Kans, Sept 5, 39; m 61; c 2. CELL BIOLOGY, DEVELOPMENTAL BIOLOGY. *Educ:* Phillips Univ, BA, 61; Wash Univ, St Louis, PhD(biol), 66. *Prof Exp:* Res asst, Wash Univ, St Louis, 65-66, asst prof biol, 66-67; from asst prof to assoc prof, 67-78, from vchmn dept to chmn, 74-83, PROF ZOOL, UNIV CALIF, 78-, DEAN BIOL SCI, 85- *Mem:* AAAS; Am Soc Cell Biol; Am Soc Zool; Soc Develop Biol. *Res:* Fertilization; gamete interactious, egg activation mechanisms. *Mailing Add:* Dept Zool Univ Calif Davis CA 95616

GREY, ROGER ALLEN, b Pittsburgh, Pa, Nov 12, 47; m 72; c 2. ORGANOTRANSITION METAL CHEMISTRY. *Educ:* Grove City Col, BS, 69; Mich State Univ, PhD(chem), 73. *Prof Exp:* Teaching fel, Univ Calif, Los Angeles, 73-75; res chemist, Allied Chem Corp, 75-80; sr res chemist, Oxirane Int, 80-81, RES ADV, ARCO CHEM CO, ATLANTIC RICHFIELD CORP, 81- *Mem:* Am Chem Soc. *Res:* Synthesis, reactions and mechanistic studies of organometallic complexes; hydrogenation of organic compounds using homogeneous and heterogeneous catalysts; synthesis of surfactants for enhanced oil recovery. *Mailing Add:* 111 Piedmont Rd West Chester PA 19382-7257

GREYSON, JEROME, b New York, NY, Nov 7, 27; m 57; c 3. PHYSICAL CHEMISTRY, CLINICAL CHEMISTRY. *Educ:* Hunter Col, AB, 50; Pa State Univ, PhD, 56. *Prof Exp:* Chemist, Sylvania Elec Corp, 51-53; mem tech staff, Bell Tel Labs, 56-57; staff chemist, Int Bus Mach Corp, NY, 57-62; group leader phys chem, Stauffer Chem Co, 62-64; res specialist, Atomics Int Div, NAm Aviation, Inc, 64-67, prin scientist, Rocketdyne Div, NAm Rockwell Corp, 67-70; sect head phys sci, Ames Res Lab, 70-76, dir, Blood Chem Res & Develop Lab, Ames Co Div, 76-81, dir res & develop, Lab Tech Div, Miles Labs, 81-82; tech dir, Precision Scientific, 82-85; OWNER, J & J G ASSOC, PROD DEVELOP CONSULT, 85- *Mem:* Am Chem Soc; Sigma Xi. *Res:* Physical chemistry of surfaces and membranes, biosensors. *Mailing Add:* 10742 Timothys Rd Conifer CO 80433

GREYSON, RICHARD IRVING, b Nelson, BC, May 26, 32; m 56; c 5. PLANT MORPHOLOGY. *Educ:* Univ BC, BA, 54; Univ Ore, MS, 60, PhD(biol), 65. *Prof Exp:* Instr biol, Univ Notre Dame, BC, 55-58, 59-61; lectr, 64-65, asst prof, 65-69, assoc prof, 69-80, PROF PLANT SCI, UNIV WESTERN ONT, 80- *Mem:* Am Bot Soc; Can Bot Asn. *Res:* Growth in intercalary meristems; initiation of floral primordia; plant organ culture; hormonal interaction between plant organs; description and analysis of plant organs of genetic strains of Lycopersicon, Zea and Nigella; experimental plant morphology; ears and tassels of zea. *Mailing Add:* Dept Plant Sci Univ Western Ont London ON N6A 5B7 Can

GREYTAK, THOMAS JOHN, b Annapolis, Md, Mar 24, 40; m 66; c 2. LIGHT SCATTERING, LOW TEMPERATURE PHYSICS. *Educ:* Mass Inst Technol, BS & MS, 63, PhD(physics), 67. *Prof Exp:* From instr to assoc prof, 67-77, PROF PHYSICS, MASS INST TECHNOL, 77-, HEAD DIV ATOMIC & CONDENSED MATTER PHYSICS, 88- *Concurrent Pos:* Alfred P Sloan res fel, 71-73,. *Mem:* Fel Am Phys Soc. *Res:* Light scattering from thermal fluctuations in matter; low temperature physics; spin-polarized atomic hydrogen. *Mailing Add:* Dept Physics Mass Inst Technol Cambridge MA 02139

GREYWALL, DENNIS STANLEY, b Detroit, Mich, Nov 16, 43; m 67; c 4. LOW TEMPERATURE PHYSICS. *Educ:* Univ Detroit, BS, 65; Ind Univ, MS, 67, PhD(physics), 70. *Prof Exp:* MEM STAFF LOW TEMPERATURE PHYSICS, BELL LABS, 71- *Mem:* Fel Am Phys Soc. *Res:* Properties of liquid and solid helium. *Mailing Add:* Bell Labs ID 152 Murray Hill NJ 07974

GREYWALL, MAHESH S, b Patiala, India, Oct 15, 34; US citizen; m 60; c 2. FLUID MECHANICS. *Educ:* Univ Calif, Berkeley, BS, 57, MS, 59, PhD, 62. *Prof Exp:* Vis prof eng, Univ Wash, Seattle, 62-63; mem tech staff, Aerospace Corp, Calif, 63-65; theoret physicist, Lawrence Radiation Lab, Univ Calif, Livermore, 65-69; assoc prof, 69-81, PROF, DEPT MECH ENG, WICHITA STATE UNIV, 81- *Mem:* Am Phys Soc; Am Soc Mech Engrs; Am Inst Aeronaut & Astronaut. *Res:* Fluid dynamics; computational methods; kinetic theory. *Mailing Add:* 2707 Rushwood Ct Wichita KS 67226

GREZLAK, JOHN HENRY, b Englewood, NJ, June 13, 45. ORGANIC CHEMISTRY, POLYMER CHEMISTRY. *Educ:* Pa State Univ, BS, 67; Princeton Univ, MA, 72, PhD(chem), 74. *Prof Exp:* Res chemist, Rohm and Haas Co, 69-71 & 74-77; asst prof, Mansfield State Col, 77-78; PROF CHEM, SHIPPENSBURG UNIV, 78- *Concurrent Pos:* Asst instr, Princeton Univ, 67-68, asst res, 68-69, Am Can fel, 71-72 & 72-73; consult, A D Little, Inc, 73; instr, Pa State Univ, 74-75. *Mem:* Sigma Xi; Am Chem Soc; AAAS. *Res:* Polymer synthesis; organometallics; block and graft copolymers; application of macromolecules to organic synthesis and organic reaction mechanisms. *Mailing Add:* 403 Springfield Rd Shippensburg PA 17257

GRIBBLE, DAVID HAROLD, b Seattle, Wash, Dec 26, 32; m 55; c 4. VETERINARY PATHOLOGY. *Educ:* Wash State Univ, DVM, 62; Univ Calif, Davis, PhD(comp path), 70. *Prof Exp:* Asst prof, 66-74, ASSOC PROF PATH, SCH VET MED, UNIV CALIF, DAVIS, 74- *Mem:* Am Vet Med Asn; Am Col Vet Path; Int Acad Path. *Res:* Infectious diseases; nervous system; endocrine and primate pathology. *Mailing Add:* 1563 Hwy 9 Mt Vernon WA 98273

GRIBBLE, GORDON W, b San Francisco, Calif, July 28, 41; m 85; c 3. ORGANIC CHEMISTRY. *Educ:* Univ Calif, Berkeley, BS, 63; Univ Ore, PhD(chem), 67. *Hon Degrees:* AM, Dartmouth Col, 81. *Prof Exp:* Nat Cancer Inst fel org chem, Univ Calif, Los Angeles, 67-68; from asst prof to assoc prof, 68-80, chmn, 88-91, PROF ORG CHEM, DARTMOUTH COL, 80- *Concurrent Pos:* Res Corp grant, 68; Petrol Res Fund grant, 68-70; Eli Lilly grant, 69-71; NSF grant, 69-71; NIH career develop award, 71-76; consult, Merck, Sharp & Dohme, 72-74; NSF fac develop award, 77-78, NIH grants, 73-89, Petrol Res Fund grant, 75-77, 83-86, 87-90. *Honors & Awards:* Am Cyanamid Acad Achievement Award, 88. *Mem:* Am Chem Soc; Chem Soc; Int Soc Heterocyclic Chem; Sigma Xi. *Res:* Synthetic organic chemistry; indole alkaloids; nuclear magnetic resonance spectroscopy; reaction mechanisms; chemical carcinogenesis; anticancer drug synthesis. *Mailing Add:* Dept Chem Dartmouth Col Hanover NH 03755-1894

GRIBOVAL, PAUL, b Paris, France, Aug 24, 25; m 52; c 1. ELECTRON OPTICS. *Educ:* Conserv, Nat Arts et Metiers, France, Eng dipl, 56; Univ Grenoble, PhD(electromagnetic separation), 66. *Prof Exp:* Tech asst physics & astron, Nat Ctr Sci Res, France, 47-56; engr, Fr Atomic Comn, 56-59; chief engr, Univ Grenoble, 59-66; spec res assoc electronographic camera, Dept Astron, Univ Tex, Austin, 66-86; RETIRED. *Mem:* AAAS; Am Astron Soc; Am Phys Soc. *Res:* Study of photomultipliers response; mass spectrometry of solids; isotopes separation and electrostatic accelerators; electronographic camera. *Mailing Add:* 18980 Third Dr Buena Vista CO 81211

GRICE, GEORGE DANIEL, JR, b Charleston, SC, Oct 9, 29; m 55; c 2. MARINE ZOOLOGY. *Educ:* Clemson Col, BS, 50; Fla State Univ, MA, 53, PhD(biol oceanog), 57. *Prof Exp:* Fishery res biologist, US Fish & Wildlife Serv, 57-58; fel, Guggenheim Mem Found, 58-59; assoc scientist, 59-73, chmn dept biol, 74-81, SR SCIENTIST, WOODS HOLE OCEANOG INST, 73-, ASSOC DIR SCI OPERS, 81- *Mem:* Ecol Soc Am. *Res:* Food chain dynamics, experimental ecosystems; zooplankton ecology, particularly the taxonomy and zoogeography of marine calanoid copepods. *Mailing Add:* Woods Hole Oceanog Inst Woods Hole MA 02543

GRICE, HARVEY H(OWARD), b Flint, Mich, Sept 25, 12; m 41; c 3. CHEMICAL ENGINEERING. *Educ:* Ohio State Univ, BChE, 37, MSc, 38, PhD(chem eng), 41. *Prof Exp:* Proj engr, Cent Labs, Gen Foods Corp, NJ, 41-42, supt processing & power, Diamond Crystal-Colonial Salt Div, 46-48, tech asst to dir mfg & eng, Gen Foods Corp, 48-52, plant mgr, Diamond Crystal-Colonial Salt Div, 52-53, mgr mfg & eng, Gaines Div, 53-58; pres, Graceland Col, 58-64; prof, 64-78, EMER PROF CHEM ENG, UNIV MO-ROLLA, 78- *Mem:* Fel Am Inst Chem Engrs; Am Soc Eng Educ; Am Chem Soc; fel Am Inst Chemists; Nat Soc Prof Engrs; Sigma Xi. *Res:* Engineering; mechanism of crystallization in the industrial evaporation of sodium chloride brine. *Mailing Add:* Eight Laird Ave Rolla MO 65401

GRIDER, JOHN RAYMOND, b Chester, Pa, Oct 12, 52; m 79; c 1. GASTROINTESTINAL PHYSIOLOGY, SMOOTH MUSCLE PHYSIOLOGY. *Educ:* Univ Pa, BA, 73; Hahnemann Univ, PhD(physiol & biophysics), 81. *Prof Exp:* Res assoc gastroenterol, Philadelphia Gen Hosp, Univ Pa, 73-74; teaching fel physiol, Hahnemann Univ, 75-81; NIH fel gastroenterol, 81-83, from instr to asst prof, 84-90, ASSOC PROF PHYSIOL, MED COL VA, 91- *Concurrent Pos:* NSF fel, Albert Einstein Med Ctr, Div Reprod Endocrine, Philadelphia, 69; microbiologist, Div Labs, Pa State Health Dept, 73; prin investr NIH grants, 83-; chmn, Abstract Selection Comt, Am Gastroenterol Asn, 87 & 88. *Mem:* Am Physiol Soc; Am Gastroenterol Asn; Am Motility Soc; Gastroenterol Res Group. *Res:* Elucidating the regulation of contraction of gastrointestinal smooth muscle by neuropeptides of the myenteric plexus; studies have been conducted on the regulation of the release of neuropeptides, on the neural pathways which coordinate peristalsis and on the intracellular mechanisms responsible for smooth muscle contraction. *Mailing Add:* Box 711 MCV Sta Med Col Va Richmond VA 23298-0711

GRIEB, MERLAND WILLIAM, b Carey, Idaho, Jan 26, 20. INORGANIC CHEMISTRY. *Educ:* Univ Idaho, BS, 42, MS, 49; Univ Ill, PhD(chem), 53. *Prof Exp:* Instr chem, Univ Idaho, 48-49; asst, Univ Ill, 49-53; asst prof, Wayne State Univ, 53-56; asst prof, 56-69, ASSOC PROF CHEM, UNIV IDAHO, 69- *Mem:* Am Chem Soc. *Res:* Complex ions. *Mailing Add:* 1532 35th Ave S Seattle WA 98144

GRIECO, MICHAEL H, INFECTIOUS DISEASES, ALLERGY. *Educ:* Columbia Univ, MD, 79. *Prof Exp:* CHIEF, DIV ALLERGY CLINICAL IMMUNOL & INFECTIOUS DISEASES, ST LUKE'S ROOSEVELT HOSP CTR, 73- *Mailing Add:* St Luke's Roosevelt Hosp Ctr 428 W 59th St New York NY 10032

GRIECO, PAUL ANTHONY, b Framingham, Mass, Oct 27, 44; m 71; c 4. SYNTHETIC ORGANIC CHEMISTRY. *Educ:* Boston Univ, BA, 66; Columbia Univ, MA, 67, PhD(org chem), 70. *Prof Exp:* NSF fel, Harvard Univ, 70-71; from asst prof to prof chem, Univ Pittsburgh, 71-80; prof, 80-85, EARL BLOUGH PROF CHEM, IND UNIV, 85-, CHMN, DEPT CHEM, 88- *Concurrent Pos:* Fel, Alfred P Sloan, 74-76 & Japan Soc Prom Sci, 78-79; mem, Med Chem A Study Sect, NIH, 78-82, Rev Panel Nat Res Coun Associateship Prog, 87- *Honors & Awards:* William P Timmie lectr, Emory Univ; Ernet Guenther Award Chem Essential Oils & Related Prod, 82. *Mem:* Am Chem Soc; Royal Soc Chem; Chem Soc Japan; Swiss Chem Soc. *Res:* Development of new synthetic methods for construction of complex natural products. *Mailing Add:* Dept Chem AO53 Ind Univ 1410-3000 Bloomington IN 47405

GRIEGO, RICHARD JEROME, b Albuquerque, NMex, June 11, 39; m 60; c 2. MATHEMATICS. *Educ:* Univ NMex, BS, 61; Univ Ill, PhD(math), 65. *Prof Exp:* Lectr math, Univ Calif, Riverside, 65-66; from asst prof to prof, Univ NMex, 66-85, coordr, Chicano Studies Prog, 70-71, dir, Minority Access Res Careers Prog, 75-85, chmn, math dept, 77-80, dir, Resource Ctr, Sci & Eng, 79-85, dean grad studies, 88-91, PRESIDENTIAL PROF MATH, UNIV NMEX, 85- *Mem:* Am Math Soc. *Res:* Markov processes; potential theory; probability theory. *Mailing Add:* Dept Math & Statist Univ NMex Albuquerque NM 87131

GRIEM, HANS RUDOLF, b Kiel, Ger, Oct 7, 28; nat US; m 57; c 4. PHYSICS. *Educ:* Univ Kiel, PhD(physics), 54. *Hon Degrees:* PhD, Ruhr Univ, 89. *Prof Exp:* Asst upper atmospheric physics, Univ Md, 54-55; asst high temperature physics, Univ Kiel, 55-57; res asst prof plasma physics, 57-61, assoc prof, 61-63, PROF PHYSICS, UNIV MD, COLLEGE PARK, 63- *Concurrent Pos:* Consult, US Naval Res Lab, 57-, Los Alamos Nat Lab, 76- & Lawrence Livermore Nat Lab, 76- *Honors & Awards:* Meggers Award, Optical Soc Am, 87. *Mem:* Fel Am Phys Soc. *Res:* High temperature and plasma physics; spectroscopy; line broadening theory. *Mailing Add:* Lab Plasma Res Univ Md Col Park MD 20742

GRIEM, MELVIN LUTHER, b Milwaukee, Wis, May 22, 25; m 51; c 3. RADIOLOGY, PHYSICS. *Educ:* Univ Wis, BS, 48, MS, 50, MD, 53. *Prof Exp:* From instr to assoc prof, 57-68, PROF RADIOL, SCH MED, UNIV CHICAGO, 68-, dir, Chicago Tumor Inst, 66-82. *Concurrent Pos:* Am Cancer Soc clin fel, 58-60, career res develop award, 63-65; mem, Med Adv Bd, Nuclear Regulatory Comn, 80-; mem, Radiation Study Sect, HEW, 84-88. *Mem:* Am Asn Cancer Res; Am Roentgen Ray Soc; Am Soc Therapeut Radiol; Radiation Res Soc; Radiol Soc NAm. *Res:* Radiobiology; radiation therapy; radiologic physics; radioactive isotopes; micro vascular imaging of radiation and drug induced vascular injury; radiation carcinogenesis. *Mailing Add:* Mitchell Hosp Univ 950 E 59th St Chicago IL 60637

GRIEM, SYLVIA F, b West Allis, Wis, Feb 24, 29; m 51; c 3. MEDICINE, DERMATOLOGY. *Educ:* Univ Wis, BS, 50, MD, 53. *Prof Exp:* asst prof med, 61-80, ASSOC PROF MED, UNIV CHICAGO, 80- *Mem:* Am Acad Dermat. *Res:* Hypersensitivity to physical agents, especially cold; microangiography of skin; diagnosis and treatment of mycosis fungoides. *Mailing Add:* Dept Med Box 409 Univ Chicago 5841 Maryland Ave Chicago IL 60637

GRIEMAN, FREDERICK JOSEPH, b Long Beach, Calif, May 19, 52; m 76; c 2. LASER SPECTROSCOPY. *Educ:* Univ Calif, Irvine, AB, 74, Berkeley, MS, 76, PhD(chem), 79. *Prof Exp:* Res fel, Phys Dept, Univ Ore, 79-81, res assoc, 81-82; ASSOC PROF, CHEM DEPT, POMONA COL, 82- *Concurrent Pos:* Vis scientist, Lab des Collinions Atomiques et Moleculaires, Univ Paris, 80-81 & Phys Chem Lab, Univ Oxford, 88-89. *Mem:* Am Chem Soc. *Res:* Laser induced fluorescence and electron impact emission spectra of molecular radicals supersonically expanded in a nozzle; spectra of molecular ions by laser induced fluorescence of ions confined in a three dimensional ion trap. *Mailing Add:* Chem Dept Pomona Col Claremont CA 91711

GRIENINGER, GERD, BIOCHEMISTRY. *Educ:* Univ Tübingen, Germany, PhD(biochem), 70. *Prof Exp:* HEAD LAB PLASMA PROTEIN REGULATION, NY BLOOD CTR, 81- *Mailing Add:* New York Blood Ctr 310 E 67th St New York NY 10021

GRIER, CHARLES CROCKER, b Pasadena, Calif, Sept 1, 38; m 60; c 1. ECOLOGY, FOREST SOILS. *Educ:* Univ Wash, BS, 68, PhD(forest soils), 72. *Prof Exp:* Res assoc ecol, Sch Forestry, Ore State Univ, 72-76; res asst prof ecol, Col Forest Res, Univ Wash, 76-78, from assoc prof to prof, 78-85; PROF ECOL, SCH FORESTRY, NORTHERN ARIZ UNIV, FLAGSTAFF, 85- *Concurrent Pos:* Affil asst prof, Univ Idaho, 78-84. *Honors & Awards:* Antarctic Serv Medal, NSF, 72. *Mem:* Ecol Soc Am; AAAS; Soil Sci Soc Am; Soc Am Foresters; Sigma Xi. *Res:* Productivity of wildland ecosystems as related to their physical environment and mineral nutrition. *Mailing Add:* Sch Forestry Northern Ariz Univ Flagstaff AZ 86011

GRIER, HERBERT E, TECHNICAL MANAGEMENT. *Educ:* Mass Inst Technol, BS & MS, 34. *Hon Degrees:* DSc, Univ Nev, 67. *Prof Exp:* Pres, EG&G, 47-76; pres, CER Geonuclear Co, 65-83; pres & chmn bd, Reynolds Elec & Eng Co, Inc, 69-71; DIR, CER CORP, 83-; CONSULT, EG&G, 83- *Concurrent Pos:* Mem, Res Group Electronic Flash Photog, Night Aerial Reconnaissance, Mass Inst Technol; atomic weapons res, Manhattan Eng Dist & AEC; dir, CER Geonuclear Corp, AUX Corp; consult, Aerospace Safety Adv Panel, NASA; mem STS-1 Readiness Adv Team, 81, Shuttle Oper Strategic Planning Group, NASA, 84. *Honors & Awards:* Commendation of the Pres, 48; Presidential Cert Appreciation, 71; Nat Medal Sci, 89. *Mem:* Nat Acad Eng. *Res:* Stroboscopic and flash lighting techniques; ultra high speed photography; atomic weapons research; development and implementation of methods of using the power of nuclear explosives in commercial applications with primary emphasis on improving methods for the extraction of underground natural resources; numerous patents and publications. *Mailing Add:* 9648 Blackgold Rd La Jolla CA 92037

GRIER, JAMES WILLIAM, b Waterloo, Iowa, Sept 15, 43; m 65; c 2. ZOOLOGY. *Educ:* Univ Northern Iowa, BA, 65; Univ Wis-Madison, MS, 68; Cornell Univ, PhD(ecol, evolutionary biol), 75. *Prof Exp:* Asst prof, 73-77, assoc prof, 77-83, PROF ZOOL, NDAK STATE UNIV, 83- *Mem:* AAAS; Wildlife Soc. *Res:* Ecology, population dynamics, reproduction and behavior of birds of prey, particularly bald and golden eagles. *Mailing Add:* Dept Zool NDak State Univ Fargo ND 58105

GRIER, NATHANIEL, b Brooklyn, NY, Mar 27, 18; m 41; c 3. MEDICINAL CHEMISTRY. *Educ:* Long Island Univ, BS, 37; Univ Mich, MS, 38, PhD(chem), 43. *Prof Exp:* Res chemist, Dept Eng Res, Univ Mich, 40-41; sr res chemist, Hoffmann-La Roche, Inc, 42-46; chief res chemist, Dar-Syn Lab, Inc, 46-57; vpres & dir res, Metalsalts Corp, 57-66; sr investr Merck Sharp

& Dohme Res Lab Div, Merck & Co, Inc, Rahway, 66-80; RETIRED. *Mem:* Fel AAAS; Am Chem Soc; NY Acad Sci; Soc Indust Microbiol. *Res:* Isolation of antibiotics; synthetic organic chemistry; process development. *Mailing Add:* 153 Morse Pl Englewood NJ 07631

GRIER, RONALD LEE, b Cedar Rapids, Iowa, May 9, 41; m 66; c 2. VETERINARY SURGERY. *Educ:* Iowa State Univ, DVM, 65; Colo State Univ, PhD(surg), 70; Am Col Vet Surgeons, dipl, 72. *Prof Exp:* Chief exp surg, Res & Develop Lab & Vivarium, Madigan Gen Hosp, US Army, 65-67; NIH fel res, Colo State Univ, 67-70; asst prof, 70-72, ASSOC PROF VET CLIN SCI, IOWA STATE UNIV & CHMN, SMALL ANIMAL TEACHING HOSP, 72- *Mem:* Am Col Vet Surgeons; Am Asn Vet Clinicians; Am Vet Med Asn; Comp Gastroenterol Soc; Sigma Xi. *Res:* Study of comparative oncology and alimentary tract diseases in the canine. *Mailing Add:* Rte 1 Arrasmith Trail Ames IA 50010

GRIERSON, JAMES DOUGLAS, b Dayton, Ohio, July 15, 31; m 53; c 4. PALEOBOTANY, PLANT MORPHOLOGY. *Educ:* Hiram Col, BA, 54; Cornell Univ, PhD(paleobot), 62. *Prof Exp:* Instr bot, Cornell Univ, 61-63; asst prof biol, 63-68, ASSOC PROF BIOL, STATE UNIV NY BINGHAMTON, 68- *Concurrent Pos:* Mem, Paleont Res Inst. *Mem:* Bot Soc Am; Paleont Soc. *Res:* Applications of anatomical and morphological techniques to the study of paleozoic fossil plants; lycopod evolution. *Mailing Add:* Dept Biol Sci State Univ NY Binghamton NY 13901

GRIES, DAVID, b Flushing, NY, Apr 26, 39; m 61; c 2. COMPUTER SCIENCE. *Educ:* Queens Col, NY, BS, 60; Univ Ill, Urbana, MS, 63; Munich Tech Univ, Dr rer nat(math), 66. *Prof Exp:* Mathematician, US Naval Weapons Lab, 60-62; res asst, Univ Ill, Urbana, 62-63; res asst math, Munich Tech Univ, 63-66; asst prof comput sci, Stanford Univ, 66-69, res assoc, Linear Accerator Ctr, 66-69; assoc prof, 69-77, chmn dept, 82-87, PROF COMPUT SCI, CORNELL UNIV, 77- *Concurrent Pos:* Guggenheim fel, 83-84; chmn, Comput Res Bd, 87-; ed, Info Processing Lett, Acta Informatica & Springer Verlag. *Mem:* Asn Comput Mach; Inst Elec & Electronics Engrs. *Res:* Programming methodology, programming languages, compiler construction. *Mailing Add:* Dept Comput Sci Upton Hall Cornell Univ Ithaca NY 14853

GRIES, GEORGE ALEXANDER, b Cambridge, Mass, May 2, 17; m 39; c 2. PLANT PATHOLOGY, HORTICULTURE. *Educ:* Miami Univ, AB, 38; Kans State Univ, MS, 40; Univ Wis, PhD(plant physiol), 42. *Prof Exp:* Asst plant pathologist, Conn Agr Exp Sta, 42-45; assoc prof bot & assoc plant pathologist, Agr Exp Sta, Purdue Univ, 45-53, prof plant physiol & plant physiologist, 53-60; plant pathologist, prof plant path & head dept, Univ Ariz, 60-66, actg head dept bot, 63-65, prof biol sci & head dept & biologist, Agr Exp Sta, 66-68; dean, Col Arts & Sci, Okla State Univ, 68-80, prof biol, 80-82; RETIRED. *Concurrent Pos:* Res demonstr, Univ Col, Swansea, Wales, 57-58; mem, Comn Educ & Natural Resources, 62-68; mem exec comt & mem, Comn Undergrad Educ Biol Sci, 68-71; consult-evaluator, NCent Asn Cols & Sec Schs; consult, Coord Coun Higher Educ, State Calif, 69-71; mem, Gov Bd, Am Inst Biol Sci, 70-78; vol diagnostician, Univ Ariz Coop Ext, 85-; consult, Ark Dept Higher Educ, 87. *Mem:* Am Inst Biol Sci (pres, 77); Am Phytopath Soc. *Res:* Physiology of parasitic fungi. *Mailing Add:* 501 W Ocotillo NBU 4603 Green Valley AZ 85614

GRIES, JOHN CHARLES, b Rapid City, SDak, May 17, 40; m 68; c 1. STRUCTURAL GEOLOGY. *Educ:* Univ Wyo, BS, 62, MS, 65; Univ Tex, Austin, PhD(geol), 70. *Prof Exp:* Instr geol, Univ Tex, Austin, 70-71; asst prof, 71-79, ASSOC PROF GEOL, WICHITA STATE UNIV, 80-, CHMN, 87- *Concurrent Pos:* Res assoc, Bur Econ Geol, Univ Tex, Austin, 71-72; mem seismicity task force, City of Wichita, Kans, 72-; Consult, 71- *Mem:* Geol Soc Am; Am Asn Petrol Geologists; Sigma Xi; Am Inst Prof Geologists. *Res:* Structural relationship between the Rio Grande Rift in New Mexico, Colorado and the evaporite tectonic region of Northern Chihuahua, Mexico; stress distribution about southern Rocky Mountain foreland thrusting. *Mailing Add:* Dept Geol Wichita State Univ Wichita KS 67208

GRIES, JOHN PAUL, b Washington, DC, June 7, 11; m 33; c 2. GEOLOGY. *Educ:* Miami Univ, Ohio, AB, 32; Univ Chicago, MS, 33, PhD(geol, paleont), 35. *Prof Exp:* Asst geologist, State Geol Surv, Ill, 35-36; from instr to assoc prof geol & mineral, SDak Sch Mines & Technol, 36-44; geologist, Magnolia Petrol Co, 44-46; assoc prof geol & mineral, SDak Sch Mines, 46-49, prof geol eng, 49-76, dean grad div, 66-76, dir grad studies, 50-76, EMER PROF GEOL ENG, SDAK SCH MINES & TECHNOL, 76-, GEOL CONSULT, 76- *Mem:* Geol Soc Am; Paleont Soc; Am Asn Petrol Geol; Am Inst Mining, Metall & Petrol Eng. *Res:* Stratigraphy of northern Great Plains and Rocky Mountain area. *Mailing Add:* 238 St Charles St Rapid City SD 57701

GRIESBACH, ROBERT ANTHONY, b Menasha, Wis, Apr 11, 24; m 54; c 5. CYTOLOGY, GENETICS. *Educ:* DePaul Univ, BS, 51, MS, 52; Univ Chicago, PhD(bot), 55. *Prof Exp:* From instr to asst prof, 55-70, chmn dept, 71-87, ASSOC PROF BIOL SCI, DEPAUL UNIV, 70- *Mem:* AAAS; Bot Soc Am; Am Genetic Asn; Genetic Soc Am. *Res:* Cytogenetics; seed dormancy; polyploidy in ornamental angiosperms, lilium, hemerocallis, phaelaenopsis:; induction methodology and hybridization, genetics of polyploids. *Mailing Add:* 1036 Belden Ave Chicago IL 60614

GRIESBACH, ROBERT JAMES, b Chicago, Ill, June 21, 55; m 84; c 2. PLANT TISSUE CULTURE, SOMATIC CELL GENETICS. *Educ:* De Paul Univ, BS, 77; Mich State Univ, PhD(genetics), 80. *Prof Exp:* Res assoc hort, Mich State Univ, 80-81; RES GENETICIST GENETICS ENG, USDA, 81- *Mem:* Am Genetics Asn; Am Soc Hort Sci; Plant Molecular Biol Asn; AAAS. *Res:* Current techniques in somatic cell genetics to plant improvement. *Mailing Add:* Florist-Nursery Crops USDA Agr Res Serv BARC-W Beltsville MD 20705

GRIESEMER, RICHARD ALLEN, b Andreas, Pa, May 8, 29; m 51; c 3. VETERINARY PATHOLOGY. *Educ:* Ohio State Univ, DVM, 53, PhD(vet path), 59. *Prof Exp:* Instr vet path, Ohio State Univ, 53-55; jr pathologist, Virol Br, Armed Forces Inst Path, 55-57; from instr to prof vet path, Ohio State Univ, 57-71, chmn dept, 67-71; assoc dir, Nat Ctr Primate Biol, Univ Calif, Davis, 71-73; sr res staff mem, Carcino-Genesis Prog, Oak Ridge Nat Lab, 73-75, prog mgr, Cancer & Toxicol Prog, 75-77; assoc dir carcinogenesis testing, Nat Cancer Inst, 77-80; dir, Biol Div, Oak Ridge Nat Lab, 80-88; DIR, DIV TOXICOL RES & TESTING, NAT INST ENVIRON HEALTH SCI, 88- *Concurrent Pos:* Mem animal resources adv comt, NIH, 69-73. *Honors & Awards:* Nat Gaines Award, Am Vet Med Asn. *Mem:* AAAS; Am Vet Med Asn; Am Col Vet Path; Int Acad Path; Am Asn Cancer Res. *Res:* Morphogenesis of cancer; environmental co-carcinogenesis. *Mailing Add:* Nat Inst Environ Health Sci PO Box 12233 Research Triangle Park NC 27709

GRIESER, DANIEL R, b Newark, Ohio, May 5, 26; m 69. OPTICAL PHYSICS. *Educ:* Ohio State Univ, BS, 53. *Prof Exp:* OPTICAL ENGR, COLUMBUS LABS, BATTELLE, 53- *Mem:* Nat Soc Prof Engrs; Soc Photo-optical Instrumentation Engrs; Optical Soc Am. *Res:* Instrumentation for environmental extremes; optical meteorology; holography and coherent optical signal processing; spectroscopy of optical transitions; laser plasma diagnostics; on-line graphical analysis by microcomputer. *Mailing Add:* Battelle Columbus 505 King Ave Columbus OH 43201

GRIESHABER, CHARLES K, b Erie, Pa, Dec 1, 41. TOXICOLOGY. *Educ:* Gannon Univ, BA, 64; Pa State Univ, MS, 66, PhD(physiol), 69. *Prof Exp:* Chief, Toxicol Br, Develop Therapeut Prog, Nat Cancer Inst, NIH, 83-90; br chief, Drug Develop Br, 90, DIR, RES & TESTING CTR, DRUG EVAL & RES, FED DRUG ADMIN, 90- *Mem:* Am Asn Cancer Res; AAAS; Am Asn Clin Chem; Sigma Xi. *Mailing Add:* Fed Drug Admin Four Research Ct Suite 314 Rockville MD 20850

GRIESHAMMER, LAWRENCE LOUIS, b Jefferson City, Mo, Aug 28, 22; m 45; c 7. ANALYTICAL CHEMISTRY, PHYSICAL CHEMISTRY. *Educ:* Univ Calif, Los Angeles, BS, 49; Univ Mo, MA, 51. *Prof Exp:* Asst, Univ Mo, 49-51; anal res chemist, Lubrizol Corp, 51-53, res supvr, Spectros Lab, 53-61, sect leader, 61-70, supvr, Anal & Instrumental Labs, 70-84; RETIRED. *Mem:* Am Chem Soc; Soc Appl Spectros; Am Soc Testing & Mat. *Res:* Emission spectroscopy; ultraviolet and infrared absorption spectroscopy; instrumental methods of analysis and analytical organic chemistry, particularly on lubricating oil additives. *Mailing Add:* 1982 Idlehurst Dr Euclid OH 44117

GRIESINGER, DAVID HADLEY, b Cleveland, Ohio, Mar 22, 44; m 69; c 1. AUDIO ENGINEERING, NUCLEAR PHYSICS. *Educ:* Harvard Univ, BA, 66, MA, 69, PhD(physics), 76. *Prof Exp:* REC ENGR MUSIC, DAVID GREISINGER REC, 66- *Concurrent Pos:* Musician, Boston Comerota Inc, 75-80; design consult, Lexicon Inc, Waltham, 77-; instr elec dept physics, Harvard Univ, 79-81. *Mem:* Audio Eng Soc. *Res:* Sound recording; microphone design, acoustics, electronics and music; digital electronics; high speed digital audio equipment and microprocessors; Mossbauer effect; nuclear magnetic resonance. *Mailing Add:* 23 Bellevue Ave Cambridge MA 02140

GRIESMER, JAMES HUGO, b Cleveland, Ohio, Dec 18, 29; m 56, 84; c 4. EXPERT SYSTEMS. *Educ:* Univ Notre Dame, BS, 51; Princeton Univ, PhD(math), 58. *Prof Exp:* Asst math, Princeton Univ, 54-57; assoc mathematician, IBM CORP, 57-58, staff mathematician, 58-60, res staff mem, 60-65, mgr, Res Comput Ctr, 64-65, mgr symbol manipulation proj, 65-76, mgr educ & develop, 76-81, RES STAFF MEM, T J WATSON RES CTR, IBM CORP, 81- *Concurrent Pos:* Vis Mackay lectr, Univ Calif, Berkeley, 70-71; adj prof, Polytech Univ, 87-88. *Mem:* Asn Comput Mach; Am Asn Artificial Intel; Math Asn Am; Inst Elec & Electronics Engrs. *Res:* Expert systems. *Mailing Add:* Seven Inningwood Rd Ossining NY 10562-2203

GRIESS, JOHN CHRISTIAN, JR, b Mt Vernon, Ind, July 13, 22; m 47; c 7. CORROSION, ELECTROCHEMISTRY. *Educ:* Ind Univ, BS, 43, AM, 47. *Prof Exp:* res assoc corrosion, Oak Ridge Nat lab, Union Carbide Corp, 47-84; res assoc corrosion, Martin Marietta Corp, 84-86; RETIRED. *Honors & Awards:* Young Authors Award, Electrochem Soc, 50. *Mem:* Nat Asn Corrosion Engrs; Sigma Xi. *Res:* Corrosion phenomena in nuclear reactor and associated technologies. *Mailing Add:* 10803 Fox Park Rd Knoxville TN 37931

GRIESS, ROBERT LOUIS, JR, b Savannah, Ga, Oct 10, 45. ALGEBRA. *Educ:* Univ Chicago, BS, 67, MS, 68, PhD(math), 71. *Prof Exp:* Hildebrandt res instr math, Univ Mich, 71-73, asst prof, 73-74; vis asst prof, Rutgers Univ, 74-75; from asst prof to assoc prof, 75-81, PROF MATH, UNIV MICH, ANN ARBOR, 81- *Concurrent Pos:* Vis prof, Yale Univ, 83-84; CNRS prof, Ecole Normale Superieure, Paris, 86-87. *Res:* Classification of finite simple groups; properties of finite simple groups; group extensions and cohomology. *Mailing Add:* Dept Math Univ Mich Ann Arbor MI 48109

GRIEVE, CATHERINE MACY, b Watertown, NY, Nov 24, 26; m 61; c 2. PLANT PHYSIOLOGY. *Educ:* St Lawrence Univ, BS, 48; Univ Calif, Riverside, PhD(bot), 78. *Prof Exp:* Res librn, Mathieson Chem Corp, Niagara Falls, NY, 48-51; res chemist, US Naval Ordnance Lab, Corona, Calif, 54-71; PLANT PHYSIOLOGIST, US SALINITY LAB, RIVERSIDE, CALIF, 79- *Mem:* Sigma Xi. *Res:* Biosystematics of Aurantiodeae (rutacae); effects of salinity on plant metabolism; synthesis of substituted phosphonitriles; polymerization kinetics of phosphonitriles. *Mailing Add:* US Salinity Labs 4500 Glenwood Ave Riverside CA 92501

GRIEVE, RICHARD ANDREW, b Aberdeen, Scotland, Sept 15, 43. METEORITE IMPACT, CRUSTAL EVOLUTION. *Educ:* Univ Aberdeen, Scotland, BSc, 65; Univ Toronto, MSc, 67, PhD(geol), 70. *Hon Degrees:* MA, Brown Univ, 83; DSc, Univ Aberdeen, 85. *Prof Exp:* From res scientist I to res scientist III geophys, Earth Physics, Energy, Mines & Resources, Can, 74-82; assoc prof geol, Brown Univ, 82-84; RES SCIENTIST IV GEOPHYS, GEOL SURV, ENERGY MINES & RESOURCES, CAN, 84- *Concurrent Pos:* Vis prof, Brown Univ, 80-81; secy, adv comt comp planetology, Int Union Geol Sci, 80-84, comn comp planetology, 84-; assoc ed, Proc Conf Multi Ring Basins & Proc 12th Lunar & Planetary Sci Conf, 81, Meteoritics, 88-; assoc ed, J Geophys Res, 82-84. *Honors & Awards:* Barringer Medal, Meteoritical Soc, 90. *Mem:* Geol Asn Can; Can Geophys Union; Am Geophys Union; fel Meteoritical Soc; Sigma Xi; AAAS. *Res:* Impact phenomena and their relation to crustal evolution and the terrestrial biosphere; early crustal evolution of the terrestrial planets; use of large spatial geophysical data bases for the interpretation of crustal structure. *Mailing Add:* Geophys Div Geol Surv Can One Observatory Crescent Ottawa ON K1A 0Y3 Can

GRIEVE, ROBERT B, b Torrington, Wyo, Oct 27, 51; m 82; c 2. IMMUNOPARASITOLOGY. *Educ:* Univ Wyo, BS, 73, MS, 75; Univ Fla, PhD(parasitol), 78. *Prof Exp:* Posdoctoral immunoparasitol, Cornell Univ, 78-79, res assoc, 79-81; asst prof parasitol, Univ Pa, 81-84; assoc prof, Univ Wis, 84-87; assoc prof parasitol, 87-90, PROF PARASITOL, COLO STATE UNIV, 90- *Concurrent Pos:* Assoc investr, USAF, 79-84; adv, WHO, 84; consult, NIH, 85 & 88; head, Lab Parasitol, Univ Pa, 82-84; reviewer, Competitive Grants Prog, USDA, 85 & Spec Constraints Grants Prog, AID-CSRS, 88; mem coun, Am Soc Parasitologists, 89- *Honors & Awards:* Henry Baldwin Ward Medal, Am Soc Parasitologists. *Mem:* AAAS; Am Asn Immunologists; Am Asn Vet Parasitologists; Am Soc Parasitologists; Am Soc Trop Med & Hyg. *Res:* The immunology and biology of host and parasite interactions in the course of parasitic nematode infections. *Mailing Add:* Dept Path Colo State Univ Ft Collins CO 80523

GRIEVES, ROBERT BELANGER, b Evanston, Ill, Oct 15, 35; m 66; c 2. CHEMICAL ENGINEERING, SANITARY ENGINEERING. *Educ:* Northwestern Univ, BA, 56, MS, 59, PhD(chem eng), 62. *Prof Exp:* Asst prof civil eng, Northwestern Univ, 61-63; from asst prof to assoc prof environ eng, Ill Inst Technol, 63-67; chmn dept, 67-69, prof chem eng Univ Ky, 67-, dir, Ky Water Resources Res Inst, 74-, ASSOC DEAN ADMIN & GRAD PROG & RES, COL ENG, 76-; DEAN ENG, UNIV TEX, EL PASO. *Mem:* Am Inst Chem Engrs; Water Pollution Control Fedn. *Res:* Critical point of multicomponent hydrocarbon mixtures; foam fractionation for industrial water and waste treatment; biological waste treatment; chemical separations; membrane processes. *Mailing Add:* Univ Tex El Paso TX 79968

GRIFFEL, MAURICE, b Brooklyn, NY, Mar 10, 19; m 64. PHYSICAL CHEMISTRY. *Educ:* City Col New York, BS, 39; Univ Mich, MS, 40; Univ Chicago, PhD(chem), 49; Yale Univ, MPH, 75. *Prof Exp:* Asst prof chem, Iowa State Col, 49-55; adv chemist, Westinghouse Elec Corp, 55-57; vis scientist, Saclay Nuclear Res Ctr, France, 57-58; prof chem, US Naval Postgrad Sch, 59-62; mem staff, Inst Defense Anal, 62-65; vis res assoc chem, Univ Pa, 65-67; dir div prof educ, NY State Educ Dept, 67-71; dean instr, Piedmont Va Community Col, 72-73; CONSULT, 73- *Concurrent Pos:* Consult, Nat Bur Standards, 51 & Lawrence Radiation Lab, Univ Calif, Berkeley, 59- *Mem:* AAAS; Am Chem Soc; NY Acad Sci; Royal Soc Chem. *Res:* Thermodynamics; biochemistry; kinetics of fast reactions; epidemiology; occupational health. *Mailing Add:* Ten Sage Hill Lane Albany NY 12204

GRIFFEN, DANA THOMAS, b Washington, DC, Sept 29, 43; m 67; c 5. CRYSTAL CHEMISTRY. *Educ:* US Naval Acad, BS, 65; Va Polytech Inst & State Univ, MS, 73, PhD(mineral), 75. *Prof Exp:* Res assoc geol, Brigham Young Univ, 75-77; res geologist, Phillips Petrol Co, 77-79; from asst prof assoc prof, 79-88, PROF GEOL, BRIGHAM YOUNG UNIV, 88-, CHAIR DEPT, 90- *Concurrent Pos:* Assoc ed, Am Mineralogist, 86- *Mem:* Mineral Soc Am; Am Crystallog Asn; Sigma Xi. *Res:* Crystallography and crystal chemistry of silicate minerals. *Mailing Add:* Dept Geol Brigham Young Univ Provo UT 84602

GRIFFEN, WARD O, JR, b New Orleans, La, July 21, 28; m 52; c 7. SURGERY. *Educ:* Princeton Univ, AB, 48; Cornell Univ, MD, 53; Univ Minn, PhD(surg), 63. *Prof Exp:* From instr to asst prof surg, Med Col, Univ Minn, Minneapolis, 62-65; assoc prof surg, physiol & biophys, 65-66, PROF SURG & CHMN DEPT, SCH MED, UNIV KY, 67- *Concurrent Pos:* USPHS fel, 61-63; Markle scholar acad med, 62-67. *Mem:* Soc Exp Biol & Med; Sigma Xi. *Res:* Gastrointestinal and hepatic physiology and surgery. *Mailing Add:* Dept Surg Univ Ky Med Ctr Lexington KY 40506

GRIFFENHAGEN, GEORGE BERNARD, b Portland, Ore, June 9, 24; m 46; c 3. PHARMACY. *Educ:* Univ Southern Calif, BS, 49, MS, 50. *Prof Exp:* Dir pharmaceut res & asst gen mgr, Nion Corp, Calif, 50-52; cur div med sci, US Nat Mus, Smithsonian Inst, 52-59; consult cur & archivist, Am Pharmaceut Asn, 53-59, managing ed, J, 59-62, ed, J, 62-76, dir div commun, 59-69, exec dir commun, 69- 86, dir int affairs, 86-89, CONSULT, AM PHARMACEUT ASN, 90- *Concurrent Pos:* Lectr, Univ Southern Calif, 50-52; del, Int Pharmaceut Fedn, London Assembly, 55, Brussels Assemby, 58, Copenhagen Assembly, 60, Vienna Assembly, 62, Amsterdam Assembly, 64, Hamburg Assembly, 68, Geneva Assembly, 70, Lisbon Assembly, 72, Rome Assembly, 74, Warsaw Assembly, 76, The Hague, 77, Montreal, 85, Helsinki, 86, Amsterdam, 87, Sydney, 88, Munich, 89, Istanbul, 90; secy gen, Pan-Am Cong Pharm & Biochem, Washington, DC, 57, vpres, Pan-Am Pharm Fedn, 63-; proj adminr, USPHS, secy organizing comn, Int Cong Pharmaceut Sci, Washington, DC, 71, secy gen, Int Cong Hist Pharmaceut, 83, secy gen, Japan-US Cong Pharmaceut Sci, Honolulu, HI, 87, secy, Pharm World Cong, Washington, DC, 91. *Honors & Awards:* Squibb Pan-Am Pharmaceut & Biochem Award, 63; Am Red Cross Meritorious Serv Citation, 64; Edward Kremers Award, 69; Nat Coord Coun on Drug Educ Award, 70; Distinguished Serv Award, Am Wholesalers Asn, 71; Hugo H Schaefer Award, 84; Distinguished Serv Award, Pharm Guild of Australia, 88; Remington Honor Medal, 91. *Mem:* Am Inst Hist Pharm (pres, 60-61); hon mem, Mex, Arg & Gt Brit Pharmaceut Asn; Int Hist Pharm (treas, 70-81, 89-); Int Acad Hist Pharm (treas, 70-81). *Res:* History of pharmacy and medical science; pharmaceutical journalism; pharmaceutical philately. *Mailing Add:* Am Pharmaceut Asn 2215 Constitution Ave NW Washington DC 20037

GRIFFIN, ALLAN, b Vancouver, BC, Feb 10, 39; m 78. CONDENSED MATTER PHYSICS. *Educ:* Univ BC, BSc, 60, MSc, 61; Cornell Univ, PhD(theoret physics), 65. *Prof Exp:* Asst prof, 67-70, assoc prof, theoret physics, 70-76, PROF PHYSICS, UNIV TORONTO, 76- *Mem:* Am Phys Soc; Can Asn Physicists. *Res:* Many-body problems in condensed matter physics; superfluid 4 helium; high temperature superconductivity. *Mailing Add:* Dept Physics Univ Toronto Toronto ON M5S 1A7 Can

GRIFFIN, ANSELM CLYDE, III, b Greenville, Miss, Dec 19, 46; m 69; c 3. PHYSICAL ORGANIC CHEMISTRY. *Educ:* Miss Col, BS, 69; Univ Tex, Austin, PhD(chem), 75. *Prof Exp:* from asst prof to assoc prof, 75-83, PROF CHEM & POLYSCI, UNIV SOUTHERN MISS, 83- *Mem:* Am Chem Soc; North Am Thermal Anal Soc; Royal Soc Chem. *Res:* Structure-property relationships in liquid crystals; solid state chemistry; liquid crystalline polymers; synthetic polymer chemistry. *Mailing Add:* Univ Southern Miss Box 5043 Southern Sta Hattiesburg MS 39406

GRIFFIN, CHARLES CAMPBELL, b Philadelphia, Pa, July 23, 38; m 60; c 2. BIOCHEMISTRY. *Educ:* Catholic Univ, AB, 60; Johns Hopkins Univ, PhD(biochem), 69. *Prof Exp:* Res assoc biochem, Armed Forces Inst Path, 60-64; asst prof biochem, 68-74, actg chmn dept chem, 84-86, ASSOC PROF BIOCHEM, MIAMI UNIV, 74- *Mem:* Am Soc Biochem & Molecular Biol; Am Chem Soc. *Res:* Mechanisms of enzyme action; membrane transport; heparin derivatives. *Mailing Add:* Dept Chem Miami Univ Oxford OH 45056

GRIFFIN, CHARLES FRANK, b Slaton, Tex, Nov 2, 35; m 58; c 3. PHYSICS. *Educ:* Tex Tech Col, BS, 59, MS, 61; Ohio State Univ, PhD(physics), 64. *Prof Exp:* Asst physics, Tex Tech Col, 59-61 & Ohio State Univ, 61-64; assoc prof, Sam Houston State Col, 64-67; from asst prof to assoc prof, 67-76, PROF PHYSICS, UNIV AKRON, 76- *Mem:* Am Phys Soc; Am Asn Physics Teachers. *Res:* Nuclear magnetic resonance; computer aided instruction. *Mailing Add:* Dept Physics Univ Akron Akron OH 44325-4001

GRIFFIN, CLAIBOURNE EUGENE, JR, b Rocky Mount, NC, Oct 15, 29; m 72; c 2. ORGANIC CHEMISTRY. *Educ:* Princeton Univ, BA, 51; Univ Va, MS, 53, PhD(chem), 55. *Prof Exp:* Instr chem, Univ Va, 53-55, res assoc biochem, Sch Med, 55; USPHS res fel chem, Cambridge Univ, 55-57; from instr to prof, Univ Pittsburgh, 57-69; prof & chmn dept, Univ Toledo, 69-74; dean grad studies & res, 74-77, PROF CHEM, UNIV AKRON, 74-, DEAN, COL ARTS & SCI, 77- *Concurrent Pos:* Consult, Stauffer Chem Co, 62-82; adj prof chem, Bowling Green State Univ, 73-74. *Res:* Synthesis and reactions of organophosphorus compounds; nuclear magnetic resonance spectroscopy. *Mailing Add:* Col Arts & Sci Univ Akron Akron OH 44325

GRIFFIN, CLAUDE LANE, b Lebanon, Mo, Nov 9, 37; m 62; c 3. PHARMACOLOGY. *Educ:* Univ Mo-Kans City, BS, 61, MS, 63; Ore State Univ, PhD(cardiovasc pharmacol), 66. *Prof Exp:* Sr scientist, Smith Kline & French Labs, 66-67, group res leader, 67-68, asst dir, 69-70; dir biomet & res support, Merrell-Dow Res Inst, 70-77; VPRES & MANAGING DIR, DEVELOP RES, MARION MERRELL DOW, 82- *Res:* Cardiovascular pharmacology, metabolism of cardiac glycosides. *Mailing Add:* Marion Merrell Dow 2110 E Galbraith Rd Cincinnati OH 45215

GRIFFIN, CLAYTON HOUSTOUN, b Atlanta, Ga, June 14, 25; m; c 3. TRANSMISSION & DISTRIBUTION SYSTEMS. *Educ:* Ga Inst Technol, BEE, 45, MS, 50. *Prof Exp:* Elec officer, USN, 51-53; tester, Ga Power Co, 49-51, test engr, 53-62, protection engr, 62-67, chief protection engr, 67-79, mgr, Syst Protection & Control Dept, 79-89; CONSULT ELEC ENGR, 89- *Concurrent Pos:* Vis prof, Sch Elec Eng, Ga Inst Technol, 75-89; dir, sem cogeneration protection, Inst Elec & Electronics Engrs, 85- *Mem:* Fel Inst Elec & Electronics Engrs. *Res:* High impedance faults on distribution systems; ground fault protection of AC turbine generators; interface protection of cogeneration installations; coordination of transmission line directional ground relays. *Mailing Add:* 221 S Chace Atlanta GA 30328

GRIFFIN, DANA GOVE, III, b Fort Worth, Tex, Nov 9, 38; m 64. BRYOLOGY. *Educ:* Tex Tech Col, BS, 61, MS, 62; Univ Tenn, PhD(bot), 65. *Prof Exp:* Res asst bot, Tex Tech Col, 60-62; from instr to asst prof, Univ Tenn, 65-67; from asst prof to assoc prof, 67-79, PROF BOT, UNIV FLA, 79- *Concurrent Pos:* Fulbright lectr, Peru, 65-66; invited lectr, Inst Nat de Pesquisas da Amazonia, Manaus, Brazil, 74. *Mem:* Bot Soc Am; Am Soc Plant Taxon; Int Asn Plant Taxon; Brit Bryol Soc; Nordic Bryol Soc; Sigma Xi. *Res:* Taxonomy and ecology of the lower plants; taxonomy of bryophytes, especially of tropical mosses; migration of floras, plant geography. *Mailing Add:* Dept Bot 206 FSM Bldg Univ Fla Gainesville FL 32611

GRIFFIN, DAVID H, b Buffalo, NY, Mar 13, 37; m 60; c 3. MYCOLOGY. *Educ:* State Univ NY, BS, 59; Univ Calif, Berkeley, MA, 60, PhD(bot), 63. *Prof Exp:* Res fel biol, Calif Inst Technol, 63-64; asst prof bot, Univ Iowa, 64-68; from asst prof to assoc prof, 68-80, PROF, COL ENVIRON SCI & FORESTRY, STATE UNIV NY, 80- *Concurrent Pos:* Air Force Off Sci Res fel, 63-64; res fel DGRST, Univ Louis Pasteur, Strasbourg, France, 74-75. *Mem:* Am Soc Microbiol; Mycol Soc Am; Am Phytopath Soc. *Res:* Physiology and biochemistry of development in fungi; mechanisms of fungal parasitism. *Mailing Add:* Col Environ Sci & Forestry State Univ NY Syracuse NY 13210

GRIFFIN, DAVID WILLIAM, b Berkeley, Calif, Nov 5, 55; m 81; c 2. ENVIRONMENTAL PHYSIOLOGY. *Educ:* Univ Calif, Davis, BS, 79 & 83, MS, 84, DVM, 85. *Prof Exp:* Physiologist, Rasor Assoc Inc, 80-81; intern med & surg, Sacramento Animal Med Group, 85-86; resident surg, Vet Med Teaching Hosp, 86-89, ASSOC VET, DEPT ANIMAL PHYSIOL, UNIV CALIF, DAVIS, 89-; STAFF SURGEON, SACRAMENTO VET SURG SERV, 89- *Mem:* Am Vet Med Asn. *Res:* Environmental physiology-effects of altered environments (light, temperature, gravity) on physiological processes. *Mailing Add:* Dept Animal Physiol Univ Calif Davis CA 95616-8519

GRIFFIN, DIANE EDMUND, b Iowa City, Iowa, May 12, 40; m 65; c 2. IMMUNOLOGY, VIROLOGY. *Educ:* Augustana Col, BA, 62; Stanford Univ, MD, 68, PhD(med microbiol), 71. *Prof Exp:* Intern med, Stanford Univ Hosp, 68-69, resident, 69-70; fel virol, 70-73, from asst prof to assoc prof, 73-86, PROF MED & NEUROL, SCH MED, JOHNS HOPKINS UNIV, 86- *Concurrent Pos:* Investr, Howard Hughes Med Inst, 75-82. *Mem:* Am Asn Immunologists; Am Soc Microbiol; AAAS; Am Fedn Clin Res; Infectious Dis Soc Am; Am Soc Clin Invest. *Res:* Immune response to viral infection; role of the immune response in recovery from or production of viral disease; neurovirulence of alphaviruses. *Mailing Add:* Meyer 6-181 Dept Med & Neurol Johns Hopkins Univ Sch Med 600 N Wolfe St Baltimore MD 21205

GRIFFIN, DONALD R(EDFIELD), b Southampton, NY, Aug 3, 15; m 41; c 4. COGNITIVE ETHOLOGY, COMPARATIVE PHYSIOLOGY. *Educ:* Harvard Col, BS, 38; Harvard Univ, MA, 40, PhD(biol), 42. *Hon Degrees:* DSc, Ripon Col, Wis, 66, Eberhard-Karls Univ, Tübingen, Ger, 88. *Prof Exp:* Teaching asst biol, Harvard Univ, 38-40, jr fel, 40-41 & 46, res assoc, 42-45; from asst prof to prof zool, Cornell Univ, 46-53; prof zool, Harvard Univ, 53-65, chmn, Dept Biol, 62-65; prof, Rockefeller Univ, 65-86; vis lectr, Dept Biol, Princeton Univ, 87-89; ASSOC, MUS COMP ZOOL, HARVARD UNIV, 89- *Concurrent Pos:* Dir, Inst Res Animal Behav, Rockefeller Univ & NY Zool Soc, 65-69; mem bd trustees, Rockefeller Univ, 73-76; pres, Harry Frank Guggenheim Found, 79-83. *Honors & Awards:* Eliot Medal, Nat Acad Sci, 61. *Mem:* Nat Acad Sci; Am Philos Soc; Am Acad Arts & Sci; Animal Behav Soc; Am Soc Zoologists; Am Physiol Soc; Ecol Soc Am. *Res:* Animal orientation; echolocation of bats, especially their use of sonar to capture flying insects; cognitive ethology; evidence of animal cognition and consciousness, especially the use of animal communication as a window on animal minds. *Mailing Add:* Concord Field Sta Harvard Univ Old Causeway Rd Bedford MA 01730

GRIFFIN, EDMOND EUGENE, b Marshall, Ark, June 5, 30; m 54; c 2. PHYSIOLOGY, RADIOBIOLOGY. *Educ:* Univ Ark, BSEd, 62; Univ Tenn, MS, 64, PhD(radiation biol), 69. *Prof Exp:* High sch teacher, Ark, 52-53; health physicist, Los Alamos Sci Lab, Univ Calif, 53-56, chem operator, 56-61; fel, Med Sch, Univ Tex, San Antonio, 69 & Sch Med, Univ Rochester, 69-73; asst prof physiol, Southwestern Med Sch, Univ Tex Health Sci Ctr Dallas, 73-77; prof biol & chmn dept, Univ Cent Ark, 77-78; assoc dir environ health, Univ Tex Health Sci Ctr, Dallas, 78-82; sci adminr, Am Heart Asn, 82-84; PROF & CHMN, BIOL DEPT, UNIV CENT ARK, 86- *Mem:* AAAS; Radiation Res Soc; Am Soc Zoologists; Sigma Xi. *Res:* Actions of hormones and effects of radiation on metabolism. *Mailing Add:* 12280 Rivercrest Dr Little Rock AR 72212-1436

GRIFFIN, EDWARD L(AWRENCE), JR, b Washington, DC, Jan 9, 19; m 41; c 2. CHEMICAL ENGINEERING. *Educ:* Cornell Univ, BCh, 40, ChE, 41. *Prof Exp:* Chem engr, James Lees & Sons, Pa, 41-42; chem engr, Eastern Regional Res Lab, Bur Agr & Indust Chem, USDA, 42-57, chief, Eng & Develop Lab, Northern Regional Res Ctr, 57-75; RETIRED. *Mem:* Am Inst Chem Engrs; Am Chem Soc; Am Oil Chem Soc; Am Asn Cereal Chemists; Inst Food Technol. *Res:* Chemical engineering pilot plant studies; rubber from guayule and Kok-Saghyz; volatile flavor from fruit juices; acrylic ester polymerization; production of allyl-sucrose; tanning material from canaigre; extraction of rutin from dried plants; cereal crop and oilseed utilization research. *Mailing Add:* 4206 Keenland Ave Peoria IL 61614

GRIFFIN, ERNEST LYLE, b Tampa, Fla, May 25, 21. MATHEMATICS. *Educ:* Emory Univ, BA, 43; Univ Chicago, MS, 47, PhD(math), 52. *Prof Exp:* From instr to asst prof math, Univ Mich, 52-62; assoc ed, Math Rev, 62-63; vis assoc prof math, Univ Pa, 63-66, assoc prof, 66-67; PROF MATH, LA STATE UNIV, BATON ROUGE, 67- *Concurrent Pos:* Vis asst prof, Columbia Univ, 57-58. *Mem:* Am Math Soc. *Res:* Theory of algebras of operators on Hilbert spaces and applications to representations of locally compact groups and quantum physics; classification and properties of von-Neumann algebras. *Mailing Add:* 940 Sanford Ave Apt 404 Baton Rouge LA 70808

GRIFFIN, FRANK M, JR, b Orangeburg, SC, May 17, 41. INTERNAL MEDICINE, INFECTIOUS DISEASES. *Educ:* Col Charleston, BS, 62; Univ SC, MD, 66. *Prof Exp:* Asst prof med, 75-83, assoc prof microbiol, 79-85, PROF MED, UNIV ALA, BIRMINGHAM, 83-, PROF MICROBIOL, 85- *Mem:* Am Fedn Clin Res; Infectious Dis Soc Am; AAAS; Am Bd Internal Med; Am Soc Microbiol. *Mailing Add:* Div Infectious Dis Univ Ala Univ Sta Med Ctr Birmingham AL 35294

GRIFFIN, GARY J, b Glen Cove, NY, Dec 23, 37; m 59; c 3. BIOLOGY. *Educ:* Colo State Univ, BS, 59, MS, 61, PhD(plant path), 62. *Prof Exp:* Plant pathologist, Agr Res Serv, USDA, 62-63; from asst prof to assoc prof biol, Morehead State Col, 63-67; asst prof plant path & physiol, 67-70, assoc prof, 70-76, PROF PLANT PATH, VA POLYTECH INST & STATE UNIV, 76- *Mem:* Am Phytopath Soc. *Res:* Soil microbiology; root diseases. *Mailing Add:* Dept Plant Path & Physiol Va Polytech Inst & State Univ Blacksburg VA 24061

GRIFFIN, GARY WALTER, b Pasadena, Calif, Nov 12, 31; m; c 4. ORGANIC CHEMISTRY. *Educ:* Pomona Col, BA, 53; Univ Ill, Urbana, PhD(org chem), 56. *Prof Exp:* Res chemist, Humble Oil & Refining Co, Tex, 56-58; from instr to asst prof chem, Yale Univ, 58-63; assoc prof, Tulane Univ, 63-66; assoc prof, 66-67, PROF CHEM, UNIV NEW ORLEANS, 67- *Concurrent Pos:* Consult, Am Cyanamid Co, 59-61, Southern New Eng Ultraviolet Co, Conn, 62-77 & Minn Mining & Mfg Co, 63-; res grants, Res Corp, 60-61, NSF, 60-62, 64-73, 78-81, Army Res Off-Durham, 61-75, George Sheffield Fund, 62-63, NIH, 63-66, 75-78, Cancer Asn Greater New Orleans, 66-67, 69-70, 74-76, Petrol Res Fund, 66-69, 71-73, 75-77, Merck Sharp & Dohme, 78-, UN Orgn Res Coun, Japan, 71 & Can, 75, Damon Runyon Mem Fund, 72-73 & 3M Co, 72-78; vis prof, Univ New Orleans, 65-66 & Inst Lipid Res, Baylor Col Med, Tex Med Ctr, Houston, 71-72; NIH sr res fel, 71-72. *Mem:* Am Chem Soc; The Chem Soc; Int Asn Heterocyclic Chemists; Am Soc Mass Spectrometry; AAAS. *Res:* Nonbenzenoid aromatic chemistry; carbene chemistry; singlet oxygen chemistry; photochemistry; small-ring chemistry; electrochemistry; plasma chromatography; chemical ionization mass spectrometry; gas phase ion molecule chemistry; biochemical and biomedical problems of a photochemical nature; photoinduced disorders. *Mailing Add:* Dept Chem Univ New Orleans Lake Front New Orleans LA 70148

GRIFFIN, GEORGE MELVIN, JR, b Baltimore, Md, Apr 14, 28; m 50; c 2. MARINE GEOLOGY, SEDIMENTOLOGY. *Educ:* Univ NC, BA, 52, MS, 54; Rice Univ, PhD(geol), 60. *Prof Exp:* Res geologist clay petrol, Explor & Prod Res Div, Shell Develop Co, 54-65; assoc prof & head dept geol, Dayton Campus, Miami-Ohio State Univ, 65-66; proj leader, World Wide Tech Serv Ctr, Gulf Oil Corp, 66-67; assoc prof geol, 67-70; PROF GEOL, UNIV FLA, 70- *Concurrent Pos:* Proj dir & chief scientist, Key Largo Lab, Harbor Br Found, 72-74; vpres explor & prod, Campbell Oil Co, 77-78; consult, minerals, oil & gas, environ probs, 70-, mem Coun Gulf Univ Res Consortium, 75- *Mem:* Am Asn Petrol Geol; Soc Econ Paleontologists & Mineralogists. *Res:* Geologic significance of mineral assemblages in sedimentary rocks, especially clay minerals; geothermal gradients; turbidity in coastal waters; and environmental effects of dredging and offshore petroleum production; sedimentation processes and products. *Mailing Add:* Dept Geol Univ Fla Gainesville FL 32611

GRIFFIN, GERALD D, b Escalante, Utah, May 31, 27; m 52; c 3. PLANT NEMATOLOGY, PLANT PATHOLOGY. *Educ:* Univ Utah, BS, 53, MS, 56; Univ Wis, PhD(plant path), 63. *Prof Exp:* Nematologist, Agr Res Serv, Utah, 56-59, Wis, 59-63, NEMATOLOGIST, AGR RES SERV, USDA, UTAH STATE UNIV, 63- *Mem:* AAAS; Am Phytopath Soc; Soc Nematol. *Res:* Biology and control of plant-parasitic nematodes associated with field crops, vegetables and stone fruit. *Mailing Add:* 1553 E 1220 N Logan UT 84321

GRIFFIN, GREGORY LEE, b Montebello, Calif. HETEROGENEOUS CATALYSIS. *Educ:* Calif Tech, BS, 75; Princeton Univ, PhD(chem eng), 79. *Prof Exp:* Fel Nat Bur Standards, 79-80, asst prof chem eng, Univ Minn, 80-87; ASSOC PROF CHEM ENG, LA STATE UNIV, 87- *Mem:* Am Inst Chem Engrs; Am Chem Soc; Am Vacuum Soc; Am Ceramics Soc. *Res:* Heterogeneous catalysis; chemical vapor deposition of ceramic coatings. *Mailing Add:* 3614 Sessions Dr Baton Rouge LA 70816-2727

GRIFFIN, GUY DAVID, b Omaha, Nebr, Aug 27, 42; m 66; c 1. CHEMICAL CARCINOGENESIS, CELL CULTURE. *Educ:* Univ Nebr at Omaha, BA, 64; Univ Nebr Med Ctr, Omaha, MS, 69, PhD(biochem), 71. *Prof Exp:* Supv cystic fibrosis res, Pediat Lab, Univ Nebr Med Ctr, 65-71, res instr, 71-72; investr cancer res, 72-76, staff scientist, Biol Div, 76-78, STAFF SCIENTIST, HEALTH EFFECTS RES, ADVAN MONITOR DEVELOP, HEALTH & SAFETY RES DIV, OAK RIDGE NAT LAB, 78- *Concurrent Pos:* Res fel, Nat Cystic Fibrosis Res Found, 71-72; fel, Am Cancer Soc 72-74 & Nat Cancer Inst, USPHS, 74-76; instr org chem, Bryan Col, 86-87, 89- *Honors & Awards:* IR-100 Award, 87. *Mem:* Am Chem Soc; Sigma Xi; Am Soc Biol Chem. *Res:* Molecular basis for the toxicity of environmental pollutants; biological indicators of extent of human exposure to chemicals; biochemial mechanisms of lymphocyte activation and control of immunoglobulin synthesis; cellular regulation by transfer RNA. *Mailing Add:* 210 Alhambra Rd Oak Ridge TN 37830

GRIFFIN, HAROLD LEE, b Canton, Ill, Nov 23, 28; m 54; c 3. STARCH, CELLULOSE. *Educ:* Univ Ill, Urbana, BS, 54. *Prof Exp:* Chemist dextran, 54-56, chemist starch, 56-59, chemist starch & enzymes, 59-63, res chemist starch enzymes & feedlot waste, 63-75, res chemist fungal cellulase, 75-85, RES CHEMIST BACTERIAL CELLULASE, NORTHERN REGIONAL RES CTR, AGR RES SERV, MIDWEST AREA, USDA, 85- *Mem:* AAAS. *Res:* Molecular parameters and dilute solution properties of native and modified macromolecular carbohydrates; enzymes of starch biosynthesis in various corn varieties; pollution control particularly feedlot waste; cellulase enzymology; Iisolation and physical enzymes and chemical characterization of fungal bacterial enzyme systems; fungal and bacterial cellulases. *Mailing Add:* Northern Regional Res Ctr 1815 N University Peoria IL 61604

GRIFFIN, HENRY CLAUDE, b Greenville, SC, Feb 14, 37; m 60; c 2. NUCLEAR SPECTROSCOPY, RADIOCHEMISTRY. *Educ:* Davidson Col, BS, 58; Mass Inst Technol, PhD(nuclear chem), 62. *Prof Exp:* Resident res assoc nuclear chem, Argonne Nat Lab, 62-64; from asst prof to assoc prof, 64-89, dir freshman studies, Dept Chem, 74-82, PROF NUCLEAR CHEM & RADIOCHEM, UNIV MICH, ANN ARBOR, 89- *Concurrent Pos:* Guest scientist, Swiss Fed Inst for Reactor Res, 71-72; res partic Nuclear Chem Div, Lawrence Livermore Lab & vis res engr Dept Nuclear Eng, Univ Calif, Berkeley, 78-79; consult, Environ Res Group Inc, Energy Data Systs Inc. *Mem:* AAAS; Am Phys Soc; Am Chem Soc. *Res:* Nuclear fission; nuclear spectroscopy; nuclear reactions; radiochemistry. *Mailing Add:* Dept Chem Univ Mich Ann Arbor MI 48109-1055

GRIFFIN, JAMES EDWARD, b Sioux City, Iowa, Oct 11, 25; m 53; c 5. ELEMENTARY PARTICLE PHYSICS. *Educ:* Iowa State Univ, BS, 51, PhD(physics), 63. *Prof Exp:* Asst prof & assoc physicist, Iowa State Univ & Ames Lab, 63-69; PHYSICIST, FERMI NAT ACCELERATOR LAB, 69- *Mem:* Am Phys Soc; AAAS. *Res:* Properties of fundamental particles and design considerations of high energy particle accelerators. *Mailing Add:* Fermi Lab MS-341 Box 500 Batavia IL 60510

GRIFFIN, JAMES EMMETT, b Kansas City, Mo, Dec 10, 44; m 68; c 3. ENDOCRINOLOGY. *Educ:* Rockhurst Col, BA, 66; Univ Kans, MD, 70. *Prof Exp:* Intern internal med, Univ Kans Med Ctr, 70-71, resident, 71-72; fel endocrinol, 74-76, from instr to prof internal med, 75-89, assoc dean med educ, 83- 86, ASSOC DEAN ACAD PLANNING, UNIV TEX SOUTHWESTERN MED SCH, 86- *Mem:* Am Col Physicians; Am Fedn Clin Res; Am Soc Clin Invest. *Res:* Disorders of sexual differentiation; androgen resistance syndromes. *Mailing Add:* Dept Internal Med Univ Tex Southwestern Med Ctr Dallas TX 75235

GRIFFIN, JAMES J, b Philadelphia, Pa, Oct 20, 30; m; c 5. THEORETICAL PHYSICS. *Educ:* Villanova Col, BS, 52; Princeton Univ, MS, 54, PhD(physics), 56. *Prof Exp:* Theoret physicist, Los Alamos Sci Lab, 56-66; from asst prof to assoc prof physics & astron, 66-74, assoc chmn dept, 68-69, PROF PHYSICS, UNIV MD, COLLEGE PARK, 74- *Concurrent Pos:* Fulbright scolar, N Bohr Inst, Copenhagen, 55-56; NSF fel, 59-60; vis lectr, Univ Wis-Madison, 65-66; Guggenheim fel, Univ Berkeley, 72-73; Alexander von Humboldt sr vis US scientist, WGer, 75-76; vis scientist, Los Alamos Nat Lab, Oak Ridge Nat Lab, Lawrence Berkeley Nat Lab, Brookhaven Nat Lab, ISN, Grenoble & GSI, Darmstadt. *Mem:* Am Phys Soc; Union Concerned Scientist; Comt Concerned Scientists; Am Civil Liberties Union; Amnesty Int. *Res:* Nuclear physics; nuclear collective model via generator coordinates; pre-equilibrium nuclear reactions; dynamical nuclear many body problem; heavy ion reactions; fission and nuclear structure; nonlinear and quantal/classical connections; hypothetical quadronium (e plus e plus e minus e minus) leptonic atom. *Mailing Add:* Dept Physics Univ Md College Park MD 20742

GRIFFIN, JAMES RICHARD, ecology, for more information see previous edition

GRIFFIN, JANE FLANIGEN, b Mar 26, 33; US citizen; m 54; c 4. PHYSICAL CHEMISTRY, BIOPHYSICS. *Educ:* D'Youville Col, BA, 54; State Univ NY, Buffalo, PhD(chem), 74. *Prof Exp:* Fel, 74-77, res scientist, 77-88, HEAD, MOLECULAR BIOPHYS DEPT, MED FEDN BUFFALO, 88- *Concurrent Pos:* Co-prin investr, NIH grant, 77-79, prin investr, 80-82. *Mem:* Am Chem Soc; Am Crystallog Asn; Royal Soc Chem; NY Acad Sci; AAAS. *Res:* Structure of steroid hormones; oproid peptides; cardioactive drugs. *Mailing Add:* 73 High St Buffalo NY 14203

GRIFFIN, JERRY HOWARD, b Miami, Fla, July 14, 45; m 67; c 2. VIBRATIONS, FRACTURE MECHANICS. *Educ:* Univ S Fla, BS & MS, 69; Calif Inst Technol, PhD(eng), 73. *Prof Exp:* Sr engr vibrations, Pratt & Whitney Aircraft, 73-74; lectr appl mech, Univ Auckland, NZ, 74-76; sr engr vibrations, Pratt & Whitney Aircraft, 76-77, tech specialist fatigue, 77-79, mgr fan struct group, 79-80; from asst prof to assoc prof, 80-87, PROF MECH ENG, CARNEGIE MELLON UNIV, 87- *Concurrent Pos:* Vis sr scientist, Air Force Wright Aeronaut Labs, 81; consult, USN, 84-85, NASA, 86-87, Rocketdyne Div, Rockwell Int, 86-, Pratt & Whitney Aircraft & AVCO Lycoming, 88-; sr partner, Griffin Consult, 83- *Mem:* Am Soc Mech Engrs; Am Inst Aeronaut & Astronaut. *Res:* Structural dynamics which include vibration of jet engines, boat drive-trains, friction effects, space structures and random response of nonlinear systems; fracture mechanics which include path independent integrals, thermo-mechanical fatigue. *Mailing Add:* Carnegie Inst Technol Carnegie Mellon Univ Pittsburgh PA 15213

GRIFFIN, JOE LEE, b Bass, Ark, Sept 8, 35; m 58; c 2. CELL BIOLOGY. *Educ:* Univ of the South, BS, 56; Princeton Univ, PhD(biol), 59. *Prof Exp:* From instr to asst prof biol, Brown Univ, 59-62; NIH spec fel anat, Harvard Univ Med Sch, 62-64; res biologist, Armed Forces Inst Path, 64-74, chief, Div Exp Neuropath, 74-83, res neuromyologist, Div Neuromuscular Path, 83- 90, RES BIOLOGIST, DIV CELLULAR PATH, ARMED FORCES INST PATH, 90- *Mem:* AAAS. *Res:* Flow cytometry; cell cycles; neuromuscular integration related to tension and health. *Mailing Add:* Armed Forces Inst Path Washington DC 20306-6000

GRIFFIN, JOHN HENRY, b Seattle, Wash, June 26, 43; m 65; c 3. BIOCHEMISTRY, EXPERIMENTAL PATHOLOGY. *Educ:* Univ Santa Clara, BS, 65; Univ Calif, Davis, PhD(biophys), 69. *Prof Exp:* Guest worker, NIH, 71-73; mem staff biochem, Ctr Nuclear Studies, Saclay, France, 73-74; asst, 74-75, assoc, 75-80, ASSOC MEM IMMUNOPATH, MOLECULAR IMMUNOL, SCRIPPS CLIN RES FOUND, 80- *Concurrent Pos:* Helen Hay Whitney Found res fel biol chem, Harvard Med Sch, 69-71. *Mem:* Am Soc Biol Chemists; Am Asn Immunol; Am Asn Path; Am Chem Soc; Sigma Xi. *Res:* Basic biochemical and clinical research on the regulation of thrombosis and hemostasis based on studies of purified blood coagulation proteins. *Mailing Add:* Scripps Clin Res FDN 10666 N Torrey Pines Rd La Jolla CA 92037

GRIFFIN, JOHN LEANDER, b Toledo, Ohio, Nov 9, 23; m 47; c 2. PHYSICAL CHEMISTRY. *Educ:* Univ Toledo, BEng, 45, MS, 53; Univ Mich, PhD(chem), 62. *Prof Exp:* Jr chemist, Chase Bag Co Lab, 45-46; teaching fel chem, Univ Toledo, 46-47, instr, 47-48, 49-50; asst anal chemist, Res Lab, Owens-Ill Glass Co, 48-49; plant chemist, Kaylo Div, 50-51; lectr chem, Univ Toledo, 51-52; sr res chemist, Gen Motors Corp, 57-69, supvry res chemist, 69-77, dept res scientist, 77-81, sr staff res scientist, electrochem dept, Res Labs, 81-84; RETIRED. *Mem:* AAAS; Am Chem Soc; Electrochem Soc (treas, 76-82); hon mem Electrochem Soc. *Res:* Electrochemistry, especially electrodeposition, secondary batteries and corrosion; industrial analytical chemistry. *Mailing Add:* 26177 Lepley Rd Howard OH 43028

GRIFFIN, JOHN R(OBERT), b Du Quoin, Ill, Apr 21, 36; m 57; c 4. CHEMICAL ENGINEERING, PLASTICS TECHNOLOGY. *Educ:* Univ Ill, BS, 59; Purdue Univ, PhD(chem eng), 63. *Prof Exp:* Res chem engr process develop, Humble Oil & Refining Co, 63-65, res chem engr, Esso Res & Eng Co, 65, supvr polymerization process develop, 65-71, supvr low density polyethylene & new prod, 71-72, mgr high performance plastics technol, 72-78, SR RES ASSOC, EXXON CHEM CO, 78- *Honors & Awards:* A M White Award, Am Inst Chem Engrs, 59. *Mem:* Am Inst Chem Engrs; Soc Plastics Engrs. *Res:* Plastic process and product research and development. *Mailing Add:* 225 Pin Oak Dr Baytown TX 77520

GRIFFIN, KATHLEEN (MARY), b Milwaukee, Wis, Oct 1, 43. SPEECH PATHOLOGY, AUDIOLOGY. *Educ:* Univ Wis, Madison, BS, 65; Stanford Univ, MA, 66; Univ Ore, PhD(speech path & audiol), 71. *Prof Exp:* Speech clinician, Holladay Ctr Crippled Children, Portland Pub Schs, 66-68; speech consult crippled childrens div, Med Sch, Univ Ore, 67; dir speech & audiol dept, Glendale Adventist Hosp, 71-74; dir clin & hosp prog, Am Speech-Lang-Hearing Asn, 74-78, dir res & prof develop dept, 78-80, dep exec dir & dir econ dept, 80-85; exec vpres, Am Col Health Care Adminr, 85-88; exec vpres opers, Rebound, 88-89, pres, 89-90; PRES & CHIEF EXEC OFFICER, AM TRANSITIONAL CARE, INC, 90- *Concurrent Pos:* Asst prof, Calif State Col, Los Angeles, 71-74. *Mem:* Fel Am Speech-Lang-Hearing Asn; NY Acad Sci; fel Am Soc Asn Execs. *Res:* Operant conditioning in stuttering and other speech and language therapy; auditory perception in cerebral palsy; quality assurance; thermography and speech functioning. *Mailing Add:* PO Box 541912 Houston TX 77254-1912

GRIFFIN, MARTIN JOHN, b Chicago, Ill, Oct 1, 33; m 63; c 4. BIOCHEMISTRY. *Educ:* Loyola Univ, Ill, SB, 55; Univ Chicago, SM, 57, PhD(org chem), 60. *Prof Exp:* Chemist, E I du Pont de Nemours & Co, Del, 60-61; from asst prof to assoc prof, 65-73, PROF BIOCHEM, SCH MED, UNIV OKLA, 73-; AT TRAVENOL LABS INC, ROUND LAKE, ILL, 85- *Concurrent Pos:* NIH fel biochem, Mass Inst Technol, 61-63; NIH fel genetics & med, NY Univ, 63-65; mem staff, Cancer Sect, Okla Med Res Found, 65-85. *Mem:* AAAS; Am Asn Cancer Res; Am Soc Biol Chemists. *Res:* Enzymology and mammalian regulatory mechanisms; membrane biochemistry and oncology; drug metabolism. *Mailing Add:* Res & Develop Dept G D Searle P-442 4701 Searle Pkwy Skokie IL 60077

GRIFFIN, MARVIN A, b Mar 28, 23; US citizen; m 49; c 4. OPERATIONS RESEARCH, INDUSTRIAL ENGINEERING. *Educ:* Auburn Univ, BS, 49; Univ Ala, MSE, 52; Johns Hopkins Univ, DrEng, 60. *Prof Exp:* Opers analyst, Army Ord Dept, Ala, 49-51; sr proj engr, Western Elec Co, NC, 52-55; chief engr, Cumberland Mfg Co, Tenn, 55-57; instr opers res, Johns Hopkins Univ, 57-60; chief indust engr, Matson Navig Co, Calif, 60-61, sr vpres, 76-80; prof indust eng, Univ Ala, Tuscaloosa, 61-76, head dept, 65-76, coordr, Comput Sci Prog, 69-76, prof comput sci, 76-79, prof & head indust eng, 84-87; RETIRED. *Concurrent Pos:* Proj dir, NASA-Univ Ala Contract, 62-65; NSF grant, 63-64; fel comput sci, Johns Hopkins Univ, 68-69; consult pvt co & States of Ala & Tenn, arbitrator-mediator, Fed Mediation & Conciliation Serv & Am Arbit Asn, expert witness safety eng. *Mem:* Nat Acad Sci; Am Soc Eng Educ; Am Inst Indust Engrs; Asn Comput Mfrs; Opers Res Soc; Maritime Transport Comn. *Res:* Industrial operations research; analysis of telemetry systems for Saturn vehicle; computer information systems. *Mailing Add:* 2640 Claymont Circle Tuscaloosa AL 35404

GRIFFIN, PATRICK J, b Warren, Ohio, Oct 26, 51; m 74, 87. NUCLEAR THEORY DEFINITION, RADIATION TRANSPORT COUPLING. *Educ:* Ohio Univ, BS, 73, PhD(physics), 79. *Prof Exp:* Res scientist, Kaman Sci Corp, 79-85, prog mgr, 85-88; SCIENTIST, SANDIA NAT LAB, 89- *Mem:* Am Phys Soc; Inst Elec & Electronics Engrs. *Res:* Reactor simulation fidelity, neutron damage to materials and radiation transport. *Mailing Add:* 2872 Tramway Circle NE Albuquerque NM 87122

GRIFFIN, PAUL JOSEPH, b Westfield, Mass, Dec 31, 54; m 83; c 1. DIELECTRIC FLUIDS. *Educ:* Am Int Col, BS, 76; Univ RI, MS, 80. *Prof Exp:* LAB MGR, DOBLE ENG CO, 78- *Mem:* Am Soc Testing & Mat. *Res:* Physical, chemical and electric properties of dielectric fluids; gassing characteristics of various liquids under electrical stresses; static electrification of flowing liquids; long term stability of dielectric liquids in-service. *Mailing Add:* 85 Walnut St Watertown MA 02172

GRIFFIN, RALPH HAWKINS, b Roanoke, Va, Feb 28, 21; m 43; c 3. SILVICULTURE, FOREST ECOLOGY. *Educ:* Va Polytech Inst, BS, 43; Yale Univ, MF, 47; Duke Univ, DF, 56. *Prof Exp:* Forester, Va Forest Serv, 47-51; prof forestry, Agr & Tech Col, NC, 53-56; from asst prof to prof, 56-86, EMER PROF FOREST RESOURCES, UNIV MAINE, 86- *Mem:* Soc Am Foresters; Ecol Soc Am; AAAS; Am Forestry Asn; Forest Hist Soc; Sigma Xi. *Mailing Add:* 14 Sylvan Rd Univ Me Orono ME 04473

GRIFFIN, RICHARD B, b Buffalo, NY, Aug 31, 42; m 65; c 2. AQUEOUS CORROSION, ENVIRONMENTAL DEGRADATION. *Educ:* Pa State Univ, BS, 64; Iowa State Univ, PhD(metall), 69. *Prof Exp:* Mat res engr, Watervliet Arsenal, NY, 69-74; mat engr, US FSTC, Charlottesville, Va, 74-77; from asst prof to assoc prof mat sci & corrosion, 77-90, ASST DEPT HEAD, UNDERGRAD PROG, MECH ENG DEPT, TEX A&M UNIV, 90- *Mem:* Nat Asn Corrosion Engrs; Am Soc Metals; Am Soc Eng Educ. *Res:* Aqueous corrosion; cathodic protection of offshore structures and corrosion of coatings. *Mailing Add:* Mech Eng Dept 3123 College Station TX 77843

GRIFFIN, RICHARD E, agricultural engineering, for more information see previous edition

GRIFFIN, RICHARD NORMAN, b Winchester, Mass, Nov 2, 29; m 55; c 3. PLASTICS ENGINEERING. *Educ:* Columbia Univ, AB, 51; Mass Inst Technol, PhD(org chem), 58. *Prof Exp:* Res chemist, E I du Pont de Nemours & Co, 57-61; res chemist, Space Div, 61-77, mgr solar progs, Energy Syst Prog Dept, 78-83, MGR ADVAN COMPOSITES, GE AIRCRAFT ENGS, GE ELEC CO, 83- *Mem:* Am Chem Soc; Soc Advan Mat & Process Eng. *Res:* Kinetics and mechanisms of organic reactions; photochemistry; energy transfer; radiation chemistry; electrophoretic purification of biologicals; plastics and composites. *Mailing Add:* 9821 Tollgate Lane Montgomery OH 45242

GRIFFIN, ROBERT ALFRED, b Long Beach, Calif, June 7, 44; m 64; c 2. GEOCHEMISTRY, SOIL CHEMISTRY. *Educ:* Univ Calif, Davis, 66, MS, 68; Utah State Univ, PhD(soil chem), 73. *Prof Exp:* Res assoc, 73-74; from asst geochemist to assoc geochemist, Ill State Geol Surv, 74-78, geochemist & head sect geochem, 78-90; CUDWORTH PROF ENVIRON ENG, UNIV ALA, 90-; DIR ENVIRON INST WASTE MGT STUDIES. *Mem:* Am Soc Agron; Soil Sci Soc Am; Am Chem Soc; Sigma Xi. *Res:* Solid and hazardous waste disposal; geochemistry of ground waters; soil physical chemistry; solution thermodynamics; adsorption by earth materials. *Mailing Add:* Chem Eng Dept Univ Ala Tuscaloosa AL 35487-0203

GRIFFIN, SUMNER ALBERT, b Ashland, NY, May 11, 22; m 51; c 3. ANIMAL HUSBANDRY. *Educ:* Cornell Univ, BS, 49; Univ Ky, MS, 50; Mich State Univ, PhD(animal husb), 55. *Prof Exp:* Instr animal husb, Mich State Univ, 52-55; mgr animal health & nutrit, Mallinckrodt Chem Works, 55-57; assoc prof animal husb & vet sci, Univ Tenn, Knoxville, 57-70; dean & prof, Sch Agr & Home Econ, Tenn Technol Univ, 70-87; CONSULT, 88- *Concurrent Pos:* Consult, Oak Ridge Nat Lab. *Mem:* Am Soc Animal Sci; Animal Nutrit Res Coun. *Res:* Swine nutrition and physiology management. *Mailing Add:* 987 Georgetown Cookville TN 38501

GRIFFIN, THOMAS SCOTT, pesticide chemistry, for more information see previous edition

GRIFFIN, TRAVIS BARTON, b Trinidad, Tex, Apr 30, 34; m 56; c 4. BIOCHEMISTRY. *Educ:* Tex A&M Univ, BS, 57, MS, 61, PhD(biochem), 66. *Prof Exp:* Asst prof biochem, Tex A&M Univ, 65-69; asst prof, 69-73, res assoc prof toxicol, Inst Comp & Human Toxicol, Albany Med Col, 73-80; EXEC VPRES, COULSTON INT, INC, 80- *Mem:* AAAS; Am Chem Soc; Sigma Xi; NY Acad Sci. *Res:* Enzymology; protein chemistry; toxicology of environmental chemicals. *Mailing Add:* 1841 Corte Del Sol Alamogordo NM 88310

GRIFFIN, VILLARD STUART, JR, b Birmingham, Ala, May 19, 37; m 66; c 4. GEOLOGY. *Educ:* Univ Va, BA, 59, MS, 61; Mich State Univ, PhD(geol), 65. *Prof Exp:* From instr to assoc prof, 64-75, actg head, 90-91, PROF GEOL, CLEMSON UNIV, 75- *Concurrent Pos:* Assoc investr, US Off Water Resources grant, 65-66; consult, Bechtel Corp, 66, John Wiley & Sons Publ, 71, E D'Appolonia Engr, 73, Fulton Nat Bank & Ga Geol Surv, 80; Chevron Corp, 76; E I duPont de Nemours & Co, Inc, 78; co-prin investr, NSF, 68-70, prin investr, 70-74; proj geologist, SC Geol Surv, 65-; mem blue ribbon geol adv panel nuclear reactor siting, SC Elec & Gas Co, 73-76; vis res investr, Geol Surv Finland, Otaniemi, 75; chmn, SC State Mapping Adv Comt, 84-85. *Mem:* AAAS; Sigma Xi; fel Geol Soc Am; Am Inst Prof Geologists. *Res:* Structural geology and petrofabrics; Appalachian geology and tectonics; metamorphic-migmatite terrane analysis; Fennoscandian migmatization; Fennoscandian geology; applied geology. *Mailing Add:* Dept Earth Sci Clemson Univ Clemson SC 29634-1905

GRIFFIN, WILLIAM DALLAS, b Plainfield, NJ, Jan 1, 25; m 47; c 6. ACADEMIC ADMINISTRATION, ORGANIC CHEMISTRY. *Educ:* Rutgers Univ, BSc, 48. *Prof Exp:* Assoc chemist, Cent Res Lab, Allied Chem & Dye Corp, 48-52, res chemist, 52-58, res chemist, Allied Chem Corp, 58-59, group leader, 59-71; sr chemist, Pavelle Corp, 72-73; sci lab coordr, 73-90, ADJ ASSOC PROF CHEM, COUNTY COL MORRIS, 90- *Res:* Process development. *Mailing Add:* 35 Terry Dr Morristown NJ 07960

GRIFFIN, WILLIAM THOMAS, b Thompson, Mo, May 13, 32; m 52; c 3. OBSTETRICS & GYNECOLOGY. *Educ:* Univ Mo-Columbia, MD, 59; Am Bd Obstet & Gynec, dipl, 67. *Prof Exp:* Resident physician, Sch Med, 60-63, from instr to assoc prof, 63-74, PROF OBSTET & GYNEC, UNIV MO-COLUMBIA, 74-, ATTEND OBSTETRICIAN & GYNECOLOGIST, MED CTR, 63- *Concurrent Pos:* Consult gynecologist, Fifth Army, Ft Leonard Wood, Mo, 69- *Mem:* Am Col Obstet & Gynec; AMA; Asn Profs Gynec & Obstet; Am Fertil Soc; Am Col Surgeons. *Res:* Urinary incontinence; cancer control. *Mailing Add:* Univ Mo Health Sci Ctr 807 Stadium St Columbia MO 65201

GRIFFING, DAVID FRANCIS, b Nanking, China, Feb 23, 26; US citizen; m 49; c 2. PHYSICS & SPORTS, PHYSICS EDUCATION. *Educ:* Miami Univ, AB, 49, MA, 50; Univ Ill, Urbana, PhD(physics), 56. *Prof Exp:* PROF PHYSICS, MIAMI UNIV, 56- *Mem:* Am Phys Soc; Am Inst Physics; Am Asn Physics Teachers; Sigma Xi. *Res:* Low temperature physics; nuclear orientation; radioactivity and nuclear spectroscopy; ultrasonic studies in metals. *Mailing Add:* Dept Physics Miami Univ 23 Culler Oxford OH 45056

GRIFFING, GEORGE WARREN, b Smith Center, Kans, Feb 28, 21; m 46; c 2. THEORETICAL PHYSICS. *Educ:* Kans State Col, Ft Hays, BA, 46; Univ Kans, MA, 48; Queen's Univ Belfast, PhD, 54. *Prof Exp:* Asst prof, Eastern Tenn State Col, 50-51; chief ionospheric reactions sect, Air Force Cambridge Res Ctr, 51-56; sr physicist, Phillips Petrol Co, 56-70; SR PHYS SCIENTIST, ENVIRON PROTECTION AGENCY, 70- *Mem:* Am Phys Soc. *Res:* Atomic and molecular collisions; reactor physics; scattering of slow neutrons by gases, liquids and solids; urban pollution modeling. *Mailing Add:* 1709 Su John Rd Raleigh NC 27607

GRIFFING, J BRUCE, b Tempe, Ariz, Feb 24, 19; m 50; c 4. GENETICS. *Educ:* Iowa State Univ, BS, 41, MS, 47, PhD(genetics), 48. *Prof Exp:* From instr to asst prof genetics, Iowa State Univ, 47-53; Nat Res Coun fel, Cambridge Univ, 53-55; prin res officer plant indust, Commonwealth Sci & Indust Res Orgn, Australia, 56-57, sr res fel, 57-59, sr prin res scientist, 59-64, chmn genetics sect, 60-62; chmn dept, 67-84, MERSHON PROF GENETICS, OHIO STATE UNIV, 65- *Mem:* Genetics Soc Am; Am Soc Naturalists. *Res:* Mathematical genetics; quantitative inheritance; selection theories. *Mailing Add:* Genetics 963 Biol Sci Bldg Ohio State Univ Main Campus Columbus OH 43210

GRIFFING, JOHN MALCOLM, chemistry, for more information see previous edition

GRIFFING, WILLIAM JAMES, b Manhattan, Kans, July 18, 22; m 45; c 4. COMPARATIVE PATHOLOGY, VETERINARY TOXICOLOGY. *Educ:* Kans State Univ, DVM, 44, MS, 60, PhD(vet path), 63. *Prof Exp:* Pvt pract, Bremen, Ind, 44-59; Nat Defense fel, Kans State Univ, 59-62, NIH fel, 62-63; sr pathologist, Eli Lilly & Co, 63-65, res pathologist, 72-74, electron microscopist, 65-85, res assoc, 74-85; RETIRED. *Mem:* Emer mem Electron Micros Soc Am; Am Vet Med Asn; fel Am Col Vet Toxicol; emer mem Int Acad Path. *Res:* Electron microscopic examination of the tissues of laboratory animals subjected to agriculture chemical compounds. *Mailing Add:* 2515 Davis Rd Indianapolis IN 46239

GRIFFIOEN, ROGER DUANE, b Grand Rapids, Mich, Sept 7, 34; m 56, 88; c 4. NUCLEAR PHYSICS, CHEMISTRY. *Educ:* Calvin Col, AB, 56; Purdue Univ, PhD(nuclear chem), 60. *Prof Exp:* Univ fel, Lawrence Radiation Lab, Univ Calif, 60-61; from instr to prof physics & chmn dept, 61-85, ACAD DEAN, CALVIN COL, 85- *Concurrent Pos:* Consult, Argonne Nat Lab, 63-67 & Lawrence Radiation Lab, 67-69; NSF sci fac fel, Fla State Univ, 70-71, sr res assoc, 71-72. *Mem:* Am Phys Soc; Am Asn Physics Teachers. *Res:* Nuclear structure; spectroscopy; radioactivity. *Mailing Add:* Acad Dean Calvin Col Grand Rapids MI 49546

GRIFFISS, JOHN MCLEOD, b Chattanooga, Tenn, July 9, 40. MEDICINE. *Educ:* Univ NC, BA, 62; Yale Univ, MD, 66; Am Bd Internal Med, cert, 72, cert infectious dis, 74. *Prof Exp:* Intern med, King County Hosp, Seattle, 66-67; asst res med, Univ Wash Affil Hosps, 67-68; prev med officer, Stuttgard Med Serv Area, Bad Cannstatt, WGer, 68-71; fel infectious dis, Walter Reed Army Inst Res, 71-73, med res officer, 73-75, actg chief, Dept Bact Dis, 76, sr med res officer, 77-79; asst prof med, Harvard Med Sch, 79-83, lectr, 83-84; assoc prof lab med & med, 83-85, PROF LAB MED & MED, UNIV CALIF, SAN FRANCISCO, 85- *Concurrent Pos:* Staff physician, Infectious Dis Serv, Walter Reed Army Med Ctr, 71-79; consult, Sch Med, George Washington Univ, 71-74, USPHS Hosp, Brighton, Mass, 80-81, Nat Acad Sci, 84-86; assoc med, Brigham & Women's Hosp, 79-82, physician, 82-83; assoc med, Dana Farber Cancer Inst, 82-83; chief microbiol, Clin Path Serv, Vet Admin Med Ctr, San Francisco & staff infectious dis, 83-; expert consult, Meningitis Trust, 87-, Wellcome Trust, 88-, WHO, 89- *Mem:* AMA; Am Soc Microbiol; Am Fedn Clin Res; Am Asn Immunologists; fel Infectious Dis Soc Am; Int Endotoxin Soc. *Res:* Immunologic function of the IgA system; epidemiology and immunochemistry of Neisseria meningitidis; immunochemistry of Neisseria gonorrhoeae; immunology of bacterial vaccines; epidemiology and immunochemistry of Gram negative sepis; bacterial outer membrane glycolipids; immunochemistry of Pneumocystis carinii; author of numerous publications. *Mailing Add:* Vet Admin Ctr/113A 4150 Clement St San Francisco CA 94121

GRIFFITH, B HEROLD, b New York, NY, Aug 24, 25; m 48; c 2. PLASTIC SURGERY, RECONSTRUCTIVE SURGERY. *Educ:* Yale Univ, MD, 48; Am Bd Plastic Surg, dipl, 59. *Prof Exp:* Intern, Grace-New Haven Hosp, Conn, 48-49; resident surg, Vet Admin Hosps, Newington, Conn, 49-50, plastic surg, Bronx, NY, 53-55; asst resident, sec surg div, Bellevue Hosp, 52-53; asst surgeon inpatients, NY Hosp-Cornell Med Ctr, 55, sr registr, Glasgow Royal Infirmary, Scotland, 55; resident plastic surg, NY Hosp-Cornell Med Ctr, 56; assoc surg, 59-62, from asst prof to assoc prof, 61-70, PROF SURG, NORTHWESTERN UNIV, 71-, CHIEF, DIV PLASTIC SURG, SCH MED, 70- *Concurrent Pos:* Res fel, Med Sch, Cornell Univ, 56-57; pvt pract, 57-; mem med adv comt, Nat Paraplegia Found, 67-79; chmn, Plastic Surg Res Coun, 68-69; dir, Am Bd Plastic Surg, 74-82, chmn, 81-82. *Mem:* AAAS; fel Am Asn Plastic Surg; Am Soc Plastic & Reconstruct Surg; fel Am Col Surgeons; NY Acad Sci; fel Royal Soc Med; Sigma Xi; Soc Head & Neck Surgeons; Am Burn Asn; Asn Mil Surgeons of US. *Res:* Experimental embryology; transplantation; cancer chemotherapy; physiology of flaps and grafts; wound-healing; decubitus ulcers; cleft lip and palate; tumors of the skin, head and neck. *Mailing Add:* 251 E Chicago Ave Chicago IL 60201

GRIFFITH, CECIL BAKER, b New Lexington, Ohio, Nov 9, 23; m 46. PHYSICAL CHEMISTRY, PROCESS METALLURGY. *Educ:* Ohio Univ, BS, 47. *Prof Exp:* Supvr gas metals, Battelle Mem Inst, 47-55; res engr phys chem, Cramet Inc, 55-58; div head, Res Ctr, Repub Steel Corp, 58-84; CONSULT, 84- *Res:* Process metallurgy of ironmaking, steelmaking and refractories. *Mailing Add:* 17856 Bennett Rd Cleveland OH 44133-6032

GRIFFITH, CECILIA GIRZ, b Cleveland Ohio, June 3, 49; m 71; c 4. SATELLITE METEOROLOGY, WEATHER FORECASTING APPLICATIONS. *Educ:* Notre Dame Col Ohio, BS, 70; Univ Dayton, MS, 73; Colo State Univ, PhD(atmospheric sci), 87. *Prof Exp:* Physicist, Nat Weather Serv, Nat Oceanic & Atmospheric Admin, 70-71 & Nat Hurricane & Exp Meteorol Lab, 72-78, meteorologist, Weather Res Prog, 78-88, dir, 88-90, ACTG CHIEF, SCI DIV, FORECAST SYSTS LAB, NAT OCEANIC & ATMOSPHERIC ADMIN, 90- *Concurrent Pos:* Counr, Am Meteorol Soc, 83-86. *Mem:* Fel Am Meteorol Soc; Am Geophys Union. *Res:* Technique to estimate convective rainfall from geostationary-satellite thermal infrared imagery that could be applied in the tropics and mid-latitudes; quartitative use of imagery in the forecasting of severe weather. *Mailing Add:* Nat Oceanic & Atmospheric Admn ERL/FSL R/E/FSI 325 Broadway Boulder CO 80303

GRIFFITH, DONAL LOUIS, b Culver City, Calif, Oct 27, 42. CELL PHYSIOLOGY. *Educ:* Univ Calif, Los Angeles, BA, 64, MA, 66, PhD(zool), 69. *Prof Exp:* NIH res grant zool, Univ Calif, Los Angeles, 69-73; lectr biol, Calif Lutheran Col, 73-80. *Mem:* Soc Protozoologists; Sigma Xi. *Res:* Locomotion of microorganisms, amoeboid glavanotaxis, cellular regeneration and growth. *Mailing Add:* 10520 Ashton Ave Los Angeles CA 90024

GRIFFITH, EDWARD JACKSON, b Atlanta, Ga, Apr 4, 25; wid; c 2. PHYSICAL CHEMISTRY. *Educ:* Howard Col, BS, 47; Univ Ky, MS, 48, PhD(chem), 51. *Prof Exp:* scientist, 51-67, adv scientist, 67-72, SR FEL, MONSANTO CO, 72- *Concurrent Pos:* lectr, Int Union Pure & Appl Chem, Chem Indust Basle Found. *Honors & Awards:* Du Boise Award. *Mem:* AAAS; Am Chem Soc; Sigma Xi; Am Inst Chemists. *Res:* Inorganic chemistry; phosphates; nitrates; tailings and mining waste disposal. *Mailing Add:* Monsanto Co 800 N Lindbergh Blvd St Louis MO 63166

GRIFFITH, ELIZABETH ANN HALL, b Washington, DC, Feb 3, 35; wid. ORGANOMETALLIC CHEMISTRY, PHYSICAL CHEMISTRY. *Educ:* Pfeiffer Col, AB, 60; Duke Univ, MA, 63; Univ SC, PhD(phys chem), 70. *Prof Exp:* Asst prof chem, Jacksonville State Univ, 62-63; instr, Pfeiffer Col, 63-64; asst prof, Campbell Col, NC, 64-65; res fel, Univ Sask, Regina, 70-73; res fel, 73-82, ASST PROF RES, DEPT CHEM, UNIV SC, 82- *Mem:* AAAS; Am Chem Soc; Am Crystallog Asn; Sigma Xi. *Res:* Structure determination by x-ray diffraction of single crystals; synthesis and structure of organometallic compounds which are models for biologically significant systems; structure of enzyme substrate model systems. *Mailing Add:* Dept Chem Univ SC Columbia SC 29208

GRIFFITH, GAIL SUSAN TUCKER, b New York, NY, Aug 30, 45; m 87; c 1. NEUROSCIENCES, SCIENCE EDUCATION. *Educ:* Mercy Col, BA, 67; Univ Kans, Lawrence, PhD(cell biol/develop biol), 73. *Prof Exp:* Res asst develop biol, 68-72, res asst mycology, Univ Kans, Lawrence, 72-73; instr, Mercy Col, 73-75; res assoc, Univ Miami, Bascom Palmer Eye Inst, Sch Med, 76-80, asst prof, 80-87; NEW WORLD SCH ARTS, 87- *Concurrent Pos:* Fel, Eye Inst, Col Physicians & Surgeons, Columbia Univ, 73-75; vis fel, Biol Labs, Harvard Univ, 80; consult, Pub Sch Syst, Dade County, Fla, 77-; Ida H Hyde grant in aid, Dept Physiol & Cell Biol, Univ Kans, 71; honoree, Am Soc Cell Biol, 84. *Mem:* Soc Neurosci; AAAS; Nat Sci Teacher Asn. *Res:* Light and electron microscopy of retinal development and cytoarchitecture of the normal and visually deprived amphibian, cat and rabbit retina; retinal aging in the human; corneal dystrophy in the cat; studies on retinal degeneration in retinitis pigmentosa in the human; development of science education strategies. *Mailing Add:* New World Sch Arts 300 NE Second Ave Miami FL 33132

GRIFFITH, GORDON LAMAR, b Bogue, Kans, Oct 12, 21; m 41; c 2. PHYSICS. *Educ:* Kans State Col, BS, 43; Univ Ill, PhD(physics), 50. *Prof Exp:* Physicist, Tenn Eastman Corp Div, Eastman Kodak Co, 44-45; res physicist, Res Lab, Westinghouse Elec Corp, Pa, 49-62; from assoc prof to prof physics, 62-85, chmn dept, Muskingum Col, 67-76, 80-85; RETIRED. *Concurrent Pos:* Consult, Stanford Res Inst, 55 & Lawrence Radiation Lab, Calif, 58; res assoc, Sci Res Coun, Daresbury Lab, Eng, 69, 77. *Mem:* Am Phys Soc; Am Asn Physics Teachers; Sigma Xi. *Res:* Elastic and inelastic scattering of neutrons; x-ray scintillation spectrometry; plasma physics; surface ion mass spectroscopy. *Mailing Add:* 1928 Adriel Ct Ft Collins CO 80524

GRIFFITH, JACK DEE, b Knox City, Tex, Aug 13, 36; m 69. PREVENTIVE MEDICINE. *Educ:* ECent State Univ, BS, 59; Okla Univ, PhD(prev med), 69. *Prof Exp:* Demonstration coordr community control epidemiol, Commun Dis Ctr, USPHS, 64-69, dep dir epidemiol, Nat Ctr Health Serv Res & Develop, 69-71, assoc dir data & eval, Off Exp Health Serv Delivery Syst, 71; asst prof epidemiol, Univ Okla, 72-73; prog dir data & eval, Div Cancer Control & Rehab, Nat Cancer Inst, NIH, 73-74; BR CHIEF EPIDEMIOL, HUMAN EFFECTS MONITORING BR, TECH SERV DIV, ENVIRON PROTECTION AGENCY, ROCKVILLE, 74- *Concurrent Pos:* Consult, Nat Ctr Health Serv Res & Develop, Off Exp Health Serv Delivery Syst, 72-74. *Mem:* Soc Epidemiol Res. *Res:* Application of epidemiologic principles to the development of an index of community and personal health status, quantifiable by aggregate indicators of community and individual health symptomatology. *Mailing Add:* Biol 203 Wilson Hall C46a Univ NC Chapel Hill Chapel Hill NC 27514

GRIFFITH, JAMES H, b Chicago, Ill, Feb 8, 36; m 59; c 3. POLYMER CHEMISTRY, TECHNICAL PRODUCTION MANAGEMENT. *Educ:* Univ Ill, BS, 59; Cornell Univ, PhD(org chem), 63; Univ Chicago, MBA, 75. *Prof Exp:* Res assoc polymer chem, Univ Ariz, 62-64; res chemist, E I du Pont de Nemours & Co, 64-67; sr scientist, 67-69, group supvr, 69, sect supvr, 69-75, resin supt, 75-76, mgr, Chicago Resin Mfg, 76-81, MGR SPECIALTY POLYMERS & CHEMICALS, SHERWIN WILLIAMS CO RES CTR, 81- *Mem:* Am Chem Soc; Fedn Socs Paint Technol; Am Asn Textile Chem & Colorists; Abrasive Eng Soc; Soc Advan Mat & Process Eng. *Res:* Emulsion and solution polymerizations of vinyl and acrylic monomers; condensation polymerization of polyesters; synthesis of nitrogen containing resins; water based protective and decorative coatings; epoxy resins and hardeners, structural polyamides, composite materials, phenolic resins, isocyanates and polyurethanes. *Mailing Add:* Sherwin-Williams Co 11541 Champlain Ave Chicago IL 60628

GRIFFITH, JOHN E(DWARD), b Easton, Pa, Mar 14, 27; m 50; c 3. ENGINEERING MECHANICS. *Educ:* Pa State Univ, BS, 50, MS, 52, PhD(eng mech), 55. *Prof Exp:* Asst prof civil eng, Yale Univ, 55-58; asst prof eng mech, Univ Fla, 58-62; assoc prof, NC State Univ, 62-64; PROF STRUCT, MAT & FLUIDS & CHMN DEPT, UNIV SFLA, 64- *Concurrent Pos:* Martin Co grant, 61-62, consult, 62-; consult, Atlantic Res Corp, 63-; exec chmn, Southeastern Conf Theoret & Appl Physics, 70-72. *Mem:* Am Soc Rheology; Am Soc Mech Engrs; Am Soc Eng Educ; Am Soc Exp Stress Anal; NY Acad Sci. *Res:* Mechanics of solid propellants and ductile fracture; creep under combined stresses; mechanical failure of materials; large deformation of circular membranes under dynamic loading; mechanics of composite materials; soil mechanics. *Mailing Add:* Dept Civil Eng Univ SFla 4202 Fowler Ave Tampa FL 33620

GRIFFITH, JOHN E(MMETT), electrical engineering, for more information see previous edition

GRIFFITH, JOHN RANDALL, b Baltimore, Md, Mar 22, 34; m 55; c 3. HOSPITAL ADMINISTRATION. *Educ:* Johns Hopkins Univ, BEng, 55; Univ Chicago, MBA, 57. *Prof Exp:* Admitting officer hosp admin, Johns Hopkins Hosp, 55; asst, Univ Rochester, 56-60; asst prof, 60-64, assoc dir, 63-70, assoc prof, 64-68, dir hosp admin, 70-82, PROF HOSP ADMIN, UNIV MICH, ANN ARBOR, 68- *Concurrent Pos:* Admin asst hosp admin, Strong Mem Hosp, 56-60; mem hosp mgt systs soc, Am Hosp Asn, 56-; assoc mem health applns sect, Opers Res Soc Am, 66-; vis assoc prof, Yale Univ, 67; consult, Health Ins Benefits Adv Coun, Social Security Admin, 68-69; chmn med care sect, Am Pub Health Asn, 70-71; mem exec comt, Asn Univ Progs in Health Admin, 72-76, pres, 74-75; dir, Medicus Systs Corp, 73-78, chmn, 74-78. *Mem:* Opers Res Soc Am; fel Am Pub Health Asn; Am Hosp Asn; fel Am Col Hosp Adminr. *Res:* Hospital management systems; health facilities planning. *Mailing Add:* Dept Pub Health Univ Mich Main Campus Ann Arbor MI 48109-2029

GRIFFITH, JOHN SIDNEY, b Hull, Eng, Dec 30, 35; Can citizen; m 63; c 3. ASTRONOMY, MATHEMATICS. *Educ:* Univ London, BSc, 57, PhD(astrophys), 62. *Prof Exp:* Asst lectr math, Univ London, 57-58; asst lectr astron, Univ Glasgow, 58-61; sci officer math, Atomic Weapons Res Estab, 61-62; sr lectr, Royal Mil Col Sci, 62-64; sr sci officer astron, Royal Greenwich Observ, 64-68; PROF MATH, LAKEHEAD UNIV, 68- *Concurrent Pos:* Consult, US Ephemeris Working Group, 71-73; prof, World Open Univ, 72-; mem, Ont Math Comn, 72-74; ed, Ont Math Gazette, 72-74; past pres, Lakehead Univ Fac Asn; past dir, Ont Confederational Univ Fac Asn; past mem exec, Confederation Ont Univs. *Mem:* Fel Royal Astron Soc; fel Brit Interplanetary Soc. *Res:* Analysis; celestial mechanics; mathematics education; education in astronomy. *Mailing Add:* Dept Math Sci Lakehead Univ Thunder Bay ON P7B 5E1 Can

GRIFFITH, JOHN SPENCER, b Utica, NY, July 18, 44. FISHERY BIOLOGY, ECOLOGY. *Educ:* Cornell Univ, BS, 66, MS, 67; Univ Idaho, PhD(fisheries), 71. *Prof Exp:* Fisheries res officer, Govt Repub Zambia, 71-74; res assoc, Environ Sci Div, Oak Ridge Nat Lab, 74-77; asst prof biol, 77-81, ASSOC PROF BIOL, IDAHO STATE UNIV, 81- *Concurrent Pos:* Assoc ed, Trans Am Fisheries Soc, 76-77. *Mem:* Am Fisheries Soc; Ecol Soc Am; AAAS; Sigma Xi. *Res:* Behavior of salmonid fishes; African fishery development; fish ecology and population dynamics. *Mailing Add:* Box 8007 Idaho State Univ Pocatello ID 83209

GRIFFITH, LINDA M(AE), b Edwards, Calif, Feb 16, 52. HEMATOPATHOLOGY. *Educ:* Tulane Univ, BS, 73; Univ Calif, Los Angeles, MA, 75; Harvard Univ, PhD(anat & cellular biol), 79; Univ Miami, MD, 87. *Prof Exp:* Fel cell biol & anat, Sch Med, Johns Hopkins Univ, 79-80; fel cell biol, Sch Med, Stanford Univ, 81-84; resident physician path, Sch Med, Yale Univ, 87-90, resident physician lab med, 90-92; FEL HEMATOPATH, LAB PATH, NAT CANCER INST, NIH, 92- *Concurrent Pos:* Am Soc Cell Biol travel award, Second Int Cong Cell Biol, Berlin, 80; Muscular Dystrophy Asn postdoctoral fel, 80-81; NIH postdoctoral fel, Dept Path, Sch Med, Stanford Univ, 83-84. *Mem:* Int Acad Path; Am Soc Clin Pathologists; Am Asn Pathologists; Am Soc Cell Biol; Biophys Soc. *Res:* Hematopathology; applications of cell biology to diagnostic surgical pathology; cell motility and the cytoskeleton. *Mailing Add:* Hematopath Sect Lab Path Nat Cancer Inst NIH Bldg 10 Rm 2N202 Bethesda MD 20892

GRIFFITH, M S, b Dec 25, 40; m; c 2. ELECTRICAL ENGINEERING. *Educ:* Univ Mich, BSSE. *Prof Exp:* TECH LECTR & AUTHOR. *Honors & Awards:* Achievement Award, Inst Elec & Electronics Engrs/I & CPS. *Mem:* Fel Inst Elec & Electronics Engrs. *Mailing Add:* Brown & Root USA Inc 03-626 PO Box 3 Houston TX 77001

GRIFFITH, MARTIN G, b Philadelphia, Pa, Dec 4, 39; m 70; c 3. ORGANIC CHEMISTRY, LUBRICATION CHEMISTRY. *Educ:* Haverford Col, BS, 61; Pa State Univ, PhD(chem), 67. *Prof Exp:* CHEMIST, EXXON RES & ENG CO, 67- *Mem:* Am Chem Soc; Am Soc Testing & Mat. *Res:* Physical and chemical properties, solution thermodynamics of petroleum and synthetic fuel fractions; petroleum lubricant development. *Mailing Add:* Box 51 Linden NJ 07036

GRIFFITH, MICHAEL GREY, b Mansfield, La, Sept 30, 41; m 64; c 2. TECHNICAL MANAGEMENT, ORGANIC CHEMISTRY. *Educ:* Northwestern Univ, BS, 63; La State Univ, Baton Rouge, PhD(chem), 67. *Prof Exp:* Res fel chem, Univ Minn, Minneapolis, 67-68; res chemist, Carothers Res Lab, Exp Sta, E I du Pont de Nemours & Co, Inc, 68-70, res chemist, Textile Res Lab, Chestnut Run, 71-73, res & develop supvr, Chattanooga Qiana Tech Lab, 73-75, res & develop supvr, Old Hickory Spunbonded Res Lab, 76-77; dir explor res, 77-80, dir insulation technol, 80-81, dir res & develop, 81-83, VPRES RES & DEVELOP, TECH CTR, OWENS-CORNING FIBERGLAS CORP, OHIO, 83- *Mem:* Am Chem Soc. *Res:* Fiber chemistry; free radical chemistry; nuclear magnetic resonance spectroscopy; chemical engineering. *Mailing Add:* 505 W Beechtree Lane Wayne PA 19087-3231

GRIFFITH, MICHAEL JAMES, b Kalamazoo, Mich, Dec 1, 48; m 69; c 2. ENZYMOLOGY, THROMBOSIS. *Educ:* Western Mich Univ, BS, 70; Univ Tex Health Sci Ctr, San Antonio, PhD(biochem), 77. *Prof Exp:* Fel, Dept Path, Univ NC, Chapel Hill, 77-80, RES ASST PROF, DEPT MED, DIV HEMAT, 80-; VPRES SCI AFFAIRS, HYLAND DIV, BAXTER HEALTH CORP. *Concurrent Pos:* Mem, Coun Thrombosis, Am Heart Asn, 80- *Mem:* Am Soc Biol Chemists; Am Chem Soc. *Res:* Thrombosis and hemostasis involving the anticoagulant, heparin and plasma protease inhibitors; the role of blood coagulation factor IX in hemostasisis. *Mailing Add:* Hyland Therapeut 1710 Flower Ave Duarte CA 91010-2923

GRIFFITH, O HAYES, b Torrance, Calif, Sept 14, 38; m 85; c 2. BIOPHYSICAL CHEMISTRY. *Educ:* Univ Calif, Riverside, BA, 60; Calif Inst Technol, PhD(chem), 64. *Prof Exp:* Nat Acad Sci-Nat Res Coun fel chem, Stanford Univ, 65-66; from asst prof to assoc prof, 66-71, PROF CHEM & RES ASSOC, INST MOLECULAR BIOL, UNIV ORE, 71- *Concurrent Pos:* Woodrow Wilson fel, 60-61, Nat Sci Found Predoctoral fel, Calif Inst Technol, 61-64; postdoctoral fel, Nat Acad Sci, Nat Res Coun, Stanford Univ, 65, Sloan Found fel, 67-69, Guggenheim Found fel, 77-78; career develop award, Nat Cancer Inst, 72-76. *Honors & Awards:* Camille & Henry Dreyfus Found Teacher - Scholar Award, 70. *Mem:* Am Chem Soc; Biophys Soc. *Res:* Physical chemistry; electron and nuclear magnetic spectroscopy; photoelectron microscopy; structure of membrane model systems and biological membranes; cell surfaces. *Mailing Add:* Dept Chem Univ Ore Eugene OR 97403-1253

GRIFFITH, OWEN MALCOLM, b Guyana, SAm, Sept 6, 28; US citizen; m 65; c 3. BIOCHEMISTRY, BIOPHYSICS. *Educ:* Morgan State Col, BS, 57; NY Univ, MS, US Int Univ, PhD, 77. *Prof Exp:* Asst res biochemist, Rockefeller Univ, 59-63; assoc appln chem, 63-65, appln chem engr, 65-74, SR APPLN RES CHEMIST, SPINCO DIV, BECKMAN INSTRUMENTS

INC, PALO ALTO, CALIF, 74- *Mem:* AAAS; Am Chem Soc; fel Am Inst Chemists. *Res:* Improvement of techniques for free boundary diffusion; advanced applications for using both preparative and analytical ultracentrifuges to stimulate faster and improved data acquisition; sedimentation coefficients; molecular weight determination and density separations of protein nucleic acids; whole cells and subcellular organelles. *Mailing Add:* Heraeus Sepatech 2706 Pruneridge Ave Santa Clara CA 95051-6238

GRIFFITH, OWEN WENDELL, b Oakland, Calif, June,19, 46. ENZYMOLOGY, METABOLISM. *Educ:* Univ Calif, Berkeley, BA, 68; Rockefeller Univ, PhD(biochem), 75. *Prof Exp:* From asst prof to assoc prof, 78-87, PROF BIOCHEM, CORNELL UNIV MED COL, 87- *Concurrent Pos:* Prin investr various NIH, March of Dimes & Juv Diabetes Found grants, 81-; career scientist, Irma Hirschl Found, 81-85. *Mem:* Am Soc Biochem & Molecular Biol; Am Soc Pharmacol & Exp Therapeut; Am Chem Soc. *Res:* Amino acid metabolism; metabolism and control of fatty acid; metabolism; nitric oxide synthesis and the design; enzyme mechanisms and the design and ;synthesis of enzyme inhibitors. *Mailing Add:* Dept Biochem Cornell Univ Med Col New York NY 10021

GRIFFITH, PETER, b London, Eng, Sept 23, 27; m 83; c 2. MECHANICAL ENGINEERING. *Educ:* NY Univ, BS, 50; Univ Mich, MSE, 52; Mass Inst Technol, ScD, 56. *Prof Exp:* Asst, Univ Mich, 50-52; asst, 52-56, from asst prof to assoc prof, 56-70, PROF MECH ENG, MASS INST TECHNOL, 70- *Mem:* Am Soc Mech Engrs; Am Nuclear Soc; Am Inst Chem Engrs. *Res:* Heat transfer with phase change and two phase flow; nuclear reactor thermal-hydraulics; reactor safety. *Mailing Add:* Rm 7-044 Mass Inst of Technol Cambridge MA 02139

GRIFFITH, PHILLIP A, b Danville, Ill, Dec 29, 40; m 60; c 2. MATHEMATICS. *Educ:* Northern Mich Univ, BS, 63; Univ Mo, MA, 65; Univ Houston, PhD(math), 68. *Prof Exp:* Instr math, Univ Houston, 67-68; instr, Univ Chicago, 68-70; from asst prof to assoc prof, 70-77, PROF MATH, UNIV ILL, URBANA, 77- *Concurrent Pos:* Alfred P Sloan fel, 71-73. *Mem:* Am Math Soc. *Res:* Infinite Abelian groups; homological algebra of locally compact Abelian groups; ring theory. *Mailing Add:* Dept Math Univ Ill 1409 W Green St Urbana IL 61801

GRIFFITH, ROBERT W, b Bristol, Eng, Nov 12, 30; m 62; c 3. RESEARCH ADMINISTRATION. *Educ:* Oxford Univ, BA, 52, BMBCh, 56. *Prof Exp:* Pathologist, Royal Air Force Hosp, Chiangi, Singapore, 58-59, Wegberg, Ger, 59-61 & Ely, Eng, 61-63; pathologist, Chelser Hosp Women, London, Eng, 63-64; toxicologist, Sandoz Ltd, Basle, Switz, 64-75, dir, Clin Res, 76-78; VPRES RES & DEVELOP, SANDOZ, INC, EAST HANOVER, NJ, 79- *Mem:* Sigma Xi. *Mailing Add:* Dept Biol Southeastern Mass Univ North Dartmouth MA 02747

GRIFFITH, THOMAS, b Minneola, Kans, June 22, 30; m 58; c 3. BIOCHEMISTRY. *Educ:* Kans State Univ, BS, 52, MS, 54; Mich State Univ, PhD(chem), 58. *Prof Exp:* Asst cereal chem, Kans State Univ, 52-54; asst chem, Mich State Univ, 54-58, instr, 58-62; assoc prof, 62-65, head dept, 65-66, assoc dean, Sch Arts & Sci, 66-67, dean, 67-72, PROF CHEM, NORTHERN MICH UNIV, 72- *Mem:* Am Chem Soc; Sigma Xi. *Res:* Plant metabolism; alkaloid biosynthesis; transmethylation; cereal chemistry; fermentation. *Mailing Add:* Dept Chem Northern Mich Univ Marquette MI 49855

GRIFFITH, VIRGIL VERNON, b Cicero, Ill, Dec 30, 28; m 48; c 5. COMPUTER SCIENCE. *Educ:* Wash Univ, AB, 56, BS, 56, MS, 58; Case Inst Technol, PhD(systems), 67. *Prof Exp:* Sr group engr, McDonnel Aircraft Corp, St Louis, Mo, 52-61; sr tech specialist, Goodyear Aerospace, 61-67; chief eng-electronics, 67-83, STAFF DIR, MCDONNEL AIRCRAFT CORP, 83- *Concurrent Pos:* Instr, Washington Univ, 58-59; lectr, St Louis Community Col, Flo Valley, 70-74. *Mem:* Am Inst Aeronaut & Astronaut; Inst Elec & Electronics Engrs. *Res:* Applicability of new digital system techniques to avionic systems for advanced fighter aircraft. *Mailing Add:* 2750 Redman Ave St Louis MO 63136

GRIFFITH, W(ILLIAM) A(LEXANDER), b Sioux Falls, SDak, Mar 28, 22; m 49; c 3. EXTRACTIVE METALLURGY. *Educ:* SDak Sch Mines & Technol, BS, 47, DBA, 86; Mass Inst Technol, SM, 50. *Hon Degrees:* DSc, Univ Idaho, 90. *Prof Exp:* Metallurgist, Santiago-Alaska Mines, Inc, 47; asst, Mass Inst Technol, 47-48, instr metall, 48-49; investr mineral dressing, Res Dept, NJ Zinc Co, 49-52, group leader, 52-55, chief milling & maintenance, 55-57; chief metallurgist, Rare Metals Corp Am, Ariz, 57-58; head res dept, Morenci br, Phelps Dodge Corp, 58-68; dir res, 68-73, vpres metall, 73-77, sr vpres, 77-78, exec vpres 78-79, pres & chief exec off, 79-85, chmn, 86-87, DIR, HECLA MINING CO, 79- *Honors & Awards:* A M Gaudin Award, Am Inst Mining, Metall & Petrol Engrs; Robert A Richards Award, Am Inst Mining, Metall & Petrol Engrs, 81. *Mem:* Hon mem Am Inst mining, metall & Petrol Engrs. *Res:* Unit operations of mineral benefication, extractive metallurgy. *Mailing Add:* 630 S 14th St Coeur D Alene ID 83814

GRIFFITH, WAYLAND C(OLEMAN), b Champaign, Ill, June 26, 25; m 61; c 2. FLUID MECHANICS, ENERGY MANAGEMENT. *Educ:* Harvard Univ, AB, 45, MS, 46, PhD(appl sci), 49. *Prof Exp:* Fel physics, Atomic Energy Comn, Princeton Univ, 49-50, from instr to asst prof, 50-57; mgr flight sci div, Missiles & Space Div, Lockheed Aircraft Corp, 57-58, assoc dir res, 58-59, asst dir res, 59-62, dir res, 62-66, vpres & asst gen mgr res & technol, Lockheed Missiles & Space Co, 66-71, asst dir new bus, 71-73; PROF MECH ENG, NC STATE UNIV, 73- *Concurrent Pos:* NSF fel, Univ Col, London, 55-56; vis prof, NC State Univ, 70-71; mem adv comt fluid mech, NASA; mem div comt math & phys sci, NSF; mem bd human resources, Nat Acad Sci; mem panels 213 & 274, Nat Bur Standards. *Mem:* Am Phys Soc; Royal Aeronaut Soc; Am Inst Aeronaut & Astronaut; Sigma Xi. *Res:* Shock waves; supersonic flow; shock tube; turbocompressors; design engineering. *Mailing Add:* Mech Aerospace Eng NC State Univ Box 7910 Raleigh NC 27695

GRIFFITH, WILLIAM KIRK, b Henry, Ill, May 25, 29; m 51; c 2. AGRONOMY. *Educ:* Western Ill Univ, BS, 51; Univ Ill, MS, 52; Purdue Univ, PhD(agron), 60. *Prof Exp:* Asst county agt, Univ Ariz Coop Exten Serv, 56-58; eastern agronomist, 60-66, asst to pres, 66-68, EASTERN DIR, POTASH & PHOSPHATE INST, 68- *Honors & Awards:* Agron Serv Award, Am Soc Agron, 74. *Mem:* Am Soc Agron; Soil Sci Soc Am; Crop Sci Soc Am; Am Forage & Grassland Coun. *Res:* Stimulation and development of sound soil fertility research in the Northeastern United States. *Mailing Add:* 865 Seneca Rd Great Falls VA 22066

GRIFFITH, WILLIAM SCHULER, b Bradford, Pa, Oct 10, 49; m 72; c 2. RELIABILITY THEORY, APPLIED PROBABILITY. *Educ:* Grove City Col, BS, 71; Univ Pittsburgh MA, 73, PhD(math statist), 79. *Prof Exp:* Teaching asst and fel, Dept Math & Statist, Univ Pittsburgh, 71-76; lectr math, Dept Exact Sci, Carlow Col, 77; casual mathematician, Systs Sci, Westinghouse Res & Develop Ctr, 77-79; asst prof, 79-85, ASSOC PROF STATIST, UNIV KY, 85- *Mem:* Inst Math Statist; Am Statist Asn; Operations Res Soc Am; Soc Indust & Appl Math; Biometric Soc. *Res:* Mathematical theory of reliability with particular emphasis on multistate reliability models, shock models, minimal repair models; applied probability models in engineering and biology. *Mailing Add:* Dept Statist Univ KY Lexington KY 40506-0027

GRIFFITH, WILLIAM THOMAS, b Palmerton, Pa, Aug 15, 40; m 67; c 3. PHYSICS. *Educ:* Johns Hopkins Univ, BA, 62; Univ NM, MS, 64, PhD(physics) 68. *Prof Exp:* NASA trainee, Univ NM, 65-67; from asst prof to assoc prof, 67-81, chmn, Dept Physics, 71-76, chmn, Div Sci & Math, 76-82, PROF PHYSICS, PAC UNIV, 81- *Concurrent Pos:* Consult, Energy Resources, Holographic Toy Design. *Mem:* Am Phys Soc; Am Asn Physics Teachers; Am Asn Crystal Growth. *Res:* Kinetics of phase transformations; plastic crystals; thermodynamics of solids; crystal growth; solar energy; science teaching. *Mailing Add:* Dept Physics Pac Univ Forest Grove OR 97116

GRIFFITHS, ANTHONY J F, b Bristol, Gr Brit, Oct 24, 40; m 63; c 4. GENETICS, MOLECULAR BIOLOGY. *Educ:* Univ Keele, BA, 63; McMaster Univ, PhD(molecular biol), 67. *Prof Exp:* Res assoc biol, Kans State Univ, 67-68; NIH fel, Oak Ridge Nat Lab, 68-70; teaching fel, 70-71, from asst prof to assoc prof, 77-82, PROF BOT, UNIV BC, 82- *Mem:* Environ Mutagen Soc; Genetics Soc Am; Genetics Soc Can. *Res:* Senescence plasmids in Neurospora. *Mailing Add:* Dept Bot Univ BC 2075 Westbrook Pl Vancouver BC V6T 1W5 Can

GRIFFITHS, CLIFFORD H, b Chester, Eng, Jan 28, 36; m 57; c 2. SOLID STATE & POLYMER PHYSICS. *Educ:* Nottingham Univ, Eng, BSc, 57; London Univ, MSc, 64, PhD(phys chem), 73. *Prof Exp:* Scientist chem, Noranda Mines Ltd, Que, 57-61; res fel physics, London Univ, 61-63; group leader, Noranda Res Ctr, Noranda Mines Ltd, 63-67; assoc scientist, Rochester Corp Res Ctr, Xerox Corp, 67-70; group leader, Multi-State Devices, Montreal, 70-72; PRIN SCIENTIST PHYSICS, WEBSTER RES CTR, XEROX CORP, 72- *Res:* Morphology structure and structure-electronic property relationships in organic and inorganic polymers; thin film structure and properties. *Mailing Add:* Xerox Corp Webster Res Ctr W114 800 Philips Rd Webster NY 14580

GRIFFITHS, DAVID, b Neath, Wales, May 2, 38; m 76, 84; c 4. PHYSICS. *Educ:* Univ Wales, BSc, 60, PhD(physics), 64. *Prof Exp:* Asst lectr physics, Univ Col Swansea, Wales, 64-65; res physicist, Eng Physics Lab, E I du Pont de Nemours & Co, 65-69, supvr, Electronics Group, 69-70, res supvr, Appl Physics Sect, 70-72, sr supvr, 72-74, field supvr & sr supvr tech serv sect, Chamber Works Chem Plant, NJ, 74-76, mgr instruments & control, Eng Physics Lab, Exp Sta, 76-79, mgr mat & mach dynamics, eng technol lab, 79-87, mgr bioeng, 87-91; RETIRED. *Mem:* Brit Inst Physics; Am Inst Physics. *Res:* Electrical breakdown and gaseous ionization; surface physics, especially conductivity, friction, static properties, polymer adhesion and highpower laser technology; color, computer technology, textile fiber technology; catalysis, coatings, wear, microscopy, metallurgy, electrochemistry; vibrational analysis, computer monitoring, robotics, micro and mini computers. *Mailing Add:* 12 Jarrell Farms Dr Newark DE 19711

GRIFFITHS, DAVID JEFFERY, b Washington, DC, Dec 5, 42; m 70; c 2. THEORETICAL PHYSICS. *Educ:* Harvard Univ, BA, 64, MA, 66, PhD(physics), 70. *Prof Exp:* Res assoc & assoc instr physics, Univ Utah, 70-72; res assoc & assoc instr, Univ Mass, 72-74; asst prof, Mt Holyoke Col, 74-77, Trinity Col, 77-78; from asst prof to assoc prof, 78-88, PROF PHYSICS, REED COL, 88- *Concurrent Pos:* Visitor, Theory Group, Stanford Linear Accelerator Ctr, 84-85. *Mem:* Am Asn Physics Teachers. *Res:* Classical and quantum theory of fields. *Mailing Add:* Dept Physics Reed Col Portland OR 97202

GRIFFITHS, DAVID JOHN, b Vancouver, BC, June 15, 38; m 62; c 3. PHYSICS. *Educ:* Univ BC, BA, 59, MSc, 60, PhD(low temperature physics), 65. *Prof Exp:* Teaching asst, Univ BC, 62-64; res assoc superconductivity & lectr, Univ Southern Calif, 65-67; asst prof, 67-71, assoc prof, 71-86, PROF PHYSICS, ORE STATE UNIV. *Concurrent Pos:* Mem tech staffs, Oak Ridge Nat Lab, 74-75, Cen Saclay, 82-83. *Mem:* Am Phys Soc. *Res:* Magnetic properties of amorphous alloys and shock-wave physics. *Mailing Add:* Dept Physics Ore State Univ Corvallis OR 97331

GRIFFITHS, DAVID WARREN, b St Louis, Mo, Oct 23, 44; m 65; c 2. PHYSICAL ORGANIC CHEMISTRY, BIO-ORGANIC CHEMISTRY. *Educ:* Wilmington Col, AB, 65; Wash Univ, St Louis, PhD(chem), 70. *Prof Exp:* Fel, Northwestern Univ, 70-73; res chemist, Petrolite Corp, 73-77, mgr water res, tretolite div, 77-83; DIR RES & TECHNOL, OLIN WATER SERV, KANS, 83- *Mem:* Am Chem Soc; Sigma Xi; AAAS. *Res:* Structure, energetics and dynamics of molecular complexes, particularly the complexes of ion-containing polymers with their substrates; principles of crystal growth inhibition; surfactants for enhanced oil recovery; efficient production of energy resources through chemistry; efficient management of water resources. *Mailing Add:* 612 E Evans St Kirkwood MO 63122

GRIFFITHS, GEORGE MOTLEY, b Thorold, Ont, Dec 12, 23; m 48; c 5. NUCLEAR PHYSICS, RESOURCE MANAGEMENT. *Educ:* Univ Toronto, BSc, 49; Univ BC, MA, 50, PhD(physics), 53. *Prof Exp:* Asst physics, Nat Res Coun Can, Chalk River, 48-49; Rutherford Mem fel, Cavendish Lab, Cambridge Univ, 53-55; from asst prof to assoc prof, 55-63, PROF PHYSICS, UNIV BC, 63- *Concurrent Pos:* Sr res fel, Calif Inst Technol, 62-63; energy adv, Ministry of State, Sci & Technol, Govt Can, 73-75. *Mem:* Am Phys Soc; Can Asn Physicists. *Res:* Low energy nuclear physics, direct radiative capture reactions; technical, economic and political factors in Canadian energy policy. *Mailing Add:* 1975 W 15th Ave Vancouver BC V6J 2L2 Can

GRIFFITHS, JAMES EDWARD, b Ft Frances, Ont, June 1, 31; m 54; c 1. INORGANIC CHEMISTRY, SPECTROSCOPY. *Educ:* Univ Man, BSc, 55, MSc, 56; McGill Univ, PhD(inorg chem), 59. *Prof Exp:* Res assoc chem, Univ Southern Calif, 58-60; mem tech staff, AT&T Bell Labs, 60-86; DIR/CONSULT, INSTRUMENTS SA INC, 85-; DISTINGUISHED CONSULT IN RESIDENCE, ARMSTRONG STATE COL, 87- *Concurrent Pos:* Vis prof physics, Univ Paris VI, 78-79. *Mem:* AAAS; Am Phys Soc; Soc Appl Spectros; fel Optical Soc Am. *Res:* Synthesis, properties, molecular structure of volatile compounds; molecular structure and interactions in gases and liquids; arc phenomena in ultraviolet and vacuum ultraviolet; Raman scattering, absolute scattering cross-sections and molecular dynamics in the liquid state; structure of glasses; thin films; semiconductor-oxide interfacial reactions; surface physics; solid state physics. *Mailing Add:* Dept Chem & Physics Armstrong State Col Savannah GA 31419

GRIFFITHS, JOAN MARTHA, b Rutherford, NJ, Apr 23, 35. BIOCHEMISTRY, CHEMISTRY. *Educ:* Univ Rochester, BS, 57; Cornell Univ, PhD(biochem), 67. *Prof Exp:* Lab technician, Sloan-Kettering Inst, 57-58; teacher sci, North Babylon Pub Sch, 58-62; res assoc biochem, Cornell Univ, 67-69, lectr, 70-76; ASST PROF CHEM & BIOCHEM, KEUKA COL, 76-; LECTR, CORNELL UNIV, 79- *Mem:* Am Chem Soc; Sigma Xi; AAAS. *Res:* Methylation reactions. *Mailing Add:* Wing Hall Cornell Univ Ithaca NY 14853

GRIFFITHS, JOHN CEDRIC, b Llanelly, Wales, Feb 29, 12; nat US; m 41; c 1. PETROLOGY, GEOLOGY STATISTICS. *Educ:* Univ Col, Swansea, BS, 33; Univ Wales, MSc, 34, PhD(sedimentary petrol), 37; Univ London, PhD(sedimentary petrol), 40, dipl, Imp Col, 40. *Prof Exp:* Petrographer, Trinidad Leaseholds, Ltd, 40-47; from asst prof to prof petrog, 47-77; head dept mineral, 55-63, head dept geochem & mineral, 63-69, dir planning res, 69-73, EMER PROF PETROG, PA STATE UNIV, 77- *Concurrent Pos:* Consult, oil & gas, ore deposits, statistics in geosci. *Honors & Awards:* WC Krumbein Award, Int Asn Math Geol, 77. *Mem:* Fel Geol Soc Am; Mineral Soc Am; Am Soc Econ Paleont & Mineral. *Res:* General systems; heavy minerals in correlation; petrography in petroleum exploration in Trinidad, British West Indies and Venezuela; texture of sediments and reservoir engineering; statistics, computers and operations research in earth sciences; exploration for and assessment of mineral resources. *Mailing Add:* 310 Deike Pa State Univ Univ Park PA 16802

GRIFFITHS, JOHN FREDERICK, b London, Eng, Feb 8, 26; m 62. METEOROLOGY. *Educ:* Univ London, BS, 47, MS, 49, dipl, Imp Col, 48. *Prof Exp:* Bioclimatologist, Brit, Colonial Sci Res Serv, 50-57; chief res, EAfrican Meteorol Dept, 57-62; from asst prof to assoc prof, 62-70, PROF METEOROL, TEX A&M UNIV, 70-, PROF GEOG, 86- *Concurrent Pos:* Mem comts instruments, agr meteorol & climat, World Meteorol Orgn, 58-63; Rockefeller & Munitalp Found travel grants, 60; pres, Appl Trop Res; chief consult, World Meteorol Orgn, 74 & 78-; prin investr climatic res, NSF, NIH & Nat Weather Serv; Key consult, Food Agr Orgn; state climatologist, Tex, 73- *Mem:* Am Meteorol Soc; World Acad Art & Sci; fel Royal Meteorol Soc; fel Royal Geog Soc. *Res:* All aspects of biometeorology, especially statistical analysis, instrumentation and agricultural climatology. *Mailing Add:* Dept Meteorol Tex A&M Univ College Station TX 77843-3146

GRIFFITHS, LLOYD JOSEPH, b Edmonton, Alta, Sept 30, 41; m 61, 84; c 3. ELECTRICAL ENGINEERING, APPLIED MATHEMATICS. *Educ:* Univ Alta, BSc, 63; Stanford Univ, MSc, 65, PhD(elec eng), 68. *Prof Exp:* Res engr, Defence Res Bd, Ottawa, 63; res asst radiosci, Stanford Univ, 63-65; circuit engr, Hewlett Packard Co, 65; res asst, Stanford Univ, 65-68; staff scientist commun, Barry Res Corp, 69-70; prof, Univ Colo, 68-69, prof elec eng, 70-84; prof elec eng, 84-86, ASSOC DEAN ENG, UNIV SOUTHERN CALIF, 86- *Concurrent Pos:* Res scientist, Nat Ctr Atmospheric Res, 71; consult, Marathon Oil Co, 71-77, IBM Corp, 73-79 Nat Oceanic & Atmospheric Admin, 76-78, SRI Int, 76-, MIT Lincoln Lab, 79-80, Ball Bros Aerospace Corp, 80-84, NCR Corp, 81-84, MRO Assoc, 81, Ford Aerospace Corp, 81-84, TRW, 84- *Honors & Awards:* B J Thompson Award, Inst Elec & Electronics Engrs, 71. *Mem:* Inst Elec & Electronics Engrs; Soc Explor Geophysicists; Sigma Xi. *Res:* Adaptive antenna arrays; baseband high density recording and equalization; seismic processing; adaptive noise cancelling; short term spectral estimation techniques; communication systems; direction finding systems; neural networks. *Mailing Add:* 4051 McLaughlin Ave Five Los Angeles CA 90066

GRIFFITHS, PHILLIP A, b Raleigh, NC, Oct 18, 38. ALGEBRAIC GEOMETRY. *Educ:* Wake Forest Univ, BS, 59; Princeton Univ, PhD(math), 63. *Hon Degrees:* Dr, Angers Univ, France, 79. *Prof Exp:* Miller fel, Univ Calif, Berkeley, 62-64, mem fac, 64-67, 75-76; Prof math, Princeton Univ, 68-72; prof, Harvard Univ, 72-83, Dwight Robinson Parker prof, 82; James B Duke prof math & provost, Duke Univ, 83-91; DIR, INST ADVAN STUDY, 91- *Concurrent Pos:* vis prof, Princeton Univ, 67-68; Guggenheim fel, 80-82; taught in China on Guggenheim fel at Peking Univ & Nanking; guest prof, Univ Peking, 83; mem, ex off, US Nat Comt Mathematicians; treas Comn Develop & Exchange, Int Math Union; mem, Comn Phys Sci, Math & Resources; bd of trustees, NC Sch Sci & Math; bd of trustees, Woodward Acad; ed, J Differential Geom, Duke Math J & Compositio; ed, J Differential Geom; mem, Bd Gov, Res Triangle Inst. *Honors & Awards:* LeRoy P Steele Prize, Am Math Soc, 72; Dinnie-Heineman Prize, Acad Sci, 79. *Mem:* Nat Acad Sci. *Mailing Add:* Inst Advan Study Princeton NJ 08540

GRIFFITHS, RAYMOND BERT, b Worcester, Mass, June 16, 15; m 51; c 1. BIOLOGY, MEDICINE. *Educ:* Univ Rochester, 37; Princeton Univ, AM, 39, PhD(biol), 40; Northwestern Univ, MD, 46. *Prof Exp:* Instr zool, Univ Ariz, 40-42; resident physician, Cushing Vet Admin Hosp, Framingham, Mass, 48-49; instr anat, Yale Univ, 52-54; consult, Med Serv Div, Ciba Pharmaceut Prods, Inc, NJ, 54-58; ed dir, med affairs dept, Am Cancer Soc, Inc, 58-60; EXEC ED, J CELL BIOL, ROCKEFELLER UNIV, 60- *Concurrent Pos:* Res fel pediat & anat, Sch Med, Yale Univ, 49-52. *Mem:* AAAS; Coun Biol Ed; Am Asn Anat. *Res:* Cytology of amphibia; human anatomy; biological and medical editing and writing. *Mailing Add:* 1700 York Ave New York NY 10021

GRIFFITHS, ROBERT BUDINGTON, b Etah, India, Feb 25, 37; US citizen. STATISTICAL MECHANICS, PHASE TRANSITIONS. *Educ:* Princeton Univ, AB, 57; Stanford Univ, MS, 58, PhD(physics), 62. *Prof Exp:* NSF fel, Univ Calif, San Diego, 62-64; from asst prof to assoc prof, 64-69, PROF PHYSICS, CARNEGIE-MELLON UNIV, 69- *Concurrent Pos:* Alfred P Sloan res fel, 66-68; vis assoc prof, State Univ NY Stony Brook, 69; Guggenheim Found, fel, 73; Otto Stern prof physics, Carnegie-Mellon, 79. *Honors & Awards:* US Sr Scientist Award, Alexander von Humboldt Found, 73; A Cressy Morrison Award, NY Acad Sci, 81; Dannie Heineman Prize, 84. *Mem:* Nat Acad Sci; Am Sci Affil; Am Phys Soc. *Res:* statistical mechanics and thermodynamics of phase transitions, especially critical points; mathematical foundations of statistical mechanics; theory of magnetism; relation of physical science and Christian theology; interpretation of quantum mechanics. *Mailing Add:* Dept Physics Carnegie-Mellon Univ Pittsburgh PA 15213

GRIFFITHS, ROLAND REDMOND, b Glen Cove, NY, July 19, 46. BEHAVIORAL PHARMACOLOGY. *Educ:* Occidental Col, BA, 68; Univ Minn, PhD(psychopharmacol), 72. *Prof Exp:* From asst prof to assoc prof, 72-86, PROF BEHAV BIOL & NEUROSCI, JOHNS HOPKINS UNIV SCH MED, 86- *Mem:* Behav Pharmacol Soc; Am Psychol Asn; Am Col Neuropsychopharmacol. *Res:* Behavioral and pharmacological analysis of human and infrahuman drug self-administration with special emphasis on benzodiazepines and caffeine. *Mailing Add:* Dept Psychiat Johns Hopkins Sch Med 720 Rutland Ave 623 Traylor Bldg Baltimore MD 21205

GRIFFITHS, THOMAS ALAN, b Lewiston, Maine, Sept 16, 51; m 73; c 2. COMPARATIVE ANATOMY, SYSTEMATICS. *Educ:* Bates Col, BS, 73; Univ Vt, MS, 76; Univ Mass, PhD(zool), 81. *Prof Exp:* Instr biol, State Univ NY Col Plattsburgh, 77-78; asst prof, 81-86, ASSOC PROF BIOL, ILL WESLEYAN UNIV, 86- *Concurrent Pos:* Res assoc, Dept Mammal, Am Mus Nat Hist, NY, 84- *Mem:* Am Soc Mammalogists; Soc Syst Zool; Soc Study Evolution; Soc Vert Paleont; Sigma Xi; Asn Tropical Biol. *Res:* Dissection and histological examination of the musculo skeletal system of bats; evolution and biogeography of bats; functional anatomy of the larynx. *Mailing Add:* Dept Biol Ill Wesleyan Univ Bloomington IL 61701

GRIFFITHS, VERNON, b Treorchy, Wales, May 4, 29; US citizen; m 54; c 5. FAILURE ANALYSIS, MATERIALS ENGINEERING. *Educ:* Univ Wales, BSc, 49, MSc, 51; Mass Inst Technol, ScD(metall), 55. *Prof Exp:* Res assoc, Univ BC, 55-59; assoc prof metall, Mont Col Mineral Sci & Technol, 59-64, dir res, 76-84, exec dir, Mont Tech Found, 76-80, dir grad sch, 77-85, PROF METALL, MONT COL MINERAL SCI & TECHNOL, 64-, HEAD DEPT, 59- *Concurrent Pos:* Res metallurgist, Sherritt Gordon Mines, Ltd, 55-57. *Mem:* Am Soc Metals; Metall Soc Am Inst Mech Engrs; Am Powder Metall Inst; Inst Metals; Am Ceramic Soc; Mat Res Soc. *Res:* Corrosion of muffle tube alloys; ceramics. *Mailing Add:* Dept Metall Mont Col Mineral Sci & Technol Butte MT 59701

GRIFFITHS, WILLIAM C, b Fall River, Mass, Nov 23, 39; m 68; c 2. CLINICAL CHEMISTRY. *Educ:* Providence Col, BS, 62, PhD(org chem), 67; Am Bd Clin Chem, dipl. *Prof Exp:* Fel, Ohio State Univ, 67-68; org chemist, RI Hosp, 68-71; asst prof, Dept Biomed Sci, 74-85, ASSOC PROF, DEPT CLIN SCI, BROWN UNIV, 85-; CLIN BIOCHEMIST, ROGER WILLIAMS HOSP, 71- *Mem:* AAAS; Am Chem Soc; Am Asn Clin Chemists; Asn Clin Scientists. *Res:* Analytical methodology in endocrinology and toxicology; protein binding of small molecules. *Mailing Add:* 825 Chalkstone Ave Providence RI 02908-4728

GRIFFITTS, JAMES JOHN, b Springfield, Ill, Dec 13, 12; wid; c 4. MEDICINE. *Educ:* Univ Va, BS, 33, MD, 37. *Prof Exp:* Intern, Lakeside Hosp, Ohio, 37-39; asst surgeon, US Marine Hosp, 39-40; from asst surgeon to surgeon, NIH, 41-49; ASSOC DIR, JOHN ELLIOTT BLOOD BANK, DADE COUNTY, 49-; PRES, DADE DIV, AM HOSP SUPPLY CO, 54- *Honors & Awards:* Elliott Award in Blood Banking, 63. *Mem:* Am Soc Clin Path; AMA. *Res:* Blood immunology; clinical pathology; nutrition. *Mailing Add:* 11000 SW 190th Ave Dunnellon FL 32630

GRIFFITTS, WALLACE RUSH, b Ann Arbor, Mich, Oct 28, 19; m 46; c 4. ECONOMIC GEOLOGY. *Educ:* Univ Mich, BS, 42, MS, 49, PhD(geol), 58. *Prof Exp:* GEOLOGIST, US GEOL SURV, 46- *Mem:* Mineral Soc Am; Geochem Soc; Am Soc Econ Geologists; Am Geophys Union; Soc Geol Appl Mineral Deposits. *Res:* Exploration geochemistry; geology of pegmatites; nonmetallic mineral deposits; geology of beryllium deposits; rock weathering. *Mailing Add:* US Geol Surv Fed Ctr Denver CO 80225

GRIFFO, JAMES VINCENT, JR, b Brooklyn, NY, Sept 17, 28; m 54; c 3. NATURAL HISTORY, PARASITOLOGY. *Educ:* Univ Ky, BS, 52, MS, 53; Univ Fla, PhD(biol), 60. *Prof Exp:* Asst zool, Univ Ky, 52-53; asst parasitol, Univ Tenn, 55-56; asst biol, Univ Fla, 56-60; instr, Farleigh Dickinson Univ, 60-61; res biologist, Patuxent Wildlife Res Ctr, 61-62; from asst prof to assoc prof, 62-69, chmn biol dept, 62-69, dean campus, 67-71, PROF BIOL, FARLEIGH DICKINSON UNIV, FLORHAM-MADISON CAMPUS, 69-, PROVOST, CAMPUS, 71- *Mem:* Am Soc Mammalogists; Wildlife Soc; Wilderness Soc. *Res:* Experimental natural history; mammals and insects; comparative parasitology; small mammal behavior. *Mailing Add:* Farleigh Dickinson Univ Off Provost 285 Madison Ave Madison NJ 07940

GRIFFO, JOSEPH SALVATORE, b Mt Morris, NY, Oct 13, 27; m 54; c 2. PHYSICAL CHEMISTRY, INORGANIC CHEMISTRY. *Educ:* Col St Bonaventure, BS, 51, MS, 54; St Louis Univ, PhD(inorg chem), 61. *Prof Exp:* Teacher high sch, NY, 51-52; res chemist, Olin Mathieson Chem Corp, 55-57; res asst boron hydrides, St Louis Univ, 57-61; sr res chemist, Mound Lab, Monsanto Res Corp, 61-66; isotopic fuels specialist, US AEC Hq, Germantown, 66-75, systs engr, Energy Res & Develop Admin, 75-77; systs engr, US Dept Energy, 77-88; RETIRED. *Mem:* AAAS; Am Chem Soc; Am Inst Physics; Sigma Xi. *Res:* Plutonium chemistry; systems analysis; chemistry of boron hydrides; high energy fuels; chemistry of rare earths; salts of barbituric acid. *Mailing Add:* 12330 Old Canal Rd Rockville MD 20854

GRIFFO, ZORA JASINCUK, b Prague, Czech, Nov 24, 28; US citizen; m 54; c 2. PHYSIOLOGY. *Educ:* St Bonaventure Univ, BS, 52; Buffalo Univ, PhD(physiol), 59. *Prof Exp:* Asst anesthesiol, Sch Med, Wash Univ, 58-61; grants assoc, Div Res Grants, 68-69, prog officer, Craniofacial Anomalies Prog, Nat Inst Dent Res, 69-70 & prog chief, 70-73, health sci adminr, Off of Dir, 73-75, SPEC PROGS OFFICER, OFF OF DIR, NIH, 75- *Res:* Cardiovascular and pulmonary physiology. *Mailing Add:* 12330 Old Canal Rd Bethesda MD 20854

GRIFFY, THOMAS ALAN, b Oklahoma City, Okla, Dec 16, 36; m 58; c 3. THEORETICAL PHYSICS. *Educ:* Rice Univ, BA, 59, MA, 60, PhD(physics), 61. *Prof Exp:* Asst prof physics, Duke Univ, 61-62; res assoc Stanford Univ, 62-65; assoc prof, 65-68, assoc dean grad sch, 70-73, chmn dept, 74-84, PROF PHYSICS, UNIV TEX, 68- *Mem:* Am Phys Soc; Acoust Soc Am. *Res:* Theoretical physics. *Mailing Add:* 6806 Pioneer Pl Austin TX 78731

GRIGAL, DAVID F, b Orr, Minn, Sept 21, 41; m 64; c 2. SOIL SCIENCE, FOREST ECOLOGY. *Educ:* Univ Minn, St Paul, BS, 63, MS, 65, PhD(soil sci), 68. *Prof Exp:* AEC fel, 68-70; res assoc, Ecol Sci Div, Oak Ridge Nat Lab, 70; from asst prof to assoc prof, 70-80, PROF SOIL SCI, UNIV MINN, ST PAUL, 80- *Mem:* AAAS; Ecol Soc Am; Am Soc Agron; Classification Soc. *Res:* Nutrient cycling; water quality; air pollution impacts; multivariate analysis of biological data. *Mailing Add:* Dept Soil Sci Univ Minn St Paul MN 55108

GRIGARICK, ALBERT ANTHONY, JR, b Redding, Calif, Dec 22, 27; m 46; c 3. ENTOMOLOGY. *Educ:* Univ Calif, BS, 53, PhD(entom) 57. *Prof Exp:* Lectr entom & asst entomologist, 57-65, assoc prof & assoc entomologist, 65-70, PROF ENTOM & ENTOMOLOGIST, UNIV CALIF, DAVIS, 70- *Mem:* Entom Soc Am; Sigma Xi. *Res:* Economic entomology; biology and control of pests in field crops; systematics; Pselaphidae; Megachilidae; Tardigrada. *Mailing Add:* Dept Entom Univ Calif Davis CA 95616

GRIGG, PETER, BIOMECHANICS, MECHANORECEPTORS. *Educ:* NY State Univ, PhD(physiol), 69. *Prof Exp:* PROF PHYSIOL, MED SCH, UNIV MASS, 72- *Mailing Add:* Dept Physiol Univ Mass Med Sch Worcester MA 01605

GRIGG, RICHARD WYMAN, b Los Angeles, Calif, Apr 12, 37; m 71; c 3. OCEANOGRAPHY. *Educ:* Stanford Univ, BS, 58; Univ Hawaii, MS, 64; Univ Calif, San Diego, PhD(oceanog), 70. *Prof Exp:* Res asst oceanog, Univ Calif, San Diego, 64-70; from asst prof to assoc prof, 70-81, PROF OCEANOG, UNIV HAWAII, 82- *Concurrent Pos:* Assoc prog dir, Nat Sea Grant Prog, 75-77; mem, WPac Fisheries Mgt Coun, 77-79; consult UNESCO & UN Develop Prog, 79-81. *Mem:* Am Soc Limnol & Oceanog; Am Inst Biol Sci; Ecol Soc Am; Sigma Xi; AAAS. *Res:* Coral reef ecology, patterns of response of marine temperate and tropical shallow water communities to stress; ecology and fishery dynamics of precious corals; paleoceanography. *Mailing Add:* Dept Oceanog Univ Hawaii 1000 Pope Rd Honolulu HI 96822

GRIGGER, DAVID JOHN, b Cleveland, May 5, 60. ELECTROCHEMICAL PROCESSES, SEMICONDUCTOR PROCESSES. *Educ:* Cleveland State Univ, BS, 82; Ohio State Univ, MS, 83. *Prof Exp:* Intern, Standard Oil Co, 82; assoc mem tech staff, RCA Corp, 84-87; CHEM ENGR, LIFE SYSTS INC, 87- *Mem:* Am Inst Chem Engrs; Soc Automotive Engrs; Int Asn Hydrogen Energy. *Res:* Development and mathematical modeling of chemical and electrochemical processes for regenerative life support systems used in space applications. *Mailing Add:* Life Systs Inc 24755 Highpoint Rd Cleveland OH 44122

GRIGGS, ALLAN BINGHAM, b Cottage Grove, Ore, June 10, 09; m 40; c 1. GEOLOGY. *Educ:* Univ Ore, BS, 32; Stanford Univ, PhD(geol), 52. *Prof Exp:* Geol field asst, US Engrs, Ore, 35-38; lab asst geol, Stanford Univ, 39-40; geologist, US Geol Surv, 40-74; RETIRED. *Concurrent Pos:* Mem staff, Inst Int Mining Res, Japan, 74-76; courtesy prof, Univ Ore, 77- *Mem:* Soc Econ Geologists; fel Geol Soc Am. *Res:* Economic geology, particularly chromite and lead-zinc deposits; areal geology of Coeur d'Alene district, Idaho. *Mailing Add:* 1121 Spyglass Dr Eugene OR 97401

GRIGGS, DOUGLAS M, JR, b Aug 14, 28; US citizen; m 56; c 2. PHYSIOLOGY, MEDICINE. *Educ:* Harvard Univ, AB, 49; Univ Va, MD, 53; Am Bd Internal Med, dipl. *Prof Exp:* From intern to asst resident, St Lukes Hosp, New York, 53-58; clin biophys, Nat Heart Inst, 61-62; from asst med to assoc prof, Hahnemann Med Col & Hosp, 62-67; assoc prof, 67-70, PROF PHYSIOL & MED, SCH MED, UNIV MO-COLUMBIA, 70- *Concurrent Pos:* NY Heart Asn fel, St Lukes Hosp, 58-60; fel, Univ Pittsburgh, 60-61; USPHS fel, 60-62, res career develop awards, 63-67 & 68-73. *Mem:* Am Fedn Clin Res; Am Physiol Soc; Soc Exp Biol & Med. *Res:* Cardiovascular physiology particularly myocardial metabolism and coronary physiology. *Mailing Add:* Dept Physiol Univ Mo Med Sci Columbia MO 65212

GRIGGS, GARY B, b Pasadena, Calif, Sept 25, 43; m 80; c 5. MARINE GEOLOGY, ENVIRONMENTAL GEOLOGY. *Educ:* Univ Calif, Santa Barbara, BA, 65; Ore State Univ, PhD(oceanog), 68. *Prof Exp:* From asst prof to assoc prof, 69-79, chmn dept, 81-84, PROF EARTH SCI, UNIV CALIF, SANTA CRUZ, 79- *Concurrent Pos:* Fulbright Fel, Greece, 74-75; vis prof, Semester at Sea. *Mem:* Geol Soc Am; Am Geophys Union; Soc Econ Paleontologist & Mineralogists; Am Shore & Beach Preserv Asn; Am Coun Environ. *Res:* Coastal geology and engineering; environmental geology; geomorphology; author and coauthor of various publications. *Mailing Add:* Div Natural Sci Univ Calif Santa Cruz CA 95064

GRIGGS, ROBERT C, b Rochester, NY, July 11, 25; m 53; c 3. MEDICINE. *Educ:* Harvard Univ, MD, 49. *Prof Exp:* From instr to assoc prof, 56-75, PROF MED, SCH MED, CASE WESTERN RESERVE UNIV, 76- *Concurrent Pos:* Webster-Underhill fel, Case Western Reserve Univ, 58-60; from asst vis physician to physician, Metrop Gen Hosp, Cleveland, 56-; consult, Marymount Hosp, 58-; attend hemat, Vet Admin Hosp, 61- *Mem:* Am Fedn Clin Res; Am Soc Hemat; Soc Exp Biol & Med. *Res:* Internal medicine; hematology; medical education. *Mailing Add:* Cleveland Mem Gen Hosp 3395 Scranton Rd Cleveland OH 41109

GRIGGS, THOMAS RUSSELL, b Lexington, NC, July 26, 43; m 66; c 1. MEDICINE, PATHOLOGY. *Educ:* Univ NC, AB, 65, MD, 69. *Prof Exp:* Med officer, Environ Protection Agency, 73-75; instr path, 73-75, asst prof, 75-78, ASSOC PROF MED & PATH, SCH MED, UNIV NC, 78- *Concurrent Pos:* Dir, Cornary Care Unit, NC Mem Hosp, 75; consult, Off Chief Med Examr, 76-; Jefferson-Pilot fel acad med, 78-82. *Mem:* Am Asn Pathologists; Int Soc Thrombosis & Haemostasis; Am Fedn Clin Res; Am Heart Asn. *Res:* Use of animal models to study hemostasis, thrombosis and atherosclerosis; study of atherogenesis in pigs with Von Willebrand's disease. *Mailing Add:* Dept Path & Med 349 Clin Sci Bldg 229 H Univ NC Sch Med Chapel Hill NC 27514

GRIGGS, WILLIAM HOLLAND, b Novelty, Mo, May 31, 16; m 44; c 3. POMOLOGY. *Educ:* Mo State Teachers Col, BS, 37; Univ Mo, MA, 39; Univ Md, PhD(pomol), 43. *Prof Exp:* Asst pomol, Univ Mo, 37-39 & Univ Md, 39-42; asst prof, Univ Conn, 46-47; from asst prof to prof, 47-77, EMER PROF POMOL, UNIV CALIF, DAVIS, 77- *Honors & Awards:* Stark Award, Am Soc Hort Sci, 64. *Mem:* Am Soc Hort Sci. *Res:* Pollination requirements of fruits and nuts; physiological problems in pear production; pear breeding. *Mailing Add:* Dept Pomol Univ Calif Davis CA 95616

GRIGORI, ARTUR, b Binab, Persia, May 1, 18; nat US. ALGEBRA. *Educ:* Univ Tübingen, PhD(philos), 50. *Prof Exp:* Asst prof math, St Bonaventure Univ, 58; asst prof, 62-74, ASSOC PROF MATH, CALIF STATE UNIV, NORTHRIDGE, 74- *Concurrent Pos:* NSF Math Award, 59. *Mem:* Math Asn Am. *Res:* Modern algebra and geometry; calculus; differential equations; theory of equations. *Mailing Add:* Box 1179 Desert Hot Springs CA 92240

GRIGORIU, MIRCEA DAN, b Bucharest, Romania, Mar 2, 43; m 69; c 1. CIVIL ENGINEERING. *Educ:* Bucharest Inst Civil Eng, dipl ing, 67; Univ Bucharest, dipl math, 72; Mass Inst Technol, PhD(civil eng), 76. *Prof Exp:* Asst prof struct anal, Bucharest Inst Civil Eng, 67-73; res asst struct reliability, Mass Inst Technol, 73-76; assoc prof eng, Univ Simon Bolivan, Caracas, 76-77; vis prof, McGill Univ & Univ Waterloo, 77-80; ASSOC PROF STRUCT RELIABILITY, CORNELL UNIV, 80- *Mem:* Am Soc Civil Engrs; Am Acad Mech. *Res:* Structural reliability analysis, especially reliability of structural systems, probabilistic models for loads and load combinations, random vibration and structural analysis. *Mailing Add:* 218 Hollister Hall Civil Eng Cornell Univ Main Campus Ithaca NY 14853

GRIGOROPOULOS, SOTIRIOS G(REGORY), b Athens, Greece, Mar 24, 33. ENVIRONMENTAL & ENGINEERING. *Educ:* Nat Univ Athens, Dipl, 55; Wash Univ, St Louis, MS, 58, ScD(sanit eng), 60. *Prof Exp:* Chemist, Sigma Chem Co, Mo, 56-57; asst civil eng, Wash Univ, 57-58, instr, 58-60; assoc prof, 60-63, Univ Mo- Rolla, prof civil eng, 63-; dir, Environ Res Ctr, 66-; PROF ENVIRON ENG, DEPT CIVIL ENG, UNIV PATRAS, GREECE. *Concurrent Pos:* Res engr, Ryckman, Edgerley, Tomlinson & Assocs, Mo, 58-60. *Mem:* Nat Soc Prof Engrs; Am Soc Civil Engrs; Am Inst Chem Engrs; Am Water Works Asn; Water Pollution Control Fedn; Sigma Xi. *Res:* Water supply and pollution control; organic chemical contaminants in water; removal of nutrients from wastewater; biological waste treatment; water treatment plant control; management of stormwater runoff. *Mailing Add:* Dept Civil Eng Univ Patras GR-26110 Patras Greece

GRIGSBY, JOHN LYNN, b Tulsa, Okla, Apr 28, 24; m 50; c 4. ELECTRICAL ENGINEERING. *Educ:* Univ Colo, BS, 49; Stanford Univ, MS, 55, PhD(elec eng), 59. *Prof Exp:* Radio operator, Fed Commun Comn, 42-43; instr elec eng, Iowa State Col, 49; engr, Gen Elec Co, 49-52; res assoc elec eng, Stanford Univ, 52-59; chief engr, Appl Technol Div, ITEK Corp, 60-61, vpres eng, 61-71, exec vpres, 71-75, pres, 75-84, corp vpres, Defense Electronics Opers, 80-84; chief exec officer, Saxpy Computer Corp, 84-86; PRES, ADVENT SYSTS, INC, 87- *Concurrent Pos:* Mem, Rientjes Comt, Dept Army, 59. *Mem:* Sr mem Inst Elec & Electronics Engrs; Sigma Xi. *Res:* Electronic countermeasures systems; reconnaissance and surveillance receiving systems; broadband low-noise amplifiers. *Mailing Add:* 729 Viol Pl Los Altos CA 94022

GRIGSBY, LEONARD LEE, b Floydada, Tex, Dec 31, 29; m 55; c 3. ELECTRICAL ENGINEERING. *Educ:* Tex Technol Col, BSEE, 57, MSEE, 62; Okla State Univ, PhD(eng), 65. *Prof Exp:* Instr elec eng, Tex Technol Col, 57-60, asst prof, 60-61; from instr to asst prof, Okla State Univ, 61-66; assoc prof, Va Polytech Inst & State Univ, 66-72, prof elec eng, 72-; G A POWER DISTINGUISHED PROF, AUBURN UNIV. *Concurrent Pos:* NASA grant, 69-71; consult, Nuclear Develop Ctr, Babcock & Wilcox, Va, 69-72; dir, Energy Res Group, 72- *Mem:* Inst Elec & Electronics Engrs; Am Soc Eng Educ. *Res:* Network and system modeling, simulation and design; control and power systems. *Mailing Add:* Dept Elec Eng Auburn Univ Auburn AL 36849

GRIGSBY, MARGARET ELIZABETH, b Prairie View, Tex, Jan 16, 23. MEDICINE. *Educ:* Prairie View State Col, BS, 43; Univ Mich, MD, 48; Univ London, DTM&H, 63. *Prof Exp:* From intern to asst resident, H G Phillips Hosp, St Louis, Mo, 48-50; asst resident, Freedmen's Hosp, Washington, DC, 50-51; from instr to assoc prof, 52-66, PROF MED, COL MED, HOWARD UNIV, 66- *Concurrent Pos:* Rockefeller Found res fel, Harvard Med Sch, 51-52; China Med Bd fel, Sch Trop Med, Univ PR, 56; attend physician, Freedmen's Hosp, 52- & DC Gen Hosp, 58-; med epidemiologist, USPHS, Ibadan, Nigeria, 66; consult, US AID, 70-71; mem anti-infective agents adv comt, Food & Drug Admin, 70-71. *Mem:* Fel Am Col Physicians; AMA; Nat Med Asn. *Res:* Internal medicine; infectious diseases antibiotic research; electrophoresis of proteins. *Mailing Add:* Dept Med Howard Univ Col Med Washington DC 20059

GRIGSBY, RONALD DAVIS, b Tulsa, Okla, Feb 28, 36; m 62; c 7. MASS SPECTROMETRY, PETROLEUM ANALYSIS. *Educ:* Univ Okla, BS, 58 & 59, PhD(phys chem), 66. *Prof Exp:* Asst chem, Univ Okla, 59-64, teaching asst Russ, 62; res chemist, Continental Oil Co, Okla, 64-68; from asst prof to assoc prof biochem & biophys, Tex A&M Univ, 68-80; res chemist, US Dept Energy, Okla, 80-82; res fel, Assoc Western Univ, 84-86; SR CHEMIST, NAT INST PETROL & ENERGY RES, NIPER, OKLA, 86- *Concurrent Pos:* Adj prof chem, Okla State Univ, 65-66; proprietor, Mass Spectrometer Accessories, 72-77; vis scientist, Nat Res Coun Can, NS, 73-74; pres, Masspec, Inc, 77-; vis prof chem, Univ Okla, 82-84, adj prof, 84-86, adj prof physics & astron, 89-; chmn, comt E-14 mass spectrometry, Am Soc Testing & Mat, 88-89, vchmn, 86-87. *Mem:* Am Soc Mass Spectrometry; Am Chem Soc; Sigma Xi; Am Soc Testing & Mat. *Res:* Analysis of mixtures from petroleum and coal by mass spectrometry; development of mass spectrometric instrumentation and computer techniques. *Mailing Add:* 275 Turkey Creek Rd Bartlesville OK 74006-8034

GRIGSBY, WILLIAM REDMAN, b Denver, Colo, Oct 15, 34; m 58; c 2. PERIODONTICS. *Educ:* Dartmouth Col, BA, 56; Univ Mo-Kansas City, DDS, 60; Med Col Va, Va Commonwealth Univ, PhD(biochem), 70; Univ Iowa, Periodont cert, 75. *Prof Exp:* Asst prof oper dent & biochem, Med Col Va, Va Commonwealth Univ, 68-73; ASSOC PROF PERIODONT, COL DENT, UNIV IOWA, 73- *Concurrent Pos:* Consult, Coun on Dent Therapeut Am Dent Asn. *Mem:* AAAS; Am Chem Soc; Am Dent Asn; Int Asn Dent Res; Am Acad Periodont. *Res:* Prevention of bacterial colonization of teeth. *Mailing Add:* Dept Periodont Univ Iowa Col Dent Iowa City IA 52242

GRILIONE, PATRICIA LOUISE, b Fresno, Calif, Apr 15, 35. MICROBIOLOGY. *Educ:* Fresno State Col, BA, 57, MA, 59; Univ Calif, Davis, PhD(microbiol), 66. *Prof Exp:* From asst prof to assoc prof, 66-76, PROF MICROBIOL, SAN JOSE STATE UNIV, 76- *Mem:* Am Soc Microbiol; Inst Food Technologists; Soc Indust Microbiologists. *Res:* General microbiology; ecology. *Mailing Add:* Dept Biol San Jose State Univ San Jose CA 95192

GRILL, HERMAN, JR, b New York, NY, June 30, 36; m 59. DAIRY SCIENCE, BIOCHEMISTRY. *Educ:* Mich State Univ, BS, 59; Univ Minn, MS, 62; Pa State Univ, PhD(dairy sci, biochem), 66. *Prof Exp:* Mgr flavor res, Carnation Res Lab, Van Nuys, Calif, 66-73, mgr fundamental res, 73-75, dir tech serv, 75-85, DIR TECH SERV & QUAL ASSURANCE, BASIC VEG PROD, VACAVILLE, CALIF, 81- *Mem:* Am Chem Soc; Inst Food Technologists; Sigma Xi; Am Soc Qual Control. *Res:* Flavor chemistry of foods, both the chemical and sensory aspects; research and development of dehydrated vegetables, primarily onions and garlic. *Mailing Add:* 1562 Rockville Rd Suisun City CA 94585

GRILLO, RAMON S, b New York, NY, May 28, 31; m 56; c 1. BIOLOGY. *Educ:* NY Univ, BA, 53; Fordham Univ, MS, 57, PhD(cytol), 60. *Prof Exp:* Instr biol, Marymount Col NY, 59-60 & Bronx Community Col, 60; from asst prof to assoc prof, Seton Hall Univ, 60-64; assoc prof, 64-69, PROF BIOL, ADELPHI UNIV, 69- *Concurrent Pos:* USPHS res grant, 63-65; US Atomic Energy Contract, 67-69. *Mem:* AAAS; Am Soc Zoologists; Am Soc Cell Biol. *Res:* Study of the dynamics of cell proliferation in normal and regenerating tissues of the newt, Triturus viridescens, using tritiated thymidine and autoradiography. *Mailing Add:* Dept Biol Adelphi Univ Garden City NY 11530

GRILLOS, STEVE JOHN, b Rock Springs, Wyo, Jan 15, 28; m 61; c 3. BOTANY, PLANT ANATOMY. *Educ:* Univ Denver, BS, 51; Univ Wyo, MS, 52; Ore State Col, PhD(bot), 56. *Prof Exp:* Asst bot, Univ Wyo, 51-52; asst, Ore State Col, 52-53, instr, 53-56; instr biol sci, Modesto Jr Col, 56-60; assoc prof bot, Univ Pac, Stockton, 60-62; prof biol, Calif State Col, Hayward, 62-65; PROF BOT, CALIF STATE UNIV, STANISLAUS, 65- *Concurrent Pos:* Vis prof, Stephen F Austin State Col, 60. *Res:* Tissue structure and development; systematics of fern and fern allies and weeds. *Mailing Add:* Dept Biol Sci Calif State Univ Stanislaus 801 W Monte Vista Ave Turlock CA 95380

GRILLOT, GERALD FRANCIS, b Versailles, Ohio, Jan 24, 14; m 41; c 2. ORGANIC CHEMISTRY. *Educ:* Ohio State Univ, BA, 36; Univ Ill, PhD(org chem), 40. *Prof Exp:* Asst chem, Univ Ill, 36-40; head dept, Blue Ridge Col, 40-41; from instr to asst prof, Univ Ky, 41-46; asst prof, 46-53, assoc prof, 53-80, EMER ASSOC PROF CHEM, SYRACUSE UNIV, 80- *Mem:* Am Chem Soc. *Res:* Mannich reaction; stereochemistry; synthesis of organic pharmaceuticals; preparation and properties of aminomethyl sulfides; Hofmann-Martius rearrangement. *Mailing Add:* 205 Laurel Ave Clarks Summit PA 18411-1004

GRILLY, EDWARD ROGERS, b Cleveland, Ohio, Dec 30, 17; m 73; c 2. THERMODYNAMICS. *Educ:* Ohio State Univ, BS, 40, PhD(phys chem), 44. *Prof Exp:* Asst chem, Univ Wis, 40-41 & Ohio State Univ, 41-44; chemist, Carbide & Carbon Chem Co, Oak Ridge, 44-46; asst prof chem, Univ NH, 46-47; mem staff, 47-80, CONSULT, LOS ALAMOS NAT LAB, UNIV CALIF, 80- *Concurrent Pos:* Mem, Los Alamos County Coun, 76-78 & NMex Legis, 67-70. *Mem:* Am Phys Soc; Sigma Xi. *Res:* Gaseous kinetic theory; cryogenics; thermodynamics; transport properties of gases; pressure-volume-temperature at low temperatures and high pressure; cryogenic laser and magnetic confinement fusion targets. *Mailing Add:* 705 43rd St Los Alamos NM 87544

GRIM, EUGENE, b Stillwater, Okla, July 19, 22; m 46; c 1. PHYSIOLOGY. *Educ:* Kans State Univ, BS, 45, MS, 46; Univ Minn, PhD(physiol chem), 50. *Prof Exp:* From instr to prof physiol, 52-89, head dept, 68-86, EMER PROF PHYSIOL, UNIV MINN, MINNEAPOLIS, 89- *Concurrent Pos:* Lederle med fac award, 54; USPHS sr res fel, Univ Minn, Minneapolis, 58-63, USPHS career develop award, 63-68; sect ed, J Physiol, 66-68; mem physiol study sect, NIH, 67-69, chmn sect, 69-71. *Mem:* Fel AAAS; Am Physiol Soc; Am Chem Soc; Am Gastroenterol Soc; Biophys Soc. *Res:* Membrane transport phenomena; intestinal absorption; visceral circulation; regional blood flow. *Mailing Add:* 60255 Millard H-Physiol Univ Minn 435 Delaware St SE Minneapolis MN 55455

GRIM, J(OHN)NORMAN, b Santa Barbara, Calif, Sept 8, 33; m 54; c 2. CELL BIOLOGY, PROTOZOOLOGY. *Educ:* Univ Calif, Santa Barbara, BA, 56; Univ Calif, Los Angeles, MA, 60; Univ Calif, Davis, PhD(zool), 67. *Prof Exp:* Lab technician, Univ Calif, Davis, 60-67; assoc prof, 67-79, PROF BIOL, NORTHERN ARIZ UNIV, 79- *Mem:* Soc Protozoologists; Am Micros Soc; Sigma Xi; Electron Microscopic Soc Am. *Res:* Ultrastructure and function of ciliated protozoan organelles; encystment and secretion; ciliate-fish gut symbioses. *Mailing Add:* Dept Biol Sci Box 5640 Northern Ariz Univ Flagstaff AZ 86011

GRIM, LARRY B, b Philadelphia, Pa, Aug 20, 45; m 69; c 2. IMAGE COMPRESSION & PROCESSING. *Educ:* Univ Del, BEE, 67; Univ Mass, MSEE, 69; Univ Pa, PhD(comput sci), 80. *Prof Exp:* res engr, E I Du Pont De Nemours Inc, 68-83; VPRES, MESH INC, 83- *Mem:* Inst Elec & Electronics Engrs. *Res:* Pattern recognition as applied to medical images and reversible image compression. *Mailing Add:* 802 Bethel Rd Oxford PA 19363

GRIM, RALPH EARLY, geology; deceased, see previous edition for last biography

GRIM, SAMUEL ORAM, b Landisburg, Pa, Mar 11, 35; m 57, 83; c 4. INORGANIC CHEMISTRY, ORGANOPHOSPHORUS CHEMISTRY. *Educ:* Franklin & Marshall Col, BS, 56; Mass Inst Technol, PhD(inorg chem), 60. *Prof Exp:* Asst prof chem, Univ Md, 60-61; res fel, Imp Col, Univ London, 61-62; from asst prof to assoc prof, 62-65, chmn Inorg Chem div, 70-77, 80-87, PROF CHEM, UNIV MD, NSF, 68- *Concurrent Pos:* Res grants, NSF, 62-81, Air Force Off Sci Res, 64-68 & NATO, 80-82; Sir John Cass's Found sr res fel, City London Polytechnic, 79-80; prof officer rotator, inorg & organometallic chem, NSF, 88-90. *Honors & Awards:* Sci Achievement Award, Sigma Xi, 83. *Mem:* AAAS; Am Chem Soc; Royal Soc Chem; NY Acad Sci; Am Inst Chemists; Sigma Xi. *Res:* Organometallic and organophosphorus chemistry; coordination compounds; phosphorus nuclear magnetic resonance studies. *Mailing Add:* Dept Chem Univ Md College Park MD 20742-2021

GRIM, WAYNE MARTIN, b York, Pa, Apr 12, 30; m 52; c 4. PHARMACEUTICAL CHEMISTRY. *Educ:* Philadelphia Col Pharm, BSc, 52, MSc, 54; Univ Mich, PhD(pharmaceut chem), 59. *Prof Exp:* Res assoc pharmaceut res, Merck & Co, Inc, 54-56, Res Labs, 59-61, unit head, 61-62, sect head, 62-64, mgr pharmaceut develop, 64-68, dir, 68-69, dir, pharmaceut res, Merck, Sharp & Dohme Res Labs, 69-75, sr dir, 75-78; dir, New Prods Develop, 78-86, VPRES, NEW PRODS DEVELOP, RORER CENT RES, WILLIAM H RORER INC, 86- *Concurrent Pos:* Fel, Am Found Pharm Educ, 56-59. *Mem:* AAAS; Am Pharmaceut Asn; Am Asn Promotion Sci; NY Acad Sci. *Res:* Rheology and the physical stabilization of suspensions; topical and inhalation aerosol development; oral and topical liquid, semisolid and solid dosage form development; particle size reduction and measurement; biopharmaceutics and bioavailability research. *Mailing Add:* William H Rorer Inc 500 Virginia Dr Ft Washington PA 19034

GRIMALDI, JOHN VINCENT, b New York, NY, Sept 6, 16; m 42; c 2. CHEMICAL & SAFETY ENGINEERING. *Educ:* NY Univ, BS, 39, MA, 41, PhD(acoust), 56; Polytech Inst Brooklyn, BChE, 51. *Prof Exp:* Apprentice, Grumman Aircraft Eng Corp, 41-42, dir safety, 42-45; res engr, Nat Conserv Bur, 45-46, dir res div, 46-47, dir indust div & asst mgr safety, 47-56; consult safety & plant protection, Gen Elec Co, 56-62, consult health, safety & plant protection, 62-67; dir, Ctr for Safety, NY Univ, 67-77; actg exec dir, Inst Safety & Systs Mgt, 77-80, dir degree progs, 77-82, EXEC DIR DEGREE PROGS, UNIV SOUTHERN CALIF, 82- *Concurrent Pos:* Mem, President's Comt Employ of Physically Handicapped, 47-56, NY State Gov Comt Occup Safety, 58-63, US Secy Labor's Nat Adv Comt Occup Safety & Health, 69-73, Nat Inst Occup Safety & Health Study Sect, 70-74, 78-, Harvard Bd Overseers Health Serv Comt, 76-81 & Bd Gov, Flight Safety Found, 78-80; consult, Creole Petrol Co, 57 & 63 & Tenn Valley Authority, 68-70. *Honors & Awards:* Medal of the Associacado National de Medicina de Trabalho, Brazil; Centennial Medal, Am Soc Mech Engrs. *Mem:* Fel Am Inst Chemists; Am Soc Mech Engrs; fel Am Soc Safety Engrs (pres, 61-62); Am Chem Soc; fel Am Acad Safety Educ. *Res:* Management performance measurement methods; employment of the physically handicapped; executive health examinations and their relative value; effect of noise on human performance. *Mailing Add:* 13044 Mindanao Way Marina Del Rey CA 90292-6463

GRIMES, CAROL JANE GALLES, b Long Beach, Calif, Sept 30, 46; m 69. BIOINORGANIC CHEMISTRY. *Educ:* Immaculate Heart Col, BA, 69; Northwestern Univ, Evanston, PhD(chem), 73. *Prof Exp:* Fel, Univ Calif, Irvine, 72-73, lectr chem, 73-75; VIS ASST PROF CHEM, UNIV SOUTHERN CALIF, 75- *Mem:* Am Chem Soc. *Res:* Electro-transfer processes in biological and bioinorganic model systems. *Mailing Add:* 3212 Tigertail Dr Los Alamitos CA 90720-4836

GRIMES, CAROLYN E, b Kansas City, Mo, Oct 21, 43. EYE RESEARCH. *Prof Exp:* Grants mgt specialist, Grants Mgt Sect, 80-85, chief, 85-87, BR CHIEF, EXTRAMURAL SERV BR, NAT EYE INST, NIH, 87- *Mailing Add:* Extramural Serv Br Nat Eye Inst NIH Bldg 31 Rm 6A52 Bethesda MD 20892

GRIMES, CHARLES CULLEN, b Norman, Okla, June 11, 31; m 57; c 3. SOLID STATE & LOW TEMPERATURE PHYSICS. *Educ:* Univ Okla, BS, 53; Stanford Univ, MS, 54; Univ Calif, Berkeley, PhD(physics), 62. *Prof Exp:* Electronics physicist, US Naval Air Missile Test Ctr, 54-55; mem tech staff, Bell Labs, 62-88 & 85-91, Head Solid State & Low Temperature Physics Res Dept, 68-84; CONSULT, 91- *Mem:* Fel Am Phys Soc. *Res:* Experimental studies of electrons in surface states outside liquid helium and electron bubbles in liquid helium. *Mailing Add:* 85 Woodland Ave Summit NJ 07901

GRIMES, CHURCHILL BRAGAW, b Washington, NC, Apr 29, 45; m 65; c 2. FISH BIOLOGY, MARINE ECOLOGY. *Educ:* ECarolina Univ, BS, 67, MS, 71; Univ NC, PhD(marine sci), 76. *Prof Exp:* Marine biologist, Fla State Dept Natural Resources, 69-72; fishery biologist, Nat Marine Fisheries Serv, 75-76; dir marine sci, NC Marine Resources Ctr, 76-77; assoc prof marine fisheries, Rutgers Univ, 77-84; RES ECOLOGIST, NAT MARINE FISHERIES SERV, 84- *Concurrent Pos:* Ecol, fishery biol & mgt deepwater grouper & snapper resources, SE US, 72-76; ecol, behavior & fisheries biol of tilefish, 78-85, ecol & fishery biol of weakfish in US Middle Atlantic waters, 79-84 & stock identity, growth & recruitment processes of coastal pelagic fishes in US S Atlantic & Gulf of Mex waters, 85-; mem, Northeastern Rep, Marine Fisheries Sect, Am Fisheries Soc, 82-84, Outstanding Publ Selection Comt, 87-88, Marine & Estaurine Resource Comt, Southern Div, 88-90; assoc ed, Trans Am Fishery Soc, Am Fisheries Soc, 84-86; chmn, Marine & Estuarine Resource Comt, Southern Div, Am Fisheries Soc, 84-86; pres-elect, Marine Fishery Sect, Am Fisheries Sect, Am Fisheries Soc, 90-92. *Honors & Awards:* Award Excellence Comt, Am Fisheries Soc, 91. *Mem:* Am Fisheries Soc; Am Soc Ichthyologists & Herpetologists; Ecol Soc Am; Am Soc Limnol & Oceanog; Estuarine Res Fedn; Sigma Xi; Am Inst Fishery Res Biologists. *Res:* Life history, population dynamics, community ecology, recruitment processes and management of marine fishes. *Mailing Add:* Nat Marine Fisheries Serv Southeast Fisheries Ctr 3500 Delwood Beach Rd Panama City FL 32408-7499

GRIMES, CRAIG ALAN, b Ann Arbor, Mich, Nov 6, 56; m 84. ELECTRICAL & COMPUTER ENGINEERING. *Educ:* Penn State Univ, BSEE & BS(physics), 84; Univ Tex, Austin, MS, 86, PhD(elec & computer eng), 90. *Prof Exp:* Res asst elec & computer eng, Univ Tex, Austin, 85-87, lectr, 87-90; RES SCIENTIST & SR MEM, LOCKHEED RES LABS, 90- *Concurrent Pos:* Pres, Crale Inc, Austin, 85- *Mem:* AAAS; Inst Elec & Electronics Engrs. *Res:* Analysis of wave propagation in materials; calculating the effective electromagnetic properties of granular media; radio frequency instrumentation and antennas. *Mailing Add:* 504 W 24th St No 137 Austin TX 78705

GRIMES, DALE M(ILLS), b Marshall Co, Iowa, Sept 7, 26; m 47; c 2. ANTENNAS, AUTOMOTIVE RADAR. *Educ:* Iowa State Univ, BS, 50, MS, 51; Univ Mich, PhD(elec eng), 56. *Prof Exp:* Lab asst physics, Iowa State Univ, 48; asst, Ames Lab, Atomic Energy Comn, 49-51; asst, Univ Mich, 51-52, res assoc, 52-54, assoc res physicist, 54-56, from assoc prof to prof elec eng, 56-76; prof elec eng & chmn dept, Univ Tex, El Paso, 76-79; prof & head dept, 79-86, PROF ELEC ENG, PA STATE UNIV, 79- *Concurrent Pos:* Consult, Boulder Labs, Nat Bur Standards, 57-68 & Gen Motors Tech Ctr, 68-; chief scientist, Conductron Corp, 60-63; chief scientist, Crale Inc, 86- *Mem:* AAAS; Inst Elec & Electronic Engrs; Am Phys Soc; Soc Automotive Engrs. *Res:* Automotive radar; antennas. *Mailing Add:* Dept Elec & Computer Eng Pa State Univ University Park PA 16802

GRIMES, DAVID, b Hull, Eng, Apr 7, 48; Brit & Can citizen; m; c 2. SMOOTH MUSCLE RECEPTOR PHARMACOLOGY, ALLERGY IMMUNOLOGY. *Educ:* Univ Bath, BSc, 74, PhD(pharmacol), 77. *Prof Exp:* Lab instr, Dept Pharmacol, Univ Bath, Eng, 74-77; sr scientist, Ayerst Labs, Montreal, 77-81, res assoc, 81-82, coordr, Isolated Tissue Sect, 82-84, SECT HEAD, ISOLATED TISSUE RECEPTOR, PHARMACOL SECT, AYERST RES, PRINCETON, NJ, 84-, SECT HEAD PULMONARY SECT, 87- *Concurrent Pos:* Mem, Res & Develop Info Syst Steering Comt, Wyeth-Ayerst Res, 88-89, co-chair, Immunopharmacol Div Subcomt, 88-90, mem, Bradykinin Discovery Team, 90- *Mem:* Am Soc Pharmacol & Exp Therapeut; Am Thoracic Soc; AAAS. *Res:* In vivo evaluation of pulmonary effects of load compounds; in vitro screening evaluation receptor profiling and mechanism studies on new compounds. *Mailing Add:* Dept Immunopharmacol Wyeth-Ayerst Res C N 8000 Princeton NJ 08543-8000

GRIMES, DONALD WILBURN, b Maysville, Okla, July 28, 32; m 57; c 3. AGRONOMY, AGRICULTURAL ECONOMICS. *Educ:* Okla State Univ, BS, 54, MS, 56; Iowa State Univ, PhD(soil fertil), 66. *Prof Exp:* Soil scientist, Soil Conserv Serv, USDA, 55-56; asst agronomist, Kans State Univ, 56-61; assoc agron, Iowa State Univ, 61-66; asst water scientist, 66-71, assoc water scientist, 71-77, WATER SCIENTIST, UNIV CALIF, DAVIS, 77- *Concurrent Pos:* USAID, Sudan, 83; consult, Dominican Repub, 86, 87. *Mem:* Am Soc Agron; Soil Sci Soc Am; Int Soc Soil Sci; Crop Sci Soc Am; Sigma Xi. *Res:* Soil, plant and water relations; shallow water table use by growing crops to meet ET demands; crop water-yield production functions. *Mailing Add:* Kearney Agr Ctr Univ Calif 9240 S Riverbend Ave Parlier CA 93648

GRIMES, GARY WAYNE, b Henderson, Ky, Oct 28, 46; m 67; c 2. DEVELOPMENTAL GENETICS. *Educ:* Ind Univ, BA, 68, MA, 72, PhD(genetics), 72. *Prof Exp:* Res assoc genetics, Ind Univ, 68-69, NIH trainee, 72-73; PROF BIOL, HOFSTRA UNIV, 73- *Mem:* Sigma Xi; Soc Protozoologists; Soc Develop Biol; Am Soc Zoologists. *Res:* Morphogenesis and cortical inheritance in ciliated protozoa; comparative structure and function of mammalian tissues; pattern formation. *Mailing Add:* Dept Biol Hofstra Univ 1000 Fulten Ave Hempstead NY 11550

GRIMES, HUBERT HENRY, b Cleveland, Ohio, Mar 11, 29; m 55; c 2. SOLID STATE PHYSICS. *Educ:* Western Reserve Univ, BS, 52, MS, 53, PhD(phys chem), 56. *Prof Exp:* RES PHYSICIST, NASA, 56-, HEAD SOLID STATE PHYSICS SECT, LEWIS RES CTR, 61- *Mem:* Am Phys Soc; Am Inst Mining, Metall & Petrol Engrs; Am Soc Metals. *Res:* Imperfections in solids, diffusion; radiation damage; composite materials; ceramics. *Mailing Add:* 31308 Aldrich Dr Bay Village OH 44140

GRIMES, L NICHOLS, b Boston, Mass. REGENERATION, WOUND HEALING. *Educ:* Harvard Univ, AB, 60; Brown Univ, MAT, 71, PhD(biol), 74. *Prof Exp:* Instr biol, Cherry Lawn Sch, 66-71; teaching asst, Brown Univ, 71-74; res assoc anat, Med Univ SC, 74-75; asst prof biol, Oakland Univ, 75-78; ASST PROF ANAT, UNIV TEX HEALTH SCI CTR, SAN ANTONIO, 78- *Concurrent Pos:* Upward Bound instr biol, Cherry Lawn Sch, 66-69, chmn sci dept, 67-71, dean students, 70-71. *Mem:* Am Asn Anatomists; Am Soc Zoologists; Sigma Xi. *Res:* Regeneration of fibrous and cartilaginous connective tissue. *Mailing Add:* Rte 5 Box 5647 Boerne TX 78006

GRIMES, ORVILLE FRANK, b San Bernardino, Calif, Jan 13, 16; m 41; c 4. THORACIC SURGERY. *Educ:* Univ Calif, AB, 37; Northwestern Univ, MD, 41. *Prof Exp:* Intern, Passavant Mem Hosp, Northwestern Univ, 41-42; from asst resident to chief resident, 42-49, assoc prof, 49-80, vchmn dept, 64-71, PROF SURG, SCH MED, UNIV CALIF, SAN FRANCISCO, 80- *Concurrent Pos:* In chg teaching surg, mem attend staff, chief thoracic surg serv, dir & chief surg consult, Outpatient Dept, San Francisco Gen Hosp, 49; consult, Hamilton AFB Hosp, 50-64. *Mem:* AMA; Am Surg Asn; Am Asn Thoracic Surg; Am Col Surg. *Res:* Management of various esophageal lesions including carcinoma, strictures, acquired and short esophagus and hiatal hernias; tissue transplantation as it reflects pulmonary changes. *Mailing Add:* Dept Surg Univ Calif San Francisco CA 94143

GRIMES, RUSSELL NEWELL, b Meridian, Miss, Dec 10, 35; m 62; c 2. BORON CHEMISTRY, CLUSTER CHEMISTRY. *Educ:* Lafayette Col, BS, 57; Univ Minn, PhD(chem), 62. *Prof Exp:* Fel, Harvard Univ, 62 & Univ Calif, Riverside, 62-63; from asst prof to assoc prof, 63-73, chmn dept, 81-84, PROF CHEM, UNIV VA, 73- *Concurrent Pos:* Fulbright res scholarship, US State Dept, NZ-US Educ Found, 74-75; pres, Va Chap Sigma Xi, 90. *Honors & Awards:* Boron USA Award, 90. *Mem:* Fel AAAS; Am Chem Soc (secy-treas, Inorgan Div, 81-84); Sigma Xi; Corp Inorg Syntheses. *Res:* Organometallic and boron chemistry, especially boranes; carboranes, metallacarboranes and carborane-metal complexes; synthesis and structural studies of electron-delocalized cage compounds; reaction mechanisms of carborane formation and cage rearrangements. *Mailing Add:* Dept Chem Univ Va Charlottesville VA 22901

GRIMES, SIDNEY RAY, JR, b Washington, DC, July 31, 47; m 69; c 3. BIOCHEMISTRY, MOLECULAR BIOLOGY. *Educ:* Univ NC, BS, 69, PhD(biochem), 73. *Prof Exp:* Res asst biochem, Univ NC, 69-73; proj investr, M D Anderson Hosp & Tumor Inst, Univ Tex Syst Cancer Ctr, 73-74, res assoc, 74-75; res assoc, Vanderbilt Univ, 75-76; asst prof biochem, 76-89, RES ASSOC PROF, LA STATE UNIV, 89-; RES CHEMIST, VET ADMIN MED CTR, SHREVEPORT, 76- *Concurrent Pos:* Merit rev grant, Vet Admin Med Ctr, 77-80, 80-82, 84-86, 86-90, 90-93; NIH grant, 78-81; asst prof, La State Univ Med Ctr, 76-88, res assoc prof, 89- *Mem:* Sigma Xi; Am Soc Cell Biol; Am Soc Microbiol. *Res:* Regulation of gene activity in eucaryotes; the role of nuclear and chromosomal proteins in cellular differentiation and development; regulation of transcription of histone genes. *Mailing Add:* Med Res Serv Vet Admin Med Ctr 510 E Stoner Ave Shreveport LA 71101-4295

GRIMES, STEVEN MUNROE, b St Louis, Mo, Aug 20, 41; m 67; c 2. NUCLEAR PHYSICS, APPLIED PHYSICS. *Educ:* Stanford Univ, BS, 63; Univ Wis-Madison, MS, 64, PhD(physics), 68. *Prof Exp:* Fel, Univ Basel, Switz, 68-69; fel, 69-71, physicist, Lawrence Livermore Lab, 71-81; vis prof, 78-79, PROF PHYSICS, OHIO UNIV, 81-, DIR, EDWARDS ACCELERATOR LAB, 85-; DIR, EDWARDS ACCELERATOR LAB, 85- *Mem:* Sigma Xi; fel Am Phys Soc. *Res:* Neutron-induced and neutron-producing reactions. *Mailing Add:* Dept Physics Ohio Univ Athens OH 45701-2979

GRIMINGER, PAUL, b Vienna, Austria, Aug 29, 20; US citizen; m 54; c 4. VITAMIN NUTRITION, GERIATRIC NUTRITION. *Educ:* Univ Ill, BS, 52, MS, 53, PhD(nutrit), 55. *Prof Exp:* Asst prof nutrit, Univ Nebr, 55-57; asst prof, 57-62, assoc prof, 62-67, chmn biol sci, 77-81, PROF NUTRIT, GRAD SCH, RUTGERS UNIV, 67-, DIR, GRAD PROG NUTRIT, 81-, SCI ADV, 87- *Concurrent Pos:* Guggenheim Found fel, 64. *Honors & Awards:* Nutrit Res Award, Am Feed Mfr Asn, 73. *Mem:* Am Inst Nutrit; Soc Exp Biol & Med; Poultry Sci Asn; fel NY Acad Sci; fel AAAS. *Res:* Nutrition and aging; nutritional factors in atherosclerosis and osteoporosis; various aspects of avian nutrition; requirements and metabolism of vitamins in man and vertebrate animals. *Mailing Add:* Dept Nutrit 103 Thompson Hall Cook Col Rutgers Univ New Brunswick NJ 08903

GRIMLEY, EUGENE BURHANS, III, b Passaic, NJ, Oct 28, 41; m 64; c 3. INORGANIC CHEMISTRY. *Educ:* Olivet Col, BA, 63; Univ Iowa, PhD(chem), 71. *Prof Exp:* ASSOC PROF CHEM, MISS STATE UNIV, 71- *Mem:* Am Chem Soc; Sigma Xi. *Res:* Studies on stoichiometry; rates and mechanisms of ligand substitution and electron transfer reactions in solution; synthesis and characterization of transition metal complexes; chlorine oxidation reduction studies; environmental studies on pollution; nuclear magnetic resonance spectroscopy. *Mailing Add:* PO Box 817 Elon College NC 27244

GRIMLEY, PHILIP M, b New York, NY, Mar 10, 35; m 62; c 3. MEDICINE, PATHOLOGY. *Educ:* City Col New York, BS, 56; Albany Med Col, MD, 61; Am Bd Path, dipl, 67, 80. *Prof Exp:* Intern, Cornell Div, Bellevue Hosp, 61-62; resident path, Univ Calif, San Francisco, 62-63;

resident, Nat Cancer Inst, 63-65, staff pathologist, 65-70, head ultrastructural path, 70-73; dir, Clin Path Lab, NY State Dept Health, 73-77; prof path, Univ Md, 78-81; PROF PATH, HEBERT MED SCH, UNIFORMED SERV UNIV HEALTH SCI, 82- *Concurrent Pos:* Nat Heart Found res fel, NY State Dept Health, 58-60; assoc ed, J Nat Cancer Inst, 70-73, J Exp Path, 84-; adj assoc prof path, Albany Med Col, 74-77. *Mem:* Am Soc Clin Path; Col Am Path; Am Asn Path. *Res:* Cellular effects of interferon; cytokine signal transduction; macrophage differentiation. *Mailing Add:* Dept Path Uniformed Serv Univ Health Sci 4301 Jones Bridge Rd Bethesda MD 20814

GRIMLEY, ROBERT THOMAS, b North Attleboro, Mass, Jan 3, 30; m 52; c 5. HIGH TEMPERATURE CHEMISTRY, MASS SPECTROMETRY. *Educ:* Univ Mass, BS, 51; Univ Wis, PhD(phys chem), 58. *Prof Exp:* Res chemist, Corning Glass Works, 57-59; res assoc physics, Univ Chicago, 59-61; from asst prof to assoc prof, 61-77, PROF CHEM, PURDUE UNIV, 77- *Mem:* Am Chem Soc; Am Phys Soc; Sigma Xi. *Res:* High temperature mass spectrometry; kinetics of vaporization; high temperature chemistry. *Mailing Add:* Dept Chem Purdue Univ 1393 Brwn Bldg West Lafayette IN 47907-1393

GRIMM, ARTHUR F, b Berwyn, Ill, June 16, 31; m 54; c 2. PHYSIOLOGY, HISTOLOGY. *Educ:* Northwestern Univ, BS, 53; Univ Ill, DDS & MS, 56, PhD(physiol), 62. *Prof Exp:* Instr, Cols Med & Dent, Univ Ill, 58-62, asst prof, Col Med, 62-66, assoc prof, Cols Dent & Med, 66-69, PROF PHYSIOL & HISTOL, COLS DENT & MED, UNIV ILL, 69- *Concurrent Pos:* Career develop awards, 64-69 & 70-74. *Mem:* Am Physiol Soc; Am Dent Asn; Sigma Xi. *Res:* Growth and development of striated muscles. *Mailing Add:* Dept Histol Univ Ill 801 S Pauline Chicago IL 60680

GRIMM, CARL ALBERT, b Cincinnati, Ohio, Apr 1, 26; m 48; c 3. MATHEMATICS. *Educ:* Univ Cincinnati, MA, 52. *Prof Exp:* From instr to assoc prof, 52-64, PROF MATH, SDAK SCH MINES & TECHNOL, 64- *Mem:* Math Asn Am; Sigma Xi. *Res:* Differential equations. *Mailing Add:* 5231 Pine Tree Dr Rapid City SD 57702

GRIMM, CHARLES HENRY, b New York, NY, Oct 28, 11; m 38; c 1. ORGANIC CHEMISTRY. *Educ:* NY Univ, BSc, 34, MSc, 38. *Prof Exp:* Res chemist, Fritzsche Bros Inc, 35-43; chief chemist & dir lab, Felton Chem Co, Inc, 43-53; dir flavor develop, 53-66, vpres, Flavor Creation Res, 66- 76, SR VPRES, FLAVOR CREATION RES, INT FLAVORS & FRAGRANCES, 76-, DIR, 67- *Concurrent Pos:* Mem food liaison panel, Nat Acad Sci-Nat Res Coun; tech consult, Navy, Army, Air Force Inst Gt Britain, Toronto, Ont, Can, 43-45; sr vpres flavor creation, Int Flavors & Fragrances, Inc, 76; res comt, Essence Oil Assoc, Inst Food Technologists. *Honors & Awards:* Merit Award for Sci Contrib, Essential Oil Asn, 53. *Mem:* AAAS; Am Chem Soc; Inst Food Technol; Am Soc Enologists; Essential Oil Asn; Am Inst Chemists; chap mem Soc Flavor Chemists. *Res:* Essential oil and flavor; development of research on flavor creation; new aroma chemicals. *Mailing Add:* 1 Lincoln Plaza 12E New York NY 10023

GRIMM, ELIZABETH ANN, b Charleston, WVa, Nov 29, 49. CELLULAR IMMUNOLOGY, LYMPHOCYTE ACTIVATION. *Educ:* Randolph-Macon Woman's Col, AB, 71; Univ Calif Sch Med, Los Angeles, PhD(microbiol), 79. *Prof Exp:* Cancer res, surg br, Nat Cancer Inst, NIH, 80-84, sr staff fel, 84-85, sr investr, immunol, Surg Neurol Br, Nat Inst of Neurol & Commun Disorders, NIH, 85-86; ASSOC PROF TUMOR BIOL & SURG, MD ANDERSON HOSP & TUMOR INST, 86- *Concurrent Pos:* Consult, Hoffman-LaRoche, NJ, 83, Bristol Lab, Rochester, 84, E I DuPont Res Facil, 84-85, Cetus Corp, Emeryville, Calif, 84, Eli Lily, Indianapolis, 83-84, Amgen Corp, Calif, 85; vis prof, Dept Microbiol, Howard Univ, 85. *Mem:* Am Asn Immunologists; Am Asn Cancer Res. *Res:* Investigation of ways to utilize cytotoxic lymphoctes for cancer therapy; studies in vitro lymphocyte activation. *Mailing Add:* 1515 Holcombe MD Anderson Hosp & Tumor Inst Box 79 Houston TX 77030

GRIMM, JAMES K, b St Paul, Va, Jan 17, 30; m 48, 75; c 4. ENTOMOLOGY, HUMAN ANATOMY. *Educ:* Concord Col, BSEd, 56; Univ Tenn, Knoxville, MS, 58, PhD(entom), 63. *Prof Exp:* PROF ENTOM, JAMES MADISON UNIV, VA, 58- *Mem:* Entom Soc Am; Am Soc Baxiatrics. *Res:* Microenvironmental studies relating to insects; gypsy moth surveys, biological control; arthropods found in birdnests and their interrelationship to certain species of birds. *Mailing Add:* Dept Biol James Madison Univ Harrisonburg VA 22807

GRIMM, LOUIS JOHN, b St Louis, Mo, Nov 30, 33; m 67; c 2. MATHEMATICS. *Educ:* St Louis Univ, BS, 54; Ga Inst Technol, MS, 60; Univ Minn, PhD(math), 65. *Prof Exp:* Chemist, Walter Reed Army Inst Res, 56-58; asst prof math, Univ Utah, 65-69; assoc prof, 69-74, chmn math & statist, 81-87, dir, Inst Appl Math, 83-87, PROF MATH, UNIV MO-ROLLA, 74- *Concurrent Pos:* Chemist, Tech Develop Labs, USPHS, 58-61; instr, Armstrong State Col, 61; vis asst prof, Univ Minn, 66; NSF res grants, 69-73 & 76-79; vis prof, Univ Nebr, Lincoln, 78-79; Nat Acad Sci exchange scientist, Polish Acad Sci, 81; vis prof, Univ Gdansk, 85, Univ Southern Calif, 87-88. *Mem:* Am Math Soc; Soc Indust & Appl Math; Ger Soc Appl Math & Mech; Polish Math Soc. *Res:* Difference and functional equations; delay-differential equations; numerical analysis; boundary value problems; singular point theory. *Mailing Add:* Dept Math & Statist Univ Mo Rolla MO 65401

GRIMM, ROBERT ARTHUR, b Two Rivers, Wis, July 25, 37; m 65; c 3. ORGANIC CHEMISTRY. *Educ:* Univ Wis, BS, 59; Stanford Univ, PhD(org chem), 63. *Prof Exp:* Res chemist, Archer Daniels Midland Co, 63-67; sr res chemist, Ashland Oil Inc, 67-73, mgr, 73, sect mgr org chem, 73-80, res mgr, 80-87; SECT MGR, PLASTICS & BONDING, EDISON WELDING INST, 87- *Mem:* Am Chem Soc; Sigma Xi; AAAS; Soc Plastics Engrs. *Res:* Fatty acid chemistry, fatty amines and salts; malonitrile; high-temperature radical reactions; catalysis; catalytic oxidation; hydroformylation; aminimides; reductive alkylation; unsym-dimethylhydrazine synthesis; chlorination; carbon suboxide; hydrometallurgy; petrochemical process discovery and development; composites; polymers joining; adhesives. *Mailing Add:* 1810 Ivanhoe Ct Upper Arlington OH 43220-3006

GRIMM, ROBERT BLAIR, b New York, NY, Nov 26, 30. PLANT PATHOLOGY. *Educ:* Univ Miami, BS, 55, MS, 57; La State Univ, PhD(bot), 60. *Prof Exp:* Asst, Univ Miami, 56-57, La State Univ, 57-59 & Nicholls State Col, 60-62; res assoc bot, Univ Miami, 62-64; asst prof, 64-77, ASSOC PROF BOT, FLA ATLANTIC UNIV, 77- *Mem:* AAAS; Phycol Soc Am; Am Inst Biol Sci; Bot Soc Am; Sigma Xi. *Res:* Cryptogamic botany; effects of environmental factors on growth and development of algae. *Mailing Add:* Dept Biol Sci Fla Atlantic Univ Boca Raton FL 33431

GRIMM, ROBERT JOHN, b Detroit, Mich, May 24, 33; m 59; c 2. NEUROLOGY, NEUROPHYSIOLOGY. *Educ:* Antioch Col, BS, 57; Univ Mich, Ann Arbor, MS, 59, MD, 61. *Prof Exp:* Intern med, Univ Mich Hosp, Ann Arbor, 61-62; Nat Inst Neurol Dis & Blindness fel physiol & biophys, Univ Wash, 62-64; Nat Inst Neurol Dis & Blindness fel, Lab Neurophysiol, 64-65, neurol residency, 65-77, ASSOC SCIENTIST, NEUROL SCI INST, MED CTR, GOOD SAMARITAN HOSP, 68-; PVT PRACT NEUROL, 77- *Concurrent Pos:* Dir, Electromyography Lab, Med Ctr, Good Samaritan Hosp, 69-71, dir neurol teaching & res, 70-73, asst dir neurol, 74-76, dir, Myasthenia Gravis Clin, 76-; asst prof neurol, Health Sci Ctr, Univ Ore, 69-74, assoc prof, 75-86, prof, 87. *Mem:* Soc Neurosci; Cooper Soc. *Res:* Disorders of movement; head injuries & labyrinth injury, bioethical issues having to do with neurological, psychiatric disorders. *Mailing Add:* Good Samaritan Hosp Med Ctr 2455 NW Marshall Suite 14 Portland OR 97210

GRIMMELL, WILLIAM C, b Brooklyn, NY, Mar 16, 41; m 66; c 1. MATHEMATICAL ANALYSIS, ELECTRONICS ENGINEERING. *Educ:* Mass Inst Technol, BS, 61; Univ Mich, Ann Arbor, MS, 62 & 64, PhD(elec eng), 65. *Prof Exp:* Mem tech staff control systs res, Bell Tel Labs, 66-68; sr mathematician, 68-74, group leader, Appl Sci Sect, 74-77, GROUP LEADER, TECH DIV, HOFFMANN-LA ROCHE, 77- *Mem:* Inst Elec & Electronics Engrs; AAAS; Sigma Xi. *Res:* Computer applications of industrial processes, pattern recognition and feature analysis, on line quality control. *Mailing Add:* 22 New England Dr Lake Hiawatha NJ 07034

GRIMMER, RONALD CALVIN, b Minneapolis, Minn, Aug 16, 41; m 63; c 3. INTEGRODIFFERENTIAL EQUATIONS. *Educ:* Carthage Col, BA, 63; Univ Iowa, MS, 65, PhD(math), 67. *Prof Exp:* from asst prof to assoc prof, 67-76, PROF MATH, SOUTHERN ILL UNIV, 76- *Concurrent Pos:* Vis assoc prof, Iowa State Univ, 73-74, vis prof Univ Warwick, 80-81. *Mem:* Am Math Soc; Soc Rheology; Soc Natural Philos. *Res:* Integrodifferential equations in the Banach spaces; application to various questions in linear viscoelasticity. *Mailing Add:* Southern Ill Univ Carbondale IL 62901-4408

GRIMM-JORGENSEN, YVONNE, b Zurich, Switz, April 19, 42; m 70; c 1. COMPARATIVE ENDOCRINOLOGY, NEUROENDOCRINOLOGY. *Educ:* Portland State Univ, BA, 69; Univ Conn, PhD(med sci), 74. *Prof Exp:* ASST PROF ANAT, UNIV CONN HEALTH CTR, 76-, MEM FAC, DEPT PHYSIOL. *Mem:* Soc Neurosci; Am Physiol Soc; Am Soc Anatomists; AAAS; Am Soc Zoologists. *Res:* Various aspects of neurosecretion; mechanism of synthesis and release of neuropeptides and the mechanism of action of neurohormones; evolution of the neurohormones and their targets; molluscan model systems. *Mailing Add:* Dept Physiol Univ Conn Health Ctr 263 Farmington Ave Farmington CT 06032

GRIMSAL, EDWARD GEORGE, b Detroit, Mich, Sept, 15, 27; div; c 2. FLUIDS, ACOUSTICS. *Educ:* Western Mich Univ, AB, 48; Iowa State Univ, MS, 50; La State Univ, Baton Rouge, PhD(physics), 55. *Prof Exp:* Instr physics, Western Mich Univ, 50-51; instr, La State Univ, 53-54; asst prof, Canisius Col, 54-56; res physicist acoustics, Armour Res Found, 56-62; chief physicist, Universal Oil Prod Inc, 62-65; ASSOC PROF PHYSICS, UNIV SOUTHWESTERN LA, 65- *Mem:* Am Phys Soc; Acoust Soc Am; Am Asn Physics Teachers. *Res:* Physical acoustics; non-linear acoustics and noise control. *Mailing Add:* 214 Longview Dr Lafayette LA 70506

GRIMSBY, F(RANK) NORMAN, b Seattle, Wash, Sept 15, 27; m 55; c 3. CHEMICAL ENGINEERING. *Educ:* Univ Wash, Seattle, BS, 48, MS, 49; Mass Inst Technol, ScD, 54. *Prof Exp:* Engr, Calif Res Corp, 48; instr indust chem, Mass Inst Technol, 51-52; engr process develop, design & eval, Shell Develop Co, 53-54, 56-69, head chem eng div, Carrington Plastics Lab, Shell Res Ltd, Eng, 69-71, SR ENGR, POLYMERS DIV, SHELL CHEM CO, 71- *Concurrent Pos:* Engr, Process Design, Chem Ctr, Md, 54-55, Phosphate Develop Works, Ala, 55 & Westvaco Res Detachment, WVa, 55-56. *Mem:* Am Chem Soc; Am Inst Chem Engrs. *Res:* Design and economics of chemical processes; epoxy resins; high-pressure polyethylene; Ziegler polyethylene and polypropylene; polymerization catalysis and kinetics. *Mailing Add:* PO Box 1380 Houston TX 77251-1380

GRIMSHAW, PAUL R, b Cedar City, Utah, Mar 17, 22; m 46; c 4. AGRICULTURAL ECONOMICS. *Educ:* Utah State Univ, BS, 48, MS, 49; Ore State Univ, PhD(agr econ & mkt), 71. *Prof Exp:* Vet farm training farm mgt, Iron County Sch Dist, 49-50; county exten agt, Utah State Univ Exten, 52-54, better farming agt farm mgt, 54-60, county agt, 60-63, mkt specialist, 63-68; grad asst, Ore State Univ, 68-70; mkt specialist, Utah State Univ Exten, 70-72; mkt specialist, Utah State Univ, 72-81, assoc dean resident instr, Col Agr, 72-81; RETIRED. *Concurrent Pos:* Chief of Party, USU-Somalia Contract, 81-83; consult, USAID Proj, Pragma Corp, Yemen, 83-84. *Res:* Interregional linear programming computer model to analyze competition in markets facing farmers in Utah and the United States; agricultural marketing. *Mailing Add:* PO Box 603 Santa Clara UT 84765-0603

GRIMSON, BAIRD SANFORD, b Durham, NC, June 23, 43. OPHTHALMOLOGY. *Educ:* Washington & Lee Univ, BA, 65; Duke Univ, MD, 69; Am Bd Ophthal, dipl, 76. *Prof Exp:* Intern, Dept Internal Med, Iowa Univ, 69-70; flight surgeon, US Army Marine Corp, Vietnam and Pa, 70-72; resident, Dept Ophthal, Univ Iowa, 73-75, assoc, 76; fel, Bascom Palmer Eye Inst, Univ Miami, 76-77; ASST PROF, DEPT OPHTHAL, UNIV NC SCH MED, 77- *Res:* Effects of serotonin and norepinephrine on carotid artery blood flow in primates; effects of head-down position during hypotension; drug testing in Horner's Syndrome; Raeder's Syndrome; fourth cranial nerve palsies following Herpes Zoster. *Mailing Add:* 617 Clin Sci Bldg CB No 7040 Univ NC Chapel Hill Chapel Hill NC 27599-7040

GRIMSRUD, DAVID T, b Minot, NDak, Aug 14, 38; m 60. LOW TEMPERATURE PHYSICS. *Educ:* Concordia Col, Minn, BA, 60; Univ Minn, MS, 63, PhD(physics), 65. *Prof Exp:* Fulbright grant, Rome, 65-66; from asst prof to assoc prof physics, Muhlenburg Col, 66-71; assoc prof physics, St Olaf Col, 71-77; staff scientist, Lawrence Berkeley Lab, Berkeley, Calif, 77-88; DIR, MINN BLDG RES CTR, GRAD SCH, UNIV MINN, 89- *Mem:* Am Phys Soc; Am Asn Physics Teachers. *Res:* Liquid helium temperature scale; transport phenomena near critical point. *Mailing Add:* Univ Minn 330 Wulling Hall 86 Pleasant St SE Minneapolis MN 55455

GRIMSTAD, PAUL ROBERT, b Dallas, Tex, Feb 24, 45; m 70; c 2. MEDICAL ENTOMOLOGY, EPIZOOTIOLOGY. *Educ:* Concordia Col, BA, 67; Univ Wis-Madison, MS, 72, PhD(entom), 73. *Prof Exp:* Prev med specialist entom, Med Serv Corps, US Army, 68-70; proj assoc entom & virol, Univ Wis-Madison, 73-74; res fel biol, 74-76, asst fac fel, 76-80, asst prof, 80-86, DIR, LAB ARBOVIRUS RES & SURVEILLANCE, UNIV NOTRE DAME, 76-, ASSOC PROF, 86- *Mem:* Entom Soc Am; Am Mosquito Control Asn; Sigma Xi; Am Soc Trop Med & Hyg; Am Soc Microbiol. *Res:* Epidemiology and ecology of Jamestown Canyon virus, its vectors & vertebrate hosts; the ecology and causes of outbreaks of LaCrosse virus encephalitis in nature; the genetics of transmission of LaCrosse virus by the mosquitoes Aedes triseriatus, Aedes hendersoni and their hybrids. *Mailing Add:* Dept Biol Scis Univ Notre Dame Notre Dame IN 46556-0369

GRIMWOOD, BRIAN GENE, b Yorkville, Ill, Aug 2, 40; m 65; c 1. BIOCHEMICAL GENETICS. *Educ:* Aurora Col, BS, 64; Northern Ill Univ, MS, 67; Univ Ariz, PhD(genetics), 72. *Prof Exp:* Res assoc, Univ Ariz, 70-72; NIH fel, Univ Tex, Austin, 72-74; RES SCIENTIST, DIV LABS & RES, DEPT HEALTH, STATE NY, 75- *Mem:* AAAS; Am Soc Microbiol. *Res:* Biochemistry and physiology of mitochondria, microbial antibiotic susceptibility testing; bacterial DNA replication; biochemistry and metabolism of parasitic protozoan cultures. *Mailing Add:* Ten Crestwood Lane Delmar NY 12054

GRIMWOOD, CHARLES, b Greenfield, Mass, July 13, 31; m 59; c 2. OPEN CHANNEL HYDRAULICS, COASTAL ENGINEERING. *Educ:* Drexel Univ, BS, 62; La State Univ, MSEE, 68; Tulane Univ, MS, 75, PhD(civil eng), 78. *Prof Exp:* Data syst engr space vehicle, Gen Elec Co, 65-66; sr engr instrument, Boeing Co, 66-70; res engr, Sch Med, Tulane Univ, 70-71; electron engr instrument, US Army Corps Engrs, 71-74, hydraulic engr, 74-77, environ engr, 77-80; ASSOC PROF HYDROL, TULANE UNIV, 80- *Mem:* Am Soc Civil Engrs; Inst Elec & Electronics Engrs; Soc Am Military Engrs; Coastal Soc. *Res:* Environmental effects of dredging operations; measurement of turbulence in open channels; effects of urbanization stormwater runoff and water supply systems. *Mailing Add:* Dept Civil Engr 210 Civil Engr Tulane Univ 6823 St Charles Ave New Orleans LA 70118

GRINA, LARRY DALE, b Seattle, Wash, June 12, 43; m 64; c 2. ORGANIC CHEMISTRY, POLYMER CHEMISTRY. *Educ:* Univ NDak, BS, 65; Univ Wash, PhD(org chem), 70. *Prof Exp:* Sr chemist, 70, res chemist, 70-78, GROUP LEADER, POLYMER RES GROUP, TEXACO INC, BEACON, 78- *Mem:* Am Chem Soc. *Res:* Synthetic and physical organic chemistry. *Mailing Add:* 21 Willow Dr Hopewell Junction NY 12533-6235

GRINBERG, ERIC L, b Galati, Romania. GEOMETRIC ANALYSIS. *Educ:* Cornell Univ, BA, 78; Cambridge Univ, 79; Harvard Univ, PhD(math), 83. *Prof Exp:* Hildebrandt A Prof Math, Univ Mich, 83-87; fel math, Inst des Hautes Etudes Sci, 85-86; PROF MATH, TEMPLE UNIV, 87- *Concurrent Pos:* Reviewer Math Rev, 83- & Nat Sci Found, 83- *Mem:* Am Math Soc. *Res:* Integral geometry; geometric analysis on homogeneous spaces; isoperimetric inequalities; several complex variables. *Mailing Add:* Dept Math Temple Univ Philadelphia PA 19122

GRINBERG, JAN, b Krakow, Poland, Nov 30, 33; US citizen; m 60; c 3. PHYSICS, ELECTRICAL ENGINEERING. *Educ:* Technion-Israel Inst Technol, BS, 60, MS, 64; Weizman Inst Sci, Israel, PhD(physics), 70. *Prof Exp:* Res & develop engr elec eng, Soreq Nuclear Res Ctr, Israel, 60-65; asst prof, Tel-Aviv Univ, 65-72; sect head device physics, 72-80, MGR, EXPLORATORY STUDIES DEPT, HUGHES RES LABS, 80- *Concurrent Pos:* Consult, Elscient, Electronic Indust, Israel, 67-72 & Soreq Nuclear Res Ctr, Israel, 70-72. *Mem:* Sigma Xi. *Res:* Solid state physics; ferroelectric phase transitions; solid state devices; liquid crystals; electrooptic devices; image processing devices; thin film technology; array computers. *Mailing Add:* 1141 Carmelina Ave Los Angeles CA 90049

GRINDALL, EMERSON LEROY, b Jackson, Mich, Sept 13, 16; m 37; c 4. UNDERWATER ACOUSTICS. *Educ:* Olivet Col, AB, 38; Mich State Col, MS, 47. *Prof Exp:* Teacher pub sch, Mich, 38-46; instr math, Mich State Col, 46-51; from asst prof to assoc prof, Pa State Univ, 51-66, prof eng res, 66-79, emer prof, 79-80; prin scientist, Tracor Applied Sci, 80-82; RETIRED. *Mem:* Am Math Asn; Acoust Soc Am; Math Asn Am; Sigma Xi. *Res:* Acoustics; physical properties of sea water; statistical analysis; design of experiments for research; reliability; maintainability; weapon system management. *Mailing Add:* 1265 Smithfield St State College PA 16801

GRINDEL, JOSEPH MICHAEL, b Kansas City, Mo, Dec 18, 46; m 70; c 3. DRUG METABOLISM, PROJECT MANAGEMENT. *Educ:* St Benedict's Col, Kans, BSc, 69; Univ Kans, PhD(med chem), 73. *Prof Exp:* Staff officer pharmacol, Walter Reed Army Inst Res, 74-75; chief, Clin Drug Metab Lab, 75-76; res scientist, 76-77, group leader, 77-78, sect head biochem, 78-80, dir, dept drug metab, 80-82, exec dir proj planning, 82-85, exec dir qual improv, McNeil Pharmaceut, 85-88; EXEC DIR PROJ MGT, JOHNSON & JOHNSON, 88- *Mem:* Am Chem Soc; Am Soc Pharmacol & Exp Therapeut; AAAS; Drug Info Asn; Proj Mgt Inst. *Res:* Project management process in new product development; management of human resources in pharmaceutical research and development; disposition and pharmacokinetics of drugs in man. *Mailing Add:* 1787 Cindy Lane Hatfield PA 19446-3218

GRINDELAND, RICHARD EDWARD, b Decorah, Iowa, Mar 11, 29; m 52; c 5. PHYSIOLOGY. *Educ:* Luther Col, Iowa, BA, 53; Univ Iowa, PhD(physiol), 58. *Prof Exp:* Asst prof, Univ Sask, 58-61; NIH fel, 60 & 61-62; vis asst prof physiol, Howard Univ, 62-63; RES SCIENTIST SPACE PHYSIOL, AMES RES CTR, NASA, 62- *Mem:* Endocrine Soc; Am Physiol Soc. *Res:* Physiology of growth hormone. *Mailing Add:* Life Sci Div Ames Res Ctr Moffett Field CA 94035

GRINDLAY, JOHN, b Glasgow, Scotland, June 24, 33; Can citizen; m 60; c 1. SOLID STATE PHYSICS. *Educ:* Glasgow Univ, BSc, 55; Oxford Univ, DPhil, 58. *Prof Exp:* Proj assoc theoret chem, Univ Wis, 58-59; from instr to asst prof theoret physics, Univ BC, 59-64; asst res officer pure chem, Nat Res Coun Can, 64-65, assoc prof, 65-69; chmn dept, 82-88, PROF PHYSICS, UNIV WATERLOO, 69- *Mem:* Am Phys Soc; Can Asn Physicists. *Res:* Elastic, dielectric and thermal properties of crystals; non-linear mechanics and chaos. *Mailing Add:* Dept Physics Univ Waterloo Waterloo ON N2L 3G1 Can

GRINDLAY, JONATHAN ELLIS, b Richmond, Va, Nov 9, 44; m 70; c 2. ASTROPHYSICS, ASTRONOMY. *Educ:* Dartmouth Col, BA, 66; Harvard Univ, MA, 69, PhD(astrophys), 71. *Prof Exp:* Jr fel astrophys, Harvard Univ, 71-74, lectr, 74-76, asst prof, 76-81, chmn dept, 85-90, PROF ASTRON, HARVARD UNIV, 81- *Concurrent Pos:* Astrophysicist, Smithsonian Astrophys Observ, 74-76. *Honors & Awards:* Bok Prize, 76. *Mem:* fel Am Phys Soc; Sigma Xi; Am Astron Soc (secy-treas, 81-84); Int Astron Union; fel AAAS. *Res:* Studies of galactic and extragalactic x-ray sources; models for x-ray and gamma ray sources; optical identification and studies of x-ray sources; studies of hard x-ray and gamma ray sources; development of imaging hard x-ray and optical detectors. *Mailing Add:* Ctr Astrophys Harvard Univ 60 Garden St Cambridge MA 02138

GRINDLEY, NIGEL DAVID FORSTER, b Leeds, UK, Nov 24, 45. DNA TRANSPOSITION, SITE-SPECIFIC RECOMBINATION. *Educ:* Cambridge Univ, UK, BA, 67; London Univ, UK, PhD(bact genetics), 74. *Prof Exp:* Sci staff bact genetics, Cent Pub Health Lab, London, UK, 67-73; postdoctoral fel molecular biol, Carnegie Mellon Univ, Pittsburgh, 73-75; asst prof, Univ Pittsburgh, 78-80; postdoctoral fel molecular biol, 75-78, from asst prof to assoc prof, 80-86, PROF, DEPT MOLECULAR BIOPHYS & BIOCHEM, YALE UNIV, 86- *Concurrent Pos:* Mem, Microbiol Physiol & Genetics NIH Study Sect, 88-92 & Immunol of Leprosy Steering Comt, WHO, 90-93. *Mem:* Am Soc Microbiol; AAAS; Am Soc Biochem & Molecular Biol. *Res:* DNA transposition and site-specific recombination; DNA polymerase I structure and function; DNA-protein interactions. *Mailing Add:* Dept Molecular Biophys & Biochem Yale Univ 333 Cedar St PO Box 3333 New Haven CT 06510

GRINDROD, PAUL (EDWARD), b Oconomowoc, Wis, Apr 5, 25; m 45; c 3. FOOD TECHNOLOGY, CHEMICAL ENGINEERING. *Educ:* Univ Wis, BS, 50, MS, 51, PhD(food tech), 54. *Prof Exp:* Asst food tech, Univ Wis, 51-54; res dir, C J Berst & Co, Wis, 54-58; CHIEF PACKAGING RES, OSCAR MAYER & CO, 58- *Mem:* Am Chem Soc; Am Dairy Sci Asn; Am Inst Chem Engrs; Sigma Xi. *Res:* Food packaging; design and packaging equipment design. *Mailing Add:* 4221 Esch Lane Madison WI 53704

GRINDSTAFF, TEDDY HODGE, b Blount Co, Tenn, Aug 15, 32; m 62; c 1. PHYSICAL CHEMISTRY, ORGANIC CHEMISTRY. *Educ:* Univ Tenn, BS, 58, MS, 61, PhD(phys chem), 63. *Prof Exp:* Proj engr prod eval, Celanese Corp Am, NC, 63-64; res chemist fiber surface, 64-69, res assoc, Dacron Res Lab, 69-86, STAFF CHEMIST, BELLE DEVELOP LAB, E I DU PONT DE NOMOURS & CO, 86- *Mem:* Fiber Soc. *Res:* Adsorption of fatty acids on copper and nickel single crystals; determination of the kinetics of radioactive fatty soil detergency from polymer materials; surface chemistry; adhesion; computer modeling. *Mailing Add:* 1964 Parkwood Rd Charleston WV 25314

GRINDSTAFF, WYMAN KEITH, b Ada, Okla, May 13, 39; m 62; c 2. INORGANIC CHEMISTRY. *Educ:* E Cent State Col, BS, 59; Univ Okla, MS, 65, PhD(inorg chem), 66. *Prof Exp:* From asst prof to assoc prof, 65-73, PROF CHEM, SOUTHWEST MO STATE UNIV, 73-, HEAD DEPT, 72- *Mem:* Am Chem Soc; Sigma Xi. *Res:* Synthesis, physical properties, structure and bonding of inorganic coordination compounds; use of computers in chemical education. *Mailing Add:* Dept Chem Southwest Mo State Univ Springfield MO 65802

GRINE, DONALD REAVILLE, b Dunkirk, NY, Aug 21, 30; m 53; c 2. GEOPHYSICS. *Educ:* Mass Inst Technol, BS, 52, MS, 54, PhD(geophys), 59. *Prof Exp:* Physicist, Stanford Res Inst, 59-61; Schlumberger Well Surv Corp, 62-64, sect head, 64-65; dept head, Stanford Res Inst, 65-71; sr res scientist, 71-72, dept head, 72-75, vpres, 75-76, sr vpres & mgr, La Jolla Div, Sci & Software, 76-81; pres, S-Cubed, 82-83; vpres & div mgr, Maxwell Labs, 83-89; CONSULT, 89- *Mem:* AAAS. *Res:* Explosions, shock waves, seismology, ultrasonics, exploration geophysics, instrumentation. *Mailing Add:* 15009 Paso Del Sol Del Mar CA 92014

GRINER, PAUL F, b Philadelphia, Pa, 33. HOSPITAL ADMINISTRATION. *Educ:* Rochester Univ, Md, 59. *Prof Exp:* PROF MED, ROCHESTER UNIV, 73- *Concurrent Pos:* Gen dir, Strong Mem Hosp, 84- *Mem:* Inst Med-Nat Acad Sci. *Mailing Add:* Strong Mem Hosp 601 Elmwood Ave Rochester NY 14642

GRINGAUZ, ALEX, b Memel, Lithuania, May 18, 34; US citizen; m 59; c 3. MEDICINAL CHEMISTRY, PHARMACY. *Educ:* Long Island Univ, BS, 56; Purdue Univ, MS, 58, PhD(pharmaceut chem), 60. *Prof Exp:* Asst org chem, Purdue Univ, 56-58; NIH fel, 60-61; from asst prof to prof org & pharmaceut chem, Brooklyn Col Pharm, Long Island 61-77, actg chmn dept, 70-72, chmn dept, 72-77, chmn, Div Pharmacotherapeut, Arnold & Marie Schwartz Col Pharm, 77-79; CONSULT. *Mem:* Am Chem Soc; Am Pharmaceut Asn. *Res:* Synthesis and stability of organic medicinal agents; synthesis and evaluation of nonsteroidal antiinflammatory drugs. *Mailing Add:* 1055 E Broadway Woodmere NY 11598

GRINKER, JOEL A, PSYCHOLOGY, NUTRITION. *Educ:* New York Univ, PhD(psychol), 67. *Prof Exp:* DIR, PROG HUMAN NUTRIT & PROF NUTRIT, SCH PUB HEALTH, UNIV MICH, 82- *Res:* Childhood obesity; dietary-induced obesity in rats. *Mailing Add:* MS170-SPH 11 Univ Mich Ann Arbor MI 48109

GRINKER, ROY RICHARD, SR, b Chicago, Ill, Aug 2, 00; m 24; c 1. PSYCHIATRY. *Educ:* Univ Chicago, BS, 19; Rush Med Col, MD, 21. *Prof Exp:* Intern, Psychopathic Hosp, Chicago, 21, Wesley Mem Hosp, 21-22 & Cook County Hosp, 22-24; instr neurol, Sch Med, Northwestern Univ, 24-27; instr, Sch Med, Univ Chicago, 27-28, asst clin prof med, 28-29, from asst prof to assoc prof neurol, 29-35, assoc prof psychiat & head dept, 35-36; chmn dept psychiat, Michael Reese Hosp, 36-76, dir inst psychosom & psychiat res & training, 51-76; RETIRED. *Concurrent Pos:* Rockefeller fel, 33-34; attend neurologist, Cook County Hosp, 26-28, 40-42; res assoc, Chicago Inst Psychoanal, 40-60; attend psychiatrist, Cook County Psychopathic Hosp, Chicago, 46-50; lectr, Social Serv Admin, Chicago, 46-50; clin prof, Col Med, Univ Ill, 51-69; prof, Dept Psychiat, Univ Chicago, 69-78. *Honors & Awards:* Gold Medal Award, Soc Biol Psychiat, 70; Salmon Medal, NY Acad Med, 70. *Mem:* AAAS; Am Psychosom Soc (pres, 52); fel Am Psychiat Asn; Am Asn Neuropath (vpres, 40); Acad Psychoanal (pres, 61). *Res:* Schizophrenia. *Mailing Add:* 910 N Lake Shore No 519 Chicago IL 60611

GRINNELL, ALAN DALE, b Minneapolis, Minn, Nov 11, 36; m 62. SYNAPTIC NEUROPHYSIOLOGY, ECHOLOCATION. *Educ:* Harvard Univ, BA, 58, PhD(biol), 62. *Prof Exp:* NSF fel biophys, Univ Col, Univ London, 62-64; from asst prof to prof biol & physiol, 64-78, PROF PHYSIOL, UNIV CALIF, LOS ANGELES, 78- *Concurrent Pos:* Guggenheim fel, 86. *Honors & Awards:* Sr Scientist Award, Alexander von Humboldt Inst, WGer, 75, 79; Javits Neurosci Investr Award, 86. *Mem:* AAAS; Soc Neurosci; NY Acad Sci; Am Physiol Soc; Biophys Soc. *Res:* Neurophysiology of audition, especially in echolocating bats; synaptic physiology; development specificity and plasticity in the nervous system. *Mailing Add:* Dept Physiol Univ Calif Los Angeles CA 90024

GRINNELL, FREDERICK, b Philadelphia, Pa, Feb 5, 45; m 69; c 3. CELL BIOLOGY, BIOCHEMISTRY. *Educ:* Clark Univ, AB, 66; Tufts Univ, PhD(biochem), 70. *Prof Exp:* NIH trainee, Biochem Sect, Vet Admin Hosp, Dallas, 70-72 & Tex Cancer Soc fel, 71-72; asst prof, 72-77, assoc prof, 77-81, PROF CELL BIOL & ANAT, SOUTHWESTERN MED SCH, UNIV TEX SOUTHWESTERN CTR, 81- *Concurrent Pos:* Vis fel dept biol, Yale Univ, 81; Meyerhoff vis prof, Weizmann Inst, Rehovoth, Israel, 84; Furcheimer vis prof, Hebrew Univ, Hadassah Med Sch, 88. *Mem:* Am Soc Cell Biol; Am Soc Biol Chem. *Res:* Wound healing; biomaterials; cell adhesion. *Mailing Add:* Dept Cell Biol Univ Tex Southwestern Med Sch Dallas TX 75235

GRINNELL, ROBIN ROY, b Palo Alto, Calif, July 21, 32; m 59; c 4. INSTRUMENTATION, BIOENGINEERING. *Educ:* Univ Minn, MS, 61; Purdue Univ, BS, 55, PhD(agr eng), 76. *Prof Exp:* Asst prof, Univ Guelph, 61-67; from asst prof to assoc prof, 67-81, PROF AGR ENG, CALIF POLYTECH STATE UNIV, 81- *Concurrent Pos:* Mem sect comt, Am Standards Inst to Int Orgn Standardization, 66- *Mem:* Am Soc Agr Engrs; Instrument Soc Am; Inst Elec & Electronics Engrs. *Res:* Mechanical properties of biological materials and the development of instrumentation techniques to obtain the measurements and information desired. *Mailing Add:* Agr Eng Dept Calif Polytech State Univ San Luis Obispo CA 93407

GRINSPOON, LESTER, b Newton, Mass, June 24, 28; m 54; c 3. PSYCHIATRY. *Educ:* Tufts Col, BS, 51; Harvard Univ, MD, 55. *Prof Exp:* SR STAFF, MASS MENT HEALTH CTR, 61-; ASSOC PROF PSYCHIAT, HARVARD MED SCH, 68- *Mem:* Am Psychiat Asn; AAAS; Group Advan Psychiat. *Res:* Drug abuse. *Mailing Add:* Dept Psychiat Med Sch Harvard Univ 74 Fernwood Rd Boston MA 02115

GRINSTEIN, REUBEN H, b Dallas, Tex, Aug 7, 35; m 64; c 2. SURFACTANT CHEMISTRY, PROCESS CHEMISTRY. *Educ:* Southern Methodist Univ, BS, 57; Rice Univ, PhD(phys org chem), 61. *Prof Exp:* Fel chem, Univ Leicester, Eng, 61-62; res chemist, Shell Chem Co, Inc, 62-72; tech mgr, PVO Int, Inc, 72-75; TECH MGR, DIAMOND SHAMROCK CORP, 75- *Mem:* Am Chem Soc; Am Oil Chemists Soc. *Res:* Development and production of surfactants; dispersants; flocculants; lubricants; functional polymers; food chemicals; solvents and catalysts. *Mailing Add:* 1070 Oak Ridge Dr Blue Bell PA 19422-3002

GRINSTEIN, SERGIO, b Mexico City, Mex, May 14, 50; Can citizen; m 76. ION TRANSPORT, MEMBRANE BIOLOGY. *Educ:* Polytech Inst, Mexico City, BSc, 72, PhD(physiol), 75. *Prof Exp:* Fel, Hosp Sick Children, Toronto, 76-78; res asst, Eidgenossische Technische Hochschule, Zurich, Switz, 78-79; asst prof, 79-84, ASSOC PROF PHYSIOL, HOSP SICK CHILDREN, 84-; PROF BIOCHEM, UNIV TORONTO, 85- *Concurrent Pos:* Scholar, Med Res Coun Can, 79-84; vis prof, Univ Montreal, 83 & Univ Coimbra, Portugal, 87; assoc prof med biophys, Univ Toronto, 85- & head, Div Cell Biol, 87-; assoc ed, Am J Physiol. *Honors & Awards:* Scientist Award, Med Res Coun, 85; Averst Award, Can Biochem Soc, 87. *Mem:* Soc Gen Physiologists; Biophys Soc; Can Biophys Soc; Am Physiol Soc. *Res:* Cytoplasmic pH regulation and control of cellular volume; Mechanisms of signal trans in cells of immune system. *Mailing Add:* Div Cell Biol Hosp Sick Children 555 Univ Ave Toronto ON M5G 1X8 Can

GRINTER, LINTON E(LIAS), b Kansas City, Mo, Aug 28, 02; m 26; c 2. STRUCTURAL ENGINEERING. *Educ:* Univ Kans, BS, 23, CE, 30; Univ Ill, MS, 24, PhD(eng), 26. *Hon Degrees:* LLD, Ariz State Univ; DSc, Univ Akron, 69. *Prof Exp:* Design engr, Standard Oil Co, Ind, 26-28; assoc prof civil eng, Agr & Mech Col Tex, 28-29, prof struct eng, 29-37; dean grad div & dir civil eng, Armour Inst Technol, 37-39, vpres & dean grad sch, 39-40; vpres & dean grad sch, Ill Inst Technol, 40-46, res prof civil eng & mech, 46-52; dean grad sch & dir res, 52-69, exec vpres univ, 69-70, CHMN SELF STUDY & FUTURE PLANNING, UNIV FLA, 70- *Concurrent Pos:* Consult, War Manpower Comn, 43, Res & Develop Bd, Off Secy Defense, 49-53, wind vibrations of missile launch towers, 60-65 & US Off Educ, 64-65; mem adv comt, NSF & Weapons Develop Ctr, Eglin AFB; chmn, Comt Ship Struct Design, Nat Res Coun, 53-59 & Comt Inter-Am Sci Coop, Nat Acad Sci, 60; dir study eng technol educ, Am Soc Eng Educ & NSF, 69-71. *Honors & Awards:* Lamme Medal, Am Soc Eng Educ, 58. *Mem:* Hon mem Am Soc Civil Engrs; Am Soc Mech Engrs; Am Soc Eng Educ (pres, 53); Int Asn Bridge & Struct Eng; Sigma Xi. *Res:* Theory and design of modern steel structures; wind stresses in skyscrapers; design of continuous frames with emphasis upon numerical analysis of plane stress problems and plasticity. *Mailing Add:* Rm 249 Grinter Hall Univ Fla Gainesville FL 32611

GRISAFE, DAVID ANTHONY, b Cincinnati, Ohio, Jan 30, 38; m 66; c 3. CERAMIC SCIENCE, MATERIALS SCIENCE. *Educ:* Univ Mo, Kansas City, BS, 61; Pa State Univ, MS, 63, PhD(ceramic sci), 68. *Prof Exp:* Res engr luminescent & ceramic mat, Sylvania Elec Prod, 67-70; RES ASSOC MAT SCI, KANS GEOL SURV, UNIV KANS, 70- *Concurrent Pos:* Consult mat & mineral resources, Kans Geol Surv, 71-; reviewer, J Am Ceramic Soc, 76- *Mem:* Am Ceramic Soc; Keramos. *Res:* Physical and chemical properties of building stone; crystal chemistry and color of selected crystal structures; physical properties of construction related materials using clays, cements and ashes; restoration of stone structures and rock art. *Mailing Add:* 2327 Bryce Dr Lawrence KS 66047

GRISAFFE, SALVATORE J, HIGH TEMPERATURE MATERIALS, PROTECTIVE COATINGS. *Educ:* Univ Ill, BS, 57; Case Inst Technol, MS, 65; Harvard Grad Sch Bus, cert, 76. *Prof Exp:* Researcher high temp mat, Nat Adv Comt Aeronaut, 57-68, head, Coating Sect, 68-72, chief, Surface Protection Br, 72-80, chief, Mat Appl & Composites Br, 80-81, chief, Metallic Mat, 81-83, CHIEF MATERIALS DIV, LEWIS RES CTR, NASA, 83- *Concurrent Pos:* Mem, Comt Coatings, Nat Mat Adv Bd, 63-65, Comt Erosion, 75-77; consult elec utilities, 70-73; mem adv panel, Lab Surface Sci, State Univ NY, 75-77; head energy conversion, Comn Mat, Nat Res Coun, 77-78; proj officer res & develop high temp mat for automotive engine, Int Energy Agency, 80-81; prog mgr, Mat Adv Turbine Engines, 80-88; mem, Comt Complex Composites, NSF. *Honors & Awards:* Coatings Award, Am Soc Metals, 73. *Mem:* Fel Am Soc Metals; Am Vacuum Soc; Am Soc Testing & Mat; Am Inst Aeronaut & Astronaut; Am Ceramic Soc. *Res:* Plasma sprayed coatings; ceramics, refractory carbides and oxides; ceramic bearings; oxidation, hot corrosion; gas turbine materials; program management; composites; power generation; aircraft propulsion; space propulsion and power. *Mailing Add:* 2771 Gibson Dr Rocky River OH 44116

GRISAMORE, NELSON THOMAS, b Sioux City, Iowa, Jan 27, 21; m 44. SCIENCE ADMINISTRATION. *Educ:* Univ Ill, BS, 48, MS, 50; George Washington Univ, PhD(physics), 54. *Prof Exp:* Res assoc rocket ballistics, George Washington Univ, 42-46; physicist, Appl Physics Lab, Johns Hopkins Univ, 47; res assoc, Electronics Res Proj, George Washington Univ, 50-55, res scientist, Opers Res Group, 56-57, from asst prof to assoc prof elec eng, 56-60, prof eng & appl sci, 60-69, exec officer, Dept Elec Eng, 60-62, asst dean res, 62-66, dir, Ctr Measurement Sci, 63-66, exec officer, Dept Elec Eng, 66-67; exec secy, ABM Data Processing Syst, Nat Acad Sci, 67-72; exec secy, Comt Fire Res, Comt Nat Disasters, Nat Acad Sci, 72-78, staff scientist, Nat Mat Adv Bd, 78-81; RETIRED. *Concurrent Pos:* Consult, US Army, 46-47. *Mem:* AAAS; Am Phys Soc; Inst Elec & Electronics Engrs; Sigma Xi. *Res:* Computer circuits and logic; physical electronics; information retrieval; operations research; systems reliability; measurement science. *Mailing Add:* 9536 E Bexhill Dr Kensington MD 20895

GRISAR, JOHANN MARTIN, b Gorlitz, Ger, July 10, 29; US citizen; m 83; c 3. ORGANIC CHEMISTRY, MEDICINAL CHEMISTRY. *Educ:* Swiss Fed Inst Technol, BS, 54; Mass Inst Technol, PhD(org chem), 59. *Prof Exp:* Res chemist, Charles Pfizer & Co, Inc, 59-63; proj leader med chem, 63- 67, sect head med chem, 67-81, SR SCIENTIST, MARION MERRELL DOW RES INST, STRASBOURG CTR, 81- *Mem:* Am Chem Soc. *Res:* Medium-sized ring transannular reactions; synthetic neuroleptic, hypolipidemic, anti-thrombotic, and cardiovascular anti-diabetic agents. *Mailing Add:* Org Chem Dept Merrell Int Strasbourg Cedex 67084 France

GRISARU, MARCUS THEODORE, b Stefanesti, Romania, May 15, 29; m 64; c 3. PHYSICS. *Educ:* Univ Toronto, BASc, 55; Princeton Univ, MA, 57, PhD(physics), 59. *Prof Exp:* Res assoc physics, Univ Ill, 58-60; asst prof, McGill Univ, 60-62; from asst prof to assoc prof, 68-74, PROF PHYSICS, BRANDEIS UNIV, 74- *Res:* Elementary particles physics and quantum field theory; supersymmetry and supergravity. *Mailing Add:* Dept Physics Brandeis Univ Waltham MA 02254

GRISCHKOWSKY, DANIEL RICHARD, b St Helens, Ore, Apr 17, 40; m; c 2. PHYSICS, OPTOELECTRONICS. *Educ:* Ore State Univ, BS, 62; Columbia Univ, AM, 65, PhD(physics), 68. *Prof Exp:* Res assoc, Columbia Radiation Lab, Columbia Univ, 68-69; res staff mem laser physics, IBM Watson Res Ctr, 69-77, sci adv, dir IBM Res Div, 78, mgr, Atomic Physics Lasers Group, 79-83, MGR, ULTRAFAST SCI LASERS GROUP, IBM WATSON RES CTR, 83- *Honors & Awards:* Boris Pregel Award, NY Acad Sci, 85; R W Wood Prize, Optical Soc Am, 89. *Mem:* Fel Am Phys Soc; fel Optical Soc Am; sr mem Inst Elec & Electronics Engrs. *Res:* Interactions of microwaves with paramagnetic spins; interactions of laser light with atomic vapors; nonlinear effects in optical fibers; ultrafast optoelectronics. *Mailing Add:* Dept Phys Sci IBM T J Watson Res Ctr PO Box 218 Yorktown Heights NY 10598

GRISCOM, ANDREW, b Boston, Mass, Oct 12, 28; m; c 2. GEOPHYSICS. *Educ:* Harvard Univ, AB, 49, MA, 56, PhD, 76. *Prof Exp:* GEOPHYSICIST, BR GEOPHYS, US GEOL SURV, 57- *Mem:* Geol Soc Am; Mineral Soc Am; Am Geophys Union. *Res:* Interpretation of magnetic and gravity data; relationship of petrology to the magnetic properties of rocks. *Mailing Add:* US Geol Surv 345 Middlefield Rd Menlo Park CA 94025

GRISCOM, DAVID LAWRENCE, b Pittsburgh, Pa, Nov 1, 38; m 70; c 2. ELECTRON SPIN RESONANCE, STRUCTURE OF GLASSES. *Educ:* Carnegie-Mellon Univ, BS, 60; Brown Univ, PhD(physics), 66. *Prof Exp:* Res assoc, Brown Univ, 66-67; Nat Res Coun fel, 67-69, head, Radiation Effects Sect, 73-79, RES PHYSICIST, NAVAL RES LAB, 69- *Concurrent Pos:* Prin investr, Lunar Sample Prog, NASA, 71-73; mem ed comt, Glass Div, J Am Ceramic Soc, 72-; vis scientist, Univ Lyon-I, France, 75-76; prog mgr, Defense Advan Res Projs Agency, Arlington, Va, 81-83; co-chmn, Mat Res Soc Symp, 85; mem, NASA Microgravity Sci & Appln Discipline Working Group, 88-; chmn, Glass & Optical Mat Div, Am Ceramic Soc, 91-92. *Mem:* Fel Am Ceramic Soc; Am Phys Soc; AAAS. *Res:* Physics of optical materials with specialization in radiation effects; electron spin resonance studies of amorphous insulators; author of over 130 articles in books and journals. *Mailing Add:* Code 6505 Naval Res Lab Washington DC 20375

GRISCOM, RICHARD WILLIAM, b Chattanooga, Tenn, Apr 15, 26; m 53; c 3. SYNTHETIC ORGANIC CHEMISTRY. *Educ:* Univ Chattanooga, BS, 45; Univ Tenn, MS, 48. *Prof Exp:* Chemist, Reilly Tar & Chem Corp, Tenn, 45-46, Reilly Labs, Ind, 46 & Phosphate Div, Monsanto Chem Co, Ala, 48-49; res chemist, Res Div, Tenn Prod & Chem Corp, Tenn, 49-63 & Tensyn Div, Velsicol Chem Corp, 63; from res chemist to sr res chemist, Rock Hill Lab, Chemetron Corp, 64-68, res assoc, Rock Hill Lab, Arapahoe Chem Inc, 68-70, sr scientist, 70-74, sr res chemist, 74-80, group leader, 80-90; RETIRED. *Mem:* Am Chem Soc. *Res:* Synthetic organic research on pharmaceuticals, pharmaceutical intermediates and fine organic chemicals which includes process development in the field of hydrogenation, esterification, oxidation and halogenation. *Mailing Add:* 2111 Collins St Morristown TN 37814

GRISHAM, CHARLES MILTON, b Minneapolis, Minn, June 29, 47; m 72; c 3. BIOLOGICAL CHEMISTRY. *Educ:* Ill Inst Technol, BS, 69; Univ Minn, PhD(chem), 72. *Prof Exp:* Res assoc biophys chem, Inst Cancer Res, 72-74; from asst prof to assoc prof, 80-87, PROF CHEM, UNIV VA, 87- *Concurrent Pos:* NIH/Nat Cancer Inst fel, Inst Cancer Res, 74. *Mem:* Am Chem Soc; Biophys Soc; Am Soc Biochem & Molecular Biol. *Res:* Molecular basis of membrane structure and function; mechanisms of enzyme catalysis; biological application of magnetic resonance spectroscopy, NMR imaging and in vivo spectroscopy. *Mailing Add:* Dept Chem Univ Va New Chem Bldg 152 Charlottesville VA 22901

GRISHAM, GENEVIEVE DWYER, b Glens Falls, NY, Apr 13, 27; m 49, 65; c 1. NUCLEAR CHEMISTRY. *Educ:* State Univ NY Col for Teachers, AB, 47, MA, 48; Univ Rochester, PhD(chem), 53. *Prof Exp:* Aeronaut res scientist, Lewis Flight Propulsion Lab, NASA, 53-56; MEM STAFF RADIOCHEM, LOS ALAMOS NAT LAB, 56- *Mem:* AAAS; Sigma Xi. *Res:* Nuclear reactions; radiochemistry. *Mailing Add:* 197 Tesuque Los Alamos NM 87544

GRISHAM, JOE WHEELER, b Brush Creek, Tenn, Dec 5, 31; m 55. MEDICINE, PATHOLOGY. *Educ:* Vanderbilt Univ, AB, 53, MD, 57. *Prof Exp:* Resident path, Sch Med, Wash Univ, 57-60, from instr to prof path & anat, 60-73; PROF PATH & CHMN DEPT, SCH MED, UNIV NC, CHAPEL HILL, 73- *Concurrent Pos:* Nat Cancer Inst fel, 58-59; Life Ins Med Res Fund fel, 59-61; Markle scholar, 64-69; mem bd sci counsellors, Nat Inst Environ Health Sci, 74-78. *Mem:* Am Asn Cancer Res; Am Asn Study Liver Dis; Int Acad Path; Am Soc Cell Biol; Am Asn Path (pres, 84-85); Tissue Cult Asn; AMA; Fedn Am Socs Exp Biol (pres, 84-85); Cell Kinetics Soc. *Res:* Liver diseases, especially cirrhosis; chemical carcinogenesis; regulation of cellular proliferation; DNA replication and repair. *Mailing Add:* Dept Path CB 7525 Univ NC Sch Med Chapel Hill NC 27599

GRISHAM, LARRY RICHARD, b Henderson, Tex, Feb 2, 49; m 72; c 3. PLASMA PHYSICS. *Educ:* Univ Tex, Austin, BS, 71; Oxford Univ, PhD(physics), 74. *Prof Exp:* Res assoc, Princeton Univ, 74-75, res staff physics, 75-82, res physicist, Plasma Physics Lab, 82-89, PRIN RES PHYSICIST, PRINCETON UNIV, 89- *Concurrent Pos:* Indust consult on ion-beams and neutral beams; Rhodes scholar, 74. *Mem:* Am Phys Soc; Am Vacuum Soc. *Res:* Heating of magnetically confined plasmas by injection of beams of energetic neutral particles; free expansion and energy confinement properties of Tokamak plasmas; physics and technology of neutral beams. *Mailing Add:* Plasma Physics Lab Forrestal Campus Princeton Univ Princeton NJ 08543

GRISKEY, R(ICHARD) G(EORGE), b Pittsburgh, Pa, Jan 9, 31; m 55; c 2. CHEMICAL ENGINEERING. *Educ:* Carnegie Inst Technol, BS, 51, MS, 55, PhD(chem eng), 58. *Prof Exp:* Sr engr textile fibers, E I du Pont de Nemours & Co, Inc, Del, 58-60; asst prof chem eng, Univ Cincinnati, 60-62; prof, Va Polytech Inst & State Univ, 62-66; prof & chmn dept, Univ Denver, 66-68; dir res & prof, Newark Col Eng, 68-71; dean, Col Appl Sci & Eng, Univ Wis-Milwaukee, 71-82, prof eng, 73-82; dean, Sch Eng, Univ Alabama, Huntsville, 82-85; exec vpres & provost, 85-88, INST PROF, STEVENS INST TECHNOL, 88- *Concurrent Pos:* Vis prof, Polish Acad Sci, 71, Monash Univ, 74, Universidade Estadual Sao Paulo, 73, Algerian Inst Petrol, 76-77; lectr, Royal Soc Chem (Gt Brit), 85-88; prin investr, grants NSF, NASA, US Dept Com & Environ Protection Agency; consult, Celanese Fibers Co, Res Co, Phillips Petrol, Monsanto, Hewlett-Packard, Litton, US Veterans Admin, Thermo-Tech Inc, Allis-Chalmers, Globe-Union, Rexnord, A O Smith, Donaldson Co, 3M, Hoechst-Celanese, Am Bowling Cong. *Mem:* Fel Am Inst Chem Engrs; fel Am Soc Mech Engrs; fel Am Inst Chemists; Soc Rheology; Am Chem Soc; Am Soc Eng Educ. *Res:* Polymer engineering; physics and chemistry (rheology, heat and mass transfer, chemical kinetics, structure and properties); chemical engineering (thermodynamics, transport processes, chemical kinetics); energy resources; applied chemistry; materials engineering; technology and society. *Mailing Add:* 88 Pine Grove Ave Summit NJ 07901

GRISMORE, ROGER, b Ann Arbor, Mich, July 12, 24; m 50; c 1. COMPUTER SCIENCE. *Educ:* Univ Mich, BS, 47, MS, 48, PhD(physics), 57; Coleman Col, BS, 79. *Prof Exp:* Asst physics, Eng Res Inst, Univ Mich, 47-56; asst physicist, Argonne Nat Lab, 56-61, assoc physicist, 61-62; assoc prof physics, Lehigh Univ, 62-67; specialist, Scripps Inst Oceanog, Univ Calif, San Diego, 67-71; prof physics, Ind State Univ, Terre Haute, 71-74; specialist physics, Scripps Inst Oceanog, Univ Calif, San Diego, 75-78; sr consult, Potomac Res Inc, 78-79; scientist, Jaycor, 79-84; LECTR "D" (PHYSICS), CALIF POLY STATE UNIV, SAN LUIS OBISPO, 84- *Mem:* Am Phys Soc; Am Geophys Union; Sigma Xi. *Res:* Nuclear physics; electronic instrumentation for nuclear measurements; environmental physics; marine radioactivity. *Mailing Add:* 535 Cameo Way Arroyo Grande CA 93420

GRISOLI, JOHN JOSEPH, b New York, NY, Mar 12, 25; m 48; c 3. MECHANICAL ENGINEERING. *Educ:* City Col New York, BME, 54. *Prof Exp:* Assoc mech engr, 54-64, mech engr, design & electromagnet design, 65-71, sr mech engr & head accelerator oper, 72-75, asst dept chmn admin, 75-77, assoc dept chmn admin, AGS Dept, 77-82, DEP DIV HEAD, ACCELERATOR DIV, BROOKHAVEN NAT LAB, 83- *Res:* High energy particle accelerator design and development. *Mailing Add:* 173 Edgewater NE Bayport NY 11705

GRISOLIA, SANTIAGO, b Jan 6, 23; US citizen; m 49; c 2. BIOCHEMISTRY. *Educ:* Univ Valencia, MD, 44; Univ Madrid, MD, 49. *Hon Degrees:* MD, Univ Salamanca, 67 & Univ Valencia, 72; PhD, Univ Barcelona, 71, Univ Madrid, 73, Univ Siena, 80, Univ Leon, 82 & Univ Florence, 88. *Prof Exp:* Asst prof phys chem, Univ Valencia, 44-45; vis asst prof biochem, Univ Chicago, 46-47 & phys chem, Univ Wis, 50-54; from assoc prof to prof biochem & med, 54-72, chmn dept, 62-72, DISTINGUISHED PROF BIOCHEM, UNIV KANS MED CTR, KANSAS CITY, 73- *Concurrent Pos:* High Res Coun fels, NY Univ, Univ Chicago & Univ Wis, 45-49; Juan de la Cierva fel, Univ Valencia, 49; fel, Univ Wis, 49, Wis Alumni fel, 50; Am Heart Asn estab investr, Univs Wis & Kans, 54-59; bd, Jimenez-Diaz Found, 67, vpres, Mediter Found, 73-; sec, Valencia Found Adv studies, 79; founding mem, Spanish Soc Emer Univ Prof, 66; dir, Inst Invest Citol Valencia, Spain, 79-; pres organizing comt, workshops human genome proj, 88 & 90; pres exec comt, Jaime I Prize Sci, 89; mem exec bd, Club Rome, Spain, 89-; mem sci comt, 1992 Seville Expos, 89-; pres, UNESCO Comn human genome proj, 90-; mem, World Energy Coun, 90-; bd mem, Found Banco de Bilbao Vizcaya, 90-; pres, Spanish Multiple Sclerosis Found, 90- *Honors & Awards:* Govt of Valencia Prize, Spain, 44; Colosus Award, Valencia, 72; Principe de Asturias Prize in Sci, 90. *Mem:* Am Soc Biol Chem; hon mem Span Soc Biochem & Span Soc Physiol; Sigma Xi; foreign hon mem Royal Acad Med Belgium. *Res:* Enzyme regulation; intermediary metabolism; protein turnover. *Mailing Add:* Inst de Ivest Citologicas Amadeo de Saboya Four Valencia 46010 Spain

GRISSELL, EDWARD ERIC FOWLER, b Washington, DC, Aug 10, 44. ETHOLOGY, TAXONOMY. *Educ:* Univ Calif, Davis, BS, 67, MS, 69, PhD(entom), 73. *Prof Exp:* Taxon entomologist, Fla Dept Agr & Consumer Serv, 73-78; RES ENTOMOLOGIST, SYST ENTOM LAB, AGR RES SERV, USDA, 78- *Mem:* Royal Entom Soc, London; Entom Soc Am; Sigma Xi. *Res:* Systematics of parasitic and aculeate Hymenoptera. *Mailing Add:* c/o US Nat Mus Syst Entom Lab Washington DC 20560

GRISSINGER, EARL H, b Lancaster, Pa, Nov 28, 31; m 59; c 4. SOIL SCIENCE, PHYSICS. *Educ:* Pa State Univ, BS, 53, MS, 55, PhD(agron), 57. *Prof Exp:* SOIL SCIENTIST, AGR RES SERV, USDA, 60- *Mem:* Am Soc Agron; Soil Conserv Soc Am; Sigma Xi. *Res:* Nature of cohesion of natural soil materials; defining soil properties which determine cohesion. *Mailing Add:* PO Box 1157 Oxford MS 38655

GRISSOM, DAVID, b Dallas, Tex, Aug 22, 35; m 60; c 1. SOLID STATE PHYSICS. *Educ:* Univ Tex, BS, 58 & 61, MS, 62, PhD, 65. *Prof Exp:* Assoc engr, Apparatus Div, Tex Instruments Inc, 58-59; res asst electronic mat res lab, Univ Tex, 61-65; admin asst res, Nat Geophys Co, 65-67; sr proj engr, Geotech Div, Teledyne Industs, 67-69; chief engr, Rogers Explor, Inc, 69-73; sr engr, Electronics Systs Div, Geosource, 73-76; consult, 76-77, sr engr, N L McCullough, 77-82; sr scientist, Gulf Appl Res, 82-84; sr proj engr, 84-86, consult, Western Res, 86-88; chief engr, Globe Universal Sci, 88-89; CONSULT, 89- *Concurrent Pos:* Consult, Chatlon Inc, Tex, 63-64. *Mem:* Am Soc Mech Engrs; Inst Elec & Electronics Engrs; Nat Soc Prof Engrs; Soc Explor Geophysicists; Am Phys Soc; Sigma Xi. *Res:* Geophysical instruments and oil exploration techniques; digital processing of seismic data; low temperature dielectric loss. *Mailing Add:* 802 Saybrook Houston TX 77024-4505

GRISSOM, ROBERT LESLIE, b Macon Co, Ill, Mar 5, 17; m 44; c 4. INTERNAL MEDICINE. *Educ:* Univ Ill, BS, 39, MS & MD, 41. *Prof Exp:* From asst to asst prof internal med, Univ Ill, 47-53; from assoc prof to prof & chmn dept, Univ Nebr Med Ctr, Omaha, 53-70, head, Div Cardiol, 70-72, prof med, 70-87; RETIRED. *Concurrent Pos:* Markle scholar, 50-55; fel, Coun Clin Cardiol, Am Heart Asn. *Mem:* Fedn Clin Res; fel Am Col Physicians; fel Am Col Cardiol; Cent Soc Clin Res; Sigma Xi. *Res:* Cardiovascular diseases. *Mailing Add:* Div Cardiol Univ Nebr Med Ctr 600 S 42nd St Omaha NE 68198-2285

GRISWOLD, BERNARD LEE, b Hastings, Mich, Apr 25, 42; m 63; c 1. FISH BIOLOGY, RESEARCH ADMINISTRATION. *Educ:* Iowa State Univ, BS, 64; Univ Maine, MS, 67; Univ Minn, PhD(fish biol), 70. *Prof Exp:* Fishery biologist, Nat Marine Fisheries Serv, 70-73; asst leader, Ohio Coop Fishery Res Unit, US Fish & Wildlife Serv, Columbus, Ohio, 73-74, leader, 75-79, supvr, Coop Fishery Res Units, Wash, DC, 79-83, dir, Great Lakes Fishery Lab, Ann Arbor, Mich, 83-86; PROG DIR, NAT SEA GRANT COL PROG, 86- *Concurrent Pos:* Fel, Univ Minn, 70; assoc ed fisheries, J Wildlife Mgt, 74-75; tech adv, Treaty Fishing Rights, Mich Dist Ct (Fed), 85-86. *Mem:* Am Inst Fishery Res Biologists; Am Fisheries Soc; Sigma Xi. *Res:* Effects of ecological stress on fish populations; Great Lakes fisheries; biological assessment of fish populations in impoundments. *Mailing Add:* 1335 East-West Hwy Silver Spring MD 20910

GRISWOLD, DANIEL H(ALSEY), b Colorado Springs, Colo, Jan 10, 09; m 31; c 2. GEOLOGICAL ENGINEERING, FIELD GEOLOGY. *Educ:* Colo Sch Mines, Geol Eng, 30. *Prof Exp:* Geophysicist & asst geol, US Smelting Refining & Mining Co, 30-31; asst engr, C T Griswold, Mining Engr, 31-32; lessee, Magnolia Petrol Co, 32-33; instrumentman, Mid Rio Grande Conserv Dist, 33; jr topog engr, Conserv Br, US Geol Surv, 33-35; jr agr engr, Soil Conserv Serv, USDA, NMex, 35-38, asst agr engr, Utah, 38-40, assoc geologist, NMex, 41-46, soil conservationist, 46-49, geologist, Eng & Watershed Planning Unit, Regional Tech Serv Ctr, 49-69; geologist, Groundwater Div, Ore State Engr Off, 69; eng geologist, Found Sci, Inc, 67-73, assoc, 73-84; RETIRED. *Concurrent Pos:* From lieutenant to lieutenant colonel civil engrs, US Army, 41-45. *Mem:* Geol Soc Am; Asn Eng Geologists. *Res:* Occurence of groundwater; geology of dam and reservoir sites; geology of tunnel sites; engineering geology of mine shaft sites; geology of landslides. *Mailing Add:* 12705 SE River Rd Apt 102T Portland OR 97222-8077

GRISWOLD, DANIEL PRATT, JR, b Birmingham, Ala, Nov 15, 28; m 53; c 2. VETERINARY MEDICINE, ONCOLOGY. *Educ:* Auburn Univ, DVM, 51. *Prof Exp:* Adv, USDA, 51; pvt practr vet med, 51-52 & 55-61; VPRES CHEMOTHER & TOXICOL RES, SOUTHERN RES INST, 61- *Concurrent Pos:* Mem, Breast Cancer Task Force, Nat Cancer Inst, 70-73, Biol Subcomt, 71-74; mem, Prog Planning Comt, Nat Large Bowel Cancer Proj, 77; mem, Prog Comt, Am Asn Cancer Res, 78, 80, 84; mem, Combined Modalities Adv Group, 78-81; mem, Ad Hoc Rev Comt for Dis-Oriented Drug Screening Proj, Nat Cancer Inst, 85- *Mem:* Am Asn Cancer Res; Cell Kinetics Soc; Sigma Xi. *Res:* Chemotherapy of experimental animal tumor systems and study of carcinogenesis of chemical agents; development of rodent tumor models for studies of their biological characteristics and response to treatment, principally chemotherapy. *Mailing Add:* Southern Res Inst PO Box 55305 Birmingham AL 35255-5305

GRISWOLD, EDWARD MANSFIELD, b New Haven, Conn, Mar 28, 05; m 39; c 2. MECHANICAL ENGINEERING. *Educ:* Carnegie Inst Technol, BS, 27; Univ Pa, MS, 39. *Prof Exp:* Design engr, Am Foundry Equip Co, 27-28; res engr, Metals Coating Co Am, 28-29; engr qual control, Weston Elec Instrument Co, NJ, 29-30; teacher, High Sch, Conn, 32-42; from instr to asst prof eng graphics, 42-52, from assoc prof to prof mech eng, 52-71, EMER PROF MECH ENG, COOPER UNION, 71- *Concurrent Pos:* Consult, Shades, Inc, 43-50 & Ardco Mfg Co, 45; eng designer, Byrne Assocs, 47-52; engr, Bell Tel Labs & Am Tel & Tel Co, 53-58; NSF res grants, 59-61; engr, Singer Co, 62-66. *Mem:* Nat Soc Prof Eng; Am Soc Eng Educ. *Res:* Engineering graphics and design; communication of complex systems and ideas clearly, precisely and easily; design of machines and systems, especially kinematics; lubrication; economics. *Mailing Add:* 141 Washington Ave Chatham NJ 07928

GRISWOLD, ERNEST, b Milan, Kans, Aug 13, 05; m 31; c 6. INORGANIC CHEMISTRY. *Educ:* Univ Kans, AB, 27, PhD(chem), 34. *Prof Exp:* From instr to prof chem, Univ SDak, 31-47; from assoc prof to prof, 47-75, EMER PROF CHEM, UNIV KANS, 75- *Mem:* Am Chem Soc. *Res:* Salt effects in nonaqueous solvents; reactions of cyano-complexes; conductivity and ion-pair equilibria in nonaqueous solvents. *Mailing Add:* 2217 Massachusetts St Lawrence KS 66046-3045

GRISWOLD, GEORGE B, b Ponca City, Okla, Dec 9, 28; m 52; c 4. GEOLOGICAL & MINING ENGINEERING. *Educ:* NMex Inst Mining & Technol, BS, 55; Univ Ariz, MS, 57, PhD, 67. *Prof Exp:* Jr engr, San Manuel Copper Corp, 55; chief engr, Blackrock Div, Wah Chang Mining Corp, Calif, 55-56; fel & mining engr, US Bur Mines, Ariz, 56-57; mining engr & fac assoc, NMex Inst Mining & Technol, 57-64; NSF trainee mining & geol eng, Univ Ariz, 64-65; assoc prof mining eng, NMex Inst Mining & Technol, 65-70, chmn dept petrol & mining eng, 68-70; mgr explor, Western Can & Alaska, Getty Oil Co, 70-73; mem tech staff, Sandia Labs, 74-77; pres, Tecolote Corp, 78-83; chmn, Dept Mining & Geol Eng, NMex Inst Mining & Technol, 84-88; RETIRED. *Concurrent Pos:* Consult, Proj Mohole, Brown & Root Co, 64-67. *Mem:* Am Inst Mining, Metall & Petrol Engrs; Soc Econ Geologists; Soc Mining Engrs. *Res:* Rock mechanics; engineering geology; ore deposits; site evaluation for radioactive waste isolation. *Mailing Add:* 11105 Malaguena Lane NE Albuquerque NM 87111

GRISWOLD, JOSEPH GARLAND, b Grand Rapids, Mich, June 15, 43; m 66; c 2. BEHAVIORAL BIOLOGY. *Educ:* Denison Univ, BS, 65; Pa State Univ, MS, 68, PhD(zool), 71. *Prof Exp:* From asst prof to assoc prof, 71-89, PROF BIOL, CITY COL NEW YORK, 90- *Concurrent Pos:* Consult. *Mem:* Am Inst Biol Sci; Sigma Xi; Animal Behav Soc; Nat Sci Teachers' Asn. *Res:* Behavioral development in dogs and the implications for the human-companion animal bond. *Mailing Add:* Dept Biol City Col New York Convent Ave & 138th St New York NY 10031

GRISWOLD, KENNETH EDWIN, JR, b Ruston, La, Oct 22, 43; m 67; c 2. CLINICAL BIOCHEMISTRY, POPULATION GENETICS. *Educ:* La Tech Univ, BS, 65, MS, 67; Univ SC, PhD(biol), 71. *Prof Exp:* Instr chem & physiol, Univ SC, 69-71; instr biochem, Med Ctr, La State Univ, Shreveport, 71-73, mem grad fac path & biochem, 73-83; PROF & HEAD, DEPT CLIN LAB SCI & BACT, LA TECH UNIV, 83- *Concurrent Pos:* Dir clin chem, Vet Admin Hosp, 71-83, co-dir, Sch Med Technol, 72-83; mem grad fac & prof, La Tech Univ, Ruston; clin assoc prof, Northeastern State Univ, Monroe, La, Centenary Col, Shreveport, La; consult forensic toxicol; clin scientist, Am Soc Clin Pathologists. *Mem:* Am Asn Clin Chem; fel Nat Acad Clin Biochemists; Sigma Xi; Am Soc Clin Pathologists. *Res:* Relationships of protein polymorphisms to population structure; methods development in clinical chemistry; relationship of nucleic acids to disease; toxicology; pathology. *Mailing Add:* PO Box 3053 Ruston LA 71272

GRISWOLD, MICHAEL DAVID, b Norman, Okla, Feb 17, 44; m 65; c 2. BIOCHEMISTRY, DEVELOPMENTAL BIOLOGY. *Educ:* Univ Wyo, BS, 66, PhD(biochem), 69. *Prof Exp:* Asst prof pharmacol, Baylor Col Med, 73-74; res assoc med res, C Best Inst Med Res, 74-76; ASST PROF CHEM, WASH STATE UNIV, 76- *Concurrent Pos:* Nat Cancer Inst fel, Univ Wis-Madison, 69-72; Europ Molecular Biol Orgn fel, Lab Cell Biol, Rome, Italy, 72-73. *Mem:* Soc Develop Biol; NY Acad Sci. *Res:* Biochemistry of metamorphosis and spermatogenesis, primarily RNA polymerase and control of hormone induced processes. *Mailing Add:* Dept Biochem Fulmer Hall Wash State Univ Pullman WA 99164

GRISWOLD, NORMAN ERNEST, b Yankton, SDak, July 17, 35; m 59; c 2. COLLEGE TEACHING. *Educ:* Univ Kans, BA, 57; Univ Nebr, MS, 61, PhD(chem), 66. *Prof Exp:* from instr to asst prof, Nebr Wesleyan Univ, 63-66, actg head dept, 67-68, assoc prof chem, 66-79, head dept, 79-89, PROF CHEM, NEBR WESLEYAN UNIV, 79- *Concurrent Pos:* Lectr, NSF Summer Inst Introd Phys Sci, 66-69 & 72-75, assoc dir, 70 & 71; vis assoc prof, Univ Ill, Urbana, 70-71 & Purdue Univ, 78-79. *Mem:* Am Chem Soc; Sigma Xi; Nat Sci Teachers Asn. *Res:* Writer - general chemistry. *Mailing Add:* Dept of Chem Nebr Wesleyan Univ Lincoln NE 68504-2796

GRISWOLD, PHILLIP DWIGHT, b Rutherfordton, NC, Feb 2, 48; m 71; c 3. POLYMER SCIENCE. *Educ:* NC State Univ, BS, 70, MS, 73, PhD(fiber & polymer sci), 76. *Prof Exp:* Res fel polymer sci & eng, Univ Mass, 76-77; res chemist polymer sci, Tenn Eastman Co, 77-79, sr res chemist, 79-87, develop assoc, 87-88, DEPT SUPT, TENN EASTMAN CO, 80- *Mem:* Am Chem Soc; Soc Rheol; Am Soc Qual Control. *Res:* Polymer rheology; fiber formation; morphological characterization of polymers; high modulus fibers; x-ray diffraction of polymers; plastics processing. *Mailing Add:* 4511 Preston Pl Kingsport TN 37764

GRITMON, TIMOTHY F, inorganic chemistry, for more information see previous edition

GRITTINGER, THOMAS FOSTER, b Milwaukee, Wis, Oct 23, 33; m 67; c 3. ECOLOGY, BOTANY. *Educ:* Univ Wis-Milwaukee, BS, 58, PhD(bot), 69; Univ Wis-Madison, MS, 62. *Prof Exp:* Instr zool, Univ Wis-Milwaukee, 62-65; from instr to assoc prof, 68-84, PROF BOT & ZOOL, UNIV WIS, SHEBOYGAN COUNTY CTR, 84- *Mem:* Ecol Soc Am; Sigma Xi. *Res:* Bog ecology and string bog development; forest ecology, hemlock relicts along Lake Michigan in Wisconsin; scent marking behavior among captive cheetahs; zoology; under water behavior; polar bears. *Mailing Add:* Sheboygan County Ctr Univ Wis One Univ Dr Sheboygan WI 53081

GRITTON, EARL THOMAS, b Tipton, Iowa, Sept 26, 33; m 52; c 3. PLANT BREEDING, HORTICULTURE. *Educ:* Iowa State Univ, BS, 60, MS, 61; NC State Univ, PhD(crop sci), 64. *Prof Exp:* From asst prof to assoc prof, 64-75, PROF AGRON, UNIV WIS-MADISON, 75- *Concurrent Pos:* Cropping systs agronomist tech adv, Nat Dept Agr Res, The Gambia, WAfrica, 88-90. *Mem:* Am Soc Agron; Crop Sci Soc Am; Pisum Genetics Asn; Soybean Genetics Asn. *Res:* Breeding, genetics and cultural practices for soybeans and processing peas, Pisum sativum; genetic resistance to plant diseases. *Mailing Add:* Dept Agron Univ Wis Madison WI 53706

GRITTON, EUGENE CHARLES, b Santa Monica, Calif, Jan 13, 41; m 80; c 2. NUCLEAR ENGINEERING. *Educ:* Univ Calif, Los Angeles, BS, 63, MS, 65, PhD(nuclear eng), 66. *Prof Exp:* Res engr, Nuclear Eng & Environ Simulation Modelling & Power Systs Eng, Rand Corp, 66-73, proj leader, Advan Undersea Technol Prog, 73-76, proj leader, Marine Technol, 74-76, head, Phys Sci Dept & Res Admin, 75-77, head, Eng & Appl Sci Dept, 77-86, dep vpres, Res Opers group, 86-90, PROG DIR APPL SCI & TECHNOL, RAND CORP, 76-, DEP VPRES, NAT SECURITY RES DIV, 86-, RESIDENT SCHOLAR TECHNOL, 90- *Concurrent Pos:* Vis lectr, Dept Mech Eng, Univ Southern Calif, 67-72 & Dept Energy Kinetics, Univ Calif, Los Angeles, 71-73. *Mem:* Am Nuclear Soc; Am Inst Astronaut & Aeronaut; AAAS. *Res:* Strategic command, control and communications technology and policy; satellite survivability issues; development of innovative weapon systems concepts; strategic force management issues; energy systems analyses. *Mailing Add:* 3616 The Strand, C Manhattan Beach CA 90266-3243

GRITZMANN, PETER, b Dortmund, WGer, Dec 17, 54; m 83. CONVEXITY MATHEMATICAL PROGRAMING, COMPLEXITY THEORY. *Educ:* Univ Dortmund, dipl, 78; Univ Siegen, PhD(math), 80, Habilitation, 84. *Prof Exp:* Prof math, Univ Siegen, WGer, 85-88; prof math, Univ Trier, WGer, 88-90, Univ Augsburg, WGer, 90-91, PROF MATH, UNIV TRIER, WGER, 91- *Concurrent Pos:* Studienstiftung des deutschen Volkes, 74-78, Fiebiger prof, Univ Augsburg, WGer, 90-91; vis assoc prof math, Univ Wash, Seattle, 86-87, vis prof, 89-90; sr res fel, Inst Math & Applications, Minneapolis, 87. *Mem:* Am Math Soc; Soc Indust & Appl Math; Math Prog Soc; Asn Comput Mach. *Res:* Mathematical programming; operations research; computational geometry; complexity theory; convexity; discrete mathematics. *Mailing Add:* Dept Math Univ Trier Postfach 3825 D-5500 Trier Germany

GRIVETTI, LOUIS EVAN, b Billings, Mont, Sept 13, 38; m 67; c 1. NUTRITION. *Educ:* Univ Calif, Berkeley, AB, 60, MA, 62; Univ Calif, Davis, PhD(geog), 76. *Prof Exp:* Res asst, Dept Biochem, Vanderbilt Univ, 64-70; admin officer, Maternal & Child Health/Family Planning Prog, Meharry Med Col, 73-75; asst prof, 76-81, assoc prof, 81-87, chmn, 83-87, PROF, DEPT GEOG & DEPT NUTRIT, UNIV CALIF, DAVIS, 87- *Concurrent Pos:* Ethel Austin Martin vis prof human nutrit, SDak State Univ, Brookings, 85; Richard Nixon vis prof, Whitter Col, 85. *Honors & Awards:* Bk Award, Nutrit Founds Europe & NAm, 77; Harding lectr, State Univ, Brookings, 85; Inst Food Technologists Sci Lectr, 87- *Mem:* Asn Am Geogrs; Soc Nutrit Educ; Am Inst Nutrit. *Res:* Food habits and nutritional implications; history and evolution of human dietary traditions; mechanisms of dietary change; cultural ecology of malnutrition; world food crisis; minority food habits. *Mailing Add:* Dept Nutrit Univ Calif Davis CA 95616

GRIVSKY, EUGENE MICHAEL, b Pskov, Russia, Dec 20, 11; US citizen; m 35; c 2. ORGANIC CHEMISTRY. *Educ:* Free Univ Brussels, BS, 36, MS, 38, DSc, 40. *Prof Exp:* Res assoc phys & org chem with Profs J Timmermans & G Chavanne, Free Univ Brussels & Int Bur Phys Chem Standards, Belgium, 39-40; res chemist & group leader, Pharmaceut Div, Union Chim, Belgium, 41-57; sr org res chemist, Wellcome Res Labs, Burroughs Wellcome Co, 57-78; chem consult, 78-83; res assoc, dept chem, Georgetown Univ, 79-83. *Concurrent Pos:* Abstractor, Chem Abstr, 58-70. *Mem:* AAAS; fel Am Inst Chem; NY Acad Sci; Am Chem Soc; Royal Chem Soc; Soc Chem France; Soc German Chemists; Royal Belg Chem Soc. *Res:* Stereochemistry; glycols oxidation by microorganisms; synthetic organic and medicinal chemistry; chemotherapy; sulfonamides; antihistamines; tranquilizers; publications in antidepressant, central muscle relaxant, hypotensive, sympatholytic, anti-inflammatory, antipyretic and anticonvulsant agents; catalysis; mechanism of reactions; patentee in field. *Mailing Add:* 4407 Eastwood Court Fairfax VA 22032

GRIZZARD, MICHAEL B, b Ft Worth, Tex, Sept 23, 45; m 72. PEDIATRICS, INFECTIOUS DISEASE. *Educ:* Austin Col, BA, 67; Univ Mich, MD, 72. *Prof Exp:* Res pediat, Sch Med, Yale Univ, 72-74; res assoc & med officer, Infectious Dis Lab, NIH, 74-77; fel infectious dis, Channing Lab, Harvard Med Sch, 77-78; asst prof pediat dis, Col Med, Univ Fla, 78-79; STAFF PEDIATRICIAN, CHMN, DEPT CLIN INVEST & CONSULT INFECTIOUS DIS, LOVELACE MED CTR, ALBUQUERQUE, 79- *Honors & Awards:* Res Serv Award, 77. *Mem:* Am Soc Microbiol; Infectious Dis Soc Am. *Res:* Pediatric infectious diseases and immunology. *Mailing Add:* Dept Pediat Lovelance Med Ctr 5400 Gibson Blvd SE Albuquerque NM 87108

GRIZZELL, ROY AMES, JR, b Sweetwater, Tenn, Mar 14, 18; m 49; c 2. FISHERIES MANAGEMENT, FORESTRY. *Educ:* Univ Ga, BS, 39; Univ Mich, MS, 47, PhD(wildlife mgt), 51. *Prof Exp:* Res biologist, US Fish & Wildlife Serv, 47-49, refuge mgr, 52-55; biologist, Soil Conserv, USDA, 55-78; PROF, SCH FOREST RESOURCES, UNIV ARK-MONTICELLO, 81-; CONSULT BIOLOGIST, 78- *Mem:* Wildlife Soc; fel Soil & Water Conserv Soc Am; Am Fisheries Soc. *Res:* Fish farming; agriculture; management of soil and water resources for fish and wildlife. *Mailing Add:* Rt One Box 496 Monticello AR 71655-9142

GRIZZLE, JAMES ENNIS, b Herald, Va, Apr 20, 30; m 51; c 3. BIOSTATISTICS. *Educ:* Berea Col, BS, 51; Va Polytech Inst, MS, 53; NC State Col, PhD(exp statist), 60. *Prof Exp:* Asst animal husb, Va Polytech Inst, 52-54; anal statistician, White Sands Proving Ground, 54; asst statist, NC State Col, 56; res assoc, Sch Pub Health, Univ NC, Chapel Hill, 57-60, from asst prof to prof biostatist, 60-89, chmn dept, 73-89; ASSOC HEAD, CANCER PREV RES PROG, PUB HEALTH SCI DIV, FRED HUTCHINSON CANCER RES CTR, 89- *Concurrent Pos:* Statistician, Med Lab, Army Chem Ctr, 55-56 & USPHS. *Mem:* Biomet Soc; Am Statist Asn; Am Pub Health Asn. *Res:* Analysis of categorical data; applications to clinical medicine and other medical research. *Mailing Add:* 10317 Lakeshore Blvd NE Seattle WA 98125

GRIZZLE, JOHN MANUEL, b Mangum, Okla, June 11, 49; m 69, 87; c 1. FISH BIOLOGY, FISH PATHOLOGY. *Educ:* Okla State Univ, BS, 71, MS, 72; Auburn Univ, PhD(fisheries), 76. *Prof Exp:* PROF FISH PATH, DEPT FISHERIES & ALLIED AQUACULT, AUBURN UNIV, 76- *Mem:* Am Fisheries Soc; Am Soc Ichthyologists & Herpetologists; Sigma Xi. *Res:* Fish histology and histopathology of fish diseases. *Mailing Add:* 145 Sunshine Rd Opelika AL 36801

GRMELA, MIROSLAV, b Trnava, Czech, May 30, 39; Can citizen; m 62; c 2. PHYSICAL MATHEMATICS. *Educ:* Czech Tech Univ, MSc, 61; Czech Acad Sci, PhD(math physics), 66. *Prof Exp:* Res assoc physics, Nuclear Res Inst, Czech Acad Sci, 61-69; res assoc dept physics, Univ BC, 69-71; res assoc math, Carleton Univ, 71-73; VIS MEM MATH FAC, MATH RES CTR, UNIV MONTREAL, 73- *Mem:* Am Math Soc; Can Math Soc. *Res:* Non equilibrium statistical physics; dynamical systems. *Mailing Add:* 4265 Edouard Montpetit Montreal PQ H3W 2Y4 Can

GROAT, CHARLES GEORGE, b Westfield, NY, Mar 25, 40; m 63; c 2. ECONOMIC GEOLOGY. *Educ:* Univ Rochester, AB, 62; Univ Mass, MS, 67; Univ Tex, Austin, PhD(geol), 70. *Prof Exp:* Res geologist, Bur Econ Geol, Univ Tex, Austin 68-71, assoc dir, 71-75, assoc prof, dept geol sci, 71-77, actg dir Bur Econ Geol, 75-77; assoc prof geol sci & chmn, Univ Tex, El Paso, 77-80; dir, LA Geol Survey, 78-90; EXEC DIR, AM GEOL INST, 90- *Mem:* Geol Soc Am; Am Asn Petrol Geologists; Soc Environ Geochem. *Res:* Geology of energy resources, especially coal; environmental aspects of resource extraction; geomorphology of coastal and arid areas. *Mailing Add:* Am Geol Inst 4220 King St Alexandria VA 22302-1507

GROAT, RICHARD ARNOLD, SURGICAL PATHOLOGY, PSYCHOLOGY. *Educ:* Bowman Gray Sch Med, MD, PhD. *Prof Exp:* PATHOLOGIST, SENTINEL LABS, 60- *Mailing Add:* 702 W Cornwallis Dr Greensboro NC 27408

GROB, DAVID, b New York, NY, Feb 23, 19; m 48; c 4. INTERNAL MEDICINE, RESEARCH & ADMINISTRATION. *Educ:* City Col New York, BS, 37; Johns Hopkins Univ, MD, 42. *Prof Exp:* From intern to asst resident, Johns Hopkins Hosp, 42-48, from instr to assoc prof, Sch Med, 48-58; dir med serv, 58-89, dir res & educ, 60-89, MED DIR, RES & DEVELOP FOUND, MAIMONIDES MED CTR, 82- *Concurrent Pos:* Fel, Johns Hopkins Hosp, 45-48; consult, Nat Cancer Inst, 51-54, Myasthenia Gravis Found, 53-, chmn, 61-63; US Army Hosp, Ft Meade, 53-58 & Surgeon Gen, US Army, 58; fel, Johns Hopkins Hosp, 45-48, physician, 51-58; consult, bur med, Food & Drug Admin, 66-78; prof med, State Univ NY Downstate Med Ctr, 58-, assoc dean, 62-89. *Mem:* Am Physiol Soc; Am Soc Clin Invest; Am Soc Pharmacol & Exp Ther; Am Neurol Asn; AMA; Sigma Xi; Asn Am Phys; fel Am Col Physicians. *Res:* Neuromuscular disease, physiology and pharmacology; clinical pharmacology, clinical immunology, clinical medicine myasthenia gravis, autoimmune diseases; polymyositis, multiple sclerosis, oncologic diseases, tissue culture. *Mailing Add:* Maimonides Med Ctr of Brooklyn 4802 Tenth Ave Brooklyn NY 11219

GROB, GERALD N, b New York, NY, Apr 25, 31; m 54; c 3. HISTORY & PHILOSOPHY OF SCIENCE. *Educ:* City Col NY, BSS, 51; Columbia Univ, MA, 52; Northwestern Univ, PhD(hist), 58. *Prof Exp:* Prof hist, Clark Univ, 57-69; SIGERIST PROF HIST MED, RUTGERS UNIV, 69- *Concurrent Pos:* Prin investr, NIMH grant, 60-92; mem, Exec Comt, Am Asn Hist Med, 78-81, Prog Comt, 86, 91 & Welch Medal Comt, 85, 89; chair, Prog Comt, Orgn Am Historians, 85. *Honors & Awards:* William H Welch Medal, Am Asn Hist Med, 86, Garrison Lectr, 86. *Mem:* Inst Med-Nat Acad Sci; Am Hist Asn; Am Asn Hist Med; Orgn Am Historians. *Res:* History of medicine; history of psychiatry; history of the care and treatment of the mentally ill in America, psychiatry, and public policy. *Mailing Add:* Inst Health Health Care Policy & Aging Res Rutgers Univ PO Box 5070 New Brunswick NJ 08903-5070

GROB, HOWARD SHEA, reproductive physiology, histophysiology; deceased, see previous edition for last biography

GROB, ROBERT LEE, b Wheeling, WVa, Feb 13, 27; m 52; c 4. ANALYTICAL CHEMISTRY, ENVIRONMENTAL SCIENCES. *Educ:* Col Steubenville, BS, 51; Univ Va, MS, 54, PhD(chem), 55. *Prof Exp:* Res analytical chemist, Esso Res & Eng Co, 55-57; from asst prof to assoc prof analytical chem, Wheeling Col, 57-63; assoc prof, 63-67, PROF ANALYTICAL CHEM, VILLANOVA UNIV, 67- *Concurrent Pos:* Consult, analytical & environ chem. *Mem:* Am Chem Soc; fel Am Inst Chemists; Asn Off Analytical Chemists. *Res:* Organic reagents for complexing of trace metals; gas chromatography; air and water pollution and control; liquid chromatography; mass spectrometry; gas and high performance liquid chromatography; environmental chemistry. *Mailing Add:* Dept Chem Villanova Univ Villanova PA 19085

GROBE, JAMES L, medicine; deceased, see previous edition for last biography

GROBECKER, ALAN J, b San Diego, Calif, Aug 6, 15; m 40; c 2. SPACE PHYSICS, GEOPHYSICS. *Educ:* Calif Inst Technol, BS, 37, MS, 41; Univ Southern Calif, MS, 49; Univ Calif, Los Angeles, PhD(planetary & space sci), 68. *Prof Exp:* Proj engr, Autonetics Div, NAm Aviation, Inc, 50-59; mem tech staff, Inst Defense Anal, 49-60; mem staff develop planning, Gen Off, NAm Aviation, Inc, 60-61, mgr, 61-62, corp res dir, 62-63, mem tech staff, Sci Ctr, 63-64, scientist, Space Info Systs Div, 64-68; mem tech staff, Inst Defense Anal, 68-71; mem tech staff, Off of Asst Secy Systs Eng & Technol, Dept Transp, 71-76; dir, Div Atmospheric Sci, NSF, 76-79; vis scholar, Univ Calif, Los Angeles, 79-88; RETIRED. *Concurrent Pos:* Res asst, E O Huburt Ctr Space Res, Naval Res Lab, DC, 65-68; consult, Univ Calif, Los Angeles, 79-86, NASA-Ames, 80- & NASA/GSEC, 84-86. *Mem:* AAAS; Am Meteorol Soc; Am Phys Soc; Am Geophys Union. *Res:* Planetary physics; investigation by rocket borne instrumentation of latitudinal and temporal variation of the homospheric boundary of the upper atmosphere; earth gas magnetism and instrumentation. *Mailing Add:* 25642 Orchard Rim Lane El Toro CA 92630

GROBLEWSKI, GERALD EUGENE, b Nanticoke, Pa, Nov 5, 26; m 53; c 2. PHARMACOLOGY, TOXICOLOGY. *Educ:* Univ Md, BA, 49; Univ Rochester, PhD(pharmacol), 64. *Prof Exp:* Pharmacol technician, Army Chem Ctr, Md, 50-56, physiologist, 56-57; admin asst chem, Distillation Prod Industs, Eastman Kodak Co, 57-59; assoc res biologist, Sterling-Winthrop Res Inst, 63-69; pharmacologist, Spec Proj Div, Woodard Res Corp, 69-73; res assoc toxicol, Inst Comp & Human Toxicol, Albany Med Col, 73-75, res asst prof, 75-76, asst prof, 76-80; mem staff, Int Ctr Environ Safety, 76-80; ADJ PROF BIOL, PARK COL, 86-; ADJ FAC CHEM & BIOL, NMEX STATE UNIV, ALAMOGORDO, 87-; vis instr biol, Chapman Col, 87- *Concurrent Pos:* Mem, Coun Thrombosis, Am Heart Asn. *Mem:* AAAS; Am Soc Pharmacol & Exp Therapeut; Int Soc biochem Pharmacol; Sigma Xi. *Res:* Cardiac automaticity; cardiotonic agents; anticholinesterase agents; thrombosis; psychopharmacology. *Mailing Add:* 1704 Crescent Dr Alamogordo NM 88310

GROBMAN, ARNOLD BRAMS, b Newark, NJ, Apr 28, 18; m 44; c 2. ZOOLOGY, ACADEMIC ADMINISTRATION. *Educ:* Univ Mich, BS, 39; Univ Rochester, MS, 41, PhD(zool), 43. *Prof Exp:* Instr zool, Univ Rochester, 43-44; res assoc, Manhattan Dist, Sch Med & Dent, 44-46; from asst prof to assoc prof biol, Univ Fla, 46-58; dir biol sci curric study, Univ Colo, 59-65; dean col arts & sci, Rutgers Univ, 65-67, dean, Rutgers Col, 67-72; vchancellor acad affairs, Univ Ill Chicago Circle, 73-75; chancellor, 75-85, RES PROF, UNIV MO-ST LOUIS, 86- *Concurrent Pos:* Res partic, Oak Ridge Inst, 50; dir, Fla State Mus, 52-58; res partic, Div Biol & Agr, Nat Res Coun, 54-70; chmn, Nat Coun Accreditation Teacher Educ, 70-71; chmn, Div Urban Affairs, Nat Asn State Univ & Land Grant Col, 78-80; adj cur, Fla Mus Natural Hist, 84- *Honors & Awards:* Stoye Prize, 40; Morrison Prize, 43; McAllister Award, 65. *Mem:* AAAS; Am Inst Biol Sci; Am Soc Ichthyologists & Herpetologists (vpres, 46, secy, 52-57, pres, 64); Nat Asn Biol Teachers (pres, 66). *Res:* Herpetology; academic administration; biological education. *Mailing Add:* Univ Mo-St Louis 8001 Natural Bridge Rd St Louis MO 63121

GROBMAN, WARREN DAVID, b Philadelphia, Pa, Sept 22, 42; m 65; c 2. SOLID STATE PHYSICS. *Educ:* Univ Pa, AB, 64; Princeton Univ, MA, 66, PhD(physics), 67. *Prof Exp:* Mem tech staff, Bellcomm Inc, Washington, DC, 67-69; mem tech staff, Appl Res Dept, 69-80, MGR X-RAY LITHOGRAPHY, SEMICONDUCTOR SCI & TECHNOL DEPT, T J WATSON RES CTR, IBM CORP, 80- *Mem:* Inst Elec & Electronics Engrs; Am Phys Soc. *Res:* Geophysical science; experimental and theoretical studies of electronic states in solids; band theory; transport measurements; ultra-high-vacuum electron spectroscopy of surfaces; electron beam and x-ray lithography; applications to fabrication of exploratory devices. *Mailing Add:* Advan Gas Technol Labs IBM Thomas J Watson Res Ctr Box 218 89-No 4 Yorktown Heights NY 10598

GROBNER, PAUL JOSEF, b Prague, Czech, July 16, 19; US citizen; m 59; c 2. MATERIALS SCIENCE, CORROSION. *Educ:* Tech Univ, Prague, Czech, MS, 41, PhD(chem eng). *Prof Exp:* Res engr metall, Chem & Metall Corp, Czech, 45-50; scientist corrosion, Res Inst Mat Protection, Czech, 50-59; develop engr mat res, Modransk Eng Works, Czech, 59-66; sr scientist, Res Inst Ferrous Metall, 66-69; sr staff metallurgist mat res, Amax Mat Res Ctr, 69-87; RETIRED. *Mem:* Metall Soc; Am Soc Metals; Nat Asn Corrosion Engrs. *Res:* Elevated temperature corrosion, stress corrosion, heat resistant steels, stainless steels. *Mailing Add:* Amax Res & Develop Ctr 5950 McIntyre St Golden CO 80403

GROBSTEIN, CLIFFORD, b New York, NY, July 20, 16; m 38; c 2. BIOLOGY, PUBLIC POLICY. *Educ:* City Col New York, BS, 36; Univ Calif, Los Angeles, MA, 38, PhD(zool), 40. *Prof Exp:* Instr zool, Ore State Col, 40-43; sr res fel, Nat Cancer Inst, 46-47; biologist, USPHS, 47-57; prof biol, Stanford Univ, 58-65; vchancellor sci & dean sch med, Univ Calif, 67-73, vchancellor univ rels, 73-77, prof biol sci & pub policy, 77-87, PROF BIOL, UNIV CALIF, SAN DIEGO, 65-, EMER PROF SCI & PUB POLICY, 87- *Honors & Awards:* Brachet Award, Belgian Royal Soc. *Mem:* Nat Acad Sci; Am Soc Zoologists, (pres, 66); Soc Develop Biol (past pres); Am Acad Arts & Sci; Am Soc Cell Biol; Inst Med. *Res:* Biomedical policy issues, particularly in heredity and development; biomedical policy issues. *Mailing Add:* HL1520 B-031 Univ Calif San Diego La Jolla CA 92093

GROBSTEIN, PAUL, b Long Beach, Calif, Mar 21, 46; m 83; c 2. INTEGRATIVE NEUROBIOLOGY & DEVELOPMENTAL NEUROBIOLOGY. *Educ:* Harvard Univ, BA, 69; Stanford Univ, MA, 70, PhD(biol), 73. *Prof Exp:* Fel neurobiol, Johns Hopkins Univ, 72-73 & Stanford Univ, 73-74; from asst prof to assoc prof neurobiol, Dept Pharmacol & Physiol Sci, Univ Chicago, 74-85; PROF & CHMN DEPT BIOL, BRYN MAWR COL, 86- *Concurrent Pos:* NIH fel, 72-74; A P Sloan Found fel, 76-80; mem, Comt Space Biol & Med Space Sci Bd, Nat Res Coun, 86-89. *Mem:* Soc Neurosci; Soc Develop Biol; AAAS; Fedn Am Scientists; Int Soc Neurobiol; Sigma Xi. *Res:* Nervous system organization and behavior; nervous system development. *Mailing Add:* Dept Biol Bryn Mawr Col Bryn Mawr PA 19010

GROCE, DAVID EIBEN, b Wilmar, Calif, July 15, 36; m 61; c 1. MEDICAL PHYSICS, NUCLEAR PHYSICS. *Educ:* Calif Inst Technol, BS, 58, PhD(nuclear physics), 63. *Prof Exp:* Res fel exp nuclear physics, Australian Nat Univ, 63-65; staff assoc res, Gen Atomic Div, Gen Dynamics Corp, 65-67; staff mem, Gulf Gen Atomic, Inc, 67-69; vpres & med consult, Res & Develop, JRB Assocs, Inc, 69-76; div mgr, Sci Appln Int Corp, 69-80, sr scientist appl med res & artificial intel, 80-87; physicist radiation physics, Avalon Inc, 87-90; RETIRED. *Concurrent Pos:* Consult, Nat Comt Radiation Protection, 69-; independent Consult weapons effects, 87-; mem, bd trustees, San Diego Natural Hist Mus, 85-90, treas, 89-90; consult, 90- *Mem:* AAAS; Am Phys Soc; Am Asn Physicist in Med; Am Pub Health Asn; Health Physics Soc. *Res:* Experimental research in low energy positrons, radioactive pharmaceuticals, fission and neutron yields; Monte Carlo radiation transport, simulation and systems analysis; consulting in cancer therapy and research, accelerators and facility design; neutron physics; radiation shielding; personnel response to external physical forces; diagnostic radiology and nuclear medicine; artificial intelligence and expert systems; decision aids; radiation effects. *Mailing Add:* 8243 Prestwick Dr La Jolla CA 92037

GROCE, JOHN WESLEY, b Stanley, NC, Nov 9, 30; m 59; c 2. BIOCHEMISTRY. *Educ:* Asbury Col, AB, 53; Purdue Univ, MS, 56, PhD, 64. *Prof Exp:* From instr to assoc prof, 57-68, PROF CHEM, HEIDELBERG COL, 68-, CHMN DEPT, 65- *Concurrent Pos:* Fel, NC State Univ, 70-71. *Mem:* Am Chem Soc. *Res:* Carbohydrate, polysaccharide chemistry; analytical organic chemistry. *Mailing Add:* Dept Chem Heidelberg Col Tiffin OH 44883

GROCE, WILLIAM HENRY (BILL), b Greer, SC, Sept 21, 40; m 64; c 1. HAZARDOUS WASTE PROCESSING. *Educ:* Newberry Col, BS, 64. *Prof Exp:* Polymer chemist, Allied Chem Co, 64-66; anal chemist, US Food & Drug Admin, 66-67; polymer lab mgr, Celanese Corp, 67-72; PRIN CHEMIST, GROCE LABS, INC, 72-, DIR RES, 87- *Concurrent Pos:* Consult, 72-; lectr, Am Inst Chem Engrs, 83- & SC State Chamber Com, 85-, Univ SC, 87, 88. *Mem:* Am Chem Soc; Am Inst Chem Engrs; fel Am Inst Chemists; Inst Hazardous Mat Mgt; Am Soc Prof Engr. *Res:* Chemical reclamation procedure; chemical detoxification; explosive chemical decomposition by wet chemical oxidation; chemical hazard classification. *Mailing Add:* 112 Montero Lane Greenville SC 29615-2750

GROCHOSKI, GREGORY T, b Grand Rapids, Mich, July 6, 46; c 2. RESEARCH ADMINISTRATION. *Educ:* Grand Valley State Col, BS(chem) & BS(econ), 68. *Prof Exp:* Technician res & develop, 64-68, chemist, 68-70, res chemist, 71-72, sr res chemist, 72-73, mgr tech serv, 73-77, mgr personal care, 77-78, admin asst policy admin, 78-79, dir corp develop, 79-82, VPRES, RES & DEVELOP, AMWAY CORP, 82- *Concurrent Pos:* US Army Med Corp, Med Field Specialties, 70-75. *Honors & Awards:* Biomed Res Award, Soap & Detergent Asn, 70. *Mem:* Soap & Detergent Asn; Chem Specialties Mfg Asn; Cosmetic Toiletry & Fragrances Asn; Soc Cosmetic Chemists; Indust Res Inst; Am Asn Poison Control Ctrs. *Res:* Corporate technical operations including quality assurance, chemical, electrical and package engineering; research and development; analytical, microbiology, technical-regulatory and toxicology departments. *Mailing Add:* 7575 E Fulton Rd Ada MI 49355

GRODBERG, MARCUS GORDON, b Worcester, Mass, Jan 27, 23; m 51; c 3. DRUGS FOR DENTAL & BONE DISEASES. *Educ:* Clark Univ, AB, 44; Univ Ill, MS, 48. *Prof Exp:* Jr res chemist, Schenley Labs, Inc, Lawrenceburg, Ind, 44-47; chemist, Marine Prod Co, Boston, 48-50 & Brewer & Co, Inc, Worcester, Mass, 50-55; tech dir, Gray Pharmaceut Co, Newton, Mass, 55-58; dir res & develop, Colgate-Hoyt Labs, Div Colgate-Palmolive Co, Canton, Mass, 58-89; RETIRED. *Concurrent Pos:* Consult, Grad Sch Bus Admin, Harvard Univ, 57-58. *Mem:* Int Asn Dent Res; Am Dent Asn; Am Pharmaceut Asn; AAAS; Am Fedn Clin Res; NY Acad Sci. *Res:* Industrial research and development of pharmaceuticals; fluoride products for dental caries and osteoporosis. *Mailing Add:* 111 Hyde St Newton MA 02161

GRODINS, FRED SHERMAN, physiology, biomedical engineering; deceased, see previous edition for last biography

GRODNER, MARY LASLIE, b Attapulgus, Ga, Jan 5, 35; m 59; c 2. EMBRYOLOGY, DEVELOPMENTAL BIOLOGY. *Educ:* Wesleyan Col, Ga, AB, 55; La State Univ, Baton Rouge, MS, 57, PhD(entom), 73. *Prof Exp:* Instr embryol cell physiol, Otterbein Col, 58-59, 61; agr res technician cotton insects develop, Cotton Insect Physiol Invest, Agr Res Serv, USDA, Baton Rouge, 69, 70; instr, 71-74, ASST PROF, DEPT ZOOL & PHYSIOL, AGR & MECH COL & LA STATE UNIV, BATON ROUGE, 74- *Mem:* Soc Develop Biol; Electron Micros Soc Am; Am Soc Zoologists; Entom Soc Am; Sigma Xi. *Res:* Developmental biology and reproductive physiology of insects, primarily the cotton boll weevil. *Mailing Add:* Coop Exten Serv La State Univ Baton Rouge LA 70803

GRODNER, ROBERT MAYNARD, b Brooklyn, NY, June 22, 25; m 59; c 2. FOOD SCIENCE, ZOOLOGY. *Educ:* Brown Univ, AB, 49; Univ Tenn, MSc, 50; La State Univ, PhD(zool, physiol), 59. *Prof Exp:* Instr biol, Berea Col, 51; from asst prof to assoc prof, Otterbein Col, 59-63; assoc prof, 63-69, PROF FOOD SCI & TECHNOL, LA STATE UNIV, BATON ROUGE, 69- *Concurrent Pos:* AEC fels, 64-66. *Mem:* AAAS; Inst Food Technol; Am Inst Biol Sci. *Res:* Autoxidation of unsaturated fatty acids; radiation pasteurization of Gulf Coast shellfish; food toxicology and enzymes; rapid microbiological methodology for assay of Gulf Coast seafood; rapid methodology for determining shrimp decomposition; bacterial survey of fresh commercial seafood. *Mailing Add:* Dept Food Sci La State Univ Baton Rouge LA 70803

GRODSKY, GEROLD MORTON, b St Louis, Mo, Jan 18, 27; m 51; c 2. BIOLOGICAL CHEMISTRY, ENDOCRINOLOGY. *Educ:* Univ Ill, BS, 47, MS, 48; Univ Calif, Berkeley, PhD(biochem), 55. *Prof Exp:* Asst res biochemist, 56-60, from asst prof to assoc prof, 60-68, PROF BIOCHEM, MED CTR, UNIV CALIF, SAN FRANCISCO, 68- *Concurrent Pos:* Nat Cancer Inst fel biochem, Cambridge Univ, 54-55; consult, Langley Porter Inst, 60-, US Naval Hosp, Oakland, 61- & USPHS & Vet Admin Hosp, San Francisco, 62-; vis prof, Univ Geneva, 68-69; Somogyi lectr, 72; Helen Martin lectr, 76; mem, Med Adv Bd, Juv Diabetes Found, 75-77 & 80-85; chmn res comt, Am Diabetes Asn, 77-79; chmn acad senate, Univ Calif San Francisco, 77-79; mem adv bd diabetes, US Secy Health, 81-85. *Honors & Awards:* David Rumbough Int Diabetes Res Award, 84; Mosenthal Hon Lectr, 86; Williams-Levine Award, 89. *Mem:* Am Soc Biol Chemists; Soc Exp Biol & Med; Am Fedn Clin Res; Am Diabetes Asn. *Res:* Metabolism and immunological aspects of insulin action and secretion; diabetes. *Mailing Add:* Metab Unit Univ Calif Med Ctr San Francisco CA 94122

GRODY, WAYNE WILLIAM, b Syracuse, NY, Feb 25, 52; m 90. DIAGNOSTIC MOLECULAR PATHOLOGY, MEDICAL GENETICS. *Educ:* Johns Hopkins Univ, BA, 74; Baylor Col Med, MD, 77, PhD(cell biol), 81. *Prof Exp:* Resident lab med, 82-85, fel med genetics, 85-86, ASST PROF MED GENETICS & MOLECULAR PATH, SCH MED, UNIV CALIF, LOS ANGELES, 87- *Concurrent Pos:* Contrib ed, MD Mag, 81-; consult DNA testing, US Food & Drug Admin, 89-; chmn, DNA Qual Assurance Code, Pac Southwest Regional Genetics Network, 89-; mem, Molecular Path Resource Ctr, Col Am Path, 89-; prin investr, March Dimes Birth Defects Found, 89- & Nat Inst Child Health & Human Develop, NIH, 89-91. *Honors & Awards:* Kleiner Mem Award, Am Soc Med Technol, 90. *Mem:* AAAS; AMA; Am Soc Clin Path; Am Soc Human Genetics; Col Am Path; Soc Inherited Metab Dis. *Res:* Molecular genetics of inborn errors of metabolism; application of recombinant DNA technology to clinical diagnosis; acquired immune deficiency syndrome; biochemical genetics. *Mailing Add:* Div Med Genetics & Molecular Path Univ Calif Sch Med Los Angeles CA 90024-1732

GRODZICKER, TERRI IRENE, b Brooklyn, NY, Nov 18, 42. MOLECULAR GENETICS, VIROLOGY. *Educ:* Wellesley Col, AB, 63; Columbia Univ, MA, 65, PhD(zool), 69. *Prof Exp:* Fel molecular genetics, Harvard Med Sch, 69-72; staff investr, 73-75, sr staff investr tumor virol, 75-79, SR SCIENTIST, COLD SPRING HARBOR LAB, 79-, ASST DIR, ACAD AFFAIRS, 86- *Concurrent Pos:* NIH fel, 69-71; adj assoc prof microbiol, Med Sch, State Univ NY Stony Brook, 81-; NCI Frederick Adv Bd, 82-88; sci adv bd, Damon Runyon, Walter Winchell Cancer Res, 87-; ed, Genes & Develop, 89- *Mem:* Am Soc Microbiol; Sigma Xi; AAAS. *Res:* Control of gene expression of viruses and their host cells; tumor viruses-SV40 and adenoviruses. *Mailing Add:* Cold Spring Harbor Lab Bungtown Rd PO Box 100 Cold Spring Harbor NY 11724

GRODZINS, LEE, b Lowell, Mass, July 10, 26; m 56; c 2. NUCLEAR PHYSICS. *Educ:* Univ NH, BS, 46; Union Col, MS, 48; Purdue Univ, PhD(physics), 54. *Prof Exp:* Res asst physics, Gen Elec Res Lab, 46-48; instr, Purdue Univ, 54-55; from res assoc to assoc physicist, Brookhaven Nat Lab, 55-59; from asst prof to assoc prof, 59-66, PROF PHYSICS, MASS INST TECHNOL, 66- *Concurrent Pos:* Consult, High Voltage Eng Co, 62-70, Avco-Everett Res Lab, 67-72, Los Alamos Nat Labs, 76-, Harvard Univ, 76-77, Brookhaven Nat Labs, 77- & Exxon Nuclear Co, 78-; Guggenheim fel, 64-65 & 71-72. *Mem:* Fel Am Phys Soc; fel Am Acad Arts & Sci. *Res:* Interaction between heavy nuclei; studies of scientific manpower; nuclear spectroscopy, beta decay and Mossbauer scattering. *Mailing Add:* 14 Stratham Rd Lexington MA 02173

GRODZINSKI, BERNARD, b Stettin, Poland, Jan 28, 46; Can citizen; m 67; c 3. PLANT SCIENCE. *Educ:* Univ Toronto, BS, 68; York Univ, MS, 71, PhD(biol), 74; Univ Cambridge, MA, 78. *Prof Exp:* Fel, Sch Bot, Oxford Univ, 74-75; demonstr, Sch Bot, Cambridge Univ, 75-79; PROF, DEPT HORT SCI, UNIV GUELPH, 80- *Mem:* Can Soc Plant Physiologist; Am Soc Plant Physiologists; Am Asn Hort Sci. *Res:* Biochemical and biophysical nature of photorespiration and glycolate metabolism; the relationship between photosynthesis and plant productivity in horticultural crops. *Mailing Add:* Hort Sci Dept Univ Guelph Guelph ON N1G 2W1 Can

GRODZINSKY, ALAN J, BIOMEDICAL ENGINEERING. *Educ:* Mass Inst Technol, SM & SB, 71, ScD, 74. *Prof Exp:* Asst prof, Elec & Dept Elec Eng & Computer Sci, Mass Inst Technol, 74-75, Esther & Harold E Edgerton asst prof, Elec & Bioeng, 75-77, assoc prof, Elec & Bioeng, Dept Elec Eng & Computer Sci, 77-84, assoc prof elec & bioeng, 79-84, PROF ELEC & BIOENG, DEPT ELEC ENG & COMPUTER SCI, MASS INST TECHNOL & HARVARD-MASS INST TECHNOL DIV HEALTH SCI & TECHNOL, 84- *Concurrent Pos:* Vis lectr & res assoc, Dept Orthop Surg, Harvard Med Sch & Children's Hosp, Boston, 76-77, vis assoc prof, Biomech Lab, Dept Mech Eng, Rensselaer Polytechnic Inst, 77; Am deleg, Peoples Repub China, Orthop Sci & Bioeng Group, 79; mem sci adv panel, NY Dept Health, 80-87, sci prog comt, Bioelec Res & Growth Soc, 81-86, mem sci coun, 84-86; assoc ed, J Orthop Res, 82-; spec reviewer, Orthop Study Sect, NIH, 85; biomed engr, Dept Biomed Eng, Mass Gen Hosp, 87-; vis prof biochem & orthop, Rush Presby St Lukes Med Ctr, Rush Univ, 89; mem sci adv bds, Nat Inst Arthritis & Musculoskeletal & Skin Dis, NIH, Sci Ctr of Res, grants Osteoarthritis & Osteop; chmn, Gordon Res Conf Bioeng & Orthop Sci, 89-90. *Honors & Awards:* Giovani Borelli Award, Am Soc Biomech, 87. *Mem:* Inst Elec & Electronics Engrs, Eng Med & Biol Soc; Bioelec Repair & Growth Soc (pres, 86-87); Bioelectromagnetics Soc; Biomed Eng Soc; Orthop Res Soc; Am Soc Biomech; NY Acad Sci; Biophys Soc; AAAS; Sigma Xi. *Res:* Influence of mechanical and electrical stresses on connective tissue growth; remodeling, repair and pathology; electromechanical and physiochemical properties of connective tissues and polymeric biomaterials; electrically controlled membrane permeability for drug delivery and separation processes; synthesis of physiologically compatible transducers for diagnosis and treatment; fundamental study and modelling of electrical, mechanical and chemical energy conversion in natural and synthetic membranes and in biological tissues; numerous technical publications; patents. *Mailing Add:* Dept Elec Eng & Computer Sci Rm 38-377 Mass Inst Technol 77 Massachusetts Ave Cambridge MA 02139

GROEGER, THEODORE OSKAR, b Gross Kunzendorf, Czech, Nov 25, 27; m 57; c 2. PHARMACOLOGICAL CHEMISTRY, MECHANICAL ENGINEERING. *Educ:* Univ Vienna, Dr & Dipl, 52; Vienna Tech Univ, PhD(chem), 57. *Prof Exp:* Prof chem, Fed Sci High Sch, Austria, 53-59; sci patent assoc pharmaceut, Ciba Corp, Switz, 59-64, mgr chem patents, Ciba-Geigy Corp, 64-80; CONSULT, 81- *Concurrent Pos:* Asst to dean, Dept Chem, Vienna Tech Univ, 51-55; prof, Bus Col & lectr, Austria, 58-59. *Honors & Awards:* Austrian Ministry Educ Achievement Award, 58; Austrian Sci Educ Sch Book Award, 62. *Mem:* AAAS; Austrian Chem Soc; Ger Chem Soc; fel Am Inst Chemists; Am Chem Soc. *Res:* Patent law; pharmaceutics; patent coverage of chemical and mechanical inventions worldwide. *Mailing Add:* Two Collamore Circle West Orange NJ 07052

GROEL, JOHN TRUEMAN, b Maplewood, NJ, Oct 5, 24; m 88. MEDICINE. *Educ:* Yale Univ, MD, 51. *Prof Exp:* Med ed, White Labs, 51-52; from asst med dir to assoc med dir, E R Squibb Corp, 52-60, assoc clin res dir, 60-71, worldwide med dir, 71-72, clin res dir, 72-79, dir, med ed serv, 79-86; RETIRED. *Mem:* AMA. *Res:* Infectious diseases; cardiovascular. *Mailing Add:* 199 Edgemere Way S Naples FL 33999

GROEMER, HELMUT (JOHANN), b Salzburg, Austria, Nov 6, 30; m 57; c 2. MATHEMATICS. *Educ:* Innsbruck Univ, PhD(math & physics), 54. *Prof Exp:* From instr to assoc prof math, Ore State Univ, 57-64; PROF MATH, UNIV ARIZ, 64- *Concurrent Pos:* NSF res grants, 60- 93. *Mem:* Am Math Soc; Math Asn Am; Austrian Math Soc. *Res:* Convex sets; packing and covering problems; geometry of numbers; integral geometry; geometric probability theory; geometric inequalities. *Mailing Add:* Dept Math Univ Ariz Tucson AZ 85721

GROENWEGHE, LEO CARL DENIS, b Antwerp, Belg, Aug 31, 25; nat US; m 83; c 5. INORGANIC CHEMISTRY, COMPUTER SCIENCES. *Educ:* Univ Ghent, Lic, 48, DSc(inorg & phys chem), 51. *Prof Exp:* Res chemist, Belg Inst Sci Res, 48-53; mem sr staff, Spencer Chem Co, Kans, 53-56; res specialist, Cent Res Dept, 56-66, sr res specialist, 66-73, MGR COMPUT SERV, CORP RES DEPT, MONSANTO CO, 73- *Mem:* Am Chem Soc; Sigma Xi; Nat Comput Graphics Asn. *Res:* Scientific computation; laboratory automation; inorganic synthetic chemistry, especially uranium, fluorine, nitrogen, and phosphorus compounds; solubility systems; nuclear magnetic resonance; organophosphorus compounds; random reorganization theory; catalysis; statistics; computer applications; computer graphics; technical forecasting; office automation. *Mailing Add:* 229 Greenbriar Estates Dr St Louis MO 63122

GROER, MAUREEN, b Cambridge, Mass, Nov 27, 44; m; c 3. PHYSIOLOGY. *Educ:* Univ Ill, PhD(physiol), 75; Univ Tenn, MSN, 81. *Prof Exp:* PROF NURSING, UNIV TENN, KNOXVILLE, 83- *Concurrent Pos:* Curriculum Consult. *Mem:* Sigma Xi; Am Nursing Asn; Am Physiol Soc. *Res:* Psychoneuroimmunology; stress. *Mailing Add:* Col Nursing Univ Tenn Blvd Knoxville TN 37996

GROER, PETER GEROLD, b Vienna, Austria, Jan 27, 41; m 68; c 3. RADIATION RISK ANALYSIS, BAYESIAN STATISTICS. *Educ:* Univ Vienna, PhD(theoret physics), 67. *Prof Exp:* Mem res staff appl nuclear physics & radiation biol, Mass Inst Technol, 67-70; mem staff theoret radiation biol, Argonne Nat Lab, 70-78; mem staff radiation biol & biostatist, Inst Energy Anal, 78-83, SR SCIENTIST, OAK RIDGE ASSOC UNIVS, 83- *Concurrent Pos:* Consult, UN Sci Comt on Effects of Atomic Radiation, 79, Nat Acad Sci, Nat Coun Radiation Protection & Measurements Comt, 80-85 & Beir IV Comt, 85-88, Comt on Interagency Radiation Res & Policy Coord, 85-; vis scientist, Radiation Effects Res Found, Hiroshima, Japan, 80 & 81. *Mem:* Radiation Res Soc; Soc Risk Anal. *Res:* Estimation of risk from ionizing radiations; probabilistic models for radiation carcinogenesis; survival (reliability) analysis; competing risk theory; Bayesian statistics. *Mailing Add:* 12009 Congressional Point Knoxville TN 37922

GROETSCH, CHARLES WILLIAM, b New Orleans, La, Feb 15, 45; m 66; c 2. NUMERICAL ANALYSIS, APPROXIMATION THEORY. *Educ:* Univ New Orleans, BS, 66, MS, 68; La State Univ, PhD(math), 71. *Prof Exp:* Assoc engr, Boeing Aerospace Corp, 66-67; from asst prof to assoc prof, 71-81, PROF MATH, UNIV CINCINNATI, 81-, HEAD DEPT, 85- *Concurrent Pos:* Vis asst prof, Univ RI, 74-75; sr vis fel, Univ Manchester, Eng, 80-81; vis prof, Univ Kaiserslautern, WGer, 83; vis fel, Australian Nat Univ, 86; vis scientist, E Ger Acad Sci, 79; res assoc, Air Force Flight Dynamics Lab, 78; assoc ed, Numerical Functional Anal & Optimization. *Mem:* Am Math Soc; Math Asn Am; Soc Indust & Appl Math. *Res:* Theory of approximate methods for inverse and ill-posed problems; methods for numerical solution of Fredholm integral equations of the first kind; theory of generalized inverses of linear operators. *Mailing Add:* Univ Cincinnati Cincinnati OH 45221-0025

GROFF, DONALD WILLIAM, b Lancaster, Pa, Apr 11, 28; m 57; c 1. GEOENVIRONMENTAL SCIENCE, WATER RESOURCES. *Educ:* Univ Redlands, BS, 52; Univ Pittsburgh, PhD(geol, geochem), 66. *Prof Exp:* Geophysicist, Western Geophys Co, 52; asst, Pa State Univ, 52-55; instr, Allegheny Col, 55-57; geophysicist, Pickands Mather Co, 57; asst seismologist & instr geol, Univ Pittsburgh, 57-61; assoc prof, Ind Univ, Pa, 61-65; chmn dept, 65-87, PROF GEOL, SPACE & ENVIRON SCI DEPT, WESTERN CONN STATE COL, 66-; PROF SOIL SCIENTIST, AM REGISTRY CERT PROF CROP, AGRONOMISTS & SOIL SCIENTISTS, UNIV WIS-MADISON, 78-, CERT PROF GEOLOGIST, 86- *Concurrent Pos:* Geol consult, Huntley & Huntley, Inc, 59-61; consult, Earth Sci Serv, Brookfield Ctr, 69-73; mem Coun Educ Geol Sci, 69-73; dir, Allegheny River Mining Co & Pittsburgh & Shawmut RR Co, 70-75; coun educ geol sci, Am Geol Inst, 70-73; mem, Health Systs Agency, Conn Region V, 75-84; geohydrol consult, Off Planning, Putnam Co, NY, 81-85; chief hydrogeologist & dir geol, Lawler, Matusky & Skelly Engrs, 87- *Mem:* Sigma Xi; Agron Soc Am; Geol Soc Am; Am Water Resources Asn; Int Asn Sedimentology; Asn Groundwater Scientists & Engrs; Nat Water Well Asn. *Res:* Radiometry; spectrochemistry as applied to sediments; soil sciences; water resources; geohydrology field mapper. *Mailing Add:* 31 Merwin Brook Rd Brookfield CT 06804

GROFF, RONALD PARKE, b Lancaster, Pa, Oct 25, 40; m 63; c 2. PHYSICS. *Educ:* Lehigh Univ, BS, 62; Univ Rochester, PhD(physics), 67. *Prof Exp:* RES PHYSICIST, E I DU PONT DE NEMOURS & CO, INC, 67- *Mem:* Am Phys Soc; Sigma Xi; AAAS. *Res:* Infrared; surface physics and chemistry, catalysis; luminescence; molecular crystals; organic conductors. *Mailing Add:* 203 Owls Nest Rd Wilmington DE 19807

GROFF, SIDNEY LAVERN, b Victor, Mont, Apr 7, 19; m 45; c 2. GEOLOGY. *Educ:* Mont State Univ, BA, 41, MA, 54; Univ Utah, PhD, 59. *Prof Exp:* Asst, Mont State Univ, 52-54; chief ground water & fuels div, 57-71, DIR & STATE GEOLOGIST, MONT BUR MINES & GEOL, 71- *Mem:* Geol Soc Am; Am Asn Petrol Geol; Am Prof Geol Scientists. *Res:* Hydrogeology; coal geology and economics; mineral deposits; reconnaissance geological mapping. *Mailing Add:* Mont Col Min Sci Technol Butte MT 59701

GROGAN, DONALD E, b Grogan, Mo, Feb 6, 38; m 62; c 2. CELL BIOLOGY. *Educ:* Univ Mo, BA, 60, MS, 62, PhD(biochem), 65. *Prof Exp:* Fel protein biochem, Col Med, Baylor Univ, 65-66, res assoc drug metab, 66-68; asst prof, 68-74, ASSOC PROF BIOL, UNIV MO-ST LOUIS, 74- *Res:* Biochemistry of wasps and bees. *Mailing Add:* Dept Biol Univ Mo 8001 Natural Bridge Rd St Louis MO 63121

GROGAN, JAMES BIGBEE, b Edwards, Miss, May 15, 32; m 56; c 2. MICROBIOLOGY. *Educ:* Miss Col, BS, 55; Univ Wis, MS, 57; Univ Miss, PhD(microbiol), 63. *Prof Exp:* Asst bact, Univ Wis, 55-57; supvr surg res bact lab, 57-63, from instr to asst prof surg & microbiol, 63-74, PROF SURG & ASSOC PROF MICROBIOL, SCH MED, UNIV MISS, 74- *Mem:* Am Asn Immunologists; Am Soc Microbiol. *Res:* Infections; hosts defense mechanisms; transplantation immunology. *Mailing Add:* Sch Med Univ Miss 2500 N State St Jackson MS 39216-4505

GROGAN, MICHAEL JOHN, b Hammond, Ind, Feb 6, 38; m 61; c 1. PHYSICAL CHEMISTRY. *Educ:* John Carroll Univ, BS, 59; Ill Inst Technol, PhD(chem), 67. *Prof Exp:* Res technician, Dept Med, Univ Chicago, 59-62; chemist, Magnaflux Corp, Ill, 62-63; chemist, Culligan Inc, 63-64; teaching asst phys chem, Ill Inst Technol, 64-67; chemist, Shell Chem Co, NJ, 67-70, sr chemist, 70-71, sr chemist, Shell Chem Co, Calif, 71-75, sr engr, 75-76, staff engr, 76-77, mgr qual assurance, Shell Chem Co, Ohio, 77-81, supvr sr staff, 81-87, RES MGR, ELASTOMERS RES & DEVELOP, SHELL DEVELOP CO, HOUSTON, 87- *Mem:* Am Chem Soc. *Res:* Spectroscopic investigations of metal-olefin complexes, structure and chemistry of titanium III complexes; removal of catalyst residues from polyolefins; determination of the structure and content of ethylene-propylene copolymers by pyrolysis gas chromotography; investigations of the chemistry of propylene polymerization and styrene-butadiene polymerization; elastomer product development; management of futures oriented research on styrenic block copolymers; chemical modification of polymers; product development of styrenic block polymers for automotive uses (SMC, functional parts); polymer modification; thermoset modification; coordinate technical recruiting for polymers. *Mailing Add:* Shell Develop Co PO Box 1380 Houston TX 77251-1380

GROGAN, PAUL J(OSEPH), b Adrian, Minn, Nov 20, 18; m 45; c 6. MECHANICAL ENGINEERING. *Educ:* Purdue Univ, BS, 43; Univ Wis, MS, 49. *Prof Exp:* From instr to asst prof mech eng, Univ Wis, 47-50; asst prof mech eng, Univ Notre Dame, 50-51; prof & chmn eng exten dept, Univ Wis, 51-66; dir, Off State Tech Serv, US Dept Com, Washington, DC, 66-68; prof eng, Univ Exten, Univ Wis-Madison, 68-85; PVT PRACT MECH ENG, 85- *Mem:* Am Soc Eng Educ; Nat Asn Power Engrs. *Res:* Power production and related problems of energy conversion; heat transfer; electrical transmission and distribution; water supply and water conditioning. *Mailing Add:* 18 Southwick Circle Madison WI 53717

GROGAN, RAYMOND GERALD, b Emma, Ga, July 22, 20; m 44; c 2. PLANT PATHOLOGY. *Educ:* Univ Ga, BSA, 41, MSA, 42; Univ Wis, PhD(plant path), 48. *Prof Exp:* Instr plant path & jr pathologist, Univ Calif, Davis, 48-50, from asst prof & asst plant pathologist to assoc prof & assoc plant pathologist, 50-61, prof plant path & plant pathologist, 61-85, chmn dept, 69-74, EMER PROF PLANT PATH, UNIV CALIF, DAVIS, 85- *Mem:* AAAS; Am Phytopath Soc. *Res:* Causes and control of diseases of vegetable crops; plant virology. *Mailing Add:* 2-1921 E Cliff Dr Santa Cruz CA 95062

GROGAN, WILLIAM MCLEAN, b Knoxville, Tenn, Feb 4, 44; m 65; c 2. LIPID METABOLISM & MEMBRANE STRUCTURAL DYNAMICS, GRADUATE & POSTDOCTORAL TRAINING. *Educ:* Belmont Col, BS, 67; Purdue Univ, PhD(biochem), 72. *Prof Exp:* Res assoc biochem, Vanderbilt Univ Sch Med, 72-75; from asst prof to assoc prof, 75-90, PROF BIOCHEM, DEPT BIOCHEM & MOLECULAR BIOPHYS, VA COMMONWEALTH UNIV, 90- *Concurrent Pos:* Co-dir, Flow Cytometry Lab, Massey Cancer Ctr, 77-; prin investr, NIH res grants, 79-; dir, Laser Cytometry Lab, Sch Basic Health Sci, 91-; assoc ed, Lipids, 91-; mem spec study sect, NIH, 91- *Mem:* Am Soc Biochem & Molecular Biol; Am Oil Chemists Soc; Am Chem Soc; AAAS. *Res:* Investigation of the role of intracellular cholesterol pools in regulation of bile acid metabolism and cholesterol homeostasis; mechanisms for regulation of those pools and effects of cholesterol and other membrane components on membrane structural dynamics and membrane-associated activities; author of 50 publications and one book. *Mailing Add:* Box 614 MCV Sta Richmond VA 23298-0614

GROGAN, WILLIAM R, ELECTRICAL ENGINEERING. *Prof Exp:* FAC MEM, WORCESTER POLYTECH INST. *Mem:* Inst Elec & Electronics Engrs. *Mailing Add:* Worcester Polytech Inst 100 Institute Rd Worcester MA 01609

GROGINSKY, HERBERT LEONARD, b Newark, NJ, July 10, 30; m 52; c 3. ELECTRONICS. *Educ:* Polytech Inst Brooklyn, BEE, 52; Columbia Univ, MS, 54, EngScD(elec eng), 59. *Prof Exp:* Instr physics, Polytech Inst Brooklyn, 51-52; staff engr, Electronics Res Lab, Columbia Univ, 52-59; dept mgr appl math, 59-70, TECH DIR, ADVAN DEVELOP LAB, RAYTHEON CO, 70- *Concurrent Pos:* Instr elec eng, Northeastern Univ, 59-65, adj prof, 65-72, instr, Sch Continuing Educ, 75-; deleg, NASA Active Microwave Workshop, 75-76, mem tech comt atmospherics, 77- *Mem:* AAAS; sr mem Inst Elec & Electronics Engrs; Am Inst Aeronaut & Astronaut; Sigma Xi. *Res:* Radar signal processing; radar meteorology; space based radars. *Mailing Add:* 25 Hillside Rd Wellesley MA 02181

GROH, HAROLD JOHN, b New Orleans, La, Jan 28, 28; m 51; c 5. PHYSICAL CHEMISTRY. *Educ:* St Louis Univ, BS, 49; Univ Rochester, PhD(phys chem), 52. *Prof Exp:* Res chemist, E I du Pont de Nemours & Co Inc, 52-59, res supvr, 59-67, res mgr, 68, sect dir, 69-78, gen supt, 78-82, prin consult, 82-89; RETIRED. *Concurrent Pos:* Consult, 90- *Mem:* Am Chem Soc; Am Nuclear Soc. *Res:* Radiochemical process development; radiation chemistry; radioisotope production; actinide chemistry. *Mailing Add:* Seven Longwood Dr Aiken SC 29803

GROHSE, EDWARD WILLIAM, b New York, NY, Dec 5, 15; m 40; c 2. MASS AND HEAT TRANSFER. *Educ:* Cooper Union, BChE, 40; Univ Del, PhD(chem eng), 48. *Hon Degrees:* ChE, Cooper Union, 45. *Prof Exp:* Analytical chemist, 37-40; chem engr, FMC Corp, WVa, 40-44 & NY, 44-45; asst res prof chem eng, Univ Del, 48-49; asst prof, Carnegie Inst Technol, 49-51; sr technologist, Monsanto Chem Co, 51-52; res assoc, Gen Elec Co Res Lab, 52-58, consult engr, Knolls Atomic Power Lab, 58-60; prof chem eng, Univ Ala, Tuscaloosa, 60-65; prof chem eng & head, Energy & Mass Transfer Lab, Univ Ala, Huntsville, 65-80; sr res engr, Brookhaven Nat Lab, 80-82; CONSULT, E W GROHSE ASSOCS, 83-; PRES, HYDROCARB CORP, 87- *Concurrent Pos:* Consult, FMC Corp, 60-65, Army Missile Command, 61-63, Brookhaven Nat Lab, 62-63 & 83-, NASA Marshall Space Flight Ctr, 70-80; mem & chmn, Huntsville Air Pollution Control Bd, 69-80. *Mem:* Am Chem Soc; fel Am Inst Chem Engrs; Asn Consult Chemists & Chem Engrs; Sigma Xi. *Res:* Mass and heat transfer; chemical process research and development; distillation, fluidization, heat transfer in nuclear power systems; fuel cell research; metal hydride slurry systems. *Mailing Add:* 75 Clifton Place Port Jefferson Station NY 11776

GROLLMAN, ARTHUR PATRICK, b Baltimore, Md, May 21, 34; m 59; c 3. PHARMACOLOGY, MEDICINE. *Educ:* Univ Calif, Berkeley, 55; Johns Hopkins Univ, MD, 59. *Prof Exp:* Resident med, Johns Hopkins Hosp, 59-61; res assoc biochem, NIH, 61-63; assoc med & molecular biol, Albert Einstein Col Med, 63-66, asst prof, 66-68, assoc prof pharmacol, molecular biol & med, 68-71, prof pharmacol, 71-74, assoc dean sci & admin affairs, 72-73; PROF PHARMACOL & MED & CHMN DEPT PHARMACOL SCI, STATE UNIV NY, STONY BROOK, 74- *Concurrent Pos:* Career scientist, Health Res Coun, New York, 63-71; consult comt on biol data handling, NIH, 66-70, mem pharmacol study sect, 70-74; attend physician, Bronx Munic Hosp Ctr, 63-74, Northport Vet Admin Hosp, 74- & St Barnabas Hosp, 74-; mem comt on drug safety, Drug Res Bd, 71- *Mem:* Am Soc Pharmacol & Exp Therapeut; Am Soc Biol Chem; Infectious Dis Soc Am; Am Soc Microbiol; Am Physiol Soc. *Res:* Molecular pharmacology; mechanism of drug action; design of chemotherapeutic agents; interaction of drugs with macromolecules. *Mailing Add:* Dept Pharmacol State Univ NY Stony Brook NY 11794

GROLLMAN, SIGMUND, b Stevensville, Md, Feb 12, 23. PHYSIOLOGY. *Educ:* Univ Md, BS, 47, MS, 49, PhD(physiol), 52. *Prof Exp:* From instr to assoc prof zool & physiol, 49-66, chmn div physiol, 66-73, prof, 66-84, dir grad studies, Zool Dept, 73-84, EMER PROF ZOOL, UNIV MD, COLLEGE PARK, 84- *Concurrent Pos:* Consult, Sci Writing & Text Bk Publ. *Mem:* AAAS; Am Soc Exp Biol & Med; NY Acad Sci; fel Am Col Sports Med. *Res:* Tissue and cellular metabolism; exercise and fatigue; aging and lipid metabolism. *Mailing Add:* Dept Zool Univ Md College Park MD 20742

GROMAN, NEAL BENJAMIN, b Chisholm, Minn, May 21, 21; m 43; c 4. MICROBIOLOGY. *Educ:* Univ Chicago, SB, 47, PhD(bact, parasitol), 50. *Prof Exp:* From instr to prof microbiol, Univ Wash, 50-89, dir, Off Biol Educ, 71-75, actg chmn microbiol, 80-81, EMER PROF MICROBIOL, UNIV WASH, 90- *Concurrent Pos:* Markle scholar, 55-60; Guggenheim fel, 58-59. *Mem:* Am Soc Microbiol; Am Acad Microbiol. *Res:* Bacteriophage; microbial genetics and physiology; medical microbiology. *Mailing Add:* Dept Microbiol Univ Wash Seattle WA 98195

GROMELSKI, STANLEY JOHN, JR, b Haydenville, Mass, Feb 26, 42; m 72; c 1. POLYMER CHEMISTRY, ORGANIC CHEMISTRY. *Educ:* Univ Mass, Amherst, BS, 65; Ohio State Univ, PhD(org chem), 71. *Prof Exp:* Chemist res training prog, Gen Elec Res & Develop Ctr, 65-67; res chemist, GAF Corp, 73-74, group leader, 74-76, mgr res & develop admin & tech info serv, 76-79, asst to vpres, 76-79, mgr latex res & develop, 79-80, mgr plastics res & develop, 81-84; dir polymer res, 84-86, VPRES TECH, REICHHOLD CHEM, DEL, 86- *Concurrent Pos:* Fel dept chem, Univ Ariz, 72-73. *Mem:* AAAS; Am Chem Soc; Tech Asn Pulp & Paper Indust; Soc Plastic Eng. *Res:* Investigations into the chemistry of acrylate, vinyl acetate and ethylene polymerizations and styrene butadiene polymerizations in aqueous systems. *Mailing Add:* 6953 Harbor Dr NW Canton OH 44718-3747

GROMKO, MARK HEDGES, b New Haven, Conn, Apr 1, 50. POPULATION GENETICS. *Educ:* Swarthmore Col, BA, 72; Ind Univ, AM, 75, PhD(pop genetics), 78. *Prof Exp:* ASST PROF GENETICS, BOWLING GREEN STATE UNIV, 78- *Mem:* Genetics Soc Am; Soc Study Evolution; Am Soc Naturalists; Sigma Xi. *Res:* Genetics of isolated populations of Drosophila in the Bowling Green area; mating patterns in natural populations; behavior genetics; the genetics of repeated mating. *Mailing Add:* 309 W Merry Ave Bowling Green OH 43402-1751

GROMME, CHARLES SHERMAN, b San Francisco, Calif, Nov 15, 33. GEOLOGY. *Educ:* Univ Calif, Berkeley, AB, 59, PhD(geol), 63. *Prof Exp:* Asst res geologist, Univ Calif, Berkeley, 63-65; GEOLOGIST, US GEOL SURV, 65- *Mem:* AAAS; Am Geophys Union; Geol Soc Am; Sigma Xi. *Res:* Rock magnetism and paleomagnetism. *Mailing Add:* US Geol Surv MS 937 345 Middlefield Rd Menlo Park CA 94025

GRON, POUL, b Skaerbaek, Denmark, Mar 26, 27; m 54; c 2. DENTISTRY. *Educ:* Royal Dent Col, Copenhagen, DDS, 50; Tufts Univ, DMD, 58. *Prof Exp:* Res fel, Sch Dent Med, Harvard Univ & Forsyth Dent Ctr, 58-60; assoc dent, 62-63, asst prof, 63-67, SR STAFF MEM, FORSYTH DENT CTR, 68-, DIR INFIRMARY DIV, 69-; ASSOC CLIN PROF ORAL BIOL & PATH PHYSIOL, HARVARD SCH DENT MED, 76- *Mem:* AAAS; Int Asn Dent Res; Nordic Odontol Soc. *Res:* Chemical and structural composition of bone, teeth and pathological calcifications; preventive dentistry for elders. *Mailing Add:* 44 Pearl St Cambridge MA 02139

GRONEMEYER, SUZANNE ALSOP, b Tulsa, Okla. MEDICAL PHYSICS, RADIATION PHYSICS. *Educ:* Washington Univ, St Louis, BA, 64, MA(nuclear chem), 67, MA(nuclear physics), 75, PhD(nuclear physics), 79. *Prof Exp:* Instr physics, Univ Mo-St Louis, 68-74; prof asst, Physics Div, NSF, 75-76; sr res aide physics, Argonne Nat Lab, 77-79; res fel radiol physics, Mass Gen Hosp, 79-80; eng physicist, Fermi Nat Accelerator Lab, 80-83; applns scientist, Siemens Med Systs, 83-88; ASSOC MEM, DIAG IMAGING DEPT, ST JUDE CHILDREN'S RES HOSP, 88-; ASSOC PROF RADIOL, COL MED, UNIV TENN, MEMPHIS, 88- *Mem:* Am Phys Soc; Am Asn Physicists Med; Soc Magnetic Resonance Med; Soc Magnetic Resonance Imaging; Radiol Soc NAm. *Res:* Clinical applications of magnetic resonance imaging and spectroscopy. *Mailing Add:* Dept Diag Imaging St Jude Childrens Res Hosp Memphis TN 38101-0318

GRONER, CARL FRED, b Wilkes-Barre, Pa, July 29, 42; m 65; c 2. ELECTROPHOTOGRAPHY. *Educ:* Pa State Univ, BS, 63, MS, 66, PhD(solid state sci), 70. *Prof Exp:* SR RES PHYSICIST ELECTROPHOTOG, EASTMAN KODAK RES LABS, 70- *Mem:* Soc Photog Scientists & Engrs. *Res:* Electrophotographic imaging systems. *Mailing Add:* 427 Mt Airy Dr Rochester NY 14617

GRONER, GABRIEL F(REDERICK), b Los Angeles, Calif, May 17, 38; m 64; c 2. ELECTRICAL ENGINEERING, SOFTWARE SYSTEMS. *Educ:* Univ Calif, Los Angeles, BS, 60; Stanford Univ, MS, 61, PhD(elec eng), 64. *Prof Exp:* Res asst elec eng, Stanford Univ, 62-64; comput scientist, Rand Corp, 64-78; sr prog mgr, Telesensory Systs Inc, 78-80, vpres & gen mgr, 80-82; vpres eng, Speech Plus Inc, 83-89; PRES, INSIGHT SOLUTIONS, 89- *Concurrent Pos:* Lectr, Ext Div, Univ Calif, Los Angeles, 69-72; consult, M D Anderson Hosp, 75-78; prin investr, NIH, 72-79. *Mem:* Inst Elec & Electronics Engrs; Comput Soc; Sigma Xi. *Res:* Pattern recognition; user-oriented interactive computer systems; medical applications of computers; electronic aids for handicapped; electronic speech synthesis. *Mailing Add:* 230 Parkside Dr Palo Alto CA 94306

GRONER, PAUL STEPHEN, b Binghamton, NY, May 23, 37; m 67; c 2. ELECTRONICS ENGINEERING. *Educ:* Polytech Inst Brooklyn, BSEE, 62. *Prof Exp:* Group head, Hughes Aircraft, 69-73; mgr, Sperry Univac, 73-82; dir, Computer Consoles, 82-87; consult engr, McDonnell Douglas, 87-88; dir, AST Res, 88-89; GEN MGR & VPRES ENG, SILICON CONTROLS, 89- *Concurrent Pos:* Vis lectr, Univ Calif Irvine, 79-84. *Mem:* Inst Elec & Electronics Engrs. *Res:* Developed key silicon including personal computer chip set and risc processor; granted several patents; author of various publications. *Mailing Add:* 2139 N Ross St Santa Ana CA 92706

GRONNER, A(LFRED) D(OUGLAS), b Vienna, Austria, Apr 30, 13; US citizen; m 48. ELECTRICAL ENGINEERING. *Educ:* Univ Vienna, EE, 38; Polytech Inst Brooklyn, MEE, 50, PhD, 55. *Prof Exp:* Proj supvr & field engr, Conatel, SAm, 42-45; chief engr, Castillo & Co, 45-46; proj engr, Liquidometer Corp, 46-50; dir res, Arma Corp, 50-51; chief proj engr, Am Mach & Foundry Co, 51-54; dir res, Liquidometer Corp, 54-55; mgr elec & comput dept, Greenwich Eng Div, Am Mach & Foundry Co, 56-61; vpres eng, Spaceonics, Inc, 61-63; chief engr, Simmonds Precision Prod Co, 63-67 & Singer Gen Precision, Little Falls, 67-78; VPRES, S&G SYSTS INC, 78- *Concurrent Pos:* Adj prof, Polytech Inst Brooklyn, 51-; vis prof, Albert Einstein Col Med, 67-; consult, 78- *Mem:* Am Inst Aeronaut & Astronaut; Inst Elec & Electronics Engrs. *Res:* Automatic control; servomechanisms; instrumentation; energy conservation. *Mailing Add:* 18 Dale St White Plains NY 10605

GRONSKY, RONALD, b Pittsburgh, Pa, July 9, 50; m 70; c 4. MATERIALS SCIENCE, ELECTRON MICROSCOPY. *Educ:* Univ Pittsburgh, BS, 72; Univ Calif, Berkeley, MS, 74, PhD(mat sci & eng), 77. *Prof Exp:* Lectr, Dept Mat Sci & Mineral Eng, 77-79, dep assoc lab dir, Energy Sci, 88-89, SCIENTIST, MAT & MOLECULAR RES DIV, LAWRENCE BERKELEY LAB, UNIV CALIF BERKELEY, 77-, PROF DEPT MAT SCI & MINERAL ENG, 81-; CHMN, DEPT MAT SCI & MINERAL ENG, UNIV CALIF, BERKELEY, 90- *Concurrent Pos:* Mgr Atomic Resolution Micros Proj, Nat Ctr Electron Micros, Berkeley, 79-86. *Honors & Awards:* First Place Phys Sci, Elec Micros Soc Sci Exhibit, 76; Exhibit Awards, Int Metallographic Soc, 76-78, 82-85; Robert Lansing Hardy Gold Medal, Metall Soc/Am Inst Mech Eng, 79; Burton Medal, Elec Micros Soc Am, 83; Bradley Stoughton Award, Am Soc Metals, 85. *Mem:* Am Inst Mining, Metall & Petrol Engrs; Electron Micros Soc Am (secy, 88-90); Am Soc Metals; AAAS; Mat Res Soc; Böhmische Phys Soc. *Res:* Atomic scale microstructural analysis and engineering; electron optics; physical metallurgy; nanostructures. *Mailing Add:* Dept Mat Sci & Mineral Eng Univ Calif Berkeley CA 94720

GRONVALL, JOHN ARNOLD, university administration, federal administration; deceased, see previous edition for last biography

GROOD, EDWARD S, b Buffalo, NY, Mar 28, 44; m 67; c 2. BIOMEDICAL ENGINEERING, ORTHOPAEDIC BIOMECHANICS. *Educ:* Rensselaer Polytech Inst, BS, 65; State Univ NY, Buffalo, MS, 68, PhD(mech eng& bioeng), 73. *Prof Exp:* Develop engr, Bell Aerospace Co, 66-72; res assoc mech eng, State Univ NY, Buffalo, 72; sr res engr, Res Inst, Univ Dayton, 72-75; from asst prof to assoc prof, 77-85, PROF OTHROP RES, MED CTR, UNIV CINCINNATI, 85- *Concurrent Pos:* Adj asst prof, Univ Dayton, 73-75; res assoc, Miami Valley Chap, Am Heart Asn, 73-75; adj prof biomech, Univ Cincinnati, 76-; dir, Giannestras Biomech Lab. *Mem:* Fel Am Soc Mech Engrs; Orthop Res Soc; Am Soc Biomech; Am Soc Testing & Mat; AAAS; Am Orthop Soc Sports Med; Am Col Sports Med; Biomed Eng Soc. *Res:* Mechanics and kinematics of biological joints; mechanical properties of collagen. *Mailing Add:* Dept Orthop Univ Cincinnati Col Med 231 Bethesda Ave Cincinnati OH 45267

GROOM, ALAN CLIFFORD, b London, UK, June 23, 26; m 52; c 4. BIOPHYSICS, PHYSIOLOGY. *Educ:* Univ London, BSc, 49, PhD(biophys), 57. *Prof Exp:* Sr physicist, St Mary's Hosp, London, 49-57, lectr med physics, Med Sch, 58-66; vis res assoc physiol, State Univ NY Buffalo, 62-63; assoc prof, Univ Western Ont, 66-72, prof biophys, 72-91, chmn dept biophys, 78-87, hon lectr anesthesia, 70-91, hon lect physiol, 80-91, EMER PROF BIOPHYS, UNIV WESTERN OHIO, 91- *Concurrent Pos:* Leverhulme res fel biophys, Univ London, 57-58; hon assoc ed, Can J Physiol & Pharmacol, 69-73 & Clin Hemorheology, 80- *Honors & Awards:* Landis Res Award, Microcirculatory Soc, 86. *Mem:* Europ Soc Microcirculation; Biophys Soc Can; Int Soc Oxygen Transport to Tissue; Biophys Soc; Microcirculatory Soc. *Res:* Microcirculation and oxygen transport to tissue in skeletal and cardiac muscle; microcirculation of spleen; role of spleen with respect to blood cells; cancer cells and the microcirculation. *Mailing Add:* Dept Biophys Univ Western Ont London ON N6A 5C1 Can

GROOM, DONALD EUGENE, b Pittsburgh, Pa, Dec 30, 34; m 62; c 2. COSMIC RAY PHYSICS. *Educ:* Princeton Univ, AB, 56; Calif Inst Technol, PhD(physics), 65. *Prof Exp:* Res assoc physics, Cornell Univ, 65-66, instr, 66-67, actg asst prof, 67-69; from asst prof to prof physics, Univ Utah, 72-87; SSC Cent Design Group, 85-89; STAFF PHYSICIST, LAWRENCE BERKELEY LAB, 89- *Concurrent Pos:* Vis res physicist, Princeton Univ, 75-76. *Mem:* Am Phys Soc; Am Astron Soc. *Res:* Interstellar cosmic ray propagation; atmospheric interactions; hadronic final states in electron-positron interactions, Stanford positron-electron asymmetric ring. *Mailing Add:* Lawrence Berkeley Lab MS 50-308 Berkeley CA 94720

GROOMS, THOMAS ALBIN, b Dayton, Ohio, Apr 6, 43; m 69; c 4. ENZYMOLOGY, FERMENTATION. *Educ:* Baldwin-Wallace Col, BS, 65; Univ Cincinnati, PhD(biochem), 73. *Prof Exp:* Sr investr, Wilson Pharmaceut & Chem Corp, 73-75; scientist, Leeds & Northrup, 75-76; proj chemist, Yellow Springs Instrument Co, 76-89; SR SCIENTIST, SHARON DR CORP, 89- *Concurrent Pos:* Lectr, Wittenberg Univ, 78-82; nutrit consult, 81-; adj assoc prof, Antioch Univ, 83- *Mem:* Am Asn Clin Chem; NY Acad Sci; Sigma Xi. *Res:* Biochemical sensors; nutritional consulting and evaluation of individuals in order to study the origins of disease and what maintains an individual's health. *Mailing Add:* 3312 US 42E PO Box 704 Cedarville OH 45314

GROOPMAN, JOHN DAVIS, b New York, NY, Nov 19, 52; m 81; c 3. TOXICOLOGY. *Educ:* Elmira Col, BA, 74; Mass Inst Technol, PhD(toxicol), 79. *Prof Exp:* Fel toxicol, Mass Inst Technol, 79-80; staff fel carcinogenesis, Nat Cancer Inst, 80-81; from asst prof to assoc prof toxicol, 81-89, assoc dir, 87-89, ASSOC PROF ENVIRON CHEM, SCH PUB HEALTH, JOHNS HOPKINS UNIV, 89- *Concurrent Pos:* Res assoc, Mass Inst Technol, 80-; consult, Sci Adv Bd, Dept Pub Health, State Mass, 81-; lectr epidemiol, Sch Med, Boston Univ, 82- *Mem:* NY Acad Sci; Am Pub Health Asn; Am Asn Cancer Res; Soc Toxicol. *Res:* Non-invasive screening techniques for measuring the exposure of people to chemical carcinogens including immunoassays, and high pressure liquid chromatography. *Mailing Add:* Sch Hyg & Pub Health Johns Hopkins Univ 615 N Wolfe St Baltimore MD 21205

GROOT, CORNELIUS, b Chicago, Ill, Nov 27, 19; m 42; c 3. CHEMISTRY. *Educ:* Univ Chicago, BS, 40, MS, 42; Univ Calif, PhD(chem), 47. *Prof Exp:* Asst chem, Univ Chicago, 41-42; jr chemist, Shell Develop Co, Calif, 42-45; asst chem, Univ Calif, 45-46; chemist, Gen Elec Co, 47-57, supvr, 57-59, mgr coolant chem, 59-64, electrochem consult, 64-66, nuclear engr, 66-70, chem engr, 70-79, engr steam generators, 79-84; RETIRED. *Honors & Awards:* Coffin Award, 51. *Mem:* Am Chem Soc; Nat Asn Corrosion Eng. *Res:* Separation processes; corrosion. *Mailing Add:* 4125 Wessex Dr San Jose CA 95136-1856

GROOTES, PIETER MEIERT, b Wieringerwaard, Holland, Jan 22, 44; m 70; c 1. PHYSICS, GEOCHRONOLOGY. *Educ:* Univ Groningen, Drs(exp physics), 70, PhD(physics), 77. *Prof Exp:* Res assoc physics, State Univ Groningen, 70-76; res assoc, 77-78, res asst prof physics, 78-83, SR RES ASSOC, QUATERNARY ISOTOPE LAB, UNIV WASH, 83- *Mem:* AAAS; Am Geophys Union; Int Glaciol Soc. *Res:* Geochronology; use of natural radioactive isotopes and stable isotopic abundances to study present and past environmental processes; paleoclimatology; glaciology and paleoclimate reconstruction. *Mailing Add:* 2539 NE 105th Pl Seattle WA 98125

GROOTES-REUVECAMP, GRADA ALIJDA, b Deventer, Netherlands, May 24, 44; m 70; c 1. MEDICAL MICROBIOLOGY, BACTERIAL GENETICS. *Educ:* State Univ Groningen, Netherlands, Drs, 68. *Prof Exp:* Res asst, med microbiol, Univ SAfrica, Pretoria, 68-69; res assoc, State Univ Groningen, 69-78; res assoc, Infectious Dis, Dept Med, Univ Wash, 78-82, researcher, 82-84. *Concurrent Pos:* Vis scientist, med microbiol, Dept Surg, Univ Wash, 77-78. *Mem:* Am Soc Microbiol; Gen Soc Microbiol; Am Soc Microbiol. *Res:* Bacterial genetics, in particular the study of drug-resistance caused by plasmids; fluorescent antibody technique, applied to urinary tract infections. *Mailing Add:* 2539 NE 105th Pl Seattle WA 98125

GROOVER, MARSHALL EUGENE, JR, b Atlanta, Ga, June 20, 10; m 61; c 2. MEDICINE. *Educ:* Univ Ga, MD, 34; Johns Hopkins Univ, MPH, 40; USAF Sch Aviation Med, dipl, 42; Am Bd Internal Med, dipl, 47. *Prof Exp:* Intern, Univ Hosp, Augusta, 34-35; with Civilian Conserv Corps, US Army Med Corps, Ft McPherson, Ga, 35-36; mem staff, Dept Pub Health, Brooks County, Ga, 37-38 & 40-41; mem staff, Rheumatic Fever Ctr, Foster Gen Hosp, Jackson, Miss, 45-46 & Trop Dis Ctr, Moore Gen Hosp, Swannanoa, NC, 46-48; chief med serv, 130th Army Hosp, Heidelberg, Ger, 48-51; chief Aviation Med Clin, Nat Defense Bldg, Washington, DC, 51-61; res scientist dept physiol, Southwest Found Res & Educ Tex, 61; assoc prof med, Sch Med, Univ Okla, 61-65; DIR HEART DIS CONTROL PROG, BUR CHRONIC DIS, FLA DIV HEALTH, 66- *Concurrent Pos:* Clin assoc prof, Univ Fla, 71-; mem endocrinol study sect, Div Res Grants, USPHS, 52-60, cardiovasc study sect, 53-59; mem coun arteriosclerosis, Am Heart Asn, 63. *Mem:* Fel Am Col Physicians; fel Am Col Cardiol; fel Am Geriat Soc; fel Am Col Angiol. *Mailing Add:* Rt 2 Box 355 Yulee FL 32097

GROOVER, MIKELL, b Kingsport, Tenn, Sept 8, 39; m 62; c 3. INDUSTRIAL ENGINEERING. *Educ:* Lehigh Univ, AB, 61, BSME, 62, MSIE, 66, PhD(indust eng), 69. *Prof Exp:* Mfg engr, Eastman Kodak Co, NY, 62-64; res asst, Inst Res, 64-66, from instr to asst prof, 66-76, assoc prof, 76-78, PROF INDUST ENG, LEHIGH UNIV, 78- *Mem:* Soc Mfg Engrs; Am Inst Indust Engrs; Am Soc Mech Engrs. *Res:* Manufacturing engineering and systems; adaptive control. *Mailing Add:* Dept Indust Eng Lehigh Univ Mohler Lab 200 Bethlehem PA 18015

GROPEN, ARTHUR LOUIS, b Huntington, NY, May 14, 32; div; c 3. TOPOLOGY. *Educ:* Univ Chicago, AB, 52, SB, 53; Duke Univ, PhD(math), 58. *Prof Exp:* Jr engr math, Sperry Gyroscope Corp, 54-55; Fulbright teaching fel, Univ Caen, 58-59; from instr to asst prof, Wellesley Col, 59-62; UNESCO specialist, Univ Satander, 62-63; from asst prof to assoc prof, Carleton Col, 63-74, prof math, 74-84; RETIRED. *Concurrent Pos:* Furniture maker. *Mem:* Am Math Soc; Math Asn Am. *Res:* Point set topology; dimension theory. *Mailing Add:* 809 St Olaf Ave Northfield MN 55057

GROPP, ARMIN HENRY, b Antigo, Wis, Sept 21, 15; m 44; c 2. PHYSICAL CHEMISTRY, ANALYTICAL CHEMISTRY. *Educ:* Univ Ore, BA, 43, MA, 45, PhD(phys chem), 47. *Prof Exp:* Res assoc chem, Univ Ore, 47; instr, Univ Fla, 47-48, from asst prof to prof, 48-64, asst chmn dept, 54-56, asst dean col arts & sci, 59-64; dean grad sch, Univ Miami, 64-65, vpres acad affairs & dean fac, 65-72, prof chem, 64-81, dir inst res, 72-81, EMER PROF CHEM, UNIV MIAMI, 81- *Concurrent Pos:* Consult chemist, Res Labs, Gen Motors Corp, 52- *Mem:* Am Chem Soc; Electrochem Soc; Nat Asn Corrosion Engrs; Am Inst Chemists. *Res:* Spectrophotometry; polarography; corrosion; electrochemistry; protective films; instrumental analysis. *Mailing Add:* 6090 SW 116th St Coral Gables FL 33124

GROSCH, CHESTER ENRIGHT, b Hoboken, NJ, Jan 13, 34; m 56; c 4. PHYSICS. *Educ:* Stevens Inst Technol, ME, 56, MS, 59, PhD(physics), 67. *Prof Exp:* Res assoc, Hudson Labs, Columbia Univ, 66-68; scientist, Teledyne-Isotopes, Inc, 68-69; assoc prof comput sci & physics, Pratt Inst, 69-73, chmn dept comput sci, 71-73; SLOVER PROF OCEANOG, DEPT OCEANOG, OLD DOMINION UNIV, 73- *Concurrent Pos:* Consult, Vitro Labs, 63-68, Nuclear Res Assocs, 69-, Advan Res Projs Agency, 70-, TRW Systs, 71-, Inst Defense Anal, 71-, Stanford Res Inst, 73-, Mantech Corp, 75- & Rand Corp, 76-; adj asst prof, Columbia Univ, 68-; pvt consult ocean & atmospheric sci, 70- *Mem:* AAAS; Soc Indust & Appl Math. *Res:* Physics of fluids; kinetic theory; rarefied gas dynamics; viscous flows; boundary layers; hydrodynamic stability; geophysical fluid dynamics; numerical methods in fluid dynamics; algorithms for parallel computing. *Mailing Add:* 1130 Manchester Ave Norfolk VA 23508

GROSCH, DANIEL SWARTWOOD, b Bethlehem, Pa, Oct 25, 18; m 44; c 5. GENETICS, ZOOLOGY. *Educ:* Moravian Col & Sem, BS, 39; Lehigh Univ, MS, 40; Univ Pa, PhD(zool), 44. *Prof Exp:* Lab instr chem, Moravian Col & Sem, 36-37, asst biol, 37-38; instr zool, Univ Pa, 41-44, asst, 41-43; asst prof zool, 46-51, assoc prof genetics, 51-57, PROF GENETICS, NC STATE UNIV, 57- *Concurrent Pos:* With Marine Biol Lab, Woods Hole. *Honors & Awards:* Comenius Award, Moravian Col. *Mem:* AAAS; Am Soc Naturalists; Entom Soc Am; Genetics Soc Am; Radiation Res Soc; Sigma Xi. *Res:* Cytology, genetics and radio-biology of habrobracon and brine shrimp; biosatellite research; altered fecundity and fertility from radioisotopes, antimetabolites and mutagenic agents. *Mailing Add:* 1222 Duplin Rd Raleigh NC 27607

GRSCHEL, DEITER HANS MAX, b Wurzburg, Ger, May 13, 31; US citizen; m 58; c 2. CLINICAL MICROBIOLOGY, INFECTIOUS DISEASES. *Educ:* Univ Cologne, Med, 57; Dr Med, 58; Am Bd Microbiol, dipl, 65. *Prof Exp:* Intern med, Univ Cologne, US Army Hosp, Landstuhl & Lutheran Hosp, Cologne, 57-59; asst neurosurg, Univ Cologne, 59-60, res assoc hyg & microbiol, 60-63; assoc microbiol, Wistar Inst, Philadelphia, Pa, 63-65; from asst prof to assoc prof, Sch Med, Temple Univ, 65-68; dir microbiol & infectious dis, Springfield Hosp Med Ctr, 68-71; assoc prof path, Univ Tex M D Anderson Hosp & Tumor Inst, 71-78, prof, 78-79; PROF PATH & INTERNAL MED, UNIV VA, CHARLOTTESVILLE, 79- *Concurrent Pos:* Clin assoc lab med, Sch Med, Univ Conn, 70-71; from assoc prof to prof path, Univ Tex Grad Sch Biomed Sci, 72-79; assoc prof med & path, Univ Tex Sch Med, Houston, 73-79; from assoc prof to prof, Univ Tex Sch Allied Health Sci, 77-79. *Mem:* Sigma Xi; fel Am Acad Microbiol; Am Soc Microbiol; Ger Soc Hyg & Microbiol; Austrian Soc Hyg Microbiol & Prev Med; fel Infectious Dis Soc Am; Soc Hosp Epidemiol Am; Fedn Am Scientists. *Res:* Clinical microbiology; hospital epidemiology; infectious diseases. *Mailing Add:* Dept Path Box 168 Univ Va Med Ctr Charlottesville VA 22908

GROSE, HERSCHEL GENE, b Clinton Co, Ind, Feb 1, 21; m 44; c 6. ORGANIC CHEMISTRY. *Educ:* Ind Cent Col, BS, 42; Ind Univ, PhD(chem), 51. *Prof Exp:* Tech supvr, US Rubber Co, 41-43; res chemist, E I du Pont de Nemours & Co, 52-53; from asst prof to prof, 53-77, ERWIN PROF CHEM, MARIETTA COL, 77-, HEAD DEPT, 57- *Mem:* AAAS; Am Chem Soc. *Res:* Chemistry of thiophenes; industrial research in high polymers. *Mailing Add:* 215 Ingleside Marietta OH 45750-3414

GROSE, THOMAS LUCIUS TROWBRIDGE, b Evanston, Ill, Dec 5, 24; m 47; c 2. GEOLOGY. *Educ:* Univ Wash, BS, 48, MS, 49; Stanford Univ, PhD(geol), 55. *Prof Exp:* Petrol geologist, Tex Co, 49-52; assoc prof geol, Colo Col, 55-64; assoc prof, 64-67, PROF GEOL, COLO SCH MINES, 67- *Concurrent Pos:* Consult geologist for indust & govt on energy & mineral explor, 54- *Mem:* Geol Soc Am; Am Asn Petrol Geol; Am Geophys Union; Soc Econ Geologists. *Res:* Structural and field geology; economic geology; oil, geothermal energy, coal, metals and nuclear facility seismic risk analysis. *Mailing Add:* 2001 Washington Circle Golden CO 80401

GROSE, WILLIAM LYMAN, b Beckley, WVa, Sept 12, 39; m 61; c 3. ATMOSPHERIC SCIENCE, FLUID MECHANICS. *Educ:* Va Polytech Inst, BS, 61, PhD(aerospace eng), 69; Col William & Mary, MS, 66. *Prof Exp:* Res scientist reentry physics, 61-69, RES SCIENTIST ATMOSPHERIC SCI, NASA, 69- *Concurrent Pos:* Asst prof lectr, George Washington Univ, 75; vis scientist, Reading Univ, Eng, 76. *Honors & Awards:* Except Sci Achievement Medal, NASA, 86. *Mem:* Am Geophys Union. *Res:* Dynamic meteorology; atmospheric chemistry and transport. *Mailing Add:* 106 Holloway Dr Williamsburg VA 23815

GROSECLOSE, BYRON CLARK, b Marion, Va, June 11, 34; m 52; c 4. NUCLEAR PHYSICS, RESEARCH ADMINISTRATION. *Educ:* Emory & Henry Col, BS, 55; Univ Va, MA, 57, PhD(physics), 59. *Prof Exp:* Asst prof physics, Okla State Univ, 59-63 & Kans State Univ, 63; physicist, 63-75, assoc div leader, 75-79, dep div leader, 79-80, head, nuclear design dept, 80-84, HEAD, DEFENSE SCI DEPT, LAWRENCE LIVERMORE NAT LAB, UNIV CALIF, 84- *Concurrent Pos:* Prin investr, US Army Res Off grant, 61-63. *Mem:* Am Phys Soc; Am Inst Aeronaut & Astronaut. *Res:* Elastic and inelastic neutron scattering; positron life-times and intensities; physics design of nuclear explosives for peaceful applications; physics design of nuclear explosives for tactical and strategic weapons applications. *Mailing Add:* Lawrence Livermore Nat Lab L-21 Livermore CA 94550

GROSEWALD, PETER, b Bronx, NY, May 26, 37; m 59; c 2. THIN FILM CRYSTAL GROWTH & SURFACE PREPARATION, ELECTRONIC PACKAGING OPTIMIZATION. *Educ:* City Col New York , BS, 59; Univ Fla, MS, 61; Pa State Univ, PhD(solid state technol), 66. *Prof Exp:* Mgr physics-engr, Airco, 66-68; IBM Res, IBM Components, 68-76, mgr thin film process engr, Components Div, 76-79, functional mgr packaging & thermal technol assurance, Systs Div, 79-81, functional mgr, Prod Mfg Ctr, 81-83, chair assignment, New Prof Educ Prog, CHQ Tech Educ Ctr, 83-85, tech ed, Strategic Planning Staff, 85-87, MGR UNIV PROGS, IBM CORP, 87- *Concurrent Pos:* Mem, Indust Exec Adv Bd, Nat Technol Univ, 88-91; home video tutorial chair, Educ Activ Bd, Inst Elec & Electronics Engrs, 89-90, financial chair, 91-; invited lectr, Frontiers Educ, Austria, Inst Elec & Electronics Engrs, 90- *Mem:* Sr mem Inst Elec & Electronics Engrs; Sigma Xi; Am Physics Soc; Am Vacuum Soc. *Res:* Design and development of thin film processes for semiconductor and magnetic film technology; thermal and packaging technologies associated with semiconductors. *Mailing Add:* IBM Corp 500 Columbus Ave Thornwood NY 10594

GROSH, DORIS LLOYD, b Kansas City, Mo, Nov 29, 24; m 50; c 3. STATISTICS, OPERATIONS RESEARCH. *Educ:* Univ Chicago, BS, 46; Kans State Univ, MS, 49, PhD(statist), 69. *Prof Exp:* Instr math, Kans State Univ, 46-49; res asst statist, Purdue Univ, 49-52; instr math, Univ Tulsa, 52-65; asst prof indust eng, 68-75, PROF INDUST ENG, KANS STATE UNIV, 75- *Concurrent Pos:* Actg dept head indust eng, Kans State Univ, 87-88. *Mem:* Am Statist Asn; Am Soc Qual Control; assoc Inst Elec & Electronics Engrs; Am Radio Relay League; Soc Women Engrs. *Res:* Reliability, such as life testing; Bayesian decision theory, hypergeometric distribution and its conjugate beta-binomial distribution. *Mailing Add:* 325 N 14th Manhattan KS 66502

GROSH, RICHARD J(OSEPH), b Ft Wayne, Ind, Oct 29, 27; m 50; c 6. MECHANICAL ENGINEERING. *Educ:* Purdue Univ, BSME, 50, MSME, 52, PhD, 53. *Hon Degrees:* DSc, Union Col, DEng, Purdue Univ. *Prof Exp:* Jr res engr, Capehart-Farnsworth Corp, 50-51; from asst prof to prof mech eng, Sch Eng Purdue Univ, 53-71, assoc dean, 65-67, dean, 67-71, head dept mech eng, 61-65; pres, Rensselaer Polytech Inst, 71-76; pres & chmn bd, Ranco Inc, 76-78, chmn & chief exec officer, 79-86; CONSULT, 86- *Concurrent Pos:* Dir indust develop, Purdue Res Park & vpres, McClure Park, 54-71; consult, Allison Div, Gen Motors Corp & Bell Labs, Sterling Drug Inc, AMF Inc, Transway Corp; Westinghouse & Graver res fels, Purdue Univ. *Honors & Awards:* Richards Mem Award, Am Soc Mech Engrs. *Mem:* Nat Acad Eng; Am Soc Eng Educ; Am Inst Aeronaut & Astronaut; fel Am Soc Mech Engrs. *Res:* Heat transfer; thermodynamics; fluid mechanics. *Mailing Add:* 18 Wexford Club Dr Hilton Head SC 29929

GROSHONG, RICHARD HUGHES, JR, b Lakewood, Ohio, Aug 10, 43; m 65; c 2. PETROLEUM GEOLOGY. *Educ:* Bucknell Univ, BS, 65; Univ Tex, Austin, MA, 67; Brown Univ, PhD(geol), 71. *Prof Exp:* Asst prof geol, Syracuse Univ, 70-73; res geologist, Cities Serv Co, 73-78, mgr struct geol res, 78-81, sr res assoc, explor & prod res, 81-83; assoc prof, 83-86, PROF GEOL, UNIV ALA, 86- *Concurrent Pos:* NSF grant, 72-73 & 84-86; adj prof geol, Tulsa Univ, 81-82; Oil & Gas Consult Int, 85-; ed, Tectonophys, 81-, geol, 81-90; vis hon chair, Univ Lausanne, 88; distinguished lectr, Am Asn Petrol Geol, 89-90. *Mem:* Fel Geol Soc Am; Am Geophys Union; Am Asn Petrol Geol. *Res:* Geometrical and Kinematic analysis of geological structures; rock deformation mechanisms; comparative structural geology of hydrocarbon accumulations. *Mailing Add:* Dept Geol Univ Ala Box 870338 Tuscaloosa AL 35487-0338

GROSKLAGS, JAMES HENRY, b Milwaukee, Wis, June 20, 29; m 57; c 2. MYCOLOGY. *Educ:* Univ Wis, BS, 51, MS, 55, PhD(bot), 60. *Prof Exp:* Asst prof, 58-68, ASSOC PROF BIOL, NORTHERN ILL UNIV, 68- *Mem:* AAAS; Mycol Soc Am. *Res:* Ecology of soil fungi. *Mailing Add:* Dept Biol Sci Northern Ill Univ De Kalb IL 60115

GROSKY, WILLIAM IRVIN, b Gulfport, Miss, Aug 4, 44; m 65; c 2. DATABASE MANAGEMENT, COMPUTER VISION. *Educ:* Mass Inst Technol, BS, 65; Brown Univ, MS, 68; Yale Univ, PhD(eng & appl sci), 71. *Prof Exp:* Asst prof computer sci, Ga Inst Technol, 71-76; assoc prof computer sci, 76-87, actg chmn, 88, PROF COMPUTER SCI, WAYNE STATE UNIV, 87- *Concurrent Pos:* Prin investr grant, IBM Corp, 87 & Ford Motor Co, 90-91; consult, USAF, 87-88, UN Indust Develop Org, 89-90, US Army, 89-; ed, Macmillan Encycl Computers, 89-; vis prof, Ford Motor Co Europe, 90. *Mem:* Asn Comput Mach; Am Asn Artificial Intel; Inst Elec & Electronics Engrs. *Res:* Database management systems, object-oriented database systems, scientific databases, heterogeneous databases, knowledge discovery in a database environment and the management of sensor- based data. *Mailing Add:* Computer Sci Dept Wayne State Univ Detroit MI 48202

GROSMAN, LOUIS HIRSCH, b Washington, DC, Nov 11, 39; m 65; c 2. COMPUTER SCIENCE, INFORMATION SYSTEMS. *Educ:* DC Teachers Col, BS, 65; Am Univ, cert mgt info systs, 72, cert sci & tech info systs, 73, MS, 73. *Prof Exp:* Mathematician, Goddard Space Flight Ctr, NASA, 61-67, NIH fel, 67-70; supvr digital comput programmer, Nat Libr Med, 70-73; sr systs analyst, 73-81, chief, Info Systs Support Sect, 81-82, sr automated systs security specialist, 82-89, CHIEF, CODES & STANDARDS, US AEC, US NUCLEAR REGULATORY COMN, 89- *Concurrent Pos:* Instr comput sci, Montgomery Col, Rockville, 74-77; first officer, Safeguards Div, Int Atomic Energy Agency, Vienna, Austria, 77-78. *Mem:* Asn Fed Comput Users. *Mailing Add:* 14809 Flintstone Lane Silver Springs MD 20905

GROSOF, MIRIAM SCHAPIRO, b New York, NY, Dec 02, 32; div; c 2. APPLIED STATISTICS, EDUCATION. *Educ:* Barnard Col, BA, 52; Columbia Univ, MA, 53; Yeshiva Univ, PhD(math), 66. *Prof Exp:* Res assoc, Yeshiva Univ, 66-68, from asst prof to assoc prof math educ, 68-83, PROF EDUC & STATIST, YESHIVA UNIV, 83-, FAC CHAIR, DIV SOCIAL & BEHAV SCI, 85- *Concurrent Pos:* Lectr, Columbia Univ, 53-58; chair, Dept Math Educ, Yeshiva Univ, 72-75; adj assoc prof, Pace Univ, 73-80, adj prof, 80- *Mem:* Math Asn Am; Am Statist Asn; Asn Women Math; Nat Coun Teachers Math. *Res:* Small college instruction in statistics; measures of olsymmetry in ratings. *Mailing Add:* 875 West End Ave Apt 6B New York NY 10025

GROSS, ALAN JOHN, b Bronx, NY, June 19, 34; m 62; c 2. MATHEMATICAL STATISTICS, BIOLOGICAL STATISTICS. *Educ:* Univ Calif, Los Angeles, BA, 56, MA, 57; Univ NC, PhD(statist), 62. *Prof Exp:* Lectr math, Univ Md, 61-62; lectr, Eindhoven Univ, 63-64; sr scientist, Booz-Allen Appl Res, Inc, 64-66; mathematician, Rand Corp, 66-69; res statistician, Sch Pub Health, Univ Calif, Los Angeles, 69-71; assoc prof pub health, Univ Mass, Amherst, 71-76; PROF BIOMED, MED UNIV SC, 76- *Concurrent Pos:* Lectr, Exten, Univ Calif, Los Angeles, 64-; consult statistician, Opportunity Systs Inc, Washington, DC, 75-76; USPHS training grant biostatist, 75-76; consult statistician, Uniworld Group, Bethesda, Md, 76-78 & Upjohn, Kalamazoo, Mich, 77-84. *Mem:* Am Statist Asn; Inst Math Statist. *Res:* Applications of statistical theory to reliability, biology and information theory problems. *Mailing Add:* Dept Biomet Med Univ SC 171 Ashley Ave Charleston SC 29425

GROSS, ARTHUR GERALD, b Amsterdam, NY, Aug 20, 35; m 78; c 4. APPLIED MATHEMATICS, COMPUTER SCIENCE. *Educ:* Rensselaer Polytech Inst, BEE, 56, MS, 59, PhD(appl math), 64. *Prof Exp:* comput-aided eng design, Comput Networking & Data Commun, 64-75, supvr, systs prog,

75-79, SUPVR, COMPUT NETWORK PLANNING & DEVELOP, BELL LABS, 79- *Mem:* Asn Comput Mach; AAAS; Sigma Xi. *Res:* Numerical analysis; application of digital computers to engineering problems; computer systems design and analysis. *Mailing Add:* 11 Darby Ct New Providence NJ 07974

GROSS, BENJAMIN HARRISON, b Chattanooga, Tenn, July 23, 30; m 64. ORGANIC CHEMISTRY. *Educ:* Univ Chattanooga, BS, 52; Univ Tenn, MS, 54, PhD(chem), 56. *Prof Exp:* Res chemist, Chattanooga Med Co, 56-64; head dept chem, 64-67, from assoc prof to prof, 67-77, GUERRY PROF CHEM & HEAD DEPT, UNIV TENN, CHATTANOOGA, 77- *Mem:* AAAS; Am Chem Soc; NY Acad Sci; Sigma Xi. *Res:* Synthetic organic chemistry; medicinal chemistry; aluminum alcoholates. *Mailing Add:* Dept of Chem Univ of Tenn Chattanooga TN 37401

GROSS, CHARLES GORDON, b New York, NY, Feb 29, 36; m; c 3. PHYSIOLOGICAL PSYCHOLOGY, NEUROPHYSIOLOGY. *Educ:* Harvard Univ, AB, 57; Cambridge Univ, PhD(psychol), 61. *Prof Exp:* Fel, Dept Psychol, Mass Inst Technol, 61-62, from lectr to asst prof, 62-65; asst prof, Harvard Univ, 65-68, lectr, 68-70; PROF PSYCHOL, PRINCETON UNIV, 70- *Concurrent Pos:* Vis lectr, Dept Psychol, Harvard Univ, 63-65; NIH fel psychol sci fel rev comt, 69-70 & exp psychol study sect, 70-74; vis scholar, Dept Physiol-Anat, Univ Calif, Berkeley, 70-72; vis prof, Dept Psychol, Mass Inst Technol, 75-76, Depts Biol & Psychol, Beijing Univ, 86; NAS exchange scientist, Shanghai Inst Physiol, 87,; Fulbright researcher/ lectr, Rio de Janeiro, 87; vis scientist, US Nat Prog, Shanghai Inst Physiol, 87, Tokyo Metrop Inst Neurosci, 88-89; vis fel, Magdalen Col, Oxford Univ, 90. *Mem:* Fel Am Psychol Asn; Soc Neurosci; fel AAAS. *Res:* Neurophysiology and neuropsychology of primate vision; psychology and physiology of perception. *Mailing Add:* 45 Woodside Lane Princeton NJ 08540

GROSS, DAVID (JONATHAN), b Washington, DC, Feb 19, 41; m 62; c 2. THEORETICAL HIGH ENERGY PHYSICS. *Educ:* Hebrew Univ Jerusalem, BSc, 62; Univ Calif, Berkeley, PhD(physics), 66. *Prof Exp:* Jr fel, Soc Fels, Harvard Univ, 66-69; vis scientist, Europ Orgn Nuclear Res, Switz, 69; from asst prof to assoc prof, 69-86, EUGENE HIGGINS PROF PHYSICS, PRINCETON UNIV, 86- *Concurrent Pos:* Sloan Found fel, 70-75. *Honors & Awards:* J Sakurai Prize, IAPS, 86; MacArthur Found Prize Fel, 87; Dirac Medal, Int Ctr Theoret Physics. *Mem:* Nat Acad Sci; Am Acad Arts & Sci; fel Am Phys Soc; fel AAAS. *Res:* High energy particle physics. *Mailing Add:* Dept Physics Jadwin Hall Princeton Univ Princeton NJ 08540

GROSS, DAVID JOHN, b Chester, Ill, June 14, 53; m 79. CELL MEMBRANES, BIOELECTRICITY. *Educ:* Univ Ill, Urbana-Champaign, BS, 75, MS, 77, PhD(physics), 82. *Prof Exp:* Fel biophys, Cornell Univ Sch Appl & Eng Physics, 82-85, res assoc biophys, 85-86; ASST PROF BIOPHYS, DEPT BIOCHEM, UNIV MASS, AMHERST, 86- *Mem:* Am Phys Soc; Am Soc Cell Biol; Biophys Soc; AAAS. *Res:* Cell surface receptor dynamics and cell signals are studied at the level of the individual cell by the quantitative optical microscopy; bioelectric effects in non-excitable cells are examined experimentally and theoretically. *Mailing Add:* Dept Biochem Univ Mass Amherst MA 01003

GROSS, DAVID LEE, b Springfield, Ill, Nov 20, 43; m 66; c 2. GEOLOGY, SCIENCE ADMINISTRATION. *Educ:* Knox Col, Ill, AB, 65; Univ Ill, MS, 67, PhD(geol), 69. *Prof Exp:* From asst geologist to assoc geologist, Ill State Geol Surv, 69-80, coordr environ geol, 79-84, geologist, Stratig & Areal Geol Sect, 80-84, geologist & head, Environ Studies & Assessment Sect, 84-89; EXEC DIR, GOV SCI ADV COMT, STATE ILL, 89- *Mem:* Fel Geol Soc Am; Am Inst Prof Geologists; Am Quaternary Asn; Int Asn Great Lakes Res; Soc Econ Paleont & Mineral; fel AAAS. *Res:* Environmental geology; limnogeology; glacial geology of the mid-continent. *Mailing Add:* Ill State Geol Surv 615 E Peabody Dr Champaign IL 61820

GROSS, DELMER FERD, b Huron, SDak, Aug 17, 46; m 69; c 1. PLANT BREEDING. *Educ:* SDak State Univ, BS, 68, PhD(agron), 74. *Prof Exp:* Instr voc agr, Redfield High Sch, SDak, 68-69; sr sorghum breeder, Trojan Brand, 74-77, WESTERN AREA RES DIR, PFIZER GENETICS, INC, 77- *Mem:* Sigma Xi; Am Soc Agron; Crop Sci Soc Am. *Res:* Development of new inbred lines and hybrids of corn through selection, breeding, testing and statistical interpretation of data. *Mailing Add:* DeKalb Pfizer Genetics Corn Res Ctr 3100 Sycamore Rd De Kalb IL 60115

GROSS, DENNIS CHARLES, b Whittemore, Iowa, Apr 2, 47. PLANT PATHOLOGY, BACTERIOLOGY. *Educ:* Iowa State Univ, BS, 70; Univ Calif, Davis, PhD(plant path), 76. *Prof Exp:* res assoc plant path, Univ Nebr, 76-79; asst prof, 79-84, ASSOC PROF PLANT PATH, WASH STATE UNIV, 84- *Honors & Awards:* Stark Award. *Mem:* Am Phytopath Soc; Am Soc Microbiol; Sigma Xi. *Res:* General biology of phytopathogenic bacteria; bacterial ice nucleation; phytotoxins and bacteriocins produced by bacteria of agronomic importance; biological control of plant pathogens. *Mailing Add:* Dept Plant Path Wash State Univ Pullman WA 99164-6430

GROSS, DENNIS MICHAEL, b Los Angeles, Calif, Sept 12, 47; m 73; c 2. CELL BIOLOGY, BIOCHEMISTRY. *Educ:* Calif State Univ, Northridge, BA, 69, MSc, 70; Univ Calif, Los Angeles, PhD(cell biol), 74. *Prof Exp:* Teaching asst cell biol, Univ Calif, Los Angeles, 71-72, teaching fel, 72-74; instr pharmacol, Med Sch, Tulane Univ, 76-77; sr res pharmacologist, Merck, Sharp & Dohme Res Labs, 77-81, res fel, 81-82, mgr, 82-87, DIR INTERN STRATEGIC & SCI PLANNING, MERCK, SHARP & DOHME RES LABS, 88- *Concurrent Pos:* Nat Heart, Lung & Blood Inst grant, Med Sch, Tulane Univ, 74-76; from adj asst prof pharmacol to adj assoc prof, Jefferson Med Col, Thomas Jefferson Univ, 77- *Mem:* Am Cell Biol Soc; NY Acad Sci; Soc Exp Biol & Med; Am Heart Asn; Soc Competitor Intel Prof; Sigma Xi. *Res:* B-blockers and hypertension, renin-angiotensin system; computer applications, computer simulation, QSAR. *Mailing Add:* 1621 Clearview Rd Blue Bell PA 19422

GROSS, DONALD, b Pittsburgh, Pa, Oct 20, 34; m 59; c 2. OPERATIONS RESEARCH. *Educ:* Carnegie Inst Technol, BS, 56; Cornell Univ, MS, 59, PhD(opers res), 62. *Prof Exp:* Opers res analyst, Atlantic Refining Co, 61-65; from asst prof to assoc prof eng & appl sci, George Washington Univ, 65-67, assoc prof opers res, 67-74, chmn dept, 77-88, PROF OPERS RES, GEORGE WASHINGTON UNIV, 74-, ACTG DEAN, SCH ENG & APPL SCI, 90- *Concurrent Pos:* Prog dir, NSF, 88-90. *Mem:* Opers Res Soc Am (pres, 89-90); Inst Mgt Sci; Am Inst Indust Engrs. *Res:* Queuing theory; inventory control theory; multi-echelon inventory design; queuing analysis of inventory models. *Mailing Add:* 3530 N Rockingham St Arlington VA 22213

GROSS, EDWARD EMANUEL, b New York, NY, July 11, 26; m 51; c 3. NUCLEAR PHYSICS, NUCLEAR REACTIONS. *Educ:* Queens Col, NY, BS, 48; Univ Calif, PhD(physics), 56. *Prof Exp:* Physicist, Radiation Lab, Univ Calif, 50-56; physicist, Oak Ridge Nat Lab, 56-70, dir Cyclotron Lab, 70-76, mgr nuclear res, 76-85, asst dir, Physics Div, 87-90; mgr, Heavy-Ion Physics, Dept Energy, 85-87; RETIRED. *Concurrent Pos:* Prog mgr Div Nuclear Physics, Dept Energy, 85-87. *Mem:* Am Phys Soc. *Res:* Investigation of the nuclear optical potential as revealed by the elastic scattering of pions, nucleons and heavy ions form complex nuclei; nuclear structure by inelastic scattering; experimental nuclear physics. *Mailing Add:* 119 Canterbury Rd Oak Ridge TN 37830

GROSS, ELIZABETH LOUISE, b 1940. PHOTOSYNTHESIS, MEMBRANE BIOCHEMISTRY. *Educ:* Univ Calif, Los Angeles, BA, 61, Berkeley, PhD(biophysics), 67. *Prof Exp:* Res assoc biochem, C F Kettening Res Lab, 67-68; asst prof, 68-73, assoc prof, 73-79, PROF BIOCHEM, OHIO STATE UNIV, 79- *Mem:* Biophys Soc; Am Soc Biol Chemists; Am Soc Plant Physiologists; Am Chem Soc; Int Solar Energy Soc; Sigma Xi. *Res:* Biophysical and biochemical studies of chloroplast membrane proteins including plastocyanin and the pigment-protein complexes; biological solar energy; chloroplast solar battery. *Mailing Add:* Dept Biochem 484 W 12th Ave Columbus OH 43210

GROSS, FLETCHER, b Colorado Springs, Colo, Nov 29, 39; m 64; c 3. MATHEMATICS. *Educ:* Calif Inst Technol, BS, 60; PhD(math), 64. *Prof Exp:* Asst prof math, Occidental Col, 63-66; asst prof, Univ Alta, 66-67; assoc prof, 67-77, PROF MATH, UNIV UTAH, 77- *Mem:* Am Math Soc; Math Asn Am. *Res:* Algebra, especially group theory. *Mailing Add:* Dept Math Univ Utah Salt Lake City UT 84112

GROSS, FRANZ LUCRETIUS, b Minneapolis, Minn, Aug 9, 37. THEORETICAL PHYSICS. *Educ:* Swarthmore Col, BA, 58; Princeton Univ, PhD(physics), 63. *Prof Exp:* Instr & res assoc physics, Cornell Univ, 63-65, actg asst prof, 65-66, asst prof, 66-69; vis assoc prof, Univ Calif, Santa Barbara, 69-70; assoc prof, Col William & Mary, 70-72; phys sci officer, US Arms Control & Disarmament Agency, 72-74; assoc prof, 74-76, PROF PHYSICS, COL WILLIAM & MARY, 76- *Concurrent Pos:* Vis prof, Carnegie-Mellon Univ, 81-82; assoc dir res, CEBAF, 84- 85, actg assoc dir, 85-86, sr staff scientist, 86- *Honors & Awards:* Sporn Award, 64. *Mem:* Am Phys Soc; Sigma Xi. *Res:* Electromagnetic structure; theoretical nuclear and partical physics; relativistic wave equations; deuteron and the nuclear force. *Mailing Add:* Dept Physics Col William & Mary Williamsburg VA 23185

GROSS, FRED, b Tokay, Hungary, Nov 11, 33; m 58; c 3. MATHEMATICS. *Educ:* Brooklyn Col, BS, 55; Columbia Univ, MA, 57; Univ Calif, Los Angeles, PhD(math), 62. *Prof Exp:* Lectr math, Brooklyn Col, 55-58; mathematician, NAm Aviation, Inc, Calif, 58-59; appl sci rep, IBM, 59-60, univ rep, IBM-Univ Calif, Los Angeles, 62-63; NSF assoc math, Nat Bur Standards, 63-64; res mathematician & consult, Naval Res Lab, 64-66; res mathematician, Bellcomm Inc, 66-68; PROF MATH, UNIV MD, BALTIMORE COUNTY, 68- *Concurrent Pos:* Consult, Naval Res Lab. *Mem:* Math Asn Am. *Res:* Theory of functions; entire and meromorphic functions; functional equations and number theory. *Mailing Add:* 1212 Noyes Dr Silver Springs MD 20910

GROSS, FRITZ A, b Ger, Oct 8, 10; m 37; c 3. ELECTRICAL ENGINEERING. *Educ:* Northeastern Univ, DE, 75. *Prof Exp:* Design engr, Samson Elec Co, 31-32 & S H Couch Co, 32-33; chief engr, Equip Eng Div & mgr, Radar & Commun Div, Waltham, 33-64, vpres & gen mgr, Equip Div, 64-68, vpres eng, Lexington, 68-75, TECH CONSULT, RAYTHEON CO, 75- *Concurrent Pos:* Naval tech adv, Int Conf Radio Aids Marine Navig, 46; tech consult, Raytheon Co, 75. *Mem:* Am Soc Naval Engrs; fel Inst Elec & Electronics Engrs. *Res:* Development and design of products involving electronics circuits; electronic test equipments; voltage and current stabilizers; radio frequency heating equipment; magnetic components; radio and audio frequency amplifiers; radar systems; sonar; electronic countermeasures. *Mailing Add:* 446 Dover Rd Westwood MA 02090

GROSS, GARRETT JOHN, b Britton, SDak, July 4, 42; m 67. PHARMACOLOGY. *Educ:* SDak State Univ, BS, 65, MS, 67; Univ Utah, PhD(pharmacol), 72. *Prof Exp:* Fel, Univ Wash, 72-73; from instr to assoc prof, 73-80, PROF PHARMACOL, MED COL WIS, 80-, ACTG CHMN PHARMACOL, 87- *Concurrent Pos:* Fac develop award, Pharmaceut Mfrs Asn, 76-78. *Mem:* AAAS; Am Heart Asn; Am Fedn Clin Res; Am Col Clin Pharmacol; Am Asn Animal Lab Sci. *Res:* Cardiovascular and autonomic pharmacology with general emphasis on drug actions on the coronary circulation. *Mailing Add:* Dept Pharmacol Med Col Wis 8701 Watertown Plank Rd Milwaukee WI 53226

GROSS, GEORGE ALVIN, b Detroit, Mich, Feb 5, 33; m 55; c 2. OSTEOPATHY. *Educ:* Chicago Col Osteop Med, DO, 61; Univ Mich, MPH, 73. *Prof Exp:* Gen pract, Detroit, Mich, 62-72; med dir, Dexter Clin, Detroit, 68-72; med dir, Barry, Eaton, Ionia Assoc Health Dept, 75; dir, Mich State Univ Col Clin, 75; assoc prof osteop, 73-77, PROF COMMUNITY HEALTH SCI, MICH STATE UNIV, 77-; DIR ALCOHOLISM UNIT, ST LAWRENCE HOSP, 73- *Mem:* Asn Teachers Prev Med; Am Pub Health Asn; Am Osteop Asn; Am Med Soc Alcoholism. *Mailing Add:* Dept Community Med Mich State Univ Col Osteop Med East Fee Hall East Lansing MI 48824

GROSS, GEORGE C(ONRAD), b Ginter Park, Va, Aug 14, 14; m 46; c 3. CHEMICAL ENGINEERING. *Educ:* Va Polytech Inst, BS, 36; Purdue Univ, MS, 38, PhD(chem eng), 41. *Prof Exp:* Lab instr chem Purdue Univ, 36-41; chem engr, J T Baker Chem Co, 46-48; CHEM ENGR, E I DU PONT DE NEMOURS & CO, INC, 48- *Mem:* AAAS; Am Chem Soc; Am Inst Chem Engrs. *Res:* Chemical engineering plant design and pilot plant work with respect to organic chemicals; preparation of sulfamide; manufacture of synthetic fibers. *Mailing Add:* 1231 Keesling Ave Waynesboro VA 22980-5217

GROSS, GERARDO WOLFGANG, b Greifswald, Ger, Sept 1, 23; m 59; c 3. GEOPHYSICS. *Educ:* Cordoba Univ, DSc, 48; Pa State Univ, PhD(geophys), 59. *Prof Exp:* Geologist, Nat Fuels Admin, 48-51; geologist, Soc Miniere Penarroya, 51-54; res fel geophys, Pa State Univ, 56-59; geophysicist, Newmont Explor Co, 59; instr sci, Dutchess Community Col, 60; from asst prof to assoc prof geophys, 60-74, PROF GEOPHYS, NMEX INST MINING & TECHNOL, 74-, GEOPHYSICIST, 60- *Mem:* AAAS; Soc Cryobiol; Am Geophys Union; Sigma Xi. *Res:* Geoelectricity; electrochemical properties of ice; groundwater hydrology. *Mailing Add:* Dept Geosci NMex Inst Mining & Technol Campus Sta Socorro NM 87801

GROSS, GUILFORD C, b Bowdle, SDak, June 8, 17; m 40; c 2. PHARMACOLOGY. *Educ:* SDak State Univ, BS, 39, MS, 40; Univ Fla, PhD(pharmacol), 52. *Prof Exp:* From instr to assoc prof pharmacol, SDak State Univ, 40-43 & 46-52, prof, 52-80, dean, 64-65; RETIRED. *Concurrent Pos:* Mem revision comt, US Pharmacopoeia, 55-60. *Mem:* Am Pharmaceut Asn. *Res:* Pharmacy. *Mailing Add:* 808 Christine Ave No 102 Brookings SD 57006-3983

GROSS, HARRY DOUGLASS, b Halifax, Pa, Mar 4, 24; m 54; c 4. AGRONOMY. *Educ:* Rutgers Univ, BS, 48, MS, 52; Iowa State Univ, PhD(agron), 56. *Prof Exp:* Teacher, Newport Union Sch Dist, Pa, 48-49; instr farm crops, Rutgers Univ, 49-52; instr agron, Iowa State Univ, 52-56; res asst prof crop sci, NC State Univ, 56-58, proj leader agr mission to Peru, 58-61, res assoc prof, 61-68, res prof crop sci, 68-86; ASST DIR, INT PROG, CALS, 86- *Concurrent Pos:* Prog coordr, Mission to Peru, 67-70. *Mem:* Am Soc Agron. *Res:* Forage crop management, particularly fertilization and cutting practices; forage crop evaluation in vitro and in vivo; sown-pasture and range grazing management; soybean and peanut physiology, particularly as related to light X temperature interactions, root growth, and symbiotic nitrogen fixation. *Mailing Add:* Box 7620 NC State Univ Raleigh NC 27695

GROSS, HERBERT MICHAEL, b Milwaukee, Wis, Mar 31, 25; m 51. PHARMACEUTICAL CHEMISTRY. *Educ:* Univ Wis, BS, 49; Univ Fla, MS, 51, PhD(pharm), 53. *Prof Exp:* Instr pharm, Univ Fla, 49-53; prod supvr, Lincoln Labs, 53; head dept pharmaceut prod develop, Com Solvents Corp, 53-56; res pharmacist, Abbott Labs, 56-57, head, Dept Pharmaceut Res, 57-61, dir, New Prod Planning & Develop, 61-68, vpres, Hosp Prod Div, 68-71, pres, 69-71, vpres, Int Sci Affairs, 71-81; CONSULT, 81- *Mem:* Am Pharmaceut Asn; Pharmaceut Mfrs Asn. *Res:* Tablets; suppositories; lyophilization; antibiotics; vitamins; flavoring; tablet coatings; research management. *Mailing Add:* N85 W15700 Ridge Rd Menomonee Falls WI 53051

GROSS, IAN, b Pretoria, SAfrica, Oct 15, 43; US citizen; m 67; c 2. DEVELOPMENTAL BIOCHEMISTRY. *Educ:* Univ Witwatersrand, BSc, 63, MB, 67; Yale, MA, 85. *Prof Exp:* Resident, Univ Witwatersrand Hosps, 68-71; resident pediat, Children's Hosp Med Ctr, Boston, 71-72; fel, Harvard Med Sch, 72-73; fel perinatal med, 73-74, from asst prof to assoc prof, 78-85, PROF PEDIAT, OBSTET & GYNEC, SCH MED, YALE UNIV, 85- *Concurrent Pos:* J H Brown Mem fel, Yale Univ, 73-74; mem, Human Embory & Develop Study Sect, NIH, 81-85. *Mem:* Soc Pediat Res; Am Physiol Soc; Am Thoracic Soc; Am Acad Pediat; Perinatal Res Soc. *Res:* Hormonal regulation of the biochemical maturation of fetal lung. *Mailing Add:* Dept Pediat 333 Cedar St New Haven CT 06510

GROSS, JAMES HARRISON, b Three Point, Ky, Apr 3, 42; m 65; c 2. MEDICAL EDUCATION, ANATOMY. *Educ:* Earlham Col, BA, 65; Univ Rochester, PhD(anat), 71. *Prof Exp:* Asst prof anat, Sch Med, Johns Hopkins Univ, 72-80; ASST PROF ANAT & MED EDUC, UNIFORMED SERV UNIV HEALTH SCI, 80- *Mem:* AAAS. *Res:* Neuroendocrinology; transmission and scanning electron microscopy. *Mailing Add:* Uniformed Serv Univ Health Sci 4301 Jones Bridge Rd Bethesda MD 20814

GROSS, JAMES RICHARD, b Rapid City, SDak, Feb 19, 46; m 67; c 2. ORGANIC CHEMISTRY. *Educ:* SDak Sch Mines & Technol, BS, 68; Univ Ariz, MS, 70. *Prof Exp:* Sr res chemist, Dow Chem Co, 70-80, proj leader, New Prod Res, 80-82, res leader, 82-85; RES FEL, KIMBERLY-CLARK, 85- *Mem:* Am Chem Soc. *Res:* Synthesis and applications of polyelectrolytes, particularly cross-linked synthetic polyelectrolytes useful in fluid absorption and retention; synthesis and characterization of new engineering thermoplastics; process design for aromatic polycarbonate. *Mailing Add:* 1419 S Fidelis Appleton WI 54915

GROSS, JEROME, b New York, NY, Feb 25, 17; m 47; c 3. DEVELOPMENTAL BIOLOGY, CELL-MATRIX RELATIONSHIP. *Educ:* Mass Inst Technol, BS, 39; NY Univ, MD, 43. *Prof Exp:* Med intern, Long Island Col Hosp, NY, 43-44; capt, Med Corps, US Army, 44-46; res assoc, Dept Biol, Mass Inst Technol, 46-55; res assoc med, Harvard Med Sch, 50-54, assoc, 54-57, from asst prof to prof, 57-87, EMER PROF MED, HARVARD MED SCH, 87-; ACTG ASSOC DIR RES, CUTANEOUS RES CTR, DEPT DERMAT, MASS GEN HOSP, 89- *Concurrent Pos:* Fel, Life Ins Med Res Fund, 46-48; clin & res fel med, Mass Gen Hosp, Boston, 48-51, assoc biologist, 51-66, biologist Med Serv, 66-, chmn comt res, 64-67 & 79-82; mem, Comt Skeletal Syst, Nat Res Coun, 53-62; estab investr, Am Heart Asn, 56-61; mem sci adv comt, Helen Hay Whitney Found, 56-91, bd trustees, 85-; mem, Adv Panel Molecular Biol, NSF, 59-62; mem, Bd Sci Counrs, Nat Inst Dent Res, 62-66, chmn, 63-66; NIH lectr, 65, mem, Molecular Biol Study Sect, 66-70, Breast Cancer Task Force, 75-78; mem bd dirs, Med Foun Boston, 74-80; Fogarty sr int fel, NIH, 76; mem, Boston Biomed Res Inst, 79-82; Wellcome vis prof, Basic Med Sci, Fed Am Soc Exp Biol, Robert Wood Johnson Med Sch, 82; mem sci affairs comt, W Alton Jones Sci Ctr, 83-, mem bd dirs, 86-; bd trustee, Helan Haywhitney Found, 84. *Honors & Awards:* Ludwig Kast Lectr, NY Acad Med, 52; Ciba Found Award for Res Relevant to the Problems of Aging, 59; Spec Award, Soc Cosmetic Chemists, 63; Kappa Delta Award, Am Acad Orthop Surg, 65; Herman Beerman Lectr, Soc Invest Dermat, 66; Harvey Soc Lectr, 73; First Paul Klemperer Award, NY Acad Med, 88. *Mem:* Nat Acad Sci; Inst Med-Nat Acad Sci; Am Soc Cell Biol; Am Soc Biol Chem; Int Soc Develop Biol; Am Acad Arts & Sci; Soc Develop Biol; Am Soc Physiologists. *Res:* Developmental and cellular biology; biomedical science; connective tissues and their diseases; aging processes. *Mailing Add:* Cutaneous Biol Res Ctr Mass Gen Hosp E Bldg 149 at 13th St Charlestown MA 02129

GROSS, JOHN BURGESS, b St Louis, Mo, Dec 26, 20; m 45; c 4. MEDICINE. *Educ:* DePauw Univ, AB, 42; Western Reserve Univ, MD, 45; Univ Minn, MS, 49. *Prof Exp:* From assoc prof to prof, Mayo Grad Sch Med, 63-74, PROF MED, MAYO MED SCH, UNIV MINN, 74-, CONSULT, MAYO CLIN & HOSPS, 50- *Mem:* Am Col Physicians; Am Gastroenterol Asn; Am Soc Human Genetics; Cent Soc Clin Res; affil mem Royal Soc Med; Sigma Xi. *Res:* Diseases of the pancreas, bowel and liver, especially the hereditary form of pancreatitis. *Mailing Add:* Mayo Found Rochester MN 55901

GROSS, JONATHAN LIGHT, b Philadelphia, Pa, June 11, 41; m; m 76; c 5. GRAPH THEORY & ALGORITHMS, MATHEMATICAL MODELING. *Educ:* Mass Inst Technol, BS, 64; Dartmouth Col, AM, 66, PhD(math), 68. *Prof Exp:* Instr math, Princeton Univ, 68-69; from asst prof to assoc prof, 69-78, PROF MATH STATIST, COLUMBIA UNIV, 78-, PROF COMPUT SCI & MATH, 79- *Concurrent Pos:* Sloan Found res fel, 72-75, IBM fel, 72-73; consult, IBM, Bell Labs, Russell Sage Found, Alfred P Sloan Found, Oak Ridge Nat Lab & Inst Defense Anal; prin investr, Off Naval Res, NSF, Exxon Found, ARCO Found, Mellon Found. *Mem:* Am Math Soc; Soc Indust & Appl Math; Asn Comput Mach. *Res:* Combinatorics; computer science; social sciences; topology. *Mailing Add:* Dept Math Columbia Univ 450 Computer Sci Bldg New York NY 10027

GROSS, JOSEPH F, b Plauen, Ger, Aug 22, 32; US citizen. BIOMEDICAL ENGINEERING, CHEMICAL ENGINEERING. *Educ:* Pratt Inst, BChE, 53; Purdue Univ, PhD(chem eng), 56. *Prof Exp:* Consult, Rand Corp, 56-57, engr, 57-70, res engr, 70-72; head dept, 75-81, PROF CHEM ENG, UNIV ARIZ, 72- *Concurrent Pos:* Fulbright scholar, Thermodyn Inst, Munich Tech & Inst Fluid Dynamics, Brunswick Tech Univ, 56-57; consult, Sch Pharmaceut Sci, 72- & Rand Corp, 72- *Honors & Awards:* Humboldt Prize. *Mem:* Am Inst Aeronaut & Astronaut; Am Inst Chem Engrs; Am Soc Mech Engrs; Microcirculatory Soc (pres, 76); fel AAAS; Int Inst Microcirculation (secy treas, 83-). *Res:* Fluid mechanics and mass transfer in the microcirculation; pharmacokinetics of anticancer drugs; boundary layer theory; cell kinetics modelling; tumor microcirculation. *Mailing Add:* Dept Chem Eng Univ Ariz Tucson AZ 85721

GROSS, KATHERINE LYNN, b Madison, Wis, Oct 28, 53; m 75; c 2. PLANT POPULATION & COMMUNITY ECOLOGY. *Educ:* Iowa State Univ, BS, 75; Mich State Univ, PhD(zool & ecol), 80. *Prof Exp:* Res asst ecol, Kellogg Biol Sta, Mich State Univ, 77-80; from asst prof to assoc prof bot & ecol, Ohio State Univ, 80-87; ASSOC PROF, KELLOGG BIOL STA, MICH STATE UNIV, 87- *Concurrent Pos:* Vis scientist, Kellog Biol Sta, 81- *Mem:* Ecol Soc Am; AAAS; Bot Soc Am; Int Soc Plant Pop Biologists; Sigma Xi; Brit Ecol Soc. *Res:* Influence of life history traits on distribution and abundance of co-occuring plant species; root morphology; population biology and community ecology; resource acquisition; competitive ability. *Mailing Add:* Kellogg Biol Sta Mich State Univ 3700 E Gull Lake Dr Hickory Corners MI 49060

GROSS, KENNETH CHARLES, b South Kingston, RI, July 20, 54; m 78. CELL WALL METABOLISM, POSTHARVEST PHYSIOLOGY. *Educ:* Pa State Univ, BS, 76, MS, 78; NC State Univ, PhD(hort), 81. *Prof Exp:* PLANT PHYSIOLOGIST, HORT CROPS QUAL LAB, AGR RES SERV, USDA, 81- *Concurrent Pos:* Res assoc plant physiol, dept crop sci, NC State Univ, 81. *Honors & Awards:* Krezdon Award, Am Soc Hort Sci, 82. *Mem:* Am Soc Plant Physiologists; Am Soc Hort Sci. *Res:* Cell wall and carbohydrate metabolism in relation to development, ripening and senescence of fruits and vegetables. *Mailing Add:* Hort Crops Qual Lab Agr Res Serv USDA Rm 114 Bldg 002 Beltsville MD 20705

GROSS, KENNETH IRWIN, b Malden, Mass, Oct 14, 38; m 64; c 1. MATHEMATICAL ANALYSIS. *Educ:* Brandeis Univ, BA, 60, MA, 62; Wash Univ, PhD(math), 66. *Prof Exp:* Asst prof math, Tulane Univ, 66-68 & Dartmouth Col, 68-73; from assoc prof to prof math, Univ NC, Chapel Hill, 73-81; head, dept math, Univ Wyo, 81-85; prog dir, NSF, 85-87; CHMN, DEPT MATH & STATISTS UNIV, VT, 89- *Concurrent Pos:* Vis lectr, Univ Calif, Irvine, 72-73; vis assoc prof, Univ Utah, 77; vis prof, Academia Sinica, Taiwan, 79, Drexel Univ, 88. *Honors & Awards:* Lester R Ford Prize, Math Asn Am, 79; Chauvenet Prize, Math Asn Am, 80. *Mem:* Math Asn Am; Am Math Soc; Nat Coun Teachers Math. *Res:* Harmonic analysis; infinite-dimensional representation theory. *Mailing Add:* Dept Math & Statist Univ VT Burlington VT 05405

GROSS, LEO, b Brooklyn, NY, Feb 13, 15; m 40; c 3. BIOPHYSICS. *Educ:* Brooklyn Col, BS, 34; Columbia Univ, MA, 36; NY Univ, PhD(biophys), 63. *Prof Exp:* Mem tech staff, Bell Tel Labs, 46-49; chief systs engr, Polarad Electronics Corp, 49-54; pres, Hub Electronics Corp, 54-58; biophysicist, Waldemar Med Res Found, 58-89; CONSULT, 89- *Concurrent Pos:* Physicist, Bur Ord, US Dept Navy, 42-43 & Los Alamos Sci Lab, Univ Calif, 43-46; Nat Acad Sci foreign exchange fel, 73. *Mem:* Am Phys Soc; Inst Elec & Electronics Engrs; Optical Soc Am. *Res:* Biophysics of cell function and change in function in neoplasia. *Mailing Add:* 220 E 67th St New York NY 10021

GROSS, LEONARD, b Brooklyn, NY, Oct 13, 41; m 63; c 4. ORGANIC CHEMISTRY, POLYMER SCIENCE. *Educ:* Pratt Inst, BS, 63; Brandeis Univ, MA, 65; Yeshiva Univ, PhD(org chem), 70. *Prof Exp:* Chemist org synthesis, Polaroid Corp, 64-65; chemist, 68-70, sr chemist polymer synthesis, 70-73, group leader polyvinyl chloride resins, 74-77, lab mgr flexible polymers, 78-79, mgr bus develop, 79-80, PROD MGR, POLYVINYL CHLORIDE HOMOPOLYMER, TENNECO CHEM INC, 80- *Mem:* Am Chem Soc; Am Soc Testing & Mat; Soc Plastics Engrs. *Res:* Applications research, polyvinyl chloride resins and compounds. *Mailing Add:* 41 Eagle Rd Marlboro NJ 07746-1810

GROSS, LEONARD, b Brooklyn, NY, Feb 24, 31; m 56; c 2. MATHEMATICS. *Educ:* Univ Chicago, MS, 54, PhD(math), 58. *Prof Exp:* Instr math, Yale Univ, 57-59, NSF fel, 59-60; from asst prof to assoc prof, 60-67, PROF MATH, CORNELL UNIV, 67- *Concurrent Pos:* Guggenheim Found fel, 74. *Mem:* Am Math Soc. *Res:* Classical analysis on Hilbert space; mathematical problems of quantum theories. *Mailing Add:* White Hall Cornell Univ Ithaca NY 14853-7901

GROSS, LUDWIK, b Krakow, Poland, Sept 11, 04; nat US; m 43; c 1. CANCER. *Educ:* Iagellon Univ, Poland, MD, 29; Am Bd Internal Med, dipl. *Hon Degrees:* DSc, Mt Sinai Sch Med, City Univ New York, 83. *Prof Exp:* Cancer res, Pasteur Inst, 32-39; res assoc, Inst Med Res, Christ Hosp, Cincinnati, 41-43; res prof, 71-73, EMER RES PROF MED, MT SINAI SCH MED, 73-; CHIEF CANCER RES, VET ADMIN MED CTR, BRONX, 46- *Concurrent Pos:* Consult, Sloan-Kettering Inst Cancer Res, 53-56, assoc scientist, 57-60; mem, Bd Dir, Am Asn Cancer Res, 73-76; distinguished physician, Vet Admin Med Ctr, Bronx, 77-81. *Honors & Awards:* Prix Chevillon, Paris Acad Med, 37; R R de Villers Found Award, Leukemia Soc, 53; Walker Prize, Royal Col Surgeons Eng, 61; James Award, James Ewing Soc, 62; UN Prize, WHO, 62; Albert Einstein Centennial Medal, 65; Spec Virus Cancer Prog Award, Nat Cancer Inst, 72; William S Middleton Award, Vet Admin, 74; Albert Lasker Basic Med Res Award, 74; Prin Paul Ehrlich-Ludwig Darmstaedter Prize, Frankfurt, 78;; Prix Griffuel, Paris, 78; Alfred Jurzykowski Found Award, NY, 85. *Mem:* Nat Acad Sci; AAAS; Am Soc Hemat; Soc Exp Biol & Med; Am Asn Cancer Res. *Res:* Experimental cancer and leukemia. *Mailing Add:* Vet Admin Med Ctr 130 W Kingsbridge Rd Bronx NY 10468

GROSS, M GRANT, JR, b Childress, Tex, Jan 5, 33; m 54; c 3. OCEANOGRAPHY. *Educ:* Princeton Univ, AB, 54; Calif Inst Technol, MS, 58, PhD(geochem), 61. *Prof Exp:* From asst prof to assoc prof oceanog, Univ Wash, 61-68; from assoc prof to prof, State Univ NY, Stony Brook, 68-72; dir, Oceanog Serv Off, Nat Oceanic & Atmospheric Admin, 72-73; head, Oceanog Sect, NSF, 73-74; dir, Chesapeake Bay Inst, Johns Hopkins Univ, 74-78; head, Int Decade Ocean Explor, 78-80, DIR DIV OCEAN SCI, NSF, 80- *Concurrent Pos:* Assoc cur sedimentol, Smithsonian Inst, 66-68. *Mem:* Am Soc Limnol & Oceanog; fel Geol Soc Am; Geochem Soc Am; Am Geophys Union; Soc Econ Paleont & Mineral. *Res:* Marine geochemistry and sedimentary processes in estuarine and coastal ocean waters, especially waste deposits from urban regions. *Mailing Add:* Div Ocean Scis NSF Washington DC 20550

GROSS, MALCOLM EDMUND, b Brownington, Vt, June 23, 15; m 37; c 2. CHEMISTRY. *Educ:* Middlebury Col, BS, 36; Western Reserve Univ, MS, 46, PhD, 50. *Prof Exp:* Res chemist, 48, sr res chemist, Res Ctr, 48-66, sect leader, 66-73, sect mgr, 73-76, MGR, RES & DEVELOP CTR, B F GOODRICH CO, 76- *Concurrent Pos:* Staff scientist, Lockheed Missiles & Space Co, Calif, 63. *Mem:* Am Chem Soc; Soc Advan Mat & Process Eng. *Res:* Structural adhesives for aerospace applications. *Mailing Add:* 5651 RS Hammock Rd C 105 Naples FL 33962

GROSS, MARK WARREN, b Euclid, Ohio, Oct 3, 56; m 86; c 2. LATTICE GAUGE THEORY, QUANTUM GRAVITY. *Educ:* Wash Univ, BS, 77; Univ Chicago, MS, 79, PhD(physics), 83. *Prof Exp:* Postdoctoral, Oxford Univ, 83-86 & Cornell Univ, 86-88; ASST PROF, DEPT PHYSICS & ASTRON, CALIF STATE UNIV, 88- *Mem:* Am Phys Soc. *Res:* Lattice gauge theory; quantum gravity. *Mailing Add:* Dept Physics & Astron Calif State Univ Long Beach CA 90840

GROSS, MARTIN, b New York, NY, Oct 18, 42. MEDICINE. *Educ:* Univ Chicago, BS, 64, MD, 69, PhD(biochem), 69. *Prof Exp:* Intern path, Univ Chicago, 69-70; res assoc, NIH, Nat Cancer Inst, 70-72; resident, 72-75, from asst prof to assoc prof, 75-86, PROF PATH UNIV CHICAGO, 86- *Mem:* AAAS; Sigma Xi; Am Soc Biochem & Molecular Biol; Am Asn Pathologists. *Res:* Regulation of protein synthesis in rabbit reticulocyto cell-free preparations. *Mailing Add:* 5841 S Maryland BH Box 299 Chicago IL 60637

GROSS, MICHAEL LAWRENCE, b St Cloud, Minn, Nov 6, 40; c 2. ANALYTICAL CHEMISTRY, ORGANIC CHEMISTRY. *Educ:* St John's Univ, Minn, BA, 62; Univ Minn, PhD(org chem), 66. *Prof Exp:* Fel org chem, Univ Pa, 66-67 & Purdue Univ, 67-68; from asst prof to prof, 68-88, C PETRUS PETERSON PROF CHEM, UNIV NEBR, LINCOLN, 88- *Concurrent Pos:* Dir, Midwest Ctr Mass Spectrometry, 78-; ed, Mass Spectrometry Revs; 3-M alumni prof chem; mem, NIH Study Sect Metallobiochem, Nat Res Coun Bd Chem Sci & Technol. *Mem:* Am Chem Soc; Am Soc Mass Spectrometry; Sigma Xi. *Res:* Mass spectrometry; structures and properties of gas-phase ions; ion-molecule reactions; fourier transform mass spectrometry; trace analysis of dioxins and related compounds; methods for structural determination of biomolecules; commonwealth of mass. *Mailing Add:* Dept Chem Univ Nebr Lincoln NE 68588-0304

GROSS, MICHAEL R, b Cuise-la-Motte, France, Dec 18, 34; US citizen; c 3. COMPUTER SCIENCES. *Educ:* Carnegie-Mellon Univ, BS, 57; Univ Pittsburgh, PhD(comput sci), 75. *Prof Exp:* Design engr mech eng, Etz-Hazait Co, Israel, 68; teaching fel comput sci, Univ Pittsburgh, 70-74; asst prof comput sci, Fla Int Univ, 74-78; assoc prof, 78-82, PROF, MATH & COMPUT SCI, CALIF UNIV PA, 82- *Mem:* Asn Comput Mach. *Res:* Knottedness in macromolecules related to entaglements forming molecular networks used to predict the mechanical properties of polymeric materials, through the development of an algorithmic treatment of topological structures; applications in logic programming and expert systems. *Mailing Add:* Dept Math & Comput Sci Calif Univ Pa California PA 15419

GROSS, MILDRED LUCILE, b Lancaster Co, Nebr, Nov 16, 20; m 42; c 3. MATHEMATICS. *Educ:* Nebr State Teachers Col, Kearney, BS, 42; Univ Nebr, MS, 59, PhD(math), 63. *Prof Exp:* From assoc prof to prof, 61-86, EMER PROF MATH, DOANE COL, 86- *Res:* Number theory; use of sieve methods in investigation of prime numbers. *Mailing Add:* 945 Longwood Dr Crete NE 68333

GROSS, MIRCEA ADRIAN, b Bucharest, Rumania, Aug 8, 23; US citizen; m 52; c 3. PATHOLOGY. *Educ:* Univ Man, BSc, 49; Univ Toronto, DVM, 54; Ont Vet Col, Guelph, Vet Surg, 54; Ohio State Univ, MSc, 56. *Prof Exp:* Res asst vet path, Ohio State Univ, 54-56; res pathologist, Agr Res Serv, Mich, 56-58, USPHS, Ohio, 59-60, Arthur D Little Co, Mass, 60-62 & Microbiol Assocs, Md, 62-64; actg chief path br, US Food & Drug Admin, 64-69, asst dir sci coord, Off Pharmaceut Res & Testing, Bur Drugs, 69-79; chief toxicol, 79-80, SR SCI ADV, ENVIRON PROTECTION AGENCY, 80- *Concurrent Pos:* Vis scientist, Biomet Br, Nat Cancer Inst, 59-60. *Honors & Awards:* Award, Am Vet Med Asn, 54. *Mem:* Am Vet Med Asn; Am Asn Cancer Res. *Res:* Pathology of cancer, virus and chemically induced; general experimental pathology; biometry and experimental statistics. *Mailing Add:* 2947 Birchtree Lane Silver Spring MD 20906

GROSS, PAUL, b Berlin, Ger, June 8, 02; nat US; m 30; c 4. PATHOLOGY. *Educ:* Western Reserve Univ, AB, 24, MD, 27, MA, 29. *Prof Exp:* Pathologist, St Vincent Charity Hosp, 31-35; pathologist, Western Pa Hosp, Pittsburgh, 35-44, Indust Hyg Found, Mellon Inst, 48-54, Indust Health Found, 54-85; from adj prof to res prof, Grad Sch Pub Health, Univ Pittsburgh, 60-71; distinguished res prof, 71-75, ADJ PROF PATH, MED UNIV SC, 75- *Concurrent Pos:* Sr fel, Indust Hyg Found, Mellon Inst, 53-68; pathologist, St Vincent's Charity Hosp, Ohio, 32-35; vol asst, Univ Vienna, 32-33; pathologist, Western Pa Hosp, 35-44, St Joseph's Hosp, 44-54; dir res lab, Indust Health Found, Pittsburgh, 54-85; consult threshold limit comt, Am Conf Govt Hygienists, 68-81. *Honors & Awards:* Adolf G Kammer Merit Authorship Award, Indust Med Asn, 67. *Mem:* AAAS; Am Col Physicians; Am Soc Clin Path; Am Col Path; Am Asn Path & Bact; AMA; Int Acad Path. *Res:* Pneumoconioses. *Mailing Add:* 28 Maui Circle Naples FL 33962

GROSS, PAUL HANS, b Berlin, Ger, Apr 17, 31; m 57; c 4. ORGANIC CHEMISTRY, PHYSICAL CHEMISTRY. *Educ:* Freie Univ Berlin, dipl, chem, 58, Dr Sci(chem), 61. *Prof Exp:* Res chemist, Schering AG, Ger, 61-62; fel amino-sugar chem, Univ of the Pac, 62-64; res fel carbohydrate chem, Mass Gen Hosp & Harvard Med Sch, 65-66; assoc prof chem, 66-70, PROF CHEM, UNIV PAC, 70- *Concurrent Pos:* grants, NSF, 68 & NIH, 87; vis prof, Frei Univ Berlin, Ger, 78, Univ Baja Calif & Mex State Univ, 80, Univ Sevilla, Spain, 88; res chemist, Hanover Med Sch, Ger, 83 & 85. *Mem:* Am Chem Soc; Ger Chem Soc; Sigma Xi. *Res:* Chemistry and properties of ethers of polyols; rheology properties of liquids; chemistry of aminosugars; chemistry of peptides. *Mailing Add:* Dept Chem Univ Pac Stockton CA 95211

GROSS, PAUL MAGNUS, JR, b Durham, NC, Jan 15, 20. PHYSICAL CHEMISTRY. *Educ:* Duke Univ, BS, 41; Brown Univ, PhD(chem), 48. *Prof Exp:* Asst dept phys chem, Harvard Med Sch, 42-46; instr chem, Univ Va, 48-51; asst prof gen & phys chem, 51-59; from assoc prof to prof phys chem & coordr honors prog, Wake Forest Univ, 59-87; RETIRED. *Concurrent Pos:* NSF fac fel, Cambridge Univ, 57-58. *Mem:* Am Chem Soc. *Res:* Physical chemistry of hydrogen peroxide-water solutions; dielectric constant measurements of hydrogen peroxide and protein solutions; fractionation and properties of proteins; solubility of strong electrolytes in hydrogen peroxide; vapor pressures of hydrogen peroxide water solutions; formation and stability of hydroperoxidates. *Mailing Add:* PO Box 21 Atlantic Beach NC 28512

GROSS, PAUL RANDOLPH, b Philadelphia, Pa, Nov 27, 28; m 49; c 1. DEVELOPMENTAL BIOLOGY, MOLECULAR BIOLOGY. *Educ:* Univ Pa, AB, 50, PhD(zool), 54; Brown Univ, MA, 63. *Hon Degrees:* MA, Brown Univ; DSC, Med Col Ohio. *Prof Exp:* From asst prof to assoc prof biol, NY Univ, 54-61; assoc prof, Brown Univ, 62-65; prof biol, Mass Inst Technol, 65-71; chmn dept, Univ Rochester, 72-76, prof biol, 72-78, dean grad studies, 74-78; pres & dir, Marine Biol Lab, Woods Hole, Mass, 78-86; prof biol, 86-89, UNIV PROF LIFE SCI, UNIV VA, 89-, DIR UNIV CTR ADVAN STUDIES, 89-, MOLECULAR BIOL INST, 90-, MARKEY CTR, 90- *Concurrent Pos:* Lalor fel, 54-55, NSF sr fel, Univ Edinburgh, 61-62; mem bd trustees, Assoc Univs, Inc, 74-83; chmn cell biol study sect, Div Res Grants, NIH; mem doctoral coun, NY State Dept Educ, 74-; mem corp, Marine Biol Lab, Woods Hole, sci adv comt, Sch Vet Med, Tufts Univ, bd trustees, Univ Rochester, adv comt, Ctr Advan Studies, Univ Virginia, sci adv comn, Doreen Grace Bezeier Ctr, Mashpee, Mass, bd trustees, Sea Educ Asn. *Mem:* Am Physiol Soc; Am Soc Zoologists; Am Soc Cell Biol; Int Soc Cell Biol; AAAS. *Res:* Chemical embryology; molecular biology; cellular physiology; control of gene expression in animal development. *Mailing Add:* Ctr Advan Study Univ Va 444 Kettle Hall Charlottesville VA 22903

GROSS, PETER A, b Newark, NJ, Nov 18, 38; m 64; c 3. INFECTIOUS DISEASES, VIROLOGY. *Educ:* Amherst Col, BA, 60; Yale Univ, MD, 64. *Prof Exp:* DIR, DEPT MED, HACKENSACK MED CTR, 80-; PROF MED, NJ MED SCH, 80- *Mem:* Fel Am Col Physicians; fel Infectious Dis Soc Am; fel Am Acad Microbiol; Am Soc Microbiol; Am Fedn Clin Res. *Res:* Clinical studies on the immune response to influenza vaccine; epidemiologic studies on hospital-acquired infections. *Mailing Add:* Dept Med Hackensack Med Ctr Hackensack NJ 07601

GROSS, PETER GEORGE, b Oradea, Rumania, Feb 6, 47; US citizen; m 68; c 4. ASTROPHYSICS, OPHTHALMOLOGY. *Educ:* Columbia Univ, BA, 68; Yale Univ, MPhil, 70, PhD(astrophys), 72; Ohio State Univ, MD, 82. *Prof Exp:* Res fel astron, Yale Univ Observ, 72; res fel astrophys, Thomas J Watson Res Ctr, IBM Corp, 72-74; from asst to assoc prof astron, Case Western Reserve Univ, 74-79; ophthal resident, Sch Med, Univ Pa, 83-86; EYE SURGEON & OPHTHALMOLOGIST, 86- *Concurrent Pos:* Founder, P G Gross Res Found, 81-; flexible resident, Bryn Mawr Hosp, Pa. *Mem:* Am Astron Soc; Int Astron Union; NY Acad Sci; AMA; AAAS; Am Acad Ophthal; Am Col Surgeons. *Res:* Stellar structure; stellar evolution; applied mathematics; computer calculations and graphics; magnetic fields and mixing in stellar plasmas; atmospheric escape rates; cosmic abundance of helium; brainstem auditory response latency and body temperature; glaucoma; choroidal melanomas; boron neutron capture therapy; rotational mechanics of eye movements and internal ocular forces; clinical measurements of the neuroretinal rim size; cardiac gaited fundus photography. *Mailing Add:* 714 Oxford Rd Bala Cynwyd PA 19004

GROSS, ROBERT ALFRED, b Philadelphia, Pa, Oct 31, 27; m 52; c 2. PLASMA PHYSICS. *Educ:* Univ Pa, BS, 49; Harvard Univ, MS, 50, PhD(appl physics), 52. *Prof Exp:* Fel & res assoc, Harvard Univ, 53; chief res engr, Fairchild Engine Div, Fairchild Engine & Airplane Corp, 54-59; NSF fel, 59-60; chmn dept mech eng, Columbia Univ, 70-76, dept appl physics & nuclear eng, 78-81, prof eng sci, 60-80, dean, Sch Eng & Appl Sci, 82-90, HUDSON PROF APPL PHYSICS, COLUMBIA UNIV, 80- *Concurrent Pos:* Guggenheim & Fulbright fels, 66-67; vis prof, Leiden Univ, 66-67; Australian Acad Sci vis fel, 67; vpres publ, Am Inst Aeronaut & Astronaut, 67-68; ed-in-chief selected reprint series, 68- *Honors & Awards:* Waverly Gold Medal Res, 60. *Mem:* Fel Am Phys Soc; fel Am Inst Aeronaut & Astronaut (vpres, 65-67). *Res:* High temperature gas dynamics; plasma physics. *Mailing Add:* Dean Appl Physics Columbia Univ Broadway & W 116th New York NY 10027

GROSS, ROBERT HENRY, b Brooklyn, NY, Mar 6, 45; m 68; c 3. RNA PROCESSING, HEAT SHOCK & THERMOTOLERANCE. *Educ:* Rensselaer Polytech Inst, Troy, NY, BES, 67, MS, 68; Johns Hopkins Univ, Baltimore, PhD(biophys), 74. *Prof Exp:* Fel molecular biol, dept biol, Johns Hopkins Univ, 74-77; asst prof biol, 77-83, asst prof, 78-83, ASSOC PROF BICHEM, DARTMOUTH COL, 83- *Concurrent Pos:* Mem staff, Norris Cotton Cancer Ctr, Hanover, NH, 79- & Arthritis Ctr, Mary Hitchcock Hosp, 80-; Molecular Genetics Ctr, Darthmouth, 85- *Mem:* AAAS; NY Acad Sci; Am Soc Cell Biol; Am Soc Microbiol. *Res:* Control of gene expression in eukaryotes; control of alternative splicing during development, utilizing liposome mediated introduction; use of computers in research training. *Mailing Add:* Dept Biol Sci Dartmouth Col Hanover NH 03755

GROSS, RUTH T, b Bryan, Tex, June 24, 20; div; c 1. MEDICINE, PEDIATRICS. *Educ:* Columbia Univ, BA, 41, MD, 44. *Prof Exp:* From instr to assoc prof pediat, Sch Med, Stanford Univ, 50-60; from assoc prof to prof, Albert Einstein Col Med, 60-66; dir dept pediat, Mt Zion Hosp & Med Ctr, 66-73; assoc dean student affairs, 73-75, prof pediat, 75-85, dir, Stanford Children's Ambulatory Care Ctr, 80-85, Dir, Gen Pediat Acad Develop Prog, 79-83, DIR, GEN AMBULATORY PEDIAT, SCH MED, STANFORD UNIV, 75-, NAT STUDY DIR, INFANT HEALTH & DEVELOP PROG, 83- *Concurrent Pos:* Commonwealth fel, 59-60; prin investr gen pediat acad develop prog, Stanford Univ & mem study youth develop, Boys Town Ctr, 78-84; Henry J Kaiser sr fel, Ctr Advan Behav Sci, Stanford, Calif, 80-81. *Mem:* Inst Med-Nat Acad Sci; Soc Res Child Develop; Soc Pediat Res; Am Pediat Soc; Ambulatory Pediat Asn. *Res:* medical education; interventions for infants at risks for developmental delay. *Mailing Add:* Bldg 460 Rm 110 Stanford Univ Stanford CA 94305-2135

GROSS, SAMSON RICHARD, b Brooklyn, NY, July 27, 26; m 52; c 3. GENETICS. *Educ:* NY Univ, BA, 49; Columbia Univ, AM, 51, PhD(genetics, zool), 53. *Prof Exp:* Res assoc biochem genetics, Stanford Univ, 53-56, asst prof genetics, 56-57; asst prof, Rockefeller Inst, 57-60; assoc prof microbiol, 60-65, dir prog genetics, 67-77, PROF GENETICS, DEPT BIOCHEM, DUKE UNIV, 65- *Concurrent Pos:* spec spec fel, Weizmann Inst, 69-70; dir undergrad training prog, Cold Spring Harbor Lab, 63 & 64, dir grad studies, Dept Microbiol & Immunol, 64-65; consult, NIH, 75-79; John Simon Guggenheim fel, 85-86. *Mem:* AAAS; Genetics Soc Am; Am Soc Biol Chem. *Res:* Molecular genetics; molecular regulatory mechanisms. *Mailing Add:* Genetics Div Dept Biochem Duke Univ Box 3711 Med Ctr Durham NC 27710

GROSS, STANLEY BURTON, b Pittsburgh, Pa, May 24, 31; m 53; c 3. TOXICOLOGY, ENVIRONMENTAL SCIENCES. *Educ:* WVa Univ, AB, 54, MS, 57, PhD(biochem), 63; Am Bd Toxicol, dipl, 82, Am Bd Indust Hyg, dipl. *Prof Exp:* Res assoc agr biochem, WVa Univ, 56-58, res assoc anesthesiol, Med Ctr, 63; res biochemist, Columbus Labs, Battelle Mem Inst, 65-67; asst prof environ health, Univ Cincinnati, 67-69, fel environ toxicol, 69-71, asst prof toxicol, 71-75; sr res inhalation scientist, Hazleton Labs Am, Inc, 75-76; consult toxicol, 76-77; SR TOXICOLOGIST & INDUST HYGIENIST, ENVIRON PROTECTION AGENCY, WASHINGTON, DC, 77- *Concurrent Pos:* Mem, Am Conf Govt Indust Hygienists. *Mem:* Soc Toxicol; Am Chem Soc; Sigma Xi; Am Indust Hyg Asn. *Res:* Environmental and occupational toxicology; trace metals; human body burden analyses; automated biomedical information systems; lipid absorption and analysis; pesticides; industrial hygienics. *Mailing Add:* Rte 1 PO Box 137 Ashburn VA 22011

GROSS, STANLEY H, b New York, NY, Apr 4, 23; m 47; c 2. SPACE PHYSICS, PLANETARY ATMOSPHERES. *Educ:* City Col New York, BS, 44; Polytech Inst Brooklyn, MS, 48, PhD(elec eng), 64. *Prof Exp:* Engr TV, Hazeltine Electronics Co, 45-46; sr develop engr servomech, Control Instruments Co, 46-49; proj engr, W L Maxson Corp, 49-51; from proj engr to engr & div chief missiles, radar & trainers, Fairchild Engine & Airplane Corp, 51-59; consult rocket & satellite payloads, Airborne Instruments Lab, 59-67; assoc prof, 67-72, PROF SPACE PHYSICS & ELECTROPHYS, POLYTECH INST NY, 72- *Concurrent Pos:* Consult, NASA, 67-70; vis lectr, Am Inst Physics, 67-72; NASA grants, 67-; NSF grants, 68-71 & 78-81; mem, Comn G & H, Int Union Radio Sci. *Mem:* Am Geophys Union; Am Phys Soc; fel AAAS; Am Astron Union; Inst Elec & Electronics Engrs; Sigma Xi. *Res:* Wave processes in atmospheres and ionospheres; physics and evolution of planetary atmospheres, solar-terrestrial interactions, dynamics and photochemistry of atmospheres, escape of gases from comets, satellites and planets. *Mailing Add:* Nine Grouse Lane Huntington NY 11743

GROSS, STEPHEN RICHARD, pharmacology, biochemistry; deceased, see previous edition for last biography

GROSS, THOMAS ALFRED OTTO, physics; deceased, see previous edition for last biography

GROSS, THOMAS LESTER, b Decatur, Ill, Aug 17, 45; m 67; c 3. MATERNAL-FETAL MEDICINE. *Educ:* Bluffton Col, BA, 67; Univ Ill Col Med, MD, 71. *Prof Exp:* Intern, St Francis Hosp, Ill, 71-72; residency obstet-gynec, Akron Gen Med Ctr, 74-77; fel maternal-fetal med, Cleveland Metropolitan Gen Hosp, Case Western Reserve Univ, 77-79; asst prof, 77-85, ASSOC PROF OBSTET & GYNEC, CASE WESTERN RESERVE UNIV, 85- *Concurrent Pos:* Asst dir, 82-85, actg prog dir, Perinatal Clin Res Ctr, 85- *Mem:* Am Col Obstetricians & Gynecologists; Physicians Social Responsibility; Soc Gynec Invest; Soc Perinatal Obstetricians; Perinatal Res Soc. *Res:* Biochemistry of amniotic fluid; effect of maternal nutrition and metabolism and how they relate to fetal growth. *Mailing Add:* Univ Ill Col Med One Illini Dr Box 1649 Peoria IL 61656

GROSS, VICTOR, electrical engineering, physics; deceased, see previous edition for last biography

GROSS, WALTER BURNHAM, b Sandusky, Ohio, Jan 17, 25; m 53; c 2. AVIAN PATHOLOGY. *Educ:* Ohio State Univ, DVM, 46; Univ Minn, MS, 52, PhD(animal path), 56. *Prof Exp:* Veterinarian, Bur Animal Indust, USDA, 47-48; ASSOC ANIMAL PATHOLOGIST, AGR EXP STA, VA POLYTECH INST & STATE UNIV, 49-, PROF VET SCI, 56- *Mem:* Am Vet Med Asn; Sigma Xi. *Res:* Diseases of poultry. *Mailing Add:* Va-Md Col Vet Med Va Polytech Inst & State Univ Blacksburg VA 24061-0442

GROSSBECK, MARTIN LESTER, b Paterson, NJ, July 5, 44; m 81. RADIATION EFFECTS, FRACTURE OF METALS. *Educ:* Rensselaer Polytech Inst, BS, 66; Cornell Univ, MS, 68; Univ Ill, PhD(metall eng), 75. *Prof Exp:* Instr reactor physics, US Naval Power Sch, 67-71; task leader, radiation effects fusion mat, 79-84, RES STAFF MEM, OAK RIDGE NAT LAB, 75- *Concurrent Pos:* US coordr, Int Prog Fusion Reactor Mats Irradiation, 81-; task leader, Radiation Effects Space Reaction Mat, Oak Ridge Nat Lab, 84-86, prin investr, Radiation Effects Tritium Prod Reactor Mat, 89-; chmn, mat sci & technol div, Am Nuclear Soc, 88-89. *Mem:* Am Nuclear Soc; Am Soc Metals; Am Vacum Soc; Sigma Xi. *Res:* Radiation effects in metals; mechanical properties; development of fusion reactor and space nuclear reactor materials; hydrogen embrittlement and gases in metals. *Mailing Add:* Oak Ridge Nat Lab PO Box 2008 Bldg 5500 Oak Ridge TN 37831-6376

GROSSBERG, ARNOLD LEWIS, b Cleveland, Ohio, June 1, 21; m 43; c 3. CHEMICAL ENGINEERING. *Educ:* Calif Inst Technol, BS, 42; Univ Mich, MS, 43. *Prof Exp:* chief design engr, Chevron Res Co, 65-72, vpres process eng dept, 72-85; asst res engr process design, Calif Res Corp, 43-44, assoc res engr, 44-46, res engr, 46-52, sr res engr, 52-56, supvr engr, 56-62, sect supvr, 62-65, chief design engr process & plant design, 65-72, VPRES PROCESS ENG DEPT, CHEVRON RES CO, 72-; LECTR CHEM ENG DEPT, UNIV CALIF, BERKELEY, 87- *Concurrent Pos:* Dir bd educ, Berkeley Unified Sch Dist, 65-71, pres, 66-68 & 70-71; mem adv bd, dept chem eng, Univ Calif, Berkeley, 80-85, Sch Eng, Stanford Univ, 82-85; lectr, Stanford Univ, 86. *Mem:* Fel Am Inst Chem Engrs. *Res:* Process design; process planning; computer and systems engineering; engineering research and development; alternate energy; technical service to refineries; chemical and natural gas plants. *Mailing Add:* Dept Chem Eng Univ Calif Berkeley CA 94720

GROSSBERG, SIDNEY EDWARD, b Miami, Fla, Nov 13, 29; m 59; c 2. VIROLOGY, ONCOLOGY. *Educ:* Emory Univ, BS, 51, MD, 54. *Prof Exp:* From intern to asst resident med, Duke Univ Hosp, 54-58; from instr to asst prof microbiol, Sch Med, Univ Minn, 59-62; asst prof, Med Col, Cornell Univ, 62-66; PROF MICROBIOL & MED & CHMN DEPT, MED COL WIS, 66-, DEP DIR, CANCER CTR, 84- *Concurrent Pos:* Virologist, US Army, 55-57; fel, Sch Med, Johns Hopkins Univ, 58-59; Nat Inst Allergy & Infectious Dis fel, 59-61; NIH Res Career Develop Award, 61-66; Markle scholar acad med, 65-70; vis investr, Pasteur Inst, Paris, 64-65 & 74-75; attend physician, NY Hosp, 64-66; vis prof, Dept Human Oncol, Univ Wis, Madison; assoc mem, Influenza Virus Comn, US Army, 67-70; sr fel, Europ Molecular Biol Orgn, 74-75; Walter Schroeder chair, Microbiol, 83; Am Cancer Soc Scholar in cancer res, 89; consult, WHO, NIH, Am Cancer Soc, Univ Tex, Med Col Pa Grad Schs. *Mem:* Am Asn Immunol; Am Soc Virol (secy-treas, 84-); Am Soc Microbiol; fel Infectious Dis Soc Am; Soc Microbiol Brit; French Soc Microbiol; Int Soc Interferon Res. *Res:* Host cell-virus relationships; cell biology; interferon; RNA viruses. *Mailing Add:* Dept Microbiol Med Col Wis Milwaukee WI 53226

GROSSBERG, STEPHEN, b New York, NY, Dec 31, 39; m 80; c 1. MATHEMATICAL BIOLOGY. *Educ:* Dartmouth Col, BA, 61; Stanford Univ, MS, 64; Rockefeller Univ, PhD(math), 67. *Prof Exp:* Asst prof appl math, Mass Inst Technol, 67-69, assoc prof, 69-75; PROF MATH, PSYCHOL, BIOMED ENG, BOSTON UNIV, 75-, DIR, CTR ADAPTIVE SYSTS, 83- *Concurrent Pos:* Ed Math Biosci, 80-, J Theoret Neurobiol, 80-, Behav Processes, 84- & J Math Psychol, 85- *Mem:* Cognitive Sci Soc; Am Math Soc; Soc Math Psychol; Soc Math Biol; Soc Neurosci. *Res:* Models of learning and perception; development, cognition, motor control; psychophysiology; global analysis of dynamical systems. *Mailing Add:* Biomed Eng Boston Univ Boston MA 02215

GROSSENBACHER, KARL A(LBERT), b Geneva, NY, Aug 24, 10; m 38, 52; c 6. PLANT PHYSIOLOGY. *Educ:* Univ Wis, PhB, 33; Univ Calif, PhD(plant physiol), 39. *Prof Exp:* Asst, Johns Hopkins Univ, 34-36 & Univ Calif, 36-38; asst in chg plant nutrit exhibit, Golden Gate Int Expos, 38-39; asst, Cabot Found, Harvard Univ, 39-43, assoc, 43-45; assoc, Exp Sta, Univ Calif, 45-46; asst prof bot, Univ Calif, Santa Barbara, 46-49; res chemist, Pabco Div, Fibreboard Paper Prod, Inc, Calif, 50-60; assoc res plant physiologist, Col Agr, Univ Calif, Berkeley, 61-69; teacher biol, Col San Mateo, 69-76; RETIRED. *Res:* Mineral nutrition; water relations; vegetative propagation; root-soil boundary zone; general biology; electron microscopy. *Mailing Add:* 3600 Richmond Pkwy Apt 7103 Richmond CA 94806

GROSSER, ARTHUR EDWARD, b New York, NY, Nov 30, 34. PHYSICAL CHEMISTRY, FOOD CHEMISTRY. *Educ:* Cornell Univ, AB, 56; Univ Wis, PhD(phys chem), 63. *Prof Exp:* Alumni Res Found fel, Univ Wis, 63-64, fel phys chem, 64-65; asst prof, 65-69, ASSOC PROF CHEM, McGILL UNIV, 69- *Mem:* Am Chem Soc; Am Phys Soc; AAAS. *Res:* Molecular beam studies of elastic and reactive collisions. *Mailing Add:* Dept Chem McGill Univ 801 Sherbrooke St W Montreal PQ H3A 2K6 Can

GROSSER, BERNARD IRVING, b Boston, Mass, Apr 19, 29; m 60; c 3. PSYCHIATRY, NEUROENDOCRINOLOGY. *Educ:* Univ Mass, BA, 50; Univ Mich, MS, 53; Western Reserve Univ, MD, 59. *Prof Exp:* Intern med, Affil Hosps, 59-60, resident psychiat, 60-65, from asst prof to assoc prof, Col Med, 67-75, PROF PSYCHIAT, COL MED, UNIV UTAH, 75-, CHMN DEPT PSYCHIAT, 79- *Concurrent Pos:* USPHS training grant, Col Med, Univ Utah, 60-62; USPHS res grants, 62-65 & 67-74; NIMH res career develop awards, 62-65 & 67-73; mem preclin psychopharmacol res rev comt, NIMH, 74-78, mem clin res rev comt, 80-85; mem sci adv bd, The Scottish Rite, 78-; sr sci adv, Alcohol & Drug Abuse, Ment Health Admin, 87-88. *Mem:* Int Soc Psychoneuroendocrinol (treas, 74-); Am Psychiat Asn; AMA; Soc Neurosci; NY Acad Sci. *Res:* Interrelationships between cortico-steroids and brain proteins, especially the role of these hormones in human behavior. *Mailing Add:* Dept Psychiat Univ Utah Sch Med Salt Lake City UT 84132

GROSSERT, JAMES STUART, b Durban, SAfrica, Jan 28, 40; m 68. ORGANIC CHEMISTRY. *Educ:* Univ Natal, BSc, 60, MSc, 61, PhD(chem), 64. *Prof Exp:* Lab asst, Dyson Perrins Lab, Oxford Univ, 64-65; fel, Brandeis Univ, 65-67; from asst prof to assoc prof, 67-90, PROF CHEM, DALHOUSIE UNIV, 90- *Concurrent Pos:* Vis fel, Org Chem Lab, Rijks Univ, Groningen, 73-74. *Mem:* Chem Inst Can; Royal Soc Chem; Am Chem Soc. *Res:* Organosulfur chemistry; heteroatom chemistry reactions in solution under high pressures; marine and other natural products chemistry. *Mailing Add:* Dept Chem Dalhousie Univ Halifax NS B3H 4J3 Can

GROSSFIELD, JOSEPH, b New York, NY, May 23, 40; m; div; c 2. BEHAVIORAL GENETICS. *Educ:* City Col New York, BS, 62; Univ Tex, MA, 64, PhD(genetics), 66. *Prof Exp:* Asst prof biol, Univ Calif, Riverside, 66; asst prof, Purdue Univ, 67; assoc prof biol, 73-78, PROF BIOL, CITY COL NEW YORK, 78- *Concurrent Pos:* Assoc ed, Am Midland Naturalist; vis prof, Latrobe Univ, 74-75; ed, Behav Genetics, 85- *Mem:* Genetics Soc Am; Behav Genetics Asn; Soc Study Evolution; Am Soc Naturalists. *Res:* Genetic control of physiological processes subserving behavior; population and behavior genetics; striped bass; mitochondrial DNA. *Mailing Add:* Dept Biol City Col Convent Ave & 138th St New York NY 10031

GROSSI, CARLO E, b Mt Vernon, NY, Sept 12, 27; c 4. SURGERY. *Educ:* Columbia Univ, AB, 45; LI Col Med, MD, 49. *Prof Exp:* Clin asst surgeon, St Vincent's Hosp & Med Ctr, NY, 59; from instr to asst prof surg, 60-65, assoc clin prof, 65-69, ASSOC PROF CLIN SURG, MED SCH, NY UNIV, 69-; PROF SURG, NY MED COL, 80- *Concurrent Pos:* Sloan Kettering Inst spec fel exp surg, Mem Hosp, New York, 53; Nat Cancer Inst res fel, Sloan Kettering Inst, 59; pres, NY Cancer Soc, 74; pres, NY Surg Soc, 85. *Honors & Awards:* Ital Star of Solidarity. *Mem:* Fel Am Col Surg; Am Fedn Clin Res; Am Surg Asn; James Ewing Soc; Am Soc Colon & Rectal Surgeons. *Res:* General surgery; cancer research; oncology. *Mailing Add:* LaGuardia Hosp St Vincent Hosp 520 E 82nd St New York NY 10028

GROSSI, MARIO DARIO, b Giuncarico, Italy, Jan 10, 25; US citizen; m 56; c 2. RADIO PHYSICS, ENGINEERING. *Educ:* Univ Pisa, Dr(radio eng), 48; Nat Res Coun, Italy, dipl microwave physics, 49. *Prof Exp:* Res engr, Magneti Marelli Cent Radio Lab, Italy, 49-50; design engr, Ital Marconi, 50-55, design supvr, 55-58; staff engr, Equip & Systs Div, 58-60, consult scientist, Space & Info Systs Div, 60-65, consult scientist, Equip Div, 65-80, CONSULT SCIENTIST, SUB SIGNAL DIV, RAYTHEON CO, 80-; RADIOPHYSICIST & ENGR, HARVARD-SMITHSONIAN CTR ASTROPHYS, 59-; RES ASSOC, HARVARD COL OBSERV, 65- *Concurrent Pos:* Asst prof, Commun Inst, Univ Genoa, 51-58. *Mem:* Sr mem Inst Elec & Electronics Engrs; assoc fel Am Inst Aeronaut & Astronaut; Am Astron Soc. *Res:* Propagation of electromagnetic waves in lithosphere, atmosphere, ionosphere, magnetosphere; remote probing by electromagnetic waves of planetary atmospheres, ionospheres and surfaces. *Mailing Add:* Submarine Signal Div Raytheon Co W Marn Rd Box 360 Portsmouth RI 02871

GROSSIE, JAMES ALLEN, b Beaumont, Tex, May 19, 36; m 64. PHYSIOLOGY, NEUROENDOCRINOLOGY. *Educ:* Sam Houston State Col, BS, 58, MA, 59; Univ Mo, PhD(endocrinol), 63. *Prof Exp:* NIH fel, Col Med, Univ Ill, 63-65; ASSOC PROF PHYSIOL, OHIO STATE UNIV, 65- *Mem:* Am Physiol Soc. *Res:* Effects of endocrines on the nervous system. *Mailing Add:* Dept Physiol Ohio State Univ Col Med Columbus OH 43210

GROSSKREUTZ, JOSEPH CHARLES, b Springfield, Mo, Jan 5, 22; m 49; c 2. ENERGY CONVERSION, SOLAR ENERGY. *Educ:* Drury Col, BS, 43; Washington Univ, MS, 48, PhD(physics), 50. *Prof Exp:* Asst physics, Univ Calif, 46-47; asst, Wash Univ, 47-48 & 49-50; res physicist, Calif Res Corp, 50-52; asst prof physics, Univ Tex, 52-56; sr physicist, 56-59, prinphysicist, 59-63, sr adv physics, 63-67, prin adv, Midwest Res Inst, 67-71; chief mech properties sect, Nat Bur Standards, 71-72; mgr solar energy progs, Black & Veatch Consult Engrs, 72-77; dir res & develop, Solar Energy Res Inst, 77-79; mgr advan technol proj, Black & Veatch, Engrs-Architects, 79-88; RES PROF PHYSICS, UNIV MO, KANSAS CITY, 89- *Mem:* fel Am Phys Soc; Fel Am Soc Testing & Mat. *Res:* Metal physics; plastic deformation; electron microscopy; small angle x-ray scattering; metal fatigue; solar energy conversion to electricity; scanning tanneling and atomic force microscopy. *Mailing Add:* 4306 W 111th Terr Leawood KS 66211

GROSSLING, BERNARDO FREUDENBURG, b Santiago, Chile, Dec 21, 18; nat US; m 53; c 5. GEOPHYSICS, OPERATIONS RESEARCH. *Educ:* Univ Chile, CE, 41, EE, 44; Calif Inst Technol, MSc, 45; Univ London, dipl & PhD(geophys), 51. *Prof Exp:* Asst math, Univ Chile, 38, asst elec eng, 40-42; asst comput, Astron Observ, Chile, 38-40; design engr, Chilian Develop Corp, 40-42, explor geophysicist, 43-44; asst soil mech, Calif Inst Technol, 45-46, Nat Oil Co, Chile, 46-49, consult oil explor, 51-53; from res geophysicist to sr res geophysicist, Calif Res Corp Div, Standard Oil Co Calif, 54-59, res assoc, 59-60; actg chief engr div, Inter-Am Develop Bank, 60-61, engr consult, 61-62, tech adv 62-64; res geophysicist, US Geol Surv, Washington, DC, 64-73, off dir, 73-79; natural resources adv, Int-Am Develop Bank, Washington, DC, 79-85; RES GEOPHYSICIST, US GEOL SURV, WASHINGTON, DC, 85- *Concurrent Pos:* Lectr, Sch Arts & Trades, Univ Chile, 42-43. *Honors & Awards:* Marcos Orrego Puelma Prize, Inst Engrs Chile, 41. *Mem:* Soc Explor Geophys; fel Royal Astron Soc; Am Econ Asn; Sigma Xi; NY Acad Sci. *Res:* Mathematical physics; theory of automata; tectonophysics; plasticity; geothermal studies; oil exploration; economics of natural resource exploration and development; unorthodox uses of man's visual system. *Mailing Add:* c/o Cosmos Club 2121 Massachusetts Ave Washington DC 20008

GROSSMAN, ALLEN S, b New York, NY, Apr 12, 38; m 61; c 2. ASTROPHYSICS. *Educ:* Hofstra Univ, BS, 61; Adelphi Univ, MS, 64, Ind Univ, PhD(astrophys), 69. *Prof Exp:* Instr electronics, Acad Aeronaut, 58-59; design engr, Brookhaven Nat Lab, 61-64; asst prof, Iowa State Univ, 68-76, assoc prof physics, 76-; AT LAWRENCE LIVERMORE NAT LAB, CALIF. *Mem:* Am Astron Soc; Royal Astron Soc. *Res:* Stellar evolution; structure of stars of mass less than one half solar mass; astronomical instrumentation. *Mailing Add:* 1020 El Capitan Dr Danville CA 94526

GROSSMAN, BURTON JAY, b Chicago, Ill, Nov 27, 24. PEDIATRICS. *Educ:* Univ Chicago, BS, 46, MD, 49; Am Bd Pediat, dipl. *Prof Exp:* Instr, Helen Hay Whitney Found, Grad Sch Med, Univ Chicago, 54-57, from asst prof to assoc prof, 57-66; med dir, La Ribida Children's Hosp & Res Ctr, 62-85; PROF PEDIAT, SCH MED, UNIV CHICAGO, 86- *Concurrent Pos:* Fel pediat, Helen Hay Whitney Found, Univ Chicago, 54-57; assoc dir med serv, LaRabida-Univ Chicago Inst, 57-62. *Honors & Awards:* Distinguished Serv Award, Arthritis Found, 85. *Mem:* Soc Pediat Res; Am Pediat Soc; Am Acad Pediat. *Res:* Pediatric rheumatology; treatment of rheumatic diseases in children; blood coagulation problems; prevention of atherosclerosis. *Mailing Add:* LaRabida Children's Hosp Res Ctr 65th & Lake Michigan Chicago IL 60649

GROSSMAN, CHARLES JEROME, b New York, NY, May 12, 45. IMMUNOENDOCRINOLOGY. *Educ:* Adelphi Univ, Long Island, NY, BA, 68; Univ Cincinnati, MS, 71, PhD(physiol), 77; Xavier Univ, Cincinnati, MBA, 84. *Prof Exp:* RES PHYSIOLOGIST RES SERV, CINCINNATI VET ADMIN MED CTR, 77-; ASSOC PROF BIOL, XAVIER UNIV, CINCINNATI, 77-; ASSOC PROF PHYSIOL & BIOPHYS, COL MED, UNIV CINCINNATI, 85- *Concurrent Pos:* Prin investr, Res Serv, Med Ctr, Vet Admin, 78- *Mem:* Soc Study Reproduction; Am Physiol Soc; AAAS; Am Asn Univ Professors. *Res:* Regulation of the cell mediated immune response by sex steroids and pituitary and thymic factors. *Mailing Add:* Vet Admin Med Ctr 3200 Vine St Cincinnati OH 45220

GROSSMAN, DAVID G, b New York, NY, Aug 20, 41; m 71. CERAMICS. *Educ:* Rutgers Univ, BS, 63, MS, 64; Univ Sheffield, PhD(glass technol), 67. *Prof Exp:* Asst lectr mat sci, Rugby Col Eng Technol, Eng, 67-69; SR CERAMIST, CORNING GLASS WORKS, 70- *Mem:* Brit Soc Glass Technol; Am Ceramic Soc. *Res:* Structure and properties of glass ceramics. *Mailing Add:* 200 Wall St Corning NY 14830

GROSSMAN, GEORGE, b New York, NY, July 1, 14; m 37, 78; c 1. MATHEMATICS. *Educ:* City Col New York, BS, 34; Columbia Univ, MA, 35. *Prof Exp:* Teacher math, Bd Educ, N Y, 35-46, chmn, dept high schs, 46-64, actg dir, 64-68, dir math , 68-78; RETIRED. *Concurrent Pos:* Suppl teacher, Brooklyn Polytech Inst, 44-61, Sci Honors Prog, Columbia Univ, 59-69, Teachers Col, Columbia Univ, 61-69 & City Col New York, 61-62; adj prof, Hunter Col, 73-76 & Webster Col, 74-; consult, IBM Corp & Olivetti Underwood Corp; mem, Nat Coun Teachers Math. *Res:* Computer mathematics; impact of automatic digital computer in mathematical education; multiplicative teacher training. *Mailing Add:* 21 Sagamore Rd E Norwich NY 11732

GROSSMAN, HERBERT H, botany, invertebrate zoology, for more information see previous edition

GROSSMAN, HERBERT JULES, b Chicago, Ill, June 15, 23; m 54; c 3. PEDIATRIC NEUROLOGY. *Educ:* Univ Ill, Chicago, BS, 44, MD, 46. *Prof Exp:* Consult, Epilepsy Clin, Univ Ill Neuropsychiat Inst, Chicago, 53-57, pediat neurol, Ctr Handicapped Children, Univ Ill Hosp 56-57, Child Study Ctr, Mt Sinai Hosp, Los Angeles, 57-62, Pac State Hosp, Pomona, 58-62, pediat & Neurol, Presby-St Luke's Hosp 65-71 & Neurol, Munic Contagious Dis Hosp, Chicago, 73-; consult, Epilepsy Clin, Univ Ill Neuropsychiat Inst, Chicago, 53-57, pediat neurol, Ctr Handicapped Children, Univ Ill Hosp, 56-57, Child Study Ctr, Mt Sinai Hosp, Los Angeles, 57-62, Pac State Hosp, Pomona, 58-62, pediat & neurol, Presby-St Luke's Hosp, 65-71 & neurol, Munic Contagious Dis Hosp, Chicago, 73-; dir pediat neurol, Presby-St Luke's Hosp, 71-76; dir, Ill State Pediat Inst, 62-76. *Mem:* Fel neurol, Univ Calif, Los Angeles, 58-61; distinguished fel Am Psychiat Asn; fel Am Asn Ment Deficiency (vpres med, 72-74); Am Acad Ment Retardation (pres, 72-74); Am Acad Pediat. *Mailing Add:* 3738 Middleton Dr Ann Arbor MI 48105

GROSSMAN, HERMAN, b Waterbury, Conn, Apr 25, 25; m 49; c 3. RADIOLOGY, PEDIATRICS. *Educ:* Univ NC, BA, 47; Wesleyan Univ, MA, 49; Columbia Univ, MD, 53; Am Bd Pediat, dipl, 58; Am Bd Radiol, dipl, 65. *Prof Exp:* Intern pediat, Bellevue Hosp, 53-54; resident, Babies Hosp, NY, 54-56; asst pediatrician, Columbia Univ, 56-61, instr pediat, 61; from asst prof to assoc prof radiol & pediat, Med Col, Cornell Univ, 65-71; PROF DIAG RADIOL & PEDIAT, MED COL, DUKE UNIV & ATTEND RADIOLOGIST & PEDIATRICIAN, MED CTR, 71- *Concurrent Pos:* Fel radiol, Columbia-Presby Med Ctr, 62-65; pvt pract, 56-62; attend physician, Hackensack Hosp, 56-62; asst pediatrician, Presby Hosp & Babies Hosp, NY, 56-62; from assoc attend radiologist to attend radiologist, NY Hosp, 65-71, assoc attend pediatrician, 65-71; from asst attend to attend radiologist & pediatrician, Mem Hosp, 67-71. *Mem:* Asn Univ Radiol; Soc Pediat Radiol. *Res:* Pediatric radiology. *Mailing Add:* Dept Radiol Duke Univ Med Ctr Box 3834 Durham NC 27710

GROSSMAN, I G, b New York City, NY, Mar 4, 17; m 56; c 3. HYDROGEOLOGY & ECONOMIC GEOLOGY. *Educ:* Brooklyn Col, BA, 44; Columbia Univ, MA, 47. *Prof Exp:* Hydrogeologist, US Geol Surv, 49-85; RES SCIENTIST, NJ GEOL SURV, TRENTON, 85- *Concurrent Pos:* Vis prof environ geol, Bejing Univ, China, 87; lectr geol, Rutgers Univ, 88. *Honors & Awards:* Geo Striders Award, US Geol Surv, 82. *Mem:* Fel Geol Soc Am; Soc Econ Geologist; Sigma Xi; Soc Econ Geol; Am Geophys Union; ASN Earth Sci Ed; fel AAAS; Asn Geoscientists for Int Develop. *Mailing Add:* 224 Flint Court Yardley PA 19067

GROSSMAN, JACK JOSEPH, b New York, NY, Jan 14, 26; m 54; c 2. PHYSICAL CHEMISTRY. *Educ:* Brooklyn Col, BS, 49; Univ Southern Calif, PhD(phys chem), 53. *Prof Exp:* Asst radiochem exchange, Univ Southern Calif, 50; fel & jr res chemist, Univ Calif, Los Angeles, 53; proj dir, Vard, Inc, 54-56; proj engr & actg mgr, Res Dept, Ralph M Parsons Co, 56-59; mem tech staff, Hughes Aircraft Co, 59-63; sr engr, Douglas Aircraft Co, 63-69; prin investr Apollo lunar sample anal prog, Space Sci Dept, McDonnell Douglas Astronaut Co, 69-74, prin scientist, environ sci, 74-78, prin scientist, Sensors Avionics, Control & Info Systs Dept, 78-85, prin scientist, Strategic Systs Div, 85- 87, MGR HEDI AERO-OPTICAL WIND TUNNEL TEST PROG, MCDONNELL DOUGLAS SPACE SYSTEMS CO, 87- *Mem:* Am Chem Soc; Inst Elec & Electronics Eng; AAAS; Sigma Xi; Am Vacuum Soc; fel Am Inst Chemists. *Res:* Electrical and optical properties of semiconductors; reaction kinetics; mass spectrometry; telemetry and range instrumentation; nucleation and growth of thin films; active and passive remote sensing; ballistic missile defense exoatmospheric infrared radiation tracking system analysis and development; optical systems analysis and development; nuclear environments. *Mailing Add:* McDonnell Douglas Astronaut Co 5301 Boisa Ave Huntington Beach CA 92647

GROSSMAN, JACOB, b NY, Aug 6, 16; m 48; c 4. NEPHROLOGY, MEDICINE. *Educ:* City Col New York, BS, 35; Columbia Univ, MA, 36; Univ Louisville, MS, 40. *Prof Exp:* Res assoc med, Montefiore Hosp, 52-66; dir med, Hosp Joint Dis & Med Ctr, 66-80; prof clin med, Mt Sinai Sch Med, 68-80; VIS PROF MED, ALBERT EINSTEIN COL MED, 81- *Concurrent Pos:* Vis prof med, Albert Einstein Col Med, 81- *Mem:* AAAS; fel Am Col Physicians; Am Physiol Soc; Am Soc Nephrology; Soc Exp Biol & Med. *Res:* Cardiovascular and renal physiology; metabolism; clinical investigation. *Mailing Add:* 64 Fayette Rd Scarsdale NY 10583

GROSSMAN, JEFFREY N, b Brooklyn, NY, Nov 6, 55; m 83; c 1. COSMOCHEMISTRY, METEORITICS. *Educ:* Brown Univ, BS, 77; Univ Calif, Los Angeles, PhD(geochem), 83. *Prof Exp:* Res assoc, Jet Propulsion Lab, 84; RES CHEMIST, US GEOL SURV, 84- *Mem:* Fel Meteoritical Soc. *Res:* Studies of meteorites and the early history of the solar system; trace element geochemistry of igneous and metamorphic rocks; development of methods in neutron activation analysis. *Mailing Add:* US Geol Surv MS 923 Reston VA 22092

GROSSMAN, JEROME H, MEDICAL ADMINISTRATION. *Prof Exp:* CHIEF EXEC OFFICER, NEW ENG MED CTR, 88-; DEPT MED, TUFTS UNIV SCH MED, 88- *Mem:* Inst Med-Nat Acad Sci. *Mailing Add:* Dept Med Tufts Univ Med Sch 136 Harrison Ave Boston MA 02111

GROSSMAN, JOHN MARK, b Santiago, Chile, Oct 13, 54; US citizen. PLASMA PHYSICS, COMPUTATIONAL PHYSICS. *Educ:* Manhattan Col, BA, 74; Univ Md, MA, 80, PhD(appl math), 82. *Prof Exp:* Fel, Nat Res Coun, Nat Acad Sci, 82-84; RES SCIENTIST, PLASMA PHYSICS DIV, NAVAL RES LAB, 84- *Mem:* Am Phys Soc. *Res:* Plasma physics; pulsed power engineering; computer simulation of experiments. *Mailing Add:* 2032 Belmont Rd NW 422 Washington DC 20009

GROSSMAN, LAURENCE ABRAHAM, b Nashville, Tenn, Sept 21, 16; m 42; c 4. INTERNAL MEDICINE, CARDIOVASCULAR DISEASES. *Educ:* Vanderbilt Univ, BA, 38, MD, 41. *Prof Exp:* From instr to assoc prof, 47-57, PROF CLIN MED, SCH MED, VANDERBILT UNIV, 57-; ASSOC PROF, MEHARRY MED COL, 48- *Concurrent Pos:* Mem bd, exec comt, Vanderbilt Univ Med Ctr, 81-87; vpres, St Thomas Hosp Develop Found, 81-83; pres, Tenn Heart Inst, 84. *Mem:* Fel Am Col Physicians; fel Am Col Cardiol; fel Am Col Chest Physicians. *Res:* Cardiology. *Mailing Add:* 4300 Lilywood Rd Nashville TN 37205

GROSSMAN, LAWRENCE, b Toronto, Ont, Feb 2, 46; m 68; c 2. COSMOCHEMISTRY, GEOCHEMISTRY. *Educ:* McMaster Univ, BSc, 68; Yale Univ, MPhil, 70, PhD(geochem), 72. *Prof Exp:* Asst prof geochem, Dept Geophys Sci, 72-76, asst prof, Enrico Fermi Inst, 74-76, assoc prof geochem 76-81, PROF GEOCHEM, DEPT GEOPHYS SCI & ENRICO FERMI INST, UNIV CHICAGO, 80-, RES ASSOC DEPT GEOL, FIELD MUS NATURAL HIST, 76- *Concurrent Pos:* Consult, Lunar Adv Comt, NASA, 73-76, Lunar Data Synthesis Rev Panel, 73-74 & Mariner Jupiter Orbiter Sci Working Group, Jet Propulsion Lab, 75; consult lunar & planetary rev panel, Lunar & Planetary Inst, 77-79; Alfred P Sloan res fel, 76-78; Lady Davis vis prof, Hebrew Univ, Jerusalem, 81-82. *Honors & Awards:* F W Clarke Medal, Geochem Soc, 74; James B Macelwane Award, Am Geophys Union, 80. *Mem:* Fel Meteoritical Soc; Geochem Soc; fel Am Geophys Union; fel Mineral Soc Am; Mineral Asn Can. *Res:* Chemical and isotopic compositions of the different petrographic components of chondritic meteorites as clues to the behavior of the elements during the condensation of the solar system; composition and origin of the planets; interstellar grains. *Mailing Add:* Dept Astron & Astrophys Univ Chicago 5640 Ellis Ave Chicago IL 60637

GROSSMAN, LAWRENCE, b New York, NY, Jan 23, 24; m 49; c 3. BIOCHEMISTRY. *Educ:* Hofstra Col, BA, 49; Univ Southern Calif, PhD(biochem), 54. *Prof Exp:* NIH fel, McCollum-Pratt Inst, Johns Hopkins Univ, 54-56; res biochemist, NIH, 56-57; from asst prof to prof, Brandies Univ, 67-75; E V McCollum prof & chmn, 75-89, DISTINGUISHED SERV PROF, DEPT BIOCHEM, SCH HYG & PUB HEALTH, JOHNS HOPKINS UNIV, 89- *Concurrent Pos:* Grants, NIH, NSF, AEC & Am Cancer Soc; Commonwealth fel, 63-64, career develop award, NIH, 64-; mem, Sci Adv Comt, Am Cancer Soc, 65-68; chmn adv comt biochem, NIH; Guggenheim fel, 73; ed, Methods Nucleic Acids & Cancer Res, Crit Rev Biochem, J Biol Chem. *Mem:* Am Soc Biol Chem; Biophys Soc. *Res:* Enzymatic mechanisms for the repair of DNA; molecular basis of mutagenesis. *Mailing Add:* Dept Biochem Johns Hopkins Sch Hyg & Pub Health Baltimore MD 21205

GROSSMAN, LAWRENCE I, b New York, NY, Nov 15, 39; m 70; c 1. MOLECULAR BIOLOGY, BIOCHEMISTRY. *Educ:* City Col New York, BS, 61; Albert Einstein Col Med, PhD(genetics & biochem), 70. *Prof Exp:* Res fel biol, Calif Inst Technol, 70-74; asst prof biochem, Sch Med, Wayne State Univ, 74-78; asst prof biol, Univ Mich, 78-85; sr ed, Science Mag, 85-86; ASSOC PROF MOLECULAR BIOL & GENETICS, WAYNE STATE UNIV, 86- *Concurrent Pos:* Fel, Jane Coffin Childs Mem Fund Med Res fel, 71-73; assoc ed, Theoret & Appl Electrophoresis, 89-, consult ed, McGraw-Hill Encycl Sci & Technol, 88-, contrib ed, Sci, 87-; mem, sci adv comt molecular biol & genetics, Am Cancer Soc, 91. *Mem:* Am Soc Biochem & Molecular Biol; Biophys Soc; Am Soc Microbiol; Am Electrophoresis Soc (treas, 89-); Sigma Xi; AAAS. *Res:* Nucleic acids; structure and function and evolution of mitochondrial DNA; cytochrome c oxidase genes. *Mailing Add:* Wayne State Univ Sch Med Molec Biol & Genetics Dept 540 East Canfield Detroit MI 48201

GROSSMAN, LAWRENCE M(ORTON), b New York, NY, Aug 2, 22; m 52; c 1. NUCLEAR ENGINEERING. *Educ:* City Col New York, BChE, 42; Univ Calif, MS, 44, PhD(eng sci), 48. *Prof Exp:* Chem engr, E I du Pont de Nemours & Co, Inc, NY, 42-43; instr mech eng, 44-46, lectr, 46-48, from asst prof to assoc prof, 48-53, chmn dept, 69-74, PROF NUCLEAR ENG, UNIV CALIF, BERKELEY, 60- CHMN DEPT, 69- *Concurrent Pos:* Fulbright lectr, Delft, 52-53; NSF sr res fel, 61-62, NATO fel, Saclay Nuclear Res Ctr, France, 74. *Mem:* AAAS; Soc Indust & Appl Math; Am Nuclear Soc. *Res:* Nuclear reactor theory; neutron transport; reactor dynamics and control. *Mailing Add:* Dept Nuclear Eng Etcherverry Hall Univ Calif Berkeley CA 94720

GROSSMAN, LEONARD N(ATHAN), b Oakland, Calif, Feb 26, 36; c 2. LIGHTING SCIENCE, NUCLEAR FUEL SCIENCE. *Educ:* Univ Calif, Berkeley, BA, 59, MS, 60. *Prof Exp:* Metallurgist & ceramist, Gen Elec Co, 60-65, sr res ceramist, 65-70, mgr ceramic develop, 70-74, mgr qual technol develop, 74-81, mgr microelectronic appls, 81-82, mgr lighting res lab, 82-88, DIR MFG DEVELOP, ABB NUCLEAR POWER, 88- *Honors & Awards:* Ross Coffin Purdy Award, Am Ceramic Soc, 60. *Mem:* AAAS; Am Nuclear Soc. *Res:* Material science and manufacturing technology for nuclear fuels. *Mailing Add:* ABB Nuclear Power 1000 Prospect Hill Rd Windsor CT 06095

GROSSMAN, MICHAEL, b New York, NY, Dec 21, 40; m 70; c 2. POPULATION GENETICS, QUANTITATIVE GENETICS. *Educ:* City Col NY, BS, 62; Va Polytech Inst & State Univ, MS, 65; Purdue Univ, PhD(genetics), 69. *Prof Exp:* From asst prof to assoc prof, 69-85, PROF, UNIV ILL, 85- *Concurrent Pos:* Vis prof, Plant Breeding Inst, Nat Inst Agr Technol, Castelar, Arg, 70, Fac Animal Husb, Gadjah Mada Univ, Yogyakarta, Indonesia, 74 & Dept Animal Breeding, Agr Univ Wageningen, Neth, 86-87; Danforth Assoc. *Mem:* Am Dairy Sci Asn; Am Soc Animal Sci; Am Genetic Asn; Biomet Soc; Genetics Soc Am; AAAS. *Res:* Theoretical and experimental population and quantitative genetics; statistical methods and design of experiments; genetic studies of dairy goat and dairy cattle lactation records. *Mailing Add:* Dept Animal Sci Univ Ill Urbana Campus 1301 W Gregory Dr Urbana IL 61801

GROSSMAN, NATHANIEL, b Chicago, Ill, Apr 17, 37; m 60; c 2. MATHEMATICS. *Educ:* Calif Inst Technol, BS, 58; Univ Minn, PhD(math), 64. *Prof Exp:* Asst, Inst Advan Study, 64-66; asst prof, 66-70, assoc prof, 70-76, PROF MATH, UNIV CALIF, LOS ANGELES, 76- *Mem:* AAAS; Am Math Soc; Math Asn Am; Am Geophys Union. *Res:* Differential geometry of manifolds; Riemannian geometry in the large; geodesy. *Mailing Add:* Dept Math Univ Calif Los Angeles CA 90024

GROSSMAN, NORMAN, b New York, NY, Nov 3, 22. AERONAUTICAL ENGINEERING. *Educ:* NY Univ, BAeroE, 43, MS, 48, PhD(math), 58. *Prof Exp:* Prin res engr, Repub Aviation, Fairchild Industs, Inc, 46-51, asst proj engr, 51-55, staff engr, Fairchild Repub Co, 55-58, chief electronics eng, 58-62, mgr res div, 62-64, chief engr, 64-68, asst gen mgr, 68-69, vpres, 69-76, pres, 76-78, chmn & chief exec officer, 78-80; MGT CONSULT, 81- *Concurrent Pos:* Lectr grad math, Adelphi Col, 51-53; adj prof, Grad Ctr, Polytech Inst Brooklyn, Farmingdale Campus, 67-75. *Mem:* Am Inst Aeronaut & Astronaut; Math Asn Am. *Res:* Design, production and test of high performance military aircraft and aircraft systems. *Mailing Add:* 70 E Tenth St Apt PHL New York NY 10003

GROSSMAN, PERRY L, b Brooklyn, NY, Jan 27, 38; m 59; c 2. MECHANICAL ENGINEERING, APPLIED MECHANICS. *Educ:* Polytech Inst Brooklyn, BME, 59, MSME, 63, PhD(mech eng), 68. *Prof Exp:* Instr mech eng, Polytech Inst Brooklyn, 62-66; asst prof, 67-70, ASSOC PROF MECH ENG, COOPER UNION, 70- *Concurrent Pos:* Nat Sci Res initiation grant, 70-71. *Mem:* AAAS. *Res:* Nonlinear vibrations of spherical shells; flexural vibration and pulse-excited response of plates elastically supported at the surface. *Mailing Add:* Dept Mech Eng Cooper Union Cooper Square NY 10003

GROSSMAN, RICHARD C, orthodontics, for more information see previous edition

GROSSMAN, ROBERT G, b New York, NY, Jan 24, 33; m 55; c 3. NEUROSURGERY, NEUROPHYSIOLOGY. *Educ:* Swarthmore Col, BA, 53; Columbia Univ, MD, 57. *Prof Exp:* From instr to assoc prof, Univ Tex Southwestern Med Sch, 63-69; from assoc prof to prof neurol surg, Albert Einstein Col Med, 69-73; prof neurosurg & chmn div, Univ Tex Med Br Galveston, 73-80; PROF & CHMN, DEPT NEUROSURG, BAYLOR COL MED, HOUSTON, 80- *Concurrent Pos:* Attend neurosurgeon, Univ Tex Med Br Hosp, 73- *Mem:* Am Asn Neurol Surg; Soc Univ Surg; Am Col Surg; Am Electroencephalog Soc; Cong Neurol Surg; Sigma Xi. *Res:* Neurophysiology of cortical neurons and neuroglial cells. *Mailing Add:* Dept Neurosurg Baylor Col Med 6501 Fannin A404 Houston TX 77030

GROSSMAN, SEBASTIAN PETER, b Coburg, Bavaria, Jan 21, 34; m 55. PSYCHOBIOLOGY. *Educ:* Univ Md, BA, 58; Yale Univ, MS, 59, PhD(psychol), 61. *Prof Exp:* Asst prof physiol psychol, Univ Iowa, 61-64; assoc prof biopsychol, 64-67, PROF BIOPSYCHOL & CHMN, UNIV CHICAGO, 67- *Concurrent Pos:* Ed, Physiol & Behav, 66-, Commun Behav Biol, 67-79, Psychopharmacologia, 69-79, J Life Sci, 70-, Biochem Psychol, 70-, Pharmacol Biochem Behav, 73- & Neurosci Biobehav Rev, 77- *Mem:* Fel AAAS; fel Am Psychol Asn; Am Physiol Soc; Royal Soc Med. *Res:* Physiology and pharmacology of brain mechanisms which mediate psychological processes related to motivation. *Mailing Add:* Dept Psychol Univ Chicago 5848 S University Ave Chicago IL 60637

GROSSMAN, STANLEY I, b New York, NY, Sept 25, 42; m 65; c 2. APPLIED MATHEMATICS. *Educ:* Cornell Univ, BA, 64; Univ Calif, Los Angeles, MA, 65; Brown Univ, PhD(appl math), 69. *Prof Exp:* Asst lectr math, Gothenburg Univ, 65-66; asst prof, McGill Univ, 69-75; ASSOC PROF MATH, UNIV MONT, 75- *Concurrent Pos:* NSF-Romanian Acad Sci res grant, Iasi, Romania, 71. *Mem:* Am Math Soc; Math Soc Can; Soc Indust & Appl Math. *Res:* Volterra integral and integrodifferential equations; existence, uniqueness, stability and properties of solutions. *Mailing Add:* 320 Keith Ave Missoula MT 59801

GROSSMAN, STEVEN HARRIS, b New York, NY, May 18, 45; m 70; c 2. PROTEIN FOLDING, FLUORESCENCE. *Educ:* NY Univ, BA, 67; Purdue Univ, PhD(biochem), 72. *Prof Exp:* Fel biochem, Univ Wis, 72-74; res scientist clin chem, Union Carbide Corp, 74-76; asst prof, Univ Southwest La, 78-81; ASSOC PROF CHEM, UNIV SFLA, 81- *Mem:* NY Acad Sci; Am Chem Soc; Am Soc Biol Chemists. *Res:* Mechanisms of protein folding, particularly multimeric proteins. *Mailing Add:* Dept Chem Univ SFla Tampa FL 33620

GROSSMAN, WILLIAM, b New York, NY, May 24, 40. ANIMAL PHYSIOLOGY. *Educ:* Columbia Univ, BA, 61; Yale Univ Sch Med, MD, 65; Am Bd Internal Med, dipl & Subspeciality Cardiovasc Dis, dipl. *Prof Exp:* From asst prof med to assoc prof med, Univ NC, Sch Med, Chapel Hill, 71-75; dir, Cardiac Catheterization Lab, Peter Bent Brigham Hosp, 75-81, from assoc prof med to prof med Harvard Med Sch, 75-84; CHIEF, CARDIOVASC DIV, BETH ISRAEL HOSP, BOSTON, MASS, 81-; HERMAN DANA PROF MED, HARVARD MED SCH, BOSTON, MASS, 84- *Concurrent Pos:* Intern Med, Peter Bent Brigham Hosp, Boston, Mass, 65-66, asst resident med, Harvard Med Sch, 68-69, res fel med, 69-71; peace corps physician, New Delhi, India, 66-68; dir, cv Richardson Cardiac Catheterization Lab, NC Memorial Hosp, 71-75; estab invest, Am Heart Asn. *Mem:* Am Soc Clin Invest; Asn Am Physicians; Am Physiol Soc; Am Heart Asn; fel Am Col Cardiol; Am Fed Clin Res; Int Soc Heart Res. *Res:* Congestive heart failure; the role of defective heart muscle relaxation. *Mailing Add:* Dept Med Beth Israel Hosp Harvard Med Sch 330 Brookline Ave Boston MA 02215

GROSSMAN, WILLIAM ELDERKIN LEFFINGWELL, b New York, NY, May 4, 38; m 69; c 3. ANALYTICAL CHEMISTRY. *Educ:* Princeton Univ, AB, 59; Cornell Univ, PhD(anal chem), 64. *Prof Exp:* Assoc spectrochem, Inst Atomic Res, Ames Lab, AEC, 63-65; assoc, Univ York, Eng, 65-66; asst prof chem, 66-74, ASSOC PROF CHEM, HUNTER COL, 74- *Mem:* Soc Appl Spectros; Optical Soc Am; Am Chem Soc. *Res:* Raman spectroscopy. *Mailing Add:* Dept Chem Hunter Col 695 Park Ave New York NY 10021

GROSSMANN, ELIHU D(AVID), b Philadelphia, Pa, Nov 29, 27; m 54; c 3. SAFETY, TOXIC WASTE DISPOSAL. *Educ:* Drexel Inst Technol, BS, 51, MS, 56; Univ Pa, PhD, 65. *Prof Exp:* Asst, 51-52, from instr to asst prof, 52-61, assoc prof, 61-81, PROF CHEM ENG, DREXEL UNIV, 81- *Concurrent Pos:* Consult, US Army, 61-83, various chem, drug & environ co, 52- *Mem:* Am Chem Soc; Am Soc Eng Educ; Am Inst Chem Engrs; Nat Fire Protection Asn; Am Soc Univ Professors. *Res:* Abatement of toxic chemical releases the processing conditions required for some hazardous chemical reactions; thermodynamics of solutions; removal of deposited carcinogens from work clothes. *Mailing Add:* Dept Chem Eng Drexel Univ 32nd & Chestnut St Philadelphia PA 19104

GROSSMANN, IGNACIO E, b Mexico City, Mex, Nov 12, 49; m 77; c 3. PROCESS OPTIMIZATION, PROCESS SYNTHESIS. *Educ:* Univ Ibero Am, Mexico City, BS, 74; Imp Col, London, MS, 75, PhD(chem eng), 77. *Prof Exp:* Res & develop engr, Mex Inst Petrol, 78-79; prof chem eng, 79-90, RUDULPH R & FLORENCE DEAN PROF CHEM ENG, CARNEGIE-MELLON UNIV, 90- *Concurrent Pos:* Acad trustee, CACHE Corp, 84-; Mary Upson vis prof eng, Cornell Univ, 86-87. *Honors & Awards:* Robert W Vaughn Lectr, 84; Presidential Young Investr Award, NSF, 84. *Mem:* Am Inst Chem Engrs; Opers Res Soc Am; Sigma Xi; Am Chem Soc. *Res:* Optimization of chemical processes with particular emphasis on aspects related to flexibility and uncertainity in design; systematic synthesis of chemical process flowsheets using mixed-integer programming techniques; planning and scheduling of process operations. *Mailing Add:* Dept Chem Eng Carnegie-Mellon Univ Pittsburgh PA 15213

GROSSMANN, WILLIAM, b Richmond, Va, Aug 25, 37; m 74; c 2. AERONAUTICAL & ASTRONAUTICAL ENGINEERING. *Educ:* Va Polytech Inst, BS, 58, MS, 61, PhD(aerospace eng), 64. *Prof Exp:* Aerospace technologist, Langley Res Ctr, NASA, 58-65; assoc res scientist, Courant Inst Math Sci, NY Univ, 64-67; asst prof appl math, Richmond Col, NY, 67-69; sr scientist, Max Planck Inst Plasma Physics, 69-74; sr res scientist, 74-80, asst dir, Magnetic Fluid Dynamics Div, 74-76, assoc dir, 76-87, RES PROF, COURANT INST MATH SCI, 80-; DIV MGR & CHIEF SCIENTIST, APPL PHYSICS OPER, SCI APPLN INT CORP, 87- *Concurrent Pos:* Lectr, Inst Electronics, Univ Padua, Italy, 71-73; vis sr fel, Ctr for Theoret Physics, Univ Md, 74-75; vis staff mem, Los Alamos Nat Lab, 73-; consult, Bell Aerosysts, 67-69, Sci Appln Inc, 78- & Princeton Plasma Physics Lab, 80-; adj prof, Dept Applied Sci, NY Univ, 78-; univ assoc & mem evaluation panel on alt fusion concepts, Dept Energy, 81; dir plasma physics, Int Ctr Theoret Physics, Trieste, Italy, 83- *Mem:* Am Phys Soc. *Res:* Problems in physics and engineering of controlled fusion research; magnetohydrodynamics; politics and history of science and technology; plasma physics and fluid dynamics. *Mailing Add:* 805 S Fairfax St Alexandria VA 22314

GROSSO, ANTHONY J, b Poughkeepsie, NY, July 19, 26; m 53; c 2. ELECTRONICS, COMPUTER SCIENCE. *Educ:* Manhattan Col, BEE, 50; NY Univ, MEE, 53. *Prof Exp:* Asst engr res & develop, Am Dist Del Co, 50, engr, 51-55, proj engr, 55-58, sr prof engr, 58-63, gen mgr tel answering serv, 63-64, gen mgr diversification & expansion opers, 64-70, gen mgr comput sci, 69-70, chief engr, 70, vpres eng & res, 70-78, vpres eng & mfg, 78-80; staff men, Dist Tel Co; SR VPRES NAT ACCTS, ADT SECURITY SYSTS. *Concurrent Pos:* Adj prof, Sch Continuing Prof Studies, Pratt Inst, 56-71; mem, Alarm Indust Comt Combating Crime, 70-, Cent Sta Indust Frequency Adv Comt, 70- & Underwriters Lab Adv Conf Burglar Alarm Systs, 70- *Mem:* Inst Elec & Electronics Engrs. *Res:* Application of computer technology to the electric protection field; development of sophisticated intrusion detection systems; signal transmission techniques; means to establish the security of transmitted signals. *Mailing Add:* ADT Security Systs Inc 300 Interpace Pkwy Parsippany NJ 07054

GROSSO, JOHN A, b Hollis, NY, Nov 2, 55; m 83; c 2. PROCESS DEVELOPMENT, PLANT START-UPS. *Educ:* New York Univ, BA, 76; Purdue Univ, PhD(med chem), 80. *Prof Exp:* Sr res chemist, Ash-Stevens Inc, 80; res investr, E R Squibb & Sons, 80-86, sr res investr, 86-90; RES GROUP LEADER, BRISTOL-MYERS SQUIBB, 90- *Concurrent Pos:* Consult. *Mem:* Am Chem Soc; AAAS. *Res:* Preparation and development of organic compounds to be administered as pharmaceuticals, cardiovascular, antimalarial sweetening; anti inflammatory and CNS agents. *Mailing Add:* Bristol-Myers Squibb Co One Squibb Dr PO Box 191 New Brunswick NJ 08903

GROSSO, LEONARD, b Brooklyn, NY, Oct 3, 22; m 54; c 6. HISTOLOGY, REPRODUCTIVE PHYSIOLOGY. *Educ:* NY Univ, AB, 47, MS, 49, PhD(zool), 58. *Prof Exp:* Res specialist, Dept Fishes & Aquatic Biol, Am Mus Natural Hist, NY, 49; instr biol, Adelphi Col, 53-54; from asst prof to prof & head dept, Col St Teresa, 54-62; dir sci div, 66-68, assoc dean, 68-72, PROF BIOL, LONG ISLAND UNIV, 62- *Concurrent Pos:* Am Physiol Soc res fel, 59; fel, Univ Minn, 61; adj prof anat, NY Col Podiatry, 73- *Mem:* Am Physiol Soc; Soc Study Reproduction; Am Soc Zool. *Res:* General and comparative endocrinology; endocrine and nutritional interrelations; steroid hormone metabolism. *Mailing Add:* Long Island Univ Zeckendorf Campus Brooklyn NY 11201

GROSSWALD, EMIL, b Bucharest, Roumania, Dec 15, 12; nat US; m 50; c 2. MATHEMATICS. *Educ:* Univ Bucharest, MS, 32; Univ Pa, PhD(math), 50. *Prof Exp:* Instr, Univ PR, 47-48; lectr, Univ Sask, 50-51; assoc math, Univ Pa, 52-54, from asst prof to prof, 54-68; prof math, Temple Univ, 68-80; RETIRED. *Concurrent Pos:* Mem, Inst Advan Study, Princeton Univ, 51-52 & 59-60; exchange prof, Univ Paris, 64-65; vis prof, Israel Technol Inst, Haifa, 72-73, 74 & 80-81. *Mem:* AAAS; Am Math Soc; Math Asn Am; NY Acad Sci; Sigma Xi. *Res:* Analytic and elementary number theory; algebraic properties of polynomials. *Mailing Add:* Dept Math Temple Univ Philadelphia PA 19122

GROSSWEINER, LEONARD IRWIN, b Atlantic City, NJ, Aug 16, 24; m 51; c 3. PHOTOBIOLOGY, LASER MEDICINE. *Educ:* City Col New York, BChE, 47; Ill Inst Technol, MS, 51, PhD(physics), 54. *Prof Exp:* Asst, Argonne Nat Lab, 47-51, from asst physicist to assoc physicist, 51-57; assoc prof, 57-62, chmn dept, 70-81, PROF PHYSICS, ILL INST TECHNOL, 62- *Concurrent Pos:* Consult, Ravenswood Hosp, 83-; mem, US Nat Comt Photobiol, 78-80, chmn, 80-81; vis prof radiol, Col Med, Stanford Univ, 83-89, physics, Col Med, Univ Ill, 83- , biomed eng, Northwestern Univ Tech Inst, 87-; secy-treas, Div Biol Physics, Am Phys Soc, 72-76, vchmn, 76-77, chmn, 77-78. *Mem:* Fel Am Phys Soc; fel NY Acad Sci; Am Soc Photobiol (secy-treas, 81-86, pres, 87-88, past pres, 88-89). *Res:* Radiation biophysics; photobiology flash photolysis; pulse radiolysis; biological photosensitization; photodynamic tumor therapy; tissue optics and laser interactions with tissues. *Mailing Add:* Dept Physics Ill Inst Technol Chicago IL 60616

GROSTIC, MARVIN FORD, b Webberville, Mich, June 16, 26; m 51; c 3. ORGANIC CHEMISTRY. *Educ:* Albion Col, AB, 50; Univ Idaho, PhD(org chem), 65; Univ Ill, 64-65. *Prof Exp:* Infrared spectroscopist, Upjohn Co, 53-61, mass spectroscopist, 64-70, mgr chem prod control, 70-85, dir prod control, 85-86; pharmaceut consult, 86-90; VPRES QUAL ASSURANCE, AVTECH LABS, INC, 90- *Concurrent Pos:* Mem rev comt, US Pharmacopeia, 75, 80 & 85. *Mem:* Am Chem Soc; Am Soc Mass Spectrometry. *Res:* Addition to bicyclo hexene-2; mass spectroscopy. *Mailing Add:* 5140 Sprinkle Rd Kalamazoo MI 49002

GROSVENOR, CLARK EDWARD, b Piqua, Ohio, May 9, 28; m 50; c 6. ZOOLOGY, ENDOCRINOLOGY. *Educ:* Otterbein Col, BS, 50; Univ Cincinnati, MS, 52, PhD(zool), 55. *Prof Exp:* Instr biol, Col Pharm, Univ Cincinnati, 53-54, instr endocrinol, 54-55; asst prof biol, Univ Tenn, 55-56; NIH fel, Univ Mo, 56-59; from asst prof to assoc prof, 59-67, PROF PHYSIOL, UNIV TENN, MEMPHIS, 67-, NIH SR RES FEL, 59- *Mem:* AAAS; Soc Exp Biol & Med; Am Physiol Soc; Endocrine Soc. *Res:* Neuroendocrinology; lactational physiology. *Mailing Add:* Dept Molecular & Cell Biol Pa State Univ 401 Altheys Labs University Park PA 16802

GROSVENOR, NILES E(ARL), b Carbondale, Pa, Feb 14, 22; m 42; c 4. MINING ENGINEERING. *Educ:* Colo Sch Mines, EM, 50, MS, 52. *Prof Exp:* From instr mining to prof, Colo Sch Mines, 52-70; sr vpres, Gates Eng Co, 70-85; PRES, GROSVENOR ENG CO, 85- *Mem:* Am Soc Eng Educ; Am Inst Mining, Metall & Petrol Engrs. *Res:* Rock mechanics; mining. *Mailing Add:* 110 N Littleton Blvd St 304 Littleton CO 80120

GROSZ, HANUS JIRI, b Brno, Czech, Feb 16, 24; US citizen; m 54; c 3. PSYCHIATRY. *Educ:* Univ Wales, BSc, 50, MB BCh, 53; Royal Col Physicians & Surgeons, DPM, 61; Am Bd Psychiat & Neurol, dipl & cert psychiat, 67; Am Col Physicians, FACP, 67; Royal Col Psychiatrists, MRCP, 74. *Prof Exp:* Intern surg, East Glamorgan Hosp, 53; intern int med, Swansea Gen Hosp, 53-54; resident, Morriston Hosp, Gt Brit, 54-55; resident psychiat, Warren State Hosp, 55-56; sr res fel, Inst Psychiat Res, Med Ctr, Ind Univ, 56-58; USPHS fel & resident neurol, Albert Einstein Col Med, 58-60; resident psychiat, Maudsley & Bethlehem Royal Hosp, London, Eng, 60-62; from asst prof to assoc prof, 62-68, PROF & CHIEF PSYCHIAT, SCH MED, IND UNIV, INDIANAPOLIS, 68-, CHIEF PSYCHOBIOL SECT, 72- *Concurrent Pos:* Consult, Vet Admin Hosp, 63, Ind Girls' Sch, 64, Div Alcoholism, Ind Dept Ment Health, 66 & Am Asn Foreign Med Grads, 68; mem bd dirs, Community Addiction Serv Agency, Inc, 70; dir alcoholism-drug addiction res, Inst Psychiat Res, 70; pres, Psychiat Clins of Ind, Inc, 73- *Honors & Awards:* Achievement Award, John Shaw Billings Hist Med Soc, 69. *Mem:* AAAS; Am Psychiat Asn; fel Am Col Physicians; Royal Col Psychiat. *Res:* Alcohol and drug addiction; psychosomatic disorders; delinquency; group process and group interaction. *Mailing Add:* 7233 Lakeside Dr Indianapolis IN 46278-1616

GROSZ, OLIVER, b Ashley, NDak, Mar 12, 06; m 31; c 3. CHEMISTRY. *Educ:* Univ Iowa, BA, 28, MS, 29, PhD(org chem), 31. *Prof Exp:* Asst, Univ Iowa, 28-31; instr chem, Cath Univ, 31-39; res chemist, Collins & Aikman Corp, Pa, 39-40, asst chief chemist, 40-42, chief chemist, 42-49; dir dyeing & finishing res, Delta Finishing Co, J P Stevens & Co, Inc, 49-55; chief chemist, Collins & Aikman Corp, Pa, 55-57; res chemist, Toms River Chem Corp, 57-71; CONSULT, 71- *Mem:* Am Chem Soc; Photog Soc Am; Am Asn Textile Chemists & Colorists. *Res:* Organic, analytical and textile chemistry; sulfonyl derivatives of amino phenols; cresidine derivatives; analytical processes; textile finishing; synthetic detergents; mothproofing; dyeing processes; vat dyeing; wood scouring; water repellants; spectrophotometric methods of analysis; chromatography. *Mailing Add:* Three River Oaks RR5 Box 32 Mahomet IL 61853-8908

GROSZMANN, ROBERTO JOSE, b Buenos Aires, Arg, Aug 17, 39; m 65; c 2. GASTROENTEROLOGY. *Educ:* Buenos Aires Univ, BS, 58, MD, 64. *Prof Exp:* Instr med, Boston Univ, 68-69, Georgetown Univ, 69-71; asst prof, Buenos Aires Univ, 71-75; ASST PROF MED, YALE UNIV, 75- *Concurrent Pos:* NIH fel, Arg, 64-65, fel hemodynamics, Georgetown Univ, 69-71 & Med Inst Nat Health, France, 73; investr, Nat Res Coun, Arg, 71-75; career develop award, Vet Admin, 76. *Mem:* Am Fedn Clin Res; Am Asn Study Liver. *Res:* Liver circulation in relation to portal hypertension and cirrhosis of the liver. *Mailing Add:* Dept Med Vet Admin Hosp Spring St West Haven CT 06516

GROSZOS, STEPHEN JOSEPH, b Trenton, NJ, July 1, 20; m 46; c 2. EDUCATIONAL & RESEARCH ADMINISTRATION. *Educ:* NY Univ, AB, 46; Johns Hopkins Univ, AM, 48, PhD(chem), 51. *Prof Exp:* Instr & consult, Johns Hopkins Univ, 46-50; sr res scientist, Stamford Labs, Am Cyanamid Co, 50-57, mgr resin develop, Formica Corp Div, 57-59; tech dir, Richardson Co, 59-63, dir res & develop, 63-65; vpres, Packer Eng Assocs, Inc, 65-67; dean sci, 67-70, dir instnl res, 70-80, dir res & planning, 80-82, EMER PROF, COL DUPAGE, 82-; CONSULT CHEMIST, 67- *Concurrent Pos:* Res chemist, Polaroid Corp, 49; vis prof, St Procopius Col, 67-68; mem, Ill Nursing Home Adminr Licensure Bd, 69-76; mem bd trustees, Col DuPage Found, 70-82, treas, 73-82; mem res adv coun, Ill Community Col Bd, 73-82; mem & secy bd dirs, Care-Tech, Inc, 73-76; mem, bd trustees, High Point Environ Ctr, Inc, 84, treas, 85-87, pres, 87-88, adv bd, 90-; adj fac chem, Guilford Tech Community Col, 89- *Mem:* Am Chem Soc; Soc Plastics Engrs; Sigma Xi. *Res:* Chemistry of rosin, fatty acid derivatives, diazocompounds, aldoketene dimers, organometallics and mold metabolites; condensation and addition polymers; synthesis and study of properties as function of structure; polymers and plastics; molding compounds, laminates, adhesives, coatings, fibers and films. *Mailing Add:* 2639-F Suffolk Ave High Point NC 27265

GROT, WALTHER GUSTAV FREDRICH, b Berlin, Ger, June 26, 29; m 57; c 3. ORGANIC POLYMER CHEMISTRY, ELECTROCHEMISTRY. *Educ:* Univ Gottingen, BC, 51; Marburg Univ, PhD(chem), 55. *Prof Exp:* Sr res chemist, Res & Develop Lab, Belle Works, 56-60 & Plastics Dept, Exp Sta, 60-72, RES ASSOC, PLASTICS DEPT, EXP STA LAB, E I DU PONT DE NEMOURS & CO, INC, 72- *Concurrent Pos:* Case Western fel, 80. *Honors & Awards:* Castner Medal, Soc Chem Indust London, 85. *Mem:* Am Chem Soc; Ger Chem Soc. *Res:* Aliphatic organic chemistry; plastic intermediates; fluorocarbons; polymers; ion exchange membranes; applied electrochemistry. *Mailing Add:* Exp Sta Lab 323 Dupont Co 21470 Wilmington DE 19880-0323

GROTA, LEE J, b Sturgeon Bay, Wis, May 12, 37; m 59; c 4. PSYCHONEUROIMMUNOLOGY. *Educ:* Marquette Univ, BS, 59; Purdue Univ, MS, 61, PhD(exp psychol), 63. *Prof Exp:* Trainee steroid biochem, Sch Med, Univ Utah, 63-65; asst prof, 65-71, ASSOC PROF PSYCHIAT & PSYCHOL, SCH MED & DENT, UNIV ROCHESTER, 71- *Concurrent Pos:* Vis assoc prof neurosci & psychiat, McMaster Univ, Hamilton, Ont, 77-87. *Mem:* Endocrine Soc; Int Soc Develop Psychobiol; Soc Neurosci. *Res:* Neuroendocrinology; psychoneuroimmunology. *Mailing Add:* Dept Psychiat Univ Rochester Med Ctr Rochester NY 14642

GROTCH, HOWARD, b Brooklyn, NY, May 29, 40; m 64; c 3. THEORETICAL PHYSICS. *Educ:* City Col New York, BS, 62; Cornell Univ, PhD(theoret physics), 67. *Prof Exp:* Mem physics fac, 67-76, PROF PHYSICS, PA STATE UNIV, 76- *Concurrent Pos:* Vis prof, Univ Sussex, Universite d'Aix Marseille & Ben Gurion Univ, Univ Calif-Los Angeles, ITP Santa Barbara. *Mem:* Fel Am Phys Soc; Am Asn Physics Teachers. *Res:* Theoretical work in quantum electrodynamics and its application to atomic systems and to radiative level shifts; research on heavy quark-antiquark spectra and decays. *Mailing Add:* Dept Physics 104 Davey Lab Pa State Univ University Park PA 16802

GROTEFEND, ALAN CHARLES, b Elmhurst, Ill, July 16, 42; m 75; c 1. PRINTING INKS, ADHESIVES. *Educ:* Ariz State Univ, BS, 64. *Prof Exp:* Chemist, Richardson Co, 66-69, Meyercord Co, 70-84, TECH DIR, MEYERCORD, 85- *Res:* Formulate inks, coatings and adhesives for gravure and screen printing heat transfer decals to decorate substrates such as glass, metal, wood, and plastics. *Mailing Add:* Meyercord Co 365 E North Ave Carol Stream IL 60188

GROTEN, BARNEY, b Brooklyn, NY, Oct 25, 33; m 55; c 3. POLYMER CHEMISTRY, ANALYTICAL CHEMISTRY. *Educ:* Brooklyn Col, BS, 54; Purdue Univ, PhD(org chem), 61. *Prof Exp:* Chemist, Nickel Processing Corp, 55-56; chemist, Reaction Motors, Inc, 56-57; instr org chem, Purdue Univ, 57-59, fel, 60-61; sr chemist & group leader, Esso Res & Eng Co, Standard Oil Co, NJ, 61-66, sect head, Esso Res SA, Belg, 66-68, res planning adv, Esso Chem SA, 68-71, planning adv chem specialties, Esso Chem Europe, Inc, 71-73, sr staff adv, Exxon Chem Co, USA, La, 74-75, prod planner, Exxon Chem Co, USA, Houston, Tex, 75-77; sr staff adv, Exxon Chem Co, USA, La, 74-75, prod planner, Houston, Tex, 75-77; dir res & bus develop, Tex Eastern Corp, 77-87; exec vpres, Tex Eastern Develop Corp, 80-87; EXEC DIR, ENERGY CTR, UNIV OKLA, 87- *Concurrent Pos:* secy & treas, Gulf Univ Res Consortium, 81- *Mem:* Am Chem Soc; AAAS. *Mailing Add:* Energy Ctr 100 E Boyd Univ Okla Norman OK 73019

GROTENHUIS, MARSHALL, b Oostburg, Wis, Oct 17, 18; m 46; c 4. PHYSICS, ENVIRONMENTAL SCIENCE. *Educ:* Milwaukee State Teachers Col, BS, 41; Marquette Univ, MS. *Prof Exp:* Instr pub sch, Mich, 41-42; asst physics, Marquette Univ, 46-48, instr, 48-49; assoc physicist, reactor eng, Argonne Nat Lab, 49-56, asst dir, Int Sch, assoc dir, Int Inst, 59-64, dir, Off Indust Coop, 65-67, supt cent shop, 67-70, supt tech serv, 70-71; proj analyst, Div Radiol & Environ Protection, Nuclear Regulatory Comn, 71-75, sr proj mgr, Div Oper Reactors, 76-88; RETIRED. *Honors & Awards:* High Qual Award, Nuclear Regulatory Comn, 75; Meritorious Serv Award, Nuclear Regulatory Comn. *Mem:* Am Nuclear Soc; Sigma Xi. *Res:* Industrial cooperation; reactor shielding and engineering; nuclear engineering education; nuclear reactor regulation; environmental statement review. *Mailing Add:* 216 Summit Hall Rd Gaithersburg MD 20877

GROTH, DONALD PAUL, b Milwaukee, Wis, Oct 2, 28; m 53; c 3. ONCOLOGY. *Educ:* Univ Wis, BS, 50, PhD(physiol chem), 54. *Prof Exp:* Asst oncol, Univ Wis, 50-52; fel biochem develop, Univ Libre de Bruxelles, 54-55; fel cellular physiol, Univ Calif, 55-56; asst prof biochem & pharmacol, 56-63, ASSOC PROF BIOCHEM, EMORY UNIV, 63-, ASST DIR CANCER CTR, 73- *Concurrent Pos:* Lederle Med Fac award, 58-60; Leukemia Soc scholar, 63-; vis prof, Karolinska Inst, Sweden, 64. *Mem:* AAAS; Am Chem Soc; Asn Cancer Res; Am Soc Biol Chem; NY Acad Sci. *Res:* Cancer; chemistry of development and morphogenesis; changes in nucleotide metabolism in differentiating and transformed cells. *Mailing Add:* Dept Biochem Emory Univ Sch Med Woodruff McAdam Bldg Atlanta GA 30322

GROTH, EDWARD JOHN, III, b St Louis, Mo, May 13, 46; m 75; c 2. ASTROPHYSICS. *Educ:* Calif Inst Technol, BS, 68; Princeton Univ, PhD(physics), 71. *Prof Exp:* From instr to assoc prof, 71-86, PROF PHYSICS, PRINCETON UNIV, 86- *Concurrent Pos:* Alfred P Sloan res fel, 73. *Mem:* AAAS; Am Astron Soc; Am Phys Soc; Int Astron Union. *Res:* Pulsars; x-ray sources; space astronomy; evolution of galaxies and clusters of galaxies; systems operations; software. *Mailing Add:* Dept Physics Princeton Univ Princeton NJ 08544

GROTH, JAMES VERNON, b Minneapolis, Minn, July 13, 45; m 67; c 2. PLANT PATHOLOGY. *Educ:* Univ Minn, BS, 67, MS, 69; Univ BC, PhD(bot), 74. *Prof Exp:* From asst prof to assoc prof, 72-85, PROF PLANT PATH, UNIV MINN, 85- *Mem:* Am Phytopath Soc; Sigma Xi. *Res:* Population genetics and quantitative descriptions of host-parasite relationships between higher plants and fungi; genetic diversity of plant pathogenic fungi; fungal pathogens of vegetables. *Mailing Add:* Dept Plant Path Univ Minn St Paul MN 55455

GROTH, JOYCE LORRAINE, b Meriden, Conn, Feb 3, 35; m 59; c 2. ANALYTICAL CHEMISTRY. *Educ:* Cent Conn State Col, BS, 57; Univ Conn, MS, 59, PhD(anal chem), 64. *Prof Exp:* Res fel, Wesleyan Univ, 64; asst prof chem, Univ Conn, 64-69; INSTR CHEM, NEW BRIT HIGH SCH, 69- *Concurrent Pos:* Eve lectr, Cent Conn State Col, 68-71. *Res:* Nonaqueous reference systems and polarography; coulometry. *Mailing Add:* 75 Coe Ave Meriden CT 06450-3854

GROTH, RICHARD HENRY, b New Britain, Conn, Oct 14, 29; m 59; c 2. ENVIRONMENTAL CHEMISTRY. *Educ:* Conn State Col, BS, 51; Ohio State Univ, PhD(org chem), 56. *Prof Exp:* Res assoc, Off Naval Res, Duke Univ, 56-57; from asst prof to assoc prof chem, Univ Hartford, 57-70; assoc prof, 70-76, chmn dept, 75-84, PROF CHEM, CENT CONN STATE COL, 76- *Concurrent Pos:* Consult, United Technol Corp, 61- *Mem:* AAAS; Am Chem Soc. *Res:* Gas analysis and chromatography; air pollution measurement technology; carbon dioxide-oxygen control in isolated atmospheres. *Mailing Add:* 75 Coe Ave Meriden CT 06450-3854

GROTHAUS, CLARENCE (EDWARD), organic chemistry; deceased, see previous edition for last biography

GROTHAUS, ROGER HARRY, b Hamilton, Ohio, Oct 13, 36; m 58; c 2. MEDICAL ENTOMOLOGY. *Educ:* Earlham Col, BA, 55; Purdue Univ, MS, 62; Okla State Univ, PhD(med entom), 67; Am Registry Prof Entomologists, cert. *Prof Exp:* Chief training div entom, Navy Dis Vector Ecol Control Ctr, Alameda, Calif, 61-64; res entomologist, USN, 64-67, chief opers div, Navy Dis Vector Ecol Control Ctr, Jacksonville, Fla, 67-69, sr med entomologist, Naval Support Activ, Vietnam, 69-70, chief entom div, Naval Med Field Res Lab, Camp Lejeune, NC, 70-75; res liaison officer, Dept Defense, USDA, 75-79, asst exec secy, Armed Forces Pest Mgt Bd, 79-82; MEM STAFF & MGR ENTOM RES, S C JOHNSON & SON, INC, 82- *Mem:* Asn Mil Surgeons US; Entom Soc Am; Am Mosquito Control Asn. *Res:* General arthropod ecology and urban entomology. *Mailing Add:* S C Johnson & Son Inc Mail Sta 402 Racine WI 53403

GROTHEER, MORRIS PAUL, b Pittsburg, Kans, Dec 15, 28; m 54; c 5. ELECTROCHEMISTRY. *Educ:* Kans State Col Pittsburg, BS, 50, MS, 51; Kans State Univ, PhD(anal chem), 57. *Prof Exp:* Res chemist, Columbia Southern Chem Co, 57-59; sr chemist, Chem Div, Pittsburgh Plate Glass Co, 59-62; supvr electrom develop, Hooker Chem & Plastics Corp, 62-70, mgr sales & mkt, Chlor-Alkali Systs, 70-73, tech mgr, 73-79, mgr com develop, 79-81, prin process engr, Electrochem Div, 81-82, mgr Electrochem Sect, 82-84, tech prog mgr, 84-88, dir, Process Technol, 88-90; MGR CHEM RES & DEVELOP, KERR MCGEE CORP, 90- *Mem:* Electrochem Soc; Am Chem Soc; Sigma Xi; Am Inst Chem Engrs. *Res:* Fundamental and applied electrochemical research, primarily in brine electrolysis for the production of chlorine, caustic, chlorates and perchlorates; electrolytic manganese dioxide and manganese metal; wide variety of inorganic processes. *Mailing Add:* 2805 Parkside Edmond OK 73034

GROTJAN, HARVEY EDWARD, JR, b Moberly, Mo, July 31, 47; m 69; c 2. ENDOCRINOLOGY, REPRODUCTIVE BIOLOGY. *Educ:* Univ Mo-Columbia, BS, 69, MS, 71; Univ Kans Med Ctr, PhD(physiol), 75. *Prof Exp:* Fel reproductive med, Med Sch, Univ Tex, Houston, 75-77, asst prof, 77-83; assoc prof physiol & pharmacol, Univ SDak, 83-88; ASSOC PROF ANIMAL SCI, UNIV NEBR, 88- *Mem:* Endocrine Soc; AAAS; Soc Study Reproduction; Am Soc Androl; Sigma Xi. *Res:* Biosynthesis of the anterior pituitary gonadotropins luteinizing hormone and follicle-simulating hormone; hormonal regulation of testosterone production by the leydig cell of the testis. *Mailing Add:* Animal Sci Dept Univ Nebraska Lincoln NE 68583-0908

GROTTA, HENRY MONROE, b New York, NY, Apr 4, 23; m 48; c 5. ORGANIC CHEMISTRY. *Educ:* NY Univ, BA, 43; Ohio State Univ, PhD(chem), 50. *Prof Exp:* Res chemist org high polymers, Rohm and Haas Co, 50-54; from res chemist to sr chemist, Battelle Mem Inst, 54-68, proj leader, 68-85; INDEPENDENT CONSULT, 85- *Mem:* Am Chem Soc. *Res:* Synthetic organic chemistry; vinyl polymerization; heterocyclic compounds; coal conversion reactions; agricultural chemicals; hazardous waste treatment. *Mailing Add:* 13238 Miller Rd Mt Vernon OH 43050-9417

GROTTE, JEFFREY HARLOW, b Youngstown, Ohio, Jan 16, 47; m 73; c 2. APPLIED MATHEMATICS, OPERATIONS RESEARCH. *Educ:* Mass Inst Technol, BS, 69; Cornell Univ, MS, 70, PhD(appl math), 74. *Prof Exp:* Mem res staff, 74-83, dep dir, Prog Anal Div, 83-84, DEP DIR, STRATEGY FORCES & RESOURCES DIV, INST DEFENSE ANALYSIS, 84- *Concurrent Pos:* Asst prof lectr, George Wash Univ, 76. *Mem:* AAAS; Soc Indust & Appl Math; Opers Res Soc Am. *Res:* Military applications of operations research; nonlinear optimization; game theory; mathematical aspects of arms control; decision support. *Mailing Add:* Inst Defense Analysis 1801 N Beauregard St Alexandria VA 22311

GROTZ, LEONARD CHARLES, b Chicago, Ill, Nov 8, 27; m 54; c 3. PHYSICAL CHEMISTRY. *Educ:* Northwestern Univ, BS, 49; Univ Calif, PhD(phys chem), 52. *Prof Exp:* Res chemist, Glidden Co, 52-54; res fel, Univ Minn, 54-55; res chemist, Union Carbide Chem Co Div, Union Carbide Corp, 55-59; from asst prof to assoc prof chem, Univ Wis-Milwaukee, 59-66; prof, Parsons Col, 66-69; assoc prof, 69-74, PROF CHEM, UNIV WIS-WAUKESHA, 74- *Concurrent Pos:* Instr, WVa State Col, 57-59. *Mem:* AAAS; Am Chem Soc; Soc Col Sci Teachers; Nat Sci Teachers Asn. *Res:* Surface and polymer chemistry; chemical education. *Mailing Add:* 2701 Lander Lane Waukesha WI 53188

GROTZINGER, PAUL JOHN, b Philadelphia, Pa, Oct 23, 18; m 52; c 3. SURGERY. *Educ:* Muhlenberg Col, BS, 39; Hahnemann Med Col, MD, 43. *Prof Exp:* From asst prof to assoc prof surg, Hahnemann Med Col, 50-61, clin prof, 61-68; med dir & chief surgeon, 60-86, EMER MED DIR & CHIEF SURGEON, AM ONCOL HOSP, 86-; PROF CLIN SURG, UNIV PA, 68- *Concurrent Pos:* Attend surgeon, Jeanes Hosp, 63- *Mem:* AMA; fel Am Col Surg. *Res:* Cancer and its clinical treatment. *Mailing Add:* 2121 Valley Rd Huntingdon Valley PA 19006

GROUPE, VINCENT, b Philadelphia, Pa, Sept 13, 18; m 42; c 2. VIROLOGY. *Educ:* Wesleyan Univ, BA, 39; Univ Pa, PhD(bact), 42. *Prof Exp:* Asst influenza lab, Children's Hosp, Pa, 39-42; from asst bacteriologist to sr bacteriologist, E R Squibb & Sons, NJ, 42-44; res assoc, Squibb Inst Med Res, 44-47; assoc prof virol in animal diseases & in-chg virus lab, Exp Sta, Univ Conn, 47-49; from assoc prof to prof microbiol, Inst Microbiol, Rutgers Univ, 49-68; vpres & sci dir, Life Sci, Inc, 68-76; expert, Nat Cancer Prog, Nat Cancer Inst, 77-80; VPRES TO PRES, NEW LIFE FOUND, 80- *Concurrent Pos:* Vis prof, Rutgers Univ, 68-70; pres, Life Sci Res Labs, 73-76. *Mem:* Soc Exp Biol & Med; NY Acad Med; Am Asn Cancer Res; Res Soc Alcohol; Am Acad Microbiol. *Res:* Cancer viruses; alcohol studies; antibiotics; chemotherapy; immunology. *Mailing Add:* New Life Found 11945 N 143 St 7202 Largo FL 34644

GROVE, ALVIN RUSSELL, JR, b Harrisburg, Pa, May 21, 14; m 36. ANATOMY, MORPHOLOGY. *Educ:* Lebanon Valley Col, BS, 36; Univ NMex, MSc, 37; Univ Chicago, PhD(bot), 40. *Prof Exp:* From asst to instr biol, Univ NMex, 36-41; from instr to assoc prof bot, Pa State Univ, University Park, 41-55, actg head dept, 61-65, asst dean col sci, 64-65, assoc dean col sci, 65-79, PROF BOT, PA STATE UNIV, UNIVERSITY PARK, 55- *Concurrent Pos:* Mem adv work group, Monongahela River Mine Drainage Remedial Proj, Dept Interior, 65-68; consult & panel mem, Comn Undergrad Educ in Biol Sci, 66-69; ed, Trout Mag, 70-78; mem citizens adv coun, Pa Dept Environ Resources, 71-84. *Mem:* Fel AAAS (past secy, Southwest Sect); Bot Soc Am; Am Inst Biol Sci; Sigma Xi. *Res:* Floral and experimental morphology. *Mailing Add:* 737 S Sparks St State College PA 16801

GROVE, ANDREW S, b Budapest, Hungary, Sept 2, 36; US citizen; m 58; c 2. ENGINEERING PHYSICS. *Educ:* City Col, New York, BS, 60; Univ Calif, Berkeley, PhD, 63. *Hon Degrees:* DSc, City Univ NY, 85; Dr Eng, Worcester Polytech Inst, 89. *Prof Exp:* Mem tech staff, Fairchild Semiconductor Res Lab, 63-66, sect head surface & device physics, 66-67, asst dir res & develop, 67-68; vpres & dir oper, Intel Corp, 68-75, exec vpres, 75-79, chief oper officer, 76-87, PRES & CHIEF EXEC OFFICER, INTEL CORP, 87- *Concurrent Pos:* Lectr, dept elec eng & comput sci, Univ Calif, Berkeley, 66-72. *Honors & Awards:* Medal Award, Am Inst Chemists, 60; J J Ebers Award, Inst Elec & Electronics Engrs, 74; George Washington Award, Am Hungarian Found, 90. *Mem:* Nat Acad Eng; fel Inst Elec & Electronics Engrs. *Res:* Metal-oxide-semiconductor devices and integrated circuits; large-scale integration technology; transport phenomena, especially fluid dynamics; author of over 40 technical publications. *Mailing Add:* Intel Corp 3065 Bowers Ave Santa Clara CA 95051

GROVE, DONALD JONES, b Pittsburgh, Pa, Oct 8, 19; m 43; c 3. PHYSICS. *Educ:* Col Wooster, AB, 41; Carnegie Inst Technol, PhD(physics), 53. *Prof Exp:* Res physicist, Res Labs, Westinghouse Elec Corp, 41-82; dep proj mgr, Princeton Univ, 54-80, prin res physicist & mgr, 82-86, dep dir, tech opers, 86-87, CONSULT PHYSICIST, PRINCETON UNIV, 54-, HEAD, TOKOMAK FUSION TEST REACTOR PROG, PLASMA PHYSICS LAB, 80- *Concurrent Pos:* Mgr, spec projs, 87- *Honors & Awards:* Arthur H Compton Prize Physics, 41. *Mem:* Am Phys Soc; Am Nuclear Soc; Sigma Xi. *Res:* Mass spectroscopy, specifically ionization and dissociation of atoms and molecules by electron impact; electronic instrumentation for above; ultra high vacuum techniques; controlled thermonuclear reactor techniques and machine design. *Mailing Add:* 191 Riverside Dr Princeton NJ 08540

GROVE, EWART LESTER, b Greensburg, Kans, May 31, 13; m 44; c 2. ANALYTICAL CHEMISTRY, PHYSICAL CHEMISTRY. *Educ:* Ohio State Univ, MA, 45; Western Reserve Univ, PhD(chem), 51. *Prof Exp:* Teacher, Pub Sch, Minn, 38-40 & Ohio, 40-47; instr chem, Minn State Teachers Col, St Cloud, 47-48; from asst prof to assoc prof anal chem, Univ Ala, 51-60; res chemist, Ill Inst Tech Res Inst, 60-63, sr chemist, 63-66, mgr anal chem res, 66-71; mem staff, Freeman Labs, Inc, 71-76; sr scientist, IIT Res Inst, 76-90; RETIRED. *Concurrent Pos:* Instr math, Fenn Col, 42-44; res partic, Oak Ridge Nat Lab, 53-54, consult, 54-58. *Honors & Awards:* Awards for Spec Instrumentation Develop, Nat Adv Comt Aeronaut. *Mem:* Am Soc Testing & Mat; Am Chem Soc; Am Inst Chemists; Soc Appl Spectros; Sigma Xi. *Res:* Spectrography; spectrophotometry; atomic absorption flame and colorimetric methods as applied to analytical methods. *Mailing Add:* 28 W 074 Gary's Mill Rd Winfield IL 60190

GROVE, HERBERT D(UNCAN), JR, chemical engineering, for more information see previous edition

GROVE, JOHN AMOS, b Youngstown, Ohio, Mar 26, 38; m 62; c 2. NUTRITIONAL BIOCHEMISTRY. *Educ:* Ohio State Univ, BS, 61, MS, 64, PhD(nutrit, biochem), 66. *Prof Exp:* Res fel biochem, Univ Minn, St Paul, 66-68; from asst prof to assoc prof, 68-80, PROF CHEM, SDAK STATE UNIV, 80- *Res:* Mechanism of vitamin E action; intermediary metabolism of amino acids. *Mailing Add:* Dept Chem SDak State Univ Box 2202 Brookings SD 57007

GROVE, LARRY CHARLES, b Madison, Minn, Jan 19, 38; m 59; c 2. MATHEMATICS. *Educ:* Univ Minn, BA, 60, MA, 61, PhD(math), 64. *Prof Exp:* J W Young res instr math, Dartmouth Col, 64-66; asst prof, Univ Ore, 66-69; assoc prof, Syracuse Univ, 69-71; assoc prof, 71-76, PROF MATH, UNIV ARIZ, 76- *Mem:* Am Math Soc; Math Asn Am. *Res:* Algebra. *Mailing Add:* Univ Ariz Tucson AZ 85721

GROVE, PATRICIA A, b Bronx, NY, Oct 3, 52; m 87. ETHOLOGY. *Educ:* Col Mt St Vincent, BS, 74; City Col New York, MA, 77; City Univ New York, MPh, 79, PhD(biol), 81. *Prof Exp:* Lectr biol, City Col New York, 74-79 & Mercy Col, 79-80; ASSOC PROF BIOL, COL MT ST VINCENT, 79-; Animal Behav Consult, 87- *Concurrent Pos:* Proj dir, Re-licensing Prog Sci, New York Bd Educ Grant, Manhattan Col, 83-89, adj assoc prof biol, 85-89; med writer, Hoffman-La Roche, Inc, 89-90; AP biol consult, 90- *Mem:* Sigma Xi; Asn Women Sci; Am Ornithologists Union; Animal Behav Soc; Am Asn Univ Prof; Audubon Soc. *Res:* Behavioral effects of aquatic toxicants on vertebrates, including dieldrin, malathion and PCBs. *Mailing Add:* Dept Biol Col Mt St Vincent Bronx NY 10471-1094

GROVE, STANLEY NEAL, b Augusta, Va, Sept 13, 40; m 62; c 2. BOTANY, CELL BIOLOGY. *Educ:* Goshen Col, BA, 65; Purdue Univ, PhD(bot), 71. *Prof Exp:* Res assoc biochem, Agr Res Serv, USDA, 70-71; fel microbiol, Univ Tex, Austin, 71-74; mem fac, 74-77, ASSOC PROF BIOL, GOSHEN COL, 77- *Mem:* Bot Soc Am; Mycol Soc Am. *Res:* Structure and function of protoplasmatic and surface components of germinating fungal spores and growing hyphal tips; cytochemical localization of macromolecules, especially enzymes, polysaccharides in fungi and in peanut seeds. *Mailing Add:* Dept Biol Goshen Col Goshen IN 46526

GROVE, THURMAN LEE, b Lewisburg, Pa, July 22, 43. SOILS AND SOIL SCIENCE. *Educ:* Wilkes Col, BA, 66; Cornell Univ, PhD(agron), 82. *Prof Exp:* Asst ecol, Cornell Univ, 66-71; proj mgr environ consult, Tex Instruments Inc, 73-75; mgr, Beak Consult Inc, 75-77; res assoc soil sci, Cornell Univ, 70-90; PROG OFFICER AGROECOL & ENVIRON, WINROCK INT, 91- *Concurrent Pos:* Fel ecol, Fed Water Pollution Control Admin, Cornell Univ, 68-70, agroecologist, US Agency Int Develop, 88- *Mem:* AAAS; NY Acad Sci; Am Soc Agron; Soil Sci Soc Am; Soc Conserv Biol; Ecol Soc Am. *Res:* Biogeochemistry, especially transport mechanisms for phosphorus and nitrogen from the landscape to surface waters; biomass energy production; international agricultural development; agroecology. *Mailing Add:* Winrock Int 1611 N Kent St Suite 600 Arlington VA 22209

GROVE, TIMOTHY LYNN, b York, Pa, July 15, 49; m; c 2. GEOLOGY, GEOCHEMISTRY. *Educ:* Univ Colo, BA, 71; Harvard Univ, AM, 75, PhD(geol), 76. *Prof Exp:* res assoc earth sci, State Univ NY, Stony Brook 75-79; asst prof, 79-84, ASSOC PROF GEOL, MASS INST TECHNOL, 84- *Concurrent Pos:* Vis asst prof, Calif Inst Technol, 79. *Mem:* Geol Soc Am; Mineral Soc Am; Am Geophys Union; Geochem Soc. *Res:* Experimental studies of magmas on the earth and terrestrial planets; study of diffusion processes in silicate minerals and melts and kinetics phase transformations in feldspars and pyroxenes. *Mailing Add:* Dept Earth Atmospheric & Planetary Sci Mass Inst Technol Cambridge MA 02139

GROVENSTEIN, ERLING, JR, b Miami, Fla, Nov 12, 24; m 54, 62; c 2. ORGANIC CHEMISTRY. *Educ:* Ga Inst Technol, BS, 44; Mass Inst Technol, PhD(org chem), 48. *Prof Exp:* From asst to res assoc chem, Mass Inst Technol, 45-48; from asst prof to prof, Ga Inst Technol, 48-65, Julius Brown Prof Chem, 65-88; RETIRED. *Mem:* Am Chem Soc; The Chem Soc; Sigma Xi. *Res:* Organoalkali metal chemistry; organic reaction mechanisms; photochemistry. *Mailing Add:* 2424 Briarmoor Rd Atlanta GA 30345

GROVER, CAROLE LEE, b Seattle, Wash, Nov 25, 48; m 77; c 2. TRANSLATOR OF RESEARCH JOURNALS IN MATH, MANAGEMENT OF COMPUTER INFORMATION SYSTEMS IN HIGHER EDUCATION. *Educ:* Univ Wash, BS, 70, MA, 71; Carnegie Mellon, PhD(math), 78, MPM, 90. *Prof Exp:* Vis asst prof math, Univ Pittsburgh, 76-78; vis instr math, Carnegie Mellon, 80-81; from asst prof to assoc prof math & computer sci, 81-87, CHAIR, DIV MATH & COMPUTER SCI, CARLOW COL, 87- *Concurrent Pos:* Translator, Allerton Press, NY, 77- *Mem:* Am Math Soc; Sigma Xi; Math Asn Am; Asn Comput Mach; Asn Women Math. *Res:* Educational reform in undergraduate education in mathematics and computer science, particularly involving women, minority, and continuing education students. *Mailing Add:* 2067 Beechwood Blvd Pittsburgh PA 15217

GROVER, GARY JAMES, b Camp Le Jeune, NC, Oct 16, 54; m 79. PHYSIOLOGY OF SHOCK. *Educ:* Rutgers Univ, BS, 76 & MS, 79; Albany Med Col, PhD(physiol), 82. *Prof Exp:* Fel physiol, Rutgers Med Sch, 82-85; SR RES INVESTR PHARMACOL, ER SQUIBB & SONS, 85- *Concurrent Pos:* Adj asst prof, Rutgers Med Sch, 85- *Mem:* Am Heart Asn; Am Physiol Soc; Int Soc Heart Res. *Res:* Efficacy and mechanism of action of novel, pharmacological agents for the treatment of myocardiol ischemia; effect of thromboxane A2 receptor antagonists and calcium antagonists in myocardiol ischemia. *Mailing Add:* Dept Pharmacol Squibb Inst Med Res Princeton NJ 08543-4000

GROVER, GEORGE MAURICE, b Garland, Utah, June 7, 15; m 47; c 1. ENGINEERING PHYSICS. *Educ:* Univ Wash, BS, 41; Univ Mich, MS, 48, PhD(physics), 50. *Prof Exp:* Meteorologist, US Weather Bur, 38-44; group leader proximity fuze, Appl Physics Lab, Johns Hopkins Univ, 44-46; consult, US Bur Mines, 48-49; staff mem weapons, Los Alamos Sci Lab, 50-53, group leader, 53-55, group leader nuclear rockets, 55-58, group leader advan concepts, 58-71; vpres res & develop, Q-Dot Corp, 71-84; CONSULT, 85. *Honors & Awards:* Holley Medal, Am Soc Mech Eng, 75; Medal, Japanese Heat Pipe Soc, 84. *Mem:* Am Phys Soc. *Res:* Development and manufacture of heat exchange equipment for energy conservation using heat pipes. *Mailing Add:* 1501 44th St Los Alamos NM 87544

GROVER, HERBERT DAVID, b Mt Holly, NJ, Feb 20, 51; m 72; c 2. ECOSYSTEMS. *Educ:* Rider Col, BA, 74; Rutgers Univ, MS, 77; Univ NMex, PhD(ecosysts ecol), 82. *Prof Exp:* Teaching asst biol & ecol, Rutgers Univ, 74-76; field ecologist, Environ Resources Comn, 76; lectr plant ecol, Rider Col, 77; res asst & teaching asst bot & biostatist, Univ NMex, 77-79; res asst, Sandia Labs, 78; res assoc, Ecosysts Res Ctr, Cornell Univ, 81-83; RES SCIENTIST, UNIV NMEX, 83- *Mem:* AAAS; Ecol Soc Am; Sigma Xi. *Res:* Identification of ecosystem properties that are corelated with the ability of ecosystems to resist, or recover from disturbance; clarification of mechanisms that affect ecosystem response to perturbation. *Mailing Add:* WT Rte Box 794A Sandia Park NM 87047

GROVER, JAMES ROBB, b Klamath Falls, Ore, Sept 16, 28; m 57; c 2. CHEMICAL PHYSICS, NUCLEAR SCIENCE. *Educ:* Univ Wash, BS, 52; Univ Calif, Berkeley, PhD, 58. *Prof Exp:* Asst chem, Univ Calif, Berkeley, 52-53, asst nuclear chem, Radiation Lab, 53-57; res assoc, 57-59, assoc chemist, 59-63, chemist, 63-77, SR CHEMIST, BROOKHAVEN NAT LAB, 78- *Concurrent Pos:* Assoc ed, Ann Rev, Inc, Calif, 68-77; vis prof, Inst Molecular Sci, Okazaki, Japan, 86-87. *Mem:* Am Chem Soc; Am Phys Soc. *Res:* Photoionization studies of gaseous molecular complexes; chemical uses of synchrotron light; reactive collisions between molecules; nuclear scattering of synchrotron light. *Mailing Add:* Brookhaven Nat Lab Upton NY 11973-5000

GROVER, JOHN HARRIS, b Rockville Centre, NY, Dec 21, 40; m 67; c 4. FISH BIOLOGY. *Educ:* Univ Utah, BS, 64; Iowa State Univ, MS, 66, PhD(zool), 69. *Prof Exp:* Lectr biol, Univ Libya, 69-71; from asst prof to assoc prof, 71-84, PROF FISHERIES, AUBURN UNIV, 84- *Mem:* Am Fisheries Soc; Am Inst Fishery Res Biologists; Am Inst Biol Sci; Sigma Xi. *Res:* Aquaculture and aquatic ecology, especially as practiced in warm fresh waters and the developing world. *Mailing Add:* Dept of Fisheries & Allied Aquacultures Auburn Univ Auburn AL 36849-5419

GROVER, M ROBERTS, JR, b Boston, Mass, Dec 5, 27; m 56; c 3. MEDICINE. *Educ:* Bowdoin Col, BA, 50; Cornell Univ, MD, 54; Univ Ore, MS, 56. *Prof Exp:* From instr to prof med, 58-68, dir continuing med educ, 65-76, actg dean, 78-79, assoc dean, 68-83, PROF MED & PSYCHIAT, SCH MED, UNIV ORE, 83- *Concurrent Pos:* Asst med dir hosps & clins, Sch Med, Univ Ore, 58-67, prog coordr, Ore Regional Med Prog, 67-68; mem, Mountain States Regional Med Prog, 67-72. *Mem:* AMA; Asn Am Med Cols. *Res:* Medical education; stress. *Mailing Add:* 9060 SW Oleson Rd Portland OR 97223

GROVER, PAUL L, JR, b Nunday, NY, Oct 2, 43; m 67; c 3. MEDICAL EDUCATION, COMMUNICATION SCIENCE. *Educ:* Univ Rochester, AB, 65, MA, 67; Syracuse Univ, PhD(instrnl technol), 71. *Prof Exp:* Asst prof med educ & commun, Sch Med & Dent, Univ Rochester, 70-75, asst prof, Sch Nursing, 74-75; PROF MED EDUC, STATE UNIV NY HEALTH SCI CTR, 75-, ASSOC DEAN, 85- *Mem:* Am Educ Res Asn; Asn Educ Commun & Technol. *Res:* Program evaluation; self-instructional techniques; instructional design; problem solving; academic counseling. *Mailing Add:* 300 Scothem Blvd Syracuse NY 13210

GROVER, RAJBANS, b Varnasi, India, Oct 18, 27; m 61; c 3. ENVIRONMENTAL SCIENCES. *Educ:* Panjab Univ, India, BSc, 49; Univ Delhi, MSc, 52; Univ Wash, PhD(phys biochem), 60. *Prof Exp:* Demonstr, Univ Delhi, 49-50, res asst plant physiol, 52-54; ed asst bot, Publ Div, Sci & Indust Res Coun, India, 54-55; res officer, Forest Nursery Sta, Res Br, Indian Head, Can Dept Agr, 60-63, Exp Farm, Regina, 63-65, res scientist, Res Sta, 65-75, sr res scientist Environ Chem Res Sta, 75-89, DIR RES STA, CAN DEPT AGR, REGINA, 89- *Concurrent Pos:* Mem, Sask Weed Adv Coun, Can, 61-76; mem, Res Appraisal Comt, Western Sect, Can Weed Comt, 62-; prog leader environ chem of herbicides, Can Dept Agr, 71- *Mem:* Am Chem Soc; Weed Sci Soc Am; Soc Environ Chem & Toxicol; Chem Inst Can; Sigma Xi; Am Soc Testing & Mat. *Res:* Environmental chemistry of herbicides. *Mailing Add:* Res Sta Agr Can Box 440 Regina SK S4P 3A2 Can

GROVER, ROBERT FREDERIC, b Rochester, NY, Feb 25, 24; m 44. MEDICINE, HIGH ALTITUDE. *Educ:* Univ Rochester, BA, 47; Univ Colo, PhD(physiol), 51, MD, 55. *Prof Exp:* Intern, St Anthony Hosp, Denver, 55-56; instr med & asst dir cardiovasc lab, Med Ctr, 58-61, from asst prof to prof med, 61-82, dir high altitude res lab, 64-82, dir Cardiovasc Pulmonary Res Lab, 65-82, EMER PROF MED, SCH MED, UNIV COLO, DENVER, 82- *Concurrent Pos:* NIH study sect, 71-75, Prin Investr res grants, 61-82; vis prof, Univ Calif, Santa Barbara, 78; consult, Environ Protection Agency, 89-91. *Honors & Awards:* D W Richards Award, Am Heart Asn, 84. *Mem:* Am Col Cardiol; Am Physiol Soc. *Res:* Cardiovascular and pulmonary physiology; high altitude; pulmonary circulation; author or coauthor 0f 170 research reports and 57 books and reviews. *Mailing Add:* 216 Mariposa Circle Arroyo Grande CA 93420

GROVES, DONALD GEORGE, b Syracuse, NY; m 49. MATERIALS SCIENCE, SCIENCE ADMINISTRATION. *Educ:* Syracuse Univ, BSc, 39, MSc, 49. *Prof Exp:* Designer, Hudson Motor Car Co, 41-43; res grant, Santo Domingo, 49-50; designer & systs engr, Gen Elec Co, 54-61; sr staff scientist mat sci & eng, Nat Acad Sci, 62-88; sr res fel, Inst Technol & Strategic Res, 88-89; SR RES CONSULT, INST DEFENSE ANAL, 88- *Concurrent Pos:* Sr staff scientist, Nat Acad Sci comts advan design criteria, 61-62, ceramic mat, 61-62, design with brittle material, 63-64, protective mat aerospace vehicles, 64-65, atomic characterization mat, 64-67, ceramic processing studies, 65-68 & fundamentals amorphous mat, ballistic missile defense hardening, 70-71, infrared, laser-glass & var mat sci & eng studies, Nat Res Coun-Nat Acad Sci, 62-88. *Honors & Awards:* Freedom Found Nat Award, 70, 73, 74, 87 & 89; Adm Dyer Award, 85. *Mem:* Mat Res Soc. *Res:* Natural resource utilization; research and development in materials, ocean science and engineering; scientific administration; author three books, published 300 articles and monographs. *Mailing Add:* 2400 Virginia Ave NW-C623 Washington DC 20037

GROVES, ERIC STEDMAN, b Seattle, Wash, Nov 14, 42. EXPERIMENTAL HIGH ENERGY PHYSICS, MEDICINE. *Educ:* Mass Inst Technol, BSc, 65; Univ Pa, PhD(physics), 70; Univ Miami, MD, 78, Am Bd Internal Med, dipl, 81. *Prof Exp:* Res assoc physics, Rutherford Lab, Sci Res Coun, Eng, 70-75; res physicist, Lawrence Berkeley Lab, 75-76; mem staff, Jackson Mem Hosp, Miami, 78-81; mem staff cellular immunol, Cancer Inst, Bethesda, MD, 81. *Mem:* Am Phys Soc; Am Med Asn; Am Col Physicians. *Res:* Fundamental structure of hadrons using electronic techniques; high transverse momentum hadron physics at the European Organization for Nuclear Research intersecting storage rings; murine cellular immunology. *Mailing Add:* NCI Immunol Br 9000 Rockville Pike RM 4 B-17 Bethesda MD 20892

GROVES, GORDON WILLIAM, b Wilmar, Calif, Jan 8, 27; m 60; c 12. PHYSICAL OCEANOGRAPHY. *Educ:* Univ Calif, Los Angeles, AB, 49, MS, 51, PhD(oceanog), 55. *Prof Exp:* Asst res oceanog, Scripps Inst, Univ Calif, 55-59; investr, Inst Geophys, Nat Univ Mex, 59-61; assoc res geophysicist, Inst Geophys & Planetary Physics, Univ Calif, San Diego, 61-64; oceanogr, Inst Geophys, Univ Hawaii, 64-80, prof oceanog, 65-80; HEAD, ATOLL RES UNIT, UNIV SOUTH PAC, 80- *Concurrent Pos:* Ford Found grant, Inst Geophys, Nat Univ Mex, 68; Smithsonian Inst grant, Lembaga Penelitian Laut, Indonesia, 70-71; vis prof geophys, Univ Sci, Malaysia, 74-75; NSF grant, Lembaga Oseanologi Nasional, Indonesia, 78-79. *Mem:* Am Geophys Union; Persatuan Sains Lautan Malaysia; Mex Geophys Union. *Res:* Physical oceanography; tides; waves; variation of sea level. *Mailing Add:* Box 414 Monterey CA 93942

GROVES, IVOR DURHAM, JR, b Bowling Green, Ky, Dec 30, 19; m 44; c 3. UNDERWATER ACOUSTICS. *Educ:* Rollins Col, BS, 48, MBA, 64. *Prof Exp:* Electronic develop engr, Oak Ridge Nat Lab, 48-51; head transducer div & physicist, Naval Res Lab, 51-69, head transducer br, 69-74, assoc supt underwater sound ref div, 74-79, head transducer br, 77-79; RETIRED. *Concurrent Pos:* Consult, Oak Harbor Marine Assoc, 80-83. *Mem:* Fel Acoust Soc Am. *Res:* Underwater standard transducers for calibration of underwater acoustic devices; underwater electroacoustic transducers. *Mailing Add:* 533 Baxter Ave Orlando FL 32806-6225

GROVES, JOHN TAYLOR, III, b New Rochelle, NY, Mar 27, 43; m 67; c 2. ORGANIC CHEMISTRY. *Educ:* Mass Inst Technol, SB, 65; Columbia Univ, PhD(org chem), 69. *Prof Exp:* From asst prof to prof org chem, Univ Mich, Ann Arbor, 69-85, dir, Ctr Catalytic Surface Sci, 80-85; PROF ORG & INORG CHEM, 85-, CHMN, DEPT CHEM, PRINCETON UNIV, 88- *Concurrent Pos:* Ed bd, Bioorg Chem, 83- *Honors & Awards:* Spec Creativity Exten Award, NSF, 91. *Mem:* Fel AAAS; Am Chem Soc; NY Acad Sci; fel Japan Soc Promotion of Sci, 87. *Res:* Physical organic chemistry; bio-organic chemistry; reaction mechanisms; metals in biology; catalysis; oxidation-reduction; free radical chemistry. *Mailing Add:* Dept Chem Princeton Univ Princeton NJ 08544

GROVES, STEVEN H, b Madison, Wis, May 2, 34. SEMICONDUCTOR PHYSICS, CRYSTAL GROWTH. *Educ:* Antioch Col, BS, 57; Harvard Univ, MS, 58 & PhD(appl physics), 63. *Prof Exp:* STAFF MEM, LINCOLN LABS, MASS INST TECHNOL, 64- *Mem:* Am Phys Soc. *Res:* Epitaxial growth of 3-4 materials. *Mailing Add:* 18 Winthrop Rd Lexington MA 02173

GROVES, THOMAS HOOPES, b Madison, Wis, May 9, 32; m 54; c 4. PHYSICS. *Educ:* Antioch Col, BS, 55; Univ Wis, MS, 59, PhD(physics), 62. *Prof Exp:* Asst prof physics, Purdue Univ, 62-66 & Univ Notre Dame, 66-68; assoc physicist, Argonne Nat Lab, Ill, 68-74; asst dir, 74-80, PHYSICIST, FERMI NAT ACCELERATOR LAB, 80- *Mem:* Am Phys Soc. *Res:* Elementary particle physics. *Mailing Add:* Fermi Nat Accelerator Lab PO Box 500 Batavia IL 60510

GROVES, WILLIAM ERNEST, b Flint, Mich, Sept 8, 35; m 62; c 3. BIOCHEMISTRY, COMPUTER SCIENCE. *Educ:* Southern Methodist Univ, BS, 57; Univ Ill, MS, 59, PhD(biochem), 62. *Prof Exp:* Asst biochem, Univ Ill, 57-62; res biochemist, Lab Biochem Genet, Nat Heart Inst, 62-64; asst prof biochem, Col Med, Univ Tenn, 64-71; asst prof clin path, Med Univ SC, 71-74, assoc prof lab med & biomet, 74-; SUPV SYSTS PROG, UNIV NC, CHAPEL HILL. *Concurrent Pos:* Res trainee, St Jude Children's Res Hosp, 64-67, asst mem, 67-71; mem grad fac, Memphis State Univ, 69-71. *Mem:* Asn Comput Mach; Inst Elec & Electronics Engrs; Sigma Xi. *Res:* Computer applications in medicine. *Mailing Add:* Off Info Technol Univ NC Chapel Hill 308 Wilson 039A Chapel Hill NC 27314

GROW, GEORGE COPERNICUS, JR, b Stafford, NY, June 10, 16; m 42; c 2. GEOLOGY. *Educ:* Lehigh Univ, AB, 38. *Prof Exp:* From jr geologist to chief geologist, Peoples Natural Gas Co, 38-50; chief geologist, Eastern Area, Transcontinental Gas Pipe Line Corp, 50-73; CONSULT GEOLOGIST, 73- *Concurrent Pos:* Mem nat gas surv, Fed Power Comn, 72- *Mem:* Geol Soc Am; Am Asn Petrol Geologists; Am Inst Mining, Metall & Petrol Engrs; Am Inst Prof Geologists. *Res:* Surface, subsurface, oil and gas geology of the Appalachian area; underground gas storage; geology of eastern United States and offshore. *Mailing Add:* 626 Shadowlawn Dr Westfield NJ 07090

GROW, RICHARD W, b Lynndyl, Utah, Oct 31, 25; m 47; c 4. MICROWAVE PHYSICS, ELECTRONICS. *Educ:* Univ Utah, BS, 48, MS, 49; Stanford Univ, PhD(elec eng), 55. *Prof Exp:* Electronic scientist, Radio Countermeasures Br, Naval Res Lab, 49-50 & Nucleonics Div, 50-51; res assoc, Stanford Electronics Lab, 53-58; assoc res prof & asst dir, High Velocity Lab, 58-59, assoc res prof & dir, 59-62, assoc res prof, 62-64, res prof, 64-66, dir, 60-80, chmn dept, 65-80, PROF ELEC ENG, MICROWAVE DEVICE & PHYS ELECTRONICS LAB, 66- *Concurrent Pos:* Part-time consult, Gen Elec Microwave Lab, 54-59, Litton Industs, 59, 68-, Eitel-McCullough, Inc, 59, Microwave Electronics Corp, 59-, Northrop Corp, 69- & Bur Radiol Health, 70- *Mem:* Sr mem Inst Elec & Electronics Engrs. *Res:* Radar countermeasures; nuclear radiation detectors; traveling-wave tubes; backward-wave oscillators; attenuation in rocket exhausts; masers; lasers; microwave dosimeters; pollution measurements; solid-state plasmas. *Mailing Add:* Dept Elec Eng Merrill Eng Bldg 2280 Univ Utah 1400 E Second St Salt Lake City UT 84112

GROWCOCK, FREDERICK BRUCE, b Monterrey, Mex, May 1, 48; US citizen; m 71; c 1. CORROSION SCIENCE, FLUID MECHANICS. *Educ:* Univ Tex, Austin, BA & BS, 69; NMex State Univ, MS, 72, PhD(phys chem), 75. *Prof Exp:* Res assoc coal res, Brookhaven Nat Lab, 74-76, from asst scientist to scientist, 76-82; sr res chemist, Dowell Div, Dow Chemical, 82-83; proj leader, Dowell Schlumberger, 83-84, res scientist, 84-88; STAFF RES SCIENTIST, AMOCO PROD CO, 88- *Concurrent Pos:* Vis scientist, Oak Ridge Nat Lab, 77; adj prof, Univ Okla, 88- *Mem:* Nat Asn Corrosion Engrs; Soc Petrol Engrs; Am Chem Soc; Am Inst Chemists; Electrochem Soc. *Res:* Characterization of coal; kinetics of coal hydrogenation; reaction kinetics of aromatic hydrocarbons; gas-phase kinetics of radical/molecule reactions; characterization and oxidation kinetics of graphite; high temperature nuclear chemistry; corrosion science; corrosion inhibition; surfactant chemistry; properties of foams; oilfield stimulation; electrochemistry; AC impedance spectroscopy; surface analysis; clay chemistry; environmental chemistry. *Mailing Add:* 8308 S Fifth St Broken Arrow OK 74011

GRUB, WALTER, agricultural engineering, for more information see previous edition

GRUBB, ALAN S, b Poteau, Okla, Apr 3, 39; m; m; c 3. PUBLIC HEALTH, COMMUNITY MEDICINE. *Educ:* Univ Okla, BA, 62, MSPH, 67, PhD(human ecol), 70. *Prof Exp:* Environmentalist, Comanche County Health Dept, Okla, 62-66; admin consult, Field Training Ctr, Sch Health, Univ Okla, 68-70, instr community health, Cols Health & Med, 70-71, from asst prof to assoc prof family pract & community med, Col Med, 71-84, from asst prof to assoc prof community dent, Col Dent, 70-84; CHIEF, PROG DEVELOP, OKLA STATE DEPT HEALTH, 84- *Mem:* Am Pub Health Asn; Asn Am Med Cols. *Res:* Patterns of health care utilization; levels of health knowledge and health attitudes; evaluation of quality of care. *Mailing Add:* Oklahoma State Dept Health 1000 NE 10th St PO Box 53551 Oklahoma City OK 73152

GRUBB, H(OMER) V(ERNON), b Rockville, Ind, July 12, 16; m 43; c 1. CHEMICAL ENGINEERING. *Educ:* Purdue Univ, BS, 38, MS, 41; Ga Inst Technol, PhD(chem eng), 51. *Prof Exp:* From instr to assoc prof, 39-58, from actg dir to dir sch, 58-76, PROF CHEM ENG, GA INST TECHNOL, 58- *Mem:* Am Inst Chem Engrs. *Res:* Heat and mass transfer. *Mailing Add:* 2440 Northside Pkwy NW Atlanta GA 30327

GRUBB, RANDALL BARTH, developmental biology; deceased, see previous edition for last biography

GRUBB, ROBERT LEE, JR, b Charlotte, NC, Aug 22, 64. NEUROLOGY. *Educ:* Univ NC, AB, 61, MD, 65. *Prof Exp:* Surg intern, Barnes Hosp, St Louis, 65-66, asst resident gen surg, 66-67, neurol surg, 69-71, sr res, 73; fel neurol surg, Sch Med, Wash Univ, 71-72, Nat Inst Neurol Dis & Stroke fel, 72-73, instr, 73-74, from asst prof to assoc prof, 74-81, PROF NEUROL SURG & RADIOL, SCH MED, WASH UNIV, 81- *Concurrent Pos:* Consult, St Louis City Hosp & County Hosp, 73-85, John Cochran Vet Admin Hosp, 73-, St Louis Regional Hosp, 85-; neurosurgeon, Barnes Hosp & St Louis Children's Hosp, 73-, Jewish Hosp, 77-; mem, NIH Study Sect Neurol Dis Prog-Proj Rev A Comt, 79-84, chmn, 82-84; mem sci adv comt, Res Found Am Asn Neurol Surgeons, 82-; ed, Neurosurg, 85-86, J Neurosurg, 86- *Mem:* Am Asn Neurol Surgeons; Soc Neurol Surg; Am Acad Neurol Surg; fel Am Col Surgeons; Soc Neurosci; Am Physiol Soc. *Res:* Cerebral blood flow; author of numerous scientific publications. *Mailing Add:* Dept Neurol Surg Med Ctr Wash Univ Barnes Hosp Plaza St Louis MO 63110

GRUBB, WILLARD THOMAS, b Springfield, Ill, Mar 14, 23; m 50; c 1. PHYSICAL CHEMISTRY. *Educ:* Harvard Univ, SB, 46, PhD(phys chem), 49. *Prof Exp:* RES ASSOC CHEM, RES LAB, GEN ELEC CO, 49- *Concurrent Pos:* Coolidge fel, Gen Elec Co, 73. *Mem:* Am Chem Soc; Electrochem Soc; Am Inst Chemists; AAAS. *Res:* Fuel cells; electroless metal plating; electrochemical sensors; ion-selective electrodes; trace element analysis. *Mailing Add:* 2271 Sweetbrier Rd Schenectady NY 12309

GRUBBS, CHARLES LESLIE, b Hodgenville, KY, Nov 29, 43; c 1. PAINTING SYSTEMS, INDUSTRIAL COATING. *Educ:* Univ Louisville, BS, 70. *Prof Exp:* Group leader, Cook Paint, 70-79; asst tech dir, Rockford Coatings, 79-84; TECH DIR, US PAINT-GROW GROUP, 84- *Res:* Environmentally safe coasts; baking enamels. *Mailing Add:* 23 Pike Trail Arnold MO 63010

GRUBBS, CLINTON JULIAN, b Shreveport, La, Mar 7, 45. MEDICAL PHYSIOLOGY, CANCER. *Educ:* Univ Miss, BA, 66, MS, 68; Univ Tenn Med Units, PhD(physiol), 73. *Prof Exp:* Res assoc biochem, Ctr Health Sci, Univ Tenn, 73-75, instr physiol, 74-75; sr physiologist, Ill Inst Technol Res Inst, 75-79; sr physiologist, Southern Res Inst, 79-82, head, Cancer Prev Sect, 82-85; DIR, RES NUTRIT & CARCINOGENESIS, UNIV ALA, BIRMINGHAM, 85- *Mem:* Am Asn Cancer Res; Soc Exp Biol & Med. *Res:* Reproductive patho-endocrinology; estrogen receptors; binding of chemical carcinogens; anticarcinogenic effect of retinoids; photo-oxidation of polycyclic hydrocarbons; nutrition and cancer. *Mailing Add:* 2705 Mountainwoods Dr Birmingham AL 35216

GRUBBS, DAVID EDWARD, b Steubenville, Ohio, July 18, 43. PHYSIOLOGICAL ECOLOGY. *Educ:* Univ Calif, Santa Barbara, BA, 65; Univ Calif, Irvine, PhD(biol), 74. *Prof Exp:* Biol engr, Life Sci Opers, Autonetics Div, NAm Rockwell Corp, 66-69; asst prof biol, 73-76, ASSOC PROF BIOL, CALIF STATE UNIV, FRESNO, 76- *Mem:* Am Soc Mammalogists. *Res:* Physiological ecology of vertebrates with particular emphasis on xeric and alpine species. *Mailing Add:* Dept Biol Calif State Univ Fresno Fresno CA 93740

GRUBBS, DONALD KEEBLE, b Pittsburgh, Pa, May 11, 38. GEOLOGY, GEOCHEMISTRY. *Educ:* Univ Va, BS, 61, MS, 63; Univ Pittsburgh, PhD(geochem), 69. *Prof Exp:* Geologist, Aluminum Co Am, 69-70, sr geologist, Australasian Minerals, Inc, Alcoa, 70-71, sr geologist, Alcoa Minerals Indonesia, 71-72, scientist geochem, 72-74, group leader, 74-81, TECH SPECIALIST, RAW MAT, ALCOA LABS, 81- *Mem:* Soc Explor Geochemists. *Res:* Raw materials evaluation; computer usage related to raw material exploration, evaluation, and mining; geochemical research; sulfur reduction in coal. *Mailing Add:* Weaver Mill Rd Rector PA 15677

GRUBBS, EDWARD, b Los Angeles, Calif, July 22, 34; m 63; c 4. PHYSICAL ORGANIC CHEMISTRY. *Educ:* Occidental Col, AB, 56; Mass Inst Technol, PhD(org chem), 59. *Prof Exp:* Fel, Univ Ill, 60-61; from asst prof to assoc prof, 61-68, PROF CHEM, SAN DIEGO STATE UNIV, 68- *Concurrent Pos:* Vis prof, Ain Shams Univ, Cairo, Egypt, 86. *Mem:* Am Chem Soc. *Res:* Kinetics of molecular rearrangements; stereochemistry; electrostatic interactions of substituents and reaction sites; electrostatic effects on the course of elimination reactions. *Mailing Add:* Dept Chem San Diego State Univ San Diego CA 92182

GRUBBS, FRANK EPHRAIM, b Montgomery, Ala, Sept 2, 13; m 37; c 1. MATHEMATICAL STATISTICS, OPERATIONS RESEARCH. *Educ:* Auburn Univ, BS, 34, MS, 35; Univ Mich, MA, 40, PhD(math statist), 49. *Prof Exp:* From instr to asst prof eng math, Auburn Univ, 34-40; chief surveillance lab, Ballistic Res Labs, US Army Aberdeen Res & Develop Ctr, 41-53, Weapons Systs Lab, 53-62, dep tech dir, Ballistic Res Labs, 62-67, chief opers res anal, 68-75; RETIRED. *Concurrent Pos:* Supvr, Auburn Printing Co, 34-36; prof, Exten Div, Univ Del, 60-61, 62-63 & 67-70; actg chmn, Panel Tracking Data Anal, Nat Acad Sci, 64; chmn, US Army Design Exp Confs. *Honors & Awards:* Samuel S Wilks Mem Medal, 64; Frank Wilcoxon & Jack Youden Technometrics Prizes, 69; Shewart Medal, Am Soc Qual Control, 72; Mil Applns Prize, Operations Res Soc Am, 73 & 77; Merit Award, Am Soc Testing & Mat, 86. *Mem:* Fel Am Soc Qual Control; Am Soc Testing & Mat; fel Am Statist Asn; fel Inst Math Statist; fel Royal Statist Soc; Int Statist Inst. *Res:* Applied statistics; weapon systems evaluations; reliability; administration of research and development. *Mailing Add:* 4109 Webster Rd Havre de Grace MD 21078

GRUBBS, ROBERT HOWARD, b Calvert City, Ky, Feb 27, 42; m 67; c 3. ORGANIC CHEMISTRY, INORGANIC CHEMISTRY. *Educ:* Univ Fla, BS, 63, MS, 65; Columbia Univ, PhD(chem), 68. *Prof Exp:* NIH fel, Stanford Univ, 68-69; from asst prof to assoc prof organometallic chem, Mich State Univ, 69-78; PROF ORG CHEM, CALIF INST TECHNOL, 78- *Concurrent Pos:* Sloan Found fel, 74; Alexander von Humboldt fel, 75-; Dreyfus Found teacher-scholar grant, 75. *Honors & Awards:* Organometallic Chem Award, Am Chem Soc, 88, Arthur C Cope Scholar Award, 89. *Mem:* Nat Acad Sci; Am Chem Soc. *Res:* Chemistry and reactions of small ring carbocycles and their metal complexes; polymer synthesis; mechanism and synthetic utility of the olefin dismutation reaction and the development of new synthetic reactions; polymer chemistry. *Mailing Add:* Div Chem & Chem Eng Calif Inst Technol 1201 E Calif Blvd Pasadena CA 91125

GRUBE, GEORGE EDWARD, b Bareville, Pa, Apr 12, 23; m 45; c 2. ECOLOGY. *Educ:* Muhlenberg Col, BS, 46; Cornell Univ, MS, 49. *Prof Exp:* Instr biol, Franklin & Marshall Col, 46-48 & Gettysburg Col, 49-54; dist rep, Quaker Oats Co, 54-56; assoc prof biol sci, Lock Haven State Col, 57-65; prof biol, 65-88, EMER PROF BIOL, DANA COL, 88- *Concurrent Pos:* NSF partic, Inst Desert Sci, Ariz State Univ, 59, Animal Ecol Inst, Univ Colo, 60 & Inst Trop Ecol & Marine Biol, Univ PR, 66; chmn, East Nebr Nature Conserv, 78- *Mem:* Fel AAAS; Ecol Soc Am; Wilson Ornith Soc; Am Ornith Union. *Res:* Ornithology; ecology of symbiotes. *Mailing Add:* Dept Biol Dana Col Blair NE 68008

GRUBE, GERALDINE JOYCE TERENZONI, b Boston, Mass, Sept 2, 42; m 72. THEORETICAL NUCLEAR PHYSICS. *Educ:* Mass Inst Technol, BS, 66; Fla State Univ, MS, 68; NC State Univ, PhD(physics), 80. *Prof Exp:* Asst physics, Fla State Univ, 66-67; teaching asst physics, NC State Univ, 68-79; asst prof, Northern Ky Univ, 79-80; vis asst prof, Univ Ill, 80; ASST PROF PHYSICS, ILL STATE UNIV, 81- *Mem:* Sigma Xi; Am Phys Soc. *Res:* Phenomenological investigations into nucleon-nucleon interactions in the region of A-100; investigation into the effect of the spin-orbit interaction in heavy ion physics with a strong emphasis on developing new formalism to simplify calculations and reveal more easily measured symmetries. *Mailing Add:* Univ High Sch 1212 W Springfield Urbana IL 61801

GRUBER, ARNOLD, b New York, NY, Feb 28, 40; m 62; c 2. SATELLITE METEOROLOGY. *Educ:* Brooklyn Col, BS, 62; Fla State Univ, MS, 64, PhD(meteorol), 68. *Prof Exp:* Instr, Meteorol Dept, Fla State Univ, 68-69; res scientist, Boeing Co, Seattle, 69-70; res meteorologist, 70-77, hq staff, 77, chief, Model Appln Br, 77-80, actg dir, 80, CHIEF, ATMOSPHERIC SCI BR, SAT RES LAB, NAT ENVIRON SATELLITE, DATA & INFO SERV, NAT OCEANIC & ATMOSPHERIC ADMIN, 80- *Concurrent Pos:* Adj assoc prof, Meteorol Dept, Univ Md, 79-; mem, Solar & Earth Radiation Working Group, Nat Climate Prog, 81-; mem, Earth Radiation Budget Environ Sci Team; vis prof, chair remote sensing, US Naval Acad, 90-91. *Mem:* Sigma Xi; fel Am Meteorol Soc; Am Geophys Union; fel Am Meteorol Soc. *Res:* Satellite meteorology; vertical temperature structure refrievals; atmospheric radiation budget and application of quantitative satellite data to problems in weather analysis; forecasting and climate problems. *Mailing Add:* 1225 Port Echo Lane Bowie MD 20716

GRUBER, B(ERNARD) A, b New York, NY, Mar 2, 25; m 68; c 2. ELECTROCHEMICAL ENGINEERING. *Educ:* Columbia Univ, BS, 45, MS, 47. *Prof Exp:* Res chemist, Harshaw Chem Co, 47-48; res engr, Battelle Mem Inst, 48-51; chief engr rare metals & hydrides, Metal Hydrides, Inc, 51-54; proj leader, Ethyl Corp, 54-56; res mgr, Monsanto Co, Mass, 55-69, res mgr, Mo, 69-70, mgr com develop, 70-79, mgr external res, 79-82; RETIRED. *Concurrent Pos:* Consult, 82- *Mem:* Am Chem Soc; Am Inst Aeronaut & Astronaut; AAAS. *Res:* Electrochemical energy conversion; high temperature materials; extractive metallurgy. *Mailing Add:* 514 Coeur de Royale Creve Coeur MO 63141

GRUBER, CARL L(AWRENCE), b Chicago, Ill, Nov 30, 38; m 59; c 2. SOLID STATE SENSORS. *Educ:* Univ Ill, Urbana, BSEE, 60, MS, 61, PhD(shock waves), 67. *Prof Exp:* From asst prof to prof elec eng, SDak Sch Mines & Technol, 64-81; sr prin eng, Honeywell Inc, 81-82, mgr, Adv Sensor Technol, 82-83; prof & prog coordr, 83-87, PROF & CHMN, DEPT ELECT ENG & ELECTRONIC ENG TECHNOL, MANKATO STATE UNIV, 87- *Concurrent Pos:* Am Soc Eng Educ fac fel, NASA-Goddard Space Flight Ctr, 67-68; exchange prof, Bergen, Norway, 72-73; consult, Minn Laser Systs, Honeywell Inc, 81- *Mem:* Inst Elec & Electronics Engrs; Nat Soc Prof Engrs; Electrochem Soc; Am Soc Eng Educ. *Res:* Laser applications development; solid state sensors; gaseous electronics research; gaseous laser development; solid state materials processing & analysis. *Mailing Add:* Dept Elect Eng & Electronic Eng Technol Mankato State Univ Box 215 Mankato MN 56002

GRUBER, CHARLES W, b Cincinnati, Ohio, Mar 5, 10; m 38; c 3. AIR POLLUTION. *Educ:* Univ Cincinnati, ME, 32; Environ Eng Intersoc, dipl, 68. *Prof Exp:* Mgr appl eng, Consolidation Coal Co, 34-36 & Bimel Co, 36-38; air pollution control & heating engr, City of Cincinnati, 38-45 & 47-69; assoc prof, 69-77, EMER ASSOC PROF ENVIRON ENG, UNIV CINCINNATI, 77- *Concurrent Pos:* Mgr eng dept, Floyd & Co, 46-47; spec mem consult, Nat Adv Comn Community Air Pollution Control, USPHS, 52-57, prin investr, res grant, 58-60 & 64-67, dir, USPHS-Nat Ctr Air Pollution Control prog, 67-69; consult, 69- *Honors & Awards:* Frank A Chambers Award, Air Pollution Control Asn, 64. *Mem:* Air Pollution Control Asn (pres, 50-51); Am Soc Mech Engrs; Nat Soc Prof Engrs; Sigma Xi. *Res:* Air pollution control in field of enforcement practices, source measurement, control and administration. *Mailing Add:* 6309 Parkman Pl Cincinnati OH 45213

GRUBER, ELBERT EGIDIUS, b Cincinnati, Ohio, Sept 11, 10; m 34; c 4. POLYMER CHEMISTRY. *Educ:* Xavier Univ, BS, 32; Univ Ill, MS, 34, PhD(org chem), 37. *Hon Degrees:* LLD, Xavier Univ, 75. *Prof Exp:* Asst chem, Univ Ill, 35-37; res chemist, B F Goodrich Co, 37-43, group leader, 43-46, res supvr res ctr, 46-50; head plastics res, Gen Tire & Rubber Co, 50-55, asst dir res, 55-62, dir res & develop, 62-71, vpres & dir res & develop div, 71-75; chem consult, 75-87; RETIRED. *Concurrent Pos:* Chmn elastomers, Gordon Res Conf, 57; mem bd dirs, Indust Res Inst, Inc, 73-76 & Textile Res Inst, 73-76; mem adv bd, Grad Sch, Univ Akron, 73-78. *Mem:* Sigma Xi; Dirs Indust Res; Am Chem Soc; Soc Plastics Engrs. *Res:* Rubber antioxidants; polymerization; plastics; vinyls; polyurethane foams and elastomers; stereoregulated polymers; solid propellants; rubber and plastics compounding and processing; energy, utilization and supply. *Mailing Add:* 1148 W Market St, Apt 422 Akron OH 44313

GRUBER, EUGENE E, JR, b Pittsburgh, Pa, Apr 13, 33; m 56; c 4. PHYSICAL METALLURGY. *Educ:* Univ Pa, BS, 59; Carnegie Inst Technol, MS, 62, PhD(metall eng), 64. *Prof Exp:* From assoc metallurgist to metallurgist, 63-88, SR SCIENTIST, ARGONNE NAT LAB, 88- *Mem:* Sigma Xi. *Res:* Pores in solids; radiation damage; surface diffusion; capillarity phenomena; computer modeling of transient behavior of nuclear fuel for the fast breeder reactor, including fission-gas effects. *Mailing Add:* Argonne Nat Lab RE/207 9700 S Cass Ave Argonne IL 60439-4841

GRUBER, GEORGE J, b Diosjano, Hungary, Mar 20, 36; US citizen; m 63. BIOENGINEERING, EPIDEMIOLOGY. *Educ:* Cath Univ, BEE, 59, MS, 60, PhD(physics), 62. *Prof Exp:* Res asst phys acoust, Ultrasonic Lab, Cath Univ, 58-62; dir ultrasonic lab, Chesapeake Instrument Corp, Md, 62-63; asst res prof underwater acoust, Cath Univ, 63-65; asst prof mech eng, Univ Tex, Austin, 65-71; sr res scientist, 71-80, STAFF SCIENTIST, SOUTHWEST RES INST, 80- *Concurrent Pos:* Am Mach & Foundry Co fel, 62-63; Off Naval Res fel, 63-65; adj assoc prof, Univ Tex Health Sci Ctr, San Antonio, 74-80. *Mem:* Am Indust Hyg Asn; Am Inst Ultrasound Med; Asn Advan Med Instrumentation. *Res:* Human factors engineering; occupational safety and health; medical ultrasonics. *Mailing Add:* 202 N Haven San Antonio TX 78729

GRUBER, GERALD WILLIAM, b LaCrosse, Wis, Sept 12, 44; m 69; c 1. PHOTOCHEMISTRY. *Educ:* Loras Col, BS, 66; Case Western Reserve Univ, PhD(org chem), 70. *Prof Exp:* Nat Cancer Inst fel, Cornell Univ, 70-71; CHEMIST, PPG INDUSTS, INC, 71- *Mem:* Am Chem Soc. *Res:* Organic synthesis; photochemistry mechanisms; reactive intermediates; radiation polymerization. *Mailing Add:* 30560 Atlanta Lane Westlake OH 44145-1881

GRUBER, H THOMAS, b Atlantic City, NJ, Nov 20, 29; m 52; c 7. PHYSICS ENGINEERING, NONDESTRUCTIVE TEST. *Educ:* Ohio Wesleyan Univ, BA, 52. *Prof Exp:* Engr, Indust Controls Div, Minneapolis-Honeywell Regulator Co, 52-54; sr physicist, Eng Physics Dept, Battelle Mem Inst, 56-68; tech adminr & proj mgr, Battelle Develop Corp, 68-71; staff physicist, 71-81, res scientist, Fabrication & Qual Assurance Sect, 81-85, RES SCIENTIST NONDESTRUCTIVE TESTING PROJ OFF, DYNAMICS SECT, BATTELLE COLUMBUS LABS, 85- *Res:* Quality assurance; process control; automation of measurements and control; application of computers and micropressors; nondestructive testing; radiographer level II. *Mailing Add:* Battelle Columbus Labs 505 King Ave Columbus OH 43201

GRUBER, HELEN ELIZABETH, b Wallace, Idaho, Nov 6, 46. QUANTITATIVE BONE HISTOMORPHOMETRY. *Educ:* Univ Idaho, BS, 69; Ore State Univ, MS, 74, PhD(cell biol), 76. *Prof Exp:* Teaching asst biol, Ore State Univ, 70-72, res asst radiation biol, 74-76; fel radiation, Northwest Col & Univ Asn Sci, Richland, Wash, 72-73, Univ Iowa, 76-78; res asst med, Inst Med, Univ Wash, 78-81; res fel med, Univ Southern Calif, 81-84, res asst prof, 84-86; ASST PROF PEDIAT, UNIV CALIF, LOS ANGELES, 86- *Mem:* AAAS; Am Soc Cell Biol; Am Soc Bone & Mineral Res; Soc Exp Biol & Med; NY Acad Sci; Am Fedn Clin Res. *Res:* Bone and mineral metabolism; metabolic bone disease; primary human bone tumors; fracture healing; bone changes during pregnancy and lactation; biology of bone. *Mailing Add:* Dept Pediat Cedars-Sinai Med Ctr 8700 Beverly Blvd ASB-3 Los Angeles CA 90048

GRUBER, JACK, b Brooklyn, NY, Apr 18, 31; m 64; c 1. MEDICAL MICROBIOLOGY, VIROLOGY. *Educ:* Brooklyn Col, City Univ NY, BS, 54; Univ Ky, PhD(microbiol), 63. *Prof Exp:* Asst microbiol, Univ Ky, 55-61; res bacteriologist, US Army Biol Labs, 62-63, microbiologist, Immunol Br, Med Invest Div, Med Sci Lab, Ft Detrick, 63-70; microbiologist, prog admin, Viral Biol Br, Nat Cancer Inst, NIH, 70-72; chief off prog resources logistics, viral oncol, 72-78, asst & dep chief, 78-84, CHIEF, BIOL CARCINOGENESIS BR, DIV CANCER ETIOLOGY, NAT CANCER INST, NIH, 84- *Mem:* AAAS; Am Soc Microbiol; Tissue Cult Asn; Sigma Xi; NY Acad Sci; Am Soc Virol. *Res:* Viral etiology of cancer; carcinogenesis; scientific administration; anthropod-borne viruses; efficacy of viral and bacterial vaccines; immunization procedures; large-scale production of tumor viruses in tissue culture systems; radioisotope serology; streptococci and rheumatic fever. *Mailing Add:* Biol Carcinogenesis Br Nat Cancer Inst Exec Plaza N Bethesda MD 20892

GRUBER, JOHN B, b Hershey, Pa, Feb 10, 35; m 61; c 3. SOLID STATE PHYSICS, OPTICAL PHYSICS. *Educ:* Haverford Col, BS, 57; Univ Calif, Berkeley, PhD(chem physics), 61. *Prof Exp:* Guest prof, Inst Tech Physics, Tech Univ Darmstadt, Repub Ger, 61-62; asst prof physics, Univ Calif, Los

Angeles, 62-66; from assoc prof to prof chem physics, Wash State Univ, 66-75, from asst dean to assoc dean, Grad Sch, 68-72; prof physics & dean, Col Sci & Math, NDak State Univ, 75-80; prof physics & chem & vpres acad affairs, Portland State Univ, 80-84; PROF PHYSICS & CHEM & VPRES ACAD AFFAIRS, SAN JOSE STATE UNIV, 84-; EXEC VPRES, CALIF LASER INST TECHNOL, 86- *Concurrent Pos:* NATO fel, 61-62; consult, Aerospace Corp, Calif, 62-65, Missile & Space Systs Div, Douglas Aircraft Co, 63-69, Space & Info Systs Div, NAm Aviation, Inc, 64-66, Pac Northwest Labs, Battelle Mem Inst, 66-69, Douglas Lab, McDonnell Douglas Astronaut Corp, 68-74 & Los Alamos Sci Lab, Univ Calif, 69-71 & 73-74; consult, Ames lab, Iowa State Univ, 76-80, US Army Harry Diamond Lab, 80-, Naval Weapons Ctr, US Navy, China Lake, Calif, 84-, Union Carbide Corp, 80-, Battelle Res Ctr, Triangle Park, NC, 87, Gen Telephone & Electronics, 87- & US Dept Defense, Washington, DC, 86-; mem, Lunar Explor Sci Apollo-NASA Task Force, 64-66; gen conf chmn, Fourteenth Rare Earth Res Conf, 79; chmn, Off Naval Technol & Panel Eval Postdoctoral Fel Appln, Am Soc Eng Educ, 87-, session meetings, Nat Meetings of Am Phys Soc. *Honors & Awards:* Willig Award, 86. *Mem:* Fel Am Phys Soc; AAAS; NY Acad Sci; Sigma Xi; Am Asn Physics Teachers; fel Polish Acad Sci; USSR Acad Sci; fel Am Soc Eng Educ. *Res:* Optical and magnetic properties of rare earth and actinide ions in solids; optical spectroscopy; high temperature superconductors, tunable vibonic lasers. *Mailing Add:* Dept Physics San Jose State Univ San Jose CA 95192-0106

GRUBER, KENNETH ALLEN, b Brooklyn, NY, June 21, 48. PATHOPHYSIOLOGY, PEPTIDE BIOCHEMISTRY. *Educ:* NY Univ, BA, 71, PhD(med sci), 74. *Prof Exp:* Fel, Dept Anat, NY Univ, 71-74; fel, Dept Physiol Chem, Roche Inst Molecular Biol, 74-76; asst prof physiol, 76-81, ASSOC PROF MED & NEPHROLOGY, BOWMAN GRAY SCH MED, 81- *Concurrent Pos:* Adj assoc prof, Dept Biol, Long Island Univ, 74-75; adj prof, Dept Biol, Middlesex County Col, 76; foreign travel award, Nat Res Coun, 80 & Am Soc Nephrology, 81; res career develop award, Nat Heart, Lung & Blood Inst, 80. *Mem:* Am Phys Soc. *Res:* Pathophysiology of hypertension as it pertains to neuro-humeral influences, especially the role of natriuretic hormone; endogenous regulators of ion transport. *Mailing Add:* Dept Med (Nephrology) Wake Forest Univ med Ctr 300 S Hawthorne Rd Winston-Salem NC 27103

GRUBER, PETER JOHANNES, b Stamford, Conn, Aug 23, 41; m 71; c 2. ELECTRON MICROSCOPY, MEMBRANES. *Educ:* Carleton Col, BA, 62; Univ Wis, PhD(bot), 71. *Prof Exp:* Fel biochem, Mich State Univ, 71-73; asst prof, 73-79, assoc prof, 79-86, PROF BIOL SCI, MOUNT HOLYOKE COL, 86- *Mem:* Am Soc Plant Physiol; AAAS; Sigma Xi. *Res:* Plant peroxisomes: ultrastructure, cytochemistry, and biogenesis; membranes of plant cells: flow, surface characteristics, and lectin-binding properties. *Mailing Add:* Dept Biol Sci Mount Holyoke Col S Hadley MA 01075

GRUBER, SAMUEL HARVEY, b Brooklyn, NY, May 13, 38; m 69; c 2. BEHAVIORAL & PHYSIOLOGY ECOLOGY, SENSORY PHYSIOLOGY. *Educ:* Univ Miami, BA, 60, MS, 66, PhD(marine sci), 69. *Prof Exp:* Res scientist visual physiol, Inst Marine Sci, Univ Miami, 69-73; from asst prof to assoc prof behav physiol, 73-84, PROF MARINE SCI, ROSENTIEL SCH MARINE & ATMOSPHERIC SCI, UNIV MIAMI, 84- *Concurrent Pos:* prin investr, Off Naval Res Contract, 69-78, 80-81 & 82-85; NSF grant biol oceanog, 79-89, US-Japan coop res, 83-88, US-Israel Binational Sci Found, 82-85; dir, Bimini Biol Field Sta, 90; pres, Elasmobranch Consult. *Mem:* Am Inst Fishery Res Scientist; Am Fisheries Soc; Am Soc Ichthyologists & Herpetologists; Soc Neurosci; Sigma Xi; distinguished fel Am Elasinobranch Soc, 88; Am Soc Zool. *Res:* Psychophysics and electrophysiology of vision in marine vertebrates and how senses, especially vision, effect the natural behavior of these creatures; bioenergetics; population dynamics and behavioral ecology of sharks; role of apex predators in the marine ecosystem; shark-human interaction especially shark repellent research. *Mailing Add:* Div Biol & Living Resources Rosentiel Sch of Marine 4600 Rickenbacker Causeway Miami FL 33149-1099

GRUBER, SHELDON, b New York, NY, Sept 9, 30; m 55; c 2. ELECTRICAL ENGINEERING. *Educ:* Purdue Univ, BSEE, 52; Mass Inst Technol, ScD(elec), 58. *Prof Exp:* Asst prof elec eng, Mass Inst Technol, 59-62; res scientist fusion, Ctr Nuclear Studies, Saclay, France, 62-63; mem res staff plasma physics, Sperry Rand Res Ctr, 63-67; PROF ELEC ENG, CASE WESTERN RESERVE UNIV, 67-, DEPT HEAD. *Concurrent Pos:* Vis scientist, Inst Res, Hydro Que, 74, Ctr Nuclear Studies, Grenoble, France, 74-76 & Inst Prod Automation, Fraun hofer Soc, Stuttgart, WGermany, 86. *Mem:* Am Phys Soc; Inst Elec & Electronic Engrs. *Res:* Robot control; industrial inspection; machine vision. *Mailing Add:* Dept Elec Eng Case Western Reserve Univ 2040 Adelbert Rd Cleveland OH 44106

GRUBER, WILHELM F, b Bischofshofen, Austria, Aug 2, 13; nat US; m 46. ORGANIC CHEMISTRY. *Educ:* Univ Vienna, PhD(org chem), 39. *Prof Exp:* Instr chem, Univ Vienna, 40-48, asst prof, 48-51; res chemist, Smith, Kline & French Labs, 51-52; res chemist, Res Coun BC, Can, 52-54; res chemist, Univ Ill, 54-56; RES CHEMIST, ELASTOMERS DEPT, E I DU PONT DE NEMOURS & CO, INC, 56- *Mem:* Am Chem Soc; Sigma Xi. *Res:* Accelerators, antioxidants and antiozonants for rubber; furocoumarins; furochromones and related aromatic derr; pyridine chemistry; sulfur compounds. *Mailing Add:* 2218 Old Orchard Rd Wilmington DE 19810

GRUBIN, HAROLD LEWIS, b Brooklyn, NY, Mar 1, 39; m 61; c 2. PHYSICS. *Educ:* Brooklyn Col, BS, 60; Polytech Inst Brooklyn, MS, 62, PhD(physics), 67. *Prof Exp:* Theoret physicist, United Aircraft Res Labs, 66-68; sr theoret physicist, United Technologies Res Ctr, East Hartford, 68-81; VPRES SOLID STATE RES, SCI RES ASSOCS, 81- *Concurrent Pos:* Adj mem fac, Univ Hartford & Univ Mass. *Mem:* AAAS; Am Phys Soc; Inst Elec & Electronics Engrs. *Res:* Solid state theory, especially on conduction phenomena and device-related physics. *Mailing Add:* 50 Porter Dr West Hartford CT 06117

GRUBMAN, MARVIN J, b Bronx, NY, Nov 4, 45; m 70; c 2. VIROLOGY. *Educ:* City Col New York, BS, 67; Univ Pittsburgh, PhD(biochem), 72. *Prof Exp:* Res assoc, Albert Einstein Col Med, 72-76; RES CHEMIST, PLUM ISLAND ANIMAL DIS CTR, 76- *Concurrent Pos:* Res fel, NIH, 72-73; US Dept Agr, Agr Res Serv, Plum Island Animal Ctr; mem, Int Comt Taxonomy Virus Picornavirus study group, 83-87; rev panel mem, Molecular Biol & Genetic Mechanisms Biotechnol Prog, USDA competitive Grants Prog, 85. *Honors & Awards:* Cleveland Prize, AAAS, 82. *Mem:* Am Soc Microbiol; AAAS; Am Soc Virol. *Res:* Replication of foot and mouth disease virus, African horse sickness virus, blue tongue virus and rinderpest virus; identifying and mapping the viral nonstructural and structural proteins and determining their function in infected cells. *Mailing Add:* PO Box 1483 Southold NY 11971

GRUBMAN, WALLACE KARL, b New York, NY, Sept 12, 28; m 50; c 2. CORPORATE GENERAL MANAGEMENT. *Educ:* Columbia Univ, BS, 50; NY Univ, MS, 54. *Prof Exp:* Chem engr, Nat Starch & Chem Co, 50, mgt, 54-68, vpres, 68-74, group vpres, 75-77, pres & chief oper officer, 77-82, pres & chief exec officer, 81-83, chmn & chief exec officer, 83-85; EXEC DIR, UNILEVER PLC & NV, 86- *Concurrent Pos:* Mem eng adv coun, Columbia Univ, 78-; dir, Courtaulds Plc, 90- *Mem:* Am Inst Chem Engrs; Soc Chem Industs; Chem Mfrs Asn; Chem Indust Asn UK. *Res:* Engineering, manufacturing, marketing and research of a wide spectrum of specialty chemicals. *Mailing Add:* Nat Starch & Chem Co Ten Finderne Ave Bridgewater NJ 08807

GRUEMER, HANNS-DIETER, b Bochum, Ger, May 25, 24; US citizen. BIOCHEMISTRY, CLINICAL CHEMISTRY. *Educ:* Univ Frankfurt, MD, 49; Am Bd Clin Chem, cert. *Prof Exp:* Res biochemist, Univ Frankfurt, 49-50; sr physician Charite, Univ Berlin, 50-54; res biochemist, Univ Frankfurt, 54-56; biochemist, Pineland Hosp, Pownal, Maine, 56-57, dir, 58-63; asst prof path 63-67, asst prof physiol chem, 64-67, from assoc prof to prof path & physiol chem, Ohio State Univ, 67-77, chief div clin chem, 63-77; PROF CLIN PATH & DIR SECT CLIN CHEM, MED COL VA, 77- *Concurrent Pos:* Instr & res asst, Boston Dispensary, Tufts Univ, 57-63; mem comt clin chem, Nat Acad Sci, 73-75; dir, Am Bd Clin Chem, 74-; mem bd ed, Clin Chem, 74-; dir, Nat Registry Clin Chem, 75- *Mem:* Am Chem Soc; Am Soc Biol Chemists; Am Asn Clin Chemists. *Res:* Membrane disease in muscular dystrophies; clinical enzymology. *Mailing Add:* Box 696 MCV Sta Med Col Va Richmond VA 23298

GRUEN, DIETER MARTIN, b Germany, Nov 21, 22; nat US; m 48; c 3. CHEMICAL PHYSICS, SURFACE CHEMISTRY. *Educ:* Northwestern Univ, BS, 44, MS, 47; Univ Chicago, PhD(chem), 51. *Prof Exp:* Chemist, Manhattan Proj, Oak Ridge, 44-46; assoc chemist, 47-59, SR CHEMIST, ARGONNE NAT LAB, 60-, ASSOC DIR MAT SCI DIV, 83- *Concurrent Pos:* Vis scientist, Lawrence Radiation Lab, Univ Calif, 54-55; US deleg, Int Conf Peaceful Uses Atomic Energy, Switz, 58; chmn, Gordon Res Conf Fused Salts, 61; vis prof, Hebrew Univ Jerusalem, 69 & Norweg Tech Univ, 73; chmn, Int Conf Inelastic Ion Surface Collisions, 86; rev comt, Mat Sci Div, Lawrence Livermore Lab, 89. *Honors & Awards:* Mat Sci Award, Dept Energy, 88. *Mem:* Am Chem Soc; Sigma Xi; Mat Res Soc. *Res:* Physical and surface chemistry; application of lasers to surface science; multiphoton resonance ionization of sputtered atoms; velocity and excited state distributions; ultrasensitive surface analysis; electronic structure of matrix isolated metal clusters; high Tc-superconductors; FTMS of laser desorbed organic molecules. *Mailing Add:* 1324 59th St Downers Grove IL 60515

GRUEN, FRED MARTIN, b Nuernberg, Ger, Feb 4, 15; m 49; c 2. CHEMISTRY. *Educ:* City Col New York, BS, 48; Ill Inst Technol, MS, 50; Mich State Univ, PhD, 64. *Prof Exp:* Res chemist, Schering Corp, NJ, 50-51; assoc prof, 51-61, PROF CHEM & HEAD DEPT, OLIVET COL, 61- *Mem:* Am Chem Soc; Sigma Xi. *Res:* Applicability of Hammett equation to thiophenecarboxylic acids; synthesis and study of decomposition rates of some bis-thenoyl peroxides. *Mailing Add:* Acad Ctr Box 46 Olivet Col Olivet MI 49076

GRUEN, H(AROLD), b Brooklyn, NY, June 4, 31; m 53; c 3. ELECTRONICS ENGINEERING, COMMUNICATIONS. *Educ:* City Col New York, BEE, 52; Univ Pa, MS, 57. *Prof Exp:* Sect head commun eng, Philco Corp, 59-61, mgr marine systs group, 61-63, dir advan commun lab, Pa, 63-68, Philco-Ford Corp, 68-69, gen mgr, Sierra Electronic Oper, 69-76; pres, Aydin Vector, 76-82; PRES, FAIRCHILD COMM & ELEC CO, 82- *Concurrent Pos:* Mem, Franklin Inst, 60- *Mem:* Sr mem Inst Elec & Electronics Engrs. *Res:* Theoretical communications; statistical detection methods; maximal information transfer; coding theory; sensory data processing, for minimum bandwidth. *Mailing Add:* 4280 Massachusets Ave NW Washington DC 20016

GRUEN, HANS EDMUND, b Berlin, Ger, Oct 20, 25; nat US. MYCOLOGY. *Educ:* Brooklyn Col, BS, 51; Harvard Univ, MA, 53, PhD(biol), 57. *Prof Exp:* Asst biol, Harvard Univ, 51, asst mycol, Farlow Herbarium, 52, asst plant physiol, 53-54, Am Cancer Soc res fel, 56-58, res fel, Farlow Herbarium, 59-64; from asst prof to assoc prof, 64-73, PROF BIOL, UNIV SASK, 73 - *Concurrent Pos:* Lalor Found fel, Univ Tokyo, 63; assoc ed, Can J Bot, 80-83; actg co-ed, Can J Bot, 82-83. *Mem:* Mycol Soc Am. *Res:* Physiology of fungi, especially growth and development in reproductive structures of fungi. *Mailing Add:* Dept Biol Univ Sask Saskatoon SK S7N 0W0 Can

GRUENBAUM, WILLIAM TOD, b Columbus, Ohio, Sept 28, 48. SYNTHETIC ORGANIC CHEMISTRY, POLYMER SYNTHESIS. *Educ:* Ohio State Univ, BSc, 70; Univ Wis, PhD(org chem), 75. *Prof Exp:* Res chemist, 75-80, SR RES CHEMIST, EASTMAN KODAK CO, 80- *Mem:* Am Chem Soc. *Res:* Synthesis of chemicals for electrophotographic applications. *Mailing Add:* Res Labs B-59 Eastman Kodak Co Kodak Park Rochester NY 14650

GRUENBERG, ERNEST MATSNER, b New York, NY, Dec 2, 15; div; c 3. EPIDEMIOLOGY, PSYCHIATRY. *Educ:* Swarthmore Col, BA, 37; Johns Hopkins Univ, MD, 41; Yale Univ, MPH, 49, DrPH, 55. *Prof Exp:* Intern, St Elizabeth's Hosp, 41-42; resident psychiat, Bellevue Hosp, 46-48; clin assoc prof, Syracuse Univ, 52-55; assoc clin prof, 58-61, prof psychiat, Col Physicians & Surgeons, Columbia Univ, 61-75, dir, Psychiat Epidemiol Res Unit, 68-75; prof ment hyg & chmn dept, 75-81, PROF PSYCHIAT, SCH MED, SCH HYG & PUB HEALTH, JOHNS HOPKINS UNIV, 76- *Concurrent Pos:* Exec dir, State Ment Health Comn, 49-54; mem tech staff, Milbank Mem Fund, 55-61, tech bd, 61-68; lectr, Sch Pub Health, Harvard Univ, 59-69; attend psychiatrist, Presbyterian Hosp, New York, 66-75 & Hudson River Psychiat Ctr, 67-75; mem adv coun ment hyg, State of Md, 76-81; mem adv bd, Sch Hyg & Pub Health, Johns Hopkins Univ, 76-81. *Honors & Awards:* Lapouse Medal, Am Pub Health Asn, 76. *Mem:* Fel Am Psychiat Asn; fel Am Pub Health Asn; Am Epidemiol Soc; fel NY Acad Med; Sigma Xi. *Res:* Epidemiology of mental disorders. *Mailing Add:* 5225 Pooks Hill Rd-S222 Bethesda MD 20814

GRUENBERG, HARRY, b Vienna, Austria, Feb 5, 21; US citizen; m 46; c 3. ELECTRICAL ENGINEERING. *Educ:* Univ BC, BASc, 44; Calif Inst Technol, PhD, 49. *Prof Exp:* Instr elec eng, Calif Inst Technol, 46-49; assoc res officer, Microwave Sect, Nat Res Coun Can, 49-56; assoc prof, Syracuse Univ, 56-61, prof elec eng, 61-87; RETIRED. *Concurrent Pos:* Lectr, McGill Univ, 52-56; Fulbright res scholar, Copenhagen, Denmark, 66-67; vis res scientist, Nat Res Coun Can, 77-78; vis prof, Univ BC, 84-85. *Mem:* Sr mem Inst Elec & Electronics Engrs. *Res:* Electromagnetic theory; antennas and microwave components; aircraft navigational aids. *Mailing Add:* 145 Tejah Ave Syracuse NY 13210

GRUENBERGER, FRED J(OSEPH), b Milwaukee, Wis, Sept 24, 18; m 42; c 1. COMPUTER SCIENCE, MATHEMATICS. *Educ:* Univ Wis, BS, 40, MS, 48. *Prof Exp:* Instr comput, Univ Wis, 48-54; computer programmer, Gen Elec Co, 54-57; assoc mathematician, Rand Corp, 57-67; prof comput sci, San Fernando Valley State Col, 67-77; PROF COMPUT SCI, CALIF STATE UNIV, NORTHRIDGE, 77- *Concurrent Pos:* Ed, Popular Comput, 73-81. *Res:* Empirical number theory; computing education. *Mailing Add:* Dept Comput Sci Calif State Univ 18111 Nordhoff St Northridge CA 91330

GRUENER, RAPHAEL P, b Jerusalem, Israel, Mar 7, 39; m 63; c 2. PHYSIOLOGY. *Educ:* Univ Calif, Berkeley, BA, 61; Univ Ill, Urbana, MSc, 63, PhD(physiol), 66. *Prof Exp:* NATO fel, 66-67; asst res physiologist, Scripps Inst, Univ Calif, 67-68; asst prof, 68-76, assoc prof, 76-80, PROF PHYSIOL, COL MED, UNIV ARIZ, 80- *Res:* Muscle physiology and biochemistry; excitation-contraction-coupling; basic mechanisms of myopathies; membrane excitability. *Mailing Add:* Dept Physiol Univ Ariz Col Med Tucson AZ 85724

GRUENER, WILLIAM B, b Springfield, Mass, 1944. COMPUTER SCIENCE PUBLISHING. *Educ:* Boston Univ, BS, 67. *Prof Exp:* Computer sci ed, Addison Wesley Publ Co Inc, 74-82, exec ed, Computer Sci Div, 82-84, publ, Electronic Educ Prod, 84-86; MKT MGR, XYVISION, 88-; PUBL, COMPUTER SCI PRESS, W H FREEMAN & CO, 89- *Concurrent Pos:* Dir sales & mkt, Avanti Educ Systs, 86; consult, 87-; chair, Publ Bd, Asn Comput Mach, 89- *Mem:* Asn Comput Mach; Inst Elec & Electronics Engrs. *Mailing Add:* W H Freeman & Co 41 Madison Ave New York NY 10010

GRUENHAGEN, RICHARD DALE, b Davenport, Iowa, June 9, 41; m 79; c 1. WEED SCIENCE, PLANT REGULATORS. *Educ:* Iowa State Univ, BS, 63; NDak State Univ, MS, 67; NC State Univ, PhD(crop sci & plant physiol), 70. *Prof Exp:* Biologist chem & plastics, Union Carbide Corp, 71-72; sr plant physiologist, Agrichem Proj, 3M Co, 72-77; sr biologist, Agr Prod Div, Union Carbide Corp, 77-79; SR PLANT PHYSIOLOGIST, FMC CORP, 79- *Concurrent Pos:* Mem, Plant Growth Regulator Working Group, Nat Acad Sci-Nat Res Coun. *Mem:* AAAS; Weed Sci Soc Am; Am Soc Plant Physiologists; Plant Growth Regulator Soc Am; Sigma Xi. *Res:* Parameters affecting herbicide activity/performance; discovery and evaluation of herbicides and plant regulators; investigation of mechanisms of action and reactors effecting efficacy. *Mailing Add:* 1610 Scenic Dr West Trenton NJ 08628

GRUENSTEIN, ERIC IAN, b New York, NY, Feb 23, 42; m 65; c 3. BIOCHEMISTRY. *Educ:* Princeton Univ, AB, 64; Duke Univ, PhD(biochem), 70. *Prof Exp:* Fel endocrinol, NIH, 70-72; fel biochem, Mass Inst Technol, 72-74; from asst prof to assoc prof, Dept Biol Chem, 74-88, PROF, DEPT MOLECULAR GEN, BIOCHEM & MICRO, MED SCH, UNIV CINCINNATI, 89- *Mem:* Biophys Soc; Am Soc Cell Biol; Sigma Xi. *Res:* Membrane biochemistry and biophysics; role of membrane transport in cell growth and in the etiology of cystic fibrosis and other genetic disorders. *Mailing Add:* Dept Molecular Gen Biochem & Microbiol Univ Cincinnati Col Med Cincinnati OH 45267-0524

GRUENWALD, GEZA, b Budapest, Hungary, Sept 13, 19; US citizen; m 49; c 1. PLASTICS ENGINEERING. *Educ:* Tech Univ Berlin, MS & PhD(org chem), 43. *Prof Exp:* Consult plastic & resins, Ger, 46-51; chemist, Farbwerke Hoechst AG, 51-57; develop chemist, Gen Elec Co, Mass, 57-60, chemist insulation res & develop, 60-68, mgr, Insulation Non-Metals Lab, 68-80; ENG CONSULT, 80- *Mem:* Am Chem Soc; German Chem Soc; Soc Plastics Eng. *Res:* Effect of plasticizers polymer properties; mechanism of crystallization of high polymers; reactivity of epoxy curing agents; fracture mechanism on polymer products; cold forming of polymers with aromatic chain links; solar energy engineering; thermorforming of plastics. *Mailing Add:* 36 W 34th St Erie PA 16508

GRUENWEDEL, DIETER WOLFGANG, MOLECULAR BIOLOGY, TOXICOLOGY. *Educ:* Eberhard Karl Univ Tübingen, Germany, PhD(biochem), 53. *Prof Exp:* PROF FOOD BIOCHEM & FOOD TOXICOL, UNIV CALIF, DAVIS, 66- *Mailing Add:* Dept Food Sci/Tech Univ Calif Davis CA 95616

GRUETT, MONTE DEANE, b Mobridge, SDak, June 2, 33; m 57; c 3. SYNTHETIC ORGANIC CHEMISTRY. *Educ:* Morningside Col, BS, 55; Univ Minn, MS, 58; Rensselaer Polytech Inst, PhD(org chem), 62. *Prof Exp:* From asst res chemist to assoc res chemist, Sterling-Winthrop Res Inst, 58-68, res chemist, 68-75, sr res chemist, 75-90, RES INVESTR, STERLING RES GROUP, 90- *Mem:* Am Chem Soc. *Res:* Synthesis of new antibacterial and anti viral agents; nitrogen heterocyclics; antiasthmatic agents; cardiotonic anti-hypertensive analgesic agents and antitumor agents. *Mailing Add:* Sterling Res Group Rensselaer NY 12144

GRUGER, EDWARD H, JR, b Murfreesboro, Tenn, Jan 21, 28; m 52; c 3. BIO-ORGANIC CHEMISTRY. *Educ:* Univ Wash, BS, 53, MS, 56; Univ Calif, Davis, PhD(agr chem), 68. *Prof Exp:* Chemist fishery technol, US Fish & Wildlife Serv, 53-54; asst gen & org chem, Univ Wash, 54-55; chemist org chem, Technol Lab, US Bur Com Fisheries, 55-59, proj leader marine oil chem, 56-65, supvry chemist, 59-62, res chemist, 62-71; res chemist, Nat Oceanic & Atmospheric Admin, 71-77, supvry res chemist, NW & Alaska Fisheries Ctr, Nat Marine Fisheries Serv, 77-83; RETIRED. *Concurrent Pos:* Res assoc, Agr Exp Sta, Univ Calif, Davis, 65-68; res prof chem, Seattle Univ, 77-82. *Mem:* AAAS; Am Chem Soc; Am Oil Chemists' Soc; Sigma Xi; Soc Environ Toxicol Chem. *Res:* Bio-organic chemistry of marine organisms and studies on effects of organic contaminants in marine life. *Mailing Add:* 3727 NE 193rd St Seattle WA 98155-2750

GRUHN, JOHN GEORGE, b Brooklyn, NY, Sept 8, 18; m 43; c 4. PATHOLOGY. *Educ:* Manhattan Col, BS, 41; Long Island Col Med, MD, 44. *Prof Exp:* Res fel med, Long Island Col Med, 47-48; resident, King's County Hosp, 48-49; asst pathologist & dir blood bank, Montefiore Hosp, 51-55; instr path, Sch Med, Univ Pittsburgh, 56-58, from asst prof to assoc prof, Chicago Med Sch, 58-66, clin prof, 66-81; dir lab, Skokie Valley Community Hosp, 64-81; ASSOC PROF PATH, NORTHWESTERN UNIV, 81- *Concurrent Pos:* Dir path lab, St Joseph's Hosp, Pittsburgh, Pa, 55-60; pathologist, Mt Sinai Hosp, Chicago, Ill, 60-64. *Mem:* Am Asn Path & Bact; Col Am Path; Am Soc Clin Path; Int Acad Path. *Res:* History of pathology. *Mailing Add:* Dept Pathol Rush Med Col 1753 W Congress Pkwy Chicago IL 60612

GRUHN, THOMAS ALBIN, b Eureka, Calif, Aug 18, 42; m 65; c 2. PHYSICAL CHEMISTRY. *Educ:* Univ San Francisco, BS, 64; Univ Calif, Berkeley, PhD(chem), 68. *Prof Exp:* From asst prof to assoc prof, 67-84, PROF CHEM, UNIV SAN FRANCISCO, 84-, CHMN DEPT, 76- *Mem:* Am Chem Soc. *Res:* Etching mechanisms of solid state particle detectors; radiation chemistry of polymers. *Mailing Add:* Dept Chem Univ San Francisco San Francisco CA 94117

GRULA, MARY MUEDEKING, b Minneapolis, Minn, Sept 26, 20; m 52; c 4. MICROBIAL ECOLOGY. *Educ:* Univ Minn, BA, 41, PhD(bact, chem), 46. *Prof Exp:* Instr bact, Univ Ky, 46-47, asst prof, 47-52; res assoc, Okla State Univ, 59-80, prof microbiol, 80-90; RETIRED. *Concurrent Pos:* NSF res grant, 63-66 & 69-71; Okla State Univ Res Found grant, 69-71; investr, US Bur Mines-Energy Res & Develop Admin contract, 74-75. *Mem:* AAAS; Am Soc Microbiol; Brit Soc Gen Microbiol. *Res:* Microbial nutrition; cell division and its control in a species of Erwinia; microbiology of deep disposal wells. *Mailing Add:* Dept Microbiol 307 Life Sci East Okla State Univ Stillwater OK 74078

GRUM, ALLEN FREDERICK, b Washington, DC, Oct 6, 31; m 57; c 5. OPERATIONS RESEARCH, DECISION ANALYSIS. *Educ:* US Mil Acad, BS, 53; Mass Inst Technol, MS, 58; Stanford Univ, PhD(decision anal), 76. *Prof Exp:* Res & develop staff officer eng equip, Hq, Dept Army, 68-70; assoc prof systs & decision anal, 74-80, PROF & HEAD, DEPT ENG, US MIL ACAD, 81- *Concurrent Pos:* Mem Educ Coun, Mass Inst Technol, 77- *Mem:* Mil Opers Res Soc; Inst Mgt Sci; Opers Res Soc Am; Am Soc Eng Educr. *Res:* Application of quantitive methods to defense decision making. *Mailing Add:* Dept Eng Mercer Univ Main Campus 1400 Coleman Ave Macon GA 31207

GRUMBACH, LEONARD, b Brooklyn, NY, July 24, 14; m 41; c 4. PHYSIOLOGY. *Educ:* Cornell Univ, AB, 34, AM, 35, PhD(physiol), 39. *Prof Exp:* Col Dent, NY Univ, intr pharmacol, 46-47; prof physiol, Des Moines Still Col Osteop, 47-52; assoc mem, Sterling-Winthrop Res Inst, 52-58, mem, 58-61; assoc prof, 61-80, PROF PHYSIOL, ALBANY MED COL, 80- *Mem:* AAAS; Am Physiol Soc. *Res:* Pharmacology of analgesics; physiology of pain. *Mailing Add:* New Scotland Ave Albany Med Col Albany NY 12208

GRUMBACH, MELVIN MALCOLM, b New York, NY, Dec 21, 25; m 51; c 3. PEDIATRICS, MEDICINE. *Educ:* Columbia Univ, MD, 48. *Hon Degrees:* Dr, Geneva, 91. *Prof Exp:* Pediatric residency, Babies Hosp, Columbia Presby Med Ctr, 49-51; asst pediat, Sch Med, Johns Hopkins Univ, 53-55; instr, Col Physicians & Surgeons, Columbia Univ, 55-56, assoc, 56-57, from asst prof to assoc prof, 57-65; chmn dept & dir Pediat Serv, 66-86, PROF SCH MED, UNIV CALIF, SAN FRANCISCO, 65-, EDWARD B SHAW PROF PEDIAT, 83- *Concurrent Pos:* Mem, Human Embryol & Develop Study Sect, 62-66, Endocrine Study Sect, 67-71, bd sci counrs, Nat Inst Child Health & Human Develop, NIH, 71-75; mem basic sci adv comt, 61-, chmn & mem, Clin Res Adv Comt, Nat Found March Dimes, 74-; consult, Letterman Gen Hosp, San Francisco, Childrens Hosp & US Naval Hosp, Oakland; mem, Int Scientific Coun, Found Princess Marie-Christine, Brussels, 80-; vis prof, Univ Wash, 79, Sch Med, Yale Univ, 81 & Univ Calif, 81, Univ NC, 82, Royal Soc Med, London, 85; Felton bequests prof, Royal Childrens Hosp, Melbourne, Australia, 83. *Honors & Awards:* Joseph Mather Smith Prize, Col Physicians & Surgeons, 62; Bicentennial Silver Medal, Col Physicians & Surgeons, 67; Borden Award, Am Acad Pediat, 71; Frederick C Moll lectr, Univ Wash, 79; Canon Eley lectr, Childrens Hosp Med Ctr, Harvard Med Sch, 79; Mali Dittman lectr, Univ Chicago, 80; Frederick M Kenny Memorial lectr, Childrens Hosp Pittsburgh, 81; Winthrop Award lectr, Am Fertility Soc, 81; Grover Powers lectr, Yale Univ Sch Med, 81; John Lind lectr, Kawlinska Inst, Stockholm, 84; Bilberbuck lectr, Ore Health Sci Univ,

86; Matthew Steiner lectr, Northwestern Univ, 89. *Mem:* Inst Med-Nat Acad Sci; fel AAAS; Am Soc Clin Invest; Soc Pediat Res; Lawson Wilkins Pediat Endocrine Soc (pres, 75-76); Endocrine Soc (pres, 81-82); Am Pediat Soc (pres, 89-90); Am Acad Pediat; Sigma Xi; hon mem Royal Soc Med; Soc Pediat Endocrin; corresp mem Europ Soc Pediat. *Res:* Metabolic and endocrine disorders of childhood and adolescence; abnormalities of sex differentiation; effects of hormones on growth and maturation; developmental neuroendocrinology, genetic and chromosome aberrations. *Mailing Add:* Dept Pediat Sch Med Univ Calif San Francisco San Francisco CA 94143-0106

GRUMBACH, ROBERT S(TEPHEN), b Morgantown, WVa, Dec 3, 25; m 51; c 2. ELECTRICAL ENGINEERING. *Educ:* Case Inst Technol, BSEE, 47; Univ WVa, MS, 51. *Prof Exp:* Distribution engr, Monongahela Power Co, 47; instr elec eng, Univ WVa, 47-53; elec engr, Elec Utility Eng Sect, Westinghouse Elec Corp, 53-57; prof elec eng, Tech Inst Bandung, Indonesia, 57-61, head dept, 60-61; assoc prof, 61-88, EMER PROF ELEC ENG, UNIV AKRON, 88- *Honors & Awards:* Centennial Medal, Inst Elec & Electronics Engrs, 84. *Mem:* Inst Elec & Electronics Engrs. *Res:* Electric utility engineering; generation; transmission and distribution of electrical energy, especially lighting protection and relaying. *Mailing Add:* Univ Akron Akron OH 44325

GRUMBLES, JIM BOB, b San Saba, Tex, Aug 26, 28; m 52; c 3. RANGE MANAGEMENT, ECOLOGY. *Educ:* Southwest Tex State Col, BS, 58; Tex A&M Univ, MS, 61, PhD(range mgt), 64. *Prof Exp:* Jr high sch instr, Tex, 58; asst range mgt, Tex A&M Univ, 58-59; from asst prof to assoc prof, Utah State Univ, 62-68; regional tech specialist, Field Res & Develop Herbicides, Dow Chem Co, Tex, 68-74; mgr res & develop, Agrovet Dept, Dow Quimica Mexicana SAm, 75-76; FIELD RES HERBICIDES, TEX DIV RES & DEVELOP, DOW CHEM CO, 77- *Mem:* Int Soc Range Mgt; Sigma Xi. *Res:* Wildlife management; grazing relationship of steers and white-tailed deer in south Texas; grazing management of seeded introduced grasses; studies involving both grazing by livestock and clipping to stimulate grazing. *Mailing Add:* PO Box 137 Montalba TX 75853-0137

GRUMER, EUGENE LAWRENCE, b New York, NY, May 25, 40; m 64, 84; c 3. ECONOMICS, PLANNING. *Educ:* Pa State Univ, BS, 61, MS, 62; Mass Inst Technol, SM, 64, ChemE, 65. *Prof Exp:* Engr, Celanese Corp, 65-69, sr engr, 69-70; econ engr, Hess Oil Virgin Islands Corp, 71-74, mgr oper planning/econ, 74-83; admin asst to vpres refining, 83-85, refinery mgr, planning/ econ, 85-90, COORDR, PLANNING & REFINERY ECON, AMERADA HESS CORP, 90- *Mem:* Am Chem Soc; Am Inst Chem Engrs; Am Asn Cost Engrs. *Res:* Planning, economic evaluation, fuels blending, budgeting and ranking of research, development and major ventures, particularly for the petrochemical process industry; combustion; explosions; fuels; safety; operations and shipment coordination and planning. *Mailing Add:* Southwyck Village Maddaket 7 Scotch Plains NJ 07076

GRUMER, JOSEPH, b New York, NY, Mar 18, 18; m 39; c 3. FUEL SCIENCE. *Educ:* City Col New York, BS, 37; NY Univ, MS, 41. *Prof Exp:* Anal chemist, Molnar Labs, NY, 37-39; org chemist, Fine Orgs, Inc, 39-41; gas analyst, US Bur Mines, 41-44, phys chemist, 44-53, from actg chief to chief flame res sect, 53-59, proj coordr flame dynamics, 59-76; CONSULT, 76- *Mem:* Am Chem Soc; Combustion Inst. *Res:* Analytical procedures; organic synthesis; flame stability; carbon formation and burning velocity; kinetics; air entrainment; combustion characteristics and exchangeability of fuels on gas burners; liquid pool burning; uncontrolled fires; air pollution by flames; coal dust explosions and extinguishment. *Mailing Add:* 1008 Farragut St Pittsburgh PA 15206-1744

GRUMET, ALEX, b New York, NY, Sept 1, 19; m 46; c 2. ELECTRONICS, PHYSICS. *Educ:* City Col New York, BEE, 42; Polytech Inst Brooklyn, MEE, 58. *Prof Exp:* Jr engr, Signal Corps, Gen Develop Lab, Ft Monmouth, NJ, 42-43; develop engr, Telicon Corp, NY, 46-47; TV instr, NY Tech Inst, NJ, 47-48; design engr, Pilot Radio Corp, NY, 48-49; sr proj engr, Edo Corp, 50-56; prog scientist, Res Div, Repub Aviation Corp, 56-65; staff scientist, Electronics Syst Res, Grumman Aerospace Corp, 65-83; PROF ENG, HOFSTRA UNIV, 83- *Mem:* NY Acad Sci; Optical Soc Am; Inst Elect & Electronics Engrs. *Res:* Investigation of nuclear magnetic resonance phenomena in liquids for application as a gyroscopic inertial sensing element; statistical communication theory for improving long-range communication and signal detection; statistical communication theory; optical matched filter processing. *Mailing Add:* 160-20 21st Ave Whitestone NY 11357

GRUMET, MARTIN, b Brooklyn, NY, June 30, 54; m 86; c 1. DEVELOPMENTAL NEUROBIOLOGY, NEURON GLIAL CELL-CELL INTERACTION. *Educ:* Cooper Union, BS, 76; Johns Hopkins Univ, PhD(biophys), 80. *Prof Exp:* Res assoc, 80-81, NIH fel, 81-84, ASST PROF NEUROBIOL, ROCKEFELLER UNIV, 84- *Concurrent Pos:* Prin investr, NIH grant, 84- & Irma T Hirschl Career Scientist Award, 86- *Mem:* AAAS; Am Soc Cell Biol; Soc Neurosci. *Res:* Molecular level transmission of information within and between cells, focusing on defining, isolating and analyzing molecules that mediate interactions at the surfaces of cells from the nervous system. *Mailing Add:* Dept Develop & Molecular Biol Rockefeller Univ 1230 York Ave New York NY 10021

GRUMMER, ROBERT HENRY, b Luzerne, Iowa, June 27, 16; m 40; c 2. ANIMAL HUSBANDRY. *Educ:* Iowa State Univ, BS, 39; Univ Wis, MS, 43, PhD(biochem, animal husb), 46. *Prof Exp:* From instr to assoc prof, 43-53, chmn dept, 54-63, prof, 53-80, EMER PROF MEAT & ANIMAL SCI, UNIV WIS-MADISON, 80- *Mem:* Am Soc Animal Sci. *Res:* Nutrition; genetics; physiology of reproduction and parasitology of swine. *Mailing Add:* Dairy Sci 266 Animal Sci Bldg Univ Wis 1675 Observatory Dr Madison WI 53706

GRUMMITT, OLIVER JOSEPH, organic chemistry; deceased, see previous edition for last biography

GRUN, PAUL, b New York, NY, May 14, 23; m; c 2. CYTOLOGY, CYTOGENETICS. *Educ:* Univ NC, BA, 44; Cornell Univ, PhD(plant breeding), 49. *Prof Exp:* Asst, Cornell Univ, 45-49; plant biologist, Carnegie Inst Washington, 49-54; from asst prof to assoc prof genetics, 54-63, prof cytol & cytogenetics, 64-89, EMER PROF CYTOL & CYTOGENETICS, PA STATE UNIV, UNIVERSITY PARK, 89- *Concurrent Pos:* Res fel, Inst Cell Res & Genetics, Karolinska Inst, Sweden, 53-54; NIH spec fel, Univ Calif, Los Angeles, 61-62; sabbatical, Univ Calif, Berkeley, 71-72, Louis Pasteur Univ, 83-84. *Mem:* Genetics Soc Am; Soc Study Evolution; Bot Soc Am; Potato Asn Am. *Res:* Cytogenetics of Solanum; cytoplasmic factors in plant evolution; protoplast manipulations of cytoplasmic factors of Solanum. *Mailing Add:* Dept Hort 102 Tyson Bldg Pa State Univ University Park PA 16802

GRUNBAUM, BRANKO, b Osijek, Yugoslavia, Oct 2, 29; US citizen; m 54; c 2. POLYHEDRA, TESSELLATIONS. *Educ:* Hebrew Univ, Jerusalem, MSc, 54, PhD(math), 58. *Prof Exp:* Mem, Inst Advan Study, 58-60; vis asst prof math, Univ Wash, 60-61; from asst prof to assoc prof, Hebrew Univ, Jerusalem, 61-65; vis prof, Mich State Univ, 65-66; prof, Univ Wash, 66-69; vis prof, Mich State Univ, 69-70; PROF MATH, UNIV WASH, 70- *Concurrent Pos:* J S Guggenheim Mem Found fel, 81-82. *Honors & Awards:* L R Ford Award, Math Asn Am, 75, C B Allendoerfer Award, 78. *Mem:* Am Math Soc; Am Math Asn. *Res:* Convexity; combinatorial geometry; graph theory. *Mailing Add:* Univ Wash GN-50 Univ Wash Seattle WA 98195

GRUNBERG, EMANUEL, b Everett, Mass, July 9, 22; m 46; c 2. CHEMOTHERAPY. *Educ:* Univ Ala, BA, 43; Yale Univ, PhD(bact), 46; Am Bd Med Microbiol, dipl. *Prof Exp:* Bacteriologist, J E Seagram & Sons, Ky, 46; sr bacteriologist, 46-57, sr scientist, 57-59, assoc dir dept, 59-60, dir dept chemother, 60-73, dir diag res, 70-77, assoc dir biol res, 73-76, dir chemother & diag res, 76-80, DIR BIOL RES, HOFFMANN-LA ROCHE INC, 80- *Concurrent Pos:* Chief div mycol, Dept Dermat, Dept Health, Newark, 51-67; guest lectr, Div Postgrad Med, Seton Hall Col Med & Dent, 59-63; adj assoc prof pharmacol, Med Col, Cornell Univ, 73-80, adj prof, 80-; adj prof med, Sch Med, Univ Calif, San Diego, 80- *Mem:* AAAS; Am Soc Microbiol; Soc Exp Biol & Med; Sigma Xi; fel Am Acad Microbiol. *Res:* Chemotherapeutic and diagnostic research. *Mailing Add:* 825 Egret Circle Apt A403 Delray Beach FL 33444

GRUNBERG, NEIL EVERETT, b Newark, NJ, Aug 9, 53; m 74; c 1. MEDICAL PSYCHOLOGY, PSYCHOENDOCRINOLOGY. *Educ:* Stanford Univ, BS, 75; Columbia Univ, MA, 77, MPhil, 79, PhD(psychol), 80. *Prof Exp:* Asst prof, 79-84, ASSOC PROF MED PSYCHOL, UNIFORMED SERVS, UNIV HEALTH SCI, 84- *Concurrent Pos:* Res assoc, Nat Cancer Inst, 82-87. *Mem:* Acad Behav Med Res; Soc Behav Med; Am Psychol Asn; Soc Psychologists in Addictive Behav. *Res:* Behavioral and biological studies of appetitive behaviors, including cigarette smoking and eating behavior; commonalities among substances of abuse and stress. *Mailing Add:* Psychol Dept Uniformed Serv Univ Sch Med 4301 Jones Bridge Rd Bethesda MD 20814

GRUNBERGER, DEZIDER, b May 29, 22; c 2. CHEMICAL CARCINOGENESIS, GENE EXPRESSION. *Educ:* Czechoslovak Acad Sci, PhD(biochem), 56. *Prof Exp:* Dept head, Inst Organic Chem & Biochem, Czech Acad Sci, Prague, 50-68; asst prof pharmacol, George Wash Univ, 64-65; PROF BIOCHEM & MOLECULAR BIOPHYS, COL PHYSICIANS & SURGEONS, COLUMBIA UNIV, 68- *Concurrent Pos:* Prin investr res grants, NIH, Environ Protection Agency & Am Cancer Soc; consult, NIH study sect; vis prof, Weizmann Inst, Israel, 83; scholar, Rockefellar Found, 82, Leukemia Soc, 72-77. *Mem:* NY Acad Sci; AAAS; Am Cancer Soc; Harvey Soc; Am Soc Biochem & Molecular Biol; Am Asn Cancer Res. *Res:* The mechanism of action of chemical carcinogens; the correlation between the structure of DNA modified by carcinogens and types of mutations induced in eukoryotic cells on the level of endogenous genes; development of antitumor drugs from propolis. *Mailing Add:* Col Physicians & Surgeons Columbia Univ 701 W 168 St New York NY 10032

GRUND, VERNON ROGER, b Baudette, Minn, Oct 27, 46. PHARMACOLOGY. *Educ:* Univ Minn, BS, 69, PhD(pharmacol), 74. *Prof Exp:* Postdoctorate clin pharmacol, Dept Internal Med, Univ Minn, 74-75; asst prof biochem pharmacol, Sch Pharm, State Univ NY, Buffalo, 75-78; ASSOC PROF PHARMACOL, COL PHARM, UNIV CINCINNATI MED CTR, 78-, CHMN, DIV PHARMACOL & MED CHEM, 85- *Concurrent Pos:* Vis assoc prof pharmacol, Univ BC, Can, 89; mem, New Investr Grant Rev Panel, Am Asn Col Pharm & Biol Sci Prog Comt, 90-91, Res & Grad Affairs Comt, 91-92. *Mem:* AAAS; Am Asn Col Pharm; Am Cancer Soc; Am Chem Soc; Am Diabetes Asn; Am Heart Asn; Am Lung Asn; Am Pharmaceut Asn; Am Soc Pharmacol & Exp Therapeut. *Res:* Immunochemical modulation of metabolic homeostasis; causes, complications and treatment of diabetes; influence of stress mediators on free fatty acid mobilization and insulin secretion; cyclic nucleotide changes in hypertension; author of numerous scientific publications. *Mailing Add:* Div Pharmacol & Med Chem Col Pharm Univ Cincinnati Med Ctr Cincinnati OH 45267-0004

GRUNDBACHER, FREDERICK JOHN, b Switz, Aug 3, 26; US citizen; m 66; c 2. HUMAN GENETICS. *Educ:* Swiss Fed Inst Technol, dipl, 53; Univ Calif, PhD(genet), 60. *Prof Exp:* Res geneticist & asst specialist, Univ Calif, 60-61; asst prof biol & genetics, Med Col Va, 65-69, assoc prof genetics, 69-74; assoc prof, 74-75, actg chmn, 75-84, PROF GENETICS, DEPT BASIC SCI, UNIV ILL COL MED, PEORIA, 74- *Concurrent Pos:* NIH fels human genet, Med Sch, Univ Mich, Ann Arbor, 61-65; NIH grants. *Mem:* AAAS; Am Soc Human Genetics; Soc Exp Biol & Med; Am Asn Blood Banks. *Res:* Immunogenetics of man; quantitative genetics of immunoglobulin and isoantibody levels; variation in erythrocyte and soluble antigens; etiology of ABO hemolytic disease of the newborn; study of a predisposition to allergies; preciptin reactions, mitogenic activity, serologic specificity and biochemistry of lectins, determinant groups of human ABH and related antigens; characterization and inheritance of human mucin antigens. *Mailing Add:* Dept Basic Sci Univ Ill Col Med Box 1649 Peoria IL 61656

GRUNDER, ALLAN ANGUS, b Kincardine, Ont, Mar 13, 35; m 72; c 1. GENETICS, ANIMAL HUSBANDRY. *Educ:* Ont Agr Col, BSA, 58; Univ Alta, MSc, 61; Univ Calif, PhD(genetics), 66. *Prof Exp:* Drainage surveyor, Ont Dept Agr, 58; salesman, Ogilvy Five Roses Co, 58-59; RES SCIENTIST GENETICS, ANIMAL RES CTR, CENT EXP FARM, AGR CAN, 66- *Mem:* Poultry Sci Asn; Sigma Xi; Genetics Soc Am; Genetics Soc Can; Can Soc Animal Sci; World's Poultry Sci Asn; Am Genetic Asn. *Res:* Elucidation of DNA restriction fragment polymorphisms associated with egg shell quality of Gallus domesticus; goose production; chicken meat quality. *Mailing Add:* Animal Res Ctr Agr Can Ottawa ON K1A 0C6 Can

GRUNDER, FRED IRWIN, b Detroit, Mich, Aug 17, 40; m 66; c 2. ENVIRONMENTAL CHEMISTRY, BIOLOGICAL MONITORING. *Educ:* Univ Mich, BSE, 63, MS, 67. *Prof Exp:* Res assoc indust health, Univ Mich, 60-69; asst lab dir environ chem, George D Clayton & Assocs, 69-72; dir environ chem & toxicol, Bethlehem Steel Corp, 72-85; DIR INDUST HYG, AM MED LABS, 85- *Concurrent Pos:* Mem, Subcomt Anal Procedures, Indust Hyg Comt, Am Iron & Steel Inst, 74 & Subcomt Anal Methods, Environ Qual Control Comt, 75 & chmn, 79-81; mem, Biol Monitoring Comt, Am Indust Hyg Asn, 84-; mem toxics planning & oversight panel, Chesapeake Res Consortium, 90. *Mem:* Am Chem Soc; Am Indust Hyg Asn; Am Soc Testing Mat. *Res:* Industrial hygiene; environmental chemistry; development of analytical methods; measurement of environmental contaminants absorbed by the body. *Mailing Add:* Am Med Labs 11091 Main St Fairfax VA 22030

GRUNDER, HERMANN AUGUST, b Basel, Switz, Dec 4, 31; US citizen; m 52; c 3. NUCLEAR STRUCTURE. *Educ:* Tech Hochschule, Ger, MS, 58; Univ Basel, Switz, PhD(physics), 67. *Prof Exp:* Bevatron mech engr, 59-62, group leader, 88-Inch Cyclotron, 62-64, Bevatron develop group leader, 68-73, group leader, Bevalac Opers, 73-79, LAB ASSOC DIR & HEAD, ACCELERATOR & FUSION RES DIV, LAWRENCE BERKELEY LAB, 79- *Concurrent Pos:* Consult, M D Anderson Med Ctr; Sr Scientist Award, Alexander von Humboldt Found, 79. *Mem:* Am Phys Soc; Europ Phys Soc; Swiss Phys Soc. *Res:* Design, construction and operation of accelerators for nuclear science and biomedical research and of neutral beam injectors for fusion energy experiments. *Mailing Add:* 12070 Jefferson Ave Newport News VA 23606

GRUNDLEGER, MELVIN LEWIS, b Brooklyn, NY, Nov 21, 47; c 2. DIABETES. *Educ:* Columbia Univ, PhD(nutrit & biochem), 79. *Prof Exp:* ASST PROF NUTRIT, COLO STATE UNIV, 82- *Mem:* Am Inst Nutrit; Sigma Xi. *Res:* Fiber; obesity. *Mailing Add:* 1806 Brookwood Dr Ft Collins CO 80525

GRUNDMANN, ALBERT WENDELL, b Salt Lake City, Utah, Sept 1, 12; m 47; c 1. PARASITOLOGY, MEDICAL ENTOMOLOGY. *Educ:* Univ Utah, BA, 37, MA, 39; Kans State Univ, PhD(med entom), 42. *Prof Exp:* Asst biol, Univ Utah, 37-39 & Kans State Col, 39-41; inspector, Salt Lake City Mosquito Abatement Dist, 42; from instr to prof biol, 46-83, head dept entom & zool, 62-68, EMER PROF BIOL, UNIV UTAH, 83- *Concurrent Pos:* Consult epizool & ecol res, Dugway Proving Grounds, 63-69; mem study comt future role lunar receiving lab, Am Inst Biol Sci-NASA, 69-70. *Mem:* Am Soc Parasitologists; Am Soc Trop Med & Hyg. *Res:* Transmission and reservoir of insect borne diseases; taxonomy and ecology of such insects; internal parasites of birds and mammals; ecology of animal parasites. *Mailing Add:* Dept Biol Univ Utah Salt Lake City UT 84112

GRUNDMANN, CHRISTOPH JOHANN, b Berlin, Ger, Dec 29, 08; nat US; m 33; c 1. CHEMISTRY. *Educ:* Univ Berlin, dipl, 31, Dr Phil, 33; Univ Heidelberg, Dr Phil Habil(org chem), 37. *Prof Exp:* Res assoc, Kaiser Wilhelm Inst, Ger, 33-39; res chemist, Deutsche Hydrierwerke, 39-45; assoc prof chem, Univ Halle, 47; prof & head dept, Univ Berlin, 50-52; assoc dir res, Olin Mathieson Chem Corp, NY, 52-58; dir res & develop, Gen Cigar Co, 58-60; prof chem, Carnegie-Mellon Univ, 67-76, sr fel, Mellon Inst, 60-85; RETIRED. *Concurrent Pos:* Privat-docent, Univ Heidelberg, 38; res assoc, Res Found, Ohio State Univ, 52-58; mem patent adv comt, US Secy Com, 60-62. *Mem:* Soc Ger Chem. *Res:* Organic chemistry. *Mailing Add:* 1518 Williamsburg Pl Pittsburgh PA 15235

GRUNDMEIER, ERNEST WINSTON, b North Mankato, Minn, Nov 15, 29; m 63. PHYSICAL CHEMISTRY. *Educ:* Mankato State Col, BA, 52; Iowa State Univ, MS, 54; Kans State Univ, PhD(chem), 64. *Prof Exp:* Asst chem, Iowa State Univ, 52-54; instr math, Univ NMex, 57; from instr to assoc prof, 58-69, PROF CHEM, MANKATO STATE UNIV, 69- *Concurrent Pos:* Consult, Mankato Stone Co, 60; fel, Univ Edinburgh, 71. *Mem:* Am Chem Soc; Sigma Xi. *Res:* Kinetics of the benzidine rearrangement and hydrogen isotope effects; kinetics of metal-catalyzed decarboxylations of amino acids; chemistry of combustion. *Mailing Add:* 1331 N 4th St Mankato MN 56001

GRUNDY, SCOTT MONTGOMERY, b Memphis, Tex, July 10, 33; m 56; c 2. PHYSIOLOGY, BIOCHEMISTRY. *Educ:* Tex Tech Col, BS, 55; Baylor Univ, MD & MS, 60; Rockefeller Univ, PhD, 66. *Prof Exp:* Instr biochem, Col Med, Baylor Univ, 60-61; asst prof, Rockefeller Univ, 68-70; chief, Phoenix Clin Res Sect, Phoenix Indian Med Ctr, Nat Inst Arthritis, Metab & Digestive Dis, 71-73; prof med in residence, Univ Calif, San Diego, 74-81; PROF INTERNAL MED & BIOCHEM & DIR, CTR HUMAN NUTRIT, UNIV TEX HEALTH SCI CTR, DALLAS, 81- *Concurrent Pos:* chief med serv, Metab Sect, Vet Admin Hosp, San Diego, 74-81. *Mem:* Am Soc Clin Invest; Asn Am Physicians. *Res:* Plasma lipoprotein structure and interconversions; cholesterol balance in man; cholesterol gallstone; atherosclerosis formation; diabetes mellitus. *Mailing Add:* Univ Tex Southwestern Med Ctr 5323 Harry Hines Blvd Dallas TX 75235

GRUNER, SOL MICHAEL, b Vineland, NJ, Oct 4, 50; m; c 2. BIOLOGICAL PHYSICS, CONDENSED MATTER. *Educ:* Mass Inst Technol, BS, 72; Princeton Univ, PhD(physics), 77. *Prof Exp:* Res assoc, 77, asst prof, 78-85, ASSOC PROF PHYSICS, PRINCETON UNIV, 85- *Concurrent Pos:* Danforth Found fel, 72-76; chmn, Nat Synchrotron Light Source Users Subgroup Time-Resolved Diffraction, 82; sci adv, Liposome Co, 84- & Enzymatics Co, 88-; co-chmn, 2nd & 3rd Biann Princeton-Liposome Res Conf, Lipid Membranes, 85 & 87; vis fel, Exxon Res & Eng, 86; vis appointment, Inst Theoret Physics, Univ Calif-Santa Barbara, 89; mem-at-large, Div Biol Physics, Am Phys Soc, 90-; organizer & mem, Princeton Mat Inst, 90- *Mem:* Biophys Soc; fel Am Phys Soc; AAAS. *Res:* Application of physics to problems in biology and condensed matter; one US patent. *Mailing Add:* Dept Physics Princeton Univ PO Box 708 Princeton NJ 08544

GRUNES, DAVID LEON, b Paterson, NJ, June 29, 21; m 49; c 3. SOIL CHEMISTRY, PLANT NUTRITION. *Educ:* Rutgers Univ, BSc, 44; Univ Calif, PhD(soil chem), 51. *Prof Exp:* Asst soil sci, Univ Calif, 48-50; soil scientist, Northern Great Plains Res Ctr, USDA, NDak, 50-59, soil phosphorus lab, Colo State Univ, 59-60, soil scientist, Northern Great Plains Res Ctr, USDA, NDak, 60-64, SOIL SCIENTIST, PLANT, SOIL & NUTRIT LAB, USDA, NY, 64- *Concurrent Pos:* Tech asst expert, Int Atomic Energy Agency, Israel & vis prof, Israel Inst Technol, 63-64; consult ed, McGraw Hill Co, 65-88; prof agron, Cornell Univ, 67- *Honors & Awards:* Cert Merit Outstanding Res, US Dept Agr, 59, 82; Agron Res Award, Am Soc Agron Northeastern Br, 88. *Mem:* Fel AAAS; fel Am Soc Agron; Int Soc Soil Sci; Sigma Xi; Coun Agr Sci & Technol; fel Soil Sci Soc Am. *Res:* Soil chemistry and fertility; plant, animal, and human nutrition; effects of soils, fertilizers and plant species and varieties on crop quality in relation to human and animal nutrition. *Mailing Add:* US Plant Soil & Nutrit Lab Tower Rd Ithaca NY 14853

GRUNES, ROBERT LEWIS, b New York, NY, Aug 15, 41; m 72; c 3. METALLURGY, MATERIALS SCIENCE. *Educ:* Polytech Inst Brooklyn, BS, 63, MS, 65, PhD(phys metall), 70. *Prof Exp:* Engr, Mat Develop Lab, Pratt & Whitney Corp, 63; res fel metall, Polytech Inst Brooklyn, 63-64; res engr, Lewis Res Ctr, NASA, 65-66; res assoc metall, Polytech Inst Brooklyn, 66-70; PRES, R L GRUNES & ASSOCS, INC, 70- *Mem:* Metall Soc; Am Soc Testing & Mat; Sigma Xi; Am Soc Mech Engrs. *Res:* Effects of interfaces upon bulk material properties; phase transformations; resource recovery. *Mailing Add:* R L Grunes & Assocs Inc 521 Fifth Ave New York NY 10017

GRUNEWALD, GARY LAWRENCE, b Spokane, Wash, Nov 11, 37. MEDICINAL CHEMISTRY, BIO-ORGANIC CHEMISTRY. *Educ:* Wash State Univ, BS & BPharm, 60; Univ Wis-Madison, PhD(org chem), 66. *Prof Exp:* From asst prof to assoc prof, 66-76, PROF MED CHEM, UNIV KANS, 76- *Concurrent Pos:* Ed, Med Res Ser, Marcel Dekker; chmn, Med Chem & Pharmacog Sect, Acad Pharmaceut Sci, 83-84; vis prof, Victorian Col Pharm, Melbourne, Australia, 84 & 86. *Honors & Awards:* Roland T Lakey Lectr, Wayne State Univ, 91. *Mem:* AAAS; Am Chem Soc; Soc Neurosci; The Royal Soc Chem; Am Pharmaceut Asn; Am Asn Pharm Scientists. *Res:* Synthesis and mechanism of action of biologically-active compounds; molecular orbital and molecular mechanics calculations and drug action; chemistry of the autonomic nervous system; computer assisted drug design. *Mailing Add:* Dept Med Chem Univ Kans Lawrence KS 66045

GRUNEWALD, KATHARINE KLEVESAHL, b Bonduel, Wis, Oct 4, 52; m 75. EXERCISE, WELLNESS. *Educ:* Univ Wis, BS, 74; Univ Ky, MS, 76, PhD, 79. *Prof Exp:* asst prof, 79-85, ASSOC PROF NUTRIT, KANS STATE UNIV, 85- *Mem:* Am Dietetic Asn; Am Col Sports Med; Am Inst Nutrit. *Res:* Nutrition and exercise; sports nutrition. *Mailing Add:* Dept Foods & Nutrit Kans State Univ Justin Hall Manhattan KS 66506

GRUNEWALD, RALPH, b Sheboygan, Wis, Nov 17, 36; m 62; c 2. BIONUCLEONICS, RADIATION SAFETY. *Educ:* Lakeland Col, BS, 58; Purdue Univ, MS, 64, PhD(health physics), 65. *Prof Exp:* Lab technician, A C Electronics Div, Gen Motors Corp, 58-61; chemist IV, Wis State Bd Health, 65-66; asst prof, 66-71, ASSOC PROF BOT, UNIV WIS-MILWAUKEE, 71- *Honors & Awards:* Kiekhofer Award, 71. *Mem:* AAAS; Health Physics Soc. *Res:* Radiotracer methodology and radioecology studies. *Mailing Add:* Dept Biol Sci PO Box 413 Univ Wis-Milwaukee Milwaukee WI 53201

GRUNSTEIN, MICHAEL, b Beclean, Romania, Aug 30, 46; Can citizen; m 67; c 2. MOLECULAR BIOLOGY, GENETICS. *Educ:* McGill Univ, BS, 67; Univ Edinburgh, PhD(molecular biol), 71. *Prof Exp:* Fel molecular biol, Sch Med, Stanford Univ, 72-74, fel, Dept Biochem, 74-75; from asst prof to assoc prof, 75-83, PROF, DEPT BIOL, UNIV CALIF, LOS ANGELES, 83- *Concurrent Pos:* Fac mem, Molecular Biol Inst, Univ Calif, Los Angeles, 75-; NIH grant, 76-79; Nat Found, March of Dimes, 78-79; comt res award, Univ Calif, Los Angeles, 78-79. *Res:* Gene organization and expression during animal cell differentiation; chromosome structure. *Mailing Add:* Molecular Biol Inst Univ Calif Los Angeles 405 Hilgard Ave Los Angeles CA 90024

GRUNSTEIN, MICHAEL M, b Beclan, Romania, Jan 16, 47. PULMONARY PHYSIOLOGY, PEDIATRIC MEDICINE. *Educ:* Sir George Williams Univ, BS, 70; McGill Univ, MD, 77, PhD(med), 74. *Prof Exp:* sr staff physician, Nat Jewish Ctr Immunol & Respiratory Med, 80-87, dir, Clin Serv, 85-87; CHIEF, DIV PULMONARY MED & ALLERGY, CHILDREN'S HOSP PHILADELPHIA, 87- *Mem:* Am Physiol Soc; Am Thoracic Soc; Soc Pediat Res. *Mailing Add:* Dept Pediat Childrens Hosp Philadelphia Univ Pa Sch Med 34th & Civic Center Blvd Philadelphia PA 19104

GRUNT, JEROME ALVIN, b Newark, NJ, Apr 6, 23; m 50; c 4. PEDIATRICS, ENDOCRINOLOGY. *Educ:* Rutgers Univ, BS, 47, MS, 48; Univ Kans, PhD, 52; Duke Univ, MD, 57. *Prof Exp:* Instr anat, Univ Kans, 48-52 & Duke Univ, 53-59; instr pediat, Harvard Univ, 61-63, from assoc to assoc prof, Yale Univ, 63-71; PROF PEDIAT, SCH MED, UNIV MO-KANS CITY, 71- *Concurrent Pos:* USPHS spec res fel, Duke Univ, 59-60; Med Found fel, Harvard Univ, 60-62, fel pediat endocrinol, 60-63. *Mem:* Lawson Wilkins Pediat Endocrinol Soc; Endocrine Soc; Am Diabetes Asn; Am Pediat Soc; Soc Pediat Res. *Res:* Pediatric endocrinology and growth. *Mailing Add:* Dept Pediat Univ Mo Sch Med Kansas City MO 64108

GRUNTFEST, IRVING JAMES, b Philadelphia, Pa, Jan 12, 17; m 48; c 2. PHYSICAL CHEMISTRY. *Educ:* Brown Univ, ScB, 37, ScM, 38; Cornell Univ, PhD(phys chem), 41. *Prof Exp:* Asst phys chemist, US Bur Mines, Md, 41 & Tenn Valley Authority, Ala, 41-43; mem staff, Bell Tel Labs, NJ, 43; res chemist, Rohm & Haas Co, 43-56; consult chemist, Gen Elec Co, 56-70; mat systs engr, Mat Res Ctr, Allied Chem Corp, Morristown, NJ, 70-74; chemist, Off Toxic Substances, US Environ Protection Agency, 74-84; RETIRED. *Concurrent Pos:* Chmn, Comt High Temp Mat, Mat Adv Bd, 61. *Mem:* Am Chem Soc; Soc Rheology; Am Inst Physics; Am Inst Chem Engrs. *Res:* Catalysis; surface activity; relationship between molecular structure and properties; high strength materials; materials for high temperature service; biomedical materials. *Mailing Add:* 140 Lake Carol Dr West Palm Beach FL 33411

GRUNTHANER, FRANK JOHN, b St Marys, Pa, Nov 4, 44; m 66; c 2. PHYSICAL INORGANIC CHEMISTRY, SOLID STATE CHEMISTRY. *Educ:* King's Col, Pa, BS, 66; Calif Inst Technol, PhD(chem), 74. *Prof Exp:* Teaching asst inorg chem, King's Col, Pa, 64-66; from res asst inorg chem to teaching asst introductory chem, Calif Inst Technol, 69-73; MEM TECH STAFF SOLID STATE CHEM, JET PROPULSION LAB, 73- *Mem:* Am Chem Soc; Am Vacuum Soc; Am Phys Soc. *Res:* Use of x-ray photoelectron spectroscopy to characterize surface chemistry of microelectronic processing technology; study of active sites of metallo-enzymes (bioinorganic chemistry) and heterogeneous catalysts. *Mailing Add:* 4800 Oak Grove 189-100 Pasadena CA 91109

GRUNWALD, CLAUS HANS, b Lakota, NDak, July 9, 34; m 58; c 2. PLANT PHYSIOLOGY. *Educ:* Univ Mo, BS, 60, PhD(bot), 65. *Prof Exp:* Staff res biologists, Yale Univ, 64-66; from asst prof to assoc prof agron, Univ Ky, 70-74; head bot & plant path, Ill Natural Hist Surv, Univ Ill, 74-85, prof plant biol, 75-90, plant physiologist, 85-90; RETIRED. *Mem:* NAm Soc Phytochem; Am Soc Plant Physiol; Scand Soc Plant Physiol; AAAS; Sigma Xi. *Res:* Physiological function and biochemistry of phytosterols, especially as related to membrane structure and function. *Mailing Add:* 800 Ben Franklin Dr No 111 Sarasota FL 34236

GRUNWALD, ERNEST MAX, b Wuppertal, Ger, Nov 2, 23; nat US; m 52; c 1. PHYSICAL ORGANIC CHEMISTRY. *Educ:* Univ Calif, Los Angeles, BS & BA, 44, PhD(chem), 47. *Prof Exp:* Instr chem, Univ Calif, Los Angeles, 47; res chemist, Portland Cement Asn, 48; Jewett fel, Columbia Univ, 49; from assoc prof to prof chem, Fla State Univ, 49-61; res chemist, Bell Tel Labs, Inc, 61-64; prof chem, 64-83, Henry F Fischbach prof, 83-89, EMER PROF CHEM, BRANDEIS UNIV, 89- *Concurrent Pos:* Weizman fel, 55; Sloan fel, 58-61; Guggenheim fel, 75-76. *Honors & Awards:* Pure Chem Award, Am Chem Soc, 59. *Mem:* Nat Acad Sci; Am Acad Arts & Sci; Am Chem Soc; Am Phys Soc. *Res:* Kinetics and equilibrium properties of solutions; thermodynamics of molecular species; interionic structure of ion pairs. *Mailing Add:* Dept Chem Brandeis Univ Waltham MA 02254

GRUNWALD, GERALD B, b Brooklyn, NY, Apr 6, 54; m 79; c 2. CELL INTERACTION IN DEVELOPMENT OF NERVOUS SYSTEM. *Educ:* Cornell Univ, BA, 76; Univ Wis-Madison, MS,; Univ Wis- Madison, PhD(zool), 81. *Prof Exp:* Sr staff fel, Lab Biochem Genetics, NIH, 81-85; asst prof, 85-89, ASSOC PROF ANAT, JEFFERSON MED COL, 89- *Concurrent Pos:* Mem, Rev Panel Develop Neurosci, NSF, 90-; Basil O'Connor scholar award, March Dimes; NIH grant awards. *Mem:* Am Soc Cell Biol; Soc Neurosci; Soc Develop Biol. *Res:* Is directed towards understanding the cellular and molecular mechanisms which guide the assembly of the nervous system during embryonic development. *Mailing Add:* Dept Anat Jefferson Med Col 1020 Locust St Philadelphia PA 19107

GRUNWALD, HUBERT PETER, b Lakota, NDak, Apr 7, 37; m 60; c 2. AMORPHOUS SEMICONDUCTORS. *Educ:* Wash Univ, BS, 59; Univ Rochester, PhD(solid state physics), 65. *Prof Exp:* Asst prof physics, Union Col, NY, 65-70; asst prof, 70-72, ASSOC PROF PHYSICS, STATE UNIV NY COL BROCKPORT, 72- *Concurrent Pos:* Consult, Xerox Corp, 83-84. *Mem:* Am Phys Soc; Am Asn Physics Teachers; Sigma Xi. *Res:* High resistivity semiconductors, particularly the study of transport mechanisms of electrical charge through these materials. *Mailing Add:* Dept Physics State Univ NY Brockport NY 14420

GRUNWALD, JOHN J, b Berehove, Czech, Aug 24, 31; US citizen; m 60; c 4. CHEMICAL MECHANISMS. *Educ:* Polytech Lausanne, Switz, BSCHE, 56; Columbia Univ, MSCHE, 58; Wayne State Univ, PhD(inorg chem), 64. *Prof Exp:* Res chemist, 58-66, dir res & develop, 66-81, VPRES RES, MAC DERMID INC, 81- *Mem:* Am Chem Soc; Am Electrochem Soc; Sigma Xi. *Res:* Electrochemistry of aqueous systems; microphotolithography; catalysis of aqueous systems; electroless deposition on conductors and nonconductors. *Mailing Add:* McDermid Israel Ltd PO Box 13011 T-A 61130 Tel Aviv Israel

GRUPEN, WILLIAM BRIGHTMAN, b Pittsburgh, Pa, Mar 30, 30; m 54; c 4. METALLURGY, SOLID STATE PHYSICS. *Educ:* Univ Calif, Los Angeles, BS, 57, MS, 59, PhD(metall), 64. *Prof Exp:* Res engr, Univ Calif, Los Angeles, 58-64; mem tech staff, 64-67, group supvr metallic films, 67-69, dept head, Mat & Process Develop, 69-81, HEAD, DEPT BATTERY RES & DEVELOP, BELL LABS, 81- *Mem:* Am Inst Elec & Electronics Engrs; Am Inst Mining, Metall & Petrol Engrs; Sigma Xi. *Res:* Preprecipitation reactions in aluminum-copper alloys; magnetic properties of 3d transition metals and their alloys; magnetic and electrical properties of thin metal films. *Mailing Add:* 25 Wolf Hill Dr Warren NJ 07060

GRUPP, GUNTER, b Esslingen, Ger, Feb 6, 20; nat US; m 58; c 6. EXPERIMENTAL MEDICINE, PHYSIOLOGY. *Educ:* Univ Freiburg, MD, 48, PhD 53. *Prof Exp:* Instr pharmacol, Univ Freiburg, 48-51, asst prof, 51-53, docent, 53-58; assoc prof, 58-65, dir AV activities, 67-69, chmn dept biomed commun, 69-77, PROF EXP MED, PHYSIOL & PHARMACOL, COL MED, UNIV CINCINNATI, 65- *Mem:* Am Soc Physiol; Am Soc Pharmacol & Exp Therapeut; Ger Pharmacol Soc; Cardiac Muscle Soc; Cent Soc Clin Res. *Res:* Pharmacology of calcium antagonists; physiology of cardiac contraction; heat-production of kidney; cardiovascular dynamics; ion-transport of heart and red blood cells; instructional methodology. *Mailing Add:* Univ Cincinnati Col Med ML 576 231 Bethesda Ave Cincinnati OH 45267-0576

GRUPP, INGRID L, b Hoechenschwand, Germany, Sept 20, 28; US citizen; c 5. PHARMACOLOGY, PHYSIOLOGY. *Educ:* Univ Freiburg, Germany, MD, 55. *Prof Exp:* Res fel physiol, Univ Freiburg, Germany, 55-56; resident, Ueberlingen Hosp, Germany, 56-57; resident pediat, Steinhoering Hosp, Germany, 57-58; RESIDENT PEDIAT, COL MED, UNIV CINCINNATI, 59-, ASSOC PROF PHARMACOL & CELL BIOPHYSICS, 84- *Concurrent Pos:* Mem, Basic Sci & Hypertension Coun, Am Heart Asn. *Mem:* Soc Exp Biol & Med; Am Heart Asn; Am Soc Pharmacol & Exp Therapeut; Cardiac Muscle Soc; Int Soc Heart Res. *Res:* Regulation of cardiac contraction; mechanism of action of cardiac glycosides and non-glycoside cardiotonic agents; comparative pharmacology of Ca; effects on vascular resistance, myocardial contraction and cardiac conduction; cardiovascular adaptation to anemia. *Mailing Add:* Dept Pharmacol & Cell Biophys Col Med Univ Cincinnati 231 Bethesda Ave Cincinnati OH 45267-0576

GRUSHKIN, BERNARD, b Brooklyn, NY, Apr 2, 32; m 61; c 3. PHYSICAL CHEMISTRY, ORGANIC CHEMISTRY. *Educ:* NY Univ, BA, 53; Univ Tex, MA, 56, PhD(chem), 58. *Prof Exp:* Res chemist, Callery Chem Co, 59-60; sr res chemist, Res Div, W R Grace & Co, 60-67; SR SCIENTIST, XEROX CORP, 67- *Mem:* Am Chem Soc; Soc Photog Scientists & Engrs. *Res:* Solution kinetics; inorganic polymers; phosphorus nitrogen compounds; chemistry of ammonia and its derivatives; organic sulfur-nitrogen and arsenic-selenium compounds and polymers; organic pigments; photoelectric pigments and polymers. *Mailing Add:* Five Old Brick Circle Pittsford NY 14534-4110

GRUSS, PETER H, b Aslfeld, Ger, June 28, 49; m 72; c 2. MOLECULAR GENETICS. *Educ:* Heidelberg Univ, PhD, 77. *Prof Exp:* Fel, Inst Virus Res, Heidelberg Univ, 77-78; fel, Nat Cancer Inst, 78-80, expert, Lab Molecular Virol, 80-81; vis scientist, Lab Molecular Virol, NIH, 81-82; prof microbiol, 82-86, mem Directorate Ctr Molecular Biol, Heidelberg Univ, 83-86; DIR, DEPT MOLECULAR CELL BIOL, MAX PLANCK INST BIOPHYS CHEM, WGER, 86- *Concurrent Pos:* Heisenberg grant, 81; hon prof, Univ Göttingen. *Mem:* NY Acad Sci; Am Soc Microbiol; Int Soc Differentiation; Europ Molecular Biol Orgn. *Res:* Elucidation of the molecular mechanisms controlling mammalian development and differentiation processes; transcription factors such as homeobox encoding genes and finger structure encoding genes. *Mailing Add:* Dept Molecular Cell Biol Max Planck Inst Biophys Chem Am Fassberg Gottingen 3400 Germany

GRUSZCZYK, JEROME HENRY, b Jersey City, NJ, Sept 11, 15. ANIMAL PHYSIOLOGY. *Educ:* Georgetown Univ, AB, 40; Woodstock Col, MD, PhL, 41, STL, 48; Fordham Univ, MS, 51. *Prof Exp:* Instr comp anat, Univ Scranton, 41-42; asst prof human anat & physiol, Le Moyne Col, 51-53; asst prof, 53-66, assoc prof, 66-79, EMER PROF BIOL, PETER'S COL, NJ, 79- *Mem:* AAAS; Bot Soc Am; Soc Protozool. *Res:* Enzyme activity in muscle tissue. *Mailing Add:* 2652 Kennedy Blvd Jersey City NJ 07306

GRUTT, EUGENE WADSWORTH, JR, b Hawthorne, Nev, Sept 29, 16; m 50; c 2. EXPLORATION GEOLOGY, NATURAL RESOURCES. *Educ:* Univ Nev, BS, 38. *Prof Exp:* Geologist, Nev Victory Mining Co, 38-42; staff geologist, Isbell Construct Co, 48-53; chief, Casper Br, AEC, 53-58, Grants Br, 58-62 & Resources Appraisal Br, 62-67, dir resource div, 67- 73, asst mgr, 73-75, mgr, Grand Junction Off, US Dept Energy, 75-78; CONSULT URANIUM RAW MAT & PRECIOUS METALS, 78- *Mem:* AAAS; Am Inst Mining, Metall & Petrol Engrs; Geol Soc Am. *Res:* Appraisal of uranium resources and supply; research and development and investigations in geology, geochemistry, exploration technology, and mining methods. *Mailing Add:* 1325 Grand Ave Grand Junction CO 81501-4511

GRUTZNER, JOHN BRANDON, b Melbourne, Australia, Aug 28, 41; m 65; c 2. ORGANIC CHEMISTRY. *Educ:* Univ Melbourne, BS, 63, PhD(chem), 67. *Prof Exp:* Fel chem, Univ Calif, Los Angeles, 67-68 & Calif Inst Technol, 68-69; from asst prof to assoc prof, 69-88, PROF CHEM, PURDUE UNIV, WEST LAFAYETTE, 88- *Concurrent Pos:* Prog officer, Nat Sci Found, 81-82. *Mem:* Am Chem Soc; AAAS; Sigma Xi; The Chem Soc; Royal Australian Chem Inst; Int Soc Magnetic Resonance. *Res:* Nuclear magnetic resonance spectroscopy; carbanions; organic reaction mechanisms; fluid dynamics; electrophoretic separations polymers. *Mailing Add:* Dept Chem Purdue Univ 1393 Brown Bldg West Lafayette IN 47907-1393

GRUVER, ROBERT MARTIN, b York, Pa, Mar 6, 21; m 46; c 1. CERAMICS ENGINEERING. *Educ:* Pa State Univ, BS, 43, MS, 48; Univ Mo, PhD(ceramics eng), 56. *Prof Exp:* Res asst ceramics, Pa State Univ, 43-53; proj dir, Linden Labs, Inc, 54-63, tech dir mat, 63-68; TECH DIR MAT, CERAMIC FINISHING CO, 68- *Mem:* AAAS; Am Ceramics Soc; Am Chem Soc. *Res:* Thermal analysis of ceramic materials; investigations of ceramics and glass including strengthening methods, fractography and local impact damage. *Mailing Add:* 1160 Oneida St State College PA 16801-4899

GRUVER, WILLIAM A, b Harrisburg, Pa, July 13, 41; m 70; c 2. ROBOTICS, MANUFACTURING AUTOMATION. *Educ:* Univ Pa, BSEE, 63, MSEE, 66, PhD(elec eng), 70; Univ London, DIC, 65. *Prof Exp:* Instr elec eng, US Naval Acad, Annapolis, 65-67; teaching fel, Moore Sch Elec Eng, Univ Pa, Philadelphia, 67-71; res engr, DFVLR Inst Dynamics Flight Systs, Munich, Ger, 71-73; asst prof elec eng & opers res, NC State Univ, Raleigh, 74-79; vpres res & develop & co-founder, Robotics Systs Div, Logistics Technol Int, Ltd, Torrance, Calif, 79-83; mgr, Indust Electronics Develop Lab, Gen Elec Co, Charlottesville, Va, 83-84 & Gen Elec Indust Automation Ctr, Frankfurt, Ger, 84-86; div pres, IRT Corp, San Diego, 86-88; DIR, CTR ROBOTICS & MFG SYSTS & RES PROF ELEC ENG, UNIV KY, LEXINGTON, 88- *Concurrent Pos:* Aerospace engr, Marshall Space Flight Ctr, NASA, 68-71; Alexander von Humboldt Found US sr scientist award, 73; Ger Acad

Exchange Serv res award, 78; assoc ed, Inst Elec & Electronics Engrs Trans Systs, Man & Cybernet, 82- & Inst Elec & Electronic Engrs Trans Robotics & Automation, 86; chmn, Robotics Comt, Inst Elec & Electronics Engrs, Systs, Man & Cybernet Soc, 82-88 & Tech Prog, Int Conf Computer Eng, Am Soc Mech Engrs, 84; mem, var tech comts, Inst Elec & Electronics Engrs, 89- *Mem:* Inst Elec & Electronics Engrs Robotics & Automation Soc (secy, 83-95); Inst Elec & Electronics Engrs Systs Man & Cybernet Soc; Am Soc Mech Engrs; Soc Mfg Engrs. *Res:* Manufacturing automation, robotics, control and optimization; sensor-based control of robotic and automated systems; robotics for hazardous environments; programming environments for automation and computer integrated manufacturing, robot hands, walking machines, robot control and path planning; author of 80 technical publications and two books. *Mailing Add:* Ctr Robotics & Mfg Systs Univ Ky Lexington KY 40506-0108

GRUY, HENRY JONES, b Victoria, Tex, June 10, 15; m 41; c 3. PETROLEUM TECHNOLOGY. *Educ:* Tex A&M Univ, BS, 37, PhD(petrol eng), 56. *Prof Exp:* Field engr, Standard Oil Co Tex, 37-38, exploit engr, Shell Oil Co, 38-45; geologist & engr, DeGolyer & MacNaughton, 45-50; chief exec officer & chmn, H J Gruy & Assocs, 50-86, CHIEF EXEC OFFICER & CHMN, GRUY ENG CORP, 86- *Concurrent Pos:* Chmn, Gruy Petrol Mgt Co, 60-88; task force mem, Fed Power Comn, Nat Gas Reserve, 71-73, Nat Gas Survey, Supply Tech Adv, 75-76, Prospective Exploit & Develop & Supply Additions to Reserves No4 Comt, Nat Gas Survey Adv Comt, 76-78; mem coordinating Sub Com Nat Petrol, Council Comt on Enhanced Oil Recovery, 82-83. *Honors & Awards:* De Golyer Medal, Soc Petrol Engrs, 83. *Mem:* Soc Petrol Engrs (treas, 65-67, pres, 68); Am Inst Mining, Metallurgical & Petrol Engrs (vpres, 69); Am Asn Petrol Geologists; Soc Petrol Eval Engrs (pres, 64); Australasian Inst Mining & Metallurgy; fel Inst Petrol. *Res:* Oil and gas resource analysis. *Mailing Add:* Suite 1090 909 Fannin Houston TX 77010

GRUZALSKI, GREG ROBERT, b Chicago, Ill, April 19, 48; m 72; c 3. MATERIALS SCIENCE, THIN FILMS. *Educ:* Western Ill Univ, BS, 70, MS, 71; Univ Nebr, PhD(physics), 77. *Prof Exp:* Res fel physics & mat sci, Harvard Univ, 77-80; STAFF MEM, OAK RIDGE NAT LAB, 80- *Mem:* Am Physics Soc; Mat Res Soc; Sigma Xi; Am Vacuum Soc. *Res:* Fabrication and study of thin-film multilayered structures with focus on ionic conductors, ferroelectrics, metals and Li-based solid-state microbatteries. *Mailing Add:* Solid State Div Oak Ridge Nat Lab PO Box 2008 Oak Ridge TN 37831

GRYC, GEORGE, b St Paul, Minn, July 27, 19; m 42; c 5. GEOLOGY. *Educ:* Univ Minn, BS, 40, MS, 41. *Prof Exp:* Field geologist, 43-60, staff geologist, 60-63, chief, Br Alaskan Geol, 63-76, regional geologist, 76-77, chief, Off Nat Petrol Reserve Alaska, 77-81, DIRS REP, WESTERN REGION, 77-, GEN CHMN, CIRCUM-PACIFIC MAP PROJ, US GEOL SURV. *Mem:* Geol Soc Am; Paleont Soc; Am Asn Petrol Geologists; Arctic Inst NAm; Cosmos Club. *Res:* Potential petroleum resources and regional geology of Circum-Pacific Region; paleontology; stratigraphy. *Mailing Add:* Dirs Rep Western Region 345 Middlefield Rd Menlo Park CA 94025

GRYDER, JOHN WILLIAM, b Los Angeles, Calif, Nov 6, 26; m 49; c 3. PHYSICAL CHEMISTRY, INORGANIC CHEMISTRY. *Educ:* Calif Inst Technol, BS, 46; Columbia Univ, PhD(chem), 50. *Prof Exp:* Jr scientist chem, Brookhaven Nat Lab, 48-49; from instr to assoc prof, 49-66, assoc dean, Fac Arts & Sci, 82-84, PROF CHEM, JOHNS HOPKINS UNIV, 66- *Mem:* Am Chem Soc. *Res:* Radiochemistry; kinetics; nature of ionic species in solution; phosphorus chemistry; surface chemistry and catalysis. *Mailing Add:* Dept Chem Johns Hopkins Univ Baltimore MD 21218

GRYDER, ROSA MEYERSBURG, b Brooklyn, NY, Aug 23, 26; m 49; c 3. BIOCHEMISTRY, TOXICOLOGY. *Educ:* Bucknell Univ, BS, 47; Yale Univ, MS, 49; Johns Hopkins Univ, PhD(biochem), 56. *Prof Exp:* Asst microbiol, Brookhaven Nat Labs, 47-48; res assoc ophthal, Johns Hopkins Hosp, 52-56; res assoc physiol chem, Sch Med, Johns Hopkins Univ, 60-63; from instr to asst prof biochem, Sch Med, Univ Md Baltimore, 66-74; staff sci adv, Off Assoc Comnr Sci, Food & Drug Admin, 74-80; toxicologist, Ctr Food Safety & Nutrit, 82-89, ASSOC DIR RES, NAT CTR TOXICOL RES, FOOD & DRUG ADMIN, 80-; CONSULT TOXICOL, 89- *Concurrent Pos:* Nat Inst Neurol Dis & Blindness fel, Wilmer Ophthal Inst, 56-59. *Mem:* NY Acad Sci; AAAS; Soc Risk Anal; Sigma Xi; Soc Toxicol. *Res:* Enzymology; neurochemistry; genetic toxicology. *Mailing Add:* 2006 W Rogers Ave Baltimore MD 21209

GRYNBAUM, BRUCE B, b Anapa, Russia, May 25, 20; nat US; m 43; c 2. MEDICINE. *Educ:* Columbia Univ, MD, 43; Am Bd Phys Med & Rehab, dipl, 52. *Prof Exp:* Intern, Metropolitan Hosp, New York, 44; resident, Mt Sinai Hosp, 47-48; asst prof, 50-58, assoc prof clin phys med & rehab, 58-68, PROF, MED CTR, NY UNIV, 68- *Concurrent Pos:* Fel, Goldwater Hosp, 49-50; dir phys med & rehab, City Dept Hosps, New York, 50-70; consult, Nat Inst Rehab & Health Serv, 64; mem adv comt auto ins & compensation study, Dept Transp, Washington, DC, 72; consult, Long Island Jewish Hosp, St Vincent's Hosp, Wycoff Heights Hosp & St Claire's Hosp, 78; vis physician, Beekman-Downtown Hosp; mem adv bd, NY State Chap, Arthritis Found; mem exec comt, Cent Labor Rehab Coun, NY, Inc; dir rehab med, Bellevue Hosp; consult dept hosp facilities, US Pub Health Serv; chmn med adv bd, Sidney Hillman Inst, mem med adv bd, St Mary's Children's Hosp, Bayside, NY; dir, Rehab Med Serv, Bellevue Hosp; mem prof adv bd, Inst Crippled & Disabled, sr med consult; mem, Health Task Force, Community Coun; dir, Rehab Med, Beekman Downtown Hosp, mem, Health & Hosp Corp Adv Comt, Amputee, Orthotic, Neuromuscular & Pediat Orthop Prog. *Mem:* AMA; Am Acad Phys Med; Am Cong Rehab Med; Int Asn Rehab Facil; Am Pub Health Asn. *Res:* Physical medicine and rehabilitation; compensation medicine. *Mailing Add:* 400 E 34th St Med Sch New York Univ 550 First Ave New York NY 10016

GRYTING, HAROLD JULIAN, b Belview, Minn, Dec 31, 19; m 54; c 2. CHEMISTRY. *Educ:* St Olaf Col, BA, 41; Purdue Univ, PhD(org chem), 47. *Prof Exp:* Grad asst chem, NDak State Col, 41-42; chemist, E I du Pont de Nemours & Co, Ill, 42-43; res chemist, Naval Weapons Ctr, 47-49, head, Properties Sect, 49-51, Properties Br, 52-54, Plastics Br, 54-56, Explosives Res Br, 56-65, tech asst to head, Explosives & Pyrotech Div, 65-72, assoc head, Appl Res & Processing Div, 75-76, res chemist, safety & technol coordr & proj mgr, 76-80; CONSULT EXPLOSIVES, PROPELLANTS & PYROTECHNICS, SOUTHWEST RES INST, 80-; PRES, GRYTING ENERGETICS SCI CO, 88-; VPRES, ENVIRON ENTERPRISES, INC, 90- *Concurrent Pos:* Mem, Tech Steering Comt, Joint Army, Navy, NASA, Air Force Safety & Environ Protection Working Group; chmn, Triserv Comt Qualification & Performance Test Methods for Explosives; mem Jannaf Propulsion Hazards Subcomt. *Mem:* Fel AAAS; fel Am Inst Chemists; Am Chem Soc; NY Acad Sci; Am Defense Preparedness Asn; Am Inst Astronaut & Aeronaut. *Res:* Alkylation, chloromethylation chlorination and fluorination; synthesis of wetting agents; germicides; high explosives compositions; explosive effects; pollution abatement; polymer chemistry, qualification, recovery and identification of explosives, propellants and pyrotechnics; synthesis of polymers that do not shrink on cure for low vulnerability ammunition propellants; analyzing and selecting explosives and other materials for numerous applications; investigation of accidental delonations/explosions; temperature insensitive explosives. *Mailing Add:* 7126 Shadow Run San Antonio TX 78250-3483

GRZYMALA-BUSSE, JERZY WITOLD, b Warsaw, Poland, April 3, 42. EXPERT SYSTEMS, MACHINE LEARNING. *Educ:* Tech Univ Poznan, Poland, MS, 64, PhD, 69; Univ Wroclaw, MS, 67; Tech Univ Warsaw, Dr habil, 72. *Prof Exp:* Instr elec eng, Tech Univ Poznan, 64-70, from asst prof to assoc prof computer sci, 70-80; PROF COMPUT SCI, UNIV KANS, 81- *Concurrent Pos:* Vis prof comput sci, Univ Kans, 77 & 80-81 & Banach Int Math Ctr, Poland, 74; assoc ed, Found Control Eng, 75-84 ed, Fundamenta Informaticae, 75-90. *Honors & Awards:* Award, Minister Sci, Higher Educ & Technol, Poland, 70, 73 & 76; Award, Polish Acad Sci, 81. *Mem:* Am Asn Artificial Intel; Asn Comput Mach; European Asn Theoret Comput Sci. *Res:* Rough set theory; machine learning; expert systems dealing with uncertainty; knowledge acquisition. *Mailing Add:* Dept Comput Sci Univ Kans Lawrence KS 66045-2192

GSCHNEIDNER, KARL A(LBERT), JR, b Detroit, Mich, Nov 16, 30; m 57; c 4. PHYSICAL METALLURGY, SOLID STATE PHYSICS. *Educ:* Univ Detroit, BS, 52; Iowa State Univ, PhD(phys chem), 57. *Prof Exp:* Staff mem phys metall, Los Alamos Sci Lab, 57-63, sect leader, 61-63; asst prof solid state physics, Dept Physics & Mat Res Lab, Univ Ill, 62-63; assoc prof phys metall, 63-67, prof phys metall, 67-79, prog dir metall & ceramics, 74-79, DIR RARE-EARTH INFO CTR, AMES LAB & DEPT MAT SCI & ENG, IOWA STATE UNIV, 66-, DISTINGUISHED PROF SCI & HUMANITIES, 79- *Honors & Awards:* William Hume-Rothery Award, Minerals, Metals & Mat Soc, 78; Co-recipient Outstanding Sci Accomplishment in Metall & Ceramics Award, US Dept Energy, 82; Frank H Spedding Prize, Rare Earth Res Conf, 91. *Mem:* AAAS; Am Chem Soc; Am Soc Metals Int; Am Crystallog Asn; fel Minerals, Metals & Mat Soc; Am Phys Soc; Iowa Acad Sci; Sigma Xi; Mat Res Soc; Inst Elect & ElectroniCs Engrs; Magnetics Soc; fel Am Soc Mat Int. *Res:* Physical metallurgy of rare-earth metals and alloys; alloy theory; electronic structures of metals; low temperature heat capacity and magnetic properties of solids; hydrogen in metals and metal hydrides; thermodynamic properties of metals. *Mailing Add:* Ames Lab Rare-earth Info Ctr Iowa State Univ Ames IA 50011-3020

GSCHWEND, HEINZ W, b Unterseen, Switz, Apr 12, 36; m 65; c 2. ORGANIC CHEMISTRY, MEDICINAL CHEMISTRY. *Educ:* Swiss Fed Inst Technol, Dipl Ing Chem, 61, Dr Sc Tech(org chem), 64. *Prof Exp:* Res assoc org chem, Swiss Fed Inst Technol, 64-65; res assoc, Harvard Univ, 65-67; sr staff scientist med chem, 67-70, mgr org chem, 70-72, dir org chem, 73-79, exec dir chem res, 80-86, VPRES DRUG DISCOVERY PHARMACEUT DIV, CIBA-GEIGY CORP, 86- *Concurrent Pos:* Gen & pharmaceut res mgt. *Mem:* Am Chem Soc; Swiss Chem Soc. *Res:* Synthesis of novel compounds with potential biological activity; general synthetic organic chemistry, in particular intramolecular cycloaddtions, heteroatom directed lithiations, new synthetic methodology; heterocyclic chemistry; drug design. *Mailing Add:* Cent Res Labs R 1060 7 22 CIBA Geigy Ltd CH 4002 Basel Switzerland

GSCHWENDTNER, ALFRED BENEDICT, b Altoona, Pa, Dec 16, 36; m 61. OPTICS. *Educ:* Pa State Univ, BS, 61, MS, 62. *Prof Exp:* Res assoc space physics, Univ Ill, 62-64; staff mem plasma physics, 64-69, asst group leader infrared systs, 69-71, GROUP LEADER OPTICAL SYSTS, LINCOLN LAB, MASS INST TECHNOL, 71- *Mem:* Sigma Xi; Am Inst Aeronaut & Astronaut; Am Inst Physics. *Res:* Application of laser and infrared optical systems to tactical problems; theory and application of artificial neural networks. *Mailing Add:* 1 Harrington Rd Lexington MA 02173

GUAL, CARLOS, endocrinology, reproductive biology, for more information see previous edition

GUALTIERI, DEVLIN MICHAEL, b Utica, NY, Dec 11, 47; m 71; c 2. CRYSTAL GROWTH, MAGNETISM. *Educ:* Syracuse Univ, BS, 70, PhD(solid state sci), 74. *Prof Exp:* Res assoc chem, Univ Pittsburgh, 74-76, res asst prof, 76-77; staff physicist, 77-81, sr res physicist, 81-83, RES ASSOC, ALLIED-SIGNAL INC, MORRISTOWN, NJ, 83- *Concurrent Pos:* Weizmann fel, Dept Chem, Univ Pittsburgh, 76-77. *Mem:* Am Phys Soc; Inst Elec & Electronics Engrs; AAAS; Am Asn Crystal Growth. *Res:* Synthesis of magnetic and optical thin films, especially epitaxy of crystal oxides for optical sensors; device structures based on thin film materials, including magneto-optical sensors. *Mailing Add:* 12 Moore St Ledgewood NJ 07852

GUARD, RAY W(ESLEY), metallurgy; deceased, see previous edition for last biography

GUARINO, ANTHONY MICHAEL, b Framingham, Mass, Dec 11, 34; m 57; c 10. RESEARCH ADMINISTRATION, TOXICOLOGY. *Educ:* Boston Col, BS, 56; Univ RI, MS, 63, PhD(pharmacol), 66. *Prof Exp:* Clin chemist, Chem Lab, Mass Gen Hosp, Boston, 54-57; teaching/res asst, Univ RI, 60-66; res assoc, Pharmacol & Toxicol Prog, Lab Chem Pharmacol, Nat Heart Inst, Md, 66-68; res pharmacologist, Lab Chem Pharmacol, Nat Cancer Inst, 68-73, chief, Lab Toxicol, 73-80; rev scientist, Bur Drugs, 80-84, CHIEF, FISHERY RES BR, FOOD & DRUG ADMIN, DAUPHIN ISLAND, ALA, 84- *Concurrent Pos:* Mem fac toxicol, NIH Grad Prog, 68-84, fac chmn, Dept Pharmacol & Toxicol, 80-84; adj prof, Dept Pharmacol, Col Med, Univ Southern Ala, Mobile, 85- *Honors & Awards:* Pub Health Serv Commendation Medal, 90. *Mem:* Am Soc Pharmacol & Exp Therapeut; Soc Toxicol; Am Chem Soc; NY Acad Sci. *Res:* Drug transport; biliary transport; drug metabolism; marine pharmacology and toxicology; fate and distribution of xenobiotics in aquatic species; regulatory pharmacology and toxicology. *Mailing Add:* Food & Drug Admin PO Box 158 Dauphin Island AL 36528

GUARINO, ARMAND JOHN, b Beverly, Mass, Jan 24, 26; m 49; c 5. BIOCHEMISTRY. *Educ:* Harvard Univ, BS, 49; Tufts Col, PhD(biochem), 53. *Prof Exp:* Instr biochem & nutrit, Med Sch, Tufts Col, 53-54; res assoc biochem, Mass Inst Technol, 54-55; from asst prof to assoc prof, Sch Med, Univ Mich, 55-63; prof & chmn dept, Woman's Med Col Pa, 63-68; prof biochem & chmn dept, 68-75, DEAN GRAD SCH BIOMED SCI & SCH ALLIED HEALTH SCI, UNIV TEX HEALTH SCI CTR, SAN ANTONIO, 73- *Mem:* Am Soc Biol Chem. *Res:* Nucleoside and nucleotide metabolism. *Mailing Add:* Dept Biochem Univ Tex Health Sci Ctr 7703 Floyd Curl Dr San Antonio TX 78284-7760

GUBBINS, KEITH E(DMUND), b Southampton, Eng, Jan 27, 37; m 60; c 2. CHEMICAL ENGINEERING, PHYSICAL CHEMISTRY. *Educ:* Univ London, BSc, 58, dipl chem eng, 59, PhD(chem eng), 62. *Prof Exp:* Vis lectr, Univ London, 59-62; fel chem eng, Univ Fla, 62-64, from asst prof to prof, 64-76; dir, 83-90, THOMAS R BRIGGS PROF ENG, SCH CHEM ENG, CORNELL UNIV, 76- *Concurrent Pos:* Vis prof, Imp Col, Univ London, 71, Oxford Univ, 79 & 86, Univ Calif, Berkeley, 82; Guggenheim fel, 86-87. *Honors & Awards:* Res Award, Alpha Chi Sigma, Am Inst Chem Eng, 86. *Mem:* Nat Acad Eng; Am Chem Soc; The Chem Soc; Am Inst Chem Engrs. *Res:* Statistical thermodynamics of liquids; transport properties of liquids; surface properties. *Mailing Add:* Sch Chem Eng Olin Hall Cornell Univ Ithaca NY 14853

GUBER, ALBERT LEE, b Heidelberg, Pa, June 17, 35; m 61; c 2. GEOLOGY. *Educ:* Univ Pittsburgh, BS, 57; Univ Ill, PhD(geol), 62. *Prof Exp:* NSF fel, Pa State Univ, 62-63, from asst prof to prof geol, 63-83, chmn, Geol Grad Prog, PROF GEOL, PA STATE UNIV, 83-; CONSULT. *Concurrent Pos:* Various offices, Marine Sci Consortium, 75- *Honors & Awards:* Lindback Award. *Mem:* Am Asn Petrol Geol; Paleont Soc; Soc Econ Paleont & Mineral; AAAS; Sigma Xi. *Res:* Coastal geochemistry; Silurian facies of the central and southern Appalachians; compaction and the migration of barrier islands; sedimentology of back-barrier fine-grained sediments. *Mailing Add:* Dept Geosci Pa State Univ 538 Deike Bldg University Park PA 16802

GUBERMAN, H(ERBERT) D(AVID), metallurgy, solar energy, for more information see previous edition

GUBERNATIS, JAMES EDWARD, b Baltimore, Md, July 22, 45; m 70. SOLID STATE PHYSICS. *Educ:* Loyola Col, Md, BS, 67; Case Western Reserve Univ, MS, 69, PhD(physics), 72. *Prof Exp:* Asst prof physics, Ohio Northern Univ, 72-73; res assoc, Lab Atomic & Solid State Physics, Cornell Univ, 73-75; staff mem physics, 75-81, dep group leader, Statist Physics & Mat Theory Group, 81-83, STAFF MEM, LOS ALAMOS NAT LAB, 83- *Concurrent Pos:* Coun Panel Math Sci Appl Mat Sci, Nat Res, 81- *Mem:* AAAS; Am Phys Soc; Soc Indust & Appl Math. *Res:* Quantum Monte Carlo simulations of strongly correlated electron systems and lower dimensional systems with electron phonon coupling; numerical solutions of phenomenlogical equations of superconductivity. *Mailing Add:* Los Alamos Sci Lab T-11 MS B262 PO Box 1663 Los Alamos NM 87545

GUBERSKY, VICTOR R, b Lamont, Alta, May 17, 35; m 57; c 3. DERMATOLOGY. *Educ:* Univ Alta, MD, 59. *Prof Exp:* From intern to resident internal med, Edmonton Gen Hosp, Alta, 59-61; resident dermat, Cleveland Clin Found, Ohio, 61-64; assoc med dir, Inst Clin Med, 64-68, head div inflammatory dis & med dir, 66-68, VPRES & GEN MGR, SYNTEX LTD, 68- *Concurrent Pos:* Fel dermat, 59-64. *Mem:* Am Acad Dermat; AMA. *Res:* Basic and clinical dermatological research, especially the effect of corticosteroids on various dermatoses; antimetabolites, antiviral and antiandrogenic agents. *Mailing Add:* 3712 Mission Ave Carmichael CA 95608

GUBLER, CLARK JOHNSON, b La Verkin, Utah, July 14, 13; m 38; c 4. BIOCHEMISTRY. *Educ:* Brigham Young Univ, AB, 39; Utah State Univ, MA, 41; Univ Calif, PhD(biochem), 45. *Prof Exp:* Asst chem, Brigham Young Univ, 38-39; asst, Utah State Univ, 39-41; asst biochem, Univ Calif, 41-45; res chemist, El Dorado Oil Works, Calif, 44-46; res assoc med, Sch Med, Univ Utah, 46-52, asst prof, 52-56; USPHS spec res fel, Univ Wis, 56-58; from assoc prof to prof, 58-78, EMER PROF CHEM, BRIGHAM YOUNG UNIV, 78- *Concurrent Pos:* Assoc prof, William Col, 44-45; instr, Univ Utah, 51-60; Am Heart Asn advan res fel, 58-60, estab investr, 60-65; vis prof, Univ Freiburg, 64-65; USPHS spec res fel, Univ San Diego, 71-72; vis prof, Univ Kuwait, 82-86, vis prof & Fulbright fel, Sultan Qaboos Univ, Oman, 89-90. *Mem:* AAAS; Am Soc Biol Chemists; Am Inst Nutrit; NY Acad Sci; Am Soc Clin Nutrit. *Res:* Iron, copper metabolism; biological functions and mechanisms of action of thiamin; alpha-keto-acid metabolism; biochemistry of learning and development. *Mailing Add:* Grad Sect Biochem Brigham Young Univ 659 W1DB Provo UT 84602

GUBLER, DUANE J, b Santa Clara, Utah, June 4, 39; m 58; c 2. VIROLOGY, MOSQUITO-BOURNE DISEASE ECOLOGY PREVENTION & CONTROL. *Educ:* Utah State Univ, BS, 63; Univ Hawaii, MS, 65; Johns Hopkins Univ, ScD, 69. *Hon Degrees:* DSc, 88. *Prof Exp:* Instr parasitol, Johns Hopkins Univ, 68-69, lectr med entom, 69-70, asst prof, 70-71; from asst prof to assoc prof med entom & parasitol, Univ Hawaii, Honolulu, 71-78; assoc prof med entom, Univ Ill, Urbana, 78-80; res microbiologist virol, Ft Collins, 80-81, dir, San Juan Labs & Chief, Dengue Br, Ctrs Dis Control, USPHS, San Juan, 81-89; DIR, DIV VECTOR-BOURNE INFECTIOUS DIS, CTRS DIS CONTROL, FT COLLINS, 89- *Concurrent Pos:* Head, med entom prog, JHU ICMRT, Calcutta, India, 69-71, guest worker virol, Pac Res Sect, Nat Inst Allergy & Infectious Dis, NIH, 71-75; consult herbicides in Vietnam, Nat Res Coun, Nat Acad Sci, 72; adv, Pan Am Health Orgn, Washington, DC, 73 & 82-86; consult, WHO, New Delhi, 74-80, 82 & 86, USAID, Jakarta, 77-, Int Develop Res Ctr, Ottawa, 79-82; head virol dept, US Naval Med Res Unit 2, Jakarta, 75-78; adv, Pan Am Health Orgn, Washington, DC, 81-; consult, US Agency Int Develop, Dominican Repub, 83-, Rockefeller Found, 87-88, govts Nepal, Indonesia, Taiwan, Mexico, Palau, Vietnam & Venezuela, 88-90; Deans lectr, Johns Hopkins Univ Sch Hygiene & Pub Health, 90. *Honors & Awards:* Cert Appreciation, President Gerald R Ford, 75; Third Lloyd E Rozeboom lectr, Johns Hopkins Univ, 79; Charles Franklin Craig lectr, Am Soc Trop Med & Hygiene, 88. *Mem:* AAAS; Am Soc Tropical Med & Hyg; Am Soc Parasitologists; Am Mosquito Control Asn; Entom Soc Am; Sigma Xi. *Res:* Ecology of mosquito-borne diseases; dengue-dengue hemorrhagic fever with primary emphasis on virology, epidemiology, entomology, prevention and control. *Mailing Add:* Ctr Dis Control Div Vector-Bourne Infectious Dis USPHS PO Box 2087 Ft Collins CO 80522

GUBNER, RICHARD S, b New York, NY, July 20, 13; m 49; c 3. INTERNAL MEDICINE, CARDIOLOGY. *Educ:* Columbia Univ, AB, 33; NY Univ, MD, 36; Am Bd Internal Med, dipl, 44; Am Bd Cardiovasc Dis, dipl, 52. *Prof Exp:* Med dir, Diag Serv Div, Equitable Life Ins Soc US, 41-69. *Concurrent Pos:* Clin prof, State Univ NY Downstate Med Ctr, 56-; consult physician, Pa RR, 57-69; dir med, Kings County Hosp, Brooklyn, 59-67. *Mem:* Sigma Xi; Fel Am Col Physicians. *Res:* Cardiovascular disease. *Mailing Add:* 2905 Mill Stream Ct Clearwater FL 34621

GUBSER, DONALD URBAN, b Alton, Ill, Dec 21, 40; m 67; c 2. SOLID STATE PHYSICS, SUPERCONDUCTIVITY. *Educ:* Univ Ill, BS, 63, MS, 64, PhD(physics), 69. *Prof Exp:* RES SCIENTIST, NAVAL RES LAB, 69-, SCI MGT & ADMINR, 81- *Concurrent Pos:* Advan grad training, Switz Fed Tech Univ, 76-77; teaching, George Washington Univ, 78-; co-ed, J Superconductivity. *Mem:* Fel Am Phys Soc; Metall Soc; Sigma Xi; Mat Res Soc. *Res:* New high temperature supeconducting oxides; superintendent of materials division with research in superconductivity, magnetism, physical metallargy, ceramics and composites and material mechanics. *Mailing Add:* Code 6300 Naval Res Lab Washington DC 20375-5000

GUCHHAIT, RAS BIHARI, b Calcutta, India, May 31, 35; m 56; c 4. BIOCHEMISTRY. *Educ:* Univ Calcutta, BS, 56, MS, 58, PhD, 62. *Prof Exp:* Fel & biochemist, Univ Wis & Vet Admin Hosp, Madison, 63-66; scientist, Univ Calcutta, India, 66-69; fel, NY Univ Med Ctr, 69 & Johns Hopkins Univ, 70-72; instr physiol chem, Sch Med, 72-73, ASST PROF EPIDEMIOL, BIOCHEM & BIOPHYS SCI, SCH HYG & PUB HEALTH, JOHNS HOPKINS UNIV, 73-; sr res scientist, Md Psychiat Res Ctr, 80-89; RETIRED. *Mem:* NY Acad Sci; Soc Neurosci. *Res:* Hormonal and metabolic regulation of enzymes; mechanism of enzyme reactions in relation to subunit structure in mammalian and bacterial systems; regulation of biogenic amine metabolism; cell culture and its application. *Mailing Add:* 1020 Circle Dr Baltimore MD 21227

GUCKEL, GUNTER, b Hamburg, Ger, Jan 30, 35; US citizen; m; c 1. FLUID DYNAMICS, FINITE ELEMENT ANALYSIS. *Educ:* Univ Ill, BS(math), 64, BS(physics), 65, MS, 67. *Prof Exp:* Sr mathematician, United Technol Res Lab, 67-73; systs mgr, Arthur D Little Systs, 73-75; vpres systs mgr, AG Becker, 75-78; mgr MIS, Norden Systs, 78-83; DIR, MIS, ESSEX GROUP INC, 83- *Concurrent Pos:* Adj fac mem, Univ Hartford, 68-73. *Mem:* Am Phys Soc; Asn Comput Mach. *Res:* Information systems and factory automation. *Mailing Add:* 2721 Sunbird Cove Ft Wayne IN 46804-3439

GUCKENHEIMER, JOHN, b Baton Rouge, La, Sept 26, 45; m 74; c 2. DYNAMICAL SYSTEMS, BIFURCATION THEORY. *Educ:* Harvard Univ, BA, 66; Univ Calif, Berkeley, PhD(math), 70. *Prof Exp:* Vis lectr, IMPA, Rio de Janeiro, 69; sr res fel, Univ Warwick, 69-70; mem, Inst Advan Study, 70-72; lectr, Mass Inst Technol, 72-73; from asst prof to prof, Univ Calif, Santa Cruz, 73-85; PROF, CORNELL UNIV, 85-, DIR, CTR APPL MATH, 89- *Concurrent Pos:* Chmn, math dept, Univ Calif, Santa Cruz, 76-78; vis mem, Courant Inst Math Sci, NY Univ, 79; mem, Inst Hautes Etudes Sci 80; mem, Math Sci Res Inst, 83-84; mem, Mittag-Leffler Inst; Guggenheim fel, 84; chmn, Univ Calif Coord Comt Nonlinear-Sci, 83-85; bd dirs, Math Sci Res Inst, 82-85. *Mem:* Am Math Soc; Soc Indust & Appl Math. *Res:* Bifurcation theory. *Mailing Add:* Dept Math Cornell Univ Ithaca NY 14853

GUCKER, GAIL, b Pa, June 22, 42. MATHEMATICS. *Educ:* State Univ Ny, BA, 73, MS, 78. *Prof Exp:* PROF MATH, ROCHESTER INST TECHNOL, 82- *Mailing Add:* Rochester Inst Technol One Lomb Memorial Dr Rochester NY 14623

GUCWA, PAUL RAMON, b Buffalo, NY, Jan 26, 47. GEOLOGY. *Educ:* Franklin & Marshall Col, BA, 69; Univ Tex, Austin, MA, 71, PhD(geol), 74. *Prof Exp:* RES GEOLOGIST, DENVER RES CTR, MARATHON OIL CO, 74- *Mem:* Geol Soc Am; Am Geophys Union; Sigma Xi. *Res:* Structural geology; tectonics; mechanics of deformation of aggregate materials; geology of continental margins. *Mailing Add:* 842 Plainwood Dr Houston TX 77079

GUDAS, LORRAINE J, DEVELOPMENTAL BIOLOGY, BIOCHEMISTRY. *Educ:* Princeton Univ, PhD(biochem), 75. *Prof Exp:* ASSOC PROF, BIOL CHEM & MOLECULAR PHARMACOL DEPT, HARVARD MED SCH, HARVARD UNIV, 86- *Mailing Add:* c/o Tumor/ Biol S Farber 44 Binney St Boston MA 02115

GUDAUSKAS, ROBERT THOMAS, b Georgetown, Ill, July 26, 34; m 56; c 3. VIROLOGY. *Educ:* Univ Eastern Ill, BS, 56; Univ Ill, MS, 58, PhD(plant path), 60. *Prof Exp:* Res asst plant path, Univ Ill, 56-60; from asst prof to assoc prof, 60-69, PROF PLANT PATH, AUBURN UNIV, 69- *Mem:* Am Phytopath Soc; Sigma Xi. *Res:* Plant virology. *Mailing Add:* Plant Path Dept Auburn Univ Auburn AL 36849-5409

GUDDER, STANLEY PHILLIP, b Centralia, Ill, Jan 6, 37; m 58; c 3. MATHEMATICS. *Educ:* Wash Univ, BS, 58; Univ Ill, MS, 60, PhD(math), 64. *Prof Exp:* Asst prof math, Army Math Res Ctr, Univ Wis-Madison, 64-65, asst prof, Univ, 65-69; assoc prof, 69-74, asst chmn dept, 73-75, chmn dept, 79-84, PROF MATH, UNIV DENVER, 74- *Concurrent Pos:* Sci Res Coun fel, Gt Brit, 75; sabbatical, Univ Nottingham, 75-76, Univ Bern, Univ Milan & Univ Genoa, 85. *Mem:* Sigma Xi. *Res:* Mathematical foundations of quantum mechanics; functional analysis; operator theory. *Mailing Add:* 1710 S Newport Way Denver CO 80224

GUDE, ARTHUR JAMES, 3RD, b Iowa, Sept 25, 17; m 42; c 1. MINERALOGY. *Educ:* Colo Sch Mines, Geol Eng, 48, MSc, 49. *Prof Exp:* Geologist, US Geol Surv, 49-52, x-ray crystallogr & mineralogist, 52-76, res geologist, 76-85; RETIRED. *Concurrent Pos:* Lectr, Colo Sch Mines; adv, Geol Surv, Pakistan, 59-61; Fulbright fel, Univ Rio Grande, Porto Alegre, 85. *Mem:* Fel Mineral Soc Am; Am Crystallog Asn; Am Asn Petrol Geologists; Am Geol Inst; Geochem Soc. *Res:* Clay mineralogy and crystallography; carbonate mineralogy; geochemistry; mineralogy and geochemistry of authigenic zeolites; scanning electron microscopy and spectrography of zeolites; industrial use of zeolites in agriculture, animal husbandry and pollution control. *Mailing Add:* 845 Dudley St Lakewood CO 80215-5410

GUDE, RICHARD HUNTER, b Eau Claire, Wis, Sept 16, 37; m 60; c 2. ZOOLOGY, ANIMAL BEHAVIOR. *Educ:* Wis State Univ, Eau Claire, BS, 60; Mich State Univ, MS, 62, PhD(zool), 65. *Prof Exp:* From asst prof to assoc prof biol, Hartwick Col, 64-68; actg chmn dept biol, 68-72, assoc prof, 68-73, chmn, Div Sci & Math, 72-78, PROF BIOL, UNIV TAMPA, 73- *Mem:* Ecol Soc Am. *Res:* Nesting behavior of Siamese fighting fish, Betta splendens; heavy metal uptake in marine invertebrates. *Mailing Add:* Dept Math & Sci Univ Tampa Tampa FL 33606

GUDEHUS, DONALD HENRY, b Jersey City, NJ, Sept 13, 39; m 68. OBSERVATIONAL COSMOLOGY, INSTRUMENTATION. *Educ:* Mass Inst Technol, BS, 61; Columbia Univ, AM, 63; Univ Calif, Los Angeles, PhD(astron), 71. *Prof Exp:* Engr & scientist, McDonnell Douglas Aerophysics Lab, 64-67; scholar, dept astron, Univ Calif, Los Angeles, 71-75; asst prof physics, dept physics, Los Angeles City Col, 74-81; asst res scientist, Dept Physics & Astron, Univ Mich, 81-89; VIS ASST PROF, DEPT PHYSICS, OKLA STATE UNIV, 89- *Mem:* Am Astron Soc. *Res:* Design of instrumentation and the taking of observations for study of structure and evolution of universe; use of CCD to observe distant clusters of galaxies. *Mailing Add:* Dept Physics Oklahoma State Univ Stillwater OK 74078

GUDERLEY, HELGA ELIZABETH, b Dayton, Ohio, Sept 24, 49; m 73; c 2. ECOLOGICAL PHYSIOLOGY, COMPARATIVE BIOCHEMISTRY. *Educ:* Earlham Col, BA, 71; Univ BC, PhD(zool), 76. *Prof Exp:* Res asst, Univ BC, 72-76; fel, Ore State Univ, 77-78; prof assoc, 78-79, asst prof, 79-84, ASSOC PROF PHYSIOL, UNIV LAVAL, 84- *Mem:* Am Physiol Soc; Can Soc Zool. *Res:* Evolution of pyruvate kinase isozymes; anaerobic metabolism in mytilus edulis and short and long term thermal adaptation of fish muscle metabolism. *Mailing Add:* Dept Biol Univ Laval Quebec City PQ G1K 7P4 Can

GUDERLEY, KARL GOTTFRIED, b Braunsdorf near Freiberg, Ger, June 15, 10; nat US; m 40; c 2. MATHEMATICS, AERODYNAMICS. *Educ:* DrIng, 38, DrIng habil, 43. *Prof Exp:* Aeronaut res engr, Air Res, Ger, 38-46; aeronaut res engr, 46-55, chief appl math br, Aeronaut Res Lab, Wright Air Develop Ctr, 55-61, sr scientist, Appl Math Res Lab, Aerospace Res Labs, 61-74; RETIRED. *Honors & Awards:* Thurman H Bane Award, 55. *Res:* Transonic flow; theoretical aerodynamics; applied mathematics. *Mailing Add:* 1645 Countryside Dr Dayton OH 45432

GUDMUNDSEN, RICHARD AUSTIN, b Salt Lake City, Utah, Dec 27, 22; m 47; c 6. PHYSICS, RESEARCH ADMINISTRATION. *Educ:* Univ Southern Calif, BE, 47, PhD(physics), 55. *Prof Exp:* Mgr semiconductor labs, Hughes Aircraft Co, 52-59; dir quantum electronics div, Quantatron, Inc, 59-60; pres, Quantum Technol Labs, 60-63; mgr lasers & electrooptics, Autonetics Div, Electronic Res Ctr, Rockwell Int Corp, Anaheim, Calif, 63-69, sci adv to vpres res & technol, 69-70, dir advan technol, 71-83; PRES, PERCEPTRIX INC, SANTA ANA, CALIF, 83- *Mem:* Am Phys Soc; Am Inst Elec & Electronics Engrs; Sigma Xi; Am Inst Physics. *Res:* Quantum and electro-optics; solid state physics; semiconductor physics and devices; invention and development of physical-electronic devices; noise theory; laser physics, devices and systems; computer memories; heat transfer; instrumentation physics; display technology; information sciences. *Mailing Add:* 12052 Larchwood Lane Santa Ana CA 92705

GUEFT, BORIS, b Cannes, France, Nov 10, 16; nat US; m 43; c 3. PATHOLOGY, CYTOCHEMISTRY. *Educ:* Columbia Univ, AB, 38; NY Univ, MD, 41. *Prof Exp:* Resident & USPHS asst, New Britain Gen Hosp, Conn, 46-47; instr, Sch Med, Yale Univ, 50-55; asst prof, Med Sch, Univ Cincinnati, 55-58; from assoc prof to prof path, Albert Einstein Col Med, 58-70, clin prof path, 70-87; DIR PATH, UNION HOSP, BRONX, 74-; adj prof path, NY Med Col, 78- *Concurrent Pos:* Fel path, Mt Sinai Hosp, New York, 47-50; pathologist, Fairfield State Hosp, Conn, 50-55 & Vet Admin Hosp, Ohio, 55-58; adj prof path, NY Med Col, 78; consult path, Bronx-Lebanon Hosp. *Mem:* Am Asn Path & Bact; AAAS; Int Acad Path; Electron Micros Soc Am. *Res:* Penicillin evaluation; experimental syphilis; cytochemistry of collagen and amyloid disease; microspectrophotometry; electron microscopy; x-ray diffraction. *Mailing Add:* 25 Vanderbilt Rd Scarsdale NY 10583

GUEHLER, PAUL FREDERICK, b Sandwich, Ill, Apr 14, 38; m 61; c 3. ORGANIC CHEMISTRY. *Educ:* Augustana Col, Ill, BA, 60; Univ Minn, Minneapolis, PhD(org chem), 65. *Prof Exp:* Sr chemist, Cent Res Labs, Minn Mining & Mfg Co, 65-67, supvr med prod div, 67-68, supvr tech mkt, New Bus Venture Div, 68-71, develop specialist, 71-74, lab mgr plant care systs, 74-77, RES & DEVELOP MGR, OCCUP HEALTH & SAFETY PROD DIV, 3M CO, 74- *Mem:* Am Chem Soc. *Res:* Efficient delivery systems of nutrients to plants. *Mailing Add:* Kasteellaan 23 Alsemberg 1641 Belgium

GUELL, DAVID LEE, b Thorp, Wis, June 29, 38; m 60; c 3. HIGHWAY ENGINEERING. *Educ:* Northwestern Univ, BS, 61, MS, 62, PhD(eng mech), 65. *Prof Exp:* ASSOC PROF CIVIL ENG, UNIV MO-COLUMBIA, 72- *Mem:* Am Soc Civil Engrs; Am Soc for Eng Educ; Transp Res Bd. *Res:* Highway Engineering. *Mailing Add:* Dept Civil Eng 1047 Eng Bldg Univ Mo Columbia Columbia MO 65211

GUENGERICH, FREDERICK PETER, b Pekin, Ill, Jan 1, 49; m 73; c 3. BIOCHEMISTRY, TOXICOLOGY. *Educ:* Univ Ill, BS, 70; Vanderbilt Univ, PhD(biochem), 73. *Prof Exp:* Res assoc biochem, Sch Med, Univ Mich, 73-75; PROF BIOCHEM, DEPT BIOCHEM, SCH MED, VANDERBILT UNIV, 75- *Concurrent Pos:* Fel, Nat Inst Gen Med Sci, 74-75; res career develop award, Nat Inst Environ Health Sci, 78-, prin investr, 77-; prin investr, Nat Cancer Inst, 78- *Honors & Awards:* J J Abel Award, Am Soc Pharmacol Exp Therapeut, 84. *Mem:* Am Soc Biol Chemists; Sigma Xi; Am Chem Soc. *Res:* Enzymic activation and detoxification of foreign chemicals of environmental interest. *Mailing Add:* Dept Biochem Vanderbilt Univ Sch Med Nashville TN 37232-0146

GUENNEL, GOTTFRIED KURT, b Oelsnitz, Ger, Dec 24, 20; nat US; m 47. PALYNOLOGY. *Educ:* Butler Univ, BS, 43, MS, 49; Ind Univ, PhD, 60. *Prof Exp:* Asst prof bot, Franklin Col, 49; paleobotanist, Geol Surv, Ind Univ, 49-61; advan res geologist, Marathon Oil Co, 61-80, sr res geologist, 80-82; RETIRED. *Mem:* AAAS; Bot Soc Am; fel Geol Soc Am; Am Asn Stratig Palynologists. *Res:* Thermal alteration of organic matter in rocks; age-dating and ecology of fossil spores and pollen. *Mailing Add:* 835 Front Range Rd Littleton CO 80120

GUENTERT, OTTO JOHANN, b Lorrach, Ger, June 6, 24; nat US; m 60; c 3. SOLID STATE PHYSICS. *Educ:* Univ Freiburg, dipl, 51; Mass Inst Technol, PhD(physics), 56. *Prof Exp:* Asst physics, Mass Inst Technol, 53-56; mem staff, 56-61, prin res scientist, 61-70, MGR MAT ANALYSIS LAB, RAYTHEON RES DIV, 70- *Mem:* Am Phys Soc; Am Crystallog Asn; Sigma Xi. *Res:* X-ray diffraction; disordered structures; x-ray optics; electron microprobe analysis. *Mailing Add:* 146 Barton Dr Sudbury MA 01776

GUENTHER, ARTHUR HENRY, b Hoboken, NJ, Apr 20, 31; m 54; c 2. CHEMISTRY, PHYSICS. *Educ:* Rutgers Univ, BS, 53; Pa State Univ, PhD(chem physics), 57. *Prof Exp:* Asst quant anal, Pa State Univ, 53-55, spectros lab, 55-56; proj officer, Physics Br, Air Force Weapons Lab, Kirtland AFB, NMex, 57-59, dir, Pulse Power Lab, 59-62, dir, Mat Dynamics Facil, 62-65, sci adv & chmn, Simulation Group, Effects Br, 65-66, sci adv, Effects Br & chmn, Sci Support Group, 66-69, chief, Technol Div, 69-70, sci dir, 70-74, chief scientist, Air Force Weapons Lab, 74-88; SCI ADV TO GOV, NMEX, 88-; CHIEF SCIENTIST, ADVAN DEFENSE TECHNOL, LOS ALAMOS NAT LAB, 88- *Concurrent Pos:* Adj prof physics, Air Force Inst Technol, 67- & elec eng & physics, Tex Tech Univ, 71-; consult, Tech Educ Res Ctr, 72-; adj prof chem, physics & elec eng, Univ NMex, 77-; consult, Lawrence Livermore Nat Lab, 78- & Los Alamos Nat Lab, 79-; mem, Int Steering Comt, 87-, chmn, 91-; mem, Int Conf Ionization Phenomena in Gases; chair, Pulsed Power Conf & numerous other major confs; Eastman lectr award, Optical Soc Am, 90-91. *Honors & Awards:* Harry Diamond Award, Inst Elec & Electronic Engrs, 71 & Peter Haas Award, 89; Arthur Schalow Medal, Laser Inst Am. *Mem:* Fel Optical Soc Am; fel Inst Elec & Electronics Engrs; Am Soc Testing & Mat; Am Chem Soc; fel Laser Inst Am; Sigma Xi. *Res:* High resolution infrared spectroscopy; fast time instrumentation; exploding wire phenomenology; nuclear weapon phenomena; molecular physics; high temperature properties and equation of state; high energy discharge systems; shock hydrodynamics; lasers; ultrasonics, optics and spectroscopy; simulation of nuclear weapon detonations; pulsed power technology; high power/energy lasers; interaction; laser damage to optical materials. *Mailing Add:* 6304 Rogers NE Albuquerque NM 87110

GUENTHER, BOBBY DEAN, b St Louis, Mo, Feb 13, 39; m 61; c 2. OPTICAL PHYSICS. *Educ:* Baylor Univ, BS, 60; Univ Mo, MS, 63, PhD(physics), 68. *Prof Exp:* Res asst, McDonnell Douglas Corp, 60-61; res physicist, US Army Missle Command, Redstone Arsenal, Ala, 68-79; physicist, 79-87, DIR PHYSICS, US ARMY RES OFF, NC, 90- *Concurrent Pos:* Rep high energy laser prog, Air Force Weapons Lab, 73-75; adj prof physics, Duke Univ. *Mem:* Am Phys Soc; Am Optical Soc. *Res:* Femto second technology; atmospheric propagation of submillimeter waves; coherent imaging problems and optical processing techniques related to imaging; manager of atomic and molecular physics, optics, lasers and opening switch technology and Army research programs. *Mailing Add:* US Army Res Off PO Box 12211 Research Triangle Park NC 27709

GUENTHER, DONNA MARIE, b Meadville, Pa, Oct 20, 38; m 72. PEDIATRICS, IMMUNOLOGY. *Educ:* Allegheny Col, BS, 60; Temple Univ, MD, 67. *Prof Exp:* Res technician immunol-transplantation, Mary Imogene Bassett Hosp, Cooperstown, NY, 60-62; res asst div exp biol, Baylor Col Med, 62-63; intern pediat, St Christopher's Hosp Children, Philadelphia, 67-68, resident, 68-69, chief resident, 69-70; staff pediatrician, Children's

Heart Hosp, Philadelphia, 70-71; fel pediat allergy and immunol, Thomas Jefferson Univ, Philadelphia, 70-71; med fel immunol, Univ Minn, Minneapolis, 71-72; resident allergy & fel immunol, Kaiser Found Hosp & Univ Calif Med Ctr, San Francisco, 72-73; CHIEF IMMUNOL & ALLERGY, CHILDREN'S HOSP MED CTR, OAKLAND, 73- *Concurrent Pos:* Asst clin prof pediat, Univ Calif, San Diego, 76-, Univ Calif, San Francisco, 77- *Mem:* Fel Am Acad Pediat; fel Am Acad Allergy; fel Am Col Allergists; Int Soc Exp Hemat; Int Col Pediat. *Res:* Clinical investigation of immunologic parameters in children with allergic disease. *Mailing Add:* Kaiser Perm Med Group Allergy 1200 El Camino Real South San Francisco CA 94080

GUENTHER, FREDERICK OLIVER, b Sheboygan, Wis, July 20, 19; m 47; c 2. CHEMISTRY. *Educ:* Univ Wis, BS, 42. *Prof Exp:* Res chemist, US Rubber Co, NJ, 42-47; res asst, Res Lab, Gen Elec Co, 48-56, chemist, Major Appliance Div, 56-64; SR PROJ CHEMIST, TEXACO CHEM CO, INC, 64- *Mem:* Am Chem Soc. *Res:* Polymerization; fluid bed coatings; cellular plastics; process development in organic chemicals; organic coatings for wire. *Mailing Add:* 6000 Reims No 2405 Houston TX 77036-3052

GUENTHER, JOHN JAMES, b Birmingham, Ala, Nov 8, 29; m 52; c 2. MEAT SCIENCE. *Educ:* La State Univ, BS, 51, MS, 53; Tex A&M Univ, PhD(meat sci), 59. *Prof Exp:* Asst, La State Univ, 51-53; asst, Tex A&M, 55-58; 58-71, PROF ANIMAL SCI, OKLA STATE UNIV, 71- *Mem:* Am Soc Animal Sci; Inst Food Technol; Am Meat Sci Asn. *Res:* Beef and pork carcass quality; muscle growth; muscle proteins; meat freezing; irradiation; freeze drying; processing and preservation. *Mailing Add:* Meats Lab Okla State Univ Stillwater OK 74075

GUENTHER, PETER T, b Chicago, Ill, June 30, 35; m 63; c 2. NUCLEAR STRUCTURE. *Educ:* Northwestern Univ, BS, 63; Univ Ill, MS, 68, PhD(physics), 77. *Prof Exp:* Sci asst, 67-78, asst physicist, 78-81, PHYSICIST, ARGONNE NAT LAB, 81- *Mem:* Am Phys Soc. *Res:* Fast-neutron physics, particularly experimental work in total, elastic, inelastic-scattering and cross sections; theoretical nuclear models. *Mailing Add:* Argonne Nat Lab Bldg 314 9700 S Cass Ave Argonne IL 60439

GUENTHER, R(ICHARD), b Lodz, Poland, Sept 9, 10; nat US; m 34. TELECOMMUNICATIONS, COMPUTER SCIENCE. *Educ:* Danzig Tech Univ, Dipl Ing, 34, Dr Ing, 37. *Prof Exp:* Asst, Danzig Tech Univ, 34-37; from engr commun lab to mgr eng, Dept Radio Commun & Missile Controls, Siemens & Halske, Inc, Ger, 37-45; consult, 45-47; consult commun instrumentation, US Sig Corps Eng Labs, NJ, 47-52; mem tech staff & systs engr, Bell Tel Labs, Inc, 52-56; mgr surface commun systs lab, Systs Eng & Adv Develop, Radio Corp Am, 56-63, chief scientist, Commun Systs Div, 63-65, mgr adv commun technol, Commun Systs Div, 65-68, staff tech adv, Commercial Electronics Systs Div, RCA Corp, 68-70, comput systs architect, Comput Systs Develop Div, 70-72; PRES, COMMUN & INFO SYSTS, INC, INT CONSULTS, 72- *Mem:* AAAS; fel Inst Elec & Electronics Engrs. *Res:* Engineering physics, particularly application in modern communication through information theory; network analysis; solid state physics; market analysis; electromagnetic theory; statistical analysis; viscosity of quartz crystals; communication systems analysis. *Mailing Add:* Commun & Info Systs Inc 42 Hunt Rd Sudbury MA 01776-1743

GUENTHER, RAYMOND A, b Chicago, Ill, Mar 13, 32; m 56; c 2. LOW TEMPERATURE PHYSICS, SOLID STATE PHYSICS. *Educ:* Ill Inst Technol, BS, 55, MS, 57, PhD(physics), 69. *Prof Exp:* Res assoc physics, Argonne Nat Lab, 60-61; instr, Ill Inst Technol, 61-62; res assoc, Argonne Nat Lab, 62-65; instr, Chicago City Jr Col, Bogan Br, 65-66; instr, Ill Inst Technol, 66-69; vis asst prof, 69, ASSOC PROF PHYSICS, UNIV NEBR, OMAHA, 69- *Res:* Heat capacity of dysprosium metal; thermal conductivity of irradiated potassium chloride crystals at temperatures between 0.4 and 4 degrees Kelvin. *Mailing Add:* 5603 Jackson Omaha NE 68106

GUENTHER, RONALD BERNARD, b North Bend, Ore, Nov 24, 37; m 64. MATHEMATICAL PHYSICS. *Educ:* Ore State Univ, BA, 59, MA, 62; Univ Colo, PhD(appl math), 64. *Prof Exp:* Adv res mathematician, Res Ctr, Marathon Oil Co, 64-66; from asst prof to assoc prof, 66-76, PROF MATH, ORE STATE UNIV, 76- *Concurrent Pos:* Gast prof, Univ Hamburg, 70-71 & 73-74; vis, Hahn-Meituer-Inst, Berlin, 78 & 81. *Mem:* Ger Math Asn. *Res:* Partial differential equations and applied mathematics. *Mailing Add:* Dept Math Ore State Univ Corvallis OR 97331-4605

GUENTHER, WILLIAM BENTON, b Dayton, Ohio, Sept 27, 28. PHYSICAL CHEMISTRY, INORGANIC CHEMISTRY. *Educ:* Oberlin Col, BA, 48; Univ Rochester, MS, 50, PhD(chem), 54. *Prof Exp:* Instr chem, Univ Alaska, 50-52; asst prof, Muhlenberg Col, 54-56; from asst prof to assoc prof, 56-68, prof chem, 68-79, F B Williams prof chem, 79-86, EMER PROF, UNIV OF THE SOUTH, 87- *Concurrent Pos:* Cotrell Res Corp grant, 58; Lilly Endowment res grant, 59. *Mem:* Am Chem Soc; Sigma Xi. *Res:* Inorganic complexes; chemical education; analytical chemistry; published three books and 24 articles. *Mailing Add:* Dept Chem Univ of the South Sewanee TN 37375-4003

GUENTHER, WILLIAM CHARLES, b Stewartville, Minn, Dec 17, 21; m 55; c 4. STATISTICS. *Educ:* Univ Iowa, BA, 43, MS, 46; Univ Wash, PhD(math), 52. *Prof Exp:* Instr math, Univ Ark, 48-49; math statistician, US Naval Ord Lab, 52-55; assoc prof, 59-63, PROF STATIST, UNIV WYO, 63- *Concurrent Pos:* NSF fac fel, 62; ed, Bulletin, The Inst Math Statist, 81-86. *Mem:* Math Asn Am; Inst Math Statist; fel Am Statist Asn. *Res:* Mathematical statistics, quality control and nonparametrics. *Mailing Add:* Dept Stat Box 3332 Univ Wyo Laramie WY 82071

GUENTHERMAN, ROBERT HENRY, b Chicago, Ill, July 9, 33; m 59; c 2. PHYSIOLOGY, PERIODONTOLOGY. *Educ:* Baylor Univ, DDS, 65, PhD(physiol), 68. *Prof Exp:* NIH training fel physiol & periodont, 65-68; asst prof physiol & periodont, 68-77, ASSOC PROF PERIODONT, BAYLOR COL DENT, 77- *Honors & Awards:* Merritt-Parks Award, 65. *Mem:* Am Dent Asn; Am Asn Dent Schs; Int Asn Dent Res; Am Acad Periodont; Sigma Xi. *Res:* Effect of oxygen and oxygenating agents on normal and diseased periodontal tissue; cardiovascular and metabolic physiology and its relationship to the periodontal structure of the oral cavity. *Mailing Add:* 10111 Estate Lane Dallas TX 75238

GUENTZEL, M NEAL, b Austin, Tex, Aug 21, 44; m 68; c 3. MICROBIOLOGY. *Educ:* Univ Tex, Austin, BA, 67, MA, 70, PhD(microbiol), 72. *Prof Exp:* Instr microbiol, Dept Microbiol, Univ Tex, Austin, 71-72, fel, 72-73; asst prof, 73-77, ASSOC PROF MICROBIOL, DIV ALLIED HEALTH & LIFE SCI, UNIV TEX, SAN ANTONIO, 77-, ASSOC PROF, DEPT MICROBIOL, HEALTH SCI CTR, 78- *Mem:* Am Soc Microbiol; Int Asn Biol Standardization. *Res:* Virulence factors of enteric bacterial pathogens with concentration on factors affecting colonization and pathogenesis of vibrio cholerae; studies of experimental candidosis and antifungal therapy. *Mailing Add:* Dept Life Sci Univ Tex San Antonio 6700 N Fm 1604 W San Antonio TX 78285

GUENZER, CHARLES S P, b Chicago, Ill, Apr 27, 43; m; c 2. ELECTRICAL ENGINEERING, PATENT LAW. *Educ:* Univ Mich, BSE, 66; Stanford Univ, MS, 68, PhD(appl phys), 72; Georgetown, JD, 80. *Prof Exp:* Res physicist solid state, Dept Navy, 72-81, prog mgr, Defense Nuclear Agency, 80-81, patent atty, Off Naval Res, US Naval Res Lab, 81-83; patent atty, Sughrue Mion et al, 83-88; PATENT ATTY, BELLCORE, 88- *Mem:* Inst Elec & Electronic Engrs. *Res:* Nuclear radiation effects; satellite vulnerability analysis; technical management. *Mailing Add:* Bellcore PO Box 7040 Red Bank NJ 07701

GUENZI, WAYNE D, b Sterling, Colo, Oct 6, 31; m 54; c 3. SOIL CHEMISTRY. *Educ:* Colo State Univ, BS, 53, MS, 59; Univ Nebr, PhD(soil sci), 64. *Prof Exp:* Chemist, 59-66, RES SOIL SCIENTIST, US DEPT AGR, 66- *Mem:* AAAS; Am Soc Agron; Soil Sci Soc Am; Sigma Xi. *Res:* Isolation of phytotoxic substances in plants and soils; interactions of chlorinated insecticides with soils, including photodegradation, microbial decomposition and volatilization; formation, release and identification of volatile organic sulfur compounds from animal feedlots and wastes applied to soil. *Mailing Add:* PO Box E Ft Collins CO 80522

GUERBER, HOWARD P(AUL), b Rifle, Colo, July 29, 26; m 62; c 2. COMPUTER SYSTEMS DESIGN, ELECTRICAL ENGINEERING. *Educ:* Univ Colo, BS, 47, MS, 54. *Prof Exp:* Develop engr, Westinghouse Elec Corp, 48-50; design & develop engr, RCA Corp, 50-51 & 53-59, supvr systs eng, 59-72; mem tech staff, 72-91, System Eng Div, Mitre Corp; RETIRED. *Mem:* Asn Comput Mach; Inst Elec & Electronic Engrs. *Res:* Specification, development and implementation of major communications networks and nodal switching for voice and data. *Mailing Add:* 1904 Abbotsford Dr Vienna VA 22182

GUERIGUIAN, JOHN LEO, b Alexandria, Egypt, Sept 20, 35; US citizen; m 69; c 2. PHARMACOGENOLOGY, HISTORY & PHILOSOPHY OF SCIENCE. *Educ:* Univ Paris, BS, 58, MS, 64, MD, 65. *Prof Exp:* Res fel biochem, Med Sch, Harvard Univ, 65-67; lectr biochem, Med Sch Univ Paris, 67-69; asst prof pharmacol, Med Sch, Univ NC, Chapel Hill, 69-73; assoc prof pharmacol, Sch Med, Univ Minn, Duluth, 73-78; MED OFFICER NEW DRUGS, CTR DRUG EVAL & RES, FOOD & DRUG ADMIN, 78- *Concurrent Pos:* Attend Staff, Peter Bent Brigham Hosp, Boston, 65-67; res scientist, Nat Inst Sci Educ & Res Med, 67-69; sci dir, Inst Develop Rationnel Agents Med Teux, 88; mem expert adv panel on biol standardization, WHO, Geneva, 88-; exec secy, endocrinol & metab drugs adv comt, Food & Drug Admin, 90. *Mem:* Endocrine Soc; Am Soc Pharmacol & Exp Therapeut. *Res:* Developing new drugs, biotechnological, reproductive, oncological and antidiabetic, in a rational and efficient manner, taking also into account historical experiences. *Mailing Add:* 14513 Woodcrest Dr Rockville MD 20853

GUERIN, MICHAEL RICHARD, b Waukegan, Ill, May 10, 41; m 65; c 2. ANALYTICAL CHEMISTRY, CANCER. *Educ:* Northern Ill Univ, BS, 63; Iowa State Univ, PhD(anal chem), 67. *Prof Exp:* res assoc, anal chem div, 67-70, group leader, anal chem div, 70-76, SECT HEAD, ORGANIC CHEM SECT, ANALYTICAL CHEM DIV, OAK RIDGE NAT LAB, 76- *Concurrent Pos:* Prin investr, Nat Inst Cigarette Smoke Chem, 68-, Coun Tobacco Res Cigarette Smoke Chem, 74-78 & 85-, Dept Energy Synfuels Chem, 76-83, Environ Protection Agency Synfuels, 76-79, Dept Defense Fuels & Lubricants, 82-, Dept Energy Complex Mixtures Chem, 83-; consult proposal rev group, Nat Cancer Inst, 70-73, 75 & 81, Nat Inst Environ Health Sci, 76 & 80, Dept Energy, 76, 80 & 84, Environ Protection Agency, 78 & 80, Nat Inst Drug Abuse, 81 & 83, Elec Power Res Inst, 81 & 82; del, Int Group Smoking & Health Res, 70; consult, Chem Carcinogenesis Contract Rev Comt, Nat Cancer Inst, 70, mem tobacco working group, 70-75. *Mem:* AAAS; NY Acad Sci; Sigma Xi; Am Chem Soc; Soc Toxicol. *Res:* Development and application of analytical chemical methodologies to the elucidation of the chemical factors responsible for environmental, fossil energy and tobacco smoke chemical carcinogenesis and mutagenesis. *Mailing Add:* Bldg 4500S E-1SS Oak Ridge Nat Lab Oak Ridge TN 37830

GUERNSEY, DUANE L, MOLECULAR CARCINOGENESIS. *Educ:* Univ Hawaii, PhD(physiol), 78. *Prof Exp:* ASST PROF PHYSIOL, COL MED, UNIV IOWA, 81- *Res:* Cellular differentiation. *Mailing Add:* Physiol-Biophys 5-53BSB Univ Iowa Iowa City IA 52242

GUERNSEY, EDWIN O(WENS), b Washington, DC, Sept 14, 20; m 43; c 3. CHEMICAL ENGINEERING, PHYSICAL CHEMISTRY. *Educ:* Univ Ill, BS, 41; Yale Univ, DEng, 49. *Prof Exp:* Sr res chem engr, Cent Res Div, Socony Mobil Oil Co, Inc, 48-54, Socony sponsorship, Oak Ridge Sch Reactor Technol, 54-55, sr res chem engr, Cent Res Div, 55-60, res assoc, 60-65, sr planning assoc, Corp Planning & Econ Dept, Mobil Oil Corp, 65-67, supvr, Mobil Res & Develop Corp, 67-69, SPEC STUDIES GROUP, MOBIL

RES & DEVELOP CORP, 69- *Mem:* Am Inst Chem Engrs; Am Chem Soc; fel Am Inst Chem; Sigma Xi. *Res:* Research planning; alternate energy sources; energy conversion; synthetic fuels; nuclear electric power generation. *Mailing Add:* 108 Mews Lane Cherry Hill NJ 08003

GUERNSEY, JANET BROWN, b Germantown, Pa, May 2, 13; m 36; c 5. PHYSICS. *Educ:* Wellesley Col, AB, 35; Harvard Univ, AM, 48; Mass Inst Technol, PhD(physics), 55. *Prof Exp:* From instr to prof physics, Wellesley Col, 43-77, Louise S McDowell Prof, 70, chmn dept, 67-77; RETIRED. *Concurrent Pos:* Mem gov bd, Am Inst Physics, 74-77. *Mem:* AAAS; Am Asn Physics Teachers (pres, 75-76); Am Inst Physics; Am Phys Soc. *Res:* Neutron physics; electrostatic accelerators. *Mailing Add:* Sabrina Farm Wellesley MA 02181

GUERNSEY, RICHARD MONTGOMERY, b Newton, Mass, Apr 10, 37. ACOUSTICS. *Educ:* Wesleyan Univ, Conn, BA, 60. *Prof Exp:* Phys sci asst, Acoust Res Br, Army Human Eng Labs, Aberdeen, Md, 60-62; sr engr, Goodfriend-Ostergaard Assocs, 63-67; contract consult, Donley, Miller & Nowikas, 68-72; consult physicist, Brown-Guernsey Co, 73-74; dir, Cedar Knolls Acoust Labs, 74-88; PRIN, R M GUERNSEY & ASSOCS, 88- *Mem:* Acoust Soc Am; Am Soc Testing & Mat. *Res:* Acoustical testing and measurement including acoustical properties of materials, generation and transmission of sound, and human response to sound. *Mailing Add:* R M Guernsey & Assocs PO Box 1517 Morristown NJ 07960-1517

GUERRANT, GORDON OWEN, b Fulton, Mo, July 11, 23; m 44; c 4. ANALYTICAL CHEMISTRY. *Educ:* Westminster Col, AB, 44; Univ Ill, PhD(anal chem), 49. *Prof Exp:* Asst anal chem, Univ Ill, 46-49; chemist, Explor & Prod Res Div, Shell Develop Co, Tex, 49-63; sr res chemist, Armour Agr Chem Co, 63-67; SUPVRY RES CHEMIST, CTR DIS CONTROL, USPHS, 67- *Mem:* AAAS; Am Chem Soc; fel Am Inst Chem; Am Asn Clin Chem; Am Soc Microbiol. *Res:* Spectrophotometry; emission spectroscopy; x-ray diffraction; micromolecular weights; gas chromatography; trace gas analyses; thermal stability of ammonium phosphates; analysis and formulation of public health pesticides; chromatographic analysis of bacterial components. *Mailing Add:* 1652 Kings Down Circle NE Danwoody GA 30338

GUERRANT, JOHN LIPPINCOTT, b Callaway, Va, Dec 28, 10; m 45; c 2. INTERNAL MEDICINE. *Educ:* Hampden-Sydney Col, BS, 33; Univ Va, MD, 37, MS, 42. *Prof Exp:* From instr to prof, 46-81, EMER PROF INTERNAL MED, SCH MED, UNIV VA, 81- *Mem:* Am Thoracic Soc; AMA; Am Col Physicians; Am Acad Allergy. *Res:* Pulmonary disease. *Mailing Add:* Univ Hosp Charlottesville VA 22908

GUERRANT, WILLIAM BARNETT, JR, b Danville, Ky, Mar 22, 22; m 46; c 1. ORGANIC CHEMISTRY. *Educ:* Austin Col, AB, 46; Univ NC, PhD, 49. *Prof Exp:* Tech dir, Southern Chem Cotton Co, Inc, 49-51; plant mgr, Celulosa Nac, SAm, 51-52; res chemist, Anderson Clayton & Co, Inc, 52-57; prof chem, Austin Col, 57-69; PROF CHEM, TEX TECH UNIV, 69- *Concurrent Pos:* Consult, Robert R King & Assocs, 58- *Mem:* AAAS; Am Chem Soc; fel Royal Soc Chem. *Res:* Fat and oil chemistry; organic mechanisms. *Mailing Add:* 10819 Crown Colony Dr No 14 Austin TX 78747-1642

GUERRERO, ARIEL HERIBERTO, b Buenos Aires, Arg, Jan 28, 22; m 55; c 2. ANALYTICAL CHEMISTRY, BIOPHYSICAL CHEMISTRY. *Educ:* Univ Buenos Aires, PhD(chem), 46. *Prof Exp:* Asst anal chem, Fac Exact & Natural Sci, Univ Buenos Aires, 42-55, lectr, 56-58, from assoc prof to prof, 58-70, dean fac, 68-69, prof, Fac Agron & Vet Med, 70-88, inorg chem, Fac Eng, 57-59; res worker, Nat Coun Res, Arg, 74-89; DIR, SCH CHEM, UNIV DEL SALVADOR, 89- *Concurrent Pos:* Anal chemist, Nat Direction Chem, 44-46; head, Anal Lab, Lever Bros, 47-49; Brit Coun fel, Univ Liverpool, 47-48; dir, APA, 50-63; indust consult, A Bonfanti Factory, 53-59; consult, Espiga Co, 60-61; tech secy, Arg Inst Oils & Fats, 62-64; consult solvent extraction eng, 65-68; dir training chem teachers, Nat Res Coun, Ministry Educ & Nat Inst, 62-68, Pan Am Union, 69-73; mem, Conf Chem Educ, Pan Am Union, Buenos Aires, 65, Int Union Pure & Appl Chem, Frascati, 69, Sao Paulo, 71, Wroclaw, 73, Madrid, 75, Liubliana, 77, Perth, 78, Lima, 78, Dublin & Davos, 79, Costa Rica, Maryland & Louvain, 81, Montpellier, 83, Sao Paulo, 85; Arg rep, Comn Teaching Chem, Int Union Pure & Appl Chem, 72-; dir, Sem Teaching Chem, UNESCO, Montevideo, 72-74, Mexico, 77; mem, Comt Expert Chem Educ, Pan-Am Union, Honduras, 74; Peruvian Sem Chem Educ, 88; Third World Acad Sci, Caracas, 90. *Mem:* Arg Chem Asn; Arg Sci Soc; Arg Asn Advan Sci; Am Chem Soc; Arg Inst Qual Control; hon mem, Peru Chem Soc. *Res:* Acetate complexes of lead; rhodamine B as reagent; buffering capacity; direct estimation of ions; spectrophotometric parameters; Liesegang rings; properties of biological membranes, before and after irradiation; methodology of science, education and creativity. *Mailing Add:* Avenida Santa Fe 2879 Buenos Aires 1425 Argentina

GUERRERO, E T, b Richmond, Tex, Nov 2, 24; m 49; c 4. PETROLEUM & MECHANICAL ENGINEERING. *Educ:* Tex A&M Univ, BS, 49, MS, 50, PhD(petrol eng), 53. *Prof Exp:* Instr petrol eng, Tex A&M Univ, 52-53; petrol reservoir engr, Seeligson Eng Comt, 53-56; trustee's prof petrol eng & head dept, 56-66, dean col eng & phys sci, 66-76, PROF ENG & PHYS SCI, UNIV TULSA, 66- *Mem:* Am Soc Eng Educ; Am Inst Mining, Metall & Petrol Engrs. *Res:* Relation between capillary pressures and relative permeabilities of large limestone cores; effect of surface and interfacial tensions on the recovery of oil by water flooding. *Mailing Add:* Col Eng/Phys Sci Univ Tulsa Tulsa OK 74104

GUERRERO, JORGE, b Andahuaylas, Peru, Oct 20, 42; m 71; c 2. PARASITOLOGY, IMMUNOLOGY. *Educ:* San Marcos Univ, Peru, BVetMed, 65, DVM, 65; Univ Ill, MS, 68, PhD(parasitol & immunol), 71. *Prof Exp:* Instr parasitol, San Marcos Univ, Lima, 65-67; res & teaching asst immunol/parasitol, Univ Ill, Urbana, 67-72, res assoc, 71-72; asst prof vet parasitol, San Marcos Univ, 72-73; prof med parasitol, Escola Paulista de Med, Sao Paulo, 73-74; mgr vet res & develop, Johnson & Johnson, Brazil, 74-76; sr parasitologist res & develop, 76-81, MGR, PRE-CLIN RES, PITMAN MOORE, INC, 81- *Concurrent Pos:* Prof parasitic dis, Univ Sao Paulo, 74-76; consult vet grad educ, CAPES, Ministry Educ, Brasilia, 75-76; health sci adv, Latin Am teaching fel prog Brazil, Tufts Univ, 75-76; prof immunol, Escola Paulista de Med, Sao Paulo, 74-76; prof & adv, Vet Parasitol Prog, Univ Fed do Rio Grande do Sul, Porto Alegre, 73-76. *Mem:* Am Soc Paratologists; Royal Soc Trop Med & Hyg; Am Vet Med Asn; Brazilian Soc Microbiol; World Asn Advan Vet Parasitol. *Res:* Development of antiparasitic compounds; in vitro cultivation of parasites, immunity to parasitic infections; development of diagnostic systems for parasitic infections; experimental models for study of host-parasite relationship. *Mailing Add:* 519 17th St Union City NJ 07087

GUERRERO-MUNOZ, FREDERICO, b Mex, Sept 26, 49; m 73; c 4. NEUROPHARMACOLOGY. *Educ:* Univ Hidalgo, BS, 67; Autonoma Univ, MD, 73. *Prof Exp:* Asst prof pharmacol, Sch Med, Univ Mex, 74-75; fel, Univ Calif, San Francisco, 76-78; PROF PHARMACOL, SCH MED, UNIV PANAMA, 79-; ASST DIR HUMAN REPROD, MINISTRY HEALTH, 84- *Concurrent Pos:* Consult & contribr, US Dispensatory, 83-84. *Mem:* Am Soc Pharmacol & Exp Therapeut; Mex Soc Pharmacol; Latin Am Soc Pharmacol; Cent Am Soc Pharmacol; Int Narcotic Res Soc. *Res:* The area of neuropharmacology with the main focus on opioid mechanism of action and the development of an epilepsy model to evaluate anti-epileptic drugs. *Mailing Add:* Dept Pharmacol Univ Panama Sch Med Panama City Panama

GUERRIERO, VINCENT, JR, b Buffalo, NY, 52; m 90. MYOSIN LIGHT CHAIN KINASE, CALMODULIN. *Educ:* Syracuse Univ, PhD(cell biol), 79. *Prof Exp:* res instr, Baylor Col Med, 79-86; ASST PROF, DEPT ANIMAL SCI, UNIV ARIZ, 86- *Mem:* Am Soc Cell Biol. *Res:* Calcium calmodulin regulation of smooth muscle contraction; function of heat shock proteins. *Mailing Add:* Dept Animal Sci Tucson AZ 85721

GUERRY, DAVENPORT, JR, analytical chemistry; deceased, see previous edition for last biography

GUERTIN, RALPH FRANCIS, b Stafford Springs, Conn, Dec 2, 38. ELASTIC WAVE PROPAGATION, SEISMIC MODELLING & INVERSION. *Educ:* Worcester Polytech Inst, BS, 61; Yale Univ, MS, 63, PhD(physics), 69. *Prof Exp:* Physicist, US Naval Ord Lab, 61-62 & Nat Bur Standards, 63; res assoc, Mid East Tech Univ, Turkey, 64-65 & Univ Nijmegen, Neth, 65-67; res assoc, Univ Calif, Berkeley & Lawrence Berkeley Lab, 69-70; asst prof physics, Rice Univ, 70-77; sr res physicist, McAdams, Roux, O'Connor Assocs, 79-81; Vpres & mgr geophys res, O'Connor Res, Inc; AT MITRE CORP. *Concurrent Pos:* Assoc Western Univ Fac fel, Los Alamos Nat Lab, 75. *Mem:* Am Phys Soc; AAAS; Soc Explor Geophysicists; Europ Asn Explor Geophysicists; Seismol Soc Am. *Res:* Seismic wave propagation; seismic modelling using finite difference, finite element, asymptotic ray theory and reflection matrix methods; seismic wave decomposition; surface waves; generalized inversion; analytic s-matrix theory; relativistic wave equations; coulomb-nuclear interference. *Mailing Add:* Mitre Corp, Dept D97 Mail Drop B115 Bedford MA 01730

GUERTIN, ROBERT POWELL, b Trenton, NJ, July 5, 39; m 66; c 2. SOLID STATE PHYSICS, MAGNETISM. *Educ:* Trinity Col, Conn, BS, 61; Wesleyan Univ, MA, 63; Univ Rochester, PhD(physics), 68. *Prof Exp:* From asst prof to assoc prof, 68-74, PROF PHYSICS, TURFTS UNIV, 74- & DEAN GRAD SCH ARTS & SCI, 85- *Concurrent Pos:* Vis scientist, Francis Bitter Nat Magnetic Lab, Mass Inst Technol, Univ Calif, San Diego. *Mem:* Am Phys Soc. *Res:* Effect of magnetic impurities on superconductivity; transport and high field magnetic properties of nearly ferromagnetic alloys; magnetic properties of superconducting and normal metals and alloys at high pressures. *Mailing Add:* 120 Packard Ave Tufts Univ Medford MA 02155

GUESS, HARRY ADELBERT, b New York, NY, Dec 24, 40; m 64; c 2. BIOSTATISTICS, EPIDEMIOLOGY. *Educ:* Ga Inst Technol, BS & MS, 64; Stanford Univ, MS & PhD(math), 72; Univ Miami, MD, 79; Am Bd Prev Med, cert; Am Bd Pediat, cert. *Prof Exp:* Asst prof math, Univ Rochester, 72-73; mem tech staff telecommun res, Bell Tel Labs, 73-75; res mathematician, Nat Inst Environ Health Sci, NIH, 75-77; res physician, NC Mem Hosp, 79-82; SR DIR EPIDEMIOL DEPT, MERCK SHARP & DOHME RES LAB, 82- *Concurrent Pos:* Consult, sci res syst, 80-82, Hoffman-La Roche Inc, 81-87; vis scientist, Mayo Clinic, 83-; auxiliary prof, epidemiol & biostatist, McGill Univ, 87-; adj assoc prof epidemiol & biostatist, Univ NC, 88-91, adj prof, 91- *Mem:* Soc Epidemiol Res Serv; AMA; Biomet Soc; fel Am Col Epidemiol; fel Am Col Prev Med; fel Am Acad Pediat. *Res:* Biostatistics; epidemiology; applications of mathematics and statistics to clinical medicine; pharmacoepidemiology; mathematical popoulation genetics; clinical trial design. *Mailing Add:* 104 Waterford Pl Chapel Hill NC 27514

GUESS, WALLACE LOUIS, b Durham, NC, July 18, 24; m 47; c 2. TOXICOLOGY. *Educ:* Univ Tex, BS, 49, MS, 51; Univ Wash, PhD(pharm), 59. *Prof Exp:* From asst prof to prof pharm, Univ Tex, Austin, 53-71, asst dir drug plastic res lab, 63-71; DEAN SCH PHARM, UNIV MISS, 71- *Concurrent Pos:* Chmn panel over-the-counter antimicrobial agents, Fed Food & Drug Admin, 74-80; chmn, Nat Ctr Toxicol Res-Assoc Univ. *Mem:* Am Asn Col Pharm; hon mem Mex Pharmaceut Asn; Soc Toxicol. *Res:* Toxicology, pharmacology and evaluation of plastics used in medical practice. *Mailing Add:* Dept Pharm Univ Miss Main Campus University MS 38677

GUEST, GARETH E, b Mobile, Ala, June 13, 33; m 54; c 3. PLASMA INSTABILITIES THEORETICAL, PLASMA THEORY-WAVE PARTICLE INTERACTIONS FOR RF HEATING. *Educ:* Vanderbilt Univ, BA, 54, MA, 56, PhD(physics/math), 60. *Prof Exp:* From asst prof to assoc prof physics, N Tex State Univ, 60-63; staff scientist, Oak Ridge Nat Lab, 63-65, group leader, fusion plasma theory, 65-75; dept mgr fusion plasma theory, Gen Atomic Co, 75-81; VPRES, APPL MICROWAVE PLASMA

CONCEPTS, 81- *Concurrent Pos:* Lectr, Oak Ridge Inst Nuclear Studies, 65-75. *Mem:* Am Phys Soc (secy/treas, 68-74). *Res:* Theoretical investigations of fusion plasma concepts, including magnetic-mirror systems, the ELMO Bumpy Torus, tokamaks, etc; high-frequency collective phenomena; the theory of electron cyclotron heating and the resulting relativistic-electron plasmas. *Mailing Add:* 5433 Caminito Rosa La Jolla CA 92037

GUEST, MARY FRANCES, b New Smyrna Beach, Fla, Oct 7, 27. PHYSIOLOGY. *Educ:* Fla State Univ, BS, 48; Tex Tech Univ, MS, 54; Tulane Univ, PhD(physiol), 64. *Prof Exp:* Instr zool, anat & physiol, Tex Tech Univ, 54-60; res asst surg, Sch Med, Tulane, 63-64; asst prof physiol, Fac Med, Univ Malaya, 64-68; from lectr to asst prof, Sch Vet Med, Univ Calif, Davis, 68-71; asst res parasitologist, G W Hooper Found, Med Ctr, Univ Calif, San Francisco, 71-72; assoc prof, 72-77, PROF BIOL, CLARKE COL, IOWA, 77-, DIR BIOL HEALTH PROG, 72-, CHMN DEPT BIOL, 81- *Mem:* Am Soc Trop Med & Hyg; Am Soc Parasitologists; Sigma Xi; NY Acad Sci. *Res:* Pulmonary physiology; pulmonary pathophysiology related to immune etiologies; immune response to filarial infections. *Mailing Add:* Dept Biol Clarke Col Dubuque IA 52001

GUEST, MAURICE MASON, b Fredonia, NY, July 30, 06; m 36; c 2. PHYSIOLOGY. *Educ:* Univ Mich, AB, 30; Columbia Univ, PhD(physiol), 41. *Prof Exp:* Field asst, Fed Bur Entom, 30-31; pub sch teacher, NY, 31-36; asst physiol, Columbia Univ, 36, instr, 36-40, res assoc, 40-42; from asst prof to assoc prof, Col Med, Wayne Univ, 46-50; prof physiol & chmn dept, 51-73, EMER ASHBEL SMITH PROF PHYSIOL & BIOPHYS, UNIV TEX MED BR GALVESTON, 73-; DIR HEMATOL DIV, SHRINERS BURNS INST, 73- *Concurrent Pos:* Mem, Subcomt Hemorrhage & Thrombosis, Nat Res Coun, 59-63, comt biol, White House Conf Aging, 60-61, comt thrombolytic agents, Nat Heart Inst, 62-67, hemat study sect, NIH, 60-64, 67-71 & med adv bd educ film prod, 67-85; assoc ed, Microvasc Res, 69-73. *Mem:* AAAS; Am Physiol Soc; Soc Exp Biol & Med; Int Soc Hemat; Microcirc Soc; Sigma Xi. *Res:* Carbohydrate metabolism; endocrinology; physiology of uterus and vagina; physiology of high altitudes; blood clotting mechanisms with special consideration of lytic and antilytic factors; effects of ions on heart and circulation; gerontology; physiology of microcirculation; physiology of decompression sickness; effects of burns on blood and microcirculation. *Mailing Add:* Div Hemat Shriners Burns Inst Univ Tex Med Br Galveston TX 77550

GUEST, RICHARD W(ILLIAM), b Oklahoma City, Okla, July 7, 32; m 59; c 3. AGRICULTURAL ENGINEERING. *Educ:* NDak State Univ, BS, 54, MS, 58. *Prof Exp:* Asst prof, 58-64, assoc prof, 64-84, PROF AGR ENG, CORNELL UNIV, 84- *Concurrent Pos:* Staff mem, Ford-New Holland, PA; consult, Alfa-Laval Co, Sweden, 74-75. *Mem:* Am Soc Agr Eng. *Res:* Electrical power and processing; materials handling; agricultural waste management; farm power and machinery. *Mailing Add:* Dept Agr & Biol Gen Riley Robb Hall Cornell Univ Ithaca NY 14853

GUEST, WILLIAM C, b Pine Bluff, Ark, July 18, 25. GENETICS. *Educ:* Emory Univ, AB, 48, MS, 50; Univ Tex, PhD(zool), 59. *Prof Exp:* Instr sci, Anderson Col, SC, 50-52; asst prof biol, Millsaps Col, 52-54; marine biologist, Marine Lab, Tex Game & Fish Comn, 54-57; res assoc genetics found, Univ Tex, 57-59; asst prof biol, Univ Ala, 59-63; from asst prof to assoc prof, 63-74, PROF ZOOL, UNIV ARK, FAYETTEVILLE, 74- *Mem:* AAAS; Genetics Soc Am; Am Genetic Asn; Soc Study Evolution; Sigma Xi. *Res:* Karyotypic studies of, and the effect of drugs on Drosophila species. *Mailing Add:* 821 E Lakeside Fayetteville AR 72701

GUETHS, JAMES E, b Shawano, Wis, Aug 18, 39; m 61; c 3. SOLID STATE PHYSICS. *Educ:* Ripon Col, BA, 61; Univ Conn, MS, 63, PhD(physics), 66. *Prof Exp:* Res assoc physics, Univ Conn, 66; from asst prof to assoc prof, 66-76, chmn dept, 69-74, PROF PHYSICS, UNIV WIS-OSHKOSH, 76-, ASST V CHANCELLOR, 74- *Mem:* Am Phys Soc; Am Asn Higher Educ. *Res:* Transport phenomena in metal and alloy single crystals; learning hierarchies in university science; organizational development. *Mailing Add:* 120 Camden Lane Oshkosh WI 54904

GUETTER, HARRY HENDRIK, b Andijk, Holland, Feb 1, 35; US citizen; m 63; c 3. ASTRONOMY, ASTROMETRY. *Educ:* Queen's Univ, Ont, BSc, 61; Univ Toronto, MA, 63. *Prof Exp:* Res asst astron, David Dunlap Observ, Richmond Hill, Ont, 63-64; ASTRONR, NAVAL OBSERV, 64- *Mem:* Am Astron Soc; Int Astron Union; Astron Soc Pac; Am Sci Affil. *Res:* Photoelectric photometry and spectroscopy of early-type stars in associations and clusters; investigations of properties of interstellar matter and distance determinations of stars with small masses. *Mailing Add:* US Naval Observ PO Box 1149 Flagstaff AZ 86002

GUEVARA, FRANCISCO A(NTONIO), b Denver, Colo, Jan 10, 24; m 51; c 5. NUCLEAR ENGINEERING, CHEMICAL ENGINEERING. *Educ:* Tex A&M Univ, BS, 50; Univ NMex, MS, 59. *Prof Exp:* Chem engr, Health & Safety Br, AEC, NMex, 50-56; staff mem, nuclear res & develop, Radiol Protection Eng, 56-90, RADIOL PROTECTION ENG CONSULT, LOS ALAMOS NAT LAB, 90- *Concurrent Pos:* Consult nuclear engr reactor licensing, Effluent Treat Systs Br, Nuclear Regulatory Comn, 72-75; reviewer, Dept Educ Fuel Cycle Facil Safety Anal Reports, 78- *Honors & Awards:* Medal Hon fel Eng, Mex Am Eng Soc. *Mem:* Am Nuclear Soc; Health Physics Soc; Mex Am Eng Soc (nat pres). *Res:* control systems for decontamination of effluents from nuclear facilities and disposal of radioactive waste. *Mailing Add:* Los Alamos Nat Lab PO Box 1663 MSM998 Los Alamos NM 87545

GUEVREMONT, ROGER M, b Thunder Bay, Ont, Jan 20, 51; m 76; c 3. ANALYTICAL CHEMISTRY. *Educ:* McGill Univ, Can, BSc, 74; Univ Alberta, Can, PhD(chem), 78. *Prof Exp:* Assoc res off anal chem, Marine Anal Chem Standards Prog, 78-90, SR RES OFF ANALYTICAL CHEM, MEASUREMENT SCI, NAT RES COUN CAN, 90- *Mem:* Chem Inst Can. *Res:* Methods for trace metal analysis of marine samples; reference materials for trace metals and organics in marine matrices; instrumentation for graphite furnace atomic absorption and inductively coupled plasma emission spectrometry and computer automation in the analytical laboratory. *Mailing Add:* Inst Environ Chem Nat Res Coun Can Montreal Rd Ottawa ON K1A 0R6 Can

GUGELOT, PIET C, b Bussum, Neth, Feb 24, 18; US citizen; m 44; c 1. PHYSICS. *Educ:* Swiss Fed Inst Technol, dipl physics, 40, PhD(nuclear physics), 45. *Prof Exp:* Res assoc nuclear physics, Swiss Fed Inst Technol, 40-47; res assoc, Palmer Physics Lab, Princeton Univ, 47-49, asst prof, 49-56; prof nuclear physics & dir, Inst Nuclear Physics Res, Univ Amsterdam, 56-66; dir nuclear physics, Space Radiation Effects Lab, 66-67; PROF PHYSICS, UNIV VA, 67- *Concurrent Pos:* Vis prof, Univ Wash, 55, Oak Ridge Nat Lab, 60, Lawrence Radiation Lab, Univ Calif, Livermore, 60-61, Stanford Univ, 63-64 & Univ Chicago, 69; vis scientist, Ctr Nuclear Physics Res, Saclay, France, 76; guest scientist, Max Planck Inst, 80-88, Tokyo Univ, 84; gus, Japanese Soc Prom Sci. *Honors & Awards:* Sr Scientist Award, Alexander von Humboldt Found, 82. *Mem:* Fel Am Phys Soc; Swiss Phys Soc; Europ Phys Soc. *Res:* Medium energy nuclear physics, particularly nuclear reactions. *Mailing Add:* Dept Physics Univ Va Charlottesville VA 22901

GUGENHEIM, VICTOR KURT ALFRED MORRIS, b Berlin, Germany, Aug 28, 23; UK citizen; m 66; c 2. TOPOLOGY. *Educ:* Oxford Univ, BA, 48, DPhil(math), 52. *Prof Exp:* Fel math, Magdalen Col, 50-54; asst lectr, Birkbeck Col, Univ London, 54-56; asst prof, Johns Hopkins Univ, 56-64; reader, Imp Col, Univ London, 64-66; PROF MATH, UNIV ILL, CHICAGO CIRCLE, 66- *Concurrent Pos:* Harkness Found fel, 52-54; vis mem, Inst Advan Study, Princeton, 72-73. *Mem:* Am Math Soc. *Res:* Algebraic topology. *Mailing Add:* Dept Math Univ Ill Chicago Chicago IL 60680

GUGER, CHARLES EDMUND, JR, b Pittsburgh, Pa, Dec 1, 42; m 66; c 2. PROCESS DEVELOPMENT, PROCESS ENGINEERING. *Educ:* Carnegie-Mellon Univ, BS, 64, MS, 65, PhD(chem eng), 69. *Prof Exp:* Jr res scientist, Koppers Co Inc, 68-70, proj scientist, 70-72, sr res engr, 72-79; ENG SPECIALIST, MOBAY CHEM CORP, 79- *Mem:* Am Chem Soc; Am Inst Chem Engrs. *Res:* Chemical process simulation; chemical reaction kinetics; chemical thermodynamics; diffusion, mass transfer, and heat transfer pertaining to chemical processing; economic evaluations, process and project engineering. *Mailing Add:* 1359 Foxboro Dr Monroeville PA 15146

GUGGENBERGER, LLOYD JOSEPH, b Cold Spring, Minn, Sept 11, 39; m 61; c 3. ORGANOMETALLIC CHEMISTRY, POLYMER CHEMISTRY. *Educ:* St Johns Univ, Minn, BA, 61; Iowa State Univ, PhD(chem), 65. *Prof Exp:* Res chemist, 66-75, GROUP LEADER CHEM, CENT RES & DEVELOP DEPT, E I DU PONT DE NEMOURS & CO, INC, 75- *Mem:* Am Chem Soc; Am Crystallog Asn. *Res:* Structure and bonding of organic, organometallic and transition metal complexes; polymer physical chemistry. *Mailing Add:* Du Pont Exp Sta PO Box 80356 Wilmington DE 19880-0356

GUGGENHEIM, STEPHEN, b New York, NY, June 4, 48; m 74; c 2. SILICATE MINERALOGY. *Educ:* Marietta Col, BS, 70; State Univ NY, Stony Brook, MS, 72; Univ Wis-Madison, PhD(geol), 76. *Prof Exp:* From asst prof to assoc prof, 76-88, PROF GEOL, UNIV ILL, CHICAGO, 88- *Concurrent Pos:* Vis fel, Australian Nat Univ, Canberra, Australia, 83-84. *Honors & Awards:* Hawley Medal, Mineralogical Asn, Can, 87. *Mem:* Am Crystallog Asn; Clay Mineral Soc Am; Mineral Soc Am; Mineral Asn Can; Mineral Soc Gt Brit. *Res:* Application of x-ray and electron optical techniques to solve geologic problems; understanding the relationship of crystal structure and crystall chemistry to layer silicate stability. *Mailing Add:* Dept Geol Sci M/C 186 Box 4348 Univ Ill Chicago IL 60680

GUGGENHEIMER, HEINRICH WALTER, b Nurnberg, Ger, July 21, 24; nat US; m 47; c 4. MATHEMATICS, ARTIFICIAL INTELLIGENCE COMPUTING. *Educ:* Swiss Fed Inst Technol, Dipl, 47, DSc(math), 50. *Prof Exp:* Lectr math, Hebrew Univ, Israel, 54-56; prof, Bar Ilan Univ, Israel, 56-59; assoc prof, Wash State Univ, 59-60; from assoc prof to prof, Univ Minn, 60-67; prof, 67-90, EMER PROF MATH, LONG ISLAND CTR, POLYTECH UNIV, 90- *Concurrent Pos:* Vis lectr, Math Asn Am, 64-65. *Mem:* Swiss Math Soc; Math Asn Am; Soc Indust & Appl Math. *Res:* Computer graphics; differential and algebraic geometry; foundations of geometry; history of mathematics; all aspects of geometry, including differential equations; Aramaic linguistics; artificial intelligence. *Mailing Add:* PO Box 401 West Hempstead NY 11552-0401

GUGGENHEIMER, JAMES, b Belgrade, Yugoslavia, Mar 4, 36; US citizen; m 69. ORAL MEDICINE. *Educ:* City Col New York, BS, 58; Columbia Univ, DDS, 62. *Prof Exp:* Intern dent, Vet Admin Hosp, Albany, NY, 62-63; resident oral surg, Strong Mem Hosp, Rochester, 63-64; asst prof, 66-70, PROF ORAL MED, SCH DENT MED, UNIV PITTSBURGH, 76- *Concurrent Pos:* USPHS fel oral med, Sch Dent Med, Univ Pa & Philadelphia Gen Hosp, 64-66; mem staff dept med, Presby Univ Hosp, 67-; clin asst dept surg, Montefiore Hosp, 68-; consult, Vet Admin Hosp, Oakland, 67-, courtesy staff, Eye & Ear Hosp. *Mem:* Am Asn Dent Schs. *Res:* Oral ulcerations; effects of cancer chemotherapeutic agents; tetracycline stain; use and effects of smokeless tobacco; epidemiology of oral cancer. *Mailing Add:* Sch Dent Med Univ Pittsburgh C137 Pittsburgh PA 15261

GUGIG, WILLIAM, b New York, NY, Aug 15, 14; m 47; c 4. ORGANIC CHEMISTRY. *Educ:* Long Island Univ, BS, 49; Polytech Inst Brooklyn, MS, 56. *Prof Exp:* From instr to prof chem, Long Island Univ, 50-81; RETIRED. *Concurrent Pos:* NSF fac fel, Long Island Univ, 57-58, adj prof chem, 81- *Mem:* Am Chem Soc; Sigma Xi. *Res:* Properties and structure of cyclopentadienones; rearrangements of aromatic hydrocarbons. *Mailing Add:* 1771 E 14 St Brooklyn NY 11229

GUIDA, VINCENT GEORGE, b Brooklyn, NY, Sept 2, 48; m 77; c 2. MARINE ECOLOGY, INVERTEBRATE SYMBIOSIS. *Educ:* Rensselaer Polytech Inst, BS, 70; NC State Univ, PhD(marine sci), 77. *Prof Exp:* RES SCIENTIST INVERTEBRATE PATHOBIOL, INST PATHOBIOL, LEHIGH UNIV, 78-, DIR MARINE ECOL, WETLANDS INST, 79- *Concurrent Pos:* Fel, Inst Pathobiol, Lehigh Univ, 78-80. *Mem:* AAAS. *Res:* Estuarine ecology including community structure and function based on predation, competition, symbiosis and toxic interactions. *Mailing Add:* 565 Brighton St Bethlehem PA 18015

GUIDA, WAYNE CHARLES, b Tampa, Fla, Mar 20, 46; div; c 1. ORGANIC CHEMISTRY. *Educ:* Univ SFla, BA, 68, PhD(chem), 76. *Prof Exp:* Fel chem, Duke Univ, 76; vis asst prof, 76-77, asst prof, 77-82, ASSOC PROF CHEM, ECKERD COL, 82- *Concurrent Pos:* Vis assoc prof, Univ SC, 83-84; sr fel, Columbia Univ, 85-86. *Mem:* Am Chem Soc; Sigma Xi. *Res:* Chemistry of oxonium ions and their utility as synthetic reagents for organic synthesis; borohydride reductions of organic compounds in non-polar solvents; chemistry of crown ethers; molecular modeling and computational chemistry. *Mailing Add:* Collegium Nat Sci Eckerd Col PO Box 12560 St Petersburg FL 33733

GUIDERI, GIANCARLO, b Siena, Italy, Oct 18, 31; US citizen; m 73; c 3. PHARMACOLOGY. *Educ:* Long Island Univ, BS, 61; NY Univ, MS, 65; NY Med Col, PhD(pharmacol), 69. *Prof Exp:* From instr to asst prof, 69-76, ASSOC PROF PHARMACOL, NY MED COL, 76- *Concurrent Pos:* Adj asst prof cell biol, Fordham Univ, 69; adj assoc prof, Inst Int Med Educ, 74-; vis prof, Cath Univ Lille, France, 86- *Mem:* Am Soc Pharmacol & Exp Therapeut; Am Soc Toxicol. *Res:* Role played by certain drug and central nervous system influences on blood pressure and the development of severe cardiac rhythm disorders; drug-drug interactions. *Mailing Add:* Dept Pharmacol NY Med Col Basic Sci Bldg Valhalla NY 10595

GUIDI, JOHN NEIL, b Chicago, Ill, July 28, 54. DATABASE & INFORMATION RETRIEVAL SYSTEMS, KNOWLEDGE REPRESENTATION. *Educ:* Purdue Univ, BS, 76. *Prof Exp:* Sci appln prog analyst, Jackson Lab, 77-82, systs prog analyst, 82-88, sr appln prog analyst, 88-89, SCI SOFTWARE ENG, JACKSON LAB, 89- *Concurrent Pos:* Software libr, Maine Local Users Group, DECUS Chap, 79-84; mem, People-to-People Computer Software Deleg, People's Repub China, 83; fac mem, med & exp mammalian genetics, Jackson Lab, 85-88; benevolent dictator, Maine Unix Users Group, 86-89; consult, med & exp mammalian genetics, Jackson Lab, 89 & 91; vis prof math, Col of the Atlantic, 90. *Mem:* Asn Comput Mach; Am Asn Artificial Intel; Math Asn Am; Inst Elec & Electronics Engrs Comput Soc. *Res:* Design and implementation of computer systems to provide support for research in genetics; research of issues in database and information retrieval systems related to access of scientific data. *Mailing Add:* Jackson Lab 600 Main St Bar Harbor ME 04609-0800

GUIDICE, DONALD ANTHONY, b Astoria, NY, Oct 12, 34; m 65; c 4. SPACE SYSTEMS TECHNOLOGY, RADIO ASTRONOMY. *Educ:* Manhattan Col, BEE, 56; Ohio State Univ, MS, 59, PhD(elec eng), 69. *Prof Exp:* Electronics engr, Air Force Avionics Lab, Ohio, 56-64; res physicist, Air Force Cambridge Res Labs, 64-71, radio astronr, 71-82, CHIEF ENGR SPACE SYST TECHNOL, AIR FORCE GEOPHYS LAB, 82- *Mem:* Am Astron Soc; Inst Elec & Electronics Engrs; Int Astron Union; Int Union Radio Sci. *Res:* Ring laser techniques for angular rotation sensing; atmospheric microwave radiometry; spectral characteristics of solar radio bursts; solar microwave instrumentation design; space environment effects on space system technologies. *Mailing Add:* Air Force Geophys Lab Hanscom AFB MA 01731

GUIDOTTI, CHARLES V, b Somerville, Mass, Sept 19, 35; m 61; c 3. GEOLOGY, PETROLOGY. *Educ:* Yale Univ, BS, 57; Harvard Univ, PhD(geol), 63. *Prof Exp:* Res asst geol, Univ Minn, 58; lectr, Univ Calif, Davis, 62-63, from asst prof to assoc prof, 63-69; assoc prof to prof geol, Univ Wis-Madison, 69-81; PROF GEOL, UNIV MAINE, ORONO, 81- *Mem:* Am Geophys Union; Mineral Asn Can; Sigma Xi; fel Mineral Soc Am. *Res:* Metamorphic petrology of high-grade pelitic and calc-silicate rocks; stratigraphy and structure of metamorphosed strata in northwest Maine and the central coast of Maine. *Mailing Add:* Dept Geol Sci Univ Maine Orono ME 04469

GUIDOTTI, GUIDO, b Florence, Italy, Nov 3, 33. BIOCHEMISTRY. *Educ:* Wash Univ, MD, 57; Rockefeller Inst, PhD(biochem), 63. *Prof Exp:* Intern, Barnes Hosp, 57-58, asst res, 58-59; asst prof biol, 63-68, assoc prof biol, biochem & molecular biol, 68-69, PROF BIOCHEM, HARVARD UNIV, 69- *Mem:* Am Soc Biol Chemists; Am Acad Arts & Sci. *Res:* Interactions of macromolecules in solution; structure and function of proteins; structure and function of biological membranes; mechanism of hormone action. *Mailing Add:* Fairchild Biochem Bldg Harvard Univ Seven Divinity Ave Cambridge MA 02138

GUIDRY, CARLTON LEVON, b Palacios, Tex, June 23, 33; m 55; c 5. ORGANIC CHEMISTRY. *Educ:* Univ Houston, BS, 58, PhD(biochem), 62. *Prof Exp:* Res chemist, Continental Oil Co, 61-63; from asst prof to assoc prof, 63-72, PROF CHEM, SAM HOUSTON STATE UNIV, 72- *Mailing Add:* 116 Southwood Dr Huntsville TX 77340

GUIDRY, CLYDE R, b Paris, Tex, Oct 4, 56. BIOCHEMISTRY. *Educ:* Univ Tex, BS, 81, PhD(cell biol), 86. *Prof Exp:* Teaching asst, Med Histol & Cell Biol, Health Sci Ctr, Univ Tex, Dallas, 84-85; postdoctoral fel, Dept Biochem, 86-88, res assoc, 88-90, RES INSTR, DEPTS BIOCHEM & OPHTHAL, UNIV ALA BIRMINGHAM, 90- *Concurrent Pos:* Arthritis Found fel, 88-91; Helen Keller Eye Res Inst investr, 90-92. *Res:* Cell biology; numerous technical publications. *Mailing Add:* Dept Biochem Univ Ala Birmingham BHSB 509 Birmingham AL 35294

GUIDRY, DIEU-DONNE JOSEPH, microbiology, for more information see previous edition

GUIDRY, MARION ANTOINE, b Lafayette, La, Oct 1, 25; m 46; c 3. BIOCHEMISTRY. *Educ:* Tulane Univ, BS, 47, MS, 49, PhD(chem), 52. *Prof Exp:* Asst prof chem, Southwestern La Inst, 52-53; asst prof ophthal, Sch Med, Tulane Univ, 53-56; asst prof biochem & path, Sch Med, La State Univ, 56-65; coordr life sci res, 65-75, PROF CHEM & BIOL, W TEX STATE UNIV, 65- *Mem:* Am Chem Soc; Am Asn Pathologists; Soc Exp Biol & Med. *Res:* Lipid chemistry; lipid metabolism. *Mailing Add:* Chem Dept WTex State Univ Canyon TX 79015

GUIDUCCI, MARIANO A, b Peckville, Pa, Nov 29, 30; m 54; c 2. ORGANIC CHEMISTRY. *Educ:* Albright Col, BS, 52; Lehigh Univ, MS, 54; Columbia Univ, PhD(org chem), 65. *Prof Exp:* Res asst, Olin Mathieson Chem Corp, 54-56, res asst, Squibb Inst Med Res, 56-61, sr res chemist, 65-70; sr res chemist, Schwarz Biores/Mann Res, 70-74; mgr process res, 74-81, SR PROCESS CHEMIST, FMC CORP, 81- *Mem:* Am Chem Soc. *Res:* Synthetic methods, especially as related to steroids and natural products; scale up of synthesis; peptide synthesis. *Mailing Add:* 68 Eden Ave Edison NJ 08817-3850

GUIER, WILLIAM HOWARD, b Wichita, Kans, July 13, 26; m 50; c 1. THEORETICAL PHYSICS. *Educ:* Northwestern Univ, BS, 48, MS, 50, PhD(physics), 51. *Prof Exp:* Sr staff physicist, 51-58, MEM PRIN STAFF, APPL PHYSICS LABS, JOHNS HOPKINS UNIV, 58-, ASSOC PROF BIOMED ENG, SCH MED, 69-, MED (CARDIOL), 88- *Honors & Awards:* Pioneer Award, Inst Elec & Electronics Engrs, 86; Magellan Prize for Navig, Am Philos Soc, 88. *Mem:* Am Heart Asn; Inst Elec & Electronics Engrs. *Res:* Random function theory and statistical mechanics; geophysics connected with artificial satellites; biomedical engineering; cardiovascular physiology; patient monitoring systems; artificial intelligence. *Mailing Add:* Appl Physics Lab Johns Hopkins Univ Laurel MD 20810

GUIHER, JOHN KENNETH, b Youngstown, Ohio, Mar 25, 17; m 46; c 1. FOREST PRODUCTS. *Educ:* Mich State Univ, BS, 47, PhD(wood technol), 53; Yale Univ, MF, 48. *Prof Exp:* Instr forest prod, Mich State Univ, 48-52; asst wood technologist, Inst Technol, Wash State Univ, 52-54; prof forestry & head dept, Stephen F Austin State Col, 54-55; asst prof, 55-70, ASSOC PROF WOOD TECHNOL & FORESTRY UTILIZATION, UNIV ILL, URBANA, 70- *Mem:* Soc Am Foresters; Forest Prod Res Soc; Soc Wood Sci & Technol; Sigma Xi. *Res:* Wood anatomy and identification; wood glues and gluing. *Mailing Add:* 3106 Meadow Brook Dr Champaign IL 61821

GUIKEMA, JAMES ALLEN, b Grand Rapids, Mich, Aug 31, 51. PHOTOSYNTHESIS. *Educ:* Calvin Col, BA, 73; Univ Mich, PhD(biol sci), 78. *Prof Exp:* Teaching asst, Univ Mich, 74-78; fel, Univ Mo, 78-81; ASST PROF BIOL, KANS STATE UNIV, 81- *Mem:* Sigma Xi. *Res:* Physiology of photosynthesis in cyanobacteria using a combination of genetic and biochemical techniques. *Mailing Add:* Div Biol Ackert Hall Kans State Univ Manhattan KS 66506

GUILARTE, TOMAS R, b Oriente Cuba, Sept 15, 53; US citizen; m 77; c 3. RADIATION HEALTH SCIENCES. *Educ:* Univ Fla, BS, 74, MS, 76; Johns Hopkins Univ, PhD(radiation health), 80. *Prof Exp:* Res assoc, 80-81, asst prof, 81-87, ASSOC PROF RADIATION HEALTH & TOXICOL SCI, JOHNS HOPKINS UNIV, 87-, ASSOC PROF, JOINT APPT DIV HUMAN NUTRIT, 90- *Concurrent Pos:* Affil, Ctr Advan Radiation Educ & Res, Johns Hopkins Univ, 86- *Mem:* Am Inst Nutrit; Soc Nuclear Med; AAAS. *Res:* Effects of maternal vitamin B-6 nutrition on the ontogenesis of neurotransmitter systems in the brain of progeny; discovery of putative endogenous convulsant and neurotoxin and investigation of its mechanism of action; the development and application of analytical methods for the analysis of B-vitamins in complex biological samples. *Mailing Add:* Dept Environ Health Sci Rm 2001 Johns Hopkins Univ 615 N Wolfe St Baltimore MD 21205

GUILBAULT, LAWRENCE JAMES, b Buffalo, NY, Jan 26, 40; m 67; c 3. POLYMER CHEMISTRY. *Educ:* Kans State Univ, BS, 64; Univ Akron, PhD(polymer sci), 70; Am Inst Chemists, PC-A, 69. *Prof Exp:* Chem technician, Gen Elec Co, 59-62, applns chemist, 64-66; fel, Univ Fla, 70-71; res assoc polymers, Calgon Corp, 71-72, group leader, 72-74, sr group leader polymers, 74-76; tech mgr res & develop, Thiokol Corp, 76-82, VPRES RES & DEVELOP, VENTRON DIV, MORTON THIOKOL INC, 82- *Concurrent Pos:* Lectr ion-containing polymers, Am Chem Soc, 75-; chmn Gordon res conf ion-containing polymers, 83. *Mem:* Am Chem Soc; Tech Asn Pulp & Paper Indust; Soc Plastics Engrs; Controlled Release Soc; Electrochem Soc. *Res:* Synthesis and development of polyelectrolytes for use in water and waste treatment and in oilfield applications; synthesis and application of boron hydrides; antimicrobial additives for plastics; structure-property correlations for water soluble polymers. *Mailing Add:* Morton Int 150 Andover St Danvers MA 01923

GUILBEAU, ERIC J, b Tullos, La, June 5, 44; m 67; c 3. BIOMEDICAL ENGINEERING, CHEMICAL ENGINEERING. *Educ:* La Tech Univ, BS, 67, MS, 68, PhD(chem eng), 71. *Prof Exp:* Res assoc chem eng, La Tech Univ, 71-72, res assoc biomed eng, 72-73, from asst prof to assoc prof, 73-77; assoc prof, 77-81, PROF CHEM & BIOMED ENG, ARIZ STATE UNIV, 81-, DIR BIOENG, 90- *Concurrent Pos:* Affil med staff, St Joseph's Hosp, Phoenix, 77- *Mem:* Am Inst Chem Engrs; Am Chem Soc; Sigma Xi; Int Soc Study Oxygen Transport to Tissue; Biomed Eng Soc; Soc Biomat. *Res:* Biomedical engineering, development of transducers for measurement of cellular biological parameters; study transport phenomena in physiological systems; investigation of myocardial protection techniques; development of pericardial substitutes. *Mailing Add:* Dept Chem & Bioeng Ariz State Univ Tempe AZ 85281

GUILBERT, JOHN M, b Newton, Mass, May 12, 31; m 54; c 3. PLATE TECTONICS, ECONOMIC PETROLOGY. *Educ:* Univ NCarolina, BS, 53; Univ Wis, MS, 55, PhD(geol), 62. *Prof Exp:* Res geologist, Geol Res Lab, Anaconda Co, 57-65; from asst prof to assoc prof, 65-73, PROF ECON

GEOL, DEPT GEOSCI, UNIV ARIZ, 73- *Concurrent Pos:* Consult, 65- *Mem:* Soc Econ Geologists; fel Geol Soc Am; fel Mineral Soc Am; Soc Geol Appl Mineral Deposits; Int Asn Genesis Ore Deposits. *Res:* Porphry base metal deposits; precambrian massive sulfide ores; ore deposit-plate tectonic interrelationships; alteration-mineralization petrology. *Mailing Add:* Dept Geosci Univ Ariz Tucson AZ 85721

GUILD, LLOYD V, b Elmira, NY, July 17, 20; m 41; c 3. CHEMISTRY. *Educ:* Univ Pittsburgh, BS, 48, MS, 61. *Prof Exp:* Vpres, Burrell Corp, 41-62; PRES, GUILD CORP, 62- *Mem:* Am Chem Soc; Am Soc Testing & Mat; Instrument Soc Am; Am Indust Hyg Asn; Sigma Xi. *Res:* Analytical instrumentation. *Mailing Add:* SKC Inc 334 Valley View Rd Eighty-Four PA 15330-9613

GUILD, PHILIP WHITE, b Ann Arbor, Mich, Oct 28, 15; m 39; c 3. ECONOMIC GEOLOGY. *Educ:* Johns Hopkins Univ, AB, 36, PhD(geol), 47. *Prof Exp:* Jr geologist, 39-42, prin geologist, 42-79, EMER PRIN GEOLOGIST, US GEOL SURV, 80- *Concurrent Pos:* Pres subcomn metallogenic map world, 72-89; adv, minerals map ser, Circum-Pac Map Proj, 73-; vchmn comn tectonics ore deposits, Int Asn Genesis Ore Deposits, 74-85; mem bd mineral & energy resources, Comn Natural Resources, Nat Res Coun, 75-78. *Honors & Awards:* Distinguished Serv Award, Dept Interior, 83. *Mem:* Soc Econ Geologists; Geol Soc Am; Am Inst Mining, Metall & Petrol Engrs (vpres, 73-75); Soc Appl Geol; Int Asn Genesis Ore Deposits. *Res:* Geology and resources of the ferrous metals; metallogenic maps and epochs; metallogeny and plate tectonics; remote sensing applied to mineral exploration. *Mailing Add:* 596 Belvedere Ct Punta Gorda FL 33950

GUILD, WALTER RUFUS, b Ann Arbor, Mich, Oct 25, 23; m 46; c 1. MOLECULAR BIOLOGY, GENETICS. *Educ:* Univ Tex, BS, 48, MA, 49; Yale Univ, PhD(biophys), 51. *Prof Exp:* Instr physics, Yale Univ, 51-54, asst prof biophys, 54-60; from assoc prof to prof biophys, Med Ctr, Duke Univ, 60-86, dir univ prog in genetics, 77-79; RETIRED. *Concurrent Pos:* Mem int fel rev comt, NIH, 68-72; USPHS sr fel, Stanford Univ, 70-71; mem, Microbial Genetics Study Sect, NIH, 81-84; emer prof biophys, Med Ctr, Duke Univ, 87- *Mem:* Am Soc Biol Chem; Biophys Soc; Am Soc Microbiol; Genetics Soc Am; Radiation Res Soc. *Res:* Molecular genetics, primarily mechanisms of gene transfer in bacteria and genetic recombination. *Mailing Add:* Dept Biochem Duke Univ Durham NC 27710

GUILE, DONALD LLOYD, b Olean, NY, Nov 12, 32; m 56; c 4. CERAMICS, MATERIALS SCIENCE. *Educ:* Alfred Univ, BS, 60, PhD(ceramic sci), 65. *Prof Exp:* RES ENGR, CORNING GLASS WORKS, 64- *Mem:* Am Ceramic Soc; Brit Ceramic Soc. *Res:* Study of high temperature refractory materials for the glass and metal industries, including particle technology and application of sintering theory. *Mailing Add:* 1245 Stonybrook E Horseheads NY 14845

GUILFORD, HARRY GARRETT, b Madison, Wis, June 20, 23; m 49; c 2. ZOOLOGY, PARASITOLOGY. *Educ:* Univ Wis, PhB, 44, PhM, 46, PhD(zool), 49. *Prof Exp:* Assoc prof biol, Mercer Univ, 49-50; from asst prof to prof bot & zool, Univ Wis-Green Bay, 50-69, secy fac, 70-77, dept chmn, 80-84, prof human biol, 69-88, EMER PROF, UNIV WIS-GREEN BAY, 88- *Mem:* Am Soc Parasitologists; Am Micros Soc; Soc Protozoologists; Sigma Xi. *Res:* Histology and life history of trematodes; myxosporida. *Mailing Add:* Dept Human Biol Univ Wis Green Bay WI 54302

GUILFOYLE, RICHARD HOWARD, b Seaford, NY, Nov 4, 39; m 61; c 3. MATHEMATICS. *Educ:* C W Post Col, BS, 61; Stevens Inst Technol, MS, 63, PhD(math), 72. *Prof Exp:* From instr to asst prof, 66-72, ASSOC PROF MATH/COMPUT SCI, MONMOUTH COL, 72- *Concurrent Pos:* NSF fel, 71-72. *Mem:* Math Asn Am; Sigma Xi. *Res:* Numerical methods for finding zeros of complex functions; variational methods applied to optimal design of switching circuits. *Mailing Add:* Dept Math Monmouth Col West Long Branch NJ 07764

GUILFOYLE, THOMAS J, b Mendota, Ill, Jan 13, 46. MOLECULAR BIOLOGY, PLANT BREEDING & GENETICS. *Educ:* Ill State Univ, BS, 68; Univ Ill, Urbana, MS, 69, PhD, 74. *Prof Exp:* NIH fel, Univ Ga, Athens, 74-76; from asst prof to prof, Dept Bot, Univ Minn, 76-86; PROF, DEPT BIOCHEM, UNIV MO, COLUMBIA, 86- *Mem:* Am Soc Biochem & Molecular Biol; Am Soc Plant Physiologists; Am Soc Microbiol; AAAS; Int Soc Plant Molecular Biol. *Res:* Molecular Plant; numerous publications. *Mailing Add:* Dept Biochem Univ Mo 117 Schweitzer Hall Columbia MO 65211

GUILLARD, ROBERT RUSSELL LOUIS, b New York, NY, Feb 5, 21; m 52, 62; c 3. BIOLOGICAL OCEANOGRAPHY, PHYCOLOGY. *Educ:* City Col New York, BS, 41; Yale Univ, MS, 51, PhD, 54. *Prof Exp:* Elec engr, Navy Yard, NY, 41-46; tutor physics, City Col New York, 46-49; res assoc bot, Univ Hawaii, 54-55; aquatic microbiologist, Oyster Inst NAm, Inc, US Fish & Wildlife Lab, 55-58; res assoc marine biol, Woods Hold Oceanog Inst, 58-61, assoc scientist, 63-73, sr scientist, 73-81; dir, 85-89, EMER DIR, PROVASOLI-GUILLARD CTR CULT MARINE PHYTOPLANKTON, 89-; PRIN INVESTR, BIGELOW LAB OCEAN SCI, 81- *Honors & Awards:* Darbaker Award, Bot Soc Am, 68. *Mem:* Fel AAAS; Am Soc Limnol & Oceanog; Phycol Soc Am; Int Phycol Soc; Sigma Xi. *Res:* Physiology, ecology and systematics of marine phytoplankton; isolation and culture of algae. *Mailing Add:* Bigelow Lab Ocean Scis West Boothbay Harbor MA 04575

GUILLEMIN, ROGER (CHARLES LOUIS), b Dijon, France, Jan 11, 24; nat US; m 51; c 6. PHYSIOLOGY, NEUROENDOCRINOLOGY. *Educ:* Dijon Univ, France, BA, 41, BSc, 42; Univ Lyons, MD, 49; Univ Montreal, PhD(exp med & surg), 53. *Hon Degrees:* Numerous from US & foreign univs, 76-89. *Prof Exp:* Asst prof & asst dir, Inst Exp Med & Surg, Univ Montreal, 51-53; asst prof physiol, Col Med, Baylor Univ, 53-57, assoc prof, 57-63, prof & dir labs neuroendocrinol, 63-70; res fel & chmn, Labs Neuroendocrinol, Salk Inst Biol Studies, 70-89, dean fac, 72 & 76; DISTINGUISHED SCIENTIST, WHITTIER INST DIABETES & ENDOCRINOL, 89- *Concurrent Pos:* Markle Found scholar, 52-56; consult, Vet Admin Hosp, Houston, 54-70; lectr, Rice Univ, 58-60; mem study sects, NIH, 59-69 & Int Comt Estab New ACTH Stand, 60; sci prog comt, Int Cong Pharmacol, Stockholm, 60; assoc dir dept exp endocrinol, Col France, 60-63; consult, Univ Tex M D Anderson Hosp & Tumor Inst Houston, 67-; vis mem grad fac, Tex A&M Univ, 68-70; mem comt endocrinol, Int Union Physiol Sci Glossary, 69; adj prof, Col Med, Baylor Univ & Univ Calif, San Diego, 70, Salk Inst, La Jolla, 89-; mem, Ctr Pop Res, NIH, 71-74, Panel Nat Pituitary Agency, Nat Acad Sci, 73-75, President's Biomed Res Panel, 75-76, Task Force Conception, Reproduction, Partuition, Contraception & Sex Differentiation, NIH, 78-, President's Comt Nat Medal Sci, 81 & Bd Sci Counselors, Nat Inst Child Health & Human Develop, 83; mem neuroendocrine comt, Fedn Am Socs Exp Biol, 71-75. *Honors & Awards:* Nobel Prize in Physiol or Med, 77; Louis Bonneau Award, French Acad Sci, 57, L LaCaze Award, 60; Int Cong Pharmacol Gold Medal, 61; Ayerst-Squibb Award, Endocrine Soc, 70; Lasker Found Award, 75; Passano Award, Borden Award, Asn Am Med Cols, 76; Nat Medal Sci, President of the United States, 77; Barren Gold Medal, 78; Dale Medal, Soc Endocrinol, UK, 78; Nobel Laureat Re-Vis Lectr, Nobel Found, 90. *Mem:* Nat Acad Sci; fel AAAS; Endocrine Soc(pres, 86); fel Am Acad Arts & Sci; Acad des Sci, Arts et Belles-Lettres; Am Physiol Soc; Int Brain Res Orgn. *Res:* Physiology and biochemistry of hypothalamic mechanisms controlling secretion of pituitary hormones; improvement of population control techniques with antagonists to hypothalamic hormones; neurochemistry of fundamental brain function. *Mailing Add:* Whittier Inst 9894 Genesee Ave La Jolla CA 92037

GUILLEMIN, VICTOR W, b Cambridge, Mass, Oct 15, 37. GLOBAL ANALYSIS. *Educ:* Harvard Univ, BA, 59, PhD(math), 62; Univ Chicago, MA, 60. *Prof Exp:* Instr math, Columbia Univ, 63-66; asst prof, 66-73, PROF MATH, MASS INST TECHNOL, 73- *Mem:* Nat Acad Sci; Am Acad Arts & Sci; Am Math Soc. *Mailing Add:* Dept Math Mass Inst Technol Cambridge MA 02139

GUILLERY, RAINER WALTER, b Greifswald, Ger, Aug 28, 29; m 54; c 4. NEUROANATOMY. *Educ:* Univ London, BSc, 51, PhD(anat), 54. *Prof Exp:* From asst lectr to lectr, Univ Col, London, 53-63, reader, 63-64; from assoc prof to prof anat, Univ Wis-Madison, 64-77; prof pharmacol & physiol sci, Univ Chicago, 77-84; DR LEE'S PROF ANAT, OXFORD, ENG, 84- *Concurrent Pos:* Rockefeller traveling fel neurophysiol, Univ Wis, 60-61. *Mem:* Fel Royal Soc. *Mailing Add:* Dept Human Anat S Parks Rd Oxford 0X1 3QX England

GUILLET, JAMES EDWIN, b Toronto, Ont, Jan 14, 27; m 53; c 4. PHYSICAL CHEMISTRY, POLYMER CHEMISTRY. *Educ:* Univ Toronto, BA, 48; Cambridge Univ, PhD(phys chem), 55, ScD, 74. *Prof Exp:* Res chemist, Eastman Kodak Co, 48-50, res chemist, Tenn Eastman Co Div, 50-52, sr res chemist, 55-62, res assoc, 63; assoc dean res & planning, Scarborough Col, 83-84; assoc prof, 63-69, PROF CHEM, UNIV TORONTO, 69- *Concurrent Pos:* Consult, Glidden Co, Ltd, 64-84, Imp Oil Enterprises, Ltd, 65-83, Royal Packaging, Holland, 72-79, Allied Canada, Inc, 82-86 & Res Labs, IBM, 85-; vis prof, Nat Ctr Sci Res, Strasbourg, France, 70-71, Kyoto Univ, 74, Univ Calif, San Diego, 78-79, Res labs, IBM, San Jose, 80, Univ Mainz, West Germany, 81, Univ Calif, Berkeley, 82 & Univ St Andrews, Scotland, 83; dir, Ecoplastics Ltd, 71-85, pres, 75-84, dir, Medipro Sci Ltd, 76-, pres, 85-90, chmn bd & chief exec officer, 91-, dir, Solarchem Corp, 83-88, pres, 84-85; Guggenheim fel, 81-82; Killam res fel, 87-89; overseas vis fel, St John's Col, Cambridge Univ. *Honors & Awards:* Dunlop Lectr, 79. *Mem:* Am Chem Soc; Chem Soc; fel Chem Inst Can; fel Royal Soc Can. *Res:* Structure-property relations in polymers; photochemical and photophysical processes in polymers; solar enrgy conversion; energy storage and migration in polymers; inverse gas chromatography; photo and biodegradable plastics. *Mailing Add:* Dept Chem Univ Toronto Toronto ON M5S 1A1 Can

GUILLORY, JACK PAUL, b Alexandria, La, Feb 28, 38; m 60; c 3. PHYSICAL CHEMISTRY. *Educ:* La State Univ, BS, 60; Iowa State Univ, PhD(phys chem), 65. *Prof Exp:* Res chem physicist, Phillips Petrol Co, 65-71, group leader, 71-73, sect supvr, 73-80, br mgr, 80-89, SR BR MGR, PHILLIPS PETROL CO, 89- *Mem:* Am Chem Soc; Catalyst Soc. *Res:* Electron diffraction study of gases; electronic spectroscopy; heterogeneous catalysis; petroleum refining processes; process modeling. *Mailing Add:* 206 CPL Phillips Petrol Co Bartlesville OK 74004

GUILLORY, JAMES KEITH, b Bunkie, La, Feb 4, 35. PHARMACY. *Educ:* Loyola Univ, La, BS, 56; Univ Wis, MS, 60, PhD(pharm), 61. *Prof Exp:* Asst prof pharm, Wash State Univ, 61-64; from asst prof to assoc prof, 64-71, PROF PHARM, UNIV IOWA, 71- *Concurrent Pos:* Vis Scientist, Burroughs Wellcome Co, 89. *Mem:* Am Chem Soc; Am Assoc Pharmaceut Scientists. *Res:* Kinetics of degradation of pharmaceuticals; polymorphism; absorption kinetics; thermal analysis; microcalorimetry. *Mailing Add:* Univ Iowa Col Pharm Iowa City IA 52242

GUILLORY, RICHARD JOHN, b San Diego, Calif, Oct 3, 30; m 56; c 3. BIOCHEMISTRY. *Educ:* Reed Col, BA, 53; Univ Calif, Los Angeles, PhD(biochem), 62. *Prof Exp:* Am Heart Asn res fel, 62-64; from asst prof to assoc prof biochem & molecular biol, Cornell Univ, 66-71; chmn dept, 73-76, PROF BIOCHEM & BIOPHYS, SCH MED, UNIV HAWAII, 71- *Concurrent Pos:* Estab investr, Am Heart Asn, 68-73. *Mem:* Am Chem Soc; Brit Biochem Soc; Biophys Soc; Am Soc Biol Chemists; Sigma Xi. *Res:* Bioenergetics; mechanism of enzyme action; oxidative phosphorylation; muscular contraction. *Mailing Add:* Dept Biochem & Biophys Univ Hawaii Sch Med Honolulu HI 96822

GUILLORY, WILLIAM ARNOLD, b New Orleans, La, Dec 4, 38; m 67; c 1. PHYSICAL CHEMISTRY. *Educ:* Dillard Univ, BA, 60; Univ Calif, Berkeley, PhD(chem), 64. *Prof Exp:* NSF fel, Univ Paris, 64-65; asst prof chem, Howard Univ, 65-69; assoc prof, Drexel Univ, 69-75; assoc prof, Univ

Utah, 75-76, prof chem & chmn dept, 76-83; PRES, INNOVATIONS CONSULT, INC, 83- *Concurrent Pos:* Res grants, Environ Protection Agency, 69-71, NSF & Dept Energy; consult, Naval Ord Lab, Md, 67-74; Sloan fel, 71-73. *Mem:* AAAS; Am Phys Soc; Am Chem Soc. *Res:* Ultraviolet and infrared spectroscopy, kinetics and photochemistry; elementary photophysical and photochemical processes; low temperature matrix-isolation spectroscopy and molecular structure; applications of infrared and visible ultraviolet lasers in chemistry. *Mailing Add:* Innovations Consult Inc 488 E 6400 S Suite 475 Salt Lake City UT 84107

GUILMETTE, RAYMOND ALFRED, b Laconia, NH, May 26, 46; m 69; c 2. RADIOLOGICAL HEALTH. *Educ:* Rensselaer Polytech Inst, BS, 68; NY Univ, MS, 71, PhD(radiol health), 75. *Prof Exp:* Nuclear engr radiation dosimetry, Consol Edison Power Co, 68-69; appointee plutonium metab, Argonne Nat Lab, 74-77; anal radiochemist, 79-82, RADIOBIOLOGIST, INHALATION TOXICOL RES INST, 77- *Concurrent Pos:* Assoc ed, Health Physics. *Mem:* Radiation Res Soc; Health Physics Soc; Sigma Xi; NY Acad Sci; Am Nuclear Soc; Am Thoracic Soc. *Res:* Biokinetics, dosimetry and dose-response relationships for plutonium, other actinides and internal emitters, particularly after inhalation exposure; dosimetry and dose-response modeling, mechanisms of lung clearance, decorporation, therapy; risk assessment, actinide radiochemistry and detection instrumentation; radon dosimetry and biological effects; nasal airway morphometry; dosimetry. *Mailing Add:* Inhalation Toxicol Res Inst PO Box 5890 Albuquerque NM 87185

GUIMARAES, ARMENIO COSTA, b Salvador, Brazil, June 17, 33; m 59; c 2. INTERNAL MEDICINE, CARDIOLOGY. *Educ:* Univ Bahia, MD, 56; Educ Coun Foreign Med Grads, cert, 62. *Prof Exp:* Intern, Prof Edgard Santos Hosp & Couto Maia Hosp, Univ Bahia, 55-56; from instr to prof internal med, 61-73, PROF CARDIOL & PROF POSTGRAD COURSE INTERNAL MED, MED SCH, FED UNIV BAHIA, 73-, VCHIEF DEPT MED, 74-, PRES RESIDENCY COMN, PROF EDGARD SANTOS HOSP, 71-, CHIEF CARDIOL, 79- *Concurrent Pos:* Fel cardiol, Dept Med, Clin Hosp, Univ Sao Paulo, 57-58; Am Col Physicians & Kellogg Found fel, 64-65; Nat Res Coun fel, 67; res assoc, C T Miller Hosp, St Paul, Minn, 65; vis prof fac med, Univ Brasilia, 66-67 & 75; pres internship comn, Santos Hosp, 68-69, vice-rector for postgrad & res activ, 76-79; vis assoc prof, Div Cardiol, Med Col, Cornell Univ, 69-70; vchief dept cardiol, Med Sch, Univ Bahia, 70-71; pres sci comt, Brazilian Soc Cardiol, 86- *Mem:* Fel Am Col Physicians; Brazilian Soc Cardiol. *Res:* Identification of primary myocardiopathies, physiology and natural history of endomyocardialfibrosis; pulmonary schistosomiasis, especially natural history of the pulmonary circulation haemodynamics abnormalities in patients with hepatosplenic schistosomiasis; Chagas myocarditis; hemodynamics and cardiac arrhythmias, arterial hypertension, hypercholesterolemia. *Mailing Add:* Dept Cardiol Hosp Prof Edgard Santos-Salvador Bahia Brazil

GUIMARAES, ROMEU CARDOSO, b Minas Gerais, Brazil, July 29, 43; m 69. PROTOZOOLOGY, NUCLEIC ACIDS. *Educ:* Fed Univ, Minas Gerais, MD, 65, DPhil(path), 70. *Prof Exp:* Asst prof path, Fed Univ Minas Gerais, 66-75; asst prof, 76-77, PROF GENETICS, STATE UNIV SAO PAULO, 78- *Concurrent Pos:* Fel res grants, USPHS, NIH, 71-73, Brazilian Nat Res Coun, 68-69 & 76- & Sao Paulo Res Found Aid, 76-; res fel assoc molecular biol, Univ Conn, Storrs, 71 & Univ Tex, Austin, 72-73 & 82; Brit Coun, Univ Kent, Canterbury, 80-81; mem gov body, Ctr Study Molecular Evolution, Austin, Tex, 84-86; Inst Biochem, Free Univ Berlin, 87-88; polymer res, Weizmann Inst, Israel, 89-90; sr res fel, Brazilian Nat Res Coun, 84- *Mem:* Brazilian Asn Advan Sci; Brazilian Soc Genetics; Brazilian Soc Path; Brazilian Soc Protozool; Brazilian Soc Hist Sci; Brazilian Soc Biochem; Int Soc Study Orig Life. *Res:* Trypanosoma cruzi and American Trypanosomiasis; cell biology; genetics; differentiation; nucleic acids; indeterminacy in biological variation; molecular evolution; trypanosomatids: cell cycle and cytotaxonomy, purine, urate metabolism and nucleotide pools; evolution of RNA's; the concept of the gene; 55 rRNA Molecular Linguistics. *Mailing Add:* Depto Genetica-Rubiao Jr IB-UNESP Botucatu Sao Paulo 18610 Brazil

GUIMOND, ROBERT WILFRID, b Fall River, Mass, Sept 4, 38; m 63; c 2. COMPARATIVE PHYSIOLOGY, MEDICAL PHYSIOLOGY. *Educ:* Univ RI, BA, 61, PhD(physiol), 70; New Eng Sch Law, JD, 78. *Prof Exp:* Teacher gen sci, E S Brown Jr High, Swansea, Mass, 61-64; asst prof, Dept Zool, Univ RI, 70-71; from assoc prof to prof, dept biol, Boston State Col, 71-82; PROF, DEPT BIOL, UNIV MASS, BOSTON, 82- *Concurrent Pos:* Atty med & environ law specialities, 78-; prof, Dept Biol, Univ Mass, Boston, 82- *Mem:* Am Physiol Soc; Am Bar Asn; Am Forestry Asn; Am Soc Law & Med. *Res:* General area of comparative respiratory physiology primarily with amphibious vertebrates; medical-legal problems in regards to the increased use of artificial mechanical life sustaining devices; military medical neglect and the Federal Torts claims act; history of biology and medicine. *Mailing Add:* Dept Biol Univ Mass Boston MA 02125

GUINAN, EDWARD F, b Philadelphia, Pa, Apr 29, 42; m 69; c 1. TROPICAL RAINFORESTS & CLIMATE, STELLAR EVOLUTION & ASTROPHYSICS. *Educ:* Villanova Univ, BS, 64; Univ Pa, PhD(astron & astrophys), 70. *Prof Exp:* Asst prof astron, Villanova Univ, 69-75; vis scientist physics, Shiraz Univ/Biruni Obs, 75-77; PROF ASTRON & ASTROPHYS, VILLANOVA UNIV, 77- *Concurrent Pos:* Vis scientist physics, Shiraz Univ, 77-78; res fel, Harvard-Smithsonian Ctr Astrophys, 82-83, vis scientist 83-91; prin investr, NASA, 80-91; sci organizing comt, Comn 42, Int Astron Union, 91- *Mem:* Am Astron Soc; Int Astron Union; Am Inst Physics; Rainforest Alliance. *Res:* Observational astrophysics; stellar structure and evolution; binary stars as tests of general relativity; magnetic activity in solar-type stars; variable stars; photoelectric photometry with robotic telescopes; ultraviolet spectroscopy. *Mailing Add:* Dept Astron & Astrophys Villanova Univ Villanova PA 19085

GUINAN, MARY ELIZABETH, b New York, NY, Sept 23, 39. MEDICINE, PHYSIOLOGY. *Educ:* Univ Tex Med Br, Galveston, PhD(physiol), 69; Johns Hopkins Univ, MD, 72; Am Bd Internal Med, dipl, 78. *Prof Exp:* Med epidemiologist hosp infections, Ctr Dis Control, 74-76; fel infectious dis, Univ Utah Med Ctr, 76-78; clin res investr, venereal dis, 78-86, asst dir sci, 86-90, SPEC ASST EVAL AIV/AIDS, CTR DIS CONTROL, 90- *Concurrent Pos:* Med epidemiologist, WHO, Delhi, India, 74-75. *Mem:* Am Med Women's Asn. *Res:* AIDS; infectious disease; epidemiology of human disease; women's health. *Mailing Add:* Center Dis Control 1600 Clifton Rd Atlanta GA 30333

GUINANE, JAMES EDWARD, b Oak Park, Ill, Mar 16, 32; m 63; c 2. PEDIATRICS. *Educ:* Brown Univ, AB, 53; Calif Inst Technol, MS, 54; Johns Hopkins Univ, MD, 63; Am Bd Pediat, dipl, 68. *Prof Exp:* Intern pediat, NY Univ-Bellevue Med Ctr, 63-64, resident, 64-66; fel pediat neurol, 69-72, asst prof pediat, 72-75, ASSOC PROF PEDIAT, TUFTS-NEW ENG MED CTR, 75- *Mem:* Fel Am Acad Pediat; Am Acad Neurol; AAAS. *Res:* Cerebrospinal fluid physiology; experimental hydrocephalus. *Mailing Add:* 40 Weston St Waltham MA 02154

GUINASSO, NORMAN LOUIS, JR, b San Francisco, Calif, June 2, 43; m 76. OCEANOGRAPHY. *Educ:* San Jose State Col, BA, 66; Tex A&M Univ, MS, 75, PhD, 84. *Prof Exp:* Assoc scientist physics, Teledyne Isotopes, 66-71; res assoc oceanog, 72-85, ASSOC RES SCIENTIST, TEX A&M UNIV, 85- *Concurrent Pos:* Proj scientist, US World Ocean Circulation Exp, 86-87. *Mem:* AAAS; Am Geophys Union; Sigma Xi. *Res:* Chemical oceanography; geochemistry. *Mailing Add:* Geochemical & Environ Res Group Tex A&M Univ College Station TX 77843-3149

GUINN, GENE, b Prairie Grove, Ark, Mar 19, 28; m 55; c 2. PLANT PHYSIOLOGY. *Educ:* Univ Ark, BS, 52, MS, 57; Tex A&M Univ, PhD(plant physiol), 61. *Prof Exp:* Agronomist, USDA, Ark, 53-57, res agronomist, Cotton & Cordage Fibers Res Br, Agr Res Serv, Exp Sta, Tex A&M Univ, 57-61, res plant physiologist, USDA & Okla Agr Exp Sta, 61-70; plant physiologist, 70-77, res leader, 77-90, RES PLANT PHYSIOLOGIST, USDA, 90- *Mem:* AAAS; Am Soc Plant Physiologists; Am Soc Agron; Sigma Xi; Crop Sci Soc Am. *Res:* Physiology of cotton, especially fruiting and fruit abscission. *Mailing Add:* Western Cotton Res Lab 4135 E Broadway Rd Phoenix AZ 85040

GUINN, THEODORE, b Fresno, Calif, Apr 29, 24; m 53; c 2. MATHEMATICS. *Educ:* Fresno State Col, BA, 49; Univ Calif, Los Angeles, MA, 59, PhD(math), 64. *Prof Exp:* Sr scientist, Douglas Aircraft Co, 53-65; asst prof math, Mich State Univ, 65-68; ASSOC PROF MATH, UNIV N MEX, 68- *Concurrent Pos:* Res asst, Univ Calif, Los Angeles, 62-63, instr, Exten, 63-65, asst res mathematician, 64-65. *Mem:* Soc Indust & Appl Math. *Res:* Calculus of variations and optimum control theory. *Mailing Add:* 1508 Roma Ave Albuquerque NM 87106

GUINN, VINCENT PERRY, b Los Angeles, Calif, Nov 9, 17; m 38, 91; c 2. RADIOCHEMISTRY. *Educ:* Univ Southern Calif, AB, 39, MS, 41; Harvard Univ, PhD(phys chem), 49. *Prof Exp:* Res chemist, Shell Develop Co, 41-46, 49-61, head radiochem group, 56-61; tech dir activation anal prog, Gen Atomic Co, 61-70; prof radiochem, Univ Calif, Irvine, 70-90; PROF RADIOCHEM, UNIV MD, COLLEGE PARK, 90- *Concurrent Pos:* Consult on activation anal; consult & expert witness on forensic activation anal; consult, Int Atomic Energy Agency. *Honors & Awards:* Spec Award Novel Applns Nuclear Energy, Am Nuclear Soc, 64; Hevesy Medal, radionanal chem, 79. *Mem:* Fel AAAS; fel Am Nuclear Soc; Am Chem Soc; fel Am Acad Forensic Sci. *Res:* Activation analysis; forensic chemistry; radio chemistry; radiotracers. *Mailing Add:* Nuclear Eng Dept Univ Md College Park MD 20742

GUION, THOMAS HYMAN, b New Bern, NC, Feb 21, 19; m 49; c 3. TEXTILE CHEMISTRY. *Educ:* Davidson Col, BS, 40; Univ NC, PhD(org chem), 49. *Prof Exp:* Assoc prof textile chem, Clemson Col, 49-52; sr res chemist, Appln Res Dept, Chemstrand Co Div, Monsanto Co, 52-55, group leader, 55-59, supvr, 59-64; assoc prof textile chem, NC State Univ, 64-84; RETIRED. *Mem:* Am Chem Soc; Am Asn Textile Chemists & Colorists; Fiber Soc; Brit Soc Dyers & Colourists. *Res:* Mechanisms of absorption of dyes and textile chemicals by man-made fibers. *Mailing Add:* 403 14th E Seattle WA 98112

GUIRARD, BEVERLY MARIE, b St Martinville, La, Dec 10, 15. BIOCHEMISTRY. *Educ:* Southwestern La Inst, BS, 36; La State Univ, MS, 38; Univ Tex, PhD(bioorg chem), 45. *Prof Exp:* Lab asst, Southwestern La Inst, 36, instr chem, 39-41; teacher pub sch, La, 38-39; tech asst biochem, Inst, Univ Tex, 41-42, asst, 42-45, res assoc, 45-48, res assoc, Dept Bact, 48-52, res scientist biochem, Inst, 52-56; asst res biochemist, Univ Calif, Berkeley, 56-76; RES SCIENTIST ASSOC II, UNIV TEX, AUSTIN, 76- *Concurrent Pos:* USPHS spec fel, Physiol Genetics Lab, Ctr Nat Res Sci, France, 63-64. *Mem:* AAAS; Am Chem Soc; Am Soc Biol Chemists; Am Soc Microbiol; Sigma Xi. *Res:* Microbial nutrition and enzymology. *Mailing Add:* 8313 Summer Place Dr Austin TX 78759

GUISELEY, KENNETH B, b Auburn, NY, July 20, 33; m 56; c 3. CARBOHYDRATE CHEMISTRY. *Educ:* Hartwick Col, BA, 55; Syracuse Univ, PhD(org chem), 60. *Prof Exp:* Sr chemist, Marine Colloids, Inc, FMC Corp, 61-75, dir res, 75-78, dir polymer res, Marine Colloids Div, 78-82, sr res assoc, 82-90, RES FEL, MARINE COLLOIDS DIV, FMC BIO PROD, 90- *Mem:* Am Chem Soc; fel Am Inst Chemists. *Res:* Extraction of polysaccharides and associated by-products from marine algae; chemical modification of polysaccharides; development of gelling agents for biochemical separations. *Mailing Add:* Marine Colloids Div FMC Bio Prod 191 Thomaston St Rockland ME 04841

GUITJENS, JOHANNES C, b Heeze, Holland, Jan 24, 35; m 67; c 2. IRRIGATION ENGINEERING. *Educ:* Univ BC, BSA, 61; Univ Calif, Davis, MS, 64, PhD(eng), 68. *Prof Exp:* From asst prof to assoc prof, 67-78, PROF IRRIG & DRAINAGE ENG, DIV PLANT SOIL & WATER SCI, UNIV NEV, RENO, 78- *Concurrent Pos:* Mem drain comn, Irrig & Drainage Div, Am Soc Civil Engrs, 76- *Mem:* Sigma Xi; Int Comn Irrig & Drainage; Am Geophys Union; Am Soc Civil Engrs. *Res:* Steady and unsteady flow in viscous flow models; mathematical simulation of ground water flow; water use efficiencies; agricultural drainage; water quality management; total water management. *Mailing Add:* Range Wildlife & Forestry Dept Univ Nev Agr Reno NV 89557

GULA, WILLIAM PETER, b Cleveland, Ohio, Dec 5, 39. THEORETICAL PLASMA PHYSICS. *Educ:* Spring Hill Col, Mobile, BS, 64; Columbia Univ, MA, 68, PhD(physics), 72. *Prof Exp:* Mem staff, inertial fusion & plasma theory group, 72-81, mem staff radiation transport group, 81-90, LAB REP/ADV DEPT ENERGY, LOS ALAMOS NAT LAB, 90- *Mem:* AAAS; Am Phys Soc; Inst Elec & Electronics Engrs; Asn Comput Mach. *Res:* Plasma physics and fluid dynamics problems associated with target designs for controlled thermonuclear fusion by means of lasers and ion beams; transport problems in plasmas. *Mailing Add:* Los Alamos Nat Lab ADNWT MS A105 Los Alamos NM 87545

GULATI, ADARSH KUMAR, b Karnal, India, July 1, 56. NEUROSCIENCE. *Educ:* Delhi Univ, BS, 75; WVa Univ, PhD(anat), 79. *Prof Exp:* Grad teaching asst neurobiol & gross anat, WVa Univ, 77, micro anat, 79, grad res asst lens regeneration, 78-79; fel Neural Develop & Regeneration, NIH, 79-81; ASSOC PROF CELLULAR BIOL & ANAT, MED COL GA, 89- *Mem:* Am Asn Anatomists. *Res:* Regeneration and repair of nerve and muscle tissues in mammal regeneration of lens and limb in amphibians, neurotrophic interaction; role of basement membrane (fibronectin and laminin) during various regenerative processes. *Mailing Add:* Dept Anat Med Col Ga Augusta GA 30912-2000

GULATI, SUBHASH CHANDER, b New Delhi, India, Dec 2, 50. CANCER RESEARCH. *Educ:* NDak State Univ, MS, 71; Columbia Univ, PhD(human genetics), 73; Univ Miami, MD, 76. *Prof Exp:* Mem adv comt, Dept Agron & Genetics, WVa State Univ, 74-76; mem fac, Dept Med, State Univ NY, Buffalo, 76-78; MEM FAC, DEPT HEMATOL & LYMPHOMA, SLOAN KETTERING CANCER INST, 78- *Concurrent Pos:* Am Cancer Soc clin oncol career develop award, 85-87. *Mem:* Am Col Physicians. *Res:* Understanding molecular changes in neoplastic growth using nucleic acid hybridization and immunological assays; bone marrow transplantation. *Mailing Add:* Mem Sloan-Kettering Cancer Ctr 1275 York Ave New York NY 10021

GULATI, SURESH T, b West Punjab, Pakistan, Nov 13, 36; m 61; c 3. MECHANICS, APPLIED MATHEMATICS. *Educ:* Univ Bombay, BS, 57; Ill Inst Technol, MS, 59; Univ Colo, PhD(mech), 67. *Prof Exp:* Stress analyst, Continental Can Co, 59-62; instr mech eng, Univ Colo, Boulder, 66-67; res scientist, Res & Develop Labs, 67-69, sr res scientist, 69-76, RES SUPVR, TECH STAFFS DIV, CORNING GLASS WORKS, 76-, SR RES ASSOC, 79-, RES FEL, 82- *Concurrent Pos:* Vis lectr dept theoret & appl mech, Cornell Univ, 69-70. *Mem:* AAAS; Am Soc Mech Engrs; Am Soc Eng Educ; Am Acad Mech; Am Ceramic Soc; Soc Automotive Engrs. *Res:* Analysis of elastic structures with particular emphasis on plates and shells; deformation and stress analysis of anisotropic elastic materials; application of the principles of mechanics to composite materials and materials science; optical waveguide design, fracture mechanics; cellular ceramics. *Mailing Add:* Corning Glass Works Sullivan Park RB4 Corning NY 14831

GULBRANSEN, EARL ALFRED, b Seattle, Wash, Jan 20, 09; m 38; c 3. PHYSICAL CHEMISTRY. *Educ:* State Col Wash, BS, 31; Univ Pittsburgh, PhD(phys chem), 34. *Prof Exp:* Asst chem, Wash State Univ, 30-31; asst, Univ Pittsburgh, 31-34; Nat Res Coun fel phys chem, Univ Calif, 34-35, res assoc, 35-36; instr chem eng & phys chem, Tufts Univ, 36-40; res engr, Res Labs, Westinghouse Elec Corp, Churchill Borough, 40-47, adv engr, 47-66, sci consult, 66-74; ADJ PROF METALL & MAT ENG, UNIV PITTSBURGH, 74- *Concurrent Pos:* US del, Int Conf Surface Reaction, Paris, 47, chmn, Pittsburgh, 48; mem, Gordon Res Conf, Chem & Physics of Metals, 50, Corrosion Eng, 55; mem, Int Union Pure & Appl Chem. *Honors & Awards:* Am Inst Mining, Metall & Petrol Engrs Award, 49; Whitney Award, Nat Asn Corrosion Engrs, 52; Acheson Medal & Prize, Electrochem Soc, 64. *Mem:* Am Chem Soc; Electrochem Soc; Am Inst Mining, Metall & Petrol Engrs. *Res:* Electron microscopy and diffraction of metal surfaces; kinetics of metal-gas reactions; corrosion high temperature alloys; high vacuum techniques; hydrogen in metals; pure metals; stress corrosion; refractory metals. *Mailing Add:* 63 Hathaway Ct Pittsburgh PA 15235-1929

GULDEN, TERRY DALE, b Seattle, Wash, May 4, 38; m 84; c 1. MATERIALS SCIENCE, CERAMIC ENGINEERING. *Educ:* Univ Wash, BS, 60; Stanford Univ, MS, 62, PhD(mat sci), 65. *Prof Exp:* Ceramist, United Technol Ctr, United Aircraft Corp, 60-61; res assoc radiation effects in mat, Berkeley Nuclear Labs, Cent Elec Generating Bd, Eng, 65-66; staff mem mat res, 67-71, mgr, Fuel Mat Br, Fuels & Mat Div, 71-80, mgr, ceramics & chem dept, 80-85, MGR, HIGH TEMPERATURE GAS COOLED REACTOR FUEL DEVELOP DEPT, GEN ATOMIC CO, 80-, DIR, DEFENSE MAT, 85- *Mem:* AAAS; Am Ceramic Soc; Am Nuclear Soc; Mat Res Soc. *Res:* Radiation effects in solids; transmission electron microscopy; internal friction; chemical vapor deposition; creep and deformation of ceramics and metals; nuclear reactor and composite materials; superconducting materials. *Mailing Add:* 1201 Virginia Way La Jolla CA 92037

GULENS, JANIS, b Rottweil, Ger, Dec 28, 44; Can citizen. ELECTROANALYTICAL CHEMISTRY, CERAMICS. *Educ:* Univ Toronto, BSc, 67; Queen's Univ, Ont, PhD(anal chem), 71. *Prof Exp:* Nat Res Coun fel, Calif Inst Technol, 71-73; from asst res officer to assoc res officer, 73-83, res officer anal chem, 83-88, BR MGR GEN CHEM BR, ATOMIC ENERGY CAN LTD, 88- *Mem:* Chem Inst Can; Am Chem Soc; Electrochem Soc. *Res:* Trace analysis, especially by application of electroanalytical techniques, of materials, solutions and effluents; ion-selective electrodes; fundamentals and applications of ion-conducting ceramics. *Mailing Add:* Chem Div AECL Res Chalk River ON K0J 1J0 Can

GULICK, SIDNEY L, III, b Oakland, Calif, July 29, 36. MATHEMATICS. *Educ:* Oberlin Col, BA, 58; Yale Univ, MA, 60, PhD(math), 63. *Prof Exp:* Instr math, Univ Pa, 63-65; from asst prof to assoc prof, 65-73, PROF MATH, UNIV MD, COLLEGE PARK, 73- *Mem:* Math Asn Am; Am Math Soc. *Res:* Theoretical mathematics; linear topological spaces and Banach algebras. *Mailing Add:* Dept Math Univ Md College Park MD 20742

GULICK, WALTER LAWRENCE, b Summit, NJ, July 4, 27; m 52; c 3. PSYCHOPHYSICS, PSYCHOPHYSIOLOGY. *Educ:* Hamilton Col, AB, 52; Univ Del, MA, 55; Princeton Univ, PhD(physiol psychol), 57. *Hon Degrees:* MA, Dartmouth Col, 68; LHD, St Lawrence Univ, 89. *Prof Exp:* From asst prof to prof psychol, Univ Del, 57-65, chmn dept, 64-65; prof, Dartmouth Col, 65-73, chmn dept, 70-72, 74-75, class of 25 prof, 73-75; prof psychol, Hamilton Col, 75-81, dean fac, 75-79, William Kenan prof, 79-81; pres, St Lawrence Univ, 81-87; VIS SCHOLAR, UNIV DEL, 88- *Concurrent Pos:* Mem coun, Nat Humanities Inst, 75-78; pres, Gulick Assocs, Inc, 88- *Mem:* NY Acad Sci; Psychonomic Soc; Sigma Xi. *Res:* Neurophysiology and electrophysiology; vision; hearing. *Mailing Add:* 205 Winslow Rd Newark DE 19711

GULICK, WILSON M, JR, b Plainfield, NJ, Mar 10, 39; m 62, 75; c 1. ANALYTICAL CHEMISTRY. *Educ:* Harvard Univ, AB, 60; Cornell Univ, PhD(chem), 65. *Prof Exp:* Asst prof chem, State Univ NY Col Cortland, 64-65; res assoc, Cornell Univ, 65-66, asst prof, 66-68; asst prof, Fla State Univ, 68-76; ASSOC PROF CHEM, MICH TECHNOL UNIV, 77- *Mem:* Am Chem Soc; Chem Soc. *Res:* Electrochemistry in non-aqueous media; magnetic resonance and optical spectroscopy of free radicals in solution. *Mailing Add:* Dept Chem & Chem Eng Mich Technol Univ Houghton MI 49931

GULKIS, SAMUEL, b West Palm Beach, Fla, Feb 3, 37; m 63; c 1. PHYSICS, AERONAUTICAL ENGINEERING. *Educ:* Univ Fla, BAeroE, 60, MS, 62, PhD(physics), 65. *Prof Exp:* NASA trainee, 63-65; res assoc radio astron, Arecibo Ionospheric Observ, Cornell Univ, 65-67, asst prof astron, 67-68; SR SCIENTIST PLANETARY & SPACE SCI, JET PROPULSION LAB, 68- *Concurrent Pos:* Co-investr, Voyage Mission Planetary Radio Astron Exp, 81-, Cosmic Background Explorer Satellite; proj scientist, Search for Extraterrestrial Intelligence, 81-; distinguished vis scientist, Radiophysics Div, NSW, Australia, Meudon Observ. *Honors & Awards:* Sci Achievement Award, NASA, 75. *Mem:* Am Astron Soc; Int Astron Union; Sigma Xi. *Res:* Radio astronomy, particularly planetary radio astronomy and planetary atmospheres; cosmic background radiation. *Mailing Add:* Space Sci Div JPL 4800 Oak Grove Dr Pasadena CA 91103

GULL, CLOYD DAKE, b Lorain, Ohio, June 17, 15; m 43; c 4. INFORMATION SCIENCE. *Educ:* Allegheny Col, AB, 36; Univ Mich, AB, 37, AM, 39. *Prof Exp:* Asst, Gen Libr, Univ Mich, 37-39; periodicals librn, NC State Col, 39-42; from spec asst processing dept to dep chief, Union Catalog Div, Libr Cong, DC, 45-52; tech analyst, Document, Inc, 52-54; admin officer, Div Eng & Indust Res, Nat Acad Sci-Nat Res Coun, 54-58; info systs consult, Gen Elec Co, DC, 58-63; prof libr sci, Ind Univ, Bloomington, 64-67; liaison officer, Nat Libr Task Force, Nat Libr Med, 67-68; rep consult, Document Systs, Inc, 68-69; pres, Cloyd Dake Gull & Assocs, Inc, 69-83; RETIRED. *Concurrent Pos:* Pres, Am Documentation Inst, Am Soc Info Sci, 59-60; chmn, US Nat Comt, Int Fedn Doc, 60-63. *Mem:* Am Library Asn; Spec Libraries Asn; Am Soc Info Sci. *Res:* Documentation; information systems; electronic systems for the selection, identification, analysis, storage, retrieval, use and dissemination of information. *Mailing Add:* Eight Pimlico Ct Silver Spring MD 20906

GULL, DWAIN D, b Meadow, Utah, Aug 21, 23; m 44; c 4. VEGETABLE CROPS. *Educ:* Utah State Univ, BS, 55; Cornell Univ, MS, 57, PhD(veg crops), 60. *Prof Exp:* PROF VEG PHYSIOL, UNIV FLA, 58-, ASST HORTICULTURIST, EXP STA, 59- *Concurrent Pos:* Adminstr, AID-Fla Contract, El Salvador, 73-75. *Mem:* Am Soc Hort Sci. *Res:* Post harvest physiology; flavor chemistry; quality control physiology; flavor and quality of fresh vegetables as affected by harvest and marketing stresses. *Mailing Add:* 2629 NW 11th Ave Gainesville FL 32605

GULL, THEODORE RAYMOND, b Hot Springs, SDak, Aug 17, 44; m 67; c 2. ASTRONOMY, SPACE INSTRUMENTATION. *Educ:* Mass Inst Technol, BS, 66; Cornell Univ, PhD(astron), 71; Loyola Col, MBA, 85. *Prof Exp:* Res asst radar astron, Lincoln Lab, Mass Inst Technol, 67; res assoc laser physics, Hughes Res Labs, 69; res assoc astron, Yerkes Observ, Univ Chicago, 70-72; asst astronr, Kitt Peak Nat Observ, 72-75; scientist astron, Lyndon B Johnson Space Ctr, Lockheed Electronics Co, Inc, 75-77; astrophysicist, Goddard Space Flight Ctr, 77-82, proj scientist, Hubble Space Telescope Second Generation Inst, 88-90, ASTRO MISSION SCIENTIST, GODDARD SPACE FLIGHT CTR, NASA, 82-, ASSOC CHIEF, LAB ASTRON & SOLAR PHYSICS, 85- *Concurrent Pos:* Spec consult, Nat Geog Soc, 75-; lectr, Am Astron Soc Shapley Ser, 84- *Mem:* Am Astron Soc; AAAS; Astron Soc Pac; Int Astron Union; Sigma Xi; Am Inst Aeronaut & Aerospace. *Res:* Interstellar medium; observations of emission nebulae especially supernova remnants and interstellar bubbles; development of astronomical instrumentation; development and operation of astronomy facilities on space shuttle; use of International Ultraviolet Explorer and x-ray satellites; management techniques applied to physical sciences. *Mailing Add:* Code 680 Goddard Space Flight Ctr Greenbelt MD 20771

GULLEN, WARREN HARTLEY, research methodology, preventive medicine; deceased, see previous edition for last biography

GULLIKSON, CHARLES WILLIAM, b El Dorado, Kans, Sept 11, 28; m 78; c 5. PHYSICS. *Educ:* Stanford Univ, BS, 51; Univ Okla, MS, 53, PhD(physics), 57. *Prof Exp:* Asst physics, Univ Okla, 52-56; res physicist, Marathon Oil Co, 56-62; dir res & treas, Cryogenic Res Co, 62-63; res physicist, Kaman Scis Corp, 63-73; lab dir, Western Environ, Inc, 74-89; lectr, Maui Community Col, 89-90; SUPVR PHYSICS LAB, EMBRY-RIDDLE AERONAUT UNIV, 90- *Mem:* Am Phys Soc; Soc Appl Spectros. *Res:* Raman, infrared, ultraviolet, absorption and atomic absorption spectroscopy. *Mailing Add:* 745 Malia Way Prescott AZ 86303

GULLINO, PIETRO M, b Saluzzo, Italy, Mar 24, 19; m 56. PATHOLOGY. *Educ:* Univ Turin, MD, 43; Ital Bd Path, dipl, 52. *Prof Exp:* Assoc prof path, Univ Turin, 52-59; vis scientist, Nat Cancer Inst, 59-63, res pathologist, 63-68, head tumor physiopath sect, 68-73, chmn exp biol comt, 71-75, chmn breast cancer task force, 75-78, chief, lab pathophysiol, 73-84; CONSULT, 88- *Concurrent Pos:* Adj prof, chem eng, Univ Del, 76-80, Carnegie-Mellon Univ, Pittsburgh, 79-; assoc ed, Cancer Res, 77-85, Cancer Letters, 78-82; mem, Int Sci Adv Comt, Danish Soc, 82- *Mem:* Europ Asn Cancer Res; Am Soc Path; Am Asn Cancer Res. *Res:* Physiopathology of cancer tissue. *Mailing Add:* Dept Biomed Sci Univ Torino Via Santena 7 Turin 10126 Italy

GULLION, GORDON W, b Eugene, Ore, Apr 16, 23; m 44; c 4. WILDLIFE MANAGEMENT. *Educ:* Univ Ore, BS, 48; Univ Calif, Berkeley, MA, 50. *Prof Exp:* Wildlife technician, Nev Fish & Game Comn, 51-52, sr game technician, 52-56, dist supvr fish & game mgt, 56-58; res assoc forest wildlife rels, 58-76, assoc prof wildlife, 76-80, PROF WILDLIFE MGT, FISH & WILDLIFE, UNIV MINN, 80- *Concurrent Pos:* Consult, Wildlife Biol, Region 2, US Forest Serv, 77-79; consult ecologist, La-Pac Corp, 84. *Mem:* Am Ornith Union; Wildlife Soc; Wilson Ornith Soc; Cert Wildlife Biol; Wildlife Soc; Ruffed Grouse Soc (res coordr); Sigma Xi; Soc Am Foresters. *Res:* Population dynamics; sex and age composition, ecology, behavior, food studies and nutrition, harvest and habitat management techniques for desert, Great Basin, Rocky Mountain and Boreal Forest small game species; ruffed grouse and aspen ecosystem research and management. *Mailing Add:* Forest Res Ctr 175 Univ Rd Cloquet MN 55720

GULLIVER, JOHN STEPHEN, b Torrence, Calif, Sept 9, 50; m 72; c 2. HYDRAULICS, FLUID MECHANICS. *Educ:* Univ Calif, Santa Barbara, BS, 74; Univ Minn, MS, 77, PhD(civil eng), 80. *Prof Exp:* Res assoc, dept civil & mineral eng, St Anthony Falls Hydraulic Lab, 80-81; asst prof, 81-87, ASSOC PROF, DEPT CIVIL & MINERAL ENG, UNIV MINN, 87-, TECH ED, HYDRO REVIEW, 87- *Concurrent Pos:* Vis res scientist, US Army Engr Waterways Exp Sta, 90. *Honors & Awards:* Rickey Medal, Am Soc Civil Engrs, 90. *Mem:* Int Asn Hydraul Res; Am Soc Civil Engrs; Am Water Resources Asn; Am Geophys Union; Nat Soc Prof Engrs. *Res:* Mass transfer at the air-water interface in rivers and lakes; hydraulic structures; hydraulics of river flow; hydropower engineering. *Mailing Add:* St Anthony Falls Hydraulic Lab Mississippi River at Third Ave SE Minneapolis MN 55414

GULLIVER, ROBERT DAVID, II, b Torrance, Calif, July 24, 45. GEOMETRY, ANALYTICAL MATHEMATICS. *Educ:* Stanford Univ, BS, 67, MS, 69, PhD(differential geom), 71. *Prof Exp:* Lectr math, Univ Calif, Berkeley, 71-73; from asst prof to assoc prof, 73-85, PROF MATH, UNIV MINN, MINNEAPOLIS, 85- *Concurrent Pos:* Max Planck Inst Organizer Nonlinear PDE, 86-87. *Mem:* Am Math Soc. *Res:* The immersion question for minimal surfaces of general topological type, solving appropriate variational problems; theory of branched immersions; rotating drops; harmonic maps. *Mailing Add:* Univ Minn Sch Math Minneapolis MN 55455

GULLO, VINCENT PHILIP, b New York, NY, Mar 30, 50; m 73; c 1. NATURAL PRODUCT DRUG DISCOVERY, PHARMACEUTICAL RESEARCH. *Educ:* City Col New York, BS, 71; Columbia Univ, PhD(org chem), 75. *Prof Exp:* sr res chemist, Merck Sharp & Dohme Res Labs, 75-80, res fel natural prod chem, 80-83; assoc dir, 83-89, DIR MICROBIOL PROD, SCHERING PLOUGH RES, 89- *Mem:* Am Chem Soc; Am Soc Microbiol; AAAS; Soc Indust Microbiol. *Res:* Natural product chemistry; chromatography; spectroscopy; structure determinations; antibiotics; pharmacological agents. *Mailing Add:* 33 Darren Woods Dr PO Box 332 Liberty Corner NJ 07938

GULLOTTA, FRANK PAUL, b Rockford, Ill, Aug 5, 42. EXPERIMENTAL NEUROPSYCHOLOGY. *Educ:* Calif State Univ, Los Angeles, BA, 70; Univ NMex, PhD(exp psychol), 76. *Prof Exp:* Vis asst prof psychol, State Univ NY, Albany, 76-77; RES SCIENTIST BEHAV RES, PHILIP MORRIS RES CTR, 77- *Mem:* Soc Neurosci; assoc Am Electroencephalog Soc; Soc Psychophysiol Res. *Res:* Experimental electroencephalography; evoked potentials. *Mailing Add:* Philip Morris Res Ctr PO Box 26583 Richmond VA 23261

GULLY, JOHN HOUSTON, b Oak Ridge, Tenn, July 27, 52; m 75; c 2. BEARINGS & SEALS, HIGH SPEED CURRENT COLLECTION. *Educ:* Univ Tex, Austin, BS, 76. *Prof Exp:* Res engr, 76-83, assoc tech dir, 83-85, DEP DIR, CTR ELECTROMECH, UNIV TEX, 85- *Concurrent Pos:* Prin investr numerous res & develop contracts, 84-; chmn, Opening Switch Panel, Tamarron Workshop, 86 & Power Conditioning Panel, Elec Energy Gun Study, Defense Advan Res Projs Agency, 88-; chmm & ed, Int Current Collector Conf, 88; ed, Superconducting Power Appln Review Comt, 89. *Res:* Use of rotating electrical generators as pulsed power supplies for powering fusion devices, high energy, high rate material processing systems, lasers and electric accelerators (railguns). *Mailing Add:* Ctr Electromechanics Balcones Res Ctr Univ Tex 10100 Burnet Rd Austin TX 78758

GULOTTA, STEPHEN JOSEPH, b Brooklyn, NY, Mar 5, 33; m 54; c 3. CARDIOLOGY. *Educ:* Brooklyn Col, BS, 54; State Univ NY Downstate Med Ctr, MD, 58. *Prof Exp:* Intern med, Montefiore Hosp, Bronx, 58-59, from jr asst resident to asst resident, 59-61; chief div cardiol, North Shore Hosp, 67-79; ASSOC PROF MED, MED COL, CORNELL UNIV, 72-; DIR CARDIAC CATHETERIZATION LAB, 82- *Concurrent Pos:* USPHS fel, NY Hosp, 61-62; Nassau & Westchester Heart Asns grant-in-aid, North Shore Hosp, 70-71; attend physician, Mt Vernon Hosp, 62 & North Shore Hosp, 67-; asst attend, NY Hosp, 69-; fel coun clin cardiol, Am Heart Asn, 71-; attend physician, St Francis Hosp, 79- *Mem:* Am Fedn Clin Res; fel Am Col Physicians; fel Am Col Cardiol; fel Am Col Chest Physicians. *Res:* Conduction disturbances and arrhythmias in man; coronary artery disease. *Mailing Add:* 100 Port Washington Blvd Roslyn NY 11576

GULRAJANI, RAMESH MULCHAND, b Patna, India, Dec 12, 44; m 76; c 2. BIOELECTRIC PHENOMENA, CARDIAC ELECTROPHYSIOLOGY. *Educ:* Univ Bombay, BE, 64; Ill Inst Technol, MS, 65; Syracuse Univ, MS, 72, PhD(elec eng), 73. *Prof Exp:* Engr FM receiver develop, Sylvania Elec Prod, Batavia, NY, 65-66; res & teaching asst, Dept Elec Eng, Syracuse Univ, 67-72; Med Res Coun Can postdoctoral fel neurophysiol, Dept Physiol, Univ Montreal, 73-75, res assoc cardiac electrophysiol, Dept Med, 76-79, res assoc biomed eng, Inst Biomed Eng, 79-87, assoc prof, 87-90, PROF BIOMED ENG, INST BIOMED ENG, UNIV MONTREAL, 90- *Concurrent Pos:* Res scholar, Can Heart Found, 78-81; annual res grants, Que Heart Found, 78-, prog grant, Med Res Coun Can, 79- & operating grant, Natural Sci & Eng Res Coun Can, 86-; sr res scholar, Fonds de la recherch en santé du Que, 86-89; session chair, World Cong Med Physics & Biomed Eng, San Antonio, Tex, 88; mem, Biomed Eng Grants Comt, Med Res Coun Can, 88-91; spec reviewer, Cardiovasc Study Sect, NIH, 90-91. *Mem:* Sr mem Inst Elec & Electronics Engrs. *Res:* Computer models of the electrical activity of the heart; forward and inverse problems of electrocardiography; bioelectric and biomagnetic phenomena. *Mailing Add:* Inst Biomed Eng Univ Montreal PO Box 6128 St A Montreal PQ H3C 3J7 Can

GULRICH, LESLIE WILLIAM, JR, b Chicago, Ill, July 31, 33; m 75; c 3. PHYSICAL CHEMISTRY. *Educ:* Kent State Univ, BS, 60, PhD(phys chem), 67. *Prof Exp:* Res chemist, 67-76, sr res chemist, 76-87, RES ASSOC, E I DU PONT DE NEMOURS & CO, INC, 87- *Mem:* Am Chem Soc. *Res:* Fiber spinning processes; fiber spinning equipment design; fiber structure; polymer molecular weight distribution; polymer solution rheology; computer aided design programming; engineered short fibers & asbestos replacement. *Mailing Add:* 10339 Ashburn Rd Richmond VA 23235

GULYAS, BELA JANOS, b Szekesfehervar, Hungary, Apr 14, 38; US citizen; m 62; c 3. CELL & DEVELOPMENTAL BIOLOGY, GAMETE PHYSIOLOGY. *Educ:* Moravian Col, BS, 62; WVa Univ, MS, 65; Univ Colo, Boulder, PhD(develop biol), 68. *Prof Exp:* Instr histol-cell biol, Georgetown Sch Med & Dent, 70-71; reproductive physiologist & sect chief, 71-84; EXEC SECY NCRR, GEN CLIN RES CTRS, 88- *Concurrent Pos:* Fel NIH, hon res fel, Mammal Develop Unit, Med Res Coun, London; vis prof, histol, George Washington Sch Med, 79-85; exec secy, Div Res Grants Reproductive Endocrinol, NIH, 84- *Mem:* Soc Develop Biol; Soc Study Reproduction; Am Soc Cell Biol; Am Asn Anatomists. *Res:* Cellular and developmental aspects of preimplantation mammalian embryos; experimental and fine structural approaches in problems concerning polyspermy, parthenogenesis, and cleavage plane and cleavage furrow formation. *Mailing Add:* NIH Westwood 10A-16 Bethesda MD 20892

GULYASSY, PAUL FRANCIS, b Bridgeport, Conn, Aug 5, 28; m 60; c 2. MEDICINE. *Educ:* Yale Univ, BA, 50; Columbia Univ, MD, 54. *Prof Exp:* From asst prof to assoc prof med, Med Ctr, Univ Calif, San Francisco, 64-72; dir, NCalif Artificial Kidney Ctr & chief, Renal & Electrolyte Sect, San Francisco Gen Hosp, 68-72; PROF INTERNAL MED & CHIEF, NEPHROLOGY DIV, UNIV CALIF, DAVIS 72- *Concurrent Pos:* Am Heart Asn adv res fel med, Renal Lab, Sch Med, Tufts Univ, 60-62 & Cardiovasc Res Inst, Univ Calif, San Francisco, 62-64; res career award, NIH, 65-72, mem, Clin Res Fel Rev Bd, 68-71 & Gen Med B Study Sect, 72- *Mem:* Am Soc Clin Invest; Am Fedn Clin Res; Am Physiol Soc; Am Soc Nephrology. *Mailing Add:* Dept Int Med Div Nephrol Univ Calif 4301 X St Sacramento CA 95817

GUM, ERNEST KEMP, JR, b Weston, WVa, Feb 1, 49; m 68; c 2. MICROBIAL BIOCHEMISTRY. *Educ:* WVa Univ, BS, 70; Va Polytech Inst & State Univ, PhD(biochem), 74. *Prof Exp:* Res assoc biochem, Va Polytech Inst & State Univ, 74-75, asst prof biochem, 75-77; res scientist, Union Carbide Corp, 77-79; proj specialist, Gen Foods Corp, 79-81, group leader, 82-85; mgr technol develop, 85-86, MGR BEVERAGE TECHNOL, WESTRECO, 87- *Mem:* AAAS; Am Chem Soc; Sigma Xi; Inst Food Technologists; Am Soc Microbiol. *Res:* Structure, function and regulation of microbial enzymes; flavor development; beverage development. *Mailing Add:* Westreco Inc Boardman Rd New Milford CT 06776

GUM, JAMES RAYMOND, JR, b Cleveland, Ohio; m 74; c 1. MOLECULAR CLONING OF TUMOR ANTIGENS, CHARACTERIZATION OF GENES. *Educ:* Univ Ky, BS, 72; Univ Tex, MS, 76, PhD(biochem & molecular biol), 80. *Prof Exp:* Postdoctoral fel biochem & molecular biol, Univ Wis, 80-84; RES BIOCHEMIST, VET ADMIN MED CTR, 84- *Honors & Awards:* Nat Res Serv Award, Nat Inst Gen Med Sci, 81. *Mem:* Am Soc Biochem & Molecular Biol. *Res:* Molecular biology of colon cancer; molecular cloning of tumor antigen and mucins; characterization of tumor suppressor genes in colon cancer cells. *Mailing Add:* GI Res VA/Med Ctr 4150 Clement St San Francisco CA 94121

GUM, MARY LOU, b Little Birch, WVa, Feb 15, 48; m 68. PHYSICAL CHEMISTRY. *Educ:* WVa Univ, BS, 70; Va Polytech Inst & State Univ, PhD(chem), 74. *Prof Exp:* Asst prof chem, Va Polytech Inst & State Univ, 74-77; sr proj scientist, 77-84, GROUP LEADER, UNION CARBIDE CORP, 84- *Mem:* Am Chem Soc; Sigma Xi. *Res:* Surface chemistry; kinetics; N-nitrosamine chemistry; microwave spectroscopy. *Mailing Add:* Silicones Bldg Union Carbide Corp Tarrytown NY 10591

GUM, OREN BERKLEY, b Gassaway, WVa, Nov 3, 19; m 46; c 2. MEDICINE. *Educ:* Ohio State Univ, BS, 41, PhD(agr chem), 44, MD, 50. *Prof Exp:* From intern to resident internal med, Colo Gen Hosp, 50-54; asst prof med, Sch Med, Univ Colo, 55-61; assoc prof, 61-68, PROF MED, SCH MED, TULANE UNIV, 68- *Concurrent Pos:* NIH fel endocrinol, Colo Gen Hosp, 54-55. *Mem:* AAAS; Am Rheumatism Asn; Sigma Xi. *Res:* Rheumatic diseases. *Mailing Add:* Tulane Sch Med 1430 Tulane New Orleans LA 70112

GUM, WILSON FRANKLIN, JR, b Pen Argyl, Pa, July 14, 39; m 59; c 4. ORGANIC CHEMISTRY, RESEARCH ADMINISTRATION. *Educ:* Muhlenberg Col, BS, 61; Univ Pa, PhD(org chem), 65. *Prof Exp:* Spec assignments chemist, Dow Chem Co, 65-67, sr res chemist, 67-68, proj leader designed prod dept, 68-70, res group leader, Org Chem Prod Res Lab, 70-71, sect head, Org Chem & Intermediates Sect, Tech Serv & Develop, 71-73, div mgr org chem, Tech Serv & Develop, Agr-Org Dept, Freeport, Tex, 74-78, lab dir oxides & intermediates, Tech Serv & Develop, Org Chem Dept, 78, RES & DEVELOP DIR, ORG CHEM DEPT, DOW CHEM USA, 78-, RES & DEVELOP DIR, POLYURETHANES, 84- *Mem:* Am Chem Soc. *Res:* New product research and development for organic chemicals, resin and polymer intermediates, alkylene oxide and derivatives; polyurethane intermediates; polyurethane products and systems; polymer chemistry and technical management. *Mailing Add:* 51 White Oak Ct Lake Jackson TX 77566-4426

GUMAN, WILLIAM J(OHN), b Greenwich, Conn, June 23, 29; m 54; c 2. AERONAUTICAL ENGINEERING. *Educ:* Rensselaer Polytech Inst, BAE, 52, MAE, 54, PhD(aeronaut eng), 65. *Prof Exp:* Res asst aeronaut eng, Rensselaer Polytech Inst, 52-54, from instr to asst prof, 54-59; sr sci res engr, Fairchild Industs, 59-60, prin sci res engr, 60-61, spec sci res engr, 61-63, staff engr, 63-65, chief propulsion, 65-79, mgr Indust Res & Develop & Res & Develop Contracts, Fairchild Repub Co, 79-82, dir res & develop, 82-84, dir technol & opers, 84-85, dir adv prod develop, 85-87; DIR CONTRACT RES & DEVELOP, TECH DEVELOP, GRUMMAN CORP, 87- *Concurrent Pos:* Prin investr 20 res & develop contracts; mem Res & Eng Comt Nat Security Indust Asn, 84-87, bd dirs, 85-87; mem Tech Comt Elec Propulsion, Am Inst Aeronaut & Astronaut, 66-68 & 78-80. *Honors & Awards:* Tech Excellence Award, Am Inst Aeronaut & Astronaut, 77; Cert Recognition, Nat Aeronaut & Space Admin, 76. *Mem:* Assoc fel Am Inst Aeronaut & Astronaut; Sigma Xi. *Res:* Theoretical and experimental gasdynamics, especially electric space propulsion; management of research and development. *Mailing Add:* 26 Gaymor Lane Commack NY 11725

GUMBINER, BARRY M, b Cleveland. Ohio. June 11, 54. PHARMACOLOGY, CELL BIOLOGY. *Educ:* Univ Cinncinnati, BS, 76; Univ Calif, PhD(neurosci), 82. *Prof Exp:* Fel, Dept Biol, Mass Inst Technol, Cambridge, 82-83, Dept Cell Biol, Europ Molecular Biol Lab, Heidelberg, WGer, 83-85; ASST PROF, DEPT PHARMACOL, UNIV CALIF, SAN FRANCISCO, 85- *Concurrent Pos:* Mem, cell biol & develop biol grad prog & prog molecular med, Univ Calif, San Francisco, joint appointment, Dept Physiol, fac mem, fel training prog digestive dis & nutrit & nephrol; Helen Hay Whitney fel, 82-85; Estab Investr Award, Am Heart Asn, 88-92. *Honors & Awards:* Estab Investr Award, Am Heart Asn, 88-92. *Mem:* Am Soc Cell Biol; AAAS; Soc Develop Biol. *Res:* Pharmacology; numerous publications. *Mailing Add:* Dept Pharmacol Sch Med Univ Calif San Francisco CA 94143

GUMBRECK, LAURENCE GABLE, endocrinology; deceased, see previous edition for last biography

GUMBS, GODFREY ANTHONY, b Georgetown, Guyana, Sept 17, 48; m 71; c 2. PHYSICS. *Educ:* Trinity Col, Cambridge, Eng, BA, 71; Univ Toronto, MSc, 73, PhD(physics), 78. *Prof Exp:* Res assoc chem, Nat Res Coun, Can, 78-82; asst prof, Dalhousie Univ, Can, 82-86; assoc prof, 86-90, PROF, UNIV LETHBRIDGE, 90- *Concurrent Pos:* Univ res fel, Natural Sci & Eng Res Coun Can, 82; Alexander von Humboldt fel, 87; vis scientist, Mass Inst Technol, 90-91. *Mem:* Am Phys Soc; Can Asn Physicists; NY Acad Sci. *Res:* Critical phenomena in systems with a surface; the effects due to a surface on a Landau quantized plasma; static shielding of impurities near a surface; collective excitations in two and three dimensions; penetration phenomena in solids, superlattices and heterostructures; transport in mesoscopic systems. *Mailing Add:* Dept Physics Univ Lethbridge Lethbridge AB T1K 3M4 Can

GUMMEL, HERMANN K, b Hannover, Ger, July 6, 23; US citizen; m 52; c 2. SOLID STATE ELECTRONICS. *Educ:* Syracuse Univ, MS, 52, PhD(physics), 57; Univ Marburg, dipl, 52. *Prof Exp:* Mem tech staff, Bell Labs, 57-67, head design anal dept, from asst dir to dir, Advan Comput Aided Design Test & Studies Lab, AT&T Bell Labs, 82-86; CONSULT, 86- *Mem:* Nat Acad Eng; fel Inst Elec & Electronics Engrs; Sigma Xi; Am Phys Soc. *Res:* Semiconductors; semiconductor devices; computer-aided design. *Mailing Add:* AT&T Bell Labs Rm 3-B 435 600 Mountain Ave Murray Hill NJ 07974-0636

GUMP, BARRY HEMPHILL, b Columbus, Ohio, Nov 12, 40; m 63; c 2. ANALYTICAL CHEMISTRY, WINE AND FOOD ANALYSIS. *Educ:* Ohio State Univ, BSc, 62; Univ Calif, Los Angeles, PhD(anal chem), 66. *Prof Exp:* From asst prof to assoc prof, 67-74, PROF CHEM, CALIF STATE UNIV, FRESNO, 74-, PROF ENOL & FOOD SCI, 81- *Concurrent Pos:* Res assoc, Bur Sci, Food & Drug Admin, Washington, DC, 66-67; consult, Cent Calif Med Labs, 69-71; vis scientist, Bioorg Standards Sect, Anal Div, Nat Bur Standards, 74-76, Danforth Assoc, 81; Fulbright lectr, Univ Repub, Montivideo, Uruguay, 83; assoc referee for sulfur dioxide in wine, AOAC, 86. *Mem:* Am Chem Soc; Am Soc Enologists; Sigma Xi; Am Inst Chemists. *Res:* Separation methods in chemistry, especially chromatographic methods; analytical methods development, trace components in foods and wine; trace hydrocarbon analysis in marine sediment, water and tissue samples. *Mailing Add:* Dept Chem Calif State Univ Fresno CA 93740-0070

GUMP, DIETER W, b Saranac Lake, NY, Mar 28, 33; c 3. INTERNAL MEDICINE, INFECTIOUS DISEASES. *Educ:* Swarthmore Col, BA, 55; Johns Hopkins Univ, MD, 60. *Prof Exp:* Intern, Johns Hopkins Hosp, 60-61; asst resident, Med Ctr, Univ Colo, 61-62 & Johns Hopkins Hosp, 62-63; from instr to asst prof med microbiol, 66-71, from asst prof to assoc prof med, 68-80, ASSOC PROF MICROBIOL, UNIV VT, 71-, PROF MED MICROBIOL, 80- *Concurrent Pos:* Fel infectious dis, Johns Hopkins Hosp, 63-66. *Mem:* Am Fedn Clin Res; Am Soc Microbiol; Sigma Xi; fel Am Col Physicians; fel Infectious Dis Soc Am. *Res:* Bacterial, mycoplasmal and chlamydial disease; antibiotics. *Mailing Add:* 226 Spear St South Burlington VT 05403

GUMP, FRANK E, b St Louis, Mo, Feb 16, 28; m 59; c 3. SURGERY. *Educ:* Harvard Univ, AB, 51; NY Univ, MD, 55; Am Bd Surg, dipl. *Prof Exp:* NIH res fel, 58-59; asst prof, 66-69, assoc prof, 69-72, PROF SURG & CHIEF BREAST SERV, COLUMBIA-PRESBY MED CTR, 72- *Mem:* Am Col Surg; Asn Acad Surg; Soc Univ Surg; Am Surg Asn; Am Asn Surg of Trauma. *Res:* Breast cancer. *Mailing Add:* Dept Surg Columbia P&S 181 Ft Washington Ave New York NY 10032

GUMP, J R, b Ann Arbor, Mich, Apr 1, 21; m 43; c 3. INORGANIC CHEMISTRY, ANALYTICAL CHEMISTRY. *Educ:* Detroit Inst Technol, BSChE, 43; Wayne State Univ, MS, 49; Univ Ky, PhD(inorg & anal chem), 52. *Prof Exp:* Instr chem, Lawrence Inst Technol, 47-49 & Univ Ky, 49-52; asst prof, Univ Cincinnati, 52-53; chemist, Inorg Res, Mich Chem Corp, 53-54, supvr, 54-55, group leader, 59-62; res chemist, Victor Chem Co, 62-63; from asst prof to prof physics, 63-87, EMER PROF, CENT MICH UNIV, 87- *Concurrent Pos:* Pres, Mitten Chem Inc, 55-67; consult, Sylvania Elec Prod, 67-71. *Mem:* AAAS; Am Chem Soc; Nat Educ Asn; Nat Sci Teachers Asn. *Res:* Brine; bromine; magnesia; rare earths; ion exchange; solvent extraction; yttrium metal; salts and oxides; scandium salts and oxides; ferrates chemistry; secondary and elementary physical science education. *Mailing Add:* Dept Physics Cent Mich Univ Mt Pleasant MI 48859

GUMPERTZ, WERNER H(ERBERT), b Berlin, Ger, Dec 26, 17; US citizen; m 49; c 2. CIVIL ENGINEERING. *Educ:* Swiss Fed Inst Technol, BCE, 39; Mass Inst Technol, SB, 48, SM, 50, BldgE, 54. *Prof Exp:* Vol engr, Amsterdam, Netherlands, 39-40; instrument man, Lockwood, Kessler & Bartlett, 40-41; field engr, M Shapiro & Sons Construct Co, Inc, 41-43; off engr shipyard, Kaiser Co, 43; struct engr, Corps Engrs, US Army, 47-48; engr, United Engrs & Constructors, 48-49; asst prof civil eng & construct mgt, Mass Inst Technol, 49-57; PRIN CONSULT ENGR & SR PRIN ENGR, SIMPSON GUMPERTZ & HEGER, INC, 57- *Concurrent Pos:* Mem comt adhesives & sealants, Bldg Res Inst, 60-63; mem adv comt, Inst Bldg Res, Pa State Univ, 64-84; mem Nat Bur Standards steering group, US Dept of Energy, 78-85; chmn comt D8 Roofing, Am Soc Testing & Mat. *Mem:* Fel Am Soc Civil Engrs; Am Concrete Inst; Am Soc Eng Educ; fel Am Soc Testing & Mat. *Res:* Investigations into physical properties of built-up roofing systems and their components; interrelation between structural building movement and the performance of nonstructural roof membranes; curtain walls; masonry; glass/glazing. *Mailing Add:* 297 Broadway Arlington MA 02174

GUMPF, DAVID JOHN, b Billings, Mont, Mar 30, 42; m 64; c 1. PLANT VIROLOGY. *Educ:* Mont State Univ, BS, 64, MS, 66; Univ Nebr, PhD(plant path), 70. *Prof Exp:* From asst prof & asst plant pathologist to assoc prof & assoc plant pathologist, 70-85, PROF PLANT PATH & PLANT PATHOLOGIST, UNIV CALIF, RIVERSIDE, 85- *Mem:* Am Phytopath Soc; Int Orgn Mycoplasmology; Int Orgn Citrus Virologists; Sigma Xi; Int Soc Citriculture. *Res:* Purification, characterization and diagnosis of citrus viruses and isolation and cultivation of plant mycoplasmas. *Mailing Add:* Dept Plant Path Univ Calif Riverside CA 92521

GUMPORT, RICHARD I, b Pocatello, Idaho, June 23, 37; m 60; c 2. BIOCHEMISTRY, ENZYMOLOGY. *Educ:* Univ Chicago, BS, 60; PhD(biochem), 68. *Prof Exp:* USPHS fel biochem, Sch Med, Stanford Univ, 68-70, fel, 70-71; from asst prof to assoc prof, 71-86, assoc head biochem, 89, PROF BIOCHEM, SCH BASIC MED SCI & SCH CHEM SCI, UNIV ILL, URBANA, 86- *Concurrent Pos:* Vis scholar biochem, dept biochem & molecular biol, Harvard Univ, 79-80; John Simon Guggenheim Mem fel, 79-80; assoc, Ctr for Advan Study, Univ Ill, 85-86. *Mem:* AAAS; Am Soc Biol Chem & Molecular Biol; Am Chem Soc; Am Soc Microbiologists; Sigma Xi. *Res:* Nucleic acid enzymology; transcription termination, site-specific genetic recombination; nucleic acid synthesis; protein-DNA interaction. *Mailing Add:* Dept Biochem Univ Ill 1209 W Calif St Urbana IL 61801

GUMPORT, STEPHEN LAWRENCE, b New York, NY, Dec 5, 13; m 49; c 4. SURGERY. *Educ:* Amherst Col, BA, 34; Cornell Univ, MD, 38; Am Bd Surg, dipl, 50. *Prof Exp:* From intern to resident, Lenox Hill Hosp, NY, 38-40; asst surg, Postgrad Med Sch, 49-51, from instr to assoc prof clin surg, 51-71, prof clin surg, 71-81, co-dir tumor serv, 66-81, CONSULT DIV ONCOL, DEPT SURG, SCH MED, NY UNIV, 81-, PROF SURG, 81- *Concurrent Pos:* From asst vis surgeon to vis surgeon, Bellevue Hosp, 49-; from asst attend surgeon to attend surgeon, Univ Hosp, 49-; hon med staff, Doctors Hosp, NY, 49-69; asst attend surgeon, NY Eye & Ear Infirmary, 58-67; from assoc attend surgeon to attend surgeon, NY Infirmary, 64-; dir, Med Amateur Radio Coun, 66-70 & NY State Cancer Progs Asn, 67-70; attend, Manhattan Vet Admin Hosp, 67-69; coordr, NY Metrop Regional Med Prog, Sch Med, NY Univ, 67-69, mem tech consult panel cancer, 68-; liaison fel, Comn Cancer, Am Col Surg, 67-70; consult, Manhattan Vet Admin Hosp, 69-, US Air Force Hosp, Mitchell AFB, 49-61, St Vincent's Hosp, 68-, NY Eye & Ear Infirmary, 67- & Doctors Hosp, 69-; chmn, Comt Cancer, Med Ctr, NY Univ, 68-73; Melanoma Immunol Study Group, 70-; dir, Cancer Rehab Ser, Risk Inst, NY Med Ctr, 80- *Honors & Awards:* 1st Biennial Award, Skin Cancer Found, 79. *Mem:* Am Asn Cancer Educ; Am Asn Cancer Res; Am Cancer Soc; AMA; Am Soc Clin Oncol; Sigma Xi. *Res:* Tumor surgery; cancer chemotherapeutic agents and their clinical application. *Mailing Add:* 530 First Ave New York NY 10016

GUMPRECHT, WILLIAM HENRY, b Potsdam, NY, Nov 16, 31; m; c 4. ORGANIC CHEMISTRY. *Educ:* Univ Ill, BS, 53; Univ Minn, PhD(chem), 57. *Prof Exp:* Res chemist, 57-66, sr res chemist, 66-85, RES ASSOC, E I DU PONT DE NEMOURS & CO, INC, 85- *Mem:* Am Chem Soc. *Res:* Dyes; fluorinated compounds; polymers; telomers. *Mailing Add:* 2606 Stephenson Dr Wilmington DE 19808

GUNASEKARAN, MUTHUKUMARAN, b Chidambaram, India, June 6, 42; m 70; c 2. CLINICAL MYCOLOGY, PLANT PATHOLOGY. *Educ:* Annamalai Univ, BS, 64; Univ Madras, MS, 66; Tex A&M Univ, PhD(plant path), 70. *Prof Exp:* Asst prof biol, Rust Col, 72-74; asst mem, Infectious Dis Dept, St Jude Children's Res Hosp, 74-81; assoc prof, 81-86, PROF BIOL, FISK UNIV, 86- *Concurrent Pos:* Fel, Brigham Young Univ, 71-72; NIH grant, 76-79, NASA grant, 86-; Fulbright scholar. *Mem:* Am Mycol Soc; Am Soc Microbiol; Indian Phytopath Soc. *Res:* Physiology and rapid diagnostic studies on fungal diseases. *Mailing Add:* Dept Biol Fisk Univ Nashville TN 37208-3051

GUNBERG, DAVID LEO, b Minneapolis, Minn, May 1, 22; m 46; c 4. HEAD & NECK ANATOMY, MAMMALIAN EMBRYO CULTURE. *Educ:* Univ Redlands, AD, 49; Univ Calif, Berkeley, MS, 52, PhD(anat), 54. *Prof Exp:* Teaching asst anat, Univ Calif, Berkeley, 50-54; instr, Med Sch, Univ Southern Calif, 54-55; from asst prof to prof, Ore Health Sci Univ Med Sch, 55-72; prof & chmn, Univ Malaya Med Fac, Malaysia, 72-75; prof & chmn anat, Ore Health Sci Univ Dent Sch, 75-88; PROF EMER, 88- *Concurrent Pos:* Asst dir, grad prog anat, Ore Health Sci Univ, 60-62 & 64-72; prog dir, postdoctoral training prog anat, AID proj med educ, Airlangga Univ, Indonesia & Univ Calif, 61-67; vis prof, Med Fac, Univ Airlangga, Indonesia, 62-64 & Med Fac, Univ WI, Jamaica, 83-84; external examiner, Univ Kebangsaan Med Fac, Malaysia, 80 & Univ WI Med Fac, Jamaica, 82, 83 & 87. *Mem:* Teratology Soc; Am Asn Anatomists; Europ Teratology Soc; Japanese Teratology Soc; Sigma Xi. *Res:* Causes of congenital defects in mammals; effects of exogenous agents on embryonic and fetal development are studied utilizing morphologic, chemical and metabolic measurements. *Mailing Add:* Anat Dept Sch Dent Ore Health Sci Univ 611 SW Campus Dr Portland OR 97201

GUNBERG, PAUL F, b Minneapolis, Minn, July 13, 17; m 45; c 5. ORGANIC CHEMISTRY. *Educ:* St Olaf Col, BA, 39; Purdue Univ, PhD(chem), 49. *Prof Exp:* Lab asst adhesives, Minn Mining & Mfg Co, 39-40; control chemist wheat prod, Washburn Crosby Milling Co, 40-41; supvr ballistics, Munitions Div, US Rubber Co, 42-43, supvr org synthesis, 44-45; asst, Purdue Univ, 46-49; res chemist rubber & plastics, US Rubber Co, 49-57; mgr tech serv, Tex-US Chem Co, 57-66, mgr appln develop, 66-77, mgr prod develop, 78-80; CONSULT RUBBER INDUST, 80- *Mem:* Am Chem Soc. *Res:* Rubber chemistry; emulsion polymerization; compounding; processing; product development. *Mailing Add:* RR 1 Box 726 Augusta NJ 07822

GUNCKEL, THOMAS L, II, b Dayton, Ohio, Jan 16, 36; m 63; c 2. ELECTRICAL ENGINEERING. *Educ:* Calif Inst Technol, BS, 58; Stanford Univ, MS, 59, PhD(elec eng), 61. *Prof Exp:* Res specialist, Autonetics Div, NAm Aviation, Inc, 61-62, supvr phys anal, 62-63, chief space syst anal, 63-66, mgr syst anal, 66-71, mgr navig systs, NAm Rockwell Corp, 71-73, mgr comput & microelectronics, 73-77, mgr systs eng, 77-80, chief engr mx eng, 80-83, chief eng, 83-87, VPRES MINUTEMAN & ADV SYSTS, STRATEGIC SYSTS DIV, 87-, VPRES & GEN MGR AUTONETICS, AUTONETIS STRATEGIC SYSTS DIV, ROCKWELL INT. *Res:* Control system theory; space guidance and navigation; optimal control; orbit determination; computer design; systems engineering. *Mailing Add:* 11242 La Vereda Dr Santa Ana CA 92705

GUND, PETER HERMAN LOURIE, b New York, NY, Feb 20, 40; m 67; c 1. MEDICINAL & COMPUTATIONAL CHEMISTRY. *Educ:* Columbia Col, AB, 61; Purdue Univ, MS, 63; Univ Mass, PhD(org chem), 67. *Prof Exp:* Res chemist antiparasitics, Am Cyanamid Agr Div, Princeton, NJ, 67-70; vis res fel, dept chem, Princeton Univ, 70-72, dept biochem, 72-75; sci res fel, 73-79, sr res fel, 79-83, dir, 83-90, SR DIR CHEM & BIOL SYSTS, MERCK SHARP & DOHME RES LABS, RAHWAY, NJ, 90- *Concurrent Pos:* treas, Div Med Chem, Chem Info & Comput Chem, Am Chem Soc, 87-90. *Mem:* Am Chem Soc; AAAS; Drug Info Asn. *Res:* Computer graphics applications in pharmaceutical chemical research; computer aided new drug design; drug conformation and mechanism of action; computer aided organic synthetic planning; quantitative structure-activity relationships; theoretical chemistry; molecular/macromolecular modeling. *Mailing Add:* Merck Sharp & Dohme Res Labs PO Box 2000 Rahway NJ 07065

GUNDERSEN, JAMES NOVOTNY, b Oak Park, Ill, Dec 20, 25; m 56; c 3. ECONOMIC GEOLOGY. *Educ:* Univ Wis, BS, 49; Univ Calif, Los Angeles, MA, 55; Univ Minn, PhD(econ geol), 58. *Prof Exp:* Geol consult, Alamos Mining Co, Mex, 51-53; raw mat engr, Columbia Iron Mining Co, US Steel Corp, Utah, 53-55; asst petrol & mineral, Univ Minn, 57-58, res assoc mining geologist, Mines Exp Sta, 58-61; from asst prof to assoc prof geol & chmn dept, Calif State Col, Los Angeles, 61-68, actg dir res & govt rels, 65; prof, Univ Ariz, 68-70; chmn dept geol, 70-76, PROF GEOL, WICHITA STATE UNIV, 70- *Mem:* Am Inst Mining, Metall & Petrol Engrs; Am Inst Prof Geologists; Geol Soc Am; Soc Econ Paleontologists & Mineralogists; Soc Archaeol Sci; Soc Am Archaeol. *Res:* Mining geology, especially taconites and iron ores; petrography and geochemistry of metamorphic and metasomatic rocks; mineral beneficiation; mineralography; hydrothermal alteration; differential thermal analysis; x-ray diffraction; spectroscopy; pre-Cambrian stratigraphy; uranium geology; general geoarchaeometry; provenances of Plains pipestones cathuite and other exotic lithics utilized by Paleoindians; provenances of obsidian raw materials. *Mailing Add:* Dept Geol Wichita State Univ Wichita KS 67208

GUNDERSEN, LARRY EDWARD, b Detroit, Mich, June 3, 40; m 63; c 4. RESEARCH ADMINISTRATION, INFORMATION SCIENCE. *Educ:* Bowling Green State Univ, BS, 62; Univ Ill, Urbana, MS, 66; Univ Iowa, PhD(biochem), 69. *Prof Exp:* Fel biochem, Scripps Clin & Res Found, 68-71; assoc dir sci info & regulatory affairs, Mead Johnson & Co, 71-76; dir, Off Sci Info & Regulatory Affairs, 76-81, dir, div sci info & regulatory affairs, Pennwalt Pharmaceut Div, 81-; mem staff, Nelson Res Regulatory Affairs, Calif; MEM STAFF, SMITHKLINE & BEECHAM PHARMACEUT, 88- *Mem:* Drug Info Asn; AAAS; Am Chem Soc. *Res:* Enzymology, protein structure; enzyme mechanisms; lactate dehydrogenase; dihydrofolate reductase; science information. *Mailing Add:* Two Woodstream Dr Wayne PA 19087

GUNDERSEN, MARTIN ADOLPH, b Glenwood, Minn, May 19, 40; m 63; c 2. LASERS, SOLID STATE PHYSICS. *Educ:* Univ Calif, Berkeley, BA, 65; Univ Southern Calif, MA, 67, PhD(physics), 73. *Prof Exp:* Engr bubble chamber, Lawrence Berkeley Lab, 65; res assoc lasers & solid state physics, Univ Southern Calif, 72-73; asst prof lasers & solid state physics, Dept Elec Eng, Tex Tech Univ, 73-77, assoc prof, 77-80; assoc prof, 80-83, PROF, DEPT ELEC ENG & PHYSICS, UNIV SOUTHERN CALIF, 83- *Mem:* Am Phys Soc; fel Inst Elec & Electronics Engrs; fel Optical Soc Am. *Res:* Applied physics; spectroscopy; solid state physics; lasers and lasers physics; pulsed power physics. *Mailing Add:* Dept Elec Eng & Physics MC 0484 Univ Southern Calif University Park Los Angeles CA 90089-0484

GUNDERSEN, RALPH WILHELM, b Chicago, Ill, Aug 22, 38; m 60; c 2. ENTOMOLOGY. *Educ:* Hamline Univ, BS, 59; Univ Minn, St Paul, MS, 62, PhD(entom), 68. *Prof Exp:* Instr, Univ Minn, Duluth, 63-64; from asst prof to assoc prof, 64-80, PROF BIOL, ST CLOUD STATE UNIV, 80- *Concurrent Pos:* Aquatic insect consult; res assoc, Minn Sci Mus, 77- *Mem:* Coleopterists Soc; Nature Conservancy. *Res:* Taxonomy and life history of aquatic insects principally Coleoptera and Hemiptera. *Mailing Add:* Dept Biol Sci St Cloud State Univ St Cloud MN 56301

GUNDERSEN, ROY MELVIN, b Ecorse, Mich, May 1, 30. MATHEMATICS. *Educ:* Ill Inst Technol, BS, 51; Univ Wis, MS, 52; Brown Univ, PhD(appl math), 56. *Prof Exp:* Asst math, Univ Wis, 51-52; asst appl math, Brown Univ, 53-55; mathematician, Univ Chicago, 55; dir res, Univ Paris, 55-56; res engr, Boeing Aircraft Co, 56-57; from asst prof to assoc prof math, Ill Inst Technol, 57-63; prof math, Univ Wis-Milwaukee, 63-85. *Concurrent Pos:* Consult, Armour Res Found, 57 & 59-60 & Bendix Systs Div, Bendix Corp, 60-61; NSF res grant, 58-60 & 68-72; vis assoc prof, Math Res Ctr, US Army, Univ Wis, 61-62. *Mem:* Soc Natural Philos; Soc Indust & Appl Math. *Res:* Applied mathematics; compressible fluid flow; partial and ordinary differential equations; magnetohydrodynamics. *Mailing Add:* 5801 N Sheridan Rd Chicago IL 60660

GUNDERSON, DONALD RAYMOND, b La Jolla, Calif, Jan 3, 42; m 85; c 2. FISH BIOLOGY, RESOURCE SURVEYS. *Educ:* Mont State Univ, BS, 63, MS, 66; Univ Wash, PhD(fisheries), 76. *Prof Exp:* Fisheries biologist, Wash State Dept Fisheries, 67-75; res biologist, Nat Marine Fisheries Serv, 75-78; asst prof, 78-81, ASSOC PROF MARINE FISHERIES, UNIV WASH, 81- *Concurrent Pos:* Consult, Int Groundfish Comn (US & Can), 71-; mem, Int N Pac Fisheries Comn, Northeast Pac Groundfish Tech Subcomt, 71-75, Pac Fishery Mgt Coun, Groundfish Plan Develop Team, 76-79, Army Corps Engrs, 84-85, Environ Protection Agency, 88 & Exxon, 90. *Mem:* Am Fisheries Soc; Am Inst Fisheries Res Biologists; Sigma Xi. *Res:* Population ecology and population dynamics of marine fish; estimation of stock abundance using trawl and hydroacoustic surveys. *Mailing Add:* Fisheries Res Inst Univ Wash Seattle WA 98195

GUNDERSON, HANS MAGELSSEN, b Walker, Minn, Oct 10, 38; m 69; c 1. BIOCHEMISTRY. *Educ:* St Olaf Col, BA, 60; Univ NH, MST, 69; Univ NDak, PhD(biochem), 74. *Prof Exp:* Teacher high schs, Minn, New Guinea & Wis, 60-68; teacher chem, Bemidji State Univ, 68-71; asst prof, 74-80, ASSOC PROF CHEM, NORTHERN ARIZ UNIV, 80- *Concurrent Pos:* Sigma Xi grad res award, 74. *Mem:* Am Chem Soc; Sigma Xi. *Res:* Biochemistry of exercise. *Mailing Add:* Box 5698 Northern Ariz Univ Flagstaff AZ 86011

GUNDERSON, HARVEY LORRAINE, b Gary, Minn, June 11, 13; m 50; c 3. ZOOLOGY, MAMMALOGY. *Educ:* Concordia Col, Moorhead, Minn, BA, 35; Univ Minn, Minneapolis, MS, 48; Univ Mich, PhD(zool), 62. *Prof Exp:* At W M Welch Sci Co, 36-39; mus asst, Mus Natural Hist, Univ Minn, 40-42 & 46-48, from asst scientist to assoc scientist, 48-64, cur mammals, 58-64; assoc dir & cur zool & mus records, State Mus, 64-78, from assoc prof zool & physiol to prof zool, Univ Nebr, Lincoln, 64-78; RETIRED. *Concurrent Pos:* Regional ed, Audubon Field Notes, 47-57. *Mem:* Am Soc Mammalogists; Wildlife Soc; Wilson Ornith Soc; Sigma Xi; Explorers Club. *Res:* Mammal distribution and ecology, particularly factors which influence distribution; population dynamics of small mammals; bison behavior, especially communication. *Mailing Add:* State Mus Morrill Hall Univ Nebr Lincoln NE 68508

GUNDERSON, LESLIE CHARLES, b Rahway, NJ, Aug 1, 35; m 57; c 3. OPTICAL FIBER TECHNOLOGY. *Educ:* Stevens Inst Technol, MechE, 57, MS, 60; Columbia Univ, MS, 64; NC State Univ, PhD(elec eng), 71; Harvard Univ, PMD, 71. *Prof Exp:* ITT Fed Labs, Int Tel & Tel Corp, NJ, 57-64; group leader, Electrooptics & Microwave Dept, Electronic Res Lab, Corning Glass Works, 64-66, supvr, Microwave Dept, 66-67, mgr, 67-72, dir phys res, 72-76, dir optical waveguides technol, 76-80, dir develop, 80-82, dir bus develop, 82-88; PRES & CHIEF EXEC OFFICER, METRICOR, 87- *Concurrent Pos:* PCO bd dir, La, 83-88, Raucom bd dir, Bolder Co, 86-88. *Mem:* Sr mem Inst Elec & Electronics Engrs; Optical Soc Am; Am Phys Soc; Am Mgt Asn. *Res:* Electromagnetics; microwave components; dielectric antennas; electronic materials and components; materials and processes for low loss optical fibers; propagation studies; optical components; communications systems. *Mailing Add:* Metricor 18800 142 Ave NE Woodinville WA 98072

GUNDERSON, NORMAN GUSTAV, b Schenectady, NY, May 29, 17; m 45; c 3. MATHEMATICS. *Educ:* NY State Teachers Col, Albany, AB, 37; Cornell Univ, AM, 41, PhD(math), 48. *Prof Exp:* Teacher, NY Pub Sch, 37-40; asst & instr math, Cornell Univ, 40-46; from instr to assoc prof, 46-66, assoc prof educ, 61-66, prof, 66-80, EMER PROF MATH & EDUC, UNIV ROCHESTER, 80- *Concurrent Pos:* Lectr math, Nazareth Col, 80-83. *Mem:* Am Math Soc; Math Asn Am. *Res:* Number theory; geometry; mathematics curriculum; teaching of mathematics. *Mailing Add:* 2085 W Henrietta Rd Rochester NY 14623

GUNDERSON, NORMAN O, b Laramie, Wyo, Sept 10, 18; m 42; c 2. ENGINEERING, URBAN STUDIES. *Educ:* Univ Wyo, BS, 39, MS, 47; Stanford Univ, CE, 55. *Prof Exp:* Jr engr, Hwy Dept, State of Wyo, 39 & US Soil Conserv Serv, 39-41; asst engr, US Bur Reclamation, 41; asst prof civil eng, Univ Hawaii, 47-48; from asst prof to prof eng, San Jose State Col, 48-56, dean, Sch Eng, 56-70, prof gen eng & cybernet syst, 77-80, dir, Cybernet Systs Grad Prog, 70-80, RETIRED. *Concurrent Pos:* Consult, City of San Jose, 55- *Mem:* Am Soc Civil Engrs; Am Soc Eng Educ; Nat Soc Prof Engrs. *Res:* Academic administration, particularly engineering role in interpretation of effect of computer controlled automatic production systems on mankind; city planning. *Mailing Add:* 1465 Weaver Dr San Jose CA 95125

GUNDLACH, ROBERT WILLIAM, b Buffalo, NY, Sept 7, 26; m 50; c 3. PHYSICS. *Educ:* Univ Buffalo, BA, 49. *Prof Exp:* Asst physics, Univ Buffalo, 49-51; physicist, Durez Plastics Co, NY, 51-52; physicist, Haloid Co, 52-55, proj leader, 55-57, res assoc, 57-59, sr res assoc, Haloid Xerox, Inc, 59-60, sr scientist, Xerox Corp, 60-63, prin scientist & consult, 63-66, res fel, 66-78, SR RES FEL, 78-, MGR, EXITE LABS, XEROX CORP, 89- *Honors & Awards:* Ives Jour Award, 64; Inventors Award, Rochester Patent Law Asn, 74; Kosar Award, Soc Photog Scientists & Engrs, 76. *Mem:* Electrostatic Soc Am (pres); Soc Photog Scientists & Engrs. *Res:* Exploratory xerographic research; physical imaging methods, electrophotography. *Mailing Add:* Exite Lab 317 Main St East Rochester NY 14445

GUNDY, SAMUEL CHARLES, b Reading, Pa, Mar 28, 18; m 43; c 2. BIOLOGY. *Educ:* Pa State Teachers Col, Kutztown, BS, 46; Cornell Univ, MS, 53. *Prof Exp:* Specialist nature, Reading Recreation Dept, 40-41; teacher entom, vert zool & ornith, Reading Pub Mus & Art Gallery, 41; teacher, Pa Pub Sch, 46-52; from asst dir to dir, Reading Pub Mus & Art Gallery, 53-67; asst prof, 67-81, EMER PROF BIOL, KUTZTOWN STATE COL, 81- *Mem:* Am Soc Ichthyologists & Herpetologists; Wilson Ornith Soc; Am Asn Mus. *Res:* Herpetology; ornithology; geology; mineralogy; paleontology. *Mailing Add:* 409 Harvard Blvd Lincoln Park Reading PA 19609

GUNION, JOHN FRANCIS, b Washington, DC, July 21, 43; m 67; c 2. HIGH ENERGY PHYSICS, THEORETICAL PHYSICS. *Educ:* Cornell Univ, BS, 65; Univ Calif, San Diego, MS, 68, PhD(physics), 70. *Prof Exp:* Res assoc, Stanford Linear Accelerator, 70-72, Mass Inst Technol, 72-73; from asst prof to assoc prof, Univ Pittsburgh, 73-75; assoc prof, 75-78, PROF HIGH ENERGY PHYSICS, UNIV CALIF, DAVIS, 78- *Concurrent Pos:* NSF res grant, 73-75; Sloan Found fel, 76-80; Dept Energy grant, 75-; NSF Int Collab grant, Univ Warsaw, 79-82; vis prof, Univ Calif, San Diego, 79, Inst Theoret Physics, 79, 88, Stanford Linear Accelerator Ctr, 80, Univ Ore, 85; ed numerous conf proc. *Mem:* Am Inst Physics. *Res:* High transverse momentum phenomena; quantum chromodynamics; bound state phenomena; unified theories of the weak and electromagnetic interactions; supersymmetry; super-conducting, super-collider phemenology; superstring phenomenology; author of more than 160 publications in scientific journals. *Mailing Add:* Dept Physics Univ Calif Davis CA 95616

GUNJI, HIROSHI, b Bangkok, Thailand, June 14, 28; m 62; c 1. MATHEMATICS. *Educ:* Univ Tokyo, BA, 54, MA, 56; Johns Hopkins Univ, PhD(math), 62. *Prof Exp:* Instr math, Cornell Univ, 62-64; asst prof, Univ Sask, 64-66; from asst prof to assoc prof, 66-77, PROF MATH, UNIV WIS, MADISON, 77- *Concurrent Pos:* NSF math res grants, 62-64 & 66-69. *Res:* Diophantine problem of eliptic curves. *Mailing Add:* 621 Van Vleck Hall Univ Wis Madison 480 Lincoln Dr Madison WI 53706

GUNKEL, RALPH D, physiological optics, for more information see previous edition

GUNKEL, ROBERT JAMES, b Chicago, Ill, May 13, 25; m 49; c 5. AERODYNAMICS. *Educ:* Univ Mich, BS, 46, MS, 47. *Prof Exp:* Design specialist, Missile & Space Systs Div, 47-53, asst supvr missile aerodyn, 53-56, chief aerodyn & astrodyn 56-61, chief res engr, 61-62, adv prog mgr space launch vehicles, 62, mgr sprint systs develop, 62-63, dir prog 437, 63-64, medium launch vehicles, 64, adv manned spacecraft systs, 64-66, dir adv spacecraft & launch systs, Missile & Space Systs Div, 66-68, dir systs develop & integration, Manned Orbiting Lab Subdiv, 68-69, adv space & launch systs, 69-70, dir space shuttle booster, Astronaut Co-West, 70-72, dep dir data processing, 72-76, dir prog eng, space sta design & advan space systs, 76-80, PROG MGR, GPS/PAM DII PROG, MCDONNELL-DOUGLAS ASTRONAUT CO, 80- *Concurrent Pos:* Mem res & tech adv comt space vehicles, NASA, 68-76; mem, AIAA Launch Vehicles Tech Comt, Spacecraft TC, Space Systs TC & Space Opers & Support TC. *Mem:* AAAS; Am Astron Soc; Am Inst Aeronaut & Astronaut; assoc fel Brit Interplanetary Soc; Sigma Xi. *Res:* Calculations of variational approach to flight mechanics; linearized supersonic wing theory; program management. *Mailing Add:* 1227 Devon Lane Newport Beach CA 92660

GUNKEL, WESLEY W, b Hope, NDak, Oct 17, 21; m 45; c 2. AGRICULTURAL ENGINEERING. *Educ:* NDak State Univ, BSc, 47; Iowa State Univ, MS, 48; Mich State Univ, PhD(agr eng), 57. *Prof Exp:* From asst prof to assoc prof, 48-59, PROF AGR ENG, NY STAT COL AGR & LIFE SCI, CORNELL UNIV, 59-; COORDR GRAD INSTR, 66-69, 88- *Concurrent Pos:* Vis prof & adv, Univ Nigeria, 62-63; consult, Ministry of Agr, Ghana, 65 & Dole Pineapple Co, Hawaii, 69-70; vis scientist, Int Rice Res Inst, Philippines, 76-77. *Mem:* AAAS; fel Am Soc Agr Engrs; Am Soc Eng Educ; Am Chem Soc; Coun for Agr Sci Technol; Nat Safety Coun. *Res:* Energy in agriculture; pest control equipment and methods; harvesting machinery; design of specialized farm and industrial equipment; farm power; agricultural machinery safety; robotics and machine vision. *Mailing Add:* Dept Agr Eng-Riley-Robb Hall Cornell Univ Ithaca NY 14853-5701

GUNN, ALBERT EDWARD, b Port Washington, NY, Oct 31, 33; m 68; c 6. MEDICAL ETHICS, MEDICAL LEGAL PROBLEMS. *Educ:* Fordham Col, BS, 55; Fordham Law Sch, LLB, 58; Nat Univ Ireland, MB, BCh, BAO, 67; LRCP(L) & MRCS(E), 67; Am Bd Int Med, cert, 77. *Prof Exp:* Asst dir govt relations, AMA, 72-74; med dir geriat servs, Suffolk County Health Servs Dept, NY, 74-75; med dir, Rehab Ctr, M D Anderson Hosp & Tumor Inst, 75-88, ASSOC PROF INTERNAL MED, MED SCH, UNIV TEX, HOUSTON, 80-, ASSOC PROF PUB HEALTH ADMIN, SCH PUB HEALTH, 81- *Concurrent Pos:* Mem, Nat Adv Health Coun, HEW, 74-77 & Adv Comt, Nat Inst Law Enforcement & Criminal Law, US Dept Justice, 74-76; assoc prof geriat, Syst Cancer Ctr, M D Anderson Hosp & Tumor Inst, Univ Tex, Houston, 79-, asst dean admis, Med Sch, 79-84, assoc dean admis, 84-; mem, Bd Regents, Nat Libr Med, Nat Insts Health, Dept Health Human Servs, 83-87, chmn, 86-87, chmn lit selection tech rev comt, 88- 91; consult, Ctrs Dis Control, 85, Legal Servs Corp, 87. *Mem:* Fel Am Col Physicians; Soc Surg Oncol. *Res:* Geriatrics, particularly in the area of medical and legal problems involved in mental impairment in the elderly. *Mailing Add:* M D Anderson Hosp Univ Tex 1515 Holcombe Blvd Houston TX 77030

GUNN, CHARLES ROBERT, b Columbus, Ohio, June 1, 27; m 48; c 2. TAXONOMY, BOTANY. *Educ:* Iowa State Univ, BS, 50, PhD(syst bot), 65; Univ Louisville, MS, 58. *Prof Exp:* Seed technologist, Ross Seed Co, 51-60; RES PLANT TAXONOMIST, AGR RES SERV, PLANT GENETICS GERMPLASM INST, USDA, 65- *Mem:* Am Soc Plant Taxon; Int Asn Plant Taxonomists; Asn Off Seed Anal. *Res:* Seed and fruit distribution by ocean currents; seed and fruits of the Fabaceous; the use of computers in the identification of seeds and fruits of all families. *Mailing Add:* US Nat Seed Herbarium Bldg 265 BARC-East Beltsville MD 20705

GUNN, CHESTERFIELD GARVIN, JR, b Bethany, Mo, Sept 2, 20; m 46; c 3. INTERNAL MEDICINE, NEUROPHYSIOLOGY. *Educ:* Univ Mo, AB, 42; Yale Univ, MD, 50. *Prof Exp:* Intern & resident med, Med Ctr, Univ Kans, 50-54; asst res anatomist, Sch Med, Univ Calif, Los Angeles, 54-56; asst prof med, physiol & prev med, 56-61, asst prof psychiat, neurol & behav sci, 57-62, assoc prof med & physiol, 61-67, PROF PHYSIOL, SCH MED, 67-, prof med, 68-, prof biol psychol, 71, Regents Prof Med, 85-88, EMER REGENTS PROF MED, UNIV OKLA, 88- *Concurrent Pos:* Res fels, Am Col Physicians, 53 & Am Heart Asn, 54-56; attend physician, Univ Hosps, 56-; consult physician, Okla VA Hosp, 61- *Mem:* Am Psychosom Soc; Am Col Cardiol; Am Acad Neurol; Cent Soc Clin Res; Sigma Xi. *Res:* Central nervous system mechanisms in normal and pathological cardiovascular function. *Mailing Add:* 5301 W Shawnee St No R3 Muskogee OK 74401

GUNN, JAMES EDWARD, b Livingstone, Tex, Oct 21, 38; m 61. ASTROPHYSICS. *Educ:* Rice Univ, BS, 61; Calif Inst Technol, PhD(astrophysics), 66. *Prof Exp:* Sr scientist lunar & planetary sci, Jet Propulsion Lab, 66-68; asst prof astrophys, Princeton Univ, 68-70; from assoc prof to prof astrophys, Calif Inst Technol, 70-77, prof physics, 77-80; EUGENE HIGGINS PROF ASTRON, PRINCETON UNIV, 80- *Concurrent Pos:* Staff mem, Hale Observ, 70-, affil, 77-; mem bd dirs, Astron Soc Pac; John & Katherine MacArthur fel, 83. *Honors & Awards:* Heinemann Prize, Am Astron Soc, 88. *Mem:* Nat Acad Sci; Am Astron Soc. *Res:* Theoretical and observational cosmology; astronomical instrumentation; structure and evaluation of galaxies and clusters; star formation. *Mailing Add:* Dept Astrophys Princeton Univ Princeton NJ 08544-1001

GUNN, JOHN BATTISCOMBE, b Cairo, Egypt, May 13, 28; wid; c 2. ENGINEERING PHYSICS. *Educ:* Cambridge Univ, BA, 48. *Prof Exp:* Res engr, Elliott Bros Ltd, 48-53; jr res fel, Radar Res Estab, 53-56; asst prof physics, Univ BC, 56-59; staff physicist, 59-71, mem corp tech comt, 71-72, IBM FEL, RES CTR, IBM CORP, 71- *Honors & Awards:* Liebmann Prize, Inst Elec & Electronics Engrs, 69; John Scott Award, 71; Valdemar Poulsen Gold Medal, Danish Acad Tech Sci, 72. *Mem:* Nat Acad Eng; fel Inst Elec & Electronics Engrs; fel Am Phys Soc; fel AAAS; fel Am Acad Arts & Sci. *Res:* Physics of semiconductors; semiconductor devices; programming language; automotive engineering. *Mailing Add:* Res Ctr IBM Corp PO Box 218 Yorktown Heights NY 10598

GUNN, JOHN MARTYN, b Lulworth Cove, Eng, June 27, 45. BIOCHEMISTRY. *Educ:* Inst Biol, MI Biol, 69; Univ Sheffield, PhD(biochem), 72. *Prof Exp:* Res technician biochem, Fison's Pharmaceut Ltd, UK, 63-66, Allen & Hanburys Ltd, UK, 66-69; from asst prof to assoc prof, 76-85, PROF BIOCHEM, TEX A&M UNIV, 85- *Concurrent Pos:* Fel, Temple Univ, 72-76; prin investr, USPHS grants, 78-86; consult, Spec Study Sect, NIH, 78- *Mem:* AAAS; Biochem Soc; Am Soc Biol Chemists; Am Soc Cell Biol; Proc Soc Exp Biol & Med. *Res:* Metabolic regulation; protein turnover; function and mechanism of polypeptide hormones. *Mailing Add:* Dept Biochem & Biophys Tex A&M Univ College Station TX 77843-2128

GUNN, JOHN WILLIAM, JR, b Lynn, Mass, Feb 23, 28; m 60; c 6. FORENSIC SCIENCE, SCIENCE ADMINISTRATION. *Educ:* Boston Col, BS, 51. *Prof Exp:* Lab shift supvr, Liberty Powder Defense Corp, 51-54; spec agent, Fed Bur Invest, Dept Justice, 54-62; lab supvr, Chas Pfizer & Co, Inc, 62-63; sr staff chemist, Appl Physics Lab, Johns Hopkins Univ, 63-66; chief investigative serv br, Bur Drug Abuse Control, Food & Drug Admin, 66-68; chief labs, Bur Narcotics & Dangerous Drugs, 68-73, actg dir Off Sci & Technol, Drug Enforcement Admin, 74-77, dir, 77-90, DEP ASST ADMINR, OFF ADMIN, DRUG ENFORCEMENT ADMIN, 90- *Concurrent Pos:* Chmn sci & technol subcomt, Cabinet Comt Int Narcotics Control, 74-76; consult, UN Div on Narcotics, 85 & 87. *Honors & Awards:* Distinguished Serv Award, Am Acad Forensic Sci, 85. *Mem:* Am Soc Crime Lab Dirs; fel Am Acad Forensic Sci; Int Asn Chiefs Police; Am Chem Soc; Int Asn Forensic Sci. *Res:* Law enforcement technology; forencic sciences; narcotics and dangerous drugs. *Mailing Add:* Drug Enforcement Admin Off Admin Washington DC 20537

GUNN, KENRICK LEWIS STUART, cloud physics; deceased, see previous edition for last biography

GUNN, ROBERT BURNS, TRANSPORT KINETICS, BIOPHYSICS. *Educ:* Harvard Univ, MD, 66. *Prof Exp:* PROF & CHMN DEPT PHYSIOL, SCH MED, EMORY UNIV, 81- *Mailing Add:* Dept Physiol Sch Med Emory Univ Physiol Bldg Rm 149 Atlanta GA 30322

GUNN, ROBERT DEWEY, b Leninakhan, Armenia, May 29, 28; US citizen; m 54; c 2. CHEMICAL ENGINEERING. *Educ:* Kans State Univ, BS, 50; Univ Calif, Berkeley, MS, 58, PhD(chem eng), 67. *Prof Exp:* Geophys engr, Robert H Ray Co, Saudi Arabia, 51-53; petrol engr, Mobil Int Oil Co, 57-62; asst prof chem eng, Univ Tex, Austin, 67-71; assoc prof, 71-76, PROF CHEM ENG, UNIV WYO, 76- *Concurrent Pos:* Distinguished guest prof, Tech Univ Aachen, WGermany. *Mem:* AAAS; fel Am Inst Chem; Soc Petrol Engrs; Am Chem Soc; Am Inst Chem Engrs. *Res:* Mass transport in porous material; thermodynamics and high pressure phase equilibria; underground coal gasification; permafrost. *Mailing Add:* Dept Chem Eng Univ Wyo Laramie WY 82070

GUNN, WALTER JOSEPH, b Harrison, NJ, Dec 9, 35; m 60; c 4. ENGINEERING, PSYCHOACOUSTICS. *Educ:* US Merchant Marine Acad, BS, 59; Univ Louisville, MA, 67, PhD(psychol), 69. *Prof Exp:* Asst prof speech & hearing, Univ Cincinnati, 69-70; res assoc occup med, Sch Pub Health, Columbia Univ, 70-71; asst acad dean, US Merchant Marine Acad, 71-73; head, Psychoacoust Sect, NASA Langley Res Ctr, 73-76; dir res & eval, bur health educ, 76-82, SR RES PSYCHOLOGIST, CID, CTR DIS CONTROL, 82- *Concurrent Pos:* Consult, Airport Residents Asn; mem, Nat Reye Syndrome Task Force; pres & owner, Arlington Assocs Inc, 86- *Honors & Awards:* Spec Recognition Award, Nat Pub Health Serv, 87. *Mem:* Nat Acad Sci; Am Psychol Asn; Acoustical Soc Am; Sigma Xi. *Res:* Human response to noise; research and evaluation of health education projects; association of aspirin and Reye Syndrome; infectious diseases in daycare centers; long latency neurological diseases associated with use of human growth hormone; chronic fatigue syndrome. *Mailing Add:* Viral Dis Div Ctr Infectious Dis Ctr for Dis Control Atlanta GA 30333

GUNNAR, ROLF MCMILLAN, b Riverside, Ill, Jan 22, 26; m; c 4. CARDIOLOGY, INTERNAL MEDICINE. *Educ:* Northwestern Univ, BS, 46, MB, 47, MS, 48, MD, 49; Am Bd Internal Med, dipl, 55; Am Bd Cardiovasc Dis, dipl, 61. *Prof Exp:* From intern to resident, Cook County Hosp, Chicago, Ill, 48-53; from clin instr to clin assoc, Univ Ill Col Med, 54-59, from asst prof to prof med, 59-72, dir cardiol sect, 68-70; Edmund F Foley prof med, Univ Ill Col Med, 71-72; clin prof, 72-73, chief, Sect Cardiol, 72-82, PROF MED, STRITCH SCH MED, LOYOLA UNIV, 73-, CHMN DEPT MED, 82- *Concurrent Pos:* Fel cardiol, New Eng Deaconess Hosp & Children's Med Ctr, 58-59; pvt pract, 54-58 & 59-62; from assoc attend physician to attend physician, Cook County Hosp, 54-, from assoc dir dept adult cardiol to dir, 59-70, dir div med, 70-71; chief, Univ Ill Med Div, 62-68, dir dept adult cardiol, 63-64; dir cardiopulmonary lab, US Vet Admin Hosp, Hines, Ill, 59, prog dir cardiol, 71-74, consult prog dir, 74-; mem bd dirs, MacNeal Mem Hosp, Berwyn, 66-; actg chief sect cardiol, Univ Ill Res & Educ Hosps, 67-68; fel coun clin cardiol, Am Heart Asn. *Mem:* AMA; Am Heart Asn; fel Am Col Physicians; fel Am Col Cardiol. *Res:* Research in the mechanisms of and hemodynamic changes in shock due to myocardial infarction and septic shock; research in vectorcardiography including development and testing of criteria for diagnosis of myocardial infarction. *Mailing Add:* Loyola Univ Sch Med 2160 S First Ave Maywood IL 60153

GUNNER, HAIM BERNARD, b Ottawa, Ont, June 18, 24; m 51; c 2. ENVIRONMENTAL BIOLOGY. *Educ:* Univ Toronto, BSA, 46; Univ Man, MSc, 48; Cornell Univ, PhD(soil sci), 62. *Prof Exp:* Field crops supvr, Sasa Coop Farm, Israel, 49-51, farm mgr, 51-53, soil conserv supvr, 53-55; asst coordr agr res, Res Coun Israel, Jerusalem, 55-56, coordr, 56-57; res asst soil microbiol, Cornell Univ, 57-61; res officer, Microbiol Res Inst, Can Dept Agr, Ont, 61-63; from asst prof to assoc prof soil microbiol, 63-71, co-dir, Tech Guid Ctr Indust Environ Control, 68-72, assoc dir res, Ctr Int Agr Studies, 69-72, PROF ENVIRON MICROBIOL & ACTG HEAD DEPT ENVIRON SCI, UNIV MASS, AMHERST, 71- *Concurrent Pos:* NSF res grant, 64-66; USPHS grant, 66-69; Off Water Resources Res, Dept Interior res grant, 67-70, 72-76 & 76-80; chmn soil biol sect, Northeast Soil Res Comt, 69; secy, Northeast Regional Res Comt Nitrogen Transformation Soil & Water, 69; past chmn, Northeast Regional Res Comt Pesticides; vis prof, Univ Tel Aviv, 71-; consult, Israel Environ Protection Serv, 71-; US Army Corps Engrs grant, 80-89; found, chmn & chief scientist, EcoSci Labs, Inc, 82-; vis prof, Harvard Univ, 87-88. *Mem:* Am Soc Microbiol; Can Soc Microbiol; Am Soc Agron; Soil Sci Soc Am; Int Orgn Biol Control Noxious Animals & Plants. *Res:* Investigations of inorganic transformations of microbial origin; biochemical and physiological aspects of microfloral interrelations with specific reference to the biological control of soil-borne pathogens; microbial degradation of industrial effluents; microbial ecosystem stress in soil and water; microbial control of insect pests; microbial control of aquatic weeds. *Mailing Add:* Dept Pub Health Univ Mass Amherst MA 01003

GUNNESS, R(OBERT) C(HARLES), b Fargo, NDak, July 28, 11; m 36; c 3. CHEMICAL ENGINEERING. *Educ:* Univ Mass, BS, 32; Mass Inst Technol, MS, 34, ScD(chem eng), 36. *Prof Exp:* From instr to asst prof chem eng, Mass Inst Technol, 35-38; group leader, Standard Oil Co, Ind, 38-43, asst dir res, 43- 45, assoc dir, 45-47, mgr, 47-52, asst gen mgr mfg, 52-54, gen mgr, Supply & Transportation Dept, 54-56, exec vpres, 56-65, pres, 65-75, mem bd dir, 53-75, corp dir, 75; RETIRED. *Concurrent Pos:* Vchmn res & develop bd, Dept Defense, 51. *Mem:* Am Chem Soc; Am Inst Chem Engrs. *Res:* Distillation; heat transfer; fluid catalytic cracking; petroleum refinery processing techniques; performance characteristics of a commercial stabilizer for absorption naphtha. *Mailing Add:* PO Box 538 Rancho Sante Fe CA 92067

GUNNING, BARBARA E, nutrition, biochemistry, for more information see previous edition

GUNNING, GERALD EUGENE, b Tamms, Ill, Apr 1, 32; m 62; c 2. FISH BIOLOGY. *Educ:* Southern Ill Univ, BA, 54, MS, 55; Ind Univ, PhD(zool), 59. *Prof Exp:* Asst zool, Southern Ill Univ, 53-55; asst, Ind Univ, 56-57; from instr to assoc prof, 59-69, PROF BIOL, TULANE UNIV, 69- *Concurrent Pos:* Biol consult, MacMillan Bloedel & Gaylord Container Corp, 63-; consult, Int Paper Co, 77, Georgia-Pacific Corp, 83. *Mem:* Am Fisheries Soc; Am Soc Ichthyologists & Herpetologists. *Res:* Fishery behavior; sensory physiology; swamp ecology; life history of fishes; freshwater fish populations; water pollution. *Mailing Add:* Dept Ecol Evolution & Organismal Biol Tulane Univ New Orleans LA 70118

GUNNING, HARRY EMMET, b Toronto, Ont, Dec 16, 16; m 43; c 1. PHYSICAL CHEMISTRY, PHOTOCHEMISTRY. *Educ:* Univ Toronto, BA, 39, MA, 40, PhD(phys chem), 42. *Hon Degrees:* DSc, Univ Guelph, 69 & Queen's Univ, 74; LLD, Univ Victoria, 78, Univ Alta, 83, Univ Sask, 84 & Simon Fraser Univ, 84. *Prof Exp:* Res fel chem, Nat Res Coun Can, 42-43, from asst res chemist to assoc res chemist, 43-46; asst prof chem, Univ Rochester, 46-48; from asst prof to prof, Ill Inst Technol, 48-57; prof & chmn dept, 57-74, Killam mem prof, 68-82, pres univ, 74-79, EMER PROF CHEM, UNIV ALTA, 82- *Concurrent Pos:* Mem, Nat Res Coun Can, 61-; consult, Syncrude Can Ltd, Imp Oil Enterprises Ltd & Dunlop Can; adv indust develop div, City of Edmonton. *Honors & Awards:* Medal, Chem Inst Can, 67; Officer of the Order of Can, 79. *Mem:* Fel AAAS; fel Royal Soc Can; fel Chem Inst Can (pres-elect, 72-74, pres, 73-74); fel NY Acad Sci; Am Chem Soc; Faraday Soc; Am Phys Soc; Can Asn Univ Teachers; Sigma Xi; corresp mem Europ Acad Sci, Arts & Lett. *Res:* Chemical kinetics; photochemistry; isotope separation; reactions of atomic sulphur; sulphur chemistry; mercury photo sensitization; flash photolysis and photo-ionization; kinetic mass spectroscopy; hydrocarbon chemistry. *Mailing Add:* Ste E-344 Chem Ctr Univ Alta Edmonton AB T6G 2E1 Can

GUNNING, ROBERT CLIFFORD, b Colo, Nov 27, 31; m 66. COMPLEX ANALYSIS, ALGEBRAIC GEOMETRY. *Educ:* Univ Colo, BA, 52; Princeton Univ, PhD(math), 55. *Prof Exp:* NSF fel, Univ Chicago, 55-56; from asst prof to assoc prof, 57-66, chmn dept, 76-79, PROF MATH, PRINCETON UNIV, 66-, DEAN FAC, 89- *Concurrent Pos:* Sloan fel, 58-61, Cambridge Univ, 59-60; vis prof, Munich, 67 & Oxford Univ, 68 & 80. *Mem:* Am Math Soc; fel AAAS. *Res:* Complex analysis, particularly functions of several complex variables, Riemann surfaces and theta functions; transcendental algebraic geometry. *Mailing Add:* Dept Math Princeton Univ Princeton NJ 08544

GUNNINK, RAYMOND, b Chandler, Minn, June 28, 32; m 54; c 5. PHYSICAL CHEMISTRY. *Educ:* Calvin Col, AB, 54; Purdue Univ, PhD(phys chem), 59. *Prof Exp:* Res scientist, Res Ctr, US Rubber Co, 59-63; GROUP LEADER, NUCLEAR CHEM DIV, LAWRENCE LIVERMORE LAB, UNIV CALIF, 63- *Mem:* Am Phys Soc; Inst Nuclear Mats Mgt. *Res:* Nuclear decay schemes; computer analysis of X- and gamma ray spectra; activation analysis; absolute counting techniques; gamma ray spectroscopy; plutonium safeguards. *Mailing Add:* Lawrence Livermore Nat Lab L233 PO Box 808 Livermore CA 94550

GUNNISON, ALBERT FARRINGTON, b Crown Point, NY, Oct 20, 39; m 73. ENVIRONMENTAL MEDICINE, TOXICOLOGY. *Educ:* Cornell Univ, BS, 64, MS, 66; Pa State Univ, PhD(entom, air pollution), 70. *Prof Exp:* Res asst entom, Cornell Univ, 64-66; USPHS fel environ med, 70-71, assoc res scientist, 71-72, asst prof environ med, 72-77, res assoc prof, 77-79, ASSOC PROF, NY UNIV MED CTR, 79- *Mem:* Sigma Xi; Soc Toxicol. *Res:* Toxicology of environmental contaminants with emphasis on food additives and air pollution. *Mailing Add:* Dept Environ Med NY Univ Med Ctr 550 First Ave New York NY 10016

GUNSALUS, IRWIN CLYDE, b Sully Co, SDak, June 29, 12; m 35, 51, 70; c 7. BIOCHEMISTRY. *Educ:* Cornell Univ, BS, 35, MS, 37, PhD(bact), 40. *Prof Exp:* From instr to prof, Cornell Univ, 37-47; prof, Ind Univ, 47-50; prof microbiol, 50-55, head dept biochem, 55-65, ctr adv study, 65-66, EMER PROF BIOCHEM, UNIV ILL, URBANA-CHAMPAIGN, 55- *Concurrent Pos:* Guggenheim fel, 49, 59, 68; asst secy general, UN; dir, Int Ctr for Genetic Eng & Biotechnol, UN Indust Develop Orgn, Trieste, Italy & New Delhi, 86- *Honors & Awards:* Mead Johnson Award, 46. *Mem:* Nat Acad Sci; AAAS; Am Soc Microbiol; Am Chem Soc; Am Soc Biol Chem. *Res:* Biological catalysis and regulation; mechanism of chemical transformations and energy transfer; formation of essential metabolites including pyridoxal phosphate and lipoic acid; oxidation and oxygenation reactions and energy transfer. *Mailing Add:* 420 Roger Adams Lab Univ Ill Urbana IL 61801

GUNSALUS, ROBERT PHILIP, b Ithaca, NY, Aug 24, 47. GENETIC REGULATION, ANAEROBIC MICROBIOLOGY. *Educ:* SDak State Univ, BS, 70; Univ Ill, MS, 72, PhD(microbiol), 77. *Prof Exp:* Fel molecular biol, Stanford Univ, 78-81; asst prof microbiol, 81-87, assoc mem, Molecular Biol Inst, 81-89, ASSOC PROF MICROBIOL, UNIV CALIF, LOS ANGELES, 87-, MEM, MOLECULAR BIOL INST, 90- *Concurrent Pos:* chmn, Gordon Conf Methanogenesis, 90. *Mem:* Am Chem Soc; Am Soc Microbiol; AAAS; Am Soc Molecular Biol. *Res:* Gene organization and regulation in the procaryotic and archaebacterial organisms; molecular aspects of adenosine triphosphatase, (unc) and fumarate reductase and operon control; control of tryptophan biosynthesis; repressor-operator interaction. *Mailing Add:* Dept Microbiol 5304 LS Univ Calif Los Angeles CA 90024

GUNSHOR, ROBERT LEWIS, b New York, NY, Oct 28, 35; m 59; c 3. ELECTRICAL ENGINEERING, SEMICONDUCTOR PHYSICS. *Educ:* New York Univ, BEE, 58; Union Col, MSE, 62; Rensselaer Polytech Inst, PhD(elec eng), 65. *Prof Exp:* Engr, Microwave Assocs, 58-60; asst prof, Union Col, 60-63 & Cornell Univ, 65-66; assoc prof, Rensselaer Polytech Inst, 66-68; prof elec eng, 68-69, DISTINGUISHED PROF ELEC ENG, PURDUE UNIV, 89- *Concurrent Pos:* Consult & vpres, Technol Assocs, 76- *Mem:* Fel Inst Elec & Electronics Engrs; fel Am Phys Soc; Sigma Xi. *Res:* Surface acoustic wave devices; solid state microwave devices; semiconductor devices; molecular beam epitaxy of semiconductors. *Mailing Add:* Sch of Elec Eng Purdue Univ West Lafayette IN 47907

GUNST, SUSAN JANE, MUSCLE & PULMONARY PHYSIOLOGY. *Educ:* Johns Hopkins Univ, PhD(physiol), 79. *Prof Exp:* ASST PROF PHYSIOL, MAYO CLIN & FOUND, 83- *Res:* Physiology of airway smooth muscle. *Mailing Add:* 1354 Woodland Dr SW Rochester MN 55902

GUNSUL, CRAIG J W, b Seattle, Wash, May 16, 37; m 60; c 2. SOLID STATE PHYSICS. *Educ:* Reed Col, BA, 63; Univ Del, MS, 66, PhD(physics), 69. *Prof Exp:* Asst prof, 69-74, ASSOC PROF PHYSICS, WHITMAN COL, 74- *Res:* Luminescence. *Mailing Add:* Dept Physics Whitman Col 345 Boyer Ave Walla Walla WA 99362

GUNTER, BOBBY J, b Wilson, Okla, July 25, 41. ANALYTICAL CHEMISTRY, EXPERIMENTAL MEDICINE. *Educ:* Southeastern State Col, BS, 63; Univ Okla, MS, 64, PhD(prev med), 67. *Prof Exp:* Chemist, City of Norman, Okla, 63-67; sr asst scientist, Nat Ctr Urban & Indust Health, USPHS, 67-72; regional indust hygienist, Region VIII, Dept Health, Educ & Welfare, 72-88; sr environ health sci, Westinghouse Corp, 89-90; REGIONAL MGR, OCCUP HEALTH & SAFETY, HARTFORD STEAM BOILER, ATLANTA, GA, 90- *Concurrent Pos:* Res asst physiol, Vet Admin Hosp, Oklahoma City, 66-67. *Mem:* Am Conf Govt Indust Hygienists; Am Indust Hyg Asn. *Res:* Atomic absorption techniques. *Mailing Add:* 200 Ashford Ctr N Suite 300 Atlanta GA 30338

GUNTER, CLAUDE RAY, b Benton, Ill, May 9, 39; c 4. BIOCHEMISTRY. *Educ:* Southern Ill Univ, BA, 61; Northwestern Univ, PhD(chem), 66. *Prof Exp:* M L Bender staff fel enzym, Lab of Chem, Nat Inst Arthritis & Metab Dis, NIH, 65-68; staff scientist biochem, US Army Chem Warfare Ctr, Edgewood Arsenal, 66-68; dir, Urine Chem Lab, Ames Div, Miles Lab Inc, 68-85, div sci serv & spec proj, 81-85,; CONSULT, MED DIAG TESTS, DRY REAGENT STRIP TESTS, CRG CONSULT SERV, 85- *Concurrent Pos:* Assoc fac chem dept, Ind Univ, South Bend, 75-76. *Mem:* Am Asn Clin Chem; Am Chem Soc. *Res:* Development of enzyme methods for clinical chemistry; dry reagent strip tests. *Mailing Add:* 29477 County Rd 16 Elkhart IN 46516

GUNTER, DEBORAH ANN, b Danville, Va. PHYSICS, MATHEMATICAL STATISTICS. *Educ:* Va State Col, Petersburg, BS, 69; Va State Univ, Petersburg, MEd, 75; Ill State Univ, Normal PhD(educ admin), 88. *Prof Exp:* Instr math, St Paul's Col, 69-70 & 80-82, coordr computer sci, 80-82; asst instr math, Va State Univ, 73-74; instr math, Trenton High Sch, 75-79; instr data processing computer sci, Adult Educ Bloomington, Ill, 85-86; instr math, Ill State Univ, 86-89; CHAIRPERSON, DEPT MATH, LANGSTON UNIV, 89- *Concurrent Pos:* Unit buying control, Sears, Roebuck & Co, Danville, Va, 71-73, point sales II, 74-75; admin secy, Mid Cent Commun Action Bloomington, Ill, 85-86; computer consult, Breedlove Agency State Farm, 89. *Mem:* Am Educ Res Asn; Math Asn Am. *Res:* Determine the level of and trends in the financial support to education by the government and to public education. *Mailing Add:* PO Box 308 Langston OK 73050

GUNTER, EDGAR JACKSON, JR, b Baltimore, Md, July 4, 34; m 55; c 3. ROTOR DYNAMICS, LUBRICATION THEORY. *Educ:* Duke Univ, BSME, 56; Univ Pa, MS, 61, PhD(eng mech), 65. *Prof Exp:* Compressor design engr, Dresser-Clark, 56-59; sr res eng, Franklin Inst, 62-65; assoc prof, 65-75, PROF MECH ENG, UNIV VA, 75- *Mem:* Sigma Xi; Am Soc Mech Engrs; Vibration Inst; Am Acad Mech. *Res:* Stability balancing and dynamic response of high speed turbomachinery; author or coauthor of over 100 publications; finite element analyses. *Mailing Add:* Dept Mech Eng Univ Va Charlottesville VA 22903

GUNTER, GORDON, b Goldonna, La, Aug 18, 09; m 32, 57; c 5. ZOOLOGY. *Educ:* La State Norm Col, BA, 29; Univ Tex, MA, 31, PhD(zool), 45. *Prof Exp:* Asst shrimp invest, US Bur Fisheries, 31-33, asst oil pollution invest, 34-35, spec oyster investr, 35-36; com oyster bus, Tex, 36-37; jr aquatic biologist, Debris Dam Fisheries Surv, Calif, US Engrs, 38; marine biologist, Game, Fish & Oyster Comn, Tex, 39-44; res assoc marine fisheries, Inst Marine Sci, Univ Tex, 45-49, actg dir, 49, dir, 54-55; dir, 55-71, EMER DIR & PROF ZOOL, GULF COAST RES LAB, 72- *Concurrent Pos:* Zoologist, La State Dept Conserv, 34; consult to fishery coordr, US Bur Fisheries, 41-45; vchmn comn on treatise on marine ecol, Nat Res Coun, 41-57; adv, Atlantic States Fisheries Comn, 46; vis prof zool, Univ Miami, 46-47; sr marine biologist, Scripps Inst, Univ Calif, 49; mem adv panel, La State Comn Wildlife & Fisheries, 53-55; mem, Gulf States Marine Fisheries Comn, 56-; mem bd sci consult, Wash Pollution Control Comn, 58-59; adv, Fla State Bd Conserv, 62-70; prof, Univ Miss & Miss State Univ, 68-; mem consult team, Impact of Corps Engrs Works on Coastal Ecol, 69-; consult comt deepwater ports, US Army Corps Engrs, 71- & Panama Canal Co, 73- *Mem:* Am Soc Limnol & Oceanog; Ecol Soc Am; Wildlife Soc; Am Soc Zool; Am Soc Naturalists. *Res:* Seasonal movements, distribution related to salinity, life histories and relative abundance of gulf animals; catastrophic mass mortalities; natural history of oysters, fishes, shrimp and marine mammals. *Mailing Add:* Gulf Coast Res Lab PO Box 7000 Ocean Springs MS 39564-7000

GUNTER, KARLENE KLAGES, b Pittsburgh, Pa, Aug 14, 39; m 61; c 3. MEMBRANE TRANSPORT, CALCIUM HOMEOSTASIS. *Educ:* Mass Inst Technol, BS, 61; Univ Calif, Berkeley, PhD(nuclear physics), 68. *Prof Exp:* ASST PROF BIOPHYS, UNIV ROCHESTER, 73- *Mem:* Biophys Soc; Sigma Xi; Am Fedn Scientists. *Res:* Calcium hemostasis; calcuim transport in mitochondria, submitochondrial particles and a reconstituted system; oxidative phosphorylation. *Mailing Add:* Dept Biophys Univ Rochester Rochester NY 14642

GUNTER, LAURIE M, GERONTOLOGY. *Educ:* Tenn A&I State Univ, BS, 48; Fisk Univ, MA, 52; Univ Chicago, PhD(human develop), 59. *Prof Exp:* Instr, 50-55, asst prof, 55-57, proj dir, Ment Health Training Grant, 57-58, actg dean, 57-58, dean, Meharry Med Col Sch Nursing, Nashville, 58-61; asst prof, nursing, 61-63, assoc prof, Univ Calif, Los Angeles, 63-65; prof, nursing, Ind Univ Med Ctr, Indianapolis, 65-66; assoc prof, 66-69, prof, Univ Wash, Seattle, 69-71; head, Dept Nursing, Pa State Univ, 71-75, prof, 71-87, emer prof, Nursing & Human Develop, 87; RETIRED. *Concurrent Pos:* Staff nurse, George W Hubbard Hosp, Nashville, 43-44, head nurse, 45-46, staff nurse, 46-47, supvr, 47-48. *Mem:* Inst Med-Nat Acad Sci; AAAS; Am Nurses Asn; Gerontol Soc; Am Asn Univ Prof; Am Pub Health Asn. *Res:* Numerous articles published in various journals. *Mailing Add:* 4008 47th Ave S Seattle WA 98118

GUNTER, THOMAS E, JR, b Montgomery, Ala, Mar 13, 38; m 61; c 3. SOLID STATE PHYSICS, BIOPHYSICS. *Educ:* Mass Inst Technol, BS, 60; Univ Calif, Berkeley, PhD(physics), 66. *Prof Exp:* Res asst med physics, Boeing Aircraft Co, 60; res fel biophys, Donner Lab, Lawrence Radiation Lab, 66-68; NSF fel, Phys-Chem Inst, Uppsala Univ, 68-70; asst prof radiation biol & biophys, 70-77, ASSOC PROF RADIATION BIOL & BIOPHYS, UNIV ROCHESTER, 77- *Mem:* Biophys Soc. *Res:* Applications of physical spectroscopic techniques to problems of interest in bioenergetics. *Mailing Add:* Dept Biophys Med Ctr Univ Rochester Rochester NY 14642

GUNTHER, GARY RICHARD, b Indianapolis, Ind, Mar 30, 48; m 69. BIOCHEMISTRY, CELL BIOLOGY. *Educ:* Wabash Col, BA, 69; Rockefeller Univ, PhD(biochem), 75. *Prof Exp:* Fel cell biol, Sch Med, Yale Univ, 75-78; ASST MEM BIOCHEM, ST JUDE CHILDREN'S RES HOSP, 78-; ASST PROF, DEPT PHYSICS & BIOPHYSICS, SCH MED, IND UNIV, 87- *Concurrent Pos:* USPHS fel, 75-77; asst prof anat, Ctr Health Sci, Univ Tenn, 80-84. *Mem:* Am Soc Cell Biol; Am Pancreatic Asn. *Res:* Mechanism of hormone-stimulated secretion in pancreatic acinar cells; role of calcium and cyclic nucleotides in mediating peptide hormone action; interaction of phorbol esters with normal and transformed pancreatic acinar cells. *Mailing Add:* Dept Physiol & Biophys Ind Univ Sch Med 545 Barnhill Dr Indianapolis IN 46205

GUNTHER, JAY KENNETH, b Chicago, Ill, Dec 19, 38; m 66; c 2. RESEARCH ADMINISTRATION, BIOCHEMICAL PHARMACOLOGY. *Educ:* Univ Notre Dame, BS, 61; Univ Ill, PhD(biochem), 66. *Prof Exp:* Trainee molecular biol, Univ Pa, 66-69, res assoc, 69-70; sr scientist, 70-73, sr investr, 73-77, ASSOC DIR DRUG REGULATORY AFFAIRS, MEAD JOHNSON & CO, 77- *Mem:* AAAS; Am Chem Soc. *Res:* Bacterial transformation; isolation and characterization of nucleases of Hemophilius influenzae; secretion of pancreatic hormones; control of lipid synthesis and degradation; regulatory affairs. *Mailing Add:* 409 Green Hill Dr Madison CT 06443-2305

GUNTHER, LEON, b New York, NY, Aug 22, 39; m 62, 79; c 4. THEORETICAL CONDENSED MATTER PHYSICS. *Educ:* City Col New York, BS, 60; Mass Inst Technol, PhD(physics), 64. *Prof Exp:* Res assoc physics, Mass Inst Technol, 64-65; from asst prof to assoc prof, 65-78, PROF PHYSICS, TUFTS UNIV, 78- *Concurrent Pos:* NATO fel, 66-67; vis prof, Technion & Tel Aviv Univs. *Mem:* Am Asn Univ Professors; Am Phys Soc. *Res:* Superconductivity; Brownian motion; mossbauer effect; phase transitions; magnetism; vacancy diffusion; solutions; macrscopic quantum tunneling. *Mailing Add:* Dept Physics Tufts Univ Medford MA 02155

GUNTHER , MARIAN W J, b Warsaw, Poland, Nov 27, 23; US citizen; m 76. THEORETICAL PHYSICS. *Educ:* Univ Warsaw, PhD(theoret physics), 48, Veniam Legendi, 51; Jagiellonian Univ, MS, 46. *Prof Exp:* Instr theoret physics, Jagiellonian Univ, 45-46; instr, Univ Warsaw, 46-48, asst prof, 50-53, assoc prof, 56-60; Polish Govt grant, Holland, 48-50 & Univ Birmingham, 50; assoc prof theoret physics, Wroclaw Univ, 53-56; Ford Found exchange visitor, Columbia Univ, Univ Calif, Berkeley & Inst Advan Study, NJ, 60-61; mem sci staff, Plasma Plysics Group, Boeing Sci Res Labs, Wash, 61-64; from assoc prof to prof theoret physics, 64-71, prof, 66-84, EMER PROF PHYSICS, UNIV CINCINNATI, 84- *Mem:* NY Acad Sci; AAAS. *Res:* Relativistic formulation of bound state problems in quantum electrodynamics; lamb-shift like corrections in helium; relativistic soluble models in quantum field theory; solving Bethe Salpeter equation; dynamical approach to composite particles in relativistic quantum field theory; symmetric algelbras, especially non-Hilbert representations of same. *Mailing Add:* 310 Bryant Ave Apt 15 Cincinnati OH 45220

GUNTHER, RONALD GEORGE, b St Paul, Minn, Jan 22, 33; m 66; c 2. ELECTROCHEMISTRY. *Educ:* Col St Thomas, BS, 55; Univ St Louis, PhL, 61; Iowa State Univ, MS, 64, PhD(phys chem), 68; Univ New Haven, MBA, 76. *Prof Exp:* Res assoc fuel cells, Pratt & Whitney Aircraft Div, United Aircraft Corp, 68-70; head fuel cell res, Catalytic Technol Div, Pioneer Systs, 70-71; mgr electrochem systs, Electrochem Res & Develop Group, Yardney Elec Corp, 71-76; STAFF RES SCIENTIST, GEN MOTORS RES LABS, 76- *Honors & Awards:* Bronze Medal, Am Inst Chemists, 55. *Mem:* Electrochem Soc; Int Soc Electrochem. *Res:* Chemical kinetics; electrolytic solutions; electrode processes; electrochemistry; coordination compounds; battery systems; electrocatalysis; Langmuir-Blodgett films; fuel cells. *Mailing Add:* Gen Motors Res Labs Physical Chem Dept Warren MI 48090-9058

GUNTHER, WOLFGANG HANS HEINRICH, b Gifhorn, Ger, Mar 26, 31; nat US; m 58; c 2. SELENIUM TELLURIUM CHEMISTRY. *Educ:* Brunswick Tech Univ, Cand Chem, 53; Univ Leeds, PhD(org chem), 58. *Prof Exp:* Demonstr org chem, Univ Leeds, 55-58; Brown Mem fel, Yale Univ, 58-59, from res asst to res assoc pharmacol, 59-66, asst prof, 66-67; scientist, Chem Res Lab, Xerox Corp, 67-68, mgr org & polymer chem, 68-80; res assoc & sr staff mem, res lab, Sterling Res Group, Eastman Kodak Co, 80-83, head sensitizing dye lab, 83-85, sr res investr, 87-90, SR FEL, STERLING RES GROUP, EASTMAN KODAK CO, 90- *Mem:* Am Chem Soc; Royal Soc Chemistry; Sigma Xi. *Res:* Sensitizing dyes; synthesis of metabolically active compounds; synthesis and study of oxygen, sulphur and selenium isologs; electrophotographic imaging materials and processes; organotellurium chemistry. *Mailing Add:* 606 John Anthony Dr West Chester PA 19382-7191

GUNTHER-MOHR, GERARD ROBERT, b Montclair, NJ, June 8, 22; m 50; c 3. PHYSICS. *Educ:* Yale Univ, BS, 43; Columbia Univ, MA, 49, PhD(physics), 53. *Prof Exp:* Researcher, Manhattan Dist, Corps Engrs, US Army & Monsanto Chem Co, 44-46; asst physics, Columbia Univ, 47-50, lectr, 50; mem staff, Watson Sci Comput Lab, 53-58, mgr semiconductor res dept, 58-61, dir solid state eng, 62-66, mgr tech develop & assurance, Components Div, 66-67, mgr tech & reliability develop, 67-69, asst to pres, Components Div, 69-74, MGR TECH DEVELOP, IBM CORP, 74- *Mem:* Fel Am Phys Soc; sr mem Inst Elec & Electronics Engrs; Sigma Xi. *Res:* Semiconductors. *Mailing Add:* 58 Constitution Hill W Princeton NJ 08540

GUNTHEROTH, WARREN G, b Hominy, Okla, July 27, 27; m 54; c 3. PEDIATRIC CARDIOLOGY, CARDIOVASCULAR PHYSIOLOGY. *Educ:* Harvard Univ, MD, 52; Am Bd Pediat, dipl, 60, cert cardiol, 61. *Prof Exp:* Intern med, Peter Bent Brigham Hosp, Boston, Mass, 52-53; resident pediat, Children's Hosp, 55-56; from instr to assoc prof pediat, 58-69, PROF PEDIAT, SCH MED, UNIV WASH, 69-, HEAD, DIV PEDIAT CARDIOL, 62- *Concurrent Pos:* Fel pediat cardiol, Children's Hosp, 53-55; spec fel physiol, Sch Med, Univ Wash, 57; consult cardiol, King County, Children's Orthop & USPHS Hosps, Seattle, 58- *Mem:* Am Physiol Soc; Biomed Eng Soc; Soc Exp Biol & Med; Am Col Cardiol; Soc Pediat Res. *Res:* Venous return; shock; crib death; physiology of cyanotic-congenital heart disease; electrocardiography; doppler echocardiography. *Mailing Add:* Dept Pediat RD-20 Univ Wash Sch Med Seattle WA 98195

GUNTON, JAMES D, b Medford, Ore, Mar 28, 37; m 62; c 3. THEORETICAL PHYSICS, STATISTICAL MECHANICS. *Educ:* Linfield Col, BA, 58; Oxford Univ, BA, 61; Stanford Univ, PhD(physics), 66. *Prof Exp:* Lectr physics, Univ Western Australia, 65-68; from vis asst prof to assoc prof physics, Temple Univ, 70-76, prof, 76-88; DEAN, COL ARTS & SCI, LEHIGH UNIV, 88- *Concurrent Pos:* Vis scientist, Kyoto Univ, US-Japan Coop Sci Prog, NSF, 74-75, Univ Geneva, 81, Kernforschungsanlage, WGer, 82, Inst Laue Langevin, Grenoble, 82. *Mem:* NY Acad Sci; fel Am Phys Soc; Sigma Xi. *Res:* Equilibrium and non-equilibrium theory of first and second order phase transitions. *Mailing Add:* Col Arts & Sci Lehigh Univ Bethlehem PA 18015

GUNTON, RAMSAY WILLIS, b Lexington, Ky, July 30, 22; Can citizen; m 51; c 4. MEDICINE. *Educ:* Univ Western Ont, MD, 45; Oxford Univ, DPhil, 49; FRCP(C), 52. *Prof Exp:* Sr med res fel, Nat Res Coun Can, 52-56; assoc in med, Fac Med, Univ Toronto, 53-63, assoc prof med, prof therapeut & head dept, 63-66; PROF MED, UNIV WESTERN ONT, 66-, ACTIVE STAFF, UNIV HOSP, 72- *Concurrent Pos:* Med res assoc, Ont Heart Found, 56-63; physician, Toronto Gen Hosp, 57-66; mem teaching staff, Victoria Hosp, 66-72; mem, Med Coun Can, 78-; mem coun, Royal Col Physicians & Surgeons; mem coun clin cardiol, Am Heart Asn, Can; pres, Royal Col Physicians & Surgeons, Can, 86-88. *Honors & Awards:* Hon Fel, Royal Australasian Col Physicians, 88. *Mem:* Am Fedn Clin Res; fel Am Col Physicians; Can Med Asn; Can Soc Clin Invest (vpres, 61); Can Cardiovasc Soc. *Res:* Clinical investigation; heart disease; therapeutics. *Mailing Add:* Dept Med Univ WOnt Fac Med London ON N6A 5C1 Can

GUNZBURGER, MAX DONALD, b Buenos Aires, Arg, Oct 11, 45; US citizen; m 71; c 4. APPLIED MATHEMATICS. *Educ:* NY Univ, BS, 66, MS, 67, PhD(math), 69. *Prof Exp:* Res scientist, NY Univ, 69-70, asst prof math, 70-71; mathematician, US Naval Ord Lab, 71-73; vis scientist, Inst Comput Applns Sci & Eng, 73-76; assoc prof math, Univ Tenn, Knoxville, 76-81; prof math, Carnegie-Mellon Univ, 81-88; PROF MATH, VA TECH, 88- *Concurrent Pos:* Consult, Inst Comput Appln Sci & Eng, 76- & Nuclear Div, Union Carbide Corp, 78-; adj assoc prof, George Washington Univ; adj prof math, Col William & Mary. *Mem:* Soc Indust & Appl Math. *Res:* Numerical and analytic methods for the solution of partial differential equations, with particular application to problems in fluid mechanics, acoustics and wave propagation. *Mailing Add:* 267 Wisteria Dr Radford VA 24141

GUO, HUA, b Sichuan, People's Repub China, Aug 20, 62; m. CHEMISTRY. *Educ:* Chengdu Inst Electronic Eng, BS, 82; Sichuan Univ, MS, 85; Sussex Univ, PhD(chem), 88. *Prof Exp:* Res assoc chem, Northwestern Univ, 88-90; ASST PROF CHEM, UNIV TOLEDO, 90- *Mem:* Am Chem Soc; Am Phys Soc. *Res:* Photodissociation and reaction dynamics calculations. *Mailing Add:* Dept Chem Univ Toledo Toledo OH 43606

GUO, SHUMEI, b Taiwan, Sept 8, 54; m 78; c 1. HUMAN GROWTH & DEVELOPMENT. *Educ:* Nat Taiwan Univ, BPH, 76; State Univ NY, Stony Brook, MS, 80; Univ Pittsburgh, PhD(biostatist), 83. *Prof Exp:* Consult statistician, cancer res, Info Mgt Servs, 83-84; RES SCIENTIST & ASSOC PROF, MATH & STATIST & DIV HUMAN BIOL, DEPT COMMUNITY HEALTH, SCH MED, WRIGHT STATE UNIV, 84- *Honors & Awards:* First Independent Res Support & Transition Award, NIH. *Mem:* Am Statist Asn; Soc Clin Trials. *Res:* Growth, body composition and risk factors; biased estimation; curve fitting and mathematical modelling; survival analysis. *Mailing Add:* Dept Community Health Sch Med Wright State Univ 1005 Xenia Ave Yellow Springs OH 45387-1698

GUO, YAN-SHI, b Jie-Yang, Kwangtung, China, Aug 16, 43; m 70; c 2. NEUROSCIENCE. *Educ:* Beijing Med Univ, MD, 64, MS(physiol), 68 & MS(endocrinol), 80. *Prof Exp:* Surgeon, Chen-Guang Hosp, Szechuan, 68-78; instr physiol, Beijing Med Univ, 81-84; VIS SCIENTIST GASTROINTESTINAL & ENDOCRINOL, DEPT SURG, MED BR, UNIV TEX, GALVESTON, 84- *Mem:* Chinese Physiol Asn. *Res:* Physiological actions of gastrointestinal hormones and their intracellular mechanisms; regulation of gastrointestinal hormones release; relationship between gastrointestinal hormones and gastrointestical cancers; relationship between gut and pancreas. *Mailing Add:* Univ Tex Med Br Galveston TX 77550

GUPTA, AJAYA KUMAR, b Allahabad, India, Sept 27, 44; US citizen; div; c 2. STRUCTURAL ENGINEERING, EARTHQUAKE ENGINEERING. *Educ:* Univ Roorkee, BE, 66, ME, 68; Univ Ill, PhD(civil eng), 71. *Prof Exp:* Supvr, structural eng, Sargent & Lundy, 71-76; sr res engr, IIT Res Inst, 76-79; assoc prof civil eng, Ill Inst Technol, 79-80; assoc prof, 80-83, PROF CIVIL ENG, NC STATE UNIV, 83- *Concurrent Pos:* Adj assoc prof, Ill Inst Technol, 77-79; vis scholar, Univ Ill, 78; consult, Zurn Indust, 79-, Sargent & Lundy, 79-, IIT Res Inst, 79- & Res Triangle Res Inst, 81-; adv, Comt Dynamic Response Mats Subjected High Strain Rate Loadings, Nat Mats Adv Bd, Nat Acad Sci, 77-78; prin investr, NSF grants, 79-81, 79-82, 82-85 & 85-; chmn, Energy Div, Am Soc Civil Engrs, 90-91. *Honors & Awards:* Huber Res Prize, Am Soc Civil Engrs, 82. *Mem:* Am Soc Civil Engrs; Am Concrete Inst; Earthquake Eng Res Inst; Am Soc Mech Engrs; Int Asn Shell & Spatial Struct; Int Asn Struct Mech Reactor Technol. *Res:* Concrete shells; seismic analysis and design: nuclear power plants, low-rise buildings; structural analysis and design; author one book. *Mailing Add:* Dept Civil Eng NC State Univ Raleigh NC 27695-7908

GUPTA, AMITAVA, b Calcutta, India, Feb 1, 47; m 73; c 2. PHOTOCHEMISTRY, PHYSICAL ORGANIC CHEMISTRY. *Educ:* Bombay Univ, BSc, 67; Indian Inst Technol, Bombay, MSc, 69; Calif Inst Technol, PhD(chem), 74. *Prof Exp:* Res fel chem, Univ Calif, Santa Cruz, 73-74; res scientist, Lockheed Aircraft Serv Co, 74-75; sr scientist chem, Jet Propulsion Lab, 75-78, res scientist, 81, tech group supvr, 79-87; VPRES, RES DEVELOP ENG, IOPTEX RES INC, 87- *Concurrent Pos:* Sr res scientist, Jet Propulsion Lab, 84-87. *Honors & Awards:* Except Scientist Achievement Medal, NASA, 86. *Mem:* Am Chem Soc; Sigma Xi. *Res:* Photochemistry of polymers and transition metal compounds in presence of polymer support systems; photochemistry of aromatics; pulse radiolysis of polymers. *Mailing Add:* Ioptex Research Inc 15715 Arrow Hwy Irwindale CA 91706-2094

GUPTA, ARJUN KUMAR, b Purkazi, India, July 10, 38; m 67; c 3. MATHEMATICAL STATISTICS. *Educ:* Benaras Hindu Univ, BSc & dipl French, 57; Univ Poona, BSc Hons, 58, MSc, 59; Purdue Univ, PhD(statist), 68. *Prof Exp:* Lectr statist, Agra Univ, 62; lectr, Benaras Hindu Univ, 62-63; teaching asst math, Purdue Univ, 63-66, res asst statist, 66-68; asst prof math, Univ Ariz, 68-71; asst prof statist, Univ Mich, Ann Arbor, 71-76; assoc prof math, 76-78, PROF MATH & STATIST, BOWLING GREEN STATE UNIV, 78- *Concurrent Pos:* Consult, Univ Ariz, 68-71, Semiconductor Prod Div, Motorola Inc, 69; UN statist consult, 81; vis lectr, Soc Indust & Appl Math, 81; consult, Editorial Math Review, 73-76; chmn, dept math & statistics, Bowling Green State Univ, 85-87; mem, Int Statist Inst. *Mem:* Inst Math Statist; fel Am Statist Asn; Royal Statist Soc; Indian Statist Asn. *Res:* Multivariate statistical analysis; classification models; central and non-central distribution problems in multivariate analysis; distribution of second elementary symmetric function and of likelihood ration criterion in the real and complex case; contingency tables; correlation tables; applied statistics. *Mailing Add:* Dept Math & Statist Bowling Green State Univ Bowling Green OH 43403-0221

GUPTA, ASHWANI KUMAR, b India, Oct 23, 48; US citizen. FUEL TECHNOLOGY & PETROLEUM ENGINEERING. *Educ:* Panjab Univ, BS, 66; Univ Southampton, UK, MS, 70; Univ Sheffield, UK, PhD(combustion), 73. *Hon Degrees:* DSc, Univ Sheffield, 86. *Prof Exp:* Res engr mech eng, Int Combustion, 67-71; res asst chem eng, Univ Sheffield, UK, 71-73, res assoc & independent res worker chem eng, 73-76; res staff chem eng, Mass Inst Technol, 77-82; assoc prof, 83-87, PROF MECH ENG, UNIV MD, 88- *Concurrent Pos:* Consult, NFK Japan, 76, various industs & orgn, 80- *Honors & Awards:* Energy Systs Award, Am Inst Aeronaut & Astronaut, 80. *Mem:* Assoc Fel Am Inst Aeronaut & Astronaut; Am Soc Mech Eng; Combustion Inst; fel Inst Energy UK. *Res:* Combustion and pollution; fuel sprays; laser diagnostics; alternate fuels; coal, gas, liquid fuels; swirl flows; gas turbine combustion; internal combustion engines; furnaces, modeling, fires, boilers; hazardous waste incineration. *Mailing Add:* Dept Mech Eng Univ Md College Park MD 20742

GUPTA, AYODHYA P, b India, July 1, 28; m 67; c 2. ENTOMOLOGY. *Educ:* St Andrew's Col, India, BSc, 50; Benares Hindu Univ, MSc, 53; Univ BC, MSc, 61; Univ Idaho, PhD(entom), 64. *Prof Exp:* Lectr zool, Besant Col, India, 56-57; asst entomologist, Pub Health Dept, Filaria Unit, 58-59; res assoc, 64-66, from asst prof to assoc prof, 66-72, dir grad prog, 81-86, PROF ENTOM, RUTGERS UNIV, 73- *Concurrent Pos:* Ed-in-chief, Int J Insect Morphol & Embryol, 71- & Recent Advances Comp Arthropod Morphol, Physiol & Develop, 86-; FAO consult, 74 & 82. *Mem:* AAAS; hon mem Arthropodan Embryol Soc Japan; fel Entom Soc India; Entom Soc Am; Entom Soc Can; Royal Entom Soc London; Am Asn Univ Prof; Electron Microscopy Soc Am. *Res:* Insect morphology and physiology with particular interest in hemocytes; hormones and pheromones, insect ultrastructure; arthropod immunity. *Mailing Add:* Dept Entom Rutgers Univ New Brunswick NJ 08903

GUPTA, BHUPENDER SINGH, b Delhi, India, Mar 29, 37; m 67; c 3. TEXTILE PHYSICS. *Educ:* Panjab Univ, India, BSc, 58; Univ Manchester, PhD(textile physics), 63. *Prof Exp:* Supvr prod textile yarns, Spinning Dept, Modi Spinning & Weaving Mills, India, 59-60; res instr textile physics, Sch Textiles, NC State Univ, 63-66, from asst to assoc prof textile eng, chem & sci, 66-79, asst head & grad adminr textile eng & sci, 84-90, PROF TEXTILE ENG CHEM & SCI, SCH TEXTILES, NC STATE UNIV, 79-, ASST HEAD UNDERGRAD PROG, 90- *Concurrent Pos:* Develop engr, Fiber Industs, 75. *Honors & Awards:* Fulbright Lectr Award, India, 85. *Mem:* Sigma Xi; Fiber Soc; Brit Textile Inst; Am Chem Soc; Am Asn Textile Technol; fel Textile Inst; Am Soc Biomat. *Res:* Mechanics of self crimping; knot characteristics of surgical sutures; fiber, yarn and fabrics structure and mechanics; melt extrusion; high energy irradiation cross-linking and curing; fiber migration; absorbent non-woven structures; arterial grafts; inter-fiber friction in fibers. *Mailing Add:* Col Textiles NC State Univ Raleigh NC 27695-8301

GUPTA, BRIJ MOHAN, b Muzaffar Nagar, India, Mar 16, 47; m 69; c 2. PHARMACY, CHEMISTRY. *Educ:* Birla Inst Technol & Sci, Pilani, India, BS, 66; Panjab Univ, MS, 68; Columbia Univ, PhD(pharm), 73. *Prof Exp:* Mfg chemist pharmaceut mfg, Allodial Chem, Meerut, India, 68-69; sr scientist, pharmaceut res & develop, Schering Corp, Bloomfield, 73-, mgr lab res serv, Kenilworth, 80-; AT RIJ PHARMACEUT CORP, NY. *Concurrent Pos:* Adj asst prof, Arnold & Marie Schwartz Col Pharm, Long Island Univ, NY, 77- *Mem:* Am Pharmaceut Asn; Am Chem Soc. *Res:* Pharmaceutical formulation research, physical pharmacy; rheology, surface chemistry. *Mailing Add:* 4890 Shasta Dr Nashville TN 37211

GUPTA, CHAITAN PRAKASH, b India, Apr 21, 39; m 65; c 3. MATHEMATICAL ANALYSIS. *Educ:* Univ Delhi, BA, 58, MA, 60; Univ Rochester, PhD(math), 67. *Prof Exp:* Lectr math, Ramjas Col, Delhi, 60-62; asst prof, Univ Va, 66-69; from asst prof to assoc prof, 69-78, PROF MATH, NORTHERN ILL UNIV, 78- *Concurrent Pos:* Instr, Univ Chicago, 67-69; NSF res grant, 71-72. *Mem:* Am Math Soc; Math Asn Am. *Res:* Nonlinear analysis; numerical methods for partial differential equations. *Mailing Add:* Dept Math Northern Ill Univ De Kalb IL 60115

GUPTA, DEVENDRA, b Nagina, India, Feb 15, 31; nat US; m 63; c 3. MATERIALS SCIENCE. *Educ:* Univ Delhi, BSc, 50; Banaras Hindu Univ, BSc, 54; NY Univ, MS, 57; Univ Ill, Urbana, PhD(metall), 61. *Prof Exp:* Sr sci officer, Nat Metall Lab, Govt India, 61-63; reader phys metall, Banaras Hindu Univ, 63; asst chief indust res develop, Planning Comn, Govt India, 63-65; fel metall & solid state physics, Univ Ill, Urbana, 65-68; STAFF SCIENTIST, T J WATSON RES CTR, IBM CORP, 68- *Concurrent Pos:* Adj prof phys & eng metall, Polytech Inst of NY, 78-79. *Mem:* Sigma Xi; fel Am Phys Soc; Minerals Metals & Mat Soc. *Res:* Internal friction in metals and alloys; diffusion in materials; ductility and fracture; mass transport in thin films for VLSI applications; epitaxy of thin films; ion implantation; materials for electronic packaging; defects in materials. *Mailing Add:* 3 Morningside Ct Ossining NY 10562

GUPTA, DHARAM VIR, b New Delhi, India, Sept 4, 45; US citizen; m 81; c 3. PROCESS DEVELOPMENT, ANTIBIOTICS. *Educ:* Univ Delhi, India, BSc(hons), 65; Indian Inst Technol, Kharagpur, BTech(hons), 68; Univ Waterloo, Ont, MS, 70; Worcester Polytech Inst, PhD(chem eng), 75. *Prof Exp:* Res engr, Am Cyanamid Co, 74-78, group leader & tech mgr, 79-83, prod mgr, 84-89, SR RES SCIENTIST, AM CYANAMID CO, 89- *Mem:* Am Soc Pharmaceut Engrs; Am Inst Chem Engrs; Am Chem Soc. *Res:* Heterogeneous catalysis; polymer engineering; reactor design; pharmaceutical plant design, trouble shooting and start up. *Mailing Add:* 1937 W Main St No 275 Stamford CT 06904-0060

GUPTA, GIAN CHAND, b Delhi, India, Oct 10, 39; m 72; c 2. SOIL CHEMISTRY, WATER & AIR POLLUTION. *Educ:* Panjab Univ, BSc, 59, BT, 60; Vikaram Univ, MSc, 62; Univ Roorkee, PhD(chem), 67. *Prof Exp:* Jr sci asst soil chem, Defense Sci Lab, Delhi, India, 62-63; jr res fel chem, Univ Roorkee, India, 63-66; sr res fel, Cent Rd Res Inst, New Delhi, India, 67-68; res assoc civil eng, Univ Miss, 68-69; assoc prof chem, Rust Col, Miss, 69-70; dir environ health, Medgar Evers Comprehensive Health Ctr, Fayette, Miss, 71-72; dir environ health & educ, Jackson Hinds Comprehensive Health Ctr, Utica, 72-77; assoc prof, 77-90, PROF, UNIV MD, EASTERN SHORE, 90- *Concurrent Pos:* Consult res soil chemist, Alcorn State Univ, 74-; guide prof colloid, soil & environ sci, World Open Univ, 75-83. *Mem:* Am Chem Soc; Am Soc Agron; Nat Environ Health Asn; Soc Environ Geochem & Health; Inst Environ Sci. *Res:* Thermal and electro-osmotic stabilization of soils; transformation of clay minerals; cation and anion exchange; x-ray, infrared, polarographic and thermal properties of natural and synthetic clay minerals; overland animal waste recycling; clay minerals and adsorbents in water purification; heavy metal and energy related metal pollution in waters; use of aquatic plants in water; air recycling in closed systems; air pollution and crop damage; acid rain; poultry litter disposal and toxicity; plant physiology. *Mailing Add:* Dept Nat Sci Univ MD Eastern Shore Princess Anne MD 21853

GUPTA, GODAVERI RAWAT, b India, Feb 1, 37; US citizen; m 67; c 2. PATHOLOGY. *Educ:* Agra Univ, India, BSc, 58, MSc, 60; Vikram Univ, India, PhD(zool), 64. *Prof Exp:* Res specialist histol, anat dept, Med Sch, Rutgers Univ, 79-80; trainee hemat, path dept, Muhlenberg Hosp, Plainfield, NJ, 81-82; HEMATOLOGIST, ROCHE BIOMED LAB, RARITAN, NJ, 86- *Mem:* Assoc Am Soc Clin Pathologists. *Res:* Morphology of blood cells. *Mailing Add:* 9 Flower Rd Somerset NJ 08873

GUPTA, GOUTAM, b India, Apr 17, 45; US citizen; m 73; c 2. COATINGS, ORGANIC-POLYMER. *Educ:* Jadaupur Univ, Calcutta, MSc, 65; Catholic Univ Am, PhD(org chem) 72. *Prof Exp:* Res assoc, Univ Va, 72-74; chemist, Res Triangle Inst, 74-76; sr res chemist, Corp Res-Ferro Corp, 76-81; sr scientist, Central Res Div, Sherman Williams, 81-84, group leader, 83-84, dir, New Prod Div, 85-88, DIR, APPL RES, CHEM COATING DIV, SHERWIN-WILLIAMS, 88- *Mem:* Am Chem Soc; Fed Soc Coating Tech. *Res:* New cure systems, polymer, oligomer synthesis for application to high solids coatings uncured coatings development. *Mailing Add:* 549 E 115th St Chicago IL 60628-5703

GUPTA, KRISHANA CHANDARA, b Jammu, Kashmir, India, Feb 22, 27; m 59; c 4. MICROBIOLOGY, IMMUNOCHEMISTRY. *Educ:* Punjab Univ, BSc, 45; Univ Montreal, MSc, 57, PhD(bact), 59. *Prof Exp:* Head dept microbiol, Drug Res Lab, Univ Jammu & Kashmir, 47-53; res officer tuberc, Patel Chest Dis Inst, Univ Delhi, 53-55; asst dir appl microbiol & antibiotics, Regional Res Lab, Univ Jammu & Kashmir, 60-68; Muskoka Hosp Fund Res Asn fel, 68-71; res assoc immunochem of tuberculo-proteins, Connaught Med Res Lab, Univ Toronto, 71-75; res mem, 75-81, RES SCIENTIST, CONNAUGHT RES INST, CONNAUGHT LABS LTD, 81-; PROF MICROBIOL & BIOTECHNOLOGY, RYERSON POLYTECHNICAL INST, TORONTO, 83- *Concurrent Pos:* Chmn session new antibiotics, Intersci Cong Antimicrobial Agents & Chemother, NYC, 64. *Mem:* Fel NY Acad Sci; Am Soc Microbiol; Can Soc Microbiol. *Res:* Infection promoting factors; genetic studies in Bacillus Calmette-Guerin; immunochemistry of tuberculo-proteins; taxonomic studies of streptomycetes; antibacterial and antifungal antibiotics; electron microscopic studies of effect of antibiotics on the cytology of tubercule bacilli; protective antigens from pathogenic microorganisms; immunology; development of marine and human monoclonal antibodies to bacterial and viral antigens; anti-idiotype antibodies as vaccines. *Mailing Add:* 21 Lowbank Ct Toronto ON M2M 3A5 Can

GUPTA, KRISHNA CHANDRA, b Sept 24, 48. MECHANISM SYNTHESIS, INDUSTRIAL ROBOTS. *Educ:* Indian Inst Technol, Kanpur, India, BTech, 69; Case Inst Technol, MS, 71; Stanford Univ, PhD(mech eng), 74. *Prof Exp:* From asst prof to assoc prof, 74-84, dir grad studies, 82-84, PROF MECH ENG, UNIV ILL, CHICAGO, 84- *Concurrent Pos:* Prin investr, Res Bd Grant, Univ Ill, NSF Grant & US Army Res Off Contracts; assoc ed, J Mech Design, Am Soc Mech Engrs, 81-82; gen conf chmn, Am Soc Mech Engrs, Design Tech Conf, Chicago, 90. *Honors & Awards:* Henry Mess Award, Am Soc Mech Engrs, 79. *Mem:* Am Soc Mech Engrs. *Res:* Mechanical engineering design; mechanisms; industrial robots; design optimization. *Mailing Add:* Dept Mech Eng Univ Ill Chicago PO Box 4348 m/c 251 Chicago IL 60680

GUPTA, KRISHNA MURARI, b Kanpur, India, Sept 11, 49; m 76; c 2. FUEL CELL ELECTRODES, POROSITY MEASUREMENT. *Educ:* Indian Inst Technol, Kanpur, BTech, 70; Mass Inst Technol, SM, 72, ScD, 76. *Prof Exp:* Assoc, Columbia Univ, 76 & Cornell Univ, 76-78; PRES, POROUS MAT INC, 78- *Mem:* Am Ceramic Soc; NY Acad Sci. *Res:* Development of novel fuel cell electrodes and porosity measurement techniques; creep resistant ceramics. *Mailing Add:* 216 Muriel St Ithaca NY 14850

GUPTA, KULDIP CHAND, b India, Oct 6, 40; m 71; c 3. COMPUTER-AIDED DESIGN OF MICROWAVE CIRCUITS & MICROSTRIP ANTENNAS. *Educ:* Punjab Univ, BSc, 58; Indian Inst Sci, Bangalore, BE, 61, ME 62; Birla Inst Sci & Technol, PhD(electronics eng), 69. *Prof Exp:* Res fel, Indian Inst Sci, Bangalore, 63-64; asst prof elec eng, Punjab Eng Col, 64-65; sr res fel, Cent Electronics Eng Res Inst, 65-68; asst prof, Birla Inst Technol & Sci, 68-69 & Indian Inst Technol, Kanpur, 69-75, prof, 75-84; PROF ELEC ENG, UNIV COLO, BOULDER, 85- *Concurrent Pos:* Coordr, Advan Ctr Electronics Systs, Indian Inst Technol, 71-79; vis assoc prof, Univ Waterloo, Can, 75-76; vis prof, Fed Polytech Univ Lausanne, Switz, 76, Tech Univ Denmark, 76-77, Eidgenossische Tech Univ, Zurich, 79, Univ Kans, 82-83 & Univ Colo, 83-84; res coordr, Ctr Microwave & Millimeter-Ware Cad, NSF, 87-; ed, Int J Microwave-Millimeter-Wave Computer-Aided Eng, 90- *Mem:* Fel Inst Elec & Electronics Engrs; Int Union Radio Sci; Electromagnetics Acad. *Res:* Computer-aided design techniques for microstrip circuits and microstrip antennas; author of five books and 140 publications. *Mailing Add:* Dept Elec & Computer Eng Univ Colo Campus Box 425 Boulder CO 80309-0425

GUPTA, MADAN MOHAN, b Lansdowne, India, Apr 10, 36; m 64; c 3. COMPUTER SCIENCE, CONTROL ENGINEERING. *Educ:* Univ Rajasthan, BE, 61, ME, 62; Univ Warwick, Eng, PhD(controls), 67. *Prof Exp:* Lectr, Univ Roorkee, 62-64; from lectr to assoc prof, 67-78, PROF CONTROLS, UNIV SASK, 78-, DIR, INTEL SYSTS RES LAB, 80- *Concurrent Pos:* Vis prof, Whiteshell Nuclear Res Plant, 68, Defense Res Estab, 71-, Univ BC, 75-76 & Europ Ctr Peace & Develop, Univ Peace, UN; sr mem, Comt Biol Signal Processing, Inst Elec & Electronics Engrs, 73-; sr indust fel, MacMillan Bloedel Res Ltd, 75-76; chmn, Fuzzy Set Symp, 76-77; assoc ed, Fuzzy Set J, 77-, guest ed, 78-; res grant, Nat Res Coun Can, 69-, Med Res Coun Can, 75-, Defence Res Bd Can, 70-76, AEC, 73-76; sr indust res fel, Taipei, 81; vis fel, Electro-Tech Lab, Tsukuba Science city, Japan, 89. *Mem:* Fel Inst Elec & Electronics Engrs; Can Biomed Eng Soc; Can Indust Comput Soc; NY Acad Sci. *Res:* Neuro-vision; neuro-control; fuzzy neural-networks; neuronal-morphology of biological systems; intelligent systems; cognitive information; new paradigms in information theory; inverse biomedical engineering. *Mailing Add:* Intel Systs Res Lab Col Eng Univ Sask Saskatoon SK S7N 0W0 Can

GUPTA, MADHU SUDAN, b Lucknow, India, June 13, 45; m 70; c 2. ELECTRONICS ENGINEERING, APPLIED PHYSICS. *Educ:* Allahabad Univ, MS, 66; Fla State Univ, MS, 67; Univ Mich, Ann Arbor, MA, 68, PhD(elec eng), 72. *Prof Exp:* Asst prof, Queen's Univ, Can, 72-73; asst prof elec eng, Mass Inst Technol, 73-78, assoc prof, 78-79; assoc prof elec eng, Univ Ill, Chicago, 79-84, dir grad studies, 80-83, prof, 84-87; SR STAFF ENGR, HUGHES AIRCRAFT CO, 77- *Concurrent Pos:* Lilly fel, Div Study & Res in Educ, Mass Inst Technol, 74-75, consult, Lincoln Lab, 77-79; vis prof, Univ Calif, Santa Barbara, 85-86. *Mem:* Fel Inst Elec & Electronics Engrs; Sigma Xi; AAAS; Asn Prof Engrs; Am Soc Eng Educ. *Res:* Microwave electronics; solid state electronic devices; noise in circuits and devices; fluctuation phenomenon; thermodynamics of electrical devices; engineering education. *Mailing Add:* Hughes Aircraft Co PO Box 2940 Torrance CA 90509-2940

GUPTA, MANJULA K, b India, Aug, 17, 42; m 86; c 2. ENDOCRINOLOGY, NEUROSCIENCES. *Educ:* Agra Univ, BS, 59, MS, 61, PhD(zool), 66. *Prof Exp:* Res fel, immunopath, Cleveland Clin Found, 69-71, res assoc, 72-73, proj scientist, 73-74, assoc staff, 74-75, STAFF & SECT HEAD, IMMUNOPATH, CLEVELAND CLIN FOUND, 75- *Concurrent Pos:* Adj clin asst prof, Cleveland State Univ, 77-; staff, endocrinol, Cleveland Clin Found, 85- *Mem:* Am Soc Clin Path; Am Soc Clin Chem; Endocrine Soc; Am Thyroid Asn; Am Asn Cancer Res; Am Asn Immunologists. *Res:* Immunology of endocrine diseases in general; role of hormones on cancer development and growth; mechanisms by which hormones initiate cancer cell proliferation and alter immune response. *Mailing Add:* Dept Immunol One Clin Ctr 9500 Euclid Ave Cleveland Clin Found Cleveland OH 44195-5131

GUPTA, NABA K, b Calcutta, India, July 2, 34; m 64. BIOCHEMISTRY. *Educ:* Univ Calcutta, BSc, 54, MSc, 57; Univ Mich, PhD(biochem), 62. *Prof Exp:* Res assoc biochem, Univ Chicago, 62-65; proj assoc, Inst Enzyme Res, Univ Wis, Madison, 65-66; from asst prof to prof chem, 66-76, mem fac chem, 76-80, PROF CHEM, UNIV NEBR, LINCOLN, 80- *Mem:* Am Soc Biol Chem. *Res:* Chemistry and biology of nucleic acids. *Mailing Add:* Dept Chem Univ Nebr Lincoln NE 68588-0304

GUPTA, NAND K, b Mathura, UP, India, June 10, 42; m 62; c 3. ENGINEERING PHYSICS, SOLID STATE PHYSICS. *Educ:* Agra Univ, BS, 61, MS, 64; Univ Mass, MS, 67, PhD(physics), 69. *Prof Exp:* Proj mgr nuclear detectors, Harshaw Chem Co, 69-74; mkt mgr instruments, Reuter-Stokes Inc, 75-76; sr scientist radation group, EMI Med Inc, 76-81; sr lab dir eng, Bio-Imaging Res, Inc, 81-87; PRES, OMEGA INTERNATIONAL TECHNOL, INC, 88- *Mem:* Am Asn Physicists Med; Soc Nuclear Med. *Res:* Program management and development of large computerized radiation based medical and industrial instruments; CAT scanners and digital radiographic machines for medical applications, large CAT scanners and filmless radiographic machines for aerospace and defense applications. *Mailing Add:* 1506 Bull Creek Dr Libertyville IL 60048

GUPTA, NARAIN DATT, b India, July 27, 36. MATHEMATICS. *Educ:* Univ Kashmir, India, BA, 56; Australian Nat Univ, Canberra, PhD(math), 65. *Prof Exp:* PROF MATH, UNIV MAN, 67- *Mem:* Math Can Soc; Am Math Soc. *Mailing Add:* Dept Math Univ Man Winnipeg MB R3T 2N2

GUPTA, OM PRAKASH, b Rupar, India, Feb 8, 26; m 57; c 3. DENTISTRY. *Educ:* Agra Univ, BS, 45; Univ Bombay, BDS, 50; Harvard Univ, MS, 54, DrPH, 59; NY Univ, MSD, 58; Univ Pittsburgh, DMD, 67. *Prof Exp:* Clin asst, CEM Dent Col, Bombay, 50 & 51; res assoc, Col Dent, NY Univ, 56-58; asst prof, Col Dent, Univ Ill, 58-60; dir, Dent Col, Trivandrum, India, 60-63; prof pub health & prev dent & head dept, Sch Dent Med, Univ Pittsburgh, 63-73, prof grad periodont & head dept, 64-73, prof, Grad Sch Pub Health, 66-73; PROF PERIODONT & CHMN DEPT, COL DENT, HOWARD UNIV, 73- *Concurrent Pos:* Res fel, Sch Dent Med, Harvard Univ, 53-56; USPHS res grants, 65-69 & NIH, 68-; consult, Vet Admin Hosp, Leech Farm, 64-73, Pittsburgh, 67-73; assoc ed, J Am Acad Oral Med, 72-78. *Mem:* Fel AAAS; Am Acad Oral Med; Int Dent Fedn; Int Asn Dent Res; Am Acad Periodontology; Am Dent Asn. *Res:* Experimental research in oral biology; epidemiologic investigations on oral problems, including dental caries and periodontal disease. *Mailing Add:* Dept Periodont Howard Univ Col Dent 600 W St NW Washington DC 20059

GUPTA, PRABODH KUMAR, b Shorkot, India, Mar 1, 37; m 69; c 1. PATHOLOGY. *Educ:* Panjab Univ, Chandigarh, India, MS & BS, 61; All India Inst Med Sci, MD, 65; Am Bd Path, AB, 75; Am Bd Cytopath, 90. *Prof Exp:* Instr, All India Inst Med Sci, New Delhi, 63-67, asst prof & head cytopath lab, 69-72, consult, dept path, 69-72; teaching fel, Harvard Med Sch, 67-78; res fel, Mass Gen Hosp, 67-68; instr, Sch Med, Johns Hopkins Univ, 68-69, asst prof cytopath, 72-79, asst pathologist, 72-74, pathologist & cytopathologist, 74-88, assoc prof path, 79-88; PROF PATH & LAB MED, UNIV PA SCH MED, 88-; DIR CYTOPATH & CYTOMETRY, PA MED CTR, PHILADELPHIA, 88- *Concurrent Pos:* Rockefeller Found fel, All India Inst Med Sci, 62-64; USPHS res fel, Mass Gen Hosp, 67-68; asst prof, Sch Health Serv, Johns Hopkins Univ, 73-75, cirric comt, 73-75; mem res rev comn, Nat Found March Dimes, New York, 75-; vis prof, Univ Mich, 80; lectr, Am Cancer Soc, 80, guest speaker, 81; prof dir, McKee Mem Sem, Univ SC; consult, WHO, 83, UN, 89. *Mem:* Fel Col Am Pathologists; fel Int Acad Cytol; Am Soc Cytol; AAAS; NY Acad Sci; AMA. *Res:* Morphologic studies of precancerous lesions; biomarker studies with special reference to cytology, automation, bacterial, viral and related infections; molecular pathology. *Mailing Add:* Six Founders Pavillion Hosp Univ Pa Philadelphia PA 19104

GUPTA, PRADEEP KUMAR, b Bandikui, India, July 9, 44; m 77; c 2. MECHANICAL ENGINEERING, TRIBOLOGY. *Educ:* Birla Inst Technol & Sci, India, BE Hons, 66; Mass Inst Technol, SM, 68, ME, 69, ScD, 71. *Prof Exp:* Res asst mech eng, Mass Inst Technol, 66-70; sr eng scientist tribology, 71-77, program advan technol, 77-79, sr consult scientist, Mech Technol Inc, 79-82; PRES, PKG INC, 83- *Honors & Awards:* Birla Inst Gold Medal, 66; Newkirk Award, Am Soc Mech Engrs, 78; Nat Acad Sci Fel, 83. *Mem:* Am Soc Mech Engrs; Sigma Xi; Soc Tribologists & Lubrication Engrs. *Res:* Rolling bearing technology; lubrication mechanics, friction and wear; applied mechanics; surface topography and mechanical interaction of rough surfaces. *Mailing Add:* 117 Southbury Rd Clifton Park NY 12065-7714

GUPTA, RADHEY SHYAM, b India, Mar 21, 47; m 73; c 2. HEAT SHOCK PROTEINS, MITO CHONDRIA-MICROTUBULE INTERACTION. *Educ:* Agra Univ, India, BSc, 65; Indian Inst Technol, MSc, 68; Univ Bombay, PhD(molecular biol), 73. *Prof Exp:* Vis mem, Tata Inst Fundamental Res, Bombay, 68-73; res assoc, Wash Univ Sch Med, 73-75, Univ Toronto, 75-78; asst prof med, 78-80, from asst prof to assoc prof biochem, 80-86, PROF BIOCHEM, MCMASTER UNIV, 86- *Concurrent Pos:* MRC scholar, 79-84, scientist, 84-89; assoc ed, Mutagenesis; ed, Drug Resistance in Mammalian Cells. *Mem:* Am Asn Can Res; Environ Mutagen Soc; Genetics Soc Can; Am Soc Microbiol; NY Acad Sci; Can Biochem Soc. *Res:* Role of mitochondria and molecular chaperone proteins in the invivo assembly/function of microtubules; cellular resistance to anticancer drugs; mechanism of action of cardiac glycosides; role of heat shock proteins in autoimmune diseases; molecular evolution of heat shock proteins; mechanism of action drugs. *Mailing Add:* McMaster Univ Health Sci Ctr 1200 Main St W Hamilton ON L8N 3Z5 Can

GUPTA, RAJ K, b India, Oct 30, 43. PHYSIOLOGY, BIOPHYSICS. *Educ:* Agra Univ, India, BS, 62; Univ Allahabad, MS, 64; Indian Inst Technol, Kanpur, PhD, 69. *Prof Exp:* Mem, IBM Watson Lab, Columbia Univ, 69-71, Thomas J Watson Res Ctr, Yorktown Hts, 71-72; vis scientist, Nat Inst Arthritis, Metab & Digestive Dis, NIH, Bethesda, Md, 72-73; sr res assoc, Inst Cancer Res, Philadelphia, 73-75, assoc mem res staff, 75-78, sr res staff, 78-82; PROF PHYSIOL, BIOPHYS & BIOCHEM, ALBERT EINSTEIN COL MED, 82- *Concurrent Pos:* Res career develop award, USPHS, 76-81; adj assoc prof biochem & biophys, Univ Pa, 78-82; mem, Biophys & Biophys Chem B Grants Study Sect, NIH, 79-83. *Mem:* Am Soc Biol Chemists; Biophys Soc; Am Phys Soc; Am Chem Soc. *Res:* Intracellular metal ions in intact cells and tissue; author of numerous scientific publications. *Mailing Add:* Dept Physiol & Biophys Albert Einstein Col Med Bronx NY 10461

GUPTA, RAJENDRA, b Mau Ranipir, India, Jan 01, 43; m; c 2. ATOMIC, MOLECULAR & OPTICAL PHYSICS. *Educ:* Agra Univ, India, BSc, 59, MSc, 61; Boston Univ, PhD, 70. *Prof Exp:* Res assoc, Columbia Univ, 70-73, lectr, 73-74, asst prof, 74-78; from asst prof to assoc prof, 78-85, PROF PHYSICS, UNIV ARK, 85-, CHAIR, 89- *Concurrent Pos:* Vis prof, Air Force Inst Technol, Wright-Patterson AFB & vis scientist, Air Force Aero Propulsion Lab, 77-78. *Mem:* Am Phys Soc. *Res:* Laser spectroscopy; atomic and molecular physics. *Mailing Add:* Dept Physics Univ Ark Fayetteville AR 72701

GUPTA, RAKESH KUMAR, b Delhi, India, Jan 15, 47; US citizen; m 70; c 2. POLYMER SCIENCE, TEXTILE TECHNOLOGY. *Educ:* Indian Inst Technol, BTech, 68; NC State Univ, MS, 70, PhD(fiber & polymer sci), 74. *Prof Exp:* Fiber physicist, Am Cyanamid Co, 70-71; res engr fibers, 74-78, prog leader Fiber Fundamentals, 78-81, group leader, 81-84, proj leader, 85-88, PROJ MGR, NON-WOVEN FIBERS RES & DEVELOP, HERCULES INC, 89- *Mem:* Fiber Soc. *Res:* Fiber and polymer science; fiber physics and mechanics; polymer characterization; fiber melt and wet spinning; polypropylene polymer and fibers; carpet yarn development; upholstery yarn development; non-wovens; thermal bonding fibers. *Mailing Add:* Hercules Inc PO Box 8 Oxford GA 30267

GUPTA, RAMESH, b India, Dec 14, 47; US citizen; m 76; c 1. REACTORS, FLUID DYNAMICS. *Educ:* Indian Inst Technol, BTech, 69; Princeton Univ, MA, 70, PhD(chem eng), 73. *Prof Exp:* Res fel, Chem Eng, Calif Inst Technol, 73-75; res engr, Air Products Inc, 75-77; engr, 77-78, proj engr, 78-79, sr engr, 79-81, staff engr, 81-82, SR STAFF ENGR, EXXON RES & ENG CO, 82- *Mem:* Am Inst Chem Engrs; Sigma Xi. *Res:* Fluid dynamics, with focus on multiphase chemical reactors for petroleum, chemicals, coal and synthesis fuels. *Mailing Add:* Dept Biochem SIll Univ Med Sch Box 19230 Springfield IL 62794

GUPTA, RAMESH C, b Vindhyachal, Uttar Pradesh, India, Feb 12, 44; US citizen; m 79; c 4. IMMUNOLOGY. *Educ:* Agra Univ, India, BSc, 61; All India Inst Med Sci, New Delhi, MB & BS, 66, MD, 71. *Prof Exp:* House officer med & surg, All India Inst Med Sci, New Delhi, 67, resident med, 68-71; resident med, Memorial Hosp, Med Sch, Univ Mass, Worcester, 71-72; fel med, rheumatology, Univ Colo, Denver, 72-74, instr med, 74-75; res fel & assoc immunol, Mayo Clin & Found, Rochester, Minn, 75-77; asst prof med, Univ Colo, Denver, 77-82; ASSOC PROF MED, UNIV TENN, MEMPHIS, 82- *Concurrent Pos:* Inst med, Mayo Med Sch, Rochester, Minn, 75-77; chief rheumatology serv, Denver Gen Hosp, 77-82; attend physician, City of Memphis Hosps, 82-; consult physician, Vet Admin Hosp, Memphis, 82- *Mem:* Am Asn Immunologists; Am Rheumatism Asn; AAAS. *Res:* Diagnosis, etiology and therapy of autoimmune diseases with particular emphasis on applying the knowledge of immunology, molecular and cell biology. *Mailing Add:* Dept Rheumatol Univ Tenn 956 Court Ave Box 3A Memphis TN 38163

GUPTA, RAMESH C, b Etah, UP, India, Aug 20, 49; US citizen; m 87. TOXICOLOGY. *Educ:* GB Point Univ Agr Technol, India, DVM, 73; Punjab Agr Univ, India, PhD(toxicol), 76. *Prof Exp:* Vis scientist res toxicol, Dept Pharmacol, Mich State Univ, 81-82; toxicologist, Med Ctr, Vanderbilt Univ, 83-86; TOXICOLOGIST, BREATHITT VET CTR, MURRAY STATE UNIV, 87- *Mem:* Int Brain Res Orgn; Soc Toxicol; Soc Pharmacol & Exp Therapeut; Soc Neurosci; Am Vet Med Asn; AAAS. *Res:* Biochemical mechanisms involved in the toxicity and tolerance to organophosphorus and carbamate insecticides and nerve agents; development of novel antidotal treatment; mechanisms involved in the weightlessness. *Mailing Add:* 3614 Stone Valley Dr Hopkinsville KY 42240

GUPTA, RAMESH K, b India, Nov 13, 53; US citizen. SATELLITE COMMUNICATIONS, MICROWAVE CIRCUIT & SUBSYSTEM DESIGN. *Educ:* Punjab Univ, Chandigarh, India, BS, 74; Univ Alta, Edmonton, Can, MS, 76, PhD(elec eng), 80; Univ Pa, MBA, 89. *Prof Exp:* Teaching assoc microwave, Univ Alta, Edmonton, Can, 74-80; mem tech staff microwaves, Comsat Labs, Clarksburg, Md, 80-85, staff scientist, 85-87, assoc mgr microwave systs, 87-90, MGR MICROWAVE COMPS, COMSAT LABS, CLARKSBURG, MD, 90- *Concurrent Pos:* Consult, Int Telecommun Satellite Orgn, 82-85; vchmn, Wash Chap, Inst Elec & Electronics Engrs Microwave Theory & Techniques Soc, 87-88, chmn, 88-89; panelist, Small Bus Innovative Res Prog, NSF, 89. *Mem:* Sr mem Inst Elec & Electronics Engrs. *Res:* Gallium arsenic microwave monolithic integrated circuits; advanced microwave hybrid circuit and subsystem development; satellite systems; author of over 40 publications; granted two patents. *Mailing Add:* 7709 Barnstable Pl Rockville MD 20855

GUPTA, RISHAB KUMAR, b Nagina, India, Apr 18, 43; m 72; c 2. IMMUNOCHEMISTRY, ONCOLOGY. *Educ:* Govind Ballabh Pant Univ, BSc Hon, 63, MSc, 65; Rutgers Univ, MS & PhD(microbiol), 68. *Prof Exp:* Lectr microbiol & immunol, Univ Calif, Los Angeles, 68-70; res assoc microbiol, Yale Univ, 70-71; res biochemist med mycol, 71-72, asst res oncologist, 72-75, assoc res oncologist, 75-79, asst prof, 79-81, ASSOC PROF IMMUNOL & MED MICROBIOL, UNIV CALIF, LOS ANGELES, 81- *Concurrent Pos:* Teaching & res asst, Govind Ballabh Pant Univ, 63-65; res asst, Rutgers Univ, 65-68; consult microbiologist, Ralph Stone & Co Engrs Inc, 68-; microbiologist, Vet Admin Med Ctr, Sepulveda, 76-; prin investr grants, Nat Cancer Inst, 81-84, Calif Inst Cancer Res, 73-75 & Cancer Res Coord Comt, Univ Calif, 73-75 & 81-82. *Mem:* Am Soc Microbiol; Am Acad Microbiol; Am Asn Cancer Res; Sigma Xi. *Res:* Identification, detection, isolation and purification of tumor antigens (human) to develop sensitive immunologic tests (radioimmunoassay, enzyme immunoassay), for diagnosis and prognostification of cancer in humans; develop hybridoma antibody to purified tumor antigens used for treatment of human cancer. *Mailing Add:* Dept Surg & Oncol Univ Calif Med Sch Los Angeles CA 90024

GUPTA, SANJEER, b Jaipur, India, Apr 23, 54; m 84; c 2. HEPATOLOGY, GASTROENTEROLOGY. *Educ:* Univ Rajasthan, India, MBBS, 76; Postgrad Inst Med Educ & Res, Chandigarh, India, MD, 80; Royal Col Physicians, UK, MRCP, 82. *Prof Exp:* Registr med, Postgrad Inst Med Educ & Res, Chandigarh, India, 80-81; registr med & gastroenterol, Royal Postgrad Med Sch, Hammersmith Hosp, 81-85; instr hepath, Univ Southern Calif Sch Med, 85-87; sr fel gastroenterol, 87-89, ASST PROF MED, ALBERT EINSTEIN COL MED, 89- *Concurrent Pos:* Lectr, Brit Postgrad Med Found, 83-85, Royal Postgrad Med Sch, 83-85; clin investr award, NIH, 89; attend physician, Montefiore Med Ctr & Bronx Munic Hosp Ctr, 89- *Mem:* Am Gastroenterol Asn; Am Asn Studies Liver Dis. *Res:* Liver growth control; hepatocyte transplantation for somatic gene therapy and liver repopulation; prognosis and therapy of alcoholic liver disease; natural history of viral hepatitis. *Mailing Add:* Liver Res Ctr 1300 Morris Park Ave Bronx NY 10461

GUPTA, SATISH CHANDER, b Poonch, India, July 29, 45. SOIL PHYSICS, WATER RESOURCES. *Educ:* Punjab Agr Univ, BSc, 68, MSc, 69; Utah State Univ, PhD(soil physics), 72. *Prof Exp:* Res fel, Dept Soil Sci, Univ Minn, 72-77; soil scientist, Agr Res Serv, USDA, 77-85; assoc prof, 85-88, PROF SOIL SCI, UNIV MINN, 88- *Mem:* Am Soc Agron; Soil Sci Soc Am; Am Geophys Union; Int Soc Soil Sci. *Res:* Modelling simultaneous flow of salt and water in soils; effect of mechanical properties of soils on plant growth; heat flow in soils; agricultural value of sewage waste and dredged materials; effect of tillage and compaction on soil physical properties and processes; rainfall effect on soil detachment; role of earthworm maesopores on water entry into soil. *Mailing Add:* Dept Soil Sci 146 Borlaug Hall Univ Minn St Paul MN 55108

GUPTA, SHANTI SWARUP, b Saunasi, India, Jan 25, 25. MATHEMATICAL STATISTICS. *Educ:* Univ Delhi, BA, 46, MA, 49; Univ NC, PhD(math statist), 56. *Prof Exp:* Lectr math, Delhi Col, 49-53; asst, Univ NC, 54-56; mem tech staff, Bell Tel Labs, 56-57, 58-62; assoc prof math, Univ Alta, 57-58; adj assoc prof, Courant Inst Math Sci, NY Univ, 59-61; vis assoc prof, Stanford Univ, 61-62; PROF MATH & STATIST, PURDUE UNIV, 62-, HEAD DEPT STATIST, 68- *Concurrent Pos:* Can Math Cong fel, Queen's Univ, Can, 58; vis prof, Dept Statist & Opers Res, Univ Calif, Berkeley, 68-69; consult, Math Dept, Gen Motors Res Labs, 70-80. *Mem:* Am Math Soc; fel Inst Math Statist; fel Am Statist Asn; fel AAAS; fel Royal Statist Soc. *Res:* Multiple decision rules; order statistics; life testing and reliability; multivariate probabilities. *Mailing Add:* Dept Statist Purdue Univ Math Bldg West Lafayette IN 47907

GUPTA, SHYAM KIRTI, b Lucknow, India, Aug 21, 42; m 68. ORGANIC CHEMISTRY, MEDICINAL CHEMISTRY. *Educ:* Univ Lucknow, BS, 58, MS, 60, PhD(org chem), 64. *Prof Exp:* Ford Found fel, Cent Drug Res Inst, Univ Lucknow, 64; res assoc, Univ Maine, 64-65; res assoc, Ariz State Univ, 65-67; vis asst prof biochem, Ind Univ, 67-68; res assoc, Purdue Univ, 68-71; from res chemist to sr res chemist process res & develop, Chas Pfizer & Co, Inc, 71-77; group leader personal care technol, Dial Corp, 77-85, mgr, New Technol, 86-90; VPRES, TECH SERV, AM SOAP CO, 90- *Concurrent Pos:* Consult, soaps & detergents. *Mem:* Fel Am Chem Soc; Am Oil Chemists' Soc. *Res:* Natural products; development of new reactions and reagents in organic synthesis; organometallic chemistry; OTC Drug technology; biochemistry of cosmetics; development of new technology for the manufacture of soaps and detergents; research, development and marketing of soaps and other personal care products. *Mailing Add:* Am Soap Co 11170 Green Valley Dr Olive Branch MS 38654

GUPTA, SOMESHWAR C, b Ludhiana, India, Apr 23, 35; US citizen; m 58; c 3. ELECTRICAL ENGINEERING. *Educ:* Punjab Univ, India, BA, 51, MA, 53; Univ Glasgow, BSc, 57; Univ Calif, Berkeley, MS & PhD(elec eng), 62. *Prof Exp:* Trainee elec eng, Theodore Kiepe, Ger, 54, Brit Elec Authority, Eng, 54 & Elec de France, Paris, 55; trainee electronics, Pye Telecommun, Eng, 55; lab asst acoust, Brunswick Tech, 56; engr, Remington Rand Corp, Glasgow, Scotland, 57; asst lectr elec eng, Univ Glasgow, 57-58; jr res officer nuclear eng, Atomic Energy Estab, Bombay, India, 58; asst prof elec eng, Calif State Polytech Col, 58-60 & Carnegie Inst Technol, 62-63; from assoc prof to prof, Ariz State Univ, 63-67; PROF ELEC ENG, SOUTHERN METHODIST UNIV, 67-, CHMN DEPT,82- *Concurrent Pos:* NSF res grants, 64-72; NASA res grants, 67-74; consult, Gen Elec Co, 63, Mil Electronics Div, Motorola Inc, 63-, Electronics Proving Ground, Ft Huachuca, US Army, 65, Collins Radio Co, 67, Tex Instruments, 69, Western Union Co, 82- & Pagenet, 84; consult ed, Int Textbook Co; assoc ed, Int J Systs Sci, 74- *Mem:* Inst Elec & Electronics Engrs; Sigma Xi. *Res:* Space communication, mobile radio communication, computers, digital signal processing, control. *Mailing Add:* Elec Eng Dept Southern Methodist Univ Dallas TX 75275

GUPTA, SUDHIR, b Bijnor, India, Apr 14, 44; m; c 2. IMMUNOLOGY, INTERNAL MEDICINE. *Educ:* Agra Univ, BS, 61; King George's Med Col, MB, BS, MD, 66, MD, PhD, 70; FRCP(C), 75; Am Bd Allergy & Immunol, dipl, 77; Am Bd Diag Lab Immunol, dipl, 90. *Prof Exp:* Intern, King George's Med Col, 66, resident med, 67-70; resident med, Sir Ganga Ram Hosp, 71 & Mem Hosp, Worcester, 71-72; fel allergy & immunol, Roosevelt Hosp, New York, 72-74; fel immunol, Mem Sloan-Kettering Cancer Ctr, 72-76, assoc, Sloan-Kettering Inst, 76-78; instr med, Med Col, Cornell Univ, 76-77, asst prof biol, 76-78, asst prof med, Med Col, 77-78, assoc prof biol, 78-82, assoc prof med, 79-82; prof med, 82-84, PROF PATH, MICROBIOL, MOLECULAR GENETICS & NEUROL, COL MED, UNIV CALIF, 84-, DIR, DIV BASIC & IMMUNOL, 82- *Concurrent Pos:* Nat Cancer Inst clin res trainee fel, 74-76; Arthur Manzel res award, Roosevelt Hosp, 75-76; clin asst physician, Immunol Serv, Mem Hosp, 76-78; asst attend physician, Mem Hosp Cancer & Allied Dis, 78-79, assoc attend physician, 79-82; staff physician, Univ Calif Irving Med Ctr, 83-; mem, Allergy & Clin Immunol Comt, NIAID, NIH, 85-89; mem, Adv Panel Med Devices, Food & Drug Admin, 89- *Honors & Awards:* Lifetime Achievement Award, J M Found, 90. *Mem:* Am Asn Immunologists; Am Asn Pathologists; fel Am Col Physicians; fel Am Fedn Clin Res; fel Am Acad Allergy & Immunol; Am Rheumatism Asn; fel Royal Col Physicians & Surgeons Can. *Res:* Lymphocyte biology; AIDS; aging; cancer. *Mailing Add:* Basic & Clin Immunol Med Sci I C-264A Univ Calif Irvine CA 92717

GUPTA, SURAJ NARAYAN, b Haryana, India, Dec 1, 24; nat US; m 48; c 2. THEORETICAL PHYSICS. *Educ:* St Stephen's Col, BS, 45, MS, 46; Cambridge Univ, PhD(theoret physics), 51. *Prof Exp:* Fel, Imp Chem Industs, Univ Manchester, 51-53; vis prof physics, Purdue Univ, 53-56; prof, 56-61, DISTINGUISHED PROF PHYSICS, WAYNE STATE UNIV, 61- *Concurrent Pos:* Vis scientist, Argonne Nat Lab, Brookhaven Nat Lab & Nat Res Coun Can. *Mem:* Fel Am Phys Soc; fel Indian Nat Acad Sci. *Res:* Relativity; gravitation; quantum electrodynamics; field theory; nuclear physics; high energy physics. *Mailing Add:* Dept Physics Wayne State Univ 5950 Cass Ave Detroit MI 48202

GUPTA, SURENDRA MOHAN, b Delhi, India, July 10, 47; US citizen; m 73; c 1. INDUSTRIAL ENGINEERING. *Educ:* Birla Inst Technol & Sci, India, BEEE, 70; Purdue Univ, MSIE, 72, PhD(indust eng), 77; Bryant Col, MBA, 77. *Prof Exp:* Teaching asst indust eng, Purdue Univ, 71-72, instr, 72-76; asst prof systs mgt, Bryant Col, 76-78; asst prof, 78-81, ASSOC PROF INDUST ENG, NORTHEASTERN UNIV, 81- *Concurrent Pos:* Investr, Nat Sci Found, 74-77; prin investr, Nat Bur Standards, 80 & Northeastern Univ, 80-81 & 89; vis prof, Jilin Univ Technol, China, 84. *Mem:* Am Inst Indust Eng; Am Prod & Inventory Control Soc; Inst Mgr Sci; Opers Res Soc Am; Am Inst Decision Sci. *Res:* Computerized manufacturing systems; applications of computers and operations research; production management; system evaluation; simulation. *Mailing Add:* Dept Indust Eng Info Sys Northeastern Univ 360 Huntington Ave Boston MA 02115

GUPTA, TAPAN KUMAR, b Barisal, Bangladesh, Aug 2, 39; US citizen; m 69; c 2. CERAMICS. *Educ:* Univ Calcutta, BSc, 58, MSc, 61; Mass Inst Technol, SM, 64, ScD(ceramics), 66. *Prof Exp:* Lectr chem, Hooghly Mohosin Col, India, 60-61; res asst ceramics, Mass Inst Technol, 61-66; sr scientist ceramics, Westinghouse Res Labs, 67-74, fel scientist, 74-82, adv scientist, 82-85; FEL, ALCOA LAB, 85- *Concurrent Pos:* Mgr eng & develop, Bharat Heavy Elec Ltd, India, 79-80. *Honors & Awards:* Fel, Am Cerm Soc. *Mem:* Am Ceramic Soc; Mat Sci Soc; Inst Elec & Electronics Engrs. *Res:* Processing of ceramics; development of high density high strength oxides; investigation on nonlinear lightning arrester and varistor materials; studies on sintering, microstructure, thermal shock and crack healing; studies on electronic materials. *Mailing Add:* Alcoa Tech Center-C Alloa Center PA 15069

GUPTA, UDAIPRAKASH I, b Alahabad, India, Feb 4, 52. DATABASE MANAGEMENT SYSTEMS. *Educ:* Indian Inst Technol, BTech, 74; Princeton Univ, MS, 75, PhD(comput sci), 78. *Prof Exp:* ASST PROF COMPUT SCI, NORTHWESTERN UNIV, EVANSTON, 78- *Concurrent Pos:* Mem tech staff, Bell Lab, 81. *Mem:* Asn Comput Mach; Inst Elec & Electronics Engrs. *Res:* Theoretical aspects of file organizations and database management systems; theory of algorithms and data structures; graph theory. *Mailing Add:* AT&T Bell Labs 1100 E Warrenville Rd Naperville IL 60566

GUPTA, UMESH C, b Kanpur, India, Oct 25, 37; Can citizen; m 57; c 3. SOIL FERTILITY, PLANT NUTRITION. *Educ:* Agr Col, Kanpur, India, BSc, 55, MSc, 57; Purdue Univ, PhD(soil biochem), 61. *Prof Exp:* Nat Res Coun Can fel, Soil Res Inst, Ont, 61-63, res officer soil fertil & plant nutrit, PEI, 63-65, RES SCIENTIST SOIL FERTIL & PLANT NUTRIT, CAN DEPT AGR, 65- *Concurrent Pos:* Ed-in-chief, Can J Soil Sci. *Mem:* Fel Soil Sci Soc Am; fel Can Soc Soil Sci; Agr Inst Can; fel Am Soc Agron. *Res:* To characterize various forms of micro-nutrients from soils and determine sufficiency, deficiency and toxicity levels and deficiency and toxicity symptoms of these nutrients in various crops grown in Eastern Canada; to recommend rates of trace element fertilization; studies on the selenium enrichment of livestock feeds. *Mailing Add:* Agr Can Res Sta Box 1210 Charlottetown PE C1A 7M9 Can

GUPTA, VAIKUNTH N, b Sohna, India, Sept 25, 51; m 80. SPEECH PROCESSING, DIGITAL ENGINEERING. *Educ:* Banaras Hindu Univ, BTech, 74; Univ Ky, MSME, 77, PhD(biomed eng), 80. *Prof Exp:* Engr applicatons signal processing, Univ Ky, 76-80; MEM TECH STAFF SPEECH PROCESSING, COMSAT LABS, COMMUN SATELLITE CORP, 80- *Concurrent Pos:* Consult signal processing, Nat Inst Drug Abuse, 79-80. *Res:* Utilize modern digital signal processing techniques for speech compression and transmission. *Mailing Add:* 945 Pointer Ridge Dr Gaithersburg MD 20878-1132

GUPTA, VIDYA SAGAR, b Jammu, India, Dec 17, 35; Can citizen; m 62; c 3. BIOCHEMICAL PHARMACOLOGY, VIROLOGY. *Educ:* Univ Jammu & Kashmir, BSc, 54; Univ Rajasthan, MSc, 57; Indian Inst Sci, Bangalore, PhD(biochem & pharmcol), 62. *Prof Exp:* Asst prof, 68-70, assoc prof, 70-75, PROF PHARMACOL, DEPT PHYSIOL SCI, UNIV SASK, 75- *Concurrent Pos:* Res assoc, Dept Biochem, Scripps Clin & Res Found, La Jolla, Calif, 62-66; vis prof, Dept Pharmacol, Sch Med, Yale Univ, 73-74; pres, Trans-Med Technol Ltd (T-Med) Res & Develop Co. *Mem:* NY Acad Sci; Can Biochem Soc; Int Soc Antiviral Res; Inter-Am Soc Chemother. *Res:* Virus and cancer chemotherapy; mode of action of antiviral and anti cancer drugs; enzymology. *Mailing Add:* 326 Auld Pl Saskatoon SK S7H 4X1 Can

GUPTA, VIJAI PRAKASH, b Banaras, India, Aug 2, 38; m 67; c 2. CHEMICAL ENGINEERING. *Educ:* Banaras Hindu Univ, BSc, 56, MS, 58; McGill Univ, PhD(chem eng), 64. *Prof Exp:* Sr sci asst petrol, Cent Fuel Res Inst, India, 58-59; res engr, E I du Pont de Nemours & Co, Inc, Del, 64-68; sr res engr, Gen Motors Corp, Mich, 69-70; res engr, Cities Serv Oil

Co, 70-72; process supvr, Oxirane Chem Co, 72-81, ADVISOR, ARCO CHEM CO, 81- *Concurrent Pos:* Holder of many US & foreign chem process patents. *Mem:* Am Chem Soc; Am Inst Chem Engrs. *Res:* Diffusion effects in heterogeneous catalysis; petroleum processing; automobile emissions control; synthetic fibers; chemical processes. *Mailing Add:* ARCO Chem Co 3801 W Chester Pike Newtown Square PA 19073

GUPTA, VIJAY KUMAR, b Ambala, India, Apr 27, 41; m 68; c 3. PHYSICAL CHEMISTRY. *Educ:* Panjab Univ, India, BSc, 61, MSc, 62, PhD(chem), 69. *Prof Exp:* Lectr chem, D A V Col, Chandigarh, 62; lectr appl chem, Punjab Eng Col, 62-64; res asst chem, Punjab Univ, 64-67; asst prof appl chem, Punjab Eng Col, 67-68; res assoc, Wright State Univ, 68-69; PROF PHYS CHEM , CENT STATE UNIV, 69- *Concurrent Pos:* Consult, Wright Patterson AFB, 81-88, EG&G Mound Labs, 89-90. *Honors & Awards:* Outstanding Sect Award, Am Chem Soc, 89. *Mem:* Am Chem Soc; Nat Inst Sci; Soc Tribology Lubrication Engrs. *Res:* Thermodynamics of binary mixtures consisting of non-electrolytes; thermodynamic and physical properties of pure components and mixtures; energy conversion and storage processes; environmental pollution; high energy density battery systems; spectroscopic methods and their application in the analysis of trace elements; development of advanced aerospace lubricants. *Mailing Add:* Dept Chem Cent State Univ Wilberforce OH 45384

GUPTA, VIRENDRA K, b Delhi, India, Mar 14, 32; m 61; c 2. TAXONOMY, FAUNISTICS. *Educ:* Univ Delhi, BSc Hons, 52, MSc, 54; Univ Mich, ScD(zool), 60. *Prof Exp:* Postdoctoral res assoc entom, Univ Mich, 60-61; lectr zool, Univ Delhi, 61-65, reader, 65-78, prof zool, 78-82; PROF ENTOM, UNIV FLA, 82- *Concurrent Pos:* Ed, Oriental Insects & Publ Am Etom Inst, 67-; prin investr, NSF grant, Am Etom Inst, 90-92. *Mem:* Entom Soc Am. *Res:* Taxonomy of the family Ichneumonidae; monographed the Indo-Australian fauna of Icheumonidae and some groups of US; revising several genera of Icheumonidae on a worldwide basis and surveying the ichneumonid of Florida. *Mailing Add:* 2716 NW 37th Terr Gainesville FL 32605

GUPTA, VIRENDRA NATH, b Delhi, India, Apr 15, 31; Can citizen; m 57; c 2. PAPER CHEMISTRY. *Educ:* Univ Delhi, India, BSc, 51, MSc, 53, PhD(chem), 57. *Prof Exp:* Res asst chem, Univ Delhi, 53-56, lectr, 56-58; chemist, All India Inst Med Sci, 58-61; lectr, Univ Pa, Philadelphia, 61-65; res chemist, CIP Res Ltd, Subsid CIP Inc, 65-70, res assoc chem, 70-77, asst mgr, Paper & Board Res, 77-83, asst mgr prod develop & testing, 83-87, mgr, Paper Develop, 87-90, DIR, PROCESS DEVELOP, CIP RES LTD, SUBSID CIP INC, 90- *Concurrent Pos:* Res fel, Univ Pa, 61-63. *Mem:* Tech Asn Pulp & Paper Indust. *Res:* Development of new products and processes for pulp and paper making. *Mailing Add:* C P Forest Prods Res Ltd 179 Main St W Hawkesbury ON K6A 2H4 Can

GUPTA, VISHNU DAS, b Kalyanpur, India, Nov 6, 31; US citizen; m 57; c 2. PHYSICAL PHARMACY, ANALYTICAL CHEMISTRY. *Educ:* Panjab Univ, India, BS, 53, MS, 57; Univ Tex, Austin, MS, 61; Univ Ga, PhD(pharm), 67. *Prof Exp:* In charge mfg & lab, Pharmaceut Mfg House, Azad Hind Chem, India, 53-55 & Janta Pharmaceut Works, 57-60; supvr qual control lab, Schlicksup Drug Co, 61-64; dir control lab, Kapco, Inc, 65-66; teaching asst pharmaceut, Univ Ga, 66-67; from asst prof to assoc prof, 67-76, PROF PHARMACEUT, UNIV HOUSTON, 77- *Concurrent Pos:* Pharm consult, Harris County Hosp Dist, Houston, 70- & numerous pharm co. *Honors & Awards:* Lunsford Richardson Pharm Res Award, Wm S Merrell Co, Ohio, 67. *Mem:* Am Asn Pharmaceut Scientists; Indian Pharmaceut Asn; fel Acad Pharmaceut Scientists. *Res:* Pharmaceutical analysis, especially development of analytical methods in multicomponent dosage forms; physical pharmacy, especially stability studies on dosage forms and distribution coefficients. *Mailing Add:* Dept Pharm Univ Houston Houston TX 77030

GUPTA, YOGENDRA M(OHAN), b New Delhi, India, July 24, 49; US citizen; m 75; c 2. HIGH PRESSURE PHYSICS, MECHANICS. *Educ:* Birla Inst Technol & Sci, India, BSc, 66, MSc, 68; Wash State Univ, PhD(physics), 73. *Prof Exp:* Res assoc physics, Wash State Univ, 73-; res assoc physics, Brown Univ, 74; from physicist to sr physicist, shock Physics Dept, SRI INT, 75-80, asst dir, 80-81; assoc prof, 81-84, PROF, DEPT PHYSICS, WASH STATE UNIV, 84- *Concurrent Pos:* Vis prof, SRI Int, BARC, India, NRL, WDC, 86-87. *Mem:* Am Phys Soc; AAAS; Am Acad Mech. *Res:* Condensed matter response to shock waves; spectroscopic studies under high dynamic pressures; mechanical properties of solids; yield and fracture of materials; shock wave instrumentation. *Mailing Add:* Physics Dept Wash State Univ Pullman WA 99164-2814

GUPTE, SHARMILA SHAILA, b Bombay, India, Oct 4, 42. BIOPHYSICS, BIOCHEMISTRY. *Educ:* Univ Bombay, India, BS, 62, MS, 64; Mich State Univ, PhD(biophys), 78. *Prof Exp:* Teaching asst, Dept Phys, Univ Bombay, India, 64-67; res & teaching asst, Dept Biophys, Mich State Univ, 72-77; postdoctoral fel, Dept Pharmacol & Cell Biophys, Col Med, Univ Cincinnati, 77-81; res assoc, Dept Anat, Lab Cell Biol, Sch Med, Univ NC, Chapel Hill, 85-88; RES ASST PROF, DEPT BIOCHEM, UNIFORMED SERV UNIV HEALTH SCI, 88- *Concurrent Pos:* Numerous grants, US orgn. *Mem:* Biophys Soc; Am Soc Cell Biol; NY Acad Sci. *Res:* G protein mediated signal transduction from hormone receptors to effectors; relationships between catalytic events and diffusion of membrane components; structure, function and kinetics of enzymes; ion translocating ATPases; bioenergetics; DNA binding proteins; numerous scientific publications. *Mailing Add:* Dept Biochem F Edward Herbert Sch Med Uniformed Serv Univ Health Sci 4301 Jones Bridge Rd Bethesda MD 20814-4799

GUPTON, CREIGHTON LEE, b Castalia, NC, Nov 12, 33; m 60; c 2. GENETICS. *Educ:* NC State Univ, BS, 57, MAgr, 63, PhD(plant breeding, bot), 67. *Prof Exp:* Teacher, NC High Sch, 57-61; res supvr, NC State Univ, 63-64; RES GENETICIST, CROPS RES DIV, AGR RES SERV, US DEPT AGR, 67- *Mem:* AAAS; Am Soc Agron; Sigma Xi. *Res:* Derivation from Nicotiana species and tobacco introductions, and the genetic analysis of resistance to tobacco leaf diseases incited by potato virus Y, etch virus and tobacco vein mottle virus. *Mailing Add:* 705 N Columia St Poplarville MS 39470

GUPTON, GUY WINFRED, JR, b Atlanta, Ga, Nov 15, 26; m 48; c 4. BUILDING SYSTEMS OPERATION & MAINTENANCE, INDOOR AIR QUALITY. *Educ:* Ga State Univ, BBA, 67. *Prof Exp:* Assoc, Newcomb & Boyd Consult Engrs, 52-73; partner, Peters-Gupton Assocs, 73-76; PRES, GUPTON ENG ASSOCS, INC, 76- *Concurrent Pos:* Lectr mech eng building systs, Ga Inst Technol, 77-85; prin investr, Control Systs Surv, US Army Construct Eng Res Lab, 86-87; investr, VE Study Design Large Aircraft Corrosion Control Facil, Vent Systs, US Army Cofe & USAF, 85. *Honors & Awards:* Distinguished Serv Award, Am Soc Heating, Refrig & Air Conditioning Engrs, 82. *Mem:* Fel Am Soc Heating, Refrig & Air Conditioning Engrs; Nat Soc Archit Engrs. *Res:* Building systems operation and maintenance with emphasis on systems commissioning, maintenance management, investigation of failures, indoor air quality investigations and system problem solving. *Mailing Add:* 2405 Woodward Way NW Atlanta GA 30305

GUPTON, JOHN, b Norfolk, Va, Jan 4, 46; m 70; c 1. ORGANIC CHEMISTRY. *Educ:* Va Mil Inst, BS, 67; Ga Inst Technol, MS, 69, PhD(chem), 75. *Prof Exp:* Sr chemist agr res, Ciba-Geigy Corp, 75-78; from asst prof to assoc prof org chem, 78-83, PROF CHEM, UNIV CENT FLA, 86- *Concurrent Pos:* Petroleum Res Fund, NSF & Res Corp, res grant, Univ Cent Fla, 79; mem, Coun on Undergrad Res. *Mem:* Am Chem Soc; Sigma Xi. *Res:* Synthesis of agricultural chemicals; new synthetic applications of vinamidinium salts; new methods for the incorporation of fluorine into organic molecules. *Mailing Add:* Dept Chem Box 25000 Orlando FL 32816

GUPTON, OSCAR WILMOT, b Raleigh, NC, Oct 15, 24; m 50; c 4. BOTANY. *Educ:* Univ NC, AB, 50, MA, 51 & 60, PhD(bot), 63. *Prof Exp:* From instr to assoc prof, 60-69, PROF BIOL, VA MIL INST, 69- *Mem:* AAAS. *Res:* Taxonomy. *Mailing Add:* 109 Rebel Rd G Lexington VA 24450

GUPTON, PAUL STEPHEN, b Houston, Tex, Dec 30, 34; m 60; c 2. CORROSION, ELEVATED TEMPERATURE METALLURGY. *Educ:* Lamar Univ, BS, 58; Tex A&M Univ, MS, 60. *Prof Exp:* Develop engr, Murphy Eng Lab, 55-57; welding engr, Hughes Tool Co, 57-58; res fel, Tex Eng Exp Sta, Tex A&M Res Ctr, 59-60; sr fel, Monsanto Co, 60-85; METAL CONSULT, PAUL S GUPTON & ASSOCS INC, 85- *Mem:* Fel Am Soc Metals (treas, 81); Am Soc Nondestructive Testing; Am Soc Testing & Mat; Am Welding Soc; Coating Soc. *Res:* Magnetic properties of the iron, nickel, zinc system not of high temperature alloys. *Mailing Add:* 1100 Deats Rd Dickinson TX 77539

GUR, DAVID, b Haifa, Israel, Apr 7, 47; m 71; c 2. RADIATION PHYSICS, HEALTH PHYSICS. *Educ:* Israel Inst Technol, BSc, 73; Univ Pittsburgh, MS, 76, ScD, 77. *Prof Exp:* Atomic physicist, Israel Inst Technol, 71-74; fel, 75-77, asst prof radiation health, 77-80, ASSOC PROF RADIATION HEALTH & RADIOL, UNIV PITTSBURGH, 80- *Mem:* Health Physics Soc; Am Asn Physicists in Med. *Res:* Development of techniques for early diagnosis and dose reduction in breast cancer detection; computerized electronic radiography; dynamic studies with contrast material using CT scanners; radiation bio-effects in human populations. *Mailing Add:* Dept Diag Radiol Scaife Hall RC 508 Univ Pittsburgh 3550 Terr St Pittsburgh PA 15261-0001

GÜR, TURGUT M, b Istanbul, Turkey; c 2. SOLID STATE ELECTROCHEMISTRY, MIXED IONIC-ELECTRONIC CONDUCTORS. *Educ:* Middle E Tech Univ, Ankara, BS, 66, MS, 69; Stanford Univ, MS, 71, MS, 73, PhD(mat sci eng), 76. *Prof Exp:* Asst prof teaching & res, Middle E Tech Univ, 76-79; staff scientist res, Energy Systs Lab, Calif, 80-82; sr scientist res, Raychem Corp, Calif, 82-85; gen mgr mfg, Tusa AS, Istanbul, Turkey, 85-87; sr res assoc, Dept Mat Sci Eng, 87-90, TECH DIR, CTR MAT RES, STANFORD UNIV, CALIF, 90- *Mem:* Electrochem Soc; Mat Res Soc; Am Chem Soc; Am Ceramic Soc; Solid State Ionics Soc. *Res:* High temperature electrosynthesis and fuel cells; solid electrolytes; chemical and humidity sensors; heterogeneous catalysis; chemical vapor deposition techniques; hi temperature oxide superconductors; diamond thin films; gas separation membranes. *Mailing Add:* 1595 Walnut Dr Palo Alto CA 94303

GURALNICK, ROBERT MICHAEL, b Los Angeles, Calif, July 10, 50; m 74. GROUP THEORY, COMMUTATIVE ALGEBRA. *Educ:* Univ Calif, Los Angeles, BA, 72, MA, 74, PhD(math), 77. *Prof Exp:* Bateman res instr math, Calif Inst Technol, 77-79; from asst prof to assoc prof, 79-88, PROF MATH, UNIV SOUTHERN CALIF, 88- *Mem:* Am Math Soc. *Res:* Finite group theory; representations of orders; matrix theory. *Mailing Add:* Univ Southern Calif Los Angeles CA 90089-1113

GURALNICK, WALTER, b Boston, Mass. MEDICINE. *Educ:* Mass State Col, BS, 37; Harvard Sch Dent Med, DMD, 41; Am Bd Oral & Maxillofacial Surgeons, dipl, 50. *Prof Exp:* Intern oral surg, Boston City Hosp, 41-42; mil serv, 42-46; instr dent med, 54-55, instr oral surg, 55-58, from asst clin prof to clin prof & chmn dept oral surg, 58-74, actg chmn dept dent med, 66-67, chmn dept oral & maxillofacial surg, 74-81, prof, 81-87, EMER PROF, DEPT ORAL & MAXILLOFACIAL SURG, HARVARD SCH DENT MED, 87- *Concurrent Pos:* Oral surgeon, Beth Israel Hosp, Boston, 47-82, Mt Auburn Hosp, Cambridge, 48-82, New Eng Deaconess Hosp, Boston, 50-82, Newton-Wellesley Hosp, Mass, 50-82, New Eng Baptist Hosp, Boston, 52-82; actg chief, Oral Surg Serv, Mass Gen Hosp, Boston, 67-68, chief, Oral & Maxillofacial Surg Serv, 68-81, vis oral & maxillofacial surgeon, 81-, exec dir, Ambulatory Care Ctr, 82-83, dir, operating rooms, 83-91; vis prof, Columbia Presby Hosp, NY, 81-82; hon prof, Shanghai Second Med Univ, People's Repub China, 85, Xian Med Univ, 86, W China Univ Med Sci, 87. *Honors & Awards:* Award for Distinguished Achievement in Oral & Maxillofacial Surg, William J Gies Found, 87. *Mem:* Sr mem Inst Med-Nat Acad Sci; Am Dent Asn; fel Int Asn Oral Surgeons; assoc mem Brit Asn Oral Surgeons; Int Asn Dent Res; Am Asn Oral & Maxillofacial Surgeons; Am Asn Cancer Educ; Am Asn Dent Schs. *Res:* Author of over 60 technical journal articles & publications. *Mailing Add:* 118 Wallis Rd Chestnut Hill MA 02167

GURAM, MALKIAT SINGH, b Ludhiana, India, Jan 4, 28; nat US; m 50; c 3. ZOOLOGY, BOTANY. *Educ:* Punjab Univ, BS, 48, MS, 55; Ohio State Univ, PhD(biol), 67. *Prof Exp:* Lectr zool & entom, Punjab Univ, India, 48-49, res asst, 49-53, lectr, Agr Col, 53-56; asst entomologist, Punjab Agr Dept, 53-60; asst prof zool & entom, Punjab Univ, 62-64; prof biol & head dept, 67-68, div chmn natural sci dept & acad dean, 68-80, PROF ZOOL, VORHEES COL, 80-, DIV CHMN, NATURAL SCI DEPT. *Mem:* AAAS; Am Inst Biol Sci; Entom Soc Am. *Res:* Toxicology; pesticide residue detection by electron capture gas chromatography. *Mailing Add:* Dept Nat Sci & Math Voorhees Col Denmark SC 29042

GURBAXANI, SHYAM HASSOMAL, b Karachi, Pakistan, Dec 28, 28; US citizen; m 59; c 3. SOLID STATE ELECTRONICS. *Educ:* Royal Inst Sci, BSc, 49; Stanford Univ, MS, 50; Rutgers Univ, MS, 63, PhD(solid state physics), 65. *Prof Exp:* Elec engr, Gen Elec Co, 50-51 & Kuljian Corp, 51-52; elec engr, Lummus Co, 52-58, chief engr, 58, budget control consult, Lummus Int, Venezuela, 58-59, cent mgt engr, Lummus Co, 59-60, process design engr, 60-64; br head res div, US Naval Air Develop Ctr, 65-66; asst prof physics, Sacramento State Col, 66-67; assoc scientist, Radiation Co, 67-68, chief scientist, 68-69, reentry physics consult, 69-74; assoc prof, 74-76, PROF ELEC ENG & COMPUT SCI & DIR, LOS ALAMOS GRAD CTR, UNIV NMEX, 76- *Concurrent Pos:* Consult, Shell Co, Venezuela, 58-59; NSF res partic laser physics, Stanford Univ, 67; asst prof elec eng, Univ NMex, 69-74; Assoc Western Univs res partic, Los Alamos Sci Lab, 71; guest scientist, Los Alamos Sci Lab, 72-76; NSF scientist, Bombay Univ, India, 75; invited vis scientist, Tata Inst Fundamental Res, Dept Atomic Energy, India, 75; vis prof physics, Inst Sci, India, 75. *Mem:* Am Phys Soc; sr mem Inst Elec & Electronics Engrs; Am Asn Physics Teachers; Sigma Xi. *Res:* Electron spin and nuclear magnetic resonance; radiation damage; cryogenics; microwave physics; electromagnetic-acoustic energy interaction, population inversion in ionized gases and laser physics; solar energy; electromagnetic pulse in nuclear detonation. *Mailing Add:* Dept Elec-Comp Eng Univ NMex Albuquerque NM 87131

GURD, FRANK ROSS NEWMAN, b Montreal, Can, Jan 20, 24; nat US; m 56; c 4. BIOCHEMISTRY. *Educ:* McGill Univ, BSc, 45, MSc, 46; Harvard Univ, PhD(biochem), 49. *Prof Exp:* Res assoc phys chem, Harvard Univ, 48-50, instr, 50-51, res assoc, 51-52, asst prof, 52-55; asst prof clin biochem, Med Col, Cornell Univ, 55-60; prof biochem med, Ind Univ, Indianapolis, 60-65; prof chem & biochem, 65-79, DISTINGUISHED PROF BIOCHEM & CHEM, IND UNIV, BLOOMINGTON, 79- *Concurrent Pos:* Guggenheim & Whitney fels, Wash Univ, 54-55; asst dir bur med res, Equitable Life Assurance Soc US, 55-59; mem biophys & biophys chem B study sect, NIH, 66-70, chmn, 68-70. *Mem:* Am Chem Soc; Am Soc Biol Chemists; Biophys Soc; Harvey Soc; NY Acad Sci. *Res:* Lipoproteins; protein purification; interactions of peptides and proteins with metals; relation of structure to function of proteins; protein modification and sequence determination; magnetic resonance spectroscopy of proteins and peptides; electrostatic interaction in proteins; semisynthesis of proteins. *Mailing Add:* Dept Chem Ind Univ Bloomington IN 47401

GURD, FRASER NEWMAN, b Montreal, Que, Mar 19, 14; m 38; c 5. SURGERY. *Educ:* McGill Univ, BA, 34, MD & CM, 39; Univ Pa, MSc, 47; FRCPS(C) & FACS, 48. *Prof Exp:* Assoc prof surg, McGill Univ, 59-63, prof & chmn dept, 63-71; assoc secy, Royal Col Physicians & Surgeons Can, 72-75, consult, 76-81; EMER PROF, FAC MED, MCGILL UNIV, 80- *Concurrent Pos:* From sr surgeon to surgeon-in-chief, Montreal Gen Hosp, 59-71; mem coun, 64-72; regent, Am Col Surgeons, 66-74; consult staff, Montreal Gen Hosp, 71- *Honors & Awards:* Duncan Graham Award, Royal Col Physicians & Surgeons Can, 85; Surg Award for Serv To Safety, Nat Safety Coun, 85; F N G Starr Award, Can Med Asn, 90. *Mem:* Can Asn Clin Surgeons (pres, 68-70); Am Asn Surg Trauma (pres, 68); Am Surg Asn (vpres, 75); James IV Asn Surgeons (pres, 81-84); Int Surg Group (treas, 72-75); Hon Mem, Canadian Asn Gen Surgeons, 87. *Res:* Surgical physiology; physiology of the liver; surgical shock; physiology of body fluids and electrolytes; gastroenterology; surgical training. *Mailing Add:* 85 Range Rd Apt 801 Ottawa ON K1N 8J6 Can

GURD, RUTH SIGHTS, b Chicago, Ill, Sept 17, 27; m 56; c 2. PROTEIN CHEMISTRY, ENDOCRINOLOGY. *Educ:* Univ Mich, Ann Arbor, BS, 49; Wash Univ, St Louis, MO, 57. *Prof Exp:* Res assoc biochem, Dept Med, Presby Hosp, Chicago, 50-51 & Dept Surg, Sch Med, Wash Univ, 51-52; fel physiol, Col Med, Cornell Univ, 57-58; med consult, Aerospace Res Appln Ctr, Ind Univ, 62-66; sr res assoc neurochem, 67-71, from asst prof to assoc prof, 72-85, PROF BIOCHEM, SCH MED, IND UNIV, 85- *Mem:* Soc Neurosci; Biophys Soc; Am Soc Biol Chemists; Am Diabetes Asn. *Res:* Structure-function relationships of peptides and proteins with particular interest in diabetes and the structure/function/mechanisms of the hormones which control metabolic fluxes; the structure/function/mechanisms of the hormones which control metabolic fluxes. *Mailing Add:* 2032 Quail Run Dr NE Albuquerque NM 87122

GUREL, DEMET, b Ankara, Turkey, July 14, 39; m 62; c 2. ORGANIC CHEMISTRY. *Educ:* Am Col Girls, BS, 59; Smith Col, MA, 61; NY, PhD(chem), 73. *Prof Exp:* Adj asst prof chem, NY Univ, 73-89; CONSULT, 89- *Concurrent Pos:* Adj asst prof, Baruch Col, City Univ New York, 79- *Honors & Awards:* Fulbright Fel. *Mem:* Am Chem Soc; NY Acad Sci; Sigma Xi. *Res:* Chemical oscillations; thyroglobulin iodination. *Mailing Add:* One Wash Sq Village New York NY 10012

GUREL, OKAN, b Kirkagac, Turkey, Aug 1, 31; m 62; c 2. BIOMATHEMATICS, BIOPHYSICS. *Educ:* Tech Univ Istanbul, YMuh, 54; Univ BC, MASc, 57; Stanford Univ, PhD(appl mech), 61. *Prof Exp:* Staff mem math, Int Bus Mach Corp, 61-62; asst prof, Mid East Tech Univ, Ankara, 64-66; staff mem, NY Sci Ctr, 66-70; staff mem, Sci Ctr HQ, IBM Corp, 70-79; staff mem, Cambridge Sci Ctr, 80-85; DISCIPLINE SPECIALIST, LIFE & HEALTH SCI, ACAD INFO SYSTS, 85- *Concurrent Pos:* Secy, Scientist Training Group, Sci & Technol Res Coun, Turkey, 64-65, mem, 65-66; consult, Ford Found Sci Lycee Proj, Ankara, 65-66; adj prof, Baruch Col, 70-; vis prof, Bosphorus Univ, Istanbul, 77; NATO sr scientist fel, 77; consult, UN Develop Prog, 78. *Honors & Awards:* Fel, NY Acad Sci. *Mem:* Am Math Soc; Int Soc Chronobiol; NY Acad Sci; Biophys Soc. *Res:* Qualitative theory of differential equations and its application; stability theory and bifurcation of dynamical systems; biological systems, particularly molecular and cellular; chemical oscillations; biomathematics and biophysics; cytology. *Mailing Add:* One Washington Sq Village 2-J New York NY 10012

GÜRER, EMIR, b Kastamonu, Turkey, Apr 6, 60; m 89. IMPURITIES & DEFECTS IN SEMICONDUCTORS, SURFACE PHYSICS. *Educ:* Middle East Tech Univ, Ankara, Turkey, BS, 82, MS, 84; Lehigh Univ, MS, 86, PhD(physics), 90. *Prof Exp:* Teaching asst introductory physics, Middle East Tech Univ, 82-84; teaching asst introductory physics, Physics Dept, 84-86, res asst, 86-90, VIS RES ASSOC, ZETTLEMOYER CTR SURFACE STUDIES, LEHIGH UNIV, 90- *Mem:* Am Phys Soc; Mat Res Soc; NAm Catalysis Soc; Sigma Xi. *Res:* Defects in semiconductors; electronic properties of materials; study of surfaces of single crystal metals using vibrational and electronic spectroscopic techniques. *Mailing Add:* Sinclair Lab No 7 Lehigh Univ Bethlehem PA 18015

GUREVITCH, MARK, b Cleveland, Ohio, Sept 11, 16; wid; c 2. PHYSICS. *Educ:* Univ Calif, AB, 38, PhD(physics), 47. *Prof Exp:* Instr physics, Univ Calif, 42-43, res physicist, Off Sci Res & Develop Proj, 43-45; from asst prof to prof physics, Univ Idaho, 47-58; prof physics & chmn dept, Portland State Univ, 58-82. *Concurrent Pos:* NIH fel biophys, 66-67. *Mem:* AAAS; Am Phys Soc; Am Asn Physics Teachers. *Res:* Spectroscopy; properties of thin films; environmental science. *Mailing Add:* Physics Dept Box 751 Portland OR 97207

GURFINKEL, GERMAN R, b La Habana, Cuba, Sept 14, 32; US citizen; m 56; c 4. STRUCTURAL ENGINEERING. *Educ:* Univ Havana, Civil Eng, 55; Univ Ill, Urbana, MS, 57, PhD(civil eng), 66. *Prof Exp:* Asst prof civil eng, Univ Havana, 59-61; vis asst prof, 62-63, from instr to assoc prof, 63-74, PROF CIVIL ENG, UNIV ILL, URBANA, 74- *Concurrent Pos:* NSF res grant, 67-68; consult, NSF-AID, India, summers 67, 68 & 69; consult struct engr. 55-; James F Licncoln Welding award, 73, 77 & 78-83. *Mem:* Am Soc Civil Engrs; Am Concrete Inst; Am Soc Testing & Mat; Am Soc Agr Engrs. *Res:* Nuclear containment structures; wood structures; structural mechanics; numerical methods; tall buildings, silos and bunkers; investigation of structural failures. *Mailing Add:* Dept Civil Eng Univ Ill Urbana Campus 208 N Romine St Urbana IL 61801

GURGANUS, KENNETH RUFUS, b Williamston, NC, Sept 9, 48; m 76; c 1. MATHEMATICS. *Educ:* Univ NC, BS, 70, MS, 72, PhD(math), 76. *Prof Exp:* Programmer, Int Bus Mach, 70; lectr, 75-76, ASST PROF MATH, UNIV NC, 76- *Mem:* Am Math Soc; Math Asn Am. *Res:* Geometric function theory; real analysis. *Mailing Add:* Dept Math Sci Univ NC 601 S College Rd Wilmington NC 28403

GURIEN, HARVEY, b Brooklyn, NY, Sept 25, 25; m 50; c 5. ORGANIC CHEMISTRY. *Educ:* Brooklyn Col, BS, 49; Univ Pa, MS, 50; Polytech Inst Brooklyn, PhD, 55. *Prof Exp:* Res chemist, Gen Aniline Works, 53-60; res chemist, S B Penick & Co, 60-62; res chemist, Hoffmann-La Roche Co, 62-84; STERLING-WINTHROP RES INST, 85- *Concurrent Pos:* Pres, Regulatory Affairs Prof, 79- *Mem:* Drug Info Asn; Am Chem Soc; Regulatory Affairs Prof Soc. *Res:* Pharmaceutical developmental research; drug regulatory affairs. *Mailing Add:* Drug Regulatory Affairs Sterling-Winthrop Res Inst Rensselaer NY 12144

GURIES, RAYMOND PAUL, b Worcester, Mass, Dec 7, 43. CONSERVATION BIOLOGY. *Educ:* Univ Mass, BS, 71; Univ Wash, PhD(forest genetics), 75. *Prof Exp:* From asst prof to assoc prof, 77-87, PROF FOREST GENETICS, UNIV WIS-MADISON, 87- *Mem:* AAAS; Soc Study Evolution; Soc Conserv Biol. *Res:* Selection and breeding of woody plants; population genetics and genecology of forest tree species. *Mailing Add:* 1630 Linden Dr Madison WI 53706

GURIN, SAMUEL, b New York, NY, July 1, 05; m 30; c 2. BIOCHEMISTRY. *Educ:* Columbia Univ, BS, 28, MS, 30, PhD(biochem), 34. *Hon Degrees:* DSc, Philadelphia Col Pharm, 64, La Salle Col, 65 & Univ Fla. *Prof Exp:* Asst physiol chem, Teachers Col, Columbia Univ, 28-32, biochemist, Col Physicians & Surgeons, 34-35; Nat Res Coun fel chem, Univ Ill, 36; instr physiol chem, Sch Med, Univ Pa, 36-42, from asst prof to assoc prof biochem, 42-47, prof physiol chem, 48-54, Benjamin Rush prof biochem, 54-65, chmn dept, 54-62, dean, 62-68, prof, 65-69; prof biol sci, Univ Fla, 69-72, prof biochem & dir, Whitney Marine Lab, 72-85; CONSULT, 85- *Concurrent Pos:* Mem panel metab, Comt Growth, Nat Res Coun; mem study sect biochem, USPHS; mem panel regulation metab, NSF; mem bd sci counr, Nat Inst Arthritis & Metab Dis & Nat Inst Gen Med Sci. *Mem:* Am Chem Soc; Am Soc Biol Chemists. *Res:* Radioisotopes in study of metabolism; metabolic studies on lipids; isolation of chorionic gonadotrophin; marine chemoreception and metabolism; marine metabolism of pyridine carboxylic acids. *Mailing Add:* Pine Run L-6 Ferry & Iron Hill Rd Doylestown PA 18901-2199

GURINSKY, DAVID H(ARRIS), b Brooklyn, NY, Apr 26, 14; m 37; c 2. METALLURGY. *Educ:* NY Univ, AB, 36, PhD(phys chem), 42. *Prof Exp:* Asst instr chem physics, NY Univ, 36-42; res assoc metall lab, Univ Chicago, 42-44, res assoc, Inst Study Metals, 46-47; res assoc, Los Alamos Sci Lab, NMex, 44-46; scientist, 47-50, sr scientist & head metall div, 50-78, DEP CHMN, DEPT ENERGY & ENVIRON, BROOKHAVEN NAT LAB, 78- *Concurrent Pos:* Head mat res on high temperature graphite reactor & consult, Gen Atomic Div, Gen Dynamics Corp, 59-60. *Honors & Awards:* Studies Award, Am Inst Chem. *Mem:* Fel Am Nuclear Soc; Am Soc Metals; Am Inst Mining, Metall & Petrol Engrs. *Res:* Nuclear fuels; graphite radiation effects; effects of radiation on structural materials; liquid metal corrosion; superconductivity; superconductor materials; superconducting power transmission. *Mailing Add:* 40 Inwood Rd Center Moriches NY 11934

GURK, HERBERT MORTON, b Philadelphia, Pa, Aug 6, 30; m 52; c 3. MATHEMATICS. *Educ:* Univ Pa, BA, 51, MA, 52, PhD(math), 56. *Prof Exp:* Instr eng res, Univ Pa, 53-56; mathematician, Systems Eng, Radio Corp Am, 56-58, leader systs anal group, 58-61, space opers anal group, 62-66, mgr imagery systs progs, 66-68, mgr earth observation systs, 68-74, STAFF SYST SCIENTIST, ADVAN PROG PLANNING, RCA CORP, 75- *Mem:* Soc Indust & Appl Math (treas, 61-65); Am Math Soc; Am Inst Aeronaut & Astronaut. *Res:* Theory of games; theory of waiting lines and stochastic processes; numerical analysis; mathematical linguistics; space systems planning and analysis; space meteorological systems. *Mailing Add:* 26 Howe Cir Princeton NJ 08540

GURKLIS, JOHN A(NTHONY), b Waterbury, Conn, July 12, 21; m 58; c 2. ENVIRONMENTAL ENGINEERING, ELECTROCHEMICAL ENGINEERING. *Educ:* Univ Md, BS, 43; Ohio State Univ, MS, 47, PhD(chem eng), 50. *Prof Exp:* Asst chem engr, Ohio State Univ, 43-44 & 49-50; prin chem engr, 50-56, sr chem engr, Electrochem Eng Dev, 56-76, sr chem engr, energy & environ systs assessment sect, 76-82, SR CHEM ENGR, ENVIRON TECHNOL & ASSESSMENT SECT, BATTELLE MEM INST, 82- *Mem:* Am Inst Chem Engrs; Am Electroplaters Soc; Soc Mfg Engrs; Am Soc Metals. *Res:* Hazardous waste management, water pollution control, wastewater treatment, industrial waste management, solid waste management, waste minimization, resource recovery, energy conservation; chemical milling; electrodeposition of metals and alloys; electrowinning; electrorefining; electropolishing; chemical engineering; metal finishing; electrochemical machining. *Mailing Add:* 2840 N Star Rd Columbus OH 43221-2959

GURLAND, JOHN, b Can, Jan 6, 17; nat US; m 48; c 2. STATISTICS, PROBABILITY. *Educ:* Univ Toronto, MA, 42; Univ Calif, PhD(math statist), 48. *Prof Exp:* Pierce instr math, Harvard Univ, 48-49; asst prof statist, Univ Chicago, 49-52; from assoc prof to prof, Iowa State Univ, 52-60; prof, Math Res Ctr, 60-63, PROF STATIST, UNIV WIS-MADISON, 63- *Mem:* Fel AAAS; fel Inst Math Statist; Biomet Soc; fel Am Statist Asn; Math Asn Am. *Res:* Mathematical and applied statistics; probability and distribution theory, estimation and test of fit; regression, biological assay, survival analysis and reliability. *Mailing Add:* Statis 4237 Comp Sci Univ Wis-Madison Madison WI 53706

GURLAND, JOSEPH, b Berlin, Ger, Jan 26, 23; nat US; m 48; c 2. PHYSICAL METALLURGY. *Educ:* NY Univ, BChE, 44, MS, 47; Mass Inst Technol, ScD, 51. *Prof Exp:* Res engr, Battelle Mem Inst, 47-48; asst metall, Mass Inst Technol, 48-51; res engr & mgr basic res, Firth-Sterling, Inc, 51-55; from asst prof to assoc prof, 55-64, prof, 64-87, EMER PROF ENG, BROWN UNIV, 88- *Concurrent Pos:* NATO fel sci, 70; NSF fac fel, 62 & foreign travel award, 78. *Honors & Awards:* Fel, Am Soc Metals, 87; Plansee Medal, Int Plansee Soc Powder Metal, 89. *Mem:* Am Soc Metals; Am Inst Mining Metall & Petrol Engrs; Am Soc Testing & Mat. *Res:* Structure of sintered carbides; composite alloys; fracture of metals; quantitative metallography. *Mailing Add:* Div Eng Brown Univ Providence RI 02912

GURLEY, LAWRENCE RAY, b Goldsboro, NC, Jan 13, 35; m 58; c 2. BIOCHEMISTRY, CHEMICAL ENGINEERING. *Educ:* NC State Univ, BChE, 58, ChE, 59; Univ NC, PhD(biochem), 64. *Prof Exp:* Prod engr, Merck & Co, Va, 58-60; staff biochemist, 64-83, LAB FEL, LOS ALAMOS NAT LAB, UNIV CALIF, 83- *Concurrent Pos:* Fel, Los Alamos Nat Lab, Univ Calif, Div leader, Life Sci, 86-88. *Honors & Awards:* Coker Award, Elisha Mitchell Soc, 65. *Mem:* Am Soc Biochem & Molecular Biol; Am Soc Cell Biol; Am Chem Soc; Protein Soc; Int Cell Cycle Soc. *Res:* Chromatin structure; histone chemistry and biology; histone metabolism; tissue culture; biochemistry of the cell life cycle; radiation biology; pulmonary biochemistry; high performance liquid chromatography. *Mailing Add:* Biochem Group Los Alamos Nat Lab M-880 Los Alamos NM 87545

GURLEY, THOMAS WOOD, b Endicott, NY, May 28, 46; m 68; c 3. ANALYTICAL CHEMISTRY. *Educ:* Houghton Col, BA, 68; Case Western Reserve Univ, PhD(chem), 76. *Prof Exp:* Sr res chemist, 76-82, SECT HEAD, GOODYEAR TIRE & RUBBER CO, 82- *Mem:* Am Chem Soc. *Res:* Development of new analytical methodology using gas chromatography, high performance liquid chromatography and wet chemistry for chemicals related to the rubber industry; polymer analysis; environmental analysis. *Mailing Add:* 1495 Limonis Ave NW Uniontown OH 44685

GURLL, NELSON, b Providence, RI, Jan 8, 42; m 65. ENDOCRINOLOGICAL SURGERY. *Educ:* Univ Calif, Berkeley, AB, 63, San Francisco, MD, 67. *Prof Exp:* Clin fel surg, Harvard Med Sch, 68-73; res assoc bioeng & mech eng, Mass Inst Technol, 70-71; teaching fel surg, Harvard Med Sch, 72-73; res investr surg gastroenterol, Walter Reed Army Inst Res, 73-76, chief dept, 74-76; asst prof, 76-80, assoc prof surg, 80-83, PROF SURG, COL MED, UNIV IOWA, 83- *Concurrent Pos:* Intern & resident, Beth Israel Hosp, Boston, 67-73; NIH acad surg trainee, Mass Inst Technol & Beth Israel Hosp, 70-73; resident physician, Clin Res Ctr, Mass Inst Technol, 70-71; attending staff physician, Vet Admin Med Ctr, Iowa City, 76-; staff surgeon, Hosp & Clin, Univ Iowa, 76-; actg chief surg, Vet Admin Med Ctr, Iowa City, 78-79; vis prof path, Univ BC, 85-86. *Honors & Awards:* AOA. *Mem:* Am Surg Asn; Am Gastroenterol Asn; Am Physiol Soc; fel Am Col Surgeons; Soc Univ Surgeons. *Res:* Investigating role of endorphins in the pathophysiology of shock and in gastrointestinal physiology. *Mailing Add:* Dept Surg Hosp & Clin Col Med Univ Iowa Iowa City IA 52240

GURNETT, DONALD ALFRED, b Cedar Rapids, Iowa, Apr 11, 40; m 64; c 2. SPACE PLASMA PHYSICS. *Educ:* Univ Iowa, BS, 62, MS, 63, PhD(physics), 65. *Prof Exp:* NASA trainee, Univ Iowa, 62-64 & Stanford Univ, 64-65; from asst prof to assoc prof, 65-72, PROF PHYSICS, UNIV IOWA, 72- *Concurrent Pos:* vis scientist, Max Planck Inst Extraterrestrial Physics, Garching, Ger, 75-76; vis prof, Inst Geophysics & Planetary Physics, Univ Calif, Los Angeles, 79-80. *Honors & Awards:* Alexander von Humboldt Sr Scientist Award, 75; J H Dellinger Gold Medal, Int Sci Radio union, 78; Distinguished Sci Achievement Award, NASA, 80; NASA Space Act Award, 86; J A Fleming Medal, Am Geophys Union, 89; Excellence in Plasma Physics, Am Phys Soc, 89. *Mem:* Am Geophys Union; Int Sci Radio Union; Am Phys Soc; Int Asn Astronaut. *Res:* Experimental and theoretical investigation of plasma waves and radio emissions in planetary magnetospheres, in the solar wind, and in the solar corona. *Mailing Add:* Dept Physics & Astron Univ Iowa Iowa City IA 52242

GURNEY, ASHLEY BUELL, entomology; deceased, see previous edition for last biography

GURNEY, CLIFFORD W, b Chicago, Ill, Apr 11, 24; m 49; c 5. INTERNAL MEDICINE. *Educ:* Univ Chicago, BS, 48, MD, 51. *Prof Exp:* From asst prof to prof med, Univ Chicago, 56-66, assoc prof physiol, 62; prof med & chmn dept, Sch Med, Rutgers Univ, 66-69; prof & chmn dept, Sch Med, Univ Kans Med Ctr, Kans City, 69-72; dep dean clin sci, 72-76, PROF MED, PRITZKER SCH MED, UNIV CHICAGO, 72- *Concurrent Pos:* Markle scholar, 59-64; staff mem, Argonne Cancer Res Hosp, 56-66. *Mem:* Am Soc Clin Invest; Am Soc Hemat; AMA; fel Am Col Physicians; Am Fedn Clin Res. *Mailing Add:* 10411 S Seeley Chicago IL 60643

GURNEY, ELIZABETH TUCKER GUICE, b Berkeley, Calif, Apr 5, 41; m 66. MOLECULAR BIOLOGY, CELL BIOLOGY. *Educ:* Univ Chicago, BA, 62; Univ Calif, Berkeley, MS, 70, PhD(genetics), 75. *Prof Exp:* Tech asst biol, Mass Inst Technol, 63-67, teaching asst animal cells & viruses, Cold Spring Harbor Lab, 66; res biochemist, Univ Calif, Berkeley, 70-73; NIH fel animal virol, 75-77, res asst prof, 76-83, RES ASSOC PROF, UNIV UTAH, 83- *Mem:* Am Soc Cell Biol; Am Soc Microbiol; AAAS. *Res:* Cellular growth control mechanisms and transformation by tumor viruses; monoclonal antibodies for tumor antigens; DNA rearrangements. *Mailing Add:* Dept Biol Univ Utah Salt Lake City UT 84112

GURNEY, RAMSDELL, b Buffalo, NY, Aug 2, 03; m 32; c 2. INTERNAL MEDICINE. *Educ:* Yale Univ, BA, 25; Univ Buffalo, MD, 29; Am Bd Internal Med, dipl, 42. *Prof Exp:* Asst prof physiol, 47-50, from asst prof to assoc prof med, 50-76, EMER ASSOC PROF MED, STATE UNIV NY BUFFALO, 76- *Concurrent Pos:* Pvt pract, 32-; attend physician, Buffalo Gen Hosp, 57-71, consult physician, 71- *Mem:* Am Col Physicians. *Res:* Obesity and sweating; cardiovascular disease. *Mailing Add:* 85 Hight St Buffalo NY 14203

GURNEY, THEODORE, JR, b Hartford, Conn, Oct 14, 38; m 66. CELL BIOLOGY. *Educ:* Harvard Col, AB, 59; Yale Univ, MS, 61, PhD(biophys), 65. *Prof Exp:* USPHS fels biol, Mass Inst Technol, 64-67; Jane Coffin Childs Mem Fund med res fel molecular biol, 67-68, asst prof, Univ Calif, Berkeley, 68-74; ASSOC PROF BIOL, UNIV UTAH, 74- *Mem:* AAAS; Am Soc Cell Biol. *Res:* Contact inhibition of growth in cell culture; cell fractionation; RNA processing; DNA rearrangements. *Mailing Add:* Dept Biol Univ Utah Salt Lake City UT 84112

GUROL, MIRAT D, b Ankara, Turkey, Sept 2, 51; m 75. WASTEWATER MANAGEMENT. *Educ:* Middle East Tech Univ, BS, 73; Univ NC, MS, 77, PhD(environ eng), 80. *Prof Exp:* Teaching asst environ eng, Middle East Tech Univ, 73-75; res asst environ eng, Univ NC, 77-79; ASST PROF ENVIRON ENG, DREXEL UNIV, 80- *Concurrent Pos:* Res scholar award, Drexel Univ, 81. *Mem:* Am Inst Chem Eng; Int Ozone Asn; Water Pollution Coun Fedn; Am Water Works Asn; Am Chem Soc. *Res:* Treatment of industrial wastewaters by chemical oxidation; reuse and recycling possibilities of industrial wastes by physical and chemical treatment; kinetics of treatment processes of water and wastewater. *Mailing Add:* Environ Studies Inst Drexel Univ Philadelphia PA 19104

GURPIDE, ERLIO, b Buenos Aires, Arg, Apr 8, 27; US citizen; m 61; c 1. BIOCHEMISTRY, BIOPHYSICS. *Educ:* Univ Buenos Aires, PhD(chem), 55. *Prof Exp:* Res assoc biochem, Columbia Univ, 59-64, asst prof, 64-69; prof biochem, obstet & gynec, Univ Minn, Minneapolis, 69-72; PROF BIOCHEM, OBSTET, GYNEC & REPROD SCI, MT SINAI SCH MED, 72- *Concurrent Pos:* Career scientist, Health Res Coun, NY, 63-69 & 73-; mem endocrinol study sect, NIH, 70-73. *Mem:* Am Soc Biol Chem; Endocrine Soc; Soc Gynec Invest. *Res:* Steroid biochemistry; use of isotopically labeled tracers in biology; endocrinology of reproduction. *Mailing Add:* Dept Biochem Ob-Gyn & Reprod Sci Mt Sinai Sch Med City Univ NY 1176 Fifth Ave New York NY 10029

GURR, HENRY S, b Ohio, Mar 9, 37; m 61; c 3. ELEMENTARY PARTICLE PHYSICS. *Educ:* Case Inst Technol, BS, 59, MS, 64 & PhD(nuclear physics), 67. *Prof Exp:* Res physicist, Univ Calif, Irvine, 67-76; asst prof, 76-78, PROF PHYSICS, UNIV SC, AIKEN, 78- *Concurrent Pos:* Vis prof physics & eng, Univ SC, Aiken, 73-76; res physicist, Univ Calif Irvine, 78- *Mem:* Am Phys Soc; Sigma Xi; AAAS. *Res:* Experimental proof of the neutrino electron scattering, at Department of Energy reactor site, which could recoil from an electron research participation continues with collaboration in neutrino deuteron neutral current experiment; author of two publications on proceedings of international conferences in cosmic rays. *Mailing Add:* Dept Sci Univ SC Aiken 171 Univ Pkwy Aiken SC 29801

GURRY, ROBERT WILTON, b Schenectady, NY, Sept 27, 13; m 39; c 2. PHYSICAL CHEMISTRY. *Educ:* Union Univ, NY, BS, 34; Yale Univ, PhD(phys chem), 37. *Prof Exp:* Phys chemist, US Steel Corp, 37-55; phys chemist, Quaker Chem Prod Corp, 55-59; tech dir, Union Steel Corp, Piscataway, 59-72; consult mat sci & failure anal, 72-88; RETIRED. *Concurrent Pos:* Instr, Stevens Inst Technol, 44-46. *Mem:* Am Chem Soc; Am Soc Metals; Nat Asn Corrosion Engrs; Am Soc Testing & Mat; Welding Res Coun, Am Soc Mech Engrs. *Res:* Surface chemistry of metals; corrosion; failure analysis; materials application; metal processing. *Mailing Add:* 52 Tropicana Dr Punta Gorda FL 33950

GÜRSEY, FEZA, b Istanbul, Turkey, Apr 7, 21; m 52; c 1. ALGEBRAIC METHODS IN QUANTUM MECHANICS. *Educ:* Istanbul Univ, BSc, 44; Imperial Col, Univ London PhD(appl math), 50. *Hon Degrees:* Dr, Istanbul Univ, 81. *Prof Exp:* Theoret physics, Istanbul Univ, 54-61; prof physics, Middle East Tech Univ, Ankara, 61-74; prof, 68-77, J W GIBBS PROF, PHYSICS, YALE UNIV, 77- *Concurrent Pos:* Res physicist, Brookhaven Nat Lab, 57-58; vis prof, Columbia Univ, New York, 60-61, Col de France, Paris, 81, Scuola Normale, Pisa, 86; ed, Il Nuovo Cimento, 68-; ed, Jour Math Physics, 87- *Honors & Awards:* Science Prize, Turkish Nat Org Sci & Tech Res, 69; J R Oppenheimer Prize, Univ Fla, 77; A Cressy Morrison Award, New York Acad of Sci; Wigner Medal, Group TH & Fund Phys Found, 86. *Mem:* Turkish Phys Soc; Am Phys Soc; fel Third World Acad Sci; fel Am Acad Arts & Scis. *Res:* Group theoretical and supersymmetric models in particle physics; Contributions to various topics in theoretical physics Lab PO Box 6666. *Mailing Add:* Ctr Theoret Physics Yale Univ Sloane Physics Lab PO Box 6666 New Haven CT 06511

GURSKI, THOMAS RICHARD, b Elizabeth, NJ, Dec 3, 40; m 64; c 6. QUANTUM OPTICS, ELECTROOPTICS. *Educ:* Loyola Univ, Los Angeles, BS, 62; Univ Mich, MS, 63; Univ Ariz, PhD(optics), 74. *Prof Exp:* Assoc physicist electrooptics, Fed Systs Div, IBM Corp, 64-68; sr physicist optics, Perkin-Elmer corp, 68-69; consult electrooptics, Kitt Peak Nat Observ, 71-74; res assoc optics, Optical Sci Ctr, Univ Ariz, 69-75; mem staff optics, Lincoln Lab, Mass Inst Technol, 75-80; mem staff electrooptics, W J Schafer Asn, 80-83; vpres res & develop, Am Surg Laser Inc, 83-84; pres & chmn, Electro-optical Technol, Inc, 85-89; PRIN ENGR, RAYTHEON CO, 89- *Mem:* Optical Soc Am; Soc Photo-Optical Instrumentation Engrs. *Res:* Coherent detection at 10.6 microns, laser oscillator frequency stability; passive infrared detection; robotics and program management; laser design and development. *Mailing Add:* 62 Martin Rd Concord MA 01742

GURSKY, HERBERT, b Bronx, NY, May 27, 30; m 58; c 2. ASTRONOMY. *Educ:* Univ Fla, BS, 51; Vanderbilt Univ, MS, 53; Princeton Univ, PhD(physics), 59. *Prof Exp:* Instr physics, Princeton Univ, 57-58; instr, Columbia Univ, 58-61; sr scientist, Am Sci & Eng, Inc, 61-66, proj dir, 66-69, vpres, Space Res Div, 69-73; astrophysicist, Smithsonian Astrophys Oberv, 73-76, assoc dir optical & infrared astron, Harvard/Smithsonian Ctr Astrophysics, 76-81; SUPT, SPACE SCI DIV, NAVAL RES LAB, 81- *Concurrent Pos:* Assoc, Harvard Univ, 71- 75, prof astron, 75-81. *Mem:* AAAS; Am Phys Soc; Am Astron Soc. *Res:* Galactic and extragalactic astronomy using x-ray and visible light observations. *Mailing Add:* Space Sci Div Naval Res Lab Code 4100 Washington DC 20375

GURSKY, MARTIN LEWIS, b New York, NY, Mar 19, 27; m 52; c 5. THEORETICAL PHYSICS. *Educ:* Ga Inst Technol, BS, 48, MS, 51; Vanderbilt Univ, PhD(physics), 58. *Prof Exp:* Instr physics, Ga Inst Technol, 48-50; mem staff, Melpar, Inc, 53-54; MEM STAFF, LOS ALAMOS SCI LAB, UNIV CALIF, 54- *Mem:* Am Phys Soc. *Res:* Nuclear theory. *Mailing Add:* 223 El Vinto Los Alamos NM 87544

GURST, JEROME E, b Atlantic City, NJ, Aug 9, 38. ORGANIC CHEMISTRY. *Educ:* Dartmouth Col, AB, 60; Stanford Univ, PhD(chem), 65. *Prof Exp:* Fel organoboranes, Purdue Univ, 64-65 & stereochem, Princeton Univ, 65-66; vis asst prof org chem, Univ Ore, 66-67; from asst prof to assoc prof org chem, 67-77, assoc prof, 77-81, PROF CHEM, UNIV WFLA, 81- *Concurrent Pos:* Vis assoc prof, Dartmouth Col, 73-80; vis res prof, Univ Nev, Reno, 80-81. *Mem:* AAAS; Am Chem Soc; Sigma Xi. *Res:* Stereochemistry; steroids; mass spectrometry; optical rotatory dispersion; circular dichroism; natural products. *Mailing Add:* Dept Chem Univ WFla Pensacola FL 32514-5751

GURTIN, MORTON EDWARD, b Jersey City, NJ, Mar 7, 34; c 2. APPLIED MATHEMATICS, MECHANICS. *Educ:* Rensselaer Polytech Inst, BME, 55; Brown Univ, PhD(appl math), 61. *Prof Exp:* Engr struct, Douglas Aircraft Co, Inc, 55-56 & Gen Elec Co, 56-59; res assoc appl math, Brown Univ, 61-62, from asst prof to assoc prof, 62-66; PROF APPL MATH, CARNEGIE-MELLON UNIV, 66- *Concurrent Pos:* Consult, Gen Elec Co, 61-63 & Mech Tech Inc, 62-64; lectr, Nat Univ Mex, 64 & Univ Wash, 65; NSF res grant, 65-; lectr, WVa Univ, 67-68; consult, Sandia Corp, 67-; lectr, Southwestern Mech Lect Series, 68, Wash State Univ, 70, Univ Rio de Janeiro, 70 & Int Ctr Mech Sci, Italy, 71, Simon Fraser Univ, 71 & Univ Pisa, 71 & 74; Sr Fulbright-Hays res fel, Univ Pisa, 74, Guggenheim fel, 74; lectr, Univ Poitiers, France, 74, Univ Glasgow, 74, Univ Strathclyde, Scotland, 74, Univ Florence, 74, Istanbul Univ, 74, Univ Bari, Italy, 76, Midwest Mechanics Lect Series, 76; consult, United Tech Ctr, 76-; Air Force Off Sci Res grant, 76-81; Army Res Office grant, 78-; consult, Los Alamos Nat Lab, 80-90; lectr, Univ Tenn, 79, Univ Paris, 81 & Univ Houston, 82. *Mem:* Am Math Soc; Soc Natural Philos; Sigma Xi. *Res:* Foundations of continuum mechanics and thermodynamics; elasticity and viscoelasticity theory; wave propagation; population dynamics; partial differential equations. *Mailing Add:* Dept Math Carnegie-Mellon Univ Pittsburgh PA 15213

GURTNER, GAIL H, b Buffalo, NY, Feb 8, 38. PULMONARY PHYSIOLOGY. *Educ:* Univ Buffalo, MD, 62. *Prof Exp:* ASSOC PROF MED, SCH MED, JOHNS HOPKINS UNIV, 76- *Mailing Add:* Dept Pulmon Med Westchester County Med Ctr Valhalla NY 10595

GURUDATA, NEVILLE, b Berbice, Guyana, Apr 23, 37; m 62; c 3. PHYSICAL ORGANIC CHEMISTRY. *Educ:* Univ St Andrews, BSc, 59, Hons, 60; Univ Western Ont, PhD(chem), 68. *Prof Exp:* Sci officer, Dept Govt Analyst, Guyana, 60-64; res fel, Univ Ottawa, 68-69; lectr & res assoc chem, Sir George Williams Univ, 69-73; chmn, Dept Chem & Chem Technol, 74-76, PROF CHEM, DAWSON COL, 74- *Concurrent Pos:* Instr, Univ Guyana, 63-64; adj asst prof, Concordia Univ, 73-75; mem, Laurenval Sch Bd, 79, chmn coun, 83-84, chmn exec, 86-87. *Mem:* Chem Inst Can. *Res:* Use of nuclear magnetic resonance spectroscopy to determine structure, stereochemistry, thermodynamic parameters; interactions between formally non-conjugated chromophores; relationships between chemical structure, spectra and chemical reactivity; approaches to synthesis of selected organic compounds. *Mailing Add:* Dawson Col 3040 Sherbrook St W Montreal PQ H3Z 1A4

GURUSIDDAIAH, SARANGAMAT, b Chitradurga, India, Mar 18, 37; US citizen; m 64; c 2. ANTIBIOTICS, NATURAL PRODUCTS. *Educ:* Bangalore Univ, India, BSc, 60, MSc, 61; Wash State Univ, Pullman, PhD(plant path), 71. *Prof Exp:* Res assoc agr chem, 70-71, mgr, bioanal ctr, 71-76, asst scientist , 76-80, assoc scientist, 80-85, ASSOC DIR, BIOANAL LAB, WASH STATE UNIV, 80-, SCIENTIST, 86- *Concurrent Pos:* Co-prin investr antibiotics, NIH grant, 79-82; prin investr antibiotics, Dow Chem Co grant, 82-86, Hoffman- LaRoche, 87-92. *Mem:* AAAS; Am Physiol Soc; Am Soc Internal Med. *Res:* Isolation and characterization of antifungal, antibacterial, antiviral and antitumor compounds from microorganisms, plants and other natural sources; natural products isolation and characterization. *Mailing Add:* Biolanal Ctr Wash State Univ Troy Hall 304-305 Pullman WA 99164-4430

GURUSWAMY, VINODHINI, b Colombo, Sri Lanka; m 67; c 2. SOLAR ENERGY CONVERSION, ELECTROCHEMISTRY. *Educ:* Univ Ceylon, BSc, 63; Univ Surrey, Dipl, 66; Univ Newcastle, PhD(chem eng), 76. *Prof Exp:* Res officer, Ceylon Inst Sci & Indust Res, 67-76, sr res officer, 76-78; res assoc phys sci, Flinders Univ, Australia, 78-79; sr res assoc chem, Tex A&M Univ, 79-81; ASST PROF CHEM ENG, VA POLYTECH & STATE UNIV, 81- *Mem:* Inst Engrs UK; Inst Chem Eng UK; Asn Am Women Sci; Am Electrochem Soc. *Res:* Electrochemistry with accent on energy conversion, namely photoelectrochemistry and fuel cells; photoelectrochemical reactors as the synthesis route of tomorrow. *Mailing Add:* 10618 Montrose Ave Bethesda MD 20814

GURVITCH, MICHAEL, physics, for more information see previous edition

GURWARA, SWEET K, medicinal chemistry, cancer, for more information see previous edition

GURWITH, MARC JOSEPH, b Paris, France, July 29, 39; US citizen; m 70; c 2. INFECTIOUS DISEASES. *Educ:* Yale Univ, BA, 61; Harvard Med Sch, MD, 65. *Prof Exp:* Intern internal med, Stanford Univ Hosp, 65-66; res assoc infectious dis, Sch Med, Univ Southern Calif, 66-67; epidemiologist, US AID Malaria Eradication Prog, 67-68; resident, Stanford Univ Hosp, 68-69, fel infectious dis, 69-70, resident, 70-71; staff physician, Kaiser Permanente, 71-72; asst prof internal med & med microbiol, Univ Man, 72-77; assoc prof med, Sch Med, Univ Kans, 77-79; assoc prof med, Univ Calif, Los Angeles, 79-80; ASSOC PROF MED & MICROBIOL & CHIEF, INFECTIOUS DIS SECT, DEPT MED, MICH STATE UNIV, 80- *Concurrent Pos:* Adj prof microbiol, Univ Man, 77-79; asst chief res, Infectious Dis Sect, Wadsworth Vet Admin Hosp, 79-80. *Mem:* Am Soc Microbiol; Am Fedn Clin Res; Can Soc Clin Invest; Royal Col Physicians & Surgeons Can; Am Col Physicians. *Mailing Add:* Wyeth Laboratories P O Box 8299 Philadelphia PA 19101

GUSBERG, SAUL BERNARD, b Newark, NJ, Aug 3, 13; m 38; c 1. MEDICINE, EDUCATIONAL & RESEARCH ADMINISTRATION. *Educ:* Harvard Univ, MD, 37; Columbia Univ, ScD, 48. *Hon Degrees:* SDc, Univ Barcelona, 87. *Prof Exp:* Fel oncol, Huntington Hosp, Harvard Univ, 38; resident obstetrician & gynecologist, Sloane Hosp Women, 46; assoc prof clin obstet & gynec, Col Physicians & Surgeons, Columbia Univ, 54-62, clin prof obstet & gynec, 62-68; prof & chmn dept, Mt Sinai Sch Med, 68-80, obstetrician & gynecologist-in-chief, 62-80, distinguished serv prof, 80-84, EMER PROF, MT SINAI SCH MED, COLUMBIA UNIV, NY, 84- *Concurrent Pos:* Assoc attend obstetrician & gynecologist, Columbia-Presby Med Ctr; consult med affairs & res, Am Cancer Soc; ed-in-chief, Gynec Oncol. *Honors & Awards:* Am Cancer Soc Medal. *Mem:* AAAS; fel Am Gynec Soc; Soc Gynec Oncol (pres, 74-75); fel Am Radium Soc (vpres, 67); Am Soc Cytol (vpres, 59-62); Sigma Xi; Asn Educ Cancer; Am Gynec & Obstet; NY Acad Med; Soc Pelvic Surg (pres, 76); Am Cancer Soc (pres, 81); Sigma Xi. *Res:* Gynecologic cancer; reproductive biology; author of textbooks on gynecological cancer. *Mailing Add:* 257 Palisade Ave Dobbs Ferry NY 10522

GUSDON, JOHN PAUL, b Cleveland, Ohio, Feb 13, 31; m 56; c 3. OBSTETRICS & GYNECOLOGY, IMMUNOLOGY. *Educ:* Univ Va, BA, 52, MD, 59. *Prof Exp:* From intern to resident, Univ Hosps Cleveland, Ohio, 59-64; from instr to asst prof, Case Western Reserve Univ, 66-67; from asst prof to assoc prof, 67-74, PROF OBSTET & GYNEC, BOWMAN GRAY SCH MED, 74-, ASSOC MICROBIOL, 67- *Concurrent Pos:* Fel immunol, Sch Med, Case Western Reserve Univ, 64-67; Josiah Macy Jr Found fac fel, 66-69. *Mem:* AAAS; Am Asn Immunol; Am Soc Immunol Reproduction (pres, 81-); Am Col Obstet & Gynec; Am Gynec & Obstet Soc. *Res:* Immunological aspects of fetal-maternal relationships. *Mailing Add:* Bowman Gray Sch Med Wake Forest Univ Winston-Salem NC 27103

GUSEMAN, LAWRENCE FRANK, JR, b New Iberia, La, Feb 16, 38; m 68; c 2. MATHEMATICS. *Educ:* Tex A&M Univ, BA, 60, MS, 62; Univ Tex, Austin, PhD(math), 68. *Prof Exp:* Appl mathematician, M D Anderson Hosp & Tumor Inst, 62; res mathematician, Theory & Anal Off, Comput & Anal Div, NASA Manned Spacecraft Ctr, 64-68; from asst prof to assoc prof, 68-81, PROF MATH, TEX A&M UNIV, 81-, DIR, OFF GRAD STUDIES, 88- *Concurrent Pos:* Teaching asst, Univ Tex, Austin, 64-66. *Mem:* Am Math Soc; Math Asn Am; Soc Indust & Appl Math; Sigma Xi. *Res:* Functional analysis, non-linear fixed point theory and optimization theory; mathematical techniques in pattern recognition & statistical decision theory. *Mailing Add:* 2801 Cherry Creek Circle Bryan TX 77802

GUSENIUS, EDWIN MAURTIZ, b Bristol, SDak, July 5, 16; m 48; c 2. INORGANIC CHEMISTRY. *Educ:* Gustavus Adolphus Col, BA, 38; Univ SDak, MA, 48; Kans State Univ, PhD(inorg chem), 63. *Prof Exp:* Instr, SDak High Sch, 38-40; chemist, E I du Pont de Nemours & Co, Inc, 41-42, supvr, 42-45; asst chem, Univ SDak, 46-47; instr, Luther Jr Col, 47-54; from asst prof to prof, Bethany Col, Kans, 54-68; prof, Kearney State Col, 68-69; prof & chmn dept, 69-81, EMER PROF, FRIENDS UNIV, 81- *Mem:* Am Chem Soc. *Res:* Metal complexes. *Mailing Add:* 321 N Second Lindsborg KS 67456-2006

GUSHEE, BEATRICE ELEANOR, b Dorchester, Mass, Sept 22, 18. INORGANIC CHEMISTRY. *Educ:* Simmons Col, BS, 42; Vassar Col, MS, 46; Univ Conn, PhD(inorg chem), 56. *Prof Exp:* Anal chemist, Celanese Corp Am, 42-44; instr chem, Hofstra Col, 46-48; asst, Univ Conn, 48-54; instr, Smith Col, 54-57; from asst prof to assoc prof, 57-89, EMER PROF CHEM, HOLLINS COL, 89- *Mem:* Am Chem Soc; Nat Sci Teachers Asn; Sigma Xi. *Res:* High temperature and solid state reactions and transformation; structural inorganic chemistry. *Mailing Add:* Dept Chem Hollins Col Hollins College VA 24020-1675

GUSKEY, LOUIS ERNEST, b Pittsburgh, Pa, Jan 13, 42; m 65; c 2. VIROLOGY, MICROBIOLOGY. *Educ:* Capital Univ, BS, 63; Ohio State Univ, MSc, 67, PhD(microbiol), 71. *Prof Exp:* Fel virol & biochem, Ciba Geigy, Basle, Switz, 71-73; res fel virol & lipid chem, Hormel Inst, 73-75; res asst prof virol, Waksman Inst Microbiol, 75-77; asst prof microbiol, Univ Wis, Milwaukee, 77-84; CHIEF, DIV VIROL, MICH DEPT PUB HEALTH, 84- *Concurrent Pos:* Res fel, Prog Projs Br Nat Heart & Lung Inst, 73-75; res asst prof, Nat Inst Allergy & Infectious Dis, 75-77. *Mem:* Am Soc Microbiol; Tissue Cult Asn; Am Soc Virol. *Res:* Virus-host cell interactions; molecular events that cause the development of viral-induced cytopathologies and lead to cell death; methods for rapid viral diagnosis. *Mailing Add:* Virol Sect Mich Dept Public Health Lansing MI 48909

GUSOVSKY, FABIAN, b Buenos Aires, Arg, Jan 15, 57; m 84; c 1. MOLECULAR PHARMACOLOGY. *Educ:* Univ Buenos Aires, dipl pharm, 78, dipl biochem, 80; Rush Med Col, PhD(pharmacol), 84. *Prof Exp:* Res assoc, Psychobiol Lab, Rush-Presby Hosp, Chicago, 81-84; vis fel, 85-87, staff fel, 87-88, SR STAFF FEL, BIOORG CHEM LAB, NAT INST DIABETES & DIGESTIVE & KIDNEY DIS, NIH, 88- *Mem:* Soc Neurosci; Am Soc Pharmacol & Exp Therapeut. *Res:* Biochemistry and pharmacology of signal transduction in mammalian systems; study of second messenger generating systems. *Mailing Add:* NIH Bldg 8 Rm 1A-15 9000 Rockville Pike Bethesda MD 20892

GUSS, CYRUS OMAR, b Wolford, NDak, Feb 1, 11; m; c 2. ORGANIC CHEMISTRY. *Educ:* Jamestown Col, BS, 32; Univ Minn, MS, 36, PhD(org chem), 40. *Prof Exp:* Res chemist, Dow Chem Co, 36-37, group leader, 40-42; res chemist, Rohm and Haas Co, 39; chemist, Forest Prod Lab, US Forest Serv, 42-46; asst prof chem, Univ Southern Calif, 46-50; assoc prof, Colo State Univ, 50-60; vis instr, Univ Ill, 60-61; chmn dept, 61-71, prof, 61-74, EMER PROF CHEM, UNIV NEV, RENO, 74- *Concurrent Pos:* Vis prof, Univ Ill, Urbana, 74-75 & Univ Calif, Irvine, 79. *Mem:* AAAS; Am Chem Soc. *Res:* Preparation and reactions of olefin oxides; rearrangements; intramolecular displacements. *Mailing Add:* 315 Ronnie Rd Golden CO 80403-9757

GUSS, MAURICE LOUIS, b Revere, Mass, June 21, 22; m 46; c 3. VIROLOGY. *Educ:* Boston Col, BS, 47; Univ Mass, MS, 49; Cornell Univ, PhD(bact), 52. *Prof Exp:* Microbiologist, US Army Biol Labs, Ft Detrick, 52-71; consult, Naval Ship Systs Command, US Navy, 71-72; microbiologist, Nat Cancer Inst, 72-83; exec officer, ABL-Basic Res Prog, Fred Cancer Res & Develop Ctr, Frederick, 83-90; CONSULT, 90- *Mem:* AAAS; Am Soc Microbiol; Sigma Xi. *Res:* Microbial physiology and structure; medical microbiology; biological waste disposal; viral oncology; research administration. *Mailing Add:* 405 Culler Ave Frederick MD 21701

GUSS, WILLIAM C, b Madison, Wis, Aug 26, 46. PLASMA DIAGNOSTICS. *Educ:* Univ Colo, BS, 68; Univ Wis, MS, 70, PhD(physics), 75. *Prof Exp:* Mem tech staff, TRW, 84-85; RES PHYSICIST, MASS INST TECHNOL, 85- *Mem:* Am Phys Soc. *Mailing Add:* NW 16-176 MIT Plasma Fusion Ctr 167 Albany St Cambridge MA 02139

GUSSIN, ARNOLD E S, b New York, NY, Dec 23, 35; m 62; c 2. PLANT PHYSIOLOGY, EDUCATION ADMINISTRATION. *Educ:* Tulane Univ La, BS, 57, MS, 59; Brown Univ, PhD(biol), 63. *Prof Exp:* NIH fel biol, Yale Univ, 62-64; asst prof zool, Butler Univ, 64-66; from asst prof to assoc prof biol sci, Smith Col, 66-73; coordr, 73-78, asst vpres & dir educ, NY Bot Garden, 78-85; DEAN, GRAD & CONTINUING STUDIES, UNION COL, 85- *Concurrent Pos:* NSF fac fel, Albert Einstein Col Med, 69-70; Oliver Cromwell Gorton Arnold biol fel, Brown Univ. *Mem:* Asn Continuing Higher Educ; Am Soc Biol Chem & Molecular Biol; Am Asn Adult & Continuing Educ; Nat Univ Continuing Educ Asn; Sigma Xi. *Res:* Educational procedures in cultural institutions; physiology and kinetics of trehalases. *Mailing Add:* Union Col One Union Ave-Wells House Schenectady NY 12308-2363

GUSSIN, GARY NATHANIEL, b Detroit, Mich, Aug 7, 39; m 86; c 2. MOLECULAR BIOLOGY. *Educ:* Univ Mich, BS, 61; Harvard Univ, PhD(biophys), 66. *Prof Exp:* NSF fels, Stanford Univ, 66-67 & Univ Geneva, 67-68; Am Cancer Soc fel, 68-69; asst prof, 69-74, assoc prof, 74-80, PROF BIOL, UNIV IOWA, 80- *Res:* Lysogeny of Escherichia coli by bacteriophage lambda; regulation of transcription initiation. *Mailing Add:* Dept Biol Univ Iowa Iowa City IA 52242

GUSSIN, ROBERT Z, b Pittsburgh, Pa, Jan 5, 38; m 86; c 2. PHARMACOLOGY, ADMINISTRATION. *Educ:* Duquesne Univ, BS, 59, MS, 61; Univ Mich, PhD(pharmacol), 65. *Prof Exp:* Res fel, State Univ NY-Upstate Med Ctr, 65-67; res pharmacologist, Lederle Labs, 67-69, group leader, 69-73, dir, 73-74; exec dir, McNeil Labs, Res Div, 74-77, vpres res & develop, 78-81, vpres sci affairs, McNeil Pharmaceut, 81-86. *Concurrent Pos:* Vpres, Res Div, McNeil Labs, 78; adj prof, Dept Pharmacol, Mich State Univ Med Sch, 88-; adv bd, David Mahoney Inst Neurol Sci, Univ Pa, 87- *Mem:* AAAS; Am Soc Nephrol; Am Fedn Clin Res; Am Soc Pharmacol & Exp Therapeut; Am Soc Clin Pharmacol & Therapeut. *Res:* Cardiovascular disease, especially hypertension. *Mailing Add:* Johnson & Johnson 410 George St Rm 1142 New Brunswick NJ 08933-2021

GUSSOW, W(ILLIA)M C(ARRUTHERS), b London, Eng, Apr 25, 08; Can citizen; m 36; c 3. PETROLEUM EXPLORATION, PETROLEUM RECOVERY. *Educ:* Queen's Univ, Kingston, Ont, BSc, 33, MSc, 35; Mass Inst Technol, PhD(econ & struct geol), 38. *Prof Exp:* Res fel geol, Royal Soc Can, 36-37; instr geol & mining, Royal Military Col, Ont, 38-39; construction engr, Foundation Co, Can, 41-43; resident construction engr, Aluminum Co, Can, 44; chief petroleum geologist, Shell Oil Co, 45-48, explor mgr, 48-50. sr geologist, 50-52; self-employed geologist, 53-55; staff petrolum geologist, Union Oil Co, Calif, 56-60, sr res assoc, 60-71; petroleum consult, Japan Nat Oil Corp, 72-74; SELF-EMPLOYED PETROL CONSULT, 75- *Concurrent Pos:* Geol mapping, Fed & Prov Geol Survs, Ont, Que & NT, 30-39; distinguished lectr, Am Asn Petrol Geologists, 55; guest lectr, Nat Acad Lincei, Milan, 57, Moscow State Univ, 60; mem, Nat Adv Comt Res Geol Sci, Ottawa, 57-59, Nat Acad Sci, Subcomt Geol, Comt Space Progs Earth Observ, Adv US Geol Surv, 66-72; United Nations expert, Res & Training Inst, Oil & Natural Gas Comn, India, 67; guest lectr, People's Rep China, 77. *Mem:* Am Asn Petroleum Geologists; emer Am Inst Prof Geologists; fel Geol Asn Can; fel Royal Soc Can; Soc Econ Geologists; fel Geol Soc Am. *Res:* Differential entrapment, explaining accumulation of gas downdip, oil updip and water only beyond limit of oil migration; Gussow method of enhanced oil recovery; patents. *Mailing Add:* 188 Dufferin Rd Ottawa ON K1M 2A6 Can

GUST, J DEVENS, JR, b Phoenix, Ariz, Nov 28, 44; m 69; c 2. ORGANIC CHEMISTRY. *Educ:* Stanford Univ, BS, 67; Princeton Univ, MA, 72, PhD(chem), 74. *Prof Exp:* Res fel chem, Calif Inst Technol, 74-75; from asst prof to assoc prof, 75-85, PROF CHEM, ARIZ STATE UNIV, 85- *Concurrent Pos:* Vis prof biophys, Museum of Natural History, Paris, France, 82, 85; vis scientist, dept phys chem, Centre d'Etudes Nucléaires de, Saclay, France, 82-84; vis scientist, Katholieke Univ Leuven, Belgium, 87; vis prof chem, Katholieke Univ Leuven, Belgium, 89. *Mem:* Am Chem Soc; Am Soc Photobiol; Biophys Soc. *Res:* Organic stereochemistry; nuclear magnetic resonance spectroscopy; solar energy conversion; photosynthesis. *Mailing Add:* Dept Chem Ariz State Univ Tempe AZ 85287-1604

GUSTA, LAWRENCE V, b Selkirk, Man, July 18, 39; m 62. PLANT PHYSIOLOGY. *Educ:* Univ Man, BSA, 63, MS, 65; Univ Minn, PhD(plant physiol), 70. *Prof Exp:* RES SCIENTIST PLANT PHYSIOL, CROP DEVELOP CTR, UNIV SASK, 71- *Concurrent Pos:* Fel, Univ Minn, 70 & Univ BC, 71. *Res:* Physiology of cold acclimation and low temperature injury in winter annuals. *Mailing Add:* 515 Copland Crescent Saskatoon SK S7H 2Z4 Can

GUSTAFSON, ALVAR WALTER, b Chicago, Ill, May 6, 46; m 67; c 1. REPRODUCTIVE BIOLOGY, ENDOCRINOLOGY. *Educ:* Gustavus Adolphus Col, BA, 68; Cornell Univ, PhD(zool), 75. *Prof Exp:* Teaching asst, Cornell Univ, 68-75; asst prof anat, 75-81, actg chmn comp med, 89-90, ASSOC PROF ANAT & CELLULAR BIOL & COMP MED, SCH DENT MED & VET MED, TUFTS UNIV, 81-, DIR HISTOL & HISTOCHEM LAB, DEPT ANAT, 76- *Concurrent Pos:* Vis assoc prof, LHRRB, Harvard Med Sch, 86-87; assoc dean acad affairs, Sch Vet Med, Tufts Univ, 89- *Mem:* AAAS; Sigma Xi; Am Asn Anatomists; Am Soc Zoologists; Am Soc Andrologists; Soc Study Reproduction. *Res:* Structure and function of the male reproductive system; endocrine control of puberty and seasonal reproductive rhythms; hibernation and the endocrines; protein binding of steroid hormones in relation to hormone action. *Mailing Add:* Dept Anat & Cell Biol Sch Med & Vet Med Tufts Univ 136 Harrison Ave Boston MA 02111

GUSTAFSON, BO AKE STURE, b Karlskrona, Sweden, Mar 28, 53; m 86. CELESTIAL MECHANICS. *Educ:* Lund Univ, Sweden, Fil Kand, 77, PhD(astron), 81. *Prof Exp:* Res asst astron, Lund Observ, Lund Univ, Sweden, 74-77, Space Astron Lab, State Univ NY, Albany, 77-80; asst res scientist astron, Space Astron Lab, 81-89, ASST RES SCIENTIST, ASTRON DEPT, UNIV FLA, 89- *Concurrent Pos:* Lectr astron, Lund Observ, Lund Univ, Sweden, 84-; consult, Devex Sa, Switz, 84-, RTS-Labs Inc, Fla; Alexander von Humboldt fel, 89-90; sci expert, Swed Coun, UNESCO, 86-89. *Mem:* Int Astron Union; Sigma Xi; Am Astron Soc; Swedish Astron Soc. *Res:* Small particles lightscattering properties; dynamical evolution of interplanetary dust and meteoroids; celestial mechanics of small bodies; evolution of and relation between comets, asteroids, meteoroids and cosmic dust. *Mailing Add:* Astron Dept 211 SSRB Univ Fla Gainsville FL 32611

GUSTAFSON, BRUCE LEROY, b Jamestown, NY, Oct 25, 54; m 75; c 2. HETEROGENEOUS CATALYSIS, SURFACE CHARACTERIZATION. *Educ:* Univ Cent Fla, BS, 76; Tex A&M Univ, PhD(chem), 81. *Prof Exp:* Res chemist, 81-84, sr res chemist, 84-89, RES ASSOC, EASTMAN CHEMICAL CO, 89- *Mem:* Am Chem Soc; NAm Catalysis Soc; Sigma Xi. *Res:* Characterization and evaluation of heterogeneous catalysts; surface science; zeolite chemistry. *Mailing Add:* 831 Sir Echo Dr Kingsport TN 37663

GUSTAFSON, CARL GUSTAF, JR, b Montclair, NJ, Apr 27, 25; m 51; c 3. ORGANIC CHEMISTRY. *Educ:* King's Col, BS, 48; Univ Del, PhD(org chem), 57; Inst NY, MS, 83. *Prof Exp:* Instr chem, King's Col, 48-50; org res chemist, Chem Corps, Med Labs, US Army Chem Ctr, Md, 52-53; asst prof chem, Roberts Wesleyan Col, 56-59; assoc prof, King's Col, NY, 59-63, actg dean, 60-62, chmn, Div Natural Sci, 65-87, prof chem, 63-90, dir acad comput, 87-90; RETIRED. *Concurrent Pos:* Spec fac appointment with Fed Water Qual Admin, Southeast Water Lab, Athens, Ga, 69-70. *Mem:* Am Chem Soc. *Res:* Sulfonic esters of hexitols; organic insecticides; fiber composition of paper; small ring heterscyolis; polyacetals. *Mailing Add:* 5205 Fielding Dr No 1 Raleigh NC 27606

GUSTAFSON, DAVID HAROLD, b Ft Wayne, Ind, Dec 7, 35; m 57, 78; c 1. ORGANIC CHEMISTRY. *Educ:* Purdue Univ, BS, 57; Univ Ill, PhD(org chem), 61. *Prof Exp:* Res chemist, Procter & Gamble Co, 61-64; SR RES CHEMIST, MARION MERRELL DOW INC, 64- *Mem:* Am Chem Soc; NY Acad Sci; Sigma Xi. *Res:* Metallocenes and terpenoid syntheses; spectroscopy; physical organic chemistry. *Mailing Add:* Marion Merrell Dow Inc 2110 E Galbraith Rd Cincinnati OH 45215

GUSTAFSON, DAVID HAROLD, b Kane, Pa, Sept 11, 40; m 62; c 3. INDUSTRIAL ENGINEERING, PREVENTIVE MEDICINE. *Educ:* Univ Mich, BS, 62, MS, 63, PhD(indust eng), 66. *Prof Exp:* Dir hosp div, Community Systs Found, 63-64; asst prof mech eng, 66-69, from asst prof to assoc prof indust eng, 69-74, assoc prof prev med, 71-74, PROF INDUST ENG & PREV MED, UNIV WIS-MADISON, 74-, DIR, CTR HEALTH SYSTS RES & ANALYSIS, 76- *Concurrent Pos:* Adv, Wis Gov Health Planning & Policy Task Force, 71-73, mem health policy coun, 73- & comn on educ & health admin, 73-74; dir, Health Systs Eng Prog; sr systs analyst, Decisions & Designs Inc, Va, 74; spec asst to dir, Bur Community Health Serv, Dept Health, Educ & Welfare, 75; consult to numerous health & social serv agencies within fed govt, univs & pvt sector. *Honors & Awards:* Ragner Onstad Award for Serv to Soc, 88; Excellence in Educ & Prev, Am Med Asn, 89. *Mem:* Opers Res Soc Am. *Res:* Applicability of behavioral decision theory and computer based decision support systems to medical and social systems; development, implementation and evaluation of computer based medical decision systems. *Mailing Add:* Indust Eng 392 Mech Eng Bldg Univ Wis Madison WI 53706

GUSTAFSON, DONALD ARVID, b Delano, Minn, Sept 11, 13; m 43. CHEMISTRY. *Educ:* Univ Wash, Seattle, BS, 37, PhD(chem), 44. *Prof Exp:* Asst chem, Univ Wash, 37-44; from instr to prof, 44-77, EMER PROF CHEM, UNIV IDAHO, 77- *Mem:* Am Chem Soc (secy-treas, 46). *Res:* Electrical conductance of nonaqueous solutions; electrical conductance of solutions with solvents from the system acetic anhydride-acetic acid-water. *Mailing Add:* 522 Eisenhower Moscow ID 83843

GUSTAFSON, DONALD PINK, b Columbus, Ohio, May 21, 20; m 49; c 5. VETERINARY VIROLOGY, PATHOBIOLOGY. *Educ:* Ohio State Univ, BSc, 41, DVM, 45; Purdue Univ, MS, 51, PhD(vet path), 53; Am Col Vet Microbiol, dipl. *Prof Exp:* Pvt practice, 45-46; from instr to assoc prof, 49-61, PROF VET MED, PURDUE UNIV, 61- *Concurrent Pos:* Consult, Com Solvents Corp, 57, Univ Fla, 58 & Eli Lilly & Co, 58-74; mem subcomt standardized methods in vet microbiol, Nat Acad Sci-Nat Res Coun, 67-72; mem, Livestock Conserv Inst, 69-, chmn swine dysentery comt, 71-76; mem, Coun Biol & Therapeut Agents, Am Vet Med Asn, 70-80, vchmn, 71, 73 & 74, chmn, 72 & 75-80, coun on res, 81-; mem res resources comt, Nat Inst Allergy & Infectious Dis, 71-75. *Mem:* Am Asn Immunologists; Am Vet Med Asn; Am Soc Microbiologists; Tissue Cult Asn; NY Acad Sci. *Res:* Infectious diseases of animals; scrapie in sheep; chronic equine diarrhea; myoclonia congenita in swine; pseudorabies. *Mailing Add:* Sch Vet Med Purdue Univ West Lafayette IN 47907

GUSTAFSON, GRANT BERNARD, b St Paul, Minn, Mar 5, 44; m 69; c 1. MATHEMATICAL ANALYSIS. *Educ:* Ariz State Univ, BS, 65, MA, 67, PhD(math), 68. *Prof Exp:* Vis prof, 68-69, from asst prof to assoc prof, 69-79, PROF MATH, UNIV UTAH, 80- *Mem:* Am Math Soc; Math Asn Am; Soc Ind & Appl Math. *Res:* Ordinary, functional and partial differential equations; nonlinear analysis. *Mailing Add:* Dept Math 233 JWB Univ Utah Salt Lake City UT 84112

GUSTAFSON, HEROLD RICHARD, b Grand Coulee, Wash, June 16, 37; m 61; c 1. QUARKS, NEUTRINOS. *Educ:* Calif Inst Technol, BS, 59; Univ Wash, MS, 61, PhD(physics), 68. *Prof Exp:* Elec engr missiles, Boeing Co, 59-61, space physics, 61-64; scholar, Univ Mich, 68-71, res assoc, 71-77, asst res scientist, 77-82, ASSOC RES SCIENTIST, UNIV MICH, 82- *Concurrent Pos:* Chmn, Fermilab, Users Orgn, 81-82; co-prin investr, NSF, 81-91. *Mem:* Am Phys Soc. *Res:* Prompt neutrino production (beam dump); searching for quarks; investigation of hadron jets; magnetic monopoles. *Mailing Add:* Randall Lab Univ Mich Ann Arbor MI 48109

GUSTAFSON, JOHN C, b Worthington, Minn, Nov 7, 44. SOLID STATE PHYSICS. *Educ:* Univ Minn, BA, 65; Northwestern Univ, PhD(physics), 72. *Prof Exp:* Res assoc physics, McMaster Univ, 72-73; MEM TECH STAFF, GTE LABS, INC, 73- *Mem:* Am Phys Soc; AAAS. *Res:* Studies of magnetic phenomena in amorphous metals; high temperature transport properties in homogenous materials and plasma-surface interactions. *Mailing Add:* GTE Labs 40 Sylvan Rd Waltham MA 02154

GUSTAFSON, JOHN PERRY, b Greeley, Colo, Aug 1, 44; m 77; c 2. PLANT BREEDING. *Educ:* Colo State Univ, BS, 67, MS, 68; Univ Calif, Davis, PhD(genetics), 72. *Prof Exp:* Res prof genetics, Univ Manitoba, Can, 72-77, assoc prof, 77-82; RES GENETICIST, AGR RES SERV, USDA, 82- *Concurrent Pos:* Res assoc, Univ Manitoba, Can, 76-77. *Mem:* Genetics Soc Am; Genetics Soc Can; Am Soc Agron; Am Genetic Asn. *Res:* Understanding and manipulation of genetic mechanisms of control of the phenotype of cereals; interspecific and intergeneric hybrids within the various grass genera. *Mailing Add:* Agr Res Serv USDA Univ Mo Columbia MO 65211

GUSTAFSON, KARL EDWIN, b Manchester, Iowa, May 7, 35; m 61; c 1. MATHEMATICS. *Educ:* Univ Colo, BS(eng) & BS(bus), 58; Univ Md, PhD(math), 65. *Prof Exp:* Instr appl math, Univ Colo, 58-60; physicist, Naval Res Lab, 60-61, mathematician, 61-63; NSF-NATO fel math, Inst Battelle, Geneva & Univ Rome, 65-66; asst prof, Univ Minn, 66-68; assoc prof, 68-74, fac fel, 71-72, PROF MATH, UNIV COLO, BOULDER, 74- *Concurrent Pos:* Vis scientist, Inst Battelle, Geneva, 67; vis prof, Fed Polytech Sch, Lausanne, Switz, 71-72. *Mem:* Am Math Soc. *Res:* Computer applications for naval intelligence problems; partial differential equations; functional analysis; operator theory; mathematical physics; nonlinear problems. *Mailing Add:* Univ Colo Boulder CO 80309-0426

GUSTAFSON, LEWIS BRIGHAM, b Timmins, Ont, Sept 4, 33; m 61; c 3. ECONOMIC GEOLOGY, MINERAL DEPOSITS. *Educ:* Princeton Univ, BSE, 55; Calif Inst Technol, MS, 58; Harvard Univ, PhD(geol), 62. *Prof Exp:* From geologist to chief geologist, El Salvador Mine, Andes Copper Mining Co, 62-69, from proj geologist to chief geologist, res & technol, Primary Metals Div, Anaconda Co, 69-75; independent consult, 75; prof econ geol, Australian Nat Univ, 75-81; head, Min Res Group, Conoco, Inc, 81-82; chief Res Geologist, Freeport Exploration Co, 82-86; PARTNER, ANNAPURNA EXPLORATION, RENO, 86- *Concurrent Pos:* T Lindsey lectr, Soc Econ Geol, 73-74. *Honors & Awards:* Distinguished lectr, Soc Econ Geol, 89. *Mem:* Soc Econ Geologists; fel Geol Soc Am; Am Inst Mining & Metallurgical Engrs. *Res:* Mineral deposits; mineral exploration; tectonic settings of ore deposits. *Mailing Add:* 3520 San Mateo Ave Reno NV 89509

GUSTAFSON, MARK EDWARD, b Cedar Rapids, Iowa, Feb 17, 52. MICROBIAL TRANSFORMATIONS. *Educ:* Col Pharm, Univ Iowa, BS, 76, PhD(pharm & med chem), 80. *Prof Exp:* Scientist I, 80-81, SCIENTIST II, FERMENTATION PROG, FREDERICK CANCER RES FACIL, NAT CANCER INST, 81- *Mem:* Am Chem Soc. *Res:* Isolation and characterization of variety of antitumor agents, including chemotherapeutics and interferon. *Mailing Add:* 1408 Navaho Trail St Charles MO 63303-7325

GUSTAFSON, PHILIP FELIX, b Ann Arbor, Mich, Apr 2, 24; m 49; c 2. ENVIRONMENTAL SCIENCES. *Educ:* Univ Mich, BS, 49; Ill Inst Technol, MS, 54, PhD(physics), 58. *Prof Exp:* Res technician physics, Argonne Nat Lab, 49-52, asst physicist, 52-59, assoc physicist, 55-66; nuclear fallout specialist, US Atomic Energy Comn, 66-68; assoc dir radiol physics div, 68-72, sr biophysicist, 71-88, mgr, 72-75, dir environ statement proj, Argonne Nat Lab, 75-80; dir, Dept Nuclear Safety, Ill, 80-83; dir, Div Environ Res, 83-85, COORDR HAZARDOUS MAT RES, 85-, DIR, ENVIRON RES PROG, ARGONNE NAT LAB, 87- *Concurrent Pos:* Mem, Ill Comn Atomic Energy, 72-, sci adv, 74- *Honors & Awards:* Glover Award, Dizlason Col, 72; Univ Geert Award, Belgium, 74. *Mem:* AAAS; Sigma Xi. *Res:* Environmental radiation; radioecology; environmental impact assessment for nuclear facilities; environmental effects and their minimization; hazardous waste management policy. *Mailing Add:* 413 Addison Rd Riverside IL 60546

GUSTAFSON, RALPH ALAN, b Denver, Colo, Apr 9, 39; m 61; c 2. MYCOLOGY, BACTERIOLOGY. *Educ:* Colo State Univ, BS, 62, MEd, 68; Univ Tex, Austin, PhD(bot), 73. *Prof Exp:* Instr biol, Univ Tex, Austin, 72-74; PROF BIOL, WINTHROP COL, 74- *Mem:* Sigma Xi; Am Soc Microbiol; Nat Environ Health Asn. *Res:* Bacteriological quality of drinking water sources. *Mailing Add:* Dept Biol Winthrop Col Rock Hill SC 29733

GUSTAFSON, STEVEN CARL, b St Paul, Minn, Mar 8, 45; m 70. OPTICS, STATISTICS. *Educ:* Univ Minn, BS, 67; Duke Univ, MA, 69, PhD(physics), 74. *Prof Exp:* Consult, Air Force Mat Lab, 75, vis scientist, 75-76; res scientist, 76-84, SR RES PHYSICIST, RES INST, UNIV DAYTON, 84- *Mem:* Am Phys Soc; Inst Elec & Electronics Engrs; Optical Soc Am; Soc Photo-Optical Instrumentation Engrs. *Res:* Optics; statistics; optical processing. *Mailing Add:* 5813 Arlmont Circle Dayton OH 45440

GUSTAFSON, TERRY LEE, b Brainerd, Minn, May 14, 53. RAMAN SPECTROSCOPY, ULTRAFAST SPECTROSCOPY. *Educ:* Moorhead State Univ, BS, 75; Purdue Univ, PhD(chem), 79. *Prof Exp:* Sr res chemist, Sohio Res & Develop, 79-82; proj leader, Standard Oil Co, 82-87; sr proj leader, BP Am Res & Develop, 87-89; ASST PROF, OHIO STATE UNIV, 89- *Honors & Awards:* Am Inst Chem Medal, 75. *Mem:* Am Chem Soc; Am Phys Soc; Soc Appl Spectros. *Res:* Molecular conformation and dynamics; time-resolved optical characterization of thin film materials; instrumental developments in ultrafast pulsed lasers for quasi-continuous and time-resolved spectroscopy. *Mailing Add:* Dept Chem Ohio State Univ 120 W 18th Ave Columbus OH 43210-1173

GUSTAFSON, WILLIAM HOWARD, b New Haven, Conn, Sept 9, 44. ALGEBRA. *Educ:* Wesleyan Univ, BA, 66; Univ Ill, MA, 67, PhD(math), 70. *Prof Exp:* Asst prof math, Ind Univ, Bloomington, 70-72; vis asst prof math, Brandeis Univ, 72-73; asst prof math, Ind Univ, Bloomington, 73-76; asst prof, 76-80, assoc prof, 80-86, PROF MATH, TEX TECH UNIV, 86- *Honors & Awards:* Lester R Ford Award, Math Asn Am, 77. *Mem:* Am Math Soc; Math Asn Am. *Res:* Representation theory and related aspects of matrix theory; number theory and algebraic geometry. *Mailing Add:* Dept Math Texas Tech Univ Lubbock TX 79409

GUSTAFSON, WINTHROP A(DOLPH), b Moline, Ill, Oct 14, 28; m 57; c 4. AERONAUTICAL ENGINEERING. *Educ:* Univ Ill, BS, 50, MS, 54, PhD(aeronaut eng), 56. *Prof Exp:* Assoc res scientist, Missiles & Space Div, Lockheed Aircraft Corp, Calif, 56-60; assoc prof aeronaut & eng sci, 60-66, actg head, Sch Aeronaut & Astronaut, 84-85, PROF AERONAUT & ASTRONAUT, PURDUE UNIV, 66-, ASSOC HEAD, 80- *Concurrent Pos:* Consult, Goodyear Aerospace Corp, Ohio, 64; vis prof, Univ Calif, San Diego, 68; fac fel, Dryden Flight Res Ctr, NASA, 76; consult, Los Alamos Sci Lab, 77-79; consult, US Army Aviation, Washington, DC, 85-86. *Mem:* Am Inst Aeronaut & Astronaut; Am Soc Eng Educ. *Res:* Aerodynamics; rarefied gas dynamics; spacecraft design. *Mailing Add:* Sch Aeronaut & Astronaut Purdue Univ W Lafayette IN 47907

GUSTAFSSON, JAN-AKE, b Stockholm, Sweden, Aug 4, 43; c 1. MOLECULAR ENDOCRINOLOGY, MOLECULAR TOXICOLOGY. *Educ:* Karolinska Inst, PhD(chem), 68, MD, 71. *Prof Exp:* Prof chem, Karolinska Inst, 76-78, Univ Gothenburg, 78-79; PROF MED NUTRIT, DEPT MED NUTRIT, KAROLINSKA INST, 79-, CTR BIOTECH, 85- & CTR NUTRIT & TOXICOL, 89- *Concurrent Pos:* Mem, Nobel Assembly, Karolinska Inst, 86; found, KaroBio, AB, 87; adj prof, Dept Cell Biol, Baylor Col Med, Houston, Tex, 87. *Honors & Awards:* Svedberg Prize in Chem, 82. *Mem:* Hon mem Am Soc Biochem & Molecular Biol. *Res:* Mechanism of action of steroid hormones and of growth hormone; structure and function of steriod receptors; regulation and function of cytochrome P-450 in the liver, brain and prostate; formation and biological effects of mutagens in fried foods. *Mailing Add:* Dept Med Nutrit Huddinge Univ Hosp F60 Novum Huddinge 141 86 Sweden

GUSTAFSSON, TORGNY DANIEL, b Halmstad, Sweden, Oct 18, 46; m 77; c 4. SURFACE PHYSICS. *Educ:* Chalmers Univ Technol, MSc, 70, PhD(physics), 73. *Prof Exp:* From res assoc to prof, Univ Pa, 74-87; PROF, RUTGERS UNIV, 87- *Concurrent Pos:* Vis scientist, Univ Wis Synchrotron Radiation Ctr, 76-77, Stanford Univ, 78 & FOM Inst, Amsterdam, 85; Sloan Found fel, 78-82. *Honors & Awards:* Nottingham Prize, Phys Electronics Conf, 74. *Mem:* Fel Am Phys Soc; Am Vacuum Soc. *Res:* Surface physics, in particular the interaction of clean metals with adsorbed species; surface structure studies with ion scattering. *Mailing Add:* Dept Physics & Astron Rutgers Univ PO Box 849 Piscataway NJ 98855-0849

GUSTAV, BONNIE LEE, b New York, NY, Feb 10, 44; m 78; c 3. PHYSICAL ANTHROPOLOGY, ANTHROPOMETRICS. *Educ:* Hunter Col, BA, 64; Univ Mass, PhD(phys anthrop), 72. *Prof Exp:* Lectr anthrop, Hunter Col, 67-68; asst prof, 72-86, ASSOC PROF ANTHROP, BROOKLYN COL, 87- *Mem:* AAAS; fel Am Asn Phys Anthrop; Am Anthrop Asn. *Res:* Skeletal growth and development; human paleoanthropology; osteology. *Mailing Add:* Dept Anthrop Brooklyn Col CUNY Brooklyn NY 11210

GUSTAVSON, FRED GEHRUNG, b New York, NY, May 29, 35; m 66; c 2. APPLIED MATHEMATICS, COMPUTER SCIENCES. *Educ:* Rensselaer Polytech Inst, BS, 57, MS, 60, PhD(appl math), 63. *Prof Exp:* Res asst math, Math Res Ctr, Univ Wis, 61-62; res assoc, Rensselaer Polytech Inst, 62-63; RES STAFF MEM MATH SCI, T J WATSON RES CTR, IBM CORP, 63- *Honors & Awards:* Outstanding Contrib Award, IBM Corp, 68, Outstanding Invention Award, 71. *Mem:* Sigma Xi; Math Asn Am. *Res:* Sparse matrix theory, especially design and implementation of computer algorithms to handle sparse matrices; analysis of algorithms; numerical analysis. *Mailing Add:* Box 218 Yorktown Heights NY 10598

GUSTAVSON, MARVIN RONALD, b Chicago, Ill, Nov 4, 27; m 53; c 2. SYSTEMS EVALUATION, RESOURCE DEVELOPMENT. *Educ:* Carnegie Inst Technol, BS, 48; Cornell Univ, PhD(phys chem), 53. *Prof Exp:* Chemist, Shell Develop Co, 53-57; div mgr, Aerojet-Gen Nucleonics Div, Gen Tire & Rubber Co, 57-60; vpres & gen mgr, Ordtech Corp, 60-61; dir, X div, 61-72, asst assoc dir, 73-78, SR ANALYST, LAWRENCE LIVERMORE NAT LAB, 78- *Concurrent Pos:* Spec lectr, Univ Calif & Exten, 54-58; consult, Govt & Indust Orgn, 60-; mem, Air Force Sci Adv Bd & Army Sci Bd. *Mem:* Am Chem Soc; Sigma Xi. *Res:* Reaction kinetics; weapon system design and evaluation; command and control; energy and resource development; research organization and administration. *Mailing Add:* Lawrence Livermore Lab PO Box 808 Livermore CA 94550

GUSTAVSON, THOMAS CARL, b Northampton, Mass, Dec 28, 36; m 59; c 4. GEOLOGY. *Educ:* Univ Mass, Amherst, BS, 61, PhD(geol), 73; Univ NDak, MS, 64. *Prof Exp:* Geologist, Humble Oil & Refining Co, 64-66 & Phillips Petrol Co, 66; instr geol, Southampton Col, Long Island Univ, 66-69; RES SCIENTIST ENVIRON GEOL, BUR ECON GEOL, UNIV TEX, AUSTIN, 73-, COORDR LAND RESOURCES LAB, 75- *Mem:* Geol Soc Am; Soc Econ Paleontologists & Mineralogists; Glaciol Soc; Sigma Xi. *Res:* Environmental geology of the Texas Coastal Plain; environmental aspects of geothermal energy development; fluvial and lacustrine sedimentary processes; geomorphology and engineering characteristics of expansive clay soil terrains; geology and geomorphology related to nuclear waste isolation. *Mailing Add:* 3406 Saddlestring Truce Austin TX 78749

GUSTIN, VAUGHN KENNETH, b Mansfield, Ohio, June 10, 36; m 57; c 2. ANALYTICAL CHEMISTRY. *Educ:* Ohio State Univ, BSc, 58, MSc, 60, PhD(anal chem), 63. *Prof Exp:* Teaching asst, Ohio State Univ, 58-63; sr anal chemist, 63-64, supvr anal lab, 65-66, mgr chem serv dept, 66-70, mgr, Control Technol, 70-, DIR MFR ENG CONSUMER & SCI PROD DIV, CORNING GLASS WORKS. *Concurrent Pos:* Partic, Prog for Mgt Develop, Harvard Bus Sch, 68. *Mem:* Am Chem Soc; Soc Appl Spectros (secy, 66); fel Am Inst Chem. *Res:* Chemical composition control of glasses, glass ceramics and associated raw materials and refractories; analytical method development in all phases of analytical chemistry; pollution control. *Mailing Add:* 9 Fox Lane W Painted Post NY 14870

GUSTINE, DAVID LAWRENCE, b Battle Creek, Mich, Jan 2, 41; m 61; c 2. BIOCHEMISTRY. *Educ:* Malone Col, BA, 64; Mich State Univ, PhD(biochem), 69. *Prof Exp:* Res fel biochem & pharmacol, Children's Hosp Res Found, Cincinnati, Ohio, 69-71; RES PLANT PHYSIOLOGIST, US REGIONAL PASTURE RES LAB, AGR RES SERV, USDA, 71- *Concurrent Pos:* Adj assoc prof crop sci, Pa State Univ. *Mem:* Am Soc Plant Physiol; Phytochem Soc NAm; AAAS. *Res:* Plant biochemical mechanisms of disease resistance; plant tissue culture; molecular biology of apomixis. *Mailing Add:* US Regional Pasture Res Lab Agr Res Serv USDA University Park PA 16802

GUSTISON, ROBERT ABDON, b Springfield, Ohio, Aug 29, 20; m 43; c 3. INORGANIC CHEMISTRY. *Educ:* Univ Chicago, BS, 42. *Prof Exp:* Res chemist, metall lab, Univ Chicago, 42-44; res engr, Int Minerals & Chem Co, 44-46; res chemist & engr, Union Carbide Chem Co Div, Union Carbide Corp, 46-54, mgr minerals & chem eng res, Union Carbide Metals Co Div, 54-61, supvr pilot plants, 61-63; mgr chem processing, Kawecki Berylco Industs, Inc, 64-; SR RES SCIENTIST, CABOT INC; CONSULT, 85- *Concurrent Pos:* Mem mat adv bd, Nat Acad Sci; metall consult, Ferro-Tech Inc, Pittsburgh, Pa, 75- *Mem:* Am Chem Soc; Am Inst Mining, Metall & Petrol Engrs; fel Am Inst Chemists. *Res:* Extractive metallurgy involving pyrometallurgy; electrometallurgy; halogenations; high pressure and vacuum; Mass transfer operations. *Mailing Add:* 720-D4 Old Mill Rd Wyomissing PA 19610

GUTAY, LASZLO J, b Fadd, Hungary, Aug 9, 35; US citizen; m 61; c 5. HIGH ENERGY NUCLEAR PHYSICS. *Educ:* Oxford Univ, BA, 59, MA, 61; Fla State Univ, PhD(physics), 64. *Prof Exp:* Res assoc, 65-67, asst prof, 67-70, assoc prof, 70-75, PROF HIGH ENERGY NUCLEAR PHYSICS, PURDUE UNIV, 75- *Concurrent Pos:* Prin investr contract res, Dept Energy, 74. *Honors & Awards:* H McCoy Award Science, Purdue Univ. *Mem:* Fel Am Phys Soc. *Res:* Pion-Pion interactions; current algebra; vector meson dominance; high energy, highly inelastic collision; hadronic temperature size; size of centrally produced hadronic matter hadronic phase transactions; super dense radronic matter; quark-gluon plasma. *Mailing Add:* Dept Physics Purdue Univ West Lafayette IN 47907

GUTBERLET, LOUIS CHARLES, b Chicago, Ill, Mar 29, 28; m 52; c 2. ORGANIC CHEMISTRY. *Educ:* Ill Inst Technol, BS, 50; Purdue Univ, MS, 54. *Prof Exp:* Asst chem, Purdue Univ, 52-54; chemist, Amoco Res Ctr, Amoco Oil Co, 54-88, res supvr, 82- 88; RETIRED. *Mem:* Am Chem Soc. *Res:* Catalytic hydrogenation; heterogeneous catalysis; hydrocarbon conversion processes. *Mailing Add:* 1520 E Wakeman Ave Wheaton IL 60187

GUTBEZAHL, BORIS, b Tientsin, China, May 1, 27; US citizen; m 52; c 2. POLYMER CHEMISTRY. *Educ:* City Col New York, BS, 47; Columbia Univ, AM, 49; Fla State Univ, PhD(chem), 52. *Prof Exp:* Asst chem, Columbia Univ, 52-53; chemist, 53-59, head process develop lab, 59-60, head plastics color lab, 60-65, supvr tech serv, 65-78, mgr anal & color serv, 78-85, QUAL COORDR, ROHM & HAAS CO, 85- *Mem:* Am Chem Soc; Soc Plastics Eng; Inter-Soc Color Coun; Am Soc Testing & Mat; Am Soc Qual Control. *Res:* Organic mechanisms; polymerization kinetics; product and process development in plastics. *Mailing Add:* 23 Rust Hill Rd Levittown PA 19056-2311

GUTCHO, SIDNEY J, b Brooklyn, NY, Oct 18, 19; m 49; c 2. IMMUNOASSAYS, IMMUNOCHEMISTRY. *Educ:* NY Univ, BA, 40, MS, 41. *Prof Exp:* Res chemist, Schwarz Labs, Inc, 41-59, head chemist, Schwarz Biores, Inc, 59-70, head chemist, 70-72, assoc dir res & develop, Schwarz/Mann, Becton- Dickinson & Co, 72-77, assoc dir res & develop, 77-87, CONSULT BECTON DICKINSON IMMUNODIAG, 87- *Mem:* Radioligand Soc; AAAS; Am Chem Soc; NY Acad Sci; Am Asn Clin Chem. *Res:* Immunochemistry; immunoassays; isolation of biochemicals from yeast; enzymatic preparation of biochemicals; biosynthesis of radioactive compounds; chemistry of nucleic acids and their derivatives. *Mailing Add:* 22 Francis Pl Monsey NY 10952

GUTELIUS, JOHN ROBERT, b Montreal, Que, Jan 18, 29; m 55; c 8. SURGERY. *Educ:* Univ Montreal, BA, 50; McGill Univ, MD, 55. *Prof Exp:* Rotating intern, Royal Victoria Hosp, 55-56; from asst resident surgeon to chief resident surgeon, Teaching Hosps, McGill Univ, 56-61, demonstr surg, McGill Univ, 61-63, from asst prof to assoc prof, 63-69, assoc dean postgrad studies & res, 68-69; prof surg & head dept & dean, Sch Med, Univ Sask, 70-73; surgeon-in-chief, Kingston Gen Hosp, Hotel Dieu Hosp, Kingston, Ont, 73-83; head dept, 73-83, PROF SURG, QUEEN'S UNIV, 73- *Concurrent Pos:* Fel exp surg, McGill Univ, 57-58; fel surg, Johns Hopkins Univ & Hosp, 59-60; R S McLaughlin Found traveling fel, 59-60; Markle Found scholar, 63-68; from asst surgeon to assoc surgeon, Royal Victoria Hosp, 63-69; mem, Grants Comt Clin Invest, Med Res Coun Can, 70-75 & comt gen surg & exam-gen surg, Royal Col Physicians & Surgeons, 70-; hon consult, Royal Victoria Hosp, Montreal, Que; mem, priorities comt, Med Res Coun Can, 75-79. *Mem:* Am Surg Asn; Soc Univ Surgeons; Int Cardiovasc Soc; Can Soc Clin Invest; Am Col Surg. *Res:* Vascular surgery; techniques; patient selection; water and electrolytes; adrenal function. *Mailing Add:* Dept Surg Queen's Univ Fac Med Kingston ON K7L 3N6 Can

GUTERMAN, MARTIN MAYR, b New York, NY, Nov 18, 41; m 64; c 2. ALGEBRA. *Educ:* Brooklyn Col, BS, 61; Cornell Univ, MS, 64, PhD(finite groups), 68. *Prof Exp:* Asst, Cornell Univ, 61-66; from instr to assoc prof 66-87, PROF MATH, TUFTS UNIV, 87- *Mem:* Am Math Soc; Math Asn Am. *Res:* Theory of finite groups. *Mailing Add:* Dept Math Tufts Univ Medford MA 02155

GUTERMAN, SONIA KOSOW, b Brooklyn, NY, June 27, 44; m 64; c 2. MOLECULAR BIOLOGY. *Educ:* Cornell Univ, BS, 64, MS, 67; Mass Inst Technol, PhD(microbiol), 71. *Prof Exp:* NIH fel microbiol, Mass Inst Technol, 66-70, res assoc, 70-71; instr biol, Brandeis Univ, 71-72; NIH fel molecular biol, Sch Med, Tufts Univ, 72-74, res assoc, 74-75, instr, 75-76; ASST PROF BIOL, BOSTON UNIV, 76- *Mem:* AAAS; Am Soc Microbiol; Genetics Soc Am; NY Acad Sci. *Res:* Studies of bacterial RNA polymerase mutants to determine functions and interactions of enzyme subunits; phage and colicin receptor genetics as a probe of inner and outer membrane structure; rho protein genetics and enzymology to study functional sites of nucleic acid interaction. *Mailing Add:* 20 Oakley Rd Belmont MA 02178

GUTFINGER, DAN ELI, b New Haven, Conn, June 17, 64. MACHINE VISION, ARTIFICIAL HEARTS & VENTRICULAR ASSIST DEVICES. *Educ:* Univ Calif, Irvine, BS, 83, MS, 86, PhD(elec & computer eng), 90. *Prof Exp:* Sports statistician sports, Univ Calif, Irvine, 83, teaching asst computer sci, 84-85, res asst image processing, 89-90, ENGR, BIOMED ENG, ARTIFICIAL HEART PROG, UNIV CALIF, IRVINE, 91-; ENG SPECIALIST SIGNAL PROCESSING, FORD AEROSPACE & AERONAUT, 84- *Concurrent Pos:* Mem, Heart Rx Subcomt, Am Heart Asn, 91- *Res:* Development of machine vision and pattern recognition systems with medical industrial and military applications; clinical evaluation of artificial hearts and ventricular assist devices; analysis of multi-sensor data with emphasis on radar signature data. *Mailing Add:* 1540 Stanford Irvine CA 92715

GUTFREUND, KURT, b Bielitz, Germany, April 18, 24; US citizen; m 54; c 3. SURFACE CHEMISTRY. *Educ:* Munich Tech Univ, Ger, dipl, 49; Univ Buffalo, NY, BA, 50; Univ Wis, MSc, 52. *Prof Exp:* Asst chemist, polymers, Union Carbide Corp, 53-55; res physicist, mat sci, Am Can Co, 55-57; sr scientist, surface chem, 57-86, SCI ADV, IIT RES INST, 86- *Concurrent Pos:* Mem, Adv Bd, Nat Heart, Blood & Lung Inst, 78-79. *Mem:* Am Chem Soc; Soc Rheology; Soc Biomat; AAAS. *Res:* Interfacial relationships in fiber-reinforced composites; interaction of polymers with solids; magnetic coatings; thermal degradation of polymers in oxidizing environments; investigation of blood coagulation by light scattering methods; compatibility of cardiovascular devices with blood constituents; effects of laser radiation on organic coatings. *Mailing Add:* 10 W 35th St Chicago IL 60616-3799

GUTH, ALAN HARVEY, b New Brunswick, NJ, Feb 27, 47; m 71; c 2. COSMOLOGY. *Educ:* Mass Inst Technol, SB & SM, 69, PhD(physics), 72. *Prof Exp:* Instr physics, Princeton Univ, 71-74; res assoc, Columbia Univ, 74-77; res assoc II, Cornell Univ, 77-79; res assoc, Stanford Linear Accelerator Ctr, Stanford Univ, 79-80; physicist, Harvard Smithsonian Ctr Astrophys, 84-90; vis assoc prof, Mass Inst Technol, 80-81, assoc prof, 81-86, prof, 86-89, JERROLD ZACHARIAS PROF PHYSICS, MASS INST TECHNOL, 89- *Concurrent Pos:* Exec comt, Astrophys Div, Am Phys Soc, 86-88, vchmn, 88-89, chmn, 89-90; vis scientist, Harvard Smithsonian Ctr Astrophys. *Honors & Awards:* Alfred P Sloan Found fel, 81; Rennie Taylor Award, Am Tentative Soc, 89-90. *Mem:* Nat Acad Sci; Fel Am Phys Soc; Am Astron Soc; fel AAAS; fel Am Acad Arts & Sci. *Res:* Theory of elementary particles; application of particle physics to the very early universe; inflationary cosmology. *Mailing Add:* Ctr Theoret Physics Mass Inst Technol Cambridge MA 02139

GUTH, LLOYD, b New York, NY, Oct 8, 29; m 55; c 2. ANATOMY. *Educ:* NY Univ, BA, 49, MD, 53. *Prof Exp:* Asst anat, Col Med, NY Univ, 50-51 & Jackson Mem Lab, 51; intern, Kings County Hosp, 53-54; neuroanatomist, Lab Neuroanat Sci, NIH, 54-61, head sect exp neurol, 61-75; prof & chem, Dept Anat, Sch Med, Univ Md, 75-90; RES PROF, DEPT BIOL, COL OF WILLIAM & MARY, 90- *Concurrent Pos:* Sen Jacob Javits Award, 89-96. *Mem:* AAAS; Am Asn Anat. *Res:* Experimental neurology; nerve regeneration; neuromuscular interrelationships, structure, function and chemistry; response of the mammalian spinal cord to traumatic in jury focusing on mechanisms that restrict or enhance axonal regeneration. *Mailing Add:* 111 Gullane Fords Colony Williamsburg VA 23188

GUTH, PAUL HENRY, b New York, NY, Mar 15, 27; m 53; c 4. INTERNAL MEDICINE, GASTROENTEROLOGY. *Educ:* NY Univ, BS, 48; Howard Univ, MD, 49; Am Bd Internal Med, dipl, 61; Am Bd Gastroenterol, dipl, 65. *Prof Exp:* Resident internal med, Kings County Hosp, 50-51; res fel gastrointestinal physiol, Fels Res Inst, Temple Univ, 51-52, Am Gastroenterol Asn res fel, 54-55, resident internal med, Temple Univ Hosp, 55-56, resident gastroenterol, 56-57; chief gastroenterol, Orange County Gen Hosp, 57-69, dir med serv, Med Ctr, 67-69; assoc prof in residence, 69-73, PROF IN RESIDENCE MED, SCH MED, UNIV CALIF, LOS ANGELES, 73- *Concurrent Pos:* From asst clin prof to assoc clin prof, Univ Calif-Calif Col Med, 66-69; asst chief gastroenterol, Wadsworth Vet Hosp, Los Angeles, 69- *Mem:* AMA; Am Fedn Clin Res; fel Am Col Physicians; Am Gastroenterol Asn; Am Physiol Soc. *Res:* Gastrointestinal physiology, gastrointestinal blood flow; pathophysiology, experimental ulcer; therapeutics. *Mailing Add:* Gastroenterol Sect Wadsworth Vet Admin Hosp Los Angeles CA 90073

GUTH, PAUL SPENCER, b May 29, 31; US citizen; m 53; c 4. PHARMACOLOGY. *Educ:* Fordham Univ, BSc, 53; Philadelphia Col Pharm, MSc, 55; Hahnemann Med Col, PhD(pharmacol), 58. *Prof Exp:* Vis researcher pharmacol, Inst Animal Physiol, 58-59; assoc, Hahnemann Med Col, 59-60; from asst prof to assoc prof, 60-66, PROF PHARMACOL, SCH MED, TULANE UNIV, 66- *Concurrent Pos:* Fel, Nat Paraplegia Found, 58-59; consult, ICI Am. *Mem:* AAAS; Am Soc Pharmacol & Exp Therapeut; NY Acad Sci. *Res:* Biochemical aspects of neuropharmacology; auditory pharmacology; acetylcholine storage and release. *Mailing Add:* Dept Pharmacol Tulane Univ Sch Med 1430 Tulane Ave New Orleans LA 70112

GUTH, S(HERMAN) LEON, b New York, NY, Dec 9, 32; div; c 2. VISION. *Educ:* Purdue Univ, BS, 59; Univ Ill, MA, 61, PhD(psychol), 63. *Prof Exp:* Lectr, 62-63, from instr to assoc prof, 63-70, PROF PSYCHOL, IND UNIV, BLOOMINGTON, 70- *Concurrent Pos:* Grants, NSF, 63-82 & NIH, 64-71; NIH spec res fel, Univ Calif, Berkeley, 71-72; NSF rotating prog dir sensory physiol & perception, 77-78; chmn, Dept Visual Sci, Ind Univ, 83-88. *Mem:* AAAS; fel Optical Soc Am; Asn Res Vision & Ophthal. *Res:* Visual psychophysics. *Mailing Add:* Dept Psychol Sch Optom Ind Univ Bloomington IN 47405

GUTH, SYLVESTER KARL, b Milwaukee, Wis, Dec 31, 08; m 31. PHYSICS, ILLUMINATION. *Educ:* Univ Wis, BS, 30, EE, 50. *Hon Degrees:* DOS, Northern Ill Col Optom, 53. *Prof Exp:* Res physicist, Lighting Res Lab, Gen Elec Co, 30-41, physicist, 46-49, head lighting res, 50-54, mgr radiant energy effects lab, 55-68, mgr appl res, Large Lamp Dept, 69-73; RETIRED. *Concurrent Pos:* Lectr, Case Inst Technol, 50-67; deleg, Int Comn Illum, Stockholm, 51, Zurich, 55, Brussels, 59, Vienna, 63, Wash, 67 & Barcelona, 71, London, 75 & Kyoto, 79, vpres comn & chmn activ comt, 71-75; consult, 74- *Honors & Awards:* Gold Medal, Illum Eng Soc, 67; Prentice Medal, Am Acad Optometry, 80. *Mem:* AAAS; fel Illum Eng Soc; Optical Soc Am; fel Am Acad Optom; Int Comn Illum (pres, 75-). *Res:* Light; vision; seeing; color; radiant energy; physical, psychological and physiological aspects of light and lighting. *Mailing Add:* 637 Quilliams Rd Cleveland OH 44121

GUTHALS, PAUL ROBERT, b Fowler, Colo, Mar 29, 29; m 58; c 2. ATMOSPHERIC CHEMISTRY & GENERAL ATMOSPHERIC SCIENCES, QUALITY PROGRAM & TECHNICAL MANAGEMENT. *Educ:* Eastern NMex Univ, BS, 57;. *Prof Exp:* Prog mgr 89, Nev Nuclear Waste Site Invest, 85-88; radio-chem air sampler, 57-59, proj leader, res, 72-75, asst group leader atmospheric studies, 80-85, PROJ LEADER, ATMOSPHERIC RES, LOS ALAMOS NAT LAB, US DEPT EDUC, 75-, QA OFFICER RES & DEVELOP. *Mem:* Am Soc Qual Control. *Res:* Active participant in atmospheric research as involves collection of samples (particulates and gaseous); sample analysis and in-situ atmospheric constituent measurements; nuclear event diagnosis; air mass movement tracing experiments. *Mailing Add:* MS 4-185 PO Box 1663 Los Alamos NM 87545

GUTHE, KARL FREDERICK, b Detroit, Mich, Aug 3, 18; m 46; c 3. MUSCULAR PHYSIOLOGY. *Educ:* Harvard Univ AB, 39, AM, 40, PhD(biol), 51. *Prof Exp:* Physicist, Naval Ord Dept, 41-45; tutor biochem sci, Harvard Univ, 49-50; from instr to prof, 50-85, EMER PROF ZOOL, UNIV MICH, ANN ARBOR, 85- *Mem:* AAAS; Biophys Soc. *Res:* Physiology of muscle. *Mailing Add:* 1407 Ferdon Ann Arbor MI 48104

GUTHEIL, THOMAS GORDON, b New York, NY, June 11, 42; m 85; c 3. FORENSIC PSYCHIATRY, FORENSIC TEACHING. *Educ:* Harvard Univ, BA, 63, MD, 67. *Prof Exp:* Actg dir, adult psychiat serv, 76-80, DIR MED STUDENT TRAINING, MASS MENT HEALTH CTR, 80-; ASSOC PROF PSYCHIAT, HARVARD MED SCH, 80- *Concurrent Pos:* Vis lectr, Harvard Law Sch, 77-; co-dir, Prog Psychiat & Law, Mass Ment Health Ctr, 80- & pres, Law & Psychiat Resource Ctr, 81-; consult, Am Bar Asn, 80- & Risk Mgt Found, Harvard Med Inst, 87- *Honors & Awards:* Guttmacher Award, Am Psychiat Asn, 83. *Mem:* Am Psychiat Asn; Am Acad Psychiat & Law; Int Acad Law & Ment Health (treas, 86-); Am Bd Forensic Psychiatry. *Res:* Liability and liability prevention; medicolegal decision making. *Mailing Add:* 74 Fenwood Rd Boston MA 02115

GUTHRIE, CHRISTINE, b Brooklyn, NY, Apr 27, 45. MOLECULAR BIOLOGY, BIOCHEMISTRY. *Educ:* Univ Mich, BS, 66; Univ Wis, PhD(genetics), 70. *Prof Exp:* Vis scientist, Max Planck Inst Molecular Genetics, Berlin, Ger, 70-71; asst prof, 73-80, ASSOC PROF BIOCHEM, UNIV CALIF, SAN FRANCISCO, 80- *Concurrent Pos:* Fel molecular biol, Univ Wis, 71-73. *Res:* Genetic and biochemical analysis of nucleic acid and protein synthesis. *Mailing Add:* Dept Biochem & Biophys Univ Calif S-964 San Francisco CA 94143

GUTHRIE, DANIEL ALBERT, b Ind, Mar 5, 39; div; c 3. VERTEBRATE ZOOLOGY. *Educ:* Amherst Col, BA, 60; Harvard Univ, MA, 62; Univ Mass, PhD(biol), 64. *Prof Exp:* Asst prof, 64-69, assoc prof, 69-76, PROF BIOL, PITZER COL, SCRIPPS COL & CLAREMONT MCKENNA COL, 77- *Mem:* AAAS; Nat Audubon Soc; Soc Vert Paleont; Am Soc Mammal; Am Ornithologists Union; Sigma Xi. *Res:* Lower Eocene mammals; carotid arteries in mammals; mammalian paleontology; evolutionary rates; environmental studies; bird population studies; pleistocene avifauna; zooarcheology. *Mailing Add:* Dept Joint Sci Claremont McKenna Scripps & Pitzer Col Claremont CA 91711

GUTHRIE, DAVID BURRELL, b Long Beach, Calif, Feb 10, 20; m 51; c 3. ORGANIC CHEMISTRY. *Educ:* Westminster Col, AB, 41; Univ Ill, PhD, 45. *Prof Exp:* Res chemist, Monsanto Chem Co, 46-59; asst chief chemist, Lucidol Div, Wallace & Tiernan, Inc, NY, 59-61, supvr & group leader, 61-63; group leader, new prod res, Petrolite Res & Develop, Petrolite Corp, 63-64, group leader, hydrocarbon additives & corrosion res, 64-66, supvr, Process Develop Eng Dept, Tretolite Div, 66-85, mgr, Pilot Plant Sect, 85-87, res chemist, 88; RETIRED. *Mem:* Am Chem Soc. *Res:* Organic peroxides; chemistry of petroleum products; synthetic organic and surfactant chemistry. *Mailing Add:* 1148 Ridgelynn Dr St Louis MO 63124-1220

GUTHRIE, DONALD, b Eureka, Calif, July 8, 33; m 54; c 2. STATISTICS. *Educ:* Stanford Univ, BS, 54, PhD(statist), 58; Columbia Univ, MA, 55. *Prof Exp:* Asst, Stanford Univ, 53-57, res assoc, 57-58; asst prof math mech, US Naval Postgrad Sch, 58-60; mathematician, Stanford Res Inst, 60-63; from assoc prof to prof statist, Ore State Univ, 63-73; PROF PSYCHIAT & BIOSTATIST, UNIV CALIF, LOS ANGELES, 73- *Concurrent Pos:* Consult, Stanford Res Inst, 58-59, 63-; vis assoc prof, dept statist, Univ NC, Chapel Hill, 69-70. *Mem:* Fel AAAS; Inst Math Statist; fel Am Statist Asn; Biomet Soc; Am Asn Ment Deficiency; Am Acad Ment Retardation; Soc Psychophysiol Res. *Res:* Application of statistical methods to mental retardation research; innovative methods of statistics instruction; statistical computing. *Mailing Add:* Univ Calif 760 Westwood Plaza Los Angeles CA 90024-1759

GUTHRIE, DONALD ARTHUR, b Winnipeg, Man, May 26, 26; US citizen; m 51; c 2. ORGANIC CHEMISTRY, RADIATION CHEMISTRY. *Educ:* Univ Man, BSc, 47; Univ Toronto, MSc, 49; McGill Univ, PhD(org chem), 52. *Prof Exp:* From res chemist to sr chemist, Esso Res & Eng Co, Standard Oil Co NJ, 51-62; mgr cent res div, Lord Corp, 62-70, vpres corp develop, 70-75; dir composite res, Ciba-Geigy Corp, 75-82; staff mem, Gulf South Res Inst, 83-88; VPRES & DIR, HELIX BIOMEDIX, 88- *Concurrent Pos:* Trustee, Edinboro Col; chmn, Gt Lakes Res Inst. *Mem:* Am Chem Soc; Am Inst Aeronaut & Astronaut; Soc Advan Mat & Process Eng; fel Am Inst Chem; Com Develop Asn. *Res:* Lubricant oxidation and detergency; radiation initiated polymerization; synthesis and application of propellant oxidizers and binders; management of basic research in solid state physics, polymer synthesis, surface and polymer physical chemistry; composite materials science. *Mailing Add:* 3232 Valcour Aime Baton Rouge LA 70820-5014

GUTHRIE, EUGENE HARDING, b Washington, DC, Apr, 9, 24; m 48; c 6. MEDICINE, PUBLIC HEALTH. *Educ:* George Washington Univ, MD, 51; Univ Mich, MPH, 55. *Prof Exp:* Intern, USPHS Hosp, Baltimore, Md, 51-52, house officer, 52-53; resident pub health, Montgomery County Health Dept, Md, 54-55; resident, State Dept Health, Calif, 55-56; chief sch health & rural health activ, Bur State Serv, USPHS, 56-59, chief prog officer, 59-62, chief, Neurol & Sensory Dis Serv Br, 62 & Div Chronic Dis, 62-66, from asst surgeon gen opers to assoc surgeon gen, 66-68; exec dir, Md Comprehensive Health Planning Agency, 68-74; consult health systs planning & develop, Am

Health Planning Asn, 74-76, assoc exec dir, 76-78; dep state health officer & health officer, Dorchester & Talbot Counties, 78-86, county health officer, Talbot County, Md, 86-87; assoc pub health admin, Sch Hyg Johns Hopkins Univ, 70-76; RETIRED. *Concurrent Pos:* Mem interdept comt health sch aged child & comt agr migrants, Dept HEW, 57-59; mem working group, President's Comt Migratory Labor & alt mem, Interagency Adv Group, President's Coun Youth Fitness, 57-59; consult, Boy Scouts Am, 58-59; mem, Commissioned Officers Awards Bd, USPHS, 63-64; mem med comt, President's Comn Employ Handicapped, 63-68; staff dir, Surgeon Gen Adv Comt Smoking & Health, 63-64; chmn bd, Am Acad Comprehensive Health Planning, 70-71; mem bd dir, Md Hosp Educ & Res Found, 70-72; regional adv group, Regional Med Prog, Md, 70-74. *Honors & Awards:* Smith-Reed-Russel Med Honor Soc, 51. *Mem:* AMA; Am Pub Health Asn. *Res:* School and rural health; chronic diseases, including heart, cancer, diabetes, arthritis, neurological and sensory; mental retardation; gerontology; preventive medicine; medical administration. *Mailing Add:* 8910 Teal Point Rd Easton MD 21601

GUTHRIE, FRANK ALBERT, b Madison, Ind, Feb 16, 27; m 55; c 4. ANALYTICAL CHEMISTRY, CHEMISTRY GENERAL. *Educ:* Hanover Col, BA, 50; Purdue Univ, MS, 52; Ind Univ, PhD(anal chem), 62. *Prof Exp:* Asst chem, Purdue Univ, 50-52; from instr to assoc prof, 52-67, chmn dept, 69-72, PROF CHEM, ROSE-HULMAN INST TECHNOL, 67-, CHIEF HEALTH PROFESSIONS ADV, 75- *Concurrent Pos:* Asst, Ind Univ, 55-56; vis asst prof, Purdue Univ, 58 & Univ Ill, 61, Charles F Kettering vis lectr, 61-62; NSF res partic, Univ Colo, 63-64, Rensselaer Polytech Inst, 66; res assoc, La State Univ, 67-, vis prof chem, US Mil Acad, W Point NY, 87-88; activ comt, Am Chem Soc, 82-86, nominations & elections, 88-; dir ed, Div Anal Chem, Am Chem Soc, 65-77, secy, 73-77, chmn elect, 78-79, chmn, 79-80. *Honors & Awards:* Alumni Achievement Award, Hanover Col, 77. *Mem:* Am Chem Soc; Sigma Xi; Coblentz Soc; fel Ind Acad Sci; Nat Asn Adv Health. *Res:* Spectroscopy; chemical instrumentation; coordination compounds. *Mailing Add:* Dept Chem Rose-Hulman Inst Technol 5500 Wabash Ave Terre Haute IN 47803-3999

GUTHRIE, FRANK EDWIN, b Louisville, Ky, Jan 14, 23; m 47; c 2. ENTOMOLOGY. *Educ:* Univ Ky, BS, 47; Univ Ill, MS, 49, PhD(entom), 52. *Prof Exp:* Asst entomologist, Northern Fla Exp Sta, Univ Fla, 52-54; from asst prof to assoc prof, 54-61, asst dean, 62-64, PROF ENTOM, NC STATE UNIV, 61- *Honors & Awards:* O Max Gardner Award, 83; Govenor's Award Sci, 83. *Mem:* Am Chem Soc; Soc Toxicol. *Res:* Metabolism of nicotine by insects; insecticide residues on tobacco; distribution and localization of insecticides; environmental aspects of pesticides. *Mailing Add:* Dept of Entom NC State Univ Raleigh NC 27650

GUTHRIE, GEORGE D, JR, b May 4, 62; m. MINERALOGY-CRYSTALLOGRAPHY, PETROLOGY-HYDROTHERMAL ALTERATION. *Educ:* Harvard Univ, AB, 84; Johns Hopkins Univ, MA, 86, PhD(geol), 89. *Prof Exp:* Res asst, Johns Hopkins Univ, 84-89; FEL, LOS ALAMOS NAT LAB, 89- *Concurrent Pos:* cur asst, Harvard Mineral Mus, 83; Asst teacher, 84, 86 & 87. *Mem:* AAAS; Am Geophys Union; Clay Minerals Soc; Mineral Soc Am; Sigma Xi. *Res:* Compositional and structural order in clay minerals; crystal-chemical aspects of turquoise structure; submicrometer fluid inclusions, using TEM and analytical electron microscopy, in turbid feldspars, turbid diamond coats and turbid garnets; hydrothermal alteration; health effects of minerals; interpreting structures with HRTEM; transmission and scanning electron microscopies/electron microprobe; x-ray diffraction techniques; ionic modeling of crystal structures. *Mailing Add:* Geol & Geochem MS D469 Los Alamos Nat Lab Los Alamos NM 87545

GUTHRIE, GEORGE DRAKE, b Indianapolis, Ind, Jan 20, 32; m 61; c 1. BIOCHEMISTRY, MOLECULAR BIOLOGY. *Educ:* Wabash Col, AB, 54; Calif Inst Technol, PhD(biophys), 62. *Prof Exp:* Res associateship, Mass Inst Technol, 62-65; from asst prof to assoc prof biochem, Ind Univ, Indianapolis, 65-71; ASSOC DIR, EVANSVILLE CTR MED EDUC, SCH MED, IND UNIV, EVANSVILLE, 72- *Mem:* AAAS; Sigma Xi; Res Soc Alcoholism; Am Soc Biochem & Molecular Biol. *Res:* Metabolic regulation, especially regulation of intermediary metabolism in liver. *Mailing Add:* 8600 Univ Blvd Evansville IN 47712

GUTHRIE, HELEN A, b Sarnia, Ont, Sept 25, 25; m 49; c 3. NUTRITION. *Educ:* Univ Western Ont, BS, 46; Mich State Univ, MS, 48; Univ Hawaii, PhD(physiol), 68. *Hon Degrees:* DSc, Univ Western Ont, 83. *Prof Exp:* From asst prof to assoc prof, 49-72, head dept, 79-89, PROF NUTRIT, PA STATE UNIV, 72-, ENDOWED PROF, 90- *Concurrent Pos:* Mem food & nutrit bd, Nat Acad Sci, 73-76, recommended dietary allowance comn, 80-85; mem joint nutrit eval & monitoring comt, Dept Health & Human Serv/USDA, 83-85. *Honors & Awards:* Borden Award, Am Home Asn, 78; Atwater Award, USDA, 89; Elvjehm Award, Am Inst Nutrit, 89. *Mem:* Am Soc Clin Nutrit; Am Dietetic Asn; Am Pub Health Asn; Am Inst Nutrit (pres, 87-88); Soc Nutrit Educ (pres, 77-78, 87-88). *Res:* Infant nutrition; nutritional evaluation; world nutrition. *Mailing Add:* Dept Nutrit S-126 Human Develop Pa State Univ University Park PA 16802

GUTHRIE, HUGH D, b Murdo, SDak, May 11, 19; m 50; c 5. CHEMICAL ENGINEERING, ECONOMICS. *Educ:* Univ Iowa, BS, 43. *Prof Exp:* Jr engr, Shell Develop Co, Calif, 43-45, engr, 45-52, sr technologist, Shell Oil Co, 52-54, asst mgr gas dept, 54-56, asst mgr catalytic cracking, 56, sr technologist, 56-57, group leader mfg econ, 57-59, mgr alkylation dept, 59-61, asst to mgr prod econ, 61-66, mgr prod econ mkt, 66-67, asst to gen mgr refineries in mfg, 67-70, sr staff assoc, 70-72, mgr indust anal, 72-74, spec assignment, 74-76; dir oil, gas & shale technol, US Energy Res & Develop Admin, 76-77; actg dir oil, gas, shale & insitu technol, Dept Energy, 77-78; dir, Energy Ctr, SRI Int, 78-80; mgr technol assessment, Occidental Res Corp, 80-82, vpres licensing, Occidental Develop Co, 82-86; DIR EXTRACTION PROJ MGT, ENERGY TECHNOL CTR, US DEPT ENERGY, MORGANTOWN, 89- *Concurrent Pos:* Mem bd, United Eng Trustees, 65-, pres, 78-79; mem bd, Eng Found, 66-80. *Honors & Awards:* Founders Award, Am Inst Chem Engrs, 74; F J Van Antwerpen Award. *Mem:* AAAS; Am Chem Soc; Am Inst Chem Engrs (vpres, 68, pres, 69); Soc Petrol Engrs; NY Acad Sci. *Res:* Separation processes, particularly distillation; distillation trays; economics, production and marketing; fossil energy technologies, particularly oil, gas, shale and insitu gasification. *Mailing Add:* Morgantown Enegry Tech Ctr PO Box 880 Morgantown WV 26507-0880

GUTHRIE, JAMES LEVERETTE, b Lawrence, Kans, July 3, 31; m 58. ADHESIVES, POLYURETHANES. *Educ:* Oberlin Col, AB, 52; Univ Mo, PhD(org chem), 56. *Prof Exp:* Res chemist, PPG Corp, 56-61; res supvr, Polymer Dept, Res Div, 61-70, RES ASSOC, W R GRACE & CO, CLARKSVILLE, 71- *Honors & Awards:* IR-100 Award, Indust Res Corp, 68 & 75. *Mem:* AAAS; Am Chem Soc; Sigma Xi; NY Acad Sci. *Res:* Chemistry and applications of organic materials; polymers; photosensitive systems; composite structures, urethanes, coatings, adhesives, sealants, foams, plastisols. *Mailing Add:* 1318 Patuxent Dr Ashton MD 20861

GUTHRIE, JAMES PETER, b Port Elgin, Ont, Feb 3, 42; m 65; c 2. BIO-ORGANIC CHEMISTRY. *Educ:* Univ Western Ont, BSc, 64; Harvard Univ, PhD(org chem), 68. *Prof Exp:* Nat Res Coun Can fel biochem, Princeton Univ, 68-69; from asst prof to assoc prof, 69-78, PROF CHEM, UNIV WESTERN ONT, 78- *Concurrent Pos:* Alfred P Sloan fel, 75 & E W R Steacie fel, 80. *Mem:* Am Chem Soc; Chem Inst Can. *Res:* Mechanisms of enzyme catalysis; enzyme model systems; thermodynamics of metastable intermediates. *Mailing Add:* Dept Chem Univ Western Ont London ON N6A 5B7 Can

GUTHRIE, JAMES WARREN, b Kansas City, Kans, Sept 7, 23; m 47; c 3. PHYTOPATHOLOGY. *Educ:* Utah State Univ, BS, 49, MS, 50; Univ Wis, PhD(plant path), 52. *Prof Exp:* From asst prof to prof plant path, 52-77, prof plant sci, Univ Idaho, 77-82; RETIRED. *Mem:* Am Phytopath Soc; assoc Asn Off Seed Analysts; Int Seed Testing Asn. *Res:* Seed pathology, particularly Fusarium disease of corn, diseases of Poa prantensis and Pisum arvense; bacterial diseases of Phaseolus vulgaris. *Mailing Add:* 2433 East F Moscow ID 83843

GUTHRIE, JOHN DAULTON, b Detroit, Mich, Mar 19, 03; m 35, 68; c 2. BIOCHEMISTRY. *Educ:* Ohio State Univ, BSc, 25, PhD(plant biochem), 29. *Prof Exp:* Asst biochemist, Boyce Thompson Inst, 25-41; from sr chemist to prin chemist, Southern Regional Res Ctr, USDA, 41-61, chief, Cotton Chem Reactions Lab, 61-73, collabr, 73-87; RETIRED. *Concurrent Pos:* Asst, Ohio State Univ, 27-29; vis prof, La State Univ Med Ctr, 73-79. *Honors & Awards:* John Scott Medal, 65. *Mem:* AAAS; Am Chem Soc; Am Inst Chemists; Sigma Xi. *Res:* Glutathione in plants; industrial utilization of agricultural commodities; analytical chemistry; chemical modification of cotton fibers; flameproofing of cotton; cellulose chemistry; iron-sulfur proteins. *Mailing Add:* 6848 Louisville St New Orleans LA 70124

GUTHRIE, JOHN ERSKINE, b Montreal, Que, Sept 25, 26; m 52; c 2. ECOLOGY, RADIATION ECOLOGY. *Educ:* McGill Univ, BSc, 50, MSc, 55; Univ Man, PhD(entom), 69. *Prof Exp:* Officer artillery, Can Army Regular, 44-59; res officer monitoring, Chalk River Nuclear Labs, 59-63; BR HEAD, ENVIRON BR, WHITESHELL NUCLEAR RES ESTAB, ATOMIC ENERGY CAN LTD, 63- *Concurrent Pos:* Adj prof, Dept Entom, Univ Man, 72- *Mem:* Ecol Soc Am; Entom Soc Can; Health Physics Soc. *Res:* Effects of chronic exposure to low dose rates of ionizing radiation on ecological systems; environmental radiation monitoring; fate of radionuclides in the environment. *Mailing Add:* PO Box 41 Pinawa MB R0E 1L0 Can

GUTHRIE, JOSEPH D, b Indianapolis, Ind, Apr 13, 42; m 65; c 4. ANALYTICAL CHEMISTRY. *Educ:* Wabash Col, AB, 64; Ohio Univ, MS, 66. *Prof Exp:* Res scientist, Alcoa Labs, 66-68, group leader, 68-76, staff scientist anal chem, 76-80, tech supvr, 80-85, SR TECH SUPRV, ALCOA LABS, 85- *Mem:* Am Chem Soc. *Res:* Analysis of thin films on aluminum, determination of hydrogen in aluminum and lubrication in aluminum fabrication processes; cleaning, pretreating and organic coating of aluminum. *Mailing Add:* Alcoa Tech Ctr Alcoa Center PA 15069

GUTHRIE, MARSHALL BECK, medicine, dermatology; deceased, see previous edition for last biography

GUTHRIE, MARY DUNCUM, neuroscience; deceased, see previous edition for last biography

GUTHRIE, RICHARD LAFAYETTE, b Union Springs, Ala, May 29, 41; m 62; c 3. INTERNATIONAL AGRICULTURE, SOIL CONSERVATION. *Educ:* Auburn Univ, BS, 62, MS, 65; Cornell Univ, PhD(soil sci), 68. *Prof Exp:* Asst soil scientist, Soil Conserv Serv, USDA, 71-72, soil correlator, 72-74; exten soil specialist, Ala Coop Exten Serv, 74-75, prof & head Dept Agron & Soils, 83-85, actg dean, Col Agr, 85-88, ASSOC DEAN, INT AGR, COL AGR, AUBURN UNIV, 88- *Concurrent Pos:* Grad res asst, Dept Agron & Soils, Auburn Univ, 63-65, Cornell Univ, 65-68; ed, J Soil Surv Horizons, 73-78. *Mem:* Am Soc Agron; Soil Sci Soc Am; Soil & Water Conserv Soc Am; Int Soil Sci Soc. *Res:* Morphology, genesis and classification of soils with particular emphasis on properties that relate to land use; investigation of soil chemistry, soil mineralogy and soil wetness to better predict soil behavior. *Mailing Add:* Col Agr Auburn Univ 103 Comer Hall Auburn AL 36849-5401

GUTHRIE, ROBERT, b Marionville, Mo, June 28, 16; m 41; c 6. MEDICAL GENETICS, MICROBIOLOGY. *Educ:* Univ Minn, BA, 41, MD, 45, PhD(bact), 46; Univ Maine, MS, 42. *Prof Exp:* Asst med bact, Univ Minn, 43-45; Nat Res Coun fel biochem, Univ Wis, 45-46; physician, Argonne Nat Lab, 46; sr asst surgeon, NIH, USPHS, 46-48, surgeon, 48-49, microbiologist, hosp, Staten Island, 50; prof bact & chem dept, Univ Kans, 49-50; asst, Sloan-Kettering Inst, 51-54; prin cancer res scientist, Roswell Park Mem Inst, State Dept Health, NY, 54-58; res assoc prof, dept pediat & dir biochem genetics sect, Children's Hosp, 58-75, PROF PEDIAT & MICROBIOL, DEPT PEDIAT, SCH MED, STATE UNIV NY, BUFFALO, 75- *Concurrent Pos:* Career res scientist award, Am Acad Ment Retardation, 81. *Honors & Awards:* Kimble Methodology Award, 65; Am Asn Ment Deficiency Sci Award, 70. *Mem:* Am Soc Microbiol; Fedn Am Scientists; Am Chem Soc. *Res:* Pseudomonas hydrophilia; nutrition of bacteria and invertebrates; biochemical genetics of Escherichia coli and Bacillus subtilus; cytology of bacterial endospores; nucleic acid metabolism; cancer chemotherapy; human biochemical genetics and biochemical individuality. *Mailing Add:* 352 Acheson Hall State Univ NY Buffalo Buffalo NY 14214

GUTHRIE, ROBERT D, b Bronxville, NY, June 27, 36; m 63; c 3. PHYSICAL ORGANIC CHEMISTRY. *Educ:* Oberlin Col, BA, 58; Univ Rochester, PhD(org chem), 63. *Prof Exp:* NSF fel, Univ Calif, Los Angeles, 63-64, univ fel, 64, lectr chem, 64-65; from asst prof to assoc prof, 65-77, PROF CHEM, UNIV KY, 77- *Mem:* Am Chem Soc. *Res:* Carbanion chemistry; electron transfer reactions. *Mailing Add:* Dept Chem Univ Ky Lexington KY 40506-0002

GUTHRIE, ROGER THACKSTON, b Spartanburg, SC, Dec 5, 24; m 50; c 3. ORGANIC CHEMISTRY. *Educ:* Wofford Col, BS, 49; Univ NC, MA, 52, PhD(chem), 53. *Prof Exp:* Asst, Univ NC, 50-53; res chemist, Am Enka Corp, 53-59, head dielec mat sect, 59-62, head textile yarn develop sect, 63-66; tech mgr, New Prod Explor, Celanese Plastics Co, 66-69, New Prod Develop, 69-70, mgr, New Venture Anal, 70; res mgr, 71-77, mgr special proj mkt, 77-79, tech mgr, mfg, Chem Div, 80-85, coordr, mfg processes, 85-86, TECH ASSOC, INDUST CHEM DIV, M&T CHEM INC, 86- *Mem:* AAAS; Am Chem Soc; NY Acad Sci; Asn Res Dirs; Soc Plastics Eng; Sigma Xi. *Res:* Polymer chemistry and physics; electrical insulating materials. *Mailing Add:* 457 Colonia Blvd Colonia NJ 07067

GUTHRIE, ROLAND L, b Charleston, WVa, Apr 5, 28; m 54; c 3. PLANT ANATOMY, DENDROLOGY. *Educ:* WVa Univ, BSF & MS, 53, PhD(bot), 68. *Prof Exp:* Forester & naturalist, Isaac W Bernheim Found, 53-56; arboretum asst, 59-61, from instr to asst prof, 61-73, ASSOC PROF BIOL, WVA UNIV, 73-, CUR ARBORETUM, 61- *Mem:* Bot Soc Am; Sigma Xi; Int Asn Wood Anatomists. *Res:* Ecological anatomy of woody plants. *Mailing Add:* Dept Biol WVa Univ PO Box 6057 Morgantown WV 26506-6057

GUTHRIE, RUFUS KENT, b Mullin, Tex, July 4, 23; m 48; c 2. MICROBIAL ECOLOGY. *Educ:* Univ Tex, BA, 48, MA, 50; Baylor Univ, PhD(microbiol), 54. *Prof Exp:* Bacteriologist, State Dept Health, Tex, 47-48 & Brownwood Mem Hosp, 49; asst microbiol, Col Med, Baylor Univ, 50-52 & 53-54; bacteriologist, Vet Admin Hosp, Houston, 52-53; from asst prof to prof biol, NTex State Univ, 54-69; prof microbiol, Clemson Univ, 69-74, dir biol div, 69-71, assoc dean, Col Physics, Math & Biol Sci, 71-74; PROF MICROBIOL & ECOL, SCH PUB HEALTH, UNIV TEX HEALTH SCI CTR, HOUSTON, 74-, CONVENER, DIS CONTROL, 83- *Concurrent Pos:* Consult, Flow Mem Hosp, Denton, Tex, 54-62, Alcon Labs, Ft Worth, 57-60 & Ctr Environ Studies, Va Polytech Inst & State Univ, 79; NSF sci fel microbiol, Col Med, Univ Iowa, 62-63. *Mem:* Soc Exp Biol & Med; fel Am Acad Microbiol; Am Water Resources Asn; Am Pub Health Asn; Am Soc Microbiol. *Res:* Immunological mechanisms; effects of chemical and physical pollutants on natural aquatic biota; medical microbiology; infectious diseases. *Mailing Add:* Sch Pub Health Univ Tex Health Sci Ctr Houston TX 77225

GUTHRIE, RUSSELL DALE, b Nebo, Ill, Oct 27, 36; m 61, 78; c 3. ZOOLOGY. *Educ:* Univ Ill, BS, 58, MS, 59; Univ Chicago, PhD(zool), 63. *Prof Exp:* Assoc prof, 63-74, PROF ZOOL, UNIV ALASKA, 74- *Concurrent Pos:* NSF instnl grant, 63-64, res grant, 65-68, grant, 70-72, 72-74, 80-81 & 85-86. *Mem:* AAAS; Soc Study Evolution; Soc Vert Paleont; Am Soc Mammal. *Res:* Evolutionary mechanics; Pleistocene vertebrate paleontology; vertebrate evolution. *Mailing Add:* Biol Zool Univ Alaska Fairbanks AK 99701

GUTHRIE, WILBUR DEAN, b Woodward, Okla, Mar 3, 24; m 46; c 5. ENTOMOLOGY. *Educ:* Okla State Univ, BS, 50, MS, 51; Ohio State Univ, PhD, 58. *Prof Exp:* ENTOMOLOGIST, AGR RES SERV, USDA, 51- *Mem:* Entom Soc Am; Crop Soc Agron. *Res:* Insect resistance in crop plants; relative degree of resistance of inbred lines and hybrid corn to the European corn borer; sources of germ plasm. *Mailing Add:* Rte 2 Box 199 Nevada IA 50021

GUTIERREZ, GUILLERMO, b Palma Soriano, Cuba, Mar 10, 46; US citizen; m 83; c 4. MAGNETIC RESONANCE SPECTROSCOPY, CRITICAL CARE MEDICINE. *Educ:* City Col New York, BS, 68; Univ Dayton, MS, 70; Case Western Univ, MD, 77, PhD(biomed eng), 78. *Prof Exp:* Develop engr, Gen Motors Res Labs, 68-71, sr scientist, 77-78; resident intern med, Univ Mich, 78-71, fel pulmonary med, 81-83; ASSOC PROF PULMONARY MED, UNIV TEX HEALTH CTR, 83-; DIR, PULMONARY & CRITICAL CARE DIV, HERMANN HOSP, 90- *Honors & Awards:* Career Investr Award, Am Lung Assoc, 90. *Mem:* Am Col Physicians; Am Thoracic Soc; Am Col Chest Physicians; Am Physiol Soc; Soc Critical Care Med; Am Soc Clin Invest. *Res:* Determination of parameters governing the transport of oxygen to the tissues and their effect on cellular bioenergetics during hypoxia; sepsis and multiple systems. *Mailing Add:* Dept Internal Med Univ Tex Health Sci Ctr PO Box 20036 Houston TX 77225

GUTIERREZ, LEONARD VILLAPANDO, JR, b Tanawan, Batangas, Philippines, Apr 2, 32; US citizen; m 57; c 3. ENVIRONMENTAL ENGINEERING. *Educ:* De La Salle Col, Manila, BS, 53; Univ Philippines, BS, 54; Purdue Univ, MS, 56. *Prof Exp:* Civil engr, Ayres Lewis Norris & May, 56-58; sanit engr, Gannett Fleming Corddry & Carpenter, 58-60, County San Diego, 60-63, Eng Sci Inc, 63-64, County San Diego, 64-66 & Ralph M Parsons, 66-68; sr engr, 68-72, VPRES ENG, CAMP DRESSER & MCKEE, 72- *Honors & Awards:* Presidential Sci Award, Repub Philippines, 77. *Mem:* Am Soc Civil Engrs; Am Acad Environ Engrs. *Mailing Add:* Camp Dresser & McKee Ten Cambridge Ct Cambridge MA 02142

GUTIERREZ, PETER LUIS, b Colombia, SAm, June 9, 39; US citizen; c 2. BIOPHYSICS, PHYSICAL PHARMACOLOGY. *Educ:* Wheaton Col, BS, 62; Calif State Univ, Los Angeles, MS, 70; Southern Ill Univ, PhD(radiation biophys), 73. *Prof Exp:* Engr asst electronics, Automatic Elec Co, 62-64; engr, Hycon Mfg Co, 66-67; instr phys biochem, Sch Med, Tulane Univ, 72-74; staff biophysicist biol electron spin resonance, Med Col Wis, 74-79; ASSOC PROF BIOCHEM & ONCOL, UNIV MD CANCER CTR, 79- *Mem:* Biophys Soc; Am Asn Cancer Res; Int Soc Free Radical Res; Am Chem Soc. *Res:* Free radical metabolites of antitumor agents; electron spin resonance in carcinogenesis-free radical and other paramagnetic changes monitored during the development of experimental tumors; spin labeled drugs; membrane fluidity changes in cancer. *Mailing Add:* Div Develop Ther Univ Md Cancer Ctr 655 W Baltimore St Baltimore MD 21201

GUTIERREZ-MIRAVETE, ERNESTO, b Mexico City, Mex, June 29, 54; m 87. MATHEMATICAL MODELING, METALS PROCESSING. *Educ:* Nat Univ Mex, Chem Metall Eng, 78; Mass Inst Technol, PhD(metall), 85. *Prof Exp:* Lectr mat sci, Nat Univ Mexico, 77-79; postdoctoral assoc metall, Mass Inst Technol, 85-87; ASST PROF METALL, HARTFORD GRAD CTR, 87- *Concurrent Pos:* Consult, Sicartsa, Hylsa, Mex, 79, Cent Bank Mus, Ecuador, 88, United Technol Corp, 89, consult, Olin Corp, 90. *Mem:* AAAS; Sigma Xi; Univ Mat Coun; Mat Res Soc; Minerals, Metals & Mat Soc; Am Soc Mat. *Res:* Apply and develop theoretical and computational tools to the analysis of materials and manufacturing processing operations; forming processes which start from molten and/or particulate matter. *Mailing Add:* Hartford Grad Ctr 275 Windsor St Hartford CT 06120

GUTJAHR, ALLAN L, b Hosmer, SDak, Mar 20, 38; m 59, 81; c 5. STOCHASTIC HYDROLOGY, APPLIED STOCHASTIC PROCESSES. *Educ:* Univ Wash, BS, 61; Johns Hopkins Univ, MS, 63; Rutgers Univ, PhD(math statist), 70. *Prof Exp:* Mem tech staff, Bell Tel Labs, 62-68; asst statist, Rutgers Univ, 68-69; mem tech staff, Bell Tel Labs, 69-71; from asst prof to assoc prof, 71-81, chmn dept, 85-88, PROF MATH, NMEX INST MINING & TECHNOL, 81-, RES MATHEMATICIAN, 78- *Concurrent Pos:* Vis scientist, Sch Mines, Paris, 79, US Geol Surv, 79 & Stanford Univ, 89; assoc vpres, acad affairs, NMex Inst Mining & Technol, 90- *Mem:* Sigma Xi; Am Geophys Union; Math Asn Am; NAm Coun Geostatist; Am Statist Asn. *Res:* Stochastic models in hydrology; geostatistics; applied statistics; statistical inference in semi-Markov processes and other stochastic processes; queueing theory. *Mailing Add:* 445 Aquina Ct Belen NM 87002

GUTKIN, EUGENE, mathematics, for more information see previous edition

GUTKNECHT, JOHN WILLIAM, b Youngstown, Ohio, Apr 13, 37; m 62; c 1. PHYSIOLOGY, MARINE BIOLOGY. *Educ:* Ohio Wesleyan Univ, BA, 59; Univ NC, PhD(zool), 64. *Prof Exp:* Res biologist, Radiobiol Lab, US Bur Com Fisheries, NC, 63-66; USPHS res fel biophys, Univ EAnglia, 66-67; res assoc, 67-69, from asst prof to assoc prof, 69-80, PROF PHYSIOL, MED CTR, DUKE UNIV, 80- *Mem:* AAAS; Am Inst Biol Sci; Am Physiol Soc; Biophys Soc; Soc Gen Physiol. *Res:* Transport of solutes and water across cell membranes; membrane permeability; bioelectricity. *Mailing Add:* Dept Physiol Marine Lab Duke Univ Beaufort NC 28516

GUTMAN, DAVID, b Isle of Malta, Nov 2, 34; US citizen; m 65; c 3. PHYSICAL CHEMISTRY. *Educ:* Univ Calif, Berkeley, BS, 60; Univ Ill, PhD(phys chem), 65. *Prof Exp:* From asst prof to prof chem, Ill Inst Technol, 64-88, PROF CHEM & CHMN, CATH UNIV AM, 88- *Concurrent Pos:* Prog Dir Chem Dynamics, NSF, Washington, DC, 76-77. *Mem:* Am Chem Soc; Am Phys Soc. *Res:* Chemical kinetics; laser induced reactions; theoretical and experimental studies of gas phase reactions. *Mailing Add:* Dept Chem Cath Univ AM Washington DC 20064

GUTMAN, GEORGE ANDRE, b Domme, France, Sept 15, 45; US citizen; m 77; c 2. MOLECULAR IMMUNOLOGY, MOLECULAR EVOLUTION. *Educ:* Columbia Col, AB, 66; Stanford Univ, PhD(biol sci), 73. *Prof Exp:* Res fel, dept path, Stanford Med Sch, 73-74 & Walter & Eliza Hall Inst Med Res, Australia, 74-76; from asst prof to assoc prof, 76-89, PROF, DEPT MICROBIOL & MOLECULAR GENETICS, UNIV CALIF, IRVINE, 89- *Concurrent Pos:* Researcher, Jackson Lab, 62, Roswell Park Mem Inst, 63 & Yale Univ, 64; Fulbright scholar, Inst Pasteur, Paris, 66-67; res fel, USPHS, 72-74 & Arthritis Found, 74-77; prin investr, USPHS-Nat Inst Allergy & Infectious Dis res grants, 78-89, Am Cancer Soc res grants, 78-79 Kroc Found grant, 80-82 & Res Can Develop Award, USPHS, 78-83. *Mem:* Am Asn Immunologists. *Res:* Organization and expression of immunoglobulin genes; evolution of closely related gene families in rodents and primates. *Mailing Add:* Dept Microbiol & Molecular Genetics Univ Calif Irvine CA 92717

GUTMAN, GEORGE GARIK, b Leningrad, USSR, Apr 22, 53; US citizen; m 81; c 2. CLIMATOLOGY OF LAND, REMOTE SENSING FROM SATELLITES. *Educ:* Leningrad Hydrometeorol Inst, MA, 76; Tel Aviv Univ, PhD(climate), 84. *Prof Exp:* Postdoctoral resident res assoc, Nat Res Coun, Nat Acad Sci, 85-87; res assoc, Coop Inst Climate Studies, 87-90, PHYS SCIENTIST, SATELLITE RES LAB, LAND SCI BR, NAT OCEANIC & ATMOSPHERIC ADMIN, NAT ENVIRON SATELLITE DATA & INFO SERV, 90- *Mem:* Am Meteorol Soc. *Res:* Remote sensing techniques for studying land surface-climate interactions; cloud detection from space; rain estimate from space; monitoring earth ecosystems using remote sensing. *Mailing Add:* 311 Lyric Lane Silver Spring MD 20901

GUTMANN, HELMUT RUDOLPH, b Strasbourg, France, July 31, 11; nat US; m 46; c 4. BIOCHEMISTRY. *Educ:* Univ Goettingen, MD, 36. *Prof Exp:* Res assoc, Univ Tenn, 48-50; asst prof cancer res, Univ Fla, 50-52; from asst prof to assoc prof physiol chem, Univ Minn, Minneapolis, 52-64; biochemist, Radioisotope Serv, 56-61, BIOCHEMIST, CANCER RES LAB, VET ADMIN HOSP, 61-; EMER PROF BIOCHEM, UNIV MINN, MINNEAPOLIS, 82- *Concurrent Pos:* Coxe mem fel biochem, Yale Univ, 46-48; USPHS fel biochem, Max-Planck Inst, Munich, 60-61; biochemist,

Radioisotope Serv, Cancer Res Lab, Vet Admin Hosp, 56-65; prof, Univ Minn, 64-82. *Mem:* AAAS; Am Chem Soc; Am Soc Biol Chem; Am Asn Cancer Res; Royal Soc Chem. *Res:* Metabolism and mechanism of action of carcinogenic and toxic compounds. *Mailing Add:* Cancer Res Lab Vet Admin Hosp Minneapolis MN 55417

GUTMANN, LUDWIG, b Frankfurt, Ger, Apr 7, 33; US citizen; m 54; c 3. NEUROLOGY. *Educ:* Princeton Univ, BA, 55; Columbia Univ, MD, 59. *Prof Exp:* From intern med to resident neurol, Med Ctr, Univ Wis-Madison, 59-63; chief neurol, US Air Force Hosp, Scott AFB, Ill, 63-65; Nat Inst Neurol Dis & Blindness fel neurophysiol, Mayo Clin, 65-66; from asst prof to assoc prof neurol, 66-70, asst prof physiol & biophys, 68-70, PROF NEUROL & CHMN DEPT, SCH MED, WVA UNIV, 70-, DIR ELECTROMYOGRAPH LAB, 66-, ASSOC PROF PHYSIOL & BIOPHYS, 70- *Concurrent Pos:* Neurol consult, Alton State Hosp, Ill, 63-65; mem, Myasthenia Gravis Found. *Mem:* Am Acad Neurol; Am Asn Electromyog & Electrodiag; Asn Res Nerv & Ment Dis; Asn Univ Prof Neurol; Am Neurol Asn (pres, 84-86). *Res:* Electromyography and neuromuscular diseases. *Mailing Add:* Dept Neurol WVa Univ Med Ctr Morgantown WV 26505

GUTMANN, RONALD JAY, b Brooklyn, NY, Nov 16, 40; m 67; c 2. SOLID STATE DEVICES & TECHNOLOGY. *Educ:* Rensselaer Polytech Inst, BEE, 62, PhD(electrophys), 70; NY Univ, MEE, 64. *Prof Exp:* Mem tech staff, Bell Tel Labs, 62-66; sr engr, Lockheed Electronics Co, 66-67; res asst, Rensselaer Polytech Inst, 67-69, engr, 69-70, from asst prof to prof, dept elec & systs eng, 70-81; prog dir, Solid State & Microstruct Eng, NSF, 81-83; AT ELEC COMP & SYST ENG, CTR INTEGRATED ELECTRONICS, RENSSELAER POLYTECH INST, 83- *Concurrent Pos:* Consult, govt & various industs, 72-81; ed, Continuing Educ Electronics, McGraw Hill Bk Co, 74-78. *Mem:* Inst Elec & Electronics Engrs; Sigma Xi; Am Asn Univ Prof. *Res:* Semiconductor devices; integrated electronics technology; microwave techniques. *Mailing Add:* Elec Comp & Syst Eng Rensselaer Polytech Inst Troy NY 12180-3590

GUTOFF, EDGAR B(ENJAMIN), b New York, NY, June 2, 30; m 56; c 2. COATING TECHNOLOGY, STATISTICAL PROCESS CONTROL. *Educ:* City Col New York, BChE, 51; Mass Inst Technol, SM, 52, ScD(chem eng), 54. *Prof Exp:* Chem engr, Fertilizers & Chem, Ltd, Israel, 52; sr process engr, Brown Co, 54-58; sr chem engr, Ionics, Inc, 58-60; sr scientist & sr prin engr, Polaroid Corp, 60-88; CONSULT CHEM ENGR, 88- *Concurrent Pos:* Instr chem, Ctr Prof Advan, 79-81; part-time lectr, adj prof & vis res prof, Grad Sch Eng, Northeastern Univ, 81- *Mem:* Am Chem Soc; fel Am Inst Chem Engrs; Soc Imaging Sci & Technol. *Res:* Coating operations drying; dynamic contact angles; surface elasticity; coagulation of silver halide suspensions; backmixing in extraction columns; efficiencies of mixing tanks; adsorption of polymers, rubber-filler interactions; photographic emulsions; crystallization of silver halides; syneresis. *Mailing Add:* 194 Clark Rd Brookline MA 02146-5824

GUTOWSKI, GERALD EDWARD, b Jackson, Mich, June 8, 41; m 64; c 4. RESEARCH & DEVELOPMENT ADMINISTRATION. *Educ:* Mich State Univ, BS, 63; Wayne State Univ, PhD(org chem), 67. *Prof Exp:* Res assoc chem, Wayne State Univ, 67-68; sr res chemist, 68-73, res scientist, Chem Res Div, 73-75, MGR, PROD INTRO COORD, LILLY RES LABS, ELI LILLY & CO, 75- *Mem:* Am Chem Soc; Pharmaceut Soc Japan. *Res:* Antibiotics; carbohydrates; medicinal chemistry; structure determination; organic synthesis; mechanisms of organic reactions; anti-tumor agents; anti-viral agents; alkaloids; nucleosides. *Mailing Add:* 7472 Galloway Ct Indianapolis IN 46250

GUTOWSKI, PAUL RAMSDEN, b Rhyll, UK, Apr 22, 47; Can citizen; m 72; c 1. SEISMOLOGY. *Educ:* Univ Alta, BSc, 69, MSc, 70, PhD(seismol), 74. *Prof Exp:* Petrol geophysicist petrol explor, Amoco Can Petrol Co Ltd, 74-75, RES SCIENTIST THEORET SEISMOL, AMOCO PROD CO, 75- *Concurrent Pos:* Adj asst prof, Univ Tulsa, 78- *Mem:* Soc Explor Geophysicists; Seismol Soc Am. *Res:* Wave propagation in the earth; numerical analysis and communication theory as applied to seismology; inverse theory. *Mailing Add:* 2171 N Vancouver Ave Tulsa OK 74127

GUTOWSKY, HERBERT S(ANDER), b Bridgman, Mich, Nov 8, 19; m; c 3. MICROWAVE ROTATIONAL SPECTROSCOPY. *Educ:* Ind Univ, AB, 40; Univ Calif, MS, 46; Harvard Univ, PhD(phys chem), 49. *Hon Degrees:* DSc, Ind Univ, 83. *Prof Exp:* From instr to prof, 48-56, head div phys chem, 56-62, head dept chem & chem eng, 67-70, prof chem, 56-83, dir, Sch Chem Sci & head, dept chem, 70-83, RES PROF CHEM, UNIV ILL, URBANA, 83-, PROF CHEM, CTR ADVAN STUDY, 83- *Concurrent Pos:* Guggenheim fel, 54-55; vis prof, Univ Calif, 56; Walker Ames vis prof, Univ Wash, 57. *Honors & Awards:* Langmuir Prize, Am Chem Soc, 66, Midwest Award, St Louis Sect, 73, Peter Debye Award, 75; Int Soc Magnetic Resonance Award, 74; Nat Medal Sci, 77; G B Kistiakowsky lectr chem, Harvard Univ, 80; Wolf Found Prize, 83; Krug lectr, Univ Ill, 84. *Mem:* Nat Acad Sci; fel AAAS; Am Chem Soc; fel Am Phys Soc; Am Acad Arts & Sci; Am Philos Soc. *Res:* Molecular and solid state structure; nuclear magnetic resonance. *Mailing Add:* Dept Chem Univ Ill 505 S Matthews Ave Urbana IL 61801

GUTSCHE, CARL DAVID, b Oak Park, Ill, Mar 21, 21; m 44; c 3. ORGANIC CHEMISTRY, BIO-ORGANIC CHEMISTRY. *Educ:* Oberlin Col, BA, 43; Univ Wis, PhD(org chem), 47. *Prof Exp:* Res assoc biochem, USDA Off Sci Res & Develop, Univ Wis, 43-44; from instr to assoc prof chem, Wash Univ, 47-59, chmn dept, 70-76, prof, 59-89, EMER PROF CHEM, WASH UNIV, 89-; ROBERT A WELCH PROF, TEX CHRISTIAN UNIV, 89- *Concurrent Pos:* Consult, Petrolite Corp, Mo, 51-89 & Monsanto Co, 59-79; mem adv bd, Petrol Res Fund, 72-74; mem med chem study sect, NIH, 77-81, chmn, 78-81; Guggenheim fel, 81. *Honors & Awards:* St Louis Award, Am Chem Soc, 72, Midwest Award, 88. *Mem:* Fel AAAS; Am Chem Soc; Royal Soc Chem. *Res:* Carbocyclic synthesis with emphasis on ring expansion processes; photochemistry of cyclic carbonyl compounds; chemistry of diazoalkanes and carbenes; polyfunctional catalysts and enzyme models; micellar catalysis; calixarenes. *Mailing Add:* Dept Chem Tex Christian Univ Ft Worth TX 76129

GUTSCHE, GRAHAM DENTON, b Oak Park, Ill, June 29, 25; m 48; c 3. PHYSICS. *Educ:* Univ Colo, BS, 50; Univ Minn, MS, 52; Catholic Univ, PhD, 60. *Prof Exp:* Instr, Northwestern Schs, Minn, 50-52; from asst prof to assoc prof, 52-63, PROF PHYSICS, US NAVAL ACAD, 63- *Concurrent Pos:* Asst, Univ Minn, 51-52; lectr, Univ Md, 59-; NSF fel, Cambridge Univ, 67-68. *Mem:* Am Asn Physics Teachers; Am Astron Soc; Am Sci Affil; Sigma Xi. *Res:* Nuclear reactions; Raman spectroscopy; stellar photometry; determination and analysis of the light curves of fast eclipsing binary star systems. *Mailing Add:* Sci Dept US Naval Acad Annapolis MD 21402

GUTSCHICK, RAYMOND CHARLES, b Chicago, Ill, Oct 3, 13; m 39; c 2. APPLIED & STRUCTURAL GEOLOGY. *Educ:* Univ Ill, BS, 38, MS, 39, PhD(geol), 42. *Prof Exp:* Geologist, Aluminum Ore Co, Ill, 42; instr geol, Univ Ill, 42-43; geologist, Magnolia Petrol Co, Okla, 43-46, Aluminum Ore Co, Ill, 46-47 & Gulf Oil Corp, 47; from asst prof to prof geol, 47-78, head dept, 56-70, lay fac award, 64, EMER PROF GEOL, UNIV NOTRE DAME, 78- *Concurrent Pos:* Vis prof geol, Ind Univ, 51-73; geologist, US Geol Surv, Denver, 75-; consult, 78- *Honors & Awards:* Niel A Miner Award, Nat Asn Geol Teachers, 77-78. *Mem:* Paleont Soc; fel Geol Soc Am; Nat Asn Geol Teachers; Am Asn Petrol Geologists. *Res:* Paleozoic stratigraphy, sedimentation and paleontology; Mississippian micropaleontology; foraminifera; geology of Great Basin and Northern Rockies; carbonate resources anomalous structure black shales basin analysis. *Mailing Add:* 2901 Leonard Medford OR 97504-5825

GUTSTADT, ALLAN MORTON, b Chicago, Ill, Jan 9, 26; m 56; c 1. GEOLOGY, STRATIGRAPHY. *Educ:* Univ Ill, BA, 49; Northwestern Univ, PhD(geol), 54. *Prof Exp:* Geologist stratig, Ind Geol Surv, 53-57 & Creole Petrol Corp, Standard Oil Co, NJ, 57-61; from asst prof to assoc prof, 62-71, prof, 71-88, EMER PROF GEOL, CALIF STATE UNIV, NORTHRIDGE, 88- *Mem:* Geol Soc Am. *Res:* Lower Paleozoic stratigraphy, central interior United States; Precambrian sedimentary rocks, southwestern United States; legal aspects of geology. *Mailing Add:* Dept Geol Sci Calif State Univ Northridge CA 91330

GUTSTEIN, WILLIAM H, b New York, NY, July 12, 22; m 45; c 2. PATHOLOGY, BIOPHYSICS. *Educ:* Univ Wis, AB, 43; NY Univ, MD, 46. *Prof Exp:* Instr path, State Univ NY Downstate Med Ctr, 52-55; pract pathologist, 55-57; asst prof path, NY Univ, 57-63; assoc prof, 63-76, PROF PATH, NY MED COL, 76- *Concurrent Pos:* Partic, Nat Conf Cardiovasc Dis, DC, 64. *Mem:* Am Heart Asn; Am Soc Exp Path. *Res:* Biophysical aspects and pathogenesis of atherosclerosis. *Mailing Add:* Dept Path Elmwood Hall Med Col NY Basic Sci Bldg Valhalla NY 10595

GUTTAY, ANDREW JOHN ROBERT, b Los Angeles, Calif, May 12, 24; m 48; c 4. SOIL ECOLOGY, MYCORRHIZAE. *Educ:* Mich State Univ, BS, 48, PhD(soil mgt), 59; Iowa State Univ, MS, 50. *Prof Exp:* Land use specialist, Mich Dept Conserv, 50-51; instr soil mgt, Mich State Univ, 51-58; dist rep soil fertil, Nat Plant Food Inst, 58-61; head dept plant sci, 61-74, PROF AGRON, UNIV CONN, 61- *Mem:* Fel AAAS; Am Soc Agron; Soil Sci Soc Am; Am Soc Hort Sci; Mycol Soc Am. *Res:* Soil environment factors and soil management practices upon the development and function of vesicular-arbuscular mycorrhizae. *Mailing Add:* Box 467 Dept Plant Sci Storrs CT 06268

GUTTENPLAN, JACK DAVID, b Baton Rouge, La, Oct 10, 25; m 49; c 3. CORROSION, ELECTROCHEMISTRY. *Educ:* Case Inst Technol, BS, 45, MS, 48. *Prof Exp:* Res asst gasoline additives, Case Inst Technol, 47-48; res assoc protective coatings, C F Prutton & Assocs, Ohio, 48-49; group leader electrochem, Chrysler Corp, Mich, 49-61; mgr develop biol electrochem ocean batteries, Magna Corp, Calif, 61-62; res specialist electrochem & corrosion, Autonetics Div, NAm Rockwell Corp, 62-74, MEM TECH STAFF, AUTONETICS DIV, ROCKWELL INT, 74-, M-X MISSILE GUID & CONTROL, CORROSION PREVENTION & CONTROL REP, 80- *Concurrent Pos:* Lectr, Chrysler Inst Eng, 52-59. *Mem:* Nat Asn Corrosion Eng; fel Am Inst Chem; Sigma Xi; fel Am Soc Metals. *Res:* Electroplating; thin film circuitry; physics of failure of microelectronic components; fuel cell development; bio-electrochemistry; surface potential difference. *Mailing Add:* 2210 W Avalon Santa Ana CA 92706

GUTTENPLAN, JOSEPH B, b Staten Island, NY; c 2. MUTAGENESIS, CARCINOGENESIS. *Educ:* Brooklyn Col, BS, 65; Brandeis Univ, MS, 70, PhD(chem), 70; Columbia Univ, MPH, 91. *Prof Exp:* Res asst prof biochem, Mt Sinai Med Ctr, 73-74; from asst prof to assoc prof, 74-86, PROF BIOCHEM, DENT CTR, NY UNIV, 86- *Concurrent Pos:* Res assoc prof environ med, Med Sch, NY Univ, 83- *Mem:* Am Asn Cancer Res; Am Soc Biol Chem & Molecular Biol; Environ Mutagen Soc. *Res:* Carcinogenesis; mutagenesis; DNA damage and repair; metabolism of carcinogens; mutational spectra; host-mediated mediated mutagenesis; site-specific mutagenesis by DNA adducts. *Mailing Add:* Dept Biochem Dent Sch NY Univ New York NY 10010

GUTTERMAN, JORDAN U, b Flandreau, SDak, Oct 15, 38; m. CLINICAL IMMUNOLOGY. *Educ:* Univ Va, BA, 60; Med Col Va, MD, 64; Am Bd Internal Med, dipl, 72, dipl hemat, 73. *Prof Exp:* Fac assoc, Dept Develop Therapeut, Univ Tex M D Anderson Cancer Ctr, 71-72, from asst prof to assoc prof, 72-81, prof med & dep head, Dept Clin Immunol & Biol Therap, 81-90, VIRGINIA H COCKRELL PROF IMMUNOL, UNIV TEX M D ANDERSON CANCER CTR, 87-, PROF & CHMN, 87- *Concurrent Pos:* NIH Health Res Ctr Develop Award, 74-79; NCI comt French-Am Agreement Therapeut Cancer Res, 77-80; co-chmn, Am Cancer Soc Interferon Comt, 78-84; mem, Lasker Award Jury, Lasker Found, 78-89, Combined Modality Comt, Div Cancer Treatment, Nat Cancer Inst, 78-81,

Citizens Comt, Rosalind Russell Med Res Ctr Arthritis, 80-83; Sci Adv Coun Cancer Res Inst, 79-; sr res investr, Clayton Found Res, 81- *Mem:* Am Asn Cancer Res; Am Chem Soc; Am Col Physicians; AAAS; Am Asn Immunologists; Am Fedn Clin Res; AMA; Am Soc Clin Oncol; Am Soc Hemat; Int Soc Exp Hemat; NY Acad Sci. *Res:* Clinical immunology. *Mailing Add:* Dept Clin Immunol & Biol Therap, Box 41 Univ Tex M D Anderson Cancer Ctr 1515 Holcombe Blvd Houston TX 77030

GUTTMAN, BURTON SAMUEL, b Minneapolis, Minn, Apr 12, 36; m 63; c 2. BIOLOGY. *Educ:* Univ Minn, BA, 58; Univ Ore, PhD(biol), 63. *Prof Exp:* NSF res fel biol, Calif Inst Technol, 63-65; asst prof cell biol, Med Ctr, Univ Ky, 65-72; MEM FAC, EVERGREEN STATE COL, 72- *Mem:* Am Soc Microbiol. *Res:* Regulation of metabolism; bacteriophage development; general biological organization; philosophy of biology; biology teaching strategies. *Mailing Add:* Evergreen State Col Olympia WA 98505

GUTTMAN, CHARLES M, b Cincinnati, Ohio, Apr 11, 39; m 60, 76; c 3. POLYMER SCIENCE. *Educ:* Earlham Col, BA, 61; Brandeis Univ, PhD(chem), 67. *Prof Exp:* Mem tech staff, Bell Tel Labs, 66-67; RES CHEMIST, NAT INST STANDARDS & TECHNOL, 67- *Concurrent Pos:* Adj prof, Univ Md & John Hopkins Univ. *Mem:* Am Phys Soc; Am Chem Soc; Am Soc Testing & Mat. *Res:* Statistical mechanics of polymeric systems and ionic solutions; properties of polymeric glasses; kinetics of polymer crystallization; polymers on surfaces; thermal analysis; diffusion of small molecules in polymers; properties of single polymer chains. *Mailing Add:* Polymer Div Nat Inst Standards & Technol Gaithersburg MD 20899

GUTTMAN, FRANK MYRON, b Montreal, Que, Feb 24, 31; m 53; c 2. SURGERY. *Educ:* McGill Univ, BSc, 52; Univ de Geneve, Switz, MD, 57; Royal Col Physicians & Surgeons, FRCS(C), 63; Am Bd Surg, dipl, 67. *Prof Exp:* PROF SURG, MCGILL UNIV, 81-; DIR PEDIAT SURG, MONTREAL CHILDREN'S HOSP, 81- *Concurrent Pos:* Mem staff, Div Surg Res, Jewish Gen Hosp, 76-81; children-consult surgeon, St Justine Hosp. *Mem:* Can Asn Pediat Surgeons; Am Pediat Surg Asn; Brit Asn Pediat Surg; Am Col Surgeons; Am Acad Pediat; Am Soc Tx Surg. *Res:* Preservation of tissues and organs; cryobiology. *Mailing Add:* 2300 Tupper St Montreal PQ H3H 1P3 Can

GUTTMAN, HELENE AUGUSTA NATHAN, b New York, NY, July 21, 30. MICROBIOLOGY, BIOCHEMISTRY. *Educ:* Brooklyn Col, BA, 51; Harvard Univ, AM, 56; Columbia Univ, MA, 58; Rutgers Univ, PhD(bact), 60. *Prof Exp:* Res technician immunol, Pub Health Res Inst New York, 51-52; control bacteriologist, Burroughs-Wellcome, Inc, 52-53; asst microbiologist, Haskins Labs, 52-56, res assoc, 56-59, staff mem, 59-64; res assoc, Goucher Col, 60-62; asst prof biochem & cell physiol, Univ Col & Grad Sch Arts & Sci, NY Univ, 62-65, assoc prof, 65-67; from assoc prof to prof biol sci, Univ Ill, Chicago Circle, 67-75, prof microbiol, Col Med, 69-75, fac assoc urban systems lab, Col Eng, 74-75, assoc dir res, 75; expert, Off Dir, Heart, Lung & Blood Inst, Bethesda, Md, 75-77, res resources coordr, Off Prog Planning & Eval, 77-79; dep dir, sci adv bd, US Environ Protection Agency, Washington, DC, 79-80; prog coordr, Sci Educ Coord Off, Sci & Educ Directorate, 80-83, assoc dir, Beltsville Human Nutrit Res Ctr, 83-89, ANIMAL CARE COORDR, NAT PROG STAFF, AGR RES SERV, USDA, 89- *Concurrent Pos:* Dazian Found fel, 56; Soc Am Bacteriologists' pres fel, 57; lectr, Queens Col, NY, 56-57; res collabr, Brookhaven Nat Labs, 58; res asst prof, Med Col Va, 60-62; res grants, NIH, 56-69, NSF, 59-60 & 61-63, Tidewater Fund, 61-62, AEC, 65, Hartford Found, 70-73, Fed Water Pollution Control Agency, 70-74, NASA, 72-74, NIH contract, 74-75 & Ill Dept Ment Health, 74-75, chairperson task force, 74-75; mem educ comt, Ill Comn Status Women, 74-75 & bd dirs, DuPage County Comprehensive Health Care Planning Agency, 74-75; consult sci adv bd, US Environ Protection Agency, 75-79, Int Joint Comn, US & Can, 87-; mem ed bd, J Am Med Women's Asn, 78-81, Nutrit Running & Fit News, 84-, J Protozool, 72-75; chmn, Educ Policy Comt, Am Soc Cell Biol, 66-69; chmn, Prof Opportunities Women Comn, Am Inst Chemists, 74-78; mem nomination comt, Tissue Culture Asn, 75-76, mem comt, 90-, educ comt, 90-; ed adv bd, Creative Woman, 77-; mem, Civil Serv Affairs Comn, Am Acad Microbiol, 79-85, mem, Status of Women Microbiol Comn, 80-85, mem, Publ Serv Adult Educ Comn, 81-90, prof, Affairs Comn, 87-, Comn on Use Microbiol in Res, 88-89; Tellers/Auditing Comn, Am Soc Clin Nutrit, 79-82; bd dirs, Montgomery Area Sci Fair Asn, 80-85; bd dirs, Am Running & Fitness Asn, 88-; pres, HNG Assocs, 82-; ed, Scientists Ctr Animal Welfare, 88-, rep to AAAS, 90- *Honors & Awards:* Thomas Jefferson Murray Award, Theobalt Smith Soc, 59. *Mem:* Fel Am Inst Chemists; Royal Soc Chem; Am Soc Neurochem; fel Am Acad Microbiol; fel NY Acad Sci; Am Soc Microbiol; Am Soc Biol Chem; Am Soc Cell Biol; fel AAAS; Am Soc Clin Nutrit. *Res:* Behavioral biochemistry; control of inducible syntheses, especially small peptides, enzymes and antibody; isolation and purification of bioactive natural products; nutritional biochemistry; drug mode of action at cellular level. *Mailing Add:* Nat Prog Staff Agr Res Serv USDA BARC-W Bldg 002 Room 105 Beltsville MD 20705-2350

GUTTMAN, LESTER, b Minneapolis, Minn, Apr 18, 19; m 55. SOLID STATE PHYSICS. *Educ:* Univ Minn, BChem, 40; Univ Calif, PhD(chem), 43. *Prof Exp:* Asst chem, Univ Calif, 40-42; assoc scientist, Manhattan Eng Dist, US War Dept, NMex, 43-46; res assoc, Inst Study Metals, Univ Chicago, 46-47, from instr to asst prof, 47-55; Guggenheim fel, UK Atomic Energy Authority, Eng, 55-56; phys chemist, res lab, Gen Elec Co, 56-60; sr chemist, Argonne Nat Lab, 60-86; GUEST SCIENTIST, 86- *Concurrent Pos:* Assoc ed, J Appl Physics & Appl Physics Lett, 65-73, ed, J Appl Physics, 74-89, consult ed, 90- *Mem:* Fel Am Phys Soc. *Res:* Statistical thermodynamics of alloys; x-ray diffraction from alloys; structure of covalent glasses. *Mailing Add:* Mat Sci Div Argonne Nat Lab 9700 S Cass Ave Argonne IL 60439

GUTTMAN, NEWMAN, b Minneapolis, Minn, Apr 4, 27; m 85. PSYCHOACOUSTICS. *Educ:* Univ Minn, BA, 49; Univ Ill, AM, 51. PhD(speech), 54. *Prof Exp:* Res assoc speech, Univ Ill, 54-55; asst speech, Mass Inst Technol, 55-56; mem tech staff, Bell Tel Labs, Naperville, 56-90; RETIRED. *Concurrent Pos:* Mem, sensory study sect, Social & Rehab Serv, Dept HEW, 65-69; mem subcomt sensory aids, Comt Prosthetics Res & Develop, Nat Res Coun, 69-77, adv, Comt on Hearing & Bioacoust, 70-90. *Mem:* Fel Acoust Soc Am; fel Am Speech & Hearing Asn. *Res:* Phonetics; speech and hearing pathology. *Mailing Add:* 2033 Sherman Ave No 501 Evanston IL 60201

GUTTMAN, SHELDON, b New York, NY, Mar 9, 41; m 63. ZOOLOGY. *Educ:* City Univ New York, BS, 62; Univ Tex, MA, 65, PhD(evolution), 67. *Prof Exp:* From prof to assoc prof, 67-76, PROF ZOOL, MIAMI UNIV, 76- *Mem:* Soc Study Evolution; Soc Study Amphibians & Reptiles. *Res:* Ecology and evolution of the amphibians and reptiles; biochemical genetics and biochemical taxonomy. *Mailing Add:* Dept Zool Miami Univ Oxford OH 45056

GUTTMANN, MARK, b Baltimore, Md, Aug 10, 24. NUCLEAR PHYSICS. *Educ:* Cath Univ, BS, 47, MS, 56; Univ Notre Dame, PhD(physics), 62. *Prof Exp:* Teacher high sch, Pa, 47-53; instr physics, La Salle Col, 53-55 & De La Salle Col, 55-56; asst prof, 62-70, ASSOC PROF PHYSICS, LA SALLE COL, 70- *Mem:* Am Phys Soc; Am Asn Physics Teachers; Sigma Xi. *Res:* Beta and gamma ray spectroscopy; optics. *Mailing Add:* Dept Physics La Salle Col Philadelphia PA 19141

GUTTMANN, RONALD DAVID, b Minneapolis, Minn, Aug 16, 36; m 64; c 3. IMMUNOGENETICS, MEDICAL DIAGNOSTICS. *Educ:* Univ Minn, BA, 58, BS & MD, 61. *Prof Exp:* Instr med, Harvard Med Sch, 69-70; dir transplantation serv, Royal Victoria Hosp, 70-75, dir histocompatibility & immunogenetics lab, 75-83, vpres res, 83-86; assoc prof, 70-74, assoc dir, Clin, 80-88, PROF MED, MCGILL UNIV, 75-, DIR UNIV CTR CLIN IMMUNOBIOL & TRANSPLANTATION, 88- *Concurrent Pos:* Consult, WHO-UN Develop Prog, 82 & Medicorp, Inc, 86-; Bethune exchange prof, China, 82; vchmn bd mgt, Royal Victoria Hosp Res Inst, 83-86. *Honors & Awards:* Med Award, Kidney Found Can, 87. *Mem:* Transplantation Soc (vpres, 81-83); Am Soc Transplant Physicians (pres, 81-83); Can Transplantation Soc (pres, 85-86); emer mem Am Soc Clin Invest; Asn Am Physicians; Am Asn Immunologists; Am Soc Pathologists; Soc Anal Cytol. *Res:* Immunogenetics and pathophysiology of transplantation and autoimmune diseases; use of monoclonal antibodies and genetic probes in disease models; expert systems and computer applications in medicine. *Mailing Add:* Royal Victoria Hosp 687 Pine Ave W Montreal PQ H3A 1A1 Can

GUTZKE, WILLIAM H N, b Richmond, Va, Feb 8, 50; m 79; c 4. ENVIRONMENTAL PHYSIOLOGY, PHYSIOLOGICAL ECOLOGY. *Educ:* Va Commonwealth Univ, BS, 75, MS, 77; Colo State Univ, PhD(zool), 84. *Prof Exp:* Instr biol, Colo State Univ, 83-84; LECTR BIOL, UNIV TEX, AUSTIN, 84- *Mem:* AAAS; Ecol Soc Am; Soc Study Evolution; Am Soc Ichthyologists & Herpetologists; Herpetologists League; Am Soc Zoologists. *Res:* Influence of the environment on embryos and hatchlings of reptiles; sex determining mechanisms. *Mailing Add:* Dept Biol Memphis State Univ Memphis TN 38152

GUTZMAN, PHILIP CHARLES, b Salmon, Idaho, June 23, 38; m 60; c 2. MAINTAINABILITY, LOGISTICS. *Educ:* Univ Ariz, BS(prod mgt) & BS(personnel mgt), 62; Univ Okla, MPA, 77. *Prof Exp:* Chief, advan develop logistics eng, Gen Dynamics Land Systs, 82-85; mgr, opers, Global Positioning Syst, 85-88, dir tech, Saudia Arabian Tank Conversion, 88-89, MGR, LOGISTICS ENG, GEN DYNAMICS SERV CO, 89- *Concurrent Pos:* Guest lectr, Polit Sci Hist & Anthrop Dept, Univ San Francisco, 68-69; mgr, Logistics Res, Gen Dynamics Land Systs Div, 84-85. *Mem:* Soc Logistics Engrs; Am Defense Preparedness Asn. *Res:* Effects of extreme climates, tropic, on complex mechanical and electronic systems; long term functional degredation based on environmental influences. *Mailing Add:* 47735 Jeffry Utica MI 48317

GUTZWILLER, MARTIN CHARLES, b Basel, Switz, Oct 12, 25; m 52; c 2. THEORETICAL PHYSICS. *Educ:* Swiss Fed Inst Technol, BS, 47, MS, 50; Univ Kans, PhD(physics), 53. *Prof Exp:* Physicist, Brown, Boveri & Co, Baden, Switz, 50-51, Explor & Prod Res Div, Shell Develop Co, Tex, 53-60 & Res Div, Int Bus Mach, Zurich, 60-63; Watson Lab, New York, dir, Gen Sci Dept, 63-70 & 74-77, PHYSICIST, T J WATSON RES CTR, IBM CORP, 70- *Concurrent Pos:* Adj prof, Columbia Univ. *Mem:* Am Phys Soc. *Res:* Propagation of waves, solid state physics; quantum and classical mechanics especially the chaotic phenomenum; ergodic theory; applied mathematics. *Mailing Add:* TJ Watson Res Ctr IBM Corp PO Box 218 Yorktown Heights NY 10598

GUVEN, NECIP, b Boyalik, Turkey, Apr 14, 36; US citizen; m 70; c 5. GEOLOGY, MINERALOGY. *Educ:* Univ Gottingen, Dr Sci, 62. *Prof Exp:* Fel mineral, Columbia Univ, 63-65; fel crystallog, Carnegie Inst Geophys Lab, 65-67; asst prof geol, Univ Ill, Urbana, 67-72; assoc prof, 72-76, PROF GEOL, TEX TECH UNIV, 76- *Concurrent Pos:* Assoc ed, Clay Minerals Soc, 75-87, Appl Clay Sci, 86- *Mem:* Mineral Soc Am; Am Crystallog Asn; Clay Minerals Soc (vpres, 87, pres, 88); Electron Micros Soc Am; Am Asn Petrol Geologists. *Res:* Mineralogy and colloid chemistry of clays; x-ray and electron diffraction of clays in reservoir rocks, shales, and in drilling fluids; rheology of clay dispersions. *Mailing Add:* Dept Geosci Tex Tech Univ Lubbock TX 79409

GUY, GEORGE ANDERSON, b Love, Miss, Nov 10, 14; m 39; c 1. METEOROLOGY. *Educ:* Memphis State Col, BS, 36; Univ Calif, Los Angeles, MA; George Wash Univ, MBA, 62. *Prof Exp:* Meteorologist, USAF, 43-49, chief equip develop, Hqs, Air Weather Serv, 49-52, chief meteorol equip develop, Hqs, Air Res & Develop Command, 52-57, chief, Weather Syst Proj Off, 57-62, chief tactical syst prog off, Syst Command, 63-67; sr engr, Hughes Aircraft Co, 67-81; CONSULT, 81- *Mem:* Am Meteorol Soc; Am Geophys Union; NY Acad Sci. *Res:* Meteorological instrumentation; electronic systems program management. *Mailing Add:* 707 Solana Circle E Solana Beach CA 92075

GUY, LEONA RUTH, b Kemp, Tex, Mar 17, 13. IMMUNOLOGY. *Educ:* Baylor Univ, AB, 34, MS, 49; Stanford Univ, PhD, 53. *Prof Exp:* Instr bact, Sch Dent, Baylor Univ, 48-49; clin asst prof microbiol, Univ Tex Health Sci Ctr, 53-62, from asst prof to prof med technol, Sch Allied Health Sci, 62-82, chmn dept, 62-78, from asst prof to prof path, 77-82, EMER PROF PATH, UNIV TEX HEALTH SCI CTR, 82- *Concurrent Pos:* Assoc dir blood bank, Parkland Mem Hosp, 53-78; consult, Vet Admin Hosps, Dallas, 62-82 & Temple Vet Admin Hosp, 64-82. *Honors & Awards:* John Elliott Award, Am Asn Blood Banks, 83; Distinguished Serv Award, Am Soc Clin Path, 89. *Mem:* Am Soc Microbiol; hon fel Am Soc Clin Path. *Res:* Immunohematology; bacteriology. *Mailing Add:* 5825 Birchbrook Dallas TX 75229

GUY, REED AUGUSTUS, b Andalusia, Ala, Sept 4, 44; m 66; c 2. NUCLEAR PHYSICS. *Educ:* Univ Ala, BS, 66; Univ Va, PhD(physics), 70. *Prof Exp:* Asst prof physics, Univ Wis-Oshkosh, 70-74; asst prof, 75-77, ASSOC PROF PHYSICS, SEATTLE UNIV, 77-, CHAIRPERSON DEPT, 78- *Mem:* Sigma Xi; Am Phys Soc; Am Asn Physics Teachers. *Res:* Nuclear structure; pion interactions in nuclei; history of physics. *Mailing Add:* Dept Physics Seattle Univ Seattle WA 98122

GUY, WILLIAM THOMAS, JR, b Abilene, Tex, Dec 11, 19; m 41; c 3. MATHEMATICS. *Educ:* Agr & Mech Col, Tex, BS, 40; Univ Tex, MA, 48; Calif Inst Technol, PhD(math), 51. *Prof Exp:* Design engr, Westinghouse Elec Corp, 40-42; asst, Calif Inst Technol, 48-51; from instr to asst prof appl math, 46-53, from asst prof to assoc prof math & astron, 53-59, actg chmn dept, Univ, 58-59, prof math & astron & chmn dept, 59-77, PROF MATH, UNIV TEX, AUSTIN, 77-, RES SCIENTIST, DEFENSE RES LAB, 53- *Mem:* AAAS; Am Math Soc; Math Asn Am. *Res:* Tensor, extensor and functional analysis; integral transforms; applied mathematics. *Mailing Add:* Dept Math RLM 8 100 Univ Tex Austin TX 78712

GUYDA, HARVEY JOHN, b Winnipeg, Man, July 5, 38; m 64; c 2. PEDIATRIC ENDOCRINOLOGY, CELL BIOLOGY. *Educ:* Univ Man, BSc & MD, 62. *Prof Exp:* Teaching fel pediat endocrinol, Johns Hopkins Univ, 66-69; teaching fel endocrinol, McGill Univ-Royal Victoria Hosp, 69-71, from asst prof to assoc prof pediat, 71-79, PROF PEDIAT & MED, MCGILL UNIV, 79-, CO-DIR, PROTEIN HORMONE LAB, MCGILL UNIV-ROYAL VICTORIA HOSP, 73-; DIR, ENDOCRINOL, MONTREAL CHILDREN'S HOSP, 80-, ENDOCRINOL & METAB, 86- *Concurrent Pos:* Asst physician pediat, Montreal Children's Hosp, 71-73, assoc physician, 73-; travel awards, Schering Ltd-Can Soc Clin Invest, Merck, Sharp & Dohme-Royal Col Physicians & Surgeons & France-Que Exchange Prog, 75; mem, panel C, Grants Rev Comt, Nat Cancer Inst Can, 80-84; vis prof, Nat Inst Endocrinol, Havana, Cuba, 83 & Hadassah, Hebrew Univ Jerusalem, 85; mem, Sci Rev Comt, JDFI, 86- *Mem:* Can Soc Endocrinol & Metab (pres, 82-84); Soc Pediat Res; Can Soc Clin Invest; Endocrine Soc; Am Pediat Soc; Am Soc Cell Biol. *Res:* Purification, characterization and biologic study of insulin-like growth factors; regulation of growth factor receptors, with correlations of morphology and receptor-mediated functions; characterization of sites of synthesis and regulation of production of insulin-like growth factor peptides; clinical studies of neuroendocrine regulation of growth hormone and prolactin secretion. *Mailing Add:* Dept Med & Pediat Montreal Children's Hosp McGill Univ 2300 Tupper St Rm E315 Montreal PQ H3H 1P3 Can

GUYER, CHERYL ANN, b Greensboro, NC, Nov 23, 57; m; c 1. BIOCHEMISTRY, MOLECULAR BIOLOGY. *Educ:* Univ NC Greensboro, BS, 79; Vanderbilt Univ, PhD(biochem), 85. *Prof Exp:* Res assoc, Dept Biochem, Sch Med, Vanderbilt Univ, 85-88, Am Heart Asn fel, Dept Pharmacol, 86-89, res instr, 89-91, RES ASST PROF, DEPT BIOCHEM, SCH MED, VANDERBILT UNIV, 91- *Mem:* Am Soc Biochem & Molecular Biol. *Res:* Epidermal growth factor receptor; structure and mechanism; biochemistry; numerous technical publications. *Mailing Add:* Dept Molecular Biol Vanderbilt Univ Sch Med Nashville TN 37232

GUYER, GORDON EARL, b Kalamazoo, Mich, May 30, 26; m 50; c 1. ENTOMOLOGY. *Educ:* Mich State Univ, BS. 50, MS, 52, PhD(entom), 54. *Prof Exp:* From asst prof to assoc prof, 54-63, chmn dept entom, 63-73, PROF ENTOM, ASST DEAN, COL AGR & NATURAL RESOURCES & DIR COOP EXTEN SERV, MICH STATE UNIV, 73- *Mem:* Am Entom Soc; Entom Soc Can; Sigma Xi. *Res:* Biology and taxonomy of aquatic insects; economic entomology; vegetable and forest insect research. *Mailing Add:* Coop Exten Serv Mich State Univ East Lansing MI 48824

GUYER, KENNETH EUGENE, JR, b Pampa, Tex, June 11, 34; m 57; c 2. BIOCHEMISTRY. *Educ:* Univ Tex, BA, 55; Ohio State Univ, MSc, 60, PhD(physiol chem), 62. *Prof Exp:* Chemist, Celanese Corp, 56-58; res assoc physiol chem, Ohio State Univ, 63; asst prof biochem, Med Col Va, 64-75; ASSOC PROF BIOCHEM, MARSHALL UNIV, 75- *Concurrent Pos:* Nat Heart Inst fel physiol chem, Sch Med, Johns Hopkins Univ, 63-64. *Mem:* Sigma Xi; Am Chem Soc; Am Oil Chemists' Soc; Soc Exp Biol & Med; AAAS. *Res:* Chemistry and metabolism of minor lipids and fat-soluble vitamins; hypocholesterolemic agents; biochemistry of autism. *Mailing Add:* Dept Biochem Marshall Univ Sch Med 1542 Spring Valley Dr Huntington WV 25704

GUYER, PAUL QUENTIN, b Linn Co, Mo, Mar 31, 23; m 48; c 3. ANIMAL SCIENCE AND NUTRITION. *Educ:* Univ Mo, BS, 48, MA, 49, PhD, 54. *Prof Exp:* Instr animal husb, Univ Mo, 48-54; from asst to assoc exten animal husbandman, 54-62, PROF ANIMAL SCI, UNIV NEBR, LINCOLN, 62- *Honors & Awards:* Exten Livestock Specialist Award, Am Soc Animal Sci, 73. *Mem:* Fel Am Soc Animal Sci; Sigma Xi. *Res:* Animal nutrition; beef, sheep and swine production. *Mailing Add:* Dept Animal Sci Univ Nebr Col Agr Lincoln NE 68583-0908

GUYKER, WILLIAM C, JR, b Donora, Pa, Aug 21, 33; US citizen; m 71; c 1. POWER & TRANSMISSION ENGINEERING, ENERGY RESEARCH. *Educ:* Mass Inst Technol, BS, 59. *Prof Exp:* VAR POSITIONS, ALLEGHENY POWER SERV CORP, 59-, PRIN ENGR, 83- *Concurrent Pos:* Consult, 68-; adj prof elec eng, Univ Pittsburgh & WVa Univ, 68- *Honors & Awards:* Centennial Medal, Inst Elec & Electronics Engrs, 84. *Mem:* Fel Inst Elec & Electronics Engrs; AAAS. *Res:* Research and development that supports power, transmission, distribution, and generation engineering; customer utilization service and power quality; nuclear engineering. *Mailing Add:* Allegheny Power Service Corp 800 Cabin Hill Dr Greensburg PA 15601

GUYNN, ROBERT WILLIAM, b Streator, Ill, Oct 27, 42. BIOCHEMISTRY. *Educ:* Mich State Univ, BA, 63; Johns Hopkins Univ, MD, 67. *Prof Exp:* Intern internal med, Univ Hosp Cleveland, Ohio, 67-68; resident psychiat, Johns Hopkins Hosp, Phipps Clin, 68-70; clin assoc, NIMH, Washington, DC, 70-73; from asst prof to assoc prof, 73-83, prof psychiat & vchmn dept, 83-87, PROF PSYCHIAT & INTERIM CHMN DEPT, MED SCH, UNIV TEX, HOUSTON, 87- *Concurrent Pos:* Surgeon, USPHS, 73; prin investr, var fed & private res grants. *Mem:* Am Soc Biol Chemists; Biochem Soc; Am Chem Soc; Res Soc Alcoholism; Am Psychiat Asn. *Res:* Basic biochemistry into the short-term regulation of metabolism with special emphasis on intermediary metabolism and the pathways of serine synthesis and degradation, using as models of focus: alcoholism. *Mailing Add:* PO Box 20708 Houston TX 77225-0708

GUYON, JOHN CARL, b Washington, Pa, Oct 16, 31; m 55; c 2. ANALYTICAL CHEMISTRY. *Educ:* Washington & Jefferson Col, BA, 55; Univ Toledo, MS, 58; Purdue Univ, PhD(anal chem), 61. *Prof Exp:* Res chemist, Thatcher Glass Mfg Co, NY, 57-58; from asst prof to assoc prof anal chem, Univ Mo, 61-71; prof & chmn dept, Memphis State Univ, 71-74; dean col sci, 74-76, assoc vpres res & dean grad sch, 76-80, VPRES ACAD AFFAIRS & RES, SOUTHERN ILL UNIV, CARBONDALE, 81- *Res:* Absorption spectroscopy; organic analytical reagents; fluorescence analysis; chromatography; heteropoly compounds. *Mailing Add:* Off of the Pres Southern Ill Univ Anthony Hall Carbondale IL 62901-4304

GUYSELMAN, JOHN BRUCE, b Albion, Mich, Mar 17, 24. ZOOLOGY. *Educ:* Albion Col, AB, 47; Northwestern Univ, PhD(zool), 52. *Prof Exp:* From instr to assoc prof zool, Carleton Col, 52-64; mem fac staff biol, 64-69, PROF BIOL, ALBION COL, 69- *Concurrent Pos:* Chief reader biol, Advan Placement Prog, Educ Testing Serv, 71-75, trustee NCent Col, Naperville, Ill, 77- *Mem:* Am Soc Zoologists; Am Soc Limnol & Oceanog; Am Physiol Soc; Soc Gen Physiol; Sigma Xi. *Res:* Comparative animal physiology; environmental physiology; biological rhythms. *Mailing Add:* Box 786-CCNC Pinehurst NC 28374

GUYTON, ARTHUR CLIFTON, b Oxford, Miss, Sept 8, 19; m 43; c 10. PHYSIOLOGY. *Educ:* Univ Miss, BA, 39; Harvard Univ, MD, 43. *Prof Exp:* Intern & resident surg, Mass Gen Hosp, 43-44 & 46; actg assoc prof physiol, Univ Tenn, 47; assoc prof pharmacol, 47-48, PROF PHYSIOL & CHMN DEPT PHYSIOL & BIOPHYS, MED CTR, UNIV MISS, 48- *Concurrent Pos:* Mem coun, Nat Heart & Lung Inst, 71-75. *Honors & Awards:* President's Citation, 56; Gould Award, AAAS, 60; ALZA Award, Biomed Eng Soc, 71; Ross McIntyre Award, Univ Nebr, 72; Circulation Group Wiggers' Award, Am Physiol Soc, 72; Annual Distinguished Res Achievement Award, Am Heart Asn, 75; Ciba Award, 80; Merck Int Award, 84. *Mem:* Am Physiol Soc (pres, 74-75); Fedn Am Soc Exp Biol (pres, 75-76); Am Heart Asn; Biomed Eng Soc; hon fel Am Col Cardiol. *Res:* Medical electronic development; circulatory physiology. *Mailing Add:* Dept Physiol & Biophys Univ Miss Med Ctr Jackson MS 39216

GUYTON, JAMES W, b New Albany, Miss, Nov 3, 32; m 54. GEOLOGY. *Educ:* Univ Calif, Berkeley, AB, 57; Univ Wyo, MA, 60, PhD(geol), 65. *Prof Exp:* Jr geologist, Pan Am Petrol Corp, 57-58; seismologist, Geotech Corp, 61-64; asst prof geol, 64-69, assoc prof geol & phys sci, 69-77, PROF GEOL, CALIF STATE UNIV, 77- *Mem:* Geol Soc Am; Seismol Soc Am. *Res:* Areal geology; microseisms of short period; recent and ancient landslides. *Mailing Add:* Dept Phys Sci Calif State Univ Chico CA 95929

GUYTON, WILLIAM F, b Oxford, Miss, Oct 15, 17; m 35, 87; c 5. GROUND WATER HYDROLOGY. *Educ:* Univ Miss, BA & BS, 38. *Prof Exp:* Hydraul engr, Ground Water Br, US Geol Surv, 39-45; sci consult ground water, US Army, 45; hydraul engr, US Geol Surv, 46-50; CONSULT GROUND WATER HYDROLOGIST, 50- *Mem:* Am Geophys Union; Am Soc Civil Engrs; Am Water Works Asn; Nat Soc Prof Engrs; Nat Water Well Asn; Soc Petrol Engrs; Geol Soc Am. *Res:* Ground water hydrology. *Mailing Add:* William F Guton Assoc Inc 3355 Bee Cave Rd Suite 401 Austin TX 78746

GUZE, CAROL (KONRAD LYDON), b St Louis, Mo, Nov 12, 35; m 59, 69, 83; c 3. CELL BIOLOGY, HUMAN GENETICS. *Educ:* Wash Univ, AB, 57; Univ Calif, Berkeley, PhD(zool), 63. *Prof Exp:* Res biologist, Naval Biol Labs, Univ Calif, Berkeley, 63-64; instr biol, Brandeis Univ, 64-65 & Univ Mass, Boston, 65-66; fel bact, Univ Calif, Los Angeles, 66-67; from asst prof to assoc prof biol, 67-78, coordr, Biol Grad Prog, 76-78, prof & chmn dept, 78-81, univ grad adv, 81-83, COODR, HUMAN CYTOGENETIC TECHNOL PROG, CALIF STATE UNIV, DOMINGUEZ HILLS, 81-, ASSOC VPRES ACAD AFFAIRS, 83- *Concurrent Pos:* Res assoc med genetics, Harbor Gen Hosp, Med Sch Campus, Univ Calif, Los Angeles, 74-76; extramural assoc, NIH, 82 & Inst Educ Mgt, Harvard Univ, 85. *Mem:* AAAS; Sigma Xi; Am Soc Human Genetics; Am Soc Cell Biol. *Res:* Physiology of mitotic cells; bacteriophage genetics; blue-green algae mutant production; human genetics. *Mailing Add:* 31 Ave 28 Venice CA 90291

GUZE, SAMUEL BARRY, b New York, NY, Oct 18, 23; m 46; c 2. INTERNAL MEDICINE, PSYCHIATRY. *Educ:* Wash Univ, MD, 45; Am Bd Internal Med, dipl, 54; Am Bd Psychiat & Neurol, dipl, 57. *Prof Exp:* Intern med, Barnes Hosp, 45-46, fel med, 46 & 48-49; resident, Vet Admin Hosp, Conn, 49-50; from instr to asst prof med, 51-64, from asst prof to prof psychiat, 55-75, asst to dean, Sch Med, 65-71, dir psychiat clin, 55-75, vchancellor med affairs & pres, Med Ctr, 71-89, Spencer T Olin prof psychiat & head dept, 75-89, ASSOC PROF MED, SCH MED, WASHINGTON UNIV, 64- *Concurrent Pos:* Fel psychiat & resident, Barnes Hosp & Wash Univ Sch Med, 50-53, asst physician, 51-, from asst psychiatrist to assoc psychiatrist, 55-75, psychiatrist-in-chief, 75-89, psychiatrist, 89-; from asst psychiatrist to assoc psychiatrist, Renard Hosp, 55-75, psychiatrist-in-chief, 75-89, psychiatrist 89-; vis physician & consult, St Louis City Hosp, 51-85; consult, H G Phillips Hosp, 54-85. *Mem:* Inst Med-Nat Acad Sci; fel Am Psychiat Asn; fel Am Col Physicians; Am Psychopath Asn; Asn Res Nerv & Ment Dis; fel Royal Col Psychiatrists; Psychiat Res Soc. *Res:* Natural history psychiatric disease; effects of drugs on behavior. *Mailing Add:* Dept Psychiat Wash Univ Sch Med St Louis MO 63110

GUZELIAN, PHILIP SAMUEL, b Milwaukee, Wis, May 21, 41; m 75; c 3. INTERNAL MEDICINE, DRUG METABOLISM. *Educ:* Univ Wis-Madison, BA, 63, MD, 67; Am Bd Internal Med, dipl, 72. *Prof Exp:* Resident internal med, Univ Wis, 70-71; USPHS clin fel liver dis, Yale Univ, 71-72; USPHS res fel gastrointestinal, Univ Calif, San Francisco, 72-74; asst prof, 74-78, ASSOC PROF MED, MED COL VA, 78- *Concurrent Pos:* Investr, NIH Liver Metab Prog Proj, 75-; prin investr, Nat Inst Environ Health Sci, 75-, Allied Chem Corp, 76- & Va Environ Endowment, 78- *Honors & Awards:* Clin Investr Award, NIH, 75. *Mem:* Am Fedn Clin Res; Am Asn Study Liver Dis; Am Asn Biol Chemists. *Res:* Hepatic drug metabolism and toxicity; collagen metabolism in the liver. *Mailing Add:* Dept Med Med Col Va Box 267 MCV Sta Richmond VA 23298-0267

GUZIEC, FRANK STANLEY, JR, b Chicago, Ill, Sept 16, 46; div; c 1. ORGANIC & MEDICINAL CHEMISTRY, BIO ORGANIC CHEMISTRY. *Educ:* Loyola Univ, BSc, 68; Mass Inst Technol, PhD(org chem), 72. *Prof Exp:* Res fel chem, Imp Col, London, 72-74; res assoc, Mass Inst Technol, 75-76 & Wesleyan Univ, 76-77; asst prof chem, Tufts Univ, 77-81; ASSOC PROF CHEM, 81-90, PROF CHEM, NMEX STATE UNIV, 90- *Mem:* Am Chem Soc; Chem UK; AAAS; Sigma Xi. *Res:* Organic synthesis; peptides, lipids, natural products; chemistry of sterically crowded molecules; synthetic methods; organo selenium chemistry; organo tellurium chemistry; conducting organic materials; medicinal chemistry. *Mailing Add:* Dept Chem NMex State Univ Las Cruces NM 88003

GUZIEC, LYNN ERIN, b Long Beach, Calif, Aug 18, 51; m 91. SYNTHESIS OF HIGHLY HINDERED COMPOUNDS, MUSTARDS & QUINONES. *Educ:* Russell Sage Col, BA, 79; NMex State Univ, PhD(org chem), 88. *Prof Exp:* ASST PROF ORG CHEM, NMEX STATE UNIV, LAS CRUCES, 88- *Concurrent Pos:* Proj leader, Nat Cancer Inst, 87-88, prin asst, 88-91. *Mem:* Am Chem Soc. *Res:* Synthesis and reactions of organo-sulfur and organo-selenium compounds, semi-conductors, hindered halogen compounds, strained organic molecules, heterocycles, mustards & quinones. *Mailing Add:* Chem Dept Box 3C NMex State Univ Las Cruces NM 88003

GUZMAN, RUBEN JOSEPH, toxicology, pharmacology, for more information see previous edition

GUZMAN, VICTOR LIONEL, b Huaraz, Peru, May 14, 14; nat US; m 46; c 5. HORTICULTURE, VEGETABLE CROPS. *Educ:* Nat Col Agr, Peru, BS, 40; Univ Fla, MS, 42; Cornell Univ, PhD, 45. *Prof Exp:* Asst prof hort, La State Univ, 45-46; head dept, La Molina Exp Sta & Nat Col Agr, Peru, 46-51; asst horticulturist, 52-56, mgr, Sullivans Farm, 56-58, assoc horticulturist, 58-66, horticulturist, 66-80, PROF VEG CROPS, UNIV FLA, 80- *Concurrent Pos:* Mem, Plant Growth Regulator Working Group & Integrated Crop Mgt Systs. *Honors & Awards:* Agro-Indust Award, Belle Glade Chamber Of Com, 83. *Mem:* Am Soc Hort Sci; Am Soc Plant Physiol; Am Inst Biol Sci; Am Soc Hort Sci Trop Region. *Res:* Nutrition of vegetable crop plants and cultural practices; lettuce breeding; growth regulators. *Mailing Add:* 1325 NW Ave L Belle Glade FL 33430

GUZMAN FORESTI, MIGUEL ANGEL, b San Salvador, El Salvador, Dec 20, 25; m 51; c 4. EXPERIMENTAL STATISTICS, PUBLIC HEALTH NUTRITION. *Educ:* Univ Tenn, BA, 49; NC State Col, MS, 56, PhD(exp statist), 61. *Prof Exp:* Chief labs Nutrit Cent Am & Panama, 53-54, chief div statist, 57-82, assoc dir, 75-82; PROF PATH & BIOMET, MED CTR, LA STATE UNIV, 82- *Concurrent Pos:* Consult, WHO, 59, 66 & 68, E I du Pont de Nemours & Co, Inc, 60 & USDA, 61; vis assoc prof, Dept Nutrit & Food Sci, Mass Inst Technol, 64-65, vis lectr, 66-88; vis res prof, Dept Path, Med Ctr, La State Univ, 71; vis prof, Int Pub Health, Univ Ala, Birmingham, 84-89; sr lectr, Int Pub Health, J J Sporkman Ctr Int Pub Health Educ, Sch Pub Health, Univ Ala, Birmingham. *Honors & Awards:* Inst Nutrit Cent Am & Panama Award, 89. *Mem:* Am Statist Asn; Biomet Soc; Am Inst Nutrit; Latin Am Nutrit Asn; Am Pub Health Asn. *Res:* Atherosclerosis--etiology and epidemiology; human growth and development; mathematical methods in biology--applications to estimation procedures; experimental design--variance components and their application. *Mailing Add:* Dept Path Med Ctr La State Univ 1901 Perdido St New Orleans LA 70112-1393

GWADZ, ROBERT WALTER, b Chicago, Ill, Nov 24, 40; m 63, 79; c 3. MEDICAL ENTOMOLOGY. *Educ:* Univ Notre Dame, BS, 62, PhD(biol), 70. *Prof Exp:* NIH fel, Sch Pub Health, Harvard Univ, 70-72; staff fel, Primate Malaria Unit, NIH, Ga, 72-74, res biologist, 74-80, sr scientist, 80-85, CAPT, USPHS & HEAD MED ENTOM UNIT, LAB PARASITIC DIS, NAT INST ALLERGY & INFECTIOUS DIS, 85- *Concurrent Pos:* Proj officer, Epidemiol & Control Arthropod-Borne dis, Egypt, Malaria Res & Training Ctr, Bamako, Mali, 90- *Mem:* Am Entom Soc; Am Mosquito Control Asn; Am Soc Trop Med & Hyg; Royal Soc Trop Med Hyg; Am Soc Parasitol. *Res:* Genetics of vector capacity; malaria; malaria vaccine development; reproductive physiology; biological control; host-parasite relationship. *Mailing Add:* Lab Parasitic Dis Bldg 5 Nat Inst Allergy & Infect Dis Bethesda MD 20892

GWALTNEY, JACK MERRIT, JR, b Norfolk, Va, Dec 24, 30; m 54; c 2. MEDICINE. *Educ:* Univ Va, BA, 52, MD, 56. *Prof Exp:* From intern to resident internal med, Univ Hosps, Cleveland, Ohio, 56-59; chief resident, Univ Va Hosp, 59-60, prof asst respiratory virus res, 62-63, from instr to asst prof, 64-70, assoc prof internal med & head epidemiol & virol, 70-75, PROF MED & WADE HAMPTON FROST PROF EPIDEMIOL, SCH MED, UNIV VA, 75-, DIR CTR PREV DIS & INJURY, 84- *Concurrent Pos:* Res fel med & prev med, Univ Va, 63-64; Trudeau fel, Am Thoracic Soc, 64-67; assoc mem comn acute respiratory dis, Armed Forces Epidemiol Bd, 68; mem adv panel infectious dis ther, US Pharmacopeia, 70 & 75-80. *Honors & Awards:* Joseph E Smadel Award, Infectious Dis Soc Am, 87. *Mem:* Am Fedn Clin Res; Am Pub Health Asn; Am Soc Microbiol; Am Thoracic Soc; fel Am Col Physicians; Sigma Xi. *Res:* Etiology, epidemiology and pathogenesis of the acute respiratory diseases; virology; rhinoviruses; antiviral compounds. *Mailing Add:* Dept Internal Med Univ Va Health Sci Ctr Charlottesville VA 22908

GWATKIN, RALPH BUCHANAN LLOYD, b Newport, UK, May 23, 29; nat US; m 54; c 3. REPRODUCTIVE DEVELOPMENTAL & CELL BIOLOGY. *Educ:* Univ Toronto, BA, 50; MA, 51; Rutgers Univ, PhD(microbiol), 54. *Prof Exp:* Fel plant path, Univ Ill, 54-55; res assoc, Connaught Med Res Labs, Univ Toronto, 56-58; assoc, Wistar Inst, Univ Pa, 59-62, asst prof reprod physiol, Sch Vet Med, 63-66; dir, Tissue Cult Lab & asst dir microbiol, Merck Sharp & Dohme Res Lab, Rahway, NJ, 67-70, dir physiol, 71-75, sr res fel & dir, Reproductive & Cell Biol, Merck Inst Therapeut Res, 76-81; prof, dept obstet & gyn, Med Fac, McMaster Univ, 82-83; DIR REPRODUCTIVE & DEVELOP & DIR CLIN IN VITRO FERTIL LAB CLEVELAND CLIN FOUND, OHIO, 83- *Concurrent Pos:* NIH career develop award, 64; vis prof biol, dept biol sci, Dartmouth Col, Hanover, NH, 81-82; ed-in-chief, Gamete Res, 78- *Honors & Awards:* Rubin Award, Am Fertil Soc; Pac Coast Wyeth Award, Soc Study Reprod. *Mem:* Soc Study Reprod; Am Soc Cell Biol; Am Fertil Soc; NY Acad Sci; Soc Develop Biol; Molecular Reproduction Develop. *Res:* Sperm-egg interaction and embryonic developments in mammals; tissue and organ culture, both basic and applied. *Mailing Add:* Dept Reproductive Biol Case Western Reserve Univ Sch Med Cleveland OH 44106

GWAZDAUSKAS, FRANCIS CHARLES, b Waterbury, Conn, July 25, 43; m 71; c 4. DAIRY SCIENCE, REPRODUCTIVE ENDOCRINOLOGY. *Educ:* Univ Conn, BS, 66; Univ Fla, MS, 72, PhD(animal sci), 74. *Prof Exp:* From asst prof to assoc prof, 74-86, PROF REPRODUCTIVE PHYSIOL, VA POLYTECH INST & STATE UNIV, 86- *Mem:* Am Dairy Sci Asn; Am Soc Animal Sci; Soc Study Reproduction; Sigma Xi; Soc Exp Biol Med. *Res:* Factors which determine the optimal time for insemination; proteins in normal and abnormal uterine secretions; adrenal responsiveness to ACTH; early bovine embryo development; embryo manipulation and gene insertion; transgenics. *Mailing Add:* Dept Dairy Sci Va Polytech Inst & State Univ Blacksburg VA 24061

GWIAZDA, S(TANLEY) J(OHN), b Philadelphia, Pa, Feb 14, 22; m 44; c 1. MECHANICAL ENGINEERING, SCIENCE EDUCATION. *Educ:* Drexel Inst, BS, 44, MS, 52. *Prof Exp:* Head counr, 51-61, ASSOC PROF MECH ENG, DREXEL UNIV, 46-, DEAN EVE COL, 63- *Mem:* Am Soc Eng Educ; Asn Cont Higher Educ (pres, 85-86). *Res:* Ordnance mechanisms; stress analysis and machine design problems. *Mailing Add:* 3360 Salmon St Philadelphia PA 19134

GWILLIAM, GILBERT FRANKLIN, b Park City, Utah, Aug 28, 25; m 50; c 2. NEUROBIOLOGY, INVERTEBRATE ZOOLOGY. *Educ:* Univ Calif, AB, 50, PhD(zool), 56. *Prof Exp:* Fel marine biol, Scripps Inst Oceanog, Univ Calif, San Diego, 56-57; from instr to assoc prof, 57-68, PROF BIOL, REED COL, 68- *Concurrent Pos:* NSF sr fel, 64-65; spec postdoctoral fel, Nat Inst Neurol Dis & Stroke, NIH, 70-71; vpres & provost, Comt Dis & Stroke, Nat Inst Neurol, 78-79 & 84-85. *Mem:* AAAS; Am Soc Zoologists; Soc Gen Physiol; Soc Neurosci. *Res:* Functional morphology of invertebrates; neuromuscular mechanisms in invertebrates; invertebrate sensory physiology; general invertebrate zoology. *Mailing Add:* Dept Biol Reed Col Portland OR 97202

GWILT, JOHN RUFF, science administration, applied chemistry, for more information see previous edition

GWIN, REGINALD, b Chattanooga, Tenn, Jan 14, 24; m 48; c 5. PHYSICS. *Educ:* Univ Chattanooga, BS, 48; Emory Univ, MS, 49; Univ Tenn, PhD(physics), 62. *Prof Exp:* Assoc prof physics, Mercer Univ, 49-50; physicist, Gaseous Diffusion Plant, Union Carbide Corp, 50-52 & Y-12 Plant, 52-55; PHYSICIST, OAK RIDGE NAT LAB, 55- *Mem:* Am Phys Soc; Am Nuclear Soc; Sigma Xi. *Res:* Measurement of capture and fission cross sections of fissile isotopes as a function of energy. *Mailing Add:* 218 Wakefield Rd Knoxville TN 37922

GWINN, JAMES E, b Harvey, WVa, Mar 30, 23; m 43; c 2. GEOLOGY. *Educ:* WVa Univ, BS, 48, MS, 50. *Prof Exp:* PROD MGR, CNG TRANSMISSION CORP, 63- *Mem:* Soc Petrol Engrs; Geol Soc Am. *Mailing Add:* 445 W Main St Clarksburg WV 26302-2450

GWINN, JOEL ALDERSON, b Quinnimont, WVa, Feb 24, 29; m 54; c 4. PHYSICS. *Educ:* WVa Univ, BSEd, 51, MA, 56, MS, 57, PhD(physics), 62. *Prof Exp:* From instr to assoc prof, 61-70, PROF PHYSICS, UNIV LOUISVILLE, 70- *Mem:* Am Asn Physics Teachers. *Mailing Add:* Dept Physics Univ Louisville Louisville KY 40292

GWINN, JOHN FREDERICK, b Dayton, Ohio, June 18, 42; wid. COMPUTER BASED TESTING. *Educ:* Manchester Col, BA, 64; Purdue Univ, MS, 67; Kent State Univ, PhD(physiol), 72. *Prof Exp:* From instr to asst prof biol, 72-81, ASSOC PROF BIOL, UNIV AKRON, 81- *Mem:* Sigma Xi. *Res:* Computer based testing; development of individualized instruction; science education. *Mailing Add:* Dept Biol Univ Akron Akron OH 44325-3908

GWINN, WILLIAM DULANEY, b Bloomington, Ill, Sept 28, 16; m 42, 53; c 3. PHYSICAL CHEMISTRY. *Educ:* Univ Mo, AB, 37, MA, 39; Univ Calif, PhD(chem), 42. *Prof Exp:* From instr to assoc prof, 42-55, res prof, Miller Res Inst, 61-62, prof 55-79, EMER PROF CHEM, UNIV CALIF, BERKELEY, 79- *Concurrent Pos:* Res assoc, div 10, Nat Defense Res Comt, 42-44 & Manhattan Dist Proj, 44-45; Guggenheim fel, 54; Sloan res fel, 55-59; vis prof, Univ Minn, Minneapolis, 69-70. *Mem:* Fel Am Phys Soc; Am Chem Soc. *Res:* Microwave spectroscopy and molecular structure; quantum mechanics; astrophysics. *Mailing Add:* 8506 Terrace Dr El Cerrito CA 94530-2721

GWINUP, GRANT, b Denver, Colo, June 28, 29; m 55; c 2. INTERNAL MEDICINE. *Educ:* Univ Colo, BA, 53, MD, 56; Am Bd Internal Med, dipl. *Prof Exp:* Instr internal med, Sch Med, Univ Tex, 60-61; asst prof internal med & endocrinol, Health Ctr, Ohio State Univ, 62-65; ASSOC PROF ENDOCRINOL & CHMN DIV METAB, UNIV CALIF, IRVINE-CALIF COL MED, 65- *Concurrent Pos:* USPHS fel, 59-60; ed, Metabolism, 66- *Mem:* Am Fedn Clin Res; Am Diabetes Asn; Endocrine Soc. *Res:* Endocrinology and metabolism. *Mailing Add:* Dept Med Univ Calif Irvine Med Orange CA 92668

GWINUP, PAUL D, b Shawnee, Okla, Oct 13, 31; m 57; c 3. PHYSICAL CHEMISTRY, INORGANIC CHEMISTRY. *Educ:* Okla State Univ, BS, 57, PhD(chem), 67. *Prof Exp:* Asst prof, 65-80, PROF CHEM, ARK STATE UNIV, 80- *Mem:* Am Chem Soc; Sigma Xi. *Res:* Thermoanalytical chemistry; single-crystal x-ray diffraction. *Mailing Add:* Box 908 State University AR 72467

GWIRTZ, PATRICIA ANN, CORONARY CIRCULATION, CARDIAC DYNAMICS. *Educ:* Thomas Jefferson Univ, PhD(physiol), 78. *Prof Exp:* ASSOC PROF CORONARY CIRCULATION & GASTROINTESTINAL PHYSIOL, TEX COL OSTEOP MED, 82- *Mem:* Am Physiol Soc; Am Heart Soc; Sigma Xi. *Res:* Neuro-regulation of cardiovascular system. *Mailing Add:* Dept Physiol Tex Col Osteop Med Camp Bowie Montgomery Ft Worth TX 76107

GWYN, CHARLES WILLIAM, b Sterling, Colo, Sept 3, 36; m 57; c 3. ELECTRICAL ENGINEERING, COMPUTER SCIENCE. *Educ:* Univ Kans, BS, 61; Univ NMex, MS, 63, PhD(elec eng), 68. *Prof Exp:* Staff mem nuclear radiation effects, Sandia Labs, 61-73, supvr comput aided design, 73-79, mgr, Integrated circuit design dept, 79-; asst gen mgr, Microelectronics Ctr, United Technol, Colo, 83, vpres eng, 85-89; MGR, SEMICONDUCTOR PROG, SANDIA LABS, 89- *Concurrent Pos:* Res assoc & instr elec eng, Univ NMex, 67-68. *Mem:* Fel Inst Elec & Electronics Engrs. *Res:* Nuclear radiation effects in semiconductor devices and electronic circuits; computer aided design of integrated circuits; computer controlled data acquisition systems; circuit and device modeling; radiation effects in solid state devices and electronic circuits; integrated circuit design and testing; modeling and design semiconductor manufacturing equipment. *Mailing Add:* Dept 2160 Sandia Nat Labs PO Box 5800 Albuquerque NM 87185

GWYN, J(OHN) E(DWARD), b La Junta, Colo, Dec 5, 27; m 47; c 2. CHEMICAL ENGINEERING. *Educ:* Univ Colo, BS, 50; Univ Wis, MS, 52, PhD, 55. *Prof Exp:* Asst chem eng, Univ Wis, 50-52, res assoc eng exp sta, 52-55; res engr, Shell Oil Co, 52-62, sr res engr, 62-67, staff res engr, 67-70, sr staff res engr, 70-77, res consult, 77-90; RETIRED. *Mem:* Am Chem Soc; Nat Soc Prof Engrs; Am Inst Chem Engrs. *Res:* Chemical and petroleum processes. *Mailing Add:* 26602 Willow Lane Katy TX 77494-5418

GWYNN, BERNARD HENRY, organic chemistry, for more information see previous edition

GWYNN, DONALD EUGENE, b Columbus, Ohio, Aug 28, 35; m 58; c 2. SYNTHETIC ORGANIC CHEMISTRY. *Educ:* Ohio State Univ, BSc, 57; Univ Ill, PhD(org chem), 62. *Prof Exp:* Res fel, Calif Inst Technol, 62-63; asst prof org chem, Univ Ark, 63-69; CHEMIST, TEX EASTMAN CO, 69- *Mem:* Am Chem Soc. *Res:* Organic synthesis. *Mailing Add:* 205 N Pope Dr Overton TX 75684-1529

GWYNN, EDGAR PERCIVAL, b Baltimore, Md, Jan 23, 23; m 50; c 1. CYTOGENETICS. *Educ:* Univ Md, BS, 50; Univ Ky, MS, 51; Johns Hopkins Univ, PhD(biol), 58. *Prof Exp:* Jr instr biol, Johns Hopkins Univ, 51-53; from lectr to assoc prof, 53-64, PROF BIOL, WASHINGTON COL, 64-, CHMN DEPT, 69- *Mem:* AAAS; Bot Soc Am; Soc Study Evolution; Am Genetic Asn; Sigma Xi. *Res:* Cytology; evolution. *Mailing Add:* Rte 3 Chestertown MD 21620

GWYNN, ROBERT H, b Dublin, Eire, July 11, 22; m 58; c 2. PHYSIOLOGY. *Educ:* Univ Dublin, BA, 46; Univ London, PhD(exp path), 57. *Prof Exp:* Res scientist cancer, London Hosp Med Col, 47-57; res fel, Roscoe B Jackson Mem Lab, Maine, 57-60; head dept chem, Benedict Col, 60-63; asst prof, 63-65, ASSOC PROF PHYSIOL & MICROBIOL, UNIV HARTFORD, 65- *Mem:* AAAS. *Res:* Co-carcinogenesis; carcinogenic activity of tobacco products; memory in rats. *Mailing Add:* 45 Lawton Rd Manchester CT 06040

GYAN, NANIK D, Can citizen. DRUGS & MEDICAL DEVICES, BIOPHARMACEUTICAL STUDIES. *Educ:* Univ Gujarat, India, BS, 57; Univ Freiburg, WGer, PhD(pharmaceut chem), 64. *Prof Exp:* Tech dir, res & develop, Beecham Labs, 71-78; tech dir, res & develop, Cooper Labs, 78-79; dir, new prod develop, Kalipharma Inc, 79-82; sr vpres, Res & Develop, Thermascam Inc, 83-89; PRES, EURO-AM PHARMA INC, 89- *Concurrent Pos:* Fel, health, protection res, Drug Res Labs, Ottawa, 67-69; sr vpres, Baors-Krey Assocs Inc, 83- *Res:* Dosage form design and development; analytical development and clinical investigation. *Mailing Add:* 64 Montrose Ave Fanwood NJ 07023

GYBERG, ARLIN ENOCH, b Luverne, Minn, Dec 9, 38; m 62; c 3. PHYSICAL CHEMISTRY. *Educ:* Mankato State Col, BS, 61; Univ Minn, PhD(anal chem), 69. *Prof Exp:* Teacher high sch, Minn, 61-63; from asst prof to assoc prof, 67-85, PROF CHEM, AUGSBURG COL, 85- *Concurrent Pos:* Consult, chem. *Honors & Awards:* Kohltoff Award. *Mem:* AAAS; Sigma Xi; Am Chem Soc. *Res:* Light scattering; optical methods; polymer solutions. *Mailing Add:* Dept Chem Augsburg Col 731 21st Ave S Minneapolis MN 55454

GYFTOPOULOS, ELIAS P(ANAYIOTIS), b Athens, Greece, July 4, 27; US citizen; m 62; c 3. MECHANICAL & NUCLEAR ENGINEERING. *Educ:* Univ Athens, dipl, 53; Mass Inst Technol, ScD(elec eng), 58. *Prof Exp:* Instr elec eng, 54-58, from asst prof to prof elec & nuclear eng, 58-70, FORD PROF ENG, MASS INST TECHNOL, 70- *Concurrent Pos:* Dir, Thermo Electron Corp, Thermo Instrument Systs Corp, Mercantile Bank; chmn, Nat Energy Coun Greece, 75-78; vchmn Trustees Anatolia Col. *Mem:* Nat Acad Eng; Fel AAAS; fel Am Nuclear Soc; fel Acad Athens; fel Am Acad Arts & Sci; fel Am Soc Mech Engrs. *Res:* Nuclear reactor dynamics; physics of ionized gases; surface physics; direct energy conversion; quantum thermodynamics; energy conservation. *Mailing Add:* Dept Nuclear Eng Mass Inst Technol Cambridge MA 02139

GYLES, C L, b Jamaica, WI, May 15, 40; m 64; c 2. VETERINARY MICROBIOLOGY. *Educ:* Univ Toronto, DVM, 64; Univ Guelph, MSc, 66, PhD, 68. *Prof Exp:* Lectr diag bact, Ont Vet Col, Univ Guelph, 64-66; Med Res Coun Can fel, Eng & Denmark, 68-69; from asst prof to assoc prof, 69-75, PROF VET BACT, ONT VET COL, UNIV GUELPH, 75- *Concurrent Pos:* Deleg cholera panel, US-Japan Coop Med Sci Prog, Japan, 70; dean grad studies, Univ Guelph, 81-86; Ont Coun Univ Affairs, 81-87. *Mem:* Can Vet Med Asn; Am Soc Microbiol. *Res:* Escherichia coli enterotoxins; R factors and other plasmids; pathogenesis of Escherichia coli infection in pigs and calves; exotoxin of corynebacterium pseudotuberculosis; pathogenesis of salmonellosis in calves; virulence of salmonella; edema disease principle; Shiga-like toxins; cloning and DNA sequencing. *Mailing Add:* Dept Vet Microbiol & Immunol Univ Guelph Guelph ON N1G 2W1 Can

GYLES, NICHOLAS ROY, b Jamaica, WI, Jan 30, 22; US citizen; m 55; c 2. GENETICS, ANIMAL BREEDING. *Educ:* Univ BC, BSA, 46, MSA, 49; Univ Mo, PhD(animal breeding), 52. *Prof Exp:* From asst prof to prof poultry genetics, 52-76, PROF ANIMAL SCI, UNIV ARK, FAYETTEVILLE, 76- *Concurrent Pos:* Consult geneticist, Peterson Poultry Breeding Farm, Ark, 56- *Mem:* Genetics Soc Am; Poultry Sci Asn; Sigma Xi. *Res:* Effectiveness of selection for economic traits in poultry; genetic mechanisms of resistance to Rous and leukosis viruses in chickens. *Mailing Add:* 1336 Northview Fayetteville AR 72701

GYLYS, JONAS ANTANAS, b Kaunas, Lithuania, June 27, 28; US citizen; m 56; c 3. PHARMACOLOGY. *Educ:* Roosevelt Univ, BS, 51; Loyola Univ, Ill, MS, 54, PhD(pharmacol), 57. *Prof Exp:* Pharmacologist, Bio-Test Labs, Warner Lambert Res Inst, 56-58, scientist gen & behav pharmacol, 58-60, sr scientist, NJ, 60-66; sr res scientist, 66-68, ASST DIR PHARMACOL RES, BRISTOL LABS, SYRACUSE, 68-; ASSOC DIR, NEUROPHARMACOL, BRISTOL MYERS SQUIBB, WALLINGFORD, CONN, 87- *Mem:* AAAS; Am Soc Pharmacol & Exp Therapeut; Sigma Xi. *Res:* Operant behavior; anoretic agents; hypnotics; tranquilizers; antidepressants; stimulants; performance and memory enhancers; muscle relaxants; investigational new drug application and new drug application work; good laboratory practice implementation; lab facility design; anticonvulsants; migraine research. *Mailing Add:* 153 Ciccio Rd Southington CT 06489-2104

GYOREY, GEZA LESLIE, b Budapest, Hungary, July 31, 33; US citizen; m 57; c 2. NUCLEAR ENGINEERING, TECHNICAL MANAGEMENT. *Educ:* Univ Mich, BSE, 55, MSE, 57, PhD(nuclear eng), 60. *Prof Exp:* Assoc prof nuclear eng, Univ Mich, 60-66; mgr nuclear methods, 66-68, mgr nuclear design, 68-70, mgr nuclear fuel strategic planning, 70-73, mgr uranium enrichment projs, 73-74, mgr reactor licensing, 74-76, mgr design reviews, 76-80, MGR ADVAN FUEL PROGRAMS, GEN ELEC NUCLEAR ENERGY GROUP, 80- *Mem:* Am Nuclear Soc; AAAS. *Res:* Nuclear power reactor and fuel technology, reliability and safety. *Mailing Add:* 19941 Winter Lane Saratoga CA 95070

GYORGY, ERNST MICHAEL, b Heidelberg, Ger, Feb 26, 26; nat US; m 51; c 1. EXPERIMENTAL SOLID STATE PHYSICS. *Educ:* Mass Inst Technol, BS, 50, PhD(physics), 53. *Prof Exp:* MEM TECH STAFF, BELL TEL LABS, 53- *Mem:* Fel Am Phys Soc. *Mailing Add:* Six Woodcliff Dr Madison NJ 07940

GYORKEY, FERENC, IMMUNOPATHOLOGY, ELECTRO-MICROSCOPY. *Educ:* United Hungarian Med Schs, MD, 45. *Prof Exp:* PROF PATH & TUMOR DIAG, BAYLOR COL MED, 65- *Res:* Tumor ogenesis; viralocology. *Mailing Add:* Dept Path Pharmacol & Virol Baylor Col Med 1913 Canterbury Houston TX 77030

GYRISCO, GEORGE GORDON, entomology; deceased, see previous edition for last biography

GYSLING, HENRY J, b Philadelphia, Pa, Dec 29, 41; m 73; c 2. INORGANIC CHEMISTRY, ORGANOMETALLIC CHEMISTRY. *Educ:* St Joseph's Col, Pa, 63; Univ Del, PhD(inorg chem), 67. *Prof Exp:* Res fel chem, NY Univ, 67-68; res fel inorg chem, Univ Newcastle upon Tyne, 68-70; sr res chemist, 70-75, RES ASSOC, EASTMAN KODAK CO, 75- *Concurrent Pos:* Adj sr res assoc, Pa State Univ, 81; assoc lectr, Univ Rochester, 80-; adj prof, Rochester Inst Technol. *Mem:* Soc Photog Scientists & Engrs; Am Chem Soc; NY Acad Sci; Mat Res Soc. *Res:* Synthetic and physical inorganic chemistry; coordination chemistry; ambidentate ligand chemistry; tellurium chemistry; organometallic chemistry. *Mailing Add:* Res Labs Eastman Kodak Co 1999 Lake Ave Rochester NY 14650-2111

GYULASSY, MIKLOS, b Szolnok, Hungary, Mar 9, 49; m 78; c 3. THEORETICAL PHYSICS. *Educ:* Univ Calif, Berkeley, AB, 70, PhD(physics), 74. *Prof Exp:* Fel physics, Gesselschaft Schwerionenforschung, 74-76; fel, 76-81, DIV FEL PHYSICS & NUCLEAR SCI & SR STAFF SCIENTIST, LAWRENCE BERKELEY LAB, 81- *Honors & Awards:* Alexander von Humboldt US Sr Scientist Award, 86; E O Lawrence Mem Award, 87. *Mem:* Am Phys Soc. *Res:* Physics of dense, highly excited nuclear matter formed in high energy nuclear collisions; nuclear phase transitions, Quark-gluon plasmas. *Mailing Add:* Nuclear Sci Div Lawrence Berkeley Lab Berkeley CA 94720

GYURE, WILLIAM LOUIS, b Chicago, Ill, Sept 26, 39. CLINICAL BIOCHEMISTRY, TOXICOLOGY. *Educ:* Aquinas Col, BS, 61; Tufts Univ, PhD(biochem), 70; Am Bd Clin Chem, dipl, 75 & 76. *Prof Exp:* Am Cancer Soc res fel biochem, Max Planck Inst Biochem, Munich, 70-71; fel clin biochem, Univ Hosp, Univ Wash, 71-73; clin biochemist & toxicologist, Cape Cod Hosp, 73-78; clin biochemist & toxicologist, Jersey City Med Ctr, 78-87; asst prof clin path, Univ Med & Dent NJ, Newark, 79-89; CONSULT. *Concurrent Pos:* Consult clin chem, Falmouth Hosp, 74-78; vis prof, Tokyo Inst Geront; chmn, Jersey City Munic Environ Comn, 86-89. *Mem:* Fel Am Asn Clin Chemists; Am Chem Soc. *Res:* Pteridine metabolism; purine and pteridine analytical techniques; clinical chemistry methodologies. *Mailing Add:* Seton Hall Univ 400 S Orange Ave South Orange NJ 07079

H

HA, TAI-YOU, b Youngam, Chonnam, Mar 5, 33; m 64; c 3. CELLULAR IMMUNOLOGY, IMMUNOPHARMACOLOGY & ALLERGY. *Educ:* Chonnam Nat Univ, MD, 60, Master Med Sci, 62, PhD(bacteriophage), 68. *Prof Exp:* Instr bact, Dept Microbiol, Med Sch, Chonnam Nat Univ, 66-68, from aast prof to assoc prof immunol, 68-74; res assoc immunol, Med Sch Yale Univ, 71-73; assoc prof & chmn microbiol & immunol, 74-79, PROF & CHMN IMMUNOL, MED SCH, CHONBUK NAT UNIV, 79- *Concurrent Pos:* Vdean, Med Sch, Chonbuk Nat Univ, 74-79, dir, Inst Med Sci, 77-83, dean, 79-80; pres, Asn Korean Med Educ, 80, Korean Soc Microbiol, 84-85 & Korean Soc Immunol, 87-89; adj prof, Immunol Div, Mont State Univ, 81-82; counr, Bact Div, Int Union Microbiol Socs, 82-86, rep, Int Comt Syst Bact, 86- *Mem:* Am Soc Microbiol; Am Asn Immunol; NY Acad Sci; Soc Leukocyte Biol; Int Soc Immunopharmacol; Int Soc Endotoxin. *Res:* Immunoregulation by T and B cells and cytokines; mechanisms of immediate hypersensitivity effects of platelet activation factor and platelet activation factor antagoinists on anaphylaxis; host-parasite relationships; alcohol immunology dealing with infection and tumor. *Mailing Add:* Dept Microbiol & Immunol Chonbuk Nat Univ Med Sch Chonju Chonbuk 560-182 Republic of Korea

HAACK, DONALD C(ARL), b Oak Park, Ill, Dec 7, 23; m 46; c 3. ENGINEERING MECHANICS. *Educ:* Ill Inst Technol, BS, 45; Univ Nebr, MS, 51; Kans State Univ, PhD, 68. *Prof Exp:* Instr civil eng, Ill Inst Technol, 44-46; instr eng drawing & mach design, Univ Colo, 46-47; from instr to prof, 47-89, EMER PROF ENG MECH, UNIV NEBR, LINCOLN, 89- *Concurrent Pos:* Sci fac fel, NSF, 59-60. *Res:* Application of engineering in dentistry. *Mailing Add:* 5011 Glade St Lincoln NE 68506

HAACKE, GOTTFRIED, b Bad Lausick, Ger, Nov 27, 30; US citizen; m 63; c 2. III-V COMPOUND SEMICONDUCTORS, DYE ABSORBER TECHNOLOGY. *Educ:* Univ Cologne, Ger, dipl physics, 54, PhD(physics), 57. *Prof Exp:* Res physicist semiconductors, AEG, Frankfurt Ger, 58-61; res physicist semiconductors, Am Cyanamid Co, 62-68, group leader electrochromism, 68-77, mgr mat res, 77-84, RES FEL MAT RES, AM CYANAMID CO, 84- *Mem:* Am Phys Soc; Inst Elec & Electronics Engrs. *Res:* Semiconductor epitaxy; wide-band gap semiconductors; thermoelectric compounds and devices; electrochromism devices; heterogeneous catalysis; laser hardening; author of various publications; granted 18 patents. *Mailing Add:* 1937 W Main St Stamford CT 06904

HAAG, FRED GEORGE, b Weehawken, NJ, May 11, 31; m 57; c 2. ENGINEERING. *Educ:* Stevens Inst Technol, ME, 53; Rensselaer Polytech Inst, MS, 58, DrEngSci(mech eng), 64. *Prof Exp:* Reactor studies, Oak Ridge Nat Lab, 53-54; supv engr, Knolls Atomic Power Lab, 54-67; assoc prof mech eng, Union Col, NY, 67-72; dir, Bur Tech Serv, Div Air Resources, 72-79, POWER GENERATION PLANNER, PUB SERV COMN, STATE NY, 79- *Concurrent Pos:* Adj prof, Union Col, 77- *Mem:* Sigma Xi. *Res:* electric power generation; systems analysis; cost engineeering; electric system planning. *Mailing Add:* Pub Serv Comn 3 Empire State Plaza Albany NY 12223-0001

HAAG, JOSEPH ROY, nutrition, for more information see previous edition

HAAG, KIM HARTZELL, biological control, insect-host plant interactions, for more information see previous edition

HAAG, ROBERT EDWIN, b New Haven, Conn, Mar 23, 54; m 76; c 2. GRINDING, PRODUCTION OF METAL COMPONENTS. *Educ:* Univ Conn, BS, 75, MS, 77. *Prof Exp:* Metallurgist, Robin Steel Co, 77-80; chief metallurgist, Super Ball Div, Lydall, Inc, 80-83, mgr eng, 83-84; mgr eng, Hoover Universal, RA Gray, 84-85; consult, 85-86; ASST PROF MFG, WATERBURY STATE TECH COL, 86- *Concurrent Pos:* Consult, 85-; prin investr, NSF/ILI grant, Waterbury State Tech Col, 90-91. *Res:* Metallurgy; precision grinding; bearing production; computers in manufacturing. *Mailing Add:* 750 Chase Pkwy Waterbury CT 06708

HAAG, ROBERT MARLAY, b Dayton, Ohio, Mar 2, 22; m 49; c 3. PHYSICAL CHEMISTRY. *Educ:* Ohio State Univ, BS, 43; Northwestern Univ, MS, 49; Ind Univ, PhD(chem), 52. *Prof Exp:* Asst, Res Found, Ohio State Univ, 43-44; lab asst, Northwestern Univ, 46-47, asst, 47-48; chemist, US Naval Ord Lab, 48; asst, Ind Univ, 48-50; res assoc, Knolls Atomic Power Lab, Gen Elec Co, 50-57, phys chemist, Res Lab, 57-60, prin engr, Aircraft Nuclear Propulsion Dept, 60-61; sr consult scientist res & develop, Avco Systs Div, Avco Corp, 61-71; mem res staff, Ampex Corp, 71-85; RETIRED. *Mem:* Am Ceramic Soc; Combustion Soc. *Res:* Thermodynamics of gases; thermodynamics of solutions; ultrasonic interferometry; gas-metal reactions; magnetic oxides. *Mailing Add:* 1300 Montclaire Way Los Altos CA 94024

HAAG, WERNER O, b Heilbronn, Ger, Jan 28, 26; US citizen; c 4. CATALYSIS, ZEOLITE CATALYSIS. *Educ:* Univ Tuebingen, Ger, dipl org chem, 54; Northwestern Univ, PhD(org chem), 58. *Prof Exp:* Res chemist, Swift & Co, 55; res chemist, 59-81, SR SCIENTIST & LAB ADV, MOBIL RES & DEVELOP CORP, 81- *Honors & Awards:* Petrol Chem Award, Am Chem Soc, 88; Ipatieff Lectr, Northwestern Univ, 92. *Mem:* Am Chem Soc; Catalysis Soc; Int Zeolite Asn; Sigma Xi. *Res:* Zeolite catalysis directed towards basic understanding of sorption, diffusion, mechanisms of hydrocarbon conversions and the nature of active sites; exploratory zeolite catalysis towards shape-selective and other novel processes in the petroleum and petrochemical industry. *Mailing Add:* Mobil Res & Develop Corp PO Box 1025 Princeton NJ 08543-1025

HAAK, FREDERIK ALBERTUS, b Amsterdam, Neth, July 7, 21; US citizen; m 70; c 2. TECHNICAL PROGRAM MANAGEMENT. *Educ:* Univ Amsterdam, Neth, BS, 41, MS, 47, PhD(phys chem), 48. *Prof Exp:* Res chemist, Royal Dutch Shell Lab, 48-52; oil res analyst, Phillips Petrol Co, 52-55; staff physicist, Thompson-Ramo-Wooldridge Corp, 55-58; sr staff physicist, Hughes Aircraft Co, 58-67; physicist, Electronic Res Ctr, NASA, 67-70; opers res analyst, US Army Aviation Syst Command, 70-72; PROJ MGR, DEFENSE ADV RES PROJ AGENCY OFF, ARMY MISSILE COMMAND, 72- *Mem:* Am Phys Soc; Optical Soc Am. *Res:* High energy laser technology; optical components with high flux density capability and electro-optical modulators; operations research in acquisition tracking pointing and rapid retargeting experiments. *Mailing Add:* 710 Clinton Ave Huntsville AL 35801

HAAK, RICHARD ARLEN, b Fairmont, Minn, Sept 14, 44; m 79; c 3. MEMBRANE BIOPHYSICS. *Educ:* MacMurray Col, BA, 66; Southern Ill Univ, MA, 68, PhD(molecular sci), 72. *Prof Exp:* Instr physics, Southern Ill Univ, 68-72; vis asst prof biophys, 72-73, asst prof med biophysics, 73-78, ASSOC PROF MICROBIOL & IMMUNOL, SCH MED, IND UNIV, INDIANAPOLIS, 78- *Mem:* Biophys Soc; Am Phys Soc; Am Soc Microbiol. *Res:* Electron spin resonance. *Mailing Add:* Dept Biophys Microbio Ind Univ Purdue Univ Med 1100 W Mich St Indianapolis IN 46223

HAAKE, EUGENE VINCENT, b Cleveland, Ohio, Sept 25, 21; m 51; c 3. NUCLEAR & CONTROL ENGINEERING. *Educ:* Western Reserve Univ, BS, 43; Univ Calif, Los Angeles, MA, 48. *Prof Exp:* Assoc physicist, Nepa Div, Fairchild Engine & Airplane Corp, Tenn, 48-50, exp physicist, 50-51; assoc physicist, Oak Ridge Nat Lab, 51-52; sr nuclear engr, Gen Dynamics/Convair, 52-55, nuclear group engr, 55-62; staff mem nuclear power syst design, Gen Atomic Div, Gen Dynamics Corp, 62-65, sect leader, High Temperature Gas Reactor Dept, 65-68; br mgr, Eng Div, Gen Atomic Co, 68-85; RETIRED. *Mem:* Am Phys Soc; Am Nuclear Soc. *Res:* Nuclear instrumentation and measurement; control and instrumentation of nuclear reactors and power plants; nuclear safety; dynamic and performance analysis of nuclear power plants. *Mailing Add:* 3703 Brandywine St San Diego CA 92117-5738

HAAKE, PAUL, b Winona, Minn, June 28, 32; div; c 2. HISTORY & PHILOSOPHY OF SCIENCE. *Educ:* Harvard Univ, AB, 54, PhD(chem), 61. *Prof Exp:* Noyes fel chem, Calif Inst Technol, 60-61; from instr to assoc prof, Univ Calif, Los Angeles, 61-68; prof chem, 68-84, PROF MOLECULAR BIOL & BIOCHEM, WESLEYAN UNIV, 84- *Concurrent Pos:* A P Sloan fel, 64-66; NATO sr fel, 69, NSF fac fel, 74, Wesleyan Ctr Humanities, 76. *Mem:* Hist Sci Soc; Brit Hist Sci Soc; Philos Sci Asn. *Res:* Reaction mechanisms; models for enzymatically catalyzed reactions; organophosphorus chemistry; ascorbic acid; history of science. *Mailing Add:* Dept Molecular Biol & Biochem Wesleyan Univ Middletown CT 06457

HAAKONSEN, HARRY OLAV, b Brooklyn, NY, Mar 12, 41; m 68; c 3. SCIENCE EDUCATION, RESEARCH ADMINISTRATION. *Educ:* Taylor Univ, BA, 62, BS, 63; Syracuse Univ, MS, 66, PhD(sci educ), 69. *Prof Exp:* Teacher, Warsaw Community Schs, 63-65, chmn sci dept, 64-65; dir audio-tutorial genetics prog, Syracuse Univ, 67-69; PROF SCI EDUC, SOUTHERN CONN STATE COL, 69-; PROF CHEM, SOUTHERN CONN STATE UNIV, 75- & DIR, CTR ENVIRON. *Concurrent Pos:* Grants, Conn State Audio-Tutorial Proj, Southern Conn State Col, Audio-tutorial Drug Educ Proj, & US Off Educ grant environ educ & Proj Disperse; sci consult, Branford, Granby, New Haven, Wallingford & Wilton, Conn Sch Systs; lectr, Univ Conn; dir audio-tutorial workshop, NSF; coordr environ studies, Southern Conn State Univ, 71-, dir, Conn Marine Studies Consortium, 86-; dir, Land Use Decision Making proj & Master Plan for Metric Educ, US Office Educ; student sci training prog on marine studies, NSF; Inst Study & Instr in Sci; co-ed, Conn J Sci Teaching; vis scientist, Marine Biol Lab, Woods Hole, 81 & Fisheries Directorate, Bergen, Norway, 81; dir, NSF Sponsored Inst Sci Instruction & Study. *Mem:* Am Sci Affil; Nat Asn Res Sci Teaching; Nat Sci Teachers Asn; Am Chem Soc; Acad Underwater Res; Acad Arts & Scis; NY Acad Sci. *Res:* The effect of individualized approaches to instruction upon student attitude and achievement; ultrasonic strategies for monitoring migratory behavior of Atlantic salmon; enhancement of adult salmon migration in the Farmington River. *Mailing Add:* Southern Conn State Univ 501 Crescent St New Haven CT 06515

HAALAND, CARSTEN MEYER, b Winnepeg, Man, Apr 9, 27; m 80; c 3. PHYSICS, SYSTEMS ANALYSIS. *Educ:* St Olaf Col, Minn, BS, 50; Univ Minn, MS, 55. *Prof Exp:* Engr klystron tubes, Sperry Gyroscope Co, 54-58; physicist physics, ITT Res Inst, 58-66; PHYSICIST CIVIL DEFENSE, OAK RIDGE NAT LAB, 66- *Mem:* Am Phys Soc; Health Physics Soc. *Res:* Civil defense; nuclear weapons effects; atmospheric physics and chemistry. *Mailing Add:* 121 Iris Circle Oak Ridge TN 37830

HAALAND, DAVID MICHAEL, b Chicago, Ill, Mar 16, 46; m 68; c 1. PHYSICAL CHEMISTRY, ANALYTICAL CHEMISTRY. *Educ:* Univ NMex, BS, 68; Univ Rochester, PhD(phys chem), 72. *Prof Exp:* MEM TECH STAFF PHYS CHEM, SANDIA LABS, ALBUQUERQUE, NMEX, 72- *Mem:* Am Chem Soc; Am Phys Soc. *Res:* High temperature carbon phase diagram; development of solid electrolyte oxygen sensors; molecular structure and spectroscopy; surface studies by infrared spectroscopy; catalysis; properties of silicon solar cell materials; fossil energy research; quantitative infrared spectroscopy, chemometrics. *Mailing Add:* Div 1823 Sandia Labs Albuquerque NM 87185

HAALAND, JOHN EDWARD, b Preeceville, Sask, Sept 12, 35; US citizen; m 58; c 3. SCIENCE ADMINISTRATION, BIOPHYSICS. *Educ:* Univ Minn, BA, 57, PhD(zool, phys chem), 68. *Prof Exp:* Res asst biophys, Univ Minn, 61-62; design engr, Apollo Proj Aeronaut Div, Honeywell, Inc, 62-65, res scientist biotech. Res Systs & Res Group, 65-67, sr res scientist, 68-69; res assoc cell biol, Univ Minn, Minneapolis, 69; vpres systs & res, health serv systs div, Food & Drug Res Labs, Inc, NY, 69-71, dir ecol sci, 70-71; vpres environ systs, Mgt & Environ Systs, Pillsbury Co, 73-76, vpres info, 76-83; PRES, BIOSYSTS, INC, 71-73 & 83-; PRES, TOTAL FITNESS SYST, INC, 86- *Concurrent Pos:* Res asst, Univ Minn, Minneapolis, 67-68; Stanford Exec Prog, 75; trustee, Midwest Res Inst, 75-, N StarRes Found; mem, Nat Res Coun Eval Panel Energy Conserv Progs, Nat Bur Standards, 77-79, Off Technol Assessment Rev Comt Five Year Plan, Environ Protection Agency, 77, trustee, Minn Acad Sci, 78-82, Solar Energy Res Inst Adv Bd & Oversight Comt, 82-; vchmn, Archaeus Proj, 86. *Mem:* AAAS; World Future Soc; NY Acad Sci. *Res:* Energy and food systems; environmental systems; health delivery systems; corporate and technical information systems; long range strategic planning; organizational design and theory; integrative fitness systems for the mind and body. *Mailing Add:* 339 1/2 Irvine Ave St Paul MN 55102

HAALAND, RONALD L, b Havre, Mont, Nov 14, 46; m 67; c 2. PLANT BREEDING. *Educ:* Mont State Univ, BS, 68, MS, 70; NMex State Univ, PhD(plant breeding), 73. *Prof Exp:* Plant breeder, NAm Plant Breeders, 73-74; asst prof forage breeding, Auburn Univ, 74-79, assoc prof, 79-80; PRES, SUN RISE INC, 80- *Mem:* Am Soc Agron; Crop Sci Soc Am. *Res:* Breeding forage grasses, tall fescue and phalaris, for improved winter productivity and nematode resistance; breeding alfalfa for improved persistence and forage production. *Mailing Add:* Sun Rise Inc PO Box 2085 Auburn AL 36830

HAAN, CHARLES THOMAS, b Randolph Co, Ind, July 10, 41; m 67; c 3. ENGINEERING, HYDROLOGY. *Educ:* Purdue Univ, BS, 63, MS, 65; Iowa State Univ, PhD(agr eng), 67. *Prof Exp:* Asst agr eng, Purdue Univ, 63-64; res asst, Iowa State Univ, 64-67; from asst prof to prof agr eng, Univ Ky, 67-78; prof & head dept, 78-84, prof, 85-88, REGENTS PROF & SARKEYS DISTINGUISHED PROF AGR ENG, OKLA STATE UNIV, 88- *Concurrent Pos:* Vis prof civil eng, Colo State Univ, 73-74; hydrol eng specialist, Steffen Robertson & Kirsten, Johannesburg, SAfrica. *Honors & Awards:* Am Soc Agr Engrs Award, 69; Young Researcher Award, Am Soc Agr Engrs, 75; Hancor Award, 90. *Mem:* Fel Am Soc Agr Engrs. *Res:* Mathematical, statistical and empirical models of various phases of the hydrologic cycle; hydrology of agricultural, surface mined and forest lands; statistical applications and mathematical modelling in hydrology; erosion and sedimentation. *Mailing Add:* Dept Agr Eng Okla State Univ Stillwater OK 74078

HAAR, JACK LUTHER, b Elmore, Ohio, July 9, 42; m 71; c 1. ANATOMY, ELECTRON MICROSCOPY. *Educ:* Capital Univ, BSc, 64; Univ Ariz, MSc, 66; Ohio State Univ, PhD(anat), 70. *Prof Exp:* Teaching asst anat, Univ Ariz, 65-66; instr biol, Buena Vista Col, 66-67; instr, Capital Univ, 67-71; asst prof, 71-77, ASSOC PROF ANAT, MED COL VA, VA COMMONWEALTH UNIV, 77- *Concurrent Pos:* NSF res grant, Col Sci Improv Prog, Ohio State Univ, 70-71; NIH grant. *Honors & Awards:* William & Wilkins Award, Am Asn Anat, 70. *Mem:* AAAS; Am Inst Biol Sci; Am Asn Anat. *Res:* Isolation of hemopoietic stem cells from yolk sac, fetal liver and adult bone marrow; histochemistry; chemotastic migration, colony-forming assay; tissue culture, immunofluorescence and electron microscopy. *Mailing Add:* Dept Anat Med Col Va Commonwealth Univ Box 163 MCV Scis 709 MCV Sta Richmond VA 23298

HAARD, NORMAN F, b Queens, NY, Dec 4, 41; m 63; c 2. POSTMORTEM PHYSIOLOGY FISH. *Educ:* Rutgers Univ, BS, 63; Univ Mass, Amherst, PhD(food biochem), 66. *Prof Exp:* Res fel, Inst Enzyme Res, Univ Wis-Madison, 66-68; assoc prof food biochem, Rutgers Univ, New Brunswick, 68-77; PROF BIOCHEM, MEM UNIV NFLD, 77- *Honors & Awards:* S C Prescott Award, Inst Food Technologists, 73; Fraser Award, Air Pollution Info & Comput Syst, 78. *Mem:* Inst Food Technologists. *Res:* Structure of biological membranes; noncatalytic proteins of mitochondria; mechanism of plant senescence; biochemistry of fruit ripening; mitochondrial-linked reactions in senescing plant tissue; role of peroxidase in plant senescence; hormonal control of plant senescence; stress metabolites in sweet potato; fishery waste utilization. *Mailing Add:* Dept Food Sci & Technol Inst Marine Resources Univ Calif Davis CA 95616

HAAS, ALBERT B, b Hungary, Oct 27, 11; US citizen; m 40; c 1. MEDICINE. *Educ:* Royal Hungarian Med Sch, MD, 37. *Prof Exp:* Asst prof rehab med, Inst Phys Med & Rehab, 58-65, assoc prof exp rehab med, 66-72, PROF EXP REHAB MED, INST REHAB MED, MED CTR, NY UNIV, 72-, DIR CARDIOPULMONARY SERV, 66- *Honors & Awards:* Legion of Honor, France, 71. *Mem:* NY Acad Sci; Am Thoracic Soc; Fr Nat Soc Phys Med. *Res:* Work physiology; pulmonary physiology. *Mailing Add:* Dept Rehab Med NY Univ Med Ctr 400 E 34th St New York NY 10016

HAAS, CAROL KRESSLER, b Reading, Pa, Mar 8, 48. ORGANOMETALLIC CHEMISTRY, CATALYSIS. *Educ:* Ursinus Col, BS, 70; Mass Inst Technol, PhD(org chem), 74. *Prof Exp:* Res chemist organometallic chem, Res Div, 74-77, mem res staff coal chem, Feedstocks Div, 77-78, group leader coal chem, Feedstocks Div, 78-80, RES SUPVR, FEEDSTOCKS DIV, DU PONT CO CENT RES & DEVELOP, E I DU PONT DE NEMOURS & CO, INC, 80- *Mem:* AAAS; Am Chem Soc; Sigma Xi. *Res:* Routes to alternative sources to chemical feedstocks; heterogeneous and homogeneous catalysis. *Mailing Add:* Birkdale 9011 Highgate Hill Dr Chesterfield VA 23832

HAAS, CHARLES GUSTAVUS, JR, b Portsmouth, Va, May 26, 23; c 2. INORGANIC CHEMISTRY. *Educ:* Va Polytech Inst, BS, 44, SM, 49, PhD(chem), 51. *Prof Exp:* Asst chem, Univ Chicago, 47-48; res assoc, 50-51, from asst prof to assoc prof, 51-66, PROF CHEM, PA STATE UNIV, 66- *Mem:* Am Chem Soc. *Res:* Extraction of metal chlorides by organic solvents; coordination compounds. *Mailing Add:* 108 Whitmore Lab Pa State Univ University Park PA 16802

HAAS, CHARLES JOHN, b Platteville, Wis, Oct 9, 35; m 62; c 3. ROCK MECHANICS. *Educ:* Wis State Col & Inst Technol, BS, 60; Colo Sch Mines, MS, 62, DSc(rock mech), 64. *Prof Exp:* Asst res prof rock mech, 64-68, assoc prof eng mech & sr investr rock mech & explosives, 68-76, PROF MINING ENG, UNIV MO, ROLLA, 76- *Concurrent Pos:* Student coop mem, Geophys Serv Inc, 60; res engr, Colo Sch Mines Res Found, 62, res assoc, 64; chmn subcomt, Am Soc Testing & Mat, 74- *Mem:* Am Soc Testing & Mat; Int Soc Rock Mech; Soc Mining Metall & Explor. *Res:* Rock mechanics research, specifically the behavior of rock materials subjected to static, dynamic and impulse loading; design and stability of underground structures. *Mailing Add:* Dept Mining Eng Univ Mo, Rolla Rolla MO 65401

HAAS, DAVID JEAN, b Buffalo, NY, Apr 10, 39; m 62; c 3. ELECTRONIC SECURITY, RADIATION SAFETY. *Educ:* State Univ NY, BS, 62, PhD(biophysics), 65. *Prof Exp:* Res physicist, US Naval Res Lab, DC, 65-66; asst prof biol, Purdue Univ, 67-70; prin scientist, x-rays & security systs, Philips Electronics, 70-83; PRES, TEMTEC, INC, 83- *Concurrent Pos:* NIH fel, Davey Faraday Lab, Royal Inst London, 65-66; consult, Radiation Safety, 71- *Mem:* AAAS; Am Soc Indust Security; Am Inst Physics; Am Crystallog Asn; Soc Photo-Optical Instrumentation Engrs. *Res:* Electronic security screening; access control; x-ray imaging for contraband; security identification and credentials. *Mailing Add:* Nine Marget Ann Lane Suffern NY 10901

HAAS, ERWIN, b Budapest, Hungary, Sept 11, 06; nat US; m 32; c 2. BIOCHEMISTRY. *Educ:* Gauss Sch, Berlin, ME, 28; Univ Chicago, PhD(chem), 42. *Prof Exp:* Mechanic, C Lorenz Co, Berlin, 21-28; asst chem, Kaiser-Wilhelm Inst Cell Physiol, Berlin, 28-38; from instr to asst prof chem, Univ Chicago, 38-44; sr mem, Worcester Found Exp Biol & State Hosp, 44-45; asst prof exp path, Western Reserve Univ, 45-46; asst dir inst med res, Cedars of Lebanon Hosp, 46-53; DIR BEAUMONT MEM LABS, MT SINAI HOSP, 53- *Concurrent Pos:* Mem coun high blood pressure res, Am Heart Asn; mem sci coun hypertension, Int Soc Cardiol. *Mem:* Am Soc Biol Chemists; Nat Soc Med Res; Am Chem Soc. *Res:* Chemistry of respiratory enzymes, defense enzymes against invasion and bacteria; biochemistry of high blood pressure. *Mailing Add:* Beaumont Mem Labs Mt Sinai Med Ctr 1800 E 105th St Cleveland OH 44106

HAAS, FELIX, b Vienna, Austria, Apr 20, 21; US citizen; m 48, 87; c 3. MATHEMATICS. *Educ:* Mass Inst Technol, BS, 48, MS, 49, PhD(math), 52. *Prof Exp:* Instr math, Lehigh Univ, 52-53; Fine instr, Princeton Univ, 53-55; asst prof math, Univ Conn, 55-56; from asst prof to prof math, Wayne State Univ, 56-62, head dept, 60-62; prof & head, Div Math Sci, 62, dean, Sch Sci, 62-74, provost, 74-75, exec vpres & provost, 75-86, ARTHUR G HANSEN DISTINGUISHED PROF MATH, PURDUE UNIV, 86- *Concurrent Pos:* Consult to indust, 52- *Mem:* Am Math Soc. *Res:* Nonlinear differential equations; transformation groups. *Mailing Add:* Dept Math Purdue Univ West Lafayette IN 47907

HAAS, FRANCIS XAVIER, JR, b Covington, Ky, June 16, 38; m 62, 86; c 5. NUCLEAR PHYSICS. *Educ:* Xavier Univ, Ohio, BS, 60; St Louis Univ, MS, 62; Univ Cincinnati, PhD(physics), 69. *Prof Exp:* Res physicist, Mound Lab, Monsanto Res Corp, 62-69, sr res physicist, 69-77; res specialist, 77-80, sr res specialist, 85-88, ASSOC SCIENTIST, EG&G ROCKY FLATS, 80- *Mem:* Am Phys Soc; Inst Nuclear Mat Mgt; Am Soc Testing & Mat. *Res:* Nuclear materials safeguards research and development; radometric nondestructive assay. *Mailing Add:* 30434 Inverness Lane Evergreen CO 80439

HAAS, FRANK C, b Hays, Kans, Sept 19, 31; m 55; c 4. INORGANIC CHEMISTRY, METALLURGY. *Educ:* Ft Hays State Col, BS, 55, MS, 56. *Prof Exp:* Anal chemist, Phillips Petrol Co, 56-58, chief chemist, 58-60; res chemist, Mallinckrodt Chem Works, 60-61; chief chemist, Hazen Res, Inc, 61-66; sr res chemist, 66-68, res group leader chem & metall, 68-81, MGR PROCESS RES, TOSCO CORP, 81- *Mem:* Am Chem Soc; Am Inst Mining, Metall & Petrol Engrs. *Res:* Minerals beneficiation; extractive metallurgy of metals from ores and concentrates; production of synthetic fuels from carbonaceous materials. *Mailing Add:* 12278 W 70th Ave Arvada CO 80004

HAAS, FREDERICK CARL, b Buffalo, NY, Feb 16, 36; m 57; c 3. CHEMICAL & PULP & PAPER ENGINEERING. *Educ:* Purdue Univ, BS, 57; Rensselaer Polytech Inst, MS, 59, PhD(chem eng), 60. *Prof Exp:* Assoc chem engr, Cornell Aeronaut Lab, Inc, 60-62, res engr, 62-63; tech asst to corp res dir, WVa Pulp & Paper, 63-64, tech serv supt, 64-66, tech dir, 66-67; asst prof, Potomac State Col, WVa Univ, 67-68; tech asst to vpres fine papers div, New York, 68-72, plant mgr, Pa, 72-74, gen mfg mgr, Kraft Div, 74-76, assoc corp res dir, 77-81, vpres, 79-81, SR VPRES, WESTVACO CORP, 82- *Mem:* Am Inst Chem Engrs; Am Chem Soc; Can Pulp & Paper Asn; Nat Soc Prof Engrs. *Res:* Heat transfer; fluid mechanics; heat transfer instrumentation; pulp and paper processes; water and air pollution abatement; coatings; research management. *Mailing Add:* Westvaco Corp Westavco Bldg 299 Park Ave New York NY 10171

HAAS, GEORGE ARTHUR, b Vienna, Austria, Sept 7, 26; nat; m 51; c 4. SURFACE PHYSICS. *Educ:* Univ Notre Dame, BS, 50, PhD(physics), 53. *Prof Exp:* PHYSICIST, SURFACE PHYSICS SECT, NAVAL RES LAB, WASHINGTON, DC, 53-, HEAD SECT, 54- *Mem:* Am Vacuum Soc; Am Phys Soc. *Res:* Electron emission and surface phenomena of semiconductors and metals; solid state properties of semiconductors; physical and chemical properties of oxide cathodes; thermionic energy converters; electron beam activated semiconductor devices; reliability physics for electron devices. *Mailing Add:* Naval Res Lab 4555 Overlook Ave SW Washington DC 20375-5000

HAAS, GERHARD JULIUS, b Munich, Ger, Apr 1, 17; nat US; m 59; c 2. MICROBIOLOGY, BIOCHEMISTRY. *Educ:* Cambridge Univ, BA, 39, MA, 43; Univ Pa, PhD(microbiol), 52. *Prof Exp:* Org chemist, Hoffmann-La Roche, Inc, 43-50; dir res, Liebmann Breweries, 52-60; chief chemist, Desitin Chem Co, 60-61; prin scientist, Gen Foods Corp, 61-87; SR RES ASSOC, NY BOT GARDENS, 87- *Concurrent Pos:* Adj prof, Lehman Col, 87- *Mem:* Am Soc Microbiol; Am Chem Soc; Soc Indust Microbiol; AAAS; Inst Food Technologists; Am Phytopath Soc. *Res:* Enzymes; food biochemistry; brewery microbiology; yeast; lactobacilli; food preservation; natural preservatives; microbial plant interface. *Mailing Add:* NY Bot Gardens Bronx NY 10458-5126

HAAS, GREGORY MENDEL, plasma physics, technical management, for more information see previous edition

HAAS, GUSTAV FREDERICK, b Vienna, Austria, June 15, 27; US citizen; div. AUDIOLOGY, BIOMEDICAL ENGINEERING. *Educ:* Yale Univ, BE, 63, ME, 64; Univ Calif, Los Angeles, PhD(sensory systs), 70; Univ Iowa, MA, 80. *Prof Exp:* Prod planner, Westinghouse Elec Corp, 52-54; field engr commun equip, RCA Serv Co, 54-56; field engr optical character recognition, Intelligent Mach Res Corp, 57-59; proj mgr electronic design & mfg, Epsco, Inc, 59-67; scholar, Univ Calif, 70-72; sr staff officer, Comt Prosthetics Res & Develop, Nat Acad Sci, 72-76; res fel, Univ Iowa, 76-78; specialist res audiol, Univ Calif, San Francisco, 78-79; DIR, AUDITORY REHAB ENG, 79- *Concurrent Pos:* Mem res adv comt, Helen Keller Nat Ctr Deaf-Blind Youths & Adults, 74-84, chmn, 78-84; mem adv comt, Gallaudet Col Rehab Eng Ctr, 82-85. *Mem:* Acoust Soc Am; Am Speech & Hearing Asn; Inst Elec & Electronics Engrs. *Res:* Speech processing schemes for the hearing-impaired; time-varying-parameter electric networks; sensory aids for the deaf and the blind; audiology and hearing aids; special devices for rehabilitation of the hearing-impaired. *Mailing Add:* Auditory Rehab Eng 1230 Hopkins Ave Redwood City CA 94062

HAAS, HERBERT, b Biel, Switz, July 7, 34; US citizen; m 63; c 1. GEOCHRONOLOGY, GEOCHEMISTRY. *Educ:* Kantonales Technikum Biel, Switz, dipl EngrHTL, 58; Southern Methodist Univ, Dallas, PhD(geol), 71. *Prof Exp:* Systs engr telecommun, Siemens-Albiswerk, Zurich, Switz, 58-59; engr-supvr comput systs, Remington Rand Univac, Switz, 59-63; engr geochronology, Grad Res Ctr Southwest, Dallas, Tex, 64-66; asst dir, 72-73, DIR CARBON-14 LAB, SOUTHERN METHODIST UNIV, DALLAS, 73- *Mem:* Mineral Soc Am; Geochem Soc; Am Geophys Union; AAAS. *Res:* Precise scintillation counting of low energy and very low level beta radiation involving equipment improvement and statistical methods; chemical pretreatment methods for reliable radiocarbon age dating of archeological bone samples and humic acids. *Mailing Add:* 8516 Thunderbird Lane Dallas TX 75238

HAAS, HERBERT FRANK, b Kansas City, Kans, Apr 4, 14; m 36; c 3. MICROBIOLOGY. *Educ:* Kans State Col, BS, 38, MS, 40, PhD(bact), 42. *Prof Exp:* Asst chemist, Sinclair Ref Co, 34-35; lab asst, Univ Wis, 38-39 & Kans State Col, 39-42; res assoc, Am Petrol Inst, Calif, 42-43; asst dir biol prod, Richardson-Merrell, Inc, 43-58, dir, Tissue Cult Prod Develop Lab, 58-63, head biol process develop, Jensen-Salsbery Labs, 63-79; RETIRED. *Mem:* Am Soc Microbiol; Soc Cryobiol; Tissue Cult Asn. *Res:* Bacterial utilization of petroleum products; veterinary biological product and process development; mass culture technique; tissue culture; freeze-drying. *Mailing Add:* 1100 Shawnee Rd Kansas City KS 66183

HAAS, HOWARD CLYDE, b New York, NY, Oct 31, 20; m 48, 71; c 4. POLYMER CHEMISTRY. *Educ:* City Col New York, BS, 41; Polytech Inst Brooklyn, PhD(chem), 49. *Prof Exp:* Res chemist, Gen Foods, Inc, 41-42, 45-46; res assoc, Polytech Inst Brooklyn, 46-48; supvr polymer res, Polaroid Corp, 49-55, mgr dept, 55-67, res fel, 67-82; RETIRED. *Concurrent Pos:* Adv asst, USAF, Off Naval Res & Nat Res Coun; consult, 82- *Mem:* Am Chem Soc. *Res:* Copolymerization; free radicals; kinetics; graft copolymers; ionic polymerization; tacticity; peroxide decomposition kinetics; novel monomers; microencapsulation; rheology; diffusion transfer photography; reversible gelation; gelatin substitutes; light sensitive systems; photoresists; duplication processes; fluoropolymers; polymer membranes; dyeing. *Mailing Add:* 140 Pleasant St Arlington MA 02174

HAAS, JERE DOUGLAS, b Lancaster, Pa, Sept 15, 45; m 68; c 2. HUMAN ADAPTABILITY, HUMAN GROWTH & DEVELOPMENT. *Educ:* Franklin & Marshall Col, BA, 67; Pa State Univ, MA, 70 & PhD(anthropol), 73. *Prof Exp:* Instr anthropol, Pa State Univ, 72-73; asst prof, Univ Mass, Amherst, 73-75; from asst prof to assoc prof, 75-87, PROF NUTRIT, CORNELL UNIV, 87-, DIR, HUMAN BIOL PROG, 83- *Concurrent Pos:* Hon res fel, dept anat, Univ Aberdeen, Scotland, 82; vis prof, Food Res Inst, Stanford Univ, 88-89; consult govt, Bolivia, Indonesia, & Philippines; mem, Comt Nutrit During Pregnancy & Lactation, Nat Acad Sci. *Mem:* AAAS; Am Asn Phys Anthropol; Am Inst Nutrit; Am Soc Clin Nutrit; Am Anthrop Asn; fel Human Biol Coun. *Res:* Effects of under-nutrition on fetal, infant and adolescent growth and development of the capacity for reproduction and physical work as adults in populations living in less developed countries. *Mailing Add:* Div Nutrit Sci Savage Hall Cornell Univ Ithaca NY 14853

HAAS, JOHN ARTHUR, KIDNEY PHYSIOLOGY, PATHOPHYSIOLOGY. *Educ:* Mo Valley Col, BS, 68. *Prof Exp:* RES ASSOC RENAL PHYSIOL, MAYO CLIN, 71- *Res:* Renal hormones. *Mailing Add:* Dept Physiol Mayo Clin 901 Guggenheim Bldg Rochester MN 55901

HAAS, JOHN WILLIAM, JR, b Philadelphia, Pa, Feb 19, 30; m 53; c 3. PHYSICAL CHEMISTRY, HISTORY OF SCIENCE. *Educ:* King's Col, NY, BS, 53; Univ Del, MS, 55, PhD(chem), 58. *Prof Exp:* Assoc prof chem, Grove City Col, 57-61; PROF CHEM, GORDON COL, 61-, CHMN, DEPT CHEM, 80- *Mem:* Am Chem Soc; Am Sci Affiliation; Hist Sci Soc. *Res:* Carbohydrate structure and reactions; chemiluminescence,; sugar complexes in seawater; science and religion. *Mailing Add:* Dept Chem Gordon Col 255 grapevine Rd Wenham MA 01984

HAAS, KENNETH BROOKS, JR, b Pittsburgh, Pa, Jan 14, 28; m 53; c 4. VETERINARY MEDICINE. *Educ:* Ohio State Univ, DVM, 49; Western Mich Univ, MA, 61. *Prof Exp:* Vet in charge, small animal hosp, Christensen Animal Hosp, Ill, 49-53; asst vet, Vet Div, Dept Vet Med, Upjohn Co, 53-63 & Int Vet Opers, 63-67, asst vet, Med Serv, 67-91; RETIRED. *Concurrent Pos:* Asst ed, Vet Med, 53. *Mem:* Am Vet Med Asn. *Mailing Add:* 2722 Carlyle Kalamazoo MI 49008

HAAS, LARRY ALFRED, b Zeeland, NDak, Nov 28, 35; m 59; c 2. PHYSICAL CHEMISTRY, CHEMICAL METALLURGY. *Educ:* Univ SDak, BS, 57; Univ Minn, MS, 64. *Prof Exp:* Res chemist, Honeywell Corp, 57-62; res chemist, 62-70, PROJ LEADER CHEM, US BUR MINES, 70- *Mem:* Am Inst Metall Engrs. *Res:* Thermodynamics, kinetics, catalysis, pyrometallurgy, hydrometallurgy, base metal sulfides and sulfur gases. *Mailing Add:* US Bur Mines 5629 Minnehaha Ave S Minneapolis MN 55417

HAAS, MARK, b New York, NY, Jan 30, 55. MEMBRANE TRANSPORT. *Educ:* Duke Univ, AB, 77, PhD(physiol) & MD, 82. *Prof Exp:* Res assoc physiol, Duke Univ Med Ctr, 83; resident physician physiol, Yale-New Haven Hosp, 83-85; fel physiol, 85-86, ASST PROF PATH & PHYSIOL, YALE UNIV SCH MED, 86- *Mem:* Biohphys Soc; Soc Gen Physiologists; Sigma Xi; Am Asn Pathologists. *Res:* The physiology and biochemistry of ion transport processes, particularly the co-transport system of red blood cell and renal epithelial cell membranes. *Mailing Add:* Dept Path Yale Univ Sch Med 310 Cedar St New Haven CT 06510

HAAS, MERRILL WILBER, b Albert, Kans, July 9, 10; m 44; c 4. PETROLEUM GEOLOGY. *Educ:* Univ Mich, BA, 32. *Prof Exp:* Paleontologist, Humble Oil & Refining Co, Tex, 33-34, Lago Petrol Corp, Venezuela, 34-38 & Standard Oil Co Venezuela, 38-41, div geologist, 41-42; div geologist, Creole Petrol Co, 42-49; area geologist, Standard Oil Co NJ, NY, 49-50; chief geologist, Carter Oil Co, Okla, 50-53, explor mgr, 53-56, dir & explor mgr, 56-57, vpres & dir, 57-58; asst coordr, Standard Oil Co NJ, NY, 58-59; vpres & dir, Carter Oil Co, Okla, 59-60; vpres, Humble Oil & Refining Co, 60-73 & Exxon USA, 73-75; PETROL CONSULT, 75- *Concurrent Pos:* Chmn manpower comt, Am Geol Inst, 60-67; chmn comt reserves & prod capacity, Am Petrol Inst, 71-74; trustee, Am Asn Petrol Geologists Found, 76- *Honors & Awards:* Sidney Powers Mem Award, Am Asn Petrol Geologists, 86. *Mem:* Hon mem Am Asn Petrol Geologists (pres, 74-75); Geol Soc Am; Paleont Res Inst (pres, 73-75). *Res:* Geology of petroleum deposits; micropaleontology; stratigraphy; structural geology. *Mailing Add:* 10910 Wickwild Houston TX 77024

HAAS, MICHAEL JOHN, b Denver, Colo, Mar 2, 64; m 86; c 2. BIOCHEMISTRY, MOLECULAR BIOLOGY. *Educ:* Univ Mo, Rolla, BS, 86; Univ Nebr, PhD(biochem), 91. *Prof Exp:* Adj fac chem, Bellevue Col, 91; GRAD STUDENT BIOCHEM, UNIV NEBR MED CTR, 86- *Mem:* AAAS. *Res:* Transcriptional and translational control of oncogene expression in both normal and tumor cells and experience in the use of RFLP's in studying the origins of cancer. *Mailing Add:* 2416 Adler Circle Middleton WI 53562

HAAS, PAUL ARNOLD, b Rolla, Mo, Aug 11, 29; m 58; c 4. CHEMICAL ENGINEERING, NUCLEAR FUEL CYCLES. *Educ:* Univ Mo, Rolla, BS, 50; Mont State Univ, MS, 51; Univ Tenn, PhD(chem eng), 65. *Prof Exp:* Develop engr, 51-57, GROUP LEADER, OAK RIDGE NAT LAB, 57- & SR RES ENGR, 70- *Mem:* Fel Am Inst Chem Engrs; Am Chem Soc; Am Nuclear Soc. *Res:* Engineering development of chemical processes for the nuclear fuel cycle, including nuclear fuel preparation, reprocessing and waste disposal. *Mailing Add:* Oak Ridge Nat Lab Bldg 4500 PO Box 2008 Oak Ridge TN 37831-6268

HAAS, PETER HERBERT, radiation physics; deceased, see previous edition for last biography

HAAS, RICHARD, b New York, NY, July 6, 29; m 52; c 2. ETHOLOGY, ZOOLOGY. *Educ:* Univ Calif, Los Angeles, AB, 50, MA, 58, PhD(zool), 69. *Prof Exp:* Teacher, pub sch, Calif, 55-64; consult, WHO, 64, 78-79, 81 & 84; from asst prof to assoc prof, 69-75, PROF BIOL, CALIF STATE UNIV, FRESNO, 75- *Concurrent Pos:* Consult, dept path, Sch Med, Univ Calif, Los Angeles, 63-64 & USAID, Sri Lanka, 83, WHO, 64-88. *Mem:* AAAS. *Res:* Immune responses to skin heterographs in fishes and amphibia; reproductive ethology of freshwater fishes; biology of cyprinodontid fishes; behavioral isolating mechanisms in sympatric fishes. *Mailing Add:* Dept Biol Calif State Univ Fresno CA 93740

HAAS, TERRY EVANS, b East Grand Rapids, Mich, Sept 25, 37; m 62; c 2. INORGANIC CHEMISTRY. *Educ:* Mich State Univ, BS, 58; Mass Inst Technol, PhD(inorg chem), 63. *Prof Exp:* Asst prof, 63-68, chmn dept, 73-79, ASSOC PROF INORG CHEM, TUFTS UNIV, 68- *Concurrent Pos:* Vis prof, Univ Alta, 69-70. *Mem:* Am Chem Soc. *Res:* Solid state ionics; thin film materials and processing; structure of coordination compounds; bioinorganic chemistry. *Mailing Add:* Dept Chem Tufts Univ Medford MA 02155

HAAS, THOMAS J, b Staten Island, NY, Mar 5, 51; m 74; c 3. HAZARDOUS MATERIALS. *Educ:* USCG Acad, BS, 73; Univ Mich, MS, 76, MS, 77; Rensselaer Polytech Inst, MS, 83; Univ Conn, PhD(chem), 87. *Prof Exp:* Br chief hazardous mat, Coast Guard Hq, Wash, DC, 77-81; from instr to asst prof, 81-88, ASSOC PROF CHEM, USCG ACAD, 88- *Concurrent Pos:* Vis prof, Univ Conn, 87-; asst dean, USCG Acad, 87-91, sect chief, 91-; Yale fel, 91-92. *Mem:* Sigma Xi; NY Acad Sci. *Res:* Synthesis, properties and reactions of 1.1 metallocenophanes which have shown to have application to solar energy production. *Mailing Add:* 15 Seabury Ave Ledyard CT 06339

HAAS, TRICE WALTER, b Dallas, Tex, July 24, 32; m 55; c 5. PHYSICAL CHEMISTRY. *Educ:* St Mary's Univ, Tex, BS, 54; Iowa State Univ, PhD(phys chem), 60. *Prof Exp:* Sr res chemist, Field Res Lab, Socony Mobil Oil Co, Tex, 60-62; sr res chemist, Nat Cash Register Co, Ohio, 62-64; res scientist, 64-71, res group leader, Aerospace Res Labs, 71-75, RES GROUP LEADER & RES FEL, WRIGHT LAB, WRIGHT-PATTERSON AFB, DAYTON, 75- *Mem:* Am Phys Soc; Sigma Xi. *Res:* Physics and chemistry of solid surfaces; low energy electron diffraction; electron spectroscopy; electronic devices and materials; secondary ion mass spectroscopy; thermionic emission; molecular beam epitaxy. *Mailing Add:* Wright Lab/MLBM Wright-Patterson AFB OH 45433-6533

HAAS, VIOLET BUSHWICK, applied mathematics, electrical engineering; deceased, see previous edition for last biography

HAAS, WARD JOHN, b New York, NY, Aug 26, 21; m 43; c 3. BIOCHEMISTRY. *Educ:* Mass Inst Technol, BS, 43, PhD, 49. *Prof Exp:* Asst biol, Mass Inst Technol, 46-49; biochemist, Grasselli Tech Dept, E I du Pont de Nemours & Co, 49-51; attache, London Embassy, US Foreign Serv, 51-54; asst to dir agr res & develop, Chas Pfizer & Co, 54-56, opers res group, 56-57, asst to pres, 57-60, dir opers, Pfizer Labs Div, 60-64; assoc prof mgt & dir space sci res ctr, Univ Mo-Columbia, 64-68; dir, Warner-Lambert Res Inst, 68-70, vpres res & develop, 69-72; vpres corp res & develop, S C Johnson & Son, Inc, 72-75; vpres res & develop, Chesebrough-Ponds Inc, 75-86; PRIN, INNOVATION MGT, 86- *Mem:* AAAS; Am Chem Soc; Asn Res Dirs; Am Inst Chem; Sigma Xi; fel AAAS; fel Am Inst Chem. *Res:* Enzymology; spectroscopy biological materials; antibiotics in nutrition; research management; administration; management of innovation. *Mailing Add:* Innovation Mgt PO Box 644 Southport CT 06490

HAAS, WERNER E L, b Magdeburg, Ger, Sept 14, 28; m 61; c 3. LIQUID CRYSTALS, ELECTROOPTICS. *Educ:* Univ Lisbon, Portugal, Licenciado, 56. *Prof Exp:* From jr eng to proj scientist, Res Div Philco-Ford Co, 58-66; assoc scientist, 66-68, scientist, 68-69, sr scientist, 69-73, prin scientist, 73-78, res fel, 78-81, MGR & SR RES FEL, CORP RES, XEROX CORP, 81- *Mem:* Fel Soc Photog Scientists & Engrs; Electrochem Soc; fel Soc Info Display. *Res:* Optical and electrooptical properties of liquid crystals; non-impact printing technologies; displays and storage devices; electrooptic effects in ferroelectrics and their applications; anodic oxidation and mechanisms of electrolytic rectification. *Mailing Add:* 768 High Tower Way Webster NY 14580

HAAS, ZYGMUNT, b Poland, Oct, 56; Israeli citizen. PERFORMANCE EVALUATION OF COMMUNICATION NETWORKS, OPTICAL NETWORKS. *Educ:* Israel Inst Technol, BSc, 79; Tel-Aviv Univ, MSc, 85; Stanford Univ, PhD(elec eng), 88. *Prof Exp:* Engr digital design, Govt Israel, 79-85; res asst computer commun, Stanford Univ, 85-88; MEM TECH STAFF, COMMUN NETWORKS, AT&T BELL LABS, 88- *Concurrent Pos:* Lectr digital design & computer commun, Tel-Aviv Univ, Israel, 82-85. *Mem:* Sr mem Inst Elec & Electronics Engrs. *Res:* Performance evaluation of computer and communication networks, especially optical networks; interconnection networks and switches; multi-media application of traffic characterization and traffic integration; very high-speed networks; high-speed and high-performance protocols, especially transport protocols. *Mailing Add:* AT&T Bell Labs Rm 4F-501 Holmdel NJ 07733-1988

HAASE, ASHLEY THOMSON, b Evanston, Ill, Dec 8, 39; m 62; c 3. VIROLOGY. *Educ:* Lawrence Col, BA, 61; Columbia Col Physicians & Surgeons, MD, 65. *Prof Exp:* Intern, Osler Serv, Johns Hopkins Hosp, 65-66, resident, 66-67; clin assoc & sr clin assoc, NIH, 67-70; vis scientist, Nat Inst Med Res, London, 70-71; from asst prof to prof, Dept Med, Univ Calif, San Francisco, 71-84, Dept Microbiol, 74-84; PROF MICROBIOL & HEAD DEPT, MED SCH, UNIV MINN, 84- *Concurrent Pos:* Chief, infectious dis, Vet Admin Med Ctr, San Francisco, 71-84, clin investr, 71-78, med investr, 78-84; prof, dept med, Univ Minn, 84-; Anna Fuller Found fel; mem adv coun, Nat Inst Allergy & Infectious Dis, NIH, 86-90; chmn AIDS panel, US-Japan Coop Med Sci Progr, 86-91; ed-in-chief, Virol, Microbiol Pathogenesis, 88-; gov coun biotechnol, Minn, 85-87; investr, Javits Neurosci, 88-; merit investr, Nat Inst Allergy & Infectious Dis, 89- *Mem:* Am Soc Microbiol; Am Soc Virol; Infectious Dis Soc Am; Am Soc Clin Investrs; Sigma Xi; Asn Am Physicians. *Res:* The molecular mechanisms of viral pathogenicity as they relate to diseases of the central nervous system (for example, Alzheimer's disease, multiple sclerosis and acquired immune deficiency syndrome). *Mailing Add:* Dept Microbiol Med Sch Univ Minn Box 196 UMHC 420 Del S E Minneapolis MN 55455

HAASE, BRUCE LEE, b Napoleon, Ohio, Dec 18, 38; m 73. AQUATIC ECOLOGY, FISHERIES. *Educ:* Concordia Teachers Col, BS, 61; Bowling Green State Univ, MA, 62; Univ Wis, PhD(zool), 69. *Prof Exp:* Teacher, St Peters Lutheran Sch, 59-60; instr biol, Milwaukee Lutheran High Sch, 62-64; assoc prof, 69-74, PROF BIOL, EAST STROUDSBURG UNIV, 74- *Concurrent Pos:* Consult, Monroe County Planning Comn, 70-; pres, Marine Sci Consortium. *Mem:* Am Fisheries Soc; Am Soc Limnol & Oceanog. *Res:* Flocking behavior of the common crow; life history of the longnose gar; ecology of selected fishes; general ecology of small, shallow lakes; ecology of acquatic insects; marine and freshwater plankton. *Mailing Add:* Dept Biol East Stroudsburg Col East Stroudsburg PA 18301

HAASE, DAVID GLEN, b Baton Rouge, La, Dec 6, 48; m 73; c 1. LOW TEMPERATURE PHYSICS. *Educ:* Rice Univ, BA, 70; Duke Univ, MA, 72, PhD(physics), 75. *Prof Exp:* Res asst physics, Duke Univ, 70-75; from asst prof to assoc prof physics, 76-87, PROF PHYSICS, NC STATE UNIV, 87- *Concurrent Pos:* Staff, Triangle Univs Nuclear Lab, 88-; vis prof physics, LA State Univ, 88; vis asst prof, NC State Univ, 75-76. *Mem:* Am Phys Soc; Sigma Xi; Am Asn Physics Teachers. *Res:* Properties of quantum solids; cryogenic polarized targets. *Mailing Add:* Dept Physics NC State Univ Raleigh NC 27695-8202

HAASE, DONALD J(AMES), b Gillett, Tex, Apr 18, 38. CHEMICAL ENGINEERING. *Educ:* Univ Tex, Austin, BS, 60, MS, 62, PhD(chem eng), 65. *Prof Exp:* Res scientist, Continental Oil Co, Okla, 64-67; sr chem engr, Tenneco Chem Inc, 67-70, group leader, 70-77, mgr cosorb technol, 77-78, dir separations technol, 80-83; pres, Conejo Chem Co, 83-86; CHIEF EXEC OFFICER, HERR HAASE INC, 86- *Mem:* Am Chem Soc; Am Inst Chem Engrs. *Res:* Reaction kinetics of high temperature partial oxidation systems; gas-solid purification processes and the regeneration of the substrates; applications of digital computers to control high temperature oxidation reactions; general separations technology; chemical separations technologies; research and technology management. *Mailing Add:* HCR 2 Box 48 Nixon TX 78140-9207

HAASE, EDWARD FRANCIS, b Milwaukee, Wis, Apr 29, 37; m 65; c 3. BOTANY, PLANT ECOLOGY. *Educ:* Marquette Univ, BS, 59; Univ Wis-Milwaukee, MS, 65; Univ Ariz, PhD(bot), 69. *Prof Exp:* Res ecologist, Southwest Watershed Res Ctr, Agr Res Serv, 69-70; res assoc & asst prof arid land resources, Off Arid Lands Studies, Univ Ariz, 70-76; head, Dept Smoke Invest, 76-79, land use & reclamation coordr, 79-85, SR ENVIRON ANALYST, PHELPS DODGE CORP, 85- *Mem:* AAAS; Air & Waste Mgt Asn. *Res:* Arid lands plant ecology; ecology; utilization of native desert plants; air pollution effects on vegetation; reclamation of surface mined areas; environmental regulation effects on copper mining, milling, smelting and refining. *Mailing Add:* 13420 N 82nd St Scottsdale AZ 85260

HAASE, GUNTER R, b Chemnitz, Ger, Sept 30, 24; US citizen; m 49; c 4. NEUROLOGY. *Educ:* Univ Munich, MD, 49. *Prof Exp:* Assoc prof neurol, Sch Med, Univ Okla, 60-64; prof, Sch Med, Temple Univ, 64-73; prof neurol, Univ Pa, 74-88; AT PENN HOSP, 88- *Concurrent Pos:* Consult, St Christopher's Hosp Children, 64-73; sr attend physician, Philadelphia Gen Hosp, 64-70; mem training comt A, Nat Inst Neurol Dis & Blindness, 67-71. *Mem:* AMA; Am Acad Neurol. *Res:* Clinical neurology; muscular diseases. *Mailing Add:* Penn Hosp Eight Spruce St Philadelphia PA 19107

HAASE, JAN RAYMOND, b Buffalo, NY, June 24, 41; m 63; c 2. PHOTOGRAPHIC CHEMISTRY. *Educ:* State Univ NY Buffalo, BA, 63; NY Univ, PhD(org chem), 68. *Prof Exp:* Chemist, Tech Serv Lab, Union Carbide Corp, 63-64; LAB HEAD ORG CHEM, EASTMAN KODAK CO RES LABS, 67- *Mem:* Am Chem Soc. *Res:* Design and synthesis of organic image forming materials for color photographic applications with emphasis on image dyes. *Mailing Add:* Nine White Birch Circle Rochester NY 14624

HAASE, OSWALD, b Berlin, Ger, Oct 4, 25. SOLID STATE PHYSICS. *Educ:* Univ Hamburg, BSc, 52, MSc, 55, PhD(physics), 60. *Prof Exp:* Mem tech staff res physics, Bell Tel Labs, Inc, 60-62; PROF PHYSICS, FAIRLEIGH DICKINSON UNIV, 64- *Res:* Electron diffraction; thin films and surfaces; epitaxy; oxidation; electron microscopy of lattice defects; x-ray; structure of liquids; diffraction contrast of dislocations in crystals; corrosion; applied optics, light scattering in glass. *Mailing Add:* Dept Physics Fairleigh Dickinson Univ Teaneck NJ 07666

HAASE, RICHARD HENRY, b Cleveland, Ohio, Feb 27, 24; m 51; c 3. APPLIED STATISTICS. *Educ:* Purdue Univ, BS, 45; Tulane Univ, MBA, 49; Univ Calif, Los Angeles, PhD(eng), 61. *Prof Exp:* Design engr, Convair Corp, Calif, 46-47; from appln engr to dist engr, Lamp Div, Gen Elec Co, 47-51; lectr & asst head eng exten, Univ Calif, Los Angeles, 51-61; mem tech staff, Rand Corp, 61-65; systs engr, Missile & Space Div, Gen Elec Co, 65-67; prof statist & chmn dept, Drexel Univ, 67-87; RETIRED. *Concurrent Pos:* Consult, Missile & Space Div, Gen Elec Co, Pa, Ford Motor Co, Campbell Soup Co, Mrs Smith's Frozen Foods, 67- *Mem:* Am Inst Decision Sci; Sigma Xi. *Res:* Applied statistics and decision theory as applied to management decision making in business; innovative education; statistical process control. *Mailing Add:* 3422 SW Bobalink Way Palm City FL 34990-2617

HAASER, NORMAN BRAY, b Hartford, Conn, July 8, 17; m 45; c 1. APPLIED MATHEMATICS. *Educ:* Univ Notre Dame, BS, 43, MS, 48; Brown Univ, PhD(appl math), 50. *Prof Exp:* Asst prof, 50-57, ASSOC PROF MATH, UNIV NOTRE DAME, 57- *Concurrent Pos:* Consult, advan develop labs, Bendix Aviation Corp. *Mem:* Am Math Soc; Math Asn Am. *Res:* Real analysis; numerical analysis; fluid dynamics. *Mailing Add:* PO Box 398 Notre Dame IN 46556

HAAVIK, CORYCE OZANNE, b St Paul, Minn, Sept 20, 33. PHARMACOLOGY. *Educ:* Bryn Mawr Col, BA, 54; Univ Wis, PhD(pharmacol), 65. *Prof Exp:* Res assoc, Sch Med, Yale Univ, 54-57; res asst, Univ Tex Southwestern Med Sch, Dallas, 57-59; instr pharmacol, Sch

Pharm, Univ Wis, 64-65, asst prof, 65-71; asst prof, 71-77, ASSOC PROF PHARMACOL, MED COL WIS, 78-, ASSOC DEAN RES & GRAD STUDIES, 78- *Mem:* AAAS; Am Soc Pharmacol & Exp Therapeut. *Res:* Cardiovascular and hypothermic effects of tetrahydracannabinol and analogs. *Mailing Add:* Med Col Wis 8701 Watertown Plank Rd Milwaukee WI 53226

HABAL, MUTAZ, b Damascus, Syria, Apr 27, 38; US citizen; m 64; c 2. SURGICAL ONCOLOGY, TRANSATOLOGY. *Educ:* Am Univ Beirut, Lebanon, BS, 59; Am Univ Med, Beirut, MD, 64; Suny Buffalo, MD, 69; Harvard Univ, MD, 72. *Hon Degrees:* Royal Col Surgeons, Can, FRCSC, 74. *Prof Exp:* Asst Prof Surg, Clin Ctr Ind, 72-74; assoc prof surg, Cin develop Fla, 74-78; prof surg, 78-80, RES PROF & CLIN PROF SURG, UNIV S FLA, 80- *Mem:* Am Col Surgeons; Am Acad Pediat; Am Cleft Palate Assoc; Soc Univ Surgeons; Am Soc Plastic Surgeons; Sigma Xi. *Res:* Bone graft induction, utilizing the basic principles for major reconstruction; clinical investigation, the long term follow up research. *Mailing Add:* 6358 MacLaurin Dr Tampa FL 33647-1164

HABASHI, FATHI, b Minia, Egypt, Oct 9, 28; m 58, 82; c 2. METALLURGY. *Educ:* Univ Cairo, BSc, 49; Vienna Tech Univ, Dr Tech(inorg chem), 59. *Prof Exp:* Supvr, Fertilizer Factory, Suez, UAR, 49-52 & Misr Cotton Mills, Mahalla Kobra, 52-53; asst dir, Munic Chem Lab, Alexandria, 53-56; res fel chem, Vienna, 59-60; fel, Dept Mines & Tech Surv, Ottawa, 60-62; asst ed, Chem Abstr, 63-64; assoc prof metall, Mont Sch Mines, 64-67; sr res engr, Extractive Metall Res Div, Anaconda Co, Ariz, 67-70; PROF METALL, LAVAL UNIV, 70- *Concurrent Pos:* vis prof, USSR Acad Sci, 76, Nat Autonomous Univ Mexico City, 80, Royal Inst Technol, Stockholm, 81, Brazilian Nat Coun Sci & Tech Develop, 82, Acad Sinica Beiging, 84, Tech Univ Oruro, Bolivia, 86, Higher Inst Chem Technol, Sfia, Bulgaria, 87, Cent SUniv Tech, Changsha, China; hon prof, Tech Univ Oruro, Bolivia, 86; guest lectr, Mining Res Ctr, Havana, 87; consult, UN develop prog, Cuban Laterite Proj, 89. *Mem:* Am Chem Soc; Can Inst Mining & Metall; Am Inst Mining, Metall & Petrol Engrs. *Res:* Extractive metallurgy; extraction of uranium and lanthanides from phosphate rock; kinetics of metallurgical processes; chalcopyrite chemistry and metallurgy; mineral resources of Arab countries; chemical reactions of asbestos; history of metallurgy. *Mailing Add:* Dept Mining & Metall Univ Laval Cite Univ Quebec PQ G1K 7P4 Can

HABASHI, WAGDI GEORGE, b Egypt, June 29, 46; Can citizen; m 67; c 3. COMPUTATIONAL FLUID DYNAMICS, TRANSONIC AERODYNAMICS & TURBOMACHINERY. *Educ:* McGill Univ, BE, 67, ME, 69; Cornell Univ, PhD(aerospace), 75. *Prof Exp:* Asst prof mech eng, Stevens Inst Technol, 74-75; from asst prof to assoc prof, 75-84, PROF MECH ENG, CONCORDIA UNIV, 84- *Concurrent Pos:* Aerodyn consult, Pratt & Whitney, Can, 77- *Honors & Awards:* E W R Steacie Mem Fel, Sci Achievement, Natural Sci & Eng Res Coun, Can, 88-89; Cray Gigaflop Award, 90. *Mem:* Am Inst Aeronaut & Astronaut; Am Soc Mech Engrs; Sigma Xi. *Res:* Numerical methods for the analysis and design of aerodynamic components in transonic flow; design of transonic turbomachinery. *Mailing Add:* Mech Eng Dept Concordia Univ 1455 de Maisonneuve Blvd W Montreal PQ H3G 1M8 Can

HABBAL, SHADIA RIFAI, b Damascus, Syria, Sept 30, 48; m 70; c 2. SPACE PLASMA PHYSICS. *Educ:* Damascus Univ, BS(math) & BS(physics), 70; Am Univ Beirut, MS, 73; Univ Cincinnati, PhD(physics), 77. *Prof Exp:* Fel solar physics, Nat Ctr Atmospheric Res, 77-78; res fel physics, Harvard Col Observ, 78-79, res assoc, 79-82; PHYSICIST, SMITHSONIAN ASTROPHYS OBSERV, 82- *Mem:* Am Phys Soc; Am Geophys Union; Asn Women Sci; Am Astron Soc; Int Astron Union. *Res:* Solar physics; space plasma physics. *Mailing Add:* Ctr Astrophys 60 Garden St Cambridge MA 02138

HABECK, DALE HERBERT, b Bonduel, Wis, Oct 21, 31; m 59; c 1. ENTOMOLOGY. *Educ:* Univ Wis, BS, 53, MS, 54; NC State Univ, PhD(entom), 59. *Prof Exp:* Asst entomologist, Exp Sta, Univ Hawaii, 59-63; from asst entomologist to assoc entomologist, 63-73, PROF ENTOM, UNIV FLA, 73- *Mem:* Entom Soc Am. *Res:* Biology and immature stages of Lepidoptera; biological control of weeds. *Mailing Add:* Dept Entom Univ Fla Gainesville FL 32611

HABECK, JAMES ROBERT, b Ashland, Wis, June 3, 32; m 57; c 2. ECOLOGY. *Educ:* Univ Wis, BS, 54, MA, 56, PhD(bot), 59. *Prof Exp:* Instr biol, Ore Col Educ, 59-60; from asst prof to assoc prof, 60-69, PROF BOT, UNIV MONT, 69- *Mem:* Ecol Soc Am. *Res:* Phytosociology and dendrochronology; fire ecology; planning uses for fire in forest and wildlife range systems; designing and fulfilling a research natural area system; ecological baseline monitoring. *Mailing Add:* Dept Bot Div Biol Sci Univ Mont Missoula MT 59812-1002

HABEEB, AHMED FATHI SAYED AHMED, b Gerga, Egypt, Apr 10, 28; US citizen. PROTEIN CHEMISTRY, IMMUNOLOGY. *Educ:* Univ Cairo, BPharm, 48; Univ London, PhD(pharmaceut), 54, DSc(protein struct & immunol), 76. *Prof Exp:* Fel chem, Yale Univ, 56-58; fel biol, Div Appl Biol, Nat Res Coun Can, 58-60; sr res asst microbiol, Connaught Med Res Labs, Univ Toronto, 60-62; sr cancer res scientist clin biochem, Roswell Park Mem Inst, 62-65; asst prof microbiol, Univ Tenn, 65-68; asst prof, Lab Immunol, St Jude Children's Hosp, 65-68; assoc prof microbiol, Med Ctr, Univ Ala, Birmingham, 68-77; PROF BIOCHEM, MED SCI CAMPUS, UNIV PR, 77- *Mem:* NY Acad Sci; Am Asn Immunol; Am Soc Biol Chemists; Am Soc Trop Med Hyg. *Res:* Protein structure, isolation and characterization; chemical modification of proteins, antibodies and toxins; Clostridium perfringens antigens; immunochemical studies on S mansoni. *Mailing Add:* Dept Biochem & Nutrit Med Sci Campus Univ PR GPO Box 5067 San Juan PR 00936

HABEL, ROBERT EARL, b Toledo, Ohio, Aug 8, 18; m 42; c 2. VETERINARY ANATOMY. *Educ:* Ohio State Univ, DVM, 41, MSc, 47; Univ Utrecht, MVD(vet anat), 56. *Prof Exp:* Vet, Bur Animal Indust, USDA, Pa, 41-42; county vet, Fulton County Bd Health, Ohio, 42; instr vet embryol, histol & anat, Ohio State Univ, 46-47; from asst prof to prof, 47-78, head dept, 60-75, EMER PROF VET ANAT, NY STATE COL VET MED, CORNELL UNIV, 78- *Concurrent Pos:* Vchmn, Int Comt Vet Anat Nomenclature, 63-80 & chmn, 80-86; fel, Nat Libr Med, NIH, Anat Inst, Vet Col, Vienna, 67-68 & 75-76; vis prof vet anat, Univ Utrecht, 79 & Univ Sydney, 81. *Mem:* Am Vet Med Asn; Am Asn Anatomists; Am Asn Vet Anatomists (pres, 64-65); World Asn Vet Anatomists (pres, 71-75). *Res:* Applied veterinary anatomy; ruminant anatomy; anatomy of the digestive system; microscopic anatomy of domestic animals; comparative medical anatomy. *Mailing Add:* Dept Anat NY State Col Vet Med Ithaca NY 14853

HABENER, JOEL FRANCIS, b Indianapolis, Ind, June 29, 37; m 62. ENDOCRINOLOGY. *Educ:* Univ Redlands, BS, 60; Univ Calif, Los Angeles, MD, 65; Am Bd Internal Med, dipl. *Prof Exp:* From intern to asst resident med & fel, Johns Hopkins Hosp, 65-67; res assoc endocrinol & surgeon, NIH, 67-69; res fel, 69-71, from instr to assoc prof, 71-89, PROF MED, HARVARD MED SCH, 89- *Concurrent Pos:* Assoc physician, Mass Gen Hosp, 76-, chief, Lab Molecular Endocrinol; USPHS res career develop award, 72-75; investr, Howard Hughes Med Inst, 76- *Mem:* Am Fedn Clin Res; Endocrine Soc; Am Soc Clin Invest; Am Soc Biol Chemists; Am Soc Bone & Bone Mineral Res. *Res:* Investigations of the molecular biochemistry of peptide hormone synthesis and secretion. *Mailing Add:* Lab Molecular Endocrinol Mass Gen Hosp Boston MA 02114

HABENSCHUSS, ANTON, b Ruma, Yugoslavia, June 1, 44; US citizen. PHYSICAL CHEMISTRY. *Educ:* Kent State Univ, BS, 66; Iowa State Univ, PhD(phys chem), 73. *Prof Exp:* Res assoc chem, Iowa State Univ, Ames, 73-78; res assoc chem eng, Univ Del & Oak Ridge Nat Lab, 78-80, RES SCIENTIST, OAK RIDGE ASSOC UNIV, 80- *Mem:* Am Chem Soc; Sigma Xi; AAAS; NY Acad Sci. *Res:* Thermodynamics; transport properties; x-ray diffraction; aqueous electrolyte solutions; rare earth solution chemistry; structure of simple fluids; synchrotron radiation. *Mailing Add:* Oak Ridge Nat Lab Box X Oak Ridge TN 37830

HABER, ALAN HOWARD, b Chicago, Ill, Apr 1, 30; m 56; c 4. RADIATION BOTANY, PLANT PHYSIOLOGY. *Educ:* Calif Inst Technol, BS, 53; Univ Wis, PhD(bot), 56. *Prof Exp:* Asst, Univ Wis, 53-56; biologist, Oak Ridge Nat Lab, Tenn, 56-73; chmn dept, 73-83, PROF BIOL SCI, STATE UNIV NY, BINGHAMTON, 73-; MASTER HINMAN COL, 85- *Concurrent Pos:* Proj mgr, Elec Power Res Inst, 79-80. *Mem:* AAAS; Radiation Res Soc. *Res:* Radiation biology. *Mailing Add:* Dept Biol Sci State Univ NY Binghamton NY 13901

HABER, BERNARD, b Lodz, Poland, July 20, 34; US citizen; m 59. BIOCHEMISTRY. *Educ:* McGill Univ, BSc, 56, MSc, 57, PhD(biochem), 62. *Prof Exp:* Res assoc neuropharmacol, Res Lab, Galesburg State Res Hosp, Ill, 62-64; assoc res scientist neurochem, Div Neurosci, City of Hope Med Ctr, Calif, 65-70; asst biochem & neurol, 70-72, ASSOC PROF HUMAN BIOL CHEM, GENETICS & NEUROL, UNIV TEX MED BR GALVESTON, 72-, CHIEF NEUROCHEM SECT, DIV COMP NEUROBIOL, MARINE BIOMED INST, 70- *Concurrent Pos:* Ed-in-chief, J Neurosci Res, 79- *Mem:* AAAS; Am Soc Neurochem; Int Soc Neurochem; Am Soc Biol Chem; Int Brain Res Orgn; Sigma Xi. *Res:* Transport of amino acids and sugars; biogenic amines, purification of brain enzymes related to neurotransmitters and their localization in situ; tissue culture of neurons and glia; cellular communication; muscular distrophy, tumor neovascularization; biochemistry of human head injury; mechanisms of central excitation and inhibition. *Mailing Add:* Marine Biomed Inst Univ Tex 200 University Blvd Galveston TX 77550

HABER, EDGAR, b Berlin, Ger, Feb 1, 32; US citizen; m 58; c 3. IMMUNOLOGY, CARDIOVASCULAR DISEASES. *Educ:* Columbia Univ, AB, 52, MD, 56; Am Bd Internal Med, dipl. *Hon Degrees:* AM, Harvard Univ, 68. *Prof Exp:* From instr to prof, 63-88, CLIN PROF MED, HARVARD MED SCH, 90-; PHYSICIAN, MASS GEN HOSP, 69-; PRES, BRISTOL-MYERS SQUIBB PHARM RES INST, 90- *Concurrent Pos:* Lab res with C Anfinsen, NIH, 58-61; hon clin asst, Cardiac Dept, St George's Hosp & Nat Heart Hosp, London, Eng, 62-63; asst, Mass Gen Hosp, Boston, 63-64 chief cardiac Unit , 64-88 & asst physician, 65-68; mem study sect allergy & immunol, NIH, 65-68; fel coun clin cardiol, Am Heart Asn, 66, mem prog , 66-68, mem exec comn, 68-76, mem res study comn physiol & pharm, 69-72, mem res comt, 70-76 & chmn adult & pediat cardiol res study comn, 70-72, vpres res & chmn res comt, 73-74; vchmn panel on heart & blood vessel dis, Nat Heart Act of 72, NIH, 72-73; mem task force on immunol & dis, Nat Inst Allergy & Infectious Dis, 72-73; mem tissue & organ biol interdisciplinary cluster, President's Biomed Res Panel, 75; mem US deleg, US-USSR Health Exchange; mem ed bd, Clin Immunol & Immunopath, 71-89, Herz, 80-, Hybridoma, 80-, J Hypertense, 82-88; vis prof, Univ Tex, Dallas, 82; Univ Queensland, Brisbane, 85; Stanford Univ, 85; Univ Calif Los Angles, 87 & State Univ NY, Syracuse, 87; mem, Working Group Arteriosclerosis, Nat Heart, Lung & Blood Inst, 78-81 & Comt Frontiers in Basic Sci, 86-; trustee, Life Sci Res Found, 88- *Honors & Awards:* Franz Volhard Award, Int Soc Hypertension, 80; Otsulca Award Outstanding Res, Int Soc Heart Res, 85; Res Achievement Award, Am Heart Asn, 86; CIBA Award Hypertension Res, Coun High Blood Pressure Res, Am Heart Asn, 89; Distinguished Scientist Award, Am Col Cardiol, 91; Various Named lectr from various univs & asns, 73-90. *Mem:* Am Soc Clin Invest; Asn Am Physicians; Asn Univ Cardiol (vpres, 79-80, pres, 80-81); Am Asn Immunol; London Royal Soc Med; fel AAAS; Am Soc Biol Chem; fel Am Acad Arts & Sci. *Res:* Protein chemistry; immunochemistry; cardiology. *Mailing Add:* Cardiac Unit Mass Gen Hosp Boston MA 02114

HABER, FRED, b New York, NY, July 1, 21; m 48; c 2. ELECTRICAL ENGINEERING. *Educ:* Pa State Univ, BS, 48; Univ Pa, MS, 53, PhD(elec eng), 60. *Prof Exp:* Elec engr, Arma Corp, 48 & Radio Corp Am, 48-51; from instr to prof, 51-87, EMER PROF ELEC ENG, MOORE SCH ELEC ENG, UNIV PA, 87- *Concurrent Pos:* Elec engr, Gen Precision, Inc, 62-63; vis prof, Pahlavi Univ, Shiraz, Iran, 68, Eindhoven Univ Technol, Neth, 83-84, Delft Univ Technol, Neth, 87-88. *Mem:* Fel Inst Elec & Electronics Engrs; Sigma Xi. *Res:* Electrical communications theory; modulation systems; statistical methods; interference; effects of electrical noise; sensor array processing. *Mailing Add:* 210 Locust St Apt 8AW Philadelphia PA 19106

HABER, HOWARD ELI, b Brooklyn, NY, Feb 3, 52. THEORETICAL ELEMENTARY PARTICLE PHYSICS. *Educ:* Mass Inst Technol, SB & SM, 73; Univ Mich, PhD(physics), 78. *Prof Exp:* Comput programmer, Math Lab, Educ Res Ctr, Mass Inst Technol, 72-73; teaching & res physics, Univ Mich, 73-78; fel theoret physics, Lawrence Berkeley Lab, 78-80 & Univ Pa, 80-82; AT DEPT PHYSICS, UNIV CALIF, SANTA CRUZ, 82- *Concurrent Pos:* Dept energy, outstanding jr investr, 85- *Mem:* Am Phys Soc; Am Asn Physics Teachers; Sigma Xi. *Res:* High energy physics, theory and phenomenology, especially models of weak interactions and grand unification phenomenology of Higgs particles; perturbative quantum chromodynamics; field theory at finite temperature and density. *Mailing Add:* Dept Physics Univ Calif Santa Cruz CA 95064

HABER, JAMES EDWARD, b Pittsburgh, Pa, Feb 23, 43; m 65; c 2. MOLECULAR GENETICS. *Educ:* Harvard Col, AB, 65; Univ Calif, Berkeley, PhD(biochem), 69. *Prof Exp:* Asst prof, 72-77, assoc prof, Dept Biol & Rosenstiel Basic Med Sci Res Ctr, 77-80, PROF BIOL, BRANDEIS UNIV, 80- *Concurrent Pos:* NSF fel, 70. *Mem:* Am Soc Microbiol; Genetics Soc Am; Am Soc Biol Chemists; Am Soc Cell Biol. *Res:* Genetic and biochemical controls of meiosis and cell differentiation, in the yeast Saccharomyces cerevisiae; control of mating type interconversion; regulation of nucleic acid and protein synthesis. *Mailing Add:* Rosenstiel Ctr Brandeis Univ Waltham MA 02254

HABER, MERYL H, b Cleveland, Ohio, Dec 28, 34; wid; c 3. PATHOLOGY. *Educ:* Northwestern Univ, BS, 56, MS, 58, MD, 59; Am Bd Path, dipl anat, path, 64 & clin path, 65. *Prof Exp:* Instr path, Med Sch, Northwestern Univ, 63-66; from assoc prof to prof, Sch Med, Univ Hawaii, Manoa, 66-73; dir continuing med educ, Sch Med, Univ Nev, Reno, 74-76, prof lab med & chmn dept, 73-80, med dir, Med Technol Prog, 73-80; consult labs, Vet Admin Hosp, 73-80; prof clin path, 81-85, UNIV LECTR, NORTHWESTERN UNIV MED SCH, 85-; PROF PATH, RUSH UNIV, 82- *Concurrent Pos:* USPHS fel path, Northwestern Univ, 60-64; USPHS fel, Univ Col, Univ London & Wellcome Assoc, Royal Soc Med, 61-62; head, dept anat path, Passavant Mem Hosp, Chicago, 64-66; dir labs, St Francis Hosp, Honolulu, 66-73; vchmn adv coun, Am Soc Clin Pathologists, 72-77, chmn, 77-, dep comnr for regional progs, 73-; dir, CME, Rush Univ, 87- *Mem:* AAAS; Col Am Path. *Res:* Renal disease; urinalysis. *Mailing Add:* Dept Path Rush Univ Med Ctr 600 S Paulina Chicago IL 60612

HABER, ROBERT MORTON, b Cleveland, Ohio, June 8, 32. MATHEMATICS. *Educ:* Ohio State Univ, BSc, 53, MA, 55, PhD(math), 58. *Prof Exp:* From instr to asst prof math, Univ Ill, 58-61; asst prof, Case Inst Technol, 61-65; ASSOC PROF MATH, WRIGHT STATE UNIV, 65- *Res:* Combinatorial analysis. *Mailing Add:* 1422 Sam Ctr Rd Apt 201 Cleveland OH 44124

HABER, SEYMOUR, b Brooklyn, NY, Oct 7, 29; m 54; c 4. NUMERICAL ANALYSIS. *Educ:* Yeshiva Univ, BA, 50; Syracuse Univ, MA, 51; Mass Inst Technol, PhD(math), 54. *Prof Exp:* Asst, Syracuse Univ, 50-51 & Mass Inst Technol, 51-54; res assoc math, Courant Inst Math Sci, NY Univ, 54-55; asst prof, Bar-Ilan Univ, Israel, 55-56; mathematician, Weizmann Inst, 56-57; from instr to asst prof math, Polytech Inst Brooklyn, 57-60; mathematician, Nat Bur Standards, 60-87, PROF, TEMPLE UNIV, PHILADELPHIA, PA, 87- *Concurrent Pos:* Vis Prof, Univ Md, Baltimore County, 70-71 & Temple Univ, Philadelphia, 83-84; Fulbright-Hayes sr fel, Hebrew Univ Jerusalem, 73-74. *Mem:* Am Math Soc; Soc Indust & Appl Math; Math Asn Am; Hist Sci Soc. *Res:* Numerical quadrature; numerical analysis; function theory. *Mailing Add:* 418 Levering Mill Rd Merion PA 19066

HABER, SONJA B, b New York, NY, Nov 7, 51. HUMAN FACTORS, EXPERIMENTAL PSYCHOLOGY. *Educ:* State Univ NY, Binghamton, BA, 72; Miami Univ, MA, 75, PhD(psychol), 76. *Prof Exp:* Res assoc pop genetics, Brookhaven Nat Lab, 76-78, asst scientist, 78-80, assoc scientist, 80-84, SCIENTIST, BROOKHAVEN NAT LAB, 84- *Concurrent Pos:* Adj asst prof, State Univ NY, Stony Brook, 78-81; adj assoc prof, Miami Univ, 81-; health scientist adminr, Nat Heart, Lung & Blood Inst, NIH, 81-82. *Mem:* Behav Genetics Asn; Soc Study Social Biol (secy, 77-81, treas, 78-81); Sigma Xi; AAAS. *Res:* Behavioral science; behavioral factors in nuclear safety; organizational behavior; methodology and quantitative analysis in experimental psychology. *Mailing Add:* Dept Nuclear Energy Brookhaven Nat Lab Bldg 130 Upton NY 11973

HABER, STEPHEN B, b Washington, DC, June 18, 50; m 76; c 2. IMMUNOPHARMACEUTICALS, MEDICINAL CHEMISTRY. *Educ:* Stanford Univ, BS, 72; Mass Inst Technol, PhD(org chem), 76. *Prof Exp:* Res fel, Syntex Res, 76-78; res chemist, 78-82, group leader, 82-84, res supvr, 84-86, RES MGR, E I DU PONT DE NEMOURS & CO, INC, 86- *Mem:* Am Chem Soc; NY Acad Sci; Sigma Xi; Soc Nuclear Med. *Res:* Radiolabeled monoclonal antibodies for diagnosis and therapy; organic synthesis; anti-inflammatories; beta-lactam antibiotics. *Mailing Add:* Du Pont Merck Pharmaceut Co 331 Treble Cove Rd 200-2 North Billerica MA 01862

HABERFELD, JOSEPH LENNARD, b Cleveland, Ohio, Jan 17, 45; div; c 2. MICROSCOPY, POLYMER CHARACTERIZATION. *Educ:* Ohio Univ, BS, 66; Univ Calif, MS, 68; Univ Conn, PhD(polymer chem), 74. *Prof Exp:* Fel, Inst Mat Sci, Univ Conn, 74-75; res chemist, Uniroyal Inc, 75-85, Akzo Coatings, 86-90; RES CHEMIST, L J BROUTMAN & ASSOC, 90- *Mem:* Am Chem Soc; Sigma Xi; NY Micros Soc. *Res:* Development of analytical techniques to characterize and improve rubber and polymeric materials. *Mailing Add:* 1143 S Plymouth Ct No 301 Chicago IL 60605

HABERFIELD, PAUL, b Carnuntum, Czech, May 29, 33; US citizen; m 66; c 4. ORGANIC CHEMISTRY. *Educ:* Mass Inst Technol, BS, 55; Univ Calif, Los Angeles, PhD(chem), 60. *Prof Exp:* Fel, Purdue Univ, 60-61; from instr to assoc prof, 61-72, PROF CHEM, BROOKLYN COL, 73- *Mem:* AAAS; Am Chem Soc. *Res:* Solute-solvent interactions in ground and excited states; proximate charge effects on mechanism and equilibria, calorimetry, phototropic molecules. *Mailing Add:* Dept Chem Brooklyn Col Brooklyn NY 11210

HABERLAND, MARGARET ELIZABETH, MACROPHAGES ATHEROGENESIS, LIPOPROTEIN METABOLISM. *Educ:* Case Western Reserve Univ, PhD(biochem), 71. *Prof Exp:* ASST PROF MED & BIOCHEM, UNIV CALIF, LOS ANGELES, 80- *Mailing Add:* Dept Med Univ Calif Med Ctr Los Angeles CA 90024

HABERMAN, CHARLES MORRIS, b Bakersfield, Calif, Dec 10, 27. MECHANICAL & AERONAUTICAL ENGINEERING. *Educ:* Univ Calif, Los Angeles, BS, 51; Univ Southern Calif, MS, 54, MechE, 57, MS, 61. *Prof Exp:* Group engr in charge comput applns, Northrop Aircraft Corp, 51-59; from asst prof to prof eng, 59-69, PROF MECH ENG, CALIF STATE UNIV, LOS ANGELES, 69- *Concurrent Pos:* Consult, Northrop Aircraft Corp, 59-61, Rockwell, 62-63, Lockheed, 66 & Aerospace Corp, 78. *Mem:* Am Soc Eng Educ; Am Inst Aeronaut & Astronaut. *Res:* Thermoheat transfer; applied mechanics; aeronautics; computer applications. *Mailing Add:* Dept Mech Eng 5151 State College Dr Los Angeles CA 90032

HABERMAN, HERBERT FREDERICK, b Toronto, Ont, Apr 29, 34; m 67; c 3. DERMATOLOGY. *Educ:* Univ Toronto, MD, 59; FRCPCan, 64; Am Acad Dermat, dipl, 64. *Prof Exp:* Intern & resident med, Univ Toronto, 59-61; resident and res fel dermat, McGill Univ, NY Univ & Montreal Gen Hosp, 61-65; res assoc dermat, 65-66, asst prof med, 67-72, ASSOC PROF MED, UNIV TORONTO, 72-; ACTIVE ATTEND STAFF DERMAT, TORONTO WESTERN HOSP, 65-, DIR DIV DERMAT, 75- *Concurrent Pos:* Grants, Med Res Coun, 66, Ont Cancer Treatment & Res Found, 65, Nat Cancer Inst Can, 70 & Can Geriat Res Found, 70; Ont Cancer Treat & Res Found; dir, Sexually Transmitted Dis Clin & Dermat, Toronto Western Hosp; consult, Addiction Res Found, 74- & Metrop Toronto Homes for Aged, 74-; assoc, Ont Cancer Treatment & Res Found, 77-, dir Photother & Vitiligo Units, 74- *Mem:* Can Dermat Asn; Soc Investigative Dermat; Am Soc Photobiol; Am Asn Cancer Res; Am Acad Dermat. *Res:* Pigmentation, especially melanin research and vitiligo phototherapy, photobiology and cutaneous carcinogenesis. *Mailing Add:* Dept Med Univ Toronto Toronto ON M5S 1A8 Can

HABERMAN, JOHN PHILLIP, b Techumseh, Nebr, June 2, 38; m 62; c 2. DRILLING & CEMENTING. *Educ:* Univ Nebr, BSc, 60; Univ Wis, PhD(anal chem), 66. *Prof Exp:* Tech staff anal chem & dent res, Procter & Gamble Co, 66-79; chief chem, Ohio River Valley Water Monitoring Comn, 79-80; SR RES CHEM, BELLAIRE RES LABS, TEXACO USA INC, 80- *Mem:* Am Chem Soc; Soc Petrol Engrs; Sigma Xi. *Res:* Drilling fluid loss, formation damage and cementing; evaluating and developing chemical systems and processes to drill oil and gas wells, to cement casing in wells and instrumental methods to model them in the laboratory. *Mailing Add:* Texaco EPTD PO Box 425 Bellaire TX 77401

HABERMAN, RICHARD, b Brooklyn, NY, June 27, 45; m 69; c 2. NONLINEAR WAVE PHENOMENA, SINGULAR PERTURBATION METHODS. *Educ:* Mass Inst Technol, BS, 67, PhD(appl math) 71. *Prof Exp:* Asst, Univ Calif, San Diego, 71-72; asst prof math, Rutgers Univ, 72-77 & Ohio State Univ, 77-78; assoc prof, 78-85, PROF MATH, SOUTHERN METHODIST UNIV, 85- *Mem:* Soc Indust & Appl Math. *Res:* Nonlinear wave phenomena; hydrodynamic stability; solitons for dispersive waves; slowly varying transitions for ordinary and partial differential equations. *Mailing Add:* Dept Math Southern Methodist Univ Dallas TX 75275-0156

HABERMAN, WARREN OTTO, b Milwaukee, Wis, Aug 19, 18; m 46; c 2. ENTOMOLOGY. *Educ:* Univ Wis, BA, 40, MA, 42, PhD(vet entom), 49. *Prof Exp:* Asst, Univ Wis, 46-48, instr vet sci, 48-49; mgr, Biol-Sanit Res Lab, Ralston Purina Co, St Louis, 49-80; RETIRED. *Mem:* AAAS; Am Soc Parasitol; Entom Soc Am; Sigma Xi. *Res:* Biology of Hypoderma bovis and Hypoderma lineatum. *Mailing Add:* 416 Algonquin Dr Ballwin MO 63011

HABERMAN, WILLIAM L(AWRENCE), b Vienna, Austria, May 4, 22; US citizen; m 57. THERMODYNAMICS, FLUID DYNAMICS. *Educ:* Cooper Union, BS, 49; Univ Md, MS, 52, PhD(viscous flow), 56. *Prof Exp:* Physicist, Bur Ships, US Dept Navy, 49-50, David Taylor Model Basin, 50-57, dep dir gas dynamics div, 57-59, chief res br, 59-61, res physicist, 61-62, dir advan planning div, Off Naval Res, 62-63; sr staff scientist, NASA, 63-71; chmn, Dept Mech Eng, Newark Col Eng, 71-73; CONSULT, 73- *Concurrent Pos:* Lectr, Univ Md, 58-67, vis prof, 67-68, prof, 68-71; prof, Montgomery Col, 73-78. *Mem:* AAAS; Am Phys Soc; Am Inst Aeronaut & Astronaut; Am Geophys Union; Soc Indust & Appl Math. *Res:* Potential theory; viscous and two-phase flows; cavitation; theoretical mechanics; thermal radiation; research administration; planetary sciences. *Mailing Add:* ERC Co PO Box 1727 Rockville MD 20850

HABERMANN, ARIE NICOLAAS, b Groningen, Neth, June 26, 32; m 56; c 4. COMPUTER SCIENCE. *Educ:* Free Univ, Amsterdam, BS, 53, MS, 58; Eindhoven Technol Univ, PhD(comput sci), 67. *Prof Exp:* Teacher, high sch, Neths, 54-62; from lectr to asst prof math prog, Eindhoven Technol Univ, 62-68; vis res scientist prog systs, 68-69, assoc prof, 69-73, actg head, comput sci dept, 79-80, PROF COMPUT SCI, CARNEGIE-MELLON UNIV, 73-, HEAD DEPT, 80- *Concurrent Pos:* Software consult, govt; ed, Acta Info &

ACM Trans on Prog Lang & Systs; vis prof comput sci, Univ Newcastle, Eng, 73; vis prof, Tech Univ, Berlin, 76-; vis scientist, Siemens Corp, 83; actg dir, Software Eng Inst, 84-85. *Mem:* Asn Comput Mach; NY Acad Sci. *Res:* Programming languages; programming systems; operating systems. *Mailing Add:* Dept Comput Sci Carnegie Mellon Univ Pittsburgh PA 15213

HABERMANN, CLARENCE E, physical chemistry, metallurgy, for more information see previous edition

HABERMANN, HELEN M, b Brooklyn, NY, Sept 13, 27. PLANT PHYSIOLOGY. *Educ:* State Univ NY, AB, 49; Univ Conn, MS, 51; Univ Minn, PhD, 56. *Prof Exp:* Asst bot, Univ Conn, 49-51; asst, Univ Minn, 51-53, asst plant physiol, 53-55, head residence counr, Comstock Hall, 55-56; res assoc, Res Inst, Univ Chicago, 56-57; res fel, Hopkins Marine Sta, Stanford Univ, 57-58; from asst prof to assoc prof, 58-70, chmn dept, 63-66, 68 & 78-79, PROF BIOL SCI, GOUCHER COL, 70- *Concurrent Pos:* NIH spec res fel, Res Inst Advan Study, Baltimore, 66-67. *Mem:* Fel AAAS; Am Soc Plant Physiol; Scand Soc Plant Physiol; Japanese Soc Plant Physiol; Phytochem Soc NAm (secy, 87-93). *Res:* Photosynthesis; albino physiology; photomorphogenesis; light control of stomatal opening. *Mailing Add:* Dept Biol Sci Goucher Col Towson MD 21204

HABER-SCHAIM, URI, b Berlin, Ger, Feb 8, 26; m 47; c 3. PHYSICS, SCIENCE EDUCATION. *Educ:* Hebrew Univ, Jerusalem, MS, 49; Univ Chicago, PhD, 51. *Prof Exp:* Res physicist, Weizmann Inst Sci, Israel, 51-53 & Univ Bern, 53-55; res assoc, Univ Ill, 55-56; asst prof physics, Mass Inst Technol, 56-61; proj dir, Phys Sci Study Comt, 63-71; dir div sponsored res & phys sci group, Newton Col Sacred Heart, 71-73; prof sci educ & dir phys sci group, Sch Educ, Boston Univ, 74-76, dir, Inst Curric Develop Sci & Math & prof phys & sci educ, 76-84, adj prof phys, 84-86; CONSULT & AUTH, 86- *Honors & Awards:* Oersted Medal, Am Asn Physics Teachers, 70. *Mem:* Am Phys Soc; Am Asn Physics Teachers; fel AAAS. *Res:* Theoretical and high energy physics; teacher training in physics; curriculum development in science and mathematics. *Mailing Add:* 24 Stone Rd Belmont MA 02178

HABERSTICH, ALBERT, b Basel, Switz, Dec 11, 27; US citizen; m 58; c 2. MAGNETIC FUSION RESEARCH. *Educ:* Univ Md, MS, 58, PhD(physics), 64. *Prof Exp:* Engr electronics, Autophon A G, Solothurn, Switz, 49-56; res asst electronics & physics, Univ Md, 56-64; res assoc, Plasma Physics Lab, Princeton Univ, 64-67; staff mem, 67-77, assoc group leader, 77-86, DEPT GROUP LEADER, LOS ALAMOS NAT LAB, 86- *Mem:* Am Phys Soc. *Res:* Fluid dynamics; experimental research in magnetic fusion. *Mailing Add:* 159 Laguna St Los Alamos NM 87544

HABERSTROH, ROBERT D, b Altoona, Pa, Feb 11, 28; div; c 1. MECHANICAL ENGINEERING. *Educ:* Carnegie Inst Technol, BS, 50; Mass Inst Technol, SM, 51, MechE, 52, ScD(mech eng), 64. *Prof Exp:* Engr, Struct, Inc, Pa, 53-56; indust liaison off admin, Mass Inst Technol, 56-59; asst prof mech eng, Colo State Univ, 59-62; instr, Mass Inst Technol, 62-64; assoc prof, 64-70, PROF MECH ENG, COLO STATE UNIV, 70- *Mem:* Am Soc Mech Engrs; Am Inst Chem Engrs. *Res:* Heat transfer and fluid mechanics; single-phase heat convection; thermodynamics; drying of solids. *Mailing Add:* Dept Mech Eng Colo State Univ Ft Collins CO 80523

HABETLER, GEORGE JOSEPH, b McKees Rocks, Pa, Oct 31, 28; m 53; c 3. MATHEMATICAL PHYSICS. *Educ:* Duquesne Univ, BA, 49; Carnegie Inst Technol, DSc(math), 52. *Prof Exp:* Asst math, Carnegie Inst Technol, 49-51; res assoc, Knolls Atomic Power Lab, Gen Elec Co, 52; PROF MATH, RENSSELAER POLYTECH INST, 52- *Mem:* Am Math Soc; Soc Indust & Appl Math. *Res:* Reactor theory; numerical and functional analysis; linear algebra. *Mailing Add:* Dept Math Rensselaer Polytech Inst Troy NY 12180

HABIB, DANIEL, b New York, NY, Sept 16, 36; m 71. PALYNOLOGY, MICROPALEONTOLOGY. *Educ:* City Col New York, BS, 58; Univ Kans, MA, 60; Pa State Univ, PhD(geol), 65. *Prof Exp:* Lectr geol, 65-66, asst prof, 66-69, assoc prof geol & grad adv dept geol & geog, 69-72, PROF & CHMN DEPT EARTH & ENVIRON SCI, QUEENS COL, NY, 72- *Mem:* AAAS; Paleont Soc; Soc Econ Paleont & Mineral; Brit Paleont Asn; Am Asn Stratig Palynologists. *Res:* Sedimentology of organic detritus; palynological age dating and source of sediments in deep sea cores; Carboniferous palynological paleoecology and distribution of palyniterous sediments; palynological age dating of Phanerozoic rocks; Mesozoic dinoflagellate stratigraphy; deep sea drilling. *Mailing Add:* Dept Geol Queens Col 65-30 Kissena Blvd Flushing NY 11367

HABIB, EDMUND J, b Dover, NH, May 24, 27; m 54; c 5. ELECTRICAL ENGINEERING, PHYSICS. *Educ:* Cath Univ Am, BEE, 49. *Prof Exp:* Group leader rocket instrumentation, Naval Res Lab, 50-56, sect head minitrack calibration sect, 56-58; br head, Systs Eval Br, Goddard Space Flight Ctr, NASA, 58-61, asst div chief, Space Data Acquisition Div, 61-63, asst chief, Data Systs Div, 63-65, assoc chief, Info Processing Div, 65-68, assoc chief, Advan Develop Div, 68-76; mgr com prog develop, Comsat Corp, Intelsat Mgt, 76-80; asst vpres eng, Am Stat Corp, 77-82; vpres, Satellite Systs Eng, 82-85; vpres, Fairchild Communs Systs Div, 86-87; PRIN, SYSTS RES & APPL CORP, 87- *Concurrent Pos:* Mem, Int Telemetry Conf. *Honors & Awards:* Vanguard Proj Outstanding Award, US Navy, 58; Aerospace Commun Award, Am Inst Aeronaut & Astronaut, 70. *Mem:* AAAS; Inst Elec & Electronics Engrs; Sigma Xi. *Res:* Rocket instrumentation; radio tracking and guidance systems; satellite tracking systems; time standards; tracking calibration systems; satellite instrumentation; radio star and sun position determination; computers; electronic data processing systems. *Mailing Add:* 7201 Dear Lake Lane Derwood MD 20855

HABIB, EDWIN EMILE, b Trinidad, WI, Dec 23, 27; Can citizen; m 58; c 2. PHYSICS. *Educ:* Univ Birmingham, BSc, 55; McMaster Univ, PhD(physics), 61. *Prof Exp:* From lectr to asst prof, 59-63, ASSOC PROF PHYSICS, UNIV WINDSOR, 63- *Concurrent Pos:* Nat Res Coun Can hon res grant, 56-58, 60- *Mem:* Am Phys Soc. *Res:* Beta and gamma ray spectroscopy; nuclear reactions using the tandem van de Graaf accelerator. *Mailing Add:* Dept Physics Univ Windsor Windsor ON N9B 3P4 Can

HABIB, IZZEDDIN SALIM, b Tripoli, Lebanon, Nov 16, 34; US citizen; m 62; c 2. HEAT TRANSFER, APPLIED MATHEMATICS. *Educ:* Am Univ Beirut, BME, 56; Va Polytech Inst, MSME, 61; Univ Calif, Berkeley, PhD(mech eng), 68. *Prof Exp:* Engr, Dept Hydraul, Lebanon, 57-60; instr eng, Univ Mich-Dearborn, 62-64; group leader, Chrysler Corp, 64-66; PROF MECH ENG, UNIV MICH, DEARBORN, 69- *Mem:* Am Soc Mech Engrs. *Res:* Heat and mass transfer in turbulent flow; heat transfer with change of phase; heat transfer in radiating gases; flow of high temperature gases. *Mailing Add:* Dept Mech Eng Univ Mich-Univ Chicago Dearborn MI 48128

HABIBI, KAMRAN, b Tehran, Iran, Dec 14, 37; m 65; c 2. PROJECT ENGINEERING, ENVIRONMENTAL RESEARCH. *Educ:* Univ Birmingham, BSc, 61, MSc, 62, PhD(chem eng), 63. *Prof Exp:* Tech officer process develop, Midland Silicones, 63-66; supvr automotive technol, 66-74, mgr proj eng, 74-76, mem staff energy & hydrocarbon planning & develop, 76-78, govt liason, 78-80, MEM STAFF CORP PURCHASING, E I DU PONT DE NEMOURS & CO, INC, 80- *Concurrent Pos:* Mem, Coord Res Coun, 70-74. *Mem:* Brit Inst Chem Engrs. *Res:* Preflame reactions, development of instrumentation and analytical techniques in high-speed diesel engines; characterization of particulate emmisions from cars; development of vehicle emmision control systems; environmental impact research and analysis. *Mailing Add:* 719 Taunton Rd Wilmington DE 19803

HABICHT, ERNST ROLLEMANN, JR, b Charleston, WVa, Dec 22, 38; m 66. ENVIRONMENTAL SCIENCES, ENERGY CONVERSION. *Educ:* Harvard Univ, AB, 60; Stanford Univ, PhD(chem), 67. *Prof Exp:* fel, asst prof biol & lectr chem, Univ Calif, San Diego, 67-71; staff scientist & dir energy prog, Environ Defense Fund, 71-80; INDEPENDENT ENERGY ANALYST & CONSULT, 80- *Concurrent Pos:* Consult res & develop Presidential report, US Atomic Energy Comn, 75; mem adv coun to chmn NY Pub Serv Comn, 75-; mem Nat Gas Surv Task Force on Rate Design, Fed Power Comn, 75-77, Nat Coal Pol Proj, 76-80, Mass Inst Tehcnol Energy Lab, 78-80 & Polit Econ Res Ctr, 84- *Mem:* AAAS; Am Chem Soc; Am Econ Asn; NY Acad Sci. *Res:* Economics of energy utilization; energy technology assessment; marginal cost-based pricing and investing policies for regulated utilities. *Mailing Add:* PO Box 65 Port Jefferson NY 11777

HABICHT, GAIL SOREM, b Oakland, Calif, Oct 1, 40; m 66; c 2. IMMUNOLOGY, PHYSIOLOGY. *Educ:* Stanford Univ, BS, 62, PhD(physiol), 65. *Prof Exp:* USPHS fel immunol, Rockefeller Univ, 65-66, fel immunophysiol, Stanford Univ, 66-67; asst prof biol, Univ San Diego, 67-68; res assoc, Scripps Clin & Res Found, 68-71; asst prof, 71-83, assoc prof, 83-88, PROF PATH, SCH MED, STATE UNIV NY, STONY BROOK, 88- *Concurrent Pos:* Lectr, San Diego State Col, 68-71; guest investr, Brookhaven Nat Lab, 71-83. *Mem:* Am Asn Immunologists; NY Acad Sci; Am Soc Microbiologists; Reticuloendothelial Soc. *Res:* Acquired immunological tolerance; autoimmunity; aging; babesiosis; parasitology; lyme disease; interleukin 1; inflammation. *Mailing Add:* Dept Path Sch Med State Univ NY Stony Brook NY 11794-8691

HABICHT, JEAN-PIERRE, b Geneva, Switz, Dec 15, 34; US citizen; div; c 3. NUTRITIONAL EPIDEMIOLOGY. *Educ:* Univs Geneva & Zurich, Switz, MD, 62; Harvard Sch Pub Health, MPH, 68; Mass Inst Technol, PhD(nutrit, biochem & metab), 69. *Prof Exp:* Med officer, Div Human Develop, World Health Orgn, Inst de Nutricion de Centro Am y Panama, Guatemala, 69-74; spec asst to dir, Div Health Exam Statist, Nat Ctr Health Statist, 74-77; PROF NUTRIT EPIDEMIOL, DIV NUTRIT SCI, CORNELL UNIV, 77- *Concurrent Pos:* Prof maternal & child health, Univ San Carlos, Guatemala, 73-74; adv, WHO, 75-, Nat Acad Sci, 75-, USDA, 82-, UN, 83-; prin investr, var grants & contracts. *Mem:* Soc Epidemiol Res; Am Inst Nutrit; fel Am Col Epidemiol; Pop Asn Am; Am Pub Health Asn; Int Epidemiol Asn. *Res:* Maternal and child health in developing countries; primary medical care in poor rural areas; nutritional influences on health in developed and developing countries; ethics of population research; organization of health and nutritional surveillance systems. *Mailing Add:* Div Nutrit Sci Savage Hall Cornell Univ Ithaca NY 14853-6301

HABIG, ROBERT L, b West Reading, Pa, July 28, 40; c 2. CLINICAL CHEMISTRY. *Educ:* Lebanon Valley Col, BS, 62; Purdue Univ, PhD(anal chem), 66. *Prof Exp:* Instr biochem & asst dir clin chem lab, 66-69, assoc biochem & actg dir clin chem lab, 69-70, asst prof biochem & dir clin chem lab, 70-77, DEP DIR, DUKE HOSP LABS, MED CTR, DUKE UNIV, 77- *Concurrent Pos:* Co-prin investr, USPHS Grant, 67-68, prin investr, 68-70. *Mem:* Am Chem Soc; Am Asn Clin Chem. *Res:* Automation of reaction-rate measurements for determination of inorganic catalysts; kinetics of catalyzed inorganic oxidation-reduction reactions; improvements in automation for clinical chemistry; application of isoenzyme studies to clinical chemistry. *Mailing Add:* Miles Inc Diag Div 511 Benedict Ave Tarrytown NY 10591

HABIG, WILLIAM HENRY, m; c 3. BACTERIAL TOXINS. *Educ:* Rutgers Univ, BS, 64; Univ Vt, PhD(biochem), 68. *Prof Exp:* Fel, NY State Dept Health, 71-72; staff fel, NIH, 72-78; CHIEF LAB BAC TOXINS, FOOD & DRUG ADMIN, 75- *Res:* Biochemistry; pharmacology. *Mailing Add:* FDA Bur Biologics Bldg 29 Rm 407 Bethesda MD 20205

HABOUSH, WILLIAM JOSEPH, b New York, NY, June 7, 42; m 66. REPRESENTATION THEORY. *Educ:* Cornell Univ, BA, 63; Columbia Univ, PhD(math), 70. *Prof Exp:* Lectr math, Univ Mich, 68-70; asst prof, Brooklyn Col, 70-71; from asst prof to prof math, State Univ NY, Albany, 71-86; PROF, UNIV ILL, URBANA, CHAMPAIGN, 86- *Concurrent Pos:* Fel math, Nagoya Univ, 72-73; vis mem staff, Inst Adv Study, 74-75; vis assoc prof, Univ Calif, Los Angeles, 75-76; vis prof, Univ Rome, 79, Univ Paris, 80. *Res:* Geometry of homogeneous spaces of semi-simple groups and the representation theory of semi-simple groups over fields of positive characteristic. *Mailing Add:* Dept Math 273 Altgeld Hall Univ Ill 1409 W Green St Urbana Champaign Urbana IL 61801

HABOWSKY, JOSEPH EDMUND JOHANNES, b Coburg, Ger, Sept 27, 28; Can citizen; m 68. CYTOLOGY, HISTOLOGY. *Educ:* Univ Munich, BS, 55; Univ Toronto, MS, 58, PhD(zool), 62. *Prof Exp:* Nat Res Coun Can fel, 62-63; lectr zool, Univ Toronto, 64; from asst prof to assoc prof, 64-74, PROF BIOL, UNIV WINDSOR, 74- *Honors & Awards:* Armour Pharmaceut Award, 73; William J Stickel Award, 74; Achievement Award, Asn Media & Technol Educ Can, 90; Iseta Award, 91. *Mem:* Electron Micros Soc Am; Micros Soc Can; Int Soc Individual Instr. *Res:* Transmission electron microscopy of differentiating cells in spermatogenesis in invertebrates; transmission electron microscopic studies of the effects of drugs in invertebrate systems; cytology of the human skin as related to function; cytological effects of vitamin A and hormones on the skin in mammals, analyzed by autoradiography, transmission and scanning electron microscopy; development and implementation of modules in individualized and self-paced instruction using audio and video tapes, variable speed tape recorders, micorcomputers and video disks; rationalizations in the design of learning centres; development and testing of the appropriate combination of media (including microcomputers, video disks) to achieve effective learning and problem solving skills. *Mailing Add:* Dept Biol Sci Univ Windsor Windsor ON N9B 3P4 Can

HABTEMARIAM, TSEGAYE, b Addis Ababa, Ethiopia, Apr 25, 42; m 74; c 3. BIOMEDICAL DECISION SUPPORT SYSTEM, SIMULATION MODELING. *Educ:* Agr & Mech Col, Ethiopia, BSc, 64; Colo State Univ, DVM, 70; Univ Calif, Davis, MPVM, 77, PhD(epidemiol), 80. *Prof Exp:* Asst prof animal sci, Agr & Mech Col, Ethiopia, 70-74; res assoc epidemiol, Univ Calif, Davis, 77-79; assoc prof, 79-84, PROF EPIDEMIOL, SCH VET MED, TUSKEGEE INST, 84-, DIR, INFO MGT, 82- *Concurrent Pos:* Prin investr, Tuskegee Inst, 82- *Mem:* Soc Med Decision Making; Am Vet Comput Soc; Am Vet Med Asn; Sigma Xi. *Res:* Computer modeling of diseases; biomedical decision-making support systems; analytic and quantitative epidemiology; biomedical information management. *Mailing Add:* Tuskegee Inst PO Box 446 Tuskegee AL 36008

HABTE-MARIAM, YITBAREK, b Asella, Ethiopia, Jan 28, 49. BIOPHYSICAL CHEMISTRY, MAGNETIC RESONANCE. *Educ:* Rutgers Univ, BA, 74, PhD(chem), 77. *Prof Exp:* Fel chem, Univ Va, 77-78 & Ga State Univ, 78-79; res assoc, 79-80, ASST PROF CHEM, ATLANTA UNIV, 80- *Concurrent Pos:* Teaching asst, Rutgers Univ, 74-76; instr chem, NJ Med Sch, 76-77; prin investr, Atlanta Univ, 80- *Mem:* Am Chem Soc. *Res:* Biophysical chemistry and biophysics; multinuclear magnetic resonance spectroscopy; ligand macromolecule interactions; metol ion ligand interactions and solution equilibria; experimental approaches to solution conformation and determination of inter-nuclear distances; motional properties of polymers and biopolymers as probed by nuclear relaxation. *Mailing Add:* Dept Chem Atlanta Univ Atlanta GA 30314

HAC, LUCILE R, b Lincoln, Nebr, May 18, 09. BIOCHEMISTRY, ORGANIC CHEMISTRY. *Educ:* Univ Nebr, BA, 30, MSc, 31; Univ Minn, PhD(org chem), 35. *Prof Exp:* Chem asst, Univ Nebr, 29-31; chem asst, Univ Minn, 31-35; assoc bacteriologist, State Dept Health, Md, 35-36; from asst to res assoc gynec, Kuppenheimer Fund, Chicago, 36-43; res chemist, Int Minerals & Chem Corp, Tex, 43-44, prin chemist & supvr sugar beet res, Calif, 44-56, supvr air pollution res, Fla, 56-58, supvr microbiol res, Ill, 58-61; assoc prof, 61-77, EMER PROF BIOCHEM, MED SCH, NORTHWESTERN UNIV, CHICAGO, 77- *Mem:* AAAS; Am Chem Soc; Am Soc Microbiol; Soc Exp Biol & Med; NY Acad Sci. *Res:* Orthoquinones: Eberthella typhi aglutinins; gonococci; chemotherapy; microbiological assay of amino acids; amino acids of sugar beet and sugar beet juices; sugar beet genetics; air pollution; microbial physiology; fermentation; minerals, vitamins and hormonal effects on bone metabolism and fluorosis in bone. *Mailing Add:* 812 Oakwood Ave Wilmette IL 60091

HACH, EDWIN E, JR, b Shippenville, Pa, Jan 5, 34; m 55; c 2. PHYSICAL CHEMISTRY. *Educ:* Clarion State Col, BS, 59; Univ NH, MS & PhD(phys chem), 67. *Prof Exp:* Instr acad chem, Bradford Area High Sch, 59-62; asst prof phys chem, 67-70, ASSOC PROF CHEM, ST BONAVENTURE UNIV, 70- *Mem:* Am Chem Soc; Sigma Xi. *Res:* Electrochemistry; potentiometric stability constant determinations; non-aqueous polarography; solute species in non-aqueous solutions; titrations in non-aqueous solutions. *Mailing Add:* Dept Chem St Bonaventure Univ St Bonaventure NY 14778

HACH, VLADIMIR, b Prague, Czech, Nov 17, 24. ORGANIC CHEMISTRY, MEDICINAL CHEMISTRY. *Educ:* Prague Tech Univ, Dipl Ing, 49, DrSc(org chem), 53. *Prof Exp:* Res scientist, Spofa Res Inst Pharm & Biochem, Czech, 53-66 & Org Div, Austrian Nitrogen Works, 66; sr res chemist & group leader, MacMillan Bloedel Res Ltd, 66-70; sr res assoc, Dept Chem, Univ BC, 70-73; group leader process develop, Delmar Chem Ltd, 73-76; TECH DIR, WHITEHALL LABS, 77- *Mem:* Am Chem Soc; Chem Inst Can. *Res:* Synthetic organic chemistry; synthesis of new compounds; development of synthetic reactions; natural compounds; analytical control methods research and development. *Mailing Add:* 260 Scarlett Rd Apt 1901 Toronto ON M6N 4X5 Can

HACHIGIAN, JACK, b Paterson, NJ, Feb 21, 29; m 64. MATHEMATICS, STATISTICS. *Educ:* Univ Mich, BS, 50; Ind Univ, PhD(math), 61. *Prof Exp:* Lectr math, Ind Univ, 61-62; asst prof, Cornell Univ, 62-64 & Univ Mass, Amherst, 64-68; assoc prof, Univ RI, 68-73; ASSOC PROF MATH, HUNTER COL, NY, 73- *Concurrent Pos:* Consult radiosterilized safety of foods, US Army, 65- *Mem:* Inst Math Statist; Am Math Soc. *Res:* Markov processes, particularly the Chapman-Kolmogorov equation and its probabalistic consequences. *Mailing Add:* Dept Math Hunter Col 695 Park Ave New York NY 10021

HACHINSKI, VLADIMIR C, b Zhitomir, Ukraine, Aug 13, 41; m 67; c 3. NEUROLOGY, CEREBROVASCULAR DISEASES. *Educ:* Univ Toronto, MD, 66; FRCP(C), 72; McMasters Univ, MSc, 87; Univ London, DSc, 88. *Hon Degrees:* McMasters Univ, MSc, 87. *Prof Exp:* Lectr neurol, Univ Toronto, 74-76, from asst prof to assoc prof, 76-83; assoc prof, 80-83, PROF NEUROL, UNIV WESTERN ONT, 83-, PROF EPIDEMIOL & PHYSIOL, 88-; DIR, STROKE & AGING GROUP, ROBARTS RES INST, 84- *Concurrent Pos:* Assoc ed, J Stroke, 84-86; J Archives Neurol; Richard & Beryl Ivey Prof & Chmn, Dept Clin Neurol Sci, London, Ont, Can; chmn, Steering Comt, AAAS; counr, Am Acad Neurol. *Honors & Awards:* Trillium Clin Scientist Award. *Mem:* Can Stroke Soc (pres, 81); World Fedn Neurol; AAAS; Int Soc Cerebral Blood Flow & Metab; Am Neurol Asn; Am Acad Neurol. *Res:* Cerebral blood flow and metabolism; multi-infarct dementia; migraine; cerebrovascular diseases; cardiac complications of stroke; Alzheimer's disease. *Mailing Add:* Univ Hosp 339 Windermere Rd London ON N6A 5A5 Can

HACK, JOHN TILTON, b Chicago, Ill, Dec 3, 13; m 42; c 2. GEOLOGY. *Educ:* Harvard Univ, AB, 35, MA, 38, PhD(geomorphol), 40. *Prof Exp:* Geologist, Awatovi Exped, Peabody Mus, Harvard Univ, 37-39; instr geol, Hofstra Col, 40-42; geologist, US Geol Surv, 42-66, asst chief geologist regional geol, 66-71, res geologist, 71-81; prof lectr, George Washington Univ, 81-83. *Honors & Awards:* Kirk Bryan Award, 60; Distinguished Serv Award, US Dept Interior, 72; G K Warren Prize, Nat Acad Sci, 82. *Mem:* AAAS; fel Geol Soc Am. *Res:* Geomorphology of nonglaciated areas; general and military geology; geology of archaeological sites; geomorphology of Appalachians and Coastal Plain; geologic environments of tree species. *Mailing Add:* Dept Mech Eng Box 2157 Yale Sta New Haven CT 06520-2157

HACKAM, REUBEN, b Baghdad, Iraq, Feb 18, 36; Can & Brit citizen; m 64; c 4. HIGH VOLTAGE INSULATION, ELECTRIC POWER ENGINEERING. *Educ:* Technion, Israel Inst Technol, BSc, 60; Univ Liverpool, Eng, PhD(elec eng), 64, DEng, 88. *Prof Exp:* Sr engr, Eng Elec-Gen Elec Co, 64-69; lectr, 69-73, sr lectr, 73-74, reader, Dept Elec Eng, Univ Sheffield, 74-78; chmn, 80-81, 84-86, PROF, ELEC EMG, DEPT ELEC ENG, UNIV WINDSOR, 78- *Concurrent Pos:* Vis staff, math, Staffordshire Polytechnic, 64-69; vis staff, math Sheffield City Polytech, 70-78 & Sheffield City Polytech, 70-78; consult, elec, British Rail, 75-78, Eng Elec Co, 75-77, Windsor Star, 81-87, Hiram Walker & Sons, 83-86, Corp City Windsor, 83-88. *Mem:* Fel Inst Elec & Electronics Engrs. *Res:* Research in gas insulated systems and devices; electromagnetics, polymer insulators, power system planning; weed eradication using electric energy and control of reactive power in power systems; author of various articles. *Mailing Add:* Dept Elec Eng Univ Windsor Windsor ON N9B 3P4 Can

HACKBARTH, WINSTON (PHILIP), b Iowa Falls, Iowa, June 30, 24; m 56; c 2. PHYSIOLOGICAL ECOLOGY. *Educ:* Univ Iowa, BA, 47; Idaho State Col, BS, 48; Univ Denver, MS, 50; Iowa State Univ, PhD, 56. *Prof Exp:* Instr biol, Drake Univ, 55-56; asst prof biol, Augustana Col, 56-59; assoc prof, 59-71, PROF BOT, LA TECH UNIV, 71- *Mem:* AAAS; Ecol Soc Am; Am Inst Biol Sci; Sigma Xi. *Res:* Acer saccharum in north central Louisiana; ecological analysis of undisturbed broadleaf forests of northeastern Louisiana; contrasts between broadleaf and coniferous forests of northern Louisiana; plant Physiology. *Mailing Add:* Dept Biol Sci Col Life Sci La Tech Univ Box 3179 Ruston LA 71272

HACKEL, DONALD BENJAMIN, b Boston, Mass, July 7, 21; m 47; c 3. PATHOLOGY. *Educ:* Harvard Univ, AB, 43, MD, 46. *Prof Exp:* From instr to assoc prof path, Sch Med, Western Reserve Univ, 49-60; PROF PATH, SCH MED, DUKE UNIV, 60- *Concurrent Pos:* USPHS career res award, 63-, mem study sect primates, 61-64, mem study sect path, 65-69, mem, Gen Med Sci Prog Proj Comn, 70- *Honors & Awards:* Parke Davis Award, Am Soc Exp Path, 61. *Mem:* Soc Exp Biol & Med; Am Asn Pathologists; Am Heart Asn. *Res:* Studies of cardiac function and pathology; physiology. *Mailing Add:* Dept Path Duke Univ Med Ctr Box 3712 Durham NC 27710

HACKEL, EMANUEL, b Brooklyn, NY, June 17, 25; m 50, 81; c 6. HUMAN GENETICS, TRANSPLANTATION IMMUNOLOGY. *Educ:* Univ Mich, BS, 48, MS, 49; Mich State Col, PhD(zool), 53. *Prof Exp:* Asst zool, Univ Mich, 48-49; instr biol sci, 49-52, instr natural sci, 52-53, from asst prof to prof natural sci, 53-74, chmn dept, 63-74, PROF, DEPTS MED & ZOOL, MICH STATE UNIV, 74. *Concurrent Pos:* Vis investr, Blood Group Res Unit, Lister Inst, Eng, 56-57; asst dean, Univ Col, Mich State Univ, 58-63; res fel, Galton Lab, Univ Col, Univ London, 70-71 & 77-78; mem, bd dirs, Am Asn Blood Banks, 83-84; consult, Mem Blood Ctr, Minneapolis, 83- & Blood Bank, Ingham Med Ctr, Lansing, 83- *Honors & Awards:* Cooley Mem Award, Am Asn Blood Banks, 69, Elliot Mem Award, 87. *Mem:* Am Asn Blood Banks; Sigma Xi; Am Soc Human Genetics; Genetics Soc Am; NY Acad Sci; Am Soc Histocompatibility & Immunogenetics. *Res:* Blood groups; immunohematology; immunogenetics; tissue typing; transplantation; genetic counseling. *Mailing Add:* Dept Med Mich State Univ B220 Life Sci E Lansing MI 48824

HACKEL, LLOYD ANTHONY, b Litte Chute, Wis, Oct 14, 49; m 71; c 2. HIGH RESOLUTION LASER SPECTROSCOPY. *Educ:* Univ Wis-Madison, BS, 71; Mass Inst Technol, MS, 73, ScD, 74. *Prof Exp:* Mem staff, Lab Electronics, Mass Inst Technol, 74-76; PHYSICIST, UNIV CALIF, LAWRENCE LIVERMORE NAT LAB, 76- *Mem:* Am Phys Soc. *Res:* Atomic vapor laser isotope separation with emphasis on separator pilot plant operation; atomic physics, spectroscopy, pulse propagation and e-beam generation of atomic vapor; hyperfine and automization structure, precision photon absorption cross sections; resonant and non-resonant pulse propagation. *Mailing Add:* 1072 Sherry Way Livermore CA 94550

HACKENBERG, ROBERT ALLAN, b Fargo, NDak, Mar 12, 28; m 63. MEDICAL ANTHROPOLOGY, APPLIED ANTHROPOLOGY. *Educ:* Univ Minn, BA, 50, MA, 51; Cornell Univ, PhD(anthrop), 61. *Prof Exp:* Asst dir, Bur Ethnic Res, Dept Anthrop, Univ Ariz, 58-62; res anthropologist, Demog Sect, Biomet Br, Nat Cancer Inst, 62-66; PROF ANTHROP, UNIV COLO, BOULDER & PROG DIR, INST BEHAV SCI, 66- *Concurrent Pos:* Dir res training prog cult change, Dept Anthrop, Univ Colo, Boulder, 66-, NIMH grant res training cult change, 68-79; Nat Inst Child Health res grants, demog transition without urbanization, Davao City, Philippines, 69-72, social mobility & fertil control, 72-74 & longitudinal fertil in develop city, 77-79; NIMH res grant, ment health epidemiol, Papago Indian Reservation, Ariz, 69-72; mem div behav sci, Nat Res Coun, 72-; mem, pop study sect, Div Res Grants, NIH, 72-, chmn, 73-76; sr fel, East West Pop Inst, East West Ctr, Honolulu, 75-79 & 80. *Mem:* AAAS; fel Am Anthrop Asn; fel Soc Appl Anthrop; Pop Asn Am; Asn Asian Studies. *Res:* Adaptations to ecological stress brought about by population growth and requiring changes in community organization; modernization and urbanization variables related to this problem in the Philippines and American Southwest. *Mailing Add:* Dept Anthrop Univ Colo Campus Box 233352 Boulder CO 80309

HACKENBROCK, CHARLES ROBERT, b Brooklyn, NY, Dec 23, 29; m 56; c 3. CELL BIOLOGY, ELECTRON MICROSCOPY. *Educ:* Wagner Col, BS, 61; Columbia Univ, PhD(anat), 65. *Prof Exp:* Asst prof anat, Sch Med, Johns Hopkins Univ, 65-68, assoc prof, 68-71; prof cell biol, Univ Tex Southwestern Med Sch, Dallas, 71-77; PROF & CHMN DEPT ANAT, UNIV NC SCH MED, 77- *Mem:* Am Asn Anat; Am Soc Cell Biol; Am Inst Biol Sci; Am Soc Biol Chemists; NY Acad Sci. *Res:* Ultrastructure and metabolism in mitochondria; membrane structure; bioenergetics. *Mailing Add:* Dept Anat Univ NC Sch Med 108 Taylor Hall Chapel Hill NC 27599

HACKER, CARL SIDNEY, b Newport News, Va, July 22, 41; c 1. POPULATION BIOLOGY. *Educ:* Col William & Mary, BS, 63; Rice Univ, PhD(biol), 68. *Prof Exp:* Res assoc biol, Univ Notre Dame, 68-69, NIH res fel, 69-71; asst prof ecol, 71-75, MEM FAC, GRAD SCH BIOMED SCI, UNIV TEX HEALTH SCI CTR, HOUSTON, 73-, ASSOC PROF ECOL, SCH PUB HEALTH, 75-, ASSOC PROF COMMUN MED, SCH MED, 80- *Concurrent Pos:* Adj prof math sci, Rice Univ, 73-; vis prof biol, Univ Houston, 77. *Mem:* AAAS; Am Bar Asn; Ecol Soc Am. *Res:* Population biology; vector biology and genetics; computer modelling and simulation in ecology; ecotoxicology; environmental and health law. *Mailing Add:* Sch Pub Health Univ Tex Health Sci Ctr PO Box 20186 Houston TX 77225

HACKER, DAVID S(OLOMON), b Brooklyn, NY, June 9, 25; m 49; c 2. CHEMICAL ENGINEERING, PHYSICAL CHEMISTRY. *Educ:* Univ Ill, BS, 49; Mass Inst Technol, MS, 50; Northwestern Univ, PhD(chem eng), 54. *Prof Exp:* Res engr, Gen Elec Co, 54-56; sr scientist, Chicago Midway Labs, 56-57; sr scientist & group leader aerochem, Ill Inst Technol Res Inst, 57-61; supvr combustion res, Inst Gas Tech, 61-65; assoc prof energy eng, Univ Ill, Chicago Circle, 65-81; SR STAFF RES ENGR, AMOCO CHEM CORP, 81- *Concurrent Pos:* Mem, US Army Missile & Rocket Adv Coun, 59-63; consult, Ill Inst Technol Res Inst, 65; mem indust staff, Amoco Oil Co, Chicago, 72; mem res staff, Argonne Nat Lab, 76-78 & 80-81. *Mem:* Am Phys Soc; fel Am Inst Chem Engrs; Combustion Inst; Sigma Xi; fel Am Inst Chem Engrs. *Res:* Aerothermochemistry; high temperature kinetics; heat transfer and fluid mechanics; multiphase flow; chemical process design and economics. *Mailing Add:* 343 Beech St Highland Park IL 60035

HACKER, HERBERT, JR, b Cleves, Ohio, June 4, 30; m 55; c 3. ELECTRICAL ENGINEERING. *Educ:* Univ Ohio, BS, 57; Princeton Univ, MS, 59; Univ Mich, PhD, 64. *Prof Exp:* Asst elec eng, Princeton Univ, 57-59; asst prof, Univ Ohio, 59-60; res asst, Univ Mich, 60-62, from instr to asst prof, 62-65; asst prof, 65-68, ASSOC PROF ELEC ENG, DUKE UNIV, 68- *Res:* Electromagnetic theory; magnetism; solid state theory; electron paramagnetic resonance studies of amorphous materials. *Mailing Add:* Dept Elec Eng 111 Eng Bldg Duke Univ Durham NC 27706

HACKER, MILES PAUL, PHARMACOLOGY. *Educ:* Univ Tenn, PhD(pharmacol), 75. *Prof Exp:* ASST PROF PHARMACOL & TOXICITY, UNIV VT, 80- *Res:* Pharmacology of anti-cancer drugs; mechanism of action; mechanisms of toxicity. *Mailing Add:* Dept Pharmacol Univ Vt Agr Col 85 S Prospect St Burlington VT 05405

HACKER, PETER WOLFGANG, b San Francisco, Calif, May 31, 42; m 82; c 4. PHYSICAL OCEANOGRAPHY, RESEARCH ADMINISTRATION. *Educ:* Univ Calif, Berkeley, BS, 64; Univ Calif, San Diego, MS, 66, PhD(phys oceanog), 73. *Prof Exp:* Asst prof earth & planetary sci, Johns Hopkins Univ, 73-77; phys oceanog prog dir, NSF, DC, 77-80; consult, 80-81; phys oceanog prog dir, NSF, Washington, DC, 81-84; RES SCIENTIST, TEX A&M UNIV, COL STA, TEX, 84- *Concurrent Pos:* Prin investr, Chesapeake Bay Inst, Johns Hopkins Univ, 73-77, staff oceanogr, 77-79; sci officer, US TOGA Proj Off, 84-87. *Mem:* Am Geophys Union; AAAS; Am Meteorol Soc; Oceanog Soc. *Res:* Dynamics of ocean currents on continental shelf regions; circulation and transport processes in estuaries; oceanic microstructure and mixing processes; equatorial circulation and dynamics. *Mailing Add:* 2422 Briarwood Rd Baltimore MD 21209-4302

HACKERMAN, NORMAN, b Baltimore, Md, Mar 2, 12; c 4. CHEMISTRY. *Educ:* Johns Hopkins Univ, BA, 32, PhD(chem), 35. *Hon Degrees:* DSc, Austin Col, 75, Tex Christian Univ, 78; LLD, Abilene Christian Univ, 78, St Edward's Univ, 72. *Prof Exp:* Asst prof phys chem, Loyola Col, Md, 35-39; asst chemist, USCG, 40-41; asst prof chem, Va Polytech Inst, 41-43; res chemist, Kellex Corp, NY, 44; from asst prof to prof chem, Univ Tex, Austin, 45-70, chmn dept, 52-62, dir corrosion res lab & dean res & sponsored progs, 61-62, vpres & provost, 62-63, vchancellor acad affairs, 63-67, pres, Univ, 67-70; prof chem & pres, 70-85, EMER PRES & DISTINGUISHED EMER PROF CHEM, RICE UNIV, 85-; EMER PROF CHEM, UNIV TEX, AUSTIN, 85- *Concurrent Pos:* Res chemist, Colloid Corp, 36-40; res consult, Univ Tex; mem environ pollution panel, President's Sci Adv Comt; chmn, Gordon Res Conf on Corrosion, 52 & Conf on Chem at Interfaces, 59; mem, Int Comt Electrochem Thermodyn & Kinetics; Matiello mem lectr, 64; mem, Nat Sci Bd, NSF, 68-80, chmn, 74-80, mem Defense Sci Bd, 78-85; ed, J Electrochem, 69-; mem, Gordon Conf Res Bd, 70-76. *Honors & Awards:* Whitney Award, Nat Asn Corrosion Engrs, 56; Palladium Medal, Electrochem Soc, 65; Gold Medal, Am Inst Chemists, 78; Edward Goodrich Acheson Award, Electrochem Soc, 84; Charles Lathrop Parsons Award, Am Chem Soc, 87; Philip Hauge Ableson Prize, AAAS, 87. *Mem:* Nat Acad Sci; AAAS; Am Chem Soc; Electrochem Soc (vpres, 54-57, pres, 57-58); Am Philos Soc. *Res:* Physical chemistry; corrosion of metals; surface chemistry of metals and oxides; passivity; electrical double layer at solid metal-solution interfaces. *Mailing Add:* Dept Chem Rice Univ PO Box 1892 Houston TX 77001

HACKERT, MARVIN LEROY, b Pella, Iowa, Sept 23, 44; div; c 2. PROTEIN CRYSTALLOGRAPHY. *Educ:* Cent Col, BA, 66; Iowa State Univ, PhD(phys chem), 70. *Prof Exp:* NIH fel biol sci, Purdue Univ, 70-74; from asst prof to assoc prof, 74-86, PROF BIOCHEM, UNIV TEX, AUSTIN, 86- *Concurrent Pos:* Sci dir, UIL, 84- *Mem:* Am Chem Soc; Am Crystallog Asn. *Res:* X-ray crystallography and the molecular structure of proteins; relation of enzyme structure and mechanisms; biophysical methods. *Mailing Add:* Dept Chem Univ Tex Austin TX 78712

HACKERT, RAYMOND L, b Anamoose, NDak, Mar 27, 27; m 53; c 6. ANALYTICAL CHEMISTRY. *Educ:* St John's Univ, Minn, BA, 49; Univ Detroit, MS, 51. *Prof Exp:* From chemist to sr chemist, 51-66, SR RES CHEMIST, E I DU PONT DE NEMOURS & CO, 66- *Mem:* AAAS; Am Chem Soc; Soc Appl Spectros. *Res:* Spectroscopy; polymer chemistry. *Mailing Add:* 221 North Blvd Salisbury MD 21801

HACKETT, ADELINE J, b Rosetown, Sask, Can, Jan 6, 23; m 44; c 1. VIROLOGY, GENETICS. *Educ:* Univ Calif, Berkeley, BS, 58, MS, 60, PhD, 65. *Prof Exp:* From assoc res virologist to res virologist Naval Biomed Res Lab, Sch Pub Health, Univ Calif Berkeley, 73-78, mem staff Biomed Div, Lawrence Berkeley Lab, 78-83; FOUNDER & DIR, PERALTA CANCER RES INST, 77- *Concurrent Pos:* Pres, Int Asn Breast Cancer Res, 83-85. *Mem:* Am Soc Microbiol; Sigma Xi. *Res:* Mechanisms of viral interference; the in vitro biology of human tumor cells. *Mailing Add:* 82 Evergreen Dr Orinda CA 94563

HACKETT, COLIN EDWIN, b Birmingham, Eng, Apr 9, 43; m 68; c 3. FLUID PHYSICS, THERMAL PHYSICS. *Educ:* Cambridge Univ, BA, 65, MA, 68; Brown Univ, PhD(eng mech & appl math), 71. *Prof Exp:* Proj supvr aeronaut, Aircraft Res Asn, Eng, 65-66; engr-scientist, Lockheed Ga Res Lab, 66-67; res assoc fluid mech, Mass Inst Technol, 71-73; mem tech staff fluid physics, Exp Fluid Physics Div, Albuquerque, 73-79, SR MEM TECH STAFF, EXP THERMAL & FLUID SCI, SANDIA NAT LABS, LIVERMORE, 80- *Concurrent Pos:* Fel, Mass Inst Technol, 71-73; tech consult, AVCO Corp, 72-73; adj prof mech eng, Univ NMex, 75-79. *Mem:* Sigma Xi; AAAS; Am Inst Aeronaut & Astronaut. *Res:* Experimental fluid physics; laser diagnostics, airborne tunable laser spectroscopy, diode, molecular and dye lasers; laser fluid velocimetry in turbulent compressible chemically reacting flows, application to high energy chemical laser devices; chemical physics; energy storage; supersonic laser assisted combustion and fuel synthesis; advanced energy conversion and storage systems. *Mailing Add:* 4318 Pomona Way Livermore CA 94550-3449

HACKETT, EARL R, b Moulmein, Burma, Feb 16, 32; US citizen; m 53; c 7. NEUROLOGY. *Educ:* Drury Col, BS, 53; Case Western Reserve Univ, MD, 57. *Prof Exp:* From instr to assoc prof neurol, 62-72, PROF NEUROL & PHYSIOL, SCH MED, LA STATE UNIV MED CTR, 72-, HEAD DEPT NEUROL, 77- *Concurrent Pos:* Consult, USPHS Hosp, Carville. *Mem:* Fel Am Acad Neurol; Am Asn Electromyography & Electrodiag. *Res:* Factors involved in nerve injury and repair; methods of evaluating nerve injury. *Mailing Add:* Dept Neurol & Physiol La State Univ Med Ctr 1542 Tulane Ave New Orleans LA 70112

HACKETT, JAMES E, b Nottingham, Eng, Feb 10, 36; m 57; c 2. FLUID MECHANICS, AERODYNAMICS. *Educ:* Hatfield Tech Col, BSc, 59; Univ London, DIC & PhD(aeronaut eng), 64. *Prof Exp:* Student apprentice, DeHavilland Aircraft Co, Eng, 54-59; lectr, Hatfield Tech Col, 59-61; from sci officer to sr sci officer aerodyn res, Nat Phys Lab, Teddington, 62-67; STAFF SPECIALIST AERODYN RES, LOCKHEED-GA CO, 67- *Concurrent Pos:* Secy, Powered Lift Comt, Aeronaut Res Coun, Eng, 63-66. *Mem:* Am Inst Aeronaut & Astronaut. *Res:* Fluid mechanics of flight at low speeds; formation, structure and consequences of trailing vortices behind wings, bodies, and lifting jets; now predominately methods for testing and interpretation of wind tunnel experiments. *Mailing Add:* Dept 7393 Zone 0644 Lockheed Aernaut Systs Co S Cobb Dr Marietta GA 30063

HACKETT, JOHN TAYLOR, b Chicago, Ill, July 24, 41; m 70; c 3. NEUROPHYSIOLOGY, NEUROPHARMACOLOGY. *Educ:* Univ Ill, Urbana, BS, 64; Univ Ill Med Ctr, PhD(pharmacol), 70. *Prof Exp:* Muscular Dystrophy Asn Can fel, Univ BC, 69-72; NIH fel, Univ Iowa, 72-73; from asst prof to assoc prof, 73-84, PROF PHYSIOL, SCH MED, UNIV VA, 84- *Concurrent Pos:* Res scientist develop award, Nat Inst Drug Abuse, 75-80. *Mem:* Soc Neurosci; Am Physiol Soc. *Res:* Physiology and pharmacology of identifiable central and peripheral synapses. *Mailing Add:* Dept Physiol Univ Va Box 395 Charlottesville VA 22908

HACKETT, JOSEPH LEO, b Springfield, Ohio, Jan 11, 37; m 63; c 4. MYCOLOGY, HEMATOLOGY. *Educ:* Ohio State Univ, BSc, 59, MSc, 63, PhD(clin path), 68. *Prof Exp:* Med lab technologist, Bact Lab, Ohio State Univ Hosp, 61-63, res technologist, Infectious Dis Lab, 63-68; dir qual control, Courtland Div, Abbott Labs 68-69, sect head med microbiol, Ref Lab, 69-72; supvry qual control microbiologist, Pfizer Diag, 72-74; supvry microbiologist, Standards Div, 74-77, IDE coordr res, 77-80, chief microbiol & immunol, Standards Br, 80-83, CHIEF MICROBIOL HEMAT, PATH CLIN LABS DIV, FOOD & DRUG ADMIN, 83- *Concurrent Pos:* Clin lab dir, Am Bd Bioanal, 76-86; adv, biol indicators panel, US Pharmacopeia, 81-82; mem, area comt immunol, Nat Comt Clin Lab Devices, 82-85, mem, antimicrobial susceptibility subcomt, 82- *Mem:* Am Soc Microbiol. *Res:* Effectiveness of clinical microbiological and hematological diagnostic devices. *Mailing Add:* Food & Drug Admin 1390 Piccard Dr Rockville MD 20850-4332

HACKETT, LE ROY HUNTINGTON, JR, b Long Beach, Calif, Dec 1, 44. MICROELECTRONIC PROCESS ENGINEERING, ELECTRONIC & METALLURGICAL FAILURE ANALYSIS. *Educ:* Calif State Univ, BS, 74. *Prof Exp:* Technician, MacDonald Douglas Aircraft, 64-65; res assoc, Sci Ctr, Rockwell Int, 65-73; mem tech staff, Hughes Aircraft Co, 73-83; engr, Gigabit Lowe, 83-84 & Plesscar Optronics, 84-86; MEM TECH STAFF, HUGHES RES LABS, 86- *Mem:* Am Ceramic Soc. *Res:* Integrating micro electro mechanical devices with microwave integrated circuits on III-V materials; process development for microwave monolithic integrated circuits; electron beam lithography on gallium-arsenide FET's. *Mailing Add:* 4238 Torreon Dr Woodland Hills CA 91364

HACKETT, NORA REED, b Abington, Pa, June 17, 43; m 68; c 3. CELL PHYSIOLOGY, CELL KINETICS. *Educ:* Pa State Univ, BS, 64; Brown Univ, ScM, 67, PhD(biol), 69. *Prof Exp:* Res fel biol chem, Harvard Med Sch, 68-69; res fel med, Miriam Hosp, Brown Univ, 69-71; assoc biologist, Inhalation Toxicol Res Inst, Lovelace Found, 75-79; consult, 80-82; patent adv, Off Pres, Univ Calif, 84-87; PATENT ADV, LAWRENCE LIVERMORE NAT LAB, 87- *Concurrent Pos:* Patent agent, Patent & Trademark Off, 89- *Mem:* Am Physiol Soc; Am Thoracic Soc; Sigma Xi. *Res:* Biotechnology patents; cell physiology; cell kinetics; biochemical regulation and ion transport; toxicology of non-nuclear fuel effluents; lung; kidney; adipose tissue. *Mailing Add:* Lawrence Livermore Nat Lab Patent Group L703 PO Box 808 Livermore CA 94550

HACKETT, ORWOLL MILTON, b Vayland, SDak, Jan 30, 20; m 46; c 4. GEOLOGY, HYDROLOGY. *Educ:* Univ Minn, BA, 48. *Prof Exp:* Geologist, US Geol Surv, 49-56, dist geologist, Boston, Mass, 56-61, chief, Ground Water Br, 61-65, chief, Off Water Data Coord, 65-68, assoc chief hydrologist, 68-79, staff hydrologist, 80-90; RETIRED. *Concurrent Pos:* UNESCO tech adv, Jordan, 65; US mem comn hydrol, World Meteorol Orgn, 72-80; alt chmn, US Nat Comt Sci Hydrol, 74-79. *Mem:* Fel Geol Soc Am; Am Inst Prof Geologists; Am Water Resources Asn; Am Geophys Union. *Res:* Groundwater geology and hydrology. *Mailing Add:* 2908 N Stafford St Arlington VA 22207

HACKETT, PETER ANDREW, b Havering, Eng, July 16, 48; m 70; c 3. LASER CHEMISTRY. *Educ:* Univ Southampton, BSc, 69, PhD(photochem), 72. *Prof Exp:* Res officer chem, 73-80, HEAD, LASER CHEM GROUP, NAT RES COUN CAN, 81- *Mem:* Chem Inst Can. *Res:* Multiphoton chemistry; laser isotope separation; photofragment spectroscopy; resonant scattering; industrial laser induced processes. *Mailing Add:* 303 Blair Rd Gloucester ON K1J 7M1 Can

HACKETT, RAYMOND LEWIS, b Hartford, Vt, Oct 30, 29; m 55; c 3. PATHOLOGY. *Educ:* Univ Maine, BA, 51; Univ Vt, MD, 55. *Prof Exp:* From instr to assoc prof, 62-72, PROF PATH, COL MED, UNIV FLA, 72-, ASSOC CHMN DEPT, 78- *Concurrent Pos:* Chief lab serv, Vet Admin Hosp, 71-78. *Mem:* Am Asn Path & Bact; Asn Exp Path; Int Acad Path. *Res:* Ultrastructure of renal lithiasis. *Mailing Add:* Dept Path Box J-275 JHMHC Univ Fla Gainesville FL 32610

HACKETT, WESLEY P, b Modesto, Calif, Apr 28, 30; m 53; c 4. HORTICULTURE. *Educ:* Univ Calif, Davis, BS, 53, MS, 59, PhD(plant physiol), 62. *Prof Exp:* Asst prof ornamental hort, Univ Calif, Los Angeles & asst plant physiologist, Agr Exp Sta, 62-68; assoc prof ornamental hort, Agr Exp Sta, Univ Calif, Davis 68-73, prof ornamental hort & plant physiologist, 73-82, chmn dept, 73-78; CONSULT, 78- *Honors & Awards:* Alex Laurie Award, Am Soc Hort Sci, 69 & 71. *Mem:* Am Soc Hort Sci. *Res:* Environmental physiology; growth and development of ornamental plants. *Mailing Add:* Hort Sci Univ Minn 252 Alderman St Paul MN 55108

HACKLEMAN, DAVID E, b Coos Bay, Ore. INTEGRATED CIRCUIT PROCESS TECHNOLOGY, MICRO FLUID MECHANICS. *Educ:* Ore State Univ, BS, 73; Univ NC, Chapel Hill, PhD(chem), 78. *Prof Exp:* Mem tech staff, Hewlett Packard Integrated Circuits, 78-81, prog mgr res & develop, 81-84, mem tech staff res & develop, 84-85, prog mgr res & develop, Hewlett Packard Ink-Jet Components Div, 85-89, fac loan, 89-90, PROG MGR RES & DEVELOP, INK-JET COMPONENTS DIV, HEWLETT PACKARD CO, 90- *Concurrent Pos:* Vis lectr, Nat Youth Sci Camp, 78-90, phys sci coordr, 90, vis scientist, 91; vis prof elec eng-solid state, City Col New York, 89-90. *Mem:* Am Chem Soc; Electrochem Soc. *Mailing Add:* Hewlett-Packard Co 1000 NE Circle Blvd Corvallis OR 97330

HACKLER, LONNIE ROSS, b Cloud Chief, Okla, Sept 20, 33; m 54; c 3. BIOCHEMISTRY, NUTRITION. *Educ:* Okla Agr & Mech Col, BS, 55; Univ Ill, MS, 57, PhD(animal nutrit), 58. *Prof Exp:* Asst, Univ Ill, 55-58; from asst prof to assoc prof biochem & nutrit, NY State Agr Exp Sta, Cornell Univ, 60-79; prof nutrit & head, dept foods & nutrit, Univ Ill, 79-83; CHMN DEPT HOME ECON, STATE UNIV COL NY, ONEONTA, 84- *Concurrent Pos:* Vis scientist, Human Nutrit Div, USDA, Md, 68-69; actg chief, Energy Metab Br & asst biol sci, Fitzsimons Gen Hosp, Denver, 59-60. *Mem:* Am Home Econ Asn; Am Inst Nutrit. *Res:* Bioassay procedures for the determination of protein quality; effect of processing procedures on utilization of proteins; metabolism and utilization of amino acids; amino acid methodology; role of dietary fiber in nutrition. *Mailing Add:* Dept Home Econ State Univ Col Oneonta NY 13820

HACKMAN, ELMER ELLSWORTH, III, b Philadelphia, Pa, Mar 22, 28; m 53; c 2. POLLUTION CONTROL, ENVIRONMENTAL TECHNOLOGY. *Educ:* Juniata Col, BS, 49; Univ Pa, MS, 57; Univ Del, PhD(chem), 67. *Prof Exp:* Proj engr, Chem Corps, US Army, 50-52; proj chemist, Am Viscose Corp, 53-54; asst mgr spec prod, ARCO Chem, 54-55; mgr res proj, Elkton Div, Thiokol Chem Corp, 58-73; PRES, NST/ENGRS, INC, 73- *Concurrent Pos:* Indust consult, Univ Del, 67-73; mem, Gov Sci Adv Bd, Md, 72-74; lectr, Chem Eng Mag & McGraw-Hill, 79-82. *Mem:* Am Chem Soc; Am Inst Chem Engrs. *Res:* Chemical process and equipment design; surfactants; energy conversion; radiation; combustion catalysis; air, water and solid waste treatment and process design; power plant emissions control; control of hazardous materials dissemination. *Mailing Add:* 505 Runnymede Rd Hockessin DE 19707

HACKMAN, JOHN CLEMENT, b Dayton, Ohio, May 16, 47; m 68; c 3. NEUROPHARMACOLOGY, NEUROPHYSIOLOGY. *Educ:* Univ Miami, BS, 69, MS, 76, PhD(biol), 79. *Prof Exp:* Adj asst prof neurol, 79-81, res asst prof, 81-82, asst prof neurol & pharmacol, 82-87, ASSOC PROF NEUROL, SCH MED, UNIV MIAMI, 87-; RES PHYSIOLOGIST NEUROL, VET ADMIN MED CTR, MIAMI, 79-, ASSOC PROF PHARMACOL, 91- *Mem:* Soc Neurosci; AAAS; Am Physiol Asn; Am Soc Pharmacol & Exp Therapeut. *Res:* Basic mechanism of actions of drugs and neurotransmitters in spinal cord function. *Mailing Add:* Dept Neurol D4-5 Sch Med Univ Miami Miami FL 33101

HACKMAN, MARTIN ROBERT, b New York, NY, Mar 20, 42; m 71; c 1. LABORATORY INFORMATION MANAGEMENT SYSTEMS. *Educ:* Brooklyn Col, BS, 65; City Univ New York, MA, 74. *Prof Exp:* Chemist electrochem, Princeton Appl Res Corp, 68-69; assoc scientist, 69-80, scientist anal chem, Hoffmann-La Roche Inc, 80-85; tech adminr, Whitehall Labs, 85-86; DATA MGT SPECIALIST, ANAQUEST INC, 86- *Concurrent Pos:* Lectr, Ctr prof advan, 77-78; mem, Subcomt LIMS, Am Soc Testing & Mat. *Mem:* Am Chem Soc; Am Asn Pharmaceut Scientists. *Res:* Method development of drugs in biological fluids incorporating analytical instrumentation such as gas and liquid chromatography, fluorescence, techniques specializing in electroanalytical chemistry and computerized data reduction; laboratory information management, system management, system analysis and robotics based on implementation of information and method development. *Mailing Add:* Nine Watchung Rd East Brunswick NJ 08816

HACKMAN, ROBERT J(OSEPH), geology, photogrammetry; deceased, see previous edition for last biography

HACKMAN, ROBERT MARK, b Pittsburgh, Pa, May 29, 53. NUTRITION EDUCATION, NUTRITION & ATHLETIC PERFORMANCE. *Educ:* Johns Hopkins Univ, BA, 75; Pa State Univ, MS, 77; Univ Calif, Davis, PhD(nutrit), 81. *Prof Exp:* ASSOC PROF NUTRIT, UNIV ORE, 81-, PARTIC PROF GERONT & PHYS EDUC, 82- *Mem:* Soc Nutrit Educ; AAAS; Am Inst Nutrit; Am Alliance Health Phys Educ Recreation & Dance; Teratol Soc. *Res:* Nutritional aspects of athletic performance; motivational strategies to help people make better dietary choices. *Mailing Add:* Dept Pub Health Univ Ore 250 Esslinger Hall Eugene OR 97403-4273

HACKNEY, CAMERON RAY, b Charleston, WVa, Oct 24, 51; m 74; c 1. FOOD MICROBIOLOGY. *Educ:* WVa Univ, BS, 73, MS, 75; NC State Univ, PhD(food sci), 80. *Prof Exp:* Res asst, NC State Univ, 76-80; from asst prof to assoc prof food microbiol, La State Univ, 80-85; ASSOC PROF, VA POLYTECH INST & STATE UNIV, 85-, SUPT SEAFOOD EXP STA, VA POLYTECH INST, 85- *Concurrent Pos:* Asst instr, WVa Univ, 73-75. *Mem:* Int Asn Milk, Food & Environ Sanitation; Am Soc Microbiol; Inst Food Technologists; Sigma Xi. *Res:* Microbial cell injury; survival, recovery and physiology of pathogenic Vibrios; food safety; food quality. *Mailing Add:* Dept Food Sci La State Univ Baton Rouge LA 70803

HACKNEY, COURTNEY THOMAS, b Mt Holly, NJ, Aug 11, 48; c 2. COMMUNITY ECOLOGY. *Educ:* Univ SAla, BS, 70; Emory Univ, MS, 72; Miss State Univ, PhD(zool), 77. *Prof Exp:* Res assoc ecol, Miss State Univ, 77; asst prof biol, Univ Southwestern La, 78-80; from asst prof to assoc prof, 80-88, PROF BIOL, UNIV NC, 88- *Concurrent Pos:* Mem NC Coastal Resources Comn, 89- *Mem:* Estuarine Res Fedn; Ecol Soc Am; Am Fisheries Soc; Soc Wetland Scientists (vpres, 84-85, pres, 86-87). *Res:* Coastal studies of shallow estuarine communities; productivity and decomposition of marsh plants; energetics of estuarine communities and the effect of human activity on estuarine systems and sea level rise. *Mailing Add:* Dept Biol Univ NC Wilmington NC 28406

HACKNEY, DAVID DANIEL, b New Orleans, La, Sept 6, 48. ENZYMOLOGY, BIOENERGETICS. *Educ:* La State Univ, New Orleans, BS, 70; Univ Calif, Berkeley, PhD(biochem), 75. *Prof Exp:* Fel, dept chem, Molecular Biol Inst, Univ Calif, Los Angeles, 75-78; asst prof, 78-83, ASSOC PROF BIOCHEM, DEPT BIOL SCI, CARNEGIE-MELLON UNIV, 83- *Mem:* Am Chem Soc; Biophys Soc; Am Soc Biol Chemists. *Res:* Mechanism, regulation, and structure of enzymes, involving investigation into the molecular basis of cooperativity and allosteric regulation, the mechanism of biological energy coupling, and the organzation of membrane proteins. *Mailing Add:* Dept Biol Sci Carnegie-Mellon Univ 4400 Fifth Ave Pittsburgh PA 15213

HACKNEY, JACK DEAN, b Marion, Ill, Oct 11, 24; m 46; c 2. INTERNAL MEDICINE, PHYSIOLOGY. *Educ:* St Louis Univ, MD, 48; Am Bd Internal Med, dipl, 56; Am Toxicol Sci, dipl. *Prof Exp:* Resident internal med, Vet Admin Hosp, Mo, 49-51; resident, White Mem Hosp, 53-54; instr, Loma Linda Univ, 54-56, asst clin prof, 56-57, asst prof, 57-64, from asst prof to assoc prof physiol, 65-69; assoc prof, 69-78, PROF MED, UNIV SOUTHERN CALIF, 78-, CHIEF ENVIRON HEALTH SECT, RANCHOS LOS AMIGOS HOSP, 70- *Concurrent Pos:* From actg dir to dir pulmonary lab, Los Angeles County Hosp, 54-63; clin physiologist, Rancho Los Amigos Hosp, 62-69, chief pulmonary function lab, 69- *Mem:* Fel Am Col Chest Physicians; fel Am Col Physicians; Am Col Toxicol; Am Physiol Soc; Sigma Xi. *Res:* Environmental health, especially air pollution health effects; experimental pathology of the lung, especially ozone, nitrogen dioxide and oxygen poisoning; pulmonary physiology, clinical application, especially non-invasive measures of important cardiopulmonary parameters. *Mailing Add:* 51 Med Sci Bldg 7601 E Imperial Hwy Downey CA 90242

HACKNEY, JOHN FRANKLIN, b Altdena, Calif, May 12, 45; m 67; c 2. ENDOCRINE PHARMACOLOGY, TERATOLOGY. *Educ:* Occidental Col, Los Angeles, BA, 67; Stanford Univ, PhD(pharmacol), 72. *Prof Exp:* NIH fel pharmacol, Univ Wis, Madison, 71-73; instr, 74-75, asst prof, 75-80, ASSOC PROF PHARMACOL, COL MED, UNIV SFLA, TAMPA, 80- *Concurrent Pos:* Res assoc, Courtauld Inst Biochem, Med Sch, Middlesex Hosp, London, 71. *Mem:* Am Chem Soc; Am Soc Pharmacol & Exp

Therapeut. *Res:* Biochemical mechanisms of gluocorticoid regulatory action particularly as it pertains to the control of cell growth; glucocorticoid receptor found in L929 mouse fibroblast tissue culture cells which are growth inhibited by glucocorticoids and which can be made resistant to growth inhibition. *Mailing Add:* Dept Pharmacol - Box 9 Univ SFla Col Med 12901 N Bruce B Downs Blvd Tampa FL 33612

HACKNEY, ROBERT WARD, b Louisville, Ky, Dec 11, 42; m 69; c 1. NEMATOLOGY & HELMINTHOLOGY, TAXONOMY & IDENTIFICATION. *Educ:* Northwestern Univ, BA, 65; Murray State Univ, MS, 69; Kans State Univ, PhD(parasitol), 73. *Prof Exp:* Instr histol, Murray State Univ, 67-68, instr zool, 68-69; instr bot, Kans State Univ, 72-73; scholar, Nema Seminars, Univ Calif, Riverside, 73-75; assoc plant nematologist, 75-85, sr plant nematologist, supvr, 85-89, SR PLANT NEMATOLOGIST, SPECIALIST, CALIF DEPT FOOD & AGR, 89- *Concurrent Pos:* Chmn, Calif Nematode Diag Adv Comn, 80-; com arbitrator, Am Arbitration Asn, 80-; prin investr & proj leader, Exotic Pest Res Grant, 86-87; chmn, Soc Nematologists Regulatory Comt, 89-91. *Mem:* Soc Nematologists; Int Coun Study Viruses & Virus Dis Grape; Sigma Xi. *Res:* Plant disease diagnosis and new pest detection; nematode taxonomy and identification using morphology, electrophoresis of proteins, cytology, cytogenetics, immunochemistry, monoclonal antibodies and nucleic acid analysis; computer modeling; computer-aided profile analysis. *Mailing Add:* Calif Dept Food & Agr PO Box 942871 Sacramento CA 94271-0001

HACKWELL, GLENN ALFRED, b Manti, Utah, Jan 20, 31; m 55; c 2. ENTOMOLOGY. *Educ:* Brigham Young Univ, BS, 57, MS, 58; Ore State Univ, PhD(entom), 67. *Prof Exp:* Assoc prof biol, 61-71, PROF BIOL, CALIF STATE UNIV, STANISLAUS, 71- *Concurrent Pos:* NSF res partic, Ore State Univ, 67; res partic, Univ Calif, Berkeley, 68-70; consult, Fennimore Chem Co, Campbell Soup Co & Food & Agr Orgn, UN. *Mem:* AAAS; Entom Soc Am. *Res:* Biology of Hymenoptera; insect ecology and behavior; eclosion and duration of larval development in the alkali bee, Nomia melanderi Cockerell; revision of bee genus Panurqinus. *Mailing Add:* Dept Biol Calif State Univ Stanislaus Turlock CA 95380

HACKWELL, JOHN ARTHUR, b Ashton-upon Lyne, Lancashire, Eng, May 31, 47; m 73. ASTROPHYSICS. *Educ:* Univ London, BSc, 68, PhD(physics), 71. *Prof Exp:* Asst prof physics & astron, 71-77, assoc prof physics, 82, PROF PHYSICS, UNIV WYO, 82-, DIR, WYO INFRARED OBSERV, 77- *Mem:* Am Astron Soc; Royal Astron Soc; Sigma Xi. *Res:* Observational infrared astronomy and the development of new infrared astronomical instrumentation. *Mailing Add:* Aerospace Corp MS M2-266 PO Box 92957 Los Angeles CA 90009

HACKWOOD, SUSAN, b Liverpool, Eng, May 23, 55. ROBOT SENSORS, ROBOT SYSTEM DESIGN. *Educ:* Leicester Polytech, UK, BSc, 76, PhD(solid state ionics), 79. *Prof Exp:* Fel solid state physics, AT&T Bell Labs, 79-81, mem tech staff, 81-82, supvr device robotics, 83, head, Dept Robotic Technol, 84; prof elec eng, 84, DIR, CTR ROBOTIC SYSTS IN MICROELECTRONICS, UNIV CALIF, SANTA BARBARA, 85- *Concurrent Pos:* Ed, J Robotic Systs, 84-; mem adv comt, Div Information, Robotics & Intelligent Systs, NSF, 85-; prin investr, NSF grants, 85-, Delco Electronics GM Robot & Effector for Vacuum Environ, 85- & SRC A Color Syst, 86-87; mem org comt, Int Workshop on Intelligent Systs, Inst Elec & Electronics Engrs, 87- *Mem:* Inst Elec & Electronics Engrs; Electrochem Soc. *Res:* Robot system design; sensory integration preception and organization. *Mailing Add:* Dept Elec & Comp Eng Univ Calif Santa Barbara CA 93106

HACQUEBARD, PETER ALBERTUS, b Rotterdam, Holland, Apr 9, 18; nat Can; m 45; c 4. GEOLOGY. *Educ:* State Univ Leiden, BSc, 37, MSc, 40; State Univ Groningen, PhD(coal petrol), 43. *Hon Degrees:* LLD, Dalhousie Univ, 80. *Prof Exp:* Geologist coal petrol, Geol Bur, Heerlen, Holland, 40-44; geologist field geol, Shell Oil Co Can, 46-48; sr geologist chg coal res sect, Coal Petrol & Palynology, Geol Surv Can, 48-83; ADJ PROF, DEPT GEOL, DALHOUSIE UNIV, 83- *Concurrent Pos:* Mem, Int Comt Coal Petrol, 53- *Honors & Awards:* G H Cady Medal, Geol Soc Am, 79; Reinhardt Thiessen Medal, Int Comt Coal Petrol, 79; Distinguished Lectr Award, Can Inst Mining & Metall, 83. *Mem:* Fel Geol Soc Am; Geol Asn Can; Can Inst Mining & Metall; Royal Soc Can. *Res:* Coal geology; coal petrology for stratigraphic, coal utilization and hydrocarbon exploration purposes. *Mailing Add:* Atlantic Geosci Ctr PO Box 1006 Dartmouth NS B2Y 4A2 Can

HADD, HARRY EARLE, b Baltimore, Md, Dec 24, 18; m 53; c 1. BIOCHEMISTRY, ENDOCRINOLOGY. *Educ:* Purdue Univ, BS, 46; Temple Univ, MA, 49; Ind Univ, PhD(biochem), 64. *Prof Exp:* Res asst metab, Lankenau Hosp Res Inst, 49-51; metab chemist, Philadelphia Gen Hosp, 51-53; instr, Div Endocrinol, Sch Med, Temple Univ, 53-57; scientist, Worcester Found Exp Biol, 57-60; res assoc, Div Endocrinol, Sch Med, Ind Univ-Purdue Univ, 60-65, asst prof urol, 65-66, asst prof obstet & gynec, 66-72, asst prof urol, 73-77, asst prof, 77-81, ASSOC PROF BIOCHEM, NORTHWEST CTR MED EDUC, SCH MED, IND UNIV, 81- *Concurrent Pos:* USPHS grant, 64-70 & 73-75; staff mem, Worcester Found Exp Biol, 72; consult to dir, Nat Ctr Toxicol Res, Food & Drug Admin, Jefferson, Ark, 77- *Mem:* AAAS; Am Chem Soc; Endocrine Soc; fel Am Inst Chemists. *Res:* Synthesis; biosynthesis and isolation of steroid hormone conjugates, including their role in the biological economy; synthesis prostate imaging agents for early detection prostate cancer in man; synthesis of steroid carboranes for BNCT. *Mailing Add:* 1051B N James Town Rd Decatur GA 30033

HADDAD, EUGENE, nuclear physics; deceased, see previous edition for last biography

HADDAD, GEORGE I, b Aindara, Lebanon, Apr 7, 35; US citizen; m 58; c 2. ELECTRICAL ENGINEERING. *Educ:* Univ Mich, BS, 56, MS, 58, PhD(elec eng), 63. *Prof Exp:* From instr to prof elec eng, Univ Mich, 60-69, dir, Electron Physics Lab, 68-75, chmn dept, 75-87, PROF ELEC ENG, UNIV MICH, ANN ARBOR, 69-, DIR, CTR HIGH FREQUENCY MICROELECTRONICS, 87-, DEPT CHMN, 91- *Concurrent Pos:* Mem tech prog comt, Int Solid-State Circuits Conf, 68-74; ed, J Transactions Microwave Theory & Techniques, 68-71; Robert J Hiller prof eng, 91- *Honors & Awards:* Curtis W McGraw Res Award, 70. *Mem:* Am Soc Eng Educ; Am Phys Soc; fel Inst Elec & Electronics Engrs. *Res:* Masers; parametric amplifiers; detectors; microwave electron beam devices; avalanche diodes; microwave solid-state devices; three-five semiconductor devices and integrated circuits; optoelectronic devices and integrated circuits. *Mailing Add:* Dept Elec Engr & Comput Sci Univ Mich 3303 EECS Ann Arbor MI 48109-2122

HADDAD, JERRIER ABDO, b New York, NY, July 17, 22; m 44; c 5. COMPUTERS, EDUCATION. *Educ:* Cornell Univ, BEE, 45. *Hon Degrees:* DSc, Union Col, 71, Clarkson Univ, 78. *Prof Exp:* Var positions, Endicott Lab, IBM Corp, 45-50, develop engr, Poughkeepsie, 50-52, mgr component develop, 52-53, engr, 53-54, dir adv mach develop, 54-56, gen mgr, Spec Eng Prod, 56-59 & Adv Systs Develop, 59-61, vpres data systs, 61-62 & data processing, 62-63, dir eng, Prog & Technol, 63-67, vpres eng prog & technol, 67-70, vpres & dir, Poughkeepsie Lab, 70-77, vpres develop systs prod div, 77-81; RETIRED. *Concurrent Pos:* Mem comput sci & eng bd, Nat Acad Sci, 67-72, vchmn, 71-72; mem adv coun col eng, Cornell Univ, 68-87; trustee, Clarkson Col Technol, 68-91, Webb Inst Naval Archit, 88-; dir, Am Dist Tel Co, 68-88; chmn Nat Res Coun Comt Educ & Utilization Engr, 81-85, mem Bd Army Sci & Technol, 78-82. *Honors & Awards:* Order of Cedars Medal, Off Rank, Lebanese Repub, 70. *Mem:* Nat Acad Eng; fel Inst Elec & Electronics Engrs; Sigma Xi. *Res:* Management in engineering, programming and technology area. *Mailing Add:* 162 Macy Rd Briarcliff Manor NY 10510

HADDAD, JOHN GEORGE, VITAMIN D METABOLISM, METABOLIC BONE DISEASES. *Educ:* Tulane Univ, MD, 62. *Prof Exp:* PROF MED & CHIEF, DEPT ENDOCRINOL, SCH MED, UNIV PA, 80- *Mailing Add:* Dept Med Univ Pa Sch Med 422 Curie Blvd CRB 611 Philadelphia PA 19106

HADDAD, LOUIS CHARLES, b Waterbury, Conn, Dec 15, 48; m 88. ANALYTICAL BIOCHEMISTRY. *Educ:* Fairfield Univ, BS, 70; Ind Univ, Bloomington, PhD(biol chem), 75. *Prof Exp:* NIH res assoc biophys chem, Univ Minneapolis, 75-77; sr res biochemist, Gen Mills Inc, Minneapolis, 77-78; SR RES SPECIALIST, ANALYTICAL BIOCHEM, 3M CO, ST PAUL, MINN, 78- *Mem:* Am Chem Soc; Am Asn Advan Sci. *Res:* Analytical biochemistry, chromatography, protein adsorption, enzymology and protein chemistry. *Mailing Add:* 2152 Juno Ave St Paul MN 55116

HADDAD, RICHARD A, b Brooklyn, NY, Nov 26, 34; m 61; c 3. ELECTRICAL ENGINEERING. *Educ:* Polytech Inst Brooklyn, BEE, 56, MEE, 58, PhD(elec eng), 62. *Prof Exp:* Res fel, Polytech Inst NY, 56-57, from instr to assoc prof elec eng, 57-78, assoc dean, 80-82, DIR, WESTCHESTER GRAD CTR, POLYTECH INST NY, 82- *Concurrent Pos:* Lectr indust, Gen Elec Co, Ford Instrument Co, Grumman Aircraft Corp & Am Bosch Arma Corp; consult, Sperry Gyroscope Co, NY, 63-64, Bell Tel Labs, NJ, 62 & 65-69 & Concord Res Corp, Mass, 71-73, Fairchild Camera, Syosset, NY, 75-76 & US Army Armaments Res & Develop Command, Dover, NJ, 74-78 & 81-; head, Div Eng, Inst Nat d'Electricite et d'Electronique, Boumerdes, Algeria, 78-80; US Nat Acad Sci export to Ministry of Educ, Lanzhou Univ, People's Repub China, 85. *Mem:* Inst Elec & Electronics Engrs. *Res:* Feedback control theory; digital and discrete-time system analysis and synthesis; simulation. *Mailing Add:* Dept Elec Eng Polytech Univ 333 Jay St Brooklyn NY 11201

HADDAD, ZACK H, b Jan 27, 38; m 64; c 2. ALLERGY, IMMUNOLOGY. *Educ:* Univ Cairo, BSc, 56; Univ Paris, MD, 61; Am Bd Allergy & Immunol, dipl, 67 & 72; Am Bd Pediat, dipl, 66. *Prof Exp:* Resident pediat, Cook County Hosp, Univ Ill, 62-63 & Bellevue Hosp, Med Sch, NY Univ, 64-65; res fel immunol, Children's Hosp Pittsburgh, Med Sch, Univ Pittsburgh, 65-66; USPHS immunol allergy fel, Med Ctr, Univ Calif, San Francisco, 66-67; from asst prof to assoc prof pediat, 67-79, PROF PEDIAT, SCH MED, UNIV SOUTHERN CALIF, 79-, DIR RES & TRAINING, 67-, CHIEF, PEDIAT ALLERGY & IMMUNOL CLIN DIV, 67- *Concurrent Pos:* Res assoc path, Children's Hosp of Los Angeles, Med Sch, Univ Pittsburgh, 63-64; attending physician, Pediat Allergy Clin, Children's Hosp Pittsburgh, 65-66; Calif Lung Asn Res & Educ Fund grant, Syntex, Inc clin res grant & Calif Lung Asn gen res fund grant, 72-73; consult various nat & int med facil, 73-; Union Bank pvt grant, 74-75; Astra Labs, Inc clin res grant, 75-76; Pvt Pharm Indust clin res grant, 77-78 & 78; Astra Labs res grant, 78; Pharmacia Int res grant, 78, Sandoz, 80. *Honors & Awards:* Bela Shick Mem Lectr Award, 81. *Mem:* Fel Am Acad Allergy; fel Am Acad Pediat; fel Am Col Allergists; Soc Pediat Res; fel Am Asn Cert Allergists. *Res:* Mechanisms of immediate hypersensitivity mediated by IgE; immunoglobulin E-IgE sensitization of heterologous mast cells; drug hypersensitivity; immunochemical studies of food allergens and pollen allergens; mechanisms of immunotherapy; cyclic nucleotides in bronchial asthma. *Mailing Add:* Dept Pediat Univ Southern Calif 2025 Zonal Ave Los Angeles CA 90033

HADDEN, CHARLES THOMAS, b Newport News, Va, May 28, 44; m 66; c 4. BACTERIAL METABOLISM, BACTERIAL GENETICS. *Educ:* Univ Chicago, BS, 66; Univ Wash, PhD(microbiol), 70. *Prof Exp:* Res assoc radiation biol, Univ Fla, 71-73; res asst prof radiation biol, Oak Ridge Grad Sch Biomed Sci, Univ Tenn, 74-83, res assoc prof radiation biol, 83-84; STAFF MICROBIOLOGIST, OAK RIDGE RES INST, TENN, 84- *Mem:* AAAS; Am Soc Microbiol. *Res:* Genetics and mechanisms of DNA repair in bacillus subtilis; physiological responses of Bacillus subtilis to chemical mutagens; DNA-mediated transformation and other genetics systems in anaerobes; microbial treatment of wastewater and soil contaminated with heavy metals. *Mailing Add:* SAIC PO Box 2501 Oak Ridge TN 37831

HADDEN, JOHN WINTHROP, b Berkeley, Calif, Oct 23, 39; m 64; c 1. INTERNAL MEDICINE, IMMUNOPHARMACOLOGY. *Educ:* Yale Univ, BA, 61; Columbia Univ, MD, 65. *Prof Exp:* Intern med, Roosevelt Hosp, NY, 65-66, resident, 66-68; USPHS-NIH spec fel, Dept Pediat & Path, Univ Minn, Minneapolis, 69-72, asst prof path, 72-73; assoc prof, Grad Sch

Med Sci, Cornell Univ, 73-; assoc mem & dir lab immunopharmacol, Sloan Kettering Mem Inst, 73-; PROF MED & DIR PROG IMMUNOPHARMACOL, MED COL, UNIV SFLA, TAMPA, 83- *Concurrent Pos:* Am Heart Asn estab investr, 72-77; assoc attend, Mem Hosp, 73-; assoc ed, Int J Immunopharmacol, 78- *Honors & Awards:* Angier Res Prize; Kellog Res Prize. *Mem:* Am Asn Pathologists; Am Asn Immunologists; NY Acad Sci; Int Soc Immunopharmacol (pres, 85-88, past pres, 88-91). *Res:* Immunopharmacology; mechanisms of regulation of cellular function by cyclic nucleotides and related cell surface events; biochemistry of lymphocyte activation; mechanisms of thymic hormone action; lymphokines and macrophage proliferation and activation; immunotherapeutic drug development and analysis. *Mailing Add:* Prog Immunopharmacol Univ SFla Med Col 12901 N 30th St Tampa FL 33612

HADDEN, STUART TRACEY, chemical engineering, mathematical statistics; deceased, see previous edition for last biography

HADDOCK, AUBURA GLEN, b Jasper, Ark, May 29, 35; m 54; c 4. TOPOLOGY. *Educ:* Ark State Teachers Col, BS, 54; Okla State Univ, MS, 58, PhD, 61. *Prof Exp:* Partic, Ark Exp Teacher Educ, 54-55; teacher, pub sch, Ark, 55-57; asst prof math, Ark Col, 58-59; asst, Okla State Univ, 59-61; prof, Ark Col, 61-63, acad dean, 63-65, dean, 65-67; assoc prof, 67-69, chmn dept, 68-81, PROF MATH, UNIV MO, ROLLA, 69- *Concurrent Pos:* NSF prog, Okla State Univ, 57-58; mathematician, Oak Ridge Nat Labs, 78-79. *Mem:* Math Asn Am; Am Math Soc. *Res:* Point set topology; fixed point theorems. *Mailing Add:* Rte 4 Box 159 Rolla MO 65401

HADDOCK, FREDERICK THEODORE, JR, b Independence, Mo, May 31, 19; div; c 2. RADIO ASTRONOMY. *Educ:* Mass Inst Technol, BS, 41; Univ Md, MS, 50. *Hon Degrees:* DSc, Southwestern at Memphis, 65 & Ripon Col, 66. *Prof Exp:* Physicist & electronic scientist, US Naval Res Lab, 41-56; assoc prof astron & elec eng, 56-59, prof elec eng, 59-67, PROF ASTRON, UNIV MICH, 59-, DIR RADIO ASTRON OBSERV, 61- *Concurrent Pos:* Trustee, Assoc Univs, Inc, 64-68; mem vis comt, Nat Radio Astron Observ, 56-58, 63-64; adv panel astron, NSF, 57-60, 63-66, adv panel, Int Year Quiet Sun, 63-66, mem facil panel, Off Instr Progs, 63-64, 67-70; mem, Space Sci Bd ad hoc comt astron, Nat Acad Sci, 58-62, radio frequency requirement sci res, 61-68, Whitford panel astron facil, 63-64; mem ad hoc panel on US Navy 600 foot radio telescope, President's Sci Adv Comt, 62, Air Force Off Sci Res Arecibo eval panel, Arecibo Observ, 62-69, adv comt planetary & interplanetary sci, Off Space Sci, NASA, 61-62, ad hoc working group on Apollo sci exp & training, 62-63, NASA hq astron subcomt, 67-69; consult, Off Space Sci & Applns, 70-, NASA astron missions bd, Radio Astron Panel, 68-71, ad hoc comt on Nat Astron Space Observ, 70, NASA outer planets grand tour mission, Radio Astron Team, 71-; vis res assoc, Calif Inst Technol, 66, mem adv comt for Owens Valley Observ, 69-; mem, US Navy Eclipse Expeds, Aleutian Islands, 50, Khartoum, 52; mem comn radio astron, Int Astron Union, 55-, exec & organizing comts, 64-67, comn space astron, 64-; chmn comn five, Int Sci Radio Union, 54-57, del to assemblies, 58-61, mem nat comt, 58-61. *Mem:* Am Astron Soc (vpres, 61-63); fel Inst Elec & Electronics Eng; Am Inst Aeronaut & Astronaut; fel Royal Astron Soc; Sigma Xi. *Res:* Radio astronomy; extragalactic sources, solar and planets; antennas and radiometers; space research. *Mailing Add:* Box 1245 Ann Arbor MI 48106

HADDOCK, GERALD HUGH, b Neosho, Mo, Mar 7, 29; m 60; c 3. GEOLOGY. *Educ:* Wheaton Col, Ill, BS, 56; State Col Wash, MS, 59; Univ Ore, PhD(geol), 67. *Prof Exp:* From instr to prof, 71-91, EMER PROF GEOL, WHEATON COL, ILL, 91- *Mem:* Geol Soc Am; Nat Asn Geol Teachers; Soc Econ Paleontologists & Mineralogists; Mineral Soc Am; Sigma Xi. *Res:* Petrology of welded tuffs. *Mailing Add:* Dept Physics/Geol Wheaton Col Wheaton IL 60187-5539

HADDOCK, JOHN R, MATHEMATICS. *Prof Exp:* FAC MEM, DEPT MATH SCI, MEMPHIS STATE UNIV. *Mailing Add:* Dept Math Sci Memphis State Univ Memphis TN 38152

HADDOCK, JORGE, b Caguas, PR, Aug 15, 55; m 80; c 2. SIMULATION, PROJECT MANAGEMENT & PRODUCTION PLANNING. *Educ:* Univ PR, BS, 78; Rensselaer Polytechnic Inst, MS, 79; Purdue Univ, PhD(indust eng), 81. *Prof Exp:* From instr to asst prof, Indust Eng, Univ PR, 78-83; asst prof indust eng, Clemson Univ, 84-86; asst prof, 86-90, ASSOC PROF INDUST ENG & OPER RES, RENSSELAER POLYTECHNIC INST, 90- *Concurrent Pos:* Consult var corp, 78-90; res chmn, Oper Res Div, Inst Indust Engrs, 85-86, newslett ed, 87-88, publ cluster leader, 88-91. *Honors & Awards:* Outstanding Young Indust Eng Award, Inst Indust Engrs, 90. *Mem:* Corresp mem Nat Acad Eng Mex; Oper Res Soc Am; Soc Computer Simulation; assoc mem Inst Mgt Sci; sr mem Inst Indust Engrs. *Res:* Modeling of manufacturing/production and inventory control systems, as well as the design and implementation of simulation modeling and analysis tools; author or co-author of over 60 technical publications. *Mailing Add:* 102 Old Coach Rd Clifton Park NY 12065

HADDOCK, LILLIAN, b Naguabo, PR, Feb 25, 29. INTERNAL MEDICINE, ENDOCRINOLOGY. *Educ:* Univ PR, BS, 50; Temple Univ, MD, 54; Am Bd Internal Med, dipl, 64; Am Bd Endocrinol & Metab, dipl, 72. *Prof Exp:* Internship, Bayamon Dist Hosp, PR, 54-55; resident internal med, San Juan City Hosp, 55-57; NIH trainee endocrinol, Johns Hopkins Hosp, 57-59; from assoc to assoc prof med, Sch Med, Univ PR, 59-70, chief sect endocrinol & diabetes, 60, dir endocrinol & diabetes, 60-76, dir med educ, Univ Hosp, 68-72, assoc dean acad affairs, Med Sci Campus, 76-78, dean, 78-85, actg dean med, 77-78, PROF MED, SCH MED, UNIV PR, 70- *Concurrent Pos:* Vis prof, Harvard Med Sch, 73-74 & 86; Nat Comn Diabetes, 75; mem, NLM Biomed Libr Rev Comt, 83-87; mem, Nat Diabetes Adv Bd, 90-93; gov, PR Chap, Am Col Physicians, 92-96. *Honors & Awards:* PR Acad Arts & Sci Prize Acomplishments Endocrinol, 89. *Mem:* Am Diabetes Asn; Endocrine Soc; Am Fedn Clin Res; fel Am Col Physicians; Pan-Am Med Asn; Am Soc Bone & Mineral Res. *Res:* Metabolism; vitamin D; calcium and phosphorus metabolism; diabetes mellitus; clinical research in all types of endocrine and metabolic diseases. *Mailing Add:* Univ PR Med Sci Campus San Juan PR 00936-5067

HADDOCK, PHILIP GEORGE, b San Diego, Calif, Mar 19, 13; m 42; c 5. FORESTRY. *Educ:* Univ Calif, BS, 34, PhD(plant physiol), 42. *Prof Exp:* Jr forester, Calif Forest & Range Exp Sta, US Forest Serv, 34-36; asst, dept forestry, Univ Calif, 36-39; asst prof forest bot, State Univ NY Col Forestry, Syracuse, 46-47; asst prof forestry, Univ Wash, 47-53; from assoc prof to prof, 53-78, EMER PROF FORESTRY, UNIV BC, 78- *Mem:* Fel AAAS; Soc Am Foresters; fel Can Inst Forestry; cor mem Ital Acad Forestry Sci. *Mailing Add:* 4620 W Second Ave Vancouver BC V6R 1L1 Can

HADDOCK, ROY P, b Nogales, Ariz, Mar 30, 28; m 52; c 5. NUCLEAR PHYSICS. *Educ:* Univ Calif, Berkeley, AB, 52, PhD(physics), 57. *Prof Exp:* Asst res physicist, Univ Calif, Berkeley, 57-59, from asst prof to assoc prof, 59-71, prof nuclear physics, 71-77, PROF PHYSICS, UNIV CALIF, LOS ANGELES, 77- *Concurrent Pos:* Consult, Lawrence Berkeley Lab, 59- *Mem:* Fel Am Phys Soc. *Res:* High energy nuclear physics research and instrumentation. *Mailing Add:* Dept Physics 3-174 Univ Calif 405 Hilgard Ave Los Angeles CA 90024

HADDON, ROBERT CORT, b Sheffield, Eng, Mar 12, 43; Australian citizen; m 83; c 1. ORGANIC CHEMISTRY, ORGANIC METALS. *Educ:* Univ Melbourne, BSc Hons, 66; Pa State Univ, PhD(org chem), 71. *Prof Exp:* Res fel theoret org chem, Univ Tex, Austin, 72-73; hon fel, Res Sch Chem, Australian Nat Univ, 73-76; vis scientist org chem, 76-78, MEM TECH STAFF CHEM PHYSICS, BELL LABS, 78- *Concurrent Pos:* Queen Elizabeth II fel, 73-75. *Mem:* Am Chem Soc; Royal Soc Chem; fel Royal Australian Inst. *Res:* Theoretical chemistry; organic chemistry; organic metals; molecular engineering; theory, design and synthesis of new materials. *Mailing Add:* MH1A-359 AT&T Bell Labs Murray Hill NJ 07974-2070

HADDON, WILLIAM F, (JR), b Denver, Colo, Jan 5, 42; c 1. MASS SPECTROMETRY. *Educ:* Harvey Mudd Col, BS, 64; Purdue Univ, PhD(chem), 68. *Prof Exp:* Res chemist, Celanese Res Co, NJ, 68-69; RES CHEMIST, WESTERN REGIONAL LAB, AGR RES SERV, USDA, 69- *Mem:* Am Soc Mass Spectrometry; AAAS. *Res:* Organic mass spectrometry applied to structure elucidation and trace analysis; computer interfacing of chemical instrumentation. *Mailing Add:* 35 Willow Ave Larkspur CA 94939-1321

HADDOX, CHARLES HUGH, JR, b Maramec, Okla, July 14, 21; m 45; c 1. CLINICAL CHEMISTRY. *Educ:* Univ Tex, BA, 48, MA, 49, PhD(zool), 51. *Prof Exp:* Res scientist, Rockefeller Found, Univ Tex, 48-52; instr & res assoc genetics & oncol, Sch Med, La State Univ, 52-56; res biochemist, Home for Jewish Aged, Pa, 56-59 & Nat Children's Cardiac Hosp, 59-66; mem staff, Univ Miami, 65-66; CHEMIST, GOOD SAMARITAN HOSP, DAYTON, 66- *Mem:* Am Asn Clin Chemists; Am Chem Soc; Asn Clin Sci; Sigma Xi. *Res:* Biochemistry of rheumatic fever and related disease states; M proteins of streptococci; isoenzyme characterization. *Mailing Add:* Good Samaritan Hosp 2222 Philadelphia Dr Dayton OH 45406

HADDY, FRANCIS JOHN, b Walters, Minn, Sept 6, 22; m 46; c 3. PHYSIOLOGY, INTERNAL MEDICINE. *Educ:* Univ Minn, BS, 43, BM, 46, MD, 47, MS, 49, PhD(physiol), 53. *Prof Exp:* Asst prof physiol, Northwestern Univ, 53-55, asst prof med & physiol, 57-61; prof physiol & chmn dept & assoc prof med, Sch Med, Univ Okla, 61-66; prof physiol & chmn dept, Mich State Univ, 66-76; chmn dept, 76-87, PROF PHYSIOL, UNIFORMED SERV UNIV, 87- *Concurrent Pos:* Sr grade physician, Vet Admin Res Hosp, Chicago, 53-55, clin investr, 57-59, chief res, 59-61; chief circulation sect, US Army Med Res Lab, Ft Knox, Ky, 55-57. *Honors & Awards:* Wiggers Award, 66. *Mem:* Am Physiol Soc (pres, 81); Am Soc Clin Invest. *Res:* Cardiovascular physiology. *Mailing Add:* Dept Physiol 4301 Jones Bridge Rd Bethesda MD 20814

HADEEN, KENNETH DOYLE, b Haxtun, Colo, Mar 8, 31; m 53; c 2. METEOROLOGY, COMPUTER SCIENCE. *Educ:* Colo State Univ, BS, 53; Tex A&M Univ, MS, 61, PhD(meteorol), 66. *Prof Exp:* Chief tech serv div, HQ Air Weather Serv, 72-73; Air Force Global Weather Ctr Opers officer, Air Weather Serv, USAF, 73-75, wing vcomdr, 75-76; assoc dir, First Global Atmospheric Res Prog Global Exp, Dept Defense, HQ USAF, 76-77; dep dir, Ctr Environ Assessment, 77-84; DIR, NAT CLIMATIC DATA CTR, 84- *Honors & Awards:* Merewether Award, Air Weather Serv, Dept Air Force, 70. *Mem:* AAAS; Am Meteorol Soc. *Res:* Numerical analysis and forecasting; physical scientist administration. *Mailing Add:* 226 Wildflower Rd Asheville NC 28804

HADEISHI, TETSUO, OPTICAL PHYSICS. *Prof Exp:* MEM STAFF, LAWRENCE BERKELEY LAB, UNIV CALIF. *Mem:* Am Phys Soc. *Mailing Add:* 70-143 Lawrence Berkeley Lab Univ Calif Berkeley CA 94720

HADEMENOS, JAMES GEORGE, b Houston, Tex, June 13, 31; m 62; c 3. BIOPHYSICS, SCIENCE EDUCATION. *Educ:* Univ Houston, BS, 58; Univ SDak, MA, 66; Syracuse Univ, PhD(biophys), 70. *Prof Exp:* Teacher, Spring Br Independent Sch Dist, Tex, 58-65; instr physics, Stephen F Austin State Univ, 66-67; teaching asst, Syracuse Univ, 67-69; assoc prof, Lock Haven State Col, 69-70; assoc prof sci educ, 70-74, PROF EDUC & HEAD DEPT, ANGELO STATE UNIV, 74- *Concurrent Pos:* NSF consult, Phys Sci Study Comt Physics, Univ Andhra, India, 67. *Mem:* AAAS. *Res:* Ozone and air ions accompanying biological applications of electric fields; Piaget-type conservation tasks. *Mailing Add:* Dept Educ Angelo State Univ San Angelo TX 76901

HADEN, CLOVIS ROLAND, b Houston, Tex, Apr 10, 40; m 56; c 3. ELECTRICAL ENGINEERING. *Educ:* Univ Tex, Arlington, BS, 61; Calif Inst Technol, MS, 62; Univ Tex, PhD, 65. *Prof Exp:* Design engr, Tex Instruments Inc, 62; res engr scientist, Electronic Mat Res Lab, Univ Tex, 63-64; asst prof elec eng, Univ Okla, 65-68; assoc prof, Tex A&M Univ, 68-71, prof, 71-72, dir inst solid-state electronics, 69-72; dir, Sch Elec Engr & Comput Sci, Univ Okla, 72-78; vpres acad affairs & provost, 87-89, DEAN COL ENG & APPL SCI, ARIZ STATE UNIV, 78- *Concurrent Pos:* Exec ed, Elec Power Systs Res, 78- *Mem:* Am Soc Eng Educ; Nat Soc Prof Engrs; fel Inst Elec & Electronics Engrs. *Res:* Application of superconductivity; semiconductor devices and physics; electric power systems. *Mailing Add:* Col Eng Ariz State Univ Tempe AZ 85287

HADEN, WALTER LINWOOD, JR, b Richmond, Va, May 26, 15. PHYSICAL INORGANIC CHEMISTRY. *Educ:* Univ Richmond, BS, 36; Univ NC, PhD(chem), 41. *Prof Exp:* Asst, Univ NC, 36-39; res chemist, Columbia Chem Div, Pittsburgh Plate Glass Co, 41-49; res chemist, Attapulgus Minerals & Chem Corp, 49-52, res supvr, 53-57, asst dir res, Minerals & Chem Corp Am, 57-58, assoc dir res, 58-71; ASSOC DIR RES, ENGELHARD MINERALS & CHEM CORP, 71- *Mem:* AAAS; Am Chem Soc; Am Ceramic Soc; Clay Minerals Soc. *Res:* Chain photolysis of acetaldehyde in intermittent light; kinetics; synthesis and properties of inorganic compounds; adsorption and catalysis; particle size analysis and pore size distribution; mineral synthesis; crystal structures and thermal transformations. *Mailing Add:* 438 Poe Ave Westfield NJ 07090-2729

HADER, ROBERT JOHN, b Hartford, Wis, Apr 13, 19; m 42; c 2. MATHEMATICAL STATISTICS. *Educ:* Univ Chicago, BS, 43; NC State Univ, PhD(statist), 49. *Prof Exp:* Instr exp statist, NC State Univ, 49; statistician, Los Alamos Sci Lab, 49-51; from assoc prof to prof, 51-82, EMER PROF EXP STATIST, NC STATE UNIV, 83- *Mem:* Fel Am Statist Asn. *Res:* Sampling acceptance inspection; experimental designs for industrial research. *Mailing Add:* 3313 Cheswick Dr Raleigh NC 27609

HADER, RODNEY N(EAL), b Paris, Mo, Sept 13, 22; m 52; c 3. SCIENTIFIC & EDUCATIONAL ADMINISTRATION. *Educ:* Univ Ill, BS, 44. *Prof Exp:* Partic, Manhattan proj, US Army Corps Engrs, 44-46; develop engr, Firestone Plastics Co, Firestone Tire & Rubber, 44 & 46-50; assoc ed, Appl Publ, 50-56, ed, J Agr & Food Chem, 56-64, asst to ed dir, 60-62, exec asst to dir publ, 62-70, actg secy, 70-76, exec asst to exec dir, 73-77, dep to exec dir, 79-80, chief operating officer, 80-82, SECY, AM CHEM SOC, 77-, DEP EXEC DIR, 83- *Concurrent Pos:* Actg ed, Chem, Am Chem Soc, 62-63, J Chem Eng & Data, 64, I&EC Prod Res & Develop, 65-68; mem bd dirs, Centcom, Ltd, 82- *Mem:* Am Chem Soc; AAAS; Am Inst Chem Engrs. *Mailing Add:* 907 Kenbrook Dr Silver Spring MD 20902-3228

HADERLIE, EUGENE CLINTON, b Thayne, Wyo, Mar 23, 21; m 45; c 2. BIOLOGICAL OCEANOGRAPHY. *Educ:* Univ Calif, AB, 43, MA, 48, PhD, 50. *Prof Exp:* From asst to lectr zool, Univ Calif, 47-50; instr biol sci, Monterey Peninsula Col, 50-54, chmn dept, 54-65; assoc prof, 65-68, prof oceanog, Naval Postgrad Sch, 68-86; RETIRED. *Concurrent Pos:* NSF fac fel, Univ Bristol, Eng & Stazione Zoologia, Naples, Italy, 58-59; liaison scientist, Off Naval Res, London, 62-63; actg assoc prof, Hopkins Marine Sta, Stanford Univ, 62, 64 & vis prof, 75; consult, US Dept Energy, Ocean Thermal Progs, 74-78. *Res:* Helminth parasites of the fishes of California; California intertidal invertebrates; marine boring and fouling organisms; marine pollution. *Mailing Add:* Dept Oceanog Naval Postgrad Sch Monterey CA 93940

HADERLIE, LLOYD CONN, b Afton, Wyo, Feb 23, 46; m 71; c 8. WEED SCIENCE. *Educ:* Utah State Univ, BS, 71; Univ Ill, PhD(agron), 75. *Prof Exp:* Asst prof agron, Univ Nebr, Lincoln, 75-81; assoc prof weed sci, Univ Idaho, 81-88; RES & CONSULT, AGRA SERV INC, 88- *Mem:* Weed Sci Soc Am; Am Soc Plant Physiologists; AAAS. *Res:* Weed biochemistry and physiology; perennial weed, rhizome; bud dormancy; translocation physiology; herbicide mode of action; weed control in potatoes, sugar beets and cereals. *Mailing Add:* Agra Serv Inc 3243 W Joanna Ct American Falls ID 83211

HADERMANN, ALBERT FELIX, b New York, NY, Mar 20, 38; m 62; c 4. PHYSICAL CHEMISTRY, POLYMER CHEMISTRY. *Educ:* City Col New York, BS, 59; Am Univ, MS, 69, PhD(chem), 70. *Prof Exp:* Sr chemist, Melpar, Inc, Va, 63-67; sr scientist, Ctr Environ Anal, Va, 70; asst prof phys chem, American Univ, 71-74; consult, Enviro-Control, Inc, 74-76; consult, JRB Assocs, Inc, 76-82; vpres res & develop, Gen Technol Appln, Inc, 82-91; PRES, HADERMANN CONSULTS, INC, 91- *Concurrent Pos:* Consult, EAS Sci Inc, Md, 70-, Appl Phys Chem Inc, 70- & Mineral Pigments Co, 71; fel, Am Univ, 71. *Mem:* AAAS; Am Chem Soc; NY Acad Sci; Am Defense Preparedness Asn. *Res:* Magnetically stabilized electrophoresis; photochemistry; electrokinetic behavior at high field strengths; fractionation of biological materials; applications of membrane theory; model biological membranes; properties of macroradicals; oil-spill removal products; composite development. *Mailing Add:* 11609 Browningsville Rd Ijamsville MD 21754

HADFIELD, MICHAEL GALE, b Seattle, Wash, Feb 8, 37; m. INVERTEBRATE ZOOLOGY, MARINE BIOLOGY. *Educ:* Univ Wash, AB, 59, MS, 61; Stanford Univ, PhD(biol sci), 67. *Prof Exp:* Asst prof zool, Pomona Col, 66-68; from asst prof to assoc prof, 68-79, PROF, DEPT ZOOL & PAC BIOMED RES CTR, KEWALO MARINE LAB, UNIV HAWAII, 79- *Concurrent Pos:* Vis prof, Friday Harbor Labs, 68, 78, 81, 84, 86, 88, Hopkins Marine Sta, 89; ed, Marine Biol; assoc ed, J Exp Zool. *Mem:* Fel AAAS; Am Soc Zool; Sigma Xi; Western Soc Naturalists; Int Soc Invertebrate Reproduction; Soc Conserv Biol. *Res:* Invertebrate development; embryology larval development and metamorphosis of marine gastropods; evolutionary and conservation biology of Hawaiian tree snails. *Mailing Add:* Kewalo Marine Lab 41 Ahui St Pac Biomed Res Ctr Univ Hawaii Honolulu HI 96813

HADIDI, AHMED FAHMY, b Damanhur, Egypt, Mar 1, 37; US citizen; m 64; c 3. MOLECULAR BIOLOGY, MICROBIOLOGY. *Educ:* Cairo Univ, BSc Hons, 58; Univ Minn, MS, 62; Kans State Univ, PhD(plant path), 67. *Prof Exp:* Res fel, Univ Ky, 67-68 & Purdue Univ, 68-69; res assoc, Baylor Col Med, 70 & Univ Calif, Berkeley, 70-73; virologist, Litton Bionetics, Inc, 73-75; MICROBIOLOGIST, USDA, 75- *Mem:* Am Soc Virol; Am Phytopath Soc; AAAS; NY Acad Sci; Sigma Xi. *Res:* Replication and structure of plant viruses and viroids. *Mailing Add:* Nat Plant Germ Plasms Quaratine Lab USDA Agr Res Serv Beltsville MD 20705

HADIDIAN, ZAREH, b Aintab, Turkey, Feb 9, 11; US citizen; m 39; c 3. PHARMACOLOGY. *Educ:* Rensselaer Polytech Inst, BS, 34, MS, 35; Clark Univ, PhD(gen physiol), 39. *Prof Exp:* Res assoc physiol, Clark Univ, 39-40, instr, 40-41; asst pharmacologist, Lederle Labs, NY, 41-43; instr pharmacol, Albany Med Col, 43-44; physiologist, Mem Found Neuro Endocrine Res, 44-46; biochemist, Worcester Found Exp Biol, 46-48; asst prof pharmacol, Med Sch, Tufts Univ, 48-50, assoc prof, 50-58; liaison officer, Off Naval Res, London, 58-60; dir pharmacol, Mason Res Inst, Pennwalt Corp, 60-66, dir off sci info, Pharmaceut Div, 66-76; CONSULT, ASTRA PHARMACEUT PRODS INC, 76- *Mem:* AAAS; Soc Toxicol; Am Physiol Soc; Am Soc Pharmacol & Exp Therapeut; Royal Soc Med. *Res:* Hyaluronic acid; hyaluronidase; snake venoms; toxicology. *Mailing Add:* PO Box 1003 Berlin MA 01503-1003

HADJIAN, RICHARD ALBERT, b Boston, Mass, Nov 2, 49; m 81. PROTEIN BIOCHEMISTRY. *Educ:* Boston Univ, AB, 71; Univ NH, PhD(biochem), 76. *Prof Exp:* Fel pharmacol, Health Ctr, Univ Conn, 76-81; SR RES ASSOC, NEW ENG NUCLEAR CORP, 81- *Mem:* Sigma Xi. *Res:* Protein biochemistry and enzymology; role of protein phosphorylation in the contracture response of plated and acetylcholine receptor function; role of antibodies as specific carriers for medical diagnostic purposes. *Mailing Add:* Dupont Merck Pharmaceut 331 Treble Cove Rd North Billerica MA 01862

HADJIMICHAEL, EVANGELOS, b Thessaloniki, Greece, Aug 16, 37; m 60; c 2. THEORETICAL NUCLEAR PHYSICS. *Educ:* City Col New York, BS, 60; Univ Calif, Berkeley, PhD(physics), 65. *Prof Exp:* Asst prof physics, Calif Polytech Col, 65-66; res assoc, Yale Univ, 66-68; from asst prof to assoc prof, 67-74, PROF PHYSICS, FAIRFIELD UNIV, 74- *Concurrent Pos:* NSF grants intermediate energy nuclear physics, nuclear defense & nuclear arms limitations. *Mem:* Am Phys Soc. *Res:* Heavy ion scattering; deuteron photodisintegration; scattering reactions. *Mailing Add:* Dept Physics Fairfield Univ Fairfield CT 06430

HADLER, HERBERT ISAAC, b Toronto, Ont, Aug 22, 20; US citizen; m 47; c 3. BIOCHEMISTRY. *Educ:* Univ Toronto, BASc, 42; Univ Wis, PhD, 52. *Prof Exp:* Demonstr chem eng, Univ Toronto, 42-43 & 44-45; supvr dryer house, Seagram's Can, 43-44; asst chem, Univ Wis, 47-51; fel, McArdle Mem Lab, 51-53; assoc surg, Chicago Med Sch, 53-54, from asst prof to assoc prof, 54-59; asst prof biochem, Inst Enzyme Res, Univ Wis, 63-66; assoc prof chem, 66-74, PROF CHEM & BIOCHEM, SOUTHERN ILL UNIV, 74- *Concurrent Pos:* Fogarty sr int fel, NIH, 79-80. *Honors & Awards:* Parker Award, Chicago Med Sch, 58. *Mem:* Am Asn Cancer Res; Am Soc Biol Chem; Sigma Xi; Am Col Toxicol; Am Chem Soc. *Res:* Carcinogenesis mitochondrial genes and oxidative phosphorylation. *Mailing Add:* Dept Microbiol Univ Minn Med Sch 420 Delaware St Minneapolis MN 55455

HADLER, NORTIN M, b New York, NY, Nov 13, 42; m 65; c 2. RHEUMATIC DISEASE. *Educ:* Yale, AB, 64; Harvard Univ, MD, 68. *Prof Exp:* PROF MED, MICROBIOL & IMMUNOL, UNIV NC, 73- *Mem:* Am Soc Clin Invest; Am Asn Immunol. *Res:* Industrial rheumatology; clinical investigations on back pain, disability and compensation. *Mailing Add:* Sch Med 932 FLO Bldg 231H Univ NC Chapel Hill NC 27599-7280

HADLEY, BRUCE ALAN, b Pittsburgh, Pa, Aug 1, 50; m 71; c 1. PLANT PATHOLOGY, PLANT BREEDING. *Educ:* Pa State Univ, BS, 72, MS, 73; NC State Univ, PhD(plant path), 78. *Prof Exp:* Res technician plant path, NC State Univ, 73-78; RES PATHOLOGIST PLANT PATH, E I DU PONT DE NEMOURS & CO, INC, 78- *Mem:* Am Phytopath Soc. *Res:* Chemical control of plant diseases. *Mailing Add:* E I du Pont Co Stien-Haskell Res Ctr Box 30 Elkton Rd Newark DE 19714

HADLEY, CHARLES F(RANKLIN), b Redlands, Calif, Nov 27, 14; m 42; c 3. ELECTRONICS ENGINEERING. *Educ:* Calif Inst Technol, BS, 37, MS, 38; Stanford Univ, PhD(elec eng), 44. *Prof Exp:* Asst develop geophys instruments, Magnolia Petrol Co, Tex, 38-40; actg instr, Stanford Univ, 40-44; res assoc radar counter measures, Radio Res Lab, Harvard Univ, 44-45; res engr geophys instruments, Stanolind Oil & Gas Co, 45-51, sr res engr, 51-58; res assoc, Amoco Prod Co, 58-70, spec res assoc, 70-79; RETIRED. *Mem:* Inst Elec & Electronics Engrs; Sigma Xi. *Res:* Microwave measurements; microwave vacuum tubes; velocity distribution of velocity modulated electron beams; electrical engineering. *Mailing Add:* 5623 S Quebec Ave Tulsa OK 74135

HADLEY, DONALD G, b Artesia, NMex, July 24, 36; m 63; c 3. SEDIMENTOLOGY, STRATIGRAPHY. *Educ:* Eastern NMex Univ, BS, 59; Univ Wis, MS, 63; Johns Hopkins Univ, PhD(geol), 68. *Prof Exp:* Technician biochem, Univ Wis Inst Enzyme Res, 61-63; GEOLOGIST, US GEOL SURV, 67- *Mem:* AAAS; Am Asn Petrol Geologists; fel Geol Soc Am; Soc Econ Paleont & Mineral. *Res:* Paleocurrents, basin analysis, stratigraphy and sedimentary petrography of Precambrian quartzites, Ontario and Quebec, Canada; anthracite, Pennsylvania, and bituminous, West Virginia and Pennsylvania, coal basins; regional geology and structure of the Arabian Shield. *Mailing Add:* Br Eastern Mineral Resources US Geol Surv 954 National Ctr Reston VA 22092

HADLEY, ELBERT HAMILTON, b Springville, NY, July 10, 13; m 41; c 2. CHEMISTRY. *Educ:* Univ Mich, BS, 36, MS, 37; Duke Univ, PhD(chem), 40. *Prof Exp:* Asst, Duke Univ, 38-40; res chemist, Electrochem Dept, E I du Pont de Nemours & Co, Inc, NY, 40-47; assoc prof chem, Col Sci, Southern Ill Univ, Carbondale, 47-55, asst chmn dept, 60-65, asst dean, 65-71, prof chem, 55-71, dean, 71-88; RETIRED. *Concurrent Pos:* Vis prof, Kabul Univ, Afghanistan, 60. *Mem:* AAAS; Am Chem Soc; Electrochem Soc. *Res:* Pharmaceutical synthetic organic chemistry; vapor phase catalytic fluorination; fluid catalytic reactions; electroplating; cyanides. *Mailing Add:* Dept Chem Col Sci Southern Ill Univ 1002 Briarwood Dr Carbondale IL 62901

HADLEY, ELMER BURTON, b Iola, Kans, Oct 24, 36. BOTANY, PLANT ECOLOGY. *Educ:* Univ Calif, Santa Barbara, BA, 58; Univ Ill, MS, 60, PhD(plant ecol), 63. *Prof Exp:* Asst bot, Univ Ill, 58-61, asst plant ecol, 61-63; fel, Brookhaven Nat Lab, 63-64; asst prof biol, Univ NDak, 64-65; from asst prof to prof, 65-70, actg head dept biol sci, 69-71, head dept, 71-74, dean, Col Lib Arts & Sci, 74-80, PROF BIOL, UNIV ILL, CHICAGO CIRCLE, 70- *Concurrent Pos:* Co-dir, Univ Ill & City Cols of Chicago Partnership Prog, 86-87. *Mem:* Ecol Soc Am; Sigma Xi. *Res:* Physiological ecology of alpine vascular plants; carbon dioxide exchange rates of plants in the field. *Mailing Add:* Dept Biol Sci Univ Ill Chicago Circle Box 4348 Chicago IL 60680

HADLEY, EVAN C, b Washington, DC, Jan 19, 47; m; c 2. GERIATRICS RESEARCH. *Educ:* Yale Univ, BA, 71; Univ Pa, MD, 78. *Prof Exp:* Staff fel, Geront Res Ctr, Nat Inst Aging, NIH, 78-80, health sci adminr, Geriat Res Sect, Biomed Res & Clin Med Prog, 80-82, actg chief, Geriat Br, 82-84, chief, 84-90, ASSOC DIR GERIAT, NAT INST AGING, NIH, 90- *Concurrent Pos:* Mem, Nutrit Coord Comt, Diabetes Mellitus Interagency Coord Comt & Digestive Dis Interagency Coord Comt, 84-87, Arthritis & Musculoskeletal Dis Interagency Coord Comt, 84-; mem, Comt Nat Res Agenda Aging, Nat Acad Sci, 89, res comt, Am Geriat Soc, 89-; chair, Res Task Force, Clin Med Sect, Geront Soc Am, 89- *Honors & Awards:* J Marion Sims Lectr, Am Uro-Gynec Soc, 88. *Mem:* Am Geriat Soc; Geront Soc Am. *Res:* Protein catabolism in human fibroblasts in vitro and effects of proteases on fibroblast proliferation; glycosylated hemoglobins in man in relation to age and other indices of glucose metabolism; formation and removal of carcinogen-DNA abducts in human cells in tissue culture; therapies for urinary incontinence; interventions to reduce physical frailty and fall injuries among older persons. *Mailing Add:* Geriat Prog Bldg 31 Rm 5C27 Nat Inst Aging 9000 Rockville Pike Bethesda MD 20892

HADLEY, FRED JUDSON, b Kansas City, Mo, July 18, 46; m; c 2. PHYSICAL CHEMISTRY. *Educ:* Univ Kans, BA, 68; Rice Univ, PhD(chem), 77. *Prof Exp:* Asst instr chem, Tex A&I Univ, 70-73; vis asst prof chem, Univ Ill, Champaign, 76-78; vis asst prof, 78-80, asst prof chem, Wabash Col, 80-84; ASSOC PROF CHEM, ROCKFORD COL, 84- *Mem:* Am Chem Soc; Sigma Xi. *Res:* Mass spectrometry and gas-phase kinetics; stopped flow solution kinetics; kinetics in non-aqueous media. *Mailing Add:* 424 N Gardiner Rockford IL 61107

HADLEY, GEORGE RONALD, b Memphis, Tenn, Nov 25, 46; m 67; c 3. LASER PHYSICS, OPTICS. *Educ:* Wichita State Univ, BA, 68; Iowa State Univ, PhD(physics), 72. *Prof Exp:* STAFF MEM PHYSICS, SANDIA CORP, 72- *Mem:* Optical Soc of Am. *Res:* Modelling of semiconductor lasers; nonlinear optics. *Mailing Add:* Orgn 1124 Sandia Labs Bldg M071 Rm G PO Box 5800 Albuquerque NM 87185-5800

HADLEY, GILBERT GORDON, b Takoma Park, Md, May 3, 21; m 44; c 3. PATHOLOGY. *Educ:* Wash Missionary Col, BS, 43; Col Med Evangelists, Loma Linda Univ, MD, 44. *Prof Exp:* From instr to assoc prof path, 49-69, from asst dean to assoc dean admin affairs, 64-77, dean sch med, 77-86, PROF PATH, SCH MED, LOMA LINDA UNIV, 69-; DIR, HEALTH & TEMPERANCE DEPT, WORLD HQS, SEVENTH DAY ADVENTISTS, 88- *Concurrent Pos:* Assoc prof & actg head dept, Christian Med Col, India, 55-58; vis prof, Kabul Univ, Afghanistan, 60-61, WHO vis prof, 63-64 & 70-72; clin prof path, Univ Southern Calif, 65- *Mem:* Am Soc Clin Pathologists; Am Med Asn; Asn Am Pathologists & Bacteriologists; Col Am Path. *Res:* Teaching medical students; author of articles in Clinical Hematology and geographic pathology. *Mailing Add:* Dept Health & Temperance Gen Conf Seventh-Day Adventist 6840 Eastern Ave NW Washington DC 20012

HADLEY, HENRY HULTMAN, b Blackstone, Ill, July 25, 17; m 44; c 3. GENETICS, PLANT BREEDING. *Educ:* Univ Ill, BS, 40, PhD(agron), 51; Tex Agr & Mech Col, MS, 42. *Prof Exp:* From asst prof to assoc prof genetics, Tex Agr & Mech Univ, 46-57; from assoc prof to prof plant genetics, 57-87, EMER PROF, UNIV ILL-URBANA, 87- *Mem:* AAAS; Sigma Xi; Am Soc Agron; Am Genetic Asn. *Res:* Cytogenetics and breeding of soybeans and sorghum. *Mailing Add:* Agron/AW 106 Turner Hall Univ Ill Urbana IL 61801

HADLEY, HENRY LEE, b Washington, DC, Aug 21, 22; m 45; c 4. UROLOGY. *Educ:* Columbia Union Col, BS, 42; Loma Linda Univ, MD, 46. *Prof Exp:* Chief, Urol Clin, White Mem Med Ctr, 51; sr attend physician, Los Angeles Co Gen Hosp, 58-70; PROF UROL, LOMA LINDA UNIV MED CTR, 66-, CHMN DEPT, 67- *Concurrent Pos:* Chief, Vet Hosp, Loma Linda, Calif, 77- & San Bernadino Co Med Ctr, 79- *Mem:* Soc Int Urol; Soc Univ Urologists; Am Urol Asn; Am Col Surgeons; Am Fertil Soc. *Res:* Author or coauthor of thirty publications. *Mailing Add:* 11370 Anderson St Suite 1100 Loma Linda CA 92354

HADLEY, JAMES WARREN, b Pasadena, Calif, Nov 15, 24; m 54; c 5. PHYSICS. *Educ:* Calif Inst Technol, BS, 45; Univ Calif, PhD, 52. *Prof Exp:* Physicist, 46-65, assoc leader, 78-83, dep leader, 83-87, ASST LEADER, SP PROJ PROG, LAWRENCE LIVERMORE LAB, 87- *Concurrent Pos:* Sci mgr seismol, Nev Opers, Atomic Energy Comn, 69-72. *Res:* Experimental nuclear physics, especially high energy particle scattering; nuclear reactions; reactor design; nuclear and aerospace propulsion; geophysics; seismology. *Mailing Add:* L-389 Lawrence Livermore Lab Livermore CA 94550

HADLEY, JEFFERY A, b St Joseph, Mich, Dec 10, 58; m 82. FOOD PROCESS MACHINE DEVELOPMENT. *Educ:* Tri-State Univ, BSc, 82; Purdue Univ, MSc, 88. *Prof Exp:* Br mgr, Am Testing & Eng, 78-81; mgr corp eng, Ball Corp, 82-88; div eng mgr, 88-90, DIR BUS CTR, FMC-FPSD, 90- *Mem:* Am Soc Civil Eng. *Res:* Computer modelling of materials and mechanical behaviors for product development; development of new processes and equipment to service the food processing industry; development of computer engineering applications. *Mailing Add:* FMC Corp 2300 Industrial Ave Madera CA 93639

HADLEY, KATE HILL, b New York, NY, Mar 24, 50; m 85. GEOLOGY. *Educ:* Mass Inst Technol, BS, 71, PhD(geophys), 75. *Prof Exp:* Res & res supv, Exxon Prod Res Co, 75-80, prod geol, 80-81, explor geol, 81-82, explor planning, 82-84, div drilling coord, 84-86, HQ WELL EVAL COORDR, EXXON CO, USA, 86- *Concurrent Pos:* Assoc ed, J Geophy Res, 76-78; mem-at-large, US Nat Comt Rock Mech, 78-81, chmn, 81-82; corp vis comt, Dept Earth, Atmospheric & Planetary Sci, Mass Inst Technol, 83-88; mem bd earth sci, NAS, 84-87; mem, SPE prog coord comt, 89-90, chmn, 91; chmn, Dept Energy, US Geol Surv NSF Coun Continental Sci Drilling, 90- *Mem:* Soc Petrol Engrs; Am Asn Petrol Geologists; Am Geophys Union. *Res:* Formation evaluation; reservoir evaluation; engineering geology. *Mailing Add:* Exxon Co USA PO Box 2180 Houston TX 77252-2180

HADLEY, LAWRENCE NATHAN, b Valley Center, Kans, Oct 14, 16; m 39, 90; c 2. PHYSICAL OPTICS. *Educ:* Friends Univ, AB, 37; Univ Okla, MS 39; Univ Mich, PhD(physics), 47. *Prof Exp:* Lab instr physics, Friends Univ, 35-37; lab instr physics, Univ Okla, 37-39; instr physics & math, Northern Okla Jr Col, 39-41; lab instr physics, Univ Mich, 41, res assoc, 42-45; asst prof, Colo Agr & Mech Col, 47; instr, Dartmouth Col, 47-48, asst prof, 48-55; from assoc prof to prof physics, Colo State Univ, 55-87, actg head, 59-61, chmn dept, 65-68, EMER PROF PHYSICS, COLO STATE UNIV, 87- *Concurrent Pos:* Fund adv ed fel, Univ Calif, 54-55; vis staff mem, Los Alamos Sci Labs, 64-65. *Mem:* Am Phys Soc; Sigma Xi; Am Asn Physics Teachers; fel Optical Soc Am. *Res:* Optical properties of metal semiconductor and dielectric films; reflection and transmission interference filters; solar energy utilization. *Mailing Add:* 3417 Canadian Pkwy Ft Collins CO 80524

HADLEY, MAC EUGENE, b San Jose, Calif, June 16, 30; m 66; c 1. COMPARATIVE ENDOCRINOLOGY. *Educ:* San Jose State Col, BA, 54; Brown Univ, MS, 64, PhD(biol), 66. *Prof Exp:* Teacher high sch, Calif, 59-62; USPHS res fel, 66-68; assoc prof cell & develop biol, 68-80, prof gen biol, 80-83, PROF ANAT, UNIV ARIZ, 83- *Mem:* AAAS; Am Inst Biol Sci. *Res:* Pineal and pituitary roles in vertebrate pigmentation; mechanisms of hormone action; regulation of pars intermedia function; regulation of vertebrate chromatophores. *Mailing Add:* Dept Anat Rm 4205 Basic Sci Bldg Univ Ariz Col Med 1501 N Campbell Tucson AZ 85724

HADLEY, NEIL F, b Dearborn, Mich, Oct 13, 41; m 75; c 2. ENVIRONMENTAL PHYSIOLOGY. *Educ:* Eastern Mich Univ, BA, 63; Univ Colo, PhD(zool), 66. *Prof Exp:* From asst prof to assoc prof, 66-71, PROF ZOOL, ARIZ STATE UNIV, 76-, ASSOC DEAN, GRAD COL, 89- *Concurrent Pos:* Asst vpres res, 88-89. *Honors & Awards:* Cert Merit for Distinguished Contrib Arid Zone Res, AAAS, 86. *Mem:* Am Arachnology Soc; Am Soc Zool; Sigma Xi. *Res:* Structure and function of arthropod cuticle; physiological adaptations of terrestrial arthropods to desert and montane environments; adaptive role of lipids in biological systems. *Mailing Add:* Dept Zool Ariz State Univ Tempe AZ 85287-1503

HADLEY, RICHARD FREDERICK, b Minneapolis, Minn, Jan 4, 24; m 52; c 2. GEOMORPHOLOGY. *Educ:* Univ Minn, BA, 48, MS, 50. *Prof Exp:* Geologist, US Geol Surv, 48-67, chief soil & moisture conserv prog, 67-74, chief, Pub Lands Hydrol Prog, 74-80, hydrologist, water resources div, 67-84; ADJ PROF, DEPT GEOG, UNIV DENVER, 84- *Concurrent Pos:* Vpres, Int Comn Continental Erosion, Int Asn Hydrol Sci, 75-79 & 80-83, secy, 83-87. *Honors & Awards:* Meritorious Serv Award, US Dept Interior, 82. *Mem:* AAAS; Int Asn Hydrol Sci; Am Water Resources Asn; fel Geol Soc Am; Am Geophys Union; Sigma Xi. *Res:* Hydrology and geomorphology of arid regions; relations of drainage basin characteristics to erosion and sedimentation; aggradation and erosion in stream channels; rehabilitation potential of surface-mined lands. *Mailing Add:* Dept Geog Univ Denver Denver CO 80208

HADLEY, STEVEN GEORGE, b Richmond, Ind, Aug 3, 42; m 66. PHYSICAL CHEMISTRY. *Educ:* Univ Ore, BA, 63; Univ Calif, Davis, PhD(chem), 66. *Prof Exp:* Res chemist, Nat Bur Standards, 66-68; asst prof chem, Univ Utah, 68-72; res chemist, US Air Force Weapons Lab, 72-77; mgr technol, 77-81, CHIEF PROJ ENGR, EXCIMER LASER PROG, ROCKETDYNE DIV, ROCKWELL INT, 81- *Concurrent Pos:* Nat Res Coun res assoc, Nat Bur Standards, 66-68. *Mem:* Am Chem Soc; Am Phys Soc. *Res:* Spectroscopy and kinetics associated with chemical lasers. *Mailing Add:* Satellite & Space Electronics Div Rockwell International 2600 Westminster Blvd PO Box 3644 Seal Beach CA 90740-7644

HADLEY, SUSAN JANE, b Madison, Wis; m 51; c 2. PULMONARY DISEASES, MICROBIOLOGY. *Educ:* Univ Wis, BA, 41; Cornell Univ, MD, 44. *Prof Exp:* Instr microbiol, Sch Med, NY Univ, 50-51; from instr to prof, 50-86, asst to chmn, 65-85, EMER PROF MED, CORNELL UNIV, 86- *Concurrent Pos:* Asst dir labs, New York Hosp-Cornell Med Ctr, 51-64, physician OPIS, 51-55, asst attend physician, 55-63, assoc attend physician & dir lab microbiol, 63-72, attend physician, 72-86; chief pulmonary clin, New York Hosp, 64-86. *Mem:* NY Acad Med (exec secy, 85-86); Am Thoracic Soc. *Res:* Immunology. *Mailing Add:* 1035 Fifth Ave New York NY 10028

HADLEY, WILLIAM KEITH, b Eugene, Ore, Nov 12, 28; m 53; c 2. CLINICAL PATHOLOGY, MICROBIOLOGY. *Educ:* Univ Calif, Berkeley, AB, 50, PhD(bact), 67; Yale Univ, MD, 59. *Prof Exp:* Teaching asst bact, Univ Calif, Berkeley, 52-55; intern med, Yale New Haven Med Ctr, 59-60, from intern to asst resident path, 60-62; jr res bacteriologist, Univ Calif, Berkeley, 62-64; resident, 64-67, lectr clin path, 67, from asst prof to assoc prof clin path & lab med, 67-76, CLIN PROF LAB MED & MICROBIOL, UNIV CALIF, SAN FRANCISCO, 73-; CHIEF, MICROBIOL DIV, SAN FRANCISCO GEN HOSP, 67- *Concurrent Pos:* Res fel path, Sch Med, Univ Calif, San Francisco, 64-67; trainee path, Sch Med, Yale Univ, 61-62; trainee bact, Univ Calif, Berkeley, 62-64. *Mem:* Acad Clin Lab Physicians & Scientists; AAAS; Am Fedn Clin Res; Am Soc Clin Path; Am Soc Microbiol; Am Venereal Dis Asn; Infectious Dis Soc Am; Soc Hosp Epidemiologists Am; Sigma Xi; Soc Anal Cytol. *Res:* Clinical microbiology; infectious disease; rapid and automated systems for microbiologic identification; antimicrobiol action of therapy; indigenous microorganisms of humans; pathogenesis of sexually transmitted disease. *Mailing Add:* 18 Reed Ranch Rd Tiburon CA 94920-2071

HADLEY, WILLIAM MELVIN, b San Antonio, Tex, June 4, 42; m 90; c 3. BIOTRANSFORMATION, NASAL MEMBRANES. *Educ:* Purdue Univ, BS, 67, MS, 71, PhD(toxicol), 72. *Prof Exp:* From asst prof to assoc prof pharmacol & toxicol, 72-82, asst dean, 84-86, DEAN, COL PHARM, UNIV NMEX, 86-, PROF PHARMACOL & TOXICOL, 82- *Concurrent Pos:* Investr, NIH Minority Schs Biomed Advan, 74-87, NIH ROI, 84-87; vis scientist toxicol, Inhalation Toxicol Res Inst, 81-82; actg dean, Col Pharm, Univ NMex, 85; mem, US Pharmacopeial Conv, 86- & bd dirs, Nat Ctr Toxicol Res Assoc Univ, 90-; adv bd mem, US Dept Energy Waste Educ & Res Consortium, 90-; pres, Rocky Mountain Chap Soc Toxicol, 90-91. *Mem:* Soc Toxicol; Southwestern Asn Toxicologists; Western Pharmacol Soc; AAAS; Am Asn Cols Pharm. *Res:* Biotransformation of xenobiotics in nasal membranes; effects of heavy metals on biotransformation; cadmium; effects of xenobiotics on the immune system. *Mailing Add:* Col Pharm Univ NMex Albuquerque NM 87131

HADLEY, WILLIAM OWEN, b Covington, Tenn, Mar 14, 39; m 64; c 3. CIVIL ENGINEERING, ENGINEERING MATERIALS. *Educ:* Univ Tenn, BS, 61; Univ Tex, MS, 68, PhD(civil eng), 72. *Prof Exp:* Asst design engr, Clark Daily & Dietz, 61; design engr, Clark Dietz Painter & Assoc, 63-64; field engr, Leavell-Kiewit, 64-65; res engr, Tippett & Gee, 65-66; res assoc civil engr, Ctr Hwy Res, Univ Tex, 67-71; assoc prof, 71-75, PROF CIVIL ENG, LA TECH UNIV, 77-, DIR, MAT RES LAB, 81- *Concurrent Pos:* Prin investr fed & state proj, Fundamental Eng Properties Construct Mat, 77- & Gap Graded Concretes, 75-77. *Mem:* Transp Res Bd; Asn Asphalt Paving Technologists; Am Soc Civil Engrs; Am Soc Testing & Mat. *Res:* Material characterization of construction materials; pavement design and analysis; geotechnical engineering. *Mailing Add:* TRDF 2602 Dellana Lane Austin TX 78746

HADLOCK, CHARLES ROBERT, b Brooklyn, NY, Apr 19, 47; m 67; c 2. RISK ANALYSIS, ENVIRONMENTAL MODELING. *Educ:* Providence Col, BS, 67; Univ Ill, Urbana, MA, 68, PhD(math), 70. *Prof Exp:* Asst prof math, Amherst Col, 70-76 & Bowdoin Col, 76-77; sr mem staff, A D Little Inc, 77-80, mgr safety & environ risk, 80-90; PROF & CHMN, MATH SCI DEPT, BENTLEY COL, 90- *Concurrent Pos:* Sr Fulbright lectr, Colombia, 76. *Mem:* Am Soc Mech Eng; Soc Indust & Appl Math; Math Asn Am; Int Asn Math Geol. *Res:* Risk analysis with emphasis on environmental applications; applied mathematics. *Mailing Add:* Dept Math Sci Bentley Col Waltham MA 02154-4705

HADLOCK, RONALD K, b Oneonta, NY, Sept 26, 34; m 61; c 1. METEOROLOGY. *Educ:* State Univ NY Albany, BS, 56, MS, 57; Fla State Univ, PhD(meteorol), 66. *Prof Exp:* Instr physics, State Univ NY Binghamton, 57-60; res assoc meteorol, Fla State Univ, 64-67, asst prof oceanog & meteorol, 67-73; sr res scientist, Batelle Pac Northwest Labs, 73-77, mgr appl meteorol, 77-83, PNL Admin PRECP Prog, 84-85, dir Erica Prog, 85-91, PROG MGR ATMOSPHERIC SCI, BATELLE PAC NORTHWEST LABS, 91- *Mem:* AAAS; Am Meteorol Soc. *Res:* Dynamics of rotating fluids as laboratory models of geophysical prototypes, especially chemical analogs to latent heat release; meteorology of a small heated tropical island; boundary layer structure; power generation meteorology; intensification of oceanic storms; meteorological field programs; meteorological emergency preparedness. *Mailing Add:* Battelle Pac Northwest Labs PO Box 999 Richland WA 99352

HADLOW, WILLIAM JOHN, b West Park, Ohio, Apr 8, 21; m 52; c 2. VETERINARY PATHOLOGY. *Educ:* Ohio State Univ, DVM, 48. *Prof Exp:* Instr vet med, Univ Minn, 48-50; vet pathologist, Rocky Mt Lab, NIH, 52-58; vet pathologist animal dis & parasite res div, USDA, 58-61; RES VET, ROCKY MT LAB, NIH, 61- *Mem:* Am Vet Med Asn; Am Asn Path; Am Col Vet Path; Int Acad Path; Wildlife Dis Asn. *Res:* Relation of viruses to chronic progressive disease in man and animals; comparative pathology. *Mailing Add:* 908 S Third St Hamilton MT 59840

HADOW, HARLO HERBERT, b Beaver Dam, Wis, Feb 10, 45; m 82. BEHAVIORAL ECOLOGY, EVOLUTION. *Educ:* Milton Col, BA, 67; Univ Colo, MA, 72, PhD(biol), 77. *Prof Exp:* Teacher sci, Crowley County High Sch, 68-70; lab asst psychol, Vet Admin Hosp, 70-72; instr biol, Ft Lewis Col, 72-76; asst prof, 77-83, ASSOC PROF BIOL, COE COL, 83- *Concurrent Pos:* Instr anat, Arapaho Jr Col, 71-72; consult, Aspen Colo Trout Unlimited, 72 & Iowa Conserv Comn, 78- *Mem:* Cooper Ornith Soc; Wilson Ornith Soc; Am Ornith Union; Ecol Soc Am; Wildlife Soc. *Res:* Growth and development of nesting woodpeckers; woodpecker ecology; audible communication by woodpeckers; mammalian and avian distribution. *Mailing Add:* Dept Biol Coe Col 1220 First Ave NE Cedar Rapids IA 52402

HADWIGER, LEE A, b Alva, Okla, Nov 2, 33; m 57; c 4. PLANT PATHOLOGY, BIOCHEMISTRY. *Educ:* Okla State Univ, BS, 55, MS, 59; Kans State Univ, PhD(hort), 62. *Prof Exp:* Fel biochem, Okla State Univ, 62-63; fel biochem & biophys, Univ Calif, Davis, 63-65; from asst prof to assoc prof, 65-73, PROF PLANT PATH, WASH STATE UNIV, 73- *Honors & Awards:* Asgrow Award, Am Soc Hort Sci, 64. *Mem:* Am Soc Plant Physiol; Am Phytopath Soc; Am Soc Hort Sci. *Res:* In vitro and in vivo biosynthesis of pyridine compounds, aromatic amino acids, asparagine, B-cyano alanine and orcylalanine in plants; biochemistry of resistant gene action and host-parasite interactions. *Mailing Add:* Dept Plant Path Wash State Univ Pullman WA 99164-6430

HADZERIGA, PABLO, b Buenos Aires, Argentina, Dec 25, 29; US citizen; m 57; c 3. INORGANIC CHEMISTRY. *Educ:* La Plata Univ, BSc, 51; Univ Buenos Aires, MS, 55, PhD(chem), 56. *Prof Exp:* Chemist, Bonneville, Ltd, Utah, 57-59, chief chemist, 59-61, dir res, 61-64; sr process engr, Hazen Res, 64-78; INDEPENDENT CONSULT, 78- *Mem:* AAAS; Am Inst Mining, Metall & Petrol Eng; Am Chem Soc; NY Acad Sci. *Res:* Phase chemistry of aqueous solutions; fertilizer industry; potash recovery from brines and ores. *Mailing Add:* 6058 Owens St Arvada CO 80004-4645

HADZIJA, BOZENA WESLEY, b Zagreb, Yugoslavia, Jan 5, 28; nat US; m 60; c 2. PHYSICAL PHARMACY, DRUG METABOLISM. *Educ:* Univ Zagreb, BS, 49, MS, 51, PhD(biochem), 60. *Prof Exp:* Lectr pharm, Univ Zagreb, 49-56; chief analyst, Pharmaceut Indust, Zagreb, 56-61; fel, Univ Conn, Storrs, 61-63, res asst, 63-64; sr lectr pharm, UST, WAfrica, 64-71; asst prof, 71-77, ASSOC PROF PHARM, UNIV NC, CHAPEL HILL, 77- *Mem:* Am Pharmaceut Asn; Am Asn Col Pharm; Asn Prof Sleep Soc; Sigma Xi. *Res:* Pharmacokinetics and bioavailability studies of drugs in vivo; design and application of new analytical methods for quantitation of drugs in biological fluids; metabolic pathways of drugs in animals and humans; stability of drug products. *Mailing Add:* 408 Highview Dr Chapel Hill NC 27514

HADZIYEV, DIMITRI, b Ak-Hisar, Turkey, Dec 10, 29; Can citizen; m 55; c 1. FOOD SCIENCE, AGRICULTURE. *Educ:* Univ Belgrade, BSc, 53, PhD(chem natural prod), 64. *Prof Exp:* Res assoc, Qual Control Lab, Albus, Yugoslavia, 55-59; res assoc appln radioisotopes in agr, Inst Agr Res, 59-63; res assoc agr chem, Univ Novi Sad, 63-65, asst prof biochem, 65-66; from asst prof to assoc prof, 69-78, PROF FOOD CHEM, UNIV ALTA, 78- *Concurrent Pos:* Bach's Inst Biochem fel, Acad Sci USSR, 66-67; fel, Univ Alta, 67-69. *Mem:* Can Inst Food Sci & Technol. *Res:* Chemistry and biochemistry of potatoes; aspects of lipids, starch, vitamins. *Mailing Add:* Dept Food Sci Fac Agr & Forestry Univ Alta Edmonton AB T6G 2P5 Can

HAEBERLE, FREDERICK ROLAND, b Philadelphia, Pa, Oct 6, 19; m 46; c 2. EXPLORATION GEOLOGY. *Educ:* Yale Univ, BS, 47, MS, 48; Columbia Univ, MBA, 62. *Prof Exp:* Geologist, Standard Oil Co, Tex, 48-52; chief geologist, J J Lynn Oil Div, 52-53; dist mgr petrol exp prod, Mayfair Minerals, Inc, 53-54; prof geol, McMurry Col, 54-57; chief subsurface geologist, Venezuelan Atlantic Refining Co, 57-60; elec data process coordr, 62-74, coordr spec studies, Mobil Oil Co, Libya, 74-77, mem staff explor statist, Explor & Prod Res Div, 77-83, CONSULT GEOLOGIST, MOBIL OIL CO, 83- *Concurrent Pos:* Asst prof, Univ Houston, 48-50. *Mem:* Fel Geol Soc Am; Am Asn Petrol Geologists; Soc Prof Well Log Analysts. *Res:* Carbonates; hydrocarbon gravity; information systems; application of computers to exploration. *Mailing Add:* 4036 Northview Lane Dallas TX 75229

HAEBERLI, WILLY, b Zurich, Switzerland, June 17, 25; div; c 3. NUCLEAR PHYSICS. *Educ:* Univ Basel, PhD(physics), 52. *Prof Exp:* Res assoc physics, Univ Wis, 52-54; vis prof physics, Duke Univ, 54-56, from asst prof to assoc prof, 56-61, PROF PHYSICS, UNIV WIS, MADISON, 61- *Honors & Awards:* Tom H Bonner Prize, Am Physics Soc, 79. *Mem:* Fel Am Phys Soc; Europ Phys Soc; fel AAAS. *Res:* Elastic scattering and polarization in nuclear reactions; design of polarized-ion sources; nuclear reactions with beams of polarized ions; development of sources of polarized ions; parity violation in nuclear interactions. *Mailing Add:* Dept Physics 1506 Sterling Hall Univ Wis Madison WI 53706

HAEDER, PAUL ALBERT, b Yale, SDak, July 31, 15; m 39; c 4. APPLIED MATHEMATICS. *Educ:* Huron Col, BA, 39; Univ SDak, MA, 50; Iowa State Univ, PhD, 68. *Prof Exp:* Supt pub schs, SDak, 41-44; from instr to asst prof math, Univ SDak, 52-59; asst, Iowa State Univ, 56-57, instr, 59-60; assoc prof, Univ SDak, 60-65; instr, Iowa State Univ, 65-68; chmn, Dept Math, Univ Nebr, Omaha, 68-74, prof math, 68-80, emer prof, 80-; RETIRED. *Concurrent Pos:* Asst sci aide, Northern Regional Res Lab, US Govt, 42; assoc ed, Am Math Monthly, 78- *Mem:* Math Asn Am. *Res:* Partial differential equations of mathematical physics. *Mailing Add:* 3060 Stephanos Dr Lincoln NE 68516

HAEDRICH, RICHARD L, b Wilmington, Del, Dec 12, 38; m 62; c 3. ICHTHYOLOGY, BIOLOGICAL OCEANOGRAPHY. *Educ:* Harvard Univ, AB, 61, AM, 63, PhD(biol), 66. *Prof Exp:* Asst scientist, Woods Hole Oceanog Inst, 66-70, assoc scientist, 70-79; PROF FISHERIES BIOL, MEM UNIV NFLD, 79-, DIR, NFLD INST COLD OCEAN SCI, 82- *Concurrent Pos:* Assoc, Harvard Univ, 65-; ed-in-chief, Biol Oceanog, 79-; mem & chmn, pop biol grant selection comt, NSERC, 81-85. *Res:* General biology and evolutionary relationships of fishes; distribution of oceanic animals, especially the establishment of high-seas faunal regions; ecology of mid-water and bottom fishes; patterns in the structure, seasonality, and long-term variation in animal communities; biology of eels; biology of northern aquatic ecosystems. *Mailing Add:* Dept Biol Mem Univ St John's NF A1C 5S7 Can

HAEFELI, ROBERT J(AMES), b Paterson, NJ, Sept 7, 26; m 50; c 2. SANITARY ENGINEERING. *Educ:* Univ Mich, BS, 48; Rutgers Univ, MS, 81; Environ Eng Intersoc Bd, dipl. *Prof Exp:* Civil engr, Bur Reclamation, US Dept Interior, Colo, 48-52, engr water resources, 52-54, actg off engr, Bur Reclamation Team, Beirut, Lebanon, 54-57; sr engr, Ebasco Serv, Inc, NY, 57-60; prin engr & proj mgr, Hydrotech Corp, 60-61, vpres & managing engr, 61-66; assoc & water resources engr, Hazen & Sawyer, 66-69; sanit consult engr, 69-72; pres, Frank & Haefeli Assoc, 72-76; pres, Haefeli Eng, 77-79; planning coordr, Havens & Emerson, Inc, 79-85; chief hydraulic engr, N H Bettigole, Pa, 85-88; regional serv mgr, Post, Buckley, Schult & Jernigan, Inc, 88-; CONSULT. *Concurrent Pos:* Mem, Int Comn Irrig, Drainage & Flood Control & NJ Dept Environ Protection, 68-; chmn NJ Clean Air Coun, 78-79, NJ Clean Water Coun, 87-; nat dir, Nat Soc Prof Engrs, 83-84 & 86-87. *Mem:* Water Pollution Control Fedn; Am Geophys Union; Am Water Works Asn; Am Soc Civil Engrs; Nat Soc Prof Engrs. *Res:* Hydraulic model studies; photoelastic laboratory studies; hydrodynamic effect of earthquakes on dams; surge suppression analyses; computer applications. *Mailing Add:* Four Caldwell Terr Marlboro NJ 07746

HAEFF, ANDREW V(ASILY), b Moscow, Russia, Dec 30, 04; nat US; m 36; c 1. ELECTRONICS. *Educ:* Russ Polytech Inst, China, EE & ME, 28; Calif Inst Technol, MS, 29, PhD(elec eng), 32. *Prof Exp:* Lab asst, Calif Inst Technol, 28-29, spec res fel, 32-33; res engr, Radio Corp Am, 34-41; consult physicist, Naval Res Lab, 41-42, prin radio engr, 43-45, consult electronics & head vacuum tube res br, 45-50; mem adv coun & head electron tube lab, Hughes Aircraft Co, 50-54, dir res labs & vpres co, 54-61, consult, 61-64; sr staff scientist, TRW Systs, 65-76; CONSULT ELECTRONICS, NAVAL RES LABS, WASHINGTON, DC, 76- *Concurrent Pos:* Consult, Naval Res Lab, 42-45, Off Sci Res & Develop, Res & Develop Bd & Dept Defense, 42; vis res assoc, Calif Inst Technol, 79-81. *Honors & Awards:* Diamond Award, Inst Elec & Electronics Engrs. *Mem:* Fel Am Phys Soc; fel Inst Elec & Electronics Engrs; Am Phys Soc. *Res:* Electron tubes, particularly ultrahigh frequency, microwave and storage tubes; radar systems; microwave signal generators; space charge effects in electron beams and interaction of electrons with traveling e-m fields. *Mailing Add:* 11134 Bellagio Rd Los Angeles CA 90049

HAEFNER, A(LBERT) J(OHN), b Cincinnati, Ohio, Oct 25, 23; m 45, 60; c 5. POLYMER CHEMISTRY, PLASTICS. *Educ:* Univ Cincinnati, ChE, 49, MS, 51, PhD(org chem), 53. *Prof Exp:* Lab asst, Firestone Tire & Rubber Co, 42-43; control chemist, Hilton-Davis Chem Co, 46-49; res chemist, Ethyl Corp, 53-58, res supvr, 58-63, asst dir chem res, 63-65, assoc dir polymer res, 65-67, dir plastics res & appln, 69-88; RETIRED. *Mem:* Am Chem Soc. *Res:* Polymerization of elastomers and resins; chlorination. *Mailing Add:* 11646 Glenhaven Dr Baton Rouge LA 70815

HAEFNER, PAUL ALOYSIUS, JR, b Lancaster, Pa, Dec 5, 35; m 62; c 3. ZOOLOGY, BIOLOGICAL OCEANOGRAPHY. *Educ:* Franklin & Marshall Col, BS, 57; Univ Del, MS, 59, PhD(biol), 62. *Prof Exp:* Asst prof biol, La State Univ, New Orleans, 62-63; fishery biologist, Maine Coop Fishery Unit, US Fish & Wildlife Serv, 63-69; assoc marine scientist, Va Inst Marine Sci, 69-77; head, 77-83, PROF, DEPT BIOL, ROCHESTER INST TECHNOL, 77- *Concurrent Pos:* Asst prof, Univ Maine, Orono, 63-68; assoc prof zool, Univ Maine, Orono, 68-69; assoc prof, Univ Va & Col William & Mary, 69-77. *Mem:* Am Soc Zool; Sigma Xi; Marine Biol Asn UK; Atlantic Estuarine Res Soc; Crustacean Soc, (secy, 80-87, pres-elect, 88-). *Res:* Reproductive biology and physiological ecology of marine and estuarine decapod Crustacea. *Mailing Add:* Dept Biol Rochester Inst Technol Rochester NY 14623

HAEFNER, RICHARD CHARLES, b Lancaster, Pa, Dec 13, 43; c 1. VOLCANOLOGY. *Educ:* Franklin & Marshall Col, AB, 65; Pa State Univ, MS, 69, PhD(geol), 72. *Prof Exp:* Consult geologist, Pa, 72-73; vis asst prof geol, State Univ NY Col New Paltz, 73-74; asst prof mineral, Col Charleston, 75-76; CONSULT, 77- *Concurrent Pos:* Lectr, Gov Speakers Bur on Energy, SC, 75- *Mem:* Geol Soc Am; Mineral Soc Am; fel Gemmological Asn Gt Brit; Mineral Asn Can; AAAS. *Res:* Physicochemical environments derived from statistical analysis of the crystal habits of minerals; petrogenesis of acid volcanic rocks; geology of the Death Valley region; identification and characterization of new mineral species. *Mailing Add:* 217 Nevin St Lancaster PA 17603

HAEGEL, NANCY M, b New Haven, Conn, Sept 7, 59; m 91. SEMICONDUCTOR CHARACTERIZATION. *Educ:* Univ Notre Dame, BS, 81; Univ Calif, Berkeley, MS, 83, PhD(mat sci), 85. *Prof Exp:* Res asst, Lawrence Berkeley Lab, 81-85; postdoctoral scientist, Siemens Res Labs, Erlangen, Fed Repub Ger, 86-87; asst prof, 87-89, ASSOC PROF MAT SCI & ENG, UNIV CALIF, LOS ANGELES, 89- *Concurrent Pos:* Trustee, Univ Notre Dame, 87-90; mem, fel panel, Nat Res Coun, 88-91; Kellogg fel, Kellogg Found, 90-93. *Honors & Awards:* Ross Tucker Award, Am Inst Mining Metall & Petrol Engrs, 85, Hardy Gold Medal, 89. *Mem:* Mat Res Soc; Am Phys Soc; Am Inst Mining Metall & Petrol Engrs Metall Soc. *Res:* Optical and electrical characterization of semiconductors and other electronic materials; photoconductivity and photoluminescence; high resistivity transport; infrared detectors. *Mailing Add:* Dept Mat Sci & Eng 5731 Boelter Hall Univ Calif Los Angeles CA 90024-1595

HAEGELE, KLAUS D, b Stuttgart, Ger, Oct 7, 41; m 64; c 2. DRUG METABOLISM, PHENOTYPE & DRUG METABOLISM. *Educ:* Univ Tubingen, Ger, BS, 64, MS, 67, PhD(org chem), 70. *Prof Exp:* Res asst prof chem, Univ Tubingen, Ger, 70-71; vis asst prof biochem, Baylor Col Med, Houston, 71-75; asst prof pharmacol, Health Sci Ctr, Univ Tex, San Antonio, 75-77, assoc prof, 78-79; dept head, drug metab, 79-90, DIR DEVELOP RESOURCES, MARION MERRELL DOW RES INST, STRASBOURG, FRANCE, 90- *Mem:* Am Soc Pharmacol & Exp Therapeut; Am Soc Clin Chem. *Res:* Pharmacokinetics and drug metabolism; pharmacodynamics; development of novel analytical techniques. *Mailing Add:* Dept Drug Metab Marion Merrell Dow Res Inst 16 rue d'Ankara Strasbourg 67009 France

HAEGER, BEVERLY JEAN, b Gainesville, Ga; m. INORGANIC CHEMISTRY. *Educ:* Bethany Nazarene Col, BS, 61; Univ Tenn, Chattanooga, MAT, 63; Univ Ga, PhD(inorg chem), 74. *Prof Exp:* Asst prof chem, Trevecca Nazarene Col, 67-71; chemist, Neutron Devices Dept, 74-79, prog mgr, Develop Planning & Control, 79-85, mgr, Power Sources Develop, 85-90, MGR, ENG OPER SURETY, GEN ELEC CO, 90- *Concurrent Pos:* Environ, health & safety mgt. *Honors & Awards:* Hoover Award, 85. *Mem:* Am Chem Soc. *Res:* Synthesis and characterization of thermal battery related materials; scale-up of synthesis processes from laboratory to pilot plant operation. *Mailing Add:* PO Box 2908 Largo FL 34649

HAELTERMAN, EDWARD OMER, b Norway, Mich, Oct 14, 18; m 46. VETERINARY VIROLOGY. *Educ:* Mich State Univ, DVM, 52; Purdue Univ, MS, 55, PhD(path), 59. *Hon Degrees:* Hon Dr, Univ Ghent, Belgium, 73. *Prof Exp:* Instr vet sci, Purdue Univ, 52-59, assoc prof vet microbiol, 59-64, asst dean, Sch Vet Sci & Med, 59-62, prof vet microbiol, 64-84; RETIRED. *Concurrent Pos:* Vis investr, Rockefeller Inst, 58; USDA res vet, East African Vet Res Orgn, Kenya, 67-68. *Mem:* AAAS; Am Vet Med Asn; Am Soc Microbiol; Conf Res Workers Animal Dis; Am Asn Swine Practitioners. *Res:* Enteric disease of swine; investigations on pathogenesis, immunology and epidemiology of transmissible gastroenteritis of swine. *Mailing Add:* 300 Six Soldier's Home Rd West Lafayette IN 47906

HAEN, PETER JOHN, b Udenhout, Neth, Aug 29, 38; m 74; c 2. IMMUNOLOGY, ENDOCRINOLOGY. *Educ:* Nat Univ Ireland, BSc(zool), & BSc(psychol), 64, PhD(biol), 67. *Prof Exp:* Chmn dept biol, Cardinal Otunga Col, Kenya, 67-71; chmn dept sci, Pius X High Sch, Downey, Calif, 71-72; from asst prof to assoc prof, 72-83, chmn dept, 76-84, PROF BIOL, LOYOLA MARYMOUNT UNIV, 83- *Concurrent Pos:* Dir, Rosecrans Chair Conserv Natural Resources, Loyola Marymount Univ, 77-83. *Mem:* AAAS; Sigma Xi. *Res:* Biochemical and immunological taxonomy of animals; relationships between science and theology. *Mailing Add:* Dept Biol Loyola Blvd at 80th St Los Angeles CA 90045

HAENDEL, RICHARD STONE, b Brooklyn, NY, Nov 10, 39; m 65; c 2. MODULATION, RADIO FREQUENCY LINK ANALYSIS. *Educ:* Rensselaer Polytech Inst, BEE, 60; NY Univ, MEE, 63. *Prof Exp:* Mgr develop eng, Collins Govt Avionics, 82-87, tech staff, Collins Govt Avionics, 87-90, MGR RES & DEVELOP, COLLINS AVIONICS DIV, ROCKWELL-INT, 90- *Concurrent Pos:* Lectr internal, Collins Avionics, Rockwell-Int, 82- *Mem:* Armed Forces Commun & Electronics Asn; Inst Elec & Electronics Engrs; Am Defense Preparedness Asn. *Res:* Microwave device development; opto-electronics electronic sensing as applied to weather detection for aircraft; radio propagation; avionics systems; receiver and transmitter design for anti-jam communications systems. *Mailing Add:* 400 Collins Rd MS 124-223 Cedar Rapids IA 52498

HAENDLER, BLANCA LOUISE, b Boston, Mass, Nov 22, 48. SURFACE CHEMISTRY. *Educ:* Goucher Col, BA, 69; Johns Hopkins Univ, PhD(phys chem), 74. *Prof Exp:* Lectr chem, Johns Hopkins Univ, 74-75; asst prof chem, Lafayette Col, 75-; AT CLOROX TECH CTR. *Mem:* Am Chem Soc; AAAS; Catalysis Soc NAm; Sigma Xi. *Res:* Study of adsorbed species and reaction mechanisms on solid surfaces and other aspects of heterogeneous catalysis. *Mailing Add:* Clorox Tech Ctr PO Box 493 Pleasanton CA 94566

HAENDLER, HELMUT MAX, b Boston, Mass, June 19, 13; m 36; c 2. INORGANIC CHEMISTRY. *Educ:* Northeastern Univ, BS, 35; Univ Wash, Seattle, PhD(inorg chem), 40. *Prof Exp:* Assoc chem, Univ Wash, Seattle, 39-40, instr, 40-42; res chemist & supvr div war res, Carbide & Carbon Chem Co & Substitute Alloy Mat Labs, Columbia Univ, 42-45; from asst prof to prof, 45-78, EMER PROF CHEM, UNIV NH, 78- *Concurrent Pos:* Fulbright Res Scholar, Max Planck Inst Solid State Res, 72-73. *Mem:* AAAS; Am Chem Soc; Am Crystallog Asn. *Res:* X-ray applications. *Mailing Add:* Dept Chem Parsons Hall Univ NH Durham NH 03824-3598

HAENER, JUAN, b Hamba, Rumania, Aug 11, 17; US__citizen; m 51; c 4. THEORETICAL PHYSICS, MECHANICAL ENGINEERING. *Educ:* Tech Univ, Berlin, MS, 43; Tech Univ, Vienna, PhD, 46. *Prof Exp:* Dynamics engr, Heinkel Aircraft, Ger, 43-45; asst, Inst Theoret Physics, Vienna, 45-48; head dynamics, Inst Aerotech, Cordoba, Arg, 49-56; res scientist in chg dynamics com, mil & helicopter div, Cessna Aircraft Co, 56-58; staff specialist & consult, Dynamics & Stress, Solar Aircraft Co, 58-60; STAFF SCIENTIST & CONSULT, RES & DEVELOP DIV, NARMCO DIV, WHITTAKER CORP, 60- *Res:* Dynamics of structures; flutter-vibration; development of composite materials. *Mailing Add:* 8215 Harton Pl San Diego CA 92123

HAENISCH, SIEGFRIED, b Dresden, Ger, Sept 20, 36; US citizen; m 61; c 2. MATHEMATICS. *Educ:* Trenton State Col, BS, 58, MA, 61; Rutgers Univ, New Brunswick, EdD, 67. *Prof Exp:* Teacher high sch, NJ, 58-62; instr math, Trenton State Col, 62-66; asst math educ, Rutgers Univ, New Brunswick, 66-67; guest prof math educ, Goethe Univ, 67-68; assoc prof, 68-73, chmn dept, 73-77, PROF MATH, TRENTON STATE COL, 73- *Concurrent Pos:* Guest instr, Rutgers Univ, 70-; Nat Endowment Humanities fel, Yale, 80. *Mem:* Math Asn Am; Nat Coun Teachers Math. *Res:* Foundations of mathematics and its application to the teaching and learning of mathematics. *Mailing Add:* Dept/Math Statist Trenton State Col Hillwood Lakes Trenton NJ 08625

HAENLEIN, GEORGE FRIEDRICH WILHELM, b Mannheim, Ger, Oct 27, 27; nat US; m 54; c 5. ANIMAL NUTRITION, ANIMAL BREEDING. *Educ:* Hohenheim Agr Univ, Dipl, 50, Dr Agr(animal nutrit & breeding), 53; Univ Del, MS, 60; Univ Wis-Madison, PhD(dairy sci, biochem), 72. *Prof Exp:* Asst animal nutrit res, Exp Sta, Hohenheim Agr Univ, 50-53; asst mgr & herdsman, Zeitler Farms, Inc, Del, 53-57; from asst prof to assoc prof animal sci, Col Agr Sci, 64-74, RES ASSOC & SUPVR DAIRY HERDS, UNIV DEL, 57-, PROF ANIMAL SCI & AGR BIOCHEM, 74- *Concurrent Pos:* US State Dept exchange student scholar, 51-52; NSF teaching fel, 65-66; res asst dairy sci, Univ Wis-Madison, 66-67; abstractor, Biol Abstr; chmn comt animal nutrit of goats, Nat Res Coun, 75-; mem, Northeastern Regional Dairy Res Steering Comt, 75-; ed in chief, J Small Ruminant Res; chmn, dairy goat handbook, US Dept Agr SE Exten, 81- *Mem:* Hon fel, Am Asn Advan Sci; Am Soc Animal Sci; Am Dairy Sci Asn; NY Acad Sci; Nat Mastitis Coun; Sigma Xi. *Res:* Milk composition; forage evaluation; dairy cattle genetics; animal behavior; dairy goat management. *Mailing Add:* 2071 S College Ave Newark DE 19702

HAENNI, DAVID RICHARD, b Peoria, Ill, June 6, 47; m 69; c 1. X-RAY SPECTROSCOPY, HIGH-SPIN STATES. *Educ:* Washington Univ, AB, 69; Tex A&M Univ, PhD(chem), 75. *Prof Exp:* Vis scientist, Nuclear Res Estab, Jülich, WGer, 77-79; res scientist, Oak Ridge Nat Lab, 79-80; fel, 76-77, RES SCIENTIST, CYCLOTRON INST, TEX A&M UNIV, 80- *Mem:* Am Chem Soc; Am Physical Soc; Inst Elec & Electronics Engrs. *Res:* In beam ray spectroscopy following massive transfer reactions; decay and reaction ray spectroscopy with both light and heavy ion accelerators; particle accelerator computer control systems. *Mailing Add:* SSC Lab 2550 Beckleymeade Ave MS 1046 Dallas TX 75237

HAENNI, EDWARD OTTO, b St Louis, Mo, May 25, 07; m 38; c 1. ANALYTICAL CHEMISTRY, FOOD CHEMISTRY. *Educ:* Wash Univ, St Louis, AB, 29, MS, 31; Univ Md, PhD(org chem), 40. *Prof Exp:* Asst instr chem, Wash Univ, St Louis, 29-31; res chemist, Food & Drug Admin, USDA,

31-43; chief solvents sect, Chem Bur, War Prod Bd, Washington, DC, 43-45; chief drugs sect, Civilian Prod Admin, 46; res chemist, Food & Drug Admin, Fed Security Agency, 47-51; res chemist, Entom Res Br, Agr Res Ctr, USDA, 52-56; res chemist, Bur Foods, US Food & Drug Admin, 57-64, chief, Additives & instrumentation Br, 64-71, Dir Chem & Physics Div, 71-73, consult, 74-79; RETIRED. *Mem:* Am Chem Soc; fel Asn Off Anal Chem; fel Washington Acad Sci. *Res:* Food additives and contaminants; polynuclear hydrocarbons; analytical research and instrumentation; foods and drugs; organic aerosol dispenser development. *Mailing Add:* 7907 Glenbrook Rd Bethesda MD 20814-2403

HAENSEL, VLADIMIR, b Freiburg, Ger, Sept 1, 14; nat US; m 39; c 2. CATALYSIS. *Educ:* Northwestern Univ, BS, 35, PhD(chem), 42; Mass Inst Tech, MS, 37. *Hon Degrees:* DSc, Northwestern Univ, 57, Univ Wis-Milwaukee, 79. *Prof Exp:* Res chemist, Universal Oil Prod Co, 37-45; tech observer, Petrol Admin for War, 45; div coordr, 45-55, dir ref res, 55-59, dir process res, 60-63, vpres & dir res, 64-72, vpres sci & technol, 72-79, CONSULT, UOP INC, 79-; PROF CHEM ENG, UNIV MASS, 80- *Concurrent Pos:* Mem, Panel Catalysts Automotive Emission Devices & Petrol Refining, Nat Mat Adv Bd, 73-74; chmn, US-USSR Technol Exchange Chem Catalysis, 76-79; consult, Exxon Res & Eng Co, 80-81, Environ Res & Technol, Inc, 81, Olin Chem, 82-84, Dow Chem Co, 84, Halcon Res, 83-85, Heico, Inc, 85-89, Catalytica, 89- & others; lectr, R A Welch Conf Chem Res, 81; chmn, Adv Comt Indust Sci & Technol Innovation, NSF, 82-85. *Honors & Awards:* Prof Progress Award, Am Inst Chem Engrs, 57; Modern Pioneer Medal Creative Chem, Nat Asn Mfrs, 65; Perkin Medal Appl Chem, 67; Nat Medal of Sci, 73; Eugene J Houdry Award Appl Catalysis, 77; Chancellor Lectr, Univ Mass, 84; Nat Acad Sci Award for Chem in Serv to Soc, 91. *Mem:* Nat Acad Sci; Nat Acad Eng; Am Chem Soc; Catalysis Soc NAm (pres, 76-80). *Res:* Process and catalysis in petroleum refining and petrochemical field; author of more than 120 scientific and technical papers, over 145 United States patents and over 450 foreign patents. *Mailing Add:* Dept Chem Eng Univ Mass 159 Goessmann Lab Amherst MA 01003

HAENSLY, WILLIAM EDWARD, b Boston, NY, Oct 14, 27; m 54; c 6. VETERINARY MEDICINE, GERONTOLOGY. *Educ:* Pa State Univ, BS, 52; Iowa State Univ, MS, 56, PhD(genetics), 62, DVM, 64. *Prof Exp:* Asst prof vet med, Col Vet Med, Iowa State Univ, 64-66, assoc prof, 66-70; PROF VET ANAT, COL VET MED, TEX A&M UNIV, 71- *Mem:* Am Asn Vet Anat; Can Asn Vet Anatomists; Am Vet Med Asn; Geront Soc; World Asn Vet Anat. *Res:* Qualitative and quantitative aspects of aging in cells, tissues and organs of animals, employing histological and histochemical procedures. *Mailing Add:* Dept Vet Anat Tex A&M Univ College Station TX 77843

HAENSZEL, WILLIAM MANNING, b Rochester, NY, June 19, 10; m 46; c 3. BIOSTATISTICS, EPIDEMIOLOGY. *Educ:* Univ Buffalo, BA, 31, MA, 32. *Hon Degrees:* Hon Dr, Univ Valle, Colombia, 70. *Prof Exp:* From jr statistician to statistician, NY State Dept Health, 34-47; dir, Conn Bur Vital Statist, 47-52; head, Biomet Sect, Nat Cancer Inst, NIH, 52-57, assoc chief, 56-60, chief, Biomet Br, 60-76; SR EPIDEMIOLOGIST, ILL CANCER COUN, 76-; PROF EPIDEMIOL, SCH PUB HEALTH, UNIV ILL, 76- *Concurrent Pos:* Mem US Nat Comt Vital Health Statistics, 64-67; adj prof, Grad Sch Pub Health, Univ Pittsburgh, 66-82. *Mem:* Biomet Soc; Am Pub Health Asn; Am Statist Asn; Popul Asn Am; Am Epidemiol Soc. *Res:* Vital statistics; epidemiology of cancer; methodology of field investigations in chronic diseases. *Mailing Add:* 341 E Hawthorne Blvd Wheaton IL 60187

HAERER, ARMIN FRIEDRICH, b Stuttgart, Ger, Mar 28, 34; US citizen; m 58; c 2. NEUROLOGY. *Educ:* Univ Mich, BS, 57, MD, 59; Am Bd Psychiat & Neurol, dipl neurol, 66. *Prof Exp:* Intern, Univ Mich, 59-60, resident neurol, 60-63; clin instr, Univ Louisville, 63-65; from asst prof to assoc prof, 65-76, PROF NEUROL, UNIV MISS, 76- *Concurrent Pos:* Consult neurol, Wayne County Gen Hosp, Mich, 62-63, US Vet Hosp, Jackson, Miss, 67- & Gulfport, 69- *Mem:* AMA; fel Am Col Physicians; Am Neurol Asn; fel Am Acad Neurol. *Res:* Clinical and basic neurology, especially cerebrospinal fluid and cerbrovascular disorders. *Mailing Add:* Dept Neurol Univ Hosp Jackson MS 39216

HAERING, EDWIN RAYMOND, b Columbus, Ohio, Dec 8, 32; m 56; c 3. CHEMICAL PROCESS INDUSTRIES, HAZARDOUS MATERIALS MANAGEMENT. *Educ:* Ohio State Univ, BChE & MS, 56, PhD(chem eng), 66. *Prof Exp:* From instr to asst prof chem eng 59-73, vchmn chmn dept 73-77, acting chmn, 77-78, assoc prof chem eng, 73-82, PROF CHEM ENG, OHIO STATE UNIV, 82- *Concurrent Pos:* Tech Consult 66- *Mem:* Am Chem Soc; Am Inst Chem Engrs. *Res:* Adsorption; catalysis; kinetics; process development and design; process safety; hazardous material management. *Mailing Add:* Dept Chem Eng Ohio State Univ 140 W 19th Ave Columbus OH 43210

HAERING, GEORGE, b Rangoon, Burma, Oct 7, 30; US citizen; m 85; c 3. AIR WARFARE, TECHNICAL MANAGEMENT. *Educ:* Princeton Univ, AB, 52. *Prof Exp:* Analyst, US Bur Budget, 55-58 & Opers Eval Group, 59-60; asst to dir, Naval Warfare Anal Group, 60-61; rep to Comdr, Seventh Fleet Opers Eval Group, 61-62, rep to Comdr in Chief Pac Fleet, 65-66, chief Naval Opers, Systs Anal Div, 67-84, DIR OPERS RES, NAVAL AIR SYSTS COMMAND, DEPT OF NAVY, 85- *Concurrent Pos:* Consult, Joint Tech Coord Group Munitions Effect, 63-65 & 68-; chmn, Joint Aircraft Attrition Prog, 71-75. *Mem:* AAAS. *Res:* Analysis of combat data from tactical air warfare; development of optimal force structures within constrained budgets; experimentation in dogfights. *Mailing Add:* 209 E Nelson Ave Alexandria VA 22301-1815

HAERING, RUDOLPH ROLAND, b Switzerland, Feb 27, 34; nat Can; m 54; c 2. EXPERIMENTAL AND THEORETICAL SOLID STATE PHYSICS. *Educ:* Univ BC, BA, 54, MA, 55; McGill Univ, PhD(theoret physics), 57. *Hon Degrees:* DSc, Mem Univ, 86, McMaster Univ, 89, Univ Waterloo, 90. *Prof Exp:* Nat Res Coun Can fel math physics, Univ Birmingham, 57-58; asst prof physics, McMaster Univ, 58-60; mem res staff, Int Bus Mach Corp, 60-63; prof physics, Univ Waterloo, 63-64; vis prof physics, Univ BC, 64-65; prof physics & head dept, Simon Fraser Univ, 65-72; head dept, 73-77, prof physics, 73-86, HON PROF PHYSICS, UNIV BC, 86- *Concurrent Pos:* Ed, Can J Physics, 68-72; mem, Nat Res Coun Can, 74- *Honors & Awards:* Officer, Order of Canada; Herzberg Medal, 70; Gold Medal, Can Asn Physicists, 82; Gold Medal, Brit Col Science Coun, 84. *Mem:* Am Phys Soc; Can Asn Physicists; fel Royal Soc Can. *Res:* Experimental and theoretical solid state physics; battery research. *Mailing Add:* Dept Physics Univ BC 2075 Westbrook Pl Vancouver BC V6T 1W5 Can

HAERTEL, JOHN DAVID, b Elgin, Ill, Aug 4, 37; m 62; c 2. AVIAN BIOLOGY, HERPETOLOGY. *Educ:* Univ Ill, Urbana, BS, 62; Ore State Univ, Corvallis, MS, 67, PhD(zool), 69. *Prof Exp:* PROF ZOOL, SDAK STATE UNIV, 69- *Mem:* Am Inst Biol Sci; Am Soc Ichthyologists; Inland Bird Banding Asn. *Res:* Biology of Charadriiformes in South Dakota; food analysis during spring migration. *Mailing Add:* Biol Dept SDak State Univ Brookings SD 57007-0595

HAERTEL, LOIS STEBEN, b Hinsdale, Ill, Dec 16, 39; m 62; c 2. AQUATIC ECOLOGY. *Educ:* Univ Ill, Urbana-Champaign, BS, 61, MS, 63; Ore State Univ, PhD(oceanog), 69. *Prof Exp:* Asst oceanog, Ore State Univ, 63-65; from asst prof to assoc prof, 69-88, PROF BIOL, SDAK STATE UNIV, 88- *Concurrent Pos:* Res grants, Off Water Resources, SDak State Univ, 70-72 & 74-79 & 87-89. *Mem:* Am Soc Limnol & Oceanog. *Res:* Aquatic ecology; algae; eutrophication; biological control of algal blooms. *Mailing Add:* Dept Biol SDak State Univ Box 2201 Brookings SD 57007

HAERTLING, GENE HENRY, b Old Appleton, Mo, Mar 15, 32; m 58; c 3. CERAMICS. *Educ:* Univ Mo-Rolla, BS, 54; Univ Ill, MS, 60, PhD(ceramics), 61. *Prof Exp:* Res ceramic engr, Ipsen Ceramics Inc, 54-55, plant engr ceramics develop, 57-58; staff mem ceramics res, Sandia Corp, 61-65, div supvr elec ceramics res & develop, 65-73; pres, Optoceram, Inc, Albuquerque, 73-74; mgr opto-ceramics dept, 74-80, OFFICER TECH STAFF, MOTOROLA INC, 80- *Honors & Awards:* Pace Award, Nat Inst Ceramic Engrs, 72. *Mem:* Fel Am Ceramic Soc; Nat Inst Ceramic Engrs; fel Inst Elec & Electronics Engrs. *Res:* Electrooptic ceramics; oxide ceramics; ferroelectric and Piezoelectric phenomena; sintering studies; hot-pressing of ferroelectrics. *Mailing Add:* Prof Sr Investr 135 Wardlaw Lake Rd Central SC 29630

HAESELER, CARL W, b Northampton, Mass, Feb 25, 29; m 54; c 3. POMOLOGY, HORTICULTURE. *Educ:* Univ Mass, BSc, 54; Cornell Univ, MSc, 58; Pa State Univ, PhD(hort), 62. *Prof Exp:* Asst prof pomol & exten pomologist, 61-66, ASSOC PROF POMOL, COL AGR, PA STATE UNIV, 66-, SUPT ERIE COUNTY FIELD RES LAB, 74- *Mem:* Am Soc Hort Sci; Am Pomol Soc; Am Soc Enol; Sigma Xi. *Res:* Plant physiology; viticulture, primarily cultural techniques, especially nutrition, root-stocks, growth regulators, training and pruning. *Mailing Add:* Erie Co Field Res Lab Pa State Univ North East PA 16428

HAEUSSERMANN, WALTER, b Kuenzelsau, Ger, Mar 2, 14; US citizen; m 40. ELECTRICAL ENGINEERING, APPLIED PHYSICS. *Educ:* Stuttgart Tech Univ, BSE, 35; Darmstadt Inst Technol, ME, 38, Dr Ing(physics, math), 44. *Prof Exp:* Asst prof elec eng, Darmstadt Tech Univ, 37-39, asst prof appl physics, 42-46, develop engr elec eng, 46-47; chief engr guid control, Peenemuende, Ger, 39-42; supvr aeronaut develop engr, Ord Dept, US Army, Ft Bliss, 47-50 & Ord Mission Lab, Redstone Arsenal, Ala, 50-56, dir guid & control, Army Ballistic Missile Agency, 56-60; dir astrionics lab, Marshall Space Flight Ctr, 60-69, dir cent syst eng, 69-72, dir science & eng, 72-75, asst syst engr, Sci & Eng Directorate, 75-78; consult, Bendix Guidance Systs, 78-83, Teledyne Brown Eng, 83-86; SR SYSTS ENGR, APPL RES INC, 86- *Concurrent Pos:* Prof elec eng, Auburn Univ, 66-; mem, Int Fedn Automatic Control Space Comt. *Honors & Awards:* Superior Achievement Award, Inst Navig, 70; Medal of Merit, Baden-Wuerttemberg, 85; Wernher von Braun Distinction, Ger Aerospace Soc, 88; Medaris Award, US Army, 90. *Mem:* Fel Am Astronaut Soc; Am Inst Navig; fel Am Inst Aeronaut & Astronaut; Sigma Xi. *Res:* Guidance, navigation and control of missiles and space vehicles; development of satellites. *Mailing Add:* 1607 Sandlin Ave SE Huntsville AL 35801

HAFELE, JOSEPH CARL, b Peoria, Ill, July 25, 33; US citizen; m 58; c 4. RELATIVITY PHYSICS. *Educ:* Univ Ill, BS, 59, MS, 60, PhD(physics), 64. *Prof Exp:* Res assoc physics, Univ Ill, 64; mem staff, Los Alamos Sci Lab, 64-66; asst prof, Wash Univ, 66-72; mem staff, res dept, 72-77, mem, eng gen off, Caterpillar Tractor Co, 77-85; asst prof, Eureka Col, 85-90; CONSULT, 90- *Concurrent Pos:* NASA Summer Res Fel, 87-90. *Mem:* AAAS; Am Phys Soc; Sigma Xi; Am Asn Univ Prof. *Res:* Relativity; mathematics. *Mailing Add:* 7721 N Galena Rd Peoria IL 61615

HAFEMAN, DEAN GARY, b Green Bay, Wis, Aug 19, 49; m 76; c 1. BIOCHEMISTRY. *Educ:* Univ Wis, PhD(biochem), 75. *Prof Exp:* RES SCIENTIST, MOLECULAR DEVICES CORP, 84- *Concurrent Pos:* Asst prof, Case Western Reserve Univ, 82-84. *Mem:* Biophys Soc; Mats Res Soc; AM Chem Soc; Am Asn Immunologists; Am Soc Cell Biol. *Res:* Analytical and electrochemical reseach. *Mailing Add:* 1350 Buckingham Way Hillsborough CA 94010

HAFEMANN, DENNIS REINHOLD, computer science, for more information see previous edition

HAFEMEISTER, DAVID WALTER, b Chicago, Ill, July 1, 34; m 61; c 3. SOLID STATE & NUCLEAR PHYSICS. *Educ:* Northwestern Univ, BSME, 57; Univ Ill, MS, 60, PhD(physics), 64. *Prof Exp:* Mech eng, Argonne Nat Lab, 57-58; physics asst, Univ Ill, 58-64; physicist & fel, Los Alamos Sci Lab, NMex, 64-66; asst prof physics, Carnegie-Mellon Univ, 66-69; PROF PHYSICS, CALIF STATE POLYTECH UNIV, SAN LUIS OBISPO, 69- *Concurrent Pos:* Sci Cong Fel with Sen John Glenn, 75-76 & sci adv, 76-77; spec asst to Under Secy of State, 77-78; expert consult, State Dept, 78-80; vis scientist, Mass Inst Technol, 83-84, Lawrence Berkeley, 85; Sci Cong fel, AAAS, 75-76. *Mem:* AAAS; Am Phys Soc. *Res:* Mossbauer effect; energy policy; electronic structure of the alkali halides; positron annihilation. *Mailing Add:* 553 Serrano San Luis Obispo CA 93401

HAFEN, ELIZABETH SUSAN SCOTT, b Springfield, Mo, July 25, 46; m 70. PARTICLE ASTROPHYSICS. *Educ:* Iowa State Univ, BS, 68, PhD(physics), 73. *Prof Exp:* From instr to assoc prof physics, 73-83, PRIN RES SCIENTIST, LAB NUCLEAR SCI, MASS INST TECHNOL, 83- *Mem:* Am Phys Soc; AAAS. *Res:* Short-lived resonances produced in hadron collisions as a probe of quark-quark interactions; large underground detectors. *Mailing Add:* Dept Physics, Rm 24-416 Mass Inst Technol, 77 Mass Ave Cambridge MA 02139

HAFER, LOUIS JAMES, b Youngstown, Ohio, Jan 21, 55; m 80. COMPUTER AIDED DESIGN. *Educ:* Carnegie-Mellon Univ, BS, 76, MS, 78, PhD(comput aided design), 81. *Prof Exp:* ASST PROF COMPUT SCI, SIMON FRASER UNIV, 81- *Mem:* Inst Elec & Electronics Engrs. *Res:* Automated synthesis of digital system hardware from behavioural descriptions; computer-aided design; hardware descriptive languages. *Mailing Add:* Dept Comput Sci Simon Fraser Univ Burnaby BC V5A 1S6 CAN

HAFF, RICHARD FRANCIS, b New York, NY, Oct 1, 29; m 53; c 2. MICROBIOLOGY. *Educ:* Univ NC, BA, 51; Western Res Univ, PhD(microbiol), 57. *Prof Exp:* Microbiologist, Stine Lab, E I du Pont de Nemours & Co, 57-63; virologist, Smith Kline & French Labs, 63-67, assoc dir microbiol, 67-72, sci dir, 72-78, dir new prod eval, Com Develop, 78-81; VPRES, SCI DEVELOP, WARNER LAMBERT CO, 81- *Mem:* Am Soc Microbiol. *Res:* Viral chemotherapy; animal virology; tissue culture. *Mailing Add:* Warner Lambert Co 2800 Plymouth Rd Ann Arbor MI 48105

HAFFLEY, PHILIP GENE, b Richmond, Ind, Oct 9, 41; m 63; c 3. ORGANIC CHEMISTRY. *Educ:* Ind Univ, BS, 63; Iowa State Univ, PhD(org chem), 67. *Prof Exp:* Assoc prof chem, 67-81, chmn, Div Biol & Phys Sci, 71-81, PROF CHEM & ASST DEAN ACAD AFFAIRS, IND UNIV, 81- *Mem:* Am Chem Soc; Nat Sci Teachers Asn. *Res:* Free radical chemistry. *Mailing Add:* Ind Univ Kokomo IN 46901

HAFFNER, ALDEN NORMAN, b Brooklyn, NY. OPTOMETRY. *Educ:* Brooklyn Col, AB, 48; Pa Col Optom, OD, 52; NY Univ, MPA, 60, PhD, 64. *Hon Degrees:* Dr Ocular Sci, Mass Col Optom, 60; LHD, Southern Col Optom, 80. *Prof Exp:* Dean, Col Optom, 71-76; exec dir, Optom Ctr, State Univ NY Col Optom, 57-80, pres, Col Optom, 76-80; vchancellor health sci, 78-88, PRES, COL OPTOM, STATE UNIV NY, ALBANY, 88- *Honors & Awards:* Carel Koch Award. *Mem:* Fel Am Acad Optom; Am Pub Health Asn; Am Sch Health Asn; Am Pub Welfare Asn; Am Soc Pub Admin. *Res:* Public administration of health and social services. *Mailing Add:* Col Optom State Univ NY 100 E 24th St New York NY 10010

HAFFNER, JAMES WILSON, b Ft Wayne, Ind, Mar 30, 29; m 55. NUCLEAR PHYSICS. *Educ:* Miami Univ, AB, 50; Mass Inst Technol, MS, 52, PhD(nuclear physics), 55. *Prof Exp:* Asst, Mass Inst Technol, 53-55; prin engr, Aircraft Nuclear Propulsion Dept, Gen Elec Co, Ohio, 55-59; dept mgr indust appln, Radiation Counter Labs, Inc, Ill, 59-61; group leader radioisotope appln, Armour Res Found, 61-63; RES PHYSICIST AEROSPACE SCI, SPACE & INFO SYSTS, NAM ROCKWELL CORP, 63- *Concurrent Pos:* Actg asst prof, Grad Sch, Univ Cincinnati, 56-59; instr, Dept Physics, Ill Inst Technol, 61-63; lectr, Calif State Univ, Long Beach. *Res:* Nuclear radiations and their interactions with matter. *Mailing Add:* Satellite & Space Div Rockwell Int 2600 Westminster Blvd PO Box 3644 Seal Beach CA 90740-7644

HAFFNER, RICHARD WILLIAM, b Bucyrus, Ohio, Feb 15, 31; wid; c 3. SURFACE CHEMISTRY, COMPOSITE MATERIALS. *Educ:* Ohio Univ, BS, 53; Purdue Univ, MS, 55; George Washington Univ, MBA, 65. *Prof Exp:* Proj officer, Air Force Aero Propulsion Lab, 55-58; proj officer, Air Force Space & Missile Orgn, 58-64; asst prof chem, Air Force Acad, 65-68; prog mgr, Air Force Off Sci Res, 68-72; inspector, Hq Air Force Systs Command, 72-74; prog mgr, Air Force Off Sci Res, 74-80; SR INDUST SPECIALIST, NAT SYSTS MGT CORP, 80- *Concurrent Pos:* Consult, tech problems related to DOD use of electro-optic night vision devices. *Mem:* Am Chem Soc; Sigma Xi; AAAS. *Res:* Air Force basic research program in surface chemistry including research in areas such as lubrication, corrosion and catalysis. *Mailing Add:* 13020 Chalfont Ave Ft Washington MD 20744

HAFFNER, RUDOLPH ERIC, b Thorun, Poland, Feb 27, 20; nat US; m 45; c 3. BIOLOGICAL OCEANOGRAPHY. *Educ:* Univ Maine, BA, 42; Yale Univ, PhD(zool), 52. *Prof Exp:* Instr biol, Colby Col, 46-47; instr biol, Wesleyan Univ, 52-53, asst prof, 53-58; assoc prof biol, Bucknell Univ, 58-60; ed dir sci pub, Am Ed Pub, Inc, 60-67; dir sci & prof biol, 67-87, EMER PROF BIOL, HARTFORD COL WOMEN, 87- *Mem:* Fel AAAS; NY Acad Sci; Sigma Xi. *Res:* Oceanography. *Mailing Add:* 60 Hickory Hill Rd Simsbury CT 06070-2833

HAFFORD, BRADFORD C, b Albion, Mich, June 2, 16; m 38; c 3. ANALYTICAL CHEMISTRY, INORGANIC CHEMISTRY. *Educ:* Univ Mich, BS, 38; Univ Wis, MS, 40, PhD(chem), 42. *Prof Exp:* Asst limnol, Univ Wis, 38-42; group leader, Carteret Sect, Tech Dept, Westvaco Chem Div, Food Mach & Chem Corp, 42-47, sect dir, 47-52, staff asst res & develop dept, 52-54; tech asst to mgr res, Gulf & Western Industs, 54-60, asst mgr res, 60-68, gen mgr res & develop, 68-70, dir, 70-73, vpres res & develop, Natural Resources Group, 73-81; CONSULT, 81- *Mem:* Am Chem Soc; Electrochem Soc; Am Ceramics Soc; Oil & Colour Chemists Asn. *Res:* Industrial chemicals; nonferrous metals. *Mailing Add:* Hemlock Farms Box 1089 Hawley PA 18428

HAFLEY, WILLIAM LEROY, b Phila, Pa, Jan 26, 30; m 54; c 3. GROWTH YIELD, MANAGEMENT SCIENCE APPLICATIONS. *Educ:* Pa State Univ, BS, 53; NC State Univ, MF, 57, PhD(exp statist), 60. *Prof Exp:* Staff consult statistician, Aluminum Co, Am, 60-63; sr consult statistician, Westinghouse Elec Corp, Pa, 63-66; prof forestry & statist, NC State Univ, 66-91; RETIRED. *Mem:* Am Statist Asn; Soc Am Foresters. *Res:* Statistical techniques for biological applications. *Mailing Add:* Sch Forest Resources NC State Univ Raleigh NC 27607

HAFNER, ERICH, b Graz, Austria, July 26, 28; US citizen; wid; c 3. SOLID STATE PHYSICS. *Educ:* Graz Univ, PhD(physics), 52. *Prof Exp:* Res asst theoret physics, Graz Univ, 50-52; res physicist, US Army Command, 53-60, leader frequency control devices, Solid State & Frequency Control Div, 60-81; PRES, XOTEC CORP. *Honors & Awards:* C B Sawyer Mem Award, 83. *Mem:* Am Phys Soc; fel Inst Elec & Electronics Engrs. *Res:* Frequency control and timekeeping; physical and electrical properties of quartz crystals; stable oscillators; noise; physical acoustics; internal friction in dielectrics; lattice theory; microwave theory. *Mailing Add:* Xotec Corp PO Box 1026 Eatontown NJ 07720

HAFNER, GARY STUART, b Greensboro, NC, Aug 25, 43; m 78. NEUROANATOMY, NEUROCYTOLOGY. *Educ:* Hanover Col, BA, 65; Drake Univ, MA, 67; Ind Univ, Bloomington, PhD(neuroanat), 72. *Prof Exp:* Asst prof biol, Hanover Col, 72; res assoc neurocytol, 74-76, asst prof, 76-80, ASSOC PROF, SCH OPTOM, IND UNIV, BLOOMINGTON, 81- *Concurrent Pos:* Fel neurocytol, Ind Univ, Bloomington, 73; trainee neuroanat, Univ Calif, Los Angeles, 73-74. *Mem:* Am Asn Anatomists; Am Soc Zoologists; Asn Res Vision & Ophthalmol; Soc Neurosci. *Res:* Combined cytological, biochemical and autoradiographic studies of photo- and chemoreceptors in vertebrates and invertebrates, and the distribution of neurofilaments and microtubules in receptors and other central nervous system neurons. *Mailing Add:* 5161 McNeely Ellettsville IN 47429

HAFNER, THEODORE, microwave electronics; deceased, see previous edition for last biography

HAFS, HAROLD DAVID, b Genoa City, Wis, July 2, 31; m 52; c 3. REPRODUCTIVE PHYSIOLOGY, ENDOCRINOLOGY. *Educ:* Univ Wis, BS, 53; Cornell Univ, MS, 57, PhD(animal physiol), 59. *Prof Exp:* From asst prof to prof reprod physiol, Mich State Univ, 59-80, chmn, dept dairy sci, 76-80; vpres, Animal Sci Res & Develop, 80-90, VPRES, ANIMAL HEALTH SCI AFFAIRS, MERCK & CO INC, 90- *Concurrent Pos:* NIH spec fel, Harvard Univ, 65-66; vis prof, Univ Nottingham, 73-74. *Honors & Awards:* Animal Physiol Award, Am Soc Animal Sci, 73; Animal Breeders Award, Am Dairy Sci Asn, 71; Jr Scientist Award, Sigma Xi, Mich State Univ, 71, Sr Scientist Award, 78. *Mem:* Soc Study Reprod (secy, 70-73); Endocrine Soc; Am Soc Animal Sci; Brit Soc Study Fertil; Am Physiol Soc. *Res:* Sperm production; sperm capacitation; seminal plasma and uterine fluid; gonadotropins and reproductive growth at puberty and during estrous cycle; utero-ovarian relationships, pituitary control of testicular function. *Mailing Add:* Merck & Co Inc PO Box 2000 Rahway NJ 07065

HAFSTAD, LAWRENCE RANDOLPH, b Minneapolis, Minn, June 18, 04. PHYSICS. *Educ:* Univ Minn, BSc, 26; Johns Hopkins Univ, PhD(physics), 33. *Prof Exp:* Engr, Northwest Bell Tel Co, 20-28; assoc physicist, Carnegie Inst, 28-33, physicist, 33-45; from res physicist to dir appl physics lab, Johns Hopkins Univ, 42-47, dir inst Coop Res, 47-49; dir Nuclear Reactor Develop div, Atomic Energy Comn, 49-55; dir Atomic Energy Div, Chase Manhattan Bank, 55; vpres res labs, Gen Motors Corp, 55-69; chmn comt undersea warfare, Nat Res Coun, 69-72; RETIRED. *Concurrent Pos:* Mem nat defense res comn, Off Sci Res & Develop, 41-46; exec secy res & develop bd, US Dept Defense, 47-49; mem gen adv comt, Atomic Energy Comn, 62-64, chmn, 64-68; consult Nat Security Coun, 72 & Energy Res & Develop Admn, 74; ad hoc consult, Exec Off President & US Dept Navy. *Honors & Awards:* Proctor Prize, Sci Res Soc Am, 56. *Mem:* Nat Acad Eng; AAAS; fel Am Phys Soc. *Res:* Radio propagation; high voltage techniques; geophysics; nuclear physics. *Mailing Add:* RD 1 Box 319 Chester MD 21619

HAFT, JACOB I, b E Orange, NJ, Jan 15, 37; m 64; c 2. INTERNAL MEDICINE, INVASIVE & NON CARDIOLOGY. *Educ:* Harvard Univ, BA, 58; Col Phys Columbia Univ, MD, 62. *Prof Exp:* Chief Cardiol, Bronx Vet Admin Hosp, 69-74; from asst prof to assoc prof med, Mt Sinai Med Sch, 69-74; CHIEF CARDIOL, ST MICHAELS HOSP, 74-; CLIN PROF MED, NJ COL MED. *Concurrent Pos:* Consult cardiol, Holy Name Hosp, 75-, Hackensack Hosp, 75-, Clara Mass Hosp, 75-; mem, Blue Shield Multispecialty Comt Cardiol, 82-; mem, NJ Gov Adv Comt Cardiol, 76-87; mem, NJ Comnr Health Cardiac Serv Comt, 87- *Mem:* Am Col Cardiol fel (gov NJ, 79-82); Am Col Chest Physicians; Am Col Physicians; Clin Coun Am Heart Assoc fel. *Res:* Clinical research on cardiology, electrophysiology, arrhythmias, platelet function, thrombosis, coronary atheroscilero and its progression, angioplasty and aortic valvuloplasty, echocardiology. *Mailing Add:* 306 King Blvd Newark NJ 07102

HAFTEL, MICHAEL IVAN, b Philadelphia, Pa, Mar 22, 43; m 67; c 4. ATOMIC & MOLECULAR COLLISION THEORY. *Educ:* Pa State Univ, BS, 65; Univ Pittsburgh, PhD(physics), 69. *Prof Exp:* Nat Res Coun resident physics, Nat Acad Sci & res assoc nuclear physics, Naval Res Lab, 69-71; PHYSICIST, NAVAL RES LAB, 71-; ASST PROF PHYSICS, GEORGE WASH UNIV, 81- *Concurrent Pos:* Fulbright lectr physics, Univ Graz, Austria, 78-79. *Mem:* Am Phys Soc. *Res:* Nuclear few body problems and nuclear forces, especially pertaining to three nucleon bound and scattering states; few body problems in atomic and molecular collisions; molecular dynamics at surfaces. *Mailing Add:* Code 4651 Naval Res Lab Washington DC 20375

HAGAN, CHARLES PATRICK, synthetic organic chemistry, for more information see previous edition

HAGAN, JAMES JOYCE, b Baltimore, Md, Sept 13, 26; m 61; c 3. CONTRAST AGENTS FOR MAGNETIC RESONANCE IMAGING, RADIOPHARMACEUTICALS. *Educ:* Upsala Col, AB, 49; Rutgers Univ, MSc, 51; Univ Louisville, Phd(biochem), 66. *Prof Exp:* Chemist, Gen Foods Corp, 50-51 & Lederle Labs-Am Cyanamid, 51-61; BIOCHEMIST, BRISTOL-MYERS SQUIBB CO, 66- *Mem:* Am Chem Soc; Soc Magnetic Resonance Med. *Res:* Evaluation and preparation of executive summaries on new diagnostic (in vivo) products or products in various stages of development. *Mailing Add:* 12 Ardmore Pl Holmdel NJ 07733

HAGAN, MELVIN ROY, b Beaver Co, Okla, Jan 20, 34; m 64; c 2. MATHEMATICS. *Educ:* Northwestern State Col, Okla, BS, 59; Okla State Univ, MS, 61, PhD(math), 64. *Prof Exp:* Asst prof math, Stephen F Austin State Col, 64-65; asst prof, 65-71, ASSOC PROF MATH, NTEX STATE UNIV, 71- *Mem:* Am Math Soc; Math Asn Am. *Res:* Topology; peripherally continuous functions and connectivity maps. *Mailing Add:* Dept Math NTex State Univ Denton TX 76202

HAGAN, ROBERT M(OWER), b Oakland, Calif, Dec 5, 16; m 39; c 2. RESOURCE MANAGEMENT. *Educ:* Univ Calif, BS, 37, MS, 42, PhD(soil sci), 48. *Prof Exp:* Assoc bot, Exp Sta, Col Agr, Univ Calif, Davis, 37-39, asst soils, 39-40, assoc irrig, 46-48, instr & jr irrigationist, 48-49, instr & asst irrig technologist, 49-50, asst prof & asst irrig technologist, 50-53, assoc prof & assoc irriagtionist, 53-57, chmn dept water sci & eng, 54-63, prof water sci & eng, 57-87, assoc dir, Int Agr Inst, 68-70, EMER PROF WATER SCI & EXTEN WATER SPECIALIST, COL AGR, UNIV CALIF, DAVIS, 87- *Concurrent Pos:* Mem Int Comn Irrig, Drainage & Flood Control, 60-64; pres, Water Educ Found, 90- *Mem:* Fel AAAS; Soil Sci Soc Am; Am Soc Hort Sci; Soc Int Develop; fel Am Soc Agron; Sigma Xi. *Res:* Water-soil-plant relations; irrigation planning and management; optimizing use of limited water supplies; antitranspirants; irrigation in relation to international agricultural development; water policy. *Mailing Add:* Land/Air/Water Resources Univ Calif Davis CA 95616

HAGAN, WALLACE WOODROW, b Griggsville, Ill, Feb 3, 13; m 40; c 2. GEOLOGY. *Educ:* Univ Ill, BS, 35, MS, 36, PhD(geol), 42. *Prof Exp:* Asst geologist, J V Wicklund Develop Co, Mich, 37-39; consult geologist, Ky & Ill, 39-40; geologist in-chg groundwater sect, Div Geol, Ind State Dept Conserv, 42-44; geologist, Sohio Petrol Co, 45-48; Ky chief geologist, Felmont Oil Corp, 48-52; mem adv bd, 52-92, dir & state geologist, 58-78, EMER GEOLOGIST, KY GEOL SURV, UNIV KY, 78- *Concurrent Pos:* Consult geologist, Ky, 52-58 & 78-; gov rep res comt, Interstate Oil Compact Comn, 59-; chmn res comt, 65-66; bd dirs natural resources comt, Ky Conserv Cong, 61-67; statistician, Asn Am State Geologists, 63-66; mem res & policy comt, Ky Water Resources Inst, 65-78; mem, Ky Water Resources Coun, 65-78; Asn Am State Geologists rep to adv comt water data for pub use, US Dept Interior, 68-78; mem mineral resources subcomt, Lower Miss Region Comprehensive Study, 70-78; mem subcomt radioactive waste disposal, Ky Sci & Technol Coun, 71-78; mem adv coun, Inst Mining & Minerals Res, Univ Ky, 71-78; mem, Ky State-Fed Water Resources Adv Coun, 73-78, Maxey Flats Nuclear Waste Disposal Site Decommisioning Plan Adv Bd, Ky, 83-84 & Ky Oil & Gas Asn State Regulatory Comt, 82-86; ex officio mem, Ky Develop Cabinet, 73-78; pres, Ky sect, Am Inst Prof Geologists, 82-83; deleg, Ky-Tenn Sect Geol Soc Ky, Am Asn Petrol Geologists, 83-86. *Honors & Awards:* John Wesley Powell Award, Geol Surv, US Dept Interior, 72; Pub Serv Award, Am Asn Petrol Geologists, 82. *Mem:* Hon mem Asn Am State Geologists (vpres, 66-67, pres elect); fel Geol Soc Am; Am Asn Petrol Geologists; Am Inst Prof Geologists; Sigma Xi. *Res:* Stratigraphy; petroleum geography. *Mailing Add:* 317 Jesselin Dr Lexington KY 40503

HAGAN, WILLIAM JOHN, JR, b Port Washington, NY, May 17, 56. SURFACE CHEMISTRY, CHEMICAL EVOLUTION. *Educ:* Bowdoin Col, BA, 78; Rensselaer Polytech Inst, PhD(organic chem), 85, MS, 86. *Prof Exp:* Fel biochem, Univ Rochester, 85-87; ASST PROF CHEM, ST ANSELM COL, 87- *Concurrent Pos:* Vis assoc prof, Rensselaer Polytech Inst, 88. *Mem:* Am Chem Soc; AAAS; Am Soc Photobiol; Int Soc Study of Origins of Life; Hist Sci Soc. *Res:* Surface-catalyzed chemical reactions; thermal and photochemical probes of mineral surfaces; history of chemistry. *Mailing Add:* Dept Chem St Anselm Col Box 1662 Manchester NH 03102

HAGAN, WILLIAM LEONARD, b Atkinson, Ill, Mar 21, 36; m 57; c 2. PLANT BREEDING, PHYTOPATHOLOGY. *Educ:* Univ Ill, BS, 59, MS, 60, PhD(plant genetics & path), 63; John F Kennedy Univ, MBA, 85. *Prof Exp:* Res asst plant breeding & path, Univ Ill, 59-63; plant breeder, 63-67, mgr agr res & serv, 67-75, mgr agr res, 75-79, mgr plant breeding, 79-83, dir agr res & seed oper, 83-88, DIR AGR SCI & BIOTECHNOL, DEL MONTE CORP, 88- *Mem:* Am Phytopath Soc; Am Soc Agron; Am Mgt Asn. *Res:* Genetics, diseases and breeding of vegetable and agronic crops; bio-control and biotechnology management. *Mailing Add:* 628 Broadmoor Blvd San Leandro CA 94577

HAGANS, JAMES ALBERT, b Cincinnati, Ohio, Nov 9, 22. BIOSTATISTICS, THERAPEUTICS. *Educ:* Marietta Col, AB, 44; Univ Cincinnati, MD, 46; Univ Okla, MS, 58, PhD, 60. *Prof Exp:* Instr med, Col Med, Univ Cincinnati, 51-52; instr, Sch Med, Emory Univ, 54-55; asst prof med & dir exp therapeut unit, Sch Med, Univ Okla, 55-57, instr med, prev med & pub health, 57-60, assoc prof med biostatist, prev med, pub health & med, 60-62; mgr med biostatist, clin pharmacol & med res, Upjohn Co, 63-71; dir epidemiol biostatist & res data systs med affairs, Merck Sharp & Dohme Res Labs, Pa, 71-73; chief coop studies prog, Dept Med & Surg, Vet Admin, 72-85; RETIRED. *Concurrent Pos:* Jr attend, Cincinnati Gen Hosp, 51-52; attend med, Grady Hosp, Atlanta, Ga, 54-55, Vet Admin Hosp, Okla City, 55-62, Cent State Hosp, 55-62 & Sch Med, Univ Okla, 55-62; prof biostat, Dept Epidemiol & Pub Health, Sch Med, Univ Miami, 73-; mem, US Comt, WHO. *Mem:* Fel Am Col Physicians; Am Statist Asn; Am Soc Pharmacol & Exp Therapeut; Am Soc Clin Pharmacol & Therapeut; Soc Clin Trials. *Res:* Clinical and epidemiological biostatistics; clinical pharmacology; experimental therapeutics; internal medicine; biomedical computer sciences. *Mailing Add:* 9201 W Broward Blvd Apt C-115 Plantation FL 33324

HAGAR, CHARLES FREDERICK, b Los Angeles, Calif, Aug 7, 30. ASTRONOMY, SCIENCE COMMUNICATIONS. *Educ:* Univ Calif, Los Angeles, BA, 54; Univ Calif, Berkeley, MA, 60. *Prof Exp:* Astronr lectr, Griffith Observ, Los Angeles, 52-57; asst dir astron, AF Morrison Planetarium, San Francisco, 57-59; PROF ASTRON, SAN FRANCISCO STATE UNIV, 59-, DIR OBSERV & PLANETARIUM, 75- *Concurrent Pos:* Planetarium consult, Carl Zeiss, WGer, 54-, Hong Kong Space Mus, 75-79; sci fac fel, NSF, 67-68; careers ed, Int Planetarium Soc, 75-77. *Mem:* Astron Soc Pacific; Int Planetarium Soc. *Res:* Planetarium design and operations; planetarium survey. *Mailing Add:* Dept Physics & Astron San Francisco State Univ 1600 Holloway Ave San Francisco CA 94132

HAGAR, LOWELL PAUL, b Girard, Kans, Aug 30, 26; m 49; c 3. ENZYMOLOGY, BIOSYNTHESIS. *Educ:* Valparaiso Univ, AB, 47; Univ Kans, MA, 50; Univ Ill, PhD(microbiol), 53. *Prof Exp:* Asst prof biochem, Harvard Univ, 55-60; assoc prof, 60-65, head dept, 67-88, PROF BIOCHEM, UNIV ILL, 65-, DIR BIOTECHNOL CTR, 88- *Concurrent Pos:* Fel Guggenheim Found, 60. *Mem:* Am Chem Soc; Am Soc Biochem & Molecular Biol. *Res:* Enzymology and enzyme mechanisms; biosynthesis of halo metabolites. *Mailing Add:* 801 W Delaware Urbana IL 61801

HAGAR, SILAS STANLEY, b Chenoa, Ill, Apr 7, 31. BOTANY-PHYTOPATHY. *Educ:* Univ Maine, Orono, BS, 67, PhD(plant sci), 71. *Prof Exp:* Res asst enzyme purification, Univ Maine, Orono, 67-71, res assoc, 71; asst prof biol, Unity Col, Maine, 71-72; dir, 81-87, RES PLANT PATHOLOGIST, BROOKLYN BOTANIC GARDEN RES CTR, 72- *Concurrent Pos:* Adj assoc prof, Pace Univ, Pleasantville, NY, 78- *Mem:* Am Phytopath Soc; Sigma Xi; Am Ivy Soc. *Res:* Disease physiology of horticultural plants; culture of fastidious, pathological micro-organisms. *Mailing Add:* Brooklyn Botanic Garden Res Ctr 712 Kitchawan Rd Ossining NY 10562

HAGBERG, ELROY CARL, b Bayfield, Wis, July 15, 19; m 46; c 2. FOOD SCIENCE. *Educ:* Univ Wis, BS, 41, MS, 47, PhD(dairy indust), 49. *Prof Exp:* Chemist, dairy & poultry prods control, Wilson & Co, Ill, 41-44; asst, Univ Wis, 46-49; assoc scientist, milk proteins, Nat Dairy Res Labs, Inc, 49-56; res chemist, Campbell Inst Food Res, Campbell Soup Co, Camden, 56-59, div head dairy prods res, 59-77, dir food sci, 77-80, sr res scientist, 80-82; RETIRED. *Mem:* Am Chem Soc; Am Dairy Sci Asn; Inst Food Technol. *Res:* Dairy, cereal and protein products in heat processed and frozen convenience foods. *Mailing Add:* 6737 Collins Ave Pennsauken NJ 08109

HAGBERG, ERIK GORDON, b Gothenburg, Sweden, Nov 12, 49. NUCLEI FAR FROM STABILITY, DELAYED PARTICLE DECAYS. *Educ:* Univ Gothenburg, BSc, 72, PhD(nuclear physics), 78, Docent(nuclear physics), 81. *Prof Exp:* Teaching asst, Chalmers Univ Technol, Gothenburg, Sweden, 72-76, res asst, 78; res fel, Europ Coun Nuclear Res, Geneva, Switz, 76-78, Atomic Energy Can Ltd, 78-80; res assoc, Queens Univ, Kingston, Can, 80-82; assoc res officer, 83-84, RES OFFICER ATOMIC ENERGY CAN, LTD, 85- *Concurrent Pos:* Adj prof, Univ Man, 90- *Mem:* Am Phys Soc. *Res:* Structure of nuclei far from stability; experimental studies of the properties of ground states, and highly excited states of far-unstable nuclei and theoretical model calculations of these properties; nuclear weak interaction. *Mailing Add:* Atomic Can Energy Ltd Chalk River Labs Chalk River ON K0J 1J0 Can

HAGE, KEITH DONALD, b Kandahar, Sask, Mar 15, 26; m 52; c 3. METEOROLOGY. *Educ:* Univ BC, BA, 49; Univ Toronto, MA, 50; Univ Chicago, PhD(meteorol), 57. *Prof Exp:* Meteorologist, Meteorol Off, Alta, 50-54; asst meteorol, Univ Chicago, 55-57; meteorologist, Suffield Exp Sta, 57-60; sr res scientist, Travelers Res Ctr, Inc, Conn, 60-67; prof, 67-85, EMER PROF GEOG, UNIV ALTA, 85- *Honors & Awards:* Patterson Medal, Atmospheric Environ Serv, Can, 89. *Mem:* Can Meteorol Soc; fel Am Meteorol Soc; Royal Meteorol Soc; fel Can Soc Agrometeorol. *Res:* Synoptic-dynamic meteorology and micrometeorology; atmospheric diffusion; mesometeorology. *Mailing Add:* Dept Geog Univ Alta Edmonton AB T6G 2H4 Can

HAGEAGE, GEORGE JOHN, JR, b Allentown, Pa, Nov 4, 35; m 60; c 2. MEDICAL MICROBIQLOGY. *Educ:* Muhlenberg Col, BS, 57; Univ Md, MS, 60, PhD(microbiol), 63; Am Bd Med Microbiol, dipl, 81. *Prof Exp:* Asst prof microbiol, Univ NH, 63-66; staff fel, Lab Histol & Path, Nat Inst Dent Res, 66-68, res microbiologist, Lab Biol Struct, 68-75; CLIN MICROBIOLOGIST, ST VINCENT HOSP & MED CTR, 75- *Concurrent Pos:* Prof microbiol, Med Col Ohio, 75-; adj prof biol sci, Bowling Green State Univ, 76- *Mem:* Am Soc Clin Pathologists; Am Soc Microbiol; NY Acad Sci. *Res:* Ultrastructure of microorganisms; oral microbiology; development of techniques for the rapid identification of pathogenic microorganisms. *Mailing Add:* St Vincent's Hosp & Med Ctr 2213 Cherry St Toledo OH 43608

HAGEDORN, ALBERT BERNER, b Salt Lake City, Utah, Jan 15, 15; m 44; c 4. MEDICINE. *Educ:* Univ Calif, AB, 38; Stanford Univ, MD, 43; Univ Minn, MS, 46. *Prof Exp:* From asst prof to assoc prof, 50-69, PROF MED, MAYO GRAD SCH MED, UNIV MINN, 69- *Concurrent Pos:* Consult, Mayo Clin, 46- & US Army Reserve, 54-55. *Mem:* Am Soc Hemat; Am Nuclear Soc; AMA; fel Am Col Physicians; Int Soc Hemat; Sigma Xi. *Res:* Hematology; iron therapy and metabolism. *Mailing Add:* 200 First St W Rochester MN 55901

HAGEDORN, CHARLES, b Washington, DC, Aug 13, 47; m 69. SOIL MICROBIOLOGY, FOREST ECOLOGY. *Educ:* Kans State Univ, BS, 70; Iowa State Univ, MS, 72, PhD(microbiol), 74. *Prof Exp:* Teaching asst microbiol, Iowa State Univ, 70-73, res asst, 73-74; asst prof, Dept Microbiol, Ore State Univ, 74-79; ASSOC PROF, DEPT AGRON, MISS STATE UNIV, 79- *Mem:* Am Soc Microbiol; Soc Indust Microbiol; Soil Sci Soc Am; Am Inst Biol Sci; Am Soc Agron. *Res:* Nitrogen fixation and enzymatic activities in agricultural soils; plant decomposition and microbial antagonisms in forest soils and use of sewage sludge as a source of fertilizer for crops. *Mailing Add:* Dept Agron Va Polytech Inst 365 Smith Hall Blacksburg VA 24061-0404

HAGEDORN, DONALD JAMES, b Moscow, Idaho, May 18, 19; m 43; c 1. PLANT BREEDING, MICROBIOLOGY. *Educ:* Univ Idaho, BS, 41; Univ Wis, MS, 43, PhD(plant path), 48. *Hon Degrees:* DSc, Univ Idaho, 79. *Prof Exp:* From asst prof to prof agron & plant path, 48-64, PROF PLANT PATH, UNIV WIS-MADISON, 64- *Concurrent Pos:* Courtesy prof plant path, Ore State Univ, 72-73; vis scientist plant path, Dept Sci & Indust Res, Lincoln Res Ctr, NZ, 80-81; consult, Hyderabad, India, 79; mem, Plant Variety Protection Bd, USDA, Washington, DC, 77-78; fel, Am Phytopath Soc, 76. *Honors & Awards:* Campbell Award, AAAS, 61; CIBA-Geigy Award, Am Phytopathol Soc, 75. *Mem:* Sigma Xi. *Res:* Diseases of peas and beans; control of these diseases through breeding for disease resistance. *Mailing Add:* Univ Wis 1630 Linden Dr Madison WI 53706

HAGEDORN, FRED BASSETT, b Boone, Iowa, June 8, 28; m 54; c 2. MAGNETISM. *Educ:* Iowa State Univ, BS, 52; Calif Inst Technol, PhD(physics), 57. *Prof Exp:* Mem tech staff, Bell Tel Labs, Inc, 57-85; dir tech staff, 85-87, CONSULT SOLID STATE PHYSICS, PARKS-JAGGERS AEROSPACE, 87- *Mem:* Am Phys Soc; Inst Elec & Electronics Engrs; Sigma Xi. *Res:* Low energy nuclear physics; magnetic flux reversal processes, superconducting and ferromagnetic thin films; magnetic materials research; magnetic bubble device research and development. *Mailing Add:* 1000 Cordova Dr Orlando FL 32804

HAGEDORN, GEORGE ALLAN, b Santa Monica, Calif, Oct 18, 53; m 77; c 1. MATHEMATICAL PHYSICS, QUANTUM CHEMISTRY. *Educ:* Cornell Univ, BA, 74; Princeton Univ, MA, 75, PhD(math), 78. *Prof Exp:* Res assoc math physics, Rochefeller Univ, 78-80; from asst prof to assoc prof, 80-88, PROF, VA POLYTECH INST & STATE UNIV, 88- *Concurrent Pos:* Mem, Am Math Soc prog comt nat meetings, 88-91; chmn, Joint Prog Comt Nat Meeting, Am Math Soc & Math Asn Am, 91. *Mem:* Am Math Soc; Int Asn Math Physics. *Res:* Rigorous results on N particle Schrodinger operators. *Mailing Add:* Dept Math Va Polytech Inst Blacksburg VA 24061

HAGEDORN, HENRY HOWARD, b Milwaukee, Wis, Apr 4, 40; m 64; c 2. INSECT REPRODUCTION, INSECT ENDOCRINOLOGY. *Educ:* Univ Wis-Madison, BS, 65, MS, 66; Univ Calif, Davis, PhD(entom), 70. *Prof Exp:* Asst prof insect physiol, Entom Dept, Univ Mass, Amherst, 73-77; from assoc prof to prof insect physiol, Entom Dept, Cornell Univ, 77-88; PROF INSECT PHYSIOL, ENTOM DEPT, UNIV ARIZ, TUCSON, 88-; DIR, CTR INSECT SCI, 89- *Concurrent Pos:* Von Humbolt sr scientist, von Humbolt Stiftung, 81. *Mem:* Fel AAAS; Am Soc Zoologists; Entom Soc Am; Royal Entom Soc London; Sigma Xi. *Res:* Endocrinology of mosquito reproduction; peptide hormones and the hormonal control of gene expression. *Mailing Add:* Dept Entom Univ Ariz 430 Forbes Bldg 36 Tucson AZ 85721

HAGEE, GEORGE RICHARD, b Cincinnati, Ohio, July 10, 25; m 50; c 5. NUCLEAR & RADIOLOGICAL PHYSICS, ENVIRONMENTAL RADIATION. *Educ:* Xavier Univ, Ohio, BS, 49, MS, 54; Univ Cincinnati, PhD(physics), 65; Am Bd Radiol, dipl, 76. *Prof Exp:* Physicist nuclear measurements, USPHS, 50-52; res physicist appl nuclear & health physics, Mound Lab, Monsanto Res Corp, 52-54; electronic physicist, Wright Air Develop Ctr, USAF, 54; res physicist, Div Radiol Health, USPHS, 54-56, chief phys & instrumental methods, 56-59; sr nuclear res physicist, Mound Lab, Monsanto Res Corp, 59-68; assoc prof physics, USAF Inst Technol, Wright-Patterson AFB, 68-80; res sci, Monsanto Mound Lab, 80-85; vis prof, Nuclear Eng, Univ Cincinnati, 86-87; CONSULT, ENVIRON RADIATION & MED PHYSICS, 87- *Concurrent Pos:* USPHS rep, Am Standards Asn Comt Nuclear Instrumentation, 58-59. *Mem:* AAAS; Am Phys Soc; Am Nuclear Soc; Health Physics Soc. *Res:* Low energy nuclear physics and instrumentation; environmental radioactivity; techniques and methodology; tracer applications; nuclear gauging; low level counting; nuclear spectroscopy; nuclear medicine. *Mailing Add:* 5481 E Kemper Rd Cincinnati OH 45241

HAGEL, ROBERT B, b Newark, NJ, Apr 15, 43; m 69. REGULATORY COMPLIANCE, ANALYTICAL CHEMISTRY. *Educ:* Rutgers Univ, Newark, BA, 65, MS, 68, PhD(inorg chem), 69. *Prof Exp:* Instr chem, Rutgers Univ, 68-70; sr chemist, 70-73, group leader, 73-75, tech fel, 75-79, mgr, 79-83, asst dir QA, 83-88, DIR QA, HOFFMANN-LA ROCHE INC, 88- *Mem:* Am Chem Soc; Parenteral Drug Asn; Sigma Xi. *Res:* Characterization and analysis of new drug substances and antibiotics. *Mailing Add:* 571 Green Hill Rd Kinnelon NJ 07405

HAGEL, WILLIAM C(ARL), b Pittsburgh, Pa, Apr 5, 27; m 54; c 3. PHYSICAL METALLURGY. *Educ:* Cornell Univ, BMetE, 51; Carnegie-Mellon Univ, PhD(metall), 54. *Prof Exp:* Metallurgist, Metall Div, Oak Ridge Nat Lab, AEC, Tenn, 51; res metallurgist, Metals Res Lab, Carnegie Inst Technol, 52-54 & turbine div lab, Gen Elec Co, NY, 54-58, mgr metall measurements lab, Mass, 58-59, metallurgist, Metall & Ceramics Res Dept, 59-66; prof metall, chmn dept & head metall div, Denver Res Inst, Univ Denver, 66-70; mgr mat develop, Mat & Process Technol Labs, Aircraft Engine Group, Gen Elec Co, 70-72; mgr res, Climax Molybdenum Co, 73-84; PRES, ARBORMET LTD, 84- *Mem:* Fel Am Soc Metals; Electron Micros Soc; Electrochem Soc; Am Ceramic Soc; Am Inst Mining, Metall & Petrol Engrs. *Res:* Structure of metals; phase transformations in metals; high-temperature alloys; thermoelectric, magnetic and semiconductor materials; surface reactions; diffusion in solids. *Mailing Add:* Arbormet Ltd 685 Skynob Dr Ann Arbor MI 48105

HAGELBARGER, DAVID WILLIAM, b Kipton, Ohio, May 3, 20; m 44; c 2. COMMUNICATIONS SCIENCE. *Educ:* Hiram Col, AB, 42; Calif Inst Technol, PhD(physics), 47. *Prof Exp:* Asst physics, Calif Inst Technol, 42-43, res physicist, Nat Defense Res Comt Proj, 44-45; instr aeronaut eng, Univ Mich, 46-49, res physicist, 46-49; mem tech staff, Bell Labs, 49-86; CONSULT, NAT OPTICAL ASTRON OBSERVS, KITT PARK, 88- *Concurrent Pos:* Consult, Study Group 3 , 60-64, eng concepts, curriculum proj, 65-70. *Mem:* AAAS; Am Phys Soc; Inst Elec & Electronics Engrs; Sigma Xi. *Res:* Computers; electron dynamics; pressure temperature and composition of upper atmosphere; astronomical instrument and telescope design; communication systems; man machine interface. *Mailing Add:* 2180 S Double O Pl Tucson AZ 85713

HAGELBERG, M(YRON) PAUL, b Manistee, Mich, Jan 9, 33; m 55, 84; c 2. EXPERIMENTAL PHYSICS. *Educ:* Mich State Univ, BS, 54, MS, 56, PhD, 61. *Prof Exp:* Asst physics, Mich State Univ, 55-56; from instr to asst prof, 58-71, chmn dept, 69-86, PROF PHYSICS, WITTENBERG UNIV, 71- *Mem:* Acoust Soc Am; Am Phys Soc; Am Asn Physics Teachers; Sigma Xi; AAAS; Optical Soc Am. *Res:* Diffraction of light by ultrasonic waves; sound propagation in liquids and solids; nonlinear acoustic. *Mailing Add:* Dept Physics Wittenberg Univ PO Box 720 Springfield OH 45501

HAGELIN, JOHN SAMUEL, b Pittsburgh, Pa, June 9, 54; m 85. TECHNOLOGIES FOR THE DEVELOPMENT OF HUMAN CONSCIOUSNESS. *Educ:* Harvard Univ, MA, 76, PhD(physics), 81. *Prof Exp:* Sci assoc, Europ Lab Particle Physics, 81-82; res assoc, Stanford Linear Accelerator Ctr, 82-83; assoc prof, 83-84, PROF & DEPT CHMN PHYSICS, MAHARISHI INT UNIV, 84- *Concurrent Pos:* Dir doctoral prog physics, Maharishi Int Univ, 84-; prin investr, NSF Grant, 89-; pres, Maharishi Int Asn Unified Field Scientists, 90- *Mem:* AAAS; Am Inst Physics; Am Phys Soc. *Res:* Supersymmetric unified quantum field theories based on the superstring and on the practical applications of the unified field to the individual and society; author of approximately 60 publications. *Mailing Add:* Dept Physics Maharishi Int Univ Box 1069 Fairfield IA 52556

HAGEMAN, DONALD HENRY, b Quincy, Ill, Nov 2, 18; m 42. PHYSICS, ELECTRICAL ENGINEERING. *Educ:* Northwestern Univ, Ill, BS, 48; Mass Inst Technol, MS, 51; Univ Southern Calif, PhD(physics), 67. *Prof Exp:* Res asst digital circuit design, servomechanisms Lab, Mass Inst Technol, 49-51; res engr, Hughes Aircraft Co, 51-55; asst proj engr, Litton Industs, Inc, 55-59; mem tech staff, Hughes Aircraft Co, 59-63; physicist, Naval Ocean Systs Ctr, 67-88; RETIRED. *Mem:* Am Phys Soc; NY Acad Sci. *Res:* Acoustic signal processing, radiation and scattering; ocean acoustics; laser velocimetry; theory and simulation of vehicle motion in fluids. *Mailing Add:* 6673 Aranda Ave La Jolla CA 92037

HAGEMAN, GILBERT ROBERT, b Covington, Ky, May 21, 47; m 68; c 5. CARDIOLOGY. *Educ:* Thomas More Col, AB, 68; Loyola Univ, Chicago, PhD(physiol), 75. *Prof Exp:* Estab investr, Am Heart Asn, 80-85; Cardiovasc Res Training Ctr, 74-84, instr physiol, 75-77, asst prof, 77-82, instr cardiol, 82-85, coursemaster, 80-83, ASSOC PROF PHYSIOL, DEPT PHYSIOL & BIOPHYS, MED CTR, UNIV ALA, BIRMINGHAM, 82-, ASSOC PROF MED, 85-, SCIENTIST, CARDIOVASC RES TRAINING CTR, 85-, VCHMN ACAD AFFAIRS, 88- *Concurrent Pos:* Fel circulation coun, Am Heart Asn, 81- & cardiovasc sect, Am Physiol Soc. *Mem:* AAAS; Am Physiol Soc; Sigma Xi. *Res:* Neural regulation of the heart during health and disease; neurogenic cardiac arrhythmias and sudden death. *Mailing Add:* Dept Physiol & Biophys Med Ctr Univ Ala Birmingham AL 35294

HAGEMAN, GREGORY SCOTT, OCULAR RESEARCH. *Educ:* Univ Southern Calif, PhD(biol), 83. *Prof Exp:* ASST PROF MICROBIOL RES, SCH MED, UNIV SOUTHERN CALIF, 83- *Mailing Add:* Bethesda Ave Inst 3655 Vista St Louis MO 63110

HAGEMAN, JAMES HOWARD, b Washington, Iowa, Nov 8, 42; m 68; c 8. BIOCHEMISTRY, BACTERIAL PHYSIOLOGY. *Educ:* Univ Bristol, Int, BSc, 61; Univ Ill, BS, 64; Univ Calif, Los Angeles, PhD(biochem), 68. *Prof Exp:* Fel, Am Cancer Soc, Yale Univ, 69-71; from asst prof to assoc prof, 71-81, PROF CHEM, NMEX STATE UNIV, 82-, DIR GRAD PROG, MOLECULAR BIOL, 88- *Concurrent Pos:* Mem exam comt biochem, Am Chem Soc, 74-; Intergovt Personnel Act fel, NIH, 78-79. *Mem:* Am Soc Biol Chemists; AAAS; Am Chem Soc; Am Soc Microbiol. *Res:* Metabolic regulation; mechanisms of enzyme regulation; bacterial proteases; protein turnover; control of bacterial sporulation; bacterial calmodulin; nutritive value of forages; applications of boronic acids. *Mailing Add:* Dept Chem NMex State Univ Box 3C Las Cruces NM 88003-0001

HAGEMAN, LOUIS ALFRED, b Danville, Ill, Oct 8, 32; m 71; c 2. MATHEMATICS. *Educ:* Rose Hulman Inst Technol, BS, 55; Univ Pittsburgh, PhD(math), 62. *Prof Exp:* ADV MATHEMATICIAN, WESTINGHOUSE ELEC CORP, 72- *Concurrent Pos:* Sr lectr, Carnegie-Mellon Univ, 64-76. *Mem:* Soc Indust & Appl Math. *Res:* Numerical analysis; iterative solution methods. *Mailing Add:* Bettis Atomic Power Lab PO Box 79 Pittsburgh PA 15122

HAGEMAN, RICHARD HARRY, b Powell, Wyo, Apr 14, 17; m 41; c 3. BIOCHEMISTRY. *Educ:* Kans State Col, BS, 38; Okla Agr & Mech Col, MS, 40; Univ Calif, Berkeley, PhD(biol), 54. *Prof Exp:* Asst, Okla Agr & Mech Col, 38-40; asst chemist, Exp Sta, Univ Ky, 40-41 & 46-47; chemist & plant physiologist, Fed Exp Sta, PR, 47-50; from asst prof to prof, 54-84, EMER PROF AGRON, UNIV ILL, URBANA, 84- *Concurrent Pos:* Rockefeller fel, Long Ashton Res Sta, Bristol; vis prof, Mich State Univ, 67-68; sr res fel, Australian-Am Scholarly Exchange (Fullbright), Melbourne, Australia; Dugger Lectr, Auburn Univ; Am Soc Bot del, Peoples Repub China. *Honors & Awards:* Crop Sci Award, Am Soc Agron, Agron Res Award, 84; Funke Award, Funke Seed Corn Co; Spencer Award, Am Chem Soc, 85; Hoagland Award, Am Soc Plant Physiologists, 85. *Mem:* Fel Am Soc Agron; Am Chem Soc; Am Soc Plant Physiologists; Fedn Am Socs Exp Biol; fel Crop Sci Soc Am. *Res:* Plant biochemistry; enzymes; minor elements; nitrogen metabolism; enzyme induction, mode of inheritance and genetic control of nitrate reductase. *Mailing Add:* 1302 E McHenry Urbana IL 61801

HAGEMAN, WILLIAM E, b Glendale, Ohio, Sept 2, 39; m 64; c 4. PHARMACOLOGY, PHYSIOLOGY. *Educ:* Univ Cincinnati, BSPh, 63; Univ Pittsburgh, MS, 66, PhD(pharmacol), 68. *Prof Exp:* Sr scientist, McNeil Labs, Inc, 68-73, group leader, 73-77, sect head cardiovasc pharmacol, res fel, 77-82, prin scientist, McNeil Pharmaceut, 82-88; prin scientist, Janssen Res Found, 88-90; PRIN SCIENTIST, RWJ PHARMACEUT RES INST, 90- *Concurrent Pos:* Mem, Pulmonary Discussion Group, Inflammation Res Asn. *Mem:* Am Soc Pharmacol & Exp Therapeut; AAAS. *Res:* Cardiovascular pulmonary & inflammation pharmacology and physiology. *Mailing Add:* Immunopharmacol Dept Exp Therapeut RWJ Pharmaceut Res Inst Raritan NJ 08869-0602

HAGEMARK, KJELL INGVAR, optical recording, for more information see previous edition

HAGEN, ARNULF PEDER, b Tacoma, Wash, June 6, 42; c 3. INORGANIC CHEMISTRY. *Educ:* Univ Wash, Seattle, BSc, 64; Univ Pa, PhD(chem), 68. *Prof Exp:* From asst prof to assoc prof, 67-82, PROF CHEM, UNIV OKLA, 82- *Mem:* Am Chem Soc; Sigma Xi; Asn Asphalt Paving Technologists. *Res:* Synthesis of inorganic and organometallic compounds containing silicon, phosphorus, sulfur, boron and fluorine; high pressure synthesis; chemistry of asphalt and coal. *Mailing Add:* Dept Chem Univ Okla 620 Parrington Oval Norman OK 73019

HAGEN, ARTHUR AINSWORTH, b Hot Springs, SDak, Oct 9, 33; m 57; c 4. PHARMACOLOGY. *Educ:* Univ SDak, BA, 55, MA, 57; Univ Tenn, PhD(pharmacol), 61. *Prof Exp:* USPHS trainee steroid biochem, Salt Lake City, Utah, 61-63; Swed Med Res Coun fel, Stockholm, 63-64; from asst prof to assoc prof pharmacol, Col Med, Univ Tenn, Memphis, 65-80, prof, 80-87; PROF, DEPT PHYSIOL & PHARMACOL, UNIV SDAK, VERMILLION, 87- *Mem:* Endocrine Soc; Soc Study Reproduction; Am Soc Pharmacol & Exp Therapeut. *Res:* Endocrinology; prostaglandins; gonadal physiology and biochemistry; cerebral vasospasm. *Mailing Add:* Dept Physiol & Pharmacol Univ SDak Vermillion SD 57069

HAGEN, CARL RICHARD, b Chicago, Ill, Feb 2, 37; m 65; c 3. PHYSICS. *Educ:* Mass Inst Technol, SB & SM, 58, PhD(physics), 62. *Prof Exp:* Res assoc, 63-64, res asst & asst prof, 64-65, asst prof, 65-68, assoc prof, 68-74, PROF PHYSICS, UNIV ROCHESTER, 74- *Concurrent Pos:* Res fel, Imp Col, Univ London, 64; vis scientist, Int Ctr Theoret Physics, Italy, 67. *Mem:* Am Phys Soc. *Res:* Field theory; particle physics; group theory. *Mailing Add:* Dept Physics Univ Rochester Wilson Blvd Rochester NY 14627

HAGEN, CHARLES ALFRED, b East Rutherford, NJ, Feb 1, 25; m 51; c 3. MICROBIOLOGY. *Educ:* Univ Chicago, AB, 52, MS, 56. *Prof Exp:* Bacteriologist, Bobs Robert Hosp, Univ Chicago Clin, 54-55; sr med technician, Inst Tuberculosis Res, 55-56; res bacteriologist, Nat Dairy Prod Corp, 56-62; assoc bacteriologist Life Sci Div, IIT Res Inst, 62-65, res bacteriologist, 65-69; lab mgr space biol, Space Systs Div, Avco Corp, 69-71; lab mgr, planetary quarantine, Jet Propulsion Lab, Bionetics Corp, 71-76; chief bacteriologist, Becton Dickinson Labware, 76-79, regulatory affairs officer, 79-86; qual assurance engr, Spectramed Inc, Oxnard, Ca, 86-87, regulatory affairs mgr, 87-90; CONSULT, 90- *Mem:* Am Soc Microbiol; Soc Indust Microbiol; Am Soc Qual Control; Asn Adv Med Instrumentation; Regulatory Affairs Prof Soc. *Res:* Anaerobic culture techniques; space biology; industrial microbiological processes; normal microflora of animals; animal disease; pathogenic microorganisms in foods; tuberculosis vaccination; clean room techniques and contamination control; diagnostic medical bacteriology; medical device quality control; radiation and ethylene oxide sterilization processes; medical device manufacturing compliance to State and Federal requirements. *Mailing Add:* Spectramed Inc 2085 Lyndhurst Ave Camarillo CA 93010

HAGEN, CHARLES WILLIAM, JR, b Spartanburg, SC, Mar 21, 18; m 42; c 3. ACADEMIC ADMINISTRATION, PLANT SCIENCE. *Educ:* Cornell Univ, AB, 39; Ind Univ, PhD(bot), 44. *Prof Exp:* Asst bot, Ind Univ, 39-43; mem war res staff, Off Sci Res & Develop, SC & Ft Benning, Ga, 43; asst, Metall Labs, Manhattan Dist, Univ Chicago, 43, res assoc & group leader, 43-45, assoc biologist & group leader, 45-46; from instr to emer prof biol, Ind Univ, Bloomington, 46-, actg chmn, Div Biol Sci, 64-65, assoc dean arts & sci, 65-66, assoc dean , Fac, 66-69, Acad Affairs, 69-72, assoc dean & dean, Resource Develop, 72-75, dir long-range planning, 75-81; RETIRED. *Concurrent Pos:* Fulbright res award & Guggenheim fel, 57-58. *Mem:* AAAS; Am Inst Biol Sci; Sigma Xi. *Res:* Cytogenetics of Oenothera; radiobiology; plant tissue culture; anthocyanins; flavonoid pigments; university administration; botany-phytopathology. *Mailing Add:* Rte 5 Box 388 Nashville IN 47448

HAGEN, DANIEL RUSSELL, b Springfield, Ill, Sept 29, 52; m 78; c 4. REPRODUCTIVE PHYSIOLOGY, ENDOCRINOLOGY. *Educ:* Univ Ill, Urbana, BS, 74, PhD(animal sci), 78. *Prof Exp:* Res asst, Univ Ill, Urbana, 74-75, fel, 75-77; asst prof, 78-84, ASSOC PROF REPRODUCTIVE PHYSIOL, PA STATE UNIV, 84- *Concurrent Pos:* Res assoc, Cornell Univ, 78. *Mem:* Soc Study Reprod; Soc Study Fertil; Am Soc Animal Sci; AAAS; Sigma Xi. *Res:* Endocrine relationships between dam and fetus affecting prenatal and postnatal growth and survival, including ovarian and placental function; role of somato tropin in reproduction. *Mailing Add:* 304 Henning Bldg University Park PA 16802

HAGEN, DONALD E, b Dayton, Ohio, Oct 29, 43; m 69; c 3. WATER MICROPHYSICS, CLOUD CHAMBER EXPERIMENT. *Educ:* Univ Dayton, BS, 65; Purdue Univ, MS, 67, PhD(physics), 70. *Prof Exp:* Fel physics, Battelle Mem Inst, 70-71; fel, 71-74, res assoc, 74-76, RES ASST PROF CLOUD PHYSICS, UNIV MO, ROLLA, 76- *Concurrent Pos:* Consult, NASA, 76-78, prin investr contracts, 78-81 & 82-84. *Mem:* Am Phys Soc. *Res:* Cloud microphysics; the nature of and evolution of small water drops or clusters; computer hardware and software. *Mailing Add:* Dept Physics Univ Mo PO Box 249 Rolla MO 65401

HAGEN, DONALD FREDERICK, b Boscobel, Wis, Sept 7, 32; m 53; c 4. ANALYTICAL CHEMISTRY, GENERAL CHEMISTRY. *Educ:* Univ Wis-Madison, BS, 57; Okla State Univ, MS, 61. *Prof Exp:* Anal res chemist, Continental Oil Co, 57-63; ANALYTICAL CORP CHEMIST, 3M CO, 63- *Concurrent Pos:* Co-founder & dir, Minn Chromatog Forum, 78- *Mem:* Am Chem Soc. *Res:* Analytical chemistry; fluorine chemistry; plasma chromatography; atomic emission detecters; derivatization; elemental tagging for chromatography; membranes for separations, purifications and reactions. *Mailing Add:* 7149 Windgate Rd Woodbury MN 55125

HAGEN, GRETCHEN, b Lake Placid, NY, July 23, 48. DEVELOPMENTAL BIOLOGY, MOLECULAR BOTANY. *Educ:* State Univ NY, Potsdam, BA, 70, State Univ NY Col Environ Sci & Forestry, Syracuse, MS, 73; Univ Ga, PhD(bot), 78. *Prof Exp:* Res asst bot, Univ Ill, 72-74; lab coord eng biol, Univ Ga, 74-75, grad res asst bot, 75-76, grad teaching asst, 76-77, grad res asst, 77-78; res specialist cell & molecular biol, 78-80, NIH FEL, BOT DEPT, UNIV MINN, 80- *Res:* Molecular mechanisms involved in plant growth and development, specifically at the nucleic acid and protein level; somatic cell senescence. *Mailing Add:* Dept Biochem Univ Mo 117 Schweitzer Hall Columbia MO 65211

HAGEN, HAROLD KOLSTOE, b Plummer, Minn, Nov 18, 24; m 48; c 1. FISH BIOLOGY. *Educ:* Univ Wyo, BS, 49; Univ Wash, Seattle, PhD(fisheries sci), 56. *Prof Exp:* Biologist fisheries, Wyo Game & Fish Comn, 49; trout water biologist, SDak Dept Game, Fish & Parks, 54-55; asst prof, 55-60, ASSOC PROF FISHERIES, COLO STATE UNIV, 60- *Concurrent Pos:* Edmund Niles Huyck fel, 63; Off Water Resources, Dept Interior grant, thermal pollution, Yellowstone Nat Park, 65; pres, Western Fisheries Consults; tech adv US Peace Corps, Peru & Venezuela, dir fishery resources of Guatemala, 75; bd dir, US Trout Farmers Asn, 73-76; fisheries eval Chile, Booz, Allen, Hamilton & World Bank, 75-76; leader tech adv team fisheries, AID/Peru, 76-80; sr consult fisheries, United Nations, Ecuador, 78; consult pvt trout cult, Mex, 78. *Mem:* Am Fisheries Soc; Am Inst Fishery Res Biol. *Res:* Fresh water fisheries; farm and ranch fish ponds in northern latitudes; aquatic weed control; in situ stream nutrient cycling. *Mailing Add:* Dept Fishery & Wildlife Biol Colo State Univ Ft Collins CO 80523

HAGEN, JACK INGVALD, b Coeur d'Alene, Idaho, Sept 21, 14. NUCLEAR PHYSICS. *Educ:* Ore State Univ, BS, 48, MS, 49. *Prof Exp:* Asst res eng extractive metall, Anaconda Copper Co, 49-51; res staff physics, Johns Hopkins Univ, 51-53; sr engr nuclear eng, Atomic Power Div, Westinghouse Elec Corp, 53-57; assoc physicist, Argonne Nat Lab, 57-65; assoc res prof, 65-80, EMER RES PROF PHYSICS & ELECTROCHEM, DEPT ELEC ENG, UNIV IDAH0, 80- *Concurrent Pos:* Prin investr, NSF, 74-78. *Mem:* Am Phys Soc; Am Nuclear Soc; Sigma Xi. *Res:* Electro catalysis and surface reaction studies in electrochemical energy conversion, concerning metallic and semiconductor electrodes. *Mailing Add:* 1408 Hall Rd Viola ID 83872

HAGEN, JOHN WILLIAM, b Minneapolis, Minn, May 11, 40. RESEARCH ADMINISTRATION, SCIENCE ADMINISTRATION. *Educ:* Univ Minn, BA, 62; Stanford Univ, PhD(psychol), 65. *Prof Exp:* FROM ASST PROF TO PROF PSYCHOL, UNIV MICH, 65-, DIR, CTR HUMAN GROWTH & DEVELOP, 82- *Concurrent Pos:* Exec officer, Soc Res Child Develop, 89-; mem, Study Sect Ment Retardation, NIH & chmn Study Sect Behav & Neurosci. *Mem:* Soc Res Child Develop; fel Am Psychol Asn; Am Psychol Soc; Int Soc Study of Behav Develop; fel Int Asn Res Learning Disabilities; Am Educ Res Asn. *Res:* Cognitive development and its relationship to academic and social behavior in children with learning problems or chronic health conditions. *Mailing Add:* 3421 Burbank Dr Ann Arbor MI 48105

HAGEN, JON BOYD, b Moscow, Idaho, Apr 18, 40; m 74. RADIOPHYSICS. *Educ:* Stanford Univ, BS, 62; Univ Idaho, MS, 64; Cornell Univ, PhD(elec eng), 74. *Prof Exp:* Instr physics, Hampton Inst, 67-69; RES ASSOC, ARECIBO OBSERV, CORNELL UNIV, PR, 72- *Mem:* Am Geophys Union; Sigma Xi; Inst Elec & Electronics Engrs. *Res:* Aeronomy research using incoherent scatter radar. *Mailing Add:* NAIC Dept Cornell Univ 124 Maple Ave Ithaca NY 14850

HAGEN, KENNETH SVERRE, b Oakland, Calif, Nov 26, 19; m 43; c 1. ENTOMOLOGY. *Educ:* Univ Calif, BS, 43, MS, 48, PhD, 52. *Prof Exp:* Lab technician biol control, Exp Sta, 47-52, jr entomologist, 52-53, asst entomologist, 53-59, assoc entomologist, 59-69, lectr, 64-69, PROF ENTOM & ENTOMOLOGIST, BIOL CONTROL, UNIV CALIF, 69- *Concurrent Pos:* Collabr, Entom Res Br, Agr Res Serv, USDA, 54; expert, Int Atomic Energy Comn, 61-63. *Mem:* AAAS; Entom Soc Am; Soc Syst Zool; Ecol Soc Am; Entom Soc Can. *Res:* Insect nutrition; biological control of insect pests; ecology of Coccinellidae; systematics of Anthicidae, Coccinellidae and Encyrtidae. *Mailing Add:* 14 Cambridge Way Piedmont CA 94611

HAGEN, LAWRENCE J, b Rugby, NDak, Mar 6, 40. WIND EROSION. *Educ:* NDak State Univ, BS, 63, MS, 67; Kans State Univ, PhD(mech eng), 80. *Prof Exp:* Agr engr, 67-88, RES LEADER, AGR RES SERV WIND EROSION RES UNIT, USDA, 88- *Concurrent Pos:* Adj prof, Kans State Univ, 70-91. *Mem:* Am Soc Agr Engrs; Soil & Water Conserv Soc. *Res:* Modeling and experimental research on wind erosion. *Mailing Add:* Kans State Univ E Waters Hall Rm 105-B USDA Wind Erosion Res Unit Manhattan KS 66506

HAGEN, OSKAR, b Heddal, Norway, Apr 5, 26; US citizen; m 58; c 2. SOLID & FLUID MECHANICS. *Educ:* Vienna Tech Univ, Dipl Ing, 54; Univ Pittsburgh, PhD(mech eng), 64. *Prof Exp:* Draftsman, Hq Air Arms Inspection, Norweg Air Force, 46-47; asst engr, Esco Armature Factory, Oslo, 47-49; engr, Norweg Hydrogen Plant, 54 & Elliott Co, Jeannette, Pa, 54-56; engr atomic equip div, Westinghouse Elec Corp, Cheswick, Pa, 56-60, sr engr, 60-64, fel engr, 64-66, lead eng & eng-anal mgr, electro-mech div, 66-90; RETIRED. *Mem:* Am Soc Mech Engrs; Sigma Xi. *Res:* Stress; dynamics; fluid flow; heat transfer. *Mailing Add:* 204 Craig Dr Greensburg PA 15601

HAGEN, PAUL BEO, b Sydney, Australia, Feb 15, 20; m 56; c 2. BIOCHEMISTRY, PHARMACOLOGY. *Educ:* Univ Sydney, MB, BS, 45. *Prof Exp:* Lectr physiol, Univ Sydney, 48-50, sr lectr, 50-51; sr lectr, Univ Queensland, 51-52; Martin fel pharmacol, Oxford Univ, 52-54; Brown fel, Yale Univ, 54-55, asst prof pharmacol, 55-56; asst prof, Harvard Med Sch, 56-59; prof biochem & head dept, Univ Man, 59-64; prof & head dept, Queen's Univ, Ont, 64-68; dean grad studies, 68-82, PROF PHARMACOL, UNIV OTTAWA, 68- *Honors & Awards:* Centennial Medal, Govt Can, 67; Jubilee Medal, 77; Fulbright Award, 54. *Mem:* Am Soc Pharmacol & Exp Therapeut; fel Chem Inst Can; Brit Pharmacol Soc; Physiol Soc Gt Brit. *Res:* Biochemistry and pharmacology of naturally occuring amines, endocrinology; whole animal metabolism; metabolic regulation. *Mailing Add:* Dept Pharmacol Univ Ottawa Med Facil Ottawa ON K1H 8M5 Can

HAGEN, RICHARD EUGENE, b Hillsboro, Kans, Apr 4, 37; m 62; c 3. BIOCHEMISTRY, FOOD SCIENCE. *Educ:* Southwestern Col, Kans, BA, 59; Univ Okla, PhD(biochem), 65. *Prof Exp:* Sr develop scientist, Pillsbury Co, Minn, 65-68; dir res, Roman Meal Co, Wash, 68-71; dir res & develop, Universal Foods Corp, 72-77; vpres & mgr, Washington Lab, Nat Food Processors Asn, 77-82; dir qual assurance, 82-87, MGR ENVIRON AFFAIRS, US PHARMACEUT & NUTRIT GROUP, BRISTOL MYERS, 87- *Mem:* Am Chem Soc; Am Asn Cereal Chemists; Inst Food Technologists; Am Soc Qual Control. *Res:* Nutrient research, ie raw and processed foods, food safety research including spoilage microorganisms, microbial toxins and indirect additives; industrial research and development of new foods and improvement of existing foods and ingredients particularly yeast, cereals, baked goods, engineered foods and citrus flavonoids; quality assurance of infant formulas, enteral foods, pharmaceuticals; environmental sciences. *Mailing Add:* 7408 E Walnut St Evansville IN 47715

HAGEN, SUSAN JAMES, b Detroit, Mich, March 30, 53. GASTROENTEROLOGY. *Educ:* Mich State Univ, BS, 75, MS, 79, PhD, 82. *Prof Exp:* Fel, Dept Cell Biol & Anat, Sch Med, Johns Hopkins Univ, 82-84; instr, 84-90, ASST PROF ANAT & CELL BIOL, DEPT MED, HARVARD MED SCH, 91- *Concurrent Pos:* Grad res award, Col Human Med, Mich State Univ, 80; assoc cell biologist, Brigham & Women's Hosp, 84- *Mem:* Am Soc Cell Biol. *Res:* Cell biology of the brush border after damage with lectins; biochemistry of cytoskeletal proteins in developing microvilli; light and EM immunocytochemical localization of proteins within cells; author of numerous scientific publications. *Mailing Add:* Dept Med Gastroenterol Harvard Med Sch Brigham & Women's Hosp 75 Francis St Boston MA 02115

HAGENAUER, FEDOR, b Zagreb, Yugoslavia, Mar 12, 20; US citizen; m 58; c 3. CHILD PSYCHIATRY, PSYCHOANALYSIS. *Educ:* Univ Zagreb, MD, 49. *Prof Exp:* Resident psychiat, State Res Hosp, Galesburg, Ill, 56-58; resident, 59-60, fel child psychiat, 60-62, from instr to asst prof, 62-69, ASSOC PROF CHILD PSYCHIAT, UNIV CINCINNATI, 69- *Concurrent Pos:* Staff child psychiatrist, Children's Psychiat Ctr, Cincinnati, 62-73, dir inpatient serv, 73-; consult, Rollman Receiving Hosp, 68-; staff, Cincinnati Inst Psychoanal, 75-, supv & training anal, 81- *Mem:* Fel Am Psychiat Asn; Am Psychoanal Asn; Am Acad Child Psychiat. *Mailing Add:* 3001 Highland Ave Cincinnati OH 45229

HAGENBACH, W(ILLIAM) P(AUL), b Rochester, NY, Sept 14, 22; m 46; c 3. CHEMICAL ENGINEERING. *Educ:* Univ Rochester, BS, 44, MS, 47; Univ Ill, PhD(chem eng), 51. *Prof Exp:* Res engr, E I du Pont de Nemours & Co, 51-56, process develop engr, 56-60; dir eng res, A E Staley Mfg Co, 60-69, dir eng res & serv, 69-71, environ sci, 72-73, environ sci & energy conserv, 74-83, environ sci & safety, 83-86; RETIRED. *Mem:* Am Chem Soc; Inst Chem Engrs. *Res:* Process development in field of polyester polymers and films; engineering research studies on corn products; plant environmental compliance; energy conservation programs; safety programs. *Mailing Add:* 4560 Williamsburg Ct Decatur IL 62521

HAGENLOCHER, ARNO KURT, b Herrenberg, Ger, May 20, 28; m 67; c 3. SOLID STATE PHYSICS. *Educ:* Stuttgart Tech Univ, BS, 50, MS, 53, PhD(physics), 58. *Prof Exp:* Res assoc semiconductors, Stuttgart Tech Univ, 54-57; physicist, Telefunken Ulm, Ger, 58-60; mem tech staff, Gen Tel Electronics Corp, 60-74; mgr elec devices, Coulter Systs Corp, 74-80; MGR RES & DEVELOP, ACTON RES CORP, 80- *Concurrent Pos:* Lectr, Polytech Inst Brooklyn, 62-66. *Mem:* Am Phys Soc: Electrochem Soc; Sigma Xi; Ger Phys Soc; Inst Elec & Electronics Engrs. *Res:* Semiconductors; ferroelectrics; crystal growth; thin films study by epitaxial growth, evaporation and r f sputtering; electrical properties of above materials; plasmas in solids; electron beam semiconductor devices; charge coupled devices, electrophotography; photovoltaics. *Mailing Add:* 2789 Northpoint Pkwy Santa Rosa CA 95401

HAGENLOCKER, EDWARD EMERSON, b Marysville, Ohio, Nov, 18, 39; m 58; c 3. PHYSICS. *Educ:* Ohio State Univ, BS & MS, 62, PhD(physics), 64. *Prof Exp:* Sr res scientist, 64-73, chief engr physics, 73-78, chief engr light trucks, 78-79, MGR LIGHT TRUCK PROD DEVELOP, FORD MOTOR CO, 79- *Mem:* Am Phys Soc; Soc Automotive Engrs; Sigma Xi. *Res:* Nonlinear optics; semiconductor devices; plasma physics; holography; automotive design and development. *Mailing Add:* 290 Lone Pine Rd Bloomfield MI 48013

HAGENMAIER, ROBERT DOLLER, b Mt Gilead, Ohio, May 26, 39; m 62; c 2. QUALITY CONTROL, BIOCHEMISTRY. *Educ:* Univ Detroit, BS, 65; Purdue Univ, PhD(phys chem), 70. *Prof Exp:* Assoc res chemist, Food Protein Res & Develop Ctr, Tex A&M Univ, 70-75; mgr, Coconut Foods Pilot Plant, Univ San Carlos, Philippines, 75-80; plant mgr, Red V Coconut Prod, Ltd, 80-85; consult, Niro Atomizer S PTE, Singapore, 85-86; qual control mgr, Holly Hill Fruit Prod, 86-88; RES CHEMIST, USDA CITRUS SUBTROP LAB, 88- *Concurrent Pos:* Lectr, Univ San Carlos, 78-79. *Mem:* Am Chem Soc; Inst Food Technologists; Am Asn Milk Food & Environ Sanitarians. *Res:* Processing technology for coconut protein and dried coconut milk, edible coatings. *Mailing Add:* 1891 18th St NW Winter Haven FL 33881

HAGER, ANTHONY WOOD, b Marshfield, Wis, Dec 16, 39; c 1. LATTICE-ORDERED GROUPS, CATEGORICAL TOPOLOGY. *Educ:* Pa State Univ, BS, 60, PhD(math), 65. *Prof Exp:* Asst scientist, Leeds & Northrup Co, Pa, 60-61; instr math, Univ Rochester, 65-67, asst prof, 67-68; from asst prof to assoc prof, 68-75, PROF MATH, 75-, CHMN, WESLEYAN UNIV, 88- *Concurrent Pos:* Vis prof, Czech Acad Sci, 73 & 75 & Univ Padua, 78. *Mem:* Math Asn Am; Am Math Soc. *Res:* Lattice-ordered algebra, uniform spaces, categorical algebra and topology. *Mailing Add:* Dept Math Wesleyan Univ Middletown CT 06457

HAGER, BRADFORD HOADLEY, b Johnstown, Pa, June 21, 50; m 72; c 1. TECTONOPHYSICS, SOLID EARTH GEOPHYSICS. *Educ:* Amherst Col, BA, 72, Harvard Univ, AM 76, PhD(geophysics), 78. *Prof Exp:* Instr physics, Cushing Acad, 72-74; Weizmann fel geophysics, Harvard Univ, 78-79; asst prof, State Univ NY, Stony Brook, 79-80; asst prof, 80-85, ASSOC PROF GEOPHYS, CALIF INST TECHNOL, 85- *Concurrent Pos:* Alfred P Sloan res fel, 82-86. *Honors & Awards:* James B Macelwane Award, Am Geophys Union, 86. *Mem:* Fel Am Geophys Union; AAAS; Sigma Xi. *Res:* Geodynamics; the driving mechanisms for plate motions; mantle convection; rheology of mantle materials; interpretation of the geoid; GPS satellite geodesy; large-scale parallel computations. *Mailing Add:* Dept Earth Atmosphere & Planetary Sci Mass Inst Technol Bldg 54-622 Cambridge MA 02139

HAGER, CHESTER BRADLEY, b Madison, WVa, Oct 15, 38; m 62; c 2. CLINICAL MEDICINE. *Educ:* WVa Univ, AB, 62, MS, 64, PhD(biochem), 66. *Prof Exp:* Radioisotope technician, Miami Valley Hosp, Dayton, Ohio, 60-61, blood bank technician, 62; asst biochem, WVa Univ, 63-64; Nat Inst Child Health & Human Develop res fel biochem regulation, Biol Div, Oak Ridge Nat Lab, 66 & Am Cancer Soc Res Fel, 66-68; res scientist, Miles Labs Inc, 68, asst dir, 68-72, dir tech serv, Ames Co Div, 72-78, dir res, develop & planning, Res Prod Div, 78-81; dir res & develop, Micromedic Systs, Inc, 81-86; DIR, HAZLETON BIOCHEM CORP, DIV HAZLETON LABS AM, 86. *Concurrent Pos:* Mem, Med Devices Technol Adv Bd, Am Nat Standards Inst, 74-78. *Mem:* AAAS; Am Asn Clin Chemists; NY Acad Sci; Am Chem Soc; Am Asn Clin Scientists; Am Soc Med Technol; Sigma Xi. *Res:* Clinical biochemistry, mechanism of protein synthesis; endocrinology; diagnostic medicine, endocrinology. *Mailing Add:* 9200 Leesburg Pike Vienna VA 22030

HAGER, DOUGLAS FRANCIS, b Plum City, Wis, July 13, 49; m 82; c 2. DRUG DELIVERY, PHARMACEUTICALS. *Educ:* Univ Wis, BS, 71; Harvard Univ, AM, 72, PhD(phys chem), 76. *Prof Exp:* Staff chemist, Procter & Gamble co, 76-81, sect head corp technol, 81-85, sect head paper technol, 85-86; dir drug delivery, Nestle-Alcon Labs, Inc, 87-89; HEAD, DRUG DELIVERY SYSTS, SANDOZ RES INST, 89- *Concurrent Pos:* Adj asst prof, dept chem & nuclear eng, Univ Cincinnati, 78-84. *Mem:* Am Chem Soc; Am Phys Soc; Am Inst Chem Engrs; Am Asn Pharm Scientists. *Res:* The transport of pharmaceuticals to their sites of action; sustaining the duration of effect of peptide drugs; transdermal delivery of drugs. *Mailing Add:* Nine Jay Dr Randolph NJ 07869-4102

HAGER, E JANT, b 1951; m 87. AVIAN MYOGENESIS. *Educ:* Univ Va, MS, 81, PhD, 87. *Prof Exp:* Res asst biol, Univ Va, 81-87; FEL, DIV ONCOL, STANFORD UNIV MED CTR, 87- *Concurrent Pos:* Comprehensive sci teacher, Glasgow, 74-78. *Mem:* Am Soc Cell Biol. *Res:* Ulrastructural analysis of gene expression in drosophila melanogastes polyterie chromation. *Mailing Add:* Div Oncol Standford Univ Med Ctr Rm M211 Stanford CA 94305-5306

HAGER, GEORGE PHILIP, JR, b Baltimore, Md, Mar 16, 16; m 38; c 3. MEDICINAL CHEMISTRY. *Educ:* Univ Md, BS, 38, MS, 40, PhD(pharmaceut chem), 42. *Prof Exp:* Res org chemist, Eli Lilly & Co, 42-44; prof pharmaceut chem, Univ Md, 44-55; sr scientist, Smith Kline & French Labs, 55-57; dean col pharm, Univ Minn, Minneapolis, 57-65; dean, 66-74, prof med chem, 66-81, EMER DEAN & PROF, SCH PHARM, UNIV NC, CHAPEL HILL, 81- *Concurrent Pos:* Mem adv bd-cardiovasc lit proj, Nat Res Coun, 56-60; adv comt antiradiation drug prog, Walter Reed Army Inst Res, 59-65; med chem study sect, NIH, 60-64; chmn comt mod methods handling chem info, Nat Acad Sci-Nat Res Coun, 61-66; mem ad hoc panel narcotic addiction & drug abuse, Off Sci & Technol, Exec Off President, 62-63; panel scientist-to-scientist commun, Surgeon Gen Conf Health Commun, USPHS, 62; panel undergrad sci educ prog, NSF, 63; consult chem info & data syst, Army Res Off, 63-65; mem, Nat Adv Comt Selection Physicians, Dentists & Allied Specialists, Selective Serv Syst, 64-68 & Nat Health Resources Adv Comt, 68-70; mem panel handling toxicol info, Off Sci & Technol, Exec Off President, 64-65; gen res support adv comt, Div Res Facil & Resources, NIH, 64-67, pharm rev comt, Bur Health Manpower Educ, 71-73; consult, Off Sci Info Serv, NSF, 65-72; nat civilian consult pharm, Off Surgeon Gen, USAF, 67-76. *Honors & Awards:* Achievement Award Advan of Pharm, Am Pharmaceut Asn, 68. *Mem:* Fel Acad Pharmaceut Sci (pres, 68-69); Am Pharmaceut Asn; Am Asn Cols Pharm (vpres, 64-65, pres, 65-66); Am Col Apothecaries. *Res:* Chemistry-synthesis and analysis of drugs and medicines; hormones; antibiotics; fluorine substituted aromatic acids; molecular structure-biological activity relationships, especially in agents with effects on the peripheral nervous system; documentation of information and use of modern equipment in handling chemical and biological data in medicinal chemistry research. *Mailing Add:* Univ NC 7360 Beard Hall Chapel Hill NC 27599

HAGER, GORDON L, b Girard, Kans, Dec 5, 42; m; c 1. ONCOLOGY. *Educ:* Univ Kans, BS, 64; Institut de Biologie Moleculaire, Geneva, Switz, PhD(genetics), 70. *Prof Exp:* Assoc res biochemist, Dept Biochem & Biophys, Univ Calif, 76; chief, Viral Immunogenetics Sect, Lab Tumor Virus Genetics, 80-83, CHIEF, HORMONE ACTION & ONCOGENESIS SECT, LAB MOLECULAR VIROL, NAT CANCER INST, NIH, 83- *Concurrent Pos:* Adv panels, Genetics & Biol, NSF, 83-87, NSF Site Visit Panel, Columbia Univ, 88, fel Study Sect, NIH, 90-93, Molecular Biol Study Sect, NIH, 91-94; NATO Int Coop fel, 87. *Honors & Awards:* Karlson Lectr, 90. *Mem:* Am Soc Biol Chemists; AAAS; Am Soc Microbiol; Int Asn Breast Cancer Res. *Res:* Regulation of gene expression in eucaroytic cells; chromatin structure and its relationship to gene expression; mechanism of oncogenic transformation; gene therapy; numerous publications. *Mailing Add:* Hormone Action & Oncogenetic Sect Nat Cancer Inst NIH Bldg 37 Rm 3C19 Bethesda MD 20892

HAGER, JEAN CAROL, b Avalon, Calif; m 81. EXPERIMENTAL ONOCOLOGY, CANCER CONTROL. *Educ:* Bates Col, BS, 65; Univ Ill, Champaign-Urbana, MS, 69; PhD(med sci), 74. *Prof Exp:* Res assoc vet microbiol, Col Vet Med, Univ Ill, 74-75; asst prof path & med, Col Med, Brown Univ, 75-79; res assoc, Roger Williams Gen Hosp, 75-79; asst prof path, Wayne State Univ, 79-81; sr scientist & dir, Extramural Prog, AMC Cancer Res Ctr, 81-89. *Concurrent Pos:* Mem, Mich Cancer Found, 79-81. *Mem:* Am Asn Cancer Res; Tissue Cult Asn; Int Asn Res Comparative Leukemia & Related Dis; Int Asn Breast Cancer Res; Tissue Cult Asn. *Res:* Cell biology and immunology of breast cancer; study of the metastatic process and hetergeneity; cancer control and cancer education. *Mailing Add:* Am Cancer Res Ctr 1600 Pierce St Denver CO 80214

HAGER, JOHN P(ATRICK), b Miles City, Mont, Oct 2, 36; m 61; c 7. METALLURGY. *Educ:* Mont Sch Mines, BS, 58; Mo Sch Mines, MS, 60; Mass Inst Technol, ScD(metall), 69. *Prof Exp:* From asst prof to assoc prof metall, 66-71, prof & head dept, 71-74, St Joe Minerals Corp prof extractive metall, 74-88, HAZEN RES PROF EXTRACTIVE METALL, COLO SCH MINES,88- *Mem:* Minerals, Metals, Mat Soc. *Res:* Physical chemistry of extractive metallurgy; extractive metallurgy process analysis and development. *Mailing Add:* Dept Metall Eng Colo Sch Mines Golden CO 80401

HAGER, JUTTA LORE, b Frankfurt, Main Ger, Feb 9, 42; c 1. GEOTECHNICAL INVESTIGATIONS, BEDROCK & SURFICIAL GEOLOGY. *Educ:* Radcliffe Col, BA, 63; Harvard Univ, MA, 73, PhD(geol sci), 78. *Prof Exp:* Postdoctoral fel geol, Harvard Univ, Cambridge, Mass, 78-79; asst prof geol, Bentley Col, Waltham, Mass, 79-80 & Wellesley Col, Mass, 83-84; prin scientist, Energy Resources Co, Inc, 80-82; assoc, S A Alsup & Assocs, Inc, 82-83; PRIN, HAGER-RICHTER GEOSCI, INC, 84- *Concurrent Pos:* Publ chmn, 1984 Ann Meeting, Asn Eng Geologists, 82-84, treas, 85-87, chmn, New Eng Sect, 87-89. *Mem:* AAAS; Geol Soc Am; Soc Econ Paleontologists & Mineralogists; Am Inst Prof Geologists; Asn Eng Geologists; Nat Water Well Asn. *Res:* Applications of geophysics to problems in engineering geology and the environmental field; surficial and bedrock geology of the Boston Basin. *Mailing Add:* Eight Industrial Way-D10 Salem NH 03079

HAGER, LOWELL PAUL, b Girard, Kans, Aug, 30, 26; m 49; c 3. BIOCHEMISTRY. *Educ:* Valparaiso Univ, AB, 47; Univ Kans, MA, 50; Univ Ill, PhD, 53. *Prof Exp:* Fel, NIH, 53-54; asst biochemist, Mass Gen Hosp, 54-55; asst prof chem, Harvard Univ, 55-60; assoc prof, 60-65, PROF CHEM, UNIV ILL, URBANA, 65-, DEPT HEAD, 67- *Concurrent Pos:* Guggenheim Mem fel, 59; mem, Physiol Chem Study Sect, NIH; vis scientist, Imp Cancer Res Fund Inst, London, 73. *Mem:* Am Chem Soc; Am Soc Biol Chem; Am Soc Microbiol. *Res:* Chemistry of heme proteins; enzymatic activation mechanisms; biochemistry of animal tumor viruses. *Mailing Add:* 413 Roger Adams Lab Univ Ill 1209 W California Urbana IL 61801

HAGER, MARY HASTINGS, b Upland, Calif, Mar 27, 48; m 82; c 2. DIETETICS. *Educ:* Univ Del, BS, 71; Univ Calif, Davis, MS, 73, PhD(nutrit), 78. *Prof Exp:* Staff scientist, Procter & Gamble Co, 78-83, clin biochem, 86-87; asst prof nutrit, Col Mt St Joseph, 83-86; asst prof nutrit & dietetics, Tex Christian Univ, 87-89; ASSOC PROF FOODS & NUTRIT, COL ST ELIZABETH, 89- *Concurrent Pos:* Adj prof chem, Col Mt St Joseph, 83-86; tech consult, Food & Pharmaceut Industs. *Mem:* AAAS; Am Physiol Soc; Am Inst Nutrit; Am Dietetic Asn. *Res:* Nutrient endocrine interrelationships; physiological and behavioral response to ingestion of nonabsorbable lipids. *Mailing Add:* Nine Jay Dr Randolph NJ 07869

HAGER, NATHANIEL ELLMAKER, JR, b Lancaster, Pa, June 3, 22; m 48; c 2. HEAT TRANSFER PHYSICS, THERMAL PROPERTIES OF MATERIALS. *Educ:* Franklin & Marshall Col, BS, 43; Lehigh Univ, MS, 48, PhD(physics), 53. *Prof Exp:* Instr physics, Lehigh Univ, 50-51; physicist, Vitro Corp, 52-54; res physicist, Res & Develop Ctr, Armstrong World Indust, Inc, 54-68, res assoc, 68-76, sr res assoc, 76-89; RETIRED. *Concurrent Pos:* Lectr & consult health effects accident, Three Mile Island, 79-; lectr & writer radon in the home, 82-; mem, Consultative Coun, Nat Inst Bldg Sci, 82-, Indoor Air Qual Proj Comt, 83-88; consult, Radon Comt, Am Lung Asn, Pa, 87- *Mem:* AAAS; Optical Soc Am; sr mem Instrument Soc Am; Int Inst Refrig; NY Acad Sci; Am Soc Heating, Refrigerating & Air Conditioning Engrs; Asn Energy Engrs; Air & Waste Mgt Asn. *Res:* Measurement of thermal properties; spectroscopy; physical optics; underwater sound; heat transfer; thermal conductivity; measurement of temperature and heat flux; industrial process heating; heating devices; experimental polymer physics; energy conservation techniques; measurement and theory of radon in residential structures; energy sciences. *Mailing Add:* 1410 Clayton Rd Lancaster PA 17603

HAGER, RICHARD ARNOLD, b Morgantown, WVa, Oct 10, 32; m 60; c 1. PLANT PATHOLOGY, MICROBIOLOGY. *Educ:* Univ Pittsburgh, BS, 59; WVa Univ, MS, 61; Pa State Univ, PhD(plant path), 66. *Prof Exp:* Dir res plant pathol & microbiol, Frangella Bros Inc, 66-81; PESTICIDE CONTROL SPECIALIST, NY DEPT ENVIRON CONSERV, 81- *Concurrent Pos:* Consult, Cambridge Valley Mushroom Farm, 84. *Mem:* Am Phytopath Soc; Mycol Soc Am; US Fedn Culture Collections; Sigma Xi. *Res:* Forest pathology; commercial mushroom spawn production; development of mushroom strains, mushroom disease development and control; commercial mushroom production; exotic mushroom production. *Mailing Add:* Biechman Rd RD1 Box 188B Ravena NY 12143

HAGER, ROBERT B, b Philadelphia, Pa, Oct 2, 37; m 59. ORGANIC CHEMISTRY. *Educ:* Rensselaer Polytech Inst, BS, 58; Yale Univ, MS, 59, PhD(org chem), 62. *Prof Exp:* NSF-NATO fel, 62-63; res chemist, 63-67, prof leader, 67-68, group leader, 68-75, assoc mgr, Cent Res & Develop, 75-81, DIR RES, ORG CHEM DIV, PENWALT CORP, 81- *Mem:* Am Chem Soc; Sigma Xi. *Res:* Photodimerization; metallic reduction of organics; fluorochemicals; surface phenomena; reaction mechanisms; flammability; pest control; controlled delivery; pvc stabilization; additives for plastics; process development; research and development organization. *Mailing Add:* 772 Evansburg Rd RD 1 Collegeville PA 19426

HAGER, STANLEY LEE, b Pittsfield, Ill, Apr 25, 46; m 75. THERMAL ANALYSIS, POLYMER CHARACTERIZATION. *Educ:* Univ Ill, BS, 68; Univ Wis, PhD(chem), 74. *Prof Exp:* Chemist, Hercules, Inc, 68-70; CHEMIST, UNION CARBIDE CORP, 74- *Mem:* Am Chem Soc; Soc Plastics Engrs; NAm Thermal Anal Soc. *Res:* Chemical reactions of polymers; thermal analysis of polymers; rheology and dynamic mechanical testing polymers. *Mailing Add:* 5310 Edgebrook Rd Cross Lanes WV 25313

HAGER, STEVEN RALPH, b Amery, Wis, June 26, 51; m 71; c 2. CELLULAR MECHANISMS. *Educ:* St Olaf Col, BA, 73; Univ Wis-Madison, MS, 75, PhD(physiol/biochem), 80. *Prof Exp:* Res assoc, Univ Ill Med Ctr, Chicago, 79-81; asst prof biol, Univ Scranton, 81-84; ASST PROF BIOL, MED COL WIS, 85- *Concurrent Pos:* Vis asst prof, Univ Wis-Madison, 84. *Mem:* Am Physiol Soc; Sigma Xi; Am Diabetes Asn; Endocrine Soc. *Res:* Cellular mechanisms associated with insulin resistance in muscle and adipose tissue; understanding of insulin action in peripheral tissues. *Mailing Add:* 916 Sunset Dr Hartford WI 53027

HAGER, WAYNE R, b Baltimore, Md, Sept 6, 41; c 3. ENGINEERING, CHEMICAL ENGINEERING. *Educ:* Univ Utah, BS, 63, Univ Idaho, MS, 70, PhD(chem eng), 72. *Prof Exp:* Res engr, Jackson Lab, E I Du Pont, 66-68, plant supvr, Chambers Works, 68-69; prof chem engr, Col Eng, Univ Idaho, 72-88, prog dir spec summer progs, 81-88, asst dean, 86-88; DEPT HEAD, COL ENG, PA STATE UNIV, 88- *Concurrent Pos:* Dir Inst Resource Mgt, Univ Idaho, 81-84, chmn eng sci dept, 76-85; fulbright sr & exchange scholar, Univ Mauritius, Reduit, 85-86, vis scholar, Sch Indust Technol; vis fac, dept earth resources, Pac Lutheran Univ, 80-81; consult, Save The Children, Mauritius & Rodrigues, 85-, Nat Energy Found, Salt Lake City, 76-85, Western Power Admin, Sacramento, 84-; NDEA fel, 69-71. *Mem:* AAAS; Am Inst Chem Engrs; Am Soc Eng Educ; Nat Sci Teachers Asn. *Res:* Alternative energy technologies, resource assessment and utilization and modelling for developing countries; utilization of bagasse for the production of energy. *Mailing Add:* Col Eng Pa State Univ 101 Hammond Bldg University Park PA 16802

HAGER, WILLIAM WARD, b Altadena, Calif, Apr 29, 48; m 80; c 2. NUMERICAL ANALYSIS. *Educ:* Harvey Mudd Col, BS, 70; Mass Inst Technol, MS, 71, PhD(math), 74. *Prof Exp:* Asst prof math, Univ SFla, 74-76; asst prof math, Carnegie Mellon Univ, 76-80; from assoc prof to prof math, Pa State Univ, 86-88; PROF MATH, UNIV FLA, 88- *Concurrent Pos:* Mem, Spec Interest Group on Numerical Anal, Am Comput Mach. *Mem:* Soc Indust & Appl Math; Am Math Soc; Math Prog Soc; Math Asn Am. *Res:* Numerical analysis; control theory; optimization. *Mailing Add:* Dept Math Univ of Fla Gainesville FL 32611

HAGERMAN, ANN ELIZABETH, US citizen. PLANT PHENOLICS, PHYTOCHEMISTRY. *Educ:* Occidental Col, AB, 76; Purdue Univ, PhD(biochem), 80. *Prof Exp:* Teaching fel bot, Purdue Univ, 80-82; ASST PROF 82-, ASSOC PROF CHEM, MIAMI UNIV, OHIO, 87- *Mem:* Am Chem Soc; Am Soc Biol Chemists; Sigma Xi. *Res:* Function of secondary products in higher plants; protective role of phenolic compounds; biochemical characterization of the activity of phenolics, lignin and tannin; development of methods for analyzing phenolics in plant tissues. *Mailing Add:* Dept Chem Miami Univ Oxford OH 45056

HAGERMAN, DONALD CHARLES, b Boulder, Colo, May 2, 29; m 51. PHYSICS. *Educ:* Univ Colo, BA, 51; Stanford Univ, PhD(physics), 55. *Prof Exp:* Asst div leader, 68-72, MEM STAFF, LOS ALAMOS NAT LAB, 55-, GROUP LEADER, 65-, ASSOC DIV LEADER, 72- *Mem:* Fel Am Phys Soc; Sigma Xi. *Res:* Nuclear physics; plasma physics; controlled thermonuclear reactors; accelerators. *Mailing Add:* 107 Dos Brazos Los Alamos NM 87544

HAGERMAN, DWAIN DOUGLAS, b Boulder, Colo, June 2, 24; m 50; c 4. BIOCHEMISTRY, OBSTETRICS & GYNECOLOGY. *Educ:* Univ Colo, AB, 45, MS, 48; Harvard Univ, MD, 50. *Prof Exp:* From instr to asst prof biol chem, Harvard Med Sch, 56-67; from assoc prof to prof, 67-86, EMER PROF BIOCHEM & OBSTET & GYNEC, SCH MED, UNIV COLO, DENVER, 86- *Concurrent Pos:* Guggenheim fel & Fulbright scholar, Australian Nat Univ, 61-62; dir, Fearing Res Lab, Boston Hosp Women, 63-67; Fogarty sr int fel, Univ Edinburgh, 76; adj prof chem, Miami Univ, Oxford, Ohio, 90- *Mem:* AAAS; Soc Gynec Invest; Am Soc Biol Chemists; Am Chem Soc; Endocrine Soc. *Res:* Biochemistry related to reproduction in mammals. *Mailing Add:* 6195 Fairfield Rd No Ten Oxford OH 45056-1509

HAGERMAN, LARRY M, b Owensboro, Ky, Oct 12, 40; m 79; c 2. PHARMACOLOGY, NUTRITIONAL BIOCHEMISTRY. *Educ:* Vanderbilt Univ, BA, 61. *Prof Exp:* Assoc scientist, Mead Johnson, 64-66, scientist, 66-72, sr scientist nutrition, 72-75; res scientist pharmacol, Warren-Teed, 75-77, RES SCIENTIST PHARMACOL, ADRIA/WARREN-TEED, 77- *Mem:* AAAS; Am Chem Soc; Am Fedn Sci. *Res:* Cholesterol and bile salt metabolism; blood lipids, lipoproteins, cardiovascular disease. *Mailing Add:* 1351 Castleton Rd N Columbus OH 43220

HAGERUP, HENRIK J(OHAN), b Horten, Norway, Nov 22, 32; US citizen; m 58; c 2. AERONAUTICS. *Educ:* Mass Inst Technol, BSc & SM, 56; Princeton Univ, PhD(aeronaut eng), 63. *Prof Exp:* Asst prof, 62-65, ASSOC PROF AERONAUT ENG, RENSSELAER POLYTECH INST, 65- *Mem:* AAAS; Am Inst Aeronaut & Astronaut. *Res:* Fluid mechanics. *Mailing Add:* Dept Mech Eng & Aeronaut Eng Rensselaer Polytech Inst Troy NY 12181

HAGESETH, GAYLORD TERRENCE, b Minot, NDak, Aug 13, 35; m 58; c 3. PHYSICS. *Educ:* Univ NC, BS, 58; Cath Univ, MS, 62, PhD(physics), 67. *Prof Exp:* From asst prof to assoc prof, 65-75, PROF PHYSICS, UNIV NC, GREENSBORO, 75- *Mem:* AAAS; Am Phys Soc; Am Asn Physics Teachers. *Res:* Low energy nuclear physics, acoustics and chemical physics; solid state physics thermoluminescense in crystals; kinetics and thermodynamics of isothermal seed germination; x-ray damage in crystals. *Mailing Add:* Dept Physics Univ NC 1000 Spring Garden St Greensboro NC 27412

HAGFORS, TOR, b Oslo, Norway, Dec 18, 30; US citizen; m 53; c 4. IONOSPHERIC PHYSICS, PLANETARY ASTRONOMY. *Educ:* Tech Univ Norway, MA, 55; Univ Oslo, Norway, PhD(physics), 59. *Prof Exp:* Scientist, Norweg Defence Res Estab, 55-59 & 61-63; res assoc, Stanford Univ, Calif, 59-60; staff mem, Lincoln Lab, Mass Inst Technol, 63-67 & 69-71; dir, Jicamarca Radio Observ, Peru, 67-69 & Europ Incoherent Scatter Asn, 76-82; dir opers, Arecibo Observ, PR, 71-73; prof elec eng, Univ Trondheim, Norway, 73-82; PROF ELEC ENG & ASTRON, CORNELL UNIV, 82-; DIR, NAT ASTRON & IONOSPHERE CTR, 82- *Honors & Awards:* Vander Pol Gold Medal, Union Radio Sci Int, 87; Alexander von Humboldt Sr Scientist Award, Alexander von Humboldt Found, 89. *Mem:* Am Astron Soc; Am Geophys Union; fel Inst Elec & Electronics Engrs; Int Union Radio Sci. *Res:* Earth's atmosphere, the surfaces of the moon and other solar system bodies by radar and by radio wave emissions. *Mailing Add:* Nat Astron & Ionosphere Ctr Cornell Univ 502 Space Sci Bldg Ithaca NY 14853-6801

HAGGARD, BRUCE WAYNE, b Muncie, Ind, Dec 26, 43; m 63; c 2. GENETICS. *Educ:* Ind Univ, Bloomington, BA, 66, MA, 70, PhD(genetics), 72. *Prof Exp:* From asst prof to assoc prof, 72-88, PROF BIOL, HENDRIX COL, 88- *Concurrent Pos:* USPHS genetics training grant, 66-72. *Mem:* Sigma Xi; AAAS; Genetics Soc Am. *Res:* Determination, differentiation and development in the protozoan Paramecium aurelia. *Mailing Add:* 925 Cadrow Square Rd Conway AR 72032

HAGGARD, J D, b Wellston, Okla, Nov 5, 17; m 42; c 2. MATHEMATICS. *Educ:* Northeast Okla Col, BS, 42; Univ Mo, MS, 48, EdD(math), 51. *Prof Exp:* Asst prof math, Delta State Teachers Col, 51-52; vis asst prof & Carnegie intern, Univ Chicago, 52-53; asst prof, Pittsburgh State Univ, 53-66, prof & dean grad studies, 66-89; RETIRED. *Mem:* Math Asn Am. *Res:* Mathematical logic; foundation of mathematics. *Mailing Add:* RR 4 Box 54B Pittsburg KS 66762

HAGGARD, JAMES HERBERT, b Macon, Ga, Apr 22, 41; m 68. BIOCHEMISTRY. *Educ:* Auburn Univ, BS, 63, MS, 66, PhD(biochem), 69. *Prof Exp:* Biochemist, Armour & Co, 69-71; asst prof, 71-74, assoc prof, 74-81, PROF CHEM, SAMFORD UNIV, 81- *Res:* Phospholipid metabolism. *Mailing Add:* 1837 Shades Crest Rd Birmingham AL 35216

HAGGARD, MARY ELLEN, b Topeka, Kans, Apr 23, 24. MEDICINE. *Educ:* Tex Woman's Univ, BA & BS, 45; Univ Tex, MD, 51; Am Bd Pediat, dipl, 56. *Prof Exp:* Intern, Univ Hosp, Ohio State Univ, 51-52; resident, Hosps, 52-54, from instr to assoc prof, 54-69, PROF PEDIAT, UNIV TEX MED BR GALVESTON, 69- & DIR, DIV PEDIAT HEMATOL-ONCOL, 69- *Concurrent Pos:* Fel hemat, Boston Children's Hosp, 61. *Mem:* AMA; fel Am Acad Pediat; Am Soc Hemat; Sigma Xi. *Res:* Pediatrics; hematology; medical education; chemotherapeutic agents in leukemia and in solid tumors of childhood; folic acid metabolism in hemolytic disease. *Mailing Add:* Dept Pediat Univ Tex Med Br Galveston TX 77550

HAGGARD, PAUL WINTZEL, b Bennington, Okla, Aug 31, 33; m 60; c 2. MATHEMATICS. *Educ:* Southeastern State Col, BS, 53; NTex State Univ, MS, 60. *Prof Exp:* Instr math, Lamar State Col, 58-61; teaching asst math, Univ Tex, 61-63; assoc prof, 63-89, PROF MATH, ECAROLINA UNIV, 89- *Mem:* Math Asn Am; Nat Coun Teachers Math. *Res:* Algebra and topology; metric spaces; number theory; special functions. *Mailing Add:* Dept Math ECarolina Univ Greenville NC 27858

HAGGARD, RICHARD ALLAN, b Pittsburgh, Pa, 36; m 58; c 3. ORGANIC CHEMISTRY, PHYSICAL ORGANIC CHEMISTRY. *Educ:* Cornell Univ, AB, 58, PhD(org chem), 65. *Prof Exp:* Res chemist & group leader, 65-74, RES SECT MGR, ROHM & HAAS CO, 75- *Mem:* Sigma Xi. *Res:* Electrophilic reactions, compounds; organic chemistry of sulfur and of nitrogen compounds; anionic polymerization; solution polymers; general physical organic chemistry; process development; solution polymers. *Mailing Add:* 1207 Nash Dr Ft Washington PA 19034

HAGGARD, WILLIAM HENRY, b Woodbridge, Conn, Nov 20, 20; m 44; c 2. CLIMATOLOGY. *Educ:* Yale Univ, BS, 42; Univ Chicago, SM, 46. *Prof Exp:* Instr physics, NC State Col, 46-47; meteorologist, US Weather Bur, 47-51 & 54-61, from dep dir to dir, Nat Weather Records Ctr, 61-70, dir nat climatic ctr, Nat Oceanic & Atmospheric Agency, 70-75; PRES, CLIMAT CONSULT CORP, 76- *Mem:* AAAS; Am Meteorol Soc; Am Geophys Union. *Res:* Climatology; extended forecasting; tropical and marine meteorology. *Mailing Add:* RR 2 Box 372 Asheville NC 28805

HAGGERTY, JAMES FRANCIS, b Andover, Mass, Oct 26, 16; m 43; c 2. BIOCHEMISTRY. *Educ:* Tufts Col, BS, 40; Georgetown Univ, MS, 52, PhD(biochem), 57. *Prof Exp:* Chemist, Firestone Rubber & Latex Prod Co, Mass, 46-48; biochemist, Food & Drug Admin, DC, 48-51; biochemist div biol & med, US AEC, 51-60; chief, Res Grants Br, Nat Cancer Inst, 60-64, chief, Res Grants Rev Br, 64-68, chief scholars & fels br, Fogarty Int Ctr, NIH, 68-75; dep assoc dir, Blood Serv, Am Red Cross, 76-85; CONSULT, 85. *Mem:* AAAS; Asn Mil Surg; Am Chem Soc; assoc Soc Nuclear Med. *Res:* Water soluble vitamins; radiobiology and biochemistry of cancer; science administration. *Mailing Add:* 9401 Singleton Dr Bethesda MD 20817

HAGGERTY, JOHN S, b Washington, DC, Oct 23, 38; m 61; c 3. PHYSICAL METALLURGY & CERAMICS. *Educ:* Mass Inst Technol, SB, 61, SM, 63, PhD, 65. *Prof Exp:* Res asst chem, Harvard Univ, 59-60; res asst, Mass Inst Technol, 61-62, 62-65; sr specialist ceramist, Arthur D Little, Inc, 65-77; SR RES SCIENTIST & PROJ MGR ADVAN TECHNOL, ENERGY LAB, MASS INST TECHNOL, 77- *Mem:* Fel Am Ceramic Soc; Am Asn Crystal Growth; Am Solar Energy Soc. *Res:* Crystal growth processes, materials characterization, optical, thermal and mechanical properties, laser processing of materials, solar energy systems and components. *Mailing Add:* Mat Processing Ctr Mass Inst Technol Bldg 12-011 77 Massachusetts Ave Cambridge MA 02139

HAGGERTY, JOHN S, b Brooklyn, NY. EXPERIMENTAL HIGH ENERGY PHYSICS. *Educ:* Manhattan Col, BS, 75; Harvard Univ, AM, 77, PhD(physics), 81. *Prof Exp:* Res assoc, Fermi Nat Accelerator Lab, 81-83, Lawrence Berkeley Lab, 83-86; ASSOC, BROOKHAVEN NAT LAB, 86- *Mem:* Am Phys Soc. *Res:* Production of bound states of charmed and bottom quarks; data analysis; electronic data acquisition; new partical detectors; rare decays. *Mailing Add:* Bldg 510A Brookhaven Nat Lab Upton NY 11973

HAGGERTY, ROBERT JOHNS, b Saranac Lake, NY, Oct 20, 25; m 49; c 4. BEHAVIORAL PEDIATRICS. *Educ:* Cornell Univ, AB, 46, MD, 49; Am Bd Pediat, dipl. *Hon Degrees:* MA, Harvard, DSc, Ind Univ. *Prof Exp:* Intern med, Strong Mem Hosp, Rochester, NY, 49-51; from jr asst res to chief med res, Children's Hosp Med Ctr, Boston, 53-55; asst prof pediat, Harvard Med Sch, 55-64; prof pediat & chmn dept, Sch Med & Dent, Univ Rochester, 64-75; Roger I Lee prof pub health & pediat & head dept health serv, Harvard Sch Pub Health & Harvard Med Sch, 75-80; WILLIAM T GRANT FOUND CLIN PROF PEDIAT, CORNELL UNIV, 80- *Concurrent Pos:* Med Dir, Boston Poison Info Ctr, 55-64; chief child health div, Children's Hosp Med Ctr, 58-64; Markle scholar acad med, 61-66; mem health serv res study sect, USPHS, 64-70, chmn 68-70; pediatrician-in-chief, Strong Mem Hosp, Rochester, 64-75; fel, Ctr Advan Study Behav Sci, Stanford, Calif, 74-75; vis prof pediat, Hard Med Sch & dir, Robert Wood Johnson Gen Pediat Acad Develop Prog, 78-; ed, Pediat in Rev, Am Acad Pediat; chmn, Health Serv Sci Res Study Sect, Nat Ctr Health Serv Res. *Honors & Awards:* Lienhard Award, Inst Med Nat Acad Sci; M M Eliot Award, Am Pub Health Asn. *Mem:* Inst Med-Nat Acad Sci; fel Am Acad Pediat; Am Asn Poison Control Ctrs (pres, 62-64); Am Pub Health Asn; AAAS. *Res:* Preventive pediatrics; health services research; growth and development; accident prevention; infectious and social disease. *Mailing Add:* William T Grant Found 515 Madison Ave New York NY 10022

HAGGERTY, WILLIAM JOSEPH, JR, b Kansas City, Mo, Nov 3, 32; m 67. ANALYTICAL CHEMISTRY. *Educ:* Rockhurst Col, BS, 54; Univ Mo, Kansas City, MA, 59, PhD(pharmaceut chem), 68. *Prof Exp:* Assoc chemist, Midwest Res Inst, 59-65; res assoc anal methods develop, Univ Mo, Kansas City, 65-66; sr chemist, 66-73, prin chemist, 73-74, HEAD BIO & PHARMACEUT ANALYSIS SECT, MIDWEST RES INST, 74- *Mem:* AAAS; Am Chem Soc; Am Pharmaceut Asn. *Res:* Organic synthesis and structure determination; synthesis and analysis of drugs used for anti-tumor investigation; analytical methods for drug dosage forms; teratogens present in environment; isolation and identification of natural products; development of bioanalytical techniques. *Mailing Add:* 425 Volker Blvd Kansas City MO 64110

HAGGIS, GEOFFREY HARVEY, b London, Eng, Sept 6, 24; Brit & Can citizen; m 55, 77; c 5. ELECTRON MICROSCOPY, MOLECULAR BIOLOGY. *Educ:* Cambridge Univ, BA, 45, MA, 50; London Univ, PhD(biophys), 51. *Prof Exp:* Lectr med physics, Middlesex Hosp Med Sch, London, 51-56; sr lectr physiol, Med Sch, Edinburgh Univ, 56-69; res scientist, Dept Agr, Can, 69-90; RETIRED. *Mem:* Royal Micros Soc; Can Micros Soc (vpres, 67-80, pres, 81-83). *Res:* Development of new preparative techniques for electron microscopy of biological material; freeze-fracture techniques suited to scanning electron microscopy and deep-etch replication. *Mailing Add:* The Cottage Frogmore South Devon TQ72NR England

HAGGITT, RODGER C, b Detroit, Mich, Aug 28, 42; m 65; c 2. GASTROINTESTINAL PATHOLOGY, SURGICAL PATHOLOGY. *Educ:* Univ Tenn, MD, 67. *Prof Exp:* Instr path, Harvard Med Sch, 74-77; clin assoc prof, Univ Tenn, 77-84; PROF PATH & DIR, HOSP PATH, UNIV WASH, 84- *Concurrent Pos:* Dir surg path, Baptist Mem Hosp, Memphis, 77-84; workshop dir, Am Soc Clin Pathologists, 77-, educ course dir, 79-, chmn, Coun Anat Path, 82-85; mem, Anat Path Test Comt, Am Bd Path, 87-92; vis prof, Univ Calif, Los Angeles, San Diego & San Francisco, Harvard Med Sch, Univ Conn, Univ Ill, Univ Tex & Univ SFla, 87- *Honors & Awards:* Distinguished Serv Award, Am Soc Clin Pathologists, 89. *Mem:* Fel AAAS; Am Soc Clin Pathologists; Gastrointestinal Path Soc (pres, 81); Am Gastroenterol Asn; Col Am Pathologists. *Res:* Biology of pre-malignant lesions of the gastrointestinal tract using histologic, flow cytometric and molecular genetic techniques. *Mailing Add:* Univ Wash Med Ctr RC-72 1959 NE Pacific St Seattle WA 98195

HAGHIRI, FAZ, b Tehran, Iran, Jan 18, 30; m 53; c 3. AGRONOMY. *Educ:* Univ Nebr, BS, 55, MS, 56, PhD(agron), 59. *Prof Exp:* Asst soil chem, Univ Nebr, 57-59; asst prof agron, Ohio Agr Exp Sta, 59-64; assoc prof agron, Ohio Agr Res & Develop Ctr, 64-68, PROF AGRON OHIO AGR RES & DEVELOP CTR & OHIO STATE UNIV, 68-, ASSOC CHMN DEPT, 78- *Mem:* Soil Sci Soc Am; Am Soc Agron; Int Soil Sci Soc. *Res:* Soil and water pollution; heavy metals in soils and plants. *Mailing Add:* 2742 Tuckahoe Rd Wooster OH 44691

HAGIHARA, HAROLD HARUO, soil science, plant nutrition, for more information see previous edition

HAGINO, NOBUYOSHI, b Hokkido, Japan, Feb 8, 32. NEUROENDOCRINOLOGY. *Educ:* Toyko-Gekei Univ, MD, 57, PhD(neuro endocrinol), 67. *Prof Exp:* PROF CELL BIOL, HEALTH SCI CTR, UNIV TEX, 76- *Mem:* Int Soc Neuroendocrinol; Int Brain Res Orgn; Am Asn Anatomists; Am Physiol Soc. *Mailing Add:* Dept Cellular & Struct Biol Health Sci Ctr Univ Tex 7703 Floyd Curl Dr San Antonio TX 78284

HAGINS, WILLIAM, BIOPHYSICS. *Prof Exp:* CHEM PHYSICS LAB, NAT INST ARTHRITIS, DIABETES, DIGESTIVE & KIDNEY DIS, NIH. *Mem:* Nat Acad Sci. *Mailing Add:* Nat Inst Arthritis Diabetes Digestive & Kidney Dis NIH Bethesda MD 20892

HAGIS, PETER, JR, b Phila, Pa, Jan 16, 26; m 53; c 2. MATHEMATICS. *Educ:* Temple Univ, BSEd, 50, MA, 52; Univ Pa, PhD(math), 59. *Prof Exp:* From instr to assoc prof math, 52-68, PROF MATH, TEMPLE UNIV, 68- *Mem:* Math Asn Am; Am Math Soc; Fibonacci Asn. *Res:* Theory of numbers. *Mailing Add:* 880 Edison Philadelphia PA 19116

HAGIWARA, SUSUMU, neurophysiology; deceased, see previous edition for last biography

HAGLER, JAMES NEIL, b Denver, Colo, July 18, 45. MATHEMATICAL ANALYSIS. *Educ:* Cornell Univ, BA, 67; Univ Calif, Berkeley, MA, 69, PhD(math), 72. *Prof Exp:* Vis asst prof math, State Univ NY, Binghamton, 72-73; asst prof math, Cath Univ Am, 73-88; ASST PROF MATH & COMP SCI, UNIV DENVER, 88- *Mem:* Am Math Soc. *Res:* Isomorphic problems for separable and nonseparable Banach spaces, especially how the structure of an arbitrary Banach space relates to those of the classical spaces. *Mailing Add:* Dept Math & Comp Sci Univ Denver University Park 2360 S Gaylord Denver CO 80208

HAGLER, MARION O(THO), b Temple, Tex, Sept 7, 39; m 62; c 3. PULSED POWER, ENGINEERING EDUCATION. *Educ:* Rice Univ, BA, 62, BS, 63; Univ Tex, Austin, MS, 64, PhD(elec eng), 67. *Prof Exp:* Res engr, Univ Tex, Austin, 66-67; from asst prof to prof, 67-81, dir, Ctr Energy Res, 77-85, chmn, Dept Elec Eng & Comput Sci, 83-86, HORN PROF ELEC ENG, TEX TECH UNIV, 81-, CHMN, DEPT ELEC ENG, 87- *Mem:* Am Phys Soc; fel Optical Soc Am; fel Inst Elec & Electronics Engrs; Am Soc Eng Educ. *Res:* Pulsed Power; energy; coherent optical systems; engineering education; fusion. *Mailing Add:* 4504 11th St Lubbock TX 79416-4816

HAGLER, THOMAS BENJAMIN, b Louisville, Ala, May 16, 13; m 39; c 3. PLANT SCIENCE. *Educ:* Auburn Univ, BS, 39, MS, 47; Univ Md, PhD(hort), 54. *Prof Exp:* Pub sch teacher, 39-45; assoc prof hort & assoc horticulturist, Auburn Univ, 45-58; prof hort & head dept, Clemson Univ, 58-60; chmn plant sci div, Coop Exten Serv, 60-76, HEAD EXTEN HORT, AUBURN UNIV, 76- *Mem:* Am Soc Hort Sci; Sigma Xi. *Res:* Coordination of program phases of educational program in horticulture. *Mailing Add:* 435 N Dean Rd Auburn AL 36830

HAGLUND, JOHN RICHARD, b Bessemer, Mich, Aug 31, 31; m 54; c 3. MICROBIOLOGY. *Educ:* Univ Wis, BS, 53, MS, 57; Iowa State Univ, PhD(food microbiol), 65. *Prof Exp:* Asst bact, Univ Wis, 55-57; dir qual control, Country Gardens, Inc, Wis, 57-59; microbiologist, Cent Qual Control Lab, Gen Mills, Inc, 59-62; res assoc food technol, Iowa State Univ, 62-65; res chemist, James Ford Bell Res Ctr, Gen Mills, Inc, Minn, 65-72; microbiologist, Hunt Wesson Foods, Inc, 72-75; consult, 75-78; SR MICROBIOLOGIST, CASTLE & COOKE, INC, 78-, MGR, FOOD & PROCESS SAFETY DEVELOP, 80- *Mem:* Am Soc Microbiol; Inst Food Technologists. *Res:* Bacterial food infections; fluorescent antibody; food processing and preservation; food plant sanitation; food spoilage; food chemistry and product development. *Mailing Add:* Castle & Cooke Inc 2102 Commerce Dr San Jose CA 95131

HAGLUND, RICHARD FORSBERG, JR, b Washington, DC, Sept 17, 42; m 68; c 5. LASER PHYSICS, NON-LINEAR OPTICS. *Educ:* Wesleyan Univ, BA, 67; State Univ NY, Stonybrook, MA, 69; Univ NC, PhD(physics), 75. *Prof Exp:* Fel, Los Alamos Nat Lab, 75-77, mem staff, 77-84; ASSOC PROF PHYSICS, DEPT PHYSICS & ASTRON, VANDERBILT UNIV, 84- *Concurrent Pos:* Alexander von Humbolt Found fel, 82-83; vis prof, Heraeus Found, 91. *Mem:* Am Phys Soc; Optical Soc Am. *Res:* Laser-surface interactions; nonlinear optical materials; photon-stimulated desorption. *Mailing Add:* Dept Physics & Astron Vanderbilt Univ Nashville TN 37235

HAGLUND, WILLIAM ARTHUR, b Minneapolis, Minn, May 3, 30; m 52; c 2. PLANT PATHOLOGY. *Educ:* Univ Minn, BS, 53, MS, 58, PhD, 60. *Prof Exp:* From asst plant pathologist to assoc plant pathologist, 60-72, PLANT PATHOLOGIST, NORTHWESTERN EXP STA, WASH STATE UNIV, 72- *Concurrent Pos:* Pvt consult plant path. *Mem:* Am Phytopath Soc; Soc Nematol; Can Pytopath Soc; Am Arbit. *Res:* Root rot complex of vegetable crops and nematodes in relation to development of root rot; plant breeding, peas. *Mailing Add:* 1229 11th Burlington WA 98233

HAGMAN, DONALD ERIC, b Boston, Mass, July 3, 45; m 65; c 2. INDUSTRIAL PHARMACY. *Educ:* Mass Col Pharm, BS Pharm, 68; Purdue Univ, MS, 70, PhD(indust pharm), 72. *Prof Exp:* Sr res pharmacist, 72-77, mgr, 77-80, DIR PROD EVAL & SPEC PHARM SERV, A H ROBINS CO, INC, 80- *Mem:* Am Pharmaceut Asn; Sigma Xi. *Res:* Application of physical chemical principles to the development and improvement of delivery systems for drug products. *Mailing Add:* 5413 Winetavern Lane Dublin OH 43017

HAGMANN, SIEGBERT JOHANN, b Neuenhaus, WGer, Apr 9, 48. CHARGE TRANSFER. *Educ:* Westfaelische Wilhelms Univ Muenster, dipl physics, 73; Univ Koln, Dr rer nat, 77. *Prof Exp:* Res scientist, GSI, Darmstadt, 77-78; res assoc, 78-79, asst prof, 79-84, ASSOC PROF, DEPT PHYSICS, KANS STATE UNIV, 84- *Mem:* Am Phys Soc. *Res:* atomic physics; collision dynamics of very heavy and light quasimolecular systems; spectroscopy of very highly excited atomic states. *Mailing Add:* Dept Physics J R MacDonald Lab Manhattan KS 66506

HAGNAUER, GARY LEE, b Highland, Ill, Oct 13, 43; m 67; c 2. POLYMER COMPOSITE MATERIALS RESEARCH & POLYMER MATERIALS TESTING, POLYMER DURABILITY-LIFE PREDICTION. *Educ:* Southern Ill Univ, BA, 65; Univ Iowa, MS, 68, PhD(phys chem), 70. *Prof Exp:* Res chemist, Army Mat & Mech Res Ctr, 69-85, supvr res chemist, 85-90, SR RES SCIENTIST, ARMY MAT TECHNOL LAB, 90- *Mem:* Am Soc Testing & Mat; Soc Plastic Engrs; Am Chem Soc; Sigma Xi. *Res:* Polymer characterization; polymerization behavior; durability, life prediction and structure-property relationships of polymers and composite materials; automation/artificial intelligence technology for processing, testing and evaluation of polymers and composite materials; advanced testing methodology for polymers. *Mailing Add:* US Army Mat Technol Lab Attn SLCMT-EMP Watertown MA 02172

HAGNI, RICHARD D, b Howell, Mich, Apr 29, 31; m 53; c 4. ECONOMIC GEOLOGY, PROCESS MINERALOGY. *Educ:* Mich State Univ, BS, 53, MS, 54; Univ Mo, PhD(geol), 62. *Prof Exp:* Instr geol, 56-60, from asst prof to prof, 60-84, GULF OIL FOUND PROF GEOL, UNIV MO, ROLLA, 84-, CHMN DEPT GEOL & GEOPHYS, 85- *Concurrent Pos:* Consult explor geologist, Tex Gulf Sulphur Corp, 67-70; consult, Superior Mining, 72-73 & Kerr-McGee, 74-75; US Bur Mines grants, 74-88; US Geol Surv grant, 78-79; dir & lectr, Appl Ore Microscopy, 72-85; vchmn prog comt, Int Cong Mineral & Appl Mineral Indust, 81-84. *Mem:* Int Geol Cong; Int Asn Genesis Ore Deposits; Geol Soc Am; Soc Econ Geol; Am Inst Mining, Metall & Petrol Eng. *Res:* Mississippi Valley zinc-lead mineral deposits; ore microscopy; paragenesis and genesis of ore deposits; ore microscopic applications to beneficiation problems; fluid inclusion geothermometry. *Mailing Add:* Dept Geol & Geophys Univ Mo Rolla MO 65401

HAGOPIAN, CHARLES LEMUEL, b Sacramento, Calif, Nov 25, 40; m 70; c 2. TOPOLOGY. *Educ:* Sacramento State Col, BA, 62; Ariz State Univ, MA, 65, PhD(math), 68. *Prof Exp:* Res fel, Calif Inst Technol, 68-69; asst prof math, Calif State Univ, Sacramento, 69-71; vis prof, Ariz State Univ, 71-72; asst prof, 72-80, PROF MATH, CALIF STATE UNIV, SACRAMENTO, 80- *Mem:* Am Math Soc; Math Asn Am. *Res:* Geometric topology, in particular, continua theory involving fixed-point properties, homogeneity and various forms of connectivity. *Mailing Add:* Dept Math Calif State Univ Sacramento CA 95819

HAGOPIAN, MIASNIG, b Providence, RI, June 29, 27; m 58; c 4. ORGANIC CHEMISTRY, BIOLOGICAL CHEMISTRY. *Educ:* Univ RI, BS, 50; Clark Univ, MA, 55, PhD(bio-org chem), 65. *Prof Exp:* Res asst biochem, Worcester Found Exp Biol, 51-56, staff scientist, 56-61; res chemist, Mem Hosp, 61-67; BIOCHEMIST, MASON RES INST, EG&G, INC, 67- *Mem:* AAAS; Am Chem Soc; fel Am Inst Chemists. *Res:* Steroid metabolism; structure determination of natural products; synthesis of radio labelled steroids and catecholamines; biogenesis of catecholamine hormones; drug disposition and metabolism; chemical carcinogenesis; tobacco and marihuana smoke analysis; applied enzymology; lipid biochemistry. *Mailing Add:* 88 Sachem Ave Worcester MA 01606

HAGOPIAN, VASKEN, b Lebanon, Apr 21, 37; US citizen; m 65; c 1. HIGH ENERGY PHYSICS. *Educ:* Am Univ, Beirut, BS, 57; Univ Pa, MS, 60, PhD(physics), 63. *Prof Exp:* Res investr physics, Univ Pa, 63-65; actg asst prof, Univ Calif, Berkeley, 65-66; asst prof, Univ Pa, 66-69; assoc prof, 70-75, PROF PHYSICS, FLA STATE UNIV, 75- *Mem:* Am Phys Soc. *Res:* Experimental high energy physics using high energy accelerators. *Mailing Add:* Dept Physics B-159 Fla State Univ Tallahassee FL 32306

HAGSTROM, GEROW RICHARD, b New York, NY, Oct 16, 31; m 60; c 2. AGRONOMY. *Educ:* Univ Conn, BS, 57, MS, 59; Univ Wis, Madison, PhD(soils), 64. *Prof Exp:* Agron serv rep, Int Minerals & Chem Corp, Ill, 64-68; mgr agron serv, Duval Sales Corp, 68-85; DIR AGRON SERV, WESTERN AG MINERALS CO. *Mem:* Am Soc Agron; Soil Sci Soc Am; Sigma Xi. *Res:* Secondary and micronutrient element requirements of agronomic crops. *Mailing Add:* 14911 Walters Rd Houston TX 77068

HAGSTROM, JACK WALTER CARL, b Rockford, Ill, Dec 2, 33. PATHOLOGY. *Educ:* Amherst Col, AB, 55; Cornell Univ, MD, 59. *Prof Exp:* Intern path, New York Hosp, Cornell Med Ctr, 59-60, resident, 60-63; from instr to asst prof, Med Col, Cornell Univ, 62-68; assoc prof, Case Western Reserve Univ & attend pathologist, Univ Hosps, Cleveland, 68-70; assoc prof, 70-75, PROF PATH, COL PHYSICIANS & SURGEONS, COLUMBIA UNIV, 75-, ASSOC DIR DEPT; ATTEND PATHOLOGIST, PRESBY HOSP, 70-; DIR DEPT PATH, HARLEM HOSP CTR, 81- *Mem:* Am Asn Path & Bact; Am Fedn Clin Res; Harvey Soc; fel Am Col Cardiol; fel Royal Soc Trop Med & Hyg. *Res:* Cardiovascular and pulmonary pathology. *Mailing Add:* Harlem Hosp Ctr Dept Path Columbia Univ Col Physicians & Surgeons New York NY 10037

HAGSTROM, RAY THEODORE, b Minneapolis, Minn, Nov 25, 47; m 68. FRACTIONAL CHARGE SEARCH, INSTRUMENT DEVELOPMENT. *Educ:* Mass Inst Technol, BS, 69; Univ Wash, MS, 70; Univ Calif, Berkeley, PhD(physics), 79. *Prof Exp:* Fel, Lawrence Berkeley Lab, 79-80; asst physicist, 80-85, PHYSICIST, ARGONNE NAT LAB, 85- *Mem:* Am Phys Soc; AAAS. *Res:* Fundamental experimental physics investigations without the use of high energy particle accelerators. *Mailing Add:* High Energy Physics Div Argonne Nat Lab 9700 S Cass Ave Argonne IL 60439

HAGSTROM, STANLEY ALAN, b Lincoln, Nebr, Nov 30, 30; m 56; c 2. PHYSICAL CHEMISTRY. *Educ:* Univ Omaha, BA, 52; Iowa State Univ, PhD(chem), 57. *Prof Exp:* Instr chem, Ind Univ, 58-59, asst dir, Res Comput Ctr, 59-60; res scientist, Lockheed Missiles & Space Co, 60-62; from asst prof to assoc prof, 62-71, actg chmn dept comput sci, 75-77, PROF CHEM & COMPUT SCI, IND UNIV BLOOMINGTON, 71- *Concurrent Pos:* Sloan fel, 57; indust consult, molecular quantum mech, theoret spectros, radiation transport. *Mem:* Am Phys Soc; Asn Comput Mach. *Res:* Molecular spectroscopy; molecular quantum mechanics; digital computers; numerical analysis. *Mailing Add:* Dept Chem Ind Univ Bloomington 708 S Woodlawn Ave Bloomington IN 47401

HAGSTROM, STIG BERNT, b Barkeryd, Sweden, Sept 21, 32; m 57; c 4. INDUSTRIAL RESEARCH MANAGEMENT. *Educ:* Univ Uppsala, Sweden, BSc, 57, MSc, 59, PhD, 61, DSc, 64. *Hon Degrees:* ScD, Univ Link06ping, Sweden, 87. *Prof Exp:* Res assoc physics, Univ Uppsala, 61-64; res assoc chem, Lawrence Berkeley Lab, 65-66; asoc prof physics, Chalmers Univ Technol, 66-69; prof, Linkoping Univ, 69-76; prin scientist, Xerox Palo Alto Res Ctr, 76-77, mgr, 77-87; PROF MAT SCI & ENG & CHMN DEPT, STANFORD UNIV, 87- *Concurrent Pos:* Vchancellor, Linkoping Univ, 70-76; vis prof, Stanford Univ, 73-74; dir, Stanford Ctr Mat Res, 89-. *Mem:* Fel Am Phys Soc; AAAS; Europ Phys Soc; Swedish Phys Soc; Royal Swed Acad Eng Sci; Royal Norweg Soc Sci & Let. *Res:* Surface studies of electron structure of semiconductors and metals using electron spectroscopic techniques; instrumentation in surface science; synthesis of diamond films; surface magnetism. *Mailing Add:* Dept Mat Sci & Eng Stanford Univ Stanford CA 94305-2205e

HAGSTRUM, HOMER DUPRE, b St Paul, Minn, Mar 11, 15; m 48; c 2. SURFACE PHYSICS. *Educ:* Univ Minn, BEE, 35, BA, 36, MS, 39, PhD(physics), 40. *Hon Degrees:* DSc, Univ Minn, 86. *Prof Exp:* Asst, Univ Minn, 35-40; res physicist, Bell Labs, 40-54, head surface physics, res, 54-78, res physicist surface physics, 78-85; RETIRED. *Concurrent Pos:* Chmn, Div Electron & Atomic Physics, Am Phys Soc, 57; mem bd fel, Nat Res Coun, 76-77; gen chmn, Phys Electronics Conf Comt, 76-80. *Honors & Awards:* Welch Award, Am Vacuum Soc, 74; Davisson & Germer Prize, Am Phys Soc, 75. *Mem:* Nat Acad Sci; fel Am Phys Soc; Am Vacuum Soc; NY Acad Sci; fel AAAS. *Res:* Ionization and dissociation of diatomic gases by electron impact; microwave magnetrons; mass spectroscopy; electron ejection from solids by ions and metastable atoms; electronic state densities at solid surfaces; particle-solid interactions; surface physics. *Mailing Add:* 30 Sweetbriar Rd Summit NJ 07901-3256

HAGY, GEORGE WASHINGTON, b San Antonio, Tex, Aug 21, 23; m 54; c 3. GENETICS. *Educ:* Univ Tex, BA, 44, MA, 48, PhD(genetics), 53. *Prof Exp:* Instr microanat, Southwestern Med Sch, Univ Tex, 53-55, asst prof, 55-57; from asst prof to assoc prof, 57-73, PROF BIOL, BROWN UNIV, 73- *Mem:* AAAS; Am Soc Human Genetics; Soc Social Biol; Am Genetic Asn. *Res:* Growth and development; allergy; tissue culture. *Mailing Add:* 18 Fireside Dr Barrington RI 02806

HAHN, ALEXANDER J, b Bielitz, Poland, Sept 9, 43; US citizen; m 72; c 1. ALGEBRA, MATHEMATICS. *Educ:* Loyola Univ, BS, 65; Univ Notre Dame, PhD(math), 70. *Prof Exp:* PROF MATH, UNIV NOTRE DAME, 72- *Concurrent Pos:* NSF fel, WGer, 70-71; vis prof, Univ Calif, Santa Barbara, 82-83, Univ Innsbruck, Austria, 88. *Mem:* Am Math Soc. *Res:* The classical groups and K-theory; co-author one book. *Mailing Add:* Univ Notre Dame Notre Dame IN 46556

HAHN, ALICE ULHEE, colloid & physical chemistry, for more information see previous edition

HAHN, ALLEN W, b St Louis, Mo, Dec 28, 33; m 57; c 3. VETERINARY MEDICINE, BIOMEDICAL ENGINEERING. *Educ:* Univ Mo, BS & DVM, 58; Drexel Inst Technol, MS, 64, PhD(chem eng), 68; Am Col Vet Int Med, dipl, 74. *Prof Exp:* Res asst vet med, Auburn Univ, 58-61; res instr med cardiol, Univ Pa, 61-62; instr biol sci, Drexel Inst Technol, 66-68, assoc prof, 68-69; PROF VET MED, SURG & BIOENG & INVESTR, DALTON RES CTR, UNIV MO, COLUMBIA, 69-, ASSOC DIR, 80- *Concurrent Pos:* Prin co investr grants & contracts, NIH, NSF, US Army, NASA & various commercial co; sr ed, Eng Med & Biol, Inst Elec & Electronics Engrs. *Mem:* AAAS; Am Vet Med Asn; Am Col Vet Int Med; Inst Elec & Electronics Engrs; Am Physiol Soc. *Res:* Comparative cardiology; computers in veterinary medicine; implantable polymeric materials. *Mailing Add:* Dalton Res Ctr Univ Mo Columbia MO 65211

HAHN, BEVRA H, b Wheeling, WVa, Dec 9, 39; c 2. RHEUMATOLOGY. *Educ:* Ohio State Univ, BSc, 60; Johns Hopkins Univ, MD, 64. *Prof Exp:* Intern med, Barnes Hosp, Wash Univ, 64-65, asst resident med, 65-66; fel med connective tissue div, Johns Hopkins Univ, 66-69; instr prev med, 69-71, asst prof, 71-78, ASSOC PROF MED, SCH MED, WASH UNIV, 78- *Concurrent Pos:* Res assoc, Vet Admin Hosp, St Louis, 69-70; consult rheumatology, Montebello State Hosp, 67-69; clin investr, Vet Admin Hosp, St Louis, 70-73. *Mem:* Am Rheumatism Asn; Am Fedn Clin Res. *Res:* DNA antibodies in human and murine lupas; cellular immunity in SLE; steroid-induced osteopenia. *Mailing Add:* 10833 Le Conte Los Angeles CA 90024

HAHN, C(HARLES) ARCHIE, JR, b Walla Walla, Wash, Apr 10, 14; m 40; c 3. MECHANICAL ENGINEERING. *Educ:* Univ Va, BS, 35, BSME, 36. *Prof Exp:* Jr indust engr, Grasselli Chem Dept, E I du Pont de Nemours & Co, 36-40, asst div engr, War Construct Div, 40-42, div engr, 42; indust engr, Remington Arms Co, NY, 42-51; develop engr, E I du Pont de Nemours & Co, 51-54, develop proj engr photoprods equip develop, 54, res supvr, Textile Res Lab, 54-56, develop supvr, Eng Develop Lab, 56-66, specialist engr, Design Div, 66-70, sr mech engr, 70-74; RETIRED. *Mem:* Am Soc Mech Engrs. *Res:* Development of specialized equipment. *Mailing Add:* 173 SW 51st St Cape Coral FL 33914

HAHN, DOWON, b Pyung Buck, Korea, Nov 20, 31; m 63; c 3. ENDOCRINOLOGY, REPRODUCTIVE PHYSIOLOGY. *Educ:* Mich State Univ, BS, 60, MS, 63; Univ Mo, PhD(endocrinol), 67. *Prof Exp:* Ford Found fel endocrinol, Worcester Found Exp Biol, 67-68; from assoc scientist to sr scientist reproductive physiol, 68-73, group leader, 73-75, sect head, 75-82, asst dir, 82-87, DIR, REPRODUCTIVE/ENDOCRINE RES, RW JOHNSON PHARMACEUT RES INST, 87- *Concurrent Pos:* Adj prof, Grad Sch, Rutgers Univ & Eastern Va Med Sch. *Honors & Awards:* Philip B Hofmann Res Scientist Award, Johnson & Johnson, 73, 85; Johnson medal, 90. *Mem:* Soc Study Reproduction; Am Fertil Soc; Am Physiol Soc; Endocrine Soc; Soc Gynec Invest. *Res:* Fertility control. *Mailing Add:* RW Johnson Pharmaceut Res Inst Raritan NJ 08869-0602

HAHN, ELLIOT F, b New York, NY, June 28, 44; m 68; c 3. OPIATE AGONISTS & ANTAGONISTS, STEROID HORMONES. *Educ:* City Col New York, BS, 66; Cornell Univ, PhD(org chem), 70. *Prof Exp:* Vis scientist, Technion-Israel Inst Technol, Haifa, 70-72; res fel, Inst Steroid Res, Montefiore Hosp, Bronx, NY, 72-74, investr, 74-77; asst prof, Albert Einstein Col Med, 76-78; asst prof, 77-83, assoc prof biochem, Rockefeller Univ, 83-88; assoc dir res & develop, Ivax Corp, 88-89, VPRES RES, BAKER CUMMINS PHARM, DIV IVAX CORP, 89-; RES ASSOC PROF, UNIV MIAMI, 88- *Concurrent Pos:* Estab Fel, NY Heart Asn. *Mem:* Am Chem Soc; AAAS; Sigma Xi; NY Acad Sci. *Res:* Chemistry, biochemistry and pharmacology of narcotic agonists and antagonists; biotransformation of androgens in the brain; central nervous system mechanisms regulating hypertension; anti-HIV agents. *Mailing Add:* Ivax Corp 8800 NW 36th St Miami FL 33166

HAHN, ERIC WALTER, b New York, NY, June 1, 32; m 56; c 3. RADIATION BIOLOGY. *Educ:* Univ Ga, BS, 54, MS, 57; Univ Ill, Urbana, PhD(physiol), 60. *Prof Exp:* Asst physiol, Univ Ga, 56-57; asst, Univ Ill, Urbana, 57-60; instr, Univ Rochester, 60-64, asst prof radiation biol & biophys & head sect exp endocrinol, 64-68; asst prof radiol & dir res, Health Sci Ctr, Univ Minn, Minneapolis, 68-69; head sect radiation ther, Sloan-Kettering Inst Cancer Res, 69-77, assoc mem, 79-81; prof dept radiother, Mt Sinai Med Ctr, 81-85; SCIENTIST, GER CANCER RES CTR, 85- *Concurrent Pos:* Fel steroid training prog, Worcester Found Exp Biol, 63-64; adj assoc prof, Cornell Univ, 69-; attend radiobiologist, Mem Hosp, 75-81, adj attend radiobiologist, 81-; chmn, biol res dept, King Faisal Specialist Hosp, Riyadh, Saudi Arabia. *Mem:* AAAS; Endocrine Soc; Radiation Res Soc; Am Physiol Soc; Soc Study Reproduction. *Res:* Basic and applied radiation biology in the treatment of cancer; mechanisms of action of hyperthermia and its application in the treatment of cancer. *Mailing Add:* Ger Cancer Res Ctr Neuenheimer Feld 280 Heidekberg 6900 Germany

HAHN, ERWIN LOUIS, b Sharon, Pa, June 9, 21; m 80; c 3. PHYSICS. *Educ:* Juniata Col, BS, 43; Univ Ill, MS, 47, PhD(physics), 49. *Hon Degrees:* DSc, Juniata Col, 66; DSc, Purdue Univ, 74. *Prof Exp:* Asst physics, Purdue Univ, 43-44; res assoc, Univ Ill, 50; Nat Res Coun fel, Stanford Univ, 50-51, instr, 51-52; res physicist, Watson Sci Comput Lab, IBM Corp, 52-55; from asst prof to assoc prof, 55-61, PROF PHYSICS, UNIV CALIF, BERKELEY, 61- *Concurrent Pos:* Consult, Off Naval Res, Stanford Univ, 50-52; assoc, Columbia Univ, 52-55; consult, Atomic Energy Comn, 55-; assoc prof, Miller Inst Basic Res, 58-59, prof, 66-67; spec consult, US Navy, 59; Guggenheim & NSF fels, 61-62; mem adv panel, Radio Standards Div, Nat Bur Standards, 61-64; mem comt basic res, Nat Acad Sci-Nat Res Coun; adv, US Army Res Off, Durham; Guggenheim fel, Brasenose Col, Oxford, 69-70; Alexander von Humboldt Award, 76 & 77; vis Eastman Prof, Balliol Col, Oxford Univ, 88-89. *Honors & Awards:* Buckley Prize, Am Phys Soc; 1971 prize, Int Soc Magnetic Resonance; Wolf Prize in Physics, 84; Hon Fel Brasenose Col, Oxford Univ, 84. *Mem:* Nat Acad Sci; AAAS; fel Am Phys Soc; foreign mem Acad Sci, Slovenia. *Res:* Nuclear and electron spin magnetic resonance; electronic instruments; electron and nuclear spin resonance coupling in molecules and solids; laser-physics. *Mailing Add:* Dept Physics Univ Calif Berkeley CA 94720

HAHN, FLETCHER FREDERICK, b Spokane, Wash, May 8, 39; m 61; c 2. INHALATION TOXICOLOGY, RADIOBIOLOGY. *Educ:* Wash State Univ, BS & DVM, 64; Univ Calif, Davis, PhD(comp path), 71. *Prof Exp:* Vet lab officer, Walter Reed Army Inst Res, 64-66; NIH fel, Univ Calif, Davis, 66-70; EXP PATHOLOGIST, 71-, PATH GROUP SUPVR, INHALATION RES INST, 84- *Concurrent Pos:* Clin assoc, Sch Med, Univ NMex, 72-; chmn, Task Group 9, Nat Coun Radiation Protect Comt, 57, 77-, 77-; mem, animal models for aging, Nat Acad Sci-Inst Lab Animal Res, 79; fac assoc, Sch Vet Med, Colo State Univ, 84-; fac assoc, Sch Vet Med, Purdue Univ, 86-; clin assoc prof, Sch Pharm, Univ NMex, 87-; assoc ed, Radiation Res, 89- *Mem:* AAAS; Am Col Vet Pathologists; Am Vet Med Asn; Health Physics Soc; Radiation Res Soc; Soc Tox Path. *Res:* Pathogenesis of early and late biologic effects and dose response relationships of inhaled environmental pollutants, especially radionuclides and metals; induced and spontaneous diseases of laboratory animals. *Mailing Add:* Inhalation Toxicol Res Inst PO Box 5890 Albuquerque NM 87185

HAHN, FRED ERNST, molecular biology, molecular pharmacology; deceased, see previous edition for last biography

HAHN, GEORGE, b Vienna, Austria, Jan 31, 26; nat US; m 49; c 3. BIOLOGY. *Educ:* Univ Calif, AB, 52, MA, 54; Stanford Univ, PhD, 65. *Prof Exp:* Mathematician, Radiation Lab, Univ Calif, Berkeley, 51-54; physicist, Dalmo Victor Co, Calif, 54-57, head systs group, 58-60; asst head, Electronic Dept, Firestone Tire & Rubber Co, 57-58; assoc prof electronics, US Naval Postgrad Sch, 60-66; res assoc, 69, from asst prof to assoc prof, 69-82, PROF RADIOL, SCH MED, STANFORD UNIV, 82- *Concurrent Pos:* Consult, Wright Air Develop Ctr, Ohio, USAF, 58, Dalmo Victor Co, 60-69, Ampex Co, 60-65, Hewlett-Packard Co, 77-80, Nat Cancer Inst & NIH, 80-83; ed, Hyperthermia, 85- *Mem:* Radiation Res Soc. *Res:* Cell kinetics; radiobiology of mammalian cell cultures; computer simulations of therapy of malignant disease; hyperthermia as modality of cancer treatment. *Mailing Add:* Dept Radiol & Oncol Stanford Univ Sch Med CBRL Rm GK103 Stanford CA 94305-5468

HAHN, GEORGE LEROY, b Muncie, Kans, Nov 12, 34; m 55; c 4. BIOMETEOROLOGY. *Educ:* Univ Mo, Columbia, BS, 57, PhD(atmospheric sci), 71; Univ Calif, Davis, MS, 61. *Prof Exp:* res engr, 57-61, RES PROJ LEADER, AGR RES SERV, US DEPT AGR, 61- *Concurrent Pos:* Res leader, 74-77, tech advisor, Agr Res Serv, US Dept Agr, 72-81; prof, Univ Mo, Columbia, 71-78 & Univ Nebr-Lincoln, 78-; vis prof, Agr Univ Norway, 82-83. *Honors & Awards:* Bioclimatol Award, Am Meteorol Soc, 76; Farm Bldg Award, Am Soc Agr Engrs, 76. *Mem:* Fel Am Soc Agr Engrs; Am Meteorol Soc; Int Soc Biometeorol; Am Soc Animal Sci. *Res:* Development and validation of models for assessing impact of environmental factors (primarily climatic) on farm animal stress, performance and well-being; application of such models in rational decisions concerning environmental modification for farm animals. *Mailing Add:* US Dept Agr PO Box 166 Clay Center NE 68933

HAHN, GERALD JOHN, b Karlsruhe, Ger, Sept 11, 30; m 56; c 3. STATISTICS. *Educ:* City Col New York, BBA, 52; Columbia Univ, MS, 53; Union Col, MS, 65; Rensselaer Polytech Inst, PhD(statist & opers res), 71. *Prof Exp:* MGR, MGT SCI & STATIST PROG, CORP RES & DEVELOP, GEN ELEC CO, 55- *Concurrent Pos:* Adj prof statist, Union Col,

Schenectady, NY, 65- *Honors & Awards:* Brumbaugh Award, Am Soc Qual Control, 74, 80 & 82, Wilcoxon Prize, 79 & 88, Shewell Prize, 75 & 89, Youden Address, 87. *Mem:* Fel Am Statist Asn; fel Am Soc Qual Control; Inst Math Statist; Am Soc Testing & Mat. *Mailing Add:* 1404 Orlga Dr Schenectady NY 12309

HAHN, HAROLD THOMAS, b New York, NY, May 31, 24; m 48; c 4. PHYSICAL INORGANIC CHEMISTRY, ELECTROMAGNETIC MATERIALS. *Educ:* Columbia Univ, BS, 44; Univ Tex, PhD, 53. *Prof Exp:* Oper foreman, Los Alamos Sci Lab, Calif, 46-47, mem staff, 47-50; sr scientist, Hanford Lab Oper, Gen Elec Co, 53-58; sect chief chem res, Phillips Petrol Co, 58-64; staff engr, Lockheed Missiles & Space Co, 64-73; staff scientist, 73-84, SR STAFF SCIENTIST, LOCKHEED PALO ALTO RES LAB, 84- *Concurrent Pos:* Prin investr, ferritic mat; Judge, Regional & Intl Sci Fair, 60-80. *Mem:* Sigma Xi; fel Am Inst Chemist; Inst Elec & Electronics Engrs. *Res:* Reactor fuel processing; heavy element chemistry; diffusion; isotope separation; pyrochemistry; plasma chemistry; ammonolysis of carbon halides; surface coatings; space materials; propulsion; ocean mineral processing; ferrites. *Mailing Add:* 661 Teresi Lane Los Altos CA 94024

HAHN, HENRY, b Brno, Czech, Feb 5, 28; US citizen; m 52; c 2. MATERIALS SCIENCE, METALLURGY. *Educ:* Mass Inst Technol, BS, 51; Rensselaer Polytech Inst, 53; Columbia Univ, Prof Engr, 56. *Prof Exp:* Chief proj engr, Wright Aero Div, Curtiss-Wright Corp, 56-63; mgr mat lab, Res Div, Melpar-Am Stand, Inc, 63-70; PRES, ARTECH CORP, 70- *Honors & Awards:* Fel, Am Soc Metals Int, 87. *Mem:* Am Soc Metals; Am Inst Mining, Metall & Petrol Engrs; Am Soc Testing & Mat; Am Welding Soc. *Res:* Aerospace and aircraft materials; basic research in metallurgy; research management; consumer protection technology. *Mailing Add:* Artech Corp 14554 Lee Rd Chantilly VA 22021

HAHN, HONG THOMAS, b Seoul, Korea, Feb 5, 42; US citizen; m 67; c 3. COMPOSITE MATERIALS, STRUCTURES. *Educ:* Seoul Nat Univ, BS, 64; Pa State Univ, MS, 68, PhD(eng mech), 71. *Prof Exp:* Fel, McMaster Univ, 71-72; res assoc, Air Force Mat Lab, 72-74; res engr, Univ Dayton Res Inst, 74-77 & Air Force Mat Lab, 77-78; mech engr, Lawrence Livermore Nat Lab, 78-79; assoc prof, Wash Univ, St Louis, 79-81, prof mech eng, 81-86; PROF MECH ENGR, PA STATE UNIV, 86- *Concurrent Pos:* Consult, Lawrence Livermore Nat Lab, 79-, Garrett, 89- & Dow, 89-; ed, J Composite Mat, 81- *Mem:* Am Soc Mech Engrs; Am Inst Aeronaut & Astronaut; Soc Mfg Engrs; Am Soc Testing & Mat; Sigma Xi; Am Ceramics Soc. *Res:* Mechanical behavior of composite materials; fracture and fatigue; reliability; processing; nondestructive testing; manufacturing. *Mailing Add:* Pa State Univ 227 Hammond Bldg Dept Engr Sci & Mech University Park PA 16802

HAHN, HWA SUK, b Pusan, Korea, Nov 1, 24; m 49. MATHEMATICS. *Educ:* Seoul Nat Univ, BS, 49; Univ Ore, MS, 57; Univ Ill, PhD(math), 61. *Prof Exp:* Instr math, Pusan Nat Univ, Korea, 49-51, asst prof, 51-55; asst prof, Pa State Univ, 61-67; assoc prof, 67-77, PROF MATH, W GA COL, 77- *Mem:* Am Math Soc; Math Asn Am. *Res:* Theory of numbers in mathematics. *Mailing Add:* WGa Col 108 Dunwoody Dr Carrollton GA 30117

HAHN, KYONG T, b Kyong Gi-Do, Korea, Apr 4, 29; m 62; c 2. MATHEMATICAL ANALYSIS. *Educ:* Seoul Nat Univ, BS, 56; Yonsei Univ, MS, 58; Stanford Univ, PhD(math), 64. *Prof Exp:* Lectr math, Yonsei Univ, Korea, 57-59; res assoc, Stanford Univ, 64; from asst prof to assoc prof, 64-76, PROF MATH, PA STATE UNIV, 76- *Concurrent Pos:* NSF grants, 68-69 & 80-81; vis assoc prof, Univ Calif, Berkeley, 72; AID vis prof, Seoul Nat Univ, Korea, 78-79. *Honors & Awards:* Sr Prof Fulbright Res Award, Ger. *Mem:* Am Math Soc. *Res:* Several complex analyses. *Mailing Add:* Dept Math Pa State Univ University Park PA 16802

HAHN, KYOUNG-DONG, b Kyungbuk, Korea, Sept 8, 54. PLASMA PHYSICS. *Educ:* Seoul Nat Univ, BS, 77; Univ Wash; MS, 80, PhD(physics), 88. *Prof Exp:* STAFF SCIENTIST, LAWRENCE BERKELEY LAB, 88- *Mem:* Am Phys Soc. *Res:* Heavy ion fusion driver and beam dynamics of the intense particle beams using analytic and numerical model; transport of particles in a magnetically confined plasma. *Mailing Add:* 4480 Fallbrook Rd Concord CA 94521

HAHN, LARRY ALAN, b Winfield, Kans, Mar 6, 40; m 64; c 2. ELECTRONICS. *Educ:* Wichita State Univ, BS, 62; Purdue Univ, MS, 64; Southern Methodist Univ, PhD, 70. *Prof Exp:* MEM TECH STAFF, SEMICONDUCTOR DEVICE RES & DEVELOP, TEX INSTRUMENTS, INC, 64- *Mem:* Inst Elec & Electronics Engrs. *Res:* Power transistor research and development; transistor modeling. *Mailing Add:* 1000 N Shiloh Garland TX 75046

HAHN, LIANG-SHIN, b Tainan, Taiwan; m 58; c 3. MATHEMATICS. *Educ:* Nat Taiwan Univ, BS, 56; Stanford Univ, PhD(math), 66. *Prof Exp:* Instr math, Johns Hopkins Univ, 66-68; asst prof, 68-72, ASSOC PROF MATH, UNIV NMEX, 72- *Concurrent Pos:* Vis scholar, Univ Wash, Seattle, 74-76, Nat Taiwan Univ, 79-80, Univ Tokyo, 81-82 & 85-87, Int Christian Univ, Tokyo, 86-87, Sophia Univ, 87. *Mem:* Am Math Soc; Math Asn Am; Math Soc Japan. *Res:* Multiplier problem in harmonic analysis. *Mailing Add:* Univ NMex 6801 Leander Ave NE Albuquerque NM 87109

HAHN, MARJORIE G, b Salt Lake City, Utah, Dec 30, 48; m 73; c 1. PROBABILITY. *Educ:* Stanford Univ, BS, 71; Mass Inst Technol, PhD(math), 75. *Prof Exp:* Lectr statist, Univ Calif, Berkeley, 75-77; from asst prof to assoc prof, 77-87, PROF MATH, TUFTS UNIV, 87- *Concurrent Pos:* Sr researcher, NSF grants, 78-; vis assoc prof, Univ Calif, Berkeley, 81-82. *Mem:* Fel Inst Math Statist; Am Math Soc; Mat Asn Am; Sigma Xi. *Res:* Probability, stochastic processes in particular, central limit theorems in function spaces, random sets, extreme values and large deviations. *Mailing Add:* Dept Math Tufts Univ Medford MA 02155

HAHN, MARTIN EARL, b Cincinnati, Ohio, Apr 8, 43; m 67; c 2. PSYCHOBIOLOGY, BEHAVIORAL GENETICS. *Educ:* Ohio State Univ, BA, 66; Miami Univ, MA, 68, PhD(exp psychol), 70. *Prof Exp:* Grad asst, Miami Univ, 68-69, res asst psychol, 69-70; res assoc behav genetics, State Univ NY Binghamton, 70-72; asst prof, 73-77, assoc prof, 78-80, PROF BIOL, WILLIAM PATERSON COL, 80- *Concurrent Pos:* NIMH trainee, State Univ NY Binghamton, 72-73; prin investr, NIMH grant, 71-72, & NSF Col Fac Prog, 74-75; NSF grant, 76-78. *Mem:* Sigma Xi; Behav Genetics Asn; Int Soc Develop Psychobiol. *Res:* Behavior genetics of social behavior; evolution of behavior and brain size. *Mailing Add:* Dept Biol William Paterson Col 300 Pompton Rd Wayne NJ 07470

HAHN, OTTFRIED J, b Berlin, Ger, June 21, 35; Can citizen; m 60; c 3. MECHANICAL & NUCLEAR ENGINEERING. *Educ:* Univ Alta, BS, 58; Univ WVa, MS, 60; Princeton Univ, MA & PhD(mech eng), 64. *Prof Exp:* Physicist, Can Gen Elec Co, 64-67; asst prof, 67-77, ASSOC PROF MECH ENG, UNIV KY, 77- *Mem:* Am Nuclear Soc; Am Soc Mech Engrs; Am Phys Soc; Am Soc Eng Educ. *Res:* Properties of nuclear reactors, including lattice measurements and two-phase flow; coal conversion, gasification and coal properties. *Mailing Add:* Mech Eng-Anderson Hall Univ Ky Lexington KY 40506

HAHN, PETER, b Berlin, Ger, Nov 8, 23; m 48; c 2. PHYSIOLOGY. *Educ:* Charles Univ, Prague, MD, 50; Czech Acad Sci, PhD(physiol), 54, DSc(physiol), 62. *Prof Exp:* Lectr physiol, Charles Univ, Prague, 50-51; researcher, Czech Acad Sci, 51-62, head dept develop nutrit, 62-68; assoc prof, 68-72, PROF PEDIAT & OBSTET, UNIV BC, 72- *Honors & Awards:* Award on Aging, Ciba Found, London, 60; Czech Acad Sci Develop Awards, Prague, 62 & 66. *Mem:* Perinatal Res Soc; Am Inst Nutrit; Europ Soc Pediat Res; Can Soc Clin Invest; Royal Col Physicians. *Res:* Human and mammalian development of lipid and carbohydrate metabolism and its regulations. *Mailing Add:* Res Ctr 950 W 28th Ave Vancouver BC V5Z 4H4 Can

HAHN, PETER MATHIAS, b Vienna, Austria, May 15, 37; m 80; c 3. COMMUNICATION SYSTEMS. *Educ:* City Col New York, BEE, 58; Univ Pa, MSEE, 62, PhD(elec eng), 68. *Prof Exp:* Jr engr, Res Div, Philco Corp, Philadelphia, Pa, 58-61; instr elec eng, Moore Sch Elec Eng, Univ Pa, 62-65; radar systs engr, Missile & Surface Radar Div, Radio Corp Am, 65-67; eng specialist, Philco-Ford Corp, 67-68, eng mgr, Commun & Electronics Div, 68-76; unit supvr, Govt Commun Systs, RCA Corp, 76-77; chief engr, Sonic Sci Corp, 77-79; staff tech adv, Govt Systs Div, RCA Corp, 79-84; sr commun engr, Space Systs Div, Gen Elec Co, 84-86; mgr commun anal & simulation, G E Fed Systs, 87-88, mgr technol develop, G E Strategic Systs, 88-89, sr staff engr, 89-90; CONSULT, SCI-TECH SERV, 91- *Concurrent Pos:* Adj prof, Drexel Univ, 80-; adj assoc prof, Univ Pa, 84-; chmn, Tech Transfer Comt, Inst Elec & Electronics Engrs. *Mem:* Sr mem Inst Elec & Electronics Engrs; Am Soc Eng Educ. *Res:* Electronics; information theory; pattern recognition; communication theory; mathematical programming; operations research; probability; statistics; systems science and engineering; signal and data processing; communication networks; satellite communications; postal systems. *Mailing Add:* 2127 Tryon St Philadelphia PA 19146

HAHN, RICHARD ALLEN, b Columbus, Ohio, Aug 30, 38; m 63; c 2. SYSTEMIC HEMODYNAMICS, MYOCARDIAL FUNCTION. *Educ:* Ohio State Univ, BS, 62, MS, 64, PhD(pharmacol), 72. *Prof Exp:* Assoc sr investr, biol res, Smith Kline & French Labs, 73-79, sr investr, 79; res scientist, 80-85, RES ASSOC, DEPT CARDIOVASC PHARMACOL, LILLY RES LABS, INDIANAPOLIS, 86- *Mem:* AAAS; Am Heart Asn; NY Acad Sci. *Res:* Cardiovascular pharmacology; autonomic pharmacology; hypertension. *Mailing Add:* 5049 Deer Ridge Dr S Carmel IN 46032

HAHN, RICHARD BALSER, b Detroit, Mich, July 6, 13; m 38; c 2. CHEMISTRY. *Educ:* Wayne State Univ, BS, 35, MS, 36; Univ Mich, PhD(chem), 48. *Prof Exp:* Teacher pub sch, Mich, 36-42; instr chem, Wayne State Univ, 42-46; instr, Univ Mich, 46-47; from asst prof to prof, 47-79, EMER PROF CHEM & CONSULT, WAYNE STATE UNIV, 79- *Concurrent Pos:* Mem res group, Oak Ridge Nat Lab, 50-51; vis scientist, Atomic Energy Res Estab, Eng, 69. *Honors & Awards:* Anachem Award, 78. *Mem:* Am Chem Soc; Sigma Xi. *Res:* Analytical chemistry of zirconium and hafnium; organic analytical reagents; radiochemistry. *Mailing Add:* 894 W Outer Dr Oak Ridge TN 37830

HAHN, RICHARD DAVID, b Baltimore, Md, Sept 5, 12; c 2. INTERNAL MEDICINE. *Educ:* Johns Hopkins Univ, AB, 32, MD, 36; Am Bd Internal Med, dipl. *Prof Exp:* Asst prof, Sch Med, 48-70, asst prof, Sch Hyg & Pub Health, 51-58, assoc prof pub health admin, 58-68, ASSOC PROF INTERNAL MED, SCH MED, JOHNS HOPKINS UNIV, 70- *Concurrent Pos:* Physician, Johns Hopkins Hosp, 47-; med consult, Social Security Admin, 61-; med dir, Baltimore Life Ins Co, 71-80. *Mailing Add:* 10533 Stevenson Rd Stevenson MD 21153

HAHN, RICHARD LEONARD, b New York, NY, May 25, 34; m 56; c 3. NUCLEAR CHEMISTRY, SOLAR NEUTRINO PHYSICS. *Educ:* Brooklyn Col, BS, 55; Columbia Univ, MA, 56, PhD(chem), 60. *Prof Exp:* Chemist, Oak Ridge Nat Lab, 62-74, dir transuranium res lab & sect head nuclear chem, chem div, 74-84, sr chemist, 84-87; res assoc nuclear chem, 60-62, CHEMIST, BROOKHAVEN NAT LAB, 87- *Concurrent Pos:* Vis scientist, Inst Nuclear Physics, Orsay, France, 72-73, Lawrence Berkeley Lab, 74-79 & 84-87, GSI Soc Heavy Ion Res, Darmstadt, Ger, 79 & Lawrence Livermore Nat Lab, 84-87; mem, Transplutonium Prog Comt, US Dept Energy, 75-80; secy, Am Chem Soc, Div Nuclear Chem & Technol, 78-80, vchmn, 81, chmn, 82; vis scholar, Southwestern Univ, Georgetown, Tex, 83; mem vis comt Isotopes & Nuclear Chem Div, Los Alamos Nat Lab, 84-86, mem int Gallex exp & vis scientist, Gran Sasso Nat Lab, Italy, 86-, vchmn Nuclear Radiochem comt, Nat Res Coun, 87- *Honors & Awards:* Radiation Indust Award, Am Nuclear Soc, 77. *Mem:* Am Phys Soc; Am Chem Soc; Sigma Xi. *Res:* Solor neutrino research; structures and interactions of ions in solutions; search for and characterization of new elements and radioactive isotopes; co-discoverer of 22 isotopes; nuclear reactions; nuclear fission; scattering with neutrons and synchrotron radiation; chemistry and physics of transuranium elements. *Mailing Add:* Chem Dept Brookhaven Nat Lab Upton NY 11973

HAHN, RICHARD RAY, food science, for more information see previous edition

HAHN, ROBERT S(IMPSON), b New York, NY, Nov 1, 16; m 41; c 3. COMPUTER SCIENCE, MECHANICAL ENGINEERING. *Educ:* Univ Cincinnati, ME, 40, MSc, 42, DSc(appl physics), 44. *Prof Exp:* RES ENGR, HEALD DIV, CINCINNATI-MILACRON, WORCESTER, 44-; PRES, HAHN ENG INC, 79- *Concurrent Pos:* Consult, 71-79. *Honors & Awards:* Am Soc Mech Engrs Medal, 81. *Mem:* Nat Acad Eng; Am Soc Mech Engrs; Soc Mfg Engrs; Int Inst Prod Eng Res. *Res:* Lubrication; metal cutting; vibration; dampers to prevent tool vibration during metal cutting; thin lubricating films; high production grinding technology; computer control of machine tools. *Mailing Add:* 160 Southbridge St PO Box 349 Auburn MA 01501

HAHN, ROGER C, b Cleveland, Ohio, Feb 20, 32; m 62, 85; c 3. ORGANIC CHEMISTRY. *Educ:* Oberlin Col, AB, 53; Ohio State Univ, PhD(org chem), 60. *Prof Exp:* From asst prof to assoc prof chem, Univ SDak, 60-63; NIH fel, Univ Wis, 63-65; asst prof, 65-71, ASSOC PROF CHEM, SYRACUSE UNIV, 71- *Mem:* Am Chem Soc. *Res:* Homogeneous nucleophile exchange; asymmetric synthesis; synthesis methods. *Mailing Add:* Dept Chem Syracuse Univ Syracuse NY 13244-1200

HAHN, SAMUEL WILFRED, b Columbia, SC, Mar 21, 21; m 47; c 3. MATHEMATICS. *Educ:* Lenoir Rhyne Col, AB, 41; Duke Univ, MA, 42, PhD(math), 48. *Prof Exp:* Instr math, Univ Mich, 47-49; asst prof, Wittenberg Univ, 49-51; prof & head dept math, Winthrop Col, 51-59; prof math, Hampden-Sydney Col, 59-60; chmn dept math, 61-67 & 76-77, assoc dean, 63-65, prof, 60-83, EMER PROF MATH, WITTENBERG UNIV, 83- *Concurrent Pos:* Bd gov, Math Asn Am, 79-82; vis prof, Wake Forest Univ, 83-84, Washington & Lee Univ, 84-85 & SDak Sch Mines & Technol, 85-86, Calif Lutheran Univ, 86-87. *Honors & Awards:* Cert Meritorious Serv, Math Asn Am. *Mem:* Am Math So; Math Asn Am. *Res:* Statistical methods; computer use in teaching undergraduate mathematics. *Mailing Add:* 1019 Redbud Lane Springfield OH 45504-1547

HAHN, THEODORE JOHN, OSTEOPOROSIS. *Educ:* Johns Hopkins Univ, MD, 64. *Prof Exp:* ASST CHIEF ENDOCRINOL, VET ADMIN MED CTR & SCH MED, UNIV CALIF, LOS ANGELES, 83- *Mailing Add:* Div Endocrinol 691/111N Wadsworth Med Ctr Wilshire & Sawtelle Blvds Los Angeles CA 90073

HAHN, W(ALTER) C(HARLES), JR, b Manasquan, NJ, Sept 21, 30; m 54; c 4. METALLURGY. *Educ:* Lafayette Col, BS, 52; Pa State Univ, MS, 58, PhD(metall), 60. *Prof Exp:* Sr res metallurgist, Olin Mathieson Chem Corp, 60; asst prof metall, Mont Sch Mines, 60-63; from asst prof to assoc prof, 63-72, PROF METALL & MAT SCI, LEHIGH UNIV, 72- *Res:* High temperature chemistry; semiconductors. *Mailing Add:* Dept Mat Sci & Eng Lehigh Univ Bethlehem PA 18015

HAHN, WALTER I, b Korea, Nov 10, 23; US citizen; m 54; c 2. NEUTRON ACTIVATION ANALYSIS. *Educ:* Seoul Nat Univ, BS, 54; Fla State Univ, MS, 60, PhD(physics), 62. *Prof Exp:* Instr physics, Seoul Nat Univ, 55-58; fel nuclear eng, Univ Fla, 63-64; asst prof physics, State Univ NY, Fredonia, 64-69; PROF PHYSICS ENG, BENEDICT COL, 70-, DIR, 76- *Concurrent Pos:* Prin investr, Minority Biomed Support Prog, NIH, 78- *Mem:* Am Phys Soc. *Res:* Neutron activation analysis of trace elements; analysis of air particulates. *Mailing Add:* Dept Physics Eng Benedict Col Columbia SC 29204

HAHN, WALTER LEOPOLD, b Duermentigen, Ger, May 1, 26; US citizen; m 50; c 4. POLYMER CHEMISTRY. *Educ:* Univ Freiburg, dipl, 50, Dr rer nat, 52. *Prof Exp:* Asst lectr chem, State Res Inst Macromolecular Chem, Freiburg, 50-52, asst prof, 52-56; sr res chemist, 56-60, RES ASSOC, BENGER LAB, E I DU PONT DE NEMOURS & CO, INC, 60- *Concurrent Pos:* Assoc prof, Univ Notre Dame, 70. *Mem:* Am Chem Soc. *Res:* Vinyl polymerization; synthesis of monomers and polymers; kinetics of polymerization; initiators; textile fibers; graft and block copolymers; inorganic and semiorganic polymers; stereo-specific polymers; high temperature polymers. *Mailing Add:* 1223 Hollins Rd Waynesboro VA 22980

HAHN, WILLIAM EUGENE, b Greeley, Colo, June 11, 37; m 57; c 2. MOLECULAR BIOLOGY, CELL BIOLOGY. *Educ:* Univ Idaho, BS, 60; Tex Tech Univ, MS, 62; Tulane Univ, PhD(cell biol), 65. *Prof Exp:* Res asst prof molecular endocrinol, Univ Wash, 65-68; from asst prof to assoc prof, 68-78, PROF ANAT, MED SCH, UNIV COLO, DENVER, 78- *Concurrent Pos:* NIH career development award, Med Sch, Univ Colo, Denver. *Mem:* AAAS; Am Soc Cell Biol; Soc Neurosci; Am Asn Anat; Sigma Xi. *Res:* Genetic expression in development of embryos; genomic structure and complexity; RNA synthesis and hormone action; genetic expression in brain and other organs; biogenesis of messenger RNA. *Mailing Add:* Cellular & Struct Biol Univ of Colo Med Ctr Denver CO 80220

HAHN, YU HAK, b Seoul, Korea, Mar 13, 34; US citizen; m 67; c 2. QUANTUM OPTICS. *Educ:* Ky Wesleyan Col, BS, 58; WVa Univ, MS, 61; Pa State Univ, PhD(physics), 67. *Prof Exp:* Asst prof physics, Slippery Rock State Col, 62-64; sr physicist, Bausch & Lomb Co, 67-69; pres, Laser Energy Inc, 70-73; PRES & CHIEF RES EXEC, CVI LASER CORP, 73- *Mem:* Am Phys Soc; Sigma Xi. *Res:* Quantum optics, especially thin film optics for high power lasers. *Mailing Add:* CVI Laser Corp 200 Dorado Pl SE Albuquerque NM 87112

HAHN, YUKAP, b Seoul, Korea, July 28, 32; m 56; c 3. THEORETICAL PHYSICS. *Educ:* Univ Southern Calif, BA, 56; Yale Univ, MS, 58, PhD(physics), 62. *Prof Exp:* Assoc res scientist physics, NY Univ, 61-65; from asst prof to assoc prof, 65-73, PROF PHYSICS, UNIV CONN, 73- *Mem:* Am Phys Soc. *Res:* Theory of hyperfine structure and corrections in one and two electron atoms; bounds on the inverse reactance matrix in reaction theory; scattering theory and rearrangement processes; asymptotic behavior in quantum field theory; plasma physics. *Mailing Add:* Dept Physics Univ Conn Storrs CT 06268

HAHNE, HENRY V, b Riga, Latvia, Jan 16, 24; US citizen; m 62. SAFETY ENGINEERING, ENGINEERING MECHANICS. *Educ:* Graz Tech Univ, Dipl Eng, 49; Stanford Univ, PhD (eng mech), 54. *Prof Exp:* Engr, Pac Car & Foundry Co, Wash, 51-52; asst prof appl mech, Wash Univ, 54-56; res engr, Lockheed Aircraft Co, 56-59; from assoc prof to prof eng mech & chmn dept, Univ Santa Clara, 59-73; PRES & CHIEF ENGR, HENRY V HAHNE, INC, CONSULT ENGRS, 73- *Mem:* Am Soc Safety Engrs; Am Soc Testing & Mat; Soc Automotive Engrs. *Res:* Industrial and traffic accident analysis and prevention research. *Mailing Add:* PO Box 396 Los Altos CA 94023-0396

HAHNE, ROLF MATHIEU AUGUST, b Seattle, Wash, Oct 8, 36; m 64; c 2. ENVIRONMENTAL MANAGEMENT, INDUSTRIAL HYGIENE. *Educ:* Stanford Univ, BS, 58; Columbia Univ, AM, 59; Univ Wis, PhD(phys chem), 64. *Prof Exp:* Sci assoc nuclear chem, Nuclear Res Ctr, Jülich, Ger, 64-65; asst prof chem, Wittenberg Univ, 65-69; asst prof chem, Kalamazoo Col, 69-72; chief chemist, Kennedy Space Ctr, Pan American World Airways, 72-74; asst dir Univ Hygienic Lab, Univ Iowa, 74-83; RES LEADER, HEALTH & ENVIRON SCI, DOW CHEM CO, 83- *Concurrent Pos:* Am Chem Soc Petrol Res Fund grant, 67-69; vis fel, Atomic Energy Comn Radiation Res Lab, 70; sci assoc, Nuclear Res Ctr, Jülich, Ger, 71-72; US Dept Energy grant, 77-80, US Environ Protection Agency grant, 83. *Mem:* AAAS; Am Ind Hygiene Asn; Am Chem Soc; Am Acad Ind Hygiene. *Res:* Industrial Hygiene chemistry. *Mailing Add:* 1803 Bldg Dow Chem Co Midland MI 48674

HAHNEL, ALWIN, b Karlsruhe, Ger, Sept 9, 12; US citizen; m 42; c 6. ELECTRONICS ENGINEERING. *Educ:* Bad Staatstechnikum, BSEE, 35; Karlsruhe Tech Univ, Dipl Eng, 40, Dr Eng, 42. *Prof Exp:* Scientist, Ferdinand Braun Inst, Berlin, 43, chief Gaisberg Br, Salzburg, 44-45; dir radio commun systs, Decimeter Labs & Decimeter Networks WGer, 45-47; consult, US Army Signal Corps Labs, Ft Monmouth, 47-57; prin engr, Stromberg-Carlson Corp, 57-67; pres, HTV Systs, 67-71; PRES, COMAMP CORP, 71-; PRES, EBH CORP, 75- *Mem:* Inst Elec & Electronics Engrs. *Res:* Capacitor microphones; electrical high tension research; radar and transmission equipment design; propagation of electromagnetic waves; radio communication and frequency control systems; electronics circuitry. *Mailing Add:* 244 Golden Rd Rochester NY 14624

HAHNERT, WILLIAM FRANKLIN, b Logansport, Ind, Oct 6, 01; m 28; c 1. INVERTEBRATE ZOOLOGY. *Educ:* DePaul Univ, AB, 27, Johns Hopkins Univ, PhD(zool), 31. *Prof Exp:* Nat Res Coun fel biol, Johns Hopkins Univ, Univ Pa & Marine Biol Lab, Woods Hole, 31-33; from asst prof to prof, 34-68, chmn dept, 41-67, EMER PROF ZOOL, OHIO WESLEYAN UNIV, 68- *Concurrent Pos:* Vis prof, Stone Lab, Ohio State Univ, 39-48 & 56-58; adminr, Ohio Biol Surv, 68-72. *Mem:* AAAS; Am Soc Zoologists; Soc Syst Zool; Am Soc Limnol & Oceanog; Sigma Xi. *Res:* Invertebrate faunistic studies. *Mailing Add:* 45 Elizabeth St No 212 Austin Manor Delaware OH 43015

HAHON, NICHOLAS, b New York, NY, Mar 24, 24; m 48; c 1. VIROLOGY. *Educ:* Davis & Elkins Col, BS, 48; Johns Hopkins Univ, ScM, 50; Am Bd Med Microbiol, dipl. *Prof Exp:* Supvry virologist, US Army Biol Labs, Ft Detrick, 51-67, chief aerobiol Ill br, 67-71; LAB DIR, MICROBIOL SECT, APPALACHIAN LAB OCCUP SAFETY & HEALTH, 71- *Concurrent Pos:* Assoc prof pediat & instr microbiol, Med Ctr, WVa Univ. *Honors & Awards:* Sigma Xi Outstanding Res Award, 65. *Mem:* Fel AAAS; NY Acad Sci; Am Soc Virol; fel Am Soc Microbiol; Sigma Xi; Int Soc Interferon Res. *Res:* Variola and related pox viruses; arbor viruses; rickettsiae; multiplication; interferon; pneumoconiosis; pathogenesis; aerosol stability and virulence; assay by fluorescent antibody; neutralization tests; mineral dusts invitro tests. *Mailing Add:* Appalachian Lab Occup Safety & Health 944 Chestnut Ridge Rd Morgantown WV 26505

HAHS, SHARON K, b Washington, Ind, Sept 12, 47; m 69; c 2. COORDINATION CHEMISTRY, SCIENCE EDUCATION. *Educ:* Ill Wesleyan Univ, BA, 70; Univ NMex, MS, 72, PhD(inorg chem), 74. *Prof Exp:* Prof chem, Metrop State Col, 74-84; DEAN HUMANITIES & SCI, UNIV SC, 84- *Mem:* Am Chem Soc; Nat Sci Teacher's Asn; AAAS. *Res:* Transition metal coordination chemistry; educational research in teaching strategies with special emphasis on Piaget's theories of cognitive development. *Mailing Add:* Sch Humanities & Sci Univ SC 800 Univ Way Spartanburg SC 29303

HAI, CHI-MING, b Hong Kong, June 30, 55; c 1. CARDIOVASCULAR PHYSIOLOGY, MATHEMATICAL MODELING IN PHYSIOLOGY. *Educ:* Univ Toronto, BSc, 78; Univ Ottawa, MSc, 80; Johns Hopkins Univ, PhD(physiol), 84. *Prof Exp:* Fel physiol, Univ Va, Charlottesville, 84-87, instr, 87-88; ASST PROF PHYSIOL, BROWN UNIV, PROVIDENCE, 88- *Mem:* Am Physiol Soc; Biophys Soc; AAAS. *Res:* Understanding the regulation of smooth muscle contraction at the levels of cytosolic calcium, crossbridge phosphorylation, crossbridge mechanics and force generation. *Mailing Add:* Sect Physiol & Biophys Div Biol & Med Brown Univ Box G Providence RI 02912

HAI, FRANCIS, b Los Angeles, Calif, July 30, 37; m 71; c 2. HIGH VOLTAGE TECHNOLOGY. *Educ:* Univ Calif, Los Angeles, BA, 60, MA, 61, PhD(physics), 69. *Prof Exp:* Physicist, Rocketdyne, NAm Rockwell Inc, 61-63; mem tech staff, 69-81, RES SCIENTIST & SECT MGR, AEROSPACE CORP, 81- *Mem:* Inst Elec & Electronics Engrs. *Res:* Partial discharge phenomena; high voltage effects in materials. *Mailing Add:* Aerospace Corp 2350 E El Segundo Blvd M2 250 El Segundo CA 90245

HAID, D(AVID) A(UGUSTUS), b New York, NY, Apr 9, 36; m 63; c 7. PROCESS DEVELOPMENT, GAS APPLICATIONS. *Educ:* Stevens Inst Technol, BS, 57; State Univ NY Buffalo, MS, 65. *Prof Exp:* Develop engr, 57-58, engr, 58-61, superconductivity, 61-64, group leader, 64-68, supvr, 68-74, mgr prod eng, 74-76, mgr, New Bus Dept, 76-79, asst mgr, Gas Prod Develop, 79-84, ASSOC DIR, INDUST GASES DEVELOP, LINDE DIV, UNION CARBIDE, 84- *Mem:* AAAS; Am Soc Mech Engrs; Sigma Xi; Am Soc Metals. *Res:* Arc welding, superconductivity, arc heating processes; applications of industrial gases to chemicals and materials processing. *Mailing Add:* Linde Div Union Carbide Old Sawmill River Rd Tarrytown NY 10591

HAIDAK, GERALD LEWIS, b Paterson, NJ, Mar 6, 17; m 37; c 4. ANATOMY. *Educ:* Chicago Med Sch, BS, 40, MD, 42; Univ Pa, MSc, 55. *Prof Exp:* Asst instr anat, Grad Sch, Univ Pa, 53-55; from asst to prof anat, Albany Med Col, 56-76; dir med educ, St Luke's Hosp, 59-76; assoc dean & assoc prof thoracic surg, 76-90; DIR MED EDUC, BERKSHIRE MED CTR, 76-; ASSOC DEAN & PROF SURG, PROF CELL BIOL & DISTINGUISHED PROF MED EDUC, UNIV MASS MED SCH, WORCESTER, MASS, 91. *Concurrent Pos:* Sr surgeon, Pittsfield Gen Hosp & St Luke's Hosp, 56-59; sr surgeon, Berkshire Med Ctr, 59-90. *Mem:* AAAS; Hist Sci Soc; Am Asn Anat; AMA; Asn Am Med Cols; fel Am Col Surgeons; Am Asn Surg Anat. *Res:* Gross anatomy; early diagnosis of pulmonary neoplasm utilizing special techniques in bronchography; experimental emphysema. *Mailing Add:* Dept Surg Med Sch Univ Mass 55 Lake Ave N Worcester MA 01605

HAIDLER, WILLIAM B(ERNARD), b Ann Arbor, Mich, July 11, 26; m 52; c 4. NUCLEAR ENGINEERING, METALLURGY. *Educ:* US Naval Acad, BS, 50; NC State Col, MS, 57; Univ Ariz, PhD(nuclear eng), 64. *Prof Exp:* USAF, 57-, nuclear res officer, Propulsion Lab, Wright Air Develop Ctr, 57-59, asst div chief, 59-60, instr mech & gen physics, Air Force Acad, 60-61, asst prof physics & course dir, 61-62, gen physics, 64-65, assoc prof gen physics & dir prescribed physics courses, 65-66, assoc prof physics & dep head dept, 66-69, prof & head dept, 69-71; asst dir res & develop, Div Mil Appln, USAEC, 71-74; dean sch systs & logistics, USAF Inst Technol, 74-76; asst to pres tech progs, 76-78, dean, Sch Bus, 78-79, asst to pres, eng progs, Southwestern Mich Col, 79-86; FAC & CONSULT, INST PROF & CAREER DEVELOP, CENT MICH UNIV, 86- *Mem:* Am Asn Physics Teachers; Am Nuclear Soc; Am Soc Eng Educ; Am Asn Univ Professors; Sigma Xi. *Res:* Achievement testing in general physics; high temperature space energy conversion systems for auxiliary power; influence of nuclear reactor radiations on ceramic material; national-international energy requirements and implications; systems analysis. *Mailing Add:* 1055 Joliet Dr Niles MI 49120

HAIG, FRANK RAWLE, b Philadelphia, Pa, Sept 11, 28. THEORETICAL PHYSICS. *Educ:* Woodstock Col, AB, 52, STB, 59, STL, 61; Bellarmine Col, PhL, 53; Cath Univ Am, PhD(physics), 59. *Hon Degrees:* LHD, State Univ NY Onondaga Commun Col. *Prof Exp:* Asst prof physics, Wheeling Col, 63-66, pres, 66-72; from asst prof to assoc prof, Loyola Col, 72-81; pres, Lemoyne Col, 81-87; PROF PHYSICS, LOYOLA COL, 87- *Concurrent Pos:* NSF fel, Univ Rochester, 62-63; vis fel, Johns Hopkins Univ, 72. *Mem:* Am Phys Soc; Am Asn Physics Teachers. *Res:* Theoretical physics; relativistic astrophysics. *Mailing Add:* Dept Physics Loyola Col 4501 N Charles St Baltimore MD 21210-2699

HAIG, JANET, b Whittier, Calif, May 9, 25. BIOSYSTEMATICS. *Educ:* Whittier Col, AB, 46; Stanford Univ, MA, 48. *Prof Exp:* Curatorial asst, Natural Hist Mus, Stanford Univ, 47-48; asst, 43-52, res assoc, 52-68, ASSOC CURATOR, ALLAN HANCOCK FOUND, UNIV SOUTHERN CALIF, 75- *Mem:* Western Soc Naturalists; Crustacean Soc. *Res:* Taxonomy, distribution and ecology of anomuran crabs, particularly Porcellanidae. *Mailing Add:* 13450 E Hadley St Whittier CA 90601

HAIG, PIERRE VAHE, b Lebanon, Sept 24, 17; US citizen; m 48; c 3. RADIATION ONCOLOGY, NUCLEAR MEDICINE. *Educ:* Occidental Col, AB, 38; Univ Southern Calif, MD, 43; Am Bd Radiol, dipl, 50. *Prof Exp:* From instr to clin prof therapeut radiol, Sch Med, Univ Southern Calif, 49-; from instr to assoc clin prof, Sch Med, Loma Linda Univ, 52-68; CLIN PROF THERAPEUT RADIOL, SCH MED, UNIV SOUTHERN CALIF, 84-; DIR RADIOTHER DEPT, ST JUDE HOSP, 70- *Concurrent Pos:* Chief therapeut radiologist, Los Angeles County Gen Hosp, 50-67; radiologist, Good Hope Med Found, 50-67; consult, Long Beach Vet Admin Hosp, 53-80; mem staff, White Mem Med Ctr, 57-71; radiation therapist, Southern Calif Permanente Med Group, 67-70; physician specialist, Orange County Med Ctr, 67-74. *Honors & Awards:* Service Award, Am Cancer Soc. *Mem:* Soc Nuclear Med; AMA; fel Am Col Radiol; Radiol Soc NAm. *Res:* Radiology; cancer. *Mailing Add:* 220 Monarch Bay 101 E Valencia Mesa Dr South Laguna CA 92677

HAIG, THOMAS O, b Ypsilanti, Mich, June 12, 21; m 46; c 5. METEOROLOGY, SPACE SCIENCE. *Educ:* Univ Ill, Urbana-Champaign, BS, 55; George Washington Univ, MS, 66. *Prof Exp:* Mgr simulation, Gen Elec Co, Valley Forge, 68-69, mgr advan manned space systs, 69-70; exec dir, Space Sci & Eng Ctr, Univ Wis-Madison, 70-79; SR SCIENTIST, SONICRAFT CORP, CHICAGO, ILL, 79- *Concurrent Pos:* Res mgt consult, 79- *Mem:* Am Meteorol Soc; Sigma Xi. *Res:* Management and direction of space related scientific and operational research programs. *Mailing Add:* 4558 Hwy 78 Black Earth WI 53515

HAIGH, WILLIAM E, b Huron, SDak, Nov 6, 37; m 56, 76; c 3. MATHEMATICS. *Educ:* Huron Col, SDak, BS, 59; Univ NDak, MS, 63; Ind Univ, EdD, 70. *Prof Exp:* Assoc prof, 63-73, PROF MATH, NORTHERN STATE UNIV, SDAK, 63-, CHMN DEPT MAT, NAT SCI & HEALTH PROF, 78- *Mem:* Math Asn Am; Nat Coun Teachers Math; Sch Sci & Math Asn. *Res:* Mathematics education; differential equations; analysis; computer education. *Mailing Add:* Dept Math Northern State Univ Aberdeen SD 57401

HAIGHT, FRANK AVERY, b Des Moines, Iowa, Sept 28, 19; m 69; c 2. APPLIED MATHEMATICS. *Educ:* Univ Iowa, BA, 40, MSc, 41; Univ New Zealand, PhD(math), 57. *Prof Exp:* Res analyst gen hqs, US Army, Tokyo, Japan, 46-48; sr lectr math, Univ Auckland, 48-57; assoc res mathematician, Univ Calif, Los Angeles, 57-63, res mathematician, 63-69; prof statist & transp, Pa State Univ, 69-88; ADJ PROF, UNIV CALIF IRVINE, 88- *Concurrent Pos:* Hon res assoc, Univ London, 56; consult, Rand Corp, 56; Fulbright prof, Royal Inst Tech, Sweden, 62-63. *Res:* Distribution theory; probability models for road traffic, queueing, transportation, accidents; stochastic processes. *Mailing Add:* Inst Transp Studies Univ Calif Irvine CA 92717

HAIGHT, GILBERT PIERCE, JR, b Seattle, Wash, June 8, 22; m 46; c 4. INORGANIC CHEMISTRY, SCIENCE EDUCATION. *Educ:* Stanford Univ, AB, 43; Princeton Univ, PhD(chem), 47. *Prof Exp:* Res assoc chem, Manhattan Proj, Princeton Univ, 43-46; Rhodes scholar, Oxford Univ, 47-48; asst prof chem, Univ Hawaii, 48-49; asst prof chem, George Washington Univ, 49-52; asst prof chem, Univ Kans, 52-54; assoc prof chem, Swarthmore Col, 54-65; prof chem, Tex A&M Univ, 65-66; prof, 66-87, EMER PROF CHEM, UNIV ILL, URBANA, 87- *Concurrent Pos:* Consult, US Bur Entom, Honolulu, 48, Naval Res Lab, 51-52 & Standard Oil Co Ind, 53; prof phys & inorg chem, Wagner Free Inst Sci, 55-59; Dodge lectr, Franklin Inst, 59; vis prof, Tech Univ Denmark, 60-61; Petrol Res Found fac fel, 65; vis prof, Colo Col, 72; vis scientist, Univ Calif, San Diego, 74-75; vis fel, Australian Nat Univ, 81-82; mem, Int Active Coop Prog-Malaysia, ITM-Midwest Univs Consortium, 86; affil prof chem, Univ Wash, 89- *Honors & Awards:* Award in Chem Educ, MCA, 76 & Am Chem Soc, 79. *Mem:* AAAS; Danish Chem Soc; Am Chem Soc; Am Asn Univ Prof; Sigma Xi. *Res:* Chemistry of solutions; mechanisms of substitution and oxidation-reduction of oxy-ions; model enzyme systems. *Mailing Add:* 10798 Manitou Park Blvd Bainbridge Island WA 98110

HAIGHT, JOHN RICHARD, b Chicago, Ill, Sept 8, 38; m 67; c 3. NEUROBIOLOGY, EVOLUTIONARY BIOLOGY. *Educ:* NCent Col, BA, 64; Mich State Univ, MS, 68, PhD(zool), 71. *Prof Exp:* Lectr, 71-74, sr lectr anat, 74-80, READER IN ANAT, UNIV TASMANIA, 81-, HEAD DEPT ANAT, 84- *Concurrent Pos:* Sr res fel, Res Sch Biol Sci, Australian Nat Univ, 80-83. *Mem:* Am Asn Anat; Am Soc Zoologists; Anat Soc Australia & NZ (secy, 75-80); Australian Mammal Soc; Australian Neurosci Soc. *Res:* Evolutionary and comparative functional morphology of mammalian nervous systems. *Mailing Add:* Dept Anat Univ Tasmania Box 252C GPO Hobart Tasmania 7001 Australia

HAIGHT, ROBERT CAMERON, b Ann Arbor, Mich, May 27, 41; m 70; c 2. NUCLEAR PHYSICS, NUCLEAR ENGINEERING. *Educ:* Yale Univ, BA, 63; Princeton Univ, MS, 65, PhD(physics), 69. *Prof Exp:* Fel & instr nuclear physics, Univ Pittsburgh, 68-70; fel, Los Alamos Sci Lab, 70-72; physicist nuclear physics, Lawrence Livermore Lab, 72-85; PHYSICIST NUCLEAR PHYSICS, LOS ALAMOS NAT LAB, 85- *Concurrent Pos:* Mem, subcomt controlled thermonuclear res, US Nuclear Data Comt, 74-76; mem, Nuclear Data Comt, US Dept Energy, 79-85; adv, Int Nuclear Data Comt, 81- *Mem:* Fel Am Phys Soc; AAAS. *Res:* Nuclear physics; neutron interactions; reaction mechanisms; few-body problems; coulomb excitation. *Mailing Add:* MS-D406 LANL Group P-15 Los Alamos NM 87545

HAIGHT, ROGER DEAN, b David City, Nebr, Mar 12, 36; m 64; c 1. MICROBIOLOGY, BIOCHEMISTRY. *Educ:* Univ Nebr, BS, 59, MS, 61; Ore State Univ, PhD(marine microbiol), 65. *Prof Exp:* Fel, NIH, Torry Res Sta, Aberdeen, Scotland, 65-66; from asst to assoc prof, 66-74, PROF MICROBIOL, SAN JOSE STATE UNIV, 74- *Concurrent Pos:* Grant, Brown-Hazen Fund, 67- & NASA educ consortium, 70- *Mem:* AAAS; Am Soc Microbiol. *Res:* Microbial physiology; thermal injury phenomena in microorganisms. *Mailing Add:* Dept Microbiol San Jose State Univ/Wash Sq San Jose CA 95192

HAIGHT, THOMAS H, b Poynette, Wis, Nov 3, 36; m 66. BOTANY, BIOMETRICS-BIOSTATISTICS. *Educ:* Carroll Col, BS, 58; Syracuse Univ, MS, 65, PhD(bot), 69. *Prof Exp:* From asst prof to assoc prof, 69-80, PROF BIOL, FRAMINGHAM STATE COL, 80- *Mem:* Can Soc Plant Physiol; Am Soc Mort Sci; AAAS. *Res:* Morphogenesis in cultured fern fronds. *Mailing Add:* Dept Biol Framingham State Col Framingham MA 01701

HAIGLER, HENRY JAMES, SR, b Columbia, SC, July 23, 41; m 64; c 2. NEUROPHARMACOLOGY. *Educ:* Wake Forest Univ, BS, 63; Bowman Gray Sch Med, Wake Forest Univ, PhD(physiol), 69. *Prof Exp:* Trainee neurophysiol, Ment Health Res Inst, Univ Mich, 69-71; res assoc neuropharmacol, Sch Med, Yale Univ, 71-74; from asst prof to assoc prof pharmacol, Emory Univ, 74-80, dir grad studies, Dept Pharmacol, 80-82; sect head, Searle Res & Develop, CNS Pharmacol, G D Searle & Co, 82-87; mgr res & develop neurodiag, Abbott Labs, 87-90; VIS ASSOC PROF, DEPT PHARMACODYNAMICS, UNIV ILL, CHICAGO, 90- *Concurrent Pos:* Ad hoc mem, Neurol Sci Study Sect, 80 & Aging Rev Comt, 80. *Honors & Awards:* Magistral lectr, Headache 80 Neurol Cong, WHO, Florence, Italy, 80. *Mem:* Soc Neurosci (pres, 81-82); Am Soc Pharmacol Exp Therapeut; AAAS. *Res:* Site and mechanism of action of narcotic analgesic drugs; effect of morphine and met-enkephalin administered directly, using the technique of microiontophoresis, on neuronal activity evoked by a nocioceptive stimulus; development and evaluation of a post-mortem diagnostic assay for Alzheimer's disease; behaviorial effects of neurotropic drugs; site and mechanism of action of lysergic acid diethylamide; analgesic effects of dimethyl sulfoxide. *Mailing Add:* Dept Pharmacodynamics Col Pharm MC 865 Univ Ill 833 S Wood St Chicago IL 60612

HAIK, GEORGE MICHEL, b New Orleans, La, Mar 24, 10; m 48; c 4. OPHTHALMOLOGY. *Educ:* Tulane Univ, MD, 34; Am Bd Ophthal, dipl; Am Acad Ophthal, dipl. *Prof Exp:* Instr ophthal, Sch Med, La State Univ, 38-40, assoc prof & head dept, 46-50, head dept, 50-77, prof, 50-77; RETIRED. *Concurrent Pos:* Intern, Charity Hosp, 34-35, from resident to sr resident, 35-38, vis surgeon, 38-42, sr vis surgeon, 46-; vis surgeon, Eye, Ear, Nose & Throat Hosp; consult, Hotel Dieu Hosp, Mercy Hosp, Baptist Hosp, De Paul Hosp & E Jefferson Gen Hosp; mem residency rev comt ophthal, Coun Med Educ, AMA, 61-67; mem vision res training comt, Nat Inst Neurol Dis & Blindness, 63-67; mem, Nat Med Found Eye Care; spec clin prof ophthal, Tulane Univ Sch Med, 77-80. *Mem:* Am Ophthal Soc; Am Asn Ophthal; Pan-Am Asn Ophthal; AMA; Asn Res Vision & Ophthal. *Mailing Add:* 812 Maison Blanche Bldg New Orleans LA 70112

HAILE, CLARENCE LEE, b Odess, Mo, Aug 9, 48; m 70. ENVIRONMENTAL CHEMISTRY, ANALYTICAL CHEMISTRY. *Educ:* Cent Mo State Univ, BS, 70; Univ Wis, MS, 72, PhD(water chem), 77. *Prof Exp:* Res asst, Univ Wis, 73-75; assoc chemist, 75-77, sr chemist, 77-80, prin chemist, 80-85, sect head, 85-88, DEPT MGR, MIDWEST RES INST, 88- *Concurrent Pos:* Treas, 85-87, chmn elect, 88-, Div Environ Chem, Am Chem Soc. *Mem:* Am Chem Soc; Sigma Xi. *Res:* Emissions and environmental transformations of hazardous organic compounds; methods for sampling and analysis of trace organic contaminants in the environment. *Mailing Add:* Pace Labs Inc 1710 Douglas Dr Minneapolis MN 55422

HAILE, JAMES MITCHELL, b Atlanta, Ga, Dec 7, 46; m 71; c 1. MOLECULAR-SCALE COMPUTER SIMULATION, STATISTICAL MECHANICS. *Educ:* Vanderbilt Univ, BS, 68; Univ Fla, ME, 74, PhD(chem eng), 76. *Prof Exp:* Dept head chem eng, Univ Tulsa, 88-89; from ast prof to assoc prof, 76-84, PROF CHEM ENG, CLEMSON UNIV, 84- *Concurrent Pos:* Vis res assoc, Physics Dept, Univ Guelph, Ont, 75; vis scientist, Physics Dept, Chalk River Nat Lab, Ont, 82 & Inst Thermo & Fluid Dynamics, Rhur Univ, Bochum, WGer, 86; presidential young investr, NSF, 84; ed, Molecular Simulation, 86-90. *Mem:* Am Inst Chem Engrs; Am Phys Soc; Am Chem Soc; Hist Sci Soc; Sigma Xi. *Res:* Molecular-scale computer simulation of dense fluids; statistical mechanics; thermodynamics of fluid mixtures; classical nonlinear dynamics applied to many-body problems. *Mailing Add:* Chem Eng Dept Clemson Univ Clemson SC 29634

HAILMAN, JACK PARKER, b St Louis, Mo, May 6, 36; m 58; c 2. ETHOLOGY, ANIMAL COMMUNICATION. *Educ:* Harvard Univ, AB, 59; Duke Univ, PhD(zool psychol), 64. *Prof Exp:* Fel, Nat Inst Ment Health, Tübingen, 64; instr animal behavior, Rutgers Univ, 64-66; hon res assoc, Vert Dept, Smithsonian Inst, 66-69; asst prof zool, Univ Md, 66-69; assoc prof, 69-73, PROF ZOOL, UNIV WIS, MADISON, 73- *Mem:* Fel AAAS; fel Animal Behavior Soc (pres, 81-82); Am Soc Naturalists; fel Am Ornith Union. *Res:* Animal behavior; communication, sensory processes, ontogeny, sociality. *Mailing Add:* Dept Zool Univ Wis Madison WI 53706

HAILPERN, RAOUL, b Alexandria, Egypt, July 19, 16; US citizen; m 38; c 2. MATHEMATICS. *Educ:* Univ London, BA, 54; Univ Buffalo, MA, 59, PhD(math), 62. *Prof Exp:* Bank exec, Barclays Bank, Egypt, 32-57, Eng, 57; from asst prof to prof lectr, Millard Fillmore Col, 67-89; asst prof, 63-67, PROF LECTR, MILLARD FILLMORE COL, 67-; HEAD MATH DEPT, PARK SCH, NY, 58- *Mem:* Math Asn Am; Am Math Soc; Nat Coun Teachers Math. *Res:* Calculus of finite differences; combinatorial analysis. *Mailing Add:* 63 Garden Ct Amherst NY 14226-3220

HAIM, ALBERT, b Paris, France, Mar 19, 31; US citizen; m 55; c 2. INORGANIC MECHANISMS, REACTION KINETICS. *Educ:* Univ Southern Calif, PhD(inorg chem), 60. *Prof Exp:* Res asst chem, Univ Southern Calif, 55-59, res assoc, 59-61; res assoc, Stanford Univ, 61-62; from asst prof to assoc prof, Pa State Univ, 62-64; assoc prof, 64-68, PROF CHEM, STATE UNIV NY STONY BROOK, 68-, ASSOC ED, INORG CHEM, 89- *Concurrent Pos:* Sloan res fel, 65-67; vis chemist, Brookhaven Nat Lab, 65 & 66; Fulbright lectr, Fac Chem, Montevideo Univ, Uruguay, 67, 69, 70 & 78; lectr, Nova Univ, Lisbon, Portugal, 78 & Univ Neuchâtel, Switz, 80 & Univ Tucuman, Arg, 89. *Mem:* Am Chem Soc; Royal Soc Chem. *Res:* Kinetic and mechanistic studies with complex ions in solution; redox, substitution and photochemical reactions; inorganic free radicals in solution. *Mailing Add:* Dept Chem State Univ NY Stony Brook NY 11794

HAIMES, FLORENCE CATHERINE, b San Jose, Calif, June 10, 17. CHEMISTRY, HISTORY OF CHEMISTRY. *Educ:* San Jose State Col, AB, 38; Stanford Univ, MA, 40, EdD, 52. *Prof Exp:* Asst Stanford Univ, 39-40; teacher pub schs, Calif, 40-47; from asst prof to assoc prof, 47-63, PROF CHEM, SAN FRANCISCO STATE UNIV, 63- *Concurrent Pos:* Mem bd dirs, CACT, 69-81, pres, 68-69. *Mem:* Fel AAAS. *Res:* History of chemistry. *Mailing Add:* Dept Chem San Francisco State Univ 1600 Holloway Ave San Francisco CA 94132

HAIMES, HOWARD B, b Middletown, NY, March 7, 50; m 72; c 2. ELECTRON MICROSCOPY, BIOCHEMICAL CYTOLOGY. *Educ:* Union Col, BS, 72; Long Island Univ, MS, 75; Albert Einstein Col Med, MS, 78, PhD(biochem cytol), 81. *Prof Exp:* Res assoc depts anat & biochem, Vanderbilt Univ Med Sch, 81-84; vis scientist biol dept, Mass Inst Technol, 84-85; STAFF SCIENTIST, ORGANOGENESIS INC, 85- *Mem:* Am Soc Cell Biol. *Res:* The nature and development of living organ equivalents for clinical uses and testing purposes; comparison to in vivo counterparts, e.g. skin and blood vessels; light, electron microscopy; immunocytochemistry; enzyme histochemistry. *Mailing Add:* Organogenesis Inc 83 Rogers St Cambridge MA 02142

HAIMES, YACOV Y, b Baghdad, Iraq, June 18, 36; m 68; c 2. WATER RESOURCES, RISK MANAGEMENT. *Educ:* Hebrew Univ, Jerusalem, Israel, BS, 64; Univ Calif, Los Angeles, MS, 67, PhD(large scale systs), 70. *Prof Exp:* From asst prof to prof, systs & civil eng, Case Western Reserve Univ, 83-86, dir Water Resources Systs Eng Prog, 72-87, dir, Ctr large scale systs, 80- 84, chmn, Systs Eng, Case Western Reserve Univ, 83-86; LAWRENCE R QUARLES PROF, SYSTS ENG & DIR, RISK MGT, UNIV VA, 87- *Concurrent Pos:* Jr petrol engr, Ministry Develop, Israel, 62-65, Res asst & res engr, Univ Calif, Los Angeles, 66-70; pres, Environ Systs Mgt, Inc, 74-, chmn, UNESCO/Int Hydrographic Prog, 80-87, mem, Bd Water Sci & Technol, Nat Res Coun, 82-84; chmn, Tech Adv Comt, Int Ground Water Modeling Ctr, Holcomb Res Inst, 85-88, Consult, Congr Off Technol Assessment, 77-89, Sci Adv Bd, Environ Protection Agency, 82- *Honors & Awards:* Sigma Xi Distinguished Res Award, 76. *Mem:* Fel Am Soc Civil Engrs; fel Inst Elec & Electronic Engrs; fel AAAS; fel Int Water Resources Asn; fel Am Water Resources Asn; Am Geophys Union; Oper Res Soc Am; Sigma Xi; Soc Risk Anal; Am Soc Eng Educ. *Res:* Study of large scale systems theory and methodology, with special emphasis on two areas, modeling and optimization involving multilevel hierarchies and multiple objectives, risk assessment and management; author or coauthor of five books. *Mailing Add:* Dept Systs Eng Univ Va Thronton Hall Charlottesville VA 22901

HAIMO, DEBORAH TEPPER, b Odessa, Ukraine, July 1, 21; US citizen; m 44; c 5. MATHEMATICAL ANALYSIS. *Educ:* Radcliffe Col, AB & AM, 43; Harvard Univ, PhD(math), 64. *Hon Degrees:* DSc, Franklin & Marshall Col, 91. *Prof Exp:* Actg head dept math, Lake Erie Col, 43-44; instr, Northeastern Univ, 44-45; lectr math, Wash Univ, 52-61; from lectr to assoc prof math, Southern Ill Univ, Edwardsville, 61-68; chmn dept, 69-72 & 73-76, PROF MATH, UNIV MO-ST LOUIS, 68- *Concurrent Pos:* Sci fac fel, NSF, 64-65; grant, NASA, 66-69, NSF, 69-71 & Air Force Off Sci Res, 71-74; mem sci team, US Support Grad Progs, Seoul Nat Univ, 74; mem Inst Advan Study, Princeton, NJ, 72-73; NSF Teacher Enhancement grant, 88-91; mem, Harvard Bd Overseers, 90-95. *Mem:* Am Math Soc; Math Asn Am (pres, 91-93); Soc Indust & Appl Math; Asn Women Math; Sigma Xi. *Res:* Harmonic analysis; integral transforms. *Mailing Add:* 7201 Cornell Ave St Louis MO 63130

HAIMO, LEAH T, CELL MOTILITY, CELL DIVISION. *Educ:* Yale Univ, PhD(cell biol), 80. *Prof Exp:* ASST PROF CELL BIOL, UNIV CALIF, RIVERSIDE, 80- *Mailing Add:* Dept Biol Univ Calif 900 University Ave Riverside CA 92521

HAIMOVICH, BEATRICE, b Romania, May 13, 58. BIOCHEMISTRY. *Educ:* Tel-Aviv Univ, Israel, BSc, 80; Univ Pa, PhD(biochem), 86. *Prof Exp:* Fel, Dept Biochem, 86-88, Dept Microbiol, 88-90, FEL, LAB JOAN BRUGGE, HOWARD HUGHES MED INST, UNIV PA, 90- *Concurrent Pos:* NIH fel award, 86-87; Muscular Dystrophy Asn fel, 87-89. *Mem:* Am Soc Biol. *Res:* Signalling pathways activated as a result of interactions between extracellular matrix ligands and their specific cell surface receptors in two model systems; PC12 cells and platelets; involvement of integrins in transformation phenotypes induced by Rous sarcoma virus; analysis of the effects of treatment with cytochalasin D and or polyHEMA substrates on the stability of adhesion plaques and generation of transformation-like phenotypes; expression of avian beta-1 integrin and truncation mutants in mouse 3T3 cells; analysis of their function and distribution; generation of antibodies specific for the voltage-dependent sodium channels from rat skeletal muscle; identification of sodium channel sub-types in adult and in vivo and in vitro developing skeletal muscle using immunocytochemistry and pharmacological probes; numerous publications. *Mailing Add:* Dept Microbiol Clin Res Bldg Rm 370 Univ Pa 422 Curie Blvd Philadelphia PA 19104

HAIN, FRED PAUL, b Milwaukee, Wis, Nov 21, 44; m 73. FOREST ENTOMOLOGY. *Educ:* Stetson Univ, BS, 69; Duke Univ, MF, 69; Mich State Univ, PhD(entom), 72. *Prof Exp:* Entomologist II, Tex Forest Serv, 72; fel forest entom, Tex A&M Univ, 73-74; res assoc, 74-76, asst prof, 76-78, ASSOC PROF ENTOM, NC STATE UNIV, 78- *Mem:* Entom Soc Am; Entom Soc Can; Sigma Xi. *Res:* Population studies on the southern pine beetle; population studies and control measures for insects and mites infesting Fraser fir. *Mailing Add:* Dept Entom NC State Univ Box 7626 Raleigh NC 27695-7626

HAINDL, MARTIN WILHELM, b Landshut, WGer, May 17, 40; US citizen; m 72; c 1. HISTORY OF PHYSICS, THERMODYNAMICS. *Educ:* Montclair State Col, BS, 63, MA, 66; NY Univ, PhD(physics, sci educ), 72. *Prof Exp:* Teacher & coordr chem, Glen Rock High Sch, 63-68; asst prof, 68-80, ASSOC PROF PHYSICS, JERSEY CITY STATE COL, 80-, CHMN DEPT, 77- *Mem:* AAAS; Am Asn Physics Teachers; Am Phys Soc; Am Inst Physics. *Res:* History of physics; thermodynamics-structure of materials; physics education and instructional strategies, computer aided instruction; educational applications of telecommunications. *Mailing Add:* Dept Physics Jersey City State Col 2039 Kennedy Memorial Blvd Jersey City NJ 07305

HAINES, BERNARD A, b Phila, Pa, Apr 8, 26; m 57; c 2. PHARMACEUTICAL CHEMISTRY. *Educ:* Phila Col Pharm, BS, 53; Purdue Univ, MS, 58, PhD(phys pharm), 60. *Prof Exp:* Formulator, Nat Drug Co, 47-49; res assoc parenteral pharmaceut, Merck Sharp & Dohme Div, 54-55; sr res scientist, E R Squibb & Sons Div, Olin Mathieson Chem Corp, 60-65, sci coordr, Olin Int, 65-67, dir prod develop, 67-71, dir int regulatory affairs group, E R Squibb & Sons, Inc, 71-89; RETIRED. *Mem:* Am Pharmaceut Asn; Am Chem Soc. *Res:* Surface and colloidal chemistry. *Mailing Add:* 2001 Pheasent Lane Charlottesville VA 22906

HAINES, CHARLES WILLS, b Phila, Pa, Apr 14, 39; m 61; c 2. APPLIED MATHEMATICS, CONTROLS. *Educ:* Earlham Col, AB, 61; Rensselaer Polytech Inst, MS, 63, PhD(appl math), 65. *Prof Exp:* Instr math, Rensselaer Polytech Inst, 65-66; asst prof math, Clarkson Col Technol, 66-71; asst provost, 73-81, assoc prof Math & Mech Eng, 71-86, ASSOC DEAN ENG, ROCHESTER INST TECHNOL, 82-, PROF MATH & MECH ENG, 86- *Concurrent Pos:* Consult, Xerox Corp, 76 & 77; rev panelist, NSF Sci Fac Prof Develop Prog, 78 & 80; panelist, Navy grad fel, 87-; projs bd, Am Soc Eng Educ. *Mem:* Soc Indust & Appl Math; Math Asn Am; Am Soc Eng Educ;

Sigma Xi; Am Soc Mech Eng. *Res:* Stochastic eigenvalue problems; boundary value and eigenvalue problems in general; ordinary and partial differential equations; stochastic aircraft control problems. *Mailing Add:* Col Eng Rochester Inst Technol Rochester NY 14623

HAINES, DANIEL WEBSTER, b Nashville, Tenn, Nov 8, 37; m 62; c 2. COMPOSITE MATERIALS, FORENSIC ENGINEERING. *Educ:* Rutgers Univ, BS, 59; Lehigh Univ, MS, 61; Columbia Univ, Eng ScD, 68. *Prof Exp:* Res asst civil eng, Lehigh Univ, 59-61; Peace Corps volunteer-teacher math & sci, Govt Col, Ibadan, Nigeria, 61-63; res asst eng mech, Columbia Univ, 64-68; Alfred P Sloan Found vis fel, Princeton Univ, 68-69; from asst prof to prof mech eng, Univ SC, 69-77; res engr, Ciba-Geigy Corp, 77-80, prod eng mgr, 80-81; prin, Midlantic Testing & Consult, 82-87; assoc dean eng, 87-89, PROF MECH ENG, MANHATTAN COL, 83- *Concurrent Pos:* Vis assoc prof, Stevens Inst Technol, 75-76; vis lectr, Yale Univ, 75-76; exec comt chmn, Tech Coun on Forensic Eng, Am Soc Civil Engr, 87-88. *Mem:* Am Soc Civil Eng; Am Soc Mech Eng; Catgut Acoust Soc. *Res:* Elastic wave propagation; vibrations of elastic solids; properties of musical instrument woods, varnishes and wood substitutes; violin and guitar acoustics; mechanics of composite materials; honeycomb sandwich panels; finite element analysis; forensic engineering. *Mailing Add:* 142 Greenridge Ave White Plains NY 10605

HAINES, DAVID CLARK, b Youngstown, Ohio, Jan 30, 42; m 79. MATHEMATICS. *Educ:* Col Wooster, BA, 64; Ohio State Univ, MSc, 67, PhD(math), 69. *Prof Exp:* Teaching asst math, Ohio State Univ, 64-67; instr, Iowa State Univ, 67-69; asst prof, 69-78, ASSOC PROF MATH, BATES COL, 78-, CHMN DEPT, 79- *Concurrent Pos:* Andrew W Mellon Found fel, 75; vis scholar, Mass Inst Technol, 77-78. *Mem:* Am Math Soc; Math Asn Am; Asn Comput Mach. *Res:* Theory of commutative rings, boolean rings and p-rings. *Mailing Add:* Dept Math Bates Col Lewiston ME 04240

HAINES, DONALD ARTHUR, b La Crosse, Wis. WILDFIRE METEOROLOGY. *Educ:* Univ Wis, BS, 58, MS, 61. *Prof Exp:* Res meteorologist satellite meteorol, Nat Environ Satellite Serv, US Dept Com, 61-65, climatologist, Nat Weather Serv, 65-68; PRIN RES METEOROLOGIST WILDFIRE CONTROL, USDA FOREST SERV, 68- *Concurrent Pos:* Mem, Ctr Appl Sci & Technol. *Honors & Awards:* Qual Res Medal, USDA Forest Serv, 84. *Mem:* Nat Weather Asn; Am Meteorol Soc (secy-treas, 69-70, vpres, 70-71 & treas, 82-84). *Res:* Assessment of the contributions of weather and climatic factors and combinations of factors that lead to large or unusual wildfire occurrence; fluid flow and heat transfer. *Mailing Add:* PO Box 6046 Stuart FL 34997

HAINES, DUANE EDWIN, b Springfield, Ohio, May 4, 43; m 83; c 2. NEUROANATOMY, EVOLUTIONARY BIOLOGY. *Educ:* Greenville Col, BA, 65; Mich State Univ, MS, 67, PhD(anat), 69. *Prof Exp:* Instr, Mich State Univ, East Lansing, 68-69; asst prof, Med Col Va, 69-73; from assoc prof to prof anat, Sch Med, WVa Univ, 73-85; PROF ANAT & CHMN, UNIV MISS MED CTR, 85- *Concurrent Pos:* Assoc, ed, J Med Primatol, 77-, Anat Rec, 85-; PI on NIH grant, 74-85; off historian, Cajal Club, 86-; auth. *Honors & Awards:* MacLachlan Award, WVa Univ, 83. *Mem:* Am Asn Anatomists; Am Asn Phys Anthropologists; Neurosci Soc; Int Bran Res Org; Fel Royal Anthrop Inst Gt Brit & Ireland; Cajal Club. *Res:* Comparative neuroanatomy of prosimian primates with an emphasis on cerebellar connections; relay nuclei and spinal pathways; research on the neurological bases for the evolution of upright locomotion in primates. *Mailing Add:* Dept Anat Univ Miss Med Ctr 2500 N State St Jackson MS 39216-4505

HAINES, HARRY CAUM, forest management; deceased, see previous edition for last biography

HAINES, HOWARD BODLEY, b Kansas City, Kans, Jan 15, 35; m 56. ZOOLOGY, PHYSIOLOGY. *Educ:* Univ Tex, BA, 58, MA, 59, PhD(zool), 61. *Prof Exp:* NIH fel, 61-63; instr zool, Duke Univ, 63-64; from asst to assoc prof, 64-73, PROF ZOOL, UNIV OKLA, 73- *Concurrent Pos:* Res grants, NSF, NIH, 65-78, NASA, 68-70, NIH, 80-83. *Mem:* Am Physiol Soc; Am Soc Zool; Sigma Xi. *Res:* Physiology of vertebrates which dwell in arid environments; water balance; metabolism. *Mailing Add:* Dept Zool Univ Okla 730 Vanvleet Oval Rm 222 Norman OK 73019

HAINES, JOHN HALDOR, b Aberdeen, Wash, Sept 17, 38; m 64; c 2. MYCOLOGY. *Educ:* Univ Wash, BSc, 64, MSc, 67; Ore State Univ, PhD(mycol), 72. *Prof Exp:* SR SCIENTIST MYCOL, NY STATE MUS & SCI SERV, 69-; MYCOLOGIST, NY STATE BIOL SURV. *Mem:* Mycol Soc Am. *Res:* Taxonomy of the Hyaloscyphaceae; identification of airborne fungus spores. *Mailing Add:* Five Burhans Pl Delmare NY 12054

HAINES, KENNETH A, b Pendleton, Ind, Mar 27, 07; m 36; c 2. ENTOMOLOGY. *Educ:* Purdue Univ, BS, 29; Ohio State Univ, MS, 31. *Prof Exp:* Field asst oriental fruit moth, Ohio Agr Exp Sta, 29-30; field asst codling moth, NJ Agr Exp Sta, 33; agent dutch elm dis eradication, Bur Entom & Plant Quarantine, NJ, Agr Res Serv, USDA, 34-38 & 39-40, entomologist pear psylla control, Wash, 41-47, entomologist, Div Gypsy Moth Control, Pa, 48 & Mass, 49-50, asst div leader, Div Cereal & Forage Insects, Washington, DC, 51-53, asst dir prog appraisal & internal audit, Agr Res Serv, 54-58,; from assoc dir to dir int prog div, Hyattsville, Md, 59-76; RETIRED. *Mem:* AAAS; Entom Soc Am. *Res:* Biology and control of cereal and forage insects; technical direction of control projects on insect vectors of plant diseases and injurious insects; appraising agricultural research; foreign research operations. *Mailing Add:* 3542 N Delaware St Arlington VA 22207

HAINES, PATRICK A, b Worcester, Mass, Feb 17, 49; m 78; c 1. MESOSCALE DYNAMICS, CUMULUS PARAMETERIZATION. *Educ:* Tufts Univ, BS, 71; Fla State Univ, MS, 76. *Prof Exp:* Sr meteorologist, SAfrican Weather Bur, 74-77; res meteorologist, 78-84, CONSULT, RES INST, UNIV DAYTON, 84- *Concurrent Pos:* Res asst, Dept Geosci, Purdue Univ, 84-87; NASA grad student res prog, 87- *Mem:* Am Meteorol Soc. *Res:* Numerical simulation of the squall line particularly in association with the dry line; cumulus parameterization for mesoscale systems; boundary layer flow in complex terrain; meteorological conditions associated with aircraft icing; heavy rain and clear air turbulence and their effect on aircraft. *Mailing Add:* 1823 N 27th St Lafayette IN 47904

HAINES, RICHARD FOSTER, b Seattle, Wash, May 19, 37; m 61; c 2. HUMAN FACTORS RESEARCH & DEVELOPMENT, PSYCHOPHYSICS RESEARCH. *Educ:* Pac Lutheran Col, BA, 60; Mich State Univ, MA, 62, PhD(exp biol), 64. *Prof Exp:* Predoctoral fel visual psychophys, NIMH, Mich State Univ, 62-63, res fel visual psychophys, Dept Psychol, 63-64; postdoctoral res fel biotechnol, NASA-Ames Res Ctr, Nat Res Ctr, 64-67, res scientist aerospace human factors, Ames Res Ctr, 67-88, chief-Space Human Factors Off, NASA-Ames Res Ctr, 86-88; assoc prof psychol, San Jose State Univ, 88-89; scientist telesci technol, Res Inst Advan Computer Sci, 88-90; CONSULT SCIENTIST TELESCI TECHNOL, FOOTHILL/DEANZA COL, 90- *Concurrent Pos:* Consult, Dept Prev Med, Stanford Univ, 66-67; mem, Tech Group on Vision, Optical Soc Am, 67-83, Subcomt Electronic Displays, SAE, 83-91 & Man/Systs Integration Adv Panel, NASA, 85-88; consult mem, Comt Vision, Nat Acad Sci-Nat Res Coun, 70-82; chmn, Advan Technol Applications Comt, NASA-CCAIA, 74-79 & Space Sta Human Factors Subcomt, Aerospace Med Asn, 87-89. *Res:* Visual psychophysics; NASA-Aerospace project research and management; anomalous aerial phenomena research and advanced telescience technology research and development. *Mailing Add:* 325 Langton Ave Los Altos CA 94022

HAINES, RICHARD FRANCIS, b Ann Arbor, Mich, Jan 15, 23; m 51; c 3. MICROBIOLOGY. *Educ:* Univ Mich, BS, 48, MS, 49 & 54, PhD(bact), 59. *Prof Exp:* Asst bact, 53-57, instr, 57-60, from instr to asst prof surg & microbiol, 60-74, ASSOC PROF MICROBIOL & SURG, UNIV MICH, ANN ARBOR, 74-, DIR TISSUE TYPING LAB, 72- *Mem:* AAAS; Am Soc Microbiol; Transplantation Soc; Sigma Xi. *Res:* Transplantation immunology; natural antitumor agents; pathogenic fungi. *Mailing Add:* 807 Bruce St Ann Arbor MI 48103

HAINES, ROBERT GORDON, b NJ, Apr 21, 29; m 52; c 6. ENTOMOLOGY. *Educ:* Rutgers Univ, BSc, 51, MSc, 54, PhD(entom), 55. *Prof Exp:* Asst prof entom, Mich State Univ, 55-60; field res specialist, Calif Chem Co, 60-62; new prod mgr, Union Carbide Corp, 62-81; consult Pesticide Litigation & Turf Mgt, 81-87; PRES, PESTICIDE CONSULT SERV, 87-; PRES, ENVIRON CONSULT SERV, 87- *Mem:* Entom Soc Am; Soc Am Florists. *Mailing Add:* 3342 Byron Rd Green Cove Springs FL 32043

HAINES, ROBERT IVOR, b Wales, Feb 13, 53; Brit & Can citizen; m 73; c 2. NUCLEAR WASTE MANAGEMENT, MINERAL SURFACE CHEMISTRY. *Educ:* Leicester Univ, BSc Hons, 74, PhD(chem), 77. *Prof Exp:* Fel inorg chem, Victoria, 77-79; asst prof inorg chem, Univ RI, 79-80, Univ Toronto, 80-81; res officer geochem, Atomic Energy Can, Ltd, 81-; PROF CHEM, SIR WILFRED GRENFELL COL, MEM UNIV NFLD. *Mem:* Am Chem Soc; Sigma Xi. *Res:* Nuclear waste mineral water interactions utilizing surface analytical techniques to study mineral alteration and radionuclide sorption onto mineral systems; inorganic reaction mechanisms in ion coordination compounds; chemistry of nickel complexes. *Mailing Add:* Sir Wilfred Grenfell Col Mem Univ Nfld University Dr Corner Brook NF A2H 6P9 Can

HAINES, ROLAND ARTHUR, b Ottawa, Ont, July 4, 39; m 64; c 4. INORGANIC CHEMISTRY. *Educ:* Dalhousie Univ, BSc, 60, MSc, 61; Univ Pittsburgh, PhD(inorg chem), 64. *Prof Exp:* Res assoc inorg chem, Univ Southern Calif, 64-65; asst prof, 65-71, ASSOC PROF INORG CHEM, UNIV WESTERN ONT, 71-, ASSOC DEAN SCI, 84- *Mem:* Am Chem Soc; Chem Inst Can. *Res:* Stability of metal complexes; studies of optically active coordination compounds. *Mailing Add:* Dept Chem Univ Western Ont London ON N6A 5B7 Can

HAINES, TERRY ALAN, b Lansdale, Pa, Feb 18, 43; m 67; c 2. AQUATIC ECOLOGY. *Educ:* Pa State Univ, BS, 65, MS, 67; Mich State Univ, PhD(fisheries, wildlife), 71. *Prof Exp:* From asst prof to assoc prof biol sci, State Univ NY Col Brockport, 72-78; assoc prof, 78-82, PROF ZOOL, UNIV MAINE, ORONO, 82-; LEADER FIELD RES UNIT, ORONO, US FISH & WILDLIFE SERV, 78- *Mem:* Am Fisheries Soc; Soc Environ Toxicol Chem. *Res:* Impact of Contaminants on fish resources; mechanisms of toxicity of metals to fish; Hg biogeochemical cycling. *Mailing Add:* Dept Zool Univ Maine Orono ME 04469

HAINES, THOMAS HENRY, b New York, NY, Aug 9, 33; c 1. MEMBRANES & LIPIDS, ION TRANSPORT. *Educ:* City Col New York, BS, 57, MA, 59; Rutgers Univ, PhD(biochem), 64. *Prof Exp:* Res biochemist, Boyce Thompson Inst Plant Res, 59-63; asst biochem, Rutgers Univ, 63; lectr, 63-64, from asst prof to assoc prof chem, 64-71, actg dir Ctr Biomed Educ, 72-74, PROF CHEM, CITY COL NEW YORK, 71-, PRES & DIR BIOCHEM, CUNY MED SCH, 74- *Concurrent Pos:* Grants, USPHS, 64-66, US Dept Interior, 68-70, US Off Educ, 70-71, Petrol Res Fund, 70-72, NSF, 76- & NIH, 77-; vis assoc prof, Univ Calif, Berkeley, 70; NATO sr fel, Inst Natural Prod Chem, France, 70; vis prof, Univ Minn, 79-80,; vis scholar, Mitsubishi Inst Life Sci, Tokyo, 86-87. *Mem:* AAAS; Am Chem Soc; Am Oil Chemists Soc; NY Acad Sci; Am Soc Biochem & Molecular Biol; Biophys Soc Am. *Res:* Structure and function of natural membranes, especially flagellar membranes; chemistry, biochemistry and function of sulfolipids, glycolipids, fatty acids, sterols and natural halogen lipids; mechanism and role of ion transport in living membranes, proton pathway in bioenergenics; origin of membranes on prebiotic earth. *Mailing Add:* Dept Chem City Col New York New York NY 10031

HAINES, THOMAS WALTON, b Xenia, Ohio, Aug 13, 17; m 44; c 3. MEDICAL ENTOMOLOGY. *Educ:* Wilmington Col, BS, 42; Univ Kans, MA, 51; Univ Md, PhD(entom), 58. *Prof Exp:* Biologist, Malaria Invests & Typhus Fever Progs, USPHS, 44-48, entomologist, Diarrheal Dis Invests, 48-54; entomologist, US Army Chem Corp, Ft Detrick, 54-58; sr scientist, Res Grants & Fels Prog, Nat Cancer Inst, 58-61, scientist dir, Res & Develop Off, Region IV, USPHS, 61-71; PROF MED ADMIN & DIR SPONSORED PROGS, VANDERBILT UNIV SCH MED, 71- *Mem:* Am Soc Trop Med & Hyg; Entom Soc Am; Am Pub Health Asn; Am Mosquito Control Asn. *Res:* Invertebrate cell biology; arthropod transmission of human diseases; ecology and control of Diptera; toxicology of crop pests; science administration. *Mailing Add:* 6621 Ellesmere Rd Nashville TN 37205

HAINES, WILLIAM C, b Albion, Mich, Aug 29, 42; m 64; c 3. FOOD SCIENCE, MICROBIOLOGY. *Educ:* Mich State Univ, BS, 64, MA, 68, PhD(food sci), 72. *Prof Exp:* Teacher chem, Montrose Twp Schs, Mich, 64-67; food technologist food sci, Post Div, Gen Foods Corp, 67-69; dir prod develop, L D Schreiber Cheese Co, Inc, 72-77; res sect leader food sci, Borden Res Ctr, 77-81; lab mgr, Stauffer Chem Co, 81-83; vpres res, Ridgewood Inc, 83-87; DIR, FOOD INDUST INST, 87- *Mem:* Inst Food Technologists; Am Dairy Sci Asn; Sigma Xi; Coun Agr & Sci Tech; AAAS. *Res:* Food technology; cultured foods. *Mailing Add:* Food Industry Institute 201 Food Sci Bldg, Michigan State Univ East Lansing MI 48824

HAINES, WILLIAM EMERSON, b Evanston, Wyo, Aug 23, 17; m 41; c 3. PETROLEUM CHEMISTRY. *Educ:* Univ Wyo, BS, 39. *Prof Exp:* Pub sch teacher, Wyo, 39-42; asst chem, Univ Wyo, 42-43; petrol chemist, Petrol & Oilshale Exp Sta, US Bur Mines, 43-51, chemist, 51-58, proj leader, 58-63, res supvr, Laramie Energy Res Ctr, 63-75, res supvr, Laramie Energy Res Ctr, US Energy Res & Develop Admin, 75-77; mgr, Div Phys Sci, Laramie Energy Technol Ctr, US Dept Energy, 77-80; RETIRED. *Concurrent Pos:* Managing ed, Petrol Div, Am Chem Soc Preprints, 81- *Mem:* Am Chem Soc; Sigma Xi. *Res:* Petroleum separation and analysis; sulfur and nitrogen compounds; asphalt. *Mailing Add:* 520 S 13th St Laramie WY 82070-4103

HAINES, WILLIAM JOSEPH, b Crawfordsville, Ind, Sept 26, 19; m 43; c 2. PHARMACEUTICAL MANAGEMENT, SCIENCE & TECHNOLOGY MANAGEMENT. *Educ:* Wabash Col, AB, 40; Univ Ill, PhD(biochem), 43. *Hon Degrees:* DSc, Wabash Col, 70. *Prof Exp:* Rockefeller asst biochem, Univ Ill, 40-43; res biochemist, Upjohn Co, 43-46, group leader endocrinol res, 46-50, head dept, 50-54; tech dir labs div, Armour & Co, Ill, 54-58; vpres & dir res, Ortho Pharmaceut Corp & Ortho Res Found, 58-65, exec vpres, Corp, 65-67; vchmn, Johnson & Johnson Int, 67-79; dir & exec comt mem, 69-79, corp vpres sci & technol, 79-82, vpres Johnson & Johnson, 81-82; RETIRED. *Concurrent Pos:* Laurentian Hormone Conf lectr, 51; mem, Indust Res Inst; mem, Joslin Diabetes Found, 74-79. *Honors & Awards:* Upjohn Prize, 52. *Mem:* Fel AAAS; Am Chem Soc; Am Soc Biol Chemists; Endocrine Soc; Soc Exp Biol & Med; Sigma Xi; NY Acad Sci; Soc Chem Indust. *Res:* Biochemistry of natural and synthetic penicillins and steroid hormones; tissue enzymatic and microbiological synthesis of adrenal cortex hormones; chromatographic analysis; automatic chromatographic fraction cutter; fluoroscopic paper-gram scanner camera; metabolism of proteins and amino acids; determination of human requirement for essential amino acids. *Mailing Add:* 5 Bedford Dr RD 2 Doylestown PA 18901

HAINING, JOSEPH LEO, b Yazoo City, Miss, Feb 18, 32; m 54; c 3. BIOCHEMISTRY. *Educ:* Miss Southern Col, BS, 54; Purdue Univ, MS, 57, PhD(biochem), 59. *Prof Exp:* Asst prof biochem, Sch Med, Univ Miss, 59-63; CHIEF BASIC SCI RES LAB, VET ADMIN CTR, 63- *Concurrent Pos:* Asst prof, Univ Miss, 63- *Mem:* AAAS; Am Chem Soc; fel Geront Soc; Am Asn Lab Animal Sci; Sigma Xi. *Res:* Enzymology; protein metabolism; gerontology. *Mailing Add:* 1663 Lelia Dr Jackson MS 39216

HAINLINE, ADRIAN, JR, b Blandinsville, Ill, Mar 16, 21; m 42; c 5. BIOCHEMISTRY & CLINICAL CHEMISTRY. *Educ:* Western Ill State Univ, BEd, 42; Univ Denver, MS, 48; Univ Mich, PhD, 52; Am Bd Clin Chem, dipl. *Prof Exp:* Control chemist, E I du Pont de Nemours & Co, 42-45; clin chemist, Cleveland Clin Found, 51-64; clin chemist, St Luke's Hosp, Kansas City, 65-67; chief coronary drug proj lab, 67-71, chief clin chem diag prod eval, 73-77, chief clin chem standardization sect, Ctr Dis Control, 77-87; CONSULT, 87- *Concurrent Pos:* Dir, Am Bd Clin Chem, 72-78; sci dir, WHO Collaborating Ctr Ref & Res in Blood Lipids, 77-; nat cholesterol educ, Prog Lab Comt, 86. *Mem:* Fel AAAS; Am Chem Soc; Am Asn Clin Chem; Nat Acad Clin Biochem; Sigma Xi. *Res:* Laboratory management; lipid laboratory standardization; quality control. *Mailing Add:* 4762 Green St Duluth GA 30136

HAINLINE, LOUISE, b New London, Conn, Apr 22, 47. PEDIATRIC VISION. *Educ:* Brown Univ, BA, 69; Harvard Univ, MA, 71, PhD(develop psychol), 73. *Prof Exp:* From asst prof to prof, 72-84, PROF PSYCHOL, BROOKLYN COL, 85- *Concurrent Pos:* Prin investr, res grants, NIH, 74-; lectr pediat, Downstate Med Ctr, State Univ NY, 79-85. *Mem:* Am Psychol Asn; Soc Res Child Develop; NY Acad Sci; Asn Res Vision & Ophthal; Sigma Xi; Int Soc Study Behav Develop. *Res:* Development of the normal human visual system in young infants and problems that can arise in vision; development of vision and impact of different environmental factors on that development; devising new methods for testing infants. *Mailing Add:* Dept Psychol Brooklyn Col Brooklyn NY 11210

HAINSKI, MARTHA BARRIONUEVO, b Buenos Aires, Arg, Feb 18, 32; US citizen; m 62; c 1. BIOLOGICAL CHEMISTRY. *Educ:* Univ Buenos Aires, PhD(biochem), 57; Pepperdine Univ, MBA, 81. *Prof Exp:* Res assoc physiol chem, Sch Med, Wayne State Univ, 57-58; res assoc biochem, Philadelphia Gen Hosp, 58-63, NIH fel, 63-65; assoc biochem, Clin Lab Med Group, 65-66; dir protein chem res, Hyland Div, Travenol Labs, 66-70; mgr protein chem, Res & Develop, Abbott Sci Prod Div, 70-76; vpres, res & develop, Alpha Therapeut Corp, 77-84, PRES, I B B PLASMA, 85- *Concurrent Pos:* Int consult biol bus, 85- *Honors & Awards:* Presidential Awards, Abbott Labs, 75, 76 & 77. *Mem:* Am Chem Soc; fel Am Inst Chemists. *Res:* Plasma protein fractionation; automation and new methods for large scale production of plasma proteins; immune globulins; intravenous gamma globulin; antihemophilic factor; prothrombin complex; hepatitis antibody; plasma expanders; platelets; blood preservation; interferon research; artificial blood research and clinical trials. *Mailing Add:* 1628 Campbell St Glendale CA 91207

HAINSWORTH, FENWICK REED, b Norfolk, Va, June 25, 41; m 70; c 2. ZOOLOGY, PHYSIOLOGY. *Educ:* Clark Univ, AB, 63; Univ Pa, PhD(zool), 68. *Prof Exp:* USPHS fel, 67-69; from asst prof to assoc prof, 69-78, PROF BIOL, SYRACUSE UNIV, 78- *Mem:* Am Soc Naturalists; Ecol Soc Am; Soc Study Evolution; Sigma Xi. *Res:* Animal behavior; comparative physiology. *Mailing Add:* 110 Harrington Rd Syracuse NY 13244-1270

HAIR, JAKIE ALEXANDER, b Williston, SC, Aug 31, 40. ENTOMOLOGY. *Educ:* Clemson Univ, BS, 62, MS, 64; Va Polytech Inst, PhD(entom), 66. *Prof Exp:* Med entomologist, Commun Dis Ctr, USPHS, Atlanta, Ga, 66-67; assoc prof, 67-75, prof ecol, biol & insect control, 75-77, PROF ENTOM, OKLA STATE UNIV, 77- *Mem:* Entom Soc Am; Am Mosquito Control Asn; Wildlife Soc. *Res:* Control, ecology, biology and bionomics of insects of medical and veterinary importance; bionomics of wildlife parasites; tick investigations. *Mailing Add:* Dept Entom Okla State Univ Stillwater OK 74078

HAIR, MICHAEL L, b North Shields, Eng, Mar 23, 34; m 58; c 3. PHYSICAL CHEMISTRY, INORGANIC CHEMISTRY. *Educ:* Univ Durham, BSc, 55, PhD(chem), 58. *Prof Exp:* Fel, Nat Res Coun Can, 58-60; tech officer, Imp Chem Industs Ltd, Eng, 60-61; res chemist, Corning Glass Works, 61-64, group leader surface chem, 64-69; mgr phys chem br, Corp Res Ctr, Xerox Corp, NY, 69-74, mgr colloid & interface sci, 74-81, mgr, synthesis & explor res, 83-87, RES FEL, XEROX RES CTR CAN LTD, 81- *Concurrent Pos:* Adj prof, Univ Toronto, 86- *Honors & Awards:* Charles E Ives Award, Soc Photographic Scientists & Engrs, 82. *Mem:* Am Chem Soc. *Res:* Surface forces; polymer adsorbtion; chemistry of surfaces; infrared spectroscopy of adsorbed molecules; colloid stability. *Mailing Add:* Xerox Res Ctr 2660 Speaknan Dr Mississauga ON L5K 2L1 Can

HAIRFIELD, HARRELL D, JR, b Oklahoma City, Okla, Jan 19, 30; m 52; c 3. ELECTRONICS. *Educ:* Okla State Univ, BSEE, 57. *Prof Exp:* Res & develop engr, Western Elec Corp, 57-64; sr engr, Atlantic Res Corp, 64-65; site mgr, Radiation Serv Co, 65-66; sr res engr, Autonetics Div, NAm Rockwell Corp, 66-67, mem tech staff, Autonetics-Navig Systs Div, 67-79; proj engr, Defense Div, Brunswick Corp, 79-80; eng specialist, Guidance & Control Systs, Litton Industs, Inc, 82-; RETIRED. *Concurrent Pos:* Develop engr, Bell Tel Labs, 57-64. *Mem:* Inst Elec & Electronic Engrs. *Res:* Radar as applied to reentry phenomenon; missile guidance for ballistic missile systems; interspace navigation and special techniques for detection of hard targets in backgrounds of clutter; confocal interferometers and applications in the field of measurements; electro-optical systems; high altitude optical observation missles, ring laser gyro applications and technology. *Mailing Add:* PO Box 32 Finley OK 74543

HAIRSTON, NELSON GEORGE, b Davie Co, NC, Oct 16, 17; m 42; c 3. ECOLOGY, ZOOLOGY. *Educ:* Univ NC, BA, 37, MA, 39; Northwestern Univ, PhD(zool), 48. *Prof Exp:* Lectr sci, Northwestern Univ, 48; from instr to prof zool, Univ Mich, Ann Arbor, 48-67, dir, Mus Zool, 67-75; W R Kenan, Jr, prof zool, Univ NC, Chapel Hill, 75-88; CONSULT, 88- *Concurrent Pos:* Adv & consult, WHO, Philippines, 54-56 & 59, Iraq, 56, Geneva, 56 & 60-, Egypt, Sudan, Tanzania, Zimbabwe, SAfrica, Ghana & Western Samoa, mem expert adv panel parasitic dis, 64-81; consult, NIH, 64-; vchmn adv comt biol & med sci, NSF, 71-72, chmn, 72. *Mem:* Sigma Xi; Brit Ecol Soc; Am Soc Naturalists; Ecol Soc Am; Soc Study Evolution; Am Soc Ichthyol & Herpet. *Res:* Population and community ecology of salamanders, soil arthropods, paramecium and freshwater snails; epidemiology of schistosomes and other helminths; sampling methods for invertebrates. *Mailing Add:* Dept Biol Univ NC Chapel Hill NC 27599-3280

HAIRSTON, NELSON GEORGE, JR, b Asheville, NC, Sept 26, 49; m 74; c 1. FRESHWATER ECOLOGY. *Educ:* Univ Mich, BS, 71; Univ Wash, PhD(zool), 77. *Prof Exp:* From asst prof to assoc prof zool, Univ RI, 77-85; assoc prof, 85-87, PROF ECOL & SYST DEPT, CORNELL UNIV, 88- *Concurrent Pos:* NSF grant, 80-82, 83-85, 86-89, 88-90 & 91-92. *Mem:* Am Soc Limnol & Oceanog; Freshwater Biol Asn; Soc Int Limnologiae; Ecol Soc Am; Sigma Xi. *Res:* Zooplankton population dynamics and community structure; adaptive significance of zooplankton pigmentation; freshwater community-ecosystem interactions. *Mailing Add:* Ecol & Syst Dept Cornell Univ Corson Hall Ithaca NY 14853-2701

HAIRSTONE, MARCUS A, b Reidsville, NC, Oct 16, 26. RESEARCH ADMINISTRATOR, ELECTRON MICROSCOPY. *Educ:* Livingstone Col, BS, 49; Duquesne Univ, MS, 50; Univ Pittsburgh, PhD (cell biol), 56. *Prof Exp:* Dir Dental Res Labs, Univ Nebr, 58-60; asst prof zool, Long Island Univ, 60-62; res assoc cytophotometry, Rockefeller Inst, 62-64; Physicians & Surgeons, Columbia Univ, 64-66; prof & dean academics, City Univ NY, 70-71; prof zool, Am Univ, Cairo, Egypt, 72-74; proj mgr, int progs, Nat Sci Found, 75-76; HEALTH SCIENCE ADMINISTRATOR, INT PROGS, NAT INST HEALTH, 77- *Concurrent Pos:* Lectr & consult math, Nat Sci Found, 60-62; Fulbright prof conserv med res Inst, Alexandria Univ, Egypt, 66-67; consult biol, India Univ Sci Imiprovement Prog, Nat Sci Found, 69; dean, Arts & Sci, Cheyney Univ, 86-87. *Mem:* Am Soc Pub Admn (chair, comt Develop Regions, 76); Am Assoc Adv Sci; NY Acad Sci; Sudan Studies Assoc (bd mem, 82-83). *Res:* Characterization of tumor viruses; administration of program planning and management, international coordination for basic research, epidemiological studies, field trials, leading to vaccine development in hepatitis, diarrheal disease, accute respiratory infections, rabies, and polio. *Mailing Add:* PO Box 30663 Bethesda MD 20814

HAISCH, BERNHARD MICHAEL, b Stuttgart-Bad Canstatt, Fed Repub Ger, Aug 23, 49; US citizen; m 77, 86; c 2. STELLAR ATMOSPHERES, SPACE ASTRONOMY. *Educ:* Ind Univ, BS, 71; Univ Wis-Madison, MS, 73, PhD(astrophysics), 75. *Prof Exp:* Res assoc, Joint Inst Lab Astrophysics, Univ Colo, Boulder, 75-79; vis scientist, Astron Inst, Univ Utrecht, Neth, 77-78; res scientist, 79-83, STAFF SCIENTIST, LOCKHEED PALO ALTO RES LAB, 83- *Concurrent Pos:* Guest investr, NASA IUE, Einstein, Exosat, Rosat progs; Chmn, IAU Colloquium no 104; principal investr, multilayer res prog; mem rev comts, NASA; ed, J Sci Explor; vis fel, Max Planck Inst Extra Physik, Garching, Ger, 91. *Mem:* Am Astron Soc; fel Royal Astron Soc; Astron Soc Pacific; Int Astron Union; Soc Sci Explor. *Res:* Investigation of solar and stellar atmospheres; space observations of ultraviolet and x-ray emission of chromospheres, coronae and flares; radiative transfer research; development of concepts for space instrumentation. *Mailing Add:* Div 91-30 Bldg 255 Lockheed Palo Alto Res Lab 3251 Hanover St Palo Alto CA 94304

HAISLER, WALTER ERVIN, b Temple, Tex, June 3, 44; m 64; c 2. AEROSPACE ENGINEERING. *Educ:* Tex A&M Univ, BS, 67, MS, 68, PhD(aerospace eng), 70. *Prof Exp:* From asst prof to assoc prof, 70-80, PROF AEROSPACE ENG, TEX A&M UNIV, 80- *Concurrent Pos:* Consult, David Ehrenpreis Engrs, 70-71. *Mem:* Am Inst Aeronaut & Astronaut. *Res:* Finite element analysis of structural and fluid mechanics problems; nonlinear mechanics; numerical analysis. *Mailing Add:* Dept Aerospace Eng Tex A&M Univ College Station TX 77843

HAISSIG, MANFRED, b Vienna, Austria, Jan 2, 42; m 66; c 2. LADLE METALLURGY. *Educ:* Montan Univ, Loeben, Austria, Dipl, 68. *Prof Exp:* Supt meltshop, Boehler Ag, Ger, 73-77, malt testing, 78-86; vpres res & develop & mkt, Steel Casting Eng, Ltd, 82-86, exec vpres, 86-90; PRES, FUCHS SYSTEMS, INC 90- *Honors & Awards:* Charles W Briggs Award, Iron & Steel Soc, 86. *Mem:* Asn Iron & Steel Engrs; Iron & Steel Soc. *Res:* Horizontol casting of steel and secondary metallurgy to improve product quality; development of near net shape casting of wire and sheet. *Mailing Add:* Steel Casting Eng Ltd 1434 W Taft Ave Orange CA 92665

HAITKO, DEBORAH ANN, b New Haven, Conn, Sept 22, 51. INORGANIC & ORGANOMETALLIC CHEMISTRY. *Educ:* Albertus Magnus Col, BA, 73; Yale Univ, MS, 74, MPhil, 75, PhD(inorg chem), 78. *Prof Exp:* Teaching asst introductory chem, Yale Univ, 73-74, teaching asst org lab, 74-75, teaching asst inorg grader, 75-77; fel, Princeton Univ, 78; fel chem, Ind Univ, 78-80; CONSULT, 80- *Mem:* Am Chem Soc; Sigma Xi. *Res:* Transition metal chemistry, specifically the activation of organic molecules by coordination to transition metals; synthesis and mechanism; asymmetric synthesis; rearrangement processes of stereochemically nonrigid organometallic species. *Mailing Add:* Gen Elec CR&D PO Box 8 K-1-5A38 Chem Synthesis Lab Schenectady NY 12345

HAITZ, ROLAND HERMANN, b Durmersheim, Ger, Oct 6, 35; m 67; c 2. SOLID STATE PHYSICS. *Educ:* Karlsruhe Tech Univ, BS, 58; Munich Tech Univ, MS, 61, PhD(physics), 63. *Prof Exp:* Sr scientist, Shockley Res Lab, Clevite Corp, 61-64; mem tech staff, Physics Res Lab, Tex Instruments Inc, 64-69; res & develop mgr, Optoelectronics Div, 69-84, RES & DEVELOP MGR, COMPONENTS GROUP, HEWLETT PACKARD, 84- *Concurrent Pos:* Assoc ed, Trans Electon Devices, 70-73, ed, 73-77. *Mem:* Inst Elec & Electronics Engrs. *Res:* Avalanche breakdown in p-n junctions; microplasmas; avalanche noise generators; noise in particle detectors; radiation damage; microwave generation; solid state image sensors; optically coupled isolators; light emitters and numeric displays; integrated photodetectors, fiber optics, optoelectronic devices and systems. *Mailing Add:* Hewlett Packard Co 370 Trimble Rd San Jose CA 95131

HAJDU, JOSEPH, b Budapest, Hungary, May 5, 41; US citizen. BIOORGANIC CHEMISTRY, LIPID BIOCHEMISTRY. *Educ:* Hebrew Univ, Jerusalem, BSc, 65, MSc, 67; State Univ NY, Stonybrook, PhD(chem), 72. *Prof Exp:* Fel bioorganic chem, Sch Med, Univ Calif, Los Angeles, 73-76; asst prof chem, Boston Col, 76-; AT DEPT CHEM, CALIF STATE UNIV, NORTHRIDGE. *Concurrent Pos:* Prin investr, NIH, 79- *Mem:* Am Chem Soc; Sigma Xi. *Res:* Mechanism of organic and biochemical reactions; modification of enzymic active siles; design and synthesis of enzyme inhibitors, phospholipases, role of metal ions in biochemistry. *Mailing Add:* Chem Dept Calif State Univ-Northridge Northridge CA 91330

HAJDU, STEPHEN, b Hungary, Jan 9, 16; nat US; m 49; c 2. PHYSIOLOGY. *Educ:* Univ Budapest, MD, 41. *Prof Exp:* Asst prof physiol, Univ Budapest, 39-41, assoc prof, 41-45; res prof, Hungarian Biol Inst, 45-57; lectr, King's Col, Univ London, 48-49; res assoc, Inst Muscle Res, Woods Hole, Mass, 49-54; from surgeon to sr surgeon, Nat Heart Inst, 54-64, med dir, 64-70, head sect exp cardiovasc dis, 65-76; RETIRED. *Concurrent Pos:* Mem, Marine Biol Lab Asn, Woods Hole, Mass; Eszterhazy Found fel, 42 & 43; Brit Coun fel, King's Col, Univ London, 47-48. *Mem:* Am Physiol Soc; Soc Gen Physiol. *Res:* Nutrition; endocrinology; nerve and muscle physiology; heart physiology. *Mailing Add:* 1061 Inlet Dr Marco Island FL 33937

HAJDUK, STEPHEN LOUIS, b Marietta, Ga, Apr 23, 52; m 75; c 2. CELL BIOLOGY, PARASITOLOGY. *Educ:* Univ Ga, BS, 77; Univ Glasgow, PhD(zool), 80. *Prof Exp:* Res asst, Univ Ga, 72-77, Univ Glasgow, 77-80; fel, Sch Med, Johns Hopkins Univ, 81-83; asst prof, 83-88, ASSOC PROF, DEPT BIOCHEM, UNIV ALA, BIRMINGHAM, 88- *Concurrent Pos:* Vis scientist, Univ Amsterdam, 77 & 79. *Honors & Awards:* S L Hutner Prize, Am Soc Protozoologists, 85. *Mem:* Am Soc Cell Biol; Am Soc Protozoologists; Royal Soc Trop Med & Hyg. *Res:* Molecular and cellular biology of protozoan parasites; mitochondrial DNA structure and function in Africa trypanosones. *Mailing Add:* Dept Biochem Univ Ala University Sta Birmingham AL 35294

HAJEK, BENJAMIN F, b Shiner, Tex, Sept 17, 31; m 55; c 4. SOIL CLASSIFICATION, SOIL GENESIS. *Educ:* Tex A&M Univ, BS, 58; Auburn Univ, MS, 62, PhD(soil sci), 64. *Prof Exp:* Soil scientist, Soil Conserv Serv, USDA, Tex, 58-60, soil scientist, Soil Serv Lab, Md, 64; chemist, Hanford Labs, Gen Elec Co, Wash, 64; res scientist, Pac Northwest Labs, Battelle Mem Inst, 65-68; from asst prof to assoc prof, 68-78, PROF AGRON & SOILS, AUBURN UNIV, 78- *Mem:* Soil Sci Soc Am; Clay Minerals Soc; Am Soc Agron; Sigma Xi. *Res:* Adsorption, migration and dispersion of ions in porous media; quantitative clay mineralogy; surface and colloidal chemistry. *Mailing Add:* Dept Agron & Soils Auburn Univ Auburn AL 36849-5412

HAJEK, OTOMAR, b Belgrade, Yugoslavia, Dec 22, 30; m 55; c 1. MATHEMATICS, CONTROL THEORY. *Educ:* Charles Univ, Prague, Dr rer nat(math anal), 53, Cand Sci, 63. *Prof Exp:* Asst math, Fac Electrotech Eng, Prague Tech Univ, 53-56, sr asst, 56-58; sci officer comput sci, Res Inst Math Mach, 58-65; sr sci officer math, Inst Math, Charles Univ, Prague, 65-66; assoc prof, 66-69, PROF MATH, CASE WESTERN RESERVE UNIV, 69-, PROF SYST ENG, 88- *Honors & Awards:* Sr US Scientist Award, Alexander von Humboldt Stiftung, 75; Fulbright Award, 90. *Mem:* Am Math Soc; Gesellschaft Angewandte Math & Mech. *Res:* Dynamical system theory; qualitative theory of differential equations; control theory; game theory. *Mailing Add:* Dept Math & Statist Case Western Reserve Univ Cleveland OH 44106

HAJELA, DAN, b Shillong, India, Apr 23, 60; US citizen. MATHEMATICAL ECONOMICS & FINANCIAL STRATEGIES, DATA COMMUNICATIONS & NETWORKS. *Educ:* Ohio State Univ, BS, 75, MS, 78, MS, 81, PhD(math), 83. *Prof Exp:* Lectr math, Ohio State Univ, 76-83, vis prof, 83-84; mgr & sr scientist, Bellcore, 84-90; sr analyst, Bear, Stearns & Co, Inc, 91; SR PARTNER, FIRST BOSTON/TECH PARTNERS, 91- *Concurrent Pos:* Consult, Bell Labs, 83; vis prof, Calif Inst Technol, 84. *Mem:* NY Acad Sci; Inst Elec & Electronics Engrs; AAAS. *Res:* Mathematical analysis and computer modeling of financial instruments and arbitrage strategies; data communications; optical communications; local area networks; distributed databases; software development; information theory; neural networks; applied mathematics. *Mailing Add:* 66 Glenbrook Rd No 2327 Stamford CT 06902

HAJELA, PRABHAT, b Kanpur, India, Dec 25, 56; US citizen; m 91. OPTIMAL STRUCTURAL DESIGN, INTELLIGENT DESIGN SYSTEMS. *Educ:* Indian Inst Technol, Kanpur, BTech, 77; Iowa State Univ, MS, 78; Stanford Univ, MS, 81, PhD(aeronaut & astronaut), 82. *Prof Exp:* Res asst, aeronaut, Stanford Univ, 79-82; res assoc, optimal struct design, Univ Calif Los Angeles, 82-83; from asst prof to assoc prof aerospace eng, Univ Fla, Gainesville, 83-90; ASSOC PROF AERONAUT ENG, RENSSELAER POLYTECHNIC INST, 90- *Concurrent Pos:* Consult, Univ Calif, Los Angeles, 83-84, RCA Astro Electronics 83-84, Occidental Petrol, Columbia, 87; prin investr numerous res grants, 83-; res fel, struct design, Eglin Air Force Armament Lab, 86; NASA Lewis Res Ctr, 89; mem, Am Inst Aeronaut & Astronaut Tech Comt, multidisciplinary design optimization; assoc ed, Am Inst Aeronaut & Astronaut Jour, 90- *Honors & Awards:* Ralph Teetor Award, Soc Automotive Eng, 87. *Mem:* Assoc fel, Am Inst Aeronaut & Astronaut; Am Asn Artificial Intel; Am Soc Eng Educ. *Res:* Development of efficient methods for optimal design; adapting genetic algorithms and neural network based optimization strategies; use of smart materials to detect structural damage. *Mailing Add:* 5020 JEC Rensselaer Polytechnic Univ Troy NY 12180-3590

HAJIAN, ARSHAG B, b Cairo, Egypt, Oct 21, 30; nat US. MATHEMATICS. *Educ:* Univ Chicago, MS, 54; Yale Univ, PhD(math), 57. *Prof Exp:* Instr math, Univ Rochester, 57-59; NSF res fel, Yale Univ, 59-61; asst prof math, Cornell Univ, 61-63; assoc prof, 63-66, PROF MATH, NORTHEASTERN UNIV, 66- *Mem:* Am Math Soc. *Res:* Analysis; measure, ergodic and information theories; topological dynamics. *Mailing Add:* Dept Math Northeastern Univ 360 Huntington Ave Boston MA 02115

HAJI-SHEIKH, ABDOLHOSSEIN, b Dezful, Iran, Nov 27, 33; m 59; c 2. MECHANICAL ENGINEERING. *Educ:* Univ Tehran, Dipl Eng, 56; Univ Mich, MS, 59, MA, 61; Univ Minn, Minneapolis, PhD(mech eng), 65. *Prof Exp:* Instr & res fel mech eng, Univ Minn, 65-66; from asst prof to assoc prof, 66-70, PROF MECH ENG, UNIV TEX, ARLINGTON, 70- *Concurrent Pos:* NSF initiation res grant, 67-68 & NSF res grant, 84-86. *Honors & Awards:* Halliburton Award. *Mem:* Am Soc Mech Engrs; Sigma Xi; Am Inst Aeronaut & Astronaut. *Res:* Various numerical methods of solution concerning diffusion equation; film cooling studies in supersonic and subsonic flow; design of space radiator for lunar missions; thermal property measurement; energy conservation; thermal storage systems. *Mailing Add:* Dept Mech Eng Univ Tex Arlington TX 76019

HAJIYANI, MEHDI HUSSAIN, b Chital, India, Oct 8, 39; m 61. PHARMACEUTICAL CHEMISTRY, BIOCHEMISTRY. *Educ:* Osmania Univ, India, BSc, 60; Univ Southern Calif, PhD(pharmaceut & biomed chem), 70. *Prof Exp:* Sr chemist, Calatomic, Calbiochem, Calif, 67-69; NIH sr res assoc natural prod, Howard Univ, 70-71; asst prof, 71-76, ASSOC PROF CHEM, UNIV DC, 76- *Mem:* AAAS; Am Chem Soc. *Res:* Chemical structure-biological activity correlations; drug design via molecular modifications; syntheses of radioisotopically labeled compounds of biomedicinal and pharmaceutical interest; isoquinoline alkaloids, isolation, identification, structure determination and biogenesis. *Mailing Add:* 1420 Gerard St Rockville MD 20850

HAJJ, IBRAHIM NASRI, b Lebanon, June 21, 42; Can citizen; m 73; c 2. COMPUTER-AIDED DESIGN, RELIABILITY. *Educ:* Am Univ Beirut, BE, 64; Univ NMex, Albuquerque, MS, 66; Univ Calif, Berkeley, PhD(elec eng), 70. *Prof Exp:* Asst prof, Univ Waterloo, Can, 73 & 75-78, assoc prof, 79-82; asst prof, Lebanese Univ, Beirut, 73-75; from asst prof to assoc prof, 79-85, PROF, UNIV ILL, URBANA-CHAMPAIGN, 85- *Concurrent Pos:*

Vis prof, Tech Univ Denmark, 87. *Mem:* Fel Inst Elec & Electronics Engrs; Sigma Xi. *Res:* Computer-aided design and simulation of very large scale integrated (VLSI) circuits; fault simulation and testing; parallel-processing algorithms; statistical analysis; optimization; design automation; reliability analysis; design for reliability. *Mailing Add:* Coord Sci Lab Univ Ill 1101 W Springfield Ave Urbana IL 61801

HAJJAR, NICOLAS PHILIPPE, b Cairo, Egypt, Nov 17, 46; US citizen; m 72; c 2. PESTICIDE TOXICOLOGY, BIOCHEMICAL TOXICOLOGY. *Educ:* Cairo Univ, BSc, 67, MSc, 71; Univ Calif, Berkeley, PhD(entom), 78. *Prof Exp:* Res asst anthrop biochem, US Naval Res Univ No 3, Cairo, Egypt, 67-72; res asst entom, Univ Calif, Berkeley, 72-78; fel biochem toxicol, NC State Univ, Raleigh, 78-80; SR SCIENTIST & DEPT MGR, HEALTH & ENVIRON RES DIV, DYNAMAC CORP, 80- *Mem:* AAAS; Am Chem Soc; Soc Toxicol. *Res:* Conduct quantitative risk assessments on the effects of industrial pollutants on human health, prepare scientific support documents; evaluate pesticide metabolism and toxicology data; determine environmental fate and transport of pollutants; pesticide toxicology and insect growth regulators, mode of action, structure-activity relationships and metabolism. *Mailing Add:* Dynamac Corp 11140 Rockville Pike Rockville MD 20852

HAJOS, ZOLTAN GEORGE, b Budapest, Hungary, Mar 3, 26; US citizen; m 55. ORGANIC CHEMISTRY. *Educ:* Budapest Tech Univ, dipl, 47, DSc, 49. *Prof Exp:* Assoc prof, Veszprem Tech Univ, 52-53; res assoc chem, Princeton Univ, 57-60; sr chemist, Res Div, Hoffmann-La Roche, Inc, 60-67, res fel, 67-70; res assoc chem, Univ Vt, 72-73; res assoc pharm, Univ Toronto, 73-74; prin scientist chem res, Ortho Pharmaceut Corp, Raritan, 75-90; RETIRED. *Mem:* Am Chem Soc. *Res:* Synthesis and stereochemical investigation of compounds with physiological interest: glycosides, hydrophenanthrenes, steroidal hormones, heterocyclic compounds, especially alkaloids, furan and dioxane derivatives; asymmetric synthesis of intermediates of natural product chemistry. *Mailing Add:* Pauler U 2 Tr 21 Budapest I 1013 Hungary

HAJRA, AMIYA KUMAR, b W Bengal, India, Apr 8, 35; m 65; c 2. BIOCHEMISTRY, NEUROSCIENCES. *Educ:* Univ Calcutta, BSc, 53, MSc, 56; Northwestern Univ, PhD(biochem), 63. *Prof Exp:* Asst res biochemist, Ment Health Res Inst, 64-68, asst prof, 69-73, assoc res biochemist, 68-77, assoc prof, 73-, PROF & RES BIOCHEMIST, NEUROSCI LAB, MENT HEALTH RES INST, UNIV MICH, ANN ARBOR. *Mem:* AAAS; Am Soc Biol Chem. *Res:* Phospholipid metabolism; brain biochemistry; biomembrane structure. *Mailing Add:* Mental Health Res Inst Univ Mich 1103 E Huron Ann Arbor MI 48109

HAJRATWALA, BHUPENDRA R, b Navsari, India, April 8, 42; US citizen; m 67; c 2. PHYSICAL-INDUSTRIAL PHARMACY. *Educ:* Gujarat Univ, India, BPharm, 62; Univ Colo, MS, 65; Univ Iowa, PhD(phys pharm), 70. *Prof Exp:* Anal chemist, aerosols, George Barr & Co, 65-67; asst instr pharmaceut, Univ Iowa, 67-70; dir res parenterals, Invenex Pharmaceut, 70-72; sr lectr pharmaceut, Univ Otago, NZ, 72-79; assoc prof pharmaceut, Wayne State Univ, 79-90; PVT TUTOR & FINANCIAL PLANNER, AM EDUC & FINANCIAL SERV, 90- *Concurrent Pos:* Vis prof, Univ Iowa, 78-79. *Mem:* Am Inst Hist Pharm; Int Asn Financial Planners; Asn Pharm Teachers India; Indian Pharmaceut Asn. *Res:* Kinetics and mechanism of degradation of drugs; color stability; dissolution of drugs and factors affecting unit operations in industrial pharmacy and drug-ethanol and drug-drug interactions; formulation and development of dosage forms; pharmacokinetics; biostatistics. *Mailing Add:* Am Educ & Financial Serv 42955 Ford Rd Canton MI 48187

HAK, LAWRENCE J, b Roanoke, Va, Oct 15, 44; m 74; c 3. NEPHROLOGY, CLINICAL DRUG RESEARCH. *Educ:* Philadelphia Col Pharm & Sci, BS, 67, PharmD(pharm), 71. *Prof Exp:* Resident, Univ Pa Hosp, 69-70; instr clin pharm, Philadelphia Col Pharm & Sci, 71-73; from instr to asst prof, 73-79, CLIN ASSOC PROF MED & ASSOC PROF PHARM, UNIV NC, 79- *Mem:* Am Soc Hosp Pharmacists; Am Soc Parenteral & Enteral Nutrit; Am Soc Nephrology; Am Col Clin Pharm; Am Asn Pharmaceut Scientists. *Res:* Total parenteral nutrition in rats; drug and nutritional therapy of human patients with chronic renal failure; nephrotoxic and ischemic acute renal failure in rats; drug development research in humans. *Mailing Add:* Sch Pharm Univ NC Chapel Hill NC 27599-7360

HAKALA, MAIRE TELLERVO, b Helsinki, Finland, Feb 15, 17; US citizen; m 56; c 1. BIOCHEMICAL PHARMACOLOGY. *Educ:* Univ Helsinki, MSc, 47; Duke Univ, PhD(biochem), 55. *Prof Exp:* Res asst pharmacol, Yale Univ, 53-56; sr scientist, 56-62, Roswell Park Mem Inst, 59-62, assoc scientist, 62-69, prin scientist, 69-80, assoc chief cancer res sci, 80-83, res prof pharmacol & exp ther, 80-83; RETIRED. *Concurrent Pos:* Jane Coffin Childs Mem Fund fel, Yale Univ, 53-55; USPHS grant, Roswell Park Mem Inst, 58-; assoc res prof, Dept Biochem, State Univ NY Buffalo, 68-80. *Mem:* Am Soc Biol Chemists; Am Chem Soc; Am Asn Cancer Res. *Res:* Cell culture studies on drugs and antimetabolites; cellular uptake, metabolism and site of action of anticancer agents; drug resistance. *Mailing Add:* 260 Lakewood Pkwy Amherst NY 14226

HAKALA, REINO WILLIAM, b Albany, NY, Aug 25, 23; m 50; c 3. CHEMICAL PHYSICS, APPLIED MATHEMATICS. *Educ:* Columbia Univ, AB, 46, MA, 47; Syracuse Univ, PhD(phys chem), 65. *Prof Exp:* Instr org chem, Assoc Cols Upper NY, 47-48; teaching asst org chem, Syracuse Univ, 48-49, phys chem, 49-51; instr chem, Pa State Univ, 53-54; instr phys chem & inorg chem, Fairfield Col, 54-57; asst prof, Earlham Col, 57-59; lectr, Howard Univ, 59-63; from asst prof to assoc prof phys chem, Mich Tech Univ, 64-67, chmn sect & assoc prof math, 65-67; prof math, Oklahoma City Univ, 67-72, chmn dept math & dir lab, 67-70, chmn dept physics, 68-70; prof math, Washington Tech Inst, 72-73; dean sci & technol, 73-77, prof chem & physics, 73-78, prof math & physics, Lake Superior State Col, 78-; AT DEPT MATH-PHYSICS & ENVIRON SCI, GOVS STATE UNIV. *Concurrent Pos:* Res Corp grant, 59; Am Chem Soc Petrol Fund grant, 59-63; consult, Am Dent Asn Res Lab, Nat Bur Stand, 63; Mich Tech Univ grant, 67; consult, Universal Kleen-Rite Chem Corp, 70-72. *Mem:* Am Chem Soc; Am Phys Soc; fel Am Inst Chemists; Am Math Soc; Math Asn Am. *Res:* Thermodynamics; statistical and quantum mechanics; coordination compounds; numerical analysis; theory of electrolytes; mathematical social science. *Mailing Add:* 2945 Chayes Park Dr Homewood IL 60430

HAKALA, WILLIAM WALTER, b Blackberry, Minn, Aug 5, 35; m 60; c 3. CIVIL ENGINEERING, SOIL MECHANICS. *Educ:* Univ Minn, BS, 58, MSCE, 60; Va Polytech Inst, PhD(civil eng), 65. *Prof Exp:* Assoc res engr, Boeing Co, 60-61; asst prof civil eng, Va Polytech Inst, 61-63 & Univ NMex, 64-68; dep dir eng, Environ Res Corp, 68-70; dir opers, John A Blume & Assocs, 70-72; prog mgr, 72-81, HEAD, EARTHQUAKE HAZARD MITIGATION PROG, NSF, 82- *Concurrent Pos:* Consult, Sandia Corp, 66-68. *Mem:* Am Soc Civil Engrs; Am Soc Eng Educ. *Res:* Soil and rock dynamics; nuclear effects. *Mailing Add:* 3800 Powell Lane No 721 Falls Church VA 22041

HAKE, CARL (LOUIS), b Hoyleton, Ill, Nov 23, 27; m 52, 80; c 4. CONSULTING. *Educ:* DePauw Univ, BA, 48; Univ Ill, PhD(chem), 56; Am Bd Toxicol, dipl, 80, 85. *Prof Exp:* Biochemist, Dow Chem Co, 56-61, head chem res dept, Pitman-Moore Div, 61-65, asst to dir, Human Health Res & Develop Labs, 65-69, admin mgr, 69-72; asst prof, Med Col Wis, 72-74, assoc prof environ med & vchmn dept, 74-78, assoc prof pharmacol & toxicol, 78-81; consult, 80-88; RETIRED. *Concurrent Pos:* sabbatical, Univ Neuchatel, Switzerland, 78; adj prof pharmacol & toxicol, 81-91. *Mem:* Am Chem Soc; Soc Toxicol; Am Indust Hyg Asn; Am Pub Health Asn. *Res:* Effect of environmental and industrial contaminants upon health. *Mailing Add:* 4969 Cave Point Dr Sturgeon Bay WI 54235

HAKE, RICHARD ROBB, b Denver, Colo, July 15, 27; m 55; c 3. PHYSICS, SOLID STATE PHYSICS. *Educ:* Univ Colo, BS, 50; Univ Ill, MS, 51, PhD(physics), 55. *Prof Exp:* Asst, Univ Ill, 51-55, res assoc, 55-56; sr physicist, Atomics Int Div, NAm Aviation, Inc, 56-64, res physicist, Sci Ctr, 64-70; PROF PHYSICS, IND UNIV, BLOOMINGTON, 70- *Concurrent Pos:* Consult, Los Alamos Sci Lab, Univ Calif, 70-73; vis prof, Univ Calif, San Diego, 87-88. *Mem:* Fel Am Phys Soc; AAAS; Am Asn Physics Teachers. *Res:* Solid state and low temperature physics; electronic properties; superconductivity; magnetism; cryogenics; science education. *Mailing Add:* Dept Physics Swain Hall W 117 Ind Univ Bloomington IN 47405

HAKEN, WOLFGANG, b Berlin, Ger, June 21, 28; m 53; c 4. MATHEMATICS. *Educ:* Univ Kiel, PhD(math), 53. *Prof Exp:* Microwave engr, Siemens & Halske AG, 54-62; vis prof math, Univ Ill, 62-63; temp mem, Inst Advan Study, 63-65; PROF MATH, UNIV ILL, URBANA, 65- *Mem:* Am Math Soc; Ger Math Asn. *Res:* Topology of 3-dimensional manifolds. *Mailing Add:* Dept Math 373 Altgeld Hall Univ Ill Urbana IL 61801

HAKES, SAMUEL D(UNCAN), b Colorado Springs, Colo, Aug 17, 30; m 51; c 2. ELECTRICAL ENGINEERING. *Educ:* Univ Wyo, BS, 57, MS, 58; Univ Iowa, PhD(elec eng), 69. *Prof Exp:* Asst prof, 59-69, PROF ELEC ENG, UNIV WYO, 69-, DEAN, 75- *Concurrent Pos:* Consult engr, Digital Comput Ctr, Univ Wyo, 59-64, Ideal Aerosmith, 61-, Wyo Hwy Patrol, 67-68, US Navy Stand Lab, 69-70 & Gates Rubber Co, 70-; NSF sci faculty fel, 64, dir grant, 70- *Mem:* Inst Elec & Electronics Engrs; Instrument Soc Am. *Res:* Active network synthesis; precision instrumentation; computer design. *Mailing Add:* Box 3321 University Station Laramie WY 82070

HAKEWILL, HENRY, JR, b Chicago, Ill, Dec 14, 18; m 46; c 4. ORGANIC CHEMISTRY, INORGANIC CHEMISTRY. *Educ:* Elmhurst Col, BS, 41; DePaul Univ, MS, 47. *Prof Exp:* Res chemist, Nubian Paint & Varnish Co, 41-43, Pure Oil Co, 43-44 & Inst Gas Technol, 44-51; chief chemist, Cent Com Co, 51-62; mgr active prod, Multigraphics Div, Addressograph-Multigraph Corp, 62-77; RETIRED. *Concurrent Pos:* Tech Consult, Duraclean Int, Inc, 85- *Mem:* Am Chem Soc; Tech Asn Pulp & Paper Indust; Am Soc Testing & Mat; Sigma Xi. *Res:* Electrophotography; pigmented coatings; resins; gas analysis. *Mailing Add:* 1355 Wilmot Rd Deerfield IL 60015

HAKIM, EDWARD BERNARD, b Jersey City, NJ, July 11, 36; m 64; c 4. SOLID STATE PHYSICS, ELECTRONICS. *Educ:* Fairleigh Dickenson Univ, BS, 59; Univ Conn, MS, 62. *Prof Exp:* Res physicist, IBM Corp, 61-62; BR CHIEF, RELIABILITY, TEST & QA BR, DEPT ARMY, FT MONMOUTH, 62- *Mem:* Inst Elec & Electronic Engrs. *Res:* Microcircuit reliability and physics of failure. *Mailing Add:* Dept the Army SLCET-RR Ft Monmouth NJ 07703

HAKIM, MARGARET HEATH, b Lansing, Mich, Nov 29, 38; m 63; c 2. CLINICAL CHEMISTRY. *Educ:* Oberlin Col, AB, 60; Wayne State Univ, MS, 62, PhD(biochem), 66. *Prof Exp:* Res assoc, Wayne County Gen Hosp, 77-84; res coordr, Henry Ford Hosp, Detroit, 84-85; SUPVR, LIGAND ASSAY, ST JOHN CLIN LABS, DETROIT, MICH, 85- *Mem:* Clin Ligand Assay Soc; Midwest Radio Assay Soc. *Mailing Add:* 21245 Prestwick Harper Woods MI 48225

HAKIM, RAZIEL SAMUEL, b New York, NY, Mar 4, 47; m 69; c 2. DEVELOPMENTAL BIOLOGY, HISTOLOGY. *Educ:* Queens Col, BA, 67; Harvard Univ, MA, 68, PhD(biol), 72. *Prof Exp:* Teaching asst biol, Harvard Univ, 68-71; asst prof biol, Univ Ky, 71-77; asst prof, 77-84, ASSOC PROF ANAT, HOWARD UNIV, 84- *Concurrent Pos:* Prin investr, Nat Inst Child Health & Human Develop grants, 73-80; NIH grant, 80-84, NSF grant, 85-88. *Mem:* AAAS; Soc Develop Biol; Am Soc Zool; Sigma Xi; Am Asn Anatomists. *Res:* Problems relating to cellular differentiation, and cellular determination, using 2 insect model systems: salivary glands/midgut. *Mailing Add:* Dept Anat Rm 1105 Howard Univ 525 West St NW Washington DC 20059

HAKIMI, JOHN, b New York, NY, Dec 4, 50. IMMUNOPHARMACOLOGY. *Educ:* Univ NY, BA, 72; Univ Rochester, PhD(pharmacol), 77. *Prof Exp:* Fel, Dept Pharmacol, Sch Med, Univ Rochester, 77; fel develop biol & cancer, Albert Einstein Col Med, 79-82; sr scientist, 82-88, RES INVESTR, DEPT IMMUNOPHARMACOL, HOFFMANN-LA ROCHE, 88- *Concurrent Pos:* Lectr pharmacol nursing students, Sch Med, Univ Rochester, 72 & 77, lab instr pharmacol med students, 74; res fel, Leukemia Soc Am, 79-81. *Mem:* AAAS; Am Chem Soc; Am Soc Cell Biol; NY Acad Sci; Sigma Xi. *Res:* Immunopharmacology; numerous publications. *Mailing Add:* Dept Immunopharmacol Hoffman-La Roche Inc Roche Res Ctr 340 Kingsland St Nutley NJ 07110

HAKIMI, S(EIFOLLAH) L(OUIS), b Meshed, Iran, Dec 16, 32; nat US; m 65; c 3. ELECTRICAL ENGINEERING, COMPUTER SCIENCE. *Educ:* Univ Ill, BS, 55, MS, 57, PhD, 59. *Prof Exp:* Asst, Univ Ill, 55-58, asst prof elec eng, 59-61; from assoc prof to prof elec eng, indust eng, mgt sci & appl math, Northwest Univ, 61-86, chmn dept, 72-77; PROF & CHMN ELEC ENG & COMPUT SCI, UNIV CALIF, DAVIS, 86- *Mem:* Fel Inst Elec & Electronics Engrs; Sigma Xi; Soc Indust & Appl Math. *Res:* Applied graph theory; networks; coding theory; discrete optimization. *Mailing Add:* Dept Elec Eng & Comput Sci Univ Calif Davis CA 95616

HAKKILA, EERO ARNOLD, b Canterbury, Conn, Aug 4, 31; m 88; c 3. ANALYTICAL CHEMISTRY. *Educ:* Cent Conn State Col, BS, 53; Ohio State Univ, PhD(anal chem), 57. *Prof Exp:* Dep group leader, Safeguards Syst, 80-82, group leader, 82-83, PROG COORDINATOR, INT SAFEGUARDS, 87-; STAFF MEM ANALYTICAL CHEM, LOS ALAMOS NAT LAB, 57- *Mem:* Am Chem Soc; Am Nuclear Soc; Inst Nuclear Mat Mgt; fel Am Inst Chem; Sigma Xi. *Res:* X-ray absorption and emission; electron microprobe; analytical chemistry of U and Pu; chemistry of irradiated nuclear fuels; analytical chemistry in nuclear safeguards; ion microprobe analysis; international safeguards. *Mailing Add:* N-4 E-541 Los Alamos Nat Lab Los Alamos NM 87545

HAKKILA, JON ERIC, b Columbus, Ohio, June 27, 57; m 86; c 1. INTERMEDIATE-MASS PECULIAR ABUNDANCE STARS, SPATIAL DISTRIBUTIONS GAMMA RAY BURSTS. *Educ:* Univ Calif, San Diego, BA(physics), 80, BA(eng/am lit), 80, NMex State Univ, MS, 85, PhD(astron), 86. *Prof Exp:* Electromech technician I & II, Los Alamos Nat Lab, 78-79; teaching asst & eng aid physics, Univ Calif, San Diego, 79-80; res & teaching asst astron, Astron Dept, NMex State Univ, 80-81 & 83-86; math & sci programmer analyst, Comsysts Div, Sci Applications Int Corp, 81-83; asst prof astron, 86-89, ASSOC PROF ASTRON, MANKATO STATE UNIV, 89- *Concurrent Pos:* Lectr physics & astron, Univ San Diego, 82-83; NASA/ASEE summer fac fel, NASA, Space Sci Lab, Huntsville, Ala, 89 & 90; prin investr, Joint Venture Prog, NASA & Mankato State Univ, 90-93. *Mem:* Am Astron Soc; Int Astron Union. *Res:* Stellar kinematics and spatial distributions; infrared and visible photometry; peculiar-abundance stars; mass transfer in evolving binary star systems; spatial distributions of gamma ray bursts. *Mailing Add:* Dept Math Astron & Statist Mankato State Univ Mankato MN 56002-8400

HAKKINEN, RAIMO JAAKKO, b Helsinki, Finland, Feb 26, 26; nat US; m 49; c 2. AERODYNAMICS, FLUID DYNAMICS. *Educ:* Helsinki Univ Technol, Dipl, 48; Calif Inst Technol, MS, 50, PhD(aeronaut), 54. *Prof Exp:* Engr, Finnish Aeronaut Asn, 48; instr aeronaut & mech eng, Tampere Tech Col, Finland, 49; engr, Aeronaut Div, Valmet Corp, Finland, 49; asst gas dynamics, Calif Inst Technol, 50-53; mem res staff, Mass Inst Technol, 53-56, res engr, Missile & Space Systs Div, Douglas Aircraft Co, Inc, Calif, 56-64, chief scientist phys sci dept, 64-70, chief scientist flight sci, 70-82, DIR RES, MCDONNELL DOUGLAS RES LABS, 82- *Concurrent Pos:* Lectr, Univ Calif, Los Angeles, 57-59; vis assoc prof, Mass Inst Technol, 63-64. *Mem:* fel Am Inst Aeronaut & Astronaut; Am Phys Soc; Sigma Xi. *Res:* Fluid physics; boundary layers; experimental gas-dynamics. *Mailing Add:* Dept 222 Bldg 110 McDonnell Douglas Res Lab PO Box 516 St Louis MO 63166

HAKOMORI, SEN-ITIROH, b Sendai City, Japan, Feb 13, 29; m 56; c 3. BIOCHEMISTRY, IMMUNOCHEMISTRY. *Educ:* Tohoku Univ, Japan, MD, 52, DMedSc(biochem), 56. *Prof Exp:* Asst, Biochem Inst, Tohoku Univ, Japan, 55-56, asst prof, Med Sch, 57-59; prof & chief dept, Inst Cancer Res, Tohoku Col Pharmaceut Sci, 59-63; res assoc, Harvard Med Sch, 63-66; res assoc, Mass Gen Hosp, 63-66; Am Cancer Soc scholar, Brandeis Univ, 66-68; assoc prof prev med, Sch Med, 68-71, PROF PATHOBIOL & MICROBIOL, MED SCH & SCH PUB HEALTH, UNIV WASH, 71-, PROF IMMUNOL, 77- *Concurrent Pos:* Res fel biochem, Harvard Med Sch, 56-57; mem, Fred Hutchinson Cancer Res Ctr, 75-87; dir, Biomembrane Inst, 87- *Mem:* Am Asn Biol Chemists; Am Asn Immunologists; Am Asn Cancer Res. *Res:* Biochemistry and immunochemistry of glycoproteins and glycolipids. *Mailing Add:* Biomembrane Inst 201 Elliott Ave W Seattle WA 98119

HALABAN, RUTH, b Tel Aviv, Israel, Nov 30, 38; US citizen; m 63; c 2. GENETICS, DEVELOPMENTAL BIOLOGY. *Educ:* Hebrew Univ, Israel, BSc, 62, MSc, 64; Princeton Univ, PhD(molecular & develop biol), 68. *Prof Exp:* Instr biol, Princeton Univ, 68-69; res assoc develop biol, Brookhaven Nat Lab, 69-71; res assoc genetics, State Univ NY Albany, 71-73; res assoc, 73-81, SR RES ASSOC GENETICS, SCH MED, YALE UNIV, 81- *Res:* Hormonal regulation of growth and differentiation in cultured somatic cells; analysis of mutants and their hybrid cells; factors involved in transformation to malignancy of normal pigment cells. *Mailing Add:* Sch Med Yale Univ 333 Cedar St New Haven CT 06510

HALABISKY, LORNE STANLEY, b Ottawa, Ont, Jan 4, 45. APPLIED MATHEMATICS. *Educ:* Univ Alta, BSc, 66; Brown Univ, PhD(appl math), 70. *Prof Exp:* Asst prof math, Univ BC, 70-76; systs analyst, 76-81, SR SYSTS ANALYST, INS CO BC, 81- *Mem:* Soc Indust & Appl Math; Fedn Am Sci. *Res:* Evolution of finite disturbances in dissipative gas dynamics; boundary valve problems in dissipative gas dynamics. *Mailing Add:* 11791 Pintail Dr Richmond BC V7E 4N7 Can

HALABY, GEORGE ANTON, b Jaffa, Palestine, Mar 12, 38; US citizen. FOOD TECHNOLOGY, NUTRITION. *Educ:* Alexandria Univ, BS, 62; Columbia Univ, MS, 66; Univ Mass, Amherst, PhD(food technol), 71. *Prof Exp:* Chief chemist chem & pharm, Govt Cent Labs, Ministry of Health, Amman, Jordan, 62-65; res scientist food sci, John Morrell & Co, Chicago, 71-72; mgr, Stokely-Van Camp, Inc, Indianapolis, 72-79, dir, Res Develop & Lab Serv Food Sci, 79-85; vpres res & develop, Quaker Oaks Co, Barrington, 85-90; VPRES RES & DEVELOP, SARA LEE, 90- *Concurrent Pos:* WHO fel, Univ Alexandria, 59-61; Fulbright fel award, Columbia Univ, 65-66; NIH fel, Univ Mass, 67-71. *Mem:* Am Chem Soc; Inst Food Technologists; Am Oil Chemists Soc; fel Col Sports Med. *Res:* Nutrient analysis of foods; nutritional biochemistry; lipid chemistry; thermal processing canning and bacteriology; research and development of special dietary foods; exercise physiology; sports nutrition. *Mailing Add:* 740 Oxbow Lane Barrington IL 60010

HALABY, SAMI ASSAD, b Jerusalem, Palestine, Feb 14, 33; US citizen; m 57; c 2. PHYSICAL CHEMISTRY, COMPUTER SCIENCE. *Educ:* Mich State Univ, BSc, 53; Univ Cincinnati, PhD(phys chem), 59; Harvard Bus Sch, PMD, 69. *Prof Exp:* Asst chem, Univ Cincinnati, 54-57; sr chemist, Corning Glass Works, 59-64, mgr electronic mat res dept, 64-70, mgr tech planning, 70-71, res assoc, Tech Staff's Div, 71-75; res physicist, Am Optical Corp, 76-77; mgr systs technol, 78-84, mgr, microprocessor applns, 85-87, DIR MICROPROCESSOR APPLN & ENG DEVELOP, WARNER LAMBERT CO, 87- *Concurrent Pos:* Consult, Nat Inst Environ Health, 75-79; Twitchell fel, Univ Cincinnati, 57-58. *Mem:* Am Chem Soc; Sigma Xi; Inst Elec & Electronics Engrs. *Res:* Aqueous electrolyte solutions; thin film research; active electronic film devices; film structure; biomedical instrumentation; computer applications in manufacturing. *Mailing Add:* Warner-Lambert Co Consum Prod Res & Develop 175 Tabor Rd Morris Plains NJ 07950

HALACSY, ANDREW A, electrical engineering; deceased, see previous edition for last biography

HALARIS, ANGELOS, b Athens, Greece, Nov 30, 42; US citizen. BIOLOGICAL PSYCHIATRY, PSYCHOPHARMACOLOGY. *Educ:* Univ Munich, MD & PhD(med), 67. *Prof Exp:* Res asst neurochem, Max Planck Inst Psychiat, Munich, 67-69; fel psychiat, Univ Chicago, 71-74, asst prof, 74-78, assoc prof, 78-; DEPT PSYCHIAT, CASE-WESTERN RESERVE UNIV & CLEVELAND METRO GEN HOSP. *Concurrent Pos:* Found Fund Res Psychiat fel, 71-73; psychiat consult, Michael Reese Hosp & Med Ctr, 77- *Honors & Awards:* Physician's Recognition Award, AMA, 77. *Mem:* Am Col Neuropsychopharmacol; Collegium Int Psychopharmacologicum; Am Psychiat Asn; Am Soc Pharmacol & Exp Therapeut; Soc Neurosci. *Res:* Biochemistry of affective disorders and mode of action of psychotropic drugs. *Mailing Add:* Dept Psychiat Case-Western Reserve Univ Metro Gen Hosp 3395 Scranton Rd Cleveland OH 44109

HALARNKAR, PREMJIT P, b Bombay, India, Aug 31, 59; m. BIOCHEMISTRY. *Educ:* Univ Bombay, BSc, 79, MSc, 81, Univ Nev, PhD(biochem), 87. *Prof Exp:* Chemist, Paper Prod Ltd, Bombay, India, 81-82; res asst, Dept Biochem, Univ Nev, Reno, 83-87; fel, Dept Entom, Univ Calif, 87-89; FEL, DEPT BIOCHEM, UNIV NEV, 89- *Mem:* Am Soc Biochem & Molecular Biol; Am Chem Soc; Entom Soc Am. *Res:* Biochemistry; numerous publications. *Mailing Add:* Dept Biochem Univ Nev Reno NV 89557-0014

HALASA, ADEL F, b Madaba, Jordan, Dec 24, 33; US citizen; m 54; c 2. POLYMER BLENDS. *Educ:* Univ Okla, BS, 55; Butler Univ, MS, 59; Purdue Univ, PhD(chem), 64. *Prof Exp:* Group leader elastomer res, Firestone Tire & Rubber Co, 65-68, res assoc, 68-74, sr res assoc, 74-79; dir petrol & petrochem, Kuwait Inst Sci Res, 79-83; RES & DEVELOP FEL, ELASTOMER RES, GOODYEAR TIRE & RUBBER CO, 83- *Concurrent Pos:* Lectr, Purdue Univ, 62-64; chmn, Akron Polymer Lect Group, 74-75; prog dir cent regional meeting, Am Chem Soc, 76-77. *Mem:* Am Chem Soc; Am Inst Chemists; NY Acad Sci; AAAS; Sigma Xi. *Res:* Anionic polymerization for the preparation of novel elastomer with controlled micro and macro structure; studies in the area of miscible and immiscible blends of various elastomers and their effect on physical properties; studies in hydrogenetics of block copolymers; 150 US patents. *Mailing Add:* 5040 Everrett Rd PO Box 825 Bath OH 44210

HALASI-KUN, GEORGE JOSEPH, b Zagreb, Austria-Hungary, July 28, 16; US citizen; wid; c 2. HYDROLOGY, CARTOGRAPHY. *Educ:* Inst Technol, Budapest, MS, 38; Univ Foreigner, Italy, cert, 38; Slovak Tech Univ, Bratislava, CE, 49; Brunswick Tech Univ, DrEngSc(geohydrol), 68. *Prof Exp:* Hwy engr, Bd of Works, Lipt Mikulas, Czech, 46-48; dir water res & prof, Tech Univ Kosice & Col Water Eng, 48-53; mgr chief engr, Pozemne Stavby Construct Co, Kosice, 54-57; CHMN HYDROL, SEM POLLUTION & WATER RES, COLUMBIA UNIV, 58-; STATE TOPOG ENGR MAPPING & HYDROL, NEW JERSEY STATE, 71- *Concurrent Pos:* Mem regional planning comt, Dept Interior, Prague, Czech, 52-53; res assoc, Dept Civil Eng, Brunswick Tech Univ, 66-69; organizer & speaker, Int Conf on Pollution & Water Resources, 70-; adj prof dept life sci, NY Inst Technol, 71-76; vis prof dept environ resources, Rutgers Univ & Fairleigh Dickinson Univ, 76-; Nat Acad Sci fel, researcher geohydrol, WGer, Yugoslavia & Hungary, 77, 82 & 84; Fulbright Scholar, Hungary, Czech, 90-91. *Mem:* Fel Am Soc Civil Engrs; fel Geol Soc Am; Int Water Resources Asn; Am Cong Surv & Mapping; Am Water Resources Asn; Am Inst Hydrol. *Res:* Interdisciplinary program studies on pollution and water resources as an environmental problem; extreme surface flow of smaller watersheds including water storage capacity based on permeability of subsurface geology. *Mailing Add:* 31 Knowles Ave Pennington NJ 08534

HALASZ, NICHOLAS ALEXIS, b Budapest, Hungary, Mar 13, 31; US citizen; m 64; c 2. SURGERY, ANATOMY. *Educ:* Trinity Col, BS, 48; Yale Univ, MD, 54. *Prof Exp:* From instr to asst prof surg, Univ Calif, Los Angeles, 62-67; assoc prof, 67-70, PROF SURG & DIR KIDNEY TRANSPLANT UNIT & HEAD DIV ANAT, UNIV CALIF, SAN DIEGO, 70- *Concurrent*

Pos: Dir Am Bd Surg, 84; Markle Scholar in Acad Med, 64-69. *Mem:* Am Col Surg; Soc Exp Biol & Med; Soc Univ Surg; Tissue Cult Asn; Transplantation Soc; Am Surg Asn. *Res:* General, pediatric and thoracic surgery; transplantation immunology; antigen pretreatment; transplacental passage of antigen; organ preservation using perfusion, cooling and freezing. *Mailing Add:* Dept Surg Univ Calif San Diego CA 92103

HALAZON, GEORGE CHRIST, b Milwaukee, Wis, June 2, 19; m 47. BIOLOGY. *Educ:* Univ Wis, PhB, 42, MS, 50; Washington State Univ, PhD, 54. *Prof Exp:* Res asst, Wash State Col, 50-54; from asst prof to assoc prof, 54-81, PROF ZOOL, KANS STATE UNIV, 81-, WILDLIFE SPECIALIST, 54- *Mem:* Am Soc Mammal; Animal Behav Soc; Ecol Soc; Wilderness Soc; Wildlife Soc. *Res:* Ecology; physiology; behavior. *Mailing Add:* Div Biol Kans State Univ Ackert Hall Manhattan KS 66505-4901

HALBACH, KLAUS, b Wuppertal, Ger, Feb 3, 25; US citizen; m 45; c 1. PHYSICS. *Educ:* Univ Basel, PhD(physics), 54. *Prof Exp:* Lectr physics, Univ Fribourg, 54-56, Privatdozent, 56-57 & 59-60; res assoc, Stanford Univ, 57-59; mem staff, 60-80, SR STAFF MEM, LAWRENCE BERKELEY LAB, 80- *Concurrent Pos:* Consult, LANL, TRW & Varian. *Mem:* Am Phys Soc. *Res:* Accelerator technology; nuclear magnetic resonance; plasma physics. *Mailing Add:* 1492 Grizzly Peak Blvd Berkeley CA 94708

HALBERG, CHARLES JOHN AUGUST, JR, b Pasadena, Calif, Sept 24, 21; m 41; c 3. MATHEMATICS. *Educ:* Pomona Col, BA, 49; Univ Calif, Los Angeles, MA, 53, PhD(math), 55. *Prof Exp:* Instr math, Pomona Col, 49-50; assoc, Univ Calif, Los Angeles, 54-55; from instr to assoc prof, 55-68, PROF MATH, UNIV CALIF, RIVERSIDE, 68- *Concurrent Pos:* Lectr, Univ Calif, Los Angeles, 56-; NSF fel, Copenhagen Univ, 61-62; docent, Gothenburg Univ, 69-70. *Mem:* Am Math Soc. *Res:* Linear operator theory; spectral theory of bounded linear operators. *Mailing Add:* Dept Math Univ Calif 900 University Ave Riverside CA 92521

HALBERSTADT, MARCEL LEON, b St Raphael, France, June 4, 37; US citizen. AUTOMOTIVE EMISSIONS, AIR POLLUTION. *Educ:* City Col New York, BS, 58; Yale Univ, MS, 60, PhD(phys chem), 64. *Prof Exp:* Nat Acad Sci-Nat Res Coun res assoc, Nat Bur Standards, 63-65; asst prof chem, Univ Mo-St Louis, 65-72; proj phys chemist, Bendix Res Labs, 72-78; sr staff scientist, 78-80, mgr, Res & Anal Dept, 81-87, DIR, ENVIRON DEPT, MOTOR VEHICLE MFRS ASN, 87- *Mem:* AAAS; Am Chem Soc; Air Pollution Control Asn; Combustion Inst; Soc Automotive Engrs. *Res:* Emissions from internal combustion engines; motor vehicle emissions and air quality. *Mailing Add:* Motor Vehicle Mfrs Asn 7430 Second Ave Suite 300 Detroit MI 48202

HALBERSTAM, HEINI, b Most, Czech, Sept 11, 26; Brit citizen; m 50, 72; c 4. DEVELOPMENT AND APPLICATION OF SIEVE METHODS. *Educ:* London Univ, Univ Col, BSc, 46, MSc, 48, PhD(math), 52. *Prof Exp:* Lectr math, Univ Exeter, Eng, 49-57; reader, Royal Holloway Col, Univ London, 57-62; Erasmus Smith prof math, Trinity Col, Dublin, 62-64; prof, Univ Nottingham, Eng, 64-80; prof & head, dept math, 80-88, PROF MATH, UNIV ILL, URBANA-CHAMPAIGN, 88- *Concurrent Pos:* Vis instr, Brown Univ, RI, 55-56; vis prof, Univ Mich, Ann Arbor, 66 & Univ Tel Aviv, Israel, 73; co-prin investr, NSF res grants in number theory, 81-89. *Honors & Awards:* Fel, Univ Col London, 87. *Mem:* Am Math Soc. *Res:* Analytic and combinational number theory, especially in sieve theory. *Mailing Add:* Math Dept 1409 W Green St Urbana IL 61801

HALBERT, EDITH CONRAD, b New York, NY, Apr 23, 31; m 51; c 3. THEORETICAL NUCLEAR PHYSICS. *Educ:* Cornell Univ, AB, 51; Univ Rochester, PhD, 57. *Prof Exp:* PHYSICIST, OAK RIDGE NAT LAB, 56- *Mem:* Fel Am Phys Soc. *Res:* Heavy ion collisions, nuclear structure, effective interactions. *Mailing Add:* Oak Ridge Nat Lab Box 2008 MS6374 Oak Ridge TN 37831

HALBERT, MELVYN LEONARD, b Philadelphia, Pa, Aug 12, 29; m 51; c 3. NUCLEAR PHYSICS. *Educ:* Cornell Univ, AB, 50; Univ Rochester, PhD(physics), 55. *Prof Exp:* PHYSICIST, OAK RIDGE NAT LAB, 55- *Concurrent Pos:* Exchange scientist, Brookhaven Nat Lab, 71-72 & Niels Bohr Inst, Copenhagen, 74-75. *Mem:* Fel Am Phys Soc; Sigma Xi. *Res:* Heavy-ion nuclear physics; mesonic x-rays; cross-section fluctuations; nuclear spectroscopy; nucleon-nucleon forces; few-nucleon problems. *Mailing Add:* Oak Ridge Nat Lab PO Box 2008 MS-6368 Oak Ridge TN 37831-6368

HALBERT, SEYMOUR PUTTERMAN, b Philadelphia, Pa, Mar 20, 17; m 52. MEDICAL MICROBIOLOGY, IMMUNOLOGY. *Educ:* Univ NC, BA, 37; Johns Hopkins Univ, MD, 41. *Prof Exp:* Intern, Long Island Col Hosp, 41-42; assoc bact, Sch Med, Univ Pa, 42-45; asst surgeon, NIH, 45-46; asst prof exp med, Sch Pub Health, Univ NC, 46-49; assoc prof microbiol, Col Physicians & Surgeons, Columbia Univ, 49-60, prof ophthal res, 60-65; prof pediat, Sch Med, Univ Miami, 65-78; vpres res & develop, Cordis Labs, 68-89; RETIRED. *Concurrent Pos:* Guggenheim fel, 56; Helen Hay Whitney fel, 56. *Mem:* AAAS; Am Soc Microbiol; Am Chem Soc; Asn Res Vision & Ophthal; Harvey Soc. *Res:* Pneumococcal hemolysin; Shigella vaccines; oil-emulsion vaccine adjuvants; antibiotic-producing intestinal bacteria and ocular flora; streptolysin O; analysis of streptococcal infections; immunology of ocular tissues; biochemical evolution of proteins; pharmacology of bacterial toxins; cystic fibrosis; cardiac auto-immune system; plasma proteins of pregnancy; enzyme labeled immunoassays. *Mailing Add:* 12450 Rock Garden Lane Miami FL 33156

HALBERT, SHERIDAN A, b Shelton, Wa, July 3, 43. REPRODUCTIVE BIOLOGY, INFERTILITY. *Educ:* Univ Wash, BS, 65, PhD(physiol & biophys), 72. *Prof Exp:* Fel, Ctr Bioeng, 73, res assoc, dept biol structure, 74, actg asst prof, 74-76, asst prof, 76-81, ASSOC PROF BIOENG & BIOL STRUCTURE, UNIV WASH, 81- *Concurrent Pos:* Res affiliate, Regional Primate Res Ctr, Univ Wash, 77- *Mem:* Am Asn Anatomists; Soc Study Reproduction. *Res:* Structure and function of the mammalian fallopian tube in normal animals and in animal models representing disease states associated with infertility. *Mailing Add:* Alden Analytical Labs Inc 1001 Kliktitat Way SW Suite 108 Seattle WA 98134

HALBERT, THOMAS RISHER, b Schenectady, NY, July 20, 50; m 73; c 3. SOLID STATE CHEMISTRY. *Educ:* Rutgers Univ, BA, 72; Stanford Univ, PhD(org chem), 77. *Prof Exp:* Res chemist, 77-79, sr chemist, 79-81, STAFF CHEMIST, SOLID STATE CHEM, CORP RES LABS, EXXON RES & ENG CO, 81-; PROJ LEADER, ADVAN CATALYST & HEAVY FEED UPGRADING, EXXON RES & DEVELOP LABS, 88- *Mem:* Am Chem Soc. *Res:* Transition metal-sulfur cluster chemistry; hydrotreating catalysis; solid state inorganic chemistry; intercalation chemistry; new materials with applications in energy related fields; organometallic chemistry and bioinorganic chemistry. *Mailing Add:* Exxon Res & Develop Labs PO Box 2226 Baton Rouge LA 70822

HALBIG, JOSEPH BENJAMIN, b Sparta, Ill, July 19, 38; m 60, 80; c 2. GEOCHEMISTRY, ENVIRONMENTAL GEOLOGY. *Educ:* Southern Ill Univ, Carbondale, BS, 62; Pa State Univ, MS, 65, PhD(geochem), 69. *Prof Exp:* Asst prof, State Univ NY Col Geneseo, 69-75; assoc prof, 80-85, PROF GEOL, UNIV HAWAII, HILO, 85- *Mem:* Sigma Xi; Int Asn Geochem & Cosmochem; Soc Environ Geochem & Health; Soc Archeol Sci. *Res:* Geochemistry of natural materials; environmental geology; geoarchaeology. *Mailing Add:* Geol Dept Col Arts & Sci Hilo HI 96720-4091

HALBLEIB, JOHN A, b Louisville, Ky, Aug 29, 36. RADIATION TRANSPORT. *Educ:* Univ Louisville, BS, 58; Carnegie-Mellon Univ, MS, 64, PhD(physics), 66. *Prof Exp:* MEM TECH STAFF, SANDIA NAT LABS, 66- *Mem:* Am Phys Soc. *Res:* Methods in Monte Carlo radiation transport. *Mailing Add:* 12428 Chelwood Pl NE Albuquerque NM 87112

HALBOUTY, MICHEL THOMAS, b Beaumont, Tex, June 21, 09; m 81; c 1. PETROLEUM GEOLOGY. *Educ:* Tex A&M Univ, BS, 30, MS, 31, Geol Eng, 56. *Hon Degrees:* DEng, Mont Col Mineral Sci & Technol, 66; PhD, USSR Acad Sci, 90. *Prof Exp:* Geologist & petrol engr, Yount-Lee Oil Co, 31-35; chief geologist & petrol engr, Glenn H McCarthy, 35-37; consult geologist & petrol engr, Halbouty Ctr, 37-81, CHMN & CHIEF EXEC OFFICER, MICHEL T HALBOUTY ENERGY CO, 81- *Concurrent Pos:* Distinguished lectr, Soc Petrol Engrs, Am Inst Mining & Petrol Engrs, 64, rep, Div Earth Sci, Nat Res Coun, Nat Acad Sci, 65-74; distinguished lectr, Am Asn Petrol Geologists, 65-66, sr ed, Geol Giant Petrol Fields, 68-69, spec educ for publ, 68-; chmn comt technol & water, Nat Acad Sci, 70-71; mem, Nat Energy Study Comt, 70-71 & Gov Energy Adv Coun, State of Tex, 74-; adj prof dept geosci, Tex Tech Univ, 72-; chmn & pres, Circum-Pac Coun Energy & Mineral Resources, 72- *Honors & Awards:* Anthony F Lucas Gold Medal, Am Inst Mining, Metall & Petrol Engrs, 75; Human Needs Award, Am Asn Petrol Geologists, 75; Sidney Powers Mem Award, Am Asn Petrol Geologists, 77; William T Pecora Award, NASA, 77; Horatio Alger Award, Am Schs & Cols Asn, 78; Hoover Medal, Am Asn Eng Soc, 82. *Mem:* Nat Acad Eng; Am Inst Mining, Metall & Petrol Engrs (vpres, 66-67); Geol Soc Am; Soc Independent Prof Earth Scientists; Am Asn Petrol Geologists (pres, 66-67). *Res:* Salt dome geology of Gulf Coast region; remote sensing. *Mailing Add:* Halbouty Ctr 5100 Westheimer Rd Houston TX 77056

HALBRENDT, JOHN MARTHON, b Chicago, Ill, Feb 3, 49; div; c 2. NEMATOLOGY, MYCOLOGY. *Educ:* Southern Ill Univ, BS, 72, MS, 74; Univ Mo, PhD(plant path), 85. *Prof Exp:* Teacher biol & advan chem, Stavanger Am Sch, Norway, 74-76; teacher biol & advan biol, Frankfurt Int Sch, WGer, 76-78 & Cairo Am Col, Egypt, 78-80; res asst, dept plant path, Univ Mo, 80-84, res assoc, 85; res scientist, dept plant path, Clemson Univ, 85-88; ASST PROF, PA STATE UNIV, 88- *Mem:* Am Inst Biol Sci; Soc Nematologists; Sigma Xi. *Res:* Fruit crop pathology-diseases caused by nematodes; nematode/virus interaction; graft-transmissible agents; biology of plant parasitic nematodes. *Mailing Add:* Pa State Fruit Res Lab PO Box 309 Biglerville PA 17307

HALDAR, DIPAK, b Bankura, India, Dec 8, 37; US citizen; m 63; c 2. CELL BIOLOGY. *Educ:* Univ Calcutta, BSc, 56, MSc, 58, DPhil(biochem), 63; Univ London, PhD(biochem), 66. *Prof Exp:* Res fel biochem, Indian Inst Biochem & Exp Med, Univ Calcutta, 59-63; mem sci staff, Nat Insts Med Res, London, 63-66; fel, McMaster Univ, 66-69; asst mem biochem, St Jude Children's Res Hosp, 69-71; lectr, Calcutta Univ, 71-73; vis res investr, Pub Health Res Inst City of New York, 74-75; from asst prof to assoc prof, 75-83, PROF BIOL SCI, ST JOHN'S UNIV, NY, 83- *Concurrent Pos:* Vis asst prof, Memphis State Univ, 70-71; vis prof cell biol, Med Ctr, NY Univ, 85. *Mem:* NY Acad Sci; AAAS; Am Chem Soc; Sigma Xi; Am Soc Biochem & Molecular Biol; Am Soc Cell Biol. *Res:* Mitochondria; structure and function of mitochondrid outer membrane; phospholipid metabolism. *Mailing Add:* Dept Biol Sci St John's Univ Jamaica NY 11439

HALDAR, JAYA, b Calcutta, India, Apr 23, 39; US citizen; m 63; c 2. NEUROHYPOPHYSICAL HORMONES. *Educ:* Nat Inst Med Res, London, Eng, PhD(physiol), 66. *Prof Exp:* Asst prof pharmacol, Columbia Univ, 77-83; assoc prof, 83-89, PROF BIOL & NEUROPHYSIOL, ST JOHNS UNIV, 89- *Mem:* NY Acad Sci; Am Physiol Soc; Soc Neurosci; Endocrine Soc; Sigma Xi; Brit Brain Res Asn; Europ Brain & Behav Soc. *Res:* Neuro-endocrinology. *Mailing Add:* Dept Biol Sci St Johns Univ Grand Central & Utopia Pkwys Jamaica NY 11439

HALDE, CARLYN JEAN, b Calif, June 16, 24. MEDICAL MYCOLOGY. *Educ:* Univ Calif, Los Angeles, BA, 45, MA, 47; Duke Univ, PhD(microbiol), 53. *Prof Exp:* Lectr bact, Univ Hawaii, 48; bacteriologist, Hawaiian Med Lab, US Dept Army, 48-50, Fulbright scholar, Philippines, 50-51; res asst, Sch Med, Stanford Univ, 53-55; jr res microbiologist, Div Dermat, Sch Med, Univ Calif, Los Angeles, 55-57, asst res microbiologist, 57-58, vis asst prof med educ, Univ Proj at Univ Indonesia, 58-60; from asst prof to assoc prof, 60-86, PROF MICROBIOL, SCH MED, UNIV CALIF, SAN FRANCISCO, 86-

Concurrent Pos: Giannini Found fel, 54; La State Univ-Inter-Am fel, 58. *Mem:* AAAS; Am Soc Microbiol; Mycol Soc Am; Int Soc Human & Animal Mycol; Soc Women Geogr. *Mailing Add:* Dept Microbiol Univ Calif San Francisco CA 94143

HALDEN, FRANK, b Marshalltown, Iowa, Mar 29, 29; m 54; c 4. CERAMICS, PHYSICS. *Educ:* Iowa State Col, BS, 51; Mass Inst Technol, ScD(ceramics), 54. *Prof Exp:* Res asst ceramics, Ames Lab, AEC, 48-51; asst metal-ceramic interactions, Mass Inst Technol, 51-54; sr ceramist, Stanford Res Inst, 56-65, dir ceramics & metall div, 65-69, assoc dir mat lab, 69-72; vpres res & develop, 72-80, DIR, CRYSTAL TECHNOL, INC, 67-, EXEC VPRES, 80- *Mem:* Am Ceramic Soc; Am Asn Crystal Growers. *Res:* High temperature materials, synthesis, fabrication and evaluation; refractory coatings; ultra-high-purity materials; growth and evaluation of refractory single crystals; ceramic materials for the electronic industry; growth, manufacturing and applications of optical acoustic, electro-optic single crystals. *Mailing Add:* 822 Cascade Dr Sunnyvale CA 94087

HALDER, NARAYAN CHANDRA, b Ranchi, India, Aug 9, 39; m 64; c 2. SOLID STATE PHYSICS. *Educ:* Univ Bihar, BS, 58, MS, 60; Indian Inst Technol, Kharagpur, PhD(physics), 63. *Prof Exp:* Assoc lectr physics, Indian Inst Technol, Kharagpur, 63-64; res assoc mat sci, Yale Univ, 65-67; asst prof physics, State Univ NY Albany, 67-72; assoc prof, 72-75, PROF PHYSICS, UNIV S FLA, TAMPA, 75- *Concurrent Pos:* NY State Res Found grant, 68-71; NSF grants, 72 & 75, USAF grants; Res Award, Sigma Xi, 78. *Mem:* Am Vacuum Soc; fel Am Phys Soc; AAAS; Sigma Xi. *Res:* Line broadening effects in x-ray patterns from solids; electronic properties and tunneling in metal-oxide-ceramics; study of defects in semiconductors and devices. *Mailing Add:* Dept Physics Univ SFla Tampa FL 33620

HALE, BARBARA NELSON, b Buffalo, NY, Apr 6, 38; m 64. PHYSICS. *Educ:* Syracuse Univ, BS, 60; Purdue Univ, MS, 64, PhD(physics), 67. *Prof Exp:* Asst prof physics, Rochester Inst Technol, 68-69; vis asst prof math, Univ Mo-Rolla, 69-71, res assoc physics, 71-73, assoc prof, 73-82, PROF PHYSICS, UNIV MO-ROLLA, 83- *Concurrent Pos:* NSF res grant, 72-79; NASA res grant, 76-79; NSF res grant, 80-83, 84-87 & 88-91. *Mem:* Am Phys Soc; Am Asn Physics Teachers; Am Meteorol Soc; Sigma Xi. *Res:* Atmospheric physics; theory of nucleation phenomena; molecular modeling of pre-nucleation water clusters; growth and nucleation of ice; computer simulations of water on substrates. *Mailing Add:* Dept Physics Univ MO PO Box 249 Rolla MO 65401

HALE, CECIL HARRISON, b Kilgore, Tex, June 12, 19; m 45; c 3. CHEMISTRY. *Educ:* Trinity Univ, Tex, BS, 38; La State Univ, MS, 40; Purdue Univ, PhD(phys chem), 48. *Prof Exp:* Asst chem, La State Univ, 38-40; asst chem, Purdue Univ, 46-48; anal chemist, Esso Labs, Standard Oil Develop Co, 40-45, res chemist, 48-50; owner, Southwestern Anal Chem Inc, 50-84; RETIRED. *Mem:* Am Chem Soc; fel Am Inst Chem. *Res:* Polarography; spectrophotometry; analytical chemical methods; electrochemistry. *Mailing Add:* Southwestern Analytical Chem Inc 821 E Woodward Austin TX 78704

HALE, CREIGHTON J, b Hardy, Nebr, Feb 18, 24. EXERCISE PHYSIOLOGY. *Educ:* Colgate Univ, BA, 48; Springfield Col, MS, 49; NY Univ, PhD(educ), 51. *Prof Exp:* Instr physiol, Springfield Col, 49-51, from asst prof to assoc prof, 51-55; dir res & vpres, 55-72, PRES, LITTLE LEAGUE BASEBALL, INC, 73- *Concurrent Pos:* Mem, Comt Helmets, Nat Res Coun-Nat Acad Sci. *Honors & Awards:* Merit Award, Am Soc Testing & Mat. *Mem:* Am Asn Health, Phys Educ & Recreation; Am Soc Testing & Mat; fel Am Col Sports Med; affil Am Orthop Soc Sports Med. *Res:* Child growth and development and various phases of athletics; development of safety equipment. *Mailing Add:* PO Box 3485 Williamsport PA 17701

HALE, EDWARD BOYD, b Washington, DC, July 16, 38; m 64. SOLID STATE PHYSICS & MATERIAL SCIENCE. *Educ:* Univ Md, BSEE, 60; Purdue Univ, PhD(physics), 68. *Prof Exp:* Res assoc spin resonance, dept physics & astron, Univ Rochester, 68-69; from asst prof to assoc prof, 69-82, res assoc, 72-82, PROF PHYSICS, UNIV MO-ROLLA, 82-, ASSOC DIR, MAT RES CTR, 88- *Mem:* AAAS; Am Phys Soc; Am Asn Physics Teachers; Am Vacuum Soc; Mat Res Soc. *Res:* Auger, ESCA and SEM surface studies in solids; ion implantation studies in solids; electron-nuclear double resonance studies; defects in semiconductors; spin resonance studies. *Mailing Add:* Dept Physics Univ Mo Rolla MO 65401

HALE, FRANCIS JOSEPH, b Manila, Philippines, Oct 24, 22; US citizen; m 49, 80; c 3. MECHANICAL & AEROSPACE ENGINEERING. *Educ:* US Mil Acad, BS, 44; Mass Inst Technol, SM, 52, ScD(aeronaut, astronaut), 63. *Prof Exp:* Dep dir, Ballistics Missile Div, US Air Force, 56-59, prof astronaut & head dept, Air Force Acad, 62-63, chief anal group, Directorate Develop Plans, Washington, DC, 63-65; from assoc prof to prof, 65-89, EMER PROF MECH & AEROSPACE ENG, NC STATE UNIV, 89- *Concurrent Pos:* Prof lectr, George Washington Univ, 63-65; vis prof mech eng, Middle East Tech Univ, Ankara, Turkey, 73-74; vis prof mech, US Mil Acad, West Point, 77-78; consult, Int Civil Aviation Orgn, Brazil; consult to indust & govt; tech dir, Wrightsville Beach Test Facil, 81-82; secy of naval fel aerospace eng, US Naval Acad, Annapolis, 90-91. *Honors & Awards:* Ralph R Teetor Educ Award, Soc Automotive Engrs, 85. *Mem:* Assoc fel Am Inst Aeronaut & Astronaut; Am Soc Eng Educ; Am Soc Mech Engrs; Soc Automotive Engrs. *Res:* Dynamic analysis and control with emphasis on applications to flight vehicles and processes. *Mailing Add:* 2853 Rue Sans Famille Raleigh NC 27607-3048

HALE, HARRY W(ILLIAM), b Winslow, Ind, Sept 15, 20; m 85; c 1. ELECTRICAL ENGINEERING. *Educ:* Purdue Univ, BS, 42, MS, 49, PhD(elec eng), 53. *Prof Exp:* From instr to assoc prof elec eng, Purdue Univ, 48-58; prof & head dept, Wayne State Univ, 58-60; PROF ELEC ENG, IOWA STATE UNIV, 60- *Mem:* Inst Elec & Electronics Engrs. *Res:* Network theory; systems. *Mailing Add:* Dept Elec Eng Iowa State Univ Ames IA 50011

HALE, HARRY W, JR, b New York, NY, Feb 3, 17; m 46; c 5. SURGERY. *Educ:* Rensselaer Polytech Inst, BS, 38; Univ Rochester, MD, 43. *Prof Exp:* From instr to prof surg, Sch Med, State Univ NY Buffalo, 52-69; chmn dept surg, 69-86, CHIEF GEN SURG, MARICOPA COUNTY HOSP, 87- *Concurrent Pos:* From asst attend surgeon to attend surgeon, E J Meyer Mem Hosp, 52-, assoc dir surg, 59-; Buswell res fel, State Univ NY Buffalo, 56-59; clin prof surg, Univ Ariz, 78- *Mem:* AMA; Am Asn Surg of Trauma; Am Col Surg; Soc Surg Alimentary Tract; Am Burn Asn; Sigma Xi. *Res:* Surgery of trauma; elderly surgery; surgical bacteriology. *Mailing Add:* Dept Surg Maricopa County Hosp PO Box 5099 Phoenix AZ 85010

HALE, JACK KENNETH, b Dudley, Ky, Oct 3, 28; m 49. MATHEMATICS. *Educ:* Berea Col, AB, 49; Purdue Univ, MS, 51, PhD(math), 53. *Hon Degrees:* DSc, Univ Ghent, Belgium, 79, Stuttgart, 88. *Prof Exp:* Instr math, Purdue Univ, 53-54; mem staff, Sandia Corp, 54-57; mem staff, Univac, Remington Rand Div, Sperry Rand Corp, Minn, 57-58; mem staff, Res Inst Advan Study, Martin Co, 58-64; prof appl math, Brown Univ, 64-89; PROF APPL MATH, GA TECH, 88- *Concurrent Pos:* Guggenheim fel, 79-80. *Honors & Awards:* Chauvenet Prize, Am Math Asn, 65. *Mem:* Am Math Soc; fel Nat Acad Mechs; Sigma Xi; corresp mem Brazilian Acad Sci. *Res:* Ordinary, functional and partial differential equations. *Mailing Add:* Sch Math Ga Tech Atlanta GA 30332

HALE, JOHN, b Chicago, Ill, Nov 13, 21; m 48, 79; c 2. RADIOLOGICAL PHYSICS. *Educ:* Duke Univ, AB, 43; Univ Pa, MS, 49, PhD(elec eng), 57; Am Bd Radiol, dipl, 51; Am Bd Health Physics, dipl, 60. *Prof Exp:* Asst physics, Duke Univ, 43; res asst eng, Motorola, Inc, 46-47; asst elec eng, Univ Pa, 48-49, radiation safety physicist, 49-50, from asst prof to prof radiol physics, Med Sch, 57-67, prof elec eng, Moore Sch Elec Eng, 67-74, prof bioeng, Sch Eng & Appl Sci, 74-87, prof, 67-87, emer prof radiol physics, Med Sch, 87-90; RETIRED. *Concurrent Pos:* Am ed, Physics in Med & Biol, 69-76. *Mem:* Am Asn Physicists Med (pres, 61-62); Radiol Soc NAm; Radiation Res Soc; Am Radium Soc; Am Col Radiol. *Res:* Radiation dosimetry and health physics; radiologic physics. *Mailing Add:* 4900 Homestead Littleton CO 80123-1529

HALE, JOHN DEWEY, b Salt Lake City, Utah, Mar 26, 37. INORGANIC & PHYSICAL CHEMISTRY. *Educ:* Brigham Young Univ, BS, 57, PhD(inorg phys), 63. *Prof Exp:* Teaching fel, Oxford Univ, Eng, 63-65; res group leader chem res & develop, Tech Ctr, Kerr-McGee Corp, 67-69, mgr bus anal corp develop & mkt res, 69-72, sr planning analyst corp planning & econ, 72-75, mgr technol develop new process & mining tech, 75-76, mgr res & develop, Indust Sci & Eng Tech Ctr, 76-79, mgr tech ctr, Corp Res & Develop Orgn, 79-82, vpres tech opers, 83-85, res, 85-88, tech eval & planning, 88-90, VPRES PROD RES & DEVELOP, KERR-MCGEE CORP, 90- *Concurrent Pos:* NIH fel, Waldham Col, Oxford Univ, 63-65. *Mem:* Am Chem Soc; Sigma Xi; Am Inst Chem Engrs. *Res:* Directing of industrial research and development through chemical, metallurgical, engineering and economic specialties; process development from ores and other raw materials to bulk and specialty chemical products. *Mailing Add:* Kerr-McGee Corp PO Box 25861 Oklahoma City OK 73125

HALE, KIRK KERMIT, JR, b Bentonville, Ark, Nov 20, 40; m 61; c 2. POULTRY SCIENCE, FOOD SCIENCE. *Educ:* Univ Ark, Fayetteville, BSA, 63, MS, 67; Purdue Univ, PhD(food sci), 70. *Prof Exp:* Asst prof poultry prod technol, Univ Ga, 70-76; ASSOC PROF POULTRY SCI, CLEMSON UNIV, 76- *Mem:* Poultry Sci Asn; Inst Food Technologists; Am Meat Sci Asn. *Res:* Development and improvement of poultry products; harvesting and processing of poultry and eggs. *Mailing Add:* Egg Technol Inc 6180 Harbor Rd Port Orange FL 32119

HALE, LEONARD ALLEN, b Troup, Tex, Feb 17, 37; m 56; c 5. MECHANICAL ENGINEERING. *Educ:* Tex Tech Col, BS, 59, MS, 61; Univ Tex, PhD(mech eng), 64. *Prof Exp:* Instr mech eng, Tex Tech Col, 59-60; asst prof, Lamar State Col, 64-66; from asst prof to assoc prof, 66-77, PROF MECH ENG, TEX A&M UNIV, 77- *Mem:* Am Soc Mech Engrs; Am Soc Eng Educ. *Res:* Energy utilization; heat transfer. *Mailing Add:* Sch Eng Gonzaga Univ E 502 Boone Ave Spokane WA 99258

HALE, MARTHA L, b San Antonio, Tex, Sept 6, 46. VIROLOGY, HEMATOLOGY. *Educ:* Va Commonwealth Univ, PhD(microbiol), 74. *Prof Exp:* Staff fel, LMG, Nat Inst Child Health & Human Develop, NIH, 74-77; sr assoc, Nat Res Coun, 80-82; PRIN INVESTR, ARMED FORCES RADIOBIOL RES INST, 82- *Mem:* Am Asn Microbiol; Am Asn Immunologists. *Res:* Isolation and characterization of the multipotent hematopoietic stem cell; investigation into the differentation of the hematopoietic stem cell. *Mailing Add:* Armed Forces Radiobiol Res Inst Bldg 42 Palmer Rd S Bethesda MD 20814-5145

HALE, MASON ELLSWORTH, JR, botany; deceased, see previous edition for last biography

HALE, MAYNARD GEORGE, b Mentor, Ohio, Apr 5, 20; m 43; c 2. PLANT PHYSIOLOGY. *Educ:* Ohio State Univ, BS, 47, MS, 49, PhD(bot), 51. *Prof Exp:* Asst bot, Ohio State Univ, 47-51; assoc prof, 51-85, EMER PROF PLANT PHYSIOL, VA POLYTECH INST & STATE UNIV, 85- *Mem:* Am Soc Plant Physiologists; Sigma Xi. *Res:* Procedures for culturing groups of plants under aseptic conditions; effects of chemicals on root exudations from aseptic plants; alleopathic potential of plants which interact through root exudations; plant growth. *Mailing Add:* 301 Apperson Dr Blacksburg VA 24060

HALE, RAYMOND JOSEPH, b Watertown, NY, Feb 23, 18; m 42; c 5. CHEMICAL ENGINEERING. *Educ:* Pratt Inst, Cert, 39. *Prof Exp:* Engr, Chemipulp Corp, 40; chemist res lab, St Regis Paper Co, 40-42; tech supvr, Ala Ord Works, E I du Pont de Nemours & Co, 42-44, chemist, Clinton Labs, 44, supvr, Hanford Works, 44-46; area supvr, Hanford Works, Gen Elec Co, 46-49, contact engr, 49-51; tech asst, Atomic Energy Div, E I du Pont de

Nemours & Co, 51-52, supvr design, Savannah River Lab, 52-63, chief supvr design develop, 63-76; exec dir, Am Eng Model Soc, 76-89; RETIRED. *Mem:* Am Eng Model Soc. *Res:* Three dimensional scale models in design and construction. *Mailing Add:* 117 Grace Circle SW Aiken SC 29801

HALE, ROBERT E, b North Vernon, Ind, Jan 4, 29; m 53; c 4. PHYSICS. *Educ:* Ball State Univ, BS, 51, EdD(educ admin), 73; Western Mich Univ, MA, 60. *Prof Exp:* Teacher, White Pigeon Community Schs, 56-60; teacher, Tucson High Sch, 60-61; from asst prof to assoc prof, 61-73, PROF PHYS SCI, HUNTINGTON COL, 73-, CHMN DIV SCI, 67- *Mem:* Am Asn Physics Teachers; Nat Sci Teachers Asn. *Res:* Aviation weather. *Mailing Add:* Dept Nat & Math Scis Huntington Col 2303 Col Ave Huntington IN 46750

HALE, RON L, b Maud, Okla, Dec 6, 42; m 71; c 2. ORGANIC CHEMISTRY, RADIOCHEMISTRY. *Educ:* Okla State Univ, BS, 64, MS, 66; Ga Inst Technol, PhD(org chem), 68. *Prof Exp:* Sr res chemist synthesis radiolabeled org compounds, Hoffmann-La Roche, Inc, 71-73; mgr radiochem synthesis radiolabeled org compounds, Dynapol, 73-80, assoc dir, chem synthesis, 80-82; sect mgr, Clin Assays Div, Baxter Healthcare Corp, 82-86; PRES, METAFLUOR, INC, 86- *Concurrent Pos:* NIH fel, Stanford Univ, 68-71. *Mem:* Am Chem Soc; AAAS; Am Asn Clin Chem. *Res:* Organic synthesis of isotopically labeled food additives; pharmaceuticals; natural products and other biologically active compounds and synthetic polymers for the study of their absorption; metabolism and excretion in biological systems; fluorescent rare earth chelates for immunodiagnostics. *Mailing Add:* 110 C Pioneer Way Mountain View CA 94041

HALE, WARREN FREDERICK, b Cambridge, Mass, Aug 1, 29; m 54; c 2. POLYMER CHEMISTRY. *Educ:* Northeastern Univ, BS, 52; Polytech Inst Brooklyn, MS, 54; Univ Md, PhD(org chem), 58. *Prof Exp:* Chemist plastics res, 58-66, proj scientist chem & plastics, 66-71, develop scientist, Plastics Div, 71-75, group leader, 75-80, asst to dir, 80-82, MGR FDA LIAISON, UNION CARBIDE CORP, 82- *Mem:* Am Chem Soc; Sigma Xi; Royal Soc Chem. *Res:* Monomer synthesis; polyethers; water miscible polymers, adhesives and sealants. *Mailing Add:* 94 Dreahook Rd Whitehouse Station NJ 08889

HALE, WILLIAM HARRIS, b Richmond, Ky, Feb 11, 20; m 45; c 5. ANIMAL NUTRITION. *Educ:* Univ Ky, BS, 46, MS, 47; Univ Wis, PhD(biochem, animal husb), 50. *Prof Exp:* Asst animal husb, Univ Wis, 47-50; asst prof, Univ Ill, 50-52 & Iowa State Univ, 52-57; nutritionist, Chas Pfizer & Co, Inc, 57-60; PROF ANIMAL SCI, UNIV ARIZ, 60- *Mem:* AAAS; Am Soc Animal Sci; Am Inst Nutrit; Soc Exp Biol & Med; Am Chem Soc; Sigma Xi. *Res:* Ruminant nutrition; cobalt and sulphur in sheep nutrition; forage utilization by ruminants; synthetic estrogens in growth of ruminants; nutrition of rumen microorganisms; purified rations for ruminants; vitamin A; grain processing for ruminants; fat utilization and metabolism by ruminants. *Mailing Add:* 7412 N Ellison Dr Tucson AZ 85704

HALE, WILLIAM HENRY, JR, b Ft Smith, Ark, Aug 18, 40; m 62; c 3. CHEMICAL ENGINEERING, INORGANIC CHEMISTRY. *Educ:* Univ Tex, BS, 63; Univ Calif, Berkeley, PhD(chem), 66. *Prof Exp:* Res chemist, E I du Pont de Nemours & Co, Inc, 66-71, sr res supvr, 71-75, res mgr, Savannah River Lab, 75-76, tech supt, Beaumont Works, 76-78, prof supt, Cape Fear Plant, 78-79, tech mgr polyester & acrylic intermediates, 79-81, dir, Data Ctr Opers Div, 81-86, dir, Appliances Syst Div, 86-88; VPRES, CONOCO INFO SYSTS, CONOCO, INC, 88- *Res:* Sulfur-fluorine chemistry; lanthanide complexes; separations chemistry and technology; pressurized ion exchange. *Mailing Add:* Conoco Info Systs Conoco Inc PO Box 1267 Ponca City OK 74603

HALEMANE, THIRUMALA RAYA, b Ednad, India, May 27, 53; m; c 2. TELECOMMUNICATIONS, SOFTWARE SYSTEMS. *Educ:* Bangalore Univ, BSc, 72; Indian Inst Technol, Madras, MSc, 74; Univ Rochester, MA, 76, MA, 80, PhD(physics), 80. *Prof Exp:* Asst prof, physics, State Univ NY, Fredonia, 81-85; MEM TECH STAFF, AT&T BELL LABS, 85- *Mem:* Am Phys Soc; Inst Elec & Electronics Engrs; Optical Soc Am. *Res:* Theoretical physics; ferroelectrics; lightwave communication system. *Mailing Add:* 62 Yellowstone Lane E Howell NJ 07731

HALES, ALFRED WASHINGTON, b Pasadena, Calif, Nov 30, 38; m 62; c 3. MATHEMATICS, ALGEBRA. *Educ:* Calif Inst Technol, BS, 60, PhD(math), 62. *Prof Exp:* NSF fel, Cambridge Univ, 62-63; Benjamin Peirce instr math, Harvard Univ, 63-66; from asst prof to assoc prof, 66-73, PROF MATH, UNIV CALIF, LOS ANGELES, 73- *Concurrent Pos:* Consult, Jet Propulsion Lab, 66-70 & Inst Defense Anal, 64-65, 76-90; vis lectr, Univ Wash, 70-71; vis fel, Univ Warwick, 77-78; mem, Math Sci Res Inst, Berkeley, 86-87. *Honors & Awards:* Polya Prize in Combinatorics, Soc Indust & Apl Math, 72. *Mem:* Am Math Soc; Math Asn Am; Soc Indust & Appl Math; Sigma Xi. *Res:* Algebra, especially structure of groups, modules and lattices; combinatorial analysis. *Mailing Add:* Dept Math Univ Calif Los Angeles CA 90024

HALES, CHARLES A, b Greeley, Colo, Apr 27, 41; m 65; c 3. PULMONARY HYPERTENSION, PULMONARY EDEMA. *Educ:* Emory Univ, BA, 62, MD, 66. *Prof Exp:* Intern & resident, Boston City Hosp, 66-68; Leutenant comdr, USNR, 68-70; med res II, Univ Calif, San Francisco, 70-71; clin res fel, Harvard Univ & Mass Gen Hosp, 71-73, from instr to asst prof med, 73-79, assoc physician, 84-90, ASSOC PROF MED, HARVARD UNIV & MASS GEN HOSP, 79-, PHYSICIAN, 90- *Concurrent Pos:* Chmn, cardiopulmonary coun, Am Heart Asn, 90- *Mem:* Am Thoracic Soc; Am Physiol Soc; Am Soc Clin Invest; Am Heart Asn. *Res:* Investigation of the mechanism of pulmonary vasoconstriction in response to hypoxia transforming the vessel on chronic basis producing pulmonary hypertension and COR pulmonale; investigation of smoke to determine the toxins producing pulmonary edema. *Mailing Add:* Pulmonary Unit Mass Gen Hosp Boston MA 02114

HALES, DONALD CALEB, b American Fork, Utah, May 10, 29; m 48; c 4. FISH BIOLOGY, AQUATIC ECOLOGY. *Educ:* Univ Utah, BS, 53, PhD(limnol), 67; Utah State Univ, MS, 55. *Prof Exp:* Fish biologist, Idaho Dept Fish & Game, 55-56; fish biologist, Utah Fish & Game Dept, 56-58, regional fishery mgr, 58-62, asst chief fishery mgt, 62-64; asst unit leader fisheries, Pa Coop Fishery Unit, Pa State Univ, 67-69; unit leader, SDak Coop Fishery Res Unit, Dept Wildlife & Fisheries, SDak State Univ, 70-77; Fishery Resources Coordr, US Fish & Wildlife Serv, Anchorage, 77-80, asst dir, Nat Fishery Res Lab, Wis, 80-83, staff fishery biologist, Washington, DC, 83-86, Fishery Res Biologist, Dexter Nat Fish Hatchery, US Fish & Wildlife Serv, NMex, 86-90; CONSULT, 90- *Concurrent Pos:* Teaching Aquatic Invert, Pa State Univ, 67-69; freshwater ecol, Fishery Mgt, SDak State Univ, 70-77. *Mem:* Am Fisheries Soc; NAm Benthological; Sigma Xi; Desert Fishes Coun. *Res:* Aquatic invertebrates; paddlefish; endangered species. *Mailing Add:* 550 Shawnee Dexter NM 88230

HALES, EVERETT BURTON, physics, for more information see previous edition

HALES, HUGH B(RADLEY), b Salt Lake City, Utah, Mar 17, 40; m 63; c 3. CHEMICAL ENGINEERING. *Educ:* Univ Utah, BS, 62, MS, 63; Mass Inst Technol, ScD, 67. *Prof Exp:* Sr res engr, Esso Prod Res Co, 67-68; asst prof chem eng, Mass Inst Technol, 68-70; SCIENTIST, MOBIL OIL CORP, 70- *Mem:* Am Inst Chem Engrs; Soc Petrol Engrs. *Res:* Fluid mechanics; petroleum resevoir simulation. *Mailing Add:* 6905 Echo Bluff Dallas TX 75248

HALES, J VERN, b Provo, Utah, July 21, 17; m 40; c 7. ATMOSPHERIC PHYSICS. *Educ:* Brigham Young Univ, AB, 38; Calif Inst Technol, MS, 41; Univ Calif, Los Angeles, PhD(meteorol), 52. *Prof Exp:* Observer meteorol, US Weather Bur, 38-41; meteorologist, Pan-Am Grace Airways, Inc, 41-42; weather officer, US Army Air Corps, 42-46; prof, Univ Utah, 46-65, head dept meteorol, 46-63; consult, Gen Elec Co, 65-71; PRES, HALES & CO, 54- *Concurrent Pos:* Mem, comt sci & arts, Franklin Inst, 73-, chmn, 85-86; mem, Nat Coun Indust Meteorologists. *Mem:* Am Meteorol Soc. *Res:* Solution of problems in atmospheric physics and chemistry; air quality, diffusion phenomena, cloud physics; meteorological statistics, measurement and analysis methods and equipment development. *Mailing Add:* Hales & Co 609 Lorient Dr West Chester PA 19382-4970

HALES, JEREMY M(ORGAN), b Seattle, Wash, Oct 15, 37; m 62; c 2. CHEMICAL ENGINEERING. *Educ:* Univ Wash, BS, 60, MS, 62; Univ Mich, Ann Arbor, PhD(chem eng), 68. *Prof Exp:* Sanit engr, Div Air Pollution, HEW, 62-64; sr res engr, 68-74, mgr, Atmos Dynamics & Chem Sect, 74-77, dir, MAP3S/Raine Prog, 79-81, assoc mgr, earth sci dept, 81-84, SR RES ASSOC, PAC NORTHWEST LABS, BATTELLE MEM INST, 77-; ASSOC PROF, DEPT CIVIL ENG, UNIV WASH, 74-, MGR ATMOSPHERIC SCI DEPT, 84-, DIR DEPT ENERGY ATMOSPHERIC CHEM PROG, 89- *Concurrent Pos:* Ed, J Appl Meteorol, 76-81; dir, Dept Energy precp prog, Univ Wash, 85-89. *Mem:* Am Inst Chem Engrs; Am Meteorol Soc. *Res:* Chemical engineering applications in the environmental sciences, particularly with respect to atmospheric sciences, with special emphasis on transport and reaction-rate phenomena. *Mailing Add:* Atmos Sci Dept Pac Northwest Labs PO Box 999 Richland WA 99352

HALES, MILTON REYNOLDS, b Springville, Utah, Aug 15, 18; m 42; c 3. MEDICINE, PATHOLOGY. *Educ:* Univ Calif, Los Angeles, AB, 40; Univ Southern Calif, MD, 44; Am Bd Path, dipl, 55. *Prof Exp:* Intern, Los Angeles County Hosp, 43-44; intern path, New Haven Community Hosp, 47-48; James Hudson Brown Mem res fel, Sch Med, Yale Univ, 48-49, instr, 49-50; asst prof, Univ Southern Calif, 50-55; from asst prof to assoc prof, Sch Med, Yale Univ, 55-68; chmn dept, 69-74, PROF PATH, SCH MED, WVA UNIV, 68- *Concurrent Pos:* From asst resident to resident, New Haven Community Hosp, 48-50. *Mem:* Am Asn Path & Bact; Int Acad Path; Sigma Xi. *Res:* Liver and pulmonary diseases, particularly associated vascular changes. *Mailing Add:* 66 Sherman Ave Morgantown WV 26505

HALES, RALEIGH STANTON, JR, b Pasadena, Calif, Mar 16, 42; m 67; c 2. MATHEMATICS, MATHEMATICAL ECONOMICS. *Educ:* Pomona Col, BA, 64; Harvard Univ, MA, 65, PhD(math), 70. *Prof Exp:* From instr to assoc prof, Pomona Col, 67-85, prof math, 85-90, assoc dean, 73-90; VPRES ACAD AFFAIRS & PROF MATH, COL WOOSTER, 90- *Concurrent Pos:* Consult, Div Savings & Loan, Calif, 68-69, Irvine Co & Univ Calif, Irvine, 71 & Develop Econ, Inc, 71-73; Wig distinguished prof, 71. *Mem:* Am Math Soc; Math Asn Am. *Res:* Graph theory, including numerical invariants of graphs, especially related to products of graphs; estimation, linear models and multi-dimensional scaling of mathematical economics. *Mailing Add:* Galpin Hall Col Wooster Wooster OH 44691

HALEVY, SIMON, b Bucharest, Romania, June 5, 29; US citizen; m 68; c 1. CARDIOVASCULAR PHARMACOLOGY, UMBILICAL VASCULATURE PHARMACOLOGY. *Educ:* Fr Inst Higher Studies, PCB-SPCN, 47; Univ Bucharest Med Sch, MD, 53; Romanian Bd Anesthesiol, cert anethesiol, 59 & Am Bd Anesthesiol, 68; Am Col Anesthesiologists, cert anesthesiol, 67. *Prof Exp:* Instr anesthesiol, Univ Hosp Coltzea, Bucharest, 55-57, chief lab, 57-60; attending anesthesiologist, Univ Hosp Fundeni, Bucharest, 60-63; asst prof anesthesiol, Mt Sinai Sch Med, 67-68, Albert Einstein Col Med, 69-74; attending anethesiologist, Bronx Munic Hosp Ctr, 71-74; assoc prof anesthesiol, Columbia Univ Col Physics, 74-75; PROF ANESTHESIOL, STATE UNIV NY, 76- *Concurrent Pos:* Consult, Astoria Gen Hosp, 71-86; chmn sci exhib, Postgrad Assembly Anesthesiol, 71-81; dir obstet anesthesiol, Nassau County Med Ctr, State Univ NY, 76-, assoc chmn, 83- *Mem:* Am Med Asn; Am Soc Anesthesiologists; Am Soc Pharmacol & Exp Therapeut; AAAS; Am Asn Clin Pharmacol & Therapeut; Shock Soc. *Res:* Pharmacology of shock and mircocirculation; endocrine pharmacology; physiology and pharmacology of microcirculation as related to anesthetics; cardiovascular actions of anesthetics and drugs related to anesthesia; reactivity of umbilical circulation to anesthetics and other agents. *Mailing Add:* Dept Anesthesiol SUNY Nassau County Med Ctr 2201 Hempstead Turnpike East Meadow NY 11554

HALEY, ALBERT JAMES, parasitology, for more information see previous edition

HALEY, BOYD EUGENE, b Greensburg, Ind, Sept 22, 40; m 65; c 2. BIOCHEMISTRY. *Educ:* Franklin Col, BA, 63; Univ Idaho, MS, 67; Wash State Univ, PhD(chem), 71. *Prof Exp:* NIH fel physiol, Med Sch, Yale Univ, 71-74; from asst prof to prof biochem, Univ Wyo, 74-85; PROF BIOCHEM, UNIV KY, 85- *Concurrent Pos:* Dreyfus Found vis researcher & lectr, Univ Wis, 75. *Mem:* Am Soc Biol Chemists; AAAS. *Res:* Use of photoaffinity analogs of nucleotides to photolabel nucleotide binding sites of membranes and complex regulatory enzymes. *Mailing Add:* Dept Biochem Univ Ky 800 Rose St Lexington KY 40536

HALEY, EDWARD EVERETT, biochemistry, for more information see previous edition

HALEY, HAROLD BERNARD, b Madison, Wis, Sept 11, 23; m 45; c 5. SURGERY. *Educ:* St Louis Univ, MD, 46. *Prof Exp:* Intern, Milwaukee County Hosp, 46-47; resident surg, Methodist Hosp, Madison, 47-48; res fel & Damon Runyon Clin fel, Harvard Med Sch, 50-51; resident, Peter Bent Brigham Hosp, Boston, 51-55; from asst prof to prof, Stritch Sch Med, Loyola Univ Chicago, 55-69; prof surg & assoc dean student affairs, Med Col Ohio, 69-72, prof social med, 70-72; prof surg & assoc dean clin serv, Sch Med, Univ Va, Roanoke, 72-80; assoc chief staff educ & clin prof surg, 80-90, EMER CLIN PROF SURG, HOUSTON VET ADMIN MED CTR, BAYLOR COL MED, 90- *Mem:* Asn Am Med Cols; Am Col Surg; Am Asn Cancer Educ (pres, 72-73); Am Anthrop Asn; Soc Am Archivists. *Res:* Physicians attitudes towards cancer; behavioral aspects of medical education; medical anthropology; medical ethics, especially in areas of cancer, death and dying, resource allocation; Precolumbian medicine and death beliefs, with emphasis on Mexican codices; education of resident physicians. *Mailing Add:* 7447 Cambridge No 119 Houston TX 77054

HALEY, KENNETH WILLIAM, b San Francisco, Calif, Sept 6, 39; m 65; c 1. CHEMICAL ENGINEERING. *Educ:* Stanford Univ, BS, 61; Univ Ill, MS, 63, PhD(chem eng), 65. *Prof Exp:* Res engr, Standard Oil Co Calif, 65-70, sr res engr, Chevron Res Co, 70-86, MGR, ENERGY FORCASTING, CHEVRON, 86- *Concurrent Pos:* Vis lectr, Univ Ariz, 72. *Mem:* Am Inst Chem Engrs. *Res:* Boiling heat transfer. *Mailing Add:* 711 Camino Ricardo Moraga CA 94556

HALEY, LESLIE ERNEST, b Windsor, NS, May 2, 38; m 63; c 2. GENETICS. *Educ:* NS Agr Col, dipl, 58; Ont Agr Col, BSA, 60, MSA, 63; Univ Calif, Davis, PhD(genetics), 67. *Prof Exp:* Asst prof biol, Univ Sask, Regina, 67-70; from asst prof to assoc prof, 70-80, chmn Dept Educ, 81-89, PROF BIOL, DALHOUSIE UNIV, 80-, ASST DEAN ARTS & SCI; PRIN, NOVA SCOTIA AGR COL, 89- *Mem:* Genetics Soc Can. *Res:* Agricultural genetics. *Mailing Add:* Nova Scotia Agr Col PO Box 550 Truro NS B2N 5E3 Can

HALEY, NANCY JEAN, b Huntington, NY, July 3, 48; m 73; c 1. BIOCHEMISTRY. *Educ:* St Joseph's Col, BA, 70; St John's Univ, MS, 72, PhD(biochem), 75. *Prof Exp:* Fel cell biol, Rockefeller Univ, 75-77, res assoc atherosclerosis, 77-80; assoc, Am Health Found, 80-84, assoc chief, Div Nutrit & Endocrinol, 85-90; ASSOC DIR, MET LIFE LAB, 90- *Concurrent Pos:* Consult, Int Asn Res Cancer, Lyon, France, Cine-Med, Woodbury, Conn; ed, Cellular & Molecular Neurobiol, 80- *Mem:* Am Soc Cell Biol; Soc Exp Med & Biol; NY Acad Sci; Am Heart Asn; Am Asn Clin Chem. *Res:* risk factor analysis including hypercholesterolemia, hypertension and cigarette smoking; nicotine metabolism and passive smoking; assay development for biomarkers of disease risk. *Mailing Add:* Metrop Life Lab Four Westchester Plaza Elmsford NY 10523

HALEY, SAMUEL RANDOLPH, b Ft Worth, Tex, July 12, 40; m 66; c 3. DEVELOPMENTAL ANATOMY. *Educ:* Univ Tex, BA, 62, MA, 64, PhD(zool), 67. *Prof Exp:* PROF ZOOL, UNIV HAWAII, MANOA, 67- *Mem:* AAAS; Soc Develop Biol; Am Soc Zool; Sigma Xi; Crustacean Soc. *Res:* Developmental and reproductive biology. *Mailing Add:* Dept Zool-2538 The Mall Univ Hawaii at Manoa Honolulu HI 96822

HALEY, THOMAS JOHN, b Crosby, Minn, Nov 4, 13; m 64; c 2. PHARMACOLOGY. *Educ:* Univ Southern Calif, BS, 38, MS, 42; Univ Fla, PhD(pharmacol), 45. *Prof Exp:* Instr org chem lab, Col Pharm, Univ Southern Calif, 40-42; asst pharmacol, Sch Pharm, Univ Fla, 42-45; med dir, E S Miller Labs, Los Angeles, 45-47; chief, Div Pharmacol & Toxicol, Atomic Energy Proj, Med Sch, Univ Calif, Los Angeles, 47-59, assoc clin prof indust med, 52-66, chief, Div Pharmacol & Toxicol, Dept & Labs Nuclear Med & Radiation Biol, 59-66; prof pharmacol, Univ Hawaii Sch Med, 66-69; group leader pharmacol & toxicol, Res Triangle Inst, 69-70; pharmacologist, Nat Ctr Toxicol Res, 70-82; prof pharmacol, Med Ctr, Univ Ark, 73-85; RETIRED. *Concurrent Pos:* Fel, Med Sch, Univ Southern Calif, 46-48; Del Amo fel, 55; vis scientist, Sandoz Labs, Switz, 56; adj prof, Univ NC Sch Med; consult, Food & Drug Admin; co-chmn symp response of nervous syst to ionizing radiation, AEC-NIH, 60 & 63; asst to dir, Nat Ctr Toxicol Res. *Honors & Awards:* Clinton H Thienes Award, Am Acad Clin Toxicol, 82. *Mem:* Am Chem Soc; Am Soc Pharmacol & Exp Therapeut; Am Pharmaceut Asn; fel Am Inst Chemists; fel Am Col Clin Pharmacol & Therapeut. *Res:* Radiobiology; toxicology; neurophysiology. *Mailing Add:* 774 Rivertree Dr Oceanside CA 92054

HALFACRE, ROBERT GORDON, b Newberry, SC, June 22, 41; m 63, 84; c 2. HORTICULTURE. *Educ:* Clemson Univ, BS, 63, MS, 65; Va Polytech Inst, PhD(hort), 68; NC State Univ, MLA, 73. *Prof Exp:* From asst prof to assoc prof hort, NC State Univ, 68-74; from assoc prof to prof, 74-90, ALUMNI PROF HORT, CLEMSON UNIV, 90- *Concurrent Pos:* Pres, Facul Sen, Clemson Univ, 89-90. *Honors & Awards:* Sigma Xi Res Award, 68; Julian C Miller Res Award, Am Soc Hort Sci, 68. *Mem:* Am Soc Hort Sci; Am Soc Landscape Architects. *Res:* Taxonomy and landscape architectural use of plants; computer graphics. *Mailing Add:* Dept Hort Clemson Univ Clemson SC 29634-0375

HALFAR, EDWIN, b Alexandria, La, Dec 10, 17; m 39; c 2. TOPOLOGY. *Educ:* Southern Ill Univ, BEd, 39; Univ Iowa, MA, 41, PhD(math), 47. *Prof Exp:* Instr math, Univ Iowa, 43-44; asst mathematician, Naval Ord Lab, DC, 44-45; instr math, Univ Iowa, 45-47; from instr to assoc prof, 47-63, chmn dept, 64-70, PROF MATH, UNIV NEBR-LINCOLN, 63- *Mem:* Am Math Soc; Math Asn Am; Am Soc Clin Pharmacol Therapeut. *Res:* General topology. *Mailing Add:* Dept Math Statist Univ Nebr 913 Oldfather Lincoln NE 68503

HALFERDAHL, LAURENCE BOWES, b Ottawa, Ont, Sept 14, 30; m 54; c 4. EXPLORATION GEOLOGY. *Educ:* Queen's Univ, Ont, BSc, 52, MSc, 54; Johns Hopkins Univ, PhD, 59. *Prof Exp:* Mineralogist & indust mineralogist, Res Coun Alta, 57-69; CONSULT GEOL ENGR, HALFERDAHL & ASSOC LTD, 69- *Mem:* Mineral Asn Can; Can Inst Mining & Metall. *Res:* Physical and economic mineralogy; exploration for metallic and industrial minerals; mining geology; coal. *Mailing Add:* 11539 73 Ave Edmonton AB T6G 0E2 Can

HALFF, ALBERT HENRY, b Midland, Tex, Aug 20, 15; m 40; c 2. SANITARY & CIVIL ENGINEERING. *Educ:* Southern Methodist Univ, BSCE, 37; Ill Inst Technol, MSCE, 42; Johns Hopkins Univ, DEng, 50. *Prof Exp:* Off asst, Tex Hwy Dept, 36-37; asst off engr, Koch & Fowler Engrs, 37-39; instr Tex Col Arts & Indust, 39-41; instr civil eng, Ill Inst Technol, 40-42; design engr, Charles DeLeuw & Co, 42; engr, Brown & Bellows, 42-43; asst prof, Southern Methodist Univ, 43-46; sanit engr, Brown & Bellows, 50-53; partner, Hundley & Halff, 53-60; owner, Albert H Halff Assoc Engrs, 60-70, pres, 70-83, CHMN, ALBERT H HALFF ASSOCS, INC, 83- *Concurrent Pos:* Indust consult on waste prevention & treatment. *Honors & Awards:* Award of Honor, Am Soc Civil Engr. *Mem:* AAAS; Am Water Works Asn; Am Soc Civil Engrs; Water Pollution Control Fedn; NY Acad Sci; Am Acad Environ Engrs; Am Water Works Asn. *Res:* Water desalinization; industrial wastes; ion exchange desalting; hydraulic mechanisms; water recycle; thermal energy conversion. *Mailing Add:* 8616 Northwest Plaza Dallas TX 75225

HALFHILL, JOHN ERIC, b Cedar Rapids, Iowa, Aug 25, 32; m 58; c 1. ENTOMOLOGY. *Educ:* San Jose State Col, BA, 54; Univ Idaho, PhD(entom), 70. *Prof Exp:* Lab technician, Agr Lab, Stauffer Chem Co, Calif, 54-56; lab technician, Citrus Exp Sta, Univ Calif, Riverside, 56-58; res entomolgist, Agr Res Serv, USDA, 63-88; RETIRED. *Mem:* Entom Soc Am; Int Orgn Biol Control. *Res:* Biology and control of vegetable insect pests. *Mailing Add:* 1208 S 28th Ave Yakima WA 98902

HALFMAN, CLARKE JOSEPH, b Chicago, Ill, Dec 19, 41; m 67; c 1. CLINICAL CHEMISTRY, ANALYTICAL BIOCHEMISTRY. *Educ:* Univ Ill, Champaign-Urbana, BS, 63, PhD(food chem), 70. *Prof Exp:* Res fel biochem, Georgetown Univ, 70-71, trainee clin chem, Med Sch, 71-72; asst prof path, Emory Univ, 72-75; clin chemist, BioSci Labs, 75-77; ASSOC PROF PATH, CHICAGO MED SCH, UNIV HEALTH SCI, 77- *Concurrent Pos:* Consult clin chemist, North Chicago Vet Admin Med Ctr, 77-, res assoc, 78-, co-prin res investr, 79-82; assoc dir clin lab, Chicago Med Sch, Univ Health Sci, 78- *Mem:* Am Asn Clin Chemists; Am Chem Soc; Nat Acad Clin Biochemists; Acad Clin Lab Physicians & Scientists; AAAS; Am Soc Clin Pathologists. *Res:* Binding of small molecules to proteins; physical chemical properties of immunoglobulins; application of fluorescent labels in immunoassays. *Mailing Add:* 1324 Hampton Lane Mundelein IL 60060-3218

HALFON, EFRAIM, b Tripoli, Libya, Mar 31, 48; Can citizen; m 71; c 2. ECOLOGICAL MODELLING, FATE OF TOXIC CONTAMINANTS. *Educ:* Univ Milan, Laurea, limnol, 71; Univ Ga, PhD(systs ecol), 75. *Prof Exp:* Fel, 75, res scientist 1, 76-78, RES SCIENTIST 2 SYSTS ECOL, NAT WATER RES INST, CAN CENTRE INLAND WATERS, 78-; PROF, DEPT MICROBIOL, MCMASTER UNIV, HAMILTON, ONT, 90- *Concurrent Pos:* Fulbright fel, 71-72; consult, Ecol Environ Adv Comt, Halton Region, Ont, Can, 76-77; guest scholar, Int Inst Appl Systs Anal, Laxenburg, Austria, 78-; adj assoc prof, State Univ NY Binghamton, 79-82; vis prof, GSF-PUC, Munich, WGer, 87, 90. *Mem:* Int Soc Ecol Modelling (vpres, 82-89); Int Asn Ecol; Sigma Xi. *Res:* Model prediction ability; fate models of toxic substance in lakes and rivers; ranking hazard of toxic contaminants. *Mailing Add:* Nat Water Res Inst 867 Lakeshore Rd Burlington ON L7R 4A6 Can

HALFON, MARC, b Brooklyn, NY, Aug 22, 45. CHEMISTRY. *Educ:* Brooklyn Col, BS, 67; Brown Univ, PhD(org chem), 74. *Prof Exp:* Res chemist, Photo Prod Div, E I du Pont de Nemours & Co, Inc, 73-75; RES CHEMIST PROC RES & DEVELOP INSECTICIDES, FMC CORP, 76- *Mem:* Am Chem Soc. *Res:* Research and development of agricultural process chemistry. *Mailing Add:* FMC Corp Box 8 Princeton NJ 08540

HALFORD, GARY ROSS, b Fayette Co, Ill, Dec 7, 37; m 59; c 3. MATERIALS SCIENCE, ENGINEERING MECHANICS. *Educ:* Univ Ill, BS, 60, MS, 61, PhD(eng mech), 66. *Prof Exp:* SR RES SCIENTIST, LEWIS RES CTR, NASA, 66- *Concurrent Pos:* Caterpillar Tractor, 55-57; Deere & Co, 60-61; vis res scientist, Aeronaut Res Lab, Melbourne, Australia, 89-90. *Honors & Awards:* Except Eng Achievement Medal, NASA, 82. *Mem:* Am Soc Testing & Mat; Am Soc Metals; Am Soc Mech Engrs; Am Inst Aeronaut & Astronaut; Soc Automotive Engrs; Metal Properties Coun; Am Acad Mech. *Res:* Analytical and experimental research of cyclic flow, fatigue and fracture of materials under high-temperature; thermal-fatigue exposure as found in gas turbine and rocket engines; developer of life prediction methods for fatigue; author 110 publications; aeronautical & astronautical. *Mailing Add:* c/o Nasa Lewis Res Ctr 21000 Brookpark Rd Mail Stop 49-7 Cleveland OH 44135

HALFORD, JAKE HALLIE, b Columbia, SC, Nov 26, 41; m 66; c 3. ELECTRICAL ENGINEERING, SOLID STATE PHYSICS. *Educ:* Univ SC, BSEE, 65; Duke Univ, MSEE, 68, PhD(elec eng), 74. *Prof Exp:* Instr math, Guilford Col, 66-67; asst prof elec eng, US Naval Acad, 73-78; assoc prof & dir, The Citadel, 78-83; ASSOC PROF ELEC ENG, COL CHARLESTON, 83-; PRES, HALFORD & ASSOC, 83- *Mem:* Inst Elec & Electronics Engrs; Am Soc Eng Educ. *Res:* Thin films in general and thin film Amorphous semiconductor nonvolatile memory devices; computer interfacing. *Mailing Add:* 15 Orange St Charleston SC 29401

HALGREN, LEE A, b Minneapolis, Minn, Nov 5, 42; m 66; c 2. ENTOMOLOGY, INVERTEBRATE ECOLOGY. *Educ:* Gustavus Adolphus Col, BS, 64; Kans State Univ, MS, 66, PhD(entom), 68. *Prof Exp:* Res asst entom, Kans State Univ, 64-68; prof biol & chmn dept, 68-79, assoc vpres acad affairs, Southwest State Univ, 79-84; VCHANCELLOR, UNIV WIS, PLATTEVILLE, 84- *Mem:* AAAS; Entom Soc Am; Sigma Xi. *Res:* Flight behavior and ecology of aphids; curriculum development in the biological sciences, environmental education. *Mailing Add:* Vchancellors Off Univ Wis Platteville WI 53818

HALGREN, THOMAS ARTHUR, b Berwyn, Ill, Nov 1, 41; m 78. QUANTUM CHEMISTRY, ORGANIC CHEMISTRY. *Educ:* Wabash Col, AB, 63; Calif Inst Technol, PhD(chem), 68. *Prof Exp:* Res fel chem, Harvard Univ, 68-71; asst prof, Rollins Col, 71-74; res fel, Harvard Univ, 74-75; asst prof chem, 75-80, ASSOC PROF CHEM, CITY COL, CITY UNIV NEW YORK, 81- *Mem:* Am Chem Soc. *Res:* Computational studies of chemical reactions; calculation of reaction pathways; development of computational methods; localized orbital studies of chemical bonding. *Mailing Add:* 566 Highland Ave Upper Montclair NJ 07043

HALIBURTON, T(RACEY) ALLAN, civil engineering; deceased, see previous edition for last biography

HALIJAK, CHARLES A(UGUST), b Milwaukee, Wis, Oct 30, 22. ELECTRICAL ENGINEERING. *Educ:* Univ Wis, BS, 47, MS, 49, PhD(elec eng), 56. *Prof Exp:* Sr res & develop engr electronics, Goodyear Aerospace Corp, 50-55; instr elec eng, Univ Wis, 55-56; from assoc prof to prof, Kans State Univ, 56-66; prof, Univ Denver, 66-69; PROF ELEC ENG, UNIV ALA, HUNTSVILLE, 69- *Concurrent Pos:* NASA grant, 65-66. *Mem:* Math Asn Am; Inst Elec & Electronics Engrs; Sigma Xi. *Res:* Analog and digital computers; carrier-frequency servos; modulation theory; network analysis and synthesis. *Mailing Add:* Dept Elec Eng Univ Ala-Huntsville Huntsville AL 35899

HALIK, RAYMOND R(ICHARD), b Kilkenny, Minn, Feb 10, 17; m 45; c 3. CHEMICAL ENGINEERING, SYNTHETIC FUELS. *Educ:* Univ Minn, BChE, 40; Univ Rochester, MS, 41; Carnegie Inst Technol, PhD(chem eng), 48. *Prof Exp:* Asst chem eng, Univ Rochester, 40-41; develop engr, Rohm and Haas Co, Pa, 41-46; instr chem eng, Carnegie Inst Technol, 47; from technologist to sr technologist, Socony Mobil Oil Co, 48-53, res assoc, 53-59, supvr process develop, 59-66, eng consult, Mobil Oil Corp, 66-81; CONSULT, 81- *Mem:* Am Chem Soc; Nat Soc Prof Engrs; Am Inst Chem Engrs. *Res:* Petroleum, petrochemicals and related energy sources, especially planning and economics of processes or entire ventures in oil, synthetic fuels or in uranium. *Mailing Add:* 9 Newcomb Dr New Providence NJ 07974

HALISKY, PHILIP MICHAEL, b Andrew, Alta, Can, Sept 11, 24; m; c 2. PHYTOPATHOLOGY. *Educ:* Univ Alta, BS, 50; Wash State Univ, PhD(plant path), 57. *Prof Exp:* Instr plant path, Univ Calif, Davis, 56-58, asst prof, 58-68; assoc prof, 68-74, PROF PLANT PATH, RUTGERS UNIV, 74- *Mem:* Am Phytopath Soc; Mycol Soc Am; Bot Soc Am. *Res:* Pathogenic fungi; agrostology; genetics; diseases of crops, especially of grain crops and turfgrasses; diseases of coastal vegetation. *Mailing Add:* Dept Plant Path Rutgers The State Univ New Brunswick NJ 08903

HALITSKY, JAMES, b New York, NY, Oct 18, 19; m 44; c 2. AIR POLLUTION, METEOROLOGY. *Educ:* City Col New York, BME, 40; NY Univ, MAE, 52, PhD(meteorol, oceanog), 70. *Prof Exp:* Aeronaut engr, Edo Aircraft Corp, 40-48; sr res scientist, NY Univ, 48-71; assoc prof civil eng, Univ Mass, Amherst, 71-75; CONSULT AIR POLLUTION CONTROL, 75- *Concurrent Pos:* USPHS grants, 58-60 & 62-65; meteorol consult air pollution problems of industry & pharmaceut res labs; adj prof appl sci, NY Univ & civil eng, City Univ NY, 79- *Mem:* AAAS; Am Meteorol Soc; Air Pollution Control Asn; Sigma Xi. *Res:* Diffusion of airborne pollutants in the atmosphere and the wind tunnel; chimney jets; modeling of atmospheric turbulence. *Mailing Add:* 122 N Highland Pl Croton-on-Hudson NY 10520

HALKERSTON, IAN D K, biochemistry, endocrinology, for more information see previous edition

HALKET, THOMAS D, b New York, NY, July 20, 48; m 77; c 4. COMPUTERS. *Educ:* Mass Inst Technol, SB, 71; Columbia Univ, JD, 74. *Prof Exp:* Assoc, Sullivan & Worcester, Boston, 74-78 & Dewey, Ballantine, Bushby, Palmer & Wood, 78-82; atty, Engelhard Corp, 82-83, asst gen counr, 83-86; counr, Summit Rovins & Feldesman, 86-88; partner, Scheffler Karlinsky & Stein, 88-89; PVT PRACTR LAW, 89- *Concurrent Pos:* Ed, Antitrust Law J, 79-81; chmn, Aerospace Law Div, Am Bar Asn, 79-83, coun mem, Sect Sci & Technol, 82-85, chmn, Div Ventures & Enterprise, 86-89 & Long Range Planning Comt, 89-90. *Mem:* Am Bar Asn (secy, 85-86); Computer Law Asn; Am Phys Soc; AAAS; NY Acad Sci; sr mem Am Inst Aeronaut & Astronaut; Sigma Xi. *Mailing Add:* Box 942 Shore Dr Larchmont NY 10538-0942

HALKIAS, CHRISTOS, b Monasterakion Doridos, Greece, Aug 23, 33; US citizen. ELECTRONICS. *Educ:* City Col New York, BEE, 57; Columbia Univ, MS, 58, PhD(elec eng), 62. *Prof Exp:* Prof & dir, Grad & Undergrad Electronics Labs, Columbia Univ, 62-73; PROF ELEC ENG & CHMN DEPT, NAT TECH UNIV ATHENS, 73-, CHAIR ELECTRONICS, 80- *Concurrent Pos:* Nat Sci Found grant, Columbia Univ, 64-70; proj engr, Underwriters Labs, NY, 57; Fulbright vis prof, Greece, 69. *Mem:* Inst Elec & Electronics Engrs. *Res:* Multiport network theory; electronics and instrumentation; digital signal processing. *Mailing Add:* Four Kosti Palama St Paleon Psyhion Athens Greece

HALKIAS, DEMETRIOS, b Kosma, Greece, Aug 6, 32; US citizen; m 62; c 4. MEDICAL MICROBIOLOGY. *Educ:* Univ Ill, BS, 57; Loyola Univ, Ill, MS, 59, PhD(microbiol), 64; Am Bd Med Microbiol, dipl. *Prof Exp:* Bacteriologist, Chicago Park Dist, Ill, 60-61; La State Univ-Inter-Am Prog Trop Med & Parasitol fel, 61; instr microbiol & virol, Creighton Univ, 63-67, from asst prof to assoc prof med microbiol, 67-70, path, 70-72; assoc prof, 72-78, PROF MICROBIOL & PATH, UNIV S FLA, 78- *Concurrent Pos:* Chief microbiol sect, Vet Admin Hosp, Tampa, Fla, 72- *Mem:* Am Soc Microbiol; NY Acad Sci. *Res:* Diagnostic microbiology; tissue culture and radiomimetic drugs. *Mailing Add:* 10413 Butia Pl Tampa FL 33618

HALKIN, HUBERT, b Liege, Belg, June 5, 36; div; c 2. MATHEMATICS. *Educ:* Univ Liege, Dipl, 60; Stanford Univ, MS, 61, PhD(math), 63. *Prof Exp:* Res assoc, Res Inst Advan Studies, Md, 61-62; mem tech staff, Bell Tel Labs, 63-65; assoc prof math & assoc res mathematician, Inst Radiation Physics & Aerodyn, 65-69, PROF MATH, UNIV CALIF, SAN DIEGO, 69-, CHAIR DEPT, 81- *Concurrent Pos:* Assoc ed, J Optimization Theory & Applns, 68-; Guggenheim Mem Found fel, Univ Calif, San Diego, 71-72; vis prof, Ctr Opers Res & Economet, Cath Univ Louvain, 71-72. *Mem:* Am Math Soc; Soc Indust & Appl Math; Economet Soc. *Res:* Functional analysis; calculus of variations; optimization theory; convexity; control theory; mathematical programming; applications of mathematics to economics. *Mailing Add:* Dept Math Univ Calif-San Diego La Jolla CA 92093

HALKO, BARBARA TOMLONOVIC, b Des Moines, Iowa, Nov 9, 40; m 76; c 2. COORDINATION CHEMISTRY. *Educ:* Marycrest Col, BA, 65; Wayne State Univ, MS, 70, PhD(chem), 75. *Prof Exp:* Teacher chem & physics, Unihan High Sch, 65-66, St Albert High Sch, 66-67, Walsh High Sch, 67-69 & Bourgade High Sch, 70-72; res assoc, Ore State Univ, 75-76; teaching fel, Univ BC, 76-78; vis asst prof, Univ Idaho, 78-79; SR LECTR, CHEM, ST MARTIN'S COL, 81- *Concurrent Pos:* Res award, Sigma Xi, 75. *Mem:* Am Chem Soc. *Res:* Synthesis, characterization and reactivities of polynuclear transition metal complexes of catalytic and biological interest. *Mailing Add:* 2903 NW Angelica Dr Corvallis OR 97330

HALKO, DAVID JOSEPH, b Great Falls, Mont, Feb 17, 45. INORGANIC CHEMISTRY, BIOINORGANIC CHEMISTRY. *Educ:* Col Great Falls, BS, 67; Univ Calif, Davis, PhD(chem), 73. *Prof Exp:* Res assoc chem, Wayne State Univ, 73-75; res assoc chem, Ore Grad Ctr, 75-76; res fel chem, Univ BC, 76-78; asst prof, Univ Idaho, 78-79; res chemist, ITT Rayonier Inc, 79-; mem staff, Rockwell Hanford Opers, Richland; RES & DEVELOP CHEMIST, HEWLETT PACKARD, 88- *Mem:* Am Chem Soc. *Res:* Bioinorganic systems and analytical. *Mailing Add:* 2903 NW Angelica Dr Corvallis OR 97330

HALL, A(LLEN) S(TRICKLAND), JR, b Greensboro, Vt, Dec 12, 17; m 40; c 1. MECHANICAL ENGINEERING. *Educ:* Univ Vt, BS, 38; Columbia Univ, MS, 39; Purdue Univ, PhD(eng), 46. *Prof Exp:* From instr to assoc prof, 39-53, PROF MECH ENG, PURDUE UNIV, 53- *Concurrent Pos:* Mem sci adv coun, Picatinny Arsenal, 55-64; vis prof, Univ Calif, Berkeley, 64-65; mem munitions command adv group, US Army, 64-68. *Honors & Awards:* Mach Design Award, Am Soc Mech Eng, 74. *Mem:* Am Soc Mech Engrs; Am Soc Eng Educ. *Res:* Kinematics of machines; design of machines. *Mailing Add:* 715 Sugar Hill Dr West Lafayette IN 47906

HALL, ALAN H, b S Bend, Ind, Jan 8, 49; m 67; c 1. CLINICAL TOXICOLOGY, AEROSPACE MEDICINE. *Educ:* Ind Univ, BA, 73, MD, 77. *Prof Exp:* ASST PROF CLIN TOXICOL, UNIV COLO HEALTH SCI CTR, 87-; CLIN TOXICOLOGIST, DEPT PEDIAT, ROCKY MOUNTAIN POISON & DRUG CTR, 87- *Concurrent Pos:* Mem US Air Force Soc Flight Surgeons. *Mem:* Am Acad Clin Toxicol; Am Col Emergency Physicians; Aerospace Med Asn; AMA. *Res:* General clinical toxicology; ibuprofen poisoning, cyanide poisoning and its antidotal treatment; author of articles on toxicity of industrial chemicals. *Mailing Add:* c/o Michael Medix's 600 Grant St Denver CO 80203-3527

HALL, ALBERT C(ARRUTHERS), b Port Arthur, Tex, June 27, 14; m 41; c 2. ELECTRICAL ENGINEERING. *Educ:* Tex A&M Col, BS, 36; Mass Inst Technol, MS, 38, ScD(elec eng), 43. *Prof Exp:* Asst elec eng, Mass Inst Technol, 37-39, instr, 39-43, asst prof, 43-46, assoc prof & lab dir, 46-50; assoc dir res labs, Bendix Aviation Corp, 50-52, tech dir, 52-54, gen mgr, 54-57; dir res, Martin Co, 58-60, vpres eng, 60-61, vpres & gen mgr space systs div, 62-63; dep dir, Off Dir Defense Res & Eng, Off Secy Defense, 63-65; vpres advan technol, Martin Co, 65-67, vpres eng, Martin Marietta Corp, 67-71; PRES, ALBERT C HALL, PA, 77- *Honors & Awards:* Intel Medal, Defense Intel Agency, 72. *Mem:* Nat Acad Eng; Int Acad Astronaut; fel Inst Elec & Electronics Engrs; fel Am Inst Aeronaut & Astronaut. *Res:* Automatic control; electronics; dynamics; basic control mechanisms; compensating circuits; analysis and synthesis of linear servomechanisms. *Mailing Add:* 4001 N Ninth St Apt 923 Arlington VA 22203

HALL, ALBERT M(ANGOLD), b Brooklyn, NY, Oct 8, 14; m 38; c 3. METALLURGY. *Educ:* Columbia Univ, AB, 35, BS, 36 & MS, 37. *Prof Exp:* Asst metall, Columbia Univ, 35 & 36; res metallurgist, Int Nickel Co, WVa, 37-45; res engr, Battelle Mem Inst, 45-46, metallographer in chg, 46-50, asst supvr, 50-53, div chief, 53-67, sr tech adv, 67-69, asst mgr, Metall Dept, Columbus Labs, 69-79; exec dir, Mat Technol Inst Chem Process Indust, 79-86; CONSULT, 86- *Concurrent Pos:* Mem staff, Atomic Enegy Comn. *Honors & Awards:* Fel Am Soc Metals. *Mem:* Am Soc Metals; Am Inst Mining, Metall & Petrol Engrs; Am Soc Testing Mat; Sigma Xi. *Res:* Use of color photography in study of non-metallic inclusions in metals; stainless and high-alloy steels; high-temperature alloys; cast iron-chromium-nickel alloys; new United States coinage; structural materials for coal conversion systems; alloys with special physical properties; physical metallurgy of iron and steel; graphite formation in steel. *Mailing Add:* 1194 Kenbrook Hills Dr Columbus OH 43220

HALL, ANTHONY ELMITT, b Tickhill, Eng, May 6, 40; m 65; c 2. CROP PHYSIOLOGY, CROP ECOLOGY. *Educ:* Univ Calif, Davis, BS, 66, PhD(plant physiol), 70. *Prof Exp:* Field officer, Agr Exten, Ministry Agr, Tanzania, 61-63; Carnegie Inst Wash fel plant biol, Stanford Univ, 70-71; lectr water sci & eng, Univ Calif, Davis, 71; asst plant physiologist, Agr Exp Sta, 71-76; asst prof, 71-76, assoc prof, 76-81, assoc plant physiologist, Agr Exp Sta, 76-81, PROF PLANT PHYSIOL, UNIV CALIF, RIVERSIDE, 81-, PLANT PHYSIOLOGIST, AGR EXP STA, 81- *Concurrent Pos:* Consult int agr develop, 74- *Mem:* Scand Soc Plant Physiol; Am Soc Plant Physiol; Am Soc Agron; Crop Sci Soc Am. *Res:* Environmental plant physiology and agronomy with emphasis on crop adaptation to semi arid environments and the development of improved cultivars and management methods for these. *Mailing Add:* Dept Bot & Plant Sci Univ Calif Riverside CA 92521

HALL, ARTHUR DAVID, III, b Lynchburg, Va, Apr 13, 24; m 47; c 3. TELECOMMUNICATIONS, SYSTEMS ENGINEERING. *Educ:* Princeton Univ, BSE, 49. *Prof Exp:* Head, TV Eng Dept, Bell Tel Labs, Inc, 50-66; vpres eng, Jerrold Corp, 66-67; vpres res, Systs Eng & Develop, Smith-Corona Div, SCM Corp, 68-70, pres, Melabs, Inc, Div, 69-70; PRES, ARTHUR D HALL, INC, 71-; PRES, ADVAN DECISION HANDLING INC, 76- *Concurrent Pos:* Vis prof, Univ Pa, 67-69, adj prof, 70-85. *Honors & Awards:* Outstanding Achievement in Systs Sci & Systs Eng Award, Inst Elec & Electronics Engrs, 74. *Mem:* Opers Res Soc Am; Inst Mgt Sci; Inst Elec & Electronic Engrs, Systs Man & Cybernet Soc. *Res:* Communication systems; operations research; telecommunication and agricultural systems engineering; education and writings in systems methodology; behavior sciences; public policy issues in telecommunications; real-time computer control of any type of farm. *Mailing Add:* 594 Liberty Grove Rd Port Deposit MD 21904

HALL, BARRY GORDON, b New York, NY, July 17, 42; m 64; c 3. MOLECULAR BIOLOGY, EVOLUTION. *Educ:* Univ Wis-Madison, BS, 68; Univ Wash, PhD(genetics), 71. *Prof Exp:* Res assoc, Inst of Molecular Biol, Univ Ore, 71-72, NIH fel, 72-73; fel genetics, Univ Minn, 73-74; asst prof molecular biol, Fac Med, Mem Univ Nfld, 74-77; from asst prof to assoc prof biol, Univ Conn, 77-85, prof molecular cell biol, 85-89; PROF BIOL, UNIV ROCHESTER, 89- *Concurrent Pos:* Res Career Develop Awards, NIH, 80-85; Fulbright sr scholar, 84-85; vis prof fel, Univ Wales, 84-85. *Mem:* Genetics Soc Am; Am Soc Microbiol; Am Soc Biol Chemists; Am Soc Naturalists. *Res:* Experimental evolution of new enzymatic functions in bacteria; molecular evolution; microbial evolution. *Mailing Add:* Biol Dept Univ Rochester Rochester NY 14627

HALL, BENJAMIN DOWNS, b Berkeley, Calif, Dec 9, 32; m 54; c 2. MOLECULAR GENETICS. *Educ:* Univ Kans, AB, 54; Harvard Univ, AM, 56, PhD(chem), 59. *Prof Exp:* From instr to assoc prof chem, Univ Ill, 58-63; from assoc prof to prof, 63-90, chmn genetics dept, 80-84, PROF GENETICS & BOT, UNIV WASH, 90- *Concurrent Pos:* Guggenheim fel, 62; mem, Microbial Chem Study Sect, NIH, 71-75. *Mem:* Genetics Soc; Am Soc Biol Chem; AAAS. *Res:* Molecular genetics of higher plants and of yeast; eukaryotic gene transcription; molecular evolution. *Mailing Add:* Dept Genetics Univ Wash Seattle WA 98195

HALL, BEVERLY FENTON, b White Plains, NY, Jan 20, 55. INTERNAL MEDICINE. *Educ:* Harvard Col, AB, 77; NY Univ, MD, 84, PhD, 85, Am Bd Internal Med, dipl, 90. *Prof Exp:* Jr asst resident, Dept Med; med staff fel, Lab Clin Invest, Nat Inst Allergy & Infectious Dis, 86-87, med staff fel, Unit Microbiol Pathogenesis, Lab Parasitic Dis, 87-90, VACCINE DEVELOP PROG OFFICER, PARASITOL & TROP DIS BR, DIV MICROBIOL & INFECTIOUS DIS, NIH, 91- *Concurrent Pos:* Fel, Infectious Dis Sect, Dept Internal Med, Sch Med, Yale Univ, 89-91. *Honors & Awards:* Bertram M Gesner Mem Award, Sch Med NY Univ, 84. *Mem:* AAAS; Am Soc Cell Biol. *Res:* Development and evaluation of vaccines for parasitic diseases; mechanisms of evasion of host immune responses by parasites; cell biology and biochemistry of intracellular parasitism; gene regulation and developmental biology of parasites; numerous publications. *Mailing Add:* Div Microbiol & Infectious Dis Parasitol & Trop Dis Br NIAID NIH Bethesda MD 20892

HALL, BRIAN KEITH, b Port Kembla, Australia, Oct 28, 41; m 66; c 2. DEVELOPMENTAL BIOLOGY. *Educ:* Univ New Eng, Australia, BSc, 63, Hons, 65, PhD(zool), 68, DSc(biol), 78. *Prof Exp:* Res asst zool, Univ New Eng, Australia, 63-64; from asst prof to assoc prof biol, 68-75, chmn dept, 78-85, PROF BIOL, DALHOUSIE UNIV, 75-, PROF PHYSIOTHER, 88-, KILLIAM RES PROF, 90- *Concurrent Pos:* Vis prof, Guelph, 75, Toronto, 80, Brisbane, 81 & Southampton, 82; Nuffield Overseas fel, 82; assoc ed, Can J Zool, 82-85; Warwick James fel, Univ London, 89; vis lectr, Alta Heritage Found Med Res, 85; ed, Anatomy & Embryol, 88- *Honors & Awards:* Turner Newall lectr, Univ Manchester, 85. *Mem:* Int Asn Dent Res; Int Soc Develop Biol; Can Soc Cell Biol; Am Soc Zool; fel Royal Soc Can; Brit Soc Develop Biol; Int Soc Differentiation. *Res:* Differentiation of bone and cartilage from common germinal cells; morphogenesis of skeletal system neural crest skeleton; avian embryology; evolution and development. *Mailing Add:* Dept Biol Dalhousie Univ Halifax NS B3H 4J1 Can

HALL, CARL ELDRIDGE, b Danville, Ark, Mar 7, 37; m 59; c 2. MATHEMATICS. *Educ:* WTex State Univ, BS, 61; NMex State Univ, MS, 63, PhD(math), 65. *Prof Exp:* Asst prof math, Va Polytech Inst, 65-69; ASSOC PROF MATH, UNIV TEX, EL PASO, 69-, CHMN DEPT, 71- *Mem:* Am Math Soc; Math Asn Am. *Res:* Topological algebra, especially topological groups. *Mailing Add:* Dept Math Univ Tex El Paso TX 79968

HALL, CARL W(ILLIAM), b Tiffin, Ohio, Nov 16, 24; m 49; c 1. AGRICULTURAL & MECHANICAL ENGINEERING. *Educ:* Ohio State Univ, BS & BAE, 48; Univ Del, MME, 50; Mich State Univ, PhD(agr eng), 52, JFK Sch, Harvard, SMG, 83. *Prof Exp:* Instr agr eng, Univ Del, 48-50, asst prof, 50-51; from asst prof to prof, Mich State Univ, 51-70, chmn dept, 64-70; prof & dean, 70-82, EMER DEAN, COL ENG & PROF MECH ENG, WASH STATE UNIV, 87- *Concurrent Pos:* Res adv, Mich State Univ, 56-64; res consult, Univ PR, 57 & 63 & Nat Univ Colombia, 60; collabr, USDA & Dept State Sci Exchange deleg to USSR, 58; dairy eng consult, Ohio State Univ-Govt India, 61; consult, UN Spec Fund Proj 80, Latin Am, 64-70, vpres, Int Comn Agr Eng, 64-74, pres sect IV, 68-74; mem bd dirs & secy, Eng Coun Prof Develop; mem, Mich State Univ Mission Ecuador, 66; Nat Acad Sci deleg, Brazil, 72, NSF, Brazil, 86 & 87, Indonesia, 78 & Peoples Repub China, 78; dir, Nat Soc Prof Eng, 75-79; founding ed, Drying Tech-Am Int J, 82-89; dep asst dir, NSF, 82-90, actg asst dir, 84 & 88; registr prof eng. *Honors & Awards:* Massey-Ferguson Gold Medal, Am Soc Agr Engrs, 76, Cyrus Hall McCormick Medal, 84; Max Eyth Medal, Ger, 79; Distinguished Serv Award & Medal, NSF, 88. *Mem:* Nat Acad Eng; fel Am Soc Agr Engrs (pres, 74-75); fel Am Soc Mech Engrs; Nat Soc Prof Engrs; Inst Food Technol; fel AAAS; Am Soc Eng Educ; fel Accreditation Bd Eng & Technol. *Res:* Application of heat and mass transfer principles to engineering of food processing and drying; administration. *Mailing Add:* Eng Info Serv 2454 N Rockingham St Arlington VA 22207-1033

HALL, CAROL KLEIN, b New York, NY, Apr 23, 46; m 67; c 3. CHEMICAL ENGINEERING, CHEMICAL PHYSICS. *Educ:* Cornell Univ, BA, 67; State Univ NY Stony Brook, MA, 69, PhD(physics), 72. *Prof Exp:* Res assoc chem physics, Cornell Univ, 73-76; mem tech staff econ anal, Bell Tel Labs, 76-77; asst prof chem eng, Princeton Univ, 77-85; assoc prof, 85-87, PROF CHEM ENG, NC STATE UNIV, 87- *Mem:* Am Phys Soc; Am Inst Chem Engrs; Sigma Xi; Am Chem Soc; AAAS. *Res:* Application of statistical mechanics to fluid and fluid mixtures containing chainlike molecules; polymers and colloids, metal hydrides, gases in zeolites, semiconductor interfaces, and bioseparations. *Mailing Add:* Chem Eng Dept NC State Univ Raleigh NC 27695-7905

HALL, CAROLE L, ENZYMES, FLAVOPROTEINS. *Educ:* Purdue Univ, PhD(molecular biol), 66. *Prof Exp:* SR RES SCIENTIST, GA INST TECHNOL, 83- *Res:* Beta-oxidation. *Mailing Add:* Sch Appl Biol Ga Tech Atlanta GA 30332

HALL, CHARLES A, b Castine, Maine, Mar 7, 20; m 44, 74; c 5. MEDICINE. *Educ:* Univ Maine, BA, 41; Yale Univ, MD, 44. *Prof Exp:* Fel pharmacol, Sch Med, Yale Univ, 47-48; from instr to assoc prof med, 51-66, PROF MED, ALBANY MED COL, 66-; DIR NUTRIT LAB CLIN ASSESSMENT & RES, VET ADMIN MED CTR. *Concurrent Pos:* Chief, Hemat Sect, Vet Admin Hosp, Albany, 53-78, chief radioisotope serv, 57-69, assoc chief staff, 62-71, med investr, 71; Fulbright lectr, Turku Univ, 60-61. *Mem:* Am Fedn Clin Res; Am Soc Hemat; Am Soc Clin Nutrit; Soc Exp Biol & Med. *Res:* Clinical and research hematology; vitamin B-12 metabolism; nutrition. *Mailing Add:* Nutrit Lab Clin Assessment & Res Vet Admin Med Ctr Holland Ave Albany NY 12208

HALL, CHARLES A(INSLEY), b Bloomington, Ill, Feb 27, 37; m 66; c 2. CERAMIC ENGINEERING. *Educ:* Univ Ill, BS, 59, MS, 60, PhD(ceramic eng), 62. *Prof Exp:* STAFF MEM CERAMIC CAPACITOR COMPONENTS, SANDIA LAB, 62- *Mem:* Am Ceramic Soc. *Res:* Ceramic and materials technology; ceramic capacitors; ferroelectric materials; electrooptic ceramics and other electrical ceramics. *Mailing Add:* Dept Math & Statist Univ Pittsburgh Thackeray Hall Rm 603 Pittsburgh PA 15260

HALL, CHARLES ADDISON SMITH, b Hingham, Mass, May 3, 43; m. SYSTEMS ECOLOGY, GENERAL COMPUTER SCIENCES. *Educ:* Colgate Univ, BA, 65; Pa State Univ, MS, 66; Univ NC, Chapel Hill, PhD(zool), 72. *Prof Exp:* Res assoc ecol, Brookhaven Nat Lab, 70-72, asst ecologist, 73-74; asst scientist, Ecosyst Ctr, Marine Biol Lab, 72-77; vis asst prof, Cornell Univ, 72-76, asst prof biol & ecol, Ecosyst Ctr, 77-85; res assoc prof, Univ Mont, 85-87; ASSOC PROF, STATE UNIV NY, 87- *Concurrent Pos:* Mem panel environ impacts natural resource mgt, Nat Acad Sci, 75-76; rep, Ecolog Soc Am to AAAS; Fulbright fel, 86. *Mem:* Ecol Soc Am; AAAS; Am Soc Limnol & Oceanog. *Res:* Nutrient cycles and productivity of ecosystems, with emphasis on limnetic and coastal systems, aquatic and global primary productivity, fish life history and migration, interactions of industrial and ecological energy patterns including the global carbon budget and energy return on investment for industrial fuels and materials; geographical modeling; modeling land use change in Central America and the Carribean. *Mailing Add:* Environ & Forest Biol State Univ NY Environ Sci & Forestry Syracuse NY 13210

HALL, CHARLES ALLAN, b Pittsburgh, Pa, Mar 19, 41; m 62; c 3. NUMERICAL ANALYSIS. *Educ:* Univ Pittsburgh, BS, 61, MS, 62, PhD(math), 64. *Prof Exp:* Instr math, Univ Pittsburgh, 63-64; mathematician, US Army, 64-66; sr mathematician, Bettis Atomic Power Lab, Westinghouse Elec Corp, 66-70; assoc prof, 70-77, PROF MATH, UNIV PITTSBURGH, 78- *Concurrent Pos:* Instr math, NMex State Univ, 65-66; lectr, Univ Pittsburgh, 67-70; consult, Gen Motors Res, 71-, Westinghouse Elec, 74-81 & Pittsburgh Corning, 80-86, Contraves Goerz, 87-88; exec dir, Inst Comput Math & Applns, 78-88. *Mem:* Soc Indust & Appl Math. *Res:* Matrix analysis; interpolation and approximation by piecewise polynomial functions; error bounds for interpolating polynomials; finite element methods; computational fluid dynamics; computational mechanics. *Mailing Add:* ICMA Math & Statist Univ Pittsburgh Pittsburgh PA 15260

HALL, CHARLES E, b Flushing, NY, June 9, 26; m 48; c 8. VETERINARY MEDICINE. *Educ:* Alfred Univ, AB, 50; Cornell Univ, DVM, 53. *Prof Exp:* Intern med & obstet, Cornell Univ, 53-54; pvt practr, NY State, 54-69; ASSOC PROF REPRODUCTIVE STUDIES, NY STATE VET COL, CORNELL UNIV, 69- *Mem:* Am Vet Med Asn; Asn Bovine Practrs; Am Asn Vet Clinicians. *Res:* Reproductive diseases of cattle. *Mailing Add:* PO Box 0 Ovid NY 14521

HALL, CHARLES ERIC, b Montreal, Que, Aug 27, 16; m 44. PHYSIOLOGY, ANATOMY. *Educ:* McGill Univ, BSc, 41, MSc, 42, PhD(anat), 46. *Prof Exp:* Asst anat, McGill Univ, 45-46; instr psychobiol, Johns Hopkins Hosp, 46-47; from asst prof to assoc prof, 47-54, PROF PHYSIOL & BIOPHYS, UNIV TEX MED BR GALVESTON, 54- *Concurrent Pos:* Sabbatical, Karolinska Inst, Stockholm, Sweeden, 54. *Honors & Awards:* Fantham Mem prize, Can, 41. *Mem:* Fel AAAS; Am Physiol Soc; Endocrine Soc; Soc Exp Biol & Med; Am Heart Asn. *Res:* Endocrinology; andrenal cortex; pituitary; cardiovascular lesions; hypertension; parabiosis disease; renal enzymes. *Mailing Add:* Dept Physiol & Biophys Univ Tex Med Br Galveston Galveston TX 77550-2781

HALL, CHARLES FREDERICK, b San Francisco, Calif, Apr 7, 20; m 42; c 3. ASTRONAUTICS, AERONAUTICS. *Educ:* Univ of Calif, BS, 42. *Prof Exp:* Res scientist aeronaut, 42-56, br chief wind tunnel, 56-59, asst div chief astronaut, Vehicle Environ Div, 59-62, PIONEER PROJ MGR SPACE ASTRONAUT, AMES RES CTR, NASA, 62- *Mem:* Sigma Xi. *Res:* Performance of wings and air inlets; stability and control of aircraft at subsonic, supersonic, and transonic speeds. *Mailing Add:* 10651 Castine Ave Cupertino CA 95014-1312

HALL, CHARLES MACK, b El Dorado, Ark, Sept 18, 41; m 67; c 3. ORGANIC CHEMISTRY. *Educ:* Univ of the South, BA, 63; Univ Minn, PhD(org chem), 67. *Prof Exp:* NIH fel, Univ Munich, 67-68; asst prof, Univ Minn, 68-69; res assoc, 69-77, res head, 78-84, DIR, HYPERSENSITIVITY DIS RES, UPJOHN CO, 84- *Mem:* Am Chem Soc; Royal Soc Chem. *Res:* Synthesis of heterocyclic compounds and nucleosides; medicinal chemistry; anti-allergy agents. *Mailing Add:* Hypersensitivity Dis Res Upjohn Co Kalamazoo MI 49001

HALL, CHARLES THOMAS, b Baltimore, Md, July 1, 29; m 62; c 3. MICROBIOLOGY, IMMUNOLOGY. *Educ:* Univ Md, BS, 54, MS, 60, PhD(microbiol), 62. *Prof Exp:* Asst bact, Univ Md, 54-57, asst immunol, 57-62; med bacteriologist, Ctr Dis Control, USPHS, 73-75, chief, Microbiol & Serol Unit, Proficiency Testing Sect, 68-86, chief sexually transmitted dis, 75-86; RETIRED. *Mem:* Sigma Xi. *Res:* Variations of antigens within bacterial populations; application of fluorescent antibody procedures; synthesis and degeneration of streptococcal M and T proteins. *Mailing Add:* 3742 Kayanne Ct Tucker GA 30084

HALL, CHARLES VIRDUS, b Ash Flat, Ark, June 18, 23; m 49; c 3. VEGETABLE CROPS, HORTICULTURE. *Educ:* Univ Ark, BS, 50, MS, 53; Kans State Univ, PhD(entom, bot), 60. *Prof Exp:* Asst, Fruit & Truck Br Exp Sta, Univ Ark, 51-53; from asst prof to prof hort, Kans State Univ, 53-74; PROF HORT & HEAD DEPT, IOWA STATE UNIV, 74- *Honors & Awards:* Asgrow Award, 64; Marion W Meadows Award, 72. *Mem:* Fel Am Soc Hort Sci; AAAS. *Res:* Vegetable breeding and genetics. *Mailing Add:* Dept Hort Iowa State Univ Ames IA 50011

HALL, CHARLES WILLIAM, b Gage, Okla, Feb 8, 22; m 43, 62, 79; c 5. EXPERIMENTAL SURGERY. *Educ:* Univ Kans, AB, 50, MA, 52, MD, 56; Am Bd Surg, dipl, 63. *Hon Degrees:* Dr Eng, Rose-Hulman Inst Technol, 85. *Prof Exp:* Intern, Med Ctr, Univ Kans, 56-57, resident gen surg, 57-60, sr resident gen surg, 61-62, fel C-V surg, Baylor Col Med, 62-64; proj officer artificial heart prog, Nat Heart Inst, 64; asst prof surg & physiol, Baylor Col Med, 64-68; mgr artificial organs res, 68-70, dir dept bioeng, 70-75, INST PHYSICIAN, SOUTHWEST RES INST, 70- INST MED SCIENTIST, 75- *Concurrent Pos:* Fel physiol, Baylor Col Med, 57, fel surg, 62-64; fel, Med Ctr, Univ Kans, 57-58, fel cardiol, 60, fel med, 60-61; consult, Nat Heart Inst, 64-, mem training grants comt; hon prof, Cath Univ Cordoba, 65-; clin asst prof, Dept Surg, Univ Tex Health Sci Ctr, San Antonio, 69-; pres, Acad Surgical Res, 85; founding chmn bd, Biomat Res Int Ctr, 86-; founding pres, Soc Biomat, 74. *Honors & Awards:* Minister Bienstar Sociale Medal of Merit, Arg, 74; Imagineer Award, Mind Sci Found, 87. *Mem:* Am Soc Artificial Internal Organs; Am Heart Asn; fel Am Col Chest Physicians; fel Am Col Cardiol; Soc Biomet (pres, 74); Acad Surg Res (pres, 85). *Res:* Comparative anatomy; artificial heart; pulmonary surgery; development of techniques and equipment in medical monitoring; biomedical materials. *Mailing Add:* Southwest Res Inst PO Drawer 28510 San Antonio TX 78228-0510

HALL, CHESLEY BARKER, b Concord, NH, Sept 5, 20; m 43; c 4. PLANT PHYSIOLOGY. *Educ:* Univ NH, BS, 42; Purdue Univ, MS, 48; Cornell Univ, PhD(veg crops), 50. *Prof Exp:* Asst veg crops, Cornell Univ, 48-50; from asst horticulturist to assoc horticulturist, 50-64, actg chmn, veg crops dept, 78-79, HORTICULTURIST, UNIV FLA, 64- *Mem:* Fel AAAS; fel Am Soc Hort Sci; Am Soc Plant Physiologists; Int Soc Hort Sci. *Res:* Physiological and biochemical aspects of vegetable quality as affected by environmental factors with emphasis on tomato. *Mailing Add:* Veg Crops Dept IFAS Univ Fla Gainesville FL 32611

HALL, CLARENCE ALBERT, JR, b Los Angeles, Calif, Jan 5, 30; div; c 2. STRATIGRAPHY, TECTONICS. *Educ:* Stanford Univ, BS, 52, MS, 53, PhD, 56. *Prof Exp:* Instr geol, Univ Ore, 54-55; instr, Stanford Univ, 56; from asst prof to assoc prof, 56-68, chmn dept, 74-78, PROF GEOL, UNIV CALIF, LOS ANGELES, 68-, DIR, UNIV CALIF WHITE MOUNTAIN RES STA, 80-, DEAN DIV PHYS SCI, 83- *Concurrent Pos:* Fulbright res scholar, 63-64 & 70-71. *Mem:* Fel Geol Soc Am; Paleont Soc. *Res:* Structural and stratigraphic geology of western California; large faults along coastal central California and in southern France; geology of the White Mountains and eastern California. *Mailing Add:* Dept Earth & Space Sci Univ Calif Los Angeles CA 90024

HALL, CLARENCE CONEY, JR, b Burkburnette, Tex, Sept 22, 21; m 46; c 2. ACAROLOGY. *Educ:* Southern Methodist Univ, BS, 49, MS, 50; Univ Kans, PhD(entom), 61. *Prof Exp:* Instr biol, Howard Col, 50-51, from asst prof to assoc prof, 55-63; assoc prof, 63-68, PROF BIOL, UNIV TEX, ARLINGTON, 68- *Mem:* Entom Soc Am. *Res:* Taxonomy and biology of phytophagous mites, especially Eriophyoidea. *Mailing Add:* 4006 Fairway Ct Arlington TX 76013

HALL, COLBY D(IXON), JR, b Ft Worth, Tex, Apr 17, 19; m 53; c 2. PHYSICAL CHEMISTRY, PETROLEUM ENGINEERING. *Educ:* Tex Christian Univ, BS, 39, MS, 41; Univ Tex, PhD(chem), 52. *Prof Exp:* Sr chemist res, Dowell Div, Dow Chem Co, 52-55, lab group leader, 55-64, supvr basic res, 64-69; staff res engr, Amoco Prod Co, 69-86; RETIRED. *Mem:* Am Chem Soc; Soc Petrol Engrs. *Res:* Oil well stimulations; rheology of slurries and polymer solutions; enhanced oil recovery. *Mailing Add:* 5005 E 38th Pl Tulsa OK 74135

HALL, DAVID ALFRED, b Warsaw, NY, Aug 12, 40; m 65; c 2. ELECTROCHEMISTRY. *Educ:* Rochester Inst Technol, BS, 63; Univ Mich, Ann Arbor, MS, 65, PhD(chem), 69. *Prof Exp:* Res scientist, 67-80, RES ASSOC ELECTROCHEM, ELI LILLY & CO, 80- *Mem:* Am Chem Soc; Sigma Xi; Electrochem Soc. *Res:* Investigation of the electrochemical redox behavior of pharmaceutically interesting organic compounds; application of the knowledge gained from these mechanistic studies to the electroorganic synthesis of the compounds. *Mailing Add:* 2506 Bluffwood Dr W Indianapolis IN 46208

HALL, DAVID GOODSELL, SR, b Port Jefferson, Ohio, Aug 7, 03; m 25; c 2. MEDICAL ENTOMOLOGY, PARASITOLOGY. *Educ:* Ohio State Univ, BSc, 26; Kans State Univ, MSc, 29. *Prof Exp:* Asst entomologist, Univ Ark, 26-28; asst entomologist, USDA, 29-35, assoc, 35-42; med entomologist, CEngrs, USAAF, 42-46; entomologist, USDA, 46-48, asst chief, Insect Pest Surv Info, Bur Entom Plant Quarantine, 48-50, chief, 50-54, chief, Publ Br, Agr Res Serv, 54-66, English ed, Agr Exp Sta, Univ PR, 66-76, coop scientist, Insect Identification & Beneficial Insect Introd Inst, Agr Res Serv, USDA, 76-80; RETIRED. *Honors & Awards:* Superior Serv Award, USDA, 54; Silver Anvil, Am Pub Rels Asn, 55. *Mem:* Entom Soc Am; Sigma Xi; Agr Communicators Educ; AAAS. *Res:* Biology and taxonomy of diptera involved in the health of man and animals; bites and stings of insects and related arthropods. *Mailing Add:* RR 1 Rocheport MO 65279

HALL, DAVID GOODSELL, III, b Fayetteville, Ark, June 11, 27; m 51; c 3. OBSTETRICS & GYNECOLOGY. *Educ:* Univ Va, MD, 53; Am Bd Obstet & Gynec, dipl. *Prof Exp:* Intern, Univ Va Hosp, 53-54, from jr asst resident to sr asst resident obstet & gynec, 54-58; asst prof, 58-64, PROF OBSTET & GYNEC & CHMN DEPT, SCH MED, UNIV MO-COLUMBIA, 64-, ATTEND OBSTETRICIAN & GYNECOLOGIST, MED CTR, 58- *Concurrent Pos:* Am Cancer Soc fel, Univ Va Hosp, 56-57; Univ Mo rep med educ, Nat Defense Comt. *Mem:* Fel Am Col Surg; Am Asn Obstet & Gynec; AMA; Am Col Obstet & Gynec; Asn Profs Gynec & Obstet. *Res:* Effects of magnesium on uterine contractions and its part in toxemia of pregnancy; diagnosis and therapy. *Mailing Add:* 807 Stadium Rd Columbia MO 65201

HALL, DAVID GOODSELL, IV, b Charlottesville, Va, Nov 7, 52; m 77; c 2. INTEGRATED PEST MANAGEMENT. *Educ:* Univ Mo, BA, 76, MS, 78; Tex A&M Univ, PhD(entom), 81. *Prof Exp:* CHIEF ENTOMOLOGIST, RES DEPT, US SUGAR CORP, 81- *Mem:* Sigma Xi; Am Soc Sugar Cane Technologists; Entom Soc Am; Am Registry Prof Entomologists. *Res:* Economic injury levels, biological control, chemical control; management strategies for insect pests of sugar cane & citrus. *Mailing Add:* US Sugar Corp PO Drawer 1207 Clewiston FL 33440

HALL, DAVID JOSEPH, b Morristown, NJ, June 21, 43; m 62; c 3. GEOPHYSICS. *Educ:* Beloit Col, BS, 65; Univ Mass, MS, 70, PhD(geol & geophys), 74. *Prof Exp:* Lectr geol, Bucknell Univ, 70-71; asst prof earth sci, Adrian Col, 71-73, chmn dept, 73-74; proj geophysicist, Houston Tech Serv Ctr, 74-78, dir, US Explor Interpretation Sect, Gulf Res & Develop Co, 78-79; dir seismic eval sect, 79-80, dir regional geophys sect, 80-81, sr staff geophysicist, Gulf Cent Explor Group, Gulf Res & Develop, 81-82; mgr geophys, Zenith Petrol, 82-83; mgr geophys, Int Oil & Gas Corp, 83-86; sr to CHIEF GEOPHYSICIST, TOTAL MINATOME CORP, HOUSTON, TX, 86- *Mem:* Am Geophys Union; Soc Explor Geophysicists; Am Assoc Petrol Geologists. *Res:* Integrated analysis of regional geophysical and geological data; development of geotectonic models consistent with multidisciplinary data sets; use of geophysical data in international and domestic exploration for accumulations of hydrocarbons. *Mailing Add:* 15 Twelve Pines Ct The Woodlands TX 77381

HALL, DAVID MICHAEL, b Birmingham, Ala, Mar 26, 36; m 64; c 4. TEXTILE CHEMISTRY & ENGINEERING. *Educ:* Auburn Univ, BSTC, 58; Clemson Univ, MSTC, 62; Univ Manchester, PhD(polymer chem), 64. *Prof Exp:* Assoc prof, 65-76, PROF TEXTILE ENG, AUBURN UNIV, 76- *Concurrent Pos:* Fel, Swiss Fed Inst, Zurich, 65; NSF grant, 65-67, Water Resources Res Inst grant, 69-71, ITT Rayonier grant, 79, Off Water Res & Technol grant, 80, Dow Chem grants, 86-88, US Army & USN grants, 90, 91; consult, Textile & Related Industs, 69-; fel & chartered technologist, Textile Inst Gt Brit, 74 & fel & chartered colorist, Soc Dyers & Colourists, Gt Brit, 74. *Mem:* Fiber Soc; sr mem Am Asn Textile Chem & Colorists; fel Soc Dyers & Colourists Gt Brit; Nat Soc Prof Engrs; fel Textile Inst; Sigma Xi. *Res:* Fiber and polymer chemistry; carbohydrate chemistry as it relates to fibers and starch; chemistry of dyes and coloring matters; chemistry of natural and man made fibers. *Mailing Add:* Dept Textile Eng Auburn Univ Auburn AL 36849-5327

HALL, DAVID WARREN, b Las Vegas, Nev, June 8, 35; div; c 3. ORGANIC CHEMISTRY. *Educ:* Univ Ariz, BS, 57; Calif Inst Technol, PhD(org chem), 63. *Prof Exp:* Res chemist, Denver Res Ctr, Marathon Oil Co, 62-65, adv res chemist, 65-67; res fel chem, Univ BC, 67-69; asst prof, 69-70, assoc prof chem, Colo Sch Mines, 70-78; sr res chemist, Loctite Corp, 78-79, scientist, 79; STAFF ANALYTICAL CHEMIST, IBM, 81- *Mem:* Am Chem Soc. *Res:* Substituent effects on reactions of ferrocene derivatives; oxidation kinetics and mechanisms. *Mailing Add:* 572 E 4050S Apt 9C Salt Lake City UT 84107-1857

HALL, DENNIS GENE, b Belleville, Ill, Mar 7, 48; m 70; c 3. GUIDED-WAVE OPTICS, THIN-FILMS. *Educ:* Univ Ill, BS, 70; Southern Ill Univ, MS, 72; Univ Tenn, PhD(physics), 76. *Prof Exp:* Asst prof physics, Southern Ill Univ, 76-78; sr engr, McDonnell Douglas Astronautics Co, McDonnell Douglas Corp, 78-80; asst prof, 80-82, assoc prof, 82-87, PROF OPTICS, INST OPTICS, UNIV ROCHESTER, 87- *Concurrent Pos:* Bd dirs, Optical Soc Am, 91- *Mem:* Am Phys Soc; fel Optical Soc Am; fel Soc Photo-Optical Instrumentation Engrs. *Res:* Integrated optics; optical properties of solids; thin-film phenomena; quantum electronics; surface phenomena; study of optical guided-waves. *Mailing Add:* Inst Optics Univ Rochester Rochester NY 14627

HALL, DIANA E, b New Haven, Conn, May 11, 38; m 60, 89; c 3. HISTORY OF MEDICINE & WOMENS STUDIES. *Educ:* Smith Col, BA, 59; Yale Univ, MA, 60, PhD(hist of sci & med), 66. *Prof Exp:* Res assoc, Yale Univ, 67-70; lectr biol, Boston Univ, 70-73, asst prof biol & hist, 73-83; dir, F C Wood Inst Hist Med, Col Physicians, Philadelphia, 83-89; DIR, WOMEN'S STUDIES, UNIV SOUTHERN MAINE, 89- *Concurrent Pos:* Fel, Radcliffe Inst, 76-77; NSF & NIH res grants, 76-78; vis sr historian, Nat Libr Med, 89-90. *Mem:* AAAS (secy, Sect L, 77-81); Hist of Sci Soc; Am Asn Hist Med. *Res:* Biomedical research in twentieth century; eighteenth century medical science; sex research-scientific and social aspects. *Mailing Add:* 94 Bedford St Portland ME 04103

HALL, DICK WICK, b Los Angeles, Calif, June 7, 12; m 48; c 2. PURE MATHEMATICS. *Educ:* Univ Va, PhD(math), 38. *Prof Exp:* Lectr math, Univ Calif, Los Angeles, 38; Nat Res Coun fel, Univ Pa & Univ Va, 38-39; from instr to asst prof, Brown Univ, 39-42; asst prof, Univ Md, 43-47, prof, 47-56; prof, 56-81, EMER PROF MATH, STATE UNIV NY BINGHAMTON, 81- *Mem:* Math Asn Am. *Res:* Structure theory of Peano spaces; continuous transformations defined on topological spaces; chromatic polynomials. *Mailing Add:* 26 Audubon Ave Binghamton NY 13903

HALL, DONALD D, b Cincinnati, Ohio, Mar 9, 33; m 56; c 3. GEOLOGY, INVERTEBRATE PALEONTOLOGY. *Educ:* Univ Cincinnati, BS, 55, MS, 60; Univ Mich, PhD(geol), 65. *Prof Exp:* Res geologist, Res Ctr, Union Oil Co, Calif, 64-66; assoc prof geol, 66-80, DIR AREA & INTERDISCIPLINARY STUDIES, CALIF STATE UNIV, CHICO, 66-, PROF GEOL & PHYS SCI, 80- *Mem:* AAAS. *Res:* Micropaleontology; Tertiary ostracoda. *Mailing Add:* Dept Geo & Phys Sci Calif State Univ Chico CA 95927

HALL, DONALD EUGENE, b Takoma Park, Md, Oct 27, 40; m 61; c 2. PHYSICS, ASTROPHYSICS. *Educ:* Southern Missionary Col, BA, 61; Univ Iowa, MA, 73; Stanford Univ, MS, 64, PhD(physics), 67. *Prof Exp:* From asst prof to assoc prof physics, Walla Walla Col, 67-73; from asst prof to assoc prof 74-83, PROF PHYSICS, CALIF STATE UNIV, SACRAMENTO, 83- *Mem:* Acoust Soc Am; Am Guild Organists; Am Asn Physics Teachers; Am Astron Soc. *Res:* Mathematical physics; musical acoustics; astrophysics; plasma theory; nonlinear problems. *Mailing Add:* Dept Physics Calif State Univ 6000 J St Sacramento CA 95819

HALL, DONALD HERBERT, b Maple Creek, Sask, Nov 23, 25; m 55; c 3. GEOPHYSICS, SCIENTOMETRICS. *Educ:* Univ Alta, BSc, 48; Univ Toronto, MA, 50; Univ BC, PhD(geophys), 59. *Prof Exp:* Lectr, Univ BC, 57-59; asst prof, Univ Sask, 59-62; from asst prof to assoc prof, 62-69, head, Dept Earth Sci, 78-87, PROF GEOPHYS, UNIV MAN, 69- *Concurrent Pos:* Regional ed, Geoexplor, 63-83. *Mem:* Am Geophys Union; Can Geophys Union; Geol Asn Can. *Res:* The Earth's lithosphere using surface and spacecraft data; history of geosciences and science policy and its application to the geosciences primarily from scientometric historical methods. *Mailing Add:* Dept Geol Sci Univ Man Winnipeg MB R3T 2N2 Can

HALL, DONALD NORMAN BLAKE, b Sydney, Australia, June 26, 44; m 67; c 2. ASTRONOMY, SPECTROSCOPY. *Educ:* Univ Sydney, BSc, 66; Harvard Univ, PhD(astron), 70. *Prof Exp:* Exp officer physics, Div Physics, Commonwealth Sci & Indust Res Orgn, Sydney, 66-67, div studentship astron, 67-70; res assoc, Kitt Peak Nat Observ, 70-72, assoc astronr, 72-76, astronr, 76-81; dep dir, Space Telescope Sci Inst, 82-84; DIR, INST ASTRON, UNIV HAWAII, 84- *Concurrent Pos:* Consult, Smithsonian Inst Astrophys Observ, 67-69; teaching fel, Dept Astron, Harvard Univ, 68; mem Space Sci Bd, Nat Acad Sci, 84-; mem Astron Adv Comt, NSF, 84-87; mem Astrophys Coun, NASA, 84-; mem, Space Sci Working Group, 84-; mem, Hubble Space Telescope Sci Working Group, 86- *Honors & Awards:* Newton Lacey Pierce Prize, Am Astron Soc, 78. *Mem:* Am Astron Soc; Int Astron Union. *Res:* High resolution infrared spectroscopy (including observations and development of instrumentation) of astronomical sources applied to the detection of interstellar molecules and investigation of both extremely young and highly evolved stars; author of numerous technical publications. *Mailing Add:* Inst Astron Univ Hawaii 2680 Woodlawn Dr Honolulu HI 96822

HALL, DONALD WILLIAM, b Muncie, Ind, Dec 11, 42; m 65; c 2. MEDICAL ENTOMOLOGY. *Educ:* Purdue Univ, BS, 64, MS, 67; Univ Fla, PhD(entom), 70. *Prof Exp:* Asst prof entom, Univ Mass, 70-74; from asst prof to assoc prof, 75-80, PROF ENTOM, UNIV FLA, 80- *Mem:* Entom Soc Am; Soc Invert Path; Am Mosquito Control Asn. *Res:* Mosquito pathology. *Mailing Add:* Dept Entom & Nemat Bldg 970 Hull Rd Univ Fla Gainesville FL 32611-0740

HALL, DOUGLAS SCOTT, b Lexington, Ky, May 30, 40; m 64, 81; c 2. ASTRONOMY. *Educ:* Swarthmore Col, BA, 62; Ind Univ, MA, 64, PhD(astron), 67. *Prof Exp:* Res assoc astron, Arthur J Dyer Observ, 67; asst prof, 68-71, assoc prof, 71-80, PROF PHYSICS & ASTRON, 80-, DIR, ARTHUR J DYER OBSERV, VANDERBILT UNIV, 86- *Concurrent Pos:* NSF grant, Vanderbilt Univ, 68-, 70, 76 & 84; Res Corp grant, 78; NASA grants, 77, 78, 79, 80 & 81; ed, Int Amateur-Prof Photoelec Photom Commun. *Honors & Awards:* US Sr Scientist Award, Alexander Von Humboldt Found, 73. *Mem:* Am Astron Soc; Astron Soc Pac; Am Asn Variable Star Observers; Int Astron Union; Int Amateur-Prof Photoelec Photom; Sigma Xi. *Res:* Eclipsing binary stars; astronomical photometry; stellar evolution; star clusters; variable stars. *Mailing Add:* Dyer Observ Vanderbilt Univ Nashville TN 37235

HALL, DWIGHT HUBERT, b Rumford, Maine, July 27, 40; m 63. MOLECULAR GENETICS, MICROBIAL GENETICS. *Educ:* Bowdoin Col, BA, 62; Purdue Univ, MS, 65, PhD(biophys), 67. *Prof Exp:* NSF fel biol, Mass Inst Technol, 66-68; asst prof biochem, Med Ctr, Duke Univ, 68-77; ASSOC PROF BIOL, GA INST TECHNOL, 77- *Concurrent Pos:* NIH res grant genetics, Med Ctr, Duke Univ, 68-77, Ga Inst Technol, 77-, res career develop award, 70-75. *Mem:* AAAS; Am Soc Biol Chemists; Genetics Soc Am; Am Soc Microbiol; Environ Mutagen Soc. *Res:* Biochemical genetics of bacterial viruses; production, isolation and characterization of mutants; organization and expression of the genome of bacteriophage T4; regulation; recombination; biosynthesis and interconversions of pyrimidine nucleotides; mechanisms of resistance to antimetabolites; genetic engineering. *Mailing Add:* Sch Biol Ga Inst Technol Atlanta GA 30332

HALL, EDWARD DALLAS, b Bedford, Ohio, June 16, 50; m 70; c 2. NEUROPHARMACOLOGY. *Educ:* Mt Union Col, BS, 72; Cornell Univ, PhD(pharmacol), 76. *Prof Exp:* Fel pharmacol, Med Col, Cornell Univ, 76-77; asst prof biol sci, Kent State Univ, 78-82; from asst prof to assoc prof pharmacol, Col Med, Northeastern Ohio Univ, 78-82; res scientist, 82-84, sr res scientist, 85-90, SR SCIENTIST, CENT NERV SYST DIS RES, UPJOHN CO, 90- *Concurrent Pos:* Prin investr, NIMH grant, 78-82, Amyotrophic Lateral Sclerosis Soc Am grant, 78-82; adj assoc prof, Western Mich Univ, Physicians Asst Prog, 83-89; mem, sci adv coun, Am Paralysis Asn, 85-88; adj assoc prof, 85-90, adj prof, Northeastern Ohio Univ Col Med, 90-; assoc ed, J Neurotrauma, 88- *Mem:* Am Soc Pharmacol & Exp Therapeut; Soc Neurosci; NY Acad Sci; Soc Neurotrauma (vpres, 89-90); Sigma Xi. *Res:* Acute treatment of central nervous system trauma and stroke; treatment of degenerative neurological diseases; role of oxygen radicals and lipid peroxidation in acute and chronic neuronal degeneration. *Mailing Add:* Cent Nerv Syst Dis Res Unit Upjohn Co Kalamazoo MI 49001

HALL, EDWARD DUNCAN, physical chemistry, chemistry of dyeing textiles, for more information see previous edition

HALL, ELIZABETH ROSE, b Sunnyside, Wash, Aug 29, 14. BACTERIOLOGY. *Educ:* Univ Wash, BS, 37; Univ Mich, MS, 41; Wash State Univ, PhD(bact), 52. *Prof Exp:* Instr bact, Med Sch, Univ Okla, 41-44; from instr to prof, 44-76, EMER PROF BACT, WASH STATE UNIV, 76- *Mem:* Am Soc Microbiol; NY Acad Sci. *Res:* Medical microbiology; immunology. *Mailing Add:* NW 412 Sunset Dr Pullman WA 99163

HALL, ELTON HAROLD, physical chemistry, for more information see previous edition

HALL, ERIC JOHN, b Abertillery, Gt Brit, July 5, 33; m 57; c 1. RADIOBIOLOGY. *Educ:* Univ London, BSc, 53; Oxford Univ, DPhil(radiobiol), 62. *Hon Degrees:* MA, Oxford Univ, 66. *Prof Exp:* Prin physicist, Churchill Hosp, Oxford Univ, 55-68; PROF RADIOL, COLUMBIA UNIV, 68- *Mem:* Radiation Res Soc; Brit Hosp Physicist's Asn; Brit Inst Radiol. *Res:* Effects of ionizing radiation on cells in culture; carcinogenesis; applications of radiobiology to cancer therapy. *Mailing Add:* Radiol Res Lab Col P-S 630 W 168th St New York NY 10032

HALL, ERNEST LENARD, b Naylor, Mo, Dec 8, 40; m 69; c 4. BIOENGINEERING, RADIOLOGY. *Educ:* Univ Mo-Columbia, BS, 65, MS, 66, PhD(bioeng), 71. *Prof Exp:* Asst prof elec eng, bioeng & radiol, Univ Mo-Columbia, 71-72; asst prof radiol, Yale Univ, 72-73; asst prof elec eng & radiol, Univ Southern Calif, 73-76; from assoc prof to prof elec eng, Univ Tenn, Knoxville, 80-88; CTR ROBOTICS, UNIV CINCINNATI, 88- *Concurrent Pos:* NSF grant, Yale Univ, 71-72; Nat Inst Occup Safety & Health grant, Univ Southern Calif, 73-; consult, Oak Ridge Nat Lab, Jet Propulsion Lab, Calif Inst Technol, 73- & Lab Radiation Biol, Univ Calif, Los Angeles, 73- *Honors & Awards:* Centennial Medal, Inst Elec & Electronics Engrs, 84. *Mem:* Nat Acad Eng; fel Inst Elec & Electronics Engrs; fel Soc Photo-Optical Instrumentation Engrs; Am Asn Physicists Med; Asn Advan Med Instrumentation; Int Soc Optical Eng. *Res:* Automated measurements; pattern recognition and picture processing in medicine; intelligent robotics. *Mailing Add:* Ctr Robotics ML-72 Univ Cincinnati Cincinnati OH 45221

HALL, ESTHER JANE WOOD, pharmacy, health care administration, for more information see previous edition

HALL, FORREST G, b La Feria, Tex, June 7, 40; c 2. PHYSICS, REMOTE SENSING. *Educ:* Univ Tex, BS, 63; Univ Houston, MS, 68, PhD(physics), 70. *Prof Exp:* Aerospace technologist, NASA-Johnson Space Ctr, 63-64, physicist, 64-74, proj scientist large area crop inventory exp, 74-78, chief scientist, Earth Observations Div, 78-80, chief, Scene Anal Br, Earth Resources Res Div, 80-85. *Mem:* AAAS; Inst Elec & Electronics Engrs. *Res:* Application problems of multispectral remote sensing to large area survey estimates of crop area, yield and production. *Mailing Add:* Goddard Space Ctr Greenbelt MD 20771

HALL, FRANCIS RAMEY, b Salmon, Idaho, Feb 25, 25; m 60. HYDROLOGY. *Educ:* Stanford Univ, BS, 49, PhD(geol), 61; Univ Calif, Los Angeles, MA, 53. *Prof Exp:* Ground-water geologist, US Geol Surv, 51-56; tech asst, Stanford Univ, 56-58; ground-water geologist, Stanford Res Inst, 58-59, res geologist in charge geohydrol, 59-61; assoc hydrologist, NMex Inst Mining & Technol, 61-64; assoc prof soil & water sci, 64-71, prof hydrol, Inst Natural & Environ Resources, 71-83, prof hydrogeol, Earth Sci Dept, 83-90, EMER PROF, UNIV NH, 90- *Mem:* AAAS; Geol Soc Am; Am Geophys Union; Am Water Resource Asn; Nat Water Well Asn. *Res:* Ground-water hydrology, applications of digital computers to fluid flow and chemical transport problems; effect of drainage basin parameters on pattern of ground-water outflow; impact of waste disposal on ground water; radon gas in ground water. *Mailing Add:* PO Box Ten Durham NH 03824

HALL, FRANK FOY, b Seymour, Tex, Apr 2, 40; m 63; c 4. BIOCHEMISTRY. *Educ:* Tex A&M Univ, BS, 62, PhD(biochem), 66; Am Bd Clin Chem, dipl, 73; Am Bd Bioanal, cert lab dir, 75. *Prof Exp:* Res assoc biochem, Tex A&M Univ, 66; chief chem br, Clin Biochem, Med Lab, Ft Sam Houston, Tex, 66-68; fel, Scott & White Mem Hosp, 68-70, clin biochemist, 70-74; tech dir, Damon Med Lab, Inc, 74-75; dir, Bio-Sci Labs, St Louis, 75-76; pres, Int Clin Labs Mo, Inc, St Louis, 76-81; sci dir & vpres Lab Opers, Dallas, 81-86, VPRES OPERS, INT CLINS LABS, INC, W REGION, IRVING, 86- *Concurrent Pos:* Adj asst prof chem, Baylor Univ; NIH grant; assoc prof biochem, Tex Tech Univ Sch Med, 86- *Mem:* AAAS; Am Chem Soc; fel Am Asn Clin Chemists; Sigma Xi; Am Inst Chem. *Res:* Clinical laboratory methodology, clinical laboratory management, medical research, clinical chemistry, clinical biochemistry. *Mailing Add:* 5705 Grand Oak Ct Colleyville TX 76034

HALL, FRANKLIN ROBERT, b Boston, Mass, Oct 30, 34; m 57; c 2. ENTOMOLOGY. *Educ:* Univ Mass, Amherst, BS, 56; State Univ NY Col Forestry, Syracuse Univ, MS, 61; Purdue Univ, PhD(entom), 67. *Prof Exp:* Field rep entom, Niagara Chem Co, 60-63; fruit crop specialist, Chevron Chem Co, Standard Oil Co Calif, 67-70; prof develop rep agr, Chemagro Corp, 70; asst prof, Ohio Agr Res & Develop Ctr, Ohio State Univ, 70-73, assoc prof, 73-78, prof entom, 73-78; HEAD, LAB PEST CONTROL APPLN TECHNOL, 81- *Mem:* Entom Soc Am. *Res:* Biology, ecology and management of insect and mite pests of fruit; crop loss assessments; pesticide application technology; computerized decision support systems. *Mailing Add:* 1535 Morgan Wooster OH 44691

HALL, FREDERICK COLUMBUS, b Milwaukee, Wis, Apr 19, 27; m 84; c 2. PLANT ECOLOGY. *Educ:* Purdue Univ, BS, 51; Ore State Univ, MS, 56, PhD(plant ecol), 65. *Prof Exp:* Range conservationist, US Forest Serv, 56-58, vis asst prof range mgt, Idaho, 58-60, plant ecologist, 60-72, REGIONAL ECOLOGIST, US FOREST SERV, 72- *Concurrent Pos:* Consult, Fish & Wildlife Comt, USMC, 51-54. *Honors & Awards:* Super Serv, USDA, 76, Cert Merit, 78. *Mem:* Soc Range Mgt; Soc Am Foresters; Ecol Soc Am; Wildlife Soc. *Res:* Ecology of forest and range areas; forest community ecology; evaluation of grazing and logging effects; prediction of vegetation reactions; evaluation of total dry matter production; interpretation of data for land management application. *Mailing Add:* USDA Forest Serv PO Box 3623 Portland OR 97208

HALL, FREDERICK KEITH, b Leeds, Eng, Jan 3, 30; m 56; c 3. CELLULOSE CHEMISTRY. *Educ:* Univ Manchester, BS, 51; Univ Leeds, PhD(textile chem), 55; Harvard Bus Sch, AMP, 79. *Prof Exp:* Res chemist, Courtaulds (Can), Ltd, 56-58, asst tech mgr, 58-60, tech mgr, 60-63, dep plant mgr, 63-66; dir tech serv, Int Pulp Sales Corp, Int Paper Co, 66-70, asst dir, Corp Res Ctr, 70-72, dir primary process, 72-75, corp dir res, 75-77; dir, 77-79, dir res, 79-82, CHIEF SCIENTIST, S & T LABS, 79-, DIR SCI & EXPLOR DEVELOP, 82- *Concurrent Pos:* Mem res adv comt, Textile Res Inst, 71-, mem bd trustees, 90-; officer & treas, Empire State Paper Res Asn, 78-; dir, Tech Asn Pulp & Paper Indust, 83-86. *Mem:* Fel Royal Soc Chem; fel Textile Inst Eng; Chem Inst Can; fel Tech Asn Pulp & Paper Indust (vpres, 89-91, pres 91-); fel Am Inst Chemists. *Res:* Pulp and paper chemistry; textile chemistry; wood chemistry; plastics; fibers; films; biological science, particularly as applied to trees; engineering development as applied to forest products and pulp industries; biomedical engineering. *Mailing Add:* Int Paper Co Box 797 Long Meadow Rd Tuxedo Park NY 10987

HALL, FREEMAN FRANKLIN, JR, b Kansas City, Mo, Sept 29, 28; m 52; c 4. ATMOSPHERIC SCIENCES, REMOTE SENSING. *Educ:* Occidental Col, BA, 50; Univ Calif, Los Angeles, MS, 57, PhD(meteorol), 67. *Prof Exp:* Physicist, US Naval Ord Test Sta, Calif, 51 & Lockheed Missile Systs Div, 54-57; sr staff mem, ITT Fed Labs Div, Int Tel & Tel Corp, 57-63, assoc lab dir, 63-66; res scientist, Douglas Advan Res Labs, 66-70; res meteorologist, Nat Oceanic & Atmospheric Admin, 70-87; CHIEF SCIENTIST, HARRIER CONSULTS, 88- *Concurrent Pos:* Expert witness, visibility & audibility. *Mem:* Optical Soc Am; Am Meteorol Soc. *Res:* Optical radiometry; atmospheric scattering, propagation and emission; infrared astronomy; optical instrument design; laser applications; atmospheric acoustics; planetary boundary layer observation and instrumentation; development, application and interpretation of infrared Doppler lidar for atmospheric dynamics studies. *Mailing Add:* 202 Ocean St Solana Beach CA 92075

HALL, GARY R, b Milledgeville, Ga, Jan 14, 44; m 68; c 1. CORROSION. *Educ:* Univ Pittsburgh, BS, 66. *Prof Exp:* Lab technician glass & ceramic, Kopp Glass, 63-64 & alloy & coatings, Alcoa Appl Res & Develop Labs, 64-68; SR CHEMIST CERAMICS & COATINGS, SAUEREISEN CEMENTS, 68- *Mem:* Nat Asn Corrosion Engrs; Am Inst Chem Engrs; Am Soc Testing & Mat; Am Ceramic Soc; Am Concrete Inst. *Res:* Corrosion of metals, concrete, ceramics, polymers and plastics; methods of controlling corrosion, including high temperature corrosion; concrete and ceramic technologies. *Mailing Add:* Sauereisen Cements Co 160 Gamma Dr Pittsburgh PA 15238

HALL, GENE STEPHEN, b Plainfield, NJ, Feb 6, 51. RADIOANALYTICAL CHEMISTRY, BIO-ENVIRONMENTAL ANALYSIS. *Educ:* Tusculum Col, BS, 73; Va Tech & State Univ, PhD(chem), 78. *Prof Exp:* Teaching asst, Va Polytech & State Univ, 73-78; ASST PROF CHEM, RUTGERS UNIV, NJ, 79- *Concurrent Pos:* Prin investr, Rutgers Univ, NJ, 81- *Mem:* Am Chem Soc; Am Optical Soc; Nat Orgn Prof Advan Black Chemists & Chem Engrs. *Res:* Analysis of biological and environmental samples for trace and toxic metals using proton-induced X-ray emission; positronium reactions of chlorophyll and other compounds. *Mailing Add:* 42 Poplar Rd Piscataway NJ 08854

HALL, GEORGE ARTHUR, JR, b Parkersburg, WVa, June 16, 20; m 63; c 1. PHYSICAL CHEMISTRY, ORNITHOLOGY. *Educ:* Univ WVa, BS, 41; Ohio State Univ, PhD(phys chem), 45. *Prof Exp:* Instr chem, Ohio State Univ, 44-46 & Univ Wis, 46-50; from asst prof to prof, 50-86, EMER PROF CHEM, WVA UNIV, 86- *Concurrent Pos:* Ed, Wilson Bul, Wilson Ornith Soc, 63-73. *Mem:* AAAS; Am Chem Soc; Wilson Ornith Soc (2nd vpres, 75-77, 1st vpres, 77-79, pres, 79-81); Am Ornith Union. *Res:* Kinetics of solution reactions; distribution and ecology of West Virginian birds. *Mailing Add:* Dept Chem WVa Univ Morgantown WV 26506-6045

HALL, GEORGE E, b New Haven, Conn, July 14, 17; m 42; c 6. ORGANIC CHEMISTRY. *Educ:* Yale Univ, BS, 38, PhD(org chem), 42. *Prof Exp:* Res chemist, Calco Chem Div, Am Cyanamid Co, NJ, 41; from instr to prof, 46-79, chmn dept, 66-72, EMER PROF CHEM, MT HOLYOKE COL, 79- *Concurrent Pos:* Advan Educ fel & vis prof, Calif Inst Technol, 53-54, vis assoc, 69-70; NSF fel & vis lectr, Bristol Univ, 61-62; hon res fel, Univ Col London, 76-77. *Mem:* AAAS; Am Chem Soc; Royal Soc Chem; Sigma Xi. *Res:* Halogenation of ethers; electrical effects of substituents; biphenylene and derivatives; nuclear magnetic resonance. *Mailing Add:* 15 Silverwood Terr South Hadley MA 01075

HALL, GEORGE LINCOLN, b Brandywine, Va, Feb 18, 26; m 55; c 3. THEORETICAL PHYSICS, SOLID STATE PHYSICS. *Educ:* Col William & Mary, BS, 49; Syracuse Univ, MS, 51; Univ Va, PhD(physics), 56. *Prof Exp:* Asst physics, Col William & Mary, 48-49; asst, Syracuse Univ, 49-50; sr engr traveling waves and electron devices, Fed Telecommun Labs, 51-53; res assoc classified proj, Elec Eng Dept, Univ Va, 53, instr physics, 56; theoret physicist info theory, Res Labs, Westinghouse Elec Corp, 56-57; solid state physicist, Res Inst Advan Study, Baltimore, 57-60; instr math statist, Loyola Col, Md, 59-60; assoc prof physics, Kans State Univ, 60-64, prof, 64-66; PROF PHYSICS, NC STATE UNIV, 66- *Mem:* Am Phys Soc; Am Asn Physics Teachers; AAAS. *Res:* Order-disorder phenomena; lattice distortion around vacancies; lattice summation methods; Wigner solids; mathematical physics. *Mailing Add:* Dept Physics NC State Univ Box 8202 Raleigh NC 27650

HALL, GEROGE FREDERICK, b Spickard, Mo, Mar 5, 31; m 58; c 2. SOILS, SOIL SCIENCE. *Educ:* Univ Ill, BS, 59, MS, 61; Iowa State Univ, PhD(soil genesis, classification), 65. *Prof Exp:* Res asst agron, Univ Ill, 60-61; res assoc, Iowa State Univ, 61-64; soil scientist, Soil Conserv Serv, 64-65; from asst prof to assoc prof, 65-79, PROF AGRON, OHIO STATE UNIV, 79- *Concurrent Pos:* Consult, US AID, India, 72-73; res leave, Commonwealth Sci & Indust Res Orgn, Australia, 86-87. *Mem:* AAAS; Am Asn Quaternary Res; Am Soc Agron; Sigma Xi. *Res:* Earth science; soil genesis and classification; strip mine reclamation; soil geomorphology; landscape evolution; Pleistocene geology; Paleosols; landslips; arctic soils; land use. *Mailing Add:* Dept Agron Ohio State Univ 2021 Coffey Rd Columbus OH 43210-1086

HALL, GLENN EUGENE, b Tiffin, Ohio, Apr 21, 31; m 84; c 2. GRAIN QUALITY. *Educ:* Mich State Univ, BS, 59, MS, 60, PhD(agr eng), 67. *Prof Exp:* Asst prof res, Ohio Agr Res & Develop Ctr, 60-68; asst prof, eng topics, Univ Ill, 68-73; RES ENG, THE ANDERSONS, 73- *Mem:* Am Asn Cereal Chemists; Am Soc Mech Engrs; Am Soc Agr Engrs. *Res:* Grain drying, handling, quality, dust control and grain storage areas; sorption product development. *Mailing Add:* 420 W William Maumee OH 43537

HALL, GRETCHEN RANDOLPH, b Laurel, Md, Jan 16, 49. ANALYTICAL CHEMISTRY, INORGANIC CHEMISTRY. *Educ:* McGill Univ, BSc, 70; Univ Ill, PhD(chem), 75. *Prof Exp:* SR RES CHEMIST ANALYTICAL CHEM, MOBIL RES & DEVELOP CORP, 75- *Mem:* Am Chem Soc; Sigma Xi; Catalysis Soc. *Res:* Structure and composition research pertaining to petroleum and coal-derived materials. *Mailing Add:* Mobil Res/Develop Corp Billingsport Rd Paulsboro NJ 08066

HALL, GUSTAV WESLEY, b Chillicothe, Ohio, Jan 14, 34; m 56; c 2. BOTANY. *Educ:* Ohio Univ, AB, 55, MS, 57; Ind Univ, PhD(bot), 67. *Prof Exp:* Asst prof, 63-68, ASSOC PROF BIOL, COL WILLIAM & MARY, 68- *Mem:* AAAS; Soc Study Evolution; Am Soc Plant Taxon. *Res:* Plant taxonomy; biosystematics, especially of the family Compositae; flora of Virginia; ornithology. *Mailing Add:* Dept Biol Col William & Mary Williamsburg VA 23185

HALL, HARLAN GLENN, INSECT GENETICS. *Educ:* Univ Calif, Berkeley, PhD(genetics), 78. *Prof Exp:* Staff scientist, Lawrence Berkeley Labs, Univ Calif, 80-88; ASST PROF & HONEYBEE GENETICIST, DEPT ANAT & HEMAT, UNIV FLA, GAINESVILLE, 88- *Res:* Analysis of African honeybee DNA. *Mailing Add:* IFAS 0740 Bldg 970 Univ Fla Hull Rd Gainesville FL 32611

HALL, HAROLD HERSHEY, b Kinsman, Ohio, July 18, 24; m 48; c 5. THEORETICAL PHYSICS. *Educ:* SDak State Col, BS, 48; Univ Ore, MS, 49; Univ Wis, PhD(physics), 53. *Prof Exp:* Sr scientist, Theoret Group, Radiation Lab, Univ Calif, 52-54; sr scientist, Missiles & Space Div, Lockheed Aircraft Corp, 54-56; asst dir reentry & space systs, Aeronutronic Div, Philco Corp, Ford Motor Co, 56-64; chief scientist remote area conflict, Advan Res Proj Agency, Dept Defense, 64-65; dir appl res lab, Aeronutronic Div, Philco Ford, 65-68; pres, HRB-Singer, Inc, 68-70, vpres & chief tech off, Aerospace Group, Singer Co, 70-72, mgr, Syst Sci Lab, actg mgr, Xerox Palo Alto Res Ctr, 73-75; vpres, Corp Res Staff, Xerox Corp, 75-88; RETIRED. *Res:* Nuclear physics. *Mailing Add:* Rte One Box 75A Fulton SD 57340

HALL, HARVEY, b Butte, Mont, Aug 18, 04; m 34; c 3. RESEARCH ADMINISTRATION, EDUCATION ADMINISTRATION. *Educ:* Occidental Col, BA, 27; Univ Calif Berkeley, MA, 30 & PhD(physics), 31. *Prof Exp:* Instr physics, Columbia Univ, 31-34; lectr math physics, NY Univ, 34-36; asst prof physics, Col City NY, 36-42; prin physicist, Navy Dept, Bur Aeronaut, 42-48; head, physics dept, Fla State Univ, 58-61; chief scientist, Off Manned Space Flight, NASA, 61-73; RETIRED. *Concurrent Pos:* Chmn, Navy Comt Study Feasibility Space Rocketry, Bur Aeronaut, 46-48 & pioneer, Early Earth Satellite Prog; mem comt upper atmosphere, NASA, 43-51; dir, Off Naval Res, 51-58. *Mem:* AAAS; fel Am Phys Soc. *Res:* Theoretical physics theory of relativistic photo effect; photo disintegration of nucleus; stopping of fast charged particles by matter; coherent scattering of gamma rays. *Mailing Add:* 9000 Belvoir Wood Pkwy Apt 211 Ft Belvoir VA 22060-2702

HALL, HENRY KINGSTON, JR, b New York, NY, Dec 7, 24; m 51; c 3. ORGANIC CHEMISTRY. *Educ:* Brooklyn Polytech Inst, BS, 44; Pa State Col, MS, 46; Univ Ill, PhD(chem), 49. *Prof Exp:* Res assoc phys chem, Cornell Univ, 49-50; res assoc phys org chem, Univ Calif, Los Angeles, 50-52; res chemist, Textile Fibers Div, E I du Pont de Nemours & Co, Inc, 52-58, sr chemist, 58-65, group leader, Cent Res Dept, 65-69; head dept, 70-73, PROF CHEM, UNIV ARIZ, 69- *Concurrent Pos:* Sr vis fel, Japan Soc Promotion Sci, 81; consult, Hoechst-Celanese Co, 78-, Eastman Kodak Co, 80-, Ethicon Co, 79-, Exxon Co, 81-, Amoco Chems Co, 77- & Chevron Co, 76-; vis prof, Imperial Col, London, 76, Nagoya Univ, Japan, 81 & Mainz, Ger, 88. *Mem:* Am Chem Soc. *Res:* Mechanisms of polymerizations; synthesis and polymerization of novel monomers and high polymers. *Mailing Add:* Dept Chem Univ Ariz Tucson AZ 85721

HALL, HERBERT JOSEPH, b Springfield, Mass, Sept 30, 16; m 49; c 3. ELECTROSTATIC PRECIPITATION, HIGH VOLTAGE POWER SUPPLIES. *Educ:* Trinity Col, Conn, BS, 39; Univ Mich, Ann Arbor, MS, 40. *Prof Exp:* Mem staff, Radiation Lab, Mass Inst Technol, 41-45; sr physicist, Research-Cottrell, Inc, 45-54, asst dir res, 55-58, dir res & develop, 62-68, consult scientist aerosol physics, electrostatic precipitation & air polution control systs & equip, 68-69; vpres, Recon Systs, Inc, 69-72; PRES, H J HALL ASSOCS, INC, 73- *Concurrent Pos:* H E Russell fel, Trinity Col, 39; physicist, Bikini Tests, AEC, Los Alamos Lab, 46; sci consult, 59-62. *Mem:* Am Phys Soc; AAAS; NY Acad Sci; Air Pollution Control Asn; int fel Inst Electrostatics Japan. *Res:* Electrostatic precipitation; aerosol physics; physical electronics; high voltage electromagnetic components. *Mailing Add:* 258 Opossum Rd Skillman NJ 08558

HALL, HOMER JAMES, b Uniontown, Pa, Dec 12, 11; m 41; c 5. INFORMATION SCIENCE. *Educ:* Marietta Col, AB, 31; Ohio State Univ, MSc, 32, PhD(org chem), 35. *Prof Exp:* Lab asst biol & chem, Marietta Col, 27-31; asst, dept chem, Ohio State Univ, 31-35; res chemist, Res Div, Esso Labs, Standard Oil Develop Co, 35-48, group head, 43-48, tech adv & expert witness, Patent Div, Esso Res, 48-59, gen secy, Patent Comt, 53-59, head indust technol, Tech Info Div, Exxon Res & Eng Co, 59-63, head chem info, 63-65, spec ed, 65-67, info analyst, 67-68, proj mgr, Govt Res Labs, Linden, 68-77; NSF proj dir, Grad Sch Libr & Info Studies, 78-82, VIS RES PROF, RUTGERS UNIV, 81- *Concurrent Pos:* Dir, Wainwright House Develop Human Resources, 70-80, trustee, 80-; mem toxic substances subcomt, Sci Adv Bd, Environ Protection Agency, 80-82; chmn, Union County Cultural & Heritage Bd, 85-90. *Mem:* Fel AAAS; Am Chem Soc; Am Inst Chemists (chmn, secy, 75-80); Am Soc Info Sci; Air Pollution Control Asn; NY Acad Sci. *Res:* Pure hydrocarbons, distillation and analysis; petroleum processing; fluidized solids; technical analysis of patents and inventions; air pollution technology United States and foreign; trace elements in coal; problems in the evaluation of information as distinct from technology producing it; training of information analysts; information strategy; multi-dimensional interactions of value systems; professional ethics and codes. *Mailing Add:* 260 Prospect St No 17 Westfield NJ 07090-4016

HALL, HOWARD TRACY, b Ogden, Utah, Oct 20, 19; m 41; c 7. HIGH PRESSURE PHYSICS. *Educ:* Univ Utah, BS, 42, MS, 43, PhD(phys chem), 48. *Hon Degrees:* DSc, Brigham Young Univ, 71; DHumanities, Weber State Univ, 87. *Prof Exp:* Asst, US Bur Mines, Utah, 42-44, res assoc, 46; res assoc, Gen Elec Co, 48-55; dir res, 55-67, fac lectr, 64, DISTINGUISHED PROF CHEM, BRIGHAM YOUNG UNIV, 67- *Concurrent Pos:* Alfred P Sloan res fel, 59-63; Olin Mathiesen lectr, Yale Univ, 64; govt & indust consult. *Honors & Awards:* Res Medal, Am Soc Tool & Mfg Eng, 62; James E Talmage Sci Achievement Award, Brigham Young Univ, 65; Pioneer Chem Award, Am Inst Chem, 70. *Mem:* Fel AAAS; Am Chem Soc; fel Am Inst Chemists. *Res:* Ultra high pressure; high temperature technique and phenomena; synthesis of diamonds. *Mailing Add:* Dept Chem Bldg ESC-141 Brigham Young Univ Provo UT 84602

HALL, HUGH DAVID, b Henryetta, Okla, May 15, 31; m 60; c 3. ORAL & MAXILLOFACIAL SURGERY, PHYSIOLOGY. *Educ:* Univ Okla, BS, 53; Harvard Univ, DMD, 57; Am Bd Oral & Maxillofacial Surg, dipl, 65; Univ Ala, MD, 77. *Prof Exp:* From instr to asst prof oral surg, Sch Dent, Univ Ala, Birmingham, 61-65, assoc prof & chmn dept, 65-68; PROF ORAL SURG & CHMN DEPT, SCH MED, VANDERBILT UNIV, 68- *Concurrent Pos:* Fel oral surg, Med Ctr, Univ Ala, Birmingham, 59-62; USPHS grants, 60-67, res career develop award, 62-64, dir training grants, 64-68; vis assoc prof physiol & biophysics, Univ Ala, 70-; consult oral surg, US Army, Ft Campbell, Ky, 71-75; examr, Am Bd Oral & Maxillofacial Surg, 80-85. *Mem:* Int Asn Dent Res; Am Physiol Soc; Am Asn Oral & Maxillofacial Surgeons; Sigma Xi; Int Asn Oral & Maxillofacial Surgeons; Soc Educ Oral & Maxillofacial Surg (founding mem); Am Soc Temporomandibular Joint Surgeons (founding mem). *Res:* Clin oral and maxillofacial surgery, neural regulation of salivary gland growth and function. *Mailing Add:* Dept Oral Surg Vanderbilt Univ Sch Med Nashville TN 37232

HALL, IAN WAVELL, b Leeds, UK. ELECTRON MICROSCOPY. *Educ:* Univ Leeds, BSc, 70, PhD(metall), 74. *Prof Exp:* Postdoctoral fel metall, Leeds Univ, 74-75; lectr metall, Tech Univ Denmark, 75-78; res fel mat sci, Technion-Israel Inst Technol, 78-80; asst prof mat sci, Univ Del, 80-86; CHMN MAT SCI, MAT SCI PROG, UNIV DEL, 88- *Concurrent Pos:* Res scientist, Univ Bordeaux, 89. *Mem:* Mat Res Soc; Am Soc Metals Int; Electron Microscope Soc Am; Metall Soc Am Inst Mech Engrs. *Res:* Studies of the relationship between structure and properties in metal matrix composites with emphasis on fiber/matrix interfaces. *Mailing Add:* Spencer Lab Univ Del Newark DE 19716

HALL, IRIS BERYL HADDON, b Richmond, Va, Nov 19, 37; m 64; c 4. CELL PHYSIOLOGY, BIOCHEMISTRY. *Educ:* James Madison Univ, Va, BS, 59; Univ Tenn, Knoxville, MS, 61; Univ NC, Chapel Hill, PhD(physiol), 65. *Prof Exp:* Asst zool & physiol, Univ Tenn, Knoxville, 60-61; asst physiol, Sch Med, Univ NC, 61-65; USPHS fel radiobiol, Med Ctr, Duke Univ, 65-66, assoc radiol, 66-67; biochemist & physiologist, Chem & Life Sci Lab, Res Triangle Inst, 67-70; from instr to asst prof, 70-77, assoc prof med chem, Sch Pharm, Univ NC, Chapel Hill, 77-83, PROF MED CHEM & NATURAL PROD, 83- *Mem:* AAAS; Am Asn Cols Pharm; Am Chem Soc; Am Inst Chemists; Sigma Xi; Am Oil Chemist. *Res:* Anoxic, hyperbaric, drug, hormonal radiation toxicological effects on cellular metabolism and enzymes; pharmacokinetics & distribution of radioactive drugs; identification of active contraceptive agents, antitumor, anti-inflammatory and hypocholesteremic agents; athersclerosis LDL & HDL receptor activity; serum lipiproteins; mechanism of action of these agents. *Mailing Add:* Med Chem Parm Sch Univ NC Chapel Hill NC 27599-7360

HALL, IVAN VICTOR, b Parrsboro, NS, Aug 24, 27; m 53; c 4. BOTANY. *Educ:* Acadia Univ, BSc, 48, MSc, 49; Cornell Univ, PhD(bot), 53. *Prof Exp:* BOTANIST, KENTVILLE RES STA, CAN DEPT AGR, 49- *Concurrent Pos:* Asst, Cornell Univ, 51-52; adv, Nat Cranberry Mag. *Mem:* Bot Soc Am; Am Soc Hort Sci; Can Bot Soc; Agr Inst Can. *Res:* Ecology of native lowbush blueberries; factors influencing cranberry production in Eastern Canada. *Mailing Add:* 59 Elm Ave Kentville NS B4N 1Z2 Can

HALL, J(AMES) A(LEXANDER), b Providence, RI, Apr 25, 20; m 45; c 3. PHYSICS, ELECTRICAL ENGINEERING. *Educ:* Brown Univ, ScB, 42; Univ RI, PhD(elec eng), 71. *Prof Exp:* Engr, Allen B DuMont Labs, Inc, NJ, 42-46; jr instr physics, Johns Hopkins Univ, 46-48, equip engr, 48, asst physics, 49-50; engr, Westinghouse Elec Corp, 50-56, mgr eng sect TV camera tubes, 56-65, mgr electron tube tech lab, 65-71; assoc prof elec eng, Univ RI, 71-76; sr adv engr, Systs Develop Div, Advan Tech Labs, Westinghouse Elec Corp, 76-80; MGR RES & DEVELOP SECT, HEWLETT-PACKARD CO, BOISE, ID. *Mem:* Am Phys Soc; Am Vacuum Soc; Optical Soc Am; fel Inst Elec & Electronics Engrs. *Res:* Solid state imaging sensors, diode arrays and charge-coupled devices; electron optical imaging systems; light amplifiers; television camera tubes, especially for imaging at very low light levels. *Mailing Add:* Hewlett-Packard Co PO Box 15 Boise ID 83707

HALL, J(OHN) GORDON, b Westview, BC, Oct 14, 25; nat US; m 50; c 2. FLUID MECHANICS, HEAT TRANSFER. *Educ:* Univ BC, BASc, 50; Univ Toronto, MASc, 51, PhD(aeronaut eng, aerophysics), 54. *Prof Exp:* Asst aerophysics, Univ Toronto, 50-54, res assoc & asst prof aeronaut eng, 54-58; res aerodynamicist, Cornell Aeronaut Lab, Inc, 58-59, prin aerodynamicist, 59-61, asst head aerodyn res dept, 61-66, head dept, 66-69; prof eng & appl sci & dir fluid & thermal sci lab, 69-76, PROF MECH ENG, STATE UNIV NY BUFFALO, 76- *Mem:* Am Inst Aeronaut & Astronaut; Am Soc Mech Engrs; Am Phys Soc. *Res:* Environmental transport phenomena. *Mailing Add:* Dept Mech Eng State Univ NY Buffalo NY 14260

HALL, J HERBERT, b Minneola, Kans, Mar 16, 31; m 53; c 2. ORGANIC CHEMISTRY. *Educ:* Univ Kans, BS, 53; Univ Mich, MS, 58, PhD(org chem), 59. *Prof Exp:* Chemist, US Naval Ord Test Sta, Calif, 53-55; fel, Pa State Univ, 59-60; asst prof org chem, Kans State Teachers Col, 60-62; from asst prof to assoc prof, 62-71, PROF CHEM, SOUTHERN ILL UNIV, CARBONDALE, 71- *Mem:* Am Chem Soc. *Res:* Small ring heterocyclic compounds; organic azides; nitrene intermediates; cycloaddition reactions; electron spin resonance; charge transfer complexes; small ring heterocyclics. *Mailing Add:* Dept Chem Southern Ill Univ Carbondale IL 62901

HALL, JAMES CONRAD, b Ont, Apr 26, 19; m 45, 86; c 2. CELL METABOLISM, CARCINOGENESIS. *Educ:* Univ Toronto, BA, 40, PhD(physiol zool), 46; Univ Western Ont, MA, 42. *Prof Exp:* Asst prof biol, Univ New Brunswick, 45-47; from instr to prof biol, Rutgers Univ, 47-62, prof physiol, 62-86, assoc dean, Grad Sch, 80-82, EMER PROF PHYSIOL, RUTGERS UNIV, 86- *Concurrent Pos:* Vis prof, Univ Nebr Col Med, 70 & Baylor Med Sch, 71; chmn dept zool & physiol, Rutgers Univ, Newark, 62-69; consult, NIH, USPHS, 66-80; prin investr, NIH grant, 60-82; dir, Grad Prog Zool, Rutgers Univ, Newark, 75-78 & 79-81. *Mem:* Am Physiol Soc; Am Soc Cell Biol; Soc Exp Biol & Med; Am Zool Soc; Can Zool Soc; Sigma Xi. *Res:* Effects of hormones, especially insulin, on cellular respiration and metabolism; liver regeneration; effects of carcinogenic agents, pollutants, drugs on cell structure and function; oncogenesis; atherosclerosis; membrane structure. *Mailing Add:* 226 Eagle Rock Ave Roseland NJ 07068

HALL, JAMES DANE, b Columbus, Ohio, Aug 31, 33; m 55; c 2. FISH BIOLOGY. *Educ:* Univ Calif, Berkeley, AB, 55; Univ Mich, MS, 60, PhD(fisheries), 63. *Prof Exp:* Res fel fisheries, Inst Fisheries Res, Mich Dept Conserv, 58-62; res instr, Fisheries Res Inst, Univ Wash, 62-63; from asst to assoc prof, 63-82, PROF FISHERIES, ORE STATE UNIV 82- *Concurrent Pos:* Teaching fel zool, Univ Mich, 59-60. *Mem:* Am Fisheries Soc; Ecol Soc Am. *Res:* Population dynamics of freshwater fish; effects of watershed practices on streams; stream ecology. *Mailing Add:* Dept Fisheries & Wildlife Ore State Univ Corvallis OR 97331-3803

HALL, JAMES EDISON, b Ft Worth, Tex, June 19, 42; m 64; c 2. NUCLEAR PHYSICS. *Educ:* Tex Christian Univ, BA, 64; Iowa State Univ, MS, 66, PhD(nuclear physics), 70. *Prof Exp:* Res asst nuclear physics, Ames Lab, AEC, 64-70; assoc, Swiss Fed Inst Technol, 70-72; develop proj engr, Schlumberger Technol Corp, 72-76, sect mgr nuclear physics, 76-79, head eng physics dept, 79-82, eng dept 83-86, DIR ENG, STATHAM TRANSDUCER DIV, SCHLUMBERGER TECHNOL CORP, 86- *Mem:* Am Phys Soc; Soc Prof Well Log Analysts; Soc Petrol Engrs; Inst ELec & Electronic Engrs; Instrument Soc Am. *Res:* Design and development of pressure transducers & transmitters for use in aerospace instrumentation, process control systems and nuclear reactor environments. *Mailing Add:* Schlumberger Indust 2230 Statham Blvd Oxnard CA 93033

HALL, JAMES EMERSON, b Berwyn, Ill, Oct 3, 36; m 59; c 3. MATHEMATICS. *Educ:* Northern Ill Univ, BS, 58; Harvard Univ, AM, 59; Univ Wis, PhD(math), 65. *Prof Exp:* Instr math, Northern Ill Univ, 59-61; from asst prof to assoc prof, Univ Wis-Madison, 65-77, prof math, 77-; PROF MATH, WESMINISTER UNIV. *Mem:* Am Math Soc; Math Asn Am; Sigma Xi. *Res:* Stability of ordinary differential equations. *Mailing Add:* Dept Math Westminister Col Box 23 New Wilmington PA 16172

HALL, JAMES EWBANK, b Sewannee, Tenn, June 2, 41. BIOPHYSICS. *Educ:* Pomona Col, BA, 63; Univ Calif, Riverside, MA, 65, PhD(physics), 68. *Prof Exp:* Research physics, US Army, 68-70; res fel biophysics, Calif Inst Technol, 70-74; asst prof biophys, Duke Univ, 74-77; assoc prof, 78-84, PROF PHYSIOL, UNIV CALIF, IRVINE, 84- *Concurrent Pos:* Physiol Study Sect, NIH, 86-90. *Mem:* Biophys Soc; Soc Gen Physiologists; AAAS. *Res:* Investigation of the molecular mechanisms of voltage-dependent conductances in membranes; gap junctions. *Mailing Add:* Dept Physiol & Biophysics Univ Calif Irvine CA 92717

HALL, JAMES LAWRENCE, b Springfield, Mass, Jan 11, 28; m 53; c 3. ANATOMY. *Educ:* Am Inat Col, AB, 51; Univ Conn, MS, 53; St Louis Univ, PhD(anat), 57. *Prof Exp:* Asst prof anat, Univ Kans, 57-61; from asst prof to assoc prof, Ohio State Univ, 61-68; dir grad studies, 70-75, PROF ANAT, COL MED, UNIV CINCINNATI, 68- *Concurrent Pos:* Consult & vis scientist, Neuroanat Vis Scientists Prog, Ohio State Univ, 64-68, regional dir, 68-70; lectr, Ann LeRoy Sante Mem Lect, 65. *Mem:* Am Asn Anat; Sigma Xi. *Res:* Autonomic nervous system, using morphological, physiological and histochemical techniques; innervation of cerebral blood vessels; peripheral nerve repair. *Mailing Add:* Anat & Cell Biol Dept Univ Cincinnati 231 Bethesda Ave Cincinnati OH 45267-0521

HALL, JAMES LESTER, b Pine Grove, WVa, Dec 26, 10; m 48; c 3. PHYSICAL CHEMISTRY. *Educ:* WVa Univ, AB, 31, MS, 32; Univ Wis, PhD(phys chem), 39. *Prof Exp:* Lilly asst phys chem insulin, Med Col, Cornell Univ, 39-40; instr chem, Tufts Col, 40-41; res chemist, WVa Geol Surv, 41-43; from asst prof to prof chem, WVa Univ, 46-76; RETIRED. *Concurrent Pos:* Emer prof chem, W Va Univ, 76- *Mem:* AAAS; Am Chem Soc. *Res:* Complex inorganic ions; dielectric constant and conductance of electrolytic solutions. *Mailing Add:* 1060 Takoma St Morgantown WV 26505

HALL, JAMES LOUIS, b Gunnison, Colo, Jan 30, 27. INORGANIC CHEMISTRY. *Educ:* Western State Col Colo, AB, 47; Univ Tex, PhD(inorg chem), 53. *Prof Exp:* Instr chem, Univ Wis-Milwaukee, 47-49; asst prof, Mich State Univ, 53-59; assoc prof, Colo Sch Mines, 59-63; ed publ, Am Chem Soc, 63-68; sr ed, W A Benjamin, Inc, 68-74; DIR, ASPEN WRITING CTR, AUTHOR-PUBLISHER SERV, 74- *Mem:* Am Chem Soc. *Mailing Add:* PO Box 3161 Aspen CO 81612

HALL, JAMES TIMOTHY, b Los Angeles, Calif, June 12, 50; m 78; c 3. THIN FILM OPTICS, IR DETECTOR MATERIAL. *Educ:* Stanford Univ, BS, 72; Univ Calif, Santa Barbara, MA, 77, PhD(physics), 78. *Prof Exp:* Res asst physics, Univ Calif, Santa Barbara, 75-78; physicist, Nat Bur Standards, 78-80; physicist, Hughes Aircraft Co, 80-87, prog mgr, 81-87; STAFF PHYSICIST & PRIN INVESTR, NORTHROP RES & TECHNOL CTR, 87- *Concurrent Pos:* Res assoc, Nat Res Coun, 78-80; mem res tech staff & group leader, Northrop Res & Technol Ctr. *Mem:* Am Phys Soc. *Res:* Thin film deposition technology; characterization spectroscopies; superconductors; semiconductors; chemical vapor; thin films for device applications; research & development scientist. *Mailing Add:* Northrop Electronic Systs Div 2301 W 120th St J412/N4-1 Hawthorne CA 90251-5032

HALL, JEFFREY CONNOR, b Brooklyn, NY, May 3, 45. BEHAVIORAL GENETICS, NEUROBIOLOGY. *Educ:* Amherst Col, AB, 67; Univ Wash, MS, 69, PhD(genetics), 71. *Prof Exp:* Fel behav genetics, Calif Inst Technol, 71-73; asst prof, 74-79, ASSOC PROF BIOL, BRANDEIS UNIV, 79- *Mem:* Genetics Soc Am; Soc Neurosci. *Res:* Behavioral genetics of Drosophila; studies of reproductive behavior in mutant and mosaic insects; developmental, physiological, and behavioral abnormalities induced by neurochemical mutants in Drosophila. *Mailing Add:* Dept Biol Brandeis Univ 415 South St Waltham MA 02254

HALL, JENNIFER DEAN, b Bethesda, Md, Dec 15, 44; m; c 3. MOLECULAR BIOLOGY. *Educ:* Harvard Univ, BA, 67; Yale Univ, PhD(molecular biophys & biochem), 74. *Prof Exp:* Fel biochem sci, Princeton Univ, 74-76; asst prof cell develop biol, 76-82, ASSOC PROF MOLECULAR CELL BIOL, UNIV ARIZ, 82- *Concurrent Pos:* Fel, Am Cancer Soc, 74-76; NIH grant, 77-; scholars grant, Am Cancer Soc, 84. *Mem:* AAAS. *Res:* DNA replication in mammalian cells; animal viruses. *Mailing Add:* Dept Molecular Cellular Biol Univ Ariz Tucson AZ 85721

HALL, JEROME WILLIAM, b Brunswick, Ga, Dec 1, 43; m 65; c 3. PHYSICS, CIVIL ENGINEERING. *Educ:* Harvey Mudd Col, BS, 65; Univ Wash, MS, 68, PhD(civil eng), 69. *Prof Exp:* Instr, Univ Wash, 69-70; from asst prof to assoc prof civil eng, Univ Md, College park, 70-77; asst dean eng, 85-88, PROF CIVIL ENG, UNIV NMEX, 80-, CHMN, CIVIL ENG DEPT, 90- *Concurrent Pos:* Vpres, Transp Planning & Eng, Seattle, 67-70; hwy safety specialist, Fed Hwy Admin, 71; safety consult, Tech Adv Serv for Attorneys, Pa, 71-87; mem, Transp Res Bd, Nat Acad Sci-Nat Res Coun, 71-; consult hwy eng, 81- *Mem:* Fel Inst Transp Engrs; Am Soc Eng Educ; Am Rd & Transp Builders Asn. *Res:* Transportation systems safety and operation. *Mailing Add:* Dept Civil Eng Univ NMex Albuquerque NM 87131-1351

HALL, JERRY DEXTER, community ecology, mammalogy, for more information see previous edition

HALL, JERRY LEE, b Boulder, Colo, Feb 2, 38; m 57; c 3. MECHANICAL & AEROSPACE ENGINEERING. *Educ:* Iowa State Univ, BS, 59, MS, 63, PhD(mech & aerospace eng, math), 67. *Prof Exp:* From instr to asst prof, 60-70, assoc prof, 70-77, PROF MECH ENG, IOWA STATE UNIV, 77- *Honors & Awards:* Ralph Teetor Award, Soc Automotive Engrs, 77. *Mem:* Am Soc Mech Engrs; Am Soc Eng Educ; Soc Automotive Engrs; Am Inst Aeronaut & Astronaut; Nat Fire Protection Asn. *Res:* Gas dynamics; instrumentation and measurements; shock tube flows; design of experiments; environmental emissions; combustion; fires. *Mailing Add:* RR 4 Ames IA 50010

HALL, JOHN B, b New York, Oct 12, 18; m 57; c 2. ORGANIC CHEMISTRY. *Educ:* Cornell Univ, PhD(chem), 50. *Prof Exp:* Org chemist, Air Reduction Co, 53-57; proj leader, Int Flavors & Fragrances Inc, 58-65, res assoc, 65-68, assoc dir, 68-73, dir, 73-75, vpres res & develop, 75-85; RETIRED. *Mem:* Am Chem Soc. *Res:* Fragrance chemistry; terpenes; relationship between structure and odor. *Mailing Add:* PO Box 404 Rumson NJ 07760

HALL, JOHN BRADLEY, b Denver, Colo, Nov 2, 33; m 63. MICROBIAL PHYSIOLOGY. *Educ:* Univ Kans, BA, 56; Univ Calif, Berkeley, PhD(biochem), 60. *Prof Exp:* NIH fel biochem, Calif Inst Technol, 60-62; from asst prof to assoc prof, 62-73, PROF MICROBIOL, UNIV HAWAII, MANOA, 73- *Concurrent Pos:* NIH res grant, 63-65; Damon Runyon res grant, 64-65. *Mem:* Am Soc Microbiol; Sigma Xi. *Res:* Energy metabolism in anaerobic bacteria. *Mailing Add:* Dept Microbiol Univ Hawaii Honolulu HI 96822

HALL, JOHN EDGAR, b Meadville, Pa, Dec 28, 29; m 57; c 3. ZOOLOGY, PARASITOLOGY. *Educ:* Univ NH, BS, 51, MS, 53; Purdue Univ, PhD(parasitol), 58. *Prof Exp:* Asst zool, Univ NH, 51-52; asst, Purdue Univ, 55-56; from instr to assoc prof, 58-70, PROF MICROBIOL, SCH MED, WVA UNIV, 70- *Concurrent Pos:* Inter-Am fel trop med & parasitol, La State Univ, 60; NIH res career develop award, 62-67. *Mem:* AAAS; Am Micros Soc; Am Soc Parasitol; Am Soc Trop Med & Hyg; Entom Soc Am; Soc Invert Path. *Res:* Medical parasitology; morphology, life history and host-parasite relationships of helminths; reactions of insects to helminth parasites; endosymbionts of pathogenic freshwater amebae. *Mailing Add:* Dept Microbiol & Immunol WVa Univ Health Sci Ctr Morgantown WV 26506

HALL, JOHN EDWARD, c 3. HYPERTENSION, BLOOD PRESSURE REGULATION. *Educ:* Kent State Univ, BS, 68; Mich State Univ, PhD(physiol), 74. *Prof Exp:* from instr to assoc prof, 75-82, PROF, DEPT PHYSIOL & BIOPHYSICS, UNIV MISS MED CTR, 82-, DIR GRAD PROG, 80- *Concurrent Pos:* Grad teaching asst, Physiol Dept, Mich State Univ, 73-74; fel, NIH nat Res Serv Award, Dept Physiol & Biophysics, Univ Miss Med Ctr, 74-75; coun high blood pressure res, Am Heart Asn; mem grad fac, Dept Physiol & Biophysics, Univ Miss Med Ctr, 77-; vchmn, Dept Physiol & Biophysics, Univ Miss Med Ctr, 88-89, chmn, 89- *Honors & Awards:* Goldblatt Award, Am Heart Asn; Marion Award, Am Soc Hypertension. *Mem:* Am Physiol Soc; Am Soc Nephrology; Int Soc Nephrology; Am Soc Hypertension; Int Soc Hypertension. *Res:* Renal and cardiovascular physiology. *Mailing Add:* Dept Physiol & Biophys Univ Miss Med Ctr 2500 N State St Jackson MS 39216

HALL, JOHN EMMETT, b Wadena, Sask, Apr 23, 25; m 52; c 6. ORTHOPEDIC SURGERY. *Educ:* Univ Sask, BA, 48; McGill Univ, MD, CM, 52; FRCS(C), 57. *Prof Exp:* Assoc surg, Univ Toronto, 58-68, asst prof, 68-71; PROF ORTHOP SURG, HARVARD MED SCH, 71-; CHIEF CLIN SERV, DEPT ORTHOP SURG, CHILDREN'S HOSP MED CTR, BOSTON, 71-, CLIN PROF ORTHOP SURG, CHIEF ORTHOP SURGEON, 86- *Concurrent Pos:* Orthop consult, Ont Crippled Children's Ctr, 58-68; asst surgeon, Hosp for Sick Children, 58-68, chief div orthop surg, 68-71. *Mem:* Can Med Asn; Can Orthop Asn; fel Am Col Surg. *Res:* Prosthetics; development of myoelectric arm; spinal curvatures. *Mailing Add:* Sch Med Harvard Univ Boston Children's Hosp 300 Longwood Ave Boston MA 02115

HALL, JOHN FREDERICK, geology; deceased, see previous edition for last biography

HALL, JOHN JAY, b Cambridge, Mass, Oct 20, 31; m 53; c 2. SOLID STATE PHYSICS. *Educ:* Columbia Univ, BS, 56, MA, 58, PhD(physics), 62. *Prof Exp:* Res assoc physics, Watson Res Labs, IBM Corp, Columbia, 62-63; res staff mem, T J Watson Res Ctr, IBM Corp, 63-80; MANAGING PARTNER, SHEARWATER CO, 80- *Mem:* Inst Elec & Electronics Engrs. *Res:* Display technologies; transport properties of semiconductors and semi-metals; ultrasonic properties of insulating and semiconducting crystals; piezoelectric and ferroelectric materials; electrical discharges in gases. *Mailing Add:* 3030 Emmons Ave 4A Shearwater Co Brooklyn NY 11235

HALL, JOHN L, b Denver, Colo, Aug 21, 34; m 58; c 3. LASERS, OPTICS. *Educ:* Carnegie Inst Technol, BS, 56, MS, 58, PhD(physics), 61. *Hon Degrees:* Dr, Univ Paris Nord, 89. *Prof Exp:* NRC postdoctoral fel, Nat Bur Standards, 61-62, physicist, 62-71, SR SCIENTIST, NAT INST SCI & TECHNOL, 71-; LECTR, PHYSICS DEPT, UNIV COLO, 67- *Concurrent Pos:* Fel, Joint Inst Lab Astrophys, 64-; mem, Comn VII, Int Union Radio Sci; deleg, Consultative Comt Definition Meter, Sevres, France, 70-; mem, Comt Recommendations, NAC-Army Res Off, 76-79; mem, Nat Res Comt Fundamental Constants, Nat Res Coun-Nat Acad Sci, 76-79; mem, prog comt, Tenth Int Conf Quantum Electronics, 78. *Honors & Awards:* Gold Medal, US Dept Com, 69 & 74; Samuel Wesley Stratton Award, Nat Bur Stand, 71; E U Condon Award, Nat Bur of Standards, 79; Charles H Townes Award, Optical Soc of Am, 84, Frederic Ives Medal, 91; Davisson-Germer Award, Am Phys Soc, 88. *Mem:* Nat Acad Sci; fel Am Phys Soc; fel Optical Soc Am; Sigma Xi. *Res:* Laser and solid state physics; negative ion photodetachment; nonlinear optics; laser stabilization; application of precision measurement techniques to fundamental physical measurements; five patents. *Mailing Add:* Univ Colo Campus Box 440 Boulder CO 80309

HALL, JOHN SYLVESTER, b Westfield, Mass, July 15, 30; m 54; c 5. ZOOLOGY. *Educ:* Univ Mass, BS, 51, MA, 56; Univ Ill, PhD(zool), 60. *Prof Exp:* Asst prof, 60-64, assoc prof, 64-71, PROF BIOL, ALBRIGHT COL, 71- *Mem:* AAAS; Am Soc Mammalogists; Nat Spleleol Soc; Animal Behav Soc. *Res:* Mammalian ecology and evolution, especially population ecology of bats; osteometric variation of mammals; Pleistocene distribution of mammals; behavior of bats and rodents. *Mailing Add:* Dept Biol Albright Col PO Box 516 Reading PA 19604

HALL, JOHN WILFRED, b Midland, Ont, May 30, 41. STATISTICS. *Educ:* Univ Toronto, BSc, 65, MSc, 69; Va Polytech Inst & State Univ, PhD(statist), 73. *Prof Exp:* Res asst statist, Connaught Biosci, 66-69; consult, Health Protection Br, Health & Welfare Can, 72-74, sect head, 74-77; statistician, 77-84, STATIST SCIENTIST, RES BR, AGR CAN, 84- *Mem:* Am Statist Asn; Biomet Soc; Can Statist Soc. *Res:* Statistical methods for agricultural and medical research. *Mailing Add:* Vancouver Res Sta 6660 NW Marine Dr Vancouver BC V6T 1X2 Can

HALL, JON K, b Derby, Conn, June 23, 37; m 60; c 4. SOIL CHEMISTRY, ENVIRONMENTAL QUALITY. *Educ:* Univ Conn, BS, 58; Pa State Univ, MS, 63, PhD(agron), 66. *Prof Exp:* Instr agron, 65-66, asst prof soil chem, 66-73, ASSOC PROF SOIL CHEM, PA STATE UNIV, UNIVERSITY PARK, 73- *Mem:* Am Soc Agron; Soil Sci Soc Am; Sigma Xi. *Res:* Chemistry, mobility and persistence of herbicides in soils and tillage systems. *Mailing Add:* Dept Agron Pa State Univ 116 Agr Sci Ind Bldg University Park PA 16802

HALL, JOSEPH GLENN, b Cincinnati, Ohio, July 22, 23; m 50; c 4. VERTEBRATE ZOOLOGY. *Educ:* Princeton Univ, AB, 48; Univ Colo, MA, 50; Univ Calif, PhD(zool), 56. *Prof Exp:* Asst prof zool, Colo State Univ, 56-57; instr, San Francisco State Univ, 57-59, from asst prof to assoc prof, 59-67, prof biol, 67-; RETIRED. *Mem:* Am Soc Mammal; Wildlife Soc; Ecol Soc Am; Cooper Ornith Soc. *Res:* Wildlife ecology. *Mailing Add:* 1907 Monument Canyon Dr Grand Junction CO 81503

HALL, JOSEPH L, b Boston, Mass, Jan 22, 36; m 61; c 4. HEARING. *Educ:* Williams Col, BA, 59; Mass Inst Tech, SB, 59, SM, 59, PhD(elec eng), 63. *Prof Exp:* Asst prof commun theory, Mass Inst Tech, 63-64 & Johns Hopkins Univ, 64-66; MEM TECH STAFF, AT&T BELL LABS, 66- *Concurrent Pos:* Chmn, Tech Comt Physiol & Psychol Acoust, Acoust Soc Am, 78-79; assoc ed, J Acoust Soc Am, 81-83; mem, Exec Coun, Acoust Soc Am, 88-91. *Mem:* Fel Acoust Soc Am. *Res:* Application of communication theory to auditory information processing; models for information processing at the auditory periphery. *Mailing Add:* Rm 2D-546 AT&T Bell Labs Murray Hill NJ 07974-2070

HALL, JUDD LEWIS, b Canton, Ohio, Sept 2, 34. POLYMER CHEMISTRY. *Educ:* Capital Univ, BS, 56; Akron Univ, MS, 57, PhD(polymer chem), 62. *Prof Exp:* Res chemist, 56-62, res dir, 62-64, exec vpres, 64-73, PRES, SURETY RUBBER CO, 73- *Mem:* Am Chem Soc. *Res:* Anionic polymerization of styrene, kinetic and synthesis studies; mechanical and physical studies of thin films of natural and synthetic rubbers; chemical resistance studies of thin polymer films. *Mailing Add:* 944 Thomas Ave NW Carrollton OH 44615

HALL, JUDY DALE (MRS RICHARD MODAFFERI), b Chicago, Ill, July 15, 43; m 77; c 1. CELL BIOLOGY, ELECTRON MICROSCOPY. *Educ:* Ill Wesleyan Univ, BA, 64; Smith Col, MA, 66; Purdue Univ, PhD(biol), 71. *Prof Exp:* Electron microscopist, Dept Path & Toxicol, Human Health Res & Develop Labs, Dow Chem Co, 71-72; res fel, Boston Biomed Res Inst, 72-74; asst prof biol, State Univ NY Binghamton, 74-81; ELECTRON MICROSCOPIST, DEPT PATH, ROBERT PACKER HOSP, SAYRE, PA, 81- *Mem:* AAAS; Am Soc Cell Biol; Sigma Xi; Electron Micros Soc Am. *Res:* biological and diagnostic electron microscopy. *Mailing Add:* Dept Path Robert Packer Hosp Sayre PA 18840-1698

HALL, KENNETH DALAND, b Durham, NC, Oct 26, 26; m 54; c 4. ANESTHESIOLOGY. *Educ:* Duke Univ, AB, 49, MD, 53. *Prof Exp:* Head neuroanesthesiol, Nat Inst Neurol Dis & Blindness, 56-58; from asst prof to assoc prof, 58-68, PROF ANESTHESIOL, DUKE UNIV, 68- *Concurrent Pos:* Consult, Vet Admin Hosp, Durham & Womack Army Hosp, Ft Bragg, NC. *Mem:* AAAS; Am Soc Anesthesiol; Aerospace Med Asn. *Res:* Physiology of respiration; clinical hypothermia; physiology of halogenated anesthetics. *Mailing Add:* Dept Anesthesiol Duke Univ Med Ctr Box 3094 Durham NC 27710

HALL, KENNETH LYNN, b Great Falls, Mont, Aug 23, 27; m 52; c 3. PHYSICAL INORGANIC CHEMISTRY. *Educ:* Reed Col, BA, 49; Univ Calif, MS, 51; Univ Mich, PhD(chem), 55. *Prof Exp:* Asst nuclear chem, Radiation Lab, Univ Calif, 49-51; asst, Univ Mich, 51-55; res chemist, Calif Res Corp, 55-64; sr res chemist, Chevron Res Co, 64-68, sr res assoc, 68-86; RETIRED. *Mem:* Am Chem Soc; Sigma Xi. *Res:* Corrosion and wear; petroleum products; internal combustion engines. *Mailing Add:* 16 Sussex Ct San Rafael CA 94903

HALL, KENNETH NOBLE, b Lawrence, Mass, Sept 18, 35; m 58; c 4. FOOD SCIENCE, NUTRITION. *Educ:* Univ Maine, BS, 57; Univ Md, MS, 60; Wash State Univ, PhD(food sci), 64. *Prof Exp:* From asst prof to assoc prof, 66-72, PROF NUTRIT SCI, UNIV CONN, 83- *Concurrent Pos:* Coord res & develop, Food Sci Lab, US Army Natick Develop Ctr, 68-79; staff officer, Office Dep Chief of Staff for Res & Develop Acquisition, Pentagon, 80-85; regional communicator, off Sci & Pub Affairs, Inst Food Technologists, 81-; mem adv bd, Nutrit Res News Letter, Lyda Assoc, 82-, corresp, Food Safety Notebk, Lyda Asn, 90; Coun Agr Sci & Technol; interim nat prog leader, Exten Serv, USDA, 90-91. *Mem:* Inst Food Technologists; Poultry Sci Asn; Sigma Xi. *Res:* Poultry products technology; egg and poultry meat quality; microwave cooking; cooked meat quality. *Mailing Add:* Dept Nutrit Sci Univ Conn Storrs CT 06269-4017

HALL, KENNETH RICHARD, b Tulsa, Okla, Nov 5, 39; m 76; c 5. CHEMICAL ENGINEERING. *Educ:* Univ Tulsa, BSChE, 62; Univ Calif, Berkeley, MS, 64; Univ Okla, PhD(chem eng), 67. *Prof Exp:* Asst prof chem eng, Univ Va, Charlottesville, 67-70 & 72-74; asst to pres, ChemShare, 70; sr res engr, Amoco Prod Co, 70-71; assoc prof, Thermodyn Res Ctr, 74-78, dir, 79-85; from asst dir to assoc dir, 85-88, DEP DIR, TEX ENG EXP STA, 89-, ASSOC DEP CHANCELLOR, 90-; PROF CHEM ENG, TEX A&M UNIV, 78- *Concurrent Pos:* NATO fel, Cath Univ Louvain, 71-72. *Mem:* Am Inst Chem Engrs; Am Chem Soc; Am Soc Eng Educ; Am Soc Testing & Mat. *Res:* Chemical engineering thermodynamics covering: precise data on PVT, enthalpy and VLE; equations of state; critical region modeling. *Mailing Add:* Dept Chem Eng Tex A&M Univ College Station TX 77843

HALL, KENT D, b Versailles, Mo, Aug 3, 37; m 61; c 2. REPRODUCTIVE PHYSIOLOGY, ENVIRONMENTAL BIOLOGY. *Educ:* Cent Mo Univ, BS, 60; Emporia Kans State Col, MS, 64; Univ Kans, PhD(zool), 69. *Prof Exp:* Teacher biol & gen sci, Jr High Sch, Kans, 60-62 & Sr High Sch, 62-63; instr biol, Univ Kans, 64-67; asst prof reproductive & ecol physiol, 68-74, ASSOC PROF BIOL, UNIV WIS-STEVENS POINT, 74- *Mem:* Am Inst Biol Sci. *Res:* Hibernation in 13-lined ground squirrel; premarital pregnancy and abortion problems. *Mailing Add:* Dept Biol Univ Wis Stevens Point WI 54481

HALL, KIMBALL PARKER, b Richmond, Va, Nov 7, 14; m 42; c 3. CHEMICAL ENGINEERING. *Educ:* Dartmouth Col, AB, 37; Cornell Univ, PhD(org chem), 42. *Prof Exp:* Asst chem, Cornell Univ, 37-40 & 41-42, Nat Defense Res Comt Proj, 40-41; res chemist, Eastman Kodak Co, NY, 42-43; res lab analyst, Lockheed Aircraft Corp, Calif, 43-44, process engr, 44-45; dir aircraft specialty develop, Coast Paint & Chem Co, 45-47; sr res engr, Jet Propulsion Lab, Calif Inst Technol, 47-55; res assoc, Guggenheim Lab Aerospace Propulsion Sci, Dept Aeronaut Eng, Princeton Univ, 57-61, consult, 62-63; tech consult, Rocket Power Inc, Ariz, 62-63; consult, Precision Devices & Eng Co, Inc, NJ, 64-67; pres, Hall Marine Corp, 67-75; sr scientist, Princeton Combustion Labs, NJ, 75-76, bus mgr, 76-77; GEN ENGR, NAVAL ORD STA, 77- *Concurrent Pos:* Consult, Greyrad Corp, 67-71; assoc fel of Am Inst Aeronaut & Astronaut; mem, Sigma Xi. *Mem:* Am Inst Aeronaut & Astronaut; Am Chem Soc; AAAS. *Res:* Sealing compounds and coatings; solid rocket propellant development; combustion processes; underwater propulsion. *Mailing Add:* 113 Wood Duck Circle La Plata MD 20646

HALL, LARRY CULLY, b Akron, Ohio, Apr 6, 30; m 53; c 4. ANALYTICAL CHEMISTRY, ELECTROCHEMISTRY. *Educ:* Bowling Green State Univ, BA, 52; Univ Ill, PhD(chem, metall), 56. *Prof Exp:* Asst chem & elec eng, Univ Ill, 52-53, chem, 53-56; ASSOC PROF CHEM, VANDERBILT UNIV, 56-, COORDR GEN CHEM, 77- *Mem:* Am Chem Soc; Am Soc Metals; Nat Asn Corrosion Engrs. *Res:* Electrochemistry in molten salts and organic solvents; methods of analyzing pharmaceutical compounds; trace analysis studies in semiconductor materials. *Mailing Add:* Box 1530 B Vanderbilt Univ Nashville TN 37235

HALL, LAURENCE STANFORD, b Winona, Minn, July 24, 29; m 51, 68; c 3. PLASMA PHYSICS. *Educ:* Univ Minn, BA & BPhysics, 53; Univ Calif, Berkeley, MA, 58, PhD(plasma physics), 61. *Prof Exp:* Physicist, Lawrence Livermore Lab, Univ Calif, Livermore, 54-88; RETIRED. *Concurrent Pos:* Lectr, Dept Appl Sci, Univ Calif, Davis. *Mem:* Am Phys Soc. *Res:* Theoretical plasma physics; microinstabilities; equilibria; arc discharges; statistical mechanics; statistical transport theory. *Mailing Add:* 7695 Crest Ave Oakland CA 94605

HALL, LAWRENCE JOHN, b London, Eng, Sept 9, 55; m 82; c 2. PARTICLE PHYSICS, COSMOLOGY. *Educ:* Oxford Univ, UK, BA, 77; Harvard Univ, PhD(physics), 81. *Prof Exp:* From asst prof to assoc prof physics, Harvard Univ, 83-86; Miller fel, Univ Calif, Berkeley, 81-83, from asst prof to assoc prof, 86-90, PROF PHYSICS, UNIV CALIF, BERKELEY, 90- *Concurrent Pos:* Sloan fel, Sloan Found, 84-89; NSF presidential young investr, 86-91. *Mem:* Am Phys Soc. *Res:* Extending the standard model of elementary particles to understand: Fermion masses, the scale of weak interactions, CP violations, neutrino masses, prospects at future collider experiments and dark matter. *Mailing Add:* Physics Dept Univ Calif Berkeley CA 94720

HALL, LEO MCALOON, b Denver, Colo, Jan 28, 29; m 53; c 6. BIOCHEMISTRY. *Educ:* Creighton Univ, BS, 51; Okla State Univ, MS, 53; Univ Wis, PhD(physiol chem), 57. *Prof Exp:* Biochemist, Res Div, Am Cyanamid Co, NY, 57-58; from asst prof to assoc prof biochem, Univ Ala, Birmingham, 58-70, prof biochem, 71-80, assoc prof path, 77-80. *Mem:* Am Chem Soc; AAAS. *Res:* Automated methods in clinical diagnostic enzymology. *Mailing Add:* Dept Biochem Univ Ala Sch Med Univ Sta Birmingham AL 35294

HALL, LEO TERRY, b Gettysburg, SDak, Apr 5, 41; m 67; c 3. MICROBIOLOGY. *Educ:* Northern State Col SDak, BS, 63; Okla State Univ, PhD(microbiol), 73. *Prof Exp:* Sci math teacher, Clarkfield High Sch, Minn, 63; NDEA fel microbiol, Okla State Univ, 63-66, NIH predoctoral fel, 66-68; from asst prof to assoc prof, 68-80, PROF BIOL, PHILLIPS UNIV, ENID, OKLA, 80- *Concurrent Pos:* Vis res assoc, Okla Med Res Found, 75; res assoc, Biochem Dept, Okla State Univ, 77, 79, 80 & 89; chmn div sci & math, Phillips Univ, Enid, Okla, 80-83, Samuel Roberts Endowed chair, Biol Dept, 86-91; consult, Hillcrest Infertility Ctr, Hillcrest Hosp, Tulsa, Okla, 83-86; vis prof, Biol Dept, Univ Md, Baltimore Co, 84-85; prin investr, NIH Acad Res Enhancement Award, 85-87. *Mem:* Am Soc Microbiol. *Res:* Regulation of fusidic acid and streptomycin resistance in Staphylococcus aureus. *Mailing Add:* Univ Sta PO Box 2000 Enid OK 73702

HALL, LEON MORRIS, JR, b Springfield, Mo, July 31, 46; m 67; c 2. MATHEMATICAL ANALYSIS. *Educ:* Univ Mo, Columbia, BS(educ), 69, Univ Mo, Rolla, BS(appl math), 69, MS, 71, PhD(math), 74. *Prof Exp:* Teacher high sch, St Louis, Mo, 69-70; from asst prof to assoc prof math, Univ Nebr, Lincoln, 74-85; ASSOC PROF MATH, UNIV MO, ROLLA, 85- *Concurrent Pos:* NSF grant, 75- *Mem:* Am Math Soc; Math Asn Am; Sigma Xi. *Res:* Solvability of ordinary and functional differential equations in the neighborhood of a singular point; functional analysis techniques applied to differential equations; functional equations; difference equations. *Mailing Add:* Dept Math & Statist Univ Mo-Rolla Rolla MO 65401

HALL, LOWELL HEADLEY, II, b Akron, Ohio, Sept 29, 37; m 60; c 2. PHYSICAL & PHARMACEUTICAL CHEMISTRY, STRUCTURE-ACTIVITY RELATIONS. *Educ:* Eastern Nazarene Col, BS, 59; Johns Hopkins Univ, MA, 61, PhD(phys chem), 63. *Prof Exp:* Res assoc x-ray crystallog, Nat Bur Standards, 63-64; asst prof chem, Fla Atlantic Univ, 64-67; assoc prof, 67-71, PROF CHEM, EASTERN NAZARENE COL, 71-, HEAD DEPT CHEM, 67-, CHMN DIV SCI & MATH, 78- *Concurrent Pos:* NSF Col Coop Sci consult, City of Quincy & Minnemast Sci Progs, 70-73; adj prof, Mass Col Pharm, 74-; mem, Acad Adv Coun, GTE Labs, 81-83; Environ Protection Agency grant, 81-84; consult, GTE Labs, Allied Signal Corp, Eastman Pharmaceut. *Mem:* AAAS; Am Chem Soc; Am Crystallog Asn; NY Acad Sci. *Res:* X-ray crystallogrphy; mass spectrometry; boron hydrides; structural inorganic chemistry; molecular structure by x-ray diffraction techniques; quantitative relation of molecular structure to properties and drug activity (SAR); development of molecular connectivity; semiempirical MO calculations; author of 3 books. *Mailing Add:* Dept Chem Eastern Nazarene Col Quincy MA 02170

HALL, LUTHER AXTELL RICHARD, b Pittsburgh, Pa, Oct 19, 24; m 51; c 2. SCIENCE WRITING, ORGANIC CHEMISTRY. *Educ:* Wesleyan Univ, BA, 44; Calif Inst Technol, MS, 46; Univ Kans, PhD(org chem), 50. *Prof Exp:* Res chemist, Gen Chem Co, 46-47; fel, Sch Pharm, Univ Kans, 50-51; res chemist, E I du Pont de Nemours & Co, 51-64; res assoc, Geigy Chem Corp, 64-66, dir polymer res, 66-69, asst to res dir, Plastics & Additives Div, 69-74, SR PATENT AGENT, CIBA-GEIGY CORP, 74- *Mem:* Am Chem Soc. *Res:* Synthesis, fabrication and analysis of vinyls, polyamides, polyolefins, polyurethanes, polyesters and polyhydrocarbons; organic phosphorus chemistry; physical chemistry of polymers; process development on polymer systems; organic chemical products and processes. *Mailing Add:* 36 Old Farms Rd Woodcliff Lake NJ 07675

HALL, LYLE CLARENCE, b Mason City, Iowa, Feb 12, 35. PHYSICAL CHEMISTRY. *Educ:* Luther Col, Iowa, BA, 56; Univ Iowa, MS, 60, PhD(phys chem), 61. *Prof Exp:* Res fel phys chem, Univ Minn, 61-64; admin asst to exec secy, Minn State Jr Col Bd, State Minn, 64-65; from asst prof to assoc prof, 65-81, PROF CHEM, UNIV WIS-RIVER FALLS, 81- *Concurrent Pos:* Vis prof, Univ Cape Town, 69, Univ Witwatersrand, 70, Monash Univ, 70, Univ New South Wales, 71 & Univ PR, Mayaguez, 84, Univ Costa Rica, 87-88. *Res:* Infrared spectroscopy; science education. *Mailing Add:* Dept Chem Univ Wis River Falls WI 54022

HALL, LYNN RAYMOND, surgery, for more information see previous edition

HALL, MADELINE MOLNAR, b Cleveland, Ohio, June 16, 36; m 77; c 2. PHARMACOLOGY. *Educ:* Ohio State Univ, BS, 64, PhD(pharmacol), 70. *Prof Exp:* Fel pharmacol, Univ Chicago, 70-72; res fel cardiovasc, Cleveland Clin Found, 72-73, staff mem, 73-77; asst prof ,73-78, ASSOC PROF, DEPT BIOL, CLEVELAND STATE UNIV 78- *Honors & Awards:* Lower Prize, 73. *Mem:* AAAS; Am Soc Pharmacol & Exp Therapeut; NY Acad Sci. *Res:* The interrelationships between angiotensin and prostaglandins in the various tissues, brain, uterus, stomach and aorta; naltrexone-treated animals; effect of adrenalectomy in obese rats. *Mailing Add:* 3793 Bushnell Rd University Heights OH 44118

HALL, MARION TRUFANT, b Gorman, Tex, Sept 6, 20; m 44; c 3. TAXONOMY, CYTOGENETICS. *Educ:* Univ Okla, BS, 43, MS, 48; Washington Univ, PhD(bot), 51. *Prof Exp:* Ranger, Nat Park Serv, USDA, 42; instr bot, Univ Okla, 46-47, asst zool, Washington Univ, 48-50; prof bot & head dept, Butler Univ, 56-62; dir, Stovall Mus, Univ Okla, 62-66; DIR, MORTON ARBORETUM, 66- *Concurrent Pos:* Collabr, Dept Bot, Univ Mich, 50-56; adj prof biol sci, Univ Ill, Chicago Circle, prof hort, Univ Ill, Urbana; adj prof biol, Northern Ill Univ. *Mem:* Am Soc Plant Taxon; Ecol Soc Am; Soc Study Evolution; Int Soc Plant Taxon. *Res:* Taxonomy and cytogenetics of flowering plants; variation and evolution in cupressaeeae and angiosperms; plant ecology; land use management. *Mailing Add:* 1415 Birchwood Box 89 Ellison Bay WI 54210

HALL, MICHAEL OAKLEY, b Pretoria, SAfrica, Dec 14, 36; US citizen; m 63; c 2. BIOCHEMISTRY, OPHTHALMOLOGY. *Educ:* Univ Natal, BSc, 57; Univ Calif, Los Angeles, PhD(biochem), 61. *Prof Exp:* Lectr biochem, Univ Natal, 61-62; asst res biol chemist, 63-67, asst prof surg, 67-69, assoc prof ophthal, 69-80, RES PROF OPHTHAL & ASSOC DIR, JULES STEIN EYE INST, UNIV CALIF, LOS ANGELES, 80- *Concurrent Pos:* NSF Univ Res Coun grant, 63-64; Nat Coun Combat Blindness grants, 64-66; Nat Inst Neurol Dis & Blindness grant, 64-67; partic investr, USPHS Prog Proj grant, Jules Stein Eye Inst, Univ Calif, Los Angeles, 66-73; Nat Eye Inst grants, 69-82. *Mem:* AAAS; Asn Res Vision & Ophthal. *Res:* Biochemistry of the retina; visual pigments; retinal degeneration; electron microscopy of the retina. *Mailing Add:* Jules Stein Eye Inst 100 Stein Plaza UCLA Med Ctr Los Angeles CA 90024

HALL, NANCY K, b Washington, DC, July 7, 47. PATHOLOGY. *Educ:* Kans State Univ, PhD(immunol), 76. *Prof Exp:* ASSOC PROF IMMUNOL, HEALTH SCI CTR, OKLA UNIV, 81- *Mem:* Med Mycol Soc Am; Am Soc Microbiol; Int Asn Pathologists; Sigma Xi. *Mailing Add:* Univ Okla Health Sci Ctr PO Box 26901 Oklahoma City OK 73190

HALL, NATHAN ALBERT, b Bozeman, Mont, July 3, 18; c 2. PHARMACOLOGY. *Educ:* Univ Wash, BSc, 39, PhD(pharmaceut chem), 48. *Prof Exp:* Instr pharmaceut chem, Univ Wash, 49; asst prof pharm, Philadelphia Col Pharm, 49-50; pharmaceut chemist, Eli Lilly & Co, 50-51; res chemist, Univ Wash, 51-52, asst prof pharmaceut chem, 52-54, asst prof pharm, 54-56, assoc prof, 56-62, prof pharm, 62-85; RETIRED. *Concurrent Pos:* Fulbright lectr, Univ Malaya, 59-60; vis prof, Univ Sydney, 68. *Mem:* Fel AAAS; Am Pharmaceut Asn; Am Asn Cols Pharm; Acad Pharmaceut Sci. *Res:* Self-medication; folk medicine. *Mailing Add:* 10553 41st Pl NE Seattle WA 98125

HALL, NEWMAN A(RNOLD), b Uniontown, Pa, June 14, 13; m 38; c 2. MECHANICAL ENGINEERING. *Educ:* Marietta Col, AB, 34; Calif Inst Technol, PhD(math), 38. *Hon Degrees:* ScD, Marietta Col, 59; MA, Yale Univ, 56. *Prof Exp:* Instr, Queens Col NY, 38-41; res mathematician, Chance Vought Div, United Aircraft Corp, Conn, 41-42, supvr, Eng Personnel, 42-43, power plant anal engr, 43-44, head, Thermodyn Group, Res Dept, 44-46, anal sect, 46-47; prof mech eng & in charge heat power div, Univ Minn, 47-55; prof mech eng & asst dean in charge grad div, Sch Eng, NY Univ, 55-56; prof mech eng, Yale Univ, 56-58, Strathcona prof, 58-64, chmn dept, 56-64; exec dir, Comn Eng Educ, 62-68 & Nat Acad Eng, 68-71; Secy M Eng, Am Asn Arts & Sci, 64-74; sci adv, USAID Mission, Korea, 72-74; consult, Ministry Sci & Technol, Repub Korea, 74-75; consult eng & educ, 76-77; RETIRED. *Concurrent Pos:* Dir, Div Eng Sci, Off Ord Res, US Army, 52-53; hon dir, Int Combustion Inst, 65- *Mem:* Am Soc Mech Engrs; Soc Automotive Engrs; Am Soc Eng Educ (vpres, 60-62); assoc fel Am Inst Aeronaut & Astronaut; Int Combustion Inst. *Res:* Thermodynamics; fluid mechanics; gas dynamics; heat transfer; combustion; administration of engineering education. *Mailing Add:* 511 Town Hill Rd New Hartford CT 06057

HALL, OTIS F, b Cleveland, Ohio, Oct 5, 21; m 58; c 3. FOREST MANAGEMENT, FOREST ECONOMICS. *Educ:* Oberlin Col, AB, 43; Yale Univ, MF, 48; Univ Minn, PhD(forest econ), 54. *Prof Exp:* Instr & asst prof forestry, Univ Minn, 48-57; prof forest mgt, Purdue Univ, 57-68; prof forestry & head dept, Univ NH, 68-74; prof & head dept, 74-84, ENDOWED PROF FORESTRY, VA POLYTECH INST, 84- *Mem:* Fel Am Foresters. *Res:* Forest management, economic analysis and impact, computer applications; expert systems. *Mailing Add:* Sch Forestry Va Polytech Inst Blacksburg VA 24061

HALL, PETER, b Whangarei, NZ, Oct 17, 26; m 84; c 1. PHYSIOLOGY, PHARMACOLOGY, ANESTHESIA. *Educ:* Univ Otago, NZ, MB, ChB, 50; Royal Col Surgeons, fel fac anesthetists, 55. *Prof Exp:* Jr house surgeon med, Auckland Hosp, Bd, NZ, 50, sr house surgeon, 51, jr registr anesthesia, 52-53; jr house officer, Univ Col Hosp, Univ London, 54-55 & Hosp for Sick Children, Eng, 55-56; locum sr registr, Royal Free Hosp, 56; intern, Bispebjerg Hosp, Copenhagen, Denmark, 57, first resident, 58; instr, Univ Pa Hosp, 59-60, instr pharmacol, Sch Med, 60-64, assoc, 64-68; assoc prof physiol & biophys, Colo State Univ, 68-78; PROF PHYSIOL, ST GEORGE SCH MED, GRENADA, WEST INDIES, 79- *Concurrent Pos:* USPHS grants, 60-62 & fel, 62-63, assoc pharmacol training grant, 63-64, training grant, 64-65; Pa Plan scholar, 65-68; consult, Coun Drugs, AMA, 61, 63 & 64; physician to Univ Pa Underwater Archaeol Exped, 62; vis prof, dept anesthesia, Univ Hosp Wales, Cardiff, UK, 82; vis scholar, Dept Res Med Educ, Univ Southern Calif, 76, anesthesiolgist, Gen Hosp Grenada, 87-90. *Mem:* Fel Royal Soc Health; NY Acad Sci. *Res:* Chemical control of respiration and cerebral blood flow in man; hyperbaric oxygen in research and therapy; chemoreceptor responses; altitude physiology; diving physiology. *Mailing Add:* PO Box Seven St George's St George Univ Sch Med Grenada West Indies

HALL, PETER FRANCIS, b Sydney, Australia, Dec 13, 24; m 69; c 2. PHYSIOLOGY, ENDOCRINOLOGY. *Educ:* Univ Sydney, MB, BS, 47, MD, 56; Univ Utah, PhD(biochem), 61. *Prof Exp:* Jr resident med officer, Royal Prince Alfred Hosp, Sydney, Australia, 47-48, sr resident med officer, 48-50; pvt pract internal med, Australia, 50-52 & Postgrad Sch Med, London, 52; registr, Maudsley Hosp, London, 53; registr, Asst lectr & asst dir student health, Guy's Hosp & Guy's Hosp Med Sch, London, 53-55; asst lectr physiol, Univ Sydney, 55-59; USPHS trainee steroid biochem, Sch Med, Univ Utah, 59-61 & res fel biochem, 61-62; asst prof physiol, Sch Med, Univ Pittsburgh, 62-66; prof biochem, Univ Melbourne, 66-71; chmn dept, 71-78, PROF PHYSIOL, UNIV CALIF, IRVINE-CALIF COL MED, 71- *Concurrent Pos:* Asst physician, Sydney Hosp, Australia, 55-59; consult endocrinologist, Prince of Wales Hosp, 55-59; pvt pract, 55-59. *Mem:* Am Physiol Soc; Australian Biochem Soc; Am Soc Biol Chemists. *Res:* Mechanisms by which trophic hormones stimulate steroidogenesis; metabolism of testis; action of follicle-stimulating hormone; biochemistry. *Mailing Add:* Dept Endocrinol Prince of Wales Hosp Avoca St Randwick Sydney NSW 2031 Australia

HALL, PETER M, b Belmont, NY, July 31, 34; m 56; c 4. PHYSICS. *Educ:* Hobart Col, BA, 54; Iowa State Univ, MS, 56, PhD(physics), 59. *Prof Exp:* Asst physics, Iowa State Univ, 54-59; distinguished mem tech staff, Bell Tel Labs, Allentown, 59-90; DISTINGUISHED PROF PHYSICS, JOHNSON C SMITH UNIV, CHARLOTTE, NC, 90- *Concurrent Pos:* Fel, Bell Labs. *Mem:* Am Phys Soc; fel Inst Elec & Electronics Engrs; Int Soc Hybrid Microelectronics; Soc Exp Mech; Am Soc Mech Eng. *Res:* Solid state physics, especially magnetic and electrical properties of matter; thin films; packaging of electronic circuits. *Mailing Add:* 10308 Katelyn Dr Charlotte NC 28269

HALL, PHILIP LAYTON, b Orange, NJ, Dec 12, 40; m 64; c 2. BIOORGANIC CHEMISTRY, LIGNIN BIODEGRADATION. *Educ:* Col Wooster, AB, 63; Univ Chicago, SM, 65, PhD(chem), 67. *Prof Exp:* NIH fel, Columbia Univ, 67-68; asst prof, 68-74, assoc prof chem, Va Polytech Inst & State Univ, 74-79, asst univ provost, 79-84; assoc univ provost, 84-85 , VPRES ACAD AFFAIRS & DEAN, MARY WASH COL, 85- *Mem:* Am Chem Soc; Sigma Xi; AAAS. *Res:* Organic reaction mechanisms of biological interest; mechanisms of enzyme action; the chemistry and biochemistry of lignin biodegradation; enzyme model system. *Mailing Add:* Off Dean Mary Washington Col Fredericksburg VA 22401-5358

HALL, PHILIP WELLS, III, b Cranford, NJ, June 11, 25; m 49; c 5. MEDICINE, PHYSIOLOGY. *Educ:* Bethany Col, WVa, BS, 48; Western Reserve Univ, MS, 52, MD, 55. *Prof Exp:* Demonstr physiol, Western Reserve Univ, 52-55; Arthritis & Rheumatism Found fel & asst med, Sch Med, Boston Univ, 56-58; from instr to assoc prof, 59-81, PROF MED, SCH MED, CASE WESTERN RESERVE UNIV, 81- *Concurrent Pos:* Estab investr, Am Heart Asn, 62-67; dir nephrology, Cleveland Metrop Gen Hosp; Porter Fel, Am Physiol Soc, 51-52. *Mem:* Soc Exp Biol & Med; Am Fedn Clin Res; Am Soc Nephrol; Int Soc Nephrol. *Res:* Kidney physiology; acid-base physiology and electrolytes; clinical nephrology; environmental toxicology and epidemiology. *Mailing Add:* 1127 Forest Rd Cleveland OH 44107

HALL, R(OYAL) GLENN, JR, b Koloa, Hawaii, June 23, 21; m 43; c 3. ASTRONOMY. *Educ:* Park Col, BA, 41; Univ Chicago, PhD(astron), 49. *Prof Exp:* Instr, Univ Chicago, 49-52, res assoc, 53; astronomer, US Naval Observ 53-81; RETIRED. *Mem:* Int Astron Union; Am Astron Soc; AAAS. *Res:* Mass ratios of binary stars; orbits of binary stars; time. *Mailing Add:* 3612 Spring St Chevy Chase MD 20815

HALL, RANDALL CLARK, b Iowa City, Iowa, Aug 30, 46; m 67; c 3. ANALYTICAL CHEMISTRY. *Educ:* Tex A&M Univ, BS, 68, PhD(chem), 71. *Prof Exp:* From asst prof to assoc prof chem, Purdue Univ, 70-75; vis chemist, Environ Protection Agency, 75-76; sr res chemist & head detector develop, Tracor, Inc, 76-79; head, Org Chem Dept & sr scientist, 79-80, PROG MGR CHEM INSTRUMENTATION, RADIAN CORP, 81- *Mem:* Am Chem Soc; Sigma Xi. *Res:* Development of analytical instrumentation and chemical methodology for the analysis of trace organics. *Mailing Add:* 410 Chimney Hill College Station TX 78720

HALL, RAYMOND G, JR, b Sherman, Tex, Mar 11, 37; m 59; c 2. CELL PHYSIOLOGY. *Educ:* Union Col, Nebr, BA, 59; Walla Walla Col, MA, 62; Loma Linda Univ, PhD(physiol, biophys), 68. *Prof Exp:* Instr physiol & biophys, 68-70, ASSOC PROF PHYSIOL, PHARMACOL & BIOPHYS, LOMA LINDA UNIV, 70- *Concurrent Pos:* Am Cancer Soc fel, Dept Molecular, Cellular & Develop Biol, Boulder, Colo, 69-70. *Mem:* Am Soc Cell Biol; Sigma Xi. *Res:* Regulatory mechanisms in bone. *Mailing Add:* Dept Physiol Loma Linda Univ Loma Linda CA 92354

HALL, RICHARD BRIAN, b Mendota, Ill, Mar 24, 47; m 67; c 2. FOREST GENETICS, SILVICULTURE. *Educ:* Iowa State Univ, BS, 69; Univ Wis-Madison, PhD(plant breeding & genetics), 74. *Prof Exp:* From asst prof to assoc prof, 74-82, PROF FORESTRY, IOWA STATE UNIV, 82- *Mem:* AAAS; Sigma Xi; Soc Am Foresters. *Res:* The development of improved selection and breeding techniques for Populus species and Alnus glutinosa; genetic and cultural improvement of nitrogen fixation in Alnus; variation in DNA content and redundancy in tree genomes. *Mailing Add:* Dept Forestry Iowa State Univ 251 Bessey Hall Ames IA 50011

HALL, RICHARD CHANDLER, b Northampton, Mass, Apr 12, 39; m 67; c 2. ASTRONOMY. *Educ:* Amherst Col, AB, 60; Ind Univ, PhD(astron), 65. *Prof Exp:* Asst prof, 65-71, ASSOC PROF PHYSICS & ASTRON, NORTHERN ARIZ UNIV, 65- *Mem:* Fel AAAS; Am Astron Soc. *Res:* Image tubes; photometry; polarization; cinematography; Orion Nebula polarization. *Mailing Add:* Dept Physics & Astron Northern Ariz Univ C328 Flagstaff AZ 86011

HALL, RICHARD EUGENE, b Sioux City, Iowa, Jan 19, 43; m 68. INORGANIC CHEMISTRY. *Educ:* Univ SDak, BA, 64; Ohio State Univ, MS, 67, PhD(inorg chem), 69. *Prof Exp:* RES CHEMIST, FMC CORP, 69- *Mem:* Am Chem Soc. *Res:* Synthesis and characterization of molecular hydride complexes containing boron and aluminum; investigation and development of reaction catalysts and process and product development in organic and peroxide chemicals. *Mailing Add:* FMC Corp PO Box 8 Princeton NJ 08543

HALL, RICHARD EUGENE, b Indianapolis, Ind, Feb 15, 36; m 64; c 3. NEUROPHYSIOLOGY. *Educ:* Ind Univ, AB, 58; Univ Calif, San Diego, PhD(neurosci), 73. *Prof Exp:* Asst prof physiol, Bowman Gray Sch Med, 73-76; ASST PROF BIOL, THOMAS NELSON COMMUNITY COL, 73- *Mem:* Am Physiol Soc; Soc Neurosci; Am Heart Asn; Am Fedn Clin Res. *Res:* Neurophysiology of emotion; the function of brain structures implicated in causing psychosomatic disorders. *Mailing Add:* Dept Biol Thomas Nelson Community Col Hampton VA 23670

HALL, RICHARD HAROLD, b Akron, Ohio, Jan 12, 27; m 54; c 2. ORGANIC CHEMISTRY, CHEMICAL ENGINEERING. *Educ:* Case Inst Technol, BSChE, 50; Univ Del, MS, 51, PhD(chem), 53. *Prof Exp:* Chemist, B F Goodrich Co, 52; lab instr org chem, Univ Del, 52-53; res chemist, 53-67, group leader, Polymer Res Lab, 67-69, sr res engr phys res lab, 69-73, proj mgr imbiber systs eng & metal prod, 73-76, proj mgr functional prod & systs, 76-79, RES LEADER, NEW MONOMERS & POLYMERS, DOW CHEM CO, 79- *Mem:* Am Chem Soc; Am Inst Chem Engrs; Sigma Xi; Royal Soc Chem. *Res:* Petrochemical processes; industrial organic chemistry; synthesis; oxidation of hydrocarbons; monomers and polymerization; reinforced plastics and laminating resins; plastic foams; t-butylstyrene, chlorostyrene, vinylbenzyl chloride; imbibing or absorbing polymers; pollution control; systems to protect the environment; economic evaluation. *Mailing Add:* Rte 2 1187 Steward Rd Midland MI 48640

HALL, RICHARD L, b 1940. MATHEMATICS, PHYSICS. *Educ:* Univ London, BC, 62, PhD(combinatorics & finite math), 67. *Prof Exp:* PROF MATH, CONCORDIA UNIV, 72- *Mem:* Am Statist Soc. *Mailing Add:* Concordia Univ 1455 de Maisonneuve Blvd W Montreal PQ H3G 1M8 Can

HALL, RICHARD LELAND, b Roseland, Nebr, June 14, 23; m 48; c 3. FLAVOR CHEMISTRY & TECHNOLOGY, NATURAL TOXICANTS. *Educ:* Harvard Univ, SB, 43, AM, 48, PhD(chem), 51. *Prof Exp:* From asst sr tutor to sr tutor, Harvard Univ, 47-50; res chemist & dir res, McCormick & Co, Inc, 50-57, dir res & develop, 57-68, vpres res & develop, 68-75, vpres sci & technol, 75-88; CONSULT 88- *Concurrent Pos:* Mem panel chem & health, President's Sci Adv Comt, 70-73; mem exec comt, Int Union Food Sci & Technol, 75-78, vpres, 78-83, pres, 83-87; pres, Int Food Biotechnol Coun, 88-91. *Mem:* Hon mem Soc Flavor Chemists; fel Inst Food Technologists (pres, 71-72); Soc Toxicol; Am Chem Soc; fel AAAS. *Res:* Flavor and odor constituents of spices and other natural products; food toxicology and safety evaluation; pharmacology; food science. *Mailing Add:* 7004 Wellington Ct Baltimore MD 21212

HALL, RICHARD TRAVIS, b Berkeley, Calif, Apr 8, 38; m 61; c 2. AUTOMOTIVE RESEARCH, AUTOMOTIVE DEVELOPMENT. *Educ:* Mass Inst Technol, SB, 59; Univ Calif, Berkeley, PhD(chem), 63. *Prof Exp:* Mem tech staff, Aerospace Corp, El Segundo, Calif, 62-75, Germantown, Md, 75-79, prog mgr, 77-80, mem tech staff, Washington, DC, 79-, syst dir, 80-; VPRES ENGR, TELEPHONICS CORP. *Mem:* Am Chem Soc; Am Phys Soc; AAAS; Sigma Xi; Soc Automotive Engrs. *Res:* Planning and management of electric, hybrid vehicle and development programs. *Mailing Add:* Telephonics Corp 790 Park Ave Huntington NY 11743

HALL, ROBERT, b Melbourne, Australia, Mar 26, 39; m 63; c 2. FIELD CROP DISEASES. *Educ:* Univ Melbourne, BAgrSc, 61, PhD(bot), 64. *Prof Exp:* Univ Adelaide fel bot, Waite Agr Res Inst, 64-66; res plant pathologist, Univ Calif, Riverside, 66-67; asst prof, 67-72, assoc prof, 72-80, PROF PLANT PATH, UNIV GUELPH, 80- *Concurrent Pos:* Ed, Can J Plant Path, 90- *Mem:* Am Phytopath Soc; fel Can Phytopath Soc (pres, 87-88). *Res:* Pathology, physiology, taxonomy of plant pathogenic fungi; chemical taxonomy of fungi; pathology of field crops; environmental plant pathology. *Mailing Add:* Dept Environ Biol Univ Guelph Guelph ON N1G 2W1 Can

HALL, ROBERT DICKINSON, b Washington, DC, Mar 6, 47; m 68, 90; c 2. MEDICAL & VETERINARY ENTOMOLOGY. *Educ:* Univ Md, College Park, BA, 73; Va Polytech Inst & State Univ, MS, 75, PhD(entom), 77. *Prof Exp:* from asst prof to assoc prof, 77-89, PROF ENTOM, UNIV MO, COLUMBIA, 89- *Concurrent Pos:* Med entomologist, US Army Reserve, 77- *Mem:* Entom Soc Am; Sigma Xi. *Res:* Taxonomy, biology and control of arthropods affecting man and animals, with particular interest in Diptera (Calliphoridae, Muscidae, Sarcophagidae). *Mailing Add:* Dept Entom 1-87 Agr Bldg Univ Mo Columbia MO 65211

HALL, ROBERT DILWYN, b Philadelphia, Pa, Jan 19, 29; m 62; c 3. PHYSIOLOGICAL PSYCHOLOGY, NEUROBIOLOGY. *Educ:* Dartmouth Col, AB, 56; Brown Univ, MA, 58, PhD(psychol), 60. *Prof Exp:* Instr psychol, Brown Univ, 59-60; res fel commun biophysics, Res Lab Electronics, Mass Inst Technol, 60-63, res assoc, 63-68, staff scientist, 68-74, staff scientist, Neurosci Res Prog, 74-76; staff scientist neurobiol, Worcester Found Exp Biol, 76-83; vis prof psychol, Bourdoin Col, 83-84; AT AUDITORY PROTHESIS RES LAB, DEPT OTOLARYNGOL, MASS EYE & EAR INFIRMARY. *Mem:* Aero Sci Soc; Soc Neurosci; AAAS; Sigma Xi. *Res:* Cochlear implants; animal auditory psychophysics; auditory psychophysics. *Mailing Add:* Dept Otolaryngol Auditory Prothesis Res Lab 243 Charles St Boston MA 02114

HALL, ROBERT EARL, b Philadelphia, Pa, Aug 1, 49. PHYSICAL OCEANOGRAPHY, GEOPHYSICAL FLUID DYNAMICS. *Educ:* Calif Inst Technol, BS, 71; Univ Calif, San Diego, PhD(eng physics), 76. *Prof Exp:* Asst prof geophys, 76-81, RES ASSOC, DEPT GEOL & GEOPHYS, YALE UNIV, 81- *Mem:* Am Geophys Union; Am Meteorol Soc. *Res:* Theoretical and observational studies in physical oceanography. *Mailing Add:* Sci Appln 1200 Prospect St PO Box 2351 La Jolla CA 92038

HALL, ROBERT EVERETT, b Sioux City, Iowa, Apr 27, 24; m 49; c 3. VETERINARY MEDICINE. *Educ:* Iowa State Univ, DVM, 50; Univ Wis, MS, 64. *Prof Exp:* Veterinarian, Mo, 50-51; field supvr regulatory progs, Wis State Dept Agr, 51-52, diagnostician, Animal Health Lab, 52-59, chief diagnostician, 59-61; exten vet, Univ Wis-Madison, 61-80. *Mem:* Am Vet Med Asn; Am Asn Swine Practitioners; Am Asn Exten Vet. *Res:* Prevention, treatment, control and eradication of livestock diseases. *Mailing Add:* 5718 Dogwood Pl Madison WI 53705

HALL, ROBERT JOSEPH, b Buffalo, NY, June 4, 26; m 48; c 5. CARDIOLOGY. *Educ:* Univ Buffalo, MD, 48; Am Bd Internal Med & Am Bd Cardiovasc Dis, dipl. *Prof Exp:* Intern, Mercy Hosp, Buffalo, NY, 48; chief med serv, Hosps, Korea & Japan, US Army, 52-55, internist, Valley Forge Hosp, 55-56, mem cardiol serv, Walter Reed Army Hosp & Inst Res, 56-61, chief Brooke Gen Hosp, 61-66, chief, Walter Reed Gen Hosp, 66-69; CHIEF CARDIOL SERV, ST LUKE'S HOSP, HOUSTON, 69-; MED DIR, TEX HEART INST, 69- *Concurrent Pos:* Clin prof med, Baylor Col Med & Univ Tex Med Sch, Houston; assoc chief med, St Luke's Hosp; consult cardiol, Vet Admin Hosp, Houston & US Army; fel coun clin cardiol, Am Heart Asn. *Mem:* Fel Am Col Physicians; fel Am Col Cardiol; AMA; Am Heart Asn. *Res:* Coronary heart disease; electrocardiography; exercise testing; antianhythemic agents; phonocardiography; prosthetic heart valves; coronary bypass surgery; coronary circulation. *Mailing Add:* Tex Heart Inst PO Box 20269 MC1-102 Houston TX 77225-0269

HALL, ROBERT LESTER, b Little Rock, Ark, Dec 17, 39; m 62. MATHEMATICAL ANALYSIS. *Educ:* Univ Ark, BS, 61; Rice Univ, MA, 63, PhD(math), 65. *Prof Exp:* Ritt instr math, Columbia Univ, 65-67; asst prof, 67-71, ASSOC PROF MATH, UNIV WIS-MILWAUKEE, 71-, CHAIRPERSON DEPT, 80-, ASSOC DEAN, 82- *Mem:* Math Asn Am. *Res:* Boundary behavior of functions analysis in the unit disc. *Mailing Add:* Dept Math Sci Univ Wis Milwaukee WI 53201

HALL, ROBERT NOEL, b New Haven, Conn, Dec 25, 19; m 41; c 2. SEMICONDUCTORS. *Educ:* Calif Inst Technol, BS, 42, PhD(physics), 48. *Prof Exp:* Lab asst, Corp Res & Develop Ctr, Gen Elec Co, 42-46, res assoc, 48-52, physicist, res lab, 52-90; RETIRED. *Honors & Awards:* David Sarnoff Award, Inst Elec & Electronics Engrs, 63; Jack A Morton Award, Inst Elec & Electronics Engrs, 76; Solid State Sci & Tech Award, Electrochem Soc, 77. *Mem:* Nat Acad Eng; Nat Acad Sci; fel Inst Elec & Electronics Engrs; fel Am Phys Soc; Electrochem Soc. *Res:* Semiconductor physics; device technology; recombination; solar cells; junction lasers; gamma detectors. *Mailing Add:* 2315 Gurenson Lane Schenectady NY 12309

HALL, RONALD HENRY, b Wheeling, WVa, Sept 13, 30; m 55; c 2. SPECTROCHEMISTRY, ANALYTICAL CHEMISTRY. *Educ:* West Liberty State Col, BS, 52; Pa State Univ, MS, 56. *Prof Exp:* Asst fuel chem, Pa State Univ, 54-56; asst spectroscopist, Vanadium Corp Am, 56-63; SR SPECTROCHEMIST, OWEN ILL CO, TOLEDO, 63- *Concurrent Pos:* Adj prof, Geol Dept, Univ Toledo. *Mem:* Am Chem Soc; Soc Appl Spectros; Am Soc Testing & Mat. *Res:* Study and development of high precision analytical methods in ICP spectroscopy utilizing the present state of the art technology in electronics, computers and systematic evaluation of analytical systems. *Mailing Add:* 4802 Wickford Dr E Sylvania OH 43560-3351

HALL, ROSS HUME, b Winnipeg, Man, Nov 22, 26; nat US; m 50; c 3. BIOCHEMISTRY, ENVIRONMENTAL MANAGEMENT. *Educ:* Univ BC, BA, 48; Toronto Univ, MA, 50; Cambridge Univ, PhD(chem), 53. *Prof Exp:* Fel, Univ BC, 53-54; res chemist, Lederle Labs Div, Am Cyanamid Co, 54-58; prin scientist cancer res, Roswell Park Mem Inst, 58-67; assoc prof biochem, State Univ NY Buffalo, 65-67; prof biochem & chmn dept, McMaster Univ, 67-88; RETIRED. *Concurrent Pos:* Mem grants rev comt, Med Res Coun Can; mem, Can Environ Adv Coun, 75-81; co-founder & dir, En-Trophy Inst Advan Study Food, Nutrit & Food Policy, 77; chmn & mem, bd dirs, Pollution Probe Found, Toronto, Ont, 83-87; chmn, ministers panel selectional priority substances, Dept Environ, Govt Can; co-chmn, health comt, Int Joint Comn, 90- *Mem:* Am Soc Plant Physiologists; Can Biochem Soc (pres, 72-73); Am Chem Soc; Am Soc Biol Chem; Am Asn Cancer Res. *Res:* Assessment of effects of biological technologies; food processing, agriculture, drug, medical technologies; study of cellular mechanisms for processing information. *Mailing Add:* Box 239 Danby VT 05739

HALL, RUSSELL P, RHEUMATOLOGY, DERMATOLOGY. *Educ:* Univ Mo, MD, 75. *Prof Exp:* ASST PROF, DEPT MED, MED CTR, DUKE UNIV, 84- *Mailing Add:* Dept Anesthesiol Duke Univ Box 3135 Durham NC 27710

HALL, SEYMOUR GERALD, b Brooklyn, NY, June 30, 40; m 63; c 2. TEXTILE CHEMISTRY. *Educ:* Elon Col, BS, 64; Appalachian State Univ, MA, 66; Univ NC, Greensboro, PhD(textiles), 75. *Prof Exp:* Process chemist, E I du Pont de Nemours & Co, Inc, 66-68; mgr res & develop, Cone Mills Corp, 68-76; vpres oper, Chemonic Indust, 76-79; PRES, PROCHEM, 79- *Mem:* Am Chem Soc; Am Asn Textile Chemists & Colorists; Air Pollution Control Asn; Synthetic Org Chem Mfrs Asn. *Res:* Developing textile finishing and dyeing chemicals; flame retardant research; water and air pollution development projects; surfactants; esterification reactions. *Mailing Add:* 119 Manchester Pl Greensboro NC 27410

HALL, STAN STANLEY, b Platteville, Wis, July 4, 38; div; c 2. ORGANIC CHEMISTRY, BIO-ORGANIC CHEMISTRY. *Educ:* Univ Wis, Madison, BS, 63; Mass Inst Technol, PhD(org chem), 67. *Prof Exp:* NIH fel & res assoc chem, Stanford Univ, 67-68; from asst prof org chem to assoc prof chem, 68-78, PROF CHEM, RUTGERS UNIV, 78- *Concurrent Pos:* NIH spec fel, Scripps Inst Oceanog, 72-73. *Honors & Awards:* Geigy Award, Chem Indust, Summit, NJ, 80. *Mem:* Am Chem Soc; Sigma Xi. *Res:* Synthetic methods; chemistry of natural products; metal-ammonia reductions; model enzyme studies; molecular rearrangements. *Mailing Add:* Dept Chem Rutgers Univ Newark NJ 07102

HALL, STANTON HARRIS, b Boise, Idaho, 1940; m 62; c 3. BIOCHEMISTRY OF EXTRACELLULAR MATRIX. *Educ:* Northwestern Univ, MS & DDS, 67; Univ Wash, PhD(exp path), 74, cert orthod, 79; Am Bd Orthod, dipl, 91. *Prof Exp:* Capt, Wilford Hall USAF Hosp, Lackland AFB, Tex, 67-69; trainee exp path, Univ Wash, 69-74; res assoc, Lab Molecular Genetics, Nat Inst Child Health & Human Develop, NIH, 74-77; asst prof, 79-85, ASSOC PROF ORTHOD, UNIV WASH, 85- *Concurrent Pos:* USPHS trainee, Univ Wash, 70, res grant & co-prin investr, dept orthod, 81-86. *Mem:* AAAS; Am Soc Bone & Mineral Res; Am Asn Dent Res; Am Asn Orthodontists. *Res:* Morphogenesis of craniofacial sutures and the modulation of phenotypic expression accompanying osteodifferentiation. *Mailing Add:* Dept Orthod Univ Wash Seattle WA 98195

HALL, STEPHEN KENNETH, b Hong Kong, Dec 15, 34; US citizen. INDUSTRIAL HYGIENE, ENVIRONMENTAL CHEMISTRY. *Educ:* Int Christian Univ, Tokyo, BS, 61; Univ Toronto, MS, 63; Univ Pittsburgh, PhD(chem), 67; Harvard Univ, MSHyg, 75. *Prof Exp:* Qual control chemist, Ford Motor Co Can, 63-64; res fel, Univ Alta, 67-69; from asst prof to prof vchem, Southern Ill Univ, 69-85; assoc dean, Med Col Ohio, Toledo, 85-89; DEAN, CHICAGO STATE UNIV, 89- *Mem:* AAAS; Am Chem Soc; The Chem Soc; Am Indust Hyg Asn; Brit Occup Hyg Soc. *Res:* Industrial hygiene chemistry; occupational health and safety; toxicological effects of chemicals; environmental pollution. *Mailing Add:* PO Box 100 Chicago Heights IL 60411-0100

HALL, TERENCE ROBERT, neuroendocrinology, comparative physiology, for more information see previous edition

HALL, THEODORE (ALVIN), b New York, NY, Oct 20, 25; m 47; c 3. BIOPHYSICS. *Educ:* Harvard Univ, BS, 44; Univ Chicago, PhD(physics), 50. *Prof Exp:* USPHS res fel radiobiol, Univ Chicago, 50-52; physicist, Sloan-Kettering Inst Cancer Res, 52-62; biophysicist, Cavendish Lab, 62-74; biophysicist, dept zool, Cambridge Univ, 74-84; RETIRED. *Mem:* Hon fel Royal Micros Soc; Cambridge Philos Soc; hon mem Microbeam Anal Soc Am. *Res:* X-ray histochemistry; electron probe microanalysis; biology of the minor elements. *Mailing Add:* 49 Owlstone Rd Cambridge CB3 9JH England

HALL, THOMAS CHRISTOPHER, b New York, NY, Nov 26, 21; m 45, 64, 78; c 8. CANCER, ONCOLOGY. *Educ:* Harvard Univ, MD, 49; Am Bd Internal Med, dipl. *Prof Exp:* Intern med, Peter Bent Brigham Hosp, 49-50; res fel, Harvard Med Sch, 50-53, asst, 53 & 55-57, instr, 57-65, asst clin prof med, 65-68; prof med & pharmacol & dir div oncol, Sch Med & Dent, Univ Rochester, 68-73; Am Cancer Soc prof med & biochem, Sch Med, Univ Southern Calif, 73-75; prof med, Univ BC & dir, Cancer Control Agency BC, 75-77; vis mem staff, Pasadena Found Med Res, 77-78; dir, Cancer Control, Educ & Community Affairs, Cancer Ctr Hawaii, 78-87; prof, 78-87, EMER PROF MED & PHARMACOL, UNIV HAWAII SCH MED, 87; DIR, CANCER PROG, UNITED HOSP MED CTR, NEWARK, 87- *Concurrent Pos:* Clin & res fel, Mass Gen Hosp, 50-55; fel, Harvard Med Sch, 54; asst resident, Mass Gen Hosp, 55-61; vis intern & consult, Pondville Hosp, 55-68; asst, Sch Med, Tufts Univ, 57-60; dir oncol div, Med Serv, Lemuel Shattuck Hosp, 57-60; consult, USPHS, 57 & Free Hosp Women, 59; assoc physician, Peter Bent Brigham & Children's Hosps, Boston, 65-68; dir, Minority Based Clin Oncol Prog, Univ Med & Dent NJ, 87- *Mem:* Radiation Res Soc; Asn Cancer Res; fel Am Col Physicians; Am Asn Cancer Educ; Am Soc Clin Oncol; hon fel Royal Col Physicians. *Res:* Internal medicine and cancer chemotherapy, especially as related to nucleic acid metabolism and hormonal therapy of breast cancer; cancer control. *Mailing Add:* Cancer Prog United Hosp Med Ctr 15 S Ninth St Newark NJ 07107

HALL, THOMAS KENNETH, b Minneapolis, Minn, Sept 28, 36; m 59; c 5. ORGANIC CHEMISTRY. *Educ:* Univ Santa Clara, BS, 61; Iowa State Univ, PhD(org chem), 65. *Prof Exp:* Asst prof, Chico State Col, 65-66; from assoc to asst prof, Univ Ga, 66-68; sr res chemist, Eastman Kodak Co, 68-70; asst prof chem, Rochester Inst Technol, 70-71; prof chem, West Valley Col, 71-78; PRES & FINANCIAL MGR, T K HALL ENTERPRISES, INC, 78- *Mem:* Am Chem Soc; Royal Soc Chem. *Res:* Organic photochemistry; oxidation mechanisms; ozonide decomposition. *Mailing Add:* T K Hall Enterprises Inc 3275 Stevens Creek Blvd San Jose CA 95117

HALL, THOMAS LIVINGSTON, b Great Barrington, Mass, Aug 14, 31; m 91; c 3. PUBLIC HEALTH, MEDICINE. *Educ:* Harvard Univ, AB, 53, MD, 57, MPH, 61; Johns Hopkins Univ, DrPh, 67; Am Bd Prev Med & Pub Health, Cert. *Prof Exp:* Med dir, Castaner Gen Hosp, PR, 58-60; dir training & res, Community Teaching Health Ctr, Guaynabo, PR, 61-62; res assoc pub health, Sch Hyg, Johns Hopkins Univ, 63-67, from asst prof to assoc prof, 67-71; dep dir, Carolina Pop Ctr & assoc prof health admin, Univ, 71-74, actg dir, 74-75, dir, Carolina Pop Ctr, Univ NC, 75-77, prof, Dept Health Admin, Sch Pub Health, 74-79; prof health admin, Univ NC, Chapel Hill, 74-79; vis prof, Sch Pub Health & Community Med, Univ Wash, 79-80; exec dir, Puget Sound Health Systs Agency, Seattle, 80-82; health care consult, 82-84, prin med officer res, Dept Health, Wellington, NZ, 85-86 & 86-88; HEALTH CARE CONSULT, CTR AIDS PREV STUDIES, UNIV CALIF, SAN FRANCISCO, 87- *Concurrent Pos:* Consult, Md State Comt Med Care, 65, Pan Am Health Orgn, 65 & 66 & WHO, 70-; AID consult, Chile, 66 & Chilean Nat Health Serv, 67-70, World Bank, 88-; comt, Int Health, Inst Med, 77-80; lectr, Dept Epidemiol & Biostatist, Sch Med, Univ Calif, San Francisco; gov coun, 87-89. *Mem:* fel Am Pub Health Asn. *Res:* Public health administration; medical care; health manpower and general health planning; population dynamics and family planning program administration; AIDS program planning; author of numerous publications & major reports. *Mailing Add:* 1515 16th Ave San Francisco CA 94122

HALL, TIMOTHY COUZENS, b Darlington, Eng, Aug 29, 37; m 60; c 3. PLANT PHYSIOLOGY, PLANT BIOCHEMISTRY. *Educ:* Univ Nottingham, BSc, 62, PhD(bot), 65. *Prof Exp:* Res fel, Dept Hort Sci, Univ Minn, St Paul, 65-66; from asst prof to prof hort, Univ Wis-Madison, 66-82; DISTINGUISHED PROF & HEAD, DEPT BIOL, TEX A&M UNIV, 84- *Concurrent Pos:* Prin investr, NSF, NIH, USDA, Sci Educ Admin & Herman Frasch Found grants; sabbatical leave, vpres & dir advan res, Agrigenetics Corp, 81-84. *Mem:* AAAS; Am Soc Microbiol; Am Soc Plant Physiologists; Brit Biochem Soc; Am Soc Virol; Agr Soc Biol Chemists; Sigma Xi. *Res:* Gene cloning, isolation and expression; biosynthesis, structure and function of legume seed proteins; in vitro systems of protein synthesis; infective role of aninoacylation of viral RNA; replication of viral RNAs. *Mailing Add:* Dept Biol Tex A&M Univ College Station TX 77843-3258

HALL, W(ILLIAM) J(OEL), b Berkeley, Calif, Apr 13, 26; m 48; c 3. EARTHQUAKE ENGINEERING, STRUCTURAL ENGINEERING. *Educ:* Univ Kans, BS, 48; Univ Ill, MS, 51, PhD(struct eng), 54. *Prof Exp:* Asst civil eng, Univ Kans, 47-48; engr, Sohio Pipe Line Co, Mo, 48-49; res asst, 49-50, from asst prof to assoc prof, 51-59, assoc mem, Inst Advan Study, 63-64, PROF CIVIL ENG, UNIV ILL, URBANA, 59-, HEAD DEPT, 84- *Concurrent Pos:* Consult, numerous indust orgn & govt agencies, 70-; mem & chmn, numerous sci & adv comts, Nat Acad Sci, Nat Acad Eng, Nat Res Coun, NSF, US Geol Surv & Am Soc Civil Engrs. *Honors & Awards:* A Epstein Mem Award, 58; Walter L Huber Res Award, Am Soc Civil Engrs, 63, Newmark Medal & E E Howard Award, 84, Martin Duke Award, 91. *Mem:* Nat Acad Eng; hon mem Am Soc Civil Engrs; Am Concrete Inst; Am Welding Soc; Earthquake Res Inst; fel AAAS; Am Soc Mech Engrs; Am Soc Testing & Mat. *Res:* Structural analysis and design; structural dynamics; plasticity; brittle fracture; earthquake and nuclear effects; author or co-author of more than 180 technical publications. *Mailing Add:* 2106 Newmark Civil Eng Lab Univ Ill 205 N Mathew Ave Urbana IL 61801

HALL, W KEITH, b McComb, Ohio, June 30, 18; m 45; c 1. PHYSICAL CHEMISTRY, CATALYSIS. *Educ:* Emory Univ, BS, 40; Carnegie Inst Technol, MS, 48; Univ Pittsburgh, PhD(chem), 56. *Hon Degrees:* DSc, Emory Univ, 74; Dr, Katholieke Univ Leuven, Belg, 77; PHU, Univ Santa Fe, Argentina, 80. *Prof Exp:* Res assoc chem, Nat Defense Res Coun, Div 8, 41-45; phys chemist, US Bur Mines, 45-51; fel, Mellon Inst, 51-56, sr fel, 56-70; sr scientist, Gulf Res & Develop Co, 70-73; distinguished prof chem, Univ Wis-Milwaukee, 73-85; DISTINGUISHED RES PROF, UNIV PITTSBURGH, 85- *Concurrent Pos:* Mem, Petrol Res Adv Bd, Am Chem Soc, 72-75; coordr, US-USSR Exchange Prog in chem catalysis, 72-78; mem, Advan Res Prop Agency Conf on res needs of Dept Defense, 73; co-organizer, NSF Workshop on Catalysis Related to Energy Prod, Houston, 74; mem, Comt Direction, Nat Ctr Sci Res, Res Ctr, Crystalline Solids, France, 75-79; ed, J Catalysis, 76-89; exec comm, Colloid & Surface Chem Div, Am Chem Soc, 74-81, chmn div, 79; bd Trustee, Gordon Res Conf, 81-87. *Honors & Awards:* Kendall Award, Am Chem Soc, 74; Von Homboldt Award, 84;

Petrol Award, Am Chem Soc, 87; Exxon Award Excellence Catalysis, 90. *Mem:* Am Chem Soc; Catalysis Soc NAm (pres, 81-85). *Res:* Chemistry; surface chemistry; energy; solar energy; petroleum chemistry; automotive emissions control. *Mailing Add:* Box 97 Mill Run PA 15464

HALL, W(ILLIAM) M(OTT), b Burlington, Vt, July 10, 06; wid; c 3. ELECTRICAL ENGINEERING. *Educ:* Mass Inst Technol, SB, 28, SM, 32, ScD(elec eng), 35. *Prof Exp:* Sound recording, Metro-Goldwyn-Mayer, NY, 28-29 & Warner Bros, 29-30; instr elec eng, Mass Inst Technol, 30-37, asst prof commun lab, 37-47; engr, Raytheon Co, 41-62, consult scientist, 62-71, dir develop eng, 68-71; CONSULT ENGR, 71- *Concurrent Pos:* Indicator sect head, Radiation Lab, Mass Inst Technol, 40-41; consult, Nat Defense Res Coun, 40-42. *Mem:* Fel Acoust Soc Am; fel Inst Elec & Electronics Engrs. *Res:* Sound measurements; infrared detection; microwave radar design. *Mailing Add:* 1357 Massachusetts Ave Lexington MA 02173

HALL, WARREN A(CKER), water resources, systems analysis; deceased, see previous edition for last biography

HALL, WARREN G, b Urbana, Ohio, July 14, 48. PSYCHOBIOLOGY, NEUROBIOLOGY. *Educ:* Univ Ariz, BA, 71; Johns Hopkins Univ, MA, 73, PhD(physiol psychiat), 75. *Prof Exp:* Fel develop psychobiol, Inst Animal Behav, Rutgers, 75-77; res scientist, NC Div Ment Health, Dorothea Dix Hosp, 77-; AT DEPT PSYCHOL, DUKE UNIV. *Mem:* Int Soc Develop Psychobiol; Animal Behav Soc; Am Psychol Asn; Soc Neurosci. *Res:* Developmental psychobiology and neurobiology of motivation and learning. *Mailing Add:* Dept Psychol Duke Univ 210 Soc-Psych Bldg Durham NC 27706

HALL, WAYNE CLARK, b Vandalia, Mont, Oct 16, 19; m 41; c 2. PLANT PHYSIOLOGY. *Educ:* Univ Iowa, BA, 41, MS, 47, PhD(plant physiol), 48. *Prof Exp:* Res assoc, Univ Iowa, 46-48; asst prof, Univ Ky, 48-49; plant physiologist, Tex A&M Univ, 49-51, from assoc prof to prof, 51-65, head dept, 58-60, vpres acad affairs & dean grad col, 65-68; dir fels & new prog adv, Off Sci Personnel, Nat Acad Sci-Nat Res Coun, 68-71; prof biol & vpres grad studies & res, State Univ NY Binghamton, 71-74; VPRES ACAD AFFAIRS & PROF BIOL, VA COMMONWEALTH UNIV, 74- *Concurrent Pos:* Consult, Union Carbide Chem Co, 58-65, Ethyl Corp, 63-65; NSF fel metab biol, 63-65. *Mem:* Am Soc Plant Physiologists; Scand Soc Plant Physiol; Japanese Soc Plant Physiol; AAAS; Am Inst Biol Sci. *Res:* Metabolism and biochemistry of plants; abscission; ethylene biogenesis and physiology. *Mailing Add:* 401 Chimney Hill Dr Col Sta TX 77840-1833

HALL, WAYNE HAWKINS, b Fairland, Okla, Apr 4, 36; m 55; c 3. MATHEMATICS EDUCATION. *Educ:* Kans State Col Pittsburg, BS, 58, MS, 64; George Peabody Col, PhD(math, educ-psychol), 70. *Prof Exp:* Teacher math, High Schs, Kans, 59-66; instr, Kans State Col Pittsburgh, 66-67; asst prof math educ, 70-73, co-dir, Ctr Safety Educ, 74, assoc prof math & educ, 73-80, PROF MATH & EDUC, EASTERN WASH UNIV, 80- *Mem:* Sigma Xi. *Res:* Mathematics education, especially the effect on performance of number of exercises; feedback and detail; investigation of two instructional variables in learning non-metric geometry. *Mailing Add:* Dept Math Eastern Wash Univ Cheney WA 99004

HALL, WENDELL HOWARD, b Redfield, SDak, May 18, 16; m 48; c 4. INTERNAL MEDICINE. *Educ:* Univ Minn, BS, 38, MB, 40, MD, 41, PhD(med), 50; Am Bd Internal Med, dipl. *Prof Exp:* Intern, Detroit Receiving Hosp, 40-41; intern internal med, 41-42, asst, 43-45, instr, 45-47, from asst prof to assoc prof internal med & microbiol, 47-61, prof internal med & microbiol, Univ Minn Hosp, 61-86; RETIRED. *Concurrent Pos:* Chief bact lab, Vet Admin Hosp, 49-52, chief clin labs, 52-60, chief med serv, 60-71, chief infectious dis serv, 71-80. *Mem:* Am Soc Clin Invest; Am Col Physicians; Infectious Dis Soc Am. *Res:* Infectious diseases; brucellosis; tuberculosis; antibiotics; immunology. *Mailing Add:* 6609 Galway Dr Minneapolis MN 55439

HALL, WILLIAM BARTLETT, b Cincinnati, Ohio, July 12, 25; m 51; c 3. GEOLOGY, PHOTOGEOLOGY. *Educ:* Princeton Univ, AB, 50; Univ Cincinnati, MS, 51; Univ Wyo, PhD, 61. *Prof Exp:* Geologist, Pure Oil Co, 51-54; asst geol, Univ Wyo, 54-57, instr, 58; asst prof, Mont Sch Mines, 58-65; assoc prof, 65-69, PROF GEOL, UNIV IDAHO, 69- *Concurrent Pos:* Mem, Fourth Int Field Inst, Italy, 64. *Mem:* Geol Soc Am; Am Soc Photogram; Am Asn Petrol Geologists; Am Inst Prof Geologists. *Res:* Montana geology; geomorphology; photogeology; photogrammetry; engineering geology; aerial photography. *Mailing Add:* Dept Geol Univ Idaho Moscow ID 83843

HALL, WILLIAM CHARLES, b Peoria, Ill, Aug 2, 40; m 62; c 1. NEUROPSYCHOLOGY, NEUROANATOMY. *Educ:* Duke Univ, BA, 62, PhD(physiol psychol), 67. *Prof Exp:* NIMH fel, Duke Univ, 67-68; Nat Inst Neurol Dis & Stroke fel, Brown Univ, 68-70; ASSOC PROF ANAT, DUKE UNIV, 70- *Concurrent Pos:* NIMH res scientist career develop award, 71-; Nat Inst Neurol Dis & Stroke res grant award, 71-; NSF res grant award, 75- *Honors & Awards:* C J Herrick Award, Am Asn Anat, 72. *Mem:* AAAS; Am Asn Anat; Neurosci Soc. *Res:* Structure and function of neocortex, especially the use of comparative technique; identification and comparison of particular structure in a variety of species to determine how structures vary in response to the particular ecological niches of species and as a function of the evolutionary history of species. *Mailing Add:* Dept Anat & Psychol Duke Univ 3011 Med Ctr Durham NC 27710

HALL, WILLIAM EARL, b Fayetteville, Ark, Mar 2, 38; m 60; c 3. PHARMACY, PHYSICAL CHEMISTRY. *Educ:* Univ Ark, BS, 61; Univ Wis, MS, 64, PhD(pharm), 67. *Prof Exp:* Instr pharm, Univ Wis, 65-66; asst prof, Sch Pharm, Univ NC, Chapel Hill, 66-73; group leader, Burroughs Wellcome Co, 73-74, sect head, 74-77, dept head, 77-84, VALIDATION COORDR, PHARMACEUT RES DEVELOP LAB, QUAL ASSURANCE DIV, BURROUGHS WELLCOME CO, 84- *Mem:* Am Asn Pharmaceut Scientists; Am Pharmaceut Asn. *Res:* Computer optimization of pharmaceutical formulations, microscopy of particulate matter, sustained release of drug dosage forms, image analysis, regulatory compliance in the pharmaceutical industry; validation of pharmaceutical processes. *Mailing Add:* Burroughs Wellcome Box 1887 Greenville NC 27834

HALL, WILLIAM FRANCIS, b Cassville, Mo, Dec 20, 28; m 58; c 2. SPEECH PATHOLOGY. *Educ:* Southwest Mo State Col, BS, 55; Univ Mo, MA, 57, PhD(speech path), 62. *Prof Exp:* Elem teacher, Barry County Pub Schs, 46-51; instr speech & supvr speech & hearing clin, Univ Mo, 57-62; prof speech path, Speech & Hearing Clin, NE Mo State Univ, 62-, head, Div Spec Progs, 66-, dir, Speech & Hearing Clin, 80-; RETIRED. *Concurrent Pos:* Fel, Northwestern Univ, 64-65. *Mem:* Am Speech & Hearing Asn; Am Acad Cerebral Palsy. *Res:* Audiology; theater. *Mailing Add:* One Grim Ct Kirksville MO 63501

HALL, WILLIAM HEINLEN, b Cairo, WVa, June 19, 10; m 38; c 2. CHEMISTRY. *Educ:* Muskingum Col, AB, 32; Ohio State Univ, PhD(phys chem), 39. *Prof Exp:* Asst chem, Ohio State Univ, 32-36; from instr to prof chem, 36-76, chmn dept chem, 54-72, EMER PROF CHEM, BOWLING GREEN STATE UNIV, 76- *Concurrent Pos:* Consult, Chem Info. *Mem:* Fel AAAS; Am Chem Soc. *Mailing Add:* 324 Pennsylvania Ave Kutztown PA 19530-1809

HALL, WILLIAM JACKSON, b Beltsville, Md, Nov 13, 29; m 54; c 4. STATISTICS. *Educ:* Johns Hopkins Univ, AB, 50; Univ Mich, MA, 51; Univ NC, PhD(statist), 55. *Prof Exp:* Mem tech staff probability & statist, Bell Tel Labs, 54-55; asst chief polio surveillance unit, Commun Dis Ctr, USPHS, 55-57; from asst prof to prof, Univ NC, Chapel Hill, 57-69; chmn, Dept Statist, 69-81, actg dir, Div Biostatist, 87-90, PROF STATIST & BIOSTATIST, UNIV ROCHESTER, 69- *Concurrent Pos:* Vis prof, Stanford Univ, Univ Calif, Berkeley, 67-69 & Univ Wash, 82. *Mem:* Fel AAAS; fel Am Statist Asn; fel Inst Math Statist; Royal Statist Soc. *Res:* Statistical inference. *Mailing Add:* 75 Chelmsford Rd Univ Rochester Rochester NY 14618

HALL, WILLIAM JOEL, b Berkeley, Calif, Apr 13, 26; m 48; c 3. EARTHQUAKE ENGINEERING, STRUCTURAL DYNAMICS. *Educ:* Univ Kans, Lawrence, BS, 48; Univ Ill, Urbana, MS, 51, PhD(civil eng), 54. *Prof Exp:* Engr, Sohio Pipe Line Co, 48-49; res asst, Univ Ill, Urbana, 49-51, res assoc, 51-54, from asst prof to assoc prof res & teaching, 54-59, PROF RES & TEACHING, UNIV ILL, URBANA, 59- *Concurrent Pos:* Consult, Alyeska Pipeline Serv Co, 70-, Yukon Pac Corp, 89-, Defense Nuclear Facil Safety Bd, 89-; dir, Earthquake Eng Res Inst, 79-82; founding dir, Civil Eng Res Found, 89-; comn mem, State Ill Low-Level Radioactive Waste Disposal Facil Siting Comn, 90- *Honors & Awards:* W L Huber Res Award, Am Soc Civil Engrs, 63, E Howard Award, 84, Newmark Medal, 84, Martin Duke Award, 91; John Parmer Award, Struct Engrs Asn, 90. *Mem:* Nat Acad Eng; Am Soc Mech Engrs; Am Concrete Inst; Nat Soc Prof Engrs; hon mem Am Soc Civil Engrs. *Res:* Structural engineering; structural dynamics; earthquake engineering; materials; fracture mechanics; protective structures; author of 180 publications. *Mailing Add:* 3105 Valley Brook Dr Champaign IL 61821

HALL, WILLIAM MYRON, JR, technical & research administration, for more information see previous edition

HALL, WILLIAM SPENCER, b Ancon, Panama, Oct 9, 35; m 63. APPLIED MATHEMATICS. *Educ:* Univ Va, BEE, 58; Cambridge Univ, BA, 65, MA, 70; Brown Univ, PhD(appl math), 68. *Prof Exp:* Asst prof, 68-74, ASSOC PROF MATH, UNIV PITTSBURGH, 75- *Concurrent Pos:* Lectr, Dept Math, Univ Col, Galway, Ireland, 73-74; Joint Int Res & Exchange Bd & Nat Acad Sci Scholar, Czech Acad Sci, 75-76 & 78-79. *Mem:* Am Math Soc; Sigma Xi. *Res:* Periodic solutions of partial differential equations. *Mailing Add:* 1140 23rd St NW Washington DC 20037

HALL, WILLIAM THOMAS, cytology, for more information see previous edition

HALL, ZACH WINTER, b Atlanta, Ga, Sept 15, 37; m; c 3. PHYSIOLOGY GENERAL. *Educ:* Harvard Univ, PhD(biochem), 66. *Prof Exp:* PROF PHYSIOL, UNIV CALIF, SAN FRANCISCO, 76- *Mem:* AAAS; Soc Neurosci; Am Soc Biol Chemists; Am Soc Cell Biol. *Mailing Add:* Dept Physiol Univ Calif Box 0444 San Francisco CA 94143

HALLADA, CALVIN JAMES, b Grand Forks, NDak, Oct 16, 33; m 57; c 3. INDUSTRIAL CHEMISTRY, METALLURGY. *Educ:* Univ NDak, BA, 56; Univ Mich, Ann Arbor, MS, 58, PhD(chem), 61. *Prof Exp:* Res chemist, Union Carbide Corp, 61-62 & Lawrence Radiation Lab, 62-63; group leader chem, Conductron, Inc, 63-65; sr res chemist, Climax Molybdenum Co, 65-66, group leader chem & extractive metall, 66-70, supvr, 70-74, mgr chem & process res, 74-84, DIR CHEM DEVELOP, CLIMAX MOLYBDENUM CO, 87- *Concurrent Pos:* Lectr, Univ Calif, Berkeley, 62-63. *Mem:* Am Chem Soc; NAm Catalyst Soc. *Res:* Chemistry of transition metals as it applies to extraction of the metals and their use in catalysis, lubrication, pigments, flame retardants and aqueous corrosion inhibitors; energy storage materials. *Mailing Add:* Climax Molybdenum Co PO Box 407 Ypsilanti MI 48197

HALLAHAN, WILLIAM LASKEY, b Philadelphia, Pa, Dec 19, 46. BIOLOGY, ETHOLOGY. *Educ:* Colo Col, BA, 69; Duke Univ, PhD(zool), 76. *Prof Exp:* ASST PROF BIOL, NAZARETH COL, ROCHESTER, 76- *Mem:* AAAS; Am Inst Biol Sci; Animal Behav Soc; Nat Sci Teachers Asn. *Res:* Ecology; endocrinology. *Mailing Add:* Biol Dept Nazarath Col 4145 East Ave Rochester NY 14610

HALLAM, THOMAS GUY, b Chicago, Ill, Sept 3, 37; m 64; c 2. MATHEMATICS. *Educ:* Univ Southern Ill, BA, 59, MA, 61; Univ Mo, PhD(math), 65. *Prof Exp:* From asst prof to prof math, Fla State Univ, 65-76; vis prof math & zool, Univ Ga, 76-77; PROF MATH & ECOL, UNIV TENN, KNOXVILLE, 77- *Concurrent Pos:* Vis assoc prof, Univ RI, 70-71; vis prof, Inst Math Sci, Univ Sao Paulo, Brazil, 74. *Mem:* Ecol Soc Am; AAAS; Soc

Indust & Appl Math. *Res:* Ordinary and functional differential equations; asymptotic behavior; stability theory; mathematical modelling; magnetohydrodynamics, rotating fluid flows, ecological systems; ecotoxicology; mathematical ecology. *Mailing Add:* Dept Math Univ Tenn Knoxville TN 37916

HALLANGER, LAWRENCE WILLIAM, b Oakland, Calif, Aug 9, 39; div; c 1. OCEAN ENGINEERING, RESEARCH ADMINISTRATION. *Educ:* Harvey Mudd Col, BS, 61; Calif Inst Technol, MS, 62, PhD(appl mech), 67. *Prof Exp:* Proj engr, US Naval Civil Eng Lab, 65-79; proj mgr, Seacoast Test Fac, Seaco, Hawaii, 79-81, actg exec dir, Natural Energy Lab, 80-82, sr prog engr, 82-84; PROG MGR, SAIC, 84- *Mem:* Am Soc Mech Engrs; Marine Technol Soc. *Res:* Diver work systems and techniques; alternate energy applications; undersea surveillance systems. *Mailing Add:* SAIC 970 N Kalaheo Suite A-203 Kailua HI 96734

HALLANGER, NORMAN LAWRENCE, b Salt Lake City, Utah, June 13, 12; m 35; c 2. METEOROLOGY. *Educ:* Calif Inst Technol, BS, 34, MS, 36. *Prof Exp:* Meteorologist, United Airlines, 36-39; meteorologist, Pan Am Airways, 39-52, asst supt meteorol, 44-50; chief meteorologist, R M Parsons Co, 52-54, prin scientist, Booz, Allen Appl Res, Inc, 57-71; sr scientist & mgr indust meteorol, Meteorol Res, Inc, 71-77; consult, 77-83; RETIRED. *Mem:* Fel Am Meteorol Soc; Nat Coun Indust Meteorol. *Res:* Micrometeorology; atmospheric diffusion; industrial meteorology; tropical meteorology; forecast system development. *Mailing Add:* 26551 Hempstead Ct Sun City CA 92381

HALLAS, LAURENCE EDWARD, b Montgomery, Ala, Feb 10, 54; m 78; c 2. ENVIRONMENTAL MICROBIOLOGY, MICROBIAL ECOLOGY. *Educ:* Miami Univ, Ohio, BA, 75; Univ Cincinnati, MS, 77; Univ Md, PhD(microbiol), 81. *Prof Exp:* Fel toxicol, Cornell Univ, 81-82; MONSANTO AGR CO, 82- *Mem:* Am Soc Microbiol; Soc Indust Microbiol; Sigma Xi. *Res:* Environmental microbiology and toxicology; heavy metal biotransformations; biodegradation of industrial and agricultural chemical. *Mailing Add:* Monsanto Agr Co T4G 800 N Lindbergh Blvd St Louis MO 63167

HALLAUER, ARNEL ROY, b Netawaka, Kans, May 4, 32; m 64; c 2. PLANT BREEDING. *Educ:* Kans State Univ, BS, 54; Iowa State Univ, MS, 58, PhD(crop breeding), 60. *Prof Exp:* Res agronomist, USDA, 58-61, res geneticist, 61-89; PROF AGRON, IOWA STATE UNIV, 77- *Honors & Awards:* Crop Sci Award, Crop Sci Soc Am, 81, DeKalb-Pfizer Crop Sci Distinguished Career Award, 90; Applied Res & Extension Award, Iowa Agr Exp Sta, 81; Res Recognition Award, Northop-King, 84; Nat Coun Com Plant Breeders Award, 84; Scientist of Year Award, Agr Res Serv, USDA, 85; Agron Achievement Award, Am Soc Agron, 89; Gov Sci Award, 89. *Mem:* Nat Acad Sci; fel Crop Sci Soc Am; Am Genetics Asn; Biometric Soc; fel Am Soc Agron. *Res:* Experimental quantitative genetic studies and their relation to corn breeding methodology. *Mailing Add:* Dept Agron Iowa State Univ Ames IA 50011

HALLBERG, CARL WILLIAM, b Detroit, Mich, July 13, 18; m 42; c 3. PARASITOLOGY, EMBRYOLOGY. *Educ:* Univ Mich, BS, 47, MS, 48, PhD(zool), 52. *Prof Exp:* From asst prof to prof biol, Bowling Green State Univ, 51-83, actg chmn biol sci, 74-75; RETIRED. *Concurrent Pos:* Adj prof, Col Health & Community Serv, Bowling Green State Univ, 84-87. *Mem:* Am Soc Parasitologists; Am Asn Adv Health; Am Micros Soc; Am Inst Biol Scientists; AAAS; Sigma Xi; Audubon Soc; Nat Geog Soc; Am Mus Natural Hist. *Res:* Studies of migration routes to kidneys of mammals and resultant pathology caused by Dioctophyma renale; giant kidney worm of man; germ cell cyclein digenetic trematodes. *Mailing Add:* 1109 Clark Bowling Green OH 43402

HALLBERG, GEORGE ROBERT, b Chicago, Ill, Nov 11, 46; m 70; c 2. ENVIRONMENTAL GEOLOGY, SUSTAINABLE AGRICULTURE. *Educ:* Augustana Col, BA, 70; Univ Iowa, PhD(geol), 75. *Prof Exp:* Res geologist, Quaternary Geol & Earth Resources Observ Satellite & coordr remote sensing, 71-73, sr res geologist quaternary geol, 73-75, chief, Res Div, 75-79, chief geol studies, 79-86, CHIEF, ENVIRON GEOL, IOWA GEOL SURV, 86- *Concurrent Pos:* Adj prof geol, Univ Iowa, 75-; assoc prof geol collabr, Iowa State Univ, 78-; liason rep to Geol Soc Am, 78-81. *Mem:* Fel Geol Soc Am; Soil Sci Soc Am; Am Quaternary Asn; Asn Ground Water Scientists & Engrs; Prof Soil Classifiers. *Res:* Soil genesis; gound water quality; agriculture and nonpoint source pollution; stratigraphy, engineering geology and hydrology of quaternary deposits; midwestern United States. *Mailing Add:* 1110 N Dubuque St Iowa City IA 52245

HALLBERG, RICHARD LAWRENCE, heat shock, mitochondrial biogenesis, for more information see previous edition

HALLE, MORRIS, b Liepaja, Latvia, July 23, 23; nat US. LINGUISTICS. *Educ:* Univ Chicago, MA, 48; Harvard Univ, PhD(ling), 55. *Hon Degrees:* DSc, Brandeis Univ, 89. *Prof Exp:* From asst prof to prof mod lang, Mass Inst Technol, 51-76, Ferrari P Ward prof mod lang & ling, 76-81, actg head, Dept Foreign Lang & Ling, 76 & Dept Ling & Philos, 76-81, INST PROF, MASS INST TECHNOL, 81- *Concurrent Pos:* J S Guggenheim Found fel, 60-61; fel, Ctr Advan Study Behav Sci, 60-61; vis prof, Col France, 87. *Honors & Awards:* James R Killian Jr Lectr, 78. *Mem:* Nat Acad Sci; Am Acad Arts & Sci; Ling Soc Am (vpres, 73, pres, 74). *Mailing Add:* Dept Ling 20D219 Mass Inst Technol 77 Massachusetts Ave Cambridge MA 02139

HALLECK, FRANK EUGENE, b Waterbury, Conn, Dec 9, 27; m 56; c 2. BIOMEDICAL ENGINEERING. *Educ:* Wesleyan Univ, BA, 48, MA, 49; Rutgers Univ, PhD(microbiol, biochem), 52. *Prof Exp:* Res assoc, Univ Md, 52-53; assoc prof microbiol & bact, Loyola Univ, Ill, 53-60; res assoc, Pillsbury Co, 60-64; res dir, Proprietaries & Hosp Prod, Chesebrough Pond's Inc, 64-72; vpres, Cytomed Labs, Inc, 73; corp dir, Sci Affair, Am Sterilizer Co, Inc, Pa, 73-84; pres, Tech Consult Ltd, 84-87; SR DIR QUAL ASSURANCE, FABERGE & ELIZABETH ARDEN, INC, 87- *Concurrent Pos:* Res consult, Wahl-Henius Inst, 53-55; prof, Worsham Col Mortuary Sci, 57-58; consult, Oppenheimer Casing Co, 57-59, Res Armour Co, 55-57 & numerous med & pharmacol co. *Mem:* Parenteral Drug Asn; Soc Indust Microbiol; AAAS; Am Soc Microbiol; NY Acad Sci; Am Chem Soc. *Res:* Microbial fermentaitons and metabolism; chemical composition of polysaccharides of microbial origin; factors influencing microbial growth; food spoilage problems; product development of hospital and proprietary ethical and over-the-counter products; business development through acquisition evaluations; regulatory liaison; microbiological deterioration of cosmetic products; health care products, hardware and disposable products; antibacteerial products; enzyme drugs. *Mailing Add:* 820 Elizabeth Lane Erie PA 16515

HALLECK, MARGARET S, b 1937. CELL BIOLOGY. *Educ:* Univ NMex, PhD(biol), 80. *Prof Exp:* RES INSTR, DEPT PHARMACOL, SCH MED, UNIV TEX, HOUSTON, 86- *Res:* Red cell-mediated injection; chromatin structure. *Mailing Add:* Dept Biochem Univ Utah Med Ctr 50 N Medical Dr Salt Lake City UT 84132

HALLECK, SEYMOUR LEON, b Chicago, Ill, Apr 16, 29; c 3. PSYCHIATRY. *Educ:* Univ Chicago, PhB, 48, BS, 50, MD, 52. *Hon Degrees:* ScD, Rockford Univ, 69. *Prof Exp:* Clin asst path, Sch Med, Univ Chicago, 49-51; intern, USPHS Hosp, San Francisco, 52-53; staff psychiatrist, Dept Justice, Med Ctr Fed Prisoners, Springfield, Mo, 53-55; resident psychiat, Menninger Found, Topeka, Kans, 55-58; asst prof psychiat, Univ Wis, 58-60, clin asst prof, 60-63, lectr sociol, 63-64, from assoc prof to prof psychiat, 63-72, actg chmn dept, 71-72; PROF PSYCHIAT & DIR RESIDENT TRAINING, SCH MED, UNIV NC, CHAPEL HILL, 72- *Concurrent Pos:* Psychiat consult, Wis Diag Ctr, 58-60 & Wis Sch Girls, 58-60; coordr inst serv, Psychiat Serv, Div Corrections, 60-61, chief psychiat serv, 61-63; chief psychiat consult, Wis Div Corrections, 62-72; mem adv comt, Cent State Hosp, 60-72; columnist, Madison Capital Times & Chapel Hill Newspaper, 71-73. *Res:* Student mental health and adolescent adjustment. *Mailing Add:* Dept Psychiat Univ NC Chapel Hill NC 27514

HALLEEN, ROBERT M(ARVIN), b Detroit, Mich, June 26, 33; m 55; c 4. MECHANICAL ENGINEERING. *Educ:* Univ Mich, BSE, 55; Wash State Univ, MSME, 60; Stanford Univ, PhD, 67. *Prof Exp:* Asst prof mech eng, Wash State Univ, 55-67; engr gas turbine div, 67-75, sr staff engr basic engine eng, 75-76, ASST CHIEF ENGR ENGINE ENG, CATERPILLAR INC, 76- *Mem:* Am Soc Mech Engrs; Am Soc Eng Educ; Soc Automotive Engrs. *Res:* Evaporation and scale formation; energy dissipation; non-conventional power plants; fluid mechanics; turbulent boundary layer; free turbulent shear flow; turbomachinery; combustion; exhaust emissions; heat exchangers; bearings; pistons; large diesel engines. *Mailing Add:* 2805 N Elmcroft Peoria IL 61604

HALLENBECK, GEORGE AARON, surgery, for more information see previous edition

HALLENBECK, PATRICK CLARK, b Hornell, NY, Apr 17, 51; c 2. NITROGEN FIXATION, ANAEROBIC METABOLISM. *Educ:* Syracuse Univ, BSc, 72; Univ Calif, Berkeley, PhD(biophys), 78. *Prof Exp:* Postdoctoral fel microbiol, Dept Sanit Eng, Univ Calif, Berkeley, 79; nuclear engr biochem, Ctr Nuclear Studies, Grenoble, 79-81; postdoctoral fel, Univ Calif, Davis, 81-87; invited prof, 87-89, ASSOC PROF MICROBIOL, UNIV MONTREAL, 89- *Concurrent Pos:* Prin investr, Med Res Coun Can & Natural Sci & Eng Res Coun, 88- *Mem:* Am Soc Microbiol; Can Soc Microbiologists; Am Soc Biochem & Molecular Biol. *Res:* Physiology and genetics of bacterial anaerobic metabolism; biochemical and physiological basis of bacterial nitrogen fixation; purification and characterization of iron-sulfur proteins. *Mailing Add:* Dept Microbiol & Immunol Univ Montreal CP 6128 Succursale A Montreal PQ H3C 3J7 Can

HALLENBECK, WILLIAM HACKETT, b Albany, NY, Sept 14, 45. ENVIRONMENTAL SCIENCE, CHEMISTRY. *Educ:* State Univ NY Albany, BS, 67, MS, 69; Univ NC, MSPH, 73; Univ Ill Med Ctr, DrPH, 77. *Prof Exp:* Chemist food chem, NY State Dept Agr & Mkt, 70-72; instr, 73-77, asst prof environ sci, 77-80, ASSOC PROF, UNIV ILL MED CTR, 80- *Concurrent Pos:* Prin investr, US Dept Interior grant, 74-77 & US Environ Protection Agency grant, 77-82. *Res:* Asbestos, food, drinking water and air pollution. *Mailing Add:* Dept Pub Health Col Med Univ Ill PO Box 6998 Chicago IL 60680

HALLER, CHARLES REGIS, b Kansas City, Kans, Nov 21, 31; m 64; c 1. PETROLEUM GEOLOGY. *Educ:* Univ Mo, Columbia, AB, 53, MA, 57; Univ Calif, Berkeley, PhD(paleont), 67. *Prof Exp:* Paleontologist, Amerada Petrol Corp, 59-60 & Oasis Oil Co of Libya, Inc, 60-62; sr paleontologist, Amoco Int Oil Co, 65-74; sr geologist petrol explor, Marathon Petrol Norte Brazil, 74-80, chief geologist, 80-81, explor mgr, 81-85; explor mgr, Clam Petrol Co, 85-88; SR PROJ COORDR, MARATHON OIL CO, 88- *Mem:* Swiss Geol Soc; Am Asn Petrol Geologists; Soc Econ Paleontologists & Mineralogists; Paleont Soc. *Mailing Add:* Marathon Oil Co PO Box 3128 Houston TX 77056

HALLER, EDWIN WOLFGANG, b Stuttgart, Ger, May 19, 36; US citizen; c 2. PHYSIOLOGY, NEUROENDOCRINOLOGY. *Educ:* Park Col, BA, 59; Western Reserve Univ, PhD(physiol), 67. *Prof Exp:* Res & develop chemist, Lucidol Div, Wallace & Tiernan, Inc, 59-61; res assoc physiol, Sch Med, Univ Md, Baltimore, 67-69, asst prof, 69-71; asst prof physiol, 71-72, ASSOC PROF PHYSIOL & BIOL, SCH MED, UNIV MINN, DULUTH, 72- *Concurrent Pos:* Vis scientist, Inst Animal Physiol, Agr Res Coun, Cambridge, Eng, 77-78; consult, Am Asn Accreditation Lab Animal Care, 78-88, Coun on Accreditation, 88-; vis prof, Health Sci Ctr, State Univ NY, Syracuse, 90. *Mem:* Sigma Xi; AAAS; Endocrine Soc; Soc Neurosci. *Res:* Central nervous system regulation of gonadotropin secretion; neurophysiological aspects of neurosecretion; regulation of secretion of vasopressin and oxytocin from the posterior pituitary. *Mailing Add:* Univ Minn Sch Med Duluth MN 55812-2487

HALLER, ELDEN D, b Chillicothe, Ohio, Oct 20, 09; c 1. CHEMICAL ENGINEERING. *Educ:* Ohio State Univ, BChE, 33, MSc, 35, PhD(chem eng), 40. *Prof Exp:* Pvt jobber, 27-34; asst chem eng, Ohio State Univ, 35, asst mach lab, 37-39; chief chemist & chem engr, Nat Lime & Stone Co, 40; tech ed, Ord Dept, Washington, DC, 41; chem engr, Nat Tech Labs, 41-52; eastern sales mgr, Beckman Instruments, Inc, 52-54; from consult to asst to the pres, Arthur H Thomas Co, 54-59, vpres & sales mgr, 59-74; RETIRED. *Concurrent Pos:* Consult, 74- *Mem:* Am Chem Soc; Instrument Soc Am; Soc Appl Spectros; Am Inst Chem Engrs. *Res:* Instrumentation; spectrophotometric pH control and radiation; corrosion of galvanized iron. *Mailing Add:* 8505 Hazelwood Dr Bethesda MD 20814

HALLER, EUGENE ERNEST, b Basel, Switz, Jan 5, 43; m 73; c 2. SEMICONDUCTOR PHYSICS, ELECTRONIC DEVICE PHYSICS. *Educ:* Univ Basel, Switz, dipl, 67, PhD(physics), 70. *Prof Exp:* Res & teaching asst physics, Inst Appl Physics, Univ Basel, 70-71; Fel solid state res, Lawrence Berkeley Lab, 71-73, staff scientist, 73-77, sen staff scientist, 77-80, assoc prof, 80-82, PROF SEMI CONDUCTORS & MAT SCI, DEPT MAT SCI & MINING ENG, UNIV CALIF, BERKELEY, 82-, FAC SR SCIENTIST, LAWRENCE BERKELEY LAB, 80- *Concurrent Pos:* Prog leader, Ctr Advan Mat, Lawrence Berkeley Lab, 84-; prin investr, Space Infrared Telescope Facil, NASA, 85- *Honors & Awards:* US Sr Scientist Award, Alexander von Humboldt Found, 86, Miller Inst Prof, 90. *Mem:* Fel Am Phys Soc; Mat Res Soc. *Res:* Spectroscopic and electronic characterization of one and two dimensional defects in semi-conductors; growth of ultra-pure and doped single crystals of germanium and GaAs; advanced far infrared, x-ray and gamma ray semiconductor detectors. *Mailing Add:* Dept Mat Sci & Mining Eng Univ Calif 286 Hearst Mining Bldg Berkeley CA 94720

HALLER, GARY LEE, b Loup City, Nebr, July 10, 41; m 62; c 3. PHYSICAL CHEMISTRY. *Educ:* Kearney State Col, BA, 62; Northwestern Univ, PhD(chem), 66. *Prof Exp:* NATO fel phys chem, Oxford Univ, 66-67; from asst prof to assoc prof eng & appl sci, 67-80, Becton prof eng & appl sci, 84-87, PROF CHEM ENG & CHEM, YALE UNIV, 80-, DEP PROVOST, PHYS SCI & ENG,87- *Mem:* Am Chem Soc; Am Inst Chem Eng; Catalysis Soc. *Res:* Heterogeneous catalysis; kinetics and mechanisms of catalyzed reactions; surface chemistry and structure of solid catalysts; environmental problems; air pollution. *Mailing Add:* Dept Chem Eng Yale Univ 2159 Yale Sta New Haven CT 06520

HALLER, IVAN, b Budapest, Hungary, June 8, 34; US citizen; m 65; c 2. PHYSICAL CHEMISTRY, PLASMA CHEMISTRY. *Educ:* Budapest Technol Univ, BS, 56; Univ Calif, Berkeley, PhD(chem), 61. *Prof Exp:* Asst, Univ Calif, Berkeley, 57-59; assoc mem staff, 60-61, MEM STAFF, T J WATSON RES CTR, IBM CORP, 61- *Mem:* Am Phys Soc; Am Chem Soc; Am Inst Chemists. *Res:* Materials science of electronic materials; photon charged particles; beam and plasma induced chemical processes; spectroscopy and molecular structure. *Mailing Add:* IBM Watson Res Ctr PO Box 218 Yorktown Heights NY 10598

HALLER, KURT, b Vienna, Austria, May 6, 28; nat US; m 52; c 2. THEORETICAL HIGH ENERGY PHYSICS. *Educ:* Columbia Univ, AB, 49, PhD(physics), 58. *Prof Exp:* Instr physics, Newark Col Arts & Sci, Rutgers Univ, 53-54; instr, Newark Col Eng, 54-56; res asst, Columbia Univ, 56-57; res assoc, Washington Univ, 57-59; from asst prof to assoc prof, NY Univ, 59-64; assoc prof, 64-70, PROF PHYSICS, UNIV CONN, 70- *Concurrent Pos:* Sr Fulbright lectr, Univ Graz, Austria, 73; vis prof, Inst Theoret Physics, 73. *Mem:* Fel Am Phys Soc. *Res:* Elementary particle theory; quantum field theory. *Mailing Add:* Dept Physics U-46 Rm 107 Univ Conn 2152 Hillside Rd Storrs CT 06269-3046

HALLER, WILLIAM T, b Watertown, NY, June 28, 47; m 68; c 3. AQUATIC BOTANY, PLANT PHYSIOLOGY. *Educ:* Cornell Univ, BS, 69; Univ Fla, MS, 71, PhD(agron), 74. *Prof Exp:* Res assoc aquatic ecol, Agr Res Ctr, 71-72, from asst prof to assoc prof aquatic bot, 74-84, dir, Ctr Aquatic Weeds, 83-85, PROF AGRON, UNIV FLA, 84- *Concurrent Pos:* Prin investr, Fla Dept Natural Resources grants, 73-, US Army Corps Engrs grants, 77-, US Environ Protection Agency grant, 78-81 & US Agency Int Develop grant, 81-85; ed, Aquatic Plant Mgt Soc. *Mem:* Weed Sci Soc Am; Int Asn Aquatic Plant Biologists; Aquatic Plant Mgt Soc. *Res:* Aquatic weed control and effects of chemical, mechanical and biological aquatic weed control on the ecology of lakes and rivers; physiology of aquatic plants and interrelationships of aquatic plants and fish. *Mailing Add:* Dept Agron Univ Fla Gainesville FL 32611

HALLER, WOLFGANG KARL, b Vienna, Austria, Dec 9, 22; nat US. PHYSICAL CHEMISTRY, GLASS TECHNOLOGY. *Educ:* Univ Vienna, PhD(chem), 50. *Prof Exp:* Res assoc, Univ Vienna. 49-51; res chemist & Fulbright fel, Univ Wash, 51-52; res assoc, Univ Calif, Berkeley, 52-53; prof phys chem, Inst Silicate Res, Univ Toledo, 53-57; physicist, Inst Mat Sci & Technol, Nat Bur Standards, washington, DC, 58-66, chief, Inorg Glass & Optical Mat Sect, 66-84; RES ASSOC & CONSULT, MAT SCI & ENG LAB, NAT INST STANDARDS & TECHNOL, WASHINGTON, DC, 84- *Concurrent Pos:* vis prof, Max Planck Inst Protein Res, Munich, Ger, 68-69 & Inst Appl Phys & Chem, Univ Heidelburg, Ger, 75. *Honors & Awards:* Gold Medal, US Dept Com, 73; Alexander von Humboldt Award, Ministry for Res & Technol, Fed Repub Ger, 75. *Mem:* Am Ceramic Soc; Brit Soc Glass Technol; Am Soc Microbiol. *Res:* Structure of glass; physics of surfaces; separation science; chromatography; microbiology. *Mailing Add:* 4620 N Park Ave Chevy Chase MD 20815-4558

HALLESY, DUANE WESLEY, b Denton, Mont, Feb 2, 28; m 56; c 3. ENVIRONMENTAL HEALTH. *Educ:* Mont State Col, BS, 53; Univ Chicago, PhD(pharmacol), 56. *Prof Exp:* Res assoc pharmacol, Univ Chicago, 56-57; res pharmacologist, Lederle Labs Div, Am Cyanamid Co, 57-63, sr res pharmacologist & group leader, 63-65; sr indust toxicologist, Standard Oil Co Calif, 65-69; toxicologist, 69-71, mgr toxicol sect, 71-73, head toxicol dept, 73-83, dir toxicol, Syntex Res Div, Syntex Corp, 83-88; RETIRED. *Mem:* AAAS; Soc Toxicol. *Res:* Preclinical safety evaluation of proposed new therapeutic agents. *Mailing Add:* 1246 Emerson St Palo Alto CA 94301

HALLET, BERNARD, b Ciney, Belg, Sept 28, 48; m 74; c 3. GEOLOGY, GLACIOLOGY. *Educ:* Univ Calif, Los Angeles, BS, 70, PhD(geol), 75. *Prof Exp:* Asst prof geol & appl earth sci, Stanford Univ, 75-80; PROF, DEPT GEOL & QUATERNARY RES CTR, UNIV WASH, SEATTLE, 80- *Mem:* Int Glaciol Soc; Am Geophys Union; Geol Soc Am. *Res:* Glacial and periglacial geomorphology and permafrost studies; processes that shape landscape in arctic and alpine areas. *Mailing Add:* Dept Geol & Quaternary Res Univ Wash AK-60 Seattle WA 98195

HALLET, RAYMON WILLIAM, JR, b Chicago, Ill, Nov 21, 20; m 44; c 3. THERMODYNAMICS, NUCLEAR ENGINEERING. *Educ:* Purdue Univ, BS, 42; Univ Calif, Los Angeles, MS, 58. *Prof Exp:* Chief nuclear eng, Douglas Aircraft Co, McDonnell Douglas Corp, 62-64, dir res & develop, 64-66, vpres & dep gen mgr, Douglas United Nuclear, 66-70, dir energy systs, 72-74, prog mgr design integration, Solar I Pilot Plant, 78-84, dir solar progs, 84-86; CONSULT, 86- *Concurrent Pos:* Consult, NSF, 73-74; mem solar energy rev bd, NSF, 73-75. *Mem:* Am Inst Aeronaut & Astronaut. *Res:* Application of solar energy to thermal electric power generation involving development of heliostats, solar boilers and thermal storage subsystems. *Mailing Add:* 1906 Holiday Rd Newport Beach CA 92660

HALLETT, FREDERICK ROSS, b High River, Alta, July 14, 42; m 67; c 2. BIOPHYSICS. *Educ:* Univ Calgary, BSc, 63, MSc, 66; Pa State Univ, PhD(biophys), 69. *Prof Exp:* From asst prof to assoc prof, 69-82, PROF PHYSICS, UNIV GUELPH, 82- *Mem:* Am Asn Physics Teachers; Biophys Soc; Can Asn Physicists. *Res:* Photon correlation spectroscopic and small angle neutron scattering of biopolymers in solution. *Mailing Add:* Dept Physics Univ Guelph Guelph ON N1G 2W1 Can

HALLETT, JOHN, b Bristol, UK, Dec 2, 29; m 60; c 4. CLOUD PHYSICS. *Educ:* Bristol Univ, BSc, 53; Imp Col, dipl, 54, Univ London, PhD(meteorol), 58. *Prof Exp:* Asst lectr meteorol, Imp Col, London, 58-60; asst prof, Univ Calif, Los Angeles, 60-62; lectr physics, Imp Col, London, 62-66; RES PROF ATMOSPHERIC PHYSICS, UNIV NEV, RENO, 66-, MARSTON CHAIR ATMOSPHERIC PHYSICS, 79-, VCHAIR ATMOSPHERIC SCI. *Concurrent Pos:* Vis prof physics, Univ Manchester, Eng, 78; vis fel, Japan Soc Promotion Sci, 85. *Mem:* Am Meteorol Soc; Glaciol Soc; fel Royal Meteorol Soc. *Res:* Laboratory studies of the mechanism of ice crystal growth from vapor and liquid; ice phase evolution in convective clouds in the atmosphere and its influence on cloud electrification; combustion aerosol physics and chemistry; low gravity crystal growth; freezing of biological materials. *Mailing Add:* Dept Physics Univ Nev Reno NV 89557

HALLETT, MARK, b Philadelphia, Pa, Oct 22, 43; m 66; c 2. HUMAN MOTOR CONTROL, CLINICAL NEUROPHYSIOLOGY. *Educ:* Harvard Univ, BA, 65, MD, 69. *Prof Exp:* Intern med, Peter Bent Brigham Hosp, Boston, 69-70; staff assoc, Lab Neurobiol, NIMH, 70-72; resident neurol, Mass Gen Hosp, 72-74, chief resident, 74-75; dir neurophysiol labs, Sect Neurol, Peter Bent Brigham Hosp, 76-84; CLIN DIR, NAT INST NEUROL & COMMUN DISORDERS & STROKE, NIH, 84-, DIR CLIN NEUROSCI PROG, DIV INTRAMURAL RES, 87-; CLIN PROF NEUROL, UNIFORMED SERVS UNIV HEALTH SCI, 87- *Concurrent Pos:* Clin fel med, Harvard Med Sch, 69-70, clin fel neurol, 72-75, from instr to assoc prof neurol, 76-84; guest lectr physiol, Howard Med Sch, 70-72; staff physician, Mass Rehab Hosp, Boston, 73-74, New Eng Baptist Hosp, 78-84, Parker Hill Hosp, 79-84, Brigham & Women's Hosp, 82-84; Moseley travelling fel-Harvard Med Sch, Dept Neurol, Inst Psychiat, London, Eng, 75-76; jr assoc med/neurol, Peter Bent Brigham Hosp, 76-77, assoc, 77-82; consult neurol, Robert Breck Brigham Hosp, 76-80 & Naval Hosp, Bethesda, Md, 84-; dir, Children's Hosp Med Ctr-Affil Hosps Ctr Inc Muscular Dystrophy Clin, 80-84, Neurol Div, Acute Rehab Unit, Brigham & Women's Hosp, 82-84; clin assoc prof neurol, Uniformed Servs Univ Health Sci, 84-86; assoc dir intramural res prog, Nat Inst Neurol & Commun Disorders & Stroke, NIH, 86-87. *Mem:* Fel Am Acad Neurol; Am Neurol Asn; Soc Neurosci; Am EEG Soc; Am Asn Electromyography & Electrodiag (secy-treas, 87-); Int Med Soc Motor Disturbances (pres, 88-). *Res:* Physiology of how the brain controls movement in humans; studies of normal physiology and of pathophysiology of both disordered voluntary movements and involuntary movements; myoclonus, bradykinesia and ataxia. *Mailing Add:* Clin Dir NINCDS NIH Bldg 10 Rm 5N-226 Bethesda MD 20892

HALLETT, PETER EDWARD, b Sliemma, Malta, Apr 30, 37; Brit & Can citizen; m 60; c 3. VISUAL PSYCHOPHYSICS, EYE MOVEMENTS & VISION THEORY. *Educ:* Univ Oxford, BA, 58, BSc, 60, BM, 62, MA, 62. *Prof Exp:* Asst lectr physiol, Univ Col London, 63-65; asst prof, Univ Alta, 65-69; PROF PHYSIOL, 75-, PROF ZOOL, UNIV TORONTO, 79- *Concurrent Pos:* Prof Biomed Eng, Univ Toronto, 85. *Mem:* Physiol Soc; Can Physiol Soc; Asn Res Vision & Ophthal; Optical Soc Am. *Res:* Behavioral and non-invasive approaches to human peripheral vision and its clinical assessment; human night vision and eye movements; optics, visual neuroanatomy and computational models. *Mailing Add:* Dept Physiol Univ Toronto Toronto ON M5S 1A8 Can

HALLETT, WILBUR Y, b La Paz, Bolivia, Feb 7, 26; US citizen; m 48; c 4. MEDICINE. *Educ:* Univ Rochester, MD, 53. *Prof Exp:* Nat Tuberc Asn teaching fel, Univ Wash, 57-58; staff physician, Firland Sanatorium, Seattle, 58-60; physician-in-chg pulmonary physiol lab, City of Hope Med Ctr, 60-66, dir respiratory dis dept, 63-66; assoc prof med, 66-77, dir, Chronic Respiratory Dis Proj, 69-77, ASSOC CLIN PROF MED, SCH MED, UNIV SOUTHERN CALIF, 77- *Concurrent Pos:* Pvt pract chest dis, 60- *Mem:* Am Thoracic Soc; Am Fedn Clin Res; fel Am Col Physicians. *Res:* Clinical applications of pulmonary physiology. *Mailing Add:* 655 N Central Ave Suite 101 Glendale CA 91203

HALLEY, JAMES WOODS, (JR), b Chicago, Ill, Nov 16, 38; m 70; c 1. THEORETICAL PHYSICS. *Educ:* Mass Inst Technol, BS, 61; Univ Calif, Berkeley, PhD(physics), 65. *Prof Exp:* NSF fel, Fac of Sci, Orsay Ctr, Univ Paris, 65-66; asst prof physics, Univ Calif, Berkeley, 66-68; assoc prof, 68-77,

PROF PHYSICS, UNIV MINN, MINNEAPOLIS, 77- *Concurrent Pos:* Prin investr, NSF, Corrosion Res Ctr, 79-; vis scientist, Harvard, 79, Mich State Univ, 80 & 81, Argonne Nat Lab, 81-90, Univ Calif, Santa Barbara, 83-84, Australian Nat Univ, 88; Bush fel, 83-84, Paul Flory IBM Sabbatical, 87. *Mem:* Am Phys Soc; AAAS. *Res:* Theory of optical and other properties of normal and superfluid liquids; percolation theory; magnetic phase transitions and optical properties of magnetic materials; physics of human motion and human powered technology; polymers; physics of electrochemistry; electronic structure of oxides; high temperature superconductivity. *Mailing Add:* Dept Physics Univ Minn Minneapolis MN 55455

HALLEY, ROBERT, b San Diego, Calif, Jan 5, 20; m 43, 63; c 3. UNDERWATER ACOUSTICS. *Educ:* San Diego State Col, AB, 41; Univ Calif, Los Angeles, MA, 49. *Prof Exp:* Jr physicist, Div War Res, Univ Calif, 45-46; instr math & physics, San Diego State Col, 46 & 49-50; physicist, US Naval Ocean Systs Ctr, 49 & 50-77; PHYSICIST, COMPUT SCI CORP, 78- *Mem:* Acoust Soc Am. *Res:* Underwater sound, especially ambient sea noise and ship noise measurement; passive sonar target classification; information processing for classification. *Mailing Add:* 1714 Malden St Pacific Beach CA 92109

HALLEY, ROBERT BRUCE, b Englewood, NJ, June 2, 47; m 68; c 1. SEDIMENTOLOGY. *Educ:* Oberlin Col, AB, 69; Brown Univ, MS, 71; State Univ NY Stony Brook, PhD(geol), 74. *Prof Exp:* Res assoc geol, State Univ NY, Binghamton, 73-74; GEOLOGIST, US GEOL SURV, 74- *Mem:* Sigma Xi; Soc Econ Paleontologists & Mineralogists; Am Asn Petrol Geologists. *Res:* Deposition and diagenesis of carbonate sediments. *Mailing Add:* 1153 Williams Dr S St Petersburg FL 33705

HALLFORD, DENNIS MURRAY, b Abilene, Tex, Feb 11, 48; m 71; c 1. REPRODUCTIVE PHYSIOLOGY, ENDOCRINOLOGY. *Educ:* Tarleton State Univ, BS, 70; Okla State Univ, MS, 73, PhD(animal breeding), 75. *Prof Exp:* Instr animal sci, Tarleton State Univ, 70-71; from asst prof to assoc prof, 75-83, PROF ANIMAL SCI, NMEX STATE UNIV, 83- *Mem:* Am Soc Animal Sci; Sigma Xi. *Res:* Improving reproductive performance in domestic animals. *Mailing Add:* Dept Animal & Range Sci NMex State Univ Box 3I Las Cruces NM 88003

HALLFRISCH, JUDITH, b Gary, Ind. NUTRITION, PUBLIC HEALTH & EPIDEMIOLOGY. *Educ:* Ind Univ, AB; Univ Md, PhD(nutrit sci). *Prof Exp:* Res nutritionist, Nutrit Inst, USDA, 74-84; ASST PROF NUTRIT BIOCHEM, UNIV MD, 84-; SR STAFF FEL, NAT INST AGING, NIH, 84- , MEM NUTRIT COORDR COMT, 87- *Concurrent Pos:* Prin investr, Geront Nutrit Study, Baltimore Longitudinal Study, 84-; vis prof nutrit biochem, Med Sch, Johns Hopkins Univ, 86-, continuing educ prog, 86-87, asst prof med, 88. *Mem:* Am Asn Cereal Chemists; Am Col Nutrit; Am Inst Nutrit; Am Soc Clin Nutrit; Am Chem Soc; AAAS; Geront Soc Am. *Res:* Effects of various dietary carbohydrates on cardiac risk factors and diabetes indicators; effects of fiber on cardiac risk factors, mortality from heart disease and copper and zinc status in chronically ill patients; diet and performance in master's athletes over 50. *Mailing Add:* Geront Res Ctr 4940 Eastern Ave Baltimore MD 21224

HALLGREN, ALVIN ROLAND, b Aug 24, 19; US citizen; m 48; c 2. FORESTRY, FOREST MANAGEMENT. *Educ:* Univ Minn, BS, 49, PhD(forestry), 67; Yale Univ, MF, 50. *Prof Exp:* Asst dist forester, Crossett Co, 50-52, conserv forester, 53-55, asst wood mgt, 56-59; assoc prof, Cloquet Forestry Ctr, Univ Minn, 59-75, coordr & prof forestry, 75-87; RETIRED. *Concurrent Pos:* Chmn, Minn Timber Law Comt, 56-; forestry consult, St Paul Bd Water Comnrs, Minn, 57-75. *Res:* Timber laws; effects of mechanized harvesting. *Mailing Add:* 909 Jasper St Cloquet MN 55720

HALLGREN, HELEN M, b Lancaster, Minn, Jan 6, 40; m 59; c 2. HUMAN AGING. *Educ:* Univ Minn, MS, 75. *Prof Exp:* Asst prof, 76-84, ASSOC PROF IMMUNOL, UNIV MINN, 84- *Honors & Awards:* Sci Creativity Award, Am Soc Med Technol, 87. *Mem:* Am Asn Immunologists; NY Acad Sci; AAAS; Acad Clin Lab Physicians & Scientists; Am Soc Med Technol. *Res:* Determination of the mechanisms underlying the decline in immune functions in aging humans. *Mailing Add:* Dept Lab Med & Path Univ Minn Box 198 Mayo Minneapolis MN 55455

HALLGREN, RICHARD E, b Kersey, Pa, Mar 15, 32; m 54; c 3. METEOROLOGY, PHYSICS. *Educ:* Pa State Univ, BS, 53, PhD(meteorol), 60. *Hon Degrees:* DSc, State Univ NY. *Prof Exp:* Res asst cloud physics, Pa State Univ, 56-60; opers res analyst, IBM Corp, 60-63, mgr meteorol systs, 63-64; sci adv to asst secy com, US Govt, 64-66; dir world weather systs, Environ Sci Serv Admin, 66-69, asst adminr environ systs, 69-71; assoc adminr environ monitoring & prediction, Nat Oceanic & Atmospheric Admin, 71-73; dep dir, Nat Weather Serv, 73-77; dep asst adminr, Oceanic & Atmospheric Servs, Nat Oceanic & Atmospheric Admin, 77-79; dir, 79-83, asst Adminr Weather Serv, Nat Weather Serv, 83-88; EXEC DIR, AM METEOROL SOC, 88- *Concurrent Pos:* US Rep to World Meteorol Orgn, 81- *Honors & Awards:* Arthur S Fleming Award, 68; Gold Medal, US Dept Com, 69; Charles F Brooks Award, Am Meteorol Soc, 86; Int Meteorol Prize, World Meteorol Orgn, 90. *Mem:* Am Meteorol Soc (pres, 82). *Res:* Cloud physics; atmospheric electricity; meteorological systems. *Mailing Add:* Am Meteorol Soc 45 Beacon St Boston MA 02108

HALLIBURTON, LARRY EUGENE, b Cairo, Mo, Oct 13, 43; m 65; c 2. SOLID STATE PHYSICS. *Educ:* Univ Mo-Columbia, BS, 65, MS, 67, PhD(physics), 71. *Prof Exp:* Engr scientist microwave standards, Douglas Aircraft Co, 65-66; from asst prof to assoc prof, 71-81, PROF PHYSICS, OKLA STATE UNIV, 81- *Mem:* Am Phys Soc; Sigma Xi. *Res:* Investigations of impurities and other point defects in electro-optic, electro-acoustic, and laser materials using electron spin resonance, electron-nuclear double resonance, optical absorption, and luminescence techniques. *Mailing Add:* Dept Physics Okla State Univ Stillwater OK 74078

HALLIDAY, IAN, b Lloydminster, Sask, Nov 10, 28; m 51; c 2. METEORITICS, ASTROPHYSICS. *Educ:* Univ Toronto, BA, 49, MA, 50, PhD(astron), 54. *Prof Exp:* Astrophysicist, Dom Observ, Can Dept Mines & Tech Surv, 52-70; RES OFFICER, NAT RES COUN CAN, 70- *Concurrent Pos:* Lectr, St Patrick's Col, 58-60; ed, Royal Astron Soc Can J, 70-75; mem steering group, Int Halley Watch, 81-, chmn, 85- *Mem:* Int Astron Union; Am Astron Soc; Royal Astron Soc Can (pres, 80-82); Meteoritical Soc; Can Astron Soc; Planetary Soc; fel Royal Soc Can. *Res:* Stellar and meteor spectra; positional astronomy; meteors; meteorite craters; diameter of Pluto; meteor spectroscopy; meteorite astronomy and recovery; comets. *Mailing Add:* Herzberg Inst of Astrophys Nat Res Coun Can Ottawa ON K1A 0R6 Can

HALLIDAY, ROBERT WILLIAM, b New York, NY, July 17, 42; m 64; c 3. INORGANIC CHEMISTRY. *Educ:* Bates Col, BS, 64; Wesleyan Univ, MA, 66; Wayne State Univ, PhD(inorg chem), 69. *Prof Exp:* Asst prof, 69-77, ASSOC PROF INORG CHEM, ADELPHI UNIV, 77- *Concurrent Pos:* Cottrell Corp res grant, Adelphi Univ, 70- *Res:* Preparation and characterization of inorganic coordination compounds; optical and magnetic properties of inorganic complexes; conformation and configuration of optically active coordination compounds. *Mailing Add:* Dept of Chem Adelphi Univ Garden City NY 11530

HALLINAN, EDWARD JOSEPH, biology, for more information see previous edition

HALLINAN, THOMAS JAMES, b Albany, NY, Sept 30, 41; m 64; c 2. AURORAL PHYSICS. *Educ:* Cornell Univ, BS, 64; Univ Alaska, MS, 69, PhD(geophysics), 76. *Prof Exp:* Engr TV, Cornell Univ, 64; engr TV, Aero-Geo-Astro, 64-65; engr TV, 65-76, asst prof, 76-81, ASSOC PROF GEOPHYSICS, UNIV ALASKA, 81- *Mem:* Am Geophys Union; AAAS; Archaeol Inst Am; Nat Speleol Soc. *Res:* Detailed morphology of the Aurora Polaris using a stereo pair of low-light-level television cameras; experiments in space, including explosive chemical releases and accelerated electron beams. *Mailing Add:* Dept Geophys Inst Univ Alaska 116 Bunnell Fairbanks AK 99701

HALLIWELL, ROBERT STANLEY, b Lovell, Wyo, Sept 8, 31; m 54; c 3. PLANT VIROLOGY. *Educ:* Univ Wyo, BS, 56, MS, 59; Ore State Univ, PhD(plant path, biochem), 62. *Prof Exp:* Instr plant path, Ore State Univ, 60-62; from asst prof to assoc prof, 62-71, PROF PLANT VIROL, TEX A&M UNIV, 71- *Concurrent Pos:* Consult, Food & Agr Orgn, UN, Korea, 74-75. *Mem:* Am Phytopath Soc. *Res:* Physiology of virus diseased plants; plant virus disease effects on individual cells. *Mailing Add:* Dept Plant Path & Microbiol Tex A&M Univ Col Sta TX 77843

HALLOCK, GILBERT VINTON, b Worcester, Mass. PSYCHIATRY. *Educ:* Harvard Univ, BA, 50; Tufts Univ, MD, 54. *Prof Exp:* Med Dir, Dir Prof Serv & Prof Info, Astra Pharmaceut Prod Inc, 56-73; dir Med Serv, Dome Labs Div, Miles Labs, 74; assoc med dir & assoc dir Prof Serv, Lederle Labs Div, Am Cyanamid Co, 74-78; staff physician, Worcester State Hosp, 80-83; RETIRED. *Concurrent Pos:* Assoc med staff, Worcester City Hosp, 69-83, consult pharmacol, 77-83; consult pharmacol, Worcester Mem Hop 65-83. *Mem:* Drug Info Assoc (charter mem & dir, 72-74); Aerospace Med Assoc; Am Med Assoc; Am Med Writers Assoc; Am Pub Health Assoc; Am Col Gen Practice. *Res:* Research and development in the field of Anesthetics, Hematinics, Dermatologicals and Antibiotics. *Mailing Add:* 16 Duncannon Ave Apt 2 Worcester MA 01604-5128

HALLOCK, JAMES A, b Patarson, NJ, Oct 28, 42; m 65; c 3. PEDIATRICS. *Educ:* Steon Hall Univ, AB, 63; Georgetown Univ, MD, 67. *Prof Exp:* Dir, Pediat Ambulatory Serv, Univ SFla, 75-77, asst med dir, Ambulatory Care Ctr, 76-77, assoc dean, Sch Med, 77-83, dep dean, 83-85, exec dean, 85-87; DEAN, SCH MED, E CAROLINA UNIV, 88-, VCHANCELLOR HEALTH SCI, 90- *Concurrent Pos:* Chmn, Med Deans Adv Comt & Prog Policy Comt Microelectronic NC, 89-, Ad Hoc Task AIDS, NC Med Soc, 90-91; rep, Accreditation Coun Continuing Med Educ, 91- *Mem:* Southern Med Asn; AMA; Asn Am Med Cols; fel Am Acad Pediat. *Res:* Medical education and health care delivery. *Mailing Add:* Sch Med ECarolina Univ Brody Bldg AD-48 Greenville NC 27858-4354

HALLOCK, ROBERT B, b Washington, DC, Dec 9, 43; m 65; c 2. LOW TEMPERATURE PHYSICS, CONDENSED MATTER PHYSICS. *Educ:* Univ Mass, BS, 65; Stanford Univ, MS, 67, PhD(physics), 69. *Prof Exp:* Air Force Off Sci Res-Nat Res Coun fel, Stanford Univ, 69-70; from asst prof to assoc prof physics, 70-79, PROF PHYSICS, UNIV MASS, AMHERST, 79-, HEAD, DEPT OF PHYSICS & ASTRON, 85- *Concurrent Pos:* A P Sloan res fel, Univ Mass, Amherst, 72-76; Fulbright fel, 77-78; adj prof polymer sci & eng, Univ Mass, 85- *Mem:* fel Am Phys Soc; Am Asn Physics Teachers. *Res:* Macroscopic quantization effects in He II; NMR in 3He-4He mixtures; properties of persistent currents and third sound in helium films; super fluid onset; conducting polymers; classical localization of helium waves; polarized quantum systems; viscosity and boundary kinetics of polymer solutions; antibody-antigen kinetics at surfaces; high Tc superconductivity; transport measurements in novel conducting systems; porous media. *Mailing Add:* Dept of Physics & Astron Univ of Mass Amherst MA 01003

HALLOCK, ZACHARIAH R, b Southampton, NY, June 27, 42; m 66; c 2. PHYSICS. *Educ:* Polytech Inst Brooklyn, BS, 70; Univ Miami, MS, 73, PhD(phys oceanog), 77. *Prof Exp:* Phys oceanogr, US Naval Oceanog Off, 77-81; PHYS OCEANOGR, NAVAL OCEANOG & ATMOSPHERIC RES LAB, 81- *Mem:* Am Geophys Union. *Res:* Design and implementation of oceanographic measurement programs; analysis and interpretation of data from oceanographic observations and evaluation of theoretical predictions. *Mailing Add:* Naval Oceanog & Atmospheric Res Lab Stennis Space Center MS 39529

HALLOCK-MULLER, PAMELA, b Pierre, SDak, June 2, 48; m 69. CARBONATE SEDIMENTOLOGY, CORAL REEF ECOLOGY. *Educ:* Univ Mont, BA, 69; Univ Hawaii, Manoa, MS, 72, PhD(oceanog), 77. *Prof Exp:* asst prof, earth sci, Univ Tex, Permian Basin, 78-83; assoc prof, 83-88, PROF MARINE SCI, UNIV S FLA, 88- *Concurrent Pos:* Environ consult, Hawaiian Elec Co Inc, 75-77; res fel, Univ Copenhagen, 78 & Kiel Univ, 79 & Gollard Space Flight Ctr, 87; consult, Cities Serv Co, 81; prin investr, NSF grants, 81-91; assoc ed, J Foraminiferal Res; consult, Continental Shelf Asn, 88-89; bd dirs, Cushman Found Foraminiferal Res, 90- *Mem:* Fel Geol Soc Am; Soc Sedimentary Geologists; Asn Women Geoscientists; Paleont Soc; fel Cushman Found Foraminiferal Res. *Res:* Role of nutrients in coral reefs, carbonate sedimentology and paleoceanography; role of algal symbiosis in carbonate production, community structure and evolution; larger foraminiferal ecology/paleoecology. *Mailing Add:* Dept Marine Sci Univ SFla St Petersburg FL 33701

HALLOIN, JOHN MCDONELL, b Green Bay, Wis, Aug 14, 38; m 67; c 2. PLANT PHYSIOLOGY, PLANT PATHOLOGY. *Educ:* Univ Wis-Madison, BS, 60; Univ Minn, St Paul, MS, 64; Mich State Univ, PhD(bot), 68. *Prof Exp:* Asst plant physiol, Univ Minn, St Paul, 60-64; res fel plant path, Univ Wis, 68-71; asst prof bot, Mich State Univ, 71; res assoc biochem, Iowa State Univ, 71-72; res plant physiologist, Nat Cotton Path Res Lab, Col Sta, Tex, 72-87, RES PLANT PHYSIOLOGIST, USDA, AGR RES SERV, E LANSING, MICH, 87- *Concurrent Pos:* Mem grad fac, Tex A&M Univ, 73-87; mem fac, Mich State Univ, 88- *Mem:* Am Phytopath Soc; Am Soc Sugar-Beet Technologists; Am Soc Agron; Crop Sci Soc Am; Sigma Xi. *Res:* Studies the nature and heritability of resistance to sugarbeet diseases; nuclear magnetic resonance imaging of plants and independent research and consulting on seed quality and the nature and heritability of resistance to seed deterioration and seedling disease. *Mailing Add:* Dept Bot-Plant Path Mich State Univ East Lansing MI 48824

HALLORAN, HOBART ROOKER, b San Francisco, Calif, Jan 8, 16; m 38; c 1. POULTRY NUTRITION, ANIMAL NUTRITION. *Educ:* Univ Calif, Berkeley, BS, 37. *Prof Exp:* Chemist & salesman, Adhesive Prod, Inc, 38-39; chemist & plant mgr, Farallone Div, Borden Co, 39-43; vpres & mgr feed prod div, Collett-Week-Nibecker, Inc, 43-48; res dir, Poultry Producers Cent Calif, 48-56; nutrit consult, 56-59; pres, Halloran Res Farm, Inc, 59-88; RETIRED. *Concurrent Pos:* Mem, Agr Adv Comt, Univ Calif, 64-67; assoc ed, Poultry Sci Jour. *Mem:* Fel Poultry Sci Asn; Am Chem Soc; Animal Nutrit Res Coun; Am Inst Nutrit. *Res:* Poultry feed ingredients; new ingredient development; ingredient improvement; computer evaluation; research planning and reviewing; technical service functions. *Mailing Add:* Halloran Res Farms 8801 Bainbridge Pl Stockton CA 95209

HALLORAN, PHILIP FRANCIS, b Ont, Can, June 44; c 3. MEDICINE, SURGERY. *Educ:* Univ Toronto, MD, 68, FRCP(C), 73; Univ London, PhD(immunol), 76. *Prof Exp:* Assoc prof med, 80-87, ASSOC PROF SURG, UNIV TORONTO, 81-, PROF MED, 87-; DIR, DIV NEPHROL/IMMUNOL, UNIV ALTA, 91- *Concurrent Pos:* Staff physician, Mt Sinai & Toronto Gen Hosp, 75-; dir renal transplantation, Tri-Hosp Nephrol Serv, 75- *Mem:* Can Soc Immunologists; Can Soc Nephrology; Am Soc Nephrol; Am Fedn Clin Res; Am Asn Immunologists; Am Soc Clin Invest. *Mailing Add:* Dept Nephrol & Immunol Univ Alberta, Blood Trans Fusion Serv Bldg 8249, 114th St Edmonton AB T6G 2E1 Can

HALLS, LOWELL KEITH, b Monticello, Utah, May 7, 18; m 46; c 5. RANGE CONSERVATION. *Educ:* Colo State Univ, BS, 47; Tex A&M Univ, MS, 48. *Prof Exp:* Instr range mgt, Colo State Univ, 48-49; range conservationist, Southeastern Forest Exp Sta, Ga, 49-57, res forester, Southern Forest & Range Exp Sta, La, 57-61, Range Conservationist, Nacogdoches Wildlife Habitat & Silvicult Lab, Southern Forest & Range Exp Sta, Forest Serv, USDA, 61-81. *Concurrent Pos:* Part-time instr, Sch Forestry, Stephen F Austin State Univ, 69-79. *Honors & Awards:* Outstanding Forest Res, Tex Forestry Asn, 73. *Mem:* Soc Range Mgt; Wildlife Soc; Soc Am Foresters. *Res:* Forest wildlife habitat; grazing on forest lands; coordination of wildlife, cattle and timber production. *Mailing Add:* 2720 Dogwood Nacogdoches TX 75961

HALLUM, CECIL RALPH, b Stamford, Tex, Jan 6, 44; m 64; c 2. APPLIED STATISTICS, MEDICAL STATISTICS. *Educ:* Tex Tech Univ, BS, 66, MS, 69, PhD(math statist), 72. *Prof Exp:* Teaching asst math, Tex Tech Univ, 66-68, consult statistician, 68-72; assoc prof math, Loyola Univ, La, 72-78; aerospace technician, earth observ div, Johnson Space Flight Ctr, NASA, 78-; AT DEPT MATH, UNIV HOUSTON, CLEAR LAKE. *Concurrent Pos:* NASA traineeship, 67-69; consult statist, La State Univ Med Ctr. *Mem:* Am Statist Asn; Int Inst Statist. *Res:* Remote sensing in the area of medical statistics, experimental designs and music therapy. *Mailing Add:* Dept Math Univ Houston Clear Lake 2700 Bay Area Blvd Houston TX 77058

HALLUM, JULES VERNE, b Minneapolis, Minn, Mar 18, 25; m 45; c 3. VIROLOGY. *Educ:* Univ Minn, BA, 48; Univ Iowa, PhD(chem), 52. *Prof Exp:* Du Pont instr, Univ Minn, 52-53; res assoc, Ind Univ, 53-54; head org chem dept, Columbian Carbon Co, 59-65; res assoc microbiol, Sch Med, Univ Pittsburgh, 65-68, asst prof, 68-70; assoc prof microbiol & immunol, Tulane Univ, 70-74; CHMN DEPT MICROBIOL, SCH MED, UNIV ORE, 74- *Concurrent Pos:* Sr fel, Mellon Inst, 54-59. *Res:* Interferon; persistent viral infections; measles viruses; mumps viruses. *Mailing Add:* Univ Ore Med Sch 3181 SW Sam Jackson Portland OR 97201

HALM, DAN ROBERT, b Fort Dodge, Iowa, Apr, 25, 55; m 86; c 1. EPITHELIAL ION TRANSPORT. *Educ:* Univ Iowa, BA, 77 & PhD(physiol), 81. *Prof Exp:* Res fel, 81-85, RES ASST PROF PHYSIOL, DEPT PHYSIOL, UNIV ALA BIRMINGHAM, 85- *Concurrent Pos:* Prin investr, Nat Inst Healt grants. *Mem:* Am Physiol Soc; Biophys Soc. *Res:* Regulation of electrolyte transport across epithelial tissues; measurement of ion flow across the epithelium and the membranes of epithelial cells during activiation of these processes. *Mailing Add:* Dept Physiol & Biophys Univ Ala Birmingham UAB Sta Birmingham AL 35294

HALM, JAMES MAURICE, b Chicago, Ill. ORGANOMETALLIC & PHYSICAL ORGANIC CHEMISTRY. *Educ:* Univ Ill, BS, 53; St Louis Univ, MS, 55; Va Polytech Inst & State Univ, PhD(organometallic chem), 72. *Prof Exp:* Instr org chem, Morton Col, 57-67; chemist, Addressograph-Multigraph Corp, 72-74; sr chemist, Org Photoconductors, A B Dick Co, 74-80; sr scientist-imaging, St Regis Corp, 81-85; sr develop chemist, Polychrome Corp, 85; chem instr, Rockland Community Col, 86-87; assoc prof org chem, Norwich Univ, 87-90; SR RES & DEVELOP CONSULT, CORP RES CTR, INT PAPER CO, 90- *Mem:* Am Chem Soc; Soc Photog Scientists & Engrs. *Res:* Theory, design, synthesis, electrochemistry and spectral properties of donor, acceptor, and charge transfer complexes of organic photoconductors. *Mailing Add:* 24 Lime Kiln Rd Suffern NY 10901

HALMI, NICHOLAS STEPHEN, b Budapest, Hungary, June 6, 22; nat US; m 63; c 2. ENDOCRINOLOGY. *Educ:* Univ Budapest, MD, 47. *Prof Exp:* Asst path, St Johns Hosp, Budapest, 45-47; asst anat, Univ Pecs, 47-49; instr, Univ Chicago, 49-50; from instr to assoc prof, 50-58, PROF ANAT, 58-, PROF PHYSIOL, 66-, PROF ENDOCRINOL, 69-; DEPT ANAT, MT SINAI SCH MED, NY. *Concurrent Pos:* NSF sr fel, 61-62; mem endocrinol study sect, NIH, 62-66. *Honors & Awards:* Ciba Award, Endocrine Soc, 57. *Mem:* AAAS; Soc Exp Biol & Med; Am Asn Anat; Am Thyroid Asn; Am Physiol Soc. *Res:* Pituitary histophysiology; hypothalamic regulation of adenohypophysis; thyroid regulation; staining techniques; morphogenesis of pituitary tumors; metabolism of iodide. *Mailing Add:* 200 Winston Dr No 708 Cliffside Park NJ 07010

HALMOS, PAUL RICHARD, b Budapest, Hungary, Mar 3, 16; nat US; m 45. PURE MATHEMATICS. *Educ:* Univ Ill, BS, 34, MS, 35, PhD(math), 38. *Hon Degrees:* DSc, Univ St Andrews, 80, DePauw Univ, 80, LHD, Kalamazoo Col, 86. *Prof Exp:* Asst math, Univ Ill, 35-36, instr, 38-39, assoc, 42-43; fel, Inst Advan Study, 39-40, asst, 40-42; asst prof math, Syracuse Univ, 43-46; from asst prof to prof, Univ Chicago, 46-61; prof, Univ Mich, 61-68; prof & chmn dept, Univ Hawaii, 68-69; prof, 69-70, distinguished prof math, Ind Univ, Bloomington, 70-84; PROF MATH, SANTA CLARA UNIV, 84- *Concurrent Pos:* Mem staff, Radiation Lab, Mass Inst Technol, 45; Guggenheim fel, Inst Advan Study, 47-48; prof, Univ Montevideo, 51-52; prof, Univ Calif, Santa Barbara, 76-78. *Honors & Awards:* Guggenheim Fel, 47. *Mem:* Am Math Soc; Math Asn Am; fel Royal Soc Edinburgh; Hungarian Acad Sci. *Res:* Measure theory; Hilbert space; algebraic logic. *Mailing Add:* Dept Math Santa Clara Univ Santa Clara CA 95053

HALONEN, MARILYN JEAN, b Duluth, Minn, July 8, 41; m 80; c 1. IMMUNOLOGY. *Educ:* Univ Minn, BS, 63; Iowa State Univ, MS, 68; Univ Ariz, PhD(molecular biol), 74. *Prof Exp:* Instr hemat, State Univ NY Buffalo, 63-64; technician org chem, Univ Chicago, 64-65; med technologist clin path, Mercy Hosp, Des Moines, Iowa, 65-66; technician biochem, 68-69, from res asst to res assoc immunol, 69-77, adj asst prof med, 77-83, RES ASSOC PROF INTERNAL MED, 83-, ASSOC PROF PHARMACOL, UNIV ARIZ, 87- *Concurrent Pos:* NIH res career develop award, 78; prin investr NIH grant, 77- *Mem:* Am Asn Immunologists; Int Soc Immunopharmacol; Am Thoracic Soc; Am Acad Allergy & Immunol; Am Soc Pharmacol & Exp Therapeut. *Res:* mechanism of IgE-induced acute allergic reactions; the capacity of IgE to alter neuronal regulation of airways; relationship of IgE to asthma. *Mailing Add:* Div Respiratory Sci Univ Ariz Col Med Tucson AZ 85724

HALOULAKOS, VASSILIOS E, b Gytheion, Greece, Jan 13, 31; US citizen; m 53; c 1. ENGINEERING, APPLIED MATHEMATICS. *Educ:* Univ Southern Calif, BSME, 59, MSAE, 62, EngrD(aeronaut eng), 65. *Prof Exp:* Sr engr, Rocketdyne Div, NAm Rockwell Corp, 59- 66, spec lectr, 65-66; sr staff mem, Mech Res, Inc, 66; supvr res & develop propulsion, 66-77, PROPULSION SPECIALIST, EXOTIC PROPULSION, MCDONNELL DOUGLAS SPACE SYSTS CO, 77- *Concurrent Pos:* Sr lectr & mem bd trustees, W Coast Univ, 66-; spec consult math, Los Angeles Sch Dist. *Mem:* Assoc fel Am Inst Aeronaut & Astronaut; fel Inst Advan Eng; Nat Space Soc. *Res:* Propulsion, flight mechanics, space nuclear power and propulsion, spacecraft control dynamics and systems analysis; uses of new mathematical techniques in solution of novel-type problems; missile propulsion; energy management systems; organized & lectured National Courses on Nuclear Propulsion. *Mailing Add:* 1031 Fairmount Rd Burbank CA 91501

HALPER, JAROSLAVA, b Prague, Czech, Aug 1, 53; Can citizen; m 79; c 3. CELL BIOLOGY. *Educ:* Univ Toronto, MD, 80; Univ Minn, PhD(path), 86. *Prof Exp:* Resident path, Albert Einstein Col Med, 80-81; fel exp path, Mayo Clin/Found, 81-82, resident path, 82-86; ASST PROF PATH, COL VET MED, UNIV GA, 86- *Concurrent Pos:* Res assoc cell biol, Vanderbilt Univ, 85-86; clinician-investr, Mayo Clin/Found, 84-86. *Mem:* AAAS; Am Asn Path; Am Soc Cell Biol; Int Acad Path. *Res:* Characterization of a novel (transforming) growth factor; cell and molecular biology; protein purification; biochemistry; growth factors in growth and development. *Mailing Add:* Dept Path Col Vet Med Univ Ga Athens GA 30602

HALPERIN, BERTRAND ISRAEL, b Brooklyn, NY, Dec 6, 41; m 62; c 2. SOLID STATE PHYSICS, STATISTICAL MECHANICS. *Educ:* Harvard Univ, AB, 61; Univ Calif, Berkeley, MA, 63, PhD(physics), 65. *Prof Exp:* NSF fel physics, Univ Paris, 65-66; mem tech staff, Bell Tel Labs, 66-76; chmn Physics Dept, 88-91, PROF PHYSICS, HARVARD UNIV, 76- *Concurrent Pos:* Lectr, Harvard Univ, 69-70; assoc ed, Rev Modern Physics, 73-80. *Honors & Awards:* Oliver Buckley Prize, Am Phys Soc, 82. *Mem:* Nat Acad Sci; Am Phys Soc; Am Acad Arts & Sci; Am Philos Soc. *Res:* Theory of disordered systems, critical phenomena, magnetism, metal-insulator transitions, two-dimensional systems and electrons in strong magnetic fields. *Mailing Add:* Dept Physics Harvard Univ Cambridge MA 02138

HALPERIN, DON A(KIBA), b Cleveland, Ohio, Jan 22, 25; m 49; c 2. ENGINEERING MECHANICS. *Educ:* Case Inst Technol, BS, 45; Univ Ill, BS, 48; Va Polytech Inst, MS, 57, PhD(eng mech), 64. *Prof Exp:* Civil engr, Bur Yards & Docks, US Dept Navy, 45-46; architect, Braverman & Halperin,

48-53; assoc prof, 53-67, dir, Sch Bldg Construct, 73-80, PROF BLDG CONSTRUCT, UNIV FLA, 67- *Concurrent Pos:* Chmn res & grad study comt, Assoc Schs Construct, 65. *Mem:* Int Asn Bridge & Struct Engrs; Am Inst Constructors (secy, 80-82); Am Coun Construct Educ (vpres, 80-81). *Res:* Structural and architectural design; construction methods. *Mailing Add:* Sch of Bldg Construct Univ of Fla Gainesville FL 32611

HALPERIN, HERMAN, electrical engineering; deceased, see previous edition for last biography

HALPERIN, JOSEPH, b Gomel, Russia, Apr 1, 23; nat US; m 51; c 4. INORGANIC CHEMISTRY, NUCLEAR CHEMISTRY. *Educ:* Univ Chicago, BS, 43, MS, 50, PhD(chem), 51. *Prof Exp:* Asst chem, Metall Lab, Univ Chicago, 43; assoc chemist, Clinton Lab, Tenn, 44-46; chemist, Oak Ridge Nat Lab, 51-87; RETIRED. *Mem:* AAAS; Am Chem Soc; Sigma Xi; Am Phys Soc; Am Nuclear Soc. *Res:* Heavy element chemistry and physics. *Mailing Add:* 109 Baker Lane Oak Ridge TN 37830

HALPERIN, MAX, mathematical statistics, clinical trials; deceased, see previous edition for last biography

HALPERIN, STEPHEN, b Kingston, Ont, Feb 1, 42; m 79; c 2. RATIONAL HOMOTOPY THEORY. *Educ:* Univ Toronto, BSc, 65, MSC, 66; Cornell Univ, PhD(math), 70. *Prof Exp:* From asst prof to assoc prof, 70-79, PROF MATH, UNIV TORONTO, 79- *Mem:* Fel Royal Soc Can; Can Math Soc; Am Math Soc. *Res:* Rational homotopy theory, with a particular interest in the loop spaces of finite complexes. *Mailing Add:* Dept Math Univ Toronto Toronto ON M5S 1A1 Can

HALPERIN, WALTER, b San Jose, Calif, Oct 20, 32; m 56; c 2. BOTANY. *Educ:* Brown Univ, AB, 54; Univ Conn, PhD(bot), 65. *Prof Exp:* Asst prof bot, Univ Mass, 66-68; from asst prof to assoc prof, 68-88, PROF BOT, UNIV WASH, 88- *Mem:* AAAS; Bot Soc Am; Am Soc Plant Physiol. *Res:* Physiology of growth and development in plants. *Mailing Add:* Dept Bot Univ Wash Seattle WA 98195

HALPERIN, WILLIAM PAUL, b Ottawa, Ont, July 16, 45; m 68; c 3. LOW TEMPERATURE PHYSICS, NUCLEAR MAGNETIC RESONANCE. *Educ:* Queen's Univ, Ont, BSc, 67; Univ Toronto, MSc, 68; Cornell Univ, PhD(physics), 75. *Prof Exp:* Fel physics, Cornell Univ, 74-75; from asst prof to assoc prof, 75-86, PROF PHYSICS, NORTHWESTERN UNIV, EVANSTON, 86-, CHMN, DEPT PHYSICS & ASTRON, 90- *Concurrent Pos:* Fac res partic, Argonne Nat Lab, 75-; Alfred P Sloan fel, 77; Yamada Found Fel, 84. *Mem:* Am Phys Soc. *Res:* Thermal and magnetic properties of liquid helium 3, solid helium 3, solutions of helium 3 in helium 4; superfluidity; superconductivity; cryogenic devices; absolute thermometry; nuclear magnetic resonance; platinum metal particles; molecular diffusion; organic; metals and magnets; porous materials; cementitious materials. *Mailing Add:* Dept Physics Northwestern Univ Evanston IL 60208

HALPERIN-MAYA, MIRIAM PATRICIA, b Brooklyn, NY, July 1, 45; m; c 2. PURE MATHEMATICS. *Educ:* Radcliffe Col, AB, 66; Brandeis Univ, MA, 68, PhD(math), 73. *Prof Exp:* Asst prof math, Univ Md, College Park, 72-74 & Tufts Univ, 74-80; syst designer, 80-88, CONSULT, CAMEX CO. 88- *Mem:* Asn Comput Mach; Math Asn Am; Asn Women in Math. *Res:* Singularities of algebraic and complex analytic varieties; equisingularities; equivalence of zerocycles on algebraic varieties; computer graphics including scan conversion, splines, image representations and manipulations. *Mailing Add:* Resat Corp 25 Marshall St Brookline MA 02146

HALPERN, ALVIN M, b New York, NY, July 17, 38; m 66; c 2. ATOMIC PHYSICS. *Educ:* Columbia Col, AB, 59; Columbia Univ, MA, 61, PhD(physics), 65. *Prof Exp:* Instr physics, Pratt Inst, 64-65; from instr to assoc prof, 65-74, chmn dept, 80-90, PROF PHYSICS, BROOKLYN COL, 75-, EXEC DIR, APPL SCI INST, 90- *Concurrent Pos:* Recipient & dir, NSF educ grants, 70-71 & 72-74, co-prin investr, NSF res grant, 78-80 & 80-82; vis guest scientist, Internat Ctr Theoret Physics, Trieste, Italy & Theoret Chem Dept, Oxford Univ, 73. *Mem:* Am Phys Soc; Am Asn Physics Teachers; AAAS; NY Acad Sci. *Res:* Theoretical studies of atomic scattering processes, especially charge transfer, charge exchange, excitation, and ionization of atoms and ions. *Mailing Add:* Dept Physics Brooklyn Col Brooklyn NY 11210

HALPERN, ARTHUR MERRILL, b Bayonne, NJ, Aug 4, 43; m 66; c 4. PHYSICAL CHEMISTRY, SPECTROSCOPY. *Educ:* Rutgers Univ, New Brunswick, BA, 64; Northeastern Univ, PhD(phys chem), 68. *Prof Exp:* Fel, Univ Minn, Minneapolis, 68-70; vis mem staff, Bell Tel Labs, 70; asst prof chem, NY Univ, Bronx, 70-73; from asst prof to prof chem, Northeastern Univ, 73-90; PROF & CHMN, IND STATE UNIV, 90- *Concurrent Pos:* Indust consult; Alfred P Sloan res fel, 74-76; sr sci fel, NATO, 80. *Mem:* Am Chem Soc; AAAS; Am Phys Soc; Europ Photochem Asn; Inter-Am Photochem Soc. *Res:* Photophysics of organic molecules especially saturated amines; measurement of fast luminescence phenomena; photokinetics/photoassociation; laser spectroscopy; chemical education. *Mailing Add:* Dept Chem Ind State Univ Terre Haute IN 47809

HALPERN, BENJAMIN DAVID, b Malden, Mass, May 19, 21; m 51; c 5. CHEMISTRY, CHEMICAL ENGINEERING. *Educ:* Mass Inst Technol, BS, 43; Univ Notre Dame, PhD(chem), 49. *Prof Exp:* Chem engr & proj leader polymerization, Rohm & Haas Co, 43-49; res dir med specialties, Dajac Labs, 49-55; res dir, Borden Chem Co, 55-63; PRES, POLYSCI INC, 63- *Concurrent Pos:* Mem staff, Monomer-Polymer, Inc, 49-51, treas, 51-53, secy & pres, 53-55; vis prof, Hahnemann Med Col; pres, Polaron Instruments, Inc. *Mem:* Am Chem Soc; NY Acad Sci; Electron Micros Soc; Asn Consult Chemists & Chem Engrs (past pres); Histochem Soc. *Res:* Polymers; monomers; organic intermediates; monomer synthesis; polymerization; medical specialties; medical application of polymers; materials and chemicals for life sciences. *Mailing Add:* Polysci Inc 400 Valley Rd Warrington PA 18976

HALPERN, BERNARD, b Chicago, Ill, Feb 13, 18; m 41; c 3. BACTERIOLOGY. *Educ:* Univ Ill, BS, 39, MS, 42; Northwestern Univ, PhD(bact), 57. *Prof Exp:* Jr bacteriologist, Michael Reese Hosp, 39-43; res bacteriologist, St Luke's Hosp, 46-51; from asst prof to assoc prof microbiol, Obstet & Gynec, Med Sch, Northwestern Univ, Chicago, 62-88; RETIRED. *Mem:* Am Soc Microbiol. *Res:* Immunology research, especially carbohydrate metabolism, choriocarcinoma (trophoblastic disease) and amebiasis; growth of mycobacteria tuberculosis. *Mailing Add:* 17725 S Stonebridge Hazel Crest IL 60429

HALPERN, BRUCE PETER, b Newark, NJ, Aug 18, 33; m 56; c 2. NEUROSCIENCES, TASTE. *Educ:* Rutgers Univ, AB, 55; Brown Univ, ScM, 57, PhD(physiol psychol), 59. *Prof Exp:* NIH vis fel physiol, Cornell Univ, 59-61, res assoc, 59-60, lectr, 60-61; asst prof, State Univ NY Upstate Med Ctr, 61-66; assoc prof psychol & biol, 66-73, field rep psychol, Grad Sch, 70-73, chmn dept psychol, 74-80, 90-95, PROF PSYCHOL, NEUROBIOL & BEHAV, CORNELL UNIV, 73- *Concurrent Pos:* Consult, Behav Sci Div, Food Sci Lab, US Army Natick Develop Ctr, 73, vis scientist, 73-74; consult, NSF Sensory Physiol & Perception Prog, 77-79; mem commun disorders panel, Nat Inst Neurol & Commun Disorders & Stroke, NIH, 77-79; Fogarty Sr Int Fel, Osaka Univ, 82; mem sensory prog adv comt, NIH, 84-87; exec ed, Chem Senses, 84-88; chmn, Gordon Conf Chem Senses, 87-90; mem, Int Comn Olfaction & Taste, 85- *Mem:* Asn Chemoreception Sci; Am Physiol Soc. *Res:* Sensory physiology; sensory function, especially in gustation; relation between environmental energy, neural responses and chemosensory behavior. *Mailing Add:* Dept Psychol Uris Hall Cornell Univ Ithaca NY 14853-7601

HALPERN, DANIEL, b New York, NY, May 28, 17; m 43; c 3. MEDICINE. *Educ:* NY Univ, BA, 36; Chicago Med Sch, MD, 43. *Prof Exp:* Pvt pract, 48-58; instr phys med & rehab, Bird S Coler Hosp, New York Med Col, 61-62; from asst prof to assoc prof, Children's Rehab Ctr, Univ Hosps, Univ Minn, 62-74, prof phys med & rehab, 74-80. *Concurrent Pos:* Mem interdisciplinary adv comt active nursing home care, Minneapolis Dept Health, 63; consult, Minneapolis Area Study of Educ Serv for Multiply Handicapped Sch Children, 68; med consult, St Paul Sch Syst, 71-; mem comt nursing home rehab & stand, Minn Dept Welfare, 72; consult educ prog for handicapped children, St Paul Sch Syst, 72-74; mem spec consult comt, Dakota's Children's Nursing Home, Hill Found, 73-74; adj prof rehab med, Univ Wis-Madison Med Sch, 80- *Mem:* Am Acad Pediat; Am Acad Phys Med & Rehab; Am Acad Cerebral Palsy; Am Cong Phys Med & Rehab. *Res:* Learning of motor skills in handicapped and normal individuals; mechanisms of abnormal muscle tone and evaluation of treatment; special learning disabilities in handicapped children. *Mailing Add:* 200 Tanglewood Dr South Chatham MA 02659

HALPERN, DAVID, b Montreal, Que, June 24, 42; m 66; c 2. PHYSICAL OCEANOGRAPHY. *Educ:* McGill Univ, BSc, 64; Mass Inst Technol, PhD(phys oceanog), 69. *Prof Exp:* Res assoc phys oceanog, Mass Inst Technol, 69; oceanogr, Pac Oceanog Labs, Nat Oceanic & Atmospheric Admin, 69-74; proj supvr, Ocean-Atmospheric Response Studies, Pac Marine Environ Lab, Nat Oceanic & Atmospheric Admin, 74-81, sr scientist, 81-85; prin res assoc, Atmospheric Sci Dept & Sch Oceanog, Univ Wash, 85-86; MEM TECH STAFF, JET PROPULSION LAB, CALIF INST TECHNOL, PASADENA, 86- *Concurrent Pos:* Affil prof oceanog & atmospheric sci, Univ Wash, 81-85; ed, Trop Ocean-Atmospheric Newsletter, 79-84, assoc ed, Tropicale Oceanologie. *Honors & Awards:* Graham Medal, 62-63; Silver Medal, Dept Com, 81. *Mem:* AAAS; Am Geophys Union; Am Meteorol Soc; Marine Technol Soc; Sigma Xi; Can Meteorol & Oceanog Soc. *Res:* Near-surface circulation; coastal and equatorial upwelling; ocean-atmosphere response studies; wind and current measurements; oceanography of tropical and equatorial waters, satellite observations of wind, ocean color and sea surface temperature; author or coauthor of over 100 publications. *Mailing Add:* Jet Propulsion Lab M/S 300-323 Calif Inst Tech 4800 Oak Grove Dr Pasadena CA 91109

HALPERN, DONALD F, b New York, NY, May 9, 36; m 65; c 1. ORGANOMETALLIC CHEMISTRY, LABORATORY COMPUTING. *Educ:* Queen's Col, BS, 59, MA, 64; City Univ New York, PhD(chem), 71. *Prof Exp:* Chemist, Sun Chem Corp, 60-64; instr chem, Queen's Col, 64-73; sr scientist, Bio-Med Sci, 73; tech serv mgr, Tuck Indust, 74; CONSULT SCIENTIST, ANAQUEST DIV, BOC HEALTHCARE INC, 75- *Mem:* Am Chem Soc; Sigma Xi. *Res:* Product oriented organic synthesis, pharmaceutical research and development; use of computers in research. *Mailing Add:* Anaquest Div BOC Healthcare Inc 100 Mountain Ave Murray Hill NJ 07974

HALPERN, EPHRIAM PHILIP, b Montreal, Que, July 8, 22; nat US; m 46; c 3. BIOCHEMISTRY. *Educ:* McGill Univ, BSc, 44, MSc, 45; Wash State Univ, PhD(biochem), 49; Am Bd Clin Chem, dipl; Am Inst Chemists, cert. *Prof Exp:* Plant chemist, Can Packers, Ltd, 45-46; res assoc biochem, Inst Cancer Res & Lankenau Hosp Res Inst, 49-51; head chem div, Northern Div, Albert Einstein Med Ctr, Philadelphia, 51-54; dir pvt med & indust lab, Park Labs, 54-70; tech dir, Philadelphia Med Labs, Inc, 70-75, dir nat qual assurance, Damon Corp, 75-79; CONSULT, 79- *Concurrent Pos:* Consult, Vet Admin Hosp, Philadelphia, 66- *Mem:* Am Chem Soc; fel Am Asn Clin Chemists; fel Am Inst Chemists. *Mailing Add:* 7185 Huntington Lane Delray Beach FL 33446-2519

HALPERN, FRANCIS ROBERT, b New York, NY, Mar 5, 29; m 51; c 2. THEORETICAL PHYSICS. *Educ:* Cornell Univ, AB, 49; Univ Chicago, MS, 49; Univ Calif, PhD(physics), 57. *Prof Exp:* Dynamics engr, Bell Aircraft Corp, 49-50; instr res engr, Armour Res Found, Ill Inst Technol, 50-52; instr physics, Princeton Univ, 57-61; from asst prof to assoc prof physics, Univ Calif, San Diego, 61-80, prof, 80-84; lectr, Univ MD, 85-87; RETIRED. *Concurrent Pos:* Fulbright fel, Italy, 59-60. *Mem:* Am Phys Soc. *Res:* Field theory; electrodynamics and pi meson physics. *Mailing Add:* 4024 Quartz Dr Santa Rosa CA 95405

HALPERN, HOWARD S, b Newark, NJ, July 7, 25; m 49; c 3. PHYSICS, SYSTEMS ENGINEERING. *Educ:* Union Col NY, BS, 46; Univ Md, MS, 53. *Prof Exp:* Physicist commun & radar, Naval Res Lab, 47-50; electronic scientist, Naval Gun Factory, 50-53; res engr radar, pinspotter, Am Mach & Foundry Co, 53-55; chief physics & chem sect, Avco Res & Advan Develop, 55-56; mem staff res & planning mil syst radar, Norden Syst Inc, Subsid United Technol Corp, 56-90; CONSULT, 90- *Mem:* Assoc fel Am Inst Aeronaut & Astronaut; Inst Elec & Electronics Engrs; Am Phys Soc. *Res:* Military electronics systems and equipments; radar; cosmology; seven patents. *Mailing Add:* 182 Clay Hill Rd Stamford CT 06905

HALPERN, ISAAC, b New York, NY, May 15, 23; m 51; c 3. PHYSICS. *Educ:* City Col New York, BS, 43; Mass Inst Technol, PhD(physics), 48. *Prof Exp:* Instr physics, Princeton Univ, 43-44; res assoc nuclear lab, Mass Inst Technol, 46-53; from asst prof to assoc prof physics, 53-60, PROF PHYSICS, UNIV WASH, 60- *Mem:* Am Phys Soc. *Res:* Nuclear physics; nuclear reactions; fission; interactions of mesons and photons with nuclei. *Mailing Add:* Dept Physics Univ Wash Seattle WA 98195

HALPERN, JACK, b Poland, Jan 19, 25; nat Can; m 49; c 2. INORGANIC CHEMISTRY, ORGANOMETALLIC CHEMISTRY. *Educ:* McGill Univ, BSc, 46, PhD(chem), 49. *Prof Exp:* Nat Res Coun Can fel, Univ Manchester, 49-50; from instr metall to prof chem, Univ BC, 50-62; prof, 62-71, LOUIS BLOCK DISTINGUISHED SERV PROF CHEM, UNIV CHICAGO, 71- *Concurrent Pos:* Nuffield Found traveling fel, Cambridge Univ, 59-60; Sloan fel, 59-63; vis prof, Univ Minn, 62, Harvard Univ, 66-67, Calif Inst Technol, 69, Princeton Univ, 70-71, Univ Copenhagen, 78 & Univ Shelfield, 81-82; mem adv panel chem, NSF, 67-70; bd trustees, Gordon Res Conf, 68-70; chem rev comt, Argonne Nat Lab, 70-72; mem adv bd, Am Chem Soc Petrol Res Fund, 72-74; mem, NIH Med Chem B Study Sect, 75-78; assoc ed, J Am Chem Soc & Inorganica Chimica Acta; guest scholar, Kyoto Univ, 81; consult, Monsanto Co, Argonne Nat Lab, IBM, Air Prods & Chem & Eni Chem, Enimont & Rohm & Haas; R B Woodward vis prof, Harvard Univ, 91. *Honors & Awards:* Inorg Chem Award, Am Chem Soc, 68 & 85; Catalysis Award, Chem Soc London, 76; Humboldt Award, 77; Kokes Award, Johns Hopkins Univ, 77; Willard Gibbs Medal, 86; Bailar Medal, Univ Ill, 86; von Hoffmann Medal, Ger Chem Soc, 88; Chem Pioneer Award, Am Inst Chemists, 91. *Mem:* Nat Acad Sci; fel Am Acad Arts & Sci; fel NY Acad Sci; fel Royal Soc London; fel AAAS; Max Planck Soc; hon fel Royal Soc Chem. *Res:* Kinetics and mechanisms of inorganic reactions; organometallic chemistry; electron transfer processes; catalytic phenomena; fast reactions; bioinorganic chemistry. *Mailing Add:* Dept Chem Univ Chicago 5735 S Ellis Ave Chicago IL 60637

HALPERN, JAMES DANIEL, b Detroit, Mich, Aug 4, 34; m 59; c 2. MATHEMATICAL LOGIC. *Educ:* Univ Mich, AB, 55, MS, 56; Univ Calif, Berkeley, PhD(math), 62. *Prof Exp:* Bateman res fel math, Calif Inst Technol, 62-64, instr, 64-65; mem, Inst Advan Study, 65-66; asst prof math, Univ Mich, Ann Arbor, 66-70; assoc prof, Univ Toledo, 69-74; PROF MATH, UNIV ALA, BIRMINGHAM, 74- *Concurrent Pos:* Consult, Math Rev, 71-74. *Mem:* Am Math Soc. *Res:* Foundations of set theory; independence problems involving axiom of choice; combinatorial set theory. *Mailing Add:* 1060 Continentals Way No 4 Belmonte CA 94002

HALPERN, JOSHUA BARUCH, b Brooklyn, NY, Jan 21, 46; m 74. PHOTOCHEMISTRY, SPECTROSCOPY. *Educ:* Johns Hopkins Univ, BA, 66; Brown Univ, PhD(physics), 72. *Prof Exp:* Teaching assoc, Notre Dame Radiation Lab, 71-73; sci assoc physics, Univ Bielefeld, 73-76; res asst prof chem, 76-79, asst prof, 79-84, ASSOC PROF CHEM, HOWARD UNIV, 84- *Mem:* Am Chem Soc; Optical Soc Am. *Res:* Photochemistry and spectroscopy of small molecules and free radicals using laser generation and detection techniques. *Mailing Add:* Dept Chem Howard Univ 2400 Sixth St NW Washington DC 20059

HALPERN, LAWRENCE MAYER, b New York, NY, July 3, 31; m 52; c 3. PHARMACOLOGY, NEUROPHARMACOLOGY. *Educ:* Brooklyn Col, BSc, 53; Albert Einstein Col Med, PhD(pharmacol), 61. *Prof Exp:* Unit head neuropharmacol, Merck Inst Therapeut Res, 63-65; asst prof, 65-68, ASSOC PROF PHARMACOL, SCH MED, UNIV WASH, 69- *Concurrent Pos:* Vis scientist, NIH & consult, Dir Nat Inst Neurol Dis & Blindness, 67-75; clin consult, Univ Wash Hosp, 70-; dir, Drug Abuse Info Serv, Univ Wash, 68-75. *Mem:* NY Acad Sci; Am Soc Pharmacol & Exp Therapeut; Soc Neurosci; Am Pain Soc. *Res:* Epileptogenesis in cerebral cortex and anticonvulsant drugs; antidepressants, analgesics and psychosedative agents; electrophysiology in pharmacology and behavior; addiction behavior; chronic pain medicines. *Mailing Add:* 1204 E Lynn Seattle WA 98102

HALPERN, LEOPOLD (ERNST), b Vienna, Austria, Feb 17, 25. GRAVITATIONAL THEORY, THEORETICAL PHYSICS. *Educ:* Brit Inst Eng Technol, AM, 47; Univ Vienna, PhD(physics), 52. *Prof Exp:* Asst theoret physics, Univ Vienna, 56-59; res assoc, Europ Orgn Nuclear Res, Geneva, 59-60 & Inst Field Physics, Univ NC, 60-61; fel, Niels Bohr Inst, Copenhagen, 62-63; vis prof, Univ Stockholm, 63-66; vis researcher, Inst H Poincare, Paris, 66-67; assoc prof, Univ Windsor, Can, 67-79; researcher, Univ Libre, Bruxelles, Belg, 70-73 & Univ Amsterdam, 73-74; SR RES ASSOC THEORET PHYSICS, FLA STATE UNIV, 74- *Concurrent Pos:* Sr res assoc, Jet Propulsion Lab, Calif Inst Technol, Nat Res Coun Grantee, 86-88. *Mem:* Am Phys Soc. *Res:* Gravitational theory and its relation to elementary particle physics and the quantum theory; foundation of generalizations of the general theory of relativity based on local covariance with respect to semisimple lie groups. *Mailing Add:* Dept Physics Fla State Univ Tallahassee FL 32306-3016

HALPERN, MARTIN, b Montreal, Que, Dec 24, 37; div; c 2. PETROLEUM EXPLORATION, GEOCHRONOLOGY. *Educ:* McGill Univ, BSc, 59; Univ Wis, MS, 61, PhD(geol), 63. *Prof Exp:* Res assoc geol, Geophys & Polar Res Ctr, Univ Wis, 63-64; res assoc geol & geochronology, Univ Tex, Dallas, 64-67, from asst prof to prof, 67-81; sr staff geologist, 81-82, DIST EXPLOR GEOLOGIST, ENSERCH EXPLOR, INC, 82- *Concurrent Pos:* NSF res grants, 64-76; vis res fel, Univ Leeds, 71-72; adj prof, Univ Tex Marine Sci Inst Galveston, 74-81; mem, Inter-Union Comn Geodynamics Study Group 1: Study Group on the Cocos Plate, 74-81; vis prof & consult, Inst Geol, Nat Univ Mex, 75; chmn US working group, Int Geol Correlation Prog Proj 120, Magmatic Evolution of Andes Proj Working Group, 75-80; vis prof geophys & planetary sci, Tel-Aviv Univ, 77-78; adj prof geol & geochronology, Univ Tex, Dallas, 81- *Honors & Awards:* Antarctic Serv Medal, 76. *Mem:* Am Asn Petrol Geologists; Geol Soc Am; Am Inst Prof Geologists. *Res:* Geochronologic and geologic investigations in Latin America, Antarctica and Middle East. *Mailing Add:* Enserch Explor Inc 4849 Greenville Ave Suite 1200 Dallas TX 75206

HALPERN, MARTIN B, b Newark, NJ, Aug 26, 39; m 62; c 1. THEORETICAL HIGH ENERGY PHYSICS. *Educ:* Univ Ariz, BS, 60; Harvard Univ, AM, 61, PhD(physics), 64. *Prof Exp:* Fel physics, Europ Orgn Nuclear Res, Switz, 64-65, Univ Calif, Berkeley, 65-66 & Inst Advan Study, Princeton Univ, 66-67; from asst prof to assoc prof, 67-73, PROF PHYSICS, UNIV CALIF, BERKELEY, 73- *Mem:* Am Phys Soc. *Res:* Theory of elementary particles; high energy theoretical physics. *Mailing Add:* Dept Physics Univ Calif 2120 Oxford St Berkeley CA 94720

HALPERN, MIMI, b Antwerp, Belg, June 19, 38; US citizen; m 61; c 2. NEUROSCIENCE, PSYCHOBIOLOGY. *Educ:* Oberlin Col, AB, 60; Adelphi Univ, PhD(psychol), 64. *Prof Exp:* From instr to asst prof anat, 69-74, assoc prof, 74-79, asst dean sch grad studies, 75-83, dir grad prog biol psychol, 76-85, assoc dean sch grad studies, 83-90, co-dir grad prog neurol behav sci, 85-90, PROF ANAT & CELL BIOL, STATE UNIV NY DOWNSTATE MED CTR, 79- *Mem:* Fel Am Psychol Asn; Am Asn Anatomists; Soc Neurosci; Am Asn Zoologists; Sigma Xi; fel AAAS. *Res:* Biological basis of behavior; comparative neuropsychology; visual and limbic systems of vertebrates; chemical communication; reptilian neuroanatomy and behavior. *Mailing Add:* Dept Anat & Cell Biol 450 Clarkson Ave Brooklyn NY 11203

HALPERN, MORDECAI JOSEPH, b Ruda Pabianica, Poland, Mar 20, 20; US citizen. ANALYTICAL CHEMISTRY, PHYSICAL CHEMISTRY. *Educ:* Hebrew Univ Jerusalem, MSc, 54, PhD(phys chem), 59. *Prof Exp:* Asst prof anal chem, Hebrew Univ Jerusalem, 54-58; asst prof gen chem, Tel-Aviv Univ, 58-59; asst dir, Dept Radioisotopes, Weizmann Inst Sci, 59-60; res chemist phys chem, Duquesne Univ, 60-61 & NY Univ, 61-65; assoc prof chem, Staten Island Community Col, 65-67; PROF CHEM, NY INST TECHNOL, 67- *Concurrent Pos:* Res chemist, Engelhard Industs, Newark, NJ, 64-65. *Mem:* Am Chem Soc: Am Asn Univ Prof; Israeli Soc Advan Sci. *Res:* Heterometric microdetermination of metals; mechanism of enzymatic sugar hydrolysis; complex formation in solvent extraction of metals; organic electrochemical oxidation-reduction. *Mailing Add:* Dept Life Sci NY Inst Technol PO Box 170 Old Westbury Campus Old Westbury NY 11568

HALPERN, MYRON HERBERT, b New York, NY, May 12, 24; m 60; c 2. ANATOMY, HISTOLOGY. *Educ:* Ind Univ, AB, 44; WVa Univ, ScM, 47; Univ Mich, PhD(anatomy), 52. *Prof Exp:* Dir biol labs, Univ Rochester, 47-49; asst anat, Med Sch, Univ Mich, 49-52; asst prof, Hahnemann Med Col, 52-57; dir info res sect & admin dir common cold & virus res prog, Nat Drug Co, 57-61; dir life sci & space med, RCA, 61-63; pres, Bio-Med Res Corp, 63-67; chmn div allied health sci, 68-72, dean instr, 69-71, PROF BIOL, CAMDEN COUNTY COL, 67- *Mem:* Am Soc Ichthyol & Herpet; Sigma Xi. *Mailing Add:* 109 Leeds Rd Mt Laurel NJ 08054

HALPERN, SALMON RECLUS, b Brooklyn, NY, June 24, 07; m 43; c 3. ANATOMY. *Educ:* Univ Mich, AB, 30; NY Univ, MS, 31; Univ Colo, PhD(anat), 34, MD, 37; Am Bd Pediat, dipl, 43, dipl pediat allergy, 48. *Prof Exp:* Asst anat, Univ Colo, 31-35; intern, Montefiore Hosp, New York, 37-38; intern, New York Hosp, 38-39, asst resident, 39-40; resident, Children's Med Ctr, 40-41, asst med dir, 41-42, from instr to assoc clin prof, 46-67, CLIN PROF PEDIAT, UNIV TEX HEALTH SCI CTR, DALLAS, 67-, DIR ALLERGY CLIN, CHILDREN'S MED CTR, 71- *Concurrent Pos:* Instr, Baylor Col Med, 41-42. *Mem:* Fel AMA; fel Am Diabetes Asn; fel Am Acad Allergy; fel Am Acad Pediat; fel Am Col Allergists. *Res:* Pediatric allergy. *Mailing Add:* 6319 Bandera Apt C Dallas TX 75225

HALPERN, TEODORO, b Buenos Aires, Arg, Sept 5, 31; m 56; c 3. SOLID STATE PHYSICS, INSTRUMENTATION. *Educ:* Nat Univ La Plata, BSc, 54; Nat Univ Cuyo, MSc, 58; Univ Stuttgart, MSc & Dr rer nat(physics), 61. *Prof Exp:* Researcher, Arg AEC, 58-64; from asst prof to assoc prof solid state physics, Nat Univ Cuyo, 58-63; prof phys metall, Nat Univ La Plata, 63-64; res assoc physics, Univ Chicago, 64-68; vis scientist, Argonne Nat Lab, 68-70; vis assoc prof solid state physics, Yeshiva Univ, 70-74; from asst prof to assoc prof, 74-78, dir, 78-79, dean of schs & actg vpres acad affairs, 79-82, PROF PHYSICS, SCH THEORET & APPL SCI, RAMAPO COL, NJ, 78- *Concurrent Pos:* Sci consult, lasers in med, Yeshiva Univ, 81-, semiconductors, Stauffer Chem, 82-; Fred & Florence Thomases fac award, 82; fel, Princeton Univ, 83 & Am Cyanamid, 84-90; educ consult, 90-; exec secy, Inter-Am Coun Physics Educ, 90- *Mem:* AAAS; Sigma Xi; Ger Soc Metall; Arg Soc Metals; NY Acad Sci; Am Asn Physics Teachers. *Res:* Optical properties of solids; electroreflectance; high pressure optical spectroscopy; semiconductor to metal transition; high voltage, high speed pulse-induced semiconductor to metal transition; quantum states in sputtered metal dispersions in semiconductor phases; lasers in medicine; international energy; bio-engineering; semiconductor crystal growth plasma enhanced chemical vapor deposition; epitaxial growth; diffusion. *Mailing Add:* Sch Theoret & Appl Sci Ramapo Col Mahwah NJ 07430-1680

HALPERN, WILLIAM, b Oct 27, 23; c 2. HYPERTENSION, CORONARY & BRAIN ISCHEMIA. *Educ:* Univ Vt, PhD(physiol & biophys), 69. *Prof Exp:* EMER PROF MED INSTRUMENTATION, DEPT PHYSIOL & BIOPHYS, COL MED, UNIV VT, 69- *Mailing Add:* 416 S Willard St Burlington VT 05401

HALPERT, JAMES ROBERT, b Los Angeles, Calif, Dec 5, 49. DRUG-METABOLIZING ENZYMES, ENZYME INHIBITORS. *Educ:* Univ Calif, BA, 71; Uppsala Univ, PhD, 77, MSc, 78. *Prof Exp:* Res assoc, Dept Pharmacol, Karolinska Inst, 77-78, Vanderbilt Univ, 78-80; asst prof, 83-87, ASSOC PROF PHARMACOL & TOXICOL, DEPT PHARM & TOXICOL, UNIV ARIZ, 87- *Mem:* Int Soc Toxicol; Am Soc Pharmacol & Exp Ther; Am Soc Biochem & Molecular Biol. *Res:* Structure and function of mammalian hepatic cytochromes P-450 with special emphasis on the use of specific irreversible inhibitors as probes and modulators of P-450 function. *Mailing Add:* Dept Pharmacol & Toxicol Col Pharmacy Univ Ariz Tucson AZ 85721

HALPIN, DANIEL WILLIAM, b Covington, Ky, Sept 29, 38; m 63; c 1. ENGINEERING MANAGEMENT, CIVIL ENGINEERING. *Educ:* US Mil Acad, BS, 61; Univ Ill, Urbana, MS, 69, PhD(civil eng), 73. *Prof Exp:* Opers res analyst, Construct Eng Res Lab, Champaign, Ill, 70-72; mem fac civil eng, Univ Ill, 72-73; prof civil eng, Ga Inst Technol, 81-85, mem fac civil eng, 73-85; Clark chair prof, Univ Md, 85-87; DIV HEAD, SCH ENG, PURDUE UNIV, 87- *Concurrent Pos:* Proj dir res proj, Dept Energy, 76-78; vis assoc prof civil eng, Univ Sydney, Australia, 81; vis scholar, Tech Univ Munich, Ger, 79; vis prof, Swiss Tech Inst (ETH), Zurich, 85; chmn const res coun, Am Soc Civil Engrs, 84-86; consult, Off Tech Assessment, 86 & Tenn Valley Authority. *Honors & Awards:* Huber Res Prize, Am Soc Civil Engrs, 79. *Mem:* Am Soc Civil Engrs; Am Soc Eng Educ; Sigma Xi. *Res:* Simulation of construction operations; data base management for complex construction projects, impact of international competition on construction technology; author or co-author of 5 engineering textbooks. *Mailing Add:* Head Div Construct Eng Mgt Purdue Univ Civil Eng Bldg West Lafayette IN 47907

HALPIN, JOSEPH JOHN, b Philadelphia, Pa, Nov 10, 39; m 61; c 4. PHYSICS, ELECTRONICS. *Educ:* St Joseph's Col, Philadelphia, BS, 61; Georgetown Univ, MS, 71. *Prof Exp:* Res asst, Georgetown Univ, 64-66; res physicist, US Naval Res Lab, 66-71; tech specialist, 71-75, BR CHIEF NUCLEAR WEAPONS EFFECTS, HARRY DIAMOND LABS, US ARMY, 75- *Mem:* Sr mem Inst Elec & Electronics Engrs; Sigma Xi. *Res:* Nuclear weapons effects to include blast, thermal radiation, electromagnetic pulse and transient radiation-induced effects on materials, components and systems; technical and managerial aspects of nuclear effects phenomena. *Mailing Add:* Harry Diamond Labs 2800 Powder Mill Rd Adelphi MD 20783-1197

HALPIN, ZULEYMA TANG, b Ciudad Bolivar, Venezuela, Mar 9, 45; US citizen; m 78. ANIMAL BEHAVIOR, BEHAVIORAL ECOLOGY. *Educ:* St Louis Univ, BS, 67; Univ Calif, Berkeley, MA, 70, PhD(zool), 74. *Prof Exp:* NIMH fel ecol, Univ BC, 74-76; ASST PROF BIOL, UNIV MO, ST LOUIS, 76- *Mem:* Animal Behav Soc; Ecol Soc Am; Am Soc Zoologists; Am Soc Mammalogists. *Res:* Social behavior; dispersal and population dynamics; physiological correlates of social behavior; animal communication; chemical communication. *Mailing Add:* Dept Biol Univ Mo 8001 Natural Bridge Rd St Louis MO 63121

HALPRIN, ARTHUR, b Portsmouth, NH, Aug 11, 35; m 58. PHYSICS. *Educ:* Univ NH, BS, 57; Worcester Polytech Inst, MS, 59; Univ Pa, PhD(physics), 65. *Prof Exp:* Asst prof, 64-69, assoc prof, 69-80, PROF PHYSICS, UNIV DEL, 80- *Mem:* Am Phys Soc. *Res:* Particle physics. *Mailing Add:* Dept Physics & Astron Univ Del Newark DE 19716

HALPRIN, KENNETH M, b Brooklyn, NY, Mar 19, 31. DERMATOLOGY, BIOCHEMISTRY. *Educ:* Univ Chicago, BA, 50, MD, 55. *Prof Exp:* Intern, Univ Chicago Clins, 55-56, resident dermat, 56-59; NSF fel, St John's Hosp, London, 61-62; asst prof dermat, Univ Chicago, 63; from asst prof to assoc prof, Med Sch, Univ Ore, 64-68; assoc prof, 68-71, PROF DERMAT, SCH MED, UNIV MIAMI, 71-; CHIEF DERMAT, MIAMI VET ADMIN HOSP, 68- *Mem:* Am Acad Dermat; Soc Invest Dermat; Am Fedn Clin Res; NY Acad Sci. *Res:* Enzymology; carbohydrate metabolism; carcinogenesis. *Mailing Add:* 1201 NW 16th St Miami FL 33125

HALPRYN, BRUCE, b New York, Mar 5, 57. CARDIOLOGY. *Educ:* State Univ NY, BS, 79, PhD(biol & pharmacol), 83. *Prof Exp:* Nat Res Coun fel, Cardiac Res Labs, NASA/Ames, 83-86; SECT HEAD/PROJ LEADER, DIV PROCTER & GAMBLE, NORWICH PHARMACEUT INC, 86- *Concurrent Pos:* Mem bd dir, Nonprofit Community Health Clin, Our Health Ctr, 82-84. *Mem:* Am Physiol Soc. *Res:* Development of novel inotropic agents for congestive heart failure. *Mailing Add:* Norwich Pharmaceut Inc PO Box 191 Norwich NY 13815-0191

HALS, FINN, b Trondheim, Norway, July 1, 24; m 60; c 3. MECHANICAL ENGINEERING. *Educ:* Tech Univ Norway, MSc, 51. *Prof Exp:* Consult engr, Asn Norweg Steam Power Generators, 51-55; mgr, G Hartman Inc, Norway, 55-57; asst power supt, WVa Pulp & Paper Co, Pa, 57-58; mech engr, Stone & Webster Eng Corp, Mass, 58-60; prin res engr, Avco-Everett Res Lab, 60-87, CONSULT, 87- *Mem:* Am Soc Mech Engrs; AAAS. *Res:* Magnetohydrodynamic electrical power generation; coal combustion and utilization; pollution control. *Mailing Add:* 14 Vine Brook Rd Lexington MA 02173

HALSALL, H(ALLEN) BRIAN, b Eng, Apr 13, 43; m 70; c 2. BIOCHEMISTRY. *Educ:* Univ Birmingham, Eng, BSc, 64, PhD(phys biochem), 67. *Prof Exp:* Scholar, Univ Calif, Los Angeles, 67-69, actg asst prof biochem, 69-70; staff mem & consult phys biochem, Oak Ridge Nat Lab, 70-74; asst prof, 74-81, assoc prof, 81-87, PROF CHEM, UNIV CINCINNATI, 87- *Mem:* Biophys Soc; Brit Biophys Soc; Am Chem Soc; Sigma Xi; Protein Soc; Am Asn Clin Chemists; Clin Ligand Assay Soc. *Res:* Biophysical studies of protein-ligand interactions, particularly therapeutic drugs with orosomucoid; application of electrochemical techniques to immunoassay methodology; glycoprotein structure and function. *Mailing Add:* Dept Chem Univ Cincinnati Cincinnati OH 45221-0172

HALSEY, BRENTON S, b Newport News, Va, Apr 8, 27; m 54; c 4. CHEMICAL ENGINEERING. *Educ:* Univ Va, BChE, 51; Inst Paper Chem, Lawrence Col, cert. *Prof Exp:* Res engr, Va-Carolina Chem Corp, 53-55; process develop engr, Albemarle Paper Mfg Co, 55-57, asst tech dir, 57-60, dir res & develop, 60-63, vpres planning, 63-70; mem staff, 70-77, CHMN BD, JAMES RIVER PAPER CORP, 77- *Mem:* Am Chem Soc; Tech Asn Pulp & Paper Indust; Am Inst Chem Engrs. *Res:* Pulp and paper; wood chemistry; plastics; heat transfer; fluid flow. *Mailing Add:* 213 Ampthill Rd Richmond VA 23226

HALSEY, GEORGE DAWSON, JR, b Washington, DC, May 28, 25. PHYSICAL CHEMISTRY. *Educ:* Univ SC, BS, 43; Princeton Univ, PhD(chem), 48. *Prof Exp:* Jr fel, Soc Fels, Harvard Univ, 48-51; from asst prof to assoc prof chem, 51-58, PROF CHEM, UNIV WASH, 58- *Honors & Awards:* Kendall Award, Am Chem Soc, 65. *Mem:* Am Chem Soc. *Res:* Adsorption; catalysis; viscous elasticity; statistical mechanics; solutions. *Mailing Add:* 5819 17th Ave NE Seattle WA 98105-2511

HALSEY, JAMES H, JR, b Paris, France, Sept 3, 33; US citizen; c 2. NEUROLOGY. *Educ:* Univ Bridgeport, BA, 55; Yale Univ, MD, 59. *Prof Exp:* From instr to assoc prof neurol, 65-69; assoc prof & dir div, 69-72, chmn dept, 73-85, PROF NEUROL, UNIV ALA MED CTR, BIRMINGHAM, 72- *Concurrent Pos:* Mem, Res Comt, Am Heart Asn, 71-75; mem, Prog Proj Comt, Nat Inst Neurol Dis & Stroke, 73-77. *Mem:* Am Neurol Asn; Int Soc Oxygen Transport Tissue; Am Acad Neurol; AMA; Am Heart Asn; Int Soc Cerebral Blood Flow & Metab. *Res:* Cerebrovascular disease and clinical neurophysiology. *Mailing Add:* Dept Neurol Univ Ala Birmingham AL 35294

HALSEY, JOHN FREDERICK, b St Petersburg, Fla, Feb 8, 42; m 67; c 2. CLINICAL IMMUNOLOGY. *Educ:* Univ Fla, BS, 65, MS, 67; Johns Hopkins Univ, PhD(biochem), 73. *Prof Exp:* Res assoc phys biochem, Univ Va, 73-75; instr molecular pharmacol, Med Univ SC, 75-76; asst prof microbiol, Univ Okla, 76-78; ASST PROF BIOCHEM, MED CTR, UNIV KANS, 78- *Concurrent Pos:* Consult, Int Diag Inc; prin investr, NIH grant. *Mem:* Am Chem Soc; Am Asn Immunologists; Am Soc Microbiol. *Res:* Immunochemistry, immunology and biochemistry. *Mailing Add:* Univ Kans Med Ctr 39th & Rainbow Blvd Kansas City KS 66103

HALSEY, JOHN JOSEPH, b Jersey City, NJ, Dec 26, 18; m 44; c 4. CHEMISTRY, OPERATIONS RESEARCH. *Educ:* St Peters Col, BS, 40; NY Univ, MS, 48, MBA, 54. *Prof Exp:* Prod chemist, Calco Chem Div, Am Cyanamid Corp, 40; anal chemist, Brooklyn Army Base, 41-42; chief, chem prods br, Qm Depot, NJ, 42-44; chief, chem sect, Merck & Co, Inc, 46-47, mgr, qual standards dept, 48-54 & testing & inspection dept, 54-57; mgr, qual servs dept, 57-63, spec proj, 63-64 & opers anal, 64-68; dir, mgt sci dept, Gen Foods Corp, 69-82; RETIRED. *Concurrent Pos:* Mem bd dir, Extrudo Film Corp, NY, 59-62; dir, Neuwirth Fund, Inc, 68- *Mem:* Am Chem Soc; Am Soc Qual Control. *Res:* Analytical chemistry; testing methods; statistics in chemistry; management of technical operations; management sciences; operations research; financial and portfolio management. *Mailing Add:* 9900 S Ocean Dr Suite 405 Jensen Beach FL 34957

HALSTEAD, BRUCE W, b San Francisco, Calif, Mar 28, 20; m 41; c 6. CHEMISTRY. *Educ:* Univ Calif, BA, 43; Loma Linda Univ, MD, 48. *Prof Exp:* Asst surgeon, USPHS, 47-48; instr path & med zoologist, Sch Trop & Prev Med, Loma Linda Univ, 48-58, asst prof prev med & pub health, Sch Med & head dept biotoxicol, Sch Trop & Prev Med, 54-58; DIR, WORLD LIFE INST, 58- *Concurrent Pos:* Mem, Great Barrier Reef Comt, Australia, 68; mem, Adv Coun, Life Bound, 71; consult, NIH, Bur Med & Surg, USN, Marine Colloids, Inc, NAm Aviation, Inc, W J Voit Rubber Co, Dow Chem Co, Tempo, Gen Elec Co, 70; mem, Joint Group Experts Sci Aspects Marine Pollution, WHO, UN, Paris, 70; partic, Fishery Develop Proj, Food & Agr Orgn, UN, Mauritius, 70; mem, Sci Adv Bd, Environ Defense Fund, 71 & Expert Adv Panel Food Additives, WHO, 71; med dir, Rancho Mediterranean Med Clin. *Mem:* AAAS; Marine Technol Soc; Am Inst Chem; Royal Soc Health; fel NY Acad Sci. *Res:* Poisonous and venomous marine animals; natural product chemistry; author or coauthor of numerous publications; preventive medicine. *Mailing Add:* World Life Res Inst 2300 Grand Terr Rd Colton CA 92324

HALSTEAD, CHARLES LEMUEL, b Norfolk, Va, Aug 24, 28; m 52; c 3. ORAL PATHOLOGY, ORAL MEDICINE. *Educ:* Univ Va, BA, 49; Med Col Va, DDS, 54; Emory Univ, MS, 66; Am Bd Oral Path, dipl, 69. *Prof Exp:* Pvt pract gen dent, Norfolk, Va, 57-64; assoc prof oral med & chmn dept, 66-71, PROF ORAL PATH & CHMN DEPT, EMORY UNIV, 71- *Concurrent Pos:* Consult, Ga Dept Human Resources, 70-; NIH grant, 73-; consult, Coun Int Exchange Scholars, 78- *Mem:* Am Acad Oral Path; Int Asn Dent Res; Soc Teachers Oral Path; Am Asn Dent Schs (secy, 74-75). *Res:* Oral cancer, particularly fluorescent antibody studies and chemical carcinogenesis; inorganic bone and dental pulps; ceramic implants; differential diagnosis of soft tissue diseases utilizing computer assistance. *Mailing Add:* 825 Emerson Dr Charlottesville VA 22901

HALSTEAD, RONALD LAWRENCE, b Alta, Can, Sept 18, 23. SOIL FERTILITY, SOIL CHEMISTRY. *Educ:* Univ Man, BSA, 50; Univ Wis-Madison, PhD(soil sci & microbiol), 54. *Prof Exp:* Res scientist, Chem Div, Can Dept Agr, 54-59, Soil Res Inst, 59-75, res coordr, Planning & Eval Directorate, Res Br & Cent Exp Farm, 75-82, dir gen, Prog Coordr Directorate, 82-85, dir gen insts, 85-87; RETIRED. *Mem:* Int Soc Soil Sci; fel Can Soc Soil Sci; Agr Inst Can. *Res:* Soil organic and inorganic phosphorous; sewage sludge disposal in soils. *Mailing Add:* 1094 Bedbrook St Ottawa ON K2C 2R7 Can

HALSTEAD, SCOTT BARKER, b Lucknow, India, Jan 23, 30; m 55; c 3. MEDICAL MICROBIOLOGY, TROPICAL MEDICINE. *Educ:* Yale Univ, BA, 51; Columbia Univ, MD, 55; Am Bd Med Microbiol, dipl, 64; Am Bd Prev Med, dipl, 73. *Hon Degrees:* DrMedS, Mahidol Univ, Bangkok. *Prof*

Exp: Intern, Bellevue Hosp, New York, 55-56, resident internal med, 56-57; chief virol, Dept Virus & Rickettsial Dis, 406th Med Gen Lab, Japan, 57-59; mem, Dept Virus Dis, Walter Reed Army Inst Res, 59-61; chief virol dept, SEATO Med Res Lab, Thailand, 61-65; res assoc virol, Yale Univ, 65-68; prof trop med, med microbiol & chmn dept, John A Burns Sch Med, Univ Hawaii, 68-83; ACTG DIR HEALTH SCI DIV, ROCKEFELLER FOUND. *Concurrent Pos:* Prof pub health, Sch Pub Health, Univ Hawaii; assoc mem, Armed Forces Epidemiol Bd, Comn Virus Dis, 68-73; consult epidemiol & control dengue hemorrhagic fever, WHO, 67-; expert consult, Southeast Asia Regional Off, WHO, 74-83; Fogarty Int Fel, 75-76; mem, Armed Forces Epidemiol Bd, 87-; mem HIV Vaccine Selection Comt, NIH, 89-; mem infectious dis adv bd, Ctr Dis Control, 89- *Honors & Awards:* Langmuir Lectr, Ctr Dis Control, 81. *Mem:* Am Asn Immunologists; Am Epidemiol Soc; Am Soc Trop Med & Hyg; AAAS; Int Dis Soc Am; Int Epidemiol Soc. *Res:* Japanese encephalitis ecology; clinical, epidemiological and virological studies of dengue hemorrhagic fever; epidemiology of primary bladder stone disease; development and evaluation of live virus vaccines; technology transfer; science-based development. *Mailing Add:* Health Sci Div Rockefeller Found 1133 Ave Americas New York NY 10036

HALSTEAD, THORA WATERS, b Chicago, Ill; c 2. GRAVITATIONAL & SPACE BIOLOGY, CELL BIOLOGY. *Educ:* Wash State Univ, BS, 50; Univ Tex, Austin, MS, 55; Univ Md, PhD(microbiol), 68. *Prof Exp:* Chair, Biol Dept, Okaloosa Walton Jr Col, Fla, 64-65; sr scientist sci anal, Army Inst Advan Study, 68-71; instr biol, Sinclair Community Col, 71-73; sr scientist sci anal, George Washington Univ, DC, 73-74; MGR SPACE BIOL, LIFE SCI DIV, NASA, 74- *Concurrent Pos:* Lectr, Univ Col WI, Jamaica, 52; ed, Am Soc Gravitational & Space Biol Publ, 88- *Honors & Awards:* Outstanding Serv Award, Am Soc Gravitational & Space Biol, 87. *Mem:* Am Soc Gravitational & Space Biol (vpres, 86, pres, 87); Am Soc Plant Physiol; Am Physiol Soc; Aerospace Med Asn. *Res:* Formulation of how life on earth has adapted to gravity, how gravity effects its function, and how microgravity can be used to explore biological mechanisms. *Mailing Add:* NASA Hq Washington DC 20546

HALSTED, A(BEL) STEVENS, b Pasadena, Calif, Jan 18, 38; m 59; c 2. ELECTRICAL ENGINEERING, APPLIED PHYSICS. *Educ:* Stanford Univ, BS, 59, MS, 60, PhD(elec eng), 65. *Prof Exp:* Res asst plasma physics, Electronics Lab, Stanford Univ, 62-65; sr mem tech staff gas lasers & head ion laser sect, Hughes Res Labs, 65-70, mgr, Laser Dept & Laser Prod Line, Electron Dynamics Div, Hughes Aircraft Co, 70-75; prog mgr, Electra-Optical & Data Systs Group, Huges Aircraft Co, 75-87; PROG MGR, SANTA BARBARA RES CTR, 87- *Concurrent Pos:* Mem comt, Int Electron Devices Conf, 67-71; mem tech prog comts, Int Electron Devices Conf, 67-70; mem, Electron Device Res Conf, 70 & Conf Laser Eng & Appln, 71. *Mem:* Inst Elec & Electronics Engrs; Sigma Xi. *Res:* Electron devices; microwave tubes and devices; plasma physics; quantum electronics; lasers, infrared systems. *Mailing Add:* Santa Barbara Res Ctr Subsid & Hughes Aircraft 75 Coromar Dr Bldg B25 MS71 Goleta CA 93117

HALSTED, CHARLES H, b Cambridge, Mass, Oct 2, 36. CLINICAL NUTRITION. *Educ:* Univ Rochester, MD, 62. *Prof Exp:* PROF INTERNAL MED, UNIV CALIF, DAVIS. *Mem:* Am Soc Clin Invest; Am Soc Clin Nutrit; Am Gastroesterol Asn. *Mailing Add:* Div Clin Nutrit Univ Calif Sch Med TB156 Davis CA 95616

HALTER, JEFFREY BRIAN, b Minneapolis, Minn, Aug 25, 45; c 2. ENDOCRINOLOGY, METABOLISM. *Educ:* Univ Minn, BA, 66, BS & MD, 69. *Prof Exp:* Res asst metab, Univ Minn Sch Med, 64-68; physician & surgeon internal med, USPHS Outpatient Clin, Washington, 71-73; staff physician, Seattle Vet Admin Med Ctr, 74-75, advan spec resident endocrinol & metab, 75-77; acting instr, Sch Med, Univ Wash, 74-77, from asst prof to assoc prof med, 77-84; assoc dir geriat res, Educ & Clin Ctr, Seattle Vet Admin Med Ctr, 78-84; CHIEF, INST GERIAT, UNIV MICH MED CTR, 84- *Concurrent Pos:* Res & educ assoc, Seattle Vet Admin Med Ctr, 77-78; NIH grants, 77- *Mem:* Am Diabetes Asn; Am Fedn Clin Res. *Res:* Autonomic nervous system function and insulin secretion in diabetes mellitus; neuroendocrine responses in aging. *Mailing Add:* 1010 Walls St Box 004 Rm 1508 Ann Arbor MI 48109

HALTERLEIN, ANTHONY J, US citizen. HORTICULTURE. *Educ:* Univ Mo, BS, 72, MS, 74; Kans State Univ, PhD(hort), 78. *Prof Exp:* Asst res horticulturist, Delta Br, Miss Agr & Forestry Exp Sta, 78-81; ASST PROF, MIDDLE TENN STATE UNIV, 81- *Mem:* Am Soc Hort Sci. *Res:* Cultural practices. *Mailing Add:* Dept Agr Box 5 Middle Tenn State Univ Murfreesboro TN 37132

HALTERMAN, JERRY J, b Parowan, Utah, May 7, 22; m 44; c 4. AGRICULTURAL ENGINEERING. *Educ:* Univ Calif, Davis, BS, 50, MEd, 55; Ohio State Univ, PhD(agr educ & eng), 64. *Prof Exp:* Instr agr, Red Bluff Union High Sch, 51-55 & Modesto Jr Col, 55-65; res asst agr educ, Ohio State Univ, 63-65; coordr agr, Chico State Col, 65-68; prof agr educ, Ohio State Univ, 68-76, prof eng educ, 76-; AT DEPT AGR, RICKS COL. *Concurrent Pos:* Consult, Ctr Voc & Tech Educ, Columbus, Ohio, 65 & 67 & Off Educ, Washington, DC, 66; mem, Conf Undergrad Educ Biol Sci Students in Agr & Natural Sci, 66; dir, Agr Tech Inst, 68- *Res:* Educational needs of vocational, technical and professional workers in the agricultural manpower force. *Mailing Add:* Dept Agr Ricks Col Rexburg ID 83440

HALTINER, GEORGE JOSEPH, b St Paul, Minn, Nov 26, 18; m 47; c 5. METEOROLOGY. *Educ:* Col St Thomas, BS, 40; Univ Wis, PhM, 42, PhD(math), 48. *Prof Exp:* Asst math, Univ Wis, 40-42; from asst prof to prof aerological eng, US Naval Postgrad Sch, 46-69, chmn dept meteorol & oceanog, 64-68, actg dean sci & eng, 77-78, distinguished prof meteorol, 69-81, chmn dept meteorol, 68-81; RETIRED. *Concurrent Pos:* Mem tech adv comt, Air Pollution Control Dist, Monterey & Santa Cruz counties. *Mem:* Fel Am Meteorol Soc; Royal Meteorol Soc. *Res:* Dynamical meteorology and numerical weather prediction. *Mailing Add:* 1134 Alta Mesa Rd Monterey CA 93940

HALTIWANGER, JOHN D(AVID), b Irmo, SC, June 10, 25; m 54; c 2. CIVIL ENGINEERING. *Educ:* Univ SC, BS, 45; Univ Ill, MS, 49, PhD(civil eng), 57. *Prof Exp:* Instr civil eng, Ala Polytech Inst, 46-48, asst prof, 49-51; from asst to assoc prof, 48-59, PROF CIVIL ENG, UNIV ILL, URBANA, 59-, ASSOC HEAD DEPT, 67-84. *Concurrent Pos:* Consult, Strategic Structures Div, Targets Div, Defense Nuclear Agency, US Dept Defense; mem adv comt, US Off Civil Defense, 66-70; mem, Ill State Tech Serv Adv Coun, 67-70 & USCG Acad Adv Comt, 80-83. *Honors & Awards:* Bliss Medal, Soc Am Military Engrs, 80. *Mem:* Am Soc Civil Engrs; Am Soc Eng Educ; Am Concrete Inst; Sigma Xi. *Res:* Structural analysis and design; design of structures for blast-induced loads. *Mailing Add:* 3129 Newmark Lab Univ Ill 205 N Mathews St Urbana IL 61801

HALTNER, ARTHUR JOHN, b Milwaukee, Wis, Aug 28, 27. PHYSICAL CHEMISTRY. *Educ:* Univ Wis, BS, 51; Univ Calif, PhD(phys chem), 55. *Prof Exp:* Asst phys chem, Univ Wis, 51; assoc chem, Univ Calif, 51-53; phys chemist, Res Lab, 55-64, surface physicist, Space Sci Lab, 64-69, CONSULT SURFACE PHYSICIST, ASTRO SPACE DIV, GEN ELEC CO, 69- *Mem:* Am Chem Soc; sr mem Am Vacuum Soc; fel Soc Tribologists & Lubrication Engrs. *Res:* Friction, wear and adhesion of solid surfaces in space environment; ultrahigh vacuum. *Mailing Add:* 251 W DeKalb Pike A 909 Valley View Apts King of Prussia PA 19406

HALTON, JOHN HENRY, b Brussels, Belg, Aug 25, 31; m 77; c 2. MATHEMATICS, COMPUTER SCIENCE. *Educ:* Cambridge Univ, BA, 53, MA, 57; Oxford Univ, BA, 56, MA, 57, DPhil(math), 60. *Prof Exp:* Physicist, Mech Eng Lab, Eng Elec Co, 54-56; res officer, Clarendon Lab, Oxford Univ, 60-62; vis lectr math & comput sci, Univ Colo, 62-64; mathematician, Brookhaven Nat Lab, 64-66; from assoc prof to prof comput sci, Univ Wis-Madison, 66-83; prin engr, Advan Technol Dept, Harris Corp, Melbourne, 83-84; PRES, TEDCO, 80-; PROF COMPUT SCI, UNIV NC, CHAPEL HILL, 84- *Concurrent Pos:* Consult, Rand Corp, 63, C-E-I-R Corp, 66, Lawrence Livermore Nat Lab, 83 & Los Alamos Nat Lab, 85; fel, Cambridge Philos Soc(math & physics), 54. *Mem:* Am Math Soc; Asn Comput Mach; Math Asn Am; Soc Indust & Appl Math; Am Phys Soc; Sigma Xi; Sr mem Inst Elect & Electronics Engrs; fel Inst Math & Applns; fel Brit Comput Soc. *Res:* Monte Carlo method; mathematical analysis; probability theory; combinatorial theory; theory of algorithms, computer networks and very large scale integration. *Mailing Add:* 108 Carolina Forest Chapel Hill NC 27514

HALUSHKA, PERRY VICTOR, b Chicago, Ill, June 4, 41; m 64; c 3. PHARMACOLOGY. *Educ:* Univ Ill, Chicago, BS, 63, PhD(pharmacol), 67, MD, 70. *Prof Exp:* Teaching asst pharmacol, Univ Chicago, 67-69; from intern to jr asst resident, Grady Med Hosp, Atlanta, Ga, 70-72; res assoc pharmacol & toxicol, Nat Heart & Lung Inst, NIH, 72-74; from asst prof to assoc prof, 74-81, PROF PHARM & MED, MED UNIV SC, 81- *Concurrent Pos:* Burroughs-Wellcome scholar clin pharmacol, 84. *Honors & Awards:* Found Fac Develop Award, Pharmaceut Mfg Asn, 75. *Mem:* Am Soc Pharmacol & Exp Therapeut; Am Soc Clin Pharmacol & Therapeut; Am Fedn Clin Res; Am Soc Clin Invest. *Res:* Studies on the role of prostaglandins in cardiovascular and renal disease. *Mailing Add:* Dept Pharmacol Med Univ SC Charleston SC 29425

HALVER, JOHN EMIL, b Woodinville, Wash, Apr 21, 22; m 44; c 5. NUTRITION. *Educ:* Wash State Univ, BSc, 44, MSc, 49; Univ Wash, PhD(biochem), 53. *Prof Exp:* Plant chemist, Asn Frozen Foods, Inc, Wash, 46; asst chem, Wash State Univ, 46-47; asst chemist, State Chemists Off, Purdue, 48-49; instr, Sch Fisheries, 49-50, dir western fish nutrit lab, Bur Sport Fisheries & Wildlife, 50-75, US fish & wildlife prof, 75-78, SR FISHERIES, UNIV WASH, 78- *Concurrent Pos:* Pres, Fisheries Develop Technol, Seattle, Wash. *Mem:* Nat Acad Sci; fel Am Inst Fisheries Res Biologists; fel Am Inst Nutrit; Am Fisheries Soc; Am Chem Soc. *Res:* Fundamental nutrition, metabolism and comparative biochemistry of fish and other experimental animals; basic nutritional requirements and specific functions of each in metabolism. *Mailing Add:* Sch Fisheries HF-15 Univ Wash Seattle WA 98195

HALVERSON, ANDREW WAYNE, b Bancroft, SDak, Aug 17, 20; m 56; c 3. AGRICULTURAL CHEMISTRY. *Educ:* SDak State Univ, BS, 43; Univ Wis, MS, 47, PhD(biochem), 49. *Prof Exp:* Prof sta biochem, SDak State Univ, 49-85; RETIRED. *Concurrent Pos:* Fel, Johns Hopkins Univ, 58-59. *Mem:* Am Chem Soc; Am Inst Nutrit; Sigma Xi. *Res:* Selenium toxicity in the albino rat; carotene stability in feeds. *Mailing Add:* 442 Harvey Dunn St Brookings SD 57006

HALVERSON, FREDERICK, b West Prairie, Wis, Sept 8, 17; m 46. CHEMISTRY. *Educ:* Luther Col, BA, 39; Johns Hopkins Univ, PhD(phys chem), 43. *Prof Exp:* Asst chem, Johns Hopkins Univ, 40-43, instr, 43-46, res chemist, USN Proj, 44; res chemist, Nat Defense Res Comt & US Army Proj, 45; res physicist, Am Cynamid Co, 46-51, theoret chemist, 51-53, group leader phys chem, 53-54, mgr basic res sect, 54-62, res fel, 62-68, sr res fel, 68-82; RETIRED. *Concurrent Pos:* Am Cyanamid Co sr res award, 57-58; vis scientist, Cambridge Univ, 57-58. *Mem:* Am Phys Soc; Faraday Soc; Am Chem Soc; Am Optical Soc; Sigma Xi. *Res:* Electronic energy transfer; structural inorganic chemistry; raman, ultraviolet and infrared spectroscopy; use of deuterium in analysis of molecular spectra, flocculation, water treating, enhanced oil recovery. *Mailing Add:* 206 Janes Lane Stamford CT 06903

HALVORSON, ARDELL DAVID, b Rugby, NDak, May 31, 45; m 66; c 2. SOIL SCIENCE, CHEMISTRY. *Educ:* NDak State Univ, BS, 67; Colo State Univ, MS, 69, PhD(soil sci), 71. *Prof Exp:* SOIL SCIENTIST, AGR RES SERV, USDA, 71- *Concurrent Pos:* Res leader, Agr Res Serv, USDA, 88- *Honors & Awards:* USDA Cert of Merit, 90. *Mem:* Sigma Xi; Am Soc Sugar Beet Technol; fel Am Soc Agron; fel Soil Sci Soc Am; Soil Conserv Soc Am; Crop Sci Soc Am. *Res:* Soil chemistry, reclamation of saline land under semiarid conditions; soil fertility and plant nutrition; nitrogen, phosphorous, potassium and trace elements for sugarbeets, corn, beans and small grains; dryland cropping systems. *Mailing Add:* Agr Res Serv USDA PO Box 400 Akron CO 80720

HALVORSON, HARLYN ODELL, b Minneapolis, Minn, May 17, 25; m 54; c 2. MICROBIOLOGY. *Educ:* Univ Minn, BS, 48, MS, 50; Univ Ill, PhD(bact), 52. *Prof Exp:* Instr bact, Univ Mich, 52-54, asst prof, 54-56; Merck sr fel, Pasteur Inst, Paris, 55-56; from assoc prof to prof bact, 56-71, Univ Wis-Madison, prof molecular biol, 64-71, actg dir molecular biol lab, 64-66, chmn, 66-71; PROF BIOL & DIR ROSENSTIEL BASIC MED SCI RES CTR, BRANDEIS UNIV, 71- *Concurrent Pos:* Mem NSF adv panel, Qm Corps, 58-64; fel panel biochem & nutrit, NIH, 61-64, mycol & microbiol, 64-; instr marine biol lab, Woods Hole Biol Inst, 62-65 & 67, investr, 68-70, trustee & mem exec comt; instr, Hebrew Univ, 65, Univ Naples, 66 & Univ Bergen, 68; mem Nat Acad Sci Res Adv Comt to USDA; consult, NASA Comt Mariner Sterilization. *Mem:* Am Soc Microbiol (vpres, 75-76, pres, 76-77); Am Chem Soc; Am Soc Biol Chemists; Am Acad Microbiol; Am Acad Arts & Sci. *Res:* Induced enzyme synthesis; sporulation, physiology and germination of aerobic spores; cell cycle in yeast, nucleic acid synthesis, genetics and physiology of bacterial spores; regulation of gene expression at the molecular level. *Mailing Add:* 26 Fay Rd Box 81 Woods Hole MA 02543

HALVORSON, HERBERT RUSSELL, b Fergus Falls, Minn, Aug 18, 40; m 68; c 1. BIOPHYSICAL CHEMISTRY. *Educ:* Univ Minn, BS, 63; Univ Va, PhD(biophys), 71. *Prof Exp:* Mgr tech commun, Medtronics, Inc, 66-68; NIH fel chem, Univ Mass, 71-73; res assoc biochem, Univ Va, 73-74; STAFF INVESTR BIOCHEM, EDSEL B FORD INST MED RES, 75- *Concurrent Pos:* NIH study sect biophys & biophys chem, 77-; adj fac biol, Wayne State Univ, 78- *Mem:* Sigma Xi; NY Acad Sci; Fedn Am Scientists; Biophys Soc; AAAS; Am Chem Soc. *Res:* Macromolecular interactions as typified by oligomeric proteins, particularly ligand-mediated subunit association; application of physical techniques to the study of biochemical systems. *Mailing Add:* 4661 Devonshire Rd Detroit MI 48224

HALVORSON, LLOYD CHESTER, b Swift County, Minn, Feb 7, 18; m 42; c 4. AGRICULTURAL ECONOMICS. *Educ:* Univ Minn, BS, 39, PhD(agr econ), 43. *Prof Exp:* Economist credit, Farm Credit Admin, 42-45; economist agr policy, Nat Grange, 45-57; PRIN AGR ECONOMIST, COOP STATE RES SERV, USDA, 57- *Mem:* Am Agr Econ Asn. *Res:* Marketing and production economics. *Mailing Add:* 931 Douglass Dr McLean VA 22101

HALZEN, FRANCIS, b Tienen, Belg, Mar 23, 44; m 68; c 1. THEORETICAL PHYSICS. *Educ:* Cath Univ Louvain, MS, 66, PhD(physics), 69. *Prof Exp:* Fel physics, Europ Ctr Nuclear Res, 69-71; MEM FAC PHYSICS, UNIV WIS-MADISON, 72- *Concurrent Pos:* Consult, Argonne Nat Lab, 73, Brookhaven Nat Lab, 75, Rutherford Lab, 77 & Europ Orgn for Nuclear Res, 77. *Res:* High energy physics (theory). *Mailing Add:* Dept Physics Univ Wis-Madison 1150 University Ave Madison WI 53706

HAM, ARTHUR WORTH, b Brantford, Ont, Feb 20, 02; m 25; c 1. HISTOLOGY. *Educ:* Univ Toronto, MB, 26. *Hon Degrees:* DSc, Univ Western Ont, 65. *Prof Exp:* Asst cytol, Wash Univ, 29-31, instr path, 31; assoc prof anat & chmn dept, 32-66, head dept med biophys, 58-61, PROF ANAT, FAC MED, UNIV TORONTO, 66- *Concurrent Pos:* Hon secy, Banting Res Found, 38-69; head div biol res, Ont Cancer Inst, 57-62. *Honors & Awards:* Starr Gold Medal, Toronto, 34. *Mem:* Am Asn Anat; fel Royal Soc Can. *Res:* Histology of bone and bone repair; cancer. *Mailing Add:* 25 Hawkridge Ave Markham ON L3P 1V8 Can

HAM, FRANK SLAGLE, b Bronxville, NY, Aug 15, 28; m 60; c 2. PHYSICS. *Educ:* Harvard Univ, AB, 50, AM, 51, PhD(physics), 55. *Prof Exp:* Nat Res Coun fel physics, Univ Ill, 54-55; RES ASSOC, CORP RES & DEVELOP CTR, GEN ELEC CO, 55- *Concurrent Pos:* Guggenheim fel theoret studies solid state physics, Oxford Univ, 71-72; Alumni vis prof physics, Clemson Univ, 76-77. *Mem:* Am Phys Soc; Fedn Am Scientists; Sigma Xi. *Res:* Solid state physics; theory. *Mailing Add:* 1445 Valencia Rd Schenectady NY 12309

HAM, GEORGE EDWARD, b Jacksboro, Tex, May 19, 31; m 59; c 3. ORGANIC CHEMISTRY. *Educ:* Baylor Univ, BS, 51; Purdue Univ, MS, 53, PhD(org chem), 56. *Prof Exp:* Res chemist, Tex Org Res Dept, 55-58, sr res chemist, 58-62, res specialist, 62-67, res specialist, Tex Amines Res Dept, 67-71, ASSOC SCIENTIST, TEX AMINES RES DEPT, DOW CHEM CO, 71- *Mem:* Am Chem Soc. *Res:* Organic synthesis; physical organic chemistry. *Mailing Add:* Box 373 Lake Jackson TX 77566

HAM, GEORGE ELDON, b Ft Dodge, Iowa, May 22, 39; m 64; c 3. SOIL MICROBIOLOGY, AGRONOMY. *Educ:* Iowa State Univ, BS, 61, MS, 63, PhD(soil microbiol), 67. *Prof Exp:* Res assoc soil microbiol, Iowa State Univ, 65-67; from asst prof to prof soil microbiol, Univ Minn, St Paul, 67-80; prof agron & head dept, 80-89, ASSOC DEAN & ASSOC DIR, AGR EXP STA, KANS STATE UNIV, MANHATTAN, 89- *Mem:* Fel AAAS; fel Soil Sci Soc Am; fel Am Soc Agron. *Res:* Ecology and significance of Rhizobium japonicum and its impact on the nitrogen nutrition of the soybean plant. *Mailing Add:* Agr Exp Sta Waters Hall Kans State Univ Manhattan KS 66506-4008

HAM, INYONG, b Korea, Dec 22, 25; US citizen; m 49; c 2. MECHANICAL & INDUSTRIAL ENGINEERING. *Educ:* Seoul Nat Univ, BEng, 48; Univ Nebr, MSc, 56; Univ Wis, PhD(mech eng), 58. *Prof Exp:* Instr mech eng, Seoul Nat Univ, 48-50; asst, Univ Nebr, 54-56; res asst & instr, Univ Wis, 56-58; asst prof indust eng, Pa State Univ, 58-59; dir indust & asst minister, Repub S Korea Ministry Com & Indust, 60-62; from asst prof to assoc prof, 63-69, PROF INDUST ENG, PA STATE UNIV, 69-, FANUC PROF, 89-, DIR MFG RES CTR, 90-, DISTINGUISHED PROF, 91- *Concurrent Pos:* Adv, Korea Inst Sci & Technol, 73-83; consult prof, Xian Jiaotong Univ, China, 81 & Beijing Inst Technol, 84; Fulbright prof, Ga Polytech Inst, USSR, 81; chair vis prof, Univ Tokyo, 89. *Honors & Awards:* CAM-I Award, Comput Aided Mfg-Int, 78; Sargent Progress Award, 90. *Mem:* Fel Am Soc Mech Engrs; fel Soc Mfg Engrs; Int Inst Prod Eng Res; fel Inst Indust Engrs; Korean Sci & Eng Asn Am (pres, 74-75); NAm Mfg Res Inst (pres, 85-86). *Res:* Analysis of manufacturing problems; metal cutting theory and experiments; machinability evaluation; optimization of manufacturing conditions; design of cutting tools, jigs, and fixtures; group technology application; computer integrated manufacturing; manufacturing systems engineering. *Mailing Add:* Dept Indust & Mgt Systs Eng 207 Hammond Bldg Pa State Univ University Park PA 16802

HAM, JAMES M(ILTON), b Coboconk, Ont, Sept 21, 20; m 55; c 3. ELECTRICAL ENGINEERING. *Educ:* Univ Toronto, BASc, 43; Mass Inst Technol, SM, 47, ScD(elec eng), 52. *Prof Exp:* Asst prof, Mass Inst Technol, 51-52; assoc prof, 52-59, head dept, 64- dean fac eng, 66-73, chmn res bd, 74-76, dean, Grad Sch, 76-78, PROF ELEC ENG, UNIV TORONTO, 59-, PRES, 78- *Concurrent Pos:* Vis scientist, Cambridge Univ, 60-61; mem, Nat Res Coun Can; bd gov, Ont Res Found; chmn comt educ, World Fedn Eng Orgn. *Honors & Awards:* Centennial Medal of Can. *Mem:* Fel Inst Elec & Electronics Engrs; fel Eng Inst Can; Sigma Xi. *Res:* Feedback control systems; computer applications in industrial process control. *Mailing Add:* Dept Elec Eng Rosebrugh Bldg Rm 210d Univ of Toronto 35 St George St Toronto ON M5S 1A4 Can

HAM, JOE STROTHER, b Okmulgee, Okla, Mar 12, 28; m 52; c 2. PHYSICS, CHEMISTRY. *Educ:* Univ Chicago, PhB, 48, MS, 51, PhD(physics), 54. *Prof Exp:* Res chemist, Jackson Lab, E I du Pont de Nemours & Co, 53-56; from asst prof to prof physics, 56-63, prof chem, 67-78, PROF PHYSICS, TEX A&M UNIV, 63- *Concurrent Pos:* Res physicist, Ford Motor Co, 61-62; vis scientist, TNO, Delft, Neth, 71; consult, Shell Develop Co, 78-80; vis scientist, Air Force Mats Labs, 80-81. *Mem:* Am Phys Soc; Am Asn Physics Teachers; Sigma Xi. *Res:* Molecular spectra; thermosetting resins; polymer solutions; composite materials. *Mailing Add:* Dept Physics Tex A&M Univ College Station TX 77843

HAM, LEE EDWARD, b San Francisco, Calif, Dec 19, 19; m 86; c 4. URBAN PLANNING, NEW TOWN DESIGN. *Educ:* Univ Calif, Berkeley, BS, 42. *Prof Exp:* PRES, WILSEY & HAM CONSULT ENG, 45- *Concurrent Pos:* Prin in charge, Design Foster City, Calif, 60-, Design of North Star, Tahoe, Calif, 74-80 & Design of Las Positas New Town, 80-84; distinguished engr, Am Pub Works Asn, 74. *Mem:* Fel Am Soc Civil Engrs; Am Pub Works Asn; fel Am Consult Engrs Coun; Urban Land Inst. *Res:* Development of new towns, determining optimum location and how to finance the infrastructure. *Mailing Add:* 355 Lakeside Dr No 200 Foster City CA 94404

HAM, RICHARD GEORGE, b Tacoma, Wash, Feb 10, 32; m 53; c 4. CELL BIOLOGY, CELL GROWTH REQUIREMENTS. *Educ:* Calif Inst Technol, BS, 53; Univ Tex, PhD(biochem), 57. *Prof Exp:* Res scientist biochem, Biochem Inst, Univ Tex, 57-58; instr biophys, Med Ctr, Univ Colo, Denver, 58-60, asst prof, 60-65; res assoc, Inst Develop Biol, 65-66, from asst prof to assoc prof, 66-76, PROF MOLECULAR, CELLULAR & DEVELOP BIOL, UNIV COLO, BOULDER, 76- *Mem:* Am Soc Cell Biol; Soc Develop Biol; Tissue Cult Asn. *Res:* Growth requirements of cultured mammalian and avian cells; mechanisms of cellular differentiation; aging at the cellular level; effects of malignancy on cellular growth requirements. *Mailing Add:* Dept Molecular Cell & Develop Biol Univ Colo Campus Box 347 Boulder CO 80309

HAM, RICHARD JOHN, b Chester, Eng, June 14, 46; m 67; c 3. GERIATRICS, FAMILY PRACTICE. *Educ:* Univ London, BS & MB, 69; Ray Col Obstetricians & Gynecologists, DObs RCOG, 72; Royal Col Gen Practitioners, MRCGP, 74; Med Col Can, LMCC, 85. *Prof Exp:* Asst prof geriat & family med & chief, Div Geriat, Dept Family Practice, Sch Med, Southern Ill Univ, 77-82; assoc prof geriat, Dept Family Med & Suncoast Geront Ctr, Univ SFla, 82-84; Mt Pleasant Legion prof community geriat & dir, Div Community Geriat, Dept Family Pract, Univ BC, 84-86; DISTINGUISHED CHAIR, GERIAT MED, HEALTH SCI CTR, STATE UNIV NY, 87- *Concurrent Pos:* Ed, Geriat Med Ann, 83-; dir, Short Term Assessment & Treatment Ctr, Vancouver Gen Hosp, 84- *Mem:* Fel Am Geriat Soc (secy, 84-85, vpres, 85-86); Am Acad Family Physicians; Soc Teachers Family Med; fel Geront Soc Am; Can Asn Geront; Can Soc Geriat Med. *Mailing Add:* Health Sci Ctr State Univ NY 750 E Adams Syracuse NY 13210

HAM, RUSSELL ALLEN, b Belvidere, Ill, March 18, 40; m 65. GEOPRESSURED GEOTHERMAL ENERGY. *Educ:* Northern Ill Univ, BS, 66; Univ Iowa, MS, 68, PhD(chem), 70. *Prof Exp:* PROF CHEM, MCNEESE STATE UNIV, 70- *Mem:* Am Chem Soc; Soc Col Sci Teachers. *Res:* Composition of gas and fluids obtained from geopressure geothermal resevoirs in Louisiana and Texas to determine potential for energy production. *Mailing Add:* Chem Dept McNeese State Univ Lake Charles LA 70609

HAM, WILLIAM TAYLOR, JR, b Norfolk, Va, Sept 20, 08; m 40; c 2. PHYSICS, PHOTOBIOLOGY. *Educ:* Univ Va, BSE, 31, MS, 33, PhD(physics), 35. *Prof Exp:* Instr physics, Columbia Univ, 36-37; res physicist, Kendall Mills, NC, 37-38; pres, W T Ham & Co, Inc, Va, 38-40; res assoc, Univ Va, 40-43; head div physics, Inst Textile Technol, 46-48; from assoc prof to prof biophysics, 48-76, head dept, 53-76, EMER PROF BIOPHYSICS, VA COMMONWEALTH UNIV, 76- *Concurrent Pos:* Mem, Radiation Cataract Comt, Nat Acad Sci-Nat Res Coun, 54-57, Atomic Bomb Casualty Comt, 56-58; NSF panel on fels, 55-58; mem, Electromagnetic Radiation Mgt Adv Coun, exec off pres, 68-75; Nat Coun Radiation Protection & Measurements; chmn, AEC spec fel bd, 66-67, chmn, Gordon Conf Lasers Med & Biol, 70. *Mem:* Fel Am Phys Soc; Biophys Soc; Health Phys Soc (pres, 63-64); Asn Res Vision Ophthal; fel AAAS; Am Soc Photobiol; Bioelectromagnetic Soc; Optical Soc Am. *Res:* Nuclear physics; separation uranium isotopes; radiation dosimetry; radiation cataract; radiobiology; health physics; thermal injury; biological effects of lasers; ultraviolet, visible and infrared radiation effects on the eye. *Mailing Add:* 8653 Cherokee Rd Richmond VA 23235-1515

HAMA, FRANCIS R(YOSUKE), b Tokyo, Japan, Dec 6, 17; nat US; c 2. AERODYNAMICS. *Educ:* Tokyo Imp Univ, ME, 40, ScD(aerodyn), 52. *Prof Exp:* Asst prof, Tokyo Imp Univ, 43-50; asst, Johns Hopkins Univ, 50-52; res engr, Inst Hydraul Res, Univ Iowa, 52-54; from asst res prof to res prof, Inst Fluid Dynamics & Appl Math, Univ Md, 54-63; scientist specialist, Jet Propulsion Lab, Calif Inst Technol, 63-68; sr res scientist & lectr, Mech & Aerospace Eng Dept, Princeton Univ, 68-81. *Concurrent Pos:* Vis prof, Princeton Univ & Stuttgart Univ, 81- *Mem:* Am Phys Soc; Am Inst Aeronaut & Astronaut; Ger Asn Appl Math & Mech; Ger Soc Air & Space Travel. *Res:* Supersonic aerodynamics; boundary layer. *Mailing Add:* Wilramster 17 Meunchen 80 8000 Germany

HAMACHER, HORST W, b Buir, WGer, Apr 21, 51; m 72; c 4. COMBINATORIAL OPTIMIZATION. *Educ:* Univ Cologne, Dipl, 77, PhD(math), 80. *Prof Exp:* Asst prof, dept math, Univ Cologne, 77-81; asst prof, 81-84, ASSOC PROF MATH, UNIV FLA, GAINESVILLE, 84-, DIR ADMIN & RES, CTR OPTIMIZATION & COMBINATORICS, 86- *Concurrent Pos:* Prin investr, res grants combinatorial optimization, NSF, 83-85 & 85-87, Off Naval Res, 85 & NATO, 85-89; guest ed, Discrete Appl Math, 85-; assoc ed, Latin-Ibero-Am J Opers Res, 85-, Europ J Opers Res. *Mem:* Math Prog Soc; Opers Res Soc Am; Soc Indust & Appl Math. *Res:* Modelling real world problems using combinatorial optimization techniques; developing efficient algorithms for solving these models. *Mailing Add:* Dept Indust & Sys Eng Univ Kaiserslautern (D-6750) Kaiserslautern Germany

HAMACHER, V(INCENT) CARL, b London, Ont, Sept 28, 39; m 65; c 2. ELECTRICAL ENGINEERING, COMPUTER SCIENCE. *Educ:* Univ Waterloo, BASc, 63; Queen's Univ, Ont, MSc, 65; Syracuse Univ, PhD(elec eng), 68. *Prof Exp:* Asst prof elec eng, Univ Toronto, 68-72, assoc prof elec eng & comput sci, 72-82, prof elec eng & computer sci, 82-90; DEAN, FAC APPL SCI, QUEEN'S UNIV, 91- *Concurrent Pos:* Nat Res Coun Can res grant, 68- *Mem:* Inst Elec & Electronics Engrs; Asn Comput Mach. *Res:* Computer organization; computer commun networks; real time systems. *Mailing Add:* Fac Appl Sci Queen's Univ Kingston ON K7L 3N6 Can

HAMADA, HAROLD SEICHI, b Honolulu, Hawaii, Nov 1, 35; m 58; c 2. CIVIL ENGINEERING, APPLIED MECHANICS. *Educ:* Univ Hawaii, BS, 57; Univ Ill, PhD, 62. *Prof Exp:* Proj off, Air Force Weapons Lab, Kirtland AFB, NMex, 62-65; engr, Theoret Physics Div, Lawrence Radiation Lab, Calif, 65-67; assoc prof, 67-76, PROF CIVIL ENG, UNIV HAWAII, 76- *Mem:* Am Soc Civil Engrs; Am Concrete Inst. *Res:* Numerical techniques for the solution of fluid or solid mechanics problems, creep and shrinkage of concrete. *Mailing Add:* Dept Civil Eng Univ Hawaii at Manoa Honolulu HI 96822

HAMADA, MOKHTAR M, b Belbase, Egypt, Sept 21, 35; m 60; c 2. CHEMICAL ENGINEERING. *Educ:* Univ Alexandria, BSc, 57; Colo Sch Mines, MSc, 63, DSc(chem eng), 65. *Prof Exp:* Oper engr, Suez Oil Processing Co, Egypt, 57-60; sr develop engr, Monsanto Textile Div, Fla, 65-66; sr engr, Suez Oil Processing Co, Egypt, 66-67; sr process engr, Bechtel Eng Corp, Gt Brit, 68-60; sr engr, Monsanto Textile Div, Fla, 69-75, gen engr, Corp Eng Dept, 75-81, PRIN ENGR, MONSANTO CO, ST LOUIS, 81- *Mem:* Am Chem Soc; Am Inst Chem Engrs. *Res:* Chemical process development and engineering; catalysis and chemical kinetics; measurement and control. *Mailing Add:* Petrokemya 8th Floor Nikko Shinkawa Bdg 2 26 3 Tokyo 104 Shinkawa Chuo Ku Japan

HAMADA, SPENCER HIROSHI, b Honolulu, Hawaii, Mar 31, 43. DEVELOPMENTAL BIOLOGY, CELL BIOLOGY. *Educ:* Univ Hawaii, BA, 68; Ore State Univ, MS, 72, PhD(zool), 75. *Prof Exp:* Instr zool, Ore State Univ, 73; lectr biol, Calif Polytech State Univ, 74; asst prof, Fordham Univ, 76; asst prof, 78-81, ASSOC PROF BIOL, WGA COL, 81- *Concurrent Pos:* Grant, Fordham Univ, 76; Ga fac grant, 78-79; Dept Energy grant; res scientist, Med Univ SC, 82. *Mem:* AAAS; Am Inst Biol Sci; Soc Develop Biol; Tissue Cult Asn; Electron Micros Soc. *Res:* Development of neurons and glial cells in culture; self-reassembly of embryos in culture. *Mailing Add:* Dept Biol WGa Col Carrollton GA 30118

HAMAI, JAMES Y, b Los Angeles, Calif, Oct 14, 26; m 54; c 1. MANUFACTURING REPRESENTATIVE. *Educ:* Univ Southern Calif, BS, 52, MS, 55. *Prof Exp:* From process engr to sr process engr, Fluor Corp Ltd, 54-64: sr proj mgr, Cent Res Dept, Monsanto Co, 64-67, mgr res, develop & eng, Graphic Systs Dept, 67-68, mgr commercial develop & mgr, Brisbane Tech Ctr, 68-69; exec vpres, corp secy & mem bd dirs, Concrete Cutting Industs, 69-72, pres, 73-79, BD CHMN, CONCRETE CUTTING INDUSTS, INC, 80- *Concurrent Pos:* Lectr chem eng, Univ Southern Calif, 63-64; consult, Fluor Corp, 70-75, Japan bus, 76- *Mem:* Am Inst Chem Engrs. *Res:* Research management and planning; technical and commercial liaison; process design of petroleum, petrochemical and chemical plant complexes; general management. *Mailing Add:* 6600 Via La Paloma Rancho Palos Verdes CA 90274

HAMAKER, JOHN C, JR, b Canton, Ohio, Apr 21, 24; m 47; c 2. AEROSPACE APPLICATIONS. *Educ:* Univ Mich, BS, 45, MS, 47, PhD(metall eng), 52. *Prof Exp:* Plant engr, Stearns-Roger Mfg Co, Denver, 51-53; dir res & eng, Basco Metals Corp, Latrobe, Pa, 53-61, vpres & dir, 61-68; pres, Teledyne Rodney Metals, New Bedford, Mass, 68-70; group pres, Whittaker Corp, Los Angeles, 70-71; group exec & chmn, Teledyne Can, 71; group vpres, Automation Industs Inc, Los Angeles, 71-75; sr vpres, SSP Industs, Burbank, 75-77; OWNER, JCH INC & PLH INC, AUTO PARTS MFRS, 81- *Concurrent Pos:* Mem mat adv bd, Nat Acad Sci; lectr, Pa State Univ, Univ Hawaii, Univ Wis; trustee, Am Soc Metals Int. *Mem:* Am Soc Metals Int; Am Soc Testing & Mat; Am Inst Mining Metall & Petrol Engrs; Am Iron & Steel Inst; Am Chem Soc. *Res:* Development and properties of tool steels; heat treatment of special steels and alloys; superstrength steels. *Mailing Add:* JCH Inc & PLH Inc Auto Parts Mfrs 795 Kumukahi Pl Honolulu HI 96825

HAMAKER, JOHN WARREN, b Montreal, Que, Oct 25, 17; m 44; c 4. AGRICULTURAL CHEMISTRY. *Educ:* Univ Calif, Los Angeles, BA, 40; Univ Calif, PhD(chem), 44. *Prof Exp:* Res assoc chem, Univ Calif, 44-45; instr, Napa Jr Col, 45-47; from asst prof to assoc prof, Whittier Col, 47-55; chemist, Dow Chem Co, 55-86; RETIRED. *Concurrent Pos:* With AEC; with Off Sci Res & Develop, 44; vis assoc prof, Univ Chicago, 53. *Mem:* Am Chem Soc; Sigma Xi. *Res:* Microchemical technique of qualitative analysis; analytical microchemistry; radioactive or tracer chemistry; environmental fate of pesticides. *Mailing Add:* 125 Conifer Lane Walnut Creek CA 94598

HAMANN, DONALD DALE, b Moline, Ill, July 24, 33; m 54; c 4. ENGINEERING MECHANICS, FOOD TECHNOLOGY. *Educ:* SDak State Univ, BS, 55, MS, 59; Va Polytech Inst, PhD(eng mech), 67. *Prof Exp:* Head topog surv party, SDak State Univ, 55, asst agr eng, 56-58, instr, 58-61; asst prof, Va Polytech Inst, 62-68; assoc prof, 69-75, PROF FOOD ENG, NC STATE UNIV, 75- *Mem:* Am Soc Agr Engrs; Inst Food Technologists; Soc Rheology; Sigma Xi. *Res:* Rheology of foods and other biomaterials; concept and design of food processing equipment; machine analysis and design. *Mailing Add:* Food Sci Dept NC State Univ Raleigh NC 27695-7624

HAMANN, DONALD ROBERT, b Valley Stream, NY, May 16, 39; m 66; c 1. PHYSICS. *Educ:* Mass Inst Technol, BS, 61, PhD(elec eng), 65. *Prof Exp:* Staff scientist theoret physics, Ford Motor Co Sci Lab, 64-65; mem tech staff, 65-78, head, Dept Theoret Physics, 78-81, HEAD, SURFACE PHYSICS RES DEPT, BELL TEL LABS, 81- *Honors & Awards:* Davisson-Germer Prize, Am Phys Soc. *Mem:* Fel Am Phys Soc; Sigma Xi. *Res:* Theoretical solid state physics; surface physics. *Mailing Add:* Bell Telephone Labs PO Box 261 Rm 1C-327 Murray Hill NJ 07974

HAMAR, DWAYNE WALTER, b Lexington, Nebr, Mar 28, 37; m 57; c 1. BIOCHEMISTRY. *Educ:* Nebr State Teachers Col, Kearney, BA, 58; Univ Nebr, MS, 61, PhD(biochem), 64. *Prof Exp:* From instr to asst prof path, 63-70, ASSOC PROF PATH, COLO STATE UNIV, 70- *Mem:* Fel Am Inst Chem. *Res:* Chemical pathology; metabolism. *Mailing Add:* Dept Path Colo State Univ Ft Collins CO 80523

HAMARNEH, SAMI KHALAF, b Madaba, Jordan, Feb 2, 25; nat US; m 48; c 1. HISTORY OF MEDICAL SCIENCES. *Educ:* Syrian Univ, Damascus, Syria, BS, 48; NDak State Univ, MS, 56; Univ Wis, PhD(hist pharm & sci), 59. *Prof Exp:* Assoc cur, Div Med Sci, Smithsonian Inst, 59-61, cur-in-chg, 61-72, res historian, dept sci & technol, 73-78 & pharm, 72-78, field expert, 79, EMER CUR, SMITHSONIAN INST, 78- *Concurrent Pos:* Vis assoc prof, George Washington Univ, 63-64; vis prof, Univ Pa, 69 & Univ Aleppo, Syria, 79; sci adv, Arab Soc Hist Pharm, 77-; prof hist med sci, King Abdulaziz Univ, Jeddah, Saudi Arabia, 82-83; prof hist med sci, Sch Pub Health, Yarmouk Univ, Irbid, Jordan, 84-90. *Honors & Awards:* Star of Jordan Medal, 65; Ed Kremers Award, 66. *Mem:* Int Acad Pharm; Am Inst Hist Pharm; Arab Soc Hist Pharm; Arab Acad. *Res:* Health field in Islam; history of Arabic medicine, pharmacy and public health and the Arab physicians who contributed to it. *Mailing Add:* 4631 Massachusetts Ave NW Washington DC 20016-2361

HAMASAKI, DUCO I, b Maui, Hawaii, July 1, 29; m 67; c 2. OPHTHALMOLOGY. *Educ:* Washington Univ, AB, 50; Univ Calif, Berkeley, MOptom, 54, PhD(physiol optics), 59. *Prof Exp:* Asst physiol optics, Univ Calif, Berkeley, 53-59, physiol opticist, 59-61; from instr to assoc prof, 62-79, PROF OPHTHAL, UNIV MIAMI, 79- *Concurrent Pos:* NIH fel physiol, Cambridge Univ, 61-62; NIH grant, 63-64; vis scientist physiol, Tohoku Univ, Sendai, Japan, 77-78. *Mem:* Asn Res Vision & Ophthal. *Res:* Visual physiology. *Mailing Add:* Dept Ophthal Univ Miami Sch Med Miami FL 33136

HAMB, FREDRICK LYNN, b Mt Gay, WVa, May 28, 37; m 60; c 2. ORGANIC POLYMER CHEMISTRY. *Educ:* WVa Univ, BS, 60, MS, 62, PhD(org chem), 64. *Prof Exp:* Sr res chemist, Eastman Kodak Co, 63-69, res assoc, 69-71, lab head, Res Labs, 71-75, asst dir, Photog Res Div, 75-78, asst dir, 78-80, dir, Black & White Photog Div, 80-89, MGR REGIONAL BUS UNIT, JAPANESE REGION, EASTMAN KODAK CO, 90- *Mem:* Am Chem Soc; NY Acad Sci; Sigma Xi. *Res:* Condensation and addition polymers; chemistry of cyclopentadienes; Diels-Alder reaction; polymers for photography; non-silver photography; silver halide photography. *Mailing Add:* Eastman Kodak Japan Ltd Gotenyama Mori Bldg 4-7-35/Kita-Shingawa Shingawa Tokyo 140 Japan

HAMBIDGE, K MICHAEL, b Oct 28, 32. PEDIATRICS. *Educ:* Cambridge Univ, Eng, BA, 56; Westminster Med Sch, MB & BChir, 60. *Hon Degrees:* ScD, Univ Cambridge, 88. *Prof Exp:* Res fel pediat, Health Sci Ctr, Univ Colo, 66-67, asst dir, Children's Clin Res Ctr, 67-68, from instr to assoc prof, 67-77, PROF PEDIAT, HEALTH SCI CTR, UNIV COLO, 78-, DIR, CHILDREN'S CLIN RES CTR, 85- & CTR HUMAN NUTRIT, 88- *Concurrent Pos:* Co-dir, Children's Clin Res Ctr, Univ Colo, 75-84. *Honors & Awards:* Borden Award, Am Inst Nutrit, 79; Nutrit Award, Am Acad Pediat, 87. *Res:* Childhood nutrition; author of numerous scientific publications. *Mailing Add:* Dept Pediat Health Sci Ctr Univ Colo 4200 E Ninth Ave Box C-233 Denver CO 80262

HAMBLEN, DAVID GORDON, b Norwalk, Conn, Apr 20, 40; m 64; c 2. EXPERIMENTAL PHYSICS. *Educ:* Williams Col, BA, 62; Univ Ill, Urbana, MS, 63, PhD(physics), 69. *Prof Exp:* Exp physicist, United Technol Res Ctr, 69-80; VPRES, ADVAN FUEL RES, INC, 80- *Mem:* Am Phys Soc; Inst Elec & Electronics Engrs; Sigma Xi. *Mailing Add:* Advan Fuel Res 87 Church St East Hartford CT 06108

HAMBLEN, DAVID PHILIP, b Chicago, Ill, Sept 12, 28; m 59; c 2. CLINICAL ANALYZERS, ION ETCHING. *Educ:* NMex Mil Inst, BS, 51; Tulane Univ La, BS, 55; Univ Tenn, MS, 60; Univ Rochester, MS, 65. *Prof Exp:* Assoc physicist, Union Carbide Nuclear Co, 56-58 & Oak Ridge Nat Lab, 58-61; sr scientist, Bausch & Lomb, Inc, 61-67; RES ASSOC SR STAFF,

EASTMAN KODAK RES LABS, 67- *Concurrent Pos:* Consult, Redstone Arsenal, asst, Tulane Univ La, 55-56; consult, Atomic Energy Proj, US AEC-Univ Rochester, 66-67. *Mem:* Am Optical Soc. *Res:* Development of radiation scanner for tumor localization; atomic reactor experimentation; thermonuclear fusion research; abiotic synthesis of biological molecules; gas laser development; production of gradient-index optical materials; electrochemical and liquid-crystal display systems; lunar photographic devices; clinical diagnostic instrumentation; emulsion research; microelectronics development. *Mailing Add:* 42 Gateway Rd Rochester NY 14624

HAMBLEN, JOHN WESLEY, b Story, Ind, Sept 25, 24; m 87; c 1. COMPUTER SCIENCE. *Educ:* Ind Univ, AB, 47; Purdue Univ, MS, 52, PhD(math), 55. *Prof Exp:* From asst prof to assoc prof math, Okla State Univ, 55-58, dir comput ctr, 57-58; assoc prof statist & dir comput ctr, Univ Ky, 58-61; prof math & technol & dir data processing & comput ctr, Southern Ill Univ, 61-65; proj dir comput sci, Southern Regional Educ Bd, 65-72; chmn comput sci, 72-81, prof, 72-87, EMER PROF, UNIV MO, ROLLA, 87- *Concurrent Pos:* Consult, D-X Sunray Oil Co, 57-58; lectr, IBM Corp, 60, Asn Comput Mach-NSF vis scientist, 57; consult, Systs Develop Corp & prof, Ga Inst Technol, 65-66; ed, J Asn Ed Data Systs, 67-68; mem, Data Base Panel, Comput Sci & Eng Bd, Nat Acad Sci, 69-71, 79-80; chmn, Educ Comt, Am Fedn Info Processing Socs, 71-72 & 79-84; consult, NSF, 75-76; vis scientist, Ctr Appl Math, Nat Bur Standards, 81-83; assoc prog dir, SEE/DMDRI/AAT, NSF, 85-86. *Honors & Awards:* Asn Educ Data Systs Award, 71. *Mem:* Fel AAAS; Asn Comput Mach (secy, 72-76); Nat Geneal Soc; Asn Educ Data Systs (pres, 68-69). *Res:* Distributions of combinations of random variables; application of digital computers; information systems design; genealogy. *Mailing Add:* 4432 Carya Sq Columbus IN 47201-8933

HAMBLETON, WILLIAM WELDON, b Lancaster, Pa, Sept 10, 21; m 46; c 2. GEOLOGY, GEOPHYSICS. *Educ:* Franklin & Marshall Col, BS, 43; Northwestern Univ, MS, 47; Univ Kans, PhD(geol), 51. *Prof Exp:* Geologist, Pa Geol Surv, 46 & 51; instr, 49, from asst prof to assoc prof, 51-62, PROF GEOL, UNIV KANS, 62-, STATE GEOLOGIST & DIR, KANS GEOL SURV, 70- *Concurrent Pos:* Chem petrographer, US Bur Mines, 49; asst dir, Kans Geol Surv, 54-56, assoc dir, 56-70, assoc dean, Grad Sch, Univ Kans, 67-68, fac, 68-70; geophysicist, Chevron Oil Co, 55; vis scientist, Lamont Geol Observ, Columbia Univ, 59-60; mem, Bd Dirs, Mid-Continent Res & Develop Coun; Bd Dirs, Kans Sch Relig, Energy Comt, Interstate Oil Compact Comn, Mineral Resources Comt, Nat Asn State Univ & Land Grant Cols, Energy Adv Coun, Kans Energy Off, Exec Adv Comt, Nat Gas Surv, Gov Task Force Water Resources; chmn, Kans Comt Midwest Gov Conf Task Energy & Natural Resources, Geol Rev Group nuclear waste disposal, Dept Energy; vchmn adv bd, Kans Univ Sch Fine Arts. *Mem:* Geol Soc Am; Am Soc Explor Geol; Am Asn Petrol Geol; Am Asn Geol Teachers; Sigma Xi. *Res:* Geophysics, gravity and magnetism; igneous petrography; economic geology. *Mailing Add:* 1312 Raintree Pl Lawrence KS 66044

HAMBLIN, WILLIAM KENNETH, b Lyman, Wyo, May 22, 28; m 52; c 4. GEOLOGY. *Educ:* Brigham Young Univ, BA, 53, MS, 54; Univ Mich, PhD(geol), 58. *Prof Exp:* Asst prof geol, Univ Kans, 57-62; assoc prof, Univ Ga, 62-63; PROF GEOL, BRIGHAM YOUNG UNIV, 63- *Mem:* Geol Soc Am; Soc Econ Paleont & Mineral; Am Asn Petrol Geol. *Res:* Stratigraphy and sedimentation. *Mailing Add:* Dept Geol Brigham Young Univ ESC 144A Provo UT 84602

HAMBOURGER, PAUL DAVID, b Cleveland, Ohio, Dec 26, 39; m 76. EXPERIMENTAL SOLID STATE PHYSICS. *Educ:* Harvard Univ, BA, 62; Northwestern Univ, PhD(physics), 69. *Prof Exp:* Res assoc physics, Univ Pa, 69-71; asst prof, 71-75, ASSOC PROF PHYSICS, CLEVELAND STATE UNIV, 75- *Mem:* Am Phys Soc. *Res:* Low temperature solid state physics and applied physics; high-pressure physics; electronic and magnetic properties of solids; reduced-dimensionality materials; solid-state phase transitions; superconductivity; conductor-insulator composite materials. *Mailing Add:* Dept Physics Cleveland State Univ Euclid Ave at E 24th St Cleveland OH 44115

HAMBRECHT, FREDERICK TERRY, b Galesburg, Ill, Aug 18, 39; m 65; c 2. BIOMEDICAL ENGINEERING. *Educ:* Purdue Univ, BS, 61; Mass Inst Technol, MS, 63; Johns Hopkins Univ, MD, 68. *Prof Exp:* Res assoc biomed eng, Mass Inst Technol, 61-63; intern surg, Med Ctr, Duke Univ, 68-69; RES ASSOC NEURAL CONTROL, 69- & PROG DIR NEURAL PROSTHESIS, NIH, 72- *Honors & Awards:* Commendation Medal, USPHS, 75, Meritorious Serv Medal, 80, Distinguished Serv Medal, 86. *Mem:* Soc Neurosci; Biomed Eng Soc. *Res:* Application of engineering to clinical medicine, neural prostheses, neural control and neurophysiology. *Mailing Add:* Fed Bldg Rm 916 NIH Bethesda MD 20892

HAMBRICK, GEORGE WALTER, JR, b Charlottesville, Va, Dec 4, 22. MEDICINE, DERMATOLOGY. *Educ:* Concord Col, BS, 44; Univ Va, MD, 46; Am Bd Dermat, dipl, 53, dipl dermatopath, 75. *Prof Exp:* Intern, Univ Iowa Hosp, 47; resident dermat, Univ Va Hosp, 48; resident, Columbia-Presby Med Ctr, 51; instr, Duke Univ Hosp, 51-52, assoc, 52-53; instr, Col Physicians & Surgeons, Columbia Univ, 53-55; from assoc to assoc prof, Sch Med, Univ Pa, 55-66, chief dermat, Clin Hosp, 56-65; from assoc prof to prof, Sch Med, Johns Hopkins Univ, 66-76, dir dermat, 67-76; prof dermat & dir dept, Univ Cincinnati, 76-81; head div dermat, Dept Med, Col Med, Cornell Univ, 81-91; chief, NY Hosp, 81-91; CONSULT, 88- *Concurrent Pos:* Fel, Duke Univ Hosp, 51-52; mem, Attend Staff, Philadelphia Gen Hosp, 55-66; consult, US Naval Hosp, Philadelphia, 59-66 & NIH, 67-75. *Mem:* Soc Invest Dermat (secy-treas, 65-69, pres, 71-); Am Dermat Asn; AMA; Am Acad Dermat; Am Col Physicians; Skin Dis Soc (pres, 88-91). *Res:* Clinical dermatology and anatomy of the skin. *Mailing Add:* Dept Med Med Col Cornell Univ 1300 York Ave New York NY 10021

HAMBURG, BEATRIX A M, b Jacksonville, Fla, Oct 19, 23; m 51; c 2. CHILD PSYCHIATRY, PEDIATRICS. *Educ:* Vassar Col, AB, 44; Yale Univ, MD, 48. *Prof Exp:* Intern, Grace-New Haven Hosp, 48-49; resident, Yale Psychiat Inst, 49-50, Children's Hosp, 50-51 & Inst Juvenile Res, 51-53; res assoc, Med Sch, Stanford Univ, 61-71, assoc prof psychiat, 76-80; assoc prof Sch Med, Harvard Univ, 80-83, exec dir, health policy res div, 81-83; PROF PSYCHIAT & PEDIAT, 83-, DIR DIV CHILD & ADOLESCENT PSYCHIAT, MT SINAI HOSP, 88- *Concurrent Pos:* Mem, Comn Behav & Social Sci, Nat Acad Sci; mem, NY State Pub Health Coun; dir, Bush Found; trustee, William T Grant Found. *Honors & Awards:* T Roswell Gallagher Award in Adolescent Med. *Mem:* Inst Med-Nat Acad Sci; Soc Adolescent Med; Am Pub Health Asn; fel Am Acad Child Psychiat; Soc Prof Child Psychiat. *Res:* Biomedical-behavioral factors in diabetes; normal development and developmental psychopathology of adolescents; health policy research. *Mailing Add:* Mt Sinai Med Ctr One Gustave Levy Pl New York NY 10029

HAMBURG, DAVID ALAN, b Evansville, Ind, Oct 1, 25; m 51; c 2. BEHAVIORAL SCIENCE. *Educ:* Ind Univ, AB, 44, MD, 47; Am Bd Psychiat & Neurol, cert, 53. *Hon Degrees:* DSc, Ind Univ, 76, Rush Univ, 77, City Univ NY, 80, Univ Rochester, 81, Univ Ill, 84, Albert Einstein Sch Med, 85, Univ Southern Calif, Hahnemann Univ, 86 & Univ Pittsburgh, 86. *Prof Exp:* Asst psychiat, Sch Med, Yale Univ, 48-49; resident psychiatrist, Michael Reese Hosp, Chicago, 49-50; staff psychiatrist, Brooke Army Hosp, Tex, 50-52; resident psychiatrist, Walter Reed Army Inst Res, DC, 52-53; assoc dir inst psychosom & psychiat res & training, Michael Reese Hosp, 53-56; chief adult psychiat br, NIMH, 58-61; exec head dept, Sch Med, Stanford Univ, 61-69, prof psychiat, 61-76, chmn dept, 69-76; pres, Inst Med, Nat Acad Sci, 75-80; J D MacArthur prof health policy & dir, div health policy res & educ, Harvard Univ, 80-82; PRES, CARNEGIE CORP NY, 83- *Concurrent Pos:* Consult, Walter Reed Army Inst Res, 54-57, Nat Adv Comt Res Psychiat, Neurol & Psychol, Vet Admin, 64-68, UNESCO, 69-70; fel, Ctr Advan Study Behav Sci, 57; mem, bd dirs, Found Fund Res Psychiat, 57-60, Comt Res Probs Sex, Nat Acad Sci, Nat Res Coun, 58-61, Res Career Award Comt, NIMH, 61-65, Behav & Social Sci Surv Comt & Comt Life Sci & Social Policy, Nat Acad Sci, 66-73, gov bd, Nat Res Coun, 75-80, Bd Ment Health & Behav Med, Inst Med-Nat Acad Sci & Comt Int Security & Arms Control, 81-; assoc ed, J Psychiat Res, 60-; chmn, var sci comts, NIMH, HEW, WHO & Nat Acad Sci, 61-; clin prof psychiat & behav sci, George Washington Univ, 76-80; adj prof pub policy & clin prof psychiat, Duke Univ, 78-80; sr adv, Ctr Social Policy Studies Israel, 83- *Honors & Awards:* Menninger Award, Am Col Physicians, 76; Vestermark Award, Am Psychiat Asn, 77; Rosenhaus Award, Am Pub Health Asn, 78; John P McGovern Award, Sigma Xi, 86; Health for All Medal, WHO, 88. *Mem:* Inst Med-Nat Acad Sci (pres, 75-80); AAAS (pres, 84-85); Am Psychosom Soc; Am Psychiat Asn; Am Soc Human Genetics; Soc Neurosci; Am Philos Soc. *Res:* Psychological stress and endocrine function; adaptive behavior under stress; psychotherapy in crisis; genetics, hormones and behavior; author of numerous technical publications. *Mailing Add:* Carnegie Corp NY 437 Madison Ave New York NY 10022

HAMBURG, JOSEPH, b Philadelphia, Pa, Sept 9, 22. HEALTH ADMINISTRATION. *Educ:* Hahnemann Med Col, MD, 51. *Hon Degrees:* DSc, Hahnemann Med Col, 79. *Prof Exp:* From asst prof to dean, 63-66, PROF HEALTH ADMIN, UNIV KY, 86- *Mem:* Inst Med-Nat Acad Sci; Am Soc Allied Health Professions (pres, 72); AMA; Am Acad Family Pract. *Mailing Add:* Col Allied Health Professions Univ Ky Lexington KY 40506

HAMBURGER, ANNE W, b Bronx, NY, May 10, 47; m 70; c 2. CELL BIOLOGY, EXPERIMENTAL HEMATOLOGY. *Educ:* Brandeis Univ, AB, 68; NY Univ, MS, 71, PhD(biol), 73. *Prof Exp:* Instr biol, St Peter's Col, 72-73; fel cell biol, Albert Einstein Col Med, 73-75; res assoc hemat, Col Med, Univ Ariz, 75-78; res scientist cell cult, Am Type Cult Collection, 78-84; ASSOC PROF CELL BIOL, UNIV MD CANCER CTR, 84- *Concurrent Pos:* Multiple Sclerosis Soc fel, 73-75; adj asst prof hemat, Col Med, Univ Ariz, 77-78; adj asst prof med oncol, Georgetown Univ Sch Med, 80-85, Prog Projs Study Sect Nat Cancer Inst, 83-88; vis prof, Japanese Found Prom Cancer Res, 85. *Mem:* AAAS; Am Soc Hemat; Int Soc Exp Hemat; Tissue Cult Asn; Sigma Xi; Am Asn Cancer Res; NY Acad Sci. *Res:* Factors controlling tumor cell growth and differentiation; kinetics of normal hematopoiesis and leukemic cell growth; production of monoclonal antibodies to differentiation antigens. *Mailing Add:* Univ Md Cancer Ctr 655 W Baltimore St Baltimore MD 21201

HAMBURGER, MICHAEL WILE, b Rochester, NY, Dec 17, 53; m 83; c 2. SEISMOLOGY, TECTONOPHYSICS. *Educ:* Wesleyan Univ, BA, 75; Cornell Univ, MS, 82 & PhD(geol), 86. *Prof Exp:* Sr res asst seismol, Lamont-Doherty Geol Observ, 76-80; grad res asst seismol, Cornell Univ, 80-86; ASST PROF GEOPHYS, IND UNIV, 86- *Concurrent Pos:* Mem, US-USSR work group on exchange in earthquake prediction, 86-88, earthquake adv panel, Ind dept civil defense. *Mem:* Am Geophysical Union; Geol Soc Am; Seismol Soc Am. *Res:* Spatial and temporal distribution of earthquakes in a collisional plate boundary, Soviet Central Asia; seismicity of the Fiji Islands; tectonics of the southwest pacific. *Mailing Add:* Dept Geol Sci Ind Univ Bloomington IN 47405

HAMBURGER, RICHARD, b Baltimore, Md, Dec 30, 15; m 43; c 1. GEOLOGY. *Educ:* Univ Mich, AB, 38. *Prof Exp:* Geologist & engr, Inspiration Consol Copper Co, 46-55; geol engr, Div Raw Mat, AEC, 55-58, mining engr, Div Mil Appln, 58-60, asst chief, Div Peaceful Nuclear Explosives, 60-61, asst dir tech opers, 61-74; CONSULT, 74- *Mem:* Am Inst Mining, Metall & Petrol Engrs; Health Physics Soc; Geol Soc Am; Soc Econ Geologists. *Res:* Occurrence of mineral deposits; evaluation of ore deposits and of exploration techniques with emphasis on statistical analysis; site investigations. *Mailing Add:* 14900 Springfield Rd Germantown MD 20874

HAMBURGER, ROBERT NEWFIELD, b New York, NY, Jan 26, 23; m 43; c 2. PEDIATRICS, ALLERGY. *Educ:* Univ NC, BA, 47; Yale Univ, MD, 51; Am Bd Pediat, dipl, 58; Am Bd Allergy & Immunol, dipl, 74. *Prof Exp:* Intern & asst resident pediat, Strong Mem Hosp, 51-53; asst resident, New

Haven Hosp, Conn, 53-54; clin instr, Sch Med, Yale Univ, 54-58, asst clin prof, 58-60; vis assoc prof, 63-64, assoc prof pediat & asst dean, Sch Med, 64-67, PROF PEDIAT, SCH MED, UNIV CALIF, SAN DIEGO, 67-, HEAD DIV PEDIAT IMMUNOL & ALLERGY, 70- *Concurrent Pos:* Spec fel, NIH, Yale Univ & Univ Calif, San Diego, 60-63; instr pediat, Univ Rochester, 51-53; practicing pediatrician, 54-60; chief pediat serv, Milford Hosp, 59-60. *Mem:* Sigma Xi. *Res:* Lupus erythematosus; immunology; genetics; pediatric growth and development. *Mailing Add:* 9485 La Jolla Shores Dr La Jolla CA 92037-1149

HAMBURGER, VIKTOR, b Landeshut, Ger, July 9, 00; nat US; m 28; c 2. ZOOLOGY. *Educ:* Univ Freiburg, PhD(zool), 25. *Hon Degrees:* PhD, Wash Univ, 76 & Univ Uppsala, Sweden, 84. *Prof Exp:* Asst, Kaiser Wilhelm Inst, Berlin, 26-27; instr zool & privatdocent, Dept Zool, Univ Freiburg, 27-32; Rockefeller res fel, Univ Chicago, 32-33, instr zool, 33-35; from asst prof to prof, 35-69, actg head dept, 42-44, chmn dept, 44-66, EMER PROF BIOL, WASHINGTON UNIV, 69- *Concurrent Pos:* Fel, NIH, 47. *Honors & Awards:* Wakeman Award, 78; Harrison Prize, 81; Louise Gross Horwitz Prize, Cell Biol & Develop Neurobiol, Columbia Univ, 83; Nat Medal Sci, 89; Karl Lashley Award, 90. *Mem:* Nat Acad Sci; AAAS; Am Soc Naturalists; Am Soc Zool (pres, 55); Soc Develop Biol (pres, 50-51); Neurosci Soc. *Res:* Experimental neuro-embryology; mode of gene action in development; embryology of behavior. *Mailing Add:* Dept Biol Washington Univ St Louis MO 63130

HAMBURGH, MAX, biology; deceased, see previous edition for last biography

HAMBY, DAME SCOTT, b Macon, Ga, July 8, 20; m 43; c 2. TEXTILE ENGINEERING. *Educ:* Auburn Univ, BS, 46. *Hon Degrees:* Dr Textiles, Philadelphia Col Textile & Sci, 84. *Prof Exp:* Prod trainee textile testing & qual control, Goodyear Tire & Rubber Co, 37-42; textile engr res & develop, Celanese Corp, 43-45, B F Goodrich Co, 47-48; prof textile technol, 48-65 & head dept, 65-72, assoc dean, Textile Exten & Continuing Educ, 75-81, dean, Sch Textiles, 81-87, EMER DEAN, NC STATE UNIV, 87- *Concurrent Pos:* Chmn, Hamby Textile Res Labs. *Mem:* Fel AAAS; fel Brit Textile Inst; fel Am Soc Qual Control; Sigma Xi; fel Am Soc Testing & Mat. *Res:* Performance of textile materials; process evaluation and control. *Mailing Add:* Hamby Textile Res Lab Box 247 Garner NC 27529

HAMBY, DRANNAN CARSON, b Duncan, Okla, Nov 16, 33; m 52; c 2. ELECTROCHEMISTRY. *Educ:* Linfield Col, BA, 55; Ore State Univ, MA, 61, PhD, 68. *Prof Exp:* Asst prof physics, 62-69, dir res inst, 62-78, assoc prof physics & chem, 69-74, chmn, div math & natural sci, 79-83, PROF PHYSICS & CHEM, LINFIELD COL, 74-, CHEMIST, LINFIELD RES INST, 56- *Concurrent Pos:* Chemist, Field Emission Corp, 61-62; vis assoc res engr, Univ Calif, Los Angeles, 74-75, Brigham Young Univ, 81. *Mem:* Am Chem Soc; Electrochem Soc; Sigma Xi. *Res:* High temperature batteries; thermodynamics of fused salts; electrochemical machining of refractory metals; alkaline zinc secondary electrode; changing of sealed lead-acid batteries. *Mailing Add:* 232 Oregan Way McMinnville OR 97128

HAMBY, ROBERT JAY, b Perry, Ore, Sept 25, 32; m 63; c 4. PHYSIOLOGICAL ECOLOGY, MARINE BIOLOGY. *Educ:* San Jose State Col, BA, 60; Univ of the Pac, MS, 65; Univ Chicago, PhD(biol), 69. *Prof Exp:* From instr to asst prof biol, Purdue Univ, 68-71; assoc prof, 71-76, PROF BIOL, CUMBERLAND COUNTY COL, 76- *Concurrent Pos:* Alternate trustee, NJ Marine Sci Consortium, 75-78. *Mem:* Am Soc Zool. *Res:* Adaptive physiological mechanisms in marine animals. *Mailing Add:* Dept Biol Cumberland County Col Vineland NJ 08360

HAMDY, MOHAMED YOUSRY, b Cairo, Egypt, Apr 17, 38; m 61; c 1. AGRICULTURAL ENGINEERING. *Educ:* Cairo Univ, BSc, 59; Mich State Univ, MS, 62; Ohio State Univ, PhD(agr eng), 65. *Prof Exp:* Instr mech eng, Cairo Univ, 59-61; from res assoc to assoc prof, Ohio State Univ, 63-74, prof agr eng, 74-89; RETIRED. *Concurrent Pos:* Consult, USAID & Pioneer Hi-Bred Int, 79- *Honors & Awards:* Am Soc Agr Eng Paper Award, 72. *Mem:* Am Soc Agr Engrs. *Res:* Mathematical modeling of agricultural systems. *Mailing Add:* 2488 Lytham Rd Columbus OH 43220

HAMDY, MOSTAFA KAMAL, b Cairo, Egypt, May 27, 21; nat US; m 54; c 2. BACTERIOLOGY. *Educ:* Cairo Univ, BSc, 44, MSc, 49; Ohio State Univ, PhD, 53. *Prof Exp:* Instr bot, Univ Alexandria, 44-47; instr bact, Cairo Univ, 47-49; asst waste treatment lab, Eng Exp Sta, Ohio State Univ, 51-52, res assoc bact, 52-53, Muellhaupt fel, 53-54, res assoc biochem, 54-58; from asst prof bact & biochem to assoc prof food sci, 58-65, PROF FOOD SCI, UNIV GA, 65- *Mem:* Fel Am Soc Microbiol; Soc Exp Biol & Med; fel Inst Food Technol; fel Am Acad Microbiol; NY Acad Sci. *Res:* Pathogenesis of salmonella; effect of radiation and radioprotectors on the lysosomal enzymes in rats; effect of staphylococcus infection on the biochemistry of bruised and infected tissues; effect of anthocyanin on bacteria; mercury biotransformation polyclorinated biphenyl pollution control, alcohol fermentation and biotechnology. *Mailing Add:* Dept Food Sci Univ Ga Athens GA 30602

HAMED, AWATEF A, b El-Mansoura, Egypt, June 17, 44; US citizen; m 66. TURBOMACHINERY, TWO-PHASE FLOWS. *Educ:* Cairo Univ, BSc, 65; Univ Cincinnati, MSc, 69, PhD(eng), 72. *Prof Exp:* Design engr, Helwan Aircraft Estab, Egyptian Gen Aero Orgn, 65-67; grad teaching & res asst, 68-72, from asst prof to assoc prof, 73-80, PROF AEROSPACE ENG, UNIV CINCINNATI, 80- *Concurrent Pos:* Prin investr, NASA, Air Force Off Sci Res, Dept Energy, Dept Navy, Army Res Off & NSF; consult to indust & govt agencies; assoc ed, J Fluids Eng, 82-85. *Honors & Awards:* Tech Innovation Award, NASA, 83; Amelia Gerhart Fel, 67, 69 & 70. *Mem:* fel Am Inst Aeronaut & Astronaut; fel Am Soc Mech Engrs; Sigma Xi. *Res:* Gas turbine engine aerothermodynamics; laser doppler velocimetry measurements in three dimensional and two phase flows; prediction of turbomachinery single and two phase flows, erosion, performance deterioration and engine life; probablistic modeling of propulsion systems. *Mailing Add:* Dept Aerospace Eng Univ Cincinnati Cincinnati OH 45221

HAMED, GARY RAY, b Jonesboro, Ark, July 3, 50; m 74; c 1. POLYMER CHEMISTRY. *Educ:* Cornell Univ, BS, 72, MS, 73; Univ Akron, PhD(polymers), 78. *Prof Exp:* Res scientist, Firestone, 76-78, sr res scientist, 78-80; ASST PROF, UNIV AKRON, 80- *Mem:* Am Chem Soc; Adhesion Soc; Tire Soc. *Res:* Mechanical properties of polymers, especially tack fracture and adhesion of elastomers. *Mailing Add:* 3301 N Dover Rd Stow OH 44224-2423

HAMEED, SULTAN, b Ambala, India, Sept 8, 41; m 63; c 3. ATMOSPHERIC SCIENCE. *Educ:* Univ Karachi, Pakistan, BSc & MS, 61; Univ Manchester, Eng, PhD(physics), 68. *Prof Exp:* Res assoc atomic physics, Belfer Grad Sch, Yeshiva Univ, NY, 68-70; res assoc, physics dept, Columbia Univ, 70-72; sr res assoc atmospheric sci, Inst Space Studies, NASA, NY, 72-76; assoc prof, 76-86, PROF ATMOSPHERIC SCI, STATE UNIV NY, STONY BROOK, 86- *Concurrent Pos:* Consult, Pub Div, Am Inst Physics, New York, 70- *Mem:* Am Geophys Union; Am Meteorol Soc. *Res:* Many-electron effects in atomic and molecular structure; changes in atmospheric composition, cycles of trace gases in the atmosphere, climate change and boundary layer phenomena. *Mailing Add:* Lab Atmosphere Res Dept Mech Eng State Univ NY Stony Brook NY 11794

HAMEEDI, MOHAMMAD JAWED, b Budaum, India, Apr 7, 44; US citizen; m 77; c 2. COASTAL RESOURCE MANAGEMENT, ECOLOGICAL MODELING. *Educ:* Univ Karachi, BSc, 61, MSc, 63; Univ Wash, MS, 70, PhD(oceanog), 74. *Prof Exp:* Asst lectr zool, Univ Karachi, 64-66; lectr oceanog, Univ Wash, 71-74, oceangr, Apply Physics Lab, Univ Wash, 74-76; sr scientist, Sci Appln Inc, 76-81; DIR, US DEPT COM, NAT OCEANIC & ATMOSPHERIC ADMIN, 81- *Concurrent Pos:* Assoc prof, Univ Alaska, Fairbanks, 86- *Mem:* Sigma Xi; Artic Inst NAm; Oceanic Soc. *Res:* Plankton distribution and productivity; analysis of coastal and marine ecosystem; development of techniques to include observational and experimental data. *Mailing Add:* Ocean Assessments Div Nat Oceanic & Atmospheric Admin 4230 University Dr Suite 300 Anchorage AK 99508

HAMEKA, HENDRIK FREDERIK, b Rotterdam, Neth, May 25, 31; US citizen; m 72; c 2. THEORETICAL CHEMISTRY. *Educ:* Univ Leiden, Drs, 53, DSc(chem), 56. *Hon Degrees:* MA, Univ Pa, 71. *Prof Exp:* Asst chem, Univ Leiden, 55-56 & Univ Rome, 56-57; fel, Carnegie Inst, 57-58; res physicist, Philips Lamps Labs, 58-60; asst prof chem, Johns Hopkins Univ, 60-62; assoc prof, 62-67, PROF CHEM, UNIV PA, 67- *Concurrent Pos:* Consult, Bombrini-Parodi Delfino, Italy, 57; Sloan Found res fel, 62-66. *Honors & Awards:* Alexander von Humboldt Prize, 82. *Res:* Quantum theory of molecules and radiation fields; interactions between molecules and electromagnetic fields. *Mailing Add:* Dept Chem Univ Pa Philadelphia PA 19104

HAMEL, COLEMAN RODNEY, b Massena, NY, June 9, 37; m 60; c 3. HETEROCYCLIC CHEMISTRY. *Educ:* Clarkson Col Technol, BChE, 57; Bucknell Univ, MS, 63; Lehigh Univ, PhD(chem), 69. *Prof Exp:* Chemist, Merck & Co, 58-65; teaching asst, Lehigh Univ, 66-67; asst prof chem, Lafayette Col, 69-70; assoc prof, 70-73, chmn dept phys sci, 77-82, PROF CHEM, KUTZTOWN UNIV, PA, 73- *Concurrent Pos:* Consult, J T Baker Chem Co, 79-84, Connaught Labs, 84-, Air Prod & Chem, Inc, 84- & Lemmon Chem Co, 85; mem, health & safety comt, Am Chem Soc, 80-86, Wright Lab, 88- *Mem:* Am Chem Soc; fel Am Inst Chem; AAAS; NY Acad Sci. *Res:* Heterocyclic chemistry; hazardous chemical safety. *Mailing Add:* Dept Phys Sci Kutztown Univ Pa Kutztown PA 19530

HAMEL, EARL GREGORY, JR, b Pensacola, Fla, Nov 30, 28; m 55; c 4. ANATOMY. *Educ:* Spring Hill Col, BS, 51; St Louis Univ, MS, 53; Univ Iowa, PhD(anat), 59. *Prof Exp:* Instr anat, Univ Tex, 51-53; instr, Univ Iowa, 58-59; instr, 59-64, assoc prof, 64-70, PROF ANAT & CHMN DEPT, MED CTR, UNIV ALA, BIRMINGHAM, 70- *Res:* Neuroanatomy; comparative neurology; motor disturbances and behavior. *Mailing Add:* Dept Anat Sch Med Univ Ala University Sta Birmingham AL 35294

HAMEL, EDWARD E, b Alta, Can, Feb 18, 26; US citizen; m 54; c 8. ORGANIC CHEMISTRY. *Educ:* Univ Notre Dame, BS, 47; Univ Calif, Berkeley, PhD(org chem), 52. *Prof Exp:* Res chemist, E I du Pont de Nemours & Co, 51-57; res supvr chem, Aerojet Solid Propulsion Co, 57-74, mgr res & develop, Cordova Chem Co, 74-81; dir chem opers, Aerojet Strategic Propulsion, Co, 81; vpres & gen mgr, chem opers, Aerojet Propulsion Co, 81-88; RETIRED. *Mem:* Am Chem Soc; Sigma Xi; NY Acad Sci. *Res:* Aliphatic polynitro compounds; process research; continuous processing; catalytic hydrogenation; monomer synthesis; antimalarial synthesis; medicinal chemistry; aziridings; anticancer synthesis. *Mailing Add:* 7592 Lakeshore Dr Roseville CA 95678

HAMEL, JAMES V(ICTOR), b Rochester, NY, Apr 21, 44; m 67; c 1. GEOTECHNICAL ENGINEERING, ROCK MECHANICS. *Educ:* Univ Pittsburgh, BS, 65, PhD(geotech eng), 70; Mass Inst Technol, SM, 66. *Prof Exp:* Res engr, Univ Pittsburgh, 67-69; asst prof civil eng, SDak Sch Mines & Technol, 69-72; proj engr, Gen Anal, Inc, Monroeville, Pa, 72-73; CONSULT ENGR, HAMEL GEOTECH CONSULTS, 73- *Concurrent Pos:* Consult, Mo River Div, US Army CEngrs, Nebr, 70-73; adj assoc prof civil engr, Univ Pittsburgh, 74-85. *Mem:* Am Soc Civil Engrs; Asn Eng Geologists; Int Soc Rock Mech; US Comt Large Dams; Am Inst Mining, Metall & Petrol Engrs; Int Soc Soil Mech & Found Engr; Int Asn Eng Geol. *Res:* Stability and long-term behavior of natural and excavated slopes in soil and rock; stability and behavior of rock foundations and abutments of dams; disposal of solid wastes from mining, industrial, and energy generation processes; river bank instability. *Mailing Add:* 1992 Butler Dr Monroeville PA 15146

HAMELIN, CLAUDE, b Montreal, Que, Aug 25, 43; m 70; c 2. VIROLOGY, MOLECULAR GENETICS. *Educ:* Univ Montreal, BPed, 65, BSc, 70, MSc, 72, PhD(genetics), 75. *Prof Exp:* Fels, Stanford Univ, 75-76, Armand-Frappier Inst, Univ Quebec, 76-77 & Pasteur Inst, 77-78; asst prof, 78-85, dir res & develop biotics, 85-86, PROF VIROL, ARMAND-FRAPPIER INST,

UNIV QUE, 87- *Mem:* Am Soc Microbiol; Can Col Microbiologists; Genetics Soc Am; Genetics Soc Can; NY Acad Sci; Can Soc Microbiologists; French-Can Asn Advan Sci. *Res:* Biochemical analysis of the human cytomegalovirus genome; researches on the persistence, latency, late reactivation and oncogenic potential of this virus in humans; development of biosensors, bioprobes and biochips. *Mailing Add:* Dept of Virol Inst Armand-Frappier 531 Blvd Des Prairies Ville de Laval PQ H7V 1B7 Can

HAMELINK, JERRY L, limnology, toxicology, for more information see previous edition

HAMER, DEAN H, b Montclair, NJ, May 29, 51; c 1. BIOCHEMISTRY, MOLECULAR GENETICS. *Educ:* Trinity Col, BA, 72; Harvard Univ, PhD(biochem), 77. *Prof Exp:* Staff fel molecular genetics, Nat Inst Child Health & Human Develop, 77-78; staff fel, Nat Inst Allergy & Infectious Dis, 78-80; sr staff fel, 80-89, CHIEF, SECT GENE STRUCTURE & REGULATION, NAT CANCER INST, 90- *Res:* Molecular basis of gene expression; recombinant DNA technology. *Mailing Add:* Gene Structure & Regulation Sect Biochem Lab NIH Nat Cancer Inst Bldg 37 Rm 4A-17 Bethesda MD 20897

HAMER, JAN, b Gombong, Indonesia, May 2, 27; m 56; c 2. CHEMISTRY. *Educ:* Univ Leyden, BS, 49, Drs, 55, Dr(chem), 56. *Prof Exp:* Res assoc chem, Tulane Univ, 53-55 & 56-57; assoc prof, Dillard Univ, 58-60; asst prof, 60-65, ASSOC PROF CHEM, TULANE UNIV, 65- *Concurrent Pos:* Consult, USDA, 62- *Mem:* AAAS; Am Chem Soc. *Res:* Cycloaddition reactions; stereochemistry and kinetics of four-center reactions; synthesis of azahexahelicene. *Mailing Add:* 299 Walnut St New Orleans LA 70118

HAMER, JUSTIN CHARLES, b Oct 2, 14; US citizen; m 41; c 5. ORGANIC CHEMISTRY, QUALITATIVE ANALYSIS. *Educ:* Pac Union Col, BA, 49, MA, 49; Univ NMex, PhD(chem), 61. *Prof Exp:* Sci teacher chem & math, Sandia View Acad, NMex, 53-56; head sci dept, Col of the Antillas, Cuba, 56-58; chmn chem dept, Columbia Union Col, 60-68; chmn div sci & chem, MidE Col, Beirut, Lebanon, 68-74; PROF CHEM, OAKWOOD COL, 75- *Concurrent Pos:* Sci teacher, Voc Col, Alajuela, Costa Rica; res asst, Univ Md, 65-68; sr res fel, Col Med, Howard Univ, 74-75. *Mem:* Sigma Xi; Am Chem Soc. *Res:* Biochemistry of lipids; lipids in body fluids; amino acid derivatives of salicylic acid. *Mailing Add:* Dept Chem Oakwood Col Huntsville AL 35806

HAMER, MARTIN, b Indianapolis, Ind, Sept 17, 28; m 54; c 2. ORGANIC CHEMISTRY. *Educ:* Ind Univ, BS, 49; Purdue Univ, MS, 51, PhD(org chem), 53. *Prof Exp:* Res chemist, Standard Oil Co Ind, 53-60; res chemist, 60-74, mgr process & prod res, 74-77, corp toxicity coordr, 77-81, MGR TECH MKT SERV, INT MINERALS & CHEM CORP, 81- *Mem:* Am Chem Soc. *Res:* Heterocyclics; fuels; pesticides; medicinals; organic phosphorus compounds; improved processes for phosphate flotation. *Mailing Add:* 8423 N Karlov Skokie IL 60076

HAMER, WALTER JAY, b Altoona, Pa, Nov 5, 07; m 41; c 1. PHYSICAL CHEMISTRY, ELECTROCHEMISTRY. *Educ:* Juniata Col, BS, 29; Yale Univ, PhD(phys chem), 32. *Hon Degrees:* DSc, Juniata Col. 66. *Prof Exp:* Asst instr, Juniata Col, 27-29; asst, Yale Univ, 29-32, USN res fel, 32-34; res assoc chem, Mass Inst Technol, 34-35; chemist, Nat Bur Standards, 35-50, chief electrochem sect, 50-70, consult electrochem & dir electrolyte ctr, 70-72; CHEM CONSULT & SCI WRITER, 72- *Concurrent Pos:* Lectr grad sch, USDA, 40-49, Nat Bur Standards, 44-45, 52-53, Cath Univ, 44-45 & Georgetown Univ, 47-50; res chemist, Nat Defense Res Comn, Off Sci Res & Develop & Manhattan Proj, Nat Bur Standards, 43-44; consult res & develop bd, US Dept Defense, 51-53; mem comn electrochem, Int Union Pure & Appl Chem, 57-71; tech adv, Int Electrotech Comn, 50-67; chmn, dry cell comt, Am Standards Asn, 50-68; pres, Yale Chem Assoc, 58-61; adv, coun electrochem, Univ Pa, 62-63. *Honors & Awards:* Manhattan Proj Award, 45; Superior Awards, 54, 62 & 65, Gold Medal Award, 65, US Dept Com. *Mem:* Fel AAAS; fel Am Inst Chem; Am Phys Soc; hon mem Electrochem Soc (vpres, 60-63, pres, 63-64); fel Inst Elec & Electronics Eng; fel NY Acad Sci; Sigma Xi. *Res:* Standard cells; primary and secondary batteries; electrolytic solutions; standardization of pH scale; fused salts and thermal batteries; electromotive series for molten electrolytes; critical evaluation of electrochemical data; Faraday determination. *Mailing Add:* 407 Russell Ave No 305 Gaithersburg MD 20877

HAMERLY, ROBERT GLENN, b Apr 4, 31; US citizen; m 55; c 2. PHYSICS. *Educ:* Western Ill Univ, BS, 55; Univ Ill, Urbana, MS, 57; Colo State Univ, PhD(physics), 69. *Prof Exp:* Mem tech staff systs anal, Hughes Aircraft Corp, Calif, 59-60; assoc prof, 60-75, PROF PHYSICS, UNIV NORTHERN COLO, 75- *Mem:* Am Asn Physics Teachers; Sigma Xi. *Res:* Solid state physics; transport properties of semiconductors. *Mailing Add:* Dept Physics Greeley CO 80631

HAMERMAN, DAVID JAY, b New York, NY, Apr 20, 25; m 53; c 3. INTERNAL MEDICINE. *Educ:* NY Univ, MD, 48. *Prof Exp:* From asst prof to assoc prof, 56-68, chmn dept, 68-81, PROF MED, MONTEFIORE HOSP & MED CTR, ALBERT EINSTEIN COL MED, 68- *Concurrent Pos:* Dazian fel chem, Col Med, NY Univ, 52-53; Markle scholar med sci, 58; Sinsheimer scholar med sci, 63; Fogarty Int fel, 81. *Mem:* Am Soc Clin Invest; Am Rheumatism Asn; Asn Am Physicians; fel Am Col Physicians. *Res:* Joint structure and function in the normal state, in aging and in osteoarthritis. *Mailing Add:* Dept Med Albert Einstein Coll Med 1300 Morris Park Ave Bronx NY 10461

HAMERMESH, BERNARD, b Brooklyn, NY, Dec 25, 19; m 41; c 3. PHYSICS. *Educ:* City Col New York, BS, 40, NY Univ, MS, 42, PhD(physics), 44. *Prof Exp:* Tutor, City Col New York, 41-42; instr, NY Univ, 43-46; Nat Res Coun fel, Calif Inst Technol, 46-48; assoc physicist, Argonne Nat Lab, 48-58, sr physicist, 58-59; sr tech adv, Res Lab, TRW Inc, 59-60 & Space Tech Labs, 60-65, sr staff physicist, TRW Systs, 65-68; prof physics & chmn dept, Cleveland State Univ, 68-; RETIRED. *Concurrent Pos:* Vis prof physics, Univ Calif, Los Angeles, 87. *Mem:* Am Phys Soc. *Res:* Electron scattering; hygrometry; cosmic ray neutrons; cosmic ray meson energy spectrum; photo-neutron reactions; gamma rays from neutron capture; micrometeoroids; micrometeoroid accelerators. *Mailing Add:* 10433 Wilshire Blvd Apt 906 Los Angeles CA 90024

HAMERMESH, MORTON, b New York, NY, Dec 27, 15; m 41; c 3. PHYSICS. *Educ:* City Col New York, BS, 36; NY Univ, PhD(physics), 40. *Prof Exp:* Tutor physics, City Col New York, 36-37; instr, 41; from asst to assoc prof, NY Univ, 37-47; instr, Stanford Univ, 41-43, res assoc, 42-43; res assoc, Radio Res Lab, Harvard Univ, 43-45; assoc prof, NY Univ, 47-48; sr physicist, Argonne Nat Lab, 48-50, assoc dir physics div, 50-59, dir, 59-63, assoc dir, Lab, 63-65; head sch physics & astron, Univ Minn, Minneapolis, 65-69; head physics dept, State Univ NY, Stony Brook, 69-70; head sch physics & astron, 70-75, prof physics & astron, Univ Minn, Minneapolis, 75-86; RETIRED. *Concurrent Pos:* Off Sci Res & Develop, 44; consult, Brookhaven Nat Lab, 47-; ed, J Math Physics, 70-79; mem bd trustees, Argonne Univs Asn, 72-76. *Honors & Awards:* Townsend Harris Medal. *Mem:* AAAS; fel Am Phys Soc; Am Asn Univ Prof. *Res:* Theoretical nuclear physics; electromagnetic theory; longwave search antenna array; passage of neutrons through crystals and polycrystals; group theory; elementary particle physics; nonlinear phenomena; theory of solitons; symmetry principles. *Mailing Add:* 120 Yale Pl No 508 Minneapolis MN 55455

HAMERSKI, JULIAN JOSEPH, b Winona, Minn, May 21, 30; m 55; c 6. INORGANIC CHEMISTRY. *Educ:* St Mary's Col, BS, 52, MA, 57; Univ of Pac, PhD(chem), 63. *Prof Exp:* Res chemist, Am Can Co, 52-53; teacher high schs, Minn, 53-56; chem instr, Worthington Jr Col, Minn, 56-61; teacher high sch, Calif, 62-63; ASSOC PROF CHEM, EASTERN ILL UNIV, 63- *Mem:* Am Chem Soc. *Res:* Iron transition metal carbonyl phosphine complexes. *Mailing Add:* Dept Chem Eastern Ill Univ Charleston IL 61920

HAMERSMA, J WARREN, b Midland Park, NJ, July 24, 40; m 63. ENVIRONMENTAL CHEMISTRY. *Educ:* Calvin Col, AB, 62; Univ Conn, PhD(chem), 66. *Prof Exp:* NASA fel, Northwestern Univ, 66-67; sr res chemist, Arco Chem Co Div, Atlantic-Richfield Co, 68-70; mgr qual control, Int Chem & Nuclear Corp, 70-71; SR MEM PROF STAFF, TRW SYSTS, 71- *Mem:* Am Chem Soc; Royal Soc Chem. *Res:* Oxidation reactions; energy research and development; evaluation of clean fuels; environmental assessment and chemistry. *Mailing Add:* 26730 Eastvale Rd Palo Verdes Peninsula CA 90274

HAMERSTROM, FRANCES, b Needham, Mass, Dec 17, 07; m 31; c 2. BIOLOGY, NATURE WRITING. *Educ:* Iowa State Col, BS, 35; Univ Wis-Madison, MS, 40. *Hon Degrees:* DSc, Carroll Col, 61. *Prof Exp:* Biologist game, Wis Dept Natural Resources, 49-72; ADJ PROF, UNIV WIS-STEVENS POINT, 82- *Concurrent Pos:* Dir, Raptor Res Found, 74-76; mem sci bd, Wis Peregrine Soc, 90- *Honors & Awards:* Awards, Wildlife Soc, 40 & 57; Josselyn Van Tyne Award, Am Ornithologist's Union, 60; Chapman Award, Am Mus Natural Hist, 64; United Peregrine Soc Conserv Award, 80; Edwards Prize, Wilson Ornith Soc, 85. *Mem:* Corresp mem Ger Ornith Soc; Wilson Ornith Soc; Wildlife Soc; hon mem, Ger Falconry Asn; fel Am Ornithologist's Union. *Res:* Ecology and behavior of raptors; hunting ethics and habits. *Mailing Add:* Rte 1 Box 448 Plainfield WI 54966

HAMERSTROM, FREDERICK NATHAN, wildlife ecology, ornithology; deceased, see previous edition for last biography

HAMERTON, JOHN LAURENCE, b Brighton, Eng, Sept 23, 29. HUMAN GENETICS, CELL BIOLOGY. *Educ:* Univ London, BSc, 51, DSc(human genetics), 68. *Prof Exp:* Sci staff mem mammal cytogenetics group, Radiobiol Res Unit, Med Res Coun, 51-56; sr sci officer, Zool Dept, Brit Mus Natural Hist, HM Sci Civil Serv, 59-60; sr lectr, Cytogenetics Sect, Pediat Res Unit, Guy's Hosp Med Sch, 60-69, head sect, 62-69; from assoc prof to prof pediat & anat, fac med, 69-84, assoc dean, 77-81, PROF & HEAD DEPT HUMAN GENETICS, UNIV MAN, 85-, DISTINGUISHED PROF, 87- *Concurrent Pos:* Consult, Psychiat Genetics Res Unit, Med Res Coun, 62-69; adv human cytogenetics, WHO, 67; chmn, Working Group Prenatal Diag Genetic Dis & mem, Grants Panel Genetics, Med Res Coun Can, 71, res prof, 81-82; chmn, Int Standing Comt Human Cytogenetics, 71-76; dir, Dept Genetics, Children's Ctr Winnipeg, 69-80. *Honors & Awards:* Robert Roessler DeVilliers Award, Leukemia Soc Am, 56; Huxley Mem Medal, Imperial Col, London Univ, 58; Teddy Award, Childrens Hosp Winnepeg Res Found, 88. *Mem:* Royal Soc Med; Soc Pediat Res; Genetics Soc Can (pres, 77-78); Am Soc Human Genetics (pres, 75). *Res:* Human & mammalian cytogenetics; somatic cell genetics; gene mapping; molecular genetics. *Mailing Add:* Dept Human Genetics Univ Man 250-770 Bannatyne Ave Winnipeg MB R3E 0W3 Can

HAMES, F(REDERICK) A(RTHUR), b Assiniboia, Sask, July 22, 19; m 44; c 2. METALLURGICAL ENGINEERING. *Educ:* Mont Sch Mines, BS, 40; Queen's Univ, Ont, MSc, 46; Univ Mo, PhD(metall eng), 48. *Prof Exp:* Asst prof metall, Univ BC, 48-50; prof & head dept, Mont Sch Mines, 50-57; prof metall eng, Queen's Univ, Ont, 57-84; RETIRED. *Mem:* Am Soc Metals; Am Inst Mining, Metall & Petrol Engrs. *Res:* Magnetic materials; constitution and structure of alloys; x-ray diffraction. *Mailing Add:* 2132 Colinwood Rd Sidney BC V8L 4H5 Can

HAMET, PAVEL, b Klatovy, Czech, June 13, 43; m; c 4. ENDROCRINOLOGY. *Educ:* Charles Univ, MD,67; McGill Univ, PhD, 72. *Hon Degrees:* Montreal Univ, CSPd; Royal Col, FRCP(C). *Prof Exp:* dir, clin res lab, 75-90; DIR, RES CTR, HOTEL DIEU DE MONTREAL, 90- *Concurrent Pos:* Chief, Endocrinol, Hotel Dieu of Montreal; prof, Montreal Univ; assoc mem, McGill Univ; Centennial Fel, Med Res Coun Can. *Honors & Awards:* Res Award, Can Cardiovasc Soc; Astra Award, Can Hypertension Soc; Can Hypertension Soc (pres); Harry Goldblatt Award, Am Heart Asn, 90. *Mem:* Asn Med Lang Francaise Can; AAAS; Can Med Asn; Can Soc

Endocrinol & Metab; Am Fedn Clin Res; Endocrine Soc; Can Soc Clin Invest; Am Heart Asn; Can Diabetes Asn; Can Health Res. *Res:* Mechanism of action of atrial natriuretic factor; studies on nutritional calcium in hypertension; studies on heat stress proteins in hypertension; growth control of vascular smooth muscle in diabetes mellitus and hypertension; platelet functions and cyclic nucleotide metabolism in hypertension and diabetes mellitus. *Mailing Add:* Res Ctr Hotel Dieu de Montreal Pavillon Marie de la Ferre 3850 St-Urbain Montreal PQ H2W 1T8 Can

HAMID, MICHAEL, b June 7, 34; Can citizen. ELECTRICAL ENGINEERING. *Educ:* McGill Univ, BEng, 60, MEng, 62; Univ Toronto, PhD(elec eng), 66. *Prof Exp:* Res asst elec eng, McGill Univ, 60-61; sr consult engr, Sinclair Radio Labs, 62-66; from asst prof to assoc prof elec eng, 65-70, PROF ELEC ENG, UNIV MAN, 70-, HEAD ANTENNA LAB, 67- *Concurrent Pos:* Treas, assoc ed & mem bd gov, Int Microwave Power Inst, 69-, pres, 71-; gen chmn, Microwave Power Symposium, Monterey, 71; pres, Indust Microwave Res Assoc, Winnipeg, 71-75; mem, Nat Res Coun Can assoc comn on bird hazards to aircraft, 72-77; mem, Man Res Coun; consult, Defense Res Bd Can, 72-; vis prof, Naval Postgrad Sch, Monterey, Calif, 79-81; mem, Can deleg to Int Union Radio Sci, 65-; chmn grad studies, Univ Man, 83-88. *Mem:* Sr mem & fel Inst Elec & Electronics Engrs; Int Union Radio Sci; Can Med & Biol Eng Soc; fel Inst Electronic Eng; Int Microwave Power Inst; Med Prod Inst. *Res:* Antennas; diffraction; scattering; inverse scattering; microwave techniques; microwave power; acoustics; electromagnetic theory; transmission lines; biological effect of electromagnetic energy. *Mailing Add:* Dept Elec Eng Univ Man Winnipeg MB R3T 2N2 Can

HAMIELEC, ALVIN EDWARD, b Cracow, Poland, Jan 10, 35; Can citizen; m 55; c 3. CHEMICAL ENGINEERING, POLYMER SCIENCE. *Educ:* Univ Toronto, BASc, 57, MASc, 58, PhD(mass transfer), 61. *Prof Exp:* Res engr chem, Cent Res Labs, Can Industs Ltd, Que, 61-63; from asst prof to assoc prof chem eng, 63-70, PROF CHEM ENG, MCMASTER UNIV, 70- *Concurrent Pos:* Dir, McMaster Inst Polymer Prod Technol, McMaster Univ. *Honors & Awards:* ERCO Award, Can Soc Chem Eng, 74; Protective Coatings Award, Chem Inst Can, 78; Dunlop Award, Macro Sci/Eng, 87. *Mem:* Am Inst Chem Engrs; Chem Inst Can; Am Chem Soc; Can Soc Chem Engrs; Fel Royal Soc Can. *Res:* Polymer production technology; polymer synthesis and characterization; polymer reactors and production technology. *Mailing Add:* 99 Rennick Rd Burlington ON L7R 3X5 Can

HAMIL, MARTHA M, b New Edinburg, Ark, Jan 14, 39. STRUCTURAL GEOLOGY, MINERALOGY. *Educ:* La State Univ, BS, 59, MS, 65; Univ Mo-Columbia, PhD(geol), 71. *Prof Exp:* Asst prof geol, Rutgers Univ, 72-79; MEM STAFF, ENGELHARD MINERALS & CHEM, 79- *Concurrent Pos:* Fel, Va Polytech Inst, 70-71, Univ Calif, Berkeley, 72. *Mem:* AAAS; Am Geophys Union; Geol Soc Am; Geol Soc London; Mineral Soc Am. *Res:* Hercynian tectonics of Central Morocco; fabric analysis; planar structures; tectonic-mineralogic interface; fracture mechanisms in rocks; distortions of metal-oxygen polyhedra in crystal structures. *Mailing Add:* 357 N Fourth Ave Highland Park NJ 08904

HAMILL, DENNIS W, b Cedar Rapids, Iowa, Dec 7, 40; m 60; c 2. SOLID STATE PHYSICS. *Educ:* Univ Wis-Madison, BS, 62, MS, 63; Boston Univ, PhD(physics), 68. *Prof Exp:* Staff physicist, Forest Prod Lab, USDA, 63; physicist semiconductors, Dow Corning Corp, 63-65, group leader, 68-70; vpres opers, High Performance Technol Inc, 70-71; res mgr physics, 71-77, tech dir, Info Mgt Div, 77-80, TECH DIR, ELEC & MECH RES DIV, 3M CO, 80- *Concurrent Pos:* Fac mem dept physics, Saginaw Valley Col, 64-69. *Mem:* Am Phys Soc; Int Soc Hybrid Microelectronics. *Res:* Semiconductor materials and devices; role of impurities in silicon and other bulk semiconductors and films; Mossbauer effect. *Mailing Add:* 3M Dynatel Systs Div PO Box 2963 Austin TX 78769

HAMILL, JAMES JUNIOR, b Griswold, Iowa, Dec 7, 21; m 47; c 2. CHEMISTRY, CHEMICAL MICROSCOPY. *Educ:* Univ Colo, BA, 50. *Prof Exp:* Microscopist, Res Div, Goodyear Tire & Rubber Co, 50-56, sr microscopist, 56-64, res scientist, 64-65, head, Micros Sect, Res Div, 65-80; RETIRED. *Honors & Awards:* R P Dinsmore Res Award, 59. *Mem:* Am Chem Soc. *Res:* Chemical and industrial microscopy; development of new or improved processes for synthetic rubber, plastics, rubber chemicals or other products of interest to the rubber and plastics industry; high impact plastics. *Mailing Add:* 105 Audubon Rd SE Winter Haven FL 33884

HAMILL, PATRICK JAMES, b Salt Lake City, Utah; c 2. CELESTIAL MECHANICS, AEROSOL PHYSICS. *Educ:* St Edward's Univ, Austin, BS, 59; Univ Ariz, MS, 68, PhD(physics), 71. *Prof Exp:* Vis prof physics, Univ Trujillo, Peru, 62-66; fel geophys, Univ Chicago, 71-72; asst prof physics, Clark Col, Atlanta, 72-74; res assoc, Ames Res Ctr, NASA, 74-78; scientist, Systs & Appl Sci Corp, 78-81; PROF PHYSICS, SAN JOSE STATE UNIV, CALIF, 81- *Honors & Awards:* Julian Allen Award, NASA. *Mem:* Am Phys Soc; Sigma Xi; Am Meteorol Soc. *Res:* Celestial mechanics and aerosol physics; author of 50 journal articles on topics ranging from the rings of Saturn to the formation of stratospheric aerosol particles and the deposition of particles in the lung. *Mailing Add:* 10300 Moretti Dr Cupertino CA 95014

HAMILL, PETER VAN VECHTEN, b Baltimore, Md, Apr 16, 26; m 52; c 4. EPIDEMIOLOGY, PREVENTIVE MEDICINE. *Educ:* Univ Mich, BA, 47, MD, 53; Johns Hopkins Univ, MPH, 62. *Prof Exp:* Med officer in charge, Alaska Native Hosp, Tanana, 55-57; med dir, USPHS, 55-78, chief epidemiol studies air pollution, 58-60, sci dir surgeon gen study smoking & health, 62-64, chief med adv, US Nat Health Exam Surv, 64-78; CONSULT EPIDEMIOLOGIST PREV MED, HAMILL ASSOCS, INC, 78- *Concurrent Pos:* Consult, Nat Ctr Health Statist, USPHS, 78-; sr med consult, SRI-Int, 78-84; consult epidemiologist, Olin Corp, 78-; chmn, sci adv comt, Aleutian-Bering Sea Inst, 66-79; vis prof, Epidemiol & Pub Health, Univ Mass, Amherst, 78-79; prof epidemiol & prev med, Med Sch, Univ Md, Baltimore, 79-81, adj prof, 81-; sr med consult, Occidental Chem, 86- *Mem:* Soc Epidemiol Res; Soc Study Human Biol; fel Am Acad Occup Med; fel Am Col Prev Med; Asn Teachers Prev Med; fel Am Col Epidemiol. *Res:* Pulmonary disease, chronic, non-infectious, malignant and nonmalignant; environmental health, especially air pollution, smoking and occupational health; epidemiology of occupational diseases; preventive medicine in occupational diseases. *Mailing Add:* 1001 Whitehall Cove Annapolis MD 21401

HAMILL, ROBERT L, b Youngstown, Ohio, Mar 13, 27; m 53, 76; c 2. NATURAL PRODUCTS CHEMISTRY, BIOCHEMISTRY. *Educ:* Ohio Univ, BS, 50; Mich State Univ, MS, 53, PhD(biochem), 55. *Prof Exp:* Res asst, Mich State Univ, 50-51 & 52-55; sr biochemist, 55-64, res scientist, 64-69, res assoc, 69-83, group leader antibiotic isolation, 70-83, RES ADVISOR, ANTIBIOTIC ISOLATION, ELI LILLY & CO, 83- *Concurrent Pos:* Chemist, Mich Dept Health Lab, 54-55; ed, J Antibiotics, 75- & Antimicriobial Agents & Chemother, 80-85, vchmn, Int Conf Antimicrobial Agents & Chemother, 84-87, chmn, 88- *Mem:* Am Chem Soc; Am Soc Microbiol; Sigma Xi; NY Acad Sci; AAAS. *Res:* Antibiotic isolation and purification; blood coagulation; transmethylation and methylation in higher plants and animals; fibrinolysis. *Mailing Add:* Fermentation Prod Res Lilly Res Labs Eli Lilly & Co Lilly Corp Ctr Indianapolis IN 46285-1533

HAMILL, ROBERT W, b Hartford, Conn, July 30, 42; m 66; c 3. NEUROLOGY, NEUROSCIENCE. *Educ:* Springfield Col, BS, 64; Wake Forest Univ, MD, 68. *Prof Exp:* Med intern & resident, Sch Med & Dent, Univ Rochester, 68-70; med officer-internist, USN, 70-73; resident neurol, Med Col, Cornell Univ, 73-75, instr, 75-76, asst prof, 76-80; asst prof, 80-85, ASSOC PROF NEUROL, SCH MED & DENT, UNIV ROCHESTER, 85-; DIR NEUROL, MONROE COMMUNITY HOSP, 80- *Concurrent Pos:* Alfred P Sloan Found fel, 76; Nat Res Serv Award, Nat Inst Neurol Commun Dis & Stroke, 76-78, teacher investr develop award, 78-80; Jordan res fel, Nat Paraplegia Found, 77. *Mem:* Soc Neurosci; Am Acad Neurol; AAAS; Asn Res Nervous & Ment Dis; NY Acad Sci. *Res:* Growth and development, function, aging, and hormonal control of the autonomic nervous system; neurodegenerative disorders and autonomic dysfunction. *Mailing Add:* 435 E Henrietta Rd Rochester NY 14603

HAMILL, WILLIAM HENRY, b Oswego, NY, June 13, 08; m 34; c 5. PHYSICAL CHEMISTRY. *Educ:* Univ Notre Dame, BS, 30, MS, 31; Columbia Univ, PhD(phys chem), 36. *Prof Exp:* Asst prof phys chem, Fordham Univ, 31-38; prof, 38-74, EMER PROF PHYS CHEM, UNIV NOTRE DAME, 74- *Mem:* Am Chem Soc; The Chem Soc. *Res:* Ionic processes in radiation chemistry; mass spectrometry; slow electron impact phenomena for thin-film dielectric solids; electron transport in disordered solids. *Mailing Add:* 17899 Edgewood Walk South Bend IN 46635

HAMILTON, ANGUS CAMERON, b Listowel, Ont, Apr 18, 22; m 49; c 5. GEODESY. *Educ:* Univ Toronto, BASc, 49, MASc, 51. *Prof Exp:* Chief shoran sect, Geod Surv Can, 54-57, electronics sect, 57-58; sr sci officer, Can Dept Mines & Tech Surv, 58-67; coordr res & training, Surv & Mapping Br, Can Dept Energy, Mines & Resources, 67-71; chmn dept, 71-85, prof, 71-86, EMER PROF SURV ENG, UNIV NB, 86- *Honors & Awards:* Earle Fernell Award, Am Cong Surv Mapping, 83; Champlain Award, Can Coun Land Surveyors. *Mem:* Hon mem Can Inst Surv & Mapping. *Res:* Application of electronic distance measuring methods to geodetic surveying; application of gravity measurements to geoidal and isostatic studies; development of land information systems; development of studies in surveying engineering economy. *Mailing Add:* Dept Surv Eng Univ NB Fredericton NB E3B 5A3 Can

HAMILTON, BRUCE KING, b Easton, Pa, May 26, 47; m 83; c 2. BIOCHEMICAL ENGINEERING, CHEMICAL PROCESS DEVELOPMENT. *Educ:* Mass Inst Technol, BS, 74, PhD(biochem eng), 74. *Prof Exp:* Res assoc biochem eng, Mass Inst Technol, 74-75; group leader, 75-77, sect head fermentation technol, Frederick Cancer Res Ctr, 78-80; dir process develop, Genex Corp, 80-81, vpres process develop, 82-85; DIR PROCESS DEVELOP, RES DIV, W R GRACE & CO, 85- *Mem:* AAAS; Am Inst Chem Engrs; Am Chem Soc; Am Soc Microbiol; Soc Indust Microbiol. *Res:* Biochemical engineering; enzyme engineering; fermentation microbiology; drug discovery and development; biotechnology process design, development, and scale-up; biotechnology plant start-up and toubleshooting; design of biotechnology plant operating and maintenance systems. *Mailing Add:* 15229 Watergate Rd Silver Spring MD 20905

HAMILTON, BRUCE M, b Hamilton, Ont, Apr 20, 20; m. METALLURGY. *Educ:* Queens Univ, BSc, 43. *Prof Exp:* Res analyst, Dept Mines & Resources, Ottawa, 43-44; chief metallurgist, supt melting & primary rolling & dir metall, Atlas Steel Corp, Welland, Ont, 44-64; dir metall, Crucible Steel Corp, Pittsburgh, 66-67, works mgr, 67-68; pres, Colt Indust, 68-71; pres & chief exec officer, Slater Steels Corp, Hamilton, Ont, 71-86; pres, Bruce Hamilton Consult Inc, Burlington, Ont, 86-87; vpres & gen mgr, Atlas Specialty Steels, Welland, Ont, 87-88, pres, 88-89; consult to chmn, Sammi Steel Co, Seoul, Korea, 89-90; CHMN BD, SYDNEY STEEL CORP, NS, 91- *Concurrent Pos:* Bd dirs, numerous corp & orgn, 75-90. *Res:* Alloy specialty steels; automotive steels; fatigue in mining and construction steels; application and selection of specialty steels. *Mailing Add:* Bruce Hamilton Consults Inc 584 North Shore Blvd E Burlington ON L7T 1X2 Can

HAMILTON, BYRON BRUCE, b New Brighton, Pa, July 6, 34; m 58; c 2. PHARMACOLOGY. *Educ:* Syracuse Univ, AB, 56; State Univ NY, MD, 59, PhD(pharmacol), 71. *Prof Exp:* Intern med, Boston City Hosp, 59-60; asst prof pharmacol & rehab, State Univ NY Upstate Med Ctr, 67-70; from asst prof to assoc prof, clin rehab, Sch Med, Northwestern Univ, 70-84; dir res, Nat Inst Handicapped Res, Rehab Res & Training Ctr, Northwestern Univ-Rehab Inst Chicago, 70-84; CLIN ASSOC PROF REHAB MED, SCH MED, STATE UNIV NY, BUFFALO, 84- *Concurrent Pos:* USPHS fels, State Univ NY Upstate Med Ctr, 60-61 & 63-67. *Honors & Awards:* Licht Award. *Mem:* AAAS; Am Rheumatism Asn; Am Cong Rehab Med. *Res:*

Physiology of cell cation transport; renal acid and base transport; rehabilitation evaluation; rehabilitation research administration. *Mailing Add:* Dept Rehab Med State Univ NY 82 Farber Hall 3435 Main St Buffalo NY 14214

HAMILTON, C HOWARD, b Pueblo, Colo, Mar 17, 35; m 68; c 3. MECHANICAL METALLURGY, SHAPING & FORMING. *Educ:* Colo Sch Mines, BS, 59; Univ Southern Calif, MS, 65; Case Western Res Univ, PhD(metall), 68. *Prof Exp:* Mem tech staff, Mat Lab, LA Div, Rockwell Int, 68-76, mem tech staff metals processing, Sci Ctr, 76-78, group mgr, 78-80, prin scientist mat, 80-81, dir, 81-84; PROF MAT SCI, WASH STATE UNIV, 84- *Concurrent Pos:* Ed, J Mat Shaping Technol, 86-88; mem Shaping-Forming Comt, Minerals, Metals & Mat Soc, 78- *Mem:* Minerals Metals & Mat Soc; fel Am Soc Metals Int; Sigma Xi; Mat Res Soc. *Res:* Deformation processing and forming, especially at elevated temperatures, including superplasticity, superplastic forming, diffusion bonding and thermomechanical processing; number of publications and patents. *Mailing Add:* Wash State Univ Pullman WA 99164-2920

HAMILTON, CAROLE LOIS, b Butte, Mont, Jan 25, 37; m 62; c 2. ENERGY CONVERSION; SYSTEMS ANALYSIS. *Educ:* Colo State Univ, BS, 58; Calif Inst Technol, PhD(chem), 63. *Prof Exp:* Res fel chem, Calif Inst Technol, 62-65; res fel, Stanford Univ, 65-68, lectr, 68-71; with Environ Qual Lab, Calif Inst Technol, 71-74; mem tech staff, 74-78, TECH GROUP SUPVR, JET PROPULSION LAB, 79- *Mem:* Am Chem Soc; Sigma Xi. *Res:* Rearrangements in organometallic compounds; analysis of specificity alpha-chymotrypsin; spin-labeled biomolecules; energy systems engineering. *Mailing Add:* 2269 Midwock Dr Altadena CA 91001

HAMILTON, CHARLES LEROY, b Nyack, NY, Feb 19, 32; m 54; c 3. GEOLOGY. *Educ:* Lehigh Univ, BA, 53; Dartmouth Col, MA, 54; Va Polytech Inst, PhD(geol), 64. *Prof Exp:* Explor geologist, NJ Zinc Co, Va, 56-58; from instr to asst prof, Rutgers Univ, 60-69; assoc prof, 69-81, PROF GEOL, MONTCLAIR STATE COL, 81- *Mem:* Nat Asn Geol Teachers; Mineral Soc Am; Sigma Xi; Geol Soc Am. *Res:* Igneous and metamorphic petrology. *Mailing Add:* Three Lafayette Ct Wayne NJ 07470

HAMILTON, CHARLES R, b Chicago, Ill, Sept 27, 35; m 64; c 2. HEMISPHERIC SPECIALIZATION, VISUAL PERCEPTION. *Educ:* Univ of the South, BS, 57; Calif Inst Technol, PhD(biol), 64. *Prof Exp:* Res fel psychobiol, Calif Inst Technol, 64-65; asst prof psychol, Stanford Univ, 65-71; sr res fel, Calif Inst Technol, 71-74, from res assoc to sr res assoc biol, 74-90; RES SCIENTIST, SCH MED, TEX A&M UNIV, 91- *Mem:* AAAS; Am Psychol Soc; Asn Res Vision & Ophthal; Soc Neurosci; Sigma Xi. *Res:* Behavioral testing of split-brain monkeys and humans; study cerebral lateralization, interhemispheric relations and visual perception. *Mailing Add:* Human Anat & Med Neurobiol Sch Med Tex A&M Univ College Station TX 77843

HAMILTON, CHARLES W(ALLACE), computer science, chemical engineering, for more information see previous edition

HAMILTON, CHARLES WILLIAM, b Chicago, Ill, Sept 6, 19; m 43; c 4. CHEMISTRY. *Educ:* Cent YMCA Col, BS, 41; Northwestern Univ, MS, 43. *Prof Exp:* Res chemist, Armour Res Found, 43-44; assoc res chemist, Columbia-Southern Chem Corp, 46-52; asst div chief rubber & plastics chem, Battelle Mem Inst, 52-60, assoc dir, New Prod Res, Continental Can Co, Inc, 60-69; mgr prod develop & tech serv, Foster Grant Co, Inc, 69-80; dir, Develop Lab Plastics Div, Am Hoechst Corp, 80-83; RETIRED. *Mem:* Emer mem Am Chem Soc. *Res:* Polymer chemistry; synthesis and processing of thermoplastics; properties of crystalline polymers; electrical applications of plastics; condensation polymers; plastics packaging. *Mailing Add:* 118 Prospect St Leominster MA 01453

HAMILTON, CLARA EDDY, b New Haven, Conn, Mar 14, 23; m 55. ZOOLOGY, PHYSIOLOGY. *Educ:* Univ Ga, BS, 42; Univ Ill, MA, 44, PhD(zool), 46. *Prof Exp:* Asst prof zool, Univ Ga, 46-55, assoc prof, 55-56; physiologist, Div res grants, NIH, 56-78; RETIRED. *Mem:* AAAS; Am Soc Zoologists; Am Physiol Soc. *Res:* Physiology of reproduction; endocrinology. *Mailing Add:* 15424 Tierra Dr Silver Spring MD 20906

HAMILTON, D(EWITT) C(LINTON), JR, b Eufaula, Okla, Dec 4, 18; m 42. MECHANICAL ENGINEERING. *Educ:* Univ Okla, BS(mech eng) & BS(petrol eng), 41; Univ Calif, MS, 46; Purdue Univ, PhD(mech eng), 49. *Prof Exp:* Res & develop engr, Power Plant Lab, Wright Field, 46-47; res asst, Purdue Univ, 47-48, instr heat transfer & thermodyn, 48-49, asst prof heat transfer, 49-51; sr develop engr, Oak Ridge Nat Lab, 51-52, prin develop engr, 52-56; lectr in chg reactor eng, Oak Ridge Sch Reactor Technol, 56-65; head dept, 65-74, prof, 65-86, EMER PROF MECH ENG, TULANE UNIV, LA, 86- *Concurrent Pos:* Spec lectr, Ga Inst Technol, 64. *Mem:* Am Soc Mech Engrs Int. *Res:* Thermodynamics; heat transfer; solar thermal systems. *Mailing Add:* Dept Mech Eng Tulane Univ New Orleans LA 70118

HAMILTON, DAVID FOSTER, physiology woody plants; deceased, see previous edition for last biography

HAMILTON, DAVID WHITMAN, b Anaconda, Mont, Nov 29, 35; m 59. DEVELOPMENTAL BIOLOGY. *Educ:* Harvard Univ, AB, 57; Univ Kans, MA, 60; Cambridge Univ, PhD(anat), 64. *Prof Exp:* Instr anat, Univ Kans, 59-60; fel, 63-65, from instr to assoc prof, Harvard Med Sch, 65-77, Lawrence J Henderson assoc prof health sci & technol, 74-77; PROF CELL BIOL & NEUROANAT & HEAD DEPT, UNIV MINN, MINNEAPOLIS, 77-, DIR GRAD STUDIES, 80- *Mem:* Am Soc Cell Biol; Soc Study Reproduction; Am Asn Anat. *Res:* Cell and reproductive biology. *Mailing Add:* Dept Cell Biol & Neuroanat Univ Minn Jackson Hall Minneapolis MN 55455

HAMILTON, DOUGLAS J(AMES), b Canton, Ohio, Dec 6, 30; m 53; c 2. ELECTRICAL ENGINEERING. *Educ:* Case Western Reserve Univ, BS, 53; Univ Calif, Los Angeles, MS, 56; Stanford Univ, PhD, 59. *Prof Exp:* Mem tech staff, Hughes Aircraft Co, 53-57; design engr comput lab, Gen Elec Co, 57-58; asst elec eng, Stanford Univ, 58-59; assoc prof, 59-62, actg dir solid state eng lab, 65, PROF ELEC ENG, UNIV ARIZ, 62- *Concurrent Pos:* Consult, Lockheed Missile Syst Div, 59 & Motorola Semiconductor Prod, Inc, 60- *Res:* Solid-state circuit techniques and devices; integrated circuits. *Mailing Add:* Dept Elec Comp Eng Univ Ariz Tucson AZ 85721

HAMILTON, DOUGLAS STUART, b Ft Collins, Colo, June 28, 49. PHYSICS. *Educ:* Univ Colo, Boulder, BA, 71; Univ Wis-Madison, PhD(physics), 76. *Prof Exp:* Res assoc physics, Univ Wis, 77; res assoc, Univ Southern Calif, 77-80; PROF PHYSICS, UNIV CONN, 80- *Mem:* Am Phys Soc; Am Asn Physics Teachers; Sigma Xi. *Res:* Dynamical interactions between light and condensed media. *Mailing Add:* Dept Physics Univ Conn Storrs CT 06269-3046

HAMILTON, EDWIN LEE, b Sherman, Tex, Dec 20, 14; m 38; c 3. MARINE GEOLOGY. *Educ:* Tex Agr & Mech Col, BS, 36; Stanford Univ, MS, 50, PhD(geol), 52. *Prof Exp:* Lectr geol, Univ Wash. 51; marine geologist, US Navy Electronics Lab, 51-77; marine geologist, 77-84, EMER MARINE GEOLOGIST, NAVAL OCEAN SYSTS CTR, 84- *Honors & Awards:* Curl Mem Award, Naval Undersea Ctr, 71. *Mem:* Geol Soc Am; Soc Econ Paleont & Mineral; Am Asn Petrol Geologists; Am Geophys Union; Sigma Xi. *Res:* Marine geology; sediments, geomorphology of the sea floor. *Mailing Add:* Naval Ocean Systs Ctr Code 541 San Diego CA 92152

HAMILTON, ERNEST SCOVELL, b Greenfield, Mass, Oct 7, 28; m 51; c 3. PLANT ECOLOGY. *Educ:* Univ Mass, BS, 51; Rutgers Univ, MS, 52, PhD(plant ecol), 56. *Prof Exp:* Asst biol, Rutgers Univ, 54-56; instr, 56-59, asst prof, 59-62, ASSOC PROF BIOL, BOWLING GREEN STATE UNIV, 62- *Concurrent Pos:* Consult, Ohio Veg Surv, 65- *Mem:* Ecol Soc Am; Sigma Xi; Torrey Bot Club. *Res:* Tree species distribution in relation to edaphic factors, especially on Lake Erie Islands and shoreline. *Mailing Add:* Dept Biol Bowling Green State Univ Bowling Green OH 43403

HAMILTON, FRANKLIN D, b Aulcila, Fla, Oct 30, 42; c 3. PROTEIN & ENZYME CHEMISTRY. *Educ:* Univ Pittsburgh, PhD(biochem); Fla A&M Univ, BS. *Prof Exp:* Assoc prof chem, Atlanta Univ, 79-; DIR, DIV SPONSORED RES, FLA A&M UNIV. *Concurrent Pos:* Asst prof biomed sci, 71-74, assoc prof bio med sci, Grad Sch Bio Med Sci, Univ Tenn-Oak Ridge, 74-79; food found fel, USPHS postdoctoral fel; mem, NIH-MARC rev comt, Nat Acad Sci-NSF; postdoctoral fel rev panel, sci & tech adv comt, Nat Asn Equal Opportunity Higher Ed, vchair; vis lectr, FASEB Minority Inst, Prague. *Mem:* Am Soc Biol Chemists; AAAS; Am Soc Cell Biol. *Mailing Add:* Div Sponsored Res Rm 404 Foote-Hilyer Admin Ctr Tallahassee FL 32307

HAMILTON, GORDON ANDREW, b Cobden, Ont, Mar 15, 35; m 68; c 1. BIO-ORGANIC CHEMISTRY. *Educ:* Queen's Univ, Ont, BA, 56; Harvard Univ, MA, 57, PhD(org chem), 59. *Prof Exp:* Res assoc org chem, Ill Inst Technol, 59-60; from instr to asst prof, Princeton Univ, 60-66; assoc prof, 66-72, PROF ORG CHEM, PA STATE UNIV, 72- *Concurrent Pos:* Sloan res fel, 67-69; mem biochem study sect, NIH, 71-74; NIH spec res fel, Kyoto, Japan, 75; ed, Bioorganic Chem, 83-; mem, comt prof training, Am Chem Soc. *Mem:* AAAS; Am Chem Soc; Am Soc Biol Chemists; Sigma Xi. *Res:* Mechanisms of organic and enzymatic reactions, especially oxidation-reduction reactions; peroxisomal oxidases and their role in metabolism; mechanism of insulin and growth factor action; role of oxalyl thidesters in controlling metabolism. *Mailing Add:* Dept Chem Pa State Univ 152 Davey University Park PA 16802

HAMILTON, GORDON WAYNE, b Salt Lake City, Utah, Nov 5, 26; m 52, 78; c 3. PLASMA PHYSICS, CIVIL ENGINEERING. *Educ:* US Naval Acad, BS, 49; Rensselaer Polytech Inst, BCE, 51; Univ Calif, Berkeley, PhD(physics), 65. *Prof Exp:* Civil eng officer, US Navy Civil Engr Corps, 49-56; mgr produce warehouse fruit & vegetable, Hamilton Fruit Co, 56-58; physicist plasma physics, Aerojet-Gen Corp, 64-68; PHYSICIST PLASMA PHYSICS, LAWRENCE LIVERMORE LAB, UNIV CALIF, 68- *Mem:* Am Phys Soc; Sigma Xi. *Res:* Development of neutral injection systems for magnetic fusion energy experiments; development of techniques for production of negative hydrogen ions and diagnostics; design of fusion reactors. *Mailing Add:* 137 El Altillo Los Gatos CA 95030

HAMILTON, HARRY LEMUEL, JR, b Charleston, SC, May 26, 38; m 81; c 2. MICROMETEOROLOGY. *Educ:* Beloit Col, BA, 60; Univ Wis, MS, 62, PhD(meteorol), 65. *Prof Exp:* Asst prof, 65-71, chmn dept, 76-83, assoc prof atmospheric sci, State Univ NY, Albany, 71-90; OFF OF PROVOST, CHAPMAN COL, 90- *Concurrent Pos:* Environ analyst, Gen Elec Gas Turbine Div, 73-75, environ consult, 75-79. *Mem:* AAAS; Sigma Xi; Am Meteorol Soc. *Res:* Micrometeorologic instrumentation; determination of diffuse solar radiation. *Mailing Add:* Off Provost Chapman Col Orange CA 92666

HAMILTON, HOBART GORDON, JR, b Washington, DC, Feb 8, 39; m 64. INORGANIC CHEMISTRY. *Educ:* Univ Tex, El Paso, BS, 61; NMex State Univ, MS, 63, PhD(chem), 68. *Prof Exp:* Radiochemist, Gen Elec Co, 63-64; res assoc anal chem, Univ Ariz, 67-68; from asst prof to assoc prof, 68-74, chmn dept, 72-73, assoc vpres acad affairs, 73-81, PROF CHEM, CALIF STATE COL, STANISLAUS, 75- *Mem:* Am Chem Soc; Sigma Xi. *Res:* Synthesis and stereochemistry of multidentate ligand transition metal complexes. *Mailing Add:* 2010 Sconyers Ct Turlock CA 95380

HAMILTON, HOWARD BRITTON, b Augusta, Kans, Oct 28, 23; m 43; c 4. ELECTRICAL ENGINEERING. *Educ:* Univ Okla, BS, 49; Univ Minn, MS, 55; Okla State Univ, PhD(elec eng), 62. *Prof Exp:* Engr, Gen Elec Co, 49-53; instr mech eng, Univ Wichita, 53-54, assoc prof elec eng & head dept, 55-58; unit chief mfg res, Boeing Co, 58-60; prof, Wichita State Univ, 60-65;

adj prof & chief at party for Univ Pittsburgh at Valparaiso Santa Maria Univ, Chile, 65-66; prof, 66-86, chmn dept, 66-73 & 83-86, EMER PROF ELEC ENG, UNIV PITTSBURGH, 86- *Mem:* Fel Inst Elec & Electronics Engrs. *Res:* Power systems; electric machinery and current limiting. *Mailing Add:* Elec Eng Dept 348 Benedum Univ Pittsburgh Pittsburgh PA 15261

HAMILTON, HOWARD LAVERNE, b Lone Tree, Iowa, July 20, 16; m 45, 75; c 5. DEVELOPMENTAL BIOLOGY. *Educ:* Univ Iowa, BA, 37, MS, 38; Johns Hopkins Univ, PhD(biol), 41. *Prof Exp:* From asst prof to prof zool, Iowa State Univ, 46-62, actg head dept zool & entom, 60-61, chmn dept, 61-62; prof, 62-82, EMER PROF BIOL, UNIV VA, 82- *Concurrent Pos:* Mem corp, Marine Biol Lab, Woods Hole, 46-; managing ed, Am Zoologist, Am Soc Zool, 66-70. *Mem:* Soc Develop Biol; Am Soc Naturalists; Am Soc Zool; Int Soc Develop Biol; Am Inst Biol Sci. *Res:* Pigmentation in birds; culture of viruses and rickettsiae; chemotherapy in rickettsial diseases; experimental embryology; developmental effects of rare earths; chemical control of organogenesis. *Mailing Add:* Jumping Branch Farm Rte 5 Box 401 Charlottesville VA 22901

HAMILTON, IAN ROBERT, b Ft Frances, Ont, July 1, 32; m 59; c 3. MICROBIOLOGY, BIOCHEMISTRY. *Educ:* Ont Agr Col, BSc, 58, MSc, 60; Univ Wis, PhD(microbiol, biochem), 63. *Prof Exp:* Asst prof biochem, 64-67, assoc prof oral biol, 67-71, prof, 71-81, HEAD ORAL BIOL, FAC DENT, UNIV MAN, 81-85 & 90- *Concurrent Pos:* Nat Res Coun Can res fel biochem, Oxford Univ, 63-64. *Mem:* Am Soc Microbiol; Can Soc Microbiol; Int Asn Dent Res; Can Asn Dent Res. *Res:* Carbohydrate formation and utilization by oral microorganisms; bioenergenetics of sugar transport by bacteria; microbial gluconeogenesis; mechanism of antibacterial effects of fluoride; growth of oral pathogens in continuous culture. *Mailing Add:* Dept Oral Biol Univ Man Fac Dent Winnipeg MB R3E 0W2 Can

HAMILTON, J(AMES) HUGH, b Los Angeles, Calif, Feb 7, 04; m 35; c 5. ELECTRICAL & CHEMICAL ENGINEERING. *Educ:* Calif Inst Technol, BS, 25, MS, 27, PhD(elec eng), 28. *Prof Exp:* From asst prof to assoc prof elec eng, Univ Utah, 28-37; res engr in charge electrostatic precipitation res, Western Precipitation Corp, Calif, 37-44; prof eng & dir, Eng Exp Sta, Univ Utah, 44-55; mem sr staff TRW Systs Group, 55-64, sr engr, TRW Inc, 64-68; CONSULT ENGR, 68- *Mem:* Fel Inst Elec & Electronics Engrs; Am Inst Mining, Metall & Petrol Engrs. *Res:* Metallurgical engineering; high voltage engineering; Cottrell precipitation; heat transfer; magnesium, gallium, indium alloys; coal carbonization; cryogenic propellants; space propulsion systems. *Mailing Add:* 11527 Venice Blvd Los Angeles CA 90066

HAMILTON, JAMES ARTHUR, b Oct 21, 47; m 83; c 1. NUCLEAR MAGNETIC RESONANCE SPECTROSCOPY. *Educ:* Juniata Col, BS, 69; Ind Univ, PhD(chem), 74. *Prof Exp:* Res asst, Eastman Kodak, 69; from assoc instr to asst prof, 71-75, fel, chem, Ind Univ, 75-78; asst prof, 78-85, ASSOC PROF, CHEM, BOSTON UNIV SCH MED, 85- *Concurrent Pos:* Vis prof, Juniata Col, 76. *Mem:* Biophys Soc. *Res:* High resolution and solid state nuclear magnetic spectroscopy of lipids in membranes, plasma lipoproteins and atherosclerotic lesions; interaction of lipids with proteins and the movement of lipids between model membranes and proteins; application of NMR methods to lipid enzymatic reactions, such as the phosphdypose A2 and C reactions. *Mailing Add:* Biophys Dept Boston Univ Sch Med 80 E Concord St Boston MA 02118

HAMILTON, JAMES ARTHUR ROY, b Eng, May 1, 19; m 46; c 3. FISH BIOLOGY. *Educ:* Univ BC, BA, 44, MA, 47; Univ Wash, PhD(fisheries), 55. *Prof Exp:* Sr biologist, Int Pac Salmon Fisheries Comn, Can, 43-56; sr biologist fisheries, 56-70, ENVIRON COORDR, PAC POWER & LIGHT CO, 71- *Concurrent Pos:* Consult, US CEngr & Wash Dept Fisheries. *Mem:* Am Fisheries Soc; Am Soc Limnol & Oceanog; Am Inst Fishery Res Biologists. *Res:* Limnology and fresh water ecology, especially reservoir and salmon populations. *Mailing Add:* 10345 Homestead Lane Beaverton OR 97005-6624

HAMILTON, JAMES BECLONE, b Chicago, Ill, Oct 22, 33; m 65; c 2. INORGANIC CHEMISTRY. *Educ:* Northwestern Univ, PhB, 61, MS, 64; Iowa State Univ, PhD(inorg chem), 68. *Prof Exp:* Res technician chem physics, IIT Res Inst, 59-61; asst chem, Northwestern Univ, 61-63; res chemist, Morton Chem Co, 63-64; asst inorg chem, Iowa State Univ, 64-68; res assoc, 68-69, asst prof, 69-74, assoc prof inorg chem, 74-80, asst provost spec proj, 77- 80, asst dean grad sch, 76-77, asst provost undergrad educ, PROF INORG CHEM, MICH STATE UNIV, 80- *Concurrent Pos:* Petrol Res Fund grant, 69-72; pres, Mich Coun Educ Opportunity Prog, 73-75; mem, Adv Comt Int Prog, NSF, 81-84, consult, NSF; consult, AAAS, Nat Coun Educ Opportunity Prog; mem, Linkages Adv Comt, AAAS, 85-, Comt Pub Understanding Sci & Technol, 88- *Honors & Awards:* Henry A Hill Lectr Award, Northeastern Sect, Am Chem Soc, 84. *Mem:* AAAS; Am Chem Soc; Nat Orgn Black Chemists & Chem Engrs; Nat Coun Educ Opportunity Prog; Mid-Am Asn Educ Opportunity Prog Personnel. *Res:* Synthesis, structure, spectra and magnetic behavior of coordination compounds of metals of the second and third row transition series, particularly zirconium, niobium, molybdenum, hafnium, tantalum and tungsten. *Mailing Add:* 444 Admin Mich State Univ East Lansing MI 48824

HAMILTON, JAMES F(RANCIS), b Edinburg, Ind, Apr 8, 27; m 48; c 3. MECHANICAL ENGINEERING. *Educ:* Purdue Univ, BSME, 51, PhD(eng), 63; Cornell Univ, MME, 54. *Prof Exp:* Instr mech eng, Cornell Univ, 51-54; sr res engr, Clevite Res Ctr, 54-55; prof mech eng, WVa Univ, 55-65; assoc prof, 65-70, PROF MECH ENG, PURDUE UNIV, 70- *Mem:* Am Soc Mech Engrs. *Res:* Experimental stress analysis; structural vibrations; analysis and design; system analysis; noise control. *Mailing Add:* Ray W Herrick Labs Sch Mech Eng Purdue Univ Lafayette IN 47907

HAMILTON, JAMES GUTHRIE, b Eagletown, Okla, Jan 30, 23; m 50; c 4. BIOCHEMISTRY. *Educ:* Okla State Univ, BS, 48, MS, 50; Univ Minn, PhD(biochem), 53. *Prof Exp:* Asst prof exp med, Univ Tex Southwestern Med Ctr, 54-57; asst prof biochem & med, Sch Med, Tulane Univ, 57-61, assoc prof, 61-68; mem staff, Hoffman-LaRoche, Inc, 68-72, group chief, Dept Biochem Nutrit, 72-88; RETIRED. *Mem:* Am Chem Soc; Am Soc Microbiol; Am Inst Nutrit; Am Oil Chem Soc. *Res:* Chemistry and metabolism of lipids and bile acids. *Mailing Add:* 2976 Heather Bow Sarasota FL 34235

HAMILTON, JAMES WILBURN, b Louisville, Ky, Sept 20, 36; m 55; c 5. BIOCHEMISTRY. *Educ:* Univ Louisville, AB, 61, PhD(biochem), 65. *Prof Exp:* Res asst biochem, Med Sch, Univ Louisville, 58-61; fel pharmacol, Baylor Med Sch, 65-66; asst prof, biochem, Sch Dent, Univ Mo-Kansas City, 66-71, assoc prof, 71-77; co dir Calcium Res Lab, 68-82, RES BIOCHEMIST, VET ADMIN HOSP, KANSAS CITY, MO, 66-, DIR CALCIUM RES LAB, 82-; ASSOC PROF BIOCHEM, MED SCH, UNIV KANS, 79- *Mem:* AAAS; Am Chem Soc; Am Soc Biol Chemists; Endocrine Soc. *Res:* Parathyroid hormone biosynthesis; structure-function studies of vitamin D-dependent calcium binding protein. *Mailing Add:* Vet Admin Hosp 4801 Linwood Blvd Kansas City MO 64128

HAMILTON, JANET V, b Decatur, Ill, Nov 11, 36; m 60; c 2. PHYSICAL CHEMISTRY. *Educ:* Millikin Univ, AB, 58; Tulane Univ, PhD(phys chem), 63. *Prof Exp:* Sr res engr thermochem, Rocketdyne Div, NAm Aviation, Inc, 63-66; PROF CHEM, TARRANT COUNTY JR COL, 66- *Mem:* AAAS; Am Chem Soc; Sigma Xi. *Res:* Heat of formation, reaction and ablation; bond energies; heat capacity; surface tension; viscosities. *Mailing Add:* Rte 1 Box 272 Sanger TX 76266

HAMILTON, JEFFERSON MERRITT, JR, b Minneapolis, Minn, Aug 31, 18; m 42; c 2. FLUORINE CHEMISTRY. *Educ:* Johns Hopkins Univ, AB, 40, PhD(phys chem), 44. *Prof Exp:* Res supvr, Manhattan Dist, Johns Hopkins Univ, 43-46; res chemist, 46-53, res supvr, 53-60, head, Res & Develop Div, 60-82, consult, E I du Pont de Nemours & Co, Inc, 82-88; RETIRED. *Concurrent Pos:* Civilian with Off Sci Res & Develop, 44. *Mem:* AAAS; Am Chem Soc. *Res:* Organic fluorine compounds; acetylene chemistry; catalytic processes. *Mailing Add:* 4031 Kennett Pike Box 56 Wilmington DE 19807

HAMILTON, JOHN FREDERICK, b Knoxville, Tenn, Mar 19, 28; m 50; c 3. SOLID STATE PHYSICS. *Educ:* Univ Tenn, BS, 50. *Prof Exp:* Sr res physicist, Eastman Kodak Co, 50-60, res assoc, 60-71, sr res assoc, 71-82, res fel, 82-86; RETIRED. *Concurrent Pos:* Instr, Inst Optics, Univ Rochester. *Mem:* Fel AAAS; Electron Micros Soc Am; fel Am Phys Soc; fel & hon mem Soc Photog Sci & Eng; Am Vacuum Soc. *Res:* Formation of photographic latent image; electrical and structural properties of silver halides, using the electron microscope and related equipment; imperfections in crystalline solids; heterogeneous catalysis; nucleation phenomena. *Mailing Add:* 211 Glenn Abbey Johns Island SC 29455

HAMILTON, JOHN MEACHAM, b Gifford, Ill, Feb 20, 12; m 39; c 2. BIOLOGY. *Educ:* Oberlin Col, AB, 35; Wesleyan Univ, MA, 37; Yale Univ, MS, 42; Univ Iowa, PhD(zool), 51. *Prof Exp:* Asst neurophysiol res, Yale Univ, 37-41; sci master, Asheville Sch for Boys, NC, 42-46; from asst prof to prof biol, 46-77, head dept, 49-64, actg chmn sci div, 51-53, chmn sci div, 59-64, actg dean col, 55-56 & 64-67, emer prof biol, Park Col, 77; RETIRED. *Concurrent Pos:* NSF fac fel, Univ Calif, Los Angeles, 58-59; vis lectr, Inter-Am Univ PR, 67-68; scholar, Univ Calif, Davis, 71-72. *Mem:* Nat Asn Biol Teachers; Sigma Xi. *Res:* Fresh water protozoology and ecology; history of evolution. *Mailing Add:* 6611 N Platte Hills Rd Parkville MO 64152

HAMILTON, JOHN ROBERT, b Atlanta, Ga, Apr 22, 25; m 46; c 2. WOOD SCIENCE & TECHNOLOGY. *Educ:* Univ Ga, BSF, 48, MSF, 49; NC State Col, PhD, 60. *Prof Exp:* County forester, Ga Forestry Comn, 49-50; assoc forester, Ga Exp Sta, 50-53, head dept forestry, 53-55; asst, NC State Col, 55-57; assoc prof, Univ Ga, 57-64; PROF WOOD SCI & WOOD SCIENTIST, WVA UNIV, 64- *Concurrent Pos:* Ga Forest Res Coun grants, 60-64, Nat Plant Food Inst grant, 61-62, US AEC grant, 61-64 & 65-; Soc Am Wood Preservers grant, 62-63. *Mem:* Soc Am Foresters; Forest Prod Res Soc; Soc Wood Sci & Technol; Sigma Xi. *Res:* Wood anatomy. *Mailing Add:* One Bates Rd Morgantown WV 26505

HAMILTON, JOSEPH H, JR, b Ferriday, La, Aug 14, 32; m 60; c 2. EXPERIMENTAL NUCLEAR PHYSICS. *Educ:* Miss Col, BS, 54; Ind Univ, MS, 56, PhD(physics), 58. *Hon Degrees:* DSc, Miss Col, 82. *Prof Exp:* From asst prof to prof physics, 58-81, chmn dept, 79-85, LANDON C GARLAND PROF PHYSICS, VANDERBILT UNIV, 81- *Concurrent Pos:* NSF fel, Univ Uppsala, 58-59; res appointment, Inst Nuclear Physics Res, Amsterdam, Neth, 62-63; chmn, Int Conf Internal Conversion Processes, 65; chmn, Int Conf Radioactivity in Nuclear Spectros, Mod Tech & Appln, 69; chmn, Int Symposium, Directions in Nuclear Structure Res, 84; mem, Nat Acad Sci Sub-Comt Nuclear Physics, Mgt & Costs, 70; chmn, Univ Isotope Separator Group, Oak Ridge Nat Lab, 71-73, 82-83, 90-92 & nat policy bd, Nat Heavy Accelerator Lab, 75-82; mem planning comt, Int Conf on Reactions between Complex Nuclei, 73-74; chmn, Int Conf Future Directions Studies Nuclei Far From Stability, 79; Alexander von Humboldt sr fel, Frankfurt, 79-80; guest prof, Univ Frankfurt, 79-80, 87; dir, Joint Inst Heavy Ion Res, 83-; policy coun, Nuclear Sci Adv Comn & Long Range Planning Comn, 89; chmn, Int Symposium Reflection & Direction in Nuclear Res, 91; adj prof, Tsing-Hua Univ, Pei Ging, 86-; hon adv prof, Fudan Univ, Shanghai, 88-; H Branscomb distinguished prof, 83. *Honors & Awards:* Jesse Beams Gold Medal, 75; Earl Southerland Award, 88; George Pegram Gold Medal, 88; Guy & Rebecca Forman Award, 90. *Mem:* Fel Am Phys Soc; Sigma Xi; fel AAAS; Am Asn Physics Teachers. *Res:* Nuclear structure studies via beta- and gamma-ray spectroscopy studies of nuclei far from stability with heavy ions and isotope separator; Coulomb excitation; in-beam spectroscopy following heavy-ion nuclear reactions. *Mailing Add:* Box 1638 - Sta B Vanderbilt Univ Nashville TN 37235

HAMILTON, KENNETH GAVIN ANDREW, b Nottingham, Eng, Mar 13, 46; Can citizen; m 77. ENTOMOLOGY. *Educ:* Univ Man, BSA Hons, 68; Univ Ga, MS, 70, Phd(entom), 72. *Prof Exp:* RES SCIENTIST ENTOM, AGR CAN, 72- *Mem:* Entom Soc Can; Sigma Xi. *Res:* Taxonomy of Homoptera-Auchenorrhyncha; morphology of Insecta; phylogeny of Insecta; palaeontology of Insecta. *Mailing Add:* 1109 Meadowlands Dr E Ottawa ON K2C 0K5 Can

HAMILTON, KEVIN, b Calgary, Alta, Can, Feb 10, 56. CLIMATOLOGY. *Educ:* Queen's Univ, BSc, 76; McMaster Univ, MSc, 77; Princeton Univ, MA, 79, PhD(geophys fluid dynamics), 81. *Prof Exp:* Postdoc fel, Nat Ctr Atmospheric Res, 81-82; NSERC Univ res fel, Dept Oceanog, Univ BC, 82-85; asst prof, Dept Meteorol, McGill Univ, Can, 85-87; res scientist, Atmospheric & Oceanic Sci Prog, 87-88, RES SCIENTIST, GEOPHYS FLUID DYNAMICS LAB, PRINCETON UNIV, 88-, ASSOC PROF, ATMOSPHERIC & OCEANIC SCI PROG, 88- *Mem:* Am Meteorol Soc; Can Meteorol & Oceanog Soc. *Res:* General circulation of the stratosphere and mesosphere; dynamics of atmospheric waves; interannual variability of the ocean and atmosphere; author several articles and publications. *Mailing Add:* Geophys Fluid Dynamics Lab Princeton Univ PO Box 308 Princeton NJ 08542

HAMILTON, LAWRENCE STANLEY, b Toronto, Ont, June 5, 25; nat US; m 47; c 4. FORESTRY, ENVIRONMENTAL SCIENCES. *Educ:* Univ Toronto, BScF, 48; State Univ NY, MF, 50; Univ Mich, PhD, 62. *Prof Exp:* Zone forester, Dept Lands & Forests, Can, 48-51; exten forester, 51-54, from asst prof to assoc prof forestry, 54-66, prof, 65-80, EMER PROF FORESTRY, CORNELL UNIV, 80-; RES ASSOC, EAST WEST CTR, HONOLULU, 80- *Concurrent Pos:* NSF fel, 64-65; mem panel nat res sci, Nat Acad Sci-Nat Res Coun, 65-67, comt soc sci educ, 68-70; Fulbright lectr, Univ New Eng, Australia, 69-70 & Univ Walkato, NZ, 78; vis lectr, Univ Queensland, 72; dir, Venezuelan Trop Rainforest Study, 74-75; consult, UNESCO, Australia, 80, Asn Rural Develop, Costa Rica, 85, Int Union Conserv Nature & Natural Resources, 88; fel, East West Ctr, 79; Comn Ecol, Int Union Conserv Nature, 78-, comn Nat Parks & Protected Areas, 88-; Counr, Int Mountain Soc, 90- *Honors & Awards:* Fulbright Awards, 69 & 78. *Mem:* Soc Am Foresters; Int Soc Trop Foresters; Int Union Conserv Nature & Natural Resources; Int Mountain Soc. *Res:* Ecological base for land and water planning; watershed land use; international natural resources programs; tropical forests and protected areas. *Mailing Add:* Environ & Policy Inst East West Ctr 1777 East West Rd Honolulu HI 96848

HAMILTON, LEONARD DERWENT, b Manchester, Eng, May 7, 21; nat US; m 45; c 3. EXPERIMENTAL MEDICINE. *Educ:* Oxford Univ, BA, 43, BM & BCh, 45, MA, 46, DM, 51; Cambridge Univ, MA, 48, PhD, 52. *Prof Exp:* Jr asst pathologist, Radcliffe Infirmary, Oxford Univ, 45-46; resident med officer, Radiotherapeut Ctr & house physician, Dept Med, Addenbrooke's Hosp, 46; res student path, Dept Radiotherapeut, Cambridge Univ, 46-49; asst, 50-53, assoc, 53-64, head isotope studies sect, 57-64, assoc scientist, Sloan-Kettering Inst, 65-79; attend physician, Hosp Med Res Ctr, 64-85, HEAD, BIOMED & ENVIRON ASSESSMENT DIV, BROOKHAVEN NAT LAB, 73- *Concurrent Pos:* Fel med, Salt Lake Gen Hosp, 49-50; spec fel, Mem Hosp, New York, 51-53, spec fel radiation ther, 53; clin asst radiation therapist & clin asst med, Chemother Serv, Mem Hosp, NY, 54-58, asst attend physician, 58-65; consult, Off Under-Secy Spec Polit Affairs, Sci Comt Effects Atomic Radiation, UN, 60-62; mem, NY Mayor's Tech Adv Comt Radiation, 63-77; head div microbiol, Brookhaven Nat Lab, 64-73; prof med, Health Sci Ctr, State Univ NY Stony Brook, 68-; vis fel, St Catherine's Col, Oxford Univ, 72-73; mem, panel fossil fuel, UN Environ Prog, 78, nuclear energy, 78-79, renewable sources, 80, comp assessment of different sources, 80; mem, NY Comnr Health Tech Adv Comt Radiation, 78; mem, expert adv panel environ hazards & focal pt on health & environ effects of energy systs, 83- *Mem:* Am Asn Pathologists; Am Soc Clin Invest; Am Asn Cancer Res; Harvey Soc; Soc Risk Anal. *Res:* Structure and functions of nucleic acids; lymphocytes; biomedical and environmental effects of energy systems; hazards evaluation; biological effects of ionizing radiations. *Mailing Add:* Biomed & Environ Assessment Div Brookhaven Nat Lab Upton NY 11973

HAMILTON, LEROY LESLIE, b Fresno, Calif, Aug 31, 34; c 4. BIOMEDICAL ENGINEERING, ELECTRICAL ENGINEERING. *Educ:* Univ Calif, Berkeley, BSEE, 58; Cath Univ Am, MEE, 60; Case Western Reserve Univ, PhD(Med eng), 67. *Prof Exp:* Electronic engr, US Dept Defense, 58-63; res assoc med eng, Highland View Hosp, Ohio, 63-68; assoc prof elec eng, Cath Univ Am, 68-74; engr, Bur Med Devices, Food & Drug Admin, Rockville, 74-76; dir med eng & electronics, Health Indust Mfrs Asn, Washington, DC, 76-83; PRES, HAMILTON MED EQUIPMENT, INC, 87- *Concurrent Pos:* Prof lectr, Med Sch, Georgetown Univ, 69-; consult comput applns, Silver Spring, MD, 84- *Mem:* Inst Elec & Electronics Engrs; Asn Advan Med Instrumentation; Bioelectromagnetics Soc. *Res:* Analysis and modeling of human cardiovascular control system, especially postural reflexes. *Mailing Add:* 13002 Autumn Dr Silver Spring MD 20904

HAMILTON, LEWIS R, b Wilmington, Ohio, Apr 19, 41; m 61; c 2. ORGANIC CHEMISTRY. *Educ:* Ohio State Univ, BS, 63, PhD(chem), 67. *Prof Exp:* Chemist, Miami Valley Labs, Procter & Gamble Co, 67-70; CHEMIST, EASTMAN KODAK CO, 70- *Mem:* Am Chem Soc; Soc Photog Scientists & Engrs. *Res:* Photochemical and excited state behavior of dyes; color photgraphic chemistry. *Mailing Add:* 215 Rogens Pkwy Rochester NY 14617

HAMILTON, LYLE HOWARD, b Superior, Nebr, June 11, 24; m 85; c 1. PHYSIOLOGY. *Educ:* Willamette Univ, AB, 50; Univ Iowa, MS, 52, PhD(physiol), 54. *Prof Exp:* Asst prof physiol, Univ Sask, 54-57, admin asst to dean, 56-57; from asst prof to assoc prof physiol, Sch Med, Marquette Univ, 57-66; chief, Physiol Sect, Zablocki Vet Admin Med Ctr, 57-61, prin scientist, 61-86; EMER PROF PHYSIOL, MED COL WIS, 87-; PRES, QUINTRON INSTR CO, 87- *Concurrent Pos:* Prof physiol & dir clin physiol, Med Col Wis, 67-87; external examr physiol, Univ West Indies, 79-81, 83-84 & 86. *Mem:* Am Physiol Soc; Soc Exp Biol & Med; Biomed Eng Soc; Can Physiol Soc; Instrument Soc Am. *Res:* Respiration (mechanics of ventilation); development of instruments for physiological studies; exercise physiology. *Mailing Add:* 3712 W Pierce St Milwaukee WI 53215

HAMILTON, MARY JANE GILL, b Buffalo, NY, Sept 9, 25; m 48. PHYSICAL BIOCHEMISTRY. *Educ:* Univ Buffalo, BA, 47; Polytech Inst Brooklyn, MS, 50; Cornell Univ, PhD(biochem), 61. *Prof Exp:* From asst to res assoc biochem, Sloan-Kettering Inst Cancer Res, 49-64; from instr to asst prof biochem, Sloan-Kettering Div, Grad Sch Med Sci, Cornell Univ, 64-71, assoc prof, 71-; ASSOC MEM, SLOAN-KETTERING INST CANCER RES, 70-; ASSOC PROF CHEM, DIV SCI & MATH, COL LINCOLN CTR, FORDHAM UNIV, 82- *Concurrent Pos:* Fel biochem, Sloan-Kettering Inst Cancer Res, 56-60; Nat Cancer Inst fel, Nat Inst Med Res, London, 61-63; assoc, Sloan-Kettering Inst Cancer Res, 64-70. *Mem:* Am Soc Cell Biol; Biophys Soc; Am Chem Soc; NY Acad Sci; Am Soc Biol Chemists; AAAS. *Res:* Structure of hemocyanins; structure and function of ribosomes; physical chemical characterization of proteins, nucleic acids and viruses. *Mailing Add:* Dept Sci & Math Fordham Univ 113 W 60th St New York NY 10023

HAMILTON, PAT BROOKS, b Haskell, Okla, Sept 25, 30; m 55; c 3. MICROBIOLOGY, MYCOTOXICOLOGY. *Educ:* Northeastern State Col, BS, 51; Univ Wis, MS, 54, PhD(bact), 62. *Hon Degrees:* Dr Hon Causa, Marcilio Ficino Free Univ Sci, Bologna, Italy. *Prof Exp:* Assoc res biochemist, Sterling-Winthrop Res Inst, 60-62; sr res microbiologist, Continental Oil Co, 62-64; sr microbiologist, Res Triangle Inst, 64-67; assoc prof, 67-71, PROF MICROBIOL & POULTRY SCI, NC STATE UNIV, 71- *Concurrent Pos:* Exchange Scholar, Hebrew Univ, Jerusalem, Israel, 81. *Honors & Awards:* Int Award, Corn Prod Coun, 76. *Mem:* Am Soc Microbiol; Soc Indust Microbiol; Am Chem Soc; fel Poultry Sci Asn; Am Acad Microbiol; Asn Off Analytical Chemists; Int Asn Milk, Food & Environ Sanitarians. *Res:* Mycotoxins; fungicides; carotenoids; microbial transformations; mycotoxicoses. *Mailing Add:* Dept Poultry Sci NC State Univ Box 7608 Raleigh NC 27650-7608

HAMILTON, PAUL BARNARD, biochemistry; deceased, see previous edition for last biography

HAMILTON, RALPH WEST, b Raleigh, NC, Aug 16, 33; m 58; c 3. PLASTIC SURGERY. *Educ:* Lehigh Univ, BS, 55; Univ Pa, MD, 59. *Prof Exp:* From asst instr to instr surg, 60-65, from instr to assoc prof plastic surg, 65-69, from asst prof to assoc prof, 69-74, PROF SURG, UNIV PA, 74- *Mem:* Am Soc Plastic & Reconstruct Surgeons; Am Asn Plastic Surgeons; Plastic Surg Res Coun; Soc Univ Surgeons; Soc Head & Neck Surgeons. *Res:* Biology of wound healing; frozen organ preservation. *Mailing Add:* 3400 Spruce St Philadelphia PA 19104

HAMILTON, ROBERT BRUCE, b Nashville, Tenn, Aug 8, 36; m 63. WILDLIFE ECOLOGY, ORNITHOLOGY. *Educ:* Univ Tenn, Knoxville, BS, 60; Univ Calif, Berkeley, PhD(zool), 69. *Prof Exp:* Asst prof biol sci, Northwestern State Univ, 69-72; asst prof, 72-77, ASSOC PROF WILDLIFE MGT, LA STATE UNIV, BATON ROUGE, 77- *Mem:* Wildlife Soc; Ecol Soc Am; Am Ornithologists Union; Wilson Ornith Soc; Cooper Ornith Soc; Sigma Xi. *Res:* Avian community ecology; population ecology of herons, rails, and shorebirds. *Mailing Add:* Sch Forestry & Wildlife Mgt La State Univ Baton Rouge LA 70803

HAMILTON, ROBERT DUNCAN, b Edmonton, Alta, May 24, 37; m 60; c 3. AQUATIC MICROBIOLOGY. *Educ:* McGill Univ, BSc, 59; Univ Miami, PhD(marine sci), 64. *Prof Exp:* Asst res microbiologist, Univ Calif, San Diego, 64-69; res biologist, 69-72, sect leader, 72-77, dir, 77-81, RES SCIENTIST, FRESHWATER INST, 81- *Concurrent Pos:* Sr sci adv to govts. *Mem:* Am Soc Microbiol; Can Soc Microbiol; Am Soc Limnol & Oceanog; Int Asn Theoret & Appl Limnol; Sigma Xi. *Res:* Occurrence and activities of heterotrophic bacteria in marine and freshwater environments; acid rain; toxic chemicals. *Mailing Add:* Freshwater Inst 501 University Crescent Winnipeg MB R3T 2N6 Can

HAMILTON, ROBERT HILLERY, JR, b Thompsonville, Ill, Apr 24. 29; m 59; c 2. PLANT PHYSIOLOGY. *Educ:* Univ Ill, BS, 50; Rutgers Univ, MS, 52; Mich State Univ, PhD, 60. *Prof Exp:* Plant physiologist, Agr Res Serv, USDA, Mich State Univ, 55-60 & NC State Col, 60-61; from asst prof to assoc prof, 61-72, PROF BOT, PA STATE UNIV, 72- *Mem:* AAAS; Am Soc Plant Physiologists; Weed Sci Soc Am; Japan Soc Plant Physiol. *Res:* Plant growth and development; biosynthetic mecahnisms in plants; plant growth substances. *Mailing Add:* 20B Muller Lab Pa State Univ University Park PA 16802

HAMILTON, ROBERT HOUSTON, b Corsicana, Tex, Sept 12, 06; m 33. BIOCHEMISTRY. *Educ:* Univ Tex, BA, 26, MA, 27; Univ Minn, PhD(physiol chem), 33, MD, 35. *Prof Exp:* Asst & tutor zool, Univ Tex, 25-27; asst physiol chem, Univ Minn, 27-28, instr, 31-34, physiol, 35-36; from asst prof to assoc prof, 36-43, chmn dept, 44-71, PROF BIOCHEM, SCH MED, TEMPLE UNIV, 43- *Mem:* AAAS; Am Chem Soc; Am Asn Clin Chem. *Res:* Clinical analytical chemistry; biochemical analysis. *Mailing Add:* 6900 Wayne Ave Philadelphia PA 19119

HAMILTON, ROBERT L, JR, b Muba City, Calif, Dec 25, 34. PLASMA LIPO PROTEIN METABOLISM. *Educ:* Vanderbilt Sch Med, PhD(anat), 64. *Prof Exp:* PROF ANAT, SCH MED, UNIV CALIF, SAN FRANCISCO, 84- *Mailing Add:* Univ Calif Med Ctr CVRI 10304 M San Francisco CA 94143

HAMILTON, ROBERT MILTON GREGORY, b Ottawa, Ont, Dec 15, 39; m 67; c 2. POULTRY NUTRITION, BIOCHEMISTRY. *Educ:* McGill Univ, BSc, 66, MSc, 68; Univ Western Ont, PhD(biochem), 72. *Prof Exp:* Fel biochem, Univ Western Ont, 72-73; RES SCIENTIST POULTRY NUTRIT,

ANIMAL RES CTR, AGR CAN, 73- *Concurrent Pos:* Scholar, Nat Res Coun Can, 66; fel, Med Res Coun Can, 68 & Ont Prov Govt, 70. *Honors & Awards:* Poultry Sci Asn Award, 81. *Mem:* Agr Inst Can; Can Soc Animal Sci; Poultry Sci Asn; World Poultry Sci Asn. *Res:* Elucidation of nutritional, physiological and biochemical factors that affect egg shell strength; effects of mycotoxins on the productive performance of chickens. *Mailing Add:* Cent Exp Farm Agr Can Bldg 12 Maple Dr Ottawa ON K1A 0C6 Can

HAMILTON, ROBERT MORRISON, b Houston, Tex, June 20, 36. GEOPHYSICS. *Educ:* Colo Sch Mines, BSc, 58; Univ Calif, Berkeley, MA, 63, PhD(geophys), 65. *Prof Exp:* Res seismologist, Geophys Div, Dept Sci & Indust Res, NZ, 65-68; geophysicist, 68-72, dep earthquake geophys, 72-73, chief off earthquake studies, 73-78, RES GEOPHYSICIST, US GEOL SURV, 78- *Mem:* Seismol Soc Am; Am Geophys Union; Earthquake Eng Res Inst; AAAS; Geol Soc Am. *Res:* Electrical properties of minerals; seismicity of geothermal areas; explosion and earthquake seismology; seismicity of eastern United States. *Mailing Add:* Nine Emeraude Pl Reston VA 22090

HAMILTON, ROBERT W, b Davenport, Iowa, Dec 6, 39; m 65, 80; c 4. ENTOMOLOGY. *Educ:* Parsons Col, BSc, 62; Ohio State Univ, MSc, 64, PhD(entom), 69. *Prof Exp:* Teaching asst zool, Ohio State Univ, 62-69; from asst prof to assoc prof, 69-82, PROF BIOL, LOYOLA UNIV CHICAGO, 82- *Concurrent Pos:* Mem, Am Mus Natural Hist-Lerner Marine Lab Exped, Bahama Islands, 65 & Loyola Univ Chicago Exped Southwest US, 71; consult pest control, 71-; NSF grants, 72 & 74; Type Species Study, London, Eng & Hamburg, W Ger, 89. *Mem:* Coleopterists Soc Am; Entom Soc Am; Am Entom Soc. *Res:* Arthropod biodiversity; entomology, taxonomy and morphology of weevils (Coleoptera: Rhynchophora). *Mailing Add:* Dept Biol Loyola Univ 6525 N Sheridan Rd Chicago IL 60626

HAMILTON, ROBERT WILLIAM, b Stanton, Tex, June 5, 30; m 72; c 3. ENVIRONMENTAL MEDICINE. *Educ:* Univ Tex, BA, 51; Tex A&M Univ, MS, 58; Univ Minn, PhD(physiol), 64. *Prof Exp:* Instr physiol, Med Ctr, Univ Kans, 61-63; res physiologist, Linde Div, Union Carbide Corp, 64-67, sr res physiologist, 67-68, res supvr, Ocean Systs Inc, Tarrytown Tech Ctr, 69-74, dir res, Tarrytown Labs, Ltd, 74-76; PRES & SR CONSULT, HAMILTON RES, LTD, 77- *Honors & Awards:* Stover-Link Award, Undersea Med Soc; Oceaneering Award, Undersea & Hyperbaric Med Soc, 88. *Mem:* Aerospace Med Asn; Am Physiol Soc; Marine Technol Soc; Undersea & Hyperbaric Med Soc (secy, 81-82). *Res:* Development of practical procedures for decompression; commercial and scientific diving, aerospace, and medical therapy aspects; life support and human performance in stress environments: pressure, gases, thermal; hyperbaric medicine; human performance; life support systems; decompression and decompression sickness; effects of various inert gas atmospheres at low, normal and high pressures; diving physiology; G-forces. *Mailing Add:* 80 Grove St Tarrytown NY 10591-4138

HAMILTON, STANLEY R, b Ft Wayne, Ind, Dec 2, 48; m 71; c 3. GASTROINTESTINAL PATHOLOGY. *Educ:* Ind Univ, AB, 70, MD, 73. *Prof Exp:* Intern & resident path & lab med, 73-79, asst prof, 79-83, ASSOC PROF PATH, SCH MED & HOSP, JOHNS HOPKINS UNIV, 83-, ASSOC PROF ONCOL, 86- *Concurrent Pos:* Sr fel, Nat Found Ileitis & Colitis, 79-81; prin investr, Nat Cancer Inst res grant, NIH, Dept Health & Human Servs, 81-; Am Cancer Soc Nat Adv Comt & Task Force Colorectal Cancer, 89- *Mem:* Int Acad Path; Am Gastroenterol Asn; Gastrointestinal Path Soc; Am Asn Cancer Res; Am Asn Pathologists. *Res:* Colorectal carcinogenesis; Barrett esophagus; inflammatory bowel diseases; pathology and pathogenesis of colorectal carcinoma in human beings and experimental models, particularly relating to dietary factors, chemoprevention and molecular and medical genetics; dysplasia and carcinoma in Barrett esophagus. *Mailing Add:* Dept Path Johns Hopkins Hosp 600 N Wolfe St Baltimore MD 21205

HAMILTON, STEVEN J, b Sacramento, Calif, Jan 8, 47; m 68; c 1. AQUATIC TOXICOLOGY. *Educ:* Humboldt State Univ, BS, 74; Univ Mo, Columbia, MS, 80, PhD(fish & wildlife), 85. *Prof Exp:* Res fishery biologist, Columbia Nat Fisheries Res Lab, 81-84, RES FISHERY BIOLOGIST, NAT FISHERIES CONTAMINANT RES CTR, US FISH & WILDLIFE SERV, COLUMBIA, MO, 84- *Concurrent Pos:* Adj assoc prof, Univ SDak, Vermillion, 87-, SDak State Univ, Brookings, 90- *Mem:* Am Fisheries Soc; Sigma Xi; Soc Environ Toxicol & Chem. *Res:* Determine the biological effects of aquatic contaminants on warm and coldwater fish and aquatic invertebrates; development of new, complex techniques and original approaches to address resource contamination problems in the United States. *Mailing Add:* Nat Fisheries Contaminant Res Ctr US Fish & Wildlife Serv Field Res Sta RR 1 Box 295 Yankton SD 57078

HAMILTON, TERRELL HUNTER, physiology, developmental biochemistry; deceased, see previous edition for last biography

HAMILTON, THOMAS ALAN, b Philadelphia, Pa, Feb 2, 50; m 73; c 2. MONONUCLEAR PHAGOCYTE BIOLOGY. *Educ:* Univ Colo, BA, 71; Univ Ore Health Sci, PhD(biochem), 76. *Prof Exp:* Fel path, Stanford Univ Med Ctr, 76-80; res assoc immunol, St Jude Childrens Res Hosp, 80-82; from asst prof to assoc prof path, Duke Univ Med Ctr, 82-87; HEAD SECT IMMUNOL, RES INST, CLEVELAND CLIN FOUND, 89- *Mem:* Am Asn Immunologists; Reticuloendothelial Soc; AAAS. *Res:* The biology and biochemistry of mononuclear phagocytes and their activation for performance of multiple functions including host defense homeostasis and inflammation; particular emphasis is placed upon defining the genes which are expressed selectively during maerophage activation and examining the transmembrane signalling mechanisms which regulate their expression. *Mailing Add:* Cleveland Clin Found NN1-06 9500 Euclid Ave Cleveland OH 44195

HAMILTON, THOMAS CHARLES, b Chicago, Ill, Apr 24, 47; m 69. ANALYTICAL PROBLEM SOLVING, DECISION ANALYSIS. *Educ:* Mich State Univ, BS, 68, MS, 71, PhD(biophys), 72. *Prof Exp:* Fel neurophys, Univ Tex, Austin, 72-73; res analyst life sci, 74-79, instr info sci, 79-81, scientist, advan concepts staff, 81-83, chief, life sci, 83-85, DEPT CHIEF, SCI & TECHNOL, US GOVT, 85- *Concurrent Pos:* Inst environ sci, Univ Va, Northern Va Exten Campus, 75. *Res:* Problem solving techniques and strategies for intelligence analysts; decision analysis; strategies for management and allocation of government resources; application of technical advances in biophysical and life sciences. *Mailing Add:* 6506 Heather Brook Ct McLean VA 22101-1607

HAMILTON, THOMAS DUDLEY, b White Plains, NY, Jan 17, 36; m 62; c 2. QUATERNARY GEOLOGY, GLACIAL GEOLOGY. *Educ:* Univ Idaho, BS, 60; Univ Wis, MS, 63; Univ Wash, PhD(geol), 66. *Prof Exp:* From asst prof to assoc prof geol, Univ Alaska, 66-75; GEOLOGIST, US GEOL SURV, 75- *Concurrent Pos:* Consult geologist, Trans-Alaska Pipeline Syst, 69-70 & 74. *Mem:* AAAS; Arctic Inst NAm; Geol Soc Am; Glaciol Soc; Am Quaternary Asn. *Res:* Arctic and alpine geomorphology; environmental geology; late Cenozoic geology, chronology and environments; environmental reconstructions, early man sites. *Mailing Add:* 4545 Reka Dr Anchorage AK 99508

HAMILTON, THOMAS REID, b Kansas City, Mo, Apr 30, 11; m 41; c 3. MICROBIOLOGY, PATHOLOGY. *Educ:* Univ Mo, AB, 32; Univ Kans, MD, 35, MS, 41; Am Bd Path, dipl & cert anat path & clin microbiol; Am Bd Med Microbiol, dipl & cert med lab & pub health. *Prof Exp:* Intern, Univ Hosp, Univ Iowa, 35-36; resident, St Joseph Hosp, Kansas City, 36-37; from asst to assoc prof path, Sch Med, Univ Kans, 37-50, prof microbiol & path & lectr hist med, 50-69; prof path, Sch Med, Wash Univ, 69-72; prof biol, microbiol & path, Med Sch, Univ Mo-Kansas City & actg dir, div microbiol, Truman Med Ctr, 81-83; prof & head, 72-78, prof, 78-81, EMER PROF MED MICROBIOL & IMMUNOL, MED SCH, UNIV MINN, DULUTH, 81- *Concurrent Pos:* Resident, Univ Hosp, Univ Kans, 37-39, clin bacteriologist, 38-42, chmn dept med microbiol, 51-61; pathologist, Providence Hosp, 39-42, dir develop prog, 56-58; mem training grant comt, Nat Inst Allergy & Infectious Dis; consult, US Vet Admin Hosp, Mo, 46- & Kans, 47; assoc prof, Univ Minn, 48-51, vis prof, 62-63; chief lab serv, St Louis Vet Admin Hosp, 69-72; consult, Children's Cardiac Ctr & Mercy Hosp, Kansas City, Mo. *Mem:* Am Soc Microbiol; Am Soc Clin Path; Soc Exp Biol & Med; Am Asn Path. *Res:* Pathogenesis and prophylaxis of rheumatic and granulomatous disease; biology of streptococcus and L-forms as slow agents; asbestos and immune response. *Mailing Add:* Truman Med Ctr Univ Mo Kansas City MO 64108

HAMILTON, WALTER S, b Hattiesburg, Miss, Dec 12, 31; m 60; c 2. PHYSICAL CHEMISTRY. *Educ:* Univ Southern Miss, BS, 54; Tulane Univ, PhD(phys chem), 63. *Prof Exp:* Sr res engr, Rocketdyne Div, NAm Aviation, Inc, 63-66; asst prof, 66-75, ASSOC PROF PHYS CHEM, TEX WOMAN'S UNIV, 75- *Mem:* AAAS; Am Chem Soc. *Res:* Solvent effect in kinetics; growth of ice crystals; electrochemical demineralization of brackish water; thermodynamics and thermochemistry. *Mailing Add:* Dept Chem Tex Woman's Univ Denton TX 76204

HAMILTON, WARREN BELL, b Los Angeles, Calif, May 13, 25; m 47; c 3. GEOLOGY, TECTONICS. *Educ:* Univ Calif, Los Angeles, AB, 45, PhD(geol), 51; Univ Southern Calif, MS, 49. *Prof Exp:* Lectr geol, Univ Calif, Los Angeles, 51; asst prof, Univ Okla, 51-52; geologist, 52-61, RES GEOLOGIST, US GEOL SURV, 61- *Concurrent Pos:* Sr exchange scientist, USSR, 67; vis prof, Scripps Inst Oceanog, 68 & 79, Calif Inst Technol, 73, Yale, 80 & Univ Amsterdam, 81 & Plate Tetonics Deleg China, 79; distinguished lectr, Am Asn Petrol Geol, 84-85; regents lectr, Univ Calif Santa Barbara, 86, Univ Calif Los Angeles, 88, Univ Calif San Diego, 90; vis scholar, Western Mich Univ, 85. *Honors & Awards:* Distinguished Serv Award, US Dept Interior, 81; Wilbert Lectr, La State Univ, 85; Penrose Medal Geol Sc Am, 89; Hookes Distinguished Lectr, McMaster Univ, 90 & Nat Acad Sci, 89. *Mem:* Fel Geol Soc Am; Am Geophys Union; hon fel Geol Soc London, 83; Geol Soc Am (chmn cordilleran Sect, 87-88); hon mem Colo Sci Soc, 85. *Res:* Structural geology and tectonics Western North America, Antarctica, Indonesia, Melanesia, Southeast Asia, USSR; plate tectonics; marine geophysics; igneous and metamorphic petrology; crustal evolution. *Mailing Add:* US Geol Surv Stop 964 Box 25046 Denver CO 80225

HAMILTON, WILLARD CHARLSON, b Auburn, NY, Sept 29, 42; m 63; c 3. SURFACE CHEMISTRY, PHYSICAL CHEMISTRY. *Educ:* Cornell Univ, BS, 64; Univ Calif, Davis, PhD(chem), 67. *Prof Exp:* NSF res assoc, Ctr Surface & Coatings Res, Lehigh Univ, 67-68; res chemist, Res Inst, Gillette Co, 68-69, proj supvr res & develop, 69-72; mgr spec mat technol ctr, Xerox Corp, 72-75; DIR RES & DEVELOP, JOHNSON & JOHNSON CO, 75- *Mem:* Sigma Xi. *Res:* Surface chemistry; rheology; wetting phenomena; product development; materials science. *Mailing Add:* 4905 Marlborough Way Durham NC 27713

HAMILTON, WILLIAM DONALD, b Cairo, Egypt, Aug 1, 36; Brit citizen; m 67; c 3. EVOLUTIONARY BIOLOGY. *Educ:* Cambridge Univ, BA, 60; Univ London, PhD(genetics), 68. *Prof Exp:* Lectr genetics, Imp Col Sci & Technol, Univ London, 64-77; prof evolutionary biol, Mus Zool, div biol sci, Univ Mich, Ann Arbor, 77-84; ROYAL SOC RES PROF, DEPT ZOOL, UNIV OXFORD, UK, 84- *Honors & Awards:* Sci Medal, Zool Soc London, 75; Newcomb Cleveland Prize, AAAS, 81. *Mem:* Brit Genetical Soc; Am Soc Naturalists; foreign hon mem Am Acad Arts & Sci; mem Royal Soc Sci Uppsala. *Res:* Evolution of social behavior, population genetics, sex ratio, evolution of sex and sexual selection. *Mailing Add:* Dept Zool Univ Oxford S Parks Rd Oxford OX1 3PS England

HAMILTON, WILLIAM EUGENE, JR, b Washington, DC, Sept 14, 42; m 73; c 2. CONTROL THEORY, OPTIMIZATION. *Educ:* Iowa State Univ, BS, 64; Purdue Univ, MS, 66, PhD(elec eng), 70. *Prof Exp:* Systs engr, Comptek Res, Inc, 72-75 & Sierra Res Corp, 75-78; sect head, Hydra-Pt Div, Maeg Inc, 78-79, mgr systs engr, 79-81, consult, 81-82, staff engr, Aircraft Controls Div, 82-83; sr res engr, 83-87, STAFF RES ENGR, GEN MOTORS RES LABS, 87- *Mem:* Inst Elec & Electronics Engrs; Am Sci Affil. *Res:* Simulation and control of electromechanical systems; optimization; digital signal processing. *Mailing Add:* 1346 W Fairview Lane Rochester MI 48306

HAMILTON, WILLIAM HOWARD, b Greenville, Pa, Apr 2, 18; m; m; c 2. ELECTRICAL & NUCLEAR ENGINEERING. *Educ:* Wash & Jefferson Col, BS, 40; Univ Pittsburgh, MS, 48. *Prof Exp:* Res engr electronics, Res Labs, Westinghouse Elec Corp, 45-50, eng mgr, Nuclear Reactor Plants, 50-65, mgr, Operating Plants, 65-70, gen mgr, Bettis Atomic Power Lab, 70-79; CONSULT ENGR, 79- *Mem:* Fel Inst Elec & Electronics Engrs; Am Nuclear Soc. *Res:* Development, design, installation and operation of nuclear reactor propulsion and power plants. *Mailing Add:* PO Box 613 Ligonier PA 15658

HAMILTON, WILLIAM JOHN, JR, zoology; deceased, see previous edition for last biography

HAMILTON, WILLIAM KENNON, b Guthrie Center, Iowa, Dec 15, 22; m 46; c 2. CLINICAL MEDICINE, ANESTHESIOLOGY. *Educ:* Univ Iowa, BA, 43, MD, 46; Am Bd Anesthesiol, dipl, 54. *Prof Exp:* Intern, St Luke's Hosp, Duluth, Minn, 46-47; resident, Div Anesthesiol, Col Med, Univ Iowa, 49-51, clin instr, 51-53, from asst prof to prof surg, 53-67, chmn div, 58-67; PROF ANESTHESIA & CHMN DEPT, UNIV CALIF, SAN FRANCISCO, 67-, ASSOC DEAN POSTDOCTORAL AFFAIRS, 78- *Concurrent Pos:* Chief anesthesiologist, Vet Admin Hosp, Iowa City, 51-53, consult, 58-; consult, Vet Admin Hosp, Des Moines, 58- & US Naval Hosp, Oakland & US Vet Admin Hosp, San Francisco; dir, Am Bd Anesthesiol, 64- *Mem:* Soc Exp Biol & Med; Am Soc Anesthesiol; AMA; Asn Univ Anesthetists. *Res:* Respiratory physiology, especially acute changes during anesthesia and surgery and in the postoperative period; care and management of acute and chronic respiratory insufficiency; venous physiology, especially role of the veins in adjusting blood volume. *Mailing Add:* Dept Anesthesia Univ Calif San Francisco CA 94143

HAMILTON, WILLIAM LANDER, b New York, NY, May 3, 43; m 65; c 2. ORGANIC CHEMISTRY, POLYMER CHEMISTRY. *Educ:* Columbia Univ, AB, 64; Yale Univ, MS, 65, PhD(chem), 69. *Prof Exp:* Res chemist, Film Dept, E I du Pont de Nemours & Co, Inc, 69-72, res chemist, 72-75, sr develop chemist, 75-79, supvr, Photo Prod Dept, 79-81, mgr, Com Systs Develop, 81-83, bus mgr, PCM Div, Du Pont Japan Ltd, 83-88, SR RES ASSOC, DU PONT ELECTRONICS, 88- *Mem:* Am Chem Soc. *Res:* Photoresists for printed circuit industry; physical structure and chemistry of natural and synthetic macromolecules; high temperature polymers; radiation chemistry; photo-initiated polymerization. *Mailing Add:* 52 Leicester Way Chesapeake City MD 21915-1808

HAMILTON, WILLIAM OLIVER, b Lawrence, Kans, Sept 5, 33; m 56; c 3. PHYSICS. *Educ:* Stanford Univ, BS, 55, PhD(physics), 63. *Prof Exp:* Res assoc physics, Stanford Univ, 63-65, actg asst prof, 65-67, asst prof, 67-70; assoc prof, 70-76, PROF PHYSICS, LA STATE UNIV, BATON ROUGE, 76- *Concurrent Pos:* NSF fel, 63-65; vis prof, Univ Rochester, 78. *Mem:* Am Phys Soc; Am Asn Physics Teachers; AAAS. *Res:* Experimental gravitational measurements; electron paramagentic resonance in organic free radicals; magnetism; cryogenics; properties of superconductors; thin film technology; Josephson effect; infrared detection. *Mailing Add:* Dept Physics & Astron La State Univ Baton Rouge LA 70803-4001

HAMILTON, WILLIAM THORNE, b Marion Center, Pa, July 26, 17; m 41; c 3. AERONAUTICAL ENGINEERING. *Educ:* Wash Univ, Seattle, BS, 41, MS, 47. *Prof Exp:* Res engr, Ames Lab, Nat Adv Comt Aeronaut, 41-48; mem aeronaut staff & aerodyn & propulsion staff engr, Boeing Co, 48-58, chief flight tech, 58-62, chief tech staff, Airplane Div, 62-65, dir develop proj, 65-71, mgr res eng, Boeing Aerospace Co, 72-74, vpres eng, 74-75, vpres-mgr YC-14 Prog, 75-76, vpres eng, 77-78, vpres res & eng, Boeing Com Airplane Co, 78-80, vpres & chief scientist, Boeing Mil Airplane Co, 81-82; CONSULT, 82- *Concurrent Pos:* Mem res adv comt aerodyn, NASA, 56-78; mem flight mech panel, Adv Group Aerospace Res & Develop, NATO. *Mem:* Nat Acad Eng; Flight Mech Panel; Am Inst Aeronaut & Astronaut; Asn Unmanned Vehicle Systs. *Res:* Aerodynamics; aeroelasticity; dynamic stability. *Mailing Add:* 9708 SW Harbor Dr Vashon WA 98070

HAMILTON-KEMP, THOMAS ROGERS, b Lebanon, Ky, May 13, 42; m 80. NATURAL COMPOUNDS, PHYTOCHEMISTRY. *Educ:* Univ Ky, BS, 64, PhD(org chem), 70. *Prof Exp:* PROF, DEPT HORT, UNIV KY, 70- *Concurrent Pos:* Asst prof, dept food sci & nutrit, Univ Ky, 70. *Mem:* Am Chem Soc; AAAS; Am Soc Hort Sci; Am Asn Univ Professors; Sigma Xi. *Res:* Isolation, purification and structure determination of natural products including volatiles involved in host plant-pathogen interactions and cytokinins and other modified nucleosides occurring in transfer RNA. *Mailing Add:* 868 Laurel Hill Rd Lexington KY 40504

HAMILTON-STEINRAUT, JEAN A, b Airdrie, Scotland, Feb 5, 38; m 67; c 2. X-RAY CRYSTALLOGRAPHY, BIOCHEMISTRY. *Educ:* Univ Glasgow, BSc, 59, PhD(chem), 62. *Prof Exp:* Instr, 66-68, asst prof, 68-80, ASSOC PROF BIOCHEM, IND UNIV SCH MED, INDIANAPOLIS, 80- *Concurrent Pos:* Fels, Dept Chem, Univ Ill, Urbana, 62-64 & Dept Biochem, Ind Univ, Indianapolis, 64-66; NIH career develop award, 66. *Mem:* Royal Soc Chem. *Res:* X-ray crystallography of biologically important substances, such as proteins, antibiotics and hormones. *Mailing Add:* Dept Biochem Ind Univ Sch Med 635 Barnhill Dr Indianapolis IN 46202-5122

HAMIT, HAROLD F, b Stockton, Kans, Dec 29, 13; m 35; c 2. SURGERY. *Educ:* NY Univ, AB, 42, MD, 45; Univ Colo, MS, 55; Am Bd Surg, dipl, 56 & 80; Am Bd Thoracic Surg, dipl, 63. *Prof Exp:* Chief obstet & gynec, Med Corps, US Army Hosp, Ft Bragg, NC, 48, resident gen surg, Oliver Gen Hosp, Augusta, Ga, 49-50, chief post-oper sect, 1st Mobile Army Surg, Korea, 50-51, comdr, Army Hosp, Camp Leroy Johnson, La, 51-53, resident gen surg, Fitzsimons Gen Hosp, Denver, 53-55, chief gen surg serv, Army Hosp, Ft Hood, Tex, 55-56, chief surg res, Army Med Res & Develop Command, 57-60, resident thoracic surg, Letterman Gen Hosp, San Francisco, 60-62, comdr, 121st Evacuation Hosp, Ascom, Korea, 62-63, res assoc prof surg, Baylor Col Med, 63-65, chief gen surg serv, Brooke Gen Hosp, San Antonio, 65-67, dir, Div Surg, Walter Reed Army Inst Res, DC, 67-68; assoc dir clin res, Baxter Labs, Inc, 68-70; assoc dir surg, Charlotte Mem Hosp, 70-85; clin prof, 75-85, EMER PROF CLIN SURG, UNIV NC, CHAPEL HILL, 85- *Concurrent Pos:* Mem, Surg Study Sect, USPHS, 58-60; vis oper surgeon, Korean Nat Res Cross Hosp, Seoul, 62-63; consult, Eighth Army Surgeon, 62-63; lectr, Dept Surg, Northwestern Univ, 68-70. *Mem:* AAAS; fel Am Col Surg; fel Am Asn Surg Trauma; Asn Mil Surgeons US; NY Acad Sci; AMA; Pan Am Med Soc. *Res:* General, thoracic and cardiovascular surgery; trauma; burns; shock; bacterial enzymes; esophageal physiology; plasma volume expanders; tissue adhesives; antibiotics; hemodynamics of brain. *Mailing Add:* 1309 Providence Rd Charlotte NC 28207

HAMJIAN, HARRY J, b Auburn, NY, July 8, 23; m 49. METALLURGY. *Educ:* Univ Mich, BSE, 44, MSE, 47. *Prof Exp:* Chem engr, Pilot Plant, Houdaille Hershey Corp, 44-45, supvr gas opers, 45-46; scientist mat res, NASA, 47-52; mgr powder metall, Utica Drop Forge & Tool Corp, 52-53, mgr opers, 53-56 & Kelsey Hayes Corp, 56-61; vpres metall, Spec Metals, Inc, 61-65; mgr process metall, Carpenter Steel Co, 65-67, asst mgr labs, Carpenter Technol Corp, 67-69, mgr process metall res, 69-77; from vpres alloy opers to sr vpres mat tech group, Howmet, 77-86; RETIRED. *Mem:* Am Inst Mining, Metall & Petrol Engrs; Am Soc Metals; Am Vacuum Soc; Newcomen Soc. *Res:* Vacuum metallurgy; high temperature alloys; powder metallurgy; high temperature alloys and tool steels. *Mailing Add:* 170 Blackberry Dr Stanford CT 06903

HAMKALO, BARBARA ANN, b New York, NY, July 4, 44. GENETICS, CELL BIOLOGY. *Educ:* Univ Mass, PhD(radiation biophys), 68. *Prof Exp:* Nat Cancer Inst fel biochem, Harvard Med Sch, 68-70; res assoc, Biol Div, Oak Ridge Nat Lab, 70-71, staff mem, 71-73; from asst prof to assoc prof, biol sci, Dept Molecular Biol & Biochem, 79-86, assoc dean grad studies & res, 81-84, PROF MOLECULAR BIOL, UNIV CALIF, IRVINE, 86- *Mem:* Am Soc Cell Biol; Am Soc Biochem & Molecular Biol. *Res:* Structure and function of chromosomes in eukaryotes; regulation of gene expression. *Mailing Add:* Dept Molecular Biol & Biochem Univ Calif Irvine CA 92717

HAMLET, RICHARD GRAHAM, b Minneapolis, Minn, Mar 27, 38; m 58, 86; c 2. COMPUTER SCIENCE. *Educ:* Univ Wis-Madison, BS, 59; Cornell Univ, MS, 64; Univ Wash, PhD(comput sci), 71. *Prof Exp:* Intern, Shimer Col, 62-64; systs supvr, Comput Ctr, Univ Wash, 66-68; dir systs prog, Comput Ctr Corp, 68-70; from asst prof to assoc prof comput sci, Univ Md, 71-83; prof dept comput sci, Ore Grad Ctr, 83-88; PROF DEPT COMPUT SCI, PORTLAND STATE UNIV, 88- *Concurrent Pos:* Defense Mapping Agency grant, 74-75; consult, Naval Res Lab, 76-78, IBM, 78-83; NSF grant, 78-80 & 89-91; Air Force Off Sci Res grant, 79-83 & 86. *Mem:* Asn Comput Mach; Inst Elec & Electronics Engrs. *Res:* Theory of program testing; software engineering; computability theory; theory of programming; Godel numberings; systems programming; programming languages. *Mailing Add:* Computer Sci Dept Portland State Univ PO Box 751 Portland OR 97207

HAMLET, ZACHARIAS, b Pulincunnu, India, Nov 19, 30; m. ORGANIC CHEMISTRY. *Educ:* Loyola Col, Madras, India, BSc, 50; Victoria Col, Agra, MSc, 52; Univ Notre Dame, PhD(org chem), 60. *Prof Exp:* Lectr chem, St Joseph's Col, Darjeeling, India, 52-56; res assoc, Univ Southern Calif, 60-61 & State Univ NY Stony Brook, 61-62; fel, Nat Res Coun Can, 62-64; res asst, Univ Montreal, 64-66; asst prof, 66-69, ASSOC PROF CHEM, CONCORDIA UNIV, 69- *Concurrent Pos:* Vis assoc, Calif Inst Technol & Stanford Univ, 72-73; vis colleague, Univ Hawaii, Manoa, 80-81. *Mem:* Am Chem Soc; Royal Soc Chem; Chem Inst Can. *Res:* Organic free radical chemistry; organic photochemistry; organophosphorus and organosulfur chemistry; synthesis and reaction mechanisms. *Mailing Add:* Concordia Univ 7141 Sherbrooke St W Montreal PQ H4G 1R6 Can

HAMLETT, WILLIAM CORNELIUS, b Henderson, NC, Jan 22, 48; m 70; c 3. DEVELOPMENT BIOLOGY, CELL ULTRASTRUCTURE. *Educ:* Univ SC, BS, 70, MS, 73; Clemson Univ, PhD(develop bid), 83. *Prof Exp:* Res assoc anat, Med Univ SC, 73-76; asst prof biol sci, Baptist Col, Charleston, 76-77; teaching fel zool, 77-80, res fel develop biol, Dept Zool, Clemson Univ, 80-82; instr anat, 83-84, ASST PROF ANAT, MED COL, OHIO, 84- *Concurrent Pos:* fel Dept Anat, Med Col Ohio, 82-83; vis investr, Va Inst Marine Sci, 80 & 90, Bermuda Biol Sta Res, 79 & 87, Inst Marine Sci, Univ NC, 79-91, Inst Oceanog, Univ Sao Paulo, Brazil, 88. *Mem:* Am Soc Cell Biol; Am Asn Anatomist; Soc Develop Biol; Electron Micros Soc Am; Soc Study Reproduction; Am Soc Zoologist. *Res:* Comparative placentation; regulation, morphogenesis and evolution of viviparity in elasmobranch fishes; developoment of the valvular apparatus in vertebrate hearts. *Mailing Add:* Dept Anat Med Col Ohio C S 10008 Toledo OH 43699

HAMLIN, DANIEL ALLEN, b Burbank, Calif, June 21, 26; m 52; c 2. ATMOSPHERIC PHYSICS. *Educ:* Univ Calif, Berkeley, AB, 47, PhD(physics), 54. *Prof Exp:* Sr res engr, Gen Dynamics/Convair, 56-59, design specialist, 59-60, staff scientist, 60-63, sr staff scientist, 63-64, chief theoret physics group, 64-69; SCIENTIST 4, SCI APPLNS, INC, LA JOLLA, CALIF, 69- *Mem:* AAAS; Am Phys Soc; Am Geophys Union; Am Optical Soc; Sigma Xi. *Res:* Elementary particle and nuclear physics; atomic properties and processes; equation of state of high temperature gases; radiation transport; opacity; penetration of atomic particles through matter; aeronomy; infrared physics; geomagnetism. *Mailing Add:* 3224 Oliphant St San Diego CA 92106-1942

HAMLIN, GRIFFITH ASKEW, JR, b Wilson, NC, Oct 31, 45; m 66; c 1. COMPUTER SCIENCES. *Educ:* Westminster Col, BA, 68; Univ NC, MS, 70, PhD(comput sci), 75. *Prof Exp:* Instr math, William Woods Col, 69-72; vis scientist comput sci, Inst Comput Applns Sci & Eng, Univs Space Res Asn, 75-77; mem staff, Los Alamos Sci Lab, 77-83; sr scientist, Unicad Inc, Boulder, Co, 83-87; CO-FOUND, DES SCI INC, 87- *Mem:* Asn Comput Mach. *Res:* Distributed computing research related to interactive satellite computer graphics; design of user interface management system software. *Mailing Add:* 6397 Pitcairn St Cypress CA 90630

HAMLIN, JAMES T, III, b Danville, Va, Feb 6, 29; m 55; c 3. INTERNAL MEDICINE. *Educ:* Va Mil Inst, AB, 51; Univ Va, MD, 55. *Prof Exp:* Intern, Peter Bent Brigham Hosp, 55-56; instr med, NY Med Col, 59-60; guest investr, Rockefeller Inst, 60-62; from asst prof to assoc prof, Med Col Ga, 62-66; assoc prof med, Sch Med, Univ Va, 66-73, from asst dean to actg dean, 70-73; PROF MED, TULANE UNIV MED CTR, 73-, DEAN SCH MED, 75- *Concurrent Pos:* Dir clin res ctr, Univ Va Hosp, 66-71. *Mem:* Am Fedn Clin Res; AMA; Am Asn Med Clin. *Res:* Lipid metabolism and atherosclerotic disease. *Mailing Add:* 199 Fairmont Circle Danville VA 24541

HAMLIN, JOYCE LIBBY, b Apr 15, 39; div. DNA REPLICATION, GENE AMPLIFICATION. *Educ:* Univ Calif, Los Angeles, PhD(molecular biol), 71. *Prof Exp:* ASSOC PROF BIOCHEM, SCH MED, UNIV VA, 84- *Concurrent Pos:* fac res award, Am Cancer Soc. *Res:* Mammalian genome organization; DNA replication. *Mailing Add:* Biochem Dept Univ Va Sch Med Charlottesville VA 22908

HAMLIN, KENNETH ELDRED, JR, b Baltimore, Md, Mar 27, 17; m 41; c 2. CHEMISTRY, RESEARCH ADMINISTRATION. *Educ:* Univ Md, BS, 38, PhD(pharmaceut chem), 41. *Prof Exp:* Pharmacist, Univ Md, 38-41; fel & spec asst chem, Univ Ill, 41-42; instr org chem, Univ Md, 42-43; res chemist, Abbott Labs, 43-54, from asst head to head org res, 54-57, from asst dir to dir chem res, 57-61, dir res, 61-66; vpres res, 66-73, vpres res & qual assurance, 73-74, sr vpres sci opers, Cutter Labs Inc, 74-81; RETIRED. *Concurrent Pos:* Mem bd dirs, Cutters Labs, 67-81, vchmn, 79-81. *Mem:* AAAS; Am Pharmaceut Asn; Am Chem Soc. *Res:* Synthesis of amino acids; monocrotaline and related alkaloids; antimalarials; synthetic organic medicinals. *Mailing Add:* 3270 Terra Granada Dr Walnut Creek CA 94595

HAMLIN, ROBERT LOUIS, b Cleveland, Ohio, Mar 18, 33; m 60; c 2. CARDIOVASCULAR PHYSIOLOGY. *Educ:* Ohio State Univ, BSc, 56, DVM, 58, MSc, 60, PhD(physiol), 62. *Prof Exp:* From asst prof to assoc prof, 62-69, PROF VET PHYSIOL, OHIO STATE UNIV, 69- *Concurrent Pos:* NIH career develop award, 62- *Mem:* AAAS; Am Heart Asn; Am Physiol Soc; Am Vet Med Asn. *Res:* Ventricular activation processes of many species; simultaneous indicator-dilution curves for detection of mitral regurgitation; electrocardiographic and hemodynamic characteristics of sinus arrhythmia in the dog. *Mailing Add:* Dept Vet Physiol & Pharm Ohio State Univ 1900 Coffey Rd Columbus OH 43210

HAMLIN, WILLIAM EARL, b Wautoma, Wis, July 3, 22; m 51; c 2. INDUSTRIAL PHARMACY. *Educ:* Lawrence Col, BA, 44; Univ Ill, MS, 47, PhD(chem), 49. *Prof Exp:* Res chemist, Schenley Labs, Inc, 49-50; res chemist, Upjohn Co, 50-84; RETIRED. *Mem:* Am Chem Soc; AAAS; Sigma Xi. *Res:* Pharmaceutical research; product development. *Mailing Add:* 2601 Hill-n-Brook Dr Kalamazoo MI 49008

HAMLOW, EUGENE EMANUEL, b Bloomington, Ill, Feb 22, 27; m 54; c 3. PHARMACY. *Educ:* Purdue Univ, BS, 50, MS, 53, PhD, 58. *Prof Exp:* Control chemist, Upjohn Co, 52-55; sect leader, 58-70, DIR PHARMACEUT RES & DEVELOP, BRISTOL-MYERS SQUIBB CO, 70- *Mem:* Am Chem Soc; Soc Cosmetic Chem; Sigma Xi; Am Pharmaceut Asn; Am Asn Pharmaceut Scientists. *Res:* Pharmaceutical research and development. *Mailing Add:* B-M Pharmaceut Prod Develop 2404 W Lloyd Expressway Evansville IN 47721

HAMM, DONALD IVAN, b Wellington, Kans, Jan 11, 28; m 50; c 5. ORGANIC CHEMISTRY. *Educ:* Univ Okla, BS, 49, PhD(chem), 56; Purdue Univ, MS, 51. *Prof Exp:* Asst prof chem, Southwestern State Col, Okla, 51-53; asst, Okla Med Res Found, 53-56; from assoc prof to prof chem, Southwestern Okla State Univ, 56-89, chmn dept, 70-79, dean, Sch Arts & Sci, 78-89; RETIRED. *Concurrent Pos:* Vis prof, Mich State Univ, 69-70. *Mem:* Am Chem Soc. *Mailing Add:* 1617 E Davis Rd Weatherford OK 73096

HAMM, FRANKLIN ALBERT, b New Tripoli, Pa, Feb 23, 18; m 41; c 2. SOLID STATE SCIENCE. *Educ:* Muhlenberg Col, BS, 39; Cornell Univ, MS, 41, PhD(micros), 43. *Prof Exp:* Microscopist, Cent Res Lab, Gen Aniline & Film Corp, 43-52; asst to dir res, Res Div, Burroughs Adding Mach Co, Pa, 52-53; mem physics div, Res Lab, Eastman Kodak Co, 53-58; sect leader, Cent Res Dept, Minn Mining & Mfg Co, 58-64, mgr photog prod lab, 64-66, tech mgr new prod develop, 66-70, res mgr photog prod div, 70-73, sr res specialist, Cent Res Lab, 73-82; RETIRED. *Honors & Awards:* Photog Prize, Am Soc Testing & Mat, 46. *Mem:* AAAS; Am Chem Soc; Electron Micros Soc Am; Am Phys Soc; Soc Photog Scientists & Engrs. *Res:* Solid state physics; photoconductivity; physical properties of macromolecules; electron microscopy; sensitization of organic semiconductors; electrophotography. *Mailing Add:* 1505 N Second St Stillwater MN 55082

HAMM, JOSEPH NICHOLAS, b Winner, SDak, Dec 6, 15; m 39; c 2. MEDICINE. *Educ:* Creighton Univ, BS, 39; George Washington Univ, MD, 42. *Prof Exp:* Asst dean clin affairs, 75-80, ASSOC DEAN CLIN AFFAIRS & ASSOC PROF FAMILY MED, SCH MED, UNIV SDAK, 80- *Mem:* AMA; Am Acad Family Pract. *Mailing Add:* 1405 Davenport St Sturgis SD 57785

HAMM, KENNETH LEE, b Princeton, Ill, Nov 12, 23; m 49; c 2. ORGANIC CHEMISTRY. *Educ:* Carthage Col, BA, 47; Univ Ill, MA, 48; Univ Iowa, PhD, 57. *Prof Exp:* Assoc prof, 48-53, PROF CHEM, CARTHAGE COL, 53- *Mem:* Am Chem Soc; Sigma Xi. *Res:* Synthesis and oxidation of unsymmetrical hydrazines. *Mailing Add:* 4210 Farmington Lane Racine WI 53403

HAMM, RANDALL EARL, b Auburn, Wash, May 9, 13; m 37; c 3. PHYSICAL CHEMISTRY. *Educ:* Univ Wash, BS, 35, MS, 37, PhD(chem), 40. *Prof Exp:* Res assoc oceanog, Univ Wash, 40-41; asst chemist, Puget Sound Navy Yard, US Dept Navy, 40-42; instr phys sci, Western Wash Col Educ, 42; from instr to prof anal chem, Univ Utah, 42-63; prof anal chem, 63-78, EMER PROF ANALYTICAL CHEM, WASH STATE UNIV, 78- *Mem:* Am Chem Soc. *Res:* Electrochemistry; dissolved gases in solution; amperometric titrations; analytical polarography; structure and reactions of complex ions; kinetics of complex ion reactions. *Mailing Add:* Dept Chem Wash State Univ Pullman WA 99164

HAMM, THOMAS EDWARD, JR, b Denver, Colo, Dec 26, 42; m 70; c 1. LABORATORY ANIMAL MEDICINE, COMPARATIVE PATHOLOGY. *Educ:* Univ Colo, Boulder, BA, 64; Colo State Univ, DVM, 68, MS, 72; Am Col Lab Animal Med, dipl, 74; Bowman Gray Med Sch, PhD, 80. *Prof Exp:* Fel lab animal med, Colo State Univ, 70-72; NIH fel comp path, Bowman Gray Sch Med, 72-75; asst prof path, Med Ctr, Univ Colo, 75-78; expert consult lab animal med, Nat Cancer Inst, 78-80, chief toxicology br, 79-80; head dept toxicol, Chem Indust Inst Toxicol, 80-84; DIR, DIV LAB ANIMAL MED, STANFORD UNIV, 84- *Concurrent Pos:* Consult animal care, Wake Forest Univ, 74-75, Colo State Univ, 75-76, Univ Colo, Boulder & Nat Jewish Hosp, 75-78; affil fac path, Colo State Univ, 75-78; consult, Am Asn Accreditations Lab Animal Care, 78-; affil fac, NC State Univ, 81- *Mem:* Am Vet Med Asn; Am Asn Lab Animal Sci; Am Soc Lab Animal Practr; Am Soc Primatologists; Am Col Lab Animal Med. *Res:* Development of animal models of human diseases, especially atherosclerosis and carcinogenesis; toxicology. *Mailing Add:* Chmn Dept Comp Med Sch Med Stanford Univ Stanford CA 94305

HAMM, WILLIAM JOSEPH, b Belleville, Ill, July 26, 10. PHYSICS. *Educ:* Univ Dayton, BSc, 31; Cath Univ Am, MSc, 35; Wash Univ, PhD(physics), 42. *Prof Exp:* Instr physics, St Mary's Univ, Tex, 35-37 & Maryhurst Norm Col, 37-40; from asst prof to assoc prof, 42-58, PROF PHYSICS, ST MARY'S UNIV, TEX, 58- *Concurrent Pos:* Extramural assoc, NIH, 78; sr res assoc, USAF Sch Aerospace Med, Brooks AFB, Tex, 79. *Mem:* Fel Inst Elec & Electronics Engrs; Am Phys Soc. *Res:* Collision processes in gases, particularly energy loss retardation of positive ions shot through gases; electronic circuitry. *Mailing Add:* Dept Physics St Mary's Univ San Antonio TX 78284

HAMMAKER, GENEVA SINQUEFIELD, b Bolivar, Mo, Jan 9, 36; m 59; c 2. INORGANIC CHEMISTRY. *Educ:* Drury Col, AB, 57; Northwestern Univ, PhD(inorg chem), 61; Univ East Anglia, dipl econ, 78. *Prof Exp:* Asst prof chem, Marymount Col, Kans, 65-68; lectr physics, Kans State Univ, 68-72, asst prof chem, 72-74; staff scientist, Develop, Planning & Res Assocs, Inc, 74-85; SR ENVIRON SCIENTIST, MIDWEST RES INST, 85- *Mem:* Am Chem Soc. *Res:* Coordination compounds; environmental science, economics; science education. *Mailing Add:* 1013 Boathouse Ct Raleigh NC 27613

HAMMAKER, ROBERT MICHAEL, b Evanston, Ill, Feb 9, 34; div; c 2. HADAMARD TRANSFORM SPECTROMETRY, VOLATILE ORGANIC COMPOUNDS IN THE ATMOSPHERE. *Educ:* Trinity Col, Conn, BS, 56; Northwestern Univ, PhD(phys chem), 60. *Prof Exp:* Sr chemist, Res Lab, Texaco, Inc, NY, 60-61; from asst prof to prof chem, 61-74, assoc head dept, 68-76, PROF CHEM, KANS STATE UNIV, 74- *Concurrent Pos:* Sr vis fel, Univ East Anglia, England, 76-77; sabbatical, Univ Calif, Riverside, 87-88. *Mem:* AAAS; Am Chem Soc; Am Phys Soc; Royal Soc Chem; Soc Appl Spectros. *Res:* Molecular spectroscopy (IR, Raman); development of Hadamard transform spectrometry; detection of volatile organic compounds (VOCs) using Fourier transform infrared (FT-IR) spectrometry. *Mailing Add:* Dept Chem Willard Hall Kans State Univ Manhattan KS 66506-3701

HAMMAM, M SHAWKY, ELECTRICAL ENGINEERING. *Prof Exp:* FAC MEM, ELEC ENG DEPT, CLARKSON UNIV. *Mem:* Inst Elec & Electronics Engrs. *Mailing Add:* Elec Eng Dept Clarkson Univ Potsdam NY 13699

HAMMAN, DONALD JAY, b Lorain, Ohio, Sept 8, 29; m 52; c 4. ELECTRONICS. *Educ:* Univ Ohio, BS, 56; Ohio State Univ, MSc, 58. *Prof Exp:* Mem tech staff, Bell Tel Labs, 56; prin physicist, Battelle Mem Inst, 56-62, res physicist, Battelle Columbus Labs, 62-64, sr physicist, 64-65, assoc chief radiation effects in electronics, 65-76, proj leader electronics, 76-77, prin res scientist, 77-80; WITH QWIP SYSTS, DIV EXXON ENTERPRISES, INC, 80-, MGR QUAL ENG,. *Mem:* Sr mem Inst Elec & Electronics Engrs. *Res:* Nuclear and space radiation effects in electronics and materials; electronic reliability; reliability physics. *Mailing Add:* McDonell Douglas Astronaut Co-Titusville Div MS 13 PO Box 600 Titusville FL 32780

HAMMANN, JOHN WILLIAM, b St Louis, Mo, Oct 21, 14; div; c 1. ELECTRICAL ENGINEERING, ELECTRONICS ENGINEERING. *Educ:* Purdue Univ, BSEE, 36; Mo Sch Mines & Metall, MSEE, 48; Univ Grenoble, dipl higher educ, 52; Wash Univ, PhD(elec eng), 60. *Prof Exp:* Instr elec eng, Mo Sch Mines & Metall, 46-48; lectr, Wash Univ, 48-51 & 52-56; engr, Emerson Elec & Mfg Co, 56-59; assoc elec engr, Argonne Nat Lab, 60-67; from asst prof to assoc prof elec eng, Calumet Campus, Purdue Univ, 67-75, assoc prof elec technol, 77-78; assoc prof, 78-85, EMER PROF ELEC TECHNOL, IND UNIV-PURDUE UNIV, INDIANAPOLIS, 85- *Mem:* Inst Elec & Electronics Engrs; Am Soc Eng Educ. *Res:* The electrical aspects of solar energy; alternate energy sources. *Mailing Add:* 5316 Acorn Lane Indianapolis IN 46254-1351

HAMMANN, WILLIAM CURL, b Little Rock, Ark, Apr 22, 25; m 55; c 3. ANIMAL GROWTH PROMOTANTS, FOOD & FEED CHEMICALS. *Educ:* Univ Minn, BChem, 47; Univ Ill, PhD(chem), 51. *Prof Exp:* Fulbright fel, Univ Paris, 51-52; Swiss-Am fel, Swiss Fed Inst Technol, 52-53; res chemist, Monsanto Co, 53-60, group leader, 60-67, res mgr new enterprises div, Res Ctr, 67-69, mgr com develop, foods & fine chem bus group, 69-75, mgr res functional prod bus group, 75-77, dir res spec chem div, 77-80, dir res & develop, Nutrit Chem Div, 80-85, CONSULT, MONSANTO CO, 86- *Concurrent Pos:* Instr, Univ Dayton, 55-61. *Mem:* AAAS; Sigma Xi; Soc Chem Indust; Am Chem Soc. *Res:* Synthetic functional fluids; thermal stability of organic compounds; polyphenyl ethers; organic syntheses; food additives; fine chemicals; water treatment chemicals; feed additives; growth promotants. *Mailing Add:* 438 Fourwynd Dr St Louis MO 63141

HAMMAR, ALLAN H, b Springfield, Mass, June 5, 23; m 48; c 2. MICROBIOLOGY, ZOOLOGY. *Educ:* Univ Conn, BS, 50, MS, 51. *Prof Exp:* Instr virus res animal dis, Univ Conn, 51-53; biologist, Lederle Labs, Am Cyanamid Co, 53-57, dept head virus vaccine prod, 57-60, supt tissue cult vaccine prod, 60-63, supt virus prod & vacuum drying, 63-70; microbiol biol res mgr, 70-76, asst to vpres, 77-86, RES & DEVELOP QUAL ASSURANCE MGR, CUTTER LABS, 86- *Mem:* Am Soc Microbiol. *Res:* Experimental embryology; virus research in animal diseases; rickettsial and human virus vaccines; tissue culture and production activities in virus vaccines; development and study of effectiveness of new hyperimmune globulins. *Mailing Add:* 120 Florence Ave Mill Valley CA 94941

HAMMAR, SHERREL L, b Caldwell, Idaho, May 21, 31; m 56; c 2. PEDIATRICS, ADOLESCENT MEDICINE. *Educ:* Col Idaho, BA, 53; Univ Wash, MD, 57. *Prof Exp:* From instr to assoc prof pediat, Univ Wash, 62-71, asst dir, Div Child Health, 65-71; assoc prof, 71-73, PROF PEDIAT & CHMN DEPT, SCH MED, UNIV HAWAII, MANOA, 73- *Concurrent Pos:* Res fel growth & develop, Univ Wash, 60-62; med consult, Luther Burbank Sch, 60-66 & Echo Glenn Children's Ctr, 67-71; dir ambulatory serv, Kauikeolani Children's Hosp, 71-73, chief pediat, 73-; dir, Am Bd Pediat. *Mem:* Fel Am Acad Pediat; Soc Adolescent Med (pres, 80); Ambulatory Pediat Asn; Am Pediat Soc; AMA. *Res:* Adolescent medicine and obesity; school learning problems; mental retardation and growth disorders. *Mailing Add:* Kapiolani-Children's Med Ctr 1319 Punahou St Honolulu HI 96826

HAMMAR, WALTON JAMES, b Waverly, Iowa, Apr 25, 41; m 63; c 4. ORGANIC CHEMISTRY. *Educ:* Iowa State Univ, BS, 63; Purdue Univ, PhD(org chem), 67. *Prof Exp:* NIH fel, Cornell Univ, 67-69; res specialist drug res, Riker Labs, 69-76, STAFF SCIENTIST, LIFE SCI LAB, 3M CO, 76- *Mem:* Am Chem Soc; Soc Biomat; Sigma Xi. *Res:* Organic cation chemistry; photochemistry; small ring chemistry; chemotherapeutic agents; central nervous system agents; biomaterials; nonthrombogenic surfaces; polymer chemistry. *Mailing Add:* Cent Res 201-2W 3M Ctr St Paul MN 55101

HAMMARLUND, EDWIN ROY, b Seattle, Wash, Aug 24, 22; m 47; c 3. PHARMACY. *Educ:* Univ Wash, BS, 43, MS, 49, PhD(pharm), 51. *Prof Exp:* From asst prof to assoc prof pharm, Wash State Univ, 51-60; assoc prof, 60-62, PROF PHARM, UNIV WASH, 62- *Concurrent Pos:* Pfeiffer Mem res fel, Copenhagen, 58-59; WHO fel drug abuse, London, Stockholm & Copenhagen, 71; US Info Serv lectr drug abuse, WGer, 71. *Mem:* Am Pharmaceut Asn; Am Asn Cols Pharm. *Res:* Surface active agents; antacids; isosmotic solutions; blood hemolysis; drug abuse education. *Mailing Add:* Sch Pharm Univ Wash BG-20 Seattle WA 98195

HAMME, JOHN VALENTINE, b Oxford, NC, May 14, 19; m 47; c 2. MINERAL & CERAMIC ENGINEERING. *Educ:* NC State Col, BS, 40; Univ Utah, MS, 42; NC State Univ, PhD(ceramic eng), 63. *Prof Exp:* Metallurgist, Columbia Steel Co Div, US Steel Corp, 42-44; engr, Tungsten Mining Corp, NC, 47-50, asst mill supt, 50-51, mill supt, 51-55, chem plant supt, 55-58; res asst, 58-63, assoc prof, 63-83, dir coop eng educ, 69-83, EMER ASSOC PROF CERAMIC ENG, NC STATE UNIV, 83- *Mem:* Am Inst Mining, Metall & Petrol Engrs; Sigma Xi. *Res:* Mineral exploration; mineral beneficiation; nuclear fuel materials research. *Mailing Add:* 1312 Onslow Rd Raleigh NC 27606-2745

HAMMEL, EDWARD FREDERIC, b New York, NY, Jan 6, 18; m 42; c 3. PHYSICAL CHEMISTRY. *Educ:* Dartmouth Col AB, 39; Princeton Univ, PhD(phys chem), 44. *Prof Exp:* Res assoc heterogeneous catalysis, Princeton Univ, 41-42, sr scientist, 42-44; sect leader plutonium remelting, alloying & casting, high vacuum, Los Alamos Nat Lab, Univ Calif, 44-45, group leader metal physics & low temperature physics, 45-70, mem staff low temperature physics, 70-72, prog mgr superconducting transmission line & energy storage, 72-73, assoc div leader, Energy Div, 73-74, asst dir energy, 74-79; CONSULT ENERGY, LOW TEMPERATURE PHYSICS, CRYOENG, 79- *Honors & Awards:* Am Chem Soc Award, 55; Samuel C Collins Award, Cryogenic Eng Conf, 73; Wilbur T Pentzer Award, Int Inst Refrig, 85. *Mem:* Sigma Xi; fel Am Phys Soc. *Res:* Low temperature physics. *Mailing Add:* 99 Rim Rd Los Alamos NM 87544

HAMMEL, EUGENE A, b New York, NY, Mar 18, 30; m 51; c 4. SOCIAL ANTHROPOLOGY, DEMOGRAPHY. *Educ:* Univ Calif, Berkeley, AB, 51, PhD(anthrop), 59. *Prof Exp:* Asst prof anthrop, Univ NMex, 59-61; from asst prof to prof anthrop, 61-78, PROF ANTHROP & DEMOG, UNIV CALIF, BERKELEY, 78- *Mem:* Nat Acad Sci; Am Ethnol Soc (pres, 83-85); Am Anthrop Asn; Pop Asn Am. *Mailing Add:* Dept Anthrop Univ Calif Berkeley CA 94720

HAMMEL, HAROLD THEODORE, b Huntington, Ind, May 8, 21; m 48; c 2. PHYSIOLOGY. *Educ:* Purdue Univ, BS, 43; Cornell Univ, MS, 50, PhD(zool), 53. *Prof Exp:* Jr physicist, Los Alamos Sci Lab, Univ Calif, Los Angeles, 44-46, staff physicist, 48-49; from instr to asst prof physiol, Sch Med, Univ Pa, 53-61; assoc prof, Yale Univ, 61-67; prof physiol, Univ Calif, San Diego, 67-88; ADJ PROF PHYSIOL & BIOPHYS, IND UNIV, BLOOMINGTON, 88- *Concurrent Pos:* Fel, John B Pierce Found Lab, 61-67; panelist regulatory biol, NSF, 68-71; foreign sci mem, Max Planck Inst Physiol & Clin Res; Alexander von Humboldt Found US sr scientist award, 81-82. *Mem:* Am Phys Soc; Am Soc Mammal; Am Physiol Soc; fel AAAS; Am Soc Plant Physiol; Norweg Acad Sci & Lett. *Res:* Osmoregulation and regulation of body temperature in vertebrates; water relations in vascular plants. *Mailing Add:* Dept Physiol Ind Univ Sch Med Meyers Hall Bloomington IN 47405-4201

HAMMEL, JAY EDWIN, b Hutchinson, Kans, June 5, 21. PLASMA PHYSICS. *Educ:* Calif Inst Technol, BS, 44; Rice Univ, MS, 51. *Prof Exp:* Staff mem, Los Alamos Nat Lab, 52-87; CONSULT PLASMA PHYSICS, UNIV CALIF, 87- *Mem:* Fel Am Phys Soc. *Res:* High temperature solid density plasma. *Mailing Add:* 40 Chaco Los Alamos NM 87544

HAMMEL, JAY MORRIS, b New York, NY, Mar 14, 38; m 70; c 1. MICROBIOLOGY, BIOCHEMISTRY. *Educ:* City Col New York, BS, 60; Pa State Univ, MS, 62, PhD(microbiol), 65. *Prof Exp:* From instr to asst prof, 65-81, ASSOC PROF MICROBIOL, HAHNEMANN MED COL, 81- *Concurrent Pos:* Vis Prof, Dept Micro & Immunol, Ore Health Sci Univ, 81-82. *Mem:* AAAS; Am Soc Microbiol; NY Acad Sci; Sigma Xi. *Res:* Microbial physiology, especially protein synthesis; control mechanisms; extracellular enzyme synthesis; legionella pneumophila; mechanisms of pathogenesis. *Mailing Add:* Dept Microbiol Hahnemann Univ Philadelphia PA 19102

HAMMEN, CARL SCHLEE, b Newark, NJ, Aug 26, 23; m 49, 62; c 4. INVERTEBRATE PHYSIOLOGY, COMPARATIVE BIOCHEMISTRY. *Educ:* St John's Col, Md, BA, 47; Columbia Univ, MA, 49; Univ Chicago, SM, 52; Duke Univ, PhD(zool), 58. *Prof Exp:* Instr biol & chem, Mitchell Col, NC, 49-51; prof biol & math, Cedarville Col, 52-53; biologist, Vet Admin Ctr, WVa, 53-54 & Army Chem Ctr, Md, 54-56; assoc prof biol, Newark State Col, 58-60 & Adelphi Col, 60-63; from asst prof to assoc prof, 63-71, PROF ZOOL, UNIV RI, 71- *Concurrent Pos:* NSF res grants, Newark State Col, 59-60, Adelphi Col, 61-63 & Univ RI, 64-66; Fulbright fel, Morocco, 84. *Mem:* Am Physiol Soc; Am Soc Zoologists; Sigma Xi. *Res:* Metabolism of marine invertebrates, particularly bivalve mollusks and brachiopods. *Mailing Add:* Dept Zool Univ RI Kingston RI 02881

HAMMEN, SUSAN LUM, b Summit, NJ, Mar 15, 36; m 62; c 3. POPULATION BIOLOGY OF FERNS. *Educ:* Mt Holyoke Col, BA, 58; NY Univ, MS, 63; Univ RI, PhD(bot), 89. *Prof Exp:* Res asst, Kidney Lab, P B Brigham Hosp, Harvard Med Sch, 58-60 & physiol, Adelphi Univ, Garden City, 60-63; instr biol, Prout Mem High Sch, Wakefield, RI, 87-88; INSTR BOT, COL CONTINUING EDUC, UNIV RI, 89- *Concurrent Pos:* Instr, Prog Excellence Teaching Sci, Univ RI, 89-91. *Mem:* Am Fern Soc; Bot Soc Am; Sigma Xi. *Res:* Population biology of the hay-scented fern, Dennstaedtia Punctilobula, including intraspecific variation, morphology, resource allocation, and defoliation and transport of mineral nutrients. *Mailing Add:* 185 Ferry Rd Saunderstown RI 02874

HAMMER, CARL, b Chicago, Ill, May 10, 14; m 44. COMPUTER SCIENCES. *Educ:* Univ Munich, dipl, 36, PhD(math statist), 38. *Prof Exp:* Statistician, Res Labs, Tex Co, NY, 38-43 & foreign div, Pillsbury Mills, Inc, 44-47; chmn div tech educ, Walter Hervey Jr Col, 47-50; res assoc, Columbia Univ, 50-52; sr res engr, Franklin Inst, 52-55; dir, Univac Europ Comput Ctr, Ger, 55-57; staff consult, Electronic Systs Div, Sylvania Elec Prod, Inc, Mass, 57-59; adminr tech proj coord, Surface Commun Eng, Radio Corp Am, 59-61, mgr sci comput appln, 61-63; dir comput sci, Sperry Univac, 63-81; CONSULT, 81- *Concurrent Pos:* Adj prof, Am Univ, 63-81; vis prof, Indust Col Armed Forces, 65-; mem, Nat Defense Exec Reserve, 70- *Honors & Awards:* Comput Sci Man-of-the-Yr Award, Data Processing Mgt Asn, 73. *Mem:* Fel AAAS; Am Math Soc; Soc Indust & Appl Math (treas, 53-55); Am Statist Asn; Am Soc Cybernet (pres, 69-71); fel Inst Elec & Electronics Engrs. *Res:* Engineering and mathematical-statistical analysis; quantal response studies; evaluation of computing systems; design of experiments; cryptology. *Mailing Add:* 3263 O Street NW Washington DC 20007-2843

HAMMER, CARL HELMAN, BIOCHEMISTRY OF COMPLEMENT. *Educ:* Pa State Univ, PhD(biochem), 72. *Prof Exp:* SR INVESTR, NAT INST ALLERGY & INFECTIOUS DIS-NIH, 82- *Mailing Add:* Nat Inst Allergy & Infectious Dis NIH Bldg 10 Rm 11N220 Bethesda MD 20892

HAMMER, CHARLES F, b Fremont, Ohio, July 22, 33; m 57. ORGANIC CHEMISTRY, CHEMICAL INSTRUMENTATION. *Educ:* Bowling Green State Univ, BA, 55; Univ Minn, PhD(org chem), 59. *Prof Exp:* NIH fel, x-ray crystallog & steroids, Brandeis Univ, 61-63; from asst prof to assoc prof, 63-82, PROF CHEM, GEORGETOWN UNIV, 82-, DIR, INST ADVAN ANALYTICAL CHEM, 63- *Concurrent Pos:* Mem, ChemTec Writing Team, Am Chem Soc, 70-; vis prof, Dept Hydrocarbon Chem, Sch Eng, Kyoto Nat Univ, Japan, 71-72; vis scholar, Dept Chem, Univ Calif, Berkeley, 78, NIDDK, NIH, 86. *Honors & Awards:* Alan Berman Res Publ Award, NRL, 87. *Mem:* AAAS; Am Chem Soc; Am Soc Mass Spectrometry; Soc Appl Spectros; Am Soc Testing & Mat; Sigma Xi. *Res:* Chemistry and mechanisms in nitrogen heterocyclics and steroids; bromination-dehydrobromination reactions; structure elucidation of natural products by instrumental methods; polypeptide conformations by nuclear magnetic resonance; isotope ratio kinetics by mass spectrometry; computer software applications to spectrometric analysis. *Mailing Add:* Dept Chem Georgetown Univ Washington DC 20057

HAMMER, CHARLES LAWRENCE, b Buffalo, NY, June 30, 22; m 48; c 4. ELEMENTARY PARTICLE PHYSICS, MATHEMATICAL PHYSICS. *Educ:* Univ Mich, BS, 48, MS, 50, PhD(physics), 54. *Prof Exp:* Instr physics, Univ Mich, 53-54; res assoc, Iowa State Univ, 54-55, from asst prof to assoc prof, 55-61, prof physics, 61-88; RETIRED. *Concurrent Pos:* Consult, Allis Chalmers Mfg Co, 57-67 & Midwestern Univs Res Asn, 59-70; mem adv coun, NSF, 78-82. *Mem:* Am Phys Soc. *Res:* Nuclear and theoretical physics; quantum field theories and applications to coherent phenomenon. *Mailing Add:* Dept Physics Iowa State Univ Ames IA 50011

HAMMER, CHARLES RANKIN, b Memphis, Tenn, Aug 7, 27; m 56; c 2. ORGANIC CHEMISTRY. *Educ:* Univ Utah, BA, 49, PhD(org chem), 57. *Prof Exp:* Asst prof chem, Westminster Col, Utah, 56-60 & Idaho State Univ, 60-64; asst prof, 64-67, ASSOC PROF, UNIV SOUTHERN COLO, 67- *Mem:* Am Chem Soc. *Res:* Condensation of phenols with formaldehyde and primary amines; Mannich reactions involving primary amines; synthesis of heterocyclic compounds. *Mailing Add:* Dept Chem Univ Southern Colo Pueblo CO 81001

HAMMER, CLARENCE FREDERICK, JR, b Toledo, Ohio, Nov 24, 19; m 42; c 3. POLYMER SCIENCE. *Educ:* Miami Univ, BA, 40; Univ Wis, PhM, 42, PhD(chem physics), 48. *Prof Exp:* Jr physicist, Naval Ord Lab, 42-44, assoc physicist, 45, elec engr, 45-46; res asst physics, Univ Wis, 46-48; physicist, 58-51, group leader, Polychem Dept, E I du Pont de Nemours & Co, Inc, 51-55, supvr, 55-61, res assoc, Electrochem Dept, 61-71, res assoc, Plastics Dept, 71-78, res fel, Plastics Prod & Resins Dept, 78-81; CONSULT, 81- *Honors & Awards:* Award for Creative Invention, Am Chem Soc, 90. *Mem:* Am Phys Soc; Sigma Xi; Am Chem Soc; AAAS. *Res:* Infrared studies of molecular structures; theoretical depolymerization mechanisms of high polymers; crystalline structure and physical properties of high polymers; physical properties of high polymers; polymer compatibility; invented "Elvaloy" non-migrating polymeric plasticizer for PVC, "Elvaloy" weather resistant impact modifier for PVC, polymeric toughener for phenolic and epoxy resins, highly novel and versatile graft copolymers. *Mailing Add:* 5349 Delano Ct Cape Coral FL 33904

HAMMER, DAVID ANDREW, b New York, NY, Apr 5, 43; m 68; c 2. PLASMA PHYSICS, INTENSE PARTICLE BEAMS. *Educ:* Calif Inst Technol, BS, 64; Cornell Univ, PhD(appl physics), 69. *Prof Exp:* Res physicist, Naval Res Lab, Washington, DC, 69-76; assoc prof elec sci & eng, Univ Calif, Los Angeles, 77; assoc prof, 77-84, PROF NUCLEAR SCI & ENG, CORNELL UNIV, 84-, DIR, LAB PLASMA STUDIES, 85- *Concurrent Pos:* Vis assoc prof, Univ Md, 73-76; sr vis fel, Imp Col, London, Eng, 77, 83-84; consult, Physical Dynamics, 77-79, Lawrence Livermore Nat Lab, 79-, Sci Applns Int Corp, 79-, SRI Int-Jason, 80-82 & Mitre-Jason, 82-; dir assoc ed phys res lett, 88-91. *Mem:* Fel Am Phys Soc; AAAS; sr mem Inst Elec & Electronics Engrs; Sigma Xi. *Res:* Intense electron and ion beams and their application to plasma physics and controlled fusion. *Mailing Add:* Lab Plasma Studies Upson Hall Cornell Univ Ithaca NY 14853

HAMMER, GARY G, b Wichita, Kans, Jan, 9, 34; m 53; c 3. PHYSICAL ORGANIC CHEMISTRY. *Educ:* Wichita State Univ, BS, 55, MS, 57; Ga Inst Technol, PhD(chem), 62. *Prof Exp:* Res chemist, Dow Badische Co, 61-67; from asst prof to assoc prof, 67-74, chmn dept, 69-76, PROF CHEM, CHRISTOPHER NEWPORT COL, 74- *Mem:* AAAS; Am Chem Soc. *Res:* Polymerization kinetics; mechanisms of addition to nitriles. *Mailing Add:* 684 Winthrop Rd Williamsburg VA 23185

HAMMER, HENRY FELIX, b Brooklyn, NY, Oct 3, 21; m 48; c 3. PHARMACEUTICAL CHEMISTRY. *Educ:* Polytech Inst Brooklyn, BS, 43; Rensselaer Polytech Inst, PhD(org chem), 51. *Prof Exp:* Res chemist, Sterling-Withrop Res Inst, 43-46; asst & instr, Rensselaer Polytech Inst, 46-51; res chemist, Pfizer Inc, 51-58, pharmaceut res supvr, 58-62, mgr pharmaceut res, 62-64, from asst dir to dir qual control, 64-72, vpres, 72-86; RETIRED. *Mem:* Sigma Xi; Am Chem Soc. *Res:* Antibiotics; synthetic medicinals; vitamins and nutritional products; pharmaceutical dosage forms. *Mailing Add:* 29 Sea Marsh Rd Amelia Island FL 32034

HAMMER, JACOB, b Bucarest, Rumania, Apr 30, 50; US citizen; m 84; c 2. MATHEMATICAL SYSTEM THEORY, CONTROL THEORY. *Educ:* Technion-Israel Inst Technol, BSc, 74, MSc, 77, DSc(elec eng), 80. *Prof Exp:* Lectr eng, Technion-Israel Inst Technol, 79-80; res fel syst theory, Univ Fla, Gainesville, 80-82; asst prof math, Case Western Reserve Univ, 82-87; ASSOC PROF ELEC ENG, UNIV FLA, 87- *Mem:* Am Math Soc; Inst Elec & Electronics Engrs. *Res:* Nonlinear control theory; linear control theory; nonlinear filtering. *Mailing Add:* Dept Elec Eng Univ Fla Gainesville FL 32611

HAMMER, JACOB MEYER, b New York, NY, Sept 14, 27; m 51; c 3. OPTICAL COMMUNICATIONS, QUANTUM PHYSICS. *Educ:* NY Univ, BS, 50, PhD(physics), 56; Univ Ill, MS, 51. *Prof Exp:* Asst physics, NY Univ, 51-55; mem tech staff, Bell Tel Labs, 56-59; mem tech staff, RCA Labs, 59-87; CONSULT, 88- *Concurrent Pos:* Sr visitor, Cavendish Lab, Cambridge, Eng, 68-69; adj prof elec eng, Polytech Inst NY, 81. *Honors & Awards:* RCA Labs Outstanding Achievement Awards, 62, 64 & 74, 86. *Mem:* Am Phys Soc; fel Inst Elec & Electronics Engrs; Optical Soc Am; AAAS. *Res:* Photonics; optoelectronics; semiconductor lasers. *Mailing Add:* Three Leavitt Lane Princeton NJ 08540

HAMMER, JOHN A, CELL MOTILITY, GENE REGULATION. *Educ:* Pa State Univ, PhD(physiol), 80. *Prof Exp:* SR STAFF FEL, CELL BIOL LAB, NIH, 82- *Mailing Add:* Cell Biol Lab Nat Heart Lung Blood Inst NIH Bldg 3 Rm B1-22 Bethesda MD 20892

HAMMER, LOWELL CLARKE, b Council Bluffs, Iowa, Jan 7, 30; m 49; c 2. SPEECH PATHOLOGY, AUDIOLOGY. *Educ:* Wichita State Univ, BA, 52; Univ Fla, PhD(speech path), 65. *Prof Exp:* Staff speech pathologist, Inst Logopedics, Wichita, Kans, 52-55; dir speech path, Children's Rehab Inst, Reisterstown, Md, 55-60; from instr speech path & audiol to asst prof, 63-70, assoc prof speech path, 70-75, PROF COMMUN DIS & SPEECH, UNIV FLA, 75- CHIEF DEPT COMMUN DIS, 65-, CHIEF SPEECH PATH, SHAND'S HOSP. *Concurrent Pos:* Consult, Fla Crippled Childrens Comn, 66-; assoc dean Col Health Related Professions, 71- *Mem:* AAAS; Am Speech & Hearing Asn. *Res:* Clinical research in neurological diseases affecting speech and language processes; physiological research, especially electromyography of the speech mechanism. *Mailing Add:* Shan's Hosp c/o Commun Dis Box J174 JHMHC Gainesville FL 32610-0174

HAMMER, MARK J(OHN), b Fond du Lac, Wis, Apr 26, 31; m 55; c 4. CIVIL ENGINEERING. *Educ:* Northwestern Univ, BS, 55, MS, 56; Univ Mich, PhD(civil eng), 64. *Prof Exp:* Proj engr, Shannon & Wilson, consult engr, Wash, 60-61; assoc prof civil eng, 64-69, acting chmn dept, 68-69, PROF CIVIL ENG, UNIV NEBR, 69- *Mem:* Water Pollution Control Fedn; Am Water Works Asn. *Res:* Biological waste treatment, nitrogen and phosphorus removal from waste water, eutrophication of reservoirs; sanitary engineering and technology. *Mailing Add:* 1810 S 77th Lincoln NE 68506

HAMMER, RICHARD BENJAMIN, b Hornell, NY, Sept 18, 43; m 86; c 1. NEW PRODUCT MANAGEMENT & DEVELOPMENT. *Educ:* Alfred Univ, BA, 66; Syracuse Univ, PhD(org chem), 72. *Prof Exp:* Res asst, Syracuse Univ, 71-72, State Univ NY Col Environ Sci & Forestry, 72-73; group leader basic res cellulose chem & technol, ITT Rayonier Inc, 73-81; DEVELOP ENG & MGR, IBM CORP, 81- *Concurrent Pos:* Res assoc supvr, ITT Rayonier Inc, 81. *Honors & Awards:* Alfred Hitchcock Award. *Mem:* Am Chem Soc; The Chem Soc; Sigma Xi. *Res:* Basic cellulose chemistry derivative, fiber spinning and processing, polymer chemistry, pulping, ceramic engineering, circuit design, photolithography; new product and process development; polymer & colloidal chemistry; computer chip packaging and development. *Mailing Add:* 522 Hilton Rd Apalachin NY 13732-9765

HAMMER, RICHARD HARTMAN, b Latrobe, Pa, May 24, 33; m 62; c 2. PHARMACEUTICAL CHEMISTRY, ORGANIC CHEMISTRY. *Educ:* Univ Ariz, BS, 59, PhD(pharmaceut chem), 63. *Prof Exp:* From asst prof to assoc prof, Univ Fla, 63-76, chmn dept, 75-79, asst dean, 79-84, PROF PHARMACEUT CHEM, UNIV FLA, 76- *Concurrent Pos:* Am Cancer Soc grant, 64-67. *Mem:* Am Pharmaceut Asn; Am Chem Soc; Am Asn Pharmaceut Scientists. *Res:* Design and synthesis of drugs; drug metabolism. *Mailing Add:* Dept Med Chem Univ Fla Box J-485 J Hills Miller Health Ctr Gainesville FL 32610

HAMMER, ROBERT NELSON, b Kansas City, Kans, Sept 15, 24. INORGANIC CHEMISTRY. *Educ:* Univ Kans, AB, 47, MA, 49; Univ Ill, PhD(chem), 54. *Prof Exp:* Instr chem, Ark State Col, 49-51; from asst prof to assoc prof, 54-77, PROF CHEM, MICH STATE UNIV, 77- *Concurrent Pos:* Assoc dir hons col, Mich State Univ, 66-70. *Mem:* AAAS; Am Chem Soc; Brit Chem Soc. *Res:* Nonaqueous solvents; voltammetry in liquid ammonia; acid-base equilibria in nonaqueous solvents; metallosiloxane synthesis. *Mailing Add:* 4953 S Okemos Rd East Lansing MI 48823-2922

HAMMER, ROBERT RUSSELL, b Red Bluff, Calif, Aug 5, 36. PHYSICAL INORGANIC CHEMISTRY. *Educ:* Chico State Col, AB, 58; Univ Wash, PhD(inorg chem), 63. *Prof Exp:* Res chemist, Phillips Petrol Co, 64-66; res chemist, Idaho Nuclear Corp, 66-70; assoc res scientist, Allied Chem Corp, 70-79, Exxon Nuclear Corp, 79-84, Westinghouse Idaho Nuclear Corp, 84-87; RETIRED. *Concurrent Pos:* Consult, 88. *Mem:* Am Chem Soc; Sigma Xi. *Res:* Fluoride complexes of various metal ions, their stability constants, rates of formation and thermodynamic properties; computer simulation of chemical processes. *Mailing Add:* 605 Tendoy Dr Idaho Falls ID 83401

HAMMER, RONALD PAGE, JR, b Philadelphia, Pa, Feb 21, 53; m 86. NEUROENDOCRINOLOGY. *Educ:* Univ Calif, Berkeley, AB, 80; Univ Calif, Los Angeles, PhD(anat), 84. *Prof Exp:* Teaching fel psychiat, Univ Calif, Los Angeles, 80; staff fel neurosci, NIMH, 80-83, sr staff fel, 83-84; ASSOC PROF ANAT, UNIV HAWAII, MANOA, HONOLULU, 84- *Concurrent Pos:* Instr, Montgomery Col, Md, 80-83; mem adv comt, Am Asn Anatomists, 84-87. *Honors & Awards:* Res career develop award, NIH, 87. *Mem:* Soc Neurosci; Am Asn Anatomists; NY Acad Sci; AAAS; Int Soc Develop Neurosci; Int Brain Res Orgn. *Res:* Brain development; modulation by neuropeptide factors; drug effects on brain metabolism; biological substrates for drug function; drugs of abuse; neuroendocrine effects on reproductive behavior. *Mailing Add:* Dept Anat & Reproductive Biol Sch Med Univ Hawaii 1960 E-W Rd Honolulu HI 96822

HAMMER, SIGMUND IMMANUEL, geophysics; deceased, see previous edition for last biography

HAMMER, ULRICH THEODORE, b Maple Creek, Sask, Mar 25, 24; m 54; c 2. LIMNOLOGY. *Educ:* Univ Sask, BEd, 50, BA, 56, PhD(limnol), 63; Mont State Univ, MS, 59. *Prof Exp:* Teacher elem schs, Sask, 46-47 & 48-49 & Alta, 50-51 & high schs, 51-58; instr, 61-62, lectr, 62-63, asst prof, 63-66, assoc prof, 66-71, chmn water studies inst, 67-69, head dept, 73-76, PROF BIOL, UNIV SASK, 71- *Concurrent Pos:* Woods Hole Oceanog Inst fel, 63; sabbatical vis lectr, Monash Univ, Australia, 69-70, San Diego State Univ, 90; Sask res coun scholar, 60-61 & 61-62; fel Rawson Acad Aquatic Sci, 79, dir. *Mem:* Am Soc Limnol & Oceanog; Int Soc Limnol. *Res:* Ecology of bloom-forming blue-green algae; eutrophication; primary productivity of inland lakes; chemical, physical and biological limnology of saline lakes; bioaccumulation and cycling of heavy metals. *Mailing Add:* Dept Biol Univ Sask Saskatoon SK S7N 0W0 Can

HAMMERLI, MARTIN, b Hasle, Switz, Mar 30, 38; Can citizen; m 63; c 3. ELECTROCHEMISTRY. *Educ:* Queen's Univ, Ont, BSc, 61, PhD(phys chem), 65. *Prof Exp:* Nat Res Coun Can overseas fel, Univ Leeds, 65-66; asst res officer, Gen Chem Br, Chalk River Nuclear Labs, Atomic Energy Can Ltd, 66-70, assoc res officer, 71-90; CONSULT, 90- *Concurrent Pos:* Consult advan alkaline water electrolysis, Nat Res Coun Can, 78-; Can mem adv bd Int Assn Hydrogen; mem adv bd Clean Air Found, US; founding chmn Can Hydrogen Assn. *Mem:* Chem Inst Can; Int Asn Hydrogen Energy; Electrochem Soc. *Res:* Electrolytic hydrogen-deuterium separation factor as a function of over potential, temperature, charge, metal and surface state; diffusion of hydrogen and deuterium in metals; new heavy water processes; new tritium recovery processes; hydrogen economy; advanced water and water vapour electrolysis systems; managing contract research and development in hydrogen production, hydrogen use and energy storage all involving electrochemistry; industrial research and development transfer policy analysis; policy analysis for federal government research and development. *Mailing Add:* 2078 Beaconwood Dr Glausser ON K1J 8M4 Can

HAMMERLING, JAMES SOLOMON, b Philadelphia, Pa, Aug 17, 07; m 36; c 2. MEDICINE, OTOLARYNGOLOGY. *Educ:* City Col New York, BSc, 29; NY Med Col, MD, 33. *Prof Exp:* Prof otolaryngol, fac med, Dalhousie Univ, 64-76; RETIRED. *Concurrent Pos:* Consult, Izaak Walton Killam Hosp Children, Camp Hill Hosp & Halifax Infirmary, 64-76. *Mem:* Am Col Surgeons; Am Acad Ophthal & Otolaryngol; Can Otolaryngol Soc; Can Med Asn. *Mailing Add:* 6777 Quinpool Rd Halifax NS B3L 1C2 Can

HAMMERLING, ULRICH, IMMUNOCHEMISTRY, IMMUNOGENETICS. *Educ:* Univ Freiburg, Ger, PhD(immunol), 65. *Prof Exp:* MEM STAFF, MEM SLOAN-KETTERING INST, 83- *Mailing Add:* Dept Immunol Sloan-Kettering Inst 1275 York Ave New York NY 10021

HAMMERMAN, DAVID LEWIS, b Brooklyn, NY, Dec 19, 35; m 63; c 3. PHYSIOLOGY, PATHOLOGY. *Educ:* City Col New York, BS, 57; NY Univ, MS, 59, PhD(biol), 62. *Prof Exp:* Asst med, 58-59, instr biol, NYU UNIV, 62; from asst prof to assoc prof, 62-71, adj prof health sci, 81-89, PROF BIOL, LONG ISLAND UNIV, 71-, ADJ PROF SPORTS SCI, 83- *Concurrent Pos:* Consult Div Natural Sci, Brooklyn Children's Mus, Inst Arts & Sci, 63, Saunders Col Publ, 89. *Mem:* AAAS; Sigma Xi; Am Soc Zool. *Res:* Vertebrate morphogenesis; sensory physiology; physiogenesis of taste. *Mailing Add:* Dept Biol Long Island Univ Brooklyn NY 11201-5372

HAMMERMAN, IRA SAUL, b New York, NY, May 15, 42; m 65; c 4. BIOPHYSICS, BIOLOGICAL STRUCTURE. *Educ:* Brandeis Univ, BA, 64; Princeton Univ, MA, 66, PhD(physics), 69. *Prof Exp:* Res assoc elem particles, Nevis Labs, 69-71; res assoc, Technion-Israel, Inst Technol, 71-72; res assoc, Dept Biol Sci, Columbia Univ, 72-74; LECTR, DEPT LIFE SCI, BAR ILAN UNIV, ISRAEL, 74- *Mem:* Am Phys Soc; Israel Electron Micros Soc. *Res:* Three-dimensional reconstruction of cell ultrastructure; sex determination in fish; ribosome structure; statistical fluctuations in polymerization. *Mailing Add:* 33 Eisenberg St Rehovot 76287 Israel

HAMMERMAN, MARC R, b St Louis, Mo, Sept 29, 47. CELL PHYSIOLOGY. *Educ:* Washington Univ, St Louis, MD, 72. *Prof Exp:* Asst prof, 79-84, ASSOC PROF INTERNAL MED, WASH UNIV SCH MED, 84- *Mem:* Am Fedn Clin Res; Am Physiol Soc; Am Heart Asn; Am Soc Clin Invest; Am Soc Nephrology; Cent Soc Clin Res. *Mailing Add:* Dept Internal Med PO Box 8121 Wash Univ Sch Med 660 S Euclid Ave St Louis MO 63110

HAMMERMEISTER, KARL E, b Baguio, Philippines, Sept 10, 39; US citizen. CARDIOLOGY. *Educ:* Univ Wash, BS, 60, MD, 64. *Prof Exp:* Asst chief med cardiol, USPHS Hosp, Baltimore, Md, 68-70; asst chief cardiol, dir cardiovasc lab, Vet Admin Hosp, Denver & asst prof med, Univ Colo Med Ctr, 70-71; ASST CHIEF CARDIOVASC DIS SERV, VET ADMIN HOSP, SEATTLE, 71- *Concurrent Pos:* Cardiol fel, Univ Wash Sch Med, 67-68, asst prof med, 71-75 & assoc prof med, 75-; fel, Coun Clin Cardiol, Am Heart Asn, 74- *Mem:* Am Fedn Clin Res; Am Heart Asn; fel Am Col Cardiol. *Res:* Effect of left ventricular function on surgical and late mortality in patients with cardiac disease. *Mailing Add:* Vet Admin Med Ctr Cardiol Denver CO 80220

HAMMERNESS, FRANCIS CARL, b Glasgow, Mont, Aug 8, 22; m 46. PHARMACY. *Educ:* Mont State Univ, BS, 47, MS, 51; Univ NC, PhD(pharm), 56. *Prof Exp:* Instr pharm, Mont State Univ, 48-51; instr, Univ NC, 51-55, lectr, 55-56, asst prof, 56-57; assoc prof pharm admin, 57-66, prof pharm admin & dir Ctr Drug Mkt Res, 66-87, EMER PROF, UNIV COLO, BOULDER, 87- *Mem:* Am Asn Cols Pharm; Am Pharmaceut Asn. *Res:* Pharmaceutical product development and marketing studies. *Mailing Add:* 373 Ridges Blvd No 209 Grand Junction CO 81503

HAMMERSCHLAG, RICHARD, b New York, NY, June 12, 39; div; c 1. NEUROCHEMISTRY, CELL BIOLOGY. *Educ:* Mass Inst Technol, BS(humanities & sci), 60 & BS(chem), 61; Brandeis Univ, PhD(biochem), 67. *Prof Exp:* Fel neurochem, Dept Biophysics, Univ Col London, 67-69; from asst res scientist to assoc res scientist, 70-84, RES SCIENTIST NEUROCHEM, DIV NEUROSCI, BECKMAN RES INST, CITY OF HOPE, 84- *Mem:* Am Soc Neurochem; Int Soc Neurochem; Soc Neurosci; Am Soc Cell Biol. *Res:* Mechanisms and functions of axonal transport; neurotrophic phenomena. *Mailing Add:* Div Neurosci Beckman Res Inst City of Hope Duarte CA 91010-0269

HAMMERSMITH, JOHN L(EO), b Massillon, Ohio, Apr 4, 29; m 51; c 2. SPACE SCIENCE. *Educ:* Univ Mich, BS, 51. *Prof Exp:* Mathematician, US Naval Res Lab, 51-60; mathematician, Theoret Div, Goddard Space Flight Ctr, NASA, 60-61, space sci & eng assignments, Off Manned Space Flight, 61-63, mission analyst, Gemini Prog Off, NASA Hq, 63-66, sr engr, Manned Space Flight, Advan Missions Prog, 66-70, chief engr payload planning, 70-72, head payload accomodations, Space Shuttle Prog, 72-79, sr systs engr, Space Transp Syst Prog, 79-83, STAFF ENGR, SPACE SHUTTLE OPERS, NASA HQ, 83- *Concurrent Pos:* Radio reportage of US Space Prog, Voice of Am. *Mem:* Assoc fel Am Inst Aeronaut & Astronaut. *Res:* Manned space flight mission planning; systems engineering. *Mailing Add:* 3460 39th St Washington DC 20016

HAMMERSTEDT, ROY H, b Duluth, Minn, June 17, 41; m 66; c 2. BIOCHEMISTRY, REPRODUCTIVE PHYSIOLOGY. *Educ:* Univ Minn, BA, 63, PhD(biochem), 68. *Prof Exp:* NIH res fel biochem, Mich State Univ, 68-70; from asst prof to assoc prof, 70-83, PROF BIOCHEM, PA STATE UNIV, UNIVERSITY PARK, 83- *Concurrent Pos:* Vis prof, Univ Wis, 78 & Cornell Univ, 86. *Mem:* Sigma Xi; Am Chem Soc; Am Soc Biol Chemists; AAAS; Am Soc Andrology; Soc Study Reproduction; Soc Cryobiol. *Res:* Biochemical characterization of spermatozoa; bioenergetics; membrane structure and function; lipid metabolism and cryobiology; effect of microgravity on cell function. *Mailing Add:* 406 Althouse Lab Pa State Univ University Park PA 16802

HAMMERSTROM, HAROLD ELMORE, b Davis, SDak, Apr 7, 27; m 49; c 4. ANALYTICAL CHEMISTRY, INORGANIC CHEMISTRY. *Educ:* SDak State Col, BS, 53; Univ SDak, MA, 55; Univ of the Pac, PhD(chem), 68. *Prof Exp:* Prin, Linn Grove Consol Sch, Iowa, 54-57; prof chem, Northwestern Col, 57-; RETIRED. *Mem:* Am Chem Soc. *Res:* Radiometric titrations. *Mailing Add:* RR 6 6841 Box Off 6877 Northwestern Col Spirit Lake IA 51360

HAMMES, GORDON G, b Fond du Lac, Wis, Aug 10, 34; m 59; c 3. BIOCHEMISTRY, BIOPHYSICS. *Educ:* Princeton Univ, BA, 56; Univ Wis, PhD(chem), 59. *Prof Exp:* NSF fel, 59-60; from instr to assoc prof chem, Mass Inst Technol, 60-65; prof chem, Cornell Univ, 65-75, chmn dept, 70-75, Horace White prof chem & biochem, 75-88, dir biotechnol prog, 83-88; PROF CHEM & VCHANCELLOR ACAD AFFAIRS, UNIV CALIF, SANTA BARBARA, 88- *Concurrent Pos:* Mem adv panel, NIH, 67-80 & 86-88; NSF sr fel, 68-69; NIH Fogarty scholar, 75-76. *Honors & Awards:* Award Biol Chem, Am Chem Soc, 67. *Mem:* Nat Acad Sci; Am Acad Arts & Sci; Am Chem Soc; Am Soc Biochem & Molecular Biol. *Res:* Biophysical chemistry, especially enzyme kinetics and mechanisms; biochemical control mechanisms and membrane structure and function. *Mailing Add:* 5105 Cheadle Hall Univ Calif Santa Barbara CA 93106

HAMMETT, MICHAEL E, b Cowpens, SC, Dec 5, 37; m 59; c 2. MATHEMATICS. *Educ:* Furman Univ, BS, 59; Auburn Univ, MS, 61, PhD(math), 67. *Prof Exp:* Asst prof math, Furman Univ, 62-65; asst prof, Auburn Univ, 65-67; asst prof, 67-70, assoc prof, 70-79, PROF MATH, FURMAN UNIV, 79- *Mem:* Math Asn Am. *Res:* Mathematical analysis; ordinary differential equations. *Mailing Add:* Dept Math Furman Univ Poinsett Hwy Greenville SC 29613

HAMMILL, TERRENCE MICHAEL, b Potsdam, NY, Dec 28, 40; m 63; c 2. MYCOLOGY, CYTOLOGY. *Educ:* State Univ NY Col Potsdam, BS, 63; Univ Ga, MEd, 68; State Univ NY Col Forestry, Syracuse, PhD(bot), 71; Syracuse Univ, PhD(biol), 71. *Prof Exp:* Teacher biol, Indian River Cent High Sch, 63-67; from asst prof to assoc prof, 71-85, prof biol, 85-90, DISTINGUISHED TEACHING PROF BIOL, COL ARTS & SCI, STATE UNIV NY, OSWEGO, 90-, CHAIR BIOL, 91- *Concurrent Pos:* Res grant, Res Corp, 73; NSF grant, 74, 80. *Mem:* AAAS; Am Inst Biol Sci; Bot Soc Am; Electron Micros Soc Am; Mycol Soc Am; NY Acad Sci; Sigma Xi; Am Soc Microbiol. *Res:* Light and electron microscopy of fungi and fungal development; human sexuality. *Mailing Add:* Dept Biol Col Arts & Sci State Univ NY Oswego NY 13126

HAMMING, KENNETH W, b 1918; US citizen. ENGINEERING. *Educ:* Univ Ill, BS, 40. *Prof Exp:* Mech engr, Sargent & Lundy, 40-56, partner, 56-66, mgr mech & nuclear dept, 64-65, dir eng, 65-66, sr partner, 66-77; RETIRED. *Mem:* Nat Acad Eng; Am Soc Mech Eng. *Res:* Company planning, organization, direction and administration. *Mailing Add:* 527 Meadow Dr W Wilmette IL 60091

HAMMING, MYNARD C, b McBain, Mich, Nov 1, 21; m 53; c 2. SPECTROCHEMISTRY. *Educ:* Mich State Univ, BS, 50. *Prof Exp:* Chemist, Dow Chem Co, Mich, 51-54, Phillips Petrol Co, 54-55 & Koppers Co, 55-60; res chemist, 60-67, SR RES SCIENTIST, CONOCO INC, 67- *Mem:* Am Chem Soc; Am Soc Testing & Mat. *Res:* Analytical applications of high and low mass spectrometry, microprobe mass spectrometry; complex computer matrices; computer programming; gas chromatography; fossil fuel chemistry. *Mailing Add:* 1410 Reveille Dr Ponca City OK 74604-4438

HAMMING, RICHARD W, b Chicago, Ill, Feb 11, 15; m 42. COMPUTER SCIENCE. *Educ:* Univ Chicago, BS, 37; Univ Nebr, MA, 39; Univ Ill, PhD(math), 42. *Prof Exp:* Instr math, Univ Ill, 42-44; asst prof, Univ Louisville, 44-45; mem staff, Manhattan Proj, Los Alamos Sci Lab, 45-46; mem tech staff, Bell Labs, Inc, 46-64, head numerical methods res dept, 64-67, head comput sci res dept, 67-77; ADJ PROF, DEPT COMPUT SCI, NAVAL POSTGRAD SCH, 77- *Honors & Awards:* Piore Prize, Inst Elec & Electronics Engrs, 49; Turing Prize lectr, 68; Pender Prize, 81. *Mem:* Nat Acad Eng; Math Asn Am; Asn Comput Mach; fel Inst Elec & Electronics Engrs. *Res:* Numerical methods; error detecting and error correcting codes; automatic coding systems; statistics; digital filters. *Mailing Add:* Dept Comput Sci Naval Postgrad Sch Monterey CA 93940

HAMMITT, FREDERICK G(NICHTEL), nuclear & mechanical engineering; deceased, see previous edition for last biography

HAMMOCK, BRUCE DUPREE, b Little Rock, Ark, Aug 13, 47; m 72; c 3. TOXICOLOGY, ENDOCRINOLOGY. *Educ:* La State Univ, Baton Rouge, BS, 69; Univ Calif, Berkeley, PhD(entom), 73. *Prof Exp:* Fel endocrinol, Dept Biol, Northwestern Univ, 73-74; from asst prof to assoc prof entom, 74-80, ASSOC PROF ENTOM & ENVIRON TOXICOL, UNIV CALIF, DAVIS, 80- *Concurrent Pos:* Fel, Rockefeller Found, 73-74; res career develop award, 78-83; NSF, USDA, NIH & Environ Protection Agency grants; prin invester, Superfund Nat Inst Environ Health Sci Proj; vis fel, Lady Margaret Hall, Oxford Univ. *Honors & Awards:* Frash Found Award; Burrough Wellcome Toxicol Scholar; Fel, Fogarty Int. *Mem:* AAAS; Am Chem Soc; Entom Soc Am; Sox Toxicol; Soc Environ Toxicol. *Res:* Biochemistry of endocrine control of insect development; effects of xenobiotics on organisms and xenobiotic metabolism; immunoassay. *Mailing Add:* Dept Entom Univ Calif Davis CA 95616

HAMMOND, ABNER M, JR, b Marks, Miss, Jan 17, 39; m 61; c 2. ENTOMOLOGY. *Educ:* Miss State Univ, BS, 61, MS, 63; La State Univ, PhD(entom), 67. *Prof Exp:* Res entomologist, Southern Res Inst, 67-68; from asst prof to prof, 68-77, PROF INSECT PHYSIOL, LA STATE UNIV, BATON ROUGE, 77- *Concurrent Pos:* Int consult. *Mem:* Entom Soc Am, Int Insect Chemoreception Workshop; Sigma Xi. *Res:* Physiological ecology of migrant noctuid moths; chemistry and behavior of lepidopterous insect pheromones; plant/insect interactions, including the role of endophytic fungi in regulating insect pest populations. *Mailing Add:* Dept Entom Life Sci Bldg 552 La State Univ Baton Rouge LA 70803-1710

HAMMOND, ALLEN LEE, b W Chicago, Ill, Sept 6, 43; m 69; c 2. PUBLIC UNDERSTANDING OF SCIENCE, SCIENCE POLICY. *Educ:* Stanford Univ, BS, 66; Harvard Univ, MA, 67, PhD(appl math), 70. *Prof Exp:* Res news ed, AAAS, 70-79, ed Sci, 80-86; pres, ALH & Assoc Inc, 86-90; ED, WORLD RESOURCES, 90- *Concurrent Pos:* Consult, Nat Acad Sci, 83-86; ed, I Sci & Tech, 84-86, Sci Impact Letter, 87-89, Off Sci & Technol Policy, 89- *Mem:* Fel AAAS; Sigma Xi. *Res:* Environmental science; energy technologies and policy; science policy. *Mailing Add:* 9612 E Bexhill Dr Kensington MD 20895

HAMMOND, ANDREW CHARLES, b Glendale, Calif, May 25, 49; m 73; c 2. RUMINANT NUTRITION. *Educ:* Ore State Univ, BS, 71; Wash State Univ, MS, 77, PhD(nutrit), 79. *Prof Exp:* Res assoc, Dept Animal Sci, Wash State Univ, 79-80; res animal scientist, Agr Res Ctr, Beltsville, Md, 80-85, RES LEADER, SUBTROP AGR RES STA, US DEPT AGR, BROOKSVILLE, FLA, 85- *Concurrent Pos:* Adj fac mem, Univ Fla, 85- *Mem:* AAAS; Am Inst Nutrit; Am Soc Animal Sci; Coun Agr Sci Technol; Am Registry Cert Animal Scientists; Am Dairy Sci Asn; Soc Exp Biol & Med; Am Forage & Grasslands Coun; Sigma Xi. *Res:* Beef cattle nutrition; growth biology; nutritional toxicology. *Mailing Add:* Subtrop Agr Res Sta USDA Agr Res Serv PO Box 46 Brooksville FL 34605-0046

HAMMOND, BENJAMIN FRANKLIN, b Austin, Tex, Feb 28, 34. MEDICAL MICROBIOLOGY. *Educ:* Univ Kans, AB, 54; Meharry Med Col, DDS, 58; Univ Pa, PhD(microbiol), 62. *Prof Exp:* USPHS fel, Univ Pa, 58-62, asst instr, 58-60, from instr to assoc prof, 60-70, chmn dept, 72-85, PROF MICROBIOL, SCH DENT MED, UNIV PA, 70-, DEAN ACAD AFFAIRS, 84- *Concurrent Pos:* USPHS career develop award, 66; consult, Coun Dent Educ & USPHS, NSF & Am Fund Dent Educ, 74-; mem study sect, Nat Inst Dent Res; mem, Nat Adv Dent Res Coun, NIH, 81-; Ralph Metcalf chair-distinguished vis prof, Marquette Univ, 85-86. *Honors & Awards:* Hatton Award, Int Asn Dent Res, 59; Medaille d'Argent, Paris, 78; Pres lectr, Univ Pa, 81; Turpin Mem lectr, Meharry Med Col, 85. *Mem:* Am Soc Microbiol; Int Asn Dent Res; Sigma Xi; Am Asn Dent Res (pres, 77-78). *Res:* Oral microbiology with emphasis on the physiology and molecular biology of periodontopathic bacteria; precise physico-chemical definition of how specific bacteria produce human periodontal disease; characterization & mechanisms of action of cytotoxic macromolecules and their effects on the microbial ecology of the gingival crevice; DNA probes. *Mailing Add:* Dept Microbiol Univ Pa Sch Dent Med Philadelphia PA 09104

HAMMOND, BRIAN RALPH, b Innisfail, Alta, Aug 20, 34; m 58; c 2. COMPARATIVE PHYSIOLOGY. *Educ:* Univ Alta, BSc, 62, MSc, 64; Cornell Univ, PhD(physiol), 69. *Prof Exp:* Asst prof physiol, Rice Univ, 69-78, assoc, Hanszen Col, 72-78; RES SECRETARIAT, ALTA ENVIRON, 78- *Mem:* Am Soc Zool. *Res:* Hormonal control of osmoregulation and kidney function in lower vertebrates; aquatic toxicology. *Mailing Add:* 11120 39th Ave Edmonton AB T6J 0M5 Can

HAMMOND, CHARLES BESSELLIEU, b Ft Leavenworth, Kans, July 24, 36; m 58; c 2. OBSTETRICS, GYNECOLOGY. *Educ:* Duke Univ, BS, 60, MD, 61. *Prof Exp:* Intern surg, 61-62, resident obstet & gynec, 62-64 & 66-69, from asst prof to prof, 66-80, E C HAMBLEN PROF OBSTET & GYNEC & CHMN DEPT, DUKE UNIV MED CTR, 80- *Concurrent Pos:* Fel reproductive endocrinol, NIH, Bethesda, Md, 64-66; dir, Am Bd Obstet & Gynec, 78-90. *Mem:* Am Fertil Soc (pres, 84-85); fel Am Col Obstet & Gynec; fel Soc Gynec Invest. *Res:* Menopause, including hormonal replacement, osteoporosis and cardiovascular disease; prolactin disorders; reproductive endocrinopathy; malignant trophoblastic disease. *Mailing Add:* PO Box 3853 Duke Univ Med Ctr Durham NC 27710

HAMMOND, CHARLES E, US citizen. DIGITAL SYSTEM DESIGN. *Educ:* Univ Md, BSEE, 75; Loyola Col, MSES, 81. *Prof Exp:* Staff engr, Univ Md, 75-77; elec engr, Nat Bur Standards, 77-79; sr engr, Gould Inc, 79-82; PRES & CONSULT, TRONEX CORP, 83- *Mem:* Inst Elec & Electronics Engrs. *Res:* Advanced computer architecture development. *Mailing Add:* PO Box 6029 Annapolis MD 21401

HAMMOND, CHARLES EUGENE, b Lithonia, Ga, Dec 21, 40; m 62; c 2. DYNAMICS, AEROELASTICITY. *Educ:* Ga Inst Technol, BAE, 62, MSAE, 68, PhD(aero eng), 70. *Prof Exp:* Aero engr, US Naval Weapons Lab, 65-70; leader rotor aero group, Res & Technol Labs, US Army, 71-80, chief aeromech, Appl Technol Lab, 80-85; PRIN ENGR, MARTIN MARIETTA ORLANDO AEROSPACE, 85- *Concurrent Pos:* Lectr math, Thomas Nelson Community Col, 73-; ad hoc fac mem, Univ Kans, 75-77. *Honors & Awards:* Spec Achievement Award, NASA, 75, Group Achievement Award, 78. *Mem:* Am Inst Aeronaut & Astronaut; Howard Hughes Award, Am Helicopter Soc, 83. *Res:* Helicopters, aeroelasticity, structural dynamics, wind tunnel testing; digital data acquisition and analysis. *Mailing Add:* 100 Cove Colony Rd Maitland FL 32751

HAMMOND, CHARLES THOMAS, b Moline, Ill, July 11, 44; m 67; c 3. BIOLOGY, CYTOLOGY. *Educ:* Western Ill Univ, BSEd, 66; Ohio State Univ, MS, 69; Ind Univ, Bloomington, PhD, 77. *Prof Exp:* Teaching assoc bot & biol, Ohio State Univ, 67-69; assoc instr bot, Ind Univ, 69-72; asst prof biol, Wabash Col, 73-77; from asst prof to assoc prof, 77-86, PROF BIOL, ST MEINRAD COL, 86- *Mem:* AAAS; Bot Soc Am. *Res:* Development, function and fine structure of the glandular secretory system in Cannabis stavia; structure and function of glandular systems in plants; developmental and regulatory aspects of leaf morphogenesis, heterophylly. *Mailing Add:* Dept Biol St Meinrad Col St Meinrad IN 47577

HAMMOND, DAVID G, b Paterson, NJ, Sept 8, 13; m 47; c 3. CIVIL ENGINEERING. *Educ:* Pa State Univ, BS, 34; Cornell Univ, MSCE, 39. *Prof Exp:* Engr, US Army CEngr, 37-64; asst gen mgr oper engrs, San Francisco Rapid Transit, 64-73; vpres, 73-85, CONSULT ENG, DANIEL, MANN, JOHNSON & MENDENHALL, 85- *Mem:* Nat Acad Eng; fel Am Soc Civil Engrs; Am Pub Transit Asn. *Mailing Add:* 364 Noren St La Canada CA 91011

HAMMOND, DONALD L, b Kansas City, Mo, Aug 7, 27. PHYSICAL SCIENCE. *Educ:* Colo State Col, BS, 50, MS, 52. *Prof Exp:* Chief crystal res sect, Frequency Control Br, US Army Signal Eng Labs, 53-57; dir res, Sci Electronic Prod, 57-59; CONSULT, CORP DEVELOP DEPT, HEWLETT PACKARD, 88- *Concurrent Pos:* Bd dirs, Colo Crystal Corp, 87-, Mid Penninsla Bank, 88-, Iris Med Corp, 90- & Nellcor Corp, 90-; chmn, New Hope & Ivy Land RR, 90-; adv bd, Idaho Nat Eng Labs, 91- *Mem:* Nat Acad Eng. *Mailing Add:* 12660 Corte Madera Lane Los Altos Hills CA 94022

HAMMOND, DOUGLAS ELLENWOOD, b New York, NY, Jan 12, 46; m 79; c 2. MARINE CHEMISTRY, GEOCHEMISTRY. *Educ:* Univ Rochester, BA, 67, MS, 70; Columbia Univ, PhD(geol), 75. *Prof Exp:* Asst prof, 75-81, assoc prof, 81-89, PROF GEOL, UNIV SOUTHERN CALIF, 89- *Mem:* AAAS; Am Geophys Union; Geochem Soc. *Res:* Aqueous geochemistry; estuarine chemistry; application of stable and radioisotopes to the study of accumulation and diagenesis of recent sediments, dissolved gases, nutrient cycles, circulation and mixing; groundwater chemistry. *Mailing Add:* Dept Geol Sci Univ Southern Calif Los Angeles CA 90089-0740

HAMMOND, EARL GULLETTE, b Terrell, Tex, Nov 21, 26; m 51; c 4. FOOD CHEMISTRY, BIOCHEMISTRY. *Educ:* Univ Tex, BS, 48, MA, 50; Univ Minn, PhD(biochem), 53. *Hon Degrees:* Dhc, Univ Agr & Biotechnology, Olsztyn Poland, 90. *Prof Exp:* Assoc prof dairy & food indust, biochem & biophys, 53-70, chmn food technol, 85-90, PROF FOOD TECHNOL, BIOCHEM & BIOPHYS, IOWA STATE UNIV, 70- *Concurrent Pos:* Chmn food technol, Iowa State Univ, 85-90. *Honors & Awards:* Pfizer Award, Am Dairy Sci Asn, 80; Hon Medal, Univ Agr & Technol, Olsztyn, Poland, 86. *Mem:* Am Chem Soc; Am Oil Chem Soc; Am Dairy Sci Asn; Inst Food Technol. *Res:* Chemistry and analysis of lipids; autoxidation of fats; glyceride structure of fats; cheese chemistry; odor pollution; fermentations. *Mailing Add:* Dept Food Sci & Human Nutrit Iowa State Univ Ames IA 50011

HAMMOND, GEORGE DENMAN, b Atlanta, Ga, Feb 5, 23; m 46; c 4. PEDIATRICS, ONCOLOGY. *Educ:* Univ NC, AB, 44; Univ Pa, MD, 48; Am Bd Pediat, dipl, 57. *Prof Exp:* Intern, Pa Hosp, 50; resident physician pediat, Children's Hosp, Philadelphia, 50 & 52-53; resident physician, Sch Med, Univ Calif, San Francisco, 53-54, asst prof pediat, 55-57; from asst prof to assoc prof, 57-65, PROF PEDIAT, SCH MED, UNIV SOUTHERN CALIF, 65-, ASSOC DEAN SCH MED, 71- *Concurrent Pos:* Res fel pediat, Sch Med, Univ Calif, San Francisco, 54-55; Gianinni Found fel, 54-56; Am Cancer Soc scholar cancer res, 64-65; attend pediatrician, Univ Hosps, Univ Calif, 55-57; asst pediatrician, San Francisco City & County Hosp, 55-57; lectr, Sch Nursing, Univ Calif, 55-57; pediatrician, Northern Calif Sch Cerebral Palsied Children, 56-57; assoc hematologist, Childrens Hosp Los Angeles, 57-60, chief div hemat/oncol & hemat res labs, 60-71, dep physician-in-chief, 70-71; chmn, Childrens Cancer Study Group, 68-; dir, Los Angeles County-Univ Southern Calif Comprehensive Ctr Cancer, 71-; consult, NIH; pres appointee, Nat Cancer Adv Bd, 74-; dir Norris Cancer Res Inst, 78- *Mem:* AAAS; Am Asn Cancer Educ; Soc Surg Oncol; Am Soc Clin Oncol; Soc Exp Biol & Med. *Res:* Pediatric hematology. *Mailing Add:* 199 N Lake Ave 3rd Floor Pasadena CA 91101

HAMMOND, GEORGE SIMMS, b Auburn, Maine, May 22, 21; m 45, 77; c 7. ORGANIC CHEMISTRY. *Educ:* Bates Col, BS, 43; Harvard Univ, MS & PhD(chem), 47. *Hon Degrees:* DSc, Wittenberg Univ, 72, Bates Col, 73; Dr State Univ Ghent, 73, Georgetown Univ, 85 & Bowling Green State Univ, 90. *Prof Exp:* Asst, Rohm and Haas Co, Pa, 43; res assoc insect repellants, Off Sci Res & Develop, Harvard Univ, 45; Off Naval Res fel, Univ Calif, Los Angeles, 48; from asst prof to prof chem, Iowa State Univ, 48-58; prof org chem, Calif Inst Technol, 58-64, Arthur Amos Noyes prof chem, 64-72, chmn div chem & chem eng, 68-72; vchancellor div natural sci, Univ Calif, Santa Cruz, 72-75, prof chem, 72-78; foreign secy, Nat Acad Sci, 74-78; assoc dir, Allied Corp, 78-79, dir, Integrated Chem Systs, 79-85, exec dir biosci, metals & ceramics, 85-88; CONSULT, 88-; DIR MAT SCI, BOWLING GREEN STATE UNIV, 91- *Honors & Awards:* Award Petrol Chem, Am Chem Soc, 61, James Flack Norris Award Phys Org Chem, 68, Award Chem Educ, 74 & Priestley Medal, 76; E Harris Harbison Award Gifted Teaching, Danforth Found, 71. *Mem:* Nat Acad Sci; Am Chem Soc; The Chem Soc; Mat Res Soc. *Res:* Mechanisms of free radical reaction; solution photochemistry; theories of reaction rates; materials science. *Mailing Add:* 2410 Southside Dr North Garden VA 22959

HAMMOND, GORDON LEON, b Portsmouth, NH, Nov 6, 31; Wid; c 3. ATOMIC PHYSICS, SPECTROSCOPY. *Educ:* Univ NH, BS, 58; Univ Md, MS, 62, PhD(astrophys), 74. *Prof Exp:* physicist, US Naval Ord Lab, 58-66, res physicist, 66-74, res physicist, Naval Surface Weapons Ctr, 74-86; PROF MATH, UNIV SFLA, 87- *Concurrent Pos:* Consult, Dept Defense, 75-86. *Mem:* Int Astronaut Union; Am Astron Soc. *Res:* Neutral atom interactions; spectral line broadening; stellar atmospheres; plasma diagnostics; explosives properties and phenomenology; shock wave physics. *Mailing Add:* 5944 Plazaview Dr Zephyr Hills FL 33541

HAMMOND, H DAVID, b Philadelphia, Pa, Feb 10, 24; div; c 1. BOTANY, PLANT PHYSIOLOGY. *Educ:* Rutgers Univ, BS, 45, MS, 47; Univ Pa, PhD(bot), 52. *Prof Exp:* Res assoc biochem, Long Island Jewish Hosp, 55-57; asst prof physiol, Univ Del, 57-58; asst prof bot, Howard Univ, 58-68; from asst prof to assoc prof biol sci, State Univ NY Col Brockport, 68-83; ASSOC ED, NY BOT GARD, 84- *Concurrent Pos:* Sigma Xi res grant, 64-65; NSF grant, 70-72; State Univ NY Found grant, 70-72; ed, Bull of the Torrey Bot Club, 76-82, 88- *Mem:* Torrey Bot Club; Bot Soc Am; Am Inst Biol Sci; Sigma Xi. *Res:* Plant morphogenesis; function of boron; herbarium development; evolution; plant growth and development; chemotaxonomy. *Mailing Add:* 2095 Cruger Ave Apt 6C Bronx NY 10462

HAMMOND, HAROLD LOGAN, b Hillsboro, Ill, Mar 18, 34; m 86; c 1. ORAL PATHOLOGY. *Educ:* Loyola Univ Chicago, DDS, 62; Univ Chicago, MS, 67; Am Bd Oral Path, dipl, 72. *Prof Exp:* Intern dent, Univ Chicago Hosps, 62-63, Zoller fel oral path, Univ Chicago Hosps & W G Zoller Mem Dent Clin, 63-65; USPHS res fel, 65-67; from asst prof to assoc prof oral diag & oral path, 67-83, head, Div Oral Diag, 73-75, PROF ORAL PATH, COL DENT, UNIV IOWA, 83-, DIR, SURG ORAL PATH SERV, 80- *Concurrent Pos:* Consult pathologist, Gen Hosp Managua, Nicaragua, 70-89; consult ed, Odontol Fedn Rev, Cent Am & Panama, 71-77, J Am Dent Asn, 79-89; consult oral path, Vet Admin Hosp, Iowa City, 77- *Mem:* Fel AAAS; Am Dent Asn; Int Asn Dent Res; fel Am Acad Oral Path; NY Acad Sci; Am Asn Dent Res. *Res:* Salivary gland disease; oral manifestations of systemic disease; odontogenic tumors. *Mailing Add:* Univ Iowa Col Dent Iowa City IA 52242-0101

HAMMOND, JAMES ALEXANDER, JR, b Trion, Ga, Mar 10, 36; m 59; c 2. ORGANIC CHEMISTRY. *Educ:* Ga Inst Technol, BS, 59, PhD(chem), 63. *Prof Exp:* Res chemist, 63-70, SR RES CHEMIST, NYLON TECH SECT, TEXTILE FIBERS DEPT, E I DU PONT DE NEMOURS & CO, INC, 70- *Mem:* Am Chem Soc. *Res:* Natural product and polymer chemistry. *Mailing Add:* PO Box 907 Redlands CA 92373

HAMMOND, JAMES B, b Salt Lake City, Utah, Sept 27, 21. INTERNAL MEDICINE. *Educ:* Univ Utah, BA, 43, MD, 46. *Prof Exp:* From assoc to instr, Sch Med, Ind Univ, 53-61; asst prof med, Chicago Med Sch, 62-63; assoc prof med, Med Sch, Northwestern Univ, Chicago, 63-81; ASSOC PROF MED, CHICAGO MED SCH, 81-; VET ADMIN MED CTR, NORTH CHICAGO. *Concurrent Pos:* Fel biochem, Mass Inst Technol, 61-62 & Chicago Med Sch, 62-63; physician, Div Clin Res, Eli Lilly & Co, 53-61; attend physician, Marion County Gen Hosp, Indianapolis, 53-61; chief gastroenterol serv, Vet Admin Lakeside Hosp, 63-70, clin investr, 71-72, physician, 72-81; chief, Gastroenterol Serv, Vet Admin Med Ctr, North Chicago, 80- *Mem:* Am Gastroenterol Asn; fel Am Col Physicians. *Res:* Gastric and small intestinal disease including electron microscopy of endocrine cells and enzyme deficiency; pancreatitis; drug effects on gastric secretion. *Mailing Add:* Vet Admin Med Ctr 111 G North Chicago IL 60064

HAMMOND, JAMES JACOB, b West Union, Ill, Dec 11, 41; m 69; c 2. AGRONOMY. *Educ:* Univ Ill, BS, 63, MS, 65; Univ Nebr, PhD(agron), 70. *Prof Exp:* Asst to assoc prof, 70-79, PROF AGRON, NDAK STATE UNIV, 79- *Mem:* Am Soc Agron; Crop Sci Soc Am. *Res:* Flax breeding and genetics. *Mailing Add:* Dept Agron NDak State Univ Univ Station Fargo ND 58105

HAMMOND, JAMES W, b Winona, Miss, Feb 2, 13; m 37; c 2. CHEMISTRY, CHEMICAL ENGINEERING. *Educ:* Miss State Univ, BS, 35, ChE, 50; La State Univ, MS, 37. *Prof Exp:* Asst chemist, Univ Tenn, 36-41 & USPHS, 41-46; asst dir indust hyg, Ga State Dept Health, 47; indust hygienist, 47-58, chief indust hygienist, 58-63, DIR INDUST HYG, HUMBLE OIL & REFINING CO, 63- *Concurrent Pos:* Asst prof, Col Med, Baylor Univ, 48-67; instr, Post-Grad Sch Med, Univ Tex, 50-67; asst prof, Sch Pub Health, Univ Tex, Houston, 67-72, assoc prof, 72- *Mem:* Am Indust Hyg Asn; Am Acad Indust Hyg (pres, 76-77). *Res:* Industrial environment science related to health conditions in work and occupation areas; industrial toxicology; chemical testing; ventilation; personal protective measures and equipment. *Mailing Add:* 3036 Lafayette St Houston TX 77005

HAMMOND, JOSEPH LANGHORNE, JR, b Birmingham, Ala, Oct 16, 27; m 49; c 3. COMPUTER NETWORKS, COMMUNICATION NETWORKS. *Educ:* Mass Inst Technol, BS & MS, 52; Ga Inst Technol, PhD(elec eng), 61. *Prof Exp:* Asst instrumentation, Air Force Cambridge Res Ctr, 49-52; elec engr, Southern Res Inst, 52-55; from asst prof to prof elec eng, GA Inst Technol, 55-84; PROF ELEC & COMPUTER ENG, CLEMSON UNIV, 85- *Concurrent Pos:* Instr, Univ Ala, 54 & Redstone Arsenal, 55; consult, Lockheed Ga Corp, 77 & 80-81 & Sci Atlanta Inc, 78-79; prin investr, NSF grant, 81-83; consult, Lockheed Ga Corp, 84-85. *Mem:* Inst Elec & Electronics Engrs; Sigma Xi. *Res:* Instrumentation; application of random processes; computer applications; systems theory; digital communication systems; design of local computer networks. *Mailing Add:* Elec & Computer Eng Dept Clemson Univ Riggs Hall Clemson SC 29634-0915

HAMMOND, LUTHER CARLISLE, b Seneca, SC, Jan 17, 21; m 47; c 4. WATER & SOLUTE TRANSPORT, CROP WATER PRODUCTION FUNCTIONS. *Educ:* Clemson Univ, BS, 42; Iowa State Univ, MS, 47, PhD, 49. *Prof Exp:* From asst prof & asst soil physicist to assoc prof & assoc soil physicist, 50-64, PROF SOILS & SOIL PHYSICIST, UNIV FLA, 64- *Mem:* Fel AAAS; Am Soc Agron; Soil Sci Soc Am; Am Geophys Union. *Res:* Soil-water-plant relationships; irrigation water management; crop-water production functions; water and solute transport in soils; modeling water balance; plant water use efficiency; soil spatial variability. *Mailing Add:* I-F-A-S McCarty 2169 Univ Fla Gainesville FL 32601

HAMMOND, MARTIN L, b Enid, Okla, Apr 19, 36; m 60; c 2. MATERIALS SCIENCE. *Educ:* Univ Calif, Berkeley, BS, 58, MS, 59; Stanford Univ, PhD(mat sci), 65. *Prof Exp:* Grad study scientist, Lockheed Palo Alto Res Labs, 59-63, scientist, 63-65, res scientist, 65-67; sect head mat res, Philco Microelectronics Div, Lockheed Aircraft Co, 67-68; mgr mat res, Union Carbide Electronics, 68-69; tech dir, UNICORP, 72-76; dir mkt, Appl Mat Inc, 76-81; DIR MKT, PLASMA THERM, INC, 81- *Concurrent Pos:* Consult, USAF, 66; div mgr, Hugle Industs, 69-72. *Mem:* Am Ceramic Soc; Electrochem Soc; Am Soc Metals. *Res:* Development and marketing semiconductor process systems. *Mailing Add:* 15945 Casa de Pino Cupertino CA 95014

HAMMOND, MARVIN H, JR, b Great Bend, Kans, Sept 28, 39; m 66; c 2. ELECTRICAL ENGINEERING. *Educ:* Kans State Univ, BSEE & BS(math), 62, MS, 63; Ohio State Univ, PhD(elec eng), 68. *Prof Exp:* Mem tech staff, NAm Rockwell Corp, Ohio, 65-69 & Inst Defense Anal, 69-76; MGR ELEC ENG, BATTELLE MEM INST, 76- *Mem:* Inst Elec & Electronics Engrs; Nat Soc Prof Engrs; Sigma Xi. *Res:* Optimal control systems; pattern recognition; guidance and control systems. *Mailing Add:* 505 Seneca Knoll Ct Great Falls VA 22066

HAMMOND, MARY ELIZABETH HALE, b Salt Lake City, Utah, Jan 5, 42; m 64; c 3. IMMUNOPATHOLOGY. *Educ:* Univ Utah, BS, 63, MD, 67. *Prof Exp:* Intern path, Univ Hosp, Univ Utah, 67-68; fel tumor biol, Karolinska Inst, Sweden, 68-69; resident path, Mass Gen Hosp, 70-73, fel immunopath, 73-74, asst in path, 74-77; assoc prof clin path, 77-82, ASSOC PROF PATH, COL MED, UNIV UTAH, 82-; DIR EM LAB, LATTER DAY SAINTS HOSP, 77-, VET ADMIN CTR, 84- *Concurrent Pos:* Teaching fel, Harvard Med Sch, 70-74, instr, 74-76, asst prof, 76-77; Am Cancer Soc res scholar, 74-77; pathologist, Latter Day Saints Hosp, 77-82. *Mem:* Am Asn Immunologists; Am Asn Pathologists; Col Am Path; Am Soc Clin Path; Int Acad Path; Am Soc Nephrology. *Res:* Role of macrophage in cell mediated and tumor immunity, especially related to cell membrane surface changes; role of clotting system in cell mediated immunity; ultrastructure of human tumors; immunopathology of tumors, hearts and kidney; ultrastructure of heart and kidney disease. *Mailing Add:* EM Lab Latter Day Saints Hosp Eighth Ave & C St Salt Lake City UT 84143

HAMMOND, PAUL B, b Cleveland, Ohio, Apr 27, 23; m 56. TOXICOLOGY. *Educ:* Colo State Univ, DVM, 49; Univ Minn, PhD(pharmacol), 55. *Prof Exp:* From instr to prof pharmacol, Univ Minn, St Paul, 51-72; PROF ENVIRON HEALTH, UNIV CINCINNATI, 72- *Concurrent Pos:* Mem toxicol study sect, NIH, 61-65; mem comt vet drug efficacy, Nat Res Coun, 67-68, mem comt biol effects of atmospheric pollutants, 70-; mem adv coun, Nat Inst Environ Health Sci, 67-69, mem training comt, 70-; mem comt hazardous trace substances, Off Sci & Technol, 70- *Mem:* Am Soc Pharmacol & Exp Therapeut; Am Vet Med Asn; Soc Toxicol. *Res:* Heavy metal toxicology, particularly lead; fluid and electrolyte physiology and pharmacology. *Mailing Add:* Dept Environ Health 323 Eden Ave Mail Location S6 Cincinnati OH 45267

HAMMOND, PAUL ELLSWORTH, b Oakland, Calif, Jan 28, 29; m 56; c 3. GEOLOGY. *Educ:* Univ Colo, AB, 52; Univ Calif, Los Angeles, MA, 58; Univ Wash, PhD(geol), 63. *Prof Exp:* Mining geologist, Northern Pac Rwy Co, 56-60; from asst prof to assoc prof, 63-90, PROF GEOL, PORTLAND STATE UNIV, 90- *Mem:* Geol Soc Am; Geol Soc London; Am Geophys Union. *Res:* Volcanic stratigraphy; structure of volcanic rock terrain; geology of the Pacific Northwest and of geothermal resources. *Mailing Add:* Geol Dept Portland State Univ PO Box 751 Portland OR 97207-0751

HAMMOND, R PHILIP, b Creston, Iowa, May 28, 16; m 41; c 4. SEAWATER DESALINATION, NUCLEAR SAFETY. *Educ:* Univ Southern Calif, BS, 38; Univ Chicago, PhD(inorg chem), 47. *Prof Exp:* Chief chemist, Lindsay Chem Co, Ill, 38-46; mem staff, Los Alamos Sci Lab, 47-58, asst dir reactor div, 58-64; dir nuclear desalination prog, Oak Ridge Nat Lab, 62-74; energy staff, R & D Assocs, 74-86; CONSULT, 86- *Concurrent Pos:* Adv, US Del, Geneva Conf Peaceful Uses Atomic Energy, 55, 64 & 71; adj prof eng, Univ Calif, Los Angeles, 85- *Mem:* Am Nuclear Soc. *Res:* Design, engineering and economics of nuclear reactors and sea water evaporators; mobile fuel reactors; nuclear fuel reprocessing; effect of energy on environment; fusion energy process from underground containment of thermonuclear reactions, storage of nuclear waste, and energy storage systems; nuclear power plant safety; ocean thermal energy; liquid nitrogen automobile; chill-vent filter system; marine structures; engineering systems. *Mailing Add:* Box 1735 Santa Monica CA 90406

HAMMOND, RAY KENNETH, b Lake Charles, La, Oct 3, 43; m 66; c 2. BIOCHEMISTRY, MICROBIOLOGY. *Educ:* Cent Col, BS, 65; Univ Ky, PhD(biochem), 69. *Prof Exp:* Res assoc biochem, Mich State Univ, 70-72; from asst prof to assoc prof, 72-86, PROF BIOL & BIOCHEM, CENT COL, 86-, VPRES & DEAN STUDENTS, 88- *Concurrent Pos:* Vis prof biol, Silliman Univ, Philippines, 80-81; vis prof biochem, Sch Med, St Georges Univ, WI, 81-83; vis lectr, Chulalongkorn Univ, Bangkok, Thailand, 86. *Mem:* Nat Asn Adv Health Professions; AAAS. *Res:* Microbial degradation of agricultural pesticides and function of lipids in insect development; metabolism of yeast phosphoinositides; exercise biochemistry. *Mailing Add:* Div Sci Cent Coll Danville KY 40422

HAMMOND, ROBERT BRUCE, b Bethesda, Md, Mar 27, 48; m 78. SOLID STATE PHYSICS, SUPERCONDUCTORS & SEMICONDUCTORS. *Educ:* Calif Inst Technol, BS, 71, MS, 72, PhD(appl physics), 75. *Prof Exp:* Staff physicist laser physics, US Air Force Weapons Lab, NMex, 75-76; dep group leader, electronics res, Los Alamos Sci Lab, Univ Calif, 76-87; VPRES, SUPERCONDUCTOR TECHNOLOGIES INC, 87- *Mem:* Am Phys Soc; Inst Elec & Electronics Engrs. *Res:* Study of excitons and electron-hole liquid in semiconductors; photoluminescence studies of impurities and defects in semiconductors; thermo-electric power of semiconductors; picosecond photoconductivity and device applications; microwave superconductivity. *Mailing Add:* Superconductor Technologies Inc 460-F Ward Dr Santa Barbara CA 93111

HAMMOND, ROBERT GRENFELL, b Portland, Ore, Feb 14, 17; m 43; c 3. MATHEMATICS. *Educ:* Utah State Univ, BS, 48, MS, 52. *Prof Exp:* ASSOC PROF MATH, UTAH STATE UNIV, 56- *Concurrent Pos:* NSF fels, Univ Wyo, 59 & Harvard Univ, 60-61. *Mem:* Math Asn Am. *Res:* Development of teaching methods of mathematics in college and secondary school; logics; fundamental concepts of mathematics. *Mailing Add:* 499 E 200 S Logan UT 84321

HAMMOND, ROBERT HUGH, b Fullerton, Calif, May 11, 30; m 55; c 3. PHYSICS. *Educ:* Univ Calif, Berkeley, BS, 53, PhD(physics), 60. *Prof Exp:* Res & develop staff mem, Gen Atomic Div, Gen Dynamics Corp, 60-68; physicist, Lawrence Radiation Lab, Univ Calif, 68-71; sr res physicist, Stanford Univ, 71-90; CONSULT, 90- *Concurrent Pos:* Lectr, Univ Calif, Berkeley, 67-71. *Mem:* Am Phys Soc. *Res:* Solid state and low temperature physics; superconductivity; nuclear magnetic resonance; thin films; synthesis of new superconducting materials. *Mailing Add:* Ginaton Physics Lab Stanford Univ Stanford CA 94305

HAMMOND, SALLY KATHARINE, b Philadelphia, Pa, Feb 6, 49; m 72. OCCUPATIONAL HEALTH, INDUSTRIAL HYGIENE. *Educ:* Oberlin Col, BA, 71; Brandeis Univ, PhD(inorg chem), 76; Harvard Sch Pub Health, MS, 81. *Prof Exp:* Instr chem, physics, meteorol & astron, Boston Univ, 75-76; asst prof chem, Wheaton Col, 76-80; asst prof, 85-89, ASSOC PROF, MED SCH, UNIV MASS, 89- *Concurrent Pos:* Vis investr, New Eng Aquarium, 75-; res assoc, Harvard Sch Pub Health, 80-84, lectr, indust hyg, 85- *Mem:* Am Chem Soc; AAAS; Am Indust Hyg Asn; Am Coun Govt Indust Hygienists; Am Pub Health Asn. *Res:* Occupational health; effects of various industrial chemicals on workers; solvents; polycyclic aramatic compounds; epidemiological studies; methods of sample collection and analysis. *Mailing Add:* Med Ctr Univ Mass 55 Lake Ave N Worcester MA 01605

HAMMOND, SEYMOUR BLAIR, electrical engineering, for more information see previous edition

HAMMOND, WILLIAM EDWARD, b Hendersonville, NC, Jan 9, 35; m 57; c 2. BIOMEDICAL ENGINEERING, COMPUTER SCIENCES. *Educ:* Duke Univ, BSEE, 57, PhD(elec eng), 67. *Prof Exp:* Instr elec eng, 60-64, res asst, 64-67, asst prof community health sci & biomed eng, 68-71, assoc prof community health sci, 71-79, PROF COMMUNITY AND FAMILY MED, MED CTR, DUKE UNIV, 79- *Concurrent Pos:* Res fel, Duke Univ, 67-68; mem, Div Res Resources, Nat Adv Coun, 79-81; exec comt, Am Col Med Informatics, 87-88, treas, 90-; bd dirs, Am Asn Med Syst Informatics, 87-90, prog chair, 88 & 89. *Mem:* Inst Elec & Electronics Engrs; Biomed Eng Soc; Sigma Xi; Asn Comput Mach; Am Soc Testing & Mat; Am Col Med Informatics; Am Med Informatics Asn (treas). *Res:* Use of computers in clinical medicine, specifically the acquisition and analysis of medical historical information, physical examination data and clinical laboratory data; information storage and retrieval systems; computer aided clinical decision and hospital information systems; networking. *Mailing Add:* Box 3054 Durham NC 27712

HAMMOND, WILLIAM MARION, b South Bend, Ind, Nov 24, 30; m 59, 88; c 3. COMMUNICATIONS, COMPUTER SYSTEMS. *Educ:* Purdue Univ, BSEE, 57, PhD(elec eng), 66; Bradley Univ, MSEE, 60. *Prof Exp:* Asst prof elec eng, Bradley Univ, 57-63; instr, Purdue Univ, 63-66; PROF ELEC ENG, BRADLEY UNIV, 66- , DIR TECHNOL CTR, 85- *Concurrent Pos:* Consult, Surface Radar Br, Apparatus Div, Tex Instruments, Inc, Tex, 66-67; prin investr, NASA contracts, 67-68; dir & res adv, Continental Mgt Co, Ill, 68-71; pres & mem, Bd Dirs, Comput Corp Am, 69-71 & ADS, Inc, 74-; NSF grant, 71-72; consult, Caterpillar Tractor Co, 73-85. *Mem:* Am Soc Eng Educ; Inst Elec & Electronics Engrs. *Res:* Statistical communications; microelectronics; computer application systems. *Mailing Add:* Dir Technol Ctr Bradley Univ Peoria IL 61625

HAMMOND, WILLIS BURDETTE, b Truman, Minn, Sept 29, 42; m 71; c 2. COMPUTATIONAL CHEMISTRY, POLYMER MODELING. *Educ:* Northwestern Univ, BA, 64; Columbia Univ, MA, 66, PhD(chem), 67. *Prof Exp:* NIH fel, Calif Inst Technol, 67-68; asst prof chem, Yale Univ, 68-75; sr res chemist, 75-81, RES ASSOC, ALLIED CORP, 81- *Mem:* AAAS; Am Chem Soc. *Res:* Organic photochemistry; organic spectroscopy; laser chemistry; polymer characterization; organic spectroscopy; polymer characterization; organic photochemistry; currently developing computer modeling of polymers for purpose of predicting polymer structures and properties; spectroscopic characterization of polymers with special emphasis on nuclear magnetic resonance, organic synthesis and organic photochemistry. *Mailing Add:* Allied Corp PO Box 1021R Morristown NJ 07962-1021

HAMMOND, WYLDA, pediatrics, for more information see previous edition

HAMMONS, JAMES HUTCHINSON, b Chicago, Ill, Aug 8, 34; m 56; c 2. PHYSICAL ORGANIC CHEMISTRY. *Educ:* Amherst Col, BA, 56; Johns Hopkins Univ, MA, 58, PhD(chem), 62. *Prof Exp:* Fel chem, Univ Southern Calif, 62-63 & Univ Calif, Berkeley, 63-64; from instr to assoc prof, 64-76, chmn dept, 76-81, PROF CHEM, SWARTHMORE COL, 76- *Concurrent Pos:* Fel, Swiss Fed Inst Technol, 68-69; vis res prof, Univ Calif, Berkeley, 72-73 & Univ Geneva, Switz, 77-78; vis fel, Yale Univ, 81-82; NSF Sci Fac Develop Grant, 81-82; vis res prof, Univ Wash, 89. *Mem:* Am Chem Soc. *Res:* Molecular orbital theory; electron spin resonance. *Mailing Add:* Dept Chem Swarthmore Col 500 College Ave Swarthmore PA 19081-1397

HAMMONS, PAUL EDWARD, dentistry; deceased, see previous edition for last biography

HAMMONS, RAY OTTO, b Miss, Oct 2, 19; m 42; c 4. GENETICS, AGRONOMY. *Educ:* Miss State Univ, BS, 47, MS, 48; NC State Univ, PhD(agron), 53. *Prof Exp:* Res instr peanut radiation genetics, NC State Univ, 49-53, assoc genetics, 52-53; asst prof forage breeding, Purdue Univ, 53-55; geneticist, Agr Res Serv, USDA, 55-63, res geneticist, 63-72, tech adv peanuts, 73-81, res leader crops, 72-84, supvry res geneticist, 75-84; collabr, Agr Res Serv, USDA, 84-87; CONSULT, UNIV GA, 89- *Concurrent Pos:* Assoc prof, Univ Ga, 55-84, mem grad fac, 59-82; plant explorer, SAm, 68; agr consult, USAID, West Pakistan, 71; ed, Peanut Res, 72-82; consult, Int Progs Div, USDA, Israel, India & Pakistan, 73; assoc ed, Agron J, Am Soc Agron, 75-77; vis scientist award, Am Soc Agron, 81-82; consult, UN Conference Trade & Develop, Geneva & Rome, 80 & Int Crops Res Inst Semi-Arid Tropics, India, 79, 80 & 84; assoc ed, Peanut Sci, Am Peanut Res & Educ Soc, 79-85 & Crop Sci, Crop Sci Soc Am, 81-84; UN Food & Agr Orgn deleg, Int Symp Groundnut, Africa, Gambia, 82. *Honors & Awards:* Golden Peanut Res Award, Nat Peanut Coun, 75. *Mem:* Fel Am Soc Agron; fel Crop Sci Soc Am; Am Genetic Asn; fel Am Peanut Res Educ Soc; Sigma Xi. *Res:* Peanut genetics, breeding, origin, history; evolutionary development in Arachis; production research on tobacco, ornamentals, tomato and peanuts. *Mailing Add:* 1203 Lake Dr Tifton GA 31794-3834

HAMNER, CHARLES EDWARD, JR, b Schuyler, Va, Mar 26, 35; m 61; c 2. BIOCHEMISTRY, VETERINARY MEDICINE. *Educ:* Va Polytech Inst, BS, 56; Univ Ga, DVM, 60, MS, 62, PhD(biochem), 64. *Prof Exp:* Asst prof exp surg & dir vivarium, 64-67, assoc prof obstet & gynec, 67-77, from asst vpres to assoc vpres Health Serv, 77-88, PROF OBSTET & GYNEC, SCH MED, UNIV VA, 77- *Concurrent Pos:* Morris Animal Found fel, Univ Ga, 60-61, Nat Inst Gen Med Sci fel, 61-64; consult, Pharmaceut Res & Develop Mgt, WHO, NIH, 71-; dir prog coord, A H Robins Co, 74-77; pres, NC Biotechnol Ctr, 88- *Honors & Awards:* Res Career Develop Award, NIH, 71. *Mem:* Soc Study Reproduction; Drug Info Asn; Brit Soc Study Fertil; AAAS. *Res:* Sperm metabolism, composition of female reproductive tract secretions; drug development; industrial management. *Mailing Add:* NC Biotechnol Ctr PO Box 13547 Res Triangle Park NC 27709

HAMNER, MARTIN E, b Castor, La, July 28, 18; m 46; c 3. PHARMACY. *Educ:* Univ Colo, BS, 49, MS, 51, PhD(pharm), 55. *Prof Exp:* Instr pharm, Univ Fla, 51-53 & Univ Colo, 53-55; assoc prof, Southwestern State Col, Okla, 55-59; prof & chmn dept, 59-63, assoc dean, Col Pharm, 63-81, PROF, UNIV TENN, MEMPHIS, 83-, EMER PROF PHARMACEUT, 88- *Concurrent Pos:* actg dean, Univ Tenn, Col Pharm, 75; Parenteral & Sterile Prod Lab, 86-88. *Mem:* Am Asn Col Pharm; Am Pharmaceut Asn. *Res:* Bacteriology and antibiotics with emphasis on synergism and drug action; larvae and larvicidal agents; tracer studies on contamination and cleaning of contact lenses; completation retinoic acid with cyclodextrims. *Mailing Add:* 4806 Craigmont Memphis TN 38128

HAMNER, WILLIAM FREDERICK, b Pharr, Tex, Sept 25, 22; m 49; c 4. POLYMER CHEMISTRY. *Educ:* Centenary Col, BS, 43; Univ Tex, MA, 47, PhD(phys chem), 50. *Prof Exp:* Asst prof phys chem, Univ Ala, 49-51; res chemist, Monsanto Co, 51-52, from asst res group leader to res group leader, 52-59, res sect leader, 59-63, group leader, Chemstrand Res Ctr, 63-65, mgr, 65-74, mgr res & develop, 74-80, MGR TECH SERV & PHYS SCI, MONSANTO TRIANGLE PARK DEVELOP CTR, 80- *Mem:* Am Chem Soc; Am Inst Chem; Sigma Xi. *Res:* Spectroscopy; instrumental analysis; textiles and fibers; synthetic turf; research and development of synthetic recreational surfaces, polyurethane chemistry, analytical chemistry. *Mailing Add:* 724 Catawba St Raleigh NC 27609

HAMOLSKY, MILTON WILLIAM, b Lynn, Mass, May 25, 21; m 45, 79; c 3. INTERNAL MEDICINE. *Educ:* Harvard Univ, AB, 43, Harvard Med Sch, MD, 46; Am Bd Internal Med, dipl, 56. *Prof Exp:* Asst med, Beth Israel Hosp, 51-52; from asst to asst prof, Harvard Med Sch, 52-63; PROF MED, BROWN UNIV, 63-, DIR, DIV MED RES, RI HOSP, 75- *Concurrent Pos:* Assoc med res, Beth Israel Hosp, 52-63, head endocrine clin, 58-63, asst vis physician, 57-59, assoc vis physician, 59-63, consult, 63-; vis res asst prof, Brandeis Univ, 58-59; tutor, Harvard Med Sch, 57-62, lectr, 63; physician-in-chief med serv, RI Hosp, 63-; consult, Miriam Hosp, Roger Williams Hosp, Bradley Hosp & Vet Admin Hosp, 63-; chief, Dept Med, Women & Infants Hosp, 81- *Mem:* AAAS; Am Physiol Soc; fel Am Col Physicians; Am Soc Clin Invest; Am Thyroid Asn; Endocrine Soc. *Res:* Endocrinology; clinical thyroid physiology; biochemistry; pathology. *Mailing Add:* Rhode Island Hosp Providence RI 02902

HAMON, AVAS BURDETTE, b Ripley WVa, Mar 8, 40; m 65. COCCOIDEA, ALEYRODIDAE. *Educ:* Morris Harvey Col, BS, 68; Marshall Univ, MS, 69; Va Polytech Inst & State Univ, PhD(entom), 76. *Prof Exp:* TAXON ENTOMOLOGIST, DEPT AGR & CONSUMER SERV, FLA, 76- *Concurrent Pos:* Adj asst prof, Dept Entom & Nematol, Univ Fla, 78- *Mem:* Entom Soc Am; Sigma Xi; Am Registry Prof Entomologists. *Res:* Taxonomy and systematics of Coccoidea and Aleyrodidae. *Mailing Add:* 10212 NW Sixth Pl Gainesville FL 32601

HAMON, J HILL, b Junction City, Ky, July 30, 31; m 51; c 3. AVIAN ANATOMY, PALEONTOLOGY. *Educ:* East Ky State Col, BS, 52; Univ Ky, MS, 53; Univ Fla, PhD(biol), 61. *Prof Exp:* Asst prof biol, Jacksonville Univ, 60-61; assoc prof, Ind State Univ, 61-68; PROF BIOL, 68-, CHMN, BIOL DEPT, TRANSYLVANIA UNIV, 87- *Mem:* Sigma Xi. *Res:* Avian anatomy and paleontology; history and philosophy of science. *Mailing Add:* Div Natural Sci & Math Transylvania Univ Lexington KY 40508

HAMOR, GLENN HERBERT, b Kootenai, Idaho, May 14, 20; m 47; c 4. PHARMACEUTICAL CHEMISTRY. *Educ:* Univ Mont, BS, 41, MS, 47; Univ Minn, PhD(pharmaceut chem), 52. *Prof Exp:* Instr pharm, Univ Mont, 47-48; asst, Univ Minn, 48-51; from asst prof to assoc prof, 52-66, PROF PHARMACEUT CHEM, UNIV SOUTHERN CALIF, 66- *Concurrent Pos:* Cotrell res grant, 53-54; Pfeiffer Mem res fel, Univ Trieste, 66-67; vis prof, Sch Pharm, Trinity Col, Dublin, Ireland, 81-82. *Mem:* AAAS; Am Chem Soc; Am Pharmaceut Asn; Am Asn Col Pharm. *Res:* Synthesis of medicinally active pharmaceutical agents; design of enzyme inhibitors; antiinflammatory, antiepileptic and sweetening agents; heterocyclic chemistry; histidine decarboxylase; inhibition of histamine biosynthesis; literary perceptions of pharmacists. *Mailing Add:* 6519 W 87th St Los Angeles CA 90045

HAMORI, EUGENE, b Gyor, Hungary, Aug 27, 33; US citizen; m 58; c 2. BIOCHEMISTRY, PHYSICAL CHEMISTRY. *Educ:* Sci Univ Budapest, dipl chem, 56; Univ Pa, PhD(phys chem), 64. *Prof Exp:* Asst prof chem, Univ Del, 66-72; PROF BIOCHEM, SCH MED, TULANE UNIV, 82- *Concurrent Pos:* NIH grant biophys chem, Cornell Univ, 64-66; vis prof, Max-Planck Inst Biophys Chem, Goettingen, WGer, 79-80. *Mem:* Am Chem Soc; NY Acad Sci; Am Soc Biol Chemists; Biophys Soc. *Res:* Nucleic acids; kinetics of conformation changes of biopolymers; computer study of DNA sequences. *Mailing Add:* Dept Biochem Tulane Univ Sch Med 1430 Tulane Ave New Orleans LA 70112

HAMOSH, MARGIT, b Dresden, Ger, Aug 13, 33; US citizen; m 54; c 3. PHYSIOLOGY, BIOCHEMISTRY. *Educ:* Hebrew Univ, Israel, MSc, 56, PhD(biochem), 59. *Prof Exp:* Fel biochem, Hadassah Med Sch, Hebrew Univ, Israel, 59-61, from instr to asst prof biochem, 61-65; sabbatical, NIMH,

Bethesda, Md, 65-67, vis scientist endocrinol, Nat Inst Arthritis & Metab Dis, 67-74; res assoc, 74-78, assoc prof pediat, 78-84, PROF PEDIAT, MED SCH, GEORGETOWN UNIV, 84-, CHIEF, DIV DEVELOP BIOL NUTRIT, 88- *Concurrent Pos:* Mem, Pulmonary Dis Adv Comt, NIH, 80-84; Pediat Coun, Am Col Nutrit, 82-84; Coun Perinatal Res Soc; Nat Res Rev Comt, Am Thoracic Soc, 84-90 & Maternal & Child Health Res Comt, Nat Inst Child Health & Human Develop, NIH, 86-90; proj hope adv, Poland, 85; heritage prof, Alta, Can, 86; chair, Subcomt Nutrit during Lactation, Nat Acad Sci, 88-90; mem, Coun Nutrit during Pregnancy & Lactation, Nat Acad Sci, 89-91. *Mem:* Am Physiol Soc; Endocrine Soc; Soc Exp Biol & Med; Am Fedn Clin Res; Am Thoracic Soc; Perinatal Res Soc; Am Inst Nutrit; Am Col Nutrit; Am Soc Clin Nutrit. *Res:* Developmental physiology; lipid transport; development of digestive functions; lung maturation; endocrine control of organ development; role of lipoprotein lipase in lipid uptake by extrahepatic tissues; lipid digestion and clearing of lipids from the circulation system; biologic and nutritional aspects of milk (human and other species); developmental biology. *Mailing Add:* Pediat 2 PHC Med Ctr Georgetown Univ 3800 Reservoir Rd NW Washington DC 20007

HAMOSH, PAUL, b Subotica, Yugoslavia, Apr 4, 31; US citizen; m 54; c 3. NEUROSCIENCES, NUTRITION. *Educ:* Hebrew Univ, Jerusalem, MD, 59. *Prof Exp:* Resident, Hadassa Univ Hosp, Jerusalem, 58-63; fel hematol, Tel Aviv Munic Univ Hosp, 63-65; fel respiration, Georgetown Univ & Vet Admin Hosp, Washington, DC, 65-68; res assoc hemodynamics, Vet Admin Hosp, Washington, DC, 68-70; asst prof med, George Washington Univ, 70-72; from asst prof to assoc prof, 72-85, PROF PHYSIOL BIOPHYS & PEDIAT, GEORGETOWN UNIV, 85- *Concurrent Pos:* Dir, Pulmonary Function Lab, Vet Admin Hosp, Washington, DC, 69-72, Dept Pediat, Georgetown Univ Hosp, 80-84. *Mem:* Am Physiol Soc; Biophys Soc; Am Fedn Clin Res; Int Asn Study Lung Cancer; fel AAAS. *Res:* Respiration physiology, control and mechanics; infant digestion: lipases in human milk and upper gastrointestinal tract; lung metabolism and hormonal control of lung maturation; neurotransmitters. *Mailing Add:* Sch Med Georgetown Univ 3900 Reservoir Rd NW Washington DC 20007

HAMPAR, BERGE, b New York, NY, Aug 20, 32; div. VIROLOGY, ADMINISTRATION. *Educ:* Columbia Univ, BA, 54, DDS, 60; Univ Baltimore, JD, 84. *Prof Exp:* Res microbiologist, NIH, 62-67, asst chief, Lab Molecular Virol & head, Microbiol Sect, 75-81, dent dir, Nat Cancer Inst, 67-86, gen mgr, Frederick Cancer Res Fac, 81-86; PRES, BIO-MOLECULAR TECH, INC, 86- *Concurrent Pos:* Res fel microbiol, Columbia Univ, 60-62, USPHS fel, 60-63; head, Solid-Tumor Virus Sect, Nat Cancer Inst, 73-75. *Mem:* Am Asn Immunologists; Am Soc Virol. *Res:* Herpes viruses with emphasis on immunology, biological properties, oncogenesis and genetics; biology of cell transformation; monoclonal antibodies to virion and non-virion antigens. *Mailing Add:* Box 888 Middletown MD 21769

HAMPARIAN, VINCENT, b New York, NY, July 30, 27; m 55; c 4. VIROLOGY. *Educ:* Wayne State Univ, BA, 50, MS, 51; Univ Pa, PhD(med microbiol), 58. *Prof Exp:* Instr pub health & prev med, Univ Pa, 58; res assoc virol, Chas Pfizer & Co, 58; res assoc, Merck Inst Therapeut Res, Pa, 58-64; PROF PEDIAT & MED MICROBIOL, COL MED, OHIO STATE UNIV, 64-; EXEC DIR, CHILDREN'S HOSP RES FOUND, 73- *Concurrent Pos:* Mem res staff, Children's Hosp, Philadelphia, 58; mem res resources comt, Nat Inst Allergy & Infectious Dis, 67-71, chmn, 70-71. *Mem:* AAAS; Am Asn Immunol; Am Soc Microbiol. *Res:* Respiratory viruses; immunology and serology, particularly laboratory diagnosis of viral infections. *Mailing Add:* Pediat Ohio State Univ Col Med 700 Childrens Dr Columbus OH 43205

HAMPEL, ARNOLD E, b Burlington, Ill, Sept 10, 39; m 62; c 3. BIOCHEMISTRY, MOLECULAR BIOLOGY. *Educ:* Northern Ill Univ, BS, 63; Univ Wis-Madison, PhD(biochem), 69. *Prof Exp:* Fel, Los Alamos Nat Lab, Univ Calif, 69-70; asst prof to assoc prof, 70-78, PROF BIOL & CHEM, NORTHERN ILL UNIV, 78- *Concurrent Pos:* NIH career develop award, 72; vis assoc prof, Salk Inst, 75-76; vis prof, Univ Calif Davis, 85-86. *Mem:* AAAS; Am Chem Soc; Biophys Soc; Am Soc Biochem & Molecular Biol. *Res:* Catalytic RNA, the "Hairpin" Ribozyme. *Mailing Add:* Dept Biol Sci & Chem Northern Ill Univ De Kalb IL 60115

HAMPEL, CLIFFORD ALLEN, b Minneapolis, Minn, Mar 15, 12; m 35; c 1. CHEMISTRY. *Educ:* Univ Minn, BChE, 34. *Prof Exp:* Res chemist, Mathieson Alkali Works, Inc, NY, 36-42 & Diamond Alkali Co, Ohio, 42-43; res scientist, Manhattan Dist Proj, SAM Labs, Columbia Univ, 43-44; res chemist, Minn Mining & Mfg Co, 44-45; asst chief chemist, Cardox Corp, Ill, 45-46; chem engr, Armour Res Found, Ill Inst Technol, 46-48, supvr inorg technol, 48-49, extraction metall, 49-52; consult chem engr, Ill, 52-53 & 55; proj engr, Morton Salt Co, 53-54; mgr, Chem Equip Div, Fansteel Metall Corp, Ill, 55-58; CONSULT CHEM ENGR, 58- *Concurrent Pos:* Consult, USN & USCG, 47-49; lectr, Milan Polytech Univ, 50. *Mem:* Fel AAAS; Am Chem Soc; Electrochem Soc; fel Am Inst Chemists; Am Soc Metals. *Res:* Production and uses of chlorine compounds; equilibria in heterogeneous salt systems; industrial chemicals; electrochemical processes and products; sea water and natural salts; economic and market surveys on inorganic chemicals and metals; fluorine compounds; heavy chemicals; rare metals. *Mailing Add:* 169 Sunnyside Ave Crystal Lake IL 60014-5253

HAMPP, EDWARD GOTTLIEB, b St Louis Co, Mo, Jan 15, 12; m 41; c 2. MICROBIOLOGY. *Educ:* Wash Univ, DDS, 36, MS, 39. *Prof Exp:* Instr path & bact, Sch Dent, Wash Univ, 39-41, instr path, Sch Med, 40-41; from res assoc to sr res assoc, Nat Inst Dent Res, 41-59; dir res div, Am Dent Asn, 59-69; prog adminr path, Nat Inst Gen Med Sci, NIH, 69-80; RETIRED. *Concurrent Pos:* Carnegie fel, Wash Univ, Univ Rochester & Univ Mich, 36-39; Am Dent Asn sr res fel, Nat Inst Dent Res, 41-58. *Mem:* Am Dent Asn; fel Am Col Dent; Int Asn Dent Res. *Res:* Histology; pathology; bacteriology and pathology of trench mouth; micro-incineration of soft tissues. *Mailing Add:* 8787 Pelican Ct Seminole FL 34647

HAMPSON, MICHAEL CHISNALL, b Lancashire, Eng, July 29, 30; m; c 2. MYCOLOGY, HORTICULTURE. *Educ:* Univ Col North Wales, BSc, 52, dipl educ, 53; McGill Univ, MSc, 60; Cornell Univ, PhD(plant path), 69. *Prof Exp:* Teacher sci, Eng, 56-58; community teacher, Govt Can, 60-61; photographic tech, McGill UNIV, 61-62; teaching specialist biol, Greater Montreal, 62-65; RES SCIENTIST PLANT PATH, AGR CAN RES BR, 69- *Concurrent Pos:* Lectr biol, Mem Univ Nfld, 71-72; mem, Atlanta Comt Crops, 72- *Mem:* Am Phytopath Soc; Can Phytopath Soc; Sigma Xi; Mycol Soc Am. *Res:* Studies on the biological control and eradication of wart disease of potatoes and studies on the life-history and biology of its causal agent Synchtrium endobioticum. *Mailing Add:* Agr Can Res Sta PO Box 7098 St John's NF A1E 3Y3 Can

HAMPSON, ROBERT F, JR, b Washington, DC, May 4, 30; m 66; c 1. PHYSICAL CHEMISTRY. *Educ:* Cath Univ Am, AB, 51, MS, 57, PhD(chem), 59. *Prof Exp:* Res chemist, US Naval Propellant Plant, Md, 51-52; res chemist, Inorg Solids Div, 59-61, RES CHEMIST, PHYS CHEM DIV, NAT BUR STANDARDS, 61- *Mem:* AAAS; Am Chem Soc; Am Phys Soc. *Res:* Vacuum-ultraviolet photochemistry; chemical kinetics; high-temperature and quantum chemistry. *Mailing Add:* Nat Bur Standards Chem Kin Ctr Bldg 222 Rm 260 Gaithersburg MD 20899

HAMPTON, CAROLYN HUTCHINS, b Burke Co, NC, Dec 11, 36; m 63; c 1. ZOOLOGY, PARASITOLOGY. *Educ:* Appalachian State Teachers Col, BS, 59; Univ Tenn, MS, 61, PhD(zool), 63. *Prof Exp:* Asst zool, Univ Tenn, 59-61, instr, Exten Sch, 62-63; asst prof biol, Charlotte Col, 63-65 & Longwood Col, 65-70; assoc prof, 70-76, PROF SCI EDUC, ECAROLINA UNIV, 76- *Concurrent Pos:* Consult, NSF In-Serv Coop Prog High Sch Biol, 64-65 & Sch Sci Inst Coop Col, 69-70, dir, NSF Inst, 69-70 & NSF Implementation Proj, 74-75, partic, NSF Leadership Confs, 72 & 74; dir, Title I Community Serv grant, 74-77, Title IV-C ESEA grant, 79-82; NC State Dept Pub Instr, Summer Sci Inst Mid Sch Teachers, 83-85. *Honors & Awards:* Gustaf-Ohaus Award, Nat Sci Teachers Asn, 74. *Mem:* Sigma Xi; Nat Sci Teachers Asn; Am Inst Biol Sci; Am Asn Univ Professors. *Res:* Environmental education; teaching materials; field equipment; biological education; marine education. *Mailing Add:* Dept Sci Educ ECarolina Univ Greenville NC 27834

HAMPTON, CHARLES ROBERT, b Weeksville, NC, July 22, 45; m 68; c 3. MATHEMATICS. *Educ:* Univ Mich, BS, 67; Univ Wis, MA, 68, PhD(math), 72. *Prof Exp:* Teaching asst math, Univ Wis, 67-72; asst prof, 72-79, assoc prof, 79-87, PROF MATH, COL WOOSTER, 87- *Concurrent Pos:* Fulbright-Hays lectr, Liberia, West Africa, 77-78; vis scholar, Cambridge Univ, Eng, 84-85. *Mem:* Am Math Soc; Math Asn Am; Asn Christians in Math Sci. *Res:* Structure theory of group rings; philosophy of mathematics, particularly epistemological questions. *Mailing Add:* Dept Math Col Wooster Wooster OH 44691

HAMPTON, DAVID CLARK, b Mason City, Iowa, Nov 25, 34; m 57; c 2. ORGANIC CHEMISTRY, BIOCHEMISTRY. *Educ:* St Olaf Col, BA, 56; Univ NDak, MS, 58; Purdue Univ, PhD(org chem), 63. *Prof Exp:* Instr chem, Luther Col, Iowa, 61-62; from asst prof to assoc prof, 62-72, PROF CHEM, WARTBURG COL, 72-, CHMN DEPT, 66- *Concurrent Pos:* Consult, North Cent Asn; researcher, Ore State Univ, 64, 65, Ind Univ, 66, NASA-Ames Res Labs, 70, 71; prin investr, NASA, sr res grant, 72-73; chair, Iowa sect, Am Chem Soc. *Mem:* AAAS; Am Chem Soc; Royal Soc Chem; Sigma Xi; Nat Asn Adv Health Professions. *Res:* Organic reaction mechanisms; organosulfur, organic oxidation, organometallic and hydrocarbon chemistry; computers in chemical education. *Mailing Add:* Dept Chem Wartburg Col Waverly IA 50677

HAMPTON, DELON, b Jefferson, Tex, Aug 23, 33. CIVIL ENGINEERING. *Educ:* Univ Ill, BSCE, 54; Purdue Univ, MSCE, 58, PhD(civil eng), 61. *Prof Exp:* Instr civil eng, Prairie View Agr & Mech Col, 54-55; asst prof, Univ Kans, 61-64; sr res engr, IIT Res Inst, 64-68; prof civil eng, Howard Univ, 68-85; PRES, DELON HAMPTON & ASSOCS, 73- *Concurrent Pos:* Actg head soil mech res, E H Wang civil eng res facil, Univ NMex, 62-63; pvt consult, 68-70; pres, Gnaedinger Baker Hampton & Assocs, 70-74; mem hwy res bd, Nat Acad Sci-Nat Res Coun; mem, US Nat Comt for tunneling technol, 76-78. *Mem:* Am Soc Eng Educ; Am Soc Civil Engrs; Am Soc Testing & Mat; Int Soc Soil Mech & Found Engrs; Am Pub Transit Asn; Am Consult Engrs Coun. *Res:* Soil dynamics; stress wave propagation; dynamic properties of soils; foundation vibrations; soil properties; tunneling; pavement design. *Mailing Add:* Delon Hampton & Assocs 800 K St NW Suite 720 N Lobby Washington DC 20001

HAMPTON, JAMES C, anatomy, cytology, for more information see previous edition

HAMPTON, JAMES WILBURN, b Durant, Okla, Sept 15, 31; m 58; c 4. HEMATOLOGY, ONCOLOGY. *Educ:* Univ Okla, BA, 52, MD, 56. *Prof Exp:* From clin asst to assoc prof, 60-71, head hemat oncol, Col Med, 72-77, prof med, Sch Med, 71-77, head, Hemat Sect, 69-72, head, Hemat Lab, Okla Med Res Found, 71-77, dir med oncol, 77-85, CLIN PROF MED, SCH MED, UNIV OKLA, 77-, MED DIR CANCER CTR, SOUTHWEST, BAPTIST MED CTR, 85- *Concurrent Pos:* NIH career develop award, 66-76; Angiol Res Found honors achievement award, 67-68; attend, Vet Admin Hosp, Oklahoma City, 63-; consult, Tinker AFB Hosp, 65- *Mem:* Asn Am Pathologists; Cent Soc Clin Res; Am Soc Hemat; Am Physiol Soc; Am Psychosom Soc; Am Soc Clin Oncol; Sigma Xi; fel Am Col Physicians. *Res:* Physiology and pathophysiology of hemostasis, thrombosis, leukemia, ,multiple myeloma and paraneoplastic syndromes; epidemiology of cancer in native America. *Mailing Add:* Dept Med Univ Okla Col Med 3435 NW 56th St Suite 200 Oklahoma City OK 73112

HAMPTON, JOHN KYLE, JR, b Okalona, Miss, Nov 9, 23; m 44, 61; c 1. PHYSIOLOGY. *Educ:* Millsaps Col, BS, 47; Tulane Univ, PhD(physiol), 49. *Prof Exp:* From instr to prof physiol, Tulane Univ, 49-66; prof & mem, Univ Tex Dent Sci Inst & prof, Univ Tex Grad Sch Biomed Sci, Houston, 66-73; prof biol & chmn dept, Adelphi Univ, 73-76; head dept, 76-83, PROF BIOL SCI DEPT, CALIF POLYTECH STATE UNIV, SAN LUIS OBISPO, CALIF, 76- *Concurrent Pos:* Markle scholar, 51-56. *Mem:* AAAS; Soc Exp Biol & Med; Am Physiol Soc; Geront Soc Am. *Res:* Primatology; reproductive biology; gerontology. *Mailing Add:* 20213 Rimrock Rd Monroe WA 98272

HAMPTON, LOYD DONALD, b Santa Anna, Tex, Aug 23, 30; m 58; c 2. PHYSICS, OCEANOGRAPHY. *Educ:* Univ Tex, BS, 52, MA, 59; Tex A&M Univ, PhD(phys oceanog), 67. *Prof Exp:* Res physicist, 51-65, head, Signal Physics Br, 65-70, asst dir, 70-75, assoc dir, 75-80, DIR, APPL RES LAB, UNIV TEX, AUSTIN, 80- *Mem:* Sigma Xi; fel Acoust Soc Am. *Res:* Acoustics; underwater sound; physical oceanography. *Mailing Add:* 10817 RR 2222 No 2 Austin TX 78730

HAMPTON, RAYMOND EARL, b Sherman, Tex, Jan 17, 34; m 57; c 2. PLANT PATHOLOGY. *Educ:* Midwestern Univ, BS, 55; Kans State Univ, MS, 57; Univ Wis, PhD(plant path), 60. *Prof Exp:* From asst prof to assoc prof plant path, Univ Ky, 60-65; PROF BIOL, CENT MICH UNIV, 70- *Mem:* AAAS; Am Soc Plant Physiol; Am Inst Biol Sci. *Res:* Plant physiology; lichen ecology and physiology. *Mailing Add:* Dept Biol Cent Mich Univ Mt Pleasant MI 48859

HAMPTON, RICHARD OWEN, b Dalhart, Tex, Feb 17, 30; m 54; c 2. PLANT PATHOLOGY. *Educ:* Univ Ark, BSAgr, 51; Iowa State Univ, MS, 54, PhD(plant path), 57. *Prof Exp:* Teaching asst, Iowa State Univ, 52-54, res asst, 54-57; asst plant pathologist, Irrig Agr Res & Exten Ctr, Wash State Univ, 57-61, res plant pathologist, Irrig Agr Res & Exten Ctr, Agr Res Serv, USDA, Wash, 61-65; DEPT BOT & PLANT PATH, ORE STATE UNIV, AGR RES SERV, USDA, 65- *Concurrent Pos:* Asst to Third World, plant path, virol. *Mem:* Am Phytopath Soc; Int Soc Plant Pathologists. *Res:* Virus-induced diseases of food legumes; virology of legume viruses. *Mailing Add:* Dept Bot & Plant Path Ore State Univ Corvallis OR 97331

HAMPTON, SUZANNE HARVEY, b Ogdensburg, NY, Oct 27, 34; m 61. DEVELOPMENTAL BIOLOGY, GENETICS. *Educ:* Drew Univ, BA, 56; Tulane Univ, MS, 59; Univ Tex, Houston, PhD(biomed sci), 70. *Prof Exp:* Res assoc physiol, Tulane Univ, 59-66; res asst, Univ Tex Dent Sci Inst Houston, 66-70, asst mem, 70-72; asst prof biol, York Col, 72-75, assoc prof, 75-77; ASST PROF, BARNARD COL, 80- *Concurrent Pos:* Prin investr, Dept Health, Educ & Welfare grant, 71-77. *Mem:* AAAS; NY Acad Sci; Int Primatol Soc; Soc Study Reproduction. *Res:* Developmental biology of marmosets; sex determination; gametogenesis; husbandry of primates; reproductive biology. *Mailing Add:* 47 Tulip Ave Ringwood NJ 07456

HAMRE, HAROLD THOMAS, b Wautoma, Wis, Jan 1, 10; m 52; c 1. BIOLOGY. *Educ:* Univ Wis, BS, 33, MS, 34; Ohio State Univ, PhD(physiol), 61. *Prof Exp:* Instr zool, Wittenburg Col, 34-35; instr biol, Shenandoah Col, 35-38; from instr to asst prof biol, Bowling Green State Univ, 46-61, assoc prof, 61-79; RETIRED. *Mem:* AAAS. *Res:* Physiology; circulation. *Mailing Add:* 227 Biddle Bowling Green OH 43402-3238

HAMRE, MELVIN L, b Tacoma, Wash, Nov 8, 32. FOOD SCIENCE, POULTRY SCIENCE. *Educ:* Wash State Univ, BS, 54, MEd, 60; Purdue Univ, MS, 63, PhD(food technol), 66. *Prof Exp:* Asst poultry sci, Purdue Univ, 60-65; from asst prof to assoc prof, 65-73, PROF POULTRY, DEPT ANIMAL SCI, UNIV MINN, ST PAUL, 73-, EXTEN SPECIALIST, 65- *Mem:* Poultry Sci Asn; Inst Food Technologists. *Res:* Maintenance of shell egg quality; poultry management and nutrition. *Mailing Add:* Dept Animal Sci Univ Minn St Paul MN 55108

HAMRICK, ANNA KATHERINE BARR, b Atlanta, Ga, Aug 29, 47; m 69; c 1. MATHEMATICS, MATHEMATICS EDUCATION. *Educ:* Univ Ga, BSEd, 69, MEd, 74, EdD(math), 76. *Prof Exp:* Res asst math, Res & Develop Ctr, Univ Ga, 69-70; teacher math, US Army Dependent Sch Syst, 71-72, Big Bend Community Col, 72-73 & Continuing Educ, Univ Ga, 75-76, teaching asst math educ, 73-76; ASST PROF MATH, AUGUSTA COL, 76- *Concurrent Pos:* Curric evaluator, Lincoln County Sch Syst, 77- *Mem:* Nat Coun Teachers Math. *Res:* How computational skills contribute to the meaningful learning of arithmetic; beginning programming languages. *Mailing Add:* Dept Math & Comput Sci Augusta Coll 2500 Walton Way Augusta GA 30910

HAMRICK, JAMES LEWIS, b Hopewell, Va, Feb 26, 42; m 74; c 1. POPULATION BIOLOGY. *Educ:* NC State Univ, BS, 64; Univ Calif, Berkeley, MS, 66, PhD(genetics), 70. *Prof Exp:* Fel genetics, Univ Calif, Davis, 70-71; asst prof bot, Univ Kans, 71-75, assoc prof, 75-79, prof bot & syst & ecol, 79-85; PROF BOT & GENETICS, UNIV GA, 85- *Mem:* Genetics Soc Am; Soc Study Evolution; Ecol Soc Am; Am Genetic Asn; Am Soc Naturalists. *Res:* Population genetics and ecology of natural plant populations. *Mailing Add:* Dept Bot Univ Ga Athens GA 30602

HAMRICK, JOSEPH THOMAS, b Meridian, Miss, Oct 4, 33; m 54; c 3. PUBLIC HEALTH, PREVENTIVE MEDICINE. *Educ:* Univ Tenn, MD, 57; Tulane Univ, MPH, 62. *Prof Exp:* Dir, Copiah Simpson County Health Dept, Miss, 59-61; res div county health work, Miss State Bd Health, 62-64, supvr tuberc control unit, 64-65, dir div tuberc control, 65-66; assoc prof pub health admin, 66-67 & health serv admin, 67-70, chmn dept, 68-71, PROF HEALTH SERV ADMIN, SCH MED, TULANE UNIV, 70-, PROF TROP MED & PUB HEALTH & COORDR COMMUNITY MED PROG, 71- *Concurrent Pos:* Vis prof, Sch Nursing, Univ Southern Miss, 67-; consult, USPHS Hosp, New Orleans, La, La State Dept Hosps, Health Educ Authority, La, La Regional Med Prog, New Orleans City Health Dept & Sch Nursing, Univ Southern Miss; dir, City Health Dept, New Orleans, 77-78 & Local Health Serv, Dept Health & Human Resources, Off Health Serv & Environ Qual, State of La, 78- *Mem:* AMA; Am Pub Health Asn. *Res:* Community day care cancer clinic; promoting quality care in nursing homes. *Mailing Add:* Dept Community Med Sch Med Tulane Univ 1430 Tulane Ave New Orleans LA 70112

HAMRICK, JOSEPH THOMAS, b Carrollton, Ga, Mar 20, 21; m 48; c 3. POWER GENERATING SYSTEMS, BIOMASS PROCESSING. *Educ:* Ga Inst Technol, BME, 46, MSME, 48. *Prof Exp:* Aeronaut res scientist, Nat Adv Comt Aeronaut, 55-84; PRES, AEROSPACE RES CORP, 61- *Concurrent Pos:* Chief res engr, Thompson Prod, 55-61. *Honors & Awards:* Tech Achievement Award, US Dept Energy, 84. *Mem:* Am Soc Mech Engrs. *Res:* Compressor and pump; biomass fueled gas turbine. *Mailing Add:* 4353 Windy Gap Dr Roanoke VA 24014

HAMRIN, CHARLES E(DWARD), JR, b Chicago, Ill, Jan 12, 34; m 56; c 3. CHEMICAL ENGINEERING. *Educ:* Northwestern Univ, BS, 56, MS, 57, PhD(chem eng), 64. *Prof Exp:* Develop specialist nuclear div, Y-12 Plant, Union Carbide Corp, 60-66; asst prof chem eng & res engr, Denver Res Inst, Univ Denver, 66-68; assoc prof, 68-78, actg chair, 88-89, PROF CHEM ENG, UNIV KY, 78-, CHAIR, 89- *Concurrent Pos:* Consult, Rocky Flats Plant Dow Chem Co, 67-68; vis prof, Chalmers Univ Technol, Sweden, 79-80. *Mem:* Am Inst Chem Engrs. *Res:* Chemical vapor deposition; thermodynamic and transport properties; fluidized beds; biomedical and enzyme engineering; wastewater treatment; catalysis; materials processing. *Mailing Add:* Dept Chem Eng Univ Ken Lexington KY 40506-0046

HAMRUM, CHARLES LOWELL, b Franklin, Minn, Oct 2, 22; m 44; c 2. ENTOMOLOGY. *Educ:* Gustavus Adolphus Col, BA, 47; Pa State Univ, MS, 49; Iowa State Col, PhD, 57. *Prof Exp:* Assoc prof, 49-59, PROF BIOL, GUSTAVUS ADOLPHUS COL, 59- *Mem:* Am Ornith Union; Entom Soc Am; Sigma Xi. *Res:* Chemical senses of birds; physiology of Diptera; taxonomy of Diptera. *Mailing Add:* Dept Biol Gustavus Adolphus Col St Peter MN 56082

HAMSA, WILLIAM RUDOLPH, b Stanton, Nebr, Dec 18, 03; m 31; c 2. ORTHOPEDIC SURGERY. *Educ:* Univ Nebr, BSc, 27, MD, 29; Am Bd Orthop Surg, dipl, 37. *Prof Exp:* Asst orthop surg, Univ Iowa, 33-36, assoc, 36-37; from instr to assoc prof, 37-49, chmn dept, 49-69, prof orthop surg, Univ Nebr Med Ctr, Omaha, 49-65, ORTHOP SURG, CACLARKSON HOSP, 65- *Concurrent Pos:* Attend orthop surgeon, Nebr Orthop Hosp, Lincoln; consult, Vet Admin Hosp, Omaha. *Mem:* Clin Orthop Soc; AMA; Am Col Surg; Am Acad Orthop Surg. *Res:* Reconstructive orthopedic surgery, especially related to crippled children. *Mailing Add:* 4239 Farnam St Omaha NE 68131

HAMSHER, JAMES J, b Nappanee, Ind, Oct 7, 40. BIO-ORGANIC CHEMISTRY. *Educ:* Wabash Col, BA, 63; Southern Ill Univ, Carbondale, MA, 65; Purdue Univ, PhD(org chem), 69. *Prof Exp:* Res chemist, 69-72, proj leader, 72-75, mgr, 75-77, ASST DIR, PFIZER, INC, 77- *Mem:* Am Chem Soc. *Res:* Enzymology; use of enzyme transformations in organic synthesis. *Mailing Add:* Pfizer Inc 235 E 42nd St New York NY 10017

HAMSHER, KERRY DE SANDOZ, b Long Br, NJ, Dec 10, 46; div. NEUROPSYCHOLOGY. *Educ:* Trinity Col, BS, 70; Univ Iowa, MA, 74 & PhD(psychol), 77. *Prof Exp:* Asst res scientist neurol, Univ Iowa, 77-79; asst prof, 79-84, ASSOC PROF NEUROL, UNIV WIS MED SCH, 84- *Concurrent Pos:* Mem, US Dept Educ Nat Invitational Conf Traumatic Brain Injury Res, 87, Vet Admin Merit Review Bd Neurobiol, 87-89, res group dementias, World Fed Neurol, 85-; chmn dict comt, Int Neuropsychol Soc, 87- *Mem:* Fel Am Psychol Asn; Sigma Xi; Am Acad Neurol; Am Psychol Asn; Int Neuropsychol Soc; Royal Soc Med; World Fed Neurol. *Res:* Aging and brain disease; attention, audition, memory sequencing, spatial and facial perception and stereopsis; development of neuropsychological asessment instruments; language disorders; neuroaratomic basis of confusion; objective research criteria for neurobehavioral syndromes. *Mailing Add:* PO Box 342 950 N 12th St Milwaukee WI 53201-0342

HAMSON, ALVIN RUSSELL, b Lava Hot Springs, Idaho, Sept 11, 24; m 46; c 4. VEGETABLE CROPS. *Educ:* Utah State Univ, BS, 48; Cornell Univ, PhD(veg crops), 52. *Prof Exp:* Asst veg crops, Cornell Univ, 48-52; from asst prof to assoc prof, Utah State Univ, 52-55, horticulturist, 55-58, prof hort, 59-90; RETIRED. *Concurrent Pos:* Actg head dept hort, Utah State Univ, 58-60 & 65. *Mem:* Am Soc Hort Sci; Weed Sci Soc Am. *Res:* Cultural research on vegetable crops; plant breeding of vegetable crops; varietal testing of vegetable crops. *Mailing Add:* 1780 N 1200 E Logan UT 84321

HAMSTROM, MARY-ELIZABETH, b Pittsburgh, Pa, May 24, 27. PURE MATHEMATICS. *Educ:* Univ Pa, AB, 48; Univ Tex, PhD(pure math), 52. *Prof Exp:* Instr pure math, Univ Tex, 49-52; from asst prof to assoc prof math, Goucher Col, 52-57; assoc prof, 57-66, PROF MATH, UNIV ILL, URBANA, 66- *Concurrent Pos:* Mem, Inst Advan Study, 56-57. *Mem:* Am Math Soc. *Res:* Set theoretic topology; regular mappings; space homeomorphisms on manifolds. *Mailing Add:* Dept Math Univ Ill 1409 W Green St Urbana IL 61801

HAMTIL, CHARLES NORBERT, b St Louis, Mo, Apr 13, 13; m 40; c 2. PHYSICS. *Educ:* St Louis Univ, BS, 37, MS, 40, PhD(physics), 45. *Prof Exp:* Instr physics, St Louis Univ, 43-44; from instr to prof physics, 44-76, chmn, Div Natural Sci & Math, 66-75, EMER PROF PHYSICS, ROCKHURST COL, 76- *Mem:* Am Phys Soc; Sigma Xi. *Res:* Theoretical physics; classical mechanics. *Mailing Add:* 3223 Bee Ridge Rd Apt 79 Sarasota FL 34239

HAMZA, MOHAMED HAMED, b Heliopolis, Egypt, Oct 17, 36. AUTOMATIC CONTROL SYSTEMS, OPERATIONS RESEARCH. *Educ:* Mass Inst Technol, BSc, 58; Swiss Fed Inst Technol, Dr tech sci, 63; Univ Zurich, PhD(math), 66. *Prof Exp:* Res asst, Inst Opers Res & Electronic Data Processing, Univ Zurich, 64-68; assoc prof, 68-75, PROF ELEC ENG,

UNIV CALGARY, 75- *Concurrent Pos:* Consult, Univ Fribourg, 66-68; ed jour, Automatic Control Theory & Appln; pres, Icord Ltd; pres, Int Asn Sci & Technol Develop. *Mem:* Inst Elec & Electronics Engrs; Int Soc Min & Micro Computers. *Res:* Expert systems; adaptive, optimum, and stochastic control systems; mathematical programming; simulation. *Mailing Add:* Dept Elec Eng Univ Calgary 2500 Univ Dr NW Calgary AB T2N 1N4 Can

HAN, BYUNG JOON, b Seoul, Korea, Aug 15, 59; m 82; c 2. CATALYSIS, POLYMER RHEOLOGY. *Educ:* HanYang Univ, Seoul, Korea, BS, 82; Tenn Tech Univ, MS, 85; Columbia Univ, MPhil, 88, PhD(chem eng), 89. *Prof Exp:* Mem tech staff, T J Watson Res Ctr, IBM, 86-88; MEM TECH STAFF, AT&T BELL LABS, 88- *Mem:* Sigma Xi; Am Inst Chem Engrs; Mat Res Soc; Soc Plastics Engrs. *Res:* Polymer science; transitions in polymers; high temperature and performance polymer; polymers in electronic applications adhesion science; packaging; thermal science; heat transfer; numerical calculation chemical engineering. *Mailing Add:* AT&T Bell Labs Rm 1A-370 600 Mountain Ave Murray Hill NJ 07974-2070

HAN, CHANG DAE, b Seoul, Korea, Sept 28, 35; US citizen; m 62; c 3. CHEMICAL & ELECTRICAL ENGINEERING. *Educ:* Seoul Nat Univ, BS, 58; Mass Inst Technol, MS, 62, ScD(chem eng), 64; Newark Col Eng, MS, 68; NY Univ, MS, 71. *Prof Exp:* Process analyst, Am Cyanamid Co, 64-66; systs engr, Esso Res & Eng Co, 66-67; assoc prof chem eng, Polytech Inst Brooklyn, 67-72; PROF CHEM ENG, POLYTECH INST NY, 72-, HEAD DEPT, 74- *Mem:* Am Inst Chem Engrs; Am Chem Soc; Soc Rheology; Soc Plastics Engrs. *Res:* Rheology; mathematical modelling; applied mathematics; polymer processing. *Mailing Add:* Dept Chem Eng Polytech Inst NY 333 Jay St Brooklyn NY 11201

HAN, CHARLES CHIH-CHAO, b Szuchuan, China, Jan 18, 44; m 70; c 3. POLYMER CHEMISTRY. *Educ:* Univ Houston, MS, 69; Univ Wis, PhD(polymer chem), 74. *Prof Exp:* RES CHEMIST, NAT BUR STANDARDS, 74- *Mem:* Am Chem Soc; Am Phys Soc. *Res:* Neutron scattering and quasielastic light scattering of polymer solutions; polymer characterization; block copolymers. *Mailing Add:* Nat Bur Standards 440 Gaithersburg MD 20899

HAN, CHIEN-PAI, b Hunan, China, Dec 17, 36; m 66; c 2. STATISTICS. *Educ:* Nat Taiwan Univ, BA, 58; Univ Minn, MA, 62; Harvard Univ, PhD(statist), 67. *Prof Exp:* From asst prof to prof statist, Iowa State Univ, 67-82; PROF MATH, UNIV TEX, ARLINGTON, 82- *Concurrent Pos:* Vis asst prof statist, Harvard Univ, 70. *Mem:* Am Statist Asn; Inst Math Statist. *Res:* Multivariate analysis; sample survey. *Mailing Add:* Dept Math Univ Tex Arlington TX 76019

HAN, CHOONG YOL, organic chemistry, for more information see previous edition

HAN, JAOK, b Chinnampo, Korea, July 16, 30; US citizen; m 61; c 3. CARDIOLOGY, PHYSIOLOGY. *Educ:* Kyung-Pook Nat Univ, Korea, MD, 51; State Univ NY, PhD(physiol), 61. *Prof Exp:* Intern med, Jersey City Med Ctr, NJ, 55-56; med resident, Mercy Hosp, Pittsburgh, 56-57; res assoc cardiol, Masonic Med Res Lab, NY, 60-66; fromn asst prof to assoc prof, 68-73, PROF MED & DIR ELECTROCARDIOGRAPHY, ALBANY MED COL, 73- *Concurrent Pos:* Res fel physiol, State Univ NY Upstate Med Ctr, 57-60; Int Soc Cardiol Found fel, 60-61; Masonic Found Med Res & Human Welfare fel, 61-63; Am Heart Asn & NIH grants, 63-; cardiol fel, Univ Rochester Med Ctr, 66-67; attend physician & cardiologist, Albany Med Ctr Hosp, 68-; mem res comt, Am Heart Asn, NY State Affil, 69-72 & 76-79; mem comt, Sudden Cardiac Death, Nat Heart, Lung & Blood Inst, 71, Ischemic Heart Dis, 74, Beta Blocker Heart Attack Trial, 78, Cardiol Adv Comt, 81-85 & Cardiac Rhythm Studies, 85. *Mem:* Am Fedn Clin Res; Am Physiol Soc; fel Am Col Cardiol; fel Am Heart Asn (pres, Northeastern NY Chap, 80-84). *Res:* Clinical and research cardiology, especially electrophysiology of cardiac arrhythmias. *Mailing Add:* Dept Med Albany Med Col Albany NY 12208

HAN, KENNETH N, b Seoul, Korea, July 3, 38; US citizen; m 67; c 3. HYDROMETALLURGY, MINERAL BENEFICIATION. *Educ:* Seoul Nat Univ, BS, 61, MS, 63; Univ Ill, Urbana, MS, 67; Univ Calif, Berkeley, PhD(metall), 71. *Prof Exp:* Res asst beneficiation, Res Inst Mining & Metall, Seoul, 61-63; res assoc mining, Seoul Nat Univ, 63-65; lectr chem eng, Monash Univ, Melbourne, Australia, 71-74, sr lectr, 74-80; assoc prof, 81-84, PROF METALL, SDAK SCH MINES, 84-, HEAD, DEPT METALL ENG & DIR, INST MINING & METALL, 87- *Concurrent Pos:* Vis lectr, Univ Calif, Berkeley, 74, vis assoc prof, 79; res investr, Korean Inst Sci & Technol, 79-80; ed-in-chief, Mineral Processing & Extractive Metall Rev, 86-; dir, Epscor State SDak, 89-; consult, Inst Mineral & Energy Res, Taiwan, 89- *Mem:* Am Inst Mining Metall & Petrol Engrs (treas, 91). *Res:* Geochemistry and extraction of marine manganese nodules; fine particle recovery; leaching and cementation mechanisms; metal recovery from solutions; transport phenomena on metal ions in solutions; treatment of gold refractory ores; solubility of gases in solutions. *Mailing Add:* Dept Metall Eng SDak Sch Mines & Technol Rapid City SD 57701

HAN, KI SUP, b Seoul, Korea, Apr 7, 29; US citizen; m 57; c 2. ELECTROOPTICS. *Educ:* Seoul Nat Univ, Korea, BS, 57; Univ Uppsala, Sweden, Dipl, 62; Mich State Univ, MS, 64; Univ Mich, Ann Arbor, PhD(particle physics), 70. *Prof Exp:* Instr physics, Korea Univ, 59-61; res asst nuclear physics, Mich State Univ, 62-64; asst prof physics, Aquinas Col, 64-67; res asst particle physics, Univ Mich, Ann Arbor, 67-70, res assoc electrooptics, 70-72; asst group leader electrooptics, Los Alamos Nat Lab, 72-79; CHIEF SCIENTIST, SPACE TRANSP SYSTS DIV, ROCKWELL INT, 79- *Concurrent Pos:* Fel, dept nuclear med, Univ Mich, Ann Arbor, 70-72; fel, Int Atomic Energy Agency, 61-62. *Mem:* Am Phys Soc. *Res:* Infrared sensor design and infrared imaging system analysis. *Mailing Add:* 990 S Jay Circle Anaheim CA 92808

HAN, L(IT) S(IEN), b Shanghai, China, May 5, 23; m 53; c 2. MECHANICAL ENGINEERING. *Educ:* Chiao Tung Univ, BSc, 45; Ohio State Univ, MS, 48, PhD, 54. *Prof Exp:* Jr engr, Shanghai Rwy Admin, 45-47; asst res found, 47-50, res assoc, 50-54, from instr to assoc prof, 54-61, PROF ENG MECH, OHIO STATE UNIV, 61- *Concurrent Pos:* NSF sr fel & vis prof, Max Planck Inst Fluid Mech Res, Gottingen, 62-63; consult, Battelle Mem Inst, 63- & Flight Dynamics Lab, USAF, 69- *Mem:* Am Soc Mech Engrs. *Res:* Applied mechanics; thermodynamics; heat transfer. *Mailing Add:* Dept Mech Eng Ohio State Univ 1075 Robinson Lab 206 W 18th Ave Columbus OH 43210

HAN, MOO-YOUNG, b Seoul, Korea, Nov 30, 34; US citizen; m 59; c 3. PARTICLE PHYSICS, THEORETICAL PHYSICS. *Educ:* Carroll Col, Wis, BS, 57; Univ Rochester, PhD(physics), 63. *Prof Exp:* Res assoc physics, Boston Univ, 63-64; res assoc, Syracuse Univ, 64-65; asst prof, Univ Pittsburgh, 65-67; from asst prof to assoc prof, 67-77, PROF PHYSICS, DUKE UNIV, 77- *Concurrent Pos:* Distinguished Foreign Scholar, Kyoto Univ, 74; vis prof, Korea Advan Inst Sci, 82. *Mem:* Am Phys Soc. *Res:* Quantum electrodynamics; symmetries of strong interactions; quark models; quantum chromodynamics; auth of one book. *Mailing Add:* Dept Physics Duke Univ Durham NC 27706

HAN, SEONG S, b Seoul, Korea, May 12, 33; m 60; c 3. CELL BIOLOGY, SCIENCE EDUCATION. *Educ:* Seoul Nat Univ, DDS, 56; Univ Mich, PhD(anat), 61. *Prof Exp:* Res assoc anat,, Univ Mich, 60-61, from asst prof to assoc prof dent, 61-68, prof dent, Sch Dent, 68-87, prof anat, Med Sch, 70-87, EMER PROF, DEPT ANAT & CELL BIOL, SCH MED, UNIV MICH, ANN ARBOR, 87-, EMER PROF, SCH DENT, 87-; CHMN BD, YURIN SOC WORLD-WIDE, INC, ANN ARBOR, 88- *Concurrent Pos:* Consult res, Vet Admin Hosp, 68-87; mem, Spec Study Sect, NIH, 71-87; consult, Khmer Repub, 71 & Ministry Sci & Technol, Repub Korea, 73; mem fac, Interdept Prog Cell & Molecular Biol, Univ Mich, Ann Arbor, 71-87, chmn, Grad Prog Anat, 72-74 & Reproductive Endocrinol Prog, 74-87, dir, Interdisciplinary Prog Biol Aging, 75-87; res scientist dir, Prog Biol Aging, 80-87; vpres & chancellor, Med Ctr, Dong-A Univ, Korea, 87-89, prof anat, 87-; dean Med Sch, Life Sci Res Inst, Dong-A Univ, 87-89, dir, 87- *Mem:* Fel Am Asn Anatomists; Am Soc Cell Biol; Geront Soc; Soc Exp Biol & Med; Int Asn Dent Res. *Res:* Cellular mechanism of exocrine and endocrine secretion; membrane receptors to protein hormones; molecular biology of membrane aging; development of educational technology. *Mailing Add:* 1225 Lincolnshire Ann Arbor MI 48103

HAN, SHU-TANG, b Ningpo, China, Jan 10, 13; nat US; m 66. CHEMICAL ENGINEERING. *Educ:* Nat Tsinghua Univ, China, BS, 37; Univ Maine, MS, 47; Lawrence Univ, MS, 78. *Prof Exp:* From tech asst to tech assoc chem eng, Inst Paper Chem, Lawrence Col, 49-52, from res asst to res assoc, 52-57; consult chemist, Beloit Iron Works, 57-58, from res assoc to sr res assoc chem eng, Inst Paper Chem, Lawrence Univ, 58-70, chmn dept, 66-70; assoc dir res, Beloit Corp, 70-76; mem fac, 76-77, EMER MEM FAC PULP & PAPER RES, INST PAPER CHEM, 78- *Concurrent Pos:* Vis prof, Nat Taiwan Univ, 66. *Mem:* Fel Tech Asn Pulp & Paper Indust. *Res:* Transport phenomena in paper technology. *Mailing Add:* 1699 Alcan Dr Menasha WI 54952

HAN, TIN, b Mergui, Burma, May 23, 33; US citizen. IMMUNOLOGY, HEMATOLOGY. *Educ:* Rangoon Univ, ISc, 53, MD, 58. *Prof Exp:* Intern, Worcester City Hosp, Mass, 60-61; first yr resident internal med, Union Mem Hosp, Baltimore, Md, 61-62; second yr resident, Roswell Park Mem Inst, Buffalo, NY, 62-63; third yr resident, French Hosp, New York, 63-64; sr cancer res clinician, Roswell Park Mem Inst, 64-72, cancer res clinician II, 72-74; res instr, 68-70, asst res prof, 70-75, ASSOC RES PROF MED, STATE UNIV NY BUFFALO, 75-; ASSOC CHIEF CANCER RES CLINICIAN, ROSWELL PARK MEM INST, 74- *Concurrent Pos:* Nat Cancer Inst grants, 72-78 & 73-76. *Mem:* AAAS; Am Asn Cancer Res; Am Soc Clin Oncol; Am Soc Hemat; Am Asn Immunologists. *Res:* Basic research in T lymphocytes, B lymphocytes and macrophages participation in cell-mediated immunity; immunological studies in patients with cancer; immunotherapy of cancer. *Mailing Add:* Dept Med Oncol Roswell Park Mem Inst 666 Elm St Buffalo NY 14263

HAN, YURI WHA-YUL, b Seoul, Korea, Oct 15, 32; US citizen; m 55; c 1. ORGANIC CHEMISTRY, POLYMER CHEMISTRY. *Educ:* Austin Col, BA, 55; Univ Tex, Austin, PhD(chem), 60. *Prof Exp:* Res fel chem, Univ Tex, 60-61; res assoc, Univ Southern Calif, 61-62; asst prof, Mt St Mary's Col, Calif, 61-64; sr res engr, Rye Canyon Res Lab, Lockheed-Calif Co, 64-68; PROF CHEM, EAST LOS ANGELES COL, 69-, CHMN DEPT, 74- *Mem:* Sigma Xi; Am Chem Soc. *Res:* Rearrangements of alkylaryl hydrocarbons; aromatic-heterocyclic polymer chemistry. *Mailing Add:* Dept Chem East Los Angeles Col 1301 Brooklyn Ave Monterey Park CA 91754

HANAFEE, JAMES EUGENE, b Chicago, Ill, Jan 8, 37; m 62; c 2. METALLURGY. *Educ:* Univ Ill, Urbana-Champaign, BS, 58, MS, 60; Case Western Reserve Univ, PhD(metall), 66. *Prof Exp:* Res metallurgist, Int Nickel Co, 60-63; sr res metallurgist, Franklin Inst, Pa, 66-71; PROJ MGR, LAWRENCE LIVERMORE LAB, 71- *Concurrent Pos:* Contrib ed, PCM-PCE, 71-73. *Mem:* Am Soc Metals. *Res:* Mechanical properties; fabrication; beryllium; aluminum; failure analysis. *Mailing Add:* Lawrence Livermore Lab Univ Calif Livermore CA 94550

HANAFEE, WILLIAM NORMAN, b Louisville, Ky, Mar 21, 26. RADIOLOGY. *Educ:* Univ Rochester, BA, 46; Univ Louisville, MD, 49. *Prof Exp:* From asst prof to assoc prof, 53-66, chmn dept, 67-72, PROF RADIOL, UNIV CALIF, LOS ANGELES, 66- *Mem:* Am Soc Head & Neck Radiol (secy-treas); Am Col Radiol. *Res:* Head and neck radiology. *Mailing Add:* Ctr Health Sci Univ Calif Sch Med Los Angeles CA 90024

HANAFUSA, HIDESABURO, b Nishinomiya, Japan, Dec 1, 29; m 58; c 1. VIRAL ONCOLOGY. *Educ:* Osaka Univ, BS, 53, PhD(biochem), 60. *Prof Exp:* Res assoc, Res Inst Microbial Dis, Osaka Univ, 58-61; fel, Univ Calif, Berkeley, 61-64; vis scientist, Col France, Paris, 64-66; assoc mem & chief dept viral oncol, Pub Health Res Inst of City of New York, Inc, 66-68, mem, 68-73; PROF, ROCKEFELLER UNIV, 73- *Honors & Awards:* Howard Taylor Ricketts Award, Univ Chicago, 81; Albert Lasker Basic Med Res Award, Lasker Found, 82; Asahi Prize, Asahi Press, 84; Clowes Mem Award, Am Asn Cancer Res, 86. *Mem:* Nat Acad Sci; Am Soc Microbiol; Am Soc Biol Chemists; Am Asn Cancer Res; NY Acad Sci; Am Soc Virol; Am Soc Cell Biol. *Res:* RNA tumor viruses; mechanism of cell transformation; function of viral and cellular oncogenes. *Mailing Add:* Rockefeller Univ 1230 York Ave New York NY 10021

HANAHAN, DONALD JAMES, b Springfield, Ill, May 13, 19; m 47; c 5. BIOCHEMISTRY. *Educ:* Univ Ill, BS, 41, PhD, 44. *Prof Exp:* Res assoc, Manhattan Proj, Univ Chicago, 44; res assoc, E I du Pont de Nemours & Co, 45; res assoc physiol, Univ Calif, 45-48; instr chem, Univ Wash, 48-49, asst prof, 49-50, from asst prof to prof biochem, 50-60; prof & chmn dept biochem, Col Med, Univ Ariz, 67-75; chmn dept biochem, 76-83, PROF BIOCHEM, UNIV TEX HEALTH SCI CTR, SAN ANTONIO, 76- *Concurrent Pos:* Guggenheim fel, 55; NIH spec fel, 65-66; Macy fac scholar, 74. *Mem:* Am Soc Biol Chemists; Am Chem Soc. *Res:* Simple and complex lipids; lipolytic action, lipid chemical mediators. *Mailing Add:* Dept Biochem Univ Tex Health Sci Ctr 7703 Floyd Curl Dr San Antonio TX 78284-7760

HANAK, JOSEPH J, b Tarrytown, NY, Mar 21, 30; m 55; c 6. SOLID STATE SCIENCE. *Educ:* Manhattan Col, BS, 53; Univ Detroit, MS, 56; Iowa State Univ, PhD(phys chem), 59. *Prof Exp:* Asst chem, Univ Detroit, 53-55; chemist, Ethyl Corp, Mich, 55; asst rare earth res, Inst Atomic Res, Iowa State Univ, 55-59; mem tech staff mat res, Radio Corp Am, 59-71, FEL, TECH STAFF, RCA LABS, 71- *Honors & Awards:* David Sarnoff Award & John A Roebling Award, Am Soc Metals, 65; Best Sem Award, Soc Info Display, 75. *Mem:* Am Vacuum Soc; Soc Info Display; Sigma Xi. *Res:* Chemistry and metallurgy of rare earth elements; superconducting materials research; chemical vapor deposition of niobium-tin used in construction of high field superconducting solenoids; radio frequency co-sputtering of multicomponent systems; microwave; acoustic delay lines; electroluminescence; magnetic recording heads; video disc development; photovoltaic amorphous silicon solar cells. *Mailing Add:* PO Box 1459 Ames IA 50010

HANAN, BARRY BENTON, b Morgantown, WVa, Mar 10, 49; m 78; c 2. HEAVY ISOTOPE GEOCHEMISTRY. *Educ:* Univ Kans, BS, 73; Va Polytech Inst & State Univ, MS, 77, PhD(geol), 80. *Prof Exp:* Fel geol & geochem, Univ Calif, Santa Barbara, 80-; AT GRAD SCH OCEANOG, UNIV RI. *Mem:* Am Geophys Union; Geol Soc Am; Mineral Soc Am; AAAS; Geochem Soc; Sigma Xi. *Res:* Investigation of the early history of the solar system by experimentation with the uranium, thorium, lead and rubidium strontium isotopic systems in meteorites and lunar rocks. *Mailing Add:* 9121 Sinsonte Lane Lakeside CA 92040

HANAN, JOE JOHN, b Buffalo, NY, Jan 22, 31; m 56; c 3. HORTICULTURE, PLANT PHYSIOLOGY. *Educ:* Univ Mo, BS, 52; Colo State Univ, MS, 59; Cornell Univ, PhD(hort), 63. *Prof Exp:* Res asst hort, Colo State Univ, 57-59; res asst floricult, Cornell Univ, 59-62, res assoc, 62-63; asst horticulturist, 63-64, from asst prof to assoc prof hort, 65-71, leader floricult invests, 74-82, prof hort, 72-89, leader & bulletin ed, 82-89, EMER PROF HORT, COLO STATE UNIV, 89- *Honors & Awards:* Alex Laurie Award, Am Soc Hort Sci, 63, 82, Kenneth Post Award, 65 and 76. *Mem:* Fel Am Soc Hort Sci; Int Hort Soc. *Res:* Effects of temperature on plant growth and development; oxygen diffusion in soils; storage of horticultural products; air pollution; environmental control and measurement; carbon dioxide utilization by plants; water relationships of plants; greenhouse management and nutrition; computer control of plant environments. *Mailing Add:* Dept Hort Colo State Univ Ft Collins CO 80523

HANAU, RICHARD, b New York, NY, Aug 1, 17; m 41; c 1. OPTICS. *Educ:* Mass Inst Technol, SB, 39; Univ Mich, MS, 40, PhD(physics), 47. *Prof Exp:* Res assoc eng res, Univ Mich, 44-46; assoc prof, Univ Ky, 47-60, prof physics, 60-83; RETIRED. *Concurrent Pos:* Vis prof, Univ PR, 53-54; prof, Univ Indonesia, 56-58; asst, Univ Rochester, 58-59; vis scholar, Opt Sci Ctr, Univ Ariz, 73-74. *Mem:* Optical Soc Am; Am Phys Soc. *Res:* Visible and ultraviolet spectrochemical analysis; geometrical optics and lens design. *Mailing Add:* RR 4-46 Stage Coach Rd Patterson NY 12563

HANAUER, RICHARD, b Brooklyn, NY, Feb 10, 43; m 68; c 2. ANALYTICAL CHEMISTRY, BIOCHEMISTRY. *Educ:* Columbia Univ, BS, 63; Univ Wis, PhD(org chem), 68. *Prof Exp:* Sr chemist fibers, 68-70, sr chemist petrol additives, 70-80, sr chemist process res, Agr Chemicals & Biocides, 80-87, SR CHEMIST AGR RES RESIDUE/METAB/ENVIRON PATE ROHM AND HAAS CO, 87- *Mem:* Am Chem Soc. *Res:* Process research of agricultural chemicals and biocides. *Mailing Add:* 3094 Cloverly Dr Furlong PA 18925

HANAUER, STEPHEN B, b Chicago, Ill, Jan 1, 52; m 72; c 3. MEDICINE. *Educ:* Univ Mich, BS, 73; Univ Ill, 77. *Prof Exp:* Resident, internal med, 77-80, gastroenterol, 80-82, asst prof, 82-88, ASSOC PROF MED, UNIV CHICAGO, 88- *Concurrent Pos:* Am Col Physicians fel, 86; mem, Food & Drug Admin Gastrointestinal Adv Panel, 87-; Am Col Gastroenterol fel, 87. *Mem:* Am Col Physicians; Am Gastroenterol Asn; Am Col Gastroenterol; AMA. *Res:* Clinical research in gastroenterology with specific interest in management and new medical therapies of inflammatory bowel diseases. *Mailing Add:* Univ Chicago Hosp Box 400 5841 S Maryland Chicago IL 60637

HANAUER, STEPHEN H(ENRY), b New York, NY, Mar 6, 27; m 48, 69; c 6. NUCLEAR ENGINEERING. *Educ:* Purdue Univ, BSEE, 48, MSEE, 49; Univ Tenn, PhD, 60. *Prof Exp:* Instr elec eng, Purdue Univ, 48-50; sr physicist, Oak Ridge Nat Lab, 50-65; prof nuclear eng, Univ Tenn, Knoxville, 65-70; tech adv to dir regulation, USAEC, 70-78; asst dir Plant Systs, US Nuclear Regulatory Comn, 78-79, dir Unresolved Safety Issues, 79-80, dir, Div Human Factors, 80-81, dir, Div Safety Technol, 81-82; SR VPRES, TECH ANALYTICAL CORP, 82- *Concurrent Pos:* Deleg, Int Conf Peaceful Uses Atomic Energy, 55, 64, 71 & 77; mem tech subcomt reactor instrumentation, Int Electrotechnol Comn, 62-78; adv comt reactor safeguards, USAEC, 65-70, chmn, 69. *Res:* Instrumentation and control of nuclear reactors; nuclear and low-temperature physics; nuclear reactor safety; power plant construction and operation. *Mailing Add:* 6723 Whittier Ave Suite 202 McLean VA 22101

HANAWALT, PHILIP COURTLAND, b Akron, Ohio, Aug 25, 31; m 57, 78; c 4. PHOTOBIOLOGY, DNA REPAIR. *Educ:* Oberlin Col, BA, 54; Yale Univ, MS, 55, PhD(biophys), 59. *Prof Exp:* USPHS fel, Microbiol Inst, Univ Copenhagen, 58-60; Am Cancer Soc fel biophys, Calif Inst Technol, 60-61; res assoc & lectr, Biophys Lab, Stanford Univ, 61-65, assoc prof biol, 65-70, dir, Biophys Grad Prog, 68-85, chmn, Dept Biol Sci, 82-89, PROF BIOL, STANFORD UNIV, 70-, DIR, GRAD STUDIES BIOL, 90- *Concurrent Pos:* Postdoctoral fel, NIH & Am Cancer Soc, 58-61; mem, Physiol Chem Study Sect, NIH, 66-70, Chem Path Study Sect, 81-84, bd sci counselors, Nat Inst Environ Health Sci, 87-90, Outstanding Investr Res Award, Nat Cancer Inst, 87-; mem adv comt, Nucleic Acids & Protein Synthesis, Am Cancer Soc, 72-76; prog dir, Cell-Molecular Biol, Training Prog, Stanford Univ, 73-84, chmn, Admin Panel Radiol Hazards, 78-80; vis prof, Dept Molecular Biol, Univ Calif, Berkeley, 76; predoctoral fel rev panel, NSF, 85; outstanding investr res award, Nat Cancer Inst, 87-; counr, Environ Mutagen Soc, 88-91. *Honors & Awards:* Hans Falk Lectr, Nat Inst Environ Health Sci, 90. *Mem:* Nat Acad Sci; fel AAAS. *Res:* Molecular mechanism and control of repair and tolerance of damaged DNA (including: ultraviolet, psoralens and chemical carcinogens), in bacteria and mammalian cells; selective repair in defined genes. *Mailing Add:* Dept Biol Sci Stanford Univ Stanford CA 94305-5020

HANBAUER, INGEBORG, b Austria, July 30, 43; nat US. PHARMACOLOGY, NEUROSCIENCE. *Educ:* Univ Vienna, PhD, 69. *Prof Exp:* Sr res fel, NY State Res Inst, Ward's Island, 69-71; vis fel, Lab Clin Sci, NIMH, NIH, 71-74, sr staff fel, Sect Biochem Pharmacol, Hypertension-Endocrine Br, Nat Heart Lung & Blood Inst, 74-79, pharmacologist, 79-89, PHARMACOLOGIST, LAB CHEM PHARMACOL, NAT HEART LUNG & BLOOD INST, NIH, 89- *Concurrent Pos:* Vis assoc, Lab Clin Sci, NIMH, NIH, 74. *Mem:* Am Soc Neurochem; Soc Neurosci; Am Soc Pharmacol & Exp Therapeut. *Res:* Pharmacology; numerous publications. *Mailing Add:* Lab Chem Pharmacol Nat Heart Lung & Blood Inst NIH Bethesda MD 20892

HANBY, JOHN ESTES, JR, b Washington, DC, May 3, 41; m 68; c 2. PAPER CHEMISTRY. *Educ:* Ga Inst Technol, BChE, 63; Lawrence Univ, MS, 65, PhD(phys inorg chem), 68. *Prof Exp:* Res eng, James River Corp, 70-75, supvr prod develop, Cent Res Div, Camas, Wash, 75-76, tech mgr, 77-78, plant mgr, 78-80, mgr div mfg, 80-81, asst resident mgr, 81-89, VPRES TECH, COMMUN PAPERS, JAMES RIVER CORP, 89- *Mem:* Tech Asn Pulp & Paper Indust; Am Inst Chem Engrs. *Res:* Development of printing and business papers; market development of synthetic pulp; reprography. *Mailing Add:* James River Corp Camas Tech Ctr 349 NW Seventh Ave Camas WA 98607

HANCE, ANTHONY JAMES, b Bournemouth, Eng, Aug 19, 32; m 54; c 3. NEUROPHARMACOLOGY. *Educ:* Univ Birmingham, BSc, 53, PhD(neuropharmacol), 56. *Prof Exp:* Jr res pharmacologist, Univ Calif, Los Angeles, 59, asst res pharmacologist, 59-62; res assoc pharmacol, Stanford Univ, 62-65, asst prof, 65-68; ASSOC PROF PHARMACOL, SCH MED, UNIV CALIF, DAVIS, 68- *Concurrent Pos:* Res fel electrophysiol, Univ Birmingham, 57-58; consult pharmacologist, Riker Labs, Inc, Calif, 59-60; consult, Stanford Res Inst, 63-65 & Ampex Corp, 64-65; mem, Comt Psychopharmacol, NIMH, 65-69. *Mem:* AAAS; Am Soc Pharmacol & Exp Therapeut; Asn Comput Mach; Biomed Eng Soc. *Res:* Electrical activity of central nervous system of mammals and its modification by drugs and during learning. *Mailing Add:* Dept Pharmacol Univ Calif Sch Med Davis CA 95616-8654

HANCE, ROBERT LEE, b El Paso, Tex, Mar 1, 43; m 63; c 5. SURFACE CHEMISTRY, ELECTRON SPECTROSCOPY. *Educ:* Abilene Christian Univ, BS, 66; Mass Inst Technol, PhD(phys chem), 70. *Prof Exp:* From asst prof to prof chem, Abiline Christian Univ, 70-83; TECH STAFF, MOTOROLA, INC, 84- *Concurrent Pos:* Vis assoc prof chem, Univ Tex, 78-79, res assoc, 81; res grants, Robert A Welch Found, Res Corp & NSF. *Mem:* Am Chem Soc; Am Phys Soc; Am Vacuum Soc. *Res:* Interaction of atoms and molecules at catalytic surfaces investigated by electron spectroscopy; programmed thermal desorption techniques; semiconductor materials analysis and characterization. *Mailing Add:* 8805 El Rey Blvd Austin TX 78737

HANCHEY, RICHARD HOWARD, floriculture; deceased, see previous edition for last biography

HANCK, KENNETH WILLIAM, b Danvers, Ill, Dec 6, 42; m 87. ANALYTICAL CHEMISTRY. *Educ:* Ill State Univ, BS, 64; Univ Ill, MS, 66, PhD(chem), 69. *Prof Exp:* From asst prof to assoc prof, 69-78, PROF CHEM, NC STATE UNIV, 78-, HEAD DEPT, 84- *Concurrent Pos:* Sr Fulbright scholar, Australia, 80. *Mem:* Am Chem Soc. *Res:* Electrochemistry of transition metal complexes; development of analytical chemical techniques for the solution of specific chemical problems. *Mailing Add:* Dept Chem NC State Univ Box 8204 Raleigh NC 27695

HANCOCK, ANTHONY JOHN, b Smethwick, Eng, May 7, 36; m 75. BIOCHEMISTRY, ORGANIC CHEMISTRY. *Educ:* Univ Nottingham, BSc, 57, Dipl Ed, 58; Univ Ottawa, Can, PhD(biochem), 72. *Prof Exp:* Lectr chem, Univ Ottawa, 69-71; res assoc biochem, Case Western Reserve Univ, Sch Med, 72-75; from asst prof to assoc prof chem & med, Univ Mo, Kansas City, 75-83; res & develop proj mgr, 83-88, DIR, SPONSORED EXTERNAL RES, MARION LABS, INC, 88- *Concurrent Pos:* Grant, Mo Heart Asn, 76-77 & 77-78, NSF, 79-81; adj prof basic life sci, Univ Mo, Kansas City, 83- *Mem:* AAAS; Am Chem Soc; NY Acad Sci. *Res:* Chemical synthesis of lipid analogs; isolation and structure elucidation of natural lipids; physical and biochemical properties of lipids; membrane lipid-protein interaction. *Mailing Add:* Marion Labs Inc 10236 Marion Park Dr Kansas City MO 64137

HANCOCK, DEANA LORI, b Sedalia, Mo, Jan 23, 61; m 81. RUMINANT NUTRITION, GROWTH & DEVELOPMENT. *Educ:* Univ Mo, Columbia, BS, 83, MS, 86; Tex Tech Univ, Lubbock, PhD(animal sci), 89. *Prof Exp:* Postdoctoral res scientist animal nutrit & physiol, Lilly Res Labs, 89-90; ASST PROF RUMINANT NUTRIT GROWTH & DEVELOP, PURDUE UNIV, 90- *Mem:* Am Soc Animal Sci; Am Dairy Sci Asn; AAAS. *Res:* Ruminant nutrition with an emphasis on cellular and molecular regulation of nutrient utilization, growth, protein accretion, lactation and their interactions; author of numerous publications. *Mailing Add:* 2-109 Lilly Hall Purdue Univ West Lafayette IN 47907

HANCOCK, GEORGE WHITMORE, JR, b Richmond, Va, Apr 25, 42; m 80. PHYSICS. *Educ:* Univ Va, BA, 63, PhD(physics), 73. *Prof Exp:* From instr to asst prof, 68-80, ASSOC PROF PHYSICS DEPT, MARIETTA COL, 80- *Mem:* Am Asn Physics Teachers; Sigma Xi. *Mailing Add:* Dept Physics Marrietta Col Marietta OH 45750

HANCOCK, HAROLD E(DWIN), b Norwood, Ohio, Apr 8, 21; m 46; c 2. ELECTRICAL ENGINEERING. *Educ:* Univ Cincinnati, EE, 43, MS, 45, DSc, 48. *Prof Exp:* Consult, Askania Regulator Co, 44-49; partner, Eng Specialties, Madeira, 49-65; OWNER, H E HANCOCK ASSOCS, 65-; PRES, HAST INDUST INC, 74- *Concurrent Pos:* Instr, Univ Cincinnati, 43-45 & 48-49, eve col, 43-49, res engr, 43-45; consult, Wright Field, 46-47; gen mgr, Environ Instruments Div,Ohmart Corp, 71-73. *Mem:* AAAS; sr mem Inst Elec & Electronics Engrs. *Res:* Design of electronic control equipment and magnetic amplifiers for control; development of underwater logistics; electrical, electronic and mechanical research, development and manufacturing; design of nuclear instrumentation for reactor control pruposes; design and manufacture of robot monitors for water quality analysis; development of arc-less contactors. *Mailing Add:* 5209 Kenridge Dr Cincinnati OH 45242

HANCOCK, JAMES FINDLEY, JR, b Cleveland, Ohio, Jan 20, 50; m 75. PLANT EVOLUTION, PLANT BREEDING. *Educ:* Baldwin-Wallace Col, BS, 72; Miami Univ, MS, 74; Univ Calif, Davis, PhD(genetics), 77. *Prof Exp:* Lab asst biol, Baldwin-Wallace Col, 70-72; teaching asst bot, Miami Univ, 72-74; res asst genetics, Univ Calif, Davis, 74-77; asst prof biol, Univ SC, 77-80; from asst prof to assoc prof, 80-89, PROF HORT, MICH STATE UNIV, 89- *Mem:* Bot Soc Am; Am Soc Hort Sci; Genetics Soc Am; Soc Study Evolution. *Res:* Artificial and natural selection in blueberries and strawberries; polyploidy; ecological genetics; gene flow; breeding for virus resistance. *Mailing Add:* Dept Hort Mich State Univ 347 Plant & Soil Sci East Lansing MI 48824-1325

HANCOCK, JAMES WILLIAM, organic polymer chemistry, for more information see previous edition

HANCOCK, JOHN C(OULTER), b Martinsville, Ind, Oct 21, 29; m 49; c 4. ELECTRICAL ENGINEERING. *Educ:* Purdue Univ, BS, 51, MS, 55, PhD, 57. *Prof Exp:* Res engr, US Naval Avionics Facil, Ind, 51-57; from asst prof to prof elec eng, Purdue Univ, 57-80, dir, Electronics Systs Res Lab, 64-65, head sch, 65-80, dir, Appl Electronics Res Lab, 66-80, dean, Sch Eng, 72-80; exec vpres, United Telecommun, 80-88; PVT CONSULT, 88- *Concurrent Pos:* Mem, Nat Sci Bd, Inst Elec & Electronics Engrs. *Honors & Awards:* Lamme Award, Inst Elec & Electronics Engrs, 80. *Mem:* Nat Acad Eng; Am Soc Eng Educ; Inst Elec & Electronics Engrs; fel Nat Asn State Univ & Land Grant Col. *Res:* Communication theory. *Mailing Add:* 4550 Warwick Blvd Suite 901 Kansas City MO 64111

HANCOCK, JOHN CHARLES, b Lockwood, Mo, Aug 20, 38. NEURO-CARDIOVASCULAR PHARMACOLOGY, ELECTRO PHYSIOLOGY. *Educ:* Univ Mo-Kansas City, BS, 62; Univ Tex, MS, 65, PhD(pharmacol), 67. *Prof Exp:* From res assoc to asst prof pharmacol, Health Ctr, Univ Conn, 67-71; assoc prof pharmacol, Med Ctr, La State Univ, 72-78; PROF & ASSOC CHMN PHARMACOL, E TENN STATE COL MED, 78- *Concurrent Pos:* Chmn curric comt, E Tenn State Col Med. *Mem:* AAAS; Sigma Xi; Am Soc Pharmacol & Exp Therapeut; Soc Neurosci. *Res:* Regulation of cardiovascular function by endogenous vasoactive peptides and the changes in blood vessel responsiveness and transmitter release in the hypertensive rat. *Mailing Add:* Dept Pharmacol E Tenn State Col Med Johnson City TN 37614

HANCOCK, JOHN EDWARD HERBERT, organic chemistry; deceased, see previous edition for last biography

HANCOCK, JOHN OGDEN, physics, for more information see previous edition

HANCOCK, JOSEPH GRISCOM, JR, b Bridgeton, NJ, Apr 8, 38; m 60, 87; c 2. PLANT PATHOLOGY. *Educ:* Rutgers Univ, BS, 60; Cornell Univ, MS, 63, PhD(plant path), 64. *Prof Exp:* From asst prof to assoc prof plant path, 64-76, asst plant pathologist, 64-70, assoc plant pathologist, 70-76, chmn dept conserv & resource studies, 74-76 & 83-84, actg chmn dept plant path, 84-85, chmn dept, 85-89, PROF PLANT PATH & PLANT PATHOLOGIST, UNIV CALIF, BERKELEY, 76- *Concurrent Pos:* Vis prof, Imp Col, Univ London, 70. *Mem:* Am Phytopath Soc; Mycol Soc Am; Soc Gen Microbiol. *Res:* Physiological aspects of plant disease; ecology of soil-borne plant pathogenic fungi; rhizosphere biology; biological control; integrated pest management. *Mailing Add:* Dept Plant Path Univ Calif 147 Hilgard Hall Berkeley CA 94720

HANCOCK, KENNETH FARRELL, b Alexander City, Ala, Apr 17, 30; m 59; c 2. BIOLOGY. *Educ:* Jacksonville State Col, BS, 50; George Peabody Col, MA, 51; Univ Ala, PhD(biol), 60. *Prof Exp:* From asst prof to assoc prof, 60-67, chmn dept, 62-85, CHARLES A DANA PROF BIOL, BERRY COL, 67- *Res:* Plant morphology; phycology. *Mailing Add:* Dept Biol Berry Col Mt Berry GA 30149

HANCOCK, KENNETH GEORGE, b St Louis, Mo, Mar 11, 42; m 64; c 3. ORGANIC CHEMISTRY, PHOTOCHEMISTRY. *Educ:* Harvard Univ, BA, 63; Univ Wis, PhD(org chem), 68. *Prof Exp:* From asst prof to assoc prof chem, Univ Calif, Davis, 68-78; prog dir chem dynamics, 78-76, actg dir chem div, 87-88, prog dir org chem, 87-88, DIR, CHEM DIV, NSF, 90- *Concurrent Pos:* Adj prof, Univ Calif, Davis, 78-80; Legis asst, US Senate, 81; sr prog mgr, div int progs, NSF, 83-85. *Mem:* Am Chem Soc; Sigma Xi; AAAS. *Res:* Organic physical-organic and theoretical organic chemistry; photochemical and thermal reactions of organometallics and bichromophores; boron chemistry; photobiology; free radicals. *Mailing Add:* Chem Div NSF Washington DC 20550

HANCOCK, MICHAEL B, b Dallas, Tex, Mar 11, 39; m 69. NEUROPHYSIOLOGY. *Educ:* Arlington State Col, BS, 62; Univ Tex Southwestern Med Sch Dallas, PhD(neurophysiol), 69. *Prof Exp:* Fel anat, Univ Tex Southwestern Med Sch Dallas, 68-69; ASST PROF ANAT, UNIV TEX MED BR GALVESTON, 69- *Res:* Spinal cord physiology; interactions of visceral and somatic pathways at the spinal level. *Mailing Add:* Dept Anat 401 Keiller Bldg FO7 Univ Tex Med Sch 301 Univ Bldg Galveston TX 77550

HANCOCK, PETER ADRIAN, b Berkley, Eng, March 9, 53; m 79; c 2. HUMAN FACTORS, STRESS. *Educ:* Loughborough Univ, Eng, BEd, 76, MSc, 78; Univ Ill, PhD(human performance), 83. *Prof Exp:* Res Assoc, Univ Ill, 82-83; asst prof safety sci, Univ Southern Calif, 83-87; ASSOC PROF, UNIV MINN, 87- *Concurrent Pos:* Res grants, Pac Telesis, 84-86 & NASA, 86-; consult, Calif Comn Police Training, 87-, Los Angeles Cty Sheriffs Dept, 87-, Shell Oil Co, 86- & Thermacor Inc, 86-; lectr, Southern Calif Safety Inst, 87- *Mem:* Human Factors Soc; Psychonomic Soc; Am Soc Safety Engrs; Inst Elec & Electronics Engrs; AAAS. *Res:* Human performance under extremes of stress; time dilation and contraction in life-threat conditions and how to engineer human-machine interfaces to cope with such demands. *Mailing Add:* HFRL 164 Norris Hall Univ Minn 172 Pillsbury Dr SE Minneapolis MN 55455

HANCOCK, ROBERT ERNEST WILLIAM, b Merton, Eng, Mar 23, 49; Can & Brit citizen; m 73; c 2. MICROBIOLOGY, BIOCHEMISTRY. *Educ:* Univ Adelaide, BSc Hons, 71 & PhD (microbiol), 75. *Prof Exp:* From asst prof to assoc prof, 78-86, PROF MICROBIOL, UNIV BC 86-, SCI DIR, CAN BACT DIS NETWORK, 90- *Concurrent Pos:* Consult, Bristol-Myers Squibb & Co, Ltd, 84-91; vis scholar, Monash Univ, 86-87; chmn exec, Med Sci Adv Comt, Can Cystic Fibrosis Found, 90- 93. *Honors & Awards:* Can Soc Microbiologists Award, 87. *Mem:* Am Soc Microbiol (br pres, 85-86); Infectious Dis Soc Am. *Res:* Role of the outer membranes of gram negative bacteria in bacterial antibiotic resistance, barrier function, pathogenesis and interaction with macrophages. *Mailing Add:* Dept Microbiol Univ BC Vancouver BC V6T 1W5 Can

HANCOCK, RONALD LEE, b St Joseph, Mo, Nov 24, 31; m 55; c 1. BIOCHEMISTRY. *Educ:* Univ Kansas City, BA, 52; Univ Kans, MD, 59. *Prof Exp:* Assoc staff scientist, Jackson Lab, Maine, 64-66, staff scientist, 66-70; ASSOC PROF MED BIOCHEM, FAC MED, UNIV CALGARY, 70- *Concurrent Pos:* Fel biochem, Ben May Lab Cancer Res, Univ Chicago, 59-63, USPHS fel, 60-63; Am Cancer Soc Janice M Blood Mem grant cancer res, Jackson Lab, Maine, 64-66, Nat Cancer Inst grant, 66-70; Med Res Coun Can grant, Fac Med, Univ Calgary, 70- *Mem:* Am Chem Soc; Biochem Soc; Can Biochem Soc; fel Royal Soc Health; NY Acad Sci. *Res:* Biochemistry of carcinogenesis; alkylation of transfer RNA. *Mailing Add:* Center de Recherche Hotel-Dieu One Rue de L'Arsenal Quebec City PQ G1R 2J6 Can

HANCOCK, V(ERNON) RAY, b Baltimore, Md, July 10, 26; m 50; c 2. ALGEBRA. *Educ:* Va Polytech Inst, BS, 49; Johns Hopkins Univ, MA, 51; Tulane Univ, PhD(math), 60. *Prof Exp:* Jr instr, Johns Hopkins Univ, 50-51; from instr to asst prof math, Va Polytech Inst, 52-56; instr, Tulane Univ, 56-60; from asst prof to assoc prof, Va Polytech Inst, 60-63; chmn dept, 63-74, prof, 63-91, EMER PROF MATH, EMORY & HENRY COL, 91- *Concurrent Pos:* Consult, US Naval Res Lab, 54-59; visitor, Tulane Univ, 70-71. *Mem:* Am Math Soc; Math Asn Am; Nat Coun Teachers Math. *Res:* Algebraic semigroups. *Mailing Add:* Box Y Emory VA 24327

HANCOX, WILLIAM THOMAS, b New Westminister, BC, Mar 19, 40; m 63; c 2. THERMAL-HYDRAULICS, NUMERICAL FLUID MECHANICS. *Educ:* Carleton Univ, Ont, BEng, 66, MEng, 67; Univ Waterloo, Ont, PhD(mech eng), 71. *Prof Exp:* Res eng, Atomic Power Div, Westinghouse Can, 67-73; sect head, Appl Math & Comput Reactor Anal Br, 73-76, br head thermalhydraul res, 76-78, dir, Appl Sci Div, 78-84, dir, Local Energy Systs Bus Unit, 84-86, VPRES, ATOMIC ENERGY CAN LTD, 86- *Concurrent Pos:* Adj prof, Univ Waterloo, 71-73. *Mem:* Can Nuclear Soc; Am Nuclear Soc. *Res:* Aspects of candu reactor safety; reprocessing of uranium and thorium fuels; disposal of radioactive waste; radioactive waste management. *Mailing Add:* 128 Dunbarton Ct Ottawa ON K1K 4L6 Can

HAND, ARTHUR RALPH, b Los Angeles, Calif, May 15, 43; m 61; c 2. CELL BIOLOGY, CYTOCHEMISTRY. *Educ:* Univ Calif, Los Angeles, DDS, 68. *Prof Exp:* res investr, 68-78, CHIEF, LAB BIOL STRUCT, NAT INST DENT RES, NIH, USPHS, 78- *Concurrent Pos:* Vis prof, Dept Anat, McGill Univ, 76-77. *Honors & Awards:* Commendation Medal, USPHS, 75; Basic Res Oral Sci Award, Int Asn Dent Res, 78. *Mem:* Am Soc Cell Biol; Int Asn Dent Res; Sigma Xi; Histochem Soc; AAAS. *Res:* Ultrastructure, cytochemistry and function of cellular organelles; mechanisms of exocrine secretion. *Mailing Add:* Dept Pediat Dent Univ Conn Health Ctr Farmington CT 06032

HAND, BRYCE MOYER, b Jersey City, NJ, Mar 22, 36; m 63; c 2. SEDIMENTOLOGY. *Educ:* Antioch Col, BA, 58; Univ Southern Calif, MS, 61; Pa State Univ, PhD(geol), 64. *Prof Exp:* Asst prof geol, Amherst Col, 64-69; PROF GEOL, SYRACUSE UNIV, 69- *Mem:* AAAS; fel Geol Soc Am; Am Asn Petrol Geol; Soc Econ Paleont & Mineral. *Res:* Sediment transport; bedform dynamics. *Mailing Add:* Dept Geol Syracuse Univ Syracuse NY 13244-1070

HAND, CADET HAMMOND, JR, b Patchogue, NY, Apr 23, 20; m 42; c 2. INVERTEBRATE ZOOLOGY. *Educ:* Univ Conn, BS, 46; Univ Calif, MA, 48, PhD(zool), 51. *Prof Exp:* From instr to asst prof zool, Mills Col, 48-51; asst res zoologist, Scripps Inst, Univ Calif, 51-53; from asst prof to prof, 53-85, dir Bodega Marine Lab, 61-85, EMER PROF ZOOL, UNIV CALIF, BERKELEY, 85- *Concurrent Pos:* Mem, Pac Sci Bd Res Exped to Kapingamarangi Atoll, Caroline Islands, 54; NSF sr fel, NZ & Australia, 59-60; consult, NIH, 63-66 & NSF, 63-67; John Simon Guggenheim Mem Found fel, 67-68; mem, Nuclear Regulatory Comn, Atomic Safety & Licensing Bd Panel, 71-; admin judge, US Nuclear Regulatory Comn, 80- *Mem:* Soc Syst Zool; Am Soc Limnol & Oceanog; Ecol Soc Am; Am Soc Zoologists. *Res:* Systematic and natural history studies of invertebrates, particularly hydroids and sea anemones; symbiosis, particularly in hydroids. *Mailing Add:* Bodega Marine Lab PO Box 247 Bodega Bay CA 94923

HAND, CLIFFORD WARREN, b Philadelphia, Pa, Jan 3, 36; m 60. PHYSICAL CHEMISTRY. *Educ:* Cornell Univ, AB, 57; Harvard Univ, PhD(phys chem), 61. *Prof Exp:* Res assoc chem, Princeton Univ, 61-62; fel, Carnegie-Mellon Univ, 63-67, asst prof, 67-69; ASSOC PROF CHEM, UNIV ALA, 69- *Mem:* Am Chem Soc; Am Phys Soc; Sigma Xi. *Res:* Gas phase kinetics; fast reactions; flash photolysis; mass spectroscopy. *Mailing Add:* Box 870336 Tuscaloosa AL 35487-0336

HAND, GEORGE SAMUEL, JR, b Perryville, Mo, Aug 22, 36; m 61; c 2. EXPERIMENTAL EMBRYOLOGY. *Educ:* Southeast Mo State Col, BS, 58; Washington Univ, MA, 61; Univ NC, PhD(zool), 67. *Prof Exp:* NIH res fel embryol, Calif Inst Technol, 67-69; asst prof, 69, ASSOC PROF ANAT, UNIV ALA, BIRMINGHAM, 75-, ASSOC DIR MED ADMIS, 85- *Concurrent Pos:* Lectr, Univ Calif, Los Angeles, 69. *Mem:* AAAS; Am Soc Zoologists; Soc Development Biol; Sigma Xi. *Res:* Macromolecular synthesis during early development; comparative histophysiology of endometrium; fertilization physiology; RNA and protein synthesis during early development; embryonic induction. *Mailing Add:* Dept Cell Biol Box 317 UAB Sta Birmingham AL 35294

HAND, JAMES HENRY, b Jersey City, NJ, Jan 2, 43. CHEMICAL ENGINEERING. *Educ:* Newark Col Eng, BS, 66; Univ Calif, Berkeley, PhD(chem eng), 71. *Prof Exp:* From asst prof to assoc prof, Univ Mich, Ann Arbor, 77-81; ASSOC PROC CONSULT, DOW CORNING CORP, 81- *Mem:* Am Inst Chem Engrs. *Res:* Mathematical modeling of polymerization reactors; applied statistical mechanics and thermodynamics. *Mailing Add:* 2622 Abbott Midland MI 48640-0995

HAND, JUDITH LATTA, sociobiology, ethology, for more information see previous edition

HAND, LOUIS NEFF, b Hollywood, Calif, Oct 16, 33; m 56; c 4. HIGH ENERGY PHYSICS. *Educ:* Swarthmore Col, BA, 55; Stanford Univ, PhD(physics), 61. *Prof Exp:* Res fel physics, Harvard Univ, 61-62, instr, 62-64, asst prof, 64-65; assoc prof, 65-71, PROF PHYSICS, CORNELL UNIV, 71- *Concurrent Pos:* Alfred P Sloan fel, 64-66; John Simon Guggenheim Mem fel, 80-81; Sci Res Coun vis fel, Oxford, 80-81; vis foreign scientist, Deutsches Elektronec-Synchrotron, 86- *Mem:* Am Phys Soc. *Res:* Experimental high energy nuclear physics. *Mailing Add:* Dept Physics Cornell Univ Clark Hall Ithaca NY 14853

HAND, PETER JAMES, b Oak Park, Ill, Jan 5, 37; m 58, 86; c 6. ANATOMY. *Educ:* Univ Pa, VMD, 61, PhD(neuroanat), 64. *Prof Exp:* Assoc, Univ Pa, 64-65, from asst prof to assoc prof, 65-79, head, Anat Labs, 73-74 & 80-87, PROF ANAT, UNIV PA, 79-, MEM GRAD GROUP ANAT, 67- *Concurrent Pos:* NSF grants, 65-66; Nat Inst Neurol Dis & Blindness grant, 66-75; res collabr, Brookhaven Nat Lab, 79-; mem, Inst Neurol Sci, Univ Pa, 68-; vis prof, Nat Yang-Ming Med Col, Taiwan, Rep China, 87-88 & 89 & Shandong Med Univ, Jinan, People's Rep China, 88; vis scientist, Kobe Univ, Sch Med, Kobe, Japan, 90- *Mem:* Int Brain Res Orgn; Am Asn Anatomists; Am Asn Vet Anatomists; World Asn Vet Anatomists; Soc Neurosci; Sigma Xi. *Res:* Brain plasticity associated with disturbances of sensory input and determination of anatomicophysiological mechanisms associated with muscle stimulation produced analgesia (i.e. acupuncture). *Mailing Add:* Dept Animal Biol Sch Vet Med 3800 Spruce St Philadelphia PA 19174

HAND, ROGER, b Brooklyn, NY, Sept 25, 38; m 86; c 2. INTERNAL MEDICINE, QUALITY ASSURANCE & UTILIZATION REVIEW. *Educ:* NY Univ, BS, 59, MD, 62. *Prof Exp:* Clin asst prof med, Med Sch, Cornell Univ, 70-73; asst attend physician, Mem Hosp Cancer & Allied Dis, McGill Univ, 72-73, from asst prof to assoc prof, dept med & microbiol, 73-78, from asst physician to assoc physician, Clin Royal Victoria Hosp, 73-78, prof med, dept med, Cancer Ctr, 78-84, dir, Cancer Ctr, 80-84; chmn, Ill Masonic Med Ctr, 84-88, ASSOC CHIEF STAFF, UNIV HOSP, VIC COL MED, CHICAGO, 88-; PROF MED, UNIV ILL, CHICAGO, 84- *Concurrent Pos:* Sr investr, NY Heart Asn, Rockefeller Univ, 72-73; transfer receipt & vis prof, Univ Bern, 77; assoc ed, Clin & Investigative Med, 81-84; vis scientist, Int Cancer Res Technol, NIH, 83; mem bd dirs, Am Cancer Soc, Ill Div, 87-; pres, Ill div, Nat Coun against Health Fraud, 86- *Mem:* Am Soc Clin Invest; Am Soc Biol Chemists; Am Soc Clin Oncol; Am Soc Cell Biol; Am Asn Cancer Res. *Res:* Regulation of DNA replication in mammalian cells; clinical investigation in delivery of medical care; basic research in biochemistry of DNA. *Mailing Add:* Sect Gen Internal Med Rm 720S Col Med Univ Chicago Clin Sci Bldg CSB MC 787 840 S Wood St Chicago IL 60612

HAND, STEVEN CRAIG, METABOLIC REGULATION, COMPARATIVE BIOCHEMISTRY. *Educ:* Ore State Univ, PhD(physiol), 80. *Prof Exp:* Asst prof biol, Univ Southwestern La, 82-86; asst prof biol, 86-89, ASSOC PROF BIOL, UNIV COLO, 89- *Mailing Add:* Dept Environ Pop & Org Biol Univ Colo Campus Box B-334 Boulder CO 80309

HAND, THOMAS, b Olean, NY, Feb 28, 43; m 66; c 1. FORTH ENVIRONMENTS, EXTENSIBLE SOFTWARE. *Educ:* Fla Southern Col, BS, 64; Okla Univ, PhD(math), 72. *Prof Exp:* Prof comput sci, Ind State Univ, 72-78; consult, Software Environ, 78-81; CHMN, GRAD COMPUT SCI, FLA INST TECHNOL, 81- *Mem:* Asn Comput Mach; Inst Elec & Electronics Engrs. *Res:* expert systems, relational data base systems, operating systems, compilers and other extensions to FORTH. *Mailing Add:* Dept Comput Sci Fla Inst Tech 150 W Univ Blvd Melbourne FL 32901

HANDA, PAUL, b Lucknow, India, Dec 9, 50; Can citizen; m 78; c 1. THERMODYNAMICS & MATERIAL PROPERTIES, PHYSICAL CHEMISTRY. *Educ:* Punjab Univ, BSc, 70, MSc, 72; Univ Otago, NZ, PhD(chem), 75. *Prof Exp:* Fel chem, Univ Calif, Los Angeles, 75-76; fel, Wright State Univ, Dayton, 76-77; res assoc, Nat Res Coun, Can, 77-81; res chemist, Allied Chem Co, Buffalo, 81-82; RES OFFICER CHEM, NAT RES COUN, CAN, 82- *Res:* Thermodynamic properties of solids and fluids; study of stability, structure and thermophysical properties of inclusion compounds especially gas hydrates; high pressure phase diagrams; pressure and thermally induced phase transitions; relaxation processes in amorphoys and disordered materials, mechanical properties, viscoelastic properties. *Mailing Add:* Inst Environ Chem Nat Res Coun Ottawa ON K1A 0R6 Can

HANDA, V(IRENDER) K(UMAR), b India, Dec 28, 31; m 62; c 2. CONSTRUCTION, CIVIL ENGINEERING. *Educ:* Univ Calcutta, BSc, 49; Univ London, BSc, 54; Queen's Univ, Ont, MASc, 58; Univ Waterloo, MSc, 62, PhD(civil eng), 64. *Prof Exp:* Staff photographer, The Spotlight, India, 49-50; eng trainee, K Hajnal Konyi, 53-54; civil engr, William H Laithwaite, 54-56; teaching asst, Queen's Univ, Ont, 56-58; asst construct engr, Eldorado Mining & Refining Ltd, 58; proj engr soils, Racey McCallum & Assocs, Ltd, 59-60; lectr civil eng, 60-64, from asst prof to assoc prof, 64-69, PROF CIVIL ENG, UNIV WATERLOO, 69- *Concurrent Pos:* Nat Res Coun Can Grants, 62-; consult, Nat House-Builders Asn Can, 64-; vis prof & chmn dept civil eng, Univ Petrol & Minerals, Dhahran, Saudi Arabia, 69-70 & Polytech Sch, Fed Univ Paraiba, Brazil, 71; vis prof InterAm Comt Agr Develop, Univ West Indies, 74-75, 77; mem comn W.65, Int Coun Building Res & Documentation, 74-, coordr, 89-; mem, steering comt indust building construct, Can Standards Asn, 74-, vchmn, 90-; head econ group, construct indust study proj group, Ministry Planning & Develop, Govt Trinidad & Tobago, 75-77; guest prof, Swiss Fed Inst Technol, Zurich, 79; adv, Cree Housing Corp, 80-87; adv, Secondary Sch Bldg Prog, Gov Trinidad, Tobago, 77-79; external examnr, Fac eng, Univ WI, 85; UN Tokten expert, Govt India, 88; mem, Fire Code Comn, Govt Ont, 90- *Mem:* Can Standards Asn; fel Inst Engrs India; Am Soc Civil Engrs; Asn Proj Mgrs UK; Asn Researchers Construct UK; Construct Mgt Asn Am. *Res:* Construction: organization, productivity, project management, planning, and control process; economics; resource allocation; operations research; econometrics; cold regions northern engineering; industrialized buildings; housing; educational planning; training in developing countries; energy, small scale low-head hydro power. *Mailing Add:* Dept Civil Eng Univ Waterloo Waterloo ON N2L 3G1 Can

HANDEL, DAVID, b Brooklyn, NY, June 20, 38; m 63, 70; c 2. TOPOLOGY. *Educ:* Calif Inst Technol, BS, 59; Univ Chicago, MS, 60, PhD(math), 65. *Prof Exp:* NSF fel, Univ Calif, Berkeley, 65-66, actg asst prof math, 66-67, lectr, 67-68; asst prof, Univ Wash, 68-72; assoc prof, 72-78, PROF MATH, WAYNE STATE UNIV, 78- *Mem:* Am Math Soc. *Res:* Algebraic topology; embeddings and immersions of manifolds in Euclidean space; nonsingular bilinear maps; K theory; topological methods in approximation theory; thom modules. *Mailing Add:* Dept Math 646 McKenzie Wayne State Univ 5950 Cass Ave Detroit MI 48202

HANDEL, MARY ANN, b New Haven, Conn, Feb 27, 43; m 67; c 1. CELL BIOLOGY. *Educ:* Goucher Col, BA, 65; Johns Hopkins Univ, MS, 67; Kans State Univ, PhD(biol), 70. *Prof Exp:* Fel biol, Oak Ridge Nat Lab, 70-73; from asst prof to assoc prof, 73-88. PROF ZOOL, UNIV TENN, 88- *Concurrent Pos:* Am Cancer Soc fel, 70. *Mem:* AAAS; Soc Study Reproduction; Am Soc Cell Biol; Soc Develop Biol; Genetics Soc Am. *Res:* Cytological and genetic aspects of cell differentiation during spermatogenesis. *Mailing Add:* Dept Zool Univ Tenn Knoxville TN 37996-0810

HANDEL, STEVEN NEIL, b Brooklyn, NY, Jan 29, 45; m 73; c 2. PLANT ECOLOGY, POPULATION BIOLOGY. *Educ:* Columbia Univ, AB, 69; Cornell Univ, MS, 74, PhD(ecol), 76. *Prof Exp:* Teaching asst ecol, Cornell Univ, 71-76; asst prof biol, Univ SC, 76-; ASST PROF BIOL, YALE UNIV, 79- *Concurrent Pos:* Vis scientist, Sch Plant Biol, Univ Col N Wales, 76; dir, Marsh Bot Garden Yale Univ, 80- *Mem:* Ecol Soc Am; Soc Study Evolution; Bot Soc Am; Brit Ecol Soc; Sigma Xi. *Res:* Plant population biology; pollination ecology; plant-animal interactions. *Mailing Add:* Dept Biol 357 OML Yale Univ PO Box 6666 New Haven CT 06520

HANDELMAN, EILEEN T, b Holyoke, Mass, Dec 11, 28; m 59; c 1. MICROWAVE SPECTROSCOPY. *Educ:* Mt Holyoke Col, BA, 50, MA, 52; Univ Calif, Berkeley, PhD(phys chem), 55. *Prof Exp:* NSF postdoctoral fel, Univ Copenhagen, Denmark, 55-56; mem tech staff, Bell Tel Labs, Murray Hill, 56-66; PROF PHYSICS & DEAN SCI, SIMON'S ROCK COL, 68- *Mem:* Am Asn Physics Teachers. *Res:* Microwave spectroscopy; molecular structure; solid state and semiconductor physics; author of numerous publications; awarded several patents. *Mailing Add:* Dept Natural Sci Simon's Rock Col Great Barrington MA 01230

HANDELMAN, GEORGE HERMAN, b Pittsburgh, Pa, Mar 24, 21; m 49; c 2. APPLIED MATHEMATICS. *Educ:* Harvard Univ, AB, 41, AM, 42; Brown Univ, PhD(appl math), 46. *Prof Exp:* Res assoc appl math, Brown Univ, 43-47, asst prof eng, 47-48; from asst prof to assoc prof math, Carnegie Inst Technol, 48-55; chmn dept, 60-72, prof appl math, 55-78, dean sch sci, 72-78, AMOS EATON PROF APPL MATH, RENSSELAER POLYTECH INST, 78- *Concurrent Pos:* Mem, US Nat Comt Theoret & Appl Mech, 71-77; book rev ed, Soc Indust & Appl Math, Rev & Soc Indust Appl Math, News. *Mem:* Am Math Soc; Soc Indust & Appl Math; fel Am Soc Mech Eng; Math Asn Am. *Res:* Elasticity; vibrations; stability; wave motion; math problems in biology. *Mailing Add:* Dept Math Sci Rensselaer Polytech Inst Amos Eaton Bldg 307 Troy NY 12180-3590

HANDELSMAN, JACOB C, b Elizabeth, NJ, Jan 20, 19; m 43; c 4. SURGERY. *Educ:* Johns Hopkins Univ, AB, 40, MD, 43. *Prof Exp:* Asst prof, 54-63, ASSOC PROF SURG, SCH MED, JOHNS HOPKINS UNIV, 63-; SURGEON & IN CHG OUTPATIENT CLIN, JOHNS HOPKINS UNIV, 50- *Concurrent Pos:* Attend surgeon, Sinai Hosp, 51- *Mem:* Soc Univ Surgeons; Am Col Surg. *Res:* Surgical teaching; pediatric and general surgery. *Mailing Add:* 220 W Cold Spring Lane Baltimore MD 21210

HANDELSMAN, JO, b New York, NY, Mar 19, 59. MOLECULAR BIOLOGY. *Educ:* Cornell Univ, BS, 79; Univ Wis, PhD(molecular biol), 83. *Prof Exp:* Postdoctoral res fel, Am Cancer Soc, 84 & NIH, 84-85; ASST PROF PLANT PATH, UNIV WIS, 85- *Mem:* Am Soc Microbiol; Am Phytopath Soc. *Res:* Molecular basis of competitiveness of Rhizobium phaseoli in nodulation of beans; elucidating the mechanism of attachment of Agrobacterium tumefaciens to its host plant; mechanisms of biocontrol of root rot pathogens of crop plants. *Mailing Add:* Dept Plant Path Univ Wis 682 Russell Labs 1630 Linden Dr Madison WI 53706

HANDELSMAN, MORRIS, electronics; deceased, see previous edition for last biography

HANDFORD, STANLEY WING, physiology; deceased, see previous edition for last biography

HANDIN, JOHN WALTER, b Salt Lake City, Utah, June 27, 19; m 47; c 2. GEOPHYSICS. *Educ:* Univ Calif, Los Angeles, AB, 42, MA, 48, PhD(geol), 49. *Prof Exp:* Geologist, Corps Eng, 47-48; res assoc inst geophys, Univ Calif, 49-50; res geologist, Shell Develop Co, 50-63, res assoc, 63-66; dir, Ctr Tectono-Physics, Tex A&M Univ, 67-78, distinguished prof, 67-84, assoc dean, Col Geosci, 73-82, dir, Earth Resources Inst, 78-82, distinguished emer prof geol & geophys & res scientist,; pres, John Handin Inc, 84-90; RETIRED. *Concurrent Pos:* Mem comt rock mech, Nat Acad Sci-Nat Res Coun, 63-66; vis prof, Columbia Univ, 64-65; consult, US Army Corps Eng, 64-66, Off Sci & Technol, Exec Off Pres, 65, Oak Ridge Nat Lab, 67 & 88, Gen Motors Tech Ctr, 67, Defense Atomic Support Agency, 67-68, Lawrence Livermore Lab, 67-80, Los Alamos Sci Lab, 72-82 & 86-89, Lake Powell Proj, NSF, 72-74, Sandia Lab, 74-80, Environ Protection Agency, 77-78, Defense Nuclear Agency, 76-77, Lawrence Berkely Lab, 79, Dept Energy, 79-82, Calif Energy Comn, 80, Intera Environ Consult, 80-82, Res & Develop Assoc, 82-84, Earth Technol, 85, Tex A&M Found, 84- & Sci Appl Int Corp, 86-88; mem US Nat Comt Rock Mech, 67-72, chmn, 69-72; consult, US Air Force, 68-69, Advan Res Projs Agency, 69-71, Terra Tek, 70-; res geophysicist, US Geol Surv, 70-75; mem panel San Fernando Earthquake, Nat Acad Sci-Nat Acad Eng, 71; consult, State La, 71, Calif, 80; mem, US Nat Comt Tunneling Technol, Nat Acad Sci, 72-75; USAF Rev Group on ICBM Survivability, 82-84. *Honors & Awards:* Distinguished Achievement Award in Rock Mech, Am Inst Mining Engrs, 70; Spec Award for Exceptional Serv to Sci of Rock Mech, Nat Acad Sci, 81; Bucher Medal, Am Geophys Union, 83; Career Contrib Award, Geol Soc Am, 88. *Mem:* Fel Am Geophys Union; Int Soc Rock Mech (vpres, 74-79); Sigma Xi; fel Geol Soc Am. *Res:* Tectono physics, especially experimental rock deformation; structural geology; engineering geology, especially rock mechanics. *Mailing Add:* Col Geosci Tex A&M Univ Col Sta TX 77843

HANDIN, ROBERT I, b New York, NY, June 20, 41; m 67; c 2. HEMATOLOGY, HEMOSTASIS. *Educ:* Univ Calif, Berkeley, AB, 63; Univ Calif Med Ctr, San Francisco, MD, 67. *Prof Exp:* Intern, Peter Bent Brigham Hosp, 67-69; Lt Comdr res, Naval Blood Res Lab, US Naval Reserve, 69-71; fel, 71-73, instr, 73-74, asst prof, 74-78, ASSOC PROF MED, HARVARD MED SCH, 78- *Concurrent Pos:* Assoc physician & dir, Hemat Div, Brigham & Womens Hosp. *Mem:* Am Soc Clin Invest; Am Soc Hemat; Am Soc Biol Chemists; Am Fedn Clin Res. *Res:* Role of platelets in hemostasis and thrombosis. *Mailing Add:* Hemat Div Brigham & Women's Hosp Boston MA 02115

HANDLER, ALFRED HARRIS, b Boston, Mass, Apr 4, 23; m 52; c 2. PATHOLOGY. *Educ:* Providence Col, BS, 43; Boston Univ, AM, 48, PhD, 51. *Prof Exp:* Instr biol, Providence Col, 46-47; lab asst, Boston Univ, 47-48, lectr, 49; sr tech investr, Pondville Hosp, Mass, 52-53; res assoc path, Children's Med Ctr & Cancer Res Found, 52-65; assoc res prof carcinogenesis, Grad Sch Pub Health, Univ Pittsburgh, 65-68; assoc prof path, Med Sch, Rutgers Univ, 68-69; AT FINDLEY INST, FALL RIVER, MASS; mem staff, Findley Res Inc, 80- *Concurrent Pos:* Nat Cancer Inst fel, Boston Univ, 51-52; res assoc path, Harvard Med Sch, 56-65 & Bio-Res Inst, Boston Univ, 72-75; ed, Abstr Sect, Transplantation, 56-, ed, Tumor Bibliog, Transplantation Bull, 59-; lectr exp embryol, Carnegie Inst Technol, 65- *Mem:* AAAS; Am Soc Exp Path; Am Asn Cancer Res; Am Asn Pathologists; Royal Soc Med. *Res:* Experimental cancer chemotherapy; carcinogenesis; experimental morphology and pathology, particularly normal and neoplastic tissue transplantation. *Mailing Add:* 1501 Beacon St-904 Brookline MA 02146

HANDLER, EVELYN ERIKA, b Budapest, Hungary, May 5, 33; US citizen; m 65; c 2. CELL BIOLOGY. *Educ:* Hunter Col, BA, 54; NY Univ, MSc, 62, PhD(biol), 63. *Hon Degrees:* LHD, Rivier Col, 82, Univ Pittsburgh, 87, Hunter Col, 88. *Prof Exp:* Res assoc, Sloan-Kettering Inst, 58-60 & Merck Inst Therapeut Res, 58-60; lectr, Hunter Col, 62-64, from asst prof to prof biol sci, 65-80, dean sci & math, 77-80; pres, Univ NH, 80-83; PRES, BRANDEIS UNIV, 83. *Concurrent Pos:* Res grants, NIH, 64-69 & 73-76, NSF, 65-67 & 70-72; vis scientist, Karolinska Inst, 71-72; res grants, City Univ New York, 72-74; evaluator, Comn Higher Educ, Middle States Asn, 72-; vchmn univ fac senate, City Univ New York, 74-76; generalist mem, Am Coun Pharmaceut Educ, 78-80; dir, NY Acad Sci, 79; dir, NE Mutual Life Ins Co, 86; sr fel, Carnegie Found Advan Teaching, 90. *Mem:* Fel AAAS; fel NY Acad Sci; Int Soc Hemat; Harvey Soc. *Res:* Blood cell production and release in normal and leukemic states. *Mailing Add:* 26 Bradley Lane North Hampton NH 03862

HANDLER, HARRY ELIAS, experimental nuclear physics, for more information see previous edition

HANDLER, JOSEPH S, b New York, NY, Apr 19, 29; m 55. INTERNAL MEDICINE. *Educ:* Univ Pa, AB, 50, MD, 54. *Prof Exp:* Instr med, Univ Pa, 57-60; sr investr renal physiol, Nat Heart Inst, 60-67, head membrane metab unit, 67-76, SECT HEAD, LAB KIDNEY & ELECTROLYTE METAB, NAT HEART LUNG & BLOOD INST, 76- *Mem:* Am Fedn Clin Res; Am Physiol Soc; Soc Gen Physiol; Am Soc Clin Invest. *Res:* Renal physiology. *Mailing Add:* Dept Med Div Nephrol Johns Hopkins Univ Sch Med 725 N Wolfe St Hunterian 217 Baltimore MD 21205

HANDLER, PAUL, b Newark, NJ, Apr 24, 29; m 52; c 3. CLIMATE & POPULATION FORECASTS. *Educ:* Univ Chicago, PhD, 54. *Prof Exp:* Res assoc, 54-56, from asst prof to assoc prof, 56-64, PROF PHYSICS, UNIV ILL, URBANA, 64- *Concurrent Pos:* Guggenheim fel, 60-61. *Mem:* Am Meteorol Soc; Am Phys Soc. *Res:* Climate and long range weather and crop forecasting; computer-assisted instruction; population and social policy studies. *Mailing Add:* 309 Loomis Lab Physics Univ Ill 1110 W Green St Urbana IL 61801

HANDLER, SHIRLEY WOLZ, b Marshall, Tex, Jan 2, 25. BIOLOGY, GENETICS. *Educ:* Univ Tex, BA, 45, MA, 47; Univ Okla, PhD(biol educ), 58. *Prof Exp:* Asst genetics, Univ Tex, 44-47; assoc prof chem, 47-49 & biol, 49-51 & 53-58, chmn, Div Sci & Math, 71-79, PROF BIOL & HEAD DEPT, E TEX BAPTIST COL, 58- *Mem:* AAAS; Am Chem Soc; Am Soc Human Genetics; Am Genetic Asn; Nat Asn Biol Teachers. *Res:* Master's chromosomal aberration in Drosophila melanogaster as produced by x-ray; teaching methods in college biology. *Mailing Add:* Dept Biol E Tex Baptist Col 1209 N Grove St Marshall TX 75670

HANDLEY, DEAN A, b Salisbury, Md, Sept 20, 49. CARDIOVASCULAR PHARMACOLOGY, ATHEROSCLEROSIS INFLAMMATION. *Educ:* Rutgers Univ, PhD(microbiol), 78. *Prof Exp:* SR RES SCIENTIST, SANDOZ RES INST, 81- *Concurrent Pos:* Mem, Coun Arteriosclerosis, Am Heart Asn. *Mem:* Am Soc Microbiol; Am Soc Cell Biol; Am Physiol Soc; Am Asn Pathologists; Soc Exp Biol & Med; Am Heart Asn. *Mailing Add:* Sandoz Res Inst Rte 10 East Hanover NJ 07936

HANDLIN, DALE L, b Clemson, SC, Sept 4, 56; m 78. POLYMER MORPHOLOGY, STRUCTURE PROPERTY RELATIONSHIPS. *Educ:* Clemson Univ, BS, 78; Univ Mass, MS, 81, PhD(polymer sci & eng), 83. *Prof Exp:* SR RES CHEMIST, SHELL DEVELOP, 82- *Mem:* Am Phys Soc; Am Chem Soc. *Res:* Structure property relationships of multi-phase composite matrix materials and block co-polymers. *Mailing Add:* 12223 Waldemar Houston TX 77077-4904

HANDMAN, STANLEY E, b New York, NY, Jan 17, 23; m 49. SAFETY PROCESSING SYSTEMS & EQUIPMENT DESIGN & OPERATION, INNOVATIVE MECHANICAL DESIGN. *Educ:* Polytech Univ, Brooklyn, BME, 44; MME, 50. *Prof Exp:* Develop test eng, Gen Elec Co, 46-47; instr mech eng, Polytech Inst Brooklyn, 47-51; Anal engr, Subsid Pullman, Inc, M W Kellogg Co, 51-59, supvr, Mech Eng Develop Lab, Res & Develop, 59-61, sect head mech eng develop, Mech Eng Develop Div, Res & Develop Dept, Div Signal & Div Wheelabrator-Fry, 64-73, chief mech eng, Div Allied Signal, 73-78, chief engr, Div Henley Inc, 78-86; OWNER, S E HANDMAN CONSULTS, 86- *Concurrent Pos:* Instr mech eng, Polytech Inst, Brooklyn, 51-54. *Honors & Awards:* Centennial Award, Am Soc Mech Engrs, 80; Distinguished Serv Award, Welding Res Coun, 87. *Mem:* Am Soc Mech Engrs; Welding Res Coun; NY Acad Sci; Sigma Xi; AAAS. *Res:* Development of equipment and systems for the chemical, petroleum and petrochemical industry with emphasis on safety and economical designs to meet industries needs. *Mailing Add:* 32 Joyce Rd Plainview NY 11803

HANDORF, CHARLES RUSSELL, b Memphis, Tenn, Jan 1, 51; m 76; c 2. NEUROPSYCHOPHARMACOLOGY, TRACE METAL RESEARCH. *Educ:* Rice Univ, BA, 73; Univ Tenn, MD, 77, PhD(chem), 81; Am Bd Path, cert, 82. *Prof Exp:* Resident path, Univ Tenn & Methodist Hosps Memphis, 78-82; sect chief toxicol, 82-91, ASSOC DIR, MED EXPRESS LABS, 85-; CHMN & DIR LABS, METHODIST HOSPS MEMPHIS, 88- *Concurrent Pos:* Marion res fel, Marion Labs, 73; adj instr, Univ Tenn, 78-80, clin instr, 83-; nat med dir, Personal Blood Storage, 91- *Mem:* Sigma Xi; AMA; Col Am Pathologists; Am Soc Clin Pathologists. *Res:* Neuropsychopharmacology; trace metal research; urine drug screening technology. *Mailing Add:* 1591 Peabody Ave Memphis TN 38104

HANDRICK, GEORGE RICHARD, organic chemistry, for more information see previous edition

HANDSCHUMACHER, ROBERT EDMUND, b Glenside, Pa, Oct 16, 27; m 49, 81; c 2. PHARMACOLOGY. *Educ:* Drexel Univ, BS, 49; Univ Wis, MS, 51, PhD(biochem), 53. *Prof Exp:* From instr to assoc prof pharmacol, 56-64, dir grad studies, 63-70, dir div biol sci, 69-71, chmn dept, 74-77, PROF PHARMACOl, SCH MED, YALE UNIV, 64- *Concurrent Pos:* Nat Found Infantile Paralysis fel, Lister Inst Prev Med, London, 53-54; Squibb fel, Sch Med, Yale Univ, 55-56, scholar cancer res, 57-62; Eleanor Roosevelt sr fel, Prague & London, 62-63; sci consult, Am Cancer Soc, 63-, career prof, 64-74; consult, Anna Fuller Fund, 65-73, sci adv, 73-88; consult, Nat Cancer Inst, 65- *Mem:* Fel AAAS; Am Chem Soc; Am Asn Cancer Res; Am Soc Biol Chemists; Am Soc Pharmacol & Exp Therapeut. *Res:* Biochemical pharmacology; nucleic acid and amino acid metabolism; development of antimetabolites and enzymes for chemotherapy; pharmacological control of the immune response. *Mailing Add:* Dept Pharmacol B228 shm Yale Univ New Haven CT 06510

HANDWERGER, BARRY S, b Baltimore, Md, Apr 25, 43. RHEUMATOLOGY. *Educ:* Johns Hopkins Hosp, BA, 64; Univ Md, MD, 68; Am Bd Internal Med, cert, 79; Am Bd Med Lab Immunol, cert, 80. *Prof Exp:* Clin assoc, Gerontol Res Ctr, NICHD, NIH, 70-72, sr staff fel, Immunol Br, Nat Cancer Inst, 72-74; asst prof med, Sect Immunol Dept Med, Sch Med, Univ Minn, 74-81, asst prof microbiol, 74-85; assoc prof med & immunol, Mayo Med Sch, Rochester, 81-85, head, Rheumatol Res Unit, Mayo Clin/Mayo Found, 81-85; HEAD, DIV RHEUMATOL & CLIN IMMUNOL & PROF MED & MICROBIOL, SCH MED, UNIV MD, BALTIMORE, 85- *Concurrent Pos:* Mem, Ctr Grants Study Sect, Nat Arthritis Found, 79-84 & 86-87, med sci comt, Arthritis Found, 75-85, res grant rev comt, Am Diabetes Asn, 83-85, legis subcomt Res Coun, Am Col Rheumatol, 89-; consult, Dept Med Rheumatol & Immunol Mayo Clin/Mayo Found, 81-85. *Mem:* Fel Am Rheumatism Asn; Am Asn Immunologists; Am Fedn Clin Res; fel Am Col Rheumatol. *Res:* Basic rheumatology and immunology; immunology of systemic lupus erthematosus and diabetes mellitus; biochemistry and cell biology of T cell activation; numerous technical publications. *Mailing Add:* Dept Med Univ Md Sch Med Baltimore MD 21201

HANDWERGER, STUART, b Baltimore, Md, Dec 10, 38; m 64; c 2. DEVELOPMENTAL & FETAL ENDOCRINOLOGY. *Educ:* Johns Hopkins Univ, AB, 60; Univ Md, MD, 64. *Prof Exp:* Intern pediat, Jacobi Hosp, Bronx, NY, 64-65; resident, Mt Sinai Hosp, NY, 65-66; clin assoc metab, NIH, Bethesda Md, 66-68; fel endocrinol, Harvard Med Sch, Childrens Hosp Med Ctr, 68-69, Beth Israel Hosp, Boston, 69-71; PROF PEDIAT & PHYSIOL, DUKE UNIV, 71- *Concurrent Pos:* Mem Nat Adv Coun, Nat Inst Child Health & Human Develop, 87-; assoc ed, J Clin Endocrinol & Metab, 84-; vis scientist & Guggenheim fel, Weizmann Inst Sci, Israel, 79-80; bd sci adv, Barbara Davis Diabetes Ctr, Univ Colo, Denver, 87-; prin investr, NIH grants, 72-, res career develop award, 74-79; mem Human Embryol & Develop Study Sect, NIH, 78-84. *Mem:* Am Soc Clin Invest; Am Pediat Soc; Soc Pediat Res; Endocrine Soc; Am Fedn Clin Res; AAAS. *Res:* Physiology of prolactin, placental lactogen and growth factors in the mother and fetus during pregnancy; regulation of the synthesis and secretion of these hormones and factors. *Mailing Add:* Div Pediat Endocrinol Childrens Hosp Med Ctr Elland & Bethesda Aves Cincinnati OH 45229-2899

HANDWERKER, THOMAS SAMUEL, b Little Rock, Ark, Dec 9, 51; m 80; c 1. TREE FRUITS, POMOLOGY. *Educ:* Univ Tenn, BS, 73; Cornell Univ, MS, 76, PhD(pomol), 79. *Prof Exp:* Exten horticulturist pomol, Tex Agr Exten Serv, Tex A&M Univ, 79-87; PROF HORT & HEAD AGR & AQUACULTURE PROGS, UNIV MD-EASTERN SHORE, 87- *Mem:* Am Soc Hort Sci; Am Pomol Soc. *Res:* Fruit crops. *Mailing Add:* Dept Agr Univ Md-Eastern Shore Princess Anne MD 21853

HANDY, CARLETON THOMAS, b Cataumet, Mass, July 16, 18; m 43; c 3. TEXTILE CHEMISTRY. *Educ:* Yale Univ, BS, 40, PhD(org chem), 43. *Prof Exp:* Res chemist, Exp Sta, E I du Pont de Nemours & Co Inc, 43-57, sr res chemist, Textile Fibers Dept, 57-60, res supvr, 60-70, asst to tech dir, Textile Fibers Dept, 70-78, asst to res & develop dir, 79-81; RETIRED. *Mem:* AAAS; Am Chem Soc; Fiber Soc (pres, 73); NY Acad Sci. *Res:* Synthetic organic chemistry; organic peroxides; polyester and polyamide fibers; fabric construction, finishing and performance characterization. *Mailing Add:* PO Box 727 Cataumet MA 02534

HANDY, LYMAN LEE, b Payette, Idaho, Aug 4, 19; m 48; c 2. RESERVOIR ENGINEERING, SURFACE CHEMISTRY. *Educ:* Univ Wash, BS, 42, PhD(chem), 51. *Prof Exp:* Res assoc, Calif Res Corp, Standard Oil Co, Calif, 51-66; prof & chmn, Petrol Eng Dept, 66-88, EMER PROF CHEM & PETROL ENG, UNIV SOUTHERN CALIF, 88- *Mem:* Am Chem Soc; Am Inst Chem Engrs; Soc Petrol Engrs. *Res:* Colloid and surface chemistry; multiphase flow of fluids in porous media; enhanced oil recovery; structural and thermodynamic properties of certain inorganic compounds. *Mailing Add:* Petrol Eng Dept Univ Southern Calif Los Angeles CA 90089-1211

HANDY, RICHARD L(INCOLN), b Chariton, Iowa, Feb 12, 29; m 64, 82; c 1. GEOTECHNICAL ENGINEERING. *Educ:* Iowa State Univ, BS, 51, MS, 53, PhD(geol, soil eng), 56. *Prof Exp:* From asst prof to assoc prof, Iowa State Univ, 56-63, prof civil eng & head geotech res lab, 63-91, Anson Marston Distinguished prof, 87-91, EMER DISTINGUISHED PROF CIVIL ENG, IOWA STATE UNIV, 91- *Concurrent Pos:* Res consult. *Honors & Awards:* T A Middlebrooks Award, Am Soc Civil Engrs, 85. *Mem:* Fel AAAS; Soil Sci Soc Am; Am Soc Civil Engrs; fel Geol Soc Am; Am Soc Testing & Mat. *Res:* Geotechnical engineering; soil mechanics; development soil and rock testing devices; Pleistocene geology and geomorphology. *Mailing Add:* RR 1 Box 63C Madrid IA 50156

HANDY, ROBERT M(AXWELL), b Buffalo, NY, Apr 1, 31; m 55; c 3. ELECTRICAL ENGINEERING, SOLID STATE PHYSICS. *Educ:* Trinity Col, BS, 53; Northwestern Univ, MS, 58, PhD(elec eng), 62. *Prof Exp:* Sr engr, Westinghouse Res Labs, 61-65, mgr oxides & surfaces res, 65-68, mgr solid state devices res, 68-69, mgr prod planning & develop, Semiconductor Div, Westinghouse Elec Corp, 69; prod mgr, Semiconductor Div, Motorola, Inc, Phoenix, Ariz, 69-72, corp dir res, 72-75; exec dir, Ariz Solar Energy Res Commn, 75-76; dir bus & tech planning, Integrated Circuits Div, 76-80, patent agt & semiconductor technol specialist, 80-84, PATENT ATTY, MOTOROLA, INC, 84- *Concurrent Pos:* Lectr, Carnegie Inst Technol, 64-65. *Mem:* Inst Elec & Electronics Engrs; Am Phys Soc. *Res:* Solid state electronics; semiconductors; thin films; tunneling; hot electron effects in solids; surface effects on semiconductors; semiconductor devices. *Mailing Add:* 4250 E Camelback Rd 300K Phoenix AZ 85018

HANE, CARL EDWARD, b Leavenworth, Kans, Mar 19, 43; m 66; c 2. METEOROLOGY. *Educ:* Univ Kans, BA & BS, 66; Fla State Univ, MS, 68, PhD(meteorol), 72. *Prof Exp:* Fel, Nat Ctr Atmospheric Res, 72-73; atmospheric res scientist, Battelle Northwest Lab, Battelle Mem Inst, 73-76; METEOROLOGIST ATMOSPHERIC RES, NAT SEVERE STORMS LAB, ENVIRON RES LABS, NAT OCEANIC & ATMOSPHERIC ADMIN, DEPT OF COM, 76- *Concurrent Pos:* Adj assoc prof meteorol, Univ Okla. *Mem:* Am Meteorol Soc; fel Cooperative Inst Mesoscale Meteorol Studies. *Res:* Numerical modelling of convective clouds; analysis of observations from severe thunderstorms; planning and execution of severe storms observational programs. *Mailing Add:* Nat Severe Storms Lab 1313 Halley Circle Norman OK 73069

HANEBRINK, EARL L, b Mar 24, 24; US citizen; m 57; c 2. ORNITHOLOGY, ANIMAL ECOLOGY. *Educ:* Southeast Mo State Col, BSE, 48; Univ Miss, MS, 55; Okla State Univ, EdD(zool), 65. *Prof Exp:* Sci instr high sch, Mo, 48-57; from instr to assoc prof, 58-69, PROF BIOL, ARK STATE UNIV, 69- *Concurrent Pos:* NSF lectr, Southeast Mo State Col, 62-67. *Mem:* Am Ornith Union; Wilson Ornith Soc; Nat Audubon Soc; Sigma Xi; Am Birding Asn. *Res:* Heronry in Mississippi County, Arkansas; bird populations and habitat selections of birds in northeastern Arkansas; pigeon milk and pigeon behavior studies; environmental surveys. *Mailing Add:* Dept Biol Sci Ark State Univ Box 67 State University AR 72467

HANEGAN, JAMES L, b Chicago, Ill, Apr 11, 44; m 65; c 2. COMPARATIVE PHYSIOLOGY. *Educ:* Northern Ill Univ, BS, 66; Univ Maine, MS, 68; Univ Ill, PhD(physiol), 70. *Prof Exp:* from asst prof to assoc prof, 70-81, chmn dept, 74-76, PROF BIOL, EASTERN WASH UNIV, 81- *Concurrent Pos:* Fel, NASA-Ames Res Ctr, 72-73. *Mem:* AAAS; Am Soc Zool; Sigma Xi. *Res:* Neurophysiology of insect flight; insect temperature regulation; central nervous system control of mammalian thermoregulation. *Mailing Add:* Dept Biol Eastern Wash Univ Cheney WA 99004

HANEL, RUDOLF A, b Krems, Austria, July 14, 22; US citizen; m 58; c 2. ATMOSPHERIC PHYSICS. *Educ:* Vienna Tech Univ, BS, 48, MS, 50, PhD(physics), 53. *Prof Exp:* Asst prof electronics, Vienna Tech Univ, 50-53; res physicist, US Army Res & Develop Lab, Ft Monmouth, NJ, 53-59; consult, Aeronomy & Meteorol Div, 59-65, chief scientist, Lab Planetary Atmospheres, 65-77, SR SCIENTIST LAB EXTRA TERRESTRIAL PHYSICS, GODDARD SPACE FLIGHT CTR, NASA, 77- *Honors & Awards:* Spec Serv Award, NASA Group Achievement Award & Sustained Super Performance Award, 63. *Res:* Ultrasonics; space electronics; physics of planetary atmospheres; infrared; Fourier spectroscopy. *Mailing Add:* 31 Brinkwood Rd Brookeville MD 20833

HANENSON, IRWIN BORIS, b New York, NY, Apr 7, 22; m 49; c 1. MEDICINE. *Educ:* NY Univ, BS, 43, MD, 46. *Prof Exp:* Intern, US Naval Hosp, St Albans, NY, 46-47; asst resident internal med, Vet Admin Hosp, Bronx, 49-51; res asst med, Montefiore Hosp, 52-53; instr, Col Med, Univ Cincinnati, 55-58, asst prof exp med, 58-65, lectr pharmacol, 65-66, from asst prof to assoc prof med, 65-74, asst prof pharmacol, 66-69, assoc prof, 70-85, actg dir clin pharmacol, Dept Internal Med, 72-73, actg asst dir, Dept Lab Med, 74-75, prof med & path, 74-85, assoc dir, Dept Lab Med, 76-81, dir qual assurance, 80-85, asst dean clin & housestaff affairs, 81-85, EMER PROF MED, COL MED, UNIV CINCINNATI, 85- *Concurrent Pos:* Fel cardiol, Montefiore Hosp, 51-52, Rosenstock Found fel, 52-53; res fel, Harvard Med Sch, 53-55; asst, Peter Bent Brigham Hosp, Mass, 53-55; asst clinician, Cincinnati Gen Hosp, Ohio, 55-66, clinician & attend physician, 67-, dir hypertension clin, 68-, dir clin toxicol, 71-72; assoc, May Inst Med Res, Jewish Hosp, Cincinnati, 55-65, attend physician, 55-, physician-in-chief spec study unit, 63-65; estab investr, Am Heart Asn, 60-65. *Mem:* AAAS; Am Fedn Clin Res; Am Soc Pharmacol & Exp Therapeut; Soc Exp Biol & Med; Am Heart Asn; Cent Soc Clin Res; Sigma Xi. *Res:* Hypertension; clinical toxicology; biochemical and physiological studies of renal function; body fluids and electrolytes. *Mailing Add:* Med Ctr ML 578 Univ Cincinnati 231 Bethesda Ave Cincinnati OH 45267

HANES, DEANNE MEREDITH, b Weehawken, NJ, Aug 7, 42. BIOLOGY, IMMUNOLOGY. *Educ:* Hunter Col, BA, 64; Univ Calif, Davis, PhD(nutrit, biochem), 70. *Prof Exp:* NIH fel, 70-73, RES IMMUNOLOGIST, DEPT SURGERY, TRANSPLANT UNIT, HISTOCOMPATABILITY LAB, MED CTR, UNIV CALIF, SAN FRANCISCO, 73- *Concurrent Pos:* Res asst lysosomal biochem, Univ Calif, Davis, 66-70. *Mem:* AAAS; Sigma Xi; Am Soc Human Genetics. *Res:* Transplantation immunology; aspects of immunology at the cellular level. *Mailing Add:* 317 Michelle Lane Dale City CA 94015-2879

HANES, HAROLD, b Thebes, Ill, Apr 10, 31; m 56; c 2. MATHEMATICS. *Educ:* Tex Christian Univ, BA, 57; Univ Kans, MA, 59, PhD(math), 67. *Prof Exp:* From asst prof to assoc prof, 62-71, PROF MATH, EARLHAM COL, 71- *Mem:* Asn Comput Mach; Math Asn Am; Am Math Soc; Soc Indust & Appl Math. *Res:* Theory of finite groups, specifically factorizability. *Mailing Add:* Dept Math Earlham Col Richmond IN 47374

HANES, N BRUCE, b Minot, NDak, Jan 24, 34; m 54; c 5. ENVIRONMENTAL ENGINEERING. *Educ:* NDak State Univ, BS, 54; Univ Wis, MS, 57, PhD(civil eng), 61. *Prof Exp:* Instr sanit eng, Univ Wis, 55-57; instr environ eng, Mont State Col, 57-59; from asst prof to assoc prof,

61-71, chmn, Dept Civil Eng, 69-81, PROF ENVIRON ENG, TUFTS UNIV, 71- *Concurrent Pos:* Consult, Nat Coun Stream Improv, 63-64, US Public Health Serv, Environ Protection Agency, & NSF. *Mem:* Am Water Works Asn; Water Pollution Control Fedn; Am Acad Environ Engrs; Asn Environ Eng Professors (past pres). *Res:* Water pollution, survival of indicator bacteria in water and oxygen demand of benthal deposits. *Mailing Add:* Dept Civil Eng Tufts Univ Medford MA 02155

HANES, RONNIE MICHAEL, b Birmingham, Ala, Apr 15, 49; m 71; c 2. ORGANIC CHEMISTRY, CATALYSIS. *Educ:* Univ Ala, BS, 72, PhD(org chem), 76. *Prof Exp:* Res assoc, Dept Chem, Univ Ga, 76-77; sr res chemist, 77-87, GROUP LEADER, US INDUST CHEM, NAT DISTILLERS & CHEM CORP, 87- *Mem:* Am Chem Soc. *Res:* Organic synthesis via homogeneous catalysis; supported homogeneous catalysts. *Mailing Add:* 1275 Section Rd Cincinnati OH 45222

HANES, TED L, b Los Angeles, Calif, Mar 31, 28; m 60; c 4. PLANT ECOLOGY. *Educ:* Univ Calif, Los Angeles, BS, 50, MS & PhD(plant ecol), 63; Claremont Grad Sch, MA, 58. *Prof Exp:* Teacher high sch, Calif, 55-57; biologist, Citrus Jr Col, 57-69; dir Arboretum, 80-85, PROF BOT, CALIF STATE UNIV, FULLERTON, 69- *Concurrent Pos:* Environ consult, NSF res grant, 66-68, Ultrasysts, Inc, 73- & numerous environ firms. *Mem:* AAAS; Ecol Soc Am; Nat Audubon Soc. *Res:* Ecology of chaparral vegetation; chaparral succession after fire in the mountains of southern California; natural areas; environmental assessment and impact studies; vegetation community structure and dynamics. *Mailing Add:* Dept Bot Calif State Univ Fullerton CA 92634

HANESIAN, DERAN, b Niagara Falls, NY, Sept 26, 27; m 86. CHEMICAL ENGINEERING. *Educ:* Cornell Univ, BChE, 52, PhD(chem eng), 61. *Prof Exp:* Engr, 52-57, res engr, E I du Pont de Nemours & Co Inc, 60-63; from asst prof to assoc prof, 63-70, chmn dept, 75-88, PROF CHEM ENG, CHEM & ENVIRON SCI, NJ INST TECHNOL, 63-, CONSULT, CTR PLASTICS RECYCLING RES, RUTGERS, 88- *Concurrent Pos:* Grant, NSF, 67-69 & 72-74, German Acad Exchange Serv, 81-82; Fulbright scholar, USSR, 82; vis prof, Univ Edinburgh, Scotland, 81, Ctr Plastics Recycling Res, Rutgers, State Univ NJ, 89-90; mem, Adv Bd J Int Chem Eng, Am Inst Chem Engrs, 73-79, Tech prog comt, 67 & 77, prof develop comt, 70-86, nat prog comt, 74-86, secy, vchmn & chmn, 74-79, educ projs comt, 69-, vchmn, 80-82, chmn, 83-85, past chmn, 86; secy, Div Exp & Lab Oriented Studies, Am Soc Eng Educ, 78- 80, prog chmn/chmn-elect, 80-81, chmn, 82-83, past chmn, 83-84, nominating comt chmn, 86; vchmn, Chem Eng Div, Am Soc Eng Educ, 83-84, chmn, 84-85, past chmn, 85-86, mem comt, 85-87, nominating comt, 85-88; consult, E I du Pont de Nemours, 64, 65 & 66, Exxon Affil, 67, NJ Inst Technol, 72, 73 & 74, Celanese Corp, 77 & 80, Ctr Plastics Recycling Res, Rutgers, State Univ NJ, 88-; invited lectr, Algerian Petrol Inst, 78. *Honors & Awards:* Mid-Atlantic AT&T Found Award, Am Soc Eng Educ. *Mem:* Am Inst Chem Eng; Am Chem Soc; Am Soc Eng Educ; Sigma Xi. *Res:* Chemical kinetics and chemical reaction engineering; process dynamics, simulation and control optimization of chemical reactor systems; plastics recycling. *Mailing Add:* Dept Chem Eng Chem & Environ Sci 323 Dr Martin Luther King Jr Blvd Newark NJ 07102

HANESSIAN, STEPHEN, b Alexandria, Egypt, Apr 25, 35; US citizen; m 57; c 3. ORGANIC CHEMISTRY. *Educ:* Univ Alexandria, BSc, 56; Ohio State Univ, PhD(chem), 60. *Prof Exp:* Res chemist, Starch Prod Co, 56-57; NIH fel, 57-59; Charles Kettering fel, 59-60; res chemist, Parke, Davis & Co, 61-68; assoc prof, 68-70, PROF CHEM, UNIV MONTREAL, 70- *Honors & Awards:* Merck Sharpe & Dohme Award, Chem Inst Can, 74. *Mem:* Am Chem Soc. *Res:* Carbohydrates; antibiotics; natural products. *Mailing Add:* Dept Chem Universite de Montreal CP 6128 Succursale A Montreal PQ H3C 3Z7 Can

HANEY, ALAN WILLIAM, b Portsmouth, Ohio, Oct 12, 41; m 79; c 1. PLANT ECOLOGY. *Educ:* Ohio State Univ, BS, 63; Yale Univ, MF, 65; State Univ NY, PhD(forest ecol), 68. *Prof Exp:* From asst prof to assoc prof bot, Univ Ill, Urbana, 68-77; PROF BIOL, WARREN WILSON COL, 77- *Concurrent Pos:* Grant ecol wild marijuana in Ill, 69-70 & grant effects wildlife & logging on bird pop. *Mem:* Ecol Soc Am; Soc Am Foresters; Am Orinthologist's Union. *Res:* Ecology of colonizing species; mechanisms of plant interactions, especially plant succession; chemical interactions between weed and crop plants; population dynamics and community structure in disturbed forest communities. *Mailing Add:* 701 Warren Wilson Rd Box 5064 Swannanoa NC 28778

HANEY, DONALD C, b Ferguson, Ky, July 2, 34; m 56; c 2. GEOLOGY, SOILS. *Educ:* Univ Ky, BS, 60, MS, 62; Univ Tenn, PhD(geol), 66. *Prof Exp:* Instr geol, Campbellsville Col, 60-62; from instr to assoc prof geol, Eastern Ky Univ, 62-76, chmn dept, 68-80, prof, 76-80; MEM FAC, DIR & STATE GEOLOGIST, GEOL SURV, UNIV KY, 80- *Mem:* Geol Soc Am. *Res:* Structural geology of east Tennessee; research with Pennsylvanian's sediments in eastern Kentucky, including structure, sedimentology and coal resources. *Mailing Add:* Ky Geol Surv Univ KY 228 Mining & Mineral Resources Bldg Lexington KY 40506-0107

HANEY, JAMES FILMORE, b Bergholz, Ohio, May 20, 38; m; c 3. LIMNOLOGY, STREAM ECOLOGY. *Educ:* Miami Univ, BA, 61, MA, 63; Univ Toronto, PhD(limnol), 70. *Prof Exp:* Instr zool, Miami Univ, 65-66; res assoc limnol, Kellog Biol Sta, Mich State Univ, 70-72; from asst prof to assoc prof, 72-86, PROF ZOOL, UNIV NH, 86- *Concurrent Pos:* Consult, Radiol & Environ Res Div, Argonne Nat Lab, 74-; co-dir, NH Lakes Lay Monitoring Prog, 79- *Mem:* Am Soc Limnol & Oceanog; Int Soc Theoret & Appl Limnol. *Res:* Interactions of zooplankton and phytoplankton communities; feeding relationships of zooplankton; factors which regulate and consequences of diel vertical migration. *Mailing Add:* Dept Zool Univ NH Durham NH 03824

HANEY, PAUL D, sanitary engineering, chemical engineering; deceased, see previous edition for last biography

HANEY, ROBERT LEE, b Newport News, Va, Oct 14, 38; m 66; c 1. METEOROLOGY, OCEANOGRAPHY. *Educ:* George Wash Univ, BA, 64; Univ Calif, Los Angeles, PhD(meteorol), 71. *Prof Exp:* From asst prof to assoc prof meteorol, 70-83, PROF METEOROL, NAVAL POSTGRAD SCH, 83- *Concurrent Pos:* Vis prof, Univ Hawaii, 83-84; co-ed, J Phys Oceanog, 83-88. *Mem:* Am Meteorol Soc; Sigma Xi; Nat Geog Soc. *Res:* Numerical modelling and analysis of coastal oceanography. *Mailing Add:* Dept Meteorol Naval Postgrad Sch Code MR/Hy Monterey CA 93943-5000

HANFF, ERNEST SALO, b Santiago, Chile, July 16, 40; Can citizen; m 64; c 2. ELECTRICAL ENGINEERING. *Educ:* Univ Toronto, BASc, 65; Univ Western Ontario, MESc, 67, PhD(elec eng), 71. *Prof Exp:* Asst res officer, 71-75, assoc res officer, 75-81, SR RES OFFICER, NAT RES COUN, 82- *Concurrent Pos:* Lectr, Algonquin Col Technol, 74-77; chmn comt instrumentation, Aerospace Simulation Facilities, Inst Elec & Electronics Engrs, 77- *Mem:* Inst Elec & Electronics Engrs; Am Inst Aeronaut & Astronaut. *Res:* Data handling; signal extraction and instrumentation particularly as applied to studies of the dynamic stability and performance of aircraft. *Mailing Add:* NRC Montreal Rd Ottawa ON K1A 0R6 Can

HANFORD, WILLIAM EDWARD, b Bristol, Pa, Dec 9, 08; m 39; c 2. CHEMISTRY. *Educ:* Philadelphia Col Pharm, BS, 30; Univ Ill, MS, 32, PhD(org chem), 35. *Hon Degrees:* DSc, Philadelphia Col Pharm, 56, Alfred Univ, 59. *Prof Exp:* Anal chemist, Rohm and Haas, Pa, 30-31; asst chem, Univ Ill, 32-35; res chemist, Exp Sta, E I du Pont de Nemours & Co, Del, 35-36, group leader, 36-42; from asst dir res to dir res, Gen Aniline & Film Corp, Pa, 42-46; tech consult, M W Kellogg Co, 46-47, dir petrol & chem res, 48-50, vpres, dir res & mem bd dirs, 50-57; asst to pres, Olin Corp, NY, 57, vpres res & develop, Conn, 57-73; consult, World Water Resources Inc, 73-85; RETIRED. *Honors & Awards:* Gold Medal, Am Inst Chemists, 74. *Mem:* AAAS; Am Chem Soc; Soc Indust Chem; Am Inst Chem Engrs. *Res:* Synthetic organic chemistry; high pressure reactions; polymerization; petroleum chemistry. *Mailing Add:* 4956 Sentinel Dr Apt 306 Bethesda MD 20816

HANFT, RUTH S, b New York, NY, July 12, 29; m 51; c 2. HEALTH POLICY, HEALTH ECONOMICS. *Educ:* Cornell Univ, BS, 49; Hunter Col, MA, 63. *Prof Exp:* Pub admin intern & consult mgt, NY State Dept Social Welfare, 51-54; res assoc health econ, Health Res Coun, 62-63; social sci analyst, Social Security Admin, 64-66; prog analyst, Health Care Off Econ Opportunity, 66-68, Health Care & Econ Dept HEW, 68-71; spec asst to asst secy, Health Financing, Dept HEW, 71-72; sr res assoc, Health Econ Inst Med, 72-76; consult, Health Policy Self Employ, 76-77; dep asst secy, Health Policy, Dept HEW, 77-78; dep asst secy health res, Statist Technol, 79-81; consult, 81-91; RES PROF, DEPT HEALTH SCI MGT & POLICY, GEORGE WASHINGTON UNIV, 88- *Concurrent Pos:* Consult health econ, DC Dept Health, 64; vis prof, Health Policy, Dartmouth Col, 76-; res prof, George Washington Univ. *Mem:* Inst Med-Nat Acad Sci; fel Hastings Inst. *Res:* Health care financing; health manpower. *Mailing Add:* Dept Health Serv Mgt & Policy George Washington Univ 600 21st NW Washington DC 20052

HANG, DANIEL F, b Cleveland, Ohio, July 17, 18; m 41; c 2. ENGINEERING ECONOMICS. *Educ:* Univ Ill, BS, 41, MS, 49. *Prof Exp:* Proj engr, Gen Elec Co, NY, 41-47; prof, 47-84, emer prof elec & nuclear eng, Univ Ill, Urbana, 84; PRES, H T H ASSOCS, 78- *Concurrent Pos:* Consult, Argonne Nat Lab, 61-80, Commonwealth Edison Co & US Army Corps Eng, 71-76, Nat Coun Examiners Eng & Surv, 84-; mem, Ill State Prof Eng Exam Comt, 71- & Ill Atomic Energy Comn, 78-85. *Honors & Awards:* Distinguished Serv Award, Nat Coun Examiners Eng & Surv, 90. *Mem:* Nat Soc Prof Engrs; Inst Elec & Electronics Engrs; Am Soc Eng Educ; Am Nuclear Soc; Nat Coun Examiners Eng & Surv. *Res:* Engineering economy; nuclear fuel cycle; nuclear fuel management; electrical engineering; power systems. *Mailing Add:* 229C Nuclear Eng Lab Univ Ill 103 South Goodwin Ave Urbana IL 61801-2984

HANG, HSUEH-MING, b Taipei, Taiwan, May 22, 56. VISUAL COMMUNICATION, SIGNAL PROCESSING. *Educ:* Nat Chiao-Tung Univ, BS, 78, MS, 80; Rensselaer Polytech Inst, PhD(elec eng), 84. *Prof Exp:* Instr elec eng, Rensselaer Polytech Inst, 83-84; MEM TECH STAFF ELEC ENG, AT&T BELL LABS, 84- *Mem:* Sr mem Inst Elec & Electronics Engrs. *Res:* Compression algorithms for still and motion video and their hardware implementation in very large scale integration; algorithm and architecture for digital image/signal processing. *Mailing Add:* 23 Southport Dr Howell NJ 07731

HANG, YONG DENG, b Kampot, Cambodia. FOOD SCIENCE, WATER POLLUTION. *Educ:* Nat Taiwan Univ, BS, 62; Univ Alta, MS, 65; McGill Univ, PhD(microbiol), 68. *Prof Exp:* Res assoc plant proteins, NY State Agr Exp Sta, 68-70, res assoc water pollution, 70-76, asst prof, 76-82, ASSOC PROF FOOD SCI, CORNELL UNIV, 82- *Mem:* Inst Food Technologists; Am Soc Microbiol; Inst Food Technologists; Sigma Xi. *Res:* Microbial degradation of phenolic compounds; development of practical methods of isolation, processing and utilization of plant proteins; effects of processing conditions on fruit and vegetable effluents, biological treatment of processing effluents; utilization of agricultural and industrial wastes; production of fuels, chemicals and biologics from waste materials; solid-state fermentation systems. *Mailing Add:* Dept Food Sci/Tech NY Agr Expt Sta Geneva NY 14456

HANIC, LOUIS A, b Secovce, Czech, Jan 20, 28; Can citizen; m 57; c 2. PHYCOLOGY. *Educ:* Univ BC, BA, 50, PhD(biol), 65. *Prof Exp:* Fel biol, Dalhousie Univ, 65-66, asst prof, 66-71; assoc prof, 71-80, PROF BIOL, UNIV PEI, 80- *Mem:* Int Phycol Soc; Brit Phycol Soc; Can Bot Asn. *Res:* Marine algae; life histories; cultivation Irish moss; cytology; structure of the cell wall. *Mailing Add:* Dept Biol Univ PEI Charlottetown PE C1A 4P3 Can

HANIG, JOSEPH PETER, b New York, NY, Apr 29, 41; m 69; c 3. NEUROPHARMACOLOGY, TOXICOLOGY. *Educ:* Rutgers Univ, BS, 62; NY Med Col, MS, 65, PhD(pharmacol), 68; Am Bd Toxicol, dipl, 80, recert, 85 & 90; Acad Toxicol Sci, dipl, 84, recert, 89. *Prof Exp:* Pharmacologist & proj leader, Div Pharmacol, Bur Sci, 68-70, pharmacologist & proj leader, 70-73, GROUP LEADER, CTR DRUG EVAL & RES, DIV RES & TESTING, US FOOD & DRUG ADMIN, 73- *Concurrent Pos:* NIH fel, 68; Nat Acad Sci-Nat Res Coun res assoc, 68-70; adj asst prof pharmacol, New York Med Col, 74-75; vis lectr toxicol, Sch Med, Howard Univ, 75-81, vis assoc prof toxicol, 81-89, vis prof toxicol, 89-; adj assoc prof pharmacol, New York Med Col, 75- *Mem:* AAAS; Am Soc Pharmacol & Exp Therapeut; Soc Exp Biol & Med; Am Chem Soc; Am Inst Chemists; Soc Toxicol; Asn Govt Toxicologists. *Res:* Blood-brain barrier permeability; drug effects, pharmacologic mechanisms; biogenic amine and ethanol-biogenic amine interactions in brain; neurochemistry, effects on behavior; studies of putative neurotransmitter substances; drug toxicity; drug interactions, adverse effects; effects of drugs on intracranial pressure; adrenergic receptor sensitivity in hypertension; effects of nutrition on drug toxicity. *Mailing Add:* 822 Eden Ct Alexandria VA 22308

HANIN, ISRAEL, b Shanghai, China, Mar 29, 37; US citizen; m 60; c 2. PSYCHOPHARMACOLOGY. *Educ:* Univ Calif, Los Angeles, BS, 62, MS, 65, PhD(pharmacol), 68. *Prof Exp:* Vis res scientist, Dept Toxicol, Karolinska Inst, Sweden, 68; Nat Inst Neurol Dis & Blindness fel & pharmacologist, Lab Preclin Pharmacol, NIMH, St Elizabeth Hosp, Washington, DC, 69-73; prog dir psychopharmacol, Dept Psychiat, Western Psychiat Inst & Clin, Sch Med, Univ Pittsburgh, 73-86, from asst prof to prof psychiat & pharmacol, 73-86; PROF & CHMN, DEPT PHARMACOL & DIR, NEUROSCI & AGR INST, STRITCH SCH MED, LOYOLA UNIV CHICAGO, 86- *Mem:* Am Chem Soc; fel Am Col Neuropsychopharmacol; Am Soc Neurochem; Am Soc Pharmacol & Exp Therapeut (pres, 90-92); Soc Neurosci; Collegium Int Psychopharmacol. *Res:* Measurement of the extent of neurotransmitter involvement, particularly of acetylcholine, in various psychiatric disorders; measurement and correlation of the neurochemical findings with effects of drug therapy on clinical course; investigation in research animals of the effect of psychoactive drugs on the neurotransmitter mechanisms under investigation; development of animal models of neuropsychiatric disease states; animal models of Alzheimer's disease; cholinotoxins; nootropic drugs. *Mailing Add:* Dept Pharmacology Loyola Univ Chicago Stritch Sch Med 2160 S First Ave Maywood IL 60153

HANING, BLANCHE COURNOYER, b Paxton, Mass, Mar 7, 43; m 70. PHYTOPATHOLOGY, EDUCATIONAL ADMINISTRATION. *Educ:* Univ Mass, BS, 65; Iowa State Univ, MS, 67, PhD(plant path), 70. *Prof Exp:* Fel plant pathol, DeKalb AgRes, Inc, 70-74; asst prof biol sci, Northern Ill Univ, 75-76; ACAD COORDR PEST MGT & ASSOC PROF PLANT PATH, NC STATE UNIV, 77-, ASSOC PROF ENTOM, 80- *Concurrent Pos:* Mem comt, Nat Constraints Work Group, Off Technol Assessment, 78-79; mem, Food & Agr Systs Anal, Col & Univ Fac Develop Workshop, 86; assoc prof biol sci, NC State Univ. *Mem:* Am Phytopath Soc; Entom Soc Am. *Mailing Add:* NC State Univ Box 7611 Raleigh NC 27695-7611

HANIS, CRAIG L, b Seattle, Wash, Apr 11, 52; m 74; c 4. GENETIC EPIDEMIOLOGY. *Educ:* Brigham Young Univ, BS, 74, MS, 77; Univ Mich, MA, 81, PhD(human genetics), 81. *Prof Exp:* Res instr human genetics, Grad Sch Biomed Sci, 81-83, from asst prof to assoc prof, 83-91, ASSOC PROF EPIDEMIOL, SCH PUB HEALTH, UNIV TEX HEALTH SCI CTR, 87-, PROF, GRAD SCH BIOMED SCI, 91- *Concurrent Pos:* Res career develop award, NIH, 87; acting dir, Ctr Demographic & Pop Genetics & Med Genetics Ctr, Grad Sch Biomed Sci, Univ Tex Health Sci Ctr, 90- *Mem:* Am Soc Human Genetics; Am Diabetes Asn. *Res:* Genetics and epidemiology of common chronic diseases, including diabetes, cardiovascular disease and gallbladder disease; chronic disease among Mexican Americans. *Mailing Add:* Genetic Marker Lab Genetic Ctr Grad Sch Biomed Sci Univ Tex Health Sci Ctr Houston PO Box 20334 Houston TX 77225

HANKA, LADISLAV JAMES, b Mnetice, Czech, July 31, 20; nat US; m 48; c 2. MICROBIOLOGY. *Educ:* Prague Tech Univ, Ing, 46; Iowa State Col, MS, 56, PhD(bact), 58. *Prof Exp:* Teacher high sch, Czech, 42-45; agr adminr, 47-50; asst bact, Iowa State Col, 56-58; res assoc anal microbiol, 58-67 & biochem res, 67-71, SR RES SCIENTIST CANCER RES, UPJOHN CO, 71- *Mem:* Am Soc Microbiol; Am Asn Cancer. *Res:* Am Chem Soc. Res: Analytical microbiology; microbiological assay; antibiotics; reproduction physiology; fermentations; cryobiology; antimetabolites in cancer chemotherapy; methods of finding new antitumor drugs; international cooperation in cancer research. *Mailing Add:* 2917 Grace Rd Kalamazoo MI 49007

HANKEL, RALPH D, nuclear engineering, thermodynamics; deceased, see previous edition for last biography

HANKEN, JAMES, b New York, NY, July 14, 52; m 85; c 2. EVOLUTIONARY MORPHOLOGY, HERPETOLOGY. *Educ:* Univ Calif, Berkeley, AB, 73, PhD(zool), 80. *Prof Exp:* Killam res fel, Dalhousie Univ, 80-83; asst prof, 83-90, ASSOC PROF BIOL, UNIV COLO, BOULDER, 90- *Concurrent Pos:* Mem bd gov, Am Soc Ichthyologists & Herpetologists, 83-88 & publ secy, 85-88; prog officer, Am Soc Zoologists, Div Vert Morphol, 88-89 & chair-elect, 91-92; assoc ed, J Morphol, 88-, Evolution, 91- *Mem:* Am Soc Zoologists; Soc Study Evolution; Sigma Xi; Am Soc Ichthyologists & Herpetologists. *Res:* Evolutionary morphology of vetebrates; developmental and evolutionary questions about the skull and limb skeleton. *Mailing Add:* Dept Environ Population & Organismic Biol Univ Col Boulder CO 80309-0334

HANKER, JACOB S, b Philadelphia, Pa, Feb 23, 25; m 51; c 3. HISTOCHEMISTRY, CELL BIOLOGY. *Educ:* St Joseph's Col, Pa, BS, 48; Univ Md, PhD(med chem), 69. *Prof Exp:* Group leader org chem, US Army Chem Welfare Labs, 52-60; supvr surg res labs, Sinai Hosp, Baltimore, Md, 60-69; prof neurobiol prog, Univ, 69-77, PROF ORAL SURG & ORAL BIOL, DENT RES CTR, SCH DENT, UNIV NC, CHAPEL HILL, 69- *Concurrent Pos:* Consult, Polysci, Inc, Pa, 65-; asst in surg, Sch Med, Johns Hopkins Univ, 67-69; consult, NIMH, 71 & Papanicolaou Cancer Res Inst, US Navy; adj assoc prof path, Duke Univ Med Ctr, 81- *Mem:* AAAS; Am Chem Soc; Histochem Soc; Sigma Xi; Am Soc Cell Biol. *Res:* Development of cytochemical staining methods for the light and electron miroscopic study of leukocytes in health and disease, especially leukemia, granuloma and malignancy. *Mailing Add:* Dept Biomed Eng CB 7455 Univ NC Chapel Hill NC 27599-7453

HANKES, GERALD H, b Aurora, Ill, Aug 27, 36; m 59; c 4. VETERINARY SURGERY. *Educ:* Univ Ill, Urbana, BS, 59, DVM, 61; Colo State Univ, MS, 67, PhD(cardiovasc surg & physiol), 69. *Prof Exp:* Gen pract large & small animals, Ill, 61-62; instr small animal surg & med, Colo State Univ, 64-65; NIH fels, Nat Inst Arthritis & Metab Dis, 65-68 & Nat Heart Inst, 68-69; PROF SMALL ANIMAL SURG & MED, AUBURN UNIV, 69- *Mem:* Am Vet Med Asn. *Res:* Cardiovascular surgery and physiology; intensive patient care; anesthesiology; malignant neoplasia. *Mailing Add:* 1220 Sanders St Auburn AL 36830

HANKES, LAWRENCE VALENTINE, b Chicago, Ill, Nov 24, 19; m 51; c 3. BIOCHEMISTRY. *Educ:* DePauw Univ, AB, 42; Mich State Col, MS, 43; Univ Wis, PhD, 49. *Prof Exp:* Biochemist, Vet Admin Hosp, 50; BIOCHEMIST & HEAD CLIN CHEM DEPT, BROOKHAVEN NAT LAB, 51-, SR CLIN SCIENTIST, 68- *Concurrent Pos:* Mem, Nat Registry Clin Chemists; vis clin scientist, SAfrican Coun Sci & Indust Res, 65, sr clin scientist, SAfrican Atomic Energy Bd & SAfrican Med Res Coun, 68, 69, 72, 73, 75, Inst Med, Kernforschungsanlage, WGer, 87 & 88; USA partic, Int Study Group for Tryptophan Res, Padova, Italy, 74, Madison, Wis, 77, Kyoto, Japan, 80 & Munich, Ger, 83 & Baltimore, Md, 89. *Mem:* Am Soc Biol Chemists; Am Chem Soc; Soc Exp Biol & Med; Am Asn Clin Chemists; Nat Acad Clin Biochem; Asn Clin Scientists. *Res:* Animal nutrition; amino acid interrelationships; vitamins; synthetic and analytical radiochemistry; radiation effects on enzymes; neoplastic diseases metabolism; Trichinella spiralis larvae metabolism; quantitative clinical chemistry; tryptophan metabolism in scurvy, pellagra, scleroderma, dementia, Alzheimers, Huntingtons, and cancer; I-inositol meabolism in Pentosuria. *Mailing Add:* PO Box 1056 Setauket NY 11733-0804

HANKEY, WILBUR LEASON, JR, b New Kensington, Pa, Oct 31, 29; m 54; c 3. AERONAUTICAL ENGINEERING. *Educ:* Pa State Univ, BS, 51; Mass Inst Technol, SM, 53; Ohio State Univ, MS, 58, PhD(aeronaut eng), 62. *Prof Exp:* Res asst fluid mech, Mass Inst Technol, 51-53; proj engr, transonic wind tunnel, US Dept Air Force, Wright-Patterson AFB, 53-58, aerothermodynamics chief, 58-60, aerodynamics chief, Dynasoar Proj, 60-64, group leader hypersonics res, Aerospace Res Labs, 64-75, group leader, Flight Dynamics Lab, 75-84; PROF, WRIGHT STATE UNIV, DAYTON, OHIO, 85- *Concurrent Pos:* Adj prof, US Air Force Inst Technol, 64- *Honors & Awards:* Primus AFSC Award. *Mem:* Am Inst Aeronaut & Astronaut; Sigma Xi; Am Soc Mech Engrs. *Res:* Optimization studies of hypersonic lifting bodies; boundary layer-shock interactions in supersonic flow; trajectory and aerothermodynamic studies of hypersonic lifting vehicles; computational fluid dynamics; unsteady flows. *Mailing Add:* 5738 Thatchwood Circle Dayton OH 45431

HANKIN, JEAN H, b Ewen, Mich, May 7, 23. NUTRITION, EPIDEMIOLOGY. *Educ:* Milwaukee-Downer Col, BS, 45; Univ Tenn, Knoxville, MS, 54; Univ Calif, Berkeley, MPH, 63, MPH & DrPH(nutrit, epidemiol), 66. *Prof Exp:* Dietetic intern, Cincinnati Gen Hosp, 46; supvry dietitian, Cook County Hosp, Chicago, 46-48; dietician, Milwaukee Children's Hosp, 48-50 & Detroit Children's Hosp, 51-52; nutrit apprentice, Mich Dept Health, 52-53; pub health nutritionist, City Milwaukee Health Dept, 54-59; nutrit consult, Heart Dis Control Prog, USPHS, 59-60 & RI Dept Health, 60-62; instr pub health & res nutritionist, Univ Calif, Berkeley, 67-69; prof pub health, Manoa, 69-88, prof family pract & community health, Sch Med, 80-88, NUTRIT RESEARCHER & PROF PUB HEALTH, EPIDEMIOL PROG, CANCER RES CTR, UNIV HAWAII, 88- *Concurrent Pos:* Mem, Coun Epidemiol, Am Heart Asn, 69; dir proj grants, Pub Health Nutrit Training Prog, 75-83; prin investr & co-prin investr, diet & cancer epidemiol studies, Cancer Res Ctr, Univ Hawaii. *Honors & Awards:* Cottes Gen Foods Lectr, Australia, 80; Plenary lectr, Nutrit Soc Australia, 83; Lenna Frances Cooper Lectr, Am Dietetic Asn, 85. *Mem:* Am Dietetic Asn; Soc Nutrit Educ; fel Am Pub Health Asn; Am Heart Asn; Am Inst Nutrit; Am Soc Clin Nutrit; Soc Epidemiol Res. *Res:* Diet and breast cancer; dietary methods in epidemiological studies; diet, ethnicity & cancer risk; use of scientific methods for planning, conducting and evaluating applied nutrition programs. *Mailing Add:* Epidemiol Prog Cancer Res Ctr Univ Hawaii 1236 Lauhala Honolulu HI 96813

HANKIN, LESTER, b Norwich, Conn, Feb 23, 26; m 49; c 3. FOOD SCIENCE & TECHNOLOGY. *Educ:* Univ Conn, BS, 49, MS, 51; NC State Col, PhD(food sci), 54. *Prof Exp:* Asst dairy bact, NC State Col, 51-54; from asst biochemist to biochemist, 54-83, HEAD, DEPT ANALYTICAL CHEM, CONN AGR EXP STA, 84- *Mem:* Am Soc Microbiol; Asn Off Anal Chem; Int Asn Milk, Food & Environ Sanit; Am Chem Soc; Asn Food & Drug Off. *Res:* Interrelationship of nutrients to growth; methods in food analysis; solid waste recycling; environmental pollutants. *Mailing Add:* Conn Agr Exp Sta PO Box 1106 New Haven CT 06504

HANKINS, B(OBBY) E(UGENE), b Wilson, Ark, Feb 22, 29; m 58; c 1. ANALYTICAL CHEMISTRY, GEOPRESSURED-GEOTHERMAL ENERGY. *Educ:* Univ Cent Ark, BS, 51; Univ Mo, MA, 53, PhD(chem), 57. *Prof Exp:* Lab asst, Univ Cent Ark, 50-51; asst, Univ Mo, 51-57; chemist, Union Carbide Nuclear Co, 57-59; from asst prof to assoc prof, 59-63, chmn dept, 70-80, dean col sci, 80-87, PROF CHEM, MCNEESE STATE UNIV, 64-, VPRES ACAD AFFAIRS, 87- *Concurrent Pos:* Consult Dept Energy, 80-88. *Mem:* Am Chem Soc; Am Inst Chemists; Royal Soc Chem. *Res:*

Instrumental analysis; chemical concentration of trace elements; geopressured-geothermal energy; analysis of fluids from geopressured-geothermal wells. *Mailing Add:* Acad Affairs PO Box 93220 McNeese State Univ Lake Charles LA 70609

HANKINS, GEORGE THOMAS, b England, Ark, Nov 10, 25; m 51; c 6. ELECTRICAL ENGINEERING. *Educ:* US Air Force Inst Technol, BS, 55; Southern Methodist Univ, MS, 61; Union Grad Sch, PhD(elec eng, educ), 77. *Prof Exp:* Pilot, US Air Force, 47-53, engr, airborne radar systs, 51-53, proj engr, long range radar syst, Air Tech Intel Ctr, 55-59, br chief instrumentation data anal, Foreign Technol Div, Air Force Systs Command, 61-63, div chief space explor systs, 63-64, chief engr, space systs, 64-68; asst prof, Wright State Univ, 69-73, assoc prof, 73-80; assoc prof, 80-85, PROF ELEC ENG, UNIV PACIFIC, 85- *Concurrent Pos:* Western Elec fund award, 78. *Mem:* AAAS; Am Inst Elec & Electronics Engrs; Am Soc Eng Educ; Soc Hist Technol. *Res:* Radar systems; spaceborne electronic systems; digital systems; social dynamics of technology; engineering education methods; computer networking. *Mailing Add:* Univ Pacific Stockton CA 95211

HANKINS, TIMOTHY HAMILTON, b Miami, Fla, Mar 13, 41; m 77; c 2. RADIO ASTRONOMY, SIGNAL PROCESSING. *Educ:* Dartmouth Col, BA, 62, MS, 67; Univ Calif, San Diego, PhD(astron), 71. *Prof Exp:* Res physicist radio astron, Univ Calif, San Diego, 71-74; res assoc radio astron, Arecibo Observ, 74-81; assoc prof, Dartmouth Col, 81-88; assoc prof, 88-90, PROF NMEX TECH, 90- *Concurrent Pos:* Lectr physics, Univ Calif, San Diego, 71-75; Alexander von Humbolt fel, 78-79. *Mem:* Am Astron Soc; Int Sci Radio Union; Int Astron Union. *Res:* Radio observations of pulsars, including high-time resolution, polarization, microstructure dispersion measurement, hardware and software dispersion removal and signal processing; electronic engineering instrumentation. *Mailing Add:* Physics Dept NMex Tech Socorro NM 87801

HANKINS, WILLIAM ALFRED, b Indianapolis, Ind, July 4, 34; m 56; c 2. VIROLOGY. *Educ:* Purdue Univ, BS, 56, MS, 59, PhD(microbiol), 63. *Prof Exp:* Microbiologist, Virus & Rickettsiae Div, Ft Detrick, 63-66; br chief, Exp Aerobiol Div, 67-71; microbiologist, Merrell Nat Labs, 71-80; MEM STAFF, CONNAUGHT LABS INC, 80- *Mem:* Am Soc Microbiol; Sigma Xi. *Res:* Bacterial, viral vaccines; biological products; research, development and manufacture. *Mailing Add:* Connaught Labs PO Box 187 Swiftwater PA 18370

HANKINSON, DENZEL J, b Morrice, Mich, June 24, 15; m 38; c 6. FOOD SCIENCE. *Educ:* Mich State Univ, BS, 37; Univ Conn, MS, 39; Pa State Univ, PhD(dairy mfg), 42. *Prof Exp:* Asst prof dairy indust, Univ Conn, 42-44; fieldman, Sealtest, Inc, NY, 44-46; assoc prof dairy husb, Agr & Mech Col Tex, 46-48; prof dairy & animal sci & head dept, 48-64, prof food sci & technol, 64-75, EMER PROF FOOD SCI & TECHNOL, UNIV MASS, AMHERST, 75- *Concurrent Pos:* Consult dairy products & processing. *Mem:* Am Dairy Sci Asn. *Res:* Detection of neutralized cream; shrinkage of ice cream; comeup time milk pasteurization; automation cleaning. *Mailing Add:* 110 Jacaranda Dr Leesburg FL 34748-8805

HANKINSON, OLIVER, b Reading, Eng, Jan 5, 46; US citizen. GENE CLONING, SOMATIC CELL GENETICS. *Educ:* Univ Edinburgh, BSc, 67; Univ Cambridge, PhD(genetics), 72. *Prof Exp:* Res fel genetics, Med Sch, Harvard Univ, 72-74, Med Ctr, Univ Colo, Denver, 74-75 & Univ Calif, Berkeley, 75-79; asst prof, 79-86, ASSOC PROF PATH, UNIV CALIF, LOS ANGELES, 86- *Concurrent Pos:* Prin investr, Nat Cancer Inst grant, Calif, 79- & Margaret E Early Med Res trust grant, 88- *Mem:* Am Soc Cell Biol; Tissue Cult Asn; Sigma Xi. *Res:* Genetic and molecular biological analysis of the process of induction of cytochrome P-450 and other xenobiotic metabolizing enzymes with emphasis on cloning the genes involved, and their relation to cancer induction by environmental pollutants. *Mailing Add:* 5950 Canterbury Dr C314 Culver City CA 90230

HANKOFF, LEON DUDLEY, b Baltimore, Md, June 17, 27; m 57; c 4. PSYCHIATRY. *Educ:* Univ Md, BS, 50, MD, 52. *Prof Exp:* From instr to assoc prof psychiat, Col Med, State Univ NY Downstate Med Ctr, 56-71; chmn psychiat, Misericordia Hosp & Misericordia-Fordham Affil, Bronx, 70-78, prof psychiat, NY Med Col, 71-78; PROF & DIR, DIV CLIN PSYCHIAT, SCH MED, STATE UNIV NY, STONY BROOK, 78-; DEPT PSYCHIAT, NY MED COL. *Concurrent Pos:* Psychiatrist, Kings County Hosp, 56, 58-64; dir psychiat, Queens Hosp Ctr, Jamaica, NY, 64-68; actg dir psychiat, Shaar Menashe Hosp, Israel, 68-69; prog planning consult, First Dept Comnr, New York Dept Ment Health & Ment Retardation Serv, 69-70. *Mem:* AAAS; Sigma Xi; NY Acad Sci; Am Psychiat Asn; Asn Orthodox Jewish Scientists. *Res:* Community psychiatry. *Mailing Add:* Elizabeth Gen Med Ctr-Psych 925 E Jersey St Elizabeth NJ 07201

HANKS, CARL THOMAS, b Cushing, Okla, Aug 10, 39; m 61; c 2. ORAL PATHOLOGY. *Educ:* Phillips Univ, BS, 61; Wash Univ, DDS, 64; State Univ NY Buffalo, PhD(exp path), 70. *Prof Exp:* USPHS training fel, Univ Pittsburgh, 64-66; Nat Inst Dent Res fel, 66-69; asst prof oral path, Sch Dent, State Univ NY Buffalo, 69-70; Nat Inst Dent Res grant, 71-76,from asst prof to assocprof,sch Dent & Dent Res Inst, 70-80, PROF ORAL PATH, SCH DENT, UNIV MICH, ANN ARBOR 79-, ASSOC PROF PATH, SCH MED, 78- *Mem:* AAAS; Am Acad Oral Path; Int Asn Dent Res; Tissue Cult Asn. *Res:* Cell differentiation in salivary glands associated with development and neoplasia; wound healing; growth and differentiation of epithelial and connective tissue in vivo and in vitro; electromagnetic stimulation of metabolic processes in mammalian cells; influence of extracellular matrix on cellular differentiation of osteoblasts and odontoblasts; in vitro testing of biocompatibility of dental materials. *Mailing Add:* 1276 Kuehnle Ct Ann Arbor MI 48103

HANKS, DAVID L, b Tetonia, Idaho, Oct 5, 25; m 49; c 7. MYCOLOGY. *Educ:* Brigham Young Univ, BS, 62, MS, 63; Univ Mich, PhD, 66. *Prof Exp:* Asst prof bot, Brigham Young Univ, 66-69; res specialist & plant physiologist, Great Basin Exp Area, 69; from asst prof to assoc prof, 69-81, PROF MICROBIOL, NORTHEAST MO STATE UNIV, 82- *Mem:* Mycol Soc Am; Bot Soc Am. *Res:* Enzyme-sugar relationships in Neurospora crassa; taxonomy of coprophilous fungi from Great Basin and vicinity; physiology of desert fungi; chemotaxonomy of fungi using mycelial extracts. *Mailing Add:* Dept Sci Northeast Mo State Univ Kirksville MO 63501

HANKS, EDGAR C, b Johnstown, Pa, Aug 23, 21; m 43; c 1. ANESTHESIOLOGY. *Educ:* Gettysburg Col, AB, 43; Jefferson Med Col, MD, 47; Am Bd Anesthesiol, dipl, 53. *Prof Exp:* Intern, Bronx Hosp, 47-48; resident anesthesiol, 48-50, from asst to assoc attend anesthesiologist, 50-66, ATTEND ANESTHESIOLOGIST, PRESBY HOSP, NEW YORK, 66-; PROF CLIN ANESTHESIOL, COL PHYSICIANS & SURGEONS, COLUMBIA UNIV, 66- *Concurrent Pos:* From instr to assoc prof clin anesthesiol, Col Physicians & Surgeons, Columbia Univ, 50-66; consult, US Army Hosp, US Mil Acad, 56-66 & Northern Westchester Hosp, Mt Kisco, NY, 62- *Mem:* Am Soc Anesthesiol; fel Am Col Anesthesiol; AMA; NY Acad Med. *Res:* Respiration; cardiac resuscitation. *Mailing Add:* Dept Anesthesiol Columbia Univ Col 622 W 168th St New York NY 10032

HANKS, JAMES ELDEN, b Augusta, Maine, Apr 24, 24; m 46; c 1. MARINE BIOLOGY. *Educ:* Univ NH, BA, 52, MS, 53, PhD(marine zool), 60. *Prof Exp:* Asst marine biol, Woods Hole Oceanog Inst, Mass, 52-53, fel, 57-60; asst zool, Univ Hawaii, 53-55; fishery res biologist, US Fish & Wildlife Serv, 55-56; asst dir, Marine Biol Lab, US Bur Com Fisheries, Oxford, Md, 60-62, lab dir, Milford, 62-69; liaison scientist marine biol, Off Naval Res, US Navy, Eng, 69-70; dir biol lab, 70-84, LIAISON SCIENTIST AQUACULTURE, NAT MARINE FISHERIES SERV, US DEPT COM, 84- *Mem:* Nat Shellfisheries Asn. *Res:* Marine ecology; life histories of benthic marine invertebrates; reproduction and larval development of pelecypod and gastropod mollusks; predator-prey relationships in marine inter-tidal communities. *Mailing Add:* Biol Lab Nat Marine Fisheries Serv US Dept Com 212 Rogers Ave Milford CT 06460

HANKS, JOHN HAROLD, b Fowlerton, Ind, Sept 16, 06; c 4. BACTERIOLOGY, NUTRITION. *Educ:* Allegheny Col, BS, 28; Yale Univ, PhD(bact), 31. *Prof Exp:* Asst biol, Allegheny Col, 27-28; asst bact, Yale Univ, 30-31; Nat Res Coun fel, Harvard Univ, 31-32; asst & assoc prof bact, Med Sch, George Washington Univ, 32-39; prof pathobiol, Sch Hyg, Johns Hopkins Univ, 62-76; MICROBIOLOGIST, LEONARD WOOD MEM FOUND, 39-, DIR, 59-; EMER PROF PATHOBIOL, SCH HYG, JOHNS HOPKINS UNIV, 76- *Concurrent Pos:* Tissue cult, Johns Hopkins Hosp, 45-46; consult, Pan Am Health Orgn & WHO. *Mem:* Am Soc Microbiol; Sigma Xi; Am Asn Immunol; Int Leprosy Asn. *Res:* Mechanisms of tuberculin types of allergies; bacteriological investigations in leprosy; physiology and growth of host dependent agents; metabolism and cytology of the mycobacteria. *Mailing Add:* Dept Immunol & Infectious Dis Johns Hopkins Univ Sch Hyg & Pub Health 501 W University Pkwy Baltimore MD 21210

HANKS, RICHARD DONALD, b Cincinnati, Ohio, Mar 9, 18; m 44; c 4. CHEMISTRY. *Educ:* Ohio State Univ, BA, 40, MSc, 46. *Prof Exp:* Chemist, McGean Chem Co, Ohio, 41-42; res engr, Battelle Mem Inst, 42-48; asst ed, Chem Abstracts, 48-53, assoc ed, 53-61, head abstracting dept, 61-65, sr assoc ed, 65-69, sr ed, 69-80; RETIRED. *Mem:* Am Chem Soc. *Res:* Organic syntheses; polarography; patent research; economic survey; separation and purification of olefins; synthesis of organo-selenium compounds; chemical literature. *Mailing Add:* 3939 Karl Rd Apt 101 Columbus OH 43224-2400

HANKS, RICHARD W(YLIE), b St Petersburg, Fla, Oct 16, 35; m 55; c 6. NON-NEWTONIAN FLUID MECHANICS. *Educ:* Yale Univ, BE, 57; Univ Utah, PhD(chem eng), 60. *Prof Exp:* Res engr, Oak Ridge Gaseous Diffusion Plant, Union Carbide Nuclear Co, 60-63; from asst prof to assoc prof chem eng, Brigham Young Univ, 63-72, Ctr Thermochem Studies, 70-71, chmn dept, 78-84, PROF CHEM ENG, BRIGHAM YOUNG UNIV, 72-; PRES, RICHARD W HANKS ASSOCS, INC, 80-; PRIN, PROCESS SIMULATION CORP, 90- *Concurrent Pos:* NSF grants, 65-71; consult, Gen Elec Co, Wash, 65-69, Douglas United Nuclear Co, 66-70, Bechtel, Inc, Calif, 71-78, Marathon Oil Co, 76, Oak Ridge Nat Lab, 78-79, Pipline Systems Inc, Orinda, Calif, 79-80, AMAX Extractive Res & Develop, Golden, Colo, 79-80, Arthur D Little, Cambridge, Mass, 80-86, Marconaflo, Salt Lake City, 80-82, CarbonFuels, Denver, 83-89, United Conveyor, Waukegan, 87-, Eastern Generation/Transmission, NH, 90- *Mem:* Am Inst Chem Engrs; Soc Rheol; Coal & Slurry Technol Asn. *Res:* Fluid mechanics and heat transfer; turbulent and transitional flow; stability, particularly non-Newtonian fluids; non-ideal solution thermodynamics; slurry hydraulics and rheology. *Mailing Add:* 350-M Clyde Bldg Brigham Young Univ Provo UT 84602

HANKS, ROBERT WILLIAM, b Hurley, NMex, July 6, 15; m 42; c 2. PLANT CHEMISTRY, PHYTOPATHOLOGY. *Educ:* Univ NMex, BS, 37, MS, 38; Univ Chicago, PhD(plant physiol), 46. *Prof Exp:* Chemist, Kennecott Copper Co, NMex, 39-40; asst prof bot, Univ Miami, 46-49 & Fla State Univ, 49-53; biologist, Fla State Plant Bd, 54-55; asst plant physiologist, 55-73, ASSOC PLANT PHYSIOLOGIST, AGR RES & EDUC CTR, UNIV FLA, 73-, ASSOC PROF, 80- *Mem:* Am Soc Plant Physiol; Am Phytopath Soc. *Res:* Mineral nutrition of plants; systemic chemicals; physiology of host-pathogen relationships. *Mailing Add:* 310 Goodman Ave Lake Alfred FL 33850

HANKS, ROBERT WILLIAM, b Springfield, Mass, May 17, 28; m 51; c 5. FISHERIES MANAGEMENT, RESEARCH ADMINISTRATION. *Educ:* Univ NH, BA, 52, MSc, 60, PhD(zool), 69. *Prof Exp:* Proj leader test mach design, Assoc Engrs, Inc, 52-54; asst biol, Univ NH, 54-55; fisheries res biologist marine invert zool, Bur Com Fisheries, US Fish & Wildlife Serv, Maine, 55-63, fisheries res biologist & leader shellfish ecol & physiol prog, Biol Lab, 63-68, prog coordr, Nat Oceanic & Atmospheric Admin, 68-71, prog analyst, Nat Marine Fisheries Serv, Washington, DC, 71-74, asst regional dir, Nat Marine Fisheries Serv, Gloucester, Mass, 74-81, NEW ENGLAND LIAISON OFFICER, NAT MARINE FISHERIES SERV, PORTLAND, MAINE, 81- *Res:* Marine invertebrate ecology; bottom fauna; invertebrate taxonomy. *Mailing Add:* PO Box 507 Winter Harbor ME 04693

HANKS, RONALD JOHN, b Salem, Utah, Aug 4, 27; m 48; c 5. SOIL PHYSICS. *Educ:* Brigham Young Univ, BS, 50; Univ Wis, MS, 52, PhD(soils), 53. *Prof Exp:* Soil scientist, USDA, 53-68; PROF SOIL & BIOMETEOROL, UTAH STATE UNIV, 68- *Concurrent Pos:* Am Soc Agron vis scientist, 68-69; Fulbright sr scholar, Australia, 78. *Mem:* Fel Am Soc Agron; Soil Sci Soc Am; Soil Conserv Soc Am. *Res:* Evaporation of water from soils and plants; water conservation; irrigation water management; evapotranspiration; soil temperature; heat flow; modelling soil-plant-atmosphere systems. *Mailing Add:* Dept Plants Sci & Biometeorology Utah State Univ Logan UT 84322-4820

HANKS, THOMAS COLGROVE, b Washington, DC, Nov 29, 44; m 68; c 2. GEOPHYSICS. *Educ:* Princeton Univ, BSE, 66; Calif Inst Technol, PhD(geophysics), 72. *Prof Exp:* Res fel, Calif Inst Technol, 72-74; GEOPHYSICIST, US GEOL SURVEY, 74- *Concurrent Pos:* Vis assoc appl sci & geophysics, Calif Inst Technol, 74-77; vis sr scientist, Lamont-Doherty Geol Observ, 79-80; G K Gilbert fel, US Geol Surv, 82-84; Bd Dir, Am Seismol Soc. *Mem:* Fel Am Geophys Union; Seismol Soc Am. *Res:* Cause and effect of earthquakes; strong ground motion; fault-scarp geomorphology. *Mailing Add:* US Geol Surv Box 7712 Menlo Park CA 94025

HANLE, PAUL ARTHUR, b Newark, NJ, Oct 27, 47. HISTORY OF PHYSICS, SPACE SCIENCES. *Educ:* Princeton Univ, AB, 69; Yale Univ, MS, 72, PhD(hist sci), 75. *Prof Exp:* Assoc curator to curator, Sci & Technol, Nat Air & Space Mus, Smithsonian Inst, 74-80, actg chmn, Space Sci & Explor, 80-81, chmn, 81-84, ASSOC DIR RES, SPACE SCI & EXPLOR, NAT AIR & SPACE MUS, SMITHSONIAN INST, 84- *Mem:* Am Phys Soc; Sigma Xi; Hist Sci Soc; Soc Hist Technol; AAAS. *Res:* History of relativity and quantum theory; German and American aerodynamics between the wars; space science. *Mailing Add:* Maryland Acad Sci 601 Light St Baltimore MD 21230

HANLEY, ARNOLD V, b Jersey City, NJ, Nov 22, 27; m 53; c 3. ANALYTICAL CHEMISTRY. *Educ:* Seton Hall Univ, BS, 51. *Prof Exp:* Anal chemist, Am Cyanamid Co, NJ, 51-56, process develop chemist, 56; anal chemist, FMC Corp, 56-64, res chemist, 64-65, supvr anal chem methods sect, 65-75, RES ASSOC, CENT RES DEPT, FMC CORP, 75- *Mem:* Am Chem Soc; Int Confedn Thermal Anal; Am Microchem Soc. *Res:* Chemical methods of analysis; gas chromatography; thermoanalytic methods; analytical chemistry of nonoxygen containing inorganic sulfur compounds. *Mailing Add:* 33 Cutler Pl Clark NJ 07066

HANLEY, DANIEL F, JR, b Portland, Maine, May 17, 49; m 82; c 3. NEUROLOGY. *Educ:* Williams Col, BA, 71; Cornell Univ, MD, 75; Am Bd Internal Med, dipl, 78; Am Bd Neurol & Psychiat, dipl, 90. *Prof Exp:* Med intern, NY Hosp, New York, 75-76, resident, 76-78; Kettering res fel, Sloan-Kettering Inst, 78; resident neurol, Johns Hopkins Hosp, 79-81, res fel, Depts Neurol & Anesthesia, Johns Hopkins Med Insts, 81-83, asst prof, Depts Neurol, Neurosurg & Anesthesia-Critical Med, 83, co-dir, Neurol Resident Training Prog, 87 & 90, DIR, NEUROSCI CRITICAL CARE UNIT, JOHNS HOPKINS MED INSTS, 83-, ASSOC PROF, DEPT NEUROL, 90- *Concurrent Pos:* Prin investr, Am Heart Asn, 83-85, Charles Dana Found, 84-86, NIH, 86-88 & 87-92; prin collabr, Nat Inst Neurol & Communicative Disorders & Stroke, 83-87; collabr, Collab Antiviral Study Group, Nat Inst Allergy & Infectious Dis, 83-89; clinician-scientist award, Johns Hopkins Med Inst, 86-88; vis prof med, Univ Hawaii, 87; vis consult neurocritical care, Univ Va, 87, Univ Rochester Med Col, 88, Mayo Clin, 90; head, Sect Emergency & Critical Care Neurol, Am Acad Neurol, 88, mem exec bd, 91, mem, Therapeut & Technol Assessment Subcomt, 91, Neurol Self Assessment Task Force, 91. *Mem:* AMA; Am Acad Neurol; Am Soc Neurol Invest (secy-treas, 87, pres, 88); Soc Neurosci; Am Physiol Soc; Soc Critical Care Med; Am Neurol Asn; Soc Cerebral Blood Flow & Metab. *Res:* Acute care of the neurologically impaired patient; mechanisms of cerebrovascular regulation; author of numerous publications. *Mailing Add:* Dept Neurol Johns Hopkins Hosp Meyer 8-139 Baltimore MD 21205

HANLEY, HOWARD JAMES MASON, b Hove, Eng, Aug 19, 37; nat US; m 64; c 1. FLUIDS, STATISTICAL MECHANICS. *Educ:* Univ London, BSc, 59, PhD(chem), 63. *Prof Exp:* Res assoc, Pa State Univ, 63-65; phys chemist, Cryogenics Div, 65-78, THERMOPHYSICS DIV, NAT INST STANDARDS & TECHNOL, 78- *Concurrent Pos:* vis fel, Australian Nat Univ, 73-74, 78-79 & 87; adj prof, Dept Chem, Univ Colo, 75-; vis scientist, Inst Laue-Langeuin, 84, 85; NBS fel, 83; fel, Wissenschaftskolleg Zu Berlin, 89-90. *Honors & Awards:* Gold Medal, Dept Com, 85. *Mem:* Royal Chem Soc. *Res:* Statistical mechanics and thermodynamics; study of fluids via computer simulation; non linear fluid behavior; structure of liquids. *Mailing Add:* Thermophysics Div Nat Inst Standards & Technol Boulder CO 80303

HANLEY, JAMES RICHARD, JR, b Dorchester, Mass, Apr 21, 29; m 51; c 1. TERPENE CHEMISTRY. *Educ:* Univ RI, BS, 56; Univ Ill, PhD(chem), 59. *Prof Exp:* Propellant chemist, US Naval Propellant Plant, Md, 51-52; org chemist, Univ RI, 55-56, lab instr, 56; asst org & polymer chem, Univ Ill, 56-59; assoc technologist, Gen Foods Res Ctr, NY, 59-60; res chemist, Fla Citrus Canners Coop, 60-61 & Nelio Chem Inc, 61-64; res chemist, Org Chem Div, Glidden Co, 64-72; PROF CHEM, FLA COMMUNITY COL, JACKSONVILLE, 72- *Mem:* AAAS; Am Chem Soc; Am Inst Chem; Royal Soc Chem; Chem Soc Japan. *Res:* Terpenes; essential oils; thermally stable polymers; waste recycling and utilization. *Mailing Add:* Dept Chem Fla Community Col S Campus 11901 Beach Blvd Jacksonville FL 32216-6624

HANLEY, JOHN HERBERT, geology; deceased, see previous edition for last biography

HANLEY, KEVIN JOSEPH, b Utica, NY, Oct 25, 52; m 78; c 1. ORTHODONTICS. *Educ:* State Univ NY, Buffalo, BA, 74, DDS, 78. *Prof Exp:* Asst prof, 80-83, ASST CLIN PROF ORTHOD, SCH DENT MED, UNIV CONN, FARMINGTON, 83-; ORTHODONTIST, PEDODONTIC-ORTHOD ASSOCS, 83- *Mem:* Am Dent Asn; Am Asn Orthodontists. *Res:* Effects of pulsating electromagnetic fields on bone cell growth, specifically its effect on the proliferation of DNA. *Mailing Add:* Seven Englewood Ave Buffalo NY 14214

HANLEY, THOMAS ANDREW, b Chicago, Ill, Jan 24, 51; m 76; c 1. FORAGING ECOLOGY. *Educ:* Ariz State Univ, BS, 73, MS, 76; Univ Wash, PhD(forest wildlife ecol), 80. *Prof Exp:* Wildlife biologist, Bur Land Mgt, US Dept Interior, 76-77; RES WILDLIFE BIOLOGIST, PAC NORTHWEST FOREST & RANGE EXP STA, FOREST SERV, USDA, 80- *Concurrent Pos:* Res asst, Div Agr, Ariz State Univ, 74-76 & Col Forest Resources, Univ Wash, 77-80; wildlife biologist, Forest Serv, US Dept Agr, 80. *Mem:* AAAS; Am Soc Naturalists; Ecol Soc Am; Soc Range Mgt; Wildlife Soc. *Res:* Effects of forest management practices on habitat quality for animals, particularly regarding ungulate-vegetation interactions; food and habitat selection by ungulates; ruminant nutrition and bioenergetics; forest community structure and productivity. *Mailing Add:* Forest Sci Lab US Forest Serv PO Box 20909 Juneau AK 99802

HANLEY, THOMAS ODONNELL, ice science; deceased, see previous edition for last biography

HANLEY, THOMAS RICHARD, b Logan, WVa, July 26, 45; m 79; c 2. POLYMER ENGINEERING, SANITARY & ENVIRONMENTAL ENGINEERING. *Educ:* Va Polytechnic Inst, BS, 67, MS, 71, PhD(chem eng), 72; Wright State Univ, MBA, 75. *Prof Exp:* Develop engr & chem engr, USAF mat lab, 72-75; asst prof, chem eng, Tulane Univ, 75-79; assoc prof, chem eng, Rose-Hulman Inst Technol, 79-83; prof & head chem eng, La Tech Univ, 83-85; prof & chmn chem eng, Fla State Univ/Fla A&M Univ, 85-91; DEAN ENG, SPEED SCI SCH, UNIV LOUISVILLE, 91- *Concurrent Pos:* Chem eng consult, Edgewood Arsenal, Nat Space & Technol Lab, 76-78, Chevron Chem Co, 78-79, Solar Energy Res Inst, 80, El Paso Polyolefins Co, 82-84, Int Minerals & Chem Corp, 82-84, Olin Corp, 86-91, chem eng consult, Kraft, Inc, 89-90; div adv, Chem, Biochem & Thermal Div, NSF, 86-89, Biol & Crit Systs Div, 89- *Honors & Awards:* Soc Am Milit Engrs Award, 66 & 67; Ralph R Teetor Educ Award, Soc Automotive Engrs, 89. *Mem:* Am Inst Chem Engrs; Am Soc Eng Educ; Soc Automotive Engrs. *Res:* reactive mixing; bioreactor analysis; heterogeneous simulation; biomedical and biochemical processing; author of over 30 publications. *Mailing Add:* Speed Sci Sch Univ Louisville Louisville KY 40292

HANLEY, WAYNE STEWART, b Edinburgh, Scotland, Oct 30, 45; US citizen; m 68; c 2. BIO-ORGANIC CHEMISTRY. *Educ:* Tarkio Col, BA, 66; Vanderbilt Univ, PhD(org chem), 71. *Prof Exp:* Fel med chem, Univ Minn, 70-71; fel, Dept Chem, Vanderbilt Univ, 71-72; asst prof chem, Georgetown Col, 72-76; dir res & develop, Conwood Corp, 76-80; plant mgr, Kilgore Corp, 80-; CONSULT, CONWOOD CORP, 74-, DIR, RES & DEVELOP. *Concurrent Pos:* Res grant, Am Chem Soc, 74. *Mem:* Am Chem Soc; Sigma Xi. *Res:* Study of synthesis, stability and reactions of organic sulfenyl iodides; preparation of penicillamine analogs as potential antiarthritic agents; synthesis of organic sulfur compounds as antihistoplasmosis agents. *Mailing Add:* Kilgore Corp Toone TN 38381

HANLIN, RICHARD THOMAS, b Hammond, Ind, May 10, 31; m 55; c 3. MYCOLOGY. *Educ:* Univ Mich, BS, 53, MS, 55, PhD(bot), 60. *Prof Exp:* Asst plant pathologist, Ga Exp Sta, 60-66; assoc prof, 67-71, PROF MYCOL, UNIV GA, 71- *Mem:* AAAS; Bot Soc Am; Mycol Soc Am; Am Phytopath Soc; Brit Mycol Soc. *Res:* Morphological development and taxonomy of the Pyrenomycetes, especially the Hypocreales. *Mailing Add:* Dept Plant Path Univ Ga Athens GA 30602

HANLON, C ROLLINS, b Baltimore, Md, Feb 8, 15; m 49; c 8. SURGERY. *Educ:* Loyola Col, Md, AB, 34; Johns Hopkins Univ, MD, 38; Am Bd Surg, dipl Am Bd Thoracic Surg, dipl. *Hon Degrees:* DSc, Georgetown Univ, 76, St Louis Univ & Univ Ill, 86. *Prof Exp:* From instr to assoc prof surg, Johns Hopkins Univ, 46-50; prof & dir dept, Sch Med, St Louis Univ, 50-69; prof, 69-85, EMER PROF SURG, MED SCH, NORTHWESTERN UNIV, CHICAGO, 85- *Concurrent Pos:* W S Halsted fel, Johns Hopkins Univ, 39-40; William H Vogt Lectr, 51; mem surg study sect, NIH, 61-65, chmn, 65-; pres, Coun Med Spec Soc, 74-75; chmn, Coord Coun Med Educ, 76-77; gov, Am Col Surgeons, 57-59, Regent, 67-69, dir, 69-86, pres elect, 86-87, pres, 87-88, exec consult, 86- *Mem:* Soc Univ Surgeons (secy, 53, pres, 58-59); Am Asn Thoracic Surg (treas, 62-68); Int Cardiovasc Soc (pres, 63-64); Am Thoracic Soc; Am Surg Asn (pres, 81-82); hon mem Am Hosp Asn; hon mem Soc Thoracic Surgeons; hon mem Am Urol Asn; hon mem Am Col Radiol. *Res:* Shock; cardiovascular and pulmonary diseases. *Mailing Add:* 556 Earlston Rd Kenilworth IL 60043

HANLON, JOHN JOSEPH, public health; deceased, see previous edition for last biography

HANLON, MARY SUE, b New Orleans, La, Sept 14, 33; m 66; c 1. BIOCHEMISTRY. *Educ:* La State Univ, BS, 54; Univ Calif, Berkeley, PhD(biochem), 61. *Prof Exp:* Instr biochem, Chicago Med Sch, 63-64, asst prof, 64-66; asst prof biol chem, 66-69, assoc prof biochem, 69-76, PROF BIOCHEM, UNIV ILL COL MED, 76- *Concurrent Pos:* NSF fel, 61-62; NIH fel, 62-63. *Mem:* Am Chem Soc; Am Soc Cell Biol; Am Soc Biol Chemists; Biophys Soc; fel AAAS. *Res:* Physical biochemistry; conformation and chemical interaction properties of biologically important macromolecules. *Mailing Add:* Dept Biochem MC 536 Univ Ill Col Med 1853 W Polk St Chicago IL 60612

HANLON, ROGER THOMAS, b Frankfurt, Germany, May 17, 47; US citizen; m 79; c 2. CEPHALOPOD BEHAVIOR. *Educ:* Fla State Univ, BS, 69; Univ Miami, MS, 75, PhD(marine sci), 78. *Prof Exp:* Res assoc, Marine Biomed Inst, Univ Tex Med Br, 75-80; NATO res fel, dept zool, Univ Cambridge, 81-82; asst prof, 82-85, ASSOC PROF, MARINE BIOMED INST, UNIV TEX MED BR, 85-, CHIEF, DIV BIOL & MARINE RESOURCES, 83- *Mem:* Am Soc Zoologists; Animal Behav Soc; Marine Biol Asn UK; Am Malacol Union; Sigma Xi. *Res:* Ethology of cephalopods, with particular emphasis on the functional morphology of the neurally controlled chromatophore system; mariculture of cephalopods. *Mailing Add:* Marine Biomed Inst Univ Tex Med Br Galveston TX 77550-2772

HANLON, THOMAS LEE, b Durbin, WVa, May 10, 37; m 62; c 2. PHYSICAL CHEMISTRY. *Educ:* Randolph-Macon Col, BS, 59; Princeton Univ, PhD(phys chem), 63. *Prof Exp:* Sr res chemist synthetic rubber, Goodyear Tire & Rubber Co, 63-75, res scientist, 75-80; SR APPLNS CHEMIST, ETHYL CORP, 80- *Mem:* Am Chem Soc. *Res:* Stereospecific polymerizations, especially dienes; ionic copolymerization; Ziegler-Natta type catalysis; emulsion polymerization and copolymerization. *Mailing Add:* Ethyl Tech Ctr 8000 GSRI Ave Baton Rouge LA 70820

HANLY, W CAREY, b Wharton, Tex, Aug 6, 36. BIOCHEMISTRY, IMMUNOLOGY. *Educ:* Rice Univ, BA, 58, PhD(biochem), 64. *Prof Exp:* Res biochemist, Bowman Gray Sch Med, 58-60; Nat Acad Sci-Nat Res Coun res associateship biochem, Aeromed Res Lab, Holloman AFB, NMex, 64-66, res biochemist, 66-70; res assoc, 70-73, asst prof, 73-77, ASSOC PROF IMMUNOL, UNIV ILL MED CTR, 77- *Concurrent Pos:* Lectr, Holloman Exten, Univ NMex, 66-67; instr, Alamogordo Community Col, NMex State Univ, 67-68. *Mem:* Am Asn Immunologists; AAAS; Sigma Xi; NY Acad Sci. *Res:* Chemistry, genetic variation and function of immunoglobulin A; comparative biochemistry of primates, hemoglobin, red cell antigens and serum proteins; dihydroxyphenylalanine-decarboxylase and functional properties. *Mailing Add:* Dept Microbiol & Immunol M/C 790 Univ Ill PO Box 6998 Chicago IL 60680

HANN, DAVID WILLIAM, b Oakland, Calif. FOREST MENSURATION, FOREST BIOMETRY. *Educ:* Ore State Univ, BS, 68, MS, 70; Univ Wash, PhD(forest mensuration), 78. *Prof Exp:* Res forester forest mensuration, Intermountain Forest & Range Exp Sta, US Forest Serv, USDA, 71-78; ASSOC PROF FOREST MENSURATION, COL FORESTRY, ORE STATE UNIV, 78- *Mem:* Soc Am Foresters; Am Statist Asn; Biomet Soc. *Res:* Development of mathematical tools useful to forest managers, especially modeling of stand and tree dynamics, and the prediction of stand yield and of the tree product potential. *Mailing Add:* Dept Forest Mgt Ore State Univ Corvallis OR 97331

HANN, G(EORGE) C(HARLES), b United States, June 25, 24; m 47; c 5. GENERAL MANAGEMENT. *Educ:* Univ Minn, BChemE, 45, MS, 51. *Prof Exp:* Asst chem, Univ Minn, 47-48; from res engr glass tech to proj mgr systs prod, Minn Mining & Mfg Co, 49-66; dir res & prod develop, 66-68, VPRES, APACHE CORP, 68- *Mem:* AAAS; fel Am Inst Chem; Am Chem Soc; Am Inst Chem Engrs; Sigma Xi. *Res:* Fluid mechanics; fluidization; high refractive index glasses; ion exchange. *Mailing Add:* Seven Timberglade Rd Minneapolis MN 55437

HANN, HIE-WON L, b Seoul, Korea, Mar 9, 36; m 66; c 2. PEDIATRICS, PEDIATRIC ONCOLOGY. *Educ:* Seoul Nat Univ, MD, 61; Am Bd Pediat, dipl, 71. *Prof Exp:* Intern, Worcester City Hosp, Mass, 63-64; resident pediat, Children's Hosp Med Ctr, Boston, 64-65, resident pediat path, 65-66, resident pediat, 66-68, fel pediat oncol, 68-71; instr to asst prof, 72-80, ASSOC PROF PEDIAT, SCH MED, UNIV PA, 81-; RES PHYSICIAN, INST CANCER RES, PHILADELPHIA, 71- *Concurrent Pos:* Asst physician oncol, Children's Hosp Philadelphia, 72-75, assoc physician, 75-; assoc staff, Am Oncol Hosp, Philadelphia, 72-; assoc med staff, Jeanes Hosp, Philadelphia, 72- *Mem:* Nat Acad Sci; Am Soc Clin Oncol; Am Fedn Clin Res; Am Acad Pediat. *Res:* Biology of isoferritins in cancer; immunology and biochemistry of neuroblastoma; hepatitis B virus and hepatoma. *Mailing Add:* Jefferson Medical Col 1025 Walnut St Philadelphia PA 19107

HANN, ROBERT A, forest products, wood technology, for more information see previous edition

HANN, ROY WILLIAM, JR, b Oklahoma City, Okla, Mar 21, 34; m 84; c 6. CIVIL ENGINEERING. *Educ:* Univ Okla, BSCE, 56, MCE, 57, PhD(eng sci), 63. *Prof Exp:* Civil engr, US Pub Health Serv, 57-59 & C H Guernsey & Co, 59-60; prof civil eng, Univ SC, 62-65; head environ eng div, 71-75, dir, Sea Grant Prog, 75-77, PROF CIVIL ENG, ENVIRON ENG DIV, TEX A&M UNIV, 65- *Concurrent Pos:* Consult civil eng, 60-; pres, Civil Eng Systs, Inc, Hann Enterprises & Int Spill Technol Corp. *Honors & Awards:* Palladium Medal, Nat Aubudon Soc, 83. *Mem:* Fel Am Soc Civil Engrs; Nat Soc Prof Engrs; Am Soc Eng Educ; Water Pollution Control Fedn; Am Water Works Asn. *Res:* Environmental engineering; prevention and control of oil spills; mathematical simulation of aquatic systems; digital computer applications in civil engineering; design, operation and management of water resources projects. *Mailing Add:* Environ Eng Div Tex A&M Univ College Station TX 77843

HANNA, ADEL, b Cairo, Egypt, May 4, 43; Can citizen; m 72; c 1. GEOTECHNICAL ENGINEERING, FOUNDATION ENGINEERING. *Educ:* Ain Shams Univ, Egypt, BEng, 66; Cairo Univ, MEng, 72; NS Tech Col, PhD(civil eng), 78. *Prof Exp:* Proj engr, design, Pub Work, Egypt, 66-74; res assoc, NS Tech Col, 74-77; sr engr consult, Maritime Testing Ltd, 77-78; assoc prof, 78-89, PROF, CONCORDIA UNIV, 89- *Concurrent Pos:* Reviewer, J Geotech Eng, Am Soc Civil Engrs, 79-, Can Geotech J & J Transp Res Records. *Mem:* Am Soc Civil Engrs; Int Soc Soil Mech & Found Engrs; Int Soc Housing Sci; Deep Found Inst. *Res:* Author over 40 papers in the field of foundation engineering. *Mailing Add:* Civil Eng Dept Concordia Univ 1455 DeMaisonneuve Blvd W Montreal PQ H3G 1M8 Can

HANNA, CALVIN, b Danville, Ill, May 1, 23; m 46; c 2. PHARMACOLOGY. *Educ:* Univ Ill, BS, 49; Univ Iowa, MS, 50, PhD(pharmacol), 53. *Prof Exp:* Asst pharmacol, Univ Iowa, 49-52; instr, Col Med, Univ Tenn, 53-54, asst prof, 54-55; from asst prof to assoc prof, Col Med, Univ Vt, 55-61; from assoc prof to prof pharmacol, Col Med, Univ Ark, Little Rock, 61-89, emer prof, 89-; RETIRED. *Concurrent Pos:* Lederle fac award, 63-66; mem, Visual Sci Study Sect, Div Res Grants, NIH, 62-65; mem, OTC Opthal Panel, Food & Drug Admin, 75-80. *Mem:* Am Chem Soc; Am Soc Pharmacol & Exp Therapeut; Soc Exp Biol & Med. *Res:* Ophthalmology; relation of chemical structure to biological action; synthesis of physiologically active materials; tachyphylaxis; mechanism of action of drugs, radiation biology, drug toxicology. *Mailing Add:* 74 Glemere Rd Little Rock AR 72204

HANNA, EDGAR ETHELBERT, JR, b Anniston, Ala, Sept 2, 33; m 59; c 3. MICROBIOLOGY, IMMUNOLOGY. *Educ:* Tuskegee Inst, BSc, 59; Univ Minn, MS, 63, PhD(microbiol, immunol), 67. *Prof Exp:* Asst microbiol, Tuskegee Inst, 57-59; asst, Med Sch, Univ Minn, 60-64; asst prof, Health Sci Div, Va Commonwealth Univ, 69-70; sr investr, NIH, 75-83, immunobiologist, Lab Molecular Genetics, 70-83, chief, immunoregulation & cellular control sect, Develop & Molecular Immunity Lab, 83-90, SR MICROBIOL, DIV SCI REV, NAT INST CHILD & HUMAN DEVELOP, NIH, 90- *Concurrent Pos:* USPHS res fel, Lab Immunol, Nat Inst Allergy & Infectious Dis, 67-69; lectr, Howard Univ, 68-69; Nat Inst Allergy & Infectious Dis grant, 70-72; sect ed, Microbiology, 80 & 82; chmn, Immunol Div, Am Soc Microbiol, 80-82 & counr, Div Group I, 84-86; lectr, Found Microbiol, 84-85; elect bd gov, Am Acad Microbiol, 85-88. *Honors & Awards:* Sustained High Qual Res Award, NIH, 77; Becton-Dickinson-Traveling Lectr Award, Am Soc Microbiol, 82. *Mem:* AAAS; Am Asn Immunologists; Am Soc Microbiol; fel Am Acad Microbiol. *Res:* Genetic, cellular and subcellular mechanisms of the antibody response; regulatory mechanisms of antibody biosynthesis; pathways of antibody biosynthesis in dispersed cell culture; cell fusion and regulatory cell hybridomas; molecular biology; functional precursor T-cell hybridomas; cloned from nude (athrogenic mice) as models for understanding development of the T-cell phenotype. *Mailing Add:* Nat Inst Child Health & Human Develop NIH EPN Rm 520 Bethesda MD 20892

HANNA, GEOFFREY CHALMERS, b Stretford, Eng, Oct 5, 20; Can citizen; m 51; c 3. NUCLEAR PHYSICS. *Educ:* Cambridge Univ, MA, 41. *Hon Degrees:* DSc, McGill Univ, 1983. *Prof Exp:* Exp officer, Brit Ministry Supply, 41-45; sci officer, Brit Mission to Can, 45-50; asst res officer, Nat Res Coun Can, 50-52; assoc res officer, Chalk River Nuclear Lab, Atomic Energy Can Ltd, 52-57, sr res officer, 57-64, prin res officer, 64-67, res dir, Physics Div, 67-71, res dir, 71-85; RETIRED. *Mem:* Fel Royal Soc Can; Can Asn Physicists. *Mailing Add:* Five Tweedsmuir Pl PO Box 194 Deep River ON K0J 1P0 Can

HANNA, GEORGE P, JR, b Manhattan, Kans, Mar 25, 18; m 44; c 2. WATER QUALITY CONTROL. *Educ:* Ill Inst Technol, BS, 40; NY Univ, MS, 42; Univ Cincinnati, PhD, 68. *Prof Exp:* Asst sanit engr, dist off, US Army Engrs, Pa, 42-43; design engr, Holmes, O'Brien & Gere, consult engrs, NY, 46-48; sanit engr & dir bur sanit, Dept Health, Syracuse, 48-50; asst prof civil eng, Syracuse Univ, 50-52; chief engr, Mat Handling Prod Corp, 52-53; design engr, Nussbaumer, Clarke & Velzy, consult engrs, 53; civil engr, Creole Petrol Corp, Venezuela, 54-59; from assoc prof to prof civil eng & dir water resources ctr, Ohio State Univ, 59-69; prof civil eng & chmn dept, Univ Nebr, Lincoln, 69-72, interim dean, Col Eng, 71-72, dean, Col Eng & Technol, 72-79; PROF CIVIL ENG, CALIF STATE UNIV, FRESNO, 79- *Concurrent Pos:* NSF sci fac fel, 68-69; mem, Nebr Environ Control Coun, 71-72; vis prof, South China Inst Technol, Guangzhou, People's Repub China, 85. *Mem:* Am Soc Civil Engrs; Am Water Works Asn; Am Acad Environ Engrs (pres, 81-82); Am Soc Eng Educ; Nat Soc Prof Engrs (vpres, 78-79). *Res:* Industrial wastes treatment; water quality studies involving aggressive water neutralization, activated carbon adsorption of contaminants from groundwaters, and agricultural drainwater desalination. *Mailing Add:* 7389 N Bond Ave Fresno CA 93720-3005

HANNA, GEORGE R, b Boonville, Mo, July 27, 31; m 56; c 3. NEUROLOGY, NEUROPHYSIOLOGY. *Educ:* Cent Methodist Col, AB, 52; Univ Mo, BSc, 54; McGill Univ, MD & CM, 56. *Prof Exp:* From asst prof to assoc prof, 63-74, PROF NEUROL, SCH MED, UNIV VA, 74- *Concurrent Pos:* Fel neuroanat, Col Physicians & Surgeons, Columbia Univ, 60-61; prin investr, Dept Health, Educ & Welfare grants, Voc Rehab Admin, 65-69, Nat Heart & Lung Inst, 72- & Nat Inst Neurol Dis & Stroke, 73- *Mem:* Fel Am Acad Neurol; Am Neurol Asn; Am Epilepsy Soc; Asn Res Nerv & Ment Dis; Soc Neurosci. *Res:* Neuroanatomy; abnormal movements and control of ocular movements; neurophysiology of auditory and vestibular systems; neural control of motor systems; mechanisms of epilepsy; cerebrovascular disease. *Mailing Add:* Dept Neurol Univ Va Sch Med Charlottesville VA 22903

HANNA, JOEL MICHAEL, b Lock Haven, Pa, Dec 6, 38; m 62; c 1. PHYSICAL ANTHROPOLOGY. *Educ:* Pa State Univ, BS, 61, MA, 65; Univ Ariz, PhD(anthrop), 68. *Prof Exp:* Field dir anthrop, Field Lab, Pa State Univ, Peru, 66-67; asst prof, 68-74, ASSOC PROF ANTHROP & PHYSIOL, DEPT PHYSIOL, UNIV HAWAII, MANOA, 74- *Concurrent Pos:* Assoc prof, Pa State Univ, 71-; coordr, Int Biol Prog, 71- *Mem:* Am Asn Phys Anthrop; Am Anthrop Asn; Brit Soc Study Human Biol. *Res:* Human biology and ecology; environmental physiology. *Mailing Add:* Dept Physiol Univ Hawaii Burns Med Sch 1960 East-West Rd Honolulu HI 96822

HANNA, LAVELLE, b La Grande, Ore, Apr 23, 18. MICROBIOLOGY. *Educ:* Ore State Col, BA, 40; Univ Calif, Berkeley, MA, 53. *Prof Exp:* Technician, Ore State Health Dept, 41-46; technician, Univ Calif, San Francisco, 47-54, assoc microbiol, 54-62, assoc specialist, 62-70, specialist microbiol, Med Ctr, 70-82; RETIRED. *Mem:* AAAS; Am Venereal Dis Asn; Am Soc Microbiol; fel Am Acad Microbiol. *Res:* Diagnosis of viral disease; viral infections of the eye; adenoviruses; herpes viruses; biology, diagnosis and control of trachoma-inclusion conjunctivitis; interferon; host-parasite interactions; immunofluorescence in the diagnosis of infection. *Mailing Add:* 620 Ortega San Francisco CA 94122-4646

HANNA, MARTIN SLAFTER, b Detroit, Mich, Aug 8, 32; m 60; c 3. MATHEMATICS. *Educ:* Harvard Univ, AB, 53; NY Univ, MS, 59; Univ Wis, PhD(math), 63. *Prof Exp:* Asst prof, 63-68, ASSOC PROF MATH, UNIV KANS, 68- *Mem:* Am Math Soc; Math Asn Am; Soc Indust & Appl Math; Asn Comput Mach. *Res:* Numerical methods; partial differential equations; functional analysis. *Mailing Add:* Dept Math/203b ST Univ Kans Lawrence KS 66044

HANNA, MELVIN WESLEY, b Glendale, Calif, Oct 1, 32; m 56; c 3. PHYSICAL CHEMISTRY. *Educ:* Univ Calif, Los Angeles, BS, 54; Univ Minn, PhD(chem), 59. *Prof Exp:* NSF fel, Calif Inst Technol, 59-60, Noyes teaching fel, 60-61; from asst prof to assoc prof, 61-66, PROF CHEM, UNIV COLO, BOULDER, 66- *Concurrent Pos:* Alfred Sloan fel, 65-67. *Mem:* AAAS; Am Chem Soc. *Res:* Electron and nuclear magnetic resonance spectroscopy; chemical education. *Mailing Add:* Univ Nations 75-5851 Kuakini Hwy Kailua-Kona HI 96740-2199

HANNA, MICHAEL G, JR, b Cleveland, Ohio, July 7, 36; m 58; c 3. IMMUNOBIOLOGY. *Educ:* Baldwin-Wallace Col, BSc, 58; Univ Notre Dame, MSc, 60; Univ Tenn, PhD(radiation biol), 64. *Prof Exp:* Res assoc chem co-carcinogenesis, Oak Ridge Nat Lab, 64-65, res biol, 65-68, dir immunol carcinogenesis group, 68-75; res psychiatrist, Lab Psychol, NIMH, 70-72, Ment Health Study Ctr, 72-74, asst chief, 74, actg chief, 74-75, chief, Ment Health Study Ctr, 75-84, CHIEF, CLIN INFANT/CHILD DEVELOP RES CTR, HEALTH RESOURCES & SERVS ADMIN & NIMH, 84-; VPRES, OTC, DIR RES, BIONETICS RES, INST, 85- *Concurrent Pos:* dir, Litton Inst Appl Biotechnol, 82-85. *Mem:* NY Acad Sci; Am Asn Pathologists; Am Asn Cancer Res; Am Asn Immunologists; AAAS. *Res:* Immunology; immunopathology; immunotherapy. *Mailing Add:* Litton Inst Applied Biotechnol 1330 Piccard Dr Rockville MD 20850

HANNA, MICHAEL ROSS, agronomy, for more information see previous edition

HANNA, MILFORD A, b West Middlesex, Pa, Feb 26, 47; m 78; c 4. FOOD ENGINEERING, PROPERTIES OF BIOLOGICAL MATERIALS. *Educ:* Pa State Univ, BS, 69, MS, 71 & PhD(agr eng), 73. *Prof Exp:* Asst prof agr eng, Calif Polytech State Univ, 73-75; from asst prof to assoc prof agr eng, 75-85, PROF BIOL SYST ENG, UNIV NEBR, LINCOLN, 85- *Concurrent Pos:* Morrison prof food eng, 90. *Mem:* Am Soc Agr Engrs; Inst Food Technologists; Sigma Xi. *Res:* High-temperature short-time extrusion cooking effects on the quality attributes of food constituents; proteins and starches; value added processing. *Mailing Add:* Univ Nebr 211 LW Chase Hall Lincoln NE 68583-0730

HANNA, OWEN TITUS, b Chicago, Ill, June 8, 35; m 57; c 5. CHEMICAL ENGINEERING. *Educ:* Purdue Univ, BSChE, 57, PhD(chem eng), 61. *Prof Exp:* Asst prof chem eng, Rensselaer Polytech Inst, 62-65; sr res engr, Boeing Co, 65-67; assoc prof chem eng, 67-74, chmn dept, 71-73, 81-84, PROF CHEM ENG, UNIV CALIF, SANTA BARBARA, 74- *Concurrent Pos:* Consult, US Naval Missile Ctr, 67-, Lawrence Livermore Nat Lab, 82- & Los Alamos Nat Lab, 86- *Mem:* Am Inst Chem Engrs. *Res:* Applied mathematics; heat, mass and momentum transfer. *Mailing Add:* Dept Chem Eng Univ Calif Santa Barbara CA 93106

HANNA, PATRICK E, b Little River, Kans, Oct 13, 40; c 1. MEDICINAL CHEMISTRY, PHARMACOLOGY. *Educ:* Creighton Univ, BS, 63; Univ Kans, PhD(med chem), 69. *Prof Exp:* From asst prof to assoc prof, 69-84, PROF MED CHEM & PHARMACOL, UNIV MINN, MINNEAPOLIS, 84- *Concurrent Pos:* Vchmn med chem div, Am Chem Soc, 85, chmn, 86. *Mem:* AAAS; Am Chem Soc; Soc Toxicol; Am Asn Cols Pharm; Am Asn Pharmaceut Scientists; Am Soc Pharmacol & Exp Therapeut. *Res:* Bioactivation processes and drug metabolism; carcinogen metabolism; enzyme inhibitors; drug design. *Mailing Add:* HSUF 8-101 Med Chem Dept Univ Minn Minneapolis MN 55455

HANNA, RALPH LYNN, b Martinsville, Tex, Jan 5, 19; m 45; c 4. ENTOMOLOGY. *Educ:* Stephen F Austin State Col, BA, 39; Agr & Mech Col Tex, PhD(entom), 51. *Prof Exp:* Instr, 49-51, asst prof, 51-56, ASSOC PROF ENTOM, TEX A&M UNIV, 56- *Mem:* Entom Soc Am. *Res:* Control of insects affecting cotton, corn, small grains and forage crops. *Mailing Add:* Dept Entom Tex A&M Univ College Station TX 77843

HANNA, SAMIR A, b Egypt, May 1, 34; US citizen; m 66; c 2. PHARMACEUTICAL & ANALYTICAL CHEMISTRY. *Educ:* Cairo Univ, BS, 56, MS, 67; Assiut Univ, PhD(pharmaceut sci), 70. *Prof Exp:* Pharmacist, Pharmacy Fikry, 56-63; anal chemist, Morgan's Chem Co, 60-63; group leader pharmaceut, Drug Res & Control Ctr, 63-70; dir control dept pharmaceut, Natcon Chem Co, 71-74; dir, Anal Res & Develop Dept Pharmaceut, Endo Lab, Inc, 74-79; dir, Anal Res & Develop Dept Pharmaceut, Bristol Lab, 79-83; dir, Anal Res Dept Pharaceut, Res & Develop Div, Bristol- Myers Co, 83-85, vpres qual assurance, Indust Div, 85-90, VPRES QUAL CONTROL, BRISTOL-MYERS SQUIBB CO, 90- *Concurrent Pos:* Teaching, Sch Pharm, NDak State Univ, 72; lectr, Ctr Prof Advan, NJ, 78. *Mem:* Am Chem Soc; Am Pharmaceut Asn; Acad Pharmaceut Sci; Parenteral Drug Asn; Asn Anal Chem. *Res:* Drug analysis; automated analysis; biopharmaceutics and pharmacodynamics. *Mailing Add:* PO Box 4755 Syracuse NY 13221

HANNA, STANLEY SWEET, b Sagaing, Burma, May 17, 20; m 42; c 3. NUCLEAR PHYSICS, SOLID STATE PHYSICS. *Educ:* Denison Univ, AB, 41; Johns Hopkins Univ, PhD(physics), 47. *Hon Degrees:* DSc, Denison Univ, 70. *Prof Exp:* Instr physics, Johns Hopkins Univ, 43-44 & 46-48, asst, 44-45, asst prof, 48-55; assoc physicist, Argonne Nat Lab, 55-60, sr physicist, 60-63; PROF PHYSICS, STANFORD UNIV, 63- *Concurrent Pos:* Guggenheim fel, 58-59; Chmn, Nuclear Physics Panel, Comt Physics, Nat Acad Sci-Nat Res Coun, 64-65; mem, Comt Intermediate Energy Physics, NSF-AEC, 74; chmn, Div Nuclear Physics, Am Physics Soc, 76-77, mem coun, 78-82. *Honors & Awards:* Humboldt Award, 77 & 89. *Mem:* Fel Am Phys Soc; Sigma Xi. *Res:* Nuclear physics and structure; giant resonances; polarizations of nuclear radiations; lifetimes of nuclear states; resonance absorption; Mossbauer effect; nuclear moments; analog states; photonuclear reactions; hyperfine interactions; magnetism; electron scattering; intermediate energy physics weak interactions. *Mailing Add:* Varian Physics Bldg Stanford Univ Stanford CA 94305

HANNA, STEVEN J(OHN), b Indianapolis, Ind, Dec 24, 37; m 60; c 5. CIVIL ENGINEERING, TRANSPORTATION. *Educ:* Purdue Univ, BS, 60, MS, 62, PhD(civil eng), 68. *Prof Exp:* Res asst, Purdue Univ, 60-61, instr civil eng, 61-67; from asst prof to assoc prof civil eng, Univ Wyo, 67-73; assoc prof, 73-77, chmn dept eng & technol, 81-83, actg asst dean eng & actg chmn civil eng, 83-84, asst dean eng, 84-87, PROF ENG, SOUTHERN ILL UNIV, EDWARDSVILLE, 77- *Concurrent Pos:* Mem, Transp Res Bd, Nat Res Coun. *Mem:* Am Soc Testing & Mat; Am Soc Civil Engrs; Am Soc Eng Educ; Nat Soc Prof Engrs. *Res:* Statistical analysis and quality control for highway construction materials and test methods; investigation of origin and growth of critical cracks in hardened portland cement paste; highway safety; engineering materials; soils. *Mailing Add:* Sch Eng Edwardsville Southern Ill Univ Edwardsville IL 62026-1800

HANNA, STEVEN ROGERS, b Rutland, Vt, June 26, 43; m 65; c 5. METEOROLOGY. *Educ:* Pa State Univ, BS, 64, MS, 66, PhD(meteorol), 67. *Prof Exp:* Res meteorologist, Nat Oceanic & Atmospheric Admin, 67-81; res meteorologist, Environ Res & Technol, Inc, 81-85; RES METEOROLOGIST, SIGMA RES CORP, 85- *Mem:* Am Meteorol Soc. *Res:* Atmospheric boundary layer turbulence and diffusion. *Mailing Add:* 696 Virginia Rd Concord MA 01742

HANNA, WAYNE WILLIAM, b Flatonia, Tex, Apr 20, 43; m 67; c 1. PLANT GENETICS. *Educ:* Tex A&M Univ, BS, 66, MS, 68, PhD(genetics), 70. *Prof Exp:* Asst prof plant breeding, Univ Fla, 70-71; RES GENETICIST, COASTAL PLAIN EXP STA, USDA, 71- *Mem:* AAAS; Am Soc Agron; Crop Sci Soc Am. *Res:* Apomixis, chromosome manipulation and reproductive behavior in plants as related to plant improvement. *Mailing Add:* Agron Dept Coastal Plain Exp Sta PO Box 748 Tifton GA 31794

HANNA, WILLIAM F, b Chicago, Ill, July 27, 38; c 2. GEOPHYSICS. *Educ:* Ind Univ, BS, 60, AM, 62, PhD(geophys), 65. *Prof Exp:* Vis asst prof geophys, Stanford Univ, 64-65; geophysicist, 65-73, dep chief, Off Geochem & Geophys, 74-77, CHIEF, BR REGIONAL GEOPHYS, US GEOL SURV, 78- *Mem:* AAAS; Am Geophys Union; Sigma Xi; Soc Explor Geophys; fel Geol Soc Am. *Res:* Solid-earth geophysics; gravity, aeromagnetic and rock magnetic interpretation. *Mailing Add:* GEO OMR BGP Mail Stop 927 12201 Sunrise Valley Dr Reston VA 22092

HANNA, WILLIAM J(OHNSON), b Longmont, Colo, Feb 7, 22; m 44; c 2. ELECTRIC POWER, TRANSMISSION & MACHINES. *Educ:* Univ Colo, BS, 43, MS, 48, EE, 50. *Prof Exp:* From instr to assoc prof, 46-63, PROF ELEC ENG, UNIV COLO, BOULDER, 63- *Concurrent Pos:* Res engr, eng labs, 44-46; consult bd adv, Electronics Assocs, Colo, 59-63 & Los Alamos Sci Lab, 65-85; mem, State Bd Regist Prof Engrs & Land Survrs, 73-84; pres, Nat Coun Eng Examrs, 77-78. *Honors & Awards:* Archimedes Award, Calif Soc Prof Engrs, 78; Distinguished Serv Award, Nat Coun Eng Examr, 79; Alfred J Ryan Award, 78; Distinguished Serv Award, Spec Commendation, Nat Coun Eng Examr Soc, 90. *Mem:* Am Soc Eng Educ; Inst Elec & Electronics Engrs; Nat Soc Prof Engrs. *Res:* Use of foils, especially aluminum, in electromagnetic applications; generation and transmission of electric power. *Mailing Add:* Dept Elec Eng Box 425 Univ of Colo Boulder CO 80309

HANNA, WILLIAM JEFFERSON, b Chesterfield, SC, Sept 14, 13; m 41. GEOCHEMISTRY. *Educ:* Clemson Univ, BS, 34; NC State Univ, MS, 48; Rutgers Univ, PhD(soils), 51. *Prof Exp:* Asst chem, SC Agr Exp Sta, 34-35, 37-38; asst county agent, NC Agr Exten Serv, 35-37; jr scientist soils, Va Truck Crop Exp Sta, 38-42; from instr to prof soils, Rutgers Univ, 47-67, assoc ed, Soil Sci, 62-65, ed, 65-67; assoc prof geol, Old Dominion Univ, 67-69, chmn dept, 69-74, prof geophys sci, 69-76; RETIRED. *Mem:* AAAS. *Res:* Chemical analysis of sediments, soils and waters. *Mailing Add:* PO Box 4660 Virginia Beach VA 23454

HANNAH, HAROLD WINFORD, b White Heath, Ill, Jan 16, 11; m 32; c 5. VETERINARY MEDICINE LAW. *Educ:* Univ Ill, BS, 32; Univ Agr & Technol, India, 74. *Hon Degrees:* JD, Univ Ill, 35. *Prof Exp:* Asst to dean, admin, Col Agr, Univ Ill, 35-39, asst prof agr, 39-41; mil officer, 101st Airborne, 41-45; dir admin, Div Spec Serv War Vet, Univ Ill, 45-47, prof agr law, 47-54. *Concurrent Pos:* Exec secy, Ill Soil Conserv Dist, 39-41, res dir, Mich "Little Hoover Comt", agr, 52, group leader, Univ Ill Contract term, India, 55-57, adv, GP Pant Univ, India, 59, team mem, comn post-sec & higher educ, Nigeria, 60, adv; Ford Found Term, Pakistan, 62, author books on struct of univs in develop countries, 64-65, adv United Nations Develop Prog, Agr Univ, Malaysia, 72. *Mem:* Hon mem, Am Vet Asn; Am Bar Asn; Am Agr Law Asn. *Res:* Laws and internal structures of universities in developing countries; veterinary law; agricultural law; water law; landlord and tenant law. *Mailing Add:* AFBI Texico IL 62889

HANNAH, JOHN, b Kilwinning, Scotland, Aug 1, 31; m 56; c 2. ORGANIC CHEMISTRY. *Educ:* Glasgow Univ, BSc, 53; Imp Col, dipl, 56; Univ London, PhD(org chem), 56. *Prof Exp:* Harwell fel org chem, Imp Col, Univ London, 56-57; Harvard fel, Harvard Univ, 57-58; Imp Chem Industs fel, Cambridge Univ, 58-60; sr res chemist, 60-66, res fel, 66-71, sr res fel, 71-82, SR INVESTR, MERCK & CO, RAHWAY, 82- *Mem:* Am Chem Soc. *Res:* Nitrogen heterocyclic compounds; steroids; antibiotics; natural products. *Mailing Add:* Strathmore 155 Idlebrook Lane Matawan NJ 07747

HANNAH, SIDNEY ALLISON, b Clifton Forge, Va, Apr 9, 30. ENVIRONMENTAL CHEMISTRY. *Educ:* Va Mil Inst, BS, 51; Univ Fla, MS, 60, PhD(chem), 62. *Prof Exp:* Chemist, Gen Chem Div, Allied Chem & Dye Corp, 51-52 & Am Viscose Corp, 56-58; res asst, Univ Fla, 58-62; res chemist, USPHS, 63-67, Fed Water Pollution Control Admin, 67-69 & Fed Water Qual Admin, 69-70, SUPVRY RES CHEMIST, ENVIRON PROTECTION AGENCY, 70- *Mem:* AAAS; Am Chem Soc; Am Water Works Asn; Fed Water Qual Asn. *Res:* Water treatment, particularly removal of dissolved and suspended solids; physical and chemical methods for wastewater purification; water quality monitoring; control of toxic substances in wastewater; pollution control research project management. *Mailing Add:* 1648 Dell Terr Cincinnati OH 45230

HANNAN, CHARLES KEVIN, virology, for more information see previous edition

HANNAN, HERBERT HERRICK, b Liberty, Maine, Apr 3, 29; m 52; c 2. ZOOLOGY, LIMNOLOGY. *Educ:* Southwest Tex State Col, BS, 57, MA, 63; Brown Univ, MAT, 61; Okla State Univ, PhD, 67. *Prof Exp:* Sci coordr, Dickinson Independent Sch Dist, 57-59; from instr to asst prof, 60-65, chmn dept, 70-76, dir, Aquatic Sta, 76-79, ASSOC PROF BIOL, SOUTHWEST TEX STATE COL, 68-,CHEM DEPT, 80- *Mem:* Am Inst Biol Sci; Am Soc Limnol & Oceanog. *Res:* Limnology, primary productivity, plant succession, eutrophication; physiology, endocrinology. *Mailing Add:* Dept Biol Southwest Tex State Col San Marcos TX 78666

HANNAN, JAMES FRANCIS, b Holyoke, Mass, Sept 14, 22; m 51. MATHEMATICAL STATISTICS. *Educ:* St Michael's Col, PhB, 43; Harvard Univ, AM, 47; Univ NC, PhD(math statist), 53. *Prof Exp:* Instr math, Cath Univ, 50-53; asst prof, 53-55, statist, 55-57, assoc prof, 57-66, PROF STATIST, MICH STATE UNIV, 66- *Mem:* Am Math Soc; Economet Soc; Math Asn Am; Am Statist Asn; fel Inst Math Statist. *Res:* Statistical decision theory; asymptotic distribution theory. *Mailing Add:* Dept Statist & Probability Mich State Univ 4024 Wells Hall East Lansing MI 48824

HANNAN, ROY BARTON, JR, analytical chemistry; deceased, see previous edition for last biography

HANNAWAY, DAVID BRYON, b Philadelphia, Pa, Sept 14, 51; m 75. NITROGEN FIXATION, FORAGE QUALITY. *Educ:* Univ Del, BS, 73; Univ Tenn, MS, 75; Univ Ky, PhD(plant physiol), 79. *Prof Exp:* FORAGE AGRONOMIST, CROP SCI DEPT, ORE STATE UNIV, 79- *Mem:* Am Soc Agron; Crop Sci Soc Am; Am Forage & Grassland Coun. *Res:* Nitrogen fixation and mineral nutrition of forage legumes; forage physiology; forage quality; hay quality standards. *Mailing Add:* 2903 NW Hayes Corvallis OR 97330

HANNAY, DAVID G, b Catskill, NY, July 2, 45; m 65; c 2. INFORMATION SYSTEMS, THEORY OF COMPUTING. *Educ:* Wheaton Col, Ill, BS, 66; State Univ NY Stony Brook, MA, 67, MS, 70; Rensselaer Polytechnic Inst, PhD(computer Sci), 73. *Prof Exp:* Asst dir, Comput Ctr, State Univ NY Albany, 67-70, lectr computer sci, 70-73; dir computer serv, Russell Sage Col, 73-76; sr software specialist, Digital Equip Corp, 76-78; assoc prof, 78-84, CHMN COMPUTER SCI, UNION COL, NY, 84- *Concurrent Pos:* Consult, 67-90, vpres info systs, Hannay Reels, 90-; exec dir, Computator Assoc, 90- *Mem:* Asn Comput Mach; Sigma Xi; Inst Elec & Electronics Engrs. *Res:* Making computer theory more accessible via such methods as hypercard animations. *Mailing Add:* Computer Sci Dept Union Col Schenectady NY 12308

HANNAY, NORMAN BRUCE, b Mt Vernon, Wash, Feb 9, 21; m 43; c 2. SOLID STATE CHEMISTRY & PHYSICS, PHYSICAL CHEMISTRY. *Educ:* Swarthmore Col, BA, 42; Princeton Univ, MA, 43, PhD(phys chem), 44. *Hon Degrees:* PhD, Tel Aviv Univ, 78; DSc, Swarthmore Col, 79 & Polytech Inst NY, 81. *Prof Exp:* Asst chem, Princeton Univ, 42-43, instr physics, 43-44, res, Manhattan Dist, 44; res chemist, Bell Tel Labs, 44-82, chem dir, 60-67, exec dir res, mat sci & eng, 67-73, vpres res & patents, 73-82; RETIRED. *Concurrent Pos:* Mem, Coun Int Inst Appl Systs Anal; chmn, sci adv coun, Atlantic Richfield Co, Gulf Appl Technologies; consult, Alexander von Humboldt Found, OECD; mem adv comts, Princeton Univ, Univ Calif, San Diego, Univ Calif, Berkley, Harvard Univ, Lehigh Univ, Yale Univ, Penn State Univ, Cornell Univ, Stanford Univ, Duke Univ, Polytech Inst NY, Ctr Sci & Technol Policy, NY Univ, Coun Sci & Technol Develop, Dept State, Dept Defense, Dept Commerce, NSF, Off Sci & Technol Policy, Brookhaven Nat Lab & Nat Bur Standards; bd dirs, Plenum Publ Corp, Rohm & Haas Co, Gen Signal Corp & Alex, Brown & Sons; mem gov bd, Nat Res Coun; Regents prof, Univ Calif, Los Angeles, 76 & Univ Calif, San Diego, 79; Case Western Centennial Scholar, 80; consult, Chrysler & United Technol Corp, SRI Int, Merck Inst, Am Cancer Soc, Sci-Tech Holdings, Paribas. *Honors & Awards:* Monie A Ferst lectr, Ga Inst Technol, 70; Almquist lectr, Univ Idaho, 74; Acheson Medal, Electrochem Soc, 76; Perkin Medal, 83; IRI Medal, 82; Gold Medal, Am Inst Chem, 86. *Mem:* Nat Acad Sci; Nat Acad Eng (foreign secy); fel Am Phys Soc; fel Am Acad Arts & Sci; Electrochem Soc; Mexican Nat Acad Eng; Indust Res Inst; Am Chem Soc; Sigma Xi. *Res:* Molecular structure; electron emission; mass spectroscopy; semiconductors; solid-state chemistry; superconductors. *Mailing Add:* 201 Condon Lane Port Ludlow WA 98365

HANNE, JOHN R, b St Louis, Mo, Aug 25, 36; m 58; c 2. COMMUNICATION SCIENCES, MATHEMATICS. *Educ:* Dartmouth Col, AB, 58; Univ Mich, MS, 59 & 61, PhD(commun sci), 64. *Prof Exp:* Engr, Bendix Systs Div, 59-61; res asst automatic speech recognition, Commun Sci Lab, Univ Mich, 61-64, res assoc, 64-65; mem corp tech staff, Tex Instruments Inc, 65-67, mgr design automation dept, 67-76, mgr prod develop, Comput Systs Div, 76-80, asst vpres & mgr advan technol res & develop, Digital Syst Group, 80-82, vpres, 82-83; vpres eng, Osborne Comput, 83; VPRES COMPUT AIDED DESIGN, MICROELECTRONICS & COMPUT TECHNOL CORP, 83- *Concurrent Pos:* Chmn design automation comt, Inst Printed Circuits, 67-70; secy & treas, spec interest group design automation, Asn Comput Mach; chmn, comput sci adv comt, Univ Tex, Austin. *Mem:* Inst Elec & Electronics Engrs; Asn Comput Mach; Sigma Xi. *Res:* Computer aided design. *Mailing Add:* 11000 Spicewood Pkwy Austin TX 78750

HANNEKEN, CLEMENS, b Ramsey, Ill, Oct 10, 23; m 55; c 6. MATHEMATICS. *Educ:* Eastern Ill Univ, BS, 45; Univ Ill, MS, 48, PhD, 52. *Prof Exp:* From asst prof to assoc prof, 53-67, chmn dept, 64-70, PROF MATH, MARQUETTE UNIV, 67- *Mem:* Am Math Soc; Math Asn Am. *Res:* Metabelian groups; classification of irreducible congruences. *Mailing Add:* Dept Math/Comput Sci Marquette Univ 1515 W Wisconsin Ave Milwaukee WI 53233

HANNELL, JOHN W(ELDALE), JR, b Troy, NY, May 30, 23; wid; c 2. MECHANICAL ENGINEERING. *Educ:* Rensselaer Polytech Inst, BS, 47. *Prof Exp:* Engr, Rayon Res Div, E I du Pont de Nemours & Co, Inc, 47-52, res supvr, 52-58, res supvr, 58-60, res mgr, 69-79, develop fel, Indust Prods Lab, 79-85; RETIRED. *Mem:* Am Soc Mech Engrs. *Res:* Industrial uses for synthetic fibers. *Mailing Add:* 704 Ambleside Dr Wilmington DE 19808

HANNEMAN, RODNEY E, b Spokane, Wash, Mar 14, 36; m 59, 81; c 2. MATERIALS SCIENCE, METALLURGY. *Educ:* Wash State Univ, BS, 59; Mass Inst Technol, MS, 61, PhD(phys metall), 64. *Prof Exp:* Metallurgist, Hanford Labs, Wash, Gen Elec Co, 59; staff assoc electronic mat res, Lincoln Lab, Mass Inst Technol, 59-63; metallurgist, Gen Elec Co, 63-68, acting mgr, Ceramics Br, 68, mgr, Inorganic Mat & Reactions Br, 68-75, mgr, BWR struct mat, 75-77, mgr, Mat Character Lab, Res & Develop Ctr, 77-80, mgr mat prog, 80-81; vpres res, develop & energy resources, 81-85, VPRES QUAL ASSURANCE & TECHNOL, REYNOLDS METAL CO, 85- *Concurrent Pos:* Dir, Environ Struct Inc, 71-73; mem, Elec Power Res Inst, Corrosion Adv Comt, 75-80; mem adv bd, Mat Process Ctr, Mass Inst Technol, 80-; chmn, Res Coord Coun, Gas Res Inst, 82-; mem bd, Metal Properties Coun, 82- *Honors & Awards:* Geisler Award, Am Soc Metals, 71; Edison Medallion, 79; Engr Nat Achievement Award, Am Soc Metals, 73. *Mem:* Am Chem Soc; Am Soc Metals; Am Inst Mining, Metall & Petrol Engrs; NY Acad Sci; AAAS. *Res:* Nuclear technology, thermodynamics; kinetics; phase stability; diffusion; x-ray; high pressure; diamonds; material removal technology; surfaces; light sources; environment effects; analytical chemistry; aluminum technology. *Mailing Add:* 3801 Old Gun Rd W Midlothian VA 23113

HANNEMAN, WALTER W, b Oak Park, Ill, Oct 17, 27; m 51; c 4. ANALYTICAL CHEMISTRY, ORGANIC CHEMISTRY. *Educ:* Univ Ill, Urbana, BS, 49; Univ Nebr, Lincoln, MS, 56, PhD(chem) 58. *Prof Exp:* Mem staff, Chrysler Corp, 49-50, Standard Oil Co Calif, 58-63 & E I du Pont de Nemours & Co, 63-65; MEM STAFF, KAISER ALUMINUM & CHEM CORP, 65- *Mem:* Am Chem Soc. *Res:* Organic analytical chemistry; classical and instrumental methods including information retrieval, ultraviolet, nuclear magnetic resonance, mass spectroscopy, as well as gas, liquid, and thin layer chromatography. *Mailing Add:* Box 877 Pleasanton CA 94566

HANNON, BRUCE MICHAEL, b Ivesdale, Ill, Aug 14, 34; m 56; c 4. ENERGY, ECONOMIC & ECOLOGICAL MODELING. *Educ:* Univ Ill, Urbana, BS, 56, MS, 66, PhD(theoret & appl mech), 70. *Prof Exp:* Proj & res engr, US Indust Chem Corp, 56-66; instr eng, 66-70, from asst prof to assoc prof Off Vchancellor Res, 70-80, PROF GEOG, UNIV ILL, URBANA, 80- *Concurrent Pos:* Consult, US Indust Chem Corp, 66-67, Inst Environ Qual, Ill, 70-77, Ford Found, 72-73, Dept Transp, Ill, 74-75, Fed Energy Agency, Washington, DC, 74-76, Off Technol Assessment, US Cong, 75-; prin investr grants, NSF, US Energy Res & Develop Admin, Dept of Energy, Coun on Environ Qual, 71-78; Bioengineering & Hon Fac, Affil Scientist, Ill; Natural Hist Surv, 82- *Honors & Awards:* Mitchell Prize Award, Club of Rome, 75. *Res:* Modeling the flow of natural resources through society and through ecosystems. *Mailing Add:* Dept Geog Univ Ill 130A Observ Urbana IL 61801

HANNON, JAMES PATRICK, b Houston, Tex, Mar 27, 40; m 66; c 2. THEORETICAL SOLID STATE PHYSICS. *Educ:* Rice Univ, BA, 62, MA, 65, PhD(physics), 67. *Prof Exp:* Fel physics, AEK Riso, Denmark, 66-67; fel, Rice Univ, 67-69; fel, Tech Univ Munich, 69-70; asst prof, 70-74, assoc prof, 74-77, PROF PHYSICS, RICE UNIV, 77-; ASSOC, WEISS COL, 76- *Concurrent Pos:* Exchange scientist to USSR, Nat Acad Sci, 78; vis res scientist, Tech Univ Munich, 78-79 & Univ Hamburg, 78-79 & 81. *Mem:* Am Phys Soc. *Res:* Gamma ray optics; theoretical Mossbauer effect studies; quantum optics; magnetism; cooperative effects. *Mailing Add:* Dept Physics Rice Univ Box 1892 Houston TX 77251

HANNON, JOHN PATRICK, b Richmond, Calif, May 12, 27; m 52; c 4. PHYSIOLOGY. *Educ:* Univ Calif, BA, 50, PhD(physiol), 54. *Prof Exp:* Jr res physiologist, Univ Calif, 54-55; res physiologist, Arctic Aeromed Lab, 55-58, chief, Physiol Dept, 58-63; asst chief, Physiol Div, US Army Med Res & Nutrit Lab, Fitzsimons Gen Hosp, 63-65, chief, 65-74; SR RES PHYSIOLOGIST, LETTERMAN ARMY INST RES, 74- *Concurrent Pos:* Assoc prof zool, Univ Alaska, 60-63; adjoint prof, Univ Colo, 68-76. *Mem:* Fel AAAS; Am Physiol Soc; Soc Exp Biol & Med. *Res:* Effects of heat, cold and altitude on human and animal physiology and biochemistry; nutritional physiology and biochemistry; cellular and work physiology; intermediary and energy metabolism. *Mailing Add:* Letterman Army Inst Res Presidio San Francisco San Francisco CA 94129

HANNON, MARTIN J, b New York, NY, Dec 20, 42; m 69; c 2. POLYMER SCIENCE. *Educ:* Manhattan Col, BChE, 64; Case Western Reserve Univ, MS, 66, PhD(polymer sci, eng), 69. *Prof Exp:* Res chemist, Celanese Res Co, 68-70, sr res chemist, 70-74, group leader, 77-80, mem tech staff planning mgt, Celanese Plastics & Specialities Co, 80-88; QUALITY, BUS & MKT MGR, HOECHST-CELANESE, ENG PLASTICS DIV, 88- *Mem:* Am Chem Soc; Am Inst Physics; Soc Rheology; Soc Plastics Eng. *Res:* Rheological studies of polymer melts; polymer solid state structure and morphology; infrared spectroscopy and theoretical studies of molecular vibrations; mechanical and physical properties of polymers; fiber physics; powder coatings. *Mailing Add:* 103 Pomeroy Rd Madison NJ 07940

HANNON, WILLARD JAMES, JR, b Jacksonville, Ill, July 9, 38; m 60; c 3. GEOPHYSICS. *Educ:* St Louis Univ, BSc, 59, PhD(geophys), 64. *Prof Exp:* Inst mech, St Louis Univ, 61-64; asst prof geophys, Wash Univ, 65-69; leader, Geophys Sect, 80-82, PHYSICIST, LAWRENCE LIVERMORE NAT LAB, 69-, PROG MGR, SEISMIC MONITORING RES, 82- *Concurrent Pos:* post-doctoral fel, Calif Inst Technol, 65. *Mem:* Seismol Soc Am; Am Geophys Union. *Res:* Elastic wave propagation; nuclear test ban monitoring. *Mailing Add:* 309 Pearl Dr Livermore CA 94550

HANNSGEN, KENNETH BRUCE, b New York, NY, May 27, 42; m 63; c 1. VOLTERRA INTEGRAL EQUATIONS. *Educ:* Dartmouth Col, BA, 64; Univ Wis-Madison, MA, 65, PhD(math), 68. *Prof Exp:* Asst prof math, Univ Calif, Los Angeles, 68-72; from asst prof to assoc prof, 72-80, PROF MATH, VA POLYTECH INST & STATE UNIV, 80- *Mem:* Am Math Soc; Soc Indust & Appl Math; Math Asn Am. *Res:* Asymptotic behavior of solutions of integral equations and applications to stabilization of mechanical systems. *Mailing Add:* Dept Math Va Polytech Inst & State Univ Blacksburg VA 24061-0123

HANNUM, STEVEN EARL, b Long Beach, Calif, July 18, 41; m 66; c 2. PHYSICAL CHEMISTRY. *Educ:* Wheaton Col, Ill, BS, 63; Univ Ky, PhD(phys chem), 69. *Prof Exp:* Res assoc, Univ SC, 69-70; from asst prof to assoc prof chem, Aurora Col, 70-78; assoc prof, Asbury Col, 78-85, chmn, sci div, 78-84; ASSOC PROF CHEM, GEORGE FOX COL, 85- *Mem:* Am Chem Soc. *Res:* Computers and chemistry; interfacing chemical instrumentation to computers. *Mailing Add:* Dept Chem George Fox Col Newberg OR 97132

HANOVER, JAMES W, b Port Huron, Mich, Dec 10, 30; m 53; c 5. FOREST GENETICS. *Educ:* Univ Wash, BS, 53; Wash State Univ, PhD(genetics), 63. *Prof Exp:* Res forester, Intermountain Exp Sta, US Forest Serv, 56-58, geneticist, 58-65; assoc prof forest genetics, Yale Univ, 65-66; assoc prof, 66-71, PROF FORESTRY, MICH STATE UNIV, 71- *Concurrent Pos:* Assoc ed, Forest Sci, 75-78; Chmn, Soc Am Forestry Genetics Working Group, 80-82. *Mem:* AAAS; Sigma Xi; Am Soc Plant Physiologists; Soc Am Foresters. *Res:* Biochemical and physiological genetics; population genetics; tree physiology; biochemical systematics. *Mailing Add:* Dept Forestry Mich State Univ East Lansing MI 48823

HANOVER, JOHN ALLAN, b Tulsa, Okla, May 19, 53; m 87; c 2. GLYCOPROTEIN RESEARCH, CELL BIOLOGY. *Educ:* Univ Tulsa, BS(chem) & BS(biol), 76; Johns Hopkins Univ, PhD(biochem), 81. *Prof Exp:* Sr staff, Nat Cancer Inst, 81-85, RES CHEMIST, NAT INST DIABETES, DIGESTIVE & KIDNEY DIS, NIH, 85- *Concurrent Pos:* Exec ed, Anal Biochem, 88-; lectr, Univ Ky, 90; ed, Arch Biochem & Biophysics, 90-, Glycobiol, 90- *Mem:* Am Soc Biochem & Molecular Biol; Am Soc Cell Biol. *Res:* Molecular characterization of nuclear pore glycoproteins. *Mailing Add:* NIDDK LBM Bldg 10 Rm 9B15 9000 Rockville Pike Bethesda MD 20892

HANRAHAN, EDWARD S, b Marietta, Ohio, Dec 8, 29; m 52; c 3. PHYSICAL CHEMISTRY. *Educ:* Univ Miss, BS, 51; WVa Univ, MS, 56, PhD(chem), 59. *Prof Exp:* Instr chem, WVa Univ, 57-58; chemist, E I Du Pont de Nemours & Co, 58-63; from asst prof to prof chem, 63-78, chmn dept, 67-78, DEAN, COL SCI, MARSHALL UNIV, 77- *Concurrent Pos:* Consult, Polan Indust, Inc, 63-65; mem, Comt Chem Educ, Am Chem Soc, 72-78. *Mem:* AAAS; Am Chem Soc; NY Acad Sci; Sigma Xi. *Res:* Decarboxylation of organic di-acids; hydrogen bonding; phase transitions in solids; surface chemistry. *Mailing Add:* Col Sci Marshall Univ Huntington WV 25701

HANRAHAN, JOHN J, b New London, Conn, Mar 19, 32; m 57; c 4. UNDERWATER ACOUSTICS. *Educ:* Univ Conn, BA, 54, MS, 62. *Prof Exp:* Actuarial trainee, Travelers Ins Co, 54-55; jr engr, US Navy Underwater Sound Lab, 57-60; math analyst, Elec Boat Div, Gen Dynamics Corp, 60-62; opers res analyst, US Naval Underwater Systs Ctr, 62-71, head, oper systs anal & assessment, 71-83, head, Oper Systs & Spec Projs Off, 83-87; BBN SYSTS & TECHNOL, 87- *Concurrent Pos:* Exchange scientist, Admiralty Underwater Weapons Estab, Eng, 64-65; adj prof, Naval Postgrad Sch, Monterey, Calif, 83. *Mem:* Fel Acoust Soc Am. *Res:* Propagation and scattering aspects of underwater sound; performance prediction studies and new system concepts. *Mailing Add:* Four Seabreeze Dr Waterford CT 06385

HANRAHAN, ROBERT JOSEPH, b Chicago, Ill, Jan 7, 32; m 57; c 4. PHYSICAL CHEMISTRY, RADIATION CHEMISTRY. *Educ:* Loyola Univ, Ill, BS, 53; Univ Wis, PhD, 58. *Prof Exp:* Asst, Univ Wis, 53-56; res fel, NSF, Univ Leeds, 57-58; from asst prof to assoc prof, 58-71, PROF CHEM, UNIV FLA, 71- *Concurrent Pos:* Vis scientist, Hahn-Meitner Inst Nuclear Res, Berlin, 76. *Mem:* AAAS; Am Chem Soc; Radiation Res Soc; Am Soc Mass Spectrometry; Am Phys Soc; Int Am Photochem Soc. *Res:* Kinetics and mechanisms of reactions; radiochemistry; radiation chemistry, especially the effects of ionizing radiation on pure organic compounds; mass spectrometry; application of small computers in chemistry; photochemistry. *Mailing Add:* Dept Chem Univ Fla Gainesville FL 32611

HANRATTY, THOMAS J(OSEPH), b Philadelphia, Pa, Nov 9, 26; m 56; c 5. CHEMICAL ENGINEERING, FLUID DYNAMICS. *Educ:* Villanova Col, BChE, 47; Ohio State Univ, MS, 50; Princeton Univ, PhD(chem eng), 53. *Hon Degrees:* Dr, Villanova Univ, 79. *Prof Exp:* Engr, Fischer & Procter Co, 47-48; res engr, Battelle Mem Inst, 48-50; from asst prof to assoc prof chem eng, 53-63, PROF CHEM ENG, UNIV ILL, URBANA, 63-, JAMES W WESTWATER PROF, 88- *Concurrent Pos:* NSF fel, 62-63; Shell distinguished chair chem eng, 81-86. *Honors & Awards:* Colburn Award, 57, William H Walker Award, 65 & Prof Prog award, 67, Am Inst Chem Engrs; Curtis W McGraw res award, Am Soc Eng Educ, 63; Sr Res Award, Am Soc Eng Educ, 79; Ernest Thiele Award, Am Inst Chem Eng, 86. *Mem:* Nat Acad Eng; Am Inst Chem Engrs; fel Am Phys Soc; Am Chem Soc; fel Am Acad Mech. *Mailing Add:* Dept Chem Eng Univ Ill 1209 W California Urbana IL 61801

HANSARD, SAMUEL L, II, b Ft Sill, Okla, Feb 20, 44; m 80; c 4. ANIMAL SCIENCES, NEW DRUG EVALUATION. *Educ:* Univ Tenn, BS, 66; Univ Fla, MS, 68, PhD(animal sci), 75. *Prof Exp:* ANIMAL SCIENTIST NUTRIT, CTR VET MED, FOOD & DRUG ADMIN, 76- *Concurrent Pos:* Consult, Nat Heart, Lung & Blood Inst & NIH, 85. *Mem:* NY Acad Sci; Am Soc Animal Sci; Sigma Xi. *Mailing Add:* Rte 2 Box 166 Charleston WV 25414-9643

HANSARD, SAMUEL LEROY, b Knoxville, Tenn, June 5, 14; m 41; c 3. NUTRITION. *Educ:* Univ Tenn, BSA, 37; Ohio State Univ, MS, 38; Univ Fla, PhD, 52. *Prof Exp:* From instr to assoc prof animal husb, 46-50, prof & sr scientist nutrit, 51-57; prof animal sci & dir nutrit radioisotope labs, La State Univ, 57-68; prof animal sci, 68-80, head dept, 72-80, EMER PROF, UNIV TENN, KNOXVILLE, 80- *Concurrent Pos:* Calcium Carbonate Co travel fel award, 65; Gamma Sigma Delta res award, 63; travel award, Am Inst Nutrit, 63, 66, 69, 75 & 78. *Honors & Awards:* Nutrit Award, Am Soc Animal Sci, 66; Trace Mineral Award, 70; Morrison Award, 76; Distinguished Serv Award, 77. *Mem:* Fel AAAS; Soc Exp Biol & Med; Am Inst Nutrit; NY Acad Sci; Am Soc Animal Sci (pres, 74). *Res:* Mineral metabolism; placental transfer and body composition studies in farm and laboratory animals; radioisotope procedures for biological availability; fission product partition; response measurements and nutritional interrelationships; nature of farm animal anemias. *Mailing Add:* FDA HFV-136 Rockville MD 20857

HANSBURG, DANIEL, T-CELL SPECIFICITY, GRAFT REJECTION. *Educ:* Washington Univ, PhD(cell biol), 79. *Prof Exp:* ASSOC PATHOLOGIST, FOX CHASE CANCER CTR, 83- *Mailing Add:* Fox Chase Cancer Ctr 7701 Burholme Ave Philadelphia PA 19111

HANSCH, CORWIN HERMAN, b Kenmare, NDak, Oct 6, 18; m 44; c 2. CHEMISTRY. *Educ:* Univ Ill, BS, 40; NY Univ, PhD(chem), 44. *Prof Exp:* Asst chem, NY Univ, 40-44; mem staff, Manhattan Proj, Univ Chicago, 44, group leader, Richland, Wash, 44-45; res chemist, E I du Pont de Nemours & Co, Del, 45-46; from asst prof to assoc prof, 46-56, PROF ORG CHEM, POMONA COL, 56- *Concurrent Pos:* Guggenheim fel, Fed Inst Technol, Zurich, 52-53; fel, Petrol Res Fund, Am Chem Soc, Munich, 59-60; Guggenheim fel, 66-67. *Honors & Awards:* Edward E Smissman-Bristol Labs Award, Am Chem Soc, 75; Res Achievement Award Pharmaceut & Med Chem, Am Pharmaceut Asn Found, 69; Medal Ital Soc Pharmaceut Sci, Ital Pharmaceut Soc, 67; Tolman Award, Am Chem Soc, 76, Undergrad Res Award, 85. *Mem:* Am Chem Soc. *Res:* Nitrogen and sulfur heterocycles; vapor phase catalysis; correlation of chemical structure and biological activity. *Mailing Add:* Dept Chem Pomona Col Claremont CA 91711

HANSCH, THEODOR WOLFGANG, b Heidelberg, WGer, Oct 30, 41. PHYSICS, LASERS. *Educ:* Univ Heidelberg, MS, 66, PhD(physics), 69. *Prof Exp:* Asst prof physics, Univ Heidelberg, 69-70; NATO fel, 70-72, assoc prof, 72-75, PROF PHYSICS, STANFORD UNIV, 75- *Concurrent Pos:* Sloan Found fel, 73-75; Alexander von Humboldt Sr US Scientist fel, 78-79; prog co-chmn, 14th Int Quantum Electronics Conf, San Francisco, 86. *Honors & Awards:* Otto Ulung Award, 80; Cyrus B Cornsde Prize, Nat Acad Sci, 83; Herbert Bvoida Prize, Am Phys Soc, 83; William F Meggers Award, Optical Soc Am, 85. *Mem:* Fel Optical Soc Am; fel Am Phys Soc. *Res:* Spectroscopy and quantum electronics; developed powerful monochromatic pulsed dye lasers; high resolution nonlinear spectroscopy of atoms and molecules. *Mailing Add:* Sektion Physik Univ Munich Schellingstr 4 8000 Munich 40 94305 Germany

HANSCOM, ROGER H, b New York, NY, Feb 9, 44; m 68, 82; c 2. GEOLOGY, COMPUTER SCIENCE. *Educ:* Franklin & Marshall Col, AB, 65; Harvard Univ, AM, 67, PhD(geol), 73. *Prof Exp:* Sr scientist, Univ Va, 72-75; analyst, Phillips Petrol Co, 76-80; res scientist, Amoco Prod Res, 80-84; sr tech analyst, Standard Oil Ohio, 84-85; COMPUT SCIENTIST, LAWRENCE LIVERMORE NAT LAB, 85- *Mem:* Sigma Xi; Inst Elec & Electronics Engrs. *Res:* Computer applications in the geological sciences; computational mathematics; computer modeling and simulation; software systems. *Mailing Add:* PO Box 2394 Livermore CA 94551-2384

HANSEBOUT, ROBERT ROGER, b Ont, Can, Jan 11, 35; m 81. SURGICAL NEUROLOGY, HOSPITAL ADMINISTRATION. *Educ:* Univ Western Ont, MD, 60; McGill Univ, MSc, 64, dipl, 66; FRCS. *Prof Exp:* Res assoc, Univ Ottawa, 67-68; asst prof neurosurg, Hahnemann Med Col, 68-69; asst prof, McGill Univ, 69-75, assoc prof, 75-79; assoc prof, 79-82, PROF SURG, MCMASTER UNIV, 82- *Concurrent Pos:* Consult neurosurg, Nat Defense Med Ctr, 67-68; neurosurgeon, Montreal Neurol Inst, 69-79; head neurosurg co-dir, St Joseph's Hosp, Ont, 79-; co-dir Spinal Injury Unit, Montreal Neurol Hosp, 76-79; affiliate surgeon, Hamilton Civic & Chedoke-McMaster Hosp, 79-; asst dir prof serv, Montreal Neurol Hosp, 78-79, vpres, Coun Physicians, 79-; vis prof, Cleveland Clinic, 81. *Mem:* Am Asn Neurol Surgeons; Congress Neurol Surgeons; Can Neurosurg Soc; Sigma Xi; Int Asn Study Pain. *Res:* Pathogenesis and treatment of severe spinal cord injuries. *Mailing Add:* St Joseph's Hosp 50 Charlton Ave E Hamilton ON L8N 1Y4 Can

HANSEL, PAUL G(EORGE), b Grand Island, Nebr, June 22, 17; m 46; c 3. ENGINEERING PHYSICS. *Educ:* Univ Kans, BS, 46. *Prof Exp:* Radio engr, US Army Signal Corps Labs, NJ, 41-47; chief radio engr, Servo Corp Am, 47-61; vpres eng, Electronic Commun, 61-70, asst gen mgr, 64-70, vpres & gen mgr aerospace electronics, 70, vpres res & eng, 71-79; CONSULT ADV TECHNOL, 79- *Honors & Awards:* US War Dept civilian serv commendation, 46; Pioneer Award, Inst Elec & Electronics Engrs, 70. *Mem:* Fel Inst Elec & Electronics Engrs; NY Acad Sci; AAAS. *Res:* Radio direction finding and navigation; frequency control; receivers; transmitters; electronic instrumentation; light modulators; communication systems; radiation health effects. *Mailing Add:* 1374 Monterey Blvd NE St Petersburg FL 33704

HANSEL, WILLIAM, b Vale Summit, Md, Sept 16, 18; m 42; c 2. ANIMAL BIOTECHNOLOGY, REPRODUCTIVE BIOLOGY. *Educ:* Univ Md, BS, 40; Cornell Univ, MS, 47, PhD(animal physiol), 49. *Prof Exp:* Asst animal husb, Cornell Univ, 46-49, from assoc prof to prof animal physiol, 49-90, chmn physiol, 78-84; GORDON D CAIN PROF ANIMAL PHYSIOL, LA STATE UNIV, 90- *Concurrent Pos:* Guggenheim fel, Univ Chicago, 58; NSF sr fel, Commonwealth Sci Res Orgn, Australia, 66-67; Liberty Hyde Bailey Professorship, 78; consult, Merck Sharpe & Dohme, 81-83, Smith Kline Buchman, 86; vis prof, Univ Guelph, Can, 84. *Honors & Awards:* Borden Award, 72; Nat Asn Animal Breeders Award, 73; Soc Animal Sci Award Physiol & Endocrinol, 62 & NY Farmers Award, 64, Am Soc Animal Sci, 65; Morrison Award, Am Soc Animal Sci, 79; Carl Hartman Award, Soc Study Reproduction, 80; W Henry Hatch Award, 86. *Mem:* AAAS; Am Soc Animal Sci; Am Dairy Sci Asn; Am Physiol Soc; Brit Soc Study Fertil; Soc Study Reproduction (pres, 75-76); Endocrinol Soc. *Res:* Dairy cattle sterility; mechanism of the control of ovulation in farm animals; pituitary hypothalamic interrelationships; corpus luteum function; early pregnancy recognition. *Mailing Add:* Dept Vet Sci La State Univ Baton Rouge LA 70803

HANSELL, MARGARET MARY, b Weston, Ont, Feb 1, 41. ANATOMY. *Educ:* Univ Toronto, BSc, 63; Univ Calif, Riverside, PhD(biol), 68. *Prof Exp:* Nat Res Coun Can fel bot, Univ Toronto, 68-69; lectr anat, 69-71, asst prof, 71-76, ASSOC PROF ANAT, DALHOUSIE UNIV, 76- *Mem:* Can Asn Anat; Can Soc Cell Biol. *Res:* Electron microscopy of mammalian respiratory system, liver and kidney; pollution effects. *Mailing Add:* Dept Anat Dalhousie Univ Halifax NS B3H 4H6 Can

HANSELMAN, RAYMOND BUSH, b New York, NY, Dec 29, 32; m 58; c 3. ANALYTICAL CHEMISTRY, PHOTOGRAPHIC CHEMISTRY. *Educ:* Amherst Col, BA, 54; Mass Inst Technol, PhD(anal chem), 59. *Prof Exp:* Group leader anal chem, Plastics Div, Union Carbide Corp, 59-64; mgr environ technol, Space Systs Div, Avco Corp, 64-67; mgr res lab, Polaroid Corp, 67-85; dir chem res, 85-90, DIR CUSTOMER SATISFACTION & QUAL ASSURANCE, CHEM PROD, WATERS CHROMATOGRAPHY DIV, MILLIPORE CORP, 90- *Mem:* AAAS; Am Chem Soc. *Res:* Photographic sciences; polymer physical chemistry; chemical instrumentation and space sciences; oceanography; chromatography. *Mailing Add:* 132 Deacon Haynes Rd Concord MA 01742

HANSEN, AFTON M, b Mayfield, Utah, Sept 25, 25. BIOLOGY, GENETICS. *Educ:* Brigham Young Univ, BS, 52, MS, 53; Utah State Univ, PhD(zool), 62. *Prof Exp:* From instr to prof & chmn, Div Agr & Life Sci, Snow Col, 53- 72, prof & chmn, Div Natural Sci, 72-75, prof biol sci, 75-89; RETIRED. *Concurrent Pos:* Forestry aid, US Forest Serv, 54-57; range conservationist, Great Basin Res Ctr, 64-54, range scientist, 65-66. *Res:* Drasophila genetics eye abnormatities. *Mailing Add:* Box 96 Mayfield UT 84643

HANSEN, ANTHONY DAVID ANDERS, b London, Eng, May 12, 51; m; c 1. ATMOSPHERIC PHYSICS, INSTRUMENTATION. *Educ:* Oxford Univ, Eng, BA, 72; Univ Calif, Berkeley, PhD(physics), 77. *Hon Degrees:* MA, Oxford Univ, Eng, 78. *Prof Exp:* Technician instrumentation, Nuclear Physics Lab, Univ Oxford, 70-72 & Europ Orgn Nuclear Res, Geneva, 72; res assoc physics, 74-77, STAFF SCIENTIST ATMOSPHERIC PHYSICS, LAWRENCE BERKELEY LAB, UNIV CALIF, BERKELEY, 77- *Mem:* Am Phys Soc; Optical Soc Am. *Res:* Particulate air pollution; physical and chemical transformation of particles; development and validation of instrumentation, experimental techniques and data analysis. *Mailing Add:* Lawrence Berkeley Lab Bldg 70A-3363 Berkeley CA 94720

HANSEN, ANTON JUERGEN, b Hamburg, Ger, Dec 23, 28; m 61; c 1. PLANT PATHOLOGY. *Educ:* Univ Gottingen, dipl, 54; Univ Wis, PhD(plant path), 58. *Prof Exp:* Asst, Max Planck Inst Res Plant Breeding, Ger, 54-55; asst plant path, Univ Wis, 55-58, res fel, 63-65; plant pathologist, Int-Am Inst Agr Sci, 58-63; PLANT PATHOLOGIST, RES BR, CAN DEPT AGR, 65- *Mem:* Am Phytopath Soc; Am Hort Soc. *Res:* Virus diseases of stone fruits, pome fruits and grapes; diseases of grapes and tropical crops. *Mailing Add:* Rural Rt No 2 No 4 Site 106 Summerland BC V0H 1Z0 Can

HANSEN, ARTHUR G(ENE), b Sturgeon Bay, Wis, Feb 28, 25; m 72; c 5. MECHANICAL ENGINEERING, APPLIED MATHEMATICS. *Educ:* Purdue Univ, BS, 46, MS, 48; Case Western Reserve Univ, PhD(math), 58. *Hon Degrees:* DEng, Purdue Univ, 70; DSc, Ind Univ, 82. *Prof Exp:* Instr math, Purdue Univ, 46-48; res engr fluid mech, Lewis Lab, Nat Adv Comt Aeronaut, 48-49, aeronaut res scientist, 50-58; head nucleonics sect, Cornell Aeronaut Lab, Inc, 58-59; prof mech eng & chmn dept, Univ Mich, 59-66; dean eng, Ga Inst Technol, 66-69, pres, 69-71; pres, Purdue Univ, 71-82; chancellor, Tex A&M Syst, 82-86; dir res, Hudson Inst, 87-88; RETIRED. *Concurrent Pos:* Instr math, Univ Md, 49-50; lectr, John Carroll Univ, 56-57 & Baldwin-Wallace Col, 57-58; consult engr, Deming Pump Co, 61-65; lectr, Ford Motor Co, 64; vis prof, Tuskegee Inst, 65; mem bd dirs, Eng Joint Coun, 70-72; subcomt Prof Sci & Technol Manpower, US Dept Labor, 71-73; consult, 86- *Honors & Awards:* Medal Public Service, Dept Defense, 86. *Mem:* Nat Acad Eng; Am Soc Eng Educ; AAAS. *Res:* Viscous fluid flow, particularly three-dimensional boundary-layer theory; theoretical fluid mechanics; partial differential equations; pump design. *Mailing Add:* 815 Suagrbush Ridge Zionsville IN 46077

HANSEN, AXEL C, b VI, Mar 4, 19; m 46. OPHTHALMOLOGY. *Educ:* Fisk Univ, BA, 41; Meharry Med Col, MD, 44; Am Bd Opthal, dipl, 53. *Hon Degrees:* DHL, Fisk Univ, 90. *Prof Exp:* From instr to prof surg ophthal, 49-74, head, Div Ophthal, 60-74, prof ophthal & chmn dept, 74-85, DISTINGUISHED EMER PROF OPHTHAL, MEHARRY MED COL, 85- *Concurrent Pos:* Insular ophthalmologist & consult to children's serv, Dept Health, VI, 56-59, consult ophthalmologist, 69-74; consult, Tenn Dept Health & Environ, Georges W Hubbard Hosp, Mecharry Med Col & Alvin York Med, Ctr Vet Admin, Murfreesboro, TN. *Mem:* Fel AAAS; Nat Med Asn; Nat Soc Prev Blindness; Am Med Asn; fel Am Acad Opthamol; fel Am Col Surgeons. *Mailing Add:* 1716 Windover Dr Nashville TN 37218

HANSEN, BARBARA CALEEN, b Boston, Mass, Nov 24, 41; m 76; c 1. ENDOCRINOLOGY. *Educ:* Univ Calif, Los Angeles, BS, 64, MS, 65; Univ Wash, Seattle, PhD(physiol & psychol), 71. *Prof Exp:* From asst prof to assoc prof physiol & nursing, Sch Med & Sch Nursing, Univ Wash, 71-76; from assoc prof to prof, Sch Med & Sch Nursing, Univ Mich, 76-82; prof physiol & psychol, assoc vpres res & dean grad sch, Southern Ill Univ, 82-85; prof physiol, psychol & vpres, 86-90, PROF PHYSIOL, SCH MED & DIR OBESITY & DIABETES RES CTR, UNIV MD, 90- *Concurrent Pos:* Prin investr, NIH res grants, 73-; mem, Dir Adv Comt & Nutrit Study Sect, NIH, 79-83; mem, Inst Med-Nat Acad Sci, 81- *Mem:* Inst Med-Nat Acad Sci; Am Physiol Soc; Nutrit Soc; NAm Asn Study Obesity (pres); Am Asn Clin Nutrit; Nat Asn Univ Res Admin; Am Inst Nutrit; Am Soc Clin Nutrit; Int Asn Study Obesity (pres). *Res:* Non-human primates in the study of obesity and the regulation of appetite, diabetes, endocrinology and behavior. *Mailing Add:* Obesity & Diabetes Res Ctr Sch Med Univ Md Ten Pine St No 600 Baltimore MD 21201

HANSEN, BERNARD LYLE, b Providence, Utah, July 27, 16; m 38; c 1. PHYSICS. *Educ:* Brigham Young Univ, BS, 40. *Hon Degrees:* PhD honoris causa, Univ Bern, 77. *Prof Exp:* Instrument engr, US Weather Bur, Washington, DC, 42-45, physicist, Instrument Div, 45-47, physicist & dir, Snow Lab, Calif, 47-49; res assoc, Univ Minn, 49-51; physicist, Snow, Ice & Permafrost Res Estab, Wilmette, 52-60; chief tech serv div, US Army Cold Regions Res & Eng Lab, Hanover, 61-73; RES ASSOC, UNIV NEBR-LINCOLN, 74- *Honors & Awards:* Seligman Crystal Award, Int Glaciol Soc, 72. *Mem:* Am Meteorol Soc; Am Geophys Union; Int Glaciol Soc. *Res:* Design of new instruments and equipment to determine physical properties of snow, ice and frozen ground; geophysical applications of infra-red radiation; deep core drilling in ice. *Mailing Add:* 530 Hazelwood Dr Lincoln NE 68510

HANSEN, CARL FREDERICK, chemical physics, for more information see previous edition

HANSEN, CARL JOHN, b Brooklyn, NY, Dec 21, 33; m 60; c 1. ASTROPHYSICS, PHYSICS. *Educ:* Queens Col, NY, BS, 56; Yale Univ, MS, 61, PhD(physics), 66. *Prof Exp:* Reactor analyst, Combustion Eng Inc, Conn, 56-60; res assoc, 66-68, from asst prof to assoc prof, 68-74, PROF ASTROPHYS, JOINT INST LAB ASTROPHYS, UNIV COLO, BOULDER, 74-, FEL, 69- *Res:* Evolution and stability of highly evolved stars; nuclear astrophysics. *Mailing Add:* Dept Astrophys & Geophys Univ Colo Jila A408 Boulder CO 80309

HANSEN, CARL TAMS, b Greeley, Colo, July 22, 29; m 62; c 2. GENETICS. *Educ:* Colo State Univ, BS, 51; SDak State Univ, MS, 59; Univ Wis-Madison, PhD(genetics), 66. *Prof Exp:* GENETICIST, NAT CTR RES RESOURCES, NIH, 64- *Mem:* AAAS; Am Genetic Asn; Genetics Soc Am. *Res:* Developing rodent models for the study of cardiovascular disorders, metabolic diseases, parasitic and infectious diseases, immunology, cancer, endocrine disorders and behavior. *Mailing Add:* Vet Res Prog Nat Ctr Res Resources NIH Bldg 14 G Rm 101 Bethesda MD 20892

HANSEN, CHARLES M, physical chemistry, chemical engineering, for more information see previous edition

HANSEN, DAVID ELLIOTT, b Brooklyn, NY, July 18, 58; m 88; c 1. ENZYMOLOGY, ANTIBODY CATALYSIS. *Educ:* Brown Univ, ScB, 79; Harvard Univ, PhD (chem), 86. *Prof Exp:* asst prof, 86-90, ASSOC PROF CHEM, AMHERST COL, 90-; ADJ PROF, PROG MOLECULAR & CELLULAR BIOL, UNIV MASS, 90- *Concurrent Pos:* Consult, Igen, Inc, Rockville, MD, 87-; prin investr, NSF, 86-89 & NIH, 88-; Amherst Col trustee fac fel, 88; NSF presidential young investr award, 89. *Mem:* Am Chem Soc; AAAS. *Res:* Strategies for the isolation of antibodies with sequence-specific protease activity; designing enzyme inhibitors. *Mailing Add:* Dept Chem Amherst Col Amherst MA 01002

HANSEN, DAVID HENRY, b Corvallis, Ore, May 8, 45; m 68; c 2. PHYSIOLOGICAL ECOLOGY, POPULATION ECOLOGY. *Educ:* Ore State Univ, BS, 68; Univ Utah, MS, 69; Univ Calif, Irvine, PhD(pop, environ biol), 74. *Prof Exp:* ASST PROF BIOL, PAC LUTHERAN UNIV, 74- *Mem:* AAAS; Bot Soc Am; Ecol Soc Am. *Res:* Physiological plant ecology and biotic effects on plant distribution with specific interests in hemiparasitic angiosperms and high energy pollinators. *Mailing Add:* Dept Biol Pac Luthern Univ Five 121st & Park Ave Tacoma WA 98447

HANSEN, DEBORAH KAY, b Springfield, Illinois, June 17, 52. TOXICOLOGY, GENETICS. *Educ:* Eastern Ill Univ, BS, 74; Iowa State Univ, MS, 76; Indiana Univ, PhD(medical genetics), 81. *Prof Exp:* RES BIOLOGIST, DIV REPRODUCTIVE & DEVELOP, 85- *Concurrent Pos:* Post doc fellow, Dept Epidemiology & Public Health, Yale Univ, 81-82; post doc fellow, Dept Pharmacol, Univ Texas, 82-85; adj asst prof, Dept Pharmacol & Interdisciplinary, Arkansas Med Sci, 86- *Mem:* Soc Toxicology; Sigma Xi; AAAS; Teratol Soc; Soc Exp Biol & Med; Asn Women Sci. *Res:* Research interest focus on mechanisms whereby drugs and or chemicals induce developmental toxicity. *Mailing Add:* 6105 Kenwood Little Rock AR 72207

HANSEN, DONALD JOSEPH, b Kansas City, Mo, June 14, 32. MATHEMATICS. *Educ:* Southern Methodist Univ, BS, 54, MS, 55; Univ Tex, PhD(math), 62. *Prof Exp:* Lectr math, Southern Methodist Univ, 55-56; spec instr, Univ Tex, 58-62; ASST PROF MATH, NC STATE UNIV, 62- *Mem:* Am Math Soc; Math Asn Am. *Res:* Ordered algebraic systems; lattice theory; functional equations. *Mailing Add:* Dept Math NC State Univ Raleigh NC 27695-8205

HANSEN, DONALD JOSEPH, physical chemistry, for more information see previous edition

HANSEN, DONALD VERNON, b Seattle, Wash, Jan 18, 31; m 58; c 3. PHYSICAL OCEANOGRAPHY. *Educ:* Univ Wash, BS, 54, MS, 61, PhD(oceanog), 64. *Prof Exp:* Engr, Boeing Airplane Co, Seattle, 56-57; teacher sci pub schs, Seattle, 57-58; asst oceanog, Univ Wash, Seattle, 59-61, res asst, 61-64, res asst prof, 64-65; res oceanogr, 66-69, actg dir, Atlantic Oceanog & Meteorol Labs, Miami, 78-80, SUPVRY OCEANOGR, ENVIRON SCI SERV ADMIN, NAT OCEANIC & ATMOSPHERIC ADMIN, US DEPT COM, 70-, DIR, PHYS OCEANOG DIV, 69- *Concurrent Pos:* Adj prof, Univ Miami, 69- *Mem:* Am Geophys Union; Am Soc Limnol & Oceanog; AAAS; Sigma Xi. *Res:* Currents and general circulation in coastal and oceanic waters; dynamics and theory of estuarine and inshore waters; ocean processes in climate. *Mailing Add:* 5900 SW 104th St Miami FL 33156

HANSEN, DONALD WILLIS, JR, b Springfield, Ill, June 12, 43; m 73. MEDICINAL CHEMISTRY, ORGANIC CHEMISTRY. *Educ:* Univ Wis, BS, 65; Pa State Univ, PhD(org chem), 71. *Prof Exp:* Scholar dept chem, Univ Mich, 71-73, scholar dept med chem, 72-73, lectr org chem, 73-74; GROUP LEADER PHARMACEUT CHEM, G D SEARLE & CO, 74- *Concurrent Pos:* Adj asst prof med chem, Univ Ill. *Mem:* Am Chem Soc; AAAS; Soc Neurosci. *Res:* Design and synthesis of new pharmacologically active substances and the development of new synthetic methods and transformations in organic chemistry. *Mailing Add:* 5250 W Brown St Skokie IL 60077-3616

HANSEN, DOUGLAS BRAYSHAW, b Neenah, Wis, Apr 30, 29; m 57; c 3. CHILD PSYCHIATRY, PSYCHOANALYSIS. *Educ:* Antioch Col, AB, 51; Univ Rochester, MD, 55; Am Bd Psychiat & Neurol, dipl & cert psychiat, 65 & cert child psychiat, 66. *Prof Exp:* Intern med, Barnes Hosp, Wash Univ, 55-56; res assoc cerebral metab, NIMH, 56-58; resident psychiat & child psychiat, Columbia Univ & NY State Psychiat Inst, 58-62; instr child psychiat, Columbia Univ, 62-64; asst prof, 64-72, dir, Child Psychiat Training Prog, 64-80, assoc prof psychiat, Child Psychiat & Pediat, 76-86, PROF CLIN PSYCHIAT, BAYLOR COL MED, 85- *Concurrent Pos:* Chief psychiatrist, Rice Univ, 64-65; chief child psychiatrist, Tex Res Inst Ment Sci, 65-74; fel psychoanal, New Orleans Psychoanal Inst, La, 71, mem fac psychoanal, 72-73, teaching analyst, 73-80; mem, Regional Coun, Am Acad Child Psychiat, 71-; Am Soc Adolescent Psychiat liaison, Am Asn Psychiat Serv Children, 72-; mem Comt Adolescent Psychiat, Am Acad Child Psychiat, 72-; training & supv psychoanalyst, Houston-Galveston Psychoanal Inst, 77-; psychiatrist-in-chief, Tex Childrens Hosp, 75-86, St Lukes Episcopal Hosp, 78-79; pres med staff, Psychiat Inst, Houston, 81-83, med dir, 85-; pres, Houston Psychoanal Soc, 73-75, 82-84. *Mem:* Fel Am Psychiat Asn; Am Orthopsychiat Asn; Am Psychoanal Asn; Am Group Psychother Asn; Am Soc Adolescent Psychiat; fel Am Col Psychiat; fel Am Acad Child Psychiat. *Res:* Child development. *Mailing Add:* 5300 San Jacinto 160 Houston TX 77004

HANSEN, EDER LINDSAY, b Melbourne, Australia, Jan 7, 14; m 44; c 2. PHARMACOLOGY. *Educ:* Univ Melbourne, BAgrSc, 35, MAgrSc, 38; Univ Calif, MS, 40, PhD, 49. *Prof Exp:* Asst bot, Univ Melbourne, 36-39; mem res staff indust hyg, Dept Labor & Nat Serv, Australia, 42-44; asst & assoc pharmacol, Univ Calif, 44-50; asst, Kaiser Found Res Inst, 56-60, assoc, 60-66; mem res staff, Clin Pharmacol Res Inst, 66-80; RETIRED. *Mem:* Am Soc Pharmacol & Exp Therapeut; Am Soc Trop Med & Hyg; Soc Exp Biol & Med; Am Soc Parasitologists. *Res:* Experimental biology; tropical medicine. *Mailing Add:* 561 Santa Barbara Rd Berkeley CA 94707

HANSEN, EVERETT MATHEW, b Portland, Ore, Sept 8, 46; m 77; c 3. FOREST PATHOLOGY. *Educ:* Ore State Univ, BS, 68; Univ Wis-Madison, MS, 70, PhD(plant path), 72. *Prof Exp:* Res assoc, 72-75, from asst prof to assoc prof, 75-88, PROF FOREST PATH, ORE STATE UNIV, 81- *Concurrent Pos:* Assoc ed, Phytopath, Am Phytopath Soc, 85-88, chair, Forest Path Comt, 88. *Mem:* Am Phytopath Soc; Sigma Xi; Brit Myological Soc. *Res:* Biology, ecology and management of fungi causing diseases of forest trees, particularly root disease; cytology and sexuality and subsequent influences on population patterns and development of host specialization. *Mailing Add:* Dept Bot & Path Ore State Univ Corvallis OR 97331

HANSEN, GERALD DELBERT, JR, b Oil City, Pa, May 27, 21; m 49; c 3. PHYSICAL CHEMISTRY. *Educ:* Thiel Col, BS, 46. *Prof Exp:* Res chemist rubber, Lord Mfg Co, Pa, 46-47; asst chem, Duquesne Univ, 47-49; res chemist rheol, Hagan Chem & Controls, Inc, 50-56, proj leader, 56-57, group leader surface chem, 57-64, Calgon Corp, 64-65; boiler res group, Betz Labs, 65-66, group leader surface chem, 66-69, res assoc surface chem, 69-73; pres, G D Hansen Assocs, 73-77; sr chemist, Arco Chem, 77-78, prod mgr, Arco Performance Chem Co, 78-79, dir cent res, 80-87; SR CONSULT, CHEMLINK CO, 87- *Mem:* AAAS; Soc Rheol; Royal Soc Chem; fel Am Inst Chem. *Res:* Thermodynamics and electrokinetics of interfaces and the relationship between these properties and the rheology of dispersed systems. *Mailing Add:* PO Box 127 Holicong PA 18928-0127

HANSEN, GRANT LEWIS, b Bancroft, Idaho, Nov 5, 21; m 45; c 5. ELECTRICAL ENGINEERING, TECHNICAL MANAGEMENT. *Educ:* Ill Inst Technol, BSEE, 48. *Hon Degrees:* DSc, Nat Univ, 78. *Prof Exp:* Mem staff, Douglas Aircraft Co, 48-60, vpres & prog dir for Centaur, Convair Div, 60-65; vpres launch vehicle progs, Convair Div, gen Dynamics Corp, 65-69, gen mgr, 73-74, corp vpres, 74-78; asst secy Airforce for res & develop, 69-73; vpres, Gen Dynamics Corp, San Diego, 74-78, vpres & gen mgr, Convair Div, 73-78; pres, Systs Develop Corp, 78-86; RETIRED. *Concurrent Pos:* US deleg, NATO Adv Group Aerospace Res & Develop, 69-73; US mem sci comt nat reps, SHAPE Tech Ctr, The Hague, Netherlands, 69-73; mem res & tech adv coun, NASA, 71-73; mem sci adv bd, Dept Air Force, 76-83. *Mem:* Nat Acad Eng; fel Am Inst Aeronaut & Astronaut (pres, 75); AAAS; fel Int Acad Astronaut; sr mem Inst Elec & Electronics Engrs. *Res:* Liquid hydrogen propulsion; space launch vehicles; missiles; knowledge based computer systems. *Mailing Add:* 10737 Fuerte Dr La Mesa CA 91941-5740

HANSEN, HANS JOHN, ANTIBODY THERAPY. *Educ:* Tulane Univ, PhD(biochem), 60. *Prof Exp:* Dir cell biol, 85-87, VPRES, EXPLOR RES, IMMUNOMEDICS, INC, 87- *Res:* In vitro and in vivo diagnoses; cancer research. *Mailing Add:* 2617 N Burgee Dr Mystic Island NJ 08087

HANSEN, HAROLD WESTBERG, plant morphology; deceased, see previous edition for last biography

HANSEN, HARRY LOUIS, b San Francisco, Calif, Dec 13, 24; m 48; c 5. ENTOMOLOGY, BIOLOGY. *Educ:* Univ Calif, BS, 51, PhD, 55. *Prof Exp:* Asst prof entom & asst entomologist, Univ WVa, 54-58; asst res prof entom & exten entomologist, Univ RI, 58-61; chief entomologist, Div Trop Res, Standard Fruit Co, Honduras, 61-64; assoc prof entom, Purdue Univ, 64-69; PROF BIOL, STEPHENS COL, 70- *Res:* Ecology. *Mailing Add:* Dept Nat Scis Stephens Col Columbia MO 65215-0001

HANSEN, HOBART RAYMOND, medical research, for more information see previous edition

HANSEN, HOLGER VICTOR, b Rahway, NJ, Sept 11, 35; m 57; c 3. ORGANIC CHEMISTRY. *Educ:* Lehigh Univ, BS, 57, PhD(chem), 61. *Prof Exp:* Wm S Merrell Co fel, Brown Univ, 61-62; sr scientist, Warner-Lambert Res Inst, 62-66 & Shulton Inc, 66-72; SR SCIENTIST, COSAN CHEM CORP, 72- *Mem:* Am Chem Soc. *Res:* Synthetic organic chemistry, mainly relating to biological systems. *Mailing Add:* Nine Seminole Trail Denville NJ 07834

HANSEN, HOWARD EDWARD, b Lincoln, Nebr, Jan 28, 23; m 62; c 2. CHILD PSYCHIATRY, PSYCHOANALYSIS. *Educ:* Univ Nebr, BS & MD, 47. *Prof Exp:* Asst pediat, Childrens Hosp, Los Angeles & Univ Southern Calif, 54-55; chief pediat serv, 7520 US Air Force Hosp, London, Eng, 56-59; pvt practr psychiat, Calif, 60-68; asst attending pediat & psychiat, Childrens Hosp, Los Angeles, 61-67; asst clin prof, Univ Southern Calif, 62-65, asst prof, 65-67; head psychiat & dir child psychiat, Children's Hosp, Los Angeles, 68-88; EMER ASSOC PROF, PEDIAT PSYCHIAT & BEHAV SCI, 88-, MED DIR, CHILD STUDY CTR, ST JOHN'S HOSP & HEALTH CTR, 90-; ASSOC PROF PEDIAT, PSYCHIAT & BEHAV SCI, SCH MED, UNIV SOUTHERN CALIF, 67- *Concurrent Pos:* Dir child psychiat, Cedars Lebanon Hosp, Los Angeles, 61-64, assoc attending psychiat & adj pediat, Med Ctr, 61-70; med dir psychiat, Julia Ann Singer Presch Ctr, 61-64, mem, Prof Adv Comt, Psychiat Clin, 65-; sr staff psychiatrist, Mt Sinai Hosp, 61-64, adj pediat, Med Ctr, 61-70; mem attending staff, Los Angeles County Gen Hosp, 61-70; mem, Prof Adv Coun, Dubnoff Sch Educ Ther, 62-; psychiat consult, Ment Retardation Community Serv Clin, Childrens Hosp, Los Angeles, 63-; clin assoc psychoanal, Southern Calif Psychoanal Inst, 64-; coordr child psychiat training, Los Angeles County-Univ Southern Calif Med Ctr, 65-68; mem, Adv Comt, Dept Child Psychiat, Cedars Sinai Med Ctr, 65-; mem, Adv Bd, Park Cent Sch, 68-; mem, Med Adv Comt, Calif State Dept Ment Hyg, 70-71, mem task force 5 yr plan, Subcomt Child & Adolescent Psychiat, 71-; mem, Citizens Comt Bd & Care, Los Angeles City Coun, 71-72; mem, Ment Health Adv Bd, Welfare Planning Coun, 71-; co-investr, NIMH grant & contract, 71-, prin investr, Grants Found, Inc grant, 71-; mem bd dirs, Ment Health Asn, Los Angeles, 74-, Edgemont Hosp, 78- *Mem:* Am Psychoanal Asn; AMA; fel Am Acad Pediat; fel Am Psychiat Asn; fel Am Aacd Child Psychiat. *Res:* Emotional reactions to natural disasters; psychological preparation and care of amputees; renal dialysis and transplant patients; cancer patients; fatally ill child and his family; extreme cases of psycho-social isolation; suicide and self endangering behaviors of children; Gilles de la Tourette Syndrome. *Mailing Add:* Emer Ctr Geront 220 MC 0191 Univ Southern Calif Los Angeles CA 90089-0191

HANSEN, HUGH J, b Thief River Falls, Minn, March 30, 23; m 49; c 3. ENERGY MANAGEMENT, AGRICULTURAL ENGINEERING. *Educ:* NDak State Univ, BSAE, 51; Cornell Univ, MSAE, 52. *Prof Exp:* Asst prof eng, Agr Eng Dept, Purdue Univ, 52-55; ed tech eng, Reuben H Donnelley Corp, NY, 55-62; publ, Dun-Donnelley Publ Corp, 62-74; mgr, Western Regional Agr Eng Serv, 74-78; PROF ENG, AGR ENG DEPT, ORE STATE UNIV, 78- *Concurrent Pos:* Exec mgr, Nat Food & Energy Coun, 72-74. *Mem:* Fel Am Soc Agr Engrs (pres, 71-72); Irrigation Asn. *Mailing Add:* Agr Eng Dept Ore State Univ Corvallis OR 97331

HANSEN, IRA BOWERS, zoology, for more information see previous edition

HANSEN, J RICHARD, b Sioux City, Iowa, Sept 21, 22; m 48; c 2. ELECTROOPTICS. *Educ:* Univ Mo, BS, 44. *Prof Exp:* Sr res engr, Res Labs, Westinghouse Elec Corp, 47-66, sr res scientist, 66-87; RETIRED. *Mem:* Sr mem Inst Elec & Electronics Engrs; Am Astron Soc. *Res:* Electronic circuitry; imaging devices, especially electronic; infrared detection; liquid crystals; x-ray and neutron activation nondestructive testing; fiber optics. *Mailing Add:* 1462 Jefferson Heights Rd Pittsburgh PA 15235

HANSEN, JAMES E, b Green Bay, Wis, Sept 4, 26; m 48; c 4. PULMONARY DISEASES, ENVIRONMENTAL PHYSIOLOGY. *Educ:* Johns Hopkins Univ, MD, 49. *Prof Exp:* From intern to resident internal med, US Army, Letterman Gen Hosp, San Francisco, 49-53, asst chief med serv, US Army Hosp, Ft Riley, Kans, 53-56, surgeon, 32 AAA Brigade, England, 56-57, chief med serv, 34th Gen Hosp, Orleans, France, 57-59, resident pulmonary dis, Fitzsimons Gen Hosp, Denver, 60-61, chief tuberc serv, 61-62, chief med training team, Jordan Arab Army, 62, chief physiol div, US Army Med Res & Nutrit Lab, 62-65, commanding officer & sci dir, US Army Res Inst Environ Med, Mass, 65-71, chief clin invest serv, Tripler Army Med Ctr, 71-75; fel pulmonary dis, Harbor-Univ Calif, Los Angeles Med Ctr, Torrance, 75-76, head clin respiratory physiol lab, 76-86; from assoc prof to prof 76-86, EMER PROF MED, UNIV CALIF, LOS ANGELES, 86- *Concurrent Pos:* Instr, Univ Colo, 62-65; lectr, Johns Hopkins Univ, 66-71; clin prof physiol, Univ Hawaii, 72-75; assoc prof med, Univ Calif, Los Angeles, 76-78, prof, 78- *Mem:* Fel Am Col Physicians; Am Thoracic Soc; Am Physiol Soc; Am Fedn Clin Res; fel Am Col Chest Physicians. *Res:* Clinical medicine; pulmonary and infectious diseases; exercise physiology; altitude physiology. *Mailing Add:* 1692 Morse Dr San Pedro CA 90732

HANSEN, JO ANN BROWN, molecular biology, research administration, for more information see previous edition

HANSEN, JOHN C, b Miami, Fla, Mar 3, 47. COMPUTER SCIENCE. *Educ:* Univ Miami, BA, 68; Mich State Univ, PhD(comput sci), 74. *Prof Exp:* Assoc prof comput sci, Xavier Univ, 81-83; PROF COMPUT SCI, CENT MICH UNIV, 83- *Mem:* Math Asn Am; Inst Elec & Electronics Engrs; Asn Comput Mach. *Mailing Add:* Cent Mich Univ Mt Pleasant MI 48859

HANSEN, JOHN FREDERICK, b Turtle Lake, Wis, Mar 21, 42. ORGANIC CHEMISTRY. *Educ:* Wis State Univ, River Falls, BA, 64; Duke Univ, AM, 67, PhD(org chem), 69. *Prof Exp:* Fel chem, Ohio State Univ, 68-69, Univ Notre Dame, 69-71 & Wayne State Univ, 71-72; ASSOC PROF CHEM, ILL STATE UNIV, 72- *Mem:* Am Chem Soc; Sigma Xi. *Res:* Chemistry of organic heterocyclic compounds. *Mailing Add:* Dept Chem Ill State Univ Normal IL 61761

HANSEN, JOHN NORMAN, b Kearney, Nebr, July 15, 42; m 66; c 2. BIOCHEMISTRY, MOLECULAR BIOLOGY. *Educ:* Drake Univ, BA, 64; Univ Calif, Los Angeles, PhD(biochem), 68. *Prof Exp:* USPHS fel, Univ Wis, Madison, 68-71; from asst prof to assoc prof, 71-78, PROF BIOCHEM, UNIV MD, COLLEGE PARK, 78- *Mem:* Am Chem Soc; Sigma Xi; AAAS; Am Soc Microbiol. *Res:* Spore outgrowth; control of development; chromosome structure and organization; molecular genetics and mechanism of action of ribosomally-synthesized peptide antibiotics. *Mailing Add:* Dept Chem & Biochem Univ Md College Pk MD 20742-2021

HANSEN, JOHN PAUL, b Bain, Minn, Feb 11, 28; m 50; c 3. MATERIALS SCIENCE ENGINEERING, CHEMICAL ENGINEERING. *Educ:* Univ Minn, BS, 54, MS, 55, PhD(metall eng), 58. *Prof Exp:* Res fel, Univ Minn, 55-58; res metallurgist, US Bur Mines, 58-63; prof metall eng, Univ Ala, Tuscaloosa, 63-67; chief, Tuscaloosa Metall Res Lab, US Bur Mines, 67-70; head, Dept Chem & Metall Eng, Univ Ala, Tuscaloosa, 70-73, prof metall eng, 70-87; RETIRED. *Concurrent Pos:* Consult, US Army Res Off, NC, Bur Mines, Ala & Ores Res Lab, Mich. *Mem:* Am Inst Mining, Metall & Petrol Engrs; Am Inst Chem Engrs. *Res:* Reaction kinetics of heterogeneous systems; thermodynamics. *Mailing Add:* 1245 Highpoint E Springfield MO 65804

HANSEN, JOHN THEODORE, b Sheboygan Falls, Wis, Oct 10, 47; m 70; c 2. ANATOMY, CELL BIOLOGY. *Educ:* Beloit Col, BA, 70; Creighton Univ, MS, 72; Tulane Univ, PhD(anat), 74. *Prof Exp:* Adj instr anat, Tulane Univ Med Sch, 74-75; instr, 75-76, asst prof anat, 76-80, ASSOC PROF, MED SCH, UNIV TEX, SAN ANTONIO, 80- *Mem:* AAAS; Am Asn Anat; Am Soc Cell Biol; Am Heart Asn; Soc Neurosci. *Res:* Ultrastructure of paraneurons, chromaffin tissues and cardiovascular system; cardiovascular neurobiology. *Mailing Add:* 50 Tilstone Pl Rochester NY 14618-2853

HANSEN, KEITH LEYTON, b Gainesville, Fla, Nov 14, 25; m 47, 83; c 2. BIOLOGY. *Educ:* Stetson Univ, BS, 49, MS, 50; Univ Fla, PhD, 55. *Prof Exp:* Instr, 50-55, from assoc prof to prof, 55-88, RES PROF BIOL, STETSON UNIV, 88- *Concurrent Pos:* Consult ecol, Ecosyst Anal. *Mem:* Am Soc Icthyol & Herpet; Ecol Soc Am. *Res:* Freshwater ecology; ecology of anurans and reptiles; radiological research of trophic relations in aquatic communities; heavy metal detection in marine coastal fauna; reproductive and endocrine physiology of anura; systems ecology; ecology of Florida. *Mailing Add:* Dept Biol Stetson Univ De Land FL 32720

HANSEN, KENT F(ORREST), b Chicago, Ill, Aug 10, 31; m 59; c 2. NUCLEAR ENGINEERING. *Educ:* Mass Inst Technol, SB, 53, ScD(nuclear eng), 59. *Prof Exp:* Res assoc, comput ctr, 59-60, Ford fel, 60-61, from asst prof to assoc prof nuclear eng, 60-69, assoc dean eng, 79-82, PROF NUCLEAR ENG, MASS INST TECHNOL, 69- *Concurrent Pos:* Consult, US Nuclear Regulatory Comn, 75-; sci adv comn, EG&G, Idaho, 77- *Honors & Awards:* A H Compton Award, Am Nuclear Soc, 78. *Mem:* Nat Acad Eng; Am Nuclear Soc; Asn Comput Mach; Am Soc Eng Educ. *Res:* Numerical methods of reactor analysis; reactor theory; numerical mathematics; radiation shielding. *Mailing Add:* Dept Nuclear Eng Mass Inst Technol Cambridge MA 02139

HANSEN, KENT W(ENDRICH), b Salt Lake City, Utah, Apr 28, 36; m 56; c 3. CERAMIC ENGINEERING, METALLURGY. *Educ:* Univ Utah, BS, 58, PhD(ceramic eng, metall), 62. *Prof Exp:* Sr engr, Corning Glass Works, 62-65, res engr, 65-67, res assoc ceramics, 67; eng specialist glass & ceramic res, Semiconductor Prod Div, 67-70, mgr glass & ceramics, 70-76, mem tech staff, 76-77, sect mgr, Semiconductor Res & Develop Labs, 77-79, mgr, Mat Tech Lab, 79-85, MGR, PHYS ELECTRONICS & PACKAGING LAB, SEMICONDUCTOR RES & DEVELOP LABS, MOTOROLA, INC, 85- *Concurrent Pos:* Assoc mem sci adv bd, Motorola, 72. *Mem:* Am Phys Soc; Am Ceramic Soc; Soc Glass Technol; Sigma Xi; Inst Elec & Electronics Engrs. *Res:* Electrical properties of materials, particularly ceramics and glasses; sintering mechanisms in oxides; surface area and pore structure of porous materials; glass technology; semiconductor materials and packaging; glass-metal reactions and sealing; electronic packaging. *Mailing Add:* Semiconductor Prod Sector 5005 E McDowell Rd Phoenix AZ 85008

HANSEN, LARRY GEORGE, b Omaha, Nebr, July 16, 41. ENVIRONMENTAL HEALTH. *Educ:* Creighton Univ, BS, 63; Univ Nebr, Omaha, MS, 66; NC State Univ, PhD(entom, toxicol), 70. *Prof Exp:* Teacher high sch, Iowa, 63-65; res assoc zool, 70-71, asst prof physiol & pharmacol, Col Vet Med, 71-76, assoc prof vet pharmacol, 76-80, PROF VET PHARMACOL & PROF ENVIRON STUDIES, UNIV ILL, URBANA, 80- *Concurrent Pos:* Res fel toxicol, Wageningen, The Netherlands, 80. *Mem:* Am Chem Soc; Soc Toxicol. *Res:* Environmental toxicology; residue transmission; chlorinated hydrocarbons; cadmium; pollution in developing countries. *Mailing Add:* Dept Vet Biosci Univ Ill 2001 S Lincoln Urbana IL 61801

HANSEN, LEE DUANE, b Brigham City, Utah, Apr 13, 40; m 60; c 9. GENERAL CHEMISTRY. *Educ:* Brigham Young Univ, BS, 62, PhD(inorg chem), 65. *Prof Exp:* From asst prof to assoc prof chem, Univ NMex, 65-72; assoc prof dept chem & Ctr Thermochem Studies, 72-78, PROF CHEM, BRIGHAM YOUNG UNIV, 78- *Concurrent Pos:* USPHS career develop award, 69-72; James Prof Chem, St Francis Xavier Univ, 85-86; vis prof, UC Davis, 79-80 & 88. *Mem:* AAAS; Sigma Xi; The Chem Soc. *Res:* Thermodynamics of reactions in solution; proton ionization; metal ion complexation; chemical speciation in air pollutants; multiple equilibria of proteins; metabolism in plants. *Mailing Add:* Dept Chem Brigham Young Univ Provo UT 84602-1022

HANSEN, LEON A, b Idaho Falls, Idaho, Sept 15, 43; div; c 1. PLANT BREEDING. *Educ:* Univ Idaho, BS, 65, MS, 67; Ore State Univ, PhD(hort plant breeding), 76. *Prof Exp:* Fieldman, Rogers Bros Seed Co, 67-69, field dept mgr, 69-72, res asst, 72-75, plant breeder, 75-79, dir sweet corn res, 79-85, dir corp res, 85-87, vpres res, 87-90; VPRES RES, ROGERS NK SEED CO, 91- *Honors & Awards:* Nat Food Processors Asn Award, 78. *Mem:* Am Soc Hort Sci; Am Soc Crop Sci; Nat Sweet Corn Breeders Asn. *Res:* Management of all company research. *Mailing Add:* Rogers NK Seed Co PO Box 4188 Boise ID 83711-4188

HANSEN, LESLIE BENNETT, b Blooming Prairie, Minn, Aug 30, 51. ANIMAL BREEDING, BIOMETRICS. *Educ:* Univ Minn, BS, 73, MS, 78; Iowa State Univ, PhD(animal breeding), 81. *Prof Exp:* asst prof, 81-87, ASSOC PROF DAIRY CATTLE BREEDING, UNIV MINN, 87- *Concurrent Pos:* Mem, Coun Agr Sci & Technol. *Mem:* Am Dairy Sci Asn; Am Soc Animal Sci. *Res:* Population genetics of dairy cattle with special emphasis on management traits, health traits, type traits and other non-production traits; linear statistical models, especially mixed models having random and fixed effects. *Mailing Add:* 898 Hunt Pl St Paul MN 55114

HANSEN, LOUIS STEPHEN, b San Diego, Calif, May 10, 18; m 42; c 3. ORAL PATHOLOGY. *Educ:* Univ Southern Calif, BS & DDS, 41; Georgetown Univ, MS, 55; Am Bd Oral Path, dipl, 58; Pepperdine Univ, MBA, 77. *Prof Exp:* Intern oral surg, Los Angeles County Hosp, 49-50, chief dent serv, US Naval Hosp, Quantico, Va, 50-52, res oral path, US Naval Dent Sch, Bethesda, Md, 53-54, instr, 54-55, chief dent serv, US Naval Hosp, USS Haven, Long Beach, Calif, 55-57, pathologist officer educ dept, US Naval Dent Sch, 57-60, chief dent & oral div, Armed Forces Inst Path, 60-63, registr, Registry Dent & Oral Path, Am Registry Path, 60-63, head educ dept, US Naval Dent Sch, 63-66, exec officer, 66-67; prof oral path & chmn dept sch dent, prof forensic path & med & vchmn dept, Sch Med, 67-87, PROF ORAL PATH, UNIV CALIF, SAN FRANCISCO, 87- *Concurrent Pos:* Mem, Bd Dirs, Am Bd Oral Path, 71-78, pres, 78; assoc clin prof, Sch Med, George Washington Univ, 63-68; lectr, Grad Sch, Georgetown Univ, 63-68; consult, Vet Admin Hosp, San Francisco, 67-, USPHS Hosp, 73-81, Am Dent Asn, 67- & Med Ctr, US Naval Regional, Oakland, 76- *Mem:* Fel Int Col Dent; fel Am Col Dent; fel Am Acad Oral Path (pres, 65); Int Acad Path; Int Asn Dent Res; Int Asn Oral Path. *Res:* Pigmentation of the oral mucosa; pathology of the salivary glands; keratotic lesions of the oral mucosa; forensic odontology; odontogenic neoplasms. *Mailing Add:* Dept Stomatology Rm S-512 Univ Calif San Francisco CA 94143-0424

HANSEN, LOWELL JOHN, b Bemidji, Minn, Oct 3, 41; m 64; c 4. MATHEMATICAL ANALYSIS. *Educ:* Bemidji State Col, BA, 65; Univ Ill, Urbana, MA, 66, PhD(math), 69. *Prof Exp:* Asst prof, 69-73, ASSOC PROF MATH, WAYNE STATE UNIV, 73- *Res:* Complex analysis; Hardy Classes; cluster sets; univalent functions; potential theory. *Mailing Add:* Dept Math Wayne State Univ 646 Mackenzie Detroit MI 48202

HANSEN, LUISA FERNANDEZ, b Santiago, Chile; US citizen; m 54; c 1. NUCLEAR PHYSICS, HIGH ENERGY PHYSICS. *Educ:* Univ Chile, BS, 49; Univ Calif, Berkeley, MS, 57, PhD(nuclear physics), 59. *Prof Exp:* Teaching asst physics, Univ Chile, 47-52, asst prof math & physics & laboratorist cosmic rays, 50-55; res asst physics, Univ Calif, Berkeley, 55-59; SR PHYSICIST, LAWRENCE LIVERMORE NAT LAB, UNIV CALIF, 59- *Concurrent Pos:* Consult res & eng, Bechtel Corp, 76-78 & Dept Fusion Power Systs, Westinghouse Elec Corp, 76-77; mem, Bay Area Women Sci Network, 75- *Honors & Awards:* Fel, Am Nuclear Soc, 89. *Mem:* Am Phys Soc; Am Nuclear Soc; Sigma Xi. *Res:* Neutron and charged particle interactions; measurements and calculations of neutron and gamma ray production for shielding applications, fusion and hybrid reactors. *Mailing Add:* 15 Avalon Ct Walnut Creek CA 94595

HANSEN, MARC F, b Marshfield, Wis, Sept 19, 30; c 2. PEDIATRICS. *Educ:* Harvard Univ, AB, 52, MD, 56; Am Bd Pediat, dipl, 61. *Prof Exp:* Intern, Boston City Hosps, 56-57; resident, Univ Hosps, 57-59, from asst prof to assoc prof pediat, Med Ctr, 63-74, asst prof, Dept Rehab Med, 67-69, asst dean clin affairs, 70, PROF FAMILY MED, MED CTR, UNIV WIS-MADISON, 74- *Concurrent Pos:* Fel, Univ & Inst Enzyme Res, Univ Wis-Madison, 61-62; Lederle award, 66; staff physician, Univ Hosps, Med Ctr, Univ Wis-Madison, 63-, dir pediat outpatient serv & univ child health serv, 66-68, dir univ family health serv, 68-69, dir prog primary care, 69; prin policy adv & mem mgt group, Gov Health Planning & Policy Task Force, 71-72; prog dir, Family Med Prog, Univ Wis, 85-89, pres & chair, Univ Care, Health Maintenance Orgn, 90-, actg chair, Dept Family Med, Wilwaukee Clin Campus, 90- *Mem:* Am Acad Pediat; AMA. *Res:* Health system study; health policy and planning; organization and curriculum of primary care; health service resources. *Mailing Add:* Dept Pediat Fam Pract Univ Wis Med Ctr Madison WI 53792

HANSEN, MERLE FREDRICK, b Minneapolis, Minn, Jan 9, 17; m 46; c 3. PARASITOLOGY. *Educ:* Univ Minn, BA, 39, MA, 41; Univ Nebr, PhD(zool), 48. *Prof Exp:* Asst zool, Univ Nebr, 46, instr, 46-47; assoc parasitologist, Dept Animal Path, Univ Ky, 48-50; prof & parasitologist, 50-82, assoc dir, Div Biol, 74-82, EMER PROF PARASITOL, AGR EXP STA, KANS STATE UNIV, 82- *Concurrent Pos:* Sr scientist, USPHS, 57-79. *Mem:* Am Soc Parasitol; Am Inst Biol Sci. *Res:* Animal parasitology. *Mailing Add:* 2025 Thackery Manhattan KS 66502

HANSEN, MIKE, b Rockville, Minn, Nov 31, 31; m 58; c 3. VETERINARY MEDICINE. *Educ:* Univ Minn, DVM, 57. *Prof Exp:* Field vet, Minn Livestock Sanit Bd, 57-58; gen practr, 58-60; res veterinarian, Land O'Lakes Creameries, Inc, Minn, 60-71 & Land O'Lakes, Inc, 71-73; TURKEY PROD MGR, KORONIS MILL & SUPPLY CO, 73- *Mem:* Am Vet Med Asn; Am Dairy Sci Asn; Am Poultry Sci Asn; Am Soc Animal Sci. *Res:* Feed management and health programs and program development for turkeys, laying hens, pullets and chicks. *Mailing Add:* RR Painesville MN 56362

HANSEN, MORRIS HOWARD, statistics; deceased, see previous edition for last biography

HANSEN, OLE, b Frederiksberg, Denmark, May 14, 34; m 58; c 4. HIGH ENERGY HEAVY ION PHYSICS. *Educ:* Univ Copenhagen, Mag Scient, 58, DrPhil(physics), 67. *Prof Exp:* Teaching asst physics, Niels Bohr Inst, Denmark, 60-62, lectr, 62-68 & 72-74, prof exp physics, 74-82; prof physics, Univ Pa, 69-70; mem staff, Los Alamos Sci Lab, 70-72; group leader, Heavy Ion Res, 84-88, SR SCIENTIST, BROOKHAVEN NAT LAB, 81- *Concurrent Pos:* Res assoc, Mass Inst Technol, 64-65 & Rutgers Univ, NJ, 65-66; mem staff, Los Alamos Sci Lab, 69-70 & 74-75, adv, P-Div, 82-88; mem, Natural Sci Coun, Denmark, 79-81; prog adv, Oak Ridge Nat Lab, 83-85; mem, nuclear sci adv comt, NSF-Dept Energy, 83-88; chmn, Danish Physics Surv, Ministry Educ, Denmark, 90- *Honors & Awards:* Prize, Ole Rohmer Found, Denmark, 68; von Humboldt Sr Sci Award, 88. *Mem:* Am Phys Soc. *Res:* Study of collisions between nuclei at high energies to search for new forms of matter. *Mailing Add:* 15 Bailey Hollow Rd Stony Brook NY 11790

HANSEN, PAUL B(ERNARD), b Mont, Nov 23, 13; m 44; c 4. CHEMICAL ENGINEERING. *Educ:* State Col Wash, BS, 37; Lawrence Col, MS, 41, PhD(chem eng), 43. *Prof Exp:* Mem res staff, Schweitzer Paper Co, 43-44; tech dir, Bergstrom Paper Co, 44-53; tech dir, 53-64, mgr tech info systs, 64-68, res chemist pioneering res, 68-70, mgr res & eng tech infor serv, Kimberly-Clark Corp, 70-81; RETIRED. *Res:* Paper and high polymer technology; electrical papers; product development and applied science. *Mailing Add:* 740 Chestnut Neenah WI 54956

HANSEN, PAUL VINCENT, JR, b Salt Lake City, Utah, May 18, 31; m 55; c 3. PHYSICAL CHEMISTRY. *Educ:* Dana Col, BS, 52; Northwestern Univ, MS, 54; Univ Nebr, PhD, 65. *Prof Exp:* Res chemist, Swift & Co, Ill, 54-56; from instr to asst prof chem, Dana Col, 56-60; from asst prof to assoc prof, 62-69, PROF CHEM, CARTHAGE COL, 69- *Mem:* Am Chem Soc; Sigma Xi. *Res:* Hydrogen bonding and association; application of phase rule and phase diagrams; ultraviolet and infrared spectroscopy. *Mailing Add:* 7840 42nd Ave Kenosha WI 53142

HANSEN, PETER GARDNER, b Curryville, Mo, Nov 2, 27; m 54; c 3. ENGINEERING MECHANICS, THEORETICAL MECHANICS. *Educ:* Mo Sch Mines, BS, 53, MS, 57; Wash Univ, ScD, 63. *Prof Exp:* From instr to asst prof mech, Mo Sch Mines, 53-60; from assoc prof to prof eng mech, 60-90, chmn dept, 71-88, EMER PROF ENG MECH, UNIV MO-ROLLA, 90- *Concurrent Pos:* Grant, Univ Mo, 69 & Bur Mines, US Dept Interior, 70-71. *Mem:* Soc Exp Mech; Am Acad Mech; Am Soc Eng Educ; Am Soc Metals. *Res:* Experimental stress analysis; structural analysis; materials. *Mailing Add:* Dept Basic Eng Univ Mo Rolla MO 65401-0249

HANSEN, PETER JACOB, b Willmar, Minn, Feb 27, 39; m 64. COMPUTATIONAL CHEMISTRY. *Educ:* St Olaf Col, BA, 61; Iowa State Univ, PhD(phys chem), 66. *Prof Exp:* Lectr phys chem, Univ Ife, Nigeria, 67-68; ASSOC PROF CHEM, NORTHWESTERN COL, 69- *Concurrent Pos:* Lectr chem, Makerere Univ, Uganda, 76-77; adj prof, Ariz State Univ, Tempe, 79-80; vis assoc prof, Penn State Univ, 86-87. *Mem:* AAAS; Am Chem Soc; Am Asn Univ Prof. *Res:* Quantitative structure property relationships. *Mailing Add:* Box 246 Orange City IA 51041-0246

HANSEN, POUL M T, b Vejle, Denmark, Sept 24, 29; m 57; c 3. FOOD SCIENCE, DAIRY TECHNOLOGY. *Educ:* Royal Vet & Agr Col, Copenhagen, BSc, 56; Univ Ill, MSc, 58, PhD(food technol), 60. *Prof Exp:* Res off dairy res, Commonwealth Sci & Indust Res Orgn, Australia, 60-63; res assoc, 64-65, from asst prof to assoc prof, 65-75, PROF DAIRY TECHNOL, OHIO STATE UNIV, 75- *Mem:* AAAS; Am Chem Soc; Inst Food Technol; Am Dairy Sci Asn; Am Asn Cereal Chem. *Res:* Milk technology; heat effects on milk proteins; hydrophilic colloids and their interactions. *Mailing Add:* Dept Food Serv & Nutrition Ohio State Univ Main Campus Columbus OH 43210

HANSEN, R(OBERT) J(OSEPH), b Tacoma, Wash, May 27, 18; m 48; c 2. CIVIL ENGINEERING. *Educ:* Univ Wash, Seattle, BS, 40; Mass Inst Technol, ScD(civil eng), 48. *Prof Exp:* Res engr, Nat Res Coun, Washington, DC, 40-43, Princeton Univ, 43-44 & Arthur D Little Co, 45; res assoc, 47-48, from asst prof to prof struct eng, 48-75, EMER PROF CIVIL ENG, MASS INST TECHNOL, 75- *Concurrent Pos:* partner, Hansen, Holley & Biggs, prin, 55-88; Mem security resources panel, Exec Off President, 57; proj harbor, Nat Acad Sci, 64. *Honors & Awards:* Moisseiff Award & Reese Prize, Am Soc Civil Engrs. *Mem:* Am Soc Civil Engrs. *Res:* Structural dynamics; experimental techniques for stuctural design; design of tall buildings; earthquake design of nuclear power reactors; effects of wind on constructed facilities; properties and behavior of glass. *Mailing Add:* 25 Cambridge St Winchester MA 13201

HANSEN, RALPH HOLM, b Brooklyn, NY, Jan 23, 23; m 45; c 4. PLASTICS CHEMISTRY & PROCESSING. *Educ:* Cornell Univ, AB, 44; NY Univ, MS, 49, PhD(org chem), 52. *Prof Exp:* Res chemist, Heyden Chem Corp, 44-47; asst chem, NY Univ, 48-52; supvr appl org res, Bell Labs, 52-68; dir chem finishing & explor technol, J P Stevens & Co, 68-69; dir develop, Raychem Corp, 69-70; supvr plastics deterioration & stabilization, Bell Labs, 70-81; sr res assoc, Chomerics, Inc, 81-83; sr res assoc, Canusa Coating Systs, 83-86; PLASTICS CONSULT, RAVE CONSULT ASSOC, 86- *Concurrent Pos:* Chmn, Conf Chem & Physics Cellular Mat, Gordon Res Conf, 66; mem vis comt chem dept, NY Univ; res prof & dir, Polymer Durability Ctr, Polytech Univ, 87- *Honors & Awards:* Mobay Award, Soc Plastics Indust, 65; Union Carbide Chem Award, Am Chem Soc, 66; Polyolefins Award, Soc Plastics Engrs, 76. *Mem:* Am Chem Soc; Soc Plastics Engrs; fel NY Acad Sci; Asn Res Dir; Asn Consult Chem & Chem Engrs. *Res:* Studies of the modification and optimization of polymer properties by physical and chemical techniques, by use of composite structures, and by development of new and improved additives for polymers; flame retardants, antioxidants, and blowing agents. *Mailing Add:* 7 Hickory Hill Rd Belle Mead NJ 08502-3801

HANSEN, RALPH W(ALDO), b NDak, July 30, 26; m 51; c 2. AGRICULTURAL ENGINEERING. *Educ:* NDak Agr Col, BS, 51, MS, 52. *Prof Exp:* Asst utilization adv, NDak Rural Elec Co-op, 50; asst, NDak Agr Col, 51-52; design engr, New Idea Div, Avco Mfg Co, Ohio, 52-53; exten agr engr, Iowa State Univ, 53-56; asst prof agr eng, 56-64, acting head dept, 64-65, ASSOC PROF AGR ENG, COLO STATE UNIV, 64- *Concurrent Pos:* Fulbright lectr, Univ Helsinki, 58-59; consult, Res Inst Agr Mach, Finland, 58-59. *Mem:* Am Soc Agr Eng; Am Soc Eng Educ. *Res:* Agricultural machinery; farmstead mechanization; farm structures; agricultural waste management; solar energy applications in agriculture. *Mailing Add:* Dept Soc Scis Augustana Col 639 38th St Rock Island IL 61201

HANSEN, RICHARD (THOMAS), astronomy, for more information see previous edition

HANSEN, RICHARD LEE, b Charles City, Iowa, June 17, 50. PHYSICAL CHEMISTRY. *Educ:* Iowa State Univ, BS, 72; Univ Wis-Madison, MS, 75, PhD(phys chem), 79. *Prof Exp:* sr res chemist, Monsanto Indust Chem, 79-84, res specialist, 84-86, sr res specialist, 86-88, SR RES GROUP LEADER, MONSANTO ELEC WATER, 88- *Mem:* Sigma Xi; Am Chem Soc; Mat Res Soc; Electrochem Soc. *Res:* Statistical mechanics; kinetic theory of fluids; molecular collisions; molecular potentials; solid state chemistry; nucleation theory; dynamics of crystal growth; dynamics of phase changes; electronic-materials; czochralski crystal growth; tagucho loss functions and measures of process efficiency; raw materials impact on electronic silicon. *Mailing Add:* 1616 Greenhill Dr St Louis MO 63146-3833

HANSEN, RICHARD M, b Goshen, Utah, Jan 11, 24; m 45; c 3. ZOOLOGY. *Educ:* Univ Utah, BS, 50, MS, 51, PhD(vert zool), 54. *Prof Exp:* Lab asst, Univ Utah, 50-51, lab instr, 51-52; asst prof zool, Exp Sta, Col State Univ, 54-55, asst biologist, Exp Sta, 55-61, assoc biologist 61-68, prof biol, 68-74, prof range sci, 74-85; res scientist, Kiboko Range Res Sta, Kenya, 83-86; RETIRED. *Mem:* Am Soc Mammal; Ecol Soc Am; Wildlife Soc; Soc Range Mgt. *Res:* Ecology and distribution of mammals and vertebrates; vertebrate management; evolution and physiology; food-habits quantification techniques; conflicts and complementary effects of herbivores and the partitioning of range plants to animals and microorganisms. *Mailing Add:* 720 E Steuart St Ft Collins CO 80525

HANSEN, RICHARD OLAF, b Ottawa, Ont, Oct 4, 46; US citizen; m 68. POTENTIAL FIELDS. *Educ:* Carleton Univ, BSc, 68; Univ Chicago, MS, 69, PhD(physics), 73. *Prof Exp:* Res assoc physics, Univ Pittsburgh, 73-75; postdoctoral res asst math, Univ Oxford, 75-76; lectr math, Univ Calif, Berkeley, 76-78; numerical analyst, EG&G Geometrics, 79-81, staff scientist, 81-85; assoc res prof, 85-88, RES PROF GEOPHYSICS, COLO SCH MINES, 88- *Concurrent Pos:* Assoc ed, Geophysics, 87-91. *Mem:* Soc Explor Geophysicists; Am Geophys Union; Europ Asn Explor Geophysicists; Am Phys Soc; Am Math Soc; Soc Indust & Appl Math. *Res:* Numerical methods for processing and interpretation of gravity and magnetic data applied to resource exploration. *Mailing Add:* Dept Geophysics Colo Sch Mines Golden CO 80401

HANSEN, ROBERT C(LINTON), b St Louis, Mo, Aug 3, 26; m 52; c 2. ELECTRICAL ENGINEERING. *Educ:* Univ Mo, BS, 49; Univ Ill, MS, 50, PhD(elec eng), 55. *Hon Degrees:* DEng, Univ Mo, Rolla, 75. *Prof Exp:* Res assoc, Antenna Lab, Univ Ill, 50-55; sr staff engr, Microwave Lab, Hughes Aircraft Co, 55-59 & telecommun lab, Space Technol Labs, 59-60; dir test mission anal off, Aerospace Corp, Calif, 60-67; head electronics div, KMS Technol Ctr, 67-71; PRES & CONSULT, R C HANSEN, INC, 71- *Concurrent Pos:* Mem Comn B, Int Sci Radio Union, chmn, Inst Elec & Electronics Engrs Antennas & Propagation Soc, 64; ed, Microwave Scanning Antennas, 64-65. *Mem:* Fel Inst Elec & Electronics Engrs; Am Phys Soc; Sigma Xi; fel, Inst Elec Eng, London. *Res:* Electromagnetic theory applied to surface waves; slot arrays; antennas and near field studies; electronic scanning antennas and systems; ferrite loop antennas; data processing antenna systems; adaptive antenna systems; satellite telemetry and command systems; software systems for satellite command and data handling; computer solutions to electromagnetic problems. *Mailing Add:* 18651 Wells Dr Tarzana CA 91356

HANSEN, ROBERT CONRAD, b Rice Lake, Wis, Nov 3, 31; m 60; c 2. ANALYTICAL CHEMISTRY, INORGANIC CHEMISTRY. *Educ:* Wis State Col, Eau Claire, BS, 53; Univ Wis, MS, 56, PhD(chem), 63. *Prof Exp:* Assoc prof, 59-67, PROF CHEM, UNIV WIS-PLATTEVILLE, 67- *Mem:* Am Chem Soc; AAAS; Sigma Xi. *Res:* Physical chemistry of electrolytic solutions; complexes and analytical methods for transition metals; design and development of instruments for teaching instrumental analysis; trace metals in water. *Mailing Add:* 607 Mitchell Hollow Rd Platteville WI 53818

HANSEN, ROBERT DOUGLAS, b Quincy, Ill, Feb 25, 29; m 50; c 3. PHYSICAL CHEMISTRY. *Educ:* Valparaiso Univ, BS, 50; Iowa State Univ, PhD(chem), 53. *Prof Exp:* Proj leader, Dow Chem Co, 53-71, sr res specialist, Phys Res Lab, 71-79, res assoc, Specialty Prod Res Lab, 79-82, RES ASSOC, MINING PRODS RES & DEVELOP LAB, DOW CHEM CO, 83- *Honors & Awards:* Vaaler Award, 80. *Mem:* Am Chem Soc; Sci Res Soc Am; Am Inst Mining, Metall & Petrol Engrs. *Res:* Ion exchange; surface chemistry; hydrometallurgy; solvent extraction; powder metallurgy; emulsion polymerization; emulsion polymers for use in dielectric papers, and as binders in nonwoven fabrics; mineral and coal flotation; frothers and collectors for mineral flotation. *Mailing Add:* PO Box 159 Waters MI 49797-0159

HANSEN, ROBERT J, b Houston, Tex, July 14, 40; m 66; c 1. RESEARCH ADMINISTRATION. *Educ:* Stanford Univ, BS, 62; Mass Inst Technol, MS, 64, ME, 65, ScD, 69. *Prof Exp:* Postdoctoral fel, Nat Res Coun, Naval Res Lab, 68-70, mech engr, 70-76, head, Boundary Layer Hydrodynamics, 76-86, dep dir, Lab Computational Physics & Fluid Dynamics, 86; prog mgr, 86-87, dir, Defense Sci Div, 87-91, ASSOC DIR, APPL RES & TECHNOL DIRECTORATE, OFF NAVAL RES, 91- *Mem:* Am Soc Mech Engrs. *Res:* Fluid dynamics; active control of systems; signal processing. *Mailing Add:* Code 120 Off Naval Res Arlington VA 22217-5000

HANSEN, ROBERT J, b Harvey, Ill, July 2, 37. PHYSIOLOGY, CHEMISTRY. *Prof Exp:* ASSOC DEAN STUDENT SERV & PROF PHYSIOL CHEM, UNIV CALIF, BERKELEY, 82- *Concurrent Pos:* Prin investr grant, Univ Calif. *Mem:* Am Physiol Soc; Am Diabetes Asn. *Mailing Add:* Dept VM Physiolsci Univ Calif Davis CA 95616

HANSEN, ROBERT JACK, b Houston, Tex, July 14, 40; m 66; c 1. HYDRODYNAMICS. *Educ:* Stanford Univ, BS, 62; Mass Inst Technol, MS, 64, ScD, 69. *Prof Exp:* Nat Res Coun fel, Off Naval Res, 68-70, res mech engr, 70-78, head boundary layer hydrodynamics, 78-86, dep dir, Lab Comput Physics & Fluid Dynamics & Naval Res Lab & prog mgr, Off Naval Res, 86-87, HEAD, DEFENSE SCI DIV, OFF NAVAL RES, 87- *Concurrent Pos:* Adj asst prof, Dept Physics, Univ Md, 71-72. *Mem:* Am Soc Mech Engrs; Sigma Xi. *Res:* Transitional and turbulent boundary layer flows and associated flow-induced vibration and acoustic phenomena. *Mailing Add:* 12718 Mac Duff Dr Ft Washington MD 20744

HANSEN, ROBERT JOHN, b Harvey, Ill, July 2, 37; m 59; c 2. PHYSIOLOGICAL CHEMISTRY, ENDOCRINOLOGY. *Educ:* George Williams Col, BS, 60, MS, 62; Univ Chicago, PhD(physiol), 69. *Prof Exp:* Instr biol & phys educ, George Williams Col, 62-63; from asst prof to assoc prof, 68-84, PROF PHYSIOL CHEM, UNIV CALIF, DAVIS, 84-; ASSOC DEAN STUDENT SERV, SCH VET MED, 82- *Concurrent Pos:* von Humboldt Found fel & guest prof, Biochem Inst, Freiburg, Ger, 75. *Mem:* AAAS; Am Physiol Soc; Am Diabetes Soc. *Res:* Action of insulin on protein turnover in animal tissues. *Mailing Add:* Dept Vet Physiol Univ Calif Davis CA 95616

HANSEN, ROBERT M(ARIUS), b Gulfport, Miss, Jan 19, 24; m 46; c 2. CHEMICAL ENGINEERING. *Educ:* La State Univ, BS, 50, PhD(chem eng), 55; Newark Col Eng, MS, 52. *Prof Exp:* Chem engr, Victor Div, Radio Corp Am, 50-52; asst, La State Univ, 52-55; sr chem engr, Res Dept, Baton Rouge, Kaiser Aluminum & Chem Corp, 55-56, group leader, Prod Control Dept, 56-57, prod control supt, 57-58, proj supvr, Develop Dept, 58-59, from asst mgr to mgr process develop, 59-61, tech supt, Gramercy, 61-68, prod supt, 63, prod mgr, 69-72, tech mgr, 73-74, environ mgr, alumina, coke & chem plants, 75-81, mgr, field develop & bauxite cord, 81-83, mgr, Lab Res & Develop, 83-84; RETIRED. *Concurrent Pos:* On site, tech asst, Hindalco Alumina Plant, India, 70; plpnt stsrt-up Euralumina, 74. *Res:* Chemical processing; production control; technical service and development; supervision of scientific personnel; production and plant management. *Mailing Add:* 9621 N Parkview Dr Baton Rouge LA 70815

HANSEN, ROBERT SUTTLE, b Salt Lake City, Utah, June 17, 18; m 39; c 1. PHYSICAL CHEMISTRY. *Educ:* Univ Mich, BS, 40, MS, 41, PhD(phys chem), 48. *Hon Degrees:* DSc, Lehigh Univ, 78. *Prof Exp:* From asst prof to prof, Iowa State Univ, 48-67, distinguished prof chem, Col Sci & Humanities, 67-88, dir, Ames Lab, 68-88, EMER DISTINGUISHED PROF, US DEPT ENERGY & INST PHYS RES & TECHNOL, 88-, ASSOC, AMES LAB, 88- *Concurrent Pos:* Assoc chemist, Ames Lab, 48-55, sr chemist, 55-65, chief chem div, 65-68, chmn dep chem, Iowa State Univ, 65-68; consult, Union Carbide & Carbon Corp, 52-86, Interchem Corp, 56-68; NSF sr fel, Univ Utrecht & Univ Southern Calif, 59-60; consult, Procter & Gamble Co, 62-86; mem NSF Adv Panels chem, 70-75 & mat sci, 76-80. *Honors & Awards:* Award Colloid & Surface Chem, Am Chem Soc, 66; Midwest Award, Am Chem Soc, 80; Iowa Award, Am Chem Soc, 87. *Mem:* AAAS; Am Chem Soc; Am Phys Soc. *Res:* Adsorption, boundary tensions; surface thermodynamics; thermodynamics of nonelectrolytic solutions; surface chemistry and electrochemistry; catalysis. *Mailing Add:* Ames Lab US Dept Energy & Inst Phys Res & Technol Iowa State Univ Ames IA 50011

HANSEN, RODNEY THOR, b Spokane, Wash, Mar 27, 40; m 64; c 3. MATHEMATICS, STATISTICS. *Educ:* Whitworth Col, BS, 62; Univ Wash, MA, 64; Wash State Univ, PhD(math), 67. *Prof Exp:* From asst prof to assoc prof math, Mont State Univ, 67-82; assoc prof, 81-85, ASSOC DEAN UNDERGRAD AFFAIRS, WHITWORTH COL, 84-, PROF MATH, 85- *Concurrent Pos:* Vis assoc prof math, Univ Ore, 76. *Res:* Combinatorial number theory; combinatorics. *Mailing Add:* Dept Math & Comput Sci Whitworth Col Spokane WA 99251

HANSEN, ROGER GAURTH, b Smithfield, Utah, Aug 18, 20; m 43; c 3. NUTRITIONAL BIOCHEMISTRY. *Educ:* Univ Wis, BS, 44, MS, 46, PhD(biochem), 48. *Prof Exp:* Asst prof biochem, Univ Utah, 48-50; from assoc prof to prof, Univ Ill, 50-57; prof & head dept, Mich State Univ, 57-68; provost, 68-85, PROF BIOCHEM, NUTRIT & FOOD SCI, UTAH STATE UNIV, 68- *Concurrent Pos:* Mem, Nutrit Study Sect, NIH, Coun Foods & Nutrit, AMA & Food & Nutrit Bd, Nat Res Coun. *Honors & Awards:* Borden Award, Am Inst Nutrit. *Mem:* Am Soc Biol Chem; Am Chem Soc; Soc Exp Biol & Med; Am Inst Nutrit. *Res:* General biochemistry and nutrition of animals; formation and utilization of galactose and hereditary disorders of metabolism; human requirements for nutrients. *Mailing Add:* 1676 E 1030 North Logan UT 84321

HANSEN, TED HOWARD, b Madison, Wis, June 6, 47. IMMUNOLOGY, GENETICS. *Educ:* Univ Mich, PhD(human genetics), 75. *Prof Exp:* Sr staff fel, Transplantation Sect, Immunol Br, Nat Cancer Inst, NIH, 78-80; res fel, Merck Inst Therapeut Res, 80-; AT DEPT GENETICS, UNIV WASHINGTON. *Mem:* Sigma Xi; Am Asn Immunol; Fedn Am Socs Exp Biol. *Res:* Immunogenetics; transplantation; immunochemistry. *Mailing Add:* Dept Genetics Wash Univ Sch Med Box 8232 4566 Scott Ave St Louis MO 63110

HANSEN, TIMOTHY RAY, b Detroit, Mich, Aug 25, 45; m 67; c 1. MEDICAL PHYSIOLOGY. *Educ:* Univ Mich, Ann Arbor, BS, 67, PhD(physiol), 73. *Prof Exp:* Teaching fel physiol, Med Sch, Univ Mich, Ann Arbor, 70-73; res fel physiol, Harvard Med Sch, 73-75; ASST PROF PHYSIOL, UNIV HEALTH SCI-CHICAGO MED SCH, 75- *Mem:* Assoc Am Physiol Soc. *Res:* Physiology of vascular smooth muscle in cardiovascular disease. *Mailing Add:* 74 S Seventh Ave La Grange IL 60525

HANSEN, TORBEN CHRISTEN, b Frederiksberg, Denmark, May 30, 33; m 56; c 1. CONCRETE, CEMENT. *Educ:* Tech Univ, Denmark, MSc, 56; Royal Inst Technol, Stockholm, Sweden, DrSci, 63. *Prof Exp:* Res engr, Royal Inst Technol, 56-60; develop engr concrete, Portland Cement Asn, Skokie, Ill, 60-62; asst prof, G4 Stokholm, Sweden, 63-65; assoc prof bldg mat, Stanford Univ, Calif, 65-67; PROF BLDG MAT, TECH UNIV, DENMARK, 67- *Concurrent Pos:* Lectr bldg mat, Univ Calif, Berkeley, 62; sr expert, UN Develop Prog, Banoung Indonesia, 76-81. *Honors & Awards:* Wason Medal, Am Concrete Inst, 66. *Mem:* Int Union Testing & Res Lab Mat & Struct (vpres, 88-91, pres); fel Am Concrete Inst; Concrete Soc; Inst Concrete Technologists. *Res:* Teaching and consultancy in cement and concrete. *Mailing Add:* Sundvaenget 12 Hellerup 2900 Denmark

HANSEN, UWE JENS, b Kiel, Ger, June 7, 33; US citizen; m 59; c 2. ACOUSTICS. *Educ:* Brigham Young Univ, BS, 54, MA, 61, PhD(physics), 66. *Prof Exp:* Nat Acad Sci-Nat Res Coun res assoc, US Naval Res Lab, 66-68, res physicist, 68; assoc prof physics, 68-77, interim chmn dept, 80-83, PROF PHYSICS, IND STATE UNIV, 77- *Mem:* Am Phys Soc; Am Soc Metals; Am Asn Physics Teachers; Acoust Soc Am. *Res:* Vibrational studies using holographic interferometry and modal analysis. *Mailing Add:* 715 S 34th St Terre Haute IN 47803

HANSEN, VAUGHN ERNEST, b Syracuse, Utah, July 26, 21; m 41; c 10. WATER RESOURCES, IRRIGATION ENGINEERING. *Educ:* Utah State Univ, BS, 43, MS, 47; State Univ Iowa, PhD(mech eng & hydrol) 49. *Prof Exp:* Prof eng, Utah State Univ, 55-66, dir res eng exp sta, 58-64; coordr resource develop, Inter-Am Ctr Land & Water, 64-66; dir water res lab, Utah State Univ, 64-66; gen consult, Vaughn Hansen Assoc, 66-74, consult engr water resources, 74-88; RETIRED. *Concurrent Pos:* Consult, numerous foreign countries, 54-, NSF, 64-68; pres, Agr Develop & Eng Serv, 54-61; vpres, Western Admixture Co, 58-66; consult, numerous foreign countries, 54- *Honors & Awards:* Collingwood Prize, Am Soc Civil Engrs, 54. *Mem:* Sigma Xi; Consult Engrs Coun; Am Water Resources Asn. *Res:* Water resource development; irrigation; ground water development. *Mailing Add:* 2046 Walker Lane Salt Lake City UT 84117

HANSEN, WAYNE RICHARD, b Oakgrove, Wis, Aug 6, 39; m 60; c 2. RADIOECOLOGY, HEALTH PHYSICS. *Educ:* Univ Wis-Eau Claire, BS, 61; Univ Kans, MS, 63; Colo State Univ, PhD(radiation biol), 70; Am Bd Health Physics, dipl, 72. *Prof Exp:* Health physicist, Univ Colo, 63-67, Health Educ & Welfare, Food & Drug Admin & Bur Radiation Health, 70-71 & Environ Protection Agency & Off Radiation Prog, 71-75; sr radiobiologist, Nuclear Regulatory Comn, 75-77; staff mem, environ surveillance, Los Alamos Nat Lab, 77-78, group leader, 78-84, dep leader, Div Health Safety & Environ, 84-87, prog mgr, 87-90, GROUP LEADER, ENVIRON SCI, LOS ALAMOS NAT LAB, 90- *Concurrent Pos:* Consult, Fed Res Biol & Health Effects of Ionizing Radiations, Nat Acad Sci, 80, EG&G, Idaho, 81-88; mem, NMex Radiation Technol Adv Coun, 83-; lectr, US Air Force Interservice Nuclear Weapons Sch, Dept Defense, 84-88; mem, NCRP Scientific Comt-64, 87-; consult, Dept Energy Nuclear Facil Safety Comt, 88- *Mem:* Fel AAAS; Int Radiation Protection Asn; Health Physics Soc; Sigma Xi. *Res:* Environmental behavior of radionuclides and hazardous chemicals; transuranic elements and polonium in ecosystems or waste sites. *Mailing Add:* Los Alamos Nat Lab M5 J495 Environ Sci Los Alamos NM 87545

HANSEN, WILFORD NELS, b Cardston, Alta, May 30, 28; US citizen; m 51; c 6. SURFACE PHYSICS, SPECTROSCOPY,. *Educ:* Brigham Young Univ, BS, 50; Iowa State Univ, PhD(chem), 56. *Prof Exp:* Sr chemist, Atomics Int, NAm Aviation, Inc, 56-57; res assoc chem, Iowa State Univ, 57-58; asst prof, Brigham Young Univ, 58-60; res specialist, Atomics Int Div, NAm Rockwell Corp, 60-62, mem tech staff, Sci Ctr, Calif, 62-68; prof physics, 68-73, PROF PHYSICS, CHEM & BIOCHEM, UTAH STATE UNIV, 73- *Concurrent Pos:* Consult, Atomics Int Div, 59-62, NAm Aviation, Inc, 62-69 & Southern Res Inst, 70-72, Los Alamos Nat Lab, 82- *Mem:* Am Chem Soc; Electrochem Soc; Am Phys Soc. *Res:* Surface physics and chemistry; optics and spectroscopy of surface regions; electrochemistry. *Mailing Add:* Dept Physics Utah State Univ 5305 Logan UT 84322

HANSEN, WILLIAM ANTHONY, b Chicago, Ill, July 14, 48; m 69; c 4. COMPUTER SCIENCES. *Educ:* Ill Inst Technol, BS, 70; Northwestern Univ, MS, 71, PhD(math), 74. *Prof Exp:* Training coordr, Comput Serv Div, Abbott Labs, 72-74; asst prof comput sci & math, Wilkes Col, 74-76; prod mgr, Deltak, Inc, 76-78, mgr qual assurance, 78-80; PRES, HANSEN TRAINING SYSTS, INC, 80- *Mem:* Asn Comput Mach. *Res:* A computer calculation of the homology of the lambda algebra; computer science education, particulary learner paced modules utilizing text, computer-based instruction and video tapes. *Mailing Add:* 1981 Abbotsford Dr Barrington IL 60010-5560

HANSEN-SMITH, FEONA MAY, MUSCLE DEVELOPMENT, EXTRA-CELLULAR MATRIX. *Educ:* Med Col Wis, PhD(physiol), 74. *Prof Exp:* Asst prof human anat & gen histol, Oakland Univ, 85-86. *Res:* Neuromuscular biology. *Mailing Add:* Dept Biol Sci Oakland Univ Rochester MI 48063

HANSFORD, RICHARD GEOFFREY, b Market Bosworth, UK, Apr 29, 44; US citizen; div; c 3. ION TRANSPORT, ENZYMOLOGY. *Educ:* Univ Bristol, UK, BSc, 65, PhD(biochem), 69. *Prof Exp:* Fel, dept physiol chem, Johns Hopkins Univ, 69-70; lectr biochem, Univ Col, Cardiff, Wales, 70-73; vis assoc & scientist, 73-79, res chemist, 79-86, SECT CHIEF ENERGY METAB & BIOENERGETICS SECT, LAB CARDIOVASC SCI, NAT

INST AGING, NIH, 86- *Concurrent Pos:* Assoc prof med, Johns Hopkins Univ, 90- *Mem:* Am Soc Biol Chemists; Biophys Soc. *Res:* Control of energy metabolism, especially by calcium ions; isolated mitochondria and cells; impact of senescence on metabolism. *Mailing Add:* Geront Res Ctr NIA NIH Francis Scott Key Med Ctr Baltimore MD 21224

HANSFORD, ROWLAND CURTIS, b Belington, WVa, Jan 26, 12; m 53. PHYSICAL CHEMISTRY. *Educ:* Davis & Elkins Col, BS, 33; George Washington Univ, MA, 37. *Hon Degrees:* DSc, Davis & Elkins Col, 71. *Prof Exp:* Works chemist, E I du Pont de Nemours & Co, WVa, 33-35; res chemist, Socony-Vacuum Oil Co, NJ, 37-44, res assoc, 44-52; res assoc, Union Oil Co Calif, 52-57, sr res assoc, 57-65, staff consult, 65-77; RETIRED. *Honors & Awards:* Tolman Medal, Am Chem Soc, 71; Philadelphia Catalysis Club Award, 74; Chem Pioneer Award, Am Inst Chemists, 76. *Mem:* Fel AAAS; Am Chem Soc; Catalysis Soc; fel Am Inst Chemists. *Res:* Catalysis in petroleum conversion; fundamentals of catalysis. *Mailing Add:* 5470 Paseo Del Lago E Laguna Hills CA 92653-2623

HANSHAW, BRUCE BUSSER, b Harrisburg, Pa, May 7, 30; m 54; c 2. GEOCHEMISTRY. *Educ:* Mass Inst Technol, ScB, 53; Univ Colo, MS, 58; Harvard Univ, PhD(geochem), 62. *Prof Exp:* Geologist, US AEC, 53-54 & 56 & Petrol Res Corp, 58-61; geologist, Water Resources Div, 61-70, staff scientist, Off Dir, Washington, DC, 70-73, GEOCHEMIST, US GEOL SURV, 73- *Concurrent Pos:* Mem comt pollution, Nat Acad Sci-Nat Res Coun, 65-67, mem US nat comt, Int Hydrol Decade, 67. *Honors & Awards:* O E Meinzer Award, Geol Soc Am, 73. *Mem:* AAAS; Geol Soc Am; Geochem Soc; Am Geophys Union; Am Chem Soc; Sigma Xi. *Res:* Isotope chemistry; membrane phenomena applied to clay minerals; mineral-solution equilibria; factors affecting natural water chemistry; paleoclimate. *Mailing Add:* US Geol surv GD & 01G Mail Stop 917 1220 Sunrise Valley Dr Reston VA 22092

HANSHAW, JAMES BARRY, b Scarsdale, NY, Dec 23, 28; c 5. PEDIATRICS, VIROLOGY. *Educ:* Syracuse Univ, AB, 50; State Univ NY Upstate Med Ctr, MD, 53; Am Bd Pediat, dipl, 61. *Hon Degrees:* DSc, Syracuse Health Sci Ctr, State Univ NY, 91. *Prof Exp:* Nat Found fel, Sch Pub Health, Harvard Univ, 58-60; USPHS career res develop award, 62-72, prof pediat & microbiol, Sch Med & Dent, Univ Rochester, 72-75; prof pediat & chmn dept, Med Sch, Univ Mass, 75-85, pediatrician-in-chief, Med Ctr, 75-85, interim vchancellor & acad dean, 85-86, dean & provost 86-89; RETIRED. *Concurrent Pos:* Pediatrician-in-chief, Genesee Hosp, 72-75, dir med educ, 73-75, chmn med bd, 74-75; vis prof pediat, Inst Child Health, Univ London; lectr pediat, Harvard Med Sch, 75- *Mem:* AAAS; Am Acad Pediat; AMA; Infectious Dis Soc Am; Am Pediat Soc; Soc Pediat Res. *Res:* Seroepidemiological studies of cytomegaloviruses and other herpes viruses; congenital infections in man and their effect on the development of the central nervous system. *Mailing Add:* Univ Mass Med Ctr 55 Lake Ave N Worcester MA 01655

HANSMA, PAUL KENNETH, b Salt Lake City, Utah, Apr 28, 46; m 68; c 2. SOLID STATE PHYSICS, SURFACE PHYSICS. *Educ:* New Col, Fla, BA, 67; Univ Calif, Berkeley, PhD(physics), 72. *Prof Exp:* From asst prof to assoc prof, 72-80, PROF PHYSICS, UNIV CALIF, SANTA BARBARA, 80- *Concurrent Pos:* Res fel, Alfred P Sloan Found, 75. *Mem:* Fel Am Phys Soc; fel AAAS. *Res:* Scanning ion conductance microscopy; atomic force microscopy. *Mailing Add:* Dept Physics Univ Calif Santa Barbara CA 93106

HANSMAN, MARGARET MARY, b Colorado Springs, Colo, Oct 16, 11. MATHEMATICS. *Educ:* Colo Col, BA, 31, MA, 36; Univ Ill, PhD(math), 41. *Prof Exp:* Asst math, Univ Ill, 36-41; asst math,Colo Col, 34-36,from instr to assoc prof math, 41-76; RETIRED. *Mem:* Am Math Soc; Math Asn Am. *Res:* Group theory; theory of numbers; geometric investigation into metabelian groups generated by four elements of order p. *Mailing Add:* 2038 Armstrong Colorado Springs CO 80904

HANSMAN, ROBERT H, b Ft Madison, Iowa, Aug 22, 28. GEOLOGY. *Educ:* Univ Iowa, BA, 52, MS, 55, PhD(geol), 58. *Prof Exp:* Fel, Univ Cincinnati, 59-61, mus specialist invert paleont, US Nat Mus, Smithsonian Inst, 62-63; res assoc, Univ Ill, 63-64 & 65-66; res assoc, Princeton Univ, 64-65, mem res staff & cur invert paleont, 66-73; tech ed earth sci, Ebasco Serv, Inc, Greensboro, NC, 76-77 & 78-79; custodian of collections, Field Mus Natural Hist, 77; ed, Mo Div Geol & Land Surv, 79-90. *Concurrent Pos:* Assoc, Clare Hall, Cambridge Univ, 69-70. *Res:* Historical research on atmospheric effects of the Lakagígar (Skaftáreldar) eruption (1783-84), Iceland. *Mailing Add:* 1134 Ave E Madison IA 32627

HANSMANN, DOUGLAS R, b Olympia, Wash, Oct 16, 44. BIOENGINEERING, APPLIED MECHANICS. *Educ:* Princeton Univ, BSE, 66; Univ Calif, Berkeley, MS, 68, PhD(mech eng), 72. *Prof Exp:* Prog mgr res, Cardio-Dynamics Lab Inc, 71-72, vpres, 72-78; mgr new prod develop, 78-80, DIR TECH DEVELOP, PHYSIO-CONTROL CORP, 80- *Mem:* Inst Elec & Electronics Engrs; Asn Advan Med Instruments. *Res:* Medical therapeutic instrumentation and devices relating to treatment of blood; cardiovascular diagnostic and monitoring instrumentation. *Mailing Add:* VP Res & Develop Cardiovasc Devices Inc 2801 Barranca Rd Irvine CA 92714

HANSMANN, EUGENE WILLIAM, b Whiting, Ind, Dec 20, 33; m 62; c 2. AQUATIC ECOLOGY, ALGOLOGY. *Educ:* Mich State Univ, BS, 56, MS, 63; Ore State Univ, PhD(bot), 69. *Prof Exp:* Biologist, Plankton Lab, USPHS, Cincinnati, 62-63; botanist, NASA, 63-65; asst prof biol, Univ Conn, 68-71; NSF fel biol, Univ NMex, 71-77; wetland coordr, 77-79, PROG MGR, ENVIRON CONTAMINANTS, FISH & WILDLIFE SERV REGION SIX, 79- *Concurrent Pos:* Grants, Inst Water Resources, Univ Conn, 69 & 70 & US Dept Interior, 71. *Mem:* AAAS. *Res:* Algal ecology of streams and reservoirs including productivity; community change; biomass accumulation as affected by watershed practices and eutrophication; algal taxonomy. *Mailing Add:* 971 S Arbutus St Lakewood CO 80228

HANSON, ALBERT L, b Gainesville, Fla, July 9, 52; m 81; c 2. NUCLEAR ANALYTICAL METHODS, X-RAY PHYSICS. *Educ:* NC State Univ, BS, 74; Univ Mich, MSE, 76, PhD(nuclear sci), 79. *Prof Exp:* Fel, Brookhaven Nat Lab, 79-81, asst scientist physics dept, 81-83, assoc physicist dept appl sci, 83-85, physicist dept appl sci, 85-89, PHYSICIST, DEPT NUCLEAR ENERGY, BROOKHAVEN NAT LAB, 89- *Honors & Awards:* R & D 100 Award, 88. *Mem:* Am Phys Soc; Am Nuclear Soc; AAAS; Int Soc Radiation Physics. *Res:* Development of analytical techniques, mostly synchrotron radiation and ion beam based, utilizing nuclear and atomic physics used in materials sciences, energy sciences and biological sciences; x-ray physics. *Mailing Add:* Bldg 130 Dept Nuclear Energy Brookhaven Nat Lab Upton NY 11973

HANSON, ALFRED OLAF, b Braddock, NDak, Sept 26, 14; m 42; c 4. EXPERIMENTAL NUCLEAR PHYSICS. *Educ:* Univ NDak, BS, 36, MA, 38; Univ Wis, PhD(physics), 42. *Prof Exp:* Teacher pub sch, 36-37; asst, Off Sci Res & Develop, Univ Wis, 42-43; scientist, Los Alamos Sci Lab, Univ Calif, NMex, 43-46; from asst prof to assoc prof, 46-51, PROF PHYSICS, UNIV ILL, URBANA, 51- *Concurrent Pos:* Fulbright scholar, Torino, Italy, 55-56; Fulbright lectr, Sao Paolo, 60 & Brookhaven Nat Lab, NY, 61-62. *Mem:* Fel Am Phys Soc; Sigma Xi. *Res:* Nuclear physics with electrostatic generator and betatron; rotating target for neutron sources; neutron detector with uniform response; electron scattering; photonuclear reactions; accelerator development. *Mailing Add:* Dept Physics Univ Ill Urbana IL 61801

HANSON, ALLEN LOUIS, b Crookston, Minn, Aug 7, 15; m 44; c 2. PHYSICAL CHEMISTRY. *Educ:* Concordia Col, BA, 35; Univ Iowa, MS, 40, PhD(phys chem), 42. *Prof Exp:* Instr pub schs, NDak, 35-37 & Minn, 37-39; res chemist, Photoprod Dept, E I du Pont de Nemours & Co, 42-43; prof chem, Concordia Col, 46-52; from assoc prof, to prof chem, 52-80, EMER PROF, ST OLAF COL, 80- *Concurrent Pos:* Chmn dept chem, St Olaf Col, 64-70 & div natural sci & math, 73-77; vis prof, Univ Lancaster, 70-71. *Mem:* Am Chem Soc. *Res:* Properties of gelatin; activity of solutes from vapor pressure lowering; electrochemistry, especially electrodeposition from nonaqueous solvents; gamma radiolysis; photolysis and x-radiolysis of alkyl halides; enzyme kinetics. *Mailing Add:* Dept Chem St Olaf Col Northfield MN 55057

HANSON, ANDREW JORGEN, b Los Alamos, NMex, Feb 22, 44; m 68; c 2. ARTIFICIAL INTELLIGENCE, THEORETICAL PHYSICS. *Educ:* Harvard Col, BA, 66; Mass Inst Technol, PhD(physics), 71. *Prof Exp:* Res assoc, Inst Advan Study, 71-73, Cornell Univ, 73-74, Stanford Linear Accelerator Ctr, 74-76 & Lawrence Berkeley Lab, 76-78; proj scientist, Inst Advan Comput, 79-80; comput scientist, Artificial Intel Ctr, SRI Int, 80-89; ASSOC PROF COMPUTER SCI, IND UNIV, BLOOMINGTON, 89- *Concurrent Pos:* NSF fel, 71-72; proj coordr & consult, Explor Sci Mus, 76-78. *Mem:* Am Phys Soc; Sigma Xi; Asn Comput Mach; Inst Elec & Electronics Engrs; Am Asn Artificial Intel. *Res:* Theoretical elementary particle physics; gauge theories and gravitation; mathematical physics; applications of artificial intelligence to machine vision; interactive computer-based tools for scientific visualization and intuition development. *Mailing Add:* Computer Sci Dept Ind Univ Bloomington IN 47405

HANSON, ANGUS ALEXANDER, b Chilliwack, BC, Jan 1, 22; US citizen; m 48; c 3. AGRONOMY. *Educ:* Univ BC, BSA, 44; McGill Univ, MSc, 46; Pa State Col, PhD(agron), 51. *Prof Exp:* Asst, McGill Univ, 44-46, lectr agron, 46-48, asst prof, 48-49; agt, US Regional Pasture Res Lab, Agr Res Serv, USDA, Pa, 49-52, agronomist, 52, sr agronomist, Forage & Range Br, Plant Sci Res Div, 53-54, prin agronomist, 54-57, head grass & turf invests, 57-65, chief br, Plant Indust Sta, 65-72, dir, Beltsville Agr Res Ctr, 72-79; dir res & vpres, 80-86, vpres, 87-88, BD DIR, W-L RES INC, 89- *Concurrent Pos:* Assoc prof, Pa State Col, 51-52; ed, Crop Sci, 62-64, Turfgrass Sci, 69, J Environ Qual, 71- 73, Alfalfa & Alfalfa Improv, 88, Practical Handbk Agr Sci, 90. *Mem:* Fel AAAS; fel Am Soc Agron; fel Crop Sci Soc Am (pres, 67); hon mem Asn Off Seed Certifying Agencies. *Res:* Inheritance of economic characters in forage grasses; cytogenetic investigations; development of proprietary alfalfa cultivars; technique studies, including methods of determining combining ability and methods of isolating disease resistant lines, research organization and management. *Mailing Add:* 10411 Sweetbriar Pkwy Silver Spring MD 20903

HANSON, AUSTIN MOE, b Roland, Iowa, Feb 26, 17; m 44; c 2. BIOCHEMISTRY. *Educ:* Luther Col, Iowa, BA, 39; Univ Wis, MS, 41, PhD(bact), 44. *Prof Exp:* Head bact sect, Res Labs, Western Condensing Co, 44-51; tech dir, Grain Processing Corp, 51-63, mgr res admin, 63-64; dir, Nutrit Res Lab, 64-78, VPRES, KENT FEEDS, INC, 78-; VPRES, GRAIN PROCESSING CORP, 79- *Mem:* Am Soc Microbiol; Am Chem Soc. *Res:* Production of microbial products and starch derivatives; enzymatic conversion of starch to syrups and dextrose; dextrose crystallization. *Mailing Add:* 2013 Burnside Dr Muscatine IA 52761

HANSON, BARBARA ANN, b San Francisco, Calif, Nov 2, 48. MICROBIAL BIOCHEMISTRY, MICROBIAL PHYSIOLOGY. *Educ:* Stanford Univ, BA, 70; Univ Calif, San Diego, PhD(biol), 75. *Prof Exp:* Fel microbial biochem, Univ Calif, Irvine, 76-77; fel microbiol biochem, Med Ctr, 77-80, ASST PROF, SCH BIOL SCI, UNIV KY, 80- *Mem:* Am Soc Microbiol; Sigma Xi. *Res:* Regulation of lipid and cell wall biosynthesis in Saccharomyces, Neurospora, and a pathogenic yeast, Candida albicans. *Mailing Add:* Dept Biol Sci Univ Ky Lexington KY 40506

HANSON, BERNOLD M, b Mayville, NDak, May 7, 28; m 51; c 3. PETROLEUM GEOLOGY. *Educ:* Univ NDak, BS 51; Univ Wyo, MA, 54. *Hon Degrees:* LLD, Univ Wyo, 87. *Prof Exp:* Geologist, Magnolia Petrol Mobil, 51-52; first lieutenent topog mapping, US Army, 52-53; lab instr geol, Univ Wyo, 53-54; geologist, Humble Oil & Refining, Exxon, 55-60; consult oil & gas, 60-66; PRES, HANSON CORP, 66- *Concurrent Pos:* Chmn adv bd, Am Asn Petrol Geologists, 87-88; chmn adv bd, geol dept, Univ NDak,

80- & Univ Wyo, 81- *Mem:* Hon mem Am Asn Petrol Geologists (pres, 86-87); Soc Econ Paleontologists & Mineralogists, Permian Basin sect (vpres, 61, pres, 62-63); Soc Independent Prof Earth Scientists; Am Inst Prof Geologists. *Res:* Economic geology of the Permian Basin describing in detail the reservoir of the oil and gas producing zones and their economic input. *Mailing Add:* PO Drawer 1269 Midland TX 79702

HANSON, CARL VEITH, b Pueblo, Colo, Nov 19, 43; m 73; c 2. MOLECULAR VIROLOGY, DIAGNOSTIC VIROLOGY. *Educ:* Harvard Univ, BA, 65; Stanford Univ, PhD(biophysics), 70. *Prof Exp:* Fel phys chem, Univ Calif, 71-76; res specialist virol, 77-82, RES SCIENTIST, VIRAL & RICKETTSIAL DIS LAB, CALIF DEPT HEALTH SERV, 82- *Concurrent Pos:* Instr, UNESCO Cell Biol Training Course, Szeged, Hungary, 74; prin investr, US Army Res Contract, 77-81. *Mem:* Am Soc Microbiol; Pan Am Group Rapid Viral Diag. *Res:* Development of inactivated viral vaccines and immunodiagnostic reagents; photochemical inactivation of viruses; structural probes for use in molecular virology; novel methods for the rapid immunodiagnosis of viral disease; development of immunoassays for the detection of mutagens, carcinogens, pesticides and toxins in humans; AIDS diagnostic methods; HIV neutralizing antibodies. *Mailing Add:* Viral & Rickettsial Dis Lab Calif Dept Health Serv 2151 Berkeley Way Berkeley CA 94704

HANSON, D(ONALD) N(ORMAN), b Minooka, Ill, Aug 3, 18; m 43; c 3. CHEMICAL ENGINEERING. *Educ:* Univ Ill, BS, 40; Univ Wis, MS, 41, PhD, 43. *Prof Exp:* Instr chem eng, Univ Wis, 43-44, chem engr, Shell Develop Co, Calif, 44-46, 47; asst prof, Kans State Col, 46-47; from asst prof to assoc prof, 47-58, chmn dept, 63-66, PROF CHEM ENG, UNIV CALIF, BERKELEY, 58- *Mem:* Am Chem Soc; Am Inst Chem Eng. *Res:* Calculation and design of distillation columns; process development for energy conservation. *Mailing Add:* Dept Chem Eng Univ Calif Berkeley CA 94720

HANSON, DANIEL JAMES, b Faribault, Minn, Oct 7, 28; m 53; c 2. PATHOLOGY. *Educ:* Univ Minn, BA, 50, BS, 51, MD, 53. *Prof Exp:* Assoc pathologist, 61-65, chief of staff, 73-75, DIR PATH SERV, MERCY HOSP, 65- *Concurrent Pos:* Pathologist, Path Labs, Toledo, Ohio, 61-; consult pathologist, Res Inst, Mercy Hosp, 61-; consult, Pharmaceut Firms, 61-; adj prof, Univ Toledo, 62-; clin assoc prof path, Med Col Ohio, 69- *Mem:* Am Soc Clin Path; Col Am Path; AMA; Pan-Am Med Asn; Int Acad Path; Am Chem Soc. *Res:* Anatomical, chemical, gynecological and clinical pathology; instrumentation techniques; drug toxicity; hemolytic disease; pharmaceutical evaluation; injection complications. *Mailing Add:* 2200 Jefferson Ave Toledo OH 43624

HANSON, DANIEL RALPH, b Fargo, NDak, Jan 26, 47; m 69; c 3. BEHAVIORAL GENETICS, EXPERIMENTAL PSYCHOPATHOLOGY. *Educ:* Univ Minn, BA, 69, PhD(psychol), 74, MD, 83. *Prof Exp:* Asst prof psychol & psychiat, McMaster Univ, 74-77; fel human genetics, Univ Minn, 77-79. *Concurrent Pos:* Sr fel human genetics, Univ Wash, 76-77. *Mem:* Am Psychol Asn; Am Soc Human Genetics; Behav Genetics Asn; Soc Study Social Biol; Am Psychiat Asn; Am Psychopath Asn. *Res:* Biological and genetic bases of abnormal behavior; antecedents of adult psychopathology. *Mailing Add:* 13277 Valley Creek Trail S Afton MN 55001

HANSON, DAVID LEE, b Minneapolis, Minn, July 14, 35; m 56; c 4. PROBABILITY STATISTICS, MATHEMATICS. *Educ:* Mass Inst Technol, BS, 56; Ind Univ, MA, 59, PhD(probability statist), 60. *Prof Exp:* Staff mathematician, Int Bus Mach Corp, 60-61 & Sandia Corp, 61-63; assoc prof statist, 63-64, from assoc prof to prof statist & math, Univ Mo, Columbia, 64-73, chmn dept statist, 71-73; PROF MATH SCI, STATE UNIV NY BINGHAMTON, 73-, CHMN DEPT, 83- *Concurrent Pos:* Assoc ed, Ann Math Statist, 67-72 & Ann Probability & Ann Statist, 72-79; prog dir statists & probability, NSF, Washington, DC, 79-80. *Mem:* Am Math Soc; fel Inst Math Statist; Am Statist Asn; Int Statist Inst. *Res:* Probability theory; convergence rates; non-parametric regression; stochastic approximation; utility theory. *Mailing Add:* Dept Math Sci State Univ NY Binghamton NY 13902-6000

HANSON, DAVID M, b Waseca, Minn, June 22, 42; m 64; c 2. CHEMICAL PHYSICS. *Educ:* Dartmouth Col, AB, 64; Calif Inst Technol, PhD(chem), 68. *Prof Exp:* From asst prof to assoc prof, 69-78, PROF CHEM, STATE UNIV NY STONY BROOK, 78- *Concurrent Pos:* NATO fel, Munich Tech Univ, 68-69; Alfred P Sloan Found res fel, 72; fel, Synchrotron Ultraviolet Radiation Facil, Nat Bur Standards, 80-81; vis scientist, Inst Molecular Sci, Okazaki, Japan, 85. *Mem:* Am Chem Soc; Am Phys Soc. *Res:* Soft x-ray spcetroscopy and molecular fragmentation processes. *Mailing Add:* Dept Chem State Univ NY Stony Brook NY 11794-3400

HANSON, DONALD FARNESS, b Urbana, Ill, Mar 5, 46. ELECTRICAL ENGINEERING. *Educ:* Univ Ill, Urbana-Champaign, BS, 69, MS, 72, PhD(elec eng), 76. *Prof Exp:* Res asst comput sci, Univ Ill, Urbana-Champaign, 69-71, teaching asst elec eng, 71-73, lectr archit, 73-74, instr elec eng, 74-76; asst prof, Iowa State Univ, 76-77; assoc prof elec eng, Syracuse Univ, 83-84; asst prof, 77-83, ASSOC PROF ELEC ENG, UNIV MISS, 84- *Concurrent Pos:* Consult, Electronics Technol & Devices Lab, US Army, Ft Monmouth, NJ, 90-91. *Mem:* Inst Elec & Electronics Engrs; Asn Comput Mach. *Res:* Electromagnetic field theory, especially numerical methods, scattering and applications; applications of digital and analog electronics and microprocessors; applications of computer science; hardware description languages. *Mailing Add:* Dept Elec Eng Univ Miss University MS 38677

HANSON, DONALD WAYNE, b Denver, Colo, Feb 9, 37; m 63; c 2. SATELLITE COMMUNICATIONS, TIME & FREQUENCY METROLOGY. *Educ:* Univ Colo, BS, 59; Stanford Univ MS(elec eng), 61. *Prof Exp:* Scientist, Lockheed Missile & Space Co, 59-61; engr, Ford Western Develop Labs, 61-63; ENGR, NAT BUR STANDARDS, 63- *Concurrent Pos:* Consult, BDM Corp, 83- *Honors & Awards:* M Barr Carlton Award, Inst Elec & Electronic Engrs, 75. *Mem:* Fel Inst Elec & Electronics Engrs. *Res:* Application of advanced communication techniques for high accuracy transfer of time and frequency information; spread spectrum signals; encryption of data; tracking of satellites; use of satellites for broadcasting and point-to-point distribution; corrections for atmospheric dispersion of radio signals. *Mailing Add:* 735 Jonquil Pl Boulder CO 80302

HANSON, EARL DORCHESTER, b Shahjahanpur, India, Feb 15, 27; m 48; c 3. CELL BIOLOGY, EVOLUTIONARY BIOLOGY. *Educ:* Bowdoin Col, AB, 49; Ind Univ, PhD(zool), 54. *Prof Exp:* From instr to asst prof zool, Yale Univ, 54-60; from assoc prof to prof biol & sci in soc, 60-72, FISK PROF NATURAL SCI, WESLEYAN UNIV, 72- *Concurrent Pos:* Guggenheim & Fulbright fels, 60-61; consult, Off Biol Educ, Am Inst Biol Sci, 68-74, mem gov bd, 70-74; mem, US Del Binational Conf Educ & Res Life Sci, India, 71; mem, Biol Advan Placement Exam Comt, Col Entrance Exam Bd, 72-75, Biol Discipline Comt, 74-76; Fulbright fel, 78-79. *Mem:* AAAS; Soc Protozool; Am Inst Biol Sci; NY Acad Sci. *Res:* Development, genetics and evolution of ciliated protozoa; bioethics, especially with regard to genetic engineering; interactions between humanities and natural sciences. *Mailing Add:* Biol Dept Wesleyan Univ Middletown CT 06457

HANSON, EARLE WILLIAM, b Wheaton, Minn, Oct 18, 10; m 41; c 2. PLANT PATHOLOGY, FIELD CROPS. *Educ:* Univ Minn, BS, 33, MS, 39, PhD(plant path), 41. *Prof Exp:* Asst plant path, Univ Minn, 34-35; pathologist, Bur Plant Indust, Soils & Agr Eng, USDA, Minn, 37-46 & Agr Res Serv, Wis, 46-51; from asst prof to assoc prof, 51-56, plant pathologist, Univ Wis-Exten, 72-76, PROF PLANT PATH, UNIV WIS-MADISON, 56- *Concurrent Pos:* Ed, Phytopath, Am Phytopath Soc, 55-58; mem bd dirs, Am Grassland Coun, 62-65; mem, Univ Wis-US AID Team, Univ Ife, Nigeria, 67-71, chief of party, 68-71. *Mem:* AAAS; Am Phytopath Soc; Am Soc Agron; Crop Sci Soc Am; Am Inst Biol Sci. *Res:* Plant disease diagnosis; legume viruses; crop seed microflora; field crop diseases; breeding for disease resistance. *Mailing Add:* 105 Frigate Dr Madison WI 53705

HANSON, FLOYD BLISS, b Brooklyn, NY, Mar 9, 39; m 62; c 1. APPLIED MATHEMATICS, COMPUTER SCIENCE. *Educ:* Antioch Col, BS, 62; Brown Univ, MS, 64, PhD(eng), 68. *Prof Exp:* Tech asst phys chem, Space Physics Dept, Convair Astronaut, 61; appl mathematician, High Temperature Physics Group, Arthur D Little, Inc, 61; physicist, Plasma Physics Lab, Aeronaut Res Lab, Wright-Patterson AFB, Dayton, Ohio, 62; assoc res scientist, Courant Inst Math Sci, NY Univ, 67-69; from asst prof to assoc prof, 69-83, PROF MATH, UNIV ILL, CHICAGO, 83- *Concurrent Pos:* Proj dir equip grant, NSF, 73, assoc investr res grant, 70-84, prin investr, 88-; fac res partic, Argonne Nat Lab, 85, 86 & 87-88. *Mem:* Soc Indust & Appl Math; Inst Elect & Electronics Engrs; Asn Comput Mach; Resource Modeling Asn. *Res:* Parallel computer computations; application of stochastic differential equations; mathematical biology; kinetic theory, separated flows, laminar boundary layer analysis; asymptotic analysis; integral equations. *Mailing Add:* Dept Math M/C 249 Univ Ill Chicago Box 4348 Chicago IL 60680-6998

HANSON, FRANK EDWIN, b Bryan, Tex, Mar 14, 38; m; c 2. NEUROSCIENCES, BEHAVIORAL PHYSIOLOGY. *Educ:* Univ Iowa, BA, 60; Univ Pa, PhD(zool), 65. *Prof Exp:* Sr asst scientist biol, NIH, 64-66; USPHS fel, Univ Pa, 66-67; asst prof zool, Univ Tex, Austin, 68-72; assoc prof, 72-81, PROF BIOL SCI, UNIV MD BALTIMORE COUNTY, 81- *Mem:* AAAS. *Res:* Insect chemoreceptor physiology; insect feeding behavior; behavioral physiology of bioluminescent communication mechanisms. *Mailing Add:* Dept Biol Sci Univ Md Baltimore County Catonsville MD 21228

HANSON, GAIL G, b Dayton, Ohio, Feb 22, 47; m 68; c 2. ELEMENTARY PARTICLE PHYSICS. *Educ:* Mass Inst Technol, BS, 68, PhD(physics), 73. *Prof Exp:* Res assoc, Stanford Univ, 73-76, staff physicist high energy physics, Stanford Linear Ctr, 76-90; PROF PHYSICS, IND UNIV, 89- *Mem:* Am Phys Soc; AAAS. *Res:* Experimental high energy physics; $e+$ $e-$ colliding beams; jet structures in final state; topologies associated with new particle production; particle detectors; wire chambers; hadron collider physics and detectors. *Mailing Add:* Physics Dept Ind Univ Swain Hall W-117 Bloomington IN 47405

HANSON, GEORGE H(ENRY), b Alpena, Mich, Jan 26, 18; m 45. CHEMICAL & NUCLEAR ENGINEERING. *Educ:* Univ Mich, BSE, 39, MS, 40, PhD(chem eng), 42. *Prof Exp:* Sr chem engr, res div, Phillips Petrol Co, 42-47; chem engr, Monsanto Chem Co & Carbide & Carbon Chem Co, Tenn, 47-48; group leader & asst sect chief, res div, Phillips Petrol Co, 48-51, sr chem & nuclear engr, atomic energy div, 51-54; mem, Rocky Mountain Nuclear Power Study Group, 54-56; sect chief reactor eng sect, atomic energy div, Phillips Petrol Co, 56-60, sr scientist, exp org cooled reactor tech br, 60-63, asst mgr, long range planning, 63-65, sr reactor engr-scientist, 65-69; assoc scientist, Idaho Nuclear Corp, 69-71; assoc scientist, Aerojet Nuclear Corp, 71-76; eng specialist, EG&G Idaho, 76-83; CONSULT, 83- *Concurrent Pos:* Phillips Petrol Co rep, Nuclear Test Ctr Study Group, Washington, DC, 57. *Mem:* AAAS; Am Nuclear Soc; Am Chem Soc; Am Inst Chem Engrs. *Res:* Nuclear reactor physics and engineering, including safety analysis; reactor hydrodynamics; heat transfer and hydraulics; thermodynamics; physical separations; industrial chemical processes. *Mailing Add:* 444 Seventh St Idaho Falls ID 83401

HANSON, GEORGE PETER, b Conde, SDak, July 20, 33; m 58, 69; c 3. PLANT BREEDING. *Educ:* SDak State Univ, BS, 56, MS, 58; Ind Univ, PhD(genetics), 65. *Prof Exp:* Asst prof biol, Thiel Col, 62-65; asst prof bot, Butler Univ, 65-67; asst prof, Eastern Ill Univ, 67-68; biologist, 68-70, SR BIOLOGIST, LOS ANGELES STATE & COUNTY ARBORETUM, 70- *Concurrent Pos:* Consult, Comn Undergrad Educ Biol Sci, 66. *Mem:* AAAS; Am Genetics Soc; Am Inst Biol Sci; Bot Soc Am. *Res:* B-chromosome effects upon crossing over and mutation in maize; ornamental plant breeding; air pollution, especially the effects on plants; guayule breeding and research. *Mailing Add:* Los Angeles Arboretum 301 N Baldwin Ave Arcadia CA 91006

HANSON, GILBERT N, b Minneapolis, Minn, Apr 30, 36; m 63; c 3. PETROLOGY, GEOCHEMISTRY. *Educ:* Univ Minn, BA, 58, MA, 62, PhD(geol), 64. *Prof Exp:* Res assoc geochem, Minn Geol Surv, 64-65; res assoc, Swiss Fed Inst Technol, 65-66; from asst prof to assoc prof, 66-75, PROF GEOCHEM, STATE UNIV NY STONY BROOK, 75-, CHMN, EARTH & SPACE SCI DEPT, 83- *Mem:* AAAS; Am Geophys Union; Geol Soc Am; Geol Soc India; Geochem Soc. *Res:* Application of geochemical data to petrogenetic problems in igneous and sedimentary systems; geochemical and geochronological studies of igneous and metamorphic terranes. *Mailing Add:* Dept Earth & Space Sci State Univ NY Stony Brook NY 11794

HANSON, HAROLD PALMER, b Virginia, Minn, Dec 27, 21; m 44; c 2. GOVERNMENT ADMINISTRATION, CONDENSED MATTER PHYSICS. *Educ:* Superior State Teachers Col, BS, 42; Univ Wis, MS, 44, PhD(physics), 48. *Prof Exp:* Res physicist, Naval Ordnance Lab, USNR, 44-46; WARF fel, Univ Wis, 46-48, vis lectr, 57; from assoc prof to prof, Univ Tex, 54-69, chmn dept physics, 61-69, dir, Ctr Structural Studies, 67-69; provost, Boston Univ, 78-80, Wayne State Univ, 82-84; from instr to assoc prof, Univ Fla, 48-54, dean grad sch, 69-71, vpres acad affairs, 71-74, exec vpres, 74-78, ADJ PROF PHYSICS & EMER EXEC VPRES, UNIV FLA, 90- *Concurrent Pos:* Res physicist, Lincoln Lab, Mass Inst Technol, 53; Fulbright res fel, Univ Oslo, 60-61, NAVF fel, 68; consult, Gen Atomic, 64; exec dir, US House Rep Comt Sci, Space & Technol, 80-82 & 84-90; chmn, Sr Res Adv Bd, Fla Inst Technol. *Honors & Awards:* St Olav Medal, 76; Franklin Medal, NSF, 90. *Mem:* Fel Am Phys Soc; Am Asn Physics Teachers; AAAS; Sigma Xi. *Mailing Add:* Dept Physics Univ Fla Gainesville FL 32611

HANSON, HARRY THOMAS, b Chicago, Ill, May 25, 39; m 65; c 2. POLYMER SCIENCE, ORGANIC CHEMISTRY. *Educ:* Univ Notre Dame, BSc, 61; Wayne State Univ, PhD(org chem), 65. *Prof Exp:* Res chemist, 65-70, SR RES CHEMIST, CELANESE RES CO, 70- *Mem:* Am Chem Soc; AAAS; NY Acad Sci. *Res:* Organic polymer synthesis, characterization and utilization, particularly in coatings field; computer applications in polymer chemistry. *Mailing Add:* 53 Maple St Millburn NJ 07041

HANSON, HARVEY MYRON, b Akron, Ohio, Mar 24, 31; m 53; c 2. THEORETICAL PHYSICS. *Educ:* Univ Akron, BS, 52; Ohio State Univ, MSc, 54, PhD(physics), 56. *Prof Exp:* Res fel physics, Ohio State Univ, 56-57; asst prof, Univ Akron, 57-60; assoc prof, Southwestern at Memphis, 60-65; PROF PHYSICS WRIGHT STATE UNIV, 65- *Concurrent Pos:* Chmn dept physics, 65-77. *Res:* Theoretical and experimental aspects of molecular spectra in the infrared region. *Mailing Add:* Dept Physics Wright State Univ Colonel Glenn Hwy Dayton OH 45435

HANSON, HENRY W A, III, b Hagerstown, Md, Aug 2, 32; m 54; c 2. SEDIMENTOLOGY. *Educ:* Univ Alaska, BS, 60; Pa State Univ, MS, 65, PhD(geol), 68. *Prof Exp:* Asst prof, 66, ASSOC PROF GEOL, DICKINSON COL, 66- *Mem:* Geol Soc Am; Soc Econ Paleontologists & Mineralogists; Sigma Xi. *Res:* Environmental system and processes of barrier islands. *Mailing Add:* Dept Geol Dickinson Col Carlisle PA 17013

HANSON, HIRAM STANLEY, b Whitesburg, Ga, July 26, 23; m 47; c 1. GEOLOGY, GEOCHEMISTRY. *Educ:* Emory Univ, BA, 46, MS, 49, MA, 59; Univ Ariz, PhD, 66. *Prof Exp:* Asst chemist, Bell Aircraft Corp, 44-45; res inorg chemist, Tenn Corp, 47-48; asst prof chem, Middle Ga Col, 48-49; instr sci, Reinhardt Col, 49-54; asst geol, Emory Univ, 54-56, instr chem & geol, 56-58; asst prof, Sul Ross State Col, 58-61; asst geol, Univ Ariz, 61-64; asst prof, 64-67, assoc prof, 67-75, PROF GEOL, GA SOUTHERN COL, 75- HEAD DEPT, 69- *Concurrent Pos:* Shell merit fel, 70. *Mem:* Geol Soc Am; Mineral Soc Am; Nat Asn Geol Teachers; Geochem Soc; Sigma Xi. *Res:* Igneous and metamorphic petrology and petrography. *Mailing Add:* Geol & Geog Dept Ga Southern Col Statesboro GA 30460

HANSON, HOWARD GRANT, b Nelson, Minn, Jan 22, 20; m 45; c 4. PHYSICS. *Educ:* St Cloud State Teachers Col, BS, 43; Univ Wis, PhD(physics), 48. *Prof Exp:* From asst prof to prof physics, Univ Minn, Duluth, 47-85, head dept, 50-85; RETIRED. *Concurrent Pos:* NSF fac fel, Univ Stockholm, 63. *Mem:* Am Phys Soc; Am Asn Physics Teachers. *Res:* Quenching and exciting of fluorescence by different gases; spectra of scintillating crystals; tesselations of random sphere aggregations. *Mailing Add:* Dept Physics Univ Minn Duluth MN 55812

HANSON, HUGH, b Lewis, Kans, Dec 16, 15; m 41. ANIMAL ECOLOGY. *Educ:* Kans State Teachers Col, BS, 39; Univ Ill, MS, 41, PhD(zool), 48. *Prof Exp:* From asst prof to assoc prof, 48-65, chmn dept, 59-62, prof, 65-78, EMER PROF ZOOL, ARIZ STATE UNIV, 78- *Mem:* Sigma Xi. *Mailing Add:* 1025 E 11th Ave Emporia KS 66801-3215

HANSON, JAMES CHARLES, b Modesto, Calif, Jan 6, 31; m 56; c 2. CELL BIOLOGY, DEVELOPMENTAL BIOLOGY. *Educ:* Univ of the Pac, BA, 55, MA, 57; Ore State Univ, PhD, 67. *Prof Exp:* Chmn, Div Sci & Math, 60-62, Dept Biol Sci, 60-66 & 71-74, assoc prof biol sci, 68-72, PROF BIOL SCI, STANISLAUS STATE COL, 72- *Res:* Early development and cell division in marine invertebrates, especially the fertilization process and the first mitotic period. *Mailing Add:* Dept Zool Calif State Col Stanislaus 801 W Monte Vista Ave Turlock CA 95380

HANSON, JAMES EDWARD, mathematics; deceased, see previous edition for last biography

HANSON, JAMES EDWARD, b Wichita, Kans, Mar 25, 62. TECHNICAL STAFF ADMINISTRATION. *Educ:* Tex Christian Univ, BS, 84; Calif Inst Technol, PhD(chem), 90. *Prof Exp:* MEM TECH STAFF, AT&T BELL LABS, 89- *Mem:* Sigma Xi; Am Chem Soc. *Res:* Design and synthesis of new materials, primarily polymers for microelectronics manufacturing. *Mailing Add:* AT&T Bell Labs Rm 1A-232 600 Mountain Ave Murray Hill NJ 07974-2070

HANSON, JERRY LEE, b Belmond, Iowa, Aug 6, 32; m 54; c 3. ELECTRICAL ENGINEERING. *Educ:* St Olaf Col, BA, 54; US Air Force Inst Technol, BS, 61; Univ Ill, MS, 62, PhD(elec eng), 70. *Prof Exp:* From instr to assoc prof, US Air Force Acad, 62-67, from asst prof to assoc prof, USAF Inst Technol, 70-74, chief, Electronics Div, Foreign Technol Div, Air Force Systs Command, 74-75; assoc prof physics, 77-83, RES ASSOC, AUGUSTANA RES INST, 75-; CHMN COMPUT SCI DEPT, AUGUSTANA COL, 83- *Mem:* Inst Elec & Electronics Engrs. *Res:* Linear systems; circuits; digital systems; microprocessor and microcomputer systems. *Mailing Add:* Dept Physics/Comput Sci Augustana Col Gilbert Sci Ctr 204 Sioux Falls SD 57105

HANSON, JOE A, b Los Angeles, Calif, July 8, 28; m 87; c 2. RESOURCE MANAGEMENT, SCIENCE WRITING. *Educ:* Univ Calif, Los Angeles, BA, 56. *Prof Exp:* Mem tech staff, Systs Develop Corp, Rand Corp, 58-59, Thomson-Ramo-Wooldridge, 59-62 & Hughes Ground Systs Group, 62-63; mem sr staff, Planning Res Corp, 63-68, dir Hawaii Opers, 68-70; sr syst scientist, Oceanic Inst, 70-77; mem tech staff, 77-86, TECH MGR, JET PROPULSION LAB, CALIF INST TECHNOL, 87- *Concurrent Pos:* Co-dir, Hawaii Environ Simulation Lab, 71-72; task force chmn, Gov Comn Alternate Energy Sources Hawaii, 74; exec secy, Seaward Advan Exec Action Group, 74-76. *Mem:* Marine Technol Soc; Sigma Xi; World Maricult Soc; NY Acad Sci; AAAS. *Res:* Systems approaches to the design and development of resource management methods and technology which recycle resources, are synergistic with natural dynamics and are solar powered; self-sustaining experimental ecosystems; ecology life support systems for extended space habitation. *Mailing Add:* 1865 Midick Altadena CA 91001

HANSON, JOHN BERNARD, b Denver, Colo, Mar 24, 18; m 43; c 3. PLANT PHYSIOLOGY. *Educ:* Univ Colo, BA, 48; State Col Wash, MS, 50, PhD(bot), 52. *Prof Exp:* Asst, State Col Wash, 48-51; Nat Res Coun fel, Calif Inst Technol, 52-53; from asst prof to prof, 53-85, head dept bot, 67-77, EMER PROF AGRON & PLANT BIOL, UNIV ILL, URBANA, 85- *Concurrent Pos:* Fulbright res scholar, Waite Agr Exp Sta, Australia, 59-60; NATO sr fel, Univ East Anglia, 68. *Honors & Awards:* Barnes Award, Am Soc Plant Physiol, 80, Gude Award, 89. *Mem:* Am Soc Plant Physiologists (pres, 73-74). *Res:* Mineral absorption; respiration. *Mailing Add:* Plant Biol Univ Ill 289 Morrill Hall Urbana IL 61801

HANSON, JOHN ELBERT, b Toledo, Ohio, Mar 5, 35; m 59; c 2. INORGANIC CHEMISTRY. *Educ:* Olivet Nazarene Col, AB, 57; Purdue Univ, PhD(chem), 64. *Prof Exp:* From asst prof to assoc prof, 61-70, PROF CHEM, OLIVET NAZARENE COL, 70-, CHMN DEPT, 71- *Concurrent Pos:* Res assoc, Univ Chicago, 74; vis prof, Univ Wis, 84-85. *Mem:* Am Chem Soc. *Res:* Organosilicon chemistry; coordination compounds; bioinorganic chemistry. *Mailing Add:* Olivet Nazarene Univ Kankakee IL 60901

HANSON, JOHN M, b Brookings, SDak, Nov 16, 32; m 60; c 4. STRUCTURAL ENGINEERING, FAILURE ANALYSIS. *Educ:* SDak State Univ, BS, 53; Iowa State Univ, MS, 57; Lehigh Univ, PhD(civil eng), 64. *Prof Exp:* Jr engr, Boeing Airplane Co, 53; designer-detailer, Sverdrup & Parcel, Inc, 55-56; engr, J T Banner & Assocs, 57-58 & Phillips-Carter-Osborn, Inc, 58-60; res instr civil eng, Lehigh Univ, 60-64, asst res prof, 64-65; sr develop engr & prin res engr, Portland Cement Asn, 65-69, asst mgr struct res sect, 69-72; vpres opers, 72-79, PRES, WISS, JANNEY, ELSTNER & ASSOCS, 79- *Honors & Awards:* State-of-the-Art Civil Eng Award, Am Soc Civil Engrs, 74, Raymond C Reese Award, 76, T Y Lin Award, 79; Distinguished Serv Award, Am Concrete Inst, 76; Martin P Korn Award, Prestressed Concrete Inst, 78. *Mem:* Am Soc Civil Engrs; Am Concrete Inst; Prestressed Concrete Inst; Earthquake Eng Res Inst; Int Asn Bridge & Struct Eng. *Res:* Behavior and ultimate strength, particularly shear strength of concrete beams; structural concrete members in general. *Mailing Add:* Wiss Janney Elstner Assocs 330 Pfingsten Rd Northbrook IL 60062-2095

HANSON, JOHN SHERWOOD, b New York, NY, May 25, 27; m 52; c 4. MEDICINE, PHYSIOLOGY. *Educ:* Yale Univ, BA, 50; NY Univ, MD, 54. *Prof Exp:* Intern, Mary Fletcher Hosp, Burlington, Vt, 54-55, resident internal med, 55-57; from instr to assoc prof, 58-71, PROF INTERNAL MED, COL MED, UNIV VT, 71- *Concurrent Pos:* Res fel, Nat Heart Inst, 57-58; Am Philos Soc Daland med res fel, 59-61; Nat Heart Inst spec res fel, 61-63, res career prog award, 63-74; mem, New Eng Regional Res Rev Comt, Am Heart Asn. *Mem:* Am Phys Soc; Am Fedn Clin Res; Am Col Physicians; NY Acad Sci; Sigma Xi. *Res:* Cardiopulmonary hemodynamics; exercise physiology; computer applications to cardiopulmonary physiology; congenital and acquired cardiac disease; pulmonary disease. *Mailing Add:* Cardiopulmonary Lab Mary Fletcher Hosp Burlington VT 05401

HANSON, JONATHAN C, b Chicago, Ill, Sept 11, 41; m 69; c 2. COMPUTER AUTOMATION, COMPUTER GRAPHICS. *Educ:* Northwestern Univ, BS, 63; Univ Mich, Ann Arbor, PhD(chem), 69. *Prof Exp:* Res assoc chem, Univ Wash, 69-70, res assoc biol structure, 70-73; res assoc biophysics, Johns Hopkins Univ, 73-77; res assoc, 77-79, COMPUT ANALYST, BROOKHAVEN NAT LAB, 79- *Mem:* Am Crystallog Asn. *Res:* Automation of physical chemical experiments, crystallographic computing and molecular graphics; molecular interactions from an experimental and theoretical standpoint. *Mailing Add:* Dept Chem Bldg 555A Brookhaven Nat Lab Upton NY 11973

HANSON, KENNETH MARVIN, b Brookings, SDak, Feb 21, 27; c 2. PHYSIOLOGY. *Educ:* Sioux Falls Col, BS, 51; Ind Univ, MS, 63, PhD(physiol), 65. *Prof Exp:* Teaching asst biol, Augustana Col, 45-49; lab technician, Univ Minn, 54-57; res asst physiol, Ind Univ, 57-63, res assoc, 63-65, instr, 65-66; asst prof, 66-71, assoc prof, 71-76, PROF PHYSIOL, OHIO STATE UNIV, 76- *Mem:* Am Physiol Soc; Sigma Xi; Soc Exp Biol & Med; Am Heart Asn. *Res:* Regional blood flow; vascular physiology; liver function and blood flow; gastrointestinal physiology. *Mailing Add:* Dept Physiol Ohio State Univ Col Med 333 W 10th Ave Columbus OH 43210

HANSON, KENNETH MERRILL, b Mt Vernon, NY, Apr 17, 40; div; c 2. RADIOGRAPHIC IMAGING, TOMOGRAPHIC RECONSTRUCTION. *Educ:* Cornell Univ, BEng Phys, 63; Harvard Univ, MS, 67, PhD(physics), 70. *Prof Exp:* Res assoc nuclear physics, Cornell Univ, 70-75; STAFF MEM IMAGE SCI, LOS ALAMOS NAT LAB, 75- *Concurrent Pos:* Consult, tomographic reconstruction. *Honors & Awards:* Nuclear Weapons Prog Award of Excellence, 87. *Mem:* Am Phys Soc; sr mem Inst Elec & Electronics Engrs. *Res.* Industrial and medical imaging; tomographic reconstruction and image analysis; detection of signals in radiographic images; interactive image processing and image display techniques. *Mailing Add:* Los Alamos Nat Lab MS-P940 Los Alamos NM 87545

HANSON, KENNETH RALPH, b Birmingham, Eng, Sept 7, 30; m 58; c 2. BIOCHEMISTRY. *Educ:* Univ Liverpool, BSc, 51, PhD(org chem), 54. *Prof Exp:* Fel enzym, Nat Res Coun Can, 56-58; Jane Coffin Childs fel biochem, Sch Med, NY Univ, 58-60; from asst biochemist to biochemist, 60-74, SR BIOCHEMIST, CONN AGR EXP STA, 74- *Concurrent Pos:* Lectr, Japan Soc Promotion Sci, 80. *Mem:* Am Soc Biol Chem; Am Soc Plant Physiol; Am Chem Soc; Phytochem Soc NAm (pres, 72-73). *Res:* Stereochemistry and mechanisms of enzymatic reactions; aromatic amino acids and plant phenols and their biosynthesis and regulation; stereochemical concepts; photosynthetic metabolism; use of stereospecifically tritiated glyceric acid to study loss during photorespiration of CO2 fixed by photosynthesis in C-3 plants (such as wheat and tobacco); isolation and characterization of a starchless mutant of Nicotiana Sylvestris in which plastid phosphoglucomutase is altered. *Mailing Add:* Dept Biochem & Genet Conn Agr Exp Sta PO Box 1106 New Haven CT 06504

HANSON, KENNETH WARREN, b Graceville, Minn, July 15, 22; m 46; c 4. HORTICULTURE. *Educ:* Univ Minn, BS, 48, MS, 51, PhD(hort), 52. *Prof Exp:* Field agent fruit res, USDA, Univ Minn, 42-43 & 46-51; asst horticulturist, Univ Ga, 52, assoc horticulturist, 52-54; asst prof pomol, Geneva Agr Exp Sta, Cornell Univ, 54-60; assoc prof hort, Am Univ, Beirut, 60-63; dir & horticulturist, Mo State Fruit Exp Sta, 63-84; RETIRED. *Concurrent Pos:* Consult, Int Exec Serv Corps, 85- *Mem:* AAAS; Am Soc Hort Sci; Am Inst Biol Sci; Sigma Xi. *Res:* Pomology; the creation of new varieties of fruits through breeding. *Mailing Add:* 227 Royal St Gulfport MS 39503

HANSON, LESTER EUGENE, b Willmar, Minn, Mar 31, 12; m 37; c 3. ANIMAL SCIENCE. *Educ:* Univ Minn, BS, 36; Cornell Univ, MSA, 37, PhD(animal husb), 40. *Prof Exp:* Asst animal husb, Cornell Univ, 36-40; from instr to prof, Univ Nebr, 40-50; prof animal husb, Univ Minn, St Paul, 50-80, head dept, 56-66. *Concurrent Pos:* Mem, Cult Res Team Animal Husb for USDA, USSR, 59; Ford Found consult, Chile, 63. *Honors & Awards:* Am Feed Mfrs Award, 55; Distinguished Nutritionist Award, Distillers Feed Res Coun, 64. *Mem:* AAAS; Am Soc Animal Sci (vpres, 61-62, pres, 62-63); Am Inst Nutrit; Am Inst Biol Sci; Am Dairy Sci Asn. *Res:* Nutrition and management of swine. *Mailing Add:* 1413 Idaho Ave St Paul MN 55108

HANSON, LOUISE I KARLE, b Washington, DC, Aug 26, 46; m 69; c 1. BIOPHYSICS, QUANTUM CHEMISTRY. *Educ:* Univ Mich, BS, 67, MS, 69; Univ Wash, PhD(chem), 73. *Prof Exp:* Fel & res chemist, NIH, 73-77; sr res assoc, 77-78, asst chemist, 78-80, ASSOC CHEMIST, BROOKHAVEN NAT LAB, 80- *Concurrent Pos:* Prof lectr, Chem Dept, The Am Univ, Washington, 77. *Mem:* Am Chem Soc; Biophys Soc. *Res:* Elucidation of the electronic structure of biological chromophores (chlorophyll, heme derivatives and flavins) both in vivo and model systems, and the role of these compounds in mediating photosynthetic and biochemical processes. *Mailing Add:* Bldg 815 Brookhaven Nat Lab Upton NY 11973

HANSON, LYLE EUGENE, b Sarona, Wis, Oct 2, 20; m 45; c 4. VETERINARY MICROBIOLOGY. *Educ:* Northland Col, PhB, 42; Mich State Univ, DVM, 50; Univ Ill, MS, 53, PhD(vet path), 57. *Prof Exp:* Veterinarian, Wis Dept Agr, 50; from instr to assoc prof, 51-61, head dept, 67-79, prof vet path & hyg, Col Vet Med, 61-, assoc dean res, 79-85, EMER PROF, UNIV ILL, URBANA, 85- *Mem:* Am Vet Med Asn; Am Asn Avian Path; US Animal Health Asn; Am Col Vet Microbiol. *Res:* Animal virology, microbiology. *Mailing Add:* Dept Vet Path Univ Ill VMBS Bldg Urbana IL 61801

HANSON, MARVIN WAYNE, b Longstreet, La, May 12, 28; m 65, 81; c 3. ANALYTICAL CHEMISTRY. *Educ:* Centenary Col, BS, 50; Univ Houston, MS, 53, PhD(chem), 64. *Prof Exp:* Asst prof chem, McNeese State Col, 54-55 & 57-59; assoc prof, Centenary Col La, 59-71, chmn dept, 68-78, prof, 71-78; prof chem, Monroe Community Col, 78-79; WITH BAYOU STATE OIL CORP, 79- *Mem:* Am Chem Soc; Am Inst Chemists; Nat Asn Corrosion Engrs. *Res:* Analytical chemistry; spectroscopy and spectrophotometry; qualitative organic and reaction mechanisms. *Mailing Add:* Bayou State Oil Co PO Box 158 Hosston LA 71043

HANSON, MERLE EDWIN, b Oakes, NDak, July 8, 34; m 58; c 2. PHYSICS, ENGINEERING. *Educ:* NDak State Univ, BS, 61, MS, 62; NMex Inst Mining & Technol, PhD(physics), 73. *Prof Exp:* Teaching asst eng, NDak State Univ, 61-62; res engr mech eng, NAm Aviation, 62-65; sr res engr continuum mech, NMex Inst Mining & Technol, 65-73; Physicist, Lawrence Livermore Lab, 73-82; dir res & develop, Hunter Geophys, 82-84; vpres, Comdisco Resources, 84-89; VPRES, DUNCAN ENERGY CO, 89- *Mem:* Soc Petrol Engrs. *Res:* Fracture mechanics and continuum mechanics; early time detonation phenomena and interaction of explosives with other material; reservoir stimulation phenomena. *Mailing Add:* 1134 Geneva Livermore CA 94550

HANSON, MERVIN PAUL, b Skellytown, Tex, Sept 14, 37; m 64; c 2. PHYSICAL CHEMISTRY. *Educ:* Humboldt State Col, BS, 61; Cornell Univ, PhD(chem), 66. *Prof Exp:* PROF CHEM, HUMBOLDT STATE COL, 65- *Mem:* Am Chem Soc. *Res:* Irreversible thermodynamics. *Mailing Add:* Dept Chem Humboldt State Col Arcata CA 95521

HANSON, MILTON PAUL, b Mitchell, SDak, Aug 22, 38; m 61; c 2. CHEMISTRY. *Educ:* Augustana Col, SDak, BA, 60; Rice Univ, PhD(org chem), 64. *Prof Exp:* from asst prof to assoc prof, 64-78, PROF CHEM, AUGUSTANA COL, SDAK, 78- *Concurrent Pos:* Vis lectr, Univ Ill, Urbana-Champaign, 69-70; vis assoc prof, Univ Ariz, 78; post doctoral assoc, Univ Ga, Athens, Ga, 85-86. *Mem:* Am Chem Soc; Sigma Xi. *Res:* Organic chemistry; reaction mechanisms; polymers; quantative structure-property relationships. *Mailing Add:* Dept Chem Augustana Col Sioux Falls SD 57197

HANSON, MORGAN A, b Wowan, Australia, Jan 28, 30; m 54; c 5. OPERATIONS RESEARCH, STATISTICS. *Educ:* Univ Queensland, BSc, 52; Univ Melbourne, MSc, 55; Univ New South Wales, PhD(statist), 64. *Prof Exp:* Physicist, Imp Chem Industs, Ltd, 55; mathematician, Lysaght's Works Ltd, Australia, 56; opers res analyst, Australian Gas Light Co, 56-58; from lectr to sr lectr math & statist, Univ New South Wales, 58-64; vis assoc prof statist, Fla State Univ, 65; from assoc prof to prof math, Queen's Univ, Ont, 66-68; PROF STATIST, FLA STATE UNIV, 68- *Concurrent Pos:* Can Nat Res Coun res grants math prog, 66-68; consult, Aerospace Res Labs, USAF, 66-; Off Naval Res grant, weather modification, 80; vis fel, La Trobe Univ, Australia, 80. *Honors & Awards:* Stevenson Prize for Maths, Queensland Univ, 60. *Mem:* Am Math Soc; Math Prog Soc. *Res:* Mathematical programming; approximation theory; economic optimization; control theory; calculus of variations. *Mailing Add:* Dept Statist Fla State Univ Tallahassee FL 32306

HANSON, NORMAN WALTER, b Manistique, Mich, Jan 18, 25; m 51; c 2. STRUCTURAL ENGINEERING, APPLIED MECHANICS. *Educ:* Univ NMex, BS, 50; Kans State Univ, MS, 53. *Prof Exp:* Test engr aircraft struct, NAm Aviation, 51-52; res asst struct, Kans State Univ, 52-53; asst to sr develop engr, 53-68, prin struct engr, 68-86, PRIN ENGR, CONSTRUCT TECHNOL LABS, PORTLAND CEMENT ASN, 86- *Mem:* Fel Am Concrete Inst; Am Soc Civil Engrs; Instrument Soc Am. *Res:* Structural reinforced concrete; seismic resistance; masonry buildings; testing methods; instrumentation. *Mailing Add:* 3026E Dodgelake Rd Manistique MI 49854

HANSON, PAUL ELIOT, b Dawson, Minn, Oct 19, 53; m 87. TAXONOMY, BIOLOGICAL CONTROL. *Educ:* Univ Minn, Morris, BA, 75; Univ Minn, St Paul, MS, 79; Ore State Univ, PhD(entom), 87. *Prof Exp:* PROF BIOL CONTROL, UNIV COSTA RICA, 87- *Mem:* Pac Coast Entom Soc. *Res:* Biological control of insects; taxonomy of parasite Hymenoptera. *Mailing Add:* Dept Biol Univ Costa Rica San Pedro San Jose Costa Rica

HANSON, PER ROLAND, b Medford, Mass, Aug 30, 12; m 40; c 2. ELECTRONICS, ELECTRICAL ENGINEERING. *Educ:* Mass Inst Technol, BS, 35; Northeastern Univ, MS, 59. *Prof Exp:* Receiving tube engr, Champion Radio Works, 35-39; receiving tube engr, Raytheon Mfg Co, Newton, 39-42, mgr eng & mfg groups, Power Tube Div, Waltham, 42-61; mgr eng & mfg groups, Bomac Div, Varian Assoc, Beverly, 61-66; prin engr, Spec Microwave Devices Oper, Raytheon Co, Burlington, 66-77, consult, 78-82; consult, 82-87; RETIRED. *Mem:* Inst Elec & Electronics Eng. *Res:* Microwave tubes and components engineering and manufacturing. *Mailing Add:* Seven Wadleigh Point Rd Kingston NH 03848

HANSON, PETER, b Appleton, Wis, Oct 9, 35. EXERCISE PHYSIOLOGY, CARDIOLOGY. *Educ:* Univ Wis-Madison, BS, 58; Univ Ill, Urbana, MS, 62; Univ NMex, MD, 71. *Prof Exp:* PROF, MED SCH, UNIV WIS, 74-, CLIN DIR, BIODYNAMICS LAB, 78- *Mem:* Am Physiol Soc; Am Col Sports Med; Am Heart Asn; Am Fedn Clin Res. *Res:* Physiological responses to exercise in cardiovascular disease. *Mailing Add:* Dept Med Univ Wis Sch & Hosp 600 Highland Ave Madison WI 53792

HANSON, RICHARD STEVEN, b Platte, SDak, Nov 14, 35; m 56; c 3. BIOCHEMISTRY, MICROBIOLOGY. *Educ:* SDak State Univ, BS, 59; Univ Ill, PhD(microbiol), 62. *Prof Exp:* Nat Acad Sci fel microbiol chem, Pioneering Labs, Northern Regional Res Labs, Agr Res Serv, USDA, 62-63; USPHS fel biochem, Enzymol Lab, Nat Ctr Sci Res, Gil Sur Yvette, France, 63-64; asst prof biochem, Med Ctr, Univ Ill, Chicago, 64-65; from asst prof to prof bact, Univ Wis-Madison, 65-81, chmn dept bact, 73-76; DIR, GRAY FRESHWATER BIOL RES INST, UNIV MINN, 81- *Honors & Awards:* Res Award, Charles Lindberg Found, 84. *Mem:* AAAS; Am Soc Microbiol; Fedn Biol Chem. *Res:* Regulation of enzyme synthesis in bacteria and the biochemistry of sporulation and its control in bacteria; chemolithotrophic bacteria; biochemistry, ecology and genetics of methanotrophic bacteria; biological degradation of xenobiotic chemicals. *Mailing Add:* 4162 Hillcrest Rd Wayzata MN 55391

HANSON, RICHARD W, b Oxford, NY, Nov 10, 35; m 61; c 3. BIOCHEMISTRY. *Educ:* Northeastern Univ, BS, 59; Brown Univ, MS, 61, PhD(biol), 63. *Prof Exp:* From asst prof to prof biochem, Fels Res Inst, Med Sch, Temple Univ, 75-77; PROF BIOCHEM, CASE WESTERN RESERVE UNIV, 78-, CHMN DEPT, 78- *Concurrent Pos:* USPHS fel, 65-66; Nat Inst Arthritis & Metab Dis career develop award, 70-74; mem, Biochem Study Sect, NIH, 74-78; mem, Comn Life Sci, Nat Res Coun; assoc ed, J Biol Chem, 85- *Honors & Awards:* Mead Johnson Award, Am Inst Nutrit, 71; Kaiser-Permanente Award, 82. *Mem:* AAAS; Biochem Soc; Am Inst Nutrit; Am Soc Biochem & Molecular Biol; Am Soc Microbiol. *Res:* Hormonal regulation of gene expression; development of metabolic processes; regulation of gluconeogenesis in mammalian liver and kidney. *Mailing Add:* Dept Biochem Case Western Reserve Univ Sch Med Cleveland OH 44106

HANSON, ROBERT BRUCE, b Minneapolis, Minn, Jan 26, 47; m 71; c 2. ASTROMETRY, STATISTICAL ASTRONOMY. *Educ:* Carleton Col, BA, 67; Univ Calif, Santa Cruz, PhD(astron), 74. *Prof Exp:* Lectr astron, Dept Astron, Yale Univ, 75-76, res staff astromr, 76-78; sr res fel, Royal Greenwich Observ, Herstmonceux, Eng, 78-80; asst res astromr, 80-87, ASSOC RES ASTRONR, LICK OBSERV, UNIV CALIF, SANTA CRUZ, 87- *Honors & Awards:* Trumpler Award, Astron Soc Pac, 76. *Mem:* Am Astron Soc; Astron Soc Pac; Sigma Xi; Royal Astron Soc; Int Astron Union. *Res:* Astrometry: proper motions and parallaxes; statistical astronomy: stellar luminosities and kinematics; stellar populations; galactic structure; cosmic distance scale. *Mailing Add:* Lick Observ Univ Calif Santa Cruz CA 95064

HANSON, ROBERT C, b Berwyn, Ill, Aug 13, 44; m; c 2. AUTONOMIC PHARMACOLOGY. *Educ:* Cornell Col, BA, 66; De Paul Univ, MS, 70; Univ Mo, PhD(physiol), 73; Univ Evansville, MBA, 83. *Prof Exp:* Sr res assoc, pharmaceut res & develop, cardiovasc pharmacol, Bristol-Myers, 78-85; sect head autonomic pharmacol, 85-87, MGR LIC & RES PLANNING, NOVA PHARMACEUT CORP, 87- *Mem:* Am Physiol Soc; Int Soc Nephrology; Am Soc Nephrology; Am Heart Asn; Am Soc Zoologists; Licensing Execs Soc. *Mailing Add:* Nova Pharmaceut Corp 6200 Freeport Ctr Baltimore MD 21224

HANSON, ROBERT D(UANE), b Albert Lea, Minn, July 27, 35; m 59; c 2. CIVIL & EARTHQUAKE ENGINEERING. *Educ:* Univ Minn, BS, 57, MSCE, 58; Calif Inst Technol, PhD(civil eng), 65. *Prof Exp:* Asst prof civil eng, Univ NDak, 59-61; asst prof eng, Univ Calif, Davis, 65-66; from asst prof to assoc prof, 66-74, chmn dept, 76-84, PROF CIVIL ENG, UNIV MICH, ANN ARBOR, 74- *Concurrent Pos:* UNESCO expert, Int Inst Seismol & Earthquake Eng, Tokyo, 70-71; div dir, NSF, Washington, DC. *Honors & Awards:* Reese Res Award, Am Soc Civil Engrs, 80. *Mem:* Nat Acad Eng; Earthquake Eng Res Inst; Am Soc Civil Engrs; Am Concrete Inst. *Res:* Structural dynamics; earthquake resistant design; building failure and repair; mechanical damping systems for buildings. *Mailing Add:* Dept Civil Eng Univ Mich Ann Arbor MI 48109-2125

HANSON, ROBERT HAROLD, b Minn, Aug 12, 18; m 49; c 2. STATISTICS, MATHEMATICS. *Educ:* Luther Col, BA, 40; Univ Iowa, MS, 41. *Prof Exp:* Statistician, Census Bur, US Dept Com, 46-78; consult, Westat Inc, Rockville, Md, 76-84; RETIRED. *Honors & Awards:* Gold Medal, US Dept Com. *Mem:* Fel Am Statist Asn; fel AAAS; Int Asn Surv Statisticians. *Res:* Probability sampling; statistical methods; survey methodology. *Mailing Add:* 6240 White Oak Dr Frederick MD 21701

HANSON, ROBERT JACK, b LaPorte, Ind, May 18, 22; m 49; c 3. MICROBIOLOGY. *Educ:* Valparaiso Univ, AB, 48; Univ Ill, MS, 49, PhD(microbiol), 55. *Prof Exp:* Bacteriologist, Stine Labs, E I du Pont de Nemours & Co, 49-50; instr microbiol, Col Med, Univ Ill, 54-56; PROF BIOL, VALPARAISO UNIV, 56- *Concurrent Pos:* Chief bacteriologist, Methodist Hosp, Ind, 56-60; partic, NSF-AEC Prog, Inst Radiation Biol, Cornell Univ, 66-67; consult, Northern Labs, Inc, Valparaiso, Ind. *Mem:* Am Soc Microbiol; Am Pub Health Asn; Am Inst Biol Sci; Sigma Xi. *Res:* Phagocytosis of myxoviruses; host-parasite relationships. *Mailing Add:* 807 McCord Rd Valparaiso IN 46383

HANSON, ROBERT PAUL, epizootiology; deceased, see previous edition for last biography

HANSON, ROGER BRIAN, b Southgate, Calif, June 12, 43; m 68; c 2. MARINE MICROBIOLOGY. *Educ:* Univ Calif, Los Angeles, BA, 67; Calif State Univ, Long Beach, MS, 70; Univ Hawaii, PhD(microbiol), 74. *Prof Exp:* Res assoc marine microbiol, Univ Ga Marine Inst, 74-77; asst prof, 76-80, assoc prof, 80-88, PROF OCEANOG, SKIDAWAY INST OCEANOG, 88- *Concurrent Pos:* NSF Adv Panel, 85, prog mgr, Div Polar Prog, 88; adj assoc prof, Univ Ga. *Mem:* Am Soc Limnol & Oceanog; Am Soc Microbiol. *Res:* Microbiology and nitrogen cycling in coastal and oceanic ecosysystems; bacterioplankton growth in oceanic waters. *Mailing Add:* Skidaway Inst Oceanog PO Box 13687 Savannah GA 31416

HANSON, ROGER JAMES, b Hutchinson, Minn, Oct 27, 27; m 50; c 4. MUSICAL ACOUSTICS, NUCLEAR SPECTROSCOPY. *Educ:* Gustavus Adolphus Col, BS, 50; Univ Nebr, MA, 53, PhD(physics), 56. *Prof Exp:* From asst prof to prof physics, Grinnell Col, 56-69; head dept physics, 69-80, PROF PHYSICS, UNIV NORTHERN IOWA, 69- *Concurrent Pos:* NSF fac fel, Harvard Univ, 61-62; res assoc, Inst Physics, Aarhus Univ, 66-67. *Mem:* Acoust Soc Am; Catgut Acoust Soc; Am Asn Physics Teachers; Sigma Xi. *Res:* Musical acoustics. *Mailing Add:* Dept Physics Univ Northern Iowa Cedar Falls IA 50614

HANSON, ROGER WAYNE, b Minn, Sept 3, 22; m 48; c 2. BIOLOGY. *Educ:* Univ Iowa, BA, 46, MA, 48; Univ Calif, Los Angeles, PhD, 52. *Prof Exp:* Asst zool, Univ Iowa, 47-48; asst, Univ Calif, Los Angeles, 48-52; lectr biol, Univ Calif, Santa Barbara, 52-53; from instr to assoc prof, Med Col, Univ Ala, Birmingham, 54-67, chmn, Div Natural Sci & Math, Univ Col, 67-73, prof pharmacol, Med Col & prof biol, Univ Col, 67-81, dean, Sch Natural Sci & Math, Univ Col, 73-81; dir & coordr curricula, 81-85, EMER PROF, BASIC ALLIED HEALTH SCI, UNIV ALA, 85- *Concurrent Pos:* NSF fel, Med Col, Univ Ala, Birmingham, 53-54, USPHS fel, 55-57; NIH fel, La State Univ Inter-Am Prog Trop Med, Cent Am, 61; vis assoc prof, Med Col, Univ Calif, San Francisco, 64-65; contract assoc prof, Sch Med, Univ Calif, San Francisco-AID Proj, Indonesia. *Mem:* AAAS; Am Physiol Soc; Nat Sci Teachers Asn; Am Asn Allied Health Prof. *Res:* Intermediary metabolism and metabolic effect of endocrines and organic salivary secretions and mechanisms of salivation. *Mailing Add:* 1734 Woodbine Dr Birmingham AL 35216

HANSON, ROLAND CLEMENTS, b Moose Lake, Minn, Mar 23, 34; m 62; c 4. EXPERIMENTAL SOLID STATE PHYSICS. *Educ:* Mich Technol Univ, BS, 55; Univ Ill, MS, 56, PhD(physics), 60. *Prof Exp:* Asst prof physics, Reed Col, 60-63; res assoc eng physics, Cornell Univ, 63-66; from asst prof to assoc prof, 66-76, PROF PHYSICS, ARIZ STATE UNIV, 77- *Concurrent Pos:* Guest scientist, Max Planck Inst Solid State Res, Stuttgart, WGer, 73-74. *Mem:* Am Phys Soc; Am Asn Physics Teachers. *Res:* Silver and copper halides; elastic constants; Raman scattering; high pressure; condensed gases. *Mailing Add:* Dept Physics Ariz State Univ Tempe AZ 85287

HANSON, RONALD LEE, b Lincoln, Nebr, Feb 2, 44. BIOCHEMISTRY. *Educ:* Univ Minn, BA, 65; Univ Wis, PhD(biochem), 70. *Prof Exp:* Asst prof biochem, Col Physicians & Surgeons, Columbia Univ, 72-79; mem sr sci staff, Sandoz Res Inst, 79-87; SR RES INVESTR, BRISTOL-MYERS SQUIBB, 87- *Concurrent Pos:* NIH fel, Harvard Med Sch, 70-72. *Mem:* Am Soc Biochem & Molecular Biol. *Res:* Biotransformation of organic compounds by enzymes and microbial cells; enzymology. *Mailing Add:* Bristol Myers Squibb One Squibb Dr New Brunswick NJ 08903

HANSON, ROY EUGENE, b Grand Forks, NDak, Dec 26, 22; m 46; c 3. GEOPHYSICS. *Educ:* Univ NDak, BS, 47; St Louis Univ, PhD(seismol), 57. *Prof Exp:* Geologist, Shell Oil Co, 48-49; prog dir geophys, NSF, Va, 59-69; liaison scientist, Off Naval Res Br Off, London, 69-71; prog dir geophys, 71-80, SR SCIENTIST, NAT ACAD SCI, NSF, 80- *Mem:* Seismol Soc Am; Soc Explor Geophys; Am Geophys Union. *Res:* Seismology; physical geodesy; physical geology; synoptic meteorology. *Mailing Add:* 208 Salesberry St Seabreeze Rehoboth Beach DE 19971

HANSON, RUSSELL FLOYD, internal medicine, biochemistry; deceased, see previous edition for last biography

HANSON, TREVOR RUSSELL, b Cambridge, Eng, Sept 24, 55; US & UK citizen. INTERACTIVE SYSTEMS, PROGRAMMING ENVIRONMENTS. *Prof Exp:* Syst programmer, Sears, Roebuck, & Co, 75-77 & Nat CSS, Inc, 77-79; CHMN, HANSON-SMITH, LTD, 80- *Concurrent Pos:* Reviewer, Asn Comput Mach Comput Reviews, 85- *Mem:* Asn Comput Mach; Inst Elec & Electronics Engrs. *Res:* How languages and environments affect problem solving; design and construction of experimental and practical support systems; operating systems; database; language; distributed computing; creating integrated applications. *Mailing Add:* 58 Martinka Dr Shelton CT 06484

HANSON, VIRGIL, b Lima, Peru, June 30, 20; US citizen; m 43; c 2. MEDICINE. *Educ:* Univ Calif, Los Angeles, BA, 42; Johns Hopkins Univ, MD, 45. *Prof Exp:* Resident pediat, Children's Hosp, Los Angeles, 50-52, chief resident, 52-53; pvt pract, Riverside, Calif, 53-56; instr pediat, Children's Hosp, Los Angeles, 56-57, asst clin prof, 57-60; from asst prof to assoc prof, 60-73, PROF PEDIAT, SCH MED, UNIV SOUTHERN CALIF, 73- *Concurrent Pos:* Nat Found Congenital Defect Ctr, 62-63; Arthritis Found grant, 63-65; consult, Calif State Dept Health, 54-60. *Mem:* AAAS; Am Rheumatism Asn; Am Acad Pediat. *Res:* Rheumatology; rheumatic diseases in childhood. *Mailing Add:* 4650 Sunset Blvd Los Angeles CA 90027

HANSON, WARREN DURWARD, b Laporte, Minn, Jan 5, 21; m 48; c 7. PLANT GENETICS, QUANTITATIVE GENETICS. *Educ:* Univ Minn, BS, 47; Purdue Univ, MS, 49, PhD(genetics), 50. *Prof Exp:* From asst prof to assoc prof agron, Univ Fla, 50-54; biometrician & statist consult, Agr Res Serv, USDA, 54-57, res geneticist, Crops Res Div, Plant Indust Sta, 57-61; PROF GENETICS & GENETICIST, NC STATE UNIV, 61- *Mem:* AAAS; Genetics Soc Am; fel Am Soc Agron; Biomet Soc; Am Genetic Asn. *Res:* Mathematical and statistical aspects of genetics through theoretical development, experimental ramifications of physiological processes as related to selection for productivity. *Mailing Add:* Dept Genetics NC State Univ Box 7614 Raleigh NC 27650

HANSON, WAYNE CARLYLE, b Kennewick, Wash, Sept 5, 23; div; c 2. WILDLIFE MANAGEMENT, RADIATION ECOLOGY. *Educ:* Wash State Univ, BS, 49; Colo State Univ, MS, 71, PhD, 73. *Prof Exp:* Biol scientist, Biol Lab, Gen Elec Co, 49-61, sr scientist, 61-64; sr res scientist, Pac Northwest Lab, Battelle Mem Inst, 65-70; res asst, Colo State Univ, 70-73; alt group leader environ study group, Los Alamos Sci Lab, 73-78; sr scientist, Pac Northwest Lab, Battelle Mem Inst, 78-81; sr ecologist, Ertec Northwest Inc, 81-82; assoc, Dames & Moore Co, 82-84; VPRES, HANSON ENVIRON RES SERV, 84- *Honors & Awards:* Arthur S Einarsen Award, Wildlife Soc, 80. *Mem:* Wildlife Soc; Ecol Soc Am; Am Soc Mammal; Am Ornith Union; Cooper Ornith Soc; Sigma Xi. *Res:* Radionuclide cycling in Arctic ecosystems; ecological implications and arctic resource development; biological monitoring. *Mailing Add:* 1902 Yew St Rd Bellingham WA 98226-8909

HANSON, WILLIAM A, ELECTRICAL ENGINEERING, SYSTEMS DESIGN. *Educ:* Rensselaer Polytech Inst, BS, 73, MS, 76. *Prof Exp:* Assoc programmer, Software Dept, Oswego, NY, 76-78, staff engr, Systs Eng Develop Dept, 78, assoc engr, 78-82, staff engr Adv Systs Eng Dept, 82, mem staff, Palo Alto Sci Ctr, 86-87, CHIEF DESIGNER, IMAGE SCI APPL DEPT, IBM CORP, 87- *Concurrent Pos:* Prin investr, NASA Landsat 4 & 5 Earth Observ Satellite Projs, 86-87, Sanford Univ, 87-, Darthmouth Col, 87- *Mem:* Inst Elec & Electronics Engrs. *Res:* Design and implementation of the Computational Network Environment System; medical image processing on IBM mainframes; information theory; analysing earth observation data to study spruce trees stress syndrome supected to be caused by acid rain; author of numerous books articles and reviews on image processing. *Mailing Add:* Drop 0200 IBM Corp Rte 17C Owego NY 13827

HANSON, WILLIAM BERT, b Warroad, Minn, Dec 30, 23; m 90; c 4. PLANETARY ATMOSPHERES, SPACE SCIENCE. *Educ:* Univ Minn, BChE, 44, MS, 49; George Washington Univ, PhD(physics), 54. *Prof Exp:* Physicist liquid helium, Nat Bur Standards, 49-56; physicist space physics, Lockheed Aircraft Corp, 56-62; PROF ATMOSPHERIC & SPACE SCI, UNIV TEX, DALLAS, 62-, DIR, CTR FOR SPACE SCI, 69- *Concurrent Pos:* Cecil H & Ida M Green Honors Chair in Natural Sci, 89- *Honors & Awards:* John Adam Fleming Medal, Am Geophys Union, 85. *Mem:* AAAS; fel Am Geophys Union. *Res:* Ionospheric physics; aeronomy. *Mailing Add:* 7831 La Sobrina Dallas TX 75248

HANSON, WILLIAM LEWIS, b Rutledge, Ga, Mar 21, 31; m 63. PARASITOLOGY, PROTOZOOLOGY. *Educ:* Univ Ga, BS, 57, MS, 60, PhD(zool), 63. *Prof Exp:* Fel parasitol, Rutgers Univ, 63-65; res assoc zool, 65-67, from asst prof to assoc prof vet path & parasitol, Sch Vet Med, 67-76, PROF PARASITOL, UNIV GA, 76-, HEAD DEPT, 74- *Mem:* Soc Protozool; Am Soc Parasitol; Am Soc Trop Med & Hyg. *Res:* Biology, morphology, physiology, immunology and taxonomy of Protozoa of the family Trypanosomatidae. *Mailing Add:* Dept Parasitol & Vet Med Univ Ga Athens GA 30602

HANSON, WILLIAM RODERICK, b Hanley, NDak, Nov 25, 18; m 46; c 3. ZOOLOGY. *Educ:* Univ Mont, BA, 43; Okla State Univ, PhD(zool), 53. *Prof Exp:* Develop res leader conserv, State Game & Fish Dept, NDak, 48-50, asst proj leader, Ariz, 53-55; assoc game specialist, Natural Hist Surv, Ill, 55-59; prof zool, Calif State Univ, Los Angeles, 60-87; RETIRED. *Concurrent Pos:* Fulbright lectr, Univ Turku, 67-68. *Mem:* Ecol Soc Am; Wildlife Soc; Brit Ecol Soc; Int Asn Ecol. *Res:* Population structure and dynamics of animals. *Mailing Add:* Dept Biol Calif State Univ Los Angeles CA 90032

HANSON, WILLIS DALE, b Omaha, Nebr, July 23, 26; m 50; c 2. FISH BIOLOGY. *Educ:* Univ Mo, AB, 54, MA, 56. *Prof Exp:* Fishery biologist, 63-75, SR FISHERIES RES BIOLOGIST, MO DEPT CONSERV, 75- *Mem:* Am Fisheries Soc. *Res:* Fish population dynamics, water quality, habitat and environmental changes in small to large reservoirs; fish distribution, movement and production of small streams; creel census; recreational use studies; channel catfish studies. *Mailing Add:* 420 Hwy 124 Hwy Hallsville MO 65255

HANSROTE, CHARLES JOHNSON, JR, b Bowling Green, Md, Nov 22, 30; m 53; c 1. PHYSICAL CHEMISTRY, ORGANIC CHEMISTRY. *Educ:* Va Mil Inst, BS, 52; Univ Richmond, MS, 55; Univ Va, PhD(chem), 58. *Prof Exp:* Asst instr chem, Va Mil Inst, 52-53; asst chem lab, Univ Va, 55-57; res chemist, E I du Pont de Nemours & Co, 58-62; from assoc prof to prof chem, Frostburg State Col, 62-65; chmn dept, 70 & 72-83, PROF CHEM, LYCHBURG COL, 65-, CHMN PHYS SCI & MATH, 82. *Concurrent Pos:* Sr res assoc, Duke Univ, 71-72. *Mem:* Am Chem Soc. *Res:* Tetrazolium compounds; schiff bases; reactions of diphenyl methylene norcamphor; amino acid uptake and release by larvae of American oyster crassostrea Virginica; detection of amino ethyl phosphoric acid in marine sediments; monomer preparation; polymer preparation; dyeing of yarns; physical testing of yarns; science education. *Mailing Add:* Lynchburg Col Lakeside Dr Lynchburg VA 24501

HANSS, ROBERT EDWARD, b St Louis, Mo, Aug 17, 33. GEOPHYSICS. *Educ:* Univ Mo, Rolla, BS, 54; Washington Univ, St Louis, PhD(geophys), 65. *Prof Exp:* Teacher high sch, Wis, 58-61; chmn dept, 65-70, ASSOC PROF GEOL, ST MARY'S UNIV, TEX, 70- *Concurrent Pos:* Vis scientist, NASA-Johnson Space Ctr, 75. *Mem:* Assoc Am Geophys Union; Nat Asn Geol Teachers. *Res:* Mineralogy; shock pressures; magnetic domain studies. *Mailing Add:* 450 Senova Dr San Antonio TX 78216

HANSSEN, GEORGE LYLE, b Monticello, Iowa, Dec 19, 27; m 54; c 3. OCEANOGRAPHY. *Educ:* Okla State Univ, BS, 52. *Prof Exp:* Meteorologist, US Navy Hydrographic Off, 52-55, oceanogr, 55-60; head forecaster, US Naval Oceanog Off, 60-71, proj mgr, Integrated Command Antisubmarine Warfare Prediction Syst & head, Environ Div, 71-78, liaison to Naval Air Systs Command, 78-80; SR SCIENTIST OCEANOG & METEOROL, SCI APPLICATIONS INT INC, 80- *Honors & Awards:* First Annual Sci & Eng Award, US Naval Oceanog Off, 74. *Mem:* Am Meteorol Asn; Am Geophys Union. *Res:* The development of the Integrated Command Antisubmarine Warfare Prediction System; the first real time numerical support system to use on-scene environmental data. *Mailing Add:* 14900 W Ridge Rd Accokeek MD 20607

HANSSON, CAROLYN M, b Hazel Grove, Cheshire, Eng, Apr 15, 41; m 83; c 1. CONCRETE REINFORCEMENT CORROSION, SURFACE TREATMENT & PROPERTIES. *Educ:* Imperial Col, London Univ, BSc, 62, PhD(metall), 66. *Prof Exp:* Res scientist, mat sci, Martin Marietta Labs, 66-70; asst prof metall, Columbia Univ, 70-71; from asst prof to assoc prof mat sci, State Univ NY, Stony Brook, 71-76; mem tech staff metall, AT&T Bell Labs, 76-80; res scientist, Danish Corrosion Ctr, 80-86, dept head metall & concrete, 86-89; DEPT HEAD MAT & METALL ENG, QUEEN'S UNIV, 90- *Concurrent Pos:* Overseas fel eng, Churchill Col Cambridge, 77-78; mem, Danish Acad Tech Sci, 87-; Guggenheim fel, John Simon Guggenheim Found. *Honors & Awards:* Hardy Gold Medal, Minerals, Metals & Mat Soc, 70; Concrete Prize, Danish Concrete Asn, 88. *Mem:* Scand Electron Microscope Soc; Minerals Metals & Mat Soc; Am Soc Testing & Mat; Am Concrete Inst; Danish Concrete Asn; Danish Metal Soc. *Res:* Reinforcement corrosion and those properties of concrete which influence it; surface treatment by laser; corrosion, erosion and wear of materials. *Mailing Add:* Dept Mat & Metall Eng Queen's Univ Nicol Hall Kingston ON K7L 3N6 Can

HANSSON, GORAN K, b Lysekil, Sweden, Nov 2, 51; m 73; c 2. CARDIOVASCULAR DISEASES, CELLULAR IMMUNOLOGY. *Educ:* Gothenburg Univ, MD, 77, PhD(cell biol), 80. *Prof Exp:* Res assoc med, Gothenburg Univ, Sweden, 80-81; fel path, Univ Wash, 81-82; resident med, Sahlgren's Hosp, Gathenburg, 83-85; asst prof, 85-89, ASSOC PROF LAB MED, GOTHENBURG UNIV, 89-; SR SCIENTIST, MED RES COUN, 90- *Concurrent Pos:* Prin investr, res vascular immunol, Heart Asn, 84- & res vascular biol, Med Res Coun, 85-; consult, Lab Med, Sahlgren's Hosp, Sweden, 89- *Mem:* Am Asn Pathologists. *Res:* Interactions between immune system and blood vessels; immunologic responses in cardiovascular diseases; cytokine effects on cell growth and metabolism; pathogenesis of artherosclerosis. *Mailing Add:* Dept Clin Chem Gothenburg Univ Sahlgren's Hosp Goteborg S-41345 Sweden

HANSSON, INGE LIEF, wear & erosion, non-destructive evaluation, for more information see previous edition

HANTHORN, HOWARD E(UGENE), b Portland, Ore, Oct 29, 09; m 35; c 3. CHEMICAL ENGINEERING. *Educ:* Ore State Col, BS, 32; Case Inst Technol, MS, 33. *Prof Exp:* Lab asst, Grasselli chem dept, E I du Pont de Nemours & Co, Ohio, 33-40; res engr, B F Goodrich Co, 40-46; sr engr, Gen Elec Co, 46-64; sr develop engr, Pac Northwest Lab, Battelle Mem Inst, 65-68, res assoc, 68-70; res assoc, Westinghouse-Hanford Corp, 70-77; RETIRED. *Mem:* Fel Am Inst Chem; Am Inst Chem Engrs. *Res:* Development and design of chemical processes; nuclear reactor safeguards. *Mailing Add:* 2108 Van Gieson St Richland WA 99352

HANTMAN, ROBERT GARY, b New York, NY, Feb 9, 41; m 70; c 2. FLUID MECHANICS, ENERGY CONVERSION. *Educ:* Case Western Reserve Univ, PhD(fluid mech), 69. *Prof Exp:* Res engr turbomach, Pratt & Whitney Aircraft, 68-74; engr energy conversion, Argonne Nat Lab, 74-77; eng mgr, STD Res Corp, 77-78; mech engr fluid mech, Res & Develop Ctr, Gen Elec Corp, 78-82, liaison scientist, Res & Develop Ctr, 82-84, MGR ADVAN TECHNOL, GE LIGHTING BUS GROUP, 84- *Mem:* Sigma Xi; Am Soc Mech Engrs; Soc Mfg Eng. *Res:* Automation and process technology; thermodynamics; energy systems analysis; turbomachinery; internal aerodynamics and combustion. *Mailing Add:* Gen Elec Corp Lighting Bus Group Nela Park Cleveland OH 44112

HANTO, DOUGLAS W, SURGERY. *Prof Exp:* ASSOC PROF SURG, UNIV CINCINNATI, 91- *Mailing Add:* Dept Surg Univ Cincinnati 231 Bethesda Ave Cincinnati OH 45267-0558

HANTON, JOHN PATRICK, b St Paul, Minn, Apr 5, 35. ELECTRICAL ENGINEERING. *Educ:* Univ Minn, BS, 57, MS, 59, PhD(elec eng), 64. *Prof Exp:* From asst prof to assoc prof, 64-77, PROF ELEC ENG, MONT STATE UNIV, 77- *Concurrent Pos:* NSF res grant, 65-67. *Res:* Field of magnetic materials. *Mailing Add:* Dept Elec Eng Mont State Univ Bozeman MT 59717

HANUKOGLU, ISRAEL, b Istanbul, Turkey, Mar 14, 52; Israeli citizen; div; c 1. CYTOCHROME SYSTEMS, STEROID HORMONES. *Educ:* Hebrew Univ Jerusalem, BA, 74; Univ Wis-Madison, MSc, 76, PhD(endocrinol-biochem), 80. *Prof Exp:* Res asst biochem, Univ Wis-Madison, 75-80; res fel & assoc molecular biol, Univ Chicago, 80-83; lectr biochem & molecular biol, Technion-Israel Inst Technol, 84-87, sr lectr, 86-87; sr scientist, 87-91, ASSOC PROF BIOCHEM & MOLECULAR BIOL, WEIZMANN INST SCI, 91- *Concurrent Pos:* Chmn, Int Symp Molecular View of Steroid Biosynthesis & Metab, 91; vchmn, Ninth Int Symp Microsomes & Drug Oxidations, 92. *Honors & Awards:* Henri Gutwirth Award, Technion-Israel Inst Technol, 84; Hans Lindner Prize in Biochem Endocrinol, Israel Endocrine Soc, 88; Lubell Prize, Weizman Inst Sci, Sci Coun, 91. *Mem:* Am Soc Biochem & Molecular Biol; Endocrine Soc; Israel Biochem Soc; Int Soc for Study Xenobiotics. *Mailing Add:* Dept Hormone Res Weizmann Inst Sci Rehovot 76100 Israel

HANUMARA, RAMACHANDRA CHOUDARY, b Peyyeru, India, Aug 28, 37; m 68. STATISTICS. *Educ:* Univ Madras, BA, 56; Guzarat Univ India, MA, 58; Mich State Univ, MS, 62; Fla State Univ, PhD(statist), 68. *Prof Exp:* Res asst statist, Govt of India, 59-61; res asst, Mich State Univ, 61-63 & Fla State Univ, 63-68; from asst prof to assoc prof, 68-88, PROF STATISTICS, UNIV RI, 88- *Mem:* Inst Math Statist; Am Statist Asn; Sigma Xi. *Res:* Theory and applications of statistics. *Mailing Add:* Dept Comp Sci Univ RI Kingston RI 02881

HANWAY, DONALD GRANT, b Broadwater, Nebr, Aug 6, 18; m 42; c 3. GENETICS. *Educ:* Univ Nebr, BSc, 42, MSc, 48; Iowa State Col, PhD, 54. *Prof Exp:* From instr to prof agron, 48-76, chmn dept, 55-76, prof agron & extension crops specialist, Univ Nebr, 76-84; RETIRED. *Concurrent Pos:* Chief of party, AID-Univ Nebr group, Ataturk Univ, Turkey, 65-67; consult, agronomists & AID-Nebr prog, 68 & 70; mem, Inst Progs Title XII Core Staff, 79-84; coordr, Morocco Dryland Farming Prog, 80-82. *Honors & Awards:* Honor Award, Soil Conserv Soc Am, 73. *Mem:* Fel Am Acad Sci; fel Am Soc Agron; fel Crop Sci Soc Am; Am Inst Biol Sci; Soil Conserv Soc Am. *Res:* Soybean breeding; genetics; production practices; administration of research in crop breeding, crop production systems, soil and water conservation and soil science. *Mailing Add:* 6025 Madison Ave Lincoln NE 68507

HANWAY, JOHN E(DGAR), JR, b Fairmont, WVa, May 14, 22; m 46. CHEMICAL ENGINEERING. *Educ:* WVa Univ, BSChE, 49, MSChE, 51. *Prof Exp:* Asst, Eng Exp Sta, WVA Univ, 49-51; prin chem engr, Battelle Mem Inst, 51-57, asst div chief, 57-61, div chief, 61-65; dir eng, Copeland Process Corp, Ill, 65-67, vpres eng, 67-68; asst dir res, Chicago Bridge & Iron Co, 68-74; PRES, JECON ENGRS, INC, 74- *Mem:* Am Inst Chem Engrs; Am Inst Mining, Metall & Petrol Engrs. *Res:* Waste effluent treatment; paper and pulp industry; non-ferrous extractive metallurgy; metallurgical and other applications of fluidized-bed reactor systems; chloride chemistry; design of chemical processing systems. *Mailing Add:* Route 4 PO Box 108 Wautoma WI 54982

HANWAY, JOHN JOSEPH, b Broadwater, Nebr, Dec 22, 20; m 43; c 7. SOILS. *Educ:* Univ Nebr, BS, 42, MS, 48; Iowa State Univ, PhD(soil fertility), 54. *Prof Exp:* Chemist & supvr explosives, E I du Pont de Nemours & Co, Okla, 42-45; asst soils, Univ Nebr, 45-50, from asst prof to assoc prof, 50-58; prof soils, Iowa State Univ, 58-85; RETIRED. *Concurrent Pos:* Sr officer, Joint FAO-IAEA Div, Int Atomic Energy Agency, Vienna, Austria, 66-68. *Mem:* Fel Am Soc Agron; Soil Sci Soc Am; Int Soc Soil Sci. *Res:* Soil chemistry and fertility; soil testing; plant analysis; soil-plant relations. *Mailing Add:* 215 Parkridge Circle Ames IA 50010

HANYSZ, EUGENE ARTHUR, b Akron, Ohio, Mar 15, 22; m 52; c 2. ELECTRICAL ENGINEERING. *Educ:* Univ Mich, BS, 45, MSE, 48. *Prof Exp:* Electronic engr, Naval Res Lab, 45-49; res engr, 49-57, supvry res engr, 58-69, asst head electronics & instrumentation dept, 69-74, ASST HEAD ELECTRONICS DEPT, GEN MOTORS CORP, 74- *Mem:* Inst Elec & Electronics Engrs. *Res:* Ferrite and dielectric materials; magnetic logic; electromagnetic applications; control systems; sensors. *Mailing Add:* 3308 Newgate Rd Troy MI 48084

HANZEL, ROBERT STEPHEN, b Johnstown, Pa, July 18, 32; m 59; c 2. ORGANIC CHEMISTRY, TEXTILE TECHNOLOGY. *Educ:* St Vincent Col, BS, 53; Univ Notre Dame, MS, 55, PhD(org chem), 60. *Prof Exp:* Res chemist, E I du Pont de Nemours & Co, 59-67, tech serv rep, 67-74, tech serv specialist, textile fibers dept, 74-84, sr tech mkt specialist, 84-90; RETIRED. *Mem:* Am Chem Soc. *Res:* Textile fibers developments; textured fibers development in spandex elastomeic fibers and applications to fabric formation. *Mailing Add:* 336 Hampton Rd Wilmington DE 19803

HANZELY, LASZLO, b Satoraljaujhely, Hungary, June 29, 39; US citizen; m 68; c 2. BOTANY, CYTOLOGY. *Educ:* Colo State Univ, BS, 62, MS, 64; Southern Ill Univ, PhD(bot), 69. *Prof Exp:* PROF BIOL, NORTHERN ILL UNIV, 69- *Mem:* Electron Micros Soc Am; Bot Soc Am; Sigma Xi. *Res:* Ultrastructural plant anatomy and cytology. *Mailing Add:* Dept Biol Sci Northern Ill Univ De Kalb IL 60115

HANZELY, STEPHEN, b Sátoraljaújhely, Hungary, Dec 30, 40; US citizen; m 62; c 2. PHYSICS. *Educ:* Kent State Univ, BS, 62; Univ Toledo, MS, 64; NMex State Univ, MS, 67, PhD(physics), 69. *Prof Exp:* Owens-Ill Glass Co asst, Univ Toledo, 62-64; asst, NMex State Univ, 64-68, NASA asst, 67; asst prof, 68-73, assoc prof, 73-80, chmn dept, 74-79, PROF, PHYSICS & ASTRON YOUNGSTOWN STATE UNIV, 80- *Mem:* Am Inst Physics; Am Asn Physics Teachers; Sigma Xi; Am Phys Soc; Nat Educ Asn. *Res:* Study of x-ray emission spectra of the first series transition elements under threshold level excitation; x-ray spectroscopy; vacuum technology. *Mailing Add:* Dept Physics & Astron Youngstown State Univ Youngstown OH 44555

HANZLIK, ROBERT PAUL, b Chicago, Ill, Nov 4, 43; m 65. MEDICINAL CHEMISTRY. *Educ:* Southern Ill Univ, BA, 66; Stanford Univ, PhD(org chem), 70. *Prof Exp:* NATO res fel & US Hon Ramsay Mem fel, Cambridge Univ, 70-71; from asst prof to assoc prof, 71-79, PROF MED CHEM, UNIV KANS, 80- *Concurrent Pos:* Mem, Bioorg & Natural Prods Study Sect, NIH, 87-91. *Honors & Awards:* Sato Mem Int Award, FAES & Pharmaceut Soc Japan, 90. *Mem:* AAAS; Royal Soc Chem; Am Chem Soc; Soc Toxicol; Int Soc Study Xenobiotics. *Res:* Drug metabolism; biochemical toxicology; enzyme mechanisms and inhibition; drug design. *Mailing Add:* Dept Med Chem Univ Kans Lawrence KS 66045-2506

HAPAI, MARLENE NACHBAR, b Honokaa, Hawaii, July 19, 48; m 72; c 2. ECOLOGY, BIOLOGICAL CONTROL. *Educ:* Gonzaga Univ, BA, 70; Univ Hawaii, Manoa, MS, 77, PhD(entom), 81. *Prof Exp:* Teacher biol, physiol & earth sci, Kohala High Sch, 70-76; teacher biol & chem, Hilo High Sch, 78-79; teacher & curric writer physiol & entom, Manoa, Univ Hawaii, 79-82, researcher ecol, Natural Energy Inst, Manoa, 85-86, instr biol & ecol, Hawaii Community Col, 82-86, asst prof biol sci educ, 86-89, ASSOC PROF BIOL SCI EDUC, UNIV HAWAII, HILO & MANOA, 89-, ASST DIR, CTR GIFTED & TALENTED NATIVE HAWAIIAN CHILDREN, 89- *Honors & Awards:* Frances Davis Mem Award, Univ Hawaii, 85. *Mem:* Entom Soc Am; Sigma Xi; AAAS; Nat Sci Teachers Asn; Nat Asn Gifted Children. *Res:* Insect cecidogenesis; biological science curriculum; acid precipitation; physiology; science education. *Mailing Add:* 523 W Lanikaula St Hilo HI 96720-4091

HAPKE, BERN, b Berlin, Ger, June 4, 43; US citizen. NUCLEAR MEDICINE. *Educ:* Univ Ill, BSc, 65; Univ Mich, MS, 68, PhD(med chem), 70. *Prof Exp:* Fel, Northwestern Univ, 70-73; SR CHEMIST NUCLEAR MED, 3M CO, 73- *Mem:* AAAS; Soc Nuclear Med. *Res:* Diagnostic procedures in medicine; nuclear medical diagnosis and in vitro radioassays. *Mailing Add:* 7590 N 62 St St Paul MN 55115

HAPKE, BRUCE W, b Racine, Wis, Feb 17, 31; m 54; c 3. PLANETARY SCIENCES. *Educ:* Univ Wis, BS, 53; Cornell Univ, PhD(eng physics), 62. *Prof Exp:* Sr res assoc astrophys, Cornell Univ, 60-67; assoc prof, 67-76, PROF GEOL & PLANETARY SCI, UNIV PITTSBURGH, 77- *Concurrent Pos:* Chair, Div Planetary Sci, Am Astron Soc. *Mem:* AAAS; Am Astron Soc; Am Geophys Union; Int Astron Union; AAAS. *Res:* Nature and origin of planets; nature of planetary surfaces; theory of light scattering. *Mailing Add:* Dept Geol & Planetary Sci Univ Pittsburgh Pittsburgh PA 15260

HAPNER, KENNETH D, b Goshen, Ind, May 7, 39; m 66; c 2. BIOCHEMISTRY. *Educ:* Ind Univ, Indianapolis, PhD(biochem), 66. *Prof Exp:* NIH fel, Sch Med, Ind Univ, Indianapolis, 65-66 & Univ Wash, 66-67; actg asst prof biochem, Univ Wash, 67-69; from asst prof to assoc prof, 69-84, PROF CHEM, MONT STATE UNIV, 84- *Concurrent Pos:* Res grants, Agr Exp Sta, 69 & 74, USDA, 76 & NSF, 78, 85. *Mem:* Am Soc Biochem & Molecular Biol; Entom Soc Am; Soc Invert Path. *Res:* Structural and functional roles of amino acids, proteins, and enzymes; isolation and characterization of naturally occurring proteinase inhibitors; protein primary structure; affinity chromatography; phytohemagglutinins; insect agglutinins, immunology; molecular cloning; DNA sequence. *Mailing Add:* Dept of Chem Mont State Univ Bozeman MT 59717

HAPP, GEORGE MOVIUS, b St Louis, Mo, May 18, 36. INSECT PHYSIOLOGY, DEVELOPMENTAL BIOLOGY. *Educ:* Principia Col, BS, 58; Cornell Univ, PhD(animal physiol), 64. *Prof Exp:* Asst prof biol, Cath Univ Am, 65-67; from asst prof to assoc prof, NY Univ, 67-72; from assoc prof to prof zool & entom, Colo State Univ, 72-78; interim vprovost, 90-91, PROF ZOOL & CHMN DEPT, UNIV VT, 78- *Concurrent Pos:* NIH fel, 64-65. *Mem:* AAAS; Entom Soc Am; Am Soc Cell Biol; Am Soc Zool; Int Soc Develop Biol. *Res:* Insect endocrinology and reproductive development; histochemistry, ultrastructure and development of exocrine glands. *Mailing Add:* Dept Zool Marsh Life Sci Bldg Univ Vt Burlington VT 05405

HAPP, HARVEY HEINZ, b Berlin, Ger, June 27, 28; US citizen; m 51; c 2. ELECTRICAL ENGINEERING, SYSTEMS THEORY. *Educ:* Ill Inst Technol, BSEE, 53; Rensselaer Polytech Inst, MEE, 58; Univ Belgrade, DSc(appl sci), 62. *Prof Exp:* Engr, High Voltage Lab, Gen Elec Co, 54-56, anal engr, Elec Utility Eng Oper, 56-68, sr engr, 68-72, mgr, Anal Eng Serv, 72-77, mgr, Advan Syst Technol, 77-82, mgr, Syst Anal, 82-87; consult, 87-88; CONSULT, NY STATE DEPT PUB SERV, 88- *Concurrent Pos:* Lectr various cols & univs; co-founder, Power Syst Computations Conf, 62. *Mem:* Fel Inst Elec & Electronics Engrs; Power Syst Computations Conf; Tensor Soc Gt Brit (vpres, 73-); Conf Int Des Grands Reseaux Electricques; Sigma Xi. *Res:* Basic understanding of new and advanced technologies for system analysis and synthesis and incorporate the use of these technologies to benefit electric utilities and consumers of electric energy; author of numerous books and contributor to numerous articles and book reviews to professional journals, chapters to technical books. *Mailing Add:* 2211 Webster Dr Schenectady NY 12309

HAPP, STAFFORD COLEMAN, b Sparrow Bush, NY, Sept 16, 05; m 35; c 1. GEOLOGY, ENGINEERING. *Educ:* Marietta Col, AB, 31; Columbia Univ, PhD(geol), 39. *Prof Exp:* Asst geol, Am Mus Natural Hist, 31-33 & Columbia Univ, 33-34; from jr engr to head stream & valley sect, Soil Conserv Serv, USDA, 35-43; head geol-soils sect, Corps Engrs, US Army, Ocala, Fla, 43-44, head geol & subsurface explor, Kansas City Dist, 44-55; chief geol reports, Grand Jct Opers Off, AEC, 55-57, asst chief geol br, 57-58, chief prod serv, 59-64; res geologist, US Geol Surv, 64-65; res geologist, 65-72, COLLBR, SEDIMENTATION LAB, USDA, 72- *Concurrent Pos:* Chmn, Engr Geol Div, Geol Soc Am, 59-60. *Mem:* AAAS; fel Geol Soc Am; Am Soc Civil Eng; Am Inst Mining, Metall & Petrol Eng. *Res:* Alluvial sedimentation; geomorphology; engineering, uranium and military geology. *Mailing Add:* 503 N 14th St Oxford MS 38655

HAPPEL, JOHN, b New York NY, April 1, 08; m 51; c 3. CATALYSIS, PROCESS ECONOMICS. *Educ:* Mass Inst Technol, BS, 29, MS, 30; Polytech Inst NY, PhD(chem), 48. *Prof Exp:* Chem engr, Mobil Oil Corp, 30-48; prof, 48-72, EMER PROF CHEM ENG, NEW YORK UNIV, 72-; PRES, CATALYSIS RES CORP, 73- *Concurrent Pos:* Spec res assoc chem eng, Columbia Univ, 73- *Honors & Awards:* Founders Award, Am Inst Chem Engrs, 87. *Mem:* Nat Acad Eng; NY Acad Sci; Am Chem Soc; Am Inst Chem Engrs. *Res:* Development of catalysts and catalytic processes, especially as related to petrochemicals and energy; chemical process economics; fluid-solid dynamics. *Mailing Add:* 69 Tompkins Ave Hastings-on-Hudson NY 10706

HAPPEL, LEO THEODORE, JR, b New Orleans, La, Aug 5, 43; m 66; c 2. NEUROPHYSIOLOGY. *Educ:* Tulane Univ, BS, 66, MS, 70; La State Univ Med Ctr, New Orleans, PhD(physiol), 72. *Prof Exp:* Instr, 72-74, ASST PROF PHYSIOL, NEUROL & NEUROSURG, LA STATE UNIV MED CTR, NEW ORLEANS, 74- *Mem:* Soc Neurosci. *Res:* Spinal cord; computer analysis of bioelectric potentials; pain; stroke. *Mailing Add:* 4716 Green Acres Ct Metairie LA 70003

HAPPER, WILLIAM, JR, b Vellore, India, July 27, 39; US citizen; m 67; c 2. ATOMIC PHYSICS. *Educ:* Univ NC, BSc, 60; Princeton Univ, PhD(physics), 64. *Prof Exp:* Res physicist Columbia Univ, 64-65, from instr to prof, Radiation Lab, 65-80, dir, 71-78; PROF PHYSICS, PRINCETON UNIV, 80- *Concurrent Pos:* Sloan fel, 66-; trustee, MITRE Corp, 88- *Honors & Awards:* Alexander von Humboldt Award, 75. *Mem:* Fel Am Phys Soc. *Res:* Atomic beams; optical double resonance and level crossing spectroscopy; optically pumped microwave masers; laser spectroscopy; spin polarized atoms and nuclei. *Mailing Add:* Physics Dept Jadwin Hall Princeton Univ PO Box 708 Princeton NJ 08540

HAQ, BILAL U, US citizen. ENVIRONMENTAL SCIENCE, EARTH & MARINE SCIENCE. *Educ:* Panjab Univ, Pakistan, BS, 61, MS, 63; Univ Stockholm, Sweden, PhD(marine geol), 67, DSc, 72. *Prof Exp:* Lectr paleont, Panjab Univ, Pakistan, 63-64; UNESCO res assoc micropaleont, Geol Surv Austria, 64-65; res fellow, Swed Int Develop Agency, Univ Stockholm, Sweden, 66-70; asst & assoc scientist, Woods Hole Oceanog Inst, 70-82; sr res specialist, 82-84, res assoc, Exxon Prod Res Co, 84-88; DIR MARINE GEOL & GEOPHYSICS PROG, NSF, 88- *Concurrent Pos:* Adj docent, Univ Stockholm, 72-; co-chief scientist, Deep Sea Drilling Proj, 73; vis res scientist, Soc Nat Elf, Aquitaine, France, 78-79; vis scholar, Univ Paris, 79; chief ed, Marine Micropaleont J, 76-90; assoc ed, Micropaleont, Am Mus Natural Hist, 80-; mem, Geophys Study Comt, Nat Res Coun, Nat Acad Sci; chmn subcomt oceans non-living resources, Int Union Geol Sci/Comn Marine Geol; mem guiding group, experts ocean resources, UNESCO, co-chief scientist, Ocean Drilling Prog, 88; distinguished lectr, Am Asn Petrol Geol, 88-89; coordr, environ issue, Exxon Prod Res Co, 86-88; vis prof, Univ Copenhagen, Denmark, 91-92; mem, UK Natural Environ Res Coun vis comn Brit Geol Surv, 91. *Mem:* Fel Geol Soc Am; Am Geophys Union; AAAS; Am Asn Petrol Geol; Geol Soc Sweden; fel Geol Soc London; Oceanog Soc. *Res:* Marine geology and geophysics; biostratigraphy; paleobiogeography; paleoceanography; petroleum geology; paleontology; global stratigraphy; global change. *Mailing Add:* Nat Sci Found Washington DC 20550

HAQ, M SAFIUL, b Bangladesh, Feb 1, 35; Can citizen. MATHEMATICAL STATISTICS. *Educ:* Univ Dhaka, BSc, 54, MSc, 55; Univ Toronto, MA, 64, PhD(math), 66. *Prof Exp:* Statistician II, Food & Agr Coun Pakistan, Karachi, 56-58, statistician I, Agr Census, 58-59; lectr statist, Univ Dacca, 59-60; dep dir statist, Small Industs Corp, Bangladesh, 60-62; sr lectr, Rajshahi Univ, 62-67; asst prof math, 67-70, assoc prof, 70-80, PROF STATIST, UNIV WESTERN ONT, 80- *Mem:* Inst Math Statist; fel Royal Statist Soc; Statist Soc Can; Int Statist Inst. *Res:* Predicitive inference; multivariate statistical analysis; marginal likelihood inference. *Mailing Add:* Dept Statist & Actuarial Sci Univ of Western Ont London ON N6A 5B9 Can

HAQ, MOHAMMAD ZAMIR-UL, b Saharanpur, India, Oct 26, 36; m 68; c 3. ANALYTICAL CHEMISTRY, FORENSIC CHEMISTRY. *Educ:* Univ Ottawa, PhD(chem), 67. *Prof Exp:* Robert Welch fel, Baylor Univ, 67-68; NSF grant, Howard Univ, 68-70; from scientist to sr scientist, Meloy Labs, Inc, Va, 71-74; forensic chemist, 74-76, TOXICOLOGIST, FORENSIC SCI LAB, OFF CHIEF MED EXAMR, DEL, 76- *Mem:* AAAS; Am Chem Soc; Int Asn Forensic Toxicologists; Forensic Sci Soc; Am Acad Forensic Sci. *Res:* Identification and quantitation of toxic substances including drugs and narcotics using gas chromatography, mass spectrometry and infrared; analysis of some carcinogens and cocarcinogens from marijuana and cigarette smoke condensates; natural product chemistry and organic synthesis of cyclohexanes, norbornanes, hydroperoxides and heterocyclics with extensive use of nuclear magnetic resonance spectroscopy. *Mailing Add:* Medlab Clin Testing 212 Cherry Lane New Castle DE 19702

HAQUE, AZEEZ C, b Bangalore, India, May 8, 33; m 67; c 3. PHYSICS, MATERIALS SCIENCE. *Educ:* Univ Mysore, BSc, Hons, 56, MSc, 58; Wash State Univ, PhD(physics), 62. *Prof Exp:* Lectr physics, Univ Mysore, 56-57; scientist electronics, Min Defense, 57-58; res assoc physics, Brown Univ, 62-64, asst prof, 64-67; DISTINGUISHED MED SCI STAFF

PHYSICS, BELL LABS, 67- *Concurrent Pos:* Fel, Brown Univ, 62-63. *Mem:* Am Phys Soc; Am Vacuum Soc; sr mem Inst Elec & Electronics Engrs. *Res:* Surface science; low energy electron diffraction; mass spectrometry; auger electron spectroscopy; electric contacts; ultra high vacuum technology; connector materials; quality engineering and manufacturing engineering. *Mailing Add:* AMHI Inc 3620 N High St Columbus OH 43214

HAQUE, RIZWANUL, b India, Dec 15, 40; m 69; c 1. ENVIRONMENTAL CHEMISTRY, ENVIRONMENTAL TOXICOLOGY. *Educ:* Aligarh Muslim Univ, India, MS, 59; Univ BC, PhD(chem), 66. *Prof Exp:* From asst prof to assoc prof agr chem, Ore State Univ, 66-74; WITH US ENVIRON PROTECTION AGENCY, 74- *Mem:* Am Chem Soc; fel Am Inst Chem; Sigma Xi; AAAS; Soc Environ Toxicol & Chem. *Res:* Physical chemistry and its application to environmental problems; transport and fate of toxic substances; modeling, regulation, exposure assessment and hazard assessment of toxic chemicals; pesticide chemistry; hazardous waste. *Mailing Add:* ASCI Corp 1365 Beverly Rd McLean VA 22101

HARA, SABURO, b Yamanashi, Japan, Feb 16, 28; US citizen; m 58; c 1. PEDIATRICS, HEMATOLOGY. *Educ:* Tokyo Med Col, MD, 53; Am Bd Pediat, dipl. *Prof Exp:* Intern, Bronson Methodist Hosp, Kalamazoo, Mich, 55; resident pediat, Res & Educ Hosp, Univ Ill, 58, asst, Col Med, 58-59; from instr to assoc prof pediat, 59-71, asst proj dir, 63-64, co-dir res lab ment retardation demonstration clin, 66-69, PROF PEDIAT, MEHARRY MED COL, 71-, DIR, DIAG & TRAINING LAB, 69- *Concurrent Pos:* Res fel pediat & hemat, Univ Ill Col Med, 59. *Mem:* Fel Am Acad Pediat; Am Soc Human Genetics. *Res:* Medical and clinical science; human genetics. *Mailing Add:* Dept Pediat Meharry Med Col 1005 18th Ave N Nashville TN 37208

HARA, TOSHIAKI J, b Kumamoto, Japan, Aug 22, 32; m 59; c 1. NEUROBIOLOGY. *Educ:* Yokohama Munic Univ, BS, 56; Univ Tokyo, MS, 58, PhD(zool), 63. *Prof Exp:* NIH fels, Columbia Univ, 63 & Univ Wash, 63-66; instr physiol, Med Sch, Kumamoto Univ, 67-69; RES SCIENTIST, FRESHWATER INST, CAN DEPT FISHERIES & OCEANS, 69- *Concurrent Pos:* Adj prof zool, Univ Man, 78- *Mem:* Soc Neurosci; Asn Chemoreception Sci; Can Soc Zool; Europ Chemorection Res Orgn. *Res:* Neurobiological study of chemosensory, olfactory and gustatory, processes in fishes and its application to aquaculture and environmental pollution researches. *Mailing Add:* Dept Fisheries & Oceans Freshwater Inst Winnipeg MB R3T 2N6 Can

HARAGAN, DONALD ROBERT, b Houston, Tex, Apr 15, 36; m 66; c 2. ATMOSPHERIC SCIENCE, HYDROLOGY & WATER RESOURCES. *Educ:* Univ Tex, Austin, BS, 59, PhD(atmospheric sci), 70; Tex A&M Univ, MS, 60. *Prof Exp:* Res scientist meteorol, Tex A&M Univ, 59-60; res scientist meteorol, Elec Eng Res Lab, Univ Tex, Austin, 60-66, instr, 62-69; from asst prof to prof atmospheric sci, Tex Tech Univ, 69-82, chmn dept, 72-82, assoc dean, A&S, 82-85, interim vpres, 85-87, vpres acad affairs & res, 86-89, EXEC VPRES & PROVOST, TEX TECH UNIV, 89- *Concurrent Pos:* Res grants, US Dept of Interior, Tex Dept Water Resources & Int Ctr Study Arid & Semi-arid Lands, 70- *Mem:* Am Meteorol Soc; AAAS; Am Water Res Asn; Am Soc Civil Eng. *Res:* Physics and dynamics of severe storms; cloud physics and atmospheric water resources; precipitation processes; clouds and precipitation in arid and semi-arid environments. *Mailing Add:* 6914 Nashville Dr Lubbock TX 79413

HARAKAL, CONCETTA, b Chieti, Italy, Nov 25, 23; US citizen; m 48. PHARMACOLOGY. *Educ:* Univ Pa, AB, 45; Temple Univ, MS, 50, PhD(pharmacol), 62. *Prof Exp:* From instr to assoc prof, 62-76, PROF PHARMACOL, SCH MED, TEMPLE UNIV, 76- *Honors & Awards:* Sowell Award, Temple Univ Sch Med, 79, 90; Lindbach Award, Temple Univ, 80. *Mem:* AAAS; Am Soc Pharmacol & Exp Therapeut; Am Heart Asn; NY Acad Sci; Sigma Xi. *Res:* Cardiovascular and renal pharmacology; hemodynamic effects of adrenalectomy; medical education; effect of drugs on vascular smooth muscle. *Mailing Add:* Dept Pharmacol Temple Univ Sch Med Philadelphia PA 19140

HARAKAS, N(ICHOLAS) KONSTANTINOS, b Karyai, Greece, July 15, 34; US citizen; m 63; c 2. CHEMICAL ENGINEERING. *Educ:* Clemson Univ, BS, 58; NC State Univ, MS, 60, PhD(chem eng), 62. *Prof Exp:* Res chem engr, Am Cyanamid Co, 62-63; res chem engr, Chemstrand Res Ctr, Inc, 63-69, group leader, Monsanto Co, 69-73, sr res specialist, 73-77, sr res group leader, 77-88, BIOTECH CONSULT, MONSANTO CO, 88- *Concurrent Pos:* Monsanto vis scientist, Med Ctr, Univ Ala, 75-76. *Mem:* Am Inst Chem Engrs; Am Chem Soc; Tissue Cult Asn; NY Acad Sci. *Res:* Heat transfer to particulate solids; synthetic fiber spinning and morphology; polymerization, fiber reinforced composites; epoxy resins; thermodynamics of steel refining; inviscid melt spinning of steel; gas carburization of steel sheet; crystallization of steel sheet; steel tire cord adhesion; phosphate conversion coatings; silane coupling agents; mammalian cell culture technology; protein fractionation; construction and operation of recombinant DNA facilities; fibrinolysis and hemostasis; cellular immunology; neovascularization; monoclonal antibodies; industrial scale mammalian cell culture facilities operation; growth factors; diagnostics; bone healing. *Mailing Add:* Monsanto Co 800 N Lindbergh Blvd St Louis MO 63167

HARALICK, ROBERT M, b Brooklyn, NY, Sept 30, 43; m 67; c 1. ELECTRICAL ENGINEERING. *Educ:* Univ Kans, BA, 64, BS, 66, MS, 67, PhD(elec eng), 69. *Prof Exp:* From asst prof to assoc prof, 69-75, prof elec eng & adj prof computer sci, Univ Kans, 75-78; prof elec eng & comput sci, Va Polytech Inst & State Univ, 79-; vpres res, Mach Vision Int, 84-86; BOEING CLAIRMONT EGTVEDT PROF ELEC ENG & ADJ PROF COMPUTER SCI, UNIV WASH, 86-, ADJ PROF CTR BIOENG, 88- *Concurrent Pos:* Assoc ed, Computer Vision, Graphics & Image Processing, 75-, Pattern Recognition, 77-, IEEE Transactions on Systs, Man & Cybernetics, 79-, Commun of ACM Image Processing, 82-; mem Computer Soc Pattern Anal Mach Intel & Tech Comt, Inst Elec & Electronic Eng, 75-, Pattern Recognition Tech Subcomt, 75-81, Data Struct & Pattern Recognition Subcomt, 75-81, Biomed Pattern Recognition Comt, 75-81, prog comt, Pattern Recognition & Image Processing Conf, 78; co-dir, Advan Study Inst Imaging Processing, NATO, 78, co-chair, 80, dir, Advan Study Inst Pictorial Data Anal, 82. *Honors & Awards:* Young outstanding fac award, Dow Chem, Am Eng Educ, 75. *Mem:* Inst Elec & Electronics Engrs; Soc Gen Systs Res; Pattern Recognition Soc; Asn Comput Mach. *Res:* Pattern recognition; clustering techniques as well as discrimination techniques, particularly as they apply to image patterns; general systems research and scene analysis; artificial intelligence; computer vision; author of four books, twenty-eight book chapters, one hundred twenty-one journal articles and one hundred twenty-six conference papers. *Mailing Add:* Dept Elec Eng FT-10 Univ Wash Seattle WA 98195

HARAMAKI, CHIKO, b Hayward, Calif, Oct 3, 25. ORNAMENTAL HORTICULTURE, WEED SCIENCE. *Educ:* Ore State Col, BS, 52; Univ Ill, MS, 53; Ohio State Univ, PhD(hort), 57. *Prof Exp:* Asst prof hort, 57-68, assoc prof ornamental hort, 68-77, PROF ORNAMENTAL HORT, PA STATE UNIV, 77- *Concurrent Pos:* Assoc exp sta, Univ Calif, Riverside, 70-71; vis prof, Univ Peradeniya, Sri Lanka, 85. *Mem:* AAAS; Am Soc Hort Sci; Weed Sci Soc Am; Int Soc Hort Sci; Int Plant Propagators Soc; Int Soc Arboriculture; Plant Growth Regulators Soc Am. *Res:* Herbicides; asexual propagation of ornamental plants; arboriculture; tissue culture. *Mailing Add:* 7012 Seventh Ave Blvd NW Bradenton FL 34209

HARAMOTO, FRANK H, entomology, for more information see previous edition

HARARY, FRANK, b New York, NY, Mar 11, 21. FAULT TOLERANCE, EDUCATING THE GIFTED. *Educ:* Brooklyn Col, BA, 41, MA, 45; Univ Calif, Berkeley, PhD(math), 48. *Hon Degrees:* DSc & MSc, Univ Aberdeen, Scotland, 75; Fil Dr, Univ Lund, Sweden, 78. *Prof Exp:* Prof math, Univ Mich, Ann Arbor, 48-86; DISTINGUISHED PROF COMPUTER SCI, NMEX STATE UNIV, LAS CRUCES, 86- *Concurrent Pos:* Vis prof & lectr, numerous foreign & US univs & socs, 57-; founding ed, J Combinatorial Theory, 66- & J Graph Theory, 77-; vis fel, Wolfsen Col, Oxford Univ, 73-74; scientist-in-resident, NY Acad Sci, 77; Humboldt sr US scientist award, Munich, 78; overseas fel, Churchill Col, Cambridge Univ, 80-81; Ulam chair math, Univ Colo, 82; res prof, Stevens Inst Technol, 84; distinguished vis prof, Indian Statist Inst, Calcutta, 86. *Honors & Awards:* C E Cullis Mem Lectr, Calcutta Math Soc, 86; R N Prasad Mem Lectr, Allahabad Math Soc, 86. *Mem:* Am Math Soc; Asn Comput Mach; Can Math Soc; Math Asn Am; Soc Indust & Appl Math. *Res:* Theory of graphs and the vast variety of subjects to which it applies effectively as an essential mathematical model for the analysis of structure; author of six books and over 500 publications. *Mailing Add:* Computer Sci Dept NMex State Univ Las Cruces NM 88003-0001

HARARY, ISAAC, b New York, NY, Mar 15, 23; m 48; c 1. BIOCHEMISTRY. *Educ:* Brooklyn Col, BA, 45; NY Univ, PhD(biochem), 52. *Prof Exp:* Asst dir radioisotope unit, Vet Admin Hosp, Calif, 55-58; asst prof biol chem, biophys & nuclear med, 58-60, assoc prof, 60-64, PROF BIOL CHEM, SCH MED, UNIV CALIF, LOS ANGELES, 64- *Concurrent Pos:* Am Cancer Soc fel, Univ Chicago, 52-54; NIH fel, 54-55. *Mem:* Am Chem Soc; Am Soc Biol Chem. *Res:* Enzymology and intermediary metabolism; carbon dioxide fixation; cholesterol synthesis; metabolism of nicotinic acid and acetoin in bacteria; control of glycolysis; acyl phosphates and thyroxin action; enzymes of cellular differentiation; control of specific function and protein synthesis in developing heart cells in culture. *Mailing Add:* Dept Biol-Chem Univ Calif Sch Med 900 Veteran Ave Los Angeles CA 90024

HARB, JOSEPH MARSHALL, b Oakland, Calif, Dec 20, 38; m 72; c 1. CELL BIOLOGY, DIAGNOSTIC ELECTRON MICROSCOPY. *Educ:* Oglethorpe Univ, BS, 61; Tulane Univ, MS, 66, PhD(biol), 69. *Prof Exp:* Tech res specialist cell biol, Tulane Univ, 63-69, instr med, Cardiovascular Lab, Med Sch, 69-72, asst prof med, Cardiovasc Lab, Med Ctr, Tulane Univ, 72-76; dir, Electron Micros Lab, Vet Admin Hosp, 76-80; DIR, ELECTRON MICROS SERV, CHILDREN'S HOSP WIS, 80-; ASSOC PROF PEDIAT, MED COL WIS-MILWAUKEE, 81- *Concurrent Pos:* Consult, Dept Path Electron Micros, Delta Regional Primate Ctr, 65; USPHS fel, Tulane Univ, 73-75; adj instr path, Wayne State Med Sch, 76-80; planning & exec comt, Vet Admin Coop Study No 147, 80-; adj instr, Anat Dept, Univ Detroit Sch Dent, 80-; ed, Electron Micros Soc Am Bull. *Mem:* Am Soc Cell Biol; Electron Micros Soc Am. *Res:* Ultrastructure of myxomycete cells; ultrastructure of pseudobranchs and gills; cardiac, hepatic, renal, pancreatic, and aortic diseases of viral cause; diagnostic electron microscopy. *Mailing Add:* Dept Pediat Children's Hosp Wis 1700 W Wisconsin Milwaukee WI 53201

HARBACH, RALPH EDWARD, b Streator, Ill, Mar 23, 48; m 70; c 2. MEDICAL ENTOMOLOGY, MOSQUITO SYSTEMATICS. *Educ:* Western Ill Univ, BS, 71, MS, 72; Univ Ill, PhD(entom), 76. *Prof Exp:* Res assoc entom, NC State Univ, 76-79; entomologist, Walter Reed Army Inst Res, 80-84; med entomologist & taxonomist, Armed Forces Res Inst Med Sci, Thailand, 85-88; MED ENTOMOLOGIST, MOSQUITO SYSTEMATIST & MGR, BIOSYSTEMATICS UNIT, WALTER REED ARMY INST RES, 88- *Concurrent Pos:* Res assoc, Smithsonian Inst. *Mem:* Am Mosquito Control Asn; Entom Soc Am; Sigma Xi; Am Registry Prof Entomologists; Am Soc Zool Nomenclature. *Res:* Biosystematic research on mosquito vectors of disease including taxonomy, morphology, bionomics and medical importance. *Mailing Add:* Dept Entom Walter Reed Army Inst Res Washington DC 20307-5100

HARBATER, DAVID, b New York, NY, Dec 19, 52. ALGEBRAIC GEOMETRY, ARITHMETIC GEOMETRY. *Educ:* Harvard Univ, AB, 74; Brandeis Univ, MA, 75; Mass Inst Technol, PhD(math), 78. *Hon Degrees:* MA, Univ Pa, 84. *Prof Exp:* Postdoctoral fel, Am Math Soc, 78-79, Nat Sci Found, 82-83; asst prof math, 78-83, ASSOC PROF MATH, UNIV PA, 83- *Concurrent Pos:* Nat Sci Found grantee, 79-82 & 83-86; Sloan Found fel, 84-87; vis mem, Math Sci Res Inst, 86-87, Nat Sci grantee, 90- *Mem:* Am Math Soc; Sigma Xi. *Res:* Algebric geometry research and its connections to number theory, Galois theory and topology. *Mailing Add:* Dept Math Univ Pa Philadelphia PA 19104-6395

HARBAUGH, ALLAN WILSON, b Houston, Tex, Oct 22, 26; m 60. OPERATIONS RESEARCH. *Educ:* Univ Tex, Austin, BS, 47, PhD(eng), 53; Univ Mich, MEng, 49; Univ Calif, Los Angeles, MBA, 61. *Prof Exp:* Geophysicist oil explor, Shell Oil Co, 49-50; res engr electromagnetics, Appl Res Lab, Univ Tex, 50-56; scientist opers res, Rand Corp, 56-59; sr staff mem space res, Space-Gen Corp, 59-62; mgr opers res, Comput Sci Corp, 62-65; mgr mgt systs, Tex Instruments Inc, 65-67; asst to vpres finance, LTV Corp, 67-72; CONSULT, OPERS RES, 72- *Mem:* Inst Mgt Sci. *Res:* Application of mathematical methods to industry and finance, options analysis, and financial management. *Mailing Add:* 7009 Meadowcreek Dr Dallas TX 75240-2712

HARBAUGH, BRENT KALEN, b Topeka, Kans, Feb 15, 48; m 69; c 4. ORNAMENTAL HORTICULTURE, ENTOMOLOGY. *Educ:* Washburn Univ, BS, 70; Kans State Univ, MS, 72, PhD(hort), 75. *Prof Exp:* From asst prof to assoc prof, 75-86, PROF ORNAMENTAL HORT, GULF COAST RES & EDUC CTR, 86- *Mem:* Am Soc Hort Sci. *Res:* Development of production systems for ornamental crops, encompassing plant nutrition, cultural practices, water management, photoperiod, media, growth regulators, pest management and post harvest techniques. *Mailing Add:* Gulf Coast Res & Educ Ctr 5007 60th St E Bradenton FL 34203

HARBAUGH, DANIEL DAVID, b Prairie du Rocher, Ill, Jan 2, 42; m 66; c 2. ANIMAL NUTRITION, PURCHASING. *Educ:* Southern Ill Univ, Carbondale, BS, 65, MS, 66; Kans State Univ, PhD(animal nutrit), 70. *Prof Exp:* Dir nutrit res & develop, Fermbionics, Inc, Mo, 70-71; field nutritionist, 71-76, bd dirs, 76, formulation & nutrit consult, 76-78, formulation & purchasing consult, 78-85, dir formulation & purchasing & mkt consult, 85-88, VPRES NUTRIT & TECH SERV, ARBIE MINERAL FEED CO, 88- *Concurrent Pos:* Lectr, Feed Ingredient Inst, 82-83; mem & chmn elect, Computer Usage Comt, Am Feed Industs Asn, 80-84. *Mem:* Poultry Sci Asn; Sigma Xi; Am Soc Animal Sci; Am Feed Industs Asn. *Res:* Recycling of animal waste; nutritional consulting; feed technology. *Mailing Add:* Arbie Mineral Feed Co 409 S Center Marshalltown IA 50158

HARBAUGH, JOHN WARVELLE, b Madison, Wis, Aug 6, 26; wid; c 3. ECONOMIC GEOLOGY. *Educ:* Univ Kans, BS, 48, MS, 50; Univ Wis, PhD(geol), 55. *Prof Exp:* Prod geologist, Carter Oil Co, Okla, 51-53; from asst prof to assoc prof, 55-66, chmn dept, 69-72, PROF APPL EARTH SCI, STANFORD UNIV, 66- *Honors & Awards:* Levorsen Award, Am Asn Petrol Geol, Distinguished Serv Award. *Mem:* Fel Geol Soc Am; Soc Econ Paleont & Mineral; Am Asn Petrol Geol. *Res:* Petroleum geology; application of digital computers in analysis and simulation of geologic processes and exploration for petroleum and coal; decision analysis in oil exploration. *Mailing Add:* Dept Appl Earth Sci Stanford Univ Stanford CA 94305

HARBECK, RONALD JOSEPH, b Nov 7, 42; m; c 2. CLINICAL IMMUNOLOGY, IMMUNOPATHOLOGY. *Educ:* Univ SDak, PhD(microbiol), 71. *Prof Exp:* DIR CLIN LABS, JEWISH CTR IMMUNOL & RESPIRATORY MED, DENVER, COLO, 83- *Concurrent Pos:* Assoc prof, Dept Microbiol, Immunol, Path & Med; chairperson, Diag & Clin Immunol Div, Am Soc Microbiol, 91. *Mem:* Am Asn Immunologists; Am Soc Microbiol; Clin Immunol Soc; Am Asn Clin Chemists. *Res:* Gamma-delta T cell receptors; diagnostic immunology. *Mailing Add:* Clin Labs Nat Jewish Ctr Immunol & Respiratory Med 1400 Jackson St Denver CO 80206

HARBER, LEONARD C, b New York, NY, June 22, 27; m 62; c 2. DERMATOLOGY. *Educ:* Johns Hopkins Univ, AB, 49; NY Univ, MD, 53, MS, 58. *Prof Exp:* From instr to prof dermat, Sch Med, NY Univ, 58-73; PROF DERMAT & CHMN DEPT, COL PHYSICIANS & SURGEONS, COLUMBIA UNIV, 73- *Concurrent Pos:* Fulbright res scholar dermat & syphil, Univ Copenhagen, 56-57; fel dermat, Post-Grad Med Sch, NY Univ, 57-58; guest investr, Rockefeller Univ, 70; consult, dermat adv comt, Food & Drug Admin, 76-79. *Honors & Awards:* Husik Award, 60. *Mem:* Soc Invest Dermat; Am Med Writers' Asn; Am Dermat Asn; Am Soc Clin Invest; Am Acad Dermat. *Res:* Clinical and research problems associated with adverse skin reactions following sun exposure, porphyria, drug photosensitivity; protoporphyria; drug photoallergy. *Mailing Add:* Dept Dermat Columbia Univ 630 W 168th St New York NY 10032

HARBERS, CAROLE ANN Z, b Ironton, Ohio, Mar 19, 43; m 78. FOOD SCIENCES, NUTRITION. *Educ:* Ohio Univ, BS, 69; Va Polytech Inst & State Univ, MS, 76; Kans State Univ, PhD(foods & nutrit), 79. *Prof Exp:* Vocational home econ teacher, Ironton City Sch, 69-74; asst prof, 79-84, ASSOC PROF FOODS & NUTRIT, KANS STATE UNIV, 84- *Mem:* Inst Food Technologists; Am Dietetics Asn; Sigma Xi; Am Asn Cereal Chemists. *Res:* Physical, chemical, sensory and nutritional aspects of corn-based products including ethnic food, products containing high fructose corn syrup and red meats; food color; food losses and conservation. *Mailing Add:* Dept Foods & Nutrit Justin Hall Kans State Univ Manhattan KS 66506

HARBERS, LENIEL H, b La Grange, Tex, Nov 11, 34; m 61; c 3. ANIMAL NUTRITION. *Educ:* Tex A&M Univ, BS, 57, MS, 58; Okla State Univ, PhD(nutrit), 61. *Prof Exp:* Univ fel pharmacol, Univ Chicago, 61-62, USPHS fel, 62-63; assoc chemist, Am Meat Inst Found, 63-64; assoc prof, 64-76, asst dean, 88-89, PROF ANIMAL NUTRIT, 76- , CHAIR, FOOD SCI GRAD PROG, GRAD SCH, KANS STATE UNIV, 90- *Concurrent Pos:* AID consult to Nigeria, 66-68; consult to Bulgaria, UN Food & Agr Org, 75; vpres, Kans-Paraguay Partners, 87-89. *Mem:* Am Inst Nutrit; Am Soc Animal Sci. *Res:* Monogastric and ruminant nutrition; scanning electron microscopy; neon infrared spectroscopical analysis of foods and feeds. *Mailing Add:* Dept Animal Sci Kans State Univ Manhattan KS 66506

HARBERT, CHARLES A, b Indianapolis, Ind, Apr 7, 40; m 61; c 2. ORGANIC CHEMISTRY, MEDICINAL CHEMISTRY. *Educ:* Univ Colo, BA, 62; Univ Mo-Columbia, PhD(org chem), 67. *Prof Exp:* NIH fel org chem, Stanford Univ, 67-69; res chemist, 69-72, proj leader, 72-76, mgr, 76-81, dir cent res, 81-84, EXEC DIR RES CENT, PFIZER INC, 84- *Concurrent Pos:* Chair, Mech Chem Gordon Res Conf, 90. *Mem:* Am Chem Soc; Sigma Xi; AAAS. *Res:* Heterocycles; natural products. *Mailing Add:* Cent Res Pfizer Inc Groton CT 06340

HARBIN, WILLIAM T(HOMAS), b Wheeler Co, Ga, July 6, 08; m 30; c 3. CHEMICAL ENGINEERING. *Educ:* Ga Inst Technol, BS, 38. *Prof Exp:* Chem engr, Hercules Powder Co, Ga, 37-38, Union Bag & Paper Co, 38-39 & Chatham Processes Co, 39-40; head, Dept Eng, E R Squibb & Sons, NJ, 40-46; asst supt, Celanese Corp of Am, 46-49, plant supt, Chem Div, 49-54; plasticizer plant supt, Kolker Chem Corp, 54-58, prod mgr, 58-61; plant mgr, Thompson Chem Co, 61-65, asst plant mgr, Thompson Apex Co Div, Continental Oil Co, 65-67, plant mgr, Teknor Apex Co, 67-73; RETIRED. *Concurrent Pos:* Consult engr, 73- *Mem:* AAAS; Am Chem Soc; fel Am Inst Chem Engrs. *Res:* Terpene compounds and chemistry; sulfa drugs and related chemicals; plasticizers; chloronated products; resins; peroxides. *Mailing Add:* 3922 Gail Dr Oakwood GA 30566

HARBISON, G RICHARD, b Miami, Fla, March 4, 41; m. MARINE BIOLOGY, BIOLOGICAL OCEANOGRAPHY. *Educ:* Columbia Col, AB, 66; Fla State Univ, PhD(biochem), 71. *Prof Exp:* From asst scientist to assoc scientist, 72-86, SR SCIENTIST, WOODS HOLE OCEANOG INST, 86- *Concurrent Pos:* Prin res scientist, Australian Inst Marine Sci, 80-82; dir, Div Marine Sci, Harbor Br Oceanog Inst, 87-89. *Mem:* Am Soc Limnol & Oceanog; Am Soc Zoologists; Crustacean Soc; Oceanog Soc; Soc Syst Zool; Systematics Asn. *Res:* Biology of gelatinous zooplankton; systematics, physiology and ecology; associations between amphipods and gelatinous zooplankton; behavior of mesopelagic organisms. *Mailing Add:* Woods Hole Oceanog Inst Woods Hole MA 02543

HARBISON, GERARD STANISLAUS, b Manchester, Eng, Jan 1, 58; m 80; c 3. PHYSICAL CHEMISTRY, BIOLOGICAL CHEMISTRY. *Educ:* Univ Dublin, BA (MOD), 77; Harvard Univ, PhD (biophys), 84. *Prof Exp:* ASST PROF CHEM, STATE UNIV NY, STONYBROOK, 86- *Concurrent Pos:* NSF presidential young investr, 90. *Res:* Nuclear magnetic resonance and nuclear quadruple resonance spectroscopy; structure and dynamics of DNA and proteins. *Mailing Add:* Dept Chem State Univ NY Stony Brook NY 11794-3400

HARBISON, JAMES PRESCOTT, b Philadelphia, Pa, Apr 5, 51; m 73; c 3. CRYSTAL GROWTH, SEMICONDUCTOR PHYSICS. *Educ:* Harvard Col, AB, 73; Harvard Univ, PhD(appl physics), 77. *Prof Exp:* IBM fel, Div Appl Sci, Harvard Univ, 77-78; mem tech staff, Bell Labs, Murray Hill, NJ, 78-83; MEM TECH STAFF, BELLCORE, 84- *Mem:* Am Phys Soc; Am Vacuum Soc; Mat Res Soc; Minerals Metals & Mat Soc. *Res:* Crystal growth of thin films of semiconductors and metals; molecular beam epitaxy. *Mailing Add:* Bellcore NVC 3X-211 Red Bank NJ 07701-7040

HARBISON, RAYMOND D, b Peru, Ill, Jan 1, 43; m 62; c 4. PHARMACOLOGY, TOXICOLOGY. *Educ:* Drake Univ, BS, 65; Univ Iowa, MS, 67, PhD(pharmacol), 69. *Prof Exp:* Instr pharmacol, Sch Med, Tulane Univ, 69-70, asst prof, 71-72; asst prof pharmacol & biochem, Sch Med, Vanderbilt Univ, 72-76, assoc prof, 77-80; PROF PHARMACOL & INTERDISCIPLINARY TOXICOL, SCH MED SCI, UNIV ARK, LITTLE ROCK, 80- *Concurrent Pos:* Mem, Narcotic Addiction & Drug Abuse Rev Comt, NIMH, 71-73; mem, Nat Inst Drug Abuse Rev Comt, 73-75, Nat Inst Occup Safety & Health Rev Comt, 80- *Honors & Awards:* Achievement Award, Soc of Toxicol, 78. *Mem:* Soc Toxicol; Am Soc Pharmacol & Exp Therapeut; AAAS; Teratology Soc. *Res:* Teratology; developmental pharmacology; hazardous materials; drug metabolism; toxicology. *Mailing Add:* Dept Pharmacol Univ Fla Sch Med Progress Ctr Alachua FL 32615

HARBISON, S P, b Pittsburgh, Pa, Jan 26, 52; m 82. SOFTWARE SYSTEMS, COMPILERS. *Educ:* Princeton Univ, AB, 74; Carnegie-Mellon Univ, Pittsburgh, PhD(comput sci), 80. *Prof Exp:* Res programmer, 74-77, asst, 78-80, RES COMPUT SCIENTIST, DEPT COMPUT SCI, CARNEGIE-MELLON UNIV, 80- *Mem:* Asn Comput Mach; AAAS; Sigma Xi. *Res:* Operating systems; software engineering; personalized computing environments. *Mailing Add:* 840 Canterbury Ln Pittsburgh PA 15232

HARBO, JOHN RUSSELL, b Tracy, Minn, Nov 20, 43; m 66; c 2. APICULTURE. *Educ:* Gustavus Adolphus Col, BA, 65; Mich Technol Univ, MS, 67; Cornell Univ, PhD(entom), 71. *Prof Exp:* RES ENTOMOLOGIST APICULTURE, BEE BREEDING & STOCK CTR LAB, AGR RES SERV, USDA, 71- *Concurrent Pos:* Mem, Am Bee Res Conf. *Mem:* Entom Soc Am; Bee Res Asn. *Res:* Reproductive physiology of honey bees. *Mailing Add:* 7240 Palmetto Dr Baton Rouge LA 70808

HARBOLD, MARY LEAH, b Millersville, Pa, Oct 26, 12. PHYSICS, ACOUSTICS. *Educ:* Goucher Col, AB, 33; Univ Pa, MS, 46; Temple Univ, MA, 55, PhD(physics), 57. *Prof Exp:* Teacher various jr & sr high schs, 36-46; asst prof physics, Pa State Teachers Col West Chester, 46-53; asst, 53-56, ASSOC PROF PHYSICS, TEMPLE UNIV, 56- *Concurrent Pos:* Off Naval Res grant, Temple Univ, 64-69. *Mem:* AAAS; Am Phys Soc; Am Asn Physics Teachers; Acoust Soc Am. *Res:* Diffraction of acoustic waves; musical acoustics. *Mailing Add:* 2411-D Delancey Pl Philadelphia PA 19103

HARBORDT, C(HARLES) MICHAEL, b Houston, Tex, Apr 8, 42; m 60; c 3. ENVIRONMENTAL ENGINEERING, PHYSICAL ORGANIC CHEMISTRY. *Educ:* Stephen F Austin State Univ, BS, 63; Southern Methodist Univ, MS, 65; Tex A&M Univ, PhD(phys chem), 70. *Prof Exp:* Assoc chemist, Texaco, Inc, 65-67, sr chemist, 70-71; dir environ control, Temple Industs, Diboll, 71-75; dir environ affairs, Temple-Eastex Inc, 75-89, dir Eviron Affairs, 80-88, VPRES TEMPLE-INLAND FPC, 90- *Mem:* AAAS; Air Pollution Control Asn; Water Pollution Control Fedn; Tech Asn Pulp & Paper Indust; fel Am Inst Chemists. *Res:* Waste water treatment technology; air pollution. *Mailing Add:* Environ Affairs Inc PO Drawer N Diboll TX 75941

HARBOTTLE, GARMAN, b Dayton, Ohio, Sept 25, 23; m 49; c 1. NUCLEAR CHEMISTRY. *Educ:* Calif Inst Technol, BS, 44; Columbia Univ, PhD(chem), 49. *Prof Exp:* From asst chemist to assoc chemist, Brookhaven Nat Lab, 49-54, chemist, 54-65; dir div res & labs, Int Atomic Energy Agency, Vienna, Austria, 65-67; SR SCIENTIST, BROOKHAVEN NAT LAB, 68- *Concurrent Pos:* Fel, AEC, 51-52; Guggenheim fel, 57-58. *Honors & Awards:* George von Hevesy Medal, 84. *Mem:* Soc Am Archaeol; AAAS; Mex Soc Archaeol; Soc Archaeol Sci (pres, 88-89). *Res:* Chemical consequences of nuclear transformations; application of nuclear techniques to archaeology; radon; forensic science. *Mailing Add:* Brookhaven Nat Lab Upton NY 11973

HARBOUR, JERRY, b Coleman, Tex, Nov 24, 27; m 51; c 3. GEOLOGY. *Educ:* Univ NMex, AB, 51, MS, 57; Univ Ariz, PhD, 66. *Prof Exp:* Instr geol, WTex State Univ, 57-59; geologist, US Geol Surv, Va, 62-69; geologist sci & technol div, Inst Defense Anal, 69-71; eng geologist, Environ Safety Br, US AEC, 71-75; br chief, Site Safety Res Br, US Nuclear Regulatory Comn, 75-81, admin judge, Atomic Safety & Licensing Bd Panel, 81-90; CONSULT, 91- *Concurrent Pos:* Consult, Mus NMex, Santa Fe, 61-62. *Mem:* Soc Vert Paleont. *Mailing Add:* 308 Poplar Dr Falls Church VA 22046

HARBOUR, JOHN RICHARD, b Portage, Wis, May 5, 44; div; c 1. COLLOIDS & INTERFACES. *Educ:* Univ Wis, Eau Claire, BS, 66; Univ Wyo, PhD(chem), 71. *Prof Exp:* Fel chem, Univ Ariz, 71-73; fel chem, Univ Western Ont, 73-74; from scientist to sr scientist chem, Xerox Res Ctr Can, 74- 84, mgr develop physics & characterization, 85-88; ADV SCIENTIST, DEFENSE WASTE PROCESSING TECHNOL, WESTINGHOUSE, 88- *Mem:* Am Chem Soc; Mat Res Soc. *Res:* High-level nuclear waste management; electron spin resonance spectroscopy. *Mailing Add:* Westinghouse Savannah River Co Bldg 773-43A PO Box 616 Aiken SC 29802

HARBOUR, ROBERT MYRON, nuclear chemistry, physical chemistry, for more information see previous edition

HARBOURT, CYRUS OSCAR, b Baton Rouge, La, June 1, 31; m 52; c 5. ELECTRICAL ENGINEERING. *Educ:* La State Univ, BS, 52; Mass Inst Technol, MS, 55; Syracuse Univ, PhD(elec eng), 61. *Prof Exp:* Asst, Mass Inst Technol, 52-54; instr elec eng, La State Univ, 54-55, Univ Del, 56-57 & Syracuse Univ, 57-61; from asst prof to assoc prof, Univ Tex, 61-67; prof elec eng, 67-77, DIR ENG EXTEN, UNIV MO, COLUMBIA, 82-86 & 87- *Concurrent Pos:* NSF grant, Univ Tex, 62-65; elec eng, Bonneville Power Admin, 77-78; interim dean eng, Univ Mo-Columbia, 86-87. *Mem:* Inst Elec & Electronics Engrs; Am Soc Eng Educ; Nat Soc Prof Engrs. *Res:* Active and nonlinear circuit analysis and design; solid state device applications; power systems. *Mailing Add:* Dept Elec Eng Univ Mo Columbia MO 65201

HARBRON, THOMAS RICHARD, b Clinton, Iowa, Dec 16, 37; m 64; c 3. COMPUTER SCIENCES. *Educ:* Iowa State Univ, BS, 60, MS, 61. *Prof Exp:* Instr physics, Anderson Univ, 61-65, asst prof, 65-70, assoc prof physics & computer sci, 70-77, chmn, Computer Sci Dept, 80-90, DIR COMPUT CTR, ANDERSON UNIV, 65-, PROF COMPUTER SCI, 77- *Concurrent Pos:* mem bd, Interex (Hewlett-Packard Users Group), 74-78, chmn bd, 78-79; consult. *Mem:* Inst Elec & Electronics Engrs; Inst Elec & Electronics Engrs Computer Soc; Asn Comput Mach. *Res:* File systems; database systems; programming methods; algorithms. *Mailing Add:* Dept Computer Sci Anderson Univ Anderson IN 46012

HARBURY, HENRY ALEXANDER, b The Hague, Neth, Dec 11, 27; nat US; m 47; c 4. BIOCHEMISTRY. *Educ:* Cornell Univ, AB, 47; Johns Hopkins Univ, PhD(biochem), 53. *Prof Exp:* From instr to assoc prof biochem, Yale Univ, 53-67; prof biochem & chmn sect biochem & molecular biol, dept biol sci, Univ Calif, Santa Barbara, 67-72, chmn dept, 69-72; chmn dept, 72-89, PROF BIOCHEM, DARTMOUTH MED SCH, 72-; PRES, DARTMOUTH-HITCHCOCK MED CTR, 80- *Concurrent Pos:* Markle scholar med sci, 56-61. *Mem:* AAAS; Am Chem Soc; Am Soc Biochem & Molecular Biol. *Res:* Structure and function of proteins; oxidative enzymes. *Mailing Add:* Dept Biochem Dartmouth Med Sch Etna NH 03750

HARCLERODE, JACK E, b Everett, Pa, June 29, 35; m 60; c 3. ZOOLOGY. *Educ:* Shippensburg State Col, BS, 57; Pa State Univ, MS, 58, PhD(zool), 62. *Prof Exp:* Asst prof zool, Univ Ohio, 62-65; from asst prof to assoc prof, 65-72, chmn dept biol, 68-78 & 86-90, PROF BIOL, BUCKNELL UNIV, 72- *Concurrent Pos:* Herbert L Spencer prof, 77. *Mem:* Endocrine Soc. *Res:* Comparative thyroid physiology in birds, reptiles and mammals; pharmacology of marihuana phencyclidine and cocaine. *Mailing Add:* Dept Biol Bucknell Univ Lewisburg PA 17837

HARCOMBE, PAUL ALBIN, b McMinnville, Ore, Oct 13, 45; m 69; c 2. PLANT ECOLOGY. *Educ:* Mich State Univ, BS, 67; Yale Univ, PhD(biol), 73. *Prof Exp:* From asst prof to assoc prof, 72-87, PROF BIOL, RICE UNIV, 87-, CHAIR, DEPT ECOL 7 EVOLUTIONARY BIOL, 89- *Concurrent Pos:* Res assoc & consult, Rice Ctr Community Design & Res, 73-76; assoc ed, Ecol & Ecol Monograph, 78-80, ed, 80-82; mem, Bd Prof Cert, Ecol Soc Am, 83-86. *Mem:* AAAS; Ecol Soc Am; Brit Ecol Soc; Asn Trop Biol. *Res:* Structure and dynamics of coastal plain forest vegetation; woody plant demography; global change. *Mailing Add:* Dept Ecol & Evolutionary Biol Rice Univ PO Box 1892 Houston TX 77251

HARCOURT, DOUGLAS GEORGE, b Toronto, Ont, Mar 23, 26; m 49; c 3. INSECT ECOLOGY, INTEGRATED PEST MANAGEMENT. *Educ:* Univ Toronto, BSA, 49; Cornell Univ, PhD(econ entom), 54. *Prof Exp:* Tech officer entom div, 49-51, res officer entom res inst, 54-65, res scientist, 65-70, chief entom sect, 70-77, SECT HEAD, OTTAWA RES STA, CAN DEPT AGR, 78-, PRIN RES SCIENTIST, 82- *Concurrent Pos:* Adj prof, Carleton Univ, 68- & Guelph Univ, 80-; vis prof, Univ Parana, Brazil, 72; consult, UN Food & Agr Orgn, Argentina, 74-79, Taiwan Nat Coun Agr, 85, Embrapa, Brazil, 81; vis prof, Univ Tucuman, Argentina, 88; study area leader, IPM, Plant Res Centre, Ottawa, 89- *Mem:* Entom Soc Am; Jap Soc Pop Ecol; fel Entom Soc Can. *Res:* Population dynamics of insects attacking forage crops; spatial pattern and sequential sampling of crop insects; pest management. *Mailing Add:* 25 Okanagan Dr Nepean ON K2H 7E9 Can

HARD, CECIL GUSTAV, b Ill, June 2, 23; m 56; c 3. HORTICULTURE. *Educ:* Mich State Univ, BS, 51, MS, 52, PhD(hort), 54. *Prof Exp:* PROF HORT & EXTEN HORTICULTURIST, 54- & PROF LANDSCAPE ARCHIT, UNIV MINN, ST PAUL, 78- *Res:* Ornamental horticulture; landscape design. *Mailing Add:* Hort & Land-Arch Univ Minn 305 Alderman St Paul MN 55108

HARD, MARGARET MCGREGOR, b Can, June 26, 19; nat US; m 48; c 2. FOOD SCIENCE, NUTRITION. *Educ:* Univ Sask, BHSc, 40; Univ Wis, MS, 42. *Prof Exp:* Res chemist, 42-45, asst prof & asst home economist, 45-51, assoc home economist, 51-60, actg dean, Col Home Econ, 73-75, CHMN DEPT HOME ECON, AGR EXP STA, WASH STATE UNIV, 51-, PROF FOODS & NUTRIT, COL HOME ECON, 60- *Concurrent Pos:* Mem, Comt of Nine, USDA, 74-77. *Mem:* AAAS; Am Chem Soc; Am Home Econ Asn; Inst Food Technol. *Res:* Food processing; availability of nutrients for humans; nutritional status of population groups in Washington state; consumer food quality. *Mailing Add:* SE 1000 Spring Pullman WA 99163

HARD, RICHARD C, JR, b May 14, 33. BONE MARROW TRANSPLANTATION, AIDS. *Educ:* St Louis Univ, MD, 58. *Prof Exp:* ASSOC PROF PATH, MED COL VA, 65- *Mem:* Am Asn Immunologists; Am Asn Pathologists; AMA. *Res:* Aids; bone marrow transplantation. *Mailing Add:* Dept Path Med Col Va Box 662 Richmond VA 23298-0662

HARD, ROBERT PAUL, b Seattle, Wash, Oct 16, 44; m 64; c 2. CELL BIOLOGY. *Educ:* Univ Wash, BS, 67, MS, 70; State Univ NY, Albany, PhD(biol), 75. *Prof Exp:* Res assoc cell biol, Dartmouth Col, 75, Univ Ore, 75-78; ASST PROF ZOOL, ORE STATE UNIV, 78- *Concurrent Pos:* Trainee cell biol, NIH, State Univ NY, 71-75. *Mem:* Am Soc Cell Biol. *Res:* Nature of microtubule-dynein interactions in the mechanism of ciliary movement and the role of microtubules in the mechanism of chromosome movement during mitosis. *Mailing Add:* Dept Zool Ore State Univ Corvallis OR 97331

HARD, THOMAS MICHAEL, b Florence, Italy, Nov 25, 37; US citizen. PHYSICAL CHEMISTRY, OPTICS. *Educ:* Harvard Univ, AB, 60; Univ Wis, Madison, PhD(chem), 65. *Prof Exp:* Res assoc chem, Mass Inst Technol, 65-67; physicist, NASA Electronics Res Ctr, 67-70; chemist, US Dept Transp 70-76; RES FEL, PORTLAND STATE UNIV, 77- *Mem:* AAAS; Optical Soc Am; Am Chem Soc; Sigma Xi. *Res:* Atmospheric chemistry; determination of trace constituents; laser spectroscopic methods; interference suppression in detection of weak light signals; chemistry of tropospheric free radicals. *Mailing Add:* 3258 SE Sherman Portland OR 97214

HARDAGE, BOB ADRIAN, b Checotah, Okla, Apr 5, 39; m 60; c 1. EARTH SCIENCES. *Educ:* Okla State Univ, BS, 61, MS, 67, PhD(physics), 67. *Prof Exp:* Res geophysicist, Phillips Petrol Co, 66-80, chief geophysicist seismic stratig, 80-85, chief geophysicist, Europe-Africa, 85-88; RES SCIENTIST, BUR ECON GEOL, AUSTIN, 91- *Mem:* Soc Explor Geophys; Am Asn Petrol Geol. *Res:* Hypervelocity impact; wave propagation in earth materials; well logging; synthetic seismograms; numerical modeling; seismic data processing; geological interpretation of seismic data. *Mailing Add:* Univ Tex Univ Sta Box X Austin TX 78713

HARDAWAY, ERNEST, II, b Columbus, Ga, Mar 3, 34. ORAL SURGERY, MAXILLOFACIAL SURGERY. *Educ:* Howard Univ, BS, 57, DDS, 66, cert(oral surg), 72; Johns Hopkins Univ, MPH, 73. *Prof Exp:* Prof staff, Comt Ways & Means, US Cong, 72; proj officer, Bur Qual Assurance, 74-75, dent off, 75-77; CHIEF, POLICY COORD BR, BUR MED SERV, DEPT HEALTH & HUMAN SERV, 77-, DEP DIR, BUR MED SERV, PUB HEALTH SERV, 81- *Concurrent Pos:* Exec asst to dir, DC Govt, Div Human Resources, 70-71; chief resident oral surg, Howard Univ Med Ctr, 71-72, asst prof, 74-75; spec asst to dir, Off Policy Planning & Eval, Off Asst Secy, Dept HEW, 72-73; comnr pub health, Washington DC, 84- *Mem:* Am Asn Oral & Maxillofacial Surgeons; Am Dent Asn; Nat Dent Asn; Am Col Dentists; Acad Dent Int. *Res:* Health care policy and administration. *Mailing Add:* 2778 Unicorn Lane NW Washington DC 20015

HARDAWAY, JOHN E(VANS), b Philadelphia, Pa, Mar 2, 36; m 63; c 2. HYDROLOGY, GEOLOGICAL ENGINEERING. *Educ:* Princeton Univ, BSE, 59, MSE, 63. *Prof Exp:* Jr geologist, NJ Geol Surv, 59; geophysicist, Geotech & Resources, Inc, 59; asst geophys, Princeton Univ, 62-63; hydrologist, Desert Res Inst, Univ Nev, 63-64; hydrologist, Isotopes, Teledyne, Inc, 64-67, geol engr-hydrologist, 67-69; hydrologist, Fed Water Pollution Control Admin, 69-71; phys sci adminr, Environ Protection Agency, 71-78, asst regional dir tech anal & res, Off Surface Mining, 78-81; mgr regional environ affairs, Tosco Corp, 81-83; scientist, Kaman Tempo, 83-84; TECH MGR ENVIRON AFFAIRS, HOMESTAKE MINING CO, 84- *Concurrent Pos:* Consult, Resource Conserv & Control Act, Superfund. *Honors & Awards:* Gold Medal, US Environ Protection Agency. *Mem:* AAAS; Am Geophys Union; Soc Petrol Engrs; Geol Soc Am; Sigma Xi. *Res:* Characteristics and control of liquid effluents; mine drainage; oil shale development; heavy metal pollution; hydrology; environmental data base management. *Mailing Add:* 3590 N Moore St Wheat Ridge CO 80033

HARDAWAY, ROBERT M, III, b Camp John Hay, Philippines, Jan 9, 16; US citizen; m 39; c 4. SURGERY. *Educ:* Univ Denver, AB, 36; Washington Univ, MD, 39; Am Bd Surg, dipl, 52. *Prof Exp:* Dir, Div Surg, Med Corps, Walter Reed Army Inst Res, Washington, DC, 60-67, commanding officer, 97th Gen Hosp, NY, 67-71, commanding gen, William Beaumont Army Med Ctr, El Paso, 70-75; PROF SURG, SCH MED, TEX TECH UNIV, 76- *Mem:* AMA; Asn Mil Surgeons US; Am Col Surg; Am Asn Surg Trauma; Am Col Angiol. *Res:* Surgical research, particularly with regard to shock and disseminated intravascular coagulation; critical care medicine. *Mailing Add:* 6121 Pinehurst El Paso TX 79912

HARDBERGER, FLORIAN MAX, b Atlanta, La, Dec 18, 14; m 44; c 3. ZOOLOGY, BOTANY. *Educ:* Northwestern State Col, La, BS, 42; La State Univ, MS, 50. *Prof Exp:* From asst prof to assoc prof, 50-71, head dept biol sci, 61-73, PROF ZOOL, NICHOLLS STATE UNIV, 71- *Res:* Affects of heat on mammalian reproduction. *Mailing Add:* 1706 Lynn Ave Thibodaux LA 70301

HARDCASTLE, DONALD LEE, b Wheeler, Tex, July 20, 38; m 59; c 4. COMPUTER SCIENCE, NUMERICAL METHODS. *Educ:* Tex Tech Col, BS, 60, MS, 62; Tex A&M Univ, PhD(physics), 67. *Prof Exp:* Instr physics, South Plains Jr Col, 61-62, Tex Tech Col, 62-63 & Tex A&M Univ, 63-65; from asst prof to assoc prof, 67-81, PROF PHYSICS & DIR COMPUT & INFO SYSTS, BAYLOR UNIV, 81- *Mem:* Am Phys Soc; Am Asn Physics Teachers; Sigma Xi. *Res:* Theoretical physics. *Mailing Add:* Ctr Comput & Info Systs Baylor Univ PO Box 97268 Waco TX 76798

HARDCASTLE, JAMES EDWARD, b San Diego, Calif, Aug 14, 32; m 70; c 6. RADIOCHEMISTRY, BIOCHEMISTRY. *Educ:* Col William & Mary, BS, 53; Univ Richmond, MS, 60; Univ Ariz, PhD(agr chem), 67. *Prof Exp:* Res chemist, Philip Morris, Inc, 58-63; res assoc radiochem, Univ Ariz, 63-67; asst prof chem & biol, Rocky Mt Col, 67-68; asst prof, 68-83, ASSOC PROF BIOCHEM, TEX WOMAN'S UNIV, 83- *Mem:* Am Chem Soc. *Res:* Analytical biochemistry; metal ions in biological systems; the uptake and metabolism of metal ions by algae and fungi; leukotriene production in macrophage cell cultures. *Mailing Add:* Dept Chem & Physics Tex Woman's Univ Box 23973 Denton TX 76204

HARDCASTLE, KENNETH IRVIN, b San Jose, Calif, Jan 2, 31; m 60; c 3. INORGANIC CHEMISTRY, STRUCTURAL CHEMISTRY. *Educ:* San Jose State Col, AB, 52; Univ Miss, MS, 54; Univ Southern Calif, PhD(chem), 61. *Prof Exp:* Chemist, Peninsula Labs, 54-56; res assoc chem, Tufts Univ, 61-63; PROF CHEM, CALIF STATE UNIV, NORTHRIDGE, 63- *Concurrent Pos:* NSF fac fel, Mass Inst Technol, 71-72; vis fac, Univ Sussex, 74, Univ Parma, 80 & 83 & Univ Col London, 85; Fulbright Grant, 83. *Mem:* Am Crystallog Asn; Sigma Xi; Am Chem Soc. *Res:* Metal hydrides; non-stoichiometric compounds; rare earths; x-ray and neutron diffraction studies; crystallographic structure determinations. *Mailing Add:* Dept Chem 18111 Nordhoff St Calif State Univ Northridge CA 91330

HARDCASTLE, WILLIS SANTFORD, b Homer, La, June 16, 21; m 55; c 5. WEED SCIENCE. *Educ:* La State Univ, BS, 50, MS, 52, PhD(plant path), 58. *Prof Exp:* Biologist, La Wildlife & Fisheries Comn, 55-61; asst agronomist, 61-80, ASSOC AGRONOMIST, GA EXP STA, 80- *Mem:* Weed Sci Soc Am; Am Soc Agron. *Res:* Laboratory, field and greenhouse evaluations of herbicides in agronomic crops, pastures, non-crop and aquatic areas; soil-herbicide interactions; physiological and life cycle studies on crops and weeds receiving herbicide applications. *Mailing Add:* 217 Hillandale Dr Griffin GA 30223

HARDEBECK, ELLEN JEAN, b Chicago, Ill, July 25, 39; m 68; c 1. RADIO ASTRONOMY. *Educ:* Univ Chicago, BS, 61; Harvard Univ, MS, 63, PhD(astron), 65. *Prof Exp:* Res assoc radio astron, Cornell Univ, 65-69; res fel, Calif Inst Technol, 69-72; AIR POLLUTION CONTROL OFFICER, GREAT BASIN AIR POLLUTION CONTROL DIST, 85- *Mem:* Am Astron Soc; Int Astron Union; Int Union Radio Sci; Air Pollution Control Asn. *Res:* Interstellar medium; interstellar molecules; interplanetary scintillations studies with radio astronomical techniques. *Mailing Add:* 3106 Tumbleweed Rd Bishop CA 93514

HARDEE, DICKY DAN, b Snyder, Tex, July 21, 38; m 59; c 2. ENTOMOLOGY, ECOLOGY. *Educ:* Tex Tech Col, BS, 60; Cornell Univ, MS, 62, PhD(entom), 64. *Prof Exp:* Res asst entom, Cornell Univ, 60-64; res entomologist, Boll Weevil Res Lab, USDA, 64-74; PRES, PEST MGT SPECIALISTS, INC, 74- *Honors & Awards:* Super Serv Award, USDA, 75. *Mem:* AAAS; Entom Soc Am. *Res:* Insecticides residues in or on raw agricultural commodities; biology, ecology and control of forage and grass insects; field and laboratory studies with plant and sex attractants for boll weevil; use of traps and pheromones in detection and management of insect pests. *Mailing Add:* 905 N Deer Creek Dr E Leland MS 38756

HARDEGREE, MARY CAROLYN, b Wichita Falls, Tex, Oct 2, 33; m 61; c 2. PEDIATRICS, IMMUNOLOGY. *Educ:* Baylor Univ, BS, 55; Univ Tex, MD, 58; Am Bd Pediat, dipl. *Prof Exp:* Rotating intern, Minneapolis Gen Hosp, Minn, 58-59; resident pediat, Sch Med, Univ Minn, Minneapolis, 59-61, med fel specialist, 61-62; med officer, Div Biologics Stand, NIH, 62-68, chief bact toxins sect, 68-72, DIR BACT TOXINS BR, BUR BIOLOGICS, FOOD & DRUG ADMIN, 72- *Mem:* Infectious Diseases Soc Am; Am Asn Immunol; Soc Pediat Res; Am Soc Microbiol; fel Am Acad Pediat. *Res:* Bacterial toxins and toxoids; adjuvants; tetanus. *Mailing Add:* Off Biologics Res Ctr Biol Eval & Res FDA 8800 Rockville Pike Bethesda MD 20892

HARDEKOPF, ROBERT ALLEN, b St Louis, Mo, Oct 14, 40; m 62; c 3. NUCLEAR PHYSICS. *Educ:* Auburn Univ, BS, 62; Duke Univ, PhD(physics), 72. *Prof Exp:* STAFF MEM PHYSICS, LOS ALAMOS NAT LAB, UNIV CALIF, 72- *Mem:* Am Phys Soc. *Res:* Polarization phenomena in nuclear reactions, including experimental research and design, development of polarized ion sources. *Mailing Add:* Group AT-3 MS H-804 Los Alamos Nat Lab Los Alamos NM 87545

HARDELL, WILLIAM JOHN, b Chicago, Ill, May 28, 28; m 56; c 1. MATHEMATICS. *Educ:* Northwestern Univ, BS, 50; Mich State Univ, MS, 53, PhD(math), 59. *Prof Exp:* Mathematician, Univac Div, Sperry-Rand Corp, Minn, 56-59; systs engr, Defense Electronic Prod Div, Radio Corp Am, NJ, 59-60; from asst prof to assoc prof, 60-67, PROF MATH, WORCESTER POLYTECH INST, 67- *Mem:* Am Math Soc; Math Asn Am; Inst Mgt Sci; Sigma Xi. *Res:* Associative algebras. *Mailing Add:* Dept Math Worcester Polytech Inst 1000 Inst Rd Worcester MA 01609

HARDEN, DARREL GROVER, b Fay, Okla, Dec 29, 30; m 58; c 3. THERMODYNAMICS. *Educ:* Univ Okla, BS, 57; Southern Methodist Univ, MS, 60; Okla State Univ, PhD(mech eng), 63. *Prof Exp:* Res assoc heat transfer, Argonne Nat Lab, 62-63; asst prof, 63-67, ASSOC PROF MECH ENG, UNIV OKLA, 67- *Mem:* Am Soc Mech Engrs; Am Soc Eng Educ. *Res:* Dynamic response of fluid flow systems with heat addition, including nuclear reactors and combustion instability. *Mailing Add:* Sch Aerospace & Mech Eng Univ Okla 209A Felgar Hill Norman OK 73019

HARDEN, PHILIP HOWARD, b SDak, May 17, 08; m 39, 59; c 1. ENTOMOLOGY. *Educ:* Greenville Col, AB, 35; Univ Minn, PhD(zool, entom), 49. *Prof Exp:* Asst zool, Univ Minn, 47-49; chmn div sci, Pasadena Col, 49-57; pres, Wessington Springs Col, 57-61; prof, 61-74, EMER PROF BIOL, ROBERTS WESLEYAN COL, 74- *Res:* Medical entomology; plecoptera. *Mailing Add:* Heritage Square Lot 0-2 Mission TX 78572

HARDENBURG, ROBERT EARLE, b Ithaca, NY, July 27, 19; m 43; c 2. HORTICULTURE STORAGE, POSTHARVEST PHYSIOLOGY. *Educ:* Cornell Univ, BS, 41, MS, 47, PhD(veg crops), 49. *Prof Exp:* Assoc horticulturist, Plant Indust Sta, USDA, 49-53, horticulturist, Biol Sci Br, Agr Mkt Serv, 53-55, sr horticulturist, 56-59, prin horticulturist, 59-67, invests leader, 67-72, chief, Hort Crops Mkt Lab, Hort Crops Inst, 72-81; RETIRED. *Concurrent Pos:* Sci adv coun, Refrig Res Found, 75-91. *Honors & Awards:* Distinguished Serv Award, Produce Mkt Asn, 63. *Mem:* AAAS; fel Am Soc Hort Sci; Produce Mkt Asn; Int Inst Refrig. *Res:* Post-harvest horticulture and physiology of fruits and vegetables; prepackaging; handling and storage. *Mailing Add:* 648 Bird Bay Dr W Venice FL 34292-4027

HARDER, DAVID RAE, b Marshfield, Wis, Jan 10, 50; m 71; c 3. PHYSIOLOGY, ELECTROPHYSIOLOGY. *Educ:* Univ Wis, BA, 72; Med Col Wis, MS, 75, PhD(physiol), 76. *Prof Exp:* Lectr physiol, Univ Wis-Milwaukee, 76-77; res assoc, Med Col Wis, 76-77; fel electrophysiol, Med Sch, Univ Va, 77-78; asst prof physiol, Sch Med, E Tenn State Univ, 78-80; res asst prof physics & biophys, Univ Vt, Burlington, 80-83; ASSOC PROF NEUROL & PHYSICS, MED COL WIS, MILWAUKEE, 83-, DIR NEUROL RES, 84- *Concurrent Pos:* Consult scientist, Cleveland Clin Found, 77; estab investr, Am Heart Asn, mem sci coun; res career scientist, Vet Admin, 85. *Mem:* Sigma Xi; Am Physiol Soc; Fedn Am Soc Exp Biol; Soc Gen Physiologists; Biophys Soc; Int Soc Heart Res. *Res:* Electrophysiology of excitable cells; ionic conductances across cell membranes; control of vascular smooth muscle. *Mailing Add:* Dept Physiol Med Col Wis 8701 Watertown Plank Rd Milwaukee WI 53226

HARDER, DONALD EDWALD, b Bassano, Alta, May 17, 39; m 65; c 2. PLANT PATHOLOGY, CELL BIOLOGY. *Educ:* Univ Alta, BSc, 62, MSc, 64; Wash State Univ, PhD(plant path), 68. *Prof Exp:* Fel plant sci, Univ Man, 68-69, asst prof agr, 69-73; RES SCIENTIST PLANT PATH, RES BR, AGR CAN, 73- *Concurrent Pos:* Adj prof, Dept Bot, Res Admin, Univ Man. *Mem:* Am Phytopath Soc; Can Phytopath Soc; Sigma Xi; Micros Soc Can. *Res:* Genetics of host parasite relations in the cereal rust diseases; physiology, ultrastructure and cytochemistry of host parasite relations in the cereal rusts. *Mailing Add:* Agr Can Res Sta 195 Dafoe Rd Winnipeg MB R3T 2M9 Can

HARDER, EDWIN L, b Buffalo, NY, 1905. ELECTRICAL ENGINEERING. *Educ:* Cornell Univ, EE, 26; Univ Pittsburgh, MS, 30, PhD, 46. *Prof Exp:* SR CONSULT, WESTINGHOUSE ELEC CO, 70- *Concurrent Pos:* Res & writing energy resources, processes & basic fundamentals. *Honors & Awards:* Lamme Award, Am Inst Elec Engrs; Distinguished Serv Award, Am Fedn Info Processing Socs; Centennial Award, Am Inst Elec Engrs, Nat Acad Engrs, 76, 90. *Mem:* Nat Acad Eng; Am Inst Elec Engrs; Am Math Soc. *Res:* Railroad electrification; electric power engineering; computers; advanced system engineering; power system problems. *Mailing Add:* 1204 Milton Ave Pittsburgh PA 15218

HARDER, HAROLD CECIL, b Lansing, Mich, Feb 27, 43; m 65; c 3. PHARMACOLOGY, CANCER. *Educ:* Alma Col, BS, 64; Mich State Univ, MS, 66, PhD(biophys), 70. *Prof Exp:* Fel pharmacol, Yale Univ, 70-72; asst prof pharmacol, George Washington Univ, 72-78; assoc prof pharmacol, Oral Roberts Univ, 78-85; ASSOC PROF PHARMACOL, BLESSINGS INT, 85- *Concurrent Pos:* Pres, Blessings Int. *Mem:* Am Asn Cancer Res; Biophys Soc; Sigma Xi; Am Soc Pharm & Exp Therapeut; NY Acad Sci. *Res:* Cancer chemotherapy, drug effects on mammalian cell cycle kinetics, drug induced DNA damage and repair, template activity of drug treated DNA and polynucleotides, heavy metal pharmacology and toxicology, drug metabolism, drug interactions, pharmacokinetics. *Mailing Add:* 5725 E 97th Pl Tulsa OK 74136

HARDER, JAMES ALBERT, b Fullerton, Calif, Dec 2, 26; div; c 3. ENGINEERING SCIENCE. *Educ:* Calif Inst Technol, BS, 48; Univ Calif, Berkeley, MS, 52, PhD(fluid mech), 57. *Prof Exp:* Design engr, soil conserv serv, USDA, 48-50; res engr, 52-57, asst prof hydraul eng, 57-62, assoc prof civil eng, 62-69, PROF HYDRAUL ENG, UNIV CALIF, BERKELEY, 69- *Concurrent Pos:* Hydrologist, Aquatechnics, San Francisco. *Mem:* Fel AAAS; Am Soc Civil Engrs; Soc Sci Explor. *Res:* Pollution travel in pipe networks; non-linear system analysis; feedback control of canal systems; flood propagation. *Mailing Add:* Dept Civil Eng Univ Calif Berkeley CA 94720

HARDER, JOHN DWIGHT, b O'Neill, Nebr, May 21, 43; m 66; c 2. REPRODUCTIVE PHYSIOLOGY, WILDLIFE BIOLOGY. *Educ:* Hastings Col, BA, 65; Colo State Univ, MS, 67; Ohio State Univ, PhD(zool), 71. *Prof Exp:* Res asst wildlife biol, Colo Coop Wildlife Res Unit, Colo State Univ, 65-67; asst prof, State Univ NY Col Oswego, 70-73; ASSOC PROF ZOOL, OHIO STATE UNIV, 73- *Mem:* Am Soc Mammal; Wildlife Soc; Soc Study Reproduction. *Res:* Reproductive endocrinology of mammals; population biology; wildlife ecology; ovariectomy, ovarian analysis; radio immunoassay of ovarian steroids; physiology regulations of estrus; ovulation and partarition. *Mailing Add:* Dept Zool Ohio State Univ 1735 Neil Ave 063 B-Z Columbus OH 43210

HARDER, JOHN JURGEN, industrial engineering; deceased, see previous edition for last biography

HARDER, ROGER WEHE, b Fond du Lac, Wis, May 26, 17; m 46; c 3. SOIL CONSERVATION, SOIL FERTILITY. *Educ:* Univ Wis, BA, 42, MS, 47. *Prof Exp:* From asst prof to assoc prof agron, 47-64, assoc prof agr biochem & soils, 64-72, assoc prof & assoc soil scientist, 72-77, prof plant & soil scientist, Soil Fertil, Univ Idaho, 77-83; RETIRED. *Mem:* Soil Conserv Soc Am; Soil Sci Soc Am; Am Soc Agron. *Res:* Soil erosion control and management. *Mailing Add:* 314 E Seventh St Moscow ID 83843

HARDESTY, BOYD A, b Cheney, Wash, May 15, 32; m 52; c 3. BIOCHEMISTRY. *Educ:* Wash State Univ, BS, 53, MS, 56; Calif Inst Technol, PhD(biochem), 60. *Prof Exp:* NSF fel, Med Sch, Yale Univ, 61; NSF fel, Med Sch, Univ Ky, 60-62, USPHS fel, 62-63; from asst prof to assoc prof, 63-74, PROF CHEM, UNIV TEX, AUSTIN, 74- *Mem:* AAAS; Am Chem Soc; Am Soc Biol Chem. *Res:* Hemoglobin biosynthesis and polyribosomes in rabbit reticulocytes; ribosome structure and functions; translational control in normal and transformed erythroid cells. *Mailing Add:* Clayton Found Dept Chem Univ Tex Austin TX 78712

HARDESTY, GEORGE K(IRWAN) C(OLLISON), b Mayo, Md, Nov 15, 09; m 32; c 1. APPLIED OPTICS, ILLUMINATING ENGINEERING. *Prof Exp:* Asst to supt Interior Commun Div, US Naval Res Lab, 41-43, asst supt, 43-47, head, Indicators Sect, 47-49, head, Interior Commun Br, 50-56, electronic scientist, 56-60, electronics engr, 60-61, res electronics engr, US Navy Marine Eng Lab, 61-65; consult & display engr, 65-69; CONSULT ENGR, 69-; VPRES, GAYSEA ENG CORP, 81- *Concurrent Pos:* Consult Polaris Proj Off, Navy Bur Weapons, 61-65, Miller Co, Conn, 62-65 & Belsinger-Signtific Display Div, Md, 64-65. *Mem:* Soc Automotive Engrs; fel Illum Eng Soc; Optical Soc Am; sr mem Instrument Soc Am. *Res:* Information display systems; instrument and display illumination; work space surveys, proposals; colored signal specifications and evaluations; spectroradiometry and colorimetry; human factors interface problems; invention development, tutorial writings; expository technical writing; approximately 40 US & foreign patents awarded; electro-optics, electro-mechanical product design; security-system keyed locks; safety systems. *Mailing Add:* PO Box 355 Secretary MD 21664

HARDESTY, PATRICK THOMAS, b Owensboro, Ky, Feb 9, 51; m 77. CHEMISTRY. *Educ:* Brescia Col, BA, 73; Purdue Univ, MS, 74; Univ Ill, PhD(anal chem), 80. *Prof Exp:* Res chemist, Monsanto Co, 75-77, RES CHEMIST, E I DU PONT DE NEMOURS & CO, INC, 80- *Mem:* Am Chem Soc; Sigma Xi. *Res:* Metabolism research to determine the environmental fate of pesticides. *Mailing Add:* 822 Starvegut Rd Kennett Square PA 19348

HARDGROVE, GEORGE LIND, JR, b Barberton, Ohio, Dec 2, 33; m 61; c 2. PHYSICAL CHEMISTRY. *Educ:* Oberlin Col, AB, 56; Univ Calif, Berkeley, PhD(chem), 59. *Prof Exp:* From asst prof to assoc prof, 59-71, PROF CHEM, ST OLAF COL, 71- *Concurrent Pos:* NIH spec fel, Oxford Univ, 65-66; vis prof, Univ Fla, 73-74; vis consult, Oak Ridge Nat Lab, 81-82; Res Fel, Univ Durham, Eng, 88-89. *Mem:* Am Crystallog Asn; Am Chem Soc. *Res:* Molecular structure determination by NMR and x-ray diffraction. *Mailing Add:* Dept Chem St Olaf Col Northfield MN 55057-1098

HARDHAM, WILLIAM MORGAN, b Philadelphia, Pa, Oct 19, 39; m 61, 71; c 3. PHOTOGRAPHIC CHEMISTRY, ORGANIC CHEMISTRY. *Educ:* Pa State Univ, BS, 61; Calif Inst Technol, PhD(chem), 65. *Prof Exp:* NIH fel, 64-65; res chemist, 65-69, res supvr, Photo Prod Dept, 69-87, ASST TO DIR RES, ELECTRONICS DEPT, E I DU PONT DE NEMOURS & CO, INC, 87- *Mem:* Am Chem Soc. *Res:* Photochemistry of maleic anhydride; ketone photoreductions; mechanism of the photoreaction of orthoquinone monoimines with arylmethanes; sensitized and direct photodecomposition of biimidazoles; silver halide emulsions. *Mailing Add:* Du Pont Electronics BMP 21 PO Box 80021 Wilmington DE 19880-0021

HARDIE, EDITH L, b Ancon, CZ, Oct 1, 31. CARDIOPULMONARY PHYSIOLOGY. *Educ:* Georgetown Univ, BSN, 54; Med Col Va, PhD(physiol), 69. *Prof Exp:* Head nurse, Med-Surg Unit, Med Col Va, 54-56, head nurse-technician, Pulmonary Function Lab, 56-61, instr, 69-71, ASST PROF PHYSIOL, MED COL VA, VA COMMONWEALTH UNIV, 71- *Concurrent Pos:* Richmond Area Heart Asn fel, Med Col Va, 70-71, A D William Fund fels, 71-72 & 73-74; Va Heart Asn fel, 72-73; vis scientist, Stritch Sch Med, Loyola Univ, Ill, 75-76. *Mem:* Sigma Xi; assoc mem Am Physiol Soc; AAAS. *Res:* Cardiac and pulmonary receptors and their reflex effects on circulation and respiration; experimental pulmonary embolism; location of atrial subsidiary pacemakers. *Mailing Add:* 300 W Franklin St Apt 508 E Richmond VA 23220

HARDIE, GERALD, b Winnipeg, Man, Feb 7, 31; US citizen. NUCLEAR PHYSICS. *Educ:* Univ Man, BS, 55, MS, 57; Univ Wis, PhD(physics), 62. *Prof Exp:* Assoc physicist, IIT Res Inst, 62-64, res physicist, 64-65; asst prof, 65-67, assoc prof, 67-72, PROF PHYSICS, WESTERN MICH UNIV, 72- *Mem:* Am Phys Soc. *Res:* Low energy nuclear physics. *Mailing Add:* Dept Physics Western Mich Univ 1128 EVR Kalamazoo MI 49008

HARDIE, ROBERT HOWIE, physics, astronomy; deceased, see previous edition for last biography

HARDIE, WILLIAM GEORGE, b Terre Haute, Ind, Mar 6, 58; m 82. LOW DIELECTRIC CONSTANT MATERIALS, HIGH TEMPERATURE POLYMERS. *Educ:* State Univ NY, Stony Brook, BS, 80. *Prof Exp:* Chemist, E I DuPont, 80-82; PROCESS DEVELOP CHEMIST, W L GORE & ASSOCS INC, 82- *Mem:* Soc Plastics Engrs. *Res:* Development of new processes and new materials for use in the production of high performance wire and cable products; high temperature, high speed products. *Mailing Add:* 750 Otts Chapel Rd PO Box 8038 Newark DE 19714

HARDIN, BOBBY OTT, b Lexington, Ky, Sept 9, 35; m 60; c 2. CIVIL ENGINEERING. *Educ:* Univ Ky, BS, 56, MS, 58; Univ Fla, PhD, 61. *Prof Exp:* Hwy engr, Ky Dept Hwys, 55-56; from instr to assoc prof, 56-67, PROF CIVIL ENG, UNIV KY, 67- *Concurrent Pos:* Struct design engr, Gregg & Assocs, 57-; asst, Univ Fla, 59; Ford Found resident engr, Tenn Valley Authority, Nickajack Proj, 65-66. *Honors & Awards:* Alfred Noble Prize, 66; Walter Huber res prize, Am Soc Civil Engrs, 68, Norman Medal 73, & Thomas A Middlebrooks Award, 79; C A Hugontogler Award, Am Soc Testing & Mat, 79; J James R Croe Medal, 90. *Mem:* Fel Am Soc Civil Engrs; Am Soc Testing & Mat. *Res:* Soil dynamics; constitutive equations. *Mailing Add:* Dept Civil Eng Univ Ky Lexington KY 40506

HARDIN, CAROLYN MYRICK, physiology, neuroendocrinology, for more information see previous edition

HARDIN, CLIFFORD MORRIS, b Knightstown, Ind, Oct 9, 15; m 39; c 5. AGRICULTURAL ECONOMICS. *Educ:* Purdue Univ, BS, 37, MS, 39, PhD(agr econ), 41. *Hon Degrees:* DSc, Purdue Univ, 53, NDak State Univ, 69, Mich State Univ, 69; Dr, Nat Univ Colombia, 68; LLD, Creighton Univ, 56, Ill State Univ, 73; DHumL, Univ Nebr, 78, Okla Christian Col, 79. *Prof Exp:* Fac mem agr econ, Univ Wis, 41-44; prof dir & dean agr, Mich State Univ Col Agr, 44-55; chancellor, Univ Nebr, 54-69; secy, USDA, 69-71; vchmn & dir corp res, Ralston Purina Co, 71-80; dir, Ctr Study Am Bus, 81-82, scholar-in-residence, Washington Univ 80-81 & 83-85; CONSULT, STIFEL NICOLAUS & CO, INC, 80- *Concurrent Pos:* Bd dir, Ralston Purina Co, 71-81 & Ralston Purina Can, 71-80; bd trustees, Rockefeller Found, 61-81, Int Agr Develop Serv, 75-85, Farm Found, 73-84, Am Assembly, 75-, Univ Nebr Found, 75-; mem bd dir, 81-; trustee, Kettering Found, 80-, Winrock Int Inst Agr Develop, 85- *Mailing Add:* Ten Roan Lane St Louis MO 63124

HARDIN, CLYDE D, b Ft Worth, Tex, May 26, 25; m 48; c 3. ELECTRONICS ENGINEERING. *Educ:* Wake Forest Col, BS, 48. *Prof Exp:* Asst staff engr electronics, Inst Coop Res, Johns Hopkins Univ, 48; res physicist & group leader, Nat Bur Stands, 48-53; group leader & sect chief, Harry Diamond Labs, 53-58, chief adv res lab, 58-65, develop lab, 65-69; spec asst on SE Asia, to Asst Secy Army, Res & Develop, 69-72, dir defense res, develop, test & eval counterpart group, Korea, 72-73; dir, US Army Electronics Warfare Lab, 73-80, tech dir, US Army Electronics Res & Develop Command, 80-81; vpres, Int Opers, J S Lee Assocs Inc, 85-86; CONSULT, 81- *Honors & Awards:* Secy Defense Meritorious Serv Medal, 73; Republic Korea Pres Cheonsu Medal, 76; Asn Old Crows Nat Medal Electronic Warfare Mgt, 76; Res & Develop Achievement Award, Dept Army, 69, Army Meritorious Serv Medal, 71. *Mem:* Fel Inst Elec & Electronics Engrs; AAAS; Am Defense Preparedness Asn; Asn Old Crows; Armed Forces Commun & Electronics Asn. *Res:* Military electronics; radar; guidance; fuzing and communications systems and techniques. *Mailing Add:* One Sorrelwood Cross Savannah GA 31411

HARDIN, CREIGHTON A, b Clinton, NC, July 20, 18; m 47; c 3. SURGERY. *Educ:* Univ Wis, BA, 40, MD, 43; Univ Kans, MS, 50; Am Bd Surg & Am Bd Plastic Surg, dipl. *Prof Exp:* From instr to assoc prof, 50-68, PROF SURG & CHIEF GEN SURG, MED CTR, UNIV KANS, 68- *Mem:* AAAS; Am Surg Asn; Soc Univ Surgeons; Am Col Surg. *Res:* Arterial grafting; hypothermia; organ and tissue transplantation; immunological factors. *Mailing Add:* Univ Kans Col Health 39th & Rainbow Blvd Kansas City KS 66103

HARDIN, EDWIN M(ILTON), b Birmingham, Ala, June 11, 26; m 48; c 2. CIVIL & STRUCTURAL ENGINEERING. *Educ:* Univ Ala, BS, 56. *Prof Exp:* Engr, Rust Eng Co, 56-58 & Schoel Eng Co, 58-63; sr design engr, Rust Int Corp, 63-68, chief struct engr, 68-70, chief design engr, 70-73, chief engr, 74-79, vpres eng, 79-88, VPRES OPERS, RUST INT CORP, 89- *Concurrent Pos:* Instr, Univ Ala, Birmingham; pres, Rust Eng Co, Va, 77-83, Mich, 79-, NY, 82- & NC, 87- *Mem:* Am Soc Civil Engrs; Am Inst Steel Construct; Nat Soc Prof Engrs; Am Soc Eng Educ; Tech Design Asn; Soc Am Military Engrs. *Mailing Add:* Rust Int Corp PO Box 101 Birmingham AL 35201-0101

HARDIN, GARRETT (JAMES), b Dallas, Tex, Apr 21, 15; m 41; c 4. BIOLOGY, HUMAN ECOLOGY. *Educ:* Univ Chicago, ScB, 36; Stanford Univ, PhD(biol), 41. *Hon Degrees:* DHumanities, Puget Sound Univ, 75; LHD, Northland Col, 76. *Prof Exp:* Asst, Stanford Univ, 36-37, 38-42; asst, Chicago City Jr Col, 38; mem staff, Div Plant Biol, Carnegie Inst, 42-46; asst prof bact, 46-50, assoc prof biol, 50-57, prof, 57-63, prof human ecol, 63-78, EMER PROF HUMAN ECOL, UNIV CALIF, SANTA BARBARA, 78- *Concurrent Pos:* Actg asst prof, Stanford Univ, 45, vis prof, 48; sr res assoc, Calif Inst Technol, 52-53; vis prof, Univ Calif, Los Angeles, 61, Berkeley, 64 & Univ Chicago, 70; pres Environ Fund, 80-81. *Honors & Awards:* Margaret Sanger Award, 80. *Mem:* AAAS; Ecol Soc Am; Am Philos Soc; Am Acad Arts & Sci; Int Soc Ecol Econ. *Res:* Evolution; human ecology. *Mailing Add:* Dept Biol Sci Univ Calif Santa Barbara CA 93106

HARDIN, GEORGE C, JR, b Oakwood, Tex, Oct 6, 20; m 42; c 2. GEOLOGY. *Educ:* Agr & Mech Col, Univ Tex, BS, 41; Univ Wis, PhM, 42. *Prof Exp:* Geologist & mining engr, Victory Fluorspar Co, Ill, 42; geologist, US Geol Surv, Washington, DC, 42-45 & Carter-Gragg Oil Co, Tex, 45-46; geologist & petrol engr, Michel T Halbouty, 46-51, explor & prod mgr, Michel T Halbouty Oil & Gas Interests, 51-59, gen mgr, 59-61; partner, Hardin & Hardin, Consult Geologists, 61-64; mgr oil & gas explor, Kerr-McGee Oil Industs, Inc, 64-65, vpres, NAm Oil & Gas Explor, Kerr-McGee Corp, 65-67, explor, 67-68; pres, Royal Resources Corp, 68-71; pres, Ashland Explor Co, 71-80, sr vpres, Ashland, Inc, 71-80; vchmn & dir, Integrated Energy, Inc, 81-84; DIR, AM GAS & OIL INVESTORS, 81- *Concurrent Pos:* Exec vpres, Halbouty Alaska Oil Co, 58-61; vis geoscientist, Am Geol Inst, 63. *Mem:* Am Asn Petrol Geol (secy-treas, 64-66); Geol Soc Am; Nat Soc Prof Eng; Soc Econ Paleont & Mineral; Soc Explor Geophys. *Res:* Petroleum geology; sedimentation; stratigraphy. *Mailing Add:* 1115 Barkdull Houston TX 77006

HARDIN, HILLIARD FRANCES, b Columbia, SC, Dec 12, 17. MICROBIOLOGY, MYCOLOGY. *Educ:* Duke Univ, AB, 39, MA, 49, PhD(microbiol), 53. *Prof Exp:* Instr microbiol, Med Ctr, Univ Ark, Little Rock, 53-58; res assoc mycol, Med Ctr, Duke Univ, 58-63; chief mycol training unit, Commun Dis Ctr, 63-68; DIR, BACT-MYCOLOGY SECT, LAB SERV, VET ADMIN HOSP, 68-; ASSOC PROF MICROBIOL, MED CTR, UNIV ARK, LITTLE ROCK, 68- *Concurrent Pos:* Microbiologist, Allied Health Prog Med Technol, 72-; guest fac, Continuing Educ Prog, Am Soc Med Technol, 72-; guest lectr mycol, Sch Med, East Tenn State Univ, 79- *Mem:* Med Mycol Soc Am; NY Acad Sci. *Res:* Mycological immunoserology. *Mailing Add:* Microbiol Sect Lab Serv Vet Admin Hosp 4300 W 7th St Little Rock AR 72205

HARDIN, IAN RUSSELL, b Glasgow, Scotland, Aug 3, 44; US citizen; m 67; c 2. TEXTILE CHEMISTRY, POLYMER SCIENCE. *Educ:* Auburn Univ, BS, 65; Inst Textile Technol, MS, 67; Clemson Univ, PhD(chem), 70. *Prof Exp:* Fel polymers, Univ Mich, 70-71; asst prof textiles, 71-76, dept head consumer affairs, 77- 82, ASSOC PROF TEXTILES, AUBURN UNIV, 76-; CONSULT. *Concurrent Pos:* Chmn, Gen Fac & Univ Senate, Auburn Univ, 85-86. *Mem:* Am Chem Soc; Am Asn Textile Chemists & Colorists; Am Home Econ Asn; Sigma Xi; Fiber Soc; AAAS. *Res:* Textile flammability; morphology of polymers; photochemical effects in fibers. *Mailing Add:* Dept Consumer Affairs Auburn Univ Auburn AL 36849

HARDIN, JAMES T, b Inez, Ky, Sept 28, 34; m 57; c 3. INDUSTRIAL CONTROLS, POWER ELECTRONICS. *Educ:* Bradley Univ, BS, 56, MS, 63. *Prof Exp:* Proj engr, Lockheed Aircraft Corp, 56-57; grad asst circuit anal, Bradley Univ, 61-63; mgr & vpres eng, Prestolite Div, Allied Signal Corp, 64-85; chief engr, Septor Electronics Corp, 85-90; PRES, TERAMAR TECHNOLOGIES INC, 91- *Concurrent Pos:* Chmn, Ignition Comt, Soc Automotive Engrs, 80-85; mem, US deleg Int Standards Orgn, 80-85. *Mem:* Inst Elec & Electronics Engrs; Soc Automotive Engrs. *Res:* Developed and patented electronic ignition systems, voltage regulators and alternators for automotive and marine engines; developed unique machine control system hardware based on digital computer techniques and logic. *Mailing Add:* Box 944 Santa Teresa NM 88008

HARDIN, JAMES W, b Baton Rouge, La, Nov 16, 46. ENDOCRINOLOGY, MOLECULAR ENDOCRINOLOGY. *Educ:* Purdue Univ, PhD(biochem), 72. *Prof Exp:* ASSOC PROF MED & BIOCHEM, UNIV ARK MED SCI, 80-; ASSOC DIR RES, ARK CANCER RES CTR, 88- *Concurrent Pos:* Reviewer, Nat Cancer Inst. *Mem:* AAAS; Am Soc Cell Biol; Sigma Xi; Am Soc Microbiol. *Res:* Molecular biology of hormone and proto oncogene activity. *Mailing Add:* Dept Med Biochem Univ Ark Med Sci 4301 W Markham St Little Rock AR 72205

HARDIN, JAMES WALKER, b Charlotte, NC, Mar 31, 29; m 57; c 3. SYSTEMATIC BOTANY, DENDROLOGY. *Educ:* Fla Southern Col, BS, 50; Univ Tenn, MS, 51; Univ Mich, PhD(bot), 57. *Prof Exp:* Instr bot, Univ Mich, 56-57; from asst prof to assoc prof, 57-68, PROF BOT, NC STATE UNIV, 68-, CUR HERBARIUM, 57- *Concurrent Pos:* Vis prof, Mt Lake Biol Sta, Va, 62, 64 & 83; pres, Highlands Biol Sta, 63-69; consult, Res Triangle Inst, 64-70; vis prof biol sta, Univ Okla, 67 & 70; ed, ASB Bull, 80-86, Systematic Bot, 85-91. *Honors & Awards:* Cooley Award, Am Soc Plant Taxon, 58. *Mem:* Am Soc Plant Taxon; Bot Soc Am; Int Asn Plant Taxon; Soc Econ Bot; Asn Southeastern Biol (pres, 79-80). *Res:* Taxonomic botany; taxonomy of woody Angiosperms; flora of the southeastern United States; endangered species; poisonous plants; scanning electron microscopy of foliar surfaces. *Mailing Add:* Dept Bot NC State Univ Raleigh NC 27695-7612

HARDIN, JAMES WILLIAM, b Paintsville, Ky, July 16, 43. VERTEBRATE ECOLOGY, ANIMAL BEHAVIOR. *Educ:* Univ Ky, BS, 65, MS, 67; Southern Ill Univ, PhD(zool), 74. *Prof Exp:* Instr biol, Memphis State Univ, 67-69; researcher, 74, asst dir, wildlife res lab, Southern Ill Univ, 74-78, asst prof zool, 74-78; from asst prof to assoc prof, 78-85, PROF WILDLIFE, COL NATURAL RESOURCES, UNIV WIS, STEVENS POINT, 85- *Mem:* Am Soc Mammalogists; Wildlife Soc; Ecol Soc Am; Wilson Ornith Soc. *Res:* Life history, population dynamics and behavior of vertebrates emphasizing the roles that these parameters play in enabling populations to exist in the wild. *Mailing Add:* 5527 Jefferson St Stevens Point WI 54481

HARDIN, JAY CHARLES, b Indianapolis, Ind, Oct 22, 42; m 64; c 3. ACOUSTICS, FLUID MECHANICS. *Educ:* Purdue Univ, BS, 64, MS, 65, PhD(eng sci), 69. *Prof Exp:* SCIENTIST ACOUST, NASA, 69- *Concurrent Pos:* Fel, Inst of Sound & Vibration Res, Southampton, Eng, 72; prof, George Washington Univ, 70- & Christopher Newport Col, 81- & Old Dominion Univ, 83- *Mem:* Sigma Xi. *Res:* Aeroacoustics; stochastic processes; computational fluid mechanics. *Mailing Add:* 116 Mistletoe Dr Newport News VA 23606

HARDIN, JOHN AVERY, b Washington, Ga, Aug 12, 43. RHEUMATOLOGY. *Educ:* Med Col Ga, MD, 69. *Prof Exp:* From asst prof to assoc prof, 76-87, PROF MED, SCH MED, YALE UNIV, 87- *Concurrent Pos:* Chief, Rheumatology Sect, Sch Med, Yale Univ. *Mem:* Am Soc Clin Invest. *Mailing Add:* Dept Med Yale Univ 333 Cedar St New Haven CT 06510

HARDIN, ROBERT CALVIN, medicine; deceased, see previous edition for last biography

HARDIN, ROBERT TOOMBS, b Dyas, Ga, Sept 14, 31; m 57; c 3. BIOMETRICS, POULTRY BREEDING. *Educ:* Univ Ga, BSA, 56; Purdue Univ, MS, 60, PhD(genetics), 62. *Prof Exp:* from asst prof to assoc prof, 62-75, PROF POULTRY GENETICS, UNIV ALTA, 75-, CHMN DEPT ANIMAL SCI, 82- *Res:* Data analysis; design and analysis of agricultural experiments. *Mailing Add:* Dept Animal Sci Univ Alta Edmonton AB T6G 2M7 Can

HARDING, BOYD W, b Provo, Utah, June 4, 26; m 55; c 4. BIOCHEMISTRY, ENDOCRINOLOGY. *Educ:* Brigham Young Univ, BS, 50; Univ Utah, MD, 54, PhD(biochem), 61. *Prof Exp:* USPHS fel biochem, Univ Utah, 57-58 & Cambridge Univ, 58-59; Am Cancer Soc fel endocrinol, 59-61, from instr med to assoc prof 60-70, PROF MED & BIOCHEM, SCH MED, UNIV SOUTHERN CALIF, 70- *Concurrent Pos:* USPHS res develop award gen med sci, 65-69; mem, Endocrine Study Sect, NIH, 73-77. *Mem:* AAAS; Am Soc Biol Chem; Am Fedn Clin Res; Endocrine Soc. *Res:* Hydroxylation reactions; control of adrenal cortex. *Mailing Add:* Dept Med & Biochem HMR 703 Sch Med Univ Southern Calif 2011 Zonal Los Angeles CA 90033

HARDING, CHARLES ENOCH, b Paris, Tenn, Nov 30, 42; m 66; c 2. PHYSICAL ORGANIC CHEMISTRY. *Educ:* Univ Tenn, Martin, BS, 64, Knoxville, PhD(chem), 69. *Prof Exp:* Alexander von Humboldt fel, Univ T06bingen, 69-70; asst prof, 71-73, assoc prof, 73-80, PROF CHEM, UNIV TENN, MARTIN, 73-, CHMN DEPT, 80- *Mem:* Am Chem Soc; Am Inst Chemists; Sigma Xi. *Res:* Experimental studies in organic reaction mechanisms. *Mailing Add:* Dept Chem Univ Tenn Martin TN 38238A

HARDING, CLIFFORD VINCENT, JR, b Cranston, RI, Apr 27, 25; m 48; c 2. CELL PHYSIOLOGY. *Educ:* Brown Univ, AB, 46; Yale Univ, MS, 48; Univ Pa, PhD(zool), 50. *Prof Exp:* Lab asst zool, Yale Univ, 46-48; asst instr, Univ Pa, 48-49; NIH fel cell physiol, Wenner-Grens Inst, Stockholm & Zool Sta, Naples, 50-52; asst prof zool, Univ Southern Calif, 52-54 & Univ Pa, 54-58; physiologist & asst to chief, Med Br, Div Biol & Med, US AEC, 56-57; from asst prof to assoc prof physiol, Col Physicians & Surgeons, Columbia Univ, 58-64; prof biol & chmn dept, Oakland Univ, 64-73; PROF OPHTHAL & DIR RES, KRESGE EYE INST, SCH MED, WAYNE STATE UNIV, 73- *Concurrent Pos:* Lalor fel, 58; mem corp, Marine Biol Lab, Woods Hole, Mass; NIH career develop award, 63-64; adj prof, Oakland Univ, 73- *Honors & Awards:* Fight-for-Sight Citation, 82. *Mem:* Am Soc Cell Biol; Tissue Cult Asn; Soc Develop Biol; Int Soc Cell Biol; Asn Res Vision & Opthal; Int Soc Eye Res. *Res:* Cell nucleus; fertilization proteins in hybrid embryos; radiation on cell division; DNA and mitosis; control of cell division and growth in ocular tissues; lens, cornea, ocular vasculature and human cataracts; energy dispersive X-ray analysis of eye tissue elements. *Mailing Add:* KRSG EYE Wayne State Univ 540 E Canfield Detroit MI 48202

HARDING, CLIFFORD VINCENT, III, b Arlington, Va, Jan 31, 57; m 83. CELL BIOLOGY. *Educ:* Harvard Col, BA, 79; Washington Univ, St Louis, PhD(cell biol) & MD, 85. *Prof Exp:* RESIDENT PATH, WASHINGTON UNIV & BARNES HOSP, 85- *Mem:* Am Soc Cell Biol. *Res:* Receptor-mediated endocytosis; intracellular protein processing and transport; roles of endosomal compartments; transferrin endocytosis and recycling. *Mailing Add:* Wash Univ Sch Med Box 8118 660 S Euclid Ave St Louis MO 63110

HARDING, DUANE DOUGLAS, b Damarascotta, Maine, Dec 14, 47. METEOROLOGY, CLOUD PHYSICS. *Educ:* Univ Mich, BS, 69, PhD(atmospheric sci), 77. *Prof Exp:* Asst res meteorol, Univ Mich, 71-78; res meteorologist, Mauna Lao Observ, Nat Oceanic & Atmospheric Admin, 78-79; asst prof, Eastern Ky Univ, 79-; AT WKYT-TV. *Mem:* AAAS; Am Meteorol Soc; Sigma Xi; Nat Weather Asn. *Res:* Precipitation scavenging; acid rain. *Mailing Add:* 5301 Stewart Dr Virginia Beach VA 23464

HARDING, FANN, b Henderson, Ky, Jan 29, 30; m 56; c 1. ANATOMY. *Educ:* Coker Col, AB, 51; Med Univ SC, MS, 54, PhD(anat), 58. *Prof Exp:* Teaching & res fel anat, Med Col SC, 53-58; pub health res prog analyst, Extramural Prog, Nat Heart Inst, 58-61, health scientist adminr, res & training grants br, 61-64, sr health scientist adminr, sect chief, Res Grant Br Sect, 64-69, sr health scientist adminr Thrombosis & Hemorrhagic Dis Br & Arteriosclerosis Dis Br, Nat Heart & Lung Inst, 69-74; ASST TO DIR DIV BLOOD DIS & RESOURCES, NAT HEART, LUNG & BLOOD INST, 74- *Concurrent Pos:* Consult, James F Mitchell Found, Washington, DC, 62-67, Wash Vet Admin Hosp, 68-70; mem, Nat Heart Inst Fel Bd, 66-68; mem, Civil Serv Qual Rev Bd, NIH, 71- Women's Action Prog adv coun, US Dept Health Educ & Welfare, 71-72; founding pres, Fedn Orgn Prof Women, 72-; mem staff, res training & res career develop & transfusion med, Prob Area US-USSR Health Exchange, Blood Dis & Res Adv Comt; prog dir, Extramural Res Training & Career Develop Blood Dis & Transfusion Med, Nat Heart, Lung & Blood Inst, exec secy, Blood Dis & Resources Adv Comt; asst coordr, US-USSR Health Exchange Prog, 74-; mem bd dirs, Lupus Found Am, 85-88 & Asn Women Sci Edn Found, 73-77; bd visitors, Coker Col, 74-78. *Honors & Awards:* Ruth Patrick Award, 51; NIH Award Sustained Performance, 73; Distinguished Serv Award, Am Asn Blood Banks, 90. *Mem:* Int Soc Thrombosis & Haemostasis; Microcirc Soc; Reticuloendothelial Soc; Am Heart Asn; Am Asn Blood Banks; Int Soc Blood Transfusion; AAAS. *Res:* Histology and physiology of liver; normal and pathologic microcirculatory physiology. *Mailing Add:* Nat Heart Lung & Blood Inst NIH Div Blood Dis & Resources Bethesda MD 20892

HARDING, GODFREY KYNARD MATTHEW, b Trinidad, W Indies, Sept 21, 42; Can citizen; m 69; c 2. INFECTIOUS DISEASES. *Educ:* Univ Man, BSc & MD, 69; FRCPS(C), 74; Am Bd Internal Med, dipl, 74; FACP, 85. *Prof Exp:* Fel infectious dis, Univ Man, 72-73; fel Univ Calif, Los Angeles, 73-74; asst microbiologist, Health Sci Ctr,75-81, asst prof med & med microbiol, 75-80, assoc prof, 80-87, PROF MED & MED MICROBIOL, UNIV MAN, 87-; HEAD INFECTIOUS DIS, ST BONIFACE HOSP, 81- *Concurrent Pos:* Consult, St Boniface Gen Hosp, 75-81, Health Sci Ctr, 75-; grants, Med Res Coun Can, 76-83, Health & Welfare Can, Nat Health Res Develop Prog, 82- *Mem:* Am Soc Microbiol; fel Royal Col Physicians Can; Can Soc Clin Invest; Can Infectious Dis Soc; fel Infectious Dis Soc Am; Fel Am Col Physicians. *Res:* Pathogenesis and management of recurrent urinary tract infections; anaerobic infections; hospital-acquired infections. *Mailing Add:* Dept Infect Dis St Boniface Gen Hosp 409 Tache Ave Winnipeg MB R2H 2A6 Can

HARDING, HOMER ROBERT, b Newark, NJ, Feb 1, 28; m 52; c 1. ENDOCRINOLOGY. *Educ:* Williams Col, Mass, BA, 48; Rutgers Univ, PhD(zool), 59. *Prof Exp:* Anal chemist, Carroll Dunham Smith Pharmacal Co, 52-54; asst endocrinol, Rutgers Univ, 55-58; scientist cancer res, Roswell Park Mem Inst, 58-63; endocrinologist, Sterling-Winthrop Res Inst, 63-83, coordr res admin, 83-90; RETIRED. *Mem:* Am Diabetes Asn; Endocrine Soc; Drug Info Asn. *Res:* Diabetes; control sterpidogenesis and steroid action. *Mailing Add:* Res Admin Sterling-Winthrop Res Inst Rensselaer NY 12144

HARDING, JAMES A, b Tustin, Calif, Jan 20, 35; m 55, 62, 72, 86; c 3. GENETICS, HORTICULTURE. *Educ:* Univ Calif, Davis, BS, 57, PhD(genetics), 65. *Prof Exp:* Lab technician agron, 57-65, asst prof landscape hort, 65-71, assoc prof environ hort, 71-76, PROF & CHAIR ENVIRON HORT, UNIV CALIF, DAVIS, 71- *Concurrent Pos:* Assoc ed, Euphytica. *Mem:* Am Soc Hort Sci. *Res:* Quantitative genetics; genetic structure of natural and artificial populations of plants; multi-trait selection in ornamental species. *Mailing Add:* Dept Environ Hort Univ Calif Davis CA 95616

HARDING, JAMES LOMBARD, b Harvey, Ill, Aug 4, 29; m 54. GEOLOGICAL OCEANOGRAPHY. *Educ:* Miss Southern Col, BS, 56; Univ Tenn, MS, 57; Tex A&M Univ, PhD(oceanog), 64. *Prof Exp:* Instr geol, Univ Tenn, 56-59; asst prof, Miss Southern Col, 59-60; res scientist, Tex A&M Univ, 63-64; vpres, Oceanonics, Inc, 64-71; oceanogr, Marine Resources Ctr, Skidway Inst Oceanog, 71-74; sr geologist, UNDP/ESCHP, Bangkok, Thailand, 88-89; assoc dir, Sea Grant Prog, Univ Ga, 74-77, dir, Marine Exten Serv, 77-85, assoc dir res, 85-87, ASSOC DIR RES, MARINE EXTEN SERV, UNIV GA, 90- *Concurrent Pos:* Prof earth sci, Nicholls State Univ, 65-68; consult, major oil co, US, 61-71, Oceaneering Int, Inc, 71-78 & Oceanonics, Inc, 78-84. *Mem:* Fel Geol Soc Am; Soc Econ Paleont & Mineral; Am Inst Prof Geol; Am Inst Mining & Metall Engrs; Sigma Xi. *Res:* Marine sedimentation; coral reef geology and oceanography; coastal oceanography; carbonate petrology and sedimentation; applied marine geophysics; applied coastal oceanography; exploration and marine mining; applied marine geotechnics and soils mechanics. *Mailing Add:* PO Box 879 Richmond Hill GA 31324-0879

HARDING, JOSEPH WARREN, JR, b Orange, NJ, Feb 9, 48; m 69; c 2. NEUROCHEMISTRY. *Educ:* Alegheny Col, BS, 70; Univ Del, PhD(biochem), 75. *Prof Exp:* Fel neurochem, Roche Inst Molecular Biol, 74-76; res assoc, Dept Vet Microbiol & Path, 76-77, asst prof path, 77-79, ASST PROF NEUROSCI & PHYSIOL, DEPT VET & COMP ANAT, PHARMACOL & PHYSIOL, WASH STATE UNIV, 79-, ASSOC PROF. *Concurrent Pos:* Adj prof, Prog Biochem & Biophysics, Wash State Univ, 78- & Dept Psychol, 80- *Mem:* Soc Neurosci. *Res:* Neural control of cardiovascular function: brain-angiotensiun interactions; neural regeneration and plasticity: the primary olfactory system as a model; mechanisms of neurotransmitter release. *Mailing Add:* Dept Vet & Comp Anat Pharmacol & Physiol Wash State Univ Pullman WA 99164

HARDING, KENN E, b Ponca City, Okla, Oct 2, 42; m 64, 76; c 1. ORGANIC CHEMISTRY. *Educ:* Okla State Univ, BS, 64; Stanford Univ, PhD(org chem), 68. *Prof Exp:* NIH fel chem, Harvard Univ, 68-69; asst prof, 69-76, assoc prof, 76-86, PROF CHEM, TEX A&M UNIV, 86- *Concurrent Pos:* Dir, Synthetic Org Chem Prog, NSF, 85-86. *Mem:* Am Chem Soc. *Res:* Organic synthesis; biogenetic-like olefin cyclizations; alkaloid synthesis; natural products synthesis; stereoselective and enantioselective synthesis. *Mailing Add:* Dept Chem Tex A&M Univ College Station TX 77843

HARDING, MATTHEW WILLIAM, immunopharmacology, tumor & cellular immunology, for more information see previous edition

HARDING, MAURICE JAMES CHARLES, b Surrey, Eng, Dec 29, 38; m 65; c 2. PESTICIDES. *Educ:* Univ London, BS, 60, PhD(org chem), 63. *Prof Exp:* Res fel, Johns Hopkins Univ, 63-65; sr chemist, Laporte Chem Ltd, Luton, Eng, 65-68; res scientist, Res & Develop Div, Union Camp Corp, 68-75; sr res chemist, 75-79, PROCESS DEVELOP MGR, AGR CHEM GROUP, FMC CORP, 79- *Mem:* Am Chem Soc; The Chem Soc. *Res:* Peroxides; peroxyacids; epoxides; nitrogen heterocycles; pyrroles; porphyrins; imidazoles; chemiluminescence; singlet oxygen; musk compounds; catalytic hydrogenation; sulfur chemistry; clay catalysis; pesticide chemistry; statistical design of experiments; process development in pyrethroids and herbicide chemistry; scale-up and commercialization of specialty chemicals, agrochemicals, etc. *Mailing Add:* 28 Fisher Ave Princeton NJ 08540

HARDING, PAUL GEORGE RICHARD, b Kitchener, Ont. REPRODUCTIVE PHYSIOLOGY, PERINATOLOGY. *Educ:* Univ Western Ont, MD, 58, MSc, 61; FRCS(C), 63. *Prof Exp:* Med Res Coun Can fel, 66-67, scholar, 67-71, ASSOC PROF PHYSIOL, OBSTET & GYNEC, UNIV WESTERN ONT, 70- *Mem:* Can Physiol Soc; Soc Obstet & Gynec Can. *Res:* Fetal and placental physiology, particularly energy metabolism in the fetus. *Mailing Add:* Dept Physiol Univ Western Ont Med Sch London ON N6A 5C1 Can

HARDING, R(ONALD) H(UGH), b Chicago, Ill, June 26, 31; m 60; c 2. NEW PRODUCT SYSTEMS DEVELOPMENT. *Educ:* Purdue Univ, BS, 53, PhD(chem eng), 58. *Prof Exp:* Asst, Purdue Univ, 52-53; asst, Lilly Varnish Co, 54-57; proj chemist, Res Dept, Chem Div, Union Carbide Corp, 57-62, group leader, Res & Develop Dept, 63-72, technol mgr, 69-72, mkt mgr, Mkt Dept, 70-72, prod mkt mgr, Latex Opers Dept, Chem & Plastics Div, 73-75, bus develop mgr, 76-77, sr develop scientist, 78-85; PRIVATE CONSULT 86- *Concurrent Pos:* Counr, Men's Residence Halls, Purdue Univ, 53-57. *Mem:* Am Chem Soc. *Res:* Applications research, product development and technical service of elastomers, urethane foams, water-soluble polymers, oil production chemicals, leather chemicals, latexes, adhesives, mining and water treatment chemicals. *Mailing Add:* Five Janson Dr Westport CT 06880

HARDING, ROY WOODROW, JR, b Arthurdale, WVa, Sept 16, 40; m 61; c 2. MOLECULAR GENETICS. *Educ:* George Washington Univ, BS, 62; Calif Inst Technol, PhD(biochem), 68. *Prof Exp:* Fel, Univ Tex, Austin, 68-70; GENETICIST, SMITHSONIAN INST, 70- *Mem:* AAAS; Am Soc Plant Physiologists; Am Soc Photobiol; Int Soc Plant Molecular Biol. *Res:* Light mediated biological responses; carotenoid biosynthesis; regulation of metabolic pathways. *Mailing Add:* 148-B Weyandt Hall Indiana Univ Pa Indiana PA 15705

HARDING, SAMUEL WILLIAM, b Salt Lake Co, Utah, Feb 18, 15; m 39, 79; c 2. PHYSICS. *Educ:* Utah State Agr Col, BS, 39; Pa State Col, MS, 42, PhD(physics), 47. *Prof Exp:* Asst physics, Utah State Agr Col, 39-40; asst, Pa State Col, 40-42, instr, 42-46, res assoc, 46-47; from asst prof to prof, 47-81, asst head, Dept Physics & Astron, 71-81, dir, Sci & Math Teaching Ctr, 74-78, EMER PROF PHYSICS, UNIV WYO, 81- *Concurrent Pos:* Consult, Optical Res Lab, Tex, 48-51. *Mem:* Am Phys Soc; Optical Soc Am; Am Asn Physics Teachers; fel AAAS. *Res:* Geometrical and physical optics with emphasis on resolution of optical instruments and thin films; science education. *Mailing Add:* 2317 Sunset Dr Lewiston ID 83501-3436

HARDING, THOMAS HAGUE, b Oxford, Miss, Mar 1, 45; m 85. NEUROPHYSIOLOGY, NEUROSCIENCE. *Educ:* Tex A&M Univ, BS, 67, MS, 71; Purdue Univ, PhD(neurophysiol), 77. *Prof Exp:* Teaching asst psychol & res asst, Tex A&M Univ, College Station, 70-71; David Ross res fel, Purdue Univ, West Lafayette, 71-74; traveling scholar, biomed eng ctr, Technol Inst, Northwestern Univ, 75-77; vis fel, Australian Nat Univ, 77-79; res neurophysiologist, 79-82, CHIEF SENSORY NEUROSCI BR, US ARMY AEROMED RES LAB, FORT RUCKER, ALA, 82- *Mem:* Asn Res in Vision & Ophthal; AAAS. *Res:* Neurophysiology of the visual system; visual psychophysics; computer vision; visual perception; expert systems. *Mailing Add:* US Army Aeromed Res Lab PO Box 577 Ft Rucker AL 36362

HARDING, WALLACE CHARLES, JR, entomology, botany; deceased, see previous edition for last biography

HARDING, WINFRED MOOD, b Huntsville, Tex, May 23, 20; m 44; c 2. BIOLOGICAL CHEMISTRY. *Educ:* Sam Houston State Col, BS, 41; Univ Tex, MA, 46, PhD(chem), 49. *Prof Exp:* Teacher pub sch, Tex, 42-43; asst chemist, Dow Chem Co, 43-44; asst, Univ Tex, 44-47; assoc prof chem, Southwest Tex State Col, 48-59; assoc prof, Sam Houston State Univ, 59-80, prof chem, 80-88; RETIRED. *Concurrent Pos:* Lab tutor, Univ Tex, 44-45. *Mem:* Am Chem Soc; Am Soc Microbiol; Sigma Xi. *Res:* Microbiological metabolism. *Mailing Add:* 2030 Ave Q Huntsville TX 77340

HARDING-BARLOW, INGEBORG, b Johannesburg, SAfrica, June 14, 38. CHEMISTRY, PATHOLOGY. *Educ:* Univ Cape Town, BSc, 57, Hons, 58, PhD(chem), 61. *Prof Exp:* Instr chem, Univ Cape Town, 59-61; res assoc physics, Univ Tenn, 61; Am Asn Univ Women int fel & res investr path, Univ Tex M D Anderson Hosp & Tumor Inst, 61-62; res assoc internal med, Washington Univ, 63, res instr, 63-65; res assoc path, Stanford Univ, 65-69; consult, Ames Res Ctr, NASA, 69-72; sr res assoc, Inst Chem Biol, Univ San Francisco, 72-80. *Mem:* Am Chem Soc; Soc Appl Spectros; assoc Royal Inst Chem; Royal Soc Health; Soc Environ Geochem & Health. *Res:* Emission spectroscopy; amounts and functions of trace metals in humans and animals, analytical chemistry; toxicity and carcinogenicity of trace elements; mass spectroscopy; laser microprobe spectroscopy. *Mailing Add:* 3717 Laguna Ave Palo Alto CA 94306

HARDINGE, MERVYN GILBERT, PHARMACOLOGY, HEALTH PROMOTION. *Educ:* Pac Union Col, BS, 39; Loma Linda Univ Sch Med, MD, 42; Harvard Univ, MPH, 49, DrPH, 51; Stanford Univ, MA, 52, PhD(pharmacol), 56. *Prof Exp:* Prof health prom, Loma Linda Univ, Calif, 43-80; RETIRED. *Mailing Add:* Sch Pub Health Loma Linda Univ Loma Linda CA 92354

HARDIS, LEONARD, b New York, NY, Sept 9, 16; m 52; c 2. MECHANICAL ENGINEERING. *Educ:* Carnegie Inst Technol, BS, 38; Univ Md, MS, 53. *Prof Exp:* Asst, Carnegie Inst Technol, 38-39; ord engr, Naval Gun Factory, 39-48; supvr mech eng, US Naval Ord Lab, 48-60, chief gen eng div, 60-61; staff engr, orbiting geophys observ proj, Goddard Space Flight Ctr, NASA, 61-69, resources technol satellite, 69-77; Systs Engr, OAO Corp, 77-84; DESIGN CONSULT, 84- *Res:* Design of naval ordnance; experimental stress analysis; rocket blast measurement; design and development of spacecraft. *Mailing Add:* 1316 Midwood Pl Silver Spring MD 20910

HARDISON, JOHN ROBERT, b Yakima, Wash, Jan 12, 18; m 37, 88; c 4. PLANT PATHOLOGY. *Educ:* State Col Wash, BS, 39; Univ Mich, MS, 40, PhD(mycol), 42. *Prof Exp:* Asst forage crop invests, Exp Sta, Univ Ky, 42-44; res pathologist, Forage & Range Res Br, Plant Sci Res Div, Agr Res Serv, USDA, 44-72, res leader, legume & grass seed prod, 72-80; PROF PLANT PATH, ORE STATE UNIV, 81-; PRES, AG TECH, INC, 85- *Mem:* Am Phytopath Soc; Mycol Soc Am; Am Soc Agron; Crop Sci Soc Am. *Res:* Biology and control of diseases of forage crops. *Mailing Add:* Bot & Plant Path Dept Ore State Univ Corvallis OR 97331

HARDISON, ROSS CAMERON, b Nashville, Tenn, Mar 29, 51. BIOCHEMISTRY, MOLECULAR GENETICS. *Educ:* Vanderbilt Univ, BA, 73; Univ Iowa, PhD(biochem), 77. *Prof Exp:* Res fel molecular biol, Div Biol, Calif Inst Technol, 77-80; asst prof, 80-86, ASSOC PROF BIOCHEM, DEPT MOLECULAR & CELL BIOL, PA STATE UNIV, 86- *Concurrent Pos:* Fel, Jane Coffin Childs Mem Fund Med Res, 77-79; vis prof, Biomed Res Centre, Unv BC, Vancouver, BC Can, 88; prin investr on NIH funded grants, 80- *Honors & Awards:* Res Career Develop Award, NIH, 86-91. *Mem:* Am Chem Soc; Am Soc Biochem & Molecular Biol; Am Soc Microbiol. *Res:* Molecular basis for control of gene expression; mammalian globin gene evolution and regulation; plastid gene expression during fruit ripening. *Mailing Add:* Dept Molecular & Cell Biol-Althouse Lab Penn State Univ Univ Park PA 16802

HARDISON, WESLEY AUREL, b Barren Co, Ky, May 24, 25; m 46; c 2. ANIMAL NUTRITION. *Educ:* Western Ky State Col, BS, 47; Univ Ky, MS, 49; Cornell Univ, PhD, 52. *Prof Exp:* Asst dairy husb, Univ Ky, 48-49; asst animal husb, Cornell Univ, 49-52; assoc prof dairy sci, Va Polytech Inst, 53-62; prod officer, Food & Agr Orgn, UN, 62-67; proj specialist fodder & animal prod, Ford Found, 67-74; livestock specialist, Int Agr Develop, World Bank, 74-87; RETIRED. *Concurrent Pos:* Consult, livestock prod, 87- *Honors & Awards:* Am Feed Mfrs Award, 60. *Mem:* Am Soc Animal Sci; Am Inst Nutrit; Am Dairy Sci Asn. *Res:* Forage utilization, including pasture herbage, with special emphasis on use of indicator techniques; dairy calif nutrition. *Mailing Add:* 4600 Connecticut Ave NW Washington DC 20008

HARDMAN, BRUCE BERTOLETTE, b Mineola, NY, Jan 18, 42; m; c 2. MATERIALS QUALITY MANAGEMENT, ENVIRONMENTAL REGULATORY COMPLIANCE. *Educ:* Union Col, NY, BS, 64; Rensselaer Polytech Inst, PhD(phys chem), 67. *Prof Exp:* Res chemist, Gen Elec Co, 67-74, chem develop specialist, 74-77, mgr, Anal Servs, 77-78, mgr RTV Qual Control, 78-79, mgr, Process Qual, 80-81, mgr, Anal Serv & Process Qual, 81-84, mgr environ testing & anal, Silicone Prod Bus Div, Gen Elec Co, 84-89; MGR, QUAL OPERS, PCR, INC, GAINESVILLE, FLA, 90- *Mem:* AAAS; Am Chem Soc; The Chem Soc; Royal Inst Chem. *Res:* Quality management systems; chemical process technology; environmental science and technology. *Mailing Add:* PO Box 23804 Gainesville FL 32602

HARDMAN, CARL CHARLES, b Atlantic City, NJ, Aug 30, 19; m 48; c 3. ELECTROCHEMISTRY. *Educ:* Pa State Univ, BS, 41; Univ Wis, MS, 47. *Prof Exp:* Engr, INCO, 41-43; res chemist, Union Carbide Co, 50-53; res chemist, Diamond Shamrock, 53-59; div scientist electrochem, Aerovox, 60-63; SR ENGR, WESTINGHOUSE RES LAB, 63- *Mem:* Am Chem Soc; Electrochem Soc. *Res:* Alkaline and acid batteries; electrolytic capacitors. *Mailing Add:* 6564 Rosemoor Dr Pittsburgh PA 15217

HARDMAN, HAROLD FRANCIS, b East Orange, NJ, Aug, 2, 27; m 50; c 4. PHARMACOLOGY. *Educ:* Rutgers Univ, BSc, 49; Univ Ill, MSc, 51; Univ Mich, PhD, 54, MD, 58. *Prof Exp:* Instr pharmacol, Med Sch, Univ Mich, 54-58, asst prof, 58-60; from assoc prof to assoc dean, 60-70, PROF PHARMACOL & CHMN DEPT, MED COL WIS, 62- *Concurrent Pos:* Vis prof, Cali, Colombia, 65-66; Markle scholar med sci, 58-62. *Mem:* AAAS; Am Soc Clin Pharmacol & Therapeut; Am Soc Pharmacol & Exp Therapeut (pres, 82-83); Fedn Am Soc Exp Biol (pres, 83-84). *Res:* Effect of pH and drug ionization on the pharmacological action of drugs affecting the heart; physiology of oxygen utilization by the heart; pharmacology of theophylline, barbiturates, epinephrine, procaine derivatives and antianginal agents; pharmacology of marijuana and derivatives. *Mailing Add:* Dept Pharmacol & Toxicol Med Col Wis PO Box 26509 Milwaukee WI 53226

HARDMAN, JOEL G, b Colbert, Ga, Nov 7, 33; m 55; c 4. MOLECULAR PHARMACOLOGY. *Educ:* Univ Ga, BS, 54, MS, 59; Emory Univ, PhD(pharmacol), 64. *Prof Exp:* Instr pharm, Univ Ga, 57-60; instr physiol, Sch Med, Vanderbilt Univ, 64-67, from asst prof to prof, 67-75, prof pharmacol & chmn dept, 75-90, ASSOC VCHANCELLOR HEALTH AFFAIRS, VANDERBILT UNIV MED CTR, 91- *Concurrent Pos:* Mem adv bd, Advan Cyclic Nucleotide Res, 73-; Francqui Foreign vis prof, Free Univ Brussels, 74; mem, pharmacol study sect, NIH, 75-77, chmn, 77-79; ed, Molecular Pharmacol, 83-86; mem, res comt, Am Heart Asn, 79-84, mem, Coun Basic Sci; chmn, Bd Publ Trustees, Am Soc Pharmacol & Exp Therapeut, 90-; mem exec comt, Int Union Pharmacol, 90- *Honors & Awards:* H B Van Dyke Award, Columbia Univ, 81. *Mem:* Am Soc Pharmacol & Exp Therapeut; Am Soc Biol Chemists; Am Heart Asn; AAAS. *Res:* Regulation of cyclic nucleotide metabolism; mechanisms of drug and hormone action. *Mailing Add:* D-3300 MCN Vanderbilt Univ Med Ctr Nashville TN 37232

HARDMAN, JOHN KEMPER, b Waynesboro, Pa, Aug 11, 34; m 67; c 3. BIOCHEMISTRY, MOLECULAR GENETICS. *Educ:* Mt St Mary's Col, Md, BS, 56; Georgetown Univ, MS, 59; Univ Md, PhD(microbiol, biochem), 62. *Prof Exp:* Chemist, NIH, 58-62; Nat Acad Sci fel, Stanford Univ, 62-63, NIH fel, 63-65; asst prof biol, Dept Biol & McCollum-Pratt Inst, Johns Hopkins Univ, 65-72; assoc prof, 72-76, PROF BIOL, UNIV ALA, 76- *Mem:* AAAS; Am Chem Soc; Sigma Xi; Am Soc Biochem & Molecular Biol. *Res:* Structure-function relationships in proteins. *Mailing Add:* Box 870344 Tuscaloosa AL 35487-0344

HARDMAN, JOHN M, b Matheson, Colo, Jan 15, 33; m 78; c 2. PATHOLOGY. *Educ:* Univ Colo, BS, 54, MD, 58; Baylor Univ, MS, 65; Am Bd Path, cert anat & clin path, 63, cert, neuropath, 67. *Prof Exp:* Intern, Walter Reed Army Med Ctr, Washington, DC, 58-59; resident anat & clin path, Brooke Gen Hosp, San Antonio, Tex, 59-63; fel neuropath, Armed Forces Inst Path, 65-67; assoc clin & clin prof, John A Burns Sch Med, Univ Hawaii, 70-75; clin prof, Georgetown Univ, 76-77; PROF & CHMN, DEPT PATH, JOHN A BURNS SCH MED, UNIV HAWAII, MANOA, 77-, PROG DIR, INTEGRATED PATH RESIDENCY PROG, 78- *Concurrent Pos:* Neuropath consult, Nat Naval Med Ctr, 66-68; vis prof neuropath, William Beaumont Gen Hosp, El Paso, Tex, 66, 68, 71 & 75, Madigan Gen Hosp, Tacoma, Wash, 70 & 74; dir-at-large, Am Cancer Soc, 71-75 & 82-; liaison mem, Army Med Dept-NIH Study Group Sect Path B, 76-77; neuropath consult, St Francis Hosp, Tripler Army Med Ctr & Med Examr City & County Honolulu, 78-; dir labs, Kapiolani-Children's Med Ctr, 78-85; dir educ & res, Dept Labs, Kapiolani Med Ctr for Women & Children, 85-; bd dirs, Alzheimer's Dis & Related Disorders Asn, Inc, 87- *Honors & Awards:* Sir Henry Wellcome Medal, Scroll & Prize, Asn Mil Surgeons US, 68. *Mem:* Col Am Pathologists; Am Soc Clin Path; Am Asn Neuropathologists; Am Asn Pathologists & Bacteriologists; Int Acad Path; Am Asn Pathologists; Sigma Xi; Soc Neurosci; AMA. *Res:* Effects of traumatic injury on the nervous system; diagnosis of central nervous system diseases, particularly brain tumors and dementias; acute nervous system decompression sickness in dogs. *Mailing Add:* Dept Path John A Burns Sch Med 1960 East-West Rd Honolulu HI 96822

HARDMAN, JOHN MICHAEL, b Exeter, Ont, Oct 31, 47; m 71; c 1. INSECT ECOLOGY. *Educ:* Dalhousie Univ, BSc Hons, 68; Imp Col, Univ London, MSc & DIC, 69; Simon Fraser Univ, PhD(insect ecol), 73. *Prof Exp:* Res scientist insect ecol, Div Entom, Commonwealth Sci & Indust Res Org, Canberra, Australia, 73-77; res assoc, Pestology Ctr, Simon Fraser Univ, 77; res scientist grasshopper ecol, Agr Can Res Sta, Lethbridge, 77-82; RES SCIENTIST, TREE FRUIT ENTOM, KENTVILLE, NS, 82- *Mem:* Entom Soc Can; Ecol Soc Am; Brit Ecol Soc. *Res:* Use of systems analysis and simulation models in management of insect pests. *Mailing Add:* Res Sta Agr Can Kentville NS B4N 1J5 Can

HARDORP, JOHANNES CHRISTFRIED, astrophysics; deceased, see previous edition for last biography

HARDRATH, HERBERT FRANK, aerospace & structural engineering; deceased, see previous edition for last biography

HARDT, ALEXANDER P, metallurgy & physical metallurgical engineering, for more information see previous edition

HARDT, ALFRED BLACK, b Fond du Lac, Wis, Nov 7, 30. PHYSIOLOGY. *Educ:* Iowa State Univ, BS, 53; San Jose State Col, MA, 60; Univ Colo, PhD(physiol), 69. *Prof Exp:* Asst prof zool, Univ Wyo, 68-72; assoc prof oral biol, Univ Nebr, Lincoln, 72-90; RETIRED. *Mem:* AAAS; Int Asn Dent Res. *Res:* Hard tissue physiology. *Mailing Add:* 11 Trout Pond Lane Bandon OR 97411

HARDT, DAVID EDGAR, b Bryn Mawr, Pa, Sept 11, 50; m 74; c 3. CONTROL SYSTEMS, MANUFACTURING. *Educ:* Lafayette Col, BSME, 72; Mass Inst Technol, MS, 74, PhD(mech eng), 78. *Prof Exp:* Res assoc, 78-79, asst prof, 79-85, ASSOC PROF MECH ENG, 85-, DIR, LAB MFG, MASS INST TECHNOL. *Mem:* Am Soc Mech Engrs; Inst Elec & Electronics Engrs; Sigma Xi. *Res:* Application of system dynamics and control to batch manufacturing processes, design/manufacturing integration. *Mailing Add:* Lab Mfg Productivity Mass Inst Technol Rm 35-234 Cambridge MA 02139

HARDT, JAMES VICTOR, b Red Wing, Minn, Feb 10, 45. BIOFEEDBACK, PSYCHOPHYSIOLOGY. *Educ:* Carnegie Inst Technol, BS, 67; Carnegie-Mellon Univ, MS, 69, PhD(psychol), 73. *Prof Exp:* Instr psychol, Carnegie-Mellon Univ, 68-71; asst res psychologist, Univ Calif, San Francisco 77-78, asst adj prof med psychol, 78-90; RES PROF, INST TRANSPERSONAL PSYCHOL, MENLO PARK, CALIF, 90- *Concurrent Pos:* Res assoc, Langley Porter Neuropsychiat Inst, Univ Calif, San Francisco, 73-74, NIMH fel, Interdisciplinary Training Prog, 74-77, co-prin investr, 74-77, prin investr, Long-Term EEG Feedback Proj, 77-90. *Mem:* AAAS; Biofeedback Soc Am; Soc Psychophysiol Res; affil mem Am Psychol Asn; Asn Appl Psychophysiol & Biofeedback; Int Soc Study Subtle Energies & Energy Med. *Res:* Altered states of consciousness; mind and body interrelationships; personality therapy; microcomputer based data acquisition, analysis, and bio- feedback instruments; individual differences. *Mailing Add:* Langley Porter Neuropsychiat Inst 401 Parnassus Ave San Francisco CA 94143

HARDTKE, FRED CHARLES, JR, b Chicago, Ill, Aug 30, 31. CHEMISTRY. *Educ:* Ill Inst Technol, BS, 53; Ore State Univ, PhD(phys chem), 59. *Prof Exp:* Anal chemist, Sci Control Lab, 53-54; instr, Ore State Univ, 57-58; appointee, Inst Nuclear Sci & Eng, Argonne Nat Lab, 58-60, asst chemist, Remote Control Div, 60-65; ASST PROF CHEM, UNIV MO-ROLLA, 65- *Mem:* Am Chem Soc; Am Nuclear Soc. *Res:* Photoconductivity; electrets; compton current; cyclic voltommetry. *Mailing Add:* Dept Chem Univ Mo Rolla MO 65401

HARDTMANN, GOETZ E, b Leipzig, Ger, Oct 4, 32; US citizen; m 61; c 3. ORGANIC CHEMISTRY, MEDICINAL CHEMISTRY. *Educ:* Brunswick Tech Univ, Dipl, 59, PhD(chem), 61. *Prof Exp:* Res assoc org chem, Brunswick Tech Univ, 61 & Univ Wis, 61-63; sr res chemist, 63-68, res group leader med chem, 69-71, res sect head med chem, 71-77, DIR CHEM DEVELOP, SANDOZ PHARMACEUT, 77- *Mem:* Am Chem Soc; Ger Chem Soc. *Res:* Total synthesis of Terramycin; synthesis of medicinal agents; heterocyclics; metallorganics. *Mailing Add:* Sandoz Res Inst Sandoz Pharmaceut Corp East Hanover NJ 07936

HARDWICK, DAVID FRANCIS, b Vancouver, BC, Jan 24, 34; m 56; c 3. PATHOLOGY, PEDIATRICS. *Educ:* Univ BC, MD, 57; FRCP(C), 65; Am Bd Path, cert anat & clin path, 65. *Prof Exp:* Res assoc physiol, Univ Southern Calif, 60-62; clin instr path, Univ BC, 63-65, from asst prof to assoc prof, 65-74, head, dept path, 76-90, PROF PATH, UNIV BC, 74-; HEAD DEPT PATH, CHILDREN'S HOSP, VANCOUVER, 69-, ASSOC DEAN, RES & PLANNING, 90- *Concurrent Pos:* Asst clin chemist, Vancouver Gen Hosp, 63-64, from asst pathologist to assoc pathologist, 65-72, pediat pathologist, 72-; consult, Ministry Health, Prov of BC, 69-; mem senate, Univ BC, 69-75; Nat Inst Neurol Dis & Blindness trainee, Children's Hosp, Los Angeles, 60-62; chief med staff, Children's Hosp, Vancouver, 70-86; US & Can Acad Pathol. *Honors & Awards:* Queen Elizabeth II Medal, 78. *Mem:* Can Asn Path; Can Med Asn; fel Col Am Path; Int Acad Path; NY Acad Sci. *Res:* Pediatric neonatal pathology; metabolic diseases; laboratory management. *Mailing Add:* Dean's Off Fac Med Univ BC Vancouver BC V6T 1W5 Can

HARDWICK, JOHN LAFAYETTE, b Atlanta, Ga, June 28, 44; m 70; c 2. PHYSICAL CHEMISTRY, SPECTROSCOPY. *Educ:* Princeton Univ, AB, 66; Ga Inst Technol, PhD(chem), 72. *Prof Exp:* Fel chem, Univ Western Ont, 72-75; res assoc spectros, Herzberg Inst Astrophys, 75-78; vis asst prof chem, Univ Colo, 78-79; asst prof specialist, Radiation Lab, Univ Notre Dame, 79-85; SR RES ASSOC, DEPT PHYSICS, UNIV ORE, 85- *Mem:* Am Chem Soc. *Mailing Add:* Dept Physics Univ Ore Eugene OR 97403-1274

HARDWICK, WILLIAM AUBREY, JR, research administration, food chemistry, for more information see previous edition

HARDWICKE, JAMES ERNEST, JR, b Winston-Salem, NC, Oct 21, 24; m 50; c 5. ORGANIC CHEMISTRY. *Educ:* Univ NC, BS, 47; Northwestern Univ, MS, 48; Univ Calif, PhD(org chem), 51. *Prof Exp:* Asst chem, Northwestern Univ, 47-48 & Univ Calif, 48-51; res chemist, Tenn Eastman Co, 51-55; mgr org res lab, Shulton Inc, 55-58; dir res & develop, Cardinal Mfg Co, 58-63, gen mgr, Cardinal Chem Co & pres, Cardinal Stabilizers, Inc, 64-67, vpres, 67; pres, Hardwicke Chem Co, 67-88; HARWICKE CONSULT, 88- *Concurrent Pos:* Mem bd dirs, McLaughlin Gormley King Co, Minneapolis, Minn, 75; mem, Chem & Specialties Mfrs Coun. *Mem:* AAAS; Am Chem Soc; Sigma Xi. *Res:* Synthetic organic chemistry in general; synthetic insecticidal synergists; perfumery synthetics; high pressure synthesis; organotin chemistry; vinyl stabilizers; fire retardants; pyrethrim substitutes intermediates for synthetic pyrethroids; surface active agents; carbamates; N-alkyl anilines; benzyl chemicals; insect repellents. *Mailing Add:* 805 Kilbourne Rd Columbia SC 29206

HARDWICKE, NORMAN LAWSON, b Winnipeg, MB, May 21, 24; Can & US citizen; m 50; c 3. POLYMER CHEMISTRY. *Educ:* Tex Col Arts & Indust, BS(chem eng) & BS(chem), 49; Univ Okla, MChE, 51, PhD(chem eng), 67. *Prof Exp:* Res engr, Monsanto Co, 51-58, group leader polymers, 58-63, res specialist, 66-67, group leader, 67-72, res specialist, 72-79, sr technol specialist, 79-82; RETIRED. *Mem:* Am Inst Chem Eng; Am Chem Soc. *Res:* Vinyl monomers; vinyl polymers; polymer stabilization; reaction simulation; analog and digital model development; liquid diffusion. *Mailing Add:* 1867 Taylor St Victoria BC V8R 3G3 Can

HARDWIDGE, EDWARD ALBERT, b Chicago, Ill, June 16, 46; m 68; c 2. PHARMACEUTICAL CHEMISTRY. *Educ:* Ill Inst Technol, BS, 67; Univ Wash, PhD(chem), 72. *Prof Exp:* Res assoc chem, Univ Alta, 72-74; scientist, UpJohn Co, 74-76, lab head quality control, 76-77, res head, 77-84, mgr packaging eng, 84-86, group mgr, exp clin prod, 86-89; DIR, QUAL CONTROL, CATER-WALLACE INC, 89- *Mem:* Parenteral Drug Asn; Am Chem Soc. *Res:* Bioavailability of oral dosage forms of drugs and correlations between bioavailability and in vitro physical tests; physical and particulate properties of parenteral drugs. *Mailing Add:* Wallace Labs 434 N Morgan St Decatur IL 62525

HARDY, CECIL ROSS, b Cleveland, Utah, Apr 4, 08; m 34; c 2. MAMMALOGY, ECOLOGY. *Educ:* Univ Utah, BS, 33, MS, 38; Univ Mich, PhD(zool), 43. *Prof Exp:* Instr schs, Utah, 33-38; head biol sci div, Dixie Jr Col, 38-46; head dept biol, Weber Col, 46-49; prof, 49-73, EMER PROF ZOOL, CALIF STATE UNIV, LONG BEACH, 73- *Concurrent Pos:* Teacher, Exten Div, Univ Utah, 44-49; vis prof, San Jose State Col, 49. *Honors & Awards:* Desert Tortoise Coun Award, 81. *Mem:* Am Soc Mammal; Wilson Ornith Soc; Herpetologists League; Wildlife Soc; Cooper Ornith Soc. *Res:* Ecology of mammals, birds and reptiles; taxonomy and distribution of mammals; influence of types of soil upon the local distribution of some mammals in southwestern Utah; studies of desert tortoise. *Mailing Add:* 5351 Las Lomas St Long Beach CA 90815

HARDY, CLYDE THOMAS, b Fremont, Ohio, Apr 23, 21. STRUCTURAL GEOLOGY. *Educ:* Ohio State Univ, BA, 43, MS, 48, PhD(geol), 49. *Prof Exp:* From asst prof to assoc prof, 50-66, head dept, 68-82, PROF GEOL, UTAH STATE UNIV, 67- *Mem:* Geol Soc Am; Am Asn Petrol Geol; Am Geophys Union. *Mailing Add:* Dept Geol Utah State Univ Logan UT 84322-4505

HARDY, D ELMO, b Lehi, Utah, Sept 3, 14; m 35; c 4. ENTOMOLOGY. *Educ:* Brigham Young Univ, AB 37; Utah State Univ, 38; Univ Kans, PhD(entom), 41. *Prof Exp:* Field entomologist, Exp Sta, Utah State Col, 37-38; asst instr entom, Univ Kans, 38-41, fel taxon, 41-42; asst state entomologist, Iowa State Col, 45-48; from prof to sr prof, 48-80, from assoc entomolgist to entomologist, 48-60, sr entomologist, 60-80, chmn dept, 58-68, EMER PROF ENTOM, UNIV HAWAII, 81- *Concurrent Pos:* State nursery inspector, Kans, 39-41; dep state entomologist, Kans Entom Comn, 42; med entomologist, US Army, 42-45; hon assoc & mem, BP Bishop Mus entom exped, New Guinea, 57; mem, State of Hawaii Natural Area Reserves Comn, 70-76 & Animal Species Adv Coun, 78-80. *Mem:* AAAS; assoc Soc Syst Zool; assoc Entom Soc Am. *Res:* Diptera taxonomy; taxonomy of the Bibionidae and Dorilaidae of the world; Diptera of Hawaii; Tephritidae of the Orient and Pacific. *Mailing Add:* Dept Entom Univ Hawaii Rm 310 3050 Maileway Honolulu HI 96822

HARDY, EDGAR ERWIN, b Charlottenburg, Ger, July 6, 13; nat US; m 41; c 4. CHEMISTRY. *Educ:* Univ Zurich, PhD(law), 35; Univ Minn, BS, 38, MS, 40. *Prof Exp:* From res chemist to res group leader, Monsanto Chem Co, 42-45, from org res supvr, Phosphate Div, to asst dir res, 45-53, asst dir, Org Div, 54; res dir, Mobay Chem Co, 54-59; from develop assoc, Plastics Div, to assoc dir res, Monsanto Res Corp, Monsanto Co, 59-63, dir develop, Plastic Prod & Resins Div, 64-65, dir, Dayton Lab, 65-78; prof chem, Wright Univ, 78-80; prof chem, Calif State Polytech Univ, 80-84; ADJ PROF, CHEM & NATURAL SCI DEPTS, SAN DIEGO STATE UNIV, 84- *Concurrent Pos:* Instr, Wright State Univ. *Mem:* Am Chem Soc. *Res:* Polyurethanes; plastic foams and coatings composites; synthetic rubber; organic isocyanates; organic phosphorus compounds; detergent synthesis and application; organic fluorine compounds. *Mailing Add:* 3655 Jackdaw St San Diego CA 92103

HARDY, FLOURNOY LANE, b Columbia, SC, Sept 10, 28; m 54; c 2. MATHEMATICS. *Educ:* Oglethorpe Univ, AB, 55; Emory Univ, MA, 56; Ohio State Univ, PhD(math), 62. *Prof Exp:* From instr to asst prof, Emory Univ, 56-65; prof & chmn dept, Armstrong State Col, 65-68 & Chicago State Univ, 68-74; PROF MATH, DEKALB COMMUNITY COL, 74- *Mem:* Math Asn Am; Am Math Soc. *Res:* Abelian group theory in abstract algebra; projective planes in geometry. *Mailing Add:* Dept Sci Dekalb Col N 2201 Womack Rd Dunwoody GA 30338

HARDY, H(ENRY) REGINALD, JR, b Ottawa, Ont, Aug 19, 31; m 54; c 2. ENGINEERING MECHANICS. *Educ:* McGill Univ, BSc, 53; Univ Ottawa, MSc, 62; Va Polytech Inst, PhD(eng mech), 65. *Prof Exp:* Sci officer rock mech, mines br, Can Dept Mines & Tech Survs, 53-60, rock physics, 60-66; assoc prof, 66-70, PROF MINING, PA STATE UNIV, 70- *Concurrent Pos:* Mem, US Nat Comt Rock Mech; Deutsche Forschungsgemeinschaft vis prof, Univ Aachem, WGer, 80; sr vis fel, Japanese Soc Promotion Sci, 85 & Tohoku Univ, Sendai, Japan, 86. *Honors & Awards:* Templin Award, Am Soc Testing & Mat, 68 & C A Hogentogler Award, 83; Gold Award, Acoustic Emission Working Group, 88. *Mem:* Can Asn Physicists; Soc Exp Stress Anal; Am Geophys Union; Am Soc Testing & Mat; Am Soc Nondestructive Testing. *Res:* Stress and strain in earth's crust particularly problems of mining at depth; deformation of geologic materials especially their viscoelastic properties; mechanics of gas storage, microseismic activity, acoustic emission, design of structures in salt. *Mailing Add:* Dept Mineral Eng Rm 110 Bldg MS Pa State Univ University Park PA 16802

HARDY, HENRY BENJAMIN, JR, b Quincy, Ill, Dec 6, 25; m 50; c 4. ELECTRICAL INSULATION, TEXTILE PHYSICS. *Educ:* Univ Ill, BS, 48; Univ Wis, MS, 50. *Prof Exp:* Physicist, Pioneering Res Lab, E I du Pont de Nemours & Co, Inc, 50-53, res physicist, Textile Res Lab, 53-57 & Indust Prod Res Lab, 57-64, mkt develop rep, Nomex Tech Div, 64-66, tech serv rep, 66-75, specialist, Spunbonded-Nomex Div, 75-82, sr specialist, Indust Fibers Div, 82-85; RETIRED. *Res:* Physics and mechanics of textile fibers and fabrics; heat and moisture transfer between human body and environment; clothing comfort and aesthetics; automobile dynamics, especially suspension and steering; mechanical properties of tires; electrical insulating materials. *Mailing Add:* 1514 Ridge Rd Wilmington DE 19809

HARDY, JAMES C, b Salina, Kans, Apr 14, 30; m 53; c 3. SPEECH PATHOLOGY. *Educ:* Northeast Mo State Col, BS, 51; Univ Iowa, MA, 57, PhD(speech path), 61. *Prof Exp:* Speech therapist, Pub Schs, Mo, 51-52, 54-55; speech therapist, Univ Hosp Sch, Univ Iowa, 56-57, supvr speech & hearing, 57-67, from asst prof to assoc prof, 61-69, PROF, DEPT SPEECH PATH, AUDIOL & DEPT PEDIAT, UNIV IOWA, 69-, DIR SPEECH & HEARING CLIN, 72-, DIR PROF SERV, DIV DEVELOP DISABILITIES, 79-, DIR, IOWA PROG ASSISTIVE TECHNOL, 90- *Concurrent Pos:* Prin investr, Nat Inst Neurol Dis & Blindness res grant, 60-73; consult, Nat Inst Dent Res grants, 63- *Mem:* AAAS; Am Speech & Hearing Asn; Am Acad Cerebral Palsy. *Res:* Speech physiology and neurology of speech; speech disorders associated with trauma and pathologies of nervous system. *Mailing Add:* Dept Pediatrics & Speech Path Univ Iowa Iowa City IA 52242

HARDY, JAMES D, b Birmingham, Ala, May 14, 18; m 49; c 4. SURGERY, BIOCHEMISTRY. *Educ:* Univ Ala, BA, 38; Univ Pa, MD, 42, MS, 51; Am Bd Surg, dipl, 50; Bd Thoracic Surg, dipl, 52. *Prof Exp:* Asst instr med, Univ Pa, 43-44, asst instr surg, 46-49, instr, 49-51; from asst prof to assoc prof, Col Med, Univ Tenn, 51-55, dir surg labs, 51-55; PROF SURG & CHMN DEPT, SCH MED, UNIV MISS, 55-, DIR SURG RES, MED CTR, 55- *Concurrent Pos:* Consult, Oak Ridge Inst Nuclear Sci, 51-54; surgeon-in-chief, Univ Hosp, Univ Miss, 55-; chief surg consult, Jackson Vet Hosp, 55-; vis prof, Univ SC, 56, Emory Univ, 60, Hartford Hosp, Conn, 61, St Louis City & Los Angeles County Hosps, 62, Univ Calif, San Francisco, 62, Tulane Univ, 64 & Honolulu Hosps, 65; mem adv comt res ther, Am Cancer Soc, 63-66; mem adv bd, Am Bd Surg, 64-70; mem anesthesiol training comt, NIH, 66-69; vchmn, Am Bd Surg, 69-79. *Honors & Awards:* Smithy Mem Lectr, Med Col SC, 56; Agnew Lectr, Sch Med, Univ Pa, 57; Elkin Lectr, Emory Univ, 60; Banks Mem Lectr, Univ Liverpool, 63; John L Madden Lectr, Hershey Med Sch, 81; Emmett B Frazier Lectr, Univ S Ala, 83; Walter Estelle Lee Lectr, Grad Hosp, Univ Pa, 84. *Mem:* AAAS; Soc Univ Surg (secy, 56-58, pres, 61-62); Am Surg Asn (pres, 75-76); Soc Surg Chmn (secy-treas, 72-74, vpres, 74-76, pres, 76-78); Int Cardiovasc Soc; Am Heart Asn; Sigma Xi; Am Asn Surg Trauma; Am Asn Thoracic Surg; Am Col Surg. *Res:* Systemic response to injury; body fluid metabolism and composition; surgical endocrinology; dynamics of the circulation. *Mailing Add:* Dept Surg Univ Miss Sch Med 2500 N St Jackson MS 39216-4505

HARDY, JAMES EDWARD, b Salmon Arm, BC, Mar 21, 32; m 73; c 2. THEORETICAL PHYSICS. *Educ:* Univ BC, BA, 55, MSc, 57; Princeton Univ, PhD(theoret physics), 62. *Prof Exp:* Asst res officer, Nat Res Coun Can, 62-63; asst prof, 63-66, ASSOC PROF PHYSICS, CARLETON UNIV, 66- *Mem:* Can Asn Physicists. *Res:* Quantum field theory. *Mailing Add:* Dept Physics Carleton Univ Colonel By Dr Ottawa ON K1S 5B6 Can

HARDY, JOHN CHRISTOPHER, b Montreal, Que, July 10, 41; m 64; c 4. NUCLEAR PHYSICS. *Educ:* McGill Univ, BSc, 61, MSc, 63, PhD(nuclear physics), 65. *Prof Exp:* Nat Res Coun Can overseas fel nuclear physics, Oxford Univ, 65-67; Miller fel, Lawrence Radiation Lab, Univ Calif, Berkeley, 67-69, physicist, 69-70; assoc res officer, 70-74, head, Nuclear Physics Br, 83-87, SR RES OFFICER, CHALK RIVER LABS, AECL RES, 75- *Concurrent Pos:* Sci assoc, CERN, 76-; chmn, Bd of Dir, Deep River Sci Acad, 87-; asst vpres, Tactical Air Support Coord Ctr, 86-88, dir, 88- *Honors & Awards:* Ambridge Prize, 65; Herzberg Medal, 76; Rutherford Medal, 81. *Mem:* Can Asn Physicists; fel Am Phys Soc; fel Royal Soc Can. *Res:* Nuclear spectroscopy; beta-decay; delayed proton radioactivity; transfer reactions; nuclear isospin; nuclei far from stability. *Mailing Add:* Phys Sci Chalk River Labs AECL Res Chalk River ON K0J 1J0 Can

HARDY, JOHN THOMAS, b Booneville, Miss, Oct 30, 38; m 66; c 2. NUMBER THEORY. *Educ:* Univ Miss, BS, 60; La State Univ, MS, 62, PhD(math), 65. *Prof Exp:* Asst prof math, Southwestern La Univ, 65-66 & Univ Ga, 66-69; vis asst prof, 69-70, asst prof, 70-72, ASSOC PROF MATH, UNIV HOUSTON, 72- *Mem:* Am Math Soc. *Res:* Algebra; rings of integers in the Cayley algebra; quadratic forms; sums of two squares in quadratic rings; quaternions. *Mailing Add:* Dept Math Univ Houston 4800 Calhoun Rd Houston TX 77004

HARDY, JOHN THOMAS, b Detroit, Mich, June 3, 41; m 66; c 1. AQUATIC TOXICOLOGY, RESEARCH MANAGEMENT. *Educ:* Univ Calif, Santa Barbara, BA, 64; Ore State Univ, MS, 66; Univ Wash, PhD(marine bot) & PhD(aquatic ecol), 71. *Prof Exp:* Res fel, Univ Wash, 69-71; asst prof marine sci, Am Univ Beirut, 72-75; dir, Environ Consult Serv, 75-79; group leader, Battelle Northwest, 79-88; ASSOC PROF, ORE STATE UNIV, 88- *Concurrent Pos:* Dir, Ctr Aquatic Surface Res, 85-88; affil prof, Sch Fisheries, Univ Wash, 84-; dir, Ctr Aquatic Surface Res, 85- *Mem:* Fel AAAS; Am Soc Limnol & Oceanog. *Res:* Biological oceanography; dynamics processes at the sediment-water and atmosphere-water interfaces; photochemistry; biological community analysis and primary productivity, effects of UV-B radiation. *Mailing Add:* Huxley Col Environ Studies ES-539 Western Wash Univ Bellingham WA 98225

HARDY, JOHN W, JR, b Tucson, Ariz, Apr 21, 27; m 51; c 2. MATHEMATICS, COMPUTER SCIENCE. *Educ:* Stanford Univ, BS, 49, MS, 52, PhD(math), 55. *Prof Exp:* Mathematician, Kaiser Electronics Lab, 55 & Lawrence Radiation Lab, 55-71; PROF MATH, CALIF STATE COL, BAKERSFIELD, 71- *Concurrent Pos:* Consult, 777 Lock & Eng Corp, 63-68 & Madsen Corp, Calif, 64-65; vpres res & develop, Resource Data Corp, 66-69. *Mem:* Am Math Soc; Sigma Xi. *Res:* Numerical analysis; analysis. *Mailing Add:* 9001 Stockdale Hwy Bakersfield CA 93309

HARDY, JOHN WILLIAM, b Murphysboro, Ill, Jan 12, 30; c 2. ORNITHOLOGY. *Educ:* Univ Southern Ill, BS, 52; Mich State Univ, MS, 54; Univ Kans, PhD, 59. *Prof Exp:* Asst zool, Mich State Univ, 52-54; asst zool, Univ Kans, 54-57; ed, Bullentin, Kans Ornith Soc, 57-59; asst res zoologist, Univ Calif, Los Angles, 60-61, from asst prof to assoc prof biol, Occidental Col & dir, Moore Lab, 61-73; chmn dept, 73-78, PROF ZOOL & CUR ORNITH, DEPT NATURAL SCI, FLA STATE MUS, UNIV FLA, 73- *Concurrent Pos:* Chapman Mem Fund res grant, Am Mus Natural Hist, 56; Kans Acad Sci fel, 55; NSF fel, Marine Lab, Duke Univ, 57; ed recent lit, Auk, Am Ornith Union, 62-64, ed, Am Ornith Union Monographs, 72-78; Nat Geog Soc res grant, 73, 78 & 81; NSF res grant, 74-78. *Honors & Awards:* Tucker Award, Am Ornith Union, 58. *Mem:* Wilson Ornith Soc; Cooper Ornith Soc (treas, 64-); Am Ornith Union; AAAS. *Res:* Ornithology; behavior; phylogeny; behavior and phylogeny of New World jays; behavior of parrots, ecology; life history studies; fossil birds; communal social behavior of neotropic jays; avian bioacoustics. *Mailing Add:* Fla State Mus Univ of Fla Gainesville FL 32611

HARDY, JUDSON, JR, b New Orleans, La, Nov 8, 31; m 54; c 3. NUCLEAR PHYSICS, REACTOR PHYSICS. *Educ:* Univ NC, BS, 53; Princeton Univ, PhD(physics), 58. *Prof Exp:* ADV SCIENTIST, BETTIS ATOMIC POWER LAB, WESTINGHOUSE ELEC CORP, 58- *Mem:* Am Phys Soc; Am Nuclear Soc. *Mailing Add:* 3226 Kennebec Rd Pittsburgh PA 15241

HARDY, KENNETH REGINALD, b Saskatoon, Sask, July 30, 29; US citizen; m 57; c 2. SATELLITE METEOROLOGY. *Educ:* Univ Sask, BA, 50, Hons, 52; Univ Toronto, MA, 53; Univ Mich, PhD(meteorol), 63. *Prof Exp:* Meteorologist, Can Meteorol Serv, 52-56 & Weather Eng Corp Can, 57-58; res assoc meteorol, Univ Mich, 58-62; res physicist, Air Force Cambridge Res Labs, 63-66, chief, Weather Radar Br, 66-70 & 71-74; div mgr, Environ Res & Technol, Inc, 74-81; chief Satellite Meteorol Br, Air Force Geophys Lab, 81-90; SR STAFF SCIENTIST, RES & DEVELOP DIV, LOCKHEED, 90- *Concurrent Pos:* Vis assoc prof, Univ Wash, 70-71; mem comn F, Int Union Radio Sci. *Honors & Awards:* Darton Prize, Royal Meteorol Soc, 63. *Mem:* Am Meteorol Soc; fel Royal Meteorol Soc; Int Union Radio Sci. *Res:* Cloud physics; growth of precipitation and raindrop size distributions; clear air turbulence and clear air atmospheric structure as deduced from radar and other meteorological observations; satellite remote sensing of the atmosphere and earth's surface. *Mailing Add:* Lockheed 0/91-01 B/255 3251 Hanover St Palo Alto CA 94304-1191

HARDY, LAURENCE MCNEIL, b Tulsa, Okla, Feb 24, 39; m 60; c 2. SYSTEMATIC HERPETOLOGY, ANATOMY. *Educ:* NMex State Univ, BS, 62; Univ Kans, MA, 65; Univ NMex, PhD(biol), 69. *Prof Exp:* Asst prof, 68-71, assoc prof, 71-78, PROF BIOL, LA STATE UNIV SHREVEPORT, 78- *Mem:* Am Soc Ichthyol & Herpet; Soc Study Amphibians & Reptiles; Soc Study Evolution; Am Soc Syst Zool; Herpetologists League (treas); Sigma Xi. *Res:* Systematic studies of colubrid snakes, especially from the New World; karyotypes of snakes; reproductive biology and population dynamics of salamanders; biogeography of amphibians and reptiles; reproductive anatomy of reptiles. *Mailing Add:* Biol Sci Dept La State Univ 8515 Youree Dr Shreveport LA 71115

HARDY, LESTER B, b Foxboro, Mass, July 4, 32; m 56; c 2. PHYSIOLOGY, TOXICOLOGY. *Educ:* Univ Maine, BS, 55; Univ Del, MS, 57; Univ Md, PhD(physiol), 60. *Prof Exp:* Asst animal physiol, Univ Del, 55-57; asst, Univ Md, 57-59; asst internal med, Sch Med, Yale Univ, 60; sr scientist endocrinol, G D Searle & Co, Ill, 60-61; sr scientist physiol, Gillette Safety Razor Co, Mass, 61-64, sr scientist, Gillette Co Res Inst, Washington, DC, 65, toxicologist, Med Eval Div, Md, 65-68, asst dir med rev, 68-70, dir, 70-71, vpres, Med Eval Labs, 71-74, PRES, GILLETTE MED EVAL LABS, 74- *Mem:* AAAS; Soc Toxicol; NY Acad Sci; Am Col Toxicol. *Res:* Medical safety evaluation of drugs and cosmetics; product toxicology and pharmacology; industrial hygiene. *Mailing Add:* Gillette Med Eval Labs 401 Prof Dr Gaithersburg MD 20879

HARDY, MATTHEW PHILLIP, b Evanston, Ill, Feb 7, 57; m 84. STEROID BIOCHEMISTRY. *Educ:* Oberlin Col, BA, 79; Univ Va, PhD(biol), 85. *Prof Exp:* Fel, Sch Hyg & Pub Health, Johns Hopkins Univ, 85-89, res assoc, 90-91; STAFF SCI, POP COUN, 91- *Mem:* Sigma Xi; Am Soc Zoologists; Am Soc Andrologists; Endocrine Soc; Soc Study Reproduction; AAAS. *Res:* Differentiation of Leydig cells during puberty. *Mailing Add:* Population Coun 1230 York Ave New York NY 10021

HARDY, PAUL WILSON, b Waukesha, Wis, Apr 18, 27; m 48; c 4. CHEMISTRY. *Educ:* Carroll Col, Wis, BA, 48. *Prof Exp:* From sr res chemist to group leader, Am Can Co, 48-65, res assoc 65-78, sr res assoc, 78-82; RETIRED. *Mem:* Am Chem Soc. *Res:* Exploratory research; color and color measurement; ceramics. *Mailing Add:* PO Box 4096 University Park NM 88003-4096

HARDY, RALPH WILBUR FREDERICK, b Lindsay, Ont, July 27, 34; m 54; c 5. SCIENCE POLICY, BIOCHEMISTRY. *Educ:* Univ Toronto, BSA, 56; Univ Wis, MS, 58, PhD(biochem), 59. *Prof Exp:* Asst prof biochem, Univ Guelph, 60-63; res biochemist, Cent Res & Develop Dept, Exp Sta, E I Du Pont De Nemours & Co, Inc, 60 & 63-67, res supvr, 67-74, assoc dir res, 74-79, dir life sci, 79-84; pres, Biotech Int Inc, 84-86, dep chmn, 86-90; PRES & CHIEF EXEC OFFICER, BOYCE THOMPSON INST, 86- *Concurrent Pos:* Exec comt mem, NRC Bd Agr, 83-88, mem NRC Comn Life Sci, 84-90, NRC Bd Basic Biol, 84-90, NRC Comt Biotechnol, 88-, NRC Bd Sci Technol Int Develop, 90-; Int Coun Sci Unions comt genetic experimentation, 81-; mem bd, Indust Biotechnol Asn, 86-89; vis prof life sci, Cornell, 84-86; mem bd, Biotech Int Inc, 85- *Honors & Awards:* Gov Gen Silver Medal, Am Chem Soc, 56, Delaware Award, 68 & Hendricks Medal & Award, 86. *Mem:* Am Soc Plant Physiologists (treas, 73-76); Am Soc Biol Chemists; Am Soc Agron; Am Chem Soc (secy, Biol Chem Div, 78-81); Am Soc Microbiol. *Res:* Plant biology and biochemistry including nitrogen and carbon inputs into major crops, nitrogen fixation, photosynthesis and partitioning of assimilates. *Mailing Add:* 330 The Parkway Ithaca NY 14850

HARDY, ROBERT J, b Port Angeles, Wash, Jan 26, 35; m 70. PHYSICS. *Educ:* Reed Col, BA, 56; Lehigh Univ, MS, 58, PhD(physics), 62. *Prof Exp:* Physicist, US Naval Radiol Defense Lab, Calif, 62; res assoc physics, Lehigh Univ, 62-63; vis prof ctr res & adv studies, Nat Polytech Inst, Mex, 64; res assoc & instr, Univ Ore, 65-67; from asst prof to assoc prof, 71-76, PROF PHYSICS, UNIV NEBR, LINCOLN, 76-, VCHMN, PHYSICS & ASTRON, 84- *Concurrent Pos:* Consult, Lawrence Livermore Lab, 70- *Mem:* Am Phys Soc; Am Asn Physics Teachers; Am Asn Univ Prof. *Res:* Theory of equilibrium and transport properties of crystal lattices; statistical mechanics; theory of conduction electrons in magnetic fields; solid state physics. *Mailing Add:* Dept Physics & Astron Univ Nebr Lincoln NE 68588

HARDY, ROBERT W, b Brantford, Ont, June 8, 52. DIABETES, PATHOLOGY. *Educ:* Univ Waterloo, BSc, 75; Univ Toronto, MSc, 82, PhD(clin biochem), 88. *Prof Exp:* Technologist chem, Can Med Labs, 76-80, Sunnybrook Med Ctr, Toronto, 80-82, Hosp for Sick Children, Toronto, 85-87; fel, Clin Chem Prog, Sch Med, Washington Univ, St Louis, Mo, 87-90; fel, 90-91, RES INSTR & ASSOC CLIN CHEM, PATH DEPT, UNIV ALA, BIRMINGHAM, 91- *Concurrent Pos:* Fel, Can Diabetes Asn, 88-; young investr award, Acad Clin Lab Physicians & Scientists, 89. *Honors & Awards:* Med-Chem Award, Can Soc Clin Chemists, 86. *Res:* Interaction of calmodulin and polylysine; regulation of glucose transport. *Mailing Add:* Dept Path Univ Ala L R Bldg Rm 506 Birmingham AL 35294

HARDY, ROLLAND L(EE), b Carthage, Ill, May 2, 20; m 44, 53, 90; c 3. GEODESY, MATHEMATICAL PHYSICS. *Educ:* Univ Ill, BS, 47; Univ Mo-Rolla, BSCE, 50, CE, 56; Karlsruhe Tech Univ, Dr Ing, 63. *Prof Exp:* Field engr, US Geol Surv, 47-51; supvr geod engr, map serv, US Army, 51-52, civil engr, Eng Res & Develop Labs, 52-55; asst prof civil eng, George Washington Univ, 56-58; hwy res engr, Bur Pub Rds, Dept Commerce, 58-59; geod engr, AID, US Opers Mission, Sudan, 59-61; gen engr adv systs off, mapping, charting & geod directorate, Dept Defense, 63-67; geodesist, Geod Res Lab, Inst Earth Sci, Environ Sci Serv Admin, 67; prof civil eng, Iowa State Univ, 67-85, prof-in-chg geod, photogram & surv, 80-88; pres, Int Inst Sci & Technol, Monterey, Calif, 90-91; CONSULT, 91- *Concurrent Pos:* Lectr, Northern Va Ctr, Univ Va, 56-58; consult, 55-58; dir geod & cartographic sci prog, George Washington Univ, 63-66; lectr, Darmstadt & Univ Stuttgart, Ger, 89. *Honors & Awards:* Performance Award Cert, US Dept Army, 59 & 60; Earle Fennel Award, Am Cong Surv & Mapping, 86. *Mem:* Fel Am Soc Civil Engrs; fel Am Cong Surv & Mapping; Am Geophys Union; Am Soc Photogram & Remote Sensing; Sigma Xi. *Res:* Design of engineering instruments; development of advanced geodetic and photogrammetric systems; data reduction; satellite geodesy and photogrammetry; multiquadric analysis; biharmonic potential theory. *Mailing Add:* 1213 Wanda Ave Seaside CA 93955

HARDY, RONALD W, b Vancouver, BC; US citizen; m 70; c 2. FISH NUTRITION. *Educ:* Univ Wash, BS, 69, PhD(fisheries), 78; Wash State Univ, MS, 73. *Prof Exp:* ASSOC PROF FISHERIES, UNIV WASH, 78-; SUPVRY RES CHEMIST, NAT MARINE FISHERIES SERV, NAT OCEANIC & ATMOSPHERIC ADMIN, 83- *Concurrent Pos:* Mem subcomt nutrient requirements warmwater fishes, Nat Res Ctr, Nat Acad Sci. *Mem:* Am Fisheries Soc; World Mariculture Soc; Am Inst Nutrit. *Res:* Diet and disease resistance, diet development for aquaculture, nutritional requirements, lipid composition of fish. *Mailing Add:* NW Fisheries Ctr 2725 Montlake Blvd E Seattle WA 98122

HARDY, VERNON E, b Oklahoma City, Okla, Apr 5, 37; m 66; c 2. ELECTRONICS ENGINEERING. *Educ:* Univ Okla, BSEE, 60, MSEE, 62. *Prof Exp:* Grad asst electronics, Univ Okla, 60-62; engr & sect mgr, Tex Instruments, 62-69; staff engr & sect mgr, Electronic Memories & Magnetics, 69-71; eng mgr & dir eng, Standard Memories-Trendata, 71-73; dir eng, MSI Data, 83-87; eng mgr, Rockwell Int, 87-89; MGR & DIR ENG, CMC, DIV ROCKWELL INT, 89- *Concurrent Pos:* Consult, ElectroCom Automation, 87. *Mem:* Inst Elec & Electronics Engrs. *Mailing Add:* 18912 Santa Mariana Fountain Valley CA 92708

HARDY, WALTER NEWBOLD, b Vancouver, BC, Mar 25, 40; m 59; c 2. SOLID STATE PHYSICS. *Educ:* Univ BC, BSc, 61, PhD(physics), 65. *Prof Exp:* Nat Res Coun Can fel, Saclay Nuclear Res Ctr, France, 64-66; mem tech staff, Sci Ctr, NAm Rockwell Corp, Calif, 66-71; assoc prof, 71-80, PROF PHYSICS, UNIV BC, 80- *Concurrent Pos:* Vis prof, Ecole Normale

Superieure, Paris, 80-81, 85. *Mem:* Am Phys Soc. *Res:* Nuclear and electron magnetic resonance; cryogenics; microwave properties of high temperature superconductors; spin aligned atomic hydrogen. *Mailing Add:* Dept Physics Univ BC Vancouver BC V6T 2A6 Can

HARDY, WILLIAM LYLE, b BC, June 18, 36; m 60. BIOPHYSICS, PHYSIOLOGY. *Educ:* Univ BC, BASc, 59, MSc, 61; Univ Wash, PhD(physiol, biophys), 69. *Prof Exp:* Asst physicist radiol physics, BC Cancer Inst, 61-63; USPHS fel, Dept Physiol & Biophys, Univ Wash, 69-71; asst prof physiol & cardiol, Sch Med, Boston Univ, 71-85. *Concurrent Pos:* Comput consult, Worchester Found Exp Biol, 72. *Mem:* AAAS. *Res:* Excitable membrane biophysics of nerve and muscle; excitation-contraction coupling; pacemaker activity in cardiac cells; digital computer applications in teaching and in clinical and biological research. *Mailing Add:* 508 Border St East Boston MA 02128-2464

HARDY, YVAN J, b Quebec City, Can, Aug 11, 41; m 67; c 2. FORESTRY ENTOMOLGY, SILVICULTURE. *Educ:* Laval Univ, BScA, 65, MSc, 68; State Univ NY, Syracuse, PhD(forest entom), 71. *Prof Exp:* From asst prof to prof forest entom, Fac Forestry & Geod, Laval Univ, 70-85, chmn dept forest mgt, 75-78, vice dean, 79-80, dean, 80-85; DIR GEN, QUE REGION, CAN FORESTRY SERV, 85- *Concurrent Pos:* Consult forest entom, Dept Lands & Forest, Que, 71-76; dir res, Forest Res & Develop Found, 76-82; working group leader Can-USA Prog, USDA Forest Serv, 77-80; vis prof, Univ Maine, Orono, 78-79; mem adv comt forest res, Que Dept Energy & Resources, 81-86; ad hoc comnr, Que Pub Hearing Bur Environ, 82-83; guest lectr, Universite de Quebec, Laval Univ, 85-; pres, Int Seminar on Private Woodlot Mgt, FAO, 87; rapporteur, Nat Forum Sustainable Forest Develop, Halifax, 89. *Mem:* Entom Soc Can; Sigma Xi; Can Inst Forestry. *Res:* Epidemiology of the spruce budworm in relation to bioregions and dynamics of forest cover; host-insect relationships; vulnerability and susceptibility of host species in various ecological environments; methodology in private woodlot management. *Mailing Add:* 1055 P E P S St PO Box 3800 Ste Foy PQ G1V 4C7 Can

HARE, CURTIS R, b Collingdale, Pa, Dec 13, 33; m 57; c 1. INORGANIC CHEMISTRY, PHYSICAL CHEMISTRY. *Educ:* Pa State Univ, BS, 55; Mich State Univ, PhD(inorg chem), 61. *Prof Exp:* Asst chem, Pa State Univ, 55 & Mich State Univ, 57-59, 60-61; NSF fel, Inst Phys Chem, Copenhagen Univ, 61-62, res assoc, 62-63; from asst prof to assoc prof, State Univ NY Buffalo, 63-69; ASSOC PROF, UNIV MIAMI, 69- *Mem:* AAAS; Am Inst Chem; Am Chem Soc; Royal Soc Chem; Am Crystallog Asn. *Res:* Spectral and magnetic properties of transition metal complexes; ligand field theory; inorganic stereochemistry. *Mailing Add:* Dept Chem Univ Miami Coral Gables FL 33124

HARE, FREDERICK KENNETH, b Wylye, Eng, Feb 5, 19; Can citizen; m 53; c 3. PUBLIC INQUIRIES. *Educ:* Univ London, 39; Univ Montreal, PhD(geog), 50. *Hon Degrees:* LLD, Queen's Univ, 64, Univ Western Ont, 68, Trent Univ, 79; DSc, McGill Univ, 69, Adelaide, 74, York Univ, 78; DS Litt, Thorneloe Col, 83; DLitt, Mem Univ, 85; LID, Toronto, 87; DSc, Windsor 88. *Prof Exp:* Asst lectr geography/meteorol, Univ Manchester, 40-41; asst prof geog & meteorol, McGill Univ, 45-49, assoc prof, 50-52, prof, 52-62, dept chmn, 50-62, dean arts & sci, 62-64; prof, King's Col, Univ London, 64-66, master, Birkbeck Col, 66-68; pres, Univ BC, 68-69; prof geog & physics, Univ Toronto, 69-84, Univ prof, 76-84; provost, 79-86, EMER UNIV PROF, TRINITY COL, 84-; CHANCELLOR, TRENT UNIV, 88- *Concurrent Pos:* Chmn, Can Climate Planning Bd, 79-90; comnr, Ontario Nuclear Safety Rev, 86-88; chair, adv bd, Int Prog, Univ Toronto, 90- *Honors & Awards:* Dawson Medal, Royal Soc Can, 87; Companion, Order of Can; Int Meteorol Orgn Prize, 88. *Mem:* Can Asn Geographers; Asn Am Geographers; Royal Geog Soc; fel Royal Soc Can; Royal Meteorol Soc (pres, 67-68); Sigma Xi. *Res:* High latitude climatology and biogeography; behavior of the stratosphere; water and energy balance of North America; aspects of climatic change; arid zone climates. *Mailing Add:* 301 Lakeshore Rd W Oakville ON L6K 1G2 Can

HARE, JAMES FREDERIC, b Philadelphia, Pa, June 4, 45; m 75; c 2. MEMBRANES, CELL BIOLOGY. *Educ:* Lafayette Col, AB, 67; Univ NH, MS, 69; Purdue Univ, PhD(biol sci), 73. *Prof Exp:* Teach asst zool & cell biol, Purdue Univ, 69-73; res assoc biochem, Harvard Med Sch, 73-75; res assoc biochem, Calif Inst Technol, 75-77; asst prof, 77-82, ASSOC PROF BIOCHEM, HEALTH SCI CTR, UNIV ORE, 82- *Mem:* Am Chem Soc; Am Soc Biochem & Molecular Biol. *Res:* Biosynthesis and degradation of the proteins of biological membranes; mechanisms of protein degradation. *Mailing Add:* Health Sci Ctr SM Univ Ore Portland OR 97201

HARE, JOAN CONWAY, b Berlin, NH, June 6, 43; m 72. PHYCOLOGY, MARINE BOTANY. *Educ:* Univ NH, BA, 67; Univ Mass, MS, 72, PhD(bot), 76. *Prof Exp:* NATO fel marine bot, Biologische Anstalt Helgoland, 78-79; chem tutor, 76-83, INSTR, CONT EDUC, UNIV MASS, 81- *Concurrent Pos:* Consult algae, Mass Inst Technol, 81. *Mem:* Bot Soc Am; Phycol Soc Am; Int Phycol Soc; Brit Phycol Soc. *Res:* Development, morphogenesis and life histories of algae; photobiology and nutrition of algae; environmental control of algal development and morphology. *Mailing Add:* One Chadwick Ct Amherst MA 01002

HARE, JOHN DANIEL, III, b Fresno, Calif, Jan 30, 48; m 82; c 2. INSECT ECOLOGY, EVOLUTIONARY ECOLOGY. *Educ:* Stanford Univ, BA, 70; State Univ NY Stony Brook, PhD(ecol & evolution), 78. *Prof Exp:* Asst entomologist, Conn Agr Exp Sta, 77-84; asst prof, 84-89, ASSOC PROF, DEPT ENTOM, UNIV CALIF, RIVERSIDE, 89- *Mem:* AAAS; Entom Soc Am; Ecol Soc Am; Soc Study Evolution. *Res:* Plant-insect interactions; integrated control of agricultural insect pests; chemical ecology; co-evolution; tritrophic interactions. *Mailing Add:* Dept Entom Univ Calif 900 University Ave Riverside CA 92521

HARE, JOHN DONALD, b Rochester, NY, Jan 23, 28; m 52; c 3. VIROLOGY. *Educ:* Harvard Univ, BA, 50; Univ Rochester, MS, 53, MD, 55. *Prof Exp:* Intern med, Harvard Serv, Boston City Hosp, Mass, 55-56, asst resident, 58-59; from instr med & bact to assoc prof microbiol, 59-71, PROF MICROBIOL, MED CTR, UNIV ROCHESTER, 71- *Mem:* Infectious Dis Soc Am; Am Soc Microbiol; Am Asn Cancer Res. *Res:* Viral oncogenesis; cellular metabolism; infectious diseases; immunology. *Mailing Add:* Dept Microbiol & Med Univ Rochester Med Ctr Rochester NY 14642

HARE, LEONARD N, b Rangoon, Burma, Dec 5, 21; US citizen; m 44; c 2. PLANT PHYSIOLOGY. *Educ:* Pac Union Col, BA, 44; Univ Md, MA, 59, PhD(plant physiol), 61. *Prof Exp:* Prin, Burma Union Training Sch, 46-48, Tenasserim Mission Sch, Burma, 48-50 & Raymond Mem Training Sch, India, 51-56; assoc prof, 61-70, PROF BIOL, ANDREWS UNIV, 70- *Res:* Effects of environment, especially light duration and quality, on differentiation, elongation, metabolism and free amino acid content of plant cells. *Mailing Add:* PO Box 43 Berrien Springs MI 49103

HARE, MARY LOUISE ECKLES, b Sardis, Miss, Nov 19, 16; m 40; c 2. CYTOLOGY, BOTANY. *Educ:* Miss State Col Women, BS, 36; Miss State Univ, MS, 38; Univ Wis, PhD(cytol), 42. *Prof Exp:* Teaching asst, Miss State Univ, 36-39; lab asst plant path, Univ Wis, 42-43; actg asst prof, 59-61, asst prof, 64-74, ASSOC PROF BOT, MISS STATE UNIV, 74- *Mem:* AAAS; Bot Soc Am. *Res:* Cytology and anatomy of angiosperms. *Mailing Add:* 215 Bridle Path Starkville MS 39759

HARE, PETER EDGAR, b Maymyo, Burma, Apr 14, 33; US citizen; m 54; c 2. ORGANIC GEOCHEMISTRY. *Educ:* Pac Union Col, BS, 54; Univ Calif, Berkeley, MS, 55; Calif Inst Technol, PhD(geochem), 62. *Prof Exp:* Instr chem, Pac Union Col, 55-58; res fel, Calif Inst Technol, 62-63; STAFF MEM, CARNEGIE INST GEOPHYS LAB, 63- *Mem:* Geol Soc Am. *Res:* High sensitivity amino acid analysis including D & L amino acid isomers; application fo geochemical systems; amino acid sequence of fossil peptides; development of chiral separations in liquid chromatography; stable isotopes of nitrogen and carbon in amino acids from fossils; paleodiets and paleoenvironments of fossils. *Mailing Add:* Carnegie Inst Washington Geophys Lab 5251 Broad Branch Rd NW Washington DC 20015-1305

HARE, ROBERT RITZINGER, JR, b Indianapolis, Ind, Feb 23, 25; m 47; c 4. APPLIED MATHEMATICS, OPERATIONS RESEARCH. *Educ:* DePauw Univ, AB, 48, MA, 49. *Prof Exp:* Mathematician & chief math anal sect, USAF Missile Test Ctr, Patrick AFB, Fla, 51-53; mem staff math, Opers Res Off, Johns Hopkins Univ, 53-57; sr scientist, Opers Res, Inc, Md, 57-69; from assoc prof to prof, 69-90, EMER PROF MATH, NDAK STATE UNIV, 90- *Concurrent Pos:* Instr, Rollins Col, 52, Univ Fla, 52-53 & Montgomery Jr Col, 56-57; consult, Army Res Off, 70. *Mem:* AAAS; Am Math Soc; Opers Res Soc Am. *Res:* Systems analysis and operations research in military and transportation fields; applied probability theory; computer applications, simulation. *Mailing Add:* Dept Math NDak State Univ Fargo ND 58105

HARE, WILLIAM CURRIE DOUGLAS, b Edinburgh, Scotland, Jan 10, 25; m 75; c 3. EMBRYOS, CYTOGENETICS. *Educ:* Univ Edinburgh, BSc, 50, PhD(vet anat), 53, DVM & S, 60. *Hon Degrees:* MA, Univ Pa, 71. *Prof Exp:* Asst lectr vet anat, Royal Sch Vet Studies, Univ Edinburgh, 50-53, lectr, 53-55; assoc prof, Ont Vet Col, 55-58; from assoc prof to prof vet anat, Sch Vet Med, Univ Pa, 58-74; RES SCIENTIST, FOOD PROD & INSPECTION BR, AGR CAN, 74- *Mem:* Can Vet Med Asn; fel Royal Col Vet Surgeons; Int Embryo Transfer Soc; Sigma Xi; US Animal Health Asn. *Res:* Veterinary cytogenetics; embryo transfer; disease control. *Mailing Add:* Animal Dis Res Inst Box 11300 Sta H Ottawa ON K2H 8P9 Can

HARE, WILLIAM RAY, JR, b Murfreesboro, Ark, June 29, 36; m 61; c 3. MATHEMATICS. *Educ:* Henderson State Teachers Col, BS, 57; Univ Fla, MS, 59, PhD(math), 61. *Prof Exp:* From instr to asst prof math, Duke Univ, 61-64; assoc prof, 64-74, PROF MATH & HEAD DEPT, CLEMSON UNIV, 74- *Mem:* Am Math Soc; Math Asn Am. *Res:* Topological and combinatorial convexity; generalizations of convexity. *Mailing Add:* Dept Math Sci Clemson Univ 201 Sikes Hall Clemson SC 29631

HAREIN, PHILLIP KEITH, b Waterville, Minn, May 17, 28; m 56; c 4. ENTOMOLOGY. *Educ:* Mankato State Col, BS, 51; Va Polytech, MS, 56; Kans State Univ, PhD(entom), 61. *Prof Exp:* Asst entomologist, Pillsbury Co, 56-57; instr entom, Kans State Univ, 57-60; entomologist, Agr Res Serv, USDA, 60-65; assoc prof, 65-67, PROF ENTOM, UNIV MINN, ST PAUL, 67-, EXTEN ENTOMOLOGIST, INST AGR, 65- *Mem:* Entom Soc Am. *Res:* General extension entomology; pesticide safety; associations between stored-product insects and micro-organisms, especially bacteria and fungi. *Mailing Add:* 226 EFW Bldg Dept Entomoly Univ Minn St Paul MN 55108

HARENDZA-HARINXMA, ALFRED JOSEF, chemical physics; deceased, see previous edition for last biography

HARES, GEORGE BIGELOW, b Corning, NY, Jan 11, 21; m 45; c 2. INORGANIC CHEMISTRY. *Educ:* Syracuse Univ, BS, 48, MS, 49; Pa State Col, PhD(chem), 52. *Prof Exp:* Sr res chemist, 52-59, res assoc chem, 59-67, SR RES ASSOC CHEM, CORNING GLASS WORKS, 67- *Mem:* Am Chem Soc; Am Ceramic Soc; Brit Soc Glass Technol; fel Am Inst Chem. *Res:* Chemistry of glass. *Mailing Add:* 31 E Fourth St Corning NY 14830

HARESIGN, THOMAS, b Sandy Creek, NY, Sept 4, 32; m 57; c 3. ZOOLOGY. *Educ:* State Univ NY Albany, BS, 57, MS, 59; Univ Mass, PhD(zool), 64. *Prof Exp:* Instr biol, Marist Col, 62-64; from asst prof to assoc prof, 64-74, PROF BIOL, SOUTHAMPTON COL, LONG ISLAND UNIV & DIR NAT SCI DIV, 74- *Concurrent Pos:* Worcester Found fel exp biol, 70-71. *Mem:* AAAS. *Res:* Endocrinology; steroid biochemistry; computer applications. *Mailing Add:* Dept Biol Southampton Col Long Island Univ Southampton NY 11968

HAREWOOD, KEN RUPERT, MOLECULAR GENETICS, CANCER RESEARCH. *Educ:* City Univ New York, PhD(biochem), 70. *Prof Exp:* PRIN RES INVESTR, MOLECULAR GENETICS RES DEPT, PFIZER INC, GROTON, CT, 71- *Concurrent Pos:* Mem, educ task force, Pfizer Inc. *Mem:* Am So Biochem & Molecular Biol; Harvey Soc. *Res:* Biochemistry. *Mailing Add:* Cent Res Div Eastern Pt Rd Groton CT 06340

HARFENIST, ELIZABETH JOYCE, fibrinogen, platelets, for more information see previous edition

HARFENIST, MORTON, b New York, NY, Dec 6, 22; div; c 3. ORGANIC CHEMISTRY, MEDICINAL CHEMISTRY. *Educ:* City Col New York, BS, 42; Polytech Inst Brooklyn, MS, 46; Univ Ill, PhD(chem), 48. *Prof Exp:* Jr chemist, Wellcome Res Labs, 43-44; res chemist, Chas Pfizer & Co, Inc, 48-52; sr org chemist, 52-70, group leader, 70-74, actg head org chem dept, 74-77, GROUP LEADER, WELLCOME RES LABS, 77- *Concurrent Pos:* Naval med res, US Navy, 45-46. *Mem:* Am Chem Soc; Royal Soc Chem UK. *Res:* Medicinal synthetic organic chemistry. *Mailing Add:* Burroughs Wellcome Co Inc Research Triangle Park NC 27709

HARFORD, AGNES GAYLER, b St Louis, Mo, June 27, 41. MOLECULAR BIOLOGY, GENETICS. *Educ:* Harvard Univ, AB, 63; Johns Hopkins Univ, PhD(biophys), 71. *Prof Exp:* Fel cell biol & genetics, Yale Univ, 71-73; asst prof, 73-80, ASSOC PROF CELL & MOLECULAR BIOL, STATE UNIV NY BUFFALO, 80- *Concurrent Pos:* Fel, NIH, 71-73; NIH grant, 74- *Mem:* Am Soc Cell Biol; Genetics Soc Am. *Res:* Molecular genetics of eukaryotes, DNA, chromosomes. *Mailing Add:* Biol Sci Cooke Hall State Univ NY North Campus Buffalo NY 14260

HARFORD, CARL GAYLER, b St Louis, Mo, June 27, 06; wid; c 3. MEDICINE. *Educ:* Amherst Col, AB, 28; Wash Univ, MD, 33. *Prof Exp:* Asst med, 33, asst med bact & immunol, 35-36, instr bact & immunol & asst clin med, 38-42, instr clin med, 42-43, from asst prof to prof, 43-74, EMER PROF MED, SCH MED, WASH UNIV, 74- *Concurrent Pos:* Fel path & bact, Rockefeller Inst, 36-38. *Mem:* Am Soc Microbiol; Am Soc Clin Invest; AMA; Asn Am Physicians; Infectious Dis Soc Am. *Res:* Methods to increase susceptibility of cells to viruses; infectious diseases. *Mailing Add:* 660 S Euclid Box 8051 St Louis MO 63110

HARFORD, JAMES J, b Jersey City, NJ, Aug 19, 24; m 52; c 4. MECHANICAL ENGINEERING. *Educ:* Yale Univ, BE, 45. *Prof Exp:* Appln engr fluid mach, Worthington Corp, 46-49; assoc ed indust prod methods, Mod Indust, 50-52; free lance writer, 52-53; exec secy astronaut, Am Rocket Soc, 53-63; exec dir, 64-89, EMER EXEC DIR, AM INST AERONAUT & ASTRONAUT, 89- *Mem:* Fel AAAS; fel Brit Interplanetary Soc; fel Am Inst Aeronaut & Astronaut; assoc fel Royal Aeronaut Soc; Inst Elec & Electronics Engrs. *Res:* Technical society administration, technical publishing and journalism. *Mailing Add:* 601 Lake Dr Princeton NJ 08540

HARGENS, CHARLES WILLIAM, III, b Philadelphia, Pa, Oct 21, 18; m 41; c 3. ELECTRICAL ENGINEERING. *Educ:* Mass Inst Technol, SB, 41. *Prof Exp:* Design engr, Lockheed Aircraft Corp, 41-42; proj engr, Gilfillan Bros, Inc, Calif, 42-45; group leader adv develop, Radio Corp Am, NJ, 46-47; sr staff engr, 48-59, head, bio-electronics br, 59-60, tech dir labs in charge elec eng, 60-68, fel, res labs, Franklin Inst, 68-88; RETIRED. *Concurrent Pos:* Vis mem staff, radiation lab, Mass Inst Technol, 42-44; res assoc, Wills Eye Hosp, 70-; adj assoc prof acoust, Temple Univ, 76 & 77 & Drexel Univ, 78-87; lectr acoust, Philadelphia Col Art, 81. *Mem:* Sigma Xi; fel Inst Elec & Electronics Engrs. *Res:* Electrical instrumentation; medical and biological applications; electrical communication engineering; acoustics. *Mailing Add:* 1006 Preston Rd Philadelphia PA 19118

HARGER, ROBERT OWENS, b Flint, Mich, Sept 15, 32; m 60; c 4. ELECTRICAL ENGINEERING. *Educ:* Univ Mich, Ann Arbor, BSE, 55, MS, 59, PhD(elec eng), 61. *Prof Exp:* Res engr, Inst Sci & Technol, Univ Mich, Ann Arbor, 61-68, asst prof elec eng, 63-66; assoc prof, 68-75, PROF ELEC ENG & CHMN DEPT, UNIV MD, COLLEGE PARK, 75- *Concurrent Pos:* Consult, NASA, 69-72, Dept Defense, 73-75 & Environ Res Inst Mich, 77- *Honors & Awards:* Barry Carlton Award, Inst Elec & Electronics Engrs, 77. *Mem:* Fel Inst Elec & Electronics Engrs; Int Radio Sci Union; AAAS. *Res:* Communication theory; radar and remote sensing systems; optical communication theory. *Mailing Add:* Dept Elec Eng Univ Md College Park MD 20742

HARGEST, THOMAS SEWELL, b Phillipsburg, NJ, Jan 3, 25; m 45; c 4. BIOMEDICAL ENGINEERING. *Educ:* Lafayette Col, BA, 50. *Prof Exp:* Geophysicist, Stand Oil & Gas Co, 51-53; opers mgr, Oil & Gas Div, Brown Container Co, 53-56; vpres, Slade Oil & Gas, Inc, 56-58; pres, Overland, Inc, 58-60; vpres, Western Oil Corp, 60-62; pres, Metcon, Inc, 62-63; dir res, Metcon Div, Chatleff Controls, 63-64; dir eng develop sect, Shriners Burn Inst, 64-65; dir, Div Clin Eng, Dept Surg, Med Univ SC, 65-76, from asst prof to assoc prof surg, 70-80, prof surg & biomet, 80-85; RETIRED. *Concurrent Pos:* Consult, Vet Admin Hosp, Charleston, SC & SC Retarded Children's Rehabilitation Ctr, Ladson, Support Syst Int, SA, Montpellier, France, Am Hosp Supply Corp, Evanston, Ill, Kum Klao Proj, Royal Thai Air Force, Bangkok, Thailand. *Mem:* Asn Advan Med Instrumentation; Am Burn Asn; Am Soc Artificial Internal Organs; Sigma Xi. *Res:* Development of systems and devices to improve the care and well-being of the patient. *Mailing Add:* PO Box 21118 Charleston SC 29413

HARGIS, BETTY JEAN, b Madison, Ind, Aug, 14, 25. BIOLOGY, IMMUNOLOGY. *Educ:* Purdue Univ, West Lafayette, BS, 47; Boston Univ, AM, 58, PhD(biol), 67. *Prof Exp:* Technician immunol, Med Sch, Northwestern Univ, 48-54; res asst protein chem, Rheumatic Fever Res Inst, 54-55; chief res asst immunol, Peter Bent Brigham Hosp, 55-62; res asst, Children's Hosp Med Ctr, Boston, 65-71, res assoc path, 71-81, res asst, Sidney Farber Cancer Inst, 63-78, res assoc, 78-81, assoc path, 81-82; sr res assoc, Dana Farber Cancer Inst, 83-86; consult, Greater Boston In-Vitro Assocs, 83-86; DIR, CYTOGEN RES & DEVELOP, NEWTON-WELLESLEY IN-VITRO LABS, 86- *Concurrent Pos:* Assoc path immunol, Harvard Med Sch, 74-82. *Mem:* AAAS; Am Asn Immunol; Sigma Xi; NY Acad Sci; Boston Cancer Res Asn. *Res:* Immunologic responses in parasitic, viral, and oncologic diseases; reproductive biology & cryogenics. *Mailing Add:* 115 Park St Boston MA 02146

HARGIS, I GLEN, b Hail, Ky, June 5, 39; m 61; c 3. PHYSICAL CHEMISTRY, POLYMER CHEMISTRY. *Educ:* Kent State Univ, BS, 61; Ohio Univ, PhD(chem), 66. *Prof Exp:* Sr res chemist, Monsanto Res Corp, 65-66; sr res chemist, Gencorp, 66-71, res scientist, Res & Develop Ctr, 71-72, group leader, Res Div, 72-81, sect head, 81-89, SR RES ASSOC, GENCORP, 89- *Mem:* Am Chem Soc. *Res:* Materials chemistry; polymer characterization; elastomer synthesis and characterization; identifying structural and morphological parameters which significantly contribute to vulcanizate performance; synthesis and evaluation of organometallic initiator catalyst systems. *Mailing Add:* 679 Atwood Dr Tallmadge OH 44278

HARGIS, J HOWARD, b Fayetteville, Ark, Nov 2, 42; m 64; c 1. ORGANIC CHEMISTRY. *Educ:* Eastern NMex Univ, BS, 64; Univ Utah, PhD(org chem), 69. *Prof Exp:* Res assoc org chem, Univ Ill, Urbana, 69-70; asst prof, 70-78, ASSOC PROF CHEM, AUBURN UNIV, 70- *Mem:* Am Chem Soc. *Res:* Free radical chemistry; mechanistic organophosphorus chemistry; conformational analysis. *Mailing Add:* Dept Chem Auburn Univ Auburn AL 36849

HARGIS, LARRY G, b Ferndale, Mich, Nov 22, 39; m 61; c 2. ANALYTICAL CHEMISTRY. *Educ:* Wayne State Univ, BS, 61, MS, 63, PhD(anal chem), 64. *Prof Exp:* Fel, Purdue Univ, 64-65; asst prof, 65-69, ASSOC PROF CHEM, UNIV NEW ORLEANS, 69- *Mem:* Am Chem Soc. *Res:* Ultraviolet and visible absorption spectroscopy; chemistry of heteropoly acids; analytical applications of solution kinetics; chemical applications of on-line computer systems. *Mailing Add:* Dept Chem Univ New Orleans New Orleans LA 70148

HARGIS, PHILIP JOSEPH, JR, b New Orleans, La, Nov 28, 44; m 64; c 3. QUANTUM OPTICS. *Educ:* Univ New Orleans, BS, 66; Univ Pa, MS, 68, PhD(physics), 72. *Prof Exp:* MEM TECH STAFF OPTICS, SANDIA LABS, 72- *Mem:* Optical Soc Am. *Res:* Application of tunable lasers to atomic and molecular spectroscopy; laser induced molecular photodissociation; laser radar; laser based diagnostics for chemical vapor deposition and plasma etching studies. *Mailing Add:* 3208 Reina Dr NE Albuquerque NM 87116

HARGIS, WILLIAM JENNINGS, JR, b Lebanon, Va, Nov 24, 23; m 79. BIOLOGICAL OCEANOGRAPHY, PARASITOLOGY. *Educ:* Univ Richmond, AB, 50, MA, 51; Fla State Univ, PhD(zool), 54. *Prof Exp:* Asst prof biol & chem, The Citadel, 54-55; assoc marine scientist, Va Inst Marine Sci, 55-59, dir, 59-69; from assoc prof to prof marine sci, Col William & Mary, 55-69; dean sch, Univ Va, 60-79, prof marine sci & chmn dept, 63-79; MEM STAFF, SCH MARINE SCI, COL WILLIAM & MARY, 79- *Concurrent Pos:* Mem nat tech adv comt on water qual criteria, Fed Water Pollution Control Admin, Dept Interior; co-chmn adv comt, Atlantic States Marine Fisheries Comn, 59-; mem, Ocean Affairs Adv Comt, US State Dept, 71-; mem adv comt, Coastal Plains Regional Comn Marine Resources, 75-; voting mem & chmn, S&S comt, Mid-Atlantic Mgt Coun, 76-; mem, Comt Offshore Technol, Nat Acad Sci, 78- *Mem:* Fel AAAS; Am Soc Limnol & Oceanog; Soc Syst Zool; Am Soc Parasitol; fel Marine Technol Soc (pres, 78-); Sigma Xi. *Res:* Resource and environmental management; information management; science administration; coastal zone management; fisheries. *Mailing Add:* Sch Marine Sci Gloucester Point VA 23062

HARGITAY, BARTHOLOMEW, b Arad, Romania, Aug 25, 24; US citizen; m 67. PHYSICAL CHEMISTRY, POLYMER CHEMISTRY. *Educ:* Pazmany Peter Univ, Budapest, BS, 46; Univ Basel, MS & PhD(phys chem), 50. *Prof Exp:* Fel, Fritz Hoffmann la Roche Found, Basel, 50-52; fel chem, Harvard Univ, 52-54; group leader polymers, Europ Res Assocs, Brussels, Belg, 54-65; sr scientist, Corp Res Lab, Union Carbide Corp, 65-81; STAFF SCIENTIST, TECHNICON INSTRUMENT CO, 81- *Res:* Physico-chemical processes that make up physiology, particularly effects due to the polymeric and colloidal nature of biological systems; contractility; kidney action; semipermeability and separations; topochemical effects in catalysis; materials used in clinical diagnostics. *Mailing Add:* 15 Edna St White Plains NY 10606

HARGRAVE, PAUL ALLAN, b Clifton Springs, NY, Nov 30, 38; m 67; c 2. BIOCHEMISTRY. *Educ:* Colgate Univ, AB, 60; Univ EAfrica, Dipl, 62; Univ Ill, MS, 66; Univ Minn, PhD(biochem), 70. *Prof Exp:* From asst prof to prof biochem, Sch Med, Southern Ill Univ, 73-84; PROF OPHTHAL & BIOCHEM, SCH MED, UNIV FLA, 85- *Concurrent Pos:* Am Cancer Soc fel, Calif Inst Technol, 70-72, NIH fel, 72-73; mem visual disorders study sect, NIH, 81-85. *Mem:* AAAS; Am Chem Soc; Asn Res Vision & Ophthal; Am Soc Photobiol; Am Soc Biol Chemists. *Res:* Protein biochemistry; structure and function of membrane proteins; rhodopsin biochemistry. *Mailing Add:* Ophthal Dept 411 Univ Fla Box J-284 JHMHC Bldg Gainesville FL 32610

HARGRAVES, PAUL E, b Providence, RI, July 27, 41; m 63; c 3. BIOLOGICAL OCEANOGRAPHY, MARINE BOTANY. *Educ:* Univ RI, BS, 63, MS, 65; Col William & Mary, PhD(marine sci), 68. *Prof Exp:* Res scientist, Lamont Geol Observ, Columbia Univ, 67-68; cur, 68-69, res assoc, 69-70, asst prof, 71-77, PROF OCEANOG & BOT, NARRAGANSETT MARINE LAB, UNIV RI, 86- *Concurrent Pos:* Vis prof, Nat Univ Costa Rica, 79. *Mem:* Int Phycol Soc; Plankton Soc Japan; Am Soc Limnol & Oceanog; Phycol Soc Am; Am Inst Biol Sci; Brit Phycol Soc. *Res:* Systematics and ecology of marine microalgae and phytoplankton, especially diatoms; ecology of marine algae. *Mailing Add:* Grad Sch Oceanog Univ RI Narragansett RI 02882-1197

HARGRAVES, ROBERT BERO, b Durban, SAfrica, Aug 11, 28; US citizen; m 55; c 3. GEOLOGY. *Educ:* Univ Natal, BSc, 48, Hons, 49, MSc, 52; Princeton Univ, PhD(geol), 59. *Prof Exp:* Geologist, Uruwira Minerals Ltd, Tanganyika, 49-50, Union Corp Ltd, SAfrica, 50-52 & Newmont Mining Corp, NY, 53-54; res fel geol, Econ Geol Res Unit, Univ Witwatersrand, 59-61; from asst prof to assoc prof, 61-71, PROF GEOL, PRINCETON UNIV, 71- *Mem:* Mineral Soc Am; Geol Soc Am; Geol Soc SAfrica; Am Geophys Union. *Res:* Igneous petrology, rock and paleomagnetism; economic geology, Precambrian geology. *Mailing Add:* Dept Geol & Geophys Sci Princeton Univ Princeton NJ 08544

HARGREAVES, GEORGE H(ENRY), b Chico, Calif, Apr 2, 16; m 51; c 4. CIVIL ENGINEERING, SOIL SCIENCE. *Educ:* Univ Calif, BS, 39; Univ Wyo, BS, 43. *Prof Exp:* Soil surveyor, Univ Calif, 40; rural rehab supvr, Farm Security Admin, Calif, 40-41; soils technologist, Bur Reclamation, US Dept Interior, 41-42, hydraul engr, 46-48; reclamation engr, NAtlantic Div, US Army CEngrs, Greece, 48-49 & Econ Coop Admin, 49-50; agr irrig engr, Inst Inter-Am Affairs, 50-51 & Haiti, Foreign Opers Admin, 51-56; Int Coop Admin, Philippines, 56-57, water resources adv, 57-60, agr irrig engr, US Agency Int Develop, Brazil, 60-62, water resources adv, 62-65, agr adv, Colombia, 65-68; chief, Civil Eng Br, Nat Resources Div, Inter-Am Geod Surv, US Army, Ft Clayton, CZ, 68-70; res engr, Dept Agr & Irrig Eng, 70-86, EMER RES PROF, UTAH STATE UNIV, 86- *Concurrent Pos:* Asst, Univ Wyo, 42-43; mem, Regional Water Resources Conf, US Del to UN, Manila, 57; US Nat Comt, Int Comn Irrig & Drainage. *Mem:* Fel Am Soc Civil Engrs; Am Soc Agr Engrs. *Res:* Water resources planning and development, especially irrigation. *Mailing Add:* Dept Agr & Irrig Eng UMC 4150 Utah State Univ Logan UT 84322

HARGREAVES, LEON ABRAHAM, JR, b Pearson, Ga, Jan 11, 21; m 46; c 3. FORESTRY. *Educ:* Univ Ga, BSF, 46, MSF, 47; Univ Mich, MPA & PhD(forestry), 53. *Prof Exp:* Forestry specialist, Agr Exten Serv, Univ Ga, 47-49, asst prof, Sch Forestry, 49-54; asst dir, Ga State Forestry Comn, 54-60; asst admin mgt lands & forests, St Regis Paper Co, 60-62; PROF FORESTRY, UNIV GA, 62-, DEAN, 80- *Mem:* Soc Am Foresters; Sigma Xi. *Res:* Operations analysis; valuation; property taxes. *Mailing Add:* Sch Forest Univ Ga Athens GA 30601

HARGREAVES, RONALD THOMAS, b Manchester, Eng, Mar 3, 46; m 72; c 2. ORGANIC CHEMISTRY. *Educ:* Univ EAnglia, BSc, 67, MSc, 68; Univ Rochester, PhD(chem), 74. *Prof Exp:* Res chemist spectros, 74-80, sr res chemist, 80-81, MGR TECH REGULATORY AFFAIRS, MED RES DIV, LEDERLE LABS, AM CYANAMID CO, 81- *Concurrent Pos:* Fel, Univ Ill, Urbana, 72-74. *Res:* Mass spectrometry, structural studies of natural products, chromatography. *Mailing Add:* Med Res Div Lederle Labs 401 N Middletown Rd Pearl River NY 10965

HARGROVE, CLIFFORD KINGSTON, b St John, NB, Nov 22, 28; m 53; c 1. EXPERIMENTAL HIGH ENERGY PHYSICS. *Educ:* Univ NB, BA, 49; McGill Univ, BSc, 55, MSc, 57, PhD(nuclear physics), 61. *Prof Exp:* SCI OFFICER, NAT RES COUN CAN, 61- *Concurrent Pos:* Adj prof, Carleton Univ, 73- *Mem:* Inst Particle Physics; Can Asn Physicists; Am Phys Soc. *Res:* Muon and pion physics with emphasis on elementary particle aspects; lepton conservation laws, and rare decays of muons; instrumentation for particle physics with emphasis on gas counters; Operation Alert detector. *Mailing Add:* Herzberg Inst Astrophys Nat Res Coun Can 100 Sussex Dr Ottawa ON K1A 0R6 Can

HARGROVE, GEORGE LYNN, b Cantril, Iowa, Sept 20, 35; m 64; c 2. QUANTITATIVE GENETICS, STATISTICS. *Educ:* Iowa State Univ, BS, 65; NC State Univ, MS, 68, PhD(animal sci), 70. *Prof Exp:* From asst prof to assoc prof, 70-83, PROF DAIRY SCI, PA STATE UNIV, UNIVERSITY PARK, 83- *Concurrent Pos:* USAID, Uruguay, 68. *Mem:* Am Dairy Sci Asn. *Res:* Dairy cattle production and genetics. *Mailing Add:* Dept Dairy & Animal Sci Pa State Univ University Park PA 16802

HARGROVE, LOGAN EZRAL, b Spiro, Okla, Mar 3, 35; div; c 2. ACOUSTICS, OPTICS. *Educ:* Okla State Univ, BS, 56, MS, 57; Mich State Univ, PhD(physics), 61. *Prof Exp:* Res assoc, Ultrasonics Lab, Mich State Univ, 61-62, mem tech staff, Mech Res Dept, Bell Tel Labs, 62-69, mem tech staff, Guided Wave Res Lab, 69-76; PHYSICIST, PHYSICS DIV, OFF NAVAL RES, 76- *Honors & Awards:* Biennial Award, Acoust Soc, 70. *Mem:* Fel Acoust Soc Am; Optical Soc Am; fel Am Phys Soc; sr mem Inst Elec & Electronics Engrs; AAAS. *Res:* Physical acoustics and optics. *Mailing Add:* Off Naval Res Phys Div (Code 1112) 800 N Quincy St Arlington VA 22217-5000

HARGROVE, ROBERT JOHN, b Willard, Ohio, June 12, 42; m 67; c 2. ORGANIC CHEMISTRY, ENVIRONMENTAL TOXICOLOGY. *Educ:* Ohio Wesleyan Univ, BA, 64; Univ Utah, PhD(org chem), 74; Univ Ala, Birmingham, MPH, 87. *Prof Exp:* Comn officer, USN, 64-68; teaching fel chem, Univ Utah, 69-74; asst prof chem, Dickinson Col, 74-75; asst prof, 75-81, chmn dept, 79-81, PROF CHEM, MERCER UNIV, 81-, CHMN DEPT CHEM, 88- *Concurrent Pos:* Sr Fulbright-Hays lectr, Cultington Univ, 81-82; title III project coordr, Mercer Univ, 82-84, dean protem, Col Liberal Arts, 84-86. *Mem:* AAAS; Am Chem Soc. *Res:* Mechanistic organic chemistry, history of chemical education, vinyl trifluoromethane-sulfonate chemistry; pesticide residues in human adipose tissue. *Mailing Add:* Dept Chem Mercer Univ Macon GA 31207

HARI, V, b Trichur, India, June 23, 36; m 67; c 1. VIROLOGY, MOLECULAR BIOLOGY. *Educ:* Annamalai Univ, Madras, BSc, 58; Madras Univ, MSc, 59, PhD(bot, virol), 64. *Prof Exp:* Fel virol, Waite Agr Res Inst, Australia, 64-65; res assoc, Univ Ariz, 65-70; officer scientist pool, Univ Madras, 70-71; asst virologist, Univ Calif, Berkeley, 71-73; res assoc virol, 73-76, asst prof biol, 76-81, ASSOC PROF BIOL, WAYNE STATE UNIV, 81- *Mem:* Am Soc Virol; Plant Molecular Biol Asn; Sigma Xi. *Res:* structure and replication of plant viruses by molecular techniques including gene cloning and sequencing; introduction of genes into plants using vectors; viral immunodiagnostics including HIV. *Mailing Add:* Biology Dept 119 Sci Wayne State University 5950 Cass Ave Detroit MI 48202

HARING, OLGA M, b Oradea, Romania, Aug 25, 17; US citizen; m 38; c 1. MEDICINE. *Educ:* Univ Vienna, MD, 38; Am Bd Internal Med, dipl, 64, 74. *Prof Exp:* Dir med, Health Ctr 1, Dept Pub Health, Nicaragua, 40-46; foreign asst cardiol, Sorbonne, 47-49; assoc, Div Cardiol, Chicago Med Sch, 50-56; assoc attend, Cook County Hosp, Chicago, 64-65; assoc prof, 70-75, prof, Dept Med & Community Health & Prev Med, Med Sch, 75-87, dir cardio-pulmonary-renal clins, 65-85, EMER PROF, NORTHWESTERN UNIV, CHICAGO, 87- *Concurrent Pos:* Fel coun clin cardiol, Am Heart Asn; UN Children's Emergency Fund fel, 48; Hektoen Inst pediat cardiol fel, 51-52. *Mem:* AAAS; fel Am Col Cardiol; fel Am Col Physicians. *Res:* Experimental production of cardiac malformations by prenatal hypercapnea in the rat and in the chicken; medical records; application of computers in medicine; clinical trials in cardiology. *Mailing Add:* 1201 Judson Evanston IL 60202

HARINGTON, CHARLES RICHARD, b Calgary, Alta, May 22, 33. MAMMALOGY. *Educ:* Univ Alta, BA, 54, BSc, 57, PhD, 77; McGill Univ, MSc, 61. *Prof Exp:* Wildlife biologist, Can Wildlife Serv, 60-65; CUR QUATERNARY ZOOL, CAN MUS NATURE, 65-, CHIEF PALEOBIOLOGY DIV, 82- *Honors & Awards:* Can Asn Geographers Prize, 57; Massey Medal, 87. *Mem:* AAAS; Am Soc Mammal; fel Arctic Inst NAm; fel Royal Geog Soc; fel Royal Can Geog Soc. *Res:* Canadian Pleistocene mammals; life history, distribution, ecology and evolution of the muskoxen and polar bear; climatic change in Canada during the Quaternary. *Mailing Add:* Paleobiol Can Mus Nature Ottawa ON K1P 6P4 Can

HARIS, STEPHEN JOSEPH, number theory, for more information see previous edition

HARITATOS, NICHOLAS JOHN, b Rome, NY, May 29, 31; m 61; c 2. CHEMICAL ENGINEERING. *Educ:* Mass Inst Technol, SB, 52, SM, 53, ScD(chem eng), 56. *Prof Exp:* Res engr, 56-61, group supvr, 61-65, sr eng assoc, 66-81, sr supv process engr, 81-87, ENG CONSULT, CHEVRON RES & TECHNOL CO, RICHMOND, CALIF, 88- *Mem:* Am Chem Soc; Am Inst Chem Engrs. *Mailing Add:* 1354 Contra Costa Dr El Cerrito CA 94530

HARJU, PHILIP HERMAN, b Bismarck, NDak, Aug 2, 30; m 54; c 5. PHYSICAL CHEMISTRY. *Educ:* NDak State Univ, BS, 56, MS, 57; Univ Pittsburgh, PhD(phys chem), 69. *Prof Exp:* Res assoc petrol, Mellon Inst Technol, 57-59, jr fel, 60-64; sr res scientist phys chem, 69-73, sr proj scientist, 73-75, GROUP MGR CATALYSIS, KOPPERS CO, INC, 75- *Mem:* Am Chem Soc. *Res:* Catalysis; chemical kinetics; interfacial chemistry; photochemistry and photopolymerization; organic geochemistry and organometallics; water pollution. *Mailing Add:* Six Maple Dr Spring Church PA 15686

HARKAVY, ALLAN ABRAHAM, b New York, NY, Mar 16, 25; m 52; c 2. MAGNETIC RESONANCE, VISUAL PHYSIOLOGY. *Educ:* City Col New York, BS, 49; New Sch Social Res, MA, 55; NY Univ, MS, 58, PhD(physics), 67. *Prof Exp:* Scientist, Westinghouse Atomic Power Div, 57-59; instr physics, Hofstra Col, 59-64; ASSOC PROF-PROF PHYSICS, STATE UNIV NY, NEW PALTZ, 64- *Mem:* Am Phys Soc. *Res:* Theory of lateral geniculate body and visual cortex; theory of physical basis of human will; magnetic resonance. *Mailing Add:* State Univ NY New Paltz NY 12561

HARKE, DOUGLAS J, b Edmonton, Alta, Apr 18, 42; m 64; c 3. PHYSICS, SCIENCE EDUCATION. *Educ:* Univ Alta, BSc, 63; Wash Univ, MAEd, 64; Purdue Univ, MS & PhD(sci educ), 69. *Prof Exp:* Teacher high sch, Leduc, Alta, 64-65 & South Bend, Ind, 65-67; dean grad studies & res, 74-88, ASST PROF PHYSICS, STATE UNIV NY COL, GENESEO, 69-, DIR RES, 88- *Mem:* Am Asn Physics Teachers; Nat Asn Res Sci Teaching. *Res:* Physics achievement testing and computerized grading of physics tests; evaluation of training programs for secondary physics teachers; student evaluation of college teachers. *Mailing Add:* Dir Res State Univ NY Col Geneseo NY 14454

HARKER, DAVID, crystallography; deceased, see previous edition for last biography

HARKER, J M, b San Francisco, Calif, June 30, 26. ENGINEERING. *Educ:* Swarthmore Col, BSME, 50; Univ Calif, Berkeley, MSE, 51; Stanford Univ, MSEE, 62. *Prof Exp:* Engr, IBM Corp, 58-87; RETIRED. *Res:* Authored six publications. *Mailing Add:* 840 Melville Ave Palo Alto CA 94301

HARKER, KENNETH JAMES, b Long Beach, Calif, July 4, 27; m 54; c 2. PHYSICS. *Educ:* Univ Calif, Los Angeles, AB, 48, MA, 50, PhD(physics), 54. *Prof Exp:* Mem tech staff, Bell Tel Labs, 54-59; res physicist, Hansen Phys Lab, 59-69, res physicist, 69-75, sr res assoc, Inst Plasma Res, 75-81, SR RES ASSOC, RADIOSCI LAB, STANFORD UNIV & SR RES PHYSICIST, SRI INT, 81- *Concurrent Pos:* Mem Comn G & vchmn, US Comn H, Int Union Radio Sci; assoc ed, Radio Sci. *Mem:* Fel Am Phys Soc; sr mem Inst Elec & Electronics Engrs; Int Asn Geomagnetism & Aeronomy. *Res:* Plasma physics; wave phenomena; and wave-particle interactions in plasmas. *Mailing Add:* 694 Camellia Way Los Altos CA 94024

HARKER, ROBERT IAN, b Glasgow, Scotland, Aug 2, 26; nat US; m 55; c 3. GEOLOGY. *Educ:* Cambridge Univ, BA, 49, MA & PhD, 54. *Prof Exp:* Res assoc geochem, Pa State Univ, 53-55, asst prof, 55-56; res chemist, Res Ctr, Johns-Manville Corp, 56-62; pres, Tem-Pres Res, Inc, 62-70; chmn dept geol, 74-79, PROF GEOL, UNIV PA, 70- *Mem:* Fel Mineral Soc Am; Brit Geol Soc. *Res:* Mineralogy and petrology; phase equilibria in silicate, oxide and carbonate systems, hydrothermal research and mineral syntheses, crystal growth; curing of cements and calcium silicates; high pressure-high temperature experimental systems; ion exchange materials; resource evaluations. *Mailing Add:* 37 Penarth Rd Bala-Cynwyd PA 19004

HARKER, YALE DEON, b Idaho Falls, Idaho, June 11, 37; m 59; c 5. REACTOR PHYSICS. *Educ:* Idaho State Univ, BS, 59; Case Western Reserve Univ, MS, 62; Colo State Univ, PhD(physics), 69. *Prof Exp:* Res scientist, Phillips Petrol Co, Idaho, 62-66 & Idaho Nuclear Corp, 66-71; res scientist, Reactor Develop Br, 71-72, res & eng supvr, Nuclear Physics Br, Aerojet Nuclear Co, 72-76; res & eng supvr, Nuclear Physics Br, 76-80, sr scientist, 80-83, SCI SPECIALIST, NUCLEAR SCIS UNIT, EG&G IDAHO INC, 83- *Mem:* Am Nuclear Soc. *Res:* Physics constants for fast reactor technology; high pressure solid state research; Raman scattering from crystalline solids; laser development research; study of the liquid and low temperature solid state; neutron spectrometry; radiation dosimetry; neutron scattering. *Mailing Add:* Nuclear Sci Mail Stop 7113 EG&G Idaho Inc Idaho Falls ID 83415

HARKEY, JACK W, b Mar 17, 21; US citizen; m 47; c 2. MECHANICAL ENGINEERING. *Educ:* Southern Methodist Univ, BSME, 44. *Prof Exp:* From instr to assoc prof, 46-70, dir coop prof, 53-67, PROF MECH ENG, SOUTHERN METHODIST UNIV, 70- *Concurrent Pos:* Consult & designer, D C Pfeiffer & Assocs, 50-53. *Mem:* Am Soc Eng Educ. *Res:* Environmental aspects of fallout shelter habitability; photoelastic analysis of sucker rod joints. *Mailing Add:* Dept Civil Eng Southern Methodist Univ Dallas TX 75275

HARKIN, JAMES C, b Fayette, Miss, Dec 9, 26; m 52; c 1. PATHOLOGY, NEUROPATHOLOGY. *Educ:* Univ Nebr, BS & MD, 51. *Prof Exp:* Intern med, Univ Hosp, Univ Cleveland, 51-52; resident path, Inst Path, Western Reserve Univ, 52-55, demonstr, 54-55; instr, Sch Med, Wash Univ, 55-57, asst prof, 57-59; asst prof, Med Col, Cornell Univ, 61-62; assoc prof, 62-69, PROF PATH & ANAT, SCH MED, TULANE UNIV, 69- *Concurrent Pos:* Asst pathologist, Barnes & St Louis Children's Hosps, 55-59; assoc dir labs, Hosp Spec Surg, New York, 59-62; asst attend pathologist, New York Hosp, 61-62; vis pathologist, Charity Hosp, 62-; vis scientist, Oxford Univ, 65; mem ed adv bd, Armed Forces Inst Path; consult, Int Ctr Med Res & Training, Univ Valle, Colombia, 68-78; fel neuropath, Montefiore Hosp, NY, 56-57; USPHS res career prog award, 63-72. *Mem:* Am Asn Neuropath; Am Asn Path; fel Col Am Path. *Res:* Pathology and experimental pathology; neuroanatomy; electron microscopy; tumors of the peripheral nervous system. *Mailing Add:* Sch Med Tulane Univ 1430 Tulane Ave New Orleans LA 70112

HARKIN, JOHN MCLAY, b Paisley, Scotland, Apr 7, 33; US citizen; m 55; c 3. ORGANIC CHEMISTRY, BIOCHEMISTRY. *Educ:* Glasgow Univ, BSc, 55; Univ Heidelberg, Dr rer nat(chem), 59. *Prof Exp:* Res asst, Res Inst for Chem Wood & Polysaccharides, Univ Heidelberg, 55-66; proj leader lignin chem, US Forest Prod Lab, Wis, 66-74; assoc prof soil sci, 74-77, PROF WATER RESOURCES & SOILS, UNIV WIS-MADISON, 77- *Mem:* Am Chem Soc; Tech Asn Pulp & Paper Indust. *Res:* Lignin structure and utilization; anatomy, physiology, chemistry and biochemistry of wood; pulping reactions and byproducts; nature and fate of natural and synthetic organic compounds in soil and water. *Mailing Add:* 3991 Plymouth Circle Madison WI 53705

HARKINS, CARL GIRVIN, b Colorado City, Tex, June 14, 39; m 60; c 3. OPTICS, PHYSICAL CHEMISTRY. *Educ:* McMurry Col, BA, 60; Johns Hopkins Univ, MA, 62, PhD(phys chem), 64. *Prof Exp:* Res assoc chem eng, Univ Calif, Berkeley, 64-65; res fel, Rice Univ, 65-66; asst prof chem, Mat Res Div, Southwest Ctr Advan Studies, 66-68, actg head dept, 67-68; sr res scientist, Rice Univ, 68-73, adj prof mat sci, 73-85; mem tech staff, 78-85, PROJ MGR, HEWLETT-PACKARD MFG RES CTR, 85- *Mem:* Am Chem Soc; Catalysis Soc; fel Am Inst Chemist; Nat Asn Corrosion Eng; Sigma Xi. *Res:* Surface chemistry; catalysis and photocatalysis; corrosion and stress corrosion; tribology; surface mount technology, laser microfabrication. *Mailing Add:* Hewlett-Packard Labs 3500 Deer Creek Rd Palo Alto CA 94304

HARKINS, KRISTI R, b Grand Island, Nebr. FLOW CYTOMETRY, PLANT CELL BIOLOGY. *Educ:* Nebr Wesleyan Univ, BS, 80; Northwestern Univ, MS, 81. *Prof Exp:* RES TECHNOLOGIST, UNIV NEBR, LINCOLN, 81- *Mem:* Soc Anal Cytol. *Res:* Maintain a multi-user flow cytometry facility; instrument modification and procedure development for application in plant cell biology. *Mailing Add:* 2723 S Mission Rd Tucson AZ 85713

HARKINS, ROBERT W, b Chester, Pa, Oct 18, 35; m 58; c 3. NUTRITION, PHYSIOLOGICAL CHEMISTRY. *Educ:* Pa State Univ, BS, 57; Wayne State Univ, PhD(chem), 61. *Prof Exp:* Assoc scientist antibiotic develop, Parke Davis & Co, 57-58, sr scientist nutrit res, Mead Johnson & Co, 61-64, group leader, 64-66; dir sect food sci, Dept Foods & Nutrit & secy food indust liaison comt, Am Med Asn, 66-68; dir nutrit res, Ross Labs, 68-70; dir sci affairs, Grocery Mfg Am, Inc, 70-79, vpres sci affairs, 74-79; dir res & develop, 79-87, VPRES, RES & DEVELOP, MCNEIL SPECIALTY PROD CO, DIV JOHNSON & JOHNSON, 87- *Mem:* NY Acad Sci; Am Inst Nutrit; fel Am Inst Chem; Am Chem Soc. *Res:* Absorption and utilization of nutrients; rate of growth and body composition-mental development; safety and regulation of food ingredients. *Mailing Add:* McNeil Specialty Prod Co 501 George St PO Box 2400 New Brunswick NJ 08903

HARKINS, ROSEMARY KNIGHTON, b Amarillo, Tex, Aug 5, 38; div. MEDICAL TECHNOLOGY, HEALTH PROFESSIONS & EDUCATION ADMINISTRATION. *Educ:* WTex State Univ, BA, 64; Univ Okla, MS, 71, PhD(anat), 72; Central State Univ, BS, 76. *Prof Exp:* Med technologist, Univ Texas Med Br, Galveston, 59-62, St Anthony's Hosp, Amarillo, Tex, 62-66, Vet Admin Hosp, Oklahoma City, 68-70 & Baptist Hosp, Oklahoma City, 69-70; asst prof, 72-77, assoc prof & actg dir, 77-81, PROF ANAT, COL MED & DIR, SCH ALLIED HEALTH PROFESSIONS, UNIV OLKA HEALTH SCI CTR, 81-, ASSOC DEAN, COL ALLIED HEALTH, 81- *Concurrent Pos:* Mem, Nat Comt Accreditation & Inst Eligibility, US Dept Educ, 81-83. *Mem:* Am Soc Med Technologists; Sigma Xi; fel Am Soc Allied Health Professions (secy, 81-83). *Res:* Comparative hematology and tissue culture of testicular and mammary tumors in androgen insensitive rats; allied health education; minorities in the health professions. *Mailing Add:* Howard Univ 2501 Kingsway Rd Ft Washington MD 20744

HARKINS, THOMAS REGIS, b McKeesport, Pa, Jan 5, 29; m 61; c 3. CHEMISTRY. *Educ:* Univ Pittsburgh, BS, 52, PhD(chem), 56. *Prof Exp:* Anal res chemist, E I du Pont de Nemours & Co, 56-59; supvr chem & mech testing sect, Res Ctr, 59-62, proj supvr stainless steels, 62-64, res assoc, 64-65, sr develop engr, Sales Develop Dept, 65-69, mgr indust sales process equip & construct, 69-71, mkt dir stainless & specialty steel plate, Allegheny Ludlum Steel Corp, 71-76, dist sales mgr, Houston, 76-77; VPRES SALES & MKT, SPEC METALS CORP, 77- *Mem:* Am Chem Soc; Am Soc Metals; Am Ord Asn. *Res:* Analytical chemistry and applied spectroscopy; infrared spectroscopy; chelates and chelating agents; organic coatings; electrodeposition; corrosion; stainless steels and specialty metals. *Mailing Add:* 302 Fox Chapel Rd Pittsburgh PA 15238

HARKLESS, LAWRENCE BERNARD, b Longview, Tex, Jan, 51; m 80; c 2. PODIATRIC MEDICINE & SURGERY, DIABETES & ITS COMPLICATIONS. *Educ:* Cal Col Podiatric Med, BS & DPM, 75. *Prof Exp:* DIR PODIATRY RESIDENCY TRAINING, UNIV TEX HEALTH SCI CTR, SAN ANTONIO, 77- *Res:* Foot disorders caused by diabetes. *Mailing Add:* 314 N Hackberry Suite 101 San Antonio TX 78202

HARKNESS, DONALD R, b Mitchell, SDak, Aug 23, 32; m 54; c 4. HEMATOLOGY, BIOCHEMISTRY. *Educ:* Univ Calif, Berkeley, BA, 54; Wash Univ, MD, 58. *Prof Exp:* Intern med, Wash Univ, 58-59, resident internal med, 59-60; res assoc, Nat Inst Arthritis & Metab Dis, 60-63; Am Cancer Soc fel, 63, NIH spIH spec res fel, 64; from asst prof to prof med, Univ Miami Sch Med, 64-80, assoc prof biochem, 69-75, mem grad fac, 70-80; chief hemat, Vet Admin Hosp, 68-80; PROF & CHMN DEPT MED, UNIV WIS-MADISON, 80- *Concurrent Pos:* John & Mary Markle scholar, 66; dir, Miami Comprehensive Sickle Ctr, 73-78; chmn, Bd-u Care, Univ HMO, 84- *Mem:* Am Soc Hematol; Am Fedn Clin Res; Am Soc Clin Invest; Asn Am Phys; Sigma Xi. *Res:* Biochemistry and enzymology of erythrocytes; control of hemoglobin function, 2, 3-diphosphoglycerate and inositol polyphosphate metabolism; hemoglobinopathy; sickle cell anemia. *Mailing Add:* 110 Standish Ct Madison WI 53705

HARKNESS, SAMUEL DACKE, b Richmond, Va, Oct 28, 40; m 63; c 3. METALLURGY, NUCLEAR ENGINEERING. *Educ:* Cornell Univ, BMetEng, 63; Univ Fla, PhD(metall eng), 67; Univ Pittsburgh, MBA, 83. *Prof Exp:* Res engr, Atomics Int Div, NAm Aviation, Inc, 63-64; assoc metalurgist & group leader radiation effects, Argonne Nat Lab, 67-73, mgr, mat technol, Nuclear Power Systs Div, combustion eng, 73-76, assoc dir, fusion power, 76-79; mgr, Fuel Develop, Bettis Atomic Power Lab, 79-86, mgr, Mat Technol Div, 86-90; GEN MGR SYSTS, PROCESS & TECHNOL DIV, WESTINGHOUSE SCI & TECHNOL CTR, 90- *Concurrent Pos:* Mem, fusion mat coord comn, Off Fusion Energy, Dept Energy, 77-79. *Honors & Awards:* Robert Lansing Hardy gold medal, Am Inst Mining, Metall & Petrol Engrs, 69. *Mem:* Am Soc Metals; Am Nuclear Soc. *Res:* Radiation damage to materials; nuclear fuel performance and corrosion of reactor cladding materials; design of fusion reactor systems; performance of steam generators. *Mailing Add:* Sci & Technol Ctr Westinghouse Elec Corp 1310 Beulah Rd Pittsburgh PA 15235

HARKNESS, WILLIAM LEONARD, b Lansing, Mich, June 25, 34; m 56, 80; c 3. STATISTICS, APPLIED STATISTICS. *Educ:* Mich State Univ, BS, 55, MA, 56, PhD(statist), 59. *Prof Exp:* Res mathematician, Inst Air Weapons Res, 57; res asst statist, Mich State Univ, 57-59, instr, 59; from asst prof to assoc prof math, Pa State Univ, University Park, 59-69, actg head dept statist, 69-70, head dept, 70-87, PROF STATIST, PA STATE UNIV, UNIVERSITY PARK, 69- *Concurrent Pos:* Vis assoc prof statist, Calif State Col, Hayward, 67; NIH consult, 77-; vis prof statist, Stanford Univ, 79-80. *Mem:* Fel Am Statist Asn; fel Inst Math Statist (prog secy, 74-80); Biomet Soc; Royal Statist Soc; Math Asn Am. *Res:* Contingency tables; distribution theory; sociometry; applied probability theory. *Mailing Add:* Dept Statist Pa State Univ University Park PA 16802

HARKRIDER, DAVID GARRISON, b Houston, Tex, Sept 25, 31; div; c 2. SEISMOLOGY. *Educ:* Rice Univ, BA, 53, MA, 57; Calif Inst Technol, PhD(geophys), 63. *Prof Exp:* Res fel geophys, Calif Inst Technol, 63-65; from asst prof to assoc prof, Brown Univ, 65-70; assoc dir, Seismol Lab, 77-79, assoc prof, 70-78, PROF GEOPHYS, CALIF INST TECHNOL, 79- *Concurrent Pos:* Elect pres Seismol Soc Am, 87-88. *Mem:* Fel Am Geophys Union; Seismol Soc Am; Soc Explor Geophys. *Res:* Elastic wave propagation and infrasonics; coupling and excitation of surface waves, especially Rayleigh, Love, acoustic-gravity waves and tsunamis. *Mailing Add:* Seismo Lab 252-21 Calif Inst Technol Pasadena CA 91125

HARLAM, RUTH (MRS FRITZ P LEHMANN), mechanical engineering; deceased, see previous edition for last biography

HARLAN, HORACE DAVID, b Austin, Tex, Feb 9, 29; m 55; c 3. PHYSICAL CHEMISTRY, ORGANIC CHEMISTRY. *Educ:* Univ Tex, BS, 50; Baylor Univ, PhD(kinetics), 61. *Prof Exp:* Instr chem, San Antonio Col, 54-55; from instr to assoc prof, Southwest Tex State Col, 55-67; head dept, 67-89, PROF CHEM, ANGELO STATE UNIV, 67- *Mem:* Am Chem Soc. *Res:* Ion exchange; water analysis. *Mailing Add:* Dept Chem Angelo St Univ 2601 West Ave N San Angelo TX 76909

HARLAN, JACK RODNEY, b Washington, DC, June 7, 17; m 39; c 4. ARCHEOBOTANY. *Educ:* George Wash Univ, BS, 38; Univ Calif, PhD(genetics), 42. *Prof Exp:* Agronomist, Div Forage Crops & Diseases, USDA, 42-51 & Agr Exp Sta, Okla State Univ, 51-66; prof, 66-84, EMER PROF AGRON, UNIV ILL, URBANA, 84- *Concurrent Pos:* Vis prof, Univ Calif, Davis, 75, Univ Calif, Riverside, 76 & Univ Nagoya, Japan, 79; John

Simon Guggenheim Mem fel, 59; Vis scientist, Am Soc Agron, 63. *Honors & Awards:* Frank N Meyer Mem Medal, Am Genetic Soc; Crop Sci Award, Am Soc Agron, 71 & Int Ser in Agron Award, 76; Meyer Medal, Am Genetic Asn Am. *Mem:* Nat Acad Sci; fel AAAS; fel Am Soc Agron; Crop Sci Soc Am (pres, 65-66); Soc Econ Bot; fel Am Acad Arts & Sci. *Res:* Origin and evolution of cultivated plants, biosystematics, germ plasm conservation, genetics and evolution. *Mailing Add:* 1016 N Hagar St New Orleans LA 70119

HARLAN, JOHN MARSHALL, b Chicago, Ill, July 18, 47; m 73; c 2. HEMATOLOGY, ONCOLOGY. *Educ:* Loyola Univ, BS, 69; Univ Chicago, MD, 73; Am Bd Internal Med, cert hemat, 78, cert oncol, 80. *Prof Exp:* From asst prof to assoc prof, 78-88, PROF MED, UNIV WASH, 89-, HEAD DIV HEMAT, 90- *Concurrent Pos:* Clin scientist, Am Heart Asn, 80-85, estab investr, 86-91; consult, Biogen Corp, 88-90, Cytel Corp, 90-; assoc ed, J Immunol, 88-, Blood J, 90; adj prof path, Univ Wash, 89-; vis prof, numerous univs & biotechnol co. *Mem:* Am Soc Clin Invest; Asn Am Physicians. *Res:* Cell biology of vessel wall focusing on interaction of leukocytes with endothelial cells. *Mailing Add:* Div Hemat Rm 10 Univ Wash Seattle WA 98195

HARLAN, PHILLIP WALKER, b McCook, Nebr, Oct 6, 44; m 70. SOIL MANAGEMENT. *Educ:* Univ Nebr, BS, 67; Purdue Univ, MS, 72, PhD(soils), 75. *Prof Exp:* INT TRAINING SPECIALIST, OFF INT COOP & DEVELOP, INT TRAINING DIV, USDA, 75- *Mem:* Soil Sci Soc Am; Am Soc Agron; Soil Conserv Soc Am; Sigma Xi. *Res:* Loess soils; defining and evaluating their natural soil characteristics which influence or are influenced by different uses and farming practices. *Mailing Add:* 1200 N Quantico St Arlington VA 22205-1736

HARLAN, RONALD A, b Mansfield, La, Dec 25, 37; m 57; c 3. NUCLEAR CHEMISTRY. *Educ:* Northwestern State Col, La, BS, 58; Fla State Univ, PhD(nuclear chem), 63. *Prof Exp:* Asst prof chem, Univ Ark, 63-66; sr nuclear chemist, Idaho Nuclear Corp, 66-67, assoc scientist, 67-69; sr res chemist, Dow Chem Co, Rocky Flats Div, 70-73, res specialist, 73-75; res specialist II, 75-81, sr res specialist, Rockwell Int Energy Systs Group, Rocky Flats Plant, 81-89, SR RES SPECIALIST, EG&G, ROCKY FLATS, INC, 90- *Concurrent Pos:* Consult, Energy Co, 83; prin investr, various instrumentation develop, 83-; standardized writing groups, Am Soc Testing & Mat, 88-; chair, Inst Nuclear Mat Mgt-9 Nondestructive Assay, Am Nat Standards Inst, 88- *Mem:* Am Chem Soc; Inst Nuclear Mat Mgt; Sigma Xi; Am Nuclear Soc. *Res:* Nuclear safeguards and accountability; nuclear spectroscopy and models; environmental chemistry and detection of low activity radionuclides. *Mailing Add:* EG&G Rocky Flats Inc MS 881 PO Box 464 Golden CO 80402-0464

HARLAN, WILLIAM R, JR, b Richmond, Va, Nov 1, 30; m 81; c 4. MEDICINE. *Educ:* Univ Va, BA, 51; Med Col Va, MD, 55. *Prof Exp:* Resident internal med, Duke Univ, 58-61; asst prof med, Med Col Va, 63-67, assoc prof, 67-70, dir clin res ctr, 63-71; prof med & assoc dean sch med, Univ Ala, Birmingham, 71-73; prof med & community health sci, Duke Univ, 73-75; prof & chmn dept post grad med, 75-82, PROF INTERNAL MED, MED SCH, UNIV MICH, 75- *Concurrent Pos:* Consult, Nat Ctr Health Statist, 75-, Atherosclerosis & Hypertension Adv Ctr, Nat Heart & Lung Inst, 75-79, NIH Epidemiol & Dis Control Study Sect, 71-74, World Bank, 81-82; dir, Physicians Asst Training Prog & Medex Training Prog, USAF sci adv bd, 79-85, Armed Forces Epidemiol Bd, 81-; NIH res fel, Dept Biochem, Duke Univ, 61-63; Markle scholar, 64. *Mem:* Sigma Xi; AAAS; Am Fedn Clin Res; Am Col Physicians; Am Heart Asn. *Res:* Medical research; lipid metabolism; blood pressure regulation; epidemiology of cardiovascular disease; health care delivery and development of paramedical personnel. *Mailing Add:* Fed Bldg Rm 212 7550 Wisconsin Ave Bethesda MD 20892

HARLAND, BARBARA FERGUSON, b Chicago, Ill; m 47; c 3. NUTRITION. *Educ:* Iowa State Univ, BS, 46; Univ Wash, MS, 49; Univ Md, PhD(nutrit), 71. *Prof Exp:* Head dietitian, Lakeview Mem Hosp, 46-47; food serv asst, Univ Northern Iowa, 47; nutrit instr, Univ Md, College Park, 64-69; RES BIOLOGIST NUTRIT, FOOD & DRUG ADMIN, WASHINGTON, DC, 68- *Mem:* Am Inst Nutrit; Am Chem Soc; Am Dietetic Asn; Soc Exp Biol & Med; Soc Nutrit Educ. *Res:* The effects of phytate and dietary fiber on human and animal nutrition. *Mailing Add:* Dept Human & Nutrit & Food Sch Human Ecol Howard Univ Washington DC 20059

HARLAND, GLEN EUGENE, JR, b Salina, Kans, Jan 30, 33; m 56; c 2. SOLID STATE PHYSICS. *Educ:* Kans State Univ, BS, 58, MS, 60, PhD(physics), 64. *Prof Exp:* Instr physics, Kans State Univ, 60-64; sr physicist, Delco Radio Div, 64-72, STAFF ENGR, DELCO ELECTRONICS, GEN MOTORS CORP, 72- *Concurrent Pos:* Mem assoc fac, Ind Univ, 64- *Mem:* Soc Automotive Eng; Inst Elec & Electronics Engrs; Int Soc Hybrid Microelectronics; Sigma Xi. *Res:* Semiconductor physics; materials research; x-ray diffraction phenomena. *Mailing Add:* 2301 Willow Spring Rd Kokomo IN 46902

HARLAND, RONALD SCOTT, b Rochester, NY, May 25, 62. CONTROLLD RELEASE, POLYELECTROLYTES. *Educ:* Purdue Univ, BS, 83, MS, 85, PhD(chem eng), 88. *Prof Exp:* Engr, IBM Inc, 83; res asst & teaching asst chem eng, Purdue Univ, 83-88; res scientist, Kimberly-Clark Corp, 88-89, sr res scientist, 90; MGR BASIC RES, KV PHARMACEUT CO, 90- *Concurrent Pos:* Session chair, Am Inst Chem Engrs, 88- *Mem:* Am Inst Chem Engrs; Am Chem Soc; Sigma Xi; NAm Membrane Soc; Controlled Release Soc; NY Acad Sci; AAAS. *Res:* Inventor of aqueous-based and thermal-based controlled- release (24 hour) oral devices; inventor of osmotic superabsorbents, proteinaceous (blood) superabsorbents, heterogeneous absorbents and absorbent microfiber absorbents. *Mailing Add:* 1944 Schottler Valley Dr Chesterfield MO 63017

HARLAP, SUSAN, medicine, for more information see previous edition

HARLE, THOMAS STANLEY, b Detroit, Mich, Aug 17, 32; m 60; c 2. RADIOLOGY. *Educ:* Northwestern Univ, BS, 54, MD, 57. *Prof Exp:* Radiologist, Ft Detrick, Md, 61-62; chief radiol, Irwin Army Hosp, Ft Riley, Kans, 62-64; chief radiol, Brooke Army Hosp, Ft Sam Houston, Tex, 64-65; from instr to assoc radiol, Baylor Col Med, 65-68; 65-69; prof, Med Sch, Duke Univ, 69-71; prof radiol, Univ Tex Med Sch Houston, 71-78; prof radiol & assoc radiologist, Univ Tex M D Anderson Hosp & Tumor Inst Houston, 71-78; prof radiol, Mich State Univ, 78-; AT DEPT RADIOL, UNIV TEX HEALTH & SCI CTR. *Concurrent Pos:* Radiologist, Kelsey-Seybold Clin, Houston, 65-66; asst radiologist, Univ Tex M D Anderson Hosp & Tumor Inst Houston, 65-66; assoc prof radiol, Baylor Col Med, 68-69; chief radiol serv, Vet Admin Hosp, Durham, 69-71; consult, Res Triangle Inst, 70- & Brooke Army Hosp, 74- *Mem:* Asn Univ Radiologists; AMA; Radiol Soc NAm; Am Col Radiol. *Res:* Xeroradiography. *Mailing Add:* Univ Tex MD Anderson Cancer Ctr 1515 Holcomb Blvd Houston TX 77030

HARLEMAN, DONALD R(OBERT) F(ERGUSSON), b Palmerton, Pa, Dec 5, 22; m 50; c 3. ENVIRONMENTAL & CIVIL ENGINEERING. *Educ:* Pa State Univ, BS, 43; Mass Inst Technol, SM, 47, ScD(civil eng), 50. *Prof Exp:* Design engr, Curtiss-Wright Corp, 44-45; asst fluid mech, 45-47, res assoc, 47-50, asst prof hydraul, 50-56, assoc prof, 56-63, dir, R M Parsons Lab Water Resources, 73-83, PROF CIVIL ENG, MASS INST TECHNOL, 63-, FORD PROF ENG, 75- *Concurrent Pos:* Vis prof, Calif Inst Technol, 62-63; sr visitor, Dept Appl Math & Theoret Physics, Cambridge Univ, 68-69; Guggenheim fel, 68-69; consult, Tenn Valley Auth, US Army Corps Engrs, E I du Pont de Nemours & Co, Inc, Arthur D Little & Consol Edison, NY, EPA. *Honors & Awards:* Hydraul Award, Boston Soc Civil Engrs, 55, Desmond Fitzgerald Medal, 67; Res Prize, Am Soc Civil Engrs, 60, Karl Emil Hilgard Prize, 71, 73; First Hunter Rouse Hydraul Eng lectr, Am Soc Civil Engrs, 80; Wesley W Horner Award, Am Soc Civil Engrs, 83; Res Scholar, Rockefeller Found, 85. *Mem:* Nat Acad Eng; Am Soc Civil Engrs; Am Geophys Union; Int Asn Hydraul Res; Water Pollution Control Fedn. *Res:* Fluid mechanics; stratified flow; heat disposal from power generation; temperature distributions in lakes and reservoirs; water quality control; tidal motion; mixing processes in lakes, estuaries and coastal waters. *Mailing Add:* R M Parsons Lab Rm 48-311 Mass Inst Technol Cambridge MA 02139

HARLETT, JOHN CHARLES, b Tiffin, Ohio, Apr 7, 36; m 63; c 3. OCEANOGRAPHY. *Educ:* Ohio State Univ, BSc, 60; US Navy Postgrad Sch, MS, 67; Ore State Univ, PhD(oceanog), 71. *Prof Exp:* Engr, US Navy, 61-68; instr oceanog, US Naval Acad, 71-74; cmndg officer, Oceanog Unit One, 74-75; oceanogr, Off Naval Res, Boston, 75-77, cmndg officer, 77-79; mem staff, Off Oceanogr Navy, Off Naval Technol, 79-82; ASST DIR PROGS, APPL PHYSICS LAB, UNIV WASH, 82- *Mem:* Sigma Xi; Am Geophys Union; US Naval Inst. *Res:* Continental shelf and deep ocean sedimentation processes; bottom currents; bottom boundary layer; shore processes and prediction of beach changes. *Mailing Add:* 11601 NE 143rd Pl Kirkland WA 98034

HARLEY, JOHN BARKER, b Baltimore, Md, Sept 13, 49; m 72. INTERNAL MEDICINE, IMMUNOLOGY. *Educ:* Dickinson Col, BS, 71; Univ Pa, MD, 75, PhD(biochem), 76. *Prof Exp:* Fel, Nat Inst Allergy & Infectious Dis, Univ Pa & Imp Cancer Res Fund Lab, London, 76-77; intern, Yale Univ, 77-78, res internal med, 78-79; clin assoc, 79-82, asst prof, 82-87, ASSOC PROF, DEPT MED, NAT INST ALLERGY & INFECTIOUS DIS, USPHS, 87- *Concurrent Pos:* Adj asst prof, Dept Microbiol & Immunol, Health Sci Ctr, Univ Okla, 82-88, adj assoc prof, 88-; affil asst prof, assoc mem, Okla Med Res Found, Okla City, 82-89; chem investr, Dept Vet Affairs, Okla City, 87- *Honors & Awards:* Baldwin-Lucke Award, Univ Pa. *Mem:* Am Fed Clin Res; Sigma Xi. *Res:* Etiology and puterogenesis of autoimmune rheumatic diseases. *Mailing Add:* 439 NW 20th Oklahoma City OK 73103

HARLEY, NAOMI HALLDEN, b NY; m 64. RADIATION CARCINOGENESIS. *Educ:* Cooper Union, BSEE, 59; Grad Sch Eng & Sci, NY Univ, MS, 67, PhD(radiation health), 71. *Prof Exp:* Phys scientist radioactivity measurement, AEC, 51-65; RES PROF RADIATION PHYSICS, MEASUREMENT, & DOSIMETRY, MED CTR, NY UNIV, 67- *Concurrent Pos:* Prin investr, Environ Protection Agency, US Dept Energy & other grants, 71-; mem, Nat Coun Radiation Protection & Measurement. *Mem:* AAAS; Health Physics Soc; NY Acad Sci; Am Statist Asn. *Res:* Environmental radon study to determine mechanisms for entry and removal in homes; radiation carcinogenesis; lung cancer risk from environmental radon daughter exposure; environmental radioactivity measurement. *Mailing Add:* Dept Environ Med NY Univ Med Ctr 550 First Ave New York NY 10016

HARLEY, PETER W, III, b St George, SC, Oct 4, 40; m 62; c 3. MATHEMATICS. *Educ:* Wofford Col, AB, 62; Univ Ga, MA, 65, PhD(math), 66. *Prof Exp:* Asst prof math, Univ Ga, 66-67; asst prof, 69-74, ASSOC PROF MATH, COMPUT SCI & STATIST, UNIV SC, 74- *Mem:* Am Math Soc. *Res:* Topology and set theory. *Mailing Add:* Dept Math & Statist Univ SC Columbia SC 29208

HARLEY, ROBISON DOOLING, b Pleasantville, NJ, Feb 27, 11; m 44; c 7. OPHTHALMOLOGY. *Educ:* Rutgers Univ, BSc, 32; Univ Pa, MD, 36; Univ Minn, PhD(neurol), 41. *Prof Exp:* Resident physician, Philadelphia Gen Hosp, 36-38; asst ophthal, Mayo Clin, 41-42; from assoc to assoc prof ophthal, Temple Univ, 47-68, prof & chmn dept, 68-78; attend surgeon & dir pediat ophthal & motility, Willis Eye Hosp, 68-80; RETIRED. *Concurrent Pos:* Asst chief ophthal, Atlantic City Hosp, 47-50, chief, 50-68; attend surgeon, St Christopher's Hosp for Children; consult, Shore Mem Hosp, 48-50, Children's Hosp, Philadelphia, Children's Seashore Home & Betty Bacharach Home for Children; mem med bd, Proj Hope, Proj Orbis; adj prof ophthal, Thomas Jefferson Univ; mem med bd, Nat Found Retinitis Pigmentosa; consult surgeon, Wills Eye Hosp. *Honors & Awards:* Antonio Navas Lectr, PR, 72; Al Morgan Lectr, Toronto, 74; Frank Costenbaker Lectr, Wash, 79; Honor & Sr Honor Awards, Am Acad Ophthal, 87; Honor Award, Am Asn Pediat Ophthal & Strabismus, 90; Vasso Nunez de Balboa-Legion of Honor, Repub

Panama. *Mem:* Am Ophthal Soc; fel Asn Res Vision & Ophthal; fel Am Col Surg; Am Acad Ophthal & Otolaryngol; Am Orthoptic Coun. *Res:* Clinical research in ophthalmology. *Mailing Add:* 1910 Academy Pl Wilmington DE 19806

HARLIN, MARILYN MILER, b Oakland, Calif, May 30, 34; wid; c 2. ALGOLOGY, ECOLOGY. *Educ:* Stanford Univ, AB, 56, MA, 57; Univ Wash, PhD(bot), 71. *Prof Exp:* Teacher biol, Leysin Am Sch, Switz, 62-66; from asst prof to assoc prof, 71-83, PROF, UNIV RI, 83-; CONSULT, APPL SCI ASN, 87-,. *Concurrent Pos:* Instr, Am Col Switz, 65-66; NSF fel, 71; NSF travel award, 74; res grants, Sea Grant, 75-82, EPA, 83-84; vis prof bot, La Trobe Univ, Australia, 84; Conservation Law Fedn, 88- *Mem:* AAAS; Phycol Soc Am; Int Phycol Soc; Aust Soc Phycol & Aquatic Bot; Am Soc Limnol & Oceanog; Int Asn Aquatic & Vascular Plant Biologist. *Res:* Physiological ecology of marine macroalgae; epiphytic algae; relationships with host plants; nitrogen uptake by marine macroalgae; algal development on artificial substrata. *Mailing Add:* Dept Bot Univ RI Kingston RI 02881

HARLIN, VIVIAN KRAUSE, b Seattle, Wash, Dec 26, 24; m 48; c 3. PUBLIC HEALTH ADMINISTRATION. *Educ:* Univ Wash, BS, 46, MS, 70; Univ Ore, MD, 50. *Prof Exp:* Pvt pract, Seattle, Wash, 52-53; exam physician, Seattle Pub Schs, 53-57, dir health serv, 57-79; dir health serv, Off Supt Pub Instr, Tumwater, 79-84; immunization consult, Wash Dept Social & Health Serv, 87-90; RETIRED. *Concurrent Pos:* Clin physician, Snohomish County Health Dept, 53; clin instr, Sch Pub Health & Community Med, Univ Wash, 67-, sr fel dept prev med, 68-69. *Honors & Awards:* Physicians Recognition Award, AMA, 70-73; William A Howe Award, Am Sch Health Asn, 79. *Mem:* Am Med Womens Asn (vpres, 78, pres, 81); Am Sch Health Asn (vpres, 73, pres-elect, 74, pres, 75). *Mailing Add:* PO Box 340 Ravensdale WA 98051

HARLING, OTTO KARL, b Staten Island, NY, Oct 1, 31; m 57; c 4. PHYSICS, NUCLEAR ENGINEERING. *Educ:* Ill Inst Technol, BS, 53; Univ Heidelberg, dipl, 55; Pa State Univ, PhD(physics), 62. *Prof Exp:* Physicist, Curtiss-Wright Res Div, 56-59; staff physicist, H R B-Singer, Inc, 59-62; sr physicist, Hanford Labs, Gen Elec Co, 62-65; staff scientist, Pac Northwest Lab, Battelle Mem Inst, 65-76; DIR NUCLEAR REACTOR LAB & PROF NUCLEAR ENG, MASS INST TECHNOL, 76- *Concurrent Pos:* Assoc prof, Joint Ctr Grad Studies, Richland, 63-; adj assoc prof, Wash State Univ. *Mem:* Am Phys Soc; Am Nuclear Soc. *Res:* Applied physics; technology of fusion reactors; experimental nuclear and solid state physics; surface science. *Mailing Add:* Nuclear Reactor Lab Mass Inst Technol Cambridge MA 02139

HARLOW, CHARLES ALTON, b New Boston, Tex, Mar 14, 40; m 61; c 2. ELECTRICAL ENGINEERING. *Educ:* Univ Tex, Austin, BS, 63, PhD(elec eng), 67. *Prof Exp:* From asst prof to prof elec eng, Univ Mo, Columbia, 67-78; PROF ELEC ENG, LA STATE UNIV, 78-, DIR REMOTE SENSING & IMAGE PROCESSING LAB, 79- *Mem:* Inst Elec & Electronics Engrs; Asn Comput Mach. *Res:* Computer engineering and logical design; image analysis; pattern recognition. *Mailing Add:* Dept Elec Eng 213 Elec Eng Bldg La State Univ Boyd Hall Baton Rouge LA 70803

HARLOW, FRANCIS HARVEY, JR, b Seattle, Wash, Jan 22, 28; m 52; c 4. ANALYTICAL FLUID DYNAMICS, NUMERICAL FLUID DYNAMICS. *Educ:* Univ Wash, BS, 49, PhD(physics), 53. *Prof Exp:* Group leader, 59-73, MEM STAFF, LOS ALAMOS NAT LAB, 53-, FEL, 81- *Res:* Multiphase flow, turbulence transport theory and numerical approximation methods for partial differential equations. *Mailing Add:* 1407 11th St Los Alamos NM 87544

HARLOW, H(ENRY) GILBERT, b Plymouth, Mass, Apr 27, 14; m 40; c 5. CIVIL ENGINEERING. *Educ:* Tufts Univ, BCE, 37; Harvard Univ, MS, 40. *Prof Exp:* From instr to prof, 40-79, chmn dept, 50-79, EMER PROF CIVIL ENG, UNION UNIV, NY, 80- *Concurrent Pos:* Consult, 40- *Mem:* Am Soc Civil Engrs. *Res:* Foundations and soil mechanics; plant genetics; nuclear shielding; plant tissue culture. *Mailing Add:* Dept Civil Eng Union Col Schenectady NY 12308

HARLOW, RICHARD FESSENDEN, b Boston, Mass, Dec 16, 19; m 42; c 3. WILDLIFE ECOLOGY, PLANT ECOLOGY. *Educ:* Univ Maine, BS, 47, MS, 48; Va Polytech Inst & State Univ, MS, 71. *Prof Exp:* Game technician, Maine Inland Fisheries & Game, 49-50; soil conservationist, Soil Conserv Serv, USDA, 50-52; asst proj leader wildlife, Fla Game & Fresh Water Fish Comn, 52-62, proj leader wildlife res, 62-64, asst chief game div, 64-65; res wildlife biologist, Forest Serv, USDA, 66-85; RETIRED. *Mem:* Wildlife Soc; Sigma Xi. *Res:* Food habits of game species; influence of land use practices and other environmental changes on wildlife habitat, especially white-tailed deer and non game and endangered wildlife species. *Mailing Add:* 100 Ft Rutledge Rd Clemson SC 29631

HARLOW, RICHARD LESLIE, b Schenectady, NY, Nov 6, 42; m 77; c 3. X-RAY CRYSTALLOGRAPHY. *Educ:* Union Col, BS, 64; Univ Ill, MS, 66; Syracuse Univ, PhD(chem), 71. *Prof Exp:* Lectr chem, Univ Zambia, 70-73; res asst, Syracuse Univ, 73; fel chem, Univ Tex, Austin, 73-77; chemist, 77-79, supvr, 79-85, PRIN INVESTR, E I DU PONT DE NEMOURS & CO, INC, 85- *Concurrent Pos:* Adj prof, Univ Del, 81- *Mem:* Am Crystallog Asn; Am Chem Soc; Sigma Xi. *Res:* Molecular structure determinations (via x-ray diffraction) of inorganic and organometallic complexes; powder diffraction software and analyses. *Mailing Add:* Cent Res & Develop Exp Sta E228/316D E I du Pont de Nemours & Co Inc Wilmington DE 19898

HARMAN, CHARLES M(ORGAN), b Canonsburg, Pa, July 25, 29; m 56; c 3. MECHANICAL ENGINEERING, THERMODYNAMICS. *Educ:* Univ Md, BS, 54; Univ NDak, MS, 57; Univ Wis, PhD(mech eng), 61. *Prof Exp:* Instr mech eng, Univ NDak, 54-57 & Univ Wis, 57-59; from asst to prof mech eng, Duke Univ, 61-86, coordr res, 68-70, assoc dean grad sch, 70-80; RETIRED. *Concurrent Pos:* Sr sci adv, Army Res Off, 64-77; assoc prog dir, NSF, 67-68; co-ed, J Advan Transp, 77-; assoc dir, Inst Prod Safety, 82-87. *Mem:* Am Soc Mech Engrs; Advan Transit Asn. *Res:* Fluid mechanics; transportation; energy systems and energy storage. *Mailing Add:* Sch Eng Duke Univ Durham NC 27706

HARMAN, DENHAM, b San Francisco, Calif, Feb 14, 16; m 43; c 4. BIOCHEMISTRY. *Educ:* Univ Calif, BS, 40, PhD(chem), 43; Stanford Univ, MD, 54. *Prof Exp:* Chemist, Shell Develop Co, 43-49; res assoc, Donner Lab, Univ Calif, instr assoc med & assoc prof biochem, 58-62, asst prof med, 62-68, PROF BIOCHEM, MED SCH, UNIV NEBR AT OMAHA, 62-, PROF MED, 68- *Mem:* AAAS; Am Chem Soc; AMA; fel Am Col Physicians; Am Aging Asn. *Res:* Use of radioactive tracers in organic reactions; free radical reactions; organic chemistry of phosphorus and sulphur; arteriosclerosis; aging and cancer. *Mailing Add:* Dept Biochem & Med Univ Nebr Med Ctr 600 S 42nd St Omaha NE 68198-4635

HARMAN, GARY ELVAN, b La Junta, Colo, Nov 13, 44; m 65; c 3. PLANT PATHOLOGY, MICROBIOLOGY. *Educ:* Colo State Univ, BS, 66; Ore State Univ, PhD(plant path & biochem), 70. *Prof Exp:* Res assoc plant path & biochem, NC State Univ, 69-70; from asst to assoc prof, 70-84, chmn, Dept Hort Sci, 83-84, PROF SEED MICROBIOL, NY STATE AGR EXP STA, CORNELL UNIV, 84- *Mem:* AAAS; Sigma Xi; Am Phytopath Soc. *Res:* Detection and control of seed-borne microorganisms; biological control of seed and root-attacking microorganisms; mechanisms of seed deterioration by aging and by storage fungus; genetics of brocontrol fungi. *Mailing Add:* PO Box 462 NY State Exp Sta Geneva NY 14456

HARMAN, GEORGE GIBSON, JR, b Norfolk, Va, Dec 7, 24; div; c 3. PHYSICS, ENGINEERING. *Educ:* Va Polytech Inst, BS, 49; Univ Md, MS, 59. *Prof Exp:* Electronic engr microwave measurements, Electron Tube Lab, Nat Inst Standards & Technol, 50-53, physicist, Semiconductor Mat Sect, 53-63, physicist hot carriers semiconductors, Electronic Technol Div, 63-70 & semiconductor packaging & assembly tech, 70-74, SR RES SCIENTIST ACOUST EMISSION SEMICONDUCTOR DEVICES, SEMICONDUCTOR, DEVICES DIV, NAT INST STANDARDS & TECHNOL, 75- *Concurrent Pos:* Res fel, Dept Physics, Univ Reading, 62-63. *Honors & Awards:* Silver Medal, US Dept Com, 72; Gold Medal, US Dept Com, 79; Technol Achievement Award, Int Soc Hybrid Microelectronics, 81, Distinguished Serv Award, 86 & 87, Daniel C Hughes Award, 89, Educ Found Distinguished Serv Award, 90; Centennial Award, Inst Elec & Electronics Engrs, 84; Lewis F Miller Award, 84; Int Electronics Packaging Soc Achievement Award, 88. *Mem:* Am Phys Soc; fel Inst Elec & Electronics Engrs; Am Soc Testing & Mat; fel Int Soc Hybrid Microelectronics; Sigma Xi. *Res:* Microelectronics reliability; microelectronics assembly technology; acoustic emission; ultrasonics; semiconductor materials; hybrid microelectronics. *Mailing Add:* Semiconductor Electronics Div 727 Nat Inst Standards & Technol Gaithersburg MD 20899

HARMAN, T(HEODORE) C(ARTER), b Ind, July 22, 29; m 57; c 4. MATERIALS SCIENCE, CRYSTAL GROWTH. *Educ:* Manchester Col, AB, 51; Purdue Univ, MS, 53. *Prof Exp:* Proj leader, Battelle Mem Inst, 53-59; asst group leader, Lincoln Lab, Mass Inst Technol, 59-65; prog mgr, Adv Proj Res Agency, US Dept Defense, 65, 66; asst group leader, 66-78, sr staff, 78-87, STAFF, LINCOLN LAB, MASS INST TECHNOL, 87- *Concurrent Pos:* Ed, J Electronic Mat, 72- *Mem:* Fel Am Phys Soc; Metall Soc; Inst Elec & Electronic Engrs. *Res:* Preparation of compound semiconductors; physical properties of semimetals and semiconductors; various solid state devices; crystal growth; diffusion and defects in semiconductors; thermoelectric and thermomagnetic materials and devices. *Mailing Add:* Lincoln Lab Mass Inst Technol Lexington MA 02173-9108

HARMAN, WALTER JAMES, b Strong, Ark, Feb 25, 28; m 54; c 2. ZOOLOGY. *Educ:* La Polytech Inst, BS, 48; Univ Ark, MA, 50; Univ Ill, PhD, 59. *Prof Exp:* From instr to assoc prof zool, La Polytech Inst, 52-61; chmn dept, La State Univ, Baton Rouge, 63-78, from asst prof to prof zool, 66-89; RETIRED. *Mem:* Am Micros Soc; Am Soc Zool; Soc Syst Zool. *Res:* Taxonomy and ecology of Oligochaeta. *Mailing Add:* 5988 S Pollard Pkwy Baton Rouge LA 70808-8867

HARMAN, WILLARD NELSON, b Geneva, NY, Apr 20, 37; m 65; c 4. FRESHWATER ECOLOGY, MALACOLOGY. *Educ:* NY State Univ, BS, 65; Cornell Univ, PhD(limnol), 68. *Prof Exp:* From asst prof to assoc prof, 68-76, chmn biol, 80-89, PROF BIOL, STATE UNIV NY ONEONTA, 76-, DIR BIOL FIELD STA. *Concurrent Pos:* Ecol Resource Adv, NY State Dept Environ Conserv, NY State, 74- *Honors & Awards:* Environ Qual Award, US Environ Protection Agency, 90. *Mem:* AAAS; Soc Limnol & Oceanog; Am Malacol Union; Soc Exp & Descriptive Malacol. *Res:* Descriptive malacology in freshwater ecosystems. *Mailing Add:* Dept Biol State Univ NY Oneonta NY 13820

HARMEL, MEREL H, b Cleveland, Ohio, May 19, 17; m 44; c 4. ANESTHESIOLOGY. *Educ:* Johns Hopkins Univ, BA, 38, MD, 43. *Prof Exp:* Asst instr surg, Johns Hopkins Univ, 44, instr, 45-47; assoc anesthesiol-in-surg, Univ Pa, 47-48; assoc prof anesthesia & assoc pharmacol, Albany Med Col, Union Univ, NY, 48-52; prof anesthesia & chmn dept, Col Med, State Univ NY Downstate Med Ctr, 52-68; chmn dept anesthesiol, Pritzker Sch Med, Univ Chicago, 68-71; CHMN DEPT ANESTHESIOL, MED CTR, DUKE UNIV, 71- *Concurrent Pos:* Vis prof, Sch Med, Yale Univ, 59; anesthesiologist-in-chief, State Univ Hosp, 66-68; Nat Res Coun fel, Johns Hopkins Univ, 46-47. *Mem:* Am Soc Anesthesiologists; AMA; Am Sci Affil; NY Acad Sci. *Res:* Action of drugs on the central nervous system. *Mailing Add:* Anesthesiol Med Ctr Duke Univ Durham NC 27710

HARMEN, RAYMOND A, b Oakland, Calif, Feb 2, 17; m 51; c 1. ENGINEERING, METEOROLOGY. *Educ:* Univ Calif, AB, 39, MS, 47; US Naval Acad, MA, 43. *Prof Exp:* Asst soil chem, Univ Calif, 39-41; soils technologist, US Bur Reclamation, 41-42; meteorologist, Landing Aids Exp Sta, Calif, 48-50; physicist, 50-52, supvry physicist, 52-54, head environ determination br, 55-58, head flight reliability br, 58-60, HEAD RELIABILITY OFF, POINT MUGU, NAVAL MISSILE CTR, 60- *Concurrent Pos:* Mem bd dir, Missile Technol Hist Asn, 80- *Res:* Reliability of missile systems; probability and stochastic processes; demonstration of flight performance capability by means of ground tests of missile equipment. *Mailing Add:* 208 Ramona Pl Camarillo CA 93010

HARMER, DAVID EDWARD, b Grand Rapids, Mich, Apr 8, 29; m 57; c 2. RADIATION CHEMISTRY, NUCLEAR DECONTAMINATION. *Educ:* Albion Col, AB, 50; Univ Mich, MS, 52, PhD(org chem), 55. *Prof Exp:* Fel chem, Univ Mich, 50-51, res asst, Eng Res Inst, 51-55; chemist org synthesis, Dow Chem Co, 55-56, head radiation chem sect, 56-71, group leader, Phys Res Lab, 71-74, proj mgr, Dow Nuclear Serv, 74-82; MGR APPL TECH, IT CORP NUCLEAR SERV, 83- *Mem:* AAAS; Am Chem Soc; Am Inst Chem Eng; Sigma Xi; Am Nuclear Soc. *Res:* Radiation chemistry, dosimetry and calculations; engineering management; radiation chemical engineering; polymer chemistry; nuclear reactor decontamination. *Mailing Add:* 11104 Windward Dr Knoxville TN 37922

HARMER, DON STUTLER, b Washington, DC, Mar 11, 28; m 52, 65; c 5. NUCLEAR CHEMISTRY, NUCLEAR PHYSICS. *Educ:* George Washington Univ, BS, 52; Univ Calif, Los Angeles, PhD(nuclear chem), 56. *Prof Exp:* Asst, George Washington Univ, 51-52 & Univ Calif, Los Angeles, 52-56; res assoc, Brookhaven Nat Lab, 56-58; spec res scientist, Eng Exp Sta, 59-60, assoc prof physics, 60-67, PROF NUCLEAR ENG, SCH NUCLEAR ENG, GA INST TECHNOL, 67- *Mem:* Am Phys Soc; Am Nuclear Soc; fel Am Inst Chem. *Res:* On-line applications of small computers in teaching and research; radiation shielding; neutron radiography; biomedical neutron radiography; bioengineering; nuclear and cosmic ray physics, neutrinos. *Mailing Add:* Dept Nuclear Eng Ga Inst Technol 225 North Ave NW Atlanta GA 30332

HARMER, MARTIN PAUL, b Hull, Eng, Aug 7, 54. CERAMIC PROCESSING, ELECTRONIC CERAMICS. *Educ:* Leeds Univ, Eng, BSc, 76, PhD(ceramics), 80. *Prof Exp:* From asst prof to assoc prof, 80-87, PROF MAT SCI, LEHIGH UNIV, 87-, DIR, CERAMICS RES LAB, MAT RES CTR, 87- *Concurrent Pos:* Presidential young investr, White House, 84; assoc ed, J Am Ceramic Soc, 84-; chmn, Kraner Award Symp, 85-; secy, Basic Sci Div, Am Ceramic Soc, 91. *Honors & Awards:* IBM Award, 84; DuPont Award, 88. *Mem:* Am Ceramic Soc. *Res:* Processing of advanced ceramics; science of sintering; microstructure and properties of advanced structural and electronic (dielectric and ferroelectric) ceramics. *Mailing Add:* Dept Mat Sci & Eng 254 Whitaker Lab Lehigh Univ Bethlehem PA 18015

HARMER, RICHARD SHARPLESS, b Whittier, Calif, Sept 14, 41; m 66; c 2. MATERIALS SCIENCE, CERAMIC ENGINEERING. *Educ:* Univ Ill, BS, 63, MS, 67, PhD(ceramic eng), 71. *Prof Exp:* Instr electron micros, Univ Ill, 66-68; asst prof mat sci, 70-77, ASSOC PROF MAT SCI, UNIV DAYTON, 77-, RES CERAMIST, RES INST, 70- *Mem:* Am Soc Metals. *Res:* Electron microprobe; scanning electron microscopy; materials characterization; x-ray diffraction; rare earth-cobalt magnetic materials. *Mailing Add:* Dept Mech Eng Univ Dayton 300 College Park Ave Dayton OH 45469

HARMET, KENNETH HERMAN, b Chicago, Ill, Aug 1, 24; m 73; c 3. ION TRANSPORT, RAPID GROWTH RESPONSES. *Educ:* Coe Col, BS, 49; Northwestern Univ, MS, 53; Univ Chicago, PhD(biol), 69. *Prof Exp:* Instr, Biol, Geneva High Sch, Ill, 50-52, Biol & Bot, East High Sch, 52-55; Instr biol, 55-61, asst prof, 61-71, ASSOC PROF BIOL & PLANT PHYSIOL, NORTHERN ILL UNIV, 71- *Concurrent Pos:* Consult, DeKalb Agres Inc, 69-73. *Mem:* Am Soc Plant Physiologists; Sigma Xi; AAAS; Am Inst Biol Sci. *Res:* Effects of mineral nutrients and rare earth ions on the rapid growth responses of AVENA coleoptile segments; membranes and ion transport; seedling physiology with emphasis on amides and cyanogenic glycosides. *Mailing Add:* 8380 Ebenezer Ovil Rd Hopkinsville KY 42240

HARMON, ALAN DALE, b Macon, Ga, Nov 11, 44; m 71; c 1. PHARMACOGNOSY, ORGANIC CHEMISTRY. *Educ:* Idaho State Univ, BS, 68; Univ Conn, MS, 75, PhD(pharmacog), 77. *Prof Exp:* Pharmacist qual control pharmaceut, USPHS, 69-72; staff fel, 77-80, ASST PROF MED CHEM & PARMACOG, NIH, 80- *Mem:* Am Soc Pharmacog; Am Chem Soc; Sigma Xi. *Res:* Biosynthesis of secondary natural products; elucidation of structures and synthesis of biologically active organic molecules; stereochemistry of enzyme reactions. *Mailing Add:* McCormormick & Co Inc Res & Dev Labs 200 Wight Ave Hunt Valley MD 21031

HARMON, BRUCE NORMAN, b Grand Rapids, Mich, 1947; m 69; c 2. COMPUTATIONAL PHYSICS. *Educ:* Ill Inst Technol, BS, 68; Northwestern Univ, MS, 69, PhD(physics), 73. *Prof Exp:* Lectr physics, Northwestern Univ, 69-71; fel physics, Ames Lab, 73-74; from instr to assoc prof, 74-81, PROF PHYSICS, IOWA STATE UNIV, 81-; PROG DIR SOLID STATE PHYSICS DIV, AMES LAB, 83- *Concurrent Pos:* Scientist, Res Estab Riso, Denmark, 76, Kernforschumgszentrum, Karlsruch, Ger, 81; ESRF, Grenoble, France, 90. *Mem:* Fel, Am Phys Soc; Am Asn Physics Teachers; Mat Res Soc. *Res:* Superconductivity, electronic, optical and magnetic properties of metals, hydrogen in metals, high temperature conductors, electron-phonon interaction, metal surfaces and rare earth metals and compounds; martensitic phase transformations. *Mailing Add:* Dept Physics Iowa State Univ Ames IA 50011

HARMON, BUD GENE, b Camden, Ind, July 2, 31; m 53; c 3. ANIMAL SCIENCE, ANIMAL NUTRITION. *Educ:* Purdue Univ, BS, 58; Mich State Univ, PhD(animal nutrit), 62. *Prof Exp:* From asst prof to prof animal nutrit, Univ Ill, Urbana, 62-75; DIR MONOGASTRIC RES, RALSTON PURINA, 75- *Concurrent Pos:* Mem comt, Nat Acad Sci Comn Educ in Agr & Natural Resources, 67- *Honors & Awards:* Am Soc Animal Sci Teaching Award, 71; Distillers Feed Res Coun Distinguished Nutrit Award, 74. *Mem:* AAAS; Am Soc Animal Sci; Am Inst Nutrit; Animal Nutrit Coun. *Res:* Mineral nutrition. *Mailing Add:* Ralson Purina One Checkerboard Sq St Louis MO 63164

HARMON, DALE JOSEPH, b Cuyahoga Falls, Ohio, Mar 1, 27; m 51; c 3. POLYMER CHEMISTRY. *Educ:* Kent State Univ, BS, 51; Univ Akron, MS, 60. *Prof Exp:* Sr res chemist, 51-68, res assoc, 68-79, sr res assoc, B F Goodrich Res Ctr, Breckville, 79-; RETIRED. *Mem:* Am Chem Soc; Am Soc Testing & Mat. *Res:* Physical structure of polymers and the relationship of the structure to processing and mechanical properties; gel permeation and liquid chromatography; solution properties of polymers. *Mailing Add:* 993 Cotswold Dr Copley OH 44321

HARMON, DAVID ELMER, JR, b Kittanning, Pa, Aug 12, 32; m 60; c 6. PETROLEUM EXPLORATION, NON-METALLIC MINERAL EXPLORATION. *Educ:* Marietta Col, Ohio, AB, 54. *Prof Exp:* Geologist, BH Putnam Oil Producer, 54-58; chief geologist FE Moran Oil Co, 58-67; chief engr, Jefferson Country, Ky, 67-69; chief geologist, Guernsey Petrol Corp, 69-73; vpres, Johnston Petrol corp, 73-75 & O'Neal Petrol, Inc, 75-83; founder & pres, Concord Energy Inc, 83-84; CONSULT, 84- *Concurrent Pos:* Chmn, Dist 3 Comt Govt Affairs, Am Asn Petrol Geologist, 77-87; mem Environ Comn, Ohio Oil & Gas Asn, 78-82. *Mem:* Am Asn Petrol Geologists; Am Inst Prof Geologists; fel Geol Soc Am; Geol Soc Am; Soc Petrol Engr. *Res:* Exploration for and exploitation of oil and gas deposits domestically and internationally. *Mailing Add:* PO Box 1 New Concord OH 43762-0001

HARMON, FREDERICK ROBERT, immunodiagnostics, enzymology, for more information see previous edition

HARMON, G LAMAR, b Baltimore, Md, Feb 21, 31; m 57; c 2. NUCLEAR PHYSICS. *Educ:* Emory Univ, AB, 52, MS, 54; Vanderbilt Univ, PhD(physics), 57. *Prof Exp:* Mgr, New Guidance Systs, 57-71, proj mgr, Helicopter Fire Control Exp, 71, div staff engr electronics, 72-74, MGR SYSTS ENG RES & TECHNOL & ENG MGR, LANTIRN NAVIGATION PAD DEVELOP, MARTIN MARIETTA CORP, ORLANDO, 74- *Honors & Awards:* Inventor of the Year, Martin Marietta Corp, 66. *Res:* Automation for nighttime pilot situational awareness enhancement; target recognition automation; precision laser target designation systems; correlation guidance systems angle; angle rate ordinance delivery; loop control and display techology. *Mailing Add:* 1362 Granville Dr Winter Park FL 32789

HARMON, GEORGE ANDREW, b Lewisville, Idaho, Aug 31, 23; m 54; c 2. BIOCHEMISTRY, MICROBIOLOGY. *Educ:* Univ Calif, Los Angeles, AB, 48; Stanford Univ, MA, 50, PhD, 58. *Prof Exp:* Instr microbiol, Univ Calif, Santa Barbara, 58-59; res biochemist, Med Res Serv, Vet Admin Ctr, 59-61; asst to exec head dept biol sci, Stanford Univ, 62-66; prof biol, Clarion State Col, 66-88; RETIRED. *Mem:* AAAS; Am Chem Soc. *Res:* Microbial intermediary metabolism; nucleic acid structure. *Mailing Add:* 709 S Brandywine St West Chester PA 19382

HARMON, GLYNN, b Hollister, Calif, Nov 4, 33; div. INFORMATION SCIENCE. *Educ:* Univ Calif, Berkeley, BA, 60, MA, 63; Case Western Reserve Univ, MS, 65, PhD(info sci), 69; Southwest Tex Univ, MBA, 73. *Prof Exp:* Researcher info, Case Western Reserve Univ, 64-65; librn & instr govt, Calif State Univ, Chico, 65-66; asst prof info, Univ Denver, 66-70; assoc prof, 70-75, PROF INFO, UNIV TEX, AUSTIN, 75- *Concurrent Pos:* Res assoc, Case Western Reserve Univ, 65-72. *Mem:* Am Soc Info Sci; Asn Comput Mach. *Res:* Education; public and private investment; systems theory, management information systems and medical information. *Mailing Add:* Grad Sch Libr & Info Sci Univ Tex EDB 564 Austin TX 78712-1276

HARMON, H JAMES, b Granite City, Ill, Sept 14, 46; m 69; c 3. BIOENERGETICS, BIOPHYSICS. *Educ:* Purdue Univ, BS, 68, MS, 71, PhD(biol), 74. *Prof Exp:* Instr biol, dept biol sci, Purdue Univ, 69-72; res bioenergetics, Johnson Res Found, Univ Pa, 74-77; from asst prof to assoc prof biol & physics, 77-88, prof zool & physics, 88-90, PROF MICROBIOL & ADJ PROF PHYSICS, OKLA STATE UNIV, 90- *Concurrent Pos:* Researcher, Johnson Res Found, Univ Pa, NIH, 75-77, Am Heart Asn, 80-82; prin investr, USAF, 84-88 & 89-90. *Mem:* Biophys Soc; AAAS; NY Acad Sci; Am Soc Biochem & Molecular Biol; Sigma Xi. *Res:* Structure-function of inner mitochondrial membrane; instrumentation/spectrophotometer design; mechanism of electron transport; membrane topography; effect of drugs on protein structure; aging of heart and brain; effects of environmental contaminants; proton interaction with cytochrome oxidase; rapid enzyme kinetics; spectroscopy and spectrometry. *Mailing Add:* Dept Microbiol Okla State Univ Stillwater OK 74078

HARMON, J FRANK, b Van Wert, Ohio, Feb 23, 39; m 63; c 3. PHYSICS, CHEMICAL PHYSICS. *Educ:* Portland State Univ, BS, 63; Univ Wyo, MS, 65, PhD(physics), 68. *Prof Exp:* Res assoc physics, Univ Utah, 68-69; from asst prof to assoc prof, 69-79, chmn dept, 71-76, PROF PHYSICS, IDAHO STATE UNIV, 80-, CHAIR DEPT, 83- *Concurrent Pos:* NSF fel, Univ Wyo, 70; assoc Western Univ fac fel, Los Alamos Sci Labs, 71 & Argonne Nat Lab W, 81; consult, Los Alamos Nat Lab, 79-84, Idaho Nat Eng Lab, 86- *Mem:* Sigma Xi; Int Soc Magnetic Resonance. *Res:* Magnetic resonance; molecular motions in condensed phases; generation and application of charged particle and photon beams to materials analysis. *Mailing Add:* Dept Physics Idaho State Univ Pocatello ID 83201

HARMON, JOAN T, b Staten Island, NY. RECEPTOR INTERACTION, PLATETE. *Educ:* Univ Rochester, PhD(biochem), 76. *Prof Exp:* RES ASSOC & SCIENTIST II, BIOMED RES & DEVELOP LAB, AM RED CROSS, 84- *Mem:* Am Soc Biol Chemists; Sigma Xi. *Mailing Add:* Five Trailridge Ct Potomac MD 20854

HARMON, JOHN W, b White Plains, NY, Apr 22, 43; m 66; c 2. SURGICAL GASTROENTEROLOGY, ENDOCRINOLOGY. *Educ:* Harvard Col, BA, 65; Columbia Univ, MD, 69. *Prof Exp:* Surg resident, Harvard Surg Serv, Boston City Hosp & New Eng Deaconess Hosp, 69-75; investr surg gastroenterol, Walter Reed Army Inst Res, 75-77, chief dept, 77-81 & dir div surg, 82-85; CHIEF SURG SERV, WASHINGTON VET ADMIN MED

CTR, 85- *Concurrent Pos:* Staff surgeon, Walter Reed Army Med Ctr, 78-85; chief acad surg, Uniformed Serv Univ Health Sci, 78-85; prof surg, Georgetown Univ, George Washington Univ & Uniformed Serv Univ; clin prof surg, Howard Univ. *Honors & Awards:* William Beaumont Prize Gastroenterol, US Army, 82. *Mem:* NY Acad Sci; Am Physiol Soc; Am Col Surgeons; Soc Univ Surgeons; Am Gastroenterol Asn. *Res:* Surgical gastroenterology and endocrinology; basic tissue mechanisms of injury and protection; peptic ulcer disease and reflux esophagitis. *Mailing Add:* Chief Surg Serv Washington VA Med Ctr Washington DC 20422

HARMON, KENNETH MILLARD, b Washington, DC, May 16, 29; m 49; c 2. INORGANIC CHEMISTRY. *Educ:* San Jose State Col, BA, 54; Univ Wash, PhD(org chem), 58. *Prof Exp:* From instr to assoc prof chem, Harvey Mudd Col, 58-69, actg chmn dept, 63-64; PROF CHEM, OAKLAND UNIV, 69- *Concurrent Pos:* Petrol Res Fund res fel, 64-65. *Mem:* Am Chem Soc; Sigma Xi. *Res:* Infrared study of strong hydrogen bonds; organoborane chemistry. *Mailing Add:* Dept Chem Oakland Univ Rochester MI 48309

HARMON, LAURENCE GEORGE, b Mountain Grove, Mo, Mar 18, 13; m 39; c 1. FOOD SCIENCE, MICROBIOLOGY. *Educ:* Kans State Univ, BS, 36; Tex Tech Univ, MS, 40; Iowa State Univ, PhD, 54. *Prof Exp:* From instr to prof dairy indust, Tex Tech Univ, 36-54; from asst prof to assoc prof dairy indust, 54-60, prof food sci, 60-70, assoc chmn dept, 71-79, prof, 70-80, EMER PROF FOOD SCI & HUMAN NUTRIT, MICH STATE UNIV, 80- *Concurrent Pos:* Consult, Dairy Res Inc, 71-82. *Honors & Awards:* Pfizer Award, Am Dairy Sci Asn, 66, Award of Hon, 79; Distinguished Serv Award, Int Asn Sanitarians, 80. *Mem:* Am Dairy Sci Asn (pres, 75-76); Inst Food Technol; Int Asn Sanitarians. *Res:* Food microbiology and processing. *Mailing Add:* Dept Food Sci & Human Nutrit Mich State Univ East Lansing MI 48824

HARMON, LEON DAVID, biomedical engineering, pattern recognition; deceased, see previous edition for last biography

HARMON, ROBERT E, b Conrad, Mont, Jan 22, 31; m 57; c 4. ORGANIC CHEMISTRY, MEDICINAL CHEMISTRY. *Educ:* Wash State Univ, BS, 54; Wayne State Univ, PhD(org chem), 59. *Prof Exp:* From asst prof to assoc prof, 61-71, PROF ORG CHEM, WESTERN MICH UNIV, 71- *Concurrent Pos:* Consult, chem notations study, Nat Acad Sci-Nat Res Coun, 61-63 & Rohm & Haas, 85- *Mem:* Am Chem Soc; Brit Chem Soc; NY Acad Sci; Ger Chem Soc. *Res:* Cancer chemotherapy; reaction mechanisms in organic chemistry, particularly in the fields of heterocyclics and carbohydrates. *Mailing Add:* Dept of Chem 4040 McK Western Mich Univ Kalamazoo MI 49008

HARMON, ROBERT WAYNE, b Winchester, Ind, Oct 22, 29; m 51; c 4. INSULATION APPLICATION, HI-VOLTAGE ENGINEERING. *Educ:* Purdue Univ, BSEE, 51, MSEE, 55. *Prof Exp:* Dir, New Prod Develop, Ohio Brass Co, 55-68; chief engr, AB Chance Co, 68-91; CONSULT, 91- *Concurrent Pos:* Lectr, Univ Belo Horizonte, Brazil, 74-76, Univ Mo, 70-85; expert witness, litigation consult, 80-91. *Mem:* Fel Inst Elec & Electronics Engrs. *Res:* Insulation, Forensic Engineering; electrical shock hazards, protective grounding, insulation design, shock protection of cranes, vehicles, insulators for power lines; over 25 United States patents in insulation, hot-line tools, protective devices. *Mailing Add:* Rte 4 Box 135 Centralia MO 65240

HARMON, WALLACE MORROW, b Colorado Springs, Colo, June 1, 33; m 58; c 4. PARASITOLOGY, PROTOZOOLOGY. *Educ:* Colo Col, BS, 55; Univ Calif, Los Angeles, PhD(parasitol), 63; Syracuse Univ, MS, 65. *Prof Exp:* NIH fel biol, Rice Univ, 63-65; from asst prof to assoc prof, 65-74, PROF ZOOL, CALIF STATE UNIV, FRESNO, 74- *Mem:* Soc Protozool. *Res:* Parasites of man in Nigeria; avian hematozoa. *Mailing Add:* Dept Biol Calif State Univ 6241 N Maple Fresno CA 93740

HARMON, WILLIAM LEWIS, b Grove City, Pa, Aug 22, 28; m 55; c 4. PHYSICS. *Educ:* Ohio State Univ, BSc, 49, MSc, 55. *Prof Exp:* Jr physicist physics, Commun & Navigation Lab, 49-56; electronic scientist, Air Force Avionics Lab, 56-61, sr physicist, 61-66; laser field engr, Cloudcroft Electro-optical Site, 66-69; sr physicist lasers, Air Force Avionics Lab, 69-90; RETIRED. *Mem:* Inst Elec & Electronics Engrs. *Res:* Management of carbon dioxide laser radar programs for strategic and tactical aircraft weapon delivery applications. *Mailing Add:* 5907 Rosebury Dr Dayton OH 45424-4355

HARMONY, JUDITH A K, Decatur, Ill, Sept 29, 43. BIOCHEMISTRY. *Educ:* Kans Univ, BA, 65, MA, 67, PhD(chem), 71. *Prof Exp:* Fel biochem, Kans Univ, 71-75, dir, Enzyme Lab, 73-74; fel chem, Ind Univ, 75-76, asst prof biochem, 76-80, adj asst prof med sci, 78-80; ASSOC PROF ANAT & CELL BIOL, COL MED, UNIV CINCINNATI, 80- *Concurrent Pos:* Estab investr, Am Heart Asn, 80-85, fel, Coun Arteriosclerosis, 81-82. *Mem:* Am Heart Asn; Am Soc Biol Chemists; Am Asn Immunologist; Am Soc Cell Biol; AAAS. *Res:* Mechanism of action of immunoregulatory human plasma lipoproteins as it relates to lipoprotein structure and to cell-cell collaboration; structures and physiological importance of lipid transfer proteins which exist in human plasma. *Mailing Add:* Pharmacol Dept ML 575 Univ Cincinnati Col Med Cincinnati OH 45267-0521

HARMONY, MARLIN D, b Lincoln, Nebr, Mar 2, 36; m 84. MICROWAVE SPECTROSCOPY, MOLECULAR SPECTROSCOPY. *Educ:* Univ Kans, BS, 58; Univ Calif, Berkeley, PhD(chem), 61. *Prof Exp:* NSF fel, Harvard Univ, 61-62; from asst prof to assoc prof, 62-67, PROF CHEM, UNIV KANS, 71-, CHMN DEPT, 80- *Concurrent Pos:* Mem comt gas-phase interatomic distances, Nat Bur Standards, 70-79; vis res assoc, Cambridge Univ, 71. *Mem:* Fel AAAS; Am Chem Soc; Am Phys Soc; Sigma Xi; Am Asn Univ Prof. *Res:* Studies in molecular structure using microwave and laser spectroscopy. *Mailing Add:* Dept Chem Univ Kans Lawrence KS 66045-0046

HARMS, ARCHIE A, b Apr 18, 34; Can citizen; m 57; c 3. NUCLEAR ENGINEERING, ENGINEERING PHYSICS. *Educ:* Univ BC, BSc, 63; Univ Wash, MSE, 65, PhD(nuclear eng), 69. *Prof Exp:* Mathematician, Int Power & Eng Consults, 63-65; lectr eng mech, Univ Wash, 65-66; assoc prof, 69-80, PROF ENG PHYSICS, MCMASTER UNIV, 80- *Concurrent Pos:* Consult, Int Atomic Energy Agency & various industs. *Mem:* Am Nuclear Soc; Am Phys Soc; Can Nuclear Soc. *Res:* Fusion-fission-spallation energy systems; nuclear reactor physics; neutron diagnostics; science & engineering education; engineering mathematics. *Mailing Add:* Dept Eng Physics Fac Eng McMaster Univ Hamilton ON L8S 4M1 Can

HARMS, BENJAMIN C, b Gladbrook, Iowa, June 1, 39; m 68; c 2. ELEMENTARY PARTICLE PHYSICS. *Educ:* Iowa State Univ, BS; Univ Vt, MS, 64; Fla State Univ, PhD(physics), 69. *Prof Exp:* Res assoc physics, Fla State Univ, 69-70; vis prof, 70-71, PROF PHYSICS, UNIV ALA, 85- *Concurrent Pos:* Cottrell Found Res Corp res grant, 71-72; NSF teaching grant, 75-77, DOE res grant, 84- *Mem:* Am Phys Soc; Sigma Xi. *Res:* Superstring theory. *Mailing Add:* Box 870324 Tuscaloosa AL 35487-0324

HARMS, CLARENCE EUGENE, b Ulysses, Kans, Jan 22, 34; m 54; c 5. PARASITOLOGY. *Educ:* Tabor Col, BA, 55; Univ Kans, MA, 57; Univ Minn, PhD(zool), 62. *Prof Exp:* From instr to prof biol, Tabor Col, 57-69; assoc prof, Westminster Col, Pa, 69-73, dean col, 85-87, chmn dept, 69-85, PROF BIOL, WESTMINSTER COL, PA, 73-, CHMN DEPT, 89- *Concurrent Pos:* NSF res grants bact transformation, 64-66 & axenic cultivation, 65-67; NIH res grant, 65-68; Fulbright Scholar, 88-89; Nat Inst Oceanog, Pakistan, 88-89. *Mem:* Sigma Xi; AAAS; Am Soc Parasitologists. *Res:* Morphogenesis and nutrition of acanthocephalans; parasites of marine fishes. *Mailing Add:* Dept Biol Westminster Col New Wilmington PA 16172-0001

HARMS, JOHN CONRAD, b Albuquerque, NMex, July 29, 30; m 52; c 3. GEOLOGY. *Educ:* Columbia Univ, BA, 51; Univ Colo, PhD(geol), 59. *Prof Exp:* Geologist, Continental Oil Co, 52-54; res geologist, Marathon Oil Co, 59-80, assoc res geologist & mgr regional explor dept, 80-82; CONSULT, HARMS & BRADY GEOL CONSULTS INC, 82- *Mem:* Geol Soc Am; Am Asn Petrol Geol; Sigma Xi; Soc Explor Paleontologists & Mineralogists; AAAS. *Res:* Detrital sedimentary rocks; structural geology. *Mailing Add:* 855 Front Range Rd Littleton CO 80120

HARMS, PAUL G, b Fairbury, Ill, Nov 21, 41; m 66; c 2. REPRODUCTIVE ENDOCRINOLOGY. *Educ:* Univ Ill, BS, 63, MS, 65; Purdue Univ, PhD(animal physiol), 69. *Prof Exp:* Res physiologist, Capt USAF, USAF Sch Aerospace Med, Brooks AFB, Tex, 68-72; fel neroendocrinol, dept physiol, Univ Tex SW Med Sch, Dallas, 72-74; from assoc prof to assoc prof, 74-82, PROF PHYSIOL, DEPT ANIMAL SCI, TEX A&M UNIV, 82- *Mem:* Am Soc Animal Sci; Am Physiol Soc; Endocrine Soc; Soc Study Reprod; Int Soc Neuroendocrinol; Soc Study Fertil. *Res:* Hypothalamic pituitary control of reproduction. *Mailing Add:* Physiol Reproduction Lab Animal Sci Texas A&M Univ College Station TX 77843

HARMS, ROBERT HENRY, b Dover, Ark, Sept 27, 23; m 44; c 2. POULTRY NUTRITION. *Educ:* Univ Ark, BS, 53, MS, 54; Agr & Mech Col Tex, PhD(poultry nutrit), 56. *Prof Exp:* Pub sch teacher, Ark, 46-51; asst poultry, Univ Ark, 52-53 & Agr & Mech Col Tex, 53-55; asst prof, Univ Tenn, 55-57; from assoc prof to prof, 57-64, chmn dept poultry sci, 64-82, GRAD RES PROF, UNIV FLA, 83- *Honors & Awards:* Am Feed Mfg Award, Poultry Sci Asn, 65, Broiler Res Award, 84. *Mem:* Fel Poultry Sci Asn; Soc Exp Biol & Med; Am Inst Nutrit. *Res:* Protein, minerals and energy requirements of broilers and laying hens; vitamin A; vitamin K; unidentified factors. *Mailing Add:* Dept Poultry Sci Univ Fla Gainesville FL 32601

HARMS, VERNON LEE, b Newton, Kans, May 31, 30; m 55; c 4. SYSTEMATIC BOTANY. *Educ:* Bethel Col, Kans, BS, 55; Kans State Teachers Col, MS, 59; Univ Kans, PhD(bot), 63. *Prof Exp:* From asst prof to assoc prof, Univ Alaska, 63-69, cur herbarium, 63-69; assoc prof, 69-75, PROF BOT & CUR, FRASER HERBARIUM, UNIV SASK, 69- *Mem:* Bot Soc Am; Am Soc Plant Taxon; Int Asn Plant Taxon; Can Bot Asn. *Res:* Plant biosystematics; arctic, subarctic, and boreal phytogeography; biosystematics of Heterotheca, Petasites and Sparganium; flora of Saskatchewan; Saskatchewan rare native plants. *Mailing Add:* Fraser Herbarium Univ Sask Dept Plant Ecol Saskatoon SK S7N 0W0 Can

HARMS, WILLIAM OTTO, b Alton, Ill, Sept 27, 23; m 41; c 3. PHYSICAL METALLURGY. *Educ:* Wayne State Univ, BS, 48; Univ Minn, MS, 50, PhD(phys metall), 53. *Prof Exp:* Metallurgist, Oak Ridge Nat Lab, 53-55; assoc prof metall eng, Univ Tenn, 55-60; head ceramics lab, Metals & Ceramics Div, Oak Ridge Nat Lab,60-65, asst sect chief res & develop, 65-66, sect chief, 66-72, mgr fast reactor progs, 69-72, dir, Breeder Reactor Prog, 72-77, dir, nuclear reactor technol progs, 77-85, dir mats & structs technol mgt ctr, 79-84; RETIRED. *Concurrent Pos:* Lectr & in-charge mat course, Oak Ridge Sch Reactor Tech, 55-56; consult, Oak Ridge Nat Lab, 55-60. *Mem:* Fel Am Soc Metals Int; Am Inst Mining, Metall & Petrol Engrs; fel Am Nuclear Soc; Sigma Xi. *Res:* Engineering mechanics; properties of materials; metallurgy and metallurgical engineering; radiation effects; reactor fuels; metallography; nuclear reactor engineering; fuel element design and development; physical chemistry; phase equilibria; thermodynamics. *Mailing Add:* 4029 Hiawatha Dr Knoxville TN 37919

HARMSEN, RUDOLF, b Medemblik, Neth, Mar 11, 33; m 60; c 2. SYSTEMS BIOLOGY. *Educ:* Univ Toronto, BA, 59, MA, 60; Cambridge Univ, PhD(insect physiol), 63. *Prof Exp:* Rockefeller sr res fel parasitol, Univ Col, Nairobi, Kenya, 63-66; asst prof, 66-68, ASSOC PROF BIOL, QUEEN'S UNIV, 68- *Mem:* Entom Soc Am; Entom Soc Can; Can Soc Zoologists. *Res:* Control systems in biology; ecological and behavior genetics; integrative population biology; community structure. *Mailing Add:* Dept Biol Queen's Univ Kingston ON K7L 3N6 Can

HARMUTH, CHARLES MOORE, b Bridgeville, Pa, Dec 30, 22; m 49; c 3. ORGANIC CHEMISTRY, POLYMER CHEMISTRY. *Educ:* Duquesne Univ, BS, 49, MS, 52, PhD(org chem), 56. *Prof Exp:* Res chemist, E I Du Pont De Nemours & Co, Inc, 55-60, chem assoc. 60-61, res supvr, 61-68, sr res supvr, 68-71, tech coordr indust finishes, 71-74, res assoc, 74-85; RETIRED. *Mem:* Am Chem Soc. *Res:* Organic reactions, particularly reaction mechanisms; fiber technology involving structure and property relationships; solution polymerization of condensation polymers; polymer properties versus properties of coating and films; powder coatings; industrial finishes; automotive finishes; high solids enamel. *Mailing Add:* 100 Edgewood Rd Wilmington DE 19803

HARMUTH, HENNING FRIEDOLF, b Vienna, Austria, July 27, 28; US citizen; m 57; c 1. ELECTRICAL ENGINEERING, PHYSICS. *Educ:* Vienna Tech Univ, dipl, 51, Dr tech Sci(elec eng), 53. *Prof Exp:* Scientist, Signal Corps Eng Labs, 53-55 & Gen Elec, 55-59; sect head commun, Gen Dynamics, 59-60; researcher physics, Nat Univ Mex & Tokyo Univ, 60-61; PROF ELEC ENG, CATH UNIV AM, 72- *Concurrent Pos:* Var positions, Battelle Inst, Frankfurt, 61-65; consult electronics, 65-71; lectr, Univ Karlsruhe, 65-68; assoc prof, Univ Md, 69-71; guest lectr, Stanford Univ, 71; consult UNESCO, Warangal Univ, India, 72; guest prof, Australian Nat Univ, 73, Northwest Telecommun Eng Inst, Xi'an, China, 81, Univ Witwatesrand, SAfrica, 83, Univ Sydney, Australia, 84, 90, Kuwait Univ, 85, 89, Beijing Inst Aeronaut & Astronaut, China, 87 & Tech Univ Dresden, Ger, 88; exchange scientist, USSR Acad Sci, 79. *Mem:* Inst Elec & Electronics Engrs. *Res:* Nonsinusoidal electromagnetic waves. *Mailing Add:* Dept Elec Eng Cath Univ Am Washington DC 20064

HARN, STANTON DOUGLAS, b Pomona, Calif, Mar 14, 45; m 87; c 5. ANATOMY. *Educ:* La Verne Col, BA, 67; Univ Utah, PhD(anat), 72. *Prof Exp:* ASSOC PROF ORAL BIOL & CUR DENT MUS, COL DENT, UNIV NEBR-LINCOLN, 72-; CUR, DENT MUS. *Mem:* Am Asn Anatomists; Int Asn Dent Res; Am Acad Hist Dent; Sigma Xi; Am Asn Clin Anatomists; Am Asn Dent Schs. *Res:* Anesthesia; mandibular nerve; history of dentistry. *Mailing Add:* Dept Oral Biol Univ Nebr Col Dent Lincoln NE 68583-0740

HARNDEN, JOHN D, JR, b Schenectady, NY, Sept 20, 28; m 74; c 2. ELECTRONICS ENGINEERING. *Educ:* Union Col, BS, 50. *Prof Exp:* Develop engr electronics, 50-59, proj engr, 60-78, MGR TECHNOL ADVAN ELECTRONICS PROD, GEN ELEC CO, 79- *Mem:* Fel Inst Elec & Electronics Engrs; Am Soc Inventors; Radio Club Am. *Res:* Developing the next generation of solid state devices, with impact on new components (senses and power control) as well as new products with emphasis on consumer markets. *Mailing Add:* Gen Elec Co Corp Res & Develop One River Rd Schenectady NY 12305

HARNED, BEN KING, TOXICOLOGY, NERVOUS SYSTEM. *Educ:* Washington Univ, St Louis, PhD(biochem & pharmacol), 29. *Prof Exp:* Dir res, The Johnson Div, 30-64; RETIRED. *Res:* Drugs that affect the nervous system. *Mailing Add:* RR 8 Petersburg Rd Evansville IN 47711-3336

HARNED, HERBERT SPENCER, JR, b Philadelphia, Pa, Feb 1, 21; m 47; c 3. MEDICINE. *Educ:* Yale Univ, BS, 42, MD, 45. *Prof Exp:* Intern pediat, Johns Hopkins Hosp, 45-46; resident, Boston Children's Hosp, 48-50; asst clin prof pediat, Sch Med, Yale Univ, 52-54, asst prof, 54-58; from asst prof to assoc prof, 58-70, PROF PEDIAT, SCH MED, UNIV NC, CHAPEL HILL, 70- *Concurrent Pos:* Consult, Wake Mem Hosp, Raleigh, NC; res fel pediat cardiol, Sch Med, Yale Univ, 50-52. *Mem:* Am Pediat Soc; Am Med Asn; Am Fedn Clin Res. *Res:* Pediatric cardiology; newborn circulation. *Mailing Add:* Spring Dell Ln Chapel Hill NC 27514

HARNER, JAMES PHILIP, b Quakertown, Pa, May 1, 43; m 65; c 2. DAIRY SCIENCE. *Educ:* Del Valley Col, BS, 65; Univ Md, MS, 68; Univ Ill, PhD(dairy sci), 72. *Prof Exp:* Herd supvr, Agron Dairy Forage Res Ctr, Univ Md, 67-68; dairy cattle supt, Cornell Univ, 71-77; mgr, Meridale Farms, 77-78; asst prof dairy, dept sci & agr, Del Valley Col, 78-86; RES FARM SUPVR, RUTGERS UNIV, 86-, ASST DIR, ANIMAL CARE PROG, 90- *Mem:* Am Dairy Sci Asn. *Res:* Dairy production. *Mailing Add:* 20 Stacey Dr Doylestown PA 18901

HARNEST, GRANT HOPKINS, b Carthage, Ill, Nov 23, 16; m 45. ORGANIC CHEMISTRY. *Educ:* Knox Col, AB, 39; Middlebury Col, MS, 41; Univ Va, PhD(org chem), 46. *Prof Exp:* Res assoc, Off Sci Res & Develop contract, Univ Va, 45-46; RETIRED. *Concurrent Pos:* Bd of Vis res award, Univ, Va, 47; bd dirs. Porter Med Cgtr, 85-87. *Honors & Awards:* James Flack Norris Award, Am Chem Soc, 74. *Mem:* Am Chem Soc. *Res:* Synthetic antimalarials and analgesics; heterocyclic amino alcohols; synthesis of antimalarials. *Mailing Add:* 125 S Main Middlebury VT 05753

HARNETT, R MICHAEL, b Winnfield, La, June 3, 44; m 68; c 2. LARGE-SCALE SYSTEMS OPTIMIZATION. *Educ:* La Polytech Inst, BS, 68; Univ Ala, Huntsville, MS, 72, PhD(indust & systs eng), 74. *Prof Exp:* Engr, Northrop Space Labs, 67-70; OR analyst, Safeguard Systs Command, 70-74; from asst prof to assoc prof systs eng, Clemson Univ, 74-83; assoc dean eng, La Tech Univ, 83-88; DEPT HEAD INDUST ENG, KANS STATE UNIV, 88- *Concurrent Pos:* Res fel, Air Force Off Sci Res, 81. *Mem:* Opers Res Soc Am; Inst Indust Engrs; Am Soc Eng ducators; Soc Mfg Engrs. *Res:* Large-scale systems optimization; methods of economic analysis; biological system modelling, especially humans in cold stress situations. *Mailing Add:* 2483 Oregon Lane Manhattan KS 66506

HARNEY, BRIAN MICHAEL, b New York, NY, Jan 20, 44; m 67; c 2. OIL SHALE TECHNOLOGY. *Educ:* Manhattan Col, BS, 65; Univ Pittsburgh, PhD(phys chem), 70. *Prof Exp:* Res chemist, US Bur of Mines, 70-72, staff res coordr, 72-74; br chief, div fossil energy res, US Energy Res & Develop Admin 75-77; dir, Div Technol Assessment Off Shale Resource Applns, 78-79, dir, Off Oil Shale, US Dept Energy, 79-81; sr assoc engr, Mobil Res & Develop Corp, 81-84, MGR, GROUNDWATER PROGS, MOBIL OIL CORP, 84-, MGR POLIT & ENVIRON AFFAIRS. *Mem:* Am Chem Soc. *Res:* Synthetic fuels engineering with emphasis on eastern and western oil shale; infrared and Raman spectroscopy to determine molecular structure and to perform vibrational analyses; remote sensing of air pollution; conversion of coal and oil shale to clean liquid fuels; hazardous waste management. *Mailing Add:* 6612 Jill Court McLean VA 22101-1613

HARNEY, PATRICIA MARIE, b New York, NY, Jan 15, 25. GENETICS, HORTICULTURE. *Educ:* McGill Univ, BSc, 50, MSc, 59, PhD(genetics), 63. *Prof Exp:* Asst horticulturist, NS Dept Agr & Mkt, 50-53; lectr hort, Macdonald Col, McGill Univ, 54-63; from asst prof to assoc prof, 63-72, PROF ORNAMENTAL HORT & PLANT BREEDING, UNIV GUELPH, 72- *Mem:* Am Soc Hort Sci; Can Soc Hort Sci; Genetics Soc Can. *Res:* Phylogeny and genetics of Pelargonium x hortorum and Pelargonium x domesticum; interspecific hybridization in Brassica, using embroyo culture, anther culture and protoplast fusion; transfer of herbicide resistance from Brassica napus to Brassica oleracea; micropropagation of woody ornamentals. *Mailing Add:* Dept Hort Sci Univ Guelph Guelph ON N1G 2W1 Can

HARNEY, ROBERT CHARLES, b Pasadena, Calif, Sept 28, 49; m 72; c 1. LASER PHYSICS, OPTICAL SCIENCES. *Educ:* Harvey Mudd Col, BS(chem) & BS(physics), 71; Univ Calif, Davis, MS, 72, PhD(appl sci), 76. *Prof Exp:* Physicist laser physics, Lawrence Livermore Lab, Univ Calif, 71-76; STAFF SCIENTIST OPTICS, LINCOLN LAB, MASS INST TECHNOL, 76- *Concurrent Pos:* Fannie & John Hertz Found fel, 72-76; consult, Lawrence Livermore Lab, Univ Calif, 76-81; lectr, Appl Physics, Univ Lowell, 80- *Mem:* Am Asn Physics Teachers; Am Chem Soc; Am Phys Soc; Astron Soc Pac; Optical Soc Am. *Res:* Application of laser techniques to investigation of problems in physics, chemistry and biology; imaging infrared radar systems and studies of forbidden raman scattering from atoms. *Mailing Add:* 6852 Parson Brown Dr Orlando FL 32819

HARNSBERGER, HUGH FRANCIS, b Taizhou, Jiangsu, China, Jan 3, 24; US citizen; m 43; c 4. PHYSICAL CHEMISTRY. *Educ:* Col William & Mary, BS, 44; Univ Calif, Berkeley, MS, 47, PhD(phys chem), 50. *Prof Exp:* From asst prof to assoc prof chem, Duquesne Univ, 49-52; res chemist, Chevron Res Co, 52-57, group supvr, 57-64, sr res assoc, head catalyst res group, 64-79; proj leader, Technol Acquisition, 79-83; CATALYST & TECHNOL CONSULT, HARNSBERGER ASSOCS, INC, 83- *Concurrent Pos:* Chmn, Gordon Conf Catalysis, 72; guest scientist microanal, Mass Inst Technol, Cambridge, 66-67, Univ Louvain, 73 & Mat Panel, Proj Rev Bd, Stanford Synchrotron Radiation Lab, 83-86; founder & pres, ICI Ebonex Technol Inc, Emeryville, Calif, 83-86, dir & consult, 87- *Mem:* AAAS; Am Chem Soc; Catalysis Soc; Microbeam Anal Soc; Am Ceramic Soc; Mat Res Soc. *Res:* Kinetics of inorganic reactions; physical, chemical and surface studies of solid microporous catalysts; electron microprobe-scanning electron microscope catalyst research. *Mailing Add:* 30 Green Valley Ct San Anselmo CA 94960

HAROIAN, ALAN JAMES, b Hartford, Conn, Sept 28, 50; c 2. NEUROANATOMY. *Educ:* Drew Univ, BA, 72; St Louis Univ, PhD(anat), 78. *Prof Exp:* Instr hist & neuroanat, 78-79, sr instr gross anat & neuroanat, 79-82, ASST PROF NEUROANAT, HAHNEMANN UNIV, 82- *Mem:* Soc Neurosci; Am Asn Anatomists. *Res:* Neuroanatomical studies of the afferent and efferent connections of the rat cerebellum using autoradiographic and horseradish peroxidase techniques; developmental plasticity of the cerebellothalamic system. *Mailing Add:* Dept Anat Hahnemann Med Col 85230 N Broad St Philadelphia PA 19102

HAROLD, FRANKLIN MARCEL, b Frankfurt-on-Main, Ger, Mar 16, 29; nat US; m 54; c 1. CELL BIOLOGY, BIOENERGETICS. *Educ:* City Col, BS, 52; Univ Calif, PhD(comp biochem), 55. *Hon Degrees:* DSc, Univ Osnabr06ck, WGer, 84. *Prof Exp:* Asst, Univ Calif, 52-55; res biochemist, Nat Jewish Hosp, Denver, Colo, 59-62, chief dept exp chem, 62-68; from asst prof to assoc prof microbiol, 63-74, PROF BIOCHEM, SCH MED, UNIV COLO, DENVER, 74-; SR BIOCHEMIST, NAT JEWISH CTR, 68- *Concurrent Pos:* NSF res fel, Calif Inst Technol, 57-58, USPHS res fel, 58-59. *Mem:* Am Soc Biol Chem; Am Soc Microbiol; Am Soc Cell Biol. *Res:* Bioenergetics and physiology of microoraganisms; ion currents in relation to growth and development. *Mailing Add:* Dept Biochem Colo State Univ Ft Collins CO 80524c

HAROLD, RUTH L, b New York, NY, July 16, 31; m 54; c 1. BIOLOGY OF FUNGI. *Educ:* Univ Ariz, BA, 52; Univ Calif, Berkeley, MA, 54. *Prof Exp:* Lab technician, Calif Inst Technol & various other univs, 54-61; res assoc, 62-72, RES ASSOC, NAT JEWISH CTR IMMUNOL & RESPIRATORY MED, 78- *Mem:* Am Soc Microbiol; Am Soc Cell Biol. *Res:* Chemotropic growth in water molds; apical growth and branching in water molds and other fungal hyphae. *Mailing Add:* Dept Biochem Colo State Univ Ft Collins CO 80523

HAROLD, STEPHEN, b Regina, Sask, Nov 10, 28; m 59; c 5. BIOCHEMISTRY. *Educ:* Univ Sask, BA, 51; Univ Alta, PhD(biochem), 64. *Prof Exp:* Biochemist, Regina Gen Hosp, 52-60, clin biochemist, 63-90; RETIRED. *Mem:* Chem Inst Can; Can Soc Clin Chemists; Can Soc Lab Technologists. *Res:* Administration and teaching; methodology studies, instrumentation and work-flow as applied to clinical chemistry; investigation of disease. *Mailing Add:* 83 Sunset Dr Regina SK S4P 2R6 Can

HAROWITZ, CHARLES LICHTENBERG, b Newport News, Va, Sept 24, 26; m 48; c 3. ORGANIC CHEMISTRY. *Educ:* Va Polytech Inst, BS, 48. *Prof Exp:* From res chemist to sr chemist, Va-Carolina Chem Corp, 48-56, sect leader, 56-58, mgr org prod, 58; mgr indust chem div, Mobil Chem Co, 58-67, mkt mgr phosphorus chem & cleaners, chem div, 67-69 , gen sales mgr, Indust Chem Div, 69-74, gen mkt mgr, 74-83, mgr bus develop, 83-84; RETIRED. *Concurrent Pos:* Independent mgt consult, 84- *Mem:* Am Chem Soc; Com Develop Asn; AAAS. *Res:* Organic phosphorus compounds; urethane polymers; agricultural chemicals; metal process chemicals. *Mailing Add:* 8207 Shelley Rd Richmond VA 23229

HARP, GEORGE LEMAUL, b Butler, Mo, Oct 21, 36; m 65; c 2. ZOOLOGY, LIMNOLOGY. *Educ:* Univ Kans, AB, 58; Univ Mo, MA, 63, PhD(zool), 69. *Prof Exp:* Instr aquatic entom, Univ Mo, 67; from asst prof to assoc prof, 67-79, PROF ZOOL, ARK STATE UNIV, 79- *Concurrent Pos:* Consult, US Army Corps Engrs, 72-74, 78-80, US Soil Conserv Serv, 74-80, Ark Eastman Co, 74, Ark Power & Light Co, 78 & Frit Indust, 83-87. *Mem:* Odonatologica; NAm Benthological Soc; Brit Freshwater Biol Asn. *Res:* Ecology of altered freshwater ecosystems; taxonomy and ecology of aquatic insects. *Mailing Add:* Dept Biol PO Box 599 Ark State Univ State University AR 72467

HARP, JAMES A, MICROBIOLOGY. *Educ:* Univ Ill, BS, 67, Southern Ill Univ, MA, 69, Mont State Univ, PhD(microbiol), 83. *Prof Exp:* Microbiologist, Ill Dept Pub Health, Chicago, 69-76, Vet Res Lab, Mont State Univ, Bozeman, 77-80; fel path-immunol, Stanford Univ Med Sch, 83-84; microbiologist, 84-88, LEAD SCIENTIST, IMMUNOL RUMINANT PERINATAL DIS PROJ, NAT ANIMAL DIS CTR, AGR RES SERV, USDA, AMES, IOWA, 88- *Concurrent Pos:* Res assoc, USDA, 85 & 89. *Mem:* Am Asn Immunologists; Am Asn Vet Immunologists; Fedn Am Socs Exp Biol; Sigma Xi; Conf Res Workers Animal Dis. *Res:* Lymphocyte trafficking and communication between lymphatic and mucosal regions of the immune system in domestic animals; immunologic mechanisms of resistance and recovery from protozoan parasites, particularly cryptosparidium parvum. *Mailing Add:* Nat Animal Dis Ctr PO Box 70 Ames IA 50010

HARP, JIMMY FRANK, b Mar 2, 33; US citizen; m 58; c 2. CIVIL ENGINEERING. *Educ:* Univ Ark, BS, 55, MS, 59; Univ Ariz, PhD(civil eng), 63. *Prof Exp:* Instr eng mech, Univ Ark, 57-60; instr civil eng, Univ Ariz, 60-63; assoc prof, 63-76, PROF CIVIL ENG, UNIV OKLA, 76- *Mem:* Am Soc Civil Engrs; Am Soc Eng Educ. *Res:* Hydraulics; hydrology. *Mailing Add:* Dept Civil Eng Univ Okla 202 W Boyd Norman OK 73019

HARP, WILLIAM R, JR, b Roanoke, Va, Nov 28, 19; m 41; c 2. SPECTROSCOPY. *Educ:* Roanoke Col, BS, 41; Tulane Univ, MS, 43. *Prof Exp:* Asst physics, Tulane Univ, 41-43; molecular spectroscopist, 43-57, supvr anal res, 57-60 & spectros res, 60-61, from asst head to head anal dept, 61-65, mgr serv div, 65-69, mgr anal dept, 69-72, mgr anal chem dept, 72-76, mgr health, safety & environ, Shell Develop Co, 76-80; RETIRED. *Mem:* Am Chem Soc. *Res:* Analytical absorption spectroscopy; molecular structure determination using spectroscopic techniques. *Mailing Add:* 2210 Sewell Lane Roanoke VA 24015

HARPAVAT, GANESH LAL, b Udaipur, India, May 13, 44; m 70. FLUID MECHANICS. *Educ:* Univ Jodhpur, BE, 65; Univ Rochester, MS, 67, PhD(fluid mech), 68; Univ Dallas, MBA, 78. *Prof Exp:* Mech engr, Birla Jute Mfg Co, India, 65; teaching & res assoc mech, Univ Rochester, 65-68; assoc scientist res div, Xerox Webster Res Ctr, 68-70, scientist, 70-74, sr scientist, 75-; VIS ASSOC PROF, DEPT BUS COMPUT INFO SYST, N TEX STATE UNIV. *Mem:* Am Soc Mech Engrs; Am Inst Aeronaut & Astronaut; Inst Elec & Electronics Engrs. *Res:* Ink jet technology; science of xerographic development; transfer, fusing, and cleaning; system simulation and modeling; fluid mechanics and hydrodynamic stability; applied math; computers. *Mailing Add:* 921 Angela Dr Lewisville TX 75067

HARPELL, GARY ALLAN, b Kingston, Ont, Sept 20, 37; m 65; c 3. POLYMER SCIENCE, PHYSICAL CHEMISTRY. *Educ:* Queen's Univ, BSc, 60, MSc, 61; Univ Leeds, PhD(phys chem), 65. *Prof Exp:* Res chemist rubber, B F Goodrich Co, 65-71; mgr appln peroxide uses, Lucidol Div, Pennwalt Corp, 71-77; RES ASSOC POLYMER SCI, ALLIED CHEM CORP, 77- *Mem:* Am Chem Soc. *Res:* Free radical polymerizations and thermoset resin systems; structure property relationships of polymeric systems; the evaluation of ballistic phenomena, with particular emphasis on optimization of ballistic resisant composites and fabric structures; polymerization processes; thermoset care systems; structure property relationships of polymeric systems. *Mailing Add:* 19 Elder Dr Morristown NJ 07960

HARPENDING, HENRY COSAD, b Penn Yan, NY, Jan 13, 44; m 66; c 1. BIOLOGICAL ANTHROPOLOGY. *Educ:* Hamilton Col, AB, 64; Harvard Univ, MA, 65, PhD(anthrop), 72. *Prof Exp:* Asst prof anthrop, Yale Univ, 71-72; from asst prof to prof anthrop, Univ NMex, 72-; PROF ANTHROP, PA STATE. *Mem:* Am Asn Phys Anthrop; Biomet Soc. *Res:* Human population genetics; numerical methods in anthropology. *Mailing Add:* Dept Anthrop 409 Carpenter Bldg Pa State Univ Main Campus University Park PA 16802

HARPER, ALEXANDER MAITLAND, b Lethbridge, Alta, Mar 10, 26; m 56; c 4. ECONOMIC ENTOMOLOGY, ECOLOGY. *Educ:* Univ Alta, BSc, 48, MSc, 51; Wash State Univ, PhD(entom), 57. *Prof Exp:* Res scientist entom, Sci Serv Lab, 48-59; res scientist, 59-86, EMER SCIENTIST ENTOM, RES STA AGR CAN, 87- *Mem:* Entom Soc Am; Entom Soc Can; Can Nature Fedn. *Res:* Biology, ecology, and control of forage insects, sugar beet insects, and aphids; insect transmission of plant diseases. *Mailing Add:* 1654 Scenic Heights S Lethbridge AB T1K 1N5 Can

HARPER, ALFRED EDWIN, b Lethbridge, Alta, Aug 14, 22; nat US; m 48; c 2. BIOCHEMISTRY. *Educ:* Univ Alta, BSc, 45, MSc, 47; Univ Wis, PhD(biochem), 53. *Prof Exp:* Instr biochem, Univ Alta, 48-49, lectr, 49-52, asst prof, 52-54, NSF-Nat Res Coun res fel, Cambridge Univ, 55-56; from asst prof to assoc prof, Univ Wis, 56-61; prof nutrit, Mass Inst Technol, 61-65; PROF BIOCHEM, UNIV WIS-MADISON, 65-, PROF NUTRIT SCI & CHMN DEPT, 68- *Concurrent Pos:* Mem, comt amino acids, Food & Nutrit Bd, Nat Res Coun, Nat Acad Sci, 56-70, chmn, 63-70, dietary allowances comt, 65-74, chmn, 69-74, dietary guidelines comt, 77-, mem bd, 61-71 & 77-, chmn, 78-81; mem metab study sect, NIH, 63-67; mem nutrit training comt, Nat Inst Gen Med Sci, 67-71, chmn, 68-71; comt dietary guidelines, 77-79; chmn Food & Nutrit Bd, 78- *Honors & Awards:* Borden Award, Am Inst Nutrit, 65; Storer Lectr, Life Sci, Univ Calif, Davis, 81. *Mem:* Fel AAAS; Am Soc Biol Chem; Biochem Soc; Am Chem Soc; Am Inst Nutrit (pres, 70). *Res:* Amino acid metabolism and interrelationships; amino acid transport; control of food intake; metabolic adaptations. *Mailing Add:* Dept Biochem Univ Wis Madison WI 53706

HARPER, C(HARLES) A(RTHUR), b Ridgeley, WVa, June 21, 26; m 50; c 3. MATERIALS SCIENCE, CHEMICAL ENGINEERING. *Educ:* Johns Hopkins Univ, BChemE, 49. *Prof Exp:* Res engr, Glidden Co, Md, 49-50; mfg engr, Point Breeze Works, Western Elec Co, 52-55; mat engr aerospace div, Westinghouse Elec Corp, 55-57, sr engr, 57-63, fel engr, 63-66, prog mgr, 66-87; RETIRED. *Concurrent Pos:* Tech adv to numerous nat & int tech conferences; lectr & keynote speaker for numerous seminars & conferences. *Mem:* Inst Elec & Electronics Engrs; Soc Plastics Engrs. *Res:* Materials engineering; plastics; electronics. *Mailing Add:* PO Box 487 Lutherville MD 10193

HARPER, CHARLES WOODS, JR, b San Antonio, Tex, Jan 10, 38; m 59; c 1. INVERTEBRATE PALEONTOLOGY. *Educ:* Mass Inst Technol, SB, 59, SM, 61; Calif Inst Technol, PhD(geol), 64. *Prof Exp:* NSF fel geol, Univ Col Swansea, Wales, 64-65; vis res assoc paleont, Smithsonian Inst, 65-66; asst prof, 67-70, PROF GEOL, UNIV OKLA, 70- *Mem:* Paleont Soc; Sigma Xi. *Res:* Evolution, taxonomy and distribution in time and space of lower paleozoic brachiopods; evolutionary paleontology; quantitative methods. *Mailing Add:* Dept Geol Univ Okla 100 East Boyd St Norman OK 73019

HARPER, CURTIS, b Auburn, Ala, May 13, 37. PHARMACOLOGY. *Educ:* Tuskegee Inst, BS, 59, MS, 61; Iowa State Univ, MS, 63; Univ Mo, PhD(biochem), 69. *Prof Exp:* Fel biochem, Yale Univ, 69-70, Univ NC, 70-71; instr pharmacol, Univ NC, 71-72; sr staff fel, NIH, 72-76; ASSOC PROF PHARMACOL, UNIV NC, 76- *Mem:* Am Chem Soc; AAAS; Sigma Xi; Soc Toxicol; Am Soc Pharmacol & Exp Therapeut. *Res:* Metabolism of drugs and other chemicals by pulmonary microsomal enzymes; pulmonary toxicity of halogenated hydrocarbons; biochemical toxicology; oxidative stress and ischemic lung injury. *Mailing Add:* Dept Pharmacol Flob CB No 7365 Sch Med Univ NC Chapel Hill NC 27599

HARPER, DEAN OWEN, b Cincinnati, Ohio, Dec 1, 34; m 59; c 3. CHEMICAL ENGINEERING, POLYMER SCIENCE & ENGINEERING. *Educ:* Purdue Univ, BSChE, 56, MS, 59; Univ Cincinnati, PhD(chem eng), 67. *Prof Exp:* Jr engr, Redstone Arsenal, Thiokol Chem Corp, 56-57; asst prof chem eng, WVa Univ, 63-69; assoc prof, 69-83, PROF CHEM ENG, UNIV LOUISVILLE, 83- *Concurrent Pos:* Consult, US Bur Mines, 67-69, E I du Pont de Nemours & Co, 69, Gen Elec, Appliance Park, Louisville, 72 & 81 & Brown & Williamson Tobacco Corp, 84 & 85. *Mem:* Am Inst Chem Engrs; Am Soc Eng Educ; Soc Plastics Engrs; Fedn Soc Coatings Technol; Soc Rheology; Nat Soc Prof Engrs. *Res:* Interfacial phenomena; properties of polymers and polymer processing; rheology; fluidization; fluid dynamics; numerical solution of partial differential equations; organic paints and coatings; interstitial composite materials. *Mailing Add:* Dept Chem Eng Univ Louisville Louisville KY 40292

HARPER, DOYAL ALEXANDER, JR, b Atlanta, Ga, Oct 9, 44; m 67; c 3. ASTRONOMY. *Educ:* Rice Univ, BA, 66, PhD(space sci), 71. *Prof Exp:* Asst prof, 71-77, assoc prof astron, Univ Chicago, 71-; AT YERKES OBSERV, WILLIAMS BAY, WI. *Mem:* Am Astron Soc. *Res:* Far infrared astronomy; investigations of regions of star formation, active galactic nuclei, interstellar medium, infrared stars and planetary atmospheres. *Mailing Add:* Astron & Astrophys Box 0258 Univ Chicago 5640 Ellis Ave Chicago IL 60637

HARPER, EDWIN T, b Chicago, Ill, May 4, 35; m 57; c 3. ORGANIC CHEMISTRY, BIOCHEMISTRY. *Educ:* Grinnell Col, BA, 55; Univ Minn, PhD(org chem), 59. *Prof Exp:* Res assoc org chem, Univ Minn, 59-60; NIH fel, 62-64; asst prof chem, Clarkson Col Technol, 64-70; asst prof biochem, 70-74, ASSOC PROF BIOCHEM, MED CTR, IND UNIV-PURDUE UNIV, INDIANAPOLIS, 74- *Mem:* AAAS; Am Chem Soc; Sigma Xi. *Res:* Reaction mechanisms; mechanism of enzyme action. *Mailing Add:* Dept Biochem Med MS418 Sch Med Ind Univ 635 Barnhill Dr Indianapolis IN 46202

HARPER, ELVIN, b New York, NY, Nov 19, 30; m 76; c 2. BIOCHEMISTRY. *Educ:* City Col New York, BS, 52; Brooklyn Col, MA, 61; Albert Einstein Col Med, PhD(biochem), 66. *Prof Exp:* From instr to assoc prof biol chem, Harvard Med Sch, 68-73; assoc prof, 73-81, PROF CHEM, UNIV CALIF, SAN DIEGO, 81- *Concurrent Pos:* NIH fel, Weizmann Inst Sci, 66-67; res fel biol chem, Harvard Med Sch, 67-68, Arthritis Found fel, 69-71; USPHS res career develop award, 71-76. *Honors & Awards:* John E Fogarty Sr Int Scholar Award, 80. *Mem:* AAAS; Am Chem Soc; NY Acad Sci; Am Rheumatism Asn; Am Soc Biol Chem; Sigma Xi. *Res:* Mechanism of action of enzymes utilizing structural proteins as substrates, in particular the structural protein collagen and the enzyme collagenase. *Mailing Add:* Dept Chem D 006 Univ Calif at San Diego La Jolla CA 92093

HARPER, FRANCIS EDWARD, b Boston, Mass, Feb 28, 36; m 58; c 2. SEMICONDUCTOR ENGINEERING, ENGINEERING SCIENCE. *Educ:* Univ Calif, Berkeley, BS, 61, MS, 62, PhD(eng sci), 66. *Prof Exp:* Mem tech staff, Bell Tel Labs, Inc, NJ, 66-69; sr physicist, KEV Electronics Corp, 69-71; staff scientist, Technol Inc, 71-72; eng mgr, Solitron 72-73; supvry engr, Fairchild Semiconductor, 73-74; eng mgr, Eurosil Intermetall Gmbh, 74-76; staff consult, ITT 76-78; mgr advan projs, Rockwell Int, 78-79; staff scientist, Hughes Aircraft, 79-80; PROJ MGR, WESTERN DIGITAL, 80- *Mem:* Am Phys Soc; Inst Elec & Electronics Engrs. *Res:* Solid state research; semiconductor processing. *Mailing Add:* Microship Technol 2355 W Chandler Blvd Chandler AZ 85224

HARPER, FRANK RICHARD, b Calgary, Alta, June 22, 29; m 56. PHYTOPATHOLOGY, MYCOLOGY. *Educ:* Univ Alta, BSc, 52, MSc, 57; Iowa State Univ, PhD(plant path), 61. *Prof Exp:* Sci officer, 52-82, HEAD, PLANT PATH SECT, LETHBRIDGE RES STA, AGR CAN, 82- *Mem:* Am Phytopath Soc; Can Phytopath Soc. *Res:* Losses caused by diseases in cereal crops; epidemiology of leaf diseases of cereals. *Mailing Add:* 3509 Tenth Ave A S Lethbridge AB T1K 0A5 Can

HARPER, HAROLD ANTHONY, biochemistry; deceased, see previous edition for last biography

HARPER, HENRY AMOS, JR, b Carlsbad, NMex, Dec 7, 42; m 66; c 3. NUCLEAR ENGINEERING, REACTOR PHYSICS. *Educ:* Univ Tex, Austin, BS, 65; Tex A&M Univ, ME, 72, PhD(nuclear eng), 76. *Prof Exp:* NUCLEAR ENGR FAST REACTOR SAFETY, ARGONNE NAT LAB, UNIV CHICAGO, 76- *Mem:* Am Nuclear Soc. *Res:* Fast reactor safety research in the area of in-pile experiments such as SLSF and TREAT-Mark III; reactor physics research. *Mailing Add:* 6330 Pine Cone Dr Idaho Falls ID 83401

HARPER, JAMES ARTHUR, b Delemar, Idaho, Nov 18, 16; m 42; c 2. POULTRY SCIENCE. *Educ:* Ore State Univ, BS, 40; Pa State Univ, MS, 42. *Prof Exp:* Instr, 42, prof in charge turkey res, poultry sci dept, 42-82, EMER PROF POULTRY SCI, ORE STATE UNIV, 82- *Concurrent Pos:* Res Award, Nat Turkey Fedn, 51. *Mem:* Poultry Sci Asn. *Res:* Physiology of reproduction in turkeys; nutrition and breeding problems of turkeys. *Mailing Add:* 1545 Dixon Corvallis OR 97330

HARPER, JAMES DOUGLAS, b Havre-de-Grace, Md, Sept 9, 42; m 60; c 3. INSECT PATHOLOGY, BIOLOGICAL CONTROL. *Educ:* Univ Ill, BS, 64, MS, 65; Ore State Univ, PhD(insect path), 69. *Prof Exp:* From asst prof to prof insect path, 69-89, HEAD, DEPT ENTOM, AUBURN UNIV, 89- *Mem:* Soc Invert Path (treas, 80-82); Entom Soc Am; Int Orgn Biol Control; Am Orchid Soc. *Res:* Theoretical and applied research with insect pathogens, especially viruses, bacteria and fungi. *Mailing Add:* Dept Entom NC State Univ Box 7613 Raleigh NC 27695-7613

HARPER, JAMES EUGENE, b Syracuse, Kans, Jan 19, 40; m 64; c 2. PLANT PHYSIOLOGY, AGRONOMY. *Educ:* Kans State Univ, BS, 62, MS, 66, PhD(plant physiol), 68. *Prof Exp:* From asst prof to assoc prof, 68-82, PLANT PHYSIOL, USDA/ARS, UNIV ILL, URBANA, 68- *Concurrent Pos:* Assoc ed, Crop Sci, 81-83, tech ed, 84-86; vis scientist, Commonwealth Sci Indust Res Org, 81-82; assoc ed, Plant Physiol, 84- *Honors & Awards:* Am Soybean Asn Res Tour. *Mem:* Am Soc Plant Physiologists; fel Am Soc Agron; fel Crop Sci Soc Am. *Res:* Mineral nutrition of soybean; fertility, nutrient uptake and metabolism; nitrogen nutrition involving nitrate reductase and symbiotic nitrogen fixation. *Mailing Add:* RR 1 Box 92 St Joseph IL 61873

HARPER, JAMES GEORGE, b Fresno, Calif, Aug 23, 34; m 55; c 5. MATERIALS SCIENCE, METALLURGY. *Educ:* Univ Calif, Berkeley, BS, 56, MS, 57; Stanford Univ, PhD(mat sci), 69. *Prof Exp:* Metallurgist, Astronaut Div, Gen Dynamics Corp, 57; mem tech staff, Cent Res Labs, Tex Instruments, Inc, 57-61 & Semiconductor Res & Develop Lab, 61-62, sect head device fabrication, 62-66 & 69-72, eng mgr semiconductor prods, 73-79, strategy mgr front end processing & equip, 79-82; gen mgr, Veeco Integrated Automation, 82-89; STRATEGIC ANALYSIS, SEMATECH, 89- *Mem:* Am Inst Mining, Metall & Petrol Engrs. *Res:* Semiconductor processes; device fabrication; crystal growth; thermoelectrics; optical properties of solids; creep of metal; process automation; research and program management. *Mailing Add:* Sematech 2706 Monotopolis Austin TX 78741

HARPER, JOHN DAVID, b Toronto, Ont, Mar 21, 39; m 62; c 2. STRATIGRAPHY, SEDIMENTOLOGY. *Educ:* Univ Toronto, BASc, 61, MASc, 64; Brown Univ, PhD(geol), 69. *Prof Exp:* Teacher, Danforth Tech Inst, 61-62; res geologist, Shell Develop Co, Calif, 68-70, sr geologist, Shell Oil Co, Midland Tex, 70-73, sr geologist, Shell Can, Ltd, 73-76; vpres, Trend Explor Ltd, 76-80; MGR GEOL, CONCEPT RESOURCES, INC, 80- *Concurrent Pos:* Consult, 80- *Mem:* Am Asn Petrol Geologists; Soc Econ Paleontologists & Mineralogists; Int Asn Sedimentologists; fel Geol Soc Am; Can Soc Petrol Geologists. *Res:* Carbonate and clastic sedimentology; related paleoecology; exploration of petroleum; computer applications to geology. *Mailing Add:* 15 Balmoral Pl St Johns NF A1A 4P4 Can

HARPER, JON JAY, b Cleveland, Ohio, May 28, 41; m 67; c 2. INDUSTRIAL ORGANIC CHEMISTRY. *Educ:* Col Wooster, BA, 63; Princeton Univ, MA, 65, PhD(chem), 68 Midwest Col of Eng, MChE, 83. *Prof Exp:* Chemist, 68-73, sr res chemist, 73-89, ASSOC SR RES CHEMIST, AMOCO CHEM CORP, 89- *Mem:* Am Chem Soc; Sigma Xi. *Res:* Study processes for the preparation of polyester intermediates, especially terephthalic acid. *Mailing Add:* Amoco Chem Corp PO Box 400 Naperville IL 60566-7011

HARPER, JON WILLIAM, physiology, neuroanatomy, for more information see previous edition

HARPER, JUDITH JEAN, b Whittier, Calif, June 29, 51; m 76. BIOCHEMISTRY. *Educ:* Univ Calif, San Diego, BA, 73, MS, 77, PhD, 79. *Prof Exp:* Lab asst, 73-75, res asst chem, 75-79, RES FEL, UNIV CALIF, SAN DIEGO, 79- *Concurrent Pos:* USPHS trainee cell biol & genetics, Univ Calif, San Diego, 76-77. *Mem:* AAAS. *Res:* The regulation of the enzyme collagenase in mammalian systems; both in normal and disease states. *Mailing Add:* Dept Chem D-006 Univ Calif San Diego La Jolla CA 92093

HARPER, JUDSON M(ORSE), b Lincoln, Nebr, Aug 25, 36; m 58; c 3. EXTRUSION, FOOD ENGINEERING. *Educ:* Iowa State Univ, BS, 58, MS, 60, PhD(food technol), 63. *Prof Exp:* Instr food technol, Iowa State Univ, 58-64; res assoc food, Gen Mills, Inc, Minn, 64-68, dept head eng sci, 68-69, venture mgr nylon polymers, 69-70; prof agr & chem eng & head dept, 70-82, int pres, 89-90, VPRES RES, COLO STATE UNIV, 82- *Concurrent Pos:* Sabbatical leave, Technion, Haifa, Israel; Fulbright-Hayes Scholar, 78. *Honors & Awards:* Food Eng Award, Dairy & Food Ind Supply Asn & Am Soc Agr Engrs, 83; Cert Merit, Off Int Coop & Develop, USDA, 83; Int Award, Inst Food Technologists, 90. *Mem:* AAAS; Am Soc Agr Engrs; Inst Food Technologists; Am Inst Chem Engrs; Am Soc Eng Educ; Am Chem Soc. *Res:* Food processing including drying, thermoprocessing and extrusion; application of operations research techniques to food processes; the effect of processing on the nutritive value of food; new feeding systems including school lunch. *Mailing Add:* VPres Res Colo State Univ Ft Collins CO 80523

HARPER, KENNARD W(ATSON), b Pittsfield, Mass, July 4, 06; m 47; c 3. MECHANICAL ENGINEERING. *Educ:* Northeastern Univ, BME, 30. *Prof Exp:* Develop engr, optical & sci instruments, Am Optical Co, 38-46, chief inspector, 46-50, chief mil instrument develop, 50-56; consult engr optical-mech, Gen Elec Co, 56-65, optical design engr, 65-71; CONSULT, ITHACO, INC, 71- *Mem:* Am Soc Mech Engrs; Optical Soc Am; Am Ord Asn. *Res:* Optics. *Mailing Add:* RR 2 Box 149 Interlaken NY 14847

HARPER, KIMBALL T, b Oakley, Idaho, Feb 15, 31; m 58; c 6. BOTANY, SOIL SCIENCE. *Educ:* Brigham Young Univ, BS, 58, MS, 60; Univ Wis, PhD(bot), 63. *Prof Exp:* Range technician, US Forest Serv, 57-60; asst prof bot, Univ Utah, 63-68, assoc prof biol, 68-73; chmn dept, 73-76, PROF BOT & RANGE SCI, BRIGHAM YOUNG UNIV, 73- *Honors & Awards:* Pres Pub Serv Award, Nature Conservancy, 87. *Mem:* Fel AAAS; Brit Ecol Soc; Am Inst Biol Sci; Ecol Soc Am; Soc Range Mgt; Bot Soc Am. *Res:* Community ecology; plant reproduction biology. *Mailing Add:* Dept Bot & Range Sci Brigham Young Univ Provo UT 84602

HARPER, LAURA JANE, b Jackson, Miss, Aug 18, 14. NUTRITION, HUMAN PHYSIOLOGY. *Educ:* Belhaven Col, BS, 34; Univ Tenn, Knoxville, MS, 49; Mich State Univ, PhD(human nutrit & physiol), 56. *Hon Degrees:* DSc, Univ Helsinki, 90. *Prof Exp:* From assoc prof to prof foods & nutrit, Va Polytech Inst & State Univ, 49-80, head, Dept Home Econ, 58-60, asst dir, Res Div, 66-71, dean, Col Home Econ, 60-80; consult, Food & Agr Organ, UN, 80-86, US Int Develop, 80-86; EMER DEAN, COL HOME ECON, 80-; CONSULT, WORLD BANK, 88- *Concurrent Pos:* Mem comn home econ, Nat Asn State Univ & Land Grant Cols, 65-69; permanent coun, Int Fedn Home Econ, 70-84. *Mem:* Am Home Econ Asn; Am Dietetic Asn; Inst Food Technologists; Am Asn Univ Women; Int Fedn Home Economics; Sigma Xi. *Res:* Infant, child and adult nutrition; published three books for Food & Agr Organ, UN. *Mailing Add:* 100 Sunset Blvd Blacksburg VA 24060

HARPER, LAURENCE RAYMOND, JR, b Atlanta, Ga, Dec 10, 29; m 56; c 4. MATHEMATICS. *Educ:* Talladega Col, AB, 50; Univ Chicago, SM, 53, PhD(math), 59. *Prof Exp:* Instr, 56-59, ASST PROF MATH, UNIV MINN, MINNEAPOLIS, 59- *Res:* Structure theory of non-associative algebras; number theory. *Mailing Add:* Dept Math Univ Minn 206 Church St SE Minneapolis MN 55455-0487

HARPER, LAWRENCE HUESTON, b Hemet, Calif, Aug 24, 38; m 90; c 3. MATHEMATICS, THEORY. *Educ:* Univ Calif, Berkeley, BA, 61; Univ Ore, MA, 63, PhD(math), 65. *Prof Exp:* Res assoc math, Rockefeller Univ, 65-66, asst prof, 66-70; assoc prof, 70-80, PROF MATH, UNIV CALIF, RIVERSIDE, 80- *Concurrent Pos:* Part time employee, Commun Res Div, Jet Propulsion Lab, 70-74. *Mem:* Am Math Soc. *Res:* Combinatorial optimization using combinatorial; algebraic and analytic methods; complexity theory in computer science. *Mailing Add:* Dept Math Univ Calif 900 University Ave Riverside CA 92521

HARPER, MICHAEL JOHN KENNEDY, b London, Eng, Feb 25, 35; US citizen; m 85; c 4. REPRODUCTIVE PHYSIOLOGY, ENDOCRINOLOGY. *Educ:* Cambridge Univ, BA, 57, MA, 61, PhD(reproductive physiol), 62, ScD, 79; Univ Reading, 57-58, dipl, 58; Univ Tex, San Antonio, MBA, 84. *Prof Exp:* Tech officer, Pharmaceut Div, Imp Chem Industs, Ltd, Eng, 61-64; staff scientist, Worcester Found Exp Biol, 64-65; tech officer, Pharmaceut Div, Imp Chem Industs, Ltd, 65-66; staff scientist, Worcester Found Exp Biol, 66-68, sr scientist, 68-70, prog dir prostaglandin res, 70-72; med officer, Human Reprod Unit, WHO, Switz, 72-75; found scientist, Southwest Found Res & Educ, San Antonio, 75-76; assoc prof, 75-81, PROF OBSTET, GYNEC & PHYSIOL, & CHIEF DIV REPRODUCTIVE BIOL, DEPT OBSTET & GYNEC, UNIV TEX HEALTH SCI CTR, SAN ANTONIO, 81- *Concurrent Pos:* Lectr, Clark Univ, 71; mem med adv comt int fertil res prog, Carolina Pop Ctr, 71; mem steering comt, 75-81, chmn Spec Prog, Human Reprod Task Force, 83, Task Force, Postovulatory Methods Birth Control, WHO, 84-87; Nat Inst Child Health & Human Develop res career develop award, 69-72; mem Nat Inst Child Health & Human Develop, Contraceptive Develop Contract Review, 77-78; consult, US Agency Int Dev, Contraceptive Dev Br, Nat Inst Child Health & Human Develop; grant reviewer, NIH, USAID, NSF, & USDA; manuscript reviewer, Biol Reprod, J Reprod & Fertil, J Med Chem, Am J Physiol, Fertil & Steril Endocrinology, Prostaglandings, J Lipid Res, J Neurochem & J Cellular Biochem; mem adv comt, Family Health Int, 86- *Honors & Awards:* Res Career Develop Award, Nat Inst Child Health & Human Develop, 69-72. *Mem:* Soc Study Reprod; Am Fertil Soc; Endocrine Soc; Am Asn Anat; Brit Soc Study Fertil; Am Physiol Soc; Soc Gynec Invest; Soc Study Reprod; Am Pub Health Asn. *Res:* Ovum development and uterine preparation for blastocyst implantation in animals and primates; prostaglandins as regulators of reproductive function; development of contraceptive agents; platelet-activating factor in reproduction; two US and one UK patent. *Mailing Add:* Dept Obstet & Gynec Univ Tex Health Sci Ctr San Antonio TX 78284-7836

HARPER, PAUL ALVA, b Watertown, Conn, Sept 18, 04; m 52; c 3. MATERNAL & CHILD HEALTH. *Educ:* Dartmouth Col, AB, 26; Yale Univ, MD, 31; Johns Hopkins Univ, MPH, 47. *Prof Exp:* Instr pediat, Sch Med, Yale Univ, 34-35, clin instr, 35-42; assoc prof pub health admin, 47-51, prof, 51-65, assoc prof pediat, Sch Med, 59-75, EMER PROF MATERNAL

& CHILD HEALTH & POP DYNAMICS, SCH HYG & PUB HEALTH, JOHNS HOPKINS UNIV, 75- *Mem:* Am Pediat Soc; Am Pub Health Asn; AMA; Am Acad Pediat. *Res:* Study of prematurely born children; population problems. *Mailing Add:* 21 Elmwood Rd Baltimore MD 21210

HARPER, PAUL VINCENT, b Chicago, Ill, July 27, 15; m 39; c 4. SURGERY. *Educ:* Harvard Univ, AB, 39, MD, 41. *Prof Exp:* From instr to assoc prof, 49-60, prof surg, 60-80, prof radiol, 72-80, PROF SURG & MEM STAFF, FRANKLIN MCLEAN MEM RES INST, UNIV CHICAGO, 80- *Concurrent Pos:* Assoc dir, Argonne Cancer Res Hosp, 63-67. *Mem:* AAAS; Soc Exp Biol & Med; Soc Nuclear Med; Am Physiol Soc; Radiation Res Soc. *Res:* Applications of the methods of nuclear medicine of problems in clinical and experimental surgery. *Mailing Add:* 950 E 59th St Chicago IL 60637

HARPER, PIERRE (PETER) PAUL, b Masson, Que, Sept 4, 42; m 67; c 1. AQUATIC ENTOMOLOGY. *Educ:* Laval Univ, BA, 63; Univ Montreal, BSc, 66, MSc, 67; Univ Waterloo, PhD(biol), 71. *Prof Exp:* From asst prof to assoc prof, 70-75, dir biol sta, 75-86, PROF BIOL, UNIV MONTREAL, 81- *Concurrent Pos:* Fel, Paul Sabatier Univ, France, 71; consult, Can Agency Int Develop, 75-77; sabbatical leave, Ore State Univ, Corvallis, 80. *Mem:* Entom Soc Can; NAm Benthological Soc. *Res:* Ecology of stream ecosystems particularly with regard to the benthic fauna; taxonomy and ecology of aquatic insects. *Mailing Add:* Dept Biol Sci Univ Montreal Montreal PQ H3C 3J7 Can

HARPER, RICHARD ALLAN, b Anderson, Ind; m 64; c 1. BIOPHYSICS, PHYSICS. *Educ:* Univ Chicago, SB, 63; NY Univ, MS, 65, PhD(physics), 70. *Prof Exp:* Physicist & asst res, NIH, 62-63, scientist & asst hosp spec surg, Res Lab, 63-71; asst prof, 71-76, ASSOC PROF PHYSICS, RENSSELAER POLYTECH INST, 76- *Mem:* AAAS; Am Crystallog Asn; Int Asn Dent Res; Sigma Xi. *Res:* X-ray scattering; biological structure; biological mineralization. *Mailing Add:* 171 N Lake Ave Troy NY 12180

HARPER, RICHARD WALTZ, b Portland, Ore, July 21, 43; m 67; c 3. SYNTHETIC ORGANIC CHEMISTRY. *Educ:* Willamette Univ, BA, 65; Ore State Univ, MS, 68; Mass Inst Technol, PhD(org chem), 73. *Prof Exp:* Org chemist, 67-69, SR ORG CHEMIST, ELI LILLY & CO, 73- *Mem:* Am Chem Soc; Royal Soc Chem. *Res:* Synthesis of novel, biologically-active heterocyclic compounds. *Mailing Add:* DROP 0444 Lilly Res Labs Lilly Corp Ctr Indianapolis IN 46285

HARPER, ROBERT JOHN, JR, b Savannah, Ga, Aug 16, 30; m 56; c 6. CELLULOSE CHEMISTRY, TEXTILE CHEMISTRY. *Educ:* Fordham Univ, BS, 52; Ohio State Univ, PhD(org chem), 57. *Prof Exp:* Res chemist organo/metallic chem, Ethyl Corp, Baton Rouge, 58-62; res chemist fluoroaromatic chem, Mat Lab, Wright-Patterson AFB, Ohio, 62-63; res chemist, 63-73, res leader, cellulose chem, 73-85, LEAD SCIENTIST, SOUTHERN REGIONAL RES CTR, 85- *Mem:* Am Chem Soc; Am Asn Textile Chemists & Colorists. *Res:* Cellulose and textile chemistry with emphasis on durable press, flame retardant, smolder resistance, speciality dyeing and weather resistance; organo metallics; organo synthesis; flouroaromatic chemistry. *Mailing Add:* 4701 Page Dr Metairie LA 70001

HARPER, VERNE LESTER, b Monroe, SDak, Aug 13, 02; m 27; c 1. FOREST MANAGEMENT, FOREST RESEARCH ADMINISTRATION. *Educ:* Univ Calif, BS, 26, MS, 27; Duke Univ, PhD(forestry), 43. *Hon Degrees:* DSc, NC State Univ, 67. *Prof Exp:* Asst, Univ Calif, 26-27; jr forester, Southern Forest Exp Sta, US Forest Serv, 27-29, leader field studies, 29-35, chief field div forest mgt res, 35-37, asst chief div silvics, US Forest Serv, Washington, DC, 37-43, chief div forest econ, 43-45, dir Forest Serv, Northeast Exp Stat, 45-51, dep chief chg res, Washington, DC, 51-66; prof, 66-73, EMER PROF FORESTRY, UNIV FLA, 73- *Concurrent Pos:* Mem permanent comt, Int Union Forest Res Orgn, 56-61, vpres, 61-67; chmn World Forestry Comt, Soc Am Foresters, 56-67; chmn Latin Am forest res comt, UN Food & Agr Orgn, 58-61; pres, Int Union Soc Foresters, 69-75. *Honors & Awards:* Distinguished Serv Award, USDA, 61; Fernow Int Forestry Award, Am Forestry Asn, 65; Forest Farmers Asn Award, 68. *Mem:* Fel Soc Am Foresters; Am Forestry Asn; fel Int Union Soc Foresters; hon mem Int Union Forestry Res Orgn. *Res:* Silviculture; forest economics; naval stores production. *Mailing Add:* 1812 SW Sixth Terr Gainesville FL 32601-8406

HARPER, WILLIS JAMES, b Lafayette, Ind, June 9, 23; m 45; c 3. DAIRYING. *Educ:* Purdue Univ, BS, 46; Univ Wis, MS, 47, PhD(dairy indust), 49. *Prof Exp:* From asst prof to prof dairy tech, Ohio State Univ, 49-67, assoc prof biochem & molecular biol, 67-71, prof biochem, 71-80; AT NEW ZEALAND DAIRY RES INST. *Honors & Awards:* Borden Award, 46, 58. *Mem:* Am Chem Soc; Am Dairy Sci Asn. *Res:* Treatment of food plant wastes, biochemistry of food fermentations; food protein characterization and functionality; membrane processing of fluid food systems. *Mailing Add:* 3563 S Old 3-C Rd Galena OH 43021-9518

HARPOLD, MICHAEL ALAN, b Charleston, WVa, June 24, 40; c 2. BIO-ORGANIC CHEMISTRY. *Educ:* WVa State Col, BS, 63; Univ NC, Chapel Hill, PhD(org chem), 67. *Prof Exp:* NIH fel, Lab Chem Biodynamics, Univ Calif, Berkeley, 67-68; res chemist, Res & Develop Dept, Union Carbide Tech Ctr, 68-73, mkt mgr biochem intermediates, 73-77, sr group leader, Corp Res Dept, 77-78, mgr systs commercialization, Med Prod Div, Union Carbide Corp, 78-79, dir Res & Develop, 79-81; plant mgr, Baker Instruments Corp, 81-82; dir, process technol, Technicon Instruments Corp, 82-86, dir, clin biochem, 86-89; DIR, QUAL ASSURANCE DEVELOP, MILES, INC, 89- *Mem:* Am Chem Soc; NY Acad Sci; AAAS; Am Asn Clin Chem; Am Soc Qual Control. *Res:* Management of clinical chemistry research and development activities. *Mailing Add:* 56526 County Rd 19 Bristol IN 46507

HARPP, DAVID NOBLE, b Albany, NY, Jan 20, 37; m 61; c 2. CHEMISTRY. *Educ:* Middlebury Col, AB, 59; Wesleyan Univ, MA, 62; Univ NC, PhD(chem), 65. *Prof Exp:* Fel org chem, Cornell Univ, 65-66; from asst prof to assoc prof, 66-75, PROF CHEM, McGILL UNIV, 75- *Honors & Awards:* Union Carbide Award Chem Educ, 82; Chem Mfrs Asn Nat Award. *Mem:* Am Chem Soc; Can Soc Chem; AAAS; The Chem Soc. *Res:* Organic synthesis; sulfur and silicon compounds; new methods; teaching innovations. *Mailing Add:* Dept Chem McGill Univ Montreal PQ H3A 2T5 Can

HARPST, JERRY ADAMS, b Glasgow, Ky, Sept 27, 36; m 61; c 2. PHYSICAL BIOCHEMISTRY. *Educ:* Wabash Col, AB, 58; Yale Univ, MS, 60, PhD(phys chem), 62. *Prof Exp:* USPHS fel, Univ Calif, San Diego, 62-63, res assoc traineeship, 63-64, NIH fel biophys chem, 62-64; asst prof, 65-71, ASSOC PROF BIOCHEM, SCH MED, CASE WESTERN RESERVE UNIV, 71- *Concurrent Pos:* USPHS career develop award, 67-77; vis scientist, Nat Inst Med Res, Mill Hill, London, 71-72; vis assoc prof biochem & biophys, Ore State Univ, 80-81. *Mem:* Am Chem Soc; Am Soc Biol Chemists; Biophys Soc. *Res:* Physical biochemistry of DNA, proteins and DNA-protein interactions; DNA-protein interactions; structure-function relationships of biological macromolecules; nucleoprotein structure and DNA replication in adenovirus. *Mailing Add:* 3077 Huntington Rd Shaker Heights OH 44120

HARPSTEAD, DALE D, b Sioux Falls, SDak, Sept 10, 26; m 48; c 5. GENETICS, PLANT BREEDING. *Educ:* SDak State Col, BSc, 50, MS, 53; Univ Nebr, PhD(genetics), 61. *Prof Exp:* Plant breeder, SDak State Col, 53-61; assoc geneticist, Rockefeller Found, 61-69; chmn dept crop & soil sci, 69-84, PROF CROP & SOIL SCI, MICH STATE UNIV, 85- *Concurrent Pos:* Mem bd, Int Food & Agr Develop, Dept State, Washington, DC, 84-86, mem, Coun Agr Sci & Technol. *Mem:* Am Soc Agron; Crop Sci Soc Am. *Res:* Improved yield; quality and adaptation of maize; world food supply, agricultural development. *Mailing Add:* Dept Crop & Soil Sci Mich State Univ East Lansing MI 48824

HARPSTEAD, MILO I, b Wilmot, SDak, Sept 28, 30; m 63. SOILS. *Educ:* SDak State Col, BSc, 53, MSc, 57; Univ Minn, PhD(soils), 62. *Prof Exp:* Instr soils, Univ Minn, 57-60; PROF SOIL SCI, UNIV WIS-STEVENS POINT, 61- *Concurrent Pos:* Prof, Univ Ife, Nigeria, 68-70. *Res:* Soil genesis, morphology and fertility. *Mailing Add:* Dept Soil Sci Univ Wis Stevens Point Stevens Point WI 54481

HARPSTER, JOSEPH, b Sewickley, Pa, June 29, 32; m 75; c 4. APPLIED PHYSICS, ELECTRICAL ENGINEERING. *Educ:* Geneva Col, BS, 59; Case Western Reserve Univ, MS, 64; Ohio State Univ, PhD(nuclear eng), 71. *Prof Exp:* Plant engr phys chem, Ashland Oil, Freedom, Pa, 54-59; proj leader solid state physics, Harshaw Chem Co, Cleveland, 60-62; res assoc physics, Horizons Inc, 62-63; mgr solid state physics, Ohio Semiconductors Inc, Columbus, 63-65; lab supvr appl physics, Ohio State Univ, 65-72; vpres res & develop appl sci, Ohio Semitronics Inc, Columbus, 72-77; PRES APPL SCI, INTEK INC, COLUMBUS, 77- *Concurrent Pos:* Comt mem, Am Soc Testing & Mat, 63-65. *Honors & Awards:* C E McCartney Sci Award, Geneva Col, 59; Apollo Soyuz Medallion Award, NASA, Marshall Space Flight Ctr, 76. *Mem:* Am Phys Soc; Am Asn Small Businesses; Sigma Xi; Instrument Soc Am. *Res:* Industrial process controls; fossil fuel heating; applied electronics; space manufacturing; experimentation hardware; semiconductor materials; applied chemistry; military equipment; ocean wave energy conversion; thermoelectric systems; electrooptics; electric power monitoring. *Mailing Add:* 11450 Overbrook Lane Galena OH 43021

HARPSTER, ROBERT E, b Olney, Ill, Sept 25, 30; m. CLAY MINERALOGY, SEISMIC GEOLOGY. *Educ:* Beloit Col, BS, 52; Univ Tex, MA, 57. *Prof Exp:* Proj geologist, Bechtel Corp, 56-57; sr proj eng geologist, Calif Dept Water Resources, 57-71; from sr res scientist to vpres & dir qual assurance, Woodward-Clyde Consult, 71-87; MGT ASSESSMENT & QUAL ASSURANCE CONSULT, MACTEC, 87- *Concurrent Pos:* Instr earthquake anal & soil eng, Antelope Valley Community Col, 70-72; mem, US Comt Large Dams; tech fel, Woodland-Clyde Consult; res grant, US Geol Survey. *Mem:* Fel Geol Soc Am; Am Soc Civil Engrs; Am Asn Petrol Geologists; Am Soc Qual Control; Clay Mineral Soc. *Res:* Study of the earth with respect to seismic events (seismic geology); study of the earth and younger materials to assess relative activity of faulting (neotectonics). *Mailing Add:* 107 Starview Court Oakland CA 94618

HARPUR, ROBERT PETER, b Marton, NZ, Dec 20, 21; nat Can; m 47; c 3. BIOCHEMISTRY. *Educ:* Univ NZ, BSc, 42; McGill Univ, MSc, 47, PhD(biochem), 49. *Prof Exp:* Asst prof, MacDonald Col, McGill Univ, 50-71, assoc prof, Inst Parasitol, 71-81; RETIRED. *Mem:* Am Chem Soc; Can Biochem Soc. *Res:* Biochemistry of parasites. *Mailing Add:* Macleod Cresent PO Box 1719 Alexandria ON K0C 1A0 Can

HARR, ROBERT DENNIS, b Astoria, Ore, June 26, 41; m 68; c 1. HYDROLOGY. *Educ:* Wash State Univ, BS, 63; Colo State Univ, PhD(watershed mgt), 67. *Prof Exp:* Sr res scientist, Pac Northwest Labs, Battelle Mem Inst, 69-71; asst prof forest hydrol, Ore State Univ, 71-88, hydrologist, US Forest Serv, 74-88; PROF FORESTRY, UNIV WASH, 88- *Mem:* Soc Am Foresters; Am Geophys Union. *Res:* Water use by phreatophytes; subsurface flow of water in steep forested areas. *Mailing Add:* Forest Resources Mail Stop AR10 Univ Wash Seattle WA 98195

HARRACH, ROBERT JAMES, b Holyoke, Colo, Mar 30, 37; m 62; c 2. LASER PHYSICS, ATOMIC & MOLECULAR PHYSICS. *Educ:* Dartmouth Col, AB, 60; Univ Colo, MS, 62, PhD(physics), 65. *Prof Exp:* Res physicist, Nat Bur Standards, Boulder, 62-66; SR PHYSICIST, LAWRENCE LIVERMORE NAT LAB, 66 - *Concurrent Pos:* Lectr, Univ Calif Davis, Livermore, 69-74; vis assoc prof physics, Dartmouth Col, 78-79. *Mem:* Am Phys Soc; Am Asn Physics Teachers. *Res:* Lasers, laser-matter interactions; atomic beam spectroscopy; fluid dynamics; radiation transport. *Mailing Add:* Lawrence Livermore Nat Lab L-297 PO Box 808 Livermore CA 94550

HARRAR, JACKSON ELWOOD, b Boulder, Colo, June 8, 30; m 61; c 4. ELECTROCHEMISTRY, INSTRUMENT DESIGN. *Educ:* Purdue Univ, BS, 52; Univ Wash, PhD(chem), 58. *Prof Exp:* Asst, Los Alamos Sci Lab, 54; CHEMIST & GROUP LEADER, LAWRENCE LIVERMORE NAT LAB, UNIV CALIF, 58- *Concurrent Pos:* US Army Arctic Test Team, 54-56. *Honors & Awards:* Weapons Award, Dept Energy, 85. *Mem:* Electrochem Soc; Am Chem Soc; Sigma Xi. *Res:* Electroanalytical chemistry including polarography, potentiometric titrations, coulometry and mechanisms of electrode processes; automated measurement instrument design; geothermal chemistry; corrosion measurements; electrosynthesis; field analytical chemistry. *Mailing Add:* Analytical Chem Sect L-310 Lawrence Livermore Nat Lab Box 808 Livermore CA 94550

HARRAWOOD, PAUL, b Akin, Ill, Aug 28, 28; m 53; c 1. CIVIL ENGINEERING, HYDRAULICS. *Educ:* Univ Mo, Rolla, BS, 51, MS, 56; NC State Univ, PhD(civil eng), 67. *Prof Exp:* Instr civil eng, Univ Mo, Rolla, 54-56; from instr to asst prof, Duke Univ, 56-67; assoc prof & assoc dean, 67-70, dir eng sci div, 67-71, actg dean, 70-71, assoc dean & dep sch eng, 71-79, PROF CIVIL ENG, VANDERBILT UNIV, 70-, DEAN SCH ENG, 79- *Concurrent Pos:* Test engr exp stress anal, McDonnell Aircraft Corp, 57; engr construct mgt, US Army Corps Eng, 58. *Mem:* AAAS; Am Soc Civil Engrs; Soc Am Mil Engrs; Am Soc Eng Educ; Am Asn Higher Educ; Sigma Xi. *Res:* Fluid mechanics, steady and unsteady flow; open channel flow; effects of construction on hydrologic environment; environmental propulsion systems. *Mailing Add:* Sta B Box 1607 Vanderbilt Univ Nashville TN 37235

HARRELL, BYRON EUGENE, b St Paul, Minn, Feb 1, 24; m 55. ZOOLOGY. *Educ:* Univ Minn, BA, 47, MA, 51, PhD(zool), 59. *Prof Exp:* Instr zool, Univ Minn, 54; from asst prof to assoc prof, 57-63, PROF ZOOL, UNIV SDAK, 63-, CUR ZOOL, W H OVER DAKOTA MUS, 70- *Concurrent Pos:* Dir, Acad Year Inst, NSF, 64-65. *Mem:* AAAS; Am Soc Zool; Ecol Soc Am; Soc Study Evolution; Soc Syst Zool. *Res:* Ornithology; ecology; biogeography of middle America. *Mailing Add:* Dept Biol Univ SDak Univ 414 Clark St Vermillion SD 57069

HARRELL, GEORGE T, b Washington, DC, June 16, 08; m; c 2. MEDICINE. *Educ:* Duke Univ, AB, 32, MD, 36. *Hon Degrees:* LLD, Duke Univ, 83; DSc, Univ Fla, 80, Georgetown Univ, 83. *Prof Exp:* Resident med & path, Duke Hosp, 36-41, instr med, Duke Med Sch, 39-41; from asst prof to res prof med, Bowman Gray Sch Med, Wake Forest Univ, 41-53, dir, Dept Internal Med, 43-52; dean & prof med, Col Med, Univ Fla, Gainesville, 54-64; dean, prof med & dir, 64-72, vpres med sci, 72-73, EMER VPRES, MILTON S HERSHEY MED CTR, PA STATE UNIV, HERSHEY, 73- *Concurrent Pos:* Assoc prof prev med, Bowman Gray Sch Med, Wake Forest Univ, 41-53; emer dean, Col Med, Univ Fla, Gainesville, 64-; mem adv coun & panels, Nat Acad Sci, USPHS, Vet Admin & others. *Honors & Awards:* Abraham Flexner Medal, Am Asn Med Clins, 73. *Mem:* AAAS; Sigma Xi; Am Soc Clin Invest. *Res:* Infectious diseases; radioisotopes; chemotherapy; fluid balance; myxedema; facilities for medical education; animal care; over 300 publications. *Mailing Add:* 2010 Eastridge Rd Timonium MD 21093

HARRELL, JAMES W, JR, b Rose Hill, NC, July 14, 42; m 64; c 2. NUCLEAR MAGNETIC RESONANCE, MAGNETIC MATERIALS. *Educ:* Univ NC, Chapel Hill, BS, 64, PhD(physics), 69. *Prof Exp:* Asst prof, Univ NDak, 69-75, assoc prof physics, 75-80; assoc prof, 80-88, PROF PHYSICS, UNIV ALA, 88- *Mem:* Am Phys Soc; Inst Elec & Electronics Engrs Magnetics Soc. *Res:* Applications of pulse nuclear magnetic resonance to study of molecular motion in solids and liquids; characterization of particulate magnetic media for information storage. *Mailing Add:* Dept Physics & Astron Univ Ala Tuscaloosa AL 35487-0324

HARRELL, JERALD RICE, b Greenup, Ky, Jan 21, 35; m 56; c 2. POLYMER CHEMISTRY. *Educ:* Univ Louisville, BS, 56, PhD(chem), 59. *Prof Exp:* Res chemist, Film Dept, 59-64, res chemist, Elastomer Chems Dept, 64-80, RES ASSOC, POLYMER PRODS DEPT, E I DU PONT DE NEMOURS & CO, 80- *Mem:* Am Chem Soc; Res Soc Am; Sigma Xi. *Res:* Radiation chemistry; films; elastomers; polymerization. *Mailing Add:* 2427 Granby Dr Wilmington DE 19810

HARRELL, RONALD EARL, b Sanford, Fla, Nov 17, 44; m 66; c 1. PURE MATHEMATICS. *Educ:* Univ Maine, BA, 66; Univ Md, MA, 69, PhD(math), 71. *Prof Exp:* ASST PROF MATH, ALLEGHENY COL, 71- *Mem:* Am Math Soc; Math Asn Am. *Res:* Geometry in Banach spaces and how it is related to the classification of Banach spaces; biomathematics, specifically, ecosystem models and population dynamics. *Mailing Add:* Dept Math Allegheny Col Meadville PA 16335

HARRELL, RUTH FLINN, mental retardation, behavioral science; deceased, see previous edition for last biography

HARRELL, T GIBSON, b Charleston, SC, June 6, 16; m 49; c 2. ANALYTICAL CHEMISTRY. *Prof Exp:* Technician chem, R J Reynolds Industs, Inc, 41-56, from jr chemist to chemist, 56-66, group leader chem, Res Dept, 66-78; RETIRED. *Mem:* Am Chem Soc. *Res:* General analytical chemistry; nonaqueous titration; spectroscopy; food chemistry; ion-selective electrodes. *Mailing Add:* 2535 Amesbury Rd Winston-Salem NC 27103

HARRELL, WILLIAM BROOMFIELD, b New Orleans, La, Sept 18, 28; m 57; c 2. MEDICINAL CHEMISTRY. *Educ:* Univ Wash, BS, 49; Univ Tex, MS, 53; Ore State Univ, PhD(pharmaceut chem), 66. *Prof Exp:* Asst prof pharm, 49-53, assoc prof pharmaceut chem, 53-68, PROF MED CHEM, TEX SOUTHERN UNIV, 68- *Mem:* Am Chem Soc; Am Pharmaceut Asn. *Res:* Synthesis and pharmacological screening of indolizines in a search for new medicinal agents with special emphasis on Mannich bases derived from indolizines. *Mailing Add:* Dept Fac Assembly & Senate Tex Southern Univ Old President's House 3100 Cleburne Ave Houston TX 77004

HARRELL, WILLIAM KNOX, bacteriology, for more information see previous edition

HARREN, RICHARD EDWARD, b New York, NY, Dec 14, 22; m 49; c 5. CHEMISTRY. *Educ:* Queen's Col, NY, BS, 47. *Prof Exp:* Chemist prod & develop, Nat Starch Prod, Inc, NY, 47-49; with tech sales, Mallinckrodt Chem Co, 49-50; chemist prod develop, Randolph Prod Co, 50-55; lab head, Rohm & Haas Co, 55-64, res supvr polymer res & develop, 64-73, dir coatings res, 73-87; RETIRED. *Mem:* Am Chem Soc. *Res:* Solution, dispersion and solution high polymers for coatings, floor polishes, cement, additives. *Mailing Add:* 68 Hillcrest Dr Doylestown PA 18901

HARRENSTIEN, HOWARD P(AUL), b Kansas City, Mo, Jan 14, 31; m 52; c 4. STRUCTURAL ENGINEERING. *Educ:* Kans State Univ, BS, 53, BArch, 54; Iowa State Univ, MS, 56, PhD(theoret & appl mech), 59. *Prof Exp:* Asst appl mech, Kans State Univ, 53-54; asst theoret & appl mech, Iowa State Univ, 54-55, instr, 55-56, from instr to asst prof civil eng, 56-60; from assoc prof to prof, Univ Ariz, 60-67; prof, assoc dean & dir, Ctr Eng Res, Univ Hawaii, 67-77; PROF, DEPT ENG, UNIV MIAMI, 77- *Mem:* Int Asn Shell Struct; Am Soc Civil Engrs; Am Concrete Inst; Am Soc Eng Educ; Sigma Xi. *Res:* Expression of structure in architecture; structural research in the areas of plate and shell structures; development of models and model testing laboratories; shelter research in civil defense. *Mailing Add:* 618 Geronimo Coral Gables FL 33146

HARRIES, HINRICH, b Berlin, Ger, June 29, 28; m 61; c 2. ECOLOGY, SOIL SCIENCE. *Educ:* Rutgers Univ, MSc, 60, PhD(soil sci), 65. *Prof Exp:* Instr biol, Dartmouth Col, 62-63; asst prof, 63-69, ASSOC PROF BIOL, MT ALLISON UNIV, 69- *Concurrent Pos:* Nat Res Coun Can operating res grant, 67-68; Can Wildlife Serv res contract, 67-68. *Mem:* Can Wildlife Fedn; Ecol Soc Am; Can Bot Asn. *Res:* Ecosystem ecology of forests and wetlands, with emphasis on soil-vegetation interrelationships, humus studies and soil genesis. *Mailing Add:* Dept Biol Mt Allison Univ Sackville NB E0A 3C0 Can

HARRIES, WYNFORD LEWIS, b Llannon, Wales, June 1, 23; nat US; m 58. PHYSICS. *Educ:* Univ Wales, BSc, 49; Oxford Univ, DPhil(physics), 53. *Prof Exp:* Res assoc biophys, Mass Inst Technol, 53-54; asst dir res, Transistor Prod, Inc, Clevite Corp, Mass, 54-56; sr proj engr, Phys Sci & Mat Lab, Int Tel & Tel Corp, 56-60; mem res staff, Plasma Physics Lab, Princeton Univ, 60-70; PROF PHYSICS, OLD DOMINION UNIV, 70- *Concurrent Pos:* Consult, Sch Ophthal, NY Col Med & Int Tel & Tel Corp. *Mem:* Fel Am Phys Soc; sr mem Inst Elec & Electronics Engrs; fel Brit Inst Physics; fel Brit Inst Elec Engrs. *Res:* Electrical discharges in gases; semiconductors; plasma physics; energy conversion; lasers. *Mailing Add:* 258 N Blake Norfolk VA 23505

HARRILL, INEZ KEMBLE, b Carrollton, Mo, May 19, 17; m 42. NUTRITION, BIOCHEMISTRY. *Educ:* Univ Mo, Columbia, BS, 49, MS, 50; Cornell Univ, PhD(nutrit), 55. *Prof Exp:* Res asst nutrit, Univ Mo, Columbia, 49-50, instr food & nutrit, 50-52; res asst nutrit, Univ Wis, Madison, 52-53; from asst prof to assoc prof, 55-70, PROF NUTRIT, COLO STATE UNIV, 70- *Mem:* Fel AAAS; Am Inst Nutrit; Am Chem Soc; Inst Food Technologists; Am Dietetic Asn. *Res:* Nutritional status of aged. *Mailing Add:* Colo State Univ PO Box 665 Ft Collins CO 80522

HARRILL, ROBERT W, b Denver, Colo, Sept 30, 41; m 75; c 2. ENVIRONMENTAL SCIENCE, GLOBAL CHANGE ISSUES. *Educ:* Grinnell Col, BA, 63; Univ Calif, Los Angeles, PhD(inorg chem), 67. *Prof Exp:* Asst prof chem, Univ Calif, Los Angeles, 67-68 & New Col, 68-70; assoc prof environ sci, Prescott Col, 70-75, vpres acad affairs, 73-74, actg pres, 74-75; acad dean, Antioch Col W, 77-78; vpres inst advan, Bryant Col, 85-87; SR ASSOC, WOODS HOLE RES CTR, 87- *Res:* Biotic aspects of global and regional environmental problems, including global warming, land use in the tropics, biotic impoverishment. *Mailing Add:* PO Box 296 Woods Hole MA 02543

HARRIMAN, BENJAMIN RAMAGE, b Monroe Bridge, Mass, Aug 9, 13; m 37; c 2. ORGANIC CHEMISTRY. *Educ:* Dartmouth Col, AB, 35; Pa State Univ, MS, 36, PhD(org chem), 39. *Prof Exp:* Asst, Pa State Univ, 36-37; res chemist, Wm S Merrell Co, Ohio, 38-41; res chemist, Ansco Div, Gen Aniline & Film Corp, 41-45, group leader, 45-46, mgr coating develop, 47-55; mgr photo prods res & develop, Haloid Xerox, Inc, 55-60; mgr chem res imaging res, 3M Co, 60-68 & Europ sci & tech liaison, 68-76, with spec assignments in corp res & develop, 76-78; RETIRED. *Concurrent Pos:* Teacher & naturalist, Dodge Nature Ctr, 78-; mem, Minn Water Resources Bd, 79-84; consult, Univ Wis-Stout, 81. *Mem:* Fel AAAS; Am Chem Soc; Royal Soc Chem; Royal Inst London. *Res:* Organic syntheses; photographic science and technology; research and development administration; scientific and technical liaison; technical planning and coordination. *Mailing Add:* 1335 Pinehurst Ave St Paul MN 55116

HARRIMAN, JOHN E, b Appleton, Wis, Dec 17, 36. CHEMICAL PHYSICS, QUANTUM CHEMISTRY. *Educ:* Univ Wis, BS, 59; Harvard Univ, PhD(chem physics), 63. *Prof Exp:* Asst prof phys chem, 64-67, assoc chmn dept chem, 70-72, ASSOC PROF PHYS CHEM, UNIV WIS-MADISON, 67- *Concurrent Pos:* NSF fel quantum chem group, Univ Uppsala, 62-64; Sloan Found fel, 68-70. *Mem:* AAAS; Am Phys Soc. *Res:* Electron spin resonance; molecular quantum mechanics; reduced density matrices. *Mailing Add:* 1307 Dept Chem Bldg Univ Wis Madison 1101 University Ave Madison WI 53706

HARRIMAN, NEIL ARTHUR, b St Louis, Mo, Aug 1, 38; m 63. PLANT TAXONOMY. *Educ:* Colo Col, BA, 60; Vanderbilt Univ, PhD(bot), 65. *Prof Exp:* Vis lectr bot, Colo Col, 61; from asst prof to assoc prof, 64-74, PROF BIOL, UNIV WIS-OSHKOSH, 74- *Concurrent Pos:* Vis lectr, Univ Wis, 67. *Mem:* Am Soc Plant Taxon; Int Asn Plant Taxon. *Res:* Taxonomy of Cruciferae and Juncaceae. *Mailing Add:* Dept Biol Univ Wis 800 Algoma Blvd Oshkosh WI 54901

HARRIMAN, PHILIP DARLING, b San Rafael, Calif, Nov 24, 37; m 59; c 1. MICROBIAL GENETICS, BIOPHYSICS. *Educ:* Calif Inst Technol, BS, 59; Univ Calif, Berkeley, PhD(biophys), 64. *Prof Exp:* Asst prof biochem, Med Ctr, Duke Univ, 68-75; from asst prof to assoc prof biol, Univ Mo, Kans City, 75-79; sr sci assoc to asst res dir biol, behav & soc sci, 81-81, PROG DIR GENETIC BIOL, NSF, WASHINGTON, DC, 77- *Concurrent Pos:* USPHS fel, Genetics Inst, Univ Cologne, 64-65 & Lab Pasteur, Paris, 65-66; Am Cancer Soc fel, Cold Spring Harbor Lab, NY, 66-68; prog dir genetic biol, NSF, Washington, DC, 77-, sr sci assoc to asst res dir biol, behav & soc sci, 81-82; Cong fel, 80-81; vis fac mem, Dept Molecular Biol & Genetics, Johns Hopkins Univ Med Sch, Baltimore, Md, 87-88. *Mem:* Genetics Soc Am; Am Soc Microbiol; AAAS. *Res:* Genetics of bacteria and bacterial viruses. *Mailing Add:* Genetics Prog Dir NSF DMB Rm 325 Washington DC 20550

HARRINGTON, DALTON, b Rice Lake, Wis, June 12, 34; m 59; c 2. BOTANY, PHYCOLOGY. *Educ:* Univ Omaha, BA, 61; Univ Mo, Kansas City, MS, 65; Univ Nebr, Lincoln, PhD(bot), 69. *Prof Exp:* Instr bot, Univ Nebr, Lincoln, 68-69; from asst prof to assoc prof, 69-77, assoc prof biol & chmn dept, 73-77, PROF BIOL, CALIF STATE COL, SAN BERNARDINO, 77-; DIR CALIF STATE UNIV & COL DESERT RES CTR, MOJAVE DESERT, 75- *Mem:* Am Inst Biol Sci; Bot Soc Am; Phycol Soc Am; Am Soc Limnol & Oceanog; Sigma Xi. *Res:* Ecology of aquatic plants, particularly members of Division Charophyta, studies of their excretion products and effect of these on other members of the aquatic biota. *Mailing Add:* Dept Biol Calif State Col 500 State College Pkwy San Bernardino CA 92407

HARRINGTON, DANIEL DALE, b Flint, Mich, Aug 12, 37; m 64; c 3. VETERINARY PATHOLOGY. *Educ:* Mich State Univ, DVM, 62, MSc, 63; Mass Inst Technol, PhD(nutrit path), 69. *Prof Exp:* NIH fel, Mass Inst Technol, 63-69; asst prof nutrit path, Univ Ky, 69-73; ASSOC PROF VET PATH, PURDUE UNIV, 73- *Mem:* Am Vet Med Asn; Asn Am Vet Col; Am Col Vet Path. *Res:* Nutritional diseases with emphasis on vitamin and mineral associated pathology, especially of the cardiovascular system. *Mailing Add:* Vet Pathobiol Purdue Univ West Lafayette IN 47907

HARRINGTON, DAVID HOLMAN, b Laconia, NH, Feb 13, 37; m 75; c 3. FARM FINANCIAL MANAGEMENT. *Educ:* Cornell Univ, BS, 59; Univ NH, MS, 64; Purdue Univ, PhD(agr econ), 73. *Prof Exp:* Agr econ, econ res serv, USDA, 64-73; agr econ, Agr Can, 73-77; exec dir, Canfarm, Agiculture, Can, 77-79; br chief, 79-86, DEP DIR, ECON RES SERV, US DEPT AGR, 86- *Mem:* Am Arg Econ Asn. *Res:* Research on structure of agriculture, farm financial management, financial conditions of American agriculture and farm rural economy linkages. *Mailing Add:* ARED/ERS/USDA Rm 328 1301 New York Ave NW Washington DC 20250

HARRINGTON, DAVID ROGERS, b North Tonawanda, NY, Sept 23, 35; m 63, 88; c 1. PHYSICS. *Educ:* Carnegie Inst Technol, BS, 57, MS, 59, PhD(physics), 61. *Prof Exp:* NSF fel, 61-62; instr & res assoc physics, Cornell Univ, 62-63, actg asst prof, 63-64; from asst prof to assoc prof, 64-84, PROF PHYSICS, RUTGERS UNIV, 84-, GRAD DIR, 89- *Mem:* Am Phys Soc; Am Asn Physics Teachers. *Res:* Theoretical nuclear physics. *Mailing Add:* Dept Physics & Astron Rutgers Univ PO Box 849 Piscataway NJ 08855-0849

HARRINGTON, DEAN BUTLER, b Schenectady, NY; m 49; c 3. ELECTRICAL ENGINEERING, ELECTROMAGNETIC PHENOMENA. *Educ:* Mass Inst Technol, BS, 44. *Prof Exp:* Engr, Gen Elec Co, 44-67, mgr generator advan eng, 67-82; RETIRED. *Concurrent Pos:* Consult, elec eng, 82- *Honors & Awards:* Charles P Steinmetz Award, Gen Elec Co, 77; Nikola Tesla Award, Inst Elec & Electronics Engrs, 81. *Mem:* Nat Acad Eng; fel Inst Elec & Electronics Engrs. *Res:* Electric power engineering; electric machine theory. *Mailing Add:* 22 Via Maria Dr Scotia NY 12302

HARRINGTON, EDMUND ALOYSIUS, JR, b Wareham, Mass, Dec 23, 43; m 70; c 1. TELECOMMUNICATIONS SCIENCE, TELECOMMUNICATIONS ENGINEERING. *Educ:* Providence Col, BS, 65; Univ Notre Dame, PhD(physics), 70. *Prof Exp:* Mem tech staff switching, Bell Tel Labs, 70-71; eng specialist telecommun, GTE Sylvania Inc, 72-; DIR, CALIF MICROWAVE LABS. *Concurrent Pos:* Lectr, Northeastern Univ, 74-75, 79- & Boston Univ, 78-; nat sem leader, Telephony Publ Corp & Info Gatekeepers Inc, 78- *Honors & Awards:* Distinguished Serv Award, Armed Forces Commun & Electronics Asn, 77. *Mem:* Inst Elec & Electronics Engrs; Am Inst Physics; Am Phys Soc; Sigma Xi. *Res:* Telecommunication networks; computer communications; fiber optics; satellite communications. *Mailing Add:* 169 North St Mattapoisett MA 02739

HARRINGTON, EDWARD JAMES, b Springfield, Mass, Feb 19, 26; m 56; c 4. VERTEBRATE ZOOLOGY. *Educ:* Tufts Univ, BS, 50, MEd, 51; Cornell Univ, PhD, 55. *Prof Exp:* From asst prof to prof biol & sci educ, San Jose State Col, 55-70; PROF BIOL & VPRES PRES ACAD AFFAIRS, CENT WASH Univ, COL, 70- *Concurrent Pos:* Assoc dean, San Jose State Col, 64-65, dean undergrad studies, 67-69, actg dean acad planning, 69-70. *Mem:* Am Soc Mammal; Sigma Xi. *Res:* Mammals. *Mailing Add:* Cent Wash Univ Ellensburg WA 98926

HARRINGTON, FRED HADDOX, b Albany, Calif, Nov 2, 47; m 72, 90; c 2. ETHOLOGY, BEHAVIORAL ECOLOGY. *Educ:* Univ Del, BA, 69; State Univ NY Stony Brook, PhD(biol), 75. *Prof Exp:* fel ethology, Dalhousie Univ, 75-77; from asst prof to assoc prof, 77-87, PROF, DEPT PSYCHOL, MT ST VINCENT UNIV, 87- *Concurrent Pos:* Consult, Nfld-Labrador Wildlife Dir, 86- & Labrador Inuit Asn, 88- *Mem:* Animal Behav Soc. *Res:* Canid behavior and communication and their relation to the animals ecology; social and environmental influences on olfactory and acoustic communication in wolves and coyotes; human attitudes toward animals; behavioral ecology of caribou and black bear. *Mailing Add:* Dept Psychol Mt St Vincent Univ Halifax NS B3M 2J6 Can

HARRINGTON, GEORGE WILLIAM, b New York, NY, Nov 13, 29; m 55; c 2. ANALYTICAL CHEMISTRY. *Educ:* NY Univ, AB, 54, PhD(chem), 59. *Prof Exp:* Asst chem, NY Univ, 57-59; proj engr, Philco Corp, 59; from instr to assoc prof chem, 59-67, assoc dean col lib arts, 68-71, chmn dept, 78-81, PROF CHEM, TEMPLE UNIV, 67- *Mem:* Am Chem Soc; Am Inst Chemists. *Res:* Electroanalytical chemistry; N-nitrosamines; instrumentation. *Mailing Add:* Dept Chem Temple Univ Broad & Montgomery Philadelphia PA 19122

HARRINGTON, GLENN WILLIAM, b Los Angeles, Calif, Oct 13, 32; m 75; c 2. PARASITOLOGY, MICROBIOLOGY. *Educ:* Univ Calif, Los Angeles, BA, 54, MA, 61; Rice Univ, PhD(parasitol, biol), 64. *Prof Exp:* Asst prof microbiol, State Univ NY Upstate Med Ctr, 66-70; asst prof, Univ Health Sci, 70-73; assoc prof microbiol, 73-85, DIV CELL BIOL & BIOPHYS, SCH BASIC LIFE SCI, UNIV MO SCH DENT, KANS CITY, 85- *Concurrent Pos:* NIH fel microbiol, State Univ NY Upstate Med Ctr, 64-66. *Mem:* AAAS; Am Soc Parasitol; Soc Protozool; Am Soc Microbiol; Am Soc Clin Path. *Res:* Lipid chemistry and metabolism of tapeworms and algae; dental microbiology. *Mailing Add:* Sch Basic Life Scis Univ Mo Sch Dent Kansas City MO 64110

HARRINGTON, JAMES FOSTER, b Newark, NJ, Nov 24, 16; m 39; c 3. OLERICULTURE. *Educ:* Ohio State Univ, BS, 39, MS, 40; Cornell Univ, PhD(veg crops), 44. *Prof Exp:* Asst res prof hort, Iowa State Col, 44-46; asst prof truck crops & asst olericulturist, 46-50, assoc prof veg crops & assoc olericulturist, 50-58, prof, 58-80, EMER PROF VEG CROPS & OLERICULTURIST, UNIV CALIF, DAVIS, 80- *Concurrent Pos:* Fels, Fulbright, 54-55 & USDA, 62-63; consult, Agency Int Develop, Brazil, 65-66, Food, Agr Org, Egypt, 69 & Rockefeller Found, India, 69. *Mem:* AAAS; Am Soc Hort Sci; Am Soc Plant Physiol; Am Soc Agron. *Res:* Seed physiology; vegetable crops; aging of seeds. *Mailing Add:* Dept Veg Crops Univ Calif Davis CA 95616

HARRINGTON, JAMES PATRICK, b Salem, Ohio, Dec 21, 39; m 66; c 3. ASTROPHYSICS. *Educ:* Univ Chicago, SB, 61; Ohio State Univ, MSc, 64, PhD(astron), 67. *Prof Exp:* Asst prof, 67-74, ASSOC PROF ASTRON, UNIV MD, COLLEGE PARK, 74- *Mem:* Am Astron Soc; Int Astron Union; Royal Astron Soc. *Res:* Gaseous nebulae; radiative transfer; stellar atmospheres. *Mailing Add:* Astron Prog Space Sci Bldg 224 Univ Md College Park MD 20742

HARRINGTON, JOHN VINCENT, b New York, NY, May 9, 19; m 43; c 5. PHYSICS, ELECTRICAL ENGINEERING. *Educ:* Cooper Union, BEE, 40; Polytech Inst Brooklyn, MEE, 48; Mass Inst Technol, ScD, 58. *Prof Exp:* Student engr, Consol Edison Co, 40-41; asst engr, Am Gas & Elec Serv Corp, NY, 41-46; chief data transmission br, USAF Cambridge Res Ctr, 46-51; leader data transmission group, Lincoln Lab, Mass Inst Technol, 53-57, assoc head commun & components div, 56-58, head radio physics div, 58-63, dir ctr space res & prof elec, aeronaut & astronaut eng, 63-73; vpres res & eng, Commun Satellite Corp, 73-79, sr vpres res & develop & dir, Comsat Labs, 79-83; RETIRED. *Concurrent Pos:* Mem bd dirs, Epsco, Inc, 64-72 & Shawmut County Bank, 64-73, Comsat Gen Corp, 73-80, Comsat Gen Telesysts, Inc, 73-81, Environ Res & Technol, 79-81. *Honors & Awards:* USAF Medal for Except Civilian Serv, 52; Citation for Distinguished Prof Achievement, Cooper Union, 65, Gano Dunn Award Eng, 82. *Mem:* Fel AAAS; fel Inst Aeronaut & Astronaut Eng; fel Inst Elec & Electronics Engrs. *Res:* Statistical detection theory; digital data transmission; storage systems; radio physics and space communications. *Mailing Add:* 1048 San Mateo Dr Punta Gorda FL 33950-6364

HARRINGTON, JOHN WILBUR, geology; deceased, see previous edition for last biography

HARRINGTON, JOSEPH ANTHONY, b Monroe, Mich, Aug 23, 39; m 65; c 1. COMBUSTION. *Educ:* Univ Mich, Ann Arbor, BS, 62; Northeastern Univ, MS, 70; York Univ, PhD(physics), 70. *Prof Exp:* Sr scientist, Systs Div, Avco Corp, 62-67, staff scientist, 69-70; prin res scientist assoc, res staff, Ford Motor Co, 70-81; SR RES PHYSICIST, ENG & ENVIRON RES, AMOCO OIL CO, 81- *Mem:* Combustion Inst; Sigma Xi; Am Asn Artificial Intel; Inst Elec & Electronics Engrs. *Res:* Determination of fundamental chemical and radiative properties of gases or their diagnostic applications; hydrocarbon fuel combustion and phase transition kinetics; combustion of solid matter; artificial intelligence-expert systems. *Mailing Add:* 1512 N Columbia Naperville IL 60540

HARRINGTON, JOSEPH D, b Butte, Mont, Aug 24, 30. ZOOLOGY, PHYSIOLOGY. *Educ:* Carroll Col, Mont, AB, 52; Cath Univ Am, MS, 58, PhD(biol), 60. *Prof Exp:* Asst prof biol, 60-69, acad dean, 65-69, pres, 69-74, PROF BIOL, CARROLL COL, MONT, 74- *Mem:* AAAS; Am Inst Biol Sci; Sigma Xi; Am Soc Microbiol. *Mailing Add:* Dept Biol Carroll Col Helena MT 59625

HARRINGTON, JOSEPH DONALD, b Washington, DC, Oct 6, 26; m 51; c 2. AGRONOMY. *Educ:* Univ Md, BS, 53; Pa State Univ, MS, 55, PhD(agron), 59. *Prof Exp:* From asst to assoc prof, 53-71, PROF AGRON, PA STATE UNIV, UNIVERSITY PARK, 71- *Mem:* Am Soc Agron; Potato Asn Am. *Res:* Cultural evaluations and ecological factors influencing the quantity and quality of white potatoes for fresh market and processing purposes. *Mailing Add:* 116 Agr Sci & Industs Bldg University Park PA 16801

HARRINGTON, MARSHALL CATHCART, b Rockford, Ohio, Aug 28, 04; m 58. OPTICS. *Educ:* Princeton Univ, AB, 26, AM, 27, PhD(physics), 32. *Prof Exp:* Instr physics, Princeton Univ, 27-29, asst, 30-31; from asst prof to prof, Drew Univ, 31-55, chmn div sci, 43-51; physicist, David Taylor Model Basin, 55-57, head fluid dynamics br, 57-60, contract res adminr, 60-62; physicist, Gen Physics Div, Air Force Off Sci Res, 62-74, consult, 74-76; RETIRED. *Concurrent Pos:* Vis prof, NY Univ, 46-52; tech asst expert,

UNESCO, Iraq, 52-54; prof lectr, George Washington Univ, 55-57 & Univ Md, 56-57. *Mem:* Fel AAAS; fel Am Phys Soc; Am Asn Physics Teachers; assoc fel Am Inst Aeronaut & Astronaut; fel Optical Soc Am. *Res:* Atomic and molecular physics; radio astronomy. *Mailing Add:* 10450 Lottsford Rd Apt 2207 Mitchellville MD 20721-2734

HARRINGTON, ROBERT D(EAN), b Philip, SDak, Apr 23, 28; m 50; c 2. METROLOGY, STANDARDS. *Educ:* Univ Colo, BS, 51, MS, 52. *Prof Exp:* Physicist, Naval Res Lab, 52-53; physicist, 53-67, asst div chief, 67-68, asst chief tech admin & coord, 68-70, PROG COORDR, NAT BUR STAND, BOULDER LABS, 70- *Mem:* AAAS; Inst Elec & Electronics Engrs; Sigma Xi. *Res:* Magnetic materials, measurements and phenomena at radio, microwave and optical frequencies; lasers; metrology and standards; administration. *Mailing Add:* 5321 Holmes Pl Boulder CO 80303

HARRINGTON, ROBERT JOSEPH, b Rochdale, Eng, July 6, 41; US citizen; m 64; c 4. POWER SYSTEMS ANALYSIS, TRANSIENT PHENOMENA ANALYSIS. *Educ:* Univ Liverpool, BEng, 62, PhD(elec eng), 67. *Prof Exp:* Res eng elec power, Nelson Res Labs, English Elec Co, 66-67; lectr elec eng, Univ Newcastle Upon Tyne, UK, 68-69; assoc prof, 80-85, PROF ELEC ENG, GEORGE WASHINGTON UNIV, WASHINGTON, DC, 85-, CHMN, 91- *Concurrent Pos:* Expert witness, elec equip. *Mem:* Sr mem Inst Elec & Electronics Engrs; fel Gt Brit Inst Elec Engrs. *Res:* Transient analysis of electrical power systems and machinery; traveling wave phenomena on power transmissions systems; corona and field effects of extra high voltage systems. *Mailing Add:* 9202 Leamington Ct Fairfax VA 22031

HARRINGTON, ROBERT SUTTON, b Newport News, Va, Oct 21, 42; m 76; c 2. ASTROMETRY, CELESTIAL MECHANICS. *Educ:* Swarthmore Col, BA, 64; Univ Tex, PhD(astron), 68. *Prof Exp:* ASTRON, US NAVAL OBSERV, 67- *Mem:* Am Astron Soc; Int Astron Union. *Res:* Photographic astrometry; multiple star dynamics; solar system dynamics. *Mailing Add:* Naval Observ Washington DC 20392-5100

HARRINGTON, RODNEY B, b Bethel, Maine, Apr 30, 31; m 59; c 2. ANIMAL SCIENCE, STATISTICS. *Educ:* Univ Maine, BS, 54; Okla State Univ, MS, 57, PhD(animal breeding, statist), 63; Univ New South Wales, MSc, 62. *Prof Exp:* Asst animal sci, Okla State Univ, 56-57, 59-62; tech officer, Univ New South Wales, 57-59; from instr to assoc prof, 62-69, PROF ANIMAL SCI, PURDUE UNIV, 69- *Concurrent Pos:* Vis sr res fel, Dept Animal Breeding, Agr Univ, Wageningen, Neth. *Mem:* Am Soc Animal Sci; Biomet Soc; AAAS. *Res:* Design of experiments and applications of computers in animal science; population genetics. *Mailing Add:* Dept Animal Sci Purdue Univ Lafayette IN 47907

HARRINGTON, RODNEY E, b Mayville, NDak, Jan 9, 32; m 79; c 2. BIOCHEMISTRY, PHYSICAL CHEMISTRY. *Educ:* Univ SDak, BA, 53; Univ Wash, PhD(phys chem), 60. *Prof Exp:* Res chemist, Ames Lab, AEC, 53-56; res assoc biophys, Univ Calif, San Diego, 60-62; asst prof chem, Univ Ariz, 62-65; from asst prof to assoc prof, Univ Calif, Davis, 65-72; prof chem, 72-81, chmn dept, 72-76, PROF BIOCHEM, UNIV NEV, RENO, 81- *Mem:* Am Chem Soc; Am Phys Soc; Sigma Xi. *Res:* Biophysical chemistry of biological macromolecules and aggregate structures including chromatin. *Mailing Add:* Dept Biochem Univ Nev Reno NV 89507

HARRINGTON, ROGER F(ULLER), b Buffalo, NY, Dec 24, 25; m 54; c 4. ELECTRICAL ENGINEERING. *Educ:* Syracuse Univ, BS, 48; Ohio State Univ, PhD(elec eng), 52. *Prof Exp:* From instr to prof, 48-68, DISTINGUISHED PROF ELEC ENG, SYRACUSE UNIV, 68- *Concurrent Pos:* Vis prof, Univ Calif, Berkeley, 64-65, E China Normal Univ, 83; guest prof, Tech Univ Denmark, 69-70; vis scientist, Yugoslavian Academies Sci, 72. *Honors & Awards:* Sigma Xi Res Award, 71; Schlesinger Award, Inst Elec & Electronics Engrs, Distinguished Achievement Award, 89. *Mem:* Fel Inst Elec & Electronics Engrs; Int Sci Radio Union; Am Asn Univ Profs. *Res:* Electromagnetic field theory; electric network theory; applied mathematics. *Mailing Add:* Dept Elec Eng Syracuse Univ Syracuse NY 13244-1240

HARRINGTON, ROY VICTOR, b Brooklyn, NY, Sept 28, 28; m 52; c 3. INORGANIC CHEMISTRY. *Educ:* Polytech Inst Brooklyn, BS, 49; Univ Colo, PhD(chem), 56. *Prof Exp:* Chemist, Gen Foods Corp, 49-52; chemist, Corning Glass Works, 55-58, supvr, 58-63, mgr appl res glass, 63-64, mgr mat res, 64-66, mgr appl chem res, 66-68; assoc dir res, 68-72, VPRES & CORP DIR RES, FERRO CORP, 72- *Concurrent Pos:* Panel mem, Nuclear Waste Disposal, Nat Acad Sci. *Mem:* Am Chem Soc; Am Ceramic Soc; Indust Res Inst. *Res:* Research management; chemistry and physics of glassy state; high temperature inorganic chemistry; frit; color; fiber glass; refractories; abrasives; composites. *Mailing Add:* 6821 Rosemont Dr Brecksville OH 44141

HARRINGTON, STEVEN JAY, b Portland, Ore, Nov 28, 47; m 69. COMPUTER GRAPHICS, SYSTEMS DESIGN. *Educ:* Ore State Univ, BS(physics) & BS(math), 68; Univ Wash, MS(physics), 69, MS(comput sci) & PhD(physics), 76. *Prof Exp:* Res assoc prof comput physics, Univ Utah, 76-78; asst prof comput sci, State Univ NY, Brockport, 78-81; PRIN SCIENTIST, XEROX CORP, 81- *Mem:* Asn Comput Mach; Inst Elec & Electronics Engrs; Soc Info Display. *Res:* Architectures, algorithms, and system design for electronic imaging. *Mailing Add:* Xerox Corp 800 Phillips Rd Bldg 128- 27E Webster NY 14580

HARRINGTON, WALTER JOEL, b Salamanca, NY, Nov 9, 16; m 41; c 2. MATHEMATICS. *Educ:* Cornell Univ, AB, 37, AM, 38, PhD(math), 41. *Prof Exp:* Instr math, Pa State Univ, 41-44; res assoc, Allegheny Ballistics Lab, Md, 44-46; vis asst prof, Cornell Univ, 46-47; from asst prof to assoc prof math, Pa State Univ, 47-57; prof math, NC State Univ, 57-82, actg head dept, 79-80, emer prof, 82-; RETIRED. *Concurrent Pos:* Dir, Dept Defense Res Proj, 67-72. *Mem:* Am Math Soc; Soc Indust & Appl Math; Math Asn Am; Sigma Xi. *Res:* Number theory; analysis of exterior ballistics; analysis and operational mathematics. *Mailing Add:* 3010 Ruffin St Raleigh NC 27607

HARRINGTON, WILLIAM FIELDS, b Seattle, Wash, Sept 25, 20; m 48; c 5. MOLECULAR BIOLOGY. *Educ:* Univ Calif, BS, 48, PhD(biochem), 52. *Prof Exp:* Res biochemist, Virus Lab, Univ Calif, 52-53; fel colloid sci, Nat Found Infantile Paralysis, Cambridge Univ, 53-54 & Nat Cancer Inst, Carlsberg Lab, Denmark, 54-55; asst prof biophys chem, Iowa State Univ, 55-56; chemist biochem, Nat Heart Inst, 57-60; chmn dept biol & dir McCollum Pratt Inst, 73-83, PROF BIOL, MCCOLLUM PRATT INST, JOHNS HOPKINS UNIV, 60-, HENRY WALTERS PROF BIOL, 76- *Concurrent Pos:* Mem adv panel physiol chem, NIH, 62-67, adv panel biophys chem, 68-72; mem sci coun, Nat Inst Arthritis & Metab Dis, 68-72, adv coun, 87-90; mem vis comt biol, Brookhaven Nat Lab, 69-73. *Mem:* Nat Acad Sci; fel Am Acad Arts & Sci; Biophys Soc; Am Chem Soc; Am Soc Biol Chemists. *Res:* Physical chemistry of proteins; protein structure and function; molecular basis of muscle contraction. *Mailing Add:* Dept Biol McCollum-Pratt Inst Johns Hopkins Univ Baltimore MD 21218

HARRIOTT, PETER, b Ithaca, NY, July 21, 27; m 53; c 5. CHEMICAL ENGINEERING. *Educ:* Cornell Univ, BCE, 49; Mass Inst Technol, ScD, 52. *Prof Exp:* From asst prof to prof, 53-66, FRED H RHODES PROF CHEM ENG, CORNELL UNIV, 75- *Mem:* Am Chem Soc; Am Inst Chem Engrs; Sigma Xi. *Res:* Chemical kinetics; reactor design; heat and mass transfer; air pollution control. *Mailing Add:* 248 Olin Hall Cornell Univ Ithaca NY 14853-5201

HARRIS, ALAN WILLIAM, b Portland, Ore, Aug 3, 44; m 70; c 3. ASTRONOMY, PLANETARY PHYSICS. *Educ:* Calif Inst Technol, BS, 66; Univ Calif, Los Angeles, MS, 67, PhD(planetary & space sci), 75. *Prof Exp:* RES SCIENTIST, 80-, SUPVR, EARTH & PLANETARY PHYSICS GROUP, JET PROPULSION LAB, 83- *Concurrent Pos:* Mem tech staff, Jet Propulsion Lab, 74-; prin investr, Lunar & Planetary Prog, NASA, 76-; vis prof physics, Univ Calif, Santa Barbara, 78; vis prof earth & space sci, Univ Calif, Los Angeles, 79; pres, Comn 15 Int Astron Union, 91-94; chmn, Div Dynamics, Am Astron Soc, 91-92. *Honors & Awards:* Asteroid 2929 Harris Named in Recognition of Planetary Res. *Mem:* Am Astron Soc; Am Geophys Union; Int Astron Union. *Res:* Dynamical evolution of the solar system, origin and evolution of satellite and ring systems; physical studies of asteroids; author of one book and over 50 technical papers in books, journals, conference proceedings, etc. *Mailing Add:* Jet Propulsion Lab MS 183-501 4800 Oak Grove Dr Pasadena CA 91109

HARRIS, ALBERT HALL, medicine; deceased, see previous edition for last biography

HARRIS, ALBERT KENNETH, JR, b Colorado Springs, Colo, Nov 5, 43; m 65; c 3. EMBRYOLOGY, CELL BIOLOGY. *Educ:* Swarthmore Col, BA, 65; Yale Univ, MPhil, 70, PhD(biol), 71. *Prof Exp:* Damon Runyon-Walter Winchell Mem Fund res fel cancer res, Strangeways Res Lab, Eng, 71-72; from asst prof to assoc prof zool, 72-83, PROF BIOL, UNIV NC, CHAPEL HILL, 83- *Mem:* Am Soc Cell Biol; Soc Study Amphibians & Reptiles; Soc Develop Biol. *Res:* Mechanism of locomotion of tissue cells in embryonic development, cancer and cell sorting; mechanism of cell adhesion and role of adhesion in cell motility; connective tissue morphogenesis. *Mailing Add:* Dept Biol Coker Hall CB 3280 Univ NC Chapel Hill NC 27599-3280

HARRIS, ALBERT ZEKE, b Gary, Ind, Mar 29, 38; m 64; c 1. ANALYTICAL CHEMISTRY, PHYSICAL CHEMISTRY. *Educ:* Tex Western Col, BS, 60; Tex A&M Univ, MS, 63, PhD(phys chem), 67. *Prof Exp:* Asst prof, 67-75, assoc prof, 75-80, PROF CHEM, UNIV MONTEVALLO, 80- *Mem:* Am Chem Soc. *Res:* Inorganic complex salts; phosphorescence of carcinogenic compounds. *Mailing Add:* 494 Overland Dr Montevallo AL 35115

HARRIS, ALEX L, b New York, NY, Sept 23, 54; m 82; c 2. BIOCHEMICAL PHARMACOLOGY, ENZYMOLOGY. *Educ:* State Univ NY, Buffalo, BS, 77, PhD(biochem pharmacol), 82. *Prof Exp:* Postdoctorate, Dept Physiol, Univ Mich, 81-83; sr res scientist, 83-88, group leader, 88-91, RES LEADER STERLING RES GROUP, 91- *Mem:* Am Soc Pharmacol. *Res:* Discovery of novel substances that inhibit or interfere with rational or cell based mechanisms. *Mailing Add:* Sterling Res Group 81 Columbia Turnpike Rensselaer NY 12144

HARRIS, ALEXANDER L, b Madison, Wis, Jan 27, 54; m 87. PHYSICAL CHEMISTRY. *Educ:* Swarthmore Col, BA, 78; Univ Calif, Berkeley, PhD(phys chem), 85. *Prof Exp:* MEM TECH STAFF, AT&T BELL LABS, 85- *Mem:* Am Phys Soc; Am Chem Soc; AAAS. *Res:* Physical chemistry of interfaces and surfaces; chemical dynamics of surfaces studied by time-resolved optical methods; nonlinear optics at interfaces; ultrafast phenomena studied by short optical laser pulses. *Mailing Add:* AT&T Bell Labs Rm 1A-364 Murray Hill NJ 07974

HARRIS, ALVA H, b Morehead City, NC, Dec 22, 28; m 54; c 3. PARASITOLOGY, MARINE BIOLOGY. *Educ:* NC State Univ, BS, 56, MS, 58, PhD(parasitol), 66. *Prof Exp:* Ecologist, Carolina Biol Supply Co, NC, 58-61; asst prof, NC State Univ, 66-67; prof marine biol, Nicholls State Univ, 67-88, dir, Marine Sci Lab, 72-88. *Concurrent Pos:* Dept Health, Educ & Welfare fel, 66-67; Nat Sea Grant Prog, NSF grant, 68-70; consult, Int Chem Corp, 68- & Inmont Corp, 68-70; Nat Sea Grant Prog res grant, US Dept Com, Nat Oceanic & Atmospheric Admin, 70-72; mem, La Adv Comn Coastal & Marine Resources, 71- *Mem:* Am Soc Parasitol; Am Fisheries Soc; Wildlife Dis Asn. *Res:* Epidemiology of wildlife parasites in fresh and estuarine environments; development of aquacultural methods for commercially important estuarine animals, especially shrimp culture; environmental studies on marine shrimp and fish. *Mailing Add:* 628 Fairway Dr Thibodaux LA 70809

HARRIS, ANDREW LEONARD, b Saranac Lake, NY, Oct 18, 51. BIOLOGY GENERAL, PHYSIOLOGY GENERAL. *Educ:* Univ Calif, San Diego, BA, 72; Stanford Univ, PhD(neurosci), 79. *Prof Exp:* Fel, dept neurosci, Albert Einstein Col Med, 79-82; res assoc, dept biol sci, Stanford Univ, 82-84; res assoc, dept anat, Harvard Med Sch, 84-85; ASST PROF, DEPT BIOPHYS, JOHNS HOPKINS UNIV, 85- *Concurrent Pos:* Staff fel, Div Comput Res & Technol, Nat Inst Arthritis, Diabetes & Digestive & Kidney Dis, NIH, 85. *Mem:* Biophys Soc; Soc Neurosci; Am Soc Cell Biol. *Res:* Cellular neurophysiology; regulation and gating of gap junction channels in native membranes and reconstituted systems. *Mailing Add:* Dept Biophys Johns Hopkins Univ 3400 N Charles St Baltimore MD 21218

HARRIS, ARLO DEAN, b Dayton, Ohio, Sept 17, 34. INORGANIC CHEMISTRY. *Educ:* Univ Dayton, BSc, 61; Tulane Univ, PhD(chem), 64. *Prof Exp:* Instr chem, Univ Dayton, 59-61; teaching asst, Tulane Univ, 61-64; fel, Univ Calif, Berkeley, 64-65; asst prof chem, Calif State Col, Fullerton, 65-67; from asst prof to assoc prof chem, 67-78, PROF INORG CHEM, CALIF STATE COL, SAN BERNARDINO, 79- *Concurrent Pos:* Vis lectr, Univ Nottingham, 69-70; res assoc, Univ London, King's Col, 78-79; vis prof, Univ Khartoum, Sudan, 80 & Univ Queensland, Brisbane Australia, 82, 84 & 87. *Mem:* Am Chem Soc; Royal Soc Chem; fel Am Int Chemists; NY Acad Sci; Am Acad Arts & Sci. *Res:* Hydride complexes of platinum metals; cancer chemotherapy of cis-platin derivatives. *Mailing Add:* 1005 Chelsea Ave Dayton OH 45420

HARRIS, ARTHUR BROOKS, b Boston, Mass, Mar 25, 35; m 58; c 3. SOLID STATE PHYSICS. *Educ:* Harvard Univ, AB, 56, AM, 59, PhD(solid state physics), 62. *Prof Exp:* Asst physics, Harvard Univ, 57-58, res asst, 58-61; res assoc, Duke Univ, 61-62, instr, 62-64; res assoc, UK Atomic Energy Auth, 64-65; from asst prof to assoc prof, 65-77, PROF PHYSICS, UNIV PA, 77- *Mem:* Am Phys Soc. *Res:* Calorimetric determination of energy levels in magnetic compounds; static and dynamic properties of the chemical shift tensor; theory of metals in the narrow band limit. *Mailing Add:* Dept Physics Univ Pa Philadelphia PA 19312

HARRIS, ARTHUR HORNE, b Middleborough, Mass, May 18, 31; div; c 3. VERTEBRATE PALEOBIOLOGY, VERTEBRATE ZOOLOGY. *Educ:* Univ NMex, BA, 58, MS, 59, PhD(vert zool), 65. *Prof Exp:* Asst prof zool, Ft Hays Kans State Col, 63-65; from asst prof to assoc prof, 65-71, PROF BIOL, UNIV TEX, EL PASO, 71- *Concurrent Pos:* Cur vert paleobiol, Lab Environ Biol, 76-, co-dir resource collections, 80-; managing ed, Southwestern Naturalist, 78-82. *Mem:* Am Soc Mammal; Soc Vert Paleont; Am Quaternary Asn; Sigma Xi. *Res:* Distribution and ecology of modern Southwestern vertebrates; past distributions and environments as interpreted from fossil and archaeological faunas, particularly of the late Pleistocene and post-Pleistocene. *Mailing Add:* Dept Biol Sci Univ Tex El Paso TX 79968

HARRIS, B(ENJAMIN) L(OUIS), b Savannah, Ga, Aug 1, 17; m 42; c 5. CHEMICAL ENGINEERING. *Educ:* Johns Hopkins Univ, BE, 38, PhD(chem eng), 41. *Prof Exp:* Res asst to Dr Frankenburg, Johns Hopkins Univ, 41, asst prof chem eng, 46-53; pres, Eng Res Co, Md, 47-53; chief plants br & dep asst chief, Toxic Chem Warfare Div, Res & Eng Command, 52-55; asst to sci dir, Chem Warfare Labs, 55-60, dep dir develop, 60-62, dir develop support, 62-66, chief systs anal div, Army Chem Res & Develop Labs, 66, dep asst dir, Off of Dir Defense Res & Eng, 66-70, tech dir, Edgewood Arsenal, 70-77; dep dir & tech dir, US Army Chem Systs Lab, 77-81; pres & chief exec officer, Eng Res Co, Glen Arm, Md, 81-87; RETIRED. *Concurrent Pos:* Mem comt hazardous mat, Nat Acad Sci-Nat Res Coun, chmn risk anal panel, 67-76; consult, 87- *Mem:* Fel AAAS; Am Chem Soc; fel Am Inst Chem Engrs; Sigma Xi; Am Defense Preparedness Asn. *Res:* Adsorption of gases by solids; adsorption from solution; permeability of organic films to water and gases; physical constants; stability and corrosion of chemical agents; operations research; engineering evaluation; mass transfer; management of research, development and engineering. *Mailing Add:* 11323 Glen Arm Rd Glen Arm MD 21057

HARRIS, BARNEY, JR, b Prescott, Ark, Dec 20, 31; m 64; c 3. DAIRY SCIENCE. *Educ:* Okla State Univ, BS, 54, PhD(dairy nutrit), 64; La State Univ, MS, 56. *Prof Exp:* Instr dairy, Southern State Col, 58-60; asst prof dairy sci & exten dairy nutritionist, Fla Coop Exten Serv, 63-69, assoc prof dairy sci, 69-74, PROF DAIRY SCI, UNIV FLA, 74- *Concurrent Pos:* Chmn, Fla Dairy Indust Tech Coun, 71- & Fla Prod Conf, 71-88. *Mem:* Am Dairy Sci Asn. *Res:* Effect of parakeratosis on ruminant absorption in dairy calves; aueromycin feeding to dairy calves; nutrient requirements for high milk production; feeding and herd arrangements; studies with roughage by-products for dairy cattle; the value of sodium bicarbonate, fat and enzymes in dairy rations; mineral needs of dairy cattle; protein sources for dairy cattle; personnel management on large dairies; feed additives for dairy cattle; feeding and management of dairy goats. *Mailing Add:* 203 Dairy Sci Dept Univ Fla Gainesville FL 32601

HARRIS, BEN GERALD, b Altus, Okla, Sept 25, 40; m 63; c 2. ENZYMOLOGY. *Educ:* Southwestern State Col, Okla, BS, 62; Okla State Univ, MS, 65, PhD(physiol), 67. *Prof Exp:* NIH & univ fels physiol, Rice Univ, 67-68; asst prof, 68-73, assoc prof, depts basic health sci & biol sci, 73-77, PROF DEPT BIOCHEM, NTEX STATE UNIV, 77- & TEX COL OSTEOP MED, 77- *Mem:* Am Chem Soc; Am Soc Biol Chemists; Am Soc Parasitologists. *Res:* Structure and function of enzymes; protein associations with cell organelles; metabolism of parasite enzymes; control mechanisms in parasites. *Mailing Add:* Res Dept Tex Col Osteo Med NTex State Univ 3500 Camp Bowie Blvd Ft Worth TX 76107-2690

HARRIS, BERNARD, b New York, NY, June 20, 26; m 49, 83; c 6. MATHEMATICAL STATISTICS, COMBINATORICS. *Educ:* City Col New York, BBA, 46; George Washington Univ, MA, 53; Stanford Univ, PhD(statist), 58. *Prof Exp:* Statistician, Richard Manville Assocs, 47; dir, Res Dept, Statist Serv Bur, NY, 48-50; math statistician, US Census Bur, 50-52; mathematician, Nat Security Agency, 52-58; from asst prof to assoc prof math, Univ Nebr, 58-64; PROF MATH, MATH RES CTR & DEPT STATIST, UNIV WIS-MADISON, 64- *Concurrent Pos:* Instr, George Washington Univ, 52-57; prof & lectr, Am Univ, 57-58; consult, Agr Res Serv, USDA, 59-63; vis prof, Eindhoven Technol Univ, 70-, Tech Univ Munich, 73-74; vchmn, comt nuclear regulatory res, Am Statist Asn, chmn, comt on AIDS, statist working group, Mil Hdbk Plastics Aerospace Vehicles. *Honors & Awards:* Wilks Mem Medal, 82. *Mem:* Economet Soc; fel Am Statist Asn; fel Inst Math Statist; Am Math Soc; Int Statist Inst; AAAS; Bernoulli Soc. *Res:* Discrete stochastic processes; moment inequalities; combinatorial methods in probability and statistics; reliability theory; statistical decision theory. *Mailing Add:* Dept Statist Univ Wis, 1210 W Dayton St Madison WI 53706

HARRIS, BERNARD, b New York, NY, Oct 13, 27; c 9. ELECTRICAL ENGINEERING. *Educ:* Cooper Union, BEE, 49; Columbia Univ, MS, 51, EngScD, 61; Pace Univ, MBA, 78. *Prof Exp:* Design engr, RCA Lab Div, 51-54; res scientist syst, NY Univ, 54-63; tech & mgt staff, Sperry Rand Corp, 63-65; res scientist oceanog, Hudson Lab, Columbia Univ, 65-68; chief engr microwaves, Polarad Electronics Corp, 68; vpres, Ocean & Atmospheric Sci, Inc, 68-79; PRES, HARRIS SCI SERV, 79- *Concurrent Pos:* Adj assoc prof, NY Univ, Pratt Inst & Manhattan Col, 60-70 assoc prof, Manhattan Col, 79- *Mem:* Acoust Soc Am; Inst Elec & Electronics Engrs; Oper Res Soc Am; AAAS; Am Soc Eng Educ; Sigma Xi. *Res:* Applications of atmospheric and oceanographic acoustics to such diverse fields as noise pollution and Navy oceanography. *Mailing Add:* 15 Overlook Rd Dobbs Ferry NY 10522

HARRIS, BEVERLY HOWARD, b Lee's Summit, Mo, Aug 22, 27; m 50; c 3. MATHEMATICAL ANALYSIS. *Educ:* Southwest Mo State Univ, BS, 49; Univ Mo, MA, 53, DEd, 63. *Prof Exp:* High sch teacher, 49-51; instr, 52-64, PROF MATH & CHMN DEPT, SOUTHWEST BAPTIST UNIV, 64- *Mem:* Math Asn Am. *Res:* Mathematics education; undergraduate mathematics, specifically elementary and intermediate analysis; summation of infinite series in intermediate analysis. *Mailing Add:* 910 E Division Bolivar MO 65613

HARRIS, BRUNO, b Ploesti, Romania, Mar 1, 32; nat US. MATHEMATICS. *Educ:* Calif Inst Technol, BS, 52; Yale Univ, PhD, 56. *Prof Exp:* NSF fel math, Yale Univ, 56-57; from instr to asst prof, Northwestern Univ, 57-60; Air Force Off Sci Res res assoc, Inst Advan Study, 60-61; assoc prof, 61-65, PROF MATH, BROWN UNIV, 65- *Concurrent Pos:* Mem, Inst Advan Study, 64-65; vis prof math, Princeton Univ, 81. *Mem:* Am Math Soc. *Res:* Algebra; geometry; homogeneous spaces; vector bundles and k-theory. *Mailing Add:* Dept Math Brown Univ Brown Sta Providence RI 02912

HARRIS, C EARL, JR, b Mineral Point, Wis, May 1, 30; c 2. GEOLOGY. *Educ:* Kent State Univ, BS, 57; Miami Univ, MS, 58; Sussex Col Technol, PhD, 70. *Hon Degrees:* DSc, Ohio Christian Col, 70. *Prof Exp:* Areal geologist, US Geol Surv, 58-59; instr geol, Wis State Univ, Superior, 59-61; chmn, Dept Geol, 61-77, ASSOC PROF GEOL, YOUNGSTOWN STATE UNIV, 75- *Concurrent Pos:* Geol consult, Columbiana County Prosecutor's Off, 74-75; mem screening comt prof cert, Am Inst Prof Geologists. *Mem:* AAAS; fel Geol Soc Am; Am Inst Prof Geologists; Am Geol Inst; Sigma Xi. *Res:* Subsurface stratigraphy of Louisiana; structural geology of Grandfather Mountain Region, North Carolina; geology of Catskill Mountains, New York; Ohio Devonian biostratigraphy; origin of salt domes; influence of industrial wastes on surface-subsurface water. *Mailing Add:* 3333 Kiwatha Rd Youngstown OH 44511

HARRIS, CARL MATTHEW, b Brooklyn, NY, Mar 29, 40; m 69; c 2. OPERATIONS RESEARCH, SYSTEMS ENGINEERING. *Educ:* Queens Col, NY, BS, 60; Polytech Inst Brooklyn, MS, 62, PhD(math), 66. *Prof Exp:* Instr math, St John's Univ, NY, 63-65; sr res mathematician, Eng Res Ctr, Western Elec Co, 65-67; mem tech staff, Adv Res Dept, Res Anal Corp, 67-70; assoc prof opers res, George Washington Univ, 70-75; prof indust eng & opers res & chmn dept, Syracuse Univ, 75-78, consult, 78-81; prof systs eng, Univ Va, 81-85; PROF OPER RES & APPL STATIST & CHMN DEPT, GEORGE MASON UNIV, 85- *Concurrent Pos:* Instr, Rutgers Univ, 66; assoc prof lectr, George Washington Univ, 67-70; consult, US Dept Energy, Washington, DC, Dept Corrwctions, Syracuse Police Dept, US Dept Justice, US Dept Com & Internal Revenue Serv, US Dept State, Nat Inst Standards & Technol. *Mem:* Sigma Xi; Inst Indust Engrs; Am Statist Asn; Am Res Soc Am; Opers Res Soc Am (pres, 90-91); Math Asn Am; Inst Mgt Sci. *Res:* Methodological research in applied probability and statistics with emphasis on queuing theory; applied research in analysis of public systems. *Mailing Add:* 12016 Whippoorwill Ln Rockville MD 20852

HARRIS, CECIL CRAIG, b Raymond, Miss, Feb 20, 25; m 47; c 3. NUCLEAR MEDICINE. *Educ:* Univ Tenn, BS, 49, MS, 51. *Prof Exp:* Instr elec eng, Miss State Univ, 49-50; jr develop engr, Oak Ridge Sch Reactor Tech, 50-52, develop engr, Physics Div, Oak Ridge 3949 Lab, 52-57, res staff mem, Thermonuclear Div, 57-67; ASSOC PROF RADIOL, DIV NUCLEAR MED, DUKE UNIV MED CTR, 71- *Concurrent Pos:* Mem task group on scanning, Int Comn Radiation Units & Measurements. 63-68. *Mem:* Soc Nuclear Med (secy, 62-65, vpres, 66-67, pres, 68-69); Am Col Nuclear Med; fel Am Col Nuclear Physicians; fel Am Col Radiol; Radiol Soc NAm. *Res:* Development of training materials and methodology in nuclear medicine science; development of instrumentation for evaluation of coronary artery hemodynamics by radionuclide means; development of multi-port clinical data processing systems for nuclear medicine. *Mailing Add:* Duke Univ Med Ctr PO Box 3808 Durham NC 27710

HARRIS, CHARLES, b New York, NY, Jan 7, 23; m 49; c 3. PATHOLOGY. *Educ:* Cornell Univ, BA, 43; Long Island Col Med, MD, 46. *Prof Exp:* NIH fel, Fels Res Inst, Sch Med, Temple Univ, 50-51, pathologist, Fels Res Inst & assoc path, Sch Med, 54-62; mem staff, Geront Res Inst, Philadelphia Geriat Ctr, 62-69, dir clin labs, 69-88; CONSULT, 88- *Concurrent Pos:* Vis asst prof, Woman's Med Col Pa, 64-69. *Mem:* AAAS; Am Soc Exp Path; Am Soc Hemat; Am Asn Path & Bact; AMA. *Res:* Breast cancer; leukemia. *Mailing Add:* Box 1489 Island Heights NJ 08732

HARRIS, CHARLES BONNER, b New York, NY, Apr 24, 40. PHYSICAL CHEMISTRY. *Educ:* Univ Mich, BS, 63; Mass Inst Technol, PhD(chem), 66. *Prof Exp:* AEC fel physics, 66-67; PROF CHEM, UNIV CALIF & PRIN INVESTR, LAWRENCE BERKELEY LAB, 67- *Concurrent Pos:* Alfred P Sloan Found fel, 70-74; Humboldt sr scientist award, 80. *Mem:* Am Chem Soc; Am Phys Soc. *Res:* General studies of coherence and coherent properties of matter and radiation; energy transfer processes in condensed phase and on metal surfaces; laser spectroscopy. *Mailing Add:* Dept Chem Univ Calif Berkeley CA 94720

HARRIS, CHARLES LAWRENCE, b Chicago, Ill, Nov 13, 42; m 63; c 3. MICROBIOLOGY. *Educ:* Univ Ill, Urbana, BS, 66, PhD(biochem), 70. *Prof Exp:* Res assoc biochem, Col Med, Univ Ill, 70; fel neurochem, Ill State Psychiat Inst, 70-72; from asst prof to assoc prof, 72-82, PROF BIOCHEM, WVA UNIV, 82- *Mem:* AAAS; Am Soc Biochem & Molecular Biol. *Res:* Structure, biosynthesis and function of transfer RNA; role and synthesis of modified nucleotides in RNA; synthetase complex, aminoacyl and RNA. *Mailing Add:* Dept Biochem Sch Med WVa Univ Morgantown WV 26506

HARRIS, CHARLES LEON, b Christiansburg, Va, Jan 3, 43; m 71. NEUROBIOLOGY. *Educ:* Va Polytech Inst, BS, 66; Pa State Univ, MS, 67, PhD(biophys), 69. *Prof Exp:* From asst prof to assoc prof, 70-81, PROF BIOL, STATE UNIV NY COL PLATTSBURGH, 81- *Mem:* AAAS; Am Soc Zool; Sigma Xi; Am Inst Biol Sci. *Res:* Behavioral role of small neuronal systems, especially giant axons of cockroach; mechanisms of learning and memory; history and philosophy of biology. *Mailing Add:* Dept Biol Sci State Univ NY Col Arts & Sci Plattsburgh NY 12901

HARRIS, CHARLES RONALD, b Kimberley, BC, Oct 3, 32; m 57; c 2. INTEGRATED PEST MANAGEMENT, ENVIRONMENTAL TECHNOLOGY. *Educ:* Univ BC, BA, 54, MA, 56; Univ Wis, PhD(entom), 61. *Prof Exp:* Res scientist, Can Dept Agr, 56-90, head, Soil Pesticide Sect, 66-90; PROF & CHAIR, UNIV GUELPH, 90- *Concurrent Pos:* Hon lectr, Univ Western Ont, 66-90; fac assoc, Univ Guelph, 69-90; mem, Ont Pesticides Adv Comt, 72-; vis res fel, New South Wales Dept Agr, 86-87. *Honors & Awards:* Bussart Award, Entom Soc Am, 68. *Mem:* Fel Entom Soc Can (pres, 74-75); Entom Soc Am. *Res:* Efficacy, behaviour and fate of insecticides in soil; insect resistance to insecticides; control of soil insect pests. *Mailing Add:* Dept Environ Biol Univ Guelph Guelph ON N1G 2W1 Can

HARRIS, CLARE I, b Canandaigua, NY, May 20, 33; m 55; c 3. PLANT SCIENCE. *Educ:* Cornell Univ, BS, 55; Purdue Univ, MS, 60, PhD(hort), 62. *Prof Exp:* Instr hort, Purdue Univ, 59-62; soil scientist, Agr Res Serv, USDA, 62-67, prin horticulturist, Coop State Res Serv, 67-70; res assoc, Univ Calif, Davis, 70-71; dir plant sci res prog, 71-74, dep adminr, 74-85, ASSOC ADMINR, COOP STATE RES SERV, USDA, 85- *Mem:* AAAS; Am Chem Soc; Weed Sci Soc Am; Am Soc Hort Sci. *Res:* Agriculture; rural development and consumer services research. *Mailing Add:* Coop State Res Serv USDA Washington DC 20250

HARRIS, COLIN C(YRIL), b Leeds, Eng, Jan 9, 28; wid. MINERAL ENGINEERING, COAL PREPARATION. *Educ:* Univ London, BSc, 52; Univ Leeds, PhD(mineral eng & coal preparation), 59. *Prof Exp:* Res asst coal preparation & mineral eng, Univ Leeds, 52-56, lectr, 57-60, 62-63; from asst prof to prof mineral eng, 60-70, PROF MINERAL ENG, COLUMBIA UNIV, 70- *Concurrent Pos:* Consult, Macmillan & Co, 63 & Envirotech, 77-83; supvr res proj, Am Iron & Steel Inst, 67-71, Cities Serv Co, 73-, Asarco, Inc, 77- & Dept of Energy, 88-; assoc ed, Int J Mineral Processing; adv on fac appointments, res & grad programs to US & foreign univs; adv on res proposals to govt funding agencies; adv & consult to mining, res & mfg co; mem organizing comts for int conf on mineral processing. *Honors & Awards:* Gaudin Award, Am Inst Mining, Metall & Petrol Engrs, 90. *Mem:* Brit Oper Res Soc; Brit Inst Mining & Metall; Am Inst Mining, Metall & Petrol Engrs. *Res:* Powder technology; fine particle statistics and measurement; fracture of brittle materials; comminution kinetics; floatation machine hydrodynamics; floatation kinetics; fluid flow and retention in porous media; sedimentation; design and scale-up of processing machinery; coal preparation and mineral processing; ceramics engineering. *Mailing Add:* 907 Eng Ctr Columbia Univ New York NY 10027

HARRIS, CURTIS C, b Anthony, Kans, Jan 9, 43. CARCINOGENESIS, CLINICAL ONCOLOGY. *Educ:* Univ Kans, MD, 69. *Prof Exp:* CHIEF, LAB HUMAN CARCINOGENESIS, NAT CANCER INST, 80- *Mem:* Am Asn Cancer Res; Am Soc Cell Biol; Am Asn Path; Am Soc Clin Invest. *Mailing Add:* Lab Human Carcinogenesis Nat Cancer Inst Bldg 37 Rm 2C09 9000 Rockville Pike Bethesda MD 20892

HARRIS, CYRIL MANTON, b Detroit, Mich; m 49; c 2. ACOUSTICS. *Educ:* Univ Calif, Los Angeles, BA, 38, MA, 40; Mass Inst Technol, PhD(physics), 45. *Hon Degrees:* ScD, NJ Inst Technol, 81; Northwestern Univ, 89. *Prof Exp:* Asst, Univ Calif, Los Angeles, 39-40; mem staff war res, Carnegie Inst of Wash, 41; mem staff war res, Nat Defense Res Comt & teaching fel, Mass Inst Technol, 41-45; res engr, Bell Tel Labs, Inc, 45-51; sci consult, London Br, Off Naval Res, 51; Fulbright lectr, Delft Univ Technol, 51-52; from assoc prof to prof elec eng, 52-76, Charles Batchelor prof, 76-87, EMER CHARLES BATCHELOR PROF ELEC ENG & EMER PROF ARCHIT, COLUMBIA UNIV, 87- *Concurrent Pos:* Fulbright vis prof, Univ Tokyo, 60; S Charles Lee vis prof, Univ Calif, Los Angeles, 91- *Honors & Awards:* Emile Berliner Maker of Microphone Award, 77; Franklin Medal, 77; Wallace Clement Sabine Medal, 79; Gold Medal, Audio Eng Soc, 84; Gold Medal, Acoust Soc Am, 87. *Mem:* Nat Acad Sci; Nat Acad Eng; fel Inst Elec & Electronics Engrs; hon mem Audio Eng Soc; fel Acoust Soc Am (vpres, 60-61, pres, 64-65); Am Soc Testing & Mat; NY Acad Sci (vpres, 88-90, pres-elect, 91); Am Philos Soc. *Res:* Architectural acoustics; noise control. *Mailing Add:* S W Mudd Bldg Columbia Univ New York NY 10027

HARRIS, D LEE, b Jenkinjones, WVa, July 11, 16; m 43; c 1. OCEANOGRAPHY, METEOROLOGY. *Educ:* Concord State Col, AB, 37; George Washington Univ, MS, 51; Univ Mich, PhD(meteorol), 65. *Prof Exp:* Res meteorologist, US Weather Bur, 49-65, res meteorologist, ESSA, 65-67, chief oceanog br, 67-78, sr scientist, Coastal Eng Res Ctr, 78-80; mem fac, Coastal & Oceanog Engr Dept, Univ Fla, 80-85; RETIRED. *Mem:* Am Meteorol Soc. *Res:* Coastal oceanography; sea surface waves; storm surges; astronomical tides. *Mailing Add:* 2516 NW 21st Ave Gainesville FL 32605

HARRIS, DANIEL CHARLES, b New York, NY, May 30, 48. BIOPHYSICAL CHEMISTRY. *Educ:* Mass Inst Technol, SB, 68; Calif Inst Technol, PhD(chem), 73. *Prof Exp:* Instr, Calif Inst Technol, 72-73; res fel, Albert Einstein Col Med, 73-75; asst prof chem, Univ Calif, Davis, 75-80; assoc prof chem, Franklin & Marshall Col, 80-83; OPTICAL & ELECTRONIC MAT, NAVAL WEAPONS CTR, 83- *Honors & Awards:* Meller Award Basic Med Res, Albert Einstein Col Med, 75. *Mem:* Am Chem Soc; Am Ceramic Soc. *Res:* Optical and electronic materials; analytical chemistry textbook writing. *Mailing Add:* Naval Weapons Ctr Code 3854 China Lake CA 93535

HARRIS, DANIEL EVERETT, b Summit, NJ, Aug 5, 34; m 67; c 3. RADIO ASTRONOMY. *Educ:* Haverford Col, BA, 56; Calif Inst Technol, MS, 57, PhD(astron & physics), 61. *Prof Exp:* Res assoc, Inst di Fisica, Bologna, Italy, 61-64 & Arecibo Observ, Cornell Univ, PR, 65-69; sci investr, Inst Argentino di Radioastron, 69-70; res assoc & lectr, Harvard Col Observ, 70-73; prin sci res officer, Radiosterrenwach, Neth Found Radioastron, 74-77; res assoc astron, Dominion Radio Astrophys Observ, Nat Res Coun, 77-80; ASTRONOMER, SMITHSONIAN ASTROPHYS OBSERV, MASS, 80- *Mem:* Am Astron Soc. *Res:* Non-thermal astrophysics. *Mailing Add:* 60 Garden St Cambridge MA 02138

HARRIS, DAVID OWEN, b Price, Utah, July 2, 39; div; c 2. CHEMICAL PHYSICS, MOLECULAR SPECTROSCOPY. *Educ:* Univ Calif, Berkeley, PhD, 65. *Prof Exp:* From asst prof to assoc prof, 65-75, chmn dept, 80-82, PROF CHEM, UNIV CALIF, SANTA BARBARA, 75- *Mem:* Am Phys Soc. *Res:* Laser spectroscopy; molecular structure. *Mailing Add:* Dept Chem Univ Calif Santa Barbara CA 93106

HARRIS, DAVID R, b Schenectady, NY, Aug 17, 32. PHYSICAL CHEMISTRY. *Educ:* Univ Colo, BA, 57, PhD(phys chem), 63. *Prof Exp:* Sr cancer res scientist biophys, Roswell Park Mem Inst, 62-65; asst prof comput sci, Utah State Univ, 65-69; assoc prof, 69-70, PROF COMPUT SCI, CALIF STATE UNIV, CHICO, 70- *Mem:* AAAS; Am Crystallog Asn. *Res:* X-ray structure analysis of organic and biological compounds; protein structure; adaptation of scientific problems to digital computers; artificial intelligence. *Mailing Add:* Dept Comput Sci Calif State Univ Chico CA 95929

HARRIS, DAVID VERNON, geology; deceased, see previous edition for last biography

HARRIS, DELBERT LINN, b Boone, Iowa, Sept 24, 43; m 61; c 4. VETERINARY MICROBIOLOGY. *Educ:* Iowa State Univ, DVM, 67, PhD(vet microbiol), 70. *Prof Exp:* Prof vet microbiol, Iowa State Univ, 70-82; VPRES, PIG IMPROV CO, INC, 85- *Concurrent Pos:* Consult, Merck, Sharp & Dohme Co, 75-76, Pig Improv Co, 76- *Honors & Awards:* H Dunne Mem, Am Asn Swine Practr. *Mem:* Am Vet Med Asn; Am Soc Microbiol; Int Pig Vet Soc; Conf Res Workers in Animal Diseases; Am Asn Swine Practr. *Res:* Bacterial diseases of the respiratory and digestive tract; swine dysentery; effect of environment on pig and human health; 3 site production of pigs. *Mailing Add:* PIG Improv Co Inc PO Box 348 Franklin KY 42134

HARRIS, DENNIS GEORGE, b Indiana, Pa. ATOMIC & MOLECULAR PHYSICS, OPTICS. *Educ:* Cornell Univ, BS, 70, PhD(appl physics), 80; Univ Ill, MS, 72. *Prof Exp:* MEM TECH STAFF, ROCKETDYNE DIV, ROCKWELL INT, 79- *Concurrent Pos:* Res assoc, Coord Sci Lab, Univ Ill, 72-74. *Mem:* Am Phys Soc. *Res:* The molecular energy transfer, as well as the visible, ultraviolet and vacuum ultraviolet spectroscopy of small molecules; visible and ultraviolet lasers. *Mailing Add:* 4193 Minnecota Thousand Oaks CA 91360

HARRIS, DENNY OLAN, b Louisville, Ky, May 2, 37; m 62; c 1. PHYCOLOGY, MICROBIOLOGY. *Educ:* Univ Louisville, BA, 61, MS, 63; Ind Univ, Bloomington, PhD(microbiol), 67. *Prof Exp:* Instr biol, Univ Louisville, 61-63; USPHS res fel microbiol, Ind Univ, Bloomington, 63-67; asst prof, 67-71, ASSOC PROF BIOL SCI, UNIV KY, 71- *Concurrent Pos:* US Dept Interior res grant, 68-; vis prof, Cambridge Univ, 73. *Mem:* Int Phycol Soc; Am Phycol Soc. *Res:* Fresh water algology, especially the study of inhibitory products produced by the algae to include the chemistry, biochemistry and ecology of algal inhibition and the effects of naturally occurring algicides upon algae, protozoans, bacteria and other forms of aquatic life. *Mailing Add:* Dept Biol Sci Univ Ky Lexington KY 40506

HARRIS, DEVERLE PORTER, b Lovell, Wyo, Jan 21, 31; m 67; c 6. MINERAL RESOURCES APPRAISAL, MINERAL SUPPLY SYSTEMS. *Educ:* Brigham Univ, BS, 56, MS, 58; Pa State Univ, PhD(mineral econ), 65. *Prof Exp:* Struct & photogeologist, Geophoto Serv, Inc, Denver, Colo, 57-59, Calgary, Can, 59-60; res asst opers res, dept mineral econ, Pa State Univ, 62-65; res geologist & geostatistician, res dept, Union Oil Co Calif, 65-66; from asst prof to prof mineral econ, dept mineral econ, Pa State Univ, 66-74; PROF & DIR MINERAL ECON, DEPT MINING & GEOL ENG, UNIV ARIZ, 74- *Concurrent Pos:* Chmn adv task force, Fed Power Comn appraisal of resource base for natural gas, 76; mem adv comt, Joint US Dept Energy & US Geol Surv prog appraisal of uranium resources, 79-81; mem, comt tech aspects of critical & strategic mat, Nat Res Coun Nat Acad sci, 79-81, panel to rev statist prog of US Bur Mines, 81-83; elected mem, US Nat Comt, Int Asn Math Geol, 85- *Mem:* Soc Mining Engrs; Am Econ Asn; Int Asn Energy Economists; Soc Econ Geologists; Int Asn Math Geol. *Res:* Development of concepts and quantitative methods for the estimation of undiscovered mineral and energy resources; design and implementation of computer systems to simulate exploration, development and production of mineral resources and provide a means of describing the potential supply stock and future supply flows. *Mailing Add:* Dept Mining & Geol Eng Col Eng Univ Ariz Tucson AZ 85721

HARRIS, DEWEY LYNN, b Red Rock, Tex, June 23, 33; m 55; c 3. ANIMAL GENETICS, STATISTICS. *Educ:* Tex A&M Univ, BS, 54, MS, 58; Iowa State Univ, PhD(animal breeding), 61. *Prof Exp:* Asst county agr agent, Tex Agr Exten Serv, 54-55; asst genetics, Tex A&M Univ, 57-58; asst animal breeding, Iowa State Univ, 58-60, from instr to asst prof statist, 60-64; biomet geneticist, De Kalb Agr Asn, Inc, 64-69, asst dir, De Kalb Agr Res, Inc, 69-71, dir poultry res, De Kalb Agr Res Inc, 71-74; chief sect animal & poultry genetics, Agr Can, 74-76; res geneticist, W Lafayette, 76-86, RES GENETICIST, US MEAT ANIMAL RES CTR, USDA AGR RES SERV, CLAY CTR, NEBR, 86- *Mem:* AAAS; Biomet Soc; Am Statist Asn; Am Soc Animal Sci; Am Genetic Asn. *Res:* Systems analysis of livestock production. *Mailing Add:* USDA Agr Res Serv US Meat Animal Res Ctr Clay Center NE 68933

HARRIS, DON NAVARRO, b New York, NY, June 17, 29; m 54; c 3. BIOCHEMISTRY, PHYSIOLOGY. *Educ:* Lincoln Univ, Pa, AB, 51; Rutgers Univ, MS, 59, PhD(biochem), 63. *Prof Exp:* Sr res chemist, Colgate Palmolive Res Ctr, 63-64; asst res specialist biochem, Rutgers Univ, 64-65; sr res investr biochem, 65-84, RES FEL, SQUIBB INST MED RES, 84- *Concurrent Pos:* Coadjutant assoc prof biochem, Univ Col, Rutgers Univ, 75-80; adj assoc prof pharmacol, Temple Univ Sch Med, 90- *Mem:* AAAS; Am Chem Soc; NY Acad Sci; Am Soc Pharmacol & Exp Therapeutics; Am Heart Asn; Sigma Xi. *Res:* Biosynthesis of cholesterol; mechanism of action of platelet aggregation, arachidonic acid metabolites and cyclic adenosinemonophosphate; isolation and purification of nucleic acids; effect of hormones on nucleic acids and protein synthesis. *Mailing Add:* Dept Cardiovasc Biochem Bristol-Myer Squibb Pharmaceut Res Inst PO Box 4000 Princeton NJ 08543-4000

HARRIS, DONALD C, b NS, Can, Jan 3, 36. MINERALOGY. *Educ:* Acadia Univ, BSc, 58; Univ Toronto, MA, 61, PhD(geol), 64. *Prof Exp:* Mine geologist, Stanrock Uranium Mine, 58-59; asst cur mineral, Royal Ont Mus, 63-67; res scientist, Mineral Sci Div, Canmet, 67-81; RES SCIENTIST, ECON GEOL MINERAL DIV, GEOL SURV CAN, 81- *Mem:* Mineral Soc Am; Mineral Soc Can. *Res:* Mineralogy of ore minerals and deposits, electron microprobe. *Mailing Add:* Econ Geol Mineral Div Geol Surv Can 601 Booth St Ottawa ON K1A 0E8 Can

HARRIS, DONALD R, JR, b Johnstown, Pa, Nov 29, 25; m 54; c 7. PHYSICS. *Educ:* Carnegie Inst Technol, BS, 48, MS, 49; Princeton Univ, MA, 52; Rensselaer Polytech Inst, PhD, 76. *Prof Exp:* Asst math, Carnegie Inst Technol, 48-49; asst physics, Princeton Univ, 50-54; sr scientist, Bettis Atomic Power Lab, Westinghouse Elec Corp, 54-61, fel scientist, 61-68; staff mem, Los Alamos Sci Lab, Univ Calif, 68-71, group leader theoret div, 71-75; res engr, 75-77, ASSOC PROF NUCLEAR ENG & DIR RPI REACTOR, RENSSELAER POLYTECH INST, 77- *Honors & Awards:* Meritorious Performance in Reactor Opers Award, Am Nuclear Soc. *Mem:* Soc Comput Simulation; Econ Hist Soc; Am Phys Soc; fel Am Nuclear Soc. *Res:* Reactor physics; nuclear cross section; transport methods; shielding; Monte Carlo; systems analysis. *Mailing Add:* Dept Nuclear Eng & Eng Physics Rensselaer Polytech Inst Troy NY 12181

HARRIS, DONALD WAYNE, b Ft Scott, Kans, Sept 23, 42; m 64; c 3. BIO-ORGANIC CHEMISTRY. *Educ:* Univ Mo, BS, 66, MS, 68, PhD(agr chem), 74. *Prof Exp:* Lab technician, Chemagro Corp, 64-65; phys sci asst, US Army, 68-70; res chemist, 74-77, MGR CARBOHYDRATE POLYMER RES, CLINTON CORN PROCESSING CO, 77- *Mem:* Am Chem Soc. *Res:* Research and development concerning new products derived from polymers associated with corn. *Mailing Add:* 3333 E Lost Bridge Rd Decatur IL 62521

HARRIS, DOROTHY VIRGINIA, b Fishersville, Va, July 19, 31. EXERCISE PHYSIOLOGY, SPORT PSYCHOLOGY. *Educ:* James Madison Univ, BS, 53; Univ NC, Greensboro, MS, 58; Univ Iowa, PhD(phys educ), 65. *Prof Exp:* Instr phys educ, Wilson Mem High Sch, 53-54, Hollins Col, 54-56, Univ NC, Greensboro, 56-63 & Univ Iowa, 63-65; asst prof, James Madison Univ, 63-64; fel exercise physiol, 64-69, assoc prof, 70-73, PROF PHYS EDUC, PA STATE UNIV, UNIVERSITY PARK, 73- *Concurrent Pos:* Oscar Mayer fel, George Williams Col, 61; res training fel, USPHS, 63-69; dir res ctr women & sport, Sports Res Inst, 74-; exec bd, Nat Asn Phys Educ, 77-, Int Soc Sport Psychol, 73- & Res Consortium, Am Asn Health Phys Educ Recreation, 76-; vpres & mem bd trustees, Women's Sports Found. *Mem:* NAm Soc Psychol Sport & Phys Activ (pres, 74-75); Int Soc Sports Psychol (treas, 77); Am Asn Health Phys Educ & Recreation; Nat Asn Girls & Women's Sports; Women's Sports Found; fel Am Acad Phys Educ; fel Am Col Sports Med; fel Assoc Adv Appl Sport Psychol; fel Res Consortium; AAHPERD. *Res:* Psychological and physiological response to physical activity and sport with specific interest in the female; role of exercise in countering depression and anxiety; interaction of behavior and performance in sport and physical acitivity environments. *Mailing Add:* Pa State Univ Rec Bldg University Park PA 16802

HARRIS, DURWARD SMITH, b Dickson, Tenn, Mar 30, 31; m 68; c 1. BIO-ORGANIC CHEMISTRY. *Educ:* Austin Peay State Col, BS, 54; Univ Tenn, MS, 61, PhD(proteins), 63. *Prof Exp:* Anal chemist, Tenn Valley Auth, 54-57; from asst prof to assoc prof, 62-69, Chmn Dept, 79-85, PROF CHEM, AUSTIN PEAY STATE UNIV, 69- *Mem:* AAAS; Am Chem Soc; Sigma Xi. *Res:* Separation and identification of protein mixtures by ion-exchange. *Mailing Add:* Dept Chem Austin Peay State Univ Clarksville TN 37040

HARRIS, EDWARD DAVID, b Chicago, Ill, Dec 1, 38; m 60; c 3. BIOCHEMISTRY. *Educ:* Univ Ill, Urbana, AB, 60, MS, 65, PhD(biochem), 68. *Prof Exp:* Res asst biochem, Univ Ill, 63-68; NIH fel, Univ Chicago, 68-70; USPHS res assoc biochem & instr, Univ Mo, Columbia, 70-73; asst prof, 73-80, PROF BIOCHEM & NUTRIT, TEX A&M UNIV, 80- *Concurrent Pos:* Sr Fulbright award, Australia, 84-85. *Mem:* Am Soc Biol Chemists; Am Inst Nutrit; Soc Exp Biol & Med. *Res:* Human nutrition; biochemical function of copper. *Mailing Add:* Dept Biochem & Biophys Tex A&M Univ College Station TX 77843

HARRIS, EDWARD DAY, JR, b Philadelphia, Pa, July 7, 37; m 58; c 3. MEDICINE, RHEUMATOLOGY. *Educ:* Dartmouth Col, AB, 58; Harvard Univ, MD, 62. *Prof Exp:* Intern med, Mass Gen Hosp, Boston, 62-64, resident, 66-67; from instr to asst prof, Harvard Med Sch, 69-70; from asst prof to assoc prof med & chief, Connective Tissue Dis Sect Dartmouth Med Sch, 70-82, dir, Multipurpose Arthritis Ctr, 77-82, Eugene W Leonard prof med, 80; PROF & CHMN, DEPT MED, UNIV MED & DENT NJ-RUTGERS MED SCH, 82- *Concurrent Pos:* Asst physician, Mass Gen Hosp, Boston, 69-70; Nat Inst Arthritis & Metab Dis spec fel, Mass Gen Hosp, Boston, 67-69 & res career develop award, Dartmouth Med Sch, 70-75. *Mem:* Am Rheumatism Asn (pres, 85-86). *Res:* Mechanisms of destruction of connective tissue in man; study of collagenase release by synovial cells. *Mailing Add:* Stanford Univ Med Ctr Med Rm S102 Stanford CA 94305-5109

HARRIS, EDWARD GRANT, b Morristown, Tenn, Mar 10, 24; m 62; c 1. THEORETICAL PHYSICS. *Educ:* Univ Tenn, BS, 48, MS, 50, PhD(physics), 53. *Prof Exp:* Physicist, Naval Res Lab, 53-57; from asst prof to assoc prof, 57-64, PROF PHYSICS, UNIV TENN, KNOXVILLE, 64- *Concurrent Pos:* Consult, Oak Ridge Nat Lab, 57-80. *Mem:* Am Phys Soc; Am Asn Physics Teachers; Sigma Xi. *Res:* Plasma physics; quantum physics; relativity neural networks. *Mailing Add:* Dept Physics Univ Tenn Knoxville TN 37996-1200

HARRIS, EDWARD LYNDOL, b Roby, Tex, Nov 12, 33; m 54; c 4. PHYSICAL CHEMISTRY. *Educ:* McMurry Col, BA, 56; La State Univ, MS, 58, PhD(chem), 61. *Prof Exp:* From asst prof to assoc prof, 61-68, PROF CHEM, MCMURRY COL, 68- *Mem:* Am Chem Soc. *Res:* Ion exchange chromatography; ion exchange equilibria of inorganic complex ions; membrane electrodes. *Mailing Add:* Dept Chem McMurry Univ S 14th & Sayles Blvd Abilene TX 79697

HARRIS, EDWIN RANDALL, b Little Rock, Ark, Aug 15, 32; m 58. ORGANIC CHEMISTRY. *Educ:* Univ Okla, BS, 53, MS, 56; Univ Calif, PhD(chem), 59. *Prof Exp:* Asst chem, Univs Okla & Calif, 53-59, assoc, Univ Calif, 58-59; from asst prof to assoc prof, 59-70, PROF CHEM, CALIF STATE COL, LONG BEACH, 70- *Concurrent Pos:* Fulbright grant, Univs Tübingen & Hamburg, 54-55. *Mem:* Am Chem Soc. *Res:* Analysis and structure elucidation of terpenoid natural products from plants; chemistry of marine animals; mechanisms of rearrangements. *Mailing Add:* Dept Chem Calif State Univ 1250 Bellflower Blvd Long Beach CA 90840

HARRIS, ELIZABETH FORSYTH, b Kilgore, Tex, Aug 14, 35; m 56; c 4. ENVIRONMENTAL MICROBIOLOGY, IMMUNOLOGY. *Educ:* Tex Wesleyan Col, BA, 56; Tex Christian Univ, MA, 64; Univ Tex Southwestern Med Sch, PhD(microbiol), 70; Am Bd Med Microbiol, cert. *Prof Exp:* Bacteriologist, Ft Worth Water Dept, 57; teacher math & sci, Hurst-Euless-Bedford Schs, 58-62; lab instr biol, Tex Christian Univ, 62-63; teacher sci, New Orleans Pub Schs, 64-65; instr biol, La State Univ, New Orleans, 65-66; chmn dept, 70-83, ASSOC PROF MICROBIOL & IMMUNOL, TEX COL OSTEOP MED, 70- *Concurrent Pos:* Microbiol consult, Nat Bd Examrs Osteop Physicians & Surgeons, 73-86; consult, air conditioning & heat pump mfrs. *Mem:* Am Soc Microbiol. *Res:* Biofilms in air conditioning systems; hormones, stress and immunity; hyperbaric oxygen effects; depression and natural killer cells. *Mailing Add:* Dept Microbiol Tex Col Osteop Med 3500 Camp Bowie Blvd Ft Worth TX 76107

HARRIS, ELIZABETH HOLDER, b Winston-Salem, NC, Oct 1, 44; m 65; c 3. GENETICS, MOLECULAR BIOLOGY. *Educ:* Swarthmore Col, BA, 65; Yale Univ, PhD(microbiol), 71. *Prof Exp:* RES ASSOC GENETICS, DEPT BOT, DUKE UNIV, 72-, SR RES SCIENTIST, 83- *Mem:* Int Soc Plant Molecular Biol; Genetics Soc Am; Asn Women Sci; Phycological Soc Am. *Res:* Genetic control of organelle biogenesis; genetics of Chlamydomonas reinhardtii; plant molecular biology. *Mailing Add:* Dept Bot Duke Univ Durham NC 27706

HARRIS, ELLIOTT STANLEY, b New York, NY, June 27, 22; m 45; c 2. RESEARCH ADMINISTRATION, OCCUPATIONAL HEALTH. *Educ:* Univ Colo, BA, 48; Univ Southern Calif, MS, 50, PhD(biochem), 54; Nat Registry Clin Chem, cert. *Prof Exp:* Instr biochem, Dent Sch, Univ Southern Calif, 48-52, res assoc, Sch Med, 54; Hodgkins res fel, Roswell Park Mem Inst, Buffalo, 54-55, sr cancer res scientist clin biochem, 55-56; sr res scientist biochem, Wyeth Inst Med Res, 56-63; chief toxicol lab, Manned Spacecraft Ctr, NASA, 63-73; dir, Div Biomed & Behav Sci, Nat Inst Occup Safety & Health, 73-81; adj assoc prof toxicol, Sch Med, Univ Cincinnati, 74-81; dep dir, Nat Inst Occup Safety & Health, 81-86; ADJ ASSOC PROF ENVIRON & OCCUP HEALTH, EMORY UNIV, 84- *Concurrent Pos:* Vis prof, Univ Ariz, 80-81; consult, 86- *Mem:* NY Acad Sci; Am Asn Clin Chem; Am Indust Hyg Asn; Soc Toxicol; Sigma Xi. *Res:* Enzyme isolation and characterization; neurochemistry; enzymology; inhalation and biochemical toxicology particularly with respect to long term low level continuous exposures. *Mailing Add:* 3215 Bolero Pass Atlanta GA 30341

HARRIS, EMMETT DEWITT, JR, b Orlando, Fla, July 27, 25; m 48; c 2. ECONOMIC ENTOMOLOGY. *Educ:* Univ Fla, BSA, 51; Cornell Univ, PhD(econ entom), 56. *Prof Exp:* Lab technician, Citrus Exp Sta, Univ Fla, 51-52; asst entom, Cornell Univ, 52-55; from asst entomologist to assoc entomologist, Everglades Exp Sta, Univ Fla, 56-66; exten entomologist & pesticides chems coordr, coop exten serv, Univ Ga, 66-67; RETIRED. *Mem:* Entom Soc Am. *Res:* Coordination of pesticide information and policies. *Mailing Add:* 178 Spruce Valley Rd Athens GA 30605

HARRIS, ERIK PRESTON, b Columbus, Ohio, Dec 16, 38; m 62; c 2. SOLID STATE PHYSICS. *Educ:* Cornell Univ, BEP, 61; Univ Ill, Urbana, MS, 63, PhD(physics), 67. *Prof Exp:* Res assoc physics, Univ Ill, Urbana, 66-67; res staff mem, Thomas J Watson Res Ctr, IBM Corp, 67-90; HARDWARE TECHNOL CONSULT, 90- *Mem:* Am Phys Soc; Sigma Xi. *Res:* Semiconductor device physics; superconductivity; applications of cryogenics in computer technology. *Mailing Add:* 3B56 Old Orchard Rd Armonk NY 10504

HARRIS, ERNEST JAMES, b North Little Rock, Ark, May 24, 28; m 54; c 3. ENTOMOLOGY. *Educ:* Univ Ark, Pine Bluff, BS, 51; Univ Minn, St Paul, MS, 59; Univ Hawaii, Manoa, PhD(entom), 75. *Prof Exp:* Teacher biol, Univ Ark, Pine Bluff, 59-62; res entomologist, Agr Res Serv, USDA, Hawaii, 62-69; leader-coordr entom, US AID, Tunis, Tunisia & Rabat, Morocco, 69-72; RES LEADER ENTOM, AGR RES SERV, USDA, HONOLULU, 72-, TECH ADV, NAT RES PROG INSECT CONTROL FRUITS & VEG, 75- *Mem:* Sigma Xi; Entom Soc Am. *Res:* Basic and applied research to develop methods for the detection, control and eradication of fruit flies to improve food production. *Mailing Add:* 45-170 Ohaha Pl Kaneohe HI 96744

HARRIS, FOREST K, b Gibson Co, Ind, Aug 26, 02. ELECTRICITY, PRECISE MEASUREMENTS. *Educ:* Univ Okla, BS, 21, MS, 23; Johns Hopkins Univ, PhD(physics), 32. *Prof Exp:* Instruments sect, US Nat Bur Standards, 23-62, chief, Absolute Measurements Sect, 62-72; RETIRED. *Mem:* Fel Inst Elec & Electronics Engrs. *Res:* Absolute measurements; relations between electrical & mechanical units in a rational system; systems for measuring power loss in dielectrics. *Mailing Add:* 9905 Wildwood Rd Kensington MD 20895

HARRIS, FRANCIS LAURIE, b Nebraska City, Nebr, July 2, 39. ORGANIC CHEMISTRY. *Educ:* Univ Tulsa, BCh, 61; Univ Calif, Los Angeles, PhD(org chem), 66. *Prof Exp:* Fel, Harvard Univ, 67-68; from asst prof to assoc prof, 68-77, PROF CHEM, CALIF STATE UNIV, NORTHRIDGE, 77- *Mem:* Am Chem Soc. *Res:* Carbonium-ion reactions and mechanisms; synthesis of spin labels; synthetic methods. *Mailing Add:* Dept Chem Calif State Univ Northridge CA 91330

HARRIS, FRANK BOWER, JR, b New York, NY, May 24, 27; m 49; c 3. PHYSICS. *Educ:* Mass Inst Technol, BS, 49, PhD(physics), 55. *Prof Exp:* Mem staff, Div Indust Coop, Mass Inst Technol, 49-50, asst, 50-53; from asst prof to assoc prof physics, Utica Col, 53-58; res engr, Stanford Res Inst, 58-62; sr supvry engr, TRG-West, 62-67; res scientist, Granger Assocs, 67-68; mgr data processing & comput applns & tech adv, TCI Inc, Mountain View, 68-87; COMPUTER SOFTWARE CONSULT, 87- *Mem:* Inst Elec & Electronics Engrs. *Res:* Electromagnetic theory; computer applications and systems; antennas; structural engineering. *Mailing Add:* 196 Meadowood Dr Portola Valley CA 94025

HARRIS, FRANK EPHRAIM, JR, b Boston, Mass, Aug 26, 29; m 52. CHEMICAL PHYSICS. *Educ:* Harvard Univ, AB, 51; Univ Calif, PhD(chem), 54. *Prof Exp:* Instr chem, Harvard Univ, 53-56; asst prof, Univ Calif, 56-59; from asst prof to assoc prof, Stanford Univ, 59-68; dean col sci, 73-75, PROF PHYSICS, UNIV UTAH, 68-, PROF CHEM, 69- *Concurrent Pos:* Sloan Found fel, Univ Calif, 57-59; mem planning comt study nat ctr comput in chem, Nat Res Coun, 72-73. *Mem:* Am Chem Soc; Am Phys Soc. *Res:* Statistical mechanics; quantum mechanics; solid state theory. *Mailing Add:* Dept Physics & Chem Univ Utah Salt Lake City UT 84112

HARRIS, FRANK WAYNE, b St Louis, Mo, Jan 28, 42; m 70; c 2. POLYMER CHEMISTRY, ORGANIC CHEMISTRY. *Educ:* Univ Mo, BS, 64; Univ Iowa, MS, 66, PhD(org chem), 68. *Prof Exp:* Captain, US Army Environ Hyg Agency, 68-70; instr org chem, Towson State Col, 69-70; prof chem, Wright State Univ, 70-83; PROF, DEPT POLYMER SCI, UNIV AKRON, 83- *Concurrent Pos:* Consult, Ethyl Corp, Goodyear Tire & Rubber Co & W H Brady Co. *Honors & Awards:* Space Act Award, NASA. *Mem:* Am Chem Soc; Controlled Release Soc. *Res:* Polymer synthesis and characterization with emphasis on polymer structure property relationships; preparation and development of controlled-release formulations. *Mailing Add:* Dept Polymer Sci Univ Akron Akron OH 44325

HARRIS, FRANKLIN DEE, agricultural engineering, for more information see previous edition

HARRIS, FRANKLIN STEWART, JR, b Logan, Utah, May 24, 12; m 36; c 11. PHYSICS. *Educ:* Brigham Young Univ, AB, 31, MA, 36; Calif Inst Technol, PhD(physics), 41. *Prof Exp:* Asst, Brigham Young Univ, 35-36 & Calif Inst Technol, 37-41; lectr physics, Univ BC, 41-43; from asst prof to prof, Univ Utah, 43-63 & 80-81; mem tech staff, Aerospace Corp, 63-70; sr scientist, Dept Chem & Inst Oceanog, Old Dominion Univ, 71-73, res prof physics, Geophys Sci & Oceanog, 73-78; CONSULT, 78- *Concurrent Pos:* Consult, Sperry Utah Co, 59-62, Intermt Weather Inc, 61-63 & Nat Oceanog Atmospheric Admin, 77-79. *Honors & Awards:* Distinguished Serv Award, Optical Soc Am, 88. *Mem:* Fel AAAS; fel Optical Soc Am; Am Meteorol Soc; fel Brit Inst Phys; Am Asn Physics Teachers. *Res:* Mass spectroscopy; oligodynamic effect of silver; diffraction of visible light; polarization by diffraction; cloud physics; infrared radiation; atmospheric optics; visibility; information retrieval; air pollution. *Mailing Add:* 81 W Main St Rockville UT 84763

HARRIS, GALE ION, b Arlington, Calif, Aug 7, 35; m 56; c 2. NUCLEAR PHYSICS, MEDICAL PHYSICS. *Educ:* Univ Kans, BS, 57, MS, 59, PhD(physics), 62; Mass Inst Technol, SM, 73; Cooley Law Sch, JD, 85. *Prof Exp:* Group leader nuclear physics, Aerospace Res Labs, USAF, 62-69, actg dir, Gen Physics Res Lab, 69-72, dep dir, Solid State Physics Lab, 73-74; dir off mgt res, Johns Hopkins Univ Med Sch, 74-75; ASSOC PROF PHYSICS & RADIOL, MICH STATE UNIV, 75-, ASSOC CHMN, DEPT RADIOL, 80- *Concurrent Pos:* Mem nuclear physics subpanel, Off Aerospace Res, 64-70; adminr & asst prof, Dept Radiol, Johns Hopkins Univ, 74-75; mem, Energy Res & Develop Admin Task Force Nuclear Med, 75-77; trustee & pres, Mich Res Ctr, Inc, East Lansing, 75-; consult, US Dept Energy, 74-82, NIH, 85- *Honors & Awards:* Res & Develop Award, USAF, 63. *Mem:* AAAS; fel Am Phys Soc; Soc Nuclear Med; Am Asn Physicists Med; Sigma Xi; Asn Univ Radiologists. *Res:* Nuclear structure physics; proton capture reactions; technology transfer from physics to medicine; application of imageing methods to medical research and diagnosis. *Mailing Add:* 1312 Basswood Circle Mich State Univ East Lansing MI 48823

HARRIS, GEORGE CHRISTE, b Thessaly, Greece, Apr 2, 16; US citizen; m 42; c 2. ORGANIC CHEMISTRY. *Educ:* Harvard Univ, AB, 38, MA, 40, PhD(org chem), 42. *Prof Exp:* Mem Nat Defense Res Comt, Harvard Univ, 41-42; sr res chemist, Hercules Res Ctr, 42-80; CONSULT, 80- *Mem:* Am Chem Soc. *Res:* Isolation and structure of rosin acids; composition of commercial rosins; composition of tall oil; Mannich reactions; analgesic action; casein derivatives; starch chemistry; emulsion polymerization; synthesis of organic specialty chemicals; pharmacentical intermediates. *Mailing Add:* 1208 Norbee Dr Wilmington DE 19803

HARRIS, GORDON MCLEOD, b China, July 23, 13; nat US; m 43; c 3. PHYSICAL INORGANIC CHEMISTRY. *Educ:* Univ Sask, BSc, 39, MSc, 40; Harvard Univ, AM, 42, PhD(phys chem), 43. *Prof Exp:* Res chemist, Nat Res Coun Can, 44-45; asst prof chem, Univ Sask, 45-48; sr lectr phys chem, Univ Melbourne, 48-52; assoc prof,State Univ NY Buffalo, 53-55, head dept, 56-69, Prof Chem, 55-85, Larkin chair chem, 61-85; RETIRED. *Concurrent Pos:* Res assoc, Univ Wis, 52-53; hon res assoc, Univ Col, Univ London, 61; vis prof, Australian Nat Univ, 69 & Univ Calif, Riverside, 76. *Mem:* AAAS; Am Chem Soc; Am Inst Chemists; Royal Soc Chem; Sigma Xi. *Res:* Chemical kinetics; isotope effects; mechanisms of inorganic complex reactions. *Mailing Add:* 23438 Villena Mission Viejo CA 92692

HARRIS, GRANT ANDERSON, b Logan, Utah, July 13, 14; m 39; c 4. RANGE ECOLOGY. *Educ:* Utah State Univ, BS, 39, PhD(range mgt), 65; Univ Idaho, MS, 41. *Prof Exp:* Res scientist, US Forest Serv, 38-40; Potlatch fel, Univ Idaho, 40-41; res scientist, US Forest Serv, 41-52; asst prof & exten forester, Utah State Univ, 52-56; assoc prof, 56-67, chmn dept & extensionist, 67-80, PROF, WASH STATE UNIV, 67-, RES SCIENTIST RANGE MGT, 56-, EMER EXTENSIONIST COOP PROGS & CONSULT, 80-; CHMN BD, DECAGON DEVICES INC, SCI CONSULT, PULLMAN WA, 82- *Concurrent Pos:* Consult, Wash Dept Natural Resources, 60-80, WPakistan Agr Univ & sr ecologist consult, Am Ecol Soc, 83; mem, Gov Tech Comt N Cascades Recreation Study Team, 63-65; arid land ecol sci comt, AEC, 64-75, chmn, 70; mem, Coun Forest Sch Exec, 67-80; chmn, Inland Empire For Adv Coun, 67-80; chmn res adv comt, Intermountain Forest & Range Exp Sta, 67-80; chmn res adv comt, Pac NW Forest & Range Exp Sta, 67-80; local contact, President's Coun Environ Qual; chmn, Coun Agr Sci & Tech, 76; team leader, US AID Eval Econ Feasibility Range Forage Seed Prod, Morocco, 77; mem agr prog, Yemen, Arab Repub, 79 & agr exten prog, Chile, 80. *Mem:* Fel AAAS; Soc Am Foresters; fel Soc Range Mgt; Sigma Xi; Ecol Soc Am. *Res:* Genecology of and competition between annual and perennial range grasses; range ecosystems; plant genecology; soil-plant-water relations. *Mailing Add:* Dept Natural Res Sci Wash State Univ Pullman WA 99164-6410

HARRIS, GROVER CLEVELAND, JR, b Belington, WVa, Mar 24, 31; m 51; c 3. ANIMAL PHYSIOLOGY, ENVIRONMENTAL PHYSIOLOGY. *Educ:* WVa Univ, BS, 52, MS, 56; Univ Md, PhD(avian physiol), 60. *Prof Exp:* Asst poultry breeding, WVa Univ, 54-56; instr poultry sci, Univ Md, 56-60; from asst prof to assoc prof, 60-70, PROF PHYSIOL, UNIV ARK, FAYETTEVILLE, 70- *Concurrent Pos:* Vis scientist, UK Poultry Res Ctr, Edinburgh, Scotland, 82; founder, Univ Teaching Acad. *Honors & Awards:* Poultry Sci Res Award, 63; Univ Ark Fac Teaching & Res Award, 69. *Mem:* Fel AAAS; fel Poultry Sci Asn; World Poultry Sci Asn; Soc Study Reprod. *Res:* Cryobiology; reproductive physiology; environmental physiology. *Mailing Add:* Dept Animal & Poultry Sci Univ Ark Fayetteville AR 72701

HARRIS, GUY H, b Calif, Oct 2, 14; m 40; c 4. ORGANIC CHEMISTRY. *Educ:* Univ Calif, BS, 37; Stanford Univ, AM, 39, PhD(chem), 41. *Prof Exp:* Lab asst, Shell Develop Co, 37-38; org chemist, William S Merrell Co, 41-45 & Fiberboard Paper Prod Inc, 45-46; org chemist, Dow Chem Co, 46-59, assoc scientist, 59-62; lectr, Univ Ghana, 62-63, sr lectr, 63-64; ASSOC SCIENTIST, DOW CHEM CO, 64- *Concurrent Pos:* Chmn dept chem, John F Kennedy Univ, 65-72. *Mem:* AAAS; Am Inst Mining, Metall & Petrol Eng; The Chem Soc; WAfrican Sci Asn; Am Chem Soc. *Res:* Mining chemicals. *Mailing Add:* Merck Sharp Dohme Res Labs PO Box 2000 R80Y-360 R80Y-355 Rahway NJ 07065

HARRIS, HAROLD H, b Council Bluffs, Iowa, Mar 12, 40; m 66; c 2. PHYSICAL CHEMISTRY. *Educ:* Harvey Mudd Col, BS, 62; Mich State Univ, PhD(phys chem), 67. *Prof Exp:* Fel phys chem, Univ Calif, Irvine, 66-67, instr chem, 67-70; asst prof, 70-75, ASSOC PROF CHEM, UNIV MO, ST LOUIS, 75- *Concurrent Pos:* Vis prof, Solar Energy Res Inst, 85; vis scientist, Univ Chicago, 78. *Mem:* AAAS; Am Chem Soc; Am Phys Soc; Fedn Am Scientists. *Res:* Chemical kinetics and dynamics; materials for high photon flux environments; modeling chemical systems; polarized spectroscopy. *Mailing Add:* Dept Chem Univ Mo St Louis MO 63121

HARRIS, HAROLD JOSEPH, b Del Rio, Tex, Aug 18, 20; m 46; c 4. PSYCHIATRY, CHILD PSYCHIATRY. *Educ:* Univ Tex, BA, 46; Long Island Col Med, MD, 49. *Prof Exp:* Intern, Mt Zion Hosp, San Francisco, 49-50; adult psychiat trainee, Menninger Found, 50-52; fel child psychiat, Mass Gen Hosp, 52-53 & NC Mem Hosp, Chapel Hill, 55-56; from instr to asst prof psychiat, Sch Med, Univ NC, 56-60; assoc prof psychiat, Med Ctr, Duke Univ, 60-87; RETIRED. *Mem:* Fel Am Psychiat Asn; Am Acad Child Psychiat. *Res:* Outcome of psychotherapy; psychological crises during pregnancy; psychopharmacotherapy with children; juvenile delinquency. *Mailing Add:* 2628 McDowell St Durham NC 27705

HARRIS, HARRY, b Manchester, Eng, Sept 30, 19; m 48; c 1. GENETICS. *Educ:* Cambridge Univ, Eng, BA, 41, MB & BChir, 43, MD, 49. *Prof Exp:* Prof & chmn biochem, Kings Col, Univ London, 60-65; Galton prof human genetics, Univ Col, Univ London, 65-76; HARNWELL PROF HUMAN GENETICS, UNIV PA, 76- *Concurrent Pos:* Dir, Med Res Coun, Human Biochem Genetics Unit, 67-76. *Mem:* Foreign mem Nat Acad Sci; fel Royal Soc; Am Soc Human Genetics; Genetics Soc. *Mailing Add:* Dept Human Genetics 181 Med/G3 Univ Pa 36th & Hamilton Walk Philadelphia PA 19104

HARRIS, HENRY EARL, b Ft Valley, Ga, Jan 28, 36; m 62; c 1. ORGANIC CHEMISTRY, BIOCHEMISTRY. *Educ:* Ga Inst Technol, BS, 59, PhD(org chem), 63. *Prof Exp:* Res chemist, Chemstrand Res Ctr, Inc, 62-66; assoc prof, 66-71, PROF CHEM & HEAD DEPT, ARMSTRONG STATE COL, 71- *Concurrent Pos:* Consult biochemist, St Josephs Hosp, Savannah, Ga, 69- *Mem:* AAAS; Am Chem Soc; Am Asn Clin Chemists. *Res:* Structure reactivity correlations; free-radical reactions; linear free energy relationships; polymer chemistry; clinical chemistry. *Mailing Add:* Dept Chem Armstrong State Col Abercom Ext Rd Savannah GA 31406

HARRIS, HENRY WILLIAM, b Catawba, NC, Jan 6, 19; m 51; c 3. MEDICINE. *Educ:* Univ NC, AB, 40; Harvard Univ, MD, 43; Am Bd Internal Med & Am Bd Pulmonary Dis, dipl, 51. *Prof Exp:* Med intern, Fourth Med Serv, Thorndike Mem Lab, Boston City Hosp, 44-45, asst resident med, 45-46; resident, Chest Serv, Bellevue Hosp, NY, 47; staff mem, Dept Med, Gundersen Clin, Wis, 48-53; asst prof med, Col Med, Univ Utah, 55-59, assoc prof, 59-60; prof & chmn dept, Woman's Med Col Pa, 60-67; chmn dept med, Cath Med Ctr, Brooklyn & Queens, 67-71; PROF CLIN MED, SCH MED, NY UNIV, 71-; dir chest serv, 83-90, DIR GRAD TRAINING PROG, BELLEVUE HOSP, 71- *Concurrent Pos:* Chief pulmonary dis serv, Vet Admin Hosp, Salt Lake City, 55-57 & 59-60, chief med serv, 58-59; chief med serv, Hosp Woman's Med Col, 60-67; consult, Philadelphia Gen Hosp, Philadelphia & Wilmington Vet Admin Hosps, Valley-Forge Army Hosp, Phoenixville, Pa & Londis State Hosp; res fel, Thorndike Mem Lab, Boston City Hosp, 44 & 46; consult, Cath Med Ctr Brooklyn-Queens, 71- & NY Vet Admin Hosp, 73-; attending physician, Univ Hosp, NY Univ Med Ctr, 71- *Mem:* Am Thoracic Soc; NY Acad Med (pres, 63-64); Am Lung Asn; Am Bur Med Advan, China (pres, 87); fel Am Col Physicians. *Res:* Pulmonary diseases; human genetics; internal medicine. *Mailing Add:* Bellevue Hosp 462 First Ave New York NY 10016

HARRIS, HENSON, b Graves Co, Ky, Dec 12, 12; m 34; c 2. MATHEMATICS, ACADEMIC ADMINISTRATION. *Educ:* Murray State Col, BS, 34; Vanderbilt Univ, MA, 41; Univ Okla, EdD(math, educ), 53. *Prof Exp:* Instr pub schs, Fla, 35-41; head dept math, Campbell Col, 41-42; instr math & physics, Maxwell AFB, 42; instr math, Univ Tenn, 46 & Univ Ill, 47; prof & chmn dept, Okla Baptist Univ, 47-54; acad dean, Wayland Col, 54-57; admin vpres & dean, Georgetown Col, 57-62; dean instr, Slippery Rock State Col, 62-65; exec vpres & dean, Union Univ, Tenn, 65-66; vpres acad affairs & prof math, 66-80, EMER DEAN, CULVER STOCKTON COL, 80- *Concurrent Pos:* Asst, Vanderbilt Univ, 40-41. *Mem:* Am Math Soc. *Res:* Mathematics and education. *Mailing Add:* Rte 1 Box 417 Mayfield KY 42066

HARRIS, HERBERT H, b Manor, Tex, June 22, 06; m 32; c 1. MEDICINE. *Educ:* Baylor Univ, AB & MD, 29; Am Bd Otolaryngol, dipl, 34. *Prof Exp:* Intern, Sta Hosp, Ft Sam Houston, Tex, 29-30; resident otolaryngol, Long Island Col Med & Hosp, 30-33; mem staff, Chevalier Jackson Clins, Temple Univ, 40; clin asst prof otolaryngol, Baylor Col Med, 43-48, clin prof, 48-51, prof, 51-67; EMER PROF & CONSULT, BAYLOR, VET ADMIN, JEFFERSON DAVIS HOSP, 67- *Concurrent Pos:* Mem staff, Univ & Grad Hosps, Univ Pa, 45-46; sr attend otolaryngologist, Ben Taub Gen & St Luke's Episcopal Hosps; chief otolaryngol serv, Hermann Hosp; sr attend, Methodist Hosp; consult, Tex Children's Hosps, San Jacinto Mem Hosp, Baytown & St Joseph's Infirmary. *Honors & Awards:* Cert of Award, Am Acad Ophthal & Otolaryngol, 68; Pres Citation, Am Acad Otolaryngol, Head & Neck Surg, 81. *Mem:* AMA; Am Laryngol, Rhinol & Otol Soc (vpres, 67); Am Acad Ophthalmol & Otolaryngol; Am Broncho-Esophagol Asn (vpres, 63); Am Laryngol Asn. *Mailing Add:* 4718 Hallmark Dr Apt 807 Houston TX 77056

HARRIS, HOLLY ANN, b Denver, Colo. GROUP MEORY LGRAPH MEORY AS APPLIED TO SOLID STATE STRUCTURES, HISTORY OF WOMEN IN SCIENCE. *Educ:* Harvey Mudd Col, Calif, BS, 82; Univ Wis-Madison, PhD(chem), 88. *Prof Exp:* Res assoc crystallog, Univ Wis-Madison, 88; asst prof inorg phys chem, Grinnell Col, Iowa, 88-90; ASST PROF PHYS CHEM, CREIGHTON UNIV, 90- *Concurrent Pos:* Clare Booth Luce prof sci, Luce Found, 90. *Mem:* Am Chem Soc; Sigma Xi. *Res:* Electronic structure and bending relationship in novel main group organo-metalic and inorganic molecules. *Mailing Add:* Chem Dept Creighton Univ Omaha NE 68178-0104

HARRIS, HUBERT ANDREW, b Durant, Okla, July 18, 09; m 30; c 1. PHYTOPATHOLOGY, MYCOLOGY. *Educ:* Southeastern State Col, AB, 30; Univ Ill, MS, 32, PhD(bot), 38. *Prof Exp:* Asst biol, Southeastern State Col, 28-30, assoc prof biol, 38-42; asst botanist, Ill State Natural Hist Surv, 30-33; asst bot, Univ Ill, 35-37; asst prof, SDak State Col, 42-43, assoc plant pathologist, Exp Sta, 43; plant pathologist, Bur Plant Indust, Soils & Agr Eng, USDA, 43-44; Prof biol, Univ PR, 44-46; asst prof biol, Univ Colo, 46-49; prof bot, San Jose State Univ, 49-76; RETIRED. *Mem:* AAAS; Bot Soc Am; Am Phytopath Soc; Mycol Soc Am. *Res:* Mycology; antibiotics. *Mailing Add:* 1145 Husted Ave San Jose CA 95125

HARRIS, HUGH COURTNEY, b New Rochelle, NY, Dec 21, 47. ASTROPHYSICS. *Educ:* Cornell Univ, BS, 70; Univ Wash, PhD(astron), 80. *Prof Exp:* Vis res fel, Dominion Astrophys Observ, 80-85; ASTRONR, US NAVAL OBSERV, 85- *Concurrent Pos:* Res fel, McMaster Univ, 82-84. *Mem:* Am Astron Soc; Astron Soc Pac. *Res:* Motions and chemical abundances in stars; star clusters; variable stars; structure of galaxy and Magellanic clouds; globular clusters. *Mailing Add:* US Naval Observ PO Box 1149 Flagstaff AZ 86002

HARRIS, ISADORE, theoretical physics; deceased, see previous edition for last biography

HARRIS, J DOUGLAS, b Dallas, Tex, Apr 13, 39; m 65; c 2. COMPUTER NETWORKING, DATABASE. *Educ:* Univ Kans, BA, 66, MA, 67, PhD(math), 69. *Prof Exp:* Chmn dept, 79, PROF MATH, MARQUETTE UNIV, 69- *Mem:* Am Math Soc; Math Asn Am; Asn Comput Mach. *Res:* Topology; mathematical logic; computer language; applying topological ideas to computer networks and distributed databases; data flow architectures and languages. *Mailing Add:* Dept Math Marquette Univ 540 N 15th St Milwaukee WI 53233

HARRIS, J(OHN) S(TERLING), b Okmulgee, Okla, Dec 20, 20; m 43; c 2. CHEMICAL ENGINEERING. *Educ:* Mo Sch Mines, BS, 42; Washington Univ, St Louis, MS, 53. *Prof Exp:* Anal chemist, 42-43, chemist, Prod Develop Lab, 43-46, proj specialist, agr chem, 46-48, admin asst to dir develop, 48-50, proj specialist pharmaceut, 50-51, proj specialist mkt res, 51-53, supvr functional fluid sales, 53-55, mgr mkt res, 55-58, MGR AGR CHEM DEVELOP, MONSANTO CO, 58- *Mem:* Am Chem Soc; Sigma Xi. *Res:* Industrial administration; analysis of techniques of chemical development; relationship of research and development to commercial products. *Mailing Add:* 15 Lochhaven Lane Manchester MO 63021

HARRIS, JACK KENYON, b Albuquerque, NMex, May 2, 45; m 67; c 1. CELL BIOLOGY. *Educ:* Columbia Univ, BA, 67; Harvard Univ, MAT, 70; State Univ NY, Albany, PhD(biol), 80. *Prof Exp:* ASSOC PROF BIOL, RUSSELL SAGE COL, 79- *Mem:* AAAS; Optical Soc Am. *Res:* Light-optical techniques to localize and quantitate forces produced by biomolecules, cells, and organisms. *Mailing Add:* Biol Dept Russell Sage Col Troy NY 12180

HARRIS, JACK R, b Saginaw, Mich, Oct 17, 30; m 53; c 3. ELECTRICAL ENGINEERING, PHYSICS. *Educ:* Univ Mich, Ann Arbor, BS, 52; US Naval Postgrad Sch, MS, 60, PhD(elec eng), 65. *Prof Exp:* US Navy pilot, 52-57, sonics engr, Naval Air Develop Squadron, VXI, 62-64, instr physics, US Naval Postgrad Sch, 64-66, spec proj mgr, Naval Air Develop Ctr, 66-68, dep dir syst anal & eng dept, 68-71, sci off, 71-74; vpres, Pinkerton Comput Const, 74-79; sr dir, Tracop Inc, 79-87; CONSULT, 87- *Concurrent Pos:* Chmn, acoust working group, Dept Defense, 67-71. *Mem:* Opers Res Soc Am; Inst Elec & Electronics Engrs; Radio Club Am; Acoust Eng Soc; Sigma Xi. *Res:* Systems engineering emphasizing computers with acoustical processing applications. *Mailing Add:* 20400 Highland Hall Dr Gaithersburg MD 20879

HARRIS, JAMES EDWARD, b Ann Arbor, Mich, Aug 25, 28. ORTHODONTICS, GENETICS. *Educ:* Univ Mich, AB, 50, DDS, 54, MS, 60 & 63. *Prof Exp:* Res assoc orthod, Univ Mich, Ann Arbor, 63-64, from asst prof to assoc prof orthod & human genetics, 64-70, prof orthod & chmn dept, 69-82; RETIRED. *Concurrent Pos:* Consult, Plymouth State Hosp, Mich, 65-; NIH career develop award, 64-82; dir field projs, Egypt, 82-; lectr & writer. *Mem:* Am Asn Orthodontists; Am Dent Asn; Am Asn Phys Anthrop; Am Soc Human Genetics. *Res:* Inheritance of the craniofacial complex; study of malocclusion; facial growth; effect of orthodontic treatment on mandibular growth; the x-ray investigation of the royal mummies of Egypt; study of ancient skeletal record and modern Nubian population; application of genetics to orthodontic diagnosis and treatment. *Mailing Add:* 1918 Scottwood Ann Arbor MI 48104

HARRIS, JAMES JOSEPH, b Chicago, Ill, Jan 27, 30; m 61; c 2. INORGANIC CHEMISTRY. *Educ:* Miami Univ Ohio, AB, 52; Univ Fla, PhD(chem), 58. *Prof Exp:* Chemist, Union Carbide Corp, WVa, 53; sr scientist, Koppers Co, Inc, 58-74; sr proj scientist, Arco Chem, 74-75, sr group mgr, 75-86; lab mgr, 87-88, CHEM SERV, CHUBB NAT FOAM, 88- *Mem:* Am Chem Soc. *Res:* Organometallic compounds; metallic fluorides; polymerization catalysis; polyolefins; organic chemistry. *Mailing Add:* 1523 High Meadow Lane West Chester PA 19380

HARRIS, JAMES RIDOUT, b Lockhart, Tex, Apr 14, 20; m 43; c 3. SYSTEMS ENGINEERING, COMPUTER COMMUNICATIONS. *Educ:* Univ Richmond, BS, 41; Polytech Inst Brooklyn, MEE, 48. *Prof Exp:* Eng asst, Chesapeake & Potomac Tel Co, Va, 41-42; mem tech staff res & develop, 42-53, supvr comput res, 53-56, head dept data processing & data commun develop, 56-61, dir data systs eng, 61-62, dir data transmission systs eng, 62-65, dir customer switching eng ctr, 65-69, dir govt & spec systs eng ctr, 69-70, dir govt commun planning ctr, 70-71, dir customer equip studies ctr, 71-81, dir data network spec studies ctr, Bell Tel Labs, 81; dir spec studies ctr, Info Systs Labs, AT&T, 82-83; RETIRED. *Concurrent Pos:* Mem admin comt, Comput Soc, Inst Elec & Electronics Engrs, 62-65. *Mem:* Inst Elec & Electronics Engrs; Am Phys Soc; Sigma Xi. *Res:* Airborne communications; solid state devices and circuits; data processing; data communication; communications networks. *Mailing Add:* Eight Dogwood Lane Rumson NJ 07760-1412

HARRIS, JAMES STEWART, JR, b Portland, Ore, Aug 22, 42; m 65; c 2. SEMICONDUCTOR PHYSICS, MOLECULAR BEAM EPITAXY. *Educ:* Stanford Univ, BS, 64, MS, 65, PhD(elec eng), 69. *Prof Exp:* Instr elec eng, Stanford Univ, 68; mem tech staff semiconductor physics, Rockwell Int Sci Ctr, 69-72, group leader infrared devices, 72-80, prin scientist, optoelectronics dept, 80-82; PROF ELEC ENG, STANFORD UNIV, 82-, DIR SOLID STATE LAB, 84- *Concurrent Pos:* Consult, Varian Asn; mem Inst Elec & Electronics Engrs Electron Device Soc, 83- *Mem:* Fel Inst Elec & Electronics Engrs; Am Phys Soc; Electrochem Soc; Am Vacuum Soc; Mat Res Soc. *Res:* Semiconductor device physics; molecular beam, epitaxial growth of III-V compounds; electro-optical devices; photodiodes; lasers; resonant tunneling devices; quantum well optical modulators and detectors; thin film high temperature superconductors. *Mailing Add:* 763 Esplanda Way Stanford CA 94305

HARRIS, JANE E, b New York, NY, Feb 26, 46. OCOGENICITY. *Educ:* Yale Univ, PhD(pharmacol), 71. *Prof Exp:* Toxicologist, Food & Drug Admin, 78-84; TOXICOLOGIST, ENVIRON PROTECTION AGENCY, 84- *Mailing Add:* Am Cyanamid Co PO Box 400 ARD Princeton NJ 08543-0400

HARRIS, JAY H, b Newark, NJ, June 3, 36; m 63; c 1. ELECTRICAL ENGINEERING, BIOMEDICAL ENGINEERING. *Educ:* Polytech Inst Brooklyn, BEE, 58; Calif Inst Technol, MS, 59; Univ Calif, Los Angeles, PhD(electromagnetics), 65. *Prof Exp:* Mem tech staff antennas & propagation, Hughes Aircraft Co, 58-65; asst prof elec eng, Univ Wash, 66-70, assoc prof, 70-76; prog dir, Nat Sci Found Eng, 76-; DIR, ELEC ENG DEPT, SAN DIEGO STATE UNIV. *Concurrent Pos:* Fulbright-Hays fel, Paris, 65-66; consult urol, Sch Med Univ Wash, 67-76. *Mem:* Inst Elec & Electronics Engrs. *Res:* Radio wave propagation; antennas; plasmas; electro-optics; biomedical ultrasonics; optical spectroscopy. *Mailing Add:* Col Eng San Diego State Univ San Diego CA 92182

HARRIS, JEAN LOUISE, b Richmond, Va, Nov 24, 31; m 55; c 3. INTERNAL MEDICINE, ALLERGY. *Educ:* Va Union Univ, BS, 51; Med Col Va, MD, 55. *Hon Degrees:* DSc, Univ Richmond, 81. *Prof Exp:* Intern, Med Col Va, 55-56, resident internal med, 56-57, fel, 57-58; fel, Strong Mem Hosp-Sch Med, Univ Rochester, 58-60; instr med, Col Med, Howard Univ, 60-68, asst prof community health pract, 69-72; prof, 73-78, CLIN PROF FAMILY PRACT, VA COMMONWEALTH UNIV, 78-; PRES & CHIEF EXEC OFFICER, RAMSEY FOUND, 88- *Concurrent Pos:* Res assoc, Walter Reed Army Inst Res, Washington, DC, 60-63; pvt pract internal med & allergy, 64-71; chief, Bur Resources Develop, DC Dept Health, 67-69; dir, Ctr Community Health Consults, Div Health Manpower Intel, HEW, 69-77; asst clin prof community med, Charles R Drew Postgrad Med Sch, Los Angeles, 70-72; exec dir, Nat Med Asn Found, Washington, DC, 70-73; lectr, Dept Med Care & Hosps, Johns Hopkins Univ, 71-73; consult, Nat Ctr Health Statist, HEW, 71-72, Div Med Care Standards, Health Serv & Ment Health Admin, 73, Vet Admin Hosp Health Care Study, Nat Acad Sci, 76 & USAID, 77; mem, Adv Coun Sickle Cell Prog, Nat Heart Lung & Blood Inst, 75-79, Recombinant DNA Adv Comt, NIH, 79-82 & President's Task Force Pvt Sector Initiatives, 81-82; secy human resources, Commowealth Va, 78-82; state liaison, US Coun Int Yr Dis Persons, 81. *Mem:* Inst Med-Nat Acad Sci; fel Royal Soc Health; Am Acad Med Adminr; Am Pub Health Asn; Am Acad Family Pract; Sigma Xi. *Mailing Add:* 640 Jackson St St Paul MN 55101

HARRIS, JEROME SYLVAN, b New York, NY, Feb 27, 09; m 58. PEDIATRICS. *Educ:* Dartmouth Col, AB, 29; Harvard Univ, MD, 33. *Prof Exp:* From instr to prof pediat & biochem, Sch Med, Duke Univ, 37-79, pediatrician, Univ Hosp, 37-79, chmn dept pediat, 50-79; RETIRED. *Mem:* Fel Soc Pediat Res; fel Am Soc Clin Invest; fel Am Pediat Soc; fel Am Acad Pediat. *Res:* Biochemistry; metabolic disturbances in children. *Mailing Add:* 1701 Pleasant Green Rd Durham NC 27705

HARRIS, JESSE RAY, b Spokane, Wash, Apr 26, 37; m 70; c 2. PHYSICAL CHEMISTRY, CERAMIC ENGINEERING. *Educ:* Univ Wash, BS, 59, MS, 62; Alfred Univ, PhD(ceramic sci), 67. *Prof Exp:* SR CHEMIST, PHILLIPS PETROL CO, 66- *Mem:* Am Ceramic Soc; Nat Inst Ceramic Eng; Am Chem Soc. *Res:* Catalytic reactions of olefins; thermodynamics of adsorption. *Mailing Add:* 3409 Willowood Dr Bartlesville OK 74006

HARRIS, JOEL MARK, b Charleston, WVa, Aug 27, 50; m 72; c 1. SPECTROSCOPY. *Educ:* Duke Univ, BS, 72; Purdue Univ, PhD(chem), 76. *Prof Exp:* From asst prof to assoc prof, 76-85, PROF CHEM, UNIV UTAH, 85- *Concurrent Pos:* From adj asst prof to adj prof bioeng, Univ Utah, 80-88; Alfred P Sloan fel, 85-87. *Honors & Awards:* Coblentz Award Molecular Spectros, 86. *Mem:* Am Chem Soc; Optical Soc Am; Soc Appl Spectros; Coblentz Soc. *Res:* Analytical molecular spectroscopy: application of lasers to chemical analysis, time-resolved fluorescence and Raman spectroscopy; spectroscopic studies of liquid-solid interfaces. *Mailing Add:* Dept Chem Univ Utah Salt Lake City UT 84112

HARRIS, JOHN FERGUSON, JR, b Stroudsburg, Pa, Apr 15, 25; m 49; c 3. ORGANIC SYNTHESIS, POLYMER SYNTHESIS. *Educ:* Univ Pa, AB, 48, MS, 50, PhD(org chem), 53. *Prof Exp:* RES CHEMIST, CENT RES & DEVELOP DEPT, E I DU PONT DE NEMOURS & CO, INC, 52- *Concurrent Pos:* Guggenheim fel, 63-64. *Mem:* Am Chem Soc; Sigma Xi; AAAS. *Res:* Organic sulfur and fluorine chemistry; free radical chemistry; photochemistry; polymer chemistry. *Mailing Add:* 333 Hampton Rd Wilmington DE 19803-2425

HARRIS, JOHN KENNETH, b Reno, Nev, Jan 7, 34; m 55; c 3. TOPOLOGY. *Educ:* Fresno State Col, AB, 58, MS, 59; Univ Ore, PhD(math), 62. *Prof Exp:* From asst prof to assoc prof, 62-73, asst dean grad studies, asst dean acad affairs, 71-74, PROF MATH, PORTLAND STATE UNIV, 73-, DIR BUDGET, 75- *Mem:* Am Math Soc. *Res:* Point set topology; number theory; algebra. *Mailing Add:* Off of the Budget Portland State Univ Portland OR 97207-0751

HARRIS, JOHN MICHAEL, b London, UK, Dec 6, 42; m 73; c 2. MAMMALIAN PALEONTOLOGY. *Educ:* Univ Leicester, BSc, 64; Univ Tex, Austin, MA, 67; Univ Bristol, PhD(geol), 70. *Prof Exp:* Lectr geol, Ahmadu Bello Univ, Nigeria, 70; sr palaeontologist, Ctr Prehistory & Palaeont, Nairobi, 71-72; dir paleont, Nat Mus Kenya, 72-79; HEAD, EARTH SCI DIV, LOS ANGELES COUNTY MUS NATURAL HIST, 80- *Concurrent Pos:* Consult paleontologist, Koobi Fora Res Proj, Nat Mus Kenya, 72-; hon lectr, Dept Zool, Univ Nairobi, 76-79. *Mem:* Geol Soc London; Linnean Soc; Soc Vert Paleont; Sigma Xi. *Res:* Systematics, biostratigraphy and functional morphology of Miocene, Pliocene and Pleistocene mammals from Africa particularly fossil ungulates from sub-Saharan Africa. *Mailing Add:* Div Earth Sci Los Angeles County Mus Natural Hist 900 Exposition Blvd Los Angeles CA 90007

HARRIS, JOHN TOM, wildlife management, for more information see previous edition

HARRIS, JOHN WALLACE, b Peoria, Ill, Aug 28, 41; m 79; c 3. BIOLOGY. *Educ:* Western Ill Univ, BSEd, 63; Ind Univ, AM, 65, PhD(genetics), 68. *Prof Exp:* From asst prof to assoc prof, 68-80, dir honors prog, 78-83, PROF BIOL, TENN TECHNOL UNIV, 80- *Mem:* Sigma Xi. *Res:* Biochemical genetics and systematics. *Mailing Add:* 1615 E 7th St Cookeville TN 38505

HARRIS, JOHN WAYNE, b Syracuse, NY, Jan 8, 37; m 60; c 4. RADIATION THERAPY, RADIATION BIOLOGY. *Educ:* Syracuse Univ, BA, 58, MS, 62; Univ Rochester, PhD(radiation biol), 65; Univ Calif, San Francisco, MD, 78. *Prof Exp:* Teacher, Baldwinsville Acad, 59-61; Australian Inst Nuclear Sci & Eng fel, 65-67; asst prof radiol, dept radiation oncol, Univ Calif, San Francisco, 67-71, assoc prof, 71-; AT DEPT RADIATION ONCOL, ST JOSEPH'S HOSP. *Mem:* AAAS; Am Soc Cell Biol; Radiation Res Soc; Biophys Soc; Am Asn Cancer Res. *Res:* Cellular sulfhydryl groups; tumor growth; radiation and drug effects; cellular immunology; radiation therapy. *Mailing Add:* Dept Radiation Oncol St Joseph's Hosp 2700 Dolbeer St Eureka CA 95501

HARRIS, JOHN WILLIAM, b Boston, Mass, Mar 30, 20; m 51; c 3. CLINICAL MEDICINE. *Educ:* Trinity Col, Conn, BS, 41; Harvard Univ, MD, 44; Am Bd Internal Med, dipl, 54. *Prof Exp:* Res assoc med, Harvard Med Sch, 51-52; from sr instr to assoc prof, 52-62, PROF MED, CASE WESTERN RESERVE UNIV, 62- *Concurrent Pos:* Res fel med, Med Sch, Harvard Univ, 48-51; Stengel res fel, Am Col Physicians, 51-52; hematologist, Cleveland Metrop Gen Hosp, 52-, assoc dir, med dept, 66-81; attend physician, Crile Vet Admin Hosp, 53-58, sr attend physician, 58-; Markle scholar, 55-60; USPHS res career award, 62. *Honors & Awards:* Martin Luther King Jr Med Award, 72. *Mem:* Am Soc Hemat (vpres, 78-79, pres-elect, 79-81, pres, 81-82); Am Soc Clin Invest (vpres, 64-65); fel Am Col Physicians; Am Fedn Clin Res; Asn Am Physicians; fel Int Soc Hemat. *Res:* Internal medicine; hematology; hemolytic anemias; sickle cell anemia; nutritional and pernicious anemia. *Mailing Add:* 3395 Scranton Rd Cleveland OH 44109

HARRIS, JOHN WILLIAM, b Brampton, Ont, 49. ORGANIC CHEMISTRY. *Educ:* Univ Toronto, BSc, 72, MSc, 74, PhD(chem), 77. *Prof Exp:* Res chemist agr chem, Uniroyal Ltd, Guelph, ON, 76-81; RES CHEMIST ORG CHEM, SHELL CAN LTD, OAKVILLE, 81- *Concurrent Pos:* Indust fel, Nat Res Coun Can, 76-78. *Res:* Synthesis of novel organic compounds for screening as agricultural chemicals; Petrochemicals research; lubricants research. *Mailing Add:* Shell Can Ltd 3415 Lakeshore Rd W Box 2100 Oakville ON L6J 5C7 Can

HARRIS, JOSEPH, b Baltimore, Md, Dec 2, 19; m 44; c 2. BIOCHEMISTRY. *Educ:* Univ Md, BS, 47; Johns Hopkins Univ, MA, 49, PhD(physiol chem), 52. *Prof Exp:* Instr biochem, Johns Hopkins Univ, 52; responsible invest, Baxter Labs, Inc, 52-54; instr, Sch Med, Univ Colo, 54-55; asst prof, Albany Med Col, 55-62; chief lab neurochem, Barrow Neurol Inst, 62-75; lectr chem, 62-65, res prof, 65-75, prof chem & assoc chmn dept, 75-89, EMER PROF, ARIZ STATE UNIV, 89- *Concurrent Pos:* Vis prof, Inst Med Res, Royal NShore Hosp, Sydney, Australia, 69; exchange scientist, US-Hungary Nat Acad Sci, 71; vis prof, Biochem Dept, Oxford Univ, 81. *Mem:* Fel AAAS; fel NY Acad Sci; Am Chem Soc; Am Asn Clin Chemists; Biochem Soc. *Res:* Neurochemistry; metal ions effect on and mechanisms of action in cellular processes; biological membrane structure and neurotoxins; mechanism of drug action; biochemistry of exercise. *Mailing Add:* 2131 E Geneva Dr Tempe AZ 85282

HARRIS, JOSEPH BELKNAP, b Thomson, Ga, Sept 11, 26; m 52; c 2. CYTOLOGY. *Educ:* Emory Univ, AB, 49; Univ Ga, MS, 54; Duke Univ, PhD(bot), 59. *Prof Exp:* Asst, Emory Univ, 49; asst bot, Univ Ga, 50-51, tutor, 51-52; forestry aide, Bur Plant Indust, USDA, 52, plant physiologist tobacco sect, Crops Res Div, Ga Coastal Plain Exp Sta, 57-60; head dept biol, Coker Col, 61-65; assoc prof, 65-69, PROF BIOL, UNIV WIS-STEVENS POINT, 69- *Concurrent Pos:* Vis prof, Univ Tenn, 72 & NC State Univ, 78-79; pres, Midwest Soc Electron Microscopists, 85-87; lectr, Temi Ryazef Inst Plant Physiol, Moscow, 90, Second Mil Med Univ, Shanghai, 90 & Inst Basic Med Sci, Beijing, 90; ed, Midwest Micros, 91-92. *Mem:* AAAS; Am Soc Plant Physiol; Electron Micros Soc Am; Geront Soc; Inst Soc Ethics & Life Sci; Sigma Xi; Int Soc Plant Molecular Biol. *Res:* Ultrastructure, metabolism; senescence; leaf aging of crop plants, using techniques of physiology, ultrastructure and molecular biology. *Mailing Add:* Dept of Biol Univ of Wis Stevens Point WI 54481

HARRIS, JUDITH ANNE VAN COUVERING, b Tulsa, Okla, Feb 20, 38; div; c 4. VERTEBRATE PALEONTOLOGY. *Educ:* Univ Calif, Berkeley, BA, 60; Univ Cambridge, PhD(geol), 72. *Prof Exp:* CUR FOSSIL VERTEBRATES & ASSOC PROF NATURAL HIST, UNIV COLO MUS, 77-; ASSOC PROF GEOL, UNIV COLO, BOULDER, 77- *Mem:* Soc Vertebrate Paleont; fel Linnean Soc. *Res:* African tertiary faunas and paleoenvironments; origin of modern terrestrial communities; biogeography; evolutionary problems cichlid fish evolution; science and philosophy; feminist science. *Mailing Add:* Geol Box 250 Univ Colo Boulder Boulder CO 80309-0250

HARRIS, JULES ELI, b Toronto, Ont, Oct 12, 34; m 72; c 6. ONCOLOGY, IMMUNOLOGY. *Educ:* Univ Toronto, MD, 59; FRCPS(C), 65. *Prof Exp:* Asst prof med, Univ Tex, Houston, 66-69; from asst prof to prof med, Univ Ottawa, 72-77; prof med & dir sect med oncol, 77-86; DIR, RUSH MED CANCER CTR, 86- *Concurrent Pos:* Consult, Ont Cancer Treat & Res Found, Nat Defense Med Ctr, Ottawa, 69, 71-77; sci officer, Med Res Coun Can, 70-72; res fel med, Ont Cancer Inst, 61-62 & Brit Med Res Coun, 62-63. *Mem:* Am Soc Clin Oncol; fel Am Col Physicians; Am Soc Hemat; Am Asn Cancer Res; NY Acad Sci; Sigma Xi. *Res:* Tumor immunology; effect of chemotherapy on immune response. *Mailing Add:* Rush Med Col 1725 W Harrison St Chicago IL 60612

HARRIS, KENNETH, b Manchester, UK, Dec 26, 43; m 66; c 2. SUPERALLOY METALLURGY, MELTING REFINING & CASTING METALLURGY. *Educ:* Royal Sch Mines, BS, 65; Univ London, ARSM, 65. *Prof Exp:* Mgr tech quality, superalloys div, Union Carbide UK Ltd, 65-75; VPRES TECHNOL, CANNON-MUSKEGON CORP, 75- *Mem:* Mettal Soc; Am Soc Metals. *Res:* Developed technology to electro-slag refine wrought nickel-base precipitation hardened superalloys; developed vacuum induction refining technology for premium quality nickel-base; gamma prime strengthened cast & PM superalloys; inventor and developer of the CMSX series of single crystal superalloys. *Mailing Add:* Cannon-Muskegon Corp PO Box 506 2875 Lincoln St Muskegon MI 49443-0506

HARRIS, KENT KARREN, atmospheric physics, nuclear physics, for more information see previous edition

HARRIS, KERRY FRANCIS PATRICK, b New Orleans, La, Aug 6, 43; m 65. VIROLOGY. *Educ:* New Orleans Univ, BS, 64; Loyola Univ, MS, 68; Mich State Univ, PhD(entomol & virol), 71. *Prof Exp:* Instr biol sci, New Orleans Sch Bd, 63-65; res assoc, Loyola Univ, 65-67; res fel entomol & virol, NSF, 68, NIH, 68-71, virus-vector interactions, Nat Res Coun, Can, 71-72 & NIH, 72-73; PROF ENTOMOL & VIROL, TEX A&M UNIV, 76- *Concurrent Pos:* Fulbright scholar & res fel, 69; prin investr & prog leader, Nat Sci Found, 72-75; res scientist, Int Sorghum-Millet Improv Prog, 77-81 & Coop Regional Res Proj Viral & Mycoplasmal Dis Corn & Sorghum, 76-; assoc ed, J Econ Entomol, 77-; res scientist, Int Working Group Legume Viruses, 77-, mem, Nat Medal Sci Comt, 79-82; lectr, Int Maize & Wheat Improv Ctr, Mex, 81-, invited lectr, People's Rep China, 88 & Int lectr, WHO, Indonesia, 88; assoc ed, Plant Dis, 81-; ed, Aphids as Virus Vectors, 77, Leafhopper Vectors & Plant Agents, 79, Vectors of Plant Pathogens, 80, Plant Dis & Vectors: Ecol & Epidemiol, 81, Current Topics Vector Res, 81, Pathogens, Vector & Plant Dis: Approaches to Control, 82, Adv in Dis Vector Res, 88- *Honors & Awards:* Sigma Xi award, 71. *Mem:* Am Phytopath Soc; Entomol Soc Am; Sigma Xi; Am Soc Microbiol; Am Soc Virol. *Res:* Pathogen-Vector-Host interactions with special emphasis on virus transmission mechanisms, determinants of virus-vector specificity phenomena and the fates of noncirculative and circulative viruses in their vectors; molecular biology of insect-transmitted plant viruses. *Mailing Add:* Dept Entomol Tex A&M Univ College Station TX 77843-2475

HARRIS, LAWRENCE DEAN, b Avoca, Iowa, Sept 20, 42; div; c 4. ANIMAL ECOLOGY, RESOURCE MANAGEMENT. *Educ:* Iowa State Univ, BSc, 64; Mich State Univ, MSc, 68, PhD(ecol), 70. *Prof Exp:* Wildlife mgt officer, Tanzania Game Div, 64-66, wildlife biologist, 66-67; res assoc systs ecol, Colo State Univ, 70-72; ASST PROF WILDLIFE ECOL, UNIV FLA, 72- *Concurrent Pos:* Fel US int biol prog, Colo State Univ, 70-72; res grant US forest serv, Southeastern Forest Exp Sta, 73-74, US Nat Park Serv, 74-76, US Dept Transp, 74-75 & Fla Game Fish Comn, NSF, 73-; task force recreation & wildlife res priorities, USDA, 73-74; long range plan comn, Wildlife Soc, 77-78. *Mem:* Am Soc Mammal; AAAS; Ecol Soc Am; Sigma Xi; Wildlife Soc. *Res:* Quantitative and conceptual aspects of renewable resource management with special emphasis on decision making and mathematical modelling. *Mailing Add:* Sch Forest Res Ziegler Bldg Univ Fla 308 Newins Gainesville FL 32611

HARRIS, LAWSON P(ARKS), b Providence, RI, Aug 28, 29; m 57; c 3. ELECTRICAL ENGINEERING. *Educ:* Mass Inst Technol, BS, 50, MS, 56, ScD(elec eng), 59. *Prof Exp:* Develop engr, RCA-Victor, 50-51; res asst, Mass Inst Technol, 54-56, res engr, 56-57, instr elec eng, 57-59; ELEC ENGR, GEN ELEC RES & DEVELOP CTR, 59- *Concurrent Pos:* Chmn, Gaseous Electronics Conf, 80-82. *Mem:* Am Soc Mech Engrs; Inst Elec & Electronics Engrs; Am Phys Soc. *Res:* Control systems theory; magneto-hydrodynamics; energy conversion; gaseous electronics; electric arcs; industrial applications of electric arcs in vacuum and high-pressure electric switchgear, lighting, arc welding and plasma jets; factory automation; aircraft engine controls. *Mailing Add:* Gen Elec Res & Develop Ctr KW-D212 PO Box 8 Schenectady NY 12301

HARRIS, LEE ERROL, b Ft Pierce, Fla, Oct 29, 53; m 75; c 2. HYDROGRAPHIC SURVEYING, COASTAL & OCEANOGRAPHIC ENGINEERING. *Educ:* Fla Atlantic Univ, BS, 74; Univ Fla, MS, 75. *Prof Exp:* Res asst, Coastal & Oceanog Eng Lab, Univ Fla, 74-75; coastal engr, Jacksonville Dist, US Army Corps Engrs, 75-77; chmn, Eng Sci Div, 80-87, ASSOC PROF OCEAN ENG & OCEANOG, FLA INST TECHNOL, JENSEN BEACH, 80- *Concurrent Pos:* Consult engr & survr, 77- *Mem:* Marine Technol Soc; Am Shore & Beach Preserv Assoc. *Res:* Coastal engineering including tidal inlet hydrodynamics; beach erosion; hydrographic surveying and coastal structures. *Mailing Add:* 310 Ormond Ave Indialantic FL 32903-3432

HARRIS, LELAND, b Warren, Ohio, Sept 9, 24; m 55; c 3. BIOORGANIC CHEMISTRY, SCIENCE EDUCATION. *Educ:* Univ Mich, BS, 50; Univ Ariz, MS, 52; Univ Iowa, PhD(org chem), 55. *Prof Exp:* Res chemist, E I du Pont de Nemours & Co, Va, 55-57; from asst prof to prof chem, Knox Col, 57-71, chmn dept, 59-77, Herbert E Griffith prof, 71-; ADJ PROF, DEPT CHEM, UNIV ARIZ. *Concurrent Pos:* NSF Found sci facul fel chem, Calif Inst Technol, 68-69 & Duke Univ, 78; vis prof biochem, Rush Med Col, 72- *Mem:* AAAS; Am Chem Soc; Sigma Xi. *Res:* Pigment variations and morphological changes in various genotypes of soybeans and peanuts using HPLC and electron microscopy; synthesis and nuclear magnetic resonance studies of substituted conjugated unsaturated acids; design and curriculum studies of first year medical programs; variable temperature nuclear magnetic resonance studies of cyclohexene systems; kinetics of transition from protochlorophyll to chlorophyll. *Mailing Add:* Dept Chem Univ Ariz Tucson AZ 85721

HARRIS, LEONARD ANDREW, b San Francisco, Calif, Apr 11, 28; m 55; c 3. CIVIL ENGINEERING, APPLIED MECHANICS. *Educ:* Stanford Univ, BS, 50; Univ Ill, MS, 53, PhD(eng), 54. *Prof Exp:* Res assoc civil eng, Univ Ill, 53-54; from engr to sr eng specialist appl mech, Space & Info Systs Div, NAm Aviation Corp, 54-58, supvr, 58-59, prin scientist struct sci, 59-61, dir, 61-67, asst mgr sci & technol, 67-68, mgr struct & design, NAm Rockwell Corp, 68-70, mgr shuttle technol, 70-72; MGR MAT & STRUCT, OFF AERONAUT & SPACE ADMIN, NASA, 72- *Concurrent Pos:* Lectr aeronaut eng, Univ Southern Calif, 56-60; guest lectr, Univ Calif, Los Angeles, 59-65; mem indust adv comt mil handbk on sandwich construct, 59-65; mem ad hoc comt design with brittle mat, Mat Adv Bd, Nat Acad Sci, 64; mem res adv comt space vehicle struct, NASA, 65-67. *Mem:* Assoc fel Inst Aeronaut & Astronaut; Am Soc Civil Engrs. *Res:* Experimental and theoretical studies of buckling of plates and shells; static and fatigue strength of welds in structural materials; brittle behavior of materials. *Mailing Add:* 5515 Trent Chevy Chase MD 20815

HARRIS, LEONCE EVERETT, b New Orleans, La, July 31, 41; m 69; c 2. PROPULSION, PROPELLANTS & EXPLOSIVES. *Educ:* La State Univ, BS, 64, PhD(chem), 68. *Prof Exp:* Teaching asst chem, La State Univ, 64-68, vis asst prof, 68-70; RES CHEMIST, US ARMY ARMAMENT RES & DEVELOP COMMAND, 70- *Concurrent Pos:* Prin investr, US Army Ardec, 70-80, group leader, 80-85, exec fel, Off Tech Dir, 86-87, proj mgr, 87-; secy Army fel, Royal Inst Gt Brit, Davy-Faraday Lab, 75-76; chmn, Joint Army, Navy, NASA, AF(JANNAF) Workshop on Combustion Diag, 89; res supvr postdoctoral students, Nat Res Coun, 83. *Mem:* Fel Roy Soc Chem; Am Chem Soc; AAAS. *Res:* Electrothermal and chemical propulsion; evaluation and identification of propellants; glucidation of the mechanism of propellant and plasma interaction using combustion diagnostics and modelling; direction of propulsion aspects of system integration. *Mailing Add:* US Army Ardec AEE Br B382 Picatinny Arsenal NJ 07806

HARRIS, LESTER EARLE, JR, b Washington, DC, June 4, 22; m 49; c 4. HERPETOLOGY. *Educ:* Columbia Union Col, BA, 49, PedD, 70; Univ Md, MS, 51. *Prof Exp:* Instr, Loma Linda Univ, 49-51, asst prof, 56-65, chmn dept, 53-76 PROF BIOL, LOMA LINDA UNIV, 79- *Mem:* Herpetologists League; Am Soc Ichthyologists & Herpetologists; Soc Study Amphibians & Reptiles. *Res:* Ecology; snake anatomy. *Mailing Add:* Dept Biol Loma Linda Univ Riverside CA 92515

HARRIS, LEWIS ELDON, b Cedar, Kans, Dec 3, 10; m 35; c 2. PHARMACEUTICAL CHEMISTRY. *Educ:* Univ Nebr, BSc, 32, MSc, 33. *Hon Degrees:* ScD, Univ Nebr, 70. *Prof Exp:* Dir lab, 33-67, PRES, HARRIS LABS, INC, 67- *Concurrent Pos:* Pres, Norden Labs, Inc, 39-69; instr, Sch Nursing, Univ Nebr, 41- *Mem:* Am Pharmaceut Asn; fel Am Inst Chem. *Res:* Stabilization of calcium gluconate solutions and calcium glycerophosphate solutions; stability and antioxidant studies in powdered dairy products; antioxidants in stabilizing vitamin A in alfalfa; soluble glucosides of sulfonamides. *Mailing Add:* 7134 South St Lincoln NE 68506

HARRIS, LEWIS PHILIP, b Avon, Wash, Nov 13, 07; m 29; c 1. BIOCHEMISTRY. *Educ:* Univ Wash, BS, 30, MS, 33. *Prof Exp:* Res chemist, Can Fishing Co, BC, 30-31; anal chemist, Wenatchee Indust Labs, Wash, 33-35; res chemist, Sherwin Williams Co, Wash, 36-41, plant supt & tech dept head, Calif, 47-48, res head, Agr Chem Div, Ohio, 48-50; mgr & plant supt, Cotton States Chem Co, La, 50-53; prod mgr, Geigy Agr Chem Div, Geigy Chem Corp, NY, 53-56; tech dir insecticides dept, Acme Qual Paints, Inc, Mich, 56-64, tech dir agr chem div, Sherwin-Williams Co, 64-71; tech adv, PBI-Gordon Corp, 71-78; RETIRED. *Concurrent Pos:* Tech consult pesticide regulations, 78- *Mem:* Am Chem Soc; Entom Soc Am. *Res:* Pesticide chemicals for insect, fungus and weed control. *Mailing Add:* 8911 W 60th St Shawnee Mission KS 66202-2862

HARRIS, LOUIS SELIG, b Boston, Mass, Mar 27, 27; m 52; c 1. NEUROPHARMACOLOGY, PSYCHOPHARMACOLOGY. *Educ:* Harvard Univ, BA, 54, MA, 56, PhD(pharmacol), 58. *Prof Exp:* Sr res biologist & sect head, Sterling-Winthrop Res Inst, 58-66; assoc prof pharmacol, Sch Med, Univ NC, Chapel Hill, 66-70, assoc prof, Sch Pharm, 67-70, prof, Sch Med & Sch Pharm, 70-72; HARVEY HAAG PROF & CHMN DEPT PHARMACOL, VA COMMONWEALTH UNIV, 72- *Concurrent Pos:* Vis lectr, Albany Med Col, 60-66; mem psychotomimetic agents adv comt, NIMH, 70-72, chmn, 71-72; trustee, Sisa Inst Res, 70-84; mem, Presidential Biomed Res Panel, Nat Acad Sci-Nat Res Coun, 75-76, Comt Clin Eval Narcotic Antagonists, 76-78, Comt Probs Drug Dependence, 76- & chmn, 90-; mem comt revision analgesics, US Pharmacopeial Conv, 80-85; comt Public Concern, US Pharmacop Conv, 80-85; Bd Scientific Counrs, Nat Inst Drug Abuse Chmn, 83-86; sr sci adv to dir, Nat Inst Drug Abuse, 87-88, mem, Extramural Sci Adv Bd, 91; hon prof pharmacol, Beijing China, 90. *Honors & Awards:* Res Achievement Award Pharmacol, Am Pharmaceut Asn Found-Acad Pharmaceut Sci, 77; Hartung Mem Award, Univ NC, 81; Nathan B Eddy Mem Award, 85; Abe Wilker Award, 91. *Mem:* AAAS; Am Soc Pharmacol & Exp Therapeut; Am Col Neuropsychopharm; Am Chem Soc; Soc Neurosci. *Res:* Central stimulants and depressants; analgesics; psychopharmacology; pharmacology of marijuana and its active constituents; drug development. *Mailing Add:* Dept Pharmacol & Toxicol Va Commonwealth Univ Richmond VA 23298-0027

HARRIS, LOWELL DEE, b Salt Lake City, Utah, June 16, 43; m 67; c 5. BIOPHYSICS, BIOENGINEERING. *Educ:* Univ Utah, BS, 68, PhD(biophys & bioeng), 74. *Prof Exp:* Res asst bioeng, Univ Utah, 73-74; res asst, Mayo Grad Sch Med, Univ Minn, 74-76, sr res fel physiol & biophys, 76-79, consult, 82-91, asst prof biophys, 79-91; DISPLAY SYST, BIO-IMAGING RES INC, 91- *Concurrent Pos:* NIH fel, 75-78; prin investr, Core IV HL-04664 Prog Proj Grant, 78- *Res:* Computed tomography; display and visual enhancement of three-dimensional reconstructions of anatomic structure. *Mailing Add:* 17513 Warren Wildwood IL 60030

HARRIS, LOYD ERVIN, b Ryan, Okla, Sept 21, 00; m 23; c 2. PHARMACEUTICAL CHEMISTRY. *Educ:* Univ Okla, BS, 22, MS, 24; Univ Wis, PhD(pharmaceut chem), 26. *Prof Exp:* Asst pharm, Univ Okla, 19-23, from asst prof to prof chem, 23-46; prof pharm, Ohio State Univ, 46-63; prof pharm & dean col pharm, 63-70, EMER PROF PHARM & EMER DEAN COL PHARM, UNIV OKLA, 70- *Mem:* Am Pharmaceut Asn. *Res:* Phytochemical studies. *Mailing Add:* 2514 S Pickard Ave Norman OK 73072

HARRIS, MARTIN H, b Pittsburgh, Pa, Oct 11, 37; m 64; c 2. PETROLEUM ENGINEERING. *Educ:* Pa State Univ, BS, 59, MS, 61. *Prof Exp:* Engr, Gulf Res & Develop Co, Gulf Oil Corp, 61-65, res engr, 65-69, sr res engr, 69-70, SECT SUPVR, GULF RES & DEVELOP CO, GULF OIL CORP, 70- *Honors & Awards:* Rossiter W Raymond Mem Award, Am Inst Mining, Metall & Petrol Engrs, 67. *Mem:* Soc Petrol Engrs. *Res:* Natural gas engineering; well stimulation, completion and logging; petrophysics and fluid flow in porous media; application of digital computers to petroleum exploration and data processing. *Mailing Add:* 5107 Queensloch St Houston TX 77036

HARRIS, MARVIN KIRK, b Genoa, Nebr, Aug 15, 43; m 64; c 3. ENTOMOLOGY. *Educ:* Dana Col, BS, 68; Cornell Univ, PhD(entom), 72. *Prof Exp:* Res asst econ entom, Cornell Univ, 68-72; ASSOC PROF ENTOM, TEX A&M UNIV, 72- *Mem:* Entom Soc Am; Sigma Xi; AAAS. *Res:* Biology and pest management of pecan insects; pecan domestication as it affects the pecan insect complex; host-plant resistance to insects in agriculture; insect-plant co-evolution. *Mailing Add:* Dept Entom Tex A&M Univ College Station TX 77843

HARRIS, MARY STYLES, b Nashville, Tenn, June 26, 49; m 71. GENETICS, PUBLIC HEALTH & EPIDEMIOLOGY. *Educ:* Lincoln Univ, BA, 71; Cornell Univ, PhD(genetics), 75. *Prof Exp:* Instr genetics, Sch Med, Morehouse Col, 78- & dir admin, Sickle Cell Found Ga, Inc, 77-; dir, Genetic Serv, State Ga, 82-85, ASST DIR SCI & PUB POLICY, ATLANTA UNIV, 81-; PRES, HARRIS & ASSOCS LTD, 86- *Concurrent Pos:* Ford Found fel, 71-75; res assoc tumor virol, Rutgers Med Sch, 75-77; NSF Residency, 79-80; scientist in residence, WGTV, 79, adj pub serv asst, 79-80; instr human genetics, Emory Univ, 82; dir genetic serv, Ga Dept Human Resources, 82-85. *Mem:* Am Pub Health Asn; Am Soc Human Genetics. *Res:* Investigation of the induction of cellular DNA synthesis by Adenovirus type 2; regulation of the synthesis of fetal hemoglobin in adult patients producing hemoglobin F; developing statewide genetic screening programs. *Mailing Add:* 2058 N Mills Ave Suite 214 Claremont CA 91711

HARRIS, MATTHEW N, b New York, NY, Dec 20, 31; m 54; c 3. SURGERY, ONCOLOGY. *Educ:* NY Univ, BA, 52; Chicago Med Sch, MD, 56; Am Bd Surgery, dipl, 64. *Prof Exp:* Intern med & surg, 3rd NY Univ Surg Div, Bellevue Hosp Ctr, 56-57, resident gen surg, 57-58, 60-63, Am Cancer Soc fel, 61-63, USPHS sr clin trainee cancer, 63-64; teaching asst surg, Sch Med, NY Univ, 64-65, from instr to assoc prof, 65-75, prof clin surg, 75-79; PROF SURG & DIR DIV ONCOL, DEPT SURG, NY UNIV SCH MED, 79- *Concurrent Pos:* Vis surgeon, Bellevue Hosp Ctr; attend surgeon, NY Univ Hosp, 76; consult, Manhattan Vet Admin Hosp; instr anat, Sch Med, NY Univ, 66-68, lectr, 68-; consult surg, NY Infirmary, Beekman Downtown Hosp, 80- *Mem:* Fel Am Col Surgeons; Am Soc Clin Oncol; Asn Acad Surg; Soc Surg Oncol; Soc Med Consults to Armed Forces; Sigma Xi; Am Radium Soc. *Res:* Surgical oncology; malignant melanoma; carcinoma of the breast and soft tissue tumors. *Mailing Add:* NY Univ Med Ctr 530 First Ave New York NY 10016

HARRIS, MAUREEN ISABELLE, b Anniston, Ala, Mar 21, 43. PUBLIC HEALTH, EPIDEMIOLOGY. *Educ:* George Washington Univ, BS, 64; Yale Univ, PhD(biochem), 68; Johns Hopkins Univ, MPH, 75. *Prof Exp:* Phys sci aide, US Army Corps Engrs, 64; staff res scientist ribosome biochem, Nat Inst Arthritis & Metab Dis, 68-70, health scientist adminr, Fogarty Int Ctr, 71-76, res prog analyst, Kidney & Urol Dis, 76-77, DIR NAT DIABETES DATA GROUP, NAT INST, DIABETES, DIGESTIVE & KIDNEY DIS, NIH, 77-, DIR, WHO COLLABORATING CTR FOR DIABETES, 85- *Honors & Awards:* Hildebrand Jr Award, Am Chem Soc, 64; Spec Recognition Award, USPHS, 85; Award of Merit, NIH, 87. *Mem:* Am Diabetes Asn; Soc Epidemiol Res. *Res:* Epidemiology of diabetes. *Mailing Add:* Nat Diabetes Data Group NIDDK NIH Bethesda MD 20892

HARRIS, MELVYN H, b Lynn, Mass, July 27, 32; m 63; c 2. DENTISTRY, ORAL SURGERY. *Educ:* Harvard Univ, AB, 53; Tufts Univ, DMD, 57. *Prof Exp:* Chief oral surg, Univ Hosp, 63-; Asst prof oral surg & oral path, 62-70, ASSOC PROF STOMATOL, BOSTON UNIV, 70- *Concurrent Pos:* Chief dent staff, Sommerville Hosp, 65- *Mem:* Am Acad Oral Path. *Res:* Vascular pathways of pain. *Mailing Add:* 665 Beacon St Boston MA 02215

HARRIS, MILES FITZGERALD, b Brunswick, Ga, Feb 2, 13; m 38; c 3. METEOROLOGY, SCIENCE WRITING. *Educ:* NY Univ, BS, 44, MS, 57. *Prof Exp:* Asst observer, US Weather Bur, 32-35 & 37-43, analyst, 43-45, hurricane forecaster, 45-47, supv meteorologist, 47-48, spec projs meteorologist, 48-51, res meteorologist, Meteorol Statist Sect, 51-61, meteorologist, Ed Sect & head, Editing & Pub Br, 61-66, phys scientist & chief, Sci Info Anal Br, Environ Sci Serv Admin, 66-68, ed, Monthly Weather Rev, 68-70; tech ed, Bull of the Am Meteorol Soc & Spec Projs Off, 70-74, tech ed, Monthly Weather Rev, 74-75; RETIRED. *Mem:* Am Meteorol Soc. *Res:* Investigating the earth; popularization of meteorology. *Mailing Add:* 40 Lothrop St Beverly MA 01915

HARRIS, MILTON, chemistry; deceased, see previous edition for last biography

HARRIS, MORGAN, b St Anthony, Idaho, May 25, 16; m 40; c 2. CELL BIOLOGY. *Educ:* Univ Calif, AB, 38, PhD(zool), 41. *Prof Exp:* Harrison fel, Univ Pa, 41-42; asst zool, Stanford Univ, 42-44; instr, Univ Wash, 44-45; from instr to prof zool, 45-83, from vchmn to chmn dept, 52-63, EMER PROF ZOOL, UNIV CALIF, BERKELEY, 83- *Concurrent Pos:* Merck sr fel, Univ Paris, 53-54; mem cell biol study sect, NIH, 58-60, 61-63, nat adv gen med sci coun, 63-65; prof, Miller Inst Basic Res, 63-65; Guggenheim fel, Cambridge Univ, 60-61. *Mem:* Am Soc Zoologists; Soc Gen Physiol; Tissue Cult Asn (pres, 58-60); Am Soc Cell Biol. *Res:* Cell growth and nutrition; somatic cell genetics; tumor biology. *Mailing Add:* LSA No Six Univ Calif Berkeley CA 94720

HARRIS, MORTON E, b Apr 27, 34; m 67. ALGEBRA. *Educ:* Yale Univ, BS, 55; Harvard Univ, MA, 56, PhD(math), 60. *Prof Exp:* Asst prof math, Clark Univ, 60-61 & Tufts Univ, 61-65; from asst prof to assoc prof, Univ Ill, Chicago Circle, 65-73; PROF MATH, UNIV MINN, 73- *Concurrent Pos:* NSF res grants, 71-72. *Mem:* Am Math Soc; Math Asn Am. *Res:* Problems in the classification of finite simple groups. *Mailing Add:* Sch Math Univ Minn 127 Vincent Hall Minneapolis MN 55455

HARRIS, NATHOLYN DALTON, b Calvary, Ga, Feb 26, 39; m 67; c 2. FOOD SCIENCE. *Educ:* Berry Col, BS, 61; Ohio State Univ, MS, 62; Univ Wis, PhD(food, nutrit), 67. *Prof Exp:* Instr food & nutrit, Berry Col, 62-63; lectr food sci, Univ Wis, 66-71; from asst prof to assoc prof, 71-86, PROF FOOD SCI, FLA STATE UNIV, 86- *Mem:* Inst Food Technologists; Am Home Econ Asn; Am Chem Soc. *Res:* Microbiological aspects of food; food flavor; quality changes in food. *Mailing Add:* 413 Sandels Bldg Fla State Univ Tallahassee FL 32306

HARRIS, NELLIE ROBBINS, b Port Norris, NJ, Sept 13, 17. HUMAN PHYSIOLOGY. *Educ:* Montclair State Col, AB, 39; Univ Pa, PhD(physiol), 59. *Prof Exp:* Asst biol, Montclair State Col, 39-40; teacher high schs, NJ, 40-50; asst instr physiol, Sch Med, Univ Pa, 53-54; asst prof biol & physics, Beaver Col, 56-60; asst prof biol, Trenton State Col, 60-62; assoc prof biol, Cascade Col, 63-69; prof sci, Judson Baptist Col, 69-82; RETIRED. *Concurrent Pos:* Vis instr, Univ Ore, 63 & 67. *Mem:* AAAS; Sigma Xi. *Mailing Add:* 808 E 19th The Dalles OR 97058

HARRIS, NORMAN OLIVER, b Shinglehouse, Pa, May 27, 17; m 55; c 1. PREVENTIVE DENTISTRY. *Educ:* Temple Univ, DDS, 39; Ohio State Univ, MSD, 53. *Prof Exp:* Intern, USPHS, 39-40; UN Relief & Rehab Agency adv, Nationalist China, 46-47; chief exp dent, USAF Sch Aviation Med, 55-61, chief res & prev dent, USAF, PR, 61-63; dir res activities, Sch Dent, Univ PR, San Juan, 63-76; prof, Dept Community Dent, Dent Sch, Univ Tex, 76-87; RETIRED. *Concurrent Pos:* Prof honoris causa, Univ Pernambuco, Brazil. *Mem:* Am Dent Asn; Int Asn Dent Res; Am Col Dentists. *Res:* Basic and applied research in preventive dentistry; research administration; interhemispheric integration of research, teaching and public health administration; professional communications; programmed learning. *Mailing Add:* Rte 1 Box 1622 Boerne TX 78006

HARRIS, P(HILIP) J(OHN), b Montreal, Que, Mar 22, 26; m 53; c 2. STRUCTURAL ENGINEERING. *Educ:* Univ Man, BSc, 48; McGill Univ, MEng, 49, PhD(struct shells), 64. *Prof Exp:* Struct designer, Dom Bridge Co, Ltd, 49-51; chief civil engr, C D Howe Co Ltd, 51-58; from asst prof to assoc prof civil eng, 58-73, chmn dept, 77-84, PROF CIVIL ENG & APPL MECH, MCGILL UNIV, 73- *Concurrent Pos:* Consult engr, 58-; mem, Univ Senate, 71 & Univ Bd Gov, 75-82. *Honors & Awards:* Univ Gold Medal, Civil Eng 48. *Mem:* Am Soc Civil Engrs; Am Concrete Inst; fel Eng Inst Can; fel Can Soc Civil Eng; Can Standards Asn. *Res:* Post buckling behavior of plates and shells; eccentrically loaded joints using high strength bolts; strength of bolted connections in timber construction. *Mailing Add:* Dept Civil Eng & Appl Mech McGill Univ 817 Sherbrooke St W Montreal PQ H3A 2K6 Can

HARRIS, PATRICIA J, b Seattle, Wash, Sept 2, 21; m 73. CELL BIOLOGY, ELECTRON MICROSCOPY. *Educ:* Univ Calif, Berkeley, BA, 54, PhD(zool), 62; Yale Univ, MS, 58. *Prof Exp:* Phys sci aide in chg electron microscope lab, Naval Radiol Defense Lab, Calif, 48-52; technician fish endocrinol, Bingham Oceanog Lab, Yale Univ, 54-58; technician cell biol & electron micros, Univ Calif, Berkeley, 58-62, asst res zoologist, 62-64; from asst prof to assoc prof zool, Ore State Univ, 64-73; adj assoc prof, 73-81, ADJ PROF BIOL, UNIV ORE, 81- *Mem:* Am Soc Cell Biol; Electron Micros Soc Am; Am Soc Develop Biol; Am Soc Zoologists; fel AAAS. *Res:* Fine structure studies of mitosis in both animal and plant cells; experimental modification of mitosis and normal development, particularly in marine invertebrate embryos. *Mailing Add:* Dept Biol Univ Ore Eugene OR 97403

HARRIS, PATRICK DONALD, b Nebraska City, Nebr, Mar 30, 40; m 59; c 3. PHYSIOLOGY, ELECTRICAL ENGINEERING. *Educ:* Univ Mo, BSEE, 62, MSEE, 63; Northwestern Univ, PhD(physiol), 67. *Prof Exp:* NIH fel physiol, Med Ctr, Ind Univ, 67-68; from asst prof to prof physiol, Univ Mo, Columbia, 68-80; assoc dir, Dalton Res Ctr, 80-81; PROF & CHMN, PHYSIOL & BIOPHYS, UNIV LOUISVILLE, 81- *Concurrent Pos:* NIH spec fel, 70-72; vis assoc biomed eng, Calif Inst Technol, 77-78. *Honors & Awards:* Microcirculatory Soc Award, 66. *Mem:* Am Physiol Soc; Inst Elec & Electronics Eng; Microcirc Soc; Am Heart Asn; Biomed Eng Soc; Shock Soc; Sigma Xi; Europ Microcirculatory Soc; US Microcirculatory Soc (pres, 86). *Res:* Hormonal control of small vessel diameters during hypertension, hemorrhagic shock, sepsis and anesthesia. *Mailing Add:* Dept Physiol Univ Louisville Louisville KY 40292

HARRIS, PAUL CHAPPELL, b Nova Scotia, Can, Oct 30, 30; US citizen; m 55; c 3. POULTRY SCIENCE. *Educ:* McGill Univ, BSc, 52; Univ Md, MS, 56, PhD(poultry husb), 60. *Prof Exp:* Poultry fieldman, NS Dept Agr, Can, 52-54; asst, Univ Md, 54-59; asst prof, 59-65, ASSOC PROF POULTRY SCI, UNIV MAINE, ORONO, 65- *Mem:* Fel AAAS; Poultry Sci Asn; World Poultry Sci Asn; Sigma Xi. *Res:* Physiology of egg production; egg quality and diet utilization; layer management. *Mailing Add:* Dept Animal Vet & Aquatic Sci Hitchner Hall Univ Maine Orono ME 04469-0131

HARRIS, PAUL JONATHAN, b Park Royal, Eng, July 8, 53. PHOSPHAZENE CHEMISTRY, ORGANIC PHOSPHORUS CHEMISTRY. *Educ:* Univ Bath, Eng, BSc, 74; Bristol Univ, PhD(chem), 78. *Prof Exp:* Res asst, Pa State Univ, 77-80; ASST PROF CHEM, VA POLYTECH INST & STATE UNIV, 80- *Mem:* Am Chem Soc; Royal Soc Chem. *Res:* Synthesis and structure determination of a series of new cyclic phosphazene molecules; potential of these compounds as precursors of new polyorganophosphazenes. *Mailing Add:* BASF Inmont PO Box 5009 Southfield MI 48086-5009

HARRIS, PAUL ROBERT, b Charlotte, NC, Nov 8, 42; m 68; c 4. CHEMICAL ENGINEERING. *Educ:* Ga Inst Technol, BChE, 63; Northwestern Univ, MS, 64; Mich State Univ, PhD(chem eng), 67. *Prof Exp:* Res engr, Eastern Res Ctr, Stauffer Chem Co, 67-68, sr res engr, 70-71, plant engr, Energy Prod Dept, 71-76; supt, Plains Co-op Oil Mill, 76-77; prog mgr, Pillsbury Co, 77-79; mgr-engr, Am Maize Prod Co, 79-84; VPRES COM DEVELOP, LEE LABS, INC, 84- *Mem:* Am Chem Soc; Am Inst Chem Engrs; Sigma Xi; Int Oil Mill Supt Asn. *Res:* Chemical reactor design; photochemical reactor design; reaction kinetics; spray drying; unusual separation techniques. *Mailing Add:* 11114 Sithean Way Richmond VA 23233

HARRIS, PETER, b Cambridge, Eng, Oct 19, 30; Can citizen; m 57; c 3. WEED SCIENCE. *Educ:* Univ BC, BSF, 55; Univ London, DIC & PhD(zool), 58. *Prof Exp:* RES SCIENTIST, AGR CAN, 59- *Mem:* Entom Soc Can; Royal Entom Soc London; Int Orgn Biol Control; Weed Sci Soc Am. *Res:* Establishing procedures for the biological control of weeds with introduced insects; control of the following weed in Canada: Hypericum perforatum, Carduus nutans, Centaurea diffusa and Centaurea maculosa, Euphorbia esula and Euphorbia cyparissias, Senecio jacobaea. *Mailing Add:* Res Sta Agr Can PO Box 440 Regina SK S4P 3A2 Can

HARRIS, RAE LAWRENCE, JR, b Laurel, Miss, Apr 8, 26; m 47; c 3. GEOLOGY. *Educ:* Ore State Col, BS, 50; Columbia Univ, PhD(geol), 57. *Prof Exp:* Lectr geol, Columbia Univ, 55-57; from asst prof to assoc prof, 57-69, PROF GEOL, TEX TECH UNIV, 69- *Mem:* AAAS; Geol Soc Am; Am Geophys Union; Sigma Xi. *Res:* Petrology and structure of granite rocks; economic and engineering geology. *Mailing Add:* Geosci Dept Tex Tech Univ Lubbock TX 79409

HARRIS, RALPH ROGERS, b Winfield, Ala, Mar 9, 29; m 51; c 3. ANIMAL NUTRITION, DAIRY SCIENCE. *Educ:* Auburn Univ, BS, 51, MS, 52; Agr & Mech Col Tex, PhD(dairy sci), 59. *Prof Exp:* Asst animal husb & nutrit, Auburn Univ, 51-52, animal husbandman, 55; asst, Agr & Mech Col Tex, 56, dairy specialist voc agr, 59-60; asst animal nutritionist, 60-64, assoc prof animal sci, 64-73, PROF ANIMAL NUTRIT, AUBURN UNIV, 73- *Mem:* Am Dairy Sci Asn; Am Soc Animal Sci. *Res:* Forage utilization by beef and dairy cattle; environmental effects upon response of beef and dairy cows. *Mailing Add:* Dept Animal Sci Auburn Univ Auburn AL 36830

HARRIS, RAY EDGAR, analytical chemistry, for more information see previous edition

HARRIS, RAYMOND, medical sciences, health sciences; deceased, see previous edition for last biography

HARRIS, REECE THOMAS, b Shreveport, La, Aug 10, 32; m 60; c 2. MATHEMATICS. *Educ:* Reed Col, BS, 55; Univ Ill, MA, 56, Phd(math), 59. *Prof Exp:* Res assoc math, Duke Univ, 59-60, asst prof, 60-65; assoc prof, Univ Md, 65-68; TUTOR, ST JOHN'S COL, NMEX, 68- *Concurrent Pos:* Vis asst prof, Univ Calif, Berkeley, 62-63; vis assoc prof, NY Univ, 67-68; consult, India Proj, NSF, 67 & IBM Corp Hq, NY; vis prof, UCSC Math Dept, 85-86 & 88. *Res:* Functional and harmonic analysis; spectral theory. *Mailing Add:* Dept Math St John's Col Camino de Cruz Blanca Santa Fe NM 87501

HARRIS, RICHARD, b London, Eng, Jan 19, 44; Brit & Can citizen. SOLID STATE PHYSICS. *Educ:* Oxford Univ, BA, 65; Univ Sussex, DPhil, 68. *Prof Exp:* Res asst physics, Imp Col, Univ London, 68-70; asst prof, 70-76, ASSOC PROF PHYSICS, MCGILL UNIV, 76- *Mem:* Am Phys Soc; Brit Inst Physics. *Res:* Magnetic and structural properties of amorphous metals and other glassy solids. *Mailing Add:* Rutherford Physics Bldg McGill Univ 3600 University St Montreal PQ H3A 2T8 Can

HARRIS, RICHARD ELGIN, b Kansas City, Mo, May 28, 41; m 67; c 1. SOLID STATE PHYSICS, LOW TEMPERATURE PHYSICS. *Educ:* Univ Rochester, BS, 63; Univ Ill, MS, 65, PhD(physics), 69. *Prof Exp:* Exp physicist, United Technol Res Ctr, 69-75; PHYSICIST, NAT INST STANDARDS & TECHNOL, 75- *Mem:* Am Phys Soc; Sigma Xi; AAAS. *Res:* Superconducting electronics; Josephson effect, experiment and theory. *Mailing Add:* US Dept Com Div 724 03 Nat Inst Standards & Technol 325 Broadway Boulder CO 80303

HARRIS, RICHARD JACOB, b Canton, Ohio, May 20, 45. OPTICAL PHYSICS. *Educ:* Miami Univ, BS, 67; Yale Univ, MS, 68. *Prof Exp:* Res assoc coherent optics, McDonnell Douglas, 68-69; assoc physicist, Syst Res Inc, 69-70; physicist, Childrens Hosp Laser Lab, 70-71; optical engr, Lynex, Inc, 71-72; assoc physicist, Univ Dayton Res Inst, 72-84; at electrooptics div, IIT Res Inst, Chicago, 84-90; CONSULT, 90- *Mem:* Optical Soc Am. *Res:* Optical properties of transparent materials, laser interactions with materials. *Mailing Add:* 4155 Fowler Dr Bellbrook OH 45305

HARRIS, RICHARD LEE, b Chicago, Ill, Nov 1, 34; m 57; c 2. ORGANIC POLYMER CHEMISTRY. *Educ:* NCent Col, Ill, BS, 56; Univ Ill, PhD(org chem), 60. *Prof Exp:* Res chemist, Electrochem Dept, E I du Pont de Nemours & Co, 60-64; sr res chemist, Diamond Alkali Co, 64-68, res supvr, T R Evans Res Ctr, 68-74, res mgr, 74-80, fossil fuels specialist, 80-83, RES SPECIALIST, TECH DEPT, DIAMOND SHAMROCK CORP, 83- *Mem:* Am Chem Soc. *Res:* Polymer chemistry; product applications; polymer synthesis and development; organic specialty chemicals research. *Mailing Add:* 27900 Bishop Park No 115F Willoughby OH 44092

HARRIS, RICHARD WILSON, b Fresno, Calif, July 26, 20; m 46; c 3. HORTICULTURE. *Educ:* Univ Calif, BS, 42, MS, 47; Cornell Univ, PhD, 50. *Prof Exp:* Asst pomol, Univ Calif, 46-47 & Cornell Univ, 47-50; instr & jr pomologist, 50-52, asst prof & asst pomologist, 52-57, assoc prof landscape hort & assoc horticulturist, 58-64, chmn, Dept Environ Hort, 58-65, prof environ hort & horticulturist, 64-86, chmn, Dept Environ Hort, 78-83, EMER PROF, ENVIRON HORT, UNIV CALIF, DAVIS, 86- *Concurrent Pos:* Consult, Tree Safety. *Honors & Awards:* Norman Jay Colman Award, Am Asn Nurserymen; Sci Award, Am Hort Soc. *Mem:* AAAS; Int Soc Arboriculture (pres, 86-87); Soc Am Foresters; Am Soc Hort Sci. *Res:* Culture of landscape plants; arboriculture; management of landscape maintenance. *Mailing Add:* Dept Environ Hort Univ Calif Davis CA 95616

HARRIS, ROBERT A, b Chicago, Ill, Aug 9, 36; m 58; c 3. THEORETICAL CHEMISTRY, CHEMICAL PHYSICS. *Educ:* Univ Ill, BS, 57; Univ Chicago, MS, 59, PhD(chem), 60. *Prof Exp:* Jr fel, Harvard Soc Fels, Harvard Univ, 60-63; from asst prof to assoc prof, 63-74, PROF CHEM, UNIV CALIF, BERKELEY, 74- *Concurrent Pos:* Sloan Found fel, 65-67; NSF sr fel, 68-69; Humboldt Sr Am Scientist Award, 77-78; J S Guggenheim Found Fel, 82-83. *Res:* Electronic properties of molecules and polymers; polymer hydrodynamics, interaction of radiation with matter, general problems of quantum theory. *Mailing Add:* Dept Chem Univ Calif 2120 Oxford St Berkeley CA 94720

HARRIS, ROBERT ALLISON, b Boone, Iowa, Nov 10, 39; m 59; c 4. BIOCHEMISTRY. *Educ:* Iowa State Univ, BS, 62; Purdue Univ, MS, 64, PhD(biochem), 65. *Prof Exp:* Asst res prof , Inst Enzyme Res, Univ Wis-Madison, 69-70; assoc prof, 70-74, prof & assoc chmn, 75-88, PROF BIOCHEM & CHMN, SCH MED, IND UNIV, INDIANAPOLIS, 88- *Concurrent Pos:* Multiple Sclerosis fel, Inst Enzyme Res, Univ Wis-Madison, 66-70; estab investr, Am Heart Asn, Sch Med, Ind Univ, Indianapolis, 70-74; vis fel, Metab Res Lab, Oxford Univ, 73-74; assoc ed, Lipids, 73-87; Res Biochemist, Nat Inst Alcohol Abuse & Alcoholism, 80. *Mem:* Am Oil Chem Soc; Brit Biochem Soc; Am Heart Asn; Am Soc Biol Chem; Am Rust Nutrit. *Res:* Regulation of metabolic processes; application of recombinant DNA techniques in the regulation of branched chain amino acid catabolism. *Mailing Add:* Dept Biochem Ind Univ Sch Med Indianapolis IN 46202

HARRIS, ROBERT HUTCHISON, b St Louis, Mo, Mar 17, 20; m 46; c 3. INORGANIC CHEMISTRY. *Educ:* Calif Inst Technol, BS, 47; Purdue Univ, PhD(chem), 52. *Prof Exp:* Chem engr, Thompson Aircraft Prods Co, 42-45; from instr to prof, 51-90, EMER PROF CHEM, UNIV NEBR, LINCOLN, 90- *Mem:* Am Chem Soc. *Res:* P-block structures and reactions. *Mailing Add:* Dept of Chem Univ of Nebr Lincoln NE 68588-0304

HARRIS, ROBERT IRA, clinical psychology, for more information see previous edition

HARRIS, ROBERT L, b Pontiac, Ill, Feb 6, 29; m 60; c 1. ORGANIC POLYMER CHEMISTRY. *Educ:* Univ Ill, BSc, 51; Mass Inst Technol, PhD(org chem), 56. *Prof Exp:* Res chemist, Visking Co Div Union Carbide Corp, 56-61, Minn Mining & Mfg Co, 61-63 & Devoe & Raynolds Co, Div Celanese Corp, 63-66; sr res scientist, Oxford Paper Co, Div Ethyl Corp, 66-68; sr res chemist, Ecusta Paper Co, Div Olin Corp, 68-74; res dir, Imperial Paper Co, 74-78; sr res chemist, Moore Bus Forms Inc, 78-80; CONSULT, W&F MFG CO, 83- *Concurrent Pos:* Lectr, Ind Univ, 65. *Mem:* Am Chem Soc; Tech Asn Pulp & Paper Indust; Soc Plastics Engrs; NY Acad Sci. *Res:* New monomers and polymers for coating applications; coatings for marking and decorative systems; flexographic and gravure inks; plastisol coatings; coatings for printing papers and wallpapers; water-based coatings. *Mailing Add:* 4480 Darcy Lane Williamsville NY 14221

HARRIS, ROBERT L, JR, b Hollister, Mo, July 18, 24; c 3. OCCUPATIONAL HEALTH. *Educ:* Univ Ark, BS, 49; Harvard Univ, MS, 54; Univ NC, Chapel Hill, PhD(environ sci & eng), 72. *Prof Exp:* Jr engr, Div Indust Hyg, USPHS, 49-50, asst engr, Environ Health Ctr, 50-52, engr, Div Spec Health Serv, 52-57, engr, Occup Health Field Sta, Div Occup Health, 57-60, chief, Eng Sect, 60-64, chief, Field Invests, Abatement Prog, Nat Air Pollution Control Admin, 64-68, dir, Bur Abatement & Control, 68-70; assoc prof, 73-76, dir, Occup Health Studies Group, 74-84, PROF ENVIRON SCI & ENG, SCH PUB HEALTH, UNIV NC, CHAPEL HILL, 76- *Honors & Awards:* Henry F Smyth Award, Am Acad Indust Hyg, 85. *Mem:* Am Indust Hyg Asn; Am Conf Govt Indust Hygienists; Sigma Xi; Am Acad Indust Hyg; Brit Occup Hyg Soc. *Mailing Add:* Dept Environ Sci & Eng CB No 7400 Rosenau Hall Univ NC Chapel Hill NC 27599-7400

HARRIS, ROBERT LAURENCE, b Melrose, Mass, May 16, 23; m 46; c 2. PHYSICAL CHEMISTRY. *Educ:* Bates Col, BS, 47; Univ Wis, PhD(chem), 51. *Prof Exp:* Res chemist, Cent Res Lab, 51-52, proj leader, 53-58, mgr lab res, 59-64, asst dir res & develop, 65-66, sr tech assoc, Cent Res Lab, 66-69, res mgr, Corp Res Div, 69-71, sr scientist, 71-76, SR RES ASSOC DEPT TOXICOL, CORP RES DIV, ALLIED CORP, 77- *Mem:* Am Chem Soc; AAAS. *Res:* Building products from plastics; x-ray diffraction; structure of salts in the liquid state; catalytic hydrogenation; liquid phase oxidation. *Mailing Add:* PO Box 11 Sunset ME 04683

HARRIS, ROBERT LEE, entomology, for more information see previous edition

HARRIS, ROBERT MARTIN, b Atlantic City, NJ, Dec 5, 21; m 44; c 4. PHYSIOLOGICAL GENETICS. *Educ:* Univ Calif, BA, 49, PhD(bot sci), 53. *Prof Exp:* Asst bot, Univ Calif, 50-53; from instr to prof biol sci, Univ Ariz, 53-84, chmn comt genetics, 66-84; RETIRED. *Concurrent Pos:* Ed, J Ariz-Nev Acad Sci, 56-78. *Mem:* Fel AAAS; Am Genetics Soc; Am Genetics Asn. *Res:* Physiological genetics of stature mutants of corn; genetics of blue-green algae. *Mailing Add:* 841 E Alta Vista St Tucson AZ 85719

HARRIS, ROGER MASON, b Berkeley, Calif, Mar 22, 46; m 70; c 2. NEUROANATOMY. *Educ:* Stanford Univ, BS, 68; Univ Wash, PhD(physics), 75. *Prof Exp:* res assoc physics, Fermi Nat Accelerator Lab, 74-77; res assoc neurobiol, Wash Univ Sch Med, 77-80; fel, Sch Med, Univ Calif, San Francisco, 80-82; AT DEPT BIOL STRUCT, SCH MED, UNIV WASH. *Mem:* AAAS; Soc Neurosci. *Res:* High energy bubble chamber physics; mouse cortical barrels; computer-assisted neuroanatomy; rat somatosensory thalamus; recovery from spinal cord injury. *Mailing Add:* Dept Biol Struct Univ Wash Sch Med SM-20 Seattle WA 98195

HARRIS, RONALD DAVID, b Norman, Okla, Apr 9, 38; m 62; c 3. FOOD SCIENCE, CHEMICAL ENGINEERING. *Educ:* Ohio State Univ, BChE & MSc, 61; Univ Cincinnati, MBA, 70. *Prof Exp:* Res engr foods, Procter & Gamble Co, 61-62 & 64-66, group leader, 66-71; sect mgr, Clorox Co, 71-72, dept mgr, 72-73, dir res & develop, 73-77; vpres res & develop, Anderson Clayton Foods, Anderson, Clayton & Co, 77-87; vpres, Kraft, Inc, 87-90; VPRES, KRAFT USA TECHNOL, KRAFT GEN FOODS, 90- *Mem:* Inst Food Technologists; Am Chem Soc; Am Oil Chemists Soc. *Res:* Food science; consumer food products; edible fats and oils; dairy products; dairy analogs; shelf stable packaged foods; fresh prepared refrigerated foods; cheese. *Mailing Add:* 1250 Cascade Ct Lake Forest IL 60045-3614

HARRIS, RONALD L, b Stockton, Calif, Feb 20, 48. METAL VAPOR SYNTHESIS, METAL AMMONIA CHEMISTRY. *Educ:* Univ Hawaii, BS, 73; Univ Tex, PhD(chem), 80. *Prof Exp:* ASSOC PROF INORG CHEM, IND UNIV & PURDUE UNIV, FT WAYNE, 83- *Mem:* Am Chem Soc; Am Asn Univ Profs. *Res:* Metal vapor synthesis. *Mailing Add:* 1320 21St St No 1 Peru IL 61354

HARRIS, RONALD WILBERT, b Houston, Tex, May 23, 38; m 56; c 2. OPTICS. *Educ:* Southwestern La Univ, BS, 63; Univ Ark, MS & PhD(physics), 67. *Prof Exp:* Physicist, Comput Graphics Res, Shell Develop Co, Tex, 66-67, proj dir optical info processing, 67-68; assoc prof physics & chmn dept, Drury Col, 68-70; from assoc prof to prof, 70-85, chmn dept, 70-76, ADJ PROF PHYSICS, LOYOLA UNIV CHICAGO, 85- *Concurrent Pos:* Lectr, Univ Houston, 67-68; chmn bd, Sladco, Inc, Eunice LA, 75- *Mem:* Am Phys Soc; Inst Elec & Electronics Engrs; Optical Soc Am; Soc Photog Scientists & Engr. *Res:* Image enhancement; optical and acoustical holography; data filtering and displays; determination of liquid structure by means of x-ray diffraction. *Mailing Add:* PO Box 1289 Eunice LA 70535

HARRIS, RONNEY D, b Tremonton, Utah, Sept 8, 32; m 55; c 4. ELECTRICAL ENGINEERING. *Educ:* Univ Utah, BS, 54, PhD(elec eng), 64. *Prof Exp:* Proj engr, Microwave Tube Div, Litton Industs, Inc, 60-63; asst prof elec eng, 64-69, ASSOC PROF ELEC ENG, UTAH STATE UNIV, 69-, DIR CTR ATOMOSPHERIC & SPACE SCI, 80-, PROF ELEC ENG. *Mem:* Inst Elec & Electronics Engrs; Am Geophys Union. *Res:* Generation and focusing of large diameter electron beams; numerical solutions of charged particle trajectories; characteristics and formation of sporadic E ionization; wave propagation in ionospheric plasmas. *Mailing Add:* Dept Elec Eng Utah State Univ Logan UT 84322

HARRIS, ROY H, b Madison, Ga, Dec 17, 28; m 51; c 2. AIRBORNE & GROUND RADAR, UNDERWATER SONAR & TELECOMMUNICATIONS. *Educ:* Ga Inst Technol, 51; Polytechnic Inst, MEE, 56. *Prof Exp:* Engr, Hazeltine Elec Corp, 51-56; mem tech staff, Bell Telephone Lab, Burlington NC, 56-60, supvr, Nike Hercules design & prod eng, 60-62, supvr, Uncom, 62-64; guidance systs Eng Mgr, Western Elec Co, 65-68, mgr, mil systs eng, 68-72, mgr, mil eng Navy, 72-82; dir, Fed Systs Div, AT&T Technol, Greensboro, NC, 82-86, vpres advan technol systs, 86-87; PRES TRIAD TECHNOL, GREENSBORO, NC, 87- *Concurrent Pos:* Eng Sch Indust Adv Group, 83-; chmn, Undersea Surveillance & Oceanog Sub Comt; mem bd trustees, NC AT&T State Univ; bd dir, Southcon, 79-82 & 90-91. *Mem:* Fel Inst Elec Electronics Engrs; Nat Security Industr Asn. *Res:* Systems for airborne and ground radar, missile guidance, underwater sonar and telecommunications. *Mailing Add:* 2025 Nottingham Lane Burlington NC 27215

HARRIS, RUDOLPH, b Shreveport, La, Nov 21, 35; m 62; c 2. CELL BIOLOGY. *Educ:* Southern Univ, BS, 56; Cath Univ Am, PhD(cell biol), 73. *Prof Exp:* Asst prof biol, Fed City Col, 71-73; staff fel virol, Nat Cancer Inst, 74-77; CONSUMER SAFETY OFFICER, FOOD & DRUG ADMIN, 77- *Mem:* Am Soc Cell Biol; Sigma Xi; AAAS. *Res:* Analysis of the transport of ions across the membrane of cells in culture which has been transformed by oncogenic viruses. *Mailing Add:* Food & Drug Admin Rockville MD 20850

HARRIS, RUTH B S, b Rugby, Eng, Dec 20, 55. FOOD INTAKE REGULATION. *Educ:* Univ Leeds, Eng, PhD(physiol), 81. *Prof Exp:* res assoc, Dept Nutrit, Univ Ga, 81-86; asst mem, Monell Chem Senses Ctr, Philadelphia, 87-88; SR RES SCIENTIST, DEPT NUTRIT, KRAFT GEN FOODS, GLENVILLE, IL, 88- *Mem:* Am Inst Nutrit. *Mailing Add:* Nutrit Dept Kraft Gen Foods 801 Waukegan Rd Glenview IL 60025

HARRIS, RUTH CAMERON, b Mt Vernon, NY, Apr 23, 16; m 52; c 2. PEDIATRICS. *Educ:* Columbia Univ, AB, 37, MD, 43. *Prof Exp:* Intern, Philadelphia Gen Hosp, 43-44; resident, Willard Parker Hosp, New York, 44-45; asst resident, Babies Hosp, 45-47; from asst to assoc pediat, Columbia Univ, 47-55, asst prof, 55-72, assoc prof clin pediat, Col Physicians & Surgeons, 72-76; prof pediat & chmn dept, Sch Med, Marshall Univ, 76-81, emer prof, 81-90; RETIRED. *Concurrent Pos:* From asst pediatrician to asst attend pediatrician, Babies Hosp, 47-72, from actg dir to dir chem lab, 53-65, attend pediatrician, 72-; from asst pediatrician to asst attend pediatrician, Vanderbilt Clin, 47-72, attend pediatrician, 72-; guest prof, Yonsei Univ, Korea, 68-70. *Mem:* Soc Pediat Res; Am Soc Human Genetics; Am Asn Study Liver Dis; Am Pediat Soc; Am Acad Pediat. *Res:* Liver disease in childhood. *Mailing Add:* 317 Sycamore Glen Dr No 412 Miamisburg OH 45342

HARRIS, S RICHARD, b Lynn, Mass, May 5, 32; m 56; c 2. BIOCHEMISTRY. *Educ:* Princeton Univ, AB, 56; Boston Col, MS, 61; Boston Univ, PhD(biochem), 66. *Prof Exp:* Res chemist, Brockton Vet Admin Hosp, 62-66; lab dir, 68-73, sr anal biochemist, Biomed Assay Labs Div, 66-74, head prod develop, 74-77, head res & develop, 77-78, sr proj leader, New Eng Nuclear Corp, 78-81; SR RES CHEMIST, BIOMED PROD, DUPONT, 82- *Mem:* AAAS; Am Chem Soc; Am Asn Clin Chemists; NY Acad Sci. *Res:* Mechanism of drug action; development of methodology for quantitating tissue levels of various drugs and their metabolites; development of in vitro diagnostic kits and reagents; development of analytical techniques for naturally occurring biological chemicals. *Mailing Add:* 104 Bird St Needham MA 02192

HARRIS, SAMUEL M(ELVIN), b Chicago, Ill, Mar 25, 33; m 60; c 2. THEORETICAL NUCLEAR PHYSICS, QUANTUM OPTICS. *Educ:* Univ Ill, BS, 55, MS, 56, PhD(physics), 61. *Prof Exp:* NSF vis scientist, Univ Bonn, 60-61; from res assoc to sr res assoc physics, Columbia Univ, 62-64; asst prof, 64-67, ASSOC PROF PHYSICS, PURDUE UNIV, 67- *Mem:* Am Phys Soc. *Res:* Theory of nuclear structure; collective states of deformed nuclei; quantum theory; quantum optics. *Mailing Add:* Dept Physics Purdue Univ Lafayette IN 47907

HARRIS, SAMUEL WILLIAM, b Chicago, Ill, Jan 27, 30; m 67; c 2. TRIBIOLOGY CHEMISTRY. *Educ:* Miami Univ Ohio, BA, 52, MS, 53; Ohio State Univ, PhD(chem), 56. *Prof Exp:* Res chemist, Whiting Labs, Standard Oil Ind, 56-62; proj chemist, Propellants Div, 62-65, proj chemist, 65-77; SR RES CHEMIST, AMOCO RES CTR, AMOCO CHEM CORP, 77- *Mem:* Am Chem Soc; Sigma Xi. *Res:* Boron hydride and nitrogen oxide chemistry; heterogeneous catalysis; catalytic petroleum and solid propellant technology; lubricant chemistry. *Mailing Add:* Amoco Petrol Additives Co PO Box 3011 Naperville IL 60566-7011

HARRIS, SIGMUND PAUL, b Buffalo, NY, Oct 12, 21; m 48; c 1. PHYSICS. *Educ:* Univ Buffalo, BA, 41, MA, 43; Ill Inst Technol, PhD(physics), 54. *Prof Exp:* Asst physics, Univ Buffalo, 41-43 & Yale Univ, 43; jr physicist, Metall Lab, Univ Chicago, 43-44; jr scientist, Los Alamos Sci Lab, 44-46; assoc physicist, Argonne Nat Lab, 46-53; proj engr, Stewart-Warner Corp, 53-54; sr physicist, Tracerlab, Inc, 54-56; sr res engr, Atomics Int Div, NAm Aviation Inc, 56-64; head physics sect, Res Div, Maremont Corp, 64-66; from instr to assoc prof, Pierce Col, 66-79, prof physics, 79-86; RETIRED. *Concurrent Pos:* Indust consult. *Mem:* Am Phys Soc; Am Nuclear Soc; Sigma Xi; Am Assoc Physics Teachers. *Res:* Neutron physics; transient radiation effects; space electrical propulsion; air pollution. *Mailing Add:* 5831 Saloma Ave Van Nuys CA 91411

HARRIS, STANLEY CYRIL, b Chicago, Ill, July 2, 16; m 42; c 2. PHYSIOLOGY, PHARMACOLOGY. *Educ:* Northwestern Univ, BS, 38, MS, 41, PhD(physiol), 46. *Hon Degrees:* MA, Univ Pa, 71. *Prof Exp:* Asst physiol, Med Sch, Northwestern Univ, 39-46, from instr to prof, Dent Sch, 46-66; prof pharmacol & chmn dept, Sch Dent Med, Univ Pa, 66-73, dir div advan dent educ, 67-73; prof therapeut & assoc dean, Dent Sch, Northwestern Univ, Chicago, 73-83; RETIRED. *Concurrent Pos:* Consult, Coun on Drugs, AMA, 49-60, Coun Dent Therapeut, Am Dent Asn, 53-83 & Coun Dent Educ, 71-76; chmn comt grad educ, Am Asn Dent Schs, 64-68; mem dent adv panel, US Pharmacopoeia & prog proj comt, Nat Inst Dent Res, 64-68. *Mem:* AAAS; Am Physiol Soc; Soc Exp Biol & Med; Am Soc Pharmacol & Exp Therapeut; Int Asn Dent Res. *Res:* Pain and analgesia; hunger and appetite; oral physiology; dental therapeutics; inflammation; research design. *Mailing Add:* 350 Sunset Ct Northbrook IL 60062

HARRIS, STANLEY EDWARDS, JR, b Basking Ridge, NJ, Mar 5, 18; m 40; c 4. GEOLOGY NATURAL AREAS. *Educ:* Princeton Univ, BA, 40; Univ Iowa, MS, 42, PhD(geol), 47. *Prof Exp:* Lab asst, Iowa Geol Surv, 40-41, jr geologist, 42-43, geologist, 43-48; asst geol, Univ Iowa, 41-42; vis asst prof, Univ Mo, 48-49; from assoc prof to prof, 49-82, chmn dept, 55-66, EMER PROF GEOL, SOUTHERN ILL UNIV, 82- *Concurrent Pos:* Guest docent, Univ Hamburg, 57-58; mem, Ill State Nature Preserves Comn, 76-78. *Mem:* AAAS; Sigma Xi. *Res:* Physiography of central United States; teaching of earth science; environmental geology; Pleistocene geology; Paleozoic stratigraphy; study of natural areas. *Mailing Add:* Woods Edge Rte 7 Box 80 Carbondale IL 62901

HARRIS, STANLEY WARREN, b Dodson, Mont, Sept 18, 28; m 50; c 2. WILDLIFE MANAGEMENT. *Educ:* Wash State Univ, BS, 50, MS, 52; Univ Minn, PhD, 57. *Prof Exp:* Asst, Wash State Univ, 50-52 & Univ Minn, 52-54; res biologist, Game & Fish Div, State of Minn, 56-59; from asst prof to assoc prof, 59-69, PROF WILDLIFE MGT, HUMBOLDT STATE UNIV, 70- *Concurrent Pos:* Res biologist III, Alaska Fish & Game Dept, 73-75 & Calif Condor, 81-85; mem adv comt shorebirds, Calif Fish & Game Dept, 73-75, adv comt Calif Condor, 81-85. *Mem:* Wildlife Soc; Am Ornith Union; Cooper Ornith Soc. *Res:* Waterfowl and seabirds, especially their population dynamics and life history; ornithology; wetland ecology and management. *Mailing Add:* Dept Wildlife Mgt Humboldt State Univ Arcata CA 95521

HARRIS, STEPHEN ERNEST, b Brooklyn, NY, Nov 29, 36; m 59; c 2. ELECTRICAL ENGINEERING. *Educ:* Rensselaer Polytech Inst, BEE, 59; Stanford Univ, MS, 61, PhD(elec eng), 63. *Prof Exp:* Mem tech staff, Bell Tel Labs, 59-60; from asst prof to prof elec eng, 63-79, dir, Edward L Ginzton Lab, 83-88, PROF ELEC ENG & APPL PHYSICS, STANFORD UNIV, CALIF, 79- *Concurrent Pos:* Consult, Optical Device Group, Sylvania Electronic Systs, Calif, 63-68 & Chromatix, Inc, Calif, 68-; Guggenheim fel, 76; dir, Joint Serv Electronics Prog, Edward L Ginzton Lab, 83-90; Barbara & Kenneth Oshman prof eng, 88. *Honors & Awards:* Nobel Prize, 65; Curtis W McGraw Res Award, 73; David Sarnoff Award, 78; Charles Hard Townes Award, Optical Soc Am, 85; Einstein Prize for Laser Sci, 91. *Mem:* Nat Acad Sci; Nat Acad Eng; fel Optical Soc Am; fel Inst Elec & Electronics Engrs; fel Am Phys Soc. *Res:* Quantum electronics; lasers; nonlinear optics; acousto-optics; XUV radiation sources; spectroscopy. *Mailing Add:* Edward L Ginzton Lab Stanford Univ Stanford CA 94305

HARRIS, STEPHEN EUBANK, b New Orleans, La, June 30, 42; m 74; c 1. MOLECULAR BIOLOGY, BIOCHEMISTRY. *Educ:* Univ Tex, Austin, BA, 65, MA, 66, PhD(zool), 70. *Prof Exp:* Res instr, Baylor Col Med, Houston, Tex, 73-76; sr staff fel, Nat Inst Environ Health Sci, NC, 76-83; SR SCIENTIST MOLECULAR BIOL, W ALTON JONES CELL SCI CTR, NY, 83- *Concurrent Pos:* Mem, Cancer Ctr, Univ NC, Chapel Hill, 77- *Mem:* Fel AAAS; Soc Develop Biol. *Res:* Molecular biology of steroid hormone gene expression in growth and differentiation; cell biology. *Mailing Add:* Health Sci Ctr Univ Tex 7730 Floyd Curl Dr San Antonio TX 78284-7877

HARRIS, STEPHEN JOEL, b Los Angeles, Calif, July 9, 49; m 77; c 2. PHYSICAL CHEMISTRY. *Educ:* Univ Calif, Los Angeles, BS, 71; Harvard Univ, MA, 72, PhD(chem), 75. *Prof Exp:* RES CHEMIST PHYS CHEM, GEN MOTORS RES LABS, 77- *Concurrent Pos:* Fel, Miller Inst Basic Res Sci, Univ Calif, Berkeley, 75-77. *Mem:* Am Phys Soc. *Res:* Laser kinetics and spectroscopy; intracavity techniques; soot formation; diamond films. *Mailing Add:* Dept Phys chem Gen Motors Res Labs Warren MI 48090

HARRIS, STEVEN H, b Cincinnati, Oh, Nov 18, 24; m 55; c 5. PETROLEUM GEOLOGY, SEDIMENTOLOGY. *Educ:* Univ Cincinnati, BA, 48, MS, 50. *Prof Exp:* Geologist, Shell Oil Co, 50-52, consult geol Harris & Brown, 52-60; instr geol, Bismarck Jr Col, 58-64, consult geol, Harris, Brown & Klemer, 60-90; PRES PETROL, CAMARGO CORP, 65-, OIL PROD, PRADERA DEL NORTE INC, 81-; VPRES, CONDOR RESOURCES, 85- *Concurrent Pos:* Pres, NDak Geol Soc, 65 & 72, rep, Interstate Oil Compact Comn, 68-72, vpres exec comt, Independent Petrol Assn Am, 81-84. *Honors & Awards:* Sigma Xi. *Mem:* Am Asn Petrol Geologists; fel Geol Soc Am; Soc Paleontologists & Mineralogists; Am Inst Prof Geologists. *Res:* Author of numerous technical articles dealing primarily with petroleum exploration of the Williston Basin. *Mailing Add:* PO Box 5006 Bismarck ND 58502

HARRIS, STEWART, b Cleveland, Ohio, Jan 11, 37; m 63; c 2. STATISTICAL MECHANICS. *Educ:* Case Inst Technol, BS, 59; Northwestern Univ, MS, 60, PhD(physics of fluids), 64. *Prof Exp:* Assoc res scientist, Courant Inst Math Sci, NY Univ, 64-66; from asst prof to assoc prof, 66-78, chmn dept, 79-81, PROF MECH, STATE UNIV, NY STONY BROOK, 78-, DEAN, COL ENG & APPL SCI, 81- *Concurrent Pos:* Sci res coun sr vis fel, Dept Physics, Univ Surrey, UK, 72-73; SRC fel, 72; vis fel, Univ Surv, 79; vis, Univ Berlin & fel, 79-80. *Mem:* Am Phys Soc; Am Soc Mech Engrs. *Res:* Kinetic theory of monatomic gases; Brownian motion; environmental physics; non-equilibrium statistical mechanics; suspensions. *Mailing Add:* Dept Mech State Univ NY Stony Brook NY 11790

HARRIS, SUSANNA, b Brooklyn, NY, May 11, 19; m 40; c 2. MICROBIOLOGY. *Educ:* Brooklyn Col, BA, 40; Drexel Inst, BS, 42; Univ Pa, PhD(med bact), 48. *Prof Exp:* Instr bact, 48-52, from asst prof to assoc prof, 58-81, EMER ASSOC PROF IMMUNOL PEDIAT, SCH MED, UNIV PA, 81- *Concurrent Pos:* Mem dept res, Children's Hosp, Philadelphia, Pa. *Mem:* Am Soc Microbiol; Am Asn Immunol. *Res:* Formation of antibodies. *Mailing Add:* 5112 Woodbine Ave Philadelphia PA 19131

HARRIS, SUZANNE STRAIGHT, b Miami, Fla, Nov 18, 44; m; c 2. NUTRITION. *Educ:* Vanderbilt Univ, BA, 66; Univ Ala, Birmingham, PhD(biochem), 76. *Prof Exp:* Assoc biochemist, Southern Res Inst, Birmingham, Ala, 66-73; investr, Inst Dental Res, Univ Ala, Birmingham, 76-85, from instr to asst prof, Dept Nutrit Sci, 80-85, asst prof, Dept Comp Med, 81-85, asst prof, Dept Int Pub Health, 84-85; adminr, Human Nutrit Info Serv, USDA, 85, dep asst secy, Food & Consumer Serv, 85-89; DIR, HUMAN NUTRIT INST, INT LIFE SCI RES FOUND, 89- *Concurrent Pos:* Fel, Nat Inst Dent Res, NIH, 74-77; mem, Long Term Care Subcomt, Nat Comt Vital & Health Statist, Dept Health & Human Serv, 83-86. *Mem:* Sigma Xi; Am Inst Nutrit. *Res:* Relationship between vitamin A deficiency and calcification of bones and teeth. *Mailing Add:* Int Life Sci Inst 1126 16th St NW Suite 100 Washington DC 20036

HARRIS, THEODORE EDWARD, b Philadelphia, Pa, Jan 11, 19; m 47; c 2. MATHEMATICS. *Educ:* Univ Tex, BA, 39; Princeton Univ, MA, 46, PhD(math), 47. *Hon Degrees:* DrTechnol, Chalmers Inst Technol, Gothenburg, Sweden. *Prof Exp:* Mathematician, Rand Corp, 47-66, head math dept, 59-66; prof, 66-89, EMER PROF MATH & ELEC ENG, UNIV SOUTHERN CALIF, 89- *Concurrent Pos:* Vis assoc prof, Columbia Univ, 53; ed, Annals Math Statist, Inst Math Statist, 55-58; vis prof, Stanford Univ, 63. *Mem:* Nat Acad Sci; Am Math Soc; fel Inst Math Statist (pres elec, 65-66, pres, 66-67); fel AAAS. *Res:* Probability, stochastic processes. *Mailing Add:* Dept Math DRB 155 Univ Southern Calif Los Angeles CA 90089-1113

HARRIS, THOMAS DAVID, b Saginaw, Mich, Dec 29, 48; m 74. ORGANIC CHEMISTRY. *Educ:* Saginaw Valley State Col, BS, 71; Univ NH, PhD(chem), 75. *Prof Exp:* Res assoc chem, Univ Ore, 75-77; ASST PROF CHEM, STATE UNIV NY, FREDONIA, 77- *Mem:* Am Chem Soc; Sigma Xi. *Res:* Development of new synthetic methods; synthesis and study of organic molecules of theoretical interest; medicinal chemistry. *Mailing Add:* 56 Zion Hill Rd Salem NH 03079

HARRIS, THOMAS MASON, b Louisville, Ky, Mar 30, 28; m 50; c 2. EMBRYOLOGY, HISTOLOGY. *Educ:* Emory Univ, BA, 49; Univ NC, PhD(zool), 62. *Prof Exp:* Asst prof biol, Richmond Univ, 61-64; from asst prof to assoc prof, 64-75, PROF ANAT, MED COL VA, 75- *Mem:* Am Soc Zool; Am Asn Anat. *Res:* Mechanisms of embryonic histogenesis analyzed by descriptive and experimental techniques; reinvestigation and documentation of normal developmental events and correlation with embryonic anamolies. *Mailing Add:* Dept Anat Va Commonwealth Univ Box 565 MCV Sci Richmond VA 23298

HARRIS, THOMAS MUNSON, b Niagara Falls, NY, May 29, 34; m 58; c 2. ORGANIC CHEMISTRY. *Educ:* Univ Rochester, BS, 55; Duke Univ, PhD(chem), 59. *Prof Exp:* Res proj chemist, Union Carbide Chem Co Div, Union Carbide Corp, 58-61; res assoc chem, Duke Univ, 61-63, USPHS spec fel biochem, Med Ctr, 63-64; from asst prof to prof chem, 70-84, CENTENNIAL PROF CHEM & ASSOC DIR, MOLECULAR TOXICOL, VANDERBILT UNIV, 84- *Concurrent Pos:* Alfred P Sloan fel, 67-69; USPHS res career development award, 67-72. *Mem:* Am Chem Soc. *Res:* polycarbonyl compounds; biogenetic-type syntheses of polyketide compounds; biosynthesis of natural products; glycopeptide antibiotics; mycotoxins; DNA-carcinogen interactions. *Mailing Add:* 1706 Graybar Lane Nashville TN 37215

HARRIS, THOMAS R(AYMOND), b San Angelo, Tex, Feb 19, 37; m 63; c 2. BIOMEDICAL ENGINEERING, CHEMICAL ENGINEERING. *Educ:* Tex A&M Univ, BS, 58, MS, 62; Tulane Univ, PhD(chem eng), 64; Vanderbilt Univ, MD, 74. *Prof Exp:* Design engr, Stand Oil Co Calif, 58-60; from asst prof to assoc prof chem eng, 64-75, asst prof med, 74-80, prof biomed & chem eng & dir, Biomed Eng Prog, 75-87, assoc prof med, 80-86, CHMN, DEPT BIOMED ENG, VANDERBILT UNIV, 88-, PROF MED, 86- *Concurrent Pos:* Res assoc sch med, Vanderbilt Univ, 65. *Mem:* Am Inst Chem Engrs; Biomed Eng Soc (pres, 86); Am Heart Asn; Am Soc Clin Res; Sigma Xi. *Res:* Cardiovascular transport phenomena; physiological systems analysis; pulmonary edema. *Mailing Add:* Box 1724 Station B Nashville TN 37235

HARRIS, TZVEE N, b Russia, Aug 31, 12; US citizen; m 40; c 2. IMMUNOLOGY. *Educ:* Univ Pa, BA, 33, MD, 36. *Prof Exp:* Instr bact, Sch Med, 38-41, from instr to asst res prof pediat, 41-48, from assoc prof to prof, 48-81, EMER PROF IMMUNOL, DEPT PEDIAT, UNIV PA, 81- *Mem:* AAAS; Am Soc Microbiol; Am Asn Immunol. *Res:* Etiology of rheumatic fever; formation of antibodies; transplantation immunology. *Mailing Add:* 5112 Woodbine Ave Philadelphia PA 19131

HARRIS, VENOIA M, range management, biology, for more information see previous edition

HARRIS, VINCENT CROCKETT, b Minneapolis, Minn, Jan 26, 13; wid; c 2. NUMBER THEORY. *Educ:* Northwestern Univ, BA, 33, MA, 35, PhD, 50. *Prof Exp:* Instr math, Northwestern Univ, 46-50; from asst prof to assoc prof, 50-57, prof, 57-76, EMER PROF MATH, SAN DIEGO STATE UNIV, 76- *Concurrent Pos:* Grad asst, Univ Wis, 35-38; summer fac, Ariz State Univ, 61, Univ Mo, 62, Northwestern Univ, 63; vis prof, Univ Alta, 76; pace instr, Cent Tex Col, 89-90. *Mem:* Math Asn Am. *Res:* Multiplicative functions. *Mailing Add:* 5054 55th St San Diego CA 92115

HARRIS, WALLACE WAYNE, b Americus, Kans, Sept 15, 25; m 50; c 3. AGRONOMY, SOIL FERTILITY. *Educ:* Kans State Univ, BS, 51, MS, 55; Iowa State Univ, PhD(soil fertil), 70. *Prof Exp:* Asst agronomist, Colby Br Exp Sta, Kans State Univ, 55-70; from asst prof to prof soils, Ft Hays Kans State Univ, 70-88, chmn Dept Agr, 73-88; RETIRED. *Mem:* Am Soc Agron; Soil Sci Soc Am; Soil Conserv Soc Am. *Res:* Dryland and irrigated soil management and fertility. *Mailing Add:* PO Box 122 Americus KS 66835

HARRIS, WALTER EDGAR, b Alta, June 9, 15; m 42; c 2. ANALYTICAL CHEMISTRY. *Educ:* Univ Alta, BSc, 38, MSc, 39; Univ Minn, PhD(chem), 44. *Hon Degrees:* DSc, Univ Waterloo, 87. *Prof Exp:* Res fel synthetic rubber, Univ Minn, 43-46; from asst prof to prof, 46-80, chmn dept, 74-79, EMER PROF CHEM, UNIV ALTA, 80- *Concurrent Pos:* Anal ed, Can J Chem, 74-79; chmn pres adv comt campus reviews, Univ Alta, 80-90. *Honors & Awards:* Fisher Award. *Mem:* Am Chem Soc; fel Chem Inst Can; Sigma Xi; fel Royal Soc Can; fel AAAS. *Res:* Polarography; modifiers in synthetic rubber; electrochemistry of uranium; amperometric titrations; chromatography; chemical effects of nuclear transformations; programmed temperature gas chromatography; sampling and other uncertainties in the analytical process; hazardous waste. *Mailing Add:* Dept Chem Univ Alta Edmonton AB T6G 2E1 Can

HARRIS, WARREN WHITMAN, chemistry, electron optics; deceased, see previous edition for last biography

HARRIS, WAYNE G, b Alexandria, Minn, July 14, 33; m 54; c 2. BIONUCLEONICS, BIOCHEMISTRY. *Educ:* Purdue Univ, BS, 61, MS, 62, PhD(bionucleonics), 64. *Prof Exp:* Asst prof pharmaceut chem, SDak State Univ, 64-67; assoc prof, NDak State Univ, 67-69; MGR ANAL DIV, NEW ENG NUCLEAR CORP, 69- *Mem:* AAAS; Am Chem Soc. *Res:* Development of analytical methods for evaluation of labeled chemicals; theory and applications in liquid scintillation counting. *Mailing Add:* 67 Overlook Dr Framingham MA 01701

HARRIS, WESLEY LAMAR, b Taylorsville, Ga, Nov 12, 31; m 53; c 4. AGRICULTURAL ENGINEERING. *Educ:* Univ Ga, BSAE, 53, MS, 58; Mich State Univ, PhD(agr eng), 60. *Prof Exp:* Instr agr eng, Univ Ga, 56-58; asst, Mich State Univ, 58-60; from asst prof to assoc prof agr eng, Univ Md, College Park, 60-69, prof, 69-76 & 87-88, chmn dept, 74-76, dir, Md Agr Exp Sta, 76-87; AGR ENGR, USDA, 88- *Mem:* Am Soc Agr Engrs; Am Soc Eng Educ. *Res:* Development of a strain-stress relationship for soils; basic relationships required for design of pneumatic handling systems for non-free flowing materials; application of thermal energy for control of insects; development of sweet potato harvester. *Mailing Add:* Coop State Res Serv USDA Washington DC 20250-2200

HARRIS, WESLEY LEROY, b Richmond, Va, Oct 29, 41; m 60; c 4. AEROSPACE ENGINEERING, COMPUTATIONAL FLUID DYNAMICS. *Educ:* Univ Va, BAeroEng, 64; Princeton Univ, MA, 66, PhD(aerospace eng), 68. *Prof Exp:* NASA trainee aerospace eng, Princeton Univ, 64-66; asst prof aerospace eng, Univ Va, 68-70; assoc prof physics, Southern Univ, 70-71; asst prof aerospace eng, Univ Va, 71-72; assoc prof aeronaut & astronaut & ocean eng, 72-81, prof aeronaut & astronaut, Mass Inst Technol, 81-85; dean, Sch Eng, Univ Conn, 85-90; VPRES & PROF MECH ENG, UNIV TENN SPACE INST, 90- *Concurrent Pos:* Consult, Res Labs Eng Sci, 70-73; nuclear div, Union Carbide Corp, 71-76; US Army Sci Bd, 79-85, 89-; eminent scholar, Norfolk Stat Univ, 81-82; NSF Eng Adv Comt, 89- *Honors & Awards:* Irwin Sizer Award, 79. *Mem:* assoc fel Am Inst Aeronaut & Astronaut; Am Phys Soc; Soc Indust & Appl Math; Sigma Xi. *Res:* Shock structure in gas mixtures; quasi-linear techniques applied to aerodynamic noise analysis; uranium separation in gas dynamic flows; hypersonic flow analysis; transonic flow analysis; helicopter rotor acoustics; computational fluid dynamics. *Mailing Add:* 403 Kingsridge Blvd Tullahoma TN 37388

HARRIS, WILLIAM B(RISBANE) D(ICK), JR, mechanical engineering, for more information see previous edition

HARRIS, WILLIAM BIRCH, b Loveland, Colo, Sept 26, 19; m 42; c 2. CHEMICAL ENGINEERING. *Educ:* Univ Colo, BS, 41; Agr & Mech Col, Tex, MS, 60 Colo State Univ, PhD, 73. *Prof Exp:* Res chemist, Nat Aniline Div, Allied Chem & Dye Corp, 41-48; asst tech dir, Apache Powder Co, 48-53; assoc prof chem eng & assoc res engr, 53-77, PROF CHEM ENG, TEX A&M UNIV, 77- *Mem:* Fel Am Inst Chem Eng. *Res:* Cottonseed products; catalysis; alternative fuels; solar energy. *Mailing Add:* Dept Chem Eng Tex A&M Univ College Station TX 77843

HARRIS, WILLIAM BURLEIGH, b Norfolk, Va, July 2, 43; m 65; c 2. PETROLOGY, STRATIGRAPHY. *Educ:* Campbell Univ, BS, 66; WVa Univ, MS, 68; Univ NC, Chapel Hill, PhD(geol), 75. *Prof Exp:* Explor geologist, Texaco, Inc, 68-70; econ geologist, Va Div Mineral Resources, 70-72; teaching asst geol, Univ NC, Chapel Hill, 72-74; from instr to assoc prof, Univ NC, Wilmington, 74-82; res specialist, Exxon Prod Res Co, 82-83; PROF GEOL, UNIV NC, WILMINGTON, 84- *Concurrent Pos:* Proj dir & prin investr grant, NC Bd Sci & Technol, 76-78; consult geologist, E I du Pont de Nemours & Co, Inc, & Westinghouse Savannah River Co, Savannah River Lab, 78, 81, 87, 88, 89 & 90; dir & prin investr, US Dept Energy grant, 82; consult geologist, Arco Oil & Gas Co, 85, 86, 87 & 88. *Mem:* European Union Geoscientists; Soc Econ Paleont & Mineral; Sigma Xi; Geol Soc Am. *Res:* Stratigraphic and structural relations of the Atlantic and Gulf Coastal Plain Province, glauconite radiometric dating, carbonate petrology and depositional sedimentary environments; carbonate cement origins, isotopic compositions and dolomitization; application of the seismic sequence concepts to outcrops. *Mailing Add:* Dept Earth Sci Univ NC 601 S Col Rd Wilmington NC 28403

HARRIS, WILLIAM CHARLES, b New York, NY, Nov 4, 44; m 67; c 2. PHYSICAL CHEMISTRY, SPECTROSCOPY. *Educ:* William & Mary Col, BS, 66; Univ SC, PhD(chem), 70. *Prof Exp:* Nat Inst Arthritis & Metab Dis fel, 70-71; from asst prof to assoc prof chem, Furman Univ, 71-79; prog dir phys chem, NSF, 87-88, prog dir struct chem & Thmodyn, 77-87, assoc dir, 87-88, DIR OFF SCI & TECHNOL CTR, 88- *Concurrent Pos:* Petrol res fund grant, Res Corp, NSF, 71-74; vis prof chem, Univ NC, Chapel Hill, 75; Dreyfus teacher scholar, 75-80; fac fel, NSF, 77; guest res, Lab Chem Physics, NIH, 78- *Mem:* Am Phys Soc; Am Chem Soc; Coblent Soc (pres, 81-83); Soc Appl Spectros. *Res:* Application of mid- and far- infrared, as well as laser Raman spectroscopy, to the study of molecular conformations; matrix isolation studies of reactive molecules; molecular dynamics and force field calculations of large molecules; lipid membranes. *Mailing Add:* Nat Sci Found Chem Div Rm 533 Washington DC 20550

HARRIS, WILLIAM EDGAR, b Edmonton, Alta, Nov 28, 47. ASTRONOMY. *Educ:* Univ Alta, BSc, 69; Univ Toronto, MSc, 70, PhD(astron), 74. *Prof Exp:* fel astron, Yale Univ Observ, 74-76; from asst prof to assoc prof, 76-84, PROF PHYSICS, MCMASTER UNIV, 84- *Concurrent Pos:* Scientific adv coun, Can-France-Hawaii Telescope, 84-86, chmn, 85-86; Grant Selection Comt, Space & Astron, Nat Sci & Eng Res Coun Can, 90-93. *Mem:* Can Astron Soc; Am Astron Soc; Astron Soc Pac. *Res:* Photometry in star clusters and nearby galaxies; observational stellar evolution; galactic structure. *Mailing Add:* McMaster Univ Dept Physics Hamilton ON L8S 4M1 Can

HARRIS, WILLIAM FRANKLIN, III, b Jacksonville, Fla, Sept 27, 42; m 66; c 2. ENVIRONMENTAL SCIENCES. *Educ:* Wabash Col, BA, 64; Univ Tenn, Knoxville, MS, 66, PhD(bot), 70. *Prof Exp:* Res ecologist, Oak Ridge Nat Lab, 70-73, res group leader, 73-76, sect head, 76-80; prog dir, 80-81, dep div dir, 81-87, EXEC OFFICER, NSF, 87- *Mem:* Ecol Soc Am; fel AAAS; Am Inst Biol Sci. *Res:* Administration of agency sponsoring fundamental research in biological behavioral and social sciences; science policy; emphasizing terrestrial and freshwater ecosystems. *Mailing Add:* 5621 Herberts Crossing Burke VA 22015

HARRIS, WILLIAM J, b South Bend, Ind, June 17, 18; m 44, 78; c 2. TRANSPORTATION, FAILURE ANALYSIS. *Educ:* Purdue Univ, BS & MS, 40; Mass Inst Technol, DSc, 48. *Hon Degrees:* DEng, Purdue Univ, 78. *Prof Exp:* Head, ferrous alloys br, metall div, Naval Res Lab, 47-51; exec secy, metall adv bd, Nat Acad Sci-Nat Res Coun, 51-54, asst secy, div eng, 60-62; asst dir, Battelle Mem Inst, 54-57, asst to vpres, 62-67; asst dir technol, Columbus Labs, 67-69; vpres res, res & test dept, Asn Am Railroads, 70-85; SNEAD DISTINGUISHED PROF TRANSP ENG, DEPT CIVIL ENG, TEX A&M UNIV, 85-, ASSOC DIR, TEX TRANSP INST, 87- *Concurrent Pos:* Pres, W J Harris, Inc, 85-; distinguished res award, Transp Res Forum, 86. *Honors & Awards:* Mathewson Medal, Am Inst Metall, Mining & Petrol Engrs, 50; Carey Award, Transp Res Bd, 78, Crum Award, 89. *Mem:* Nat Acad Eng; fel Metall Soc (pres, 69-70); fel Am Soc Metals; fel Am Soc Mech Eng; hon mem Nat Security Indust Asn. *Res:* Type, thickness and arrangement of aircraft armor; shop failure analysis; brittle fracture; rocket motor failure analysis; re-entry materials; advanced structural materials; railroad wheel failures; track train dynamic interactions; labor management productivity; transportation planning and analysis. *Mailing Add:* 415 Chimney Hill Dr College Station TX 77840-1833

HARRIS, WILLIAM M, b Fresno, Calif, Mar 23, 40; m 64; c 2. PLANT ANATOMY. *Educ:* Fresno State Col, BA, 63; Univ Calif, Davis, PhD(bot), 68. *Prof Exp:* asst prof, 68-73, ASSOC PROF BOT, UNIV ARK, FAYETTEVILLE, 73- *Mem:* AAAS; Am Inst Biol Sci; Bot Soc Am. *Res:* Morphogenesis of plastids in ripening solanaceous fruits, primarily at the ultrastructure level; plastid mutants; ultrastructure of gymnosperm leaves; root hair structure and function. *Mailing Add:* Botany & MBIO Univ Ark Scen 401 Fayetteville AR 72701

HARRIS, WILLIAM MERL, b Los Angeles, Calif, Feb 23, 31; m 57. ORGANIC CHEMISTRY. *Educ:* Univ Calif, Los Angeles, BS, 56, PhD(chem), 65; Loyola Law Sch, Los Angeles, JD, 87. *Prof Exp:* Mem tech staff, Hughes Aircraft Co, 56-59; chemist, US Food & Drug Admin, 64-65; res chemist, Univ Calif, Los Angeles, 65-66; from instr to assoc prof, 66-77, coordr instr admin, 76-77, PROF CHEM, LOS ANGELES VALLEY COL, 77-, CHMN DEPT, 70- *Concurrent Pos:* Lectr chem, Univ Calif, Santa Barbara, 68 & vis prof chem, Los Angeles, 79. *Mem:* Royal Soc Chem; Am Chem Soc; Sigma Xi. *Res:* Structure of natural plant products (alkaloids, gibberellins, cyanogenetic glycosides); phosphazene chemistry. *Mailing Add:* Dept Chem Los Angeles Valley Col 5800 Fulton Ave Van Nuys CA 91401-4096

HARRIS, ZELLIG S, b Balta, Russia, Oct 23, 09. INFORMATION STRUCTURE, MATHEMATICAL LINGUISTICS. *Educ:* Univ Pa, BA, 30, MA, 32, PhD(ling), 34. *Prof Exp:* Benjamin Franklin prof, 31-79, EMER PROF, UNIV PA, 79-; SR RES SCIENTIST, CTR SOCIAL SCI, COLUMBIA UNIV, 80- *Mem:* Nat Acad Sci; Am Acad Arts & Sci; Am Philos Soc. *Res:* Science information, mathematical linguistics. *Mailing Add:* Dept Linguistics Univ Pa Philadelphia PA 19104-6305

HARRISBERGER, EDGAR LEE, b Denver, Colo, Sept 24, 24; m 69; c 5. MECHANICAL ENGINEERING. *Educ:* Univ Okla, BS, 45; Univ Colo, MS, 50; Purdue Univ, PhD(mech eng), 63. *Prof Exp:* Instr eng, Murray State Agr Col, 46-49; asst prof mech eng, Univ Utah, 50-54; assoc prof, NC State Univ, 54-60; NSF fel, 60-61; instr, Purdue Univ, 61-62; assoc prof mech eng, Okla State Univ, 63-65, Halliburton prof & dir ctr for teaching, 65-71, head sch mech & aerospace eng, 66-71; dean col sci & eng, Univ Tex of the Permian Basin, 71-75; Andrew Carnegie vis prof, Cooper Union, 76-77; prof mech eng & dir, Mech Eng Design Clin, 77-88, head mech eng dept, 83-88, co-dir Venture Clin, Univ Ala, 83-86; DISTINGUISHED VIS PROF ENG EDUC, UNIV CINCINNATI, 88- *Concurrent Pos:* Ed, MEN Mag, 65-71, ERM Mag, 68-71. *Honors & Awards:* Fred Merry Field Award Excellence Eng Design, Am Soc Eng Educ, 84; Chester F Carlson Award, Am Soc Eng Educ, 86. *Mem:* Fel Am Soc Mech Engrs; hon mem & fel Am Soc Eng Educ (vpres, pres). *Res:* Mechanism and machine design; kinematics; engineering design. *Mailing Add:* 5813 Kugler Mill Rd Cincinnati OH 45236

HARRISES, ANTONIO EFTHEMIOS, b Manchester, NH, Sept 12, 26; m 57; c 5. PARASITOLOGY, ZOOLOGY. *Educ:* St Anselm's Col, AB, 50; Univ NH, MS, 52; Univ Notre Dame, PhD, 57. *Prof Exp:* From assoc prof to prof biol, Univ Southern Miss, 57-70; PROF BIOL, SALEM STATE COL, 70- *Mem:* Sigma Xi; Am Soc Parasitol. *Res:* Taxonomy of freshwater monogenetic trematodes; culture methods. *Mailing Add:* 447 Laydon St Manchester NH 03103

HARRIS-NOEL, ANN (F H) GRAETSCH, b Cleveland, Ohio, Sept 6, 34; m 53, 80; c 2. ENVIRONMENTAL GEOLOGY. *Educ:* Kent State Univ, BS, 56; Miami Univ, MS, 58. *Prof Exp:* Res engr micros anal, Ferro Corp, 58-59; geol aide field mapping & res, US Geol Surv, 59-60; asst chem, Summer Inst, Wis State Col, Superior, 60-61; PROF GEOL, YOUNGSTOWN STATE UNIV, 61-; CONSULT, 77- *Concurrent Pos:* Adv, Struthers Total Environ Educ Prog, 73-, Environ Rev Comt, 75-; mem, adv comt energy, app by Gov Richard Celeste, Ohio, 84-87. *Mem:* NAm Thermal Anal Soc; Nat Asn Geol Teachers; Geol Soc Am; Am Inst Prof Geologist. *Res:* The location, history and layout of abandoned deep coal and/or clay mines; engineering methods best suited for the stabilization of these mines when subsidence occurs or the sealing of the dangerous drift, slope or shaft openings. *Mailing Add:* Dept Geol Youngstown State Univ 410 Wick Ave Youngstown OH 44555

HARRISON, AIX B, b Zearing, Iowa, Feb 14, 25; m 50; c 2. EXERCISE PHYSIOLOGY. *Educ:* Univ Ill, BS, 49, MS, 50; Mich State Univ, PhD(phys educ), 59. *Prof Exp:* Teacher, Galva Ill Community Schs, 49-50; from asst prof to assoc prof, 50-68, PROF PHYS EDUC & PROG DIR HEALTH & FITNESS CTR, OKLA STATE UNIV, 58- *Mem:* Am Col Sports Med; Sigma Xi. *Res:* Training programs for producing maximal oxygen consumption; training for and measuring maximal oxygen debts; logitudinal study of adult physical fitness; fitness and aging. *Mailing Add:* Health Phys Educ & Leisure Sci Okla State Univ Stillwater OK 74078

HARRISON, ALEXANDER GEORGE, b Peterborough, Ont, Apr 1, 31; m 55; c 2. PHYSICAL CHEMISTRY, ANALYTICAL CHEMISTRY. *Educ:* Univ Western Ont, BSc, 52, MSc, 53; McMaster Univ, PhD(chem), 56. *Prof Exp:* Fel, McMaster Univ, 56-57; Nat Res Coun Can fel chem, 57-59; lectr chem, 59-60, from asst prof to assoc prof, 60-67, assoc chmn dept, 71-74, PROF CHEM, UNIV TORONTO, 67- *Concurrent Pos:* Alfred P Sloan res fel, 62-64; Killam res fel, 85-87. *Honors & Awards:* Noranda Lect Award, Chem Inst Can, 71. *Mem:* Chem Inst Can; Am Chem Soc; Am Soc Mass Spectrometry. *Res:* Mechanistic studies in electron impact and chemical ionization mass spectrometry; collisional spectroscopy of gaseous ions; analytical applications of mass spectrometry. *Mailing Add:* Dept Chem Univ Toronto Toronto ON M5S 1A1 Can

HARRISON, ALINE MARGARET, b Lincoln, Nebr, Aug 30, 40; m 66; c 1. CHEMISTRY. *Educ:* Univ Mich, BS, 62; Univ Md, MS, 67, PhD(org chem), 81. *Prof Exp:* Res org chemist, Agr Res Serv, US Dept Agr, 62-63; instr chem, Northern Mich Univ, 66-68; lab instr & admin coun, Pa State Univ, York Campus, 70-72; teaching asst, Univ Md, 72-76, res asst, 76-78; asst prof, Dickinson Col, 78-79; asst instr, 79-81, ASST PROF CHEM, YORK COL PA, 81- *Mem:* Am Chem Soc; Sigma Xi; AAAS. *Res:* Synthesis of potential analgesics for replacement of morphine; immunocompetence of the human immune system under stress, in conjunction with psychological state and of cancer patients having various psychological states. *Mailing Add:* 413 Rushmore Dr York PA 17402

HARRISON, ANNA JANE, b Benton City, Mo, Dec 23, 12. PHYSICAL CHEMISTRY. *Educ:* Univ Mo, AB, 33, MA, 37, PhD(phys chem), 40. *Hon Degrees:* DSc, Tulane Univ Y Smith Col, 75, Vincennes Univ, Am Int Col & Williams Col, 78, Suffolk Univ, Worcester Polytech Inst, Hartford Univ, Hood Col & Leigh Univ, 79, Univ Mo & Eastern Mich Univ, 83, Russell Sage Col & Mt Holyoke Col, 84, Mills Col, 85; LHD, Lindenwood Col, 77, Emmanuel Col, 83, St Joseph Col & Elms Col, 85. *Prof Exp:* From instr to asst prof chem, Newcomb Col, Tulane Univ, 40-45; from asst prof to assoc prof, 45-50, chmn dept, 60-66, prof chem, 50-76, William R Kenan, Jr, prof, 76-79, EMER PROF CHEM, MT HOLYOKE COL, 79- *Concurrent Pos:* Berliner res fel, Am Asn Univ Women, Cambridge Univ, 52-53; mem, Nat Sci Bd, 72-78; distinguished vis prof, US Naval Acad, 80; chmn bd, AAAS, 83; bd dirs, Sigma Xi, 88-; bd, Volunteers in Tech Assistance. *Honors & Awards:* Frank Forrest Award, 49; Petrol Res Fund Int Award, Nat Res Coun Can, 59-60; Award, Mfg Chem Asn, 69; James Flack Norris Award, Am Chem Soc, 77 & Award Chem Educ, 82. *Mem:* AAAS (pres 83-84); Am Chem Soc (pres, 78); Sigma Xi; Int Union Pure & Appl Chem; Nat Sci Teachers Asn. *Res:* Vacuum ultraviolet spectroscopy. *Mailing Add:* Dept Chem Mt Holyoke Col South Hadley MA 01075

HARRISON, ARNOLD MYRON, b Pittsburgh, Pa, Sept 25, 46. PHYSICAL ORGANIC CHEMISTRY. *Educ:* Princeton Univ, AB, 68; Univ Chicago, MS, 70, PhD(org chem), 75. *Prof Exp:* SR RES SCIENTIST, RES & DEVELOP, UNION CARBIDE TECH CTR, 75- *Mem:* Am Chem Soc; AAAS; Sigma Xi. *Res:* Physical organic chemistry; nuclear magnetic resonance and catalysis. *Mailing Add:* PO Box 8361 South Charleston WV 25303

HARRISON, ARTHUR DESMOND, b Cape Town, SAfrica, Dec 24, 21; m 63; c 2. FRESH WATER BIOLOGY. *Educ:* Univ Cape Town, BSc, 41, MSc, 43, BEd, 45, PhD(zool), 58. *Prof Exp:* Jr lectr zool, Univ Cape Town, 46-48; res officer, Nat Inst Water Res, SAfrican Coun Sci & Indust Res, 50-55, sr res officer, 55-58, prin res officer, 58-61; sr res fel ecol freshwater snails, Univ Col Rhodesia, 61-66; prof zool, Univ Natal, 66-67; PROF BIOL, UNIV WATERLOO, 67- *Mem:* Brit Ecol Soc; Royal Soc SAfrica; Int Asn Theoret & Appl Limnol; Entomol Soc SAfrica. *Res:* Hydrobiology of rivers and streams; ecology and taxonomy of Chironomidae. *Mailing Add:* Dept Biol Univ Waterloo Waterloo ON N2L 3G1 Can

HARRISON, ARTHUR PENNOYER, JR, b Takoma Park, Md, Oct 15, 22; m 58. MICROBIOLOGY. *Educ:* Univ Md, BS, 48, MS, 49, PhD(bact), 52. *Prof Exp:* Res assoc, Univ Md, 52-53; from asst prof to assoc prof biol, Vanderbilt Univ, 53-63; mem staff, Biol Div, Oak Ridge Nat Lab, 63-68; prof bot, Univ Mo, Columbia, 68-90; RETIRED. *Concurrent Pos:* Vis investr, Oxford Univ, 57-58. *Res:* Bacterial physiology and taxonomy. *Mailing Add:* Div Biol Sci 105 Tucker Hall Univ Mo Columbia MO 65211

HARRISON, BENJAMIN KEITH, b Macon, Ga, July 29, 47; m 71; c 1. CHEMICAL PROCESS SIMULATION, THERMODYNAMICS. *Educ:* Ga Inst Technol, BS, 69, MS, 71, MS, 77; Univ Mo, Rolla, PhD(chem eng), 84. *Prof Exp:* Eng specialist, Monsanto Co, 77-86; ASSOC PROF CHEM ENG, UNIV SALA, 86- *Concurrent Pos:* Consult, Ciba Geigy Corp, 87-; prin investr, Cray Res, 88-; mem, Sci Adv Comt, Gulf Coast Hazardous Substance Res Ctr, 88-; vchmn comt, Am Soc Testing & Mat, 90- *Honors & Awards:* Halliburton Educ Found Award of Excellence, 89. *Mem:* Am Inst Chem Eng; Am Soc Testing & Mat. *Res:* Chemical process simulation, often involving supercomputers; thermodynamic physical property prediction, often in connection with hazard predictions. *Mailing Add:* Dept Chem Eng EGLB 248 Univ SAla Mobile AL 36688

HARRISON, BERTRAND FEREDAY, b Springville, Utah, Feb 20, 08; m 31; c 4. PLANT PHYSIOLOGY, AGROSTOLOGY. *Educ:* Brigham Young Univ, BS, 30, MS, 31; Univ Chicago, PhD(plant physiol), 37. *Prof Exp:* Ranger naturalist, Yellowstone Nat Park, 31; from instr to prof, 31-74, EMER PROF BOT, BRIGHAM YOUNG UNIV, 74-; CONSULT INTERMOUNTAIN RES, NAT RESOURCES RES, MGT CONSULT, PROVO, 76- *Concurrent Pos:* Asst, Univ Chicago, 36-37; res assoc, Agr Res Lab, Am Smelting & Ref Co, 42 & 43. *Mem:* Am Soc Plant Physiol; Bot Soc Am. *Res:* Physiology of range grasses, especially water and mineral requirements. *Mailing Add:* Bot Dept Brigham Young Univ Provo UT 84601

HARRISON, BERTRAND KENT, b Provo, Utah, July 21, 34; m 54; c 4. THEORETICAL PHYSICS. *Educ:* Brigham Young Univ, BS, 55; Princeton Univ, AM, 57, PhD(theoret physics), 59. *Prof Exp:* Asst physics, Princeton Univ, 56-59; staff mem, Los Alamos Sci Lab, Calif, 59-64; from asst prof to assoc prof, 64-71, PROF PHYSICS & ASTRON, BRIGHAM YOUNG UNIV, 71- *Concurrent Pos:* Nat Sci Found grant, res res grant gravitation & electromagnetism, 65-67; Nat Acad Sci-Nat Res Coun sr resident res assoc, 68-70; res affil, Jet Propulsion Lab Calif Inst Technol, 77- Pres, Fac Adv Coun, Brigham Young Univ, 75-76. *Mem:* Am Phys Soc; Sigma Xi; NY Acad Sci. *Res:* General relativity and gravitation with particular emphasis on exact solutions of field equations, and superdense stars; plasma physics; applied mathematics. *Mailing Add:* Dept Physics & Astron Brigham Young Univ Provo UT 84602

HARRISON, BETTINA HALL, b Foxboro, Mass; m 41; c 3. CYTOLOGY, IMMUNOLOGY. *Educ:* Univ Mass, BS; Radcliffe Col, AM; Boston Univ, PhD(cytol), 68. *Prof Exp:* Fac mem med technol, Lasell Jr Col, Mass, 40-41; res microscopist, Maine Mills Labs, 41-43 & 50-51; fac mem med technol, Lasell Jr Col, 59-65; asst prof, 67-74, ASSOC PROF BIOL, UNIV MASS, BOSTON, 74- *Concurrent Pos:* Fac res grants, 68-70, 71-72 & 72-74; res fel, Tufts-New Eng Med Ctr Hosp, 76-77; res grant, Nat Leukemia Asn, 77, Healy grant, Univ Mass, Boston, 86-87. *Mem:* AAAS; Am Soc Cell Biol; Soc Develop Biol; NY Acad Sci; Int Soc Immunol. *Res:* Cytological development of lymphocytes; functional aspects of palatine tonsils; ultrastructure of ozonated pollen; regeneration of bone marrow; cytological effects of acid rain on leaf structures; various publications. *Mailing Add:* Dept Biol Univ Mass Harbor Campus Boston MA 02125

HARRISON, CHARLES WAGNER, JR, b Farmville, Va, Sept 15, 13; m 40; c 2. APPLIED PHYSICS, ELECTRICAL ENGINEERING. *Educ:* Univ Va, BSE, 39, EE, 40; Harvard Univ, SM, 42, ME, 52, PhD(appl physics), 54. *Prof Exp:* Instr, Univ Va, 39-40; res engr antenna res, Bur Ships, Dept Navy, 39-41; physicist, US Naval Res Lab, 44-46; electronics officer repair & installation, Philadelphia Naval Ship Yard, 46-48; asst dir electronics res, Electronics Design & Develop Div, Bur Ships, 48-50; prog officer radio div I, US Naval Res Lab, 50-51; electronics officer electronics equip eval, Comdr Oper Develop Force, US Navy, 53-55, electronics officer res & heat studies, Staff, Comndg Gen, Armed Forces Spec Weapons Proj, 55-57; res physicist antenna & electromagnetics compatibility studies, Sandia Labs, 57-73; CONSULT, 73- *Concurrent Pos:* Lectr, Harvard Univ, 42-43, Princeton Univ, 43-44; consult, Navy Proj Sanguine, Nat Acad Sci, 71-73; vis prof, Christian Heritage Col, El Cajon, Calif, 76; auth; mem, Int Union Radio Sci, comn B & H. *Honors & Awards:* Electronics Achievement Award, Inst Elec & Electronics Engrs, 66. *Mem:* Fel Inst Elec & Electronics Engrs; Int Union Radio Sci; Sigma Xi. *Res:* Electromagnetics; acoustics; astronomy; contributed numerous articles to professional journals; author. *Mailing Add:* 2808 Alcazar St NE Albuquerque NM 87110

HARRISON, CHRISTOPHER GEORGE ALICK, b Oxford, Eng, Dec 29, 36; m 64; c 2. GEOPHYSICS. *Educ:* Cambridge Univ, BA, 60, MA, 64, PhD(geophys), 64, ScD,89. *Prof Exp:* Res geophysicist, Scripps Inst Oceanog, 61-67; from asst prof to assoc prof, 67-74, chmn div marine geol & geophys, 76-83, interim dean, 86-89, PROF GEOPHYS, ROSENSTIEL SCH MARINE & ATMOSPHERIC SCI, UNIV MIAMI, 74- *Concurrent Pos:* Assoc ed, J Geophys Res, 73-75; mem panel, NSF, 75-77; ed, Earth & Planetary Sci Lett, 78-86; mem, sea level change panel, Nat Res Coun, 84-86, panel on solid waste aspects of global change, 90-91; mem, US Nat Comt, Int Union Geod & Geophys, 84-91. *Honors & Awards:* Group Achievement Award, Magnetic Field Satellite Sci Invest Team, NASA, 83. *Mem:* fel AAAS; fel Am Geophys Union; fel Royal Astron Soc; Sigma Xi. *Res:* Paleomagnetism of deep sea sediments and other rocks; magnetic anomalies; plate tectonics; remote sensing. *Mailing Add:* Rosenstiel Sch Mar & Atmos Sci Univ Miami 4600 Rickenbacker Causwy Miami FL 33149

HARRISON, DALTON S(IDNEY), b Darlington, Fla, Oct 10, 20; m 49; c 2. AGRICULTURAL ENGINEERING. *Educ:* Univ Fla, BSA, 50, MSA, 53. *Prof Exp:* Asst agr engr, Inst Food & Agr Sci, Univ Fla, 53-60, from asst prof & assoc agr exten serv to prof agr eng & agr engr, 60-88; RETIRED. *Mem:* Am Soc Agr Engrs; sr mem Nat Soc Prof Engrs. *Res:* Drainage and irrigation. *Mailing Add:* 300 SW 41st St Gainesville FL 32607

HARRISON, DAVID ELLSWORTH, b New Haven, Conn, June 26, 42; m 67; c 3. GERONTOBIOLOGY, IMMUNOHEMATOLOGY. *Educ:* Bates Col, BS, 64; Stanford Univ, PhD(inorg chem), 69. *Prof Exp:* Fel, 69-70, assoc staff scientist, 70-73, staff scientist, 73-81, SR STAFF SCIENTIST PHYSIOL GENETICS, JACKSON LAB, 81- *Concurrent Pos:* Mem mouse subcomt, Animal Models Res, Aging Comn, Nat Res Coun, 78-80; site evaluator, Aging Rev Comn, Nat Inst Aging & Vet Admin; ed biol sci, J Gerontol, 84-87, Growth, 87-; Chmn, Gerontol Soc Am, 89-91; dir training, Jackson Lab, 90- *Mem:* Soc Develop Biol; Int Soc Exp Hemat; Gerontol Soc Am; Am Asn Immunologists; Am Aging Asn; Am Soc Hemat. *Res:* Hemopoietic and immunopoietic early precursor cells; immunohematology of mutant and aging mice; marrow grafts; aging tissue transplantations; physiological age measurements evaluating antiaging treatments; causes of senescence. *Mailing Add:* Jackson Lab Bar Harbor ME 04609

HARRISON, DAVID KENT, b Boston, Mass, Apr 6, 31; m 55; c 2. MATHEMATICS. *Educ:* Williams Col, BA, 53; Princeton Univ, PhD(math), 56. *Prof Exp:* Intern, Brown Univ, 56-58; asst prof math, Haverford Col, 58-59; asst prof, Univ Pa, 59-62, assoc prof, 62-63; assoc prof, 63-65, PROF MATH, UNIV ORE, 65- *Mem:* Am Math Soc. *Res:* Algebra. *Mailing Add:* Dept Math Univ Ore Eugene OR 97403

HARRISON, DON EDMUNDS, b 1950; US citizen. PHYSICAL OCEANOGRAPHY, APPLIED MATHEMATICS. *Educ:* Reed Col, BA, 72; Harvard Univ, MS, 73, PhD(appl math), 77. *Prof Exp:* Res fel, Harvard Univ, 77-78, lectr phys oceanog, 78; res scientist phys oceanog, 78-80, vis asst prof, 80-84, VIS ASSOC PROF, MASS INST TECHNOL, 84-; OCEANOGR, NAT OCEANIC & ATMOSPHERIC ADMIN, 85- *Concurrent Pos:* Affil prof, Univ Wash, 85- *Mem:* Am Meteorol Soc. *Res:* Numerical and analytical solution of PDE's; statistical analysis of data; long-time averaged climate of the ocean and atmosphere; roles of geophysical turbulence in the climate; equatorial oceanography; air-sea interaction. *Mailing Add:* NOAA/PMEL/R/E/PM Bldg 3 7600 Sand Point Way NE Seattle WA 98115

HARRISON, DON EDWARD, b Dennison, Ohio, Oct 29, 28; m 61; c 1. CERAMICS, MATERIALS SCIENCE. *Educ:* Pa State Univ, BS, 52, MS, 53, PhD(ceramics), 55. *Prof Exp:* Asst lectr ceramics, Univ Leeds, 55-56; assoc chem luminescence, Lamp Div, Gen Elec Co, 56-59; fel engr crystal growth, Res & Develop Ctr, Westinghouse Elec Corp, 59-65; coordr adv group tech assessment, Regional Indust Develop Corp Southwestern Pa, 65-66; mgr mat sci, Res & Develop Ctr, Westinghouse Elec Corp, 66-71, lab mgr, Westinghouse Res Lab-Europe, Brussels, Belg, 72-74, res adv, Iran-Westinghouse Prog Ctr, Tehran, 75, mgr, Nuclear Mat Dept, Res & Develop Ctr, 76-84, mgr, mat sci div, Res & Devel Ctr, 84-87, dir res & develop, Energy & Utility Syst Group, 88-89; RETIRED. *Mem:* Am Ceramic Soc; Am Chem Soc; Sigma Xi; Am Nuclear Soc. *Res:* High temperature ceramics, crystal growth, solidification, luminescence, phase equilibria; nuclear waste management. *Mailing Add:* 3809 Edinburg Dr Murrysville PA 15668-1013

HARRISON, DON EDWARD, JR, b Detroit, Mich, Aug 9, 27; m 48; c 3. SURFACE PHYSICS, COMPUTER SIMULATION. *Educ:* Col of William & Mary, BS, 49; Yale Univ, MS, 50, PhD(physics), 53. *Prof Exp:* Asst & asst instr, Yale Univ, 49-53; asst prof physics, Univ Louisville, 53-57; assoc prof eng physics, Univ Toledo, 57-61; assoc prof physics, 61-68, fac chmn, 76-77, PROF PHYSICS, NAVAL POSTGRAD SCH, 68- *Mem:* AAAS; Am Phys Soc. *Res:* Ion implantation; computer simulations of surface physics, chemistry, atomic collisions and operations analysis. *Mailing Add:* Dept Physics Naval Postgrad Sch Monterey CA 93943

HARRISON, DONALD C, b Blount Co, Ala, Feb 24, 34; m 55; c 3. CARDIOLOGY, CARDIOVASCULAR PHYSIOLOGY. *Educ:* Birmingham Southern Col, BS, 54; Med Col Ala, MD, 58. *Prof Exp:* Res fel med, Harvard Med Sch, 60-61; clin & res assoc, Nat Heart Inst, 61-63; instr med & chief resident, 63-64, from asst prof to assoc prof, 65-72, PROF MED, SCH MED, STANFORD UNIV, 72-, WILLIAM G IRWIN PROF CARDIOL, 73-, CHIEF CARDIOL, UNIV HOSP, 67- *Concurrent Pos:* Asst prof dir, Clin Res Ctr, Univ Hosp, Stanford Univ, 64-69; pres elect, Am Heart Asn, 80-81, pres, 81-82. *Mem:* Am Soc Pharmacol & Exp Therapeut; fel Am Col Cardiol; Am Asn Physicians; Asn Univ Cardiol. *Res:* Pharmacology. *Mailing Add:* Univ Cincinnati Med Ctr 141 Health Prof Bldg Cincinnati OH 45267-0663

HARRISON, DOROTHY LUCILE, food science, for more information see previous edition

HARRISON, DOUGLAS P, b Frost, Tex, Nov 19, 37; m 65; c 2. CHEMICAL ENGINEERING. *Educ:* Univ Tex, Austin, BSChE, 61, PhD(chem eng), 66. *Prof Exp:* Res engr, Monsanto Co, 66-69; from asst prof to assoc prof chem engr, 69-77, chmn dept, 77-80, PROF CHEM ENGR, LA STATE UNIV, 77- *Mem:* Am Inst Chem Engrs; Am Soc Eng Educ; Air Pollution Control Asn. *Res:* Reactor design; catalysis; pollution control. *Mailing Add:* Dept Chem Eng La State Univ 312 Chem Engr Bldg Baton Rouge LA 70803

HARRISON, EDWARD ROBERT, b London, Eng, Jan 8, 19; m 45; c 2. PHYSICS, ASTROPHYSICS. *Prof Exp:* Prin scientist, Atomic Energy Res Estab, Eng, 47-65; Nat Acad res assoc, Goddard Space Flight Ctr, Md, 65-66; prof astrophys, 66-87, DISTINGUISHED PROF PHYSICS & ASTRON, UNIV MASS, AMHERST, 87- *Concurrent Pos:* Vis scientist, Europ Orgn Nuclear Res, 59-60, Trieste Inst Theoret Physics, 73, Nat Radio Astron Observ, 76; vis prof, Woods Hole Oceanog Inst, 69, Inst Astron, Cambridge, Eng, 70, 71, 76, 84 & 90, Univ Sussex, Eng, 73, Univ Va, 76, Univ Wellington, NZ, 81 & Univ NC, 87; Chancellor's lect award, 78. *Honors & Awards:* Melcher Award, 85; Melville S Green lectr, 87. *Mem:* Fel AAAS; fel Am Inst Physics; Am Astron Soc; Int Astron Union; fel Royal Astron Soc. *Res:* Cosmology, particle physics, history of science and author of books on cosmology, history of science and the riddle of cosmic darkness; cosmology. *Mailing Add:* Dept Physics & Astron Univ Mass Amherst MA 01003

HARRISON, ERNEST AUGUSTUS, JR, b Boston, Mass, May 25, 34; m 66; c 1. ORGANIC CHEMISTRY. *Educ:* Boston Univ, BA, 57; Univ Md, MS, 62, PhD(org chem), 66. *Prof Exp:* Instr chem, Univ Md, 65; asst prof, Northern Mich Univ, 65-68; from asst prof to assoc prof, 68-81, PROF CHEM, PA STATE UNIV, 81- *Concurrent Pos:* Vis chemist, NIH-Nat Inst Diabetes, Digestive & Kidney Dis, Lab Chem , Med Chem Sect, 75-76 & 82-83, spec expert, med chem sect, 83. *Mem:* Am Chem Soc; NY Acad Sci. *Res:* Synthesis of heterocyclic compounds and molecules of potential medicinal interest; chemistry and photochemistry of bridge-carbonyl systems; development of experiments for use in the undergraduate organic laboratory. *Mailing Add:* Dept Chem York Campus Pa State Univ York PA 17403

HARRISON, FLORENCE LOUISE, b Oil City, Pa, Sept 6, 26; m 45; c 2. COMPARATIVE PHYSIOLOGY, MARINE BIOLOGY. *Educ:* Greenville Col, BS, 47; Univ Wash, MS, 53, PhD(zool), 54. *Prof Exp:* Instr zool & chem, Seattle Pac Col, 54-55; res assoc zool, Univ Wash, 55-56; NIH fel, Stanford Univ, 56-57; ENVIRON SCIENTIST, ENVIRON SCI DIV, LAWRENCE LIVERMORE NAT LAB, 64- *Mem:* Am Soc Zool; Am Physiol Soc; Health Physics Soc; Am Chem Soc. *Res:* Evaluation of the quantities present, the rates of turnover, and the potential toxic effects of energy-related pollutants, particularly radionuclides, heavy metals and hydrocarbons in marine and freshwater molluscs, crustaceans, and fishes. *Mailing Add:* Environ Sci Div L Lawrence Livermore Nat Lab PO Box 5507 Livermore CA 94550

HARRISON, FLOYD PERRY, b Picayune, Miss, Oct 18, 27; m 51; c 3. ECONOMIC ENTOMOLOGY. *Educ:* La State Univ, BS, 51, MS, 53; Univ Md, PhD, 55. *Prof Exp:* From instr to assoc prof, 55-71, PROF ENTOM, UNIV MD, 71- *Mem:* Entom Soc Am; Sigma Xi. *Res:* Ecology, bionomics and control of insects affecting agricultural crops, especially corn and tobacco pests. *Mailing Add:* Dept Entom Univ Md College Park MD 20742

HARRISON, FRANK, b Dallas, Tex, Nov 21, 13; m 46; c 3. ANATOMY. *Educ:* Southern Methodist Univ, BS, 35; Northwestern Univ, MS, 36, PhD(neuroanat), 38; Univ Tex Southwestern Med Sch, Dallas, MD, 56. *Prof Exp:* From instr to prof & chief div anat, Univ Tenn, 38-51; prof, Univ Tex Southwestern Med Sch, Dallas, 52-68, assoc dean, 56-68, assoc dean grad studies, 58-68; actg pres, Univ Tex, Arlington, 68-69, pres, 69-72; pres, Univ Tex Health Sci Ctr, San Antonio, 72-85; RETIRED. *Concurrent Pos:* Assoc dean grad studies, Univ Tex, Arlington, 65-68; adj prof, Southern Methodist Univ, 64-70. *Mem:* AAAS; Am Physiol Soc; Soc Exp Biol & Med; Biophys Soc; Am Asn Anat. *Res:* Neuroanatomy; neurophysiology. *Mailing Add:* 4168 Valley Ridge Dallas TX 75220

HARRISON, FREDERICK WILLIAMS, b Macon, Ga, Apr 14, 38; m 65; c 2. BIOLOGY, ANATOMY. *Educ:* Univ SC, BS, 60, MS, 62, PhD(biol), 69. *Prof Exp:* Assoc prof biol, Presby Col, 66-73; from asst to assoc prof anat, Albany Med Col, 73-77; head biol dept, 77-89, PROF BIOL, WESTERN CAROLINA UNIV, 77- *Concurrent Pos:* Vis res scholar, Sch Biol Sci, Univ Sydney, 73-74; assoc ed, J Morphology; vis scientist, Yunnan Univ, Peoples Repub China, 83; proj adv, Sponges in Space, Space Shuttle 5; ed, Micros Anat Invert. *Mem:* Am Soc Zool; Am Micros Soc (secy, 83-88, pres, 90); Sigma Xi. *Res:* Developmental cytology; systematics, paleolimnology of Porifera; invertebrate microscopic anatomy. *Mailing Add:* Dept Biol Western Carolina Univ Cullowhee NC 28723

HARRISON, GAIL GRIGSBY, b Denver, Colo, June 8, 43; c 2. HUMAN NUTRITION, PHYSICAL ANTHROPOLOGY. *Educ:* Univ Calif, BA, 65; Cornell Univ, MNS, 67; Univ Ariz, PhD(anthrop), 76. *Prof Exp:* Exten assoc nutrit, Cornell Univ, 67-70; instr med dietetics, Ohio State Univ, 70-72; from lectr to assoc prof community med, 76-86, PROF COMMUNITY MED, UNIV ARIZ, 86- *Concurrent Pos:* Mem, World Food & Nutrit Study, Nat Acad Sci, 76-77 & comt food consumption patterns, Food & Nutrit Bd, 78-81; comt Int Nutrit Progs NAS/NRS, 87- *Honors & Awards:* Future Leaders Award, Nutrit Found, Inc, 78. *Mem:* Am Inst Nutrit; Am Pub Health Asn; Am Dietetic Asn; fel Am Anthrop Asn; Soc Pediat Res. *Res:* Growth and development of children; international nutrition; dietary methodology. *Mailing Add:* Dept Community Med Univ Ariz 1501 N Campbell Tucson AZ 85724

HARRISON, GEORGE CONRAD, JR, b Pittsburgh, Pa, Dec 13, 29; m 52; c 3. SURFACE CHEMISTRY, MARKETING. *Educ:* Univ Pittsburgh, BS, 51, PhD(chem), 56. *Prof Exp:* Res asst chem, Mellon Inst, 51-52 & Univ Pittsburgh, 52-55; res chemist, Pennsalt Chem Corp, 59-60; sr res chemist, Amchem Prod Inc, 60-63; proj leader inorg lab, Interchem Corp, 63-65; group leader prod res, Pennsalt Chem Corp, 65-68; asst mgr chem res, Matthey Bishop Inc, Malvern, Pa, 68-70, mgr res & develop, 70-75; dir res & develop, Nicolet, Inc, 76; chem consult & vpres, Grezes, Inc, 77-78; dir chem mfg, 78-79, dir advan progs, 80-83, VPRES, CHEM OPERS, TELEDYNE MCCORMICK SELPH, 83- *Mem:* Am Chem Soc; Sigma Xi; Am Inst Chem Engrs; Chem Mkt Res Asn. *Res:* Theoretical and coordination chemistry; physical chemistry of solutions; surface and interfacial phenomena; catalysis. *Mailing Add:* Teledyne McCormick Selph PO Box 6 Hollister CA 95024

HARRISON, GEORGE H, b Newark, NJ, Mar 13, 43; m 67; c 1. RADIOBIOLOGY, RADIATION PHYSICS. *Educ:* Tufts Univ, BA, 65; Univ Md, MS, 69, PhD(nuclear physics), 72. *Prof Exp:* Comput systs analyst, Goddard Space Flight Ctr, NASA, 66-69; res asst nuclear physics, Cyclotron Lab, Univ Md, 69-72, res assoc radiobiol, 72-73, asst prof radiobiol, 73-78, ASSOC PROF RADIATION THER, SCH MED, UNIV MD, 78- *Concurrent Pos:* Res assoc nuclear physics, Cyclotron Lab, Univ Md, 72-; tutor technol, Univ Col, 73-; NSF Presidential internship, 73. *Mem:* Radiation Res Soc; Inst Elec & Electronics Engrs. *Res:* Physics and bioeffects of high energy neutrons; neutron production and dosimetry; study of microwaves and hyperthermia for use in cancer therapy; ultrasonic transducer design and bio-effect studies. *Mailing Add:* Div Radiation Res Univ Md Sch Med Baltimore MD 21201

HARRISON, GORDON R, b Wister, Okla, Dec 14, 31; m 57; c 4. PHYSICS, ELECTRICAL ENGINEERING. *Educ:* Ark State Teachers Col, BS, 52, Vanderbilt Univ, MS, 54, PhD(physics), 58. *Prof Exp:* Sr staff engr, Sperry Microwave Electronics Div, 57-62, eng staff consult, 62-64, res sect head, 64-68, eng mgr, 68-71; prin res scientist, Ga Inst Technol, 71-83, dir, Appl Sci Lab, 72-78, sr staff, Dir Off, 78-83; VPRES MAT & NEW TECHNOL, ELECTROMAGNETIC SCI, 83- *Concurrent Pos:* Health physicist, Oak Ridge Nat Lab, 53, physicist, 54 & Gen Dynamics/Convair, 55. *Mem:* Inst Elect & Electronics Engrs. *Res:* Ferrimagnetic materials; microwave components; solid state and nuclear physics. *Mailing Add:* 2915 Greenrock Trail Doraville GA 30340

HARRISON, GUNYON M, b Fredericksburg, Va, Mar 6, 21; div; c 2. PEDIATRICS. *Educ:* Va Mil Inst, BS, 43; Univ Va, MD, 46; Am Bd Pediat, dipl, 56. *Prof Exp:* Intern, St Joseph's Hosp, Baltimore, 46-47; pediat intern, Duke Univ Hosp, Durham, NC, 51-52, resident, 52-53, fel, 53; resident, Jefferson Davis Hosp, Houston, 53-54; resident, Baylor Col Med & Tex Children's Hosp, 54-55; instr pediat, 55-57, asst prof pediat & rehab, 57-64, assoc prof, 64-77, PROF PEDIAT & REHAB, BAYLOR COL MED, 77- *Concurrent Pos:* Asst instr, Baylor Col Med, 53-55; fel, Polio Respiratory Ctr, Nat Found Infantile Paralysis, Houston, 55-56; dir cystic fibrosis clin, Tex Children's Hosp, 55-61, chief respiratory ther, 68-, mem active med staff; mem med staff, Tex Children's Hosp, 58-; dir, Cystic Fibrosis & Related Pulmonary Dis Ctr, Baylor Col Med, Inst Rehab & Res & Tex Children's Hosp, 61-; chief chest clin, Tex Children's Hosp, 69-, pulmonary med, 72-; mem attend staff, Ben Taub Gen Hosp, 77-, courtesy staff, Methodist Hosp, 71- & St Luke's Hosp, 72-; consult physician, Pediat Chest Clin, Ben Taub Gen Hosp, assoc attend staff; assoc consult, Methodist Hosp & Cystic Fibrosis Clin, Baptist Mem Hosp, San Antonio; assoc attend staff, Harris County Hosp Dist. *Mem:* Am Acad Pediat; Am Thoracic Soc; Am Asn Respiratory Ther; Asn Advan Med Instrumentation. *Res:* Pulmonary pediatric diseases; cystic fibrosis of the pancreas. *Mailing Add:* Dept Rehab/Peds Baylor Col Med One Baylor Plaza Houston TX 77030

HARRISON, H KEITH, b Springville, Utah, Dec 10, 19; c 3. PLANT ANATOMY. *Educ:* Univ Calif, Berkeley, AB, 53, PhD(bot), 59. *Prof Exp:* Res botanist, Univ Calif, 59-60; asst prof bot, Mont State Univ, 60-67; assoc prof, 67-74, PROF BOT & DIR MUS NATURAL HIST, WEBER STATE COL, 74- *Concurrent Pos:* Vis asst prof, Univ Wis, 63. *Mem:* Bot Soc Am; Torrey Bot Club; Am Soc Plant Taxonomists; Sigma Xi; NY Acad Sci. *Mailing Add:* Dept Bot Weber State Col 3750 Harrison Blvd Ogden UT 84408

HARRISON, HALSTEAD, b Annapolis, Md, Apr 4, 31; m 60; c 3. ATMOSPHERIC CHEMISTRY. *Educ:* Stanford Univ, BS, 55, PhD(chem), 60. *Prof Exp:* Staff scientist, Gen Atomic Div, Gen Dynamics Corp, 60-62; res assoc atomic & molecular beams, Univ Mich, 62; NSF fel, Inst Appl Physics, Univ Bonn, 63; staff scientist, Boeing Sci Res Labs, Wash, 63-71; sr res assoc air chem, 71-73, ASSOC PROF ATMOSPHERIC SCI, UNIV WASH, 73- *Mem:* Am Chem Soc; Am Phys Soc; Am Meteorol Soc; AAAS; Am Geophys Union. *Mailing Add:* Dept Atmospheric Sci Univ Wash Seattle WA 98195

HARRISON, HAROLD EDWARD, b New Haven, Conn, July 23, 08; m 36; c 2. PEDIATRICS. *Educ:* Yale Univ, BS, 28, MD, 31. *Prof Exp:* Instr pediat, Sch Med, Yale Univ, 35-38; from instr to asst prof, Med Col, Cornell Univ, 38-42; asst prof path & investr, Off Sci Res & Develop, Sch Med, Yale Univ, 42-45; from asst prof to prof, 45-75, EMER PROF PEDIAT, SCH MED, JOHNS HOPKINS UNIV, 75- *Concurrent Pos:* Pediatrician-in-chief, Baltimore City Hosp, 45-; mem food & nutrit bd, Nat Acad Sci-Nat Res Coun.

Honors & Awards: Mead-Johnson Prize, Am Acad Pediat, 42, Borden Award, 60; Howland Award, Am Pediat Soc, 83. *Mem:* Am Soc Clin Invest; Am Pediat Soc (vpres, 73); Soc Pediat Res (pres, 53); Endocrine Soc; Am Inst Nutrit. *Res:* Nutrition; renal physiology; electrolyte and water metabolism; physiology of vitamin D; interaction of parathyroid hormone and vitamin D on calcium and phosphate homeostasis; the effect of vitamin D in increasing the diffusibility of calcium across the intestinal mucosal epithelium. *Mailing Add:* Johns Hopkins Univ Sch Med 601 N Broadway Baltimore MD 21205

HARRISON, HELEN CONNOLLY, b Annapolis, Md, April 15, 49; m 77; c 3. HORTICULTURAL SCIENCE. *Educ:* NC State Univ, BA, 71; Ohio State Univ, MS, 76; Penn State Univ, PhD (hort), 79. *Prof Exp:* ASSOC PROF HORT, UNIV WIS, 79- *Mem:* Int Soc Hort Sci; Am Soc Hort Sci; Master Gardeners Int; Nat Agr Plastics Assoc; Nat Jr Hort Assoc. *Res:* Environmental horticulture; major program areas such as mulching alternatives for small-scale food producers, row covers and other non-chemical alternatives for pest protection, heavy metal toxic problems relating to growing vegetables on contaminated soils, hydrogels, and specialty vegetables cultivarevaluation and container plantings of trees and shrubs. *Mailing Add:* Dept Hort Univ Wis Madison WI 53706

HARRISON, HELEN COPLAN, b Baltimore, Md, Sept 29, 11; m 36; c 2. BIOCHEMISTRY. *Educ:* Goucher Col, AB, 31; Smith Col, MA, 34; Yale Univ, PhD(physiol chem), 39. *Prof Exp:* Asst biol, Goucher Col, 31-33; asst physiol chem, Yale Univ, 35-37; res asst pediat, Med Col, Cornell Univ, 38-42; instr physiol chem, Yale Univ, 44-45; res asst, 47-58, res assoc, 58-59, from asst prof to assoc prof, 59-78, EMER ASSOC PROF PEDIAT, MED SCH, JOHNS HOPKINS UNIV, 78- *Concurrent Pos:* Coxe fel biochem, Yale Univ, 42; scholar, Woods Hole Marine Biol Lab, Mass. *Honors & Awards:* Mead-Johnson Award, Am Acad Pediat, 42; Howland Award, Am Pediat Soc, 83. *Mem:* Am Physiol Soc; Soc Exp Biol & Med. *Res:* Nutrition; endocrinology; renal physiology; physiology of calcium and phosphorus. *Mailing Add:* 5500 N Charles St Baltimore MD 21210

HARRISON, HUGH THOMAS, industrial chemistry, for more information see previous edition

HARRISON, IAN ROLAND, b Manchester, Eng, Apr 3, 43; US citizen; m 66; c 2. POLYMER SCIENCE. *Educ:* Univ Leeds, BSc, 64; Case Western Reserve Univ, MS, 69, PhD(macromolecular sci), 71. *Prof Exp:* Chemist, Dow Chem Co, 64-66; res grant, Case Western Reserve Univ, 71; from asst prof to assoc prof, 71-81, PROF POLYMER SCI, PA STATE UNIV, UNIV PARK, 81- *Mem:* NAm Thermal Anal Soc; Am Inst Physics; Am Phys Soc; Am Chem Soc; Sigma Xi; Soc Plastics Engrs. *Res:* Fold surface reactions of polymers; polymer characterization; thermal analysis of polymers; influence of polymer morphology on carbon structure; polymer crystallite size; small angle x-ray diffraction of polymers; deformation mechanisms; morphological control of diffusion polymer films. *Mailing Add:* 1219 Deerfield Dr State College PA 16803

HARRISON, IRENE R, b Bronx, NY, Mar 20, 52. SOFTWARE SYSTEMS. *Educ:* Univ NY, Binghamton, BA, 74; Univ Ariz, MS, 76; Rensselaer Polytech Inst, MS, 86. *Prof Exp:* Res asst optical sci, Univ Ariz, 74-76; asst res engr, United Technologies Res Ctr, 76-80, assoc res engr, 80-84, assoc anal engr, 84-86, PROG ANALYST, UNITED TECHNOLOGIES RES CTR, 86- *Mem:* Optical Soc Am; Asn Women Sci; Asn Comput Mach. *Res:* Computer programming in the sciences; simulations of processes and controls; graphics and implementation of theoretical equations; recipient of one patent. *Mailing Add:* United Technologies Res Ctr East Hartford CT 06108

HARRISON, J(OHN) D(AVID), b Pittsburgh, Pa, Oct 11, 30; m 55; c 3. METALLURGY. *Educ:* Pa State Univ, BS, 52, MS, 53; Mass Inst Technol, ScD(metall), 58. *Prof Exp:* Res assoc, Fritz Haber Inst, Max Planck Inst, Berlin, Ger, 58-59; res metallurgist, Westinghouse Elec Corp, 59-66; res staff mem, Raychem Corp, 66-77, tech mgr, Metal Prod Group, 77-81; CONSULT METALL, 81- *Mem:* Am Chem Soc; Am Inst Mining, Metall & Petrol Engrs; Am Soc Metals. *Res:* Corrosion of solid alloys by liquid metals; ice interface morphology during freezing; high pressure seeded growth and etch-pit studies of hexagonal selenium single crystals; shape memory in TiNi and other alloys. *Mailing Add:* 7 Nunes Rd Watsonville CA 95076

HARRISON, JACK EDWARD, b Tipton, Ind, Apr 30, 24; m 47; c 3. REGIONAL GEOLOGY. *Educ:* DePauw Univ, AB, 48; Univ Ill, PhD(geol), 51. *Prof Exp:* Geologist, 51-63, asst chief geologist, 63-66, RES GEOLOGIST, US GEOL SURV, 67- *Concurrent Pos:* Comnr, Am Comn Strat Nomenclature, 74-76; chmn, Int Union Geol Sci Working Group Precambrian for US & Mex, 75-, US mem, Int Union Geol Sci Subcomn Precambrian Stratigraphy, 84-88. *Honors & Awards:* Distinguished Serv Medal, Dept Interior, 78. *Mem:* Geol Soc Am; Soc Econ Geologists; Geol Asn Can; hon mem Geol Soc Belgium. *Res:* Metamorphic petrology of Precambrian rocks; geology, geochemistry and stratabound copper deposits of Belt Supergroup, Idaho-Montana. *Mailing Add:* 150 Field St Lakewood CO 80226

HARRISON, JACK LAMAR, b South Bend, Ind, Dec 2, 27; m 51; c 3. GEOLOGY. *Educ:* Ind Univ, BS, 54, MA, 55, PhD(geol), 58. *Prof Exp:* Clay mineralogist, Ind Geol Surv, 57-65, x-ray mineralogist, 65-66; assoc prof geol, Univ SDak, 66-68; HEAD PHYS ANALYTICAL LAB, GEORGIA KAOLIN CO, 68- *Mem:* Clay Minerals Soc; Electron Microscopy Soc Am; Soc Appl Spectroscopy. *Res:* Mineralogy and geochemistry. *Mailing Add:* Georgia Kaolin Res 25 Rte 22 E Springfield NJ 07081

HARRISON, JAMES BECKMAN, b Pittsburgh, Pa, May 25, 23; m 46; c 3. ORGANIC PEROXIDES, CATALYSTS. *Educ:* Allegheny Col, BS, 43; Carnegie Inst Technol, MS, 46; Univ Pittsburgh, PhD, 52. *Prof Exp:* Chemist, Lucidol Corp, 43-45, res chemist, Lucidol Div, Novadel-Agene Corp, 46-47; asst, Mellon Inst, 48-50 & Univ Pittsburgh, 50-52; from asst chief chemist to chief chemist, Wallace & Tiernan, Inc, 52-58, mgr res & develop, 58-60, dir, 60-62; pres, Aztec Chem Inc, 63-68, Aztec Chem, Div Rexall Drug & Chem Co, 69-70 & Div Dart Industs Inc, 70-74, vpres com develop, Chem Group, Dart Industs Inc, 74-79; PRES, R & A SPECIALTY CHEM CO INC, 81- *Mem:* AAAS; Com Develop Asn; Am Chem Soc; Soc Plastics Eng. *Res:* Organic peroxides and peroxy compounds; polymerization initiators; fats and oils. *Mailing Add:* 812 E 43rd St Brooklyn NY 11210

HARRISON, JAMES FRANCIS, b Philadelphia, Pa, Jan 19, 40; c 2. THEORETICAL CHEMISTRY, QUANTUM CHEMISTRY. *Educ:* Drexel Univ, BS, 62; Princeton Univ, MA, 64, PhD(chem), 66. *Prof Exp:* NSF fel, Ind Univ, Bloomington, 66-67, res assoc chem, 66-68; from asst prof to assoc prof, 68-81, PROF CHEM, MICH STATE UNIV, 81- *Concurrent Pos:* Resident scientist, Argonne Nat Lab, 80-81. *Honors & Awards:* Camille & Henry Dreyfus Award, The Camille & Henry Dreyfus Found, NY, 72. *Mem:* Am Chem Soc; Am Phys Soc. *Res:* Molecular electronic structure theory. *Mailing Add:* Dept Chem Mich State Univ East Lansing MI 48824-1322

HARRISON, JAMES MERRITT, geology, science administration; deceased, see previous edition for last biography

HARRISON, JAMES OSTELLE, b Harrison, Ga, June 17, 20; m 42; c 2. ENTOMOLOGY. *Educ:* Mercer Univ, BA, 49; Univ Ga, MA, 53; Cornell Univ, PhD(ecol), 62. *Prof Exp:* From asst entomologist to assoc entomologist, United Fruit Co, 56-62; from asst prof to prof, 62-85, EMER PROF BIOL, MERCER UNIV, 85- *Concurrent Pos:* Chmn energy adv comt, City of Macon, Ga; chmn, Water Qual Adv Comt, Mid Ga Area Planning & Develop Comn. *Mem:* Entom Soc Am; Ecol Soc Am. *Res:* Biology of banana insect pests and natural control of insect populations. *Mailing Add:* Dept Biol Mercer Univ Macon GA 31207

HARRISON, JAMES WILLIAM, JR, b Pensacola, Fla, Sept 16, 32; m 58; c 3. SOLID STATE ELECTRONICS. *Educ:* Univ Fla, BEE, 59; NC State Univ, MEE, 66, PhD(elec eng), 72. *Prof Exp:* Instrument engr, Chemstrand Corp Nylon Plant, 59-62, elec engr, Res Ctr, 62-66; res & teaching asst elec & elecronic circuits, NC State Univ, 66-71; engr, 71-74, mgr, Eng Sci Dept, 74-77, sr scientist, 77-81, DIR, ENERGY & ENVIRON RES DIV, RES TRIANGLE INST, 81- *Concurrent Pos:* Adj assoc prof, NC State Univ, 80- *Mem:* Nat Soc Prof Engrs; Sigma Xi. *Res:* Semiconductor device technology; alternative energy technology; sensing and measurement technology. *Mailing Add:* PO Box 1466 Clemson SC 29633

HARRISON, JOHN CHRISTOPHER, b Co Durham, Eng, May 20, 29; US citizen; m 60; c 3. GEOPHYSICS. *Educ:* Cambridge Univ, BA, 50, PhD(geophys), 53. *Prof Exp:* Jr res geophysicist, Inst Geophys & Planetary Physics, Univ Calif, Los Angeles, 53-55, from asst res geophysicist to assoc res geophysicist, 57-61; mem tech staff geophys & oceanog, Hughes Res Labs, 61-65; from assoc prof to prof geol, Univ Colo, Boulder, 65-83, dir, Coop Inst Res Environ Sci, 69-72, assoc dir, 72-83; SR STAFF, GEODYNAMICS CORP, 83- *Concurrent Pos:* Sr US scientist Award, Alexander von Humboldt Found, WGer, 75-76; fel, Japanese Soc Prom Sci, 82. *Mem:* Am Geophys Union; Royal Astron Soc; Soc Explor Geophysicists. *Res:* Measurement of gravity, especially at sea; earth tides and free oscillations of the earth; physical geodesy; earth tilt. *Mailing Add:* Geodynamics Corp 5520 Ekwill St Suite A Santa Barbara CA 93111

HARRISON, JOHN HENRY, IV, b Pittsburgh, Pa, Aug 8, 36; m 62; c 2. INFORMATION SCIENCE & SYSTEMS. *Educ:* Univ Tex, BS, 58, PhD(biochem), 64. *Prof Exp:* Res fel biochem, Med Sch, Harvard Univ, 64-67; from asst prof to prof chem & zool, Univ NC, Chapel Hill, 67-89, assoc provost, 85-89; DIR RES COMPUT, GLAXO INC, RESEARCH TRIANGLE PARK, NC. *Concurrent Pos:* Jane Coffin Childs fel med res, 65-67; res career develop award, NIH, 74-79; vis scientist, Oxford Univ, Eng, 78-79. *Mem:* AAAS; NY Acad Sci; Am Chem Soc; Am Soc Biol Chemists. *Res:* Physiochemical investigations of mechanism of action; subunit structure of pyridine nucleotide dependent dehydrogenases; utilization of computers in chemical research. *Mailing Add:* Glaxo Inc Five Moore Dr Research Triangle Park NC 27709

HARRISON, JOHN MICHAEL, b London, Feb 2, 15; US citizen; m 46. BEHAVIOR-ETHOLOGY. *Educ:* Univ London, dipl psychol, 47. *Prof Exp:* Prof, 62-80, EMER PROF PSYCHOL, BOSTON UNIV, 80-; SR RES FEL, CAMBRIDGE CTR BEHAV STUDIES, 85- *Concurrent Pos:* Mem, Int Brain Res Orgn, UNESCO; univ lectr, Boston Univ, 79; mem, Bd Trustees Cambridge Ctr for Behav Studies, 86-90. *Mem:* Am Asn Anat; fel Am Psychol Asn; fel Brit Psychol Soc. *Res:* Physiological psychology; investigation of relation between neural structure and behavior; comparative neuroanatomy of the auditory system and comparative study of hearing in mammals. *Mailing Add:* Dept Psychol Boston Univ 64 Cummington St Boston MA 02215

HARRISON, JOHN PATRICK, b Watford, Eng, May 5, 40; Can citizen; m 62; c 4. LOW TEMPERATURE PHYSICS. *Educ:* Univ Leeds, BSc, 61, PhD(physics), 64. *Prof Exp:* Res assoc, Cornell Univ, 64-67; fel, Univ Sussex, 67-69; from asst prof to assoc prof, 69-78, PROF PHYSICS, QUEEN'S UNIV, ONT, 78- *Concurrent Pos:* ed, J Low Temperature Physics; Killam Sr Res Fel, 83-85. *Mem:* Can Asn Physicists; Am Phys Soc. *Res:* Quantum fluids at very low temperatures; thermal properties of solids at low temperatures. *Mailing Add:* Dept Physics Queen's Univ Kingston ON K7L 3N6 Can

HARRISON, JOHN WILLIAM, b Johnstown, Pa, Apr 21, 29; m 57; c 2. PETROLEUM CHEMISTRY. *Educ:* Univ Pittsburgh, BS, 58; Carnegie-Mellon Univ, MS, 61, PhD(chem), 62. *Prof Exp:* Chemist, 62-66, from sect head to sr sect head anal chem, 66-71, dir anal & info div, 71-75, dir, Energy & Environ Eng Lab, Govt Res Labs, Exxon Res & Eng Co, 75-78, sr environ adv, 78-81, COORDR & PLANNING MGR, RES & ENVIRON HEALTH DIV, EXXON CORP, 81- *Mem:* Am Chem Soc. *Res:* Fossil fuel conversion and utilization, combustion emissions and environmental systems. *Mailing Add:* 3601 Canterbury Ct Apt E16 Bethlehem PA 18017-1329

HARRISON, JONAS P, b New York, NY, July 26, 28; m 56; c 2. PETROLEUM CHEMISTRY, CHEMICAL ENGINEERING. *Educ:* Cooper Union, BChE, 50; Univ Paris, Dr Univ(phys chem), 56. *Prof Exp:* Chem engr, Chem Corps Eng Agency, 51-52 & Schenley Indust, 52-53; res engr, Res Lab, Mobil Oil Co, 56-59 & French Inst Petrol, 59-62; sr res assoc, 62-86, PETROCHEM CONSULT, CHEVRON RES LAB, STANDARD OIL CO CALIF, 87- *Mem:* Am Chem Soc; Catalysis Soc. *Res:* Petrochemical research; catalyst research and development; pilot plant design; processes evaluation; waste treatment; petrochemicals planning. *Mailing Add:* 1799 Le Febvre Way Pinole CA 94564

HARRISON, JULIAN R, III, b Charleston, SC, Aug 23, 34; m 60; c 2. HERPETOLOGY, MALACOLOGY. *Educ:* Col Charleston, BS, 56; Duke Univ, AM, 59; Notre Dame Univ, PhD(zool), 64. *Prof Exp:* Instr biol, Western Carolina Col, 60-61; from asst prof to assoc prof, 63-74, PROF BIOL, COL CHARLESTON, 74- *Concurrent Pos:* Res assoc herpetol, The Charleston Mus, 72- *Mem:* Am Soc Ichthyologists & Herpetologists; Soc Study Amphibians & Reptiles; Herpetologists' League; Am Malacological Union. *Res:* Biology, taxonomy and evolution of reptiles and amphibians in the southeastern United States; systematics and ecology of prosobranch and pulmonate mollusks of the southeastern United States. *Mailing Add:* Dept Biol Col Charleston Charleston SC 29424

HARRISON, LIONEL GEORGE, b Liverpool, Eng, May 29, 29; m 53; c 1. PHYSICAL CHEMISTRY. *Educ:* Univ Liverpool, BSc, 49, PhD(chem), 52. *Prof Exp:* Tech officer res, Nobel Div, Imp Chem Industs, Ltd, Scotland, 52-55; Nat Res Coun Can fel pure chem, 55-57; instr, 57-59, from asst prof to assoc prof, 59-67, PROF CHEM, UNIV BC, 67- *Mem:* Faraday Soc; Brit Chem Soc. *Res:* Surface structure and bulk imperfections of solids; adsorption; gas-solid exchange reactions; diffusion; conductivity; thermal decomposition. *Mailing Add:* Dept Chem Univ BC 2075 Wesbrook Pl Vancouver BC V6T 1W5 Can

HARRISON, LURA ANN, b Enid, Okla, June 26, 42; m 69; c 2. PHYSIOLOGY. *Educ:* Univ Okla, BS, 65, PhD(physiol), 69. *Prof Exp:* NIH fel, Med Ctr, Univ Okla, 69-70; NIH training grant cardiol, 70-72; res assoc, Nat Heart & Lung Inst, NIH, 72-74; res assoc med, Med Ctr, Duke Univ, 74-79; asst prof, Okla Univ Health Sci Ctr, 79-; RETIRED. *Mem:* Am Physiol Soc; Am Col Cardiol; NY Acad Sci. *Res:* Effects of endotoxin on the cardiovascular system; cardiac electrophysiology and mechanisms of cardiac arrhythmias; influence of sympathetic neurohormones on electrophysiology of the myocardium; hormonal control of gastric secretions. *Mailing Add:* 48 Colony Rd Gretna LA 70056

HARRISON, MALCOLM CHARLES, b St Helens, Eng, July 17, 37; m 62; c 3. COMPUTER SCIENCE. *Educ:* Cambridge Univ, BA, 59; Leeds Univ, PhD(math), 62. *Prof Exp:* Div sponsored res staff mem, Solid State & Molecular Theory Group, Mass Inst Technol, 62-63, staff mem res lab electronics, 63-64; from adj asst prof to assoc prof, 65-73, dir grad studies,76-84, PROF COMPUT SCI, NY UNIV, 73- *Mem:* Asn Comput Mach. *Res:* Artificial intelligence; parallel computer systems; programming languages; operating systems. *Mailing Add:* Dept Comp Sci NY Univ 251 Mercer St New York NY 10003

HARRISON, MARJORIE HALL, b Nottingham, Eng, Sept 14, 18; US citizen; m 42; c 2. ASTROPHYSICS, ASTRONOMY. *Educ:* Univ Alta, BS, 36; Univ Chicago, PhD(astron, astrophys), 47. *Prof Exp:* Off Naval Res grant, 50-52; adj prof physics, Tex Christian Univ, 62-64; from asst prof to assoc prof, 64-72, PROF PHYSICS, SAM HOUSTON STATE UNIV, 72- *Mem:* AAAS; Am Astron Soc; Royal Astron Soc. *Res:* Stellar models; models for the sun. *Mailing Add:* 2405 Ave S Huntsville TX 77340

HARRISON, MARK, b Paris, Mo, Nov 21, 19; m 42; c 2. PHYSICS. *Educ:* Northeast Mo State Col, BS, 42; Cath Univ Am, PhD(physics), 52. *Prof Exp:* Res assoc, Columbia Univ, 42-45; res physicist, David Taylor Model Basin, 46-60; PROF PHYSICS & CHMN DEPT, AM UNIV, 60- *Mem:* Am Phys Soc; Acoust Soc Am. *Res:* Theoretical physics; acoustics; quantum and fluid mechanics; scattering; cavitation; underwater sound; turbulence. *Mailing Add:* 1511 Kingsmill Dr Salem VA 24153

HARRISON, MELVIN ARNOLD, b Seattle, Wash, Nov 16, 24; m 45; c 2. PHYSICS. *Educ:* Greenville Col, BA, 47; Univ Wash, PhD(physics), 53. *Prof Exp:* Asst prof physics, Seattle Pac Col, 53-56; assoc prof physics, Western Wash State Col, 61-62; physicist, Lawrence Livermore Lab, Univ Calif, 56-61 & 62-86; RETIRED. *Concurrent Pos:* Sci officer, AEC, 68-75. *Mem:* Am Phys Soc; Asn Comput Mach. *Res:* Conduction of electricity through gases; nuclear explosive physics; plasma physics. *Mailing Add:* PO Box 2419 Friday Harbor WA 98250

HARRISON, MERLE E(DWARD), b Durango, Colo, May 4, 33; m 67. BIOCHEMISTRY, CHEMISTRY. *Educ:* Brigham Young Univ, BS, 55; Colo State Univ, MS, 57, PhD(biochem), 60. *Prof Exp:* Instr chem, Colo State Univ, 60-61; asst prof pharm, Am Univ Beirut, 61-64; dir lab biochem, Crops Res Div, Agr Res Serv, USDA, 65-67; head dept biochem, Univ Nangrahar, Afghanistan, 67-68; asst prof, Ft Lewis Col, 68-70, chmn, Dept Chem, 69-73, assoc prof biochem & gen chem, 70-88; RETIRED. *Concurrent Pos:* Fulbright lectr biochem, 67-68. *Res:* Resistance of plants to attacks of fungal diseases; clinical tests for cysts; analysis for trace elements. *Mailing Add:* 1539 W Third Ave Durango CO 81301

HARRISON, MICHAEL A, b Philadelphia, Pa, Apr 11, 36; m; m 71; c 1. COMPUTER SCIENCE. *Educ:* Case Inst Technol, BS, 58, MS, 59; Univ Mich, PhD(commun sci), 63. *Prof Exp:* Lectr elec eng, Univ Mich, 62-63; from asst prof to assoc prof elec eng, 63-66, assoc prof comput sci, 66-71, PROF COMPUT SCI, UNIV CALIF, BERKELEY, 71- *Concurrent Pos:* Vis prof, Mass Inst Technol, 69, Hebrew Univ, Jerusalem, 70 & Stanford Univ; Guggenheim fel, 69-70; mem, Asn Comput Mach coun, 78-82 & 85- *Mem:* Asn Comput Mach (vpres, 80-82); Inst Elec & Electronics Eng; fel AAAS. *Res:* Research in programming environments user interfaces, and electronic publishing; protection in operating systems; theoretical computer science. *Mailing Add:* Dept Comput Sci Univ Calif Berkeley CA 94720

HARRISON, MICHAEL JAY, b Chicago, Ill, Aug 20, 32; m 70. THEORETICAL CONDENSED MATTER PHYSICS, THEORETICAL EPIDEMIOLOGY & IMMUNOLOGY. *Educ:* Harvard Univ, AB, 54; Univ Chicago, MSc, 56, PhD(physics), 60. *Prof Exp:* Res fel math physics, Univ Birmingham, 59-61; from asst prof to assoc prof, 61-68, fac grievance officer, 72-73, dean, Lyman Briggs Col, 73-81, PROF PHYSICS, MICH STATE UNIV, 68- *Concurrent Pos:* NSF fel, Univ Chicago, 57-59; consult, UK Atomic Energy Authority, Harwell Lab, 60, Sarnoff Labs, Radio Corp Am, NJ, 62-64 & United Aircraft Res Labs, Conn, 64-66; Am Coun Educ fel, Univ Calif, Los Angeles, 70-71; vis res physicist, Inst Theoret Physics, Univ Calif, Santa Barbara, 80-81; res affil, Theoret Biol & Biophysics, Los Alamos Nat Lab, 87-88. *Mem:* fel Am Phys Soc; Sigma Xi. *Res:* Quantum theory of condensed matter; transport phenomena in solids; plasma physics and non-equilibrium processes in many-particle systems; quantum optics and retarded Van der Waals interactions in biological tissues, surface physics, theoretical epidemiology and immunology. *Mailing Add:* Dept Physics & Astron Mich State Univ East Lansing MI 48824-1116

HARRISON, MICHAEL R, b Portland, Ore, May 5, 43; m 66; c 4. PEDIATRIC SURGERY, FETAL SURGERY. *Educ:* Yale Univ, BA, 65; Harvard Med Sch, MD, 69. *Prof Exp:* Internship surg, Mass Gen Hosp, 69-70, resident II, 70-71; res assoc, Lab Immunol, Nat Inst Allergy & Infections Dis, NIH, 71-73; resident II-IV, Mass Gen Hosp, 73-75; fel pediat surg, Children's Hosp, Los Angeles, 76-77; from asst prof to assoc prof, 78-88, PROF, UNIV CALIF, SAN FRANCISCO, 88- *Concurrent Pos:* Attending surgeon, Univ Calif Hosp, San Francisco Gen Hosp, Children's Hosp, Mt Zion Hosp & Kaiser-Permenente Med Ctr, 78- *Mem:* Am Col Surgeons; Am Pediat Surg Asn; Am Med Asn; Am Acad Pediat. *Res:* Research on physiology of fetal malformations and their correction before birth; animal models of congenital diaphragmatic hernia, hydronephrosis, hydrocephalus, and intrauterine growth retardation; clinical applications of fetal surgery; transplantation of fetal tissues and organs. *Mailing Add:* 406 Pacheco St San Francisco CA 94116

HARRISON, MONTY DEVERL, b Afton, Wyo, June 18, 34; m 53; c 4. PLANT PATHOLOGY. *Educ:* Univ Wyo, BS, 57, MS, 58; Univ Minn, PhD(plant path), 62. *Prof Exp:* From asst plant pathologist to assoc plant pathologist, Colo State Univ, 62-68, assoc prof bot & plant path, 68-73, plant pathologist, Col Natural Sci, 68-84, prof plant path, 73-90, assoc dean, 79-85, actg assoc dean, Col Agr Sci, 85, EMER PROF PLANT PATH, COLO STATE UNIV, 90- *Concurrent Pos:* Consult, PRECODEPA, CID; prin investr, fungus & bacterial dis of potatoes, 64-90; legal expert-plant dis. *Mem:* Am Phytopath Soc; Potato Asn Am. *Res:* Bacteria and air diseases of potatoes; bacterial diseases; epidemiology, bacterial ecology soil borne fungal pathogens; soil microbiology; disease control. *Mailing Add:* 1818 Manchester Dr Ft Collins CO 80526

HARRISON, NANCY EVELYN KING, b Haverhill, Mass, Apr 26, 43; m 70; c 1. ALGEBRA. *Educ:* Smith Col, BA, 64; Univ Mich, Ann Arbor, MA, 65, PhD(math), 69. *Prof Exp:* Asst prof math, Univ Mo-Columbia, 69-70; asst prof math, Brooklyn Col, 70-75 & John Jay Col Criminal Justice, City Univ New York, 75-79; from asst prof to assoc prof, 79-83, chairperson, 86-89, PROF MATH & COMPUT SCI, MERCY COL, 88- *Concurrent Pos:* Prof Res Fund grant, Univ Mo-Columbia, 70. *Mem:* Math Asn Am; Nat Coun Teachers Math. *Res:* Structure theorems for groups with length functions into the real numbers; fitting classes. *Mailing Add:* 2455 Bound Brook Lane Yorktown Heights NY 10598

HARRISON, PAUL C, b Houston, Tex, Oct 1, 36; m 60; c 4. ENVIRONMENTAL HEALTH. *Educ:* Sam Houston State Univ, BS, 61; Univ Mo, MS, 64, PhD(agr), 66. *Prof Exp:* Res asst poultry sci, Univ Mo, 61-63; from asst prof to assoc prof animal physiol, Wash State Univ, 66-73; PROF ANIMAL PHYSIOL, UNIV ILL, 73- *Honors & Awards:* Nat Poultry Res Award, Poultry Sci Asn, 70. *Mem:* Fel Poultry Sci Asn; Soc Exp Biol & Med. *Res:* Incorporate the beneficial innate physiological and behavioral responses of various farm animal species into new management equipment and practices. *Mailing Add:* Dept Animal Sci Univ Ill Urbana IL 61801

HARRISON, R(OLAND) H(ENRY), b Austin, Tex, July 30, 27; m 51; c 7. PHYSICAL CHEMISTRY, CHEMICAL ENGINEERING. *Educ:* Univ Tex, BS, 49, MS, 52, PhD(chem eng), 55. *Prof Exp:* Res engr, Univ Tex, 54-55; chem engr res & develop, Esso Stand Oil Co, 55-58; phys chemist, US Bur Mines, 58-75, Energy Res & Develop Admin, 75-77 & Dept Energy, Bartlesville Energy Technol Ctr, 77-83; sr engr, Nat Inst Petrol & Energy Res, 83-86; CONSULT, 87- *Mem:* Am Chem Soc; Am Inst Chem Engrs. *Res:* Compressibility of gases and thermodynamic properties; vapor-liquid equilibrium in coal liquefaction processes. *Mailing Add:* 2312 Parkway St Bartlesville OK 74003

HARRISON, RALPH JOSEPH, b New York, NY, Oct 3, 24; m 50, 72; c 3. SOLID STATE THEORY, COMPUTER MOLECULAR DYNAMICS. *Educ:* City Col New York, BS, 43; Mass Inst Technol, PhD(physics), 50. *Prof Exp:* Mem staff, Radiation Lab, Mass Inst Technol, 43-45; asst dept physics, 47-50; physicist, Battelle Mem Inst, 50-58; PHYSICIST, MAT SCI DIV, ARMY MAT TECH LAB, 58- *Concurrent Pos:* Res & study fel, Secy Army, 63-64. *Mem:* Am Phys Soc; sr mem Inst Elec & Electronics Eng; AAAS; Am Soc Metals; Sigma Xi. *Res:* Theoretical investigation of interatomic structures in solids and atomistic computer simulation of defect structures; artificial intelligence applications to materials science. *Mailing Add:* 48 Fellsmere Rd Newton MA 02159

HARRISON, RICHARD GERALD, b Baltimore, Md, Nov 19, 45; m 71; c 2. EVOLUTIONARY BIOLOGY. *Educ:* Harvard Univ, BA, 67; Cornell Univ, PhD(ecol & evolution), 77. *Prof Exp:* asst prof, 77-82, ASSOC PROF BIOL, YALE UNIV, 82- *Mem:* Soc Study Evolution; Genetics Soc Am; Am Soc Naturalists. *Res:* Speciation in insects; evolution of barriers to gene exchange between closely related species; evolutionary dynamics of animal m; molecular evolution. *Mailing Add:* Sect Ecol Systemology Carson Hall Cornell Univ Ithaca NY 14853

HARRISON, RICHARD MILLER, b Pineville, Ky, Apr 8, 39; m 62; c 2. REPRODUCTIVE PHYSIOLOGY, PRIMATOLOGY. *Educ:* Univ Ky, BA, 62; Mich State Univ, MS, 71, PhD(physiol), 73. *Prof Exp:* Assoc scientist biochem, Mead Johnson Res Ctr, 63-67, scientist pharmacol, 67-69; res asst physiol, Mich State Univ, 69-73; res fel reproductive physiol, 73-75, assoc scientist reproductive physiol, 75, RES SCIENTIST UROL & REPRODUCTIVE PHYSIOL, DELTA REGIONAL PRIMATE RES CTR, TULANE UNIV, 78- *Concurrent Pos:* Adj asst prof urol, Tulane Univ, 73-78, adj asst prof physiol, 75-80, from asst prof to assoc prof urol, 80-86, adj assoc prof physiol, 80-; historian, Am Soc Primatol, 85- *Honors & Awards:* Sigma Xi Res Award, 73. *Mem:* Am Soc Primatol; Soc Exp Biol & Med; Sigma Xi; Am Fertil Soc; Am Soc Andrology (pres & vpres, 80-82). *Res:* Male fertility problems; pregnancies from limited mating sessions; in vitro fertilization of non-human primate ova; utilizing laparoscopic technique for ovulation detection and diagnosis of abdominal pathology. *Mailing Add:* Delta Regional Primate Res Ctr Three Rivers Rd Covington LA 70433

HARRISON, ROBERT CAMERON, b Lamont, Alta, Aug 2, 20; m 47; c 5. SURGERY. *Educ:* Univ Alta, MD, 43; Univ Toronto, MS, 51; FRCPS(C), 50. *Prof Exp:* From assoc prof to prof surg, Univ Alta, 60-67, dir surg res, 51-67; PROF SURG & HEAD DEPT, UNIV BC, 66- *Concurrent Pos:* Consult, Dept Vet Affairs, Can. *Mem:* Fel Am Col Surg; Soc Univ Surg; Am Surg Asn. *Res:* Gastrointestinal and related research. *Mailing Add:* Fac Med Univ BC 2211 Wesbrook Mall Vancouver BC V6T 2B5 Can

HARRISON, ROBERT EDWIN, b Shreveport, La, Mar 11, 47. NEMATOLOGY, BIOLOGICAL CONTROL. *Educ:* La Tech Univ, BS, 70, MS, 72; Univ Fla, PhD(nematol), 75. *Prof Exp:* nematologist, Div Plant Industs, Tenn Dept Agr, 75-87; PROF, DEPT AGR SCI, TENN STATE UNIV, 87- *Res:* Taxonomy of entomopathogenic nematodes. *Mailing Add:* PO Box 110973 Nashville TN 37211

HARRISON, ROBERT J, b Anthony, Kans, Nov 22, 28; m 52; c 3. AUDIOLOGY, SPEECH PATHOLOGY. *Educ:* Univ Wichita, BA, 51; Northwestern Univ, PhD(audiol), 62. *Prof Exp:* Logopedist, Inst Logopedics, 51-53; dir, Hearing & Speech Ctr, Kans, 55-56; instr audiol, Med Col Ala, 56-58; asst prof audiol, 61-72, PROF OTOLARYNGOL, SCH MED, UNIV MIAMI, 71-, CHIEF, AUDIOL-SPEECH PATH, 61- *Concurrent Pos:* Nat Inst Neurol Dis & Stroke grant audiol, 63-; Int Audiol Cong grant, HEW, 64 & Voc Rehab Admin Workshop, 64-65. *Honors & Awards:* Fel Am Speech & Hearing Asn. *Mem:* Am Speech & Hearing Asn; Am Acad Otolaryngol; Am Acad Audiol. *Res:* Meniere's disease; cleft palate; retrocochlear neoplasms; middle ear impedance; licensing of professionals; cochlear otospongiosis. *Mailing Add:* Audiol-Speech Path Div R-56 Univ Miami Miami FL 33101

HARRISON, ROBERT LOUIS, b Poteau, Okla, July 29, 29; m 50; c 4. AGRONOMY. *Educ:* Okla State Univ, BSc, 54, MSc, 57; Univ Ill, PhD(agron), 61. *Prof Exp:* Asst secy, Okla Crop Improv Assoc, 55-56; secy mgr, Okla Found Seed Stocks, Inc, 56-58; asst prof agron, NDak State Univ, 61-64; mem staff, Va Polytech Inst & State Univ, 64-89, prof agron, 74-89; RETIRED. *Res:* Seed production. *Mailing Add:* 621 Watson Lane Blacksburg VA 24060

HARRISON, ROBERT W, b Brooklyn, NY, Feb 29, 36. METALLURGY. *Educ:* Xavier Univ, BS, 57, MS, 59, MBA, 75. *Prof Exp:* Mgr mat compatibility space power & propulsion, 64-70, mgr technol transfer, 70-91, CONSULT TECHNOL TRANSFER MAT & PROCESSES, GEN ELEC AIRCRAFT ENGINES, 91- *Concurrent Pos:* Dir, Technol Transfer Soc, 90- *Mem:* Fel Am Soc Metals; Technol Transfer Soc; Metall Soc. *Mailing Add:* Eng Mat Technol Labs Gen Electric Aircraft Engines One Neumann Way Mail Drop H-85 Cincinnati OH 45215

HARRISON, ROBERT WALKER, III, b Natchez, Miss, Oct 13, 41; m 63; c 2. ENDOCRINOLOGY, BIOCHEMISTRY. *Educ:* Tougaloo Col, BS, 61; Northwestern Univ, MD, 66. *Prof Exp:* Asst prof, 74-81, ASSOC PROF MED, SCH MED, VANDERBILT UNIV, 81- *Concurrent Pos:* Investr, Howard Hughes Med Inst, 77- *Mem:* Endocrine Soc; Am Col Physicians; Am Fedn Clin Res. *Res:* Biochemistry of steroid hormone effects on sensitive tissues. *Mailing Add:* 4301 W Markham Slot 587 Little Rock AR 72205

HARRISON, ROBERT WILLIAM, b Napoleon, Ohio, Nov 3, 15; m 43, 74; c 3. ZOOLOGY, PHYSIOLOGY. *Educ:* Oberlin Col, AB, 38; Wesleyan Univ, MA, 41; Yale Univ, MS, 42, PhD(zool), 49. *Prof Exp:* Asst biol, Springfield Col, 38-39 & Wesleyan Univ, 39-41; asst zool, Yale Univ, 41-42 & 46-48; from instr to prof, Univ RI, 49-77, assoc dean, Div Univ Exten, 68-69, actg dean, 69-70, actg chmn, Dept Zool, 74-75, EMER PROF ZOOL, UNIV RI, 77- *Concurrent Pos:* Officer aviation physiol, Us Naval Reserve, 42-46; vis assoc prof, Wesleyan Univ, 57; vis spec instr, Brown Univ, 58; res fel, Univ Ill, 59. *Mem:* AAAS; Am Soc Zoologists; Am Inst Biol Sci; Am Col Sports Med; Sigma Xi. *Res:* Genetics of Lebistes; survival and rescue in aviation emergencies; physiology of growth and development; chemical mediators; forensic science; physiology of exercise. *Mailing Add:* 40 Dockray Wakefield RI 02879-3915

HARRISON, SAMUEL S, b Union City, Pa, Feb 19, 41; m 63; c 3. HYDROLOGY & WATER RESOURCES. *Educ:* Allegheny Col, BS, 63; Univ NDak, Grand Forks, MS, 65, PhD(geol), 68. *Prof Exp:* Asst prof geol, Wis State Univ-Oshkosh, 68-70; from asst prof to assoc prof geol, 70-81, PROF GEOL & ENVIRON SCI, ALLEGHENY COL, 81-. CHAIRPERSON ENVIRON SCI DEPT, 75- *Concurrent Pos:* Wis State Univ, prin investr, Off Water Resources Res grant, 68-69; res assoc, Limnol Lab, 68-70; consult soils & ground water, 74- *Mem:* Soc Econ Paleont & Mineral; Soil Conserv Soc Am; Am Geol Inst; Am Inst Prof Geologists; Asn Prof Groundwater Scientists & Engrs. *Res:* Fluvial processes; groundwater monitoring; groundwater contamination; hydrogeology of hazardous waste sites' groundwater; soils; environmental geology; hydrogeologic investigations of contaminated groundwater and hazardous-waste disposal sites. *Mailing Add:* RFD No 1 Saegertown PA 16433

HARRISON, SAUL I, b New York, NY, Nov 4, 25; c 3. PSYCHIATRY, CHILD & ADOLESCENT PSYCHIATRY. *Educ:* Univ Mich, MD, 48. *Prof Exp:* Resident psychiat & child psychiat, Hosp & Med Sch, Temple Univ, 50-52 & 54-55, instr psychiat, 54-56; from asst prof to prof, 56-84, dir child & adolescent psychiat educ, Med Ctr, 56-83, EMER PROF PSYCHIAT, UNIV MICH, ANN ARBOR, 84-; PROF & DIR CHILD & ADOLESCENT PSYCHIAT, HARBOR-UNIV CALIF LOS ANGELES MED CTR, TORRANCE, 84- *Concurrent Pos:* Asst attend neuropsychiatrist, St Christopher's Hosp for Children, Philadelphia; Commonwealth Fund fel & Grant Found award, Hampstead Clin, London & Western Europ Child Psychiat Ctrs, 66. *Mem:* Am Psychoanal Asn; fel Am Psychiat Asn; fel Am Acad Child Psychiat; fel Am Orthopsychiat Asn; fel Am Col Psychiatrists. *Res:* Psychopathology, psychiatric treatment. *Mailing Add:* Harbor-Univ Calif Los Angeles Med Ctr 1000 W Carson St Torrance CA 90509

HARRISON, SHIRLEY WANDA, b Patchogue, NY, Oct 18, 23; m 46; c 3. CHEMICAL PHYSICS. *Educ:* Barnard Col, AB, 44, Columbia Univ, AM, 46; City Univ New York, PhD(physics), 70. *Prof Exp:* Lectr math, Barnard Col, Columbia Univ, 44-45; res physicist, Gen Tel & Electronics Lab, 46-61; instr physics, Queens Col, NY, 70-71; res asst chem, Hunter Col, 71-72; asst prof sci, Mahattan Community Col, 72-75; res assoc, Hunter Col, 75-77; asst prof physics, US Merchant Marine Acad, 78; instr, 78-80, from asst prof to assoc prof, 81-89, chair dept, 86-89, EMER PROF PHYS SCI DEPT, NASSAU COMMUNITY COL, 90- *Concurrent Pos:* Guest jr res assoc, Brookhaven Nat Lab, 67-71; City Univ New York Res Found new fac res award, 72; mem doctoral fac physics, City Univ New York, 75-76; res assoc, Hunter Col, 84-85 & Queens Col, 90- *Mem:* Am Phys Soc; Am Asn Univ Women. *Res:* Theoretical studies of environmental effects on structure, properties and chemical reactions in liquids and solids; quantum mechanical calculations of structure and properties of molecules, particularly hydrazoic acid and lithium fluoride; contributions of women to physics, astronomy and space science. *Mailing Add:* 42-40 208 St Bayside NY 11361-2743

HARRISON, STANLEY L, b Philadelphia, Pa, May 12, 35; m 58; c 1. ANALYTICAL CHEMISTRY, ENVIRONMENTAL SCIENCES. *Educ:* Drexel Univ, BS, 60. *Prof Exp:* Assoc scientist anal chem, Rohm & Haas Co, 60-67; assoc scientist anal chem, 67-73, mgr, Anal Res Lab, Amchem Prod, Inc,73-78; group leader, Union Carbide Agr Prod Co, Inc, 78-84, mgr regist anal chem, 84-86; mgr environ chem, 87-89, MGR, FOOD SAFETY & RESIDUE PROGS, RHONE-POULENC AG CO, 90- *Mem:* Am Chem Soc; AAAS. *Res:* Metabolism and environmental chemistry of and development of residue analytical procedures for insecticides, fungicides, herbicides, plant regulators in plant, animal and environmental matrices. *Mailing Add:* Rhone-Poulenc Ag Co PO Box 12014 Research Triangle Park NC 27709

HARRISON, STEADMAN DARNELL, JR, b New Albany, Miss, Apr 13, 47; m 69; c 1. TOXICOLOGY, PHARMACOLOGY. *Educ:* Miss State Univ, BS, 69; Ind Univ, MS, 72, PhD(pharmacol), 73; Am Bd Toxicol, dipl, 80, 86, 91. *Prof Exp:* Southern Res Inst, 73-80; assoc prof toxicol, Grad Ctr Toxicol, Univ Ky, Lexington, 80-84; SOUTHERN RES INST, 84-, HEAD CHEMOTHER DIV, 85- *Concurrent Pos:* Adj assoc prof, Sch Med, Univ Ala, Birmingham, 84- *Mem:* Royal Soc Chem; Am Asn Cancer Res; Soc Toxicol. *Res:* New drug development; preclinical project coordination; management liaison; anticancer drugs; antirheumatoid drugs. *Mailing Add:* Southern Res Inst PO Box 55305 Birmingham AL 35255-5305

HARRISON, STEPHEN COPLAN, b New Haven, Conn, June 4, 43. BIOPHYSICAL CHEMISTRY. *Educ:* Harvard Univ, AB, 63, PhD(biophys), 68. *Prof Exp:* Res assoc struct molecular biol, Children's Cancer Res Found, Boston, 68-71; from asst prof to assoc prof biochem, 71-77, PROF & CHAIR, DEPT BIOCHEM & MOLECULAR BIOL, HARVARD UNIV, 77-; INVESTR, HOWARD HUGHES MED INST, 87- *Concurrent Pos:* Harvard Univ Soc of Fels jr fel, 68-71; vis scientist biophysics, Max Planck Inst for Med Res, 71. *Honors & Awards:* George Ledlie Prize, Harvard Univ, 82; Wallace P Rowe Award, Nat Inst Allergy Infectious Dis, 88. *Mem:* Nat Acad Sci; Am Crystallog Asn; Am Soc Microbiol; AAAS. *Res:* Structure and assembly of viruses and membranes; protein-nucleic acid interactions; applications of x-ray diffraction and electron microscopy to biomolecular structure. *Mailing Add:* Fairchild Biochem Bldg Harvard Univ Cambridge MA 02138

HARRISON, STUART AMOS, b Carver Co, Minn, Jan 21, 12; m 40; c 6. CHEMISTRY. *Educ:* Univ Minn, BChE, 35, PhD(org chem), 39. *Prof Exp:* Res chemist, B F Goodrich Co, Ohio, 40-47; res chemist, Gen Mills, Inc, 47-53, head process develop sect, Chem Res Dept, 53-61, res assoc, 61-77; RETIRED. *Mem:* Am Chem Soc. *Res:* Fatty nitrogen compounds; organic coatings; polymerization; detergents; polyamides; radiation curing; acrylic resins; azophenols; plasticizers; vinyl stabilizers. *Mailing Add:* HC74 Box 2580 Hackensack MN 56452

HARRISON, THOMAS J, b Wausau, Wis, May 13, 35; m; c 2. ELECTRICAL ENGINEERING EDUCATION, COMPUTER ENGINEERING. *Educ:* Carnegie-Mellon Univ, BS, 57, MS, 58;Stanford Univ, PhD(elec eng), 64. *Prof Exp:* Consult, acad specialist, IBM Corp, 58-87; PROF ELEC ENG, FLA A&M UNIV/FLA STATE UNIV, COL ENG, 87- *Mem:* Fel Inst Elec & Electronic Engrs; fel Instruments Soc Am (pres, 86); Am Soc Eng Educ. *Res:* Small computer architecture; real time computer systems; computer process control; small computers in electrical engineering education. *Mailing Add:* Dept Elec Eng Florida State Univ Col Eng PO Box 2175 Tallahassee FL 32316

HARRISON, TIMOTHY STONE, b Kodaikanal, India, July 13, 27; US citizen; m 61; c 2. SURGERY. *Educ:* Hope Col, AB, 49; Johns Hopkins Univ, MD, 53. *Prof Exp:* Instr surg, Sch Med, Yale Univ, 61-62; from asst prof to prof surg, Med Sch, Univ Mich, Ann Arbor, 62-75; PROF SURG & PHYSIOL, COL MED, MILTON S HERSHEY MED CTR, PA STATE UNIV, 75- *Concurrent Pos:* Consult, Ann Arbor Vet Admin Hosp, Mich, 64-;

consult ed med sci, Blaisdell Publ Co, NY, 64-70; prof & chmn dept surg, Am Univ Beirut, 68-71; USPHS fel, Nat Inst Arthritis & Metab Dis, 56-57, spec fel, 59-60. *Mem:* AAAS; Am Fedn Clin Res; fel Am Col Surg; NY Acad Sci; Am Thyroid Asn. *Res:* Magnesium and calcium metabolism; adrenal medullary and catechol amine physiology; central nervous system control of endocrine function; adrenal medullary-thyroid relationships. *Mailing Add:* Milton S Hershey Med Ctr Hershey PA 17033

HARRISON, W CRAIG, b Chickasha, Okla, June 5, 40. PARTICLE PHYSICS. *Educ:* Tex Tech Col, BS, 62; Fla State Univ, PhD(physics), 69; Temple Univ, MBA, 82. *Prof Exp:* Res fel particle physics, Harvard Univ, 70-71 & Rutgers Univ, 71-75; systs analyst, 75-79, supvr technol, 79-84, MGR, SCI COMPUT, BOEING COMPUT SERV, 84- *Mem:* Am Phys Soc; Asn Comput Mach. *Res:* Experimental elementary particle physics. *Mailing Add:* 9615 California Ave SW Seattle WA 98136

HARRISON, WALTER ASHLEY, b Flushing, NY, Apr 26, 30; m 54; c 4. SOLID STATE PHYSICS. *Educ:* Cornell Univ, BEng Phys, 53; Univ Ill, MS, 54, PhD(physics), 56. *Prof Exp:* Physicist, Res Lab, Gen Elec Co, NY, 56-65; PROF APPL PHYSICS, STANFORD UNIV, 65- *Honors & Awards:* Sr US Scientist Award, von Humboldt Found, 82. *Mem:* Fel Am Phys Soc. *Res:* Quantum theory of solids; theory of molecular bonding and structures. *Mailing Add:* Dept Appl Physics Stanford Univ Stanford CA 94305

HARRISON, WILKS DOUGLAS, b Tyler, Tex, Nov 19, 32; m 56; c 4. GEOMORPHOLOGY, METEOROLOGY. *Educ:* NTex State Univ, BA, 57; Stephen F Austin State Univ, MA, 60; Univ NC, Chapel Hill, PhD(geog), 70. *Prof Exp:* Asst prof geog, Pembroke State Univ, 61-65; instr, Univ NC, Chapel Hill, 67-68; assoc prof geog, Moorhead State Univ, 68-83, assoc prof geol, 73-83; RETIRED. *Mem:* Asn Am Geogr; Am Soc Photogram; Am Meteorol Soc; Geol Soc Am. *Res:* Geography of iron ore, iron and steel; storm development; tropical climates; origin of mound topography. *Mailing Add:* 1420 Pettit Dr Tyler TX 75701

HARRISON, WILLARD WAYNE, b McLeansboro, Ill, July 28, 37; m 59; c 4. ANALYTICAL CHEMISTRY. *Educ:* Southern Ill Univ, BA, 58, MA, 60; Univ Ill, PhD(atomic absorption), 64. *Prof Exp:* From asst prof to prof anal chem, 64-88, chmn dept chem, 78-88, DEAN LIBERAL ARTS & SCI & PROF, UNIV VA, 88- *Mem:* Am Chem Soc. *Res:* Atomic absorption; mass spectrometry; hollow cathode discharge. *Mailing Add:* 2014 Turlington Univ Fla Gainesville FL 32611

HARRISON, WILLIAM ASHLEY, b Kentville, NS, Apr 26, 33. ORGANIC CHEMISTRY. *Educ:* Acadia Univ, BSc, 54, Hons, 55; McMaster Univ, MSc, 57, PhD(org chem), 60. *Prof Exp:* Nat Res Coun Can overseas fel org chem, Oxford Univ, 60-62; res chemist, Res Labs, Dominion Rubber Co, 63-66; res chemist, 66-76, SR RES SCIENTIST, RES LABS, UNIROYAL LTD, 76- *Mem:* Am Chem Soc; Chem Inst Can. *Res:* Lycopodium alkaloids; steroids; organic synthesis; agricultural chemicals. *Mailing Add:* Uniroyal Chem Ltd PO Box 1120 Guelph ON N1H 6N3 Can

HARRISON, WILLIAM EARL, b Galveston, Tex, Apr 7, 42; m 85; c 2. ORGANIC GEOCHEMISTRY. *Educ:* Lamar State Col Technol, BS, 66; Univ Okla, MS, 68; La State Univ, PhD, 76. *Prof Exp:* Geologist, Shell Oil Co, 68-71; sr res geochemist, Atlantic Richfield, 74-75; geologist, Okla Geol Surv, 75-84; RES DIR, ATLANTIC RICHFIELD CO, 84- *Mem:* Am Asn Petrol Geologists; Geochem Soc; Sigma Xi; Am Inst Prof Geologists; Soc Econ Paleontologists & Mineralogists. *Res:* Paleothermometry of sediments utilizing organic compounds; thermally-controlled diagenesis of organic matter in sediments; lipid geochemistry of ancient and modern sediments; quantitative basin analysis and modeling. *Mailing Add:* E & G Idaho Inc PO Box 1625 MS 2107 Idaho Falls ID 83404

HARRISON, WILLIAM HENRY, b Aberdeen, SDak, Feb 24, 24; m 60; c 4. BIOCHEMISTRY. *Educ:* Univ Minn, BA, 48, MS, 51, PhD(biochem), 54. *Prof Exp:* Sr biochemist antibiotics, Res Lab, Eli Lilly Co, Ind, 54-58; asst prof, Chicago Med Sch, 63-64; from asst prof to assoc prof biochem & neurol, Univ Ill Col Med, 64-71; assoc prof, 71-73, PROF BIOCHEM & NEUROSCI, MED COL, RUSH UNIV, 73- *Concurrent Pos:* Neurobiologist, Rush-Presby-St Luke's Med Ctr, 64-68, dir neurochem unit, 68-; spec fel, Columbia Univ, 58-63. *Mem:* AAAS; Am Soc Biol Chemists; Soc Neurosci; Am Chem Soc. *Res:* Vitamin E; enzymatic phosphate transfer mechanism; antibiotic isolation and characterization; biochemistry of nerve action; catecholamine metabolism and function. *Mailing Add:* Presby-St Luke's Med Ctr 1753 W Congress Pkwy Chicago IL 60612

HARRISON, WILLIAM PAUL, b Hebden Bridge, Eng, Feb 1, 20; US citizen; m 46; c 5. SOIL MECHANICS. *Educ:* Dartmouth Col, AB, 44; Clarkson Col Technol, BCE, 57; Purdue Univ, MSCE, 59. *Prof Exp:* Teacher pub sch, Mass, 46-48; from instr to asst prof civil eng, 48-60, ASSOC PROF CIVIL ENG, CLARKSON COL TECHNOL, 60-, EXEC OFFICER, CIVIL & ENVIRON ENG DEPT, 74- *Concurrent Pos:* NSF fel, 58-59. *Mem:* Am Soc Civil Engrs. *Res:* Thixotrophy and structure of sensitive clay; foundation engineering; groundwater and seepage model study and correlation with theory in earth and masonry dams; frost action; lateral earth pressure. *Mailing Add:* RR 4 Box 208 Wheeler Rd Potsdam NY 13676

HARRISON, WYMAN, b Chicago, Ill, Mar 18, 31; m 55; c 2. APPLIED GEOSCIENCE, APPLIED GEOENGINEERING. *Educ:* Univ Chicago, SB, 53, SM, 54, PhD(geol), 56. *Prof Exp:* Geologist, Ind Geol Surv, 56-59; asst prof geol, Dartmouth Col, 59-60; assoc prof & chmn dept, Norfolk Div, Col William & Mary, 60-62; assoc prof, Va Inst Marine Sci, Univ Va, 62-64; oceanogr & dir, land & sea interaction lab, Inst Oceanog, Environ Sci Serv Admin, Va, 64-69; asst prof marine sci, Univ Va & chmn, dept phys chem & geol oceanog, Va Inst Marine Sci, 69-71; prof geog, Univ Toronto, 71-74; dir, Water Resources Res Prog, 75-76, assoc dir geosci & eng, Energy & Environ Systs Div, 78-87, SPEC PROJS GROUP, ENERGY SYSTS DIV, ARGONNE NAT LAB, 89- *Concurrent Pos:* Assoc prof, Col William & Mary, 69-71; pres, Environ Res Assocs, Inc, 69-, sr prog mgr, 87- *Honors & Awards:* Distinguished Authorship Award, US Environ Sci Serv Admin, 69. *Mem:* Am Asn Petrol Geologists; Am Geophys Union; Am Inst Prof Geologists; Asn Eng Geologists. *Res:* Fate of energy-related environmental contaminants; environmental aspects of coal-conversion processes; man as a geological agent; geotechnology; beach process, responses, and protection; geologic isolation of high level nuclear wastes. *Mailing Add:* Energy Systs Div Argonne Nat Lab Argonne IL 60439

HARRISON-JOHNSON, YVONNE E, b Norfolk, Va; m 75. BIOCHEMICAL PHARMACOLOGY, DRUG METABOLISM. *Educ:* Howard Univ, BS, 59, MS, 70, PhD(pharmacol), 72. *Prof Exp:* Jr res asst, Wellcome Res Labs, Burroughs Wellcome & Co, New York, 64-69; res assoc pharmacol, Col Med, Howard Univ, 70-72; biol res coordr, 72-73, asst dir exp therapeut, 74-79, ASST DIR PHARM RES & DEVELOP, HOFFMANN-LA ROCHE INC, 79- *Mem:* NY Acad Sci; Int Soc Ecotoxicol & Environ Safety; Sigma Xi; AAAS. *Res:* Drug development responsibilities for preclinical and clinical research programs; evaluation of research projects with the objective of developing novel drug entities. *Mailing Add:* Hoffmann-La Roche Inc 340 Kingsland St Bldg 76 Nutley NJ 07110

HARRISS, DONALD K, b Zeigler, Ill, Sept 24, 34; m 54; c 4. PHYSICAL CHEMISTRY. *Educ:* Southern Ill Univ, BA, 59; Northwestern Univ, PhD(phys chem), 63. *Prof Exp:* From asst prof to assoc prof, 63-68, PROF CHEM, UNIV MINN, DULUTH, 75-, VCHANCELLOR ACAD ADMIN, 84- *Mem:* Am Chem Soc. *Res:* Quantum chemistry; non-integration methods; electron density distributions. *Mailing Add:* Dept Chem Univ Minn Duluth MN 55812

HARRISS, THOMAS T, b Memphis, Tenn, Feb 14, 18; m 43; c 2. ZOOLOGY, ENTOMOLOGY. *Educ:* Univ Ariz, BS, 42; Univ Wis, PhM, 43, PhD(zool), 49. *Prof Exp:* Asst prof biol, Whittier Col, 49-51 & 55-62, assoc prof, 62-69; from assoc prof to prof, 69-83, emer prof zool, Western State Col Colo, 83-; RETIRED. *Mem:* Am Soc Zoologists; Am Soc Parasitol; Am Inst Biol Sci. *Res:* Parasitology; acarology. *Mailing Add:* 119 Floresta Gunnison CO 81230

HARRIST, RONALD BAXTER, b Dec 24, 36; US citizen; m 59; c 3. STATISTICS, MATHEMATICS. *Educ:* Tex Tech Univ, BS, 59, MS, 63; Southern Methodist Univ, PhD(statist), 71. *Prof Exp:* Instr math, Lubbock Christian Col, 60-66; mem tech staff, TRW Systs, 66-67; instr math, Angelo State Univ, 67-68; instr comput sci, Baylor Col Med, 71-72; ASSOC PROF BIOMET, SCH PUB HEALTH, UNIV TEX, HOUSTON, 72- *Concurrent Pos:* NSF acad year inst math teachers, La State Univ, 64-65; NSF fel, Angelo State Univ, 68-69; NIH trainee statist, Southern Methodist Univ, 68-71; res asst prof biomet, Prog in Neurol & asst prof, Dept Biomed Commun, Med Sch, Univ Tex, Houston, 73- *Mem:* Math Asn Am; Biomet Soc; Am Statist Asn. *Res:* Applications of statistics in public health and medicine. *Mailing Add:* UTHSC-SPH 1200 Herman Pressler Suite 801 Houston TX 77030

HARRIS-WARRICK, RONALD MORGAN, b Berkeley, Calif, July 28, 49; m 75; c 2. NEUROCHEMISTRY, NEUROPHYSIOLOGY. *Educ:* Stanford Univ, BA, 70, Med Sch, PhD(genetics), 76. *Prof Exp:* Fel, Dept Neurobiol, Med Sch, Stanford Univ, 76-78 & Harvard Med Sch, 78-80; asst prof, 80-86, ASSOC PROF NEUROBIOL, SECT NEUROBIOL & BEHAV, CORNELL UNIV, 86-, ASSOC CHMN, 87- *Concurrent Pos:* Guggenheim fel, 86-87. *Honors & Awards:* Stephen Fox Award, Stanford Univ, 70. *Mem:* Soc Neurosci; AAAS. *Res:* Cellular and molecular mechanisms of behavior; biochemical action of neuromodulators in simple invertebrate nervous systems; control of locomotion; neuropharmacology; cloning of ion channel genes. *Mailing Add:* Sect Neurobiol & Behav Cornell Univ Ithaca NY 14853

HARROD, JOHN FRANK, b Coventry, Eng, July 15, 34; m 58. INORGANIC SYNTHESIS, PHYSICAL CHEMISTRY. *Educ:* Univ Birmingham, BSc, 55, PhD(chem), 58. *Prof Exp:* Fel chem, Univ BC, 58-60; mem staff, Res Lab, Gen Elec Co, NY, 60-66; from asst prof to assoc prof, 66-73, PROF CHEM, MCGILL UNIV, 73- *Concurrent Pos:* Consult, Dow Chem Co, 67-80. *Mem:* AAAS; Can Inst Chem; Am Chem Soc. *Res:* Coordination chemistry; catalytic activation of simple, inert molecules; heterogeneous catalysis; organometalic synthesis. *Mailing Add:* Dept Chem 801 Sherbrooke St W Montreal PQ H3A 2K6 Can

HARROLD, GORDON COLESON, b Mt Jewett, Pa, July 5, 06; m 27, 86; c 3. CHEMISTRY. *Educ:* Antioch Col, BS, 30; Univ Cincinnati, MA, 31, PhD(chem org structure), 34; Am Bd Indust Hyg, dipl, 63. *Prof Exp:* Asst chem, Univ Cincinnati, 30-34; indust hygienist, Indust Health Conserv Lab, Ohio, 35; chief chemist, Indust Hyg Dept, Chrysler Corp, 35-45; INDUST HYG CONSULT, INDUST HEALTH, HYG & SAFETY SERV, 45- *Concurrent Pos:* Co-chmn, Am Conf Indust Dis, Ohio, 39. *Honors & Awards:* Cummings Mem Award, 60, Borden Award, 80, Am Indust Hyg Asn. *Mem:* AAAS; Am Chem Soc; fel Am Pub Health Asn; hon mem Am Indust Hyg Asn (secy-treas, 39, 40-42); fel Am Inst Chem. *Res:* Air analysis and atmospheric pollution; gas fluoride detector; corrosion inhibitors; gas washing devices; lead poisoning; noise abatement; earth and airborne vibration studies; toxicity of germanium, indium, palladium, platinum, gallium. *Mailing Add:* Box 1116 Boca Raton FL 33432

HARROLD, ROBERT LEE, b Newcastle, Ind, Oct 5, 40; m 65; c 3. ANIMAL NUTRITION. *Educ:* Purdue Univ, BS, 62, MS, 64, PhD(animal nutrit), 67. *Prof Exp:* Fel nutrit & physiol, Ohio Agr Res & Develop Ctr, Ohio State Univ, 67-68; assoc prof, 68-74, PROF ANIMAL NUTRIT, NORTH DAK STATE UNIV, 80- *Mem:* Am Soc Animal Sci; Sigma Xi; Am Asn Lab Animal Sci; Animal Nutrit Res Coun. *Res:* Physiology of nutrition; nutrition of monogastric animals; nutrient requirements; utilization and nutritive value of atypical feedstuffs; analysis of amino acids in biological specimens. *Mailing Add:* Dept Animal Sci NDak State Univ Fargo ND 58105

HARROLD, RONALD THOMAS, b Fulham, London, Eng, Apr 4, 33; US citizen; m 55; c 2. ACOUSTIC WAVEGUIDE TECHNOLOGY, VAPOR-MIST DIELECTRICS. *Educ:* Twickenham Col Technol, Eng, BS, 55; Chelmsford Col Technol, Eng, BS, 62. *Prof Exp:* Student apprentice high voltage switchgear, Brit Thomson-Houston Co, 50-55; lectr radar technol, Army Sch Electronics, 55-57; develop engr cathode ray tubes, Eng Elec Valve Co, 57-61; res engr color TV, Sylvania Thorn Color TV Lab, 61-63; sr res engr, 63-75, fel res engr, 75-87, ADV SCIENTIST HIGH VOLTAGE CORONA, WESTINGHOUSE SCI & TECHNOL CTR, 75- *Concurrent Pos:* Mem, conf elec insulation & dielectric phenomena, Inst Elec & Electronics Engrs, 74-; mem comt radio-elec coord, Am Nat Standards Inst, Inc, 75- *Honors & Awards:* Indust Res-100 Award, 81, 83 & 84; Inst Elec & Electronics Engrs fel, 82. *Mem:* Brit Inst Elec Engrs; Nat Elec Mfg Asn; fel Inst Elec & Electronics Engrs; Brit Coun Eng Inst. *Res:* The detection, measurement, location and interpretation of high voltage corona and electrical discharges; acoustic waveguide technology; cure and stress monitoring of composite materials; vapor-mist dielectrics for high voltage insulation. *Mailing Add:* Westinghouse Sci & Technol Ctr Beulah Rd Pittsburgh PA 15235

HARROP, RONALD, b Manchester, Eng, May 3, 26; Can citizen; m 52; c 2. COMPUTER APPLICATIONS IN MEDICAL IMAGING. *Educ:* Cambridge Univ, BA, 46, MA, 50, PhD(math), 53. *Prof Exp:* Jr sci officer, Royal Aircraft Estab, Farnborough, 45-48; lectr math, King's Col, Univ Durham, 51-61, sr lectr, 61-65; head dept, 64-68, PROF MATH, SIMON FRASER UNIV, 64-, PROF COMPUT SCI, 74- *Concurrent Pos:* Vis asst prof, Pa state Univ, 57-58; vis prof, Univ Munich, 69 & Univ Leicester, 70; vis prof radiol, Univ Southern Calif, 78-; hon prof pharmaceut sci, Univ BC, 83-87, adj prof (hon) pharmaceut sci, 88- *Mem:* Am Math Soc; Soc Nuclear Med; Asn Symbolic Logic; Asn Comput Mach; Can Info Processing Soc; Inst Elect & Electronics Engrs. *Res:* Medical applications of computing; positron emission tomography; mathematical logic; automata theory. *Mailing Add:* Dept Math Simon Fraser Univ Burnaby BC V5A 1S6 Can

HARROW, LEE SALEM, b Washington, DC, Oct 20, 26; m 46; c 4. PHYSICAL CHEMISTRY, CHEMICAL ENGINEERING. *Educ:* George Washington Univ, BS, 46; Georgetown Univ, MS, 52, PhD(phys chem), 53. *Prof Exp:* Engr plastics lab, US Naval Ord Lab, 46; chemist & engr org dyes spectros, Spec Proj Br, US Food & Drug Admin, 46-55; supvr & mgr instruments mass spectros, Philip Morris Res Lab, 55-60; chief engr & tech dir metals & metal coatings, Am Safety Razor, 60-64; lab mgr instr org chem phys chem eng, Gen Food Corp, 64-67; vpres & tech dir, Coco-Cola Co, 68-73; vpres & corp tech dir, World Hq, H J Heinz Co, 73-88; FOOD INDUST CONSULT, 88- *Concurrent Pos:* Instr, US Dept Agr Grad Sch, 48-55, Georgetown Univ, 53; vis prof & dir, NSF, 58-68. *Mem:* Am Chem Soc; Inst Food Technol; Am Inst Chem Engrs; NY Acad Sci. *Res:* Absorption and mass spectroscopy, structure versus spectra; ion exchange, chemical equilibrium; waste utilization, enzymatic conversions of pectins, carbohydrates; technical management systems, systems analysis; hygienic practice and construction. *Mailing Add:* 412 Cezanne Dr Osprey FL 34229

HARROWER, GEORGE ALEXANDER, b Flesherton, Ont, May 15, 24; m 48, 69; c 5. PHYSICS. *Educ:* Univ Western Ont, BSc, 49; McGill Univ, MSc, 50, PhD(physics), 52. *Prof Exp:* Mem tech staff, Bell Labs, NJ, 52-55; prof physics, 55-64, dean fac arts & sci, 64-69, acad vprin, Queen's Univ, Ont, 69-76; pres, Lakehead Univ, Ont, 78-84, prof physics, 84-89; PVT PRACT, 89- *Mem:* Int Astron Union. *Res:* Radio astronomy; scintillations; distribution of sources; cosmology; instrumentation. *Mailing Add:* 204-1033 Belmont Ave Victoria BC V8S 3T4 Can

HARRY, GEORGE YOST, b Portland, Ore, Jan 14, 19; m 42; c 2. MAMMALOGY, FISHERIES. *Educ:* Ore State Col, BS, 40; Univ Mich, MS, 41; Univ Wash, PhD(fisheries), 56. *Prof Exp:* Marine fisheries biologist, State Fish Comn, Ore, 47-52, dir res, 52-57, asst state fisheries dir, 58; dir Auke Bay Biol Lab, US Bur Com Fisheries, 58-67 & Great Lakes Fishery Lab, 67-70; dir, Marine Mammal Div, Northwest Fisheries Ctr, Nat Marine Fisheries Serv, 70-80; pres, Aqua Sci Inc, 81-80; RETIRED. *Concurrent Pos:* Res assoc, Univ Mich, 67-70. *Mem:* Am Fisheries Soc; Soc Marine Mammal; Am Inst Fishery Res Biol. *Res:* Conservation of marine mammals and fish. *Mailing Add:* 4259 - 133th Ave Southeast Bellevue WA 98006

HARRY, HAROLD WILLIAM, b La, Feb 17, 21; m. INVERTEBRATE ZOOLOGY. *Educ:* La State Univ, BS, 41, MS, 42; Univ Mich, PhD(zool), 51. *Prof Exp:* Asst, La State Univ, 40-42, instr zool, 47-48; asst, Univ Mich, 48-51; asst prof, Univ Mo, 51-52; asst limnol, Philadelphia Acad Natural Sci, 52; malacologist, Army Med Serv Grad Sch, Walter Reed Med Ctr, 53-58; asst prof biol, Univ Ala, 58-60; assoc prof, NGa Col, 60-62; res assoc, Rice Univ, 62-64; asst prof, Marine Lab, Tex A&M Univ, 64-68, assoc prof biol, 66-78; RETIRED. *Mem:* Am Malacol Union. *Res:* Biology of the Mollusca; aquatic biology. *Mailing Add:* 4612 Evergreen St Bellaire TX 77401

HARSCH, HAROLD H, b Bachnung, WGer, Oct 15, 50; US citizen. PSYCHOPHARMACOLOGY, BEHAVIORAL MEDICINE. *Educ:* Marquette Univ, BS, 72; Med Col Wis, MD, 76. *Prof Exp:* NIMH fel, Stanford Univ Med Ctr, 79-80; PSYCHIATRIST, MILWAUKEE COUNTY MED COMPLEX, 82-, MED DIR, 83- *Concurrent Pos:* Assoc staff, Milwaukee Psychiat Hosp, 82-; Human Res & Rev Comt, MCMC, 84-88, Med Audit Comt, Froedert Mem Lutheran, 84-88, Impaired Residents Comt, Med Col Wis, 84-, chmn, 88-90; asst prof med, Med Col Wis, 82-88, assoc prof, 89- *Mem:* Am Med Asn; Am Psychiat Asn; fel Acad Psychosomat Med; Physicians Soc Responsibility; Asn Acad Psychiat. *Res:* Develop of medical-psychiatry units; use of functional brain imaging (single photon emission computerized tomography) in psychiatry; clinical use of novel antidepressants; psychiatric teaching in medical school. *Mailing Add:* Milwaukee County Med Complex 8700 W Wisconsin Ave Box 175 Milwaukee WI 53226

HARSHBARGER, BOYD, b Weyers Cave, Va, Feb 15, 06; m 35; c 2. MATHEMATICAL STATISTICS. *Educ:* Bridgewater Col, BA, 28; Va Polytech Inst, MS, 31; Univ Ill, MA, 35; George Washington Univ, PhD(math statist), 42. *Hon Degrees:* DSc, Bridgewater Col, 50. *Prof Exp:* Teacher, Miller Sch, Va, 28-29; from instr to assoc prof math, Va Polytech Inst & State Univ, 31-39, assoc prof statist, 41-48, in charge statist lab, 48-72, prof statist & head dept, 48-76, emer prof, 76-85; RETIRED. *Concurrent Pos:* Statist consult adv, Va Legis Comt, 47-53; founder & ed, Va J Sci, 50-55; NATO sr sci fel, 69. *Honors & Awards:* Wilks Award, 89. *Mem:* AAAS; Biomet Soc; fel Am Statist Asn; fel Inst Math Statist. *Res:* Mathematics involved in developing of new statistical designs; mathematical theory of area sampling; application of statistical designs to agriculture and engineering research. *Mailing Add:* 213 Country Club Dr SE Blacksburg VA 24060

HARSHBARGER, JOHN CARL, JR, b Weyers Cave, Va, May 9, 36. PATHOBIOLOGY. *Educ:* Bridgewater Col, BA, 57; Va Polytech Inst, MS, 59; Rutgers Univ, PhD(entom), 62. *Prof Exp:* NSF res assoc, Insect Path Pioneering Res Lab, USDA, Md, 62-64; asst res pathobiologist invert path, Univ Calif, Irvine, 64-67; DIR, REGISTRY TUMORS IN LOWER ANIMALS, MUS NATURAL HIST, SMITHSONIAN INST, 67- *Mem:* Fel AAAS; Sigma Xi; Soc Invert Path (secy, 74-76, vpres, 84-86, pres, 86-88); Am Asn Cancer Res; NY Acad Sci; Am Soc Zoologists; Am Fisheries Soc. *Res:* Etiology, epidemiology and comparative pathology of neoplasms and related diseases of invertebrate and cold blooded vertebrate animals. *Mailing Add:* Nat Mus Natural Hist Smithsonian Inst Rm W216-A Washington DC 20560

HARSHBARGER, KENNETH E, b Arcola, Ill, Nov 12, 14; m 41; c 4. DAIRY SCIENCE. *Educ:* Univ Ill, BS, 37, MS, 39, PhD, 60. *Prof Exp:* Teacher pub sch, Ill, 37-38; asst dairy prod, Univ Ill, 38-41, asst prof, 41-54; asst prof animal husb, Cornell Univ, 54-55; from asst prof to prof, 55-80, dept head, 69-79, EMER PROF DAIRY SCI, UNIV ILL, URBANA, 80- *Concurrent Pos:* Consult, Inter-Dept Comt Nutrit for Nat Defense, 57-71; dir, Higher Educ Proj Indonesia, Midwest Univ Consortium, 79-82. *Mem:* Am Dairy Sci Asn; Am Inst Nutrit; Sigma Xi. *Res:* Dairy cattle nutrition and feeding; automation of dairy cattle feeding; human nutrition. *Mailing Add:* 502 E Pennsylvania Ave Urbana IL 61801

HARSHFIELD, GREGORY ALAN, psychophysiology, behavioral genetics, for more information see previous edition

HARSHMAN, ELBERT NELSON, b Los Angeles, Calif, July 10, 10; m 39; c 5. ECONOMIC GEOLOGY. *Educ:* Calif Inst Technol, BS, 32, MS, 33; Univ Ariz, PhD(geol), 40. *Prof Exp:* Chief geologist, Cadiz Mining Co, Calif, 34; teaching fel, Univ Ariz, 34-36; chief geologist, Nielson & Co, Inc, Philippines, 36-45; liaison geologist, US Bur Reclamation, 45-52; geologist, US Geol Surv, 52-75; CONSULT MINING GEOL, 75- *Concurrent Pos:* Civilian internee, Philippines, 42-45; partic panel uranium explor geol, Int Atomic Energy Agency, Vienna, 70 & Athens, 74; mem comt sedimentary basins & sandstone-type uranium deposits, Int Atomic Energy Agency, 70- *Mem:* Geol Soc Am; Soc Econ Geologists. *Res:* Economic and engineering geology; exploration for ore deposits. *Mailing Add:* 360 Brentwood St Denver CO 80226-1354

HARSHMAN, RICHARD C(ALVERT), b Columbus, Ohio, July 31, 22; m 44; c 5. CHEMICAL ENGINEERING. *Educ:* Ohio Wesleyan Univ, BA, 47; Ohio State Univ, MS, 49, PhD(chem eng), 51. *Prof Exp:* Tech asst, Inst Paper Chem, 43-46; asst chem eng, Ohio State Univ, 47-50, res assoc, 50-51; sr res engr, Mathieson Chem Corp, 51-56, sect chief process develop eng, Indust Chem Div, Olin Mathieson Chem Corp, 56-60; assoc prof chem eng, 60-71, prof, 71-84, EMER PROF CHEM ENG, CLEMSON UNIV, 84- *Mem:* Am Chem Soc; Am Inst Chem Engrs. *Res:* Process development; reaction kinetics; design. *Mailing Add:* Chem Eng Dept Clemson Univ Clemson SC 29631

HARSHMAN, SIDNEY, b Youngstown, Ohio, Nov 14, 30; m 50; c 3. BIOCHEMISTRY, MICROBIOLOGY. *Educ:* Western Reserve Univ, BS, 50; Johns Hopkins Univ, ScD, 59. *Prof Exp:* from instr to assoc prof, 59-78, assoc dean Grad Sch, 75-81, PROF MICROBIOL, SCH MED, VANDERBILT UNIV, 78- *Mem:* AAAS; Am Chem Soc; Am Soc Microbiol; Am Soc Biol Chem; Sigma Xi. *Res:* Chemistry of normal and transformed cell membranes; biological and chemical properties of staphylococcal and alpha toxin; immunochemistry. *Mailing Add:* Dept Microbiol Vanderbilt Univ Sch Med Nashville TN 37232

HART, BENJAMIN LESLIE, b Kansas City, Mo, Aug 12. 35; m 81; c 2. ANIMAL BEHAVIOR. *Educ:* Univ Minn, BS, 58, DVM, 60, PhD(physiol psychol), 64. *Prof Exp:* Prof anat, 75-78, asst prof anat & psychol, 64-69, assoc prof anat, 69-75, PROF PHYSIOL BEHAV, UNIV CALIF, DAVIS, 78- *Concurrent Pos:* Assoc ed, Hormones & Behavior, 71-; consult ed, Appl Animal Ethology, 73- *Mem:* Animal Behav Soc; Am Physiol Soc; Am Vet Med Asn. *Res:* Reproductive behavior; animal behavior and disease management. *Mailing Add:* Dept Vet Med Physiol Univ Calif Davis CA 95616

HART, CHARLES RICHARD, b Eau Claire, Wis, Nov 11, 43; m 67; c 2. BOTANY, PLANT TAXONOMY. *Educ:* Univ Wis-Eau Claire, BS, 67; Univ Iowa, MS, 69, PhD(bot), 73. *Prof Exp:* Asst prof biol, 74-75, chmn dept, 75-76, chmn, Div Nat Sci & Health Areas, 76-78, assoc prof, 78-79, PROF BIOL & CHMN, DIV NATURAL & SOC SCI, HURON COL, 79- *Concurrent Pos:* Mem, SDak Gov's Coun Environ Educ, 74-77. *Mem:* Am Soc Plant Taxonomists; Phytochem Soc. *Res:* Taxonomic investigations dealing with classical floristic studies and studies involving the determination of systematic relationships within the Compositae, genus Bidensand Primulaceae, genus Dodecatheon. *Mailing Add:* 2021 Ohio Ave No 1 Huron SD 57350

HART, CHARLES WILLARD, JR, b Farmville, Va, Jan 30, 28; m 62. AQUATIC BIOLOGY. *Educ:* Hampden-Sydney Col, BA & BS, 49; Univ Va, MA, 51. *Prof Exp:* Asst physiol, Fla State Univ, 51-52; asst instr, Kirksville Col Osteop & Surg, 52-53; asst physiol, Fla State Univ, 53-54; instr biol, Wash Col, Md, 54-55 & Randolph-Macon Woman's Col, 55-56; med ed, Smith Kline & French Labs, 56-58; ed sci pubs, Acad Natural Sci, 58-71, invert zoologist, 62-71, adminr, Consult Progs, 69-72; dir consult progs, Acad Natural Sci Philadelphia, 72-74; prog limnologist, Smithsonian Inst, 74, asst to dir, 74-79, chmn, Dept Invert Zool, 88-91, CUR, DEPT INVERT ZOOL, NAT MUS NATURAL HIST, SMITHSONIAN INST, 79- *Concurrent Pos:* Coun Educ Biol (treas, 69-71); consult, Mediter Marine Sorting Ctr, Tunisia, 70; ed, Proc Biol Soc Wash, 78-79. *Mem:* Fel AAAS; Coun Biol Ed (treas, 69-71); Am Soc Zool; Asn Southeastern Biol (pres, 70-71); Sigma Xi; Crustacean Soc (treas, 81-84). *Res:* Administration; freshwater and marine crustacea; limnology of Jamaica, Puerto Rico and Dominica; pollution ecology of freshwater and estuarine invertebrates. *Mailing Add:* NHB W 100 Smithsonian Inst Washington DC 20560

HART, CLARENCE ARTHUR, b Norton, Va, Nov 14, 27. FORESTRY. *Educ:* Va Polytech Inst, BS, 52; NC State Col, MS, 54, PhD(wood technol), 57. *Prof Exp:* Wood technologist, 54-57, from asst prof to assoc prof, 57-69, prof wood physics, 69-80, PROF WOOD & PAPER SCI, NC STATE UNIV, 80- *Mem:* Soc Wood Sci & Technol; Forest Prod Res Soc. *Res:* Wood technology; wood physics, especially wood moisture relations. *Mailing Add:* Dept Wood & Paper Sci NC State Univ Box 8005 Raleigh NC 27695

HART, COLIN PATRICK, cardiology, orthopaedics, for more information see previous edition

HART, DABNEY GARDNER, b Jackson, Miss, Dec 3, 40; m 62. HAZARDOUS WASTE MANAGEMENT, ENVIRONMENTAL POLICY. *Educ:* Bryn Mawr Col, AB, 62, MA, 70; Am Univ, MS, 84, PhD, 89. *Prof Exp:* Biologist, Acad Natural Sci Philadelphia, 62-63, res assoc ostracod taxon, 63-67, co investr, 67-73, spec proj ed, 73-75; sr writer-ed, 75-76, mem tech staff, Dept Environ Assessment & Planning, Metrek Div, 76-80, GROUP LEADER ENERGY, RESOURCES & ENVIRON SYSTS, MITRE CORP, 80- *Concurrent Pos:* Partic, Bredin-Archbold-Smithsonian Exped to Dominica, 64, 66, Invests in Marine Shallow Water Ecosystems Prog, Smithsonian Inst, Carrie Bow Cay, Belize, 76; consult, Mediter Marine Sorting Ctr, Khereddine, Tunisia, 70; mem, Sci Adv Bd, US Environ Protection Agency, 78-82. *Mem:* Sigma Xi; AAAS; Asn Women Sci. *Res:* Environmental regulation, legislation, and policy, especially as related to hazardous waste; computer security. *Mailing Add:* MITRE Corp Civil Systs Div 7525 Colshire Dr McLean VA 22102-3481

HART, DAVID ARTHUR, b Marquette, Mich, Aug 6, 42. IMMUNOLOGY, BIOCHEMISTRY. *Educ:* Northern Mich Univ, BA, 64; Mich State Univ, PhD(biochem), 69. *Prof Exp:* Res assoc, Univ Ill Med Ctr, 69-72; from asst prof to assoc prof immunol, Univ Tex Health Sci Ctr, Dallas, 72-78; PROF MICROBIOL & INJURY & DIS & MED, UNIV CALGARY HEALTH SCI CTR. *Mem:* Am Asn Immunologists; Am Soc Microbiol; Am Soc Biochem & Microbiol Res Soc. *Res:* Plasminogen activators and their inhibitors; role of proteinases in inflammation processes. *Mailing Add:* Dept Microbiol & Infect Dis Univ Calgary 3330 Hosp Dr NW Calgary AB T2N 4N1 Can

HART, DAVID CHARLES, b Cincinnati, Ohio, July 8, 47; m 72; c 2. DYNAMICAL SYSTEMS, GRID GENERATION. *Educ:* Purdue Univ, BS, 72; Univ Calif, Berkeley, PhD(math), 80. *Prof Exp:* Asst prof math, Univ Fla, 80-86; ASST PROF MATH, UNIV CINCINNATI, 86- *Concurrent Pos:* Consult, US Air Force, 85 & 87, NASA, 86. *Mem:* Am Math Soc; Soc Indust & Appl Math. *Res:* Dynamical systems; grid generation. *Mailing Add:* Math Sci Dept Univ Cincinnati ML25 Old Chem Bldg Cincinnati OH 45221

HART, DAVID DICKINSON, b Santa Rosa, Calif, Sept 17, 52. STREAM ECOLOGY. *Educ:* Univ Calif, Santa Cruz, BA, 74; Univ Calif, Davis, PhD(ecol), 79. *Prof Exp:* Fel ecol, W K Kellogg Biol Sta, Mich State Univ, 79-80, asst vis prof, 80-83; Ruth Patrick scholar, Acad Natural Sci, 83-85, sr scientist ecol, 85-86, assoc curator, 86-90, CURATOR, ACAD NATURAL SCI, PHILADELPHIA, 91- *Concurrent Pos:* Adj asst prof, dept biol, Univ Pa, 84-90, adj prof, 91-; vis assoc prof, Swarthmore Col, 89-90. *Mem:* AAAS; NAm Benthological Soc; Am Soc Limnol & Oceanog; Ecol Soc Am. *Res:* Behavioral, population, community and ecosystem-level processes in freshwater plants and animals; watershed management. *Mailing Add:* Academy Natural Sci 19th & Pkwy Philadelphia PA 19103

HART, DAVID JOEL, b Lansing, Mich, May 15, 48. ORGANIC CHEMISTRY. *Educ:* Univ Mich, Ann Arbor, BS, 72; Univ Calif, Berkeley, PhD(chem), 76. *Prof Exp:* NIH fel chem, Calif Inst Technol, 76-78; asst prof, 78-83, ASSOC PROF CHEM, OHIO STATE UNIV, 83- *Concurrent Pos:* Eli Lilly fel, Eli Lilly Co, 82-84; Alfred P Sloan Found fel, 83-87. *Honors & Awards:* Stuart Pharmaceut Award, 86. *Mem:* Am Chem Soc; AAAS. *Res:* Synthetic organic chemistry including the development of synthetic methods, reaction mechanisms, stereochemistry, asymmetric induction, free radical chemistry and natural products total synthesis. *Mailing Add:* Dept Chem Ohio State Univ Columbus OH 43210

HART, DAVID R, b Denbo, Pa, July 26, 26; m 50; c 2. CHEMICAL ENGINEERING, CHEMISTRY. *Educ:* Auburn Univ, BS, 51; Univ Ala, Tuscaloosa, MS, 70, PhD(chem eng), 75. *Prof Exp:* Res chemist, Monsanto Chem Co, 51-55; area supvr, Liberty Powder Defense Corp, 55-56; proj engr, Cramet Inc, 56-58; chem engr, Allied Chem Co, 58-59; res chem engr, US Pipe & Foundry Co, 59-68, dir chem res, 70-76; proj mgr & secy, Harmon Eng, 76-77; proj engr, Birmingham, 77-78, gen mgr, Albany, Ore, 78-80, prog mgr, Allentown, PA, 80-81, PROJ MGR, RUST INT CORP, BIRMINGHAM, ALA, 81- *Concurrent Pos:* Adj assoc prof chem eng, Auburn Univ, 73-77 & 83-, mech eng, Univ Ala, Birmingham, 83- *Mem:* Fel Am Inst Chem Engrs; Am Chem Soc; Sigma Xi. *Mailing Add:* 2630 Greenmont Dr Birmingham AL 35226

HART, DONALD JOHN, b Montreal, Que, Mar 23, 48; m 75, 88; c 1. INDUSTRIAL HYGIENE, TOXICOLOGY. *Educ:* Oakland Univ, BS, 69; Univ Ill, Urbana, PhD(org chem), 74; Wayne State Univ, MS, 85. *Prof Exp:* Assoc sr res chemist, 73-80, sr res scientist, res labs, 80-83, staff indust hygiene, chemist, 83-87, MGR, INDUST HYGIENE LAB, GEN MOTORS CORP, 87- *Mem:* Am Chem Soc; Am Indust Hygiene Asn. *Res:* polyurethane coatings; flexible coatings; low-energy curing coating systems. *Mailing Add:* 35279 Davison Sterling Heights MI 48310

HART, EARL W, b Los Angeles, Calif, Oct 13, 27. ENVIRONMENTAL ENGINEERING. *Educ:* Univ Calif, Los Angeles, BA, 50; Univ Calif, Berkeley, MA, 71. *Prof Exp:* Field geologist, mud logging & well drilling, Litho Log Inc, 50-54; geologist, econ geol, Calif Div Mines & Geol, 54-65, field mapping, 65-73, SR GEOLOGIST & MGR, FAULT EVAL PROG, CALIF DIV MINES & GEOL, 73- *Mem:* Fel Geol Soc Am; Asn Eng Geologists; Seismol Soc Am; Am Geophys Union. *Res:* Environmental engineering, evaluation of active faults; regulatory zoning. *Mailing Add:* Six Vista Court Corte Madera CA 94925

HART, EDWARD LEON, b Brooklyn, NY, Nov 10, 30; m 65; c 2. HIGH ENERGY PHYSICS. *Educ:* City Col New York, BS, 52; Cornell Univ, PhD(physics), 59. *Prof Exp:* Instr physics, Cornell Univ, 58-59; from asst physicist to assoc physicist, Bubble Chamber Group, Brookhaven Nat Lab, 59-66; lectr physics, Univ Calif, Riverside, 66-69; assoc prof, 69-77, PROF PHYSICS, UNIV TENN, KNOXVILLE, 77- *Concurrent Pos:* Res fel, Univ Pisa, 64-65; consult, Oak Ridge Nat Lab, 69-76, part-time res staff mem, 76-82. *Mem:* Am Phys Soc. *Res:* High energy, particle physics; bubble chamber research and data processing. *Mailing Add:* Dept Physics Univ Tenn Knoxville TN 37996-1200

HART, EDWARD WALTER, b Easton, Pa, Jan 14, 18; m 40, 78; c 2. THEORETICAL PHYSICS, MATERIALS SCIENCE. *Educ:* City Col New York, BS, 38; Univ Calif, PhD(physics), 50. *Prof Exp:* Physicist degaussing & magnetic compass, US Navy Dept, 40-45; asst physics, Univ Calif, 46-47, physicist theoret physics, Radiation Lab, 47-51; physicist, Gen Elec Res & Develop Ctr, 51-76; PROF MECH & MAT SCI, CORNELL UNIV, 76- *Concurrent Pos:* Vis lectr, Rensselaer Polytech Inst, 52-63, adj prof, 63-64; consult, Adv Comt Perspectives Mat Res, Nat Acad Sci-Nat Res Coun, 59; Battelle vis prof, Ohio State Univ, 73; vis prof mech & mat sci, Cornell Univ, 75-76; vis scientist, Nat Bur Standards, 81, Tech Univ, Braunschweig, 82, Tech Univ Darmstadt, 86. *Honors & Awards:* Sr US Scientist Award, Alexander von Humboldt Found, 82. *Mem:* Fel Am Phys Soc; Am Soc Mech Engrs; Am Inst Mining & Metall Eng. *Res:* Magnetic compass; nuclear forces; field theory; plastic deformation of metals; metal physics; thermodynamics of inhomogeneous systems; fracture and failure of materials. *Mailing Add:* 330 Thurston Hall Cornell Univ Ithaca NY 14853

HART, ELWOOD ROY, b Sioux City, Iowa, Mar 6, 38; m 79; c 1. FOREST ENTOMOLOGY. *Educ:* Cornell Col, BA, 59; Tex A&M Univ, MEd, 65, PhD(entom), 72. *Prof Exp:* Fel forest entom, Tex A&M Univ, 72-74; from asst prof to assoc prof, 74-86, PROF ENTOM, IOWA STATE UNIV, 86-, PROF FORESTRY, 89- *Concurrent Pos:* vis scientist, Univ Calif, Berkeley, 85. *Mem:* Entom Soc Am; Sigma Xi; Am Registry Prof Entomologists; Int Orgn Biol Control. *Res:* Biology and management of pests of forest and shade trees; pest management in short-rotation woody crops systems. *Mailing Add:* Dept Entom 403 Sci II Iowa State Univ Ames IA 50011

HART, FREDERICK LEWIS, b New Britain, Conn, Jan 6, 45; m 71. ENVIRONMENTAL ENGINEERING. *Educ:* Univ Conn, BS, 69, MS, 71, PhD(environ), 74. *Prof Exp:* Sanit engr wastewater treat, Conn Health Dept, 69-71; asst prof, McGill Univ, 75-78; asst prof, 74-75 & 78-80, ASSOC PROF ENVIRON ENG, WORCESTER POLYTECH INST, 80- *Concurrent Pos:* Consult, Pluritec Consults, 76, SNC Group, 77; prin investr, Nat Res Coun Can, 76- & Am Soc Civil Engrs, 78- *Mem:* Water Pollution Control Fedn; Can Soc Civil Engrs. *Mailing Add:* Dept Civil Eng Worcester Plytech Inst Worcester MA 01609

HART, GARRY DEWAINE, b Los Angeles, Calif, June 18, 44; m 65. MATHEMATICS. *Educ:* Univ Calif, Riverside, BA, 66; Univ Ore, MA, 68; Kans State Univ, PhD(math), 70. *Prof Exp:* assoc prof math & coordr acad adv, 70-85, PROF MATH & CHMN MATH DEPT, CALIF STATE UNIV, DOMINGUEZ HILLS, 85- *Mem:* Am Math Soc. *Res:* Functional analysis and measure theory. *Mailing Add:* Dept Math Cal State Univ Dominquez Hills Carson CA 90747

HART, GARY ELWOOD, b Langdon, NDak, Jan 12, 34; m 62; c 2. GENETICS. *Educ:* NDak State Univ, BS, 55; Univ Calif, Berkeley, PhD(genetics), 65. *Prof Exp:* Res assoc genetics, Brookhaven Nat Lab, NY, 65-66; asst prof, 66-71, assoc prof, 71-79, PROF GENETICS, TEX A&M UNIV, 79- *Mem:* AAAS; Genetics Soc Am; Soc Study Evolution; Crop Sci Soc; Am Inst Biol Sci; Sigma Xi. *Res:* Evolutionary biology; developmental genetics; cytogenetics. *Mailing Add:* Soil & Crop Sci Dept Tex A&M Univ College Station TX 77843

HART, GEORGE EMERSON, JR, b Cleveland, Ohio, Feb 6, 29; m 54; c 2. HYDROLOGY, FORESTRY. *Educ:* Yale Univ, BA, 51; Univ Mich, Ann Arbor, BS & MF, 56, PhD(forestry), 66. *Prof Exp:* Res forester, Northeastern Forest Exp Sta, Upper Darby, Pa, 56-66; ASSOC PROF FORESTRY & OUTDOOR RECREATION, UTAH STATE UNIV, 66- *Mem:* Am Geophys Union; Soc Am Foresters; Soil Conserv Soc Am. *Res:* Water yield improvement; tree-soil-water relationships. *Mailing Add:* Dept Forest Sci Utah State Univ Logan UT 84322-5215

HART, GERALD WARREN, b Topeka, Kans, July 16, 49; m 73; c 2. BIOLOGICAL CHEMISTRY. *Educ:* Washburn Univ, BS, 71; Kans State Univ, PhD(develop biol), 77. *Prof Exp:* Postdoctoral, 77-79, from asst prof to assoc prof, 79-88, PROF BIOL CHEM, SCH MED, JOHNS HOPKINS UNIV, 880, DIR MED SCH BIOCHEM COURSE, 85- *Concurrent Pos:*

Estab investr, Am Heart Asn, 83-88; ed-in-chief, J Glycobiol, 87-89. *Honors & Awards:* H H Haymaker Award, 75; Winzler Mem Lectr, Univ Fla, 89. *Mem:* Soc Develop Biol; Am Soc Biol Chem & Molecular Biol; Sigma Xi; Soc Complex Carbohydrates; Am Soc Cell Biol; AAAS. *Res:* Biological chemistry. *Mailing Add:* Dept Biol Chem Johns Hopkins Sch Med 725 N Wolfe St Baltimore MD 21205

HART, HAROLD, b New York, NY, May 14, 22; m 42; c 4. ORGANIC CHEMISTRY, PHOTOCHEMISTRY. *Educ:* Univ Ill, BS, 41; Pa State Col, MS, 43, PhD(org chem), 47. *Prof Exp:* Asst, Petrol Refining Lab, Pa State Col, 42-46; from instr to assoc prof, 46-57, prof, 57-87, EMER PROF CHEM, MICH STATE UNIV, 87- *Concurrent Pos:* Guggenheim fel, Harvard Univ, 55-56; NSF sr fel, Cambridge Univ, 62-63; ed-in-chief, Chem Rev, 67-76. *Honors & Awards:* Petrol Chem Award, Am Chem Soc, 62. *Mem:* Am Chem Soc; The Chem Soc; AAAS. *Res:* Mechanisms of organic reactions; molecular rearrangements; small ring compounds; carbonium ions; organic synthesis; organic photochemistry; host-guest chemistry. *Mailing Add:* Dept Chem Mich State Univ East Lansing MI 48824

HART, HAROLD BIRD, b Washington, DC, July 19, 40; m 64; c 4. THEORETICAL PHYSICS. *Educ:* Univ Utah, BA, 65; Brigham Young Univ, PhD(theoret physics), 69. *Prof Exp:* Asst prof, 69-74, ASSOC PROF PHYSICS, WESTERN ILL UNIV, 74- *Res:* Conservation laws and symmetry properties of scalar-tensor gravitational theories. *Mailing Add:* Dept Physics Western Ill Univ Macomb IL 61455

HART, HAROLD M(ARTIN), b Cherokee, Okla, Nov 9, 13; m 38; c 2. ELECTRONICS. *Educ:* Univ Buffalo, AB, 34; Univ Okla, MS, 36. *Prof Exp:* Res assoc, Submarine Signal Co, 38-39, chief engr radar, Raytheon Co, 39-49, head radar systs eng, 49-55, mgr radar dept, Wayland Lab, 55-57, dir adv develop & asst mgr, 57-59, gen mgr systs mgt sub-div, 59-60, gen mgr surface radar & navig oper, 60-66, vpres & asst gen mgr, Equip Div, 66-75, vpres eng, 75-77; CONSULT ELECTRONICS, 77- *Mem:* Fel & sr mem Inst Elec & Electronics Engrs; Am Inst Aeronaut & Astronaut. *Res:* Electronic military equipment, especially radio, radar and direction finding equipment. *Mailing Add:* 5 Marvin Rd Wellesley MA 02181

HART, HASKELL VINCENT, b Chicago, Ill, Apr 17, 43; div; c 2. SPECTROSCOPY. *Educ:* Univ Ill, BS, 65; Harvard Univ, AM, 66, PhD(chem), 73. *Prof Exp:* asst prof chem, Univ NC, Wilmington, 71-79, assoc prof, 79-81; sr res chemist, 81-85, supvr, 85-87, RES MGR, SHELL DEVELOP CO, 87- *Mem:* Am Chem Soc; Soc Appl Spectros. *Res:* Spectroscopic analysis. *Mailing Add:* Shell Develop Co PO Box 1380 Houston TX 77251-1380

HART, HERBERT DORLAN, other chemistry, for more information see previous edition

HART, HIRAM, b Brooklyn, NY, May 29, 24. PHYSICS. *Educ:* City Col New York, BS, 43; NY Univ, PhD(physics), 52. *Prof Exp:* Physicist optical design, Universal Camera Corp, 43-46; instr, Bd Educ, NY, 46-53; from instr to assoc prof physics, City Col New York, 53-68; physicist, 52-61, chief med physics lab, 62-68, CHIEF PHYSICIST, MONTEFIORE HOSP, 68-; PROF PHYSICS, CITY COL NEW YORK, 68- *Concurrent Pos:* NSF fel, Yale Univ, 59-60; consult, Walter Reed Inst Res, 60-62; res assoc, Brookhaven Nat Lab, 64-67 & 71-75; assoc ed, Bull Math Biol, 73-; mem bd dirs, Soc Math Biol, 76- *Mem:* AAAS; fel Am Phys Soc; Biophys Soc; Am Asn Physicists in Med; fel NY Acad Sci; Sigma Xi. *Res:* Meson theory and scattering problems; radioisotopes in tracer and therapeutic applications; theory of perturbation-tracer experiments; x-ray and optical diffraction; circulatory physiology; radiobiology; radioisotope scanning; diffusion-interaction phenomena; scintillation proximity assay. *Mailing Add:* CCNY Dept Physics Convent Ave/138th St New York NY 10031

HART, HOWARD ROSCOE, JR, b Fayetteville, NC, Dec 6, 29; m 58; c 3. SOLID STATE PHYSICS, NUCLEAR MAGNETIC RESONANCE. *Educ:* Cornell Univ, BEng, 52;Univ Ill, MS, 55, PhD(physics), 60. *Prof Exp:* Physicist, E I du Pont de Nemours & Co, 52-54; res assoc, Univ Ill, 60; PHYSICIST, GEN ELEC RES & DEVELOP CTR, 60- *Mem:* Nat Acad Eng; Fel Am Phys Soc; Am Geophys Union. *Res:* Low temperature physics; high field superconductivity; cosmic rays; geochronology; mineral exploration; nuclear magnetic resonance; medical imaging. *Mailing Add:* Gen Elec Res & Develop Ctr K-1 1C41 PO Box 8 Schenectady NY 12301

HART, JAYNE THOMPSON, b Aurora, Ill, Apr 30, 42; m 71. PHYSIOLOGY, PHARMACOLOGY. *Educ:* NCent Col, Ill, BA, 64; Univ Wis-Madison, MS, 66, PhD(physiol), 69. *Prof Exp:* Instr physiol, Univ Wis-Madison, 68-70; res pharmacologist, Med Res Labs, Pfizer, Inc, Conn, 70-71; asst prof physiol & pharmacol, Univ Hawaii, Manoa, 72-73; asst prof biol, 74-78, ASSOC PROF BIOL, GEORGE MASON UNIV, 78- *Concurrent Pos:* Guest scientist, Naval Med Res Inst, Bethesda, 75-77; vis assoc prof, Uniformed Serv Univ Health Sci, Bethesda, MD, 85- *Mem:* Am Physiol Soc; Am Inst Biol Sci; Sigma Xi. *Res:* Animal physiology under hyperbaric conditions; physiology of pregnancy; cardiovascular physiology; vascular reactivity; neurovascular interactions. *Mailing Add:* Dept Biol George Mason Univ 4400 Univ Dr Fairfax VA 22030

HART, JOHN BIRDSALL, b Hamilton, Ohio, Aug 24, 24; m 48; c 1. PHYSICS. *Educ:* Xavier Univ, Ohio, BS, 48, MS, 50. *Prof Exp:* From instr to assoc prof math & physics, 50-68, actg chmn, dept physics, 58-61, chmn, 61-71 & 82-83, PROF MATH & PHYSICS, XAVIER UNIV, OHIO, 68- *Concurrent Pos:* Mgr Conf Found Quantum Mech, 62; lectr, Univ Cincinnati, 55-56; vis prof, Fla State Univ, 67-68; consult prof, Ohio Univ, 70-71; educ & com TV lectr sci. *Mem:* Am Asn Physics Teachers. *Res:* Pedagogy of physics; classical mechanics. *Mailing Add:* Dept Physics Xavier Univ Cincinnati OH 45207-1096

HART, JOHN HENDERSON, b Kansas City, Mo, June 18, 36; m 59; c 3. TREE PATHOLOGY, FOREST ECOLOGY. *Educ:* Dartmouth Col, BA, 58; Iowa State Univ, MS, 60, PhD(plant path), 63; Mich State Univ, MS, 85. *Prof Exp:* From asst prof to assoc prof, 63-75, PROF PLANT PATH & FORESTRY, MICH STATE UNIV, 75- *Concurrent Pos:* Vis res fel, Div Appl Chem, Commonwealth Sci & Indust Res Orgn, Melbourne, Australia, 70-71; vis scientist, Pac Forest Res Ctr, Environ Can, Victoria, BC, 77-78; vis prof, Rocky Mt Forest & Range Exp Sta, US Forest Serv, Fort Collins, Colo, 85. *Mem:* Am Phytopath Soc; Wildlife Soc; Soc Am Foresters; Ecol Soc Am; Walnut Coun (pres, 91-92). *Res:* Diseases of woody plants; effect of herbivory on forest ecosystems. *Mailing Add:* Dept Bot 237 Plant Biol Mich State Univ East Lansing MI 48824

HART, JOHN ROBERT, b New Castle, Pa, Aug 27, 35; m 66; c 3. CHEMICAL ENGINEERING, COLLOID CHEMISTRY. *Educ:* Univ Notre Dame, BS, 58; Univ Maine, MS, 68, PhD(chem eng), 70. *Prof Exp:* Instr chem eng, Univ Maine, 68-69; sect leader, Champion Papers, 69-78; asst dir, 78-80, DIR, HERTY FOUND, 80-, KUFFERATH INC, SAVANNAH, GA. *Concurrent Pos:* Fel, Paint Res Inst, 67-69. *Mem:* Am Chem Soc; Am Inst Chem Engrs; Tech Asn Pulp & Paper Indust; Sigma Xi. *Res:* Pulp and paper research; cellulose chemistry; coatings rheology of pigmented slurries; surfactant technology; synthetic fibers. *Mailing Add:* Eight Kent Ct Savannah GA 31406

HART, JOSEPHINE FRANCES LAVINIA, marine biology, for more information see previous edition

HART, KATHLEEN THERESE, b Cleveland, Ohio, July 18, 22. PHARMACOLOGY. *Educ:* Ursuline Col, BS, 44; St Louis Univ, PhD(biochem), 53. *Prof Exp:* Teacher pub sch, Ohio, 44-47; instr biochem, Univ Mich, 53-56; res assoc & sr instr internal med, St Louis Univ, 58-62; biochemist, Gen Med Res Lab, Vet Admin Ctr, Ohio, 62-66; biochemist, Pharmacol Div, Food & Drug Admin, Washington, DC, 66-69; Food & Drug Admin coordr, Strong Cobb Arner Inc, Int Chem & Nuclear Corp, 69-72; mgr regulatory affairs, Bluline Labs, Inc, 73-75; NHM FEL, WASHINGTON UNIV, ST LOUIS, 85- *Concurrent Pos:* Biochemist, Radioisotope Lab, Vet Admin Hosp, St Louis, Mo, 56-62; instr, Univ Dayton, 64-66; consult, 75- *Mem:* Am Chem Soc; Am Thyroid Asn; Am Fedn Clin Res; NY Acad Sci. *Res:* Iodine metabolism; thyroid; neurobiochemistry. *Mailing Add:* 18 S Kingshighway Apt 3H St Louis MO 63108

HART, KENNETH HOWELL, b Edmonton, Alta, Apr 5, 24; m 48; c 4. PHYSICS. *Educ:* Univ Alta, BSc, 46, MSc, 48; Univ Toronto, PhD(physics), 53. *Prof Exp:* Lectr physics, Univ Alta, 47-48; asst prof, Acadia Univ, 48-51; assoc res officer, Nat Res Coun, 53-80, sr res officer physics, 80-84; RETIRED. *Res:* Low temperature physics; properties of liquid helium; optical physics; metrology. *Mailing Add:* 141 Somerset Ottawa ON K2P 0J1 Can

HART, LARRY GLEN, b Los Angeles, Calif, Dec 6, 32; m 57; c 3. PHARMACOLOGY. *Educ:* Univ Iowa, BS, 60, MS, 62, PhD(pharmacol), 64. *Prof Exp:* Res pharmacologist, Lab Chem Pharmacol, Nat Cancer Inst, 66-68; supvr biochem pharmacol, Bio-Med Res Labs, ICI Am Inc, 68-72; asst chief pharmacol br, 72-77, actg chief lab pharmacokinetics, 78-80, ASST TO SCI DIR, NAT INST ENVIRON HEALTH SCI, 77-, ASST DIR, 80- *Concurrent Pos:* Staff fel pharmacol, Nat Heart Inst, 64-66. *Mem:* Soc Exp Biol & Med; Am Soc Pharmacol & Exp Therapeut; Sigma Xi. *Res:* Biochemical pharmacology and drug metabolism; toxicology; drug transport, primarily foreign compounds, insecticides on hepatic drug metabolizing enzymes; stimulation of drug metabolism in newborn animals. *Mailing Add:* Off Dir Hew-Pub Health Serv PO Box 12233 Research Triangle Park NC 27709

HART, LAWRENCE ALAN, b Eagle Grove, Iowa, Sept 22, 30. PURE MATHEMATICS. *Educ:* Loras Col, BS, 53. *Prof Exp:* Sec teacher, Iowa, 53-55; instr math, 55-58, from asst prof to assoc prof, 58-70, PROF MATH, LORAS COL, 70- *Concurrent Pos:* Nat Sci Found fel, Univ Wyo, 59; instr & dir, Nat Sci Found Inst, dir, Allied Health Prog, 77- study, Univ Iowa, 65-66; Nat Sci Found faculty fel, 66-67. *Mem:* Math Asn Am. *Res:* Undergraduate mathematics; analysis; topology; algebra. *Mailing Add:* Dept Math LORAS Col 1450 Alta Vista Dubuque IA 52001

HART, LEWIS THOMAS, b Chesbrough, La, Feb 2, 33; m 58; c 2. MICROBIOLOGY, BIOCHEMISTRY. *Educ:* Southeastern La Col, BS, 59; La State Univ, Baton Rouge, MS, 64, PhD(microbiol), 67. *Prof Exp:* Fel, Univ Iowa, 67-68; assoc prof, 68-80, PROF VET SCI, LA STATE UNIV, BATON ROUGE, 80- *Mem:* Am Soc Microbiol; Sigma Xi. *Res:* Microbial physiology; nitrate reduction; effect of herbicides on microbial organisms; hydrocarbon oxidation by microbes; alteration of lipide biosynthesis in anemic diseases of animals; fatty acid metabolism. *Mailing Add:* Dept Vet Sci La State Univ Baton Rouge LA 70803

HART, LYMAN HERBERT, geology, for more information see previous edition

HART, LYNN PATRICK, b Kalamazoo, Mich, Jan 2, 47; m 68, 86; c 4. PLANT PATHOLOGY. *Educ:* Calif State Polytech Univ, Pomona, BS, 73; Univ Calif, Riverside, PhD(plant path), 78. *Prof Exp:* assoc prof, 84-88, PROF PLANT PATH, MICH STATE UNIV, 88- *Mem:* Am Phytopath Soc. *Res:* Genetic and physiological bases of host-pathogen interactions. *Mailing Add:* Dept Bot & Plant Path 105 Pest Res Ctr Mich State Univ East Lansing MI 48824

HART, LYNN W, b Logan, Utah, Oct 25, 42; div; c 5. PHYSICS. *Educ:* Brigham Young Univ, BS, 67; Iowa State Univ, PhD(physics), 71. *Prof Exp:* Fel physics, Kent State Univ, 71-72; PHYSICIST, APPL PHYSICS LAB, JOHNS HOPKINS UNIV, 72- *Mem:* Am Phys Soc. *Res:* Magnetism; electromagnetic propagation; biomedical engineering. *Mailing Add:* 6177 Prophecy Pl Columbia MD 21045

HART, MAURICE I, JR, b New York, NY, Feb 3, 34; m 59; c 7. PHYSICAL CHEMISTRY. *Educ:* Maryknoll Col, AB, 55; Fordham Univ, MS, 59, PhD(phys chem), 62. *Prof Exp:* Sr chemist, Melpar Inc, Va, 62-63; from asst prof to assoc prof, 63-71, PROF PHYS CHEM, UNIV SCRANTON, 71- *Mem:* Am Chem Soc; Electrochem Soc; Sigma Xi. *Res:* Thermodynamic properties of clathrate compounds; electrochemistry in non-aqueous solvent systems; irreversible thermodynamics-thermocells. *Mailing Add:* Dept Chem Univ Scranton Scranton PA 18510

HART, MICHAEL H, b New York, NY, Apr 28, 32; m 63; c 2. ASTRONOMY, ATMOSPHERIC SCIENCE. *Educ:* Cornell Univ, AB, 52; Adelphi Univ, MS, 69; Princeton Univ, PhD(astron), 72. *Prof Exp:* Res fel astron, Hale Observ, 72-74; res fel planetary atmospheres, Nat Ctr Atmospheric Res, 74-75 & NASA Goddard Space Flight Ctr, 75-77; sr staff scientist climat, Syst & Appl Sci Corp, 77; asst prof, Dept Meteorol, Univ Md, 78; asst prof physics & astron, Dept Physics, Trinity Univ, 78-82; ASST PROF ASTRON, SCI DIV, ANNE ARUNDEL CC, 87- *Concurrent Pos:* Fel Carnegie Inst Wash, 72-74; Nat Res Coun fel, NASA, 75-77. *Mem:* Int Astron Union. *Res:* Planetary atmospheres and their evolution; early atmosphere of the earth; history of science and its impact on society; possible abundance of life in the galaxy; climate modeling. *Mailing Add:* 7301 Masonville Dr Annandale VA 22003

HART, NATHAN HOULT, b Torrington, Conn, Jan 29, 36; m 60; c 2. EXPERIMENTAL MORPHOLOGY & DEVELOPMENTAL BIOLOGY. *Educ:* Clark Univ, AB, 58; Harvard Univ, MA, 59, PhD(biol), 63. *Prof Exp:* Nat Heart Inst fel, 63-64; asst prof biol, Union Col, NY, 64-66; from asst prof to assoc prof zool, Rutgers Univ, New Brunswick, 66-71; PROF BIOL SCI, RUTGERS UNIV, 88- *Concurrent Pos:* NIH res grant, 65-68 & 83-86, NSF, 87-88. *Mem:* Soc Develop Biol; Am Soc Zoolgists; Sigma Xi; Am Soc Cell Biol. *Res:* Fish embryology; fertilization and the cortical reaction; ultrastructure and freeze fracture; developmental biology; sperm-egg interactions; egg cytoskeleton. *Mailing Add:* Dept Biol Sci Nelson Biol Labs Rutgers Univ New Brunswick NJ 08901

HART, PATRICK E(UGENE), b Portland, Ore, Jan 8, 40; m 63; c 1. CERAMIC ENGINEERING. *Educ:* Univ Wash, BS, 62; Univ Calif, Berkeley, MS, 64, PhD(ceramic eng), 67. *Prof Exp:* Sr res scientist, 67-78, tech leader nuclear ceramics, 78-80, mgr, ceramics & polymers develop sect, 80-85, assoc, Mat Sci & Technol Dept, Pac Northwest Div, 80-87, OPER MGR, APPL PHYSICS CTR, BATTELLE MEM INST, 87- *Mem:* Fel Am Ceramic Soc. *Res:* Thermal-mechanical behavior of ceramic systems including oxides and composites; fabrication techniques. *Mailing Add:* 2101 Torbett St Richland WA 99352

HART, PEMBROKE J, b Chicago, Ill, Jan 20, 29; m 76. GEOPHYSICS, SEISMOLOGY. *Educ:* Harvard Univ, AB, 50, AM, 52, PhD(geophys), 55. *Prof Exp:* Instr geol, Vanderbilt Univ, 56-58; mem staff geophys, 58-89, CONSULT, NAT ACAD SCI, 90- *Concurrent Pos:* Guest investr, Carnegie Inst Dept Terrestrial Magnetism, 65 & 68; prog officer geophys, NSF, 69-70. *Mem:* Fel AAAS; Am Geophys Union; Seismol Soc Am; fel Geol Soc Am. *Res:* Seismic refraction studies of earth's crust and upper mantle; propagation of seismic waves; structure and composition of the earth's interior. *Mailing Add:* 3252 O St NW Washington DC 20007-2847

HART, PETER E(LLIOT), b Brooklyn, NY, Feb 27, 41; m 64. COMPUTER SCIENCE. *Educ:* Rensselaer Polytech Inst, BEE, 62; Stanford Univ, MS, 63, PhD(elec eng), 66. *Prof Exp:* Res engr artificial intel, 66-74, dir, Artificial Intel Ctr, SRI Int, 74-80; mgr, Artificial Intel Res Lab, Fairchild Camera & Instrument Corp, 80-83; sr vpres, Syntelligence Inc, 83-90; SR VPRES, RICOH CORP, 91- *Mem:* AAAS; Asn Comput Mach; fel Inst Elec & Electronics Engrs; fel Am Asn Artificial Intel. *Res:* Artificial intelligence; pattern recognition; computer modelling; statistical decision theory. *Mailing Add:* 301 Arbor Rd Menlo Park CA 94025

HART, PHILLIP A, b San Jose, Calif, Mar 2, 33; m 55; c 2. ORGANIC CHEMISTRY. *Educ:* San Jose State Col, BA, 55; Univ Wash, MS, 57; Stanford Univ, PhD(org chem), 64. *Prof Exp:* Chemist, US Fish & Wildlife Serv, 57-58 & Stanford Res Inst, 58-61; from asst prof to assoc prof, 63-79, PROF PHARMACEUT CHEM, UNIV WIS-MADISON, 79- *Mem:* Am Chem Soc; AAAS. *Res:* Nucleic acid structure and function as studied by instrumental and biological techniques. *Mailing Add:* 754 Miami Pass Madison WI 53711

HART, RAYMOND KENNETH, b Newcastle, Australia, Feb 15, 28; US citizen; m 52; c 2. METALLURGICAL FAILURE ANALYSIS, ELECTRON MICROSCOPY. *Educ:* Sydney Tech Col, Australia, ASTC, 49; Univ, London, DIC, 51; Univ Cambridge, England, PhD(metall), 55; Kennedy Western Univ, JD, 91. *Prof Exp:* Sci officer metall, Aeronaut Res Labs, Australia, 55-58; assoc scientist, 59-67, sr scientist metall, Argonne Nat Lab, Ill, 68-69; prin res scientist phys sci, Ga Tech, 70-75; dir forensic metall, Pasat Res Assocs, Inc, Ga, 75-91; DIR FORENSIC METALL, RAYMOND K HART LTD, GA, 91- *Concurrent Pos:* Dir phys sci, Electron Micros Am, 69-72; prin investr, NASA, NSG-9040, 74-75; consult, Jet Propulsion Lab, Pasadena, 77-82; exec officer, Nat Defense Exec Reserve, 90- *Mem:* Fel Am Acad Forensic Sci; Am Phys Soc; Am Soc Metals Int; Am Council Independent Labs; Sigma Xi. *Res:* Composition, structure and strength of metal alloys, and the relationship of these properties to their performance characteristics in service environments. *Mailing Add:* Raymond K Hart Ltd 145 Grogan's Lake Dr Atlanta GA 30350-3115

HART, RICHARD ALLEN, b Nora Springs, Iowa, Dec 6, 30; m 58; c 3. ENTOMOLOGY, BIOLOGY. *Educ:* Univ Mo, BS, 58, MS, 60, PhD(entom), 65. *Prof Exp:* Asst entom, Univ Mo, 58-62; from instr to prof, 62-90, EMER PROF BIOL, NORTHWEST MO STATE UNIV, 90- *Concurrent Pos:* Entomologist, Fruit Fly Lab, Agr Res Serv, USDA, 66-68; biologist, IPA appointee, Environ Protection Agency region VIII, 78-80; consult, Nine-Patch Software, 90- *Mem:* Nat Sci Teachers Asn; Entom Soc Am; Am Inst Biol Sci; Am Mosquito Control Asn; Sigma Xi. *Res:* Computer supplemented science courses; mosquitos. *Mailing Add:* 315 Alco Maryville MO 64468

HART, RICHARD CULLEN, b Brooklyn, NY, Nov 7, 45; m 71; c 2. SCIENCE POLICY, ASTRONOMY. *Educ:* Wagner Col, BS, 67; Boston Univ, MS, 69, PhD(physics & astron), 73. *Prof Exp:* Lectr astron, Bentley Col, 72-74; lectr astron & physics, Boston Univ, 72-74; STAFF OFFICER, SPACE SCI BD, NAT ACAD SCI, 74- *Mem:* Am Astron Soc; Am Geophys Union; AAAS. *Res:* Solar and space physics; space astronomy; data management and computation. *Mailing Add:* Space Studies Bd Nat Res Coun 2101 Constitution Ave Washington DC 20418

HART, RICHARD HAROLD, b Villisca, Iowa, July 23, 33; m 60; c 2. RANGE SCIENCE, AGRONOMY. *Educ:* Iowa State Univ, BS, 54, MS, 58; Ore State Univ, PhD, 61. *Prof Exp:* Asst agronomist, Coastal Plain Exp Sta, Univ Ga, 61-62; res agronomist, 62-86, RANGE SCIENTIST, AGR RES SERV, USDA, 86- *Concurrent Pos:* Mem grad fac, Univ Wyo & Colo State Univ, 74- *Honors & Awards:* Am Forage & Grassland Coun Merit Cert, 82; Outstanding Achievement Award, Soc Range Mgt, 89. *Mem:* Am Soc Agron; Soc Range Mgt; Coun Agr Sci & Technol; Am Forage & Grassland Coun. *Res:* Forage and range management systems; stocking rate theory; plant responses to grazing; crop ecology; computer modeling. *Mailing Add:* High Plains Grasslands Res Sta 8408 Hildreth Rd Cheyenne WY 82009

HART, RICHARD ROYCE, b Chicago, Ill, Nov 30, 33; m 56; c 2. GEOLOGY, STRATIGRAPHY. *Educ:* Cornell Col, BA, 56; Univ Iowa, MS, 59, PhD(geol), 63. *Prof Exp:* Asst geol, Univ Iowa, 56-61; from asst prof to assoc prof geol, Ill State Univ, 61-88; RETIRED. *Mem:* AAAS; Geol Soc Am; Paleont Soc; Nat Asn Geol Teachers. *Res:* Lower Paleozoic conodonts and biostratigraphy of mid-continent. *Mailing Add:* 1406 Searle Dr Normal IL 61761

HART, ROBERT G, surface analysis, corrosion and coatings, for more information see previous edition

HART, ROBERT GERALD, b Cumberland, Md, Feb 20, 37; m 59; c 4. REPRODUCTIVE PHYSIOLOGY. *Educ:* Duquesne Univ, BS, 63, MS, 65; Univ Ill, PhD(reproductive physiol), 67. *Prof Exp:* ASSOC PROF, 67-77, PROF BIOL & CHMN DEPT, SLIPPERY ROCK STATE COL, 77- *Concurrent Pos:* Nat Sci Found travel grant, 68; Sigma Xi res grant, 68-69; res assoc reproductive physiol, Univ Ill, 69. *Mem:* AAAS; Am Soc Zool; Soc Study Reproduction; Brit Soc Study Fertil. *Res:* Biochemistry of semen coagulation; effect of age on spermatozoa. *Mailing Add:* Dept Biol Slippery Rock Univ Slippery Rock PA 16057

HART, ROBERT JOHN, b Utica, NY, Aug 16, 23; m 47; c 3. PHYSICS. *Educ:* Syracuse Univ, BA, 44, MA, 51, PhD, 55. *Prof Exp:* From asst prof to assoc prof physics, Col of William & Mary, 51-56; assoc prof, 56-63, PROF PHYSICS, STATE UNIV NY BINGHAMTON, 63- *Concurrent Pos:* Consult, Int Bus Mach, 60. *Res:* Science in general education; energy transfer. *Mailing Add:* Dept Physics & Astron State Univ NY Binghamton NY 13901

HART, ROBERT WARREN, b Yankton, SDak, Aug 17, 22; m 44; c 1. THEORETICAL PHYSICS. *Educ:* Univ Iowa, BS, 44, MS, 46; Univ Pittsburgh, PhD(physics), 49. *Prof Exp:* Instr physics, Cath Univ Am, 49-50; sr physicist, Appl Physics Lab, 50-54, group supvr, 54-75, chmn res ctr, 72-83, asst dir explor develop, 78-83, MEM PRIN STAFF, APPL PHYSICS LAB, JOHNS HOPKINS UNIV, 55-, ASST DIR, RES & EXPLOR DEVELOP, 83- *Concurrent Pos:* Assoc prof ophthal, Sch Med, Johns Hopkins Univ, 71- *Mem:* Am Phys Soc; Combustion Inst. *Res:* Scattering theory; medical physics; statistical mechanics. *Mailing Add:* 4205 SW 20th Ave Ocala FL 32674

HART, RONALD WILSON, b Syracuse, NY, Mar 23, 42; m 74; c 2. RADIOBIOLOGY, TOXICOLOGY. *Educ:* Syracuse Univ, BS, 67; Univ Ill, MS, 69, PhD(physiol), 71. *Prof Exp:* Prof radiol & dir chem & biomed environ res group, Ohio State Univ, 71-82, joint appointment biophys, pharmacol & pathobiol & dir div radiol res, 72-82; DIR, NAT CTR TOXICOL RES, 80- *Concurrent Pos:* Mem, Argonne Univ Assoc-Argonne Nat Labs Biomed Comt, 75, chmn, Argonne Univ Assoc-Biol Comt, 76-78; consult, Brookhaven Nat Lab, 75-78; mem bd toxicol & environ health, Nat Acad Sci, 76-83; chmn, Interagency Staff Group, Off Sci & Technol Policy, 83-85, Comt Coord Environ & Health Related Progs, Dept Health & Human Serv, 85-, Task Force Risk Assessment & Risk Mgt Toxic Substances, 85; chmn bd dirs, Ark Sci & Technol Authority, 83-84, mem, 85-; mem Res Bd Visitors, Memphis State Univ, 84-, Res Ctr Adv Comt, Fla A&M Univ, 85-; chmn, Agt Orange Working Group Sci Panel; vchmn, Exec Working Group Interagency Comt, Fed Lab Tech. *Honors & Awards:* Karl-August-Forster Award, Germany, 80; FDA Award, 82, 85 & 86. *Mem:* AAAS; Radiation Res Soc; fel Geront Soc; fel Am Col Toxicol (pres, 81); Sr Exec Asn; Soc Toxicol. *Res:* Genetic toxicology; mineral fiber toxicity; carcinogenesis; gerontology; management, economic development and science philosophy; molecular biology. *Mailing Add:* Dir Nat Ctr Toxicol Res Jefferson AR 72079

HART, STANLEY ROBERT, b Swampscott, Mass, June 20, 35; m 80; c 3. GEOLOGY, GEOCHEMISTRY. *Educ:* Mass Inst Technol, SB, 56, PhD(geol), 60; Calif Inst Technol, MS, 57. *Prof Exp:* Fel geochem, Carnegie Inst Dept Terrestrial Magnetism, 60-61, staff mem geochem & geophys, 61-75; prof geol & geochem, Mass Inst Technol, 75-89; SR SCIENTIST, WOODS HOLE OCEANOG INST, 89- *Concurrent Pos:* Vis assoc prof earth sci, Univ Calif, San Diego, 66-67; assoc ed, Rev of Geophys, 70-72 & Geochimica et Cosmochimica Acta, 71-76; mem earth sci adv panel, NSF, 70-73; chmn 75-76 US Nat Comt Geochem, 73-76; mem, US Nat Comt for Int Geol Correls Prog, 74-76; mem ocean crust panel, Int Phase of Ocean Drilling, 74-76; assoc ed, Physics of Earth & Planetary Interiors, 75-; adv ed, Earth & Planetary Sci Letters, 78-87. *Mem:* Nat Acad Sci; fel Geol Soc Am; fel Am Geophys Union; Geochem Soc (vpres, 83-85, pres, 85-87). *Res:* Applications of radioactive age determination to geologic problems; Sr, Nd and Pb isotope geochemistry of volcanic rocks; geochemical evolution of earth's mantle, and the oceanic lithosphere. *Mailing Add:* Woods Hole Oceanog Inst Woods Hole MA 02543

HART, WILLIAM FORRIS, b Hopewell, NJ, Jan 8, 06; m 30; c 1. ORGANIC CHEMISTRY. *Educ:* Lafayette Col, AB, 27; Princeton Univ, MA, 28; NY Univ, PhD(chem), 36. *Prof Exp:* Asst, Princeton Univ, 27-28; instr chem, Lafayette Col, 28-29; res chemist, Reed & Carnrick, 29-35 & Endo Prod, Inc, 35-37; from instr to prof, 37-57, Larkin prof, 57-71, head dept, 57-69, EMER PROF CHEM, LAFAYETTE COL, 71- *Mem:* Am Chem Soc. *Res:* Estrogenic hormones; heterocyclic phenols; quaternary morpholinium and thiamorpholinium compounds; diaryl sulfones. *Mailing Add:* 732 Burke St Easton PA 18042

HART, WILLIAM JAMES, JR, b Holyoke, Mass, June 20, 23; m 45, 79; c 7. FOOD SCIENCE. *Educ:* Univ Mass, BS, 44, MS, 47. *Prof Exp:* Dir res & qual control, Pan Am Foods, Inc, Tex, 47-48; assoc horticulturist, Va Polytech Inst, 48-49; chief food technologist, Dulany Foods, Inc, Md, 49-63, vpres, 63-67; vpres res, Ibec Foods Inc, NY, 67-71; dir res, Am Home Foods Inc, 71-73 vpres res & develop, 73-85, sr vpres technol, 85-87; CONSULT, 87- *Mem:* AAAS; Am Chem Soc; Inst Food Technol; fel Am Inst Chem. *Res:* Fruits, vegetables, meats and seafoods as they apply to processed foods; water for processing foods and industrial waste disposal. *Mailing Add:* Two Riverfield Dr Westport CT 06880

HARTBERG, WARREN KEITH, b Watseka, Ill, Jan 24, 41; m 64; c 3. ENTOMOLOGY, GENETICS. *Educ:* Wabash Col, AB, 63; Univ Notre Dame, MSc, 65, PhD(biol), 68. *Prof Exp:* Teaching asst, Univ Notre Dame, 63-66, res asst, 66-68; entomologist-geneticist, EAfrican Aedes Res Unit, WHO, 69-70; coordr inst arthopodology & parasitol, Ga Southern Col, 80-87, from asst prof to prof biol, 70-87; CHMN DEPT BIOL & PROF BIOL, BAYLOR UNIV. *Concurrent Pos:* Fac res fund grants, Ga Southern Col, 70-71 & 73-80; grant-in-aid, Sigma Xi, 71, NSF grant, 80-83; mem bd dirs, Biol Res Inst Am, Inc, 73- *Mem:* Am Genetic Asn; Genetics Soc Can; Am Mosquito Control Asn; Entom Soc Am; AAAS; Sigma Xi. *Res:* Culicidae; genetics, cytogenetics, reproductive biology, behavior, physiology, evolution, bionomics and systematics; genetic control of insects; vector-borne diseases; vector genetics. *Mailing Add:* Dept Biol Baylor Univ PO Box 94388 Waco TX 76798

HARTE, JOHN, b July 8, 39; US citizen; m 62; c 3. MATHEMATICS MODELING. *Educ:* Harvard Col, BA, 61; Univ Wis, PhD(physics), 65. *Prof Exp:* NSF fel physics, Europ Orgn Nuclear Res, Geneva, Switz, 65-66; AEC fel, Univ Calif, Berkeley, 66-68; asst prof, Yale Univ, 68-74; MEM STAFF, LAWRENCE BERKELEY LAB, UNIV CALIF, 74- *Concurrent Pos:* Pew Scholar's Prize, conservation & environ. *Mem:* Am Phys Soc. *Res:* Elementary particle physics; environmental impacts of energy; ecology; meteorology. *Mailing Add:* Energy & Resources Group T-4 Rm 100 Univ Calif Berkeley CA 94720

HARTE, KENNETH J, b New York, NY, June 16, 35; m 56, 63; c 3. ELECTRON OPTICS, ULTRAVIOLET OPTICS. *Educ:* Rensselaer Polytech Inst, BS, 58; Harvard Univ, AM, 60, PhD(physics), 65. *Prof Exp:* Staff mem, Lincoln Lab, Mass Inst Technol, 62-69; founder & vpres, Micro-Bit Corp, 69-78, mgr theoret design dept, Micro-Bit Div Control Data Corp, 78-84; chief scientist, Lithography Prods Div, 84-85, CONSULT ELECTRON ION & PHOTON BEAM TECHNOL, VARIAN ASSOC, 85- *Mem:* Am Phys Soc; Inst Elec & Electronic Engrs; Int Soc Optical Eng; Am Optical Soc; Am Ornith Union. *Res:* Theoretical electron optics; development of electron beam accessed memory; design of electron-beam lithography systems; design of far-ultraviolet optical systems. *Mailing Add:* 64 Estabrook Rd Carlisle MA 01741

HARTENBERG, RICHARD S(CHEUNEMANN), b Chicago, Ill, Feb 27, 07; m 33; c 2. MECHANICAL ENGINEERING, HISTORY OF TECHNOLOGY. *Educ:* Univ Wis, BS, 28, MS, 33, PhD(eng mech), 41. *Prof Exp:* Instr mech, Univ Wis, 30-41; from asst prof to prof, 41-75, EMER PROF MECH ENG, NORTHWESTERN UNIV, 75- *Concurrent Pos:* Ground sch instr, civilian pilot training prog, Univ Wis, 40-41; sr consult engr, Forest Prod Lab, US Dept Agr, 42-44; asst dir, Off Nat Defense Res & Develop, 43-45; staff engr & ed, Bur Aeronaut Proj, 45-50; tech rep, USAF, Wright Field, 45; vis prof, La State Univ, 76. *Honors & Awards:* Mechanisms Award, Am Soc Mech Engrs, 74; Centennial Lectr, Am Soc Mech Engrs, 80. *Mem:* Fel Am Soc Mech Engrs; Sigma Xi; Soc Hist Technol; fel Brit Inst Mech Engrs; Asn Ger Engrs. *Res:* Kinematics; machine design; history of technology. *Mailing Add:* 726 Laurel Ave Wilmette IL 60091

HARTENSTEIN, ROY, b Buffalo, NY, Jan 5, 32; m 54; c 2. PHYSIOLOGY, BIOCHEMISTRY. *Educ:* State Univ NY Col Buffalo, BSc, 53; Syracuse Univ, MSc, 57; State Univ NY Col Forestry, Syracuse, PhD(zool), 59. *Prof Exp:* Asst prof zool, State Univ NY Col Forestry, Syracuse, 59-65; Nat Cancer Inst spec fel biochem, Sch Med, Duke Univ, 65-66; consult biol sci, Smithsonian Inst, 66-67; from assoc prof to prof environ & forest biol, State Univ NY Col Forestry, 67-89; RETIRED. *Mem:* Fel AAAS; Soc Gen Physiol; Am Soc Zool. *Mailing Add:* 106 Jean Ave Syracuse NY 13210

HARTER, DONALD HARRY, b Breslau, Ger, May 16, 33; US citizen; m; c 4. NEUROLOGY, VIROLOGY. *Educ:* Univ Pa, AB, 53; Columbia Univ, MD, 57. *Prof Exp:* Intern med, Grace-New Haven Community Hosp, 57-58; from asst resident to resident neurol, New York Neurol Inst, 58-61; guest investr, Rockefeller Univ, 63-66; from asst prof to prof neurol & microbiol, Col Physicians & Surgeons, Columbia Univ, 66-75; Charles L Mix prof, 75-85, chmn dept, 75-87, Benjamin & Virginia T Boshes prof neurol, Med Sch, Northwestern Univ, 85-87; CLIN PROF NEUROL, SCH MED & HEALTH SCI, GEORGE WASH UNIV, 87-; SR SCI OFF, HOWARD HUGHES MED INST, 87-, DIR, NAT INST HEALTH RES SCHOLAR PROG, 89- *Concurrent Pos:* Mem nat comt venereal dis, Dept Health, Educ & Welfare, 70-72; mem adv comt on fels, Nat Multiple Sclerosis Soc, 76-78, chmn, 77-78, mem, Res Prog Adv Comt, 89- & mem, Sci Adv Coun, Amyotrophic Lateral Sclerosis Asn, 78, chmn, 89-91; mem sci adv bd, Am Parkinson Dis Asn, 76-90; Am Cancer Soc scholar, 73-74; Guggenheim fel, 73; vis fel, Clare Hall, Cambridge Univ, 73-74; mem, Med Adv Bd, Myasthenia Gravis Found, 80-87, NINCDS Prog Proj Comt A, 81-85 & Grants & Res Rev Bd, Nat Amyotrophic Lateral Sclerosis Found, 81-85; vis sci officer, Howard Hughes Med Inst, 86-87; mem, Bd Sci Coun, Nat Inst Dental Res, NIH, Dept Health & Human Serv, 90- *Mem:* Soc Exp Biol & Med; Infectious Dis Soc; Am Soc Virol; Am Asn Immunologists; Am Soc Microbiol; Soc Neurosci; Am Neurol Asn; fel Am Acad Neurol; Am Soc Clin Invest. *Res:* Viral infections of the nervous system; virus-induced demylination; virus-cell interaction; motor neuron disease. *Mailing Add:* Howard Hughes Med Inst 6701 Rockledge Dr Bethesda MD 20817

HARTER, GEORGE A, c 2. DEFENSE & AEROSPACE SYSTEMS. *Educ:* Purdue Univ, BS & MS; Ohio State Univ, MS. *Prof Exp:* Var tech, mgt & exec positions, TRW Inc, 57-85, vpres & asst gen mgr electronics & defense sector, 85; RETIRED. *Mem:* Nat Acad Eng; Sigma Xi; assoc fel Am Inst Aeronaut. *Mailing Add:* 38-654 Wisteria Dr Palm Desert CA 92260

HARTER, H(ARMAN) LEON, b Keokuk, Iowa, Aug 15, 19; m 43. MATHEMATICAL STATISTICS. *Educ:* Carthage Col, AB, 40; Univ Ill, AM, 41; Purdue Univ, PhD(math statist), 49. *Prof Exp:* Asst math, Univ Ill, 41-43; prof physics, Mo Valley Col, 43-44; instr math, Purdue Univ, 46-48; asst prof, Mich State Col, 49-52; math statistician, Appl Math Res Lab, Aerospace Res Labs, Wright-Patterson AFB, 52-64, sr scientist math statist, 64-75, sr scientist math statist, Appl Math Group Air Force Flight Dynamics Lab, 75-76, mathematician, 76-78; res prof, Wright State Univ, 79-84; consult, 84-88; RETIRED. *Concurrent Pos:* Co-ed, Selected Tables, Inst Math Statist, 68-75; vis prof, Air Force Inst Technol, 82-84; chmn, Sect Phys & Eng Sci, Am Statist Asn, 64; writer, 84- *Mem:* fel Am Statist Asn; fel Inst Math Statist; Int Statist Inst. *Res:* Order statistics and their use in testing and estimation; computation and compilation of statistical tables; reliability and life testing; chronological annotated bibliography on order statistics. *Mailing Add:* 203 N McKinley Ave Champaign IL 61821-3251

HARTER, JAMES A(NDREW), b Bellefonte, Pa, Sept 13, 22; m 49; c 2. ELECTRICAL ENGINEERING. *Educ:* Pa State Col, BS, 43. *Prof Exp:* Test engr, Gen Elec Co, NY, 43-44; elec engr, Tenn Eastman Corp, 44-47; health physicist, Health Physics Div, Oak Ridge Nat Lab, 47- 56, eng leader, 56-72, develop specialist, Instrumentation & Controls Div, 72-85; RETIRED. *Mem:* Health Physics Soc; sr mem Inst Elec & Electronics Engrs. *Res:* Radiation instrument design and development; radiation physics; radiation dosimetry instrumentation; basic measurement science; nuclear medicine instrumentation. *Mailing Add:* 6004 Grove Park Rd Knoxville TN 37918

HARTER, ROBERT DUANE, b Muskegon, Mich, July 6, 36; m 69; c 3. SOIL CHEMISTRY. *Educ:* Ohio State Univ, BS, 61, MS, 62; Purdue Univ, PhD(soil chem), 66. *Prof Exp:* Asst soil scientist, Conn Agr Exp Sta, 66-68; assoc res scientist clay surface chem, NY Univ, 68-69; from asst prof to assoc prof, 69-83, PROF SOIL CHEM, UNIV NH, 83- *Concurrent Pos:* Vis prof, Pa State Univ, 76, Agr Univ, Netherlands, 83-84; chmn soil chem div, Soil Sci Soc Am, 82; assoc ed, Soil Sci Soc Am J, 88- *Mem:* Fel Soil Sci Soc Am; Am Soc Agron; Int Soc Soil Sci. *Res:* Clay surface chemistry and the kinetics and thermodynamics of heavy metal adsorption by soils and clays; soil inorganic chemistry, especially phosphorous interactions in soil; nutrient cycling in forest soils. *Mailing Add:* Dept Natural Resources James Hall Univ of NH Durham NH 03824

HARTER, WILLIAM GEORGE, b Lancaster, Pa, July 18, 43; m 74; c 3. QUANTUM ELECTRONICS. *Educ:* Hiram Col, BA, 64; Univ Calif, Irvine, PhD(physics), 67. *Prof Exp:* Asst prof physics, Univ Southern Calif, 69-74; assoc prof physics, Univ Estadual de Campinas, Brazil, 74-76; vis fel physics, Joint Inst Lab Astrophysics, Univ Colo, 76-78; from asst prof to assoc prof physics, Ga Inst Technol, 78- 84; PROF PHYSICS, UNIV ARK, 84- *Concurrent Pos:* VPres, Earth-ings Corp, 80-; consult, Los Alamos Nat Lab, 81- *Mem:* Am Phys Soc. *Res:* Group theory for symmetry analysis of laser spectroscopy; new methods for studying the dynamics of molecular rotations and vibrations; high resolution spectroscopy; computer graphics. *Mailing Add:* Dept Physics Univ Arkansas Fayetteville AR 72701

HARTFIEL, DARALD JOE, b Ray Point, Tex, Oct 10, 39; m 62; c 2. MATHEMATICS. *Educ:* Southwest Tex State Univ, BS, 62; Univ Houston, MS, 66, PhD(math), 69. *Prof Exp:* Asst prof, 69-74, assoc prof, 74-81, PROF MATH, TEX A&M UNIV, 81- *Mem:* Am Math Soc; Soc Indust & Appl Math. *Res:* Matrix theory, especially non-negative matrix theory. *Mailing Add:* Dept Math Tex A&M Univ College Station TX 77843

HARTFORD, WINSLOW H, b Newton, Mass, June 1, 10; m 39; c 2. CHEMISTRY & TECHNOLOGY OF CHROMIUM. *Educ:* Boston Univ, AB, 28; Mass Inst Technol, SB, 30, PhD(inorg chem), 33. *Prof Exp:* Asst, Mass Inst Technol, 30-33; res assoc, Mutual Chem Co Am, Mass, 33-34, res chemist, Md, 34-45, res supvr, 45-54, res supvr, Mutual Chem Div, Allied Chem & Dye Corp, 55-57, res supvr, Solvay Process Div, Allied Chem Corp, 58-63, sr scientist, 63-66, sr scientist, Indust Chem Div, 66-69; chmn dept, 72-76, assoc prof, 70-78, EMER PROF ENVIRON SCI, BELMONT ABBEY COL, 79- *Concurrent Pos:* Gen consult, 69-; adj prof environ sci, Queens Col, 81-82. *Mem:* AAAS; Am Chem Soc; Am Inst Chemists; Int Res Group Wood Preserv; Sigma Xi; Air Pollution Control Asn; Int Soc Ecotoxicology. *Res:* Chemistry and technology of chromium and its compounds; inorganic wood preservatives; statistical evaluation of wood preservatives and treating methods; relation of thermodynamics to economics and environment; biochemical causes of acid rain; energy conservation; biospecificity of inorganic compounds; structural inorganic chemistry and biological activity; application of computers to process control. *Mailing Add:* 1413 Redcoat Dr Charlotte NC 28211

HARTGERINK, RONALD LEE, b Zeeland, Mich, Mar 21, 42; m 64; c 3. CATALYSIS, PETROLEUM CHEMISTRY. *Educ:* Hope Col, AB, 64; Univ Calif, Berkeley, PhD(chem), 68. *Prof Exp:* Lectr chem, Univ Calif, Berkeley, 68-69; res chemist, 69-71, sr res chemist, 71-77, res assoc & sect head, 77-82, LAB DIR, EXXON RES & ENG CO, 82- *Mem:* Am Chem Soc. *Res:* Rates and mechanisms of organic and organometallic reactions; organic synthesis; petroleum chemistry; synthetic fuels. *Mailing Add:* c/o Wyckoff Chem Co 1421 Kalamazoo St South Haven MI 49090

HARTH, ERICH MARTIN, b Vienna, Austria, Nov 16, 19; nat US; m 51; c 2. NEUROSCIENCES, MATHEMATICAL BIOPHYSICS. *Educ:* Syracuse Univ, AB, 43, MS, 48, PhD(physics), 51. *Prof Exp:* Physicist, Nucleonics Div, Naval Res Lab, 51-54; res assoc, Duke Univ, 54-57; from asst prof to prof physics, 57-90, EMER PROF PHYSICS, SYRACUSE UNIV, 90- *Mem:* AAAS; Biophys Soc; Am Phys Soc; Biophys Soc; Soc Neurosci. *Res:* Elementary particles; bubble chambers; dynamics and information processing in neural systems; visual perception; neural mechanisms in higher brain functions, optimization processes. *Mailing Add:* Dept Physics 201 Physics Bldg Syracuse Univ Syracuse NY 13244-1130

HARTIG, ELMER OTTO, b Evansville, Ind, Jan 28, 23; m 49; c 4. PHYSICS. *Educ:* Univ NH, BS, 46; Harvard Univ, MS, 47, PhD(appl physics), 50. *Prof Exp:* Group leader microwave & antennas, Goodyear Aerospace Corp, Ohio, 50-53, sect head radiation systs, 54-55, asst mgr, Aerophys Dept, 55-63, mgr, Electronics Eng Div, 63-66, chief engr, Ariz Div, 66-76, vpres res & eng, 76-81, vpres opers, defense & energy, 81-85, vpres eng & res, 85-88; RETIRED. *Concurrent Pos:* Consult, Sci Adv Bd Guidance & Control Panel, USAF, 64-78; mem, Ariz AEC, 70-75; mem, US Army Sci Bd, 80-; mem adv comt, Univ Akron, 82- *Mem:* Fel Inst Elec & Electronics Engrs; Am Inst Aeronaut & Astronaut; Sigma Xi; Aerospace Industs Asn Am. *Res:* Microwaves; antennas; radar systems; radomes; guidance systems; data processing. *Mailing Add:* 826 E Meadow Lane Phoenix AZ 85022

HARTIG, PAUL RICHARD, b St Louis, Mo, Oct 19, 49. NEUROCHEMISTRY. *Educ:* St Louis Univ, AB, 71; Univ Calif, Berkeley, PhD(biophys), 76. *Prof Exp:* Fel neurochem, Calif Inst Technol, 76-79; ASST PROF BIOL, JOHNS HOPKINS UNIV, 79- *Mem:* Soc Neurosci; Am Soc Biol Chemists; Am Soc Neurochem. *Res:* Molecular level characterization of serotonin. *Mailing Add:* Neurogenetics Corp 215 College Rd Paramus NJ 07652

HARTIGAN, JOHN A, b Sydney, Austrialia, July 2, 37; m 59; c 3. MATHEMATICAL STATISTICS. *Educ:* Sydney Univ, BSc, 59, MSc, 60; Princeton Univ, PhD(math statist), 62. *Prof Exp:* Instr math, Princeton Univ, 62-63; vis lectr statist, Cambridge Univ, 63-64; asst prof, Princeton Univ, 64-66, statistician, 66-67, assoc prof statist & asst chmn dept, 67-70; assoc prof, 70-74, chmn dept, 74-80, PROF STATIST, YALE UNIV, 74- *Mem:* Inst Math Statist; Royal Statist Soc. *Res:* Statistical inference, similarity principles, error analysis. *Mailing Add:* Dept Statist PO Box 2179 Yale Univ New Haven CT 06520

HARTIGAN, MARTIN JOSEPH, b Providence, RI, Apr 20, 43; m 66; c 2. ANALYTICAL CHEMISTRY. *Educ:* Univ RI, BS, 66, PhD(chem), 71. *Prof Exp:* Sr res chemist anal chem, Ethyl Corp, 70-72; sr appl chemist gas chromatography, Perkin-Elmer Corp, Norwalk, 72-78; MKT MGR, SPECTRA-PHYSICS CORP, SAN JOSE, 78- *Mem:* Am Chem Soc; Am Soc Testing & Mat. *Res:* High resolution glass capillary columns and specific detectors for gas chromatography; microprocessor based gas chromatographs. *Mailing Add:* Spectra-Physics Corp 3333 N First St San Jose CA 95134

HARTILL, DONALD L, b Chewelah, Wash, Mar 25, 39; m 66. PHYSICS. *Educ:* Mass Inst Technol, BS, 61; Calif Inst Technol, PhD(physics), 67. *Prof Exp:* Res assoc physics, CERN, 67-68; asst prof, 68-74, assoc prof, 74-80, PROF PHYSICS, CORNELL UNIV, 80- *Concurrent Pos:* Alfred P Sloan Found res fel, 72-76. *Res:* Experimental high energy particle physics. *Mailing Add:* Dept Physics Cornell Univ Ithaca NY 14853

HARTKE, JEROME L, b Wichita, Kans, Oct 26, 32; m 56; c 3. SOLID STATE PHYSICS. *Educ:* Kans State Univ, BS, 55, MS, 56; Univ Ill, PhD(solid state physics), 61. *Prof Exp:* Scientist photoconductor res, Xerox Corp, 61-67; group leader solid state physics, Ion Physics Corp, 67-68; mgr advan develop, KEV Electronics Corp, 68-71, gen mgr, Semiconductor Div, 71-72; pres, KSW Electronics, Inc 72-75, pres, KSW Electronics Corp, 76-81; gen mgr, Semiconductor Div, Frequency Sources Inc, 81-82; CONSULT, 83- *Res:* Semiconductor devices. *Mailing Add:* 119 Fairbank Rd Sudbury MA 01776

HARTKOPF, ARLEIGH VAN, b Glendale, Calif, Apr 3, 42; m 68; c 3. POLYMER ANALYTICAL CHEMISTRY. *Educ:* Harvey Mudd Col, BS, 64; Northeastern Univ, MS, 68, PhD(anal chem), 70. *Prof Exp:* Asst prof chem, Weber State Col, 70-71 & Wayne State Univ, 71-74; staff scientist, Northeastern Univ, 74-76; sr res chemist, 76-80, ASSOC, MOBIL CHEM CO, 80- *Concurrent Pos:* Petrol Res Fund grant, Weber State Col & Wayne State Univ, 70-73. *Mem:* Am Chem Soc; Sigma Xi. *Res:* Chromatography; spectroscopy; polymer analysis. *Mailing Add:* Mobil Chem Co PO Box 3029 Edison NJ 08818-3029

HARTL, DANIEL L, b Marshfield, Wis, Jan 1, 43; m 64, 80; c 3. POPULATION GENETICS, EVOLUTIONARY BIOLOGY. *Educ:* Univ Wis, BS, 65, PhD(genet), 68. *Prof Exp:* Trainee, Univ Calif, Berkeley, 68-69; from asst prof to assoc prof genet, Univ Minn, St Paul, 69-74; from assoc prof to prof biol, Purdue Univ, Lafayette, Ind, 74-81; PROF GENET, SCH MED, WASH UNIV, ST LOUIS, MO, 81-, DEPT HEAD, 84- *Concurrent Pos:* NIH res career develop award, 75-79; assoc ed, BioSci, Am Inst Biol Sci, 75-78, Theoret Pop Biol, 76-80, Brazilian J Genet, 77-82, Genet, 80-88. *Mem:* Genetics Soc Am (pres, 89); Am Soc Study Evolution. *Res:* Genetics and population genetics; evolutionary biology. *Mailing Add:* Dept Genetics Sch Med Wash Univ 660 S Euclid Ave St Louis MO 63110

HARTLAGE, JAMES ALBERT, b New Albany, Ind, Jan 22, 38; m 64; c 3. ORGANIC CHEMISTRY, INDUSTRIAL CHEMISTRY. *Educ:* Bellarmine Col, BS, 59; Xavier Univ, MS, 61; Mich State Univ, PhD(org chem), 65. *Prof Exp:* Res chemist, ADM Chem Co, 65-69; sr res chemist, Ashland Chem Co, 69-71, mgr indust chem, 71-77; dir res & develop, 77-80, VPRES TECHNOL, STEPAN CHEM CO, 80- *Concurrent Pos:* Steering comt mem, Soap & Detergent Assoc, 78- *Mem:* Am Inst Mining, Metall & Petrol Engrs; Am Chem Soc. *Res:* Synthesis and application of anionic and cationic surfactants; a broad range of industrially useful surface active agents, especially amines and detergent sulfonates; materials having mildness and efficacy in personal case products. *Mailing Add:* 3926 Brittany Dr Northbrook IL 60062

HARTLAGE, LAWRENCE CLIFTON, b Portsmouth, Ohio, May 11, 34; m 67; c 2. NEUROPSYCHOLOGY. *Educ:* Ohio State Univ, BSc, 59; Univ Louisville, MA, 62, PhD(psychol), 68. *Prof Exp:* Prof neurol, Med Col Ga, 72-85; Marie Wilson Howell prof, Univ Ark, 85-86; MED ADV & VOC EXPERT, BUR HEARINGS & APPEALS, SSA, 73- *Concurrent Pos:* Consult, HEW, 71-, Vet Admin Hosp, 74-85, Eisenhower Army Med Ctr, 75 & US Surgeon Gen, 73-80. *Honors & Awards:* Outstanding Contrib Award, Nat Acad Neuropsychologists, 81, 85 & Am Psychol Asn, 85. *Mem:* Nat Acad Neuropsychol (pres); Sigma Xi. *Res:* Human clinical neuropsychology. *Mailing Add:* 4227 Evans Lock Rd Evans GA 30809

HARTLAND-ROWE, RICHARD C B, aquatic ecology; deceased, see previous edition for last biography

HARTLE, JAMES BURKETT, b Baltimore, Md, Aug 20, 39; m 84. PHYSICS. *Educ:* Princeton Univ, AB, 60; Calif Inst Technol, PhD(physics), 64. *Prof Exp:* Mem, Inst Advan Study, 63-64; instr physics, Princeton Univ, 64-66; from asst prof to assoc prof, 66-71, PROF PHYSICS, UNIV CALIF, SANTA BARBARA, 71- *Concurrent Pos:* Alfred P Sloan fel, 70-72. *Mem:* Nat Acad Sci; Am Astron Soc; Am Phys Soc. *Res:* Relativity; astrophysics. *Mailing Add:* Dept Physics Univ Calif Santa Barbara CA 93106

HARTLE, RICHARD EASTHAM, b Royal Oak, Mich, May 17, 36; m 59; c 2. EARTH SCIENCES, ASTROPHYSICS. *Educ:* Univ Mich, Ann Arbor, BSE, 59; Pa State Univ, PhD(physics), 64. *Prof Exp:* Nat Acad Sci-Nat Res Coun resident res assoc, Ames Res Ctr, NASA, 64-67, head, Planetary Aeronomy Br, Goddard Space Flight Ctr, 76-85, RES SCIENTIST, LAB ATMOSPHERES, GODDARD SPACE FLIGHT CTR, NASA, 67-, ASST CHIEF, 85- *Concurrent Pos:* Proj scientist, Earth Observing Syst, Goddard Space Flight Ctr, NASA, 83-87. *Mem:* Am Phys Soc; Am Geophys Union. *Res:* Electromagnetic theory; high pressure physics; solar wind; interaction of solar wind with planetary atmospheres; planetary ionospheres; atmospheric escape and evolution; plasma physics. *Mailing Add:* 11756 Morning Mist Lane Columbia MD 21044

HARTLEY, ARNOLD MANCHESTER, b Cranston, RI, Oct 14, 26; m 61; c 3. ANALYTICAL CHEMISTRY. *Educ:* Univ RI, BS, 51; Harvard Univ, AM, 53, PhD(anal chem), 55. *Prof Exp:* From instr to assoc prof anal chem, Univ Ill, Urbana, 55-78, dir environ res lab, 71-78; DIR CHEM, ENVIRON QUAL LAB INC, 78- *Mem:* Am Chem Soc. *Res:* Analytical chemistry; atomic spectrometry. *Mailing Add:* 5407 Bay Path Lane Tampa FL 33615

HARTLEY, CHARLES LEROY, b Oregon City, Ore, Sept 9, 44; m 72. FLUID PHYSICS. *Educ:* Portland State Col, BS, 66; Univ Colo, PhD(physics), 74. *Prof Exp:* Asst prof physics, Middlebury Col, 74-75; ASST PROF PHYSICS, HARTWICK COL, 75- *Mem:* Am Asn Physics Teachers. *Res:* Critical point phenomena; dynamics of the bursting of gas bubbles at a liquid gas interface. *Mailing Add:* Dept Physics Hartwick Col Oneonta NY 13820

HARTLEY, CRAIG JAY, b Spokane, Wash, Apr 2, 44; m 68; c 3. ULTRASONIC INSTRUMENTATION, IMPLANTABLE SENSORS. *Educ:* Univ Wash, Seattle, BS, 66, PhD(elec eng), 70. *Prof Exp:* Postdoctoral bioeng, Rice Univ, Houston, 70-71, res engr, 71-73; from instr to assoc prof, 73-86, PROF MED, DEPT MED, SECT CARDIOVASC SCI, BAYLOR COL MED, HOUSTON, 86- *Concurrent Pos:* Adj asst prof, Elec Eng Dept, Rice Univ, Houston, 72-73; prin investr, NIH, 76-; adj prof elec & mech eng, Univ Houston, 81- *Mem:* Inst Elec & Electronics Engrs; Am Inst Ultrasound Med; Am Physiol Soc; Cardiovasc Syst Dynamics Soc; Am Heart Asn. *Res:* Design, development, and application of ultrasonic instrumentation and sensors for cardiovascular research especially intravascular and implantable devices; methods for measuring blood flow and dimensions in animals and in man. *Mailing Add:* Dept Med Sect Cardiovasc Sci Baylor Col Med Houston TX 77030

HARTLEY, CRAIG SHERIDAN, b Quantico, Va, Dec 9, 37; m 58; c 3. MECHANICAL BEHAVIOR METALS, DEFORMATION PROCESSING METALLIC MATERIALS. *Educ:* Rensselaer Polytech Inst, BMetE, 58; Ohio State Univ, MS, 60, PhD(metall eng), 65; Univ Fla, MFA, 80. *Prof Exp:* Metallurgist, Nuclear Metals, Inc, Concord, Mass, 58-59; proj engr lieutenant, Mat Lab, USAF Ohio, 59-62, res phys metallurgist, 62-66; from asst prof to prof mat sci & eng, Univ Fla, 66-80; prof & chair mat sci & eng, State Univ NY, Stony Brook, NY, 80-82; assoc dean & prof mech eng, La State Univ & A&M Col, 82-87; prof & chair mat sci & eng, Univ Ala, Birmingham, 87-90; PROF & DEAN MECH ENG, FLA ATLANTIC UNIV, 90- *Concurrent Pos:* Postdoctoral fel, NSF-Birmingham Univ, Eng, 65-66; acad visitor, UKAEA, AERE, Harwell, 69, 73 & 79 & Imp Col Sci & Technol, 72-73; grad fel, Savannah River Lab, 70; NSF fel, Lockheed Palo Alto Res Ctr, 75 & Westinghouse Res & Develop Ctr, 76; fac fel, Sandia Lab, 80; prog dir, NSF, 86-87; chair, E&PD Comt, Mining, Metall & Mat Soc, 90-92. *Mem:* Fel Am Soc Metals; Metall Soc; Soc Exp Mech; Am Soc Eng Educ; fel AAAS. *Res:* Author or co-author of approximately 75 articles in technical journals in the areas of dislocation theory, materials processing, mechanical behavior of materials and diffusion in metals; co-edited two volumes of conference proceedings. *Mailing Add:* 788 St Albans Dr Boca Raton FL 33486-1519

HARTLEY, DANNY L, b North Little Rock, Ark, Sept 10, 41; c 2. COMBUSTION SCIENCE, ENERGY RESEARCH. *Educ:* Ga Inst Technol, BS, 63, PhD(aeroeng), 67. *Prof Exp:* Assoc engr res, Lockheed Missile & Space Corp, 63-64; res & teaching fel, Von Karman Inst, 67-68; staff mem res, 68-72, supvr aero res, 72-76, mgr combustion res, 76-81, DIR APPL RES, SANDIA NAT LABS, 81- *Concurrent Pos:* Tech rep, US Combustion Group, Int Energy Agency, 77-; mem rev comt, Lawrence Livermore Nat Labs, 79-82, NASA, 81, Lawrence Berkeley Lab, 82, Agena Control Systs, 85 & US Dept Treas, 86; mem bd dirs, Combustion Inst, 82-; mem, Cont High Temperature Sci & Tech Res Coun, 83-85. *Mem:* Am Inst Aeronaut & Astronaut; Combustion Inst. *Res:* Fluid mechanics, gasdynamics, diagnostics, computational methods, combustion and energy systems; materials and applied physics. *Mailing Add:* Organ 6000 Sandia Nat Lab PO Box 5800 Albuquerque NM 87185

HARTLEY, FRED L, chemical engineering; deceased, see previous edition for last biography

HARTLEY, HAROLD V, JR, b Lenoxville, Pa, Feb 5, 31; m 56; c 3. AUDIOLOGY, SPEECH PATHOLOGY. *Educ:* Bloomsburg State Col, BS, 52; Pa State Univ, MEd, 58; Kent State Univ, PhD(audiol), 72. *Prof Exp:* Clinician speech & hearing, Mercer County Crippled Children's Soc, 54-55 & Sharon Pub Sch, 55-62; audiologist, Pa State Univ, 62-63; chmn dept speech & hearing, 73-78, PROF AUDIOL, CLARION UNIV, 63- *Concurrent Pos:* Clinician speech & hearing, Sharon Med Ctr, 55-60; consult, Sch Med, Univ PR, San Juan, 69; consult audiol, Polk State Sch & Hosp, 69-87, Lawrence County Crippled Children's Soc, 69-71 & Easter Seal Soc, 74-75. *Mem:* Acoust Soc Am; Am Speech & Hearing Asn; Am Auditory Soc. *Res:* Physiological response to auditory signals. *Mailing Add:* Dept Speech Path & Audiol Clarion Univ Clarion PA 16214

HARTLEY, JANET WILSON, b Washington, DC, Mar 25, 28. VIROLOGY. *Educ:* Univ Md, BS, 49; George Washington Univ, MS, 51, PhD(virol), 57. *Prof Exp:* Asst bacteriologist, Am Type Cult Collection, 52-53; med bacteriologist, Lab Infectious Dis, 53-61, res med bacteriologist, 61-63, microbiologist, 63-68, RES MICROBIOLOGIST, LAB VIRAL DIS, NAT INST ALLERGY & INFECTIOUS DIS, 68- *Concurrent Pos:* Mem virol & cell biol comt, Am Cancer Soc, 72-76. *Mem:* Sigma Xi; AAAS. *Res:* Tumor virology, especially characterization of C-type murine RNA viruses, studies of their natural history, transmission, and role in disease. *Mailing Add:* 3513 Idaha Ave NW Washington DC 20016

HARTLEY, MARSHALL WENDELL, b Madison, Wis, Sept 16, 25; m 48; c 4. ZOOLOGY, PHYSIOLOGY. *Educ:* Univ Wis, BS, 50, MS, 54, PhD(zool), 59. *Prof Exp:* Asst zool, Univ Wis, 52-56, sr asst, 56-57, instr med physiol, 57-59; instr anat & path, 59-62, from asst prof to assoc prof path, 62-72, PROF PATH, MED CTR, UNIV ALA, BIRMINGHAM, 72- *Mem:* AAAS; Electron Micros Soc Am; Am Soc Cell Biol; AMA; NY Acad Sci; Sigma Xi. *Res:* Functional ultrastructural relationships; pathogenesis of renal diseases; pituitary cytology. *Mailing Add:* Dept Path Univ Sta Univ Ala Birmingham AL 35294

HARTLEY, ROBERT WILLIAM, JR, b Memphis, Tenn, Apr 9, 27; m 56; c 4. BIOPHYSICS. *Educ:* Yale Univ, BS, 49; Mass Inst Technol, PhD(biophys), 58. *Prof Exp:* Res assoc biol, Mass Inst Technol, 58-59; physicist, Nat Cancer Inst, 59-69, PHYSICIST, NAT INST ARTHRITIS & METAB DIS, 69- *Concurrent Pos:* Wellcome fel, Med Res Coun Lab Molecular Biol, Cambridge Univ, 71-72. *Mem:* Biophys Soc; Am Chem Soc; Am Soc Biol Chem; Sigma Xi. *Res:* Structure, function and synthesis of proteins and other biological macromolecules; fractionation and characterization; protein-protein interaction; protein denaturation; protein heterogeneity. *Mailing Add:* 9311 Kingsley Ave Bethesda MD 20814

HARTLINE, BEVERLY KARPLUS, b Princeton, NJ, June 13, 50; m 72; c 2. TECHNICAL MANAGEMENT. *Educ:* Reed Col, BA, 71; Univ Wash, PhD(geophys), 78. *Prof Exp:* Vis asst prof geophysics, Hampshire Col, 77-78; sci writer, Sci Mag, 78-80; phys scientist, NASA Goddard Space Flight Ctr, 80-82; sr tech writer, Lawrence Berkeley Lab, 83, sci asst planning & develop, 83-85; sci asst to dir, 85-87, asst dir info & anal, 87-89, ASSOC DIR & PROJ MGR, CONTINUOUS ELECTRON BEAM ACCELERATOR FACIL, 89- *Mem:* AAAS; Am Phys Soc; Nat Sci Teachers Asn; Proj Mgt Inst; Am Women Sci. *Res:* Project management of major scientific user facility; precollege science and math education. *Mailing Add:* Continuous Electron Beam Accelerator Facil 12000 Jefferson Ave Newport News VA 23606

HARTLINE, DANIEL KEFFER, b Philadelphia, Pa, Dec 1, 39. NEUROPHYSIOLOGY, BIOPHYSICS. *Educ:* Swarthmore Col, BA, 61; Harvard Univ, MA, 66, PhD(biol), 67. *Prof Exp:* NIH fel, Stanford Univ, 68-70; asst prof biol, Univ Calif, San Diego, 70-78; ASSOC RESEARCHER, UNIV HAWAII, 78- *Concurrent Pos:* NIH-NSF res grants, Univ Calif, San Diego, 71- *Mem:* AAAS; Soc Neurosci. *Res:* Quantitative analysis and modeling of integration in simple nerve nets, especially crustacean cardiac and stomatogastric ganglia; computer techniques in neuroscience. *Mailing Add:* Bekesy Lab Neurobiol Univ Hawaii 1993 East-West Rd Honolulu HI 96822

HARTLINE, PETER HALDAN, b Philadelphia, Pa, Jan 26, 42; m; c 2. NEUROPHYSIOLOGY, ANIMAL BEHAVIOR. *Educ:* Swarthmore Col, BA, 64; Harvard Univ, MA, 66; Univ Calif, San Diego, PhD(neurosci), 69. *Prof Exp:* Asst res neuroscientist, Univ Calif, San Diego, 69-74; asst prof physiol, Univ Ill, Urbana, 74-77; assoc scientist, Eye Res Inst Retina Found, Boston, 77-85, sr scientist, 85-91. *Concurrent Pos:* Consult, Commun Res Mach, Inc, 71-72; NIH fel, Univ Calif, San Diego, 71-74; grants, NIH & NSF, 74-89; mem, Sensory Physiol & Perception adv panel, NSF, 83-85. *Mem:* Sigma Xi; Soc Neurosci; Int Soc Neuroethol. *Res:* Information processing by neurons in sensory systems of vertebrates and invertebrates; control of animal behavior by the nervous system; cephalopods vision; infrared and visual senses in rattlesnakes; especially auditory visual integration in cats; automated control of experiments. *Mailing Add:* 28 Neptune St Beverly MA 01915

HARTLINE, RICHARD, b Reading, Pa, July 21, 32; m. BIOCHEMISTRY, ORGANIC CHEMISTRY. *Educ:* Kutztown State Col, BS, 58; Univ Ariz, MS, 61; Univ Calif, PhD(biochem), 66. *Prof Exp:* Fel biochem, Univ Ill, 66-67; ASST PROF BIOCHEM, IND UNIV, PA, 67- *Mem:* Am Chem Soc; Am Soc Microbiol; Fedn Am Soc Exp Biol; Am Soc Biol Chem. *Res:* Microbial metabolism and metabolic control; bacterial transport. *Mailing Add:* Dept Chem Ind Univ Indiana PA 15705

HARTMAN, ALBERT WILLIAM, surgery; deceased, see previous edition for last biography

HARTMAN, ARTHUR DALTON, b Mar 22, 41; m; c 1. PHARMACOLOGY, CLINICAL SCIENCE. *Educ:* Univ Tenn, PhD(physiol), 68. *Prof Exp:* Prof physiol, La State Univ Med Sch, 85-87; MED RES DEPT, CONTROLLED THERAPEUT CORP, 88- *Mem:* Am Physiol Soc; Am Med Writers Asn; NY Acad Sci. *Res:* Metabolism. *Mailing Add:* Med Res Dept Controlled Therapeut Corp One Technology Ct Malvern PA 19355

HARTMAN, BOYD KENT, b Oct 21, 39; US citizen; m 60; c 2. PSYCHIATRY, NEUROCHEMISTRY. *Educ:* Univ Kans, AB, 62, MD, 66. *Prof Exp:* Intern path, Sch Med, Wash Univ, 66-67, res psychologist, 69-72, from asst prof to prof psychol, 72-87, from assoc prof to prof neurobiol,75-87; PROF CELL BIOL & NEUROANAT, DONALD W HASTING PROF & DIR NEUROSCI PSYCHOL, UNIV MINN, 87- *Concurrent Pos:* USPHS res assoc pharmacol & toxicol, Lab Clin Biochem, Nat Heart Inst, 67-69. *Honors & Awards:* Shernd-Stanford Award, Am Soc Clin Path, 66; A E Bennett Award, Soc Biol Psychiat, 71. *Mem:* Am Psychiat Asn; Soc Neurosci; Am Soc Neurochem; Am Soc Biol Chemists; Am Soc Pharmacol & Exp Therapeut; Int Soc Cerebral Blood Flow & Metab. *Res:* Enzymes regulating catecholamine synthesis and degradation using biochemical and immunochemical methods. *Mailing Add:* Dept Psychiat Univ Minn Box 392 Mayo 628 Diehl Hall 505 Essex St SE Minneapolis MN 55455

HARTMAN, CHARLES WILLIAM, b Denver, Colo, June 30, 32. PLASMA PHYSICS, CONTROLLED FUSION RESEARCH. *Educ:* Univ Calif, Berkeley, BS, 54, MS, 55, PhD(elec eng), 60. *Prof Exp:* Engr, 60-84, PHYSICIST, LAWRENCE LIVERMORE LABS, 84- *Mem:* Fel Am Phys Soc; AAAS. *Res:* Experimental research in plasma applications on controlled magnetic fusion and acceleration. *Mailing Add:* Lawrence Rad Lab Bldg 131 Livermore CA 94550

HARTMAN, DAVID ROBERT, b Streator, Ill, Sept 17, 40; m 62; c 2. CHEMICAL EDUCATION, ENZYMOLOGY. *Educ:* NCent Col, Ill, BS, 62; Va Polytech Inst, MS, 64, PhD(biochem), 67. *Prof Exp:* PROF CHEM, WESTERN KY UNIV, 66- *Mem:* Am Chem Soc; Sigma Xi. *Res:* Economical electrophoresis apparatus-design and construction; studies on the esterase activity of the greater wax moth larvae. *Mailing Add:* Dept Chem Western Ky Univ Bowling Green KY 42101

HARTMAN, EMILY LOU, b Kansas City, Mo, Dec 19, 30. BOTANY. *Educ:* Univ Kans, BA, 53, MA, 55, PhD(bot), 57. *Prof Exp:* Instr biol sci, Calif State Polytech Col, 57-58; asst prof biol, Kans State Teachers Col, 58-61 & Southwest Mo State Col, 61-67; from asst prof to assoc prof, 67-71, PROF BIOL, UNIV COLO, DENVER, 71- *Mem:* Am Bryol & Lichenological Soc; Am Soc Plant Taxonomists; Nat Asn Biol Teachers. *Res:* Alpine tundra floristics. *Mailing Add:* Dept Biol Univ Colo 1200 Larimer Denver CO 80204

HARTMAN, FRED OSCAR, b Toledo, Ohio, Oct 20, 15; m 43, 59; c 1. POMOLOGY. *Educ:* Univ Toledo, BS, 37; Ohio State Univ, MSc, 41, PhD(hort), 51. *Prof Exp:* Asst, Ohio State Univ, 40-42, 46-48, instr, 48-51, secy plant inst, 50-53, from asst prof to assoc prof, 51-62, prof, 62-83, emer prof hort & forestry 83-; RETIRED. *Concurrent Pos:* Partic fruit prod study, Mex, 68. *Mem:* Am Soc Hort Sci; Am Pomol Soc; Int Dwarf Fruit Tree Asn; Interam Soc Tropical Hort. *Res:* Fruit setting; anatomy and morphology; propagation; rootstocks. *Mailing Add:* Dept Hort Ohio State Univ 2001 Fyffe Ct Columbus OH 43210-1096

HARTMAN, FREDERICK COOPER, b Memphis, Tenn, Aug 17, 39; m 61; c 2. BIOCHEMISTRY, PROTEIN CHEMISTRY. *Educ:* Memphis State Univ, BS, 60; Univ Tenn, Memphis, PhD(biochem), 64. *Prof Exp:* NIH fel, Univ Ill, 64-66; sect head, Molecular & Cellular Sci, 80-88, SR RES BIOCHEMIST, OAK RIDGE NAT LAB, 66-, DIR, BIOL DIV, 88- *Honors & Awards:* Pfizer Award Enzyme Chem, 79. *Mem:* AAAS; Am Soc Biol Chem; NY Acad Sci; Am Chem Soc; Sigma Xi. *Res:* Chemical modification of proteins as method of correlating structure and function; design and use of active-site specific reagents; site-directed mutagenesis. *Mailing Add:* 103 Dansworth Lane Oak Ridge TN 37830

HARTMAN, GRANT HENRY, SR, dairy science, for more information see previous edition

HARTMAN, HERMAN BERNARD, b Baltimore, Md, Jan 1, 34; m 57, 72; c 2. COMPARATIVE PHYSIOLOGY. *Educ:* Univ Md, BS, 60; Am Univ, MS, 62; Univ Conn, PhD(physiol), 66. *Prof Exp:* Asst physiol, Am Univ, 60-62 & Univ Conn, 62-64; fel entom, US Army Natick Labs, 65-67; asst prof zool, Univ Iowa, 67-72; asst prof biol, Univ Maine, 73-75; assoc prof, Tex Tech Univ, 75-86, prof biol, 87-89; CHMN, DUQUESNE UNIV, 90- *Concurrent Pos:* Mem, Corp Marine Biol Lab, Woodshole, Mass; fel, Inst Cellular Biol, Univ Conn, 64-65; vis scientist, Ore Inst Marine Biol, 73-75, 84- *Mem:* AAAS; Am Soc Zool; Am Physiol Soc; Sigma Xi; Soc Neurosci; Crustacean Soc. *Res:* Stridulation by cockroaches; physiology of arthropod movement, position, and tension receptors; neurophysiology of simple behaviors; equilibrium reception insects; regeneration of limbs. *Mailing Add:* Dept Biol Sci Dequesne Univ Pittsburgh PA 15282

HARTMAN, ICLAL SIREL, b Elazig, Turkey, Dec 22, 30; US citizen; m 57; c 2. BIOCHEMISTRY. *Educ:* Mt Holyoke Col, AB, 50, MA, 51; Univ Fla, PhD(biochem), 63. *Prof Exp:* Head qual control chem anal, E R Squibb & Sons, 51-55; instr chem, Regis Col, Mass, 57-58 & Wellesley Col, 58-59; from instr to assoc prof, 59-74, PROF CHEM, SIMMONS COL, 74- *Concurrent Pos:* Chair, Chem Dept, Simmons Col, 85- *Mem:* Am Chem Soc. *Res:* Steroids, gas chromatographic analysis; biochemical basis of folk medicine. *Mailing Add:* Dept Chem Simmons Col 300 Fenway Boston MA 02115

HARTMAN, JAMES AUSTIN, b Lanark, Ill, Jan 29, 28; m 51; c 2. GEOLOGY. *Educ:* Beloit Col, BS, 51; Univ Wis, MS, 55, PhD(geol), 57. *Prof Exp:* Resident geologist, Reynolds Jamaica Mines, Ltd, 51-53; geologist, Union Carbide Ore Co, 56-57; exploitation engr, Shell Oil Co, 57-64, Sr prod geologist, Shell Oil Co, 64-65 & Shell Develop Co, 65-66, staff prod geologist, Shell Oil Co, 66-68, div exploitation engr, 68-70, regional geol engr, southern Region, 70-76, sr staff geol engr, 71-80, geol engr adv, 80-81, geol engr consult, 81-86, field syudies task force leader, 76-86; INDEPENDENT GEOLOGIST, 86- *Honors & Awards:* Distinguished Serv Award, Am Asn Petrol Geologists, 85. *Mem:* Sigma Xi; Am Asn Petrol Geologists (secy, 81-83); Soc Petrol Engrs; fel Geol Soc Am; Am Inst Prof Geologists. *Res:* Origin of Jamaican bauxite; titanium mineralogy of bauxites; salt dome research; Gulf Coast oil and gas fields. *Mailing Add:* 4512 Newlands St Metairie LA 70006

HARTMAN, JAMES KEITH, b Norristown, Pa, Feb 13, 43; div; c 2. ELECTROMAGNETISM, MODELING & SIMULATION. *Educ:* Cornell Univ, BS, 65, ME, 67, PhD(appl physics), 70. *Prof Exp:* Mem tech staff electronic eng, Hughes Aircraft Co, 65-66; staff engr, IBM Res Lab, 70-71; asst prof physics, Broome Community Col, 71-74, chmn dept physics & eng sci, 74-75; coordr pre-eng prog, Canisius Col, 75-78; mem tech staff, Sandia Labs, 78-81; dir freshman eng, Univ NMex, 81-83; assoc prof, State Univ NY, Binghamton, 83-85; CONSULT, 85- *Mem:* Am Phys Soc; Inst Elec Electronics Engrs. *Res:* Materials science and energy conversion; electromagnetism; software design. *Mailing Add:* PO Box 13916 Albuquerque NM 87192-3916

HARTMAN, JAMES KERN, operations research, applied mathematics; deceased, see previous edition for last biography

HARTMAN, JAMES T, b DeRidder, La, June 13, 25; m 54; c 3. MEDICINE, ORTHOPEDIC SURGERY. *Educ:* Iowa State Univ, BS, 49; Northwestern Univ, BM, 49, MD, 52. *Prof Exp:* Resident & instr orthop surg, Med Sch, Univ Mich, 53-57; registr, Nuffield Orthop Ctr, Oxford Univ, 57-58; instr, Univ Mich, 58-61; staff mem, Cleveland Clin, 61-68; assoc prof, Northwestern Univ, 68-71; prof orthop surg & chmn dept, Sch Med, Tex Tech Univ Health Sci, 71-81, dean, Sch Med, 81-88; RETIRED. *Concurrent Pos:* Chmn, Dept Orthop Surg, Cook County Hosp, Chicago, 68-71. *Mem:* Am Orthop Asn; Am Acad Orthop Surg; Am Col Surg; Asn Bone & Joint Surg; Clin Orthop Soc. *Res:* Correlation of soft tissue injury of the cervical spine and abnormalities of movement with cineradiography. *Mailing Add:* Med Sch Tex Tech Univ Health Sci Lubbock TX 79430

HARTMAN, JAMES XAVIER, b Chicago, Ill, Sept 28, 42; m 66; c 2. MICROBIOLOGY, PHYTOPATHOLOGY. *Educ:* Aquinas Col, BS, 64; Mich State Univ, MS, 67, PhD(phytopath), 71. *Prof Exp:* Asst prof, 71-77, assoc prof, 77-80, PROF MICROBIOL, FLA ATLANTIC UNIV, 80- *Mem:* Am Phytopath Soc. *Res:* Plant viruses-infection of plant tissue culture cells and protoplasts. *Mailing Add:* Dept Biol Sci Fla Atlantic Univ Boca Raton FL 33431

HARTMAN, JOHN ALAN, b Toledo, Ohio, Feb 4, 20; m 44. ORGANIC CHEMISTRY. *Educ:* Univ Toledo, BS, 46; Wayne State Univ, MS, 55, PhD(org chem), 63. *Prof Exp:* Sr technician, Henry Ford Hosp, Detroit, Mich, 46-48; res asst org chem, Detroit Inst Cancer Res, 49-56; sr chemist, Grass Lakes Platers, Mich, 56-57; res chemist, Metal & Thermit Corp, Mich, 57-60; asst org chem, Wayne State Univ, 60-61; asst prof, Wheeling Col, 63-67; assoc prof, 67-70, chmn dept, 72-77, PROF ORG CHEM, MANSFIELD STATE COL, 70- *Concurrent Pos:* NSF res fel, 65-67. *Mem:* Am Chem Soc; Sigma Xi. *Res:* Synthesis of terpeneoid natural products; electro organic chemistry. *Mailing Add:* 806 William Penn Ct Pittsburgh PA 15221

HARTMAN, JOHN DAVID, b Peninsula, Ohio, 22; m 44; c 3. NEUROPATHOLOGY, CARDIOVASCULAR PATHOLOGY. *Educ:* Case Western Reserve Univ, MD, 46. *Prof Exp:* SR ATTEND PATHOLOGIST, ALBERT EINSTEIN MED CTR, 81- *Mem:* Am Asn Path; Am Heart Asn. *Res:* Cardiovascular pathology. *Mailing Add:* Dept Path Albert Einstein Med Ctr Northern Div Philadelphia PA 19141

HARTMAN, JOHN L(OUIS), b Elgin, Ill, Aug 7, 30; m 57; c 4. FLUID MECHANICS, HEAT TRANSFER. *Educ:* Purdue Univ, BS, 52, MS, 53, PhD, 56. *Prof Exp:* Sr engr, Aircraft Nuclear Propulsion Dept, Gen Elec Co, 55-57; mem tech staff rocket propulsion, Ramo-Wooldridge Corp, 57-58; staff mem nuclear propulsion, Los Alamos Sci Lab, 58-59; systs planning engr, Allison Div, 59, asst nuclear eng, 60-63, chief advan power systs, 63-67, head elec power plants dept, Res Labs, 67-69, head electrochem dept, 69-85, FACIL MGR, RES LABS, GEN MOTORS CORP, 85- *Mem:* Am Mgt Asn; Am Inst Plant Engrs; Int Facil Mgt Asn. *Res:* Energy conversion research and development of power propulsion systems; specific emphasis directed upon advanced electrochemical energy conversion for future vehicular applications. *Mailing Add:* 688 Balfour Rd Grosse Pointe Park MI 48230

HARTMAN, JOHN PAUL, b Glasgow, Mont, Aug 15, 36; m 61; c 2. ENGINEERING EDUCATION, ENGINEERING HISTORY. *Educ:* Principia Col, BS, 57; Washington Univ, St Louis, BSCE, 59; Harvard Univ, SM, 60; Univ Fla, PhD(civil eng), 74. *Prof Exp:* Instr heat transfer, US Naval Nuclear Power Sch, 61-65; soils engr civil eng, Ardaman & Assocs, 65; sr engr nuclear effects, Martin-Marietta Co, 65-68; from asst prof to assoc prof eng, 68-76, asst dean eng, 78-89, PROF ENG, UNIV CENT FLA, 76- *Concurrent Pos:* NSF res grant, 75-77; bd dir, Am Soc Eng Educ, 81-83, vpres Prof Interest Couns, 82-83, Nat Hist & Heritage Comt, Am Soc Mech Engrs, 76- *Honors & Awards:* Western Elec Fund Award, Am Soc Eng Educ, 81. *Mem:* Fel Am Soc Civil Engrs; fel Am Soc Mech Engrs; Nat Soc Prof Engrs; Soc Hist Technol; Am Soc Eng Educ; Soc Indust Archeol. *Res:* Engineering and industrial history; impact of engineering and technology on society; geotechnical engineering. *Mailing Add:* Col Eng Univ Cent Fla Orlando FL 32816-0450

HARTMAN, JOHN STEPHEN, b Toronto, Ont, July 18, 38; m 71; c 2. INORGANIC CHEMISTRY. *Educ:* Queen's Univ, Ont, BSc, 61; Univ Ottawa, MSc, 63; McMaster Univ, PhD(inorg chem), 67. *Prof Exp:* Nat Res Coun fel, Phys Chem Lab, Oxford Univ, 67-68; from asst prof to assoc prof 68-81, dept chmn, 85-88, PROF INORG CHEM, BROCK UNIV, 81- *Mem:* Am Chem Soc; Chem Inst Can; Royal Soc Chem. *Res:* Nuclear magnetic resonance spectroscopy applied to coordination chemistry; high-resolution solid-state nuclear magnetic resonance spectroscopy with magic angle spinning, applied to inorganic solids and minerals. *Mailing Add:* Dept Chem Brock Univ St Catharines ON L2S 3A1 Can

HARTMAN, KARL AUGUST, b Wilmington, Del, Nov 30, 35; m 61; c 3. PHYSICAL BIOCHEMISTRY. *Educ:* Lehigh Univ, BS, 58; Mass Inst Technol, PhD(phys chem), 62. *Prof Exp:* Res assoc chem, Mass Inst Technol, 62-63; res chemist, E I du Pont de Nemours & Co, Inc, Del, 63-67; from asst prof to assoc prof, 67-76, PROF BIOPHYS, UNIV RI, 76- *Mem:* Am Chem Soc; Biophys Soc; Am Asn Univ Professors. *Res:* Structure, stability and interaction of nucleic acids, proteins and viruses; infrared and Raman spectroscopy. *Mailing Add:* Dept Biochem & Biophys Univ RI Kingston RI 02881-0812

HARTMAN, KENNETH EUGENE, b Johnstown, Pa, Dec 6, 42; m 67. ORGANIC CHEMISTRY, MEDICINAL CHEMISTRY. *Educ:* Geneva Col, BS, 63; Univ Pa, PhD(chem), 69. *Prof Exp:* From instr to asst prof, 68-74, assoc prof, 74-79, PROF CHEM, GENEVA COL, 80- *Mem:* Am Chem Soc. *Res:* Synthesis of isoquinoline antimalarials; group shift effects in nuclear magnetic resonance spectrometry; aromatic substitution patterns by infrared spectroscopy; reactions of quinuclidine and triethylenediammine. *Mailing Add:* Dept Chem Geneva Col College Ave Beaver Falls PA 15010

HARTMAN, KENNETH OWEN, b Philadelphia, Pa, Oct 20, 39; m 65; c 2. PHYSICAL CHEMISTRY. *Educ:* Lehigh Univ, BS, 61; Pa State Univ, PhD(chem), 65. *Prof Exp:* Fel chem, Mellon Inst, 65-67; sr res chemist, Technol Mkt, 67-73, supvr, 73-78, supt, 78-84, sr scientist, 85-88, mgr, 88-90, TECH DIR, TECHNOL MKT, 91- *Mem:* Am Chem Soc. *Res:* Kinetics and mechanisms of solid state and combustion reactions; matrix isolation of transient species; urethane polymerization kinetics; thermal stability of propellants; explosives development; jet plasma characterization; infrared and Raman spectroscopy; transport properties; polymer and slurry rheology; synthesis of energetic materials; insensitive munitions. *Mailing Add:* 609 N Second St Lavale MD 21502

HARTMAN, MARVIS EDGAR, b Oshkosh, Wis, Sept 30, 40; m 61; c 4. POLYMER CHEMISTRY, ORGANIC CHEMISTRY. *Educ:* Univ Wis-Madison, BS, 63; Carnegie-Mellon Univ, PhD(org chem), 72. *Prof Exp:* Scientist, PPG Indust Inc, 63-81. *Mem:* Am Chem Soc. *Res:* Synthesis of new polymers for coatings; functional group manipulations on polymers, particularly acrylics, polyesters and polyarethanes; synthesis of polymers containing unique functionality. *Mailing Add:* 144 Rawlins Run Rd Pittsburgh PA 15238

HARTMAN, MARY ELLEN, histology, for more information see previous edition

HARTMAN, NILE FRANKLIN, b New Marshfield, Ohio, Nov 24, 39; m 65; c 1. INTEGRATED OPTICS, FIBER OPTICS. *Educ:* Ohio Univ, BS, 61. *Prof Exp:* Sr res scientist, Columbus Labs, Battelle Mem Inst, 61-86, SR RES SCIENTIST, GA TECH RES INST, 86- *Honors & Awards:* IR-100 Award, 82. *Mem:* Optical Soc Am; Soc Photo-Optical Instrumentation Engrs. *Res:* Sensor systems; optical inspect and sensing systems. *Mailing Add:* 5595 Southern Pines Ct Stone Mountain GA 30087-5262

HARTMAN, PATRICK JAMES, b Ann Arbor, Mich, Dec 5, 44; m 68; c 2. STOCHASTIC SIMULATION, SYSTEMS DESIGN & SYSTEMS SCIENCE. *Educ:* Marquette Univ, BME, 68; Univ RI, MS, 74, PhD(eng), 76. *Prof Exp:* Mech engr, US Coast Guard, Washington, DC, 68-72; researcher eng, Univ RI, 72-76; res engr, E I du Pont de Nemours, 76-79; sr ocean engr, Gould, Inc, 79-80; SR MECH ENGR, US NAVY, WASHINGTON, DC, 80-; ADJ PROF, GEORGE WASHINGTON UNIV, FAIRFAX, VA, 91- *Concurrent Pos:* Officer, US Coast Guard Reserve, 69-72; bd gov cert, Am Soc Mech Engrs, 85. *Honors & Awards:* Bausch & Lomb Sci Award, 63; Gold Cert, Am Soc Mech Engrs, Ocean Eng Div, 85. *Mem:* Am Soc Mech Engrs; Nat Soc Prof Engrs; USN Asn Scientists & Engrs; Soc Reliability Engrs. *Res:* Efficient software tailored to increased speed of simulation of very large reliability and maintainability systems; artificial intelligence and expert systems; increased precision in stochastic simulations. *Mailing Add:* Naval Sea Systs Command Code 5121 Washington DC 70367

HARTMAN, PAUL ARTHUR, b Baltimore, Md, Nov 23, 26; m 50; c 3. FOOD & AGRICULTURAL MICROBIOLOGY. *Educ:* Univ Ill, BS, 49; Univ Ala, MS, 51; Purdue Univ, PhD(bact), 54. *Prof Exp:* From asst prof to prof bact, 54-72, chmn, 74-81, distinguished prof sci & humanities, 72-90, EMER DISTINGUISHED PROF SCI & HUMANITIES, IOWA STATE UNIV, 90- *Concurrent Pos:* Chmn, ann meeting prog comt, Am Soc Microbiol, 85-88, Meetings Bd, Am Soc Microbiol, 88-91; consult. *Mem:* Am Soc Microbiol; Inst Food Technol; Soc Indust Microbiol; fel Am Acad Microbiol; Int Asn Milk, Food & Environ Sanit; Soc Gen Microbiol. *Res:* Agricultural and industrial microbiology; microbiology of foods; coliforms; enterococci; salmonellae. *Mailing Add:* Dept Microbiol Iowa State Univ 205 Sci I Ames IA 50011-3211

HARTMAN, PAUL LEON, b Reno, Nev, July 13, 13; m 41; c 3. EXPERIMENTAL PHYSICS. *Educ:* Univ Nev, BS, 34; Cornell Univ, PhD(physics), 38. *Prof Exp:* Instr physics, Cornell Univ, 38-39; mem tech staff, Bell Tel Lab, 39-46; prof physics, 46-83, assoc dir appl & eng physics, 72-78, EMER PROF PHYSICS, CORNELL UNIV, 83- *Concurrent Pos:* Vis staff mem, Los Alamos Sci Lab, NMex, 60-61, 66-67 & 73-74. *Mem:* Am Phys Soc. *Res:* Physics; magnetrons; vacuum ultraviolet; synchrotron radiation optics; alkali halides; air fluorescence. *Mailing Add:* Dept Physics & Eng Physics Cornell Univ Ithaca NY 14853

HARTMAN, PHILIP EMIL, b Baltimore, Md, Nov 23, 26; m 55; c 3. MICROBIOLOGY. *Educ:* Univ Ill, BS, 49; Univ Pa, PhD(med microbiol), 53. *Prof Exp:* Instr med microbiol, Univ Pa, 53-54; NIH fel, Carnegie Inst, 54-55; at dept bact & immunol, Harvard Med Sch, 55-56; Am Cancer Soc fel, Animal Morphol Lab, Univ Brussels, 56-57; from asst prof to assoc prof, 57-65, PROF BIOL, JOHNS HOPKINS UNIV, 65-, WILLIAM D GILL PROF BIOL, 75- *Honors & Awards:* Ann Award, Environ Mutagen Soc, 85. *Mem:* Am Soc Photobiol; Environ Mutagen Soc. *Res:* Bacterial genetics; gene action; chemical mutagenesis; antimutagens. *Mailing Add:* Dept Biol Johns Hopkins Univ Baltimore MD 21218

HARTMAN, RICHARD LEON, b Pittsburgh, Pa, July 5, 37; m 64; c 2. OPTICAL PHYSICS, SOLID STATE PHYSICS. *Educ:* Carnegie Inst Technol, BS, 58, MS, 61, PhD(physics), 65; Mass Inst Technol, SM, 73. *Prof Exp:* Res physicist, Phys Sci Lab, 65-74, actg dir, Aeroballistics Directorate, 74-75, dir, Phys Sci Directorate, 76-77, DIR, ARMY MISSILE LAB, REDSTONE ARSENAL, 77- *Concurrent Pos:* Sloan Found fel, Mass Inst Technol, 72. *Honors & Awards:* Army Res & Develop Achievement Award, 69. *Mem:* AAAS; Am Phys Soc; Optical Soc Am; Soc Photo-Optical Instrumentation Engrs; Sigma Xi. *Res:* Spin lattice relaxation; electron paramagnetic resonance; nuclear quadrupole and magnetic resonance; optical spectroscopy; semiconductors; holography; optical data processing; laser radar; submillimeter physics. *Mailing Add:* Bell Tel Labs Rm 2C-313 Murray Hill NJ 07974

HARTMAN, RICHARD THOMAS, b Sligo, Pa, Feb 10, 21; m 44; c 2. PLANT ECOLOGY. *Educ:* Clarion State Col, BS, 42; Pittsburgh Univ, MS, 47, PhD(bot), 52. *Prof Exp:* Teacher pub sch, Pa, 42-43, 45-46; asst biol sci, 47-48, instr bot, 48-52, from asst prof to assoc prof, 52-66, PROF BIOL, UNIV PITTSBURGH, 66-, DIR PYMATUNING LAB ECOL, 73- *Concurrent Pos:* Res assoc, Carnegie Mus Natural Hist, 65-; acad dean, semester at sea, 85. *Mem:* AAAS; Ecol Soc Am; Orgn Biol Field Stas; Am Soc Limnol & Oceanog; Am Inst Biol Sci. *Res:* Physiological ecology of algae and vascular aquatic plants; fresh water algal communities; eutrophication processes in lakes. *Mailing Add:* Dept Biol Sci Crawford 162 Univ Pittsburgh 4200 Fifth Ave Pittsburgh PA 15260

HARTMAN, ROBERT CHARLES, b Bartlesville, Okla, Aug 24, 38; m 60; c 3. HIGH ENERGY ASTROPHYSICS. *Educ:* Rice Inst, BA, 60; Univ Chicago, SM, 62, PhD(physics), 67. *Prof Exp:* Res assoc space sci, Enrico Fermi Inst, Univ Chicago, 67-68; Nat Acad Sci res assoc, 68-69, ASTROPHYSICIST, GODDARD SPACE FLIGHT CTR, NASA, 69- *Mem:* Am Astron Soc; Am Phys Soc. *Res:* Primary cosmic ray electron-positron ratio and fluxes; high energy gamma ray astronomy. *Mailing Add:* Code 662 Goddard Space Flight Ctr NASA Greenbelt MD 20771

HARTMAN, ROBERT JOHN, b Ft Wayne, Ind, July 20, 26; m 54; c 5. INDUSTRIAL ORGANIC CHEMISTRY. *Educ:* De Paul Univ, BS, 52; Wayne State Univ, MS, 54. *Prof Exp:* Staff chemist, Basf Wyandotte Corp, 54-55, res chemist, 55-64, sr res chemist, 64-83, res assoc, Eurethane Chem Res, 83-88. *Mem:* Am Chem Soc. *Res:* Synthesis of high energy compounds for rocket propellant and explosives; carbohydrate chemistry; synthesis of organo-phosphorus compounds and polymers; polymerization of alkylene oxides, ethylenimine and polymers; polyether polyol synthesis. *Mailing Add:* 14676 Yorkshire Southgate MI 48195

HARTMAN, ROGER DUANE, b Kansas City, Mo, Nov 4, 35; m 55; c 1. SOLID STATE PHYSICS, MEDICAL PHYSICS. *Educ:* William Jewell Col, AB, 58; Univ Ark, MS, 60; Okla State Univ, PhD(physics), 67. *Prof Exp:* Asst physics, Univ Ark, 58-60; instr phys sci, Evangel Col, 60-61; instr physics, Univ Ark, 61-62; from asst prof to assoc prof, Univ Tulsa, 62-72; vpres med & regulatory affairs, Tech Eval & Mgt Systs, Inc, 86-88; chmn, Nat Sci Div, Sch Med, Oral Roberts Univ, 73-77, prof physics, 72-78, dir, Off Res & Grants, 73-78, dir, Res & Develop, 78-81, prof radiol & assoc dean res grants, 81-86, PROF PHYSICS, SCH MED, ORAL ROBERTS UNIV, 88-; PRES, ESTHART RES & DEVELOP CORP, 88- *Concurrent Pos:* consult, Los Alamos Sci Lab, 62-66; NSF Sci Fac fel, 65-66. *Mem:* Am Phys Soc; Optical Soc Am; Nat Coun Univ Res Adminr; Am Asn Physicists Med; Soc Res Adminr (pres elect, 79-80, pres, 80-81). *Res:* Molecular solids; optical and electrical properties of materials; radiation biology; biomedical engineering; medical ethics; medical physics; radiation therapy; medical and regulatory clinical trials. *Mailing Add:* Eshart Enterprises 2112 S Norfolk Tulsa OK 74114

HARTMAN, STANDISH CHARD, b Philadelphia, Pa, Nov 24, 31; m 57, 78; c 4. BIOCHEMISTRY. *Educ:* Mass Inst Technol, SB & SM, 54, PhD(biochem), 57. *Prof Exp:* Instr biochem, Mass Inst Technol, 57-59; asst prof, Harvard Med Sch, 59-68; assoc prof, 68-73, PROF CHEM, BOSTON UNIV, 73- *Concurrent Pos:* USPHS develop award, 60-68. *Mem:* Am Soc Biol Chem. *Res:* Enzymic reactions of purine biosynthesis; mechanisms of enzyme action. *Mailing Add:* Dept Chem Boston Univ 590 Commonwealth Ave Boston MA 02215

HARTMAN, WARREN EMERY, b Hawthorne, Calif, May 12, 14; m 40; c 3. FOOD TECHNOLOGY, PROTEIN STRUCTURING. *Educ:* Andrews Univ, BA, 40; Univ Mich, Ann Arbor, MS, 45. *Prof Exp:* Bacteriologist, Mich State Dept Health, 40-42, anal chemist, 42-44; instr lab pract, USPHS, 45-46; dir res, 46-60, VPRES RES & DEVELOP, WORTHINGTON FOODS, INC, 60- *Concurrent Pos:* Consult, USAID, 60-70; mem, Protein Technol Adv Comt, White House Conf Food & Nutrit, 70. *Mem:* AAAS; Inst Food Technol; Am Asn Cereal Chemists; AMA. *Mailing Add:* 1046 Morning St Worthington OH 43085

HARTMAN, WILBUR LEE, b Northampton, Mass, June 22, 31; m 68; c 4. FISHERY SCIENCE, LIMNOLOGY. *Educ:* Amherst Col, BA, 52, MA, 53; Cornell Univ, PhD(fishery biol), 58. *Prof Exp:* Fishery aide biol, NY Dept Conserv, Cornell Univ, 54-58; fishery scientist, US Bur Com Fisheries, 58-68, supvry fishery scientist, 68-71; supvry fishery scientist biol, US Fish & Wildlife Serv, 71-89. *Concurrent Pos:* Adj assoc prof, Ohio State Univ, 69-74; sci adv, Ohio Gov's Task Force Fishery Mgt, 73-74. *Mem:* Am Fisheries Soc; Am Soc Limnol & Oceanog. *Res:* Dynamics Great Lakes fish resources; biology and behavior of fishes; fishery management; environmental stresses on fish stocks; limnology oligotrophic lakes. *Mailing Add:* 1150 Dunlavy Lane Dexter MI 48130

HARTMAN, WILLARD DANIEL, b Cincinnati, Ohio, Sept 26, 21; m 60. INVERTEBRATE ZOOLOGY. *Educ:* Yale Univ, BS, 43, MS, 44, PhD(zool), 50. *Prof Exp:* Lab asst biol, Yale Univ, 43-44 & 46-49; instr zool, Univ Calif, 50-53; from asst prof & assoc cur to assoc prof & cur, 53-73, dir biol grad studies, 76-79, PROF BIOL & CUR INVERT ZOOL, YALE UNIV, 73- *Concurrent Pos:* Mem, Yale Univ Seychelles Exped, 57-58; actg dir, Peabody Museum, 87-89. *Mem:* Am Soc Zool; Soc Syst Zool; Am Soc Limnol & Oceanog; Marine Biol Asn UK; Sigma Xi; Int Soc Reef Studies. *Res:* Systematics, ecology and evolution of sponges. *Mailing Add:* Peabody Mus Natural Hist Yale Univ PO Box 6666 New Haven CT 06511-8161

HARTMANIS, JURIS, b Riga, Latvia, July 5, 28; nat US; m 59; c 3. MATHEMATICS, COMPUTER SCIENCE. *Educ:* Univ Marburg, Cand Phil, 49; Univ Kansas City, MA, 51; Calif Inst Technol, PhD(math), 55. *Prof Exp:* Asst math, Calif Inst Technol, 51-55; instr, Cornell Univ, 55-57; asst prof, Ohio State Univ, 57-58; res scientist, Res Lab, Gen Elec Co, 58-65; chmn dept, 65-71, prof comput sci, 65-70, WALTER R READ PROF ENG, CORNELL UNIV, 81- *Concurrent Pos:* Ed, J Comput & Systs Sci, 69 & Siam J Comput, 72- *Mem:* Nat Acad Eng; Math Asn Am; Asn Comput Mach; Am Math Soc; NY Acad Sci. *Res:* Electronic computers; automata theory; theory of computing. *Mailing Add:* Upson Hall Cornell Univ Ithaca NY 14853

HARTMANN, ALOIS J(OSEPH), b Cincinnati, Ohio, Mar 16, 36; m 62; c 2. CIVIL ENGINEERING. *Educ:* Manhattan Col, BCE, 58; Lehigh Univ, MS, 60; Univ Ill, PhD(struct), 64. *Prof Exp:* Ford of Can fel, Univ Western Ont, 64-65, asst prof eng sci, 65-66; asst prof, 66-71, assoc prof civil eng, Marquette Univ, 71-74; sr engr, 74-84, FEL ENG, WESTINGHOUSE ELEC CORP, 84- *Mem:* Am Soc Civil Engrs; Am Concrete Inst; Am Soc Mech Engrs; Sigma Xi; Struct Stability Res Coun. *Res:* Stability of structures; dynamics of structures; structural design; computer code development. *Mailing Add:* 1161 Colgate Dr Monroeville PA 15146

HARTMANN, BRUCE, b St Louis, Mo, June 30, 38; m 62; c 3. POLYMER PHYSICS, MATERIALS SCIENCE ENGINEERING. *Educ:* Cath Univ Am, AB, 60; Univ Md, College Park, MS, 66; Am Univ, PhD(physics), 71. *Prof Exp:* Mem staff physics, 60-75, HEAD POLYMER PHYSICS GROUP, NAVAL SURFACE WARFARE CTR, 75- *Mem:* Am Phys Soc; Soc Rheology; Acoust Soc Am. *Res:* Experimental and theoretical structure-property relations in polymers. *Mailing Add:* 10614 Dunkirk Dr Silver Spring MD 20902

HARTMANN, DENNIS LEE, b Salem, Ore, Apr 23, 49; m; c 2. ATMOSPHERIC SCIENCES, CLIMATOLOGY. *Educ:* Univ Portland, BS, 71; Princeton Univ, MA, 73, PhD(geophys & fluid dynamics), 75. *Prof Exp:* Res assoc meteorol, McGill Univ, 75-76; vis scientist, Nat Ctr Atmospheric Res, 76-77; asst prof, 77-83, ASSOC PROF ATMOSPHERIC SCI, UNIV WASH, 83- *Concurrent Pos:* Assoc ed, J Atmospheric Sci & J Geophys Res: Athmospheres. *Mem:* Am Meteorol Soc; AAAS. *Res:* Stratospheric dynamics and chemistry; theoretical climatology. *Mailing Add:* Dept Atmospheric Sci AK-40 Univ Wash Seattle WA 98195

HARTMANN, ERNEST LOUIS, b Vienna, Austria, Feb 25, 34; US citizen; m 61; c 2. SLEEP & SLEEP DISORDERS, DREAMING. *Educ:* Univ Chicago, AB, 52; Yale Univ, MD, 58. *Prof Exp:* Intern med, Bronx Munic Hosp Ctr, Albert Einstein Col Med, 59-60; resident psychiat, Mass Ment Health Ctr, Boston, 60-62; clin assoc, NIMH, 62-64, career investr, 64-69; dir sleep lab, Boston State Hosp, 64-80; PROF PSYCHIAT, SCH MED, TUFTS UNIV, 75-; SR PSYCHIATRIST & DIR SLEEP LAB, WEST-ROS-PARK MENT HEALTH CTR, 78-; DIR SLEEP RES LAB, LEMUEL SHATTUCK HOSP, 80-; DIR, SLEEP DIS CTR, NEWTON WELLESLEY HOSP, 84- *Concurrent Pos:* Am Cancer Soc fel, Gustave Roussy Res Inst, Univ Paris, 58-59; teaching fel, Harvard Med Sch, 60-61 & res fel, 61-62; consult, Mass Dept Prisons, 61-62; prin investr, grants on sleep, 64-69, 66-72, 74-79, 79-82, 82-85 & 85- & grants on schizophrenia, 79-82; chmn clin res rev comt, Boston State Hosp, 67-76; lectr psychiat, Sch Med, Boston Univ, 70-; pvt pract, psychiat & sleep dis, 70-; consult, panel mem, Food & Drug Admin, 72-76; consult ed, Psychopharmacol, McLean J & assoc ed, Sem Psychiat; assoc dir sleep clinic, Peter Bent Brigham Hosp, 73-; fac mem, Boston Psychoanal Inst, 75-; dir, Sleep Dis Ctr, Newton-Wellesley Hosp, 84-; ed-in-chief, Dreaming. *Honors & Awards:* Holt Bk Prize, Yale Univ, 56; A E Bennett Award Psychiat Res, 66; First Prize Psychopharmacol, Am Psychol Asn, 71. *Mem:* AAAS; fel Am Psychiat Asn; Sleep Res Soc; Soc Neurosci; fel Am Col Neuropsychopharmacol; Psychiat Res Soc; Asn Study Dreams (pres); Am Sleep Dis Asn; AAAS. *Res:* Sleep; dreaming; sleep disorders; schizophrenia; mind-brain relationships; author of 8 books and over 260 scientific articles and abstracts. *Mailing Add:* Sleep Res Lab Tufts Lemuel Shattuck Hosp 170 Morton St Boston MA 02130

HARTMANN, FLOYD WELLINGTON, bacteriology; deceased, see previous edition for last biography

HARTMANN, FRANCIS XAVIER, b New York, NY, Nov 5, 53. NEUTRINO DETECTION, LASER EFFECTS IN NUCLEAR SYSTEMS. *Educ:* US Naval Acad, BS, 75; Princeton Univ, MA, 83, PhD(chem), 85. *Prof Exp:* Res assoc physics, Princeton Univ, 85; staff scientist physics, Inst Defense Analyses, 85-89; asst chemist, 89-91, ASSOC CHEMIST, BROOKHAVEN NAT LAB, 91- *Concurrent Pos:* Vis scientist, Brookhaven Nat Lab, 83-85, Gran Sasso Nat Lab, INFN, Italy, 89-, Ben-Gurion Univ, Negev, Israel, 90-; prin investr, Solar Neutrino Proj, Brookhaven Nat Lab, 89- *Mem:* Sigma Xi; Am Phys Soc. *Res:* Experimental work in neutrino physics; low energy nuclear reactions; nuclear chemistry; laser-nuclear interactions;

nuclear structure; solid state energy transfer processes and investigations of nuclear superradiance, search for solar neutrinos; underground low level nuclear detection. *Mailing Add:* Chem Dept Brookhaven Nat Lab Upton NY 11973

HARTMANN, FREDERICK W, b Repton, Ala, Mar 29, 40; m 67; c 2. MATHEMATICS. *Educ:* Lehigh Univ, AB, 62, PhD(math), 68; Univ Pa, MA, 64. *Prof Exp:* From asst prof to assoc prof, 65-83, PROF MATH, VILLANOVA UNIV, 83-, CHMN DEPT, 81- *Concurrent Pos:* Vis lectr, Univ Western Australia, 71. *Mem:* Am Math Soc. *Res:* Mathematical analysis, summability and complex function theory. *Mailing Add:* Dept Math Villanova Univ Villanova PA 19085

HARTMANN, GEORGE CHARLES, b New Bedford, Mass, Mar 16, 27; m 49; c 3. MYCOLOGY. *Educ:* Harvard Univ, AB, 50, MA, 51; Univ RI, PhD(mycol), 63. *Prof Exp:* Prof, 58-91, EMER PROF BIOL, RI COL, 91- *Concurrent Pos:* NSF sci fac fel bot, Univ Mich, 68-69. *Mem:* Mycol Soc Am. *Res:* Entomogenous fungi; fungal cytology. *Mailing Add:* 11 Powder Mill Lane Greenville RI 02828

HARTMANN, GEORGE COLE, b Exeter, NH, Mar 19, 40; m 65; c 3. PHYSICS, ELECTROPHOTOGRAPHY. *Educ:* Mass Inst Technol, BS, 62, PhD(physics), 67. *Prof Exp:* Res assoc elem particles, Mass Inst Technol & Stanford Linear Accelerator Ctr, 67-68; scientist electrophotog, Xerox Webster Res Ctr, 68-73, mgr imaging sci, 73-74 & Xerox Res Ctr Can, 74-78, mgr mat develop, 78-83, mgr process technol sect, 83-88, MGR TECHNOL STRAG, XEROX WILSON CTR TECHNOL, WEBSTER, NY, 88- *Mem:* Am Phys Soc; Sigma Xi. *Res:* Novel electrophotographic systems including photoactive-pigment electrophotography; photoconductivity and charge exchange phenomena at interfaces; xerographic systems and materials. *Mailing Add:* Xerox Wilson Ctr Technol W147 800 Phillips Rd Webster NY 14580

HARTMANN, GREGORY KEMENYI, b Buffalo, NY, May 25, 11; m 39; c 4. PHYSICS. *Educ:* Calif Inst Technol, BS, 33; Oxford Univ, BA, 36, MA, 43; Brown Univ, PhD(physics), 39. *Prof Exp:* Asst physics, Brown Univ, 36-38; asst prof, Univ NH, 39-41; contract employee, Bur Ord, USN Dept, 41-42, physicist, 42-43, head res group appl explosives, 43-45, assoc chief explosives res div, Naval Ord Lab, 46-48, chief, 48-50, chief explosives res dept, 50-52, assoc tech dir res, 52-55, tech dir, White Oak, 55-73; CONSULT MGT, HIST & MINE WARFARE, 73- *Concurrent Pos:* Tech dir, Bur Ord Instrumentation Group, Bikini Atom Bomb Tests, 46, Eninetok Atom Tests, 48 & 50; sr visitor, Dept Appl Math & Theoret Physics, Cambridge Univ, 64-65. *Honors & Awards:* Distinguished Civilian Serv Award, US Dept Navy, 45 & US Dept Defense, 58; Nat Civil Serv League Award, 60. *Mem:* Assoc mem Am Phys Soc; fel Acoust Soc Am; Fed Prof Asn (pres, 64). *Res:* Underwater acoustics; shock waves; explosives; instrumentation for blast and underwater explosions; weapon effects; administration of research and development; antisubmarine warfare; sea based deterrence. *Mailing Add:* 10701 Keswick St Box 317 Garrett Park MD 20896

HARTMANN, HANS S(IEGFRIED), b Hohenleipisch, Ger, Feb 24, 31; US citizen; c 2. CERAMICS. *Educ:* Clausthal Tech Univ, Dipl Ing, 57; Mass Inst Technol, ScD(ceramics), 64. *Prof Exp:* Supvr customer serv lab, Kerabedarf KG, Ger, 57-59; res asst metall, Mass Inst Technol, 59-64; sr res scientist, Tech Ctr, Owens-Ill, Inc, 64-70, mgr ceramic technol, Electronics Div, 70-73; mem staff, Pioneering Res Lab, 73-82, RES ASSOC, DU PONT ELECTRONICS, 83- *Mem:* Am Ceramic Soc; Int Soc Hybrid Microelectronics; Sigma Xi. *Res:* Glass and ceramic materials for packaging, interconnection and hermetic sealing of semiconductors and similar electronic devices; precision alumina ceramics; development of alumina fibers for reinforcement of ceramic and metal-matrix composites, fabrication technology and nondestructive evaluation; fiber reinforced metals. *Mailing Add:* Du Pont Electronics PO Box 80336 Wilmington DE 19880-0336

HARTMANN, HENRIK ANTON, b Norway, Mar 20, 20; m 52; c 5. PATHOLOGY. *Educ:* Univ Oslo, MD, 49. *Prof Exp:* From asst prof to assoc prof, 54-68, PROF PATH, UNIV WIS-MADISON, 68- *Mem:* Am Asn Path; Am Asn Neuropath; Int Soc Neurochem. *Res:* Cytology of nerve cells in relation to hyperactivity; historadiography; cytochemistry of single cells; RNA and protein changes in aging neurons and environmental factors; preceding neuronal lesions; RNA changes in amyotrophic lateral sclerosis. *Mailing Add:* Dept Path Univ Wis Madison WI 53706

HARTMANN, HUDSON THOMAS, b Kansas City, Kans, Dec 6, 14; m 40, 78; c 4. HORTICULTURE. *Educ:* Univ Mo, BS, 39, MA, 40; Univ Calif, PhD(plant physiol), 47. *Prof Exp:* Asst hort, Univ Mo, 39-40; asst, Univ Calif, Davis, 40-42, assoc, Exp Sta, 42-47, from instr to prof pomol, 47-80; RETIRED. *Concurrent Pos:* Fulbright Award, Australia, 60-61, Italy, 64-65 & Greece, 68-69; Ed, Int Plant Propagators Soc, 72. *Honors & Awards:* Norman Jay Colman Award, 70; Int Award of Honor, Int Plant Propagators Soc, 90. *Mem:* Am Hort Soc; Int Plant Propagators Soc; fel Am Soc Hort Sci. *Res:* Production of the olive; plant propagation. *Mailing Add:* 35 Parkside Dr Davis CA 95616

HARTMANN, JUDITH BROWN, b Corning, Iowa, Feb 26, 46. ADVANCED RECORDING TECHNOLOGY, CONDENSED MATTER PHYSICS. *Educ:* Iowa State Univ, BS, 67; Cornell Univ, PhD(physics), 74. *Prof Exp:* Sr physicist, 77-80, res scientist, 80-82, res supvr, 82-84, RES MGR, 3M, 84- *Mem:* Am Phys Soc; Sigma Xi. *Res:* Advanced recording technology; optical and magnetic thin film materials. *Mailing Add:* 1895 Upper Afton Rd St Paul MN 55119

HARTMANN, RICHARD W, b Jersey City, NJ, Jan 5, 35; m 65; c 1. PLANT BREEDING. *Educ:* Rutgers Univ, BA, 56; Va Polytech Inst, MS, 57; Univ Calif, Los Angeles, PhD(plant sci), 62. *Prof Exp:* NIH fel genetics, Univ Calif, Davis, 62-63; asst prof, 63-70, ASSOC PROF HORT, UNIV HAWAII, 70- *Mem:* Am Soc Hort Sci; Soc Advan Breeding Res Asia & Oceania. *Res:* Plant breeding and genetics of vegetables in the tropics, especially beans and lettuce for root-knot nematode and tomato spotted wilt virus resistance. *Mailing Add:* Dept Hort Univ Hawaii 3190 Maile Way Honolulu HI 96822

HARTMANN, ROBERT CARL, b Everett, Wash, July 23, 19; m 77; c 6. HEMATOLOGY. *Educ:* Johns Hopkins Univ, AB, 41, MD, 44; Am Bd Internal Med, dipl, 56. *Prof Exp:* Intern, Pa Hosp, Philadelphia, 44-45, sr resident med, 45-46; fel med, Div Hemat, Johns Hopkins Hosp, Baltimore, 48-49, asst resident, 49-50, USPHS fel, Div Hemat, Johns Hopkins Univ, 50-51, AEC fel, 51-52; from asst prof to prof med, Sch Med, Vanderbilt Univ, 52-74; PROF MED, SCH MED, UNIV SFLA, TAMPA, 74-; 74-80, STAFF HEMATOLOGIST, SECT HEMAT & ONCOL, VET ADMIN HOSP, 80- *Concurrent Pos:* Asst vis physician, Outpatient Dept, Johns Hopkins Hosp, 48-49, 50-52; from asst vis physician to vis physician, Outpatient Dept, Vanderbilt Univ Hosp, 52-74, hematologist in-chg, 52-74; consult hematologist, Vet Admin Hosp, Nashville, Tenn, 52-74 & George Hubbard Hosp & Meharry Med Col, 67-74; consult, Inst Nutrit Cent Am & Panama Anemia & Nutrit Surv, Guatemala, Cent Am, 64-68, subcomt platelet-glass adhesion, Int Comt Haemostasis & Thrombosis, 66-70, hemat study sect, NIH, 67-71, nat nutrit surv, Nutrit Off, USPHS, 68-70 & hemophilia adv comt, Tenn Pub Health Dept, 73-74, adv coun, Nat Heart, Lung & Blood Inst, 80-84. *Mem:* Am Fedn Clin Res; Am Soc Clin Invest; Am Soc Hemat; Asn Am Physicians; Int Soc Hemat. *Res:* Blood coagulation; physiology of heparin; function of blood platelets; hemolytic anemias; endotoxin; hepatic venous thrombosis. *Mailing Add:* Dept Med-Hemat Univ SFla Col Med 12901 N 30th Tampa FL 33612

HARTMANN, SVEN RICHARD, b New York, NY, Feb 22, 32; m. PHYSICS. *Educ:* Union Col, NY, BS, 54; Univ Calif, Berkeley, PhD(physics), 61. *Prof Exp:* Res physicist, Univ Calif, Berkeley, 61-62; from asst prof to assoc prof, 62-68, dir, 68-71, co-dir radiation lab, 71-76, PROF PHYSICS, COLUMBIA UNIV, 76- *Concurrent Pos:* Sloan res fel, 63-; John Simon Gugenheim fel, 78. *Honors & Awards:* R W Wood Prize, Am Opt Soc, 83. *Mem:* Fel Am Phys Soc; fel Am Optical Soc; Sigma Xi; AAAS. *Res:* Magnetic resonance and relaxation; spin and photon echoes; interaction between light and matter. *Mailing Add:* 404 Riverside Dr Apt 3D New York NY 10025

HARTMANN, WILLIAM HERMAN, b New York, NY, Mar 13, 31; m 54; c 3. PATHOLOGY, SURGICAL PATHOLOGY. *Educ:* Syracuse Univ, BA, 51; State Univ NY, MD, 55. *Prof Exp:* From assoc to asst prof path, Johns Hopkins Hosp, Baltimore, 62-67; prof, Sch Med, Univ Tenn, 68; assoc pathologist, El Camino Hosp, Mountain View, Calif, 68-71; dir surg path, 71-73, PROF PATH & CHMN DEPT, SCH MED, VANDERBILT UNIV, 73- *Concurrent Pos:* Exchange prof path, Cayetano-Heredia Sch Med, Lima, Peru, 65; consult, St Thomas Hosp, Nashville, 72- & Vet Admin Hosp, 73-; Nat Cancer Inst trainee, Mem Hosp, NY, 58-60. *Mem:* Int Acad Path; Am Soc Clin Path. *Res:* Breast cancer. *Mailing Add:* Mem Med Ctr Dept Path 2801 Atlantic Ave Long Beach CA 90801-1428

HARTMANN, WILLIAM K, b New Kensington, Pa, June 6, 39; m 71; c 1. PLANETARY SCIENCE, PLANETARY ASTRONOMY. *Educ:* Pa State Univ, BS, 61; Univ Ariz, MS, 65, PhD(astron), 66. *Prof Exp:* Asst prof astron, Univ Ariz, 66-69; scientist, Ill Inst Technol, Res Inst, 69-70; SR SCIENTIST, PLANETARY SCI INST, SCI APPLN, INC, 70- *Concurrent Pos:* Co-investr, Mariner 9 Mars Mapping Mission, NASA, 69-72, prin investr, co-investr & consult, var NASA progs, 71-88; consult, House Select Comt Assasinations, 79. *Honors & Awards:* Nininger Meteorite Award, 65. *Mem:* Am Astron Soc; Meteoritical Soc; Int Astron Union. *Res:* Origin and evolution of planets; author of 3 textbooks in astronomy. *Mailing Add:* Planetary Sci 2421 E Sixth St Tucson AZ 85719-5234

HARTMANN, WILLIAM MORRIS, b Elgin, Ill, July 28, 39; m 67; c 2. PSYCHOACOUSTICS, ACOUSTICS. *Educ:* Iowa State Univ, BS, 61; Oxford Univ, PhD(physics), 65. *Prof Exp:* Res fel solid state theory, Argonne Nat Lab, 65-68; from asst prof to assoc prof, 68-77, PROF PHYSICS, MICH STATE UNIV, 77- *Concurrent Pos:* Vis scientist, Inst Res & Coord Acoust Mus, Paris, 81-86, actg dir Acoust, 82-83; NSF & NIH res grants; chmn, Tech Comt on Musical Acoust, Acoust Soc of Am, 80-84; working group 100, Nat Res Coun-CHABA, 87-88; assoc ed, Music Perception J, 88- *Mem:* Fel Acoust Soc Am; Am Phys Soc. *Res:* Human pitch perception; human binaural hearing system; acoustic signal processing; theory of the electron-phonon interaction; disordered solids. *Mailing Add:* Dept Physics Mich State Univ East Lansing MI 48824-1116

HARTNER, WILLIAM CHRISTOPHER, b West Rock Hill, Pa, Dec 24, 39; m 64; c 2. TRANSPLANTATION IMMUNOLOGY, ANIMAL PHYSIOLOGY. *Educ:* Pa State Univ, BS, 63; Univ Mo, MA, 68, PhD(physiol), 71. *Prof Exp:* Trainee physiol, Univ Ill, Urbana-Champaign, 71-73, res assoc, 74; asst prof biol, Wright State Univ, 74-75; asst prof biol, 75-82, STAFF SCIENTIST & LECTR, NORTHEASTERN UNIV, 83- *Mem:* AAAS. *Res:* Temperature regulation during hibernation cycles; temperature regulation and cold acclimation; control of sleep-wakefulness; mechanisms of allograft prolongation after treatment with antilymphocyte serum and donor bone marrow. *Mailing Add:* Med Lab Sci 206 Mngar Bldg Northeastern Univ Boston MA 02115

HARTNETT, JAMES P(ATRICK), b Lynn, Mass, Mar 19, 24; m 45; c 5. MECHANICAL ENGINEERING. *Educ:* Ill Inst Technol, BS, 47; Mass Inst Technol, MS, 48; Univ Calif, Berkeley, PhD(mech eng), 54. *Prof Exp:* Engr, Aircraft Gas Turbine Div, Gen Elec Co, 48-49; res engr, Univ Calif, 50-61; from asst prof to prof mech eng, Univ Minn, 54-61; prof & chmn dept, Univ Del, 61-65; prof energy eng & head dept, 65-74, DIR, ENERGY RESOURCES CTR, UNIV ILL, CHICAGO CIRCLE, 74- *Concurrent Pos:* Guggenheim fel & vis prof, Univ Tokyo, 60-61; Fulbright prof & lectr, Univ Alexandria, 61; consult, Seoul Nat Univ, 65, Rand Corp, Calif, Kaiser Eng Co, Calif & Asian Inst Technol, Bangkok, Thailand, 77; Nat Acad Sci exchange scientist, Rumania, 69; mem orgn comt, Int Centre Heat & Mass Transfer, Belgrade, Yugoslavia, 69-77; vis prof, Israel Inst Technol, 70; mem, Adv Comt USSR & Europe-Nat Acad Sci, 74-78, Ill Energy Resources Comn, 74- & Sci Coun Regional Centre Energy, Heat & Transfer Asia & Pac, 76-; co-ed, Int J Heat & Mass Transfer, Letters in Heat & Mass Transfer, Heat Transfer-Japanese Res & Previews Heat & Mass Transfer. *Honors & Awards:* Mem

Award, Heat Transfer Div, Am Soc Mech Engrs, 69; Prof Achievement Award, Ill Inst Technol, 77. *Mem:* Fel Am Soc Mech Engrs; Am Inst Aeronaut & Astronaut; Am Soc Eng Educ; Am Inst Chem Engrs; Sigma Xi. *Res:* Heat and mass transfer; fluid mechanics; energy policy. *Mailing Add:* Univ Ill Energy Resources Ctr Box 4348 257-RRB Chicago IL 60680

HARTNETT, JOHN (CONRAD), b Astoria, NY, Feb 2, 22; m 44; c 4. BIOCHEMISTRY. *Educ:* St Michael's Col, Vt, BS, 43; Univ Vt, MS, 47, PhD, 63. *Prof Exp:* From instr to assoc prof, 43-63, chmn dept, 68-74, prof, 63-87, EMER PROF BIOL, ST MICHAEL'S COL, 87- *Concurrent Pos:* Res asst, Univ Vt, 61-62, abstractor, Chem Abstr, 65-75; Am Cancer Soc res grants, 65-66, 67-68; res, Shelburne Mus, Vt, 70, assoc trustee & mus lectr, 71; co-dir TV ser human sexuality, Vt, 71; mem & vchmn, Vt Develop Dis Planning & Adv Coun, 72-81; mem bd trustees, Vt United Cerebral Palsy, 75-78; mem prof adv bd, Vt Epilepsy Asn, 75-77; mem, Vt Archit Barriers Compliance Bd, 76- & Vt Adv Coun Spec Educ, 78-; dir, Vt Asn Blind, 85-, reviewer, AAAS Sci Books & Films, 87- *Mem:* Am Inst Biol Sci; Am Inst Hist Pharm; Inst Society, Ethics & Life Sci. *Res:* Isolation and characterization of an esterproteolytic pancreatic enzyme; enzyme activity towards synthetic homopolypeptides; etiological agent in renal uremia; bacterial growth problems; history of pharmacy and therapeutics. *Mailing Add:* Nine Florida Ave Winooski VT 05404

HARTNETT, WILLIAM EDWARD, b Bartlesville, Okla, Oct 25, 25; m 50; c 6. LOGIC ANALYSIS, FUNCTIONAL ANALYSIS. *Educ:* Rockurst Col, BS, 50; Univ Mo, Kansas City, Mo, 51; Univ Kans, PhD(math), 57. *Prof Exp:* Instr, Univ Kans, 52-57; from asst prof to assoc prof math, Col Holy Cross, 57-63; asst prof math biol, Harvard Univ, 63-65; res assoc, Parke Math Labs, Inc, Mass, 65-68; chmn dept, State Univ NY Col Plattsburgh, 69-72, prof math, 68-89; RETIRED. *Concurrent Pos:* Vis prof, Sch Educ, Boston Univ, 67-68 & Nat Fac Exchange, State Univ NY, Binghamton, 85-86; hon res assoc, Founds & Philos Sci Unit, McGill Univ, 76-77. *Mem:* AAAS; Am Math Soc; Math Asn Am; Asn Symbolic Logic; Soc Math Biol; Hist Sci Soc. *Res:* Modern applied mathematics; mathematical education; measure and coding theories; general topology; functional analysis; mathematical biology. *Mailing Add:* 59 Beinkerhoff St Plattsburgh NY 12901

HARTOCOLLIS, PETER, b Greece, Nov 29, 22; US citizen; m 53; c 3. PSYCHIATRY, PSYCHOANALYSIS. *Educ:* Clark Univ, BA, 49; Mich State Univ, MA, 51, PhD(clin psychol), 54; Univ Lausanne, MD, 55; Menninger Sch Psychiat, cert psychol, 60; Topeka Inst Psychoanal, cert, 68. *Prof Exp:* Intern med & surg, St Mary's Hosp, Waterbury, Conn, 56-57; resident psychiat, Med Ctr, Univ Kans, 57; resident, Vet Admin Hosp, Topeka, Kans, 57-59; fel hosp psychiat & psychother, Menninger Found, 60-62; consult, Topeka State Hosp, 64-69, dir res hosp psychiat, C F Menninger Mem Hosp, 69-78, med dir, 73-81; PROF & CHMN PSYCHIAT DEPT, UNIV PATRAS MED SCH, GREECE, 81- *Concurrent Pos:* Mem pac Menninger Sch Psychiat, 63-; chief clin serv psychiat, C F Menninger Mem Hosp, 66-73; consult alcoholism prog, Vet Admin Hosp, Topeka, Kans, 66-; psychother supvr, Menninger Found, 67-; mem pac, Topeka Inst Psychoanal, 69-, training & supv analyst, 73-81, training & supv analyst Greek Study Group, 83- *Mem:* AMA; Am Col Physicians; Am Psychol Asn; Am Psychiat Asn; Am Psychoanal Asn. *Res:* Hospital psychiatry; psychoanalysis. *Mailing Add:* Univ Patras Sch Med Dept Psychat PO Box 1045 Patras 26110 Greece

HARTOP, WILLIAM LIONEL, JR, b Derry, NH, Sept 21, 21; m 48; c 3. RESEARCH ADMINISTRATION. *Educ:* Univ NH, BS, 44, MS, 48; Northwestern Univ, PhD(chem), 50. *Prof Exp:* Instr, Army Spec Training Prog, Univ NH, 43-44; sr res chemist, Abbott Labs, 50-57, head, Spec Proj Res Dept, 57-61, mgr, Mat Res Dept, 61-62, area mgr anal & mat res, 62-64, dir res & develop, Ross Labs Div, Ohio, 64-67; from asst dir to dir res, Ortho Pharmaceut Corp, 67-72, dir res planning, 72- & 72-83; vpres dir admin, Johnson & Johnson Biotech Ctr, LaJolla, Calif, 83-85; CONSULT, 85- *Mem:* AAAS. *Res:* Pharmaceuticals; reproduction and contraception; nutritional products; materials research; biotechnology. *Mailing Add:* 142 Seney Dr Bernardsville NJ 07924

HARTRANFT, GEORGE ROBERT, b Leola, Pa, Mar 29, 33; m 52; c 3. ORGANIC POLYMER CHEMISTRY. *Educ:* Franklin & Marshall Col, BS, 54; Univ Del, PhD(org chem), 59. *Prof Exp:* Res chemist, 59-78, res scientist, 78-83, SR RES SCIENTIST, ARMSTRONG WORLD IND INC, 83- *Mem:* Am Chem Soc. *Res:* Condensation polymers; vinyl polymers; fiber dyeing and treatment; photopolymerization. *Mailing Add:* 1516 Springside Dr Lancaster PA 17603-6321

HARTROFT, PHYLLIS MERRITT, b Detroit, Mich, Feb 1, 28; div. EXPERIMENTAL PATHOLOGY. *Educ:* Univ Mich, BS, 49; Univ Toronto, MA, 51, PhD, 54. *Prof Exp:* Asst, Banting & Best Dept Med Res, Univ Toronto, 49-54; asst path, Sch Med, Wash Univ, 54-58, asst prof, 58-61; res assoc, Hosp for Sick Children, Toronto, 61-63; asst prof path, Ind Univ, 63-66; assoc prof path, Sch Med, Wash Univ, 66-84; RETIRED. *Mem:* AAAS; Am Asn Anat; NY Acad Sci; Am Soc Nephrology; Am Asn Lab Animal Sci; Am Asn Pathologists; Am Inst Nutrit; Int Acad Path. *Res:* Kidney, renal juxtaglomerular cells; renin; cardiovascular diseases; pollution biology; laboratory animal science. *Mailing Add:* 5516 Lancaster Rd Hebron OH 43025

HARTRUM, THOMAS CHARLES, b Cambridge, Ohio, May 19, 45; m 67; c 2. DATABASE, COMPUTER PERFORMANCE EVALUATION. *Educ:* Ohio State Univ, BEE & MS, 69, PhD(elec eng), 73. *Prof Exp:* Elec engr, US Army Electronics Command, 71-75; elec engr, Aerospace Med Res Lab, 75-80; ASST PROF ELEC ENG, AIR FORCE INST TECHNOL, 80- *Mem:* Inst Elec & Electronics Engrs; Asn Comput Mach. *Res:* Analog models of neurons and their application as new communication techniques; hardware for processing speech signals for recognition and hardware for synthesis of speech. *Mailing Add:* Dept Elec Eng Air Force Inst Technol Eng Wright Patterson AFB OH 45433

HARTSAW, WILLIAM O, b Tell City, Ind, Oct 17, 21; m 46; c 1. MECHANICAL ENGINEERING. *Educ:* Purdue Univ, BS, 46, MS, 53; Univ Ill, PhD(theoret & appl mech), 66. *Prof Exp:* Instr eng, 46-52, from asst prof to prof, 52-85, head, Dept Eng, 58-60, dir, Sch Eng, 60-68, dean eng, 68-77, head, Dept Mech Eng, 77-85, DISTINGUISHED PROF MECH ENG, UNIV EVANSVILLE, IND, 85- *Concurrent Pos:* Lilly Found fel, 60; NSF fac fel, 60-62; mem, Buffalo Trace Coun, Boy Scouts of Am, 68-, mem coun, Big Bend Dist, 86-; mem Evansville Urban Transp adv comt, 75-; vchmn, Evansville Environ Protection Agency, 77-79; chmn, Evansville Sect, Am Soc Mech Engrs, 81-82, Nat Agenda Conf Deleg, 81-83, vpres, Region VI, 83-85, nat bd issues mgt, 86-, nat bd prof develop, 88- *Honors & Awards:* Centennial Metal, Am Soc Mech Engrs, 80. *Mem:* AAAS; Am Soc Mech Engrs; Am Soc Eng Educ; Am Soc Heat, Refrig & Air-Conditioning Engrs; Am Soc Testing Mat; Am Asn Univ Professors. *Res:* Stress analysis; photoelasticity; refrigeration; design; mechanics of solids; materials; the Peltier Effect; a low cycle fatigue strength investigation of a high strength steel. *Mailing Add:* Dept Mech Eng Univ Evansville 1800 Lincoln Ave Evansville IN 47722

HARTSFIELD, SANDEE MORRIS, b Bryan, Tex, May 18, 48; m 72; c 2. VETERINARY ANESTHESIOLOGY. *Educ:* Tex A&M Univ, BS, 70, DVM, 71; Mich State Univ, MS, 73. *Prof Exp:* Postdoctoral fel vet anesthesia, Mich State Univ, 71-73; asst prof, Univ Ill, 74-77; asst prof physiol, 73-74, assoc prof, 77-82, PROF VET ADMIN ANESTHESIA, TEX A&M UNIV, 82- *Concurrent Pos:* Vis prof, Dept Vet Clin Med, Univ Ill, 84-85; chief, Sect Anesthesia, Vet Teaching Hosp, Tex A&M Univ, 86- *Mem:* Am Col Vet Anesthesiologists (pres, 84); Am Soc Vet Anesthesiol (pres, 81-82); Am Vet Med Asn; Am Animal Hosp Asn; Am Asn Vet Clinicians. *Res:* Use of and cardiopulmonary effects of opioids and alpha-2 agonists and the antagonists to these pharmacologic agents in various domestic and laboratory species. *Mailing Add:* Dept VSAMS Col Vet Med Tex A&M Univ College Station TX 77843

HARTSHORN, JOSEPH HAROLD, b Cleveland, Ohio, June 23, 22; m 42, 80; c 2. GLACIAL GEOLOGY, PHYSICAL GEOGRAPHY. *Educ:* Harvard Univ, BS, 47, MA, 50, PhD(geomorphol), 55. *Prof Exp:* Geologist, US Geol Surv, 50-67; assoc prof, 67-69, head dept, 70-77, PROF GEOL, UNIV MASS, AMHERST, 69- *Concurrent Pos:* Res geologist, US Geol Surv, 67- *Mem:* AAAS; Geol Soc Am; Arctic Inst NAm; Am Quaternary Asn; Nat Asn Geol Teachers. *Res:* Glacial geology in New England, especially proglacial lakes and landforms; geomorphology and engineering geology of the East Coast of Greenland; glacial geology of the Malaspina Glacier, Alaska; Pleistocene archaeology. *Mailing Add:* Dept Geol/Geog Univ Mass Amherst MA 01003

HARTSHORNE, ROBERT (ROBIN) COPE, b Boston, Mass, Mar 15, 38; m 69; c 3. MATHEMATICS. *Educ:* Harvard Univ, AB, 59; Princeton Univ, PhD(math), 63. *Prof Exp:* Jr fel, Harvard Univ, 63-66, from asst prof to assoc prof math, 66-72; assoc prof, 72-74, PROF MATH, UNIV CALIF, BERKELEY, 74- *Concurrent Pos:* Vis prof, Tata Inst Fundamental Res, Bombay, 69-70 & Kyoto Univ, Japan, 75-76, 82-84; A P Sloan Found fel, 70-72. *Honors & Awards:* Steele Prize, Am Math Soc, 79. *Mem:* Am Math Soc. *Res:* Algebraic geometry. *Mailing Add:* Dept Math Univ Calif Berkeley CA 94720

HARTSOOK, JOSEPH THURMAN, dentistry; deceased, see previous edition for last biography

HARTSOUGH, LARRY DOWD, b Wadsworth, Ohio, Jan 23, 42; m 65; c 1. MATERIALS SCIENCE. *Educ:* Univ Calif, Berkeley, BS, 65, MS, 67, PhD(mat sci), 71. *Prof Exp:* Res metallurgist, Optical Coating Lab, Inc, 71-75; res engr, Temescal Div, Airco Inc, 75-77; sr staff scientist, Ultek Div, Perkin-Elmer, Inc, 77-81; co-founder vp res eng & dir, Technol Gryphon Prod, 81-88; SR STAFF SCIENTIST & PROJ MGR, GEN SIGNAL THIN FILM CO, 88- *Mem:* Am Soc Metals; Am Vacuum Soc; Sigma Xi; Mat Res Soc. *Res:* Synthesis of thin film materials and their characterization; development of new sputter-deposition sources, processes and systems. *Mailing Add:* 3007 Benvenue Berkeley CA 94705

HARTSOUGH, ROBERT RAY, b Salem, Ohio, Aug 11, 57; m 83; c 2. COATINGS FORMULATION, CERAMIC & GLASS CHEMISTRY. *Educ:* Youngstown State Univ, BS, 79, MS, 85, MBA, 90. *Prof Exp:* Teaching asst chem, Youngstown State Univ, 79-81; chemist, 81-86, chief chemist, 86-89, TECH DIR DISPERSION CHEM, GRAPHITE PROD CORP, 89- *Mem:* Am Chem Soc; Am Powdered Metal Inst. *Res:* Synthesis of a control release system for an anticancer drug; improving properties and stability of suspension and dispersion of forging and extrusion lubricants. *Mailing Add:* 4210 Nottingham Dr Austintown OH 44511

HARTSOUGH, WALTER DOUGLAS, b Merced, Calif, Sept 17, 24; m 45; c 4. PARTICLE PHYSICS. *Educ:* Univ Calif, AB, 44. *Prof Exp:* Physicist, physics res, 46-50, accelerator res & develop, 50-54, Bevatron opers head, 54-73, group leader, 73, assoc dir div head, eng & tech serv div, 73-75, assoc dir div head, facil mgt & tech serv div, 75-76, dir off minority outreach prog, 87, EMER ASSOC DIR, UNIV CALIF, LAWRENCE BERKELEY LAB, 87-; ASSOC, NATIONWIDE TECHNOL INC, 87- *Concurrent Pos:* Mem, Joint Comt High Energy Physics, US-Peoples Repub China, 79-87; mem, adv comt White House, 86-; Initiative Sci & Technol Adv Comm HBCV, 86- *Res:* Particle physics research and high energy particle accelerator research and development; counter and cloud chamber techniques; design and construction of prototype neutral beam injection systems for tokamaks. *Mailing Add:* 649 Ironbark Circle Orindo CA 94563

HARTSTEIN, ALLAN MARK, b New York, NY, Oct 5, 47; c 2. SOLID STATE PHYSICS. *Educ:* Calif Inst Technol, BS, 69; Univ Pa, PhD(physics), 73. *Prof Exp:* Asst instr, Univ Pa, 69-73; fel, 74-76, RES STAFF MEM PHYSICS, THOMAS J WATSON RES CTR, IBM CORP, 76- *Mem:* Am Inst Physics; fel Am Phys Soc; sr mem Inst Elec & Electronics Engrs. *Res:*

Interface phenomena; metal oxide semiconductor structures; transport in inversion layers; tunneling; physics of disordered systems; low dimensional systems; neural networks. *Mailing Add:* IBM Thomas J Watson Res Ctr Yorktown Heights NY 10598

HARTSTEIN, ARTHUR M, b Englewood, NJ, Nov 12, 40; m 64; c 2. PHYSICAL CHEMISTRY, ANALYTICAL CHEMISTRY. *Educ:* Polytech Inst Brooklyn, BS, 62; Adelphi Univ, PhD(phys chem), 70. *Prof Exp:* Res scientist, Power Sources Div, Gulton Industs, 69-70; res chemist, Anal Res Lab, Pittsburgh Mining & Safety Res Ctr, US Bur Mines, 70-75, phys chemist mining res, US Bur Mines, Washington, DC, 75-77; PHYS CHEMIST MINING RES, US DEPT ENERGY, WASHINGTON, DC, 77- *Mem:* Am Chem Soc. *Res:* Improving the nations energy picture by developing and demonstrating methods for obtaining oil from oil shale rock. *Mailing Add:* 13811 Dowlais Dr Rockville MD 20853

HARTSTIRN, WALTER, b New York, NY, Apr 16, 28; m 54; c 3. PLANT PATHOLOGY. *Educ:* Mont State Univ, BS, 51; Univ Wis, MS, 53; Ore State Col, PhD(plant path), 58. *Prof Exp:* Asst plant pathologist, Ill Natural Hist Surv, 58-77; asst prof plant path, Univ Ill, Urbana-Champaign, 73-77; mem staff anal chem, Ill Power Co, 81-87; RETIRED. *Mailing Add:* 907 Burkwood Dr Urbana IL 61801

HARTSUCK, JEAN ANN, b Enid, Okla, July 18, 39; m 62; c 3. BIOCHEMISTRY, X-RAY CRYSTALLOGRAPHY. *Educ:* Univ Okla, BS, 60; Radcliffe Col, AM, 62; Harvard Univ, PhD(chem), 64. *Prof Exp:* Asst prof, 70-74, ASSOC PROF BIOCHEM, UNIV OKLA, 74- *Concurrent Pos:* Asst mem, Okla Med Res Found, 70-73, assoc mem, 73-; NIH fel, Harvard Univ, 64-65; Career Develop Award, Okla Med Res Found, 71-76. *Mem:* Am Crystallog Asn; Am Chem Soc; Am Soc Biol Chem. *Res:* X-ray crystallography of pepsinogen; mechanisms of activation and catalysis of pepsin and homologous enzymes. *Mailing Add:* Okla Med Res Found 825 NE 13th St Oklahoma City OK 73104

HARTT, WILLIAM HANDY, b Long Beach, Calif, Apr 28, 39; m 62; c 2. METALLURGICAL & MATERIALS ENGINEERING. *Educ:* Va Polytech Inst, BS, 61; Univ Fla, PhD(metall eng), 66. *Prof Exp:* From asst prof to assoc prof, 68-75, PROF OCEAN ENG, FLA ATLANTIC UNIV, 75- *Res:* Physical and mechanical metallurgy; marine materials and corrosion; environmental cracking of materials. *Mailing Add:* Dept Ocean Eng Fla Atlantic Univ Boca Raton FL 33431

HARTUNG, G(EORGE) HARLEY, b Houston, Tex, Dec 15, 33; m 67; c 3. EXERCISE PHYSIOLOGY, PREVENTIVE MEDICINE. *Educ:* Tex A&M Univ, BS, 56; NTex State Univ, MEd, 65; Univ Tex, Austin, PhD(exercise physiol), 70. *Prof Exp:* Asst prof phys educ, Southeastern La State Univ, 69-70; asst prof, Cent Mo State Univ, 70-75; asst prof phys med, Baylor Col Med, 75-84, assoc prof med & phys med, 84-87; PROF PHYSIOL, SCH MED, UNIV HAWAII, 87- *Concurrent Pos:* Adj assoc prof, Sch Pub Health, Univ Tex Health Sci Ctr, 76-, vis assoc prof, 82-86; consult, Rehab Med Serv, Vet Admin Med Ctr, Houston, 76-87; vis grad fac, Tex A&M Univ, 77-87; adj grad fac, Univ Houston, 79- *Mem:* Fel Am Col Sports Med; Am Heart Asn; Am Physiol Asn; Am Alliance Health, Phys Educ, Recreation & Dance. *Res:* Exercise physiology, especially as it relates to plasma lipid levels, cardiovascular disease and environmental factors; role of exercise in preventive medicine; effect of alcohol on plasma lipids and lipoproteins. *Mailing Add:* Dept Physiol Sch Med Univ Hawaii Manoa 1960 E West Rd Honolulu HI 96822

HARTUNG, HOMER ARTHUR, b Baltimore, Md, Apr 20, 27; m 53; c 3. PHYSICAL CHEMISTRY. *Educ:* Univ Minn, AB, 49; Univ NC, PhD(phys chem), 55. *Prof Exp:* Res chemist, Silicones Div, Union Carbide Corp, 55-63; group leader, Texaco Exp, Inc, 63-66; SR SCIENTIST, RES CTR, PHILIP MORRIS INC, 66- *Mem:* AAAS; Am Chem Soc; Asn Comput Mach. *Res:* Thermodynamics; mathematical modelling; smoke formation, filtration, sorption. *Mailing Add:* Phillip Morris USA Res Ctr PO Box 1583 Richmond VA 23261

HARTUNG, JACK BURDAIR, SR, b Des Moines, Iowa, Mar 10, 37; m 62; c 2. PLANETARY SCIENCES, GEOCHRONOLOGY. *Educ:* Iowa State Univ, BS, 59; Rice Univ, PhD, 68. *Prof Exp:* Aerospace engr, Opers Div, Space Task Group, NASA, 59-62 & Apollo Spacecraft Proj Off, 62-64, scientist, Planetary & Earth Sci Div, Manned Spacecraft Ctr, 67-72; guest scientist, Max Planck Inst Nuclear Physics, Heidelberg, 72-74; adj assoc prof, Dept Earth & Space Sci, State Univ NY Stony Brook, 74-79; vis assoc prof, Hunter Col, City Univ NY, 79-80; lectr, Dept Geol, George Mason Univ, 80-83; geologist, US Geol Surv, Reston, Va, 82-84; field fel, Am Geophys Union, 84-85; vis scientist, Lunar & Planetary Sci Inst, Houston, 86-87; CONSULT, IOWA DEPT NATURAL RESOURCES, GEOL SURV BUR, 87- *Concurrent Pos:* vis scientist, Max Planck Inst for Nuclear Physics, Heidelberg, 85- *Mem:* Meteoritical Soc; Am Geophys Union; Am Astron Soc; Explorers Club. *Res:* Analysis of lunar rocks; meteorite impact cratering; shock metamorphism; radiometric dating; dynamics and history of the earth-moon system; impact structures; small solar system objects. *Mailing Add:* 600 E Fifth St No 607 Des Moines IA 50309

HARTUNG, JOHN, b Passaic, NJ, Nov 24, 51; m 83; c 1. DIGITAL SIGNAL PROCESSING, COMPUTER ARCHITECTURE. *Educ:* Stevens Inst Technol, BS, 73; Rutgers Univ, MS, 75 & 78, PhD(elec eng), 80. *Prof Exp:* Mem tech staff, David Sarnoff Res Ctr, RCA, 79-80; asst prof elec eng, Rutgers Univ, 79-80; mem tech staff, AT&T-Bell Labs, 80-84, Bell Commun Res, 84-85, Info Systs Lab, 85-86, supvr, Bell Labs, 86-90, MEM TECH STAFF & PRIN INVESTR, AT&T-BELL LABS, 90- *Mem:* Inst Elec & Electronics Engrs. *Res:* Digital signal processing; design and implementation of video coding algorithms and multiprocessor digital signal processing architectures. *Mailing Add:* 22 Red Hill Rd Warren Township NJ 07059

HARTUNG, JOHN DAVID, b 1947. NEUROANESTHESIOLOGY, SOCIOBIOLOGY. *Educ:* Univ Pa, BA, 73; Harvard Univ, PhD(anthropol), 81. *Prof Exp:* RES ASSOC PROF, HEALTH SCI CTR STATE UNIV NY, 81- *Concurrent Pos:* Assoc ed, J Neurosurg Anesthesiol, 88- *Mem:* Am Asn Anthropology; Am Soc Anesthesiol. *Res:* Neuroanesthesiology cerebral protection from hypoxia, ischemia and hypothermia; evolutionary theory. *Mailing Add:* Dept Anesthesiol State Univ NY Health Sci 450 Clarkson Ave Brooklyn NY 11203

HARTUNG, ROLF, b Bremen, Ger, Mar 1, 35; US citizen; m 59; c 3. TOXICOLOGY. *Educ:* Univ Mich, BS, 60, MWM, 62, PhD(wildlife mgt), 64; Am Bd Toxicol, dipl, 80. *Prof Exp:* Lectr indust health, 64, from asst prof to assoc prof, 65-74, PROF ENVIRON TOXICOL & RES SCIENTIST, UNIV MICH, ANN ARBOR, 74- *Concurrent Pos:* Mem comt health effects water pollution, Int Joint Comn (US & Can); mem comt geochem & health effects & water qual criteria, Nat Acad Sci; chair, Comt Rediation of Irrigation Induced Water Qual, Nat Acad Sci & Comt Environ, Effects, Transport & Fate, US Environ Protection Agency, Sci Adv Bd; Comn Environ, Effects, Transport &. *Mem:* AAAS; Wildlife Soc; Am Indust Hyg Asn; Soc Toxicol; Soc Environ Toxicol & Chem. *Res:* Effects of polluting oils on waterfowl; toxicity of aminoethanols; coactions between chlorinated hydrocarbon pesticides and aquatic pollutants; environmental dynamics of heavy metals; risk assessment. *Mailing Add:* Dept Environ & Indust Health Univ Mich Ann Arbor MI 48109-2029

HARTUNG, THEODORE EUGENE, b Denver, Colo, Jan 28, 29; m 51; c 3. FOOD SCIENCE. *Educ:* Colo State Univ, BS, 51, MS, 53; Purdue Univ, PhD(food tech), 62. *Prof Exp:* Instr poultry, Colo State Univ, 51-53, exten specialist, 53-61, assoc prof, 61-65; prof poultry, 65-74, dean Agr Col, 74-88, ASSOC VCHMN, AGR & NATURAL RESOURCES, UNIV NEBR, 88- *Mem:* Inst Food Technol; Poultry Sci Asn; Sigma Xi. *Res:* Microbiological, chemical and physical changes of food subjected to storage and handling variations; nutrient composition of food and alteration in processing. *Mailing Add:* Inst Agr & Natural Resources Univ Nebr Lincoln NE 68583-0708

HARTWELL, GEORGE E, JR, b Port Jervis, NY, Aug 1, 40; m 65. CATALYSIS, ORGANOMETALLIC CHEMISTRY. *Educ:* Union Col, NY, BS, 62; Univ Ill, Urbana, MS, 64, PhD, 66. *Prof Exp:* From asst prof to assoc prof chem, Ind Univ, Bloomington, 66-74; group leader, 74-90, ASSOC SCIENTIST, DOW CHEM USA, 90- *Concurrent Pos:* NATO fel, Imp Col, Univ London, 66-67. *Mem:* Am Chem Soc. *Res:* Homogeneous and supported homogeneous catalysis, especially reduction chemistry; metal clusters; heterogeneous catalysts; catalytic processes and research in developing heterogeneous and homogeous catalysts. *Mailing Add:* Cent Res Bldg 1776 Dow Chem USA Midland MI 48674

HARTWELL, JONATHAN LUTTON, b Boston, Mass, Feb 21, 06; m 33, 43. CHEMISTRY. *Educ:* Harvard Univ, AB, 28, MA, 33, PhD(chem), 35. *Prof Exp:* Asst chem, Harvard Univ, 28-29; res chemist, E I du Pont de Nemours & Co, Del, 29-32; asst chem, Harvard Univ, 32-34; res chemist, Int Printing Ink Div, Interchem Corp, NY, 35-38; fel, Nat Cancer Inst, 38-42, chemist, 42-47, head org chem sect, 47-58, asst chief, Cancer Chemother Nat Serv Ctr, 58-63, head natural prod sect, 63-75; RETIRED. *Mem:* Hon mem Am Soc Pharmacog; Am Chem Soc. *Res:* Organic chemistry; reactions of quinones, dyes, pigments, chemical carinogenesis; lignans; chemotherapy of cancer. *Mailing Add:* 8700 Jones Rd Chevy Chase MD 20815

HARTWELL, LELAND HARRISON, b Los Angeles, Calif, Oct 30, 39; div; c 3. GENETICS. *Educ:* Calif Inst Technol, BS, 61; Mass Inst Technol, PhD(microbiol), 64. *Prof Exp:* Nat Acad Sci-Nat Res Coun fel animal virol, Salk Inst Biol Studies, 64-65; from asst prof to assoc prof biochem, Univ Calif, Irvine, 65-68; assoc prof, 68-74, PROF GENETICS, UNIV WASH, 74- *Honors & Awards:* Eli Lilly Award in Microbiol & Immunol, Am Soc Microbiol, 74. *Mem:* Nat Acad Sci; AAAS; Am Soc Microbiol; Genetic Soc Am (pres, 91). *Res:* Genetic analysis of cell division and cell communication in yeast. *Mailing Add:* Dept Genetics Univ Wash Seattle WA 98105

HARTWICK, EARL BRIAN, b Toronto, Ont, Feb 19, 42; m 64; c 3. INVERTEBRATE FISHERIES ECOLOGY. *Educ:* Univ Toronto, BSc, 66, MSc, 68; Univ BC, PhD(ecol), 73. *Prof Exp:* Asst prof pop ecol, Univ Guelph, 73-75; asst prof, 75-83, ASSOC PROF MARINE BIOL, SIMON FRASER UNIV, 83- *Mem:* Can Soc Zoologists; Can Asn Underwater Sci (pres, 90-91); Inst Fisheries Anal. *Res:* Marine invertebrate ecology; biology and ecology of octopuses; population processes in marine environments; applied marine ecology. *Mailing Add:* Dept Biol Sci Simon Fraser Univ Burnaby BC V5A 1S6 Can

HARTWICK, FREDERICK DAVID ALFRED, b Arvida, Que, May 27, 41. ASTRONOMY. *Educ:* McGill Univ, BEng, 62; Univ Toronto, MA, 64, PhD(astron), 66. *Prof Exp:* NATO vis fel, Princeton Univ, 66-67; NATO res fel, Mt Wilson & Palomar Observ & Calif Inst Technol, 67-68; from asst prof to assoc prof, 68-78, PROF ASTRON, UNIV VICTORIA, BC, 78- *Concurrent Pos:* Steacie fel, 78-79. *Mem:* Royal Astron Soc Can; fel Royal Astron Soc; Am Astron Soc; Can Astron Soc; Int Astron Union. *Res:* Kinematics of galaxy halos; space distribution of quasors. *Mailing Add:* Dept Physics & Astron Univ Victoria Victoria BC V8W 2Y2 Can

HARTWICK, RICHARD ALLEN, b Hackensack, NJ, Mar 26, 51; c 3. BIOMEDICAL ANALYSES. *Educ:* Barrington Col, BA, 73; Univ RI, MS, 75, PhD(chem), 78. *Prof Exp:* NATO fel, Univ Edinburgh, Scotland, 79-80; from asst prof to assoc prof anal chem, Rutgers Univ, 80-90; CONSULT, 90- *Mem:* Am Chem Soc; Am Asn Clin Chemists; Sigma Xi. *Res:* High performance liquid chematography; applications to biomedical analyses; capillary electrophoresis. *Mailing Add:* Dept Chem State Univ NY Binghamton NJ 13902

HARTWICK, THOMAS STANLEY, b Vandalia, Ill, Mar 19, 34; m 61; c 3. QUANTUM ELECTRONICS, SOLID STATE PHYSICS. *Educ:* Univ Ill, BS, 56; Univ Calif, Los Angeles, MS, 58; Univ Southern Calif, PhD, 69. *Prof Exp:* Mem tech staff, Hughes Aircraft Co, 56-59, group leader spin wave excitation in ferrites, 59-61; sect mgr quantum electronics, Labs Div, Aerospace Corp, 61-63 & 65-73, dept head, 73-75, asst dir, Electronics Res Lab, 75-78; mgr, E-O Develop Labs, Hughes Aircraft Co, 78-83; mgr, E-O Res Ctr, TRW, 83-86, MGR MICROELECTRONICS CTR, TRW, 86- *Concurrent Pos:* Mem bd dirs, Laser Tech Inc, 67-; consult mem, Adv Group Electron Devices, Dept Defense, 78-88; chmn Working Group C, 88- *Mem:* Am Phys Soc; Optical Soc Am. *Res:* Magnetic resonance in ferrites; molecular lasers; optical communications and optical properties of ferro electrics; laser heterodyne radiometry; far infrared lasers and applications. *Mailing Add:* One Space Park Bldg 10 Rm 2830 Redondo Beach CA 90278

HARTWIG, CURTIS P, b Boston, Mass, July 13, 39; m 62, 86; c 2. COMPUTER LANGUAGES & SOFTWARE SUPPORT TOOLS. *Educ:* Mass Inst Technol, SB & SM, 62, PhD(elec sci), 66. *Prof Exp:* Tech aide sta apparatus, Bell Tel Labs, Inc, 59-60, mil res, 60, sr tech aide solid state devices, 61-62; asst elec eng, Mass Inst Technol, 62-66; res scientist, Raytheon Co, 64, sr res scientist microwave physics, Res Div, 66-74; staff scientist, Microwave Div, Sander Assoc, Inc, 75-76; adv engr, Fed Systs Div, IBM Corp, 76-77; DEVELOP ENGR, AIRCRAFT CONTROL SYST DIV, GEN ELEC, BINGHAMTON, NY, 79- *Res:* Electrical science; microwave magnetics; electromagnetic properties of matter with emphasis on application to microwave devices; exploration of phenomena in materials which may lead to new types of devices; investigating techniques for developing software for digital control systems. *Mailing Add:* Gen Elec PO Box 5000 Mail Drop 786B Binghamton NY 13902

HARTWIG, JOHN, b Marshall, Mich, Sept 19, 49. LEUKOCYTE MOTILITY. *Educ:* Harvard Univ, PhD(cell develop biol), 80. *Prof Exp:* ASST PROF MED, MED SCH, HARVARD UNIV, 83-; ASST PROF BIOL, MASS GEN HOSP, 83-; ASSOC PROF MED, BRIGHAM-WOMEN'S HOSP, 91- *Mailing Add:* Exp Med Div Brigham-Women's Hosp Longwood Med Res Ctr 221 Longwood Ave Boston MA 02115

HARTWIG, NATHAN LEROY, b Monroe, Wis, Aug 10, 37; m 63; c 1. WEED SCIENCE, SOIL CONSERVATION. *Educ:* Univ Wis-Madison, BS, 59, PhD(agron), 70; Univ Ariz, MS, 65. *Prof Exp:* ASSOC PROF WEED SCI, PA STATE UNIV, UNIVERSITY PARK, 69- *Mem:* Am Soc Agron; Weed Sci Soc Am. *Res:* Methods of controlling annual and perennial weeds in no-tillage corn; cover crop management in agronomic crop rotations; use of crownvetch as a living mulch for min- and no-tillage crop production. *Mailing Add:* Dept Agron Pa State Univ University Park PA 16802

HARTWIG, ROBERT EDUARD, b Soerabaja, Indonesia, Mar 21, 41; m 66; c 2. APPLIED MATHEMATICS. *Educ:* Univ Adelaide, BSc, 62, Hons, 63, PhD(math), 66. *Prof Exp:* NSF res fel math physics, Cornell Univ, 66-68; from asst prof to assoc prof, 68-79, PROF MATH, NC STATE UNIV, 79- *Mem:* Soc Indust & Appl Math; Math Asn Am. *Res:* Non-negative matrices; cryptography; linear algebra; matrices; asymptotics; generalized inverses. *Mailing Add:* Dept Math NC State Univ Raleigh NC 27650

HARTY, MICHAEL, b Waterford, Ireland, July 29, 12; m 54; c 2. ANATOMY, ORTHOPEDIC SURGERY. *Educ:* Univ Dublin, MB, 37, MCh, 52; Cambridge Univ, MA, 47; FRCS. *Prof Exp:* Surgeon, Emergency Med Serv, London, 39-41; demonstr anat, Cambridge Univ, 42-44 & 46-47, lectr, 49-53; vis lectr, 53-54, from asst prof to assoc prof, 54-74, PROF ANAT & ORTHOP SURG, SCH MED, UNIV PA, 74- *Honors & Awards:* Lindback Award, 80. *Mem:* Am Asn Anat; Anat Soc Gt Brit & Ireland; Brit Med Asn. *Res:* Surgical and functional anatomy of the locomotor system. *Mailing Add:* 41 Penn Blvd E Lansdowne PA 19050

HARTZ, BILLY J, b Cavalier, NDak, July 9, 26; div; c 4. ENGINEERING MECHANICS, CIVIL ENGINEERING. *Educ:* Univ Calif, Berkeley, BS, 52, MS, 54, PhD(civil eng), 55. *Prof Exp:* From asst prof to assoc prof, 55-65, prof, 65-82, EMER PROF CIVIL ENG, UNIV WASH, 82-; PRES, BILLY BOARDS, INC, 85- *Concurrent Pos:* NSF sr fel, 63-64; NTH, Zürich, Switz, Trondheim, Norway, 81-82. *Mem:* Am Soc Civil Engrs. *Res:* Matrix methods in structural mechanics; dynamic stability of elastic systems; dynamic response of structures; elastic wave propagation; author of over 55 publications. *Mailing Add:* Dept Civil Eng Univ Wash 201 More Hall Mail Stop FX10 Seattle WA 98195

HARTZ, JOHN WILLIAM, b Detroit, Mich. NEOPLASMS OF BIOCHEMISTRY, HEMATOPATHOLOGY. *Educ:* Albion Col, AB, 58; Harvard Med Sch, MD, 62; Univ Wis, PhD(physiol chem), 68. *Prof Exp:* Lieutenant, USN Chem Lab, Naval Med Res, 68-69; fel biochem, Univ Wis, 69-71; resident path, Univ Chicago Hosps & Clins, 71-74; ASST PROF PATH, BOWMAN GRAY SCH MED, 74- *Mem:* Sigma Xi. *Res:* Redcell proteins; membrane chemistry of neoplasms; hematopathology. *Mailing Add:* 230 Flintshire Rd Winston-Salem NC 27104

HARTZ, ROY EUGENE, b Duncannon, Pa, Apr 20, 40; m 65; c 1. ORGANIC POLYMER CHEMISTRY. *Educ:* Pa State Univ, BS, 62; Univ Md, PhD(org chem), 67. *Prof Exp:* Chemist, US Bur Mines, 63; res chemist, Res & Develop Ctr, Wayne, NJ, 66-69, res scientist, 69-71; res scientist, Textile & Fiber Div, Uniroyal, Inc, 71-76, tire cord res & develop mgr, 76-85, tech mgr, Tire Textiles, 87-91, TIRE CORD & ROYAL CORD RES & DEVELOP MGR, UNIROYAL-GOODRICH TIRE CO, WINNSBORO, SC, 85- *Mem:* Am Chem Soc; Sigma Xi. *Res:* Pyrolysis of allylamines; reactions of trifluoracetic acid with unsaturated compounds; free radical and anionic initiated polymerizations in the areas of block copolymers and polar monomers; adhesion; adhesion mechanism problems; pollution areas; new adhesive system. *Mailing Add:* 101 Linsbury Circle Columbia SC 29210

HARTZELL, CHARLES ROSS, III, b Butler, Pa, Aug 12, 41; m 63; c 2. BIOPHYSICAL CHEMISTRY, PHYSIOLOGICAL CHEMISTRY. *Educ:* Geneva Col, BSc, 63; Ind Univ, Bloomington, PhD(biochem), 67. *Prof Exp:* Fel & res assoc chem, Ind Univ, Bloomington, 67; res scientist, Div Protein Chem, Commonwealth Sci & Indust Res Orgn, Australia, 67-68; Nat Insts Gen Med Sci fel & res assoc, Inst Enzyme Res, Univ Wis, Madison, 68-70, asst res prof, 70-71; from asst prof to assoc prof biochem, Pa State Univ, 71-78; sr res scientist, 78-80, DIR RES, ALFRED I DUPONT INST, NEMOUR FOUND, 80- *Concurrent Pos:* Mem coun basic sci & estab investr, Am Heart Asn, 70-75. *Mem:* Biophys Soc; Am Chem Soc; Am Heart Asn; Am Soc Biochem & Molecular Biol; Am Soc Cell Biol. *Res:* Perinatal growth and development of skeletal and cardiac muscle; biochemistry of neurotransmitter synthesis, storage and release; enzyme activities and protein turnover of mammalian skeletal muscle regeneration in cell culture. *Mailing Add:* Alfred I DuPont Inst Res Dept PO Box 269 Wilmington DE 19899

HARTZELL, KENNETH R, JR, b Philadelphia, Pa, June 15, 54. LASER PHYSICS. *Educ:* Villanova Univ, BS, 76; Drexel Univ, MS, 78 & PhD, 86. *Prof Exp:* ASST PROF PHYSICS, VILLANOVA UNIV, 83- *Concurrent Pos:* Adj inst & asst prof, Drexel Univ, 78- *Mem:* Am Phys Soc; Am Asn Physics Teachers; Optical Soc Am; Inst Elec & Electronics Engrs. *Res:* Investigation of theoretical models of free electron lasers for the purpose of gain enhancement and improvement of outdoor power and efficiency. *Mailing Add:* 521 Mercer Rd Merion PA 19066

HARTZELL, THOMAS H, b Passaic, NJ, Dec 12, 32; m 56; c 2. FOOD SCIENCE. *Educ:* Pa State Univ, BS, 56, MS, 57. *Prof Exp:* Anal chemist, Res Ctr, Nat Biscuit Co, 57-60; chief chemist, 60-67, tech dir, 67-75, vpres prod & tech dir, Minn Malting Co, 74-85; vpres qual control, Minn Grain Pearling Co, 70-83; vpres malting, 85-89, SR VPRES TECH SERV, SCHREIER MALTING CO, 89- *Mem:* Master Brewers Asn Am; Am Chem Soc; Am Soc Brewing Chem; Am Asn Cereal Chem; Inst Food Technol. *Res:* Cereal chemistry; grain malting technology; food technology; new techniques in malting technology; process and product development in malting technology. *Mailing Add:* PO Box 59 Schreier Malting Co Sheboygan WI 53082-0059

HARTZEMA, ABRAHAM GIJSBERT, b The Hague, Netherlands, May 10, 47. HEALTH SERVICES, PHARMACY. *Educ:* Univ Utrecht, BSc, 73; Univ Wash, MSPH, 78; Univ Minn, PhD(social pharm), 82. *Prof Exp:* ASSOC PROF & RES ASSOC, HEALTH SERV CTR, SCH PUB HEALTH & SCH PHARM, UNIV NC, CHAPEL HILL, 81- *Concurrent Pos:* Res assoc, Vet Admin, 82-86. *Mem:* Am Pub Health Asn; Am Pharm Asn; Am Asn Col Pharm; Royal Dutch Asn Advan Pharm; Asn Health Serv. *Res:* Program evalvation; pharmaceutical services; health care organization. *Mailing Add:* PO Box 697 Chapel Hill NC 27514-0697

HARTZFELD, HOWARD ALEXANDER, b Bronaugh, Mo, Jan 7, 28; m 65; c 1. ORGANIC CHEMISTRY. *Educ:* Pittsburg State Univ, AB, 49; Iowa State Univ, PhD(org chem), 53. *Prof Exp:* Res chemist, 53-62, develop engr, 62-75, sr patent develop chemist, 75-79, SR PATENT DEVELOP SPECIALIST, PHILLIPS PETROL CO, 79- *Concurrent Pos:* Adj prof, Okla State Univ, 56-57. *Mem:* Am Chem Soc. *Res:* Organosilicon chemistry; chemical constitution and plant hormone action; petrochemistry. *Mailing Add:* 1409 SE Meadowcrest Ct Bartlesville OK 74004

HARTZLER, ALFRED JAMES, b Manhattan, Kans, Apr 17, 22; m 50; c 1. MILITARY OPERATIONS. *Educ:* Univ Chicago, BS, 43, MS, 44, PhD(physics), 51. *Prof Exp:* Res physicist, Gen Elec Co, 45-46, Carnegie Inst Technol, 50-55 & Opers Eval Group, Mass Inst Technol, 55-62; scientist, Ctr for Naval Anal, Franklin Inst, 62-65; chief opers anal div, 65-73, tech adv, 73-77, SR ANALYST OPERS ANALYSIS OFF, US ARMS CONTROL & DISARMAMENT AGENCY, 77- *Mem:* AAAS; Opers Res Soc Am. *Res:* Cosmic rays; nuclear physics; operations research; strategic and peace research. *Mailing Add:* Opers Analysis Off Rm 5726 Arms Control Disarmament Agency 320 21st St NW Washington DC 20451

HARTZLER, EVA RUTH, b Longreen, Md, Jan 24, 11. BIOCHEMISTRY. *Educ:* Juniata Col, BS, 32; Pa State Col, MS, 33, PhD(biochem), 50. *Prof Exp:* Asst vitamin A, Oberlin Col, 33-35; res technician, Mayo Clin, 35-37; instr gen chem, Juniata Col, 37-38; jr chemist B Vitamins, Bur Home Econ, USDA, 38-41; jr res chemist, Parke, Davis Co, 41-43; assoc nutritionist, Univ Hawaii, 43-47; instr biochem, Pa State Col, 47-50; asst prof, 50-57, prof, 57-76, EMER PROF CHEM, JUNIATA COL, 76- *Mem:* AAAS; Am Chem Soc. *Res:* Vitamin A utilization, purification and oxidation; vitamin-content of foods and methods of assay; utilization of ascorbic acid; ascorbic acid and enzymes in frozen vegetables. *Mailing Add:* RD 4 Box 41 Huntington PA 16652

HARTZLER, HARRIS DALE, b Ill, July 29, 32; m 55; c 3. ORGANIC CHEMISTRY. *Educ:* Univ Chicago, BA, 53, MS, 54, PhD(chem), 57. *Prof Exp:* Instr, Univ Mich, 57-59; res chemist, 59-74, group leader, 74-76, res supvr, 76-79, RES MGR, CENT RES DEPT, E I DU PONT DE NEMOURS & CO, INC, 79- *Mem:* Am Chem Soc. *Res:* Synthetic organic chemistry. *Mailing Add:* 112 S Spring Valley Rd Wilmington DE 19807

HARTZLER, HARROD HAROLD, b Ft Wayne, Ind, Apr 7, 08; m 33; c 3. PHYSICS. *Educ:* Juniata Col, AB, 30; Rutgers Univ, PhD(physics), 34. *Prof Exp:* Prof math & physics, Elizabethtown Col, 35-37; from asst prof to prof math, Goshen Col, 37-58; prof, Bluffton Col, 45-46; Stewart Observ fel astron, Univ Ariz, 48-49; from assoc prof to prof physics, Mankato State Col, 58-64, prof math & astron, 64-76, emer prof math & astron, 76-; RETIRED. *Mem:* AAAS; Am Phys Soc; Am Astron Soc; Am Math Soc; Am Sci Affil (secy-treas, 50-54, pres, 55-60, exec secy, 61-76). *Res:* Transparency of thin metallic films. *Mailing Add:* 901 College Ave Goshen IN 46526

HARTZLER, JON DAVID, b Peoria, Ill, Oct 15, 41; m 67. ORGANIC POLYMER CHEMISTRY. *Educ:* Goshen Col, BA, 63; Univ Del, PhD(org chem), 68. *Prof Exp:* Res chemist, Pioneering Res Lab, Textile Fibers Dept, 67-73, sr res chemist, 73-74, sr res chemist, Dacron Plant Tech, 74-77, sr res chemist, Indust Tech, 77-81, RES ASSOC, SPRUANCE RES LAB, TEXTILE FIBERS DEPT, E I DU PONT DE NEMOURS & CO, INC, 81- *Mem:* Am Chem Soc. *Res:* Polymer chemistry and fiber technology. *Mailing Add:* 3222 Fortunes Ridge Rd Midlothian VA 23113

HARTZOG, DAVID G(EORGE), b St Louis, Mo, June 23, 41. CHEMICAL ENGINEERING, APPLIED MATHEMATICS. *Educ:* Wash Univ, BSChE, 63, DSc(chem eng), 68. *Prof Exp:* Staff engr, 70-73, systs analyst, 73-76, systs mgr environ eng, 76-81, SYSTS MGR PROCESS DYNAMICS & ENVIRON ENG, AIR PROD & CHEM, INC, 81- *Mem:* Am Inst Chem Engrs; Am Chem Soc. *Res:* Development of deterministic mathematical models and computer programs for the simulation and design of various chemical processes, especially biochemical reactors and fixed-bed, pressure-swing adsorption units; transport of atmospheric pollutants; LNG hazard analysis; mass transfer with chemical reaction and numerical analysis, particularly parameter estimation algorithms. *Mailing Add:* Air Prod & Chem Inc PO Box 538 Allentown PA 18195

HARTZOG, JAMES VICTOR, b Reevesville, SC, Dec 3, 37; m 65; c 2. PHYSICAL CHEMISTRY. *Educ:* Clemson Col, BSc, 61, PhD(chem), 67. *Prof Exp:* Res chemist, 67-74, SR RES CHEMIST, DACRON RES LAB, E I DU PONT DE NEMOURS & CO, INC, 74- *Mem:* Am Chem Soc. *Res:* Electrical properties of polymers; surface chemistry; adsorption; polymer structure. *Mailing Add:* 1707 Sabra Dr Kinston NC 28501

HARUKI, HIROSHI, b Tokyo, Japan, Feb 18, 18; m 50; c 2. MATHEMATICS. *Educ:* Osaka Univ, MSc, 40, PhD(math), 65. *Prof Exp:* Prof math, Kobe Merchantile Marine Col, 41-45; prof, Osaka High Sch, 45-49; assoc prof, Osaka Univ, 49-66; PROF MATH, UNIV WATERLOO, 66- *Mem:* Am Math Soc. *Res:* Functional equations for analytic functions; partial differential equations. *Mailing Add:* Dept Pure Math Univ Waterloo Waterloo ON N2L 3G1 Can

HARUN, JOSEPH STANLEY, b Shenandoah, Pa, Jan 3, 25; m 52; c 5. CLINICAL RESEARCH. *Educ:* Mt St Mary Col, Md, BS, 51; Jefferson Med Col, MD, 55. *Prof Exp:* Staff physician, Merck Sharp & Dohme, 59-60; pvt practr dermat, Pa, 60-62; head, Sci Info Dept, McNeil Labs, Inc, 62-64; dir med serv, Wallace Labs, 64-66; med dir, Madison Labs Div, Ciba Pharmaceut Co, 66-69, exec dir regulatory affairs, Ciba-Geigy Pharmaceut Div, 69-72, vpres drug regulatory affairs, 72-73; vpres & med dir, 73-74, sr vpres & med dir, Wallace Labs Div, 74-76, CORP VPRES, MED & SCI AFFAIRS, CARTER-WALLACE, INC, 76- *Concurrent Pos:* Mem, NJ State Panel Sci Adv, 81-83. *Mem:* Soc Invest Dermat; fel Am Acad Dermat; Int Acad Law & Sci; AMA; Int Soc Trop Dermat. *Mailing Add:* Roxiticus Rd Mendham NJ 07945

HARUTA, KYOICHI, b Kamakura, Japan, Jan 3, 31; m 60; c 3. PHYSICS. *Educ:* Colby Col, BA, 57; Mass Inst Technol, PhD(physics), 63. *Prof Exp:* mem tech staff, 63-83, DISTINGUISHED MEM TECH STAFF, AT&T BELL LABS, 83- *Mem:* Am Phys Soc. *Res:* X-ray diffraction study of single crystals and polycrystalline materials; acoustic vibrations in bounded media; numerical analysis; computer applications; very-large-scale intergation process simulation; integrated circuit interconnection modeling. *Mailing Add:* AT&T Bell Labs 1247 S Cedar Crest Blvd Allentown PA 18103

HARVAIS, GAETAN HUGUES, mycology; deceased, see previous edition for last biography

HARVALIK, ZABOJ VINCENT, b Sinj, Yugoslavia, Feb 4, 07; US citizen; m 36; c 2. PHYSICS. *Educ:* State Col, Prague, BA, 26; Prague Univ, MS, 30, ScD(phys chem), 33, AM, 34. *Prof Exp:* Instr physics, Prague Teachers Col, 30-33; from asst prof to prof-in-chg physics & chem, State Tech Col, Prague, 36-39; res chemist, Dept Dermat, Columbia Univ, 39-40; prof physics & head dept, St Ambrose Col, 40-42 & Minn State Teachers Col, Duluth, 42-46; assoc prof, Univ Mo, 46-48; prof, Univ Ark, 48-56; dir basic res lab, Res & Develop Labs, US Army, 56-68, chief sci consult staff, Advan Mat Concepts Agency, 68-71, sci adv, 71-73. *Concurrent Pos:* Fel, Prague Univ, 36-38; res chemist, Royal Yugoslavian Factory, Obilicevo, 39; chg eng, sci & mgt defense training, St Ambrose Col, 40-41; Minn State Teachers Col, Duluth, 42-46; consult, 73- *Mem:* AAAS; Am Chem Soc; Am Phys Soc; Am Asn Physics Teachers. *Res:* Solid state physics; chemistry; biophysics; magnetism; ferro-proteins in protein metabolism. *Mailing Add:* 5901 River Dr Lorton VA 22079

HARVARTH, LIANA, b Saginaw, Mich, Apr 27, 51. INFLAMMATION, CELL BIOLOGY. *Educ:* Univ Ill, Chicago, PhD(path), 76. *Prof Exp:* Sr staff fel, Nat Cancer Inst, 79-83; RES SCIENTIST & MICROBIOLOGIST, DIV BLOOD & BLOOD PROD, FOOD & DRUG ADMIN, 83- *Mem:* Am Asn Immunologists; Am Fedn Clin Res; Reticuloendothelial Soc; Am Soc Hemat. *Mailing Add:* 8200 Wisconsin Ave 1704 Bethesda MD 20814

HARVATH, LIANA, b Saginaw, Mich, Apr 27, 51. IMMUNOLOGY. *Educ:* Mich State Univ, BS, 73; Univ Ill, PhD(path), 76. *Prof Exp:* From instr to asst prof microbiol & immunol, Univ Ill Med Ctr, Chicago, 76-79, res asst prof, Dept Med, 79; sr staff fel, Nat Cancer Inst, NIH, 79-83; SR INVESTR-MICROBIOLOGIST, DIV HEMAT, CTR BIOL EVAL & RES, FOOD & DRUG ADMIN, BETHESDA, MD, 83- *Concurrent Pos:* Immunopath res assoc, West Side Vet Admin Med Ctr, Chicago, Ill, 76-79; grad res grand award, Univ Ill Med Ctr, Chicago, 76-77; young investr res grant, Nat Heart, Lung & Blood Inst, NIH, 77-79; vis scientist, Semmelweis Med Univ, Budapest, Hungary, 87; mem, Subcomt on Flow Cytometry, Nat Comt Clin Lab Standards, 87- *Mem:* Am Asn Immunologists; Am Fedn Clin Res; Am Soc Hemat. *Res:* Neutrophil chemotactic factors; regulation of neutrophil chemotaxis; quality control in flow cytometry. *Mailing Add:* Ctr Biol Eval & Res Hemat Dept Food & Drug Admin 8800 Rockville Pike Bldg 29 Rm 331 Bethesda MD 20892

HARVEL, CHRISTOPHER ALVIN, b Columbia, SC, Dec 7, 44. ASTRONOMY. *Educ:* Georgetown Univ, BS, 67; Univ SFla, MA, 73; Univ Fla, PhD(astron), 74. *Prof Exp:* Fel, Univ Fla, 74-75; res assoc astron, Nat Res Coun/Nat Acad Sci, 75-77 & NASA, Johnson Space Ctr, 77-78; resident astronr, Int Ultraviolet Explor Observ, 78-81, ASTRONR, SPACE TELESCOPE SCI INST, COMPUT SCI CORP, 81- *Mem:* Am Astron Soc. *Res:* Photographic and electrographic studies of diffuse nebulae, star clusters and galaxies; radiative transfer in circumstellar material; spectrophotometry, including visual and ultraviolet. *Mailing Add:* 6161 Stevens Forest Rd Columbia MD 21045

HARVEY, A(LEXANDER), b Stirling, Scotland, Sept 19, 30; m 67; c 2. ELECTRICAL ENGINEERING. *Educ:* Glasgow Univ, BSc, 52; Univ NMex, MS, 74. *Prof Exp:* Engr, Can Gen Elec Co, 54-56; design engr, Atomic Energy Can, Ltd, 56-58, res officer, 58-70; mem staff, Los Alamos Sci Lab, Univ Calif, 70-78; mem staff, Swiss Inst Nuclear Res, 78-80; staff mem, Los Alamos Nat Lab, Univ Calif, 80-88; HEAD, MECH ENG DEPT, SLAC, STANFORD, CALIF, 88- *Res:* In-reactor instrumentation; high temperature thermometry; radiation-hardened components, especially accelerator magnets; radiation effects and radiation hardening of components, especially electrical engineering accelerator magnets; ceramic vacuum chambers for accelerator particle beams. *Mailing Add:* Head Mech Eng Dept Bin 12 SLAC PO Box 4349 Stanford CA 94309

HARVEY, ABNER MCGEHEE, b Little Rock, Ark, July 30, 11; m 41; c 4. MEDICINE. *Educ:* Washington & Lee Univ, AB, 30; Johns Hopkins Univ, MD, 34. *Prof Exp:* Asst, 34-37, instr, 40-41, prof, 46-73, PROF MED, SCH MED, JOHNS HOPKINS UNIV, 73- *Concurrent Pos:* Intern & asst resident, Johns Hopkins Hosp, 34-37, resident physician, 40-41, physician-in-chief, 46-73; asst prof, Vanderbilt Univ, 41-42; fel, Nat Inst Med Res, London, 37-39; fel, Johnson Found Biophys, 39-40. *Honors & Awards:* Kober Medal, Asn Am Physicians, 81. *Mem:* Am Soc Clin Invest (pres, 56); fel Am Acad Arts & Sci; Asn Am Physicians (pres, 68); Am Clin & Climat Asn (pres, 71); master Am Col Physicians; Am Philos Soc. *Res:* Neurophysiology; clinical therapeutics; history of medicine. *Mailing Add:* Dept Med Johns Hopkins Hosp 1830 E Monument St Suite 7200 Baltimore MD 21205

HARVEY, ALAN ERIC, b Pittsburgh, Pa, Aug 2, 38; m 60; c 3. FORESTRY. *Educ:* Col Idaho, BS, 60; Univ Idaho, MS, 62; Wash State Univ, PhD(plant path), 68. *Prof Exp:* PLANT PATHOLOGIST, US FOREST SERV, 65- *Concurrent Pos:* Adj prof, Univ Idaho, 65-73 & 79-, Univ Mont, 74-79 & Mich Tech Univ, 78- *Mem:* AAAS; Am Phytopath Soc; Am Inst Biol Sci; Sigma Xi. *Res:* Host parasite physiology as a means of developing selective plant disease control agents; microbial nitrogen transformations; mycorrhizae and forest tree root diseases. *Mailing Add:* Forestry Sci Lab USDA 1221 S Main Moscow ID 83843

HARVEY, ALBERT BIGELOW, b Methuen, Mass, Dec 14, 38; m 63; c 2. ELECTRONICS, OPTICAL ENGINEERING. *Educ:* Lowell Tech Inst, BS, 60; Mass Inst Technol, SM, 64; Tufts Univ, PhD(phys chem), 66. *Prof Exp:* Head chem diag br, Naval Res Lab, 66-85; PROG DIR, LIGHTWAVE TECHNOL & NEW TECHNOLOGIES, NSF, 85- *Concurrent Pos:* Vis res assoc, Microwave Lab, Stanford Univ, 73-74; rotator, NSF, 85-86, permanent, 86- *Mem:* AAAS; Am Chem Soc; Soc Appl Spectros; Sigma Xi; Coblentz Soc. *Res:* Optical spectroscopy; lasers; photonics; optoelectronics; semiconductor physics; non linear processes. *Mailing Add:* 12300 Old Colony Dr Upper Marlboro MD 20772

HARVEY, ALBERT RAYMOND, b Lewiston, Maine, Jan 12, 21; m 50; c 1. MATHEMATICS. *Educ:* Bates Col, BS, 42; Harvard Univ, AM, 43, PhD(math), 47. *Prof Exp:* From instr to asst prof math, Univ NH, 46-48; res fel, Calif Inst Technol, 48-49; from asst prof to prof & chmn dept, 49-56, EMER PROF MATH, SAN DIEGO STATE UNIV, 83- *Concurrent Pos:* Fulbright lectr, Univ Baghdad, 57-58; vis prof, Col Virgin Islands, 63. *Mem:* Am Math Soc; Math Asn Am. *Res:* Analysis; mean of a function of exponential type. *Mailing Add:* 2350 Calle De La Garza La Jolla CA 92037

HARVEY, ALEXANDER LOUIS, b New York, NY, Oct 10, 17; m 53. PHYSICS, POWER GENERATION & DISTRIBUTION. *Educ:* City Col New York, BEE, 40; Polytech Inst Brooklyn, PhD(physics), 59. *Prof Exp:* From instr to asst prof physics, Polytech Inst Brooklyn, 55-62; assoc prof, Queens Col, NY, 62-66, chmn dept, 67-70, prof, 67-89, EMER PROF PHYSICS, QUEENS COL, NY, 89- *Mem:* Sigma Xi; NY Acad Sci; fel Am Phys Soc; Am Asn Physics Teachers. *Res:* Special relativity; general relativity. *Mailing Add:* Dept Physics Queens Col Flushing NY 11367-0904

HARVEY, BIRT, b Teheran, Iran, Nov 24, 28; m; c 5. PEDIATRICS. *Educ:* Johns Hopkins Univ, BA, 52; NY Univ, MD, 52. *Prof Exp:* CLIN PROF PEDIAT, STANFORD & UNIV CALIF SAN FRANCISCO. *Concurrent Pos:* Sr fel, Inst Health, Policy Studies, Univ Calif, San Francisco. *Mem:* Inst Med-Nat Acad Sci; Am Acad Pediat (pres). *Res:* Child health policy; child health financing. *Mailing Add:* 3885 Magnolia Dr Palo Alto CA 94306

HARVEY, BRYAN LAURENCE, b Newport, Wales, Nov 1, 37; Can citizen; m 60; c 2. GENETICS, PLANT BREEDING. *Educ:* Univ Sask, BSA, 60, MSc, 61; Univ Calif, Davis, PhD(genetics), 64. *Prof Exp:* Asst prof crop sci, Univ Guelph, 64-66; PROF CROP SCI, UNIV SASK, 66-, DIR, CROP DEVELOP CTR, 83- *Concurrent Pos:* Chmn, Prairie Regist Recommending Comt Grain, Expert Comt Plant Gene Resources, Genetics Group, Univ Sask, 77-81; vis prof crop sci, Univ Nairobi, 75; asst dean agr, Univ Sask, 80-83, head crop sci & plant ecol, 83-; dir, SeCan, 89-, chmn adv comt variety regist, 87- *Mem:* Fel Am Soc Agron; fel Crop Sci Soc Am; fel Agr Inst Can; Can Soc Agron; Genetics Soc Can; Master Brewers Am. *Res:* Salinity tolerance in barley; genetics of malting quality in barley; sprouting resistance in cereals; tissue culture cyto-genetics. *Mailing Add:* Crop Sci Dept Univ Sask Saskatoon SK S7N 0W0 Can

HARVEY, CHARLES ARTHUR, b Gering, Nebr, Aug 14, 29; m 52; c 3. CONTROL THEORY, MECHANICS. *Educ:* Nebr Wesleyan Univ, AB, 51; Univ Nebr, MA, 53; Univ Minn, PhD(math), 60. *Prof Exp:* Develop engr, Aero Div, Honeywell, Inc, Minneapolis, 55-57, sr res scientist, Res Dept, St Paul, 60-64, prin scientist, 64-65, res staff scientist, 65-69; sr resident res assoc, Marshall Space Flight Ctr, NASA, 69-71; res staff scientist, Systs & Res Ctr, Honeywell, Inc, 71-80, prin res staff scientist, 80-87; vis res fel, Elec Eng Dept, Univ Newcastle, Australia, 87-88; prof, Aerospace Eng & Mech Dept, Univ Minn, Minneapolis, 89-90; VIS PROF, ELEC ENG DEPT, NAT UNIV SINGAPORE, 88-89, 91- *Mem:* Inst Elec & Electronics Engrs; Am Math Soc; Soc Indust & Appl Math; Am Inst Aeronaut & Astronaut. *Res:* Ordinary differential equations and application to control theory. *Mailing Add:* 3843 Zenith Ave S Minneapolis MN 55410

HARVEY, CLARENCE CHARLES, (JR), b Winona, Miss, June 29, 18; m 43; c 2. CHEMISTRY. *Educ:* Univ Miss, BA, 39, MS, 41. *Prof Exp:* Develop chemist, Ethyl Corp, 41-46, corrosion chemist, 47- 50, develop assoc, 51-52, supvr, 53-57, mkt analyst, 57-64, mkt res mgr, 64-84; RETIRED. *Mem:* Am Chem Soc; Chem Mkt Res Asn (secy, 79-80). *Res:* Chemical market analysis; plastics development; corrosion. *Mailing Add:* 6168 Chandler Dr Baton Rouge LA 70808

HARVEY, DOUGLAS J, b Utica, Mich, Apr 25, 24; m 50; c 3. METALLURGICAL ENGINEERING. *Educ:* Mich State Univ, BS, 49, MS, 50, PhD(metall eng), 55. *Prof Exp:* Instr mech eng, Mich State Univ, 51-56; sr res metallurgist, Gen Motors Corp, 56-80, sr staff engr, Res Labs, 80-85. *Mem:* Am Welding Soc; Am Soc Metals; Am Inst Mining, Metall & Petrol Engrs. *Res:* Physical metallurgy; cast metals; welding metallurgy. *Mailing Add:* 11185 Clinton River Sterling Heights MI 48314-3512

HARVEY, DOUGLASS COATE, b Batavia, NY, Aug 28, 17. MECHANICAL ENGINEERING. *Educ:* Purdue Univ, BS & ME, 39. *Prof Exp:* Engr, Eastman Kodak, 39-82, dir corp prod mgt, 70-73, vpres & gen mgr, Apparatus Div, 73-77, exec vpres & gen mgr, 77-82; dir, Tex Instruments Inc, 82-89; CONSULT, 82- *Concurrent Pos:* Mem, Indust Mgt Coun. *Mem:* Nat Acad Eng; Optical Soc Am. *Mailing Add:* 3155 E Ave Rochester NY 14618

HARVEY, EVERETT H, REAL TIME SYSTEMS. *Educ:* Univ Ill, BS, 67; Univ Minn, PhD(physics), 71. *Prof Exp:* Res assoc, Univ Pa, 71-72; res assoc, Univ Wis, 72-78; COMPUT SCIENTIST, LAWRENCE BERKELEY LAB, 78- *Concurrent Pos:* Consult, Gailep Asn, 85- *Mem:* Am Phys Soc. *Mailing Add:* 1238 Josephine St Berkeley CA 94703

HARVEY, F REESE, b Atlantic Beach, Fla, Feb 7, 41; m 69; c 2. GEOMETRY & ANALYSIS. *Educ:* Carnegie-Mellon Univ, BS & MS, 63; Stanford Univ, PhD(math), 66. *Prof Exp:* Instr math, Univ Calif, Berkeley, 66-68; from asst prof to assoc prof, 68-73, PROF MATH, RICE UNIV, 73- *Concurrent Pos:* Alfred P Sloan fel, 72-74; assoc ed, Ind Math Jour, 76-; mem, US Nat Comt Math, 80-83; mem bd govs, Inst Math & Appln, 81-83; mem bd trustees, Math Sci Res Inst, 83- *Mem:* Am Math Soc. *Res:* Complex analysis, partial differential equations and differential geometry. *Mailing Add:* Dept Math Rice Univ PO Box 1892 Houston TX 77001

HARVEY, FLOYD KALLUM, audio-visual communications, for more information see previous edition

HARVEY, FRANCES J, II, b Latrobe, Pa, July 8, 43; m 65; c 2. METALLURGY, MATERIALS SCIENCE. *Educ:* Univ Notre Dame, BS, 65; Univ Pa, PhD(metall), 69. *Prof Exp:* Sr engr, Westinghouse Elec Co, 69-75, fel engr, 75-78; White House fel, Dept Defense, 78-79; prog mgr, 79-82, ENG MGR, WESTINGHOUSE ELEC CORP, 82- *Mem:* Am Inst Mining, Metall & Petrol Engrs; Am Chem Soc. *Res:* Chemical metallurgy including experimental and theoretical studies on the high temperature chemical behavior of energy conversion materials and the thermodynamic and kinetics of reactions in arc heated gas streams. *Mailing Add:* 445 Roosevelt Ave No A Sunnyvale CA 94086

HARVEY, GALE ALLEN, b New York, NY, Nov 17, 38; div; c 2. METEOR PHYSICS. *Educ:* NMex State Univ, BS, 62; Va Polytech Inst, MS, 66. *Prof Exp:* PHYSICIST, LANGLEY RES CTR, NASA, 62- *Honors & Awards:* Spec Achievement Award, Langley Res Ctr, NASA, 74. *Res:* Directs and operates FTIR and mass spectroscopy lab for gas cell, optical filter characterization, and contamination analyses, develops contamination analysis programs for space flight hardware and thermal-vacuum test chambers; analyze and report data. *Mailing Add:* Two Donald St Hampton VA 23669

HARVEY, GEORGE RANSON, b Boston, Mass, Dec 30, 37; m 60; c 3. ORGANIC CHEMISTRY, ATMOSPHERIC SCIENCES GENERAL. *Educ:* Boston Col, BS, 60; Mass Inst Technol, PhD(org chem), 64. *Prof Exp:* Sr res chemist, Monsanto Co, 64-69; asst scientist, Woods Hole Oceanog Inst, 70-72, assoc scientist, 72-77; OCEANOGR, ATLANTIC OCEANOG & METEOROL LABS, NAT OCEANIC & ATMOSPHERIC ADMIN, US DEPT COM, 77- *Mem:* Am Chem Soc; Oceanog Soc; Am Geophys Union. *Res:* Chemical oceanography; organic geochemistry; humicacid chemistry, atmospheric chemistry. *Mailing Add:* Atlantic Oceanog Lab Nat Oceanog & Atmospheric Admin 4301 Rickenbacker Causeway Miami FL 33149

HARVEY, HAROLD H, b Winnipeg, Man, Nov 20, 30; m 55; c 2. LIMNOLOGY, FISHERIES. *Educ:* Univ Man, BSc, 53, MSc, 56; Univ BC, PhD(fish physiol), 63. *Prof Exp:* Dir, Cultus Lake Field Sta, Int Pac Salmon Fisheries Comn, 61-63; from asst prof to assoc prof, 63-70, PROF ZOOL, UNIV TORONTO, 71- *Concurrent Pos:* Sci adv, Environ Can, 74-75. *Mem:* Am Fisheries Soc; Can Soc Zoologists; Int Asn Theoret & Appl Limnol; Am Inst Fishery Res Biol; Can Soc Environ Biol. *Res:* Environmental physiology of fishes, particularly kinds and abundance of fishes in relation to water quality; fish and dissolved gases; lake acidification. *Mailing Add:* Dept Zool Univ Toronto St George Campus 25 Harbord St Toronto ON M5S 1A1 Can

HARVEY, HARRY THOMAS, biology, for more information see previous edition

HARVEY, J PAUL, JR, b Youngstown, Ohio, Feb 28, 22; m 58; c 5. ORTHOPEDIC SURGERY. *Educ:* Harvard Univ, MD, 45. *Prof Exp:* Assoc attend surgeon, NY Hosp, 54-62; assoc prof, 62-67, chmn sect, 67-78, PROF ORTHOP SURG, SCH MED, UNIV SOUTHERN CALIF, 67-; CHIEF PHYSICIAN II, LOS ANGELES COUNTY GEN HOSP, 62- *Concurrent Pos:* Res fel, Hosp Spec Surg, 54-60, assoc attend surgeon orthop, 54-62, res assoc, 60-62; exchange fel, Royal Acad Hosp, Uppsala, Sweden, 57; mem bd dirs, Arthritis & Rheumatism Found, Southern Calif, 63-77. *Mem:* Am Orthop Asn; Am Acad Orthop Surg; Am Rheumatism Asn; Int Soc Orthop Surg & Traumatol. *Mailing Add:* 1200 N State St Box 302 Los Angeles CA 90033

HARVEY, JEFFREY ALAN, b San Antonio, Tex, Feb 15, 55. SUPERSTRING THEORY. *Educ:* Univ Minn, BS(physics) & BS(math), 77; Calif Inst Tecnol, PhD(physics), 81. *Prof Exp:* From asst prof to prof physics, Princeton Univ, 83-90; PROF PHYSICS, UNIV CHICAGO, 91- *Mem:* Am Phys Soc. *Res:* Superstring theory including heterotic string and study of string compuitification; cosmological implications of particle physics. *Mailing Add:* Enmco Fermi Inst Univ Chicago 5640 Ellis Ave Chicago IL 60637

HARVEY, JOHN, JR, b Philadelphia, Pa, July 7, 25; m 47; c 2. TECHNICAL MANAGEMENT ENVIRONMENTAL SCIENCES GENERAL. *Educ:* Univ Pa, AB, 49, MS, 51, PhD(org chem), 53. *Prof Exp:* Res assoc, 52-70, res chemist, 70-72, sr res chemist biochem dept, 72-80, sr res assoc agr prod div, E I du Pont de Nemours & Co Inc, 80-85; pres, Greenleaf Environ Inc, 86-89; RETIRED. *Concurrent Pos:* Chmn & fel, Agrochem Div, Am Chem Soc, 85- *Mem:* Am Chem Soc. *Res:* Determination of structure of natural products; agricultural chemicals; metabolism of pesticides; environmental fate of pesticides. *Mailing Add:* 203 W Pembrey Dr Wilmington DE 19803-2008

HARVEY, JOHN ADRIANCE, b New York, NY, Oct 14, 30; m 58; c 3. PSYCHOPHARMACOLOGY, NEUROSCIENCE. *Educ:* Univ Chicago, AB, 55, PhD(biopsychol), 59. *Prof Exp:* Res assoc biopsychol, Univ Chicago, 59-61, from asst prof to assoc prof psychol & pharmacol, 61-68; prof psychol & pharmacol, Univ Iowa, 68-88; PROF PHARMACOL, MED COL PA, 88- *Concurrent Pos:* USPHS res scientist award, 69-74; mem neuropsychol res rev comt, NIMH, 70-74, mem preclin psychopharmacol res rev comt, 75-79; consult ed, J Comp & Physiol Psychol, 71-; regional ed pharmacol, Biochem & Behav, 72-; chmn, Biopsychol Res Rev Comt, NIH, 81-85; ed, J Pharmacol & Exp Therapeut, 91- *Honors & Awards:* Merit Award, NIMH, 88. *Mem:* AAAS; Am Psychol Asn (pres); Am Soc Pharmacol & Exp Therapeut; Soc Neurosci; Am Soc Neurochem; fel, Am Psychol Asn; fel, Am Col Neuropsychopharmacol. *Res:* Effects of lesions on drug action and on central biochemistry; effects of lesions, drugs and biochemical substrates on operant and respondent behavior; biological basis of learning. *Mailing Add:* Dept Pharmacol Med Col PA EPPI 3200 Henry Ave Philadelphia PA 19129

HARVEY, JOHN ARTHUR, b Saskatoon, Sask, Dec 14, 21; US citizen; m 49; c 2. NUCLEAR PHYSICS. *Educ:* Queen's Univ, Ont, BSc, 45; Mass Inst Technol, PhD(physics), 50. *Prof Exp:* Physicist, Atomic Energy Can, Ltd, 45-46; asst, Mass Inst Technol, 46-50; assoc physicist, Brookhaven Nat Lab, 51-55; PHYSICIST, OAK RIDGE NAT LAB, 55- *Mem:* Am Phys Soc. *Res:* Neutron physics; neutron time-of-flight spectroscopy; electron linear accelerators. *Mailing Add:* Oak Ridge Nat Lab PO Box 2008-MS 6356 Oak Ridge TN 37830

HARVEY, JOHN COLLINS, b Youngstown, Ohio, Sept 11, 23; m 49; c 5. INTERNAL MEDICINE, GERIATRICS. *Educ:* Yale Univ, BS, 44; Johns Hopkins Univ, MD, 47; St Mary's Univ, PhD (theol), 88. *Prof Exp:* From asst to prof med, Johns Hopkins Univ, 47-72; PROF MED, GEORGETOWN UNIV, 72- *Concurrent Pos:* House officer & asst resident, Osler Med Serv, Johns Hopkins Hosp, 47-51, resident physician, 51-52, physician, 52-, dir med care clin, 54-57, dir med clins, 57-62, dir outpatient serv, 62-68; A Blaine Brower traveling fel from Am Col Physicians, Guys Hosp, Eng, 56. *Mem:* AAAS; Biophys Soc; fel Am Col Physicians; Am Fedn Clin Res; Am Clin & Climat Asn; fel Am Pub Health Asn. *Res:* Diseases of muscle and muscle physiology; medical ethics. *Mailing Add:* Kennedy Inst Ethics Georgetown Univ 209 Poulton Hall Washington DC 20057

HARVEY, JOHN GROVER, b Waco, Tex, Aug 10, 34. MATHEMATICS. *Educ:* Baylor Univ, BS, 55; Fla State Univ, MS, 57; Tulane Univ, PhD(math), 61. *Prof Exp:* From instr to asst prof math, Univ Ill, Urbana, 61-66; assoc prof, 66-75, PROF MATH & MATH EDUC, UNIV WIS-MADISON, 75- *Concurrent Pos:* Prin investr, Wis Res & Develop Ctr Cognitive Learning, Univ Wis-Madison, 68-78; co-ed, Am Math Monthly, 68-73; Killam vis scholar, Univ Calgary, 85; mem Comt on Testing, Math Asn Am, 84-, chair, 89- *Mem:* Math Asn Am; Am Educ Res Asn. *Res:* Mathematics education; uses of technologies in mathematics instruction and testing. *Mailing Add:* Dept Math Univ Wis Madison WI 53706

HARVEY, JOHN MARSHALL, b Fresno, Calif, Jan 13, 21; m 68. PLANT PATHOLOGY. *Educ:* Fresno State Col, AB, 42; Stanford Univ, MA, 48; Univ Calif, PhD, 50. *Prof Exp:* Asst plant path, Univ Calif, 48-50; from assoc plant pathologist to sr plant pathologist, USDA, 50-60, prin plant pathologist, Mkt Qual Res Div, Hort Field Sta, 60-64, invests leader, Western Fruit & Vegetable Invest, 64-72, res leader, Mkt Qual & Transp Res Lab, Agr Res Serv, 72-87; RETIRED. *Concurrent Pos:* Assoc plant path, Univ Calif, Davis, 56-87. *Mem:* AAAS; Am Phytopath Soc; Am Soc Heating, Refrig & Air-Conditioning Engrs; Am Inst Biol Sci. *Res:* Post-harvest diseases of horticultural crops; cold storage problems in California table grapes; effects of transit environment on market quality of fruits, vegetables and cut flowers; quarantine treatments for fruits & vegetables. *Mailing Add:* 5461 E Heaton Ave Fresno CA 93727

HARVEY, JOHN WARREN, b Los Angeles, Calif, Sept 13, 40; m 68; c 1. ASTROPHYSICS. *Educ:* Univ Calif, BA, 63, MA, 64; Univ Colo, PhD(astrogeophys), 69. *Prof Exp:* Asst astronr, Kitt Peak Nat Observ, 69-74, astronr, 74-83; ASTRONR, NAT SOLAR OBSERV, 84- *Concurrent Pos:*

Mem, solar physics working group, astron surv comt, Nat Acad Sci, 79-80 & solar & space physics comt, 81-84; chmn, Solar Physics Div, Am Astron Soc, 80-81; mem, astron adv comt, NSF, 82-85; mem, space & earth sci adv comt, NASA, 84-87; vpres, Comn 12, IAU, 85-88, pres, 89-91; mem, solar physics comt, 87-88, astron surv comt solar panel, Nat Acad Sci, 89-90. *Honors & Awards:* James Arthur Lectr, Ctr Astrophys, 81. *Mem:* Int Astron Union; Am Astron Soc; Am Geophys Union; Inst Elec & Electronics Engrs. *Res:* Solar magnetic and velocity fields, photosphere, chromosphere, prominences and corona; stellar magnetic and velocity fields; helioseismology, instrumentation. *Mailing Add:* Nat Solar Observ PO Box 26732 Tucson AZ 85726

HARVEY, JOHN WILCOX, b Hamilton, Ont, June 27, 38; m 62; c 2. HEALTH PHYSICS. *Educ:* McMaster Univ, BSc, 60, PhD(nuclear chem), 65. *Prof Exp:* Res assoc nuclear chem, Mass Inst Technol, 65-66; asst prof chem, Univ Man, 66-67; radiation protection chemist, Mass Inst Technol, 67-68; SR HEALTH PHYSICIST, MCMASTER UNIV, 68- *Concurrent Pos:* Spec lectr physics & eng physics, 75- *Mem:* Health Physics Soc. *Res:* Dosimetry. *Mailing Add:* 32 Throndale Crescent Hamilton ON L8S 3K2 Can

HARVEY, JOSEPH ELDON, b Pensacola, Fla, July 10, 27; m 52; c 6. ANATOMY, OTOLARYNGOLOGY. *Educ:* Univ Calif, Berkeley, BS, 54, MA, 64, PhD(anat), 68. *Prof Exp:* Phys therapist, San Francisco Bd Educ, Calif, 54-65 & Providence Hosp, Oakland, 66-68; asst prof anat, Univ Calif, Berkeley, 68; asst prof anat & otolaryngol, 68-72, ASSOC PROF EXP OTOLARYNGOL, SCH MED, WASH UNIV, 72- *Concurrent Pos:* Phys therapist, Shriner's Hosp for Crippled Children, San Francisco, 60-63. *Mem:* AAAS; Am Asn Anat; NY Acad Sci; Am Phys Ther Asn; Asn Res Otolaryngol. *Res:* Nerve regeneration; transplantation physiology; induced phonation; electromyography of laryngeal muscles; laryngeal physiology. *Mailing Add:* Otolaryngol Dept Wash Univ 4911 Barnes Hosp PLZ St Louis MO 63110

HARVEY, KENNETH C, b Shreveport, La, Sept 3, 47. TRAPPED ATOMS, LASER SPECTROSCOPY. *Educ:* Duke Univ, BSc, 69; Stanford Univ, MSc, 70, PhD(physics), 75. *Prof Exp:* Res assoc physics, Univ Toronto, 75-77; physicist, Nat Bur Standards, 77-81; res scientist, Univ Mich, Ann Arbor, 81-86; ASSOC PROF, SOUTHERN METHODIST UNIV, DALLAS, 86- *Honors & Awards:* IR 100 Award. *Mem:* Am Phys Soc; Optical Soc Am; AAAS; Sigma Xi. *Res:* Laser spectroscopy; trapped atoms; computers in physics. *Mailing Add:* Dept Physics Southern Methodist Univ Dallas TX 75275

HARVEY, LAWRENCE HARMON, b Jacksonville, Fla, Aug 6, 30; m 55; c 2. COTTON BREEDING & PRODUCTION. *Educ:* Univ Ga, BSA, 52, MS, 59, PhD(agron), 69. *Prof Exp:* Asst breeder, Greenwood Seed Co, 56-57; cotton breeder, DeKalb Agr Assoc, Inc, 57-64; dir res, Cotton Hybrid Res, Inc, 65-66; EXTEN AGRONOMIST, CLEMSON UNIV, 66- *Mem:* Crop Sci Soc Am; Agron Soc Am; Soil Sci Soc Am. *Res:* Conducting cotton tests-demonstrations on production problems. *Mailing Add:* 275 Poole Agr Ctr Clemson Univ Clemson SC 29634-0359

HARVEY, MACK CREEDE, b Barnsdall, Okla, Dec 25, 29; m 55, 71, 76; c 2. ANALYTICAL CHEMISTRY. *Educ:* Ind Univ, BS, 52, PhD(inorg chem), 56. *Prof Exp:* Res chemist, Celanese Chem Co, 56-58; sr res chemist, W R Grace & Co, 58-61; res chemist, El Paso Natural Gas Prod Co, 61-62, anal sect leader, 62-67; res chemist, 67-68, inorg & spectros sect leader, 69-72, staff res chemist, 72-80, INORG ANALYSIS SECT HEAD, SHELL DEVELOP CO, 68-; SCIENTIST, VALCO INSTRUMENTS CO, 80- *Mem:* Am Chem Soc; Soc Appl Spectros. *Res:* Infrared, x-ray and nuclear magnetic resonance spectrometry; organic functional group analysis; phosphate chemistry; gel permeation chromatography; high performance liquid chromatography; characterization of plastics and elastomers. *Mailing Add:* 2331 Greyburn Lane Houston TX 77080

HARVEY, MALCOLM, b London, Eng, Oct 9, 36; m 59; c 3. THEORETICAL NUCLEAR PHYSICS. *Educ:* Univ Southampton, BSc, 58, PhD(appl math), 61. *Prof Exp:* Nat Res Coun Can fel physics, Chalk River Nuclear Labs, Atomic Energy Can, Ltd, 61-62, from asst res officer to assoc res officer, 62-67, sr res officer, 67-89, DIR PHYSICS, CHALK RIVER LABS, AECL RES, 89- *Concurrent Pos:* Ford Found fel physics, Niels Bohr Inst, Copenhagen, Denmark, 64-65; vis sr fel, Manchester Univ, 78; vis prof, State Univ NY Stonybrook, 81; dir, Third Summer Inst Theoret Physics, Queens Univ, 85; chmn, Gordon Res Conf on Nuclear Structure, 86; chmn, Sci Teacher's Seminars, 87-91, TASCC Workshop Heavy Ion Physics, Chalk River, 88. *Mem:* Fel Am Phys Soc; Can Asn Physicists; Inst Particle Physics; Can Nuclear Soc; fel Royal Soc Can. *Res:* Theoretical nuclear physics, particularly groups in the shell model; self-consistent field theory; foundations of the spherical shell model; hadronic quark structure and the nucleon-nucleon force; soliton models of the hadrons. *Mailing Add:* Phys Sci Chalk River Labs AECL Res Chalk River ON K0J 1J0 Can

HARVEY, MICHAEL JOHN, b Doncaster, Eng, Dec 10, 35; m 65. BOTANY, AGROSTOLOGY. *Educ:* Univ Durham, BSc, 58, PhD(bot), 62. *Prof Exp:* Dept Sci Indust Res fel bot, Univ Birmingham, 61-63; from asst prof to assoc prof bot, Dalhousie Univ, 63-77, assoc prof biol, 77-90; RETIRED. *Res:* Grasses of North America; chromosome studies associated with speciation and the origin of polyploids. *Mailing Add:* 5061 Sooke Rd RR 2 Victoria BC V9B 5B4 Can

HARVEY, RALPH CLAYTON, b Knoxville, Tenn, July 12, 52; m 74; c 1. VETERINARY MEDICINE. *Educ:* Tenn Technol Univ, BS, 74; Univ Tenn, DVM, 81; Univ NC, Chapel Hill, MS, 81; Am Col Vet Anesthesiol, dipl, 87. *Prof Exp:* Geneticist, genetic toxicol, Nat Inst Environ Health Sci, 74-78; intern vet anesthesiol, Intensive-Care Unit, Cornell Univ, NY State Col Vet Med, 81-82, resident, 83-85; vet clin pract, Concord Vet Hosp, 82-83; fel anesthesiol & lab animal med, Cornell Univ Med Ctr, 83-84; ASST PROF VET ANESTHESIOL, UNIV TENN, 85- *Mem:* Am Vet Med Asn; Am Asn Vet Clinicians; AAAS. *Res:* Veterinary and comparative anesthesiology; anesthesia and analgesia in biomedical research. *Mailing Add:* Dept Urban Pract Col Vet Med Univ Tenn PO Box 1071 Knoxville TN 37901-1071

HARVEY, RICHARD ALEXANDER, b Salt Lake City, Utah, Nov 21, 36; m 58; c 2. BIOCHEMISTRY. *Educ:* Univ Utah, BS, 59, PhD(biochem), 64. *Prof Exp:* Fel, Inst Biophys & Biochem, Paris, 64-66; from asst prof to assoc prof, 66-78, PROF BIOCHEM, RUTGERS MED SCH, 78- *Mem:* Sigma Xi. *Res:* Fluorescence polarization spectroscopy: transient-state kinetics of enzymes and other biological macromolecules; mechanism of biosynthesis of riboflavin. *Mailing Add:* Rutgers Univ Med Sch PO Box 101 Univ Heights Piscataway NJ 08854

HARVEY, RICHARD DAVID, b Evansville, Ind, Dec 30, 28; m 52; c 3. MINERALOGY, PETROLOGY. *Educ:* Ind Univ, BS, 56, AM, 58, PhD(geol), 60. *Prof Exp:* Asst geologist, 60-65, assoc geologist, 65-70, GEOLOGIST, ILL STATE GEOL SURV, 70- *Mem:* Geol Soc Am; Mineral Soc Am; Geochem Soc. *Res:* Physical properties of rocks; economic geology. *Mailing Add:* 403 W Indiana Ave Urbana IL 61801

HARVEY, ROBERT GORDON, JR, b Upland, Calif, Dec 10, 45; m 65; c 2. WEED SCIENCE, AGRONOMY. *Educ:* Wash State Univ, BS, 67, PhD(agron), 70. *Prof Exp:* From asst prof to assoc prof, 70-78, PROF AGRON, UNIV WIS-MADISON, 78- *Mem:* Weed Sci Soc Am; Agron Soc Am. *Res:* Physiology and biochemistry of herbicide action on crops and weeds; interactions between herbicides and soil constituents; weed control in field and sweet corn, alfalfa, canning peas, lupines and soybeans. *Mailing Add:* Dept Agron 254 Moore Hall Univ Wis Madison WI 53706

HARVEY, ROBERT JOSEPH, b Staten Island, NY, Nov 3, 38; m 62; c 1. ORGANIC CHEMISTRY. *Educ:* Wagner Col, BS, 61; NY Univ, PhD(org chem), 65. *Prof Exp:* Res chemist, Halcon Res & Develop Corp, 65-85; PRES, VENTU-TECH INT, INC, 85- *Concurrent Pos:* Int Sales & Technol Transfer. *Mem:* Am Chem Soc; NY Acad Sci; Sigma Xi. *Res:* Orientation and reactivity in aromatic free radical substitution reactions; instrumental analysis; development of new processes; heterogeneous and homogeneous liquid phase oxidations; olefin dimerization; organometallic chemistry. *Mailing Add:* Three Canterbury Lane New Milford NJ 07646

HARVEY, RONALD GILBERT, b Ottawa, Ont, Sept 9, 27; US citizen; m 52; c 1. ORGANIC CHEMISTRY. *Educ:* Univ Calif, Los Angeles, AB, 52; Univ Chicago, MS, 55, PhD(chem), 60. *Prof Exp:* Proj leader org chem, Sinclair Res Labs, Ill, 56-58; instr & res assoc, 60-61, from instr to assoc prof org chem, 61-68, PROF ORG CHEM, BEN MAY INST, UNIV CHICAGO, 75- *Concurrent Pos:* USPHS spec fel, Imp Col, Univ London, 63-64; consult, Nat Cancer Inst, Nat Inst Environ, Health Sci, Israel Fedn Labor. *Mem:* AAAS; Am Chem Soc; Am Asn Cancer Res; Chem Soc Brit; Am Inst Chemists. *Res:* Polycyclic aromatic hydrocarbons; chemical carcinogenesis; metal ammonia reduction, novel synthetic methods; author of more than 260 scientific papers and book chapters. *Mailing Add:* Ben May Lab Univ Chicago 950 E 59th St Chicago IL 60637

HARVEY, ROSS BUSCHLEN, b Regina, Sask, Mar 27, 17. PHYSICAL CHEMISTRY, INFORMATION SCIENCE. *Educ:* Univ Sask, BSc, 35, MSc, 38; McGill Univ, PhD, 40. *Prof Exp:* Res chemist, Nat Res Coun Can, 40-41; mem, Dept Nat Defence, 47-68, asst dir res, Defence Res Estab, Suffield, 68-72, spec proj res, Defence Sci Info Serv, 72-77, head proj support off, 78-83; RETIRED. *Mem:* Am Chem Soc; Chem Inst Can; Sigma Xi. *Res:* Structure of gases by electron diffraction; airblast and shocks from chemical explosions; analysis of random processes; information analysis centres; energy research. *Mailing Add:* 340 Fourth Ave S W Medicine Hat AB T1A 4Z7 Can

HARVEY, STEPHEN CRAIG, b Bakersfield, Calif, Apr 13, 40; m 87; c 1. PHYSICS. *Educ:* Univ Calif, Berkeley, AB, 63; Dartmouth Col, PhD(physics), 71. *Prof Exp:* Engr, Aerojet Gen Corp, 63-65; volunteer, Peace Corps, 65-67; fel, 71-73, from asst prof to assoc prof, 73-86, PROF, UNIV ALA BIRMINGHAM, 86- *Mem:* Biophys Soc; AAAS; Sigma Xi; Fedn Am Scientists; Am Soc Biol Chemists; Am Chem Soc; Am Phys Soc. *Res:* Dynamic aspects of the structure of biological macromolecules. *Mailing Add:* Dept Biochem Univ Ala Birmingham AL 35294

HARVEY, STEWART CLYDE, b Denver, Colo, Feb 16, 21; m 42, 65; c 2. PHARMACOLOGY. *Educ:* Univ Colo, BA, 43; Univ Chicago, PhD(pharmacol), 48. *Prof Exp:* Asst chem & lab coordr, Univ Colo, 43-44, instr, 44-46; instr & dir pharmacol, Dent Sch, Univ Tex, 48-49; instr, Univ Utah, 49-50, asst res prof, 50-53, from assoc prof to prof, 53-88, EMER PROF PHARMACOL, COL MED, UNIV UTAH, 88- *Concurrent Pos:* Markle scholar, 52-57. *Mem:* Am Soc Pharmacol & Exp Therapeut; Sigma Xi. *Res:* Autonomic and cardiovascular pharmacology. *Mailing Add:* 2C219 Univ Med Ctr Univ Utah Salt Lake City UT 84132

HARVEY, THOMAS G, b Racine, Wis, Mar 18, 07. SCIENCE ADMINISTRATION. *Educ:* Mass Inst Technol, BS, 29; Univ Wis, MS, 40. *Prof Exp:* Mat engr, USN, 51-77; RETIRED. *Mem:* Fel Am Soc Metals. *Mailing Add:* 5685 N Delaware St Indianapolis IN 46220

HARVEY, THOMAS LARKIN, b Ulysses, Nebr, Jan 17, 26; m 52; c 7. ENTOMOLOGY. *Educ:* Kans State Col, MS, 51; Okla State Univ, PhD(entom), 63. *Prof Exp:* Asst entomologist, NMex Col, 52-53; asst entomologist, 54-57, from asst prof to assoc prof entom, 57-70, PROF ENTOM, FT HAYS EXP STA, KANS STATE UNIV, 70- *Honors & Awards:* Entom Soc Am Agr Recognition Award for Res, 74. *Mem:* Entom Soc Am. *Res:* Host plant resistance to insects; medical and veterinary entomology. *Mailing Add:* Ft Hays Exp Sta Hays KS 67601

HARVEY, THOMAS STOLTZ, b Louisville, Ky, Oct 10, 12; m 41; c 5. PATHOLOGY. *Educ:* Yale Univ, BS, 34, MD, 41. *Prof Exp:* Pathologist, Meriden Hosp, Conn, 43-45; pathologist, Med Res Lab, Army Chem Corps, 45-47; instr path, Sch Med, Univ Pa, 48-50, asst dir, Wm Pepper Lab Clin Med, 50-52; dir labs, Princeton Hosp, 52-60; pathologist, Vet Admin Hosp, Lyons, NJ, 61-75; med dir, Statlabs, 75-80. *Concurrent Pos:* Consult, NJ Bd Med Examr, 56-61; dir, Princeton Med Labs, 60-64. *Mem:* Am Soc Clin

Pathologists; Am Chem Soc; Asn Clin Scientists. *Res:* Role of liquid chromatography in the clinical lab; cytological, morphological changes produced by viruses. *Mailing Add:* 2401 W 25th No 5-A-8 Lawrence KS 66047

HARVEY, WALTER ROBERT, b Tucumcari, NMex, June 19, 19; m 40; c 3. ANIMAL BREEDING. *Educ:* Okla Agr & Mech Col, BS, 42; Iowa State Col, MS, 47, PhD, 49. *Prof Exp:* Instr animal husb, Iowa State Col, 47-49; assoc prof dairy husb, Univ Idaho, 50-54; biometrician in chg livestock res staff, Biomet Serv, Agr Res Serv, USDA, 54-64; Prof, 64-87, EMER PROF DAIRY, ANIMAL & POULTRY SCI & GENETICS, OHIO STATE UNIV, 87- *Concurrent Pos:* Ed-in-chief, J Animal Sci, 73-76. *Honors & Awards:* Animal Breeding, Am Soc Animal Sci, 68, Morrison Award, 87. *Mem:* Fel Am Soc Animal Sci; Am Dairy Sci Asn; Am Statist Asn; Biomet Soc; Sigma Xi. *Res:* Applied statistics. *Mailing Add:* Dept Dairy Sci Ohio State Univ 2027 Coffey Rd Columbus OH 43210

HARVEY, WALTER WILLIAM, b St John's, Nfld, Dec 18, 25; US citizen; div; c 3. EXTRACTIVE METALLURGY, WASTE GAS CLEANUP. *Educ:* Bowdoin Col, BA, 47; Mass Inst Technol, PhD(phys chem), 52. *Prof Exp:* Asst metall, Mass Inst Technol, 47-50; instr chem, Bowdoin Col, 50-51; from asst prof to assoc prof, St Lawrence Univ, 51-55, actg head dept, 53-55; asst prof, Williams Col, 55-56; mem res staff, Lincoln Lab, Mass Inst Technol, 56-63; staff scientist, Ledgemont Lab, Kennecott Copper Corp, 63-67, group leader chem & surface sci, 67-75; vpres & dir process develop, EIC Corp, 75-79; CONSULT EXTRACTIVE METALL, ENERGY & ENVIRON TECHNOL, 79- *Concurrent Pos:* Regional ed, Surface Sci, 63-68; guest lectr, Mass Inst Technol, 63 & 67; chmn, Boston Sect, Electrochem Soc, 64-65, local counr & rep to bd dirs, 65-68; vis res assoc, Nat Bur Standards, 69-70; chmn, electrolytic processes comt, Metall Soc of Am Inst Mining, Metall & Petrol Engrs, 74-75 & 80-82, & hydrometallurgy subcomt, electrolytic technol adv comt, Dept Energy, 77-80; prin investr, NSF res grants, 77-78 & 82-83 & Navy res contract, 90- *Mem:* Am Chem Soc; fel Inst Mining & Metall. *Res:* Surface properties and electrochemistry of semiconductors; physical chemistry of solid state; leaching chemistry of sulfide minerals; electrowinning and electrorefining; extractive metallurgy of manganese nodules, phosphorites, nickeliferous laterites and ilmenite and chromite ores; geothermal energy utilization; electrochemical energy storage; ocean hard minerals exploitation; circuit foil etching; gas desulfurization; electrolytic corrosion. *Mailing Add:* 35 Chester St Malden MA 02148-5402

HARVEY, WILLIAM ROSS, b Clarksville, Tenn, Apr 14, 30; m 53; c 4. AUTOMATED ANALYTICAL CHEMISTRY, PROCESS MONITORING. *Educ:* Austin Peay State Col, BS, 52; Univ Richmond, MH, 83. *Prof Exp:* Chemist, Barrow-Agee Labs, 56-60, Chemetron Corp, 60-64; chemist & proj leader anal chem, Phillip Morris Res Ctr, 64-90; RETIRED. *Mem:* Am Chem Soc. *Res:* Analytical chemistry; automated analytical methods development; ion chromatography, high-performance liquid chromatography, ultra violet visible spectroscopy; on stream process analysis; classical wet analytical chemistry. *Mailing Add:* Rte 1 Box 189 Cumberland VA 23040-9502

HARVILLE, DAVID ARTHUR, b Cleveland, Ohio Sept 6, 40; m 87. STATISTICS. *Educ:* Iowa State Univ, BS, 62; Cornell Univ, MS, 64, PhD(animal breeding), 65. *Prof Exp:* Res math statistician, Aerospace Res Labs, Wright-Patterson AFB, Ohio, 65-75; PROF STATIST, IOWA STATE UNIV, 75- *Concurrent Pos:* Assoc ed, Biomet, 71-76 & J Am Statist Asn, 83-85. *Mem:* Fel Am Statist Asn; fel Inst Math Statist; Biomet Soc; Int Statist Inst. *Res:* Development of improved statistical methodology for analyzing data sets on the basis of linear statistical models. *Mailing Add:* Statist Lab Snedecor Hall Iowa State Univ Ames IA 50011

HARVILLE, JOHN PATRICK, b Eureka, Calif, Jan 13, 18; m 39; c 2. FISHERIES MANAGEMENT. *Educ:* San Jose State Col, AB, 40; Stanford Univ, MA, 50, PhD, 56. *Prof Exp:* Lab asst entom, Stanford Univ, 41; teacher pub sch, 41-45; instr biol, San Mateo Jr Col, 46-48; instr biol, San Jose State Col, 48-51, asst prof biol & sci educ, 51-56, assoc prof, 56-61, prof, 61-71, dir, Moss Landing Marine Labs, 65-71; exec dir, 71-83, CONSULT, PAC MARINE FISHERIES COMN, 83- *Mem:* Ecol Soc Am; Am Fisheries Soc; Am Inst Fisheries Res Biol. *Res:* Bionomics of the California oak moth; population dynamics; biology of fishes and fresh water insects; local fauna. *Mailing Add:* 2430 SW Boundary St Portland OR 97201

HARVIN, JAMES SHAND, b Sumter, SC, Dec 19, 29; m; c 4. PLASTIC SURGERY. *Educ:* Duke Univ, BA, 51; Med Col SC, MD, 53. *Prof Exp:* Staff surgeon, Gen Surg Serv, Wright-Patterson Air Force Hosp, 61-63, asst chief plastic surg serv, 61-62, chief, 62-63; asst chief plastic surg serv, Wilford Hall Air Force Hosp, 63-64, chief, 64-67; ASSOC PROF PLASTIC & MAXILLOFACIAL SURG & CHIEF DIV, MED UNIV SC, 67- *Concurrent Pos:* Staff mem, Charleston County Hosp, 67-; consult, Vet Admin Hosp, Charleston, 67- *Mem:* Fel Am Col Surg; Am Soc Plastic & Reconstruct Surg; Soc Head & Neck Surg; Am Cleft Palate Asn; Aerospace Med Asn. *Res:* Renal transplantation. *Mailing Add:* 2nd Gen Hosp - LARMC APO NY NY 09180

HARWELL, KENNETH EDWIN, b Kellyton, Ala, Nov 22, 36; c 3. AEROSPACE ENGINEERING, MECHANICS. *Educ:* Univ Ala, BS, 59; Calif Inst Technol, MS, 60, PhD(aeronaut), 63. *Prof Exp:* Res asst aeronaut, Calif Inst Technol, 60-63; from assoc prof to prof aerospace eng, Auburn Univ, 63-76; dir, Gas Diagnostics Res Div, 76-82, PROF AEROSPACE & MECH, SPACE INST, UNIV TENN, 76-, DEAN, 82- *Concurrent Pos:* Res initiation grant, Auburn Univ, 63-65; consult, Hayes Int Corp, 65-68, US Army Ballistic Res Lab, 69, USAF, AFB, Fla, 70-73, USN & US Army, & Auburn Res & Eng Assocs, 76-; asst to dir, US Army Missile Res & Develop & Elec Eng Lab, Huntsville, 73-74; mem, US Army Missile Sci Adv Group, 74-77; spec asst to vpres & dean, Univ Tenn Space Inst, 80-81. *Honors & Awards:* Gen H H Arnold Res Award, Am Inst Aeronaut & Astronaut, 81. *Mem:* Am Inst Aeronaut & Astronaut; Am Phys Soc; Am Soc Eng Educ. *Res:* Diagnostics; missile aerodynamics; exhaust plume gas dynamics; radiation; laser-gas interactions; laser velocimetry; laser scattering from particles. *Mailing Add:* Univ Tenn Space Inst 1916 Country Club Dr Tullahoma TN 37388-8897

HARWELL, KENNETH ELZER, b Hillsboro, Tex, Sept 11, 21; m 61. INDUSTRIAL ORGANIC CHEMISTRY, POLYMER CHEMISTRY. *Educ:* Baylor Univ, BS, 45; Univ Tex, MA, 47, PhD(org chem), 51. *Prof Exp:* Chemist, Spec Probs, Union Carbide Chem Corp, 47-48; res scientist, Tex Cotton Res Comt, 48-50; res chemist, Celanese Corp Am, 51; res chemist, Res Found, Tex A&M Univ, 52-54; sr res chemist, Jefferson Chem Co, 54-57; chief exec, Tex Fine Chem Co, 57-61; res chemist, Continental Oil Co, 61-65; sr res chemist, Gulf Res & Develop Co, 65-72; sr res chemist, Cook Paint & Varnish Co, 73-79; MGR, MERRIAM CHEM DEVELOP CO, 79- *Mem:* Am Chem Soc. *Res:* Organic synthesis; plastics technology; process development; electronics and instrumentation; synthesis of new monomers. *Mailing Add:* 7532 Abendeen Kansas City KS 66102

HARWELL, MARK ALAN, b Lewisburg, Tenn, Nov 25, 47; m 68; c 2. SYSTEMS ECOLOGY, ECOLOGY. *Educ:* Emory Univ, BS, 69, PhD(ecol), 78. *Prof Exp:* Teaching asst, Emory Univ, 75-77; res asst & comput consult, Emory Univ & AGNS Nuclear Fuel Recycling Plant, 77-78; res scientist environ assessment, Water & Land Resources Dept, Wash, 78-80; consult environ, Cannon Beach, Ore, 80-81; assoc dir, Ecosyst Res Ctr, Cornell Univ, 81-91, dir, Global Environ Prog, 87-91; CONSULT, 91- *Concurrent Pos:* Fel, NSF, 69-75 & Emory Univ, 76-77; mem sci adv bd, Environ Protection Agency. *Mem:* AAAS; Sigma Xi; Ecol Soc Am; Am Inst Biol Sci. *Res:* Econsystem model of air pollutant effects on forests; cross-systems comparison of biogeochemical cycles; long term environmental consequences of nuclear waste disposal; indirect effects of nuclear war; global environment stresses; ecosystems modeling. *Mailing Add:* 301 Redwood Lane Key Biscayne FL 33149

HARWIT, MARTIN OTTO, b Prague, Czech, Mar 9, 31; US citizen; m 57; c 3. ASTROPHYSICS, HISTORY OF ASTRONOMY. *Educ:* Oberlin Col, BA, 51; Univ Mich, MA, 53; Mass Inst Technol, PhD(physics), 60. *Prof Exp:* Physicist, Univ Mich, 54-55; NATO fel, Cambridge Univ, 60-61; NSF fel, 61-62, res assoc astrophys, 62, from asst prof to prof astron, 62-87, chmn dept, 71-76, co-dir, Prog Hist & Philos Sci & Technol, 85-87, EMER PROF ASTRON, CORNELL UNIV, 88-; DIR, NAT AIR & SPACE MUS, SMITHSONIAN INST, 87- *Concurrent Pos:* Vis res assoc, E O Hulburt Space Ctr, 63-64; NSF res grant, 63-; Nat Acad Sci exchange visitor, Czech Acad Sci, 69-70; vpres, Spectral Imaging, Inc, 70-77; Alexander von Humboldt Found Sr US Scientist fel, 76-77; Alexander von Humboldt Award, WGer, 76; external mem, Max Planck Inst Radioastron, WGer, 79- *Mem:* Am Phys Soc; Am Astron Soc; Royal Astron Soc; Hist of Sci Soc; AAAS. *Res:* Galaxy and star formation: cosmic dust; comets; infrared optics and astronomy; rocket astronomy; science policy; history and philosophy in science; educational astronomy. *Mailing Add:* Nat Air & Space Mus Smithsonian Inst Washington DC 20560

HARWOOD, CLARE THERESA, b New Bedford, Mass, Apr 5, 20. PHARMACOLOGY. *Educ:* Seton Hill Col, AB, 41; Georgetown Univ, MS, 52, PhD(pharmacol), 55. *Prof Exp:* Lab technician hemat, Goodyear Fabric Corp, Mass, 41-42; anal chemist, US Naval Torpedo Sta, RI, 42-45; asst chem, Syracuse Univ, 45-46; instr sci, Sch Nursing, Union Hosp, Mass, 47-48; biochemist neurochem, Med Labs, US Army Chem Ctr, Md, 48-51; biochemist neuroendocrinol, Walter Reed Army Inst Res, DC, 51-58; biochemist clin neuropharmacol res ctr, NIMH, 58-61; chief pharmacol & biochem lab, Clin Res Inst & asst prof pharmacol, Med Sch, Georgetown Univ, 61-68, lectr, 56-61; chief div drug sci, Bur Narcotics & Dangerous Drugs, Drug Enforcement Admin, Dept Justice, 68-70, actg chief drug control div, 70-71, chief biol res br, 71-73, chief pharmacologist, 73-76; FAC MEM, OBLATE COL, 76- *Mem:* Am Soc Pharmacol & Exp Therapeut; Am Chem Soc; Sigma Xi. *Res:* Central regulation of anterior pituitary activity. *Mailing Add:* 3506 Dunlop St Chevy Chase MD 20815

HARWOOD, COLIN FREDERICK, b Manchester, Eng, July 17, 37; US citizen; m 58. PHYSICAL CHEMISTRY, CHEMICAL ENGINEERING. *Educ:* London Univ, BSc, 63, PhD(phys chem), 69. *Prof Exp:* Sci teacher chem & physics, St John's Col, Nassau, 63-66; sci adv fine particles, ITT Res Inst, Chicago, 69-77; ASST VPRES RES & DEVELOP FINE FILTRATION, PALL CORP, 77- *Concurrent Pos:* Lectr, Ill Inst Technol, 69-71 & Ctr Prof Advan, 70-; mem, Powder Adv Ctr. *Mem:* Fine Particle Soc (secy, 77); AAAS. *Res:* Measurement and characterization of particulate properties; control of particulate emissions by filtration; development of novel monitoring concepts; development of new filter medium. *Mailing Add:* 17 Southland Dr Glen Cove NY 11542

HARWOOD, DAVID SMITH, b Manchester, Vt, Jan 19, 36; m 58; c 2. GEOLOGY, PETROLOGY. *Educ:* Dartmouth Col, AB, 58; Harvard Univ, PhD(geol), 67. *Prof Exp:* GEOLOGIST, US GEOL SURV, 64- *Mem:* Mineral Soc Am; Geol Soc Am. *Res:* Geologic mapping of early Paleozoic metamorphic rocks in west-central Maine; tectonics of southwestern Berkshire Highlands, Massachusetts and Connecticut. *Mailing Add:* 210 Royal Palm Ave Half Moon Bay CA 94019

HARWOOD, HAROLD JAMES, b Streator, Ill, Mar 2, 31; m 52; c 3. POLYMER CHEMISTRY. *Educ:* Univ Akron, BS, 52; Yale Univ, PhD(org chem), 56. *Prof Exp:* Res chemist, Monsanto Chem Co, 55-59; from asst prof to assoc prof chem, 59-69, PROF POLYMER SCI & CHEM & RES ASSOC INST POLYMER SCI, UNIV AKRON, 69- *Concurrent Pos:* Res assoc, Inst Rubber Res, 59-69; consult, Chem Industs. *Mem:* Am Chem Soc. *Res:* Organic polymer chemistry; characterization of sequence distribution in copolymers; organophosphorus chemistry; synthesis and characterization of macromolecules; polymer reactivity; synthetic polypeptides. *Mailing Add:* Dept Chem & Polymer Sci Univ Akron Akron OH 44325

HARWOOD, JULIUS J, b New York, NY, Dec 3, 18; m 41; c 3. METALLURGY, CERAMICS. *Educ:* City Col New York, BS, 39; Univ Md, MS, 53. *Hon Degrees:* DEng, Mich Tech Univ, 86. *Prof Exp:* Mat engr, US Naval Gun Factory, 40-46; metallurgist, Off Naval Res, 46-60, head metall br, 53-60; mgr metall dept, Sci Lab, Ford Motor Co, 60-69, mgr, Res Planning Eng & Res Staff, 69-71, dir, Mat Sci Lab, Eng & Res Staff, 71-83; prof eng, Wayne State Univ, 84; pres, Ovonic Synthetic Mat Co, 84-87; PRES, HARWOOD CONSULTS, 88- *Concurrent Pos:* Chmn, Gordon Res Conf Corrosion, 56; chmn corrosion res coun, Eng Found, 64-66; mem bd control, Mich Technol Univ, 66-86, chmn bd, 70-71 & 86-87; nat mat adv bd, Nat Acad Sci-Nat Res Coun; chmn, Vis Comt, Sch Eng, Rensselaer Polytech Inst; mem, Mat Adv Comt, Off Technol Assessment, US Cong; mem, Mat Adv Comt, Nat Sci Fedn, 74-76; mem, Mich Asn Gov Bds of Univs; adj prof, Wayne State Univ, 75-; chmn, Nat Mat Adv Bd, Nat Acad Sci-Nat Res Coun, 77-79; Am Ceramic Soc Orton Lectr, 78; vpres Metro Ctr High Tech, 84. *Honors & Awards:* John H Shoemaker Award, Am Soc Metals, 77. *Mem:* Nat Acad Engrs; Am Inst Mining, Metall & Petrol Engrs (pres, 76); fel Am Soc Metals; fel Metall Soc (pres, 73); fel AAAS; hon mem Am Inst Mech Eng; Am Ceramic Soc. *Res:* Physical metallurgy; stress corrosion cracking; advanced materials, research management; technology transfer; materials policy and management; ceramics surface modification technology; commercialization of materials technology. *Mailing Add:* 5272 Wrightway E West Bloomfield MI 48322

HARWOOD, RICHARD ROLAND, agronomy, for more information see previous edition

HARWOOD, THOMAS RIEGEL, b Knoxville, Tenn, Dec 9, 26; m 49, 76; c 3. SURGICAL PATHOLOGY, IMMUNOHEMATOLOGY. *Educ:* Georgetown Univ, BS, 49; Vanderbilt Univ, MD, 53; Am Bd Path, cert, 58. *Prof Exp:* Asst path, Sch Med, Vanderbilt Univ, 54-55; instr, Med Col Va, 55-56; from instr to asst prof, Sch Med, 56-73, clin assoc prof, Sch Med & Sch Dent, 73-90, EMER ASSOC PROF PATH, NORTHWESTERN UNIV, 90-; MED DIR, BLOOD SERVS CHICAGO, 73- *Concurrent Pos:* Asst pathologist, Wesley Mem Hosp, Chicago, 58-59; assoc chief lab serv, Vet Admin Lakeside Med Ctr, 69-73, actg chief, 73-76 & 80-81; assoc attend staff, Northwestern Mem Hosp, Chicago, 80-90; pathologist, Vet Admin Lakeside Med Ctr, 76-87 & Ref Path Labs, Inc. *Mem:* Am Asn Pathologists; Am Soc Clin Pathologists; Am Asn Blood Banks; AMA; AAAS; NY Acad Sci. *Res:* Neoplasia; Mallory bodies in liver disease; salivary gland disease; cardiovascular disease; blood banking and immunohematology. *Mailing Add:* 916 Williamsburg Village Dr Jackson TN 38305

HARWOOD, WILLIAM H, b San Antonio, Tex, Sept 14, 22; m 43; c 4. PHYSICAL CHEMISTRY. *Educ:* Trinity Univ, Tex, BS, 48; Univ Tex, MA, 49, PhD(chem), 54. *Prof Exp:* Res chemist, Continental Oil Co, 51-57, sr res chemist, 57-60, res assoc chem, 60-69; from asst prof to prof chem, Cameron Univ, 69-87; RETIRED. *Mem:* AAAS; fel Am Inst Chem; Soc Am Archaeol; Am Phys Soc; Am Chem Soc. *Res:* Organic electrochemistry; metallo-organic chemistry; specific heats; spectroscopy; physical chemistry of surface films; colloids, especially foams, emulsions, dispersions, solid lubricants; theoretical and experimental thermodynamics; chemical kinetics; archaeology of Oklahoma and the Southwest. *Mailing Add:* PO Box 6416 Lawton OK 73506

HARWOOD-NASH, DEREK CLIVE, b Bulawayo, Rhodesia, Feb 11, 36; m 63; c 3. PEDIATRIC RADIOLOGY, PEDIATRIC NEURORADIOLOGY. *Educ:* Univ Cape Town, MB & ChB, 60; Univ Toronto, DMR, 65; FRCPC, 67. *Prof Exp:* PROF RADIOL, UNIV TORONTO, 76-; PEDIAT NEURORADIOLOGIST, HOSP SICK CHILDREN, 68-, RADIOLOGIST-IN-CHIEF, 78- *Concurrent Pos:* Co-pres, Int Pediat Radiol Cong, 87. *Mem:* Am Soc Neuroradiol (pres, 87-); Radiol Soc NAm; Soc Pediat Radiol; Am Asn Neurol Surgeons; Can Asn Radiologists; hon mem Swed Med Soc Radiol; hon mem Europ Soc Pediat Radiol (pres, 87); hon mem Brazilian Radiol Soc. *Res:* Development and practice of pediatric neuroradiology towards disease of the brain, skull, spine and spinal cord in children and the application to neurology and to neurosurgery and the development of imaging techniques. *Mailing Add:* Dept Radiol Hosp Sick Children 555 Univ Ave Toronto ON M5G 1X8 Can

HARYETT, ROWLAND D, b Bancroft, Ont, Aug 18, 23; m 48; c 1. ORTHODONTICS. *Educ:* Univ Toronto, DDS, 51, MSD, 53; FRCD(C). *Prof Exp:* Assoc prof, 57-68, dir div, 58-74, PROF ORTHOD, FAC DENT, UNIV ALTA, 68- *Mem:* Am Asn Orthod; Can Dent Asn; fel Int Col Dent; Can Asn Orthodontists (pres, 81). *Mailing Add:* 516 Med Bldg 11010 Jasper Ave Edmonton AB T5K 0K9 Can

HASAN, ABU RASHID, b Bangladesh, May 25, 49; m 73; c 2. CHEMICAL ENGINEERING, PETROLEUM ENGINEERING. *Educ:* Bangladesh Univ Eng & Technol, BSc Eng, 72; Univ Waterloo, Ont, MASc, 75, PhD(chem), 79. *Prof Exp:* From asst prof to assoc prof, 79-90, PROF STOICH & KINETICS, UNIV NDAK, 90- *Concurrent Pos:* Assoc Western Univ fel, Idaho Nat Eng Lab, 87-88; Can Commonwealth fel, 73-78. *Mem:* Am Inst Chem Engrs; Soc Petrol Engrs; Sigma Xi. *Res:* Multiphase flow; production logging, slurry rheology; moisture removal from low-rank coal (steam drying of lignite, dewatering of peat by solvent extraction); alternatives to diesel fuel; pressure transient analysis in oil wells. *Mailing Add:* Chem Eng Dept Box 8101 Univ NDak Grand Forks ND 58202

HASAN, MAZHAR, b India, May 20, 27; m 57; c 2. NUCLEAR PHYSICS, PLASMA PHYSICS. *Educ:* Aligarh Muslim Univ, India, BSc, 48, MSc, 50; Ill Inst Technol, PhD(nuclear physics), 59. *Prof Exp:* Lab instr, Aligarh Muslim Univ, India, 48-50; lectr physics, Forman Christian Col, Pakistan, 50-51; lectr, Loyola Univ, Ill, 54; asst physicist biol sci, Chicago Univ, 54-55; instr math & physics, Northern Ill Univ, 55-56, asst prof physics, 58-60; instr, Univ Ill, 56-58; sr physicist, Res Div, Gen Dynamics/Electronics, 60-61; assoc prof, 61-67, PROF PHYSICS, NORTHERN ILL UNIV, 67- *Concurrent Pos:* Fulbright lectr, Univ Alexandria, 67-68, Tehran Pahlavi Univ & Gondishapour Univ, Iran, 68-69. *Mem:* Am Phys Soc; Am Asn Physics Teachers; Sci Res Soc Am; Sigma Xi (vpres, 64-65, pres elect, 65-66). *Res:* Nuclear reactions and nuclear structure; principal methods of mathematical physics; plasma physics; controlled thermonuclear fusion. *Mailing Add:* Dept Physics Northern Ill Univ De Kalb IL 60115

HASAN, SYED EQBAL, b Patna, India, Apr 15, 39; US citizen; m 68; c 3. GEOTECHNICS, UNDERGROUND SPACE DEVELOPMENT & UTILIZATION. *Educ:* Patna Univ, BS, 60; Rookee Univ, MS, 63; Purdue Univ, PhD(eng geol), 78. *Prof Exp:* Jr res fel geol, Rookee Univ, 63-64; jr geologist, Geol Surv India, 65-70, sr eng geologist, 70-73; res asst geol, Purdue Univ, 73-78; asst prof, 79-83, ASSOC PROF GEOL, UNIV MO, 84- *Concurrent Pos:* Vis asst prof, Univ Ariz, 78-79, Mich Technol Univ, 78, Kuwait Univ, 85-86. *Mem:* Asn Eng Geologists; Int Asn Eng Geol; Sigma Xi; Geol Soc India; Int Soc Rock Mech. *Res:* Environmental geology and geotechnics: waste management; development of underground space for human use and occupancy; landslides. *Mailing Add:* Dept Geosci Univ Mo 5100 Rockhill Rd Kansas City MO 64110-2499

HASBROOK, ARTHUR F(ERDINAND), b Wichita, Kans, Jan 9, 13; m 43. ELECTRONICS, COMPUTERS. *Educ:* San Antonio Col, 30-33. *Prof Exp:* Seismic observer-operator, Petty Geophys Eng Co, Tex, 33-35; head elec dept, Petty Labs, Inc, Petty Geophys Eng Co, 35-38, head gravity res, gravity survs, 38-42, in charge electronic surveying & navig, 46-57, res supvr, 58-66, res assoc, 66-67, mem tech staff, 67-72; off, US Air & Signals Corp, Wrightfield, OH, 42-46; mem advan tech staff, Geosource, Inc, 73-75; TECH CONSULT, AHLEGRE ALBS, 76- *Mem:* AAAS; sr mem Inst Elec & Electronics Engrs; Am Geophys Union. *Res:* Research and design of geophysical and exploration equipment; research and development of electronic surveying equipment; geophysical data processing and analysis. *Mailing Add:* 9590 Oakland Rd San Antonio TX 78240

HASCALL, GRETCHEN KATHARINE, b Cleveland, Ohio, Nov 4, 41. CELL BIOLOGY. *Educ:* DePauw Univ, BA, 64; Rockefeller Univ, PhD(cell biol), 71. *Prof Exp:* BIOLOGIST, NAT INST DENT RES, NIH, 75- *Res:* Electron microscopy of cartilage matrix proteoglycans. *Mailing Add:* 5600 Fishers Lane Rm 8B45 Rockville MD 20857

HASCALL, VINCENT CHARLES, JR, b Burwell, Nebr, May 26, 40; div; c 2. BIOCHEMISTRY. *Educ:* Calif Inst Technol, BS, 62; Rockefeller Univ, PhD(biol sci), 69. *Hon Degrees:* DMed Univ Lund, Sweden, 86. *Prof Exp:* From asst prof to assoc prof biol chem & oral biol, Sch Dent, Univ Mich, Ann Arbor, 69-75; sr staff fel, 75-76, RES CHEMIST, NAT INST DENT RES, NIH, 76-, CHIEF PROTEOGLYCAN CHEM SECT, 78- *Concurrent Pos:* Swed Med Coun vis scientist, Univ Lund, 73; mem, Pathobiol Study Sect, NIH, 76-80; foreign work study fel, Univ Monash, Australia, 81; mem res coun, Juv Diabetes Found, 84; chmn, Gordon Conf Proteoglycans, 86; vis prof, biochem, Rush-Presby-St Lukes Med Ctr, 89- *Mem:* Am Soc Biol Chemists & Molecular Biologists; Orthop Res Soc; Am Soc Cell Biologists; NAm Soc Complex Carbohydrates (pres, 87); Sigma Xi. *Res:* Biochemistry and biophysics of proteoglycans and glycoproteins of connective tissues. *Mailing Add:* Nat Inst Dent Res NIH Bethesda MD 20892

HASCHEMEYER, AUDREY ELIZABETH VEAZIE, b Chicago, Ill, Oct 31, 36; c 2. BIOCHEMISTRY, ENVIRONMENTAL PHYSIOLOGY. *Educ:* Univ Ill, Urbana, BS, 57; Univ Calif, Berkeley, PhD(phys chem), 61. *Prof Exp:* Res assoc biol, Mass Inst Technol, 61-64; asst biologist, Mass Gen Hosp, 65-69; assoc prof biol sci, 69-74, PROF BIOL & BIOCHEM, HUNTER COL, 74-, CHMN, DEPT BIOL SCI, 80- *Concurrent Pos:* USPHS fel, Mass Inst Technol, 62-64; Helen Hay Whitney Found fel, Mass Gen Hosp, 65-67; assoc, Harvard Med Sch, 67-69; mem grad fac, City Univ New York, 69-; bd fel & associateships, Nat Res Coun, 74-76; Am Cancer Soc Res Scholar, King's Col, London, 76; univ comt res, 77-78; chief scientist, R/V Alpha Helix, Caribbean-Pac, 78; proj dir, US Antarctic Res Prog, 78-, chief scientist, US Coast Guard Cutter Polar Star, Ross Sea, 81; mem corp, Marine Biol Lab, Woods Hole; grants, NSF, 69-, NIH, 70-77 & NATO, 78- *Mem:* Am Physiol Soc; Am Soc Biol Chemists; Biophys Soc; fel AAAS; French Soc Biol Chemists. *Res:* Regulation of protein synthesis in higher organisms; cold adaptation of fish; mechanism of action of thyroid hormone; biological reaction rates in vivo. *Mailing Add:* Dept Biol Sci Hunter Col-City Univ New York 695 Park Ave New York NY 10021

HASCHKE, FERDINAND, b Linz, Austria, Sept 16, 48; m 72; c 4. PEDIATRIC NUTRITION. *Educ:* Univ Vienna, MD, 72; Austrian Bd Physicians, License Pediat, 78. *Prof Exp:* Asst prof pediat, Univ Vienna, 77-79; res fel nutrit, Univ Iowa, 79-80; from asst prof to assoc prof, 81-88, PROF PEDIAT, UNIV VIENNA, 88- *Concurrent Pos:* Lectr, dept nutrit, Univ Vienna, 86-; adv, dept toxicol, WHO/Euro Copenhagen, 85; vis prof, Allama Iqbal Med Col, Pakistan, 86. *Honors & Awards:* Milupa Award, Pediat Soc WGer, 84; V Pirquet Award, Pediat Soc Austria, 86. *Mem:* Am Soc Clin Nutrit; Am Inst Nutrit; Soc Pediat Res. *Res:* Pediatric nutrition, mainly infant nutrition; growth studies of infants; body composition; trace elements; environmental contamination. *Mailing Add:* Dept Pediat Univ Vienna 18-20 Wahringer Gurtel Vienna A-1097 Austria

HASCHKE, JOHN MAURICE, b San Antonio, Tex, Aug 5, 41; m 68; c 2. SOLID STATE CHEMISTRY, HIGH TEMPERATURE CHEMISTRY. *Educ:* Tex Tech Univ, BS, 64; Mich State Univ, PhD(inorg chem), 69. *Prof Exp:* Res assoc chem, Ariz State Univ, 69-70; asst prof chem, Univ Mich, Ann Arbor, 70-76; res chemist, Rockwell Int, Golden, Co, 77-89; MEM STAFF, LOS ALAMOS NAT LAB, NMEX, 89- *Mem:* AAAS; Am Chem Soc; Sigma Xi. *Res:* Vaporization and thermodynamics of solids; phase equilibria; nonstoichiometric compounds; hydrothermal equilibria and reactions; lanthanide actinide chemistry. *Mailing Add:* Los Alamos Nat Lab MS E506 Los Alamos NM 87545

HASDAL, JOHN ALLAN, b Oak Park, Ill, Mar 17, 42; m 88. NUCLEAR WEAPON EFFECTS. *Educ:* Univ Denver, BS, 64, PhD(physics), 69. *Prof Exp:* Scientist, BDM Corp, 69-71; assoc physicist, Cornell Aeronaut Lab, 71-73; mem tech staff, Hughes Aircraft Co, 73-75; sr analyst, R/M Systs, Inc, 75-78; staff scientist, Sci Appln Inc, 78-82; sr analyst, Sci & Eng Assocs, Inc, 82-85; sr scientist, Phys Res, Inc, 85-89; PRES, ENTERPRISE SOFTWARE CONSULT, INC, 89- *Mem:* Am Phys Soc; Asn Comput Mach. *Res:* Computer code design; weapon systems operational reliability and maintainability; nuclear weapon effects; semiconductor junction damage and electronic circuit protection. *Mailing Add:* 220 41st St Downers Grove IL 60515

HASE, DONALD HENRY, resource management, resource administration, for more information see previous edition

HASE, WILLIAM LOUIS, b Washington, Mo, Mar 22, 45; m 67; c 1. PHYSICAL CHEMISTRY, CHEMICAL KINETICS. *Educ:* Univ Mo, Columbia, BS, 67; NMex State Univ, PhD(chem), 70. *Prof Exp:* Res assoc chem, NMex State Univ, 70-71 & Univ Calif, Irvine, 71-73; from asst prof to assoc prof, 73-81, PROF CHEM, WAYNE STATE UNIV, 81- *Concurrent Pos:* Fel, NMex State Univ, 68-70; prin investr, Petrol Res Found, 73-76, 81-84 & 91-93, Res Corp, 74-76 & NSF, 75-; prog officer, NSF, 83-84. *Mem:* AAAS; Am Phys Soc; Am Chem Soc. *Res:* Computer simulation of chemical reactions; Monte Carlo trajectory studies of molecular dynamics; theories of chemical kinetics; co-author graduate text. *Mailing Add:* Dept Chem 435 Chem Wayne State Univ 5950 Cass Ave Detroit MI 48202

HASEGAWA, ICHIRO, b Seattle, Wash, Aug 2, 15; m 46; c 2. ELECTRON MICROSCOPY. *Educ:* Univ Wash, BS, 43. *Prof Exp:* From jr chemist to res chemist, Atlantic-Richfield Co, 44-65; res scientist, Philip Morris Inc, 65-69, sr scientist electron micros, 69-81; RETIRED. *Concurrent Pos:* Mem adv bd, Va State Lab Servs, 82-85. *Mem:* Am Chem Soc; AAAS. *Res:* Aerosol research; microstructure of materials utilizing electron microscopy and x-ray diffraction techniques. *Mailing Add:* 2000 Riverside Dr PH1 Richmond VA 23225

HASEGAWA, JUNJI, b Stockton, Calif, June 26, 17; m 50; c 2. DERMATOLOGY, HISTOCHEMISTRY. *Educ:* Univ Calif, Berkeley, AB, 39; Univ Calif, San Francisco, MD, 42. *Prof Exp:* Instr, 60-62, from asst prof to prof, 62-84, EMER PROF DERMAT, MED SCH, NORTHWESTERN UNIV, CHICAGO, 84- *Concurrent Pos:* Chief dermat, Vet Admin Lakeside Hosp, 60-84. *Mem:* Am Acad Dermat. *Res:* Enzyme studies in human cutaneous diseases. *Mailing Add:* 608 W Arlington Pl Chicago IL 60614

HASEGAWA, RYUSUKE, b Nagoya, Japan, Feb 7, 40; m 67; c 2. SOLID STATE PHYSICS, METAL PHYSICS. *Educ:* Nagoya Univ, BEng, 62; Calif Inst Technol, MS, 68, PhD(mat sci), 69. *Prof Exp:* Res fel mat sci, Calif Inst Technol, 69-72; mem staff phys sci, Thomas J Watson Res Ctr, IBM Corp, 73-75; sr staff physicist, Mat Res Ctr, Corp Res, 75-78, group leader, 78-80, res assoc, 80-84, sr res assoc, 84-85; vpres, Nipper Amorphous Metals, 85-89; DIR, FAR EAST OPER METGLAS PROD, ALLIED-SIGNAL INC, 89- *Concurrent Pos:* Co-chmn, Amorphous Magnetism Conf, 79, Annual Magnetism & Magnetic Mat Conf, 81, chairperson, Magnetics Soc Amorphous Magnetic Mat Tech Comt, Inst Elec & Electronics Engrs, 82-; Fulbright Scholar, 64. *Honors & Awards:* Distinguished lectr, Inst Elec & Electronics Engrs Magnetics Soc, 90. *Mem:* Am Phys Soc; Inst Elec & Electronics Engrs; Mat Res Soc. *Res:* Magnetism and electron trasport properties of noncrystalline metallic solids; development of new amorphous alloys for electromagnetic and structural applications. *Mailing Add:* Metglas Prod Six Eastman Rd Parsippany NJ 07054

HASEK, ROBERT HALL, b State College, Pa, June 25, 18; m 42; c 1. ORGANIC CHEMISTRY. *Educ:* Pa State Col, BS, 39; Univ Ill, PhD(org chem), 43. *Prof Exp:* Lab asst, Charles Lennig, Pa, 37, Eastman Kodak Co, NY, 39-40 & Gen Elec Co, 42; from res chemist to sr res chemist, Res Labs, Tenn Eastman Co, 43-55, from res assoc to sr res assoc, 55-63, res dir, Eastman Res AG, Switz, 63-67, head chem res div, 67-76, res fel, 76-78; RETIRED. *Mem:* Fel AAAS; Am Chem Soc. *Res:* Polymers; oxidized cellulose; oxo reaction; hydrogenation; ketones. *Mailing Add:* 46 Crown Colony Kingsport TN 37660-9703

HASELEY, EDWARD ALBERT, b Cleveland, Ohio, Nov 23, 30; m 53; c 4. PHYSICAL CHEMISTRY, POLYMER SCIENCE. *Educ:* Kenyon Col, AB, 53; Ohio State Univ, PhD(phys chem), 56. *Prof Exp:* From res chemist to sr res chemist, 63-65, res supvr, 65-69, sr res chemist, Dacron Res Lab, 69-78, ORGN DEVELOP SPECIALIST, TEXTILE FIBERS DEPT, E I DU PONT DE NEMOURS & CO, INC, 78- *Mem:* Am Chem Soc; Sigma Xi. *Res:* Boron hydride chemistry; polyester fiber technology. *Mailing Add:* Box 147 Gifton NC 28530

HASELKORN, ROBERT, b New York, NY, Nov 7, 34; m 57; c 2. PHYSICAL BIOCHEMISTRY, VIROLOGY. *Educ:* Princeton Univ, AB, 56; Harvard Univ, PhD(biochem), 59. *Prof Exp:* Am Cancer Soc fel, Virus Res Unit, Agr Res Coun, Eng, 59-61; asst prof biophys, chem & biochem, 61-64, assoc prof, 64-69, chmn dept biophys, 69-84, FL PRITZKER DIST SERV PROF, DEPT MOLECULAR GENETICS & CELL BIOL, UNIV CHICAGO, 84- *Concurrent Pos:* Ed, Virology; chmn, Virology Study Sect, NIH, 78-80; Sackler fel, Tel Aviv Univ, 87; mem, Panel Sci Adv, Int Ctr Genetic Eng & Biotechnol, UNIDO; mem, rec DNA adv comt, NIH. *Honors & Awards:* Darbaker Prize, Bot Soc Am, 82. *Mem:* Nat Acad Sci; Am Soc Biol Chem; fel AAAS; Am Soc Virol; Int Soc Plant Molecular Biol (pres, 87-88); fel Am Acad Arts & Sci, 87. *Res:* Nitrogen fixation in cyanobacteria and in photosynthetic bacteria; function of nucleic acids in viruses and cellular organelles. *Mailing Add:* Dept Molecular Genetics & Cell Biol 920 E 58th St Univ Chicago Chicago IL 60637

HASELTINE, FLORENCE PAT, b Philadelphia, Pa, Aug 17, 42; m 70; c 2. OBSTETRICS & GYNECOLOGY. *Educ:* Univ Calif, BA, 64; Mass Inst Technol, PhD(biophys), 69; Albert Einstein Col Med, MD, 72. *Prof Exp:* From asst prof to assoc prof obstet & gynec, Yale Univ Sch Med, 76-85; DIR, CTR POP RES, NAT INST CHILD HEALTH & HUMAN DEVELOP, NIH, 85- *Concurrent Pos:* Consult, Pediat Endocrine Clin, 78-85 & Behav Med Clin, Yale Univ, 83-85. *Mem:* Am Col Obstetricians & Gynecologists; Am Fertil Soc; Soc Gynec Invest; Soc Study Reproduction; Soc Reproductive Endocrinologists; Endocrine Soc. *Res:* Human reproduction; biomedical research on problems of human fertility and infertility; development of safe and efficacious fertility regulating methods; evaluation of the medical effects and efficacy of contraceptive methods. *Mailing Add:* NIH Exec Plaza N Rm 604 6130 Executive Blvd Bethesda MD 20892

HASELTON, GEORGE MONTGOMERY, b Worcester, Mass, Feb 28, 28; m 55; c 2. GEOLOGY, GEOMORPHOLOGY. *Educ:* Colby Col, BA, 51; Univ Boston, MA, 58; Ohio State Univ, PhD(geol), 67. *Prof Exp:* Field geologist, Raw Mat Br, US AEC, Calif, 55-56; instr geol, Marietta Col, 58-61; res assoc, Inst Polar Studies, Ohio State Univ, 62-67; from asst prof to assoc prof, 67-78, PROF GEOL, CLEMSON UNIV, 78- *Concurrent Pos:* Partic, Am Geol Inst Int field inst, Spain, 71; res & teaching, Tromso Univ, Norway, 76-77; vis prof, Ore State Univ, 81-83. *Mem:* Arctic Inst NAm; Geol Soc Am; Int Asn Quaternary Res; Int Qual Asn; Nat Asn Geol Teachers; Am Qual Asn. *Res:* Glacial geology and geomorphological studies and research in Antarctica, Alaska, the Canadian Arctic and Subarctic and New England; late Pleistocene and Holocene geology of Alaska and special areas in New England. *Mailing Add:* Dept Geol Clemson Univ Clemson SC 29634

HASEMAN, JOSEPH FISH, b Columbia, Mo, June 12, 14; m 41; c 3. METALLURGY. *Educ:* Univ Mo, AB, 36, PhD(soil chem), 43; Cornell Univ, MA, 38. *Prof Exp:* Lab asst geol, Cornell Univ, 36-37, lab instr, 37-38; jr geologist, Shell Oil Co, Tex, 38-40; res chemist, Tenn Valley Authority, 42-52; group leader, Minerals Separation Group, Foote Minerals Co, 52-56; prin res engr, Int Minerals & Chem Corp, 56-62; res scientist, Armour Agr Chem Co, Fla, 62-66; metallurgist, Agrico Chem Co, Pierce, Fl, 66-75; CONSULT, 75- *Mem:* Sigma Xi. *Res:* Use of heavy minerals in studying soil origin and development; concentration of non-metallic ores by flotation; physical chemistry of phosphate fixation by clay minerals. *Mailing Add:* 1905 Camphor Dr Lakeland FL 33803

HASENFUS, HAROLD J(OSEPH), b New York, NY, Apr 9, 21; m 45; c 4. MECHANICAL ENGINEERING, ELECTRICAL ENGINEERING. *Educ:* City Col New York, BS, 43; Va Polytech Inst, MS, 76, MS, 84. *Prof Exp:* Asst plutonium technol, Metall Lab, Univ Chicago, 44; proj engr, Fercleve Corp, Tenn, 45; res engr rockets, Ballistic Res Labs, Aberdeen Proving Ground, 46-52, chief, Rocket Br, 52-60; head, Satellite Appln Div, Naval Weapons Lab, 60-61, tech dir, 61-86, EMER TECH DIR, US NAVAL SPACE SURVEILLANCE, 86- *Concurrent Pos:* US Army deleg, Tripartite Conf Armaments, Explosives & Propellants, US Dept Defense, Que, Can, 59, USN deleg, Tripartite Conf Artificial Earth Satellites, Bermuda, 71; consult, Missiles & Astronaut Div, Am Defense Preparedness Asn. *Mem:* AAAS; Am Soc Mech Engrs; Am Inst Aeronaut & Astronaut; Am Math Soc; Am Defense Preparedness Asn; Asn Comput Mach. *Res:* Stability and control of missiles and rockets; ignition and combustion in solid-fuel rockets; instrumentation for rocket research, including radio telemetry, optics and transducer development; satellite orbit theory and computation. *Mailing Add:* 311 Ingleside Dr Fredericksburg VA 22405-2344

HASH, JOHN H, b Ferrum, Va, Feb 23, 29; m 53; c 3. BIOCHEMISTRY. *Educ:* Roanoke Col, BS, 49; Va Polytech Inst, MS, 55, PhD(biochem), 57. *Prof Exp:* Res assoc biochem, Columbia Univ, 57-58; res biochemist, Lederle Labs, Am Cyanamid Corp, 58-64; asst prof, 64-67, assoc prof, 67-70, actg chmn dept, 68-72, PROF MICROBIOL, SCH MED, VANDERBILT UNIV, 70-, ASSOC DEAN BIOMED SCI, 76- *Mem:* AAAS; Am Soc Biol Chem; Am Chem Soc; Sigma Xi. *Res:* Enzymology; protein chemistry; carbohydrate metabolism. *Mailing Add:* Dept Microbiol Sch Med Vanderbilt Univ Nashville TN 37232

HASHIM, SAMI A, b Lebanon, Sept 21, 29; US citizen; m 56; c 2. METABOLISM, NUTRITION. *Educ:* Am Univ Beirut, BA, 50 MS, 52; Univ Buffalo Sch Med, MD, 55. *Prof Exp:* Intern & resident med, Peter Bent Brigham Hosp, Boston, 55-57; res fel, Dept Nutrit, Harvard Univ, 57-58, med & nutrit, St Luke's Hosp & Columbia Univ, 58-61; DIR, DIV METAB & NUTRIT, ST LUKE'S HOSP, 61-, ATTEND PHYSICIAN, 71-; PROF MED, COLUMBIA UNIV COL PHYSICIANS & SURGEONS, 78- *Concurrent Pos:* Mem, Nutrit Study Sect, NIH, 68-72, Type Two Intervention Study Monitoring Comt, 73-82, prin investr, 71- *Mem:* Am Soc Clin Invest; Am Soc Clin Nutrit; Am Inst Nutrit; Am Soc Clin Res; AAAS. *Res:* Lipid metabolism and transport, mechanisms of lipid absorption; adipose tissue metabolism; gastric factors in obesity; development of non-caloric fats; methods of improving malnutrition. *Mailing Add:* St Luke's Hosp Ctr 114th St & Amsterdam Ave New York NY 10025

HASHIMOTO, KEN, b Niigata City, Japan, June 19, 31; US citizen; m 62; c 4. ELECTRON MICROSCOPY, IMMUNOHISTOCHEMISTRY. *Educ:* Niigata Univ Sch Med, MD, 55. *Prof Exp:* Res assoc, dept dermat, Tufts Univ, 63-65, asst prof, Sch Med, 65-68; from assoc prof to prof med & anat, Univ Tenn, Memphis, 68-77, dir dermatopath, 75-77; prof anat & dermat, dir dermat & chief dermat sect, Wright State Univ, 77-80; PROF DERMAT & CHMN DEPT DERMAT & SYPHIL, WAYNE STATE UNIV, 80- *Concurrent Pos:* Res & clin fel, Boston City Hosp, 63-64; med investr, Vet Admin career develop prog, Vet Admin Med Ctr, Memphis, 68-70, chief dermat sect & dir Electron Micros Lab, 70-77; Vet Admin rep mem, Path B Study Sect, NIH, USPHS Dept Health, Educ & Welfare, 74-77; mem, biol & climate effects res task group, Environ Protection Agency, 76-77; chief, dept dermat, Harper-Grace Hosp & Detroit Receiving Hosp, 80- *Mem:* Am Soc Dermatopath (pres-elect, 86, pres, 87); Soc Investigative Dermat (vpres, 80-81); AAAS; Am Dermat Asn. *Res:* Dermatological oncology; structure and function of the skin; monoclonal antibodies related to the skin structures; cutaneous amyloidoses. *Mailing Add:* Dept Dermat Wayne State Univ Detroit MI 48202

HASHIMOTO, PAULO HITONARI, b Amagasaki, Japan, Mar 23, 30; m 60; c 5. CEREBROSPINAL FLUID ABSORPTION, ELECTRON MICROSCOPY. *Educ:* Osaka Univ, MD, 53, DMSc(med sci), 60. *Prof Exp:* From instr to assoc prof, 57-74, PROF ANAT, MED SCH, OSAKA UNIV, 74- *Concurrent Pos:* Postdoctoral fel anat, Harvard Med Sch, 63-65; counr, Japanese Asn Anatomists, 74- *Mem:* Am Asn Anatomists; Am Soc Cell Biol; Int Brain Res Orgn. *Res:* Fine structure of the central nervous system; vascular fine structure of circumventricular organs in relation to venous absorption of the cerebrospinal fluid. *Mailing Add:* Dept Anat Osaka Univ Med Sch 2-2 Yamadaoka Osaka 565 Japan

HASHIN, ZVI, b Danzig, June 24, 29; m 59; c 3. APPLIED MECHANICS, MATERIALS ENGINEERING. *Educ:* Israel Inst Technol, BSc, 53, & MSc 55; DSc, Sorbonne, 57. *Prof Exp:* Lectr mech, Israel Inst Technol, 57-58, sr lectr, 58-59; res fel struct mech, Harvard Univ, 59-60; from assoc prof to prof eng mech, Univ Pa, 60-71; vis prof, Israel Inst Technol, 68-69, prof mat eng, 71-73, chmn dept, 73-77 & 79-81, PROF SOLID MECH, MAT & STRUCT, TEL AVIV UNIV, 73-, NATHAN CUMMINGS PROF MECHS SOLIDS, 80- *Concurrent Pos:* Consult, Scott Paper Co, 61-62, Gen Elec Co, 62-66, Armstrong Cork Co, 65-66, Franklin Inst, 66-68 & Monsanto Res Corp, 66-67; sci adv, Mats Sci Corp, 70-; vis prof mat sci & eng, Univ Pa, 77-79; Israel deleg, Int Union Theoret & Appl Mech; adj prof, Univ Pa, 78-; co-chmn, Int Union Theoret Appl Mech Symp, Composite Mat, Blacksburg, Va, 82, Damage & Fatigue Symp, Haifa & Tel Aviv, Israel, 85. *Honors & Awards:* Landau Prize, 72; Medal Excellence, Ctr Composite Mat, Univ Del, 84. *Mem:* Fel Am Soc Mech Engrs; Soc Rheology; Soc Eng Sci; Israeli Soc Theoret & Appl Mech (pres, 77-). *Res:* Mechanics of solids; heterogeneous media; composite materials; fatigue. *Mailing Add:* Dept Solid Mech Mat & Struct Sch Eng Tel Aviv Univ Tel Aviv Israel

HASHMALL, JOSEPH ALAN, b New York, NY, Oct 15, 43; m 64; c 1. QUANTUM CHEMISTRY, PHYSICAL CHEMISTRY. *Educ:* Univ Chicago, BS, 64; Univ Tex, Austin, PhD(chem), 69. *Prof Exp:* Part-time instr chem, Huston-Tillotsen Col, 65-68; Swiss Nat Fund grant, Phys Chem Inst, Univ Basel, 69-70; NSF grant, Univ Calif, Berkeley, 71; asst prof chem, Georgetown Univ, 71-79; sr anal, RDS, Inc, 79-83; STAFF SCIENTIST, COMPUT SCI CORP, 83- *Mem:* Am Chem Soc; Royal Soc Chem; AAAS. *Res:* Spacecraft ground support system development. *Mailing Add:* 8116 Triple Crown Rd Bowie MD 20715-4535

HASINOFF, BRIAN BRENNEN, b Mannville, Alta, Nov 14, 44; m 70; c 2. BIOPHYSICAL CHEMISTRY, MEDICINAL CHEMISTRY. *Educ:* Univ Alta, BSc, 66, PhD(chem), 70. *Prof Exp:* Res assoc chem, Univ Kent, Canterbury, 70-72; asst prof, Notre Dame Univ, Nelson, 73-75; from asst prof to prof chem & med, Mem Univ Nfld, 75-90; ASSOC PROF, FAC PHARM, UNIV MAN, WINNIPEG, 90- *Concurrent Pos:* Chmn Biol Chem Div; rep Can Nat Comt, Int Union Biochem. *Mem:* Am Chem Soc; Can Biochem Soc; Chem Inst Can. *Res:* Fast reaction kinetics of proteins; oxy radicals, drug induced oxy radical production, anthracyline cardioprotective agents. *Mailing Add:* Fac Pharm Univ Man Winnipeg MB R3T 2N2 Can

HASKE, BERNARD JOSEPH, b Baltimore, Md, Nov 14, 30; m 58; c 6. ORGANIC CHEMISTRY. *Educ:* Loyola Col, Md, BS, 52; Univ Va, MS, 55, PhD(chem), 59. *Prof Exp:* Asst prof chem, Wheeling Col, 58-61; assoc prof, Washington Col, 61-67; assoc prof, 67-71, PROF CHEM, DOWLING COL, 71- *Mem:* Am Chem Soc. *Res:* Structural and synthetic studies of organic complexing agents. *Mailing Add:* Dept Chem Dowling Col Idle Hour Blvd Oakdale NY 11769

HASKELL, ALBERT RUSSELL, pharmacology, toxicology, for more information see previous edition

HASKELL, BARRY G, m; c 2. ENGINEERING. *Educ:* Univ Calif, Berkeley, BS, 64, MS, 65 & PhD(elec eng), 68. *Prof Exp:* Res asst, Univ Calif Electronics Res Lab, 64-68; HEAD, VISUAL COMMUN RES DEPT, AT&T BELL LABS, 87- *Concurrent Pos:* Res asst, Lawrence Livermore Lab; Instr grad courses, Rutgers Univ, City Col New York & Columbia Univ. *Mem:* Sigma Xi; fel Inst Elec & Electronics Engrs. *Res:* Digital transmission and coding of images; videotelephone; satellite television transmission; medical imaging; digital image processing; author of over 30 publications and 20 patents. *Mailing Add:* AT&T Bell Labs Visual Commun Res Ho-4c538 Holmdel NJ 07733

HASKELL, BETTY ECHTERNACH, b Lewiston, Idaho, June 1, 25. NUTRITIONAL BIOCHEMISTRY. *Educ:* Univ Idaho, BA, 46; Univ Chicago, MA, 47; Univ Calif, Berkeley, BS, 56, PhD(nutrit), 64. *Prof Exp:* Asst prof nutrit, Univ Calif, Davis, 64-69; from assoc prof to prof nutrit, Univ Ill, Urbana, 69-78; PROF NUTRIT, UNIV TEX, 78- *Concurrent Pos:* Vis prof, Med Br, Univ Tex, Galveston, 82. *Mem:* Am Inst Nutrit; Am Soc Biol Chemists; Am Chem Soc. *Res:* Vitamin metabolism; ascorbic acid. *Mailing Add:* 117 Gearing Hall Univ Tex Austin TX 78712

HASKELL, CHARLES THOMSON, b Wenatchee, Wash, Jan 27, 24; m 47; c 3. MATHEMATICS. *Educ:* Univ Wash, BA, 46; Univ Ariz, MS, 61, PhD(math), 65. *Prof Exp:* Teacher, Churchill County High Sch, Nev, 46-48; trust clerk, Peoples Nat Bank, Wash, 48-49; teller & trust clerk, First Nat Bank Nev, 49-55, asst trust officer, 55-59; asst math, Univ Ariz, 60-63; from asst prof to assoc prof, 63-71, PROF MATH, CALIF STATE POLYTECH COL, SAN LUIS OBISPO, 71- *Mem:* Math Asn Am; Am Math Soc. *Res:* Analytic number theory. *Mailing Add:* PO Box 706 Newport OR 97365

HASKELL, DAVID ANDREW, b Lakewood, Ohio, Mar 26, 28; m 58; c 2. PLANT MORPHOLOGY. *Educ:* Ohio State Univ, BSc, 51; Purdue Univ, MS, 57, PhD(bot), 60. *Prof Exp:* Asst prof bot, 60-70, chmn dept. 62-65, assoc prof, 70-79, PROF BIOL SCI, SMITH COL, 79- *Mem:* AAAS; Bot Soc Am; Am Inst Biol Sci; Sigma Xi. *Res:* Developmental morphology of angiosperm embryos; origin, structure and growth of plant apical meristems. *Mailing Add:* Dept Biol Sci Smith Col Northampton MA 01063

HASKELL, THEODORE HERBERT, JR, b Los Angeles, Calif, June 16, 21; m 47; c 4. NATURAL PRODUCTS CHEMISTRY. *Educ:* Dartmouth Col, AB, 43; Ohio State Univ, PhD(physiol chem), 49. *Prof Exp:* Res chemist, Winthrop Chem Co, 43-46; sect dir, Parke, Davis & Co, 49-81; RETIRED. *Concurrent Pos:* Instr bio-org chem, Univ Mich & instr med chem, Washtenaw Community Col; consult. *Mem:* Am Chem Soc; Japan Antibiotics Res Asn; NY Acad Sci. *Res:* Chemistry of antibiotics; natural products; author of over 62 chemical publications. *Mailing Add:* Box 303 Rte Four Anderson SC 29621

HASKELL, VERNON CHARLES, b Can, June 20, 19; US citizen; m 46; c 3. PHYSICAL CHEMISTRY. *Educ:* Univ Sask, BEng, 44, MSc, 46; Columbia Univ, PhD(chem), 48. *Prof Exp:* Asst chem, Columbia Univ, 46-48; res chemist, E I du Pont de Neumours & Co, 48-51, develop supvr, 51-56, res assoc, 56-66, res fel, Polymer Prod Dept, 66-82; RETIRED. *Concurrent Pos:* Fel, Ctr Advan Eng Study, Mass Inst Technol, 69-70. *Mem:* AAAS; Am Inst Chem Engrs; Am Chem Soc. *Res:* Reaction kinetics and mechanism; cellulose chemistry; moisture proof coatings; fine structure of regenerated cellulose; polymer chemistry; colloids. *Mailing Add:* 95 Burlwood Dr San Francisco CA 94127

HASKILL, JOHN STEPHEN, b Toronto, Ont, Nov 10, 39; m 65; c 3. IMMUNOPATHOLOGY. *Educ:* Univ Toronto, BSc, 62, MA, 64, PhD(med biophys & radiobiol), 66. *Prof Exp:* Asst prof path, Queen's Univ, Ont, 69-72; assoc prof path, McGill Univ, 72-74; assoc prof path, Dept Basic & Clin Immunol & Microbiol, Med Univ SC, Charleston, 74-77; ASSOC PROF OBSTET & GYNEC & BACT & IMMUNOLOGY, UNIV NC, CHAPEL HILL, 77- *Concurrent Pos:* Nat Cancer Inst Can res fel med biophys, Walter & Eliza Hall Inst, Melbourne, Australia, 66-68. *Mem:* Can Soc Immunol; Can Asn Path; Am Soc Exp Path. *Res:* Cellular immunology and cancer biology. *Mailing Add:* Cancer Res Ctr Univ NC CB 7295 218 Lineberger Chapel Hill NC 27599-7295

HASKIN, HAROLD H, b Niagara Falls, NY, Jan 3, 15; m 42; c 5. MARINE BIOLOGY. *Educ:* Rutgers Univ, BSc, 36; Harvard Univ, MA, 38, PhD(biol), 41. *Prof Exp:* Asst, Oceanog Inst, Woods Hole, 46; from asst prof to assoc prof, 46-58, prof zool, Rutgers Univ, 58-84; dir, Oyster Res Lab, NJ Agr Exp Sta, 50-84, EMER PROF, 84- *Concurrent Pos:* Mem corp, Bermuda Biol Sta, 38- *Mem:* AAAS; Am Soc Limnol & Oceanog; Sigma Xi; Soc Invert Path; Am Soc Zool; Nat Shellfisheries Asn. *Res:* Shellfish biology; estuarine ecology; invertebrate pathology and physiology; continuing in Shellfish biology & pathology primarily with American oyster. *Mailing Add:* Rutgers Shellfish Res Lab PO Box 687 Pt Norris NJ 08349

HASKIN, LARRY A, b Olathe, Kans, Aug 17, 34; m 63; c 3. GEOCHEMISTRY. *Educ:* Baker Univ, BA, 55; Univ Kans, PhD(radiochem), 60. *Prof Exp:* Asst prof chem, Ga Inst Technol, 59-60; from instr to prof, Univ Wis-Madison, 60-74; chief planetary & earth sci div, NASA Johnson Space Ctr, 73-76; chmn dept, 76-90, PROF CHEM & EARTH & PLANETARY SCI, WASH UNIV, 76- *Concurrent Pos:* Consult, NASA Johnson Space Ctr, 70-73. *Honors & Awards:* NASA Except Sci Achievement Medal, 71. *Mem:* AAAS; Am Chem Soc; Am Geophys Union; Geochem Soc; Meteoritical Soc. *Res:* Trace element geochemistry, especially rare earths; neutron activation analysis; lunar sample analysis; physical chemistry of trace ions in silicate liquids. *Mailing Add:* EPSc Campus Box 1169 Wash Univ St Louis MO 63130-4899

HASKIN, MARVIN EDWARD, b Ardmore, Pa, May 28, 30; m 59; c 2. RADIOLOGY. *Educ:* Temple Univ, BA, 51, MD, 55. *Prof Exp:* Chief radiologist, USAF Hosp, Andrews AFB, Washington, DC, 57-58; chief diag, Philadelphia Gen Hosp, Pa, 61-63; radiologist, Haverford Gen Hosp, Havertown, Pa, 63-69; res asst prof physiol & biophys, 69-71, PROF DIAG RADIOL & CHMN DEPT, HAHNEMANN MED COL & HOSP, 71- *Concurrent Pos:* Res fel radiol, Philadelphia Gen Hosp, Pa, 59-60; instr, Sch Med, Temple Univ, 61-63; assoc radiol, Hahnemann Med Col, 63-66, clin asst prof, 67- *Mem:* AAAS; fel Am Col Physicians; fel Am Col Radiol; Am Nat Standards Inst; Int Sci Orgn. *Res:* Biomedical engineering; computer applications in medicine. *Mailing Add:* Dept Diag Radiol Hahnemann Med Col 85230 N Broad St Philadelphia PA 19102

HASKIN, MYRA RUTH SINGER, b Philadelphia, Pa, Feb 12, 35; m 58; c 2. PHYSICAL MEDICINE & REHABILITATION. *Educ:* Univ Pa, BA, 56; Temple Univ, MD, 60. *Prof Exp:* Intern, Presby Hosp, Philadelphia, Pa, 60-61, res & asst instr phys med & rehab, Hosp Univ Pa, 61-64; chief phys med & rehab, Albert Einstein Med Ctr, South Div, 64-65; asst prof, Sch Med, Temple Univ, 66-67; psychiatrist, Philadelphia Gen Hosp, 67, assoc phys med & rehab, Hosp & instr, Sch Med, Univ Pa, 67-80; RETIRED. *Concurrent Pos:* Fel physiol, Grad Sch Med, Univ Pa, 63-64; res psychiatrist, Temple-Moss Philco Biomed Eng Dept, 66. *Mem:* AMA; Am Cong Rehab Med; Am Med Women's Asn; Am Acad Phys Med & Rehab. *Res:* Electromyology; psychiatry. *Mailing Add:* 495 E Abington Ave Philadelphia PA 19118

HASKINS, ARTHUR L, JR, b Philadelphia, Pa, Mar 31, 17; m 43; c 4. OBSTETRICS & GYNECOLOGY. *Educ:* Univ Rochester, BA, 38, MD, 43. *Prof Exp:* Instr obstet & gynec, Sch Med, Wash Univ, 52-54, asst prof, 54-55; prof obstet & gynec, Sch Med, Univ Md, Baltimore, 55-80; CLIN PROF OBSTET & GYNEC, MED COL GA, 80- *Concurrent Pos:* Consult, US Dept Army, 56- *Mem:* Am Asn Obstet & Gynec; Am Col Surg; Soc Gynec Invest; Endocrine Soc; Am Gynec Soc. *Res:* Gynecologic endocrinology; infertility. *Mailing Add:* Three Cotesby Lane Savannah GA 31411

HASKINS, CARYL PARKER, b Schenectady, NY, Aug 12, 08; m 40. GENETICS, PHYSIOLOGY. *Educ:* Yale Univ, PhB, 30; Harvard Univ, PhD(physiol), 35. *Hon Degrees:* ScD, Tufts Col, 51, Union Univ, NY, 55, Northeastern Univ, 55, Yale Univ, 58, Hamilton Col, NY, 59, George Washington Univ, 63; LLD, Cincinnati Univ, 60, Carnegie Inst Technol, 60, Boston Col, 60, Washington & Jefferson Col, 61, Univ Del, 65 & Pace Univ, 75. *Prof Exp:* Mem staff, Res Lab, Gen Elec Co, 31-35; dir, pres & chmn bd,

Haskins Labs Inc, 35-55; CONSULT, 35- *Concurrent Pos:* Dir, Schenectady Trust Co, 34-; res assoc, Mass Inst Technol, 35-45; res prof, Union Col, NY, 37-55; pres & dir, Nat Photocolor Corp, NY, 39-55; asst liaison officer, Nat Defense Comt-Off Sci Res & Develop, 40-42, sr liaison officer, 42-43; exec asst to chmn, Nat Defense Res Comt, 43-44, dep exec officer, 44-45; sci adv policy coun, Army & Navy Joint Res & Develop Bd, 47-48; res & develop bd, Nat Mil Estab, 47-51; chmn adv comt spec weapons, Secy Defense, 48-49, consult, 50-56; mem, pres sci adv comt, 55-59, consult, 59-70; trustee, 49-, pres, Carnegie Inst, Washington, 55-71, trustee, Carnegie Corp, NY, 55, chmn bd, 76-; trustee Nat Geog Soc, Woods Hole, Mass; regent, World Wildlife, Smithsonian Inst; dir, Coun Foreign Rels, Yale Univ. *Honors & Awards:* Joseph Henry Medal, Smithsonian Inst. *Mem:* Nat Acad Sci; fel AAAS; Am Phys Soc; Entom Soc Am; Genetics Soc Am; Royal Soc Arts. *Res:* Mechanism of speciation and evolution; radiation biophysics; cellular physiology, especially nutritional requirements of microorganisms. *Mailing Add:* 1545 18th St NW Suite 810 Washington DC 20036

HASKINS, EDWARD FREDERICK, b Minneapolis, Minn, Apr 10, 37. BOTANY. *Educ:* Univ Minn, BA, 59, MS, 62, PhD(protozool), 65. *Prof Exp:* Instr bot, Univ Minn, 62; NSF fel electron micros, Biol Labs, Harvard Univ, 65-66; from asst prof to assoc prof, 66-83, PROF BOT, UNIV WASH, 83- *Mem:* AAAS; Bot Soc Am; Mycol Soc Am; Am Soc Protozool; Am Soc Zool. *Res:* Developmental biology of the Myxomycetes; electron microscopy; microbial ecology. *Mailing Add:* Dept Bot Univ Wash Seattle WA 98195

HASKINS, FRANCIS ARTHUR, b Omaha, Nebr, Aug 20, 22; m 51; c 4. PLANT GENETICS. *Educ:* Univ Nebr, BSc, 43, MSc, 48; Calif Inst Technol, PhD(chem genetics), 51. *Prof Exp:* Res fel chem genetics, Calif Inst Technol, 51-52; res scientist, Univ Tex, 52-53; asst agronomist, Univ Nebr-Lincoln, 53-55, from assoc prof to prof agron, 55-58, George Holmes prof agron, 67-88, EMER PROF, UNIV NEBR-LINCOLN, 88- *Mem:* Fel AAAS; Genetics Soc Am; Am Soc Plant Physiol; fel Am Soc Agron; Phytochem Soc NAm; fel Crop Sci Soc Am. *Res:* Aromatic metabolism in neurospora; enzyme studies in maize; chemical genetics of Melilotus, sorghum and other forage species. *Mailing Add:* Dept Agron Univ Nebr Linc oln NE 68583-0915

HASKINS, JOSEPH RICHARD, b Wooster, Ohio, Mar 27, 26; m 54; c 4. NUCLEAR PHYSICS. *Educ:* Univ Tex, BS, 46; Ohio State Univ, PhD(physics), 52. *Prof Exp:* Res physicist, Ord Missile Lab, Res Lab, Redstone Arsenal, 54-59; from asst prof to assoc prof, 59-65, PROF PHYSICS & CHMN DEPT, GETTYSBURG COL, 65- *Concurrent Pos:* Asst prof, Univ Ala, 58-59. *Mem:* Am Phys Soc; Am Asn Physics Teachers. *Res:* Nuclear spectra; Mossbauer effect. *Mailing Add:* Dept Physics Gettysburg Col Gettysburg PA 17325

HASKINS, MARK, b Philadelphia, Pa, Dec 27, 44. ANIMAL MODELS, METABOLIC DISEASE. *Educ:* Pa State Univ, BS, 66; Univ Pa, VMD, 69, PhD(path), 77; Drexel Univ, MS, 73. *Prof Exp:* Asst prof, 81-84, ASSOC PROF, SCH VET MED, UNIV PA, 84- *Honors & Awards:* Ralston Purina Small Animal Res Award. *Mem:* Am Vet Med Asn; AAAS; Int Acad Path; Am Asn Pathologists; Am Soc Human Genetics. *Res:* Pathogenesis and therapy of animals models of human genetic disease. *Mailing Add:* Path Lab Univ Pa Sch Vet Med 3800 Spruce St Philadelphia PA 19104-6051

HASKINS, REGINALD HINTON, b North Bay, Ont, July 16, 16; m 49; c 3. MYCOLOGY, MICROBIOLOGY. *Educ:* Univ Western Ont, BA, 38, MA, 40; Harvard Univ, PhD(biol), 48. *Prof Exp:* Demonstr bot, Univ Western Ont, 38-40, sr demonstr, 45-46; tutor, Dunster House, Harvard Univ, 47-48; from asst res officer to sr res officer, Nat Res Coun Can, 48-67, head physiol & biochem of fungi sect, 56-67, prin res officer, 67-81, head microbial physiol & biochem, Prairie Regional Lab, 70-81; RETIRED. *Mem:* Bot Soc Am; Mycol Soc Am; Am Soc Plant Physiol; Can Soc Microbiol; Brit Mycol Soc. *Res:* Physiology of biochemistry of fungi; industrial mycological fermentations; taxonomy of fungi; sexual reproduction and ultrastructure of phycomycetes. *Mailing Add:* 222 Lake Crescent Saskatoon SK S7H 3A2 Can

HASLAM, JOHN LEE, b Salt Lake City, Utah, Jan 4, 39; m 66; c 2. PHARMACEUTICAL CHEMISTRY. *Educ:* Univ Utah, BA, 63, PhD(chem), 66. *Prof Exp:* NIH fel, Cornell Univ, 66-68; asst prof chem, Univ Kans, 68-73; ASSOC DIR DEVELOP, INTERX RES CORP, SUBSID MERCK & CO, INC, 73- *Mem:* Sigma Xi; Am Chem Soc; Am Asn Pharmaceut Scientists. *Res:* Pharmaceutical research and product development. *Mailing Add:* Interx Res Corp 2201 W 21st St Lawrence KS 66046

HASLANGER, MARTIN FREDERICK, b Dayton, Ohio, Mar 27, 47; m 69; c 2. MEDICINAL CHEMISTRY, RATIONAL DRUG DISCOVERY. *Educ:* Denison Univ, BS, 69; Univ Mich, PhD(org chem), 74. *Prof Exp:* Postdoctoral fel chem, Dept Chem, Harvard Univ, 74-76; res investr chem, Squibb Inst Med Res, 76-80, sr res investr, 80-81, group leader, 81-85; assoc dir, 85-88, DIR CHEM, SCHERING-PLOUGH RES, 88- *Concurrent Pos:* Res chemist, Eastman Kodak, 69-70; NIH fel, 74-76. *Mem:* Am Chem Soc; AAAS; NY Acad Sci; Am Heart Asn. *Res:* New drug discovery; design and synthesis of enzyme inhibitors and receptor antagonists; peptide mimetics; use of data and computational chemistry to design and optimize small molecule-large molecule interactions. *Mailing Add:* Schering-Plough Res 60 Orange St Bloomfield NJ 07003-4799

HASLEM, WILLIAM JOSHUA, b Roosevelt, Utah, Nov 28, 36; m 57; c 3. TECHNICAL MANAGEMENT. *Educ:* Univ Utah, BS, 60; Cent Mich Univ, MA, 74. *Prof Exp:* Proj officer, Dugway Proving Ground, US Army, 60-72, opers analyst, Tank-Automotive Command, 72-75, tech dir, Cold Regions Test Ctr, 75-91, TECH DIR, US ARMY DUGWAY PROVING GROUND, 91- *Concurrent Pos:* Instr, Univ Alaska, 77-85. *Mem:* Int Test & Eval Asn; Am Defense Preparedness Asn. *Res:* Test and evaluation of defense equipment. *Mailing Add:* Two Armitage Rd Dugway UT 84022

HASLER, ARTHUR DAVIS, b Lehi, Utah, Jan 5, 08; m 32; c 6. ZOOLOGY. *Educ:* Brigham Young Univ, AB, 32; Univ Wis, Madison, PhD(zool), 37. *Hon Degrees:* DSc, Mem Univ Nfld, 67, Miami Univ, Oxford, Oh, 88. *Prof Exp:* Asst biologist, US Fish Wildlife Serv, 35-37; from instr to assoc prof zool, 37-48, dir inst ecol, 71-74, chmn dept zool, 53-54 & 55-57, dir lab limnol, 63-78, prof, 68-78, EMER PROF ZOOL, UNIV WIS-MADISON, 78- *Concurrent Pos:* Mem adv comt biol, Off Naval Res, 51-54, hydrobiol comt, 52-58; Fulbright res scholar, Ger, 54-55; mem comt environ biol, NSF, 55-59, comt biol sta & facil, 55-60; Nat Acad Sci-Nat Res Coun, 55-59; comt educ & recruitment oceanog, Am Soc Limnol & Oceanog-Nat Acad Sci, 60; chmn, Int Cong Limnol, 62; scholar & exchange prof, Univ Helsinki, 63-64; convenor comt freshwater productivity, Int Biol Prog, Int Coun Sci Union, 66-67; mem US Nat Comt, Int Union Biol Sci & US Comt Int Biol Prog, Nat Acad Sci-Nat Res Coun. *Honors & Awards:* Sea Grant Award, Sea Grant Asn, 80. *Mem:* Nat Acad Sci; AAAS; Am Soc Limnol & Oceanog (vpres, 47, pres, 51); Ecol Soc Am (pres, 61); Int Asn Ecol (pres, 67-74); Am Soc Zool (pres, 71); Royal Neth Acad Sci; Soc Zool Bot Fennica. *Res:* General and experimental limnology; migration of fishes; sense of smell in orientation to parent stream; sun-orientation in fishes; radioisotopes in hydrobiology. *Mailing Add:* Ctr Limnol 680 N Park St Univ Wis Madison WI 53706

HASLER, ARTHUR FREDERICK, b Madison, Wis, Aug 21, 40; m 65; c 4. METEOROLOGY. *Educ:* Univ Wis, BS, 63, MS, 65, PhD(meteorol), 71. *Prof Exp:* Res asst satellite meteorol, Univ Wis, 63-71; sr scientist, Nat Ctr Atmospheric Res, 71-74; RES METEOROLOGIST, GODDARD SPACE FLIGHT CTR, NASA, 74- *Concurrent Pos:* Vis researcher, Dynamic Meteorol Lab, Paris, 75-76; mem, COSPAR Working Group VI, Panel A Weather & Climate, 63-79. *Mem:* Am Meteorol Soc. *Res:* Extraction of meteorological parameters from geostationary satellite data; in situ verification of the parameters using aircraft; dynamical studies of severe storms and tropical cloud clusters using geostationary satellite data; stereoscopy using geosynchronous satellites; synthetic stereo using multichannel radiometric data. *Mailing Add:* Code G12 Goddard Lab Atmospheres Goddard Space Flight Ctr Greenbelt MD 20771

HASLER, MARILYN JEAN, b Chicago, Ill, July 3, 43; m 69; c 2. REPRODUCTIVE PHYSIOLOGY. *Educ:* Swarthmore Col, BA, 65; Tulane Univ La, MS, 67; Univ Mo-Columbia, PhD(zool), 70. *Prof Exp:* Ford Found fel reproductive endocrinol, Univ Ill, Urbana, NIH fel, 72-74; qual control chemist & develop chemist, Micromedic Diagnostics, Inc, 74-78; vis asst prof biol, Franklin & Marshall Col & Lancaster Gen Hosp, 78-79, coordr assoc sci degree prog, 79-86; ADMIN ASST, EM TRAN INC, 86- *Mem:* Am Soc Zool; Soc Study Reproduction; Sigma Xi; Brit Soc Study Fertil. *Res:* Reproductive physiology and behavior of mammals; radioimmunoassay. *Mailing Add:* Em Tran Inc 197 Bossler Rd Elizabethtown PA 17022

HASLETT, JAMES WILLIAM, b North Battleford, Sask, Sept 27, 44; m 70. ELECTRONICS, SOLID STATE PHYSICS. *Educ:* Univ Sask, BS, 66; Univ Calgary, MS, 68, PhD(elec eng), 70. *Prof Exp:* From assoc prof to prof, Univ Calgary, 70-86, head dept, elec eng, 86-91; CONSULT, 91- *Mem:* Inst Elec & Electronics Engrs. *Res:* Electrical characteristics and noise in solid state devices and circuits; design and construction of instrumentation for oilfield exploration and oilwell testing. *Mailing Add:* Dept Elec Eng Univ Calgary Calgary AB T2N 1N4 Can

HASLING, JILL FREEMAN, b Bryan, Tex, June 8, 52; m 73; c 1. MARINE METEOROLOGY, TROPICAL METEOROLOGY. *Educ:* Univ St Thomas, BA, 75. *Prof Exp:* VPRES METEOROL, INST STORM RES, 74-, ASST RESEARCHER, 78-; DIR OPERS, WEATHER RES CTR, 88- *Mem:* Am Meteorol Soc. *Res:* Mechanics of hurricanes, hurricane waves, currents and storm surges; hurricane awareness; mitigation of damage and loss of life; real-time marine weather forecasting programs to aid in world-wide forecast. *Mailing Add:* Weather Res Ctr 3710 Mt Vernon Houston TX 77006

HASLUND, R L, b Brattleboro, Vt, Oct 3, 32; m 57; c 2. AERO-OPTICS. *Educ:* Univ Vt, BA, 54; Pa State Univ, MS, 56, PhD(physics), 61. *Prof Exp:* SR PRIN ENG, BOEING AEROSPACE, 64- *Mem:* Am Phys Soc; Am Inst Aeronaut & Astronaut. *Res:* Working on developing the technical base for aero-optics. *Mailing Add:* 8345 Avalon Dr Mercer Island WA 98040-5614

HASPEL, MARTIN VICTOR, b New York, NY, Aug 24, 45; m 67; c 1. HYBRIDOMA RESEARCH, CANCER RESEARCH. *Educ:* Yeshiva Univ, BA, 67; Pa State Univ, PhD(microbiol), 74. *Prof Exp:* Fel virol, Col Med, Pa State Univ, 74-75; fel immunopath, Scripps Clin Res Found, 75-77, asst mem, 77-80; sr staff fel immunol, NIH, 80-82; DIR, HYBRIDOMA RES & DEVELOP, BIONETICS RES, 82- *Concurrent Pos:* Adj assoc prof, Dept Microbiol, Col Med, Univ Md, 84-87, adj prof, 89- *Honors & Awards:* C B Thornton Adv Technol Award, Litton Industs, 84. *Mem:* Am Soc Microbiol; Sigma Xi; Am Asn Neuropathologists; Am Asn Immunologists; Am Asn Pathologists; Am Asn Cancer Res. *Res:* Development of monoclonal antibodies for cancer diagnosis and therapy; molecular basis of mechanisms of viral diseases. *Mailing Add:* Organon-Teknika/Biotechnol Res Inst 1330 Piccard Dr Rockville MD 20850-4373

HASS, ALVIN A, b New York, NY, Apr 5, 28; m 58; c 2. PLASTICS & CHEMICAL ENGINEERING. *Educ:* Polytech Inst Brooklyn, BChE, 49. *Prof Exp:* Plastics technologist, US Army Plastics Lab, Picatinny Arsenal, 49-54; mgr mat develop, Polymer Chem Div, W R Grace & Co, NJ, 56-63, asst dir tech serv, 63-66; mgr, appln res, Norchem Co, Rolling Meadows, 66-86; mgr tech serv, VSI Chem, 87; RETIRED. *Mem:* Soc Plastics Engrs. *Res:* Plastic mold and product design; coloring and compounding of polyolefin polymers for specific end uses; injection and blow molding and thermoforming of polyolefin polymers. *Mailing Add:* 908 Willow Lane Sleepy Hollow IL 60118

HASS, GEORG, b Hanau, Ger, Aug 8, 13; US citizen; m 39. OPTICAL PHYSICS, SOLID STATE PHYSICS. *Educ:* Inst Tech Danzig, Ger, Dr rer tech (physics), 37, Dr habil, 43. *Prof Exp:* Res asst physics, Inst Tech Danzig, Ger, 36-43, asst prof, 43-45; consult, US Army Eng Res & Develop Labs, Va, 46-52, chief physics res sect, 52-54, supvry physicist & chief physics res lab, 54-65, dir physics res tech area, Night Vision Lab, Electronic Command, 65-71; consult, US Army Night Vision Lab, Goddard Space Flight Ctr NASA & US Naval Res Lab, 71-80. *Concurrent Pos:* Mem bd dirs, Optical Soc Am, 62-65. *Honors & Awards:* Army Res Develop Achievement Award, 61 & 69; Sky Lab Achievement Award, 74; Welch Award, Am Vacuum Soc, 78; Frederic Ives Medal, Optical Soc Am, 81. *Mem:* Am Phys Soc; Optical Soc Am; Am Vacuum Soc. *Res:* Optical properties of metals; oxidation phenomena of metal surfaces; electron diffraction; structure and behavior of evaporated films; temperature control of satellites. *Mailing Add:* 7728 Lee Ave Alexandria VA 22308

HASS, GEORGE MARVIN, b Tingley, Iowa, Apr 29, 07; m 42; c 5. PATHOLOGY. *Educ:* Harvard Univ, MD, 29. *Prof Exp:* Instr path, Harvard Med Sch, 30-39, assoc, 39; asst prof, Med Col, Cornell Univ, 39-42; chmn dept path, Rush-Presby-St Luke's Med Ctr, 46-75, prof path & head dept, Rush Med Col, 71-75, EMER PROF PATH, RUSH-PRESBY-ST LUKE'S MED CTR, 75- *Concurrent Pos:* Prof path, Univ Ill, 46-71. *Mem:* Am Soc Exp Path; Am Asn Path & Bact; Int Acad Hist Sci; Sigma Xi. *Res:* Rickettsial diseases; aircraft accidents and safety; histochemistry; general pathology. *Mailing Add:* 53 S Lombard Ave 1753 W Congress St Lombard IL 60148

HASS, GEORGE MICHAEL, b San Antonio, Tex, Aug 20, 43; m 65; c 2. BIOCHEMISTRY. *Educ:* Northwestern Univ, BA, 65; Duke Univ, PhD(biochem), 69. *Prof Exp:* Res asst prof biochem, Univ Wash, 72-74; assoc prof biochem, Univ Idaho, 74-79, prof, 79-81; SR QUAL CONSULT, ABBOTT LAB, 81- *Concurrent Pos:* Consult, Presby St Luke's Hosp, 65 & Frederick Cancer Res Ctr, 75; investr, Howard Hughes Med Inst, 72-74. *Mem:* AAAS; Am Soc Biol Chem; Sigma Xi. *Res:* Structure, function and evolution of proteases and protease inhibitors; chalone structure and function. *Mailing Add:* Abbott Diagnostics Div Abbott Labs Abbott Park AP6C-4N D-917 Abbott Park IL 60064

HASS, JAMES RONALD, b Statesville, NC, Sept 17, 45; m 65. ANALYTICAL CHEMISTRY, ENVIRONMENTAL HEALTH. *Educ:* Appalachian State Univ, BA, 67; Univ NC, Chapel Hill, PhD(anal chem), 72. *Prof Exp:* Res fel chem, Univ Warwick, 72-73; asst prof, NC State Univ, 73-74; chem staff fel, 74-78, res chemist, Nat Inst Environ Health Sci, 78-84; PRES, TRIANGLE LAB, 84- *Concurrent Pos:* Adj asst prof environ sci & eng, Univ NC, 75-83, adj assoc prof environ sci, engr & chem, 83- *Mem:* Am Chem Soc; Am Soc Mass Spectrometry; Royal Soc Chem. *Res:* Chemical ionization mass spectrometry using novel reagent gases; application of gas chromatography and mass spectrometry in environmental and health problems; development of new gas chromatography and mass spectrometry techniques; development and application of field desorption mass spectrometry. *Mailing Add:* PO Box 13485 Research Triangle Park NC 27709

HASS, KENNETH PHILIP, b Minneapolis, Minn, Sept 17, 34; m 60; c 2. METALLURGICAL ENGINEERING. *Educ:* Univ Minn, BS, 56, MS, 59. *Prof Exp:* Trainee, 59, res engr, Res & Develop Dept, 59-61, sr res engr, 61-63, supvr raw mat, 63-74, supvr process metal, 74-84, supvr chem & coal res, 84-85, MGR PROD & PROCESS TECH, NAT STEEL CORP, 85- *Mem:* Am Inst Mining, Metall & Petrol Engrs; Am Soc Metals; Am Soc Testing & Mat; Asn Iron & Steel Engrs. *Res:* Agglomeration and beneficiation of ironbearing materials; coal and coke technology; direct reduction processes; ironmaking; economic evaluation of process metallurgy processes; steelmaking. *Mailing Add:* Nat Steel Corp 20 Stanwix St Pittsburgh PA 15222

HASS, LOUIS F, b Scranton, Pa, Aug 23, 26; m 60; c 3. BIOCHEMISTRY. *Educ:* Univ Scranton, BS, 50; Bucknell Univ, MS, 54; Duke Univ, PhD(biochem), 59. *Prof Exp:* Chemist, Armstrong Cork Co, 52-55; Nat Inst Cancer fel, Univ Minn, 59-61; chemist, NIH, 61-63; asst prof, Sch Med, State Univ NY Buffalo, 63-68; ASSOC PROF BIOL CHEM, MILTON S HERSHEY MED CTR, PA STATE UNIV, 68- *Concurrent Pos:* Prof & hon vis fel, Univ Aberdeen, Scotland, 80-81. *Mem:* Am Chem Soc; Am Soc Biol Chemists; Biochem Soc UK. *Res:* Correlation of enzyme structure and function; physicochemical characterization of native and modified proteins; mechanisms of enzyme action; erythrocyte metabolism. *Mailing Add:* 230 Bittersweet Dr Hershey PA 17033

HASS, MARVIN, b New York, NY, June 5, 30. SOLID STATE PHYSICS. *Educ:* City Col New York, BS, 50; Syracuse Univ, MS, 51; Univ Mich, PhD(physics), 55. *Prof Exp:* Asst chem, Syracuse Univ, 50-52 & Univ Mich, 52-53; physicist, US Naval Res Lab, 55-82; CONSULT, 82- *Concurrent Pos:* Vis prof, Univ Nebr, 70. *Mem:* Fel Am Phys Soc. *Res:* Light scattering; infrared spectroscopy; lattice vibrations. *Mailing Add:* 309 Yoakum Pkwy Apt 605 Alexandria VA 22304-3923

HASS, MICHAEL A, b St Louis, Mo, Oct 3, 50. CRIMINOLOGY, DNA ANALYSIS. *Educ:* Wash Univ, BA, 72; St Louis Univ, PhD(biochem), 78. *Prof Exp:* NIH res fel, Pulmonary Div, Sch Med, Univ Miami, 78-81, res asst prof, 81-89; FORENSIC DNA SPECIALIST, METRO-DADE POLICE DEPT, CRIME LAB BUR, MIAMI, FLA, 89- *Concurrent Pos:* Prin investr, Fla Lung Asn, 82-83 & Am Lung Asn, 83-85. *Mem:* Am Soc Biochem & Molecular Biol; AAAS. *Res:* Forensic DNA fingerprinting; molecular biology; protein purification and characterization; regulation of metabolism; characterization of antibody specificity and use in immunoassays; mechanisms of oxidant injury and superoxide dismutase regulation; author of 18 technical publications. *Mailing Add:* Crime Lab Bur Metro-Dade Police Dept 9105 NW 25th St Miami FL 33125

HASS, ROBERT HENRY, b Columbus, Ohio, Apr 30, 22; m 49; c 2. FUEL TECHNOLOGY, PETROLEUM ENGINEERING. *Educ:* Purdue Univ, BSc, 43; Case Inst Technol, MSc, 49, PhD(chem eng), 54. *Prof Exp:* Chem engr, B F Goodrich Res, 43-47; instr, Case Inst Technol, 48-54; CHEM ENGR, UNION OIL CO, CALIF, 54- *Mem:* Am Inst Chem Engrs; Am Chem Soc; Instrument Soc Am; AAAS. *Res:* Development of new catalysts and processes useful in petroleum refining and environmental pollution control; hydrogen sulfide recovery and abatement; flue gas treating. *Mailing Add:* 3507 Sunnywood Dr Fullerton CA 92635-1664

HASS, WILLIAM K, b Detroit, Mich, Nov 20, 29; m 53; c 3. MEDICINE, NEUROLOGY. *Educ:* Kenyon Col, AB, 50; Univ Mich, MD, 54. *Prof Exp:* Fel neurol, Neurol Inst, Columbia Univ, 59-60, asst neurol, 60; asst prof, 60-64, assoc prof, 65-71, PROF NEUROL, SCH MED, NY UNIV, 71- *Concurrent Pos:* Mem exec comt, Nat Joint Study Extracranial Arterial Occlusion, 65-68; mem exec comt, Stroke Coun, Am Heart Asn, 68-, mem publ comt, 71-75, vchmn & chmn, Stroke Coun, 76-80; spec consult to dir, Nat Inst Neurol Dis & Stroke, 70-74; ed, Current Concepts Cerebrovasc Dis-Stoke, 71-75; mem bd trustees, Princeton Conf Cerebrovasc Dis, 78- *Mem:* Fel Am Acad Neurol; Asn Res Nerv & Ment Dis; Am Neurol Asn (vpres, 79-80); fel NY Acad Med; Harvey Soc. *Res:* Cerebral blood flow and metabolism; cerebrovascular disease; clinical mass spectrometry. *Mailing Add:* Dept Neurol NY Univ Sch Med 530 First Ave New York NY 10016

HASSAN, ASLAM SULTAN, BILE AND METABOLISM. *Educ:* Univ Ore, PhD(physiol), 79. *Prof Exp:* ASST PROF PHYSIOL, UNIV ILL, 82- *Mailing Add:* Dept Vet Biosci 3641 Vet Med Basic Sci Bldg Univ Ill Urbana IL 61801

HASSAN, AWATIF E, b Alexandria, Egypt, Dec 11, 37; US citizen; div; c 2. SOIL-MACHINE SYSTEMS, MACHINE DESIGN & EQUIPMENT DEVELOPMENT. *Educ:* Univ Alexandria, Egypt, BS, 59; Univ Calif, Davis, MS, 64, PhD(eng), 68. *Prof Exp:* Asst prof mech eng, Cairo High Polytech Inst, 68-71; asst prof agr eng, Univ Alexandria, 71-73; asst prof agr & forest eng, Univ Maine, 74-75; assoc prof, 75-80, PROF FORESTRY BIOL AGR ENG, NC STATE UNIV, 80- *Concurrent Pos:* Postdoctoral Fel, McGill Univ, Que, Can, 72-74; fac adv, Soc Women Engrs, 74-75; assoc ed, J Terra Mech, Int Soc Terrain-Vehicle Systs, 84-; vis prof, Land Tech Univ, Ger Nat Inst Agr Eng, 85-86; co-chair, Coun Forest Eng, 89-90. *Mem:* Am Soc Agr Engrs; Int Soc Terrain-Vehicle Systs; Coun Forest Eng; Int Union Forestry Res Orgn; Soc Women Engrs. *Res:* Design and development of all terrain vehicles, semi-automatic tree planter, precision seeder and several other devices; 2 patents; effect of machine traffic on soil physical and mechanical properties and vehicle instrumentation and capabilities under extremely wet conditions; tested several high flotation tires used in forestry and agriculture in England. *Mailing Add:* Forestry Dept NC State Univ Box 8002 Raleigh NC 27695-8002

HASSAN, HASSAN AHMAD, b Tamra, Palestine, Feb 26, 31; m 58; c 2. AERONAUTICAL ENGINEERING. *Educ:* Univ London, BSc, 52; Univ Ill, Urbana, MS, 53, PhD(aeronaut eng), 56. *Prof Exp:* Res assoc, Univ Ill, Urbana, 53-55, instr aeronaut eng, 55-56; sr lectr mech eng, Univ Raghdad, Baghdad, Iraq, 56-59; from assoc prof to prof aeronaut eng, Va Polytech Inst, 59-62; PROF AERONAUT ENG, NC STATE UNIV, RALEIGH, 62- *Concurrent Pos:* Aerodyn engr, Douglas Aircraft Co, 61; aerospace technologist, NASA, 62; consult, Douglas Aircraft Co, 62-64, Kaman Nuclear, 63-64, Res Triangle Inst, 64-65, Defense Atomic Support Agency, 70-71 & Forge Aerospace, Inc, 73-74; contractor, Langley Res Ctr, NASA, 79- *Honors & Awards:* Western Elec Fund Award, Am Soc Eng Educ, 74; Charles Russ Mem Award, Am Soc Mech Engrs, 77. *Mem:* Am Inst Aeronaut & Astronaut; Am Soc Mech Engrs; Am Phys Soc; Am Soc Eng Educ; Sigma Xi. *Res:* Computational fluid dynamics; reacting flows; hypersonic aerodynamics. *Mailing Add:* Dept Mech Eng NC State Univ Raleigh NC 27650

HASSAN, HOSNI MOUSTAFA, b Alexandria, Egypt, Sept 3, 37; US citizen; c 2. MOLECULAR BIOLOGY OF OXYGEN TOXICITY, NATURAL PRODUCTS MUTAGENICITY. *Educ:* Univ Ain Shams, Egypt, BSc, 59; Univ Calif, Davis, PhD(microbiol), 67. *Prof Exp:* From asst to assoc prof microbiol & food sci, Univ Alexandria, Egypt, 68-72; res assoc biochem, Med Ctr, Duke Univ, 75-79; assoc prof microbiol, McGill Univ, Montreal, 79-80; assoc prof, 80-84, PROF MICROBIOL, TOXICOL & FOOD SCI, NC STATE UNIV, 84- *Concurrent Pos:* Sci Consult, Nat Res Coun, Egypt, 68-72; vis prof, McGill Univ, Montreal, 72-74; asst microbiologist, Univ Maine, 74-75. *Honors & Awards:* Fulbright Sr Res Scholar, Paris, France, 87-88. *Mem:* Am Soc Biol Chemists; Am Soc Microbiol; NY Acad Sci; Can Soc Microbiologists; Sigma Xi; fel Am Inst Chemists. *Res:* Toxicity and the mutagenicity of oxygen free radicals and the protective roles of superoxide dismutases and catalases; regulation of the biosynthesis of superoxide dismutase and catalases; food microbiology. *Mailing Add:* 1309 Swallow Dr Box 7624 Raleigh NC 27606

HASSAN, MOHAMMAD ZIA, b Gurgaon, Brit India, Apr 2, 33; m 59; c 3. MANAGEMENT SCIENCES. *Educ:* Punjab Univ, BSc, 54; Ill Inst Technol, MS, 58, PhD(indust eng), 65. *Prof Exp:* Indust engr, Dukane Corp, Ill, 56-58; sr indust engr, Webcor, Inc, Ill, 58-59, dir indust eng, 59-60; from instr to assoc prof indust eng, 60-77, assoc prof & chmn dept mgt sci, 77-83, PROF MGT SCI & INDUST MGT & DEAN, STUART SCH BUS, ILL INST TECHNOL, 83- *Concurrent Pos:* Consult, Webcor, Inc, 60-62, Warwick Electronics, Inc, 62-74, H K Porter & Co, Inc, 65-67, Standard Kollsman Indust, Inc, Melrose Park, Ill, 67, Hamilton Indust, Chicago, 69-71, GRI Inc, Chicago, 74-75, Comput Peripheral Inc, Mich, 75-76, BRK Electronics, 78, 83, 85, UN Develop Prog, 81 & Fibre Craft Mat Corp, 82, Am Generator & Armature, 86-87, Kable News, 88- *Mem:* Inst Mgr Sci; Sigma Xi; fel Am Soc Qual Control. *Res:* Strategic planning; organizational effectiveness; productivity and quality management. *Mailing Add:* Sch Bus Admin Ill Inst Technol Chicago IL 60616

HASSAN, WILLIAM EPHRIAM, JR, b Brockton, Mass, Oct 13, 23; m 51; c 2. PHYTOCHEMISTRY, PHARMACOLOGY. *Educ:* Mass Col Pharm, BS, 45, MS, 47, PhD(pharmacol), 51; Suffolk Univ, LLB, 65. *Prof Exp:* Asst pharmacog & biol, Mass Col Pharm, 47-49, instr pharmacol & biol, 49-51, asst prof, 51-55; asst dir, 55-59, assoc dir, 59-67, Dir, Peter Bent Brigham Hosp, 67-76; exec vpres, Brigham Women's Hosp, 80-82, vpres & gen counsel, 82-86; RETIRED. *Concurrent Pos:* Consult, Mass Col Pharm, 51-, prof lectr, 55-67, adj prof, 67-, trustee, 72-, vchmn bd trustees, Mass Col Pharm, 75-; pharmacist-in-chief, Peter Bent Brigham Hosp, 52-55; vpres admin, Affiliated Hosps Ctr Inc, 76-78, exec vpres, 78-80. *Mem:* Fel Am Col Hosp Adminr; Am Soc Hosp Pharmacists; Am Soc Law & Med. *Res:* Complete study of anatomy, chemistry and pharmacology of six plants in the genus Asclepias; anatomical and chemical study of Indian belladonna and the camphor basil plant; study of various fractions of licorice for electrolytic activity in the adrenalectomized rat. *Mailing Add:* 535 Boylston St Boston MA 02116-3720

HASSE, RAYMOND WILLIAM, JR, b Detroit, Mich, Aug 7, 24; m 50; c 3. UNDERWATER ACOUSTICS, COMPUTER MODELING. *Educ:* Mass Inst Technol, BS, 47; Univ Conn, MS, 58. *Prof Exp:* Prog officer acoust & oceanog, 49-53, electronic scientist, 55-59, head acoust res br, 59-62, head acoust res div, 62-66, head ocean sci div, 66-71, asst dir sonar res, 71-76, head spec projs dept, Naval Underwater Systs Ctr, New London, Conn, 76-81; FAC MEM, THAMES VALLEY STATE TECH COL, 81- *Concurrent Pos:* Res dir, Tudor Hill Lab, Bermuda Res Detachment, 60- *Honors & Awards:* Naval Underwater Sound Lab Sci Achievement Award, 69. *Mem:* Fel Acoust Soc Am. *Res:* Environmental systems and technology; deep ocean engineering; acoustic propagation loss and backscatter; ocean noise. *Mailing Add:* Thames Valley State Tech Col 574 New London Turnpike Norwich CT 06360

HASSELGREN, PER-OLOF J, b Mulndal, Sweden, May 4, 47; m 70; c 2. SURGERY, BIOCHEMISTRY. *Educ:* Univ Goteborg, MD, 73, PhD(surg), 79. *Prof Exp:* From asst prof to assoc prof surg, Univ Goteborg, 79-84; res clin fel dept surg, Univ Cincinnati, 84-86; assoc prof, Univ Lund, Sweden, 86-87; asst prof, 87-89, ASSOC PROF SURG, UNIV CINCINNATI, 89- *Mem:* AMA; Swed Med Asn; Asn Acad Surg; Surg Infection Soc; Soc Univ Surgeons; fel Am Col Surgeons. *Res:* Protein metabolism in liver and skeletal muscle during sepsis shock and ischemia. *Mailing Add:* Dept Surg Univ Cincinnati Med Ctr 231 Bethesda Ave Cincinnati OH 45267-0558

HASSELKUS, EDWARD R, b Dousman, Wis, July 18, 32; m 60; c 2. ORNAMENTAL HORTICULTURE. *Educ:* Univ Wis, BS, 54, MS, 58, PhD(hort & plant physiol), 62. *Prof Exp:* Res asst hort, 57-61, proj asst, 61, asst prof hort, 61-64, from asst prof to assoc prof hort & land archit, 64-74, PROF HORT & LAND ARCHIT, UNIV WIS, MADISON, 74-, EXTEN HORTICULTURIST, 65- *Honors & Awards:* L C Chadwick Award. *Mem:* Am Hort Soc; Am Soc Hort Sci; Am Asn Bot Gardens & Arboreta; Int Lilac Soc; Am Conifer Soc; Int Ornam Crabapple Soc. *Res:* Landscape plants; evaluation of woody ornamental plants. *Mailing Add:* Dept Hort Univ Wis 1575 Linden Dr Madison WI 53706

HASSELL, CLINTON ALTON, b Seagraves, Tex, Oct 26, 45; m 68; c 2. CLAY PROPERTIES, THERMOLUMINESCENCE. *Educ:* Baylor Univ, BS, 69; Tex A&M Univ, PhD(chem), 75. *Prof Exp:* Welsh Found res fel, Cyclotron Inst, 75-76, lectr, 76-79, asst first year prog dir, Tex A&M Univ, 77-82; lectr math, physics & chem, 82-85, INSTR CHEM, BAYLOR UNIV, 85- *Concurrent Pos:* Instr nuclear power plant technicians, 81-82. *Mem:* Am Chem Soc; Clay Minerals Soc. *Res:* Supplemental materials and large program coordination in chemical education; writing, problem solving for general chemistry; liberal arts chemistry; properties of clay especially related to thermoluminescence and electron spin resonance. *Mailing Add:* Dept Chem Baylor Univ Waco TX 76706

HASSELL, H(ORACE) PAUL, JR, industrial engineering; deceased, see previous edition for last biography

HASSELL, JOHN ALLEN, b Erie, Pa, July 27, 37; m 59; c 2. PHYSICAL CHEMISTRY. *Educ:* Pa State Univ, BS, 59; Carnegie Inst Technol, MS, 65, PhD(phys chem), 66. *Prof Exp:* Res asst heterogeneous catalysis, Mellon Inst Indust Res, Pa, 59-62; sr chemist, Tex US Chem Co, 65-68; sr res chemist, Battelle Mem Inst-Columbus, 68-90; REHAB COMPUTER CONSULT, 89- *Concurrent Pos:* Voice input disabled chem students, ECU, Greenville, NC, 81-82, instr, 82. *Honors & Awards:* IR-100 Award. *Mem:* Am Chem Soc. *Res:* Adsorption and kinetics on metals and solid hydrocarbons; Langmuir film balance studies; solution and melt rheology with Weissenberg rheogonimeter on polymer systems; molecular characterization; process development and control; development of microcomputer acquisition and analysis of polymer molecular characteristics(GPC) and polymer rheology; microcomputer use in modeling solution and molecular properties of predicting useful life of materials; product properties and service life predictions of polymers, using rheological measurements and analysis; computer modelling of diffusion and rheological processes; software development of voice entry computer to aid handicapped students perform chemistry laboratory courses; development of computer analysis of gel permeation chromatography and thermoanalysis including graphic analysis; computer applications in rehabilitation technology for vision and physical disabilities; interfacing of adaptive equipment and technology to microcomputer through mainframe applications. *Mailing Add:* 2229 Northwest Blvd Columbus OH 43221

HASSELL, JOHN ROBERT, b Cambridge, Mass, Oct 27, 43. DEVELOPMENTAL BIOLOGY, BIOCHEMISTRY. *Educ:* Cent Conn State Col, BA, 66; Univ Conn, PhD(develop biol & biochem), 72. *Prof Exp:* INVESTR DEVELOP BIOL & BIRTH DEFECTS, LAB BIOL STRUCT, NIH, 72- *Res:* Biochemical mechanisms regulating early embryonic growth; embryo nutrition; effects of pharmacological drugs on development; chick embryo culture; cytochemistry. *Mailing Add:* Dept Opthamol Univ Pittsburgh Sch Med Eye & Ear Inst Pittsburgh 203 Lothrop St Rm 1025 Pittsburgh PA 15213

HASSELL, THOMAS MICHAEL, b Gary, Ind, Jan 20, 45; m 71; c 1. DENTAL RESEARCH, PATHOLOGY. *Educ:* Ind Univ, Bloomington, BS, 67; Ind Univ, Indianapolis, DDS, 69; Univ Zurich, Dr med dent, 72; Univ Wash, PhD(path), 78. *Prof Exp:* Res assoc biotelemetry, Dent Inst, Univ Zurich, 69-72; asst prof dent auxiliary, Univ Guam, 72-73; asst prof periodont, Sch Dent, Univ NC, Chapel Hill, 78-; AT UNIV FLA, GAINESVILLE. *Mem:* Int Asn Dent Res. *Res:* Connective tissue pathology, especially fibrosis and hyperplasia; side-effects of antiepileptic drugs; pathogenesis of periodontal disease. *Mailing Add:* 222 SW 27th St Gainesville FL 32607

HASSELMAN, DIDERICUS PETRUS HERMANNUS, b The Hague, Neth, Aug 1, 31; US citizen; m 67. MATERIALS SCIENCE, CERAMIC ENGINEERING. *Educ:* Queen's Univ Ont, BSc, 57; Univ BC, MASc, 60; Univ Calif, Berkeley, PhD(ceramic eng), 66. *Prof Exp:* Physicist, Carborundum Co, 60-62; res asst, Lawrence Radiation Lab, Univ Calif, 62-66; sr ceramist, Stanford Res Inst, 66-67; sr res assoc, Mat Res Ctr, Allied Chem Corp, 68-70; prof metall & mat sci & dir ceramics res lab, Lehigh Univ, 70-76; mgr, Mont MHD Inst, Butte, Mont, 76-77; WHITTEMORE PROF MAT ENG, VA POLYTECH INST & STATE UNIV, 77- *Concurrent Pos:* Humboldt Prize. *Honors & Awards:* John Jeppson Gold Medal. *Mem:* Am Ceramic Soc; Am Carbon Soc; Soc Eng Sci. *Res:* Physical properties of refractory materials; elasticity; strength; creep; thermo-elastic fracture; crack growth and propagation; static and cyclic fatigue of single-phase and composite glasses and ceramics; thermophysical properties of high temperature materials. *Mailing Add:* Dept Mat Eng Va Polytech Inst & State Univ Blacksburg VA 24061

HASSELMEYER, EILEEN GRACE, b Brooklyn, NY, May 23, 24. PERINATAL BIOLOGY. *Educ:* NY Univ, BS, 54, MA, 56, PhD(nursing sci, human develop), 63. *Prof Exp:* Head nurse, Children's Med Serv, Bellevue Hosp, NY Univ, 46-50, supvr, 50-56; consult div nursing, USPHS, 56-59, spec asst prematurity, Nat Inst Child Health & Human Develop, 63-67, actg dir perinatal biol & infant mortality prog, 67-68; Annie Goodrich prof perinatal res, Sch Nursing, Yale Univ, 68-69; dir perinatal biol & infant mortality prog, Nat Inst Child Health & Human Develop, 69-74, chief, Pregnancy & Infancy Br, 74-79; assoc dir sci rev, USPHS, 79, asst surg gen, 81; Nat Inst Child Health & Human Develop, NIH, 81-89; EXEC DIR, FED COL NURSING TASK FORCE, UNIFORM SERVS HEALTH SCI, 89- *Concurrent Pos:* NIH spec fel, NY Univ, 62-63; Am Nurses Found res grant, 62-64; Sigma Theta Tau res grant, 68; Yale fac res develop grant, 69; Univ Conn maternal & child health grant, 69-70; lectr, Sch Nursing, Univ Tex Med Br Galveston, 53-53; vis prof, Univ BC, 66-71. *Mem:* Perinatal Res Soc. *Res:* Nursing science; relationship between early and later development for low birth weight infants; indices of fetal and neonatal maturation; influence of nursing care upon survival of high risk newborns; sudden infant death syndrome. *Mailing Add:* 8901 Paddock Lane Potomoc MD 20854

HASSERT, G(EORGE) LEE, JR, b Passaic, NJ, Dec 15, 20; m 44; c 2. PHYSIOLOGY, PHARMACOLOGY. *Educ:* Rutgers Univ, BS, 43, PhD(physiol, biochem), 63. *Prof Exp:* Sr res scientist, Pharmacol Sect, E R Squibb & Sons, Inc, 46-70, sr res investr, 70-72, sect head, dept toxicol, Squibb Inst Med Res, 72-77, admin mgr, qual control, 77-87; RETIRED. *Mem:* NY Acad Sci. *Mailing Add:* 21 Sylvan Ave Metuchen NJ 08840

HASSETT, CHARLES CLIFFORD, b New Haven, Conn, July 20, 05; m 39; c 2. PHYSIOLOGY. *Educ:* Univ Buffalo, AB, 38; Johns Hopkins Univ, PhD(physiol), 41. *Prof Exp:* Asst zool, Johns Hopkins Univ, 39-41; from asst prof to assoc prof physiol, Kirksville Col Osteop & Surg, Mo, 41-45, head dept, 43-45; physiologist, Med Res Lab, Edgewood Arsenal, 45-56, chief entom br, 56-59, asst chief physiol div, 59-66, asst to lab chief, 66-69, chief exp med dept, 70-75; RETIRED. *Concurrent Pos:* Res assoc, Still Mem Res Lab, 41-43; liaison scientist, Neth, 62-63; consult, Biomed Lab, Edgewood Arsenal, 75-76, Nat Inst occup Safety & Health, 75-79 & Nat Res Coun, 82-83. *Mem:* AAAS; Am Entom Soc; Am Physiol Soc; Soc Gen Physiol. *Res:* Pharmacology related to chemical poisons. *Mailing Add:* Four Cedar Ave Towson MD 21204

HASSIALIS, MENELAOS D(IMITRI), b New York, NY, Dec 25, 09; m 31; c 2. MINING, MINERAL ECONOMICS. *Educ:* Columbia Univ, BA, 31. *Hon Degrees:* DSc, Columbia Univ, 53. *Prof Exp:* From lectr to prof mineral eng, 37-59, exec off, 52-67, Henry Krumb prof mining, Sch Mines, 59-78, HENRY KRUMB EMER PROF MINING, SCH MINES, COLUMBIA UNIV, 78- *Concurrent Pos:* Consult, Mobil Oil Corp, 44-, Chem Warfare Serv, 42-44 & Manhattan Proj, 43-45; dir, Minerals Beneficiation Lab, AEC, 52-58; pres, Pac Uranium Mines Co; dir, Kermac Nuclear Fuels Corp & Ambrosia Lake Uranium Corp; chmn phosphate slimes panel, Minerals & Metals Adv Bd, Nat Res Coun, 53; mem, Coun Atomic Age Studies; mem minerals sci & technol comt, Nat Acad Sci, 66-69; dir, Tudor Indust, 66-69; dir, Sandvik Steel, 67-85, chmn bd, Sandvik, Inc, 74-86; trustee, Valley Hosp, Ridgewood, NJ, 68-, vpres hosp & chmn capital budget planning & control comt, 69-75; dir & pres, TIC Mining Co, 69-70; US del Geneva Confs Peaceful Uses Atomic Energy, 55 & 58; US rep, Meeting Peaceful Uses Atomic Energy, UNESCO, 58; head, UN Spec Fund Mission, Turkey, 64 & Spain, 65; chmn bd, Disston Corp, 76-82 & Strata Bit Inc, 82-; vpres, Sammine Explor Co, 81; USAEC Rep to Sweden, 52-60; US Negotiator with Sweden, re nuclear agreement, 57; co-chmn, Bioeng Comt, Col Physicians & Surgeons. *Honors & Awards:* Award, Am Chem Soc, 58; Bicentennial Medal, Freiburg Bergakademie, EGer, 65; Bicentennial Scroll, WBerlin Tech Univ, WGer, 67; Knight of Malta & St John, 85; Citation from Dept of State. *Mem:* Am Inst Mining, Metall & Petrol Engrs; Am Soc Mining Engrs; Am Chem Soc; AAAS; fel Explorers Club, Ex TBTT. *Res:* Mining, ore dressing; surface chemistry and physics; petroleum research; mineral economics and mine management. *Mailing Add:* 122 Phelps Rd Ridgewood NJ 07450

HASSINGER, MARY COLLEEN, b Minneapolis, Minn, Mar 3, 53; m 80. LASER SPECTROSCOPY. *Educ:* Univ Minn, Duluth, BS, 75; Purdue Univ, MS, 79, PhD(anal chem), 83. *Prof Exp:* ASST PROF ANALYTICAL CHEM, VITERBO COL, 83- *Concurrent Pos:* Vis asst prof, Purdue Univ, 85;

prin investr, NSF Col Sci Instrumentation Prog, 85-87. *Mem:* Am Chem Soc; Soc Appl Spectros. *Res:* Develop and test two-laser instruments for fundamental studies in molecular spectroscopy and combustion diagnostics; measurement of two-photon absorption spectra of metal chelate complexes in solution. *Mailing Add:* Dept Chem Viterbo Col 815 S Ninth St La Crosse WI 54601-8802

HASSLER, CRAIG REINHOLD, b Chicago, Ill, Mar 19, 42; m 69; c 2. PHYSIOLOGY, BIOMATERIALS. *Educ:* Northwestern Univ, Evanston, BA, 63; Depauw Univ, MA, 65; Loyola Univ, Chicago, PhD(physiol), 69. *Prof Exp:* Instr physiol, Sch Med, Loyola Univ, 69-70; physiologist, 70-74, sr physiologist, 74-80, RES LEADER, COLUMBUS LABS, BATTELLE MEM INST, 80- *Concurrent Pos:* Prin investr, Battelle Inst, 71-73 & NIH, 74- *Mem:* Am Physiol Soc; Inst Elec & Electronics Engrs; Biomat Soc; Orthop Res Soc. *Res:* Basic understanding of hard tissue; the use of biomaterials to replace hard tissue; tooth replacement systems; electrical augmentation of bone formation; cardiovascular toxicology; toxicology of biomaterials; Pharmacodynamics. *Mailing Add:* Battelle-Columbus 505 King Ave Columbus OH 43201

HASSLER, F(RANCIS) J(EFFERSON), b Cooper Hill, Mo, Aug 2, 21; m 42; c 5. AGRICULTURAL ENGINEERING. *Educ:* Univ Mo, BS, 46; Mich State Col, MS, 48, PhD(agr eng), 50. *Prof Exp:* Asst, Mich State Col, 46-47, instr, 47-48, asst, 48-50; from res asst prof to res assoc prof agr eng, 50-54, in chg grad studies, 53-61, prof, 54-61, REYNOLDS PROF BIOL & AGR ENG & HEAD DEPT, NC STATE UNIV, 61- *Mem:* AAAS; Am Soc Agr Engrs; Am Soc Eng Educ; Sigma Xi. *Res:* Frost control with infrared; fundamentals of tobacco curing; bulk curing of tobacco; patents. *Mailing Add:* Dept Biol & Agr Eng NC State Univ PO Box 7625 Raleigh NC 27695

HASSLER, THOMAS J, b McCook, Nebr, Jan 6, 34; m 56; c 4. FISHERIES, FRESH WATER ECOLOGY. *Educ:* Univ Nebr, BS, 57; Utah State Univ, MS, 60, PhD(fishery biol), 65. *Prof Exp:* Fishery biologist, NCent Reservoirs Invest, US Fish & Wildlife Serv, 64-72; ASST LEADER CALIF COOP FISHERY RES UNIT, FISH & WILDLIFE SERV, HUMBOLDT STATE UNIV, 73- *Mem:* Am Fisheries Soc; Am Inst Fishery Res Biol. *Res:* Reservoir fisheries management; toxicology; ecology of anadronous salmonids; fishery management. *Mailing Add:* Calif Coop Fishery Res Unit Humboldt State Univ Arcata CA 95521

HASSLER, WILLIAM WOODS, b Clearfield, Pa, Sept 6, 17; m 41; c 3. ANALYTICAL CHEMISTRY, ORGANIC CHEMISTRY. *Educ:* Juniata Col, BS, 39; Univ Pa, MS, 41, PhD(chem), 51. *Prof Exp:* Asst chem, Univ Pa, 39-42; res chemist plastics, Rohm and Haas Co, Pa, 42-46; asst prof chem, Drexel Inst, 46-51; prof & head dept, Beaver Col, 51-63; dean, Sch Lib Arts, Indiana Univ Pa, 63-69, pres, Univ, 69-75; pres, Wesley Col, 75-77; RETIRED. *Concurrent Pos:* Consult, Houdry Process Corp, 52-, Smith Kline & French, 54-, Harris D McKinney Co, 57, Selas Corp & Capital Systs, 60-; lectr, Eastern Baptist Col, 56-61. *Mem:* Am Chem Soc; fel Am Inst Chemists. *Res:* Technical writing and communication; potentiometric study of reactions between cupric salts and various bases; taste and odor control in water purification; synthesis of ion exchange resins; inorganic analytical chemistry. *Mailing Add:* 448 N Branddock St Winchester VA 22601-3922

HASSNER, ALFRED, b Czernowitz, Romania, Nov 11, 30; US citizen; m 57; c 2. STEREO CHEMISTRY, HETEROCYCLES. *Educ:* Inst Technol Vienna, BS, 52; Univ Nebr, MS, 54, PhD(chem), 56. *Prof Exp:* Res chemist, E I du Pont de Nemours & Co, Inc, 54; res fel chem, Harvard Univ, 56-57; from asst prof to prof chem, Univ Colo, Boulder, 57-75; leading prof chem, State Univ NY Binghamton, 75-84; PROF CHEM, BAR-ILAN UNIV, RAMAT-GAN, 84- *Concurrent Pos:* Hon res assoc, Harvard Univ, 66; vis scientist, Swiss Fed Inst Technol, 67; vis prof, Weizmann Inst, 69, 72, Univ Wurzburg, 71, Univ Lyons, 80, Kyshu Tech Univ, 85, & Univ Calif, Berkeley, 89; consult, Univ Colo Med Sch, Chem Phys Eng Corp, Med Chem Panel, NIH, NSF-URP Panels, 3M Corp & SKB; Humbolt fel, 71, NIH spec fel, 72-73, Lady Davis fel, 79 & Meyerhoff fel, 81. *Mem:* Am Chem Soc; Chem Soc; Ger Chem Soc. *Res:* Synthesis and stereochemistry of organic nitrogen compounds; stereospecific additions to olefins; synthesis and reactions of nitrogen heterocycles; steroids; azides; design of new synthetic reactions; stereoselective cycloadditions; potential anticancer agents. *Mailing Add:* Dept Chem Bar-Ilan Univ Ramat-Gan 52100 Israel

HASSOLD, GREGORY NAHMEN, b Oakland, Calif, Dec 12, 56. CONDENSED MATTER PHYSICS. *Educ:* Harvey Mudd Col, BS, 79, Univ Colo, MS, 81, PhD(physics), 85. *Prof Exp:* Res asst fel, Harvard Univ, 86-87; RES ASST FEL, UNIV MICH, 87- *Mem:* Am Phys Soc; Am Metall Soc; Sigma Xi. *Mailing Add:* 5454 Deland Rd Flushing MI 48433-8801

HASSOLD, TERRY JON, b Flint, Mich, Aug 31, 46. HUMAN CYTOGENETICS. *Educ:* Mich State Univ, BS, 68, PhD(genetics), 77. *Prof Exp:* Instr genetics, Dept Zool, Mich State Univ, 73-75, res asst cytogenetics, Human Develop Dept, 75-76; asst researcher, 76-80, ASSOC RESEARCHER CYTOGENETICS, DEPT ANAT, UNIV HAWAII, 80- *Concurrent Pos:* NIH fel, 78-81. *Mem:* Am Soc Human Genetics. *Res:* Human population cytogenetics; cytogenetics of human spontaneous abortions. *Mailing Add:* Dept Reproductive Biol Univ Hawaii 2500 Campus Road Honolulu HI 96822

HASSON, DENNIS FRANCIS, b Baltimore, Md, June 1, 34; m 61; c 2. MATERIALS SCIENCE, MECHANICAL ENGINEERING. *Educ:* Johns Hopkins Univ, BES, 55; Va Polytech Inst, MS, 58; Univ Md, College Park, PhD(chem eng), 70. *Prof Exp:* Aerospace res engr, Langley Res Ctr, NASA, 55-59, Proj Mercury, 59-60; appl physics lab, Johns Hopkins Univ, 60-61; aerospace technologist, Goddard Space Flight Ctr, NASA, 61-67; instr mech eng, Univ Md, College Park, 69-71, fel, 71; assoc prof mech eng, Univ DC, 71-73; PROF MECH ENG, US NAVAL ACAD, 73- *Concurrent Pos:* Consult, Comsat Labs, Md, 70-71, David Taylor Naval Ship Res & Develop Ctr & White Oak Lab, Naval Surface & Weapons Ctr; lectr, Univ Southern Calif, 76- & Cath Univ Am, 78- *Mem:* Am Soc Mech Engrs; Am Soc Metals; Am Inst Mining, Metall & Petrol Engrs; Sigma Xi. *Res:* Fracture mechanics and stress corrosion cracking of marine materials, corrosion fatique, welding, fractography; mechanical and fracture behavior of metal, ceramic and polymer matrix composite materials; relaxation processes and mechanical properties of bone materials. *Mailing Add:* 20 East St Annapolis MD 21401

HASSON, JACK, Jan 26, 25; m 48; c 2. AUTOPSIES. *Educ:* City Col New York, BS, 47; Downstate Med Ctr-State Univ NY, MD, 51. *Prof Exp:* CHIEF DEPT PATH, MT SINAI HOSP; ASSOC PROF NEUROPATH, UNIV CONN. *Honors & Awards:* Henry Moses Res Prize, 57. *Mem:* Asn Am Physicians; AMA; Am Soc Clin Pathologists; Can Asn Pathologists; Int Acad Path. *Mailing Add:* Dept Path Mt Sinai Hosp 500 Blue Hills Ave Hartford CT 06112

HASSOUN, GHAZI QASIM, b Haifa, Palestine, June 21, 35; US citizen; m 58. THEORETICAL PHYSICS. *Educ:* Am Univ Beirut, BS, 56; Univ Minn, Minneapolis, MS, 59, PhD(physics), 63. *Prof Exp:* Res assoc theoret physics, Univ Mich, 63-65; asst prof physics, Alfred Univ, 65-66; ASSOC PROF PHYSICS, NDAK STATE UNIV, 66- *Mem:* Am Phys Soc; Am Asn Physics Teachers; Sigma Xi. *Res:* Quantum field theory; elementary particles; electromagnetic theory; physics education. *Mailing Add:* 3720 Fairway Rd Fargo ND 58102-1223

HASSOUN, MOHAMAD HUSSEIN, b Gazieh, Lebanon, Jan 1, 61; m; c 1. ARTIFICIAL NEURAL SYSTEMS, MACHINE INTELLIGENCE. *Educ:* Wayne State Univ, BS, 81, MS, 82, PhD(elec eng), 86. *Prof Exp:* Res assoc, Digital Optics Inc, 86-87; asst prof, 87-91, ASSOC PROF, ELEC & COMPUTER ENG DEPT, WAYNE STATE UNIV, 91- *Concurrent Pos:* NSF res initiation award, 88 & presidential young investr award, 90-95; consult, Digital Optics Inc, 87-88, Federated Monetary Inc, 90; assoc ed, Inst Elec & Electronics Engrs Trans Neural Networks, 91- *Mem:* Soc Photo-Optical Instrumentation Engrs; Int Neural Networks Soc; Inst Elec & Electronics Engrs; Am Soc Eng Educ; Sigma Xi. *Res:* Massively parallel, highly interconnected, and collective computations; artificial neural networks; learning machines; associative memories; photonic computing; search and optimization utilizing highly parallel architectures; genetic algorithms search and optimization; author of numerous publications; one US patent. *Mailing Add:* 7518 Oakman Blvd Dearborn MI 48126

HAST, MALCOLM HOWARD, b New York, NY, May 28, 31; m 53; c 2. ANATOMY, PHYSIOLOGY. *Educ:* Brooklyn Col, BA, 53; Ohio State Univ, MA, 58, PhD, 61. *Prof Exp:* Instr, Univ Iowa, 61-63, res asst prof otolaryngol & maxillofacial surg, Col Med, 65-69; assoc prof, 69-74, PROF OTOLARYNGOL-HEAD & NECK SURG, MED SCH, NORTHWESTERN UNIV, CHICAGO, 74-, PROF CELL, MOLECULAR & STRUCT BIOL (ANAT), MED & DENT SCH, 77-, DIR RES, 69- *Concurrent Pos:* NIH fel otolaryngol, Univ Iowa, 63-65; NIH res grants, 64-75, 78-84; assoc staff, Northwestern Mem Hosp, 69-; mem bd dirs, Ill Soc Med Res, 73-78; NSF res grant, 75-77; guest scientist, Max Planck Inst, Munich, 76; NATO sr fel sci, Oxford Univ, 78; vis prof, Royal Col Surgeons, Eng, 80-86, Univ Edinburgh, 87; guest sci zool, Forshunginstitut U Mus Koenig, 88. *Honors & Awards:* Gould Int Award, 71; Arnott Dem, Royal Col Surgeons, Eng, 85. *Mem:* Fel AAAS; fel Am Speech & Hearing Asn; Am Asn Hist Med; Am Physiol Soc; fel Am Acad Otolaryngol; fel Linnean Soc London; fel Royal Soc Med; Anat Soc Gt Brit & Ireland; Am Asn Anatomists; Am Asn Clin Anatomists; Am Soc Mammalogists; Sigma Xi. *Res:* Neuromuscular physiology and embryology of the larynx; comparative anatomy of larynx; human anatomy. *Mailing Add:* Dept Otolaryngol-Head & Neck Surg Northwestern Univ Med Sch Chicago IL 60611

HASTENRATH, STEFAN LUDWIG, b Budapest, Hungary, July 10, 34. METEOROLOGY, CLIMATOLOGY. *Educ:* Univ Bonn, Dr. *Prof Exp:* Asst prof meteorol, Univ Cologne, 59-60; chief climat div, Nat Meteorol Serv, El Salvador, Cent Am, 60-63; proj assoc meteorol, Univ Wis-Madison, 63-65; assoc prof, Univ Wis-Milwaukee, 65-67; assoc prof, 67-70, PROF METEOROL, UNIV WIS-MADISON, 70- *Concurrent Pos:* Vis prof, Univ Witwatersrand, 71; World Meteorol Orgn/UN Develop Prog prof & head dept meteorol, Univ Nairobi, Kenya, 73-74. *Mem:* Am Meteorol Soc; Ger Meteorol Soc; Meteorol Soc Japan; Royal Meteorol Soc; Int Glaciol Soc; Soc Brasil Meteorol. *Res:* Climate; tropics; glaciers. *Mailing Add:* Dept Meteorol Univ Wis Madison WI 53706

HASTERLIK, ROBERT JOSEPH, b Chicago, Ill, Mar 17, 15. MEDICINE. *Educ:* Univ Chicago, SB, 34; Rush Med Sch, MD, 38; Am Bd Internal Med, dipl. *Prof Exp:* Clin asst, Med Sch, Northwestern Univ, 47-48; from asst prof to prof med, Univ Chicago, 48-70; clin prof, Univ Calif, San Diego, 70-72; prof med, Sch Med, Univ Southern Calif, 73-75; dir, Regional Cancer Control Prog, Cancer Ctr, Hawaii, 77-78; assoc dir, La Jolla Cancer Res Found, 79-81; CONSULT, 81- *Concurrent Pos:* Jr attend physician, Evanston Hosp, 47-48; dir health serv div, Argonne Nat Lab, US AEC, 48-53, sr scientist, Div Biol & Med Res, 50-53, consult, 53-70; assoc dir, Argonne Cancer Res Hosp, 52-63; vpres, Enviro-Med, Inc, 70-73; assoc dir, Los Angeles County-Univ Southern Calif Cancer Ctr, 73-75; assoc dir, Cancer Ctr Hawaii & dir, Comn Cancer Control Prog Hawaii, 77-78; mem, Ill Legis Comn Atomic Energy; mem, Radiation Protection Adv Coun, State of Ill; mem, Subcomt 14, Nat Comt Radiation Protection, 71-77; mem, Nat Coun Radiation Protection & Measurements; mem res comt, Zool Soc San Diego, 79-87; bd trustees, Rush-Presby-St Luke's Med Ctr, 83- *Mem:* Fel AAAS; Radiol Soc NAm; Radiation Res Soc; fel Am Col Physicians. *Res:* Radiation effects and studies on radium poisoning; biological and biochemical effects of total body irradiation; planning studies of cancer research and treatment facilities and delivery of health services. *Mailing Add:* 7722 Ludington Place La Jolla CA 92037

HASTIE, JOHN WILLIAM, b Ilford, Eng, Mar 29, 41; US citizen; m 66; c 2. INORGANIC CHEMISTRY, CHEMICAL PHYSICS. *Educ:* Univ Tasmania, BS, 63, PhD, 67. *Hon Degrees:* DSc, Univ Tasmania, 73. *Prof Exp:* Fel chem, Rice Univ, 66-69; RES CHEMIST, NAT INST STANDARDS &

TECHNOL, 69- *Honors & Awards:* Indust Res-100 Award, 80. *Mem:* Am Chem Soc; Soc Appl Spectros; Am Ceramic Soc; Int Union Pure & Appl Chem. *Res:* Thermochemical properties of inorganic materials, ie, refractory solids, molten salts and slags, vapors and gases, at high temperatures; thermodynamics; kinetics and spectroscopy of high temperature vapors; chemistry of flames, inhibition and combustion. *Mailing Add:* Nat Inst Standards & Technol Gaithersburg MD 20899

HASTINGS, DAVID FRANK, physiology, biophysics, for more information see previous edition

HASTINGS, EARL L, b Hampton, Va, June 15, 28; m 51; c 2. MINING GEOLOGY. *Educ:* Univ Ala, BS, 53; Univ Mich, MS, 54. *Prof Exp:* Res mineralogist, Int Minerals & Chem Corp, 55; geologist, Geol Surv Ala, 56-61; consult geologist, 61-65; vpres mining, R E Wilson Mining Co, 65-68; pres, Eufaula Bauxite Mining Co, 68-73; pres, Spec Minerals Corp, 74-75; opers mgr, Eufaula Minerals, Div Combustion Eng, 76-78; PRES, HASTINGS MINING CORP, 80- *Res:* Economic geology; mineralogy. *Mailing Add:* 735 Holleman Dr Eufaula AL 36027

HASTINGS, ELLSWORTH (BERNARD), b Billings, Mont, Aug 9, 10; m 35; c 3. ENTOMOLOGY. *Educ:* Mont State Col, BS, 34, MS, 39. *Prof Exp:* Field supvr, Farm Security Admin, Mont, 34-37; asst, Exp Sta, 37-39, from asst prof to assoc prof, 39-53, prof, 53-77, EMER PROF ENTOM, MONT STATE UNIV, 77- *Mem:* Entom Soc Am. *Res:* Biology and physiology of Mormon cricket; various species of grasshoppers; sugarbeet webworm; codling moth; absorption of fumigants by grain and grain products; alfalfa weevil and animal ecology. *Mailing Add:* 416 S Sixth Ave Bozeman MT 59715

HASTINGS, FELTON LEO, b Port Arthur, Tex, Jan 15, 38; m 61. INSECT TOXICOLOGY. *Educ:* Auburn Univ, BS, 60, MS, 62; NC State Univ, PhD(entom), 67. *Prof Exp:* Nat Inst Environ Health Sci res grant biochem, NC State Univ, 67-68; res entomologist, 68-69, SUPVRY RES ENTOMOLOGIST, SOUTHEASTERN FOREST EXP STA, US FOREST SERV, 69- *Mem:* Am Chem Soc; Entom Soc Am. *Res:* Develop state, effective and selective chemicals for controlling destructive forest insect populations; determine strategies which minimize adverse effects on parasites, predators and other non-target organisms. *Mailing Add:* Forestry Sci Lab PO Box 12254 Research Triangle Park NC 27709

HASTINGS, HAROLD MORRIS, b Dayton, Ohio, Nov 21, 46; m 68; c 2. MATHEMATICAL BIOLOGY. *Educ:* Yale Univ, BS, 67; Princeton Univ, MA, 69, PhD(math), 72. *Prof Exp:* From instr to asst prof math, Hofstra Univ, 68-74; vis assoc prof, State Univ NY, Binghamton, 74-75; assoc prof, 75-81, chmn dept, 85-90, PROF MATH, HOFSTRA UNIV, 81-, ASSOC DEAN, HOFSTRA COL, 90- *Concurrent Pos:* NSF grant, 77-79 & 81-83 & Woodrow Wilson grant, 80; lectr, Univ Ga, 78-79; consult comput models, 80-; NASA grant, 85-89; founder, Hastings Saalboch Assocs Inc, 83- *Mem:* Am Math Soc; Asn Comput Mach; NY Acad Sci. *Res:* Stability theory of dynamical systems and applications to mathematical ecology; artificial intelligence and brain modelling; patentee in computerized medical monitoring. *Mailing Add:* Hofstra Univ Dept Math Hempstead NY 11550

HASTINGS, IAN JAMES, b Brisbane, Australia, July 21, 43; m 68; c 2. PHYSICAL METALLURGY. *Educ:* Univ Queensland, BSc, 64, PhD(metall), 68. *Prof Exp:* Lectr, Queensland Inst Technol, 66-68; res off, Chalk River Nuclear Labs, 68-79, head, fuel properties & behav group, 79-88, MGR, FUSION BLANKET PROG, AECL RES, 86-, MGR FUEL MAT BR, 88- *Mem:* fel Am Ceramic Soc; Am Nuclear Soc; Can Nuclear Soc. *Res:* Irradiation effects in nuclear fuels; fission gas effects; fuel fabrication and design; computer modelling fuel performance; advanced fuel cycles; fusion breeder blanket materials. *Mailing Add:* AECL RES Chalk River Labs Chalk River ON K0J 1J0 Can

HASTINGS, JOHN WOODLAND, b Salisbury, Md, Mar 24, 27; m 53; c 4. BIOLUMINESCENCE, BIOLOGICAL RHYTHMS. *Educ:* Swarthmore Col, BA, 47; Princeton Univ, MA, 50, PhD(biol), 51; Harvard Univ, Ma, 66. *Prof Exp:* Instr natural sci, Col Cevenol, Le Chambonsur-Lignon, France, 47-48; AEC fel biol sci, Johns Hopkins Univ, 51-53; from instr to asst prof, Northwestern Univ, 53-57; from asst prof to prof biochem, Univ Ill, Urbana, 57-66; PROF BIOL, HARVARD UNIV, 66-, PAUL C MANGELSBERG PROF NATURAL HIS, 87- *Concurrent Pos:* Asst instr, Princeton Univ, 48-51; vis lectr, Univ Sheffield, 61-62; instr, Physiol Training Prog, Marine Biol Lab, Woods Hole, Mass, 61-62, dir course, 62-66; mem Panel Molecular Biol, NSF, 63-66, Div Biol & Sci Adv Comt, 68-71, chmn, 70-71; mem, comt postdoctoral fels chem, Nat Acad Sci, 65-68, comt photobiol, 65-72; mem, space biol subcomt, NASA, 66-70, mem, biochem training comt, Mat Inst Gen Med Sci, 68-72; mem Int Adv Comt, Red Sea Marine Res Sta, Eilat, Israel, 68-; NIH fel, Inst Phys Biochem, Paris, 72-73, Guggenheim fel, 65-66; vis prof, Rockefeller Univ, 65-66, Nat Inst Basic Biol, Okaza, Japan, 86 & Univ Konstaz, Ger, 79-80. *Honors & Awards:* Yamada Found Award, 86; Alexander von Humboldt Prize, 79. *Mem:* AAAS; Am Soc Microbiol; Biophys Soc; Am Soc Biol Chemists; Soc Gen Physiol (vpres to pres, 62-65). *Res:* Marine biology, especially microorganisms; biochemical mechanism of bioluminescence and its biological roles; studies concerning molecular mechanisms of circadian daily rhythms; biological clocks. *Mailing Add:* 16 Divinity Ave Harvard Univ Cambridge MA 02138

HASTINGS, JULIUS MITCHELL, b New York, NY, Aug 24, 20; m 43; c 3. SOLID STATE CHEMISTRY. *Educ:* NY Univ, AB, 40; Cornell Univ, PhD(phys chem), 45. *Prof Exp:* Sr physicist, Monsanto Chem Co, Ohio, 45-47; CHEMIST, BROOKHAVEN NAT LAB, 47- *Mem:* Fel Am Phys Soc. *Res:* Neutron scattering; critical phenomena. *Mailing Add:* 65 Ketcham Ave Patchogue NY 11772

HASTINGS, ROBERT CLYDE, b Tenn, Apr 23, 38; m 81; c 3. LEPROSY. *Educ:* Univ Tenn, Memphis, MD, 62; Tulane Univ, PhD(pharmacol), 71. *Prof Exp:* CHIEF PHARMACOL RES DEPT, NAT HANSEN'S DIS CTR, 71-, CHIEF LAB RES BR, 83- *Concurrent Pos:* Ed, Int J Leprosy, 79-; adj prof pharmacol & clin prof med, Tulane Med Ctr, 83- *Mem:* Int Leprosy Asn; Am Soc Pharmacol & Exp Therapeut; Am Soc Clin Pharmacol & Therapeut; Am Soc Microbiol; AAAS. *Res:* Growth of M Leprae in athymic mice; growth of M Leprae in armadillos for the supply of lepromin and bacilli; microbiology, biochemistry, pathology, pharmacology and immunology of leprosy. *Mailing Add:* Nat Hansen's Dis Ctr Carville LA 70721

HASTINGS, ROBERT WAYNE, b Memphis, Tenn, Nov 24, 43; m 69; c 2. ICHTHYOLOGY, MARINE BIOLOGY. *Educ:* Univ Fla, BS, 65; Fla State Univ, MS, 67, PhD(biol), 72. *Prof Exp:* From asst prof to assoc prof zool, Rutgers Univ, 72-88; PROF, DEPT BIOL, SOUTHEASTERN LA UNIV, 88- *Concurrent Pos:* Dir, Turtle Care Biol Res Sta, 84- *Mem:* Am Soc Ichthyologists & Herpetologists; Am Fisheries Soc; Ecol Soc Am; Estuarine Res Fedn. *Res:* Ecology and distribution of fishes; microhabitat selection in fishes; biology of artificial reefs; biology of coral reef fishes; estuarine ecology. *Mailing Add:* Dept Biol Southeastern La Univ PO Box 814 Hammond LA 70402

HASTINGS, STUART, b Glen Cove, NY, May 12, 37; m 64; c 1. MATHEMATICS. *Educ:* Brown Univ, ScB, 59; Mass Inst Technol, PhD(math), 64. *Prof Exp:* Asst prof math, Case Western Reserve Univ, 64-70; from assoc prof to prof math, State Univ NY, Buffalo, 70-87; PROF MATH, UNIV PITTSBURGH, 87- *Mem:* Am Math Soc; Soc Indust & Appl Math; Math Asn Am. *Res:* Nonlinear differential equations from science and engineering. *Mailing Add:* Dept Math Univ Pittsburgh Pittsburgh PA 15260-4114

HASTREITER, ALOIS RUDOLF, b Rio de Janeiro, Brazil, Mar 22, 27; m 58; c 1. PEDIATRIC CARDIOLOGY. *Educ:* Univ Brazil, MD, 54; Am Bd Pediat, dipl & cert cardiol, 61. *Prof Exp:* Intern, St Luke's Hosp, New Bedford, Mass, 55-56; pediat resident, Philadelphia Gen Hosp, 56-59; instr pediat cardiol, Children's Mem Hosp, Chicago, 61-63; asst prof pediat cardiol, Univ & dir cardiovasc lab, Univ Hosp, 63-67, assoc prof, 67-70, dir sect pediat cardiol, 67-82; PROF PEDIAT CARDIOL, UNIV ILL COL MED, 70- *Concurrent Pos:* Fel pediat cardiol, Children's Mem Hosp, Chicago, 59-61; instr, Northwestern Univ, 61-63; dir cardiovasc lab, Cook County Children's Hosp, Ill, 63-67. *Mem:* Fel Am Acad Pediat; fel Am Col Cardiol; Am Heart Asn; Asn Advan Med Instrumentation; Am Pediat Soc. *Res:* Digitalis, antiarrhythmics. *Mailing Add:* Dept Pediat Cardiol M C 856 Univ Ill Coll Med PO Box 6998 Chicago IL 60680

HASTY, DAVID LONG, b Atlanta, Ga, Sept 1, 47. CELL BIOLOGY, HYBRIDOMA RESEARCH. *Educ:* Carson-Newman Col, BS, 69; Univ Tenn, PhD(anat), 74. *Prof Exp:* Fel anat, Harvard Med Sch, 74-77; asst prof, 77-83, ASSOC PROF ANAT, UNIV TENN CTR HEALTH SCI, 83- *Mem:* Am Soc Cell Biol; Am Soc Develop Biol; NY Acad Sci; AAAS; Sigma Xi. *Res:* Analyzing the role of fibronectin and other extracellular macromolecule in development; adherence of bacteria to oral epithelial cells and other tissue. *Mailing Add:* Univ Tenn Ctr Health Sci Dept Anat 800 Madison Ave Memphis TN 38163

HASTY, NOEL MARION, JR, b Los Angeles, Calif, Oct 22, 44; m 68; c 2. ORGANIC CHEMISTRY. *Educ:* Univ Southern Calif, BS, 66; Univ Wis-Madison, PhD(org chem), 71. *Prof Exp:* Res asst org chem, Univ Calif, Riverside, 70-72; instr, Wesleyan Univ, 72-73; res chemist, Elastomer Chem Dept, 73-79, sr res chemist, Polymer Prod Dept, 79-83, RES ASSOC, DU PONT FIBERS, E I DU PONT DE NEMOURS & CO, INC, 83- *Mem:* Am Chem Soc. *Res:* Polyurethane elastomers; polyesters; thermoplastic elastomers. *Mailing Add:* 1414 N Van Buren St Wilmington DE 19806

HASTY, ROBERT ARMISTEAD, b Dallas, Tex, July 28, 36; div; c 3. ANALYTICAL CHEMISTRY, INDUSTRIAL CHEMISTRY. *Educ:* Univ Tex, Austin, BSChE, 58, PhD(chem), 62. *Prof Exp:* Chemist, Lawrence Radiation Lab, Univ Calif, 61-63; sr scientist chem, Hanford Atomic Prod Oper, Gen Elec Co, 63-65; sr res scientist, Pac Northwest Lab, Battelle Mem Inst, 65-68; asst prof, Mont State Univ, 68-72; sr lectr, 72-78, assoc prof chem, 78-85, Prof anal chem, Univ Witwatersrand, Johannesburg, SAfrica, 85-86; PROF, UNISA, PRETORIA, S AFRICA, 86- *Concurrent Pos:* Lectr, Univ Calif, Berkeley, 62-63; Assoc Western Univs fel, Nat Reactor Testing Sta, Idaho Falls, 71; vis res fel, Univ Strathclyde, Glasgow, 76-77; vis res scientist, Mintek, Johannesburg, 83; vis prof, Univ Wash, 90. *Mem:* Royal Soc Chem; SAfrican Chem Inst. *Res:* Kinetic and catalytic methods of analysis; continuous flow analysis; halogen chemistry; separation methods; ion chromatography. *Mailing Add:* Dept Chem UNISA PO Box 392 Pretoria 0001 South Africa

HASTY, TURNER ELILAH, b Linden, Ala, June 18, 31; m 59; c 2. PHYSICS. *Educ:* Univ Ala, BS, 53, MS, 56, PhD(physics), 59. *Prof Exp:* Instr, Univ Ala, 58-59; mem tech staff, Tex Instruments, 59-69, mgr microwave physics br, 69-73, dir advan technol lab, 73-75, dir semiconductor res & eng labs, 75-80, mgr complementaty metal-oxide semiconductor develop, 80-83, eng mgr, Semi-Custom Design Div, 83-87; mem mgt team, 87, EXEC VPRES & CHIEF OPER OFFICER, SEMATECH, 89- *Concurrent Pos:* Consult mem, Adv Group Electron Devices, 73-79. *Mem:* Am Phys Soc; Sigma Xi. *Res:* Solid state physics; hot electron transport; Gunn effect; Impatt diodes; microwave engineering; complementary metal-oxide semiconductors. *Mailing Add:* 6109 Bon Terra Dr Austin TX 78731

HATANAKA, MASAKAZU, biochemistry, virology, for more information see previous edition

HATCH, ALBERT JEROLD, b Little Rock, Ill, Apr 23, 16; m 46; c 3. NUCLEAR STRUCTURE, PLASMA PHYSICS. *Educ:* Univ Ill, BS, 39, MS, 47. *Prof Exp:* Asst physics, Univ Ill, 43-47; instr & asst physicist, NMex State Univ, 47-49, asst prof & assoc physicist, 49-56; assoc physicist, Argonne Nat Lab, 56-72, physicist, 72-73, asst dir physics div, 73-81; LECTR, ROOSEVELT UNIV, 82- *Concurrent Pos:* Lectr, Ill Inst Technol, 63-71; consult, Ultek Corp, Calif, 63-64; lectr, DePaul Univ, 80-81; instr, Triton Col, 82-83, Moraine Valley Community Col, 82- *Mem:* Fel Am Phys Soc; Sigma Xi. *Res:* High-frequency breakdown and gaseous discharges; controlled thermonuclear processes; electromagnetic levitation; nuclear structure. *Mailing Add:* 5460 S Cornell Ave Chicago IL 60615

HATCH, CHARLES ELDRIDGE, III, b Richmond, Va, Sept 30, 48; m 76; c 2. ORGANIC CHEMISTRY, TECHNICAL MANAGEMENT. *Educ:* Old Dominion Univ, BS, 71; Johns Hopkins Univ, MS, 73, PhD(org chem), 75. *Prof Exp:* From res chemist to sr res assoc, 80-81, mgr process res group, Agr Chem Group, 81-88, DIR PROCESS & FORMULATIONS, RES & DEVELOP, FMC CORP, 88- *Mem:* Am Chem Soc; Sigma Xi. *Res:* Organic photochemistry with synthetic utility resulting in a novel route to several unusual penicillin analogs; carbamoyl fluoride approach to carbamate insecticides; asymmetric synthesis of pyrethroid insecticides; transition metal catalyzed coupling of Grignard reagents to produce substituted biphenyls. *Mailing Add:* 284 Wargo Rd Pennington NJ 08534

HATCH, DORIAN MAURICE, b Burley, Idaho, Sept 25, 40; m 60; c 5. THEORETICAL PHYSICS, SOLID STATE PHYSICS. *Educ:* Utah State Univ, BS, 62; State Univ NY, Stony Brook, MA, 66, PhD(physics), 68. *Prof Exp:* Vis prof physics, State Univ NY, Farmingdale, 68; from asst prof to assoc prof, 68-80, PROF PHYSICS, BRIGHAM YOUNG UNIV, 80- *Concurrent Pos:* Fac res fel, Brigham Young Univ, 69-70; vis prof physics, Fed Inst Technol, Zurich, Switz, 72, 73, Univ Del, Newark, 80 & 81, Univ Wash, Seattle, 90. *Mem:* Sigma Xi; Am Phys Soc. *Res:* Inequivalent representations in field theory and the removal of cutoffs; field theory applied to solid state physics; second quantization of parastatistics; logic of quantum mechanics; phase transitions in solids, renormalization group; group theory. *Mailing Add:* Dept Physics & Astron 277 ESC Brigham Young Univ Provo UT 84602

HATCH, EASTMAN NIBLEY, b Salt Lake City, Utah, June 14, 27; m 52; c 3. NUCLEAR PHYSICS. *Educ:* Stanford Univ, BS, 50; Calif Inst Technol, PhD(nuclear physics), 56. *Prof Exp:* Asst physics, Calif Inst Technol, 54-56, res fel, 56-57; res assoc, Brookhaven Nat Lab, 57-58; sci liaison officer, Sci & Tech Unit, USN, Ger, 58-60; guest physicist, Heidelberg, 60-61; from assoc prof to prof physics, Iowa State Univ, 61-69, sr physicist, 66-69, asst dean grad col, 67-69; prof, 69-89, head dept physics, 72-74, dean, sch Grad Studies, 74-79, EMER PROF PHYSICS, UTAH STATE UNIV, 89- *Concurrent Pos:* Vis prof phys, Univ Freiburg, WGer, 79-80, Univ Cologne, WGer, 87. *Mem:* Fel Am Phys Soc. *Res:* Precision gamma ray spectroscopy with bent crystal spectrometers; measurement of nuclear life-times. *Mailing Add:* Dept Physics UMC 4415 Utah State Univ Logan UT 84322

HATCH, FREDERICK TASKER, b Boston, Mass, Aug 27, 24; m 46; c 4. TOXICOLOGY, FOOD SCIENCE. *Educ:* Dartmouth Col, AB, 44; Harvard Univ, MD, 48; Mass Inst Technol, PhD, 60. *Prof Exp:* Intern med, Roosevelt Hosp, NY, 48-49; res fel & instr res serv, Goldwater Mem Hosp, Columbia Univ, 49-52; assoc med, Harvard Med Sch, 60-65; biochemist, Lawrence Livermore Nat Lab, Univ Calif, 65-73, sect leader cell biol & mutagenesis, 73-80, asst assoc dir biomed & environ res prog, 80-87, CONSULT, LAWRENCE LIVERMORE NAT LAB, UNIV CALIF, 87- *Concurrent Pos:* Asst physician, Mass Gen Hosp, 60-65; investr, Am Heart Asn, 60-65; consult, Vet Admin Hosp, Livermore, Calif, 70-80; mem, lipid metab adv comt, Nat Heart & Lung Inst, NIH, 72-76, Human Subjects Inst Rev Bd, 74-, chmn, 83-, animal res comt, 79- & lab safety comt, 82-; chmn, Melanoma Invest Task Group, 84-87. *Mem:* Am Heart Asn; Am Chem Soc; Am Soc Biochem & Molecular Biol; Environ Mutagen Soc; Am Inst Chemists. *Res:* Food mutagens; genetic toxicology; satellite DNAs, chromosome structure; chemical mutagenesis; tritium metabolism; quantitative lipoprotein electrophoresis; nutrition, genetics and lipoproteins in cardiovascular disease; chemical taxonomy of proteins. *Mailing Add:* RR 2 Box 813 Meredith NH 03253

HATCH, GARY EPHRAIM, b Provo, Utah, Sept 28, 47; m 72; c 6. TOXICOLOGY, NUCLEAR CHEMISTRY. *Educ:* Brigham Young Univ, BS, 72, MS, 74; Univ Utah, PhD(pharmacol), 77. *Prof Exp:* Res assoc, dept pharmacol & med, Duke Univ, 77-79; RES PHARMACOLOGIST, INHALATION TOXICOL DIV, HEALTH EFFECTS RES LAB, US ENVIRON PROTECTION AGENCY, 79- *Concurrent Pos:* Proj officer, numerous extramural coop agreements, US Environ Protection Agency, 80-; prin investr, US Army Interagency Agreement, 84- & Innovative Res award, US Environ Protection Agency, 85- *Honors & Awards:* Sci & Technol Achievement Award, US Environ Protection Agency, 82 & 83. *Mem:* Am Soc Pharmacol & Exp Therapeut. *Res:* Inhalation toxicology of indoor and outdoor air pollutants in animals and extrapolation of results to human risk assessment; dosimetry of inhaled substances; oxygen-18 analysis in tissues; measurement of tissue oxidation and natural protective agents found in tissues. *Mailing Add:* Environ Protection Agency MD-82 ERC Research Triangle Park NC 27711

HATCH, JOHN PHILLIP, b Highland Park, Ill, May 18, 46; m 69; c 2. PSYCHOPHYSIOLOGY, BEHAVIORAL MEDICINE. *Educ:* Bradley Univ, BS, 68, MA, 72; Univ Tex, Arlington, PhD(psychol), 77. *Prof Exp:* Res fel psychol, State Univ NY, Stony Brook, 77-79; res instr, 79-81, asst prof, 81-84, ASSOC PROF PSYCHIAT, UNIV TEX HEALTH SCI CTR, SAN ANTONIO, 84-, ASSOC PROF DENT, 87- *Concurrent Pos:* Chmn, Task Force Comt, Biofeedback Soc Am, 83-87; grant reveiwer, psychobiol prog, NSF, 85-86; consult, Johnson Space Ctr, NASA, 86-87; bd dir, Asn Appl Psychophysiol, 88- *Mem:* Sigma Xi; Soc Psychophysiol Res; Am Asn Dent Res; Am Psychol Asn; Asn Appl Psychophysiol Biofeedback. *Res:* Investigating the psychological and physiological mechanisms that underlie stress related disorders including headache, hypertension, and sexual dysfunction. *Mailing Add:* Dept Psychiat Univ Tex Health Sci Ctr 7703 Floyd Curl Dr San Antonio TX 78284-7792

HATCH, MELVIN (JAY), b Winslow, Ariz, Oct 20, 26; m 58; c 3. POLYMER CHEMISTRY. *Educ:* Univ Ariz, BS, 48; Univ Calif, Los Angeles, PhD(chem), 52. *Prof Exp:* Chemist, Dow Chem Co, 52-56, proj leader, 56-58, assoc scientist, 58-66; assoc prof chem, 66-73, PROF CHEM, NMEX INST MINING & TECHNOL, 73- *Concurrent Pos:* Consult, Dow Chem Co, 66-88, Dionex Corp, 88- *Mem:* Am Chem Soc; fel Am Inst Chemists. *Res:* Ion exchange and ion retardation resins; water soluble polyelectrolytes; sulfonium compounds; halonium compounds; organic reaction mechanisms. *Mailing Add:* 1217 Apache Dr Socorro NM 87801

HATCH, NORMAN LOWRIE, JR, b Boston, Mass, May 27, 32; m 57; c 2. STRUCTURAL GEOLOGY. *Educ:* Harvard Univ, AB & AM, 58, PhD(geol), 61. *Prof Exp:* Chief br eastern regional geol, 73-78, GEOLOGIST, US GEOL SURV, 61- *Concurrent Pos:* Chmn, NE Sect, Geol Soc Am, 76-77, pres Geol Soc Wash, 85. *Mem:* fel Geol Soc Am. *Res:* Structural geology, petrology and stratigraphy of the complexly deformed metamorphic rocks of New Hampshire, Vermont and western Massachusetts. *Mailing Add:* US Geol Surv Stop 926 Reston VA 22092

HATCH, RANDOLPH THOMAS, b Milwaukee, Wis, Apr 27, 45; m 68; c 3. CHEMICAL ENGINEERING. *Educ:* Univ Calif, Berkeley, BS, 67; Mass Inst Technol, MS, 69, PhD(biochem eng), 73. *Prof Exp:* From asst prof to assoc prof chem eng, Univ Md, 72-84; DIR FERMENTATION & BIOCHEM ENG, BIOTECHNICA INT, 84- *Concurrent Pos:* Traineeship, NIH, 67-72; prog dir, NSF, 77-78, prin investr, NSF grant, 76-84. *Mem:* Am Soc Microbiol; Am Inst Chem Engrs; Am Chem Soc. *Res:* Biochemical engineering; membrane separations; reverse osmosis; fermentation technology; oxygen transfer computer control; mixed cultures. *Mailing Add:* 12 Falmouth Rd Wellesley MA 02181-1239

HATCH, RICHARD C, b Aug 20, 36; US citizen; m 62; c 1. PHYSICAL CHEMISTRY. *Educ:* Brown Univ, BS, 58; Univ NH, PhD(chem), 63. *Prof Exp:* Asst chem, Univ NH, 59-62; from asst prof to assoc prof, 62-74, PROF CHEM, MUHLENBERG COL, 74- *Mem:* Am Chem Soc. *Res:* Chemical kinetics; mechanisms of oxidation; reduction reactions in aqueous and nonaqueous media; radiation chemistry. *Mailing Add:* Dept Chem Geol Muhlenberg Col 24 & Chew St Allentown PA 18104

HATCH, RICHARD WALLACE, fish biology, for more information see previous edition

HATCH, ROBERT ALCHIN, b Kalamazoo, Mich, Feb 20, 14; m 42; c 3. MINERALOGY. *Educ:* Univ Mich, BS, 37, MS, 38, ScD(mineral), 42. *Prof Exp:* Res microscopist & phys chemist, Corning Glass Works, NY, 42-47; mineralogist, Electrotech Lab, US Bur Mines, 47-50, leader synthetic mineral sect, 50-55; sr chemist, Minn Mining & Mfg Co, 55-62, supvr, High Temp Mat Res, 62-65, mgr, 65-79; RETIRED. *Mem:* Fel Mineral Soc Am; fel Am Ceramic Soc. *Res:* Phase equilibria of silicate systems; silicate crystal chemistry; mineral synthesis; ceramics; synthetic mica. *Mailing Add:* 2223 E Maple Lane St Paul MN 55109

HATCH, ROBERT HAROLD, physiology, for more information see previous edition

HATCH, ROGER CONANT, b St Joseph, Mich, Jan 23, 35; m 56; c 2. PHARMACOLOGY, TOXICOLOGY. *Educ:* Mich State Univ, BS, 57, DVM, 59; Purdue Univ, MS, 64, PhD(pharmacol), 66. *Prof Exp:* Practitioner vet med, Berwyn Animal Hosp, Ill, 59-60; US Army Vet Corps, 60-62; instr vet pharmacol & toxicol, Purdue Univ, 62-66; assoc prof pharmacol, Ont Vet Col, Univ Guelph, 66-73. sect head pharmacol, 67-70, supvr toxicol testing lab, 66-68, assoc prof, 70-73; assoc prof & supvr, 73-79, PROF TOXICOL, DIAG ASSISTANCE LAB, COL VET MED, UNIV GA, 79-, PROF TOXICOL & PHARMACOL, DEPT VET PHYSIOL & PHARMACOL, 82- *Mem:* Am Acad Vet & Comp Toxicol; Soc Toxicol; Am Acad Vet Pharmacol & Therapeut; NY Acad Sci. *Res:* Veterinary pharmacology and toxicology; therapy and antidotes for poisoning in companion and farm animals; pathologic changes in poisoned animals; poison effects on biochemical functions; poison levels in tissues; poisonous plants and plant poisons; biology; medicines. *Mailing Add:* 2385 White Rd Conyers GA 30207

HATCH, STEPHAN LAVOR, b Logan, Utah, July 22, 45; m 67; c 7. SYSTEMATIC BOTANY. *Educ:* Utah State Univ, BS, 70, MS, 72; Tex A&M Univ, PhD(range sci), 75. *Prof Exp:* Asst prof range sci, Tex A&M Univ, 74-75, fel, 75-76; asst prof range sci, NMex State Univ, 76-79; from asst prof & cur to assoc prof & cur, 79-90, PROF & CUR S M TRACY HERBARIUM, DEPT RANGE SCI, TEX A&M UNIV, 90- *Concurrent Pos:* Tom Slick Res fel, Tex A&M Univ, 73-74, mem fac syst, 74-76. *Mem:* Am Soc Plant Taxonomists; Southwestern Asn Naturalists. *Res:* Grass taxonomy and systematics; Schizachyrium; Sporobolus; digitaria; Pennisetum and Cenchrus; grass flora of New Mexico; grasses of the Southwestern United States; Carex. *Mailing Add:* Dept Rangeland Ecol & Mgt Tex A&M Univ College Station TX 77843-2126

HATCH, WILLIAM JAMES, physiology, for more information see previous edition

HATCHARD, WILLIAM REGINALD, b Sheerness, Eng, Oct 20, 19; nat US; m 44; c 3. POLYMER CHEMISTRY. *Educ:* Univ Portland, BS, 41; Univ Ill, PhD(org chem), 44. *Prof Exp:* Lab instr gen chem, Univ Portland, 40-41; asst, Univ Ill, 41-43, res chemist, 43-45; res chemist, Off Sci Res & Develop, Mass Inst Technol, 45; res chemist, Cent Res Dept, 45-63 & Textile Fibers Dept, 63-69, RES ASSOC, EXP STA, TEXTILE FIBERS DEPT, E I DU PONT DE NEMOURS & CO, 69- *Mem:* Am Chem Soc. *Res:* Free radicals; high energy radiation reactions; isothiazole chemistry; adhesives. *Mailing Add:* 120 Meriden Pl Hockessin DE 19707-1702

HATCHER, CHARLES RICHARD, b Paris, Tex, June 3, 32; m 60; c 2. PHYSICS. *Educ:* Univ Tex, BS, 53, PhD(physics), 58. *Prof Exp:* Scientist, Univ Tex, 58; sr physicist, Lawrence Radiation Lab, 58-61; mgr exp physics dept, Santa Barbara Div, 61-66, dir res & develop, 66-69, gen mgr Los Alamos Div, 69-72, prog exec, Los Alamos Div, EG&G, Inc, 72-77, STAFF MEM, LOS ALAMOS SCI LAB, 77- *Concurrent Pos:* Cost free expert, Assigned Int Atomic Energy Agency, Vienna, Austria, 79 & 84-86. *Res:* Positron annihilation and positronium formation in organic liquids; high speed electronics instrumentation directed toward atomic weapons testing; nondestructive assay instrumentation for safeguarding special nuclear material. *Mailing Add:* Los Alamos Sci Lab PO Box 1553 Los Alamos NM 87545

HATCHER, HERBERT JOHN, b Minneapolis, Minn, Dec 18, 26; m 52; c 6. MICROBIOLOGY, BIOCHEMISTRY. *Educ:* Univ Minn, BA, 53, MS, 64, PhD(microbiol), 66. *Prof Exp:* Asst microbiologist, Glen Lake Sanatorium, Minneapolis, Minn, 53-56; microbiologist, Vet Admin Hosp, Wilmington, Del, 56-57, Smith Kline & French Labs, 57-61 & Clinton Corn Processing Co Div, Standard Brands, Inc, Iowa, 66-67; microbiologist, Econ Lab, Inc, 67-73, mgr int prod develop, 73-77, scientist, 77-78, mgr, Fermentation Res, 80-84; RETIRED. *Mem:* Am Chem Soc; Am Soc Brewing Chem; Henrici Soc. *Res:* Production of microbiological products, such as enzymes, organic acids and antibiotics; physiological studies on microorganisms; detergents; germicides. *Mailing Add:* 13761 Horizon View Rd McCall ID 83638

HATCHER, JAMES DONALD, b St Thomas, Ont, June 22, 23; m 46; c 2. PHYSIOLOGY. *Educ:* Univ Western Ont, MD, 46, PhD(physiol), 51; FRCP(C), 77. *Hon Degrees:* LLD, Queen's Univ, 85. *Prof Exp:* Instr med, Sch Med, Boston Univ, 50-52; from asst prof to assoc prof physiol, Queen's Univ, Ont, 52-59, Nat Heart Found Can sr res assoc, 59-60, assoc dean fac med, 68-71, prof physiol & chmn dept, 59-76; dean Fac Med, 76-85, PROF PHYSIOL & BIOPHYSICS, DALHOUSIE UNIV, 76-, SPEC ADV TO PRES & TECHNOL TRANSFER, 85- *Concurrent Pos:* Markle scholar med sci, Queen's Univ, Ont, 52-57; Nuffield travel fel, 56; Ont Heart Found res fel, 57-59; chmn panel arctic med & climate physiol, Defense Res Bd, 62-67; mem comt care of exp animals, Nat Res Coun, 64-66; mem med adv comt, Ont Heart Found, 66-71, dir, 67-71; mem, Med Res Coun Can, 67-70, chmn fel comt, 67-73; vis prof & vis res assoc, Cardiovasc Res Inst & Sch Med, Univ Calif, 71-72; mem bd dirs, Dalhousie Med Res Found, 80-, Connaugh Labs, 81-87; chmn Tertiary Care Adv Comt, 81-88, Task Force on Chem in Environ & Their Rel to Birth Defects, Dept Health, New Brunswick, 82-85. *Honors & Awards:* Her Majesty Queen Elizabeth Silver Jubilee Anniversary Medal, 77; Guest Hon, Atlantic Chap Weizman Inst, 85. *Mem:* Can Physiol Soc (treas, 65-68, vpres, 69-70, pres, 70-71); hon treas Asn Can Med Col (vpres). *Res:* Cardiovascular adjustments and mechanisms of hypoxic states; electrolyte and water metabolism. *Mailing Add:* Res & Technol Transfer Rm-11-12 Sir Charles Tupper Med Bldg Dalhousie Univ Halifax NS B3H 4H6 Can

HATCHER, JOHN BURTON, chemistry, physics; deceased, see previous edition for last biography

HATCHER, ROBERT DEAN, JR, b Madison, Tenn, Oct 22, 40; m 65; c 2. STRUCTURAL GEOLOGY, TECTONICS. *Educ:* Vanderbilt Univ, BA, 61, MS, 62; Univ Tenn, PhD(geol), 65. *Prof Exp:* Geologist, Humble Oil & Ref Co, 65-66; from asst prof to prof geol, Clemson Univ, 66-78; prof geol, Fla State Univ, 78-80; prof geol, Univ SC, 80-86; DISTINGUISHED SCIENTIST, OAK RIDGE NAT LAB, UNIV TENN, 86- *Concurrent Pos:* Proj dir, Appalachian Ultradeep Proj; ed, Geol Soc Am Bull, 81-88. *Honors & Awards:* Distinguished Serv Award, Geol Soc Am, 88. *Mem:* AAAS; Geol Soc Am; Am Geophys Union; Geol Asn Can; Am Asn Petrol Geologists; Sigma Xi. *Res:* Structural geology; regional tectonics; regional geophysics; stratigraphy in medium to high grade metamorphic rocks; evolution of mountain chains. *Mailing Add:* Dept Geol Sci Univ Tenn Knoxville TN 37996-1410

HATCHER, ROBERT DOUGLAS, b St John's, Nfld, June 26, 24; nat US; m 51; c 4. SOLID STATE PHYSICS. *Educ:* Dalhousie Univ, BSc, 45, MSc, 47; Yale Univ, MS, 48, PhD(physics), 49. *Prof Exp:* From instr to assoc prof physics, NY Univ, 49-62; PROF PHYSICS, QUEEN'S COL, NY, 62- *Concurrent Pos:* Vis res assoc, Brookhaven Nat Labs, 57-58, consult, 58-; vis res assoc, Atomic Energy Res Estab, Harwell, Eng, 69-70; guest scientist, Kernforschungsanlage, Jülich, WGer, 76-77, Univ Munich, 83, Univ Trento, 84 & Univ Amsterdam, 84. *Mem:* Fel Am Inst Physics. *Res:* Proton-proton scattering; relativistic equations of elementary particles; echelette diffraction gratings; far infra properties of alkali halides; defects in alkali halide crystals, ionic crystals, metals and superconductors. *Mailing Add:* Dept Physics Queen's Col 65-30 Kissena Blvd Flushing NY 11367

HATCHER, S(TANLEY) RONALD, b Salisbury, Eng, Aug 20, 32; Can citizen; m 55; c 4. CHEMICAL & NUCLEAR ENGINEERING. *Educ:* Univ Birmingham, BSc, 53, MSc, 55; Univ Toronto, PhD(chem eng), 58. *Prof Exp:* Chem engr reactor chem, Chalk River Nuclear Labs, 58-63, sr chem engr, 63-65, head chem technol br, 65-74, dir appl sci div, 74-78, vpres & mgr, Whiteshell Nuclear Res Estab, Atomic Energy Can Ltd, 78-81, vpres, marketing & sales, 81-86, PRES, ATOMIC ENERGY CAN LTD, 86- *Mem:* Can Soc Chem Eng; Chem Inst Can; Am Nuclear Soc; Can Nuclear Soc; Can Nuclear Asn. *Res:* Chemistry of nuclear reactor coolants and moderators, including mass transfer of corrosion products, behavior of impurities, decomposition and purification in water and organic coolants; radioactive waste management; nuclear fuel resources and fuel recycle; technology driven and customer oriented nuclear research and development. *Mailing Add:* AECL Corp 344 Slater St Ottawa ON K1A 0S4 Can

HATCHER, VICTOR BERNARD, b Ormstown, Que, Apr 1, 43; m 66; c 2. BIOCHEMISTRY, CELL BIOLOGY. *Educ:* Bishop's Univ, BSc, 64; McGill Univ, MSc, 66, PhD(biochem), 69. *Prof Exp:* Res asst immunol, Univ Helsinki, 69-70; res assoc biol chem, Harvard Med Sch, Mass Gen Hosp, 70-73; from asst prof to assoc prof, 73-88, PROF, ALBERT EINSTEIN COL MED, 88- *Concurrent Pos:* Estab investr award, NY Heart Asn,. *Mem:* Am Fedn Clin Res; Harvey Soc; NY Acad Sci; Soc Complex Carbohydrates; Sigma Xi; Fedn Am Soc Exp Biol. *Res:* Structure and function of human endothelial cells; bacterial interaction with human endothelial cells. *Mailing Add:* Albert Einstein Col Med 111 E 210th St Bronx NY 10467

HATCHER, WILLIAM JULIAN, JR, b Augusta, Ga, July 21, 35; m 58, 85; c 3. CHEMICAL ENGINEERING. *Educ:* Ga Inst Technol, BChE, 57; La State Univ, Baton Rouge, MSChE, 64, PhD(chem eng), 68. *Prof Exp:* Res engr, Esso Res Labs, Humble Oil & Ref Co, 60-66, sr res engr, 68-69; res assoc chem eng, La State Univ, Baton Rouge, 66-68; from asst prof to prof chem eng, Univ Ala, Tuscaloosa, 69-76, head dept, 73-83, actg dean, Col Eng, 81-83, ACTG HEAD, DEPT COMPUTER SCI, UNIV ALA, TUSCALOOSA, 90- *Concurrent Pos:* Prof engr, State Ala, 71-; chmn, State Hazardous Waste Adv Comt, 78-82; adj fac, USMC Command Staff Col, 78-84. *Mem:* Am Inst Chem Engrs; Am Inst Chemists. *Res:* Chemical reactor design; chemical kinetics; transport phenomena; catalysis. *Mailing Add:* Box 870290 Tuscaloosa AL 35487-0203

HATCHER, WILLIAM S, b Charlotte, NC, Sept 20, 35; m 59; c 3. MATHEMATICAL LOGIC. *Educ:* Vanderbilt Univ, BA, 57, MAT, 58; Univ Neuchatel, Dr es Sc(math), 64. *Prof Exp:* Res assoc, Univ Neuchatel, 61-64, prof charge de cours, 65; assoc prof math, Univ Toledo, 65-68; assoc prof, 68-72, PROF MATH, LAVAL UNIV, 72- *Concurrent Pos:* Invited prof EPFL, Lausanne, Switz, 72-73. *Mem:* Asn Symbolic Logic; Am Math Soc; Math Asn Am. *Res:* Universal algebra; category theory; logic; computer science. *Mailing Add:* Univ Laval Quebec Quebec PQ G1K 7P4 Can

HATCHETT, JIMMY HOWELL, b Mangum, Okla, June 2, 35; m 57; c 2. ENTOMOLOGY. *Educ:* Okla State Univ, BS, 59, MS, 61; Purdue Univ, PhD(entom), 69. *Prof Exp:* Res entomologist, Agr Res Serv, USDA, 61-81; MEM STAFF, DEPT ENTOM, KANS STATE UNIV, 75- *Mem:* Entom Soc Am; Am Soc Agron; Sigma Xi. *Res:* Development of wheat varieties resistant to the Hessian fly; genetics of Hessian fly biotypes; biology, ecology and control of wheat insects. *Mailing Add:* Dept Entom Kans State Univ Manhattan KS 66506

HATEFI, YOUSSEF, b Tehran, Iran, Aug 9, 29; m 61; c 2. BIOCHEMISTRY. *Educ:* Davis & Elkins Col, BSc, 52; Univ Wash, PhD(biochem), 56. *Prof Exp:* Asst prof biochem, Inst Enzyme Res, Univ Wis, 59-61; assoc prof & head dept, Pahlavi Univ, Iran, 61-63, prof, provost & dept chancellor, 64-66; PROF IN RESIDENCE BIOCHEM, UNIV CALIF, SAN DIEGO, 67- *Concurrent Pos:* Assoc mem, Scripps Clin & Res Found, 63-64 & 66-67, mem, 67-; vis prof, Sch Med, Univ Calif, San Francisco, 66, & Univ Wash, 69; ed, Arch Biochem & Biophys, 69-74; adv ed, Bioenergetics, 70-; vis prof, Univ Stockholm, 71. *Honors & Awards:* Sci Medal Honor, Ministry of Educ, Iran, 52. *Mem:* Am Soc Biol Chem; Biophys Soc; Am Chem Soc. *Res:* One-carbon metabolism; electron transport; oxidative phosphorylation; structure and function of biological membranes; mechanisms of enzyme action; water structure in biology. *Mailing Add:* Scripps Clin & Res Found 10666 N Torrey Pines Rd La Jolla CA 92037-1093

HATFIELD, CHARLES, JR, b Flemingsburg, Ky, July 30, 20; m 43; c 3. MATHEMATICS. *Educ:* Georgetown Col, AB, 40; Univ Ky, AM, 41; Cornell Univ, PhD(math), 44. *Prof Exp:* Asst prof math, Univ Minn, 46-60; chmn dept, Univ NDak, 60-64; PROF MATH, UNIV MO, ROLLA, 64- *Mem:* Am Math Soc; Math Asn Am; Sigma Xi. *Res:* Wiener integration of non-linear functionals; mathematical analysis; probability theory; average number of roots of certain random functions. *Mailing Add:* One Lovers Lane Rolla MO 65401

HATFIELD, CRAIG, b Greenfield, Ind, Jan 24, 35; m 62. GEOLOGY. *Educ:* Univ Ind, BS, 57, MA, 61, PhD(geol), 64. *Prof Exp:* From asst prof to assoc prof, 64-74, PROF GEOL, UNIV TOLEDO, 74- *Res:* Richmondian stratigraphy and paleoecology in Indiana and Kentucky; upper Cretaceous stratigraphy and paleoecology in Kansas; Devonian stratigraphy in Michigan and Ohio. *Mailing Add:* Dept Geol Univ Toledo 2801 W Rancroft Toledo OH 43606

HATFIELD, DOLPH LEE, b El Paso, Tex, Oct 3, 37; m 58, 83; c 3. MOLECULAR BIOLOGY. *Educ:* Univ Tex, BA, 58, MA, 60, PhD(biol, Chem), 62. *Prof Exp:* Fel protein chem, Duke Univ, 62-64; Am Cancer Soc fel protein synthesis, NIH, 64-66; Nat Cancer Inst spec fel bact genetics, Pasteur Inst, Paris, 66-67; RES BIOLOGIST, NAT CANCER INST, 68- *Concurrent Pos:* USPHS fel, 63. *Mem:* Am Soc Biol Chem. *Res:* Transcriptional and translational control mechanisms of protein biosynthesis. *Mailing Add:* NIH Bldg 37 Rm 3C23 Bethesda MD 20892

HATFIELD, EFTON EVERETT, b Mindenmines, Mo, Jan 25, 19; m 42; c 3. ANIMAL NUTRITION, PHYSIOLOGY. *Educ:* Univ Ark, BS, 42; Okla Agr & Mech Col, MS, 49; Univ Ill, PhD(nutrit), 55. *Prof Exp:* Instr animal indust, Univ Ark, 46-47; asst prof animal husb, Panhandle Agr & Mech Col, 49-51; from instr to assoc prof, 54-70, prof, 70-78, EMER PROF ANIMAL SCI, UNIV ILL, URBANA, 78- *Concurrent Pos:* Res partic, Oak Ridge Inst Nuclear Studies, 62. *Mem:* AAAS; Am Soc Animal Sci; Poultry Sci Asn; Am Inst Nutrit; Am Dairy Sci Asn. *Res:* Nonprotein nitrogen in animal nutrition; energy-protein relationships; vitamin and mineral nutriton; effect of antibiotics and certain drugs on nutrition and health; amino acid nutrition of ruminants. *Mailing Add:* Dept Animal Sci 124 ASL Univ Ill 1207 W Gregory Dr Urbana IL 61801

HATFIELD, G WESLEY, b Avant, Okla, Aug 2, 40; m 59; c 2. BIOCHEMISTRY, MICROBIOLOGY. *Educ:* Univ Calif, Santa Barbara, BS, 64; Purdue Univ, PhD(molecular biol), 68. *Prof Exp:* From asst prof to assoc prof, 70-78, PROF MICROBIOL, COL MED, UNIV CALIF, IRVINE, 78-, DIR, GENE RES & BIOTECHNOL PROG, 83- *Concurrent Pos:* NIH fel, Duke Univ, 68-70; Am Cancer Soc, NIH & NSF grants, Univ Calif, Irvine,

70-; USPHS career develop award, 71-76; IPA vis scientist, Lab Biochem, Nat Cancer Inst, 80-81; founder, dir, vpres res & chief scientist, Am Biogenetics Corp, Irvine, Calif, 83- *Honors & Awards:* Eli Lilly Res Award Microbiol & Immunol, Am Soc Microbiol, 75. *Mem:* AAAS; Am Soc Microbiol; Soc Exp Biol & Med. *Res:* Molecular mechanisms of biological control systems; interrelationships of tRNA and gene expression in bacteria and cultured animal cells; recombinant DNA and gene cloning; biotechnology. *Mailing Add:* Dept Microbiol & Molecular Genetics Col Med Univ Calif Irvine CA 92717

HATFIELD, GEORGE MICHAEL, pharmacognosy, for more information see previous edition

HATFIELD, JERRY LEE, b Wamego, Kans, May 1, 49; m 68; c 2. BIOMETEOROLOGY. *Educ:* Kans State Univ, BS, 71; Univ Ky, MS, 72; Iowa State Univ, PhD(agron), 75. *Prof Exp:* From asst prof to assoc prof biometeorol & biometeorologist, Univ Calif, 75-83; vis scientist, Water Conserv Lab, USDA-Agr Res Serv, 82, plant physiologist, 83-85, supvr plant physiologist, 85-89, LAB DIR, NAT SOIL TILTH LAB, USDA-AGR RES SERV, AMES, IOWA, 89- *Concurrent Pos:* Prin investr, grants, USDA, NASA, Dept Energy & Off Water Res & Technol, 76-; tech ed, Agron J, Am Soc Agron, 81-89, ed, 90- *Mem:* Fel Am Soc Agron; Crop Sci Soc Am; Soil Sci Soc Am; Am Meteorol Soc; Am Soc Photogram; Am Geophys Union; hon mem Indian Agrometerol Soc. *Res:* Evaluation of the response of soil microclimate and energy exchange to changes in the soil surface; quantification of the impact of farming systems on water on atmospheric quality. *Mailing Add:* Nat Soil Tilth Lab 2150 Pammel Dr Ames IA 50011-3250

HATFIELD, JOHN DEMPSEY, b Sneedville, Tenn, Aug 18, 19; m 43; c 4. PHYSICAL CHEMISTRY, APPLIED MATHEMATICS. *Educ:* Univ Tenn, BA, 38, MS, 39; Purdue Univ, PhD(agr chem), 42. *Prof Exp:* Control chemist, E I du Pont de Nemours & Co, 39; asst animal nutrit, Purdue Univ, 40-41, asst chemist, 41-42; res chemist, Tenn Valley Authority, 43-81; CONSULT, 81- *Honors & Awards:* Charles H Stone Award, 77. *Mem:* Am Chem Soc. *Res:* Thermodynamics of solutions; diffusion; electrolytic equilibria; complexes in solution; polyphosphate chemistry; corrosion; design and analysis of tests; statistics and regression analysis; sulfur dioxide chemistry. *Mailing Add:* 1224 Sorrento Rd Florence AL 35630-5934

HATFIELD, LYNN LAMAR, b Kansas City, Mo, July 6, 37; m 56; c 3. ATOMIC PHYSICS, PULSED POWER PHYSICS. *Educ:* Ark Polytech Col, BS, 60; Univ Ark, MS, 64, PhD(physics), 66. *Prof Exp:* Instr physics, Little Rock Univ, 60-62; res assoc, Rice Univ, 66-68; from asst prof to assoc prof, 68-86, PROF PHYSICS, TEX TECH UNIV, 86-, DIR ENG PHYSICS, 79- *Mem:* Am Phys Soc; Inst Elec & Electronics Engrs; Insulation Soc. *Res:* Atomic physics applied to pulsed power; surface effects on insulators and metals used in spark gaps; control of spark gaps and discharges used as switches in high voltage systems. *Mailing Add:* 3815 42nd St Lubbock TX 79413

HATFIELD, MARCUS RANKIN, b Chicago, Ill, Dec 21, 09; m 57. ELECTROCHEMISTRY. *Educ:* Univ Md, BS, 31, MS, 32, PhD(phys chem), 35. *Prof Exp:* Asst, State Feed & Fertilizer Control Lab, Md, 30; asst, Univ Md, 31-35; res chemist, Nat Carbon Co Div, Union Carbide Corp, 35-48, head develop & control labs, 48-52, head develop lab & mach develop dept, 52-54, battery develop mgr, 54-56, tech dir consumer prod, 56-59, dir develop, Consumer Prod Div, 60-62, dist works mgr, 62-63, dir prod technol, 64-74; CONSULT, 75- *Concurrent Pos:* Mem staff, Battery Standards Work, Am Nat Standards Inst-Int Electrotech Comn, 67-75. *Mem:* Am Chem Soc; Electrochem Soc; Am Soc Qual Control. *Res:* Batteries; development and uses of carbon; flow of fluids and heat; resins and plastics; powder metallurgy; antifreeze. *Mailing Add:* Renaissance 35 Delft Circle Olmsted Falls OH 44138-1265

HATFIELD, WILLIAM E, b Ransom, Ky, May 31, 37; div; c 5. INORGANIC CHEMISTRY. *Educ:* Marshall Univ, BS, 58, MS, 59; Univ Ariz, PhD(chem), 62. *Prof Exp:* Res assoc chem, Univ Ill, 62-63; from asst prof to prof chem, 63-88, ADJ PROF APPL SCI, UNIV NC, CHAPEL HILL, 87-, MARY ANN SMITH PROF CHEM, 88- *Concurrent Pos:* Consult, overseas liaison comt, Am Coun Educ, 76-78; alt counr, Inorg Div, Am Chem Soc, 76-80; Nat Acad Sci exchange scientist to Soviet Union, 77; Am Specialist, US State Dept, 77; dir, NATO Advan Res Inst on Molecular Metals, 78; vis scholar, Univ Cambridge, 78; mem, Acad Adv Coun, GTE, 79-81; vis prof, Univ Petrol & Minerals, Dhahran, Saudi Arabia, 80, Tata Inst for Fundamental Res, Bombay, India, 81 & Johannes-Gutenburg Univ, Mainz, WGer, 82 & Univ SParis, Orsay, 87; vis lectr, Univ PR, Mayaguez, 82; mem grants comt, NC Bd Sci & Technol, 84-86; Guggenheim fel, 85-86; chmn, NC sect, Am Chem Soc, 88; vchmn, dept chem, Univ NC, Chapel Hill, 87- *Honors & Awards:* G O Doak lectr, NC State Univ, 81; Waite Philip Fishel lectr, Vanderbilt Univ, 81; Charles H Stone Award, Carolina-Piedmont Sect, Am Chem Soc, 85. *Mem:* Am Chem Soc; AAAS; Sigma Xi; Mat Res Soc. *Res:* Structure, magnetic properties and bonding in transition metal compounds. *Mailing Add:* CB No 3290 Venble Hall Univ NC Dept Chem Chapel Hill NC 27599-3290

HATHAWAY, CHARLES EDWARD, b Corpus Christi, Tex, Feb 29, 36; m 75; c 3. PHYSICS. *Educ:* Tex A&M Univ, BS, 58; Univ Okla, PhD(physics), 65. *Prof Exp:* Jr prof physicist, US Naval Ord Test Sta, 58-59; res asst physics, Res Inst, Univ Okla, 59-65; from asst prof to prof physics, Kans State Univ, 66-81, head dept, 71-81; dean sci & math, Univ Tex, San Antonio, 81-86; VPRES ACAD AFFAIRS, WRIGHT STATE UNIV, 86- *Concurrent Pos:* Dir, Moiser Inc. *Mem:* Am Phys Soc; Optical Soc Am; AAAS; Am Asn Physics Teachers. *Res:* Light scattering in gases, liquids and solids; infrared spectroscopy; lattice dynamics; physics education. *Mailing Add:* Acad Affairs Wright State Univ Dayton OH 45435

HATHAWAY, DAVID HENRY, b Bangor, Maine, Aug 29, 51; m 73; c 2. GEOPHYSICAL FLUID DYNAMICS. *Educ:* Univ Mass, Amherst, BS, 73; Univ Colo, Boulder, MS, 75, PhD(astrophys), 79. *Prof Exp:* Teaching asst physics, Univ Colo, Boulder, 73-75; res asst, Sacramento Peak Observ, 75, res assoc, 81-82; res asst, Lab Atmospheric & Space Physics, 75-76 & dept astrogeophys, UNiv Colo, 76-79; fel, Advan Study Prog, Nat Ctr Atmospheric Res, 79-81; asst astron, Nat Solar Observ, 82-84; SCIENTIST, MARSHALL SPACE FLIGHT CTR, NASA, 84- *Mem:* Am Astron Soc; Int Astron Union; Sigma Xi. *Res:* Fluid flows in the sun, stars and planets, including small scale turbulence, large scale convection, global circulation and magnetohydrodynamical dynamos. *Mailing Add:* Space Sci Lab ES52 NASA Marshall Space Flight Ctr Huntsville AL 35812

HATHAWAY, GARY MICHAEL, b Los Angeles, Calif, Mar 6, 37. PURIFICATION OF IMMUNOGLOBULINS. *Educ:* Calif State Univ, Long Beach, BS, 64; Univ Calif, Davis, PhD(biochem), 67. *Prof Exp:* Asst res biochemist, Scripps Clin & Res Found, 67-71; asst res biochemist, 71-76, res assoc, 76-85, ACAD COORDR, UNIV CALIF, RIVERSIDE, 85- *Concurrent Pos:* Lectr, Univ Calif, Riverside, 71-72 & Calif State Col, San Bernadino, 75-76. *Mem:* Am Chem Soc; AAAS; Fedn Am Socs Exp Biol. *Res:* Effects of phosphorylation on the regulation of eukaryotic protein synthesis and the role of cyclic adenosine monophosphate-independent casein kinases and their regulation by cellular factors using kinetic and physical/biochemical techniques. *Mailing Add:* Dept Biochem Univ Calif Riverside CA 92521

HATHAWAY, RALPH ROBERT, b Chicago, Ill, Dec 20, 28; m 56; c 2. EXPERIMENTAL BIOLOGY. *Educ:* Chicago Univ, PhB, 49; Duke Univ, BA, 55; Fla State Univ, MS, 57, PhD(biol), 61. *Prof Exp:* Marine scientist, Va Inst Marine Sci, 63-64; asst prof, 64-69, ASSOC PROF MOLECULAR BIOL, UNIV UTAH, 69- *Concurrent Pos:* USPHS fel, 61-63. *Mem:* Am Soc Zool; Brit Soc Study Fertil. *Res:* Animal fertilization and gamete physiology; biochemistry of steroid hormones. *Mailing Add:* Dept Biol 201 S Biol Bldg Univ Utah Salt Lake City UT 84112

HATHAWAY, ROBERT J, b Glendive, Mont, Dec 2, 21; m 45; c 6. CARBOHYDRATE CHEMISTRY, BIOCHEMISTRY. *Educ:* Univ Ill, BS, 43; Mich State Univ, PhD(chem), 51. *Prof Exp:* Chemist res & develop, Abbott Lab, 43-46 & Miles Lab, 51-57; sr res chemist, A E Staley Co, 57-83; RETIRED. *Mem:* Am Chem Soc; Am Asn Cereal Chemists; Tech Asn Pulp & Paper Indust; Sigma Xi. *Res:* Organic reactions of starch; rheology; application to papermaking, coating, textiles, detergents and other specialties; chemicals from saccharides, cellulose and renewable resources through organic and enzymatic reactions. *Mailing Add:* 2515 Ill Circle Decatur IL 62526

HATHAWAY, RONALD PHILIP, b Denver, Colo, Aug 22, 43; m 63; c 2. PARASITOLOGY. *Educ:* Ft Lewis Col, BS, 65; Univ NMex, MS, 66; Univ Ill, Urbana, PhD(zool), 70. *Prof Exp:* Asst biol, Univ NMex, 64-66; asst zool, Univ Ill, 66-70; asst prof, 70-78, ASSOC PROF BIOL, COLO COL, 78- *Mem:* Am Soc Parasitol. *Res:* Ultrastructural studies of parasitic helminthes; reproductive physiology of trematodes. *Mailing Add:* Dept Biol Colo Col Colorado Springs CO 80903

HATHAWAY, SUSAN JANE, b Potsdam, NY, Aug 8, 50. ORGANIC CHEMISTRY. *Educ:* State Univ NY, Potsdam, BA, 71, MA, 73; Univ NH, PhD(chem), 78. *Prof Exp:* Asst prof chem, Gettysburg Col, 78-; AT GEN ELEC RES & DEVELOP CTR, SCHENECTADY, NY. *Concurrent Pos:* Fel, Univ NH, 77-78. *Mem:* Am Chem Soc. *Res:* Stereochemistry; asymmetric synthesis, in particular asymmetric homogeneous hydrogenation. *Mailing Add:* Gen Elec Co Noryl Ave Selkirk NY 12158

HATHAWAY, WILFRED BOSTOCK, b Salem, Mass, Dec 13, 19; m 48; c 3. BIOLOGY. *Educ:* Mass State Col, BS, 41; Univ Mass, MS, 47; Cornell Univ, PhD(ornamental hort, floricult & entom), 50. *Prof Exp:* Dean, Grad Studies, 67-80, prof, 50-80, EMER PROF BIOL, TOWSON STATE UNIV, 80- *Concurrent Pos:* Mem, Harford Coun Forest Conservancy Dist Bd, 76- & chmn, 81-, Harford Coun Environ Adv Bd, 84-, Md Asn Forest Conservancy Dist Bds, 76-, vpres, 85-86, pres, 91-, Hartford Hist Preserv Comn, 87- *Mem:* Sigma Xi. *Res:* Zoology, particularly isopod crustacea; economic entomology. *Mailing Add:* PO Box 185 Churchville MD 21028

HATHCOCK, BOBBY RAY, b Ripley, Tenn, Aug 29, 42; m 64; c 2. AGRICULTURE, PLANT BREEDING. *Educ:* Univ Tenn, Martin, BS, 64; Univ Tenn, Knoxville, MS, 66; Tex A&M Univ, PhD(plant breeding), 70. *Prof Exp:* ASST PROF AGRON, UNIV TENN, MARTIN, 71- *Mem:* Am Soc Agron; Soil Conserv Soc Am; Crop Sci Soc Am. *Res:* Grain and forage crops. *Mailing Add:* Dept Agr Univ Tenn Martin TN 38238

HATHCOCK, JOHN NATHAN, b Marshville, NC, Aug 13, 40; m 59; c 2. NUTRITION, PHARMACOLOGY. *Educ:* NC State Univ, BS, 62, MS, 64; Cornell Univ, PhD(animal nutrit), 67. *Prof Exp:* Res assoc biochem, Sch Med, St Louis Univ, 67-69; asst prof, Cornell Univ-Univ Philippines Exchange Prog, 69-71; asst prof biochem & biophys, Iowa State Univ, 71-72; asst prof nutrit, Div Biol Sci, Pa State Univ, 72-73; from assoc prof to prof, Iowa State Univ, 73-85; CHIEF, EXP NUTRIT BR, FOOD & DRUG ADMIN, 85-, MGR, DIET-TOXICITY INTERACTIONS PROG, 87- *Mem:* Am Inst Nutrit; Soc Toxicol; Am Soc Clin Nutrit; Inst Food Technologists. *Res:* Nutritional toxicology; diet and glutathione-mediated detoxifications; nutrient toxicities; safety of natural anticarcinogens in foods. *Mailing Add:* Exp Nutrit Br Food & Drug Admin 200 C St SW Rm 1874 HFF-268 Washington DC 20204

HATHCOX, KYLE LEE, b Gilmer, Tex, Feb 22, 43; m 66; c 3. SOLID STATE PHYSICS. *Educ:* NTex State Univ, BS, 65, MS, 68, PhD(physics), 72. *Prof Exp:* Instr physics, NTex State Univ, 72-73; instr physics & math, Tarrant Community Col, 72-74; assoc prof physics, Union Univ, 74-88, chmn, chem/physics dept, 80-88; PROF PHYSICS, GORDON COL, 88-, CHMN, SCI/MATH DIV, 88- *Mem:* Am Asn Physics Teachers; Am Chem Soc. *Res:* Solid state, transport properties and stress studies on III-V materials. *Mailing Add:* Sci/Math Div Gordon Col Barnesville GA 30204

HATHEWAY, ALLEN WAYNE, b Los Angeles, Calif, Sept 30, 37; div; c 3. HAZARDOUS WASTE CLEANUP, ENGINEERING GEOLOGY. *Educ:* Univ Calif, Los Angeles, AB, 61; Univ Ariz, MS, 66, PhD(geol eng), 71. *Prof Exp:* Res assoc, Lunar & Planetary Lab, Univ Ariz, 66-69; staff engr, LeRoy Crandall Assoc, Los Angeles, 69-71; proj engr, Geotech Br, US Forest Serv, 71-72; sr proj engr, Woodward-Clyde Consult, 72-74; proj geologist, Shannon & Wilson, Inc, San Francisco, 74-76; vpres & chief geologist, Haley & Aldrich, Inc, Cambridge, Mass, 76-81; PROF GEOL ENG, UNIV MO, ROLLA, 81- *Concurrent Pos:* Colonel, Corps Engrs, US Army Reserve, 61-; adj asst prof civil eng, Univ Southern Calif, 71-74; adj assoc prof geol, Boston Univ, 79-81; US Nat Comt, Eng Geol, 80-81 & 85-86, Tunneling Technol, 83-86; chmn, Eng Geol Div, Geol Soc Am, 80; mem, US Nat Res Coun Bd on Earth Sci, 87-90; chmn, Eng Geol Div, Geol Soc Am, 80. *Honors & Awards:* Daniel W Mead Prize, Am Soc Civil Engrs, 75; E B Burwell Mem Award, Geol Soc Am, 81. *Mem:* Geol Soc Am; Asn Eng Geologists (pres, 85); Am Soc Civil Engrs; Am Inst Mining, Metal & Petrol Engrs; Geol Soc London; Soc Am Mil Engrs; Am Geophys Union. *Res:* Waste management facility siting and design; remedial treatment of uncontrolled waste disposal sites; critical facility siting; seismic risk assessment; tunnels and underground construction; urban geology in reconstruction of cities. *Mailing Add:* Dept Geol Eng 129 V H McNutt Hall Univ Morolla Rolla MO 65401-0249

HATHEWAY, CHARLES LOUIS, b Barberton, Ohio, July 27, 32; m 60; c 2. MICROBIOLOGY, PUBLIC HEALTH. *Educ:* Ohio State Univ, BSc, 57, MSc, 61, PhD(dairy sci & immunogenetics), 64. *Prof Exp:* Res assoc immunochem, Evanston Hosp, Ill, 64-65; res supvr blood derivatives res, div labs, Mich Dept Health, 65-67; asst prof vet surg & med, Mich State Univ, 67-69; biochemist bact vaccines, Mich Dept Pub Health, 69-75; adj assoc prof, Dept Parasitol & Lab Pract, 80-88, RES CHIEF, BOTULISM LAB, DIV BACT DIS, CTR INFECT DIS, CTRS DIS CONTROL, UNIV NC, 75-; ADJ ASSOC PROF, DEPT PARASITOL & LAB PRACT, UNIV NC, 80- *Concurrent Pos:* Training fel, Med Sch, Northwestern Univ, 64-65. *Mem:* Am Soc Microbiol; Sigma Xi. *Res:* Genetics, serology and immunochemistry of blood group antigens; biologic products from blood; cancer research; bacterial vaccines and toxoids; botulism, clostridial foodborne illnesses; diagnostic procedures; epidemiology. *Mailing Add:* Botulism Lab DBD/CID Ctrs Dis Control Atlanta GA 30333

HATHEWAY, RICHARD BRACKETT, b Melrose, Mass, Oct 27, 39; m 63; c 3. GEOLOGY, PETROLOGY. *Educ:* Bowdoin Col, BA, 61; Univ Mo, Columbia, MA, 64; Cornell Univ, PhD(geol), 69. *Prof Exp:* Teaching asst geol, Univ Mo, Columbia, 62-64 & Cornell Univ, 64-68; from asst prof to assoc prof, 68-87 CHAIR, DEPT GEOL SCI, STATE UNIV COL, GENESEO, 87- *Concurrent Pos:* NSF grant, Basin & Range Field Conf, 69; State Univ NY Res Found grant, 70-71 & 74; NATO Advan Study Inst Grant, 72. *Mem:* Mineral Soc Am; Sigma Xi; Geol Soc Am. *Res:* Igneous and metamorphic petrology including origin and significance of mylonites. *Mailing Add:* Dept Geol Sci State Univ NY Col Geneseo NY 14454

HATHEWAY, WILLIAM HOWELL, b Hartford, Conn, Nov 28, 23; m 53; c 3. BOTANY. *Educ:* Chicago Univ, BS, 48, MS, 51; Harvard Univ, PhD(biol), 56. *Prof Exp:* Botanist, Atoll Res Proj, Pac Sci Bd, Nat Res Coun, 52; asst statistician, Field Staff Agr, Rockefeller Found, Colombia, 56-60, assoc statistician, Mex, 61-64; exec dir, Orgn Trop Studies, Costa Rica, 64-65; botanist, Trop Sci Ctr, 65-68; assoc prof, 69-75, PROF FORESTRY, UNIV WASH, 75- *Concurrent Pos:* Collab bot, Smithsonian Inst, 65- *Mem:* AAAS; Ecol Soc; Soc Study Evolution; Sigma Xi. *Res:* Quantitative ecology; tropical forestry; economic botany. *Mailing Add:* Col Forest Resources Univ Wash Seattle WA 98195

HATKIN, LEONARD, electronics engineering, physics; deceased, see previous edition for last biography

HATSELL, CHARLES PROCTOR, b Alexandria, Va, Sept 18, 44; m 67. ELECTRICAL ENGINEERING. *Educ:* Va Polytech Inst, BS, 67; Duke Univ, MS, 68, PhD(elec eng), 70. *Prof Exp:* Res asst, Adaptive Signal Detection Lab, Duke Univ, 69-70, res assoc, 70; asst prof elec eng, USAF Inst Technol, 70-76; res aerospace med physician, Med Crew Technol, USAF Sch Aerospace, 76-; chief med, USAF Aeromed Res Lab, Wright Patterson AFB, Dayton, Ohio; CHIEF SCIENTIST, AEROSPACE MED DIRECTORATE, ARMSTRONG LAB, BROOKS AFB, TEX. *Mem:* Inst Elec & Electronics Engrs. *Res:* Signal processing, information theory; biomedical sciences. *Mailing Add:* Aerospace Med Directorate Armstrong Lab Brooks AFB TX 78235-5000

HATSOPOULOS, GEORGE NICHOLAS, b Athens, Greece, Jan 7, 27; US citizen; m 59; c 2. THERMODYNAMICS. *Educ:* Mass Inst Technol, BS & MS, 51, ME, 54, ScD(mech eng), 56. *Hon Degrees:* DSc, NJ Inst Technol, 82. *Prof Exp:* From instr to assoc prof mech eng, Mass Inst Technol, 54-62, sr lectr, 62-90; PRES & CHMN BD, THERMO ELECTRON CORP, 58- *Concurrent Pos:* Founder, Thermo Electron Corp, 56; US rep to liaison group on thermionic elec power generation, Europ Nuclear Energy Agency, 66-; mem, NASA Res & Technol Comt Space Powers & Elec Propulsion, 68- & gov coun, Nat Acad Eng. *Honors & Awards:* Golden Plate Award. *Mem:* Nat Acad Eng; fel Inst Elec & Electronics Engrs; Am Inst Aeronaut & Astronaut; Am Soc Mech Engrs; fel Am Acad Arts & Sci. *Res:* Thermodynamics; direct energy conversion; author of over 60 technical publications. *Mailing Add:* Thermo Electron Corp 101 First Ave PO Box 9046 Waltham MA 02254-9046

HATT, ROBERT TORRENS, mammalogy, ethnography; deceased, see previous edition for last biography

HATTAN, DAVID GENE, b Independence, Kans, Feb 21, 42; m 67; c 1. PHARMACOLOGY, TOXICOLOGY. *Educ:* Univ Kans, BS, 65; Ohio State Univ, PhD(pharmacol), 73. *Prof Exp:* Pharmacist, Indian Health Div, USPHS, 65-68; asst prof pharmacol, Sch Pharm, Univ Md, 73-78; TOXICOLOGIST FOOD ADDITIVES, FOOD & DRUG ADMIN, FOOD ADDITIVES EVAL, BUR FOODS, 78- *Mem:* AAAS; Acad Pharmaceut Sci; Am Col Toxicol. *Res:* Behavioral and electrophysiological correlates of alcohol in animals; temperature regulation. *Mailing Add:* 5126 Beaverbrook Rd Columbia MD 21044

HATTEMER, JIMMIE RAY, b St Louis, Mo, Jan 4, 39; m 60; c 2. COMPUTER SCIENCE. *Educ:* Wash Univ, AB, 59, AM, 63, PhD(math), 64. *Prof Exp:* Instr math, Princeton Univ, 64-66; from asst prof to assoc prof, 66-81, PROF COMPUT SCI, SOUTHERN ILL UNIV, 81- *Mem:* Asn Comput Mach; Inst Elec & Electronics Engrs Comput Soc. *Res:* Local boundary behavior of solutions to certain partial differential equations. *Mailing Add:* Dept Computer Sci Southern Ill Univ Edwardsville IL 62026

HATTEN, BETTY ARLENE, b Oxford, Kans, Dec 14, 29. MEDICAL MICROBIOLOGY, IMMUNOLOGY. *Educ:* Drury Col, BA, 52; Univ Tex Southwestern Med Sch Dallas, MA, 62, PhD(microbiol), 65. *Prof Exp:* USPHS training grant microbiol, Univ Tex Southwestern Med Sch Dallas, 65-66, from instr to asst prof microbiol, 68-73; asst prof basic health & biol sci, Southwestern Med Col, NTex State Univ, 73-80; chair, Dept Clin Lab Sci, 83-90, ASSOC PROF CLIN LAB SCI & DIR MED TECHNOL PROG, OKLA CITY HEALTH SCI CTR, UNIV OKLA, 80- *Concurrent Pos:* Res dir, Med Res Found Tex grant, 67-68; prin investr, Nat Inst Allergy & Infectious Dis grant, 69-72. *Mem:* Am Soc Microbiol. *Res:* Animal reactions to Klebsiella antigens; ultrastructure, viability, filterability, reverting ability, antigenicity, toxicity, and pathogenicity of Brucella L-forms; immune capabilities of bats and their response to bacteriophage X 174; antigenic characterization of the mycoplasmas; age-related and leukemia-related changes in immune capabilities of mice. *Mailing Add:* Health Sci Ctr Univ Okla Oklahoma City OK 73106

HATTERSLEY-SMITH, GEOFFREY FRANCIS, b London, Eng, Apr 22, 23; Eng & Can citizen; m 55; c 2. TOPONYMY, POLAR HISTORY. *Educ:* Oxford Univ, MA, 51, DPhil(geol), 56. *Prof Exp:* Base leader glaciol, Falkland Islands Dependencies Surv, 48-50; defence sci staff officer geophys, Defence Res Bd Can, 51-71; head geotech, Defence Res Estab Ottawa, 71-73; prin sci officer, Brit Antarctic Surv, 73-91; RETIRED. *Concurrent Pos:* Mem assoc comt geod & geophys & chmn subcomt glaciers, Nat Res Coun Can, 59-66, mem assoc comt meteorites, 63-73; gov, Arctic Inst NAm, 63-66; secy, Antarctic Place Names Comt, UK, 75-91. *Honors & Awards:* Royal Geog Soc Founder's Medal, 66. *Mem:* Fel Arctic Inst NAm; fel Royal Geog Soc; Royal Soc Can. *Res:* Glaciology and glacial history of northern Ellesmere Island, Northwest Territories; toponymy of Antarctic. *Mailing Add:* Crossways Cranbrook England

HATTIN, DONALD EDWARD, b Cohasset, Mass, Nov 16, 28; m 50; c 3. GEOLOGY. *Educ:* Univ Mass, BS, 50; Univ Kans, MS, 52, PhD(geol), 54. *Prof Exp:* Asst instr geol, Univ Kans, 50-52, instr, 53-54; from asst prof to assoc prof, 54-67, PROF GEOL, IND UNIV, BLOOMINGTON, 67- *Concurrent Pos:* NSF sci fac fel, Univ Reading, 69, Univ Tex, Arlington, 78; grants, NSF, 75-77, 87-90 & Am Chem Soc, 78-80, 84-86; vis prof, Ernst-Moritz-Arndt-Univ, Greifswald, GDR, 85. *Mem:* Geol Soc Am; Soc Econ Paleontologists & Mineralogists; Am Asn Petrol Geologists; Paleont Soc. *Res:* Regional stratigraphy; depositional environments; paleontology of Upper Cretaceous formations in the US Western Interior region; paleoecology; petrology of chalk. *Mailing Add:* Dept Geol Sci Ind Univ Bloomington IN 47405

HATTIS, DALE B, b Santa Monica, Calif, Oct 31, 46; m 80; c 1. QUANTITATIVE RISK ASSESSMENT, PHARMACOKINETIC & PHARMACODYNAMIC MODELING. *Educ:* Univ Calif, Berkeley, BA, 67; Stanford Univ, PhD(genetics), 74. *Prof Exp:* Sr res assoc, Complex Systs Inst, Case Western Reserve Univ, 73-74; res assoc, Ctr Policy Alternatives, Mass Inst Technol, 75-81, prin res assoc, Ctr Technol, Policy & Indust Develop, 81-90; RES ASSOC PROF, CTR TECHNOL, ENVIRON & DEVELOP, CLARK UNIV, 90- *Concurrent Pos:* Vis sr lectr, Prog Social Ecol, Univ Calif, Irvine, 86. *Mem:* Soc Risk Anal; AAAS; Soc Occup & Environ Health. *Res:* Quantitative risk assessment for both cancer and non-cancer effects; pharmacokinetic modeling; human interindividual variability, interspecies comparison and Monte Carlo simulation analysis of uncertainties. *Mailing Add:* Ctr Technol Environ & Develop Clark Univ Worcester MA 01610

HATTMAN, STANLEY, b Brooklyn, NY, July 19, 38; m 63; c 3. MICROBIOLOGY, VIROLOGY. *Educ:* City Col New York, BS, 60; Mass Inst Technol, PhD(microbiol), 65. *Prof Exp:* Helen Hay Whitney Found fel, 65-68; from asst prof to assoc prof, 74-81, PROF BIOL, UNIV ROCHESTER, 81- *Concurrent Pos:* Prin investr, USPHS, 68-71 & 72-91 & Am Cancer Soc, 69-70; Res Career Develop Award, NIH, 72-76. *Mem:* Am Soc Biol Chemists; Am Soc Microbiol. *Res:* Genetics and biochemistry of DNA-modification and restriction; regulation of gene expression. *Mailing Add:* Dept Biol Univ Rochester Wilson Blvd Rochester NY 14627

HATTON, GLENN IRWIN, b Chicago, Ill, Dec 12, 34; m 54; c 5. NEUROPHYSIOLOGY, NEUROENDOCRINOLOGY. *Educ:* North Cent Col, BA, 60; Univ Ill, MA, 62, PhD(physiol psychol), 64. *Prof Exp:* From asst prof to assoc prof, 65-73, PROF PSYCHOL, MICH STATE UNIV, 73-, PROF PHYSIOL, 84- *Concurrent Pos:* NIH res career develop award, 70-75; prin investr res grants, Nat Inst Neurol, Commun Dis & Stroke, 70-; neurobiol adv panel, NSF, 74-77; dir, Neurosci Prog, Mich State Univ, 78-; prin investr, 81-; assoc ed, Brain Res Bull, 80-; mem Neurol Disorders prog, proj rev comt B, Nat Inst Neurol, Commun Dis & Stroke, 84-88, chmn, 86-88; Fogart sr Int fel, 82-83; sr res scholar, Univ Cambridge, UK, 82-83 Guggenheim fel, Cambridge, UK, 89-90. *Mem:* Soc Neurosci; Am Asn Anatomists; Physiol Soc London; Am Physiol Soc. *Res:* Neurophysiology, neuroanatomy and immunocytochemistry of neuropeptide-containing cells of the hypothalamus of the brain. *Mailing Add:* Psychol Res Bldg Mich State Univ East Lansing MI 48824-1117

HATTON, JOHN VICTOR, b Yorkshire, Eng, Apr 9, 34; m 61; c 3. WOOD PULPING, FOREST PRODUCTS. *Educ:* Oxford Univ, BA, 59, MA & DPhil, 61. *Prof Exp:* Fel chem, Nat Res Coun Can, 61-63; scientist org chem, Domtar Ltd, 64-65; head pulp bleaching pulp & paper, BC Res Coun, 65-67; res scientist fiber prod, Western Forest Prod Lab, 67-80, head, Wood Pulping Sect, 76-80; HEAD, RESOURCE EVAL SECT, PAPRICAN, 80- *Honors & Awards:* Pulp Manufacture Div Award, Tech Asn Pulp & Paper Indust, 83; Johann Richter Prize, Tech Asn Pulp & Paper Indust, 83. *Mem:* Can Pulp & Paper Asn; fel Tech Asn Pulp & Paper Indust; fel Chem Inst Can; Int Union Forest Res Org; Poplar Coun Can. *Res:* Complete tree utilization; chip quality; outside chip storage; process control; fibre resources for conversion to pulp; use of low quality materials in pulping; second-growth forests. *Mailing Add:* 3337 Quesnel Dr Vancouver BC V6S 1Z7 Can

HATTON, THURMAN TIMBROOK, JR, b Bartow, Fla, Feb 4, 22; m 47; c 4. HORTICULTURE. *Educ:* Univ Fla, BS, 43, MS, 49; Wash State Univ, PhD, 53. *Prof Exp:* Asst plant physiol, Univ Fla, 48-49; asst prof hort & chmn dept, Col Agr & Mech Arts, Univ PR, 49-50; instr, Wash State Univ, 50-53; exten hort specialist, NC State Col, 53-55; res leader export & qual improvement res, 72-89, RES HORTICULTURIST, AGR RES SERV, USDA, 55-; CONSULT, POSTHARVEST PHYSIOL FRUITS & VEGETABLES, 89- *Concurrent Pos:* Sr horticulturist in chg, Mkt Qual & Transp Res, USDA, 55-64, invests leader, Subtrop & Trop Fruits Invest, 64-68, invests leader, Southeast Citrus & Veg Invests, 68-72; prof, Fruit Crops Dept, Univ Fla, 72-89. *Mem:* Fel Am Soc Hort Sci; Int Hort Soc; Int Soc Citricult; Sigma Xi. *Res:* Tropical horticulture and plant physiology; post-harvest studies of fruits and vegetables, particularly maturity, storage, handling and transportation; quarantine treatment research. *Mailing Add:* PO Box 68 Chuluota FL 32766

HATTORI, TOSHIAKI, b Japan, Dec 17, 31; m 69; c 1. NEUROCYTOLOGY. *Educ:* Kyoto Univ, MS, 59, PhD, 76; Mie Univ, Japan, MD, 67. *Prof Exp:* Instr, Dept Physiol, Fac Med, Mie Univ, 59-62, lectr, 62-67; res assoc biochem, Ore State Univ, 67-70; res assoc neurol sci, Dept Psychiat, Univ BC, 70-78; ASST PROF, DEPT ANAT, UNIV TORONTO, 78- *Mem:* Sigma Xi; Soc Neurosci; Am Asn Anat; AAAS. *Res:* Functional neuroanatomy of the extrapyramidal system of rodents by combined technique of electron microscopic autoradiography and immunohistochemistry. *Mailing Add:* Dept Anat Med Sci Bldg Univ Toronto Toronto ON M5S 1A8 Can

HATZAKIS, MICHAEL, b Charia, Crete, Greece, Jan 1, 28; nat US. RESEARCH ADMINISTRATION. *Educ:* NY Univ, BS, 64, MS, 67. *Prof Exp:* Technician, Radio Eng Labs, Long Island City, New York, NY, 58-61; technician, IBM T J Watson Res Ctr, Yorktown Heights, NY, 61-67, staff mem, 67-76, mgr, 76-88, IBM FEL, IBM T J WATSON RES CTR, YORKTOWN HEIGHTS, NY, 88- *Concurrent Pos:* Vis prof & dir, Microelectronics Inst, Nat Res Ctr, Athens, Greece. *Honors & Awards:* Cledo Brunetti Award, Inst Elec & Electronics Engrs, 87. *Mem:* Nat Acad Eng; Inst Elec & Electronics Engrs; Am Vacuum Soc; Electrochem Soc; Mat Res Soc. *Res:* Lithographic material, tools and processes; early development of electron beams for device fabrication; development of polymethyl methacrylate resist; lift-off metallization process; first micron and sub-micron sized bipolar and field effect transistors; photo and electron resists and processes; scanning electron microscopy; author of various publications; granted 19 patents. *Mailing Add:* IBM Thomas J Watson Res Ctr PO Box 218 Yorktown Heights NY 10598

HATZENBUHLER, DOUGLAS ALBERT, b Amsterdam, NY, Sept 23, 45; m 88. PHARMACEUTICAL FORMULATION, PHYSICAL CHEMISTRY. *Educ:* Univ Rochester, BS, 67; Mich State Univ, PhD(phys chem), 70. *Prof Exp:* Instr chem, Univ Va, 70-71; phys chemist, Aerochem Res Labs, Sybron Corp, 71-74; phys chemist, Procter & Gamble Miami Valley Res Labs, 74-76; staff scientist, 76-78, res head, 78-84, SR RES SCIENTIST, UPJOHN CO, 84- *Mem:* Am Chem Soc. *Res:* Laser and molecular spectroscopy; laser Raman spectroscopy; pharmaceutical analysis; vibrational spectroscopy; quality control; formulation of pharmaceutical products; topical formulations; transdermal studies. *Mailing Add:* Upjohn Co Unit 7233-209-7 301 Henrietta St Kalamazoo MI 49007

HATZIOS, KRITON KLEANTHIS, b Florina, Greece, Aug 6, 49; US citizen; m 79; c 2. PLANT PHYSIOLOGY, HERBICIDE TECHNOLOGY. *Educ:* Aristotelian Univ, Thessaloniki, Greece, BS, 72; Mich State Univ, MS, 77, PhD(plant physiol), 79. *Prof Exp:* Res asst plant physiol, Mich State Univ, 76-79; from asst prof to assoc prof, 79-88, PROF PLANT PHYSIOL, VA POLYTECH INST & STATE UNIV, 88- *Honors & Awards:* Res Award, Agr Honor Soc, 85; Outstanding Young Scientist Award, Weed Sci Soc Am, 86. *Mem:* Am Soc Plant Physiologists; AAAS; Weed Sci Soc Am; Am Chem Soc; Am Soc Photobiol; Scand Soc Plant Physiologists. *Res:* Mechanisms of action of herbicides and other plant growth regulators; metabolism of herbicides in higher plants; chemical and genetic manipulation of crop tolerance to herbicides; interactions of herbicides with other agrochemicals. *Mailing Add:* Dept Plant Path Va Polytech Inst & State Univ Blacksburg VA 24061-0330

HAUBEIN, ALBERT HOWARD, b Joplin, Mo, May 1, 14; m 40; c 3. ORGANIC CHEMISTRY. *Educ:* Cent Methodist Col, AB, 37; Iowa State Col, PhD(org chem), 42. *Prof Exp:* Asst chem, Iowa State Col, 37-42; res chemist, 42-65, SR RES CHEMIST, RES CTR, HERCULES INC, 65- *Mem:* Am Chem Soc; Entom Soc Am. *Res:* Organometallic chemistry; Freidel-Crafts reaction; chemical derivatives of rosin and terpenes; synthetic organic chemistry; agricultural chemicals. *Mailing Add:* 443 Nottingham Rd Newark DE 19711

HAUBER, JANET ELAINE, b Milwaukee, Wis, July 21, 37; m 74. MATERIALS COMPATIBILITY. *Educ:* Marquette Univ, BS, 65; Stanford Univ, MS, 67, PhD(mat sci), 70. *Prof Exp:* Metallurgist, 70-74, sect leader, 74-76, dep div leader, 76-86, dep assoc dept head, 86-87, ENGR AT LARGE, LAWRENCE LIVERMORE NAT LAB, 87- *Mem:* Soc Women Engrs; Nat Asn Female Execs. *Res:* Automation of engineering record retrieval. *Mailing Add:* 4717 Sorani Way Castro Valley CA 94546

HAUBRICH, DEAN ROBERT, b Pittsburgh, Pa, Jan 4, 43; m 64; c 2. PHARMACOLOGY, NEUROCHEMISTRY. *Educ:* Bethany Col, WVa, BS, 64; Purdue Univ, MS, 66, PhD(pharmacol), 69. *Prof Exp:* Res assoc pharmacol toxicol, NIH, 69-71; res investr pharmacol & neurochem, 71-73, sr res investr pharmacol & neurochem, Squibb Inst Med Res, 74-76; res fel, 76-81, SECT HEAD NEUROPSYCHOPHARMACOL, MERCK INST THERAPEUT RES, WEST POINT, PA, 81- *Concurrent Pos:* Vis lectr neurochem, Princeton Univ, 72- *Mem:* Am Soc Pharmacol & Exp Therapeut; Soc Neurosci. *Res:* Biochemical mechanisms of drug action; biochemical control of neurotransmitter metabolism. *Mailing Add:* Dept Biol Rorer Central Res 680 Valley Forge Rd King of Prussia PA 19406

HAUBRICH, RICHARD AUGUST, geophysics, for more information see previous edition

HAUBRICH, ROBERT RICE, b Claremont, NH, May 4, 23. BIOLOGY. *Educ:* Mich State Univ, BS, 49, MS, 52; Univ Fla, PhD(biol, psychol), 57. *Prof Exp:* Asst prof biol, ECarolina Col, 57-61 & Oberlin Col, 61-62; from asst prof to prof, 62-88, EMER PROF BIOL, DENISON UNIV, 88- *Mem:* AAAS; Sigma Xi; NY Acad Sci; Int Soc Hist, Philos & Social Studies Biol. *Res:* Relationships among aggression, hierarchical behavior and learning ability in the South African frog; behavior, development and population analysis of the topminnow; conceptual structure of biology. *Mailing Add:* Dept Biol Denison Univ Granville OH 43023

HAUCK, GEORGE F(REDERICK) W(OLFGANG), b Kassel, Ger, Sept 7, 32; US citizen; m 56, 74; c 4. CIVIL ENGINEERING, STRUCTURAL ENGINEERING. *Educ:* Okla State Univ, BArchE, 59, MArchE, 60; Northwestern Univ, PhD(struct eng), 64. *Prof Exp:* Res asst soil mech, Okla Exp Sta, 58-59; designer struct eng, Am Bridge Div, US Steel Corp, 59-60; res asst, Northwestern Tech Inst, 60-63; prof civil eng & chmn dept, Tri-State Univ, 63-75; assoc prof civil eng, 75-88, PROF, UNIV MO, COLUMBIA, 88-, DIR, CIVIL ENG DEPT, 91-, EXEC DIR, COORD ENG PROG, 91- *Honors & Awards:* Harry S Truman Award, 86. *Mem:* Am Soc Civil Engrs; Am Soc Eng Educ; Soc Hist Tech; Am Soc Testing & Mat. *Res:* Structural mechanics; stability of columns and frames; history of civil engineering; technological history; forensic engineering. *Mailing Add:* UMC/UMKC Eng 600 W Mechanic St Independence MO 64050

HAUDENSCHILD, CHRISTIAN C, b St Gallen, Switz, May 5, 39. PATHOLOGY, CYTOLOGY. *Educ:* Swiss Fed Inst Technol, Zurich, Switz, 59; Univ Basel, Switz, MD, 68; Am Bd Path, cert anat path, 77. *Prof Exp:* Res fel exp med, F Hoffman-LaRoche, Basel, Switz, 68-69, res assoc, 69-72; res assoc surg & path, Children's Hosp Med Ctr, Boston, 73-75; jr res asst path, Boston City Hosp, 74-75, sr res asst, 75-76; assoc pathologist, Malloryt Inst Path, Boston, 77; from asst prof to assoc prof, 79-82, PROF PATH, SCH MED, BOSTON UNIV, 82- *Concurrent Pos:* Clin instr path, Harvard Med Sch, 76-80; consult pathologist, Boston Veterans Admin Hosp, 78-; mem, ad hoc study sects, NIH, Nat Heart Lung & Blood Inst, 80-; mem, Coun Arteriosclerosis, prog comt, Am Heart Asn, 83-84; hon consult prof, Fac med, Univ Siena, Italy, 85-; prin invest endothelium, Nat Heart Lung & Blood Inst grant, 79-; fel, Coun Arteriosclerosis, Am Heart Asn. *Mem:* Am Heart Asn; Am Soc Cell Biol; Am Asn Pathologists; Int Acad Path; AAAS; AMA. *Res:* Endothelium: motility and related functions; cardiovascular disease: structure and function of vascular cells in diabetes, hypertension, and vascularization of tumors; pathology; cell biology. *Mailing Add:* Mallory Inst Path 784 Massachusetts Ave Boston MA 02118

HAUEISEN, DONALD CARL, b Columbus, Ohio, June 14, 45; m 68; c 2. OPTICAL PHYSICS, LASERS. *Educ:* Col Wooster, BA, 67; Cornell Univ, PhD(physics), 72. *Prof Exp:* Asst prof physics, Univ Dallas, 72-75; res assoc physics, Cornell Univ, 75-77; asst prof, 77-80, ASSOC PROF PHYSICS, PAC LUTHERAN UNIV, 80- *Mem:* Am Phys Soc; Am Asn Physics Teachers; Astron Soc Pac. *Res:* Nonlinear optical effects. *Mailing Add:* MS 2C41 Nat Semiconductor 1111 39th Ave SE Puyallup WA 98374

HAUENSTEIN, JACK DAVID, b Oskaloosa, Iowa, June 10, 29; m 50; c 3. POLYMER CHEMISTRY, PHYSICAL CHEMISTRY. *Educ:* Iowa Univ, BA, 50, PhD(phys chem), 55. *Prof Exp:* Sr res chemist, 55-60, proj leader olefin plastics, 60-64, prof leader polyesters, 64-65, SECT HEAD POLYESTERS, GOODYEAR TIRE & RUBBER CO, 65- *Mem:* AAAS; Am Chem Soc; Sigma Xi. *Res:* Synthesis and characterization of polymers, especially on condensation polymers; ethylene terephthalate based plastics and fibers. *Mailing Add:* RR1 Tracy IA 50256

HAUFLER, CHRISTOPHER HARDIN, b Niskayuna, NY, Apr 20, 50. PLANT BIOSYSTEMATICS, PTERIDOLOGY. *Educ:* Hiram Col, BA, 72; Ind Univ, MS, 74, PhD(bot), 77. *Prof Exp:* Fel, Gray Herbarium Harvard Univ, 77-78 & Mo Bot Garden, 78-79; ASSOC PROF & CHAIRPERSON BOT DEPT, UNIV KANS, 79- *Mem:* AAAS; Am Soc Plant Taxonomists; Int Asn Pteridologists; Soc Study Evolution; Am Inst Biol Sci. *Res:* Application of data from such diverse sources as micromorphology, karyotype analysis, enzyme variability, and flavonoid biochemistry in examining patterns and processes of speciation in plants, especially ferns. *Mailing Add:* Dept Bot Univ Kans Lawrence KS 66045

HAUFLER, JONATHAN B, b Schenectady, NY, July 29, 52; m 75. HABITAT MANAGEMENT. *Educ:* Univ NH, BS, 74; Va Polytech Inst & State Univ, MS, 76; Colo State Univ, PhD(wildlife biol), 79. *Prof Exp:* Asst prof, 79-84, ASSOC PROF WILDLIFE ECOL, MICH STATE UNIV, 84- *Mem:* Wildlife Soc; Wildlife Dis Asn. *Res:* Wildlife responses to habitat manipulations and impacts; wildlife nutrition. *Mailing Add:* Dept Fisheries & Wildlife Mich State Univ East Lansing MI 48824-1222

HAUG, ARTHUR JOHN, b Milwaukee, Wis, Sept 27, 19; m 53; c 5. CHEMISTRY. *Educ:* Marquette Univ, BS, 41; Lawrence Col, MS, 43, PhD(pulp & paper chem & eng), 47. *Prof Exp:* Lab asst, Hummel & Downing Co, Wis, 42; supvr, Tenn Eastman Corp, Tenn, 44-45; asst, 43, process control engr, 47-48, tech dir, 49-50, supt, Paper Mill, 51-52, prod mgr, 53-58, mgr, Paper Mill, Wash, 58-60, prod serv, Pa, 60-65, plant mgr, Paper Mill, NY, 65-67, gen mgr, Northeast Opers, Me, 67-68, div vpres, Northeast Opers, Scott Paper Co, Winslow, Maine, 68-78; pres, Forster Mfg Co, Inc, Wilton, Maine, 78-86; RETIRED. *Mem:* Tech Asn Pulp & Paper Indust. *Res:* Pulp and paper problems: chemical structure and information on guar mannogalactan; alkaline hypochlorite oxidation of guar mannogalactan with respect to certain variables. *Mailing Add:* 25188 Marion Ave A-106 Punta Gorda FL 33950

HAUG, EDWARD J, JR, b Bonne Terre, Mo, Sept 15, 40; div; c 1. MECHANICAL DESIGN, APPLIED MATHEMATICS. *Educ:* Univ Mo-Rolla, BS, 62; Kans State Univ, MS, 64, PhD(appl mech), 66. *Prof Exp:* Mech engr, Systs Anal Directorate, Hq, US Army Weapons Command, 66-69, chief, 69-71, chief systs res eng, 71-73; chief concepts & technol eng, US Army Armament Command, 73-76; prof mech eng, 76-81, PROF & DIR, CTR FOR COMPUT AIDED DESIGN, UNIV IOWA, 81- *Concurrent Pos:* Adj asst prof, Univ Iowa, 68-71, adj assoc prof, 71-75, adj prof, 75-76; consult, Battelle Mem Inst & Intertech Corp, 77- *Mem:* Am Soc Mech Engrs; Sigma Xi. *Res:* Development of methods of optimal design of mechanical systems; applied functional analysis; mechanism and machine dynamics. *Mailing Add:* Col Eng Univ Iowa Iowa City IA 52242

HAUGAARD, NIELS, b Copenhagen, Denmark, Feb 25, 20; US citizen; m; c 2. BIOCHEMISTRY, PHARMACOLOGY. *Educ:* Swarthmore Col, AB, 42; Univ Pa, PhD, 49. *Prof Exp:* Instr res med, Sch Med, Univ Pa, 49-51, assoc, 51-52, asst prof physiol chem, 52-53, from asst prof to prof, 54-87, EMER PROF, SCH MED, UNIV PA, 87- *Concurrent Pos:* Guggenheim Found fel, Carlsberg Lab, Denmark, 53-54; Commonwealth Found fel, Amsterdam, 65-66; mem, Exp Cardiovasc Sci Study Sect, NIH, 78-82; vis prof, Univ Oslo, 84; mem adv bd, Molecular & Cellular Biochem, 91- *Mem:* Am Soc Biol Chem; Soc Exp Biol & Med; Am Soc Pharmacol & Exp Therapeut. *Res:* Intermediary metabolism; mechanism of hormone and drug action; carbohydrate metabolism; oxygen toxicity. *Mailing Add:* Dept Pharmacol G3 Univ Pa Sch Med Philadelphia PA 19104

HAUGE, PAUL STEPHEN, b Brookings, SDak, May 1, 45; m 67; c 4. GEOPHYSICS. *Educ:* SDak State Univ, BS, 67; Iowa State Univ, PhD(nuclear physics), 71. *Prof Exp:* NDEA fel & asst, Dept Physics, Iowa State Univ, 67-71; Nat Acad Sci-Nat Res Coun res assoc nuclear physics, Aerospace Res Lab, Wright-Patterson AFB, 71-72; res assoc, Cyclotron Dept, Mich State Univ, 72-74; geophysicist, Exxon Prod Res Co, 74-81; geophysicist, Phillips Petrol Co, 81-88; GEOPHYSICIST, CONOCO INC, 88- *Mem:* Soc Explor Geophys; Am Phys Soc. *Res:* Theoretical investigation of numerical methods for analyzing seismic data. *Mailing Add:* Conoco Inc 2342 EB PO Box 1267 Ponca City OK 74603

HAUGEN, DAVID ALLEN, b Ft Dodge, Iowa, Nov 28, 45; m 67; c 2. ENZYMOLOGY, CHEMICAL CARCINOGENESIS. *Educ:* Augsburg Col, BA, 67; Univ Wis-Madison, PhD(biochem), 73. *Prof Exp:* Scholar, Dept Biol Chem, Univ Mich, 73-75, lectr biochem, 75; asst biochemist, 75-81, BIOCHEMIST, BIOL ENVIRON & MED RES DIV, ARGONNE NAT LAB, 81- *Concurrent Pos:* NIH fel, Univ Mich, 73-75; vis scientist, Dept Biochem & Biophysics, Univ Hawaii, 77. *Mem:* Am Chem Soc; Am Soc Biochem & Molecular Biol. *Res:* Metabolism of drugs and carcinogens, especially the enzymology of monooxygenases; formation and fate of reactive metabolites, analysis of environmental exposure to carcinogens. *Mailing Add:* Energy Environ & Biol Res Bldg 202 Argonne Nat Lab Argonne IL 60439-4833

HAUGEN, GILBERT R, b Hollywood, Calif, Apr 10, 30; m 61; c 4. PHYSICAL CHEMISTRY. *Educ:* Univ Southern Calif, BS, 55, PhD(chem), 62. *Prof Exp:* Res assoc photochem, Univ Calif, Los Angeles, 60-62; proj leader phys chem, Stanford Res Inst, 62-70; SR SCIENTIST, CHEM & NATURAL SCI DEPT, LAWRENCE LIVERMORE LAB, 70- *Mem:* Am Chem Soc; fel Am Inst Chem; The Chem Soc; InterAm Photochem Soc; Soc Appl Spectros; Optical Soc Am; Sigma Xi; Western Spectros Asn. *Res:* Thermodynamics; photochemistry; chemical kinetics; spectroscopy; photoluminescence; chemluminescence; applications of computers to laboratory experiments of lasers to analytical chemistry and fiber optics to chemical and physical measurements; application of linear response theory to chemical measurement. *Mailing Add:* 3845 Pinot Ct Pleasanton CA 94566

HAUGEN, ROBERT KENNETH, b Detroit, Mich, July 12, 47; m 85; c 1. INDUSTRIAL & MANUFACTURING ENGINEERING. *Educ:* Univ Ill-Urbana-Champaign, BS, 69, MS, 73, PhD(sci educ), 78. *Prof Exp:* Chemist, Lawrence Livermore Labs, 68-69; anal chemist, Tri-Met Sanitary Dist Sludge Proj, 70-74; assoc dir environ educ, Technol & Environ Proj, LaSalle, Ill, 74-84; instr physics, math, chem, Kishwaukee Col, Malta, Ill, 85-87; INSTNL PROD MGR, ST CHARLES MFG, 87- *Concurrent Pos:* Instr chem, Univ High Sch Gifted, 70-74; subcomt chmn, Am Soc Testing & Mat, 88-91; chief consult, Glogon Eng, 90-91. *Mem:* Am Soc Testing & Mat; Am Chem Soc. *Res:* Fume hood testing and product development, including flow pattern, containment studies, baffle arrangements, velocity alarms, and associated technologies. *Mailing Add:* St Charles Mfg 1611 E Main St Charles IL 60174

HAUGH, C(LARENCE) GENE, b Spring Mills, Pa, Oct 11, 36; m 62; c 2. AGRICULTURAL ENGINEERING. *Educ:* Pa State Univ, BS, 58; Univ Ill, MS, 59; Purdue Univ, PhD(agr eng), 64. *Prof Exp:* Instr agr eng, Purdue Univ, 61-64; asst prof, Univ Fla, 64-65; from asst prof to assoc prof agr eng, Purdue Univ, 65-72, prof, 72-; CHMN, DEPT AGR ENG, VA POLYTECH INST & STATE UNIV. *Concurrent Pos:* Regist prof engr, State of Fla, 64-; lead analyst, World Food & Nutrit Study, Nat Acad Sci, 76-77. *Honors & Awards:* Young Researcher Award, Am Soc Agr Engrs, 76. *Mem:* Am Soc Agr Engrs; Inst Food Tech; sr mem Nat Soc Prof Engrs; Am Soc Heating, Refrig & Air-Conditioning Engrs. *Res:* Heat and mass transfer; physical properties of biological materials; food processing; food losses. *Mailing Add:* Dept Agr Eng Va Polytech Inst & State Univ Blacksburg VA 24061

HAUGH, EUGENE (FREDERICK), b Reedsburg, Wis, Mar 23, 29; m 57; c 4. PHYSICAL CHEMISTRY. *Educ:* Univ Wis, BS, 51, PhD(phys chem), 54. *Prof Exp:* From res chemist to sr res chemist, E I du Pont de Nemours & Co, 54-63, res supvr, 63-73, res mgr, 73-85, sr res fel, Imaging Systs Dept, 85-88; RETIRED. *Mem:* Soc Photog Sci & Eng. *Res:* Photographic chemistry and theory. *Mailing Add:* 4648 Dartmoor Dr Liftwood Estates Wilmington DE 19803

HAUGH, JOHN RICHARD, b East Liverpool, Ohio, May 13, 40. ECOLOGY, VERTEBRATE ZOOLOGY. *Educ:* Westminster Col, Pa, BS, 62; Syracuse Univ, MS, 64; Cornell Univ, PhD(ecol, evolutionary biol), 70. *Prof Exp:* Asst prof biol sci, State Univ NY, Binghamton, 70-80; CONSULT, 80- *Mem:* Ecol Soc Am; Am Ornith Union; Am Soc Mammal; Cooper Ornith Soc; Am Inst Biol Sci. *Res:* Arctic and boreal ecology; ecology of predatory birds in Alaska. *Mailing Add:* 169 Savitch Rd Binghamton NY 13900

HAUGH, LARRY DOUGLAS, b Gary, Ind, June 11, 44; m 66; c 3. APPLIED STATISTICS, QUALITY CONTROL. *Educ:* Wabash Col, BA, 66; Univ Wis-Madison, MA, 67, MS, 70, PhD(statist), 72. *Prof Exp:* Asst prof, Univ Fla, 72-75; from asst prof to assoc prof, 75-88, PROF, STATIST, UNIV VT, 88-, DIR STATIST PROG, 90- *Concurrent Pos:* Acad assoc, IBM, Burlington, 78-81; statistician, Shell Res, Amsterdam, 81-82; assoc ed, Technometrics, 81-86; lectr, indust short courses; consult statist; mem, bd dirs, Time Series Anal & Forecasting Soc, 81-85; mem, ed rev bd, Qual Progress, 87- *Mem:* Am Statist Asn; Royal Statist Soc; Inst Statisticians; Int Asn Statist Comput; Biometric Soc; sr mem Am Soc Qual Control. *Res:* Time series analysis with cross correlation identification; extremal value estimation via censoring and truncation; reliability estimation via the renewal function; dental statistics and medical statistics; low back pain research. *Mailing Add:* Statist Prog Univ Vt 16 Colchester Ave Burlington VT 05405

HAUGHEY, FRANCIS JAMES, b Brooklyn, NY, Apr 6, 30; m 51; c 6. HEALTH PHYSICS. *Educ:* Hofstra Col, BS, 59; Rutgers Univ, MS, 62, PhD(aerosol physics), 66; Am Bd Health Physics, dipl, 61. *Prof Exp:* Health physicist, Brookhaven Nat Lab, 55-58; univ health physicist, 58-63, assoc prof environ sci, 67-76, from asst prof to assoc prof, 63-83, PROF RADIATION SCI, RUTGERS UNIV, PISCATAWAY, 83- *Concurrent Pos:* Consult, NJ Comn Radiation Protection, 63-; mem, NJ Adv Comt Nuclear Med, 63-; mem gov sci adv comm, NJ, 79- *Mem:* Health Physics Soc; Am Meteorol Soc; Am Geophys Union; Air Pollution Control Asn; Am Phys Soc. *Res:* Aerosol physics; surface properties of aerosol particles and interactions with gases, other particles including radioactive particles; external and internal radiation dosimetry. *Mailing Add:* Dept Environ Sci Rutgers Univ New Brunswick NJ 08903

HAUGHT, ALAN F, b Buckhannon, WVa, Nov 5, 36; m 59; c 4. PLASMA PHYSICS. *Educ:* Amherst Col, AB, 58; Princeton Univ, MA, 60, PhD(physics), 62. *Prof Exp:* From res scientist to sr res scientist plasma physics, 62-64, from prin scientist to sr prin scientist, 64-74, MGR FUSION RES LAB, UNITED TECHNOL CORP, 74- *Mem:* Am Phys Soc; Sigma Xi. *Res:* Shock tube studies of thermal ionization rates; plasmas produced by high-intensity laser beam interaction with matter; magnetic field containment of high temperature plasmas. *Mailing Add:* 39 Fox Den Rd Glastonbury CT 06033

HAUGHTON, GEOFFREY, b Leeds, Eng, June 10, 32; m 65; c 1. IMMUNOGENETICS, IMMUNOLOGY. *Educ:* Univ Southampton, BSc, 55, PhD(bact enzymol), 58. *Prof Exp:* Demonstr physiol & biochem, Univ Southampton, 55-58; sr sci officer, Microbiol Res Estab, Eng, 61-64; vis scientist, Karolinska Inst, Sweden, 64-65; from asst prof to assoc prof, 66-71, PROF IMMUNOL & GENETICS, SCH MED, UNIV NC, CHAPEL HILL, 71- *Concurrent Pos:* Civil Serv sr res fel microbiol, Microbiol Res Estab, Eng, 58-61; Int Union Control Cancer Eleanor Roosevelt res fel, 64-65; USPHS res career develop award, 68; res asst, Royal South Hants Hosp, Eng, 55-58. *Mem:* Transplantation Soc; Am Asn Cancer Res; Am Asn Immunol. *Res:* Transplantation immunology and genetics; tumor immunobiology; basic immunology. *Mailing Add:* Dept Bacteriol & Immunol Univ NC 614 F10B CB No 7290 Chapel Hill NC 27514

HAUGHTON, JAMES GRAY, b Panama, Repub Panama, Mar 30, 25; US citizen; div; c 2. PUBLIC HEALTH ADMINISTRATION. *Educ:* Pac Union Col, BA, 47; Loma Linda Univ, MD, 50; Columbia Univ, MPH, 62; Am Bd Prev Med, dipl, 68. *Hon Degrees:* DSc, Chicago Med Sch-Univ Health Sci, 71. *Prof Exp:* Intern, Unity Hosp, Brooklyn, 49-50, fel gen surg, 51-55; USNR Med Corps, 56-58; child health clinician, New York City Health Dept, 58-60, resident pub health physician, 60-62, dir med care, 63-65, exec dir med care serv, 65-66; first dep comnr, New York City Dept Hosps, 66; first dep adminr, Health Serv Admin, New York City, 66-70; exec dir, Health & Hosps Gov Comn, Chicago, Ill, 70-79; assoc prof, Dept Community Health & Prev Med, Northwestern Univ, 77-80; exec vpres, Charles R Drew Postgrad Med Sch, 80-82, vpres finance & admin, 82-83; dir, City Houston Health Dept, 83-87; MED DIR, KING/DREW MED CTR, LOS ANGELES, 87- *Concurrent Pos:* Pvt pract, obstet & abdominal surg, 52-66; adj asst prof admin med, Sch Pub Health & Admin Med, Columbia Univ, 63-70; dep med welfare adminr, New York City Welfare Dept, 63-65, med welfare adminr & coordr welfare serv, New York City Hosp Dept, 65-66; lectr pub health, Sch Med, Yale Univ, 64-65; non-resident lectr med care orgn, Sch Pub Health, Univ Mich, 66-79; lectr admin health serv, Dept Polit Sci, Roosevelt Univ, 70-72; fac mem, Interagency Inst Fed Health Care Exec, 70-82; adj prof, Carnegie Mellon Univ, 77-78 & Univ Tex Health Sci Ctr, Houston, 84-87; prof, Dept Family Med, Drew Univ Med & Sci, 87. *Honors & Awards:* Dr Mary McLeod Bethune Award of Merit, Nat Coun Negro Women, 72; Humanitarian Award, Nat Asn Health Serv Exec, 72; Mathew B Rosenhaus Lectr, Am Pub Health Asn, 74. *Mem:* Inst Med-Nat Acad Sci; fel Am Pub Health Asn; fel Am Col Prev Med; fel NY Acad Sci; fel Royal Soc Health; AMA; Nat Med Asn. *Mailing Add:* King/Drew Med Ctr 12021 Wilmington Ave Los Angeles CA 90059

HAUGHTON, KENNETH E, b Myrtle Point, Ore, Jan 8, 28. MECHANICAL ENGINEERING. *Educ:* Univ Calif, Berkeley, BS, 52, PhD(mech eng), 64; Iowa State Col, MS, 55. *Prof Exp:* Design engr, Gen Elec Co, 52-53; instr eng graphics, Iowa State Col, 53-56; instr eng mech, Cornell Univ, 56-57; res engr, IBM Corp, San Jose, Calif, 57-77, prod develop engr, 80-82, dir eng lab, Lexington, Ky, 77-80; dean eng, Univ Santa Clara, 80-89; VPRES ENG, D A VINCI GRAPHICS, 90- *Mem:* Nat Acad Eng; Am Soc Eng; Am Soc Mech Eng. *Mailing Add:* D A Vinci Graphics 870 Hermosa Dr Sunnyvale CA 94086

HAUGHTON, VICTOR MELLET, b Willimantic, Conn, July 9, 39; m 65; c 3. RADIOLOGY. *Educ:* Harvard Col, Mass, BA, 61; Yale Univ Med Sch, MD, 67. *Prof Exp:* Instr radiol, Harvard Med Sch, MA, 73; clin instr radiol, Tufts Univ, CT, 73; radiologist, Peter Bent Brigham Boston, 73; radiologist, Beth Israel Hosp, 73; consult radiol, Vet Admin Hosp, Boston, 73 & Vet Admin Hosp Milwaukee, 74-81; asst prof, Radiol Med Col Wis, 74; radiologist & chief neuroradiol, Milwaukee County Med Complex, 74; PROF RADIOL, 81- & CHIEF NEURORADIOL RES, MED COL WIS, 82- *Concurrent Pos:* Int Travel fel, NIH, Oslo, Norway, 80; mem sci exhibits comt, Radiol Soc N Am, 85-; mem Comt MR Appln, Comn Magnetic Resonance, 87-88; mem, res overview comt, Am Soc Neuroradiol, 87-88; mem radiopharmaceut drug adv comt, Food & Drug Admn, Dept Health & Human Serv, 88- *Mem:* Radiol Soc N Am; Am Col Radiol; Am Soc Neuroradiol; Asn Univ Radiologist. *Mailing Add:* 8700 W Wisconsin Ave Milwaukee WI 53226

HAUGLAND, RICHARD PAUL, b Huron, SDak, July 17, 43; m 72; c 2. ORGANIC CHEMISTRY, BIOPHYSICS. *Educ:* Hamline Univ, BS, 65; Stanford Univ, PhD(chem), 70. *Prof Exp:* Chemist, Syntex Corp, 69; asst prof chem, Hamline Univ, 75-78; PRES, MOLECULAR PROBES, INC, 75- *Concurrent Pos:* Fel cardiovasc res inst, Univ Calif, San Francisco, 72-74, NIH fel, 74-75. *Mem:* Biophys Soc; Am Chem Soc. *Res:* Fluorescent probes, spin labels, spectroscopy, organic syntheses, protein and membrane structure. *Mailing Add:* Molecular Probes Inc 4849 Pitchford Ave Eugene OR 97402

HAUGSJAA, PAUL O, b Las Vegas, Nev, Apr 3, 42; m 65; c 3. ATOMIC & MOLECULAR PHYSICS. *Educ:* Concordia Col, Moorhead, Minn, BS, 64; Univ Denver, PhD(physics), 69. *Prof Exp:* Mem tech staff, Explor Res Lab, 69-81, res mgr, Lighting Technol Ctr, 81-82, RES MGR, COMPONENTS RES LAB GTE LABS, INC, 82- *Honors & Awards:* Leslie H Warner Award, 84. *Mem:* Am Phys Soc; Sigma Xi. *Res:* Solid state device design and development; semiconductor devices development; collisional and radiative processes pertinent to gaseous electronics and plasma processes; seismic and reproductive acoustics; semiconductor device design and process development; transportation control; electrochemical kinetics; microwave transmission effects; charge storage and piezo electric effects; electrodeless lighting and bioeffects of light and microwaves; high speed IC development; heterostructural epitaxy of III-V materials; design and fabrication of integrated optical and optoelectronic circuits; optics. *Mailing Add:* GTE Labs 40 Sylvan Rd Waltham MA 02254

HAUK, PETER, b Dallas, Tex, Jan 19, 37; m 58; c 4. PHYSICAL CHEMISTRY. *Educ:* Univ Md, BS, 58; Carnegie Inst Technol, MS, 61, PhD(chem), 63. *Prof Exp:* Nat Acad Sci-Nat Res Coun fel chem, Calif Inst Technol, 63-64, NSF fel, 64-65, res fel, 65; res chemist, Beckman Instruments Inc, 65-66; sr res chemist, Bell & Howell Res Lab, 66-75; prin res scientist, Newport Res Corp, 74-84; CONSULT, 84- *Mem:* Am Chem Soc; Am Phys Soc. *Res:* Quantum chemical modeling of small molecules and molecular electronic interactions, electrochemical sensors, photoconductor processes; optics and electro-optics instrumentation. *Mailing Add:* 10181 Beverly Dr Huntington Beach CA 92646-7199

HAUKE, RICHARD LOUIS, b Detroit, Mich, Apr 28, 30; m 58; c 4. STRUCTURAL BOTANY. *Educ:* Univ Mich, BS, 52, PhD(bot), 60; Univ Calif, MA, 54. *Prof Exp:* Asst ecol, Fla State Univ, 52-53; from instr to assoc prof bot, 59-69, PROF BOT, UNIV RI, 69- *Mem:* Bot Soc Am; Am Fern Soc (treas, 61-65, secy, 71-73, vpres, 76-77, pres, 78-79). *Res:* Morphology and anatomy of the genus equisetum. *Mailing Add:* Dept Bot Univ RI Kingston RI 02881

HAUN, CHARLES KENNETH, b Los Angeles, Calif, June 13, 30; m 56; c 3. ANATOMY. *Educ:* Pomona Col, BA, 52; Univ Calif, Los Angeles, PhD(anat), 60. *Prof Exp:* From instr to asst prof anat, Hahnemann Med Col, 59-62; asst prof, 62-68, ASSOC PROF ANAT, SCH MED, UNIV SOUTHERN CALIF, 68- *Mem:* Am Asn Anat; Int Soc Neuroendocrinol; Soc Neurosci. *Res:* Neuro-endocrine regulation; neuroanatomy; reproduction; integration of anterior and posterior pituitary secretion; acupuncture-mechanisms; the role of thyrotropin-releasing hormon (TRH), along with other neuro-peptides and transmitters in control of spinal motoneurons; the application of TRH and analogs as antispasticity agents. *Mailing Add:* Dept Anat & Cell Biol Sch Med Univ Southern Calif Los Angeles CA 90033

HAUN, J(AMES) W(ILLIAM), b Birmingham, Ala, Sept 8, 24; m 46; c 4. CHEMICAL ENGINEERING, REGULATORY MANAGEMENT. *Educ:* Univ Tex, BS, 46, MS, 48, PhD(chem eng & chem), 50. *Prof Exp:* Instr chem eng, Univ Tex, 48-49; res engr, Plastics Div, Monsanto Chem Co, 50-56, res group leader, 56; head dept chem eng, Cent Res Labs, 56-60, dir phys res, 60-63, dir & vpres corporate eng, 63-75, vpres eng policy, 75-85, CONSULT, GEN MILLS INC, 85- *Concurrent Pos:* Mem, Environ Qual Comt, Nat Asn Mfrs, 69-85; mem, Food Indust Adv Comt, Dept Energy, 78-81; mem comn environ, US Chamber Com, 78-84; mem sci Adv Bd, Environ Protection Agency, 82-90, consult, 90- *Mem:* Fel Am Inst Chem Engrs; Sigma Xi; Air & Waste Mgt Asn. *Res:* Applications of advanced engineering technology to food processing; environmental engineering; interaction between technology and public policy. *Mailing Add:* 6912 E Fish Lake Rd Minneapolis MN 55369

HAUN, JOHN DANIEL, b Old Hickory, Tenn, Mar 7, 21; m 42. PETROLEUM GEOLOGY. *Educ:* Berea Col, AB, 48; Univ Wyo, MA, 49, PhD(geol), 53. *Prof Exp:* Dir geol res, Petrol Res Co, 52-55; from asst prof to prof, 55-83, EMER PROF GEOL, COLO SCH MINES, 83- *Concurrent Pos:* Mem consult firm, Barlow & Haun, Inc, 57-90; ed, J of Am Asn Petrol Geologists, 67-71; mem comt explor, Am Petrol Inst, 71-73, 78-88; US rep, Int Comt Petrol Reserves Classification, UN, 76-77; consult, Off Technol Assessment, US Cong, 76-80; comnr, Colo Oil & Gas Conserv Comn, 77-87, chmn, 85-87; mem, energy resources comt, Interstate Oil Compact Comn, 78-90 & Nat Petrol Coun, 79-90; mem, US Nat Comt Geol, 82-89, chmn, 85-87. *Honors & Awards:* Distinguished Serv Award, Am Asn Petrol Geologists, 73; Ben H Parker Mem Award, Am Inst Prof Geologists, 83; Halliburton Award, Colo Sch Mines, 85. *Mem:* Am Inst Prof Geologists (pres, 76); fel Geol Soc Am; hon mem Am Asn Petrol Geologists (pres-elect, 78-79, pres, 79-80); Soc Econ Paleontologists & Mineralogists; Geochem Soc; Am Geol Inst (pres, 81-82). *Res:* Cretaceous stratigraphy; regional structural geology; migration and accumulation of petroleum; methods of estimating undiscovered petroleum resources. *Mailing Add:* 1238 County Rd 23 Evergreen CO 80439

HAUN, JOSEPH RHODES, b Edinburgh, Va, May 13, 22; m 45; c 5. PLANT PHYSIOLOGY. *Educ:* Berea Col, AB, 46; Univ Md, MS, 50, PhD(plant physiol), 51. *Prof Exp:* Res plant physiologist, E I du Pont de Nemours & Co, 51-58; horticulturist, New Crops Res Br, USDA, 58-61, invests leader, 61-64; tech dir, Cherry Hill Trust, Ga, 64-65; agr consult crop prod anal, 65; PROF HORT, CLEMSON UNIV, 65- *Mem:* AAAS; Soc Econ Bot; Am Soc Plant Physiol; Am Soc Hort Sci; Am Soc Agronomists. *Res:* Influence of environmental factors on plant growth; crop production analysis. *Mailing Add:* Rte 2 Box 430-D Asheville NC 28805

HAUN, RANDY S, b Ft Carson, Colo, Mar 14, 56. MOLECULAR BIOLOGY. *Educ:* Univ Iowa, BS, 78; Purdue Univ, PhD(biochem), 88. *Prof Exp:* Res chemist, Uniformed Serv Univ Health Sci, Bethesda, Md, 79-82; res asst, Dept Biochem, Purdue Univ, 82-88, teaching asst, 84, fel, 89; PRES, MOLECULAR SOFTWARE, INC, 85-; STAFF FEL, NAT HEART LUNG & BLOOD INST, NIH, 89- *Mem:* Assoc mem Am Soc Biochem & Molecular Biol. *Res:* Cholecystokinin gene expression; cloning and expression of protein-tyrosine-phosphatase. *Mailing Add:* Lab Cellular Metab NIH Nat Heart Lung & Blood Inst Bldg 10 Rm 5N307 Bethesda MD 20892

HAUN, ROBERT DEE, JR, b Lexington, Ky, Apr 3, 30; m 53, 77; c 3. ATOMIC & MOLECULAR PHYSICS. *Educ:* Univ Ky, BS, 52; Mass Inst Technol, PhD(physics), 57. *Prof Exp:* Asst physics, Mass Inst Technol, 53-57; res physicist, 57-59, fel physicist, 59-62, sect mgr, 62-63, mgr quantum electronics res & develop, 63-69, dir appl physics & math res, 69-74, mgr appl sci div, 74-81, dir indust prod res & develop, 81-83, res dir, energy & adv technol, 83-84, res dir, Bus Unit Technol Progs, 84-87, DEP CHIEF SCIENTIST, WESTINGHOUSE ELEC CORP, 87- *Mem:* Am Phys Soc; Inst Elec & Electronics Engrs; Optical Soc Am; fel AAAS. *Res:* Lasers; quantum electronics; parametric amplification; ferromagnetic resonance; atomic frequency standards; atomic beam magnetic resonance spectroscopy. *Mailing Add:* Westinghouse Sci & Technol Pittsburgh PA 15235

HAUNERLAND, NORBERT HEINRICH, b Essen, WGer, Feb 23, 55; m 88; c 2. LIPID BIOCHEMISTRY, INSECT BIOCHEMISTRY. *Educ:* Univ Münster, Ger, dipl chem, 80, PhD(biochem), 82. *Prof Exp:* Assoc insect physiol, Cornell Univ, 82-84; res assoc insert biochem, Univ Ariz, 84-86, res asst prof, 87-89; ASST PROF BIOCHEM, SIMON FRASER UNIV, 89- *Honors & Awards:* H P Kaufmann Award, Deutsche Gesellschaft Für Fettwissenschaft, 82; Young Chemist Award, Agrochem Div, Am Chem Soc, 88. *Mem:* Entom Soc Am; Am Chem Soc; Can Fedn Biol Sci; Gesellschaft Für Biol Chem; Ger Soc Chem. *Res:* Lipid transport and metabolism in insects; structure and function of lipoproteins; intracellular fatty acid binding proteins and receptors; insects as biochemical model systems. *Mailing Add:* Dept Biol Sci Simon Fraser Univ Burnaby BC V5A 1S6 Can

HAUNOLD, ALFRED, b Hollabrunn, Austria, Oct 7, 29; US citizen; m 59; c 7. PLANT GENETICS. *Educ:* State Univ Agr & Forestry, Austria, Dipl Ing, 51, Dr Agr, 52; Univ Nebr, PhD(plant breeding & genetics), 60. *Prof Exp:* Res assoc plant breeding, State Univ Agr & Forestry, Austria, 52-53; revision sect, Agr Bd Lower Austria & Vienna, 54-55; asst agron, Univ Nebr, 55-60, asst prof corn genetics & starch biochem, 60-64; prof analyst, Sci Info Exchange, Smithsonian Inst, 64-65; RES PLANT GENETICIST, AGR RES SERV, USDA, ORE STATE UNIV, 65- *Concurrent Pos:* Mem sci comn, Int Hop Prod Bur; Int Order Hop, Hop Res Coun, 83. *Mem:* Am Soc Agron; Crop Sci Soc Am; Am Soc Brewing Chemists. *Res:* Wheat genetics and breeding; corn genetics and starch biochemistry as related to starch biosynthesis; breeding, genetics and cytogenetics of hops. *Mailing Add:* USDA Crop Sci Bldg 451 B Ore State Univ Corvallis OR 97331-3002

HAUNZ, EDGAR ALFRED, b London, Eng, Dec 12, 10; nat US; m 48. ENDOCRINOLOGY, DIABETES. *Educ:* Canisius Col, BS, 37; Univ Buffalo, MD, 43; Univ Minn, MSc, 47. *Hon Degrees:* DSc, Univ NDak, 78. *Prof Exp:* From assoc to prof med & asst prof clin med, 58-75, chmn dept, 58-75, EMER PROF MED, SCH MED, UNIV NDAK, 75- *Concurrent Pos:* Mem adv bd & attend staff, United Hosp, 47-; mem staff, Grand Forks Clin, 47-; spec consult, Diabetes Sect, USPHS, 52-58, nat lectr diabetes; founder & dir, Camp Sioux for Diabetic Children, 53-; mem bd dirs, Am Diabetes Asn, 63-75; past chmn, bd gov, Am Diabetes Asn. *Honors & Awards:* Serv to Humanity Award, Sertoma Club; Gov Award Outstanding Serv to Handicapped; Pfizer Outstanding Clinician in Diabetes Award, Am Diabetes Asn, 75. *Mem:* Am Med Assn; fel Am Col Physicians; AMA. *Res:* Diabetes mellitus. *Mailing Add:* 1029 Lincoln Dr Grand Forks ND 58201

HAUPT, L(EWIS) M(CDOWELL), JR, b Kyle, Tex, May 27, 06; m 31; c 4. ELECTRICAL ENGINEERING. *Educ:* Agr & Mech Col, Tex, BS, 27, MS, 35. *Prof Exp:* Student employee, Westinghouse Elec Corp, 27-28; asst dispatcher, Operating Dept, WTex Utilities Co, 28-30; from instr to prof elec eng, Tex A&M Univ, 30-72, supvr calculator lab, 47-72, res engr, Data

Processing Ctr, 70-72; RETIRED. *Mem:* Am Soc Eng Educ; Inst Elec & Electronics Engrs; Nat Soc Prof Engrs; Sigma Xi. *Res:* Electric power generation and transmission. *Mailing Add:* Box 284 Rte 2 College Station TX 77840

HAUPT, RALPH FREEMAN, b Peabody, Kans, Mar 4, 06; m 34; c 1. ASTRONOMY. *Educ:* George Washington Univ, BS, 33, MA, 38. *Prof Exp:* Lab apprentice physics, Nat Bur Stand, 27-28; from jr sci aide to sci aide astron, Nautical Almanac Off, US Naval Observ, 28-35 & from jr astronomer to astronomer, 35-59, asst dir prod, 59-63, asst dir Off, 63-73; PVT CONSULT CELESTIAL NAVIG, 73- *Concurrent Pos:* Chmn ed bd, US Naval Observ, 64-70; adv US, UK-Can Navig Working Party for Standardization, 61- & proj off Sight Reduction Tables & Air Almanac, 64-73; mem US Civil Serv Comt Sci Exam, 63-73; mem US Study Group 7, Consultative Comt Int Radio, 70- *Honors & Awards:* Super Achievement Award, Inst Navig, 73. *Mem:* Int Astron Union; Am Astron Soc; Am Inst Navig; Sigma Xi. *Res:* Dynamical and theoretical astronomy; star catalogs; problems in optical phenomena of the earth's atmosphere; automatic celestial navigation systems; automatic electronic printing of astronomical ephemerides and navigation tables; transmission of time and frequency. *Mailing Add:* 3701 Dulwick Dr Silver Spring MD 20906

HAUPTMAN, HERBERT AARON, b New York, NY, Feb 14, 17; m 40; c 2. CRYSTALLOGRAPHY. *Educ:* City Col New York, BS, 37; Columbia Univ, MA, 39; Univ Md, PhD(math), 55. *Hon Degrees:* Univ Md, 85; DSc, City Col New York,86. *Prof Exp:* Statistician, Census Bur, 40-42; radar instr, USAF, 42-43 & 46-47; physicist, Naval Res Lab, 47-61, mathematician, 62-70; mathematician, 70-72, res dir & vpres, Med Found Buffalo, 72-85; res prof biophys sci, State Univ NY Buffalo, 70-; pres & res dir, 86-87, PRES, MED FOUND BUFFALO, INC, 88- *Concurrent Pos:* Lectr, Univ Md, 56-70. *Honors & Awards:* Nobel Prize in Chem (co-recipient), 85; Schoellkopf Award, Am Chem Soc, 86; Patterson Award, 84. *Mem:* Nat Acad Sci; Endocrine Soc; Am Crystallog Asn; Math Asn Am; Am Math Soc. *Res:* Phase problem in x-ray crystallography. *Mailing Add:* Med Found Buffalo 73 High St Buffalo NY 14203

HAUPTMAN, STEPHEN PHILLIP, IMMUNOLOGY, HEMATOLOGY. *Educ:* Col Osteop Med & Surg, DO, 68. *Prof Exp:* PROF MED, THOMAS JEFFERSON UNIV, 84- *Mailing Add:* Dept Med Thomas Jefferson Univ 1025 Walnut St Philadelphia PA 19107

HAUPTMANN, RANDAL MARK, b Hot Springs, SDak, July 6, 56; m; c 1. PLANT TISSUE CULTURE, CRYOPRESERVATION. *Educ:* SDak State Univ, BS(biol & bot) & BS(chem), 79; Univ Ill, Urbana, MS, 82, PhD(plant physiol), 84. *Prof Exp:* Postdoctoral, Monsanto Corp Res, 84-86; vis res scientist, Univ Fla, 86-88; ASST PROF, NORTHWESTERN ILL UNIV, 88-, DIR, PLANT MOLECULAR BIOL CTR, 89- *Concurrent Pos:* Consult, Amoco Life Sci Technologies, 90-91. *Mem:* Sigma Xi; Am Soc Plant Physiologists; Tissue Cult Asn; Int Asn Plant Tissue Cult; AAAS. *Res:* Identification, isolation, and transfer of agronomically important genes to crop species through the development of DNA delivery systems, isolation and characterization of stable transformants, and the study of gene structure and function relationships. *Mailing Add:* Plant Molecular Biol Ctr Northern Ill Univ 325 Montgomery Hall DeKalb IL 60115-2861

HAUPTSCHEIN, MURRAY, b New York, NY, Mar 15, 23; m 47; c 2. ORGANIC CHEMISTRY, RESEARCH ADMINISTRATION. *Educ:* City Col New York, BS, 43; Duke Univ, PhD(chem), 50. *Prof Exp:* Res chemist, Manhattan Proj, SAM Labs, Columbia Univ, 43-45; res chemist, Carbide & Carbon Chem Corp, 45-46; asst, Duke Univ, 46-47; res assoc, Res Inst, Temple Univ, 50-52, dir org chem res, 52-55; sr res chemist & proj leader, Pennwalt Corp, 55-57, group leader, 57-63, dir org res dept, 63-73, assoc mgr res & develop, 73-74, mgr cent res develop, 74-79, mgr res & develop, 79-83; CONSULT, 83- *Concurrent Pos:* Chmn, Gordon Res Conf Fluorine Chem, 65; mem, Exec Comt Div Fluorine Chem, Am Chem Soc, 67-69 & 72-74; pres, Res Mgt Group Philadelphia, 78; mem comt, Nat Am Chem Soc Corp Assocs, 78-80; mem, Indust Res Inst, 79- *Honors & Awards:* Am Chem Soc Award, 62. *Mem:* Am Chem Soc; Sigma Xi. *Res:* Organic synthesis; fluorine chemistry; polymerizations; lubricants; plastics; stain repellent finishes; surfactants for fire-fighting foams; sulfur chemistry; additives for plastics and rubber; pesticide and pharmaceutical research. *Mailing Add:* 513 Patricia Dr Glenside PA 19038

HAURANI, FARID I, b Gaza, Palestine, May 28, 28; US citizen. MEDICINE. *Educ:* Am Univ, Beirut, BA, 49, MD, 53. *Prof Exp:* From asst to assoc prof, 57-74, PROF MED, JEFFERSON MED COL HOSP, 74- *Concurrent Pos:* Nat Cancer Inst res grants. *Mem:* Am Soc Hemat; Am Col Physicians; Sigma Xi. *Res:* Anemia of defective iron reutilization; metabolites and anti-metabolite and folate coenzymes in the stimulated lymphocyte and the leukemic cell. *Mailing Add:* Cardeza Found Suite 807 Curtis Bldg 1015 Walnut St Philadelphia PA 19107

HAURWITZ, BERNHARD, meteorology; deceased, see previous edition for last biography

HAURY, LOREN RICHARD, b Tucson, Ariz, June 25, 39. BIOLOGICAL OCEANOGRAPHY. *Educ:* Yale Univ, BE, 62; Univ Calif, San Diego, PhD(biol oceanog), 73. *Prof Exp:* Investr, Woods Hole Oceanog Inst, 74-76, asst scientist, 76-78; from asst to assoc res oceanogr, 78-88, RES OCEANOGR, SCRIPPS INST, 88- *Mem:* Am Soc Limnol & Oceanog; AAAS; Am Geophys Union. *Res:* Zooplankton ecology. *Mailing Add:* Marine Life Res Group Scripps Inst Oceanog La Jolla CA 92093-0218

HAUS, HERMANN A(NTON), b Ljubljana, Yugoslavia, Aug 8, 25; nat US; m 52; c 4. ELECTRICAL ENGINEERING. *Educ:* Union Col, NY, BSc, 49; Rensselaer Polytech Inst, MEE, 51; Mass Inst Technol, ScD(elec eng), 54. *Prof Exp:* Instr elec eng, Rensselaer Polytech Inst, 49-51; asst, 51-53, instr, 53-54, from asst prof to prof, 54-87, INST PROF, MASS INST TECHNOL, 87- *Concurrent Pos:* Consult, Sylvania Elec Prod Corp, 53-56 & Raytheon Mfg Co, 56-; Guggenheim fel, Vienna Tech Univ, 59-60; Fulbright csholar, Vienna Tech Univ, 85. *Honors & Awards:* Quantum Electronics & Applns Soc Award, Inst Elec & Electronics Engrs, 84, Educ Medal, 91; C H Townes Award, Am Opt Soc, 87. *Mem:* Nat Acad Sci; Nat Acad Eng; Am Phys Soc; fel Am Acad Arts & sci; fel Inst Elec & Electronics, Engrs. *Res:* Noise in electron devices; circuit theory of linear noisy networks; fluctuation phenomena; electromagnetic theory and microwaves; electron dynamics; quantum electronics. *Mailing Add:* Dept Elec Eng Mass Inst Technol Cambridge MA 02139

HAUS, JOSEPH WENDEL, b Cleveland, Ohio, Dec 21, 48; m 70; c 6. QUANTUM FLUCTUATIONS & DIFFRACTION IN OPTICAL WAVE PROPAGATION, ELECTROMAGNETICALLY ENHANCED OPTICAL NONLINEAR RESPONSE. *Educ:* John Carroll Univ, BS, 71, MS, 72; Cath Univ Am, PhD(physics), 75. *Prof Exp:* Postdoctoral res assoc, Nat Bur Standards, 75-77; vis scientist, Kernforschungsanlage-Juelich, 77-78; scientist asst physics, Universitaet Essen, 78-83; res assoc, US Army Missile Command, 83-85; ASSOC PROF PHYSICS, RENSSELAER POLYTECH INST, 85- *Concurrent Pos:* Consult, US Army Missile Command, 83-91 & Dove Electronics, 86-87. *Mem:* Optical Soc Am. *Res:* Theoretical research in quantum optics nonlinear optics and diffusion in disordered materials. *Mailing Add:* Physics Dept Rensselaer Polytech Inst Troy NY 12180-3590

HAUS, THILO ENOCH, b Rochester, Wis, Aug 16, 18; m 45; c 2. AGRONOMY. *Educ:* Univ Wis, BS, 41, MS, 45; Purdue Univ, PhD, 55. *Prof Exp:* Asst prof genetics & asst agronomist, 45-56, assoc prof & assoc agronomist, 56-62, PROF GENETICS & AGRONOMIST, COLO STATE UNIV, 62- *Mem:* Am Soc Agron; Genetics Soc Am; Am Genetic Asn; Sigma Xi. *Res:* Small grain genetics and breeding. *Mailing Add:* Agron Dept Colo State Univ Ft Collins CO 80523

HAUSBERGER, FRANZ X, b Mühldorf, Ger, Feb, 8, 08; m 49; c 2. ANATOMY, PHYSIOLOGY. *Educ:* Univ Munich, MD, 35. *Prof Exp:* Resident med, Gertrauden Hosp, Berlin, Ger, 36-39; head internal div, Salem Hosp, Koeslin, 39-47; head endocrine div, Univ Hosp, Erlangen, 47-48; from asst prof to assoc prof, 50-63, prof, 63-76, EMER PROF ANAT, JEFFERSON MED COL, 76- *Mem:* AAAS; Am Physiol Soc; Endocrine Soc; Am Diabetes Asn; Am Asn Anat. *Res:* Adipose tissue physiology and metabolism; obesity; fat embolism; gross anatomy. *Mailing Add:* Temple Univ Sch Dent Philadelphia PA 19140

HAUSCH, H GEORGE, b Lodz, Poland, Mar 31, 41. DENTISTRY ADMINISTRATION. *Educ:* Rensselaer Polytech Inst, BAE, 63; Va Commonwealth Univ, PhD(physiol), 70. *Prof Exp:* CHIEF, SCI REV BR, NAT INST DENT RES, NIH, 81- *Mailing Add:* Sci Rev Br Nat Inst Dent Res NIH Westwood Bldg Rm 519 5333 Westbard Ave Bethesda MD 20892

HAUSCH, WALTER RICHARD, b Akron, Ohio, Oct 26, 17; m 43; c 4. RUBBER CHEMISTRY. *Educ:* Univ Akron, BS, 40; Univ Cincinnati, MS, 42; Purdue Univ, PhD(org chem), 46. *Prof Exp:* Chemist, Gen Metals Powder Corp, Ohio, 40; asst chem, Univ Cincinnati, 40-41; res chemist, Columbia Chem Div, Pittsburgh Plate Glass Co, Ohio, 41-42; asst chem, Purdue Univ, 42-43; res chemist, J T Baker Chem Co, NJ, 45-47; res chemist, Firestone Tire & Rubber Co, 47-59, head adhesives dept, 59-66, plant mgr, Permalastic Prod Co Div, 66-69, sr res chemist, Firestone Synthetic Rubber & Latex Co, 69-71, res assoc, 71-78, sr res assoc, Cent Res Lab, Firestone Tire & Rubber Co, 78-82; RETIRED. *Mem:* Am Chem Soc; Adhesion Soc. *Res:* Chlorination; fluorination; preparation of certain organic fluorides; silica pigments; polyurethane rubbers and adhesives. *Mailing Add:* 327 Caladonia Ave Akron OH 44333-3711

HAUSCHILD, ANDREAS H W, b Wense, Ger, Oct 13, 29; Can citizen; m 59; c 3. BACTERIOLOGY, FOOD MICROBIOLOGY. *Educ:* Univ Mainz, Dipl, 54; Univ Toronto, MA, 57; Queen's Univ, PhD(plant biochem), 62. *Prof Exp:* Teacher high schs, Ger, 54-56; res asst bact, Connaught Med Res Labs, Univ Toronto, 57-59; lectr, Queen's Univ, 60-61; res assoc, Connaught Med Res Labs, Univ Toronto, 63-65; res scientist, Microbiol Div, Food & Drug Directorate, Health Protection Br, 65-71, head bact ecol sect, Dept Nat Health & Welfare Can, 71-89; RETIRED. *Concurrent Pos:* Chmn, Can Botulism Ref Ctr, 74-; consult, Pan Am Health Orgn, 74-75, Int Comn Microbiol Spec Foods, 74-77, Int Standards Orgn, 77-79, Can Int Develop Agency, 82. *Mem:* Am Soc Microbiol; Can Soc Microbiol; Can Inst Food Sci & Technol. *Res:* Toxins of food-borne microorganisms; bacterial food ecology with particular reference to the control of clostridium botulinum in foods. *Mailing Add:* RR 1 Stilttsville ON K2A 8G0 Can

HAUSCHKA, STEPHEN D, b Philadelphia, Pa, Apr 18, 40; m 64; c 2. DEVELOPMENTAL BIOLOGY, NEUROBIOLOGY. *Educ:* Amherst Col, BA, 62; Johns Hopkins Univ, PhD(biol), 66. *Prof Exp:* Fel, 66-67, from asst prof to assoc prof, 67-80, PROF BIOCHEM, UNIV WASH, 74- *Concurrent Pos:* Res career develop award, NIH, 73-77; mem, Cell Biol Sect, NIH, 75-76, Develop Biol Study Sect, NSF, 77-78, Sci Adv Comt, Muscular Dystrophy Asn, 90-; assoc ed, J Develop Biol, 87 & J Growth Factors, 88- *Mem:* Soc Develop Biol; Am Soc Cell Biol; Am Soc Biol Chemists; Am Soc Zool; Am Soc Microbiol; Soc Neurosci. *Res:* In vitro studies of skeletal and heart muscle development; interaction of cells with extracellular macromolecules and growth factors; molecular basis of cellular differentiation; cell lineage analysis of human, chick and mouse muscle development. *Mailing Add:* Dept Biochem Univ Wash Seattle WA 98195

HAUSDOERFFER, WILLIAM H, b Weehawken, NJ, May 26, 13; m 39; c 1. MATHEMATICS. *Educ:* Trenton State Col, BS, 36; Columbia Univ, MA, 39; Rutgers Univ, EdD(math educ), 50. *Prof Exp:* Teacher elem sch, NJ, 36-40; demonstration teacher math, Trenton State Col, 40-42, from instr to asst prof astron & physics, 46-52, assoc prof math, 52-56, chmn dept, 56-74, prof math, 57-79; RETIRED. *Mem:* Nat Coun Teachers Math; Math Asn Am. *Res:* Differential equations. *Mailing Add:* 220 King George Rd Pennington NJ 08534

HAUSE, NORMAN LAURANCE, b Ft Lupton, Colo, July 26, 22; m 46; c 3. ORGANIC & POLYMER CHEMISTRY. *Educ:* Univ Colo, AB, 47, PhD(chem), 50. *Prof Exp:* Asst, Univ Colo, 47; res chemist, Electrochem Dept, 50-56, res supvr, Exp Sta, 56-60, lab mgr, 60-65, lab dir, Chestnut Run Lab, 65-70, lab dir & mgr polymer res sect, 70-72, res mgr polymer div, Plastic Prod & Resins Dept, 72-79, tech mgr, Ethylene Polymers Div, Polymer Prod Dept, E I du Pont de Nemours & Co Inc, 79-82; RETIRED. *Mem:* Am Chem Soc; Sigma Xi. *Res:* Mechanisms of dehydrohalogenation reactions; furfural chemistry; Diels-Alder reactions; polymers; adhesives; polymeric coatings, plasticizers and modifiers; plastics; thermoplastic elastomers. *Mailing Add:* 275 E Ridge Rd Sedona AZ 86336

HAUSEN, JUTTA, b Berlin, Ger, Jan 6, 43. MATHEMATICS. *Educ:* Univ Frankfurt, Dipl Math & PhD(math), 67. *Prof Exp:* Fel, NMex State Univ, 67-68; from asst prof to assoc prof, 68-77, PROF MATH, UNIV HOUSTON, 77- *Concurrent Pos:* Prin investr res grant, 69; NSF res grant, 72; assoc managing ed, Houston J Math, 74-80, ed, 80-90; res enabling grants & fac develop grants, Univ Houston. *Mem:* Am Math Soc. *Res:* Abelian groups. *Mailing Add:* Dept Math Univ Houston University Park Houston TX 77204-3476

HAUSENBUILLER, ROBERT LEE, b St Joseph, Mo, Sept 20, 18; m 41; c 2. SOILS, CHEMISTRY. *Educ:* Colo State Univ, BS, 41; Wash State Univ, MS, 48, PhD(soil sci), 51. *Prof Exp:* Asst soil scientist, Irrig Exp Sta, Wash State Univ, 51-56, asst prof soils, Univ Inter-col Exchange Team, Univ Panjab, W Pakistan, 56-58, asst soil scientist, Irrig Exp Sta, 58-60, from assoc prof to prof soils, 60-83; RETIRED. *Honors & Awards:* Teaching Award, R M Wade Found, 64. *Res:* Soil fertility. *Mailing Add:* NE 1015 B Pullman WA 99163

HAUSER, EDWARD J P, b Cleveland, Ohio, Apr, 14, 38; div; c 2. BIOLOGICAL SCIENCES. *Educ:* Kent State Univ, BS, 60, MA, 62, PhD(biol sci educ), 75. *Prof Exp:* NSF fel bot, Univ Ga, 62; coun fel, Mich State Univ, 62-63; instr biol sci, Kent State Univ, 62-64; assoc prof, Lorain County Community Col, 65-67; PROF BIOL SCI, LAKELAND COMMUNITY COL, 67- *Concurrent Pos:* Adj prof environ studies, Cleveland State Univ, 66- *Mem:* Nat Audubon Soc; Am Asn Environ Professionals. *Res:* Environmental assessment; plant taxonomy. *Mailing Add:* 6450 Conley Rd Painesville OH 44077

HAUSER, EDWARD RUSSELL, b Newark, NJ, July 15, 42; m 63; c 2. PHYSICAL CHEMISTRY. *Educ:* Macalester Col, BA, 64; Case Western Reserve Univ, PhD(phys chem), 69. *Prof Exp:* Sr res chemist, 68-76, res specialist, Bldg Serv & Cleaning Prod Div, 76-85, PROTECTIVE PROD DIV, 3M CO, 85- *Mem:* Am Chem Soc. *Res:* Nucleation of liquids; properties of nonaqueous dispersions with emphasis on the mechanism of surface potential; nonwoven web, fiber production and characterization; polymer crystallization phenomena; blown microfilm production and process. *Mailing Add:* 1140 34th St Hudson WI 54016

HAUSER, FRANK MARION, b Washington, DC, July 31, 43; m 65; c 2. ANTHRACYCLINE ANTIBIOTICS. *Educ:* Univ NC, BS, 65, PhD(org chem), 69. *Prof Exp:* Fel, Iowa State Univ, 69-70; fel, Res Triangle Inst, 70-71, chemist, 71-74; assoc prof, 74-80, PROF, ORE GRAD CTR, 80- *Concurrent Pos:* Res career develop award, NIH, 78-83. *Mem:* Am Chem Soc. *Res:* Total synthesis of biologically important natural products. *Mailing Add:* 1030 NE 24th Gresham OR 97030

HAUSER, GEORGE, b Vienna, Austria, Dec 13, 22; nat US; m 55. NEUROCHEMISTRY, LIPID METABOLISM. *Educ:* Ohio State Univ, BS, 49; Harvard Univ, PhD(biochem), 55. *Prof Exp:* From res assoc to sr res assoc biol chem, 55-85, MEM FAC MED, HARVARD MED SCH, 70-, PROF PSYCHIAT BIOCHEM & NEUROSCI, 85- *Concurrent Pos:* USPHS fel, 55-57; asst biochemist, McLean Hosp, 57-61, assoc biochemist, 61-78; NIH grants, 65-; NSF grant, 80-82; ed, Neurochem Res, 81; interim dir, Ralph Lowell Labs, 83-; dep chief ed, J Neurochem, 86-; consult, NSF, Initial Rev Groups, NIH; biochemist, McLean Hosp, 78-; mem coun, Am Soc Neurochem, 83-87. *Mem:* Am Soc Biochem & Molecular Biol; Brit Biochem Soc; Am & Int Soc Neurochem; Soc Neurosci. *Res:* Metabolism of the nervous system; carbohydrate and lipid synthesis and function; properties of biological membranes; effects of drugs and cell surface receptor activation on phospholipid metabolism and signal transmission; protein kinase C. *Mailing Add:* Ralph Lowell Labs McLean Hosp Belmont MA 02178-9106

HAUSER, ISIDORE, physics, for more information see previous edition

HAUSER, JOHN REID, b Mocksville, NC, Sept 19, 38; m 62; c 3. SEMICONDUCTOR DEVICES, MICROELECTRONICS. *Educ:* NC State Univ, BS, 60; Duke Univ, MS, 62, PhD(elec eng), 64. *Prof Exp:* Mem tech staff, Bell Tel Labs, 60-62; res engr, Res Triangle Inst, 63-66; from asst prof to assoc prof, 66-72, PROF ELEC ENG, NC STATE UNIV, 72- *Concurrent Pos:* Vpres, Microelectronics Ctr, NC, 81-82. *Honors & Awards:* Western Elec Fund Award, Am Soc Elec Engrs, 75. *Mem:* Inst Elec & Electronics Engrs; Am Phys Soc. *Res:* Solid state electronic materials and devices; microelectronic fabrication techniques; electronic properties of devices and materials. *Mailing Add:* 432 Daniels Hall NC State Univ Raleigh NC 27695

HAUSER, MARTIN, b Berlin, Ger, Jan 6, 34; US citizen; m 55; c 3. POLYMER CHEMISTRY. *Educ:* Union Col, BS, 55; Univ Buffalo, MA, 58, PhD(org chem), 60. *Prof Exp:* Chemist, Stein-Hall, Inc, 55-56; res chemist, Am Cyanamid Co, 60-66, sr res chemist, 66-68; sr res & anal chemist, 68-70, mgr technol, 70-75, vpres res & develop, 75-81, vpres environ health & safety, 81-87, VPRES NEW BUSINESS DEVELOP N AM, LOCTITE CORP, 87- *Mem:* Am Chem Soc. *Res:* Heterocyclic synthesis and properties of heteroatom polymers; polyester chemistry; anaerobic adhesives; cyanoacrylate adhesives. *Mailing Add:* Loctite Corp 705 N Mountain Rd Newington CT 06111

HAUSER, MICHAEL GEORGE, b Chicago, Ill, Dec 3, 39; m 60, 81; c 5. SPACE INFRARED ASTRONOMY. *Educ:* Cornell Univ, BEP, 62; Calif Inst Technol, PhD(physics), 67. *Prof Exp:* Res assoc physics, Princeton Univ, 67, instr, 67-70, asst prof, 70-72; sr res fel physics, Calif Inst Technol, 72-74; head infrared astron group, Lab High Energy Astrophys, Goddard Space Flight Ctr, 74-77, head infrared astrophys sect, 77-85, head infrared astrophyshys br, Lab Extraterrestrial Physics, 85-87, assoc chief, Lab Astron & Solar Physics, 87-88, CHIEF, LAB ASTRON & SOLAR PHYSICS, GODDARD SPACE FLIGHT CTR, 88- *Concurrent Pos:* Mem sci working group, Infrared Astron Satellite, 75-88; prin investr, Cosmic Background Explorer Mission, 76- *Honors & Awards:* Except Sci Achievement Medal, NASA, 84, Pub Serv Group Achievement Award, 84. *Mem:* AAAS; fel Am Phys Soc; Am Astron Soc; Sigma Xi; Int Astron Union. *Res:* Far infrared astronomy of galactic and extragalactic objects and diffuse cosmic infrared radiation; physics of cosmic infrared sources; sensitive detectors for submillimeter and millimeter wavelength radiation. *Mailing Add:* Astron & Solar Physics Lab Goddard Space Flight Ctr Code 680 Greenbelt MD 20771

HAUSER, NORBERT, b Poland, Aug 13, 24; US citizen. INDUSTRIAL ENGINEERING, MANAGEMENT SCIENCE. *Educ:* Cooper Union, BME, 50; NY Univ, MIE, 55, EngScD(qual control), 62. *Prof Exp:* Supvr qual control, Gen Elec Co, 50-55; assoc prof indust eng, NY Univ, 55-66; head dept opers res & syst anal, 68-72, dean mgt, 80-84, PROF INDUST ENG & MGT SCI, POLYTECH INST NY, 66- *Concurrent Pos:* Vis assoc prof, Univ Mich, 65; mem, NSF Proj, Use of Comput in Eng Design, 65; adj prof opers res, Naval Postgrad Sch, Calif, 71 & 79. *Mem:* Am Inst Indust Engrs; Inst Mgt Sci; Am Asn Univ Prof. *Res:* Computer simulation; economic analysis; quality control; modeling of social systems. *Mailing Add:* Polytech Univ 333 Jay St Brooklyn NY 11201

HAUSER, RAY LOUIS, b Litchfield, Ill, Apr 16, 27; m 51; c 4. MATERIALS ENGINEERING, CHEMICAL ENGINEERING. *Educ:* Univ Ill, Urbana, BS, 50; Yale Univ, MEng, 52; Univ Colo, PhD(chem eng), 57. *Prof Exp:* Sr engr, Conn Hard Rubber Co, 51-52; mem res staff, Univ Colo, 55-57; group engr, Martin Co, Colo, 57-58, unit head mat eng, 58-59, staff engr, 59-61; PARTNER & RES DIR, HAUSER LABS, 61- *Concurrent Pos:* Vis lectr, Univ Colo, 57-66. *Honors & Awards:* Allied Chem Co Plaskon Design Award, 67. *Mem:* AAAS; Soc Plastics Engrs; Am Inst Chem Engrs; Am Soc Test & Mat. *Res:* Applied research and development with plastics, adhesives, coatings and products; development of ultrasonic bonding techniques; medical applications and instrumentation with polymeric materials; forensic science and products liability testimony. *Mailing Add:* Hauser Labs 5680 Central Ave PO Box G Boulder CO 80306

HAUSER, RICHARD SCOTT, b Chile, Mar 15, 19; US citizen; m 44; c 5. ECOLOGY. *Educ:* Oberlin Col, AB, 41; Mich State Univ, MS, 47, PhD(bot), 53. *Prof Exp:* From instr to assoc prof biol, State Univ NY Albany, 48-63, prof, 63-82; RETIRED. *Res:* Biogeography. *Mailing Add:* 245 McCormack Rd Slingerlands NY 12159

HAUSER, ROLLAND KEITH, b Marshalltown, Iowa, Aug 30, 37; m 63; c 3. METEOROLOGY. *Educ:* Iowa State Univ, BS, 60; Univ Chicago, SM, 64, PhD(geophys sci), 67. *Prof Exp:* From asst prof to assoc prof, 67-75, chmn dept geol & phys sci, 70-72, PROF PHYS SCI, CALIF STATE UNIV, CHICO, 75- *Concurrent Pos:* Vis res prof, Environ Field Sta, Royal Mil Col Sci, England, 75; res meteorologist, Upper Air Br, Nat Meteorol Ctr, Marlow Heights, Md, 76; dir, Aerospace Educ Serv Proj, NASA, 76-78; tech dir, Nowcasting, Inc, Chico, Ca, 80-85; tech consult, Software Illus, Inc, Pleasant Hills, Calif, 84-85; sci consult, Weather Network, Inc, Sunnyvale, Calif, 85-87; meteorological consult, NOAA, 86, USDA Forest Serv, 85-, US DOD Atmospheric Res Lab, 87-88, Sonoma Technol Inc, Santa Rosa, Calif, 89-, Aerovironment Inc, Monrovia, Calif, 90; chair, Unidata McIdas Broadcast Eval Comt, 87-88. *Honors & Awards:* Clean Air Award, Lung Asn. *Mem:* Am Meteorol Soc; Am Geophys Union; AAAS; Am Asn State Climatologists; Coun Agr Sci & Technol. *Res:* Television presentation of subject matter; weather services via microcomputers; agricultural meteorology; air pollution meteorology; meteorology field studies. *Mailing Add:* 1075 Tracy Lane Chico CA 95926-1446

HAUSER, VICTOR LA VERN, b Hitchcock, Okla, July 1, 29; m 58; c 2. AGRICULTURAL ENGINEERING, APPLIED MATHEMATICS. *Educ:* Okla State Univ, BS, 52; Univ Calif, Davis, MS, 57; Tex A&M Univ, PhD(agr eng), 73. *Prof Exp:* Asst prof agr eng, Tex Tech Univ, 57; res engr water conserv, 57-70, res engr water qual, 70-76, RES ENGR WATER CONSERV, USDA, 76- *Concurrent Pos:* Adv water qual, Nat Acad Sci, 70-72. *Mem:* Fel Am Soc Agr Engrs; Am Soc Civil Engrs; Am Geophys Union; Soc Range Mgt; Soil Conserv Soc. *Res:* Hydrology; dryland water conservation; ground water recharge; water quality; irrigation; range and grassland management. *Mailing Add:* 3309 Cordova Circle Temple TX 76502

HAUSER, WILLARD ALLEN, b Cleveland, Ohio, Mar 6, 37; div; c 3. NEUROEPIDEMIOLOGY, ELECTROENCEPHALOGRAPHY. *Educ:* Case Western Reserve Univ, AB, 58; St Louis Univ, MD, 62. *Prof Exp:* Resident neurol, Northwestern Univ, 63-66; staff neurologist, US Army, 66-68; res asst, Mayo Clin, 68-69, res assoc epidemiol, 69-70; from asst prof to assoc prof neurol, Univ Minn, 70-78; assoc prof neurol, 78-81, assoc prof neurol & epidemiol, 81-84, PROF NEUROL & EPIDEMIOL, COLUMBIA UNIV, 84- *Concurrent Pos:* Spec proj assoc, Dept Epidemiol, Mayo Clin, 70-; examr, Am Bd Qual Electroencephal, 71- & Am Bd Psychiat & Neurol, 76-; co-investr, Epidemiol & Genetics Sect Comprehensive Epilepsy Prog State of Minn, 75-; consult, Biomet & Epidemiol Br, 76-77 & Nat Adv Bd Epilepsy, 78-81. *Mem:* Am Acad Neurol; Am EEG Soc; Soc Epidemiol Res; Am Epilepsy Soc; Am Neurol Soc; Am Epidemiol Soc. *Res:* Epidemiology of neurologic diseases. *Mailing Add:* Columbia Univ G H Sergievsky Ctr 630 W 168th St New York NY 10032

HAUSER, WILLIAM JOSEPH, b Davenport, Iowa, Aug 4, 42; m 69. FISH BIOLOGY, VERTEBRATE ZOOLOGY. *Educ:* Univ Wis-Madison, BS, 65; Mont State Univ, MS, 68; Univ Maine, Orono, PhD(zool), 73. *Prof Exp:* Asst res biologist, Univ Calif, Riverside, 73-77; mem staff, San Francisco Regional Lab, Nalco Environ Sci, 77-78, fisheries sect head, Hazelton Environ Sci, 78-80; REGIONAL BIOLOGIST-COOK INLET & PRINCE WILLIAM SAOUN ALASKA DEPT FISH & GAME, DIV FISHERIES REHAB, ENHANCEMENT & DEVELOP, 80- *Mem:* Am Fisheries Soc; AAAS; Sigma Xi. *Res:* Biolgoical control of aquatic weeds in irrigation systems in the lower Colorado River basin by fish, especially Tilapia zillii. *Mailing Add:* 3621 Hazen Circle Anchorage AK 99502

HAUSER, WILLIAM P, b Cincinnati, Ohio, Apr, 28, 34; m 66; c 3. PHOTOGRAPHIC CHEMISTRY. *Educ:* Univ Notre Dame, BS, 56; Univ Rochester, PhD(phys chem), 61. *Prof Exp:* Aerospace technologist, Goddard Space Flight Ctr, NASA, 61-62; fel, Radiation Lab, Mellon Inst, 62-63; fel Univ Notre Dame, 63-66; mem staff, 66-72, sr patent chemist, 72-81, PATENT ASSOC, PHOTO PROD DEPT, E I DU PONT DE NEMOURS & CO INC, 81- *Mem:* Am Chem Soc; Sigma Xi. *Res:* Thermal reaction kinetics; chemistry of upper atmosphere; radiation chemistry of gases; electron impact phenomena. *Mailing Add:* Dupont Co Photo & Elec Prod P21-2356 Barley Mill Plaza Wilmington DE 19880

HAUSERMANN, MAX, food science & technology, for more information see previous edition

HAUSFATER, GLENN, b St Louis, Mo, Feb 10, 47. PRIMATE BEHAVIOR. *Educ:* Northwestern Univ, BA, 70; Univ Chicago, PhD(evolutionary biol), 74. *Prof Exp:* Asst prof psychol & biol, Univ Va, 73-77; assoc prof neurobiol & behav, Cornell Univ, 78-83; PROF BIOL, UNIV MO, 83- *Concurrent Pos:* Co-dir, Longitudinal Study Amboseli Baboons, Kenya, 74-83; prin investr grants, Nat Sci Found, 74-; assoc ed, Am J Primatol, Am J Phys Anthropol, The Behav & Brain Sci & Behav Ecol Sociobiol, 77-; dir, Liddell Lab Animal Behav, Cornell Univ, 78-82; bd dir, NCountry Inst Nat Philos; res career develop award, Nat Inst Child Health & Human Develop, 76-81; Lady Davis vis prof, Inst Life Sci, Hebrew Univ Jerusalem, Israel. *Mem:* Am Soc Primatologists; Animal Behav Soc; AAAS; Int Primatol Soc; East Africa Wildlife Soc. *Res:* Evolutionary causes and consequences of group-living in animals, especially nonhuman primates; dominance and aggression in primates; behavioral factors in the transmission of parasites and diseases. *Mailing Add:* 1842 N Burling Chicago IL 60614

HAUSLER, CARL LOUIS, b Schenectady, NY, Aug 18, 41. ANIMAL SCIENCE, REPRODUCTIVE PHYSIOLOGY. *Educ:* Univ Vt, BS, 63, MS, 65; Purdue Univ, PhD(reproductive physiol), 70. *Prof Exp:* Asst prof, 70-77, ASSOC PROF REPRODUCTIVE PHYSIOL, SOUTHERN ILL UNIV, CARBONDALE, 77- *Concurrent Pos:* Educ specialist, Food & Agr Orgn, Southern Ill Univ, 70-72; Consult, 75, Portugal, 82-84 & Liberia, 84. *Mem:* Am Soc Animal Sci; Am Dairy Sci Asn; Soc Study Reproduction; AAAS; Sigma Xi. *Res:* Reproductive physiology applied to increased efficiency of animal production; inducing ovulation in lactating sows; estrus detection in cattle; breeding management. *Mailing Add:* Dept Animal Indust Southern Ill Univ Agr Bldg 0121 Carbondale IL 62901

HAUSLER, RUDOLF H, b Zurich, Switz, Apr 9, 34; US citizen; m; c 1. CORROSION, PETROLEUM CHEMISTRY. *Educ:* Swiss Fed Inst Tech, MS, 58, PhD(chem), 61. *Prof Exp:* Proj leader, Battelle Mem Inst, Switz, 61-63; res scientist, Universal Oil Prod Co, 63-67, assoc res coordr, 67-72, res assoc, 72-76; tech dir, Gordon Lab, Inc, Great Bend, Kans, 76-78; sr res chemist, Petrolite Corp, 79-82, prin investr, 82-86, res fel, 86-90; RES ASSOC, MOBIL RES & DEVELOP CORP, 90- *Concurrent Pos:* Lectr, Ill Inst Technol, 75-76, Univ Aachen, Ger, Univ Ferrara, Italy & Univ Manchester, Eng, 85. *Honors & Awards:* Tech Achievement Award, Nat Asn Corrosion Engrs, 89; Plenary lectr, Singapore, 89, Italy, 90, Bahrain, 91. *Mem:* Am Chem Soc; Electrochem Soc; Nat Asn Corrosion Engrs; Am Soc Metals; Soc Petrol Engrs. *Res:* Technology of nickel-cadmium batteries; organic electrochemistry; electrocatalysis; corrosion and corrosion inhibition in chemical process industry; crude oil production chemicals; industrial heat transfer; anti-foulants; corrosion inhibitors for oil and gas production; corrosion inhibition in chemical cleaning of nuclear steam generators; author of 3 books and 46 publications. *Mailing Add:* Mobil Res & Develop Corp PO Box 819047 Dallas TX 75381-9047

HAUSLER, WILLIAM JOHN, JR, b Kansas City, Kans, Aug 31, 26; m 49; c 4. PUBLIC HEALTH ADMINISTRATION. *Educ:* Univ Kans, AB, 51, MA, 53, PhD(bact), 58; Am Bd Med Microbiol, dipl. *Prof Exp:* Asst instr bact, Univ Kans, 51-57; asst sanit, Kans State Bd Health, 57-58; after asst dir, Iowa State Hyg Lab, 59-65, PROF PREV MED & ENVIRON HEALTH, COLS MED, UNIV IOWA, 90-, PROF ORAL DIAG, COL DENT, 90- *Concurrent Pos:* From asst prof to assoc prof hyg & prev med, Univ Iowa, 59-90, assoc prof oral diag, 66-90; WHO consult to Iran, 69, China, 90; comnr, Iowa Air Pollution Control Comn; ed, Standard Methods for Exam of Dairy Prods; dir Iowa State Hyg Lab, Univ Iowa, 65-; lectr, Hebei Med Col, China, 87- *Honors & Awards:* Henry Albert Award. *Mem:* Am Soc Microbiol; fel Am Pub Health Asn; Sigma Xi; fel Am Acad Microbiol; NY Acad Sci; AAAS. *Res:* Public health microbiology; physical-microbiological methods; Brucellosis; AIDS. *Mailing Add:* State Hyg Lab Univ Iowa Iowa City IA 52242

HAUSMAN, ARTHUR HERBERT, b Chicago, Ill, Nov 24, 23; m 46; c 3. ELECTROMAGNETISM. *Educ:* Univ Tex, BSEE, 44; Harvard, SM, 48. *Prof Exp:* Vpres res, Ampex Corp, 60-63, opers, 63-65, group vpres, 65-67, exec vpres, 67-71, pres & chief exec officer, 71-83, chmn bd, 81-87; RETIRED. *Concurrent Pos:* Lectr, Dept Appl Math, Univ Tex, 42-44; chmn, Tech Adv Comt, Dept Com, 73-75; mem vis comt, Dept Math, Mass Inst Technol, 83-86; mem, Pres Export Coun, 84-88, chmn, Export Admin Sub-Comt, 84-88; consult pvt indust & US Govt, 88- *Mem:* Inst Elec & Electronics Engrs. *Res:* Magnetic materials, magnetic tape and magnetic tape/disk recording; electromagnetic wave propagation in the radio high frequency spectrum. *Mailing Add:* 55 Flood Circle Atherton CA 94027

HAUSMAN, GARY J, b Carol, Iowa, Dec 4, 48. FAT CELL DIFFERENTIATION & DEVELOPMENT. *Educ:* Univ Wis-Madison, PhD(animal sci), 77. *Prof Exp:* RES PHYSIOLOGIST, RICHARD RUSSELL RES CTR, USDA, 81- *Mem:* Am Soc Animal Sci; Am Inst Nutrit. *Mailing Add:* Richard Russell Res Ctr USDA PO Box 5677 Athens GA 30613

HAUSMAN, HERSHEL J, b Pittsburgh, Pa, Aug 19, 23; c 3. NUCLEAR PHYSICS. *Educ:* Carnegie Inst Technol, BS, 48, MS, 49; Univ Pittsburgh, PhD(physics), 52. *Prof Exp:* Instr physics, Univ Pittsburgh, 49-50, res assoc, 50-52; from asst prof to prof physics, 52-89, supvr, Van De Graaff Lab, 62-89, EMER PROF PHYSICS, OHIO STATE UNIV, 89- *Concurrent Pos:* USAID consult, India, 64, 65. *Mem:* Fel Am Phys Soc; Am Asn Physics Teachers. *Res:* Nuclear scattering problems; nuclear reaction mechanisms; nuclear capture reactions. *Mailing Add:* Dept Physics Ohio State Univ 174 W 18th Ave Columbus OH 43210

HAUSMAN, ROBERT, b Takengon, Indonesia, June 19, 14; US citizen; m 44; c 2. PATHOLOGY. *Educ:* State Univ Leiden, DrMed, 36; Univ Amsterdam, MD, 38; Baylor Univ, MS, 52; Free Univ, Amsterdam, PhD(fac med), 77. *Prof Exp:* Med adv health serv, Neth Merchant Marine, NY, 42-45; dir path lab, Govt Hosp, Indonesian Pub Health Serv, Surabaja, 46-49; chief asst, Inst Criminol, Univ Indonesia, 49; res path, Baptist Mem Hosp, San Antonio, Tex, 50-52; asst res, Med Col, Cornell Univ, 52-53; clin asst prof, Sch Med, Emory Univ, 53-56; med exam, Bexar County, 56-68; dep chief med exam, New York, 68-70; consult path, Ft Worth, 70-71; pathologist, St Antonius Hosp, Utrecht, 71-74; prosector, Path Inst, Free Univ, Amsterdam, 74-78; pathologist, Inst Path, Univ Amsterdam, 78-83. *Concurrent Pos:* Dir labs, Grady Mem Hosp, Atlanta, Ga, 54-56; vis prof forensic med, Affiliation Proj, Univ Calif-Airlangga Univ Med Sch, Indonesia, 63-65; consult, USAF Hosp, Lackland AFB, Tex; consult, San Antonio State Tuberculosis Hosp, Med C, Neth Merchant Marine, 40-42; consulting pathologist, 84- *Mem:* Acad Forensic Sci; Am Soc Clin Path; Int Soc Dermatopath; Am Soc Dermatopath. *Res:* General, forensic pathology, dermatopathology. *Mailing Add:* 12 E 86th Apt 1526 New York NY 10028

HAUSMAN, ROBERT EDWARD, b New York, NY, Feb 15, 47; m 87; c 1. CELL-CELL INTERACTIONS. *Educ:* Case Western Reserve Univ, AB, 69, MA, 69; Northwestern Univ, PhD(biol), 71. *Prof Exp:* Postdoctoral fel develop biol, Dept Biol, Univ Chicago, 71-74, res assoc, Comt Develop Biol, 74-78; asst prof, 78-87, ASSOC PROF DEVELOP BIOL & DIR GRAD STUDIES BIOL, DEPT BIOL, BOSTON UNIV, 87- *Concurrent Pos:* Investr, Cancer Ctr, Univ Chicago, 76-78; mem interdept biochem prog, Boston Univ, 78- *Mem:* AAAS; Asn Res Vision & Ophthal; Int Soc Develop Biologists; Soc Neurosci; Soc Develop Biol. *Res:* Control of gene expression during embryonic development especially of nerve and muscle; specific interest in the roles of cell-cell signaling and cell interactions in such gene control. *Mailing Add:* Biol Dept Boston Univ Five Cummington St Boston MA 02215

HAUSMAN, STEVEN J, b Philadelphia, Pa, May 20, 45. IMMUNOGENETICS, CELL BIOLOGY. *Educ:* Univ Pa, BA, 67, MS, 68, PhD(biol), 72. *Prof Exp:* Assoc, Inst Cancer Res, 72-75; staff fel, Nat Inst Aging, NIH, 75-77; spec asst to assoc dir, Nat Inst Arthritis, Metab & Digestive Dis, 77-78; prog dir arthritis ctr, 78-87, DEP DIR, EXTRAMURAL ACTIVITIES, NAT INST ARTHRITIS DIABETES, DIGESTIVE & KIDNEY DIS, 86- *Concurrent Pos:* Fel, Children's Hosp Pa, 75. *Mem:* AAAS; Am Asn Immunologists; Am Chem Soc; Am Soc Cell Biol; Tissue Cult Asn. *Mailing Add:* PO Box 3819 Gaithersburg MD 20885

HAUSMAN, WARREN H, m; c 2. PRODUCTION & OPERATIONS MANAGEMENT. *Educ:* Yale Univ, BA, 61; Mass Inst Technol, PhD(indust mgt), 66. *Prof Exp:* From asst prof to assoc prof, Grad Sch Bus & Pub Admin, Cornell Univ, 65-70; assoc prof, Sloan Sch Mgt, Mass Inst Technol, 70-73; from assoc prof to prof, Grad Sch Mgt, Univ Rochester, 73-77; prof, Dept Indust Eng & Eng Mgt, Stanford Univ, 77-82; PROF & CHMN, DEPT INDUST ENG & ENG MGT, STANFORD UNIV, 82- *Concurrent Pos:* Vis prof, Inst Admin Gestion, Univ Cath Louvain, Belg, 81; vis scholar, Europ Inst Advan Studies Mgt, Brussels, Belg, 81; consult, various corp & bus, 62-; author of various publications; adv panel mem, Decision, Risk & Mgt Sci Progs, NSF, 87-90. *Mem:* Inst Mgt Sci; Inst Indust Engrs; Opers Res Soc Am. *Res:* Productions and operations management. *Mailing Add:* Dept Indust Eng Stanford Univ Stanford CA 94305

HAUSMAN, WILLIAM, b Brooklyn, NY, July 25, 25; m 47; c 4. PSYCHIATRY, SCIENCE EDUCATION. *Educ:* Wash Univ, MD, 47; Am Bd Neurol & Psychiat, dipl, 53. *Prof Exp:* Instr, Med Sch, Univ Pa, 50-52; chief psychiat serv, Letterman Gen Hosp, US Army, 61-62, chief behav sci res br, Off Surgeon Gen, 62-65, dep dir, Div Neuropsychiat, Walter Reed Army Inst, 65-66, assoc prof psychiat, Student Ment Health Serv & dir, Curriculum Options Study, Sch Med, Johns Hopkins Univ, 66-69; prof psychiat, Univ Minn, Minneapolis, 69-87, chmn dept, 69-80; RETIRED. *Concurrent Pos:* Mem res coun, US Army, 65-66; dir group rels conf, A K Rice Inst, 66-, mem bd dir, 67-74, pres bd, 70-73; mem fac, Wash Sch Psychiat, 67-69; consult, Univ Man, Winnipeg, 75-76 & Levinson Inst, 75- *Mem:* AAAS; fel Am Psychiat Asn; Int Asn Social Psychiat; fel Am Col Psychiat. *Res:* Selection and adaptation; group process; concepts basic to social psychiatry; psychiatric education. *Mailing Add:* 3785 Ranch Crest Dr Reno NV 89509

HAUSMANN, ERNEST, b Heidelberg, Ger, June 19, 29; US citizen; m 51; c 4. BIOCHEMISTRY. *Educ:* NY Univ, BA, 51; Harvard Univ, DMD, 56; Univ Rochester, PhD(biochem), 60. *Prof Exp:* Nat Inst Dent Res fel, 57-59; asst prof pharmacol, State Univ NY, Buffalo, 60-64; assoc prof periodont, Univ Ky, 64-66; assoc prof, 66-70, PROF ORAL BIOL, STATE UNIV NY BUFFALO, 70-, ASST DEAN SCH DENT, 66- *Concurrent Pos:* Nat Inst Dent Res career develop award, 61-66. *Mem:* Int Asn Dent Res. *Res:* Dental research; collagen; calcium metabolism. *Mailing Add:* 82 High Park Blvd Buffalo NY 14226

HAUSMANN, WERNER KARL, b Edigheim, Ger, Mar 9, 21; US citizen; m 49; c 1. SEPARATION TECHNIQUES. *Educ:* Swiss Fed Inst Technol, MS, 45, DSc, 47. *Prof Exp:* Res fel biochem, Univ London, Eng, 47-48; res assoc, Rockefeller Inst Med Res, 49-57; group leader antibiotics res, Lederle Labs, 57-66; assoc dir qual control, Ayerst Labs, 66-71; dir, Stuart Pharmaceut, 71-74; corp dir qual assurance & anal res & develop, Adria Labs, 74-84; CONSULT QUAL ASSURANCE & ANALYTIC RES & DEVELOP, QUAL-EX CO, 85- *Concurrent Pos:* Lecturer Am Mgt Asn, 68. *Mem:* Fel Am Soc Qual Control; fel Am Inst Chemists; fel Royal Soc Chem; Am Soc Biol Chemists; Acad Pharmaceut Sci; Am Chem Soc; fel NY Acad Sci; fel AAAS; fel Chem Soc London. *Res:* Analytical chemistry; pharmaceutical compounds as such and in dosage forms and biological fluids; purification, isolation and structure determination of biologically active compounds; new antibiotics; high performance liquid chromatography. *Mailing Add:* 4332 Post Rd San Diego CA 92117

HAUSNER, HENRY H(ERMAN), b Vienna, Austria, June 1, 01; US citizen; m 27. METALLURGY. *Educ:* Vienna Tech Univ, EE, 25, DrEng, 33. *Prof Exp:* Elec engr, Elin A G, Vienna, 25-38, res mgr, 38-40; res engr powder metall, Am Electro Metal Corp, 40-41; chief elec eng, Nichols Eng & Res Corp, 42-43; chief res engr, Gen Ceramics & Steatite Co, 43-45; adj prof illum & metall eng & res assoc metall ceramics, NY Univ, 46-48; sect head, Metall Labs, Sylvania Elec Prod Inc, 48-51, mgr eng, Atomic Energy Div, 51-56; vpres, Penn-Tex Corp, 56-58; consult engr, 58-70; mgr, Franklin Inst Res Labs, Philadelphia, Pa, 70-77; HON PROF, POLYTECH INST NY, 77- *Concurrent Pos:* Adj prof, Polytech Inst Brooklyn, 51-; consult, Sylvania Corning Nuclear Corp & Nuclear Develop Corp Am, 58- & Gen Aniline & Film Corp, 62-; lectr, Univ Calif, Los Angeles, 63-; consult, Am Standard Corp, 64- & Olivetti Co, Italy, 66-; ed chief, Int J Powder Metall, 65-; guest prof, Max-Planck Inst, Ger, 71; lectr, NY Univ, 73- *Mem:* Am Soc Metals; Am Ceramic Soc; Am Ord Asn; Am Inst Mining, Metall & Petrol Engrs; fel NY Acad Sci. *Res:* Physical and powder metallurgy; ceramics. *Mailing Add:* 69 Red Brook Rd Kings Point NY 11024

HAUSNER, MELVIN, b USA, Apr 11, 28; m; c 2. GEOMETRY, COMBINATORICS. *Educ:* Brooklyn Col, BSc, 48; Princeton Univ, MA, 49, PhD(math), 51. *Prof Exp:* Res mathematician, Rand Corp, 51-52; instr math, Brooklyn Col, 52-53, 55-56; asst prof, Stevens Inst Technol, 56-60; assoc prof, 60-66, actg chmn dept, 65-66, chmn dept, 70-73, PROF MATH, WASH SQUARE COL, NY UNIV, 66- *Concurrent Pos:* Ed, Encycl Am. *Mem:* Am Math Soc; Math Asn Am. *Res:* Geometry and combinatorics. *Mailing Add:* Dept Math NY Univ 251 Mercer St New York NY 10012

HAUSPURG, ARTHUR, b New York, NY, 1925. POWER SYSTEMS, ULTRAHIGH VOLTAGE TRANSMISSION SYSTEMS. *Educ:* Columbia Univ, BS, 45, MS, 47. *Prof Exp:* Chmn & chief exec officer, Consolidated Edison Co, 82-90; RETIRED. *Mem:* Nat Acad Eng; fel Inst Elec & Electronics Engrs. *Mailing Add:* Consolidated Edison 14 Irving Place New York NY 10003

HAUSRATH, ALAN RICHARD, b E Cleveland, Ohio, Sept 12, 45; m 71; c 3. QUALITATIVE THEORY OF DIFFERENTIAL EQUATIONS, PERIODIC SOLUTIONS OF DIFFERENTIAL EQUATIONS. *Educ:* Mass Inst Technol, ScB, 67; Brown Univ, PhD(appl math), 72; Wash State Univ, MEd, 77. *Prof Exp:* Asst prof math, Univ Pittsburgh, 71-76; PROF MATH, BOISE STATE UNIV, 76- *Concurrent Pos:* Vis prof, Univ Andes, Venezuela, 78-79; vis scholar, Univ Chile, 86. *Mem:* Am Math Soc; Math Asn Am; Soc Indust & Appl Math. *Res:* Qualitative theory of differential equations; differential arising as models of biological systems. *Mailing Add:* Dept Math Boise State Univ Boise ID 83725

HAUSSER, JACK W, b Cleveland, Ohio, Jan 6, 35; m 59; c 3. ORGANIC CHEMISTRY. *Educ:* Case Western Reserve Univ, BS, 56; Univ Ill, PhD(org chem), 60. *Prof Exp:* Res assoc org chem, Iowa State Univ, 60-62; from asst prof to assoc prof, 62-72, chmn dept, 72-75, PROF CHEM, DUQUESNE UNIV, 72- *Mem:* AAAS; Am Chem Soc; Sigma Xi. *Res:* Mechanisms of organic reactions; chemistry of small ring compounds; solvolytic displacement reactions. *Mailing Add:* Dept Chem Duquesne Univ Pittsburgh PA 15219

HÄUSSER, OTTO FRIEDRICH, b Schwabach, WGer, Dec 9, 37; m 64; c 2. NUCLEAR PHYSICS. *Educ:* Univ Erlangen, dipl, 62, PhD(physics), 64. *Prof Exp:* NATO fel, Oxford Univ, 64-66; Nat Res Coun Can fel, Chalk River Nuclear Labs, Atomic Energy Can Ltd, 66-67, res officer, 67-83; PROF PHYSICS, SIMON FRASER UNIV, BURNABY, BC, 83- *Mem:* Fel Am Phys Soc. *Res:* Electromagnetic and particle decay of nuclear states; higher order effects in coulomb excitation; hyperfine interactions; nuclear moments; intermediate energy physics; polarized targets; nucleon structure; spin physics. *Mailing Add:* TRIUMF 4004 Wesbrook Mall Vancouver BC V6T 2A3 Can

HAUSSLEIN, ROBERT WILLIAM, b New York, NY, Sept 17, 37; m 60; c 3. CHEMICAL ENGINEERING. *Educ:* Mass Inst Technol, SB, 58, PhD(chem eng), 65. *Prof Exp:* Asst prof chem eng, Mass Inst Technol, 65-66; asst dir res, Amicon Corp, 66-71; res group leader, Polaroid Corp, 71-73, sr res group leader, 73-77, mgr photosysts, 77-80, sr res lab mgr, 80-88; VPRES, HYPERIAN CATALYSIS INT. *Concurrent Pos:* Ford fel, Mass Inst Technol, 65-66. *Mem:* Electrochem Soc. *Res:* Photographic chemistry; coating technology; colloid technology. *Mailing Add:* 20 Slocum Rd Lexington MA 02173

HAUSSLING, HENRY JACOB, b Newark, NJ, June 18, 45; m 68; c 7. FLUID DYNAMICS. *Educ:* Mass Inst Technol, BS, 67; Univ Md, PhD(meteorol), 74. *Prof Exp:* MATHEMATICIAN FLUID DYNAMICS, DAVID W TAYLOR NAVAL SHIP RES & DEVELOP CTR, 68- *Mem:* Sigma Xi. *Res:* The use of numerical methods for solving partial differential equations; computational fluid dynamics; numerical ship hydrodynamics. *Mailing Add:* RR 2 Box 343H Lovettsville VA 22080

HAUSSMANN, ULRICH GUNTHER, Can citizen. STOCHASTIC OPTIMAL CONTROL, FILTERING THEORY. *Educ:* Univ Toronto, BSc, 66; Brown Univ, PhD(appl math), 70. *Prof Exp:* Fel & res assoc, 70, from asst prof to assoc prof, 70-82, PROF MATH, UNIV BC, 82- *Concurrent Pos:* Vis prof, Univ Grenoble, 75-76, Univ Paris, 83 & Univ Provence, 84. *Mem:* Soc Indust & Appl Math. *Res:* Optimal control of stochastic systems, non-linear filtering theory, theory of diffusions. *Mailing Add:* Dept Math Univ BC Vancouver BC V6T 1Y4 Can

HAUST, M DARIA, Can citizen; m; c 2. PATHOLOGY, PEDIATRICS. *Educ:* Queen's Univ, Ont, MSc, 59. *Prof Exp:* From asst prof to assoc prof path, Queen's Univ, Ont, 60-67; assoc prof, 67-68, PROF PATH, UNIV WESTERN ONT, 68-, PROF PEDIAT, 72-, PROF OBSTET & GYNEC, 77- *Concurrent Pos:* Res assoc, Ont Heart Found, 60-63, sr res assoc, 63-71; mem coun arteriosclerosis, Am Heart Asn, 65-68; vis scientist, Oxford Univ, 72-73; mem educ comt, Intl Acad Path, 69-72, coun, 70-73; task force & rec secy, Int Atherosclerosis Soc, 76-78, Chmn Comt Const, 77-79; coordr, Adv Expert Panel Pediat Path, Int Pediat Asn, 80-87. *Honors & Awards:* Alexander von Humboldt & Distinguished Serv Award, 82; Serv Award, Am Heart Asn, 86; William Boyd Lectr, Can Asn Pathologists, 90. *Mem:* Am Asn Path & Bact; Am Soc Exp Path; Can Asn Path; Int Acad Path; Int Atherosclerosis Soc (secy-treas, 78-85, treas, 85-); Int Pediat Asn; Am Heart Asn; Int Acad Path. *Res:* Basic structural and biochemical nature of fibrous connective tissue in health and disease, especially inborn errors of metabolism, selected collagen diseases and atherosclerosis; obstetrics and gynecology. *Mailing Add:* Dept Path Univ Western Ont London ON N6A 5B7 Can

HAUSTEIN, PETER EUGENE, b Detroit, Mich, Jan 17, 44. NUCLEAR STRUCTURE. *Educ:* Univ Calif, BS, 66; Iowa State Univ, MS & PhD(nuclear chem), 70. *Prof Exp:* Fel nuclear chem, Brookhaven Nat Lab, 70-72; asst prof chem, Yale Univ, 72-75; assoc chemist, Brookhaven Nat Lab, NY, 75-78, chemist, 78-90, asst chmn chem dept, 88-90, ASSOC CHMN CHEM DEPT & SR CHEMIST, BROOKHAVEN NAT LAB, NY, 90- *Concurrent Pos:* Vis scientist, Lawrence Berkeley Lab, 79, 89-90, Europ Coun Nuclear Res, 83-84; Nat Acad Sci-Comt on Nuclear & Radiochem, 80-86. *Mem:* Am Chem Soc; Am Phys Soc; Sigma Xi. *Res:* Nuclear structure and reaction studies; nuclear resonance spectroscopy; relativistic heavy-ion reactions; application of computers to nuclear science. *Mailing Add:* Dept Chem Brookhaven Nat Lab Bldg 555A Upton NY 11973-5000

HAUSWIRTH, WILLIAM WALTER, b San Francisco, Calif, Jan 1, 45. GENE TRANSMISSION, GENE EXPRESSION. *Educ:* Stanford Univ, BS, 66; Ore State Univ, PhD(chem), 71. *Prof Exp:* Asst prof biochem, Johns Hopkins Univ, 74-75; from asst prof to assoc prof, 76-85, PROF MOLECULAR GENETICS, COL MED, UNIV FLA, 85-, RYBACZKI-BALLARD PROF MOLECULAR GENETICS, 86- *Concurrent Pos:* Mem, Molecular Biol Study Sect, NIH, 86- *Mem:* Am Chem Soc; Am Soc Biol Chemists; Am Soc Microbiol; Sigma Xi; Int Soc Plant Molecular Biol. *Res:* Animal and plant mitochondrial DNA molecular biology; transmission genetics; evoluton; developmental biology. *Mailing Add:* Box J-266 JHMHC Col Med Univ Fla Gainesville FL 32610

HAUT, ARTHUR, b New York, NY, Oct 1, 27; m 53; c 4. HEMATOLOGY, INTERNAL MEDICINE. *Educ:* Columbia Univ, AB, 46, MD, 50; Am Bd Internal Med, dipl, 57. *Prof Exp:* Instr med, Col Med, Univ Utah, 57-60, asst prof, 60-63; assoc prof, 63-67, head sect hemat, 63-72, PROF MED, MED CTR, UNIV ARK, LITTLE ROCK, 67-, DIR, DIV HEMAT & ONCOL, 72- *Concurrent Pos:* Teaching fel med, Harvard Med Sch, 52-53; Am Cancer Soc fel, Col Med, Univ Utah, 53-55; Markle scholar acad med, 59-64; consult hemat, Surgeon Gen, US Army, Europe, 55-57 & Little Rock Vet Hosp, Ark, 63-; mem hemat study sect, NIH, 67-71; mem hemat test comt, Am Bd Internal Med, 70-73; mem merit rev bd hemat, Vet Admin, 74- *Mem:* Am Fedn Clin Res; fel Am Col Physicians; Am Soc Hemat. *Res:* Chemotherapy of leukemia; erythrocytic nonhemoglobin proteins; cancer chemotherapy. *Mailing Add:* Dept Med Univ Ark Med Sch 4301 W Markham St Little Rock AR 72205

HAUTALA, RICHARD ROY, b Rock Springs, Wyo, July 12, 43; m 75. ORGANIC CHEMISTRY, PHOTOCHEMISTRY. *Educ:* Colo Col, BS, 65; Northwestern Univ, PhD(org chem), 69. *Prof Exp:* Instr chem, Northwestern Univ, 69-70; res assoc, Columbia Univ, 70-72; asst prof chem, Univ Ga, 72-78, assoc prof, 78-; AT DISCOVER-DEPT CHEM RES DIV, AM CYANAMID CO. *Concurrent Pos:* Vis staff, Los Alamos Nat Labs, Univ Calif, 78- *Mem:* Am Chem Soc; Inter-Am Photochem Soc; The Chem Soc. *Res:* Mechanistic organic photochemistry; catalysis by molecular organization; properties of Micelles; time-resolved spectroscopy; energy transfer mechanisms; solar energy storage; surface properties. *Mailing Add:* Discover-Dept Chem Res Div American Cyanamid Co 1937 W Main St Stamford CT 06904-0060

HAUTH, WILLARD E(LLSWORTH), III, b Boston, Mass, May 20, 48; m 83; c 2. CERAMIC COMPOSITES, ADVANCED CERAMICS. *Educ:* State Univ NY, Alfred Univ, BS, 70, MS, 74. *Prof Exp:* Ceramic engr, Babcock & Wilcox Co, 70-74 & Corning Glass Works, 74-75; res scientist, Los Alamos Nat Lab, 75-80; pres ceramics, Mat Technol Assoc, 77-80; res scientist, Battelle Columbus Labs, 80-83; res scientist ceramics, Dow Corning Corp, 83-89; SR VPRES, RES, DEV & ENG, ADVAN REFRACTORY TECHNOLOGIES. *Concurrent Pos:* Gov, Am Asn Eng Soc, 84-86; comt mem, Nat Mat Adv Bd, 84-87; tech reviewer, Am Ceramic Soc, 84-; comt mem, Mat Tech Adv Comt, US Dept of Com, 86. *Honors & Awards:* Schwartzwalder-PACE Award, Nat Inst Ceramic Engrs, 84. *Mem:* Nat Inst Ceramic Engrs (secy/treas, 81, vpres, 82, pres-elect, 83, pres, 84); fel Am Ceramic Soc; Am Soc Testing Mat. *Res:* Applied development of oxide and non-oxide ceramic materials; fabrication technologies and advanced product forms, including chemically-derived ceramics, ceramic matrix composites and high-performance ceramic powders and composite reinforcements. *Mailing Add:* 482 Fillmore Ave East Aurora NY 14052

HAUXWELL, DONALD LAWRENCE, b Indianola, Nebr, June 25, 38; m 62; c 2. FOREST SOILS. *Educ:* Univ Idaho, BS, 60, PhD(forest soils), 67. *Prof Exp:* From asst prof to prof soil sci, 66-76, NSF instnl grant, 69, prog leader natural resources, Sch Natural Resources, 70-73, PROF NATURAL RESOURCES, HUMBOLDT STATE UNIV, 76- *Concurrent Pos:* Soil conservationist, US Soil Conserv Serv, 68; US Forest Serv admin study, Calif State Univ, Humboldt, 72-73; soil scientist, Bur Land Mgt, 79-80; vis prof, Univ Idaho, 75-76, NC State Univ, 83, Egerton Univ, Kenya, 89-90. *Mem:* Am Registry Prof Agron, Soils & Crop Sci; Am Soc Agron; Soil Sci Soc Am. *Res:* Tissue analysis in the management of soil nitrogen in a Douglas-fir seedling nursery; soils associated with various sagebrush species in Idaho; soil treatment to enhance erosion control vegetation on forest road cutbanks; moisture and temperature regimes of forest soils; agroforestry & soil fertility. *Mailing Add:* Dept Natural Resources Humboldt State Univ Arcata CA 95521

HAUXWELL, GERALD DEAN, b Indianola, Nebr, Sept 24, 35; m 64. COMPETITOR ASSESSMENT, MATHEMATICS. *Educ:* Univ Colo, BS, 58; Univ Idaho, MS, 60; Ore State Univ, PhD(chem eng), 71. *Prof Exp:* Design engr, Atomic Power Equip Dept, Gen Elec Co, 62-64; res engr, Sontara develop, Pioneering Res Lab, 64-68, res assoc, Nomex & Kevlar develop, Spruance Res Lab, 80-84, MGR, TECHNOL ASSESSMENT-FIBERS INDUST, E I DUPONT DE NEMOURS & CO, INC, 84- *Concurrent Pos:* Res fel, Dow Chem Co, 68-70 & Shell Oil Co, 70-71; pvt consult var sponsoring orgn, 74-; adj prof, Va Commonwealth Univ, 75-, Univ Va, 77- & Va State Univ, 80- *Mem:* Am Inst Chem Engrs; Nat Soc Prof Engrs; Am Chem Soc; Tech Asn Paper Pulp Ind; Soc Competitor Intel Profs. *Res:* Heat, mass and momentum transfer; related to oxygen absorption, industrial processes, environmental control and energy conservation; expertise in competitor assessment including database construction competitive strategy development. *Mailing Add:* 11400 Edenberry Dr Richmond VA 23236-4032

HAUXWELL, RONALD EARL, b Flint, Mich, May 19, 46; m 68; c 3. ORGANIC CHEMISTRY, BIOCHEMISTRY. *Educ:* Western Mich Univ, BA, 69, MA, 71. *Prof Exp:* Res chemist, Burdick & Jackson Labs, Subsid Hoffmann-La Roche, Inc, 71-72, prod mgr org chem, 72-80; MEM STAFF, CORDOVA CHEM CO, 80- *Concurrent Pos:* Mem fac, Dept Chem, Muskegon Community Col, 78- *Mem:* Am Chem Soc. *Res:* Synthesis of phosphorescent enzyme inhibitors for use in structure elucidation; innovative drug synthesis and production process improvement and maximization. *Mailing Add:* PO Box 5240 North Muskegon MI 49445-0240

HAVA, MILOS, b Prague, Czech, Oct 15, 27; m 51; c 1. CLINICAL PHARMACOL. *Educ:* Charles Univ, MD, 52, PhD(pharmacol), 55, Educ Coun Foreign Med Grads, cert, 74. *Prof Exp:* Sr res worker pharmacol, Czech Acad Sci, 55-58; head dept, Res Inst Pharm & Biochem, 58-68; from asst prof to assoc prof, Med Sch, Univ Kans, 68-73; assoc prof, Peoria Sch Med, Univ Ill Med Col, 73-75; asst med dir, Marion Labs, Inc, 75-79; med dir, 79-80, exec dir, clin drug develop, Wallace Labs, 80-; CLIN RESEARCHER, WYETH-AYERST RES. *Concurrent Pos:* Consult, Czech Ministry Health, 58-68 & Ill Valley Ment Health Asn, 74-75; adj prof, Med Sch, Univ Kans, 77-80; clin prof, Med Sch, Temple Univ, 80- *Honors & Awards:* Res Award, Czech Acad Sci, 56. *Mem:* AMA; AAAS; Am Soc Pharmacol & Exp Therapeut; Am Soc Clin Pharm Therapeut. *Res:* Clinical research and drug evaluation; pharmacology of gastrointestinal smooth muscle; pharmacology of steroids. *Mailing Add:* Clin Res Wyeth-Ayerst Res PO Box 8299 Philadelphia PA 19101

HAVARD, JOHN FRANCIS, b Helena, Mont, Mar 15, 09; m 43; c 4. MINING ENGINEERING, ECONOMIC GEOLOGY. *Educ:* Univ Wis, PhB, 34, PhM & BS, 35, EM, 43; Harvard Univ, cert, 57. *Prof Exp:* Mine supt, US Gypsum Co, 35-38, engr, 38-40, works mgr, 40-48, chief engr mines, 48-52; asst res mgr, Potash Corp Am, 52-53; vpres, Fibreboard Corp, 53-62; vpres, Kaiser Engrs, 62-74, sr vpres, 74-77. *Concurrent Pos:* Mem, US Nat Comt Geol, 75-79; consult mining, 77- *Honors & Awards:* Henry Krumb lectr, Am Inst Mining & Metall Engrs, 79; Hardinge Award, Am Inst Mining Metall & Petroleum Engrs, 82. *Mem:* Hon mem Am Inst Mining Metall & Petroleum Engrs; Soc Mining Engrs (pres, 76); Soc Econ Geologists; fel AAAS. *Mailing Add:* 18552 Augustine Rd Nevada City CA 95959

HAVAS, HELGA FRANCIS, b Vienna, Austria, Nov 26, 15; US citizen; c 2. MICROBIOLOGY. *Educ:* Columbia Univ, MA, 44; Lehigh Univ, PhD(bact), 50. *Prof Exp:* Res chemist, Sch Nutrit, Cornell Univ, 45-46; res fel chemother, Inst Cancer Res, 50-51, res assoc, 51-63; from asst prof to assoc prof microbiol, 59-72, PROF MICROBIOL, SCH MED, TEMPLE UNIV, 72- *Concurrent Pos:* USPHS fel, NIH, 64-66, career develop award, 66-70. *Mem:* Am Asn Cancer Res; Am Asn Immunol. *Res:* Role of suppressive factors and suppressor cells in tumor-induced immunosuppression; immunochemistry of carcinogen-protein conjugates; oncolytic and immunochemical properties of polysaccharides and bacterial toxins; immunochemistry; tolerance to hapten conjugates; immunochemical aspects of tolerance; effect of carrier on anti-hapten response; mechanism of immunosuppression by plasmacytomas to hapten-conjugates and pneumococcal polysaccharides; immunomodulation by bacterial vaccine in tumor bearing host. *Mailing Add:* Dept Microbiol & Immunol Temple Univ & Health Sci Campus Broad & Ont Philadelphia PA 19140

HAVAS, PETER, b Budapest, Hungary, Mar 29, 16; US citizen. THEORETICAL PHYSICS. *Educ:* Vienna Tech Univ, Absolutorium, 38; Columbia Univ, PhD(physics), 44. *Prof Exp:* Lectr physics, Columbia Univ, 41-45; instr, Cornell Univ, 45-46; from asst prof to prof, Lehigh Univ, 46-65; prof, 65-81, EMER PROF PHYSICS, TEMPLE UNIV, 81- *Concurrent Pos:* Guggenheim Mem Found fel & mem, Inst Adv Study, 53-54; vis prof, Univ Gottingen, 73; adj prof, Univ Pa & Utah State Univ, 81- *Mem:* Fel AAAS; Am Phys Soc; Soc Gen Relativity & Gravity. *Res:* Theory of radiation; elementary particle theory; mathematical physics; history of physics; theory of relativity; foundation problems. *Mailing Add:* Dept Physics Temple Univ Broad & Montgomery Philadelphia PA 19122

HAVEL, HENRY ACKEN, b Palmerton, Pa, May 23, 54; m 80. BIOPHYSICS. *Educ:* Univ Rochester, BS, 76; Univ Minn, PhD(phys chem), 81. *Prof Exp:* Scientist, Control Develop, Upjohn Co, 81-83, res scientist, Spectros Develop, 83-90, qual assurance consult, Specif Develop, 90; RES SCIENTIST, PARENTERAL RES & DEVELOP, ELI LILLY & CO, 90- *Mem:* AAAS; Am Chem Soc; Biophys Soc; Coblentz Soc. *Res:* Biophysical characterization of proteins; protein structure and dynamics; dynamic light scattering studies of protein aggregation; protein folding; antigen-antibody binding reactions. *Mailing Add:* Eli Lilly & Co Lilly Corp Ctr Indianapolis IN 46285-0835

HAVEL, JAMES JOSEPH, b Urbana, Ill, July 22, 47; m 69; c 2. ORGANIC CHEMISTRY. *Educ:* Lewis Univ, BA, 68; Pa State Univ, PhD(chem), 72. *Prof Exp:* Asst prof chem, Rice Univ, 72-77; asst prof chem, Pa State Univ, University Park, 77-80; TECH MGR ANALYTICAL SCI, DOW CHEM CO, 80- *Mem:* Am Chem Soc; Sigma Xi. *Res:* Reactions of atomic species with organic compounds; reactive organic intermediates; organometallic chemistry. *Mailing Add:* 1208 Foxwood Dr Midland MI 48640

HAVEL, RICHARD JOSEPH, b Seattle, Wash, Feb 20, 25; m 47; c 4. MEDICINE. *Educ:* Reed Col, BA, 46; Univ Ore, MS & MD, 49. *Prof Exp:* Asst biochem, Med Sch, Univ Ore, 47-49; from intern to resident physician, New York Hosp, 49-53; instr med, Med Col, Cornell Univ, 52-53; clin assoc, Nat Heart Inst, 53-56; from asst prof to assoc prof, 56-64, PROF MED, SCH MED, UNIV CALIF, SAN FRANCISCO, 64-, DIR, CARDIOVASC RES INST, 73- *Concurrent Pos:* Estab investr, Am Heart Asn, 56-61, chmn coun arteriosclerosis, 77-79; ed, J Lipids Res, 72-75. *Honors & Awards:* Bristol-Myers Squibb Award for Distinguished Achievement Nutrit Res, 89. *Mem:* Nat Acad Sci; Inst Med-Nat Acad Sci; Am Physiol Soc; Am Soc Clin Invest; Am Fedn Clin Res (pres, 64-65); Asn Am Physicians; fel AAAS; Am Physiol Soc. *Res:* Intermediary and lipoprotein metabolism. *Mailing Add:* Cardiovasc Res Inst Univ Calif San Francisco CA 94143-0130

HAVELKA, ULYSSES D, agronomy, for more information see previous edition

HAVEMEYER, RUTH NAOMI, b New York, NY, July 28, 32; m 68. PHARMACY, PROJECT MANAGEMENT. *Educ:* Columbia Univ, BS, 53; Purdue Univ, MS, 55; Univ Wis, PhD(pharm), 60. *Prof Exp:* Res pharmacist, Nepera Chem Co, NY, 55-56; sr res scientist, Squibb Inst Med Res, 60-66; group leader, 66-72, mgr prod develop, 72-80, DIR, RES ADMIN, SYNTEX CORP, 80- *Mem:* Fel AAAS; Am Asn Pharmaceut Sci; Sigma Xi. *Res:* Vitamin stability; chemical kinetics in pharmaceutical systems; biopharmaceutics. *Mailing Add:* Syntex Res Stanford Indust Park 3401 Hillview Ave Palo Alto CA 94304

HAVEN, DEXTER STEARNS, b Chicago, Ill, Nov 2, 18; m 51; c 1. FISH BIOLOGY. *Educ:* RI State Col, BS, 42, MS, 48. *Prof Exp:* From asst biologist to sr biologist, 49-68, assoc prof, 74-77, HEAD DEPT APPL BIOL, VA INST MARINE SCI, 68-, PROF MARINE SCI, COL WILLIAM & MARY, GLOUCESTER POINT, VA, 77- *Mem:* Ecol Soc Am; Nat Shellfisheries Asn. *Res:* Fisheries biology; ecology of oysters and biodeposition. *Mailing Add:* Dept Marine Sci Col William & Mary Williamsburg VA 23185

HAVENER, ROBERT D, b London, Ohio, July 24, 30; c 1. INTERNATIONAL AGRICULTURAL DEVELOPMENT. *Educ:* Ohio State Univ, BS, 52, MSc, 58; Harvard Univ, MPA, 72. *Hon Degrees:* LLD, Univ Ark. *Prof Exp:* Chmn, Coop Exten Serv, Ohio State Univ, 54-61; dist gen mgr, Ohio Farm Bur Coops, 61-63; merchandising mgr, Sugardale Provision Co, 63-64; agr prog officer, Ford Found, 66-78, prog adv, Pakistan, 66-71, dir, Arid Lands Agr Develop Prog, Lebanon, 72-76, prog develop officer, Int Ctr Agr Res in Dry Areas, Mid East, 75-76, prog adv, Asia & Pac Region, 76-78; dir gen, Int Maize & Wheat Improv Ctr, Mexico, 78-85; PRES & CHIEF EXEC OFFICER, WINROCK INT INST AGR DEVELOP, 85- *Concurrent Pos:* Vis guest lectr, Am Univ Beirut, 75-76; mem bd trustees, Int Ctr Agr Res Dry Areas, Beirut, 76-78, Int Rice Res Inst, Manila, 77-78, Int Agr Develop Serv, 80-85 & Honduras Agr Res Found, 84-85; mem bd trustees-ex officio, Int Maize & Wheat Improv Ctr, Mexcio, 78-85; mem Nat Planning Asn Food & Agr Comt. *Honors & Awards:* Distinguished Serv Award, Am Agr Ed Asn. *Mem:* Fel AAAS; Am Agr Econ Asn. *Res:* Agricultural development issues; development of agricultural production programs. *Mailing Add:* Winrock Int Rte 3 Morrilton AR 72110-9537

HAVENER, WILLIAM H, b Portsmouth, Ohio, 1927;; c 7. OPHTHALMOLOGY. *Educ:* Col Wooster, BA, 44; Western Reserve Univ, MD, 48; Univ Mich, MS, 53. *Prof Exp:* Intern med, Univ Hosps, Cleveland, Ohio, 48-50; intern neuropsychiat, Cleveland State Receiving Hosp, 50; resident ophthal, Univ Mich, 51-53, instr, 53-54; asst prof, 54-56, assoc prof & actg chmn dept, 56-59, chmn dept, 59-61, PROF OPHTHAL, OHIO STATE UNIV, 59- *Concurrent Pos:* Chmn dept, Ohio State Univ, 72-88. *Mailing Add:* Dept Ophthal 456 W Tenth Ave Ohio State Univ Columbus OH 43210

HAVENOR, KAY CHARLES, b Denver, Colo, Aug 16, 31; div; c 6. GEOLOGY. *Educ:* Colo Col, BS, 57; Univ Ariz, MS, 58. *Prof Exp:* Geologist, Pure Oil Co, 58-62; CONSULT GEOLOGIST, EXPLOR & FIELD RES, 62- *Mem:* Am Asn Petrol Geol; fel Geol Soc Am; Am Inst Prof Geol. *Res:* Regional and local geology; petroleum and economic geology; groundwater; geohydrology of West Texas, New Mexico and Arizona. *Mailing Add:* 609 E Tierra Berrenda Roswell NM 88201-7865

HAVENS, ABRAM VAUGHN, b Fairfield, Conn, Sept 10, 22; m 43; c 2. METEOROLOGY, CLIMATOLOGY. *Educ:* NY Univ, BSc, 44; Rutgers Univ, MSc, 48. *Prof Exp:* From instr to assoc prof, 48-60, chmn dept, 63-78, PROF METEOROL, RUTGERS UNIV, NEW BRUNSWICK, 60- *Concurrent Pos:* Meteorol consult, Environ Data Serv, Nat Oceanic & Atmospheric Agency, US Dept Com, 56-61, NJ Dept Environ Protection, 65-, Woodward-Envicon, Inc, NJ, 70-73 & Enviroplan, Inc, NJ, 76- *Mem:*

Am Meteorol Soc; Air Pollution Control Asn; Am Geophys Union; fel NY Acad Sci. *Res:* Micrometeorology, hydrometeorology and biometeorology; meteorological aspects of air pollution, environmental impact, water resources and coastal zone management; climate of New Jersey pine barrens. *Mailing Add:* Dept Meteorol Rutgers State Univ New Brunswick NJ 08903

HAVENS, BYRON L(UTHER), electronics; deceased, see previous edition for last biography

HAVENS, JAMES MERYLE, b Olean, NY, Aug 21, 31. METEOROLOGY, CLIMATOLOGY. *Educ:* Middlebury Col, AB, 53; Fla State Univ, MS, 56; Univ London, MSc, 62, PhD(geog), 69. *Prof Exp:* Res asst, Axel Heiberg Island res exped, McGill Univ, 60, 62; res meteorologist, US Army Natick Labs, 63-65; asst prof geog, Univ Western Ont, 66-70; asst prof geog, Univ RI, 70-72, assoc prof, 72-80; RI State climatologist, 78-80; RETIRED. *Mem:* Am Geophys Union; Am Meteorol Soc; Asn Am Geog. *Res:* Heat energy and water budgets at the surface of the earth; climatic change; glacial climatology. *Mailing Add:* PO Box 260 Treadwell NY 13846

HAVENS, JERRY ARNOLD, b Lonoke, Ark, Nov 24, 39; m 61; c 2. FIRE RESEARCH, RISK ASSESSMENT. *Educ:* Univ Ark, BS, 62; Univ Colo, MS, 62; Univ Okla, PhD(chem eng), 69. *Prof Exp:* Proj engr, Procter & Gamble Co, 62-65; res engr, Res Inst, Univ Okla, 69-70; asst prof, 70-75, assoc prof, 75-79, PROF, UNIV ARK, FAYETTEVILLE, 79- *Mem:* Am Soc Testing & Mat; Am Inst Chem Engrs; Am Asn Eng Educ; Sigma Xi. *Res:* Fire and explosion risk assessment; irreversible thermodynamics. *Mailing Add:* 809 Lighton Trl Fayetteville AR 72701-4439

HAVENS, LESTON LAYCOCK, b Brooklyn, NY, July 31, 24; m 73; c 4. PSYCHIATRY. *Educ:* Williams Col, Mass, BA, 47; Cornell Univ, MD, 52; Am Bd Psychiat & Neurol, dipl, 59; Harvard Univ, MA, 90. *Prof Exp:* Intern & asst resident med, New York Hosp, 52-54; resident psychiat, Boston Psychopath Hosp, 54-57, chief serv, 57-58; from instr to assoc clin prof, 58-70, PROF PSYCHIAT, HARVARD MED SCH, 71- *Concurrent Pos:* Examr, Am Bd Psychiat & Neurol, 59; sr psychiatrist, Mass Ment Health Ctr, Boston Psychopath Hosp, 59-64, prin psychiatrist, 64-; chief psychiat consult, Mass Rehab Comn; Carnegie vis prof humanities, Mass Inst Technol, 68; H B Williams travelling prof, Psychiat Soc Australia & NZ, 75; prin psychiatrist, Cambridge Hosp, 84- *Honors & Awards:* Bennett Award, Soc Biol Psychiat, 58; McCurdy Prize, Mass Soc Res Psychiat. *Mem:* Am Psychiat Asn; AAAS. *Res:* Schizophrenic psychopathology; perceptual processes; properties of psychiatric drugs; history of psychiatric ideas; methods of schools. *Mailing Add:* Dept Psych Havard Med Sch 25 Shattuck St Boston MA 02115

HAVENS, WILLIAM WESTERFIELD, JR, b New York, NY, Mar 31, 20; m 44; c 2. NUCLEAR PHYSICS. *Educ:* City Col New York, BS, 39; Columbia Univ, MA, 41, PhD(physics), 46. *Prof Exp:* From asst to assoc prof physics, 40-55, RES SCIENTIST, COLUMBIA UNIV, 41-, PROF PHYSICS, 55-, DIR, DIV NUCLEAR SCI & ENG, 61- *Concurrent Pos:* NSF fel, 59-60; mem nuclear cross sect adv group, AEC. *Mem:* AAAS (vpres, 67-68); fel Am Phys Soc (dep secy, exec secy, 67-); Sigma Xi. *Res:* Electronics; neutron spectroscopy; meson physics; electronuclear machines. *Mailing Add:* 335 E 45th St New York NY 10017

HAVENSTEIN, GERALD B, b Manhattan, Kans, Sept 2, 39; m 63; c 2. ANIMAL GENETICS. *Educ:* Kans State Univ, BS, 61; Univ Wis, MS, 65, PhD(genetics), 66. *Prof Exp:* Instr genetics, Univ Wis, 66-67; geneticist, H & N Inc, 67-86; chmn, Dept Poultry Sci, Ohio State Univ, 86-89; HEAD, DEPT POULTRY SCI, NC STATE UNIV, 89- *Mem:* World Poultry Sci Asn; Poultry Sci Asn; Am Inst Biol Sci; Am Asn Univ Prof. *Res:* Quantitative genetics and animal breeding. *Mailing Add:* Dept Poultry Sci NC State Univ Box 7608 Raleigh NC 27695-7608

HAVER, WILLIAM EMERY, b Somerville, NJ, Sept 19, 42; m 67; c 2. TOPOLOGY. *Educ:* Bates Col, BS, 64; Rutgers Univ, MS, 67; State Univ NY Binghamton, PhD(math), 70. *Prof Exp:* Instr math, Bates Col, 67-68; from asst prof to assoc prof, Univ Tenn, Knoxville, 70-77; ASSOC PROF MATH SCI & CHMN DEPT, VA COMMONWEALTH UNIV, 77- *Concurrent Pos:* Mem staff, Inst Advan Studies, 75-76. *Mem:* Am Math Soc; Math Asn Am. *Res:* infinite dimensional topology with particular emphasis on the study of spaces of homeomorphisms on manifolds, cellular mappings and absolute neighborhood retracts. *Mailing Add:* Dept Math Sci Va Commonwealth Univ 923 W Franklin St Richmond VA 23284

HAVERS, JOHN ALAN, b Broadview, Sask, May 15, 25; m 76; c 2. CIVIL ENGINEERING. *Educ:* Univ Sask, BSCE, 47; Purdue Univ, MSCE, 52, PhD(civil eng, soil mech), 56. *Prof Exp:* Resident engr, Sask Dept Hwy & Transp, Can, 47-49, supv engr, 49-51, prin design engr, 52-53; res engr, Purdue Univ, 53-56; dir eng & assoc partner, Harland Bartholomew & Assocs, Hawaii, 56-62; res engr, IIT Res Inst, 62-64, sr res engr, 64-65; prof civil eng, Purdue Univ, 65-86, head, Div Construct Eng, 80-86; RETIRED. *Concurrent Pos:* Indust consult, 65- *Mem:* Am Soc Civil Engrs; Eng Inst Can. *Res:* Engineering management; construction engineering and management; systems engineering. *Mailing Add:* 1930 Ashbrook Dr W Tucson AZ 85704

HAVERTY, MICHAEL IRVING, b Fresno, Calif, Oct 24, 46; m 68; c 1. INSECT ECOLOGY. *Educ:* Univ Calif, Davis, BS, 68; Univ Ariz, MS, 70, PhD(entom), 74. *Prof Exp:* Res assoc entom, Univ Ariz, 73-75; RES ENTOMOLOGIST, USDA FOREST SERV, 75- *Mem:* Sigma Xi; AAAS; Int Union Study Social Insects; Entom Soc Am; Ecol Soc Am. *Res:* Ecology of subterranean termites, especially the influence of physical and biological parameters on flight, foraging behavior, caste composition and nutrition. *Mailing Add:* 941 Carol Lane USDA Forest Serv POB 245 Lafayette CA 94549

HAVERTZ, DAVID S, b Salt Lake City, Utah, Feb 4, 31; m 57; c 4. MEDICAL ENTOMOLOGY, ECOLOGY. *Educ:* Univ Utah, BS, 56, MS, 57, PhD(invert zool), 62. *Prof Exp:* Parasitologist epidemiol lab, USAF, Tex, 62-65; asst prof, 65-68, PROF ZOOL & DIR CTR ENVIRON STUDIES, WEBER STATE COL, 68- *Mem:* Am Mosquito Control Asn. *Res:* Nematode distribution; mosquito ecology and resistance to insecticides; seasonal distribution of Triatoma Reduviidae; brime fly ecology. *Mailing Add:* Dept Zool Weber State Col 3750 Harrison Blvd Ogden UT 84403

HAVILAND, JAMES WEST, b Glenns Falls, NY, 11. MEDICINE, INTERNAL MEDICINE. *Educ:* Johns Hopkins Univ, MD, 36. *Prof Exp:* Intern med, Johns Hopkins Hosp, 36-37, intern pediat, 37, asst resident pediat, 37-38; asst resident med, New Haven Hosp, 38-39; Attend staff physician, Swed Doctors Hosp, Harilos View Hosp, Univ Hosp, Seattle; clin prof med, Sch Med, Univ Wash, emer assoc dean clin affairs, 42-46; RETIRED. *Mem:* Inst Med-Nat Acad Sci; Am Med Asn. *Mailing Add:* 1229 Madison Suite 860 Seattle WA 98104

HAVILAND, JOHN KENNETH, b Mt Kisco, NY, Jan 19, 21; m 43; c 5. AEROSPACE ENGINEERING. *Educ:* Univ London, BSc, 46; Mass Inst Technol, PhD(astronaut & aeronaut eng), 61. *Prof Exp:* Stress analyst, Bristol Aeroplane Co, Ltd, 46-47; jr res officer, Struct Lab, Nat Res Coun Can, 47-48; group leader struct & adv design, Canadair Ltd, 48-51; group supvr dynamics, Chance-Vought Corp, 51-58; res staff mem, Mass Inst Technol, 59-61, teaching asst aeronaut & astronaut, 60-61; sr tech engr, Astronaut Div, Ling-Temco Vought, 61-64, mgr struct & mat sect, 64-67; PROF AEROSPACE ENG, UNIV VA, 67- *Concurrent Pos:* Dir Div Space Sci, Va Assoc Res Ctr, 67-69. *Mem:* Assoc fel Am Inst Aeronaut & Astronaut. *Res:* Aerospace dynamics and flutter; acoustics; applications of Monte Carlo methods. *Mailing Add:* Mech Eng Thornton Hall Univ Va Charlottesville VA 22903

HAVILAND, MERRILL L, b Ridgeville, Ind, Jan 31, 33; m 52; c 3. FUEL TECHNOLOGY & PETROLEUM ENGINEERING. *Educ:* Purdue Univ, BS, 54, MS, 56. *Prof Exp:* From jr res engr to sr res engr, Gen Motors Corp, 56-76, dept res engr, res BS, 76-86, res assessment mgr, 86-87; AUTOMOTIVE DEVELOP SPECIALIST, EXXON CHEM CO, 87- *Concurrent Pos:* Mem staff, Mech Eng Sch, Wayne State Univ, 58-61. *Honors & Awards:* Henry Ford Mem Award, 61. *Mem:* Soc Automotive Engrs; Soc Tribiol & Lubrication Engrs; Sigma Xi. *Res:* Measurement of flame and end gas temperatures in automotive engines and jet engine combustors; lubrication of sliding systems; power transmission fluids; vehicle fuel economy improvement; marketing automotive lubricant additives. *Mailing Add:* 1432 Bates Birmingham MI 48009

HAVILAND, ROBERT P(AUL), b Cleveland, Ohio, Oct 18, 13; m 37; c 3. ELECTRICAL ENGINEERING, ENVIRONMENTAL SCIENCES. *Educ:* Mo Sch Mines, BS, 39. *Prof Exp:* Res engr, Schlumberger Well Surv Corp, 39-42; engr, Gen Elec Co, 47-72; CHIEF ENGR, MINILAB INSTRUMENTS CO, 73- *Concurrent Pos:* Mem, Test Equip Group, Res & Develop Bd, 51; mem spec comt space technol, Nat Adv Comt Aeronaut, 58; mem comt sci & arts, Franklin Inst, 59-; mem US deleg, Int Radio Consult Comt & Int Telecommun Union; mem commun adv sub-comt, NASA, 68-73; mem, Fed Commun Comn/World Admin Radio Conf Adv Comt, 74. *Mem:* Fel Am Astronaut Soc; fel Inst Elec & Electronics Engrs; assoc fel Am Inst Aeronat & Astronaut; fel Brit Interplanetary Soc. *Res:* Astronautics; space vehicles; communications and electronics; author of 15 books. *Mailing Add:* Minilab Instruments Co PO Box 21086 Daytona Beach FL 32121-1086

HAVIR, EVELYN A, b Scranton, Pa, Sept 5, 33; m 74. PLANT PHYSIOLOGY. *Educ:* Beaver Col, BA, 55; Temple Univ, MA, 57; Cornell Univ, PhD(biochem), 62. *Prof Exp:* Asst biochemist, Pub Health Res Inst, NY, 62-64; RES ASSOC, CONN AGR EXP STA, 64- *Mem:* Am Asn Biol Chemists; Am Soc Plant Physiologists. *Res:* Enzyme mechanisms; enzymology; regulation of photorespiration. *Mailing Add:* Conn Agr Exp Sta PO Box 1106 New Haven CT 06504

HAVIS, JOHN RALPH, b Roaring Springs, Tex, Sept 5, 20; m 49; c 3. HORTICULTURE. *Educ:* Tex Tech Col, BS, 42; Cornell Univ, MS, 47, PhD(veg crops), 49. *Prof Exp:* Asst, NY State Col Agr, Cornell Univ, 46-49; assoc prof hort, Va Polytech Inst, 49-51; horticulturist, Inter-Am Inst Agr Sci, 51-54; head, Waltham Field Sta, Univ Mass, 54-61, prof plant physiol, 61-85; RETIRED. *Mem:* Soc Hort Sci; Int Plant Propagation Soc. *Res:* Physiology of woody ornamentals. *Mailing Add:* 229 Dolomite Dr Colorado Springs CO 80919

HAVLICEK, STEPHEN, b Englewood, NJ, Aug 19, 41; m 61; c 3. ORGANIC & WATER CHEMISTRY. *Educ:* Hope Col, AB, 63; Wayne State Univ, PhD(chem), 67. *Prof Exp:* Fel marine natural prods chem, Univ Hawaii, 67-70; asst prof chem, Bemidji State Col, 70-71; sr res chemist, Campbell Soup Co, 71-76; sr res scientist, Ga Inst Technol, 76-80; tech dir, Brown & Caldwell, 80-84; VPRES, CENT COAST ANALYTICAL SERV, 84- *Mem:* AAAS; Am Chem Soc; Inst Food Technologists; Am Water Works Asn; Air Pollution Control Asn. *Res:* Trace organic chemicals in wastewaters, surface waters and drinking water; structural elucidation by instrumental methods; environmental problems, particularly those related to public health; energy; waste product utilization. *Mailing Add:* Brown/Caldwell PO Box 8045 Walnut Creek CA 94596

HAVLICK, SPENSER WOODWORTH, b Oak Park, Ill, June 21, 35; m 58; c 3. NATURAL HAZARDS & URBANIZATION IMPACTS, URBAN PLANNING BASED ON ECOLOGICAL PRINCIPLES. *Educ:* Beloit Col, BA, 57; Univ Colo, MS, 61; Univ Mich, PhD(environ planning), 67. *Prof Exp:* Asst prof natural resources, Univ Mich, 67-73; dean, Col Environ Design, 85-88, PROF ENVIRON DESIGN, UNIV COLO, 74- *Concurrent Pos:* Ed, J Ekistics, Athens, Greece, 72-; prof geog, Nat Taiwan Univ, Taipei, 90-91. *Mem:* AAAS; Am Planning Asn; Sigma Xi; Inst Behav Sci. *Res:* Urbanization and impacts on natural ecosystems including natural hazard mitigation. *Mailing Add:* C B 314 Univ Colo Boulder CO 80309

HAVNER, KERRY S(HUFORD), b Huntington, WVa, Feb 20, 34; m 54; c 3. SOLID MECHANICS, MATERIALS SCIENCE. *Educ:* Okla State Univ, BS, 55, MS, 56, PhD(appl mech), 59. *Prof Exp:* Assoc engr, Douglas Aircraft Co, Okla, 56; from instr to asst prof civil eng, Okla State Univ, 57-62; sr stress & vibration engr, Aires Mfg Div, Garrett Corp, Ariz, 62-63; sect chief solid mech res, Douglas Missile & Space Systs Div, McDonnell Douglas Corp, Calif, 63-68; assoc prof, 68-75, PROF CIVIL ENG, NC STATE UNIV, 75-, PROF MAT SCI & ENG, 82- *Concurrent Pos:* Am Inst Steel Construct scholar, 51-55; lectr, Univ Southern Calif, 65-68; sr vis, Dept Appl Math & Theoret Physics, Univ Cambridge, 81 & 89; vis fel, Clare Hall, Cambridge, Eng, 81, 89; chmn Comt Inelastic Behav, 81-83, vchmn, 80-81 & 83-84; chmn, Eng Mech div, Am Soc Civil Engrs, 87-88; NSF res grants, theoretical plasticity metal, 71-85, 87-90. *Honors & Awards:* Alco Found Distinguished Eng Res Award, 82. *Mem:* Fel Am Acad Mech; Fel Am Soc Civil Engrs; Am Soc Mech Engrs; Soc Eng Sci; Soc Indust Appl Math. *Res:* Mechanics of materials with emphasis on theoretical plasticity of metals; crystal plasticity. *Mailing Add:* 3331 Thomas Rd Raleigh NC 27607

HAVRAN, WENDY LYNN, b Houston, Tex, Sept 1, 55. BIOLOGY. *Educ:* Duke Univ, BS, 77; Univ Chicago, PhD(immunol), 86. *Prof Exp:* Sr res technician, Dept Med, Div Hemat, Duke Univ Med Ctr, 77-80, res analyst, Dept Microbiol & Immunol, 80-82; POSTDOCTORAL FEL, CANCER RES LAB, UNIV CALIF, BERKELEY, 86- *Concurrent Pos:* NIH grant, 87-89; Lucille P Markey scholar biomed sci, 89-94. *Mem:* Am Asn Immunologists. *Res:* Immunology. *Mailing Add:* Cancer Res Lab Univ Calif 415 LSA Berkeley CA 94720

HAWBECKER, BYRON L, b Freeport, Ill, Oct 2, 35; m 61; c 2. ORGANIC CHEMISTRY. *Educ:* Manchester Col, BA, 57; Univ Ariz, MS, 62; Kent State Univ, PhD(org chem), 68. *Prof Exp:* Res chemist, Ethyl Corp, Mich, 57; A E Staley Mfg Co, Ill, 58, 62-63; asst prof chem, Monmouth Col, 61-62; from instr to assoc prof, 63-78, PROF CHEM & CHMN DEPT, OHIO NORTHERN UNIV, 78- *Concurrent Pos:* Cent Ohio Heart Asn supvr undergrad res students, 65-70, 71-; ed consult, Houghton-Mifflin Co, Boston & W B Saunders Co, Philadelphia, West Publishing Co, Amesbury, Mass; consult, BF Goodrich Tire & Rubber Co; fel, NSF, DuPont & Continental Oil. *Honors & Awards:* Continental Oil Fel; DuPont Fel. *Mem:* Am Chem Soc; Sigma Xi. *Res:* Chemical education; organic synthesis, particularly heterocycles of potential biological activity; molecular rearrangements; organic reaction mechanisms. *Mailing Add:* Dept Chem Ohio Northern Univ Ada OH 45810

HAWES, ROBERT OSCAR, b Bangor, Maine, Jan 4, 35; m 58; c 4. POULTRY BREEDING & AQUACULTURE GENETICS. *Educ:* Univ Maine, BS, 56; Univ Mass, MS, 58; Pa State Univ, PhD(poultry genetics), 62. *Prof Exp:* From asst prof to assoc prof animal sci, MacDonald Col, McGill Univ, 62-71; geneticist & physiologist, Res Dept, Hy-Line Int, Des Moines, 71-78; ASSOC PROF ANIMAL SCI, UNIV MAINE, ORONO, 78- *Mem:* Poultry Sci Asn; Am Genetics Asn; World Poultry Soc Asn; Int Asn Genetics Aquacult. *Res:* Selection for improved growth rate in the American oyster (Crassostrea virginica). *Mailing Add:* Dept Animal & Vet Sci Hitchner Hall Univ Maine Orono ME 04469

HAWIGER, JACK JACEK, b Cracow, Poland, May 30, 38; US citizen; div; c 2. HEMATOLOGY, LABORATORY MEDICINE. *Educ:* Copernicus Sch Med, MD, 62; Nat Inst Hyg, PhD(med microbiol), 67. *Hon Degrees:* MA, Harvard Univ. *Prof Exp:* Instr microbiol med, Nat Inst Hyg, Warsaw, 66-67; res assoc med, Vanderbilt Univ Sch Med, 67-69, asst prof, 69-74, assoc prof, 74-78, assoc prof path, 75-78; prof med, Boston Univ, 78-79; prof path & med, Vanderbilt Univ Sch Med, 79-82; assoc prof, 83-85, PROF MED, HARVARD MED SCH, 85- *Concurrent Pos:* Prin investr, Pub Health Serv res grants, NIH, 73-; res & educ assoc, Nashville Vet Hosp, 74-75; dir, Hemostasis & Thrombosis Lab, Vanderbilt Univ Hosp, 75-78; vis assoc prof, Univ Calif, San Diego, 77; chief, Hemostasis Sect, Boston Vet Admin Med Ctr, 78-79; lectr, Tufts Univ Sch Med, 79-80; dir, Div Exp Med, New Eng Deaconess Hosp, Boston, 83-; mem, Nat Adv Comt, Commonwealth Fund Fel Prog, 83-; mem, Hemat Study Sect, NIH, 83-87; chmn, subcomt on platelets, Am Soc Hemat, 87-88, coun on thrombosis, Am Heart Asn, 87-89. *Honors & Awards:* Sci Coun Award, Polish Ministry Health & Social Welfare, 67; Edward Kowalski Mem Award & lectr, Int Cong Thrombosis & Haemostasis, 85. *Mem:* Asn Am Physicians; Am Soc Clin Invest; Am Asn Pathologists; Am Soc Hematol; Infectious Dis Soc Am; Am Soc Biochem & Molecular Biol. *Res:* Mechanism of formation and prevention of blood clots; regulation of platelet receptors for adhesive molecules; pathobiology of thrombosis; interactions of biological agents and immune complexes with platelets and the vessel wall. *Mailing Add:* New Eng Deaconess Hosp Harvard Med Sch 185 Pilgrim Rd Boston MA 02215

HAWIRKO, ROMA ZENOVEA, b Winnipeg, Man, July 27, 19. MICROBIOLOGY. *Educ:* Univ Man, BSc, 40; McGill Univ, MSc, 49, PhD(bact), 51. *Prof Exp:* From asst prof to prof microbiol, Univ Man, 51-87; RETIRED. *Mem:* AAAS; Am Soc Microbiol; Can Soc Microbiol; World Fedn Cult Collections; Soc Gen Microbiol. *Res:* Cytology and metabolism of spore formation in Clostridium species spore germination; bacteriocins and plasmids of Rhizobium species. *Mailing Add:* 250 Douglas Apt 602 Victoria BC V8V 2P4 Can

HAWK, VIRGIL BROWN, b What Cheer, Iowa, Sept 8, 08; m 36; c 4. AGRONOMY. *Educ:* Iowa State Univ, BS, 33, PhD(crop breeding), 48; State Univ Wash, MS, 38. *Prof Exp:* Surv agent, USDA, Iowa, 29-32, forage crops res agent, Ohio, 32-33, nursery mgr, Soil Conserv Serv, 35-45 & 48-54, plant mat technician, 54-65, mgr plant mat ctr, 65-70; ecol consult, Soil Conserv Serv, 70-75; environ consult, 75-85; RETIRED. *Concurrent Pos:* Prof, Okla State Univ, 47-48. *Mem:* Emer mem Am Soc Agron; Asn Retired Soil Conserv Serv Employees. *Res:* Breeding of forage crops; vegetative control of soil erosion. *Mailing Add:* PO Box 50 No 7-205 Sun City AZ 85372-0050

HAWK, WILLIAM ANDREW, THYROID, BOWEL DISEASES. *Educ:* Univ Pa, MD. *Prof Exp:* Sr consult, Dept Path, Cleveland Clin Found, 83-89; RETIRED. *Mailing Add:* 1918 Princess Ct Naples FL 33942

HAWKE, SCOTT DRANSFIELD, b Gary, Ind, May 30, 42; m 70; c 1. PHYSIOLOGICAL ECOLOGY. *Educ:* San Diego State Univ, BS, 64; Univ BC, MS, 66; Univ Calif, Riverside, PhD(biol), 70. *Prof Exp:* From asst prof to assoc prof, 71-82, PROF BIOL, WILLAMETTE UNIV, 82- *Concurrent Pos:* Instr, Ore State Corrections, 77-; adj fac, Am Col Naturopathic Med, 81- *Mem:* Am Inst Biol Sci; Am Soc Zoologists; AAAS. *Res:* Physiology and morphology of insect sense organs. *Mailing Add:* Dept Biol Willamette Univ Salem OR 97301

HAWKE, W A, b Galt, Ont, July 9, 06; c 1. PEDIATRICS, NEUROLOGY. *Educ:* Univ Toronto, MD, 30; FRCP, 30; FRCP(C), 42. *Prof Exp:* Dir neurol & psychiat, Hosp Sick Children, Toronto, 68, dir develop pediat, 68-72; prof pediat & assoc prof psychiat, 68-72, EMER PROF PEDIAT, UNIV TORONTO, 72- *Mem:* Can Pediat Soc; Can Neurol Soc; Can Psychiat Asn. *Res:* Specific educational disabilities; assessment of potential learning disabilities in preschool children; development of Jamaican children from birth onward; development of the Canadian Indian as it relates to education and ultimate employment. *Mailing Add:* 92 Garfield Ave Toronto ON M4T 1G1 Can

HAWKES, H BOWMAN, physical geography; deceased, see previous edition for last biography

HAWKES, HERBERT EDWIN, JR, b New York, NY, Dec 11, 12; m 39; c 4. ECONOMIC GEOLOGY. *Educ:* Dartmouth Col, AB, 34; Mass Inst Technol, PhD(geol), 40. *Prof Exp:* Field party chief, Hans Lundberg, Toronto, 36-37; from jr geologist to geologist, US Geol Surv, 40-53; lectr geol & geophys, Mass Inst Technol, 53-57; prof mineral explor, Univ Calif, Berkeley, 57-65; ed, Geol Soc Am, 65-66; independent consult geologist, 66-84; RETIRED. *Honors & Awards:* Jackling Award, Soc Mining Engrs, 80. *Mem:* Soc Econ Geol; Asn Explor Geochemists. *Res:* Exploration for mineral deposits. *Mailing Add:* 6120 S Camino Deoste Tucson AZ 85746

HAWKES, JOYCE W, b Portland, Ore. MARINE BIOLOGY. *Educ:* Lewis & Clark Col, BA, 61; Wake Forest Univ, MA, 65; Pa State Univ, PhD(biophys), 71. *Prof Exp:* Asst scientist cell biol, Ore Regional Primate Res Ctr, 71-76; FISHERIES RES BIOLOGIST CELL BIO, NORTHWEST & ALASKA FISHERIES RES CTR, 76- *Concurrent Pos:* NIH fel, Ore Regional Primate Res Ctr, 71-75; affil assoc prof, Univ Wash, 78-; reviewer for grants, NSF, 78-; prin investr contracts, Environ Protection Agency, 77- *Honors & Awards:* Spec Achievement Award, Dept Com, 77. *Mem:* Soc Cell Biol; Soc Invert Path; Am Soc Zool; Int Pigment Cell Soc; fel AAAS. *Res:* Ultrastructure and cell biology of the effects of environmental contaminants on marine organisms. *Mailing Add:* 2020 NE 65th Seattle WA 98115

HAWKES, ROBERT LEWIS, b Moncton, NB, Sept 12, 51; m; c 2. METEOR PHYSICS, ATMOSPHERIC SCIENCE. *Educ:* Mt Allison Univ, BSc, 72, BEd, 78; Univ Western Ont, MSc, 74, PhD(physics), 79. *Prof Exp:* Sci Teacher sci & math, Pub Sch NB, 78-80; asst prof phys educ, 80-86, ASSOC PROF PHYSICS, MT ALLISON UNIV, 86- *Mem:* Am Astron Soc; Can Astron Soc; fel Royal Astron Soc. *Res:* Development of optical observing systems for faint meteors; numerical modeling of ablation of meteoroids in planetary atmospheres. *Mailing Add:* Physics Dept Mt Allison Univ Sackville NB E0A 3C0 Can

HAWKES, STEPHEN J, b London, Eng, May 30, 28; m 65; c 2. ANALYTICAL CHEMISTRY. *Educ:* Univ London, BSc, 53, PhD(phys org chem), 63. *Prof Exp:* Lectr & res assoc chem, Univ Utah, 63-64; from asst prof to assoc prof, Brigham Young Univ, 64-68; assoc prof, 68-77, PROF CHEM, ORE STATE UNIV, 77- *Mem:* Am Chem Soc. *Res:* Variation of gas chromatographic retention indices with column conditions; diffusion in liquid polymers and in gases. *Mailing Add:* Chem Dept Ore State Univ Corvallis OR 97331

HAWKES, SUSAN PATRICIA, b May 23, 46; Brit citizen. CYTOLOGY. *Educ:* Queen Mary Col, London Univ, Eng, BSc, 69, PhD(virol), 72. *Prof Exp:* Res asst, Dept Tissue Immunol, London Hosp Med Col, 73; fel, Univ Calif, Berkeley, 74-76, staff scientist lab chem biodynamics, 76-78; assoc prof, Mich Molecular Inst, 78-85; ASSOC ADJ PROF, UNIV CALIF, SAN FRANCISCO, 85- *Concurrent Pos:* Vis researcher, Tumor Immunol Lab, St Louis Hosp, Paris, 73; fel, Am Asn Univ Women, 74-75 & Elsa Pardee Found, 75-; adj asst prof, Mich State Univ, 78-85. *Mem:* Am Soc Cell Biol. *Res:* Cell-extracellular matrix interactions; oncogenic transformation; metalloproteinases and their inhibitors; antisense oligodeoxynucleotides as potential therapeutic agents. *Mailing Add:* Dept Pharm Univ Calif S-926 Sch Phrm Parnassus Ave San Francisco CA 94143-0446

HAWKES, WAYNE CHRISTIAN, b Oakland, Calif, Apr 23, 51; m 85. PROTEIN CHEMISTRY, NUTRITIONAL BIOCHEMISTRY. *Educ:* Univ Calif, Berkeley, BS, 74; Univ Calif, Davis, PhD(biochem), 80. *Prof Exp:* Prod supvr, Medi Physics, Roche, 72-76; res asst, Dept Food Sci & Technol, Univ Calif, Davis, 76-80, postgrad res biochemist, 80-84; RES CHEMIST, WESTERN HUMAN NUTRIT RES CTR, AGR RES SERV, USDA, 84- *Mem:* AAAS; Am Chem Soc; Am Inst Nutrit; Fedn Am Socs Exp Biol. *Res:* Identify, purify, characterize and deduce the biological functions of human selenocysteine containing proteins; metabolic bases of the human selenium requirement. *Mailing Add:* Western Human Nutrit Res Ctr USDA PO Box 29997 San Francisco CA 94129

HAWKINS, A(LBERT) W(ILLIAM), b Portland, Ore, July 3, 12; m 36; c 2. CHEMICAL ENGINEERING. *Educ:* Univ Wash, Seattle, BS, 35, PhD(chem), 41. *Prof Exp:* Instr chem eng, Kans State Col, 40-41; develop chem engr, E I du Pont de Nemours & Co, Inc, 41-51, tech asst res div, Explosives Dept, Del, 51-61, sect head appl math, Eastern Lab, NJ, 61-68, sr systs analyst, 68-70, sr res chem engr planning & eval, 70-71, sr res chem engr, Polymer Intermediates Dept, 72-80; RETIRED. *Concurrent Pos:* Vis prof, Univ Va, 67-68. *Mem:* Am Chem Soc; Am Inst Chem Engrs. *Res:* Planning and technical and economic evaluation for research guidance. *Mailing Add:* 12 Devon Dr W Orange NJ 07052

HAWKINS, BRUCE, b Rochester, NY, Sept 8, 30; m 57; c 2. HANDICAPPED COMPUTER AIDS. *Educ:* Amherst Col, BA, 51; Princeton Univ, PhD(physics), 54. *Prof Exp:* NSF fel physics, Swiss Fed Inst Technol, 54-55; instr, Yale Univ, 55-57; asst prof, Oberlin Col, 57-61; from asst prof to assoc prof, 61-87, PROF PHYSICS, SMITH COL, 87- *Mem:* AAAS; Am Phys Soc; Am Asn Physics Teachers; Asn Comput Mach. *Res:* Computer data acquisition in the teaching laboratory. *Mailing Add:* Dept Physics Smith Col Northampton MA 01063

HAWKINS, C MORTON, b Memphis, Tex, Jan 17, 38; m 61. BIOSTATISTICS. *Educ:* Univ Okla, BS, 60; Univ Mich, MPH, 62; Tulane Univ, ScD(biostatist), 69. *Prof Exp:* Statistician, Nat Inst Neurol Dis, NIH, 62-66; res assoc biostatist, Sch Pub Health, Tulane Univ, 66-69; from asst prof to assoc prof, 69-77, PROF BIOMET, UNIV TEX SCH PUB HEALTH, HOUSTON, 77-, ASSOC DEAN RES, 87- *Concurrent Pos:* Mem opers comt, Vet Admin, 73- *Honors & Awards:* Albert Lasker Spec Pub Health Award, 80. *Mem:* Am Statist Asn; Biomet Soc; Am Pub Health Asn; Am Heart Asn. *Res:* Hypertension; cooperative clinic trial; propranolol heart trial; hypertension detection and followup program; systolic hypertension in the elderly program. *Mailing Add:* 5232 Arboles Dr Houston TX 77035

HAWKINS, CHARLES EDWARD, b Pontiac, Mich, Nov 29, 41; m 64, 83; c 3. PLASMA PHYSICS, ASTROPHYSICS. *Educ:* Greenville Col, AB, 64; Dartmouth Col, PhD, 71. *Prof Exp:* Asst prof physics, Spring Arbor Col, 69-74, assoc prof, 74-80; assoc prof, 80-86, PROF PHYSICS, NORTHERN KY UNIV, 86-, DIR ACAD COMPUT, 88- *Concurrent Pos:* Mem staff, NASA Lewis Res Ctr, 79-81. *Mem:* AAAS; Am Phys Soc; Am Asn Physics Teachers. *Res:* Plasma waves and instabilities; applications of plasma physics to astrophysical problems; spacecraft propulsion. *Mailing Add:* Acad Comput Northern Ky Univ Highland Heights KY 41076

HAWKINS, DANIEL BALLOU, b Dillon, Mont, Jan 24, 34; m 54, 73; c 3. GEOCHEMISTRY. *Educ:* Mont State Col, BSc, 56, MSc, 57; Pa State Univ, PhD(geochem), 61. *Prof Exp:* Anal chemist, US Geol Surv, 57-58; geochemist waste mgt br, US Atomic Energy Comn, Idaho, 61-67; assoc prof, Univ Alaska, 67-71, head dept geol, 72-75, head solid earth sci prog, 75-76, actg dir, Grad Progs, 85-86, prof geol & chem, 85-90, PROF GEOCHEM, INST WATER RESOURCES, UNIV ALASKA, FAIRBANKS, 71-, EMER PROF GEOL & CHEM, 90-; HYDROLOGIST, US GEOL SURV, 90- *Concurrent Pos:* Consult, J R Simplot Co, Idaho, 62-65, Geochem Serv, 72- & Los Alamos Sci Lab, 75-80. *Mem:* AAAS; Geochem Soc; Mineral Soc Am; Int Asn Math Geol; Int Asn Geochem & Cosmochem. *Res:* Mineral synthesis and hydrothermal alteration of rocks; radioactive waste disposal and ion-exchange studies; geochemical prospecting; environmental geochemistry of arsenic and other trace metals; study of Alaskan zeolites; hydrology and water resources; geochemistry of watersheds; statistical applications to geology. *Mailing Add:* Col Natural Sci Univ Alaska Fairbanks AK 99775

HAWKINS, DAVID GEOFFREY, b St John's, Nfld, June 21, 37. IMMUNOLOGY. *Educ:* Mem Univ Nfld, dipl, 55; Dalhousie Univ, MD, 60. *Prof Exp:* From asst prof to prof med, 68-80, dir, Div Rheumatol, 71-80, sr physician, 74-80, RES ASSOC IMMUNOL, MED CLIN, MONTREAL GEN HOSP, MCGILL UNIV, 68-; PROF & CHMN MED & CONSULT, AFFIL HOSP, MEM UNIV NFLD, 80- *Concurrent Pos:* Royal Col Physicians & Surgeons Can fel internal med, 64; res fel exp path, Scripps Clin & Res Found, 65-67, res fel allergy, immunol & rheumatol, 67-68; Med Res Coun Can scholar, 68-73; consult physician, Montreal Children's Hosp, 80. *Mem:* Am Rheumatism Asn; Am Fedn Clin Res; Can Soc Immunol; Can Fedn Biol Sci; Can Soc Clin Invest. *Res:* Mechanisms of immunologic tissue injury; role of neutrophilic leukocytes in inflammatory response; immune complex diseases; anti-flammatory agents. *Mailing Add:* Dept Med Mem Univ Nfld Fac Med Prince Philip Dr St John's NF A1B 3V6 Can

HAWKINS, DAVID ROGER, b Springfield, Ohio, Apr 10, 41; m 67; c 2. RUMINANT NUTRITION. *Educ:* Ohio State Univ, BSc, 63, MSc, 65; Mich State Univ, PhD(animal husb), 69. *Prof Exp:* From asst prof to assoc prof, 69-78, PROF ANIMAL SCI, MICH STATE UNIV, 78- *Mem:* Am Soc Animal Sci. *Res:* Ruminant nutrition; live animal growth, composition and evaluation. *Mailing Add:* 124 Anthony Hall Mich State Univ East Lansing MI 48824

HAWKINS, GEORGE ELLIOTT, JR, b Princeton, Ky, July 26, 19; m 46; c 1. DAIRY NUTRITION. *Educ:* Western Ky State Univ, BS, 41; Univ Ga, MS, 47; NC State Univ, PhD, 52. *Prof Exp:* Instr animal husb, Univ Ga, 48, asst prof dairy husb, 48-49; asst nutrit, NC State Col, 49-51; from asst prof to prof, 52-82, EMER PROF DAIRY SCI, AUBURN UNIV, 82- *Honors & Awards:* Honors Award, Am Dairy Sci Asn, 81. *Mem:* Am Soc Animal Sci; Am Dairy Sci Asn; Am Inst Nutrition. *Res:* Metabolism of volatile fatty acids by ruminants; role of saliva in ruminant nutrition; trace minerals in calf nutrition; calf management; pasture and harvested forage utilization; lipid metabolism; feeding management of dairy cattle; protein and gossypol in dairy cattle nutrition. *Mailing Add:* Dept Animal & Dairy Sci Auburn Univ Auburn AL 36830

HAWKINS, GERALD STANLEY, b Gt Yarmouth, Eng, Apr 20, 28; US citizen; m 79; c 2. ASTRONOMY. *Educ:* London Univ, BSc, 49; Manchester Univ, PhD(physics), 52, DSc, 63. *Prof Exp:* Astronr, Harvard-Smithsonian Observ, Cambridge, Mass, 54-74; prof astron, Boston Univ, 57-69; dean, Dickinson Col, Pa, 69-71; sci adv & ed, US Info Agency, 75-89; RETIRED. *Mem:* Int Astron Union. *Res:* Astronomy of Stonehenge and other ancient sites; archaeoastronomy; meteors; cosmology; mathematics of crop circle patterns. *Mailing Add:* Cosmos Club 2121 Mass Ave NW Box 1 Washington DC 20008

HAWKINS, GILBERT ALLAN, b Wichita, Kans, Dec 10, 46; m 69. SOLID STATE PHYSICS. *Educ:* Stanford Univ, BS, 69; Mass Inst Technol, PhD(physics), 73. *Prof Exp:* Miller fel basic res, Univ Calif, Berkeley, 73-75; RES PHYSICIST, PHYSICS DIV, EASTMAN KODAK CO, 75- *Mem:* Am Phys Soc; Am Inst Physics. *Res:* Magnetic and optical properties of thin films; superconductivity; dielectric phenomena, electrical conduction in polymer films; cooperative phenomena. *Mailing Add:* Eastman Kodak Co Bldg 81 Kodak Park Rochester NY 14650

HAWKINS, ISAAC KINNEY, b Johnson Co, Ark, Jan 17, 37; c 1. RESTORATIVE DENTISTRY, HUMAN ANATOMY. *Educ:* Univ Tex, BA, 58; Univ Tex Dent Br Houston, DDS, 62; Colo State Univ, PhD(anat), 69. *Prof Exp:* ASST PROF ANAT & ORAL BIOL, MED COL GA, 69-, ASST PROF RESTORATIVE DENT, 73- *Concurrent Pos:* Pvt pract dent, 64-65 & 72- *Mem:* Asn Am Dentists; Aerospace Med Asn; Am Acad Oral Med. *Res:* Inert gases and their morphophysiologic and histochemical effects; in vivo effects on argentum amalgam; in vivo cavosurface margin leakage studies; neuroscience; science education; technical management. *Mailing Add:* PO Box 3225 Augusta GA 30914-3225

HAWKINS, JAMES WILBUR, JR, b Westerly, RI, Sept 10, 32; m 59; c 1. GEOLOGY. *Educ:* Univ Conn, BA, 54; Univ Wash, Seattle, MS, 60, PhD(geol), 63. *Prof Exp:* Jr scientist glaciol, Int Geophys Year Prog, Univ Wash, Seattle, 57-58, asst geol, 59-63, asst prof, 64-65; asst prof, Univ Alaska, 63-64; vis asst prof, 65-66, from asst prof to assoc prof, 66-76, PROF GEOL, SCRIPPS INST OCEANOG, UNIV CALIF, SAN DIEGO, 76- *Concurrent Pos:* NSF instnl grant, Univ Alaska & Geol Soc Am Penrose bequest, 64; NSF grants, 66, 67, 69 & 71-78. *Honors & Awards:* Newcomb-Cleveland Award, AAAS, 80. *Mem:* Fel Geol Soc Am; Am Mineral Soc. *Res:* Petrology, geochemistry and tectonics of island arcs, oceanic ridges and their remnants in fold mountain belts of continental margins. *Mailing Add:* Scripps Inst Oceanog 1145 Sverdrup Hall Box 109 A-020 La Jolla CA 92093

HAWKINS, JOSEPH ELMER, b Waco, Tex, Mar 4, 14; m 39; c 5. SPECIAL SENSES, PHYSIOLOGICAL. *Educ:* Baylor Univ, AB, 33; Oxford Univ, BA, 37, MA, 66, DSc, 79; Harvard Univ, PhD(med sci), 41. *Prof Exp:* Instr, Harvard Med Sch, 41-45; asst investr, Nat Defense Res Comt Proj, Harvard Univ, 41-43, spec res assoc sci, Res & Develop Off Psychoacoust Lab, 43-45; asst prof physiol, Bowman Gray Sch Med, 45-46; res assoc & head, dept neurophysiol, Merck Inst Therapeut Res, 46-56; assoc prof otolaryngol, Sch Med, NY Univ, 56-63; Prof, Physiol Accoust and Chmn Grad Prog, 63-84, EMER PROF OTOLARYNGOL, MED SCH, UNIV MICH, 84- *Concurrent Pos:* USPHS spec fel, Univ Göteborg, 61-63; mem, Study Sect Sensory Dis, NIH, Commun Sci Study Sect 75-79, commun dis res training comt, Nat Inst Neurol Dis & Stroke, 65-69; chercheur e'tranger, Lab Audiol Exp, Univ Bordeaux II, 78; Vis distinguished prof biol, Baylor Univ, 85-; Humboldt Award fel, Univ Würzburg, 91. *Honors & Awards:* Award of Merit Asn Res Otolaryngol, 85. *Mem:* Asn Res Otolaryngol; Am Physiol Soc; assoc mem Am Oral Soc; fel Acoust Soc Am; fel AAAS; Sigma Xi. *Res:* Anatomy, physiology and experimental pathology of auditory and vestibular systems; medical history otology. *Mailing Add:* Kresge Hearing Res Inst Univ Mich Med Sch Ann Arbor MI 48109

HAWKINS, LINDA LOUISE, b Chicago Heights, Ill, Feb 25, 46. PSYCHIATRY, MEDICAL MICROBIOLOGY. *Educ:* Univ Mich, BS, 68; Temple Univ, PhD(microbiol), 74; Vanderbilt Univ, MD, 80. *Prof Exp:* Dir clin microbiol & asst prof path & microbiol, Meharry Med Col, 74-76; PSYCHIATRIST, HARRIET COHN MENT HEALTH CTR, CLARKSVILLE, TENN, 84- *Concurrent Pos:* Asst prof psychiat, Psychiat Dept, State Univ NY, Buffalo, 86. *Mem:* Am Soc Microbiol; Am Psychiat Asn. *Res:* Cellular immune response in histoplasmosis. *Mailing Add:* 400 Forest St Buffalo NY 14213

HAWKINS, MICHAEL JOHN, b Cleveland, Ohio, Mar 24, 47. DRUG RESEARCH. *Educ:* Washington & Lee Univ, BA, 69; Univ Va, MD, 76; Am Bd Internal Med, cert, 79, cert oncol, 81. *Prof Exp:* Intern, Dept Internal Med, Univ Wis Hosp, 76-77, resident, 77-79, fel med oncol, Wis Clin Cancer Ctr, 79-81, res assoc, Dept Human Oncol, 81-82, from instr to asst prof, 82-84; head, Biol Eval Sect, 84-88, CHIEF, INVESTIGATIONAL DRUG BR, CANCER THER EVAL PROG, DIV CANCER TREATMENT, NAT CANCER INST, 88- *Concurrent Pos:* Clin fel, Am Cancer Soc, 79-81. *Mem:* Am Col Physicians; AAAS; Am Soc Clin Oncol. *Res:* Cancer therapy; internal medicine; oncology. *Mailing Add:* Cancer Ther Eval Prog Div Cancer Treatment Nat Cancer Inst Exec Plaza N Rm 715 6130 Exec Blvd Bethesda MD 20892

HAWKINS, MORRIS, JR, b Shreveport, La, Feb 13, 44. SOMATIC CELL GENETICS, NEUROGENETICS. *Educ:* Southern Univ, BS, 65; Howard Univ, MS, 69, PhD(genetics), 71. *Prof Exp:* Tech asst life sci, Nat Res Coun, Nat Acad Sci, 69; lectr zool, Howard Univ, 69-70; prof asst neurobiol, Div Biol & Med Sci, NSF, 71; assoc prof & chmn dept biol, WVa State Col, 71-75; vis assoc prof human genetics, Sch Med, Yale Univ, 75-78; asst to dean, 82-89, ASSOC PROF MICROBIOL, HOWARD UNIV COL MED, 78- *Concurrent Pos:* Sci fac fel, WVa Univ, 74; consult, Inst Serv Educ, 74-75; NSF sci fac fel, 75-76; Nat Inst Gen Med Sci, sci fac fel, 76-78. *Mem:* Soc Neurosci; Sigma Xi; Nat Inst Sci; Am Soc Human Genetics; Am Soc Cell Biol; Am Soc Microbiol. *Res:* Biochemical and genetic analysis of enzymes involved in neurotransmitter metabolism; using techniques of cell culture, biochemistry, immunochemistry, and somatic cell genetics and molecular biology techniques to determine the chromosomal location of human genes; factors controlling the expression and regulation of genes. *Mailing Add:* Dept Microbiol Col Med Howard Univ 054278240n DC 20059

HAWKINS, NEIL MIDDLETON, b Sydney, Australia, Jan 31, 35; m 61; c 2. STRUCTURAL ENGINEERING. *Educ:* Univ Sydney, BS, 55, BE, 57; Univ Ill, MS, 59, PhD(civil eng), 61. *Prof Exp:* Lectr & sr lectr structures, Univ Sydney, Australia, 61-68; prof & chmn civil eng, 78-87, PROF STRUCT ENG, UNIV WASH, 68-, ASSOC DEAN RES, 87- *Concurrent Pos:* Design engr, Portland Cement Asn, 66-67; consult, Boeing Aircraft Co, Post-Tensioning Inst, US Navy & Bethlehem Steel Co; prin investr, NSF, 69- *Honors & Awards:* Edward Noyes Prize, Inst Engrs, Australia, 65; Wason Medal, Am Concrete Inst, 68; Reese Award, Am Concrete Inst, 76, 80 & 91 & Am Soc Civil Engrs, 79; T Y Lin Award, Am Soc Civil Engrs, 89. *Mem:* Am Concrete Inst; Am Soc Civil Engrs; Earthquake Eng Res Inst; Masonry Soc. *Res:* Strength and behavior of reinforced, precast, prestressed and composite steel and concrete structures with particular emphasis on their dynamic loading response. *Mailing Add:* 371 Loew Hall Univ Wash Seattle WA 98195

HAWKINS, RICHARD ALBERT, b Greenwich, Conn, Mar 27, 40; m 64; c 2. PHYSIOLOGY, BIOCHEMISTRY. *Educ:* San Diego State Univ, BS, 63; Harvard Univ, PhD(physiol), 69. *Prof Exp:* Instr biol, San Diego State Univ, 63-64; res asst biochem, Scripps Metab Clin, 64; fel, Oxford Univ, 69-71; staff fel neurochem, NIMH, 71-73, sr staff fel, 73-74; assoc prof physiol, NY Univ, 75-78; chief phys sci, Bur Med Devices, Food & Drug Admin, 74-76; prof physiol, Dept Anesthesia & Physiol & chief, Div Anesthesia & Metab Res, Hershey Med Ctr, Pa State Univ, 77-78; PROF & CHMN PHYSIOL, UNIV HEALTH SCI, CHICAGO MED SCH, 88- *Concurrent Pos:* Parker B Francis investr. *Mem:* Am Physiol Soc; Biochem Soc; Soc Neurosci. *Res:* Cellular physiology-biochemistry with special emphasis on the study of brain metabolism and physiology. *Mailing Add:* Dept Physiol & Biophys Univ Health Sci/Chicago Med Sch 3333 Green Bay Rd North Chicago IL 60064

HAWKINS, RICHARD HOLMES, b St Louis, Mo, Dec 16, 34; m 59; c 3. HYDROLOGY, WATERSHED SCIENCE. *Educ:* Univ Mo-Columbia, BS, 57, BS, 59; Colo State Univ, MS, 61, PhD(watershed mgt), 68. *Prof Exp:* Forester, US Forest Serv, 59-60; assoc engr, State Calif Dept Water Resources, 61-66; instr watershed mgt, Colo State Univ, Ft Collins, 67-68; asst dir, Water Resources Ctr, State Univ NY, 68-71; assoc prof, 71-80, prof watershed sci, Utah State Univ, 80-88; PROF & PROG LEADER WATERSHED SCI, SCH RENEWABLE NATURAL RESOURCES, UNIV ARIZ, 88- *Concurrent Pos:* NDEA fel, Colo State Univ, 66-68; assoc prof forest engr, State Univ NY Col Forestry, Syracuse, 69-71; hydrologist, USDA, Agr Res Serv, Tucson, 77-78, 85; vis prof, Sch Renewable Natural Resources, Univ Ariz, 77-78; hydrologist, USDA, Agr Res Serv, Beltsville, Md, 84-85; distinguished vis scientist, US Environ Protection Agency, Corvallis, Ore, 86-88; prof engr, Calif & NY; prof hydrologist, Am Inst Hydrol. *Mem:* AAAS; Am Soc Civil Eng; Am Water Resources Asn; Am Geophys Union; Asn Univ Watershed Sci; Sigma Xi; Am Inst Hydrol. *Res:* Small watershed hydrology, processes and water quality; watershed modeling and applied hydrology; runoff, erosion; sedimentation; land use effects. *Mailing Add:* SRNR 325 Biol Sci E Univ Ariz Tucson AZ 85721

HAWKINS, RICHARD HORACE, b Columbia, Mo, Nov 6, 22; m 53; c 2. MANAGEMENT RADIOACTIVE WASTE. *Educ:* Univ Mo, BS, 47, MS, 50; Texas A&M Univ, PhD(mineral), 60. *Prof Exp:* Assoc agronomist, Clemson Univ, 50-55; staff chemist, Savannah River Plant, E I du Pont de Nemours & Co, 60-89; RETIRED. *Mem:* Soil Sci Soc Am; Clay Minerals Soc. *Res:* Soil mineralogy applied to improved radioactive waste management. *Mailing Add:* 1027 Clark Rd Aiken SC 29803

HAWKINS, RICHARD L, b Mt Pleasant, Iowa, Nov 21, 54. PROTEIN ISOLATION & CHARACTERIZATION, BEHAVIORAL NEUROPHARMACOLOGY. *Educ:* Reed Col, BA, 76; Univ Calif, San Diego, PhD(physiol & pharmacol), 84. *Prof Exp:* Postdoctoral fel biochem, Health Sci Ctr, Univ Colo, 84-86; postdoctoral fel pharmacol, Univ Calif, San Diego, 86-87; res assoc zool, 87-91, RES ASSOC BIOCHEM, UNIV WASH, 91- *Mem:* Am Soc Cell Biol. *Res:* Extracellular matrix in development and growth; effect of proteases on neuronal outgrowth; neuropeptides and serotonin in behavior. *Mailing Add:* Dept Biochem SJ-7-O Univ Wash Seattle WA 98195

HAWKINS, RICHARD THOMAS, b Provo, Utah, Sept 9, 29; m 55; c 3. ORGANIC CHEMISTRY. *Educ:* Brigham Young Univ, BA, 51; Univ Ill, PhD(org chem), 59. *Prof Exp:* Res chemist, Phillips Petrol Co, 51-54; Nat Petrochem Corp, 57; fel Univ Ill, 59; assoc prof, 59-69, chmn dept, 71-74, PROF CHEM, BRIGHAM YOUNG UNIV, 69- *Concurrent Pos:* Vis prof, Du Pont, 65-66 & Phillips Petrol Co, 74-75 & 80. *Mem:* Am Chem Soc. *Res:* Synthesis of calixarenes and organoboron compounds; polymer chemistry. *Mailing Add:* Dept Chem Brigham Young Univ Provo UT 84602

HAWKINS, ROBERT C, ENGINEERING, GAS TURBINE TECHNOLOGY. *Prof Exp:* Gen mgr, Advan Technol Oper, Lynn Prod Div, 88, CONSULT, GEN ELEC, 88- *Mem:* Nat Acad Eng. *Mailing Add:* 130 Spy Glass Lane Jupiter FL 33477

HAWKINS, ROBERT DRAKE, b Washington, DC, May 2, 46. NEUROPHYSIOLOGY, EXPERIMENTAL PSYCHOLOGY. *Educ:* Stanford Univ, BA, 68; Univ Calif, San Diego, MA, 69, PhD(psychol), 73. *Prof Exp:* Res assoc psychol, Univ Calif, San Diego, 74; res fel, Nat Ctr Sci Res, Gif-sur-Yvette, France, 74-75; res fel, 75-79, staff assoc, 79-82, ASST PROF, CTR NEUROBIOL & BEHAV, COLUMBIA UNIV, 82-; RES SCIENTIST, NY STATE PSYCHIAT INST, 83- *Mem:* AAAS; Soc Neurosci. *Res:* Neural basis of behavior, neural mechanisms of sensitization and classical conditioning in Aplysia californica. *Mailing Add:* Dept Psychiat Columbia Univ 722 W 168th St Box 25 New York NY 10032

HAWKINS, THEO M, b Frederick, Okla, Mar 7, 28; m 48; c 3. PUBLIC HEALTH. *Educ:* Univ Calif, Berkeley, BS, 50, MPH, 58, DPH, 64. *Prof Exp:* Pub health lab dir, Sonoma County Health Dept Calif, 50-55; assoc pub health, Univ Calif, Berkeley, 55-62; microbiologist, Ctr Dis Control, USPHS, 64-69, chief bacteriol training br, 69-86, training consult, 87-89, PROJ MGR, NAT LAB TRAINING NETWORK PROJ, CTR DIS CONTROL, USPHS, 89- *Mem:* Am Soc Microbiol; Am Pub Health Asn. *Res:* Microbiology. *Mailing Add:* Ctr Dis Control Atlanta GA 30333

HAWKINS, THOMAS WILLIAM, JR, b Flushing, NY, Jan 10, 38; m 59; c 3. MATHEMATICS, HISTORY OF SCIENCE. *Educ:* Houghton Col, BA, 59; Univ Rochester, MS, 61; Univ Wis, PhD(hist math), 68. *Prof Exp:* From instr to asst prof math, Swarthmore Col, 66-72; assoc prof, 72-80, PROF MATH, BOSTON UNIV, 80- *Concurrent Pos:* Am Coun Learned Socs fel, Swiss Fed Inst Technol, 69-70; vis assoc prof, Yale Univ, 75; assoc ed, Historia Mathematica, 75-; vis scholar, Harvard Univ, 80-81. *Mem:* Hist Sci Soc; Am Math Soc. *Res:* History of mathematics, especially in 19th century; development of modern integration theory; history of group representation theory; history of matrix theory, lie groups and algebras. *Mailing Add:* Dept Math Boston Univ 111 Cummington St Boston MA 02215

HAWKINS, W(ILLIAM) BRUCE, b Rochester, NY, Sept 8, 30; m 57; c 2. SOFTWARE DEVELOPMENT, VIOLIN ACOUSTICS. *Educ:* Amherst Col, BA, 51; Princeton Univ, PhD(physics), 54. *Prof Exp:* Instr physics, Yale Univ, 55-57; asst prof, Oberlin Col, 57-61; from asst prof to assoc prof, 61-87, PROF PHYSICS, SMITH COL, 87- *Mem:* Am Phys Soc; Am Asn Physics Teachers; AAAS; Asn Comput Mach. *Res:* Data acquisition software for introductory laboratories; software for teaching upper level physics courses; violin acoustics. *Mailing Add:* Dept Physics Smith Col Northampton MA 01063

HAWKINS, WALTER LINCOLN, b Washington, DC, Mar 21, 11; m 39; c 2. POLYMER CHEMISTRY. *Educ:* Rensselaer Polytech Inst, ChE, 32; Howard Univ, MS, 34; McGill Univ, PhD(cellulose chem), 38. *Hon Degrees:* LLD, Montclair State Col, 74; DE, Stevens Inst Technol; DSc, Howard Univ, 86, Kean Col, 86. *Prof Exp:* Res assoc org chem, Howard Univ, 34-35; sessional lectr cellulose & lignin chem, McGill Univ, 38-41; Nat Res Coun fel alkaloid chem, Columbia Univ, 41; mem tech staff, Bell Labs, 42-63, supvr, Applied Res, 63-72, dept head, 72-74, asst dir, Chem Res Lab, 74-75; CONSULT MAT ENG, 75- *Honors & Awards:* Honor Scroll, Am Inst Chemists, 70; Int Award, Soc Plastic Engrs, 86. *Mem:* Nat Acad Eng; fel NY Acad Sci; fel Am Inst Chemist; Am Chem Soc. *Res:* The degradation and stabilization of high polymers; plastics for telecommunications. *Mailing Add:* 26 High St Montclair NJ 07042

HAWKINS, WILLARD ROYCE, b Ft Worth, Tex, July 31, 22; m; c 3. AEROSPACE MEDICINE. *Educ:* Tex Christian Univ, BS, 48; Baylor Univ, MD, 52; Johns Hopkins Univ, MPH, 56; Am Bd Prev Med, dipl. *Prof Exp:* Intern, Hermann Hosp, Houston, Tex, 52-53; USAF, 53-63, comdr, 1500th Med Group & base surgeon, Hickman AFB, Hawaii, 53-55, resident aerospace med, Johns Hopkins Univ, 55-58, chief biodynamics sect, Dept Space Med, USAF Sch Aerospace Med, 58-59, chief aeromed monitor flight opers, Proj Mercury, 59-60, mem staff bioastronaut div, Surgeon Gen, Washington, DC, 60-63; chief environ med br, 63-67, chief med opers div, 67-72, DEP DIR MED OPERS, LIFE SCI DIRECTORATE, MANNED SPACECRAFT CTR, NASA, 72- *Concurrent Pos:* Clin asst prof prev med, Ohio State Univ; med dir, Skylab Prog & Apollo-Soyuz Test Proj. *Honors & Awards:* Exceptional Serv Medal, NASA; Melbourne W Boynton Award, Am Astronaut Soc, 72; Louis H Bauer Founders Award, Aerospace Med Asn, 73. *Mem:* Fel Aerospace Med Asn. *Mailing Add:* 2303 College Green Dr Houston TX 77058

HAWKINS, WILLIAM M(ADISON), JR, b Hannibal, Mo, Aug 7, 09; m 38; c 3. NUCLEAR ENGINEERING, MECHANICAL ENGINEERING. *Educ:* Harvard Univ, AB, 31, MSE, 33; Chrysler Inst Eng, MME, 35. *Prof Exp:* Res engr, Chrysler Corp, 34-42; proj engr, Continental Aviation & Eng Corp, 42-44; exp engr, Packard Motor Car Co, 44-48; res engr, chief engr & mgr res & develop, ACF Indust, Inc, 48-59; mgr res develop, Atomic Energy Div, Allis-Chalmers Mfg Co, 59-63, appln eng, 63-69; vpres, 69-76, PRES, NUTEC, INC, 76- *Mem:* Am Nuclear Soc. *Res:* Peaceful applications of nuclear energy. *Mailing Add:* 9414 Thrush Lane Rockville MD 20854

HAWKINS, WILLIAM MAX, b Brokenhead, Man, May 19, 26; m 55; c 2. ECONOMIC GEOLOGY, MINERALOGY. *Educ:* Queen's Univ, BSc, 60; McGill Univ, MSc, 58, PhD(geol), 60. *Prof Exp:* Instr geol, Lawrence Col, 61-63; from asst prof to assoc prof, 63-69, chmn dept, 64-77, PROF GEOL, STATE UNIV NY COL CORTLAND, 69-, CHMN DEPT, 80- *Concurrent Pos:* State Univ NY Res Found grant-in-aid, 67. *Mem:* Geol Asn Can; Geochem Soc; Nat Asn Geol Teachers; Can Inst Mining & Metall; Spectros Soc Can. *Res:* Spectrochemical studies of wall rock alterations around sulfide ore deposits; significance of trace elements in ore deposits. *Mailing Add:* 538 Madison St Buffalo NY 14211

HAWKINS, WILLIS M(OORE), b Kansas City, Mo, Dec 1, 13; wid; c 3. AERONAUTICAL ENGINEERING. *Educ:* Univ Mich, BS, 37. *Hon Degrees:* DEng, Univ Mich, 65; DSc, Ill Col, 66. *Prof Exp:* Struct designer, fighter aircraft, Lockheed Aircraft Corp, 37-40 & preliminary design investr, 40-41, supvr, wind tunnel test group, 41, mgr, Preliminary Design Dept, 41-47, dir, Preliminary Design Div, 47-52, dir, preliminary design, Missiles & Space Div, 52-53, dir eng, 54-57, asst gen mgr, 57-60, vpres & gen mgr, Space Div, 60-62, vpres sci & eng, 62-63; asst secy, res & develop, US Army, 63-66; corp sr vpres, 66-74, dir, 72-80, corp sr vpres & pres, Calif Co, 76-79, sr vpres aircraft, Lockheed Aircraft Corp, 79-80, SR ADV, LOCKHEED CORP, 80- *Concurrent Pos:* Vis lectr, Univ Calif, Los Angeles; chmn, Aerospace Safety Adv Panel & mem adv coun, NASA; dir, Wackenhut Corp & Avemco. *Honors & Awards:* NASA Distinguished Civilian Serv Medal; Founders Lectr, Nat Acad Eng; Nat Medal Sci, 88; Gardner Lectr, Mass Inst Technol, 91. *Mem:* Nat Acad Eng; fel Royal Aeronaut Soc; hon fel Am Inst Aeronaut & Astronaut. *Res:* Research and development direction; weapon system management; system design and analysis; development. *Mailing Add:* Lockheed Corp 4500 Park Granada Blvd Calabasas CA 91399

HAWKINSON, STUART WINFIELD, b Harris, Minn, Feb 3, 43; m 65; c 2. CRYSTALLOGRAPHY, BIOCHEMISTRY. *Educ:* Wash State Univ, BS, 65; Univ Chicago, PhD(chem physics), 68. *Prof Exp:* USPHS fel, Dept Anat, Univ Chicago, 69; Univ Tenn investr, Biol Div, Oak Ridge Nat Lab, 69-71; asst prof biochem, Univ Tenn, Knoxville, 72-77, assoc prof, 77-80; ENGR SOFTWARE APPLN, FLOATING POINT SYSTS INC, PORTLAND, 80- *Concurrent Pos:* Guest scientist, Oak Ridge Nat Lab, 76- *Mem:* Am Chem Soc; Am Crystallog Asn. *Res:* X-ray crystallographic study of proteins, nucleic acids and interacting drugs. *Mailing Add:* 6695 SW 203rd St Beaverton OR 97997

HAWKRIDGE, FRED MARTIN, b Ft Gordon, Ga, June 20, 44; m 69; c 2. ANALYTICAL CHEMISTRY. *Educ:* Univ Ga, BS, 66; Univ Ky, PhD(chem), 71. *Prof Exp:* Fel anal chem, Case Western Reserve Univ & Ohio State Univ, 71-72; asst prof chem, Univ Southern Miss, 72-76; from asst prof to assoc prof, 76-81, PROF CHEM, VA COMMONWEALTH UNIV, 81- *Concurrent Pos:* Vis prof, Univ Del, 81-82; prog officer, Chem Div, NSF,

89-90. *Mem:* Am Chem Soc; AAAS; Sigma Xi; Electrochem Soc; Soc Electroanal Chem; fel Japan Soc Prom Sci. *Res:* Heterogeneous electron transfer properties of biological molecules; biological molecule electron transfer mechanisms; trace organic analyses. *Mailing Add:* Dept Chem Va Commonwealth Univ Richmond VA 23284

HAWKS, BYRON LOVEJOY, b New York, Nov 23, 09; m 41; c 3. OBSTETRICS & GYNECOLOGY. *Educ:* Duke Univ, BA, 36; NY Univ, MD, 39; Am Bd Obstet & Gynec, dipl, 56. *Prof Exp:* Asst chief obstet & gynec, US Naval Hosps, Portsmouth, Va, 49-50, chief, Port Lyautey, Morocco, 50-52 & US Marine Hosp, Cherry Point, NC, 52-54; asst chief, US Naval Hosps, San Diego, Calif, 54-57, chief, Yokuska, Japan, 57-59 & Great Lakes, Ill, 59-63; asoc prof, 63-71, actg chmn dept, 69-71, prof, 71-78, EMER PROF OBSTET & GYNEC, SCH MED, UNIV ARK, LITTLE ROCK, 78- *Concurrent Pos:* Commandants rep, US Navy, 64-; dir, State Ark Maternity & Infant Care Proj, 65-78; assoc dir, Ark State Dept Health, 78- *Mem:* Fel Am Col Obstet & Gynec. *Res:* Preinvasive carcinoma of cervix uteri; origin, disposal and prognostic potential of amniotic fluid; operative obstetrics. *Mailing Add:* 105 N Plaza Dr Little Rock AR 72205

HAWKS, GEORGE H, III, b Rochester, NY, Sept 5, 38; m 63; c 1. ORGANIC CHEMISTRY. *Educ:* Princeton Univ, AB, 64, MS, 67, PhD(chem), 68. *Prof Exp:* Alexander von Humboldt stipend, Univ Munich, Ger, 68-70; RES ASSOC, EASTMAN KODAK RES LABS, 70- *Mem:* AAAS; Am Chem Soc; Sigma Xi. *Res:* Synthesis and manipulation of reactive intermediates; synthesis of di-substituted sulfur ylids. *Mailing Add:* 3211 E Ave Rochester NY 14618

HAWKS, KEITH HAROLD, b New Castle, Ind, Nov 9, 41; m 62; c 3. MECHANICAL ENGINEERING. *Educ:* Purdue Univ, BS, 64, MS, 66, PhD(mech eng), 69. *Prof Exp:* Asst prof, 69-76, ASSOC PROF MECH ENG, PURDUE UNIV, 76- *Mem:* Am Soc Mech Engrs; Am Soc Eng Educ. *Res:* Power systems; cryogenics; thermal design; mathematical modeling. *Mailing Add:* Sch Mech Eng Purdue Univ West Lafayette IN 47907

HAWKSLEY, OSCAR, b Kingston, NY, June 30, 20; m 42; c 2. ECOLOGY, QUATERNARY PALEONTOLOGY. *Educ:* Principia Col, BS, 42; Cornell Univ, MS, 47, PhD, 50. *Prof Exp:* From asst prof to assoc prof, Cent Mo State Col, 47-56, prof biol, 56-79; RETIRED. *Mem:* fel Nat Speleol Soc (vpres, 57-58). *Res:* Behavior and ecology of vertebrates; speleology; Pleistocene ecology. *Mailing Add:* Rte 1 1104 S Holden Warrensburg MO 64093

HAWKSWORTH, FRANK GOODE, b Fresno, Calif, Apr 30, 26; c 3. FOREST PATHOLOGY. *Educ:* Univ Idaho, BS, 49; Yale Univ, MF, 52, PhD, 58. *Prof Exp:* From jr pathologist to asst pathologist, Div Forest Path, Bur Plant Indust, Soils & Agr Eng, USDA, 49-53; FOREST PATHOLOGIST, US FOREST SERV, 54-; FAC AFFIL, COLO STATE UNIV, 67- *Concurrent Pos:* Forest path, Mex. *Honors & Awards:* Moore Award, Soc Am Foresters. *Mem:* Soc Am Foresters; Am Phytopath Soc; Sigma Xi. *Res:* Dwarf mistletoes; tree diseases of the Rocky Mountains and Southwest. *Mailing Add:* US Forest Serv 240 W Prospect Colo State Univ Ft Collins CO 80526

HAWLEY, JOHN WILLIAM, b Evansville, Ind, Oct 7, 32; m 61; c 3. GEOLOGY, GEOMORPHOLOGY. *Educ:* Hanover Col, BA, 54; Univ Ill, PhD(geol), 62. *Prof Exp:* Part-time geologist, Nev State Dept Conserv & Natural Resources, 59-61; geologist, Ground-Water Br, US Geol Surv, 61; geologist, Soil Conserv Serv, USDA, 62-77; SR ENVIRON GEOLOGIST, NMEX BUR MINES & MINERAL RESOURCES, 77- *Concurrent Pos:* Adj prof geol, NMex Inst Mining & Tech, 80-; collabr, Los Alamos Nat Lab, 85- *Honors & Awards:* Kirk Bryan Award, Geol Soc Am, 83. *Mem:* AAAS; Geol Soc Am; Am Quaternary Assoc; Am Inst Prof Geologists; Soil Sci Soc Am; Soil Conserv Soc Am. *Res:* Geology and hydrogeology of Cenozoic deposits in the Basin and Range and southern Great Plains provinces; geomorphology and environmental geology of New Mexico; quaternary geology of Winnemucca area. *Mailing Add:* NMex Bur of Mines & Mineral Resources Socorro NM 87801

HAWLEY, MARTIN C, b Adrian, Mich, Sept 24, 39; m 59; c 1. CHEMICAL ENGINEERING. *Educ:* Mich State Univ, BS, 61, PhD(chem eng), 64. *Prof Exp:* From asst prof to assoc prof, 64-75, PROF CHEM ENG, MICH STATE UNIV, 75- *Honors & Awards:* Chem Engr Award, Am Inst Chem Engrs, 76. *Mem:* Am Inst Chem Engrs; Am Soc Eng Educ; Am Chem Soc; Sigma Xi. *Res:* Scale up of liquid-solid chromatographic columns; study of axial dispersion in packed beds. *Mailing Add:* 144 Lexington Ave Lansing MI 48823

HAWLEY, MERLE DALE, b Olin, Iowa, Apr 19, 39; m 59; c 3. ANALYTICAL CHEMISTRY. *Educ:* Northern Iowa Univ, BA, 60, MA, 62; Univ Kans, PhD(chem), 65. *Prof Exp:* Res chemist, Dow Chem Co, 65-66; asst prof, 66-70, assoc prof, 70-76, PROF ANALYTICAL CHEM, KANS STATE UNIV, 76- *Mem:* Am Chem Soc; Electrochem Soc; Soc Electroanal Chem. *Res:* Electroanalytical chemistry; electrode processes of organic compounds. *Mailing Add:* Dept Chem Kans State Univ Manhattan KS 66506-3701

HAWLEY, NEWTON SEYMOUR, JR, mathematics; deceased, see previous edition for last biography

HAWLEY, PAUL F(REDERICK), b Cananea, Mex, Apr 26, 10; m 38; c 2. ELECTRONICS. *Educ:* Univ Ariz, BS, 32, EE, 48; Calif Inst Technol, MS, 33, PhD(elec eng), 37. *Prof Exp:* Res engr, Western Geophys Co, Calif, 35-38; res engr, Stanolind Oil & Gas Co, Okla, 38-39, patent adv, 39-41; instr elec eng, Ill Inst Technol, 42-43, head instr eng, sci & mgt war training, 42-44; vpres in charge eng, Consol Eng Corp, Calif, 44-47; patent dir, Amoco Prod Co, 47-75; adj prof physics, Univ Tulsa, 70-87; RETIRED. *Concurrent Pos:* Patent solicitor, Standard Oil Co Ind, Ill, 41-44; mem fac, Okla State Univ, 50-60. *Mem:* Assoc Inst Elec & Electronics Engrs. *Res:* Electrical and seismic geophysical prospecting; design of vibration measuring and recording instruments; cathode-ray oscillography; mass spectrometry; electrical well logging to determine strata, Lenard photography with cathode-ray oscillograph. *Mailing Add:* 5536 E 35th St Tulsa OK 74135

HAWLEY, PHILIP LINES, b Hartford, Conn, Nov 8, 28; m 57; c 3. PHYSIOLOGY. *Educ:* Bowdoin Col, AB, 52; Wesleyan Univ, MA, 54; Univ Iowa, PhD(physiol), 61. *Prof Exp:* Instr biol, Am Univ, Beirut, 56-59; assoc prof, Haile Selassie Univ, 61-64; asst prof, Loyola Univ, Ill, 64-66; from asst prof to assoc prof, 66-77, PROF PHYSIOL, UNIV ILL COL MED, 78-, ASSOC DEAN, GRAD COL, 80- *Concurrent Pos:* Res fel biochem, Inst Living, Conn, 55-56. *Mem:* AAAS; Am Physiol Soc; Sigma Xi. *Res:* Intermediary metabolism; hypothermia; hibernation. *Mailing Add:* Dept Physiol M/C 901 Univ Ill Chicago IL 60680

HAWLEY, ROBERT JOHN, b Astoria, NY, July 13, 40; c 3. CLINICAL MICROBIOLOGY, INFECTIOUS DISEASES. *Educ:* Pa Mil Col, BS, 62; Catholic Univ Am, MS, 66; Col Med & Dent NJ, PhD(microbiol), 74. *Prof Exp:* Diag virol asst, Sch Med, WVa Univ, 62-64; chief, Bacteriol Sect, 249th Gen Hosp, APO San Francisco, 67-70; fel, Sch Med, Georgetown Univ, 74-75, instr to asst prof microbiol, 75-81; dir microbiol immunol, Holy Cross Hosp, Silver Spring, Md, 81-82; instr, Univ Md, College Park, 82-85; US ARMY MED MATERIAL DEVELOP ACTIVE, 85- *Concurrent Pos:* Res microbiologist, NIH & Nat Inst Dent Res, 77-79. *Mem:* Am Soc Microbiol; Sigma Xi; Electron Micros Soc Am; NY Acad Sci. *Res:* Antibiotic resistance mechanisms and recombination in streptococci; rapid identification methods. *Mailing Add:* SGRD-UIZ USAMRIID Ft Detrick Frederick MD 21701-5009

HAWLEY, ROBERT W(ILLIAM), chemical engineering, for more information see previous edition

HAWORTH, D(ONALD) R(OBERT), b Steubenville, Ohio, Jan 26, 28; m 77; c 3. MECHANICAL ENGINEERING. *Educ:* Purdue Univ, BSME, 52, MSME, 55; Okla State Univ, PhD, 61. *Prof Exp:* Helicopter designer, Bell Helicopter Corp, 52-54; instr thermodyn, Purdue Univ, 54-56; propulsion engr, Chance Vought Aircraft, 56-58; asst prof heat transfer, Okla State Univ, 58-61; sr scientist, Ling-Temco-Vought Res Ctr, 61-62; assoc prof heat transfer, Okla State Univ, 62-66; prof & chmn dept mech eng, Univ Nebr, 66-75; assoc exec dir, Am Soc Mech Engrs, 76-90; RETIRED. *Concurrent Pos:* NSF grant, 59; NASA grant, 65. *Mem:* Fel Am Soc Mech Engrs; Am Soc Eng Educ. *Res:* Heat transfer problems related to high velocity and space flight; high temperature; partially ionized gas phenomena associated with space propulsion and power generation. *Mailing Add:* PO Box 612 Sheffield MA 01257-0612

HAWORTH, DANIEL THOMAS, b Fond du Lac, Wis, June 27, 28; m 52; c 3. INORGANIC CHEMISTRY. *Educ:* Univ Wis-Oshkosh, BS, 50; Marquette Univ, MS, 52; St Louis Univ, PhD(chem), 59. *Prof Exp:* Asst chem, Marquette Univ, 50-52, instr, 55; res chemist, Allis-Chalmers Mfg Co, 58-60; from asst prof to assoc prof, 60-68, PROF CHEM, MARQUETTE UNIV, 68- *Concurrent Pos:* Chem, Bur of Ships, 52-53. *Mem:* Am Chem Soc; Sigma Xi; NY Acad Sci. *Res:* Boron hydrides and related compounds; inorganic syntheses; high vacuum techniques; thin-layer and high pressure liquid chromatography; nuclear magnetic resonance spectroscopy. *Mailing Add:* Dept Chem 535 N 14th St Milwaukee WI 53233

HAWORTH, JAMES C, b Gosforth, Eng, May 29, 23; m 51; c 4. PEDIATRICS. *Educ:* Univ Birmingham, MB, ChB, 45, MD, 60. *Prof Exp:* House physician med, Gen Hosp, Birmingham, Eng, 46; house physician pediat, Children's Hosp, 46-47; registr, Alder Hey Hosp, Liverpool, 51-52; sr registr, Children's Hosp, Sheffield, 53-57; pediatrician, Winnipeg Clin, Man, 57-64; from asst prof to assoc prof, 63-70, head dept, 79-85, PROF PEDIAT, UNIV MAN, 70- *Concurrent Pos:* Pediatrician, Winnipeg Children's Hosp, 57- *Mem:* Soc Pediat Res; Am Pediat Soc; Can Pediat Soc; fel Am Acad Pediat; fel Royal Col Physicians Can. *Res:* Metabolic diseases and nutrition in infants and children. *Mailing Add:* Dept Pediat & Child Health Childrens Hosp 840 Sherbrook St Winnipeg MB R3A 1S1 Can

HAWORTH, RICHARD THOMAS, b Wirksworth, Eng, May 24, 44; m 69; c 2. MARINE GEOPHYSICS, DEEP GEOLOGY. *Educ:* Univ Durham, BSc, 65; Cambridge Univ, PhD(geophys), 68. *Prof Exp:* Res scientist, Atlantic Geosci Ctr, Beford Inst Oceonog, Geol Surv Can, 68-83; chief geophysicist, Brit Geol Surv, 83-90; DIR GEN, GEOPHYS & MARINE GEOSCI, GEOL SURV CAN, 90- *Concurrent Pos:* Res Coun Can, 69-74; Brit Nat Comt for Geodesy & Geophys, 84-; mem, Brit Nat Lithosphere Comt, 84-; hon lectr, Nottingham Univ, 85- *Mem:* Am Geol Inst; fel Geol Asn Can; Can Geophys Union; Am Geophys Union; fel Geol Soc London; fel Royal Astron Soc. *Res:* Investigation of techniques and errors in gravity measurements at sea; marine geological and geophysical surveys of the continental shelves and oceanic areas; information storage and retrieval systems; geological and geophysical correlations of Appalachian/Caledonide structures and their offshore subsurface extension. *Mailing Add:* Geol Surv Can 601 Booth St Ottawa ON K1A 0E8 Can

HAWORTH, WILLIAM LANCELOT, b Rossendale, Eng, May 18, 41; m 68; c 2. MATERIALS SCIENCE, METALLURGY. *Educ:* Univ Liverpool, BSc, 62; Univ Alta, MSc, 64; Yale Univ, MS, 66, MPhil, 68, PhD(mat sci), 69. *Prof Exp:* Res assoc metall, Univ Ill, 69-71; from asst prof to prof metall eng, Wayne State Univ, 72-85; prog dir, Mat Res Group, 84-85, PROG DIR, MAT RES LAB, NSF, 85- *Concurrent Pos:* Mem, Res Inst Eng Sci, Wayne State Univ, 75-85, assoc chmn dept metall eng, 75-85; sr staff mem, Cent Solar Energy Res Corp, Mich Energy & Resource Res Asn, 77-78, vpres, 78-79; metall consult; assoc prog dir, Metall Prog, NSF, 82-83. *Mem:* AAAS; Am Inst Metall Eng; Sigma Xi; Am Phys Soc; Mat Res Soc. *Res:* Plastic deformation, fatigue and fracture of materials; nondestructive evaluation. *Mailing Add:* Div Mat Res NSF Washington DC 20550

HAWRYLEWICZ, ERVIN J, b Chicago, Ill. NEUROENDOCRINOLOGY, NUTRITION. *Educ:* Ill Inst Technol, BS, 50, MS, 53, PhD(biochem), 60. *Prof Exp:* Asst biochemist, Armour Pharmaceut Labs, 50-53; from asst chemist to assoc chemist, IIT Res Inst, 53-58, res biochemist, 58-61, mgr life sci div, 61-63, asst dir div, 63-69; CHMN DEPT RES & DIR RES, MERCY HOSP & MED CTR, 69-; PROF BIOCHEM, COL BASIC MED SCI, UNIV ILL, 75- *Concurrent Pos:* Mem res comt & chmn, Mercy Hosp Med Ctr, 69-; mem, Am Cancer Soc Res Comt, 73-78; chmn, Cancer Prevention Comt, vpres, Am Cancer Soc, Ill Div, 84-; mem, Inst Rev Bd, Ill Cancer Coun, 80- *Mem:* AAAS; Am Soc Neurochem; Sigma Xi; Am Inst Nutrit; Soc Neurosci. *Res:* Neonatal development with emphasis on brain as affected by malnutrition; enzyme alterations in breast tissue as a function of tumor growth; effect of diet on neuroendocrine system and breast tumors; mechanisms of disease process at subcellular level. *Mailing Add:* Dept Res Mercy Hosp & Med Ctr Stevenson Expwy King Dr Chicago IL 60616

HAWRYLKO, EUGENIA ANNA, b New York, NY, Feb 7, 42; m 74; c 2. CELLULAR IMMUNOLOGY, TUMOR IMMUNOLOGY. *Educ:* Vassar Col, AB, 62; NY Univ, MD, 66; Am Bd Allergy & Immunol, dipl, 79. *Prof Exp:* Intern & resident internal med, Bellevue Hosp, 66-68; res fel med, Harvard Med Sch, Peter Bent Brigham Hosp, 68-69; res fel & asst mem cellular immunol, Trudeau Inst, 69-74; res assoc cellular immunobiol, Sloan-Kettering Inst, 74-75, assoc, 75-81; CHIEF, IMMUNOL LABS, DEPT ALLERGY & IMMUNOL, LONG ISLAND COL HOSP, 82- *Concurrent Pos:* Fac res award, Am Cancer Soc, 76-81; clin fel allergy & immunol, NY Hosp Cornell Med Ctr, 78-80; mem, Am Cancer Soc Adv Comt, 78-82. *Mem:* Am Asn Immunologists; Am Asn Cancer Res; Am Acad Allergy & Immunol; AAAS; NY Acad Sci; Am Col Allergists; Am Soc Microbiol. *Res:* Cell mediated mechanisms in antitumor immunity and delayed type hypersensitivity and allergy. *Mailing Add:* Dept Med NY Univ Sch Med 160 E 32nd St New York NY 10025

HAWRYLUK, ANDREW MICHAEL, b Brooklyn, NY, May 24, 54; m 77; c 2. ELECTRICAL ENGINEERING. *Educ:* Mass Inst Technol, BS, 76, MS, 77, PhD(elec eng), 81. *Prof Exp:* Postdoctoral, Univ Fla, 81-83; scientist 83-89, GROUP LEADER, LAWRENCE LIVERMORE NAT LAB, 89- *Honors & Awards:* I R 100 Award, 86, 87, 88, 89, 90; Award of Excellence, Dept Energy, 86. *Res:* Microfabrication technology; semiconductor devices; semiconductor fabrication; artificial human vision. *Mailing Add:* Lawrence Livermore Lab PO Box 5508 L-487 Livermore CA 94550

HAWRYLUK, RICHARD JANUSZ, b Mansfield, Eng, June 7, 50; US citizen; m 76; c 2. PLASMA PHYSICS, SOLID STATE PHYSICS. *Educ:* Mass Inst Technol, SB & MS, 72, PhD(physics), 74. *Prof Exp:* Res assoc plasma physics, 74-75, staff physicist, 75-79, res physicist, 79-83, head Tokamak Fusion Test Reactor Oper Div, 86, DEP PROJ MGR, 90-, PRIN RES PHYSICIST, PLASMA PHYSICS LAB, PRINCETON UNIV, 83- *Concurrent Pos:* Consult, Lincoln Lab, Mass Inst Technol, 70-73 & 80. *Honors & Awards:* Excellence Plasma Physics Award, Am Phys Soc. *Mem:* Fel Am Phys Soc. *Res:* Plasma confinement and heating in toroidal confinement devices as part of the fusion program; microcircuit fabrication, electron beam lithography. *Mailing Add:* Plasma Physics Lab PO Box 451 Princeton NJ 08540

HAWRYSH, ZENIA JEAN, b Edmonton, Alta, Nov 4, 38. FOOD SCIENCE. *Educ:* Univ Alta, BSc, 59; Mich State Univ, MS, 60, PhD(food sci), 70. *Prof Exp:* From lectr to asst prof, Sch Home Econ, Univ Alta, 60-66; fel food sci, Mich State Univ, 66-68, res asst, 68-70; assoc prof, 71-76, PROF, DEPT FOODS & NUTRIT, FAC HOME ECON, UNIV ALTA, 76-, CHMN, 80- *Concurrent Pos:* Mem, Can Comt Meats, 76-77; Can sensory adv comt, Int Stds Orgn. *Mem:* Can Inst Food Sci & Technol; Inst Food Technologists; Am Dietetic Asn; Am Oil Chem Soc; Am Meat Sci Asn; Chemoreceptor Sci Asn. *Res:* Subjective (sensory) and objective (instrumental, chemical) evaluations of the eating quality characteristics of beef, pork, canola oils and canola oil products; foods for food service operations and for special diets; sensory methodology; taste perception measurements. *Mailing Add:* Foods & Nutrit Dept Univ Alta Edmonton AB T6G 2M8 Can

HAWS, BYRON AUSTIN, b Vernal, Utah, July 1, 21; m 45; c 4. AGRICULTURE, ECONOMIC ENTOMOLOGY, ENTOMOLOGY. *Educ:* Utah State Agr Col, BS, 48, MS, 49; Iowa State Col, PhD(zool, entom), 55. *Prof Exp:* Res fel entom, Univ Minn, 52-57; assoc prof, 57-71, dir int progs, 65-68, PROF ENTOM, UTAH STATE UNIV, 71-, ASST TO PRES, LATIN AM AFFAIRS, 64- *Concurrent Pos:* Chief party, Utah State Univ Tech Assistance Team, Bolivia, 68-71; consult, Honduran export stas, cocaine studies in Bolivia, African bees & bee industs in Ecuador. *Mem:* Entom Soc Am; Soc Range Mgt; Am Registry Prof Entomologists; Sigma Xi. *Res:* Range entomology; legume seed production, pollination and injurious insects. *Mailing Add:* Dept Biol Utah State Univ Logan UT 84322

HAWTHORNE, BETTY EILEEN, b Seattle, Wash, Nov 22, 20. NUTRITION, EDUCATION ADMINISTRATION. *Educ:* Univ Wash, Seattle, BS, 41, MS, 44; Mich State Col, PhD(nutrit), 54. *Prof Exp:* Asst, Univ Wash, Seattle, 41-43; field rep nutrit, Am Red Cross, Pac Area, 43-44; from instr to prof foods & nutrit, Ore State Univ, 46-83, dean home econ, 65-83; RETIRED. *Concurrent Pos:* Chmn, Asn Admin Home Econ, Nat Asn State Univ & Land Grant Cols, 79-80; mem corp board, Pacificorp, Portland, Ore; Curtice-Burns Foods, Rochester, NY. *Mem:* AAAS; Am Inst Nutrit; Am Home Econ Asn; Am Dietetic Asn. *Res:* Ascorbic acid metabolism; metabolic interrelationships in human nutrition; lipid metabolism; nutrition in aging. *Mailing Add:* 144 NW 29th St Corvallis OR 97330

HAWTHORNE, DONALD CLAIR, b Olympia, Wash, Jan 23, 26. MICROBIOLOGY, GENETICS. *Educ:* Univ Wash, Seattle, BS, 50, MS, 53, PhD(microbiol), 55. *Prof Exp:* Asst microbiol, Univ Wash, 52-55; NSF res fel microbial genetics, Calif Inst Technol, 56; USPHS res fel, Gen Physiol Lab, Nat Ctr Sci Res, Ministry Ed, France, 57-58; res instr, 59-65, res assoc prof, 65-70, PROF GENETICS, UNIV WASH, 70- *Concurrent Pos:* USPHS spec fel, France, 68; assoc prof, Univ Paris VI, 79. *Mem:* Genetics Soc Am. *Res:* Genetics of yeast; tetrad analysis; gene interactions; mutant characterization. *Mailing Add:* 2321 NE 55th St Seattle WA 98105

HAWTHORNE, EDWARD I, b Poland, Sept 25, 21; nat US; m 47; c 2. ELECTRICAL ENGINEERING. *Educ:* Cooper Union, BS, 43; Univ Pa, MS, 48, PhD(elec eng), 53. *Prof Exp:* Elec engr, Victor Div, Radio Corp Am, NJ, 43-47; assoc prof elec eng, Univ Pa, 47-58; mgr electronics & elec mech dept, Res & Adv Develop Div, Avco Corp, 58-61; mgr defense systs lab, Space & Commun Group, El Segundo, 61-68, sr spacecraft mgr, 68-70, prog mgr space systs, 70-75, CHIEF SCIENTIST SPACE TECHNOL, HUGHES AIRCRAFT CO, EL SEGUNDO, 75- *Mem:* Am Soc Eng Educ; Inst Elec & Electronics Engrs. *Res:* Electromagnetic field theory, especially electromechanical energy conversion; electromagnetic shielding; physical and space electronics; missiles and weapons systems; spacecraft systems; ion propulsion. *Mailing Add:* 42121 Village 42 Camarillo CA 93012

HAWTHORNE, FRANK CHRISTOPHER, b Bristol, Eng, Jan 8, 46; m 70; c 2. CRYSTALLOGRAPHY, MINERALOGY. *Educ:* Univ London, BSc Hons, 68, Royal Sch Mines, ARSM, 68; McMaster Univ, PhD(geol), 73. *Prof Exp:* Fel, 73-75, RES ASSOC GEOL, UNIV MAN, 75-, PROF, 84- *Honors & Awards:* Hawley Medal, Mineral Asn Can, 84; Rh Inst Award, 84. *Mem:* Mineral Asn Can; Mineral Soc Am; Am Geophys Union; Am Crystallog Asn; Mineral Soc (London); fel Royal Soc Can. *Res:* Structure and chemistry of amphiboles and pyroxenes, hierarchical architecture of inorganic crystal structures; spectroscopy of minerals. *Mailing Add:* Dept Geol Sci Univ Man Winnipeg MB R3T 2N2 Can

HAWTHORNE, MARION FREDERICK, b Ft Scott, Kans, Aug 24, 28; m 51, 77; c 2. ORGANIC CHEMISTRY, ORGANOMETALLIC CHEMISTRY. *Educ:* Pomona Col, BA, 49; Univ Calif, Los Angeles, PhD(chem), 53. *Hon Degrees:* DSc, Pomona Col, 74. *Prof Exp:* Res assoc, Iowa State Col, 53-54; sr res chemist, Redstone Arsenal Div, Rohm and Haas Co, Ala, 54-55, head metallo-org chem group, 56-60, lab head, Pa, 61; prof chem, Univ Calif, Riverside, 62-68; PROF CHEM, UNIV CALIF, LOS ANGELES, 69- *Concurrent Pos:* Vis lectr, Harvard Univ, 60, Sloan fel, 66, vis prof, 68; spec lectr, Queen Mary Col, Univ London, 63; Corn Prod lectr, Pa State Univ, 68; Wooley Lectr, Ga Inst Technol, 69; 3M Co lectr, Univ Minn, 69; ed, Inorg Chem J, Am Chem Soc, 69-; consult, Union Carbide Corp, 70-76, McMillan Sci Assoc, 72-, Adv Res Proj Agency, US Dept Defense, 74-76, Intelcom Radio Technol Corp, 75-76 & Vacuum Atmospheres Corp, 75-77; mem mat res coun, Dept Defense, 73-77; vis prof, Univ Tex, Austin, 74; Reilly lectr, Notre Dame, 74; distinguished scientist lectureship, Mich State Univ, 74; mem chem, Kaysen Panel, Nat Acad Sci, 75-77; consult, Army Sci Adv Panel, 75-78; rev panel, Stanford Synchrotron Radiation Proj, 75-77 & Nat Acad Sci US-Soviet Sci Exchange, 75-78; mem, US Air Force Sci Adv Bd, 79-86, bd Army Sci & Technol, Nat Res Coun, 86-; consult, Callery Chem Co, 85-90, Xoma Inc, 87-; chmn, Gordon Res Conf Inorg Chem, 74; Humboldt Sr Scientist Res Award, 90- *Honors & Awards:* McCoy Award, 73; Award Inorg Chem, Am Chem Soc, 74, Tolman Medal, 86, Distinguished Achievements Boron Sci Award, 88, Distinguished Serv to Inorg Chem, 88; Frontiers Sci lectr, Wayne State Univ, 74; Castle lectr chem, Univ S Fla, 86. *Mem:* Nat Acad Sci; fel AAAS; Am Acad Arts & Sci; NY Acad Sci; Am Chem Soc; Chem Soc, London. *Res:* Organometallic and inorganic chemistry; synthesis, structural and mechanism studies with boron hydrides, carboranes, and their transition metal derivatives; homogeneous catalysis; application of organometallic chemistry to immunology; cancer diagnosis and therapy. *Mailing Add:* Dept Chem & Biochem Univ Calif Los Angeles CA 90024

HAWTHORNE, ROBERT MONTGOMERY, JR, b Akron, Ohio, Nov 1, 29; m 55; c 4. ORGANIC CHEMISTRY, HISTORY OF CHEMISTRY. *Educ:* Columbia Univ, BS, 56; Rutgers Univ, PhD(org chem), 63; Univ Notre Dame, MA, 78. *Prof Exp:* Asst radiation physicist, Bellevue Hosp, New York, 54-56; chemist, Nat Starch & Chem Co, 56-59 & Am Cyanamid Co, NJ, 59-60; teaching asst chem, Rutgers Univ, 60-61, instr, 61-63; teacher, Marlboro Col, 63-68; from asst prof to assoc prof chem, NCent Campus, Purdue Univ, 68-81; chem consult, Manley Bros, 79-80; chmn, environ sci ctr, 81-83, dean fac, 82-85, PROF, UNITY COL, 86- *Concurrent Pos:* Lectr, Bloomfield Col, 62-63; secy, Maine Sect, Am Chem Soc, 82-84, chmn-elect, 88-89, chmn, 90-91, chmn/secy-treas, Div Hist of Chem, 76-80. *Mem:* AAAS; Am Chem Soc (secy-treas, 76-79); Hist Sci Soc; Soc Hist Alchemy & Chem; fel Am Inst Chemists. *Res:* Chemical history, particularly the Nineteenth Century. *Mailing Add:* Unity Col Unity ME 04988

HAWTHORNE, VICTOR MORRISON, b Glasgow, Scotland, June 19, 21; m 48; c 3. PSYCHIATRIC EPIDEMIOLOGY. *Educ:* Univ Glasgow, MD, 62; FRCP(G), 73. *Prof Exp:* Sr lectr, 66-78, SR RES FEL EPIDEMIOL, UNIV GLASGOW, 78- *Concurrent Pos:* Consult chest med, Western Regional Hosp Bd, 66-78; chmn, EDC study sect, NIH, 80-84; chmn kidney dis comt, Dept Health, State of Mich, 80- & Worksite Health Reserve, 85-; Victor M Hawthorne res fel award prog; prof epidemiol & chmn dept, Sch Pub Health, Univ Mich, 78-87. *Mem:* Fel Am Col Epidemiol; Soc Epidemiol Res; Am Epidemiol Asn. *Res:* Epidemiological studies in west of Scotland and Tecumseh, Michigan of cardio-respiratory, diabetic, hypertensive, osteoarthritic & nutritional diseases. *Mailing Add:* Dept Epidemiol Sch Pub Health Univ Mich 109 Observ St Ann Arbor MI 48109-2029

HAWTON, LARRY DAVID, physical chemistry; deceased, see previous edition for last biography

HAWTON, MARGARET H, b Jan 26, 42; div. BIOLOGICAL PHYSICS, CHEMICAL PHYSICS. *Educ:* Univ New Brunswick, HBSc, 64, MSc, 66; Univ Waterloo, PhD (physics), 81. *Prof Exp:* ASSOC PROF, LAKEHEED UNIV, 66- *Mem:* Am Phys Soc. *Res:* Applications for biological materials, rocks and semiconductora; trasnsport in and polarization of heterogeneous materials and interfaces. *Mailing Add:* Lakehead Univ Postal Station P Thunder Bay ON D7B 5E1 Can

HAWXBY, KEITH WILLIAM, b Nemaha, Nebr, Nov 6, 40; m 61; c 4. PLANT PHYSIOLOGY, AQUATIC BIOLOGY. *Educ:* Peru State Col, BS, 61; Okla Univ, MNS, 66; Okla State Univ, PhD(bot), 74. *Prof Exp:* Sci instr, Tecumseh High Sch, Tecumseh, Nebr, 61-63 & Pawnee High Sch, Pawnee City, Nebr, 63-65; asst prof biol, Robert Morris Col, Carthage, Ill, 65-74; RES ASST PROF, LANGSTON UNIV, 74- *Mem:* Am Soc Plant Physiologists; Weed Sci Soc Am. *Res:* Accumulation and degradation of various herbicides by algae species common in Oklahoma. *Mailing Add:* Rte 2 Box 350 Perkins OK 74059

HAXBY, B(ERNARD) V(AN LOAN), b Minneapolis, Minn, Nov 30, 21; m 44; c 4. ELECTRICAL ENGINEERING. *Educ:* Univ Minn, Minneapolis, BEE, 48, MS, 49, PhD(elec eng), 57. *Prof Exp:* From instr to asst prof, 49-60, assoc prof elec eng & assoc head dept, 60-86, prof elec eng & assoc head, 86-88, EMER PROF ELEC ENG, UNIV MINN, MINNEAPOLIS, 88- *Mem:* AAAS; Inst Elec & Electronics Engrs; Sigma Xi. *Res:* Physical electronics; microwave circuits. *Mailing Add:* Dept Elec Eng Univ Minn 200 Union St SE Minneapolis MN 55455

HAXHIU, MUSA A, b Peje, Yugoslavia, Mar 15, 39; m 64; c 3. MEDICINE, ANIMAL PHYSIOLOGY. *Educ:* Univ Belgrade Med Sch, MD, 63; Univ Zagreb, PhD(med), 73. *Prof Exp:* PROF PHYSIOL, MED SCH, UNIV PRISHTINA, 75- *Concurrent Pos:* Dean, Univ Prishtina Med Sch, 73-75; pres, Yugoslav Physiol Soc, 78-82; vis prof, Case Western Reserve Univ, 82- *Mem:* NY Acad Sci; Yugoslav Physiol Soc; Am Thoracic Soc; Am Asn Advan Sci. *Res:* Integrative aspects of the neuronal control of upper airway and chest wall muscles activity in health and disease; brain peptides and peptidergic pathways controlling airway tone and secretion; air pollution and chronic pulmonary diseases. *Mailing Add:* Pulmonary Div Dept Med Case Western Reserve Univ Cleveland OH 44106

HAXO, FRANCIS THEODORE, b Grand Forks, NDak, Mar 9, 21; m 47, 61; c 5. OCEANOGRAPHY. *Educ:* Univ NDak, BA, 41; Stanford Univ, PhD(biol), 47. *Prof Exp:* Asst biol, Univ NDak, 40-41; asst biol, Stanford Univ, 41-44, actg instr, 43, res assoc, Hopkins Marine Sta, 46-47; asst, Calif Inst Technol, 46; instr plant physiol, Johns Hopkins Univ, 47-48, asst prof, 49-52; from asst prof to assoc prof, Scripps Inst Oceanog, Univ Calif, San Diego, 52-63, chmn, Dept Marine Biol, 60-65, chmn marine biol res div, 60-65 & 70-77, prof marine biol, 63-68, EMER PROF MARINE BIOL, SCRIPPS INST OCEANOG, UNIV CALIF, SAN DIEGO, 88- *Concurrent Pos:* Instr, Marine Biol Lab, Woods Hole, 49-53 & 70; mem vis fac, Univ Calif, Berkeley, 57 & Univ Wash, 63; US Nat Comn Int Biol Prog, 66; pres's sci adv comt, Oceanog & Marine Biol, 66. *Mem:* Fel AAAS; Phycol Soc Am; Int Phycol Soc; Am Soc Photobiologists. *Res:* Algal physiology; photosynthesis; plant pigments. *Mailing Add:* Scripps Inst Oceanog A002 Univ Calif at San Diego La Jolla CA 92093

HAXO, HENRY EMILE, JR, b Missoula, Mont, Aug 20, 18; m 44; c 4. RUBBER CHEMISTRY, PLASTICS CHEMISTRY. *Educ:* Univ NDak, BA, 38; Yale Univ, PhD(phys chem), 41. *Prof Exp:* Res chemist, Res Ctr, US Rubber Co, 41-42 & 46-67 & Uniroyal Inc, 67-72; assoc dir res, Mat Res & Develop Div, Woodward-Lundgren Assocs, 72-73; PRES, MATRECON, INC, 73- *Concurrent Pos:* Consult, Army-Navy Munitions Bd, 51. *Mem:* AAAS; Am Chem Soc; NY Acad Sci; Soc Plastics Engrs; Am Soc Testing & Mat. *Res:* Hazardous waste management; asphalt technology; molecular spectroscopy; polymerization, dienes and vinyls; evaluation and physics of synthetic rubbers; polymer-to-polymer adhesion; filler reinforcement of rubber; permeability of polymers to gases and liquids. *Mailing Add:* Matrecon Inc 815 Altantic Ave Alameda CA 94501

HAXTON, WICK CHRISTOPHER, b Santa Cruz, Calif, Aug 21, 49; m 75; c 2. THEORETICAL PHYSICS. *Educ:* Univ Calif, Santa Cruz, BA, 71; Stanford Univ, BS, 73, PhD(physics), 75. *Prof Exp:* Res assoc, Universitat Mainz, 75-77; res assoc, 77-79, Jr Oppenheimer fel, Los Alamos Nat Lab, 79-; asst prof physics, Purdue Univ, 80-; AT DEPT PHYSICS, UNIV WASH, SEATTLE. *Mem:* Am Phys Soc. *Res:* Nuclear and particle physics; theory of weak and electromagnetic interactions; astrophysics. *Mailing Add:* Dept Physics Univ Wash Seattle WA 98195

HAY, ALLAN STUART, b Edmonton, Alta, Can, July 23, 29; m 56; c 4. ORGANIC CHEMISTRY, POLYMER CHEMISTRY. *Educ:* Univ Alta, BSc, 50, MSc, 52; Univ Ill, PhD(chem), 55. *Hon Degrees:* DSc, Univ Alta, 87. *Prof Exp:* Instr, Univ Alta, 50-52; org chemist, 55-68, mgr chem lab, Res & Develop Ctr, 68-80, RES & DEVELOP MGR, CHEM LABS, GEN ELEC CO, 80-; PROF POLYMER CHEM, MCGILL UNIV, 87- *Concurrent Pos:* Adj prof, Polymer Sci & Eng Dept, Univ Mass, 75- *Honors & Awards:* Int Gold Medal, Soc Plastics Eng, 75; Carothers Award, 85; Chem Pioneer Award, Am Inst Chemists, 85. *Mem:* Am Chem Soc; Royal Soc Chem; fel NY Acad Sci; fel Royal Soc London. *Res:* Homogenous catalytic oxidations; oxidative coupling reactions; organic and polymer synthesis. *Mailing Add:* Dept Chem McGill Univ 801 Sherbrooke St W Montreal PQ H3A 2K6 Can

HAY, DONALD IAN, b Peterborough, Eng, Nov 5, 33; m 68; c 1. BIOCHEMISTRY. *Educ:* London Univ, BSc, 59, PhD(biochem), 72. *Prof Exp:* Res scientist chem, Unilever Res Labs, Colworth House, Bedford, Eng, 59-65; from asst staff mem to assoc staff mem, 65-79, SR STAFF MEM BIOCHEM, FORSYTH DENT CTR, BOSTON, 79- *Concurrent Pos:* Asst clin prof, 76-83, assoc clin prof, Harvard Sch Dent Med, 83- *Mem:* AAAS; Am Chem Soc; Royal Soc Chem; Int Asn Dent Res. *Res:* Oral biochemistry; salivary protein structure and function; biochemical mechanisms of dental disease. *Mailing Add:* Forsyth Dent Ctr 140 Fenway Boston MA 02115

HAY, DONALD ROBERT, b Ottawa, Ont, Apr 1, 39; m 61; c 5. OPERATIONS RESEARCH, RESEARCH MANAGEMENT. *Educ:* McGill Univ, BEng, 61; Cornell Univ, MS, 64, PhD(mat sci), 66. *Prof Exp:* From asst prof to assoc prof metall eng, Drexel Univ, 66-72; invited prof, Ecole Poytechnique, 72-74, dir, Technol Ctr Develop, 74-79, dir, Indust Innovation Ctr, 79-81; tech dir, Hydrobien Ind Coun, 81-87; PRES, TEKTREND INT, INC, 73- *Concurrent Pos:* Adj prof computer sci, Concordia Univ, 81- & mining eng, McGill Univ, 90- *Mem:* Am Soc Metals; Am Soc Nondestructive Testing; Sigma Xi; Am Soc Testing & Mat. *Res:* Fracture mechanics; pattern recognition; non-destructive testing; acoustic emission; systems and reliability engineering; neural networks; artificial intelligence. *Mailing Add:* Tektrend Int Inc PO Box 322 Lachute PQ J8H 3W8 Can

HAY, ELIZABETH DEXTER, b St Augustine, Fla, Apr 2, 27. ANATOMY, CYTOLOGY. *Educ:* Smith Col, AB, 48; Johns Hopkins Univ, MD, 52. *Hon Degrees:* DSc, Smith Col, 73. *Prof Exp:* Intern, Osler Serv Hosp, Johns Hopkins Univ, 52-53, instr anat, Sch Med, 53-56, asst prof, 56-57; asst prof, Med Col, Cornell Univ, 57-60; asst prof, 60-64, Louise Foote Pfeiffer assoc prof embryol, 64-69, LOUISE FOOTE PFEIFFER PROF EMBRYOL, HARVARD MED SCH, 69-, CHMN DEPT ANAT, 75- *Concurrent Pos:* Ed-in-chief, Develop Biol, 71-75; mem, Nat Adv Coun, Nat Inst Gen Med Sci, 78-81; mem, Bd Sci Counselors, Nat Inst Dent Res, 83- *Honors & Awards:* Alcon Award for Vision Res, 88. *Mem:* Nat Acad Sci; Am Asn Anat (pres, 81-82); Am Soc Cell Biol; Am Soc Cevelop Biol; Int Soc Develop Biologists. *Res:* Origin of cells in amphibian limb regeneration; fine structure of dedifferentiating and differentiating tissues; autoradiographic studies of protein and nucleic acid synthesis in embryos and regenerates; fine structure of developing muscle, cartilage, skin and eye; collagen synthesis by epithelium; tissue interaction in the developing cornea; immunohistochemistry of collagen; gene expression during epithelial-mesenchymal transformation. *Mailing Add:* Harvard Med Sch Boston MA 02115

HAY, GEORGE EDWARD, b Durham, Ont, June 11, 14; nat US; m 43; c 3. MATHEMATICS. *Educ:* Univ Toronto, BA, 35, MA, 36, PhD(appl math), 39. *Prof Exp:* Instr math, Ill Inst Technol, 39-40; from instr to assoc prof, 40-56, chmn dept math, 57-67, assoc dean grad sch, 67-76, PROF MATH, UNIV MICH, ANN ARBOR, 56- *Mem:* Am Math Soc; Soc Indust & Appl Math; Indust Math Soc; Math Asn Am; London Math Soc; Sigma Xi. *Res:* Mathematical theory of elasticity; mechanics; method of images applied to the problem of torsion; equilibrium of a thin compressible membrane. *Mailing Add:* Dept Math Univ Mich Ann Arbor MI 48109

HAY, GEORGE WILLIAM, b Winnipeg, Man, May 19, 29; m 55; c 2. CARBOHYDRATE CHEMISTRY. *Educ:* Univ Man, BSc, 51, MSc, 52; Univ Minn, PhD(biochem), 59. *Prof Exp:* Civilian biochemist in chg, Serol Sect, Western Regional Crime Detection Lab, Royal Can Mounted Police, 52-55; fel enzymol, Nat Res Coun Can, 59-60 & biochem, Univ Minn, 60-63; asst prof, 63-68, assoc dir, Carbohydrate Res Inst, 77-85, ASSOC PROF CHEM, QUEEN'S UNIV, ONT, 68- *Concurrent Pos:* Vis prof, Inst Forestry, Univ Rural Rio De Janeiro, 79. *Mem:* Am Chem Soc; Soc Nuclear Med. *Res:* Carbohydrate chemistry; polysaccharide chemistry and polysaccharidases; synthetic sweeteners; synthesis of sugars with S or N in the ring; synthesis of radiopharmaceuticals; carbohydrates of soil. *Mailing Add:* Dept Chem Queen's Univ Kingston ON K7L 3N6 Can

HAY, IAN LESLIE, b London, Eng, Dec 6, 34; m 64; c 3. POLYMER PHYSICS, ELECTRON MICROSCOPY. *Educ:* Univ London, BSc, 57; Bristol Univ, PhD(physics), 68. *Prof Exp:* Res asst semiconductor physics, Assoc Elec Industs, 57-63; RES ASSOC, CELANESE RES CO, 68- *Mem:* Brit Inst Physics & Phys Soc; Am Phys Soc. *Res:* Morphological characterization of solid polymers; relationships between processing, morphology and properties of polymers. *Mailing Add:* 129 Sorrento Dr Moore SC 29369

HAY, JAMES ROBERT, b Can, Oct 19, 25; m 74; c 2. WEED SCIENCE. *Educ:* Ont Agr Col, Guelph, BSA, 49; SDak State Col, MS, 51; Harvard Univ, PhD(plant physiol), 55. *Prof Exp:* Res officer weed control, Plant Res Inst, Can Dept Agr, 55- 62, dir, Exp Farm, 62-68, dir, Res Sta, 68-90; RETIRED. *Concurrent Pos:* Chmn, Can Weed Comt, 69-78. *Mem:* Fel Weed Sci Soc Am (pres, 79); Agr Inst Can; Can Soc Agron; Can Soc Pest Mgt (pres, 78). *Res:* Weed control. *Mailing Add:* 238 Christopher Crescent Saskatoon SK S7J 3R6 Can

HAY, LOUISE (SCHMIR), mathematical logic; deceased, see previous edition for last biography

HAY, PETER MARSLAND, b Ft Riley, Kans, Oct 5, 21; m 52; c 1. INDUSTRIAL CHEMISTRY. *Educ:* Antioch Col, BS, 44; Ohio State Univ, PhD(org chem), 51. *Prof Exp:* Asst instr chem, Ohio State Univ, 46-51; res chemist, Olin Mathieson Chem Corp, 52-56, res group leader, 56-59; group leader prod appln, Celanese Plastics Co, 59-63; group leader prod develop, J P Stevens & Co, 63-67; chem develop mgr, Colors & Chem Div, Sandoz-Wander, Inc, 67-72, mgr environ affairs, Colors & Chem Div, Sandoz Inc, East Hanover, 72-84; mem, gov sci adv comt, NJ, 80-87; RETIRED. *Concurrent Pos:* Independent consult, 85-; pres, ARIA Chem Serv, 85- *Mem:* NY Acad Sci; AAAS; Am Chem Soc. *Res:* Polymer physical chemistry; emulsion polymerization; physical properties of plastics; textile chemistry; environmental and toxicological properties of industrial chemicals. *Mailing Add:* 24 Bedford Rd Summit NJ 07901

HAY, PHILIP JEFFREY, b Philadelphia, Pa, Oct 4, 45; m 66; c 2. THEORETICAL CHEMISTRY. *Educ:* Franklin & Marshall Col, Lancaster, Pa, 67; Calif Inst Technol, Pasadena, PhD(chem), 72. *Prof Exp:* Res fel chem, Battelle Inst, Columbus, Ohio, 71-73 & Cornell Univ, Ithaca, NY, 73-74; staff mem chem, 74-81, group leader, Theoret Chem Group, 81-88 STAFF MEM, LOS ALAMOS SCI LAB, 88- *Mem:* Am Chem Soc; AAAS. *Res:* Electronic structure of molecules; potential energy surfaces; effective core potential; relativistic effects; excited states; transition-metal; actinide compounds; simulation of materials. *Mailing Add:* Los Alamos Sci Lab T12 Mail Stop B268 Los Alamos NM 87545

HAY, RICHARD LE ROY, b Goshen, Ind, Apr 29, 26; m 73; c 2. SEDIMENTARY PETROLOGY. *Educ:* Northwestern Univ, BS, 47, MS, 49; Princeton Univ, PhD(geol), 52. *Prof Exp:* Geologist, US Geol Surv, 48-49, 52 & 54-55; asst prof geol, La State Univ, 55-57; asst prof, 57-69, prof geol, Univ Calif, Berkeley, 69-83; RALPH E GRIU PROF GEOL, UNIV ILL, URBANA, 83- *Honors & Awards:* Kirk Bryan Award, Geol Soc Am; Arnold Guyot Mem Award, Nat Geog Soc. *Mem:* Geol Soc Am; Soc Econ Paleontologists & Mineralogists; Clay Mineral Soc. *Res:* Clay mineralogy, zeolites; silicate diagenesis; archeologic geology in northern Tanzania. *Mailing Add:* 245 Natural Hist Bldg Urbana IL 61801

HAY, ROBERT E, b Kingston, Ont, Sept 3, 37; m 62; c 1. ECONOMIC GEOLOGY. *Educ:* Queen's Univ, Ont, BSc, 58; McGill Univ, MSc, 59, PhD(geol), 64. *Prof Exp:* Tech off geologist, Geol Surv Can, 59-61; instr geol, Queen's Univ, Ont, 61-63; party chief geologist, Steep Rock Iron Mines, Ltd, 63; assoc prof geol, Catawba Col, 63-66; asst prof, 66-70, assoc prof, 70-77, PROF GEOL, STATE UNIV NY COL CORTLAND, 78- *Mem:* Geol Soc Am. *Res:* Igneous and metamorphic geology; mapping on Canadian Shield; Huronian stratigraphy. *Mailing Add:* Dept Geol State Univ NY PO Box 2000 Cortland NY 13045

HAY, ROBERT J, b Winnipeg, Man, Jan 23, 38; m 59; c 2. CELL PHYSIOLOGY, DEVELOPMENTAL BIOLOGY. *Educ:* Univ Man, BSc, 60, MSc, 61; Glasgow Univ, PhD(biochem), 65. *Prof Exp:* Mem staff geront, Vet Admin, 64-67; develop biol, Carnegie Inst, 67-70; from asst prof to assoc prof biol sci, Wright State Univ, 70-75; res assoc 75-76, HEAD CELL CULT DEPT, AM TYPE CULT COLLECTION, 76- *Mem:* Am Soc Cell Biol; Soc Exp Biol Med; Am Tissue Cult Asn; AAAS. *Res:* Cellular biology of aging; mechanisms governing compensatory hyperplasia and hypertrophy; differentiated cells in culture; cell standardization and banking for distribution. *Mailing Add:* Cell Cult Dept ATCC 12301 Parklawn Dr Rockville MD 20852

HAY, RUSSELL EARL, JR, b Dayton, Ohio, Jan 5, 18; m 43; c 3. AGRONOMY. *Educ:* Miami Univ, AB, 40; Univ Nebr, MS, 42; Univ Ill, PhD(agron), 48. *Prof Exp:* Asst bot, Miami Univ, 38-40; asst bot, Univ Nebr, 41; spec asst soil fertility, Univ Ill, 46-48; plant physiologist, Biol Labs, Chem Corps, Camp Detrick, 48-50; asst supvr agr res, Battelle Mem Inst, 50-53; plant physiologist, C F Kettering Found, 53-55; assoc dir res found, Ohio State Univ, 55-67; dir res develop, Wright State Univ, 67-76; RETIRED. *Concurrent Pos:* Mem, Nat Coun Univ Res Adminrs. *Mem:* AAAS; Am Chem Soc; Am Inst Biol Sci; Am Soc Plant Physiol; Am Soc Agron. *Res:* Nitrogen metabolism; plant growth regulators; research management; university development. *Mailing Add:* Rte 2 Box 314 Carthage NC 28327

HAY, WILLIAM WALTER, b Bay City, Mich, Dec 10, 08; m 43; c 2. CIVIL ENGINEERING. *Educ:* Carnegie Inst Technol, BS, 31, MgtEngr, 48; Univ Ill, MS, 48, PhD, 56. *Prof Exp:* From track supvr to proj engr, RR Co, Pa, NY, Iowa, Wis & Alaska, 34-47; from asst prof to prof, 47-77, EMER PROF RWY CIVIL ENG, UNIV ILL, URBANA, 77- *Concurrent Pos:* Engr & supt, Maintenance of Way, Europ Theater Opers, 43-45, chief engr, Korea RR, 45-46 & Transp Co, 43-46; consult, rail & track res, various US & Can rwy, 50-, Knappen Tippets Abbett Co, NY, 51, Govts Southern Rhodesia & Mozambique, 52-58, SAfrica, 55, Sangamon County Planning Comn, Ill, 58-59, Northwestern Univ, 59-62 & Harland Bartholomew & Assocs; mem, Ill Prof Engrs Exam Comt, 59-69; adv panel, Res, Nat Res Coun; consult, CVG Ferromimera Orinoco, Venezuela, 76, Royal Comn on BC Rwy, 77-78 & Can Pac Consult Servs, 77-; annual short courses in railway eng, 75- *Mem:* Hon mem, Am Rwy Eng Asn; hon mem, Roadmasters & Maintenance of Way Asn Am. *Res:* Effects of weather on construction, maintenance and operation of transport agencies; conservation of transportation resources; engineering costs for track and roadway maintenance especially materials, lateral stability and economics of ballast studies. *Mailing Add:* Dept Civil Eng 310 Eng Hall Univ Ill 1308 W Green St Urbana IL 61801

HAY, WILLIAM WINN, b Dallas, Tex, Oct 12, 34. SEDIMENTARY MASS BALANCE, PALEOCEANOGRAPHY. *Educ:* Southern Methodist Univ, BS, 55; Univ Ill, Urbana, MS, 58; Stanford Univ, PhD(geol), 60. *Prof Exp:* From asst prof to prof geol, Univ Ill, Urbana, 60-73; prof geol & geophys, Rosenstiel Sch Marine & Atmospheric Sci, Univ Miami, 68-82, dean, 76-80; dir mus, 82-87, PROF GEOL, UNIV COLO, 82- *Concurrent Pos:* Hon fel, Univ Col, Univ London, 72-; pres, Joint Oceanog Inst, 79-82; fel, Coop Inst Res Environ Sci, 82; adj prof, Rosenstiel Sch Marine & Atmospheric Sci, Univ Miami, 82- *Honors & Awards:* Francis P Shepard Medal, Soc Econ Paleontologists & Mineralogists, 81. *Mem:* Fel AAAS; Geol Soc Am; Int Nannoplankton Asn; Swiss Geol Soc; Soc Econ Paleontologists & Mineralogists (pres, 87-88). *Res:* Paleoceanography; paleoclimatology; exogene and endogene dynamics of the earths erosion sedimentation system; taxonomy and biostratigraphy of calcareous nannofossils. *Mailing Add:* Univ Colo Campus Box 218 Boulder CO 80309

HAYA, KATSUJI, b Ashcroft, BC, Nov 22, 43; m 69; c 3. SUBLETHAL EFFECTS, INTERMEDIARY METABOLISM. *Educ:* Univ BC, BSP, 67, PhD(pharm & med chem), 73. *Prof Exp:* Fel drug metab, Dept Pharm, Chelsea Col, Univ London, 73-76; res assoc toxicol, Dept Pharm & Toxicol, Univ Kans, 76-78; RES SCIENTIST BIOCHEM & TOXICOL, DEPT FISHERIES & OCEANS, GOVT CAN, ST ANDREWS, 78- *Mem:* Soc Environ Toxicol & Chem; Soc Toxicol Can; NY Acad Sci. *Res:* Biochemical toxicology; biochemical and physiological responses of aquatic animals to sublethal concentrations of xenobiotics (pollutants); occurence, source and effects of natural marine foxins in the aquatic biosphere. *Mailing Add:* Dept Fisheries & Oceans Biol Sta St Andrews NB E0G 2X0 Can

HAYAKAWA, KAN-ICHI, b Shibukawa, Japan, Aug 12, 31; m 67. CHEMICAL ENGINEERING. *Educ:* Tokyo Univ Fisheries, BS, 55; Rutgers Univ, PhD(food sci), 64. *Prof Exp:* Res fel, Canners Asn Japan, 55-60; from asst prof to prof food sci, 64-82, DISTINGUISHED PROF FOOD ENG, RUTGERS UNIV, 82- *Concurrent Pos:* Invited lectr, Univ Campinas, Brazil, 72 & 73, Nat Taiwan Univ, Taipei, 82, WUXI Inst Light Indust, China, 86; prin investr, 64- *Mem:* AAAS; fel Inst Food Technol; Am Inst Chem Eng; Am Soc Heat, Refrig & Air-Conditioning Eng; Am Soc Agr Eng. *Res:* Biochemical engineering analysis of thermal processing of food; heat and/or mass transfer in food; heat sterilization of food; thermalphysical properties of food. *Mailing Add:* Dept Food Sci Cook Col Rutgers Univ PO Box 231 New Brunswick NJ 08903

HAYASAKA, STEVEN S, b Philadelphia, Pa, Feb 9, 47; m 72; c 2. MARINE MICROBIOLOGY. *Educ:* Pa State Univ, BS, 69; Ore State Univ, MS, 72, PhD(microbiol), 75. *Prof Exp:* Res assoc microbiol, Ore State Univ, 75; asst prof, 75-80, ASSOC PROF MICROBIOL, CLEMENSON UNIV, 80- *Mem:* Sigma Xi; Am Soc Microbiol. *Res:* Effect and interaction of marine environmental parameters on growth and function of marine bacteria; microbiology of seagrasses; diseases of maricultured animals. *Mailing Add:* Dept Microbiol Clemson Univ Clemson SC 29634-1909

HAYASHI, FUMIHIKO, b Tokyo, Japan, Mar 28, 30; US citizen; m 70; c 3. PLANT PHYSIOLOGY. *Educ:* Univ Kagoshima, BS, 55; Univ Tokyo, PhD(agr biochem), 65. *Prof Exp:* Fel, Dept Veg Crops, Univ Calif, Davis, 60-63; lectr, Dept Agrobiol, Univ Tokyo, 63-66; sr biochemist, Woodard Res Corp, 68-70; leader, Plant Studies Criteria & Eval Div, 72-77, sr environ biochemist, Hazard Eval Div, Off Pesticide Progs, 77-81, SR ENVIRON BIOCHEMIST, HEALTH & ENVIRON REV DIV, OFF TOXIC SUBSTANCES, US ENVIRON PROTECTION AGENCY, WASHINGTON, DC, 81- *Concurrent Pos:* Environ Protection Agency rep, Man & Biosphere Prog, UNESCO. *Honors & Awards:* Herman Flash Found Award; Scholar, NIH & Ministry Educ, Japan. *Mem:* Am Chem Soc; Am Soc Hort Sci; Am Soc Plant Physiologists; Weed Sci Soc Am; NY Acad Sci; Sigma Xi. *Res:* Phytotoxicity caused by pesticide and toxic substance application on crop plants, and biochemical mode-of-action of herbicides and plant growth regulators. *Mailing Add:* 6421 Berkshire Dr Alexandria VA 22310

HAYASHI, IZUMI, b Tokyo, Japan, Feb 20, 48. CELL BIOLOGY, DEVELOPMENTAL BIOLOGY. *Educ:* Int Christian Univ, Tokyo, BA, 71; Univ Calif, San Diego, PhD(biol), 77. *Prof Exp:* Fel res assoc cell biol, Dept Biol, Univ Calif, San Diego, 77-78; res assoc, Dept Biol, Mass Inst Technol, 78-80; ASST RES SCIENTIST, DIV CYTOL & CYTOGENETICS, CITY OF HOPE NAT MED CTR, 80- *Mem:* AAAS. *Res:* Myoblasts derived from normal and genetically dystrophic mice which are cultured in hormonally defined media and used to study how hormones regulate growth and differentiation. *Mailing Add:* Dept Molecular Genetics Beckman Res Inst City Hope 1450 E Duarte Rd Duarte CA 91010

HAYASHI, JAMES AKIRA, b Alameda, Calif, Dec 18, 26; m 55; c 3. BIOCHEMISTRY, TOXICOLOGY. *Educ:* Northern Ill State Teachers Col, BS, 48; Univ Wis, MS, 53, PhD(biochem), 56. *Prof Exp:* From res assoc to prof biochem, Univ Ill Col Med, 55-71; asst dean instruction, 75-79, PROF BIOCHEM, MED COL, RUSH UNIV, 71-; SR BIOCHEMIST, PRESBY-ST LUKE'S HOSP, 61- *Mem:* Am Chem Soc; Am Soc Biol Chem; Sigma Xi. *Res:* Cell walls; bacterial enzymes. *Mailing Add:* Dept Biochem Rush Med Col Chicago IL 60612

HAYASHI, MASAO, b Yokohama, Japan, Jan 4, 47; m 72. CELL BIOLOGY. *Educ:* Saitama Univ, BSci, 69; Nagoya Univ, PhD(molecular biol), 74. *Prof Exp:* Asst prof biochem, Inst Biol Sci, Univ Tsukuba, 76-85; ASSOC PROF BIOCHEM, DEPT BIOL, OCHANOMIZU UNIV, 85- *Concurrent Pos:* Counr, Japanese Soc Cell Biol, 87-91; ed, Cell Structure & Function, 91-93. *Mem:* Am Soc Cell Biol; Japanese Soc Cell Biol; Japanese Soc Biochem. *Res:* Cell adhesion and extracellular matrix, especially on a cell-adhesive glycoprotein vitronectin; transmembrane control of cellular activities by adhesive- proteins and extracellular matrix. *Mailing Add:* Dept Biol Ochanomizu Univ Bunkyo-ku Tokyo 112 Japan

HAYASHI, TERU, b Atlantic City, NJ, Feb 12, 14; m 43, 70; c 4. PHYSIOLOGY, CELL PHYSIOLOGY. *Educ:* Ursinus Col, BS, 38; Univ Mo, PhD(cell physiol), 43. *Prof Exp:* Instr physics, US Army Air Force, 43-44; instr zool, Univ Mo, 44-45, res assoc, 45-46; instr, Columbia Univ, 46-47, from asst prof to prof, 47-67, chmn dept, 62-67; prof, Biol & Chem Dept, Ill Inst Technol, 67-79; sr scientist, Papanicolaou Cancer Res Inst, 80-84; EMER TRUSTEE, MARINE BIOL LAB, 81- *Concurrent Pos:* Guggenheim & Fulbright fels, Denmark, 54-55; trustee, Marine Biol Lab, Woods Hole, 60-; vis prof, Japan Soc Adv Sci, Japan, 74-75; Fulbright fel Ger, 75; Humboldt Award, Ger, 79. *Mem:* AAAS; Soc Gen Physiol (pres, 62); Am Physiol Soc; Biophys Soc; Am Soc Cell Biol. *Res:* Cell physiology; physiology and biochemistry of muscle. *Mailing Add:* 7105 SW 112th Pl Miami FL 33173

HAYASHI, TERUO TERRY, b Sacramento, Calif, July 23, 21; m 53; c 5. MEDICINE, BIOCHEMISTRY. *Educ:* Temple Univ, AB, 44, MD, 48, MS, 54. *Prof Exp:* From instr to assoc prof obstet & gynec, Sch Med, Temple Univ, 54-65; PROF OBSTET & GYNEC, SCH MED, UNIV PITTSBURGH, 65-, CHMN DEPT, 74- *Concurrent Pos:* NIH grant, 59-; mem study sect human embryol & develop, NIH, 64-68; mem comt perinatal biol & infant mortality br, Nat Inst Child Health & Human Develop, 70-74 ad hoc consult comt, Maternal & Child Health Res, 74-, subcomt, Res Training, 78- *Mem:* Soc Gynec Invest (secy-treas, 65-71, pres, 71-72); Am Soc Biol Chemists. *Res:* Placental metabolism; nucleic acid metabolism of human placenta. *Mailing Add:* Magee-Womens Hosp Univ Pittsburgh Pittsburgh PA 15213

HAYASHI, YOSHIKAZU, b Tokyo, Japan, Feb 9, 43; m 74. DYNAMIC METEOROLOGY. *Educ:* Univ Tokyo, BSc, 67, MSc, 69, PhD(geophys), 72. *Prof Exp:* Res assoc, 72-75, mem res staff dynamic meteorol, 75-76, RES METEOROLOGIST, GEOPHYS FLUID DYNAMICS LAB, PRINCETON UNIV, 76- *Mem:* Meteorol Soc Japan; Am Meteorol Soc. *Res:* Space-time spectral analysis; dynamics of atmospheric waves simulated by a general circulation model. *Mailing Add:* Geophys Fluid Dynamics Lab/NOAA Princeton Univ Princeton NJ 08544

HAYASHIDA, TETSUO, b Berkeley, Calif, July 31, 17. ANATOMY, ENDOCRINOLOGY. *Educ:* Univ Calif, Berkeley, BA, 48, PhD(anat), 56. *Prof Exp:* Instr anat, Univ Calif, Berkeley, 56-58; from asst prof to assoc prof, 58-69, PROF ANAT, UNIV CALIF, SAN FRANCISCO, 69- *Concurrent Pos:* USPHS sr res fel, Univ Calif, San Francisco, 58-61 & res career develop award, 62-68; assoc res endocrinologist, Hormone Res Lab, Univ Calif, San Francisco, 64- *Mem:* Endocrine Soc; Am Asn Anat; NY Acad Sci. *Res:* Comparative immunochemical and biological studies with pituitary growth hormones and prolactins from various species of vertebrates. *Mailing Add:* Dept Anat Univ Calif Sch Med 513 Parnassus Ave San Francisco CA 94143

HAYAT, M A, b Iran, Jan 31, 38. BIOLOGY. *Educ:* Univ Sind, Pakistan, BS, 56; Univ Tex, MA, 59; Ind Univ, PhD, 62. *Prof Exp:* Asst prof biol, Loyola Univ Chicago, 62-63, NDak State Univ, 63-67 & Univ Dayton, 67-71; assoc prof, 71-77, PROF BIOL, KEAN COL, 78- *Concurrent Pos:* Res grants, NSF, Heart Inst & Am Philos Soc. *Mem:* Am Inst Biol Sci; Electron Micros Soc Am. *Res:* Principles and techniques of electron microscopy; practical methods for electron microscopy; biological scanning electron microscopy; positive staining for electron microscopy; fixation for electron microscopy. *Mailing Add:* Dept Biol Sci Kean Col NJ Union NJ 07083

HAYATSU, RYOICHI, b Tokyo, Japan, May 2, 28; m 60; c 1. ORGANIC CHEMISTRY. *Educ:* Univ Tokyo, BS, 51, PhD(org chem), 58. *Prof Exp:* Fel org chem, Univ Alta, 60-63; from res assoc to sr res assoc org cosmochem, Enrico Fermi Inst, Univ Chicago, 63-74; RES STAFF, ARGONNE NAT LAB, 74- *Concurrent Pos:* Vis prof, Univ Tokyo, 70. *Mem:* AAAS; Am Chem Soc. *Mailing Add:* Argonne Nat Lab 9700 S Cass Ave Bldg 200 Argonne IL 60439

HAYBRON, RONALD M, b Zanesville, Ohio, June 6, 34; m 60; c 2. PHYSICS. *Educ:* Case Western Reserve Univ, BS, 56, MS, 58, PhD(physics), 61. *Prof Exp:* Res assoc physics, Mich State Univ, 61-63, asst prof res physics, 63-65; mem staff, Oak Ridge Nat Lab, 65-68; asst prof, 68-71, asst dean, Col Arts & Sci, 74-77, ASSOC PROF PHYSICS, CLEVELAND STATE UNIV, 71- *Concurrent Pos:* Adj assoc prof, Okla State Univ, NASA aerospace specialist, 81- *Mem:* AAAS; Am Phys Soc. *Res:* Scattering of electrons by nuclei; nuclear reactions produced by protons, deuterons and alphas, especially direct reaction domain, particularly at high energy; structure of nuclei as related to scattering properties. *Mailing Add:* Dept Physics Cleveland State Univ Euclid Ave Cleveland OH 44115

HAYCOCK, DEAN A, b Atlantic City, NJ, Mar 20, 52. NEUROBIOLOGY. *Educ:* Brown Univ, AB, 76, PhD(neurobiol), 85. *Prof Exp:* Lab res asst, Neurobiol Sect, Brown Univ, 78-81; postdoctoral fel, Rockefeller Univ, 85-87, res assoc, Lab Molecular & Cellular Neurosci, 87; sr res biologist, Dept Pharmacol, 87-90, SR RES INVESTR, DEPT NEUROSCI, STERLING RES GROUP, 90- *Concurrent Pos:* Res training fel, Nat Res Serv Award, NIMH, 87. *Mem:* Am Soc Pharmacol & Exp Therapeut; Soc Neurosci; Am Soc Neurochem; Int Soc Neurochem. *Res:* Neuropharmacology of mental disorders; synaptic transmission. *Mailing Add:* Dept Pharmacol Sterling Res Group Columbia Turnpike Rensselaer NY 12144

HAYCOCK, ERNEST WILLIAM, b Kansas City, Mo, May 20, 27; m 50; c 5. MATERIALS SCIENCE. *Educ:* Bristol Univ, Eng, BSc, 48, PhD(phys chem), 51. *Prof Exp:* AEC fel chem, Univ Calif, Berkeley, 51-53; chemist, Shell Develop Co, 53-64, res supvr Mat Eng & Corrosion Dept, 64-77, mgr Mat Sci & Eng, 77-84; RETIRED. *Honors & Awards:* Frank Newman Speller Award, Nat Asn Corrosion Engrs, 84. *Mem:* Am Chem Soc; Nat Asn Corrosion Engrs. *Res:* Chemistry of solid interfaces; reaction kinetics; electrochemistry; corrosion phenomena. *Mailing Add:* 314 Panorama Ct Benicia CA 94510

HAYCOCK, JOHN WINTHROP, b Washington, DC, Jan 16, 49. BIOCHEMISTRY, NEUROCHEMISTRY. *Educ:* Mich State Univ, BS(physiol) & BS(psychol), 71; Univ Calif, Irvine, PhD(biol sci), 75. *Prof Exp:* Res assoc neurochem, Dept of Psychobiol, Univ Calif, Irvine, 75-76, Riverside, 76-77, Irvine, 77-78; instr, dept neurobiol & anat, Univ Tex Med Sch, 78-83; res assoc, Rockfeller Univ, 83-86; ASSOC PROF, DEPT BIOCHEM, LA STATE UNIV MED CTR-NEW ORLEANS, 86- *Mem:* Soc Neurosci; Am Soc Neurochem; Am Soc Biol Chem Molecular Biol. *Res:* Regulation of cutecholamine metabolism, tyrosine hydroxyluse; phosphoxylation; schizophrenia. *Mailing Add:* Dept Biochem LA State Univ Med Ctr 1100 Florida Ave New Orleans LA 70119

HAYDEN, EDGAR C(LAY), b Wooster, Ohio, Aug 23, 22; m 51; c 3. RADIOLOCATION, RADIO PROPAGATION. *Educ:* Ohio State Univ, BEE, 43; Univ Ill, MS, 52, PhD(elec eng), 58. *Prof Exp:* Res engr, Columbia Broadcasting Syst, 43-46; res asst, Univ Ill, 46-52, from res assoc to assoc prof radio propagation, 52-67; consult ionospheric telecommun, Inst Telecommun Sci, Environ Sci Serv Admin, US Dept Commerce, 67-70, consult radio telecommun, Off Telecommun, 70-73; INST SCIENTIST, SOUTHWEST RES INST, 73- *Mem:* Am Inst Elec & Electronics Engrs; Am Geophys Union; AAAS. *Res:* Propagation of radio waves through the ionosphere, and use of such information in the radiolocation process; antennas and antenna arrays; radiolocation systems. *Mailing Add:* Southwest Res Inst PO Drawer 28510 San Antonio TX 78228-0510

HAYDEN, GEORGE A, b Kansas City, Mo, Feb 25, 28; m 57; c 1. BIOCHEMISTRY. *Educ:* Univ Kans, BA, 50; Howard Univ, MS, 60, PhD(biochem), 66. *Prof Exp:* Res technician, Children's Hosp Philadelphia, 54-58; res asst biochem, Col Med, Howard Univ, 58-62, res assoc, 62-66; post-doctoral fel, Sloan-Kettering Inst Cancer Res, 66-67; res scientist blood res, Am Red Cross, 67-71, dept dir, Nat Fractionation Ctr, 71; health scientist adminr, Nat Inst Gen Med Sci, 71-72; staff mem, Nat Heart, Lung & Blood Inst, 72-89; RETIRED. *Concurrent Pos:* Dir, Minority Hypertension Res Training Prog, 76-89. *Honors & Awards:* Dirs Award, NIH, 85. *Mem:* AAAS; Sigma Xi; Am Chem Soc. *Res:* Gamma irradiation effects on proteins and polypeptides; anti-tumor drug effects on nucleic acids from mouse liver and sarcoma; glycoprotein biosynthesis in phytohemagglutinin stimulated human lymphocytes in culture. *Mailing Add:* 1312 Juniper St NW Washington DC 20012

HAYDEN, H(OWARD) WAYNE, JR, metallurgy, for more information see previous edition

HAYDEN, HOWARD CORWIN, b Pueblo, Colo, June 20, 40; m 83; c 2. ATOMIC PHYSICS. *Educ:* Univ Denver, BS, 62, MS, 64, PhD(physics), 67. *Prof Exp:* Res asst physics, Univ Denver, 63-67; fel, 67-68, asst prof, 68-76, ASSOC PROF PHYSICS, UNIV CONN, 76- *Mem:* Am Asn Physics Teachers; Am Phys Soc. *Res:* Atomic collisions; cross-section measurements of low energy charge transfer and ionization processes; energy loss measurements in inelastic scattering experiments; ion implantation; beam-foil spectroscopy; x-ray angular distributions. *Mailing Add:* Dept Physics Univ Conn Storrs CT 06268

HAYDEN, JESS, JR, b Eugene, Ore; div; c 4. ANATOMY, ANESTHESIOLOGY. *Educ:* Univ Ore, DMD, 47; Univ Mich, MS, 55; Loma Linda Univ, PhD(anat), 62; Am Bd Pedodontics, dipl, 58. *Prof Exp:* Dentist pvt pract, Eugene, Ore, 47-51; instr pedodontics, Sch Dent, Loma Linda Univ, 55-57, asst prof oral surg, 57-60, asst prof anesthesia sect, 60-62, from asst prof to assoc prof anat, Sch Med, 62-66; prof dent, coordr res & assoc dean, Col Dent, Univ Iowa, 66-74, prof pedodontics, 74-76; clin prof surg dent, Col Dent, Univ Colo Med Ctr, Denver, 76-80, clin asst prof anesthesiol, Col Med, 80-85; CAPT, DENT CORP, DEPT ANESTHESIOL, US NAVAL HOSP, USNR, SAN DIEGO, 90- *Concurrent Pos:* Am Scand Found travel grant-in-aid fel, 63; Fulbright-Hays vis prof, Royal Dent Col, Aarhus, 63-64; spec res fel, Nat Inst Dent Res, 57-60 & 73; pvt pract dent, 79-90; Am Asn Dent Educ Scholar, 91. *Mem:* Fel AAAS; Am Dent Asn; fel Am Col Dent; Sigma Xi; fel Am Dent Soc Anesthesiol. *Res:* Pathways and chronology of human tooth innervation by laboratory and clinical investigation; neuroanatomical-physiological correlates of sedation, local and general anesthesia in dentistry. *Mailing Add:* Anesthesia Dept Staff US Naval Hosp San Diego CA 92134-5000

HAYDEN, MARY VICTORIA, systematic botany, for more information see previous edition

HAYDEN, RICHARD AMHERST, b Wooster, Ohio, June 14, 28; m 55; c 4. HORTICULTURE, POMOLOGY. *Educ:* Ohio State Univ, BSc, 52, PhD, 56. *Prof Exp:* Asst horticulturist, Agr Exp Sta, Univ Ga, 56-66; assoc prof, 66-75, PROF HORT, PURDUE UNIV, 75- *Honors & Awards:* Stark Award, Am Soc Hort Sci, 71. *Mem:* Am Soc Hort Sci; Am Pomol Soc; Int Dwarf Fruit Tree Asn. *Res:* High density fruit production systems, apple and peach rootstocks pruning and training: strawberry production systems, varieties; peach tree longevity, tree fruit production and related studies. *Mailing Add:* Dept Hort Purdue Univ West Lafayette IN 47907

HAYDEN, RICHARD JOHN, b Whitefish, Mont, Feb 25, 22; m 48. PHYSICS. *Educ:* Oberlin Col, AB, 43; Univ Chicago, MS & PhD(physics), 48. *Prof Exp:* Jr physicist, Metall Lab, Univ Chicago, 43-46; assoc physicist, Argonne Nat Lab, 48-49 & 50-55; asst prof physics, 49-50 & 55-56, assoc prof, 56-58, PROF PHYSICS, UNIV MONT, 58- *Mem:* Fel Am Phys Soc. *Res:* Mass spectroscopy; radioactivity measurements; mass spectrographic mass assignment of radioactive isotopes; radiogenic geological dating by argon-potassium method. *Mailing Add:* 618 W Crestline Dr Missoula MT 59803

HAYDEN, THOMAS DAY, b Boston, Mass, Oct 23, 44; m 78; c 2. MATERIALS SCIENCE. *Educ:* Bates Col, BS, 66; Boston Univ, PhD(chem), 76. *Prof Exp:* Res scientist, Martin Marietta Labs, 77-78; RES ASSOC, W R GRACE & CO, 78- *Concurrent Pos:* Lectr, Eastern Nazarene Col, 75. *Mem:* Am Chem Soc; Am Ceramic Soc; Am Crystallog Asn; Sigma Xi; Mat Res Soc; Transportation Res Bd. *Res:* Chemical and physical reactions of portland cement hydration; experimental design methods for the development of chemical additives which control, modify or accelerate these hydration reactions and thereby extend the technical usefulness of concrete as a building material. *Mailing Add:* W R Grace & Co 62 Whittemore Ave Cambridge MA 02140

HAYDEN, THOMAS LEE, b Abilene, Tex, Feb 8, 32; m 57; c 1. MATHEMATICS. *Educ:* Univ Tex, BS, 54, PhD(math), 61. *Prof Exp:* Asst prof math, Univ Ky, 61-62; vis mem, US Army Math Res Ctr, Univ Wis, 62-63; asst prof, 63-65, assoc prof, 65-77, PROF MATH, UNIV KY, 78- *Concurrent Pos:* Consult biomath, Oak Ridge Nat Labs. *Mem:* Soc Indust & Appl Math. *Res:* Distance matrices; linear algebra; non-linear optimization. *Mailing Add:* Dept Math Univ Ky Lexington KY 40506

HAYDEN-WING, LARRY DEAN, b Webster City, Iowa, Aug 13, 35; m; c 3. WILDLIFE ECOLOGY. *Educ:* Univ Idaho, BS, 58, MS, 62, PhD(forestry), 70. *Prof Exp:* Res biologist, Wash State Univ, 62-65; asst prof zool & entom, Iowa State Univ, 69-72, assoc prof wildlife ecol, 72-77; assoc prof wildlife ecol, Univ Wyo, 77-78; dir, Wildlife Div, Land Inventory & Develop Inc, 78-80; DIR, HAYDEN-WING ASSOC, ENVIRON CONSULT, 80- *Mem:* Wildlife Soc; EAfrican Wildlife Soc. *Res:* Wild ungulate ecology and management; wildlife-habitat relationships; mechanisms controlling the selection of plants by grazing herbivores; conservation of African wildlife; environmental impacts and mitigation of energy extraction activities; other natural resource perterbations. *Mailing Add:* PO Box 6083 Sheridan WY 82801

HAYDOCK, PAUL VINCENT, GENE EXPRESSION. *Educ:* Purdue Univ, PhD(biochem), 83. *Prof Exp:* RES ASSOC, UNIV WASH, 83- *Res:* Epithelial differentiation. *Mailing Add:* Dept Periodont SM-44 Rm D664 Univ Wash Seattle WA 98195

HAYDON, GEORGE WILLIAM, b Alma Center, Wis, July 10, 14; m 44. RADIO ENGINEERING. *Prof Exp:* Engr, Dept Army, 42-54, scientist, Off Chief Signal Off, 54-59; radio consult, Cent Radio Propagation Lab, Dept Com, 59-65, RADIO CONSULT, INST TELECOMMUN SCI, OFF TELECOMMUN, DEPT COM, 65- *Concurrent Pos:* Mem, Int Radio Consult Comts Ionosopheric Propagation, 48-68 & Space Systs, 62-68; US del, Int Telecommun Radio Conf, 59 & Int Telecommun Aeronaut Radio Conf, 64 & 66. *Mem:* Inst Elec & Electronics Engrs; Sigma Xi; Int Union Radio Sci. *Res:* Radio spectrum utilization; prediction of performance of high frequency radio circuits using sky waves; international agreements concerning radio usage; selection of frequencies for satellite communication systems. *Mailing Add:* 350 Ponca Pl Boulder CO 80303-3828

HAYDU, JUAN B, b Budapest, Hungary, Apr 26, 30; US citizen; m 55; c 2. TECHNOLOGY MANAGEMENT. *Educ:* Univ Buenos Aires, Argentina, PhD(chem), 56; New Haven Univ, MBA, 79. *Prof Exp:* VPRES TECHNOL, SUBSID ASARCO INC, ENTHONE INC, 62- *Honors & Awards:* Heussner Award, Am Electroplaters' Soc, 65. *Mem:* Am Chem Soc; Electrochem Soc; Am Electroplaters' & Surface Finishers Soc. *Res:* Electronic chemicals; corrosion; protection of materials. *Mailing Add:* PO Box 1900 New Haven CT 06508

HAYDUK, WALTER, b Beauvallon, Alta, Sept 1, 31; m 57; c 3. CHEMICAL ENGINEERING. *Educ:* Univ BC, BA, 54, MA, 56, PhD(chem eng), 64. *Prof Exp:* Process engr, Polymer Corp Ltd, Ont, 55-61, tech supvr, Rubber Dept, 64-66; from asst prof to assoc prof chem eng, 66-73, PROF CHEM ENG, UNIV OTTAWA, 74- *Concurrent Pos:* Chem Inst Can fel, 78. *Mem:* Can Soc Chem Eng; Asn Prof Engrs Ont. *Res:* Solubilities and diffusivities of gases in liquids; solar energy related studies; emulsion polymerization and molecular weight distribution of vinyl chloride polYmers. *Mailing Add:* Dept Chem Eng Univ Ottawa Ottawa ON K1N 6N5 Can

HAYE, KEITH R, b Bryan, Ohio, June 13, 53. DEVELOPMENTAL BIOLOGY, MOLECULAR CELL BIOLOGY. *Educ:* Pa State Univ, PhD(microbiol), 83. *Prof Exp:* INSTR & FEL IMMUNOL, UNIV ILL, 83- *Concurrent Pos:* Muscular Dystrophy Asn fel, 84-85. *Mem:* Am Soc Cell Biol; Am Inst Biol Sci; Am Soc Microbiol. *Res:* Cell differentiation; groundwater microbiota. *Mailing Add:* Incarnate Word Col 4301 Broadway San Antonio TX 78209-6397

HAYEK, DEAN HARRISON, b Waukesha, Wis, Jan 1, 31; m 55; c 3. PHYSIOLOGY, EDUCATION ADMINISTRATION. *Educ:* Univ Wis, BS, 54; Univ Tenn, MS, 60, PhD(radiation biol), 64. *Prof Exp:* Instr physiol, Med Col Va, 64-65, asst prof, 65-70; asst dean, 70-78, ASSOC PROF PHYSIOL, SCH MED, E CAROLINA UNIV, 70-, ASSOC DEAN, 78- *Res:* Medical education. *Mailing Add:* Admis Off Sch Med E Carolina Univ Greenville NC 27858-4354

HAYEK, LEE-ANN COLLINS, b Boston, Mass, Dec 27, 43; m 70; c 4. APPLIED STATISTICS. *Educ:* Emmanuel Col, AB, 65; Catholic Univ Am, MS, 67; Univ Md, PhD(statist), 78. *Prof Exp:* Math analyst, Physics Res Div, Cambridge Res Labs, US Air Force, 62-65; chief analyst, Blue Cross Western Pa, 67-68; chief statistician, Consolidated Nat Gas Serv Co, 68-69; prof math & statist, Beaver Col, 69-73; CHIEF MATH STATISTICIAN, SMITHSONIAN INST, 73- *Concurrent Pos:* Lectr math & statist, Point Park Col, 67-71; consult, EEO Serv, 79-80, Justice Dept, Civil Rights Div, 81; res adj, Sch Med, Georgetown Univ, 81-84. *Mem:* Am Statist Asn; Biomet Soc; Am Educ Res Asn; Psychomet Soc; Inst Math Statist. *Res:* Development of statistical and mathematical models and research methodology; mathematical and statistical research activities published as solutions to problems of concern in paleontology, biological sciences, anthropology, nutrition and education. *Mailing Add:* Rm E 116 Mus Natural Hist Smithsonian Inst Washington DC 20560

HAYEK, MASON, b St Paul, Minn, Mar 28, 20; m 44; c 1. TEXTILE & PAPER CHEMICALS, TECHNICAL WRITING. *Educ:* Univ Minn, BChem, 41; Ind Univ, MA, 43, PhD(org chem), 47. *Prof Exp:* Control chemist, Joseph E Seagram & Sons, Inc, 41-42, res chemist, 43-45; asst org chem, Ind Univ, 46-47; res chemist, E I du Pont de Nemours & Co, 47-51, group leader, 51-63, res assoc, 63-82; RETIRED. *Concurrent Pos:* Consult, 82-; tech writing consult, 84-90. *Mem:* Am Chem Soc. *Res:* Chemical auxiliaries including antistats, surfactants and fluorochemicals for textiles, plastics and paper; product development; market development and writing; organic chemistry. *Mailing Add:* 113 Rockingham Dr Wilmington DE 19803

HAYEK, SABIH I, b Baghdad, Iraq, Mar 25, 38; US citizen; m 60; c 2. ENGINEERING MECHANICS, ACOUSTICS. *Educ:* Robert Col, Istanbul, BSc, 59, MSc, 60; Columbia Univ, DEngSc, 65. *Prof Exp:* From asst prof to assoc prof, Pa State Univ, 65-70, scientist, Appl Res Lab, 65-70, head, Vibration & Radiation Group, 70-83, assoc dir, Res Ctr Acoust & Vibration, 86-90, PROF ENG MECH, PA STATE UNIV, 76- *Concurrent Pos:* Sr vis fel, Inst Sound Vibration, Univ Southhampton, Eng, 72; sr scientist, Naval Ocean Systs Ctr, San Diego, 80-81. *Mem:* Fel Acoust Soc Am; fel Am Soc Mech Engrs; Inst Noise Control; Sigma Xi; founding mem Acad Mech; Soc Eng Sci. *Res:* Acoustic radiation from structures; wave propagation in elastic media; acoustic properties of composite materials; acoustic theory of diffraction and scattering; noise barriers; machinery noise. *Mailing Add:* Dept Eng Sci & Mech Pa State Univ University Park PA 16802

HAYES, ALICE BOURKE, b Chicago, Ill, Dec 31, 37; wid. BOTANY. *Educ:* Mundelein Col, BS, 59; Univ Ill, Urbana, MS, 60; Northwestern Univ, PhD(biol), 71. *Prof Exp:* Lab asst bot, Univ Ill, 59-60; microbiologist, Munic Tuberc Sanitarium, Chicago, 60-62; from instr to prof natural sci, Loyola Univ, Chicago, 62-88, dept chairperson, 68-69 & 71-77, dean, 77-80, assoc acad vpres, 80-86, vpres acad affairs, 87-89; EXEC VPRES, PROVOST & PROF BIOL, ST LOUIS UNIV, 89- *Concurrent Pos:* NSF fel, Northwestern Univ, 69-71; mem, proposal develop & proj team, Inst Renewal through Improv Teaching, HEW, 76-78, adv panels, NSF, 77-81 & Space Biol Prog, NASA, 80-84; NASA res grant, 80-84. *Mem:* Am Inst Biol Sci; Am Soc Plant Physiologists; Bot Soc Am; AAAS; Sigma Xi; Am Asn Univ Profs; Am Asn Univ Women; Am Coun Educ; Am Soc Gravitational & Space Biol. *Res:* Morphogenesis; planar form and plagiotropism of the leaf blade and sporulation in imperfect fungi. *Mailing Add:* 825 S Bemiston Ave Clayton MO 63105

HAYES, ANDREW WALLACE, b Corning, Ark, Aug 21, 39; m 63; c 3. PHARMACOLOGY, TOXICOLOGY. *Educ:* Emory Univ, AB, 61; Auburn Univ, MS, 64, PhD(biochem), 67; dipl, Am Bd Toxicol. *Prof Exp:* Res assoc biochem, Vanderbilt Univ, 66-68, from asst prof to assoc prof microbiol, 68-71; prof microbiol, Univ Ala, 75; assoc prof, Univ Miss Med Ctr, 75-76, prog dir, training prog toxicol, 77-80; group dir & corp toxicologist, 84-87, VPRES & CORP TOXICOLOGIST, RJR NABISCO, INC, 87-; PROF PHARMACOL & TOXICOL, UNIV MISS MED CTR, 76- *Concurrent Pos:* USPHS fel, 67-69; adj prof, Univ Ala, 75- & Duke Univ, 85-; res prof, Tex A&M Univ, 79-, Temple Univ, 81-; dir toxicol res, Rohm & Haas, 80-84; ed, Toxicol & Appl Pharmacol, 81-, Comments Toxicol, 85-; Res Career Develop Award, NIH, 74; NATO sr fel, 77. *Mem:* Am Soc Microbiol; Soc Toxicol; Am Inst Nutrit; Am Chem Soc; Am Soc Pharmacol & Exp Therapeut. *Res:* Carcinogenicity, teratogenecity and mode of action of mycotoxins. *Mailing Add:* Bowman Gray Tech Ctr RJR Nabisco Inc Winston-Salem NC 27102

HAYES, CAROL J, b New York, NY, Dec 13, 40. BIOLOGICAL RHYTHMS. *Educ:* St Joseph Col, NY, BA, 61; NY Univ, PhD(physiol), 75. *Prof Exp:* Res asst immunol, Col Med, Cornell Univ, 64-65; from instr to assoc prof, 65-80, PROF BIOL, ST JOSEPHS COL, NY, 80-, CHMN DEPT, 76- *Concurrent Pos:* Adj asst prof, Medgar Evers Col, City Univ New York, 70- *Mem:* Sigma Xi; AAAS; NY Acad Sci. *Res:* The effects of chemicals on rhythmic processes in vertebrates. *Mailing Add:* Dept Biol St Josephs Col 245 Clinton Ave Brooklyn NY 11205

HAYES, CHARLES AMOS, JR, b Winnipeg, Man, Apr 9, 16; nat US; m 42; c 2. MATHEMATICS. *Educ:* Univ Calif, AB, 37, MA, 38, PhD(math), 42. *Prof Exp:* Instr math, Univ Calif, Berkeley, 46-47; from asst prof to prof, 47-78, chmn dept, 59-64, EMER PROF MATH, UNIV CALIF, DAVIS, 78- *Mem:* Am Math Soc; Math Asn Am; Sigma Xi. *Res:* Theory of functions of real variables; boundary value problems for elliptic systems of linear partial differential equations; theory of differentiation and integration of set functions. *Mailing Add:* Dept Math Univ Calif Davis CA 95616

HAYES, CHARLES FRANKLIN, b Huntington, WVa, Nov 7, 41; div; c 2. PHYSICS. *Educ:* Wheaton Col, BA, 63; WVa Univ, MS, 65, PhD, 67. *Prof Exp:* Asst prof physics, 67-74, assoc prof 74-79, PROF PHYSICS & ASTRON, UNIV HAWAII, 79-, CHMN DEPT, 85- *Mem:* Am Phys Soc; Illum Eng Soc. *Res:* Acoustics; entomology. *Mailing Add:* Dept Physics Univ Hawaii 2505 Correa Rd Honolulu HI 96822

HAYES, CLAUDE Q C, b New York, Nov 15, 45. ELECTRO STATICS. *Educ:* Columbia Univ, BA, 71, MBA, 73; Western State Law Sch, JD, 78. *Prof Exp:* Design, sales & mgt, Cyberton Eng, 72-75; instr phys sci & bus law, City Col, 76-79, instr phys sci & chem, bus law, phys geog & geol, Miramar Col, 76-82, instr phys sci, Nat Univ, 80-81; sr systs analyst, Gen Dynamics/ Western Data Systs & specif analyst, Convair Div, 79-80; CONSULT, SCI & TECHNOL, 81- *Concurrent Pos:* Consult, Hitech & Inforama, Paris, France & USN, 82; instr phys sci & phys geog, San Diego Community Col Dist & Mira Costa Col, 85-90; adj asst prof phys chem, San Diego State Univ, 86; adj prof int bus mgt, Grad Sch, Univ Redlands, 86-88; defense contractor/ consult, Defense Nuclear Agency, NOSC, Defense Advan Res Proj Agency, Strategic Defense Initiative Orgn, USAF, US Army & Ballistic Missile Orgn, 86- *Mem:* AAAS; sr mem Am Inst Aeronaut & Astronaut; NY Acad Sci; Am Chem Soc. *Res:* Endothermies; physical chemistry. *Mailing Add:* Hayes & Assocs 7980 Linda Vista Rd No 49 San Diego CA 92111

HAYES, DALLAS T, b Easton, Md, Aug 15, 31; m 60; c 3. PHYSICS, MATHEMATICS. *Educ:* Mass Inst Technol, SB, 54; Univ Gottingen, dipl physics, 61, Dr rer nat(physics), 62. *Prof Exp:* Physicist systs eng, Lincoln Lab, Mass, 56-57; staff scientist plasma physics, Res & Adv Develop Div, Avco Corp, 63-67; RES PHYSICIST, AIR FORCE CAMBRIDGE RES CTR, 67- *Mem:* AAAS; Am Phys Soc. *Res:* Many body problem as applied to nuclear physics; transport properties of dense plasmas; wave propagation in plasma; turbulent media. *Mailing Add:* 583 Peakham Rd Sudbury MA 01776

HAYES, DAVID R, b Raleigh, NC, July 14, 37; div; c 3. NUMBER THEORY. *Educ:* Duke Univ, AB, 59, PhD(math), 63. *Prof Exp:* From asst prof to assoc prof math, Univ Tenn, 63-67; assoc prof, 67-72, PROF MATH, UNIV MASS, AMHERST, 72- *Concurrent Pos:* NSF fel, 66; vis prof, Oxford Univ, 74-75, Harvard Univ, 81, Univ Calif, San Diego, 83, Imp Col, London, 89. *Mem:* Am Math Soc; Math Asn Am. *Res:* Algebraic and analytic number theory; elliptic modules. *Mailing Add:* Dept Math Univ Mass Amherst MA 01003

HAYES, DAVID WAYNE, b Mineola, Tex, Nov 26, 36; m 59; c 2. OCEANOGRAPHY, CHEMISTRY. *Educ:* NTex State Univ, BS, 61, MS, 63; Tex A&M Univ, PhD(oceanog), 69. *Prof Exp:* STAFF OCEANOGR, SAVANNAH RIVER LAB, E I DU PONT DE NEMOURS & CO, INC, 66- *Mem:* Am Geophys Union; Am Chem Soc; Am Soc Limnol & Oceanog; Geochem Soc. *Res:* Marine chemistry; diffusion of natural waters; coastal oceanography; radiochemistry; model development for natural water systems. *Mailing Add:* Box 2253 Aiken SC 29801

HAYES, DENNIS E, b Saint Joseph, Mo, Oct 03, 38; m 78; c 3. MARINE GEOLOGY, GEOPHYSICS. *Educ:* Univ Kans, BSE, 61; Columbia Univ, PhD(marine geophysics), 66. *Prof Exp:* Res assoc, 66-70, assoc prof marine geophysics, 74-77, assoc dir, 78-81, SR RES SCIENTIST, LAMONT-DOHERTY GEOL OBSERV, 71-, ASSOC DIR, 84-, PROF GEOPHYSICS, COLUMBIA UNIV, 77- *Concurrent Pos:* NSF fel, 61-65; Guggenheim fel, 80-81. *Mem:* Fel Am Geophys Union; fel Soc Explor Geophysicists; fel Geol Soc Am; Am Asn Petrol Geologists; Sigma Xi. *Mailing Add:* Lamont-Daherty Geol Observatory Columbia Univ Palisades NY 10964

HAYES, DONALD CHARLES, b Oakland City, Ind, June 29, 32; m 55; c 4. PHARMACOLOGY, BIOCHEMISTRY. *Educ:* Indiana Univ, BS, 58. *Prof Exp:* Sr scientist, Mead Johnson & Co, Bristol-Myers Co, 58-87; RETIRED. *Mem:* Am Chem Soc; AAAS. *Res:* Natural products isolation; tissue culture; lysosomes; protein chemistry; enzyme separations; lipid metabolism; cardiovascular pharmacology. *Mailing Add:* 18601 Old Princeton Rd Evansville IN 47711-9116

HAYES, DONALD H, b Washington, DC, Apr 13, 37; m 64; c 2. PHYSIOLOGY, GENETICS. *Educ:* Univ Mo, BA, 59, MA, 62; La State Univ, PhD(zool), 68. *Prof Exp:* Assoc prof zool, 62-77, ASSOC PROF BIOL, SOUTHEASTERN LA UNIV, 78- *Mem:* Am Inst Biol Sci; AAAS. *Res:* Molecular physiology and genetics; history of biology, especially genetics. *Mailing Add:* PO Box 756 Hammond LA 70401

HAYES, DONALD M, b Kings Mountain, NC, Nov 6, 28; m 51; c 3. INTERNAL MEDICINE, OCCUPATIONAL MEDICINE. *Educ:* Wake Forest Col, BS, 51, MD, 54. *Prof Exp:* From instr to asst prof med, Bowman Gray Sch Med, 59-66, asst dean in charge admis, 60-61, assoc prof med & prev med, 66-70, assoc med, prof community med & chmn dept, 70-76, assoc dean community health sci, 75-76; prof community med, chmn dept & prof med, Med Sch, Univ Tex, 76-79, prof health admin, Sch Pub Health, 76-79; med dir, Burlington Indust, 79-81, dir, health & hygiene, 81-87; CORP MED DIR, SARA LEE CORP, 87- *Concurrent Pos:* USPHS fel psychiat, 55-56; partic, Acute Leukemia Coop Group B, 61-71; consult, Vet Admin Hosp, Salisbury, NC, 62-76; adj prof med sociol, Univ Houston, 76-79; prog dir, Coun Accreditation Occup Hearing Conserv, 81; clin prof community & family med, Duke Univ Med Ctr, 83-; clin assoc prof prev med & community health, Sch Med, Univ SC, 84-; adj prof epidemiol, Sch Public Health, Univ NC, 84-; mem bd sci counselors, Div Cancer Prev & Control, Nat Cancer Inst, 85-89; mem, NC Inst Med, 90- *Mem:* Fel Am Col Prev Med; Asn Teachers Prev Med; Am Asn Cancer Educ; fel Am Col Physicians; Am Col Occup Med. *Res:* Health promotion, epidemiology. *Mailing Add:* 100 Whitlock Court Winston-Salem NC 27106

HAYES, DONALD SCOTT, b Los Angeles, Calif, Oct 7, 39; m 83. PHOTOMETRY, TELESCOPE AUTOMATION. *Educ:* Pomona Col, BA, 61; Univ Calif, Los Angeles, MS, 66, PhD(astron), 67. *Prof Exp:* Asst prof astron, Rensselaer Polytech Inst, 67-75; vis assoc prof physics, Ariz State Univ, 75-79; asst support scientist, Kitt Peak Nat Observ, 79-86; ASTRONOMER, FAIRBORN OBSERV, 86- *Concurrent Pos:* Vis scientist, Mt Hopkins Observ, Smithsonian Inst, 74-75; mem, Inst Space Observations, 86- *Mem:* Int Astron Union; Am Astron Soc; Int Amateur-Prof Photoelec Photom. *Res:* Stellar photometry; automatic telescopes; full automation of telescopes and observatories. *Mailing Add:* PO Box 1907 Scottsdale AZ 85252-1907

HAYES, DORA KRUSE, b Kindred, NDak, June 19, 31; m 53; c 2. BIOCHEMISTRY. *Educ:* Hamline Univ, BS, 52; Univ Wis, MS, 53; Univ Minn, PhD(biochem), 61. *Prof Exp:* Chemist, Gen Mills, Inc, Minn, 53-54; jr scientist, Univ Minn, 54-57, teaching asst, 57-61; biochemist, Dugway Proving Ground, 61-62, res biochemist, 62-65; res biochemist, 65-76, chief, Chem Biophys Control Lab, 76-79, RES LEADER, LIVESTOCK INSECTS LAB, USDA, 79- *Concurrent Pos:* Abstractor, Chem Abstr, 61-74; equal employment officer, Fed Women's Prog Coord, Agr Res Serv, USDA, 74-77 & arranger res leaders mgt; mgt leader, ISC, 75 & 89; sabbatical, Pharmacol Dept, Johns Hopkins Univ, 85. *Mem:* AAAS; Am Chem Soc; Am Entom Soc; Am Soc Biol Chem; life mem Int Soc Chronobiol; Sigma Xi. *Res:* Insect chronobiology; insect biochemistry; membrane transport; Pasteurella tularensis metabolism; bacterial spore metabolism; biological aspects of aerosols; insect diapause; receptors; peptides. *Mailing Add:* 9105 Shasta Ct Fairfax VA 22031

HAYES, EARL T, b Wallace, Idaho, Apr 1, 12; m 35; c 3. METALLURGY. *Educ:* Univ Idaho, BS, 35, MS, 36; Univ Md, PhD(chem eng), 40. *Hon Degrees:* DSc, Univ Idaho, 71. *Prof Exp:* Mine leaser, Mullan, Idaho, 36-38; metallurgist, US Bur Mines, Utah, 40-48, chief phys metall br, Ore, 48-56, asst chief metallurgist, Washington, DC, 56-57, chief metallurgist, 57-62; asst dir mat, off dir, defense res & eng, US Dept Defense, 62-66; dep dir, US Bur Mines, 67-70, actg dir, 68 & 70, chief scientist, 70-73; consult, Mat Assocs, 74-78; RETIRED. *Mem:* Fel Am Soc Metals; Am Inst Mining, Metall & Petrol Engrs. *Res:* Metallurgy of zirconium, titanium, chromium and manganese metals and alloys; high purity and refractory metals; metallurgy and materials research and development programming direction, administration and management. *Mailing Add:* 517 Gilmore Dr Silver Springs MD 20901

HAYES, EDWARD FRANCIS, b Baltimore, Md, Sept 8, 41; m 64; c 3. QUANTUM CHEMISTRY. *Educ:* Univ Rochester, BS, 63; Johns Hopkins Univ, MA, 65, PhD(chem), 66. *Prof Exp:* Res asst chem, IBM, 62; res assoc chem, Univ Rochester, 63; res fel chem, IBM, 64; grad fel chem, Johns Hopkins Univ, 64-67; fel chem, Princeton Univ, 67-68; prof chem, Rice Univ, 68-78; prog officer, NSF, 75-76, sect head, chem div, 76-80, budget analyst, Exec Off of the Pres, 80-82, dir, chem div, 82-83, 85-87, controller, 83-85; assoc provost & vpres info systs, 87-89, VPRES GRAD STUDIES, RES & INFO SYSTS, RICE UNIV, 89- *Concurrent Pos:* Lectr, Georgetown Univ, 79-80. *Honors & Awards:* Distinguished Serv Award, NSF, 85; Meritorious Exec Award, US Govt, 86. *Mem:* Am Chem Soc; fel Am Phys Soc; fel AAAS; Sigma Xi. *Res:* Quantum theory of molecular structure; application of electronic computers in quantum chemistry; molecular potential functions and intermolecular interactions; scattering theory. *Mailing Add:* Rice Univ PO Box 1892 Houston TX 77251-1892

HAYES, EDWARD J(AMES), b Brooklyn, NY, Apr 8, 24; m 50; c 3. MECHANICAL ENGINEERING. *Educ:* Mass Inst Technol, BS, 50; Univ Md, MS, 55. *Prof Exp:* Engr design & develop, E I du Pont de Nemours & Co, NY, 50-51, dir, Electronics Lab, 51-52; proj supvr appl mech & hydraul, Johns Hopkins Univ, 52-58; dir res & develop, Kelsey-Hayes Co, 58-64, vpres res, eng & qual control, 64-65, vpres corp res & develop, 65-67, pres & gen mgr wheel, Drum & Brake Div, 68-76, CORP VPRES RES & DEVELOP & ENG, KELSEY-HAYES CO, 76-, EXEC VPRES & SR ADV TO CHMN, 76-; CORP VPRES RES & DEVELOP & ENG, FRUEHAUF CORP, 80- *Concurrent Pos:* Hom mem sr staff, Johns Hopkins Univ; vpres, Nat Friction Prods, 65-; mem bd, E&E Eng & Brembo SpA, Italy. *Mem:* AAAS; Am Soc Mech Engrs; Am Inst Aeronaut & Astronaut; Am Ord Asn; NY Acad Sci; Sigma Xi. *Res:* Research management; rocketry, thrust vector controls; applied mechanics; mechanisms; automotive control systems. *Mailing Add:* 1801 Arlington Blvd Ann Arbor MI 48104

HAYES, EVERETT RUSSELL, b Pomeroy, Ohio, Feb 5, 17; m 46; c 3. ANATOMY. *Educ:* Ohio Univ, AB, 38; Ohio State Univ, PhD(anat), 47. *Prof Exp:* Instr anat, Univ Buffalo, 45-48; from asst prof to assoc prof, Ohio State Univ, 48-57; from assoc prof to prof, 57-85, EMER PROF ANAT, STATE UNIV NY, BUFFALO, 85- *Mem:* Histochem Soc; Am Soc Zool; Am Asn Anat; Am Soc Cell Biol; Biol Stain Comn. *Res:* Plasmalogens; comparative histology and histochemistry. *Mailing Add:* 6599 Bear Ridge Rd Lockport NY 14094-9212

HAYES, GEORGE J, b Washington, DC, 1918. NEUROSURGERY. *Educ:* Johns Hopkins Univ, MD, 43; Am Bd Neurol Surg, dipl, 52. *Prof Exp:* Intern, Johns Hopkins Hosp, 44; fel neurosurg, Lahey Clin Boston, 44-46; chief neurosurg serv, Walter Reed Gen Hosp, Washington, DC, 47-49, 50-51, 55-66; fel neurosurg, Duke Hosp, 49-50; chief neurosurg serv, Brooke Army Hosp, Ft Sam Houston, Tex, 53-55; dir prof serv, Off Serv Gen, Dept Army, 66-74; RETIRED. *Concurrent Pos:* Prin dep asst secy defense health & environ, Off Secy Defense, 71-74; clin prof neurosurg, George Washington Univ. *Mem:* Inst Med-Nat Acad Sci; AMA; Am Asn Neurol Surgeons; Am Col Surgeons. *Mailing Add:* 303 Skyhill Rd Alexandria VA 22314

HAYES, GUY SCULL, b Andover, Mass, Mar 10, 12; m 47; c 6. MEDICINE, PUBLIC HEALTH ADMINISTRATION. *Educ:* Harvard Univ, AB, 34, MD, 39; Johns Hopkins Univ, MPH, 47; Am Bd Prev Med, dipl, 55. *Prof Exp:* Intern med serv, Boston City Hosp, 40-41; instr med, Med Sch, Johns Hopkins Univ, 42-43; mem staff, Int Health Div, Rockefeller Found, 43-56, asst dir med & natural sci, 57-68, asst dir, 68-69, assoc dir health sci, 69-77; RETIRED. *Mem:* Royal Soc Trop Med & Hyg; Int Epidemiol Asn; Am Pub Health Asn. *Res:* Medical education, particularly the role of departments of preventive medicine and public health. *Mailing Add:* 111 W 50th St Rm 4200 North Brooklin ME 04661

HAYES, J EDMUND, b Arcola, Saskatchewan, Can, Feb 14, 10; m 37; c 3. ELECTRICAL ENGINEERING. *Educ:* Queens Univ, BSc, 35; McGill Univ, MEng, 45. *Prof Exp:* VPres eng, Can Broadcasting Corp, 36-70; RETIRED. *Mem:* Fel Inst Elec & Electronics Engrs. *Mailing Add:* 2030 Featherston Dr Ottawa ON K1H 6P9 Can

HAYES, J SCOTT, b June 21, 46; m 78; c 1. CARDIO-VASCULAR HOMODYNAMICS, CARDIOVASCULAR PHARMACOLOGY. *Educ:* Univ Ariz, PhD(pharmacol), 76. *Prof Exp:* Res scientist, 79-86, HEAD CARDIOVASC PHARMACOL, ELI LILLY & CO, 86- *Mem:* Am Soc Pharmacol & Exp Therapeut; Am Heart Asn. *Res:* Cardio-vascular biochemistry. *Mailing Add:* Dept Pharmacol 307 E McCarty St Lilly Res Lab Indianapolis IN 46285

HAYES, JANAN MARY, b Los Angeles, Calif, Dec 10, 42. ANALYTICAL CHEMISTRY, CHEMICAL EDUCATION. *Educ:* Ore State Univ, BS, 64, MS, 65; Brigham Young Univ, PhD(inorg & anal chem), 71. *Prof Exp:* Instr chem, Fortuna Union High Sch, 65-67; assoc prof chem, Am River Col, 71-81; ASST DEAN SCI & AGR, COSUMNES RIVER COL, 81- *Concurrent Pos:* Consult, Sumar Corp, 75-81; NSF proj dir sci educ, Am River Col, 78-79. *Mem:* Am Chem Soc. *Res:* Applications of clinical chemistry procedures to non-clinical problems; analytical chemistry and public mandated concerns; chemical education for an informed public. *Mailing Add:* Merced Col 3600 M St Merced CA 95384

HAYES, JEFFREY CHARLES, b San Bernardino, Calif, Aug 28, 54. METAL REAGENTS & CATALYSTS. *Educ:* Cornell Univ, AB, 76; Harvard Univ, MA, 79, PhD(organometallic chem), 83. *Prof Exp:* NIH res fel, dept chem, Univ Calif, Berkeley, 83-85; ASST PROF ORG CHEM, DEPT CHEM & BIOCHEM, UNIV COLO, BOULDER, 85- *Mem:* Am Chem Soc. *Res:* Synthesis and reactivity of bimetallic transition metal oxo and alkylidene complexes; reactivity of metal radicals; transannular reactions in metallacycles. *Mailing Add:* Dept Chem & Biochem Box 215 Univ Colo Boulder CO 80309-0215

HAYES, JEREMIAH FRANCIS, b New York, NY, July 8, 34; m 62; c 4. COMPUTER COMMUNICATIONS, COMMUNICATION THEORY. *Educ:* Manhattan Col, BEE, 56; New York Univ, MS, 61; Univ Calif, Berkeley, PhD(elec eng), 66. *Prof Exp:* Mem tech staff, Bell Labs, 56-60; res engr, Electronics Res Labs, 60-62; actg instr elec eng, Univ Calif, 62-66; asst prof, Purdue Univ, 66-69; mem tech staff, Bell Labs, 69-78; assoc prof elec eng, McGill Univ, 78-84; PROF ELEC & COMPUT ENG & CHMN DEPT, CONCORDIA UNIV, 84- *Concurrent Pos:* Ed comput communications, Trans Communications, Inst Elec & Electronics Engrs, 81-; tech rep, Int Telegraph & Telephone Consult Comt, Int Telecom Union, 72-78; Adj prof, Polytech Inst NY, 76-78. *Mem:* Fel Inst Elec & Electronics Engrs. *Res:* Computer communications with special emphasis on local area networks; application of informaton theory to computer communications; source encoding; communication theory. *Mailing Add:* Dept Elec & Comput Eng Concordia Univ Montreal PQ H3G 1M8 Can

HAYES, JOHN A, b Maesteg, Wales, July 16, 29. PULMONARY DISEASE. *Educ:* Univ Bristol, MD, 56, PhD(path), 68. *Prof Exp:* PROF PATH, SCH MED, BOSTON UNIV, 76-, ASSOC CHMN PATH, 80-; LAB DIR LAB SERV, VA MED CTR, 84- *Mem:* Am Asn Pathologists; Int Acad Pathologists; Royal Col Path. *Res:* Toxic lung disease. *Mailing Add:* 150 South Huntington Ave Lab Serv VA Med Ctr Boston MA 02130

HAYES, JOHN BERNARD, b Omaha, Nebr, Nov 30, 34; m 61; c 2. GEOLOGY, MINERALOGY. *Educ:* Iowa State Univ, BS, 56, MS, 57; Univ Wis, PhD(geol), 61. *Prof Exp:* From instr to assoc prof geol, Univ Iowa, 60-68; sr res geologist, Denver Res Ctr, Marathon Oil Co, 68-86; RETIRED. *Concurrent Pos:* Mineral consult, Iowa Geol Surv, 62-68; vis prof geol, Univ Colo, 73 & Colo Sch Mines, 76; lectr, Am Asn Petrol Geol Schs, 78-85. *Mem:* Geol Soc Am; Soc Econ Paleont & Mineral; Clay Minerals Soc (pres, 77-78); Am Asn Petrol Geol. *Res:* Sedimentary petrology and mineralogy, especially clay mineralogy and sandstone petrology; x-ray diffraction and crystallography; sedimentary basin analysis; Alaska and California geology. *Mailing Add:* 1517 W Briarwood Ave Littleton CO 80120

HAYES, JOHN MICHAEL, b Seattle, Wash, Sept 6, 40; m 62; c 3. BIOGEOCHEMISTRY, MASS SPECTROMETRY. *Educ:* Iowa State Univ, SB, 62; Mass Inst Technol, PhD(anal chem), 66. *Prof Exp:* Res assoc cosmochem, Enrico Fermi Inst, Univ Chicago, 66; sr scientist, Ames Res Ctr, NASA, 67-68; NATO fel org chem, Org Geochem Res Unit, Bristol, Eng, 69; from asst prof to prof chem & geol, 70-84, prof, 84-90, DISTINGUISHED PROF BIOGEOCHEM, IND UNIV, BLOOMINGTON, 90- *Concurrent Pos:* Ed, Geochimica et Cosmochimica Acta, 71-74; consult, Hq, NASA, 71-79; mem precambrian paleobiol res group, Univ Calif, Los Angeles, 79-80; Guggenheim fel, 87-88. *Mem:* Geochem Soc; Am Soc Mass Spectrometry; Am Asn Petrol Geologists. *Res:* Natural chemistry of the stable isotopes of the volatile elements; development of relevant new techniques. *Mailing Add:* Biogeochem Labs Ind Univ Geol Bldg Bloomington IN 47405-5101

HAYES, JOHN TERRENCE, b Howard City, Mich, Dec 8, 28; m 69; c 3. ORTHOPEDIC SURGERY. *Educ:* Univ Mich, BS, 47, MD, 51. *Prof Exp:* Instr surg, Univ Mich, 56-58, from asst prof to assoc prof, 58-66; PROF ORTHOP SURG & CHIEF DEPT ORTHOP SURG & REHAB, BOWMAN GRAY SCH MED, 66- *Concurrent Pos:* Chief orthop surg serv, Wayne County Gen Hosp, Eloise, 56-66; attend surgeon, US Vet Hosp, Ann Arbor, 56-66; Fel ABC Traveling, 63. *Mem:* Am Med Asn; Am Col Surg; Am Acad Orthop Surg. *Res:* Bone growth and repair; infectious diseases of bones and joints. *Mailing Add:* 1342 Westgate Center Dr Winston-Salem NC 27103-2933

HAYES, JOHN THOMPSON, b Newton, Mass, Sept 10, 40; m 65, 80; c 3. INSECT ECOLOGY, MINORITY SCIENCE EDUCATION. *Educ:* Amherst Col, BA, 62; Cornell Univ, MS, 66, PhD(insect ecol), 68. *Prof Exp:* NIH fel, Savannah River Ecol Lab, Univ Ga, 67-69, US AEC res assoc, 69-73; dir pre-prof sci progs, Paine Col, 79-84 & 85-87, coordr, biol dept, 73-74 & 79, chairperson Div Natural Sci Math, 75-78, dir Text Skills Develop Ctr, 87-90, BIOL FAC, PAINE COL, 69-, DIR PRE-PROF SCI PROGS & CHAIRPERSON, DIV NATURAL SCI MATH, 90- *Concurrent Pos:* NIH Extramural Assoc, 78-79; mgr, computer based educ, Med Col Ga, 84-85. *Mem:* Asn Educ Commun & Technol; Sigma Xi; NY Acad Sci. *Res:* Insect ecology; federal agencies and historically black institutions; plant succession; microcomputers in education; affective aspects of educational success and personal health. *Mailing Add:* Biol Dept Paine Col 1235 15th St Augusta GA 30910-2799

HAYES, JOHN WILLIAM, b Johnstown, Pa, Sept 8, 44; m 67; c 2. BIOCHEMISTRY, SOLAR PHYSICS. *Educ:* Ill Benedictine Col, BS, 66; Purdue Univ, PhD(biochem), 71. *Prof Exp:* Fel & res assoc biol, Johns Hopkins Univ, 71-73; dean fac, 85-89, PROF CHEM, MARLBORO COL, 73- *Concurrent Pos:* Assoc ed & book rev ed, Passive Solar J, 81-86 & ed-in-chief, Progress in Passive Solar Energy, 82-84; mem bd dirs, Am Solar Energy Soc, 81-84, vchmn, 84. *Mem:* Am Chem Soc; AAAS; Int Solar Energy Soc; Am Solar Energy Soc (secy, 83). *Res:* Solar heating of buildings. *Mailing Add:* Marlboro Col Marlboro VT 05344

HAYES, JOHNNIE RAY, b Winston-Salem, NC, June 18, 42; m 64; c 1. NUTRITIONAL BIOCHEMISTRY. *Educ:* Pfeiffer Col, BS, 66; Appalachian State Univ, MS, 68; Va Polytech Inst & State Univ, PhD(biochem), 73. *Prof Exp:* Res assoc biochem, Va Polytech Inst & State Univ, 73-75; sr res assoc nutrit biochem, Cornell Univ, 75-80; asst prof, Dept Pharmacol, Med Col Va, 80-85; MASTER TOXICOLOGIST, RJR NABISCO, INC, 85- *Mem:* Am Chem Soc; AAAS; Am Inst Biol Sci; Sigma Xi; Soc Toxicol. *Res:* Microsomal metabolism of carcinogens; interactions between dietary insufficiencies and metabolism of foreign compounds; mode of action of the hepatocarcinogen aflatoxin. *Mailing Add:* RJR Nabisco Inc Bowman Grey Tech Ctr Winston-Salem NC 27102

HAYES, JOSEPH EDWARD, JR, b El Dorado, Ark, Nov 9, 27; m 61; c 2. BIOCHEMISTRY, RESEARCH ADMINISTRATION. *Educ:* Univ Wash, AB, 48, MA, 50, PhD(biochem), 53. *Prof Exp:* Res fel biophys, Univ Pa, 53-54; biochemist, Walter Reed Army Inst, Res, 54-57; biochemist, Nat Heart Inst, NIH, 57-65, scientist adminstr, Extramural Prog, Res Grants Br, 65-69, Nat Ctr Health Serv Res & Develop, 70-71, exec secy, Hemat Study Sect, Div Res Grants, 71-80; RETIRED. *Mem:* AAAS; Am Chem Soc. *Res:* Mechanisms of enzymic reactions; fluorescence and phosphorescence phenomena; general hematology. *Mailing Add:* 11017 Madison St Kensington MD 20895

HAYES, KENNETH CRONISE, b San Francisco, Calif, Sept 24, 39; m 61; c 2. NUTRITIONAL PATHOLOGY. *Educ:* Wesleyan Univ, BA, 61; Cornell Univ, DVM, 65; Univ Conn, PhD(nutrit path), 68. *Prof Exp:* NIH res fel nutrit path, Harvard Univ, 68-69, res assoc, 69-70, from asst prof to assoc prof nutrit, Sch Pub Health, 70-83, chmn nutrit div, New Eng Regional Primate Res Ctr, 78-83; PROF BIOL, BRANDEIS UNIV, 83-, DIR, FOSTER BIOMED RES LAB, 83- *Concurrent Pos:* NIH career develop grant, Sch Pub Health, Harvard Univ, 72-77; mem nutrit study sect, NIH, 74-78 & human nutrit adv coun, USDA, 84- *Mem:* Am Inst Nutrit; Am Soc Clin Nutrit; Int Acad Path; Soc Exp Biol & Med; Am Heart Asn. *Res:* Nutritional regulation of serum lipids and associated ultrastructure of atherogenesis in nonhuman primates; taurine and bile acid metabolism. *Mailing Add:* Foster Biomed Res Lab Brandeis Univ Waltham MA 02254

HAYES, KENNETH EDWARD, b Southampton, Eng, Jan 7, 28; m 52; c 2. PHYSICAL CHEMISTRY. *Educ:* Univ Southampton, BSc, 49; Univ Ore, PhD(chem), 52. *Prof Exp:* Fel chem, Princeton Univ, 52-54; fel & asst res officer, Nat Res Coun, 54-56; from asst prof to assoc prof, 58-71, PROF CHEM, DALHOUSIE UNIV, 71- *Res:* Physics and chemistry of surfaces, especially adsorption and catalysis and especially studies of the gas-metal interface by computer-enhanced multiple reflectance infrared spectroscopy. *Mailing Add:* Dept Chem Dalhousie Univ Halifax NS B3H 4H7 Can

HAYES, KIRBY MAXWELL, b Bourne, Mass, Aug 26, 22; m 49; c 2. FOOD TECHNOLOGY. *Educ:* Univ Mass, BS, 47, MS, 48. *Prof Exp:* Food technologist, USDA, 48-51; from assoc prof to prof, 51-85, food technologist, 51-80, & assoc head, 78-85, EMER PROF, UNIV MASS, AMHERST, 85- *Concurrent Pos:* Fulbright lectr, Univ WI, 68 & 73; vis prof Orgn Am States, Brazil, 70 & Trinidad, 70 & 75; Consult, 60- *Mem:* AAAS; fel Inst Food Technologists. *Res:* Apple and cranberry products; quality control of egg and poultry products; freezing of vegetables; nutritive values and grading of processed foods; frozen food quality; food packaging; frozen dietetic meals; thermal processing of home-canned foods; usage of under-utilized shellfish varieties. *Mailing Add:* PO Box 244 W Shore Rd Grantham NH 03753

HAYES, MARK ALLAN, b Bay City, Mich, Oct 19, 14; m 48; c 1. SURGERY. *Educ:* Univ Mich, AB, 37, MD, 40, PhD(anat), 48, MS, 51. *Hon Degrees:* MA, Yale Univ. *Prof Exp:* From intern to resident surg, Univ Mich Hosp, 40-43, instr anat, Med Sch, 46-48, instr surg, Univ Hosp & res assoc, AEC Lab, 49-51, asst prof surg, 51-52; from assoc prof to prof, 52-80, EMER PROF SURG, SCH MED, YALE UNIV, 80- *Concurrent Pos:* Kellogg fel, Univ Mich. *Mem:* Sigma Xi. *Res:* Gastrointestinal surgery; surgical anatomy; surgical metabolism and endocrinology. *Mailing Add:* 163 Ridgewood Ave N Haven CT 06473

HAYES, MILES O, b Asheville, NC, Oct 20, 34; m 58. GEOLOGY, OCEANOGRAPHY. *Educ:* Berea Col, AB, 57; Washington Univ, MA, 60; Univ Tex, PhD(geol), 65. *Prof Exp:* From instr to assoc prof geol, Univ Mass, Amherst, 64-73, dir, Coastal Res Ctr, 70-73; head dept geol, 73-77, PROF GEOL & MARINE SCI, UNIV SC, 73- *Concurrent Pos:* Res scientist, Defense Res Lab, Univ Tex, 60-64. *Mem:* Geol Soc Am; Soc Econ Paleontologists & Mineralogists. *Res:* Sedimentology of recent sediments; coastal processes of New England shoreline and southern Alaska. *Mailing Add:* Res Planning Inst 925 Gervais St Columbia SC 29201

HAYES, MILES V(AN VALZAH), b Ithaca, NY, Nov 20, 11; m 39; c 1. ENGINEERING. *Educ:* Yale Univ, AB, 32; Mass Inst Technol, BS, 34; Harvard Univ, MA, 47, PhD(eng sci), 50. *Hon Degrees:* MA, Dartmouth Col, 69. *Prof Exp:* Operating supvr, Tide Water Assoc, Oil Co, 33-41; lectr eng sci & appl physics, Harvard Univ, 50-52; res engr, Hughes Aircraft Corp, 52-53, United Aircraft Corp, 53-56 & Jones & Lamson Mach Co, 56-60; assoc prof eng sci, 60-70, prof eng, 70-77, EMER PROF ENG, THAYER SCH ENG, DARTMOUTH COL, 77- *Res:* Design and use of digital computers and numerical controls; numerical analysis; helicopter theory and design; engineering science. *Mailing Add:* 318 Davis Ave Clifton Heights PA 19018

HAYES, MURRAY LAWRENCE, b Ft Collins, Colo, July 17, 29; m 52; c 6. ZOOLOGY, MARINE FISHERIES. *Educ:* Univ Colo, BA, 51; Colo State Univ, MS, 57, PhD(zool), 59. *Prof Exp:* Fishery res biologist, US Bur Com Fisheries, Juneau, Alaska, 59-62, lab dir, Tech Lab, 62-70, assoc regional dir fisheries res, 70-71, dir, Kodiak Lab, 71-73; coordr energy related res, 73-77, dir, Resource Assessment, 77-86, CONSULT FISHERIES SCIENTIST, NAT MARINE FISH SERV, SEATTLE, WASH, 86- *Concurrent Pos:* Affil prof, Sch Fisheries, Univ Wash, 79-; assoc ed, Am Fisheries Soc, 82-84. *Mem:* Am Fisheries Soc; Am Inst Fisheries Res Biol; Sigma Xi. *Res:* Composition, distribution, abundance and condition of commercial fish stocks of eastern North Pacific; king crab biology; commercial crustaceans; fisheries management; performance of commercial fishing gear. *Mailing Add:* 20145 25th Ave NW Seattle WA 98177-2451

HAYES, RAYMOND L, JR, b Washington, DC, Feb 1, 38; m 61; c 3. ANATOMY, EMBRYOLOGY. *Educ:* Amherst Col, AB, 59; Univ Mich, MS, 61, PhD(anat), 63. *Prof Exp:* Instr anat, Harvard Med Sch, 63-65; asst prof, Sch Med, Univ Pittsburgh, 65-69, assoc prof anat & Cell Biol, 69-88; CHMN DEPT ANAT, HOWARD UNIV SCH MED, 88- *Mem:* AAAS; Am Asn Anatomists; Tissue Cult Asn. *Res:* Application of in vitro techniques to problems of embryogenesis; myological organogenesis; differentiation of embryonic duodenum as affected by cortisone; reaggregation of dissociated embryonic cells. *Mailing Add:* Dept Anat & Cell Biol Howard Univ Sch Med 520 W St NW Washington DC 20059

HAYES, RAYMOND LEROY, b Tucson, Ariz, Oct 12, 23; m 46; c 3. NUCLEAR MEDICINE. *Educ:* Univ NC, BS, 44, PhD(chem), 52. *Prof Exp:* From asst scientist to sr scientist, 50-76, CHIEF SCIENTIST, OAK RIDGE ASSOCIATED UNIVS, 76- *Mem:* AAAS; Am Chem Soc; Soc Nuclear Med; Reticuloendothelial Soc. *Res:* Radiopharmaceutical development; radiation dosimetry. *Mailing Add:* Tansi Dr Andersonville TN 37705

HAYES, RICHARD J, venture capital; deceased, see previous edition for last biography

HAYES, ROBERT ARTHUR, b Kendallville, Ind, June 29, 20; m 46; c 1. ORGANIC CHEMISTRY. *Educ:* Col Wooster, BA, 42; Univ Md, PhD(org chem), 48. *Prof Exp:* Asst chem, Univ Md, 42-44, res chemist, 46-48; chemist, Bur Entom & Plant Quarantine, USDA, Md, 44-45; res assoc, Allegheny Ballistics Lab, Md, 45; chemist, Firestone Tire & Rubber Co, 48-65, group leader plastics res, 65-67, mgr plastics & fibers res, 67-75, res assoc, 75-81, sr res assoc, 81-84; RETIRED. *Mem:* Am Chem Soc. *Res:* High polymers. *Mailing Add:* 2980 Cedar Hill Rd Cuyahoga Falls OH 44223-1497

HAYES, ROBERT GREEN, b Philadelphia, Pa, Oct 23, 36; m 60; c 5. PHYSICAL CHEMISTRY. *Educ:* Univ Pittsburgh, BS, 58; Univ Calif, Berkeley, PhD(phys chem), 62. *Prof Exp:* From instr to assoc prof, 61-74, PROF CHEM, UNIV NOTRE DAME, 74- *Mem:* Am Chem Soc. *Res:* Experimental molecular structure and dynamics, primarily utilizing photoelectron spectroscopy. *Mailing Add:* Dept Chem Univ Notre Dame Notre Dame IN 46556-5600

HAYES, ROBERT M, b Parsons, Tenn, Aug 7, 45; m 73; c 3. WEED SCIENCE, ECOLOGY. *Educ:* Univ Tenn, BS, 68; Univ Ill, PhD(agron), 74. *Prof Exp:* Asst prof agron, Univ Ky, 74-78; PROF PLANT & SOIL SCI, UNIV TENN, 78- *Mem:* Weed Sci Soc Am; Am Soc Agron. *Res:* Development of effective weed control systems for soil and energy conserving methods of crop production; crop-weed ecosystems and their interactions; biology of individual weed species. *Mailing Add:* 605 Airways Blvd Jackson TN 38301

HAYES, ROBERT MAYO, b New York, NY, Dec 3, 26; m 52; c 1. MATHEMATICS, INFORMATION SCIENCE. *Educ:* Univ Calif, Los Angeles, BA, 47, MA, 49, PhD(math), 52. *Prof Exp:* Mathematician, Nat Bur Standards, 49-52; res mathematician, Hughes Aircraft Co, 52-53; appln specialist, Nat Cash Register Co, 53-55; head bus systs dept, Res Labs, Magnavox Co, 55-59; vpres & sci dir, Electrada Corp, 59-64; pres, Adv Info Systs, Inc, 61-64; dean, Grad Sch Libr Serv, 74-89, PROF SCH LIBR SERV, UNIV CALIF, LOS ANGELES, 64- *Concurrent Pos:* Dir, Inst Library Res, Univ Calif, 64-70; vpres, Becker & Hayes, Inc, 70-73. *Mem:* AAAS (vpres); Am Math Soc; Asn Comput Mach; Am Soc Info Sci (pres, 63). *Res:* Hilbert space; numerical analysis; data processing systems. *Mailing Add:* Grad Sch Libr Info Sci Univ Calif 405 Hilgard Ave Los Angeles CA 90024-1520

HAYES, RUSSELL E, b Wichita, Kans, Nov 14, 35; m 57, 83; c 4. ELECTRICAL ENGINEERING. *Educ:* Univ Kans, BS, 58, MS, 59; Stanford Univ, PhD(elec eng), 63. *Prof Exp:* PROF ELEC ENG, UNIV COLO, 63- *Concurrent Pos:* Vis asst prof, Cornell Univ, 66-67. *Mem:* Inst Elec & Electronics Engrs. *Res:* Physics of semiconductor materials and devices. *Mailing Add:* Dept Elec & Comput Eng Univ Colo Boulder CO 80309-0425

HAYES, SHELDON P, b Provo, Utah, May 2, 13; m 35; c 2. IMMUNOLOGY, BACTERIOLOGY. *Educ:* Brigham Young Univ, BA, 34, MA, 35; Univ Utah, PhD(bact), 51. *Prof Exp:* Instr, Gila Col, 37-41; chmn div life sci & head dept bact, 41-68, PROF MICROBIOL, WEBER STATE COL, 68- *Concurrent Pos:* Res immunologist, Dept Anat, Cancer Labs, Sch Med, Univ Utah. *Mem:* Am Soc Microbiol; Am Asn Immunol; Soc Am Bacteriologists (vpres, 48). *Res:* Cellular aspects of immunity; water pollution. *Mailing Add:* 1300 25 St Ogden UT 84401

HAYES, TERENCE JAMES, b Weehawken, NJ, Feb 25, 41; m 68; c 2. IMMUNOLOGY, PATHOLOGY. *Educ:* Univ Scranton, BS, 62; Univ Pa, VMD, 66, PhD(parasitol), 70; Am Bd Toxicol, cert, 80. *Prof Exp:* Res fel, Lab Parasitol, Univ Pa, 66-70; sr scientist parasitol, Hoffmann-La Roche Inc, 71-74, asst res group chief chemother & parasitol, Animal Health Res Dept, 74-78, res group chief, 78-86, res leader, Dept Toxicol & Path, 86-89, DIR INVESTIGATIVE TOXICOL, DEPT TOXICOL & PATH, HOFFMANN-LA ROCHE INC, 89- *Mem:* Am Vet Med Asn; Am Asn Vet Parasitologists; Soc Toxicol; Am Asn Immunologist. *Res:* Toxicologic, immunologic and padhologic effects of drugs and cytokines in laboratory animals. *Mailing Add:* 525 Park Upper Montclair NJ 07043

HAYES, THOMAS B, HYDRAULIC & HYDROELECTRIC UTILITY SYSTEMS. *Educ:* Ore State Univ, BS; Mass Inst Technol, SM. *Prof Exp:* FOUNDING PARTNER & CONSULT, CH2M HILL. *Concurrent Pos:* Mem, past pres, State Ore Bd Eng Examiners; mem, Govnr Alternative Energy Develop Comn. *Mem:* Fel Am Soc Mech Engrs; fel Inst Elec & Electronics Engrs; Nat Soc Prof Engrs. *Res:* Effects of HVDC power transmission on buried structures; hydraulic systems; inventor developer of FLOmatcher scheme; conduit entrances and pumping station arrangements. *Mailing Add:* CH2M Hill Inc PO Box 428 Corvallis OR 97333

HAYES, THOMAS G, b Canonsburg, Pa, July 3, 36; m 65; c 1. HISTOLOGY, CYTOGENETICS. *Educ:* Washington & Jefferson Col, BA, 58; Case Western Reserve Univ, MA, 60; Ohio State Univ, PhD(anat), 65. *Prof Exp:* Grad asst biol, Case Western Reserve Univ, 58-60; from instr to assoc prof, 65-78, PROF ANAT, OHIO STATE UNIV, 78- *Concurrent Pos:* Fel, Tissue Cult Div, Nat Cancer Inst, 67. *Mem:* Am Asn Anat; Sigma Xi. *Res:* Cytogenetic and teratogenic effects of anticonvulsant drugs. *Mailing Add:* Dept Cell Biol Neurobiol/Anat Ohio State Univ Col Med 4058 Graves Hall 333 W Tenth Ave Columbus OH 43210

HAYES, THOMAS JAY, III, b Omaha, Nebr, Aug 26, 14; m 42; c 2. CIVIL ENGINEERING. *Educ:* US Mil Acad, BS, 36; Mass Inst Technol, MS, 39, Engr, 40. *Prof Exp:* Engr officer, US, Greenland, Bahamas, Can & Alaska, Corps Engrs, US Army, 36-43, mem fac & asst commandant, Eng Sch, Ft Belvoir, Va, 44-45, eng liaison officer & asst mil attache, London, Eng, 46-49, asst eng comnr, Washington DC govt, 49-52, dist engr, Little Rock, Ark, 52-53 & Omaha, 53-57, engr, I Corps, Korea, 58-59, chief, Los Angeles Field Off, Off Chief Engrs, 59-60, vcomdr & comdr, Corps Engrs Ballistic Missile Construct Off, Los Angeles, 60-62, asst chief engrs NASA support & dep dir mil construct for space progs, Off Chief Engrs, 62-64, dir topography & mil eng, 64-67, Comdr, SAtlantic Div, CE, Atlanta, Ga, 67-69; pres, Int Eng Co, Inc, 69-79, chmn, 79-80; dir, Burke Co, 79-87; RETIRED. *Concurrent Pos:* Mem, Permanent Int Asn Navig Cong, 51-80, US Comt Large Dams, 53-80; mem fed exec bd, Bd Engrs Rivers & Harbors, Coastal Eng Res Bd, 67-69; vchmn, SE Basin Inter Agency Comn, 67-69; pres, Cia Int de Ingenieria, Santiago, Chile, Asuncion, Paraguay, 69-80, & Int Eng Co, Kinshasa, Zaire, Cent Africa, 72-80; dir, Int Road Fedn, 78-79; vpres, Morrison-Knudson Co, Inc, Boise, Idaho, 71-76; consult, 80-88. *Honors & Awards:* George W Goethals Medal, Soc Am Mil Engrs, 61. *Mem:* Nat Acad Eng; fel Am Soc Civil Engrs; Soc Am Mil Engrs; Am Pub Works Asn. *Res:* engineering experience in connection with projects in 60 countries plus Arctic and Antarctic. *Mailing Add:* 2646 Chestnut St San Francisco CA 94123

HAYES, THOMAS L, b Oakland, Calif, Sept 12, 27; m 52; c 5. BIOPHYSICS. *Educ:* Univ Calif, Berkeley, AB, 49, PhD(biophys), 55. *Prof Exp:* BIOPHYSICIST, DONNER LAB & LAWRENCE BERKELEY LAB, UNIV CALIF, 55-, DEP DIR, DONNER LAB, 80- *Concurrent Pos:* Adj prof biophysics, Univ Calif, Berkeley, 59- *Mem:* AAAS; Biophys Soc; Electron Micros Soc Am. *Res:* Scanning electron microscopy. *Mailing Add:* 6666 Colton Blvd Berkeley CA 94611

HAYES, TIMOTHY MITCHELL, b Springfield, Mass, Nov 1, 41; m 62; c 2. SOLID STATE PHYSICS. *Educ:* Bowdoin Col, BA, 63; Harvard Univ, AM, 65, PhD(appl physics), 68. *Prof Exp:* Res assoc, Stanford Univ, 67-69; scientist, Palo Alto Res Ctr, Xerox Corp, 69-70, mem res staff, 70-86; prof physics, Colo Sch Mines, 86-89; PROF & CHAIR PHYSICS, RENSSELAER POLYTECH INST, 89- *Concurrent Pos:* Vis scientist, Atomic Energy Res Estab, Harwell, UK, 70; sr vis fel sci & eng res coun, UK, Leicester Univ, 81-82. *Mem:* Am Phys Soc. *Res:* Electron states in solids, especially eigenstates, cohesive energy, effects of disorder; thermal and electronic transport in disordered solids; atomic-scale structure of materials, and its relationship to electronic properties; x-ray absorption spectroscopy. *Mailing Add:* Physics Dept Rensselaer Polytech Inst Troy NY 12180-3590

HAYES, WALLACE D(EAN), b Peking, China, Sept 4, 18; m 48; c 3. AERONAUTICS. *Educ:* Calif Inst Technol, BS, 41, AeE, 43, PhD(physics), 47. *Prof Exp:* Jr stress analyst, Consol Vultee Aircraft Corp, Calif, 39; jr stress engr, Lockheed Aircraft Corp, Calif, 40, aerodynamicist, 43-45; Jewett fel, Princeton Univ, 47-48; asst prof appl math, Brown Univ, 48-49, assoc prof, 49-51; sci liaison officer, Off Naval Res, London, 52-54; from assoc prof to prof, 54-89, EMER PROF AERONAUT ENG, PRINCETON UNIV, 89- *Concurrent Pos:* Consult, NAm Aviation, Inc, Calif, 46-48, Space Technol Labs, Thompson-Ramo-Wooldridge, Inc, 54-58 & Aeronaut Res Assocs Princeton, Inc, 58-; Fulbright vis lectr, Delft Univ, 51-52; NSF fel & prof assoc, Univ Paris, 64-65. *Mem:* Nat Acad Eng; Am Phys Soc; Am Inst Aeronaut & Astronaut. *Res:* Gasdynamics; aerodynamics; hydrodynamics; applied mathematics. *Mailing Add:* 14 Cleveland Lane RD 4 Princeton NJ 08540

HAYES, WAYLAND JACKSON, JR, b Charlottesville, Va, Apr 29, 17; m 42; c 5. TOXICOLOGY. *Educ:* Univ Va, BS, 38, MD, 46; Univ Wis, MA, 40, PhD(zool), 42; Acad Toxicol Sci, dipl, 88. *Prof Exp:* Intern, US Marine Hosp, NY, 46-47; in charge vector-transmission invests br, Tech Develop Div, Commun Dis Ctr, USPHS, 47-49, chief toxicol sect, Tech Br, 49-67, chief toxicologist, Pesticides Prog, 67-68; prof, 68-82, EMER PROF TOXICOL, CTR TOXICOL, DEPT BIOCHEM, SCH MED, VANDERBILT UNIV, 82- *Concurrent Pos:* Consult, WHO, 50-, mem interdept comt pest control, 53-64, secy, 54-55, chmn, 59-60; mem threshold limits comt, Am Conf Govt Indust Hygienists, 61-71; vis lectr, Emory Univ, 62-68; mem res subcomt, Fed Comt Pest Control, 64-68; mem res subcomt plant & animal pests, Agr Bd, Nat Res Coun, 64-70; mem comt food protection, Food & Nutrit Bd & mem toxicol info prog comt, Div Med Sci, Nat Res Coun, 69-76; mem toxicol study sect, NIH, 68-72, chmn, 69-72; mem sci adv bd, Nat Ctr Toxicol Res, 72-76; mem stands adv comt agr, US Dept Labor, 73-76; ed, Essays Toxicol, 71-77. *Honors & Awards:* Meritorious Serv Medal, USPHS, 64; Merit Award, Soc Toxicol, 89. *Mem:* AAAS; Soc Toxicol (pres, 71-72); Am Soc Pharmacol & Exp Therapeut; Am Soc Trop Med & Hyg; Am Conf Govt Hygienists. *Res:* Toxicology of newer pesticides. *Mailing Add:* 607 Med Arts Bldg Sch Med Vanderbilt Univ Nashville TN 37212

HAYES, WILBUR FRANK, b Rhinelander, Wis, Nov 10, 36; m 79; c 6. ZOOLOGY. *Educ:* Colby Col, BA, 59; Lehigh Univ, MS, 61, PhD(biol), 65. *Prof Exp:* Asst biol, Lehigh Univ, 60-65; Nat Inst Gen Med Sci res fel & lectr biol, Yale Univ, 65-67; asst prof, 67-71, ASSOC PROF BIOL, WILKES COL, 71- *Concurrent Pos:* Vis assoc prof, Northeastern Univ, 87-88. *Mem:* Am Soc Zool; Sigma Xi; Electron Microscope Soc Am. *Res:* Comparative physiology of arthropod sense organs, especially proprioceptors and chemoreceptors in Limulus. *Mailing Add:* Dept Biol Stark Learning Ctr Wilkes Univ Wilkes-Barre PA 18766

HAYES, WILLIS B, b Long Beach, Calif, Aug 23, 42; div. GEOLOGY, COMPUTER-ASSISTED INSTRUCTION. *Educ:* Stanford Univ, BA, 63; Scripps Inst Oceanog, PhD(oceanog), 69. *Prof Exp:* Res ecologist, marine ecol, Inst Marine Resources, 69-70; asst prof biol, Am Univ Beirut, 70-73; assoc zool, 73-76, RES ASSOC GEOL, UNIV GA, 77- *Concurrent Pos:* Vis asst prof biol, Univ Victoria, 75. *Mem:* Nat Asn Geol Teachers; AAAS. *Res:* Geological modeling; geostatistics; computer-assisted instruction. *Mailing Add:* Dept Geol Univ Ga Athens GA 30602

HAYFLICK, LEONARD, b Philadelphia, Pa, May 20, 28; m 55; c 5. CELL BIOLOGY, GERONTOLOGY. *Educ:* Univ Pa, BA, 51, MS, 53, PhD(med microbiol & chem), 56. *Prof Exp:* Asst bact, Merck Sharp & Dohme, Inc, 51-52; asst instr med microbiol, Univ Pa, 55-56, asst prof, 65-68; assoc med microbiol & mem staff, Wistar Inst Anat & Biol, 58-68; prof med microbiol, Sch Med, Stanford Univ, 68-76; sr res cell biologist, Childrens Hosp Med Ctr, Bruce Lyon Mem Res Lab, Oakland, Calif, 76-81; prof zool, immunol & microbiol, Ctr Geront Studies, Univ Fla, 81-88; PROF ANAT, UNIV CALIF, SCH MED, DEPT ANAT, 88- *Concurrent Pos:* Vis scientist, Weizmann Inst Sci, Rehovoth, Israel, 80; ed-in-chief, Experimental Gerontology, 85- *Honors & Awards:* Brookdale Award, Geront Soc Am, 80; Robert W Kleemeier Award, Geront Soc Am, 82-83; Stuart Mudd Mem Award Lectr, 85; Leadership Award, Am Fedn Aging Res, 83; Pres Award, Int Orgn Mycoplasmology, 84; Samuel Roberts Noble Res Award, 84; Hon mem, Tissue Cult Asn, 89. *Mem:* AAAS; Am Soc Microbiol; Tissue Cult Asn (vpres); Am Soc Cell Biol; fel Geront Soc (vpres). *Res:* Gerontology; cell culture; human diploid cell strains; viral oncogenesis; mycoplasmas; vaccine development. *Mailing Add:* Cell Biol & Aging Sect 151E Univ Calif 4150 Clement St San Francisco CA 94121

HAYGARTH, JOHN CHARLES, b Keighley, Yorks, Sept 4, 40; US citizen; m 75. MATERIALS SCIENCE. *Educ:* Univ Leeds, BSc, 62, PhD(phys chem), 65. *Prof Exp:* Res geophysicist, Inst Geophysics, Univ Calif, Los Angeles, 65-68; res scientist, E I du Pont de Nemours & Co, Inc, 68-69; res geophysicist, Inst Geophysics, Univ Calif, Los Angeles, 69-70; sr res scientist, 70-79, chief process engr, 79-80, prin res scientist, 80-84, DIR RES & DEVELOP, TELEDYNE WAH CHANG, ALBANY, ORE, 85- *Mem:* Am Chem Soc; Am Inst Mining Metall & Petrol Engrs; AAAS; Am Ceramics Soc; Vacuum Soc. *Res:* Extractive metallurgy of group IVA, VA and VIA metals; halide chemistry of groups III, IV, V and VI; chemical thermodynamics; material properties of metals and refractory hard materials; solar energy materials. *Mailing Add:* 37699 Govier Pl Corvallis OR 97330-9317

HAYGOOD, MARGO GENEVIEVE, b Chicago, Ill, Oct 20, 54; m 82. CONTRACT RESEACH, MARINE MICROBIOLOGY. *Educ:* Harvard Univ, BA, 76; Scripps Inst Oceanog, Univ Calif, San Diego, PhD(marine biol), 84. *Prof Exp:* Sr res fel, Univ Wash, 84-85; sci officer admin, 85-88, ASST PROF UCSD, OFF NAVAL RES, 88- *Concurrent Pos:* Mem adv comt to dir, NIH, 76; Monbusho scholar, Univ Tokyo, 79-80; assoc res scientist, Johns Hopkins Univ & Chesapeake Bay Inst, 85-88. *Mem:* Am Soc Microbiol. *Res:* Marine microbiology; microbial physiology; molecular genetics; iron regulation of bacterial bioluminescene; molecular biology of bacterial symbiosis; genetics of methylotrophic bacteria. *Mailing Add:* Scripps Inst Oceanog MBRD A-002 Univ Calif La Jolla CA 92093-0202

HAYGREEN, JOHN G, b Champaign, Ill, Oct 10, 30; m 52; c 2. WOOD TECHNOLOGY. *Educ:* Iowa State Univ, BS, 52; Mich State Univ, MS, 58, PhD(forest prod), 61. *Prof Exp:* Asst prof forest prod, Mich State Univ, 58-61; asst prof Colo State Univ, 61-63; from assoc prof to prof forest prod, Univ Minn, St Paul, 63-86, head dept, Col Forestry, 71-84; ASSOC DEAN, SCH FORESTRY, AUBURN UNIV, 86- *Concurrent Pos:* Consult wood sci serv; dir, Viking Eng & Develop. *Mem:* Soc Wood Sci & Tech (pres, 69); Forest Prod Res Soc (pres, 77-78); Am Soc Test & Mat; Brit Inst Wood Sci; Soc Am Foresters; fel Int Acad Wood Sci. *Res:* Wood mechanics; mechanical behavior of fiber and particle products; applications to building design; author of 50 papers on wood utilization. *Mailing Add:* 38 Dogwood Trail Dadeville AL 36853

HAYLES, WILLIAM JOSEPH, b Dunellen, NJ, Jan 10, 27; m 48; c 3. PHYSICAL CHEMISTRY, INORGANIC CHEMISTRY. *Educ:* Wesleyan Univ, BA, 50; Iowa State Univ, PhD(phys chem), 59. *Prof Exp:* From asst prof to assoc prof chem, Rochester Inst Tehcnol, 54-64, prof, 64-80; AT AT & T BELL LABS, 80- *Concurrent Pos:* Vis prof, Mich State Univ, 69-70. *Mem:* Am Chem Soc. *Res:* Applications of computers in chemical instrumentation. *Mailing Add:* Rm 7P530 AT&T Bell Labs 600 Mountain Ave Murray Hill NJ 07974-2008

HAYMAKER, RICHARD WEBB, b San Francisco, Calif, Feb 13, 40; m 70; c 2. THEORETICAL PHYSICS. *Educ:* Carleton Col, BA, 61; Univ Calif, Berkeley, PhD(physics), 67. *Prof Exp:* Asst res physicist, Univ Calif, Santa Barbara, 67-69, lectr physics, 69; instr & res assoc, Cornell Univ, 69-70, actg asst prof, 70-71; asst prof, 71-76, assoc prof, 76-82, PROF PHYSICS, LA STATE UNIV, BATON ROUGE, 82- *Concurrent Pos:* Vis staff mem, Los Alamos Sci Lab, NMex, 77-78; vis scientist, KEK Nat Lab, Tsukuba, Japan, 85, Tata Inst, Bombay, India, 86. *Mem:* Am Phys Soc. *Res:* Theoretical physics; study of elementary particle dynamics and symmetries. *Mailing Add:* Dept Physics & Astron La State Univ Baton Rouge LA 70803

HAYMAN, ALAN CONRAD, b Wilmington, Del, Nov 19, 47; m 69. LIQUID CHROMATOGRAPHY. *Educ:* Univ Del, BS, 69; Ga Inst Technol, PhD(anal chem), 75. *Prof Exp:* Chemist, Union Carbide Corp, 75-77; appln chemist, DuPont Instruments, 77-80, sr engr, 80, ANAL CHEMIST, CENT RES & DEVELOP DEPT, E I DU PONT DE NEMOURS & CO, INC, 80- *Mem:* Am Chem Soc. *Res:* Analysis of surfactants by high performance liquid chromatography; laboratory automation in the high performance liquid chromatography laboratory; preparative high performance liquid chromatography; preparative isolation of proteins by high performance liquid chromatography. *Mailing Add:* 21 Cobble Creek Curve Newark DE 19702-2412

HAYMAN, ERNEST PAUL, b Providence, RI, Aug 37, 46. AGRICULTURAL BIOCHEMISTRY. *Educ:* Univ RI, BS, 68, PhD(agr chem), 73. *Prof Exp:* Nat Res Coun res assoc, 73-75, RES CHEMIST, FRUIT & VEG CHEM LAB, AGR RES SERV, USDA, 75- *Res:* Study and control of carotenoid, rubber and terpenoid biosynthesis and other related pathways in plants and microorganisms. *Mailing Add:* 2118 E Crary St Pasadena CA 01104

HAYMAN, SELMA, b New York, NY, May 18, 31. ENZYMOLOGY. *Educ:* Antioch Col, BS, 53; Univ Wis, MS, 56, PhD(biochem), 61. *Prof Exp:* Proj asst, Univ Wis, 57-61; res fel ophthal biochem, Howe Lab, Harvard Med Sch, 61-65; res assoc, Inst Cancer Res, 65-73; RES ASSOC CHEM, UNIV DEL, 73- *Mem:* Am Chem Soc. *Res:* Isolation and characterization of enzymes; study of enzyme active sites; study of relationship of enzyme structure to function. *Mailing Add:* 32 Rose Circle Newark DE 19711

HAYMES, ROBERT C, b New York, NY, July 3, 31; m 65; c 3. ASTROPHYSICS. *Educ:* NY Univ, BA, 52, MS, 53, PhD(physics), 59. *Prof Exp:* Asst, NY Univ, 52-58, from instr to asst prof physics, 58-62; res fel, Jet Propulsion Lab, Calif Inst Technol, 62-64; from asst prof to assoc prof, 64-72, chmn dept, 82-87, PROF SPACE PHYSICS & ASTRON, RICE UNIV, 72- *Concurrent Pos:* Chmn adv panel sci ballooning, Univ Corp Atmospheric Res, 72-74; assoc, Baker Col, 78-82,; master, Will Rice Col, 82-87; secy-treas, High Energy Astrophys Div, Am Astron Soc, 87-90; chief scientist, Astrophys Div, NASA Hq, Wash, DC, 88-90. *Honors & Awards:* Humboldt Award, Alexander von Humboldt Soc, 76. *Mem:* Am Astron Soc; Int Astron Union; Am Asn Univ Professors; Sigma Xi. *Res:* Cosmic rays; gamma ray astronomy; conducts measurements of nuclear gamma rays from various cosmic and atmospheric sources; cosmic ray neutrons. *Mailing Add:* Space Physics & Astron Dept Rice Univ Houston TX 77251-1892

HAYMET, ANTHONY DOUGLAS-JOHN, b Sydney, Australia, Feb 5, 56; m 84. STATISTICAL MECHANICS, THEORY OF LIQUIDS. *Educ:* Sydney Univ, BSc, 78; Univ Chicago, PhD(chem), 81. *Prof Exp:* Res fel physics, Harvard Univ, 81-83; ASST PROF CHEM, UNIV CALIF, BERKELEY, 83- *Concurrent Pos:* Presidential Young Investr Award, NSF, 85-90; Alfred P Sloan res fel, 86. *Mem:* Am Chem Soc; Am Phys Soc; Royal Australian Chem Inst. *Res:* Theoretical physical chemistry; theory and applications of statistical mechanics and quantum mechanics; proton transfer in solution; melting, freezing, glass formation nucleation and crystal growth; theory of liquids, liquid crystals and plastic crystals. *Mailing Add:* Dept Chem Uni Utah Salt Lake City UT 84112-1102

HAYMOND, HERMAN RALPH, b Salt Lake City, Utah, Aug 29, 24; m 49; c 6. MEDICAL PHYSICS, MEDICAL IMAGING. *Educ:* Univ Calif, AB, 44, PhD(biophys), 55. *Prof Exp:* Physicist, Radiation Lab, Univ Calif, 47-55; from asst prof to assoc prof, 55-68, PROF RADIOL, SCH MED, UNIV SOUTHERN CALIF, 68- *Concurrent Pos:* Radiation physicist, Los Angeles County-Univ Southern Calif Med Ctr, 55-70, head radiation physicist, 70-74, head med radiation physicist, 74-86. *Mem:* Radiol Soc NAm; Biophys Soc; Radiation Res Soc; Am Chem Soc; Am Asn Physicists Med. *Res:* Radiologic physics; radiation chemistry; radiation biology; medical imaging. *Mailing Add:* Sch Med Box 304 Univ Southern Calif 1200 N State St Los Angeles CA 90033

HAYMORE, BARRY LANT, b Melbourne, Australia, Nov 1, 45; US citizen; m 69; c 7. ORGANOMETALLIC CHEMISTRY, HOMOGENEOUS CATALYSIS. *Educ:* Univ Ariz, BSc, 68; Brigham Young Univ, MSc, 70; Northwestern Univ, PhD(inorg chem), 74. *Prof Exp:* Asst prof inorg chem, Ind Univ, Bloomington, 75-80; sr res scientist, 80-90, SCI FEL, CORP RES LABS, MONSANTO CO, ST LOUIS, MO, 90- *Mem:* Am Chem Soc; Sigma Xi. *Res:* Coordination chemistry and reactions of pi-bonded nitrogen containing ligands; mechanisms of homogeneous catalysts; chemistry and structures of macrocyclic ligands; interactions of proteins with transition metal ions; inorganic chemistry. *Mailing Add:* Monsanto Co Corp Res Labs 800 N Lindbergh Blvd St Louis MO 63166

HAYMOVITS, ASHER, b Jerusalem, Israel, Dec 3, 33. MEDICINE, BIOCHEMISTRY. *Educ:* Hebrew Univ, Israel, BA, 54, MD, 58. *Prof Exp:* Intern med, Hadassah Hosp, Israel, 58-59; resident, Montefiore Med Ctr, Bronx, NY, 59-60; asst, New York Hosp-Cornell Med Ctr, 60-61; resident, Med Col, Cornell, Bellevue & Mem Hosps, 61-62; asst med & endocrinol, Sloan-Kettering Div, Cornell Univ, 62-64; res assoc & asst physician, 64-67, asst prof med & assoc physician, Rockefeller Univ, 67-76; ASSOC PROF MED, DOWNSTATE MED CTR, NEW YORK, 77-; ASSOC PROF MED, ROCKEFELLER UNIV, 80- *Concurrent Pos:* Res fel, New York Hosp-Cornell Med Ctr, 60-61 & Sloan-Kettering Div, Cornell Univ, 62-64; NIH res grants, 67-69. *Mem:* AAAS; Am Chem Soc; NY Acad Sci; Endocrine Soc. *Res:* Control of metabolic processes that involve skeletal tissue and skeletal mineral; regulation of mineral concentrations in biological fluids; biochemistry of calcification. *Mailing Add:* Dept Med 1230 York Ave New York NY 10621-6399

HAYN, CARL HUGO, b Los Angeles, Calif, July 13, 16. PHYSICS. *Educ:* Gonzaga Univ, AB, 39, AM, 40; Alma Col, STL, 48; St Louis Univ, PhD(physics), 55. *Prof Exp:* Teacher pvt sch, Calif, 40-43; instr physics, Loyola Univ, Calif, 43-44; from instr to assoc prof, 55-74, PROF PHYSICS, UNIV SANTA CLARA, 74- *Mem:* Am Asn Physics Teachers. *Res:* Solid state; photoconductivity and photoemission; nuclear spectroscopy. *Mailing Add:* Dept Physics Univ Santa Clara Santa Clara CA 95053

HAYNE, DON WILLIAM, b Elgin, Ill, Apr 27, 11; m 60; c 3. BIOMETRICS. *Educ:* Kalamazoo Col, AB, 32; Univ Mich, MA, 33, PhD(zool), 47. *Prof Exp:* Asst mus zool & lab vert genetics, Univ Mich, 33-37; asst econ vert zool, Agr Exp Sta, Mich State Col, 37-45, from asst prof to assoc prof zool, 45-57; biometrician, Inst Fisheries Res, State Dept Conserv, Mich, 57-60; staff biometrician, Br Wildlife Res, US Fish & Wildlife Serv, 60-62; tech dir, 62-81, EMER PROF, SOUTHEASTERN COOP FISH & GAME STATIST PROJ, INST STATIST & PROF ZOOL & STATIST, NC STATE UNIV, 81- *Concurrent Pos:* Fel, Univ Chicago, 53-54. *Mem:* AAAS; Am Soc Mammal; Wildlife Soc; Ecol Soc Am; Am Fisheries Soc; Sigma Xi. *Res:* Quantitative ecology; fish and wildlife statistics; pesticide ecology; variation of the oldfield mouse in northwestern Florida. *Mailing Add:* Inst Statist NC State Univ Box 8203 Raleigh NC 27695-8203

HAYNES, ALFRED, b Guyana, SAm, Nov 17, 21. PREVENTIVE MEDICINE. *Educ:* Down State Med Sch, MD, 54; Harvard Univ, MPh(pub health), 63. *Prof Exp:* Asst prof pre med, Univ Vt, 64-69; prof, Drew Med Sch, 74-77, assoc dean, Community Med, Drew Med Sch, 79-86; planning dir, Drew-Meharry-Morehouse Cancer Consortium Ctr, 86-90; CONSULT, 86- *Concurrent Pos:* Mem, US Task Force Pre Health Serv, 85-; chmn, Steering Count, Nat Cancer Res Network, 86-; mem, Comt Health & Human Rights, Inst Med, 86- *Mem:* Inst Med-Nat Acad Sci; AAAS. *Res:* Published numerous articles in various journals. *Mailing Add:* 29249 First Ridge Rd Rancho Palos Verdes CA 90274

HAYNES, BOYD W, JR, b Brandenburg, Ky, July 5, 17; m 55; c 6. SURGERY. *Educ:* Univ Louisville, AB, 38, MD, 41; Am Bd Surg, dipl. *Prof Exp:* Instr surg, Med Col Va, 48-49; asst prof, Baylor Col Med, 50-53; from asst prof to assoc prof, 53-66, chmn, div trauma surg, 72-82, DIR BURN UNIT, MED COL VA, COMMONWEALTH UNIV, 54-, PROF SURG, 66- *Mem:* Fel Am Col Surg; Am Surg Asn; Soc Univ Surg; Am Asn Surg of Trauma; Am Burn Asn (past pres); Int Soc Burn Injury; Int Soc Surg; Surg Infection Soc. *Res:* Treatment of burns; author of 77 publications on burns and trauma. *Mailing Add:* Dept Surg Va Commonwealth Univ Box 661 MCV Sta Richmond VA 23298

HAYNES, CALEB VANCE, JR, b Spokane, Wash, Feb 29, 28; m 54; c 1. QUATERNARY GEOLOGY, RADIOCARBON DATING. *Educ:* Colo Sch Mines, Geol Engr, 56; Univ Ariz, PhD(geol), 65. *Prof Exp:* Sr engr, Am Inst Res, 56-59; sr engr, Martin Co, 60-62; asst prof geochronol, Univ Ariz, 65-68; from assoc prof to prof geol, Southern Methodist Univ, 68-74; PROF GEOL & ANTHROP, UNIV ARIZ, 74- *Concurrent Pos:* Consult, Nat Geog Soc, 62- *Honors & Awards:* Roald Fryxell Award, 78; Archaeol Geol Award, Geol Soc Am. *Mem:* Nat Acad Sci; Am Quaternary Asn (pres, 76-78); AAAS; Geol Soc Am; Soc Am Archaeol; Sigma Xi. *Res:* Pleistocene geology; Paleo-Indian archaeology; lunar geology; geochronology of late Quaternary time; shuttle imaging radar and geomorphology of the eastern Sahara, Egypt and Sudan; archaeological geology of Custer Battlefield National Monument. *Mailing Add:* Dept Anthrop Univ Ariz Tucson AZ 85721

HAYNES, CHARLES W(ILLARD), metallurgical engineering, for more information see previous edition

HAYNES, DEAN L, b Mich, Sept 27, 32; m 56; c 3. PLANT PROTECTION, POPULATION DYNAMICS. *Educ:* Mich State Univ, BS, 54, MS, 57, PhD(entom), 60. *Prof Exp:* Forest biologist, Can Dept Forestry, 60-65; asst prof entom, NDak State Univ, 65-66; from asst prof to assoc prof, 66-74, PROF ENTOM, MICH STATE UNIV, 74- *Res:* Population ecology of insect pests of agriculture crops; design and management of environmental systems for pest control; pest management and the analysis of agroecosystems. *Mailing Add:* Dept Entom 49 Nat Sci Bldg Mich State Univ East Lansing MI 48824

HAYNES, DOUGLAS MARTIN, b New York, NY, Jan 25, 22; m 61; c 2. OBSTETRICS & GYNECOLOGY. *Educ:* Southern Methodist Univ, BA & BS, 43; Univ Tex, MD, 46; Am Bd Obstet & Gynec, dipl. *Prof Exp:* Asst prof obstet & gynec, Univ Tex Southwestern Med Sch, 52-55; assoc prof, 55-57, chmn dept, 57-69, interim dean, 69-70, dean 70-72, PROF OBSTET & GYNEC, SCH MED, UNIV LOUISVILLE, 57- *Mem:* Fel Am Col Obstet & Gynec; fel AMA; fel Am Gynec & Obstet Soc; Am Col Surgeons. *Res:* Experimental abruptio placentae in the rabbit; uterine blood volume; medical complications of pregnancy; cesarean hysterectomy. *Mailing Add:* Univ Louisville Sch Med ACB 550 S Jackson St Louisville KY 40292

HAYNES, DUNCAN HAROLD, b Owosso, Mich, May 27, 45; m 74; c 3. DRUG-CALCIUM INTERACTION, SUSTAINED-RELEASE DRUG DELIVERY. *Educ:* Butler Univ, BS, 66; Univ Pa, PhD(molecular biol), 70. *Prof Exp:* Postdoctoral fel biophys, Max-Planck Inst, Göttingen, Fed Repub Ger, 70-73; from asst prof to assoc prof, 73-82, PROF PHARMACOL, SCH MED, UNIV MIAMI, 82- *Concurrent Pos:* Consult, 79-; vis prof, Inst Path, Univ Hamburg, Fed Repub Ger, 85; pres, Pharma-Logic, Inc, 85- *Mem:* Biophys Soc; Am Soc Pharmacol & Exp Therapeut; Soc Gen Pharmacologists; Soc Clin Invest; fel Am Col Clin Pharmacol; Am Physiol Soc. *Res:* Calcium signalling and handling in the human blood platelet; use of calcium channel blocking drugs as anti-platelet drugs in thrombosis; inventor of the patented phospholipid-coated microdroplet and microcrystal serving as injectable delivery system for water- insoluble drugs; commercial development. *Mailing Add:* Dept Pharmacol Univ Miami Med Sch PO Box 016189 Miami FL 33101

HAYNES, EMANUEL, b New York, NY, Oct 17, 16; m 42; c 3. AERONAUTICAL & ASTRONAUTICAL ENGINEERING. *Educ:* City Col New York, BME, 37; Polytech Inst Brooklyn, MME, 40. *Prof Exp:* Sales engr, Niagara Mach & Tool Works, 37-39; jr naval architect, Philadelphia Naval Shipyard, 39-40; chief eng draftsman, NY Naval Shipyard, 40-43; mech engr, Franque A Dickins, Inc, 43-45, Bur Ships, US Navy, 45-46 & Off Naval Res, 46-51; physicist, Off Chief of Ord, US Army, 51-52; dep dir aeronaut sci, Off Sci Res, US Air Force, 52-59; prog mgr, Adv Res Projs Agency, Dept Defense, 59-64; dir adv planning div, Off Naval Res, 64-66; sr staff assoc, Nat Sci Found, 69-80; consult, Nat Acad Sci, George Washington Univ, 80-83; RETIRED. *Concurrent Pos:* Mem panel on hydrol, Res & Develop Bd, Dept Defense, 49-51; mem subcomt internal flow, Nat Adv Comt Aeronaut, 51, fluid mech, 55-59; asst sec defense, Interserv Comn on Shock & Vibration, 56-59; chmn coord reps mech, Dir Defense Res & Eng, 56-59. *Res:* Fluid mechanics; aerodynamics; aeroelasticity; structures; propulsion; research management. *Mailing Add:* 6404 Earlham Dr Bethesda MD 20817

HAYNES, EMMIT HOWARD, b Irvington, Ky, Sept 21, 26. ANIMAL NUTRITION. *Educ:* Univ Ky, BSA, 51, MSA, 53; Cornell Univ, PhD(animal nutrit), 59. *Prof Exp:* asst prof animal husb & exten specialist, Iowa State Univ, 56-59, from assoc prof to prof animal sci, 59-88, leader resident instr, 59-88; RETIRED. *Mem:* Am Soc Animal Sci. *Res:* Ruminant nutrition. *Mailing Add:* 12415 Nassau Lane Middletown KY 40243

HAYNES, FRANK LLOYD, JR, b Ala, July 29, 20; m 79; c 3. PLANT BREEDING, GENETICS. *Educ:* Ala Polytech Inst, BS, 46; Cornell Univ, PhD(plant breeding), 51. *Prof Exp:* From asst prof to assoc prof, PROF HORT SCI, NC STATE UNIV, 51-, COORDR, DEPT GRAD PROGS, 78- *Concurrent Pos:* NSF grant genetics, 58-63; mem, Agency Int Develop team, in charge potato res, NC State Univ mission, Peru, 70. *Mem:* Am Genetic Asn; Am Soc Hort Sci. *Res:* Cytogenetics; evolution in solanum; adaptation of highland tropical species to temperate zone. *Mailing Add:* Dept Hort Sci NC State Univ Raleigh NC 27650

HAYNES, GEORGE RUFUS, b Nashville, Ark, Oct 30, 28; m 57; c 4. SYNTHETIC ORGANIC CHEMISTRY. *Educ:* Ark State Teachers Col, BS, 50; Univ Ark, MS, 51; Univ Tex, PhD(org chem), 57. *Prof Exp:* Chemist, Agr Res Div, 56-63, group leader synthesis vet drugs & pharmaceut, 63-68, SUPVR ORG CHEM, SHELL DEVELOP CO, 68- *Mem:* Am Chem Soc; Am Inst Chemists. *Res:* Divalent organosulphur compounds and their activity as nematocides and fungicides; synthesis of human and veterinary drugs; synthesis of animal growth improvers; scale-up synthesis of development materials. *Mailing Add:* 11706 Gardenglen Dr Houston TX 77070-2808

HAYNES, HENRY WILLIAM, JR, b Memphis, Tenn, Aug 15, 42; m 67; c 3. CHEMICAL ENGINEERING. *Educ:* Univ Miss, BS, 64; Univ Colo, MS, 66, PhD(chem eng), 69. *Prof Exp:* Res engr synthetic fuels, Esso Res & Eng Co, 68-71; assoc prof, 71-80, PROF CHEM ENG, UNIV MISS, 80- *Mem:* Am Inst Chem Eng; Am Chem Soc; Catalysis Soc; Sigma Xi. *Res:* Chemical reaction engineering, heterogeneous catalysis, synthetic fuels. *Mailing Add:* Dept Chem Eng Univ Wyo Laramie WY 82071

HAYNES, JAMES MITCHELL, b Forest Grove, Ore, May 8, 51; m 73. AQUATIC ECOLOGY. *Educ:* Carleton Col, BA, 73; Univ Minn, MS, 75, PhD(ecol), 78. *Prof Exp:* Res fel ecosyst, Battelle Pacific NW Lab, 75-77, asst prof, 78-83, ASSOC PROF BIOL, STATE UNIV NY, BROCKPORT, 84- *Concurrent Pos:* Prin investr grants, 78- *Mem:* Am Fisheries Soc; Ecol Soc Am; Int Asn Great Lakes Res. *Res:* Aquatic ecology salmonid ecology, biotelemetry and impacts of costal development. *Mailing Add:* Dept Biol Sci State Univ NY Brockport NY 14420

HAYNES, JOHN J, b Dallas, Tex, Sept 13, 25; m 48. CIVIL & HIGHWAY ENGINEERING. *Educ:* Tex Tech Col, BS, 49; Tex A&M Univ, MEng, 59, PhD(civil eng), 64. *Prof Exp:* Asst resident engr, Tex Hwy Dept, 49-51; assoc prof, 51-62, head dept, 64-72, PROF CIVIL ENG, UNIV TEX, ARLINGTON, 64- *Mem:* Am Soc Civil Engrs; Inst Transp Engrs; Am Soc Eng Educ; Nat Soc Prof Engrs. *Res:* Transportation and traffic engineering, particularly traffic flow theory and public transportation. *Mailing Add:* 3508 Rainer Dr Arlington TX 76016-3906

HAYNES, JOHN KERMIT, b Monroe, La, Oct 30, 43; m 69. CELL BIOLOGY, DEVELOPMENTAL BIOLOGY. *Educ:* Morehouse Col, BS, 64; Brown Univ, PhD(develop biol), 70. *Prof Exp:* Asst prof molecular med, Meharry Med Col, 73-78; ASSOC PROF & DIR HEALTH PROFESSIONS PROG BIOL, MOREHOUSE COL, 78- *Concurrent Pos:* Res fel molecular biol, Brown Univ, 70-71; res assoc biochem, Mass Inst Technol, 71-73; NSF grant, 76-78; Nat Found March Dimes grant, 77-78; NIH grant, 78- *Mem:* AAAS; Nat Inst Sci; Red Cell Club. *Res:* Molecular and cellular aspects of sickle cell disease; structure of cell membranes; molecular aspects of differentiation. *Mailing Add:* Dept Biol Morehouse Col Atlanta GA 30314

HAYNES, JOHN LENNEIS, b Washington, DC, Mar 25, 34; m 55; c 3. DRUG DELIVERY, PERSONAL COMPUTER SYSTEMS. *Educ:* Cornell Univ, BEE, 56; Stanford Univ, MSEE, 58; Cornell Univ, MSEE, 89. *Prof Exp:* Res engr, SRI Int, 56-61; chief engr, Pac Commun & Elctronics, 62-65, BD Electronics Lab, 65-77, gen mgr, 77-79; ASSOC DIR, BECTON DICKINSON RES CTR, 79- *Honors & Awards:* Grand Award, Circuit Design, Electronic Equip Eng, 60. *Mem:* Sr mem Inst Elec & Electronics Engrs; Asn Advan Med Instrumentation. *Res:* Health care device and instrument research and development; high performance particle analysis and sorting; electrically-assisted transdermal drug delivery; computer-aided engineering analysis in non-Newtonian systems. *Mailing Add:* 20 Kendall Dr Chapel Hill NC 27514

HAYNES, LEROY WILBUR, b Queens, NY, Jan 31, 34; m 58; c 2. ORGANIC CHEMISTRY. *Educ:* Drew Univ, BA, 56; Univ Ill, Urbana, PhD(org chem), 61. *Prof Exp:* From instr to assoc prof, 61-74, PROF CHEM, COL WOOSTER, 74- *Concurrent Pos:* NSF exten grant, 66-68; Am Chem Soc-Petrol Res Found grant, 70-71; NSF res grant, 76-77. *Mem:* Am Chem Soc; Sigma Xi. *Res:* Synthesis and reactions of various heterocyclic systems; isolation and characterization of natural products. *Mailing Add:* Dept Chem Col Wooster Wooster OH 44691

HAYNES, MARTHA PATRICIA, b Boston, Mass, April 24, 51. EXTRAGALACTIC & SPECTRAL LINE RADIO ASTRONOMY. *Educ:* Wellesley Col, Mass, BA, 73; Ind Univ, MA, 75, PhD(astron), 78. *Prof Exp:* Fel res assoc, Nat Astron & Ionosphere Ctr, Arecibo Observ, 78-80; ASST SCIENTIST & ASST DIR, GREEN BANKS OPERS, NAT RADIO ASTRON OBSERV, WVA, 81-; ASSOC PROF ASTRON, DEPT ASTRON, CORNELL UNIV, 81- *Mem:* Am Astron Soc; Sigma Xi; Int Astron Union; Int Union Radio Sci; NY Acad Sci; AAAS. *Res:* Space distribution of galaxies and the effect of intergalactic environment on the gas content of galaxies, primarily via the 21 centimeter line of neutral hydrogen. *Mailing Add:* Dept Astron Space Sci Bldg Cornell Univ Ithaca NY 14853

HAYNES, MUNRO K, b Elmira, NY, Dec 10, 23. MAGNETIC RECORDING, MAGNETO-OPTIC RECORDING. *Educ:* Univ Rochester, BS, 47; Univ Ill, MS, 48, PhD(elec eng), 50. *Prof Exp:* Res assoc, IBM, Tucson, Ariz, 50-60, mem tech staff, mgr res planning staff, dir eng planning & dir storage technol, 60-71, res eng, 71-78, res scientist, 78-88; RETIRED. *Mem:* Fel Inst Elec & Electronics Engrs; Sigma Xi. *Res:* Magnetic and magneto-optic media; recording techniques for digital information storage; media evaluation; channel characterization; coding; test methods. *Mailing Add:* 3311 E Terra Alta Blvd Tucson AZ 85716

HAYNES, N BRUCE, b Ossining, NY, Sept 26, 26. VETERINARY MEDICINE. *Educ:* Baldwin-Wallace Col, BS, 48; Cornell Univ, DVM, 52. *Prof Exp:* Pvt practice, 52-64; exten vet, NY State Col Vet Med, Cornell Univ, 65-79, assoc prof vet sci, 67-79; vet consult, 79-87; RETIRED. *Concurrent Pos:* Chmn bd dirs, Cornell Vet, Inc, 72-77; dir continuing educ, NY State Col Vet Med, 76-79. *Mem:* Am Vet Med Asn; Am Asn Vet Clinicians; Am Asn Vet Nutritionists; Am Asn Bovine Practr; US Animal Health Asn. *Res:* Cattle diseases and nutrition. *Mailing Add:* Cottage rd PO Box 9 Winthrop ME 04364-0009

HAYNES, RALPH EDWARDS, b Freeman, WVa. PEDIATRICS, MEDICAL MICROBIOLOGY. *Educ:* Med Col Va, MD, 49; Am Bd Pediat, dipl, 58. *Prof Exp:* Intern, Cincinnati Gen Hosp, 49-50; resident, Cincinnati Children's Hosp, 55-56, chief resident, 56-57; asst dir infectious dis div/sect, Columbus Children's Hosp, 65-73; from asst prof to assoc prof, 65-73, prof pediat & med microbiol, Col Med, Ohio State Univ, 73-78; dir infectious dis div/sect, Columbus Children's Hosp, 74-78; PROF & V CHMN DEPT

PEDIAT & PROF MICROBIOL & IMMUNOL, WRIGHT STATE UNIV SCH MED, 78-; DIR INFECTIOUS DISEASES, CHILDREN'S MED CTR, 78- *Concurrent Pos:* Fel, Columbus Children's Hosp, 65-66. *Honors & Awards:* Basic Sci Teaching Award, 74. *Mem:* Infectious Diseases Soc Am; Am Pediat Soc; Am Soc Microbiol; Fel Am Acad Pediat. *Res:* Viral central nervous system infections in children; chemotherapy of herpes virus infections; epidemiologic study of immunizations; evaluation of bivalent intranasal influenza vaccine; infectious diseases. *Mailing Add:* 2949 Cunnington Lane Dayton OH 45420

HAYNES, ROBERT BRIAN, b Calgary, Alta, Mar 1, 47; m 75; c 2. CLINICAL EPIDEMIOLOGY, MEDICAL INFORMATION. *Educ:* Univ Alta, BSc, 71, MD, 71; McMaster Univ, MSc, 73 & PhD(med sci), 75. *Prof Exp:* Resident internal med, Toronto Gen Hosp, 71-72 & 75-76; registr internal med & hon lectr clin epidemiol, St Thomas Hosp, 76-77; from asst prof to assoc prof, 77-85, PROF CLIN & EPIDEMIOL & MED, MCMASTER UNIV, 85- *Concurrent Pos:* Vis prof med, Univ Toronto, 84-85; chmn, Ad Hoc Working Group Patient Compliance, Us Nat Heart, Lung & Blood Inst, 84-85; dir, Prog Educ Develop, McMaster Univ, 81-87, chief, Health Info Res Unit, 87-; nat health scientist, Nat Health Res & Develop Prog, Health & Welfare Can, 87- *Honors & Awards:* Young Invest Award, Can Hypertension Soc, 80; Nat Health Scientist Award, Nat Health Res & Develop Prog, Can, 87. *Mem:* Am Fed Clin Res; fel Am Heart Asn; Can Soc Clin Invest; Can Hypertension Soc(secy & treas, 85-88, pres, 88-89); Acad Behav Med Res. *Res:* Clinical trials of treatment for vascular disorders such as stroke and hypertension; development and testing of information systems designed to reduce the gap between knowledge and practice in health care. *Mailing Add:* Dept Clin Epidemiol McMaster Univ Med Ctr 1200 Main St W Hamilton ON L8N 3Z5 Can

HAYNES, ROBERT C, b Boston, Mass, May 14, 38; m 61; c 2. LIMNOLOGY. *Educ:* Mass State Col Bridgewater, BSEd, 66; Univ NH, MS, 68, PhD(zool), 71. *Prof Exp:* Fel biol, Univ Sask, 71-73, asst prof, 73-74; asst prof, 74-81, ASSOC PROF BIOL, SANGAMON STATE UNIV, 81- *Mem:* Am Soc Limnol & Oceanog; Int Asn Theoret & Appl Limnol. *Res:* Ecology of eutrohic and saline lakes. *Mailing Add:* 1009 14 1/2 St Rock Island IL 61201

HAYNES, ROBERT CLARK, JR, b Springfield, Mo, Aug 17, 25; m 53; c 3. PHARMACOLOGY. *Educ:* Wash Univ, MD, 48; Harvard Univ, PhD(biochem), 53. *Prof Exp:* Biochemist, Worcester Found Exp Biol, 51-53; from sr instr to assoc prof pharmacol, Western Reserve Univ, 54-69; PROF PHARMACOL, SCH MED, UNIV VA, 69- *Mem:* AAAS; Endocrine Soc; Am Soc Pharmacol & Exp Therapeut; Am Soc Biol Chem. *Res:* Mechanism of action of hormones. *Mailing Add:* Dept Pharmacol Univ Va Sch Med Charlottesville VA 22908

HAYNES, ROBERT HALL, b London, Ont, Aug 27, 31; m 54; c 3. DNA REPAIR, MUTAGENESIS. *Educ:* Western Ont Univ, BSc, 53, PhD(biophys), 57. *Prof Exp:* Brit Empire Cancer Campaign fel physics, St Bartholomew's Hosp Med Col, Univ London, 57-58; from instr to asst prof biophys, Univ Chicago, 59-64; assoc prof biophys & med physics, Univ Calif, Berkeley, 64-68; chmn dept, 68-73, PROF BIOL, YORK UNIV, ONT, 68-, DISTINGUISHED RES PROF, 86- *Concurrent Pos:* Mem subcomt radiobiol, Nat Acad Sci-Nat Res Coun, 63-73; biophysicist, Lawrence Berkeley Lab, 64-68; assoc prof, dept instrnl biol, Univ Calif, Berkeley, 66-68; mem exp adv comt, Atomic Energy Can Ltd, 69-72, tech adv comt nuclear fuel waste mgt, 79-83; exchange visitor, USSR Acad Sci, 72, Japan Soc Promot Sci, Kyoto, 79 & Acad Sinica, Beijing, China, 80; Brit Coun Asn Commonwealth Univ exchange fel, Bot Sch, Oxford Univ, 73; vis fel molecular biophys, Yale Univ, 74-75; bd mem, Nat Res Coun Can, 75-82; chmn ministerial adv comt environ mutagenesis, Can Dept Nat Health & Welfare, 78-85; assoc fel & mem res coun, Can Inst Advan Res, 82-87; mem, Int Comn Protection against Environ Mutagens & Carcinogens, 87-; pres, Int Cong Genetics, Toronto, 88; pres, Int Asn Environ Mutagen Socs; assoc fel, Third World Acad Sci, 90- *Honors & Awards:* Ann Res Award, Environ Mutagen Soc, 84; Gold Medal, Biol Coun Can, 84; Flavelle Medal, Royal Soc Can, 88; Presidential Citation, Genetics Soc Can, 89; Officer of Order of Can, 90. *Mem:* Fel AAAS; Radiation Res Soc; Genetics Soc Can, (vpres, 81-82, pres, 83-85); Genetics Soc Am; Environ Mutagen Soc; fel Royal Soc Can; Indian Sci Cong Asn. *Res:* Molecular biology; radiation genetics; environmental mutagenesis; photobiology; DNA repair; blood rheology; radiological physics; yeast genetics; mathematical analysis of mutagen dose-response relations; deoxynucleotide metabolism. *Mailing Add:* Dept Biol York Univ Toronto ON M3J 1P3 Can

HAYNES, ROBERT RALPH, b Minden, La, Feb 24, 45; m 67; c 2. PLANT TAXONOMY. *Educ:* La Polytech Inst, BS, 67; Univ Southwestern La, MS, 69; Ohio State Univ, PhD(bot), 73. *Prof Exp:* Spec lectr bot, Ohio State Univ, 73-74; asst prof biol, La State Univ, Shreveport, 74-76; from asst prof to assoc prof, 76-85, PROF BIOL, UNIV ALA, 85- *Mem:* Am Soc Plant Taxonomists; Int Asn Plant Taxonomists; Bot Soc Am; Linnean Soc London. *Res:* Revisionary and phytogeographical investigations of the Alismatidae with special interests in the genera Potamogeton, Najas, Sagittaria, and Echinodorus. *Mailing Add:* Dept Biol Univ Ala Tuscaloosa AL 35487-0344

HAYNES, RONNIE J, b Trumann, Ark, May 12, 44; m 66; c 3. WILDLIFE ECOLOGY. *Educ:* Ark State Univ, BS, 66; Univ Ark, MS, 69; Southern Ill Univ, Carbondale, PhD(zool), 76. *Prof Exp:* Teacher biol & physiol, Merritt Island High Sch, Fla, 68-70; researcher ecol mined lands, Coop Wildlife Res Lab, Southern Ill Univ, Carbondale, 75-76; res assoc environ sci-terrestrial, Oak Ridge Nat Lab, Union Carbide Corp, 76-79; ENVIRON SPECIALIST, US FISH & WILDLIFE SERV, 79- *Mem:* Sigma Xi; Wildlife Soc; Nat Wildlife Fedn; Soc Wetland Scientists; Ecol Soc; Am Soc Surface Mining & Reclamation. *Res:* Reclamation of mined lands such as surface and underground; studies of plant-growth media; plant succession and utilization of mined lands by wildlife; habitat evaluation methods. *Mailing Add:* R B Russell Fed Bldg US Dept Interior 75 Spring St SW Atlanta GA 30303

HAYNES, SHERWOOD KIMBALL, b Dorchester, Mass, Apr 7, 10; m 43; c 4. PHYSICS. *Educ:* Williams Col, AB, 32; Calif Inst Technol, PhD(physics), 36. *Prof Exp:* Instr physics, Williams Col, 36-39; French sci fel, Inst Int Ed, 39-40; asst prof physics, Brown Univ, 40-42; instr elec commun, Radar Sch, Mass Inst Technol, 42-44, asst dir, 44-45; from assoc prof to prof physics, Vanderbilt Univ, 45-57; chmn dept, 57-69, prof, 57-80, EMER PROF PHYSICS & ASTRON, MICH STATE UNIV, 80- *Concurrent Pos:* Trainee & sr physicist, Clinton Labs, Oak Ridge, Tenn, 46-47; Fulbright lectr, Univ Paris, 54-55; vis physicist, Inst Nuclear Physics Res, Amsterdam, 65, Orsay, France, 71-72. *Mem:* Fel AAAS; fel Am Phys Soc; Am Asn Physics Teachers (vpres, 72-73, pres-elect, 73-74, pres, 74-75). *Res:* Cosmic ray intensities at high altitudes; low leakage condensers; decay of fission products as a function of neutron energy; decay schemes of radioactive nuclei; auger effect in radioactive nuclei; disc auger-vacancy satellites. *Mailing Add:* 2821 Mt Hope Rd Okemos MI 48864

HAYNES, SIMON JOHN, b Chelmsford, Eng, Jan 1, 44; Can citizen; m 71; c 4. ECONOMIC GEOLOGY,STRUCTURE. *Educ:* Manchester Univ, Eng, BSc, 65; Carleton Univ, MSc, 69; Queen's Univ, Ont, PhD(geol), 75. *Prof Exp:* Asst geol, Carleton Univ, 65-68 & Queen's Univ, Ont, 68-72; asst prof, Pahlavi Univ, Iran, 72-74; from asst prof to assoc prof, 74-89, PROF GEOL, BROCK UNIV, 89-, CHAIR & VCHMN COUN, 90- *Concurrent Pos:* Consult asbestos, Magcobar Iran Ltd & Fars & Khuzestan Cement Co, 73-74, chloride in cement, St Lawrence Seaway Authority, 76, copper, Nat Iranian Copper Industs, Corp, 76-78, uranium, Golden Goose Mine Ltd, 78-80, Madawaska Mines Ltd, 79 & silver-lead-zinc Ngate Explor Ltd, 80- 82, gold deposits, NS Dept Mines & Energy, 82-84, Geol Surv Can, 84-85; vis scientist, Nanjing Univ, China, 86; gypsum deposits, Ont Min Northern Develop & Mines, 86-90; gold deposits, S China, China Nat Non-Ferrous Metal Corp, China, 87-89; D Bell & Assoc, 87-90; mem, chair res comt, Geol Soc, China Inland Mission. *Honors & Awards:* World Decoration Excellence Medallion, Am Biol Inst, 89. *Mem:* Fel Geol Asn Can; Mineral Asn Can; fel Soc Econ Geologists; AAAS; Can Inst Mining & Metall. *Res:* Processes of ore deposit formation; plate tectonics and metallogenesis; regional structure; gold deposits; gypsum deposits; crushed stone aggregates. *Mailing Add:* Dept Geol Sci Brock Univ St Catharines ON L2S 3A1 Can

HAYNES, SUZANNE G, b Huntington Park, Calif; c 1. SOCIAL EPIDEMIOLOGY, GERONTOLOGY. *Educ:* Univ Tenn, Knoxville, BA, 69; Univ Tex, Austin, MA, 70; Univ Tex, Sch Public Health, Houston, MPH, 72; Univ NC, Chapel Hill, PhD(epidemiol), 75. *Prof Exp:* Res economist, Mayor's Manpower Planning Coun, Houston, 70-71; res assoc epidemiol, Univ NC, Chapel Hill, 74-76; epidemiologist, Nat Heart, Lung & Blood Inst, 75-80; res asst prof epidemiol, Univ NC, Chapel Hill, 80-84; chief, Med Statist Br, Nat Ctr Health Statist, 84-87, CHIEF, HEALTH PROMOTION SCI BR, NAT CANCER INST, 87-; PROJ COORDR, COMMUNITY SURVEILLANCE CARDIOVASC DIS, SOUTHWESTERN US, 81- *Concurrent Pos:* Ed, J Gerontol, 78-81; mem, Biomet & Epidemiol Contract Rev Comt, Nat Cancer Inst, 79-83, exec comt, Am Heart Asn Council Epidemiol, 82-85 & adv comt, Western Ctr Behav & Preventive Med, 80-87; consult, Epidemiol Br, Nat Heart, Lung & Blood Inst, 80-; estab investr, Am Heart Asn, 82-84; exec comt, Soc Epidemiol Res, 84-87; pres, nat Ctr Health Statist Womens Coun, 86; cardiovasc working group chair, Carter Ctr Closing Gap Proj, 86-87; mem, gov coun, Am Pub Health Asn, 87-89 & 90-92; NIH rep, Dept Health & Human Serv, 89-90. *Honors & Awards:* Dix Award, Nat Ctr Health Statist, 88; Ctr Dis Control Statist Award, 89. *Mem:* Fel Acad Behav Med; fel Am Col Epidemiol; fel Soc Behav Med; fel Am Heart Asn. *Res:* Epidemiology of coronary heart diease; type A behavior and risk factors among women; social, psychological, functional characteristics of aging; cancer prevention and control research in smoking, nutrition and breast cancer screening; cardiovasculary epidemiology; cancer epidemiology. *Mailing Add:* Nat Cancer Inst 9000 Rockville Pike Bethesda MD 20892

HAYNES, TONY EUGENE, b Concord, NC, Sept 8, 60; m 83. RADIATION EFFECT IN SEMICONDUCTORS, CRYSTALLINE FILMS. *Educ:* Wake Forest Univ, BS, 82; Univ NC, Chapel Hill, PhD(physics), 87. *Prof Exp:* Fel, Particle-Solid Interactions Div, Sandia Nat Lab, 85-86; MEM RES STAFF, SOLID STATE DIV, OAK RIDGE NAT LAB, 87- *Mem:* Am Phys Soc; Mat Res Soc; Am Vacuum Soc. *Res:* Materials modification by ion beams; direct ion beam deposition of thin films; ion implantation; radiation damage and rapid annealing in compound semiconductors; ion beam analysis. *Mailing Add:* Oak Ridge Nat Lab Bldg 3003-MS6048 PO Box 2008 Oak Ridge TN 37831-6048

HAYNES, WILLIAM MILLER, b Bartlesville, Okla, July 26, 36; m 56; c 4. ANALYTICAL CHEMISTRY. *Educ:* Okla State Univ, BA, 59, PhD(chem), 66. *Prof Exp:* Asst prof chem, Southeast Mo State Col, 65-68; sr res chemist, 68-73, sr res specialist anal chem, corp res dept, Monsanto Co, 73-77; sr res group leader, Environ Anal Sci Ctr, Monsanto Res Corp, 77-80, mgr, process technol, 80-85, DIR, PHYS SCI CTR, MONSANTO CO, 85- *Concurrent Pos:* Instr chem, Meramec Community Col, 70-; adv chem technol adv comt, Jr Col Dist, 75. *Mem:* Am Indust Hyg Asn; Am Soc Test & Mat; Am Chem Soc. *Res:* Polarographic analysis; ion selective electrodes; industrial hygiene sampling and analysis. *Mailing Add:* Monsanto Co 800 N Lindbergh St Louis MO 63167-0001

HAYNES, WILLIAM P, b Omaha, Nebr, Apr 7, 21; m 43; c 2. CHEMICAL ENGINEERING. *Educ:* Univ Pittsburgh, BS, 42, MS, 51. *Prof Exp:* Chem engr, Bur Mines, US Dept Interior, 45-49, supvry chem res engr, 59-76; DIV MGR, US DEPT ENERGY, 76- *Mem:* Am Inst Chem Engrs; Am Chem Soc; Catalyst Soc. *Res:* Catalytic conversion of synthesis gas to substitute natural gas and liquid hydrocarbon; production of hydrogen by steam-iron reaction; hot carbonate gas absorption process; coal gasification; processes for removal of sulfur oxides from flue gases. *Mailing Add:* 2345 Orlando Pl Pittsburgh PA 15235

HAYNIE, FRED HOLLIS, b Anniston, Ala, Feb 8, 32; m 55; c 4. CORROSION SCIENCE, ENGINEERING ECONOMICS. *Educ:* Auburn Univ, BS, 54, MS, 59; Ohio State Univ, MS, 67. *Prof Exp:* Aviator, US Navy, 54-57; chief chem metall eng, Naval Air Mat Ctr, 61-64; sr chem eng, Battelle Mem Inst, 64-69; supvr environ eng, US Environ Protection Agency, 69-90; RETIRED. *Concurrent Pos:* Capt, Naval Air Reserve, 64-81. *Honors & Awards:* Sam Tour Award, Am Soc Testing & Mat. *Mem:* Am Soc Testing & Mat; Nat Asn Corrosion Engrs. *Res:* Determining the effects of air pollution and acid deposition on man-made materials. *Mailing Add:* 300 Oakridge Rd Cary NC 27511

HAYNIE, THOMAS POWELL, III, b Hearne, Tex, Aug 9, 32; m 56; c 3. NUCLEAR MEDICINE. *Educ:* Baylor Univ, MD, 56. *Prof Exp:* Intern med, Univ Mich Hosp, 56-57, resident internal med, 57-60, instr, Med Sch, Univ, 60-62, asst prof, 62; asst prof, Med Br, Univ Tex, 62-65; assoc prof med, M D Anderson Hosp & Tumor Inst, 65-75, prof med, Syst Cancer Ctr, 75-88, JAMES E ANDERSON PROF NUCLEAR MED, M D ANDERSON CANCER CTR, UNIV TEX, 88- *Concurrent Pos:* Tech expert, Int Atomic Energy Agency, Morocco, 64; consult, Wilford Hall US Air Force Hosp, Tex, 67- & Johnson Space Ctr, NASA, 68-; consult, Los Alamos Nat Labs; prof med & physiology, Univ Tex Grad Sch & Baylor Col; ed-in-chief, J Nuclear Med, 85-89. *Mem:* AAAS; Am Soc Clin Oncol; AMA; fel Am Col Physicians; Soc Nuclear Med; Am Col Radiol; Asn Univ Radiologists; Radiol Soc NAm. *Res:* Radioactive nuclides in medicine; applications of radionuclide imaging in cancer; radioisotope therapy of cancer. *Mailing Add:* 1515 Holcombe Blvd Houston TX 77030

HAYON, ELIE M, b Cairo, Egypt, May 15, 32; m 82; c 1. PHOTOCHEMISTRY, RADIATION CHEMISTRY. *Educ:* Univ Strathclyde, Glasgow, BSc, 54; Univ Durham, PhD(phys chem), 57. *Prof Exp:* Brit Empire res fel, King's Col, Eng, 57-58; res chemist, Brookhaven Nat Lab, 58-60; sr mem, Phys Chem Dept, Churchill Col, Cambridge Univ, 60-62; res scientist, French Atomic Energy Comt, France, 63-65; head, Phys Chem Lab, US Army Natick Labs, 66-75; DEAN GRAD STUDIES & RES & PROF CHEM, QUEEN'S COL, CITY UNIV NEW YORK, 78- *Concurrent Pos:* Adj prof, Brandeis Univ, 67-75; vis prof, Hebrew Univ, Jerusalem, 72-73. *Mem:* Am Chem Soc; Royal Soc Chem; Am Soc Photobiol; Biophys Soc; NY Acad Sci. *Res:* Fast-reaction studies of the mechanism and dynamics of chemical and biological reactions; laser spectroscopy; photophysics; chemistry of excited states; free radical chemistry; electron transfer process; free radical aspects of auto-oxidation. *Mailing Add:* Dept Chem Queen's Col Flushing NY 11367

HAYRE, HARBHAJAN SINGH, b Littran, India, July 12, 29; US citizen; m 57; c 3. ELECTRICAL ENGINEERING, ACOUSTICS. *Educ:* Punjab Univ, India, BA, 49; Univ Calif, Berkeley, 52, MS, 54; Univ NMex, DSc, 62. *Prof Exp:* Asst, Univ Calif, Berkeley, 52-53; elec engr, Niagara Mohawk Power Corp, 53-55; exchange elec engr, Nordsjaellands Elektricitets Og, Sporvejs & Hamburgische Electricitata, Werkes, 55; proj engr, Standard Vacuum Oil Co, India, 56-57; instr elec eng, Univ NMex, 57-58, res assoc, 60-62; assoc prof, New Bedford Inst Technol, 58-60; chmn grad fac, Cullen Col Eng, 67-68, mem & chmn univ governance comt, 70-72, PROF ELEC ENG, UNIV HOUSTON, 65-, DIR WAVE PROPAGATION LABS, DEPT ELEC ENG, 67- *Concurrent Pos:* Asst prof & lectr, Univ Buffalo, 54-55; sr engr, Curtiss-Wright Corp, NMex, 61; mem comn 2, US Nat Comt, Int Sci Radio Union, 62-; assoc prof, Kans State Univ, 62-65; consult, US Naval Res Lab, DC, 64-71; mem undergrad res panel, Nat Sci Found, 67-68; accreditation team elec eng, Eng Coun on Prof Develop, 70-73; vpres & dir res & develop, Specific Offshore Equip Co, Tex, 70-72; Presby coordr, Proj Equality, Tex; chmn, Tex Intersoc Legis Adv Comn, 77- & Houston Sect, Inst Elec & Electronics Engrs, 78-79; gen chmn, 1980 Nat Telecommun Conf, 77-80. *Honors & Awards:* Hon Consul of India, Govt of India, 77. *Mem:* Acoustic Soc Am; sr mem Inst Elec & Electronics Engrs; Am Soc Eng Educ; Am Astron Soc; sr mem Instrument Soc Am. *Res:* Pattern recognition; electromagnetic and acoustic wave propagation; human stress and endurance; sensors-instrumentation; non-destructive testing; biomedical; geophysical systems; communications; speech analysis-synthesis and speaker identification. *Mailing Add:* 10 Legend Lane Houston TX 77024-6003

HAY-ROE, HUGH, b Edmonton, Can, Dec 7, 28; m; c 3. GEOLOGY. *Educ:* Univ Alta, BSc, 49; Univ Tex, MA, 52, PhD(geol), 58. *Prof Exp:* Asst gen mgr & geol supvr, Belco Petrol Corp, Lima, Peru, 68-71, asst vpres, NY, 71-74, vpres explor, Lima, Peru, 74-76; far east explor mgr, Superior Oil Co, Calgary, 76-78, gen mgr Dominican Opers, Santo Domingo, 78-80; CONSULT PETROL GEOL, GLOBAL ENERGY OPER, HOUSTON, 80- *Res:* Petroleum geology. *Mailing Add:* 1606 Burning Tree Rd Kingwood TX 77339-3922

HÄYRY, PEKKA J, b Vihti, Finland, Dec 13, 39; c 1. TRANSPLANTATION SURGERY, CANCER. *Educ:* Univ Helsinki, MD, 65, ScD(exp path), 66. *Prof Exp:* Postdoctoral fel immunol, Wistar Inst, Philadelphia, 67-70; docent surg, Univ Oulu, 74-80; docent immunol, 70-79, PROF TRANSPLANT SURG & IMMUNOL, UNIV HELSINKI, 79-; res surg, 70-73, assoc chief, 73-79, CHIEF TRANSPLANT IMMUNOL, UNIV HELSINKI HOSP, 79- *Mem:* Am Soc Transplant Surgeons; Am Asn Imunologists; Transplantation Soc. *Res:* Transplantation surgery and immunology, particularly the study of the mechanism of acute and chronic rejection and their prevention. *Mailing Add:* Transplantation Lab Univ Helsinki Helsinki 00290 Finland

HAYS, BYRON G, b Kansas City, Mo, Jan 1, 37; m 62; c 1. ORGANIC CHEMISTRY, PIGMENT CHEMISTRY. *Educ:* Wichita State Univ, BS, 58; Mass Inst Technol, PhD(phys org chem), 64. *Prof Exp:* Res chemist, Esso Res & Eng Co, 64-66 & Interchem Co, 66-69; mgr, Prod Develop, Porvair Ltd, 69-71; res assoc, Inmont Co, 72-76, mgr prod develop & tech mgr, 77-86; RES & DEVELOP DIR SPEC PROJS, BASF CORP, 86- *Honors & Awards:* Super Varnish & Drier Co Lectr, Nat Asn Printing Ink Mfrs, 90. *Mem:* Am Chem Soc; Tech Asn Graphic Arts. *Res:* Organic pigments; surface treatments; surfactants; dispersants; dispersions; printing inks; disperse dyes; ultraviolet-cure coatings and inks; synthetic leather; polyurethanes and resins. *Mailing Add:* 151 Park Ave Verona NJ 07044

HAYS, DAN ANDREW, b Dallas Ctr, Iowa, July 12, 39; m 64; c 2. PHYSICS. *Educ:* Iowa State Univ, BS, 61; Rutgers Univ, MS, 63, PhD(physics), 66. *Prof Exp:* Fel physics, Univ Pittsburgh, 66-68; PRIN SCIENTIST PHYSICS, XEROX CORP, 68- *Concurrent Pos:* RCA fel, 64-66, Andrew Mellon fel, 66-68. *Mem:* Am Phys Soc; Electrostatics Soc Am; Sigma Xi. *Res:* Contact electrification studies of insulating materials, electrostatic properties of small particles, characteristics of xerographic development systems. *Mailing Add:* Xerox Corp 114-220 800 Phillips Rd Webster NY 14580

HAYS, DANIEL MAUGER, b Reading, Pa, Mar 9, 19; m 51; c 4. PEDIATRIC SURGERY. *Educ:* Stanford Univ, AB, 41; Cornell Univ, MD, 44. *Prof Exp:* Asst surg, Harvard Univ, 45-46; asst, Med Col, Cornell Univ, 50-51; from clin instr to clin assoc prof, Univ Calif, Los Angeles, 55-64; clin assoc prof, 64-66, assoc prof, 66-73, prof surg, 73-81, PROF SURG & PEDIAT, SCH MED, UNIV SOUTHERN CALIF, 80- *Concurrent Pos:* Investr, NIH grant, Clin Res Ctr, Children's Hosp, Los Angeles, 61-64; prin investr, NIH grant, 65-, prin investr cancer training grant, 67-; consult, Univ Calif, Los Angeles, 54-64; clin prof, Am Cancer Soc, 84- *Mem:* Am Col Surg; Am Acad Pediat. *Res:* Hepatic regeneration; pediatric oncology; pediatric surgery; cancer: late effects; cancer survivors: economic problems. *Mailing Add:* Los Angeles Children's Hospital 4650 Sunset Blvd Los Angeles CA 90027

HAYS, DONALD BROOKS, b Bradford, Tenn, Nov 4, 39; m 64; c 2. ENTOMOLOGY. *Educ:* Memphis State Univ, BS, 65, MS, 67; Miss State Univ, PhD(entom), 70. *Prof Exp:* Res entomologist, Vero Beach, Fla, 70-75, tech planning specialist, Greensboro, NC, 75-79, sr residue specialist, Greensboro, NC, 79-80, STA MGR, DELTA RES STA, AGR DIV, CIBA-GEIGY CORP, GREENVILLE, MISS, 80- *Mem:* Entom Soc Am; Sigma Xi; Southern Weed Sci Soc Am. *Res:* Physiology of diapause in the boll weevil; physiology of behavior in Cardiochiles nigriceps, a parasitoid of Heliothis virescens; development research for insecticides, fungicides, nematicides, bactericides, herbicides, plant growth regulators. *Mailing Add:* 1100 Cloverdale Greenville MS 38701

HAYS, DONALD F(RANK), b Portland, Ore, Mar 29, 29; m 55; c 3. MECHANICAL ENGINEERING, ENGINEERING MECHANICS. *Educ:* Ore State Univ, BS, 51, MS, 52. *Prof Exp:* Sr res engr, Res Lab, Gen Motors, 52-62, asst dept head, Mech Develop Dept, 62-75, asst dept head, Engr Mech Dept, 75-79, head, Mech Res Dept, 79-83, head, Fluid Mech Dept, 83-89; RETIRED. *Concurrent Pos:* Tech ed, J Lubrication Technol, Am Soc Mech Engrs, 74-80; mem, Am Soc Lubrication Handbook Adv Bd. *Honors & Awards:* Centennial Medallions, Am Soc Mech Engrs, 80; Mayo D Hersey Award, Am Soc Mech Engrs, 83. *Mem:* Soc Tribologists & Lubrication Engrs; fel Am Soc Mech Engrs; Soc Automotive Engrs. *Res:* Hydrodynamics; mechanics; applied mathematics; fluid film bearings; seals; mensuration; fluid mechanics. *Mailing Add:* 639 Alpine Ct Rochester Hills MI 48309

HAYS, DONALD R, b Quincy, Ill, Mar 29, 29; m 53; c 2. ORGANIC CHEMISTRY. *Educ:* Bradley Univ, BS, 50, MS, 52. *Prof Exp:* Chemist, Sales Develop Lab, Fabrics & Finishes Dept, E I du Pont de Nemours & Co, Mich, 54-59; res chemist, Polymers Dept Res Labs, 59-62, sr res chemist, 62-69, sr res chemist coating processes dept, mfg develop, 69-76, sr tech engr, Basic Coatings Lab, Fisher Body, 76-85, sr tech engr, CPC Group, Paint Systs, Gen Motors Corp, 85-86; CONSULT, CHEMCO CONSULTS, INC, 85- *Concurrent Pos:* Observ demonstr, Cranbrook Inst Sci, 63-; adj instr, Eastern Mich Univ, 85-; mem, Colour Coun. *Mem:* Am Chem Soc; Fedn Socs Coatings Technol. *Res:* Solvents; thinners; solvents retention; polymers for surface coatings, applications of surface coatings and appearance assessment. *Mailing Add:* Chemco Consults Inc 2733 Buckingham Birmingham MI 48009

HAYS, ELIZABETH TEUBER, b Greenport, NY, Apr 29, 40; m 67; c 2. PHYSIOLOGY, ZOOLOGY. *Educ:* Keuka Col, BA, 62; Univ Md, PhD(zool), 68. *Prof Exp:* Assoc physiol, Univ Rochester, 70-72, asst prof, 72-78; ASSOC PROF BIOL, NAZARETH COL, ROCHESTER, 78-, CHMN DEPT, 79- *Concurrent Pos:* Res asst, Nat Resources Inst, Univ Md, 65-67; fel, Duke Univ, 67-69; fel, Univ Rochester, 69-70. *Mem:* AAAS; Biophys Soc; Am Physiol Soc; Nat Sci Teachers Asn; Soc Col Sci Teachers. *Res:* Membrane transport; muscle physiology; bioenergetics; muscle metabolism. *Mailing Add:* Biomedical Science Physiol Dept Barry Univ 11300 N E Second Ave Miami FL 33161

HAYS, ESTHER FINCHER, b Lexington, Ky, Apr 18, 27; m 51; c 4. ONCOLOGY, HEMATOLOGY. *Educ:* Cornell Univ, AB, 48, MD, 51. *Prof Exp:* Intern med, New York Hosp, 51-52, asst resident, 52-54; from instr to assoc prof med, Sch Med, 55-72, from asst res physician to assoc res physician, Lab Nuclear Med & Radiation Biol, 57-72, PROF MED, SCH MED & RES PHYSICIAN, LAB NUCLEAR MED & RADIATION BIOL, UNIV CALIF, LOS ANGELES, 72-, ASSOC DEAN MED, 84- *Mem:* Am Asn Cancer Res; Am Soc Hemat; Am Col Physicians; Int Soc Exp Hemat. *Res:* Studies of etiological factors in mouse leukemia; immune responses in leukemia and pre-leukemic mice; mechanisms of viral leukemic genesis. *Mailing Add:* Lab Biomed & Environ Sci 900 Veteran Ave Los Angeles CA 90024

HAYS, GEORGE E(DGAR), b Alton, Ill, Apr 25, 21; m 46; c 12. CHEMICAL ENGINEERING. *Educ:* Blackburn Col, AB, 41; Univ Iowa, BS, 43. *Prof Exp:* Chem engr, Phillips Petrol Co, 43-51, sr engr, 51-55, supvr fundamental eng, 56-58, mgr advan eng sect, 59-60, chem eng assoc, Res & Develop Dept, 61-68, sr eng assoc, Res & Develop, 68-85; RETIRED. *Concurrent Pos:* Secy & treas, Washington County Elder Care, Inc, 87- *Mem:* Am Chem Soc; fel Am Inst Chem Engrs. *Res:* Process engineering and economic evaluation; natural gas, petroleum and petro-chemicals technology; low temperature gas processing; fractionation; phase equilibrium; separations and rate processes; analysis and elimination of technical and economic process risk. *Mailing Add:* 1906 SE Moonlight Dr Bartlesville OK 74006

HAYS, HORACE ALBENNIE, b Kisatchie, La, Nov 25, 14; m 43; c 1. MAMMALOGY. *Educ:* Northwestern State Univ, La, AB, 36; La State Univ, MS, 41; Univ Okla, PhD, 54. *Prof Exp:* Teacher, Pub Schs, La, 37-42; from voc adv to chief, Guid Ctr, US Vet Admin, Lake Charles, La, 46-47; asst prof biol, Centenary Col, 47-51; from assoc prof to prof, 54-85, EMER PROF BIOL, PITTSBURG STATE UNIV, 85- *Mem:* Am Inst Biol Sci; Am Soc Mammal; Sigma Xi. *Res:* Mammalian ecology; vertebrate zoology; distribution and movement of small mammals; effect of microclimate on distribution of small mammals; bat research. *Mailing Add:* 1408 S Olive St Pittsburg KS 66762

HAYS, JAMES D, b Johnstown, NY, Dec 26, 33; m 65. GEOLOGY. *Educ:* Harvard Univ, AB, 56; Ohio State Univ, MS, 60; Columbia Univ, PhD(geol), 64. *Prof Exp:* RES ASSOC, LAMONT-DOHERTY GEOL OBSERV, 65-, DIR, DEEP-SEA CORE LAB, 67-; PROF GEOL, COLUMBIA UNIV, 74- *Concurrent Pos:* From asst prof to assoc prof geol, Columbia Univ, 67-74; res assoc, Am Mus Natural Hist, 70-; exec dir, Climap Proj, 71-77. *Mem:* Sigma Xi; AAAS; Geol Soc Am; Am Geophys Union. *Res:* Radiolarian extinctions and magnetic reversals; deep-sea sediment research; paleo climatic research. *Mailing Add:* Lamont Geol Observ Palisades NY 10964-0190

HAYS, JAMES FRED, b Little Rock, Ark, July 10, 33; m 56; c 1. PETROLOGY, GEOPHYSICS. *Educ:* Columbia Univ, AB, 54; Calif Inst Technol, MS, 61; Harvard Univ, PhD(geol), 66. *Prof Exp:* Geologist, US Geol Surv, 61; guest investr, Carnegie Inst Geophys Lab, 65; from asst prof to prof geol, Harvard Univ, 66-82, chmn, dept geol sci, 81-82; DIR, DIV EARTH SCI, NSF, 82-87, 89-, SR SCI ADV, 87-89, 90- *Concurrent Pos:* Mem, Lunar Sample Anal Planning Team, NASA Johnson Space Ctr, 74-76; chmn lunar & planetary rev panel, 79-81; assoc ed, J Geophys Res, 78-80, 83-85; mem, scientific adv bd, Mt St Helens Nat Volcanic Monument, 83-87; vis prof chem & geol, Ariz State Univ, 77-79; counr, Geol Soc Am, 88-, AAAS, 89- *Mem:* Fel Geol Soc Am; Am Geophys Union; fel Mineral Soc Am; Geochem Soc; AAAS; Sigma Xi. *Res:* Experimental petrology; phase equilibria in silicates at high pressures and high temperatures; kinetics of crystallization; interior of the earth, moon and planets; principal investigator of Apollo lunar samples. *Mailing Add:* Off Dir Nat Sci Found Washington DC 20550

HAYS, JOHN BRUCE, b Springfield, Ill, June 21, 37; m 61; c 3. MOLECULAR GENETICS, DNA BIOCHEMISTRY. *Educ:* Univ NMex, BS, 60; Univ Calif, San Diego, PhD(chem), 68. *Prof Exp:* Fel biol, Johns Hopkins Univ, 69-72; asst prof chem, Univ Md, Baltimore County, 72-77, assoc prof chem, 77-87; PROF, DEPT AGR CHEM, ORE STATE UNIV, CORVALLIS, OR, 87- *Mem:* Am Soc Biol Chemists. *Res:* Biochemistry and genetics of recombination, especially DNA structures that stimulate homologous recombination; roles of ultraviolet repair; DNA-cytosine methylation. *Mailing Add:* Dept Agr Chem Ore State Univ Weniger Hall Rm 339 Corvallis OR 97331

HAYS, JOHN THOMAS, b Ozark, Mo, Nov 25, 14; m 43; c 2. ORGANIC CHEMISTRY, INFORMATION SCIENCE. *Educ:* Mont State Col, BS, 35; Oxford Univ, BA & BSc, 38; Calif Inst Technol, PhD(org chem), 42. *Prof Exp:* Res chemist, Exp Sta, 42-43, asst group leader, 43-45, tech asst to dir res, 45-49, spec asst, 49-51, mgr scouting res div, 51-53, tech asst to dir exp sta, 53-56, sr res chemist, Exp Sta, 56-81, SR RES CHEMIST, RES CTR, HERCULES INC, 81- *Res:* Carbohydrates; amino acids; pyridine compounds; polymers; synthetic rubber; rosin derivatives; nitrogen chemicals; ureaform fertilizers; technical information. *Mailing Add:* 209 Sterling Ave New Castle DE 19720-4729

HAYS, KIRBY LEE, b Cullman, Ala, Aug 11, 28; m 58; c 2. INSECT ECOLOGY. *Educ:* Auburn Univ, BS, 48, MS, 54; Univ Mich, PhD(wildlife ecol), 58. *Prof Exp:* Inspector, Bur Plant Quarantine, USDA, 49-50; asst, Tularemia Proj, Univ Mich, 54-57; from asst prof to assoc prof, 57-64, PROF ZOOL & ENTOM, AUBURN UNIV, 64-, HEAD DEPT, 75- *Mem:* Entom Soc Am; Ecol Soc Am (treas, 59-63). *Res:* Ecology of medically important insects and Acarina; taxonomy of the Tabanidae. *Mailing Add:* 1044 Terrace Acres Dr Auburn AL 36830

HAYS, RICHARD MORTIMER, b Far Rockaway, NY, July 1, 27. RENAL DISEASES. *Educ:* Columbia Univ, MD, 54. *Prof Exp:* PROF MED, ALBERT EINSTIEN COL MED, 73-, DIR, DIV NEPHROL, 79- *Mailing Add:* Dept Med Albert Einstein Col Med 1300 Morris Park Ave Bronx NY 10461

HAYS, RUTH LANIER, b Cartersville, Ga, July 22, 40; m 65; c 2. ENDOCRINOLOGY, HISTOLOGY. *Educ:* Berea Col, BA, 62; Auburn Univ, PhD(physiol), 66. *Prof Exp:* From asst prof to assoc prof physiol, 65-81, PROF ZOOL, CLEMSON UNIV, 81- *Mem:* Sigma Xi; Nat Sci Teachers Asn; Asn Southeastern Biologist; Electron Micros Soc Am; Am Soc Zool. *Res:* Endocrinology; neuroendocrinology; reproductive physiology; enterochromaffin cells; age changes in mammals; seasonal variations; disease state; nutritional state. *Mailing Add:* Dept Biol Scis Clemson Univ Clemson SC 29634-1903

HAYS, SIDNEY BROOKS, b Arab, Ala, May 31, 31; m 65; c 2. ENTOMOLOGY, ANIMAL PHYSIOLOGY. *Educ:* Auburn Univ, BS, 53, MS, 58; Clemson Univ, PhD(entom), 62. *Prof Exp:* Asst entom, Auburn Univ, 58-60; asst entomologist, Univ Ga, 62-63; asst prof entom, Auburn Univ, 63-64; from asst prof to assoc prof entom res, 64-72, HEAD DEPT ENTOM, CLEMSON UNIV, 69-, PROF, 72- *Honors & Awards:* Sigma Xi Res Award, Auburn Univ, 60. *Mem:* Am Entom Soc; Entom Soc Am. *Res:* Economic entomology; insect physiology of reproduction; insect ecology. *Mailing Add:* Dept Entom 114 Long Hall Clemson Univ Clemson SC 29631

HAYS, VIRGIL WILFORD, b Waurika, Okla, Oct 1, 28; m 52; c 3. ANIMAL NUTRITION. *Educ:* Okla State Univ, BS, 54; Iowa State Univ, PhD(animal nutrit), 58. *Prof Exp:* From asst prof to prof animal nutrit, Iowa State Univ, 58-67, asst dir, Agr & Home Econ Exp Sta, 66-67; prof, 67-90, dept chair, 74-90, SCOVILL DISTINGUISHED PROF ANIMAL NUTRIT, UNIV KY, 90- *Honors & Awards:* Am Feed Mfr Nutrit Award, Am Soc Animal Sci, 74, Distinguished Serv Award, 90. *Mem:* AAAS; Am Soc Animal Sci (pres, 78-79); Am Instr Nutrit; Am Dairy Sci Asn; Poultry Sci Asn; Coun Agr Sci & Tech (pres, 89-90); Am Registry Prof Animal Scientists (pres, 85-86). *Res:* Nutrition; minerals; amino acids; drugs; hormones. *Mailing Add:* Dept Animal Sci Univ Ky Lexington KY 40546-0215

HAYS, WILLIAM HENRY, b Livingston, Mont, Mar 15, 22; m 61; c 2. GEOLOGY. *Educ:* Stanford Univ, BS, 48; Yale Univ, MS, 50, PhD(geol), 58. *Prof Exp:* GEOLOGIST, US GEOL SURV, 58- *Mem:* Fel Geol Soc Am. *Res:* Geologic hazards; stratigraphy; structural geology. *Mailing Add:* 55 S Flower St Lakewood CO 80226

HAYSLETT, JOHN P, b Greenwich, Conn, Jan 6, 35; m 60; c 2. PHYSIOLOGY. *Educ:* Col Holy Cross, AB, 56; Cornell Univ, MD, 60. *Prof Exp:* From instr to assoc prof, 67-77, PROF MED & PEDIAT, SCH MED, YALE UNIV, 77- *Concurrent Pos:* Res fel metab, Sch Med, Yale Univ, 65-67; estab investr, Am Heart Asn, 71-76. *Mem:* Am Fedn Clin; Am Soc Nephrology; Am Soc Clin Invest; Am Physiol Soc; Am Col Physicians. *Res:* Clinical problems in renal disease; mechanisms for the renal control of sodium readsorption. *Mailing Add:* Dept Int Med Yale Univ Sch Med 333 Cedar St New Haven CT 06520

HAYT, WILLIAM H(ART), JR, b Wilmette, Ill, July 1, 20; m 46; c 3. ELECTRICAL ENGINEERING. *Educ:* Purdue Univ, BS, 42, MS, 48; Univ Ill, PhD(elec eng), 54. *Prof Exp:* From asst prof to assoc prof, 48-58, head sch elec eng, 62-65, asst to vpres acad affairs, 65-73, prof, 58-86, EMER PROF ELEC ENG, PURDUE UNIV, 86- *Concurrent Pos:* Consult, Martin Co, Colo, 57-59, Int Bus Mach Corp, 61-63 & Midwest Appl Sci Corp, 63-67; auth. *Mem:* Am Soc Eng Educ; fel Inst Elec & Electronics Engrs. *Mailing Add:* 705 Sugar Hill Dr West Lafayette IN 47906

HAYTER, JOHN BINGLEY, b Auckland, NZ, Nov 27, 45; m 70; c 3. COLLOIDAL STRUCTURE, NEUTRON OPTICS. *Educ:* Univ Sydney, Australia, BSc Hons, 66, PhD(chem phys), 70. *Prof Exp:* Res assoc phys chem, Univ Oxford, UK, 71-73; staff scientist, Inst Laue-Langevin, France, 73-84; SR RES STAFF MEM, SOLID STATE DIV, OAK RIDGE NAT LAB, 84- *Concurrent Pos:* Sci dir, Advan Neutron Source Proj, Oak Ridge Nat Lab, 90- *Honors & Awards:* Inventor's Award, Martin Marietta Energy Systs, Inc, 88; R & D 100 Award, Res & Develop Mag, 89. *Mem:* Am Phys Soc; Mat Res Soc. *Res:* Theoretical and experimental investigations of complex fluids and colloidal systemms, primarily by neutron scattering; development of neutron optical techniques for surface studies. *Mailing Add:* Oak Ridge Nat Lab-FEDC PO Box 2009 Oak Ridge TN 37831-8218

HAYTER, ROY G, b Bristol, Eng, May 23, 32; m 57; c 3. INDUSTRIAL CHEMISTRY. *Educ:* Bristol Univ, BSc, 52, PhD(chem), 55. *Prof Exp:* Fel inorg chem, Cornell Univ, 55-57; tech officer chem, Imp Chem Industs, Ltd, Eng, 57-60; fel, Mellon Inst, 60-63; chemist, Shell Develop Co, Calif, 63-68, res supv, 68-70, res dir, Shell Develop Co, Houston, Tex, 70-72, venture mgr, Shell Chem Co, 72-76; mgr, Plastics Dept, Shell Develop Co, 76-83, mgr, Polymer Systs Res & Develop Dept, 83-89; sr consult, SRI Int, 89-90. *Mem:* Am Chem Soc; The Chem Soc; Soc Plastic Engrs. *Res:* Coordination, organometallic and polymer chemistry. *Mailing Add:* 1691 Yale St Mountain View CA 94040

HAYTER, WALTER R, JR, b Stamford, Conn, Mar 9, 23; m 45; c 5. ELECTRICAL ENGINEERING. *Educ:* Univ Conn, BS, 44; Stevens Inst Technol, MS, 50. *Prof Exp:* Design engr, 45-56, supv engr, 56-57, eng sect mgr, 57-70, SR ENGR, WESTINGHOUSE ELEC CORP, 70- *Honors & Awards:* Westinghouse Most Meritorious Patent Award, 55. *Mem:* Inst Elec & Electronics Engrs. *Res:* Microwave tubes and devices; power tubes and vacuum interrupters. *Mailing Add:* 512 Underwood Ave Elmira NY 14905

HAYTHORNTHWAITE, ROBERT M(ORPHET), b Whitley Bay, Eng, May 5, 22; m 52; c 4. ENGINEERING MECHANICS, CIVIL ENGINEERING. *Educ:* Univ Durham, BSc, 42; Univ London, PhD(eng), 52; Brown Univ, ScM, 53. *Prof Exp:* Jr sci off, Bldg Res Sta, Eng, 42-46; sci off, Dept Sci & Indust Res, 46-47; lectr civil eng, Univ Sheffield, 47-53; from instr to assoc prof eng, Brown Univ, 53-59; prof eng sci, Univ Mich, Ann Arbor, 59-67; head, Dept Eng Mech, Pa State Univ, 67-74, prof, 67-79; dean, Col Eng Technol, 79-81, PROF ENG SCI, TEMPLE UNIV, 79- *Concurrent Pos:* Mem res comt, Column Res Coun, 56-64; chmn, Ed Comt, Nat Cong Appl Mech, 58; consult, Land Locomotion Res Br, Ord Corps, US Army, 60-69; ed, Mechanics, 71-72 & 89- *Honors & Awards:* Walter L Huber Civil Eng Res Prize, 63. *Mem:* Am Soc Civil Engrs; Am Soc Mech Engrs; Am Soc Eng Educ; Am Acad Mech (pres, 69-71). *Res:* Mechanics of solids; framed structures, plates and shells; theory of plasticity; soil mechanics. *Mailing Add:* Dept Civil Eng Temple Univ 084-53 Philadelphia PA 19122

HAYTON, WILLIAM LEROY, b Mt Vernon, Wash, June 16, 44; m 67; c 2. PHARMACOKINETICS, DRUG ABSORPTION. *Educ:* Univ Wash, BS, 67; State Univ NY Buffalo, PhD(pharmaceut), 71. *Prof Exp:* From asst prof to prof pharm, Wash State Univ, 71-90, chmn, Grad Prog Pharmacol-Toxicol, 82-90; CHMN & PROF PHARMACEUT & PHARMACEUT CHEM DEPT, OHIO STATE UNIV, 90- *Mem:* Am Asn Pharmaceut Sci; Am Soc Pharmacol & Exp Therapeut; AAAS; Sigma Xi; Soc Toxicol. *Res:* Kinetics of drug absorption, distribution, and elimination; intestinal permeability and xenobiotic accumulation by fish. *Mailing Add:* Col Pharm Ohio State Univ 500 W 12th Ave Columbus OH 43210-1291

HAYWARD, ANTHONY R, PEDIATRIC IMMUNOLOGY, IMMUNODEFICIENCY. *Educ:* Univ London, MD, 67, PhD(immunol), 74. *Prof Exp:* PROF PEDIAT, MICROBIOL & IMMUNOL, UNIV COLO HEALTH SCI CTR, 84- *Mailing Add:* Dept Pediat Univ Colo Med Ctr 4200 E Ninth Ave Denver CO 80262

HAYWARD, BRUCE JOLLIFFE, b Pine Island, Minn, Apr 13, 28. ZOOLOGY. *Educ:* Univ Minn, BS, 50; Univ Mich, MS, 52; Univ Ariz, PhD(zool), 61. *Prof Exp:* From asst prof to assoc prof, 61-75, PROF BIOL SCI, WESTERN NMEX UNIV, 75- *Mem:* Am Soc Mammal. *Res:* Natural history studies of bats in southwestern United States, Central America and South America. *Mailing Add:* Dept Natural Sci Western NMex Univ Silver City NM 88062

HAYWARD, CHARLES LYNN, b Paris, Idaho, July 16, 03; m 30; c 2. ANIMAL ECOLOGY. *Educ:* Brigham Young Univ, BS, 27, MS, 31; Univ Ill, PhD(animal ecol), 42. *Prof Exp:* From instr to prof, 31-74, cur life sci, Mus, 69-74, chmn dept zool, 58-61, EMER PROF ZOOL, BRIGHAM YOUNG UNIV, 74- *Concurrent Pos:* Asst, Univ Ill, 38-39. *Mem:* Fel AAAS; Ecol Soc Am; Am Ornith Union; Cooper Ornith Soc. *Res:* Birds and mammals of Utah. *Mailing Add:* 959 Cedar Ave Provo UT 84604

HAYWARD, EVANS VAUGHAN, b Camp Dix, NJ, Feb 17, 22; m 47; c 2. PHYSICS. *Educ:* Smith Col, BA, 42; Univ Calif, MA, 45, PhD(physics), 47. *Prof Exp:* Physicist, Radiation Lab, Univ Calif, 47-50 & Nat Bur Standards, 50-90; PHYSICIST, DUKE UNIV, 90- *Concurrent Pos:* Guggenheim fel, Nordic Inst Theoret Atomic Physics, Denmark, 61-62; guest prof, Inst Nuclear Physics, Univ Frankfurt, 66; Sir Thomas Lyle fel, Univ Melbourne, 69; mem, gen adv comt, US Atomic Energy Comn, 72-75, chmn, 74-75; mem, Md gov sci adv coun, 72-84; vis prof physics, Univ Toronto, 75, Duke Univ, 87; guest prof exp physics, Max Planck Inst, Mainz, Ger, 82-86; chmn div nuclear physics, Am Phys Soc, 83-84; sr vis scientist, Nuclear Physics Lab, Oxford, UK, 85-86; vis scientist, Physics Inst, Univ Lund, Sweden, 88- *Honors & Awards:* Gold Medal, US Dept Com, 71; Federal Women's Award Winner, 75. *Mem:* Am Phys Soc. *Res:* Interactions of high energy radiations with matter; photonuclear reactions. *Mailing Add:* 8400 Westmont Ct Bethesda MD 20817

HAYWARD, JAMES LLOYD, b Melrose, Mass, Sept 30, 48; m 74; c 1. BEHAVIORAL ECOLOGY, TAPHONOMY OF EGGS. *Educ:* Walla Walla Col, BS, 72; Andrews Univ, MA, 75; Wash State Univ, PhD(zool), 82. *Prof Exp:* Instr biol, Andrews Univ, 75-76, Southwestern Union Col, 76-77, Walla Walla Col, 80-81; from asst prof to assoc prof biol, Union Col, 81-86; PROF BIOL, ANDREWS UNIV, 86- *Concurrent Pos:* Vis prof biol, Walla Walla Col Marine Sta, 77, 79, 83 & 85. *Mem:* AAAS; Am Ornithologists Union. *Res:* Visual communication; temporal and spatial patterning of behavior; organisms response to volcanic eruptions; factors influencing fossilization of eggs; human genetic diversity and potential. *Mailing Add:* Dept Biol Andrews Univ Berrien Springs MI 49104

HAYWARD, JAMES ROGERS, b Detroit, Mich, Dec 16, 20; m 43; c 3. ORAL SURGERY. *Educ:* Univ Mich, BS, 41, DDS, 44, MS, 46; Am Bd Oral Surg, dipl, 49. *Prof Exp:* Instr oral surg, W K Kellogg Found Inst, 47-50, assoc prof, 52-55, dir sect oral surg, Univ Hosp, 52, prof oral surg & chmn dept, W K Kellogg Found Inst, Univ Mich, Ann Arbor, 56- & prof dent & dent in surg, 78-, emer prof dent, oral & maxillofacial surg & surg, Sch Med; RETIRED. *Concurrent Pos:* Consult, Ann Arbor Vet Admin Hosp, 53-; ed, J Oral Surg. *Mem:* Am Soc Oral Surg; fel Am Col Dent; fel Am Col Oral Path. *Res:* Local anesthesia; jaw fractures; cleft lip and palate; tumors of mouth and jaw region. *Mailing Add:* 249 Second St Bonita Springs FL 33923

HAYWARD, JOHN S, b Vancouver, BC, Feb 14, 37; m 61; c 3. ENVIRONMENTAL PHYSIOLOGY, AEROSPACE MEDICINE. *Educ:* Univ BC, BSc, 58, PhD(zool), 64. *Prof Exp:* Exp officer animal physiol, Div Wildlife Res, Commonwealth Sci & Indust Res Orgn, Australia, 58-61; NATO fel, 64-65; asst prof zool, Univ Alta, 65-69; asst prof, 69-70, assoc prof zool, 70-78, PROF BIOL, UNIV VICTORIA, 79- *Mem:* Can Physiol Soc. *Res:* Physiological responses to cold; thermogenic mechanisms; brown fat and nonshivering heat production; man in cold water and cold air, his physiological responses and survival techniques; treatment of hypothermia. *Mailing Add:* Dept Biol Univ Victoria Box 1700 Victoria BC V8W 2Y2 Can

HAYWARD, JOHN T(UCKER), b New York, NY, Nov 15, 10; m 32; c 5. PHYSICS, AERONAUTICAL ENGINEERING. *Educ:* US Naval Acad, BS, 30; Univ Portland, DSc, 65. *Prof Exp:* Asst chief engr, Instruments Naval Aircraft Factory, 39-40, test pilot, Test Ctr, 40, exp officer, Manhattan Dist, Calif Inst Technol, 44-46, head plans & opers, Atomic Warfare Armed Forces, Sandia AFB, 48-49 & Group Atomic Capability Aircraft Carriers, 49-51, dir weapons res div, Atomic Energy Comn, 51-53, comdr & tech dir, Naval Ord Lab, 54-56, dep chief naval opers res & develop, 56-62, comdr, Carrier Div 2, Flagship USS Enterprise, 62-63, comdr antisubmarine warfare forces Pac, 63-66; pres, Naval War Col, Newport, 66-68; vpres, Int, Gen Dynamics Corp, 68-73; PRES, HAYWARD ASSOCS, 73- *Concurrent Pos:* Consult, Lawrence Livermore Lab & Gen Dynamics, 73-, Charles Stark Draper Lab, Inc; bd dir, Hertz Found. *Honors & Awards:* Robert D Conrad Sci Award, Secy Navy. *Mem:* AAAS; Am Inst Aeronaut & Astronaut; Am Phys Soc; Newcomen Soc N Am; fel Royal Aero Soc. *Res:* Aeronautics; ordnance engineering, especially modern underwater weapons; application of the atom to its official use by the Navy. *Mailing Add:* Three Barclay Sq Newport RI 02840

HAYWARD, LLOYL DOUGLAS, b Cupar, Sask, June 10, 19; m 45; c 2. PHYSICAL ORGANIC CHEMISTRY. *Educ:* Univ Sask, BA, 43; McGill Univ, PhD(chem), 49. *Prof Exp:* Lectr anal chem, Khaki Univ Can, Eng, 45-46; lectr, Sir George Williams Col, 47-48; sessional lectr org chem, McGill Univ, 49-50; asst prof chem, Regina Col, 50-51; from asst prof to prof org chem, Univ BC, 51-84; RETIRED. *Concurrent Pos:* Guest prof, Inst Phys Chem, Univ Uppsala, 64-65; consult, Ives Lab, 79-84, Syndel Lab, 80-, DH Stereochem, 80-; founder of Vol Prog, Do-It-Yourself Chem Exp in Elementary Sch, 86. *Mem:* NY Acad Sci; fel Chem Inst Can; Royal Chem Soc. *Res:* Induced optical activity; circular dichroism studies; chemistry and pharmacology of the nitrate esters. *Mailing Add:* 2041 W 29th Ave Vancouver BC V6J 2Z9 Can

HAYWARD, OLIVER THOMAS, b Las Cruces, NMex, Sept 26, 21; m 46; c 3. GEOLOGY. *Educ:* Univ Kans, BS, 45; Stanford Univ, MS, 51; Univ Wis, PhD(geol), 57. *Prof Exp:* Explor geologist, Phillips Venezuelan Oil Co, 45-50, consult, 50-53; area geologist, Calif Co, La, 53-55; from asst prof to assoc prof, 55-71, PROF GEOL, BAYLOR UNIV, 71- *Concurrent Pos:* Dir, Coun Ed Geol Sci, Am Geol Inst, 64-65. *Mem:* Geol Soc Am; Nat Asn Geol Teachers (pres, 80); Am Asn Petrol Geologists. *Res:* Geomorphic evolution in Central Texas; drainage history Eastward drainage American Southwest; Central Texas; sedimentary geology; structural geology of sedimentary rocks; landscape evolution in central Texas. *Mailing Add:* Dept Geol Baylor Univ Waco TX 76703

HAYWARD, RAYMOND (W)EBSTER, b Omaha, Nebr, July 28, 21; m 47; c 2. NUCLEAR PHYSICS. *Educ:* Iowa State Col, BS, 43; Univ Calif, PhD(physics), 50. *Prof Exp:* Electronic scientist, US Naval Res Lab, 43-44; physicist, Nat Bur Standards, 50-86; RETIRED. *Concurrent Pos:* Prof, Univ Md, 61- *Honors & Awards:* John Price Wetherill Medal, Franklin Inst, 66; Gold Medal, US Dept Com, 58. *Mem:* Fel Am Phys Soc. *Res:* Electromagnetic and weak interactions; relativistic quantum mechanics; Lagrangian field theory; gravitation interaction. *Mailing Add:* 8400 Westmont Ct Bethesda MD 20817-6811

HAYWARD, THOMAS DOYLE, b Hood River, Ore, June 19, 40; m 62; c 2. EXPERIMENTAL NUCLEAR PHYSICS, GENERAL PHYSICS. *Educ:* Univ Wash, BS, 62, PhD(physics), 69. *Prof Exp:* Fel nuclear physics, Duke Univ, 69-72, from instr to asst prof, 70-72; mem staff, Los Alamos Sci Lab, Univ Calif, 72-80, asst group leader physics, 74-76, alternate group leader, 77-78, group leader accelerator technol, 78-80; sr scientist, Sci Appln Int Corp, 80-83; SR PRIN ENGR, BOEING AEROSPACE, 83- *Mem:* Am Phys Soc. *Res:* Negative ion beam generation, acceleration, and high energy negative ion beam stripping processes related to stripping injection into storage rings; granted one patent; accelerators for newtron radiography; hyperconducting accelerator development; free electron laser development; radio frequency quadrupols accelerator development. *Mailing Add:* 5409 Norpoint Way Tacoma WA 98422-4239

HAYWOOD, ANNE MOWBRAY, b Baltimore, Md, Feb 5, 35. VIROLOGY, BIOCHEMISTRY. *Educ:* Bryn Mawr, BA, 55; Harvard Univ, MD, 59; Am Bd Pediat, dipl, 81. *Prof Exp:* Res fel biol, Calif Inst Technol, 60-61; res fel biochem, Columbia Univ, 61-62; res fel biol, Calif Inst Technol, 62-64; asst prof microbiol, Med Sch, Northwest Univ, 64-66; asst prof microbiol, Sch Med, Yale Univ, 66-73, resident pediat, 74-75; infectious dis fel, Univ Wash, 75-76 & Vanderbilt Univ, 76-77; ASSOC PROF PEDIAT MED & MICROBIOL, MED SCH, UNIV ROCHESTER, 77- *Concurrent Pos:* Am Cancer Soc fel, 60-62; vis asst prof, Rockefeller Univ, 71-72; vis scientist, Inst Animal Physiol, Cambridge, 72-74; Nat Inst Health spec fel, 71-73; Europ Molecular Biol Orgn fel, 73-74; vis assoc prof, Univ Calif Davis, 86; vis scientist, Univ Zürich, 87; Fogarty Sr Int Fel, 87. *Honors & Awards:* Fogarty Sr Int Fel, 87. *Mem:* Am Soc Biochem & Molecular Biol; Biophys Soc; Infectious Dis Soc Am. *Res:* Biochemistry of viruses; host-virus interaction; viral entry viral persistence; membranes; infectious diseases; pediatrics. *Mailing Add:* Dept Pediat Univ Rochester Box 777 Med Ctr 601 Elmwood Ave Rochester NY 14642

HAYWOOD, FREDERICK F, b Lumberton, NC, Sept 26, 36. HEALTH PHYSICS, TECHNICAL WRITING. *Educ:* Lynchburg Col, BS, 58; Vanderbilt Univ, MS, 64. *Prof Exp:* Health physicist, Oak Ridge Nat Lab, 59-72, res staff mem, 72-75, proj mgr, 75-81; proj mgr, 81-85, TECH DIR, EBERLINE ANALYSIS CORP, 85- *Concurrent Pos:* Consult, Int Atomic Energy Agency, 70-79; appointee, subcomt radiation shielding, adv comt civil defense, Nat Acad Sci, 73-75. *Mem:* Am Nuclear Soc; Int Radiation Protection Asn; Health Physics Soc. *Res:* Direct planning and execution of radiological and chemical investigations at hazardous waste sites; evaluates field data to characterize contaminant boundaries; identifies principal pathways to humans; assesses potential health effects and recommends measures to mitigate human exposure. *Mailing Add:* 119 Whipporwill Dr Oak Ridge TN 37830

HAYWOOD, H CARL, b Taylor County, Ga, July 2, 31; div; c 4. COGNITIVE EDUCATION. *Educ:* San Diego State Col, BA, 56, MA, 57; Univ Ill, PhD(psychol), 61. *Prof Exp:* Psychologist, US Vet Admin, 61-62; PROF PSYCHOL, GEORGE PEABODY COL, 62-; PROF NEUROL, VANDERBILT UNIV SCH MED, 71-, PROF PSYCHOL, VANDERBILT UNIV, 79- *Concurrent Pos:* Vis prof psychiatry, Univ Toronto, 65-66; dir, Inst Mental Retardation & Intellectual Develop, George Peabody Col, 70-73, Mental Retardation Res Training Prog, 68-73, John F Kennedy Ctr, Peabody Col/Vanderbilt Univ, 71-83; ed, Am Journ Mental Deficiency; presidential appointee, Nat Adv Child Health & Human Develop Coun, NIH, 84-88. *Honors & Awards:* Edgar A Doll Award, Am Psychol Assoc, 88; Nat Leadership Award, Am Asn Mental Deficiency, 85 & Res Award, 89. *Mem:* Nat Acad Sci; Am Psychol Assoc; Am Asn Mental Deficiency (pres, 80-81); Soc Res in Child Develop; Psychonomic Soc; Nat Mental Retardation Res Ctr Dirs (chmn, 79-82); Int Asn Cognitive Educ (pres, 88-92). *Res:* Psychological research on mental retardation and intellectual development, individual differences in motivation to learn and achieve, and development and modifiability of cognitive processes.; psychoeducational assessment and cognitive approaches to psychotherapy. *Mailing Add:* Box 9 Peabody Sta Vanderbilt Univ Nashville TN 37203

HAYWOOD, L JULIAN, b Reidsville, NC, Apr 13, 27; m; c 1. INTERNAL MEDICINE, CARDIOLOGY. *Educ:* Hampton Inst, BS, 48; Howard Univ, MD, 52. *Prof Exp:* from instr to asst prof, Loma Linda Univ, 60-63; from asst to asst prof, 63-66, ASSOC PROF MED, UNIV SOUTHERN CALIF, 66-; DIR, CORONARY CARE UNIT, LOS ANGLES COUNTY, UNIV SOUTHERN CALIF MED CTR, 66-, DIR, COMPREHENSIVE SICKLE CELL CTR, 72- *Concurrent Pos:* Fel cardiol, White Mem Hosp, 59-61; Am Col Physicians traveling fel, Oxford Univ, 63; mem consult staff, White Mem Hosp, 61-; sr consult cardiol, Vet Admin Hosp, Long Beach, Calif, 61-70;

consult, White Mem Hosp & St Vincent's Hosp, 68-, Calif State Div Indust Accidents & Health Resources Admin, USPHS; consult, Calif State Div Indust Accidents; fel, Coun Clin Cardiol, Epidemiol & Arteriosclerosis, Am Heart Asn past pres, greater Los Angeles Affil; consult, Health Resources Admin, USPHS; mem bd dirs & pres, Sickle Cell Dis Res Found; mem rev comt, Nat Heart, Lung & Blood Inst. *Honors & Awards:* Russell Award, Am Heart Asn, 88. *Mem:* AAAS; Am Fedn Clin Res; NY Acad Sci; fel Am Col Physicians; fel Am Col Cardiol. *Res:* Cardiovascular disease; sickle cell anemia. *Mailing Add:* Univ Southern Calif Med Ctr 1200 N State St Los Angeles CA 90033

HAYWOOD, THEODORE J, b Monroe, NC, Feb 13, 29; m 59; c 3. MEDICINE, ALLERGY. *Educ:* The Citadel, BS, 48; Vanderbilt Univ, Md, 52; Am Bd Allergy & Immunol, cert. *Prof Exp:* Asst pediat, Wash Univ, 52-53; teaching asst, Univ London, 53-54; asst, Tulane Univ, 54-55; asst, Harvard Univ, 55; assoc, McGovern Allergy Clin, 58-74; assoc prof, 70-84, ADJ ASSOC PROF ALLERGY, UNIV TEX GRAD SCH BIOMED SCI, HOUSTON, 84- *Concurrent Pos:* House physician, Hosp for Sick Children, London, Eng, 53-54; resident pediat, Charity Hosp, La, Tulane Univ, 54-55; clin assoc prof pediat allergy, Baylor Col Med, 77- *Mem:* Fel Am Col Allergists; fel Am Acad Allergy; fel Am Acad Pediat; fel Acad Psychosom Med; Sigma Xi. *Res:* Immunology of hypersensitivity; psychophysiology of allergic disease; hypothalamus and autonomic nervous system in allergic disease; insect and drug hypersensitivity; longitudinal studies on relationship of bacterial infections in asthma. *Mailing Add:* 6969 Brompton Rd Houston TX 77025-9980

HAYWORTH, CURTIS B, b Vienna, Austria, Dec 1, 20; US citizen; wid; c 2. CHEMICAL ENGINEERING. *Educ:* City Col New York, BChE, 44; NY Univ, MChE, 47, ScD(chem eng), 49, PE 49. *Prof Exp:* Lab asst, Gen Chem Div, Allied Chem Corp, 44-45, analyst, 45, tech supvr lab, 45-48, res chem engr, 48-53, asst mgr develop res, 53-54, asst dir, 54-58, dir, 58-61, asst tech dir, 61-63, asst dir, Cent Res Lab, 63-67, sr res tech assoc, Corp Res & Develop, 67-70; vpres, World Patent Develop Corp, 70-71, pres, 71-75; VAL ENG, US TREAS, 75- *Mem:* Am Chem Soc; Am Inst Chem Engrs; Sigma Xi. *Res:* Chemical engineering unit processes and economics; liquid-liquid extraction; numerous US and foreign patents. *Mailing Add:* Five Egbert Ave POB 1447 Morristown NJ 07962

HAZARD, EVAN BRANDAO, b Montclair, NJ, Nov 17, 29; m 52; c 3. MAMMMALOGY, SCIENCE WRITING. *Educ:* Cornell Univ, BS, 51; Univ Mich, MA, 55, PhD(zool), 60. *Prof Exp:* From asst prof to assoc prof biol, 58-65, head div, sci & math, 69-71, PROF BIOL, BEMIDJI STATE UNIV, 65- *Concurrent Pos:* Mem, Minnesota Writing Proj, 82-84; sci-writing & ed consult, 88-; reviewer, Choice, 83-89. *Mem:* AAAS; Am Inst Biol Sci; Am Soc Mammal; Wildlife Soc; Soc Study Evolution. *Res:* Geographic distribution of Minnesota mammals; wildlife conservation; vertebrate evolution; writing for biology majors and across-the-curriculum. *Mailing Add:* Dept Biol Bemidji State Univ 1500 Birchmont Ave NE Bemidji MN 56601-2699

HAZARD, HERBERT RAY, b Johnson City, NY, Aug 4, 17; m 42; c 3. MECHANICAL ENGINEERING, FUEL TECHNOLOGY. *Educ:* Pa State Univ, BS, 39. *Prof Exp:* Student engr, Babcock & Wilcox Co, NY, 39-40, anal engr, 40-44; proj engr, Bendix Aviation Co, NY, 44-45; res engr, Battelle Mem Inst, 45-48, asst supvr, 48-53, asst div chief, 53-60, div chief, 60-62, fel, 62-76, prin mech engr, 76-81; consult, 82-83; PRES, HERBERT R HAZARD, INC, 84- *Concurrent Pos:* Lectr, Univ Mich, 53, 55 & 58. *Mem:* Fel Am Soc Mech Engrs. *Res:* Combustion; heat transfer; fluid dynamics; gas turbines; nuclear reactors; deep-sea diving; low-emission combustion; advanced power generating concepts; boiler slagging and fouling; synthetic liquid fuels. *Mailing Add:* 2770 North Star Rd Columbus OH 43221-2960

HAZARD, JOHN BEACH, b White Horse, Pa, Jan 7, 05; m 31. PATHOLOGY. *Educ:* Univ Fla, BS, 24, MS, 25; Harvard Med Sch, MD, 30. *Prof Exp:* Instr path, Med Sch, Univ Boston, 31; from instr to asst prof path & bact, Med Sch, Tufts Univ, 32-46; chmn div & head dept path, Cleveland Clin Found, 46-70; from assoc to clin prof, 58-70, EMER CLIN PROF PATH, CASE WESTERN RESERVE UNIV, 70-; EMER CONSULT PATH, CLEVELAND CLIN FOUND, 70- *Concurrent Pos:* Researcher & asst, Mallory Inst Path, 30-34; pathologist, Faulkner Hosp, 34-46; consult, Robert B Brigham Hosp, 36-46; chief lab serv, 5th Gen Hosp, 42-44, comdr, Cent Lab, UK Base Lab, 45; mem path study sect, USPHS, 62-65. *Mem:* AAAS; Am Soc Cytol; Am Thyroid Asn (1st vpres, 61-62); Am Soc Clin Path; Int Acad Path (pres, 62-63). *Res:* Diseases of the thyroid; pathology of Beryllium diseases; cytology. *Mailing Add:* 200 Ocean Lane Dr Key Biscayne FL 33149

HAZARD, KATHARINE ELIZABETH, b Lafayette, Ind, July 15, 15. MATHEMATICS. *Educ:* Purdue Univ, BS, 36, MS, 37; Univ Chicago, PhD(math), 40. *Prof Exp:* Instr math, Winthrop Col, 40-41 & Wellesley Col, 41-45; from instr to assoc prof math, Douglass Col, Rutgers Univ, 45-65, prof math, 65-80; RETIRED. *Mem:* Am Math Soc; Math Asn Am. *Res:* Calculus of variations. *Mailing Add:* Pennswood Village J110 Newton NJ 18940

HAZDRA, JAMES JOSEPH, b Chicago, Ill, Aug 30, 33; div; c 5. HEALTH CARE EDUCATION. *Educ:* St Procopius Col, BS, 55; Purdue Univ, PhD(chem), 59. *Prof Exp:* Res chemist, Blockson Div, Olin Mathieson Chem Corp, 59-60, group leader, 60-61; chief prod res glass div, Continental Can Co, Inc, 61-63; assoc prof, 63-65, chmn dept, 64-77, Health Care Educ, 76-87, PROF CHEM,ILL BENEDICTINE COL, 69-, DIR, GRANTS & PLANNED GIVING, 87- *Concurrent Pos:* Asst prof, St Francis Col, Ill, 59; vis prof chem, Ill Benedictine Col, 60-63 & distinguished educr, 76. *Mem:* AAAS; Am Chem Soc; fel Am Col Nutrit. *Res:* Cancer and trace metal research; chemical education; allied health education, geriatric education. *Mailing Add:* Healthcare Educ Ill Benedictine Col Lisle IL 60532

HAZEL, CHARLES RICHARD, b Indianapolis, Ind, May 19, 29; m 51; c 3. AQUATIC ECOLOGY, ENVIRONMENTAL IMPACT ANALYSIS. *Educ:* Humboldt State Univ, BS, 60, MS, 63; Ore State Univ, PhD(fish biol), 69. *Prof Exp:* Res assoc water qual, Humboldt State Univ, 62-63; assoc water qual biol, Calif Dept Fish & Game, 63-64; teaching asst fisheries, Ore State Univ, 64-67; lab dir water qual, Calif Dept Fish & Game, 67-71; consult environ scientist, Jones & Stokes Assocs, Inc, 71-73, vpres environ consult, 73-82, pres, 83-90; RETIRED. *Concurrent Pos:* Phoenix Field Comt, USAF , Sacramento County, 47-56. *Mem:* Am Fisheries Soc; Ecol Soc Am; Water Pollution Control Fedn; Asn Environ Prof; Sigma Xi. *Res:* Aquatic ecology in reference to resource planning, management and policy for decisions in the use of streams, lakes, and estuaries. *Mailing Add:* 14590 Orchard Knob Rd Dallas OR 97338

HAZEL, JEFFREY RONALD, b Youngstown, Ohio, Mar 4, 45; m 66; c 2. COMPARATIVE PHYSIOLOGY, CELL PHYSIOLOGY. *Educ:* Col of Wooster, BA, 67; Univ Ill, Urbana, MS, 69, PhD(physiol), 71. *Prof Exp:* Asst prof zool, Univ Nebr-Lincoln, 71-75; from asst prof to assoc prof, 75-84, PROF ZOOL, ARIZ STATE UNIV, 84- *Concurrent Pos:* Prin investr, NSF grant, 73-75, 76-92. *Mem:* Fel AAAS; Am Physiol Soc; Am Soc Zool; Am Chem Soc. *Res:* Mechanisms of thermal adaptation in poikilotherms; lipid metabolism, membrane structure and function in cold-blooded animals. *Mailing Add:* Dept Zool Ariz State Univ Tempe AZ 85287

HAZEL, JOSEPH ERNEST, b Caruthersville, Mo, July 7, 33; m 56; c 3. INVERTEBRATE PALEONTOLOGY. *Educ:* Univ Mo, BA, 56, MA, 60; La State Univ, PhD(paleont), 63. *Prof Exp:* NSF fel, Mus Comp Zool, Harvard Univ, 63-64; chief br paleontol & stratig, US Geol Surv, 73-78, res geologist, 64-83; RES ASSOC, AMOCO PROD CO, 83- *Concurrent Pos:* Assoc prof lectr, George Washington Univ, 67-78. *Mem:* Paleontol Soc; Int Paleont Asn; Brit Micropaleont Soc; Paleont Res Inst; Soc Econ Paleontologists & Mineralogists; NAm Micropaleont Soc. *Res:* Taxonomy, biostratigraphy and biogeography of Cretaceous, Tertiary and Quaternary microfossils; computer techniques in geology and paleontology; biostratigraphical models; time scales. *Mailing Add:* 108 W Ellery Number 4 Flagstaff AZ 86001

HAZELBAUER, GERALD LEE, b Chicago, Ill, Sept 27, 44; m 70. MEMBRANE BIOLOGY, BACTERIAL BEHAVIOR. *Educ:* Williams Col, BA, 66; Case Western Reserve Univ, MS, 68; Univ Wis-Madison, PhD(genetics), 71. *Prof Exp:* Fel, Univ Wis-Madison, 71 & Inst Pasteur, Paris, 71-73; res fel molecular biol, Univ Uppsala, 73-75, asst prof, 75-81; assoc scientist biochem, Wash State Univ, 81-82, assoc prof, 82-85, actg chair, 89-90, PROF BIOCHEM, WASH STATE UNIV, 85- *Concurrent Pos:* Fel, NSF, 71-72 & Muscular Dystrophy Asn Am, 72-73; res fel, Alfred P Sloan Found, 73-75; fac res award, Am Cancer Soc, 85-90. *Mem:* Am Soc Biol Chem; Am Soc Mirobiol; Protein Soc. *Res:* Molecular biology of bacterial chemotaxis with emphasis on membrane proteins; characterization of receptors and signal transduction proteins; covalent protein modifications involved in sensory adaptation. *Mailing Add:* Biochem & Biophysics Prog Wash State Univ Pullman WA 99164-4660

HAZELRIG, JANE, b Chattanooga, Tenn, May 29, 37; m 58; c 2. MODELING BIOLOGICAL SYSTEMS, SIMULATION. *Educ:* Univ Ala, BS, 58; Univ Minn, MS, 65; Univ Ala, Birmingham, PhD(biophys sci), 76. *Prof Exp:* Assoc physicist, Southern Res Inst, 58-62; biophysics technician, Mayo Clin, 62-67; fel biophysics, Mayo Grad Sch Med, Univ Minn, 64-65; instr math, 67-77, from asst prof to assoc prof biomath, 77-82, ASST PROF PHYSIOL & BIOPHYSICS, UNIV ALA, BIRMINGHAM, 79-, ASSOC PROF BIOSTAT & BIOMATH, 82- *Concurrent Pos:* Fac mem, Interdisciplinary Grad Prog Biophys Sci, Univ Ala, Birmingham, 79- *Mem:* Sigma Xi; Soc Math Biol; AAAS; Soc Indust Appl Math. *Res:* Formulating, testing (via optimization methodology) and modifying mathematical models of biological systems to facilitate the testing of hypotheses and the design of experiments. *Mailing Add:* Dept Biomath Univ Sta Univ Ala Birmingham AL 35294

HAZELRIGG, GEORGE ARTHUR, JR, b Summit, NJ, Oct 28, 39; m 68; c 2. RESEARCH GRANT ADMINISTRATION. *Educ:* NJ Inst Technol, BS, 61, MS, 63; Princeton Univ, MA, 66, MSE, 68, PhD(aerospace eng), 69. *Prof Exp:* Engr, Curtiss-Wright Corp, 61-63; instr, NJ Inst Technol, 62-63; engr, Jet Propulsion Lab, 66-67; staff scientist, Gen Dynamics, 68-71; res staff mem, Princeton Univ, 71-75; dir syst eng, Econ, Inc, 76-82; DEP DIV DIR, DIV ELEC & COMMUN SYST, NSF, 82- *Concurrent Pos:* Assoc ed, J Spacecraft & Rockets, 77-82; consult, Mathamatica, Inc, 71-73, Princeton Univ, 76-84, Econ, Inc, 82-84, Princeton Synergetics, 84-; chair, Emerging Technol Comt, Am Soc Mech Engrs, 89- *Honors & Awards:* Outstanding Contrib to Aerospace Eng, Am Inst Aeronaut & Astronaut, 69. *Mem:* Am Inst Aeronaut & Astronaut; Am Soc Eng Educ; AAAS; Am Soc Mech Engrs. *Res:* Development of the field of microelectromechanical systems. *Mailing Add:* Div Elec & Commun Syst NSF Rm 1151 1800 G St NW Washington DC 20550

HAZELTINE, BARRETT, b Paris, France, Nov 7, 31; US citizen; m 56; c 3. ELECTRICAL ENGINEERING. *Educ:* Princeton Univ, BSE, 53, MSE, 57; Univ Mich, PhD(elec eng), 62. *Hon Degrees:* ScD, SUNY Stony Brook, 88. *Prof Exp:* From asst prof to assoc prof, 59-71, asst dean col, 61-63 & 67-72, PROF ELEC ENG, BROWN UNIV, 71-, ASSOC DEAN COL, 72- *Concurrent Pos:* Asst to mgr adv develop lab, Space & Info Systs Div, Raytheon Co, 64-65; sr lectr, Univ Zambia, 70, vis prof, 76; mem, Adv Comt Nuclear Power, Pub Utilities Comn, State of RI, 77-, NJ Dept Higher Educ & Coun for Understanding Technol in Human Affairs; vis prof, Univ Malaw, 81, 84, 89; trustee, Stevens Inst Technol. *Honors & Awards:* Fulbright lectr, Malawi, 89. *Mem:* Inst Elec & Electronic Engrs; Sigma Xi; Am Soc Eng Educ. *Res:* Engineering management; switching theory; appropriate technology. *Mailing Add:* Div Eng Brown Univ Providence RI 02912

HAZELTINE, JAMES EZRA, JR, b Warren, Pa, Sept 3, 16; m 41; c 3. PHYSICAL CHEMISTRY. *Educ:* Lafayette Col, BS, 37; Franklin & Marshall Col, MS, 50. *Prof Exp:* Asst mgr surface coatings dept, Armstrong Cork Co, 47-53, mgr plastic flooring res dept, 53-59, asst dir, Div Res, 59-61, asst dir res, 61-70, vpres & dir res, Res & Develop Ctr, 70-81; RETIRED. *Mem:* Am Chem Soc. *Res:* Research direction and administration. *Mailing Add:* 1200 Country Club Dr Lancaster PA 17601-5208

HAZELTINE, RICHARD DEIMEL, b Jersey City, NJ, June 12, 42; m 64; c 2. PLASMA PHYSICS. *Educ:* Harvard Col, AB, 64; Univ Mich, MS, 66, PhD(physics), 68. *Prof Exp:* Lectr physics, Univ Mich, 69; vis mem, Inst Advan Study, 69-71; asst dir, Inst Fusion Studies, 82-86, actg dir, 87-88; res scientist theoret physics, 71-86, PROF PHYSICS, UNIV TEX, 86- *Concurrent Pos:* Fel comt, Div Plasma Physics, Am Phys Soc, 84 & 88; Simon Ramo Award Comt, Am Phys Soc, 86; rev panel, Magnetic Fusion Sci Fel Prog; adv comt, Fusion Energy Postdoctoral Res & Prof Dev Progs; consult, Austin Res Assocs, Los Alamos Nat Lab, GA Technol, Lawrence Livermore Nat Lab & Sci Applns Inc. *Mem:* Fel Am Phys Soc; AAAS; Sigma Xi. *Res:* Theoretical plasma physics, as applied to the problem of controlled thermonuclear fusion. *Mailing Add:* 7115 Sungate Dr Austin TX 78731

HAZELTON, BONNI JANE, HUMAN RHABDOMYOSARCOMA, TUBULIAN EXPRESSION. *Educ:* Syracuse Univ, PhD(biol), 79. *Prof Exp:* RES ASSOC, ST JUDE CHILDREN'S RES HOSP, 85- *Mailing Add:* Response Technologies Inc 1775 Moriah Woods Blvd Suite 9 Memphis TN 38117

HAZELTON, RUSSELL FRANK, b Detroit, Mich, Mar 31, 14; m 41; c 3. CHEMICAL ENGINEERING. *Educ:* Wayne State Univ, BS, 35; Univ Mich, MS, 37, PhD(chem eng), 43. *Prof Exp:* Asst chemist, Kimberly-Clark Corp, NY, 37-38, Wis, 38-39; asst chem engr, Solvay Process Co, Va, 41-45; chem engr, Dow Chem Co, 45-48; assoc prof chem eng, NC State Col, 48-52; chem engr & process consult develop, Agr Div, Allied Chem Corp, 52-67; prof chem eng, W Va Inst Technol, 67-82; RETIRED. *Mem:* Am Chem Soc; Am Inst Chem Engrs; Am Soc Eng Educ; Sigma Xi. *Res:* Design of chemical processes and plants; ethanolamines; distillation steps in phenol and chlorobenzene manufacture; chlorine from salt and nitric acid; melamine crystal from urea. *Mailing Add:* PO Box 364 Charlton Heights WV 25040

HAZELWOOD, DONALD HILL, b La Crosse, Wis, May 27, 31; m 54, 71; c 2. ZOOLOGY. *Educ:* Univ Wis, BS, 53, MS, 54; Wash State Univ, PhD(zool), 61. *Prof Exp:* Asst prof, 61-67, ASSOC PROF ZOOL, UNIV MO-COLUMBIA, 67- *Concurrent Pos:* NIH res grant, 62-66; US Corps Engrs contract, 73- *Mem:* Am Soc Limnol & Oceanog. *Res:* Invertebrate zoology. *Mailing Add:* Div Biol Sci Univ Mo 105 Tucker Hall Columbia MO 65211

HAZELWOOD, ROBERT LEONARD, b Oakland, Calif, July 11, 27; m 55. ENDOCRINOLOGY. *Educ:* Univ Calif, MA, 53, PhD(physiol), 58. *Prof Exp:* From instr to asst prof physiol, Sch Med, Boston Univ, 58-63; assoc prof biol, 63-71, PROF PHYSIOL, UNIV HOUSTON, 71- *Mem:* Soc Exp Biol & Med; Am Physiol Soc; Endocrinol Soc; Am Diabetes Assoc. *Res:* Comparative studies of carbohydrate metabolism; hormonal control of growth and metabolism; regulation of pancreatic endocrine secretion; pituitary-pancreatic interrelationships. *Mailing Add:* Dept Biol Univ Houston Houston TX 77204-5513

HAZELWOOD, ROBERT NICHOLS, b Milwaukee, Wis, Jan 22, 28; m 58; c 3. ENVIRONMENTAL AUDITING, RISK ASSESSMENT. *Educ:* Haverford Col, AB, 49; Marquette Univ, MS, 52; Univ Calif, Berkeley, PhD(biophys), 57. *Prof Exp:* Sect head chem, Line Mat Co Div, McGraw-Edison, Inc, 50-54; instr gen sci, Calif Col Arts & Crafts, 55-56; staff mem, Opers Res Group, Arthur D Little, Inc, 56-63; chief scientist, Hq, US Strike Command, 63-66; tech mgr, Data Systs Div, Litton Industs, Inc, 67-70; sr mem staff, Los Angeles Tech Serv Corp, 71-74; vpres & dir, Socio-Econ Systs, Inc, 75-77; chief scientist, Global Marine Develop, Inc, 78-81; DISTINGUISHED TECH ASSOC, IT CORP, 81- *Concurrent Pos:* Assoc, Sch Med, Marquette Univ, 49-54; dir, Drake Steel Supply Co, 60-64; teacher, Hazardous Mats Prog, Univ Calif, Irvine Exten, 84-; bd ed, Environmental Auditor, 87-; regist environ assessor, REA #137, Calif, 87-; dir, Air & Waste Mgt Asn, 87-90; bd ed, J Air & Waste Mgt Asn, 89- *Honors & Awards:* Dean's Distinguished Serv Award, Univ Calif, Irvine Exten, 91. *Mem:* Fel AAAS; Air & Waste Mgt Asn (vpres, 89-90); Soc Risk Anal; Environ Auditing Roundtable. *Res:* Macromolecules; bellometrics; risk assessment; environmental impact analysis; energy systems; waste disposal; environmental carcinogenesis; air pollution. *Mailing Add:* 18861 Via Messina Irvine CA 92715

HAZEN, ANNETTE, b Pittsburgh, Pa, July 5, 29. PHYSIOLOGY, NUTRITION. *Educ:* Goucher Col, BA, 50; Vassar Col, MA, 53; Yale Univ, PhD(physiol), 61; Gen Theol Seminary, MDiv, 75. *Prof Exp:* Asst bact, Goucher Col, 50-51; asst physiol, Vassar Col, 51-54 & Yale Univ, 54; from instr to asst prof, Vassar Col, 59-64; sr res aid, Sloan-Kettering Inst Cancer Res, 63-65; asst prof, Cazenovia Col, 65-67; asst prof, 67-72, ASSOC PROF PHYSIOL, WAGNER COL, 72-, CHMN BIOL DEPT, 81- *Concurrent Pos:* Adj assoc prof, Seton Hall Univ, 71-73. *Res:* Neurophysiology; effect of nicotine upon frog reflexes; endocrinology; effect of certain hormones and vitamins upon diphosphopyridine nucleotide concentrations in rat liver; role of certain hormones in mammary cancer; feeding patterns and obesity. *Mailing Add:* Dept Biol Wagner Col 631 Howard Ave Staten Island NY 10301

HAZEN, DAVID COMSTOCK, b Greenburg, NY, July 3, 27; m 48; c 3. AERONAUTICAL ENGINEERING. *Educ:* Princeton Univ, BSE, 48, MSE, 49. *Prof Exp:* From instr to prof aeronaut eng, Princeton Univ, 49-82, assoc dean fac, 66-69; exec dir, Assembly Eng, Nat Res Coun, 80-82 & Comn Eng & Tech Systs, 82-85. *Concurrent Pos:* Assoc chmn, Aerospace & Mech Sci, 69-74; vchmn, Naval Res Adv Comt, 73-75, chmn, 75-77. *Mem:* Am Inst Aeronaut & Astronaut; Am Soc Eng Educ. *Res:* Subsonic aerodynamics; vertical and short take-off and landing stability control. *Mailing Add:* PO Box 426 Oxford MD 21654

HAZEN, GARY ALAN, b Mobile, Ala, July 18, 50; m 76. ELECTRICAL ENGINEERING. *Educ:* Univ S Ala, BS, 73; Ga Inst Technol, MS, 74. *Prof Exp:* Sr develop engr design, 75-80, DEVELOP PROJ ENGR DESIGN, SCHLUMBERGER WELL SERV, 80- *Mem:* Inst Elec & Electronics Engrs. *Res:* Hostile environment electronics; communication electronics. *Mailing Add:* Schlumberger Technol Corp PO Box 2175 Houston TX 77252

HAZEN, GEORGE GUSTAVE, b Ft Benton, Mont, Jan 1, 23; m 48; c 6. ORGANIC CHEMISTRY. *Educ:* Mont State Col, BS, 44; Univ Mich, MS, 48, PhD, 51. *Prof Exp:* Instr org chem, Univ Mich, 49-51; sr chemist process develop, 51-57, sect leader, 57-64, sect leader proc res, 64-69, mgr, 69-80, ASSOC DIR, MERCK & CO, INC, 80- *Mem:* Am Chem Soc; NY Acad Sci. *Res:* Steroids; polyhydric alcohols; substituted pyridines; development of complex organic and biosynthetic processes. *Mailing Add:* 205 Water Street Perth Amboy NJ 08861-4426

HAZEN, MARTHA L(OCKE), b Cambridge, Mass, July 15, 31; div; c 2. ASTRONOMY. *Educ:* Mt Holyoke Col, AB, 53; Univ Mich, MA, 55, PhD(astron), 58. *Prof Exp:* Instr astron, Mt Holyoke Col, 57-59; lectr & res assoc, Univ Mich, 59-60; res fel, 57-69 & 60-69, CUR ASTRON PHOTOGS, COL OBSERV, HARVARD UNIV, 69-, LECTR, 83- *Concurrent Pos:* Lectr, Wellesley Col, 61-63 & 66-67; adj assoc prof, Boston Univ, 79. *Mem:* Am Astron Soc; Int Astron Union. *Res:* Photometry; globular clusters; variable stars. *Mailing Add:* Harvard Col Observ 60 Garden St Harvard Univ 60 Garden St Cambridge MA 02138

HAZEN, RICHARD, civil engineering, sanitary engineering; deceased, see previous edition for last biography

HAZEN, RICHARD R(AY), b Zanesville, Ohio, Aug 26, 25; m 50; c 2. ELECTRICAL ENGINEERING. *Educ:* Univ Dayton, BEE, 53; Univ Cincinnati, MSEE, 62. *Prof Exp:* From inst to assoc prof, Univ Dayton, 53-71, prof elec technol, 71-86, chmn, Dept Electronic Eng Technol, 58-86; RETIRED. *Concurrent Pos:* Chmn, Eng Technol Comt, Engrs Coun Prof Develop, 74-75; pron investr, electro-mech devices, Air Forces Logistics Command, 81- *Mem:* Am Soc Eng Educ; Inst Elec & Electronics Engrs; Armed Forces Commun & Electronics Asn. *Res:* Environmental noise. *Mailing Add:* 9672 Ferry Rd Waynesville OH 45068

HAZEN, ROBERT MILLER, b Rockville Centre, NY, Nov 1, 48; m 69; c 2. MINERALOGY & CRYSTAL CHEMISTRY, SCIENTIFIC LITERACY. *Educ:* Mass Inst Technol, BA & SM, 71; Harvard Univ, PhD(mineral), 75. *Prof Exp:* NATO fel mineral, Cambridge Univ, 75-76; RES FEL & MEM STAFF, GEOPHYS LAB, CARNEGIE INST WASHINGTON, 76-; ROBINSON PROF, GEORGE MASON UNIV, 89- *Concurrent Pos:* Author of various publications. *Honors & Awards:* Mineral Soc Am Award; Ipatief Prize, Am Chem Soc; Deems Taylor Award. *Mem:* Mineral Soc Am; Am Geophys Union; Sigma Xi; Am Chem Soc; AAAS. *Res:* Development of high-pressure and high-temperature single crystal x-ray diffraction techniques and their application to silicate mineral crystal chemistry; relationships between science and society. *Mailing Add:* Carnegie Inst Washington 5251 Broad Branch Rd NW Washington DC 20015-1305

HAZEN, TERRY CLYDE, b Pontiac, Mich, Feb 7, 51; m 72; c 2. AQUATIC ECOLOGY, PARASITOLOGY. *Educ:* Mich State Univ, BS, 73, MS, 74; Wake Forest Univ, PhD(parasite ecol), 78. *Prof Exp:* Res assoc biol, Wake Forest Univ, 75-79; from asst prof biol to prof biol, Univ PR, 79-88; res scientist, E I du Pont de Nemours & Co Inc, 87-89; PRIN SCIENTIST, WESTINGHOUSE SAVANNAH RIVER CO, 89- *Concurrent Pos:* Adj prof marine sci, Univ PR, Mayagaez, 80-88 & adj prof biol, Univ SC, Columbia, 88- *Honors & Awards:* George Westinghouse Gold Signature Award, 89. *Mem:* Ecol Soc Am; Sigma Xi; AAAS; Am Soc Parasitol; Am Soc Microbiol. *Res:* Bioremediation, environmental biotechnology, microbial ecology, fish pathology and computer applications; survival, activity and distribution of bacteria in fresh and marine waters; effect of physical, chemical and biological factors on bacteria in natural and polluted systems, and mathematical models that describe those relationships. *Mailing Add:* Environ Sci Div Savannah River Lab Aiken SC 29808-0001

HAZEN, WAYNE ESKETT, b Three Rivers, Mich, Feb 8, 14; m 39; c 4. PHYSICS. *Educ:* Mass Inst Technol, BS, 36; Univ Calif, PhD(physics), 41. *Prof Exp:* From instr to asst prof physics, Univ Calif, 41-47; from asst prof to prof, 47-84, EMER PROF PHYSICS, UNIV MICH, ANN ARBOR, 84- *Concurrent Pos:* Guggenheim fel, Mass Inst Technol, 47 & Imp Col, Univ London, 54; Fulbright scholar, Polytech Sch, Univ Paris, 53-54; Smith-Mundt prof, Am Univ Beirut, 58-59; researcher, Univ Hong Kong, High Energy Physics Inst, Beijing & Cosmic Ray Inst, Univ Tokyo. *Mem:* Fel Am Phys Soc. *Res:* Cosmic-ray air showers. *Mailing Add:* Dept Physics Univ Mich Ann Arbor MI 48103

HAZEN, WILLIAM EUGENE, b Canton, NY, June 4, 25. ANIMAL ECOLOGY. *Educ:* St Lawrence Univ, BS, 47; Univ Mich, MA, 48, PhD(zool), 54. *Prof Exp:* From instr to asst prof biol, Univ Chicago, 53-62; from asst prof to assoc prof, 62-68, PROF BIOL, SAN DIEGO STATE UNIV, 68- *Concurrent Pos:* Prog dir, Ecosystem Studies, NSF, 72 & 73. *Mem:* Ecol Soc Am; Am Soc Limnol & Oceanog; Brit Ecol Soc. *Res:* Biology of aquatic invertebrates; ecology of temporary ponds. *Mailing Add:* 4173 Hilldale Rd San Diego CA 92116

HAZEYAMA, YUJI, b Tokyo, Aug 5, 43; m 82. PHYSIOLOGY. *Educ:* Musah Inst Technol, Tokyo, BS, 66, MS, 69; Univ Calif, Berkeley, PhD(elec eng & comput sci), 75. *Prof Exp:* Fel physiol, Univ Mich, 75-78; res fel, Childrens Hosp Northern Calif, Oakland, 78-80; asst prof chem, Lehigh Univ, 80-82; gen mgr, ed serv, Elsevier Sci Publ, Japan, 83-87; PRES, MYU PUBL GROUP, 87- *Mem:* AAAS; Inst Elec & Electronics Engrs. *Res:* Identification of plasma borne vasconstruction which is closely related to the genesis of basal vascular tone; biomedical instrumentation (measurement of intracranial pressure). *Mailing Add:* MYU KK 32-3-303 Sendagi 2 chrome Bunkyo-ku Tokyo 113 Japan

HAZI, ANDREW UDVAR, b Budapest, Hungary, Sept 28, 41; US citizen. CHEMICAL PHYSICS, THEORETICAL CHEMISTRY. *Educ:* Univ Calif, Los Angeles, BS, 64; Univ Chicago, PhD(chem physics), 67. *Prof Exp:* Res assoc chem, Univ Southern Calif, 67-68; asst prof, Univ Calif, Los Angeles, 68-75; lectr, Calif Polytech State Univ, San Luis Obispo, 75-76; PHYSICIST LASER PROG, LAWRENCE LIVERMORE LAB, UNIV CALIF, 76- *Concurrent Pos:* Fel, Air Force Off Sci Res, 67-68; consult, McDonald-Douglas, 68 & McMillan Assocs, 74-75. *Mem:* Am Phys Soc. *Res:* Quantum theory of excited states of atoms and molecules; electron scattering theory; resonance phenomena in collisions; fundamental processes in gas lasers. *Mailing Add:* Lawrence Livermore Lab L296 Univ Calif PO Box 808 Livermore CA 94550

HAZLEGROVE, LEVEN S, b Birmingham, Ala, Dec 1, 24; m 51; c 3. PHYSICAL CHEMISTRY, OTHER ENVIRONMENTAL. *Educ:* Howard Col, BS, 47; Emory Univ, MS, 49; Univ Ala & Univ Ala Med Sch, PhD(chem), 65. *Prof Exp:* Lab instr chem, Emory Univ, 48-49; asst prof & athletic dir, WGa Col, 49-56; res chemist, FDA, 56-57; asst prof chem, Howard Col, 57-63; res assoc biophys chem, Med Ctr, Univ Ala, 63-65; assoc prof, Samford Univ, 65-67, head dept, 78-84, prof chem, 67-90; EXEC DIR, ALA ACAD SCI, 92- *Concurrent Pos:* Instr, Exten Div, Univ Ga, 52-56; Ford Found lectr, WGa Col, 55-56; sr Danforth Assoc, 60-; NIH fel, 63-64, NSF fel, 63-65; vis prof, Med Ctr, Univ Ala, 65-, US AID, Univ Dacca & Rajshahi Univ, 67-69; chmn & trustee, Gorgas Scholarship Found, 76-; vis bioenergy ambassador, People-to-People, China, 85; vis prof, Anhui Univ, Wuhu, China, Dacca Univ. *Mem:* AAAS; Am Chem Soc; Sigma Xi; Nat Hon Soc; Am Indust Hyg Asn. *Res:* Electrodeposition manganese using mercury cathode; spray residues on foods; polydispersity of hyaluronic acid, molecular weight determination by light scattering; computer program for molecular weight determinations; bio-polymers; gel-electrophoresis; macromolecular biochemistry; immunochemistry. *Mailing Add:* 208 Rockaway Rd Birmingham AL 35209-6626

HAZLEHURST, DAVID ANTHONY, b Chester, Eng, Dec 9, 30; m 60; c 2. INORGANIC CHEMISTRY, POLYMER CHEMISTRY. *Educ:* Univ Liverpool, BSc, 52, PhD(inorg phys chem), 55. *Prof Exp:* Res fel inorg chem, Univ Tex, 55-56; res chemist, Carrington Res Lab, Shell Chem Co, Eng, 58-63; res chemist, 63-66, sr res chemist, 66-67, supvr, 67-69, sr supvr nylon prod develop, 69-82, qual systs coordr, Nylon Textile, 82-83, SR RES ASSOC, NYLON INDUST, E I DU PONT DE NEMOURS & CO, INC, 83- *Mem:* Am Chem Soc. *Res:* Polymer chemistry; product development in plastics, textile fibers and foamed materials. *Mailing Add:* 102 N Plum St Richmond VA 23220

HAZLETT, DAVID RICHARD, b Natrona Height, Pa, Apr 3, 32; m 59; c 2. PULMONARY PHYSIOLOGY. *Educ:* Univ Pittsburgh, BS, 54, MD, 58. *Prof Exp:* Resident internal med, Brooke Gen Hosp, 61-64; asst chief pulmonary dis serv, Valley Forge Gen Hosp, 64-65; cmndg officer, Ninth Field Hosp, Vietnam, 65-66; chief pulmonary function lab, Fitzsimons Army Med Ctr, 68-78; asst prof med, Univ Colo, Denver, 71-78; chief pulmonary function lab, Brooke Army Med Ctr, 78-; AT PULMONARY FUNCTION LAB, SAN ANTONIO STATE CHEST HOSP. *Concurrent Pos:* Fel pulmonary physiol, Sch Med, Univ Colo, 67-68. *Honors & Awards:* Physicians Recognition Award, AMA, 69. *Mem:* AMA; Biomed Eng Soc; fel Am Col Chest Physicians. *Res:* Lung mechanics, gas transport and computer application. *Mailing Add:* Pulmonary Function Lab San Antonio State Chest Hosp PO Box 23340 San Antonio TX 78223

HAZLETT, ROBERT NEIL, b Sterling, Kans, Oct 4, 24; m 53; c 2. PHYSICAL ORGANIC CHEMISTRY, HYDROCARBON FUELS. *Educ:* Sterling Col, BS, 47; Univ Kans, PhD(chem), 50. *Hon Degrees:* ScD, Sterling Col, 70. *Prof Exp:* Instr chem, St Ambrose Col, 50-51; chemist, Fuels Br, US Naval Res Lab, Washington, DC, 51-71, head, Fuels Sect, Chem Dynamics Br, 72-76, head fuels sect, Navy Technol Ctr for Safety & Survivability, Chem Div, 76-87; FUEL CONSULT, ALEXANDRIA, VA, 88- *Concurrent Pos:* Vis scientist, Mat Res Labs, Melbourne, Australia, 84-85. *Mem:* Am Chem Soc; Sigma Xi. *Res:* Jet aircraft fuels; hydrocarbon oxidation; fuel stability; separation of water from fuels; liqud rocket fuels; synthetic fuels; alternate energy sources; shale oil; shale derived fuels; diesel fuel; marine fuels. *Mailing Add:* 5205 Chippewa Pl Alexandria VA 22312

HAZLEWOOD, CARLTON FRANK, b Perrin, Tex, July 25, 35; m 56; c 3. PHYSIOLOGY, BIOPHYSICS. *Educ:* Tex A&M Univ, BS, 57; Univ Tenn, PhD(physiol, anat), 62. *Prof Exp:* Res physiologist, Children's Clin Res Ctr, 67-78; from instr to assoc prof physiol & pediat, 64-76; dir grad studies, dept physiol, 78-86, PROF PHYSIOL & PEDIAT, BAYLOR COL MED, 76- *Concurrent Pos:* Fel med, Sch Med, Johns Hopkins Univ, 62-64; consult, Westinghouse Elec Corp, 62-64; mem expert comt sci coop bur Eur & NAm reg (UNESCO), 83-89; sr res assoc, Dept Physics, Rice Univ, 70-75, adj assoc prof physics, 75-78, prof, 78- *Mem:* AAAS; Biophys Soc; Soc Magnetic Resonance Imaging; NY Acad Sci; Am Physiol Soc. *Res:* Mechanism of action of insulin; muscular dystrophy; ion accumulation in muscle; growth and development of muscle; physical state of ions and water in living tissues, cells and subcellular organelles; role of water in cell division; in vivo nuclear magnetic resonance scanning; scanning tunneling microscopy. *Mailing Add:* 4714 O'Meara Houston TX 77035

HAZLEWOOD, DONALD GENE, b Abilene, Tex. NUMBER THEORY. *Educ:* Univ Tex, Austin, BA, 63; Syracuse Univ, MS, 66, MA, 69, PhD(math), 71. *Prof Exp:* Sec teacher math, Independent Sch Dists, Tex, NMex & NY, 62-67; asst prof, 71-77, ASSOC PROF MATH & COMPUT SCI, SOUTHWEST TEX STATE UNIV, 71- *Mem:* Am Math Soc; Math Asn Am; Sigma Xi. *Res:* Elementary methods in analytic number theory. *Mailing Add:* Rte 3 Box 83M San Marcos TX 78666

HAZUDA, DARIA JEAN, EXPERIMENTAL BIOLOGY. *Educ:* Rutgers Univ, BA, 81; State Univ NY, Stony Brook, PhD(biochem), 86. *Prof Exp:* Grad teaching asst, State Univ NY, 81-83, grad res asst, 83-86; postdoctoral fel, Dept Molecular Genetics, Smith, Kline & French, Swedeland, Pa, 86-89; SR RES BIOCHEMIST, DEPT VIRUS & CELL BIOL, MERCK SHARP & DOHME, WEST POINT, PA, 89- *Mem:* Am Soc Biochem & Molecular Biol; Protein Soc; Sigma Xi. *Res:* Author of numerous publications. *Mailing Add:* Dept 874 Merck Sharp & Dohme WP16-215 West Point PA 19486

HAZY, ANDREW CHRISTOPHER, b Windber, Pa, Sept 10, 38; 79. ORGANIC CHEMISTRY. *Educ:* St Vincent Col, BS, 59; Univ Notre Dame, PhD(org chem), 65. *Prof Exp:* Instr chem, St Vincent Col, 59-61; sr res chemist, 64-68, group leader chem res, Olin Corp, 68-69, photochem, Horizons Res Inc, 69-71, gen mgr, Photohorizons, 71-75, vpres mfg, 76-79, VPRES TECHNOL, METALPHOTO DIV, HORIZONS RES INC, 79- *Mem:* Am Chem Soc; Soc Photog Sci & Eng. *Res:* Organic chemistry, especially related to photographic systems. *Mailing Add:* Metalphoto Div Horizons Res Inc 18531 S Miles Rd Cleveland OH 44128-4237

HAZZARD, DEWITT GEORGE, b Scranton, Pa, May 23, 32; m 53; c 5. BIOCHEMISTRY, RESOURCE MANAGEMENT. *Educ:* Pa State Univ, BS, 54, MS, 57; Univ Conn, PhD(animal nutrit), 63. *Prof Exp:* Res asst vitamin A res, Animal Nutrit Sect, Dept Animal Industs, Univ Conn, 57-62; biochemist, Nutrit Sect, Div Radiol Health, 62-63; chemist, Radiochem Sect, USPHS, 63-66, chemist, Physiol-Biophys Sect, Radiation Bio-Effects Prog, Nat Ctr Radiol Health, Rockville, 66-69, chief, Metab Studies Sect, Bur Radiol Health, Div Biol Effects, 69-72, dir, Off Extramural Res, Bur Radiol Health, Food & Drug Admin, USPHS, 78-82; DIR, CELL BIOL & ANIMAL MODELS PROGS, NAT INST AGING, NIH, 82-, HEAD, BIOL RESOURCES & RESOURCE DEVELOP PROG & BIOMED RES & CLIN MED, 85- *Concurrent Pos:* Exec secy radiol health res & training grant rev comt, Bur Radiol Health, FDA, 72-73, grants officer, 72-78. *Honors & Awards:* Comnr's Special Citation, Food & Drug Admin, 81. *Res:* Direct extramural programs in cell biology and animal models; development as related to understanding the aging process; direct development of biological and animal resources for use in aging research. *Mailing Add:* 2505 Seibel Dr Silver Spring MD 20905

HAZZARD, WILLIAM RUSSELL, b Ann Arbor, Mich, Sept 5, 36; m 61; c 4. METABOLISM. *Educ:* Cornell Univ, AB, 58, MD, 62. *Prof Exp:* Intern med, New York Hosp, 62-63; res fel metab, Sch Med, 65-66, resident med, 66-67, res fel metab, 67-69, from instr to asst prof, 69-73, res assoc, Vet Admin Hosp, Seattle, 69-70, clin investr, 70-71; dep dir, Northwest Lipid Res Clin, Univ Wash, 70-73, dir, 73-77, assoc prof metab, 73-77, investr metab, Howard Hughes Med Inst, Sch Med, 71-, prof med & assoc dir northwest lipid res clin, 77-, chief div geront & geriat med, 78-; AT DEPT MED, JOHNS HOPKINS UNIV; PROF & CHAIR INTERNAL MED, BOWMAN GRAY SCH MED, WAKE FOREST UNIV. *Concurrent Pos:* Fel, Coun Arteriosclerosis, Am Heart Asn, 74-; mem adv bd, Washington-Alaska Regional Med Prog, 75- *Mem:* Inst Med-Nat Acad Sci; Sigma Xi; Am Fedn Clin Res; Am Diabetes Asn; Am Soc Clin Invest. *Res:* Epidemiological, clinical and biochemical investigation of hyperlipidemia in man with emphasis upon endocrinological and genetic aspects and relationship to atherosclerosis in man and animal models. *Mailing Add:* Bowman Gray Sch Med 300 S Hawthorne Rd Winston-Salem NC 27103

HEACOCK, CRAIG S, BIOCHEMISTRY. *Educ:* Univ Wash, Seattle, BS, 76; Colo State Univ, Ft Collins, PhD(biochem cell & molecular biol), 83. *Prof Exp:* Fac biochem, Colo State Univ, Ft Collins, 81-83; sr instr exp therapeut, Cancer Ctr, Univ Rochester, 83-87; sr res chemist biochem, 87-89, PRIN RES SCIENTIST ANALYTICAL CHEM, CYTOGEN CORP, PRINCETON, NJ, 89- *Res:* Author of numerous publications. *Mailing Add:* Admin Off Cytogen Corp 600 College Rd E Princeton NJ 08540

HEACOCK, RAYMOND L(EROY), electrical engineering, for more information see previous edition

HEACOCK, RICHARD RALPH, b Kalama, Wash, Apr 22, 20; m 57; c 4. GEOPHYSICS, MAGNETOSPHERIC PHYSICS. *Educ:* Ore State Univ, BS, 44; Univ Wis, MPh, 46. *Prof Exp:* Instr math, Wash State Univ, 47-48; fel, Univ Calif, Berkeley, 48-49; instr, Wash State Univ, 50-51; res assoc geophys, Geophys Inst, Univ Alaska, Fairbanks, 61-63, from asst geophysicist to assoc geophysicist, 63-73, geophysicist, 73-75, prof, 75-79, consult, 79-80; RETIRED. *Mem:* Am Geophys Union. *Res:* Pulsations in the earth's magnetic field, and very low frequency electromagnetic waves in the earth's magnetosphere. *Mailing Add:* Rte 1 Box 389 Amboy WA 98601

HEACOCK, RONALD A, b London, Eng, Apr 14, 28; Can citizen; m 51; c 3. RESEARCH FUNDING, HEALTH RESEARCH POLICY. *Educ:* Univ London, BSc, 49, PhD(org chem), 52, DSc(org chem, biochem), 65. *Prof Exp:* Sci officer, Explosive Res & Develop Estab, UK, 51-54; res fel org chem, Nat Res Coun Can, 54-56; sr sci officer, Wye Col, Univ London, 56-57; chief res biochemist, Psychiat Res Unit, Univ Sask Hosp, 57-65; sr res officer & head physiol chem sect, Atlantic Regional Lab, Nat Res Coun Can, 65-74; dir, Contrib & Awards Div & res prog directorate, Health Prog Br, 74-80, DIR GEN RES PROGS, HEALTH SERV & PROM BR, DEPT NAT HEALTH & WELFARE, CAN, 80- *Concurrent Pos:* Lectr, Northampton Eng Col, London, 50-54; sessional lectr, Univ Sask, 58-64; assoc prof grad studies, Dalhousie Univ, 65-72; assoc mem, Med Res Coun Can, 80-; mem, Nat Cancer Inst, Can, 85- *Mem:* Fel Can Inst Chem; fel Royal Soc Chem. *Res:* Explosives chemistry; chemistry of heterocyclic organic compounds, including alkaloids; chemistry and biochemistry related to brain function; chemistry of biologically important substances; paper and thin-layer chromatography; gas chromatography; mass spectrometry; research programs administration; research proposal selection; research priority development. *Mailing Add:* 531 Brierwood Ave Ottawa ON K2A 2H4 Can

HEACOX, WILLIAM DALE, b Pipestone, Minn, Mar 26, 42; m 66. ASTRONOMY, MATHEMATICS. *Educ:* Whitman Col, BA, 64; Wash State Univ, MA, 72; Univ Hawaii, MS, 75, PhD(astron), 77. *Prof Exp:* Res assoc, Nat Acad Sci & Nat Res Coun, Goddard Space Flight Ctr, NASA, 77-78; res assoc astron, Lunar & Planetary Lab, Univ Ariz, 78-; AT NATURAL SCI DIV, UNIV HAWAII, HILO. *Mem:* Am Astron Soc; Astron Soc Pac; Soc Indust & Appl Math; Brit Interplanetary Soc. *Res:* Stellar atmospheres; extrasolar planetary systems; numerical smoothing; splines. *Mailing Add:* Natural Sci Lab Univ Hawaii Hilo HI 96720-4091

HEAD, CHARLES EVERETT, b Ark, Sept 6, 41; m 62; c 3. ATOMIC SPECTROSCOPY, PARTICLE THEORY. *Educ:* Univ Ark, Fayetteville, BS, 63, PhD(physics), 65. *Prof Exp:* From asst prof to assoc prof, 65-73, PROF PHYSICS, UNIV NEW ORLEANS, 73- *Concurrent Pos:* Consult, Wadsworth Publ Co, 76- *Mem:* Am Phys Soc; Optical Soc Am; Am Asn Physics Teachers. *Res:* Measurement of radiative lifetimes of excited states of neutral and ionized atoms and molecules; accelerator-based atomic and molecular physics; fundamental particle structure; laser specctroscopy. *Mailing Add:* Dept Physics Univ New Orleans New Orleans LA 70148

HEAD, H HERBERT, b Jersey City, NJ, Oct 28, 35; m 60; c 4. DAIRY SCIENCE. *Educ:* Rutgers Univ, BS, 57, MS, 59; Univ Md, PhD(animal physiol), 63. *Prof Exp:* Asst dairy sci, Rutgers Univ, 57-59 & Univ Md, 59-63; asst prof, 63-71, ASSOC PROF DAIRY SCI, UNIV FLA, 71- *Mem:* Am Dairy Sci Asn; Am Soc Animal Sci. *Res:* Ruminant metabolism studies; metabolic regulation imposed by insulin growth hormone; bovine fetal growth patterns; factors affecting insulin secretion in the ruminant; hormonal requirements and mechanisms for lactogenesis and maintenance of lactation; induced lactation. *Mailing Add:* Dept Dairy Sci Col Agr Univ Fla Gainesville FL 32611

HEAD, JAMES WILLIAM, III, b Richmond, Va, Aug 4, 41; m; c 2. STRATIGRAPHY, PLANETARY GEOLOGY. *Educ:* Washington & Lee Univ, BS, 64; Brown Univ, PhD(geol), 69. *Prof Exp:* Geologist, Bellcomm, Inc, 68-73; asst prof, 73-74, assoc prof, 74-75, assoc prof, 75-80, PROF GEOL SCI, BROWN UNIV, 80- *Concurrent Pos:* Interim dir, Lunar Sci Inst, Houston, Tex, 73-74; chmn NASA solar syst explor mgt coun, 86- *Honors & Awards:* Medal, NASA, 71. *Mem:* AAAS; Geol Soc Am; Am Geophys Union; Am Asn Petrol Geol; Am Astron Soc. *Res:* Geology and tectonics of planetology surfaces; comparative planetology. *Mailing Add:* Dept Geol Sci Brown Univ Providence RI 02912

HEAD, MARTHA E MOORE, b Little Rock, Ark, Dec 3, 41; m 62; c 3. OCEAN ACOUSTICS, ATOMIC SPECTROSCOPY. *Educ:* Univ Ark, BS, 63, MS, 64; Tulane Univ, PhD(physics), 69. *Prof Exp:* Asst prof physics & math, Southern Univ, New Orleans, 69-73; assoc prof physics, West Tex State Univ, 73-79; adj assc prof, 80-89, ADJ PROF PHYSICS, UNIV NEW ORLEANS, 89- *Concurrent Pos:* Manuscript reviewer & consult, Wadsworth Publ Co, 76-; mem staff, Acoust Div, Naval Oceanog Off, 83- *Mem:* Am Phys Soc; Optical Soc Am; Acoustical Soc Am. *Res:* Fundamental particle structure calculations; acoustic modeling; determination of mean lifetimes of excited electronic states of atoms. *Mailing Add:* Univ New Orleans PO Box 1362 Lake Front New Orleans LA 70148

HEAD, RONALD ALAN, b Birmingham, Ala, May 12, 30; m 55; c 2. CHEMISTRY. *Educ:* Birmingham-Southern Col, BS, 52; Univ Ala, MA, 57; Univ of the Pac, PhD(chem), 64. *Prof Exp:* Teacher chem, Pensacola Jr Col, 57-65; chmn, Dept Phys Sci, 65-88, CHMN, MATH/SCI DIV, OKALOOSA-WALTON COMMUNITY COL, 88- *Mem:* AAAS; Am Chem Soc; fel Am Inst Chem; Int Union Pure & Appl Chem. *Res:* Chemical education; inorganic halfwave potentials in liquid ammonia. *Mailing Add:* 403 Jame St Valparaiso FL 32580-1115

HEAD, THOMAS JAMES, b Tonkawa, Okla, Jan 6, 34; m 69; c 4. MATHEMATICS. *Educ:* Univ Okla, BS, 54, MA, 55; Univ Kans, PhD(math), 62. *Prof Exp:* Instr math, Univ Okla, 56-57 & Washburn Univ, 59-61; asst prof, Iowa State Univ, 62-65; prof math, Univ Alaska, 65-88; PROF MATH SCI, STATE UNIV NY, BINGHAMTON, 88- *Concurrent Pos:* Vis res prof, NMex State Univ, 70-71; vis prof, Univ Tex, El Paso, 72-74 & Calif Polytech State Univ, San Luis Obispo, 75; vis prof math sci, Rice Univ, 75-76; prog dir, Theoret Computer Sci, NSF, Wash, DC, 85-87. *Mem:* NY Acad Sci; European Asn Theoretical Comput Sci; Am Math Soc; Asn Comput Mach. *Res:* Theoretical computer science; abstract algebra. *Mailing Add:* Dept Math Sci State Univ NY Binghamton NY 13901

HEAD, WILLIAM FRANCIS, JR, pharmaceutical & analytical chemistry, for more information see previous edition

HEAD-GORDON, MARTIN PAUL, b Canberra, Australia, Mar 17, 62; m 85; c 1. MOLECULAR ORBITAL THEORY, GAS-SURFACE INTERACTIONS. *Educ:* Monash Univ, BSc, 83, MSc, 85; Carnegie-Mellon Univ, PhD(chem), 89. *Prof Exp:* RESEARCHER, AT&T BELL LABS, 89- *Mem:* Am Chem Soc; Am Phys Soc; AAAS. *Res:* Theoretical chemistry, including quantum chemistry and the development of molecular orbital theory, with application to problems in surface science and molecular structure and properties. *Mailing Add:* AT&T Bell Labs Rm 1A-360 600 Mountain Ave Murray Hill NJ 07974-2070

HEAD-GORDON, TERESA LYN, b Akron, Ohio, Sept 28, 60; m 85; c 1. BIOPHYSICS, PHYSICAL CHEMISTRY. *Educ:* Case Western Reserve Univ, BS, 83; Carnegie-Mellon Univ, PhD(chem), 89. *Prof Exp:* Postdoctoral res, Rutgers Univ, 89-90; POSTDOCTORAL RES, AT&T BELL LABS, 90- *Concurrent Pos:* Consult, Pittsburgh Supercomput Ctr, 87-88. *Mem:* Am Chem Soc; Am Phys Soc; AAAS. *Res:* Theoretical methods applied to biopolymers and oligopeptides; heatbath modelling; electrostatics and dielecttucs; neural networks; ab initio methods; research on protein folding. *Mailing Add:* AT&T Bell Labs Rm 1A-365 600 Mountain Ave Murray Hill NJ 07974-2070

HEADINGS, VERLE EMERY, b Hubbard, Ore, July 14, 35; m 62; c 3. MEDICINE, HUMAN GENETICS. *Educ:* Goshen Col, BA, 58; Univ Mich, Ann Arbor, MS, 62, MD, 64, PhD(human genetics), 70. *Prof Exp:* Asst prof, 69-73, assoc prof, 73-78, PROF PEDIAT, COL MED, HOWARD UNIV, 78-, MED GENETICIST, 69- *Concurrent Pos:* USPHS res grant, Col Med, Howard Univ, 70-; Am Heart Asn res grant, 71-; consult, Ctr Sickle Cell Anemia, 71 & Nat Job Corps Health Off, 71-; mem coun arteriosclerosis, Am Heart Asn. *Mem:* AAAS; NY Acad Sci; Am Soc Human Genetics; Soc Health & Human Values; Soc Pediat Res; Sigma Xi. *Res:* Gene regulation in primates; genetic susceptibility to atherosclerosis; relationship between serum alpha-antitrypsin deficiency and emphysema; diagnosis and treatment of sickle cell anemia; immunochemical detection of carriers for Duchenne muscular dystrophy; models of counseling for genetic disorders. *Mailing Add:* Rte 1 Box 682 Shepherdstown WV 25443

HEADINGTON, JOHN TERRENCE, b Grand Rapids, Mich, June 15, 30; m 57; c 2. PATHOLOGY. *Educ:* Univ Mich, BA, 52, MD, 57. *Prof Exp:* Intern, Virginia Mason Hosp, 57-58; asst resident path, 58-59, resident, 59-60, from jr clin asst to assoc prof, 60-71, PROF PATH, MED SCH, UNIV MICH, ANN ARBOR, 71-, PROF DERMAT, 77- *Mem:* AAAS; Int Acad Path. *Res:* Dermatopathology; geographic pathology. *Mailing Add:* Dept Path Univ Mich Ann Arbor MI 48109

HEADLEE, RAYMOND, b Shelby Co, Ind, July 27, 17; m 41; c 3. PSYCHIATRY, PSYCHOLOGY. *Educ:* Ind Univ, Bloomington, AB, 39, AM, 41; Ind Univ, Indianapolis, MD, 44; Am Bd Psychiat & Neurol, dipl & cert psychiat, 51. *Prof Exp:* Intern, St Elizabeth's Hosp, Washington, DC, 44-45, resident psychiat, 45; resident, Milwaukee Psychiat Hosp, 47-48; from asst prof to prof, 58-62, chmn dept, 63-70, PROF PSYCHIAT, MED COL WIS, 62- *Concurrent Pos:* Pvt pract psychiat & psychoanal, 47-; prof psychol, Marquette Univ, 65-77. *Honors & Awards:* Cert of Honor, Am Bd Psychiat & Neurol, 70. *Mem:* Am Med Asn; fel Am Psychiat Asn; fel Am Col Psychiat; fel Am Col Psychoanalysts; Am Psychol Asn. *Res:* Experimental and in vivo research into the nature of charisma. *Mailing Add:* 1055 Legion PO Box 125 Elm Grove WI 53122-0125

HEADLEE, WILLIAM HUGH, medical parasitology, tropical medicine; deceased, see previous edition for last biography

HEADLEY, ALLAN DAVE, b Jamaica, May 10, 55; m 87; c 1. PHYSICAL ORGANIC CHEMISTRY. *Educ:* Columbia Union Col, BA, 76; Howard Univ, PhD(chem), 82. *Prof Exp:* Postdoctoral chem, Univ Calif, Irvine, 82-83, lectr chem, 87-89; lectr chem, Univ WIndies, Mona, 83-87; ASST PROF CHEM, TEX TECH UNIV, 89- *Concurrent Pos:* Vis prof, Univ Calif, Irvine, 87. *Mem:* Am Chem Soc. *Res:* Analysis of structural effects on proton-transfer reactions in the gas phase; solvation effects on the reactivity and solubility of amino acids and other molecules of biological interest. *Mailing Add:* Dept Chem Tex Tech Univ Lubbock TX 79409-4260

HEADLEY, JOSEPH CHARLES, b Terre Haute, Ill, June 23, 30; m 64; c 2. NATURAL RESOURCE ECONOMICS, PRODUCTION ECONOMICS. *Educ:* Univ Ill, BS, 52, MS, 55; Purdue Univ, PhD(agr econ), 60. *Prof Exp:* Asst agr economist farm mgt, Univ Ariz, 55-57; grad asst, Purdue Univ, 57-59; assoc specialist, Univ Calif, Davis, 59-60; asst prof farm mgt, Univ Ill, 60-66; from assoc prof to prof econ, Univ Mo, 66-89; PROF & HEAD AGR POLICY, UNIV ARK, 89- *Concurrent Pos:* Vis scholar, Resources for the Future, 65-66; vis prof, Norwegian Agr Univ, 72, Swed Agr Univ, 72-73, Danish Agr & Vet Univ, 73. *Mem:* Am Agr Econ Asn. *Res:* Economics of controlling agricultural pests; developed an economic interpretation of the economic threshold for initiating pest control. *Mailing Add:* Dept Agr Econ & Rur Soc Univ Ark Fayetteville AR 72701

HEADLEY, ROBERT N, b Boyd, Md, Mar 29, 32; m 55; c 4. CARDIOLOGY. *Educ:* Univ Md, BS, 54, MD, 56. *Prof Exp:* Intern med, Univ Va Hosp, 56-57; asst resident, NC Baptist Hosp, 61-62, resident, 62-63; instr, 63-64, asst prof internal med, 64-68, assoc prof, 68-74, chief prof serv, 76-81, assoc chief, Prof serv, 81-82, actg dir, Cardiol Sect, 81-84, PROF INTERNAL MED, BOWMAN GRAY SCH MED, 74- *Concurrent Pos:* Fel cardiol, Bowman Gray Sch Med, 57-58, trainee cardiovasc dis, 60-61; Am Col Physicians Meade-Johnson scholar, 62-63; consult biomed aspects of landing impact, USAF, 60-61; attend physician, Vet Admin Regional Off, 62-63, consult, 71-; dir outpatient serv, NC Baptist Hosp, 63-72, dir, Cardiac Care Unit, 71-80; consult, Div Regional Med Prog, USPHS, 68-69; consult, Coronary Care Units, NC Med Care Comn, 68-69; dir, Heart Dis Div, NC Regional Med Prog, 69-72; dir, Dist Med Consult, Voc Rehab Admin. *Mem:* Fel Am Col Physicians; fel Am Col Cardiol. *Res:* Feasibility of the establishment of coronary care units in small community hospitals; system for specially equipped coronary care ambulances in a rural community. *Mailing Add:* Dept Med Bowman Gray Sch Med Winston-Salem NC 27103

HEADLEY, VELMER BENTLEY, b Barbados, WI, Sept 7, 34; m 65; c 2. MATHEMATICS. *Educ:* Univ Col WI & Univ London, BSc, 62; Univ BC, MA, 66, PhD(math), 68. *Prof Exp:* Asst lectr math, Univ WI, 62-63; assst prof, 68-73, chmn dept, 81-84, assoc prof, 73-88, PROF MATH, BROCK UNIV, 88- *Mem:* Am Math Soc; Math Asn Am; Can Math Soc. *Res:* Oscillation theory of ordinary and partial differential equations; turning-point problems and Gronwall-Bellman-Wendroff inequalities; approximation theory. *Mailing Add:* Dept Math Brock Univ St Catharines ON L2S 3A1 Can

HEADRICK, RANDALL L, b Lawrence, Kans, Apr 6, 60. ELECTRONIC MATERIALS, SURFACES & INTERFACES. *Educ:* Carnegie-Mellon Univ, BS, 82; Univ Pa, PhD(mat sci), 88. *Prof Exp:* Postdoctoral mem tech staff, AT&T Bell Labs, 88-90; STAFF SCIENTIST, CORNELL HIGH ENERGY SYNCHROTRON SOURCE, 90- *Concurrent Pos:* Consult, AT&T Bell Labs, 91- *Mem:* Am Vacuum Soc; Mat Res Soc; Am Phys Soc. *Res:* Surfaces and interfaces of semiconductors and metals and the growth of crystalline films of these materials; use of x-rays produced by synchrotron radiation to study structure of artificially grown thin films. *Mailing Add:* Cornell High Energy Synchrotron Source Cornell Univ Ithaca NY 14853

HEADY, HAROLD FRANKLIN, b Buhl, Idaho, Mar 29, 16; m 40, 82; c 2. PLANT ECOLOGY, RANGE SCIENCE & MANAGEMENT. *Educ:* Univ Idaho, BS, 38; Syracuse Univ, MS, 40; Univ Nebr, PhD, 49. *Prof Exp:* Teaching asst, NY State Col Forestry, Syracuse Univ, 38-40; range conservationist, Soil Conserv Serv, USDA, 41; asst prof plant ecol, NY State Col Forestry, Syracuse Univ, 42; asst prof range mgt, Col & Exp Sta, Mont State Col, 42-47; assoc prof, Agr & Mech Col Tex, 47-51; asst prof & asst plant ecologist, Univ Calif, Berkeley, 51-56, assoc prof & assoc plant ecologist, 56-62, prof forestry & plant ecologist, Exp Sta, 62-84, dean, Col Natural Resources, 74-77, asst vpres, Agr & Univ Serv & assoc dir, Agr Exp Sta, 77-80; RETIRED. *Concurrent Pos:* Fulbright res scholar, 58-59 & 66; Guggenheim fel, 58-59; consult range mgt, Food & Agr Orgn, Saudi Arabia, 62 & Malawi, 70-71; consult, Ralph M Parsons Co, 65. *Honors & Awards:* Renner Award, 80. *Mem:* Soc Range Mgt (secy-treas, 47 & pres, 80); Ecol Soc Am. *Res:* Utilization of forage by livestock; methods of sampling grassland; grassland ecology; range management; international rangeland conservation. *Mailing Add:* 1864 Capistrano Ave Berkeley CA 94707

HEADY, JUDITH E, b Cedar Rapids, Iowa, Dec 11, 39; c 2. DEVELOPMENTAL BIOLOGY, SCIENCE EDUCATION. *Educ:* Cornell Col, BA, 62; Univ Iowa, MS, 63; Univ Colo, PhD(develop biol), 70. *Prof Exp:* Res technician, Med Ctr, Univ Colo, 64-66, res assoc, 70-74; asst prof, 74-79, ASSOC PROF BIOL, UNIV MICH, 79- *Concurrent Pos:* Jr investr, Woods Hole Marine Biol Labs, 74; investr, Friday Harbor Labs, 76; vis asst prof biochem, Med Ctr, Univ Colo, 77; sabbatical res, Univ Minn, St Paul, 80-81; vis assoc prof biol, Univ Colo, Boulder, 83; vis res scientist, Marine Biol Lab, Woods Hole, Mass, 87-88. *Mem:* Soc Develop Biol; AAAS; Coun Undergrad Res. *Res:* Molecular cloning of muscle cell proteins of ascidians; writing for non-specialist audience on cancer, development, women in science; science education in areas of curricular/course reform. *Mailing Add:* Dept Natural Sci Univ Mich 4901 Evergreen Rd Dearborn MI 48128

HEAGARTY, MARGARET CAROLINE, b Charleston, WVa, Sept 8, 34. PEDIATRICS. *Educ:* Seton Hill Col, BA, 57; WVa Sch Med, BS, 59; Univ Pa, MD, 61. *Prof Exp:* Rotating intern, Philadelphia Gen Hosp, 61-62; resident pediat, St Christophers Hosp Children, 62-64; res fel, Sch Med, Harvard Univ, 64-66, asst dir family health care, 66-68; assoc, Beth Israel Hosp, Boston, 67-69; dir ambulatory pediat care, NY Hosp, Cornell Univ, 69-78; DIR PEDIAT, HARLEM HOSP CTR, 78-; PROF PEDIAT, COL PHYSICIANS & SURGEONS, COLUMBIA UNIV, 78- *Concurrent Pos:* Fel med, Children's Hosp Med Ctr, Boston, 64-66; fel, Robert Wood Johnson Found, 75; consult, Dept Health, Educ & Welfare, Nat Ctr Health Serv Res, 78-82 & 79-80; mem gov coun, Inst Med-Nat Acad Sci. *Mem:* Inst Med-Nat Acad Sci; Soc Pediat Res; Am Pediat Asn; Am Acad Pediat; Am Pub Health Asn; Ambulatory Pediat Asn. *Mailing Add:* Dept Pediat Harlem Hosp Ctr 506 Lenox Ave New York NY 10037

HEAGLE, ALLEN STREETER, b Stanley, Wis, Apr 20, 38; m 60; c 2. PLANT PATHOLOGY. *Educ:* Hamline Univ, BS, 60; St Cloud State Col, MEd, 64; Univ Minn, PhD(plant path), 68. *Prof Exp:* RES PLANT PATHOLOGIST, USDA, 68- *Honors & Awards:* Agr Res Serv Outstanding Res Award, USDA, 80. *Mem:* Am Phytopath Soc. *Res:* Effects of air pollution on agronomic crops yield; effects of air pollution on parasitism of plants by fungi. *Mailing Add:* 1216 Scott Pl Cary NC 27511

HEAGLER, JOHN B(AY), JR, b Cape Girardeau, Mo, Sept 8, 24; m 46; c 3. CIVIL ENGINEERING. *Educ:* Univ Mo, BS, 51, MS, 54. *Prof Exp:* Instr appl mech, Mo Sch Mines, Univ Mo-Rolla, 51-55, from asst prof to prof civil eng, 55-67, prof civil eng, 67-; RETIRED. *Res:* Soil mechanics, stabilization and foundations. *Mailing Add:* 1812 Independence Rolla MO 65401

HEAGY, FRED CLARK, b Stratford, Ont, Mar 17, 19; m 43; c 8. NUCLEAR MEDICINE. *Educ:* Univ Western Ont, MD, 43, MSc, 47, PhD(biochem & bact), 50; FRCP(C), 76. *Prof Exp:* Res assoc biochem, Glasgow Univ, 50-52; biochemist, London Clin, Ont Cancer Treatment & Res Found, 55-71, sr physician, 71-80; teaching staff nuclear med & diag radiol, Victoria Hosp, London, 80-84; clin prof, 78-85, PROF EMER NUCLEAR MED & DIAG RADIOL, UNIV WESTERN ONT, 85- *Concurrent Pos:* Nuffield Found Dom traveling fel, 50-51; Brit Empire Cancer Campaign exchange fel, 51-52; sr med res fel, Nat Res Coun Can, 52-55; asst prof biochem & therapeut radiol, Univ Western Ont, 61-70, assoc prof, 70-78; mem, Exec Coun, Adv Comt, Clin Uses of Radioisotopes, Dept Nat Health & Welfare, Can, 61-71, chmn, 71-73; mem teaching staff, Victoria Hosp, London; consult nuclear med, St Joseph's Hosp & Univ Western Ont Hosp; mem, Bd Govrs, Toronto Inst Med Technol, 74-77; hon staff, Victoria Hosp, 84- *Mem:* AAAS; Can Physiol Soc; Can Asn Nuclear Med (pres, 77-79); Am Soc Biol Chemists; Soc Nuclear Med. *Res:* Magnesium and temperature regulation; bacteriophage; nucleic acids; thyroid; cancer. *Mailing Add:* 1037 Brough St London ON N6A 3N5 Can

HEALD, CHARLES WILLIAM, b West Grove Pa, May 1, 42; m 61; c 2. DAIRY SCIENCE, CYTOLOGY. *Educ:* Pa State Univ, BS, 64; Univ NH, MS, 66; Va Polytech Inst & State Univ, PhD(dairy sci), 69. *Prof Exp:* Res assoc dairy sci, 69-71, asst prof, 71-78, assoc prof dairy sci, Va Polytech Inst & State Univ, 78-; AT DEPT DAIRY & ANIMAL SCI, PA STATE UNIV. *Mem:* Am Dairy Sci Asn; Am Soc Animal Sci; AAAS; Nat Mastitis Coun. *Res:* Structure and function of mammary tissue during development, differentiation, lactogenesis, galactopoiesis, involution and pathology of domestic animals with emphasis on control of lactogenesis and mammary bacterial infections. *Mailing Add:* Dept Dairy & Animal Sci Pa State Univ University Park PA 16802

HEALD, EMERSON FRANCIS, b Newport, RI, Dec 8, 34; m 59; c 3. PHYSICAL CHEMISTRY. *Educ:* Univ RI, BS, 56; Univ Hawaii, PhD(phys chem), 61. *Prof Exp:* From instr to asst prof, Univ Hawaii, 60-62; res assoc geochem, Pa State Univ, 62-64; from asst prof to assoc prof, 64-74, PROF CHEM, THIEL COL, 74- *Concurrent Pos:* Vis res fel chem, Victoria Univ, Wellington, 68-69; vis fel chem, Western Australian Inst Technol, 7S-76. *Mem:* AAAS; Am Chem Soc; Sigma Xi. *Res:* Use of computers in chemical education; interfacing computers and laboratory instruments; solid-source mass spectrometry. *Mailing Add:* 88 Chambers Ave Greenville PA 16125

HEALD, FELIX PIERPONT, JR, b Philadelphia, Pa, Dec 3, 21; m 48; c 5. MEDICINE, PEDIATRICS. *Educ:* Colo Col, AB, 46; Univ Pa, MD, 46. *Prof Exp:* Instr pediat, Harvard Med Sch, 53-59; asst prof, Sch Med, Georgetown Univ, 60-63, assoc prof, 63-66; prof & chmn dept, Sch Med, George Washington Univ, 67-70; PROF PEDIAT & HEAD DIV ADOLESCENT MED, SCH MED, UNIV MD, BALTIMORE CITY, 70- *Concurrent Pos:* Mem staff adolescent med, Children's Med Ctr, Boston, 52-59; chief adolescent med, Children's Hosp, Washington, DC, 60-70. *Mem:* Soc Pediat Res; Am Acad Pediat; Soc Res Child Develop. *Res:* Adolescent medicine; physiological, metabolic and nutritional aspects of adolescence. *Mailing Add:* Univ Med Hosp 22 S Green St Baltimore MD 21201

HEALD, MARK AIKEN, b Princeton, NJ, Jan 27, 29; m 52; c 3. PLASMA PHYSICS. *Educ:* Oberlin Col, 50; Yale Univ, MS, 51, PhD(exp physics), 54. *Prof Exp:* Res staff mem microwave diagnostics, Proj Matterhorn, Plasma Physics Lab, Princeton Univ, 54-59; from asst prof to assoc prof, 59-70, chmn dept, 68-78, PROF PHYSICS, SWARTHMORE COL, 70- *Concurrent Pos:* Tech deleg, UN Conf Peaceful Uses of Atomic Energy, Geneva, 58; NSF fac fel, Culham Lab, UK Atomic Energy Authority, 63-64; NSF sci fac fel, Plasma Physics Lab, Princeton, 69-70, vis res physicist, 74-75; vis scientist, Nat Magnet Lab, Mass Inst Technol, 78-79. *Mem:* Am Phys Soc; Am Asn Physics Teachers; Fedn Am Sci; Sigma Xi. *Res:* Plasma physics; microwave diagnostics of plasmas; RF heating; author. *Mailing Add:* Dept Physics Swarthmore Col Swarthmore PA 19081-1397

HEALD, MILTON TIDD, b Woburn, Mass, Feb 19, 19; m 41; c 3. SEDIMENTARY PETROLOGY. *Educ:* Wesleyan Univ, AB, 40; Harvard Univ, AM, 47, PhD(geol), 49. *Prof Exp:* From instr to prof 48-84, EMER PROF GEOL, WVA UNIV, 84- *Mem:* Fel Geol Soc Am; Soc Econ Paleont & Mineral; Am Asn Petrol Geol; Nat Asn Geol Teachers (vpres, 58). *Res:* Diagenesis of clastic sediments. *Mailing Add:* Dept Geol WVa Univ Morgantown WV 26506

HEALD, PAMELA, b New York, NY, June 15, 50. ECONOMIC GEOLOGY, REMOTE SENSING. *Educ:* Vassar Col, BA, 71; George Washington Univ, MS, 77. *Prof Exp:* Res asst radioactive nuclides, Yale Univ, 71-72; res geologist, remote sensing metalliferous ore deposits, 72-74, res geologist, environ ore deposition, 74-80, proj chief, comt studies, active geothermal systs & ore deposits, 80-83, RES GEOLOGIST, STUDY PRECIOUS METAL & BASE-METAL ORE DEPOSITS, US GEOL SURV, 83- *Mem:* Soc Econ Geol; Int Asn Genesis Ore Deposits. *Res:* Study of physical and chemical parameters involved in ore formation and integration with geologic factors to deduce models of ore genesis; geologic analysis and interpretation of satellite imagery and aerial photography for delineation of patterns of ore deposit distribution; study of active geothermal systems as model for precious-metal deposits. *Mailing Add:* US Geol Surv Nat Ctr MS 959 Reston VA 22092

HEALD, WALTER ROLAND, b Denver, Colo, Oct 27, 20; m 48; c 2. SOIL CHEMISTRY. *Educ:* Colo State Univ, BS, 47; State Col Wash, MS, 49; Purdue Univ, PhD(soils), 54. *Prof Exp:* Assoc prof agron, Colo Agr & Mech Col, 49-51; soil scientist, Plant Indust Sta, USDA, 53-70, res invest leader, Water Qual Mgt Invest, Northeast Br, Soil & Water Conserv Res Div, Agr Res Serv, 71-72, location & res leader, Plant, Soil & Water Lab, 72-74, res soil scientist, Northeast Watershed Res Ctr, Sci & Educ Admin-Agr Res, 74-81; RETIRED. *Mem:* Am Soc Agron. *Res:* Effect of land use on the quality of ground and surface waters. *Mailing Add:* 18 W Forest Feezor St Rte 7 Box 410 Y Tucson AZ 85747

HEALEY, ANTHONY J, b London, Eng, Sept 10, 40; m 60; c 3. MECHANICAL, OFFSHORE PIPELINE & SUB-SEA ENGINEERING. *Educ:* Univ London, BSc, 61; Univ Sheffield, PhD(mech eng), 66. *Prof Exp:* Engr, Bristol Aircraft, 61-63 & Gen Elec Co, 66-67; asst prof mech eng, Pa State Univ, 67-71; assoc prof mech eng, Univ Tex, Austin, 71-77, prof, 77-81; PROJ ENGR MGR, BROWN & ROOT, INC, 81- *Concurrent Pos:* Asst vis prof, Mass Inst Technol, 70; consult, gov & indust. *Mem:* Am Soc Mech Engrs. *Res:* Fluidic control systems and components modelling; fluid power control; system dynamics; finite element; structural dynamics; pipelines; offshore vessels, moorings. *Mailing Add:* 1109 Circle Rd Pebble Beach CA 93953

HEALEY, FRANK HENRY, b Worcester, Mass, Oct 5, 24; m 48; c 3. PHYSICAL CHEMISTRY. *Educ:* Clark Univ, AB, 47, PhD(chem), 49. *Prof Exp:* From instr to asst prof chem, Lehigh Univ, 49-56; sect chief physics & phys chem, Lever Bros Co, 56-58, sect chief detergent process develop, 58-60, dir res & develop, 60-64, vpres res & develop, 64-73, res vpres, 73-88, mem bd dirs, 68-88; RETIRED. *Concurrent Pos:* Mem bd dirs, Indust Res Inst, 70-79, pres, 78. *Mem:* Asn Res Dirs; Sigma Xi. *Res:* Surface chemistry; soap and detergent product and processing development. *Mailing Add:* 255 W Ridgewood Ave Ridgewood NJ 07450

HEALEY, JOHN EDWARD, JR, oncology, rehabilitation medicine; deceased, see previous edition for last biography

HEALEY, MARK CALVIN, b Salt Lake City, Utah, Mar 7, 47; m 69; c 3. PRODUCTION & USE OF MONOCLONAL ANTIBODIES, SUBUNIT VACCINE DEVELOPMENT FOR SELECT ANIMAL DISEASES. *Educ:* Univ Utah, BS, 71, MS, 73; Purdue Univ, PhD(immunoparasitol), 76; Miss State Univ, DVM, 81. *Prof Exp:* Teaching fel, Univ Utah, 70-73; grad instr parasitol, Purdue Univ, 73-76; instr parasitol, Tex A&M Univ, 76-77; resident parasitol, Miss State Univ, 78-81; res asst prof, 81-83, from asst prof to assoc prof, 83-91, PROF & ASST HEAD VET SCI, UTAH STATE UNIV, 91- *Mem:* Am Soc Parasitologists; Am Asn Vet Parasitologists; Am Vet Med Asn; Am Soc Microbiol; Conf Res Workers Animal Dis. *Res:* Immunoparasitology; developing subunit vaccines against avian coccidiosis (Eimeria tenella), cryptosporidiosis (Cryptosporidium parvum) and ram lamb epididymitis (Actinobacillus seminis and Haemophilus somnus). *Mailing Add:* Utah State Univ UMC-5600 Logan UT 84322-5600

HEALEY, MICHAEL CHARLES, b Prince Rupert, BC, Mar 31, 42. POPULATION ECOLOGY, FISHERIES ECOLOGY. *Educ:* Univ BC, BSc, 64, MSc, 66; Aberdeen Univ, PhD(natural hist), 69. *Prof Exp:* Nat Res Coun Can fel, Pac Biol Sta, Dept Fisheries & Oceans, 69-70, res scientist, Dept Environ, Freshwater Inst, Winnipeg, 70-74, res scientist & prog leader, Pac Biol Sta, 74-90; DIR, WESTWATER RES CTR, UNIV BC, 90- *Concurrent Pos:* Sr fel, Marine Policy & Ocean Mgt Prog, Woods Hole, Oceanog Inst, Mass, 82-83; chmn bd dir, Rawson Acad Aquatic Sci, 82-85, dir, 82-89; vis scientist, Univ BC, 88-89. *Mem:* Int Asn Theoret & Appl Limnol; Am Fisheries Soc; Pac Fishery Biologists (pres, 80-81); Rawson Acad Aquatic Sci. *Res:* Factors which influence the numbers of animals in natural populations and evolution of life history strategies; resource management strategy and application of decision analytic technology to resource management planning. *Mailing Add:* Westwater Res Ctr Univ BC Vancouver BC V6T 1W5 Can

HEALEY, PATRICK LEONARD, b Long Beach, Calif, Feb 20, 36; m 56; c 2. DEVELOPMENTAL & CELLULAR BIOLOGY. *Educ:* Univ Calif, Berkeley, AB, 60, PhD(bot), 64. *Prof Exp:* USPHS trainee & res assoc develop biol, Brown Univ, 64-65, asst prof biol sci, 65-66; asst prof, 66-72, assoc dean sch, 67-69 & 70-72, acad asst to vpres, 72-75, ASSOC PROF BIOL SCI, UNIV CALIF, IRVINE, 72-, EXEC ASSOC DEAN, SCH BIOL SCI, 80- *Mem:* AAAS; Am Soc Plant Physiologists; Bot Soc Am; Am Soc Cell Biol. *Res:* Cellular and histological changes in the shoot apex accompanying floral induction; ultrastructural and histochemical differentiation and development of plant secretory cells. *Mailing Add:* Dept Develop & Cell Biol Univ Calif Irvine CA 92717

HEALY, BERNADINE P, b New York, NY. CARDIOLOGY. *Educ:* Harvard Univ, MD, 70. *Prof Exp:* Dir, Coronary Care Unit, prof med & asst dean, postdoctoral progs & fac develop, Sch Med, Johns Hopkins Univ, 77-84; dep dir, Off Sci & Technol Policy, White House, 84-85; chmn, Res Inst, Cleveland Clin Found, 85-91; DIR, NIH, 91- *Concurrent Pos:* Chmn, White House Cabinet Working Group Biotechnol & exec secy, Panel Health Univs, White House Sci Coun; mem adv groups, Coun Nat Heart, Lung & Blood Inst, Nat Cancer Inst & White House Working Group Health Policy & Econ; bd dirs, Am Heart Asn, 83-; bd overseers, Harvard Col, 89-; chmn, Pub Policy Comt, Am Fedn Clin Res; bd gov, Am Col Cardiol; mem, adv comt to dir, NIH, White House Sci Coun, Life Sci Strategic Planning Study Comt, NASA & Spec Med Adv Comt, Dept Vet Affairs; chmn adv panel, US Cong; vchmn, President's Coun Advisors Sci & Technol; chair, Adv Panel Basic Res, Off Technol Assessment. *Mem:* Inst Med-Nat Acad Sci; Am Fedn Clin Res (pres, 83-84); Am Heart Asn (pres, 88-89). *Res:* Cardiovascular research and medicine; neurobiology; immunology; cancer; artificial organs; atherosclerosis; musculoskeletal disorders; molecular biology. *Mailing Add:* NIH Bldg 1 Rm 126 9000 Rockeville Pike Bethesda MD 20892

HEALY, GEORGE RICHARD, b Springfield, Mass, Oct 30, 24; m 52; c 2. PARASITOLOGY. *Educ:* Providence Col, BS, 49; Univ Ky, MS, 52; Rice Inst, PhD(parasitol), 56. *Prof Exp:* Asst zool, Univ Ky, 49-51; investr dove dis, Ky Fish & Game Div, 51-52; asst biol, Rice Inst, 52-54; parasitologist & chief, Protozoan Dis Br, Parasitic Dis Div, Ctr Dis Control, USPHS, 57-87; ASSOC PROF PHARMACOL, SCH MED, EMORY UNIV, 60- *Mem:* Am Soc Parasitol; Am Soc Trop Med & Hyg; Sigma Xi; AAAS; Am Soc Microbiol. *Res:* Amebiasis; Babesiosis; Amebic meningoencephalitis; host-parasite relationships; protozoan and helminthic zoonoses. *Mailing Add:* 905 Vistavia Decatur GA 30033

HEALY, GEORGE W(ILLIAM), b Muhlan, Austria, May 18, 09; US citizen; m 41; c 6. METALLURGY. *Educ:* Univ Paris, Cert de Math Gen, 30; Yale Univ, BS, 33. *Prof Exp:* Res metallurgist, Union Carbide Metals Co Div, Union Carbide Corp, 40-63; assoc prof metall, Pa State Univ, 63-74; res prof, 74-80, ADI PROF METALL, UNIV UTAH, 80- *Concurrent Pos:* Int Hofmann Prize Consortium, Spec Prize, 80-81. *Honors & Awards:* Hunt Award, Am Inst Mining, Metall & Petrol Engrs, 58. *Mem:* Am Inst Mining, Metall & Petrol Engrs; Can Inst Metallurgists; Am Ceramic Soc. *Res:* Chemical metallurgy of steel, ferroalloys and non-ferrous metals. *Mailing Add:* Dept Metall Eng Univ Utah Salt Lake City UT 84112

HEALY, JAMES C, synthetic rubber, polymer engineering, for more information see previous edition

HEALY, JOHN H, b Chicago, Ill, Aug 7, 29; m 51; c 5. GEOPHYSICS. *Educ:* Mass Inst Technol, BS, 51; Calif Inst Technol, MS, 57, PhD(geophys), 61. *Prof Exp:* Physicist, Air Force Cambridge Res Ctr, 51-53; seismologist, Chevron Oil Co, 54-56; res asst geol, Calif Inst Technol, 56-61; res geophysicist, Colo, 61-68, RES GEOPHYSICIST, US GEOL SURV, CALIF, 68- *Mem:* Am Geophys Union; Soc Explor Geophys; Seismol Soc Am. *Res:* Crustal structure; earthquake seismology. *Mailing Add:* 306 Diablo Ct Palo Alto CA 94306

HEALY, JOHN JOSEPH, b Vancouver, Wash, Apr 12, 43; m 70; c 1. STATISTICAL ENGINEERING, DIMENSIONAL ANALYSIS. *Educ:* Carnegie-Mellon Univ, BS, 65; George Washington Univ, MS, 78. *Prof Exp:* Prod engr, Elec Boat Div, Gen Dynamics Corp, 65-67; process engr, Combustion Eng Inc, 67-69; ENGR, MCDERMOTT INC CO, BABCOCK & WILCOX CO, 69-, SR ENGR, 76- *Res:* Applied statistical inference; product dimensioning and tolerancing; cleanliness of nuclear components. *Mailing Add:* Babcock & Wilcox Co PO Box 785 MC 32 Lynchburg VA 24505

HEALY, MICHAEL L, b Pocatello, Idaho, Nov 17, 36; m 58; c 2. OCEANOGRAPHY, ANALYTICAL CHEMISTRY. *Educ:* Ore State Univ, BS, 58, PhD(anal chem), 65. *Prof Exp:* PRIN OCEANOGR, UNIV WASH, 65- *Mem:* Am Chem Soc; Sigma Xi; Am Soc Limnol & Oceanog; Sigma Xi. *Res:* Chemical oceanography; trace metal analysis; ocean nutrient analysis; coulometric methods; polarography. *Mailing Add:* 12908 84th NE Kirkland WA 98034

HEALY, PAUL WILLIAM, b Phoenix, Ariz, Jan 22, 15; m 37; c 2. MATHEMATICS. *Educ:* Ohio State Univ, AB & BS, 36, MA, 39; Univ Ky, PhD(math), 50. *Prof Exp:* Teacher, Pub Sch, Ohio, 36-41; instr math, Case Inst Technol, 42-45; teacher, Pub Sch, Ariz, 45-46; assoc prof math, Southwestern Col, Kans, 48-51; asst prof math & astron, Univ NMex, 51-56; mathematician, Atomic Energy Div, Phillips Petrol Co, Idaho, 56-58; sr nuclear physicist, Aerojet-Gen Nucleonics, 58-60; head tech data systs mgt div, US Naval Weapons Sta, Weapons Aval Eng Ctr, Concord, 61-80; RETIRED. *Concurrent Pos:* Chemist, Cleveland Graphite Bronze, 44-45. *Mem:* Fel AAAS; fel Meteoritical Soc; Math Asn; Planetary Soc. *Res:* Computer mathematics. *Mailing Add:* 2448 Westcliffe Lane Walnut Creek CA 94596

HEALY, WILLIAM CARLETON, JR, b Rochester, NY, Feb 14, 25; m 49; c 3. MATHEMATICAL STATISTICS. *Educ:* Univ Mich, BSE, 48, MS, 49; Univ Ill, PhD(statist), 55. *Prof Exp:* Statistician, Gen Elec Co, 49-52; staff statistician, Ethyl Corp, Mich, 55-57, supvr math anal, 57-62; mgr, Opers Res Div, 62-69, Systs & Appln Div, 69-71 & Mgt Sci, 71-73, MGR, COMPUT SCI, MARATHON OIL CO, 73- *Mem:* Am Statist Asn; Asn Comput Mach. *Res:* Applied statistics; mathematical programming; operations research in petroleum industry; computing. *Mailing Add:* Marathon Oil Co Box 269 Littleton CO 80160

HEALY, WILLIAM RYDER, b Lynn, Mass, May 10, 38; m 63; c 3. EVOLUTIONARY ECOLOGY, HERPETOLOGY. *Educ:* Boston Col, BS, 61; Univ Mich, Ann Arbor, MS, 63, PhD(zool), 66. *Prof Exp:* From instr to assoc prof, 64-80, chmn dept, 76-81, PROF BIOL, COL HOLY CROSS, 80- *Mem:* AAAS; Am Soc Ichthyol & Herpet; Sigma Xi. *Res:* Population ecology of salamanders with emphasis on life history strategies. *Mailing Add:* Dept Biol Col Holy Cross Worcester MA 01610

HEANEY, DAVID PAUL, b Choteau, Mont, Dec 25, 27; m 52; c 1. RUMINANT NUTRITION, INTENSIVE SHEEP MANAGEMENT. *Educ:* Mont State Univ, BS, 51, MS, 56; Mich State Univ, PhD(animal nutrit), 60. *Prof Exp:* Res Scientist ruminant nutrit, Animal Res Ctr, Res Br, Agr Can, 60-89; RETIRED. *Honors & Awards:* Cert of Merit, Can Soc Animal Sci, 85. *Res:* Feeding regimes, management regimes and nutritional requirements for artificially reared lambs and totally confined, intensively managed sheep. *Mailing Add:* 2383 Baseline Rd Ottawa ON K2C 0E2 Can

HEANEY, JAMES PATRICK, b Chicago, Ill, Jan 12, 40; m 62; c 2. SANITARY ENGINEERING, OPERATIONS RESEARCH. *Educ:* Ill Inst Technol, BS, 62; Northwestern Univ, MS, 65, PhD(civil eng), 68. *Prof Exp:* Engr trainee, Metro Sanit Dist of Chicago, 57-62; consult, Am Pub Works Asn, 67; sr res engr water resources, Pac Northwest Labs, Battelle Mem Inst, 67-68; from asst prof to assoc prof, 68-80, PROF ENVIRON ENG SCI, UNIV FLA, 80-, DIR, WATER RESOURCES RES CTR, 80- *Mem:* Am Water Works Asn. *Res:* Application of operations research and mathematical economics to problems in environmental engineering. *Mailing Add:* Dept Environ Eng Univ Fla Gainesville FL 32611

HEANEY, LAWRENCE R, b Dec 2, 52. EVOLUTION, SYSTEMATICS. *Educ:* Univ Minn, BSci, 74; Univ Kans, PhD(biol), 79. *Prof Exp:* Asst prof biol & cur mammals, Univ Mich, 79-86; RES FEL, SMITHSONIAN INST, 86- *Concurrent Pos:* Vis syst specialist, Field Mus Nat Hist, 81. *Mem:* Am Soc Mammalogists; Soc Study Evolution; Soc Syst Zoologists; Sigma Xi; Soc Conserv Biol. *Res:* Island biogeography, especially Southeast Asian mammals; speciation, approached through studies of hybridization; systematics of rodents and bats; conservation biology. *Mailing Add:* Field Mus Natural Hist Div Mammals Roosevelt Rd at Lake Shore Dr Chicago IL 60605

HEANEY, ROBERT JOHN, b East Rochester, NH, Aug 31, 22; m 46; c 4. ATMOSPHERIC CHEMISTRY, WATER CHEMISTRY. *Educ:* Univ NH, BS, 43; Univ Colo, PhD, 50. *Prof Exp:* Jr chemist, Nat Bur Standards, 43-44, chemist, 46; chief chemist, Vitro Chem Co, 51-52; sr scientist, Res Ctr, 52-59, from actg dir to dir agr & meteorol res dept, 59-64, agr & meteorol res engr, Qual Control Dept, 64-70, PROCESS CONTROL & ENVIRON ENGR, KENNECOTT COPPER CORP, 70- *Concurrent Pos:* Spec engr detachment, Manhattan Proj, Colo & Washington, DC, 44-46; mem, State Adv Comt Sci & Technol, 73-; pres, Great Salt Lake Health Planning Coun, 74- *Mem:* Am Chem Soc; Sigma Xi. *Res:* Biochemical effects of air contaminants on plants and animals; effects of water pollutants on humans and animals; effects of impurities on copper quality. *Mailing Add:* 1889 Wasatch Dr Salt Lake City UT 84108-3323

HEANEY, ROBERT PROULX, b Omaha, Nebr, Nov 10, 27; m 52; c 7. HUMAN BONE & CALCIUM METABOLISM. *Educ:* Creighton Univ, BS, 47, MD, 51. *Prof Exp:* Instr, Sch Med, Univ Okla, 54-55; clin instr, Sch Med, George Washington Univ, 55-57; actg chmn, dept med, 60-61, chmn, 61-69, head, endocrinol & metab sect, 69-71, vpres, div health sci, 71-84, from asst prof to prof med, 57-84, JOHN A CREIGHTON UNIV PROF, CREIGHTON UNIV, 84- *Concurrent Pos:* Chmn, Nat Inst Dent-Res Spec Grants Rev, 82-86; mem Nutrit Res Sci Adv Comt of Nat Dairy Coun, 86-89, chmn, 87-89; mem Nat Osteoporosis Found, Sci Adv Comt, 86-91, bd dirs, 90; mem sci adv comt, Osteoporosis Found, 86. *Mem:* Fel Am Col Physicians; hon mem Am Dietetic Asn; Am Inst Nutrit; Am Soc Bone & Mineral Res; Cent Soc Clin Res. *Res:* Human bone and calcium metabolism; long term study of human osteoporosis; research and statistics for health professionals; author of over 200 scientific papers. *Mailing Add:* Creighton Univ Omaha NE 68178-0650

HEAPS, MELVIN GEORGE, b Salt Lake City, Utah. PHYSICS, ATMOSPHERIC SCIENCES. *Educ:* Brigham Young Univ, BS, 67; Utah State Univ, PhD(physics), 72. *Prof Exp:* Res assoc physics, Univ Fla, 72-75; Nat Res Coun res assoc, US Army Ballistic Res Lab, 75-76; physicist, US Army Atmospheric Sci Lab, 76-90. *Mem:* Am Geophys Union; Sigma Xi. *Res:* Atmospheric effects on propagation and electro-optical sensors; smokes and obscurants; target detection. *Mailing Add:* 2389 Sandpiper St Tallahassee FL 32303

HEARD, HARRY GORDON, b Raines, Tenn, Sept 23, 22; m 47; c 2. PHYSICS, COMPUTER SCIENCE. *Educ:* Univ Calif, BS, 49, MS, 51. *Prof Exp:* Res engr, Lawrence Berkeley Labs, 51-59; chief engr, Levinthal Electronics Prod Inc, 59-60; vpres, Radiation Inc, 59-60, Energy Systs, Inc, 60-64 & HNU Systs Div, Ohio Steel, 64-68; pres, Resalab, Inc, 68-70; gen mgr, MBA Info Systs, 70-74; scientist, Inst Advan Comput, 75-77; dir, Technol Develop Calif, 78-88; SR SYSTS DESIGNER, FORD AEROSPACE, 88- *Concurrent Pos:* Dir, Intronex, Inc, Menlo Park, 75- *Mem:* Am Phys Soc; Inst Elec & Electronics Engrs; Optical Soc Am; Sigma Xi; Asn Comput Mach; Res Soc Am. *Res:* Optical data processing; cartographic systems; computer sciences; hollow-cathode lasers; quantum electronics; high vacuum physics; high voltage systems; satellite data systems; structure software systems design; class IX (super) computer system design; Lan-Wan design; computer simulation. *Mailing Add:* 50 Skywood Way Woodside CA 94062

HEARD, HUGH COREY, rock mechanics, high pressure physics, for more information see previous edition

HEARD, JOHN THIBAUT, JR, b Houston, Tex, Sept 4, 40; m 81. MICROBIAL PHYSIOLOGY. *Educ:* Lamar State Col Technol, BS, 64; Sam Houston State Col, MS, 66; Univ Tex, Houston, PhD(biomed sci), 78. *Prof Exp:* RES ASSOC, KIRKSVILLE COL OSTEOP MED, 78- *Concurrent Pos:* Consult, Bio-Diesel Refiners Inc, Iowa, 81- *Mem:* Sigma Xi. *Res:* Biosynthesis and metabolism of nicotinamide adenine dinuclotide and its metabolites as a function of age or perturbations of the system under study. *Mailing Add:* RR 6 Kirksville MO 63501

HEARD, WILLIAM HERMAN, b Hart, Mich, Mar 4, 35; m 55; c 3. MALACOLOGY. *Educ:* Univ Mich, BS, 57, MS, 59, PhD(zool), 63. *Prof Exp:* From instr to asst prof zool, 62-67, assoc prof, 67-80, PROF BIOL SCI, FLA STATE UNIV, 80- *Mem:* Am Malacol Union; Am Micros Soc; Soc Syst Zool; Am Soc Zool; Ecol Soc Am. *Res:* Systematics, distribution and life histories of freshwater mollusks of North America. *Mailing Add:* Dept Biol Sci MS-B142 Fla State Univ Tallahassee FL 32306-2043

HEARING, VINCENT JOSEPH, JR, b Washington, DC, Aug 22, 45; m 68; c 3. BIOCHEMISTRY, MOLECULAR BIOLOGY. *Educ:* Georgetown Univ, BS, 67; Cath Univ Am, PhD(cell biol), 71. *Prof Exp:* Electron microscopist, 69-71, NIH fel, 71-72, sr staff fel biochem, 72-77, RES BIOLOGIST, NAT CANCER INST, 77- *Mem:* Am Soc Cell Biol; Am Asn Cancer Res; Sigma Xi; Am Soc Biol Chemists. *Res:* Control mechanisms involved in enzymology of melanin formation in mammals and aberrant proteins synthesized in the malignant melanocyte. *Mailing Add:* Bldg 37 Rm 1B22 NIH Bethesda MD 20892

HEARN, ANTHONY CLEM, b Adelaide, Australia, Apr 13, 37; m 70; c 2. SYMBOLIC COMPUTATION. *Educ:* Univ Adelaide, BSc, 59; Cambridge Univ, PhD(theoret physics), 62. *Prof Exp:* Res assoc physics, Stanford Univ, 62-64; sr sci officer, Rutherford High Energy Lab, Eng, 64-65; asst prof physics, Stanford Univ, 65-69; from assoc prof to prof physics, Univ Utah, 69-78, adj prof elec eng, 71-78, prof comput sci & chmn dept, 73-80; AT RAND CORP, SANTA MONICA, 80- *Concurrent Pos:* Vis scientist, Europ Ctr Nuclear Res, 62; Alfred P Sloan Found fel, 67-69; consult, Hewlett-Packard Co, 74-79 & Burroughs Corp, Detroit, 75-78. *Mem:* Am Phys Soc; Asn Comput Mach. *Res:* Algebraic simplification; computational physics. *Mailing Add:* Rand Corp 1700 Main St Santa Monica CA 90407

HEARN, BERNARD CARTER, JR, b Baltimore, Md, Feb 8, 33; m 55; c 3. GEOLOGY. *Educ:* Wesleyan Univ, BA, 54; Johns Hopkins Univ, PhD(geol), 59. *Prof Exp:* GEOLOGIST, US GEOL SURV, 57- *Mem:* AAAS; Am Geophys Union; Geol Soc Am; Mineral Soc Am. *Res:* Structure and petrology of igneous rocks; kimberlites; geothermal volcanology. *Mailing Add:* US Geol Surv 959 Nat Ctr Reston VA 22092

HEARN, CHARLES JACKSON, b Bellville, Ga, June 17, 36; m 58; c 3. PLANT BREEDING. *Educ:* Univ Ga, BSA, 57, MS, 59; Tex A&M Univ, PhD(plant breeding), 63. *Prof Exp:* Asst agronomist, Ga Coastal Plain Exp Sta, 57-58; RES GENETICIST, USDA, 62-, RES LEADER, 85- *Concurrent Pos:* Res admin trainee, USDA, Beltsville, Md, 71-72, actg lab dir, 84; adj prof, Fruit Crops Dept, Univ Fla, 79- *Mem:* Am Soc Hort Sci; Int Soc Citricult; Am Pomol Soc. *Res:* Breeding of citrus; citrus varieties; fruit breeding; fruit physiology; cytology of plants; development of improved varieties of oranges, grapefruit and tangerines. *Mailing Add:* 2120 Camden Rd Orlando FL 32803

HEARN, DAVID RUSSELL, b Tucson, Ariz, Jan 21, 42; m 77. ASTROPHYSICS. *Educ:* Calif Inst Technol, BS, 64; Harvard Univ, MA, 66, PhD(physics), 68. *Prof Exp:* Nat Acad Sci vis res assoc gamma-ray astron, Smithsonian Astrophys Observ, 68-70; res staff, Ctr Space Res, Mass Inst Technol, 70-79; sr res physicist, 79-80, physics group leader, Elscint, Inc, 81-84; STAFF MEM, LINCOLN LAB, MASS INST TECHNOL, 84- *Mem:* Am Phys Soc. *Res:* Analysis and publication of data from the soft x-ray telescope aboard the SAS-3 satellite, which was designed and built between 1970 and 1975; high energy laser system technology, adaptive optics, integrated optics and infrared imaging systems. *Mailing Add:* 18 Maurice Rd Wellesley MA 02181

HEARN, DWIGHT D, b Detroit, Mich, Apr 30, 33. APPLIED MATHEMATICS. *Educ:* Wayne State Univ, BS, 62; Univ Mich, PhD(physics), 68. *Prof Exp:* Asst prof math & physics, Lawrence Inst Technol, 64-66; asst prof physics, Eastern Mich Univ, 66-68; res assoc & assoc instr, Univ Utah, 68-70; sr analyst, Lockheed Electronics Co, 70-72; sr scientist, Braddock, Dunn & McDonald, Inc, 72-73; assoc prof appl math & comput sci, Univ Southwestern La, 73-76; mem fac, Western Ill Univ, 76-80, chmn computer sci, 80-85; vis prof, 85-90, LECTR, DEPT COMPUTER SCI, UNIV ILL, 90- *Concurrent Pos:* Consult, Petrol Assocs, Inc, 75. *Mem:* Asn Comput Mech. *Res:* Applied mathematics; computer applications in artificial intelligence; heuristic programming; numerical analysis; computer uses in education; computer graphics. *Mailing Add:* Univ Ill 2413 Digital Computer Lab 1304 W Springfield Ave Urbana IL 61801

HEARN, HENRY JAMES, JR, b Buffalo, NY, Feb 24, 27; m 51; c 3. MICROBIOLOGY, CANCER. *Educ:* Univ Buffalo, MA, 51, PhD, 55; Am Bd Microbiol, dipl pub health virol. *Prof Exp:* Bacteriologist, State Dept Health, NY, 52-55; med bacteriologist, Virol Div, Chem Corps, US Army, 55-62, supv microbiologist, Virus & Rickettsia Div, 62-67, chief, Exp Aerobiol Div, 67-71, chief, Microbiol Div, US Army Biol Defence Prog, 71-72; sci coordr viral oncol, Div Cancer Cause & Prev, Nat Cancer Inst, Frederick Cancer Res Facil, 72-81, asst proj officer basic res, Oper & Tech Support, Div Cancer Etiology, 81-86, dep gen mgr support, off dir, Nat Cancer Inst, 86-91; RETIRED. *Concurrent Pos:* Instr, Univ Md, 55-57; lectr, Hood Col, 73- *Mem:* Am Asn Immunologists; Soc Exp Biol & Med; fel Am Acad Microbiol; Am Soc Microbiol; Sigma Xi. *Res:* Viruses and rickettsiae; genetics and immunology of viral encephalitides; viruses in tissue culture; medical bacteriology and serology. *Mailing Add:* Frederick Cancer Res Facil Bldg 427 Nat Cancer Inst Frederick MD 21701

HEARN, MICHAEL JOSEPH, b Bangor, Maine, Mar 4, 49. ORGANIC CHEMISTRY. *Educ:* Rutgers Col, BA, 71; Yale Univ, MS, 73, MPhil, 75, PhD(org chem), 76. *Prof Exp:* Instr chem, Yale Univ, 76-77; from asst prof to assoc prof, 77-90, PROF CHEM, WELLESLEY COL, 90- *Concurrent Pos:* Assoc, Yale Univ, 76-77. *Mem:* Nat Sci Teachers Asn; Am Chem Soc; Sigma Xi; NY Acad Sci; Royal Inst Gt Brit; Am Inst Chemists. *Res:* Synthetic organic chemistry; new synthetic methods in the chemistry of hydrazines. *Mailing Add:* Dept Chem Wellesley Col Wellesley MA 02181

HEARN, ROBERT HENDERSON, b Phoenix, Ariz, Apr 9, 40; m 62; c 2. ELECTRONICS ENGINEERING, ENGINEERING MANAGEMENT. *Educ:* Calif Inst Technol, BS, 62,. *Prof Exp:* Engr electronics, Naval Ord Test Sta, 63-70, Naval Undersea Res & Develop Lab, 70-71, Naval Undersea Warfare Ctr, 71-72 & Naval Undersea Ctr, 72-76; ENGR ELECTRONICS, NAVAL OCEAN SYST CTR, 76- *Res:* Adaptive signal processing; ocean acoustics; real time acoustic simulation. *Mailing Add:* 6115 Radcliffe Dr San Diego CA 92122

HEARN, RUBY PURYEAR, b Winston-Salem, NC, Apr 13, 40; m 61; c 2. MATERNAL HEALTH, HEALTH POLICY. *Educ:* Skidmore Col, BA, 60; Yale Grad Sch, MS, 64, PhD(biophys), 69. *Prof Exp:* Res assoc fel, Yale Univ, 68-69; dir, Content Devolop Health Show, Children's TV Workshop, Future Works Div, 72-76; prog officer, Robert Wood Johnson Found, 76-80, sr prog officer, 80, asst vpres, 80-82, VPRES, ROBERT WOOD JOHNSON FOUND, 83- *Concurrent Pos:* Mem, Adv Coun Tele-Commun Elderly, Mt Sinai Sch Med, NY, 74-76; bd trustees, Meharry Med Col, Nashville, Tenn, 81-86; bd overseers, Dartmouth Med Sch, Hanover, NH, 86-; mem, New York City Mayoral Child Health Comn, 88- *Mem:* Inst Med-Nat Acad Sci; AAAS; Soc Res & Child Develop; Ambulatory Pediat Asn. *Res:* Development of programs to improve the health and functioning of children and youth including programs to reduce infant mortality in isolated areas, consolidate health services for high-risk young people and develop methods for assessing and treating developmental failure. *Mailing Add:* Robert Wood Johnson Found PO Box 2316 Princeton NJ 08543-2316

HEARN, WALTER RUSSELL, b Shreveport, La, Feb 20, 26; m 47, 66; c 2. BIOCHEMISTRY. *Educ:* Rice Inst, BA, 48; Univ Ill, PhD(biochem), 51. *Prof Exp:* Instr biochem, Sch Med, Yale Univ, 51-52; instr biochem, Col Med, Baylor Univ, 52-54, asst prof, 54-55; asst prof chem, Iowa State Univ, 55-60, assoc prof biochem, 60-73; SCI WRITER, ED AM SCI AFFIL NEWS, 69- *Concurrent Pos:* Am Inst Biol Sci vis biologist to cols, 61-66; res assoc, Univ Calif, Berkeley, 68-69; sr res fel, NIH, 68-69; vis assoc prof biochem, Univ Calif, Berkeley, 72-73. *Mem:* Fel AAAS; Am Chem Soc; fel Am Sci Affil. *Res:* Isolation and characterization of natural products; chemistry and metabolism of amino acids, peptides and proteins; endocrinology; bacterial pigments. *Mailing Add:* 762 Arlington Ave Berkeley CA 94707

HEARNE, HORACE CLARK, JR, b Shreveport, La, Dec 21, 30. MATHEMATICS, OPERATIONS RESEARCH. *Educ:* La State Univ, BS, 51; Mass Inst Technol, MS, 53. *Prof Exp:* Mathematician, DuPont of Can, Ltd, 57-62; mem tech staff, Res Anal Corp, 62-65; field off mgr, Booz-Allen Appl Res Inc, 65-68; partner, Roellke & Hearne, 69-70; pres, Hearne Technics Co, 70-76; pres, Crago Gear & Mach Works, Inc, 76-79; vpres, Newspaper Electronics Corp, 76-82; PRES, HUGH MATHEWS MACH WORKS, INC, 84- *Res:* Industrial operations research, especially computer applications; operations research in ground combat. *Mailing Add:* Box 336 Shawnee Mission KS 66201

HEARNEY, ELAINE SCHMIDT, neurosciences; deceased, see previous edition for last biography

HEARON, WILLIAM MONTGOMERY, b Kankakee, Ill, Feb 20, 14; m 44; c 3. BIOCHEMISTRY. *Educ:* Univ Denver, BS, 35, MS, 37; Mass Inst Technol, PhD(org chem), 40. *Prof Exp:* Asst chem, Univ Denver, 35-37; chemist, Eastman Kodak Co, 40-43; asst chief gas officer, Off Civilian Defense, Washington, DC, 43; asst prof chem, Mass Inst Technol, 46-47; dir basic res div, Cent Res Dept, Crown Zellerbach Corp, 47-49, asst dir res, 49-55, gen mgr chem prod div, 55-60, vpres res & develop, 60-67, mgt consult, 67-69; asst to vpres paper mgr, 69-77, dir chem opers, 77-79, consult, Boise Cascade Corp, 79-84; RETIRED. *Concurrent Pos:* Maj, Manhattan Dist Corp Engrs, US Army; mem vis comt, Col Forestry, Univ Wash; mem tech adv coun, Forest Prod Lab, Univ Calif & Indust Res Inst; res & develop award, Tech Asn Pulp & Paper Indust, 75. *Honors & Awards:* Com Develop Asn Honor Award, 65. *Mem:* AAAS; Am Chem Soc; Am Forestry Asn; Forest Prod Res Soc; fel Tech Asn Pulp & Paper Indust. *Res:* Research and development; wood chemistry; paper technology. *Mailing Add:* 5337 SW 34th Pl Portland OR 97201

HEARSEY, BRYAN VANDIVER, b Bellingham, Wash, Aug 2, 42; m 65; c 1. MATHEMATICS. *Educ:* Western Wash State Col, BA, 64; Wash State Univ, MA, 66, PhD(math), 68. *Prof Exp:* Teaching asst math, Wash State Univ, 64-68; asst prof, Univ Fla, 68-71; asst prof, 71-77, assoc prof, 77-81, PROF

MATH, LEBANON VALLEY COL, 81- *Mem:* Am Math Soc; Math Asn Am; Soc Actuaries. *Res:* General topology, particularly the study of convergence spaces. *Mailing Add:* Dept Math Lebanon Valley Col Annville PA 17003

HEARST, JOHN EUGENE, b Vienna, Austria, July 2, 35; US citizen; m 58; c 2. BIOPHYSICAL CHEMISTRY. *Educ:* Yale Univ, BE, 57; Calif Inst Technol, PhD(chem & physics), 61. *Prof Exp:* NSF res assoc chem, Dartmouth Col, 61-62; from asst prof to assoc prof, 62-73, PROF CHEM, UNIV CALIF, BERKELEY, 73- *Concurrent Pos:* Exec ed, Analytical Biochem, 77-82. *Mem:* AAAS; Am Chem Soc; Biophys Soc; NY Acad Sci; Am Soc Biol Chemists. *Res:* Polymer statistics; physical chemistry of DNA, hydrodynamic properties of stiff chain macromolecules; psoralen photochemistry with nucleic acids; structure of nucleic acids in viruses, ribosomes and chromosomes; nucleic acid structure-function relationships; molecular genetics of photosynthetic processes. *Mailing Add:* Dept Chem Univ Calif Berkeley CA 94720

HEARST, JOSEPH R, b Chicago, Ill, Sept 9, 31; m 57; c 3. GEOPHYSICAL WELL LOGGING. *Educ:* Reed Col, BA, 54; Mass Inst Technol, BS, 54; Boston Univ, MA, 55; Northwestern Univ, PhD(physics), 60. *Prof Exp:* PHYSICIST, LAWRENCE LIVERMORE LAB, UNIV CALIF, 59- *Concurrent Pos:* Fulbright award, Australia, 79; vis scientist, Brit Petrol Res Lab, 86. *Mem:* Soc Prof Well Log Analysts; Soc Explor Geophysicists; Minerals & Geotech Logging Soc (vpres, 85-87). *Res:* Development of improved methods of well logging, especially nuclear logging and borehole gravimetry; quality control and computer programming for well log analysis. *Mailing Add:* Lawrence Livermore Lab Livermore CA 94550

HEARST, PETER JACOB, b Stuttgart, Ger, Mar 31, 23; nat US; m 49; c 3. ORGANIC CHEMISTRY. *Educ:* Harvard Univ, BS, 43; Stanford Univ, MS, 48, PhD(chem), 51. *Prof Exp:* Asst res chemist, Calif Res Corp, Standard Oil Co Calif, 43-48; RES CHEMIST, NAVAL CIVIL ENG LAB, 50- *Mem:* AAAS; Am Chem Soc; Sigma Xi. *Res:* Pollution analysis and instrumentation; environmental chemistry; organic coatings and plastics degradation; chemical heat sources; electrical properties of materials. *Mailing Add:* 673 Devonshire Dr Oxnard CA 93030

HEARTH, DONALD PAYNE, b Fall River, Mass, Aug 13, 28; m 50; c 4. AERONAUTICAL ENGINEERING. *Educ:* Northeastern Univ, BSME, 51; Fed Exec Inst, grad, 73. *Prof Exp:* Mem staff, NASA, 51-57; dept Mgt, Marquardt Corp, Van Nuys, Calif, 57-62; mem staff, NASA, 62-67, dir planetary progs, Washington, DC, 67-70, dep dir, Goddard Space Flight Ctr, Greenbelt, Md, 70-75, Dir, Langley Res Ctr, Nasa, 75-; prof, Dept Eng, George Washington Univ, 75-85; PROF ENG, UNIV COLO, 85- *Mem:* Nat Acad Eng; fel Am Astronaut Soc; Am Soc Pub Admin; fel Am Inst Aeronaut & Astronaut. *Mailing Add:* 3815 Birchwood Dr Boulder CO 80304

HEASELL, E(DWIN) L(OVELL), b Bognor Regis, Eng, Jan 29, 31; m 52; c 2. PHYSICS, ELECTRICAL ENGINEERING. *Educ:* Imp Col, Univ London, BSc, 54, ARCS, 54, PhD(physics), 57. *Prof Exp:* Physicist, A E I Res Lab, Eng, 57-59; lectr elec eng, Imp Col, Univ London, 59-65; assoc prof, 65-67, PROF ELEC ENG, UNIV WATERLOO, 67- *Mem:* Assoc Brit Inst Physics; Inst Elec & Electronics Engrs. *Res:* Ultrasonics and its applications to chemical structure determination; the properties of semiconductors, particularly the III-V compounds. *Mailing Add:* Dept Elec Eng Univ Waterloo Waterloo ON N2L 3G1 Can

HEASLEY, GENE, b Burnips, Mich, July 21, 32; m 55; c 4. CHEMISTRY. *Educ:* Hope Col, AB, 55; Univ Kans, PhD(chem), 61. *Prof Exp:* From asst prof to assoc prof, 60-69, head dept, 75-80, PROF CHEM, BETHANY-NAZARENE COL, 69-, CHMN, DIV NATURAL SCI, 80- *Mem:* Am Chem Soc. *Res:* Mechanisms of reaction of halogens and alkyl hypohalites with conjugated dienes. *Mailing Add:* 6710 NW 33rd St Bethany OK 73008-3914

HEASLEY, JAMES HENRY, industrial materials; deceased, see previous edition for last biography

HEASLEY, VICTOR LEE, b Burnips, Mich, Mar 30, 37; m 61, 83; c 4. ORGANIC CHEMISTRY, BIOCHEMISTRY. *Educ:* Hope Col, BA, 59; Univ Kans, PhD(org chem), 63. *Prof Exp:* PROF CHEM & HEAD DEPT, POINT LOMA COL, 63- *Concurrent Pos:* Grants, Res Corp, Petrol Res Fund, Union Oil Co, NSF & US Geol Surv, 63-90. *Mem:* Am Chem Soc. *Res:* Addition of electrophiles to unsaturated hydrocarbons; author or co-author of 50 publications in the area of Electrophilic Additions. *Mailing Add:* Dept Chem Pt Loma Col 3900 Lomaland Dr San Diego CA 92106

HEASLIP, MARGARET BARKLEY, plant ecology, plant genetics, for more information see previous edition

HEASLIP, RICHARD JOSEPH, b New York, NY, Dec 7, 55. PULMONARY PHARMACOLOGY, ALLERGIC INFLAMMATION. *Educ:* Univ Pa, BA, 77; Ohio State Univ, PhD(pharmacol), 82. *Prof Exp:* Postdoctoral fel, Univ Pa, 82-84; sr scientist, Wyeth-Ayerst Res, 84-87, res scientist, 87-88, prin scientist, 89-90, RES FEL, WYETH-AYERST RES, 90- *Mem:* Am Soc Pharmacol & Exp Therapeut; Am Thoracic Soc; AAAS. *Res:* Pulmonary pharmacology; molecular pharmacological approaches to drug design; biochemical regulation of cellular reactivity; functional interaction of multiple drug-effector systems; cellular signal transduction; receptor pharmacology and physiology. *Mailing Add:* Wyeth-Ayerst Res CN 8000 Princeton NJ 08543-8000

HEASLIP, WILLIAM GRAHAM, b Brooklyn, NY, Mar 26, 28; m 50; c 5. INVERTEBRATE PALEONTOLOGY. *Educ:* Columbia Univ, BS, 53, MA, 55, PhD, 63. *Prof Exp:* Lectr geol, Brooklyn Col, 57-58; instr, Hunter Col, 58-59; asst prof, Syracuse Univ, 59-63; assoc prof, 63-70, PROF GEOL, STATE UNIV NY COL CORTLAND, 70- *Concurrent Pos:* Mem, Paleont Res Inst. *Mem:* Paleont Soc. *Res:* Genus Venericardia evolution and taxonomy; Cenozoic mollusks; sexual dimorphism; bivalves. *Mailing Add:* Dept Geol State Univ NY Cortland PO Box 2000 Cortland NY 13045

HEASTON, ROBERT JOSEPH, b Kansas City, Kans, Mar 3, 31; m 56; c 3. CHEMICAL ENGINEERING. *Educ:* Univ Ark, BS, 52, MS, 54; Ohio State Univ, PhD(chem eng), 64, Nat War Col, 75. *Prof Exp:* Sr proj engr, Propulsion Lab, Wright-Patterson AFB, Ohio, 55-58; prog mgr chem off, Adv Res Projs Agency, DC, 61-64; org chemist, Chem & Mat Br, Phys Sci Div, Off Chief Res & Develop, Hq, Dept Army, DC, 64-66, chief, Chem Br, Res & Develop Group, Europe, 66-70, chief, Technol Overview Team, Technol Div, 70-74, technol mgr weapon systs, Off Dep Chief Staff Res Develop & Acquisition, 75-87; MGR, GUIDANCE & CONTROL ANALYTICAL CTR, ILL INST TECH RES INST, CHICAGO, 87- *Concurrent Pos:* Mem solid & liquid subgroups, Interagency Chem Rocket Propulsion Group, 62-64; solid propellant instability comt, Dept Defense, 62-64; Army coord mem, panel P-3 org mat, subgroup P mat, Tripartite Tech Coop Prog, 64-66, alternate mem panel O-2 explosives & panel O-3 propellants, subgroup O ord, 65-66; alternate mem, Interagency Adv Power Group, 65-66. *Mem:* AAAS. *Res:* Technical administration of government contracts in combustion, electrochemistry and organic chemistry; theoretical physics breakthroughs: redefined four fundamental forces, published the Heaston force, speed of light to fourth power divided by Newton's gravitational constant, and derived generic field theory. *Mailing Add:* Ill Inst Tech Res Inst GACIAC 10 W 35th St Chicago IL 60616-3799

HEATH, ALAN GARD, b Wichita, Kans, July 30, 35; m 61; c 2. ECOTOXICOLOGY, COMPARATIVE PHYSIOLOGY. *Educ:* San Jose State Col, BA, 58; Ore State Univ, MS, 61, PhD(physiol), 63. *Prof Exp:* Fishery res aide, US Fish & Wildlife Serv, 58; instr physiol, Ore State Univ, 62; NIH fel, 63-64; asst prof zool, 64-69, ASSOC PROF ZOOL, VA POLYTECH INST & STATE UNIV, 69- *Concurrent Pos:* Fed Water Qual Admin fel, 70. *Mem:* AAAS; Am Soc Zool; Soc Environ Toxicol Chem; Am Inst Biol Sci; Sigma Xi. *Res:* Influence of environmental hypoxia or industrial pollutants on metabolism, cardiovascular and respiratory function of fish; physiological monitoring of pollution; temperature stress effects in fish; temperature and hypoxia effects on salamanders. *Mailing Add:* Dept Biol Va Polytech Inst & State Univ Blacksburg VA 24061

HEATH, CARL E(RNEST), JR, b Washington, DC, Jan 5, 30; m 54; c 3. CHEMICAL ENGINEERING. *Educ:* Johns Hopkins Univ, BE, 52; Univ Wis, PhD(appl reaction kinetics), 56. *Prof Exp:* Res engr, Process Res Div, Esso Res & Eng Co, 56-57, proj leader, 57-61, sect head, 61-66, govt res lab, 66-68, dir, 68-70, proj mgr, Exxon Enterprises, Inc, 70-77; res mgr, Esso Chem Res Ctr, 77-85; site mgr, Linden Tech Ctr, Exxon Chem, 85-90; PRES, CORP TRANSFORMATIONS INT, 90- *Concurrent Pos:* Consult govt adv panel on fuel cells, Off Sci & Tech, 63; mem panel on electrically powered vehicles, Dept Com, 67- *Mem:* Am Chem Soc; Am Inst Chem Engrs; Sigma Xi. *Res:* Radiation chemistry and partial oxidation of hydrocarbons; chemical reactions in shock tubes; hydrocarbon fuel cells; electrocatalysis petroleum process research and development; fuel and lubricant additives. *Mailing Add:* 48 Hawthorne Pl Summit NJ 07901

HEATH, CLARK WRIGHT, JR, b Leipzig, Ger, Jan 24, 33; US citizen; m 69; c 3. EPIDEMIOLOGY, INTERNAL MEDICINE. *Educ:* Oberlin Col, AB, 54; Johns Hopkins Univ, MD, 58. *Prof Exp:* Intern internal med, Boston City Hosp & Harvard Med Sch, 58-60, resident, 62-63; dir, Chronic Dis Div, Ctr Environ Health, Ctr Dis Control, Atlanta, 65-85; CONSULT, 85- *Concurrent Pos:* Fel hemat, Boston City Hosp & Tufts Univ, 63-65. *Mem:* Soc Epidemiol Res; Am Epidemiol Soc; Am Pub Health Asn; Am Col Epidemiol; AMA; AAAS. *Res:* Environmental health; cancer and birth defects epidemiology. *Mailing Add:* 1714 Vickers Circle Decatur GA 30030

HEATH, D(ONALD) P, b Claytonville, Ill, Jan 8, 19; m 40; c 4. CHEMICAL ENGINEERING. *Educ:* Purdue Univ, BS, 40. *Prof Exp:* Engr, Tech Serv Lab, Socony-Vacuum Oil Co, 40-45, res engr petrol fuels, Res & Develop Lab, 45-48, res assoc, 48-52; chief aviation fuels br, Petrol Admin Defense, US Dept Interior, 52-53; asst supvr petrol fuels develop, Res & Develop Lab, Socony Mobil Oil Co, 53-55, supvr petrol fuels appl res & develop, 55-63, mgr tech serv dept, Mobil Petrol Co, Inc, 63-65, mgr fuels, asphalt & spec prod, Tech Serv Dept, Mobil Int Oil Co, 65-77, coordr environ conserv, 77-80; consult, 80-91; RETIRED. *Concurrent Pos:* Mem refining comt, Mil Petrol Adv Bd, 48-54 & Coord Res Coun, 54-63. *Mem:* Am Chem Soc; Soc Automotive Engrs. *Res:* Technology of petroleum fuels including motor and aviation gasoline, jet fuel, heating oil, kerosene, diesel and industrial fuels; fuel additives; motor gasoline performance correlations; stability of supersonic jet fuels. *Mailing Add:* 11 Euclid Ave Apt 5B Summit NJ 07901-2114

HEATH, DAVID CLAY, b Oak Park, Ill, Dec 23, 42; m 64; c 3. OPERATIONS RESEARCH. *Educ:* Kalamazoo Col, BA, 64; Univ Ill, Urbana, MA, 65, PhD(math), 69. *Prof Exp:* Asst prof math, Univ Minn, Minneapolis, 69-75; from asst prof to assoc prof, 75-88, PROF OPER RES, SCH OPERS RES, CORNELL UNIV, 88- *Concurrent Pos:* Researcher, US-France Exchange Scientists Prog, NSF & Ctr Nat Sci Res, France, 73-74; consult appl probability finance & mfg. *Mem:* Am Math Soc; Opers Res Soc. *Res:* Probability theory; stochastic processes; stochastic control; game theory. *Mailing Add:* 111 Burleigh Dr Ithaca NY 14850

HEATH, EUGENE CARTMILL, b Elk City, Okla, Mar 2, 43; m 64; c 5. PHARMACOLOGY, ANALYTICAL CHEMISTRY. *Educ:* Univ Mo, BS, 65, MS, 68; Vanderbilt Univ, PhD(pharmacol), 72. *Prof Exp:* Fel, Vanderbilt Univ, 72-74, instr pharmacol, 74-75; sr scientist drug metab, Schering-Plough Corp, 75-80; DIR ANALYTICAL SERV, HARRIS LABS, 80- *Concurrent Pos:* NIH fel, 72-74; chmn, North Jersey Mass Spectrometry Discussion, 78-80. *Mem:* Am Chem Soc; Am Soc Mass Spectrometry; Am Col Clin Pharmacol; Acad Pharmacuet Sci. *Res:* Pharmacology, drug metabolism and analytical chemistry. *Mailing Add:* Harris Labs 624 Peach St Lincoln NE 68501

HEATH, EVERETT, b Boston, Mass, Jan 24, 35; m 57; c 3. ANATOMY, REPRODUCTIVE BIOLOGY. *Educ:* Swarthmore Col, BA, 58; Univ Pa, VMD, 62, PhD(anat), 69; Purdue Univ, MS, 64. *Prof Exp:* Instr vet anat, Purdue Univ, 62-64; lectr, Univ Ibadan, 64-66; res fel anat, Univ Pa, 66-69; asst prof vet anat, Univ Minn, St Paul, 69-74; sr lectr & actg head dept vet anat & physiol, Univ Ibadan, 74-78; assoc prof, 78-87, ADJ PROF, VET BIOSCI, UNIV ILL, URBANA, 87- *Concurrent Pos:* Pvt vet pract, home vet care, 87- *Mem:* Am Asn Vet Anat; World Asn Vet Anat. *Res:* Reproductive biology; veterinary anatomy. *Mailing Add:* 339 Sumpter St Lynchburg VA 24503

HEATH, G ROSS, b Adelaide, SAustralia, Mar 10, 39; c 2. GEOCHEMISTRY, MARINE GEOLOGY. *Educ:* Univ Adelaide, BSc, 60, Hons, 61; Univ Calif, San Diego, PhD(oceanog), 68. *Prof Exp:* Geologist, SAustralian Geol Surv, 61-63; res asst oceanog, Scripps Inst Oceanog, Univ Calif, San Diego, 65-67; res assoc, Ore State Univ, 68-69, asst prof, 69-72, assoc prof oceanog, 72-75; from assoc prof to prof, Univ RI, 75-78; prof oceanog & dean Col Oceanog, Ore State Univ, 78-84; PROF OCEANOG & DEAN COL OCEANOG & FISHERY SCI, UNIV WASH, 84- *Concurrent Pos:* Lithologist, Deep-Sea Drilling Proj Leg 7 & co-chief scientist, Leg 16, 84. *Mem:* Fel Geol Soc Am; fel Am Geophys Union; fel AAAS; Clay Minerals Soc; Oceanog Soc. *Res:* Mineralogy and geochemistry of deep-sea sediments; processes of deep-sea sedimentation, paleoceanography; sub-seabed disposal of nuclear wastes; deep-sea manganese nodules; tectonics of South Atlantic and equatorial Pacific. *Mailing Add:* Col Oceanog & Fishery Sci Univ Wash HN-15 Seattle WA 98195

HEATH, GEORGE A(UGUSTINE), b Concord, NH, June 10, 27; m 52; c 9. TEXTILE DYEING & FINISHING, ENVIRONMENTAL REGULATIONS. *Educ:* Univ NH, BS, 48. *Prof Exp:* Chemist textiles, Waumbec Dyeing & Finishing Co, NH, 49-51; res chemist dyestuffs, Geigy Dyestuffs Div, Geigy Chem Corp, NY, 51-56; lab mgr textiles, Ca-Vel Div, Collins & Aikman Corp, NC, 56-58, tech dir, 58-61, mgr dyeing & finishing, 61-63; dyeing engr fibers & platics, Enjay Div, Esso Res & Eng Co, Linden, NJ, 63-66; mgr dyeing tech dept, Vectra Corp, Odenton, MD, 66-80; mgr tech serv dept, Chevron Fibers Co, Odenton, Md, 80-82; CHEM ENGR, ENVIRON PROTECTION AGENCY, WASHINGTON, DC, 82- *Mem:* Am Chem Soc; Am Asn Textile Chemists & Colorists; Tech Asn Pulp & Paper Indust. *Res:* Dyeing of experimental and commercial fibers; printing of textiles; introduction of new fibers into textile processing; environmental regulations relative to pulp, paper and paperboard. *Mailing Add:* Four Elkwood Ct Catonsville MD 21228

HEATH, GEORGE L, b Toledo, Ohio, Feb 12, 24; m 49; c 3. ENERGY CONVERSION, FORENSIC ENGINEERING. *Educ:* Marquette Univ, BS, 45; Univ Mich, MS, 50. *Prof Exp:* Jr engr, Aircraft Eng Div, Packard Motorcar Co, 46-47; from instr to prof, 47-84, EMER PROF MECH ENG, UNIV TOLEDO, 84- *Concurrent Pos:* NSF sci fac fel, 62-63. *Mem:* Am Soc Mech Engrs; Am Soc Eng Educ; Nat Soc Prof Eng; Org Spare Part Equip. *Res:* Energy conversion; instrumentation and controls. *Mailing Add:* Dept Mech Eng Univ Toledo 2801 W Bancroft Toledo OH 43606

HEATH, GORDON GLENN, b Sultan, Wash, Sept 22, 22; m 54; c 5. OPTOMETRY, PHYSIOLOGY. *Educ:* Los Angeles Col Optom, BS, 50, OD, 51; Univ Calif, Berkeley, MS, 54, PhD, 60. *Prof Exp:* Am Optom Found fel, Univ Calif, 52-55; from asst prof to assoc prof, 55-64, dean, Sch Optom, 70-88, PROF OPTOM, IND UNIV, BLOOMINGTON, 64- *Concurrent Pos:* Consult, Nat Bd Examrs Optom, 55, 57, 59 & 61 & Mo Comn Higher Educ, 69-70; assoc res physiol opticist, Univ Calif, Berkeley, 62-64, vis prof, 67-68; res consult, Surgeon Gen, US Army, 64-79; mem comt on vision, Nat Res Coun-Nat Acad Sci, 64-71; mem, Nat Adv Coun Health Professions, 67-71, Nat Adv Eye Coun, 77-80 & Policy Adv Group, Nat Surv Visual Impairment & Causes, Nat Eye Inst, 80-84; mem, Nat Adv Eye Coun, 77-80; Consult dean, Univ Mo, St Louis, 78-80. *Mem:* AAAS; Am Acad Optom (pres-elect, 81-82, pres, 83-84); Optical Soc Am; Asn Schs & Cols Optom (pres, 63-65). *Res:* Physiological optics; color vision; accomodation and convergence; dark adaptation; electrophysiology of visual processes. *Mailing Add:* Sch Optom Ind Univ Bloomington IN 47405

HEATH, HARRISON DUANE, b Burlington, Wash, June 3, 23; m 56; c 2. DEVELOPMENTAL BIOLOGY. *Educ:* Stanford Univ, AB, 44, AM, 46, PhD(biol), 51. *Prof Exp:* Asst biol, Stanford Univ, 44-47 & 49-50; instr zool, Wash Univ, 51-52; Am Cancer Soc fel, Stanford Univ, 52-53; asst prof biol, Univ Fla, 53-54; asst res embryologist, Sch Med, Univ Calif, Los Angeles, 54-56; asst prof biol sci, Stanford Univ, 56-58; assoc prof zool, Univ Miami, 58-60; from assoc prof to prof, 60-90, EMER PROF BIOL SCI, CALIF STATE UNIV, HAYWARD, 90- *Concurrent Pos:* Asst clin prof pediat, Sch Med, Univ Calif, Los Angeles, 55-56; post-doctoral fel, Wash Univ, 50-51. *Mem:* Am Soc Zool; Soc Develop Biol. *Res:* Experimental embryology; growth of transplanted amphibian organs; developmental phenomena in Hydra. *Mailing Add:* Dept Biol Sci Calif State Univ Hayward CA 94542

HEATH, IAN BRENT, b Winchester, Eng, Feb 4, 45; m 67; c 1. CELL BIOLOGY, MYCOLOGY. *Educ:* Univ London, BSc & ARCS, 66, PhD(analytical cytol) & DIC, 69. *Prof Exp:* NSF fel, Univ Ga, 69-71; from asst prof to assoc prof, 71-80, PROF BIOL, YORK UNIV, 80- *Honors & Awards:* Huxley Mem Medal, Imp Col, London, 79. *Mem:* Mycol Soc Am; Soc Evolutionary Protistology (pres-elect, 81-83, pres, 83-85); Brit Soc Exp Biol; Can Soc Cell Biol (secy, 77-81, pres-elect, 81-82, pres, 82-83); Am Soc Cell Biol. *Res:* Cell ultrastructure; fungi; nuclear division; electron microscopy; morphogenesis; microtubules cytoskeleton meiosis. *Mailing Add:* Dept Biol York Univ North York ON M3J 1P3 Can

HEATH, JAMES EDWARD, b Evansville, Ind, May 3, 35; m 55; c 3. ZOOLOGY, PHYSIOLOGY. *Educ:* Univ Calif, Los Angeles, BA, 57, MA, 58, PhD(zool), 62. *Prof Exp:* Instr biol sci, Univ Calif, Santa Barbara, 61-62; NIH fel zool, Univ Calif, Los Angeles, 62-64; from asst prof to assoc prof, 64-72, head dept, 76-83, PROF PHYSIOL, UNIV ILL, URBANA, 72- *Concurrent Pos:* Lectr, Calif Lutheran Col, 63-64; Am Midland Naturalist, 72-76; prof zool & chmn dept, Univ Fla, 74-75; co-ed, Physiol Zool & J Thermal Biol, 75-; vis prof biol, Univ Nac del Sur, Bahia Blanca, Arg, 82 & Univ La Plata, Arg, 86; Fulbright sr investr, Arg, 86-87. *Mem:* Am Physiol Soc; Am Soc Zool; Am Soc Ichthyol & Herpet; Ecol Soc Am; Soc Study Evolution; Sigma Xi. *Res:* Temperature regulation and energetics; circulatory physiology. *Mailing Add:* Dept Physiol/Biophys 524 Burrill Univ Ill Urbana IL 61801

HEATH, JAMES EUGENE, b Duck Hill, Miss, Feb 7, 42; m 66; c 2. VETERINARY PATHOLOGY, LABORATORY ANIMAL ONCOLOGY. *Educ:* Auburn Univ, DVM, 66; Am Col Vet Pathologists, dipl. *Prof Exp:* Area Vet, Animal & Plant Health Inspection Serv, USDA, 66-67; pract vet, 67-70; vet diagnostician, Kord Animal Dis Lab, Tenn Dept Agr, 70-75; vet pathologist, Int Res & Develop Corp, 75-79; vet pathologist, Nat Ctr Toxicol Res, Food & Drug Admin & Univ Ark Med Sci, 79-81; VET PATHOLOGIST LAB ANIMAL PATH, SOUTHERN RES INST, 81- *Concurrent Pos:* Lectr & consult lab animal path, Univ Ark Med Sci, 79-81; consult, Int Res & Develop Corp, 79-80. *Mem:* Am Vet Med Asn; Soc Toxicol Pathologists. *Res:* Safety assessment of a wide variety of drugs and chemicals via nonclinical toxicology experiments; toxicity studies of cancer chemotherapeutic agents; publications in lab animal oncology and toxicological pathology. *Mailing Add:* Southern Res Inst 2000 Ninth Ave S Birmingham AL 35255

HEATH, JAMES LEE, b Monroe, La, Dec 6, 39; m 60; c 1. POULTRY SCIENCE, FOOD SCIENCE. *Educ:* La State Univ, BS, 63, MS, 68, PhD(prod technol), 70. *Prof Exp:* From asst prof to assoc prof poultry prod technol, 70-80, assoc dean, Col Agr, 80-82, PROF POULTRY SCI, UNIV MD, COLLEGE PARK, 80- *Honors & Awards:* Res Award, Poultry & Egg Inst Am. *Mem:* Poultry Sci Asn; Inst Food Technologists; Sigma Xi (secy, 76-77, pres, 77-78). *Res:* Chemistry and quality of poultry and egg products; development of new poultry products and processing techniques; physical changes in poultry and egg products. *Mailing Add:* Dept Poultry Sci Univ Md Animal Sci Ctr College Park MD 20742

HEATH, LARMAN JEFFERSON, b Poughkeepsie, Ark, Dec 25, 16; m 41; c 3. MECHANICAL ENGINEERING. *Educ:* Univ Okla, BSME, 40. *Prof Exp:* Draftsman, Oklahoma City Wilcox Pool Eng Asn, 40-42; engr, W Edmond Field Eng Asn, 45-47; jr petrol engr, Sohio Petrol Co, 47-48; engr, Witcher Field Eng Co, 48-49; self employed, 49-53; engr, Bur Mines, Dept Interior, US Dept Energy, 53-55, Bur Indian Affairs, 56-60 & Bur Mines, 60-65, proj leader petrol eng res, 65-77, proj leader, 77-79; RETIRED. *Mem:* Soc Petrol Engrs. *Res:* Permeability and fluid deliverability of oil and gas formations. *Mailing Add:* 515 SE Crestland Dr Bartlesville OK 74003

HEATH, LARRY FRANCIS, b Independence, Kans, Apr 18, 38; m 64; c 4. MATHEMATICAL ANALYSIS. *Educ:* Washburn Univ, BS, 60; Univ Kans, MS, 62, PhD(math), 65. *Prof Exp:* Asst prof, 65-68, ASSOC PROF MATH, UNIV TEX, ARLINGTON, 68- *Mem:* Am Math Soc; Math Asn Am; Soc Indust & Appl Math; Sigma Xi; Asn Comput Mach. *Res:* Entire and meromorphic functions; complex analysis; computational geometry. *Mailing Add:* 1804 Park Hill Dr Arlington TX 76012

HEATH, LENWOOD S, b Greenville, NC, May 23, 53; m 84; c 2. GRAPH ALGORITHMS, COMPUTATIONAL GEOMETRY. *Educ:* Univ NC, Chapel Hill, BS, 75, PhD(computer sci), 85; Univ Chicago, MS, 76. *Prof Exp:* Software engr, Telex Terminal Commun, 77-81; instr appl math computer sci, Mass Inst Technol, 85-87; ASST PROF COMPUTER SCI, VA POLYTECH INST & STATE UNIV, 87- *Concurrent Pos:* Prin investr, NSF grant, 90-92. *Mem:* Asn Comput Mach; Inst Elec & Electronics Engrs Computer Soc; Inst Elec & Electronics Engrs; Soc Appl & Indust Math. *Res:* Graph embeddings, both the traditional kind and more unusual embeddings based on, for example, stacks or queues; computational geometry; general combinatorial algorithms. *Mailing Add:* Dept Computer Sci Va Polytech Inst & State Univ Blacksburg VA 24061

HEATH, MARTHA ELLEN, b Whittier, Calif, July 28, 52. REGULATORY PHYSIOLOGY, THERMOREGULATORY PHYSIOLOGY. *Educ:* Calif State Univ, BS, 73, MA, 75; Cambridge Univ, PhD(animal physiol), 79. *Prof Exp:* Fel, Columbia Univ, 79 & Yale Med Sch, 79-80; fel, 80-83, ASST RES PHYSIOLOGIST, SCRIPPS INST OCEANOG, 83- *Concurrent Pos:* Vis asst fel, John B Pierce Found Lab, 79-80; vis scholar, Justus-Liebig Univ, WGer, 85. *Mem:* Am Soc Mammalogists; Am Soc Ichthyologists & Herpetologists; AAAS; Am Physiol Soc. *Res:* Thermoregulatory mechanism of animals and man; central integrating and controlling mechanism of body temperature; autonomic nervous system. *Mailing Add:* Dept Rehab Med Box 38 Columbia Univ 630 W 168 St New York NY 10032

HEATH, MICHELE CHRISTINE, b Bournemouth, Eng, Sept 22, 45; m 67; c 1. PHYTOPATHOLOGY, ELECTRON MICROSCOPY. *Educ:* Univ London, BS, 66, PhD(plant path) & DIC, 69. *Prof Exp:* Fel plant path, Univ Ga, 69-71; fel, Univ Toronto, 71-72, lectr, 72-73, from asst prof to assoc prof, 73-81, PROF BOT, UNIV TORONTO, 81- *Concurrent Pos:* Sr ed, J Physiol Molecular Plant Path, 82-89, Am Phytopath Soc Press, 88-91; Steacie Mem Fel, 82. *Honors & Awards:* Huxley Mem Medal, 79; Gordon Green Award, 84. *Mem:* Am Soc Plant Physiologists; fel Am Phytopath Soc; Can Phytopath Soc; Am Mycol Soc. *Res:* Electron microscopy of plant-parasite interactions; evolution and basis of plant-parasite specificity. *Mailing Add:* Dept Bot Univ Toronto Toronto ON M5S 1A1 Can

HEATH, RALPH CARR, b La Grange, NC, July 10, 25; m 47; c 2. GEOLOGY. *Educ:* Univ NC, BS, 48. *Prof Exp:* Ground water geologist, Fla, 48-53, actg dist geologist, 53-55, geologist in charge, NY, 55-60, dist geologist, NY, Conn & RI, 60-65, dist chief, NY, 65-67, dist chief, NC, 67-81, staff hydrol, US Geol Surv, NC, 81-82; CONSULT HYDROGEOLOGIST, 87- *Concurrent Pos:* Adj assoc prof geol, Rensselaer Polytech Inst, 64-67; chmn, bd regist, Am Inst Hydrol, 83-84, mem, 84-; adj prof civil eng, NC

State Univ, 83-; lectr, Univ NC, 84, Duke Univ, 85- *Honors & Awards:* Meritorious Serv Award, US Dept Interior, 81; Commendation, Nat Water Well Asn, 82, Distinguished Lectr Ground-Water Hydrol, 86; Distinguished Serv In Hydrogeol Award, Geol Soc Am, 86; Spec Award, Asn Ground-Water Scientists & Engrs, 88, Henry Darcy Distinguished Lectr, 90. *Mem:* Fel Geol Soc Am; Nat Water Well Asn; Am Inst Hydrol. *Res:* Selection and evaluation of waste-disposal sites, effect of land use on ground water, regional appraisals of ground-water resources, hydrology of barrier islands, evaluation and design of ground-water-level observation programs. *Mailing Add:* 4821 Kilkenny Pl Raleigh NC 27612

HEATH, ROBERT BRUCE, b Billings, Mont, Oct 8, 36; m 62; c 4. VETERINARY ANESTHESIOLOGY. *Educ:* Iowa State Univ, DVM, 62; Ohio State Univ, MSc, 67; Col Vet Anesthesia, Dipl, 72. *Prof Exp:* Instr vet surg, Ohio State Univ, 64-68; assoc prof vet anesthesia, 68-77, PROF CLIN SCI, COLO STATE UNIV, 77- *Concurrent Pos:* Ed, Vet Anesthesiol, 74- *Mem:* Am Soc Vet Anesthesia (vpres, 73-74, pres, 76-77); Am Soc Vet Clinicians (pres, 75-77); Am Vet Med Asn. *Res:* The effects of anesthetic drugs in the domestic species of animals, especially inhalation vapors in the horse and intramuscular agents in the cat; epidurals in animals; comparative endotracheal intubation. *Mailing Add:* Vet Teaching Hosp Colo State Univ Ft Collins CO 80523

HEATH, ROBERT GALBRAITH, b Pittsburgh, Pa, May 9, 15; m 40; c 5. PSYCHIATRY, NEUROLOGY. *Educ:* Univ Pittsburgh, BS, 37, MD, 38; Am Bd Psychiat & Neurol, dipl, 46; Columbia Univ, cert psychoanal & DMSc, 49. *Hon Degrees:* DSc, Tulane Univ, 85. *Prof Exp:* Intern, Mercy Hosp, Pittsburgh, 38-39; instr med, Univ Pittsburgh, 39-40; asst resident neurol, Neurol Inst, NY, 40-41, chief resident, 41-42; demonstr, Jefferson Med Col, 42-43; inst neurol, Col Physicians & Surgeons, Columbia Univ, 46-49; chmn dept & chief psychiat & neurol serv, 49-80, PROF PSYCHIAT & NEUROL, SCH MED, TULANE UNIV, 49-, ROBERT G HEATH, MD PROF PSYCHIAT & NEUROL, 85- *Concurrent Pos:* Sr vis physician, Charity Hosp La, New Orleans, 49-; sr consult, Southeast La Hosp & E La State Hosp. *Honors & Awards:* Gold Medal Award, Soc Biol Psychiat, 72; Frieda Fromm-Reichmann Award, Am Acad Psychoanal, 74. *Mem:* Fel AAAS; fel Am Col Physicians; fel Am Psychiat Asn; fel Am Acad Neurol; Soc Biol Psychiat (sr vpres, 66-68, pres, 68-69). *Res:* Schizophrenia; biochemical physiologic aberrations; author or coauthor of over 425 scientific publications. *Mailing Add:* 311 Rue St Peter Metairie LA 70005

HEATH, ROBERT GARDNER, b Detroit, Mich, June 15, 24; m 48; c 2. BIOMETRICS. *Educ:* Univ Mich, BS, 51; Mich State Univ, MS, 61. *Prof Exp:* Wildlife res biologist, Mich DNR, 52-60, asst biometrician, 60-61; chief mail surv, Patuxent Wildlife Res Ctr, US Dept Interior, 61-63, ctr biometrician & chief toxicol sect, 63-74; chief ecol monitoring br, Off Pesticide Progs, US Environ Protection Agency, 74-77, sr biometrician, Health Effects Br, 77-83 & Design & Develop Br, Off Toxic Substances, 83-86; BIOMET CONSULT, 86- *Honors & Awards:* Bronze Medal, US Environ Protection Agency; Cash Awards. *Mem:* Biomet Soc. *Res:* National study of underground motor fuel tanks for leakage; human pesticide exposure and associated health effects; method validation and quality assurance for analytical methods; statistical design and analysis in environmental monitoring; avian toxicology. *Mailing Add:* 11708 Eden Rd Silver Spring MD 20904

HEATH, ROBERT LOUIS, b Hermosa Beach, Calif, Feb 27, 40; m 66; c 3. BIOPHYSICS, PLANT PHYSIOLOGY. *Educ:* Calif Inst Technol, BSc, 61; Univ Mich, Ann Arbor, MSc, 63; Univ Calif, Berkeley, PhD(biophys), 67. *Prof Exp:* Asst physicist, Bell & Howell Res Ctr, Calif, 61-62; res assoc photosynthesis, Brookhaven Nat Lab, 67-69; lectr, 69-70, asst prof, 70-73, asst biologist, 69-73, assoc biologist, 73-79, ASSOC PROF PHOTOSYNTHESIS, UNIV CALIF, RIVERSIDE, 73-, PROF PLANT PHYSIOL, 79-, PLANT PHYSIOLOGIST & BIOPHYSICIST, 79-, ASSOC DEAN, COL NAT & AGR SCI, 79- *Mem:* Biophys Soc; Am Bot Soc; Am Soc Plant Physiol. *Res:* Membrane alterations by oxidants (air and water pollutants and radiation products) resulting in disruption of cellular ionic and metabolic homostasis; ionic control of photosynthesis. *Mailing Add:* Dept Bot & Plant Sci Univ Calif Riverside CA 92521

HEATH, ROBERT THORNTON, b Chicago, Ill, Mar 12, 42; m 65; c 2. THEORETICAL ECOLOGY. *Educ:* Univ Mich, Ann Arbor, BS, 63; Univ Southern Calif, PhD(biophys), 68. *Prof Exp:* NIH fel, Calif Inst Technol, 68-70; asst prof, 70-77, ASSOC PROF BIOL SCI, KENT STATE UNIV, 77- *Concurrent Pos:* Vis asst prof, Inst Ecol, Univ Ga, 75-76. *Mem:* AAAS; Am Physiol Soc; Ecol Soc Am; Am Soc Limnol & Oceanog; Am Soc Microbiol. *Res:* General system theory applied to ecosystem biology; computer simulation of model systems; holistic investigation of laboratory ecosystems; succession; phosphorus dynamics in freshwater ecosystems; microbiol ecology. *Mailing Add:* Dept Biol Sci Kent State Univ Kent OH 44242

HEATH, ROBERT WINSHIP, b Durham, NC, May 14, 33; m 55; c 2. MATHEMATICS. *Educ:* Univ NC, BS, 53, PhD(math), 59. *Prof Exp:* Asst math, Univ NC, 54 & 56-58; instr, Woman's Col NC, 58-60; from asst prof to assoc prof, Univ Ga, 60-65, NSF res grant, 64; prof, Ariz State Univ, 65-70, NSF res grant, 66-69; NSF res grant, 71-72, PROF MATH, UNIV PITTSBURGH, 70- *Concurrent Pos:* Vis assoc prof, Ariz State Univ, 64-65; vis lectr, Univ Wash, 67-69. *Mem:* Am Math Soc; Math Asn Am; Math Soc Belg; London Math Soc; Norweg Math Soc. *Res:* General topology; abstract spaces, especially metrization and generalizations of metric spaces; product topologies; ordered topological spaces; continuous selections and set-valued functions. *Mailing Add:* Dept Math William Pitt Union 911 Univ Pittsburgh 4200 Fifth Ave Pittsburgh PA 15260

HEATH, ROY ELMER, b Hastings, Mich, Feb 6, 15; m 39; c 2. INORGANIC CHEMISTRY. *Educ:* Albion Col, AB, 36; Western Reserve Univ, PhD(inorg chem), 40. *Prof Exp:* Lectr chem, Western Reserve Univ, 39-40; instr, Univ Wis, 40-42; mgr mkt res, Mich Alkali Co, Mich, 42-43; res supvr, Manhattan Proj, Univ Chicago, 43-44; mgr mkt res, Wyandotte Chem Corp, 44-45, mgr indust, railroad & aircraft div, 45-52; from assoc prof to prof chem, Mich Col Mining & Technol, 53-60; prof, Northern Mich Univ, 60-66, res coordr, 60-66, head dept, 65-66, dir res & develop, 65-66; dir res & develop, Bd Regents, Wis State Univ Syst, 66-74 & Univ Wis, Oshkosh, 74-75; DEAN, SCH GRAD STUDIES & DIR RES & DEVELOP, NORTHERN MICH UNIV, 75-, PROF CHEM, 80- *Mem:* Am Chem Soc. *Res:* Chemistry of less familiar element fluorides. *Mailing Add:* 9726 Stagecoach Ct Sun City AZ 85373

HEATH, RUSSELL LA VERNE, b Denver, Colo, June 13, 26; m 49; c 2. NUCLEAR PHYSICS. *Educ:* Colo State Univ, BS, 49; Vanderbilt Univ, MS, 56. *Hon Degrees:* DSc, Colo State Univ, 84. *Prof Exp:* Asst physics, Rutgers Univ, 49-50; sr health physicist, Am Cyanamid Co, 51-53; physicist, Atomic Energy Div, Phillips Petrol Co, 53-60, chief physics sect, Reactor Phys Br, 60-66; mgr nuclear physics br, Aerojet Nuclear Co, 66-76; mem staff, 76-80, mgr, Physics Div, 80-84, SR SCI FEL, PHYSICS DIV, IDAHO NAT ENG LAB, 84- *Concurrent Pos:* Mem bd ed, Nuclear Instr & Methods, 70-84; mem comt nuclear sci, Nat Acad Sci, 72-; mem, US Nuclear Data Comt, 75-80. *Honors & Awards:* Radiation Indust Award, 82. *Mem:* Fel Am Phys Soc; fel Am Nuclear Soc; Inst Elec & Electronics Engrs. *Res:* Nuclear structure physics; gamma-ray spectrometry; development of electronic instrumentation for nuclear research; nuclear level schemes; solid state detectors; development of field-effect low-noise preamplifiers for nuclear spectroscopy; application of on-line computer systems to experimental nuclear physics; nuclear decay data compilation; nuclear safeguards research. *Mailing Add:* Idaho Nat Energy Lab PO Box 1625 Idaho Falls ID 83401

HEATH, TIMOTHY DOUGLAS, b 52; UK citizen. CELL BIOLOGY, DRUG DELIVERY. *Educ:* Univ London, UK, BSc, 73, PhD(biochem), 76. *Prof Exp:* Asst res biochemist, Cancer Res Inst, Univ Calif, San Francisco, 81-85; asst prof, 85-90, ASSOC PROF PHARMACEUT, SCH PHARM, UNIV WIS-MADISON, 90- *Mem:* Asn Res Vision & Ophthal. *Res:* Liposome research; drug targeting. *Mailing Add:* Sch Pharm Univ Wis 425 N Charter St Madison WI 53706

HEATHCOCK, CLAYTON HOWELL, b San Antonio, Tex, July 21, 36; m 57, 80; c 4. SYNTHETIC ORGANIC CHEMISTRY. *Educ:* Abilene Christian Col, BSc, 58; Univ Colo, PhD(chem), 63. *Prof Exp:* Supvr chem tests, Champion Paper & Fibre Co, 58-60; NSF fel, Columbia Univ, 63-64; from asst prof to assoc prof, 64-75, Miller res prof, 82, 91, PROF ORG CHEM, UNIV CALIF, BERKELEY, 75- *Concurrent Pos:* Sloan Found fel, 67-69; chmn, Med Chem Study Sect, NIH, 82-84 & Div Org Chem, Am Chem Soc, 85; ed-in-chief, Org Synthesis, 85-86, J Org Chem, 88- *Honors & Awards:* Alexander von Humboldt US Sr Scientist Award, 78; Ernest Guenther Award, Am Chem Soc, 86; Creative Work in Org Synthesis Award, Am Chem Soc, 90; A C Cope Scholar, Am Chem Soc, 90. *Mem:* Am Chem Soc; Royal Soc Chem; fel AAAS. *Res:* Chemistry of natural products; total synthesis of steroids, terpenes, alkaloids and polyketides; development of new synthetic methodology; stereochemistry. *Mailing Add:* Dept Chem Univ Calif Berkeley CA 94720

HEATHER, JAMES BRIAN, b Glendale, Calif, July 24, 44. SYNTHETIC ORGANIC CHEMISTRY. *Educ:* Univ Calif, Los Angeles, BS, 67; Univ Wis-Madison, PhD(org chem), 72. *Prof Exp:* Res assoc natural prod synthesis, Sch Pharm Univ Wis-Madison, 72-74; res scientist chem process res & develop, Upjohn Co, 74-79; gen mgr, USANCO, 79-80; SR RES CHEMIST, PROCESS RES, STAUFFER CHEM CO, 80- *Mem:* Am Chem Soc; Royal Soc Chem. *Res:* New synthetic methods and reagents; chemical process design. *Mailing Add:* ICI America's Inc 1200 S 47th St Richmond CA 94804-4683

HEATHERLY, HENRY EDWARD, b Galveston, Tex, Dec 22, 36; m 56; c 1. MATHEMATICS. *Educ:* Tex A&M Univ, BA, 60, MS, 62, PhD(math), 68. *Prof Exp:* Instr math, Tex A&M Univ, 63-68; from asst prof to assoc prof, 68-76, DISTINGUISHED PROF MATH, UNIV SOUTHWESTERN LA, 76- *Concurrent Pos:* Consult, Indian Inst Technol, India, 73-74; vis assoc prof, Texas A&M Univ, 76; distinguished prof, Univ Southwestern La, 76; NASA res fel, 80; consult, Madurai Univ, 81, Bar Ilan Univ, 82. *Mem:* Am Math Soc; Math Asn Am; Sigma Xi; Soc Indust & Appl Math; London Math Soc. *Res:* Near-rings; ring theory; operational calculus; applications to difference equations; commutators of operators; general algebraic systems; applications of integral and discrete transforms. *Mailing Add:* 529 Alonda Dr Lafayette LA 70503

HEATON, CHARLES DANIEL, b Detroit, Mich, Feb 5, 21; m 48; c 3. ORGANIC CHEMISTRY. *Educ:* Univ Calif, Los Angeles, BA, 41; Stanford Univ, PhD(chem), 50. *Prof Exp:* Asst chem, Stanford Univ, 42-47; instr, Univ Wyo, 48-51; res chemist, Calif Res Corp Div, Stand Oil Co Calif, 51-60; assoc prof chem, Cent Mo State Col, 60-64; assoc prof, 64-74, PROF CHEM, NORTHERN ARIZ UNIV, 74- *Mem:* Am Chem Soc; Sigma Xi. *Res:* Natural pigments; stereoisomerism. *Mailing Add:* Box 5698 Northern Ariz Univ Flagstaff AZ 86011

HEATON, HOWARD S(PRING), b Kaysville, Utah, May 24, 35; m 57; c 5. MECHANICAL ENGINEERING. *Educ:* Univ Southern Calif, BE, 57; Stanford Univ, MS, 59, PhD(heat transfer), 63. *Prof Exp:* Sr thermodynamicist, Lockheed Missiles & Space Co, 57-59; sr develop engr, Hercules Powder Co, 62-63; from asst prof to assoc prof, 63-77, PROF MECH ENG, BRIGHAM YOUNG UNIV, 77- *Mem:* Am Soc Mech Engrs; Soc Automotive Engr; Sigma Xi. *Res:* Heat transfer and fluid mechanics. *Mailing Add:* Dept Mech Eng 242 CB Brigham Young Univ Provo UT 84602

HEATON, LEROY, b Holden, Mo, June 25, 24; m 50; c 4. PHYSICS. *Educ:* William Jewell Col, AB, 48; Univ Okla, MS, 50; Univ Mo, PhD(physics), 54. *Prof Exp:* Asst physics, Univ Okla, 48-50; instr, Univ Mo, 50-52; from asst physicist to assoc physicist, Argonne Nat Lab, 54-70; COORDR PHYSICS & ASTRON, PARKLAND COL, 70- *Concurrent Pos:* Vis scientist, Ames

Lab, Iowa State Univ, 68-69; adj prof, S F Austin State Univ, 78- *Mem:* AAAS; Am Phys Soc; Sigma Xi; Am Crystallog Asn. *Res:* Solid state physics; x-ray and neutron diffraction; structure of solids and liquids; magnetic structures. *Mailing Add:* 2202 Glen Oak Dr Champaign IL 61820

HEATON, MARIA MALACHOWSKI, b Shreveport, La, Nov 1, 32; m 57; c 2. QUANTUM CHEMISTRY. *Educ:* Chestnut Hill Col, BS, 53; Univ Vt, PhD(chem), 69. *Prof Exp:* Res chemist, Eastern Lab, E I du Pont de Nemours & Co, Inc, 53-58; res assoc appl math, Queen's Univ Belfast, 69-70; res assoc chem, Johns Hopkins Univ, 70-72; asst prof, 72-77, ASSOC PROF CHEM, NMEX STATE UNIV, 77- *Mem:* Am Chem Soc; Am Phys Soc. *Res:* Atomic and molecular structure via self-consistant-field and configuration-interaction methods; atomic and molecular properties calculations; classical and variational approaches to scattering theory. *Mailing Add:* 4120 Senna Dr Las Cruces NM 88001

HEATON, MARIETA BARROW, b Hazard, Ky. NEUROEMBRYOLOGY. *Educ:* Fla State Univ, BS, 66; NC State Univ, PhD(psychol), 71. *Prof Exp:* Fel neuroembryol, Res Div, NC Dept Ment Health, 70-72, res assoc, 72-75; from asst prof to assoc prof, 75-88, PROF NEUROSCI, COL MED, UNIV FLA, 88- *Mem:* AAAS; Soc Neurosci. *Res:* Trophic & tropic factors critical in the development of the nervous systems; mechanisms of nervous system development. *Mailing Add:* Dept Neurosci Col Med Univ Fla Gainesville FL 32610

HEATON, RICHARD CLAWSON, b Danville, Ill, July 18, 46; m 70; c 3. ANALYTICAL CHEMISTRY. *Educ:* Univ Mich, Ann Arbor, BS, 68; Univ Ill, Urbana, MS, 71, PhD(chem), 73. *Prof Exp:* Res chemist, Hercules Inc, 73-79; mem staff chem, Los Alamos Sci Lab, 73, MEM STAFF, LOS ALAMOS NAT LAB, UNIV CALIF, 79- *Mem:* AAAS. *Res:* Applied chemical research; actinide chemistry; process analytical chemistry. *Mailing Add:* 223 Rover Blvd Los Alamos NM 87544-3561

HEATON, WILLIAM ANDREW LAMBERT, b London, Eng, Nov 30, 47; m 71; c 2. HAEMATOLOGY. *Educ:* Dublin Univ, BA, 69, MB, 71, MA, 72; Univ Cape Town, MMed, 76, Col Med SAfrica, FF(path), 76; Am Bd Path, dipl & cert clin path, 81, blood banking, 83. *Prof Exp:* Intern med surg, Sir Patrick Duns Hosp, Dublin, 71-72; casualty officer, accidents, Lancaster Royal Infirmary, 72; registr path, Groote Schuur Hosp, Univ Cape Town Med Sch, 72-76, dir haematol, Red Cross Hosp, 76-77; fel blood bank, Wash Univ Med Sch, 77-78; assoc dir, Blood Bank, Barnes Hosp, 78-79; DIR, MID ATLANTIC REGION AM RED CROSS BLOOD SERV, 79-; ASSOC PROF PATH, EASTERN VA MED SCH, 82- *Concurrent Pos:* Med officer haematopathologist, Red Cross War Mem Childrens Hosp, 76-77; fel, Mo-Ill Regional Red Cross Blood Serv, 77-78, asst prof internal med & path, 78-79; asst prof path, Eastern Va Med Sch, 79-82. *Mem:* Am Asn Blood Banks; Am Soc Haematol; Am Asn Clin Path; Int Soc Hemat; Int Soc Blood Bank; Col Am Path. *Res:* Platelet kinetics and platelet involvement in thromboembolic disease; granulocyte kinetics and methods of transfusion support; red blood cell preservation and hereditary defects in red blood cell metabolism; techniques of red cell kinetic measurements; prolonged platelet storage and platelet storage solutions. *Mailing Add:* 257 Sir Oliver Rd Norfolk VA 23505

HEATWOLE, HAROLD FRANKLIN, b Waynesboro, Va, Dec 2, 34; m 55; c 2. ZOOLOGY. *Educ:* Goshen Col, BA, 55; Univ Mich, MS, 58, PhD(zool), 59, PhD(bot), 87; Univ New Eng, DSc, 81. *Prof Exp:* Instr zool, Univ Mich, 59-60; from asst prof to assoc prof, Univ PR, 60-66; sr lectr zool, 66-71, ASSOC PROF ZOOL, UNIV NEW ENG, AUSTRALIA, 71- *Concurrent Pos:* Ed, Australian J Ecol, 85-87. *Honors & Awards:* Fel Explorer's Club, Fel Inst Biol. *Mem:* Am Soc Ichthyol & Herpet; Ecol Soc Am; Australian Coral Reef Soc. *Res:* Behavior and ecology of reptiles and amphibians; ecology of islands, deserts and Antarctica. *Mailing Add:* Dept Zool Univ New Eng Armidale NSW Australia

HEAVNER, JAMES E, b Cumberland, Md, Apr 25, 44; m 67; c 3. NEUROPHARMACOLOGY, ANESTHESIA RESEARCH. *Educ:* Univ Ga, DVM, 68; Univ Wash, PhD(pharmacol), 71. *Prof Exp:* From asst prof to assoc prof anesthesiol, Med Sch, Univ Wash, 71-80; vis scientist neurophysiol, Dept Physiol, Univ Edinburgh, 78; br chief pharmacol & toxicol, Div Vet Med Res, Bur Vet Med, Food & Drug Admin, 80-82; assoc prof, 83-87, PROF ANESTHESIOL & PHYS & DIR ANESTHESIOL RES, TEX TECH UNIV HEALTH SCI CTR, 87- *Concurrent Pos:* NIH special fel, NIH, 71-74; Reviewer, Am J Vet Res, Sci, J Pharm Exp Therapeuts, Anesthesiol & J Am Vet Med Asn, 72-; prin investr grants, NIH, 73-80; assoc ed, Vet Anesthesia, 74-78; consult, Green Lake Animal Hosp, Seattle, 74-80; vis prof, Sch Med, Univ Calif, Irvine & Mich State Univ, 76; relief vet, Univ Wash, 76-77; vis prof vet med, Dept Physics & Biophys, Colo State Univ, 79; adj prof pharm, Va-Md Regional Col Vet Med, 80-82. *Mem:* Am Soc Pharmacol & Exp Therapeuts; Int Asn Study Pain; Soc Neurosci; Am Col Vet Anesthesia; AAAS. *Res:* Applied research in toxicology and pharmacology; development of new methodolgy for the evaluation of chemicals, drugs, and other toxins; human and veterinarian anesthesiology. *Mailing Add:* Dept Anesthesiol Tex Tech Univ Health Sci Ctr 3601 4th St Lubbock TX 79430

HEBARD, ARTHUR FOSTER, b New York, NY, Mar 2, 40; m 68; c 2. LOW TEMPERATURE PHYSICS. *Educ:* Yale Univ, BA, 62; Stanford Univ, MS, 64, PhD(physics), 70. *Prof Exp:* Res assoc physics, High Energy Physics Lab, Stanford Univ, 70-72; MEM TECH STAFF PHYSICS, BELL TEL LABS, 72- *Mem:* Am Phys Soc; Am Inst Physics; AAAS; Sigma Xi. *Res:* Low temperature properties of thin films; superconductors; Josephson effect; flux flow, pinning and phase transitions in superconducting films; disordered superconductors; reactive ion beam sputter deposited thin films; high Tc superconductors; thin-film dielectrics. *Mailing Add:* Bell Tel Labs Rm 1D 460 600 Mountain Ave Murray Hill NJ 07974

HEBB, MAURICE F, JR, b Tampa, Fla, Oct 19, 24; c 4. ELECTRICAL ENGINEERING. *Educ:* Univ Fla, BEE, 51. *Prof Exp:* Engr, 51-61, chief syst planning engr, 61-64, chief engr elec eng, 64-67, vpres eng, 67-71, V PRES SYST RES & DEVELOP DEPT, FLA POWER CORP, 71- *Concurrent Pos:* Mem, syst & equip comt, Edison Elec Inst, 62-65 & coord area planning comt, 65-69; tech adv comt, Nat Elec Reliability Coun, 69-; chmn tech adv comt, Southeastern Elec Reliability Coun, 69. *Mem:* Inst Elec & Electronics Engrs. *Res:* Electrical power system coordination with reference to reliability; management activities of utility system engineering; power option. *Mailing Add:* 6330 Bahama Shores Dr S 3201 34th St Box 14042 St Petersburg FL 33705

HEBBARD, FREDERICK WORTHMAN, b Eureka, Utah, Aug 6, 23. OPTOMETRY, PHYSIOLOGICAL OPTICS. *Educ:* Univ Calif, Berkeley, BS, 49, MS, 51, PhD(physiol optics), 57. *Prof Exp:* Asst optom, Univ Calif, Berkeley, 49, clin instr, 50-56; from asst prof to assoc prof, 57-67, assoc dir sch optom, 62-66, dir, 66-68, dean, Col Optom, 68-88, PROF PHYSIOL OPTICS & OPTOM, OHIO STATE UNIV, 67- *Concurrent Pos:* Dir, Nat Bd Examr Optom, 64-74, vpres, 70-72, pres, 72-74; consult, USPHS, 66-70; mem armed forces vision comt, Nat Res Coun, 67- *Honors & Awards:* Sigma Xi. *Mem:* Fel Am Acad Optom; Optical Soc Am; Am Optom Asn; Asn Schs & Cols Optom (vpres, 61-63, secy-treas, 75-79). *Res:* Binocular vision; accommodation and convergence; eye movements; clinical optometry; physiological optics; illumination; vision in aeronautics and space science. *Mailing Add:* 2100 Haverford Rd Columbus OH 43220

HEBBEL, ROBERT P, MEDICINE. *Prof Exp:* PROF MED, UNIV MINN, 88- *Mailing Add:* Dept Med Univ Minn Box 480 UMHC Harvard St at E River Rd Minneapolis MN 55455

HEBBEN, NANCY, b Detroit, Mich. NEUROPSYCHOLOGY, PAIN. *Educ:* Wayne State Univ, BA, 75, MA, 77, PhD(clin psychol), 79. *Prof Exp:* Intern neuropsychol, Boston Vet Admin Med Ctr, 78-79; res fel & res affil, Clin Res Ctr, Mass Inst Technol, 79-82; NEUROPHYSIOLOGIST, MCLEAN HOSP, BELMONT, MASS, 82- *Concurrent Pos:* Consult psychol, Mass Gen Hosp, 78-79, neuropsychol, Vet Admin Med Ctr, West Roxbury, 79-80; NIH nat res serv award, 79-82; consult, Peter Bent Brigham Hosp, Boston, Mass, 80-82; lectr, Brandeis Univ, 81; instr, Harvard Univ, 85. *Mem:* Sigma Xi; Am Psychol Asn; Int Neuropsychol Soc. *Res:* Effect of psychiatric surgery on the perception and report of pain and on the perception and expression of emotion; neuropsychological pattern associated with DSM-III psychiatric disorders. *Mailing Add:* 227 Summit Ave Apt W205 Brookline MA 02146

HEBBORN, PETER, b Leigh, Eng, Mar 6, 32; m 54; c 3. BIOCHEMICAL PHARMACOLOGY. *Educ:* Univ London, BSc, 52, PhD(cancer chemother), 55. *Prof Exp:* Sr pharmacologist, Res Dept, Boots Pure Drug Co, 58-62; from asst prof to prof biochem pharmacol, Sch Pharm, State Univ NY, Buffalo, 62-78; vpres res & develop, 78-87, CONSULT, OWEN LABS, FT WORTH, 87-; ASSOC DIR, CIRD, VALBONNE, FRANCE, 87- *Concurrent Pos:* Vpres res & develop, Westwood Pharmaceut, Buffalo, 71-78. *Mem:* Am Cancer Soc; Am Soc Pharmacol & Exp Therapeut; Soc Invest Dermat; Am Acad Dermat; Skin Pharmacol Soc. *Res:* Cancer chemotherapy; hormone antagonists; cutaneous pharmacology. *Mailing Add:* CIRD GALDERMA Sophia Antipolis Rt Lucioles 06565 Valbonne France

HEBDA, RICHARD JOSEPH, b Hamilton, Ont, Apr 6, 50; m 85; c 1. PALYNOLOGY, ETHNOBOTANY. *Educ:* McMaster Univ, BSc, 73; Univ BC, PhD(bot), 77. *Prof Exp:* Asst prof biol & earth sci, Univ Waterloo, 77-80; asst cur archaeol, BC Prov Mus, 80-86; HEAD, BOT & EARTH HIST, ROYAL BC MUS, 86- *Concurrent Pos:* Adj prof biol, Univ Victoria, 83-, Cent Earth & Ocean Res, 90- *Mem:* Can Bot Asn (vpres 87-88); Geol Asn Can; Can Asn Palynologists (pres, 82); Am Quaternary Asn; Am Asn Stratig Palynologists. *Res:* Relationship between human cultural evolution and environmental history; changing landforms (sea level change); vegetation and climate in the late quaternary of British Columbia and Central America. *Mailing Add:* Bot Unit Royal BC Mus Victoria BC V8V 1X4 Can

HEBEL, JOHN RICHARD, b Lancaster, Pa, Oct 7, 35; m 66; c 2. BIOSTATISTICS. *Educ:* Va Polytech Inst & State Univ, BS, 62, PhD(statist), 66. *Prof Exp:* Asst prof, 66-72, assoc prof biostatist, Sch Med, 72-80, assoc prof, 80-, PROF, DEPT EPIDEMIOL & PREV MED, UNIV MD. *Concurrent Pos:* Consult, Bur Biostatist, Baltimore City Health Dept, 67-72; hon res fel, Univ Birmingham, Eng, 74. *Mem:* Soc Epidemiol Res. *Res:* Epidemiologic methodology, specifically in the areas of perinatal investigation and hypertensive disease. *Mailing Add:* Dept Epidemiol & Prev Med Univ Md Sch Med 655 W Baltimore St Baltimore MD 21201

HEBEL, LOUIS CHARLES, b Oak Park, Ill, Oct 1, 30; m 53; c 3. EXPERIMENTAL PHYSICS. *Educ:* DePauw Univ, BA, 52; Univ Ill, MS, 54, PhD(physics), 57. *Prof Exp:* Asst, Univ Ill, 52-58; res physicist, Bell Tel Labs, Inc, 58-65, dept head reentry & plasma physics, 65-68; dir phys res, Sandia Labs, 68-73; mgr, Phys & Chem Sci Lab, 73-76, MGR RES PLANNING, XEROX CORP, 76- *Mem:* Fel Am Phys Soc. *Res:* Plasma physics; electronics in solids; magnetic resonance. *Mailing Add:* Xerox Parc 3333 Coyote Hill Rd Palo Alto CA 94304

HEBELER, HENRY K, b St Louis, Mo, Aug 12, 33; m 78; c 2. AERONAUTICAL & ASTRONAUTICAL ENGINEERING, STRUCTURAL ENGINEERING. *Educ:* Mass Inst Technol, SB, 56, AerE, 56, MS, 70. *Prof Exp:* Pres, Energy, Boeing Engr & Construct Co, 75-80, Missiles & Space, Boeing Aerospace Co, 80-85 & Electronics, Boeing Electronics Co, 85-87; struct engr, Boeing Co, 56-57, corp vpres planning, 87-89; RETIRED. *Concurrent Pos:* Lectr, Col Tech, 60; mem, Bd Gov, Mass Inst Technol Sloan Sch, 74-78, House Rep, Fusion Col Cemt, 76-78, Tech Adv Bd, Dept Com, 80-84, bd vis, Defense Systs Mgt, 80-84, Sci Adv Bd, Dept Energy, 84-86 & Consult Crit Mat, Dept Interior, 86- *Mem:* Fel Am Inst Aeronaut & Astronaut. *Res:* Two patents. *Mailing Add:* 24600 140th Ave SE Kent WA 98042

HEBER, DAVID, b Celle, Ger, Apr 26, 48; US citizen; m 70; c 1. ENDOCRINOLOGY, PHYSIOLOGY. *Educ:* Univ Calif, Los Angeles, BS, 69, PhD(physiol), 78; Harvard Med Sch, MD, 73. *Prof Exp:* Intern, Beth Israel Hosp, Boston, 73-74; resident, 74-75, res assoc, 75-78, ASST PROF MED, HARBOR GEN HOSP, LOS ANGELES MED CTR, UNIV CALIF, LOS ANGELES, 78- *Concurrent Pos:* NIH fel, 75-77. *Mem:* Endocrine Soc; Am Fed Clin Res; Am Col Physicians. *Res:* Control of protein metabolism in disease states; gonadotropin releasing hormone receptor physiology. *Mailing Add:* 29368 Castlehill Dr Agoura Hills CA 91301

HEBERGER, JOHN M, b Rochester, NY, Apr 23, 44; m 69; c 3. PHYSICAL CHEMISTRY. *Educ:* St John Fisher Col, BS, 66; Rensselaer Polytech Inst, PhD(phys chem), 71. *Prof Exp:* Sr res & develop chemist, Celanese Plastics Co, 70-78, staff res chemist, 78-79; staff res chemist, Am Hoechst Corp, 79-82, RES ASSOC, AM HOECHST CELANESE CORP, 82- *Mem:* Am Chem Soc. *Res:* Surface-adhesion properties of polyester film; surface modification; coatings technology; diazo microfilm; drafting film; photographic film, video tape, computer tape, floppy disks, films. *Mailing Add:* 938 River Road Greer SC 29651

HEBERLE, JUERGEN, b Koenigsberg, Ger, May 9, 25; nat US; m 57; c 2. PHYSICS. *Educ:* Swarthmore Col, AB, 44; Columbia Univ, PhD(physics), 55. *Prof Exp:* Res asst, La State Univ, 46-47; aeronaut res scientist, Nat Adv Comt Aeronaut, 48-49; res asst, Columbia Univ, 49-55; instr physics, Yale Univ, 56-57; assoc physicist, Argonne Nat Lab, 57-65; assoc prof physics, Clark Univ, 65-67; assoc prof, 68-89, EMER PROF PHYSICS, STATE UNIV NY, BUFFALO, 89- *Concurrent Pos:* Vis prof, Univ Wash, 65, Phys Lab, Univ Groningen, 66 & Physics Fac, Freiburg, Ger, 74-75; vis prof, Munich Tech Univ, 67-68. *Mem:* Am Phys Soc; Am Asn Physics Teachers. *Res:* Hydrogen atom; Moessbauer spectroscopy; molecular structure; superconducting magnets; classical electrodynamics. *Mailing Add:* 54 High Park Blvd Buffalo NY 14226

HEBERLEIN, DOUGLAS GARAVEL, b Portage, Wis, Dec 6, 16; m 38; c 2. RESEARCH ADMINISTRATION, INDUSTRIAL CHEMISTRY. *Educ:* Univ Wis, BS, 38. *Prof Exp:* Anal chemist res dept, Continental Can Co, 39-41, res chemist, 42-52; dept supt, Rocky Flats Plant, Dow Chem Co, 52-60, sect asst supt, 60-62, supt plutonium chem, 62, supt mfg tech, 62-67, patent officer, 67-75; patent officer, Rocky Flats Plant, Rockwell Int, 75-82; RETIRED. *Mem:* AAAS; Am Chem Soc; Am Nuclear Soc; Sigma Xi; fel Am Inst Chem. *Res:* Uranium, plutonium and food chemistry; process equipment design; vitamin chemistry with emphasis on methods of analysis; storage of dehydrated foods. *Mailing Add:* 2070 Neher Lane Boulder CO 80304-1606

HEBERLEIN, GARY T, b Milwaukee, Wis, Apr 11, 39; m 66; c 2. MICROBIOLOGY, PLANT PHYSIOLOGY. *Educ:* Ohio Wesleyan Univ, AB, 61; Northwestern Univ, MS, 63, PhD(plant physiol, microbiol), 65. *Prof Exp:* Jane Coffin Childs Mem Fund Med Res fel microbiol, Ghent, Belg, 66-67; asst prof microbiol & plant physiol, NY Univ, 67-70, assoc prof biol, 70-72, chmn dept, 71-72; assoc prof, Univ Mo-St Louis, 72-75, chmn dept, 73-75 & 76-80; prof biol, Bowling Green State Univ, 76-86, dean grad col & vprovost res, 80-86; VPRES RES, WAYNE STATE UNIV, DETROIT, 86- *Concurrent Pos:* Jane Coffin Childs Mem Fund res grant, 67-60; NIH res grants, 67-, NSF grants, 76-80. *Honors & Awards:* Distinguished Young Scientist Award, Sigma Xi, 71. *Mem:* AAAS; Am Soc Microbiol; Am Soc Plant Physiol; Am Inst Biol Sci; NY Acad Sci; Sigma Xi. *Res:* Mechanism by which bacteria induce plant tumors; genetics and biochemistry of crown gall tumor induction by Agrobactrium tumefaciens; genetic engineering of crop plants. *Mailing Add:* 404 Fac Admin Bldg Wayne State Univ 656 W Kirby Ave Detroit MI 48202

HEBERLEIN, JOACHIM VIKTOR RUDOLF, b Berlin, Ger, Aug 19, 39; US citizen; m 70; c 2. ARC TECHNOLOGY, PLASMA CHEMISTRY. *Educ:* Univ Stuttgart, Ger, Diplom, Physics, 66; Univ Minn, PhD(mech eng), 75. *Prof Exp:* Res fel, Heat Transfer Lab, Univ Minn, 67-75; sr engr arc res, Westinghouse Elec Corp, 75-79, mgr lamp res, 80-82, mgr plasma res, Res & Develop Ctr, 82-89, mgr nuclear & radiation technol, 84-89; ASSOC PROF, DEPT MECH ENG, UNIV MINN, 89- *Concurrent Pos:* Mem, subcomt plasma chem, Int Union Pure & Appl Chem, 84-; mem panel plasma processing, NRC, 90-91. *Honors & Awards:* Rosemount Instrumentation Award, Rosemount Inc, 74. *Mem:* Inst Elec & Electronics Engrs; Am Phys Soc; Mat Res Soc; Sigma Xi. *Res:* Arc research; plasma heat transfer; electrode effects; vacuum arcs; plasma chemistry. *Mailing Add:* Dept Mech Eng Univ Minn 111 Church St SE Minneapolis MN 55455

HEBERLING, JACK WAUGH, JR, b Chicago, Ill, Apr 22, 28; m 59. ORGANIC CHEMISTRY. *Educ:* Univ Ill, BS, 50; Univ Minn, PhD(org chem), 54. *Prof Exp:* Asst, Univ Minn, 50-53; res chemist, E I du Pont de Nemours Co, 54-59, patient liaison, Jackson Lab, Org Chem Dept, 59-77, patents & licensing, 78-85; RETIRED. *Mem:* Fel AAAS; Sigma Xi. *Res:* Sulfur and fluorine compounds; patents. *Mailing Add:* 422 Cypress Way E Naples FL 33942-1108

HEBERLING, RICHARD LEON, b Mt Penn, Pa, May 5, 26; m 58, 90; c 3. VIROLOGY. *Educ:* Albright Col, BS, 49; Univ Pa, MS, 53; Pa State Univ, PhD(bact), 57. *Prof Exp:* Res assoc microbiol, Sch Med, Univ Pittsburgh, 57-65; res microbiologist, Cancer Virol Sect, Nat Cancer Inst, 65-67; actg chmn virol dept, Southwest Found Biomed Res, 67-70, assoc found scientist, 70-75, found scientist, virol dept, Div Microbiol & Infectious Dis, 75-88. *Concurrent Pos:* Vpres, Virus Ref Lab Inc; consult, J Med Primate, Lab Animal Sci, NIH, Dept Justice. *Mem:* AAAS. *Res:* Simian viruses; human viruses; viral diagnostics. *Mailing Add:* Virus Ref Lab Inc 7540 Louis Pasteur San Antonio TX 78229-4018

HEBERT, CLARENCE LOUIS, b Brunswick, Maine, Feb 6, 12; m 42; c 2. ANESTHESIOLOGY. *Educ:* Bates Col, BS, 35; Albany Med Col, MD, 39; Am Bd Anesthesiol, dipl, 50. *Prof Exp:* Intern, USPHS Hosp, Chicago, 39-40; resident anesthesiol, Mayo Clin, 43-44; chief anesthesiol & inhalation ther, USPHS Hosp, Staten Island, NY, 44-53; chief anesthesiol dept, clin ctr, 53-77, CONSULT, NIH, 75-, CONSULT, SUBURBAN HOSP, BETHESDA, MD, 53- *Concurrent Pos:* Med dir, USPHS, 51. *Mem:* AMA; fel Am Col Anesthesiol; Int Anesthesia Res Soc. *Res:* Anesthesia. *Mailing Add:* 1604 N Neponsit Dr Venice FL 34293

HEBERT, GERARD ROSAIRE, b South Lancaster, Ont, Sept 29, 24. PHYSICS. *Educ:* Univ Ottawa, Can, BSc, 45; Univ BC, MSc(physics) & Catholic Univ, MSc(math), 56; Univ Western Ont, PhD(physics), 60. *Prof Exp:* Sr demonstr physics, Univ Ottawa, Can, 45-46, lectr physics & math, 50-53; teacher high sch, 47-50; asst prof physics, St Francis Xavier Univ, 60-64 & Univ Western Ont, 64-65; from assoc prof to prof, 65-90, EMER PROF PHYSICS & SR RESEARCHER, 90- *Honors & Awards:* Gov Gen Silver Medal, 45. *Mem:* AAAS; Am Asn Physics Teachers; Can Asn Physicists. *Res:* Properties of superfluid liquid helium; some properties of the nine point hyperbola; intensity measurements of diatomic molecular species of astrophysical interest; remote sensing. *Mailing Add:* Dept Physics York Univ 4000 Keele St North York ON M3J 1P3 Can

HEBERT, JOEL J, b Melville, La, June 23, 39; m; c 2. FAILURE ANALYSIS, FORENSIC ENGINEERING. *Educ:* La Polytech Inst, BSME, 61; Ohio State Univ, MSc, 62; Southern Methodist Univ, PhD, 70. *Prof Exp:* Construct engr, Calif Co, La, 62-64; aerothermodyn engr, Gen Dynamics, Tex, 66-67; instr mech eng, Southern Methodist Univ, 64-69; asst prof, Rice Univ, 69-74; sr mech engr, Fluor Corp, 74-77; sr mech engr, Western Geophys Co, 78-80; sect mgr, Schlumberger Technol Corp, 80-86; SR MECH ENGR, TRIODYNE INC, 87- *Concurrent Pos:* Sr consult engr, CH&A Corp, 91- *Honors & Awards:* R C Baker Found Award. *Mem:* Am Soc Mech Engrs; Sigma Xi. *Res:* Heat transfer; thermodynamics; energy conversion; seismic energy systems; mechanical safety. *Mailing Add:* 1802 Misty Hill Lane Kingwood TX 77345

HEBERT, NORMAND CLAUDE, b Bedford, Que, Jan 8, 30; US citizen; m 53; c 4. ELECTROANALYTICAL CHEMISTRY. *Educ:* St Anselm's Col, AB, 53; Univ Detroit, MS, 58; Purdue Univ, PhD(inorg chem), 61. *Prof Exp:* Chemist, Res Labs, Ethyl Corp, 56-57; res chemist, Res & Develop Div, Corning Glass Works, 61-70; PRES, MICROELECTRODES INC, 70- *Concurrent Pos:* Invited prof, Sch Med, Univ Sherbrooke, 67-72; adj prof, Univ Tex Med Br, Galveston, 73-83. *Mem:* AAAS; Am Chem Soc; Chem Soc France; Sigma Xi. *Res:* Chemistry of alkylboranes, metal carbonyls, benzene and cyclopentadienyl complexes; electrochemistry of glassy materials; hydrogen ion and ion-selective microelectrodes. *Mailing Add:* Wood Circle Goffstown NH 03045

HEBERT, TEDDY T, b Lafayette, La, Nov 24, 14; m 89; c 4. PLANT PATHOLOGY. *Educ:* Univ Southwest La, BS, 38; La State Univ, MS, 39; NC State Univ, PhD(plant path), 46. *Prof Exp:* Res asst prof, 45-54, from assoc prof to prof, 54-80, EMER PROF PLANT PATH, EXP STA, NC STATE UNIV, 80- *Concurrent Pos:* Fulbright lectr plant virol, Egypt, 63; consult, USAID Agr Prog, Peru, 67-72, & Africa, 72-84. *Mem:* Am Phytopath Soc. *Res:* Control of diseases on cereal crops; virus diseases on other crops. *Mailing Add:* 804 Ellynn Dr Cary NC 27511-4619

HECHEMY, KARIM E, b El Mansourah, Egypt, Dec 4, 38; US citizen; m 65; c 2. MICROBIAL PHYSIOLOGY. *Educ:* Cairo Univ, BSc, 60; Mass Col Pharm, MS, 65, PhD(microbiol), 68. *Prof Exp:* Trainee, Sch Med, Univ Pa, 68-71; RES SCIENTIST, NY STATE DEPT HEALTH, 71- *Mem:* Am Soc Microbiol. *Res:* Physiology of growth of bacterial cell in relation to phospholipid biosynthesis. *Mailing Add:* 29 Pico Rd Clifton Park NY 12065

HECHENBLEIKNER, INGENUIN ALBIN, b Innsbruck, Austria, Jan 4, 11; nat US; m 42; c 4. ORGANIC CHEMISTRY. *Educ:* Davidson Col, BS, 32; Mass Inst Technol, PhD(chem), 38. *Prof Exp:* Res chemist, Am Cyanamid Corp, 37-55; res dir, Shea Chem Corp, 55-58; res dir, Carlisle Chem Works, Inc, 58-63, vpres, 63-70, bd dirs, 65-70; pres, Brookfield Chem Inc, 71-83; RETIRED. *Mem:* AAAS; Am Chem Soc; NY Acad Sci. *Res:* Organic insecticides; sodium alkyls; nitrogen and phosphorus organic compounds; vinyl stabilizers; chemical additives; organometallic compounds; plastic additives including antioxidants; exploratory chemistry on novel phosphorus compounds; photochemistry. *Mailing Add:* 12115 Quail Ridge Rd Spring Hill FL 34610

HECHLER, STEPHEN HERMAN, b New York, NY, June 7, 39; m 69; c 2. MATHEMATICAL LOGIC, TOPOLOGY. *Educ:* Calif Inst Technol, BS, 61; Univ Calif, Berkeley, MA, 64, PhD(math), 67. *Prof Exp:* Asst prof math, Case Western Reserve Univ, 67-73; from asst prof to assoc prof, 73-78, PROF MATH, QUEENS COL, NY, 78- *Mem:* Sigma Xi; Asn Symbolic Logic; Am Math Soc; Math Asn Am. *Res:* Applications of set theory to combinatorial and topological structures, particularly the study of various models of set theory with the aim of proving that certain statements can be neither proven nor disproven from the axioms of set theory. *Mailing Add:* Nine Greendale Lane East Northport NY 11731

HECHT, ADOLPH, b Chicago, Ill, July 25, 14; m 42; c 2. BOTANY. *Educ:* Univ Chicago, BS, 36, MS, 37; Ind Univ, PhD(bot), 42. *Prof Exp:* Asst, Northwestern Rocky Mountain Forest & Range Exp Sta, US Forest Serv, 37-38; asst, Ind Univ, 38-42; instr bot, Univ Chicago, 46-47; from asst prof to prof, 47-79, chmn dept, 55-70, EMER PROF BOT, WASH STATE UNIV, 79- *Concurrent Pos:* Mem, Comt Examiners for Advan Biol Exam, Grad Record Exam, 65-70; ed, Plant Sci Bull, Bot Soc Am, 65-71; mem, Comn Undergrad Educ Biol Sci, 66-69; NIH spec fel, 68; ed, Northwest Sci, 76-79. *Mem:* Bot Soc Am; Am Inst Biol Sci. *Res:* Cytology of Rudbeckia and of Sarracenia; cytogenetics of Oenothera; physiology of genetic self-incompatibility in plants; warm water treatments of Oenothera styles to reduce the incompatability reaction and find out what changes have occurred in the stylar tissues. *Mailing Add:* 1409 Blvd Park Lacey WA 98503-2523

HECHT, ALAN DAVID, b New York, NY, July 23, 44; m 66; c 1. PALEOCLIMATOLOGY. *Educ:* Brooklyn Col, BS, 66; Case Western Reserve Univ, PhD(geol), 71. *Prof Exp:* Cur, Brooklyn Children's Mus, 64-66; from asst prof to assoc prof geol, W Ga Col, 70-76; prog dir, NSF, 76-; AT CLIMATE PROG OFF, NAT OCEANIC & ATMOSPHERIC ADMIN. *Concurrent Pos:* Res grants, Res Corp, 71-73 & NSF, 74-76. *Mem:* Geol Soc Am; Soc Econ Paleontologists & Mineralogists; Am Meteoritical Soc; Sigma Xi; AAAS. *Res:* Reconstruction of paleoclimates; planktonic foraminiferal ecology and paleoecology; quantitative analysis of fossil data. *Mailing Add:* Off Int Activ Environ Protection Agency 401 M St SW Washington DC 20440

HECHT, CHARLES EDWARD, b Brooklyn, NY, June 26, 30; m 55; c 4. STATISTICAL MECHANICS. *Educ:* Mass Inst Technol, SB, 52; Univ Chicago, MS, 54, PhD(phys chem), 56. *Prof Exp:* Res assoc, US Naval Res Lab, Univ Wis, 56-57; NSF res fel, Univ Amsterdam, 57-58; res assoc, Enrico Fermi Inst Nuclear Studies, Univ Chicago, 58-60; asst prof chem, Am Univ, Beirut, 60-63; from asst prof to assoc prof, Brooklyn Col, 63-67; assoc prof, 67-72, PROF CHEM, HUNTER COL, 72- *Concurrent Pos:* Vis prof physics, Delft Technol Univ, Neth, 78. *Mem:* Sigma Xi; Am Phys Soc. *Res:* Statistical mechanics and quantum mechanics applied especially to the many body problem; phase transitions; applications of fractal geometry; renormalization group transformations. *Mailing Add:* Hunter Col Dept Chem 695 Park Ave New York NY 10021

HECHT, ELIZABETH ANNE, b Spokane, Wash, June 23, 39; m 60; c 3. MEDICAL SCIENCES. *Educ:* Boston Univ, BA, 61; Harvard Univ, EdM, 62; Univ Calif, Los Angeles, MA, 74, PhD(psychol), 80. *Prof Exp:* Guest lectr, dept psychol, 84, 90, ASST PROF RES SER, DEPT PSYCHIAT, UNIV CALIF, LOS ANGELES, 80-, ASST PROF RES SER, DEPT PSYCHOL, 85- *Honors & Awards:* Danforth Award, Danforth Found, 57. *Mem:* AAAS; Am Psychol Asn; NY Acad Sci. *Res:* Effects of marijuana on cognition; tobacco cessation; depressed (and anxious) womens' cogniting; factors contributing to drug and alcohol abuse among teenagers. *Mailing Add:* 311 Amalfi Dr Santa Monica CA 90402

HECHT, EUGENE, b New York, NY, Dec 2, 38; m 60; c 3. PHYSICS. *Educ:* NY Univ, BS, 60; Rutgers Univ, MS, 63; Adelphi Univ, PhD(physics), 67. *Prof Exp:* Physicist, Astro Electronics Div, RCA Corp, 60-63; NASA fel, 67, from asst prof to assoc prof, 67-78, PROF PHYSICS, ADELPHI UNIV, 78- *Mem:* Am Phys Soc. *Res:* Low temperature solid state physics; optics. *Mailing Add:* Dept Physics Adelphi Univ Garden City NY 11530

HECHT, FREDERICK, b Baltimore, Md, July 11, 30; m 77; c 6. GENETICS, PEDIATRICS. *Educ:* Dartmouth Col, BA, 52; Univ Rochester, MD, 60. *Prof Exp:* Pediat intern & resident, Strong Mem Hosp, 60-62; from asst prof to prof pediat, med genetics & perinatal med, Med Sch, Univ Ore, 65-78; PRES, SOUTHWEST BIOMED RES INST & DIR, GENETICS CTR, 78- *Concurrent Pos:* Res fel med genetics, Univ Wash, 62-64, Nat Inst Child Health, Genetics & Pediat spec fel, 64-65; Nat Inst Child Health & Human Develop spec res fel, Genetics Unit, Mass Gen Hosp; Royal Soc Med traveling fel, Gt Brit, 71-72; vis assoc prof pediat, Harvard Med Sch; adj prof zool, Ariz State Univ, 78-; adj prof pediat, Univ Ariz Health Sci Ctr, 80-81; mem pediat staff, Ariz Childrens Hosp, St Josephs Hosp, Good Samaritan Hosp & Desert Samaritan Hosp; mem med staff, Maricopa County Hosp. *Honors & Awards:* Ross Pediat Res Award. *Mem:* Am Soc Human Genetics; Am Pediat Soc; Am Fedn Clin Res; Soc Pediat Res; Am Acad Pediat; Am Soc Pediat Hemat & Oncol; AAAS. *Mailing Add:* 4134 McGirts Blvd Jacksonville FL 32210-4362

HECHT, GERALD, b St Louis, Mo, Aug 26, 34; m 53; c 3. INDUSTRIAL PHARMACY, BIOPHARMACEUTICS. *Educ:* St Louis Col Pharm, BS, 55, MS, 57; Purdue Univ, PhD(indust pharm, bionucleonics), 65. *Prof Exp:* Chemist, Sigma Chem Co, Mo, 57-59; sr scientist pharmaceut prod develop, Mead Johnson & Co, Ind, 59-63; sect head, 65-72, dir pharmaceut sci, 72-79, dir res & develop, 80-83, DIR PHARMACEUT SCI, ALCON LABS INC, 83- *Honors & Awards:* Bristol Award, Infectious Dis Soc Am, 55. *Mem:* AAAS; NY Acad Sci; Am Pharmaceut Asn; Acad Pharmaceut Sci; Sigma Xi; Am Asn Pharmaceut Sci. *Res:* Pharmacodynamic properties of drugs as influenced by dosage form design; stabilization of pharmaceutical products and development of new dosage forms. *Mailing Add:* 6201 Wheaton Dr Ft Worth TX 76133

HECHT, HARRY GEORGE, b Powell, Wyo, May 6, 36; m 59; c 3. PHYSICAL CHEMISTRY. *Educ:* Brigham Young Univ, BS, 58, MS, 59; Univ Utah, PhD(phys chem), 61. *Prof Exp:* Asst, Argonne Nat Lab, 61-62; asst prof chem, Tex Tech Col, 62-66; mem staff, Los Alamos Sci Lab, 66-73; head dept, 73-80, PROF CHEM, SDAK STATE UNIV, 73- *Concurrent Pos:* Fulbright res grant, Commonwealth Sci & Indust Res Orgn, Australia, 65-66; Alexander von Humboldt-Stiftung fel, Univ Tubingen, 71-72; consult, Los Alamos Nat Lab, 73-78; fac res leave, Argonne Nat Lab, 83-84; exchange prof, People's Repub China, 88. *Mem:* Sigma Xi; Am Chem Soc. *Res:* Theoretical and experimental molecular structure and valence studies; actinide spectroscopy, reflectance spectroscopy, surface studies and photochemistry. *Mailing Add:* Dept Chem SDak State Univ PO Box 2202 Brookings SD 57007

HECHT, HERBERT, b Frankfort, Ger, Mar 6, 22; nat US; m 45; c 2. COMPUTER SCIENCE, CONTROL SYSTEMS. *Educ:* City Col New York, BEE; Polytech Inst Brooklyn, MEE, 49; Univ Calif, Los Angeles, PhD(eng), 67. *Prof Exp:* Engr, Mat Lab, NY Naval Shipyard, 46-48; proj engr, Flight Control Systs, Sperry Gyroscope Co, 48-53, eng sect head, Servo Syst Develop Div, 53-55, dept head, Helicopter Flight Controls, 55-58; eng supvr, Light Aircraft Dept, Sperry Phoenix Co Div, Sperry Rand Corp, 58-62; sr staff engr, Aerospace Corp, 62-64, head electromech dept, 64-70, dir comput & guid technol progs, 70-77; PRES, SOHAR INC, LOS ANGELES, 78- *Concurrent Pos:* Lectr, Univ Calif, Los Angeles Exten. *Mem:* Inst Elec & Electronics Engrs; AAAS; Asn Comput Mach. *Res:* Space vehicle systems; computer design; software and hardware reliability; economics of reliability; fault tolerant computing. *Mailing Add:* 1040 S La Jolla Ave Los Angeles CA 90035

HECHT, J(AMES) L(EE), b New York, NY, Dec 21, 26; m 53; c 2. CHEMICAL ENGINEERING. *Educ:* Cornell Univ, BChE, 49; Ga Inst Technol, MS, 51; Yale Univ, PhD(chem eng), 55. *Prof Exp:* Res engr, Film Dept, Yerkes Lab, E I du Pont de Nemours & Co, Inc, 53-59, res supvr, 59-68, res assoc, 68-70, Spruance Res & Develop Lab, 70-76, Plastic Prod & Resins Dept, 76-84, sr res assoc, 84-85; ADJ PROF & DIR, PROJ STUDY AM FUTURE, UNIV DEL, 86- *Honors & Awards:* Pub Serv Award, US Dept Interior. *Mem:* Am Chem Soc; Sigma Xi. *Res:* US and USSR cooperation. *Mailing Add:* 111 S Spring Valley Rd Wilmington DE 19807

HECHT, MAX KNOBLER, b New York, NY, Feb 15, 25; m 47; c 3. VERTEBRATE BIOLOGY. *Educ:* Cornell Univ, BS, 44, MS, 47, PhD, 52. *Prof Exp:* Lectr biol, Hunter Col, 48-51; from instr to assoc prof, 52-65, PROF BIOL, QUEENS COL, NY, 65- *Concurrent Pos:* Res assoc dept vert paleont, Am Mus Natural Hist, 58- *Mem:* Soc Study Evolution; Am Soc Ichthyol & Herpet; Am Soc Naturalists; Soc Vert Paleont; Soc Syst Zool. *Res:* Paleontology and morphology of recent orders of reptiles and amphibians; biogeography; vertebrate systematics and phylogeny. *Mailing Add:* Dept Biol Queens Col 63-19 137th St Flushing NY 11367

HECHT, MYRON J, b Oceanside, NY, May 11, 54; m 85; c 3. FAULT TOLERANT COMPUTING, RELIABILITY ENGINEERING. *Educ:* Univ Calif, Los Angeles, BS, 75, MS, 76, MBA, 82. *Prof Exp:* Assoc engr, Westinghouse, 76-78; staff scientist, Sci Appln Int Corp, 78-80; VPRES, SOHAR INC, 80- *Concurrent Pos:* Instr, Santa Monica Col, 87-88. *Mem:* Inst Elec & Electronics Engrs Computer Soc. *Res:* Development of techniques for dependable computing including fault tolerance, verification and validation, reliability prediction, safety analysis, and failure analysis. *Mailing Add:* Sohar Inc 8421 Wilshire Suite 201 Beverly Hills CA 90211

HECHT, NORMAN B, b Newark, NJ, Dec 14, 40; m 68; c 2. MOLECULAR BIOLOGY. *Educ:* Rensselaer Polytech Inst, BS, 62; Univ Ill, PhD(microbiol), 67. *Prof Exp:* Fel biol, Univ Calif, San Diego, 67-70; from asst prof to assoc prof, 70-83, PROF BIOL, TUFTS UNIV, 83- *Concurrent Pos:* USPHS-NIH res grant, 71-74, 76-79, 79-82, 80-83, 83-86, 85-88, 86-91 & 87-92; mem, clin sci study sect, NIH, steering comt, task force methods regulation of male fertility, WHO. *Mem:* AAAS; Am Soc Cell Biol; Soc Study Reproduction; Am Soc Andrology. *Res:* Biochemistry of meiosis and mitosis; chromosome structure and cellular differentiation; molecular biology of mammalian spermatogenesis, regulation of DNA and RNA synthesis. *Mailing Add:* Dept Biol Tufts Univ Medford MA 02155

HECHT, RALPH J, b St Louis, Mo, Nov 10, 42. MATERIALS SCIENCE ENGINEERING. *Educ:* Univ Mo, Rolla, BS, 64; Iowa State Univ, MS, 66. *Prof Exp:* Supvr coding develop & processing, 70-89, mgr fiber & intermetallic technol, 89-90, MGR HIGH SPEED CIVIL TRANSP MAT, PRATT & WHITNEY, 90- *Mem:* Am Soc Metals. *Mailing Add:* Pratt & Whitney PO Box 109600 West Palm Beach FL 33410-9600

HECHT, RALPH MARTIN, b New York, NY, Feb 28, 43; div; c 2. DEVELOPMENTAL GENETICS, MOLECULAR BIOLOGY. *Educ:* Wash Square Col, NY Univ, BS, 67; Univ Edinburgh, dipl animal genetics, 68, PhD(nucleic acids), 71. *Prof Exp:* Fel chromosome struct, Dept of Biophys & Genetics, 72-76, fel develop genetics, Dept Molecular Develop & Cellular Biol, 76-77; asst prof, Dept Biophys Sci, 77-83, prog dir develop biol prog, NSF, 87-88, ASSOC PROF DEVELOP GENETICS, DEPT BIOCHEM & BIOPHYS SCI, UNIV HOUSTON, 83- *Concurrent Pos:* NIH fel, 76-77, grant, 78-81. *Mem:* AAAS; Am Soc Cell Biol; Am Soc Biol Chem. *Res:* Biochemical and genetic studies of developing systems and cell-cycle genes. *Mailing Add:* Dept Biochem & Biophys Sci 4800 Calhoun Houston TX 77204-5500

HECHT, SIDNEY MICHAEL, b New York, NY, July 27, 44; m 66. CHEMISTRY, BIOLOGICAL CHEMISTRY. *Educ:* Univ Rochester, AB, 66; Univ Ill, PhD(chem), 70. *Prof Exp:* USPHS fel, Univ Wis, 70-71; from asst prof to assoc prof, Mass Inst Technol, 71-79; JOHN W MALLET PROF CHEM & PROF BIOL, UNIV VA, 78- *Concurrent Pos:* Alfred P Sloan res fel, 75-79; NIH res career develop award, 75-80; John Simon Guggenheim fel, 77-78. *Mem:* AAAS; Am Chem Soc; Royal Soc Chem; Am Soc Biol Chemists; Sigma Xi. *Res:* Site-specific transfer RNA modifications; synthesis of bleomycin; isolation of biologically active natural products. *Mailing Add:* Dept Chem/Biol Chem Bldg 237 Univ Va Charlottesville VA 22901

HECHT, STEPHEN SAMUEL, b Newark, NJ, Dec 10, 42; m 68. ORGANIC & ENVIRONMENTAL CHEMISTRY. *Educ:* Duke Univ, BS, 64; Mass Inst Technol, PhD(org chem), 68. *Prof Exp:* NIH trainee chem, Mass Inst Technol, 68-69; asst prof, Haverford Col, 69-71; Nat Res Coun fel, Eastern Mkt & Nutrit Res Div, Agr Res Serv, 71-72; head, sect org chem, Div Environ Carcinogenesis, 73-80, CHIEF, DIV CHEM CARCINOGENESIS, NAYLOR DANA INST DIS PREV, AM HEALTH FOUND, 80-, DIR RES, 87- *Concurrent Pos:* Nat Cancer Inst res career develop award, 75; lectr, Am Chem Soc Short Courses, 79-83; mem, Chem Path Study Sect, NIH, 81-85; assoc ed, Cancer Res, 81-; Nat Cancer Inst outstanding investr grant, 87- *Mem:* Am Chem Soc; Royal Soc Chem; Am Asn Cancer Res; AAAS; NY Acad Sci. *Res:* Human uptake and metabolism of environmental carcinogens; mechanism of chemical carcinogenesis. *Mailing Add:* Naylor Dana Inst Dis Prev Am Health Found Valhalla NY 10595

HECHT, TOBY T, EXPERIMENTAL BIOLOGY. *Prof Exp:* PROG DIR, BIOL RESOURCES BR, BIOL RESPONSE MODIFIERS PROG, DCT NCI-FCRDC, 87- *Mailing Add:* Biol Resources Br-Biol Response Modifiers Prog DCT NCI-FCRDC PO Box B Bldg 1052 Rm 253 Frederick MD 21702-1201

HECHTER, OSCAR MILTON, b Chicago, Ill, Sept 29, 16; m 40; c 1. PHYSIOLOGY. *Educ:* Univ Chicago, BS, 38; Univ Southern Calif, MS, 42, PhD(biochem), 43. *Prof Exp:* Asst dept metab & endocrinol, Michael Reese Hosp, Chicago, 32-37, res assoc, 37-40; res assoc, Cedars of Lebanon Hosp,

Los Angeles, 40-44; sr scientist, Worcester Found Exp Biol, 44-66; mem, Inst Biomed Res, 66-70; Nathan Smith Davis prof physiol & chmn dept, 70-78, PROF PHYSIOL, MED SCH, NORTHWESTERN UNIV, CHICAGO, 78- *Concurrent Pos:* Res assoc prof physiol, Boston Univ, 51-58; consult, Schenley Pharmaceut, Inc, 56-58 & Ayerst Labs, 58-66; vis prof biol, Brandeis Univ, 61-62; prof lectr, dept physiol, Univ Chicago, 66-70; pres, Group For Study Biol & Social Behav, 75-80; mem sci adv, Worchester Indust Exp Biol, 68-73, Lab Reproductive Biol, Univ NC, 69-73, Elau Pharmaceut Res Corp, 81-83 & Progenics, Inc, 84- *Honors & Awards:* Ciba Award, Endocrine Soc, 50. *Mem:* Am Physiol Soc; fel Am Acad Arts & Sci. *Res:* Endocrinology; mechanisms of hormone action. *Mailing Add:* Dept Physiol Northwestern Univ Med Sch 303 E Chicago Ave Chicago IL 60611

HECK, EDWARD TIMMEL, b Spencer, WVa, Aug 4, 09; m 35; c 3. GEOLOGY. *Educ:* WVa Univ, AB, 32, MS, 37, PhD(geol), 42. *Prof Exp:* Asst geologist, State Geol Surv, WVa, 35-43 & 45; assoc geologist, State Sci Serv, NY, 46; chief geologist in charge dept geol & eng, Quaker State Oil Refining Corp, 46-51; asst to vpres prod & chief engr, Minard Run Oil Co, 51-60, vpres prod & eng, 60-75, exec vpres, 75-79; CONSULT, 79- *Mem:* Fel Geol Soc Am; Am Asn Petrol Geol; Am Inst Mining, Metall & Petrol Eng; Am Geophys Union. *Res:* Migration and accumulation of oil; methods of recovery of oil; origin of oil and connate water; metamorphism of coal; methods of finding oil; conditions of deposition of oil field sediments. *Mailing Add:* 50 Williams St Bradford PA 16701

HECK, FRED CARL, b Wilkes-Barre, Pa, Nov 1, 30; m 52; c 3. VIROLOGY. *Educ:* Tex A&M Univ, BS, 59, MS, 62; Univ Tex, PhD(microbiol), 65. *Prof Exp:* Instr microbiol, Sch Vet Med, Tex A&M Univ, 59-62; asst res microbiologist, Parke, Davis & Co, 65-69; assoc prof vet microbiol, 69-80, PROF VET MICROBIOL, PARASITOL & VET PUB HEALTH, TEX A&M UNIV, 80- *Mem:* Am Soc Microbiol. *Res:* Animal virology immunology and electron microscopy; immunotheraphy of neoplastic disease. *Mailing Add:* Dept Vet Microbiol & Parasitol Tex A&M Univ College Station TX 77843

HECK, HENRY D'ARCY, b Bryn Mawr, Pa, Apr 18, 39; m 84; c 4. BIOCHEMICAL TOXICOLOGY. *Educ:* Princeton Univ, AB, 61; Northwestern Univ, PhD(chem), 66. *Prof Exp:* Fel chem, Max-Planck Inst Phys Chem, 66-68; asst prof chem, Univ Calif, Berkeley, 68-72; scientist, Stanford Res Inst, 73-77; SR SCIENTIST, CHEM INDUST INST TOXICOL, 77- *Concurrent Pos:* Adj assoc prof, Duke Univ & Univ NC, Chapel Hill; assoc ed, Fundamental & Appl Toxicol. *Honors & Awards:* Frank Blood Award, Soc Toxicol, 83. *Mem:* Am Chem Soc; AAAS; Soc Toxicol. *Res:* Inhalation toxicology; carcinogenesis; molecular dosimetry. *Mailing Add:* Chem Indust Inst Toxicol PO Box 12137 Research Triangle Park NC 27709

HECK, JAMES VIRGIL, b Louisville, Ky, Mar 14, 52; m 72; c 1. ANTIBIOTICS, CARBOHYDRATE CHEMISTRY. *Educ:* Bellarmine Col, BA, 72; Harvard Univ, MA, 74, PhD(org chem), 76. *Prof Exp:* asst dir, Synthetic Chem, 82-90, SR RES CHEMIST, MERCK, SHARP & DOHME RES LABS, 76-, SR DIR SYNTHETIC CHEM, 90- *Mem:* Am Chem Soc; AAAS; Am Soc Microbiol. *Res:* Design and synthesis of new classes of antibiotics. *Mailing Add:* Nepawin Lane Scotch Plains NJ 07076

HECK, JONATHAN DANIEL, b Vienna, Austria, July 28, 52; US citizen; m 81; c 1. GENETIC TOXICOLOGY. *Educ:* Univ NC, Greensboro, BA, 75; Univ Tex Health Sci Ctr, Houston, PhD(pharmacol & toxicol), 83; Am Bd Toxicology, dipl, 87. *Prof Exp:* Biologist, 76-78, toxicologist, 84-85, MGR LIFE SCI, LORILLARD, INC, 86- *Mem:* Soc Toxicol; Environ Mutagen Soc; Am Col Toxicol; Genetic Toxicol Asn; AAAS. *Res:* Carcinogenesis; DNA damage and repair; inhalation toxicology; application of in vitro mammalian cell culture methodologies in toxicology. *Mailing Add:* Lorillard Res Ctr PO Box 21688 420 English St Greensboro NC 27420

HECK, JOSEPH GERARD, b Philadelphia, Pa, Aug 5, 26; m 82; c 6. BACTERIOLOGY. *Educ:* Villanova Univ, BS, 49; Pa State Univ, MS, 53, PhD(bact), 57. *Prof Exp:* Res bacteriologist, Am Inst Baking, 57-58; sect head bact, Food Res Div, Armour Food Co, Armour & Co, 58-69, mgr microbiol, Qual Assurance Dept, 71-73, mgr qual assurance, 73-80; mgr vendor qual assurance, Cudahy Foods Co, 80-81; corp dir qual assurance, Knudsen Corp, 81-87; dir qual assurance, Fernando's Foods, 87-89; RETIRED. *Mem:* Inst Food Technol. *Res:* Endoproteases of cheese yeasts; food poisoning bacteria; control of spoilage of fresh and processed meats; evaluation of selective media. *Mailing Add:* 4834 Maytime Lane Culver City CA 90230

HECK, MARGARET MATHILDE SOPHIE, b Munich, WGer, May 5, 59; US citizen; m 88. EMBRYOLOGY. *Educ:* State Univ NY, Plattsburg, BA, 81; Johns Hopkins Univ, PhD(cell biol), 88. *Prof Exp:* Lab instr biol, State Univ NY, Plattsburg, 81, teaching asst, 82; lab instr cell & tissues, Sch Med, Johns Hopkins Univ, 84-87; POSTDOCTORAL TRAINING MOLECULAR GENETICS, DEPT EMBRYOL, CARNEGIE INST WASH, BALTIMORE, MD, 88- *Concurrent Pos:* Jane Coffin Childs Mem Fund med res, 88-91. *Mem:* Am Soc Cell Biol. *Mailing Add:* Dept Embryol Carnegie Inst Wash 115 W University Pkwy Baltimore MD 21210

HECK, OSCAR BENJAMIN, b Bartlesville, Okla, Jan 13, 27; m 51. PARASITOLOGY. *Educ:* Univ Kans, AB, 49, MA, 51; State Col Wash, PhD(zool), 58. *Prof Exp:* Asst prof biol, Coe Col, 56-61; asst prof, 61-69, ASSOC PROF ZOOL, OHIO UNIV, 69- *Mem:* Am Soc Parasitol; Soc Syst Zool. *Res:* Cestodes of waterfowl. *Mailing Add:* Dept Biomed Sci Ohio Univ Col Osteopath Med Grosvenor Hall Athens OH 45701

HECK, RICHARD FRED, b Springfield, Mass, Aug 15, 31. ORGANIC CHEMISTRY. *Educ:* Univ Calif, Los Angeles, BS, 52, PhD(org chem), 54. *Prof Exp:* Fel chem, Eidgenoische Tech Hoch55-56 & Univ Calif, Los Angeles, 56-57; res chemist, Hercules Inc, 57-71; PROF CHEM, UNIV DEL, 71- *Res:* Organometallic chemistry, homogeneous catalysis, reaction mechanisms. *Mailing Add:* Dept Chem Univ Del Newark DE 19716

HECK, ROBERT SKINROOD, b Mitchell, SDak, Jan 14, 29; c 4. FAMILY MEDICINE. *Educ:* Univ SDak, AB, 51; Northwestern Univ, MD, 54. *Prof Exp:* Prof family pract, Rush Med Col, 72-73; prog dir family pract residency, Christ Community Hosp, Chicago, 73-88; DEPT FAMILY PRACT, RUSH UNIV, 88- *Mem:* Soc Teachers Family Med; Am Acad Family Physicians; AMA. *Res:* Continuing medical education. *Mailing Add:* 8625 S Cicero Chicago IL 60652

HECK, RONALD MARSHALL, b Baltimore, Md, Sept 23, 43; div; c 3. CHEMICAL ENGINEERING. *Educ:* Univ Md, BSc, 65, PhD(chem eng), 69. *Prof Exp:* Res chem engr catalysis, Celanese Chem Co, 69-72; sr res chem engr, 72-76, sect head chem process develop, 76 -84, mgr, Catalytic Eng & Technol Syst Dept, 85-86, GROUP LEADER, ENVIRON CATALYST RES & DEVELOP, ENGELHARD CORP, 86- *Mem:* Am Inst Chem Eng; Air Pollution Control Asn; Soc Automotive Eng. *Res:* Catalysis, reaction kinetics and chemical reaction engineering; general research and development in chemical engineering; fluid dynamics and mass transfer; process development and scale up; development and design of catalytic air pollution control equipment; automotive catalyst; air pollution control catalysts. *Mailing Add:* Engelhard Corp 2655 US Rte 22 Union NJ 07083

HECK, WALTER WEBB, b Columbus, Ohio, May 28, 26; m 59; c 6. GROWTH & DEVELOPMENT, AIR POLLUTION EFFECTS. *Educ:* Ohio State Univ, BS, 47; Univ Tenn, MS, 50; Univ Ill, PhD, 54. *Prof Exp:* Teacher, Pub Schs, Ohio, 47-48; asst bot, Univ Tenn, 48-50; teacher, Pub Schs, Ohio, 50-51; asst bot, Univ Ill, 52-54, res biochem, 54-55; from asst prof to assoc prof biol, Ferris Inst, 55-59; assoc prof plant physiol, Tex A&M Univ, 59-63; res plant physiologist, Taft Sanit Eng Ctr, USDA, 63-68, leader coop air pollution prog, Plant Sci Res Div, Agr Res Serv, assigned to div ecol res, Environ Protection Agency, 68-72, RES LEADER AIR QUAL PROG, S ATLANTIC AREA, AGR RES SERV, USDA, 72-; PROF BOT, NC STATE UNIV, 72- *Honors & Awards:* Frank A Chambers Award, Air Pollution Control Asn, 81. *Mem:* AAAS; Air Waste Mgt Asn; Bot Soc Am; Am Soc Plant Physiol; Am Inst Biol Sci; Sigma Xi. *Res:* Physiological and biochemical effects of air pollutants on plants of horticultural and agronomic importance, using controlled growth chambers, greenhouses and field chambers. *Mailing Add:* 1509 Varsity Dr NC State Univ Raleigh NC 27606

HECKARD, DAVID CUSTER, b Steuben Co, Ind, Nov 20, 22; m 44; c 5. METALLURGICAL ENGINEERING. *Educ:* Purdue Univ, BSMetE, 43, MSMetE, 48. *Prof Exp:* Instr metall, Purdue Univ, 46-48; lab asst metallog, Armco Steel Corp, 48-50, res engr patent liaison, 50-68, mgr patents, 68-75, mgr patents & tech info, 75-86; RETIRED. *Mem:* Fel Am Soc Metals. *Res:* Metallurgy of basic steel and metallic coatings on steel. *Mailing Add:* 3100 Shadow Hill Middletown OH 45042

HECKARD, LAWRENCE RAY, b Long Beach, Wash, Apr 9, 23. BOTANY. *Educ:* Ore State Col, BS, 48; Univ Calif, PhD(bot), 55. *Prof Exp:* From instr to asst prof bot, Univ Ill, 55-60; sr herbarium botanist & lectr, Jepson Herbarium, 60-68, RES BOTANIST & CUR, JEPSON HERBARIUM & LIBR, UNIV CALIF, BERKELEY, 68- *Mem:* Am Inst Biol Sci; Soc Study Evolution; Bot Soc Am; Am Soc Plant Taxon (secy, 60-65); Int Soc Plant Taxon. *Res:* California flora; biosystematics. *Mailing Add:* Jepson Herbarium & Libr Univ Calif Berkeley CA 94720

HECKATHORN, HARRY MERVIN, III, b Warren, Ohio, Aug 30, 44; m 68; c 2. ASTRONOMY, ELECTROOPTICS. *Educ:* Carleton Col, BA, 66; Northwestern Univ, Ill, MS, 71, PhD(astrophys), 71. *Prof Exp:* Nat Res Coun res associateship, Astrophys Sect, Manned Spacecraft Ctr, NASA, 70-72; res associateship phys, Univ Houston, 72-74; prin scientist, Physics Dept, Lockheed Electronics Co, Johnson Space Ctr, NASA, 74-75; res scientist, Physics Dept, Johns Hopkins Univ, 76-77; ASTROPHYSICIST, ASTROPHYS SECT, SPACE SCI DIV, US NAVAL RES LAB, 78- *Mem:* Am Astron Soc; Soc Photo-Optical Instrumentation Engrs; Sigma Xi. *Res:* Space astronomy (unvisible to visible); galactic and extragalactic astronomy; photometry; spectroscopy; image intensification and electrography; microphotometry and computerized data analysis and display. *Mailing Add:* 7662 Sweet Hours Way Columbia MD 21046

HECKEL, EDGAR, b Plauen, Ger, July 13, 36; m 62; c 2. PHYSICAL CHEMISTRY. *Educ:* Tech Univ Berlin, BS, 58, MS, 62, PhD(chem eng), 64. *Prof Exp:* Res assoc radiation chem, Hahn-Meitner Inst Nuclear Res, Ger, 64-66; fel chem, Univ Fla, 66-67; from asst prof to assoc prof, 67-76, prof chem, ECarolina Univ, 76-80; dir IWL, WGer, 80-87, DIR IBUS, 88-, GROUP DIR LGA BAVARIA, 89-, LECTR, UNIV ERLANGEN-NURENBERG, 91- *Mem:* Am Chem Soc. *Res:* Chemical kinetics of reactions of free radicals and or ions formed by ionizing radiation in fluorocarbons or fluorocarbon-hydrocarbon mixtures; environmental research; mass spectrometry and analytical chemistry. *Mailing Add:* IBUS Asternweg 34 Koeln 90 5000 Germany

HECKEL, PHILIP HENRY, b Rochester, NY, Nov 24, 38; m 69; c 2. GEOLOGY. *Educ:* Amherst Col, BA, 60; Rice Univ, PhD(geol), 66. *Prof Exp:* Geologist, Kans Geol Surv, Univ Kans, 65-67, res assoc, 67-71; assoc prof, 71-78, PROF GEOL, UNIV IOWA, 78- *Concurrent Pos:* Vis prof, Wichita State Univ, 68-71; res assoc Kans Geol Surv, 71- *Mem:* Soc Econ Paleont & Mineral. *Res:* Sedimentary geology; stratigraphy; carbonate petrography; paleoecology. *Mailing Add:* Dept Geol Univ Iowa Iowa City IA 52242

HECKEL, RICHARD W(AYNE), b Pittsburgh, Pa, Jan 25, 34; m 59; c 2. PHYSICAL METALLURGY. *Educ:* Carnegie-Mellon Univ, BS, 55, MS, 58, PhD, 59. *Prof Exp:* Sr res metallurgist, Eng Res Lab, Exp Sta, E I du Pont de Nemours & Co, Inc, 59-63; assoc prof metall eng, Drexel Univ, 63-68, prof, 68-72; prof metall & mat sci & head dept, Carnegie-Mellon Univ, 72-76; PROF METALL ENG, MICH TECHNOL UNIV, 76- *Concurrent Pos:* With US Dept Energy, 87-90. *Honors & Awards:* Bradley Stoughton Young Teacher Award, Am Soc Metall Int, 69. *Mem:* Fel Am Soc Metals Int; Metall Soc; Am Soc Eng Educ. *Res:* Oxidation; diffusion phenomena; powder metallurgy; composite materials; coatings. *Mailing Add:* Mich Technol Univ Houghton MI 49931

HECKELSBERG, LOUIS FRED, b Indianapolis, Ind, Mar 10, 22; m 64. PHYSICAL CHEMISTRY. *Educ:* Rose Polytech Inst, BS, 48; Univ Wis, MS, 50, PhD(phys chem), 51. *Prof Exp:* Phys chemist, Phillips Petrol Co, 51-84; RETIRED. *Mem:* Am Chem Soc; Am Phys Soc; AAAS. *Res:* Thermoluminescence; catalysis; olefin disproportionation. *Mailing Add:* 2012 S Madison Blvd Bartlesville OK 74006-6940

HECKER, ART L, b Forsyth, Mont, Mar 15, 44. NUTRITION. *Educ:* Colo State Univ, PhD(nutrit biochem), 72. *Prof Exp:* DIR MED NUTRIT RES, ROSS LABS, 80-; PROF EXERCISE PHYSIOL, OHIO STATE UNIV, 85- *Mailing Add:* Ross Labs Div Abbot Labs 625 Cleveland Ave Columbus OH 43215

HECKER, GEORGE ERNST, b Hamburg, Ger, Sept 10, 39; US citizen; m 62; c 2. HYDRAULICS, FLUID MECHANICS. *Educ:* Yale Univ, BS, 61; Mass Inst Technol, MS, 62. *Prof Exp:* Res engr hydral, Tenn Valley Authority, 62-68; sr engr, Stone & Webster Eng Corp, 68-70; asst dir & asst prof, 71-75, dir, Alden Res Labs, Worcester Polytech Inst & Res Prof Civil Eng, 75-86; PRES, ALDEN RES LAB, INC, 86- *Concurrent Pos:* Chmn, exec comt, Hydraul Div, Am Soc Civil Engrs, 80-84. *Honors & Awards:* Centennial Medal, Am Soc Mech Engrs. *Mem:* Am Soc Civil Engrs; Int Asn Hydraul Res; Sigma Xi. *Res:* Hydraulic structures; mixing and dispersion; stratified flow; thermal discharges; closed conduit flow. *Mailing Add:* Alden Res Lab Inc 30 Shrewsbury St Holden MA 01520

HECKER, RICHARD JACOB, b Miles City, Mont, Mar 26, 28; m 58; c 4. PLANT BREEDING. *Educ:* Mont State Col, BS, 58; Colo State Univ, PhD(plant genetics), 64. *Prof Exp:* Res geneticist, Colo, 59-64 & Calif, 64-65, RES GENETICIST & RES LEADER, CROPS RES LAB, USDA, COLO STATE UNIV, 65- *Honors & Awards:* Merit Success Award, Am Soc Sugar Beet Technol. *Mem:* Fel Am Soc Agron; Am Soc Sugar Beet Technol; fel Crop Sci Soc Am. *Res:* Genetics of chemical and quality characters and disease resistance in sugar beets, also breeding methodology and population genetics. *Mailing Add:* Crops Res Lab USDA Colo State Univ Ft Collins CO 80523

HECKER, SIEGFRIED STEPHEN, b Tomasow, Poland, Oct 2, 43; US citizen; m 65; c 4. METALLURGY, MECHANICS. *Educ:* Case Western Reserve Univ, BS, 65, MS, 67, PhD(metall), 68. *Hon Degrees:* DSC, Col Santa Fe. *Prof Exp:* Appointee metall, Los Alamos Sci Lab, 68-70; sr res metallurgist, Phys Dept, Gen Motors Corp Res Labs, 70-73; supvr, Chem Mat Sci, 73-80, assoc div leader, 80-81, dep div leader, 81-83, div leader, 83-85, actg chmn, Ctr Mat Sci, 81-83, chmn, 85-86, DIR, LOS ALAMOS NAT LABS, 86- *Concurrent Pos:* Bd Regents, Univ NMex; bd mem, Carrie Tingley Hosp. *Honors & Awards:* Marcus Grossman Young Auth Award, Am Soc Metals, 76; E O Lawrence Award, 84. *Mem:* Nat Acad Eng; Am Inst Mining, Metall & Petrol Engrs; Metall Soc; Am Soc Metals. *Res:* Nuclear materials; high temperature and high strain rate behavior; metal-matrix fiber composite materials; plastic behavior of metals and materials; sheet metal formability. *Mailing Add:* MS A100 Los Alamos Nat Lab Los Alamos NM 87545

HECKERMAN, RAYMOND OTTO, b Rochester, Pa, Sept 16, 24; m 47; c 8. VERTEBRATE MORPHOLOGY, CYTOLOGY. *Educ:* Geneva Col, BS, 47; Univ Pittsburgh, MS, 49, PhD(zool), 55. *Prof Exp:* Asst zool, Duquesne Univ, 47-49, lectr, 49-50, instr zool & embryol, 50-55, asst prof zool, embryol & genetics, 55-58, assoc prof, 58-59; assoc prof & chmn div, Quincy Col, 59-64; assoc prof anat, physiol & biol, 64-65, assoc prof biol, 65-68, PROF BIOL, MIAMI-DADE JR COL, 68-, DIR, DIV NATURAL SCI, 65- *Mem:* Inst Environ Sci; Sigma Xi. *Res:* Environmental control technology and general education man and environment; cytology and biochemistry of cell differentiation; reticuloendothelial response to infection. *Mailing Add:* 3741 NW 108th Ct Gainesville FL 32606-4941

HECKERT, DAVID CLINTON, b Daily, WVa, Dec 31, 39; m 64; c 2. RESEARCH ADMINISTRATION, BEHAVIORAL PHARMACOLOGY. *Educ:* Manchester Col, BS, 61; Ohio State Univ, PhD(org chem), 65. *Prof Exp:* Res assoc org photochem, Iowa State Univ, 65-66; res chemist, Miami Valley Labs, 66-74, sect head, Ivorydale Tech Ctr, 74-77, sect head, Winton Hill Tech Ctr, 77-80, SECT HEAD, MIAMI VALLEY LABS, PROCTER & GAMBLE, 80- *Mem:* Am Chem Soc; Asn Chemoreception Sci; Soc Study Intake Behav. *Res:* Food chemistry; behavioral pharmacology. *Mailing Add:* 3561 Kehr Rd Oxford OH 45056

HECKLER, GEORGE EARL, b Marietta, Ohio, Dec 20, 20; m 45; c 4. PHYSICAL CHEMISTRY. *Educ:* Marietta Col, BA, 47; Univ Wis, PhD(chem), 52. *Prof Exp:* Res chemist, E I du Pont de Nemours & Co, 52-56; from asst prof to prof chem, 56-63, chmn dept, 61-83, EMER PROF CHEM, IDAHO STATE UNIV, 88- *Mem:* AAAS; Am Chem Soc; Sigma Xi. *Res:* Sensitized photo-decomposition of aqueous organic acid solutions; electrical conductivity of solutions. *Mailing Add:* Idaho State Univ Box 8023 Pocatello ID 83209

HECKLY, ROBERT JOSEPH, b Santa Barbara, Calif, May 6, 20; m 44; c 2. MICROBIOLOGY. *Educ:* Univ Calif, AB, 42; Univ Wis, MS, 49, PhD(bact), 51. *Prof Exp:* Biochemist, Camp Detrick, Md, 43-47; asst, Univ Wis, 47-51; from asst res bacteriologist to assoc res bacteriologist, 51-62, asst dir naval biomed res lab, 70-75, assoc dir, Naval Biosci Lab, 75-82, RES BACTERIOLOGIST, NAVAL BIOL LAB, 62- *Concurrent Pos:* Biochemist, USDA, Denmark, 52-53. *Mem:* AAAS; Am Chem Soc; Am Acad Microbiol; Am Soc Microbiol; Soc Cryobiol; Sigma Xi. *Res:* Lyophilization and preservation of microorganisms; chemistry and immunology of anthrax and tubercle bacilli; Pseudomonas pseudomallei and botulinum toxins; microbial degradation of wastes. *Mailing Add:* 1156 Oxford St Berkeley CA 94707

HECKMAN, CAROL A, b E Stroudsburg, Pa, Oct 18, 44. ELECTRON MICROSCOPY, IMAGE ANALYSIS. *Educ:* Beloit Col, BA, 66; Univ Mass, PhD(cell biol), 72. *Prof Exp:* Res assoc biochem, Sch Med, Yale Univ, 73-75; staff mem, Oak Ridge Nat Lab, Union Carbide Corp, 75-82; ASSOC PROF CELL BIOL, BOWLING GREEN STATE UNIV, 82- *Concurrent Pos:* Lectr cell biol, Grad Sch Biomed Sci, Univ Tenn-Oak Ridge, 76-80, adj assoc prof, 80-82; mem, NSF adv comt cell biol, 77-80; Int Cancer Res Technol Transfer fel, Int Union Against Cancer, 80, Heritage Found fel, Alta, Can, 82; dir electron micros facil, Bowling Green State Univ, 82-; guest fel, Uppsala Univ, 88-89. *Mem:* AAAS; Am Soc Cell Biol; Electron Micros Soc Am; Sigma Xi; Tissue Cult Asn. *Res:* Cell shape and adhesion mechanisms; cell responses to inducers of differentiation and growth; spatial signals modulating cell cycle. *Mailing Add:* Dept Biol Sci Bowling Green State Univ Bowling Green OH 43403-0212

HECKMAN, HARRY HUGHES, b Long Beach, Calif, Nov 25, 23; m 54; c 2. NUCLEAR PHYSICS. *Educ:* Univ Calif, PhD(physics), 53. *Prof Exp:* PHYSICIST, LAWRENCE BERKELEY LAB, UNIV CALIF, 53- *Concurrent Pos:* Consult, Los Alamos Sci Lab, 59-60. *Mem:* Am Phys Soc; Am Geophys Union. *Res:* High energy physics; geomagnetically trapped radiation. *Mailing Add:* 144 Hill Rd Berkeley CA 94708

HECKMAN, RICHARD AINSWORTH, b Phoenix, Ariz, July 15, 29; m 50; c 2. SANITARY & ENVIRONMENTAL ENGINEERING, NUCLEAR CHEMICAL ENGINEERING. *Educ:* Univ Calif, Berkeley, BS, 50. *Prof Exp:* Chem engr, Radiation Lab, Univ Calif, Berkeley, 50-51 & Calif Res & Develop Co, Standard Oil Calif, 51-53; chem engr, Lawrence Livermore Nat Lab, Univ Calif, 53-61, group leader, 61-64, sect leader, 64-68, assoc div leader, 68-74, containment scientist, 74-76, prog leader, 76-78, energy policy analyst, Atomic Vapor Laser Isotope Separation Prog, 83-84, prog engr Hazardous Waste Mgt, 84-86, proj leader, Hazardous Waste Minimization, 86-90; DIV DIR, HAZARDOUS WASTE MGT, NATIONWIDE TECH INC, 90- *Mem:* Am Nuclear Soc; AAAS; NY Acad Sci; Am Inst Chem Engrs; Am Chem Soc; Acad Hazardous Mat Mgt; fel Am Inst Chemists. *Res:* Systems studies on impacts of national energy policy on energy technologies; system studies on alternative waste management systems for the nuclear fuel cycle; process research on new technologies for hazardous waste minimization treatment and disposal. *Mailing Add:* 5683 Greenridge Rd Castro Valley CA 94552-2625

HECKMAN, RICHARD COOPER, b Richmond, Ind, Mar 28, 28; m 55; c 3. PHYSICS. *Educ:* Antioch Col, BS, 51; Duke Univ, MA, 53, PhD(physics), 56. *Prof Exp:* Res physicist, Charles F Kettering Found, 55-61; staff mem, Sandia Corp, 61-65, DIV SUPV, SANDIA NAT LAB, 65- *Concurrent Pos:* Consult, G & C Merriam Co, 57-62; from asst prof to assoc prof, Antioch Col, 56-61. *Mem:* Am Phys Soc; Sigma Xi. *Res:* Electronics engineering; thermal physics; solid state physics; atomic and molecular physics. *Mailing Add:* 7408 Pickard Ave Albuquerque NM 87110

HECKMAN, ROBERT ARTHUR, b Dobbs Ferry, NY, Sept 4, 37; m 63; c 2. ORGANIC CHEMISTRY, ANALYTIC CHEMISTRY. *Educ:* Ga Inst Technol, 59, PhD(org chem), 65. *Prof Exp:* Res chemist, 65-80, proj mgr Appl Res & Develop, 80-83, SR RES & DEVELOP, R J REYNOLDS INDUST, 83- *Mem:* Am Chem Soc; Am Inst Chemists; Sigma Xi. *Res:* Synthetic organic chemistry; isolation and characterization of natural products; medicinal, flavor and agricultural chemistry; trace organic analysis; development of analytical methods. *Mailing Add:* 847 Wellington Rd Winston Salem NC 27106-5514

HECKMAN, TIMOTHY MARTIN, b Toledo, Ohio, Oct 11, 51; m 78; c 1. EXTRAGALACTIC ASTRONOMY. *Educ:* Harvard Univ, BA, 73; Univ Wash, PhD(astron), 78. *Prof Exp:* Res fel, Leiden Observ, Leiden Univ, Neth, 78-80; Bart Bok res fel, Steward Observ, Univ Ariz, 80-82; from asst prof to assoc prof astron, Astron Prog, Univ Md, 82-89; PROF ASTRON, JOHNS HOPKINS UNIV, BALTIMORE, 89- *Concurrent Pos:* Vis prof, dept physics & astron, Johns Hopkins Univ, 84-85; chmn, users comt, K H Peak Nat Observ, 85-89. *Mem:* Am Astron Soc; Int Astron Union. *Res:* Observational studies (particulary at optical and radio wavelengths) of active galactic nuclei and quasars and their environments; observational extragalactic astronomy. *Mailing Add:* Dept Physics & Astron Johns Hopkins Univ 34th & Charles Baltimore MD 21218

HECKMANN, RICHARD ANDERSON, b Salt Lake City, Utah, Dec 7, 31; m 62; c 5. PARASITOLOGY. *Educ:* Utah State Univ, BS, 54, MS, 58; Mont State Univ, PhD(zool), 70. *Prof Exp:* Instr biol, Contra Costa Col, 62-67; lectr, Fresno State Col, 70-71, asst prof & NSF res grant, 70-72; assoc prof, 72-78, PROF ZOOL, BRIGHAM YOUNG UNIV, 78- *Concurrent Pos:* Instr, Fresno City Col, 70-71. *Mem:* Am Soc Parasitologists; Soc Protozoologists; Am Fisheries Soc; Wildlife Disease Asn; World Maricult Soc. *Res:* Diseases of fishes; ultrastructure of fish and mammalian tissue; host-parasite relationships; taxonomy and control of fish parasites; free living and parasitic ciliated protozoa; blood parasites of birds; aquaculture and mariculture. *Mailing Add:* Dept Zool Brigham Young Univ Provo UT 84602

HECKROTTE, CARLTON, b Toledo, Ohio, Nov 7, 29; m 62, 78; c 2. PHYSIOLOGICAL ECOLOGY, HERPETOLOGY. *Educ:* Univ Toledo, BS, 52; Univ Ill, MS, 56, PhD, 60. *Prof Exp:* Field asst aquatic biol, Ill Natural Hist Surv, 53-54, asst zool, 54-59; instr, Eastern Ky State Col, 59-61; asst prof, La State Univ, New Orleans, 61-64 & Harpur Col, 64-68; ASSOC PROF BIOL, ECAROLINA UNIV, 68- *Mem:* Herpetologists League; Am Soc Ichthyol & Herpet; Ecol Soc Am; Am Soc Zool; Soc Study Amphibians & Reptiles. *Res:* Physiological ecology; activity rhythms of reptiles; circadian rhythms. *Mailing Add:* Dept Biol ECarolina Univ Greenville NC 27858

HECKSCHER, STEVENS, b Philadelphia, Pa, Aug 21, 30; m 52; c 5. FUNCTIONAL ANALYSIS CONSERVATION BIOLOGY. *Educ:* Harvard Univ, AB, 52, AM, 54, PhD(math), 60. *Prof Exp:* Instr math, Rutgers Univ, 59-60; from instr to prof, math, Swarthmore Col, 60-80, prof math, 74-80; NATURALIST, NATURAL LANDS TRUST, 81- *Concurrent Pos:* NSF fac fel, Cambridge Univ, 66-67. *Mem:* Soc Conserv Biol. *Res:* Functional analysis, including Banach and topological vector spaces, especially spaces of measurable functions; applications of functional analysis, plant ecology, ecological management, mathematical ecology. *Mailing Add:* Ten Ridley Dr Wallingford PA 19086

HECKY, ROBERT EUGENE, b Akron, Ohio, June 21, 44. LIMNOLOGY, PALEOLIMNOLOGY. *Educ:* Kent State Univ, BS, 66; Duke Univ, PhD(zool), 71. *Prof Exp:* Fel & res assoc paleoecol, Woods Hole Oceanog Inst, 72-73; RES SCIENTIST LIMNOL, DEPT FISHERIES & OCEANS, FRESH WATER INST, 73- *Mem:* Am Soc Limnol & Oceanog; Int Asn Theoret & Appl Limnol; AAAS. *Res:* Description and quantification of the effects of man's use of land and water on aquatic ecosystems; reservoir ecology; limnology and paleolimnology of the African Great Lakes; phytoplankton ecology; biogeochemistry of silicon. *Mailing Add:* Freshwater Inst 501 University Crescent Winnipeg MB R3T 2N6 Can

HECTOR, DAVID LAWRENCE, b Royal Oak, Mich, Apr 12, 39; m 61; c 2. APPLIED MATHEMATICS, FLUID DYNAMICS. *Educ:* Northwestern Univ, Ill, BSSE, 62, PhD(mech eng), 66. *Prof Exp:* Asst prof, Univ Denver, 66-70, assoc prof math, 70-80; mem staff, Stearns-Roger Eng, 80-82; staff mem, 82-87, mgr, Soft Eng, 87-90, MGR, RES & DEVELOP, MELCO INDUSTS, 90- *Mem:* AAAS; Soc Indust & Appl Math. *Res:* Asymptotic methods; partial differential equations; computer applications in mathematics education; computer solutions of differential equations. *Mailing Add:* 10305 E Crestridge Lane Englewood CO 80111

HECTOR, MINA FISHER, b Oak Park, Ill, Oct 15, 47; m 75; c 1. BIOCHEMISTRY. *Educ:* Lake Forest Col, BA, 69; Univ Colo, Boulder, PhD(chem), 75. *Prof Exp:* Teacher, Chicago Bd Educ, 69-70; ASST PROF CHEM, CALIF STATE UNIV, CHICO, 75- *Concurrent Pos:* Investr, Univ Found, Calif State Univ, 76- *Mem:* Am Chem Soc; Am Soc Microbiol; Int Union Biochem. *Res:* Biodegradation of petroleum products and plant neoplastic natural products. *Mailing Add:* Dept Chem Calif State Univ Chico CA 95929

HEDAYAT, A SAMAD, b Jahrom, Iran, July 11, 37; m 70; c 2. MATHEMATICS, STATISTICS. *Educ:* Univ Tehran, BS, 61, MS, 62; Cornell Univ, MS, 66, PhD(statist), 69. *Prof Exp:* Jr statistician, Inst Econ Res, Tehran, 60-64; asst prof statist, Cornell Univ, 69-72; assoc prof, Fla State Univ, 72-74; res scientist, Ctr Drug Eval & Res, FDA, 89-90; PROF MATH, UNIV ILL, CHICAGO, 74- *Concurrent Pos:* Instr, Univ Tehran, 62-64; vis asst prof, Mich State Univ, 69, Univ Guelph, 70 & Univ Calif, Berkeley, 80-81; NSF res grant, 70, 74 & 75; Air Force Off Sci Res grants, 72-90; assoc ed, Inst Math Statist, 73-80; bd mem, Commun Statist; coord ed, J Statist Planning & Inference. *Mem:* Fel Inst Math Statist; fel Am Statist Asn; Int Statist Inst. *Res:* Theory of linear models; theory and construction of F-square designs; sum composition of orthogonal Latin square designs; successive experiment designs; multi-stage experiments; balanced designs; fractional factorial designs; optimal experimental designs; survey designs; statistical inference; repeated measurements designs. *Mailing Add:* Dept Math Statist & Comput Sci MC/ 249 Univ Ill PO Box 4348 Chicago IL 60680

HEDBERG, HOLLIS DOW, geology; deceased, see previous edition for last biography

HEDBERG, KAREN K, MONOCLONAL ANTIBODIES, IMMUNO FLUORESCENTS. *Educ:* Univ Ore, PhD(biol). *Prof Exp:* RES ASSOC, INST MOLECULAR BIOL, UNIV ORE, 84- *Mailing Add:* Inst Molecular Biol Univ Ore Eugene OR 97403

HEDBERG, KENNETH WAYNE, b Portland, Ore, Feb 2, 20; m 54; c 2. PHYSICAL CHEMISTRY. *Educ:* Ore State Col, BS, 43; Calif Inst Technol, PhD(chem), 48. *Prof Exp:* Noyes fel chem, Calif Inst Technol, 48-49, res fel, 49-52, sr res fel, 53-56; Guggenheim & Fulbright fel, Univ Norway, 52-53; from asst prof to assoc prof, 56-65, prof, 65-87, EMER PROF CHEM, ORE STATE UNIV, 87- *Concurrent Pos:* Sloan fel, 56-60; Royal Norweg Coun Sci & Indust res fel, 62-63; res prof, Royal Norweg Coun Sci & Humanities, 69-70; chmn, Div Chem Physics, Am Phys Soc, 72-73; mem comn eletron diffraction, Int Union Crystallog, 75-84; sr vis res fel, Univ Reading, England, 76-77; Norweg Marshal Plan fel, 82-83. *Mem:* Norweg Acad Sci & Lett; Sigma Xi; Fel Am Phys Soc; Am Chem Soc; Am Crystallog Asn; fel AAAS. *Res:* Electron diffraction in vapor state; molecular structure. *Mailing Add:* Dept Chem Ore State Univ Corvallis OR 97331

HEDBERG, MARGUERITE ZEIGEL, b Kirksville, Mo, Aug 27, 07; m 36. MATHEMATICS. *Educ:* Delta State Teachers Col, BS, 28; Univ Mo, MA, 29, PhD(math), 32. *Prof Exp:* Asst prof math, Delta State Teachers Col, 31-32; from asst prof to prof, Lander Col, 32-36; teacher, Baylor Univ, 38-40; prof & actg head dept, Delta State Teachers Col, 42-43; asst prof, Univ Ga, 43-44; adj prof, 46-49, assoc prof, 49-76, EMER ASSOC PROF MATH, UNIV SC, 76- *Mem:* Math Asn Am. *Res:* Geometry; invariant properties of a two-dimensional surface in hyperspace. *Mailing Add:* 738 Poinsettia St Columbia SC 29205

HEDDE, RICHARD DUANE, b Grand Forks, NDak, Dec 5, 45; m 71; c 5. ANIMAL NUTRITION. *Educ:* NDak State Univ, BS, 67, MS, 69; Colo State Univ, PhD(animal nutrit), 73. *Prof Exp:* Rumen physiologist, Syntex Res, 73-76, feedlot nutritionist, 76-78; mgr nutrit res, 78-81, Smith Kline Animal Health, 78-81, mgr nutrit & microbiol res, 81-84, dir pre clin res, 84-89, VPRES, PHARMACEUT DEVELOP, SMITH KLINE BEECHAM ANIMAL HEALTH, 89- *Mem:* Am Soc Animal Sci; Am Dairy Sci Asn; Sigma Xi; Am Inst Nutrit. *Res:* Function of microbial fermentation with emphasis on improving the utilization of protein and energy by gastrointestinal microorganisms and subsequently by the host animal; discovery of therapeutic antibiotics for livestock; development of active animal health research compounds for FDA and EEC registration. *Mailing Add:* Smith Kline Beecham Animal Health 1600 Paoli Pike West Chester PA 19380

HEDDEN, GREGORY DEXTER, b Louisville, Ky, Sept 13, 19; m 50; c 2. ORGANIC CHEMISTRY, METEOROLOGY. *Educ:* Univ Chicago, BS, 42, MS, 50, PhD(org chem), 51. *Prof Exp:* Chemist & asst to dir opers res, Inst Air Weapons Res, Univ Chicago, 51-54; res chemist org chem res & develop, E I du Pont de Nemours & Co, 54-59; tech dir, Trionics Corp, 59-62; pres & tech dir, Madison Res & Develop Labs, Inc, 62-66; prof environ resources & dir state tech serv & sea grant adv serv, Exten Div, Univ Wis-Madison, 66-83; DIR, RADIATION MEASUREMENTS, INC, MIDDLETON, WIS, 83- *Concurrent Pos:* Consult, Active Corps Execs, Small Bus Admin, 84-; pres, Gregory D Hedden & Assoc Ltd, Madison, Wis, 87-; dir, John R Cameron Med Physics Found, Madison, Wis, 87-89. *Honors & Awards:* New Prod Award, Indust Res, 63; Award of Hon, Underwater Mining Inst, 81. *Mem:* Fel AAAS; fel Am Inst Chemists; Int Asn Great Lakes Res; Am Chem Soc; Technol Transfer Soc. *Res:* Ultraviolet spectroscopy; marine resources; organometallics; dielectric fluids; meteorology; operations research; thermoluminescence radiation dosimetry. *Mailing Add:* 3805 Council Crest Madison WI 53711-2922

HEDDEN, KENNETH FORSYTHE, b Glendale, Calif, Aug 13, 41; m 63, 90; c 3. MATHEMATICAL MODELING, ENVIRONMENTAL ASSESSMENT. *Educ:* Univ Calif, Berkeley, BS, 63; Univ Calif, Davis, PhD(microbiol), 68; Univ Ga, Athens, MPA, 80. *Prof Exp:* USPHS fel microbial biochem, Sch Med, Tufts Univ, 68-70; Res assoc, Purdue Univ, West Lafayette, 70-72; bacteriologist-lab supt, Anheuser-Busch, Inc, Ind, 72-75; sanit engr, US Army Environ Hyg Agency, 75-78; sanit engr, Environ Res Lab, Athens, Ga, US Environ Protection Agency, 78-83, chem engr, Environ Monitoring Systs Lab, 83-88; ENVIRON ENGR, ENVIRON COMPLIANCE DIV, WARNER ROBINS AIR LOGISTICS CTR, ROBINS AFB, 88- *Mem:* Sigma Xi; Soc Indust Microbiol; Conf Fed Environ Engrs. *Res:* Kinetics of growth of individual cells; phospholipid metabolism in bacteria and in vitro protein synthesis; multimedia mathematical modeling of fate and transport of toxic substances. *Mailing Add:* Warner Robins Air Logistics Ctr Environ Compliance Div Robins AFB GA 31098-5990

HEDDLE, JOHN A M, b Oakville, Ont, Nov 9, 38; m 62; c 2. CYTOGENETICS, RADIOBIOLOGY. *Educ:* Univ Toronto, BSc, 61; Univ Tenn, PhD(radiation biol), 64. *Prof Exp:* Res asst, Univ Tenn, 61-62; James Picker Found fel radiol res, Oak Ridge Nat Lab, 64-65 & Med Res Coun Radiobiol Res Unit, Eng, 65-66; asst res cytogeneticist, Lab Radiobiol, Univ Calif, San Francisco, 66-67, from asst prof to assoc prof radiol in residence, 67-71; assoc prof biol, 71-76, PROF BIOL, ATKINSON COL, YORK UNIV, 76- *Concurrent Pos:* At Toronto Br, Ludwig Inst Cancer Res, 81-84 & Cell Mutation Unit, Univ Sussex; NAm ed, Mutagenesis. *Mem:* Radiation Res Soc; Environ Mutagen Soc (vpres & pres-elect, 85-86, pres, 86-87); Genetics Soc Can; Can Soc Toxicol; Biol Coun Can; Fed Can Biol Soc. *Res:* Chromosome aberrations induced by radiations or chemicals; chromosome structure; environmental mutagenesis; human chromosome breakage syndromes; carcinogenesis. *Mailing Add:* Dept Biol York Univ 4700 Keele St Downsview ON M3J 1P6 Can

HEDDLESON, MILFORD RAYNORD, b Caldwell, Ohio, Dec 20, 21; m 45; c 2. SOIL CHEMISTRY, CONSERVATION. *Educ:* Ohio State Univ, BS, 51, MS, 54, PhD, 57. *Prof Exp:* From asst to asst prof agron, Ohio State Univ, 53-62; from asst prof to prof, 62-72, COORDR ENVIRON AFFAIRS, PA STATE UNIV, UNIVERSITY PARK, 72- *Mem:* Am Soc Agron; Soil Sci Soc Am; Soil Conserv Soc Am. *Res:* Land use; waste disposal; micro-nutrients. *Mailing Add:* Dept Agron Pa State Univ University Park PA 16802

HEDDLESTON, KENNETH LUTHER, b New Matamoras, Ohio, Dec 1, 16; m 45; c 2. MICROBIOLOGY. *Educ:* Univ Md, BS, 51. *Prof Exp:* Bacteriologist, 51-53, vet bacteriologist, Animal Dis & Parasite Res Div, 53-59, sr res microbiologist, Nat Animal Dis Lab, 59-61, PRIN RES MICROBIOLOGIST & PROJ LEADER, ANIMAL DIS & PARASITE RES DIV, NAT ANIMAL DIS LAB, USDA, 61- *Concurrent Pos:* Consult, Nat Acad Sci-Nat Res Coun. *Mem:* Fel AAAS; Am Soc Microbiol; Conf Res Workers Animal Dis; Am Col Vet Microbiol; Am Asn Avian Pathologists. *Res:* Pasteurella and their significance in animals. *Mailing Add:* Rte 1 Box 895 Lakeview AR 72642

HEDENBURG, JOHN FREDERICK, b Pittsburgh, Pa, Oct 23, 24; m 46; c 3. ORGANIC CHEMISTRY. *Educ:* Univ Pittsburgh, BS, 48, MS, 52, PhD(chem), 58. *Prof Exp:* Asst petrol, Mellon Inst, 48-50, fel insecticides, 50-58; from res chemist to sr res chemist, 58-69, SECT SUPVR, GULF RES & DEVELOP CO, 70- *Mem:* AAAS; Am Chem Soc; Soc Automotive Engrs. *Res:* Automotive fuels and lubricants; synthetic lubricants. *Mailing Add:* 189 Woodshire Dr Pittsburgh PA 15215

HEDGCOCK, FREDERICK THOMAS, b Toronto, Ont, May 18, 24; m 57; c 1. PHYSICS. *Educ:* Univ Man, BSc, 49; Univ Western Ont, MSc, 50, PhD, 54. *Prof Exp:* Lectr physics, Royal Mil Col Can, 50-52; Nat Res Coun Can res fel, 54-56; from asst prof to assoc prof physics, Univ Ottawa, 56-61; group leader, Franklin Inst Philadelphia, 61-64; PROF PHYSICS, McGILL UNIV, 64- *Res:* Low temperature and solid state physics; magnetic and electrical properties of metals, alloys and semiconductors. *Mailing Add:* Dept Physics McGill Univ 843 Sherbrooke St W Montreal PQ H3A 2T8 Can

HEDGCOTH, CHARLIE, JR, b Eliasville, Tex, Jan 29, 36; m 56; c 3. BIOCHEMISTRY. *Educ:* Univ Tex, BS, 61, PhD(chem), 65. *Prof Exp:* From asst prof to assoc prof, 65-76, PROF BIOCHEM, KANS STATE UNIV, 76- *Concurrent Pos:* Vis assoc prof biochem, Univ BC, 75; NATO sr fel, 75. *Mem:* Am Chem Soc; Am Soc Biol Chemists; AAAS. *Res:* Transfer RNA; differences in tRNA from normal and virus-transformed mammalian cells; aminoacyl-tRNA synthetases; characterization of genes for plant storage proteins; plant mitochondrial DNA. *Mailing Add:* Dept Biochem Kans State Univ Willard Hall Manhattan KS 66506-3702

HEDGE, GEORGE ALBERT, b St Louis, Mo, June 7, 39; m 63; c 2. PHYSIOLOGY. *Educ:* Univ Mo, BS, 61, MA, 63; Stanford Univ, PhD(physiol), 66. *Prof Exp:* Assoc prof physiol, Col Med, Univ Ariz, 68-77; prof & chmn physiol, 77-90, ASSOC DEAN RES, WVA UNIV MED CTR, MORGANTOWN, 90- *Concurrent Pos:* NIH res fel physiol, Fac Med, Univ Utrecht, 66-68; prog dir, NIH training grant; USPHS & NSF res grants. *Mem:*

Endocrine Soc; Am Physiol Soc; Am Thyroid Asn; Int Soc Neuroendocrinol. *Res:* Neural and endocrine regulation of thyroid secretion and blood flow. *Mailing Add:* Off Assoc Dean Res WVa Univ Med Ctr Morgantown WV 26506

HEDGECOCK, DENNIS, b Torrance, Calif, Apr 23, 49; m 88. POPULATION, EVOLUTIONARY GENETICS. *Educ:* St Mary's Col, Calif, BS, 70; Univ Calif, Davis, PhD(genetics), 74. *Prof Exp:* Res geneticist, Aquacult Proj, Bodega Marine Lab, 74-78, asst prof, 78-83, ASSOC PROF, DEPT ANIMAL SCI, UNIV CALIF, DAVIS, 83- *Concurrent Pos:* Fulbright-Hayes researcher, Univ Belgrade, Yugoslavia, 76 & 82. *Mem:* Am Genetics Soc; Genetic Soc Am; Soc Study Evolution; fel AAAS; Am Soc Naturalists; Nat Shellfisheries Asn; Sigma Xi. *Res:* Population, evolutionary and quantitative genetics of marine organisms, particularly those having aquacultural or fisheries importance. *Mailing Add:* Bodega Marine Lab Univ Calif PO Box 247 Bodega Bay CA 94923

HEDGECOCK, LE ROY DARIEN, b Roy, NMex, Mar 7, 13; m 41; c 3. AUDIOLOGY, SPEECH PATHOLOGY. *Educ:* Washington Univ, BS, 36; Colo State Col Educ, MA, 39; Univ Wis, PhD, 49. *Prof Exp:* Teacher speech & lip reading, Ind State Sch for Deaf, 36-38 & NJ Sch for Deaf, 38-43; supvr speech clin, Univ Wis, 43-44; instr & speech clinician, Ind Univ, 44-46; asst prof & sr speech clinician, Univ Minn, 46-49; consult audiologist, Mayo Clin, Mayo Grad Sch Med, Univ Minn, 49-69, assoc prof audiol, 69-78; RETIRED. *Mem:* Am Speech & Hearing Asn; Speech Commun Asn. *Res:* Selection and fitting of hearing aids; hearing therapy; clinical psychology; physics of sound. *Mailing Add:* 121 NE 14th St Rochester MN 55904

HEDGECOCK, NIGEL EDWARD, b Birmingham, Eng, June 2, 34; Can citizen; m 61; c 1. SOLID STATE PHYSICS. *Educ:* Univ BC, BA, 54, MA, 56; McMaster Univ, PhD(physics), 59. *Prof Exp:* Res asst, Max Planck Inst Chem, WGer, 59-61; asst prof, 61-65, ASSOC PROF PHYSICS, UNIV WINDSOR, 65- *Mem:* Can Asn Physicists. *Res:* Electron spin resonance of transition ion impurities in crystals. *Mailing Add:* Dept Physics Univ Windsor Windsor ON N9B 3P4 Can

HEDGEPETH, JOHN M(ILLS), b Southern Pines, NC, June 29, 26; m 48, 76, 84; c 3. AEROSPACE ENGINEERING, APPLIED MATHEMATICS. *Educ:* Purdue Univ, BS, 47 & 48; Va Polytech Inst, MS, 58; Harvard Univ, PhD(appl math), 62. *Prof Exp:* Res engr, Nat Adv Comt Aeronaut, 48-52, head dynamics & aeroelasticity div, 52-57, struct mech, 57-60; dept mgr struct & mat, Martin Marietta Corp, 61-63, dept dir eng & res, 63-67; from vpres to pres, 67-83, CONSULT, ASTRO RES CORP, 83- *Concurrent Pos:* Instr, Hampton Grad Exten, Univ Va, 50-60; adj prof, Martin Exten, Drexel Inst Technol, 61-67; mem, Space Systs & Technol Adv Comt, NASA, 78-87; pres, Digisim Corp, 83- *Mem:* Fel Am Inst Aeronaut & Astronaut; Int Aerospace Acad. *Res:* Structures and aeroelasticity; magnetohydrodynamics; large space systems; deployable space structures. *Mailing Add:* 202 E Pedregosa Santa Barbara CA 93101

HEDGES, DOROTHEA HUSEBY, b Chicago, Ill, Feb 27, 28; m 49; c 3. BIOCHEMISTRY, ENZYMOLOGY. *Educ:* Southern Methodist Univ, BS, 49; Tex A&M Univ, PhD(biochem), 75. *Prof Exp:* Chemist, Tracor, Inc, 56-60, res assoc, Univ NC, Chapel Hill, 74-75; chemist, Tex A&M Univ, 61-66, instr math, 66-67, instr biochem, 71-74, asst prof biochem, 75-80, LECTR & ASST VIS PROF CHEM, TEX A&M UNIV, 80- *Mem:* Am Chem Soc; AAAS; Sigma Xi; Asn Women in Sci. *Res:* The role of proteases in biological control; the chemistry of initiation and inhibition of mechanisms of endocytosis and exocytosis in cells. *Mailing Add:* 203 Timber College Station TX 77840

HEDGES, HARRY G, b Lansing, Mich, Oct 7, 23; m 44; c 2. ELECTRICAL ENGINEERING. *Educ:* Mich State Univ, BS, 49, PhD(elec eng), 60; Univ Mich, MS, 54. *Prof Exp:* Electronics engr, Wright Air Develop Ctr, 49-51; res assoc, Willow Run Res Ctr, Univ Mich, 51-54; from instr to assoc prof elec eng, Mich State Univ, 54-69, prof & chmn, Dept Comput Sci, 69-84; AT NSF, 84- *Mem:* Inst Elec & Electronics Engrs; Asn Comput Mach. *Res:* General system theory, including analysis, design and optimization; power system economic studies. *Mailing Add:* NSF 1800 G St NW Rm 436 Washington DC 20550

HEDGES, JOHN IVAN, b Radnor, Ohio, Feb 10, 46; m 73; c 2. GEOCHEMISTRY. *Educ:* Capital Univ, BS, 68; Univ Tex, PhD(chem), 75. *Prof Exp:* Fel, Carnegie Geophys Lab, 75-77; res asst prof, 77-80, ASST PROF OCEANOG, UNIV WASH, 80- *Mem:* Geochem Soc; AAAS. *Res:* Origin, pathways and fates of organic substances in marine and fresh water environments with emphasis of the geochemistries of lignins, carbohydrates, and humic substances. *Mailing Add:* Sch Oceanog Univ Wash Seattle WA 98195

HEDGES, RICHARD MARION, b Dallas, Tex, May 27, 27; m 49; c 3. PHYSICAL CHEMISTRY. *Educ:* Southern Methodist Univ, BS, 50; Iowa State Univ, PhD(phys chem), 55. *Prof Exp:* Asst chem, Iowa State Univ, 50-55; from assoc to instr, Univ Tex, 55-60; from asst prof to assoc prof, 60-67, PROF CHEM, TEX A&M UNIV, 67- *Mem:* Am Phys Soc; Am Chem Soc. *Res:* Molecular spectroscopy and molecular quantum mechanics. *Mailing Add:* Dept Chem Tex A&M Univ College Station TX 77843

HEDGES, THOMAS REED, JR, b Sandusky, Ohio, Oct 19, 23; m 46; c 3. OPHTHALMOLOGY. *Educ:* Ohio State Univ, AB, 44; Cornell Univ, MD, 47; Univ Pa, MMedSci, 51. *Prof Exp:* Chief ophthal, Beaumont Army Hosp, El Paso, Tex, 52-54; from asst prof to assoc prof, 57-72, PROF OPHTHAL, SCH MED, UNIV PA, 72-, ASST CHIEF OPHTHAL, 67- *Concurrent Pos:* Fel neurol, Cleveland Clin, 48-49; fel neuro-ophthal, Wilmer Inst, Johns Hopkins Univ, 49-50; chief of ophthal, Burlington County Hosp, Mt Holly, 58- & Pa Hosp, 72-; sr consult ophthal, Philadelphia Gen Hosp, 67- *Mem:* Am Acad Ophthal & Otolaryngol; fel Am Col Surg; Am Ophthal Asn. *Res:* Neuro-ophthalmology; headache. *Mailing Add:* Pa Hosp 700 Spruce St Suite 508 Philadelphia PA 19106

HEDGLIN, WALTER L, b Oakland, Calif, Dec 12, 42; m 64, 85; c 4. PHARMACEUTICAL DEVELOPMENT. *Educ:* Calif State Univ, Chico, BS, 67; Univ Calif, Davis, PhD(physiol), 72. *Prof Exp:* Staff scientist, Procter & Gamble Co, 72-78, head, Sect Pharmacol & Toxicol, 78-81, assoc dir radiodiagnostics, dermat, spec prod & pharmacol, 81-85; DIR, EUROP RES & DEVELOP, NORWICH EATON PHARMACEUT INC, 85- *Mem:* Am Physiol Soc; Am Soc Bone & Mineral Res; Am Pharmaceut Asn. *Mailing Add:* Europ Res & Develop Norwich Eaton Rusham Park Whitehall Lane Egham Surrey GU21 5AP England

HEDIN, ALAN EDGAR, b St Paul, Minn, Dec 24, 35; m 69; c 1. ATMOSPHERIC PHYSICS, AERONOMY. *Educ:* Univ Minn, BS, 58, MS, 61, PhD(physics), 66. *Prof Exp:* Mem tech staff, Bellcommun, Inc, 66-67; SPACE SCIENTIST, GODDARD SPACE FLIGHT CTR, NASA, 67- *Mem:* Am Geophys Union; Sigma Xi; Am Inst Aeronaut & Astronaut. *Res:* Structure and variations of the upper atmosphere. *Mailing Add:* Goddard Space Flight Ctr Code 914 Greenbelt MD 20771

HEDIN, DAVID ROBERT, b Chicago, Ill, Nov 11, 54; m 82; c 1. ELEMENTARY PARTICLE PHYSICS. *Educ:* Southern Ill Univ, BS, 75; Univ Wis, PhD(physics), 80. *Prof Exp:* Res assoc, physics dept, State Univ NY, Stony Brook, 81-86; asst prof, 87-90, ASSOC PROF, PHYSICS DEPT, NORTHERN ILL UNIV, 90- *Mem:* Am Phys Soc. *Res:* Experimental high energy physics; construction of a detector which studies 2 TeV protein-antiproton collisions. *Mailing Add:* Physics Dept Northern Ill Univ Faraday Hall Dekalb IL 60115

HEDIN, PAUL A, b Maple Plain, Minn, July 9, 26; m 51; c 6. NATURAL PRODUCTS CHEMISTRY. *Educ:* Univ Minn, BA, 48, BS, 50, MS, 53, PhD(biol chem), 58. *Prof Exp:* Chemist, Qm Food & Container Inst, 58-60, head, Biochem Lab, 60-62; head chemist, Boll Weevil Res Lab, 62-75, HEAD SCIENTISTS, CROP SCI RES LAB, USDA, 75- *Concurrent Pos:* Adj prof, Miss State Univ, 63-; fel, Div Pest Chem, Am Chem Soc, 78. *Honors & Awards:* Outstanding Chemist, Am Chem Soc, 75; Super Serv Honor Award, USDA, 75. *Mem:* Am Chem Soc; Am Soc Agron; Entom Soc Am. *Res:* Constituents in plants which stimulate insects; sex attractants; food flavors amino acid analysis; vitamins; insect hormones; chemistry of host plant resistance to pests. *Mailing Add:* Crop Sci Res Lab USDA PO Box 5367 Mississippi State MS 39762-5367

HEDLEY, WILLIAM H(ENBY), b St Louis, Mo, June 11, 30; m 57; c 4. PROCESS DEVELOPMENT, TEACHING. *Educ:* Wash Univ, BSc, 53, DSc(chem eng), 57. *Prof Exp:* Develop engr, Uranium Div, Mallinckrodt Chem Works, 56-61; res chem engr, Monsanto Res Corp, 61-63, sr res chem engr, 63-65, res group leader phys chem & eng, 65-67, res mgr, 67-69, dir res, Monsanto Environ-Chem Systs, 69-72, mgr, 72-83, sr res eng specialist, 83-88; SR ENG SPECIALIST, EG&G MOUND APPL TECHNOLOGIES, 88- *Concurrent Pos:* Adj prof, Dept Chem Eng, Univ Dayton, 85- *Mem:* Am Inst Chem Engrs; Sigma Xi. *Res:* Process developoment (tritium control, uranium processing, pollution abatement studies); prediction and measurement of physical properties of compounds; applied physical chemistry; materials development; gas transport (flow and diffusion through fine capillaries) and materials characterization (selection of simulants for toxic materials). *Mailing Add:* EG&G Mound Appl Technologies PO Box 3000 Miamisburg OH 45343-3000

HEDLEY-WHYTE, ELIZABETH TESSA, b London, Eng, Jan 17, 37; m 59. NEUROPATHOLOGY, CELL BIOLOGY. *Educ:* Univ Durham, MB & BS, 60; Am Bd Path, dipl, 66, neuropath dipl, 73; Univ Newcastle Tyne, MD, 76. *Prof Exp:* Neuropathologist, Children's Hosp Med Ctr, 73-77; from asst prof neuropath to assoc prof, 70-78, ASSOC PROF PATH, HARVARD MED SCH, 78-; NEUROPATHOLOGIST, MASS GEN HOSP, 84- *Concurrent Pos:* Consult, Neurol Dis Prog-Proj Rev Comt, NIH & USPH, 76-81; mem bd adv, Harvard Med Sch, 76-88, fac coun, 88-; consult, Children's Hosp Med Ctr, 77-; pathologist, New Eng Deaconess Hosp, 77-81; assoc neuropathologist, Mass Gen Hosp, 81-84, pathol training prog dir, 85-, dir neuropath, 89- *Mem:* Am Soc Cell Biol; Soc Neurosci; Am Asn Pathologists; Soc Pediat Res; Am Asn Neuropathologists (vpres, 90-91). *Res:* Developmental neurobiology; electron microscopy; brain tumors; Alzheimers disease. *Mailing Add:* Dept Path Mass Gen Hosp 25 Fruit St Boston MA 02114

HEDLEY-WHYTE, JOHN, b Newcastle-on-Tyne, Eng, Nov 25, 33; US citizen; m 59. ANESTHESIOLOGY, PHYSIOLOGY. *Educ:* Cambridge Univ, BA, 55, MB, 58, MA, 59, MD, 72; Harvard Univ, AM, 67. *Prof Exp:* House physician, St Bartholomew's Hosp, 58-59, house surgeon & resident anesthetist, 59-60, chief resident, 60; resident anesthesia, Mass Gen Hosp, 60-61, chief resident, 61-62; from asst to prof, 61-76, DAVID S SHERIDAN PROF ANESTHESIA & RESPIRATORY THER, HARVARD UNIV, 76-, PROF, DEPT HEALTH POLICY & MGT, HARVARD SCH PUB HEALTH, 88- *Concurrent Pos:* Sec to Fac Med, Harvard Univ; chmn comt ventilators, Int Stands Inst & Am Nat Standards Inst, 67-; anaesthetist-in-chief, Beth Israel Hosp, Boston, Mass, 67-87. *Honors & Awards:* Litchfield Lectr, Oxford Univ, 71; Bishop Lectr, NY Anesthesia; Egan Lectr, Am Soc Respiratory Ther, 78. *Mem:* Am Physiol Soc; Asn Univ Anesthetists; Am Soc Anesthesiol; Int Anesthesia Res Soc; Am Col Physicians; Am Soc Testing & Mat. *Res:* Interaction between gases and proteins; chemistry of lung; effect of anesthetics on macromolecules; management of respiratory failure; high resolution electron microscopic radioautography; cold preservation of tissues for scanning electron microscopy. *Mailing Add:* Dept Vet Affairs Med Ctr 1400 VFW Pkwy Boston MA 02132-4927

HEDLIN, CHARLES P(ETER), b Renown, Sask, Can, Apr 18, 27; m 51; c 6. MECHANICAL ENGINEERING. *Educ:* Univ Sask, BSc, 50; Univ Minn, MSc, 52; Univ Toronto, PhD(mech eng), 57. *Prof Exp:* Assoc prof eng, Ont Agr Col, 52-60; res officer, Inst Res Construct, Prairie Regional Sta, Nat Res Coun, 60-75, officer-in-chg, 75-88; AT HEDLIN CONSULT, INC, 88- *Mem:* Fel Am Soc Heating, Refrig & Air Conditioning Engrs. *Res:* Properties of moist materials; humidity measurement; field performance of thermal insulations; energy consumption in buildings; heating and air conditioning; heat and mass transfer. *Mailing Add:* Hedlin Consult Inc 26 Weir Crescent Saskatoon SK S7H 3A9 Can

HEDLIN, ROBERT ARTHUR, b Saskatoon, Sask, Jan 5, 21; c 3. SOILS. *Educ:* Univ Sask, BSA, 45; Univ Alta, MS, 47; Univ Wis, PhD(soils), 50. *Prof Exp:* From asst prof to prof soils, Univ Man, 50-86; RETIRED. *Mem:* Soil Sci Soc Am; Am Soc Agron; Agr Inst Can. *Res:* Soil fertility. *Mailing Add:* 910 Riverwood Ave Winnipeg MB R3T 1L1 Can

HEDLUND, DONALD A, b US, Oct 10, 24; m 48; c 3. IONOSPHERIC PHYSICS, RADIO ENGINEERING. *Educ:* Yale Univ, BE, 45, ME, 47. *Prof Exp:* Engr, 47-53, sr scientist, 53-56, mgr high frequency sect, 56-61 & propagation syst & res dept, 61-69, MGR EQUIP DEPT, RAYTHEON CO, 69- *Concurrent Pos:* Mem Comn III, Int Sci Radio Union, 56-, US del, XII gen assembly, 57; mem study group 6, Int Radio Consult Comt, 57-58. *Mem:* Inst Elec & Electronics Engrs. *Res:* Characteristics of the ionosphere as a medium for propagating high frequency radio communication signals; radio sounding techniques, analytical studies to interpret these soundings. *Mailing Add:* Raytheon Co Equip Div 528 Boston Post Rd Sudbury MA 01776

HEDLUND, GUSTAV ARNOLD, b Somerville, Mass, May 7, 04; m 31; c 2. MATHEMATICS. *Educ:* Harvard Univ, AB, 25, PhD(math), 30; Columbia Univ, AM, 27. *Hon Degrees:* MA, Yale Univ, 48. *Prof Exp:* Instr math, Hunter Col, 25-27; from asst prof to assoc prof, Bryn Mawr Col, 30-39; prof, Univ Va, 39-48; prof, 48-72, EMER PROF MATH, YALE UNIV, 72- *Concurrent Pos:* Nat Res Coun fel, Princeton Univ, 33-34, mem staff, Inst Defense Anal, 59-60 & 62-63; mem, Inst Advan Study, 38-39 & 53-54; res mathematician, Off Sci Res & Develop, Columbia Univ, 44-45; vis prof, Wesleyan Univ, 72-80. *Mem:* AAAS; Am Math Soc (vpres, 51); Math Asn Am. *Res:* Differential geometry; topological dynamics. *Mailing Add:* 309 McKinley Ave New Haven CT 06515

HEDLUND, JAMES HOWARD, b Ithaca, NY, July 24, 41; m 64; c 2. MATHEMATICS, STATISTICS. *Educ:* Cornell Univ, BA, 63; Univ Mich, Ann Arbor, MA, 65, PhD(math), 68. *Prof Exp:* Asst prof math, Univ Mass, Amherst, 68-74 & Smith Col, 74-76; math statistician, 76-87, chief, Math Anal Div, 82-87, dir, driver & pedestrian res, 87-90, DIR, ALCOHOL & STATE PROGS, NAT HWY TRAFFIC SAFETY ADMIN, US DEPT TRANSP, 90- *Mem:* Am Math Soc; Am Statist Asn; Math Asn Am; Soc Automobile Engrs. *Res:* Traffic safety; applied statistics. *Mailing Add:* Nat Hwy Traffic Safety Admin NTS-20 2100 Second St SW Washington DC 20590

HEDLUND, JAMES L, b Los Angeles, Calif, Aug 1, 28; m 50; c 2. COMPUTER APPLICATIONS IN MENTAL HEALTH. *Educ:* State Univ Iowa, BA, 50, MA, 51, PhD(clin psychol), 53. *Prof Exp:* Instr psychol, US Army Med Field Serv Sch, 54-55; chief clin psychol, US Army Hosp, Frankfurt, Ger, 56-59; dir clin psychol, Intern Training, Walter Reed Army Med Ctr, 59-63; psychol consult, Off US Army Surgeon Gen, 63-66; chief biomed stress res div, US Army Med Res & Develop Command, 66-69; dir, comput support in mil psychiat, Walter Reed Army Med Ctr, 69-71; prof psychiat, 71-90, dir, Mo Inst Psychiat, 80-90, EMER PROF PSYCHIAT, UNIV MO, COLUMBIA, 91- *Concurrent Pos:* Mem exec coun, Nat Acad Sci-Nat Res Coun Comt on Hearing, Bioacoustics, Biomechanics, 66-69; mem exec coun, Armed Forces Comt on Vision, 66-69; surgeon gen rep, Behav Sci Study Sect, Nat Inst Mental Health, 66-69, & Army Human Factors Adv Comt, 66-69. *Mem:* Am Psychol Asn. *Res:* Research and development of computer applications in mental health, including sophisticated clinical data collection instruments, clinical prediction models, assessment instruments involving expert systems technology, and empirical research associated with large state-wide patient data base. *Mailing Add:* Mo Inst Psychiat 5400 Arsenal St St Louis MO 63139

HEDLUND, LAURENCE WILLIAM, b Detroit, Michigan, Mar 16, 37; m 65; c 4. IMAGING, MAGNET RESONANCE IMAGING. *Educ:* Wayne State Univ, BS, 60; Univ Pittsburgh, PhD(animal physiol), 68. *Prof Exp:* Asst prof, Univ Missouri, 70-78; asst prof, 78-89, ASSOC PROF, DUKE UNIV MED CTR, 90- *Mem:* Am Thoracic Soc; Am Asn Lab Animal Sci; Radiol Soc NAm; Soc Magnetic Resonance Imaging; Soc Magnetic Resonance Med; Asn Univ Radiologists. *Res:* Development of ventilation anesthesia and physiological motioning methods for magnetic resonance microscopy of small animals; evaluation analysis of toxic injury and tumorigenesis in lungs, liver, and kidney with magnetic resonance imaging. *Mailing Add:* Dept Radiol Duke Univ Med Ctr Box 3808 Durham NC 27710

HEDLUND, RICHARD WARREN, b Lowell, Mass, Dec 11, 35; m 60; c 3. GEOLOGY, PALYNOLOGY. *Educ:* Univ Mass, BSc, 57; Univ Okla, MSc, 60, PhD(geol), 63. *Prof Exp:* Sr res scientist, Geol Res Div, Pan Am Petrol Corp, 62-68; proj supvr, Prod Res Ctr, Geol Sci Group, Atlantic-Richfield Co, 68-71; sr res scientist, 71-74, staff res scientist, & res group supvr, 74-82, SPEC RES ASSOC, GEOL RES DIV, AMOCO PROD CO, 82- *Honors & Awards:* Distinguished Serv Award, Am Asn Stratig Palynologists, 83. *Mem:* Soc Econ Paleontologists & Mineralogists; Tulsa Geol Soc; Am Asn Stratig Palynologists (pres, 74-75). *Res:* Mesozoic and Tertiary palynology. *Mailing Add:* Amoco Prod Co PO Box 3385 Tulsa OK 74102

HEDLUND, RONALD DAVID, b Joliet, Ill, June 16, 41; m 64; c 2. RESEARCH DEVELOPMENT, MANAGEMENT OF CAMPUS-WIDE RESEARCH. *Educ:* Augustana Col, Ill, BA, 63, Univ Iowa, MA, 64, PhD(polit sci), 67. *Prof Exp:* From asst prof to prof polit sci, Univ Wis-Milwaukee, 67-89, assoc dean, Grad Sch, 80-89; VPROVOST RES, UNIV RI, 89- *Concurrent Pos:* Prin investr, Nat Conf State Legis Leaders & NSF, 66-67, NSF, 77-78 & 84-89 & Ford Found, 85-87; managing partner, Wis Pub Opinion & Mkt Res, 76-89; co-chair resource, Network RI Partnership Sci & Technol, 90- *Mem:* Nat Coun Univ Res Adminr; Soc Res Adminr; Am Polit Sci Asn; Southern Polit Sci Asn. *Res:* Effects of structural and organizational changes in state legislatures sought by studying system-level performance variables, policy output data, and individual-level perceptions. *Mailing Add:* Res Off Univ RI Kingston RI 02881

HEDMAN, DALE E, b Albert City, Iowa, Aug 22, 35; m 62; c 3. INSULATION COORDINATION. *Educ:* Univ Nebr, BS, 58, MS, 60. *Prof Exp:* Engr, Gen Elec Co, 60-69; prin engr, 69-76, treas, 76-83, VPRES, POWER TECHNOL, INC, 83- *Concurrent Pos:* Bd dir, Power Technol, Inc, 69, Pond Hill Homes, 82-, Empire Info Serv, 87-; secy, 37 Tech Comt. *Mem:* Fel Inst Elec & Electronics Engrs; Int Elec Comn; Int Conf on Large High Voltage Elec Systs. *Res:* Specialized in the general area of electric power system insulation coordination having worked on major projects including the ITAIPU hydro project in Brazil; propagation on three phase lines. *Mailing Add:* PO Box 1058 Schenectady NY 12301

HEDMAN, FRITZ ALGOT, b Arnas, Sweden, Sept 10, 06; nat US; m 36; c 1. PHYSICAL CHEMISTRY. *Educ:* State Col Wash, BS, 31, BA, 33, PhD(chem), 34. *Prof Exp:* Chemist, Exp Sta, State Col Wash, 34-36; jr seafood inspector, Food & Drug Admin, USDA, 36-37; refrig engr, Servel, Inc, Ind, 37-40; res chemist, Am Viscose Corp, 40-49; chemist & phys scientist, Chem Corps, Dept Army, 49-62; design specialist, Fairchild Stratos Corp, 62-63; consult analyst, Gen Elec Co, 63-64; nuclear physicist, Defense Commun Agency, Nat Mil Command Systs, 64-67; opers res analyst, USPHS Hosp, 67-72; CONSULT, AIR POLLUTION CONTROL, INC, BALTIMORE, 72- *Mem:* AAAS; NY Acad Sci; Am Chem Soc. *Res:* Interfacial tension of mercury and hydrocarbon interfaces; military aerosols; radiological defense. *Mailing Add:* 218 Fulford Ave Bel Air MD 21014

HEDMAN, STEPHEN CLIFFORD, b Duluth, Minn, Apr 13, 41; m 66; c 3. GENETICS, MOLECULAR BIOLOGY. *Educ:* Univ Minn, BA, 63; Stanford Univ, PhD(genetics), 68. *Prof Exp:* NIH fel, Univ Minn, St Paul, 68; asst prof biol, 68-72, assoc prof biol & biochem, 72-90, PROF BIOL & BIOCHEM/MOLECULAR BIOL, UNIV MINN, DULUTH, 90-, ASSOC GRAD SCH DEAN, 90- *Concurrent Pos:* NIH res grant, 69-72; NSF grant, 85-88. *Mem:* Genetics Soc Am; Sigma Xi. *Res:* Mitochondrial biochemistry; ctyoplasmic inheritance in Neurospora crassa; chromosome staining and identification; fish cytogenetics. *Mailing Add:* Dept Biol/Biochem Univ Minn 2400 Oakland Ave Duluth MN 55812

HEDRICH, LOREN WESLEY, b Nampa, Idaho, Dec 10, 29; m 56; c 3. MEDICINAL & ORGANIC CHEMISTRY. *Educ:* Idaho State Univ, BS, 52, MS, 64; Univ Kans, PhD(med chem), 69. *Prof Exp:* Res chemist, Gulf Res & Develop Co, 68-72; SR RES CHEMIST, GULF OIL CHEM CO, CHEVRON CORP, 72- *Mem:* Am Chem Soc; AAAS; Sigma Xi. *Res:* Design and synthesis of pharmaceutically and agriculturally active chemicals. *Mailing Add:* Chevron Chem Co 1862 Kingwood Dr Kingwood TX 77339

HEDRICK, CLYDE LEWIS, JR, b Los Angeles, Calif, Oct 14, 39; m 63; c 2. PLASMA PHYSICS. *Educ:* Univ Calif, Los Angeles, BA, 62, MS, 65, PhD(plasma physics), 70. *Prof Exp:* Physicist, US Naval Ord Test Sta, Pasadena, 63-65; MEM STAFF PLASMA PHYSICS, OAK RIDGE NAT LAB, 70- *Mem:* Am Phys Soc. *Res:* Magnetic equilibria; macroscopic stability; particle orbits; transport coefficients; fluid and kinetic models of transport and energy balance. *Mailing Add:* 144 Montana Ave Oak Ridge TN 37831

HEDRICK, GEORGE ELLWOOD, III, b Columbus, Ohio, Apr 22, 43; m 76; c 4. COMPUTER SCIENCE. *Educ:* Adams State Col, BA, 64; Iowa State Univ, MS, 68, PhD(comput sci), 70. *Prof Exp:* Programmer, Kaman Nuclear, 64-65; mathematician, US Army Air Defense Command, 65-66; sr systs analyst, Ames Lab, 70; asst prof, 70-74, assoc prof, 74-79, PROF COMPUT SCI, OKLA STATE UNIV, 80- *Concurrent Pos:* Prof math, Univ Nottingham, Gt Brit, 79. *Mem:* Asn Comput Mach; Math Asn Am. *Res:* Programming languages; operating systems; data structures. *Mailing Add:* Dept Comput & Info Sci Okla State Univ Stillwater OK 74078

HEDRICK, HAROLD BURDETTE, b Sinks Grove, WVa, May 11, 24; m 48; c 2. ANIMAL HUSBANDRY. *Educ:* WVa Univ, BS, 51; Univ Mo, MS, 55, PhD(animal husb), 57. *Prof Exp:* From instr to assoc prof, 54-69, PROF, FOOD SCI & NUTRIT DEPT, UNIV MO-COLUMBIA, 69- *Honors & Awards:* Meats Res Award, Am Soc Animal Sci, 80; R C Pollock Award, Am Meat Sci Asn, 90. *Mem:* Am Soc Animal Sci; Inst Food Technologists; Am Meat Sci Asn. *Res:* Meat technology; food science. *Mailing Add:* Dept Food Sci & Nutrit Univ Mo 21 Agr Bldg Columbia MO 65211

HEDRICK, HAROLD GILMAN, b Youngstown, Ohio, Nov 15, 24; m; c 3. ENVIRONMENTAL MICROBIOLOGY. *Educ:* Centre Col, AB, 50; Marshall Col, MA, 52, WVa Univ, PhD(agr microbiol), 57. *Prof Exp:* Instr sci, St Mary's Sch Nursing, 52-53; asst, WVa Univ, 53-55, instr agr bact, 55-57; plant pathologist, USDA, 57-58; dir microbiol res, Woodard Res Corp, 58-61; proj res scientist, Appl Sci Labs, Gen Dynamics-Ft Worth, 61-69; prof, La Tech Univ, 69-87; RETIRED. *Concurrent Pos:* Fulbright award. *Mem:* AAAS; Am Soc Microbiol; Soc Indust Microbiol; Sigma Xi. *Res:* Environmental microbiology. *Mailing Add:* 809 Sandy Lane Ruston LA 71270

HEDRICK, IRA GRANT, b Kansas City, Mo, Feb 10, 13; m 34; c 2. AEROSPACE TECHNOLOGY. *Educ:* Univ Ark, Fayetteville, BS, 36; Princeton Univ, CE, 37. *Prof Exp:* Designer, Waddell & Hardesty, Consult Engrs, 37-42; chief struct engr, Eritrean Proj, Johnson, Drake & Piper, 42-43; struct engr, US Army Eng Dept, 43; proj stress analyst, Grumman Aircraft Eng Corp, 43-46, chief of struct, 46-57, chief tech engr, 57-63, vpres eng, 63-70, sr vpres & dir tech opers, 70-73, sr vpres & dir advan systs technol, 73-75, sr vpres & presidential asst corp technol, 75-80, SR MGT CONSULT, GRUMMAN CORP, 80- *Concurrent Pos:* Mem vis comt, dept physics, Lehigh Univ, 66-74; dept aeronaut & astronaut, Mass Inst Technol, 67-73; mem res & technol adv coun & chmn res & technol adv comt mat & struct, NASA, 71-77; mem, Aeronaut & Space Eng Bd, Nat Res Coun, 77-81; mem, USAF Sci Adv Bd, 76-84, NASA Aerospace Safety Adv Panel, 79-84. *Honors & Awards:* Spirit of St Louis Award, Am Soc Mech Engrs, 67; Sylvanus Albert Reed Award, Am Inst Aeronaut & Astronaut, 71. *Mem:* Nat Acad Eng; hon fel Am Inst Aeronaut & Astronaut. *Res:* Aerospace structures and materials. *Mailing Add:* 250 Mount Joy Ave Freeport NY 11520

HEDRICK, JACK LEGRANDE, b Glen Rock, Pa, July 23, 37; m 60; c 2. ANALYTICAL CHEMISTRY. *Educ:* Elizabethtown Col, BS, 59; Univ Pittsburgh, MS, 62. *Prof Exp:* Instr chem, Albright Col, 62-63; from instr to assoc prof, 63-73, PROF CHEM, ELIZABETHTOWN COL, 73- *Mem:* AAAS; Sigma Xi; Am Chem Soc. *Res:* Chromatography; organic synthesis. *Mailing Add:* Elizabethtown Col One Alpha Dr Elizabethtown PA 17022-2298

HEDRICK, JERRY LEO, b Knoxville, Iowa, Mar 11, 36; m 57; c 4. BIOCHEMISTRY. *Educ:* Iowa State Univ, BS, 58; Univ Wis, PhD(physiol chem), 61. *Prof Exp:* Res assoc physiol chem, Univ Wis, 61-62; res assoc biochem, Univ Wash, 62-65; from asst prof to assoc prof, 65-74, PROF BIOCHEM, UNIV CALIF, DAVIS, 74- *Concurrent Pos:* NIH fel, 62-64; Guggenheim Found fel, 71-72; sabbatical Hokkaido Univ, Sapporo, Japan, 85-86 & 89; vis scientist, Mitsubishi-Kasei Inst Life Sci, Japan, 89. *Mem:* AAAS; Am Chem Soc; Am Soc Biol Chemists; Soc Study Reproduction; Sigma Xi; Am Soc Cell Biol. *Res:* Structure-function relations of proteins and glycoproteins; biochemistry of fertilization and development. *Mailing Add:* Dept Biochem & Biophys Univ Calif Davis CA 95616

HEDRICK, PHILIP WILLIAM, b Nov 21, 42; US citizen; m 65; c 2. GENETICS. *Educ:* Hanover Col, BA, 64; Univ Minn, MS, 66, PhD(genetics), 69. *Prof Exp:* Fel genetics, Univ Chicago, 68-69; asst prof systs & ecol, Univ Kans, 69-73, assoc prof genetics, 73-76, assoc prof syst & ecol, 76-87; PROF BIOL, PA STATE UNIV, 87- *Mem:* Genetics Soc Am; Am Soc Nat; Soc Study Evolution. *Res:* Population and quantitative genetics; population ecology and competition in drosophila. *Mailing Add:* Dept Biol 514 Mueller Lab Pa State Univ University Park PA 16802

HEDRICK, RONALD PAUL, b Tacoma, Wash, Aug 2, 50; m 73; c 2. VIROLOGY. *Educ:* Ore State Univ, BS, 75, PhD(microbiol), 80; Univ Ore, BS, 76. *Prof Exp:* ASSOC PROF INFECTIOUS DIS FISH, DEPT MED VET MED, UNIV CALIF, DAVIS, 82- *Honors & Awards:* Beechum Res Excellence Award, Pres Fish Health Sect, AFS, 88. *Mem:* Sigma Xi; Am Soc Microbiol; Soc Gen Microbiol; Europ Asn Fish Pathologists; Am Fish Soc; Fish Health Soc. *Res:* Study of infectious diseases of fish and shellfish and their detection, prevention and treatment. *Mailing Add:* Dept Med Vet Med Univ Calif Davis CA 95616

HEDRICK, ROSS MELVIN, b West Salem, Ill, Apr 27, 21; m 49; c 3. ORGANIC CHEMISTRY. *Educ:* Univ Ill, BS, 43; Ind Univ, AM, 44, PhD(org chem), 47. *Prof Exp:* Res chemist, Monsanto Co, 47-52, res group leader, 52-63, proj mgr & scientist, 63-69, sr sci fel, 69-72, distinguished sci fel, 72-86; RETIRED. *Mem:* AAAS; Am Chem Soc. *Res:* Organic fluorine compounds; isoflavones; herbicides; synthetic soil conditioners; polyelectrolytes; condensation polymers; ionic polymerization; reinforced plastics; block copolymers; metal/plastic laminates; nylon block copolymer reaction injection molding. *Mailing Add:* 300 Chasselle Lane St Louis MO 63141-7335

HEDRICK, THEODORE ISAAC, b Polson, Mont, Sept 11, 12; m 42; c 2. FOOD SCIENCE. *Educ:* Mont State Univ, BS, 35, MS, 37; Iowa State Univ, PhD(dairy bact), 41. *Prof Exp:* From asst to instr dairy indust, Mont State Univ, 35-39; asst prof, Iowa State Univ, 42-44; dir labs, North Star Dairy, 46-49; nat qual coordr & chemist chg dairy & poultry labs, Dairy Div, Agr Mkt Serv, USDA, 49-56; assoc prof dairy, Mich State Univ, 56-60, plant mgr, 56-67, prof, 60-78, emer prof food sci, 79-88; RETIRED. *Concurrent Pos:* Mem, bd dirs, Am Dairy Sci Asn, 73-76; food technol consult, Mich State Univ Int Progs in Indonesia & Brazil, 75-78; consult in packaging, UN Food & Agr Orgn, 79. *Honors & Awards:* USDA Merit Award, 55; Dairy Res Award, Am Dairy Sci Asn, 72. *Mem:* AAAS; Am Dairy Sci Asn; Inst Food Technologists; Int Asn Milk, Food & Environ Sanit. *Res:* Drying of milk products, ultra high temperature sterilization and aseptic packaging; new dairy foods and processes; airborne microbiology. *Mailing Add:* 1546 Sherman St SE Grand Rapids MI 49506

HEDSTROM, GERALD WALTER, b Kenosha, Wis, July 31, 33; m 57; c 1. MATHEMATICS. *Educ:* Univ Wis, BS, 55, MS, 56, PhD(math), 59. *Prof Exp:* From instr to asst prof math, Univ Mich, Ann Arbor, 59-68; from assoc prof to prof math, Case Western Reserve Univ, 68-77; MATHEMATICIAN, LAWRENCE LIVERMORE LAB, 77- *Mem:* Am Math Soc; Soc Indust & Appl Math. *Res:* Numerican analysis. *Mailing Add:* Lawrence Livermore Labs PO Box 808 Livermore CA 94550

HEDSTROM, JOHN RICHARD, b Kansas City, Kans, Oct 14, 37; m 67; c 1. MATHEMATICS, INTELLIGENT SYSTEMS. *Educ:* Univ Kans, BA, 59, MA, 66, PhD(algebra), 70; Southern Methodist Univ, MS, 90. *Prof Exp:* assoc prof math, Univ NC, Charlotte, 70-83, assoc prof comput sci, 81-83; SOFTWARE SYSTS ENGR, TEX INST, 83- *Concurrent Pos:* Vis prof, Comput Sci Dept, Univ NC Chapel Hill, 80-81. *Mem:* Inst Elec & Electronics Engrs; Math Asn Am. *Res:* Pseudo-valuation domains; signal processing Ada Compiler; Ada-OODB Interface. *Mailing Add:* 622 Perdido Garland TX 75043

HEDTKE, JAMES LEE, b Clintonville, Wis, Nov 22, 43; m 65; c 1. AQUATIC TOXICOLOGY, FISH PHYSIOLOGY. *Educ:* Univ Wis, Oshkosh, BS, 67; Univ Minn, PhD(pharmacol), 73. *Prof Exp:* Res assoc, 73-74 & 76-77, NIH res assoc, 74-76, ASST PROF FISHERIES, ORE STATE UNIV, 78-; ASSOC PROF PHARMACOL, ORE HEALTH SCI UNIV, 80- *Concurrent Pos:* Consult, Ore State Bd Dent Examr, 78- & Wash State Bd Dent Examr, 81-; asst prof pharmacol, Ore Health Sci Univ, 77-80. *Mem:* Sigma Xi. *Res:* Aquatic toxicology; comparative physiology and pharmacology with emphasis on the cardiovascular system of fish; interaction of aquatic toxicants. *Mailing Add:* PO Box 19021 Portland OR 97219

HEE, CHRISTOPHER EDWARD, b Olean, NY, Oct 2, 39; m 63; c 3. MATHEMATICS, CHEMICAL ENGINEERING. *Educ:* Univ Detroit, BChE, 61; Univ Notre Dame, PhD(math), 71. *Prof Exp:* Eng aide, Air Preheater Corp, 58-61; teacher, Penn Yan Cent Sch, NY, 63-64; lectr, 69-70, asst prof, 70-81, ASSOC PROF MATH, EASTERN MICH UNIV, 81- *Mem:* Math Asn Am. *Res:* Algebraic topology; Steenrod algebra; characteristic classes; relations over Steenrod algebra between characteristic classes; mathematical modeling; catastrophe theory. *Mailing Add:* Dept Math Eastern Mich Univ Ypsilanti MI 48197

HEEBNER, CHARLES FREDERICK, b Norwich, Conn, Apr 30, 38; m 59; c 1. PLANT PHYSIOLOGY, FORESTRY. *Educ:* Univ Conn, BA, 60, MS, 63; Wash State Univ, PhD(bot), 70. *Prof Exp:* Plant physiologist, Forest-Animal Unit, Bur Sport Fisheries & Wildlife, US Fish & Wildlife Serv, 68-74; res assoc, Western Wash Res & Ext Ctr, Wash State Univ, 75-76; natural resource res technician, State Dept Natural Resources, Olympia, 77-83; RETIRED. *Mem:* Sigma Xi; Am Soc Plant Physiologists. *Res:* Forest tree physiology and its relation to animal damage; soil biochemistry; forest genetics; Douglas-fir seed physiology. *Mailing Add:* 5503 110th Ave SW Olympia WA 98502

HEEBNER, DAVID RICHARD, b Hackensack, NJ, Feb 27, 27; m 50; c 4. SCIENCE ADMINISTRATION, TECHNICAL MANAGEMENT. *Educ:* Newark Col Eng, BSEE, 50; Univ Southern Calif, MSEE, 55. *Prof Exp:* Systs engr, Hughes Aircraft Co, 53-60; consult, Nat Acad Sci, 60-61; mgr Navy Systs Lab, Hughes Aircraft Co, 61-68; asst dir, Off Secy Defense, 68-70, dep dir defense res & eng, 70-75; VCHMN BD & EXEC VPRES, SCI APPLNS INT CORP, 75- *Concurrent Pos:* Mem, Defense Sci Bd, 87-90, Naval Studies Bd, Nat Acad Sci, 79-82; chmn, Naval Res Adv Comt, 82-86, mem, 86-87. *Mem:* Asn Unmanned Vehicle Systs (pres, 76-77); fel Inst Elec & Electronics Engrs. *Res:* Conceived initial design and led team that built the first successful towed line-array sonar, a breakthrough leading to important developments in surface and submarine surveillance systems; development of a display system that was the first use of integrated circuits in the US Navy. *Mailing Add:* Sci Applns Int Corp 1710 Goodridge Dr PO Box 1303 McLean VA 22102

HEED, JOSEPH JAMES, b New York, NY, Sept 22, 31; m 57; c 3. MATHEMATICS. *Educ:* US Mil Acad, BS, 54; St John's Univ, NY, MS, 61; Univ Nancy, DSc(math), 65. *Prof Exp:* Instr math, St Joseph's Col, NY, 58-60 & St John's Univ, NY, 60-62; from asst prof to assoc prof, 62-74, head dept, 68-78, PROF MATH, NORWICH UNIV, 74- *Mailing Add:* Dept Math Norwich Univ S Main St Northfield VT 05663

HEED, WILLIAM BATTLES, b West Chester, Pa, June 9, 26; m 54; c 3. ZOOLOGY. *Educ:* Pa State Univ, BS, 50; Univ Tex, MS, 52, PhD(zool), 55. *Prof Exp:* Res assoc genetics, Univ Tex, 55-56; instr, Univ Pa, 56-58; from asst prof to prof genetics, 58-80, PROF ECOL & EVOLUTIONARY BIOL, UNIV ARIZ, 80- *Concurrent Pos:* Prog dir, Syst Biol, Ecol & Pop Biol Sect, Biol & Med Sci Div, NSF, 75-76. *Mem:* AAAS; Am Soc Naturalists (vpres, 78); Soc Study Evolution; Genetics Soc Am; Sigma Xi. *Res:* Evolutionary genetics; population biology of Drosophila. *Mailing Add:* Dept Ecol-Evol Biol Univ Ariz Tucson AZ 85721

HEEDE, BURCHARD HEINRICH, b Riga, Latvia, Dec 3, 18; nat US; m 44; c 3. WATERSHED MANAGEMENT, EARTH SCIENCES. *Educ:* Univ Gottingen, BS, 47, MF, 49; Colo State Univ, PhD(watershed resources), 67. *Prof Exp:* Res forester, Watershed Mgt, 68-73, prin hydraul eng, 74-77, RES HYDROLOGIST, ROCKY MT FOREST & RANGE EXP STA, US FOREST SERV, 78- *Concurrent Pos:* Forestry off, Food & Agr Orgn, UN, Greece, 68-70. *Mem:* Am Geophys Union; Sigma Xi; Int Asn Hydrol Sci. *Res:* Hydrology; hydraulics; erosion control; fluvial geomorphology; development of gully control systems; dynamics of stream systems, interactions between streams and stream-side forests, overland flow and sediment delivery from forests and wood lands. *Mailing Add:* Forest Hydrol Lab Ariz State Univ Tempe AZ 85287-1304

HEEGER, ALAN J, b Sioux City, Iowa, Jan 22, 36; m 57; c 2. POLYMERCHEMISTRY. *Educ:* Univ Nebr, BS, 53; Univ Calif, PhD(physics), 61. *Hon Degrees:* MA, Univ Pa, 71. *Prof Exp:* Res assoc physics, Univ Calif, 61-62; from asst prof to assoc prof, 62-67, dir, Lab Res Struct Matter, 74-81, prof physics, 67-82, actg vprovost res, DEPT PHYSICS, UNIV PA, PHILADELPHIA, 81-82; PROF MAT, UNIV CALIF, SANTA BARBARA, 82-, PROF PHYSICS, 84-, DIR, INST POLYMERS & ORG SOLIDS, 82- *Concurrent Pos:* Sloan Found fel, 63-66; Guggenheim fel & vis prof, Univ Geneva, 68-69; bd trustees, Aspen Ctr for Physics, 73-76. *Honors & Awards:* Oliver Buckley Prize Solid State Physics, 83; John Scott Award & Medal, 89. *Mem:* Fel Am Phys Soc; Am Chem Soc. *Res:* Fundamental physics of materials synthesized by chemical techniques; one-dimensional solids. *Mailing Add:* Dept Physics Univ Calif Santa Barbara CA 93106

HEEKS, R(OBERT) E(UGENE), b Rochester, NY, Jan 19, 28; m 55; c 5. CHEMICAL ENGINEERING, MATERIALS SCIENCE. *Educ:* Univ Rochester, AB, 52, MS, 54, PhD(chem eng), 57. *Prof Exp:* Res assoc rheol, Univ Rochester, 54-55; processing engr photog papers & films, Xerox Corp, 55-60, sr chemist, 60-62, scientist, 62-66, mgr res & eng subsect, 66-73, mgr photoreceptor eng, 73-77, mgr mfg develop eng, 77-87; RETIRED. *Concurrent Pos:* Consult, Robert Christian, Inc, 78-84. *Mem:* Am Chem Soc; Am Inst Chem Engrs; Soc Photog Sci & Eng; Sigma Xi. *Res:* Photographic emulsion; process engineering of photo products plant; rheology of suspensions; cellulose chemistry as related to properties of paper; selenium and organic photoreceptors; coating technology. *Mailing Add:* 87 Hillcrest Dr Penfield NY 14526

HEELIS, RODERICK ANTONY, b Luton, Eng, June 13, 48; m 69. IONOSPHERIC DYNAMICS. *Educ:* Univ Sheffield, BSc, 69, PhD(math), 73. *Prof Exp:* RES SCIENTIST SPACE PHYSICS, UNIV TEX, DALLAS, 73- *Concurrent Pos:* Prin investr, Dynamics Explorer Mission, NASA, 76- *Mem:* Am Geophys Union. *Res:* Studies of neutral and charged particle dynamics in the upper atmospheres of the earth and planets; design and operation of satellite borne instrumentation for obtaining experimental data. *Mailing Add:* Ctr Space Sci Univ Tex PO Box 688 Richardson TX 75080

HEENAN, WILLIAM A, b Lansdale, Pa, Apr 17, 41; m 86; c 2. CHEMICAL ENGINEERING, PROCESS CONTROL. *Educ:* Univ Detroit, BChE, 64, MS, 67, DEng(chem eng), 69. *Prof Exp:* Coop engr, Ford Motor Co, 62-64; process engr, Monsanto Co, 64-67; res engr, Atomic Power Develop Assocs, 67-69; prof chem eng, Univ PR, 69-75; engr mgt, Celanese Chem Corp, 80-82; PROF CHEM ENG, TEX A&I UNIV, 75-80 & 82- *Mem:* Am Inst Chem Engrs; Am Chem Soc. *Res:* Fuel and control rod design for liquid metal fast breeder reactor; chemical processing of products from the sea, agar production; computation of chemical equilibria and chemical process optimization; computer process control, error detection & data reconsiliation. *Mailing Add:* Dept Chem Eng Tex A&I Univ Santa Gertrudis Kingscille TX 78363

HEER, CLIFFORD V, b Archbold, Ohio, May 31, 20; m 49; c 3. LASERS, QUANTUM OPTICS. *Educ:* Ohio State Univ, BSc, 42, PhD(physics), 49. *Prof Exp:* Mem Signal Corps, US Army, 42-46; from asst prof to prof physics, 49-90, PROJ SUPVR, UNIV RES FOUND, OHIO STATE UNIV, 55-, EMER PROF PHYSICS, 90- *Concurrent Pos:* Consult, Ramo-Wooldridge Corp, 56-58, Space Tech Labs, 58-65, Honeywell, Inc, 64-65, TRW Inc, 65-70 & Army Res Off, 70-71. *Honors & Awards:* William A Fowler Award, 90. *Mem:* Fel Am Phys Soc. *Res:* Low temperature solid state physics; laser physics; physics related to space technology; laser gyro. *Mailing Add:* 4174 Fairfax Dr Columbus OH 43220

HEER, EWALD, b Friedensfeld, Ger, July 28, 30; US citizen; m 52; c 2. SYSTEMS ENGINEERING, ENGINEERING PHYSICS. *Educ:* Univ Hamburg, CE, 53; City Col New York, BS, 59; Columbia Univ, MS, 60, CE, 62; Hannover Tech Univ, DESc, 64. *Prof Exp:* Engr struct, Hinz Architects, Ger, 51-54, Dixon & Evans, Can, 54-55 & Hewitt-Robins, Inc, NY, 56-59; sr res eng, Paul Weidlinger Consult Engrs, 60-64; tech specialist struct dynamics, MacDonnell Aircraft Corp, 64-65; scientist, space sci lab, Gen Elec Co, 65-66, res group leader struct mech, 66; supvr struct & dynamics res, Jet Propulsion Lab, Calif Inst Technol, 66-70; prog mgr, lunar explor off, NASA Hq, 70-71; mgr advan tech studies off, 71-77, dir, autonomous systs & space mech, Jet Propulsion Lab, Calif Inst Technol, 77-84; PRES, HEER ASSOCS INC, 84- *Concurrent Pos:* Lectr, Univ Hamburg, 63, Pa State Univ, 65-66, Univ Southern Calif, 69 & Univ Calif, Los Angeles, 69; adj prof indust & systs eng, Univ Southern Calif, 73-84; ed robotics, Mechanisms & Mach Theory J, 74-; dir, Inst Technoecon Systs, Univ Southern Calif, 78-84; prof indust & systs eng, Univ Southern Calif, 78-84; chmn, Comput Eng Div, Am Soc Mech Engrs, 84. *Mem:* Am Soc Civil Engrs; fel Am Soc Mech Engrs; assoc fel Am Inst Aeronaut & Astronaut; NY Acad Sci; Inst Elec & Electronics Engrs; Opers Res Soc Am; Am Asn Artificial Intel. *Res:* Aerospace engineering systems; automation; engineering design; cybernetics; man-machine systems; robotics; machine intelligence. *Mailing Add:* 5329 Crown Ave La Canada CA 91011

HEEREMA, NICKOLAS, b Hospers, Iowa, Oct 5, 22; m 47; c 4. MATHEMATICS. *Educ:* Univ Mich, BS, 44; Univ Tenn, MS, 49, PhD, 51. *Prof Exp:* Res engr, Union Carbide Chem Co, 46-48; prof math, Fla State Univ, 51-; RETIRED. *Concurrent Pos:* NSF grants, 60-71. *Mem:* Am Math Soc; Math Asn Am. *Res:* Pure mathematics; group theory; ring theory. *Mailing Add:* 3204 Enterprise Rd Tallahassee FL 32312

HEEREMA, RUURD HERRE, b The Hague, Holland, Aug 30, 43; m 78; c 3. MINERAL ENGINEERING. *Educ:* Delft Univ Technol, Eng, 75, Univ Minn, PhD(mineral eng), 78. *Prof Exp:* Res engr, raw mat & ironmaking res, Inland Steel Res Lab, 78-83; PROF MINERAL & PETROL ENG DEPT, DELFT UNIV TECHNOL, NETH, 86- *Concurrent Pos:* Consult, ironbearing raw mat. *Mem:* Am Inst Mining, Metall & Petrol Eng; Royal Neth Geol & Mining Soc. *Res:* Concentration and pelletizing of iron ores; selective flocculation of iron exide slimes; reduction properties of ironbearing raw materials. *Mailing Add:* Heerema Eng Serv Br Spec Proj Vondellaan 47 Leiden 2332AA Netherlands

HEEREN, JAMES KENNETH, b Fitchburg, Mass, Mar 22, 29; m 56. ORGANIC CHEMISTRY. *Educ:* Tufts Univ, BS, 51, MS, 52; Mass Inst Technol, PhD(chem), 60. *Prof Exp:* Chemist, Am Cyanamid Co, 54-56, res chemist, 60-61; fel chem, Sch Advan Study, Mass Inst Technol, 61-62; asst prof, 62-67, ASSOC PROF CHEM, TRINITY COL, CONN, 67- *Concurrent Pos:* Guest prof, Univ Heidelberg, Ger, 69. *Mem:* Am Inst Chem; Am Chem Soc. *Res:* Epoxide reactions; organometallic compounds. *Mailing Add:* Dept Chem Trinity Col 300 Summit St Hartford CT 06106

HEERMANN, DALE F(RANK), b Scribner, Nebr, Mar 2, 37; m 57; c 3. AGRICULTURAL ENGINEERING. *Educ:* Univ Nebr, BS, 59; Colo State Univ, MS, 64, PhD(agr eng), 68. *Prof Exp:* AGR ENGR, AGR RES SERV, USDA, 68- *Mem:* Am Soc Agr Engrs; Am Soc Agron; Soil Sci Soc Am; Irrig Asn; US Comt Irrig & Drainage. *Res:* Irrigation management; sprinkler and surface irrigation design; soil-plant-water relationships. *Mailing Add:* Agr Res Serv USDA Colo State Univ Foothills Campus Ft Collins CO 80523

HEERMANN, RUBEN MARTIN, b Pilger, Nebr, July 18, 21; m 44; c 3. PLANT BREEDING. *Educ:* Univ Nebr, BSc, 43, MSc, 48; Univ Minn, PhD(plant genetics), 54. *Prof Exp:* Res agronomist, Bur Plant Indust, Soils & Agr Eng, USDA, 48-53, res agronomist, Cereal Crops Br, Agr Res Serv, 53-56, prin agronomist, Coop State Res Serv, 56-70; assoc dir res, NY State Col Agr & Life Sci & assoc dir agr exp sta, Cornell Univ, 70-76; asst dir, Dakotas-Alaska Area, Agr Res Serv, USDA, Fargo, 76-80, assoc dir, 80-82, prog analyst, NCent Regional Off, Peoria, 82-84; RETIRED. *Mem:* Fel AAAS; Crop Sci Soc Am; Am Soc Agron. *Res:* Field crops; breeding and genetics of tetraploid wheats; research administration. *Mailing Add:* 739 Wonderview Dr Dunlap IL 61525

HEESCH, CHERYL MILLER, b Kermit, Tex, Dec 28, 48. NEURAL CONTROL OF CIRCULATION, CARDIOVASCULAR PHYSIOLOGY. *Educ:* NMex State Univ, BS, 71; Univ Tex, San Antonio, PhD(pharmacol), 81. *Prof Exp:* Postdoctoral researcher, Col Med, Cardiovasc Ctr, Univ Iowa, 81-83, res scientist, 83-84; asst prof neurophysiol, Dept Physiol & Biophys, La State Univ Med Ctr, Shreveport, 84-85; asst res prof cardiovasc physiol, Dept Physiol & Biophys, Univ Ky, 85-90; ASST PROF CARDIOVASC PHYSIOL, DEPT PHYSIOL, OHIO STATE UNIV, 90- *Concurrent Pos:* Prin investr, NIH grants, 84- & Am Heart Asn grant-in-aid, 84-87; ad-hoc grant reviewer, NIH, 87, 89 & 90; mem, Cardiovasc Res Study comt, Am Heart Asn, 90- *Mem:* Fel Am Heart Asn; Am Physiol Soc (secy, 87-91); Soc Neurosci. *Res:* Neural control of the circulation in hypertension and in pregnancy; central nervous system mechanisms involved in cardiovascular reflex control of the circulation. *Mailing Add:* Dept Physiol 4114 Graves Hall 333 W Tenth Ave Columbus OH 43210

HEESCHEN, DAVID SUTPHIN, b Davenport, Iowa, Mar 12, 26; m 50; c 3. RADIO ASTRONOMY. *Educ:* Univ Ill, BS, 49; Harvard Univ, PhD(astron), 55. *Hon Degrees:* ScD, WVa Inst Technol, 74, NMex Inst Mining & Technology, 89. *Prof Exp:* Instr astron, Wesleyan Univ, 54-55; lectr, Harvard Univ, 55-56; astronr, 56-58, chmn astron dept, 58-62, dir, 62-78, SR SCIENTIST, NAT RADIO ASTRON OBSERV, 77-; RES PROF, UNIV VA, 80- *Concurrent Pos:* Pres comn 40, Int Astron Union, 70-73; consult, NASA, 61, 68-72, 75 & Max-Planck Inst Radioastron, 69-76; assoc ed, Astron J, 69-72. *Honors & Awards:* Distinguished Pub Serv Award, NSF, 80; Alexander Von Humboldt Sr Scientist Award, 85. *Mem:* Nat Acad Sci; Am Acad Arts & Sci; Am Philos Soc; Int Astron Union (vpres, 76-82); Am Astron Soc (vpres, 69-71, pres, 80-82). *Res:* Radio astronomy; galactic structure; extragalactic studies. *Mailing Add:* Nat Radio Astron Observ Edgemont Rd Charlottesville VA 22901

HEESCHEN, JERRY PARKER, b Apr 14, 32; US citizen; m 56; c 3. PHYSICAL CHEMISTRY, NUCLEAR MAGNETIC RESONANCE. *Educ:* Western Reserve Univ, BS, 53; Univ Ill, PhD(phys chem), 59. *Prof Exp:* Res chemist, 58-63, sr res chemist, 63-75, RES ASSOC, DOW CHEM CO, 75- *Mem:* AAAS; Am Chem Soc; Sci Res Soc Am. *Res:* Nuclear magnetic resonance; molecular structure studies by spectroscopic techniques; molecular configuration and conformation; chemical analysis. *Mailing Add:* Dow Chem Co Analytical Sci Midland MI 48667

HEESTAND, GLENN MARTIN, b Canton, Ohio, June 17, 42. ATOMIC PHYSICS. *Educ:* Case Inst Technol, BS, 64; Univ Wis, MS, 66, PhD(physics), 69. *Prof Exp:* Res assoc physics, Aarhus Univ, Denmark, 69-70; res assoc, Univ Wis-Madison, 70-72; asst prof, Univ Wis-Superior, 72-74; ENG PHYSICIST, EXXON NUCLEAR INC, 75-; SEC LEADER, LIVERMORE NAT LAB, 88- *Mem:* AAAS; Am Phys Soc. *Res:* Development of laser isotope separation on industrial scale. *Mailing Add:* 2412 Vialos Milagros Pleasanton CA 94566

HEETDERKS, WILLIAM JOHN, b Grand Rapids, Mich, Feb 15, 48; m 69; c 5. NEURAL PROSTHES IS. *Educ:* Univ Mich, BSEE, 71 PhD(bioeng), 75; Univ Miami, MD, 83. *Prof Exp:* From asst prof to assoc prof elec eng, Cornell Univ, 76-83; med resident internal med, Butterworth Hosp, 83-86; MED OFFICER NEURAL PROSTHESES, NIH, 86- *Mem:* Inst Elec & Electronics Engrs; Am Col Physicians. *Res:* Promote the development of techniques to selectively record & stimulate small groups of neurons for the purpose of rehabilitation of neurologically impaired individuals. *Mailing Add:* NIH Fed Bldg Rm 916 Bethesda MD 20892

HEFFELFINGER, CARL JOHN, b Rochester, NY, Aug 9, 24; m 50; c 3. POLYMER PHYSICS. *Educ:* Univ Buffalo, AB, 50, PhD(chem), 53. *Prof Exp:* Asst chem, Univ Buffalo, 50-51; res chemist, 52-59, staff scientist, 59-65, res assoc, film dept, 65-80, RES FEL, POLYMER PROD DEPT, E I DU PONT DE NEMOURS & CO, INC, 80- *Mem:* Sigma Xi. *Res:* Kinetics of addition polymerization; polymer structure. *Mailing Add:* 124 Walnut Circle-PK5 124 Walnut Circle-PKS Morehead City NC 28557-9724

HEFFERLIN, RAY (ALDEN), b Paris, France, May 2, 29; US citizen; m 54; c 4. PERIODIC CHARTS FOR MOLECULES. *Educ:* Pac Union Col, BA, 51; Calif Inst Technol, PhD(physics), 55. *Prof Exp:* PROF PHYSICS & HEAD DEPT, SOUTHERN COL, 55- *Concurrent Pos:* Res grants, Res Corp, Tenn Acad Sci & NSF; consult, Oak Ridge Nat Lab, Nat Bur Standards, McDonnell Douglas Corp & Arnold Air Force Sta; Nat Acad Sci exchange scholar to USSR, 78-79 & 81; vis prof, Univ Denver, 84-85. *Mem:* Am Phys Soc; Int Astron Union; Am Chem Soc; Europ Phys Soc. *Res:* Representations of periodic law for molecules based on tabulated data; systematics of nuclides, molecules, fundamental particles and other entities. *Mailing Add:* Dept Physics Southern Col Collegedale TN 37315-0370

HEFFERNAN, GERALD R, b Edmonton, Alta, July 12, 19. METALLURGY. *Educ:* Univ Toronto, BaSc, 43. *Hon Degrees:* LLD, Queens Univ, 79. *Prof Exp:* Staff mem, Dept Metall, Univ BC, 45-46; metallurgist & asst gen mgr, Westland Iron & Steel Foundries, 46-48; supt, Western Can Steel, 48-54; managing dir, secy & vpres, Premier Steel Mills Ltd, 54-62; pres, Lake Ont Steel Co Ltd, 63-70; pres, 70-86, chmn, 87-90, DIR, CO-STEEL INC, 90- *Concurrent Pos:* Dir, Lake Ont Steel Co, Sheerness Steel Co plc, Raritan River Steel Co, Chaparral Steel Co, Tex Indust Inc, Corod Indust Inc, Harbour Petrol Co Ltd, Can Inst Advan Res, Roy L Merchant Group Inc, Nat Rubber Co Ltd & Innovations Found, Univ Toronto; vchmn, Can Inst Advan Res. *Honors & Awards:* Noranda Award, Can Inst Mining & Metall, 75; Benjamin F Fairless Award, Am Inst Mining, Metall & Petrol Engrs, 83; Gold Medal, Asn Prof Engrs, 85; Bessemer Gold Medal Award, Inst Metals, 89. *Mem:* Fel Am Soc Metals; Iron & Steel Soc; fel Can Inst Mining & Metall. *Res:* Metallurgy of steelmaking; commercial application of continuous casting; ferrous metallurgy. *Mailing Add:* Co-Steel Inc 1601 Hopkins St Whitby ON L1N 5R6 Can

HEFFERNAN, LAUREL GRACE, b San Francisco, Calif, Sept 22, 45. MOLECULAR BIOLOGY. *Educ:* San Francisco State Univ, BA, 68; Univ Calif, Santa Barbara, PhD(biol), 75. *Prof Exp:* NIH trainee genetics, Stanford Univ, 74-80; adj asst prof microbiol, Univ Calif, Los Angeles, 80-86; ASST PROF BIOL SCI, CALIF STATE UNIV, SACRAMENTO, 86- *Mem:* Sigma Xi; Am Soc Microbiol; AAAS. *Res:* Genic regulation in procaryotic organisms and their viruses. *Mailing Add:* Dept Biol Sci Calif State Univ 6000 J St Sacramento CA 95819-2694

HEFFERREN, JOHN JAMES, b Chicago, Ill, Aug 12, 28; m 65. CHEMISTRY. *Educ:* Loyola Univ, BS, 50; Univ Wis, MS, 52, PhD(pharmaceut chem), 54. *Prof Exp:* Chemist, AMA, 53-59; dir, Div Chem, Coun Dent Therapeut, Am Dent Asn, 59-66, dir, Div Biochem, 66-75, dir, Res Inst, 75-85; assoc biochem med & dent schs, Northwestern Univ, 66-80, coordr pres dent, 77-80, prof oral biol, 80-86; PRES, J J HEFFERREN RESOURCES INC, 85-; PRES, ODONTEX, INC, 87-; ADJ PROF PHARM CHEM, UNIV KANS, 87-, SCIENTIST & RES PROF, CTR BIOMED RES, 89- *Concurrent Pos:* Am Fedn Pharmaceut Educ fel, 50-53; lectr, Sch Pharm, Univ Mich, 74- *Mem:* Am Chem Soc; Am Pharmaceut Asn; AAAS; Int Asn Dent Res; NY Acad Sci; Europ Orgn Caries Res. *Res:* Organic medicinal and bioanalytical organic chemistry; laboratory, preclinical and clinical studies to assess the dental caries producing potential of foods and the function of dentifrices and other oral hygiene agents. *Mailing Add:* 3030 Campfire Dr Lawrence KS 66049

HEFFES, HARRY, b Atlantic City, NJ, Sept 1, 39; div; c 2. ELECTRICAL ENGINEERING. *Educ:* City Col New York, BEE, 62; NY Univ, MEE, 64, PhD(elec eng), 68. *Prof Exp:* Mem tech staff, AT&T Bell Labs, 62-89; PROF & DEPT HEAD, ELEC ENG & COMPUT SCI, STEVENS INST TECHNOL, HOBOKEN, NJ, 90-91. *Honors & Awards:* S O Rice Award, Inst Elec & Electronics Engrs, 87. *Mem:* Fel Inst Elec & Electronics Engrs; Opers Res Soc Am; Asn Comput Mach; Soc Indust & Appl Math. *Res:* Teletraffic and queueing theory; modeling and analysis of teletraffic systems. *Mailing Add:* Dept Elec Eng & Computer Sci Stevens Inst Technol Castle Point on the Hudson Hoboken NJ 07030

HEFFLEY, JAMES D, b Ethel, Tex, Jan 12, 41; m 63; c 5. DIET & FOOD SUPPLEMENTS. *Educ:* Abilene Christian Univ, BS, 64; Univ Tex Austin, PhD(chem), 70. *Prof Exp:* Res asst, Clayton Found Biochem Inst, 65-70, res assoc, 70-74; DIR, NUTRIT COUN SERV, 74- *Concurrent Pos:* Consult, Tex Sch Blind, 72-74; lab dir, Ctr Better Health, 84-87. *Mem:* Int Acad Nutrit & Prev Med; Pan Am Allergy Asn; Acad Orthomolecular Med. *Res:* Effects of alcohol on amino acid metabolism and on assessing total nutrient value of single foods; writing on uses of food supplements and dietary manipulation to uncover food intolerance. *Mailing Add:* 3913 Medical Pkwy No 101 Austin TX 78756

HEFFNER, REID RUSSELL, JR, b Philadelphia, Pa, Apr 16, 38; m 65; c 2. PATHOLOGY, NEUROPATHOLOGY. *Educ:* Yale Univ, BA, 60, MD, 65. *Prof Exp:* Instr path, Yale Sch Med, 68-69 & Cornell Sch Med, 69-70; neuropathologist, Armed Forces Inst Path, 70-72, chief neuromuscular div, 72-74; assoc prof, 74-84, PROF PATH, SCH MED, STATE UNIV NY, BUFFALO, 84-, ASSOC CHAIR DEPT, 85- *Concurrent Pos:* NIH fel neuropath, 68-69; consult, Vet Admin Hosp, Buffalo, NY, 75-, Roswell Park Mem Inst, 76- & Millard Fillmore Hosp, 76-; dir dept path, Erie Co Med Ctr, Buffalo, 79- *Mem:* Am Acad Neurol; Am Asn Neuropathologists; Am Soc Clin Pathologists; Int Acad Path; Soc Neurosci. *Res:* Neuromuscular disease; muscular dystrophy. *Mailing Add:* Dept Path 204 Farber Hall Buffalo NY 14214

HEFFNER, THOMAS G, b Salem, Ohio, Sept 20, 49. PSYCHOPHARMACOLOGY, ANTIPSYCHOTICS. *Educ:* Univ Pittsburgh, BS, 71, PhD(psychobiol), 76. *Prof Exp:* Teaching asst biol, Univ Pittsburgh, 71-74, teaching fel physiol, 74-76; NIMH fel pharmacol, Univ Chicago, 76-79, res asst prof, 79-83; group leader, 83-86, sr group leader, 86-90, SECT DIR, WARNER-LAMBERT-PARKE-DAVIS, 90- *Concurrent Pos:* Lectr psychopharmacol, Pritzker Sch Med, Univ Chicago, 79-83; prin investr, Brain Res Found grant, Chicago, 81-83; co-prin investr, NIMH grant, 79-83; investr, Nat Inst Neurol Commun Dis & Stroke grant, 79-83. *Mem:* Am Soc Pharmacol & Exp Therapeut; Soc Neurosci; Am Psychol Asn. *Res:* Psychopharmacology; development and discovery of novel drugs to treat psychiatric disorders; etiology of central nervous system disorders; neurochemical basis of drug action; role of neurochemical changes accompanying environmental or behavioral stimuli in drug action. *Mailing Add:* Dept Pharmacol Warner-Lambert-Parke-Davis 2800 Plymouth Rd Ann Arbor MI 48105

HEFFRON, PETER JOHN, b Flushing, NY, Oct 2, 43; m 68; c 1. INDUSTRIAL ORGANIC & SYNTHETIC ORGANIC CHEMISTRY. *Educ:* St Francis Xavier Univ, BS, 65; State Univ NY Binghamton, MA, 67; State Univ NY Stony Brook, PhD(org chem), 72. *Prof Exp:* Res fel entom, Univ Ky, 72-73; GROUP LEADER, DYES, AM CYANAMID CO, 73- *Mem:* Am Chem Soc. *Res:* Novel synthetic routes in organic chemistry; process development in synthesis of chemical intermediates and dyes; stereochemistry and stereochemically controlled reactions. *Mailing Add:* 43 Capner St Flemington NJ 08822-1313

HEFFRON, W(ALTER) GORDON, b New Orleans, La, July 25, 25; m 47; c 4. COMMUNICATIONS SCIENCE, SYSTEMS ENGINEERING. *Educ:* Tulane Univ, BE, 47; Purdue Univ, MS, 50; George Washington Univ, DSc(appl sci), 69. *Prof Exp:* Instr elec eng, Tulane Univ, 47-48 & Purdue Univ, 48-49; anal engr, apparatus div, Gen Elec Co, 45-55; sr dynamics engr, Convair Div, Gen Dynamics Corp, 55-56; proj engr, Melpar Inc, 56-60; sr engr, Chance Vought Aircraft, Inc, 60; lab mgr, Melpar Inc, 60-64; dept head aerospace guid & navig, Bellcomm, Inc, 64-72; head, Dept Data Network Planning, 72-80, HEAD DEPT VISUAL COMMUN, BELL TEL LABS, 80- *Mem:* Soc Info Display; Inst Elec & Electronics Engrs. *Res:* Systems and detailed analysis in electric utility systems; aircraft and helicopter simulation; railway train control, space navigation and guidance; celestial navigation; telephone system design; visual and audio communications systems. *Mailing Add:* Atlantic Systs R & E PO Box 2417 Springfield VA 22152

HEFFTER, JEROME L, b Minneapolis, Minn. METEOROLOGY. *Educ:* Univ Minn, BS, 55; Mass Inst Technol, MS, 60. *Prof Exp:* RES METEOROLOGIST, AIR RESOURCES LAB, NAT OCEANIC & ATMOSPHERIC ADMIN, 60- *Mem:* Am Meteorol Soc. *Res:* Atmospheric transport and dispersion of pollutants. *Mailing Add:* Air Resouces Lab Nat Oceanic & Atmospheric Admin 1325 East West Hwy Suite 9358 Silver Spring MD 20910

HEFLEY, ALTA JEAN, b Rifle, Colo, Jan 5, 41. ANALYTICAL CHEMISTRY, FOOD CHEMISTRY. *Educ:* Colo State Univ, BS, 63; Iowa State Univ, MS, 65, PhD(anal chem), 67. *Prof Exp:* From asst prof to assoc prof chem, Mundelein Col, 67-74; ANALYTICAL CHEMIST, MILES LABS, 74- *Mem:* Am Chem Soc; Sigma Xi; Inst Food Technologists. *Mailing Add:* 421 E Orchard Lane Arlington Heights IL 60005

HEFLICH, ROBERT HENRY, b Cairo, NY, Nov 10, 46; m 71. MUTAGENESIS, CARCINOGENESIS. *Educ:* Rutgers Univ, BA, 68, MS, 70, PhD(microbiol), 76. *Prof Exp:* Res asst oral microbiol, Letterman Army Inst Res, Calif, 70-72; res assoc mutagenesis res, Carcinogenesis Lab, Mich State Univ, 76-79; RES MICROBIOLOGIST MUTAGENESIS RES, DIV GENETIC TOXICOL, NAT CTR TOXICOL RES, 79- *Mem:* Am Soc Microbiol; Sigma Xi; Environ Mutagen Soc. *Res:* Mechanisms of the induction of genotoxic events in cultural mammalian cells and bacteria by physical and chemical toxicants. *Mailing Add:* Div Genetic Toxicol HFT 120 Nat Ctr Toxicol Res Jefferson AR 72079

HEFLINGER, LEE OPERT, b Pasadena, Calif, May 23, 27; m 49; c 2. APPLIED MATHEMATICS, EXPERIMENTAL PHYSICS. *Educ:* San Jose Col, BA, 50; Univ Calif, MA, 52, PhD(math), 56. *Prof Exp:* Asst, Univ Calif, 52-56; mathematician, Thompson-Ramo-Wooldridge, Inc, 56-58, Space Tech Labs, Inc, Los Angeles, 58-60 & Gen Tech Corp, 60-64; mathematician, 64-77, SR SCIENTIST, TRW SYSTS GROUP, 77- *Mem:* Am Math Soc; Am Phys Soc; Sigma Xi. *Res:* Asymptotic expansions; numerical analysis; experimental plasma research; pulsed laser holography. *Mailing Add:* 5001 Paseo de Pablo Torrance CA 90505

HEFNER, LLOYD LEE, b Bradenton, Fla, Nov 19, 23; c 4. INTERNAL MEDICINE. *Educ:* Vanderbilt Univ, BA, 46, MD, 49. *Prof Exp:* Intern, Vanderbilt Univ Hosp, 49-50, fel psychiat, 50-51; resident internal med, 53-55, fel cardiovasc dis, 55-56, from instr to prof med, 56-69, Ala heart prof cardiovasc res, 69-80, prof med, physiol & biophys, Univ Ala, Birmingham, 80-88; RETIRED. *Concurrent Pos:* Estab investr, Am Heart Asn, 60-65, chmn, Coun Basic Sci, 70-72, mem, Coun Clin Cardiol, 70- *Mem:* Am Fedn Clin Res; Am Col Physicians; AMA; Am Heart Asn; Sigma Xi. *Res:* Biophysics of ventricular contraction. *Mailing Add:* 2835 Berwick Rd Birmingham AL 35213

HEFTI, FRANZ F, b Zurich, Switz, Dec 22, 47; m 74; c 2. NEUROLOGY. *Educ:* Univ Zurich, Switz, MS, 72, PhD(biol), 76. *Prof Exp:* Res asst pharmacol, Univ Zurich, 74-77; res assoc neuroendocrinol, Mass Inst Technol, 78-80; res assoc neurobiol, Max Planck Inst, Munich, 81-82; head res lab pharmacol, Sandoz Ltd, Switz, 82-85; assoc prof pharmacol, Dept Neurol, Univ Miami, Fla, 85-89; PROF, ANDRUS GERONT CTR, UNIV SOUTHERN CALIF, LOS ANGELES, 89- *Concurrent Pos:* Parkinson res scholar, Nat Parkinson Found, 88; James E Birren Prof, 90. *Honors & Awards:* Robert Bing Prize, 90. *Mem:* Soc Neurosci; Int Soc Neurochem; Europ Neurosci Asn; Swiss Soc Pharmacol & Toxicol. *Res:* Basic research relevant to human neurodegenerative diseases, in particular Parkinson's and Alzheimer's disease, role of trophic factors and hormones in these diseases. *Mailing Add:* Geront Ctr Univ Southern Calif Los Angeles CA 90089-0191

HEFTMANN, ERICH, b Vienna, Austria, Mar 9, 18; nat US; m 42; c 3. BIOCHEMISTRY. *Educ:* NY Univ, BA, 42; Univ Rochester, PhD(biochem), 47; Am Bd Clin Chem, dipl. *Prof Exp:* Instr chem, Univ Md, 43-44; asst biochem & pharmacol, Univ Rochester, 44-47; biochemist & dir labs, Diabetes Sect, USPHS, 47-48, biochemist, Nat Cancer Inst, 48-50 & Nat Inst Arthritis & Metab Dis, 50-63, res leader, Western Regional Res Ctr, Agr Res Sci & Educ Admin, USDA, 63-83; ED, J CHROMATOGRAPHY SYMP VOLS, 83- *Concurrent Pos:* Lectr, Grad Sch, USDA, 54-62 & Georgetown Univ, 60; res fel, Calif Inst Technol, 59 & 61-64, res assoc, 64-69; vis assoc prof, Univ Southern Calif, 65-70. *Honors & Awards:* Humboldt Award, 75. *Mem:* Fel AAAS; Am Soc Biol Chemists; Am Chem Soc. *Res:* Lipids; screening method for blood glucose; biochemistry of steroids; chromatography; plant biochemistry. *Mailing Add:* PO Box 928 Orinda CA 94563

HEGARTY, PATRICK VINCENT, b Cork, Ireland, July 19, 39; m 67; c 2. NUTRITION. *Educ:* Univ Col, Cork, BSc, 61, MSc, 63; Univ London, PhD(human nutrit), 66; Univ Col, Dublin, BA, 70. *Prof Exp:* From res officer to sr res officer muscle biol, Agr Inst, Dublin, Ireland, 66-70; fel, 70-72, assoc prof, 72-76, prof human nutrit, Univ Minn, St Paul, 76-; AT DEPT HUMAN DEVELOP & CONSUMER SCI, UNIV HOUSTON. *Mem:* Am Inst Nutrit; Brit Nutrit Soc; Inst Food Technologists. *Res:* Morphological and biochemical changes during various dietary deficiencies, and the subsequent rehabilitation in skeletal muscles, adipose tissue and bones. *Mailing Add:* Dept Food Sci & Human Nutrit Mich State Univ 204 Food Sci East Lansing MI 48824

HEGE, E K, ASTRONOMY. *Prof Exp:* MEM STAFF, STEWARD OBSERV, UNIV ARIZ. *Mailing Add:* Steward Observ Univ Ariz Tucson AZ 85721

HEGEDUS, L LOUIS, b Budapest, Hungary, Apr 13, 41; m 68; c 2. CHEMICAL ENGINEERING, SPECIALTY CHEMICALS. *Educ:* Tech Univ Budapest, MS, 64; Univ Calif, Berkeley, PhD (chem eng), 72. *Prof Exp:* Res engr process develop, Res Inst Org Chem Indust, Budapest, 64-65; group leader qual control, Daimler-Ben, AG, Mannheim, WGer, 56-68; res supvr catalysis & kinetics, Gen Motors Res Lab, 72-80; dir, 80-84, VPRES INORG RES, W R GRACE & CO, COLUMBIA, MD, 84- *Concurrent Pos:* Consult ed, Am Inst Chem Engrs J, 85-88; gov bd, Coun Chem Res, 87-90; bd Chem Sci & Technol, Nat Res Coun, 90-; adv bd mem, Berkeley, Princeton, Wis & Northwestern Univs; regent's lectr, Univ Calif, Los Angeles, 91. *Honors & Awards:* Allan P Colburn Lectr, Univ Del, 75, J A Gerster Lectr, 88; B F Dodge lectr, Yale Univ, 88; R H Wilhelm Award, Am Inst Chem Eng, 88; D M Mason Lectr, Stanford Univ, 91. *Mem:* Nat Acad Eng; Am Chem Soc; Catalysis Soc; Sigma Xi; Mat Res Soc; Am Inst Chem Eng. *Res:* Specialty chemicals; applied catalysis; mathematical modeling; design and preparation of catalysts; process research; emission control. *Mailing Add:* 6625 Paxton Rd Rockville MD 20852

HEGEDUS, LOUIS STEVENSON, b Cleveland, Ohio, May 6, 43; m 67. ORGANIC CHEMISTRY, ORGANOMETALLIC CHEMISTRY. *Educ:* Pa State Univ, BS, 65; Harvard Univ, AM, 66, PhD(org chem), 70. *Prof Exp:* NIH fel org chem, Stanford Univ, 70-71; from asst prof to assoc prof, 71-79, PROF CHEM, COLO STATE UNIV, 79- *Mem:* Am Chem Soc. *Res:* Organic synthetic methods involving transition metal organometallic intermediates; homogeneous catalysis. *Mailing Add:* Dept Chem Colo State Univ Ft Collins CO 80523

HEGEDUS, STEVEN SCOTT, b Cleveland, Ohio, May 26, 55; m; c 2. SOLAR CELL DEVICE PHYSICS. *Educ:* Case Western Reserve Univ, BS, 77; Cornell Univ, MS, 81; Univ Del, PhD, 90. *Prof Exp:* Engr, Int Bus Mach Corp, 77-82; RES ASSOC, INST ENERGY CONVERSION, UNIV DEL, 82- *Mem:* Am Phys Soc; Union Concerned Scientist. *Res:* Developing low cost photo voltaic solar cells and studying amorphous silicon and CdTe devices. *Mailing Add:* Inst Energy Conversion Univ Del Newark DE 19716

HEGELE, ROBERT A, b Toronto, Ont, Nov 21, 57; m 87; c 2. ENDOCRINOLOGY & METABOLISM, ATHEROSCIEROSIS RESEARCH. *Educ:* Univ Toronto, MD, 81; FRCPC, 85 & 87. *Prof Exp:* Resident internal med, Univ Toronto, 81-84, fel endocrinol, 84-85; res fel genetics, Rockefeller Univ, 85-87 & Howard Hughes Med Inst, Univ Utah, 87-89; ASST PROF MED, UNIV TORONTO, 89- *Concurrent Pos:* Staff doctor, St Michael's Hosp, Toronto, 89-; res scholar, Heart & Stroke Found, Can, 90- *Honors & Awards:* McDonald Award, Heart & Stroke Found, Can, 90; Doupe Award, Royal Col Physicians & Surgeons, Can, 90. *Mem:* Am Heart Asn; Am Fedn Clin Res; Am Soc Human Genetics; Am Col Nutrit; fel Royal Col Physicians & Surgeons. *Res:* Genomic basis of biochemical abnormalities predisposing to premature cardiovascular disease. *Mailing Add:* 30 Bond St Toronto ON M5B 1W8

HEGEMAN, GEORGE D, b Glen Cove, NY, Aug 31, 38; m 61; c 2. MICROBIOLOGY, BIOCHEMISTRY. *Educ:* Harvard Univ, AB, 60; Univ Calif, Berkeley, PhD(comp biochem), 65. *Prof Exp:* Instr bact, Univ Calif, Berkeley, 65; NIH fel, Lab Enzym, Nat Ctr Sci Res, Gif-sur-Yvette, France, 65-66; asst prof bact, Univ Calif, Berkeley, 66-72; assoc prof, 72-78, PROF MICROBIOL, 78-, HEAD, MICROBIOL PROG, IND UNIV, BLOOMINGTON,84- *Concurrent Pos:* Mem, Microbial Biochem Study Sect, NIH, 83-; mem, sci adv bd, BioTrol, Inc. *Mem:* AAAS; Am Soc Microbiol; Brit Soc Gen Microbiol; Genetics Soc Am; Am Soc Biol Chem; Am Acad Microbiol. *Res:* Biology and chemistry of bacteria; enzymology and regulation of enzyme synthesis; bacterial physiology, genetics and biochemistry, especially the comparative control of catabolism in bacteria and evolution of catabolic pathways. *Mailing Add:* Dept Biol Jordan Hall 142 Ind Univ Bloomington IN 47405

HEGENAUER, JACK C, b Bay City, Mich, Nov, 26, 39; m 60; c 3. BIOINORGANIC CHEMISTRY. *Educ:* Univ Mich, BS, 61, MS, 68, PhD(zool), 70. *Prof Exp:* Fel, 70-72, asst res biologist, 72-77, ASSOC RES BIOLOGIST, UNIV CALIF, SAN DIEGO, 77- *Mem:* Am Inst Nutrit; Am Col Toxicol; Int Asn Bioinorg Scientists; Soc Develop Biol; Sigma Xi. *Res:* Human and animal trace element nutrition; metabolism of trace elements (iron, copper, zinc and manganese); physical chemistry of metalloproteins (ferritin, transferrin, phosphoproteins); trace element fortification of foods; epidemiology of human trace element deficiencies. *Mailing Add:* Dept Biol Univ Calif San Diego B-022 La Jolla CA 92093

HEGER, JAMES J, b Jenkinstown, Pa, Feb 9, 18. PHYSICAL METALLURGY & PROCESSING. *Educ:* Carnegie Inst Technol, BS, 40 MS, 48. *Prof Exp:* Chief staff engr stainless steel & metal processing, US Steel Res, 70-80; RETIRED. *Concurrent Pos:* Mem, Boiler & Pressure Vessel Code, Am Soc Mech Engrs. *Mem:* Am Soc Metals; Am Inst Mining Metall & Petrol Engrs; Nat Asn Corrosion Engrs; Am Soc Testing & Mat; Am Soc Mech Engrs. *Mailing Add:* 2610 Strathmore Lane Bethel PA 15102

HEGGERS, JOHN PAUL, b Brooklyn, NY, Feb 8, 33; m 77; c 6. MICROBIOLOGY, IMMUNOLOGY. *Educ:* Mont State Univ, BA, 58; Univ Md, MSc, 65; Wash State Univ, PhD(microbiol), 72; Am Bd Bioanal, dipl. *Prof Exp:* Asst to dir, Armed Forces Inst Path, 73-74; asst chief, Clin Invest Serv, Madigan Army Med Ctr, 74-77; assoc prof, Univ Chicago, 77-80, prof, dept surg, 80-83, res assoc, plastic & reconstructive surg & dir res & labs, Burn Ctr, 77-83; prof surg, Univ Health Ctr, Wayne State Univ, 83-88; PROG SURG, UNIV TEX MED BR, 88-; DIR CLIN MICROBIOL, SHRINERS BURN INST, 88- *Concurrent Pos:* Assoc prof, 77-80, prof, dept surg, Univ Chicago & res assoc, Plastic & Reconstructive Surg, 77-83; assoc lab dir, Moross Clin Lab, Detroit, 84-88. *Honors & Awards:* Fisher Award, Am Med Technol, 68; Lambert Award, Warner-Lambert, 73; Award Basic Sci Res, Am Soc Plastic & Reconstructive Surgeons, 79; Robert B Lindberg Award, Am Burn Asn, 86; Dr Stanley Reitman Mem Award, Int Soc Clin Lab Technol, 87. *Mem:* Plastic Surg Res Coun; Am Burn Asn; Nat Registry Microbiologists; Sigma Xi; Am Soc Med Technologists; Soc Exp Biol & Med; Univ Asn Emergency Med. *Res:* Pathophysiology of dermal ischemia; burn and surgical wound sepsis. *Mailing Add:* Shriners Burns Inst 610 Texas Ave Galveston TX 77550

HEGGESTAD, CARL B, b Starbuck, Minn, July 19, 30; m 53; c 3. ANATOMY. *Educ:* Univ Minn, Minneapolis, BA, 52, MD, 57, PhD(anat), 60. *Prof Exp:* From instr to prof anat, Univ Minn, Minneapolis, 59-; RETIRED. *Concurrent Pos:* Consult, Med Electronic Indust. *Mem:* AAAS; Am Asn Anat; Asn Am Med Cols. *Res:* Fetal endocrinology; placental permeability. *Mailing Add:* 1258 NE Skywood Lane Fridley MN 55421

HEGGESTAD, HOWARD EDWIN, b Stoughton, Wis, July 24, 15; m 39; c 3. PHYTOPATHOLOGY. *Educ:* Univ Wis, PhB, 40, PhD(genetics, plant path), 44. *Prof Exp:* Asst hort, Univ Wis, 38-44, instr, 44-46; from asst agronomist to agronomist, Bur Plant Indust, Soils & Agr Eng, Tobacco Exp Sta, USDA, Tenn, 46-55, agronomist, Field Crops Res Br, Agr Res Serv, 54-55, sr agronomist & breeder, Plant Indust Sta, 55-57, prin agronomist, 57-64, prin pathologist, 64, leader, Tobacco Breeding & Dis Invests, 64-66, chief air pollution lab, 66-75, plant pathologist, Plant Stress Lab, Plant Physiol Inst, Sci & Educ Admin, 75-84; RETIRED. *Concurrent Pos:* Consult, Elec Power Res Inst, Palo Alto, Calif, 79-; Biol Sci Collabr, USDA Agr Res Serv, 90-91. *Honors & Awards:* Res Award, Am Soc Hort Sci, 73; Res Award, Environ Protection Agency, 84. *Mem:* AAAS; Am Phytopath Soc; Am Inst Biol Sci; Air Pollution Control Asn. *Res:* Tobacco breeding; tobacco diseases; tobacco production; plant injury from air pollutants; effects of air pollutants, especially ozone and sulfur dioxide, singularly and in combinations, on crop productivity; technology for minimizing losses such as the identification of tolerant cultivars; interaction soil moisture stress and ozone stress on yields of soybean cultivars under field and greenhouse conditions. *Mailing Add:* 3112 Castleleigh Rd Silver Spring MD 20904

HEGGIE, ROBERT, b Glasgow, Scotland, Jan 19, 09; nat US; m 39; c 2. ORGANIC CHEMISTRY. *Educ:* Mass Inst Technol, SB, 33, PhD(org chem), 36. *Prof Exp:* From asst to res assoc chem, Mass Inst Technol, 33-39; chemist & dir res & develop, Am Chicle Co Div, Warner-Lambert Pharmaceut Co, 39-57, vpres, 57-61, exec vpres, 61-70; vpres appl sci, Warner-Lambert Res Inst, 62-65, vpres consumer prod res, 65-69, vpres tech develop & control, 69-70; CONSULT, 71- *Mem:* AAAS; Am Chem Soc; Am Inst Chemists. *Res:* Asymmetric synthesis; reaction rates; vitamin chemistry; high polymers; chemistry of flavors. *Mailing Add:* 3570 S Ocean Blvd Palm Beach FL 33480

HEGGIE, ROBERT MURRAY, b Paisley, Scotland, Mar 14, 27; Can citizen; m 51; c 4. ORGANIC CHEMISTRY. *Educ:* Univ Glasgow, BSc, 48, PhD(org chem), 51. *Prof Exp:* Fel chem, Nat Res Coun Can, 51-53, Univ Sask, 53-54 & Univ Ottawa, 54-55; sci officer, Defence Res Bd Can, 55-74 & Dept Nat Defence, Can, 74-75; chief, Defence Res Estab Suffield, 75-79, chief, Defence & Civil Inst Environ Med, 79-84, chief, Defence Res Estab Ottawa, 84-90; RETIRED. *Mem:* Fel Chem Inst Can; assoc Royal Inst Chem. *Res:* Chemistry and toxicology of biologically active materials. *Mailing Add:* 2839 Flannery Dr Ottawa ON K1V 9S8 Can

HEGGTVEIT, HALVOR ALEXANDER, b Montreal, PQ, Mar 3, 33; m 60; c 2. PATHOLOGY. *Educ:* Univ Ottawa, MD, 57; Am Bd Path, dipl, 62; FRCP(C); FRCPath. *Prof Exp:* Resident, State Univ NY Downstate Med Ctr & Kings County Hosp, 59-61, instr, Med Ctr, 61-62; from lectr to prof pathol, fac med, Univ Ottawa, 62-82, actg head dept, 73-75; pathologist, Ottawa Civic Hosp, 75-82; PROF PATH, FAC HEALTH SCI, MCMASTER UNIV, 82-; PATHOLOGIST, CHEDOKE-MCMASTER & HAMILTON GEN HOSPS, HAMILTON, ONT, 82- *Concurrent Pos:* Fel path, Univ Ottawa & Ottawa Gen Hosp, Ont, 57-59; sr res fel, Ont Heart Found, 66-70; pathologist, Ottawa Gen Hosp, 62-75, actg dir labs, 73-75; regional pathologist, Ont Atty Gen Dept, 63-82; res assoc, Ont Heart Found, 70-78; dep registr, Med Coun Can, 67-71; co-registr, Registry for Tissue Reactions to Drugs, 70-76; mem sci subcomt, Can Heart Found, 70-73; consult pathologist, Nat Defense Med Ctr, Fed Food & Drug Directorate, Ottawa, Children's Hosp of Eastern Ont, Ottawa Gen Hosp & Nat Res Coun Can; adv, Sect Cardiovasc Dis, WHO; mem coun, Can Asn Pathologists, 75-78. *Honors & Awards:* Carveth Sci Award, Can Asn Pathol, 67. *Mem:* Am Asn Path; fel Am Col Cardiol; Int Acad Path; fel Am Soc Clin Path; fel Col Am Path. *Res:* Cardiovascular pathology and electron microscopy of myocardium. *Mailing Add:* Dept Pathol McMaster Univ Med Ctr 1200 Main St W Hamilton ON L8N 3Z5 Can

HEGLUND, NORMAN C, b Lynn, Mass, Jan 1, 49. MUSCLE PHYSIOLOGY. *Educ:* Harvard Univ, PhD(biol), 79. *Prof Exp:* RES ASSOC, MUS IMPERATIVE ZOOL & DIV APPL SCI, HARVARD UNIV, 79- *Mem:* Am Physiol Soc. *Mailing Add:* 12 Northview Ave South Chelmsford MA 01824

HEGMANN, JOSEPH PAUL, b Kansas City, Mo, Sept 26, 40; m 60; c 3. QUANTITATIVE GENETICS. *Educ:* Univ Ill, BS, 62, MS, 66, PhD(quant genetics), 68. *Prof Exp:* From asst prof to assoc prof, 68-76, PROF ZOOL, UNIV IOWA, 73- *Mem:* AAAS; Soc Multivariate Exp Psychol; Genetics Soc Am; Am Genetic Asn; Sigma Xi. *Res:* Behavioral genetics; genetic variation affecting differences in complexly varying behaviors in laboratory and wild populations of rodents and insects. *Mailing Add:* Dept Biol Univ Iowa Iowa City IA 52242

HEGRE, CARMAN STANFORD, b Columbus, Mont, Dec 5, 37; m 59; c 3. BIOCHEMISTRY. *Educ:* Va Polytech Inst, BS, 59, MS, 61, PhD(biochem), 63. *Prof Exp:* Res assoc & fel pharmacol, Western Reserve Univ, 63-66; supvry res prog coordr, Nat Marine Water Qual Lab, 66-80, ENVIRON SCIENTIST, ENVIRON RES LAB, ENVIRON PROTECTION AGENCY, 80- *Concurrent Pos:* Mem group of experts on sci aspects of marine pollution, UN, 73- *Mem:* AAAS; Am Chem Soc; Am Soc Limnol & Oceanog; Water Pollution Control Fedn. *Res:* Enzymology; non-photosynthetic carbon dioxide fixation; physiology, toxicology and intermediary metabolism of marine plankton. *Mailing Add:* Environ Res Lab Narragansett S Ferry Rd Narragansett RI 02882

HEGRE, ORION DONALD, b Havre, Mont, Oct 19, 43; m 65; c 3. ANATOMY, TISSUE CULTURE. *Educ:* Univ Minn, Minneapolis, BA, 64, PhD(anat), 68. *Prof Exp:* Am Diabetes Asn fel, 68-70, asst prof, 70-77, ASSOC PROF ANAT, UNIV MINN, MINNEAPOLIS, 77- *Mem:* Tissue Cult Asn; Am Anatomists; Am Diabetes Asn; Endro Soc; Sigma Xi. *Res:* Experimental embryology and diabetes; organ and tissue culture of the islets of Langerhans; insulin synthesis and secretion. *Mailing Add:* Dept of Anat-4-112 Owre Hall Univ of Minn Minneapolis MN 55455

HEGSTED, DAVID MARK, b Rexburg, Idaho, Mar 25, 14; m 42; c 2. NUTRITION, BIOCHEMISTRY. *Educ:* Univ Idaho, BS, 36; Univ Wis, MS, 38, PhD(biochem), 40. *Hon Degrees:* AM, Harvard Univ, 62; DSc, Univ Idaho, 86. *Prof Exp:* Asst biochem, Univ Wis, 36-41; res chemist, Abbott Labs, 41-42; from instr to assoc prof, 42-62, prof nutrit, Sch Pub Health,

Harvard Univ, 62-78; adminr, Human Nutrit Ctr, USDA, 78-82; assoc dir res, New Eng Regional Primate Res Ctr, 82-86, EMER PROF NUTRIT, 82- *Concurrent Pos:* Nutrit consult, Columbian Govt, 46 & Inst Inter-Am Affairs, Peru, 50; chmn food & nutrit bd, Nat Acad Sci-Nat Res Coun; mem var exp comts, WHO & Food & Agr Orgn, UN, 60-; ed, Nutrit Revs, 68-78. *Honors & Awards:* Osborne-Mendel Award, Am Inst Nutrit, 65; Conrad Elvejhem Award, Am Inst Nutrit, 79; Eleanor Naylor Dana Award, Am Health Found, 80. *Mem:* Nat Acad Sci; AAAS; hon mem Am Dietetic Asn; Am Chem Soc; Am Inst Nutrit (pres, 72-73). *Res:* Comparative nutrition; protein and calorie requirements; calcium requirements and metabolism; iron metabolism; nutrition problems of under-developed areas; experimental atherosclerosis. *Mailing Add:* One Pine Hill Dr Southborough MA 01772

HEGSTED, MAREN, b Salt Lake City, Utah, Sept 16, 50. MINERAL METABOLISM, INTERNATIONAL NUTRITION. *Educ:* Univ Utah, BS, 73; Univ Wis-Madison, MS, 78, PhD(nutrit), 80. *Prof Exp:* Volunteer nutrit, Peace Corps, ACTION, 73-75; ASSOC PROF NUTRIT, LA STATE UNIV, 80- *Concurrent Pos:* Consult, AID, 84; sect head, Human Nutrit & Food Sect, Human Ecol, La State Univ, 83- *Mem:* Am Inst Nutrit; Inst Food Technologists; Asn Women Develop. *Res:* Diet, exercise and bone density; rice bran and blood cholesterol; trace and ultratrace elements including zinc, copper and boron. *Mailing Add:* Human Nutrit & Food Sch Human Ecol La State Univ Baton Rouge LA 70803-4300

HEGSTROM, ROGER ALLEN, b New Ulm, Minn, July 28, 41; m 67. THEORETICAL CHEMISTRY. *Educ:* St Olaf Col, BA, 63; Harvard Univ, AM, 64, PhD(chem), 68. *Prof Exp:* Res assoc chem, Nat Res Coun-Nat Bur Standards, 68-69; from asst prof to assoc prof, 69-80, PROF CHEM, WAKE FOREST UNIV, 80- *Concurrent Pos:* Guggenheim fel, 78-79; Sr vis fel, Oxford Univ, 78-79 & Univ Mich, 82; vis prof, Univ Ill, 90. *Mem:* Am Phys Soc; Am Chem Soc; Sigma Xi. *Res:* Electromagnetic and weak interactions in atoms and molecules; quantum mechanics. *Mailing Add:* Dept Chem Wake Forest Univ Winston-Salem NC 27109

HEGWOOD, DONALD AUGUSTINE, b Sumrall, Miss, Aug 28, 31; m 57; c 3. HORTICULTURE. *Educ:* Miss State Col, BS, 54, MS, 59; La State Univ, Baton Rouge, PhD(hort), 65. *Prof Exp:* Horticulturist, Coastal Plain Exp Sta, Univ Ga, 59-62 & Agr Exp Sta, La State Univ, Baton Rouge, 65-67; horticulturist, Coop Exten Serv, Univ Ga, 67-70, Coastal Plain Exp Sta, 70-72 & Dept Hort, 72-75; hort res & teaching, 75-77, assoc prof hort, Miss State Univ, 77-; AT COL AGR, UNIV MD, COLLEGE PARK. *Concurrent Pos:* USDA res grant, 66-67. *Mem:* Am Soc Hort Sci; Am Soc Agron; Soil Sci Soc Am; Int Soc Hort Sci. *Res:* Mineral nutrition, water relations and growing media development for woody ornamental nursery crops. *Mailing Add:* Hort Dept Miss State Univ PO Drawer T Mississippi State MS 39762

HEGYELI, RUTH I E J, b Stockholm, Sweden, Aug 14, 31; US citizen; wid. CANCER & CARDIOVASCULAR DISEASES. *Educ:* Univ Toronto, BA, 58, MD, 62. *Prof Exp:* Res asst anat, Univ Toronto, 59; intern med, Queen Charlotte Gen Hosp, Can, 61 & Toronto Gen Hosp, 62-63; res assoc tissue cult, Inst Muscle Res, Woods Hole, Mass, 63-65; head Tissue Cult Lab, Battelle Mem Inst, 65-67, sr res pathologist & head, Cell Biol Lab, 67-69; med officer, artificial heart res, Nat Heart Inst, 69-71 & prog planning, Off of Dir, 71-73, chief prog planning, Prog Develop & Eval Br, 73-76, actg dir, Off Prog Planning & Eval, 75-76; asst dir int relations, 76-86, ASSOC DIR INT RELATIONS, OFF OF DIR, NAT HEART, LUNG & BLOOD INST, 87- *Concurrent Pos:* Sr invest, res projs, NIH, 65-69; ed, J Soviet Res Cardiovasc Dis, 79-85; mem, sci adv bd, Giovanni Lorenzini Found, 82-; mem, bd dirs, Coun Geriat Cardiol, 88- *Honors & Awards:* Super Serv Award, Dept Health, Educ & Welfare; Ger Friendship Award, 88; Copernicus Award, 88. *Mem:* NY Acad Sci; Am Soc Cell Biol; Am Soc Artificial Internal Organs; Coun Epidemiol & Prev; Acad Med, Toronto, Can. *Res:* Cancer; cardiovascular diseases. *Mailing Add:* 24301 Hanson Ct Gaithersburg MD 20879

HEGYI, DENNIS, b Reading, Pa, Dec 23, 42; c 2. ASTROPHYSICS. *Educ:* Mass Inst Technol, BS, 63; Princeton Univ, PhD(physics), 68. *Prof Exp:* Nat Res Coun resident res assoc astrophys, Goddard Inst Space Studies, NASA, 68-70; staff mem, Dept Physics & Astron, Boston Univ, 70-73; asst prof, Bartol Res Found, 73-75; from asst prof to assoc prof, 75-86, PROF, DEPT PHYSICS, UNIV MICH, ANN ARBOR, 86- *Concurrent Pos:* Indust consult, sensor design. *Mem:* Am Phys Soc; Am Astron Soc. *Res:* Primordial helium and deuterium, cosmic blockbody radiation, maximum mass of neutron stars, galactic halos; cosmological dark matter. *Mailing Add:* Dept Physics Univ Mich Ann Arbor MI 48104

HEGYVARY, CSABA, b Debrecen, Hungary, Feb 14, 38; US citizen; m 71; c 2. PSYCHIATRY. *Educ:* Med Univ Budapest, MD, 62; Hungarian Stat Exam Bd, cert, 66. *Prof Exp:* Asst prof pathophysiol, Inst Pathophysiol, Med Univ Budapest, 62-66; res assoc virol, Krankenhaus Isar, Munich, 66-67; res assoc physiol, Med Sch, Vanderbilt Univ, 67-70; asst prof, 70-72, attend psychiat, 84, ASSOC PROF PHYSIOL, MED SCH, RUSH UNIV, 72-, ASSOC PROF CARDIOL, 75-, PROF INTERNAL MED, 78- *Honors & Awards:* Brainard Award, 76, 83. *Mem:* Biophys Soc; Physiol Soc; Am Heart Asn; AAAS. *Res:* Mechanism of ion-transport across cell membranes, mode of action of digitalis-glycosides on the heart; experimental induction of cardiac hypertrophy. *Mailing Add:* 901 Boren No 1020 Seattle WA 98104

HEGYVARY, SUE THOMAS, b Dry Ridge, Ky, Nov 28, 43; m 71; c 2. MEDICAL SOCIOLOGY, HEALTH SYSTEMS MANAGEMENT. *Educ:* Univ Ky, BSN, 65; Emory Univ, MN, 66; Vanderbilt Univ, PhD(sociol), 74. *Prof Exp:* Staff nurse, Univ Ky Med Ctr, 65; instr med-surg nursing, Col Nursing, Univ Fla, 66-67, asst prof & chmn dept, 67-69; asst prof nursing & social, Rush Univ, 72-74, assoc prof mednursing & chairperson dept, 74-77, asst prof social, Rush Univ Med Col, 77-80, PROF NURSING, ASSOC VPRES & ASSOC DEAN NURSING, COL NURSING, RUSH UNIV, RUSHY-PRESBY-ST LUKE'S MED CTR, 77- ASSOC PROF SOCIAL, MED COL, RUSH UNIV, 80- *Concurrent Pos:* Consult, Vet Admin Hosp, Miami, Fla, 68-69 & Student Health Coalition, Vanderbilt Univ, 71; investr, Dept Health, Educ & Welfare contract, Rush-Presby-St Luke's Med Ctr Chicago, 72-; sci staff, Nat Hosp Inst, Netherlands, 79-80. *Mem:* Am Sociol Asn; Am Nurses' Asn; Am Acad Nursing. *Res:* Assessment of the quality of patient care and the influence of organizational, demographic and professional variables on the quality of care. *Mailing Add:* Dean/Dir Nursing Univ Wash Seattle WA 98195

HEHEMANN, ROBERT F(REDERICK), metallurgy; deceased, see previous edition for last biography

HEHRE, EDWARD JAMES, b New York, NY, Dec 14, 12; m 38; c 3. ENZYMOLOGY, CARBOHYDRATE CHEMISTRY. *Educ:* Cornell Univ, BA, 34, MD, 37. *Prof Exp:* Intern path, New York Hosp, 37-38; from asst to assoc prof bact & immunol, Med Col, Cornell Univ, 38-56; head dept, 56-78, PROF MICROBIOL & IMMUNOL, ALBERT EINSTEIN COL MED, 56- *Concurrent Pos:* John Simon Guggenheim fel, 64-65; John Polachek fel, 70-71; vis prof, Grad Fac Sci, Tokyo Kyoiku Univ, 64-65; vis prof, Shizuoka Univ, 71. *Honors & Awards:* John Fogarty Sr Int Res Fel, NIH, 88; Medal of Merit, Japan Soc Starch Sci, 89. *Mem:* Am Chem Soc; Am Soc Biol Chem. *Res:* Enzymic synthesis of polysaccharides; glycosyl transfer concept; structural basis of dextran serotypes; glycosyl-proton interchange basis of unified class of glycosylases; use of nonglycosidic substrates to gain new understanding of catalysis by glycosylases. *Mailing Add:* Dept Microbiol & Immunol Albert Einstein Col Med New York NY 10461

HEHRE, EDWARD JAMES, JR, b New York, NY, Feb 22, 40; m 63; c 2. PHYCOLOGY, PLANT TAXONOMY. *Educ:* New Eng Col, BS, 63; Univ NH, PhD(bot), 69. *Prof Exp:* Asst prof bot, Southhampton Col, Long Island Univ, 69-77; mem fac bot, Univ NH, 77-78; CONSULT MARINE BIOL, UNIV AZORES, 78- *Mem:* Phycol Soc Am; Int Phycol Soc. *Res:* Occurrence, distribution, seasonality and reproductive periodicity of marine red algae in New Hampshire; algal and vascular plant flora of Gardiners Island, New York; floristic studies and ecology of marine algae of New Hampshire; marine flora of Azores Islands. *Mailing Add:* 26 Park St South Berwick ME 03908

HEIBA, EL-AHMADI IBRAHIM, b Egypt, May 7, 26; m 55; c 2. ORGANIC & PHYSICAL ORGANIC CHEMISTRY. *Educ:* Alexandria Univ, BSc, 48; Univ London, PhD, 52. *Prof Exp:* Asst prof, Univ Cairo, 52-54; sr res chemist, Sch Pharm, Univ London, 54-55; fel & supvr radiation chem, Phoenix Mem Lab, Univ Mich, 55-58; sr radiation chemist, Arthur D Little, Inc, Mass, 58-62; dir chem res, Mobil Res & Develop Corp, 69-75; DIR CROP CHEM RES, MOBIL CHEM CO, 75- *Mem:* Am Chem Soc. *Res:* Free radicals chemistry and reaction mechanisms; chemical factor influencing plant photosynthesis and crop productivity; homogeneous and heterogeneous catalysis. *Mailing Add:* 11 Balsam Lane Princeton NJ 08540

HEIBEL, JOHN T(HOMAS), b Brooklyn, NY, Mar 22, 43; m 67; c 1. CHEMICAL & ELECTRICAL ENGINEERING. *Educ:* Univ Calif, Berkeley, BS, 65; Univ Ariz, MS, 67, PhD(chem eng), 69. *Prof Exp:* Chem engr, Naval Biol Lab, Calif, 65; chem & systs eng process simulation & control, Indust Nucleonics Corp, Ohio, 67-69; vis asst prof chem eng, 69-70, asst prof, 70-76, ASSOC PROF CHEM ENG, OHIO STATE UNIV, 76- *Concurrent Pos:* Consult, Indust Nucleonics Corp, Ohio, 70- *Mem:* Am Inst Chem Engrs; Am Chem Soc; Instrument Soc Am; Inst Elec & Electronics Engrs; Am Soc Eng Educ. *Res:* Chemical process dynamics, especially simulation, optimization and control of chemical and environmental process; application of instrumentation and real-time computation for increased process visibility and understanding; environmental systems analysis. *Mailing Add:* 4114 Winfield Columbus OH 43220

HEIBERGER, PHILIP, b New York, NY, Mar 15, 19; m 42; c 3. PHYSICAL CHEMISTRY, ORGANIC CHEMISTRY. *Educ:* Polytech Inst Brooklyn, BSc, 39; Cornell Univ, MA, 41; Univ Tex, PhD(org chem), 53. *Prof Exp:* Asst, Cornell Univ, 39-41; chemist, Colgate-Palmolive Co, 41-42; group leader, Ralph L Evans Assoc, 42-46; chemist new prod develop, Interchem Corp, 47-49; chemist thermoplastics & lacquers, Atlas Powder Co, 49-51; res scientist, Univ Tex, 51-53; group leader oil & resin develop, Nat Lead Co, 53-57; res chemist, E I Du Pont De Nemours & Co, Inc, 57-66, develop specialist, 66-70, licensing coordr, 71-74, mgr technol licensing, 74-76, res assoc, 76-83; RETIRED. *Concurrent Pos:* Engr, Kellex Corp, 45; trustee, 74-76, pres, Paint Res Inst, 76-78; vol exec, Int Exec Serv Corp, 84 - *Mem:* Am Chem Soc. *Res:* Synthetic latices; condensation resins; organic coatings; utilization of drying oils; industrial paint systems; dispersion techniques; organic reaction mechanisms. *Mailing Add:* 88 Cherry Hill Lane Broomall PA 19008

HEIBLUM, MORDEHAI, b Tel Aviv, Israel, May 25, 47; US citizen; m 70; c 4. MOLECULAR BEAM EPITAXY GROWTH OF III-V SEMI-CONDUCTORS, BALLISTIC TRANSPORT OF CHARGED CARRIERS IN SOLIDS. *Educ:* Israeli Inst Technol, BSc, 73; Carnegie Mellon Univ, MSc, 74; Univ Calif, Berkeley, PhD(elect eng), 78. *Prof Exp:* Res staff mem, IBM, T J Watson Res Ctr, 78-86, res staff mem & group mgr, 86-90; PROF, DEPT PHYSICS, WEIZMANN INST SCI, ISRAEL, 90- *Mem:* Fel Inst Elec & Electronics Engrs; fel Am Physics Soc; Am Vacuum Soc. *Res:* In the physics and invention of new ultra fast devices, based primarily on III-V semi-conductor compounds. *Mailing Add:* Dept Physics Weizmann Inst Sci Rehovot 76100 Israel

HEICHEL, GARY HAROLD, b Park Falls, Wis, Nov 9, 40; m 58, 88; c 2. PLANT PHYSIOLOGY, PLANT BIOCHEMISTRY. *Educ:* Iowa State Univ, BS, 62; Cornell Univ, MS, 64, PhD(agron), 68. *Prof Exp:* Asst agron, Cornell Univ, 62-64, 66-67 & Univ Philippines-Cornell Univ Grad Educ Prog, 64-66; asst crop physiologist, 68-73, assoc plant physiologist, 73-76, plant physiologist, Conn Agr Exp Sta, 76; plant physiologist, USDA Agr Res Serv, 76-90; HEAD, DEPT AGRON, UNIV ILL, URBANA, 90- *Concurrent Pos:* Prog Mgr, USDA Competitive Grants Off, 81; mem, Energy Eng Bd, Nat Res Coun, 83-87; consult ed, Am Scientist, 83-90; ed-in chief, Am Soc Agron,

88-90; adj prof agron, Univ Minn, 76-90; mem, bd trustees, Am Soc Plant Physiol, 88-90; Kellogg fel, Food Agr Policy, Resources Future, 89. *Honors & Awards:* Outstanding Scientist of Yr, USDA-Agr Res Serv, 86; Crop Sci Res Award, Crop Sci Soc Am, 87. *Mem:* Fel Am Soc Agron; fel Crop Sci Soc Am (pres-elect, 90-91); Am Soc Plant Physiologists; Am Forage & Grassland Coun; Sigma Xi. *Res:* Symbiotic nitrogen fixation; photosynthesis; productivity of forage legumes; nitrogen cycling; physiology; physiological, morphogenetic and environmental limitations to forage yield; energy use in agricultural ecosystems. *Mailing Add:* Dept Agron Univ Ill 1102 S Goodwin Ave Urbana IL 61801-4798

HEICHELHEIM, HUBERT REED, b McAlester, Okla, Jan 5, 31; m 56. CHEMICAL ENGINEERING. *Educ:* Univ Notre Dame, BS, 53, MS, 56; Univ Tex, PhD(chem eng), 62. *Prof Exp:* Asst prof, 61-67, assoc prof, 67-80, ASST CHMN CHEM ENG, TEX TECH UNIV, 80- *Concurrent Pos:* Process engr, Phillips Petrol & Suntide Refining; Gastdozent, Fachhachschule Wilhelmshaven, Ger. *Mem:* Am Inst Chem Engrs; Am Chem Soc; Am Soc Engr Educ. *Res:* Volumetric behavior and thermodynamic properties of gases. *Mailing Add:* Dept Chem Eng Tex Tech Univ Lubbock TX 79406

HEICKLEN, JULIAN PHILLIP, b Rochester, NY, Mar 9, 32; m 59; c 3. ATMOSPHERIC CHEMISTRY. *Educ:* Cornell Univ, BChE, 54; Univ Rochester, PhD(chem), 58. *Prof Exp:* Res fel chem, Univ Minn, 59-60 & Univ Calif, 60-62; mem tech staff, Aerospace Corp, 62-65, mgr pyrolytic mat & kinetics sect, 65-67; assoc prof, 67-71, PROF CHEM, PA STATE UNIV, UNIVERSITY PARK, 71-, MEM TECH STAFF, CTR AIR ENVIRON STUDIES & IONOSPHERE RES LAB, 67- *Concurrent Pos:* Sigma Xi regional lectr, 71-72; consult panel biol effects hydrocarbon air pollutants, Nat Res Coun, 72; vis prof, Hebrew Univ Jerusalem, 73-74; pres, Heicklen Assocs, 75- *Honors & Awards:* Am Chem Soc Award, Creative Advan Environ Sci & Technol, 84; Boris Pregel Award Appl Sci & Technol, NY Acad Sci, 84; Frank A Chambers Award, Air Pollution Control Asn, 85. *Mem:* Fel AAAS; Am Chem Soc; fel Am Phys Soc; Royal Soc Chem; fel NY Acad Sci. *Res:* Photochemistry; reaction kinetics; vibrational spectroscopy; gerontology; carcinogenesis; mutogenesis; photochemistry; combustion chemistry. *Mailing Add:* Dept Chem Pa State Univ 222 Davey Lab University Park PA 16802

HEID, ROLAND LEO, b Erie, Pa, July 7, 14. PHYSICS. *Educ:* St Vincent Col, AB, 37; Johns Hopkins Univ, PhD(physics), 50. *Prof Exp:* From instr to asst prof math, St Vincent Col, 50-54, chmn dept, 50-54, from assoc prof to prof physics, 54-85, chmn dept, 54-68 & 72-84, EMER PROF, ST VINCENT COL, 85- *Mem:* Am Phys Soc; Optical Soc Am; Am Asn Physics Teachers. *Res:* Spectroscopy; optics. *Mailing Add:* Dept Physics St Vincent Col Latrobe PA 15650

HEIDBREDER, GLENN R, b Gerald, Mo, July 22, 29; div; c 1. ELECTRICAL ENGINEERING. *Educ:* Yale Univ, BE, 51, MEng, 56, DEng(elec eng), 59. *Prof Exp:* Instr elec eng, Yale Univ, 57-59; mem tech staff, Space Tech Labs, 59-62 & Aerospace Corp, 62-63 & 65-67; asst prof, 63-67, ASSOC PROF ELEC ENG, UNIV CALIF, SANTA BARBARA, 67- *Concurrent Pos:* Lectr, Univ Calif, Los Angeles, 60-63; sr scientist, Tech Serv Corp, 70-71. *Mem:* Inst Elec & Electronics Engrs. *Res:* Radar systems; communications theory; radio propagation in random media. *Mailing Add:* Dept Elec Eng Univ Calif Santa Barbara CA 93106

HEIDCAMP, WILLIAM H, b Saugerties, NY, Oct 26, 44; m 66; c 2. DEVELOPMENTAL BIOLOGY, CELL BIOLOGY. *Educ:* Siena Col NY, BS, 66; Univ Pittsburgh, PhD(biol), 71. *Prof Exp:* Instr biol, Carlow Col, 68-69, Carnegie-Mellon Univ, 69-71, Concordia Univ, 71-73; asst prof, 73-80, PROF BIOL, GUSTAVUS ADOLPHUS COL, 80-, CHMN DEPT, 78- *Concurrent Pos:* Fel, Montreal Cancer Inst, 71-73; res fel, Dept Path, Univ Montreal, 73-74. *Mem:* AAAS; Am Inst Biol Sci; Nat Sci Teachers Asn; Sigma Xi. *Res:* Interaction of cation metabolism and Em expression of neoplasia. *Mailing Add:* Biol Dept Gustavus Adolphus Col St Peter MN 56082

HEIDEGER, WILLIAM J(OSEPH), b Beaver, Pa, Sept 17, 32; m 58; c 3. CHEMICAL ENGINEERING, MASS TRANSFER. *Educ:* Carnegie Inst Technol, BS, 54; Princeton Univ, MSE, 56, PhD(chem eng), 59. *Prof Exp:* From asst prof to assoc prof, 57-70, PROF CHEM ENG, UNIV WASH, 70- *Concurrent Pos:* Consult, Puget Sound Div, Pac Corp, 63; Atomic Energy Comn fel, 67. *Mem:* Am Inst Chem Engrs; Am Chem Soc; Am Soc Eng Educ. *Res:* Mass transfer and interfacial phenomena; biological mass transferr. *Mailing Add:* Dept Chem Eng Univ Wash Seattle WA 98195

HEIDELBAUGH, NORMAN DALE, b Philadelphia, Pa, July 29, 27; m 63; c 3. FOOD TOXICOLOGY, PUBLIC HEALTH. *Educ:* Univ Pa, VMD, 54; Tulane Univ, MPH, 58; Mass Inst Technol, MS, 63, PhD(food sci), 70. *Prof Exp:* Vet food technologist, USAF, 54-62; food technologist, Nat Acad Sci-USAF Mission, Arg, 63-64, vet officer & lectr food sci, US Air Force Sch Aerospace Med, 64-67, chief food sci, NASA Manned Spacecraft Ctr, Houston, 70-74, mgr, USAF Food Res & Develop Prog, US Army Natick Lab, 74-76; prof Food Sci & Technol, 76-90, DIR FOOD SAFETY PROG DEVELOP, TEX A&M UNIV, 90- *Honors & Awards:* Underwood-Prescott Award, 74. *Mem:* Am Vet Med Asn; Inst Food Technol. *Res:* Food science and technology; food safety; public health; food and drug law; nutrition under stress conditions; veterinary medicine; zoonoses control; food deterioration effects on health. *Mailing Add:* Off Dean Col Vet Med Tex A&M Univ College Station TX 77843

HEIDELBERGER, MICHAEL, immunochemistry; deceased, see previous edition for last biography

HEIDELBERGER, PHILIP, b Madison, Wis, Nov 25, 51. DISCRETE EVENT SIMULATION. *Educ:* Oberlin Col, BA, 74; Stanford Univ, PhD(opers res), 78. *Prof Exp:* RES STAFF MEM, THOMAS J WATSON RES CTR, IBM, 78- *Concurrent Pos:* Assoc ed, Opers Res, 82-90; prog chmn, 1989 Winter Simulation Conf, 89; area ed, Asn Comput Mach Trans on Computer Simulation, 90- *Mem:* Asn Comput Mach; Inst Elec & Electronics Engrs; Opers Res Soc Am. *Res:* Discrete event simulation, including efficient simulation techniques and algorithms; probabilistic aspects of simulations. *Mailing Add:* Thomas J Watson Res Ctr IBM PO Box 704 Yorktown Heights NY 10598

HEIDEMANN, STEVEN RICHARD, b New York, NY, Sept 9, 49; m 73; c 2. CELL BIOLOGY, DEVELOPMENTAL BIOLOGY. *Educ:* State Univ NY, Stony Brook, BS, 71; Princeton Univ, MA, 73, PhD(biol), 76. *Prof Exp:* Res assoc cell biol, Dept Molecular, Cellular & Develop Biol, Univ Colo, 77-78; asst prof, 78-82, assoc prof physiol & biol sci prog, 82-87, PROF, DEPT PHYSIOL, MICROBIOL & PUB HEALTH, MICH STATE UNIV, 87- *Concurrent Pos:* NSF fel, 77-78; Res Career Develop Award, NIH, 81-86. *Mem:* Am Soc Cell Biol; Am Physiol Soc. *Res:* Role of extoskeleton in neuronal growth; cytomechanics of neural development. *Mailing Add:* Dept Physiol 219 Giltner Hall Mich State Univ East Lansing MI 48824

HEIDENREICH, CHARLES JOHN, b Berwyn, Ill, Nov 27, 27; m 51; c 1. ANIMAL HUSBANDRY. *Educ:* Univ Ill, BS, 51; SDak State Col, MS, 52; Univ Mo, PhD(animal physiol), 57. *Prof Exp:* Asst animal husb, SDak State Col, 51-52; from asst to instr animal sci, Univ Mo, 53-56; from asst prof to assoc prof animal physiol, Purdue Univ, 56-66; assoc prof, 66-67, PROF ANIMAL SCI, UNIV WIS-PLATTEVILLE, 67- *Concurrent Pos:* Consult livestock & feed mfg indust. *Mem:* Am Soc Animal Sci; Am Reg Prof Animal Sci. *Res:* Animal physiology, especially environmental effects upon productive efficiency of meat producing animals; non-ruminant Nutrit. *Mailing Add:* Dept Agr Sci Univ Wis-Platteville Platteville WI 53818

HEIDENTHAL, GERTRUDE ANTOINETTE, b Carbondale, Pa, June 9, 08. ZOOLOGY. *Educ:* Mt Holyoke Col, AB, 30, AM, 32; Univ Chicago, PhD(zool), 38. *Prof Exp:* Asst genetics, Rockefeller Inst, 33-37, Univ Rochester, 38-41 & Univ Pa, 41-43; instr zool, Wellesley Col, 43-45; from asst prof to prof, 45-72, EMER PROF BIOL, RUSSELL SAGE COL, 72- *Concurrent Pos:* Res assoc, Union Univ, NY, 51-53; assoc geneticist, Brookhaven Nat Lab, 53-54. *Res:* Genetics of coat color of guinea pig; genetics of Drosophila and Habrobracon. *Mailing Add:* Beechwood Apt 214 2218 Bierdett Ave Troy NY 12180

HEIDER, SHIRLEY (SCOTT) A(MBORN), b West Salem, Wis, Apr 11, 12; m 49; c 1. MECHANICAL ENGINEERING. *Educ:* Univ Wis, BS, 34. *Prof Exp:* Res & develop engr, Trane Co, 34-41; engr naval archit, Bur Ships, US Dept Navy, 41-46; underwriting supvr, Fed Housing Admin, 46-47; partner, Stem & Heider Co, 47-49; consult engr, Wilberding Co, Inc, 49-51; res engr, bldg res adv bd, Nat Acad Sci, 51-57, Fed Housing Admin, 57-64 & NSF, 64-72; gen engr, Dept Housing & Urban Develop, 72-80; CONSULT. *Concurrent Pos:* Ed consult, Nat Bur Standards, 80-82; consult, 80-83. *Mem:* Nat Soc Prof Engrs; Am Soc Testing & Mat. *Res:* Research facilities; building research; mechanical and electrical equipment; building materials and components. *Mailing Add:* 403 Russell Ave Apt G-5 Gaithersburg MD 20877-2811

HEIDGER, PAUL MCCLAY, JR, b St Johnsbury, Vt, Sept 13, 41; m 67; c 2. ANATOMY. *Educ:* Univ Northern Colo, AB, 63; Tulane Univ, PhD(anat), 67. *Prof Exp:* NIH res fel anat, Harvard Med Sch, 67-69; from asst prof to assoc prof, Med Sch, Tulane Univ, 69-74; assoc prof, 74-80, PROF ANAT, MED SCH, UNIV IOWA, 80- *Concurrent Pos:* NIH res grant, Med Sch, Tulane Univ, 71-74; NIH res contract, Med Sch, Univ Iowa, 74-76; NIH res grant, 78- *Mem:* Am Asn Anatomists; Electron Micros Soc Am; Soc Study Reproduction; Sigma Xi. *Res:* Cytochemistry and fine structure of reproductive system; tissue culture of normal and malignant prostate gland; electron microscopy of bacterial L-forms. *Mailing Add:* Dept Anat Univ Iowa Col Med Iowa City IA 52242

HEIDNER, ROBERT HUBBARD, b Holyoke, Mass, Mar 14, 19; m 44; c 3. ANALYTICAL CHEMISTRY. *Educ:* Hamilton Col, BS, 40, MA, 42. *Prof Exp:* Asst chem, Hamilton Col, 40-42; res chemist, Monsanto Chem Co, Ohio, 42-45 & Mass, 45-52; res chemist, Chemstrand Corp, 52-55, group leader, Wet Methods Anal Group, Res Ctr, Ala, 55-60; anal group leader, Chemstrand Res Ctr, Inc, 60-68; supvr anal & test methods develop, Monsanto Textiles Co, 68-78, supvr, Anal & Control Lab, Tech Ctr, 78-80, supvr, Control Lab Standards, 80-82; RETIRED. *Mem:* Am Chem Soc. *Res:* Analytical chemistry of organic monomers, intermediates, high polymers and chemical fibers; wet methods; ultraviolet, visible and infrared spectroscopy; atomic absorption; tristimulus colorimetry; polymer and fiber characterization techniques. *Mailing Add:* Six Sea Swallow Terr Ormond Beach FL 32176-2238

HEIDRICK, LEE E, b Little Valley, NY, June 23, 21; m 52. PLANT PATHOLOGY. *Educ:* Cornell Univ, BS, 43, MS, 51; WVa Univ, PhD(plant path), 61. *Prof Exp:* Res assoc plant path, Rockefeller Found, Mex, 51-52, plant pathologist, Colombia, 52-63; res specialist, Chevron Chem Co, 63-84; RETIRED. *Mem:* AAAS; Potato Asn Am; Am Phytopathological Soc; Sigma Xi. *Res:* Development of agricultural research in Latin America; organization, administration and training of agricultural technical workers; agricultural fungicides. *Mailing Add:* 100 Winthrop Dr Ithaca NY 14850-1733

HEIDRICK, MARGARET LOUISE, b Beloit, Kans, June 23, 38. BIOCHEMISTRY, IMMUNOLOGY. *Educ:* Marymount Col, BS, 59; Univ Mo, 64; Univ Nebr, MS, 68, PhD(biochem), 70. *Prof Exp:* Chemist, US Fed Food & Drug Admin, Kansas City, Mo, 59-61; res technologist, Eppley Inst Cancer Res, Omaha, Nebr, 64-68; asst prof, 73-78, ASSOC PROF BIOCHEM, UNIV NEBR MED CTR, OMAHA, 78- *Concurrent Pos:* NIH fel, Oak Ridge Nat Lab, 70-72; NIH staff fel, Geront Res Ctr, Baltimore, Md, 72-73; mem, Nat Adv Coun, Nat Inst Allergy & Infectious Dis, 75-77. *Mem:* Am Asn Immunologists; Am Soc Cell Biologists; Geront Soc; Sigma Xi. *Res:* Cancer; aging of immune system; cytochrome P-450 isozymes. *Mailing Add:* 600 S 42nd St Univ Nebr Col Med Omaha NE 68198-4525

HEIDT, GARY A, b South Bend, Ind, May 20, 42; div; c 3. MAMMALOGY. *Educ:* Manchester Col, BS, 64; Mich State Univ, MS, 68, PhD(zool), 69. *Prof Exp:* Asst prof biol sci, Mich State Univ, 69-70; from asst prof to assoc prof, 70-79, PROF BIOL & DIR BASIC ANIMAL SERV UNIT, UNIV ARK, LITTLE ROCK, 79- *Concurrent Pos:* Chmn bd dirs, Little Rock Mus Sci & Hist, 74-75; adj grad fac, Memphis State Univ, 80- *Mem:* Am Soc Mammal; Wildlife Soc; Am Asn Lab Animal Sci; Wildlife Dis Asn. *Res:* Disease, ecological and behavioral interactions of mammals; primarily dealing with carnivores and rodents. *Mailing Add:* Dept Biol Univ Ark Little Rock AR 72204

HEIFETZ, CARL LOUIS, b Somerville, NJ, Mar 9, 35; m 59; c 2. BACTERIOLOGY, CHEMOTHERAPY. *Educ:* Univ Md, BS, 57, MS, 60, PhD(microbiol), 65. *Prof Exp:* Asst microbiol, Univ Md, 57-58, from asst to instr pharmacol, 58-64; assoc res bacteriologist, Microbiol Dept, Res Div, Parke Davis & Co, 64-68, res microbiologist, 68-71, res assoc, Div Res & Develop, 71-78, sr res assoc, 78-83, sect dir, 83-89, DIR, PARKE-DAVIS PHARMACEUT RES DIV, WARNER LAMBERT CO, 83- *Concurrent Pos:* Res trainee dent microbiol, 62-64; adj prof & clin prof, Eastern Mich Univ; lectr, Sch Pub Health & Sch Pharm, Univ Mich, biol, Marquette Univ; chair-elect, Div A, Am Soc Microbiol, 90-, counr, 90- *Mem:* AAAS; Am Soc Microbiol; Brit Soc Antimicrobiol Chemother; Int Am Soc Chemother; fel Am Acad Microbiol. *Res:* Staphylococcal physiology and bacteriophages; antibacterial chemotherapy; immunomodulation; mechanism of action/toxicity of quinolones; bacterial mutations (Ames Test). *Mailing Add:* Infectious Dis Parke-Davis Pharmaceut Res Div Warner Lambert Co 2800 Plymouth Rd Ann Arbor MI 48105

HEIFETZ, JONATHAN, b Long Branch, NJ, June 28, 56; m 88; c 1. BIOMETRICS, ECOLOGY. *Educ:* Emory Univ, BS, 78; Humboldt State Univ, MS, 82. *Prof Exp:* FISHERY RES BIOLOGIST, AUKE BAY LAB, NAT MARINE FISHERIES SERV, 82- *Mem:* Am Fisheries Soc; Am Inst Fisheries Res Biologists. *Res:* Provide analyses and assessments of population dynamics of Alaskan groundfish; develop experimental designs, computer models and prepare manuscripts. *Mailing Add:* PO Box 210155 Auke Bay AK 99821

HEIFETZ, MILTON DAVID, b Hartford, Conn, Feb 7, 21; m 43; c 4. NEUROSURGERY. *Educ:* Univ Ill, BS, 41, MD, 45; Am Bd Neurosurg, dipl, 55. *Prof Exp:* From instr to asst prof, 53-72, assoc prof neurosurg, Sch Med, Loma Linda Univ, 73-; RETIRED. *Concurrent Pos:* Mem neurosurg staff, Cedars-Sinai & Med Ctr, Los Angeles; clin prof neurosurg, Sch Med, Univ Southern Calif. *Mem:* Cong Neurol Surg; assoc Am Acad Neurologists; Am Asn Neurol Surgeons. *Res:* Stereotactic radiosurgery; instrumentation design. *Mailing Add:* 704 N Bedford Dr Beverly Hills CA 90210

HEIFFER, MELVIN HAROLD, b Norfolk, Va, Sept 30, 27. PHYSIOLOGY, PHARMACOLOGY. *Educ:* Col William & Mary, BS, 49; George Washington Univ, MS, 51, PhD(physiol), 53. *Prof Exp:* Lab asst physiol, Sch Med, George Washington Univ, 51-52, asst, 52-53, fel, 53-54, asst res prof, 54-56; pharmacologist, Dept Med Chem, 56-66, CHIEF DEPT PHARMACOL, WALTER REED ARMY INST RES, 66- *Mem:* Am Soc Pharmacol & Exp Therapeut; fel Am Col Clin Pharmacol; Soc Exp Biol & Med; Radiation Res Soc; NY Acad Sci; Sigma Xi. *Res:* Drug development; autonomic pharmacology; radioprotectants; adrenergic mechanisms; antimalarials. *Mailing Add:* 11107 Whisperwood Lane Rockville MD 20852

HEIGOLD, PAUL C, b St Louis, Mo, Apr 9, 36; m 67. GROUNDWATER GEOLOGY. *Educ:* St Louis Univ, BS, 57, MS, 61; Univ Ill, Urbana, MS, 64, PhD(geol), 69. *Prof Exp:* Geophysicist, Carter Oil Co, Okla, 57-58 & Desert Res Inst, Nev, 62; asst geophysicist, 62-74, ASSOC GEOPHYSICIST, ILL STATE GEOL SURV, 74- *Concurrent Pos:* Mem, Ill-Princeton Univ Archeol Exped, Morgantina, Sicily, 70. *Mem:* Am Geophys Union. *Res:* Shallow exploration and detailed gravity surveying of the entire state of Illinois; theoretical ground-water flow system analysis. *Mailing Add:* Ill State Geol Surv 615 E Peabody Champaign IL 61820

HEIKEN, GRANT HARVEY, b Monticello, Iowa, Oct 11, 42; m 67. GEOLOGY, PLANETOLOGY. *Educ:* Univ Calif, Berkeley, BA, 64; Univ Tex, Austin, MA, 66; Univ Calif, Santa Barbara, PhD(geol), 72. *Prof Exp:* Geologist planetology, Johnson Space Ctr, NASA, 69-75; GEOLOGIST VOLCANOLOGY, LOS ALAMOS SCI LAB, UNIV CALIF, 75- *Concurrent Pos:* Mem, Lunar Sample Preliminary Exam Teams, Apollo, 69-72; mem, Visual Observations Team, Skylab, 73. *Honors & Awards:* Cert Spec Commend, Geol Soc Am, 73; Cert Commend, Johnson Space Ctr, NASA, 74. *Mem:* Geol Soc Am; Int Asn Volcanology & Chem Earth's Interior; Sigma Xi; Brit Interplanetary Soc. *Res:* Volcanology and igneous petrology; eruption phenomena and sources for geothermal energy; origins and utilization of pyroclastic rocks; planetology, in particular the formation of planetary regoliths and volcanic phenomena. *Mailing Add:* 119 Piedra Loop Los Alamos NM 87544

HEIKKENEN, HERMAN JOHN, b Detroit, Mich, June 20, 30; m 52; c 2. FOREST ENTOMOLOGY. *Educ:* Univ Mich, BSF, 53, MF, 57, PhD(forestry), 63. *Prof Exp:* Res asst forest entom, Weyerhaeuser Co, 57; entomologist, Lake States Forest Exp Sta, US Forest Serv, 58-61, proj leader forest entom, Southeastern Forest Exp Sta, 61-62; asst prof, Univ Wash, 62-67; ASSOC PROF FOREST ENTOM, VA POLYTECH INST & STATE UNIV, 67- *Mem:* AAAS; Soc Am Foresters; Entom Soc Am; Sigma Xi. *Res:* Host relations and interactions with insect pests. *Mailing Add:* 802 Preston Ave Blacksburg VA 24060

HEIKKILA, JOHN J, b Alharma, Finland, Nov 22, 50. MOLECULAR BIOLOGY. *Educ:* Toronto Univ, PhD(zool), 80. *Prof Exp:* ASST PROF BIOL, UNIV WATERLOO, 84- *Mailing Add:* Dept Biol Univ Waterloo Waterloo ON N2L 3G1 Can

HEIKKILA, RICHARD ELMER, b Painesville, Ohio, July 30, 42; m 64; c 2. BIOCHEMISTRY. *Educ:* Ohio State Univ, BA, 65, PhD(physiol chem), 69. *Prof Exp:* Teaching asst physiol chem, Ohio State Univ, 65-66, res asst, 66-68; res assoc psychiat, Columbia Univ, 70-71, res assoc neurol, 71-73; res asst prof, dept neurol, Mt Sinai Sch Med, 73-77, res assoc prof, 77-79; assoc prof, 79-85, PROF DEPT NEUROL, RUTGERS MED SCH, UNIV MED & DENT, NJ, 85- *Mem:* Am Soc Biol Chem; Soc Neurosci; NY Acad Sci; AAAS; Am Soc Pharmacol Exp Ther; Am Soc Neurochem; Int Soc Neurochem. *Res:* Studies on the mechanism of action of neurotoxins; drugs affecting dopamine systems and on the effects of oxidation-reduction reactions in biological systems. *Mailing Add:* Univ Med & Dent NJ Robert Wood Johnson Med Sch Piscataway NJ 08854-5635

HEIKKILA, WALTER JOHN, b South Porcupine, Ont, Feb 22, 28; div; c 2. SPACE PHYSICS. *Educ:* Univ Toronto, BASc, 50, PhD(low temperature physics), 54. *Prof Exp:* Group leader tropospheric physics, Radio Physics Lab, Defence Res Bd, Ottawa, Ont, 54-58, sect leader, Rocket Sect, 58-63; from assoc prof to prof, Southwest Ctr Advan Studies, 63-68, PROF PHYSICS, UNIV TEX, DALLAS, 68- *Concurrent Pos:* Adj prof elec eng, Southern Methodist Univ, 64-72; mem US comn four, Int Sci Radio Union; mem working group four, Comt Space Res. *Mem:* Am Geophys Union; Can Asn Physicists; Int Asn Geomagnetism & Aeronomy. *Res:* Rocket and satellite research on ionospheric, auroral and magnetospheric phenomena, especially soft particle fluxes on ISIS-1 and 2 satellites and laboratory studies of plasma probes; problems of population growth. *Mailing Add:* Dept Physics Univ Tex PO Box 688 Richardson TX 75083

HEIKKINEN, DALE WILLIAM, b Virginia, Minn, Nov 14, 38; m 59; c 2. NUCLEAR PHYSICS. *Educ:* Univ Minn, Duluth, BA, 60; Univ Iowa, MS, 62, PhD(physics), 65. *Prof Exp:* Res assoc nuclear physics, Stanford Univ, 65-68; SR PHYSICIST NUCLEAR PHYSICS, LAWRENCE LIVERMORE LAB, 68- *Concurrent Pos:* Vis scholar, Stanford Univ, 70-71; vis assoc prof, Univ Jyvaskyla, 72-73. *Mem:* Am Phys Soc; Am Nuclear Soc. *Res:* Basic and applied research in nuclear physics. *Mailing Add:* L-397 Lawrence Livermore Lab PO Box 808 Livermore CA 94550

HEIKKINEN, DONALD D, b Ramsay, Mich, Sept 10, 32; m 56; c 1. MATHEMATICS. *Educ:* Univ Mich, BA, 58, MA, 60, PhD(math educ), 64. *Prof Exp:* Teacher, Lee M Thurston High Sch, 58-59; asst prof math, Eastern Mich Univ, 60-61; res asst math educ, Univ Mich, 61-62; assoc prof math, Northern Iowa Univ, 63-68, assoc dir, NSF Summer Inst, 65-67; assoc prof math, Northern Mich Univ, 68-71, dir, NSF In-Serv Inst, 68-69, head, Dept Math, 68-73, chmn acad senate, 73-74, dean arts & sci, 74-89, PROF MATH, NORTHERN MICH UNIV, 68- *Concurrent Pos:* Math ed, Sch Sci & Math, 63-74. *Res:* Factors related to acceleration in the study of mathematics. *Mailing Add:* Math & Computer Sci Dept Northern Mich Univ Marquette MI 49855

HEIKKINEN, HENRY WENDELL, b Minneapolis, Minn, May 18, 35; m 64; c 2. CHEMICAL EDUCATION, CHEMISTRY. *Educ:* Yale Univ, BEng, 56; Columbia Univ, MA, 62; Univ Md, PhD(chem educ), 73. *Prof Exp:* Food engr res & develop, Gen Mills, Inc, 56-61; teacher chem, Richfield Pub Schs, Minn, 63-69; teaching assoc chem, 69-70, from instr to asst prof, 70-78, ASSOC PROF CHEM, UNIV MD, 78- *Concurrent Pos:* Adv Bd mem, Chem Mag, Am Chem Soc, 72-76 & Sci Teacher, Nat Sci Teachers Asn, 74-77; article consult, Sci Teacher, 73- *Mem:* Am Chem Soc; Nat Asn Res Sci Teaching; AAAS; Nat Sci Teachers Asn; Asn Educ Teachers Sci. *Res:* Teaching methods; curriculum development; evaluation at secondary and university levels. *Mailing Add:* 2163 Buena Vista Dr Greeley CO 80631

HEIL, JOHN F, JR, b San Francisco, Calif, Feb 29, 36. CHEMICAL ENGINEERING. *Educ:* Univ Calif, Berkeley, BS, 57, PhD(chem eng), 65. *Prof Exp:* Mgr prod develop sect, Western Res Ctr, Stauffer Chem Co Wyo, 57-76, mgr energy resourses, 76-79, dir planning, Agr Chem Div, 79-81; corp prod dir, AG/Drug Intermediates, 81-83, corp prod dir, Agr Chem Intermediates, 83-85, dir, 85-87, DIR, WESTERN RES CTR, ICI AMERICAS INC, 87- *Mem:* Am Chem Soc; Am Inst Chem Engrs. *Res:* Thermodynamics of phase equilibria. *Mailing Add:* 16 Kinross Dr San Rafael CA 94901-2420

HEIL, RICHARD WENDELL, b Chicago, Ill, Mar 16, 26; m 47; c 3. BIO-REMEDIATION OF TOXIC ORGANIC SEDIMENTS. *Educ:* Univ Ill, BS, 48. *Prof Exp:* Mining engr, US Steel Corp, 48-55; struct designer, Hazelet & Erdal, Consult Engrs, 55-58; struct engr, Metrop Water Recl Dist Greater Chicago, 58-64, prin civil engr, 64-70 & 71-79, asst chief engr, 70-71, engr treatment opers, 79-88; SR ENVIRON ENGR, HNTB ARCHITECTS, ENGRS & PLANNERS, 88- *Concurrent Pos:* Vpres, Arthur S Darr & Assocs, 62-78, Kucaba & Heil Assocs, 64-66; prin, R W Heil, Struct Engr, 78-; mem archit adv bd, Clarendon Hills, 83-; struct engr, Kudrna & Assocs, Ltd, 91- *Mem:* Fel Am Soc Civil Engrs; Am Inst Chem Engrs; Water Pollution Control Fedn. *Res:* Environmental and structural engineering; utilization of remediated sediments; three US patents. *Mailing Add:* 30 Arthur Ave Clarendon Hills IL 60514

HEIL, ROBERT DEAN, b White Butte, SDak, Aug 14, 32; m 59; c 7. SOIL CONSERVATION. *Educ:* SDak State Univ, BS, 60, MS, 64, PhD(soil sci), 72. *Prof Exp:* Soil scientist soil fertil, SDak State Univ, 60-61; soil scientist soil surv, Soil Conserv Serv, USDA, 61-63; soil scientist soil fertil & mgt, SDak State Univ, 63-65; soils exten assoc prof, Colo State Univ, 65-69; soil scientist remote sensing, SDak State Remote Sensing Inst, 69-70; assoc prof soil surv, ocnserv & mgt, 70-77, PROF AGRON, COLO STATE UNIV, 77- *Concurrent Pos:* Soil scientist, Cameron Consult Engr, 71-72, Wallace, McHary & Todd Assoc, 72, Meehen Environ Eng, 72-73 & Jones, Jones & Humkins, Attys Law, 73. *Mem:* Soil Sci Soc Am; Soil Conserv Soc Am; Am Soc Agron; Sigma Xi. *Res:* Determining the behavior and suitability of soils for application of soils data to land use and environmental planning. *Mailing Add:* 1055 Duna Laramie WY 82070-5016

HEIL, WOLFGANG HEINRICH, b Frankfurt, Ger, Nov 3, 40; m 66; c 2. TOPOLOGY. *Educ:* Univ Frankfurt, vordiplom, 63, diplom, 67; Rice Univ, MA & PhD(math), 70. *Prof Exp:* Instr math, Rice Univ, 69-70; from asst prof to assoc prof, 70-83, PROF MATH, FLA STATE UNIV, 83- *Concurrent Pos:* Vis prof, Univ Ljubljana, Yugoslavia, 77-78, J W Goethe Univ, Frankfurt, WGer, 80-81, 88, F W Alexander Univ, Erlangen, WGer, 81, 84-85. *Mem:* Am Math Soc. *Res:* Topology of 3-manifolds. *Mailing Add:* Dept Math Fla State Univ Tallahassee FL 32306

HEILENDAY, FRANK W (TOD), b Jersey City, NJ, Dec 31, 27; m 51. OPERATIONS RESEARCH. *Educ:* Mass Inst Technol, BS & MS, 49. *Prof Exp:* Aerophys engr, Convair Div, Gen Dynamics Corp, Tex, 49-51, sr dynamics engr, Calif, 53-55; assoc flight test engr, Cornell Aeronaut Lab, NY, 51-53; chief opers anal, Hq 8th Air Force, Westover AFB, Mass, 55-65, chief syst eval div, Sci & Res, Hq Strategic Air Command, Offutt AFB, 66-84; CONSULT, 84- *Concurrent Pos:* Adj prof, George Washington Univ. *Mem:* Mil Opers Res Soc. *Res:* Systems analysis studies on aircraft offensive and defensive avionics requirements. *Mailing Add:* 720 Sherman St NW Olympia WA 98502-9901

HEILES, CARL, b Toledo, Ohio, Sept 22, 39; c 2. INTERSTELLAR MATTER, RADIO ASTRONOMY. *Educ:* Cornell Univ, B Eng Physics, 62; Princeton Univ, PhD(astrophys), 66. *Prof Exp:* PROF ASTRON, UNIV CALIF, BERKELEY, 66- *Concurrent Pos:* Res astronr, Cornell Univ, 69-70; vis prof, Sci & Med Univ Grenoble, France, 83; vis fel, Joint Inst Lab Astrophys, Univ Colo, Boulder, 90. *Honors & Awards:* Heineman Prize, Am Phys Soc, 89. *Mem:* Nat Acad Sci; Am Astron Soc; Int Astron Union; Int Sci Radio Union; Astron Soc Pac. *Res:* Diffuse interstellar matter with concentration on atomic hydrogen; radio astronomy. *Mailing Add:* Astron Dept Univ Calif Berkeley CA 94720

HEILIGMAN, HAROLD A, furnace design, for more information see previous edition

HEILMAN, ALAN SMITH, b Pittsburgh, Pa, Dec 23, 27. BOTANY. *Educ:* Univ Pittsburgh, BS, 49, MS, 51; Ohio State Univ, PhD(bot), 60. *Prof Exp:* Asst bot, Univ Pittsburgh, 49-52; asst bot, Ohio State Univ, 52-56, asst instr, 55-57, instr, 57, tech asst, 57-60; from instr to asst prof, 60-70, ASSOC PROF BOT, UNIV TENN, 70- *Mem:* AAAS; Bot Soc Am; Sigma Xi. *Res:* Developmental floral and pollen morphology; pollination; microtechnique; scientific photography. *Mailing Add:* Dept Bot Univ Tenn Knoxville TN 37996-1100

HEILMAN, CAROL A, RESPIRATORY RESEARCH. *Educ:* Boston Univ, BA, 72; Rutgers Univ, MS, 76, PhD(microbiol), 79. *Prof Exp:* Asst scientist, Hoffman-La Roche, 73-74; postdoctoral res assoc, Nat Cancer Inst, 78-81, sr staff fel, 81-86; prog officer, Influenza & Viral Respiratory Dis, 86-88, actg chief, 88-89, CHIEF RESPIRATORY DIS BR, NAT INST ALLERGY & INFECTIOUS DIS, NIH, 89- *Concurrent Pos:* Grad teaching fel, Rutgers Med Sch, 76-85; mem, Childcare Comt, NIH; rep, Women's Adv Comt, NIAID, EEO Comt, HCFA Influenza Demon Proj, NIH; reviewer J Nat Cancer Inst, J Infectious Dis. *Mem:* Am Soc Microbiol; Am Soc Virol; Int Soc Antiviral Res; AAAS. *Res:* Influenza research; acute viral respiratory diseases; Reyes Syndrome; infectious disease of the elderly; regulation of gene expression during differentiation and neoplastic transformation; numerous technical publications. *Mailing Add:* Nat Inst Allergy & Infectious Dis Respiratory Dis Br NIH Westwood Bldg Rm 750 5333 Westbard Ave Bethesda MD 20892

HEILMAN, PAUL E, b Seattle, Wash, Feb 13, 31; m 54; c 2. FOREST NUTRITION, BIOMASS PRODUCTION. *Educ:* Ore State Univ, BS, 57; Univ Wash, PhD(forestry), 61. *Prof Exp:* Instr agr, Skagit Valley Col, 61-63; asst horticulturist, Wash State Univ, 63-64; assoc prof forest soils res, Univ Alaska, 64-67; exten specialist hort, 67-69, FOREST SCIENTIST, WESTERN WASH RES & EXTEN CTR, WASH STATE UNIV, 69- *Mem:* Soc Am Foresters; Am Soc Agron; Soil Sci Soc Am; Sigma Xi. *Res:* Plant productivity in relationship to soils and fertility; ecology; nitrogen nutrition of forests and sustainability; maximumizing production of intensive plantations of populus hybrids for pulp and energy. *Mailing Add:* Dept Natural Resource Sci Wash State Univ Puyallup Res & Exten Ctr Puyallup WA 98371-4998

HEILMAN, RICHARD DEAN, b Sparta, Wis, Sept 8, 37; m 67; c 4. CARDIOVASCULAR PHARMACOLOGY. *Educ:* Wis State Univ, La Crosse, BS, 59; Marquette Univ, PhD(pharmacol), 69. *Prof Exp:* From assoc scientist to scientist, 69-73, group leader pharmaceut, Ortho Pharmaceut Corp, 73-77; group leader pharmacol, Diamond Shamrock Corp, 77-80; SECT HEAD CARDIOVASC RES, G D SEARLE & CO, 80- *Mem:* Am Chem Soc; Inflammation Res Asn; Cardiovasc Pharmacol Discussion Group; Am Soc Pharmacol & Exp Therapeut. *Res:* Interactions of the autonomic nervous system and the reproductive endocrine hormones in regulation of oviducal egg transport; effect of androgen on aggressive behavior; pharmacology of narcotic antagonists; cardiovascular and central nervous system toxicology of vehicles. *Mailing Add:* Pharmaco Dynamics Res Inc 4009 Banister Lane Austin TX 78704

HEILMAN, WILLIAM JOSEPH, b Pittsburgh, Pa, Aug 31, 30; m 57; c 3. POLYMER CHEMISTRY. *Educ:* Univ Pittsburgh, BS, 52; State Univ NY Col Forestry, Syracuse, MS, 56; Tex Tech Col, PhD(phys org chem), 62. *Prof Exp:* Chemist, Res Ctr, Koppers Co, 52-54; from res chemist to sr res chemist, Gulf Oil Chem Co, 62-72, res assoc, Gulf Res & Develop Corp, 72-76, sect supvr, 76-81, dir polymer catalysis res, 81-83; DIR PROD TECHNOL, PENNZOIL, 83- *Mem:* Am Chem Soc. *Res:* Free radical chemistry; mechanistic organic chemistry; polymer chemistry; synthetic lubricants. *Mailing Add:* 14826 La Quinta Lane Houston TX 77079-4506

HEILMAN, WILLIAM PAUL, b Cheyenne, Wyo, Jan 7, 48; m 72; c 1. MEDICINAL CHEMISTRY, PHARMACOLOGY. *Educ:* Muskingum Col, BS, 70; Ohio State Univ, MS, 72, PhD(med chem), 74; Case Western Reserve Univ, MBA, 79. *Prof Exp:* Mgr New Bus Develop, Diamond Shamrock Corp, 78-81; mgr licensing, FMC Corp, 81-86; DIR TECH ACQUISITION, AM CYANAMID, 86- *Concurrent Pos:* Res fel, NIH, 71-74, Eidgenossische Technische Hochschule, Zurich, 74-75. *Mem:* Am Chem Soc; Licensing Execs Soc; Asn Univ Technol Mgr. *Res:* Design, synthesis and biological evaluation of hypolipemic agents, antihypertensive agents, anti-inflammatory agents, antiarrhythmic agents and analgetic agents. *Mailing Add:* Am Cyanamid Co Agr Res & Develop PO Box 400 Princeton NJ 08540

HEILMEIER, GEORGE HARRY, b Philadelphia, Pa, May 22, 36; m 61; c 1. SOLID STATE ELECTRONICS. *Educ:* Univ Pa, BSEE, 58; Princeton Univ, MSE, 60, AM, 61, PhD(physics, elec eng), 62. *Prof Exp:* Res scientist, RCA Labs, Inc, 58-68, head solid state & liquid state device res, David Sarnoff Res Ctr, 68-69, dir device concepts res, 69-70; White House fel, spec asst to Secy Defense, 70-71; asst dir defense res & eng electronics, Dept Defense, 71-75; dir, Defense Advan Res Proj Agency, 75-77; SR VPRES & CHIEF TECH OFFICER, CORP RES, DEVELOP & ENG, TEX INSTRUMENTS, INC, 77- *Honors & Awards:* Nat Medal of Sci, 91; David Sarnoff Award, RCA, 68 & Inst Elec & Electronics Engrs, 76; Indust Res-100 Award, 68 & 69; Centennial Medal, 84, Philips Award, 85, Founder's Medal, 86, Inst Elec & Electronics Engrs, 90; C & C Prize, Japan. *Mem:* Nat Acad Eng; fel Inst Elec & Electronics Engrs. *Res:* Low noise microwave amplifiers; solid state harmonic generation; organic semiconductors; thin film devices; liquid crystals; electro-optics; integrated circuits; computer science; artificial intelligence. *Mailing Add:* Tex Instruments Inc PO Box 655474 MS 400 Dallas TX 75265

HEILPRIN, LAURENCE BEDFORD, b New York, NY, May 26, 06; m 53; c 2. CYBERNETICS. *Educ:* Univ Pa, BS, 28, MA, 31; Harvard Univ, PhD(physics), 41. *Prof Exp:* Instr physics & math, Sch Eng & Sch Lib Arts, Northeastern Univ, 35-40; asst thermodyn, Harvard Univ, 40-41; from asst physicist to sr physicist, Nat Bur Stand, 41-51; consult physicist, Taub Eng Co, 52-54; systs engr, Melpar, Inc, 54-55; analyst opers eval group, Mass Inst Technol, Washington, DC, 55-56; physicist, Coun Libr Resources, 58-67; prof, 67-76, EMER PROF, COL LIBR & INFO SCI & DEPT COMPUT SCI, UNIV MD, COLLEGE PARK, 76- *Concurrent Pos:* Chief, Control Testing Lab, Nat Bur Stand, 43-46, lectr, Grad Sch, 49-52; lectr, George Washington Univ, 43-45; dir, Am Fedn Info Processing Socs, 65; chmn sect T, AAAS, 76-78. *Mem:* Am Phys Soc; Optical Soc Am; Am Asn Physics Teachers; Am Soc Info Sci (pres, 64, 65). *Res:* Foundations of information science; the nature of information and communication. *Mailing Add:* Col Libr & Info Sci Univ Md College Park MD 20742

HEILWEIL, ISRAEL JOEL, b Poland, May 23, 24; nat US; m 48; c 3. PHYSICAL CHEMISTRY. *Educ:* City Col New York, BS, 48; Ohio State Univ, MS, 51, PhD(phys chem), 54. *Prof Exp:* Chemist, Res Ctr, Texaco, Inc, 54-60; sr res chemist, Mobil Oil Corp, 60-68, RES ASSOC, CENT RES LAB, MOBIL OIL RES & DEVELOP CORP, 68- *Mem:* Am Chem Soc; Soc Petrol Engrs. *Res:* Physical chemistry and theology of polymers and surfactants; oil recovery and liquid deep well drilling fluids; coal/water dispersions as fuels; oil and gasoline additives; petro-proteins; gel permeation chromatography; light scattering; surface phenomena. *Mailing Add:* 47 Linwood Circle Princeton NJ 08540-3623

HEIM, LYLE RAYMOND, b Lignite, NDak, Apr 30, 33; m 57; c 3. IMMUNOLOGY, MICROBIOLOGY. *Educ:* Univ Minn, Minneapolis, BA, 63, MS, 66, PhD(immunol, microbiol), 69. *Prof Exp:* USPHS fel infectious dis, Baylor Col Med, 69-71, USPHS fel exp biol & instr immunol, 71-72; clin asst prof lab med, Med Col Wis, 72-78; clin immunologist, Columbia Hosp, 72-78; assoc prof & dir res, Dept Pediat, Tex Tech Univ Sch Med, 78-83; DIR, PHARMACEUT MED WRITING DEPT, BRISTOL-MYERS CO, 84- *Concurrent Pos:* Ed-in-chief, Exp Hemat, 70-83. *Mem:* AAAS; Coun Biol Ed; Am Soc Microbiol; Int Soc Exp Hemat; Am Med Writers Asn. *Res:* Transplantation immunology; pediatric immunology; cancer immunology; experimental hematology. *Mailing Add:* Bristol-Myers Co Five Research Pkwy Wallingford CT 06492

HEIM, WERNER GEORGE, b Mulheim an der Ruhr, Ger, Apr 7, 29; nat US; m 61, 73; c 4. HUMAN GENETICS, VERTEBRATE EMBRYOLOGY. *Educ:* Univ Calif, Los Angeles, BA, 50, MA, 52, PhD, 54. *Prof Exp:* Jr res zoologist, Col Agr, Univ Calif, Los Angeles, 54-56; instr biol, Brown Univ, 56-57; from asst prof to assoc prof, Wayne State Univ, 57-67; PROF BIOL, COLO COL,67- *Concurrent Pos:* Assoc ed, Am Biol Teacher, 70-74; vis prof biophys & genetics, Sch Med, Univ Colo, 78 & 86; chmn, dept biol, Colo Col, 71-76 & 86-89. *Mem:* Fel AAAS; Am Soc Zool; Soc Develop Biol; Int Soc Develop Biologists; Nat Soc Genetic Coun; Am Soc Human Genetics. *Res:* Methodology of karyotyping relation between maternal proteins and embryonic development. *Mailing Add:* Dept Biol Colo Col 14 E Cache La Poudre Colorado Springs CO 80903

HEIMAN, DONALD EUGENE, b Los Angeles, Calif, Nov 4, 47. CONDENSED MATTER PHYSICS, OPTICS. *Educ:* Calif State Univ, Los Angeles, BS, 67; Univ Calif, Irvine, PhD(physics), 75. *Prof Exp:* Res assoc physics, Univ Southern Calif, 75-80; MEM STAFF, FRANCIS BITTER NAT MAGNET LAB, MASS INST TECHNOL, 81- *Mem:* Am Phys Soc. *Res:* Optics and nonlinear optics as applied to condensed matter physics, with emphasis on semiconductors. *Mailing Add:* Francis Bitter Nat Magnet Lab Mass Inst Technol Cambridge MA 02139

HEIMAN, MARK LOUIS, b Fort Belvoir, Va, Oct 9, 52; m 81; c 1. ENDOCRINOLOGY. *Educ:* Univ New Orleans, BA, 74; La State Univ, PhD(physiol), 78. *Prof Exp:* Teaching asst physiol, Med Ctr, La State Univ, 76-78; res assoc physiol, Med Sch, Ind Univ, 78-; AT ELI LILLY & CO. *Concurrent Pos:* Prin investr, Am Cancer Soc, 78-79, NIH, 80-81. *Mem:* AAAS. *Res:* Regulation of anterior pituitary hormone receptors; mechanisms responsible for receptor changes; interactions between the pineal gland and the hypothalamic-pituitary-adrenal axis. *Mailing Add:* Lilly Res Lab PO Box 708 Greenfield IN 46140

HEIMANN, PETER AARON, b New York, NY, Dec 6, 49; m 72; c 2. STATISTICS, QUALITY CONTROL. *Educ:* City Col NY, BS, 71; Univ Md, MS, 73, PhD(physics), 77. *Prof Exp:* Grad res asst physics, Dept Physics, Univ Md, 74-77; staff scientist, Singer Co, Kearfott Div, 78-80; MEM TECH STAFF, BELL LABS, 80- *Mem:* Am Phys Soc; Am Soc Qual Control. *Res:* Statistical, quality control and process improvement, as applied to manufacturing and service industries. *Mailing Add:* Bell Labs Rm 7C-104 Whippany Rd Whippany NJ 07981

HEIMANN, ROBERT B(ERTRAM), b Goerlitz, Germany, Dec 31, 38; m 71. ROCK-WATER INTERACTION, ANCIENT CERAMICS TECHNOLOGY. *Educ:* Free Univ, Berlin, BSc, 61, MSc, 63, PhD(mineral), 66. *Prof Exp:* Sci asst mineral, Free Univ, Berlin, 66-71, asst prof mineral & geochem, 71-77; assoc prof, Univ Karlsruhe, 77-79; res assoc, semiconductor res, McMaster Univ, 79-81; sr res scientist, diamond synthesis, 3M Can Inc, 81-82; res geochemist, Waste Mgt, Whiteshell Nuclear Res Estab, Atomic Energy Can Ltd, 82-86; RES MGR, MFG TECHNOLOGIES DEPT, ALTA RES COUN, 87- *Concurrent Pos:* Vis prof ancient ceramics, Univ Toronto, 79-81; consult, diamond synthesis, 3M Can Inc, 80-81; adj prof mat sci, Univ Alta, 88-; adj prof, Heilongjiang Acad Sci, Harbin, Heilongjiang, China, 90-; pres, Oceangate Consult, 90- *Mem:* Mat Res Soc; Can Univ-Ind Coun Advan Ceramics; Can Advan Indust Mat Forum. *Res:* Solid state reactions in inorganic materials, and reactions catalyzed by taking place on solid surfaces and their relationship to thermodynamic and crystalline factors; interaction of hydrothermal fluids with minerals and rocks; advanced materials (ceramics) properties. *Mailing Add:* Mfg Technologies Dept Alta Res Coun 250 Karl Clark Rd Edmonton AB T6H 5X2 Can

HEIMANN, ROBERT L, b St Louis, Mo, Dec 7, 46; m 67; c 5. PLASTIC & CABLE DESIGN, THERMOPLASTIC COMPOSITES & FAILURE ANALYSIS. *Educ:* Univ Idaho, BSME, 71. *Prof Exp:* Designer aircraft & spacecraft, McDonnell-Douglas, 66-72; eng mgr prod design, Narragansett, 75-77; MGR RES & DEVELOP, ORSCHELN CO, 77- *Concurrent Pos:* Guest lectr, Univ Mo, Rolla, 90-91. *Honors & Awards:* Arch T Colwell Award, Soc Automotive Engrs, 89. *Mem:* Soc Automotive Engrs. *Res:* Plastic and ceramic coatings; seal and lube designs for automotive and industrial parking brake, transhift, and release cables; ceramic coatings for corrosion protection on rebar, prestressed and bridge stay cables; composite cables. *Mailing Add:* 1354 Heritage Pl Moberly MO 65270

HEIMBACH, DAVID M, SURGERY. *Prof Exp:* PROG SURG, UNIV WASH, 74-, DIR BURN CTR, HARBORVIEW. *Mailing Add:* Harborview Med Ctr Univ Wash 325 Ninth Ave Seattle WA 98104

HEIMBACH, RICHARD DEAN, b Chicago, Ill, Apr 5, 35; m 58; c 6. AEROSPACE MEDICINE, RADIATION BIOLOGY. *Educ:* Univ Chicago, AB, 56, BS, 57, MD, 60; NY Univ, PhD(radiation biol), 67; Harvard Univ, MPH, 72. *Prof Exp:* Chief, outpatient clin, Ellsworth AFB Hosp, 62-63, adv radiation biol, Kirtland AFB, 67-71, chief med opers aerospace med, USAF Sch Aerospace Med, 74-79, chief hyperbaric med div, 79-82; assoc dir, 82-89, DIR, HYPERBARIC MED DEPT, SOUTHWEST TEX METHODIST HOSP, 89- *Concurrent Pos:* Vis prof, Univ NMex, 69-71; vis fac, Grad Sch, Tex A&M Univ, 77-; consult, USAF Surgeon Gen, 79- *Mem:* Fel Aerospace Med Asn; fel Royal Soc Health; fel Am Col Prev Med; Undersea Med Soc (vpres, 84-, pres, 88); AMA. *Res:* Effects of hyperbaric oxygen in reversing underlying pathology associated with radiation therapy; effects of pressure changes on pulmonary physiology. *Mailing Add:* 4499 Medical Dr Sublevel 2 San Antonio TX 78229

HEIMBECKER, RAYMOND OLIVER, b Calgary, Alta, Nov 29, 22; m 50; c 5. EXPERIMENTAL SURGERY, CARDIOVASCULAR SURGERY. *Educ:* Univ Sask, BA, 44; Univ Toronto, MD, 47, MA, 48, MS, 55; FRCS(C), 55. *Prof Exp:* Res assoc, Ont Heart Found, 55-70, sr res assoc surg, 70-88; prof surg, Univ Western Ont, 73-88; chief cardiothoracic surg, Univ Hosp, London, 73-88; RETIRED. *Honors & Awards:* George Armstrong Peters Award, 50; Lister Award, 55; Gold Medal, Royal Col Surgeons Can, 67; Res & Teaching Award, Rose Found India, 76; Gordon Murray Mem Lect, Univ Toronto, 81; Medal Jeddah, Saudi Arabia, 83; P K Sen Mem Lectr, Bombay, India, 85; Wilfred Bigelow Lectr, Royal Col Physicians & Surgeons, Can, 85. *Mem:* Fel Am Col Surg; fel Am Asn Thoracic Surg; Int Cardiovasc Soc; Am Surg Asn; Can Soc Microcirc (pres, 71). *Res:* Cardiovascular physiology; experimental cardiovascular surgery. *Mailing Add:* RR 1 Collingwood ON L9Y 3Y9 Can

HEIMBERG, MURRAY, b Brooklyn, NJ, Jan 5, 25; m 47, 64; c 4. PHARMACOLOGY, BIOCHEMISTRY. *Educ:* Cornell Univ, BS, 48, MNS, 49; Duke Univ, PhD(biochem), 52; Vanderbilt Univ, MD, 59. *Prof Exp:* NIH res fel biol chem, Wash Univ, 52-54; res assoc, Sch Med, Vanderbilt Univ, 54-59, from asst prof to prof pharmacol, 59-74, asst prof med, 71-74; prof pharmacol, chmn dept pharmacol & prof med, Sch Med, Univ Mo, Columbia, 74-80; PROF PHARMACOL, CHMN DEPT PHARMACOL & PROF MED, UNIV TENN CTR HEALTH SCI, MEMPHIS, 81-, VAN VLEET PROF, 86- *Concurrent Pos:* Lederle Med Fac award, 59-62; estab investr, Am Heart Asn, 62-67; fel, Coun Arteriosclerosis & Basic Sci, Am Heart Asn. *Mem:* Am Soc Pharmacol & Exp Therapeut; Am Col Clin Pharmacol; Endocrine Soc; fel AAAS; Am Soc Biol Chem & Molecular Biol; Am Soc Study Liver Dis; Sigma Xi; Am Oil Chem Soc. *Res:* Hormonal regulation of lipid/lipoprotein metabolism; lipid absorption and transport; interrelationships between carbohydrate and lipid metabolism; biosynthesis of proteins; enzymatic oxidation of sulfur compounds; endocrine pharmacology; atherosclerosis; hyperlipidemias and hyperlipoproteinemia; regulation of hepatic lipid metabolism; mechanisms of hepatotoxicity; hepatic metabolism of drugs; endocrinology. *Mailing Add:* Dept Pharmacol Univ Tenn-Memphis Health Sci Ctr 874 Union Ave Memphis TN 38163

HEIMBURG, RICHARD W, b Silver Creek, NY, Feb 19, 38; m 59; c 2. MECHANICAL ENGINEERING. *Educ:* Syracuse Univ, BME, 59; Univ Ill, PhD(surface phenomena), 64. *Prof Exp:* Asst prof thermodyn, 64-69, ASSOC PROF MECH & AEROSPACE ENG, SYRACUSE UNIV, 69- *Res:* Cryogenic thermal regenerators; condensation and thermal accommodation coefficients in high vacuum at low temperatures. *Mailing Add:* Mech/Aerosp Eng 151 Link Hall Syracuse Univ Syracuse NY 13244-1240

HEIMER, EDGAR P, b Bayate, Oriente, Cuba, July 8, 37; nat US; c 3. BIOLOGY. *Educ:* Fla State Univ, BS, 60; Montclair State Col, MA, 71; Columbia Univ, MSc, 75, DSc Ed, 80. *Prof Exp:* Res asst, Inst Molecular Biol & dept chem, Fla State Univ, 60-66; asst scientist II, Hoffmann-LaRoche Inc, 66-68, assoc scientist, 68-74, supvr bio-org preparations, 74-80, sr scientist, 80-84, RES INVESTR, DEPT PEPTIDE RES, HOFFMANN-LAROCHE INC, 84- *Concurrent Pos:* Assoc prof sci educ, Teacher's Col, Columbia Univ, 80- *Mem:* Am Chem Soc; Res Soc Am; Sigma Xi; fel Am Inst Chemists; NY Acad Sci. *Mailing Add:* Dept Peptide Res Hoffman La Roche Inc Bldg 86 Nutley NJ 07110-1199

HEIMER, NORMAN EUGENE, b Holyoke, Colo, Sept 24, 40; m 62; c 3. ORGANIC CHEMISTRY. *Educ:* Univ Northern Colo, BA, 62; Iowa State Univ, PhD(chem), 68. *Prof Exp:* Res assoc chem, Vanderbilt Univ, 68-69; asst prof, 69-73, assoc prof chem, Univ Miss, 73-; AT FJSRL NC, USAF ACAD, CO. *Mem:* Am Chem Soc. *Res:* Organosulfur chemistry, especially the chemistry of sulfenamides; natural products chemistry; alkaloids. *Mailing Add:* Dept Chem Univ Miss University MS 38677

HEIMER, RALPH, b Vienna, Austria, Nov 4, 21; US citizen; m 43, 74; c 2. BIOCHEMISTRY. *Educ:* City Col New York, BS, 48; Columbia Univ, AM, 51, PhD(biochem), 57. *Prof Exp:* Asst prof biochem, Hosp Spec Surg, Med Col, Cornell Univ, 56-62; assoc prof biochem, NJ Col Med, 63-66; PROF BIOCHEM, JEFFERSON MED COL, 67- *Concurrent Pos:* NIH res awards, 59-76 & 80-; Am Cancer Soc res awards, 68-70; sr investr, Arthritis Found, 63-67. *Mem:* Am Rheumatism Asn; Am Soc Biol Chemists; Am Asn Immunologists; Sigma Xi. *Res:* Chemistry of glycosaminoglycans; immunochemistry; tumor immunology; technology of glycosaminoglan and proteoglycan detection; polyacrylamide gel electrophoresis and electrotransblot technology. *Mailing Add:* 235 S Third St Philadelphia PA 19106

HEIMERL, JOSEPH MARK, b Woodbury, NJ, Feb 25, 40; m 68; c 2. COMBUSTION MODELLING, CHEMICAL KINETICS. *Educ:* St Joseph's Col, BS, 62; Univ Pittsburgh, PhD(physics), 68. *Prof Exp:* Res assoc spectroscopy, Ames Ctr, NASA, 68-70; sr scientist exp aeronomy, 70-74, sr scientist modelling aeronomy, 74-78, SR SCIENTIST MODELLING COMBUSTION, BALLISTIC RES LAB, 78- *Concurrent Pos:* Nat Res Coun res assoc, Ames Ctr, NASA, 68-70. *Mem:* Combustion Inst; Am Phys Soc. *Res:* Flame inhibition; muzzle flash suppression; flame modelling; elementary gas phase chemical kinetics; aeronomy modelling; D-region ion chemistry; ion molecule reactions. *Mailing Add:* 108 Paradise Dr Havre De Grace MD 21078

HEIMLICH, RICHARD ALLEN, b Elizabeth, NJ, Aug 8, 32; m 61; c 2. GEOLOGY. *Educ:* Rutgers Univ, BS, 54; Yale Univ, MS, 55, PhD(petrol), 59. *Prof Exp:* From instr to assoc prof, 61-70, PROF GEOL, KENT STATE UNIV, 70-, CHMN, DEPT GEOL, 76- *Concurrent Pos:* Res grants, NSF, Los Alamos Nat Lab & US Dept Educ. *Mem:* Fel Geol Soc Am; Am Inst Prof Geologists. *Res:* Igneous and metamorphic petrology. *Mailing Add:* Dept Geol Kent State Univ Kent OH 44242

HEIMSCH, CHARLES W, b Dayton, Ohio, May 4, 14; m 74; c 3. PLANT ANATOMY, MORPHOLOGY. *Educ:* Miami Univ, Ohio, AB, 36; Harvard Univ, MA, 39, PhD(bot), 41. *Prof Exp:* Asst biol, Harvard Univ, 36-37 & 39-40, tech asst wood collection, 38-39, asst biol, Radcliffe Col, 37-38; Sheldon traveling fel from Harvard Univ, Univ Calif, 41-42; instr bot, Swarthmore Col, 42-46; asst prof biol, Amherst Col, 46-47; from asst prof to prof, Univ Tex, 47-59; chmn dept, 59-77, prof, 59-80, EMER PROF BOT, MIAMI UNIV, 80- *Concurrent Pos:* Res assoc, Plant Res Inst, Univ Tex, 47-50, anatomist, 50-59; ed, Am J Bot, Bot Soc Am, 65-69; sr fel, NATO, 73; prog dir, Bot Soc Am, 77-80. *Mem:* AAAS; Bot Soc Am (treas, 63-64, vpres, 71, pres, 72); Torrey Bot Club; Am Inst Biol Sci; Asn Trop Biol; Int Asn Wood Anatomists. *Res:* Wood anatomy; developmental anatomy of roots. *Mailing Add:* 316 Biol Sci Bldg Miami Univ Oxford OH 45056

HEIMSCH, RICHARD CHARLES, b Philadelphia, Pa, Dec 20, 42; m 65; c 3. FOOD MICROBIOLOGY, FERMENTATION TECHNOLOGY. *Educ:* Miami Univ, BA, 65; Univ Wis, MS, 71, PhD(bact), 73. *Prof Exp:* Instr microbiol, Ore State Univ, 71-72; from asst prof to assoc prof bact, 72-83, asst dir agr res, 86-90, PROF BACT, UNIV IDAHO, 83-, ASSOC DIR AGR RES, 90- *Concurrent Pos:* Interim head, Dept Bact & Biochem, Univ Idaho, 90- *Mem:* Am Soc Microbiol; AAAS; Sigma Xi; Inst Food Technologists; Soc Indust Microbiol. *Res:* Discovery, detection and control of microbial toxins in the food chain; production of single-cell protein and biochemicals by bioconversion of agricultural and industrial wastes; food borne disease hazards; microbial biotechnology. *Mailing Add:* Dept Bact & Biochem Col Agr Univ Idaho Moscow ID 83843

HEIN, DALE ARTHUR, b Redmond, Ore, Apr 21, 33; m 65. WILDLIFE MANAGEMENT, VERTEBRATE ECOLOGY. *Educ:* Ore State Univ, BS, 59; Iowa State Univ, MS, 62, PhD(wildlife mgt), 65. *Prof Exp:* Asst prof biol, Wake Forest Univ, 65-68; assoc prof, 68-77, PROF WILDLIFE BIOL, COLO STATE UNIV, 77- *Mem:* AAAS; Wildlife Soc; Wilson Ornith Soc; Am Ornith Union; Am Inst Biol Sci; Sigma Xi. *Res:* Vertebrate ecology; population dynamics and habitat management, especially research applicable to management of wildlife. *Mailing Add:* Dept Fish & Wildlife Biol Colo State Univ Ft Collins CO 80523

HEIN, DAVID WILLIAM, b Faith, SDak, Aug 17, 55; m 78; c 2. PHARMACOGENETICS, CHEMICAL CARCINOGENESIS. *Educ:* Univ Wis-Eau Claire, BS, 77; Univ Mich, PhD(pharmacol), 82. *Prof Exp:* From asst prof to prof pharmacol, Morehouse Sch Med, 82-89, actg chair, 83-85, chmn, 85-89, prog dir res, 86-89; PROF & CHAIR PHARMACOL, UNIV NDAK SCH MED, 89- *Concurrent Pos:* Grad fac, Atlanta Univ, 83-89; prin investr, Nat Cancer Inst grant, NIH, 83-, Div Res Resources grant, 87-89; adj asst prof, Sch Med, Emory Univ, 86-89; adj assoc prof, Ga State Univ, 89-; peer reviewer, Nat Cancer Inst Study Sect, 90-; deleg, US Pharmacopeial Conv, 91- *Mem:* Am Asn Cancer Res; Am Soc Pharmacol & Exp Therapeut; Asn Med Sch Pharmacol; Int Soc Study Xenobiotics; Sigma Xi. *Res:* Studies to understand and predict genetic predisposition to cancer and other toxicities following exposures to drugs and environmental chemicals. *Mailing Add:* Dept Pharmacol Univ NDak Sch Med Grand Forks ND 58203

HEIN, JAMES R, b Santa Barbara, Calif, Mar 15, 47. GEOLOGY, MARINE MINERAL DEPOSITS. *Educ:* Ore State Univ, BSc, 69; Univ Calif, Santa Cruz, PhD(earth sci), 73. *Prof Exp:* Lectr earth sci geol, Univ Calif, Santa Cruz, 72-73 & 80; GEOLOGIST, US GEOL SURV PAC MARINE GEOL, 73- *Concurrent Pos:* Res geologist, Nat Res Coun, 74-75; int group leader, Int Geol Correlation Proj, UNESCO, 76-86 & 91-; convenor, Penrose Conf, Geol Soc Am, 78; mem, Int Conf Siliceous Deposits, Japan, 81, Yugoslavia, 86, Marine Mineral Deposits, Hawaii, 90, Marine Mining Develop task force, US Minerals Mgt Serv, 84; ed, Elsevier, 83, Van Nostrand Reinhold, 87, Springer-Verlag, 89; guest scientist, Japanese Nat Oil Corp, 85, Chinese Ministry Petrol & Marine Geol, 86, Indian Ministry Oceanog, 86, Korea Ocean Res & Develop Inst, 88,90, Ger Res Cruises, 84, 85,86; expert witness, US Cong Comt Deep Sea Mining, 86; leader, USA-USSR prog Geochem Marine Sediments, 88-, USA-Korea prog Marine Mineral Deposits, 88-; marine geol panel, USA-Japan prog Nat Res, 88-; assoc ed, Geo-Marine Letters, 90- *Honors & Awards:* Fel, Geol Soc Am, 80. *Mem:* AAAS; Geol Soc Am; Am Geophys Union; Int Asn Sedimentol; Paleont & Mineral; Geochem Soc. *Res:* Marine geology, low temperature geochemistry, authigenic mineralogy, clay mineralogy, field geology, siliceous and calcareous deposits, ore deposits at constructive and destructive plate boundaries, island arcs, marine ferromanganese deposits; seamounts, marine mineral deposits. *Mailing Add:* US Geol Surv MS 999 345 Middlefield Rd Menlo Park CA 94025

HEIN, JOHN WILLIAM, b Chester, Mass, Sept 29, 20; m 44. DENTISTRY. *Educ:* Am Int Col, BS, 41; Tufts Col, DMD, 44; Univ Rochester, PhD(pharm), 52. *Hon Degrees:* AM, Harvard Univ, 62; DSc, Am Int Col, 79. *Prof Exp:* Instr oral path, Dent Sch, Tufts Col, 43-44; instr, Univ Rochester, 51-53, asst prof dent res & chmn dept dent & dent res, 52-55; dent dir, Colgate-Palmolive Co, 55-59; prof prev dent, lectr pharmacol & dean, Sch Dent Med, Tufts Univ, 59-62; prof dent, Sch Dent Med, Harvard Univ, 62-69; DIR, FORSYTH DENT CTR, 62- *Concurrent Pos:* Instr anat & physiol, Eastman Sch Dent Hyg, 50-55 & dent res, 53-55; assoc res specialist, Bur Biol Res, Rutgers Univ, 55-59. *Mem:* AAAS; Am Dent Asn; Int Asn Dent Res; Int Col Dentists (pres, 83-84); Sigma Xi; hon mem Royal Soc Med. *Res:* Chemical therapy for dental caries prevention; oral hygiene; experimental caries in Syrian Hamster; clinical studies of dental caries and periodontal disease. *Mailing Add:* Forsyth Dent Ctr 140 The Fenway Boston MA 02115

HEIN, PETER LEO, JR, b Chicago, Ill, Feb 12, 30. PSYCHIATRY. *Educ:* Georgetown Univ, BS, 51, MD, 55; Am Bd Psychiat & Neurol, dipl, 62. *Prof Exp:* Intern psychiat, Univ Chicago Clins, 55-56; resident, DC Gen Hosp, 56-57; chief open ward serv & outpatient dept, Sheppard AFB Hosp, Wichita Falls, Tex, 57-59; resident, Med Ctr, Georgetown Univ, 59-61; from instr to assoc prof, Duke Univ, 61-71, mem, Steering Comt Post Doctoral Res Training Prog, 65-71, head psychophysiol lab, Clin Res Unit Comt & Human Experimentation Comt, 68-71; PROF PSYCHIAT & DIR RESIDENCY TRAINING, MED CTR, WVA UNIV, 71-, PROF BEHAV MED, 80- *Concurrent Pos:* Res fel, Duke Univ, 61-63; clin investr, Vet Admin Hosp, Durham, NC, 63-65. *Mem:* Am Psychiat Asn; Soc Psychophysiol Res; Am Electroencephalog Soc. *Res:* Electroencephalography; psychophysiology. *Mailing Add:* Dept Psychiat WVa Univ Hosp Morgantown WV 26506

HEIN, R(OWLAND) F(RANK), b Minneapolis, Minn, Aug 5, 23; m 47; c 3. CHEMICAL ENGINEERING. *Educ:* Univ Minn, BChE, 44, PhD(chem eng), 52. *Prof Exp:* Res engr, E I Du Pont de Nemours & Co, 52-60, res supvr, 60-77, res assoc, 78-82; RETIRED. *Mem:* AAAS; Am Chem Soc; Am Inst Chem Engrs. *Res:* Mass transfer; process development; reactor design. *Mailing Add:* Box 603 Mendenhall PA 19357-0603

HEIN, RICHARD EARL, b Erie, Ill, May 25, 19; m 44; c 4. CHEMISTRY. *Educ:* Univ Iowa, BS, 42; Iowa State Univ, PhD(chem), 50. *Prof Exp:* Jr chemist, Manhattan Metall Proj, AEC, Iowa State Univ, 42-50; from asst prof to assoc prof chem, Kans State Col, 50-57; admin fel, Mellon Inst, 57-65; mgr food res, H J Heinz Co, 65-71, sr mgr prod res & develop, 71-73; gen mgr res & develop, H J Heinz Co, 73-77, gen mgr res & qual assurance, Tomato Prod & Condiments Div, 77-81, gen mgr Res & Qual Assurance, 81-83, SPEC TECH CONSULT, HEINZ USA DIV, H J HEINZ CO, 83- *Concurrent Pos:* Int Exec Service Corp, 83. *Mem:* AAAS; Am Chem Soc; Inst Food Technologists. *Res:* Fission product and hot-atom chemistry; application of tracers to biochemical problems; physical aspects of food technology; rheology of suspensions; gas chromatography; food product development. *Mailing Add:* 2104 Mazatlan Rd Port Charlotte FL 33983-2633

HEIN, RICHARD WILLIAM, b Cleveland, Ohio, Nov 29, 34; m 58; c 3. INDUSTRIAL ORGANIC CHEMISTRY, WOOD PRESERVATION. *Educ:* Capital Univ, BS, 56; Case Inst Technol, MS, 58, PhD(org chem), 61. *Prof Exp:* Chemist, Escambia Chem Corp, 61-68; develop scientist, BF Goodrich Chem Co, 68-73, tech mgr, 73-80, sr res & develop assoc, 80-84; MGR, MOONEY CHEMICALS, INC, 84- *Mem:* Am Wood Preservers Asn; Am Chem Soc; Sigma Xi; Am Soc Testing & Mat. *Res:* Petrochemicals; monomer synthesis; oxidations; heterogeneous catalysis; separation processes; water treatment; wood preservatives. *Mailing Add:* 51 Nantucket Dr Hudson OH 44236

HEIN, ROSEMARY RUTH, b Chicago, Ill, Sept 12, 24. DEVELOPMENTAL ANATOMY, DEVELOPMENTAL BIOLOGY. *Educ:* Carleton Col, BA, 46; George Washington Univ, MS, 51; Northwestern Univ, PhD(biol), 54. *Prof Exp:* Biologist, Nat Cancer Inst, NIH, 46-50; assoc prof anat, physiol & embryol, Keuka Col, 54-62; assoc prof anat & biol, Winthrop Col, 62-65; assoc prof anat & embryol, 65-67, actg dean, 72-73, prof anat & embryol, Upsala Col, 67-77; PROF BIOL & CHMN DIV NATURAL SCI & MATH, ST MARY'S COL, MD, 77- *Mem:* Am Soc Zool; Am Inst Biol Sci; Sigma Xi. *Res:* Nutritional effects of cancer; developmental physiology; regeneration. *Mailing Add:* Dept Biol/Math St Mary's Col St Mary's City MD 20686

HEIN, WARREN WALTER, b Plymouth, Wis, May 13, 44; m 65, 85; c 4. NUCLEAR PHYSICS. *Educ:* Wis State Univ-Whitewater, BS, 66; Iowa State Univ, PhD(nuclear physics), 70. *Prof Exp:* From asst prof to assoc prof, 70-79, PROF PHYSICS, NORTHERN STATE COL, 79-; assoc prof, 79-84, PROF PHYSICS, SDAK STATE UNIV, 84- *Mem:* Am Asn Physics Teachers; Sigma Xi; Am Soc Eng Educ; Optical Soc Am. *Res:* Optics; alignment of large x-ray optical systems; low level radioactivity of groundwater; remote sensing of soil moisture. *Mailing Add:* Physics Dept Box 2219 SDak State Univ Brookings SD 57007

HEINBERG, MILTON, b Birmingham, Ala, Apr 3, 28; m 56; c 2. PHYSICS, OPERATIONS RESEARCH. *Educ:* Oberlin Col, AB, 52; Univ Pittsburgh, PhD(physics), 56. *Prof Exp:* Res assoc physics, Univ Pittsburgh, 56-57 & Cornell Univ, 57-59; exp physicist, Lawrence Livermore Lab, Univ Calif, 59-71; sr scientist, Nuclear Defense Res Corp, Albuquerque, 71-73; staff scientist, RCA Corp, 73-74; sr test & eval scientist, EG & G Co, Albuquerque, 74-75; sr scientist, The Dikewood Corp, Albuquerque, NMex, 75-76; staff mem, Los Alamos Nat Lab, 76-90; CONSULT, DEPT ENERGY, 90- *Mem:* Am Phys Soc. *Res:* Positron annihilation; beta decay; high energy nuclear physics; high speed weapon diagnostics; military operational tests and evaluations; nuclear materials safeguards and security classification. *Mailing Add:* 8416 Harron Valley Ct Gaithersburg MD 20879

HEINDEL, NED DUANE, b Red Lion, Pa, Sept 4, 37; m 59. ORGANIC CHEMISTRY, MEDICINAL CHEMISTRY. *Educ:* Lebanon Valley Col, BS, 59; Univ Del, MS, 61, PhD(org chem), 63. *Hon Degrees:* DSc, Lebanon Valley Col, 85. *Prof Exp:* From asst to instr chem, Univ Del, 61-62; NSF fel, Princeton Univ, 63-64; asst prof, Marshall Univ, 64-66; from asst prof to prof chem, 66-76, H S BUNN CHAIR PROF, LEHIGH UNIV, 76- *Concurrent Pos:* Benedum Found fac fel, Marshall Univ, 65; grants, Petrol Res Fund, 64-65, Res Corp, 64-66, Sigma Xi Res Fund, 65, US Army, 66-72, NIH, 66-, Milheim Fund for Cancer Res, 73-74 & 81-84, Am Cancer Soc, 75-78 & Pardee Fund, 75-77 & 81-84; asst prof, Univ Ohio, 65; adj assoc prof nuclear med, Hahnemann Med Col, 73-; dir div biol chem & biophys, Ctr Health Sci, Lehigh Univ, 73-80, Ctr Health Sci, 80-87; trustee, Keystone Jr Col, LaPlume, Pa, 75-88 & Ctr Hist Chem, Philadelphia, 82-; dir, Am Chem Soc, 85- *Mem:* Am Chem Soc; Soc Nuclear Med. *Res:* Heterocyclic and medicinal syntheses; history of chemistry; nuclear medicine and diagnostic radioactive pharmaceuticals; anti-tumor compounds and central nervous system depressants; monoclonal antibody drug conjugates. *Mailing Add:* Dept Chem Lehigh Univ Bethlehem PA 18015

HEINDL, CLIFFORD JOSEPH, b Chicago, Ill, Feb 4, 26. NUCLEAR PHYSICS. *Educ:* Northwestern Univ, BS, 47, MS, 48; Columbia Univ, AM, 50 & PhD(physics), 59. *Prof Exp:* Physicist, Bendix Aviation Corp, 53-54; asst sect chief, Babcock & Wilcox Co, 56-58; res specialist physics, 59-61, group supvr nuclear physics, 61-65, tech mgr res, 65-78, DEP, DIV MGR SCI, JET PROPULSION LAB, CALIF INST TECHNOL, 78- *Mem:* Am Phys Soc; Am Nuclear Soc; Health Physics Soc; Am Inst Aeronaut & Astronaut. *Res:* Reactor physics. *Mailing Add:* 179 Mockingbird Lane South Pasadena CA 91030

HEINDSMANN, T(HEODORE) E(DWARD), b Aug 2, 25; US citizen; m 49; c 3. MARINE SCIENCES, ACOUSTICS. *Educ:* Rensselaer Polytech Inst, BEE, 44; Univ Calif, Berkeley, MS, 50. *Prof Exp:* Electronic scientist, Navy Underwater Sound Lab, Conn, 45-47, sect head acoust res, 49-55; res engr, inst transp & traffic, microwave lab, Univ Calif, Berkeley, 47-49; res specialist, tech staff, Boeing Co, 55-58, antisubmarine prog chief, 58-61, space tech mgr, 61-64, sr tech mgr, AWACS staff, 64-84, sr tech mgr, Acousts & ASW Systs, 84-90; MANAGING DIR & CONSULT, RIS ASSOCS, 91- *Concurrent Pos:* Sr lectr & ed, Ocean Tech Ser, 65-68; gen chmn, Int Ocean Eng Conf, 80, adv comt, 81, 89. *Mem:* AAAS; Acoust Soc Am; Inst Elec & Electronics Engrs; Marine Technol Soc; Ocean Eng Soc. *Res:* Propagation and reception of acoustic and electromagnetic energy; classification of signal and background sources; instrumentation for remote sensing; bioacoustics. *Mailing Add:* Sea Frost Farm 18003 Westside Hwy SW Vashon Island WA 98070

HEINE, GEORGE WINFIELD, III, b Oakland, Calif, Sept 22, 49; m 87. SIMULATED ANNEALING. *Educ:* Reed Col, BA, 71; Univ Colo, MS, 89. *Prof Exp:* MATH ANALYST, BUR LAND MGT, 84- *Mem:* Soc Indust & Appl Math; Math Asn Am; Am Math Soc. *Res:* Investigating necessary and sufficient conditions for effectiveness of simulated annealing and other combinatorial optimization algorithms; developing provably correct polygon overlay software for use in geographic information systems. *Mailing Add:* PO Box 11053 Pueblo CO 81001

HEINE, HAROLD WARREN, b Highland Park, NJ, Sept 14, 22; m 53; c 3. ORGANIC CHEMISTRY. *Educ:* Rutgers Univ, BSc, 44, PhD(org chem), 48. *Prof Exp:* Asst, Rutgers Univ, 47-48; from asst prof to assoc prof, 48-53, Bucknell presidential prof, 72-78, PROF CHEM, BUCKNELL UNIV, 54-, chair, 70-86. *Concurrent Pos:* Res chemist, Dow Chem Co, 56-57; NSF sr fac fel, Ger, 61-62; consult to Surgeon Gen, 65-; vis prof, Univ Heidelberg, 69-70 & Univ Auckland, NZ, 86; pres, Chem Div, Coun Undergrad Res, 88-90. *Honors & Awards:* Harbison Prize for Distinguished Teaching & Res, Danforth Found, 70; Catalyst Award, Chem Mfrs Asn, 80; Res Award, Am

Chem Soc, 87. *Mem:* Am Chem Soc. *Res:* Bromoamides; chlorohydrins; ethylenimines; kinetics; chemistry of diaziridines and cycloadditions; nitrones; benzodiazepines; o-quinone monoimides. *Mailing Add:* Dept Chem Bucknell Univ Lewisburg PA 17837

HEINE, MELVIN WAYNE, b Ellendale, NDak, Mar 21, 33; m 58; c 2. OBSTETRICS & GYNECOLOGY, ENDOCRINOLOGY. *Educ:* Duke Univ, MD, 58. *Prof Exp:* Intern med, Med Col Va, 58-59; asst resident obstet & gynec, Teaching Hosp, Univ Fla, 59-61, chief resident & instr, 61-62, asst prof, Col Med, 63-64, asst prof & dir endocrinol, 66-70; prof obstet & gynec & chief endocrine serv, Col Med, Univ Ariz, 70-77; PROF & CHMN DEPT OBSTET & GYNEC, TEX TECH SCH MED, 77- *Concurrent Pos:* Fel, Worcester Found Exp Biol, 62-63; Ford Found grant, Univ Fla & Univ Ariz, 67-70; Fed Drug Admin grant. *Mem:* Am Fertil Soc; Int Soc Res Reproduction; AMA; Am Col Obstet & Gynec; Am Gynec Club. *Res:* Gynecologic endocrinology; female infertility; contraception; fetal endocrinology; menopause. *Mailing Add:* Univ Ariz Health Sci Ctr Obstet/Gynec 1501 N Campbell Tucson AZ 85718

HEINE, RICHARD W, b Detroit, Mich, July 22, 18; m 40; c 2. METALLURGICAL ENGINEERING. *Educ:* Wayne State Univ, BS, 40; Univ Wis, MS, 48. *Prof Exp:* Instr metall, Gen Motors Inst, 40-43 & 46-47; from instr to assoc prof metall eng, 47-59, PROF METALL ENG, UNIV WIS-MADISON, 59-, CHMN DEPT METALL & MINERAL ENG, 64- *Concurrent Pos:* Metal casting consult. *Honors & Awards:* Sci Merit Award, Am Foundrymen's Soc, 57, Gold Medal, 66. *Mem:* Am Inst Mining, Metall & Petrol Engrs; Am Soc Metals; Am Foundrymen's Soc; Brit Inst Metals. *Res:* Solidification of metals; elimination of casting defects; molding aggregates and processes. *Mailing Add:* 7510 Widgeon Way Madison WI 53717

HEINE, URSULA INGRID, b Berlin, Ger, Feb 19, 26. CYTOLOGY, BIOLOGY. *Educ:* Univ Berlin, MS, 50, PhD(biol), 53. *Prof Exp:* Assoc, Inst Cancer Res, Ger Acad Sci, 50-59; assoc, Med Ctr, Duke Univ, 59-68; microbiologist, 68-71, HEAD ULTRASTRUCTURAL STUDIES SECT, NAT CANCER INST, 71- *Mem:* AAAS; Electron Micros Soc Am; Am Asn Cancer Res; Am Soc Cell Biol. *Res:* Electronmicroscopic studies on tumor viruses and their relation to the host cell, as well as the response of nontransformed cells to chemical carcinogens. *Mailing Add:* Frederick Cancer Res Facility Bldg 538 Rm 205-E Frederick MD 21701-1013

HEINEKEN, FREDERICK GEORGE, b Chicago, Ill, Oct 22, 39; div; c 1. CHEMICAL ENGINEERING. *Educ:* Northwestern Univ, BSChE, 62; Univ Minn, PhD(chem eng), 66. *Prof Exp:* Sr res chem engr, Monsanto Co, St Louis, 66-72; instr, Univ Colo, 75-76; sr proj eng, Cobe Labs, Inc, 76-78, dept mgr, 78-81, ther scientist, 81-85; PROG DIR, NSF, 85- *Concurrent Pos:* Fel, Univ Colo Med Ctr, 72-74, Young Pulmonary res investr, 74-76. *Mem:* Am Chem Soc; AAAS; Am Inst Chem Engrs; Am Soc Artificial Internal Organs; Asn Advan Med Instrumentation. *Res:* Mathematical aspects of enzyme kinetics; oxygen transfer; microbial enzyme synthesis; continuous fermentation; membrane separation techniques; enzyme recovery techniques; blood gas exchange; artificial kidney design; kidney machine design; physiological modelling of dialysis patients. *Mailing Add:* NSF 1800 G St NW Washington DC 20550

HEINEMAN, WILLIAM RICHARD, b Lubbock, Tex, Oct 15, 42; m 69; c 2. ANALYTICAL CHEMISTRY, ELECTROCHEMISTRY. *Educ:* Tex Tech Univ, BSc, 64; Univ NC, Chapel Hill, PhD(chem), 68. *Prof Exp:* Res chemist, Hercules, 68-70; res assoc chem, Case Western Reserve Univ, 70-71, Ohio State Univ, 71-72; from asst prof to prof chem, 72-80, DISTINGUISHED PROF, UNIV CINCINNATI, 88- *Concurrent Pos:* Ed, Cintacs, Am Chem Soc, 74-75, secy, 75-76, first vchmn & chmn elect, 76-77, chmn, 77-78, trustee, 78-81 & counr, 84-86; sci adv, Food & Drug Admin, 75-80; treas, Div Anal Chem, Am Chem Soc, 83-85; mem bd dirs, Soc Electroanal Chem, 84-; mem adv bd, Anal Chem, 84-86, analyst, 87-, Selective Electrode Rev, 87-, Biosensors, 86- & Analytica Cliica Acta, 91-93. *Honors & Awards:* Sr Humboldt Award, 89. *Mem:* Am Chem Soc; Electrochem Soc; Sigma Xi; Soc Electroanal Chem (pres, 84-85); Am Asn Clin Chem; AAAS. *Res:* Analytical chemistry; electroanalytical chemistry; optically transparent thin layer electrodes; bioelectrochemistry; thin layer differential pulse voltammetry; stripping voltammetry; development of Tc-99m radiopharmaceuticals; immunoassay by electrochemical techniques; polymer modified electrodes; extended X-ray absorption fine structure spectroelectrochemistry. *Mailing Add:* Dept Chem Univ Cincinnati Cincinnati OH 45221-0172

HEINEMANN, EDWARD H, b Saginaw, Mich, Mar 14, 08; c 2. AERONAUTICAL & MARINE ENGINEERING. *Hon Degrees:* DSc, Northrup Univ, 76. *Prof Exp:* Engr, Douglas Aircraft Co, 26-32, designer, Northrop Div, Douglas El Segundo, 32-36, chief engr, 36-58, corp vpres combat aircraft, Douglas Aircraft Co, 58-60, exec vpres guidance technol, 60-62; vpres, Gen Dynamics Corp, 62-73; pres, Heinemann Assocs, 75-; RETIRED. *Concurrent Pos:* Mem adv bd, Naval Ord Test Sta, 54-58 & 61-65; Naval Res Adv Comn, 56-, chmn, 69-71, mem lab adv bds navy ships & air warfare, 67-; mem limited war panel, President's Sci Adv Comn, 59-61, ad hoc mercury panel, 61; Defense Sci Bd, 59-64; Army Tactical Mobility Requirements Bd, 62-65; Mass Inst Technol vis comt naval archit & marine eng, 64-; adv bd, Dept Defense Joint Task Force Two, 65-67; marine bd, Nat Acad Eng, 67-71; oceanog adv comt, US Navy, 68- *Honors & Awards:* Sylvanus Albert Reed Award, 52; Collier Trophy, 53; Paul Tissandier Award, Int Aeronaut Fedn, 55; Wilbur Wright Mem Lectr, 64; Guggenheim Award, 78; Elder Statesman Aviation, Nat Aeronaut Asn, 79; Nat Medal Sci, 83. *Mem:* Nat Acad Eng; hon fel Am Inst Aeronaut & Astronaut; fel Am Astronaut Soc; fel Soc Automotive Engrs; Nat Aeronaut Asn. *Res:* Design and development of airplanes. *Mailing Add:* PO Box 1795 Rancho Santa Fe CA 92067

HEINEMANN, HEINZ, b Berlin, Ger, Aug 21, 13; nat US; m 48; c 2. PHYSICAL CHEMISTRY. *Educ:* Univ & Tech Hochschule, Berlin, Ger, BS, 35; Univ Basel, PhD(chem), 37. *Prof Exp:* Chief res chemist, Rodessa Oil & Ref Corp, 38-39; res chemist, Danciger Oil & Refining, 39-41; res fel, Carnegie Inst Technol, 41; lab supvr, Attapulgus Clay Co, 41-48; sect chief process res, Houdry Process Corp, 48-57; asst to vpres res & develop & assoc dir res, M W Kellogg Co, 57-61, mgr, 61-67, dir chem & eng res, 67-69; sr res assoc, Cent Res Lab, Mobil Res & Develop Corp, 69-70, mgr catalysis res, 70-76, mgr res contracts, 76-78; SR SCIENTIST, LAWRENCE BERKELEY LAB, UNIV CALIF, BERKELEY, 78-, LECTR CHEM ENG, 79- *Concurrent Pos:* Pres, Int Cong Catalysis, 56-60; mem, Coun Sci Res, Spain, 64-; lectr chem eng, Univ Calif, Berkeley, 79- *Honors & Awards:* E V Murphree Medallist, Am Chem Soc, 71; E J Houdry Award, Catalysis Soc NAm, 75. *Mem:* Nat Acad Eng; Am Chem Soc; fel Am Inst Chemists; fel Royal Soc; Catalysis Soc NAm. *Res:* Catalysis; petroleum processing; coal conversion and synthetic fuels; clays and adsorbents; carbohydrate to hydrocarbon conversion. *Mailing Add:* 1588 Campus Dr Berkeley CA 94708

HEINEMANN, RICHARD LESLIE, b Jan 14, 31; US citizen; m 64; c 1. CYTOGENETICS. *Educ:* Champlain Col, NY, BA, 53; Univ Rochester, MS, 58; Med Col Va, PhD(biol & genetics), 68. *Prof Exp:* Asst prof, 63-68, ASSOC PROF BIOL, LONGWOOD COL, 68- *Mem:* Soc Study Evolution; Sigma Xi. *Res:* Genetics, cytogenetics and evolution of the sex determining mechanisms in arrhenotokous parthenogenetic species. *Mailing Add:* Dept Biol Longwood Col Farmville VA 23901

HEINEMANN, WILTON WALTER, b Ritzville, Wash, Mar 9, 20; m 80; c 3. ANIMAL NUTRITION. *Educ:* Wash State Univ, BS, 42, MS, 45; Ore State Univ, PhD(animal nutrit), 54. *Prof Exp:* Exten agt & asst county agt, Kittitas County, 42-43; asst animal husb, Irrig Exp Sta, Wash State Univ, 43-45, asst animal husbandman, 45-54, assoc animal scientist animal nutrit, 54-61, prof animal nutrit & animal scientist, 61-85; NUTRIT CONSULT, 86-; TEXTBK WRITER, 87- *Concurrent Pos:* Consult, Bur Sport Fisheries, US Dept Interior, 60-70 & Battelle Mem Inst, 72-; lectr, Polish Acad Sci, Krakow, 79, Japanese farm coops, Japan & Korean farm coops, Korea, 83. *Mem:* Fel AAAS; fel Am Soc Animal Sci; Am Forage & Grassland Coun; Sigma Xi; Int Grassland Cong. *Res:* Vitamins in pork production; soil-plant-animal relationships; irrigated pastures; forage utilization by ruminants; net energy and metabolism research with ruminants; energy, metabolism and feeding research on industrial and other by-products and wastes. *Mailing Add:* 4406 Terrace Heights Dr Yakima WA 98901

HEINEN, JAMES ALBIN, b Milwaukee, Wis, June 23, 43. DIGITAL SIGNAL PROCESSING, SPEECH PROCESSING. *Educ:* Marquette Univ, BEE, 64, MS, 67, PhD(elec eng), 69. *Prof Exp:* lectr elec eng, Marquette Univ, 67-68, from asst prof to assoc prof 69-80, grad adminr, 71-73, chmn dept, 73-76, PROF, ELEC & COMP ENG, MARQUETTE UNIV, 80-, DIR GRAD STUDIES, 87- *Mem:* Inst Elec & Electronics Engrs; Am Soc Eng Educ. *Res:* Digital signal processing; speech processing; analysis and design of digital filters; stability and analysis of control systems. *Mailing Add:* Marquette Univ 1515 W Wisconsin Ave Milwaukee WI 53233

HEINER, DOUGLAS C, b Salt Lake City, Utah, July 27, 25; m 46; c 9. PEDIATRICS, ALLERGY. *Educ:* Idaho State Col, BS, 46; Univ Pa, MD, 50; McGill Univ, PhD(immunol), 69; Am Bd Pediat, cert allergy, immunol & cardiol. *Prof Exp:* Intern, Hosp Univ Pa, 50-51; resident, Boston Children's Med Ctr, 53-55, res fel cardiol, 55-56; from instr to asst prof pediat, Univ Ark, 56-59, actg head dept, 57-58; from asst prof to assoc prof, Col Med, Univ Utah, 60-69; PROF PEDIAT, MED SCH, UNIV CALIF, LOS ANGELES, 69-; DIR DIV IMMUNOL & ALLERGY, DEPT PEDIAT, HARBOR GEN HOSP, 69- *Concurrent Pos:* Asst, Harvard Med Sch, 54-55; USPHS spec res fel, Royal Victoria Hosp, 66-69; vis prof, Univ Bern, Inst Clin Immunol, Switz, 78-79; res assoc, Div Biol, Cal Tech, 89-90; Fogarty Int Fel, 78-79. *Honors & Awards:* Western Soc Pediat Res & Ross Labs Res Award, 61. *Mem:* Am Acad Allergy & Immunol; Am Acad Pediat; Am Asn Immunologists; Soc Pediat Res; Am Pediat Soc; Am Col Allergy & Immunol. *Res:* Pediatric allergy and immunology; IgE; IgD; food sensitivity; pulmonary hemosiderosis; celiac disease; pediatric AIDS; immunology. *Mailing Add:* Univ Calif-Los Angeles Sch Med Harbor-UCLA Med Ctr 1000 W Carson St J-4 Torrance CA 90509

HEINER, ROBERT E, genetics, plant breeding, for more information see previous edition

HEINER, TERRY CHARLES, b Smoot, Wyo, Feb 1, 41; m 66; c 3. BOTANY, FORESTRY. *Educ:* Utah State Univ, BS, 66; Ore State Univ, MS, 67; Iowa State Univ, PhD(bot), 70. *Prof Exp:* Asst prof, 70-80, assoc prof, 80-87, PROF BIOL SCI, WESTERN NMEX UNIV, 87- *Concurrent Pos:* Adv, NMex Environ Inst, 71- *Mem:* Bot Soc Am. *Res:* Plant competition, population ecology. *Mailing Add:* Dept Math & Sci Western NMex Univ Silver City NM 88062

HEINES, THOMAS SAMUEL, b Grand Rapids, Mich, Oct 20, 27; m 51; c 2. COMPUTER SCIENCE, CHEMICAL ENGINEERING. *Educ:* Univ Mich, Ann Arbor, BSE, 49, MSE, 50, PhD(chem eng), 54. *Prof Exp:* Res engr, Union Carbide Corp, 54-57; res assoc chem eng, Nat Distillers, 57-65; from assoc prof to prof comput sci, 65-85, CHMN COMPUT SCI, CLEVELAND STATE UNIV, 85- *Mem:* Asn Comput Mach; Inst Elec & Electronics Engrs. *Res:* Computer linguistics; simulation; modeling; control. *Mailing Add:* Dept Comp Sci Cleveland State Univ Euclid at 24th Cleveland OH 44115

HEINICKE, HERBERT RAYMOND, b Elgin, Ill, June 7, 27; m 55; c 5. FOOD SCIENCE & TECHNOLOGY. *Educ:* Northwestern Univ, BS, 49; Univ Wis, MS, 52, PhD(biochem), 55. *Prof Exp:* Res assoc leukemia, Med Sch, Univ Wis, 55-56; proj leader cat food nutrit, Res Labs, Quaker Oats Co, 56-59, proj leader pet food nutrit, 59-62, mgr pet food nutrit, 62-67, mgr pet food nutrit & eval, 67-71; TECH CONSULT, PET FOOD, HUMAN & AUTOGENOUS EXTRUSION, 71- *Mem:* Fel AAAS; Inst Food

Technologists; Asn Vitamin Chemists (pres, 71-72). *Res:* Cat and dog nutrition; care of cats and dogs as laboratory animals; pet food technology and evaluation; human nutrition; environmental sciences; technology assessment. *Mailing Add:* 1302 W Boston Ave Indianola IN 50125

HEINICKE, PETER HART, b Madison, Wis, Mar 26, 56. DATA ACQUISITION SOFTWARE, SIMULATION SOFTWARE. *Educ:* Washington Univ, BA, 77, MA, 77; Princeton Univ, MA, 79; Ill Inst Technol, MS, 85. *Prof Exp:* Engr, Int Harvester Corp, 80; SYSTS ANALYST, FERMI NAT LAB, 80- *Concurrent Pos:* Consult, Comp Method, Inc, 79- *Mem:* Asn Comput Mach; Am Phys Soc. *Res:* Design and implimation of VME data acquisition systems which will read out KMAC using Motorola 68,000 and which will feed into vax. *Mailing Add:* 25779 S Winchester Ctr No 3 Warrenville IL 60555

HEINICKE, RALPH MARTIN, b Hickory, NC, Sept 3, 14; m 44; c 1. PHARMACOLOGY. *Educ:* Cornell Univ, BS, 36; Univ Minn, PhD(biochem), 50. *Prof Exp:* Agr chemist, Shell Oil Co, 39-42; biochemist, Pineapple Res Inst, 50-55; dir chem & food res, Dole Co, 55-71, dir chem & food res, Jintan Dole Co, 64-70; mem tech staff, Kuakini Med Res Inst, 71-72; dir chem & new prod res, ARESCO, 80-84; tech pharmacol consult, Jintan Dolph Co, 74-87; vpres res & develop, BCC Systs, Inc, 85-87; CHMN, BIOTECHNOL RESOURCES INC, 87- *Concurrent Pos:* Assoc fac mem, Univ Hawaii, 50-; consult, Drug & Food Co. *Mem:* AAAS; Am Chem Soc; Am Inst Chemists; Inst Food Technologists. *Res:* Role of proteases in animal and plant physiology and their control by antagonists and hormones; medical and industrial application of plant chemicals and alkaloids. *Mailing Add:* 1124 Rostrevor Circle Louisville KY 40205

HEINIG, HANS PAUL, b Dortmund, Ger, June 30, 31; Can citizen; m 60; c 2. MATHEMATICS. *Educ:* McMaster Univ, BS, 61; Univ Western Ont, MA, 62; Univ Toronto, PhD(math), 65. *Prof Exp:* Asst prof, 65-71, assoc prof, 71-78, PROF MATH, MCMASTER UNIV, 78- *Mem:* Am Math Soc; Math Asn Am. *Res:* Classical analysis; Laplace and Fourier transforms; Banach spaces; linear operators and interpolation in Orlicz spaces; functions of bounded mean oscillation. *Mailing Add:* Dept Math McMaster Univ Hamilton ON L8S 4K1 Can

HEINIGER, HANS-JORG, b Burgdorf, Switz, Aug 30, 38; m 62; c 2. CELL BIOLOGY. *Educ:* Univ Geneva, BA, 59; Univ Bern, MA, 61, DVM, 65. *Prof Exp:* Asst prof neuropath, Inst Comp Neurol, Sch Vet Med, Univ Bern, 64, asst prof histochem, Inst path, Sch Med, 64-67; staff scientist & head, Lab Cell Biol, Inst Med, Nuclear Res Ctr, Julich, WGer, 67-70; vis scientist, 70-72, staff scientist cell biol, 72-80, dir animal resources, 81-83, sr staff scientist, Jackson Lab, 80-88; PRES & CHIEF EXEC OFFICER, CENT LAB, BLOOD TRANSFUSION SERV, 88- *Concurrent Pos:* Res collabr, Swiss Inst Exp Cancer Res, 67- & Brookhaven Nat Lab, Upton, NY, 74-; coop prof zool, Univ Maine, Orono, 1981; mem gov bd, Int Coun Lab Animal Sci. *Honors & Awards:* Edward A Stein Award, Univ Bern, 64. *Mem:* Soc Exp Med & Biol; Swiss Soc Molecular Biol; Soc Nuclear Med; Int Union Against Cancer; Am Asn Lab Animal Practitioners; Am Asn Immunologists; Fedn Am Socs Exp Biol; Int Soc Blood Transfusion. *Res:* Function of lipids, especially cholesterol in mammalian cell membranes. *Mailing Add:* Cent Lab Blood Transfusions Serv SRC Wankdorfstrasse 10 CH-3000 Bern 22 Switzerland

HEININGER, CLARENCE GEORGE, JR, b Rochester, NY, July 30, 28; m 52; c 5. PHYSICAL CHEMISTRY, ANALYTICAL CHEMISTRY. *Educ:* Villanova Univ, BS, 50; Univ Rochester, PhD(phys chem), 54. *Prof Exp:* Asst chem, Princeton Univ, 53-55; asst prof chem, Villanova Univ, 55-58; from asst prof to assoc prof, 58-64, chmn, 65-70 & 79-90, dean fac & instruction, 69-72, PROF CHEM, ST JOHN FISHER COL, 64- *Concurrent Pos:* Sect ed, Chem Abstr, 61-69; vis prof, Univ Lyon, France, 67-68 & Univ Del, 82-83. *Mem:* AAAS; Am Chem Soc; Sigma Xi. *Mailing Add:* Dept Chem St John Fisher Col 3690 East Ave Rochester NY 14618

HEININGER, S(AMUEL) ALLEN, b New Britain, Conn, June 13, 25; m 48; c 4. ORGANIC CHEMISTRY. *Educ:* Oberlin Col, BA, 48; Carnegie Inst Technol, MS, 51, DSc, 52. *Hon Degrees:* DBA, Adrian Col, 84. *Prof Exp:* Asst, Carnegie Inst Technol, 48-51; res chemist, Cent Res Dept, Monsanto Co, 52-56, group leader, 56-58, proj mgr, Develop Dept, 58-59, sect mgr, 59-64, dir develop, 65-68, dir food & fine chem bus group, Org Chem Div, 68-71, dir corp plans & develop, 71-74, gen mgr, Plasticizers Div, 74-76, dir corp res, 76-77, vpres res & develop, 77-79, vpres technol develop, 79-80, vpres corp plans & bus develop, Indust Chem Co, Monsanto Co, 80-84, corp vpres res planning, 84-90; RETIRED. *Concurrent Pos:* Mem, Indust Res Inst, 76-, vis comt sponsored res, MIT, 80, dir, Indust Res Inst, 83-85, Nat Technol Medal Select Comt, 84- 90, vpres, Indust Res Inst, 85-86, pres elect, 86-87, pres, 87-88. *Mem:* Am Chem Soc (pres elect, 90, pres, 91); Sigma Xi; Commercial Develop Asn; Chem Mkt Res Asn; NY Acad Sci. *Res:* Acrylonitrile chemistry; sulfenyl chlorides; organophosphorus; petrochemicals; food acidulants; sweeteners; food ingredients; drug intermediates; flavors and fragrances. *Mailing Add:* 11110 Hermitage Hill St Louis MO 63131

HEINIS, JULIUS LEO, b Liestal, Switz, Oct 2, 26; nat US; m 52; c 4. HORTICULTURE. *Educ:* Swiss Fed Inst Technol, IngAgr, 51; Ore State Col, PhD, 54. *Prof Exp:* Asst, NY Exp Sta, Geneva, 51; plant pathologist, Ore Dept Agr, 54-60; asst prof plant path, Citrus Ctr, Tex Col Arts & Indust, 61-62; asst prof biol, Concord Col, 62-65; assoc prof, Quincy Col, 65-67; assoc prof, 67-81, PROF BOT, FLA A&M UNIV, 81- *Concurrent Pos:* Consult, trop fruit. *Mem:* AAAS; Am Inst Biol Sci; Am Soc Microbiol; Caribbean Food Crops Soc. *Res:* Peanut proteins; tree fruit diseases. *Mailing Add:* Div Agr Sci Fla A&M Univ Tallahassee FL 32307

HEINISCH, ROGER PAUL, b St Paul, Minn, May 17, 38; m 64; c 3. OPTICS, PHYSICS. *Educ:* Marquette Univ, BS, 60, MS, 64; Purdue Univ, PhD, 68. *Prof Exp:* Instr, Purdue Univ, 64-65; res assoc, Argonne Nat Lab, 66-68; prin res scientist, Honeywell Inc, 68-73, dir res, 78-79, dir, 80-81, vpres, 81-85, vpres Mil Avionics, 85-87, vpres corp, 88-90, SR PRIN RES SCIENTIST, SYSTS & RES CTR, HONEYWELL INC, 73-, MGR OPTICS TECHNOL PROGS, BALLISTIC MISSILE DEFENSE ADVAN TECHNOL CTR, 75-, VPRES ALLIANT TECHNOL, 90- *Mem:* Am Soc Mech Eng; Am Inst Aeronaut & Astronaut; Optical Soc Am; Sigma Xi. *Res:* Optics and radiative transfer with efforts split between theoretical and experimental work; fluid mechanics. *Mailing Add:* Honeywell Inc 2700 Ridgeway Pkwy Minneapolis MN 55413

HEINKE, CLARENCE HENRY, b Rozellville, Wis, July 13, 12; m 44; c 3. MATHEMATICS EDUCATION. *Educ:* Capital Univ, BSc, 38; Ohio State Univ, MA, 42, PhD(educ), 53. *Prof Exp:* Teacher pub schs, Ohio, 38-41; asst math, Ohio State Univ, 41-42; instr, Wis State Col, Eau Clair, 45-46; from instr to assoc prof, 46-70, head dept, 56-70, prof, 70-77, EMER PROF MATH, CAPITAL UNIV, 77- *Concurrent Pos:* Am Consult, Inst Sec Math Teachers, Delhi, India, 66; mem, Entebbe Math Writing Group, Mombassa, Kenya, 67. *Mem:* Math Asn Am; Nat Coun Teachers Math. *Res:* Geometry; mathematics curriculum; development of study materials in mathematics for elementary teachers, pre-service and in-service. *Mailing Add:* 688 S Remington Columbus OH 43209

HEINKE, GERHARD WILLIAM, b Korneuburg, Austria, Dec 11, 32; Can citizen; m 52; c 2. ENVIRONMENTAL ENGINEERING, MUNICIPAL ENGINEERING. *Educ:* Univ Toronto, BASc, 56, MASc, 61; McMaster Univ, PhD(chem eng), 69. *Prof Exp:* Proj engr, G E Hanson Assocs, Consult Engr, Ont, 56-59, chief engr, 59-61; proj engr, Fischer & Porter Inc, WGer, 61-62; from lectr to assoc prof civil eng, 62-74, PROF CIVIL ENG & CHMN DEPT, UNIV TORONTO, 74- *Mem:* Am Water Works Asn; Can Inst Pollution Control; Eng Inst Can; Am Asn Prof Sanit Engrs. *Res:* Water pollution control studies in the Great Lakes and the Canadian Arctic; eutrophication studies; nutrient removal from waste water; servicing of artic communities; fire protection for northern communities; municipal engineering. *Mailing Add:* Dept Civil Eng Univ Toronto 135 St George St Toronto ON M5S 1A4 Can

HEINLE, DONALD ROGER, b New Salem, NDak. ECOLOGY, PHYSIOLOGY. *Educ:* Univ Wash, BS, 59; Univ Md, MS, 65, PhD(zool), 69. *Prof Exp:* Biologist, Fisheries Res Inst, Univ Wash, 59-60; res asst to assoc prof ecol, Chesapeake Biol Lab, Univ MD, 69-80; MGR, ENVIRON SCI, CH2M HILL, INC, NORTHWEST, 80- *Mem:* Am Soc Limnol & Oceanog; Estuarine Res Fedn. *Res:* Ecology of zooplankton, cycling of materials in estuaries, food chains, effects of pollutants; fisheries; water quality. *Mailing Add:* 10630 181st Ave NE Bellevue WA 98052

HEINLE, PRESTON JOSEPH, b Clarkdale, Ariz, Dec 5, 24; m 53; c 4. ORGANIC CHEMISTRY. *Educ:* Univ Ariz, BS, 47. *Prof Exp:* Res chemist, Hardesty Chem, Inc, 47-49 & United Verde Br, Phelps Dodge Corp, 49-53; sr chemist, Mat Test Lab, Motorola Inc, 53-58, proj chemist, 58-59, mgr mat test & inspection dept, 59-62, mgr, Plastics Lab, Semiconductor Prod Div, 62-89; CONSULT, 89- *Mem:* Am Chem Soc (treas, 61-63, secy, 64); Soc Plastics Eng; Electrochem Soc; Am Soc Testing & Mat. *Res:* Application of polymeric materials to semiconductor devices; thermoset rheology. *Mailing Add:* 5548 N 19th Pl Phoenix AZ 85016

HEINO, WALDEN LEO, b Eveleth, Minn, Apr 13, 30; m 53; c 3. INORGANIC CHEMISTRY. *Educ:* Univ Minn, BS, 53, PhD(inorg chem), 57. *Prof Exp:* Res chemist, E I du Pont de Nemours & Co, Niagara Falls, 57-63, res chemist, Wilmington, 63-64; asst prof, 64-73, ASSOC PROF CHEM, LUTHER COL, IOWA, 73- *Mem:* Am Chem Soc. *Res:* Coordination chemistry; non-aqueous systems. *Mailing Add:* Dept Chem Luther Col Decorah IA 52101-1042

HEINOLD, ROBERT H, b East Orange, NJ, June 28, 31; m 56; c 3. CHEMICAL ENGINEERING. *Educ:* Lafayette Col, Easton, Pa, ChE, 53. *Prof Exp:* Var, Hercules Inc, Wilmington, Del, 53-79; TECH DIR, A SCHULMAN, INC, AKRON, OH, 79- *Res:* chemicals; aerospace; plastics; modification of polypropylene to broaden its structural and environmental capabilities. *Mailing Add:* 3824 Chickasaw Trail Uniontown OH 44685

HEINRICH, BERND, b Ger, Apr 19, 40; US citizen; m 68; c 2. PHYSIOLOGY, ECOLOGY. *Educ:* Univ Maine, BA, 64, MS, 66; Univ Calif, Los Angeles, PhD(zool), 70. *Prof Exp:* Fel zool, Univ Calif, Los Angeles, 70-71; from asst prof to prof entom, Univ Calif, Berkeley, 71-80; PROF ZOOL, UNIV VT, BURLINGTON, 80- *Honors & Awards:* Von Humboldt Prize. *Mem:* Fel AAAS; Am Soc Zool; Am Ornith Soc; Ecol Soc Am. *Res:* Physiology of temperature regulation in insects; ecology and energetics of foragers. *Mailing Add:* Dept Zool Univ Vt Burlington VT 05405

HEINRICH, EBERHARDT WILLIAM, b Brunswick, Ger, Feb 10, 18; nat US; wid. GEOLOGY. *Educ:* Iowa State Col, BS, 40; Harvard Univ, MA, 42, PhD(geol), 47. *Prof Exp:* Asst to dir mineral, Mus, Harvard Univ, 40-42; jr geologist, US Geol Surv, 42-45; instr mineral, Mont Sch Mines, 46-47; from asst prof to prof mineral & petrol, 47-80, prof, 80-83, EMER PROF GEOL SCI, UNIV MICH, ANN ARBOR, 83- *Concurrent Pos:* Cur, mineral collection; ed, Am Mineralogist, 61-67 & Geochem News, 61-65; Fulbright Fel, 54. *Mem:* Fel Geol Soc Am; fel Mineral Soc Am; Geochem Soc; Am Inst Mining, Metall & Petrol Eng; fel Mineral Soc Gt Brit & Ireland. *Res:* Mineral paragenesis; geochemistry; economic geology; radioactive raw materials; petrography; carbonatites. *Mailing Add:* Dept Geol & Mineral Univ Mich Ann Arbor MI 48104

HEINRICH, GERHARD, b Lauban, Ger. ENDOCRINOLOGY. *Educ:* Oberlin Col, Ohio, BA, 66; Case Western Res Univ, MD, 74. *Prof Exp:* Instr med, Univ Wis-Madison, 78-79; instr med, 80-, ASST PROF, SCH MED, HARVARD UNIV. *Mem:* AAAS; Soc Neurosci. *Res:* Regulation of Nerve Growth Factor Gene Expression, and role of neurotrophic factors in neurodegenerative disease (Alzheimer's) and therapeutic applications. *Mailing Add:* Dept Med & Biochem Univ Hosp Boston Univ Med Ctr 88 E Newton St Boston MA 02118

HEINRICH, JANET, NURSING RESEARCH. *Prof Exp:* DEP DIR & DIR DIV EXTRAMURAL PROGS, NAT CTR NURSING RES, NIH, 85- *Mailing Add:* NIH Nat Ctr Nursing Res Off Dir Bldg 31 Rm 5B03 Bethesda MD 20892

HEINRICH, KURT FRANCIS JOSEPH, b Vienna, Austria, May 31, 21; US citizen; m 56; c 2. ANALYTICAL CHEMISTRY. *Educ:* Univ Buenos Aires, Dr Chem, 48. *Prof Exp:* Chemist, Nat Lead Co, Arg, 41-42, A S W Borinski, 42-46 & E Teubal, 46-50; sect chief indust chem, E Lix Klett & Co, 50-56; chemist, E I du Pont de Nemours & Co, 56-64; chemist, Spectrochem Anal Sect, Nat Bur Standards, 64-71, sect chief, Microanal Sect, Anal Chem Div, 71-82, chief, Off Int Rels, 82-89; RETIRED. *Concurrent Pos:* From lab instr to chief lab instr, Univ Buenos Aires, 48-56; lectr, Univ Del, 63-64 & Phila Col Pharm, 64. *Mem:* Am Chem Soc; Am Soc Appl Spectros; Electron Probe Anal Soc Am (pres, 69). *Res:* Inorganic analytical chemistry; trace analysis; x-ray spectrometry as an analytical tool; electron probe microanalysis. *Mailing Add:* 804 Blossom Dr Rockville MD 20850

HEINRICH, MAX ALFRED, JR, b Elmira, NY, May 22, 24; m 49; c 4. PHARMACOLOGY. *Educ:* Philadelphia Col Pharm, BSc, 47, MSc, 48; Jefferson Med Col, PhD(pharmacol), 53. *Prof Exp:* Instr physiol & pharmacol, Philadelphia Col Pharm, 48-50; instr pharmacol, Jefferson Med Col, 50-52; asst prof, Med Sch, Univ SDak, 52-59; from assoc prof to prof & head dept, NDak State Univ, 60-63; exec secy career develop rev br, NIH, 63-65, head res career sect, Nat Inst Gen Med Sci, 65-74, SCIENTIST ADMINR, MANPOWER BR, DIV HEART & VASCULAR DIS, NAT HEART, LUNG & BLOOD INST, 74- *Mem:* AAAS; Am Soc Pharmacol & Exp Therapeut; NY Acad Sci. *Res:* Central nervous system; toxicology; pulmonary pharmacology. *Mailing Add:* 5755 Box Elder Ct Frederick MD 21701

HEINRICH, MILTON ROLLIN, b Linton, NDak, Nov 25, 19; m 66. SPACE LIFE SCIENCES. *Educ:* Univ SDak, AB, 41; Univ Iowa, MS, 42, PhD(biochem), 44. *Prof Exp:* Instr biochem, Univ Iowa, 44; NIH fel, Univ Pa, 47-49; res assoc, Amherst Col, 49-58; NIH spec fel, Univ Calif, 58-60; asst prof, Univ Southern Calif, 60-63; res biochemist, Exobiol Div, 63-67, chief biol adaptation br, 67-76, sr res scientist, 76-84, space sta study scientst, Ames Res Ctr, NASA, 83-84; CONSULT, SPACE LIFE SCI, ZEROG CORP, 85- *Concurrent Pos:* USNR, 44-47. *Mem:* Am Chem Soc; Am Soc Biochem & Molecular Biol; Am Soc Gravitational & Space Biol; fel, Explorers Club. *Res:* Enzyme structure; microbial adaptation. *Mailing Add:* 27200 Deer Springs Way Los Altos Hills CA 94022-4325

HEINRICH, R(AYMOND) L(AWRENCE), b Galveston, Tex, Nov 1, 11; m 44; c 5. CHEMICAL ENGINEERING. *Educ:* Rice Univ, BS, 35. *Prof Exp:* Chemist, Humble Oil & Refining Co, 42-43, res chemist, 43-54, res chem engr, 54-57, sr res chem engr, 57-65; sr res chem engr, Esso Res & Eng Co, 65-71; pres, Eng Res Assoc, 71-77; CONSULT, 77- *Mem:* Am Chem Soc. *Res:* Separation processes; polymerization; hydrogenation; oxidation; petrochemicals production, applications and economics; organic acids, alcohols and polyols; aromatic condensation products; coal conversion to fuels and chemicals; plant start up and operation; design and implementation of research programs; air and water pollution control. *Mailing Add:* 112 Lakewood Baytown TX 77520-1510

HEINRICH, ROSS RAYMOND, b St Louis, Mo, Dec 12, 15; m 48; c 4. ATMOSPHERIC SCIENCES. *Educ:* Univ Mo, AB, 36; St Louis Univ, MS, 38, PhD(geophys), 44. *Prof Exp:* From instr to assoc prof, St Louis Univ, 38-51, prof geophys, 51-80, dir dept, 56-63, actg dean, 68-71, chmn, Dept Earth & Atmospheric Sci, 75-80, EMER PROF GEOPHYS, ST LOUIS UNIV, 81- *Concurrent Pos:* Trustee, Univ Corp Atmospheric Res, St Louis Univ, 60-71; consult ground vibrations; mem explosives control adv bd, St Louis County; seismol ed, Am Geophys Union, 46. *Mem:* Seismol Soc Am; Am Meteorol Soc; Am Soc Eng Educ; fel Geol Soc Am; Sigma Xi. *Res:* Seismicity of the Mississippi Valley; microbarographic oscillations; the moisture potential factor in precipitation and climate. *Mailing Add:* 21 Larkin Lane St Louis MO 63128

HEINRICHS, DONALD FREDERICK, b Shafter, Calif, Nov 8, 38; m 62; c 2. GEOPHYSICS, OCEANOGRAPHY. *Educ:* Stanford Univ, BS, 60, PhD(geophys), 66. *Prof Exp:* Res asst geophys, Stanford Univ, 61-66; instr physics, Menlo Col, 64-66; asst prof geophys oceanog, Ore State Univ, 66-75; prog mgr submarine geol & geophysics, 75-85, HEAD, OCEANOG CTR & FAC SECT, NSF, 85- *Concurrent Pos:* Mem staff, Off Naval Res, Arlington, Va, 74-75. *Mem:* Am Geophys Union; Soc Explor Geophys; Geol Soc Am; AAAS; Sigma Xi. *Res:* Paleomagnetics; marine magnetics; marine and land gravity. *Mailing Add:* Oceanog Sect NSF Washington DC 20550

HEINRICHS, W LEROY, b Collinsville, Okla, Aug 14, 32; m 54; c 2. BIOCHEMISTRY, OBSTETRICS & GYNECOLOGY. *Educ:* Southwestern State Col, BS, 54; Univ Okla, MD, 58; Univ Ore, MS, 65, PhD(biochem), 67; Am Bd Obstet & Gynec, dipl, 67. *Prof Exp:* Resident obstet & gynec, Harper Hosp, Detroit, 59-62; from asst prof to prof obstet & gynec, Sch Med, Univ Wash, 67-76; chmn, 76-83, PROF GYNEC & OBSTET, SCH MED, STANFORD UNIV, 76- *Mem:* AAAS; fel Am Col Obstet & Gynec; Am Soc Biol Chemists; Endocrine Soc; Soc Gyn Invest. *Res:* Steroid biochemistry; gynecological endocrinology. *Mailing Add:* Dept Gynec & Obstet Boswell A354 Stanford Univ Med Ctr Stanford CA 94305

HEINRIKSON, ROBERT L, b Sioux City, Iowa, Dec 31, 35; m 63; c 2. VIROLOGY. *Educ:* Augustana Col, Rock Island, Ill, BA, 58; Univ Chicago, PhD(biochem), 63. *Prof Exp:* Res assoc protein chem, Rockefeller Univ, New York, 63-65; NATO postdoctoral, Lab Molecular Biol, Med Res Coun, Cambridge, Eng, 65-66; from asst prof to prof biochem, Univ Chicago, 66-85; DISTINGUISHED SR SCIENTIST, UPJOHN CO, 85- *Concurrent Pos:* Ed, J Protein Chem, 75- & J Biol Chem, 88-93. *Mem:* Sigma Xi; Protein Soc. *Res:* Relationship between protein structure and function; protein chemistry and enzymology. *Mailing Add:* 2526 Pineridge Rd Kalamazoo MI 49008

HEINS, ALBERT EDWARD, b Boston, Mass, Sept 7, 12; m 39; c 1. MATHEMATICS. *Educ:* Mass Inst Technol, BS, 34, MS, 35, PhD(math), 37. *Prof Exp:* From instr to asst prof math, Purdue Univ, 35-42; res assoc, Radiation Lab, Nat Defense Res Comt, Mass Inst Technol, 42-46; from assoc prof to prof math, Carnegie Inst Technol, 46-59; PROF MATH, UNIV MICH, ANN ARBOR, 59- *Concurrent Pos:* Fel, Purdue Res Found, Brown Univ, 41; Guggenheim fel, Univ Copenhagen, 53-54; guest prof, Technische Hochschule, Darmstadt WGer, 80. *Mem:* Am Math Soc. *Res:* Representation theorems in partial differential equations; boundary value problems; wave motion; diffraction theory; electromagnetic theory; dynamics and statics of continuous media. *Mailing Add:* Dept Math Univ Mich Ann Arbor MI 48109

HEINS, CONRAD F, b Kolar, India, Apr 25, 39; US citizen; m 65; c 2. ORGANIC POLYMER CHEMISTRY. *Educ:* Drew Univ, AB, 59; Univ Ill, MS, 61, PhD(org chem), 62. *Prof Exp:* NSF fel, Cornell Univ, 62-64; org chemist, E I du Pont de Nemours & Co, Inc, 64-68; res scientist & adj prof, Denver Res Inst, Univ Denver, 68-75; head, Sci Dept, Scattergood Sch, West Branch, Iowa, 75-80; DIV HEAD SCI & TECHNOL, JORDAN COL, CEDAR SPRINGS, MICH, 80- *Mem:* Am Chem Soc; Am Solar Energy Soc; Int Solar Energy Soc. *Res:* Thermoplastic polymer composites; polymer concretes; technology transfer and solar technology; applied research in the alternative energy field with a focus on photovoltaics. *Mailing Add:* Div Sci & Technol Jordan Col Cedar Springs MI 49319

HEINS, CONRAD P, JR, structural & civil engineering; deceased, see previous edition for last biography

HEINS, MAURICE HASKELL, b Boston, Mass, Nov 19, 15; m 40; c 2. MATHEMATICAL ANALYSIS. *Educ:* Harvard Univ, AB, 37, AM, 39, PhD(math), 40. *Hon Degrees:* AM, Brown Univ, 47. *Prof Exp:* Instr & tutor math, Harvard Univ, 39-40; asst, Inst Advan Study, 40-42; asst prof math, Ill Inst Technol, 42-44; mathematician, Off Chief Ord, US Army, 44-45; from actg asst prof to prof math, Brown Univ, 45-58; prof math, Univ Ill, Urbana, 58-74; distinguished prof complex anal, 74-86, EMER PROF MATH, UNIV MD, COL PARK, 87-; EMER PROF MATH, UNIV ILL, URBANA, 74- *Concurrent Pos:* Pres fel, Brown & Fulbright res fel, Univ Paris, 52-53; vis prof, Univ Calif, Berkeley, 63-64; exchange prof, Univ Paris VI, 79. *Mem:* Am Math Soc; fel Am Acad Arts & Sci; London Math Soc. *Res:* Functions of a complex variable; conformal mapping; Riemann surfaces. *Mailing Add:* Dept Math Univ Md College Park MD 20742

HEINS, ROBERT W, petroleum & mining engineering; deceased, see previous edition for last biography

HEINSELMAN, MIRON L, b Duluth, Minn, Feb 7, 20; m 42; c 2. FOREST ECOLOGY. *Educ:* Univ Minn, BA, 42, BS, 48, MF, 51, PhD(forest ecol), 61. *Prof Exp:* Res forester silvicult, Lake States Forest Exp Sta, US Forest Serv, 48-61, forest ecologist, 61-69, forest ecologist, NCent Forest Exp Sta, 69-74; CONSULT, 74- *Concurrent Pos:* Adj prof ecol & behav biol, Dept Ecol & Behav Biol, Univ Minn, St Paul, 74- *Mem:* AAAS; Soc Am Foresters; Ecol Soc Am; Am Inst Biol Sci. *Res:* Peatlands ecology, including factors affecting growth of trees on peatlands; peatland genesis and development; factors controlling floristics and vegetation development; patterned organic terrain; silviculture of Picea mariana on peatlands; natural role of fire in northern conifer forest ecosystems. *Mailing Add:* 1783 Lindig St St Paul MN 55113

HEINSOHN, ROBERT J(ENNINGS), b Brooklyn, NY, Aug 28, 32; m 55; c 2. MECHANICAL ENGINEERING. *Educ:* Rensselaer Polytech Inst, BSME, 54; Mass Inst Technol, SMME, 55; Mich State Univ, PhD(eng), 63. *Prof Exp:* Anal engr, Pratt & Whitney Aircraft, 54-59; instr eng, eve div, Johns Hopkins Univ, 56-58; instr mech eng, Mich State Univ, 58-63; from asst prof to assoc prof, 63-70, PROF MECH ENG, PA STATE UNIV, 70- *Concurrent Pos:* NSF fac fel, 62-63, res grants, 64-72; Environ Protection Agency Serv res grant, 67-77; Pub Health Serv grants, 78-84. *Honors & Awards:* Ralph A Teetor Award, Soc Automotive Engrs, 67. *Mem:* Am Soc Mech Engrs; Air Pollution Control Asn; Am Soc Heating, Refrig & Air-Conditioning Engrs. *Res:* Control contaminants in workplace and out of doors; combustion; author of one book. *Mailing Add:* Dept Mech Eng Pa State Univ University Park PA 16802

HEINSTEIN, PETER, b Heidelberg, Ger, Apr 14, 35; US citizen; m 57; c 6. BIOCHEMISTRY. *Educ:* Strickhof Col, Switz, dipl agr, 54; NC State Univ, MS, 63, PhD(biochem), 67. *Prof Exp:* Res technician, dept animal nutrit, NC State Univ, 59-63; NIH fel, Univ Calif, Davis, 67-69; asst prof, 69-74, assoc prof biochem & chem, 74-83, PROF MED CHEM & BIOCHEM PROG, SCH PHARMACOL, PURDUE UNIV, WEST LAFAYETTE, 83- *Concurrent Pos:* Lederle fac res award, 71; sr Fulbright res fel 78-79. *Mem:* AAAS; Am Chem Soc; Am Soc Biol Chemists; Sigma Xi. *Res:* Biosynthesis of natural products; enzyme function and properties; signal transduction across plant cell membranes; mechanism of plant cell pathogen interaction. *Mailing Add:* Dept Med Chem & Biochem Prog Purdue Univ Sch Pharm West Lafayette IN 47907

HEINTZ, EDWARD ALLEIN, b Buffalo, NY, Apr 8, 31; m 56; c 2. ANALYTICAL CHEMISTRY. *Educ:* Univ Buffalo, BA, 53; Mass Inst Technol, PhD(anal chem), 57. *Prof Exp:* Asst, Mass Inst Technol, 55-57; asst, Union Carbide Metals Co, 57-61; instr, Millard Fillmore Col, State Univ NY, Buffalo, 59-61; res chemist, Cornell Aeronaut Lab, 61-63, res supvr, 63-64; mgr carbon & graphite res, Res Lab, 64-70, mgr chem & phys measurement res, 70-71, mgr res, Carbon-Graphite Tech Dept, Airco Carbon, 71-88; INDEPENDENT CONSULT, CARBON & GRAPHITE TECH, 88- *Concurrent Pos:* Guest mem, Northern Carbon Res Labs, Univ Newcastle-on-Tyne, Eng, 82 & 90. *Honors & Awards:* J F Schoellkopf Medal, Am Chem Soc, 80. *Mem:* Am Chem Soc; Am Carbon Soc; Combustion Inst; Am Sec Testing & Mat. *Res:* Reactions in non-aqueous solvents; electrodeposition of radioactive tracers; analytical chemistry of the less familiar transition elements; corrosion of graphite in aqueous solutions; gas phase reactions of graphite; carbon and graphite technology; microscopy of carbon and graphite. *Mailing Add:* Carbon & Graphite Technol 67 Red Oak Dr Buffalo NY 14221-2303

HEINTZ, ROGER LEWIS, b Jackson Center, Ohio, Mar 15, 37; m 62; c 4. BIOCHEMISTRY. *Educ:* Ohio Northern Univ, BS, 59; Ohio State Univ, MS, 61; Univ Wis-Madison, PhD(biochem), 64. *Prof Exp:* NIH fel biochem, Univ Ky, 64-66, Am Heart Asn adv res fel, 66-68; assoc prof biochem, Iowa State Univ, 68-75; assoc prof, 75-80, PROF CHEM, STATE UNIV NY, PLATTSBURGH, 80- *Concurrent Pos:* NIH res grants, 67-68 & 68-75, 75-76 & 78-; State Univ NY Res Found grant, 76 & 77, grant, 82. *Mem:* AAAS; Am Chem Soc; Am Soc Biol Chem; Sigma Xi. *Res:* Mechanism of action of thyroid hormone receptor. *Mailing Add:* Dept Chem State Univ NY Plattsburgh NY 12901

HEINTZ, WULFF DIETER, b Wurzburg, Ger, June 3, 30; m 57; c 2. ASTRONOMY. *Educ:* Univ Munich, Dr rer nat(astron), 53; Munich Tech Univ, Priv Doz, 67. *Prof Exp:* Res asst, Univ Munich Observ, 54-66, observator, 66-69; assoc prof, 69-73, chmn dept, 72-82, PROF ASTRON, SWARTHMORE COL, 73- *Mem:* Int Astron Union; Royal Astron Soc; Am Astron Soc. *Res:* Double and multiple stars; positional astronomy. *Mailing Add:* Dept Physics & Astron Swarthmore Col Swarthmore PA 19081

HEINTZELMAN, RICHARD WAYNE, b Danville, Pa, Feb 3, 47; m 67; c 2. ENVIRONMENTAL CHEMISTRY, METABOLISM. *Educ:* Lycoming Col, BA, 70; Univ Va, PhD(org chem), 74. *Prof Exp:* Res assoc sulfur-nitrogen compounds, Fels Res Inst & Dept Chem, Temple Univ, 74-75; res chemist, Amchem Prod Inc, Subsid Union Carbide Corp, 75-77, group leader, synthesis group, 77-79; group leader, synthesis group, Union Carbide Agr Prod Co, Inc, 79- 80, group leader, metab & environ chem, 80-86; GROUP LEADER, METAB & ENVIRON FATE, RHONE-POULERC AGR CO, 87- *Concurrent Pos:* NSF res fel, Univ Va, 72-74. *Mem:* Am Chem Soc; Asn Off Anal Chemists. *Res:* Organic synthesis; heterocylces; sulfur-nitrogen compounds; herbicides; plant growth regulators; insecticides; environmental fate; metabolism; radiochemistry. *Mailing Add:* 104 Crimmons Circle Cary NC 27511

HEINY, ROBERT LOWELL, b Washington, DC, June 10, 42; m 64; c 3. MATHEMATICAL STATISTICS. *Educ:* Colo Col, BS, 64; Colo State Univ, MS, 66, PhD(statist), 68. *Prof Exp:* Assoc prof, 68-77, PROF STATIST, UNIV NORTHERN COLO, 77- *Mem:* Am Statist Asn. *Res:* Educational research; estimation; statistics education; stochastic processes. *Mailing Add:* Dept Math & Appl Statist Univ Northern Colo Greeley CO 80639

HEINZ, DON J, b Rexburg, Idaho, Oct 29, 31; m 56; c 6. PLANT BREEDING, CYTOGENETICS. *Educ:* Utah State Univ, BS, 58, MS, 59; Mich State Univ, PhD(farm crops), 61. *Prof Exp:* Assoc geneticist, 61-66, head dept genetics & path, 66-78, asst dir, 77-78, vpres & dir, 79-85, PRES & DIR, EXP STA, HAWAIIAN SUGAR PLANTERS ASN, 86- *Concurrent Pos:* Affil mem, Grad Fac, Univ Hawaii, 63-; chmn germplasm comn, Int Soc Sugar Cane Technologists, 75-87; President Reagan's Coun Agr. *Mem:* Int Soc Sugar Cane Technologists; AAAS; Sigma Xi. *Res:* Sugarcane improvement; genetics; cytogenetics and disease control; agronomy; development of cell and tissue culture procedures in crop improvement. *Mailing Add:* 224 Ilihau St Kailua HI 96734

HEINZ, ERICH, b Essex, Ger; US citizen. PHYSIOLOGY, BIOPHYSICS. *Educ:* Univ Kiel, Ger, MD, 41, DrHabil, 49. *Prof Exp:* Assoc prof biochem, Med Sch, Tufts Univ, 55-58; res prof physiol, George Washington Univ, 58-59; prof biochem, T W Goethe Univ, Ger, 59-78; prof physiol & biophys, Med Sch, Cornell Univ, 78-; PROF DEPT PHYSIOL, MAX-PLANCK INST, FED REPUB GER. *Concurrent Pos:* Res assoc biophys, Med Sch, Harvard Univ, 56-58; consult, Harvard Biophys Lab, 56-58; vis prof, Nat Heart Inst, NIH, 62 & Hunter Col, City Univ NY, 68; prin investr, NIH, 78-; sponsor comt, Int Conf Biol Membranes, 70- *Mem:* Am Soc Biol Chemists; Biophys Soc. *Res:* Biological membrane transport; electrolyte metabolism; amino acid transport; ion pumps. *Mailing Add:* Dept Physiol Max-Planck Inst Rheinlanddamm 201 Dortmund 4600 Germany

HEINZ, JOHN MICHAEL, b Atlanta, Ga, Jan 9, 33; div; c 1. BIOMEDICAL ENGINEERING. *Educ:* Univ SC, BS, 55; Mass Inst Technol, SM, 58, EE, 59, ScD(elec eng), 62. *Prof Exp:* Engr, WCOS Radio & TV Sta, SC, 50-55; res asst, Mass Inst Technol, 55-61, from instr to asst prof elec eng, 61-65; res assoc speech commun, Royal Inst Technol, Stockholm, 65-67, guest res, 59; asst prof larynol & otol, Sch Med, 67-71, health serv admin, 71-76, PROF BIOMED ENG, JOHNS HOPKINS UNIV, 71-, ASST NEUROL & DIR COMMUN SCI RES LAB, KENNEDY INST, 79- *Concurrent Pos:* Consult, Bolt, Beranek & Newman, Inc, Mass, 61-65 & Sci Serv, Inc, DC, 61-65; Ford fel, 63-64; NIH spec fel, 65-66, res grants, 68-; consult, Nat Acad Engr, 79-, Nat Acad Sci, 71- & NIH, 71-; dir, Commun Sci Res Lab, Kennedy Inst, 79-, actg dir, Commun Sci & Dis Dept, 87- *Mem:* AAAS; Inst Elec & Electronics Engrs; Am Speech Lang Hearing Assoc; Sigma Xi; fel Acoust Soc Am. *Res:* Speech communication; acoustics; digital signal processing. *Mailing Add:* Commun Sci Res Lab Kennedy Inst Handicapped Children Johns Hopkins Univ Sch Med Baltimore MD 21205

HEINZ, OTTO, b Vienna, Austria, Sept 18, 24; nat US; m 48; c 3. PHYSICS. *Educ:* Univ Calif, BA, 48, PhD(physics), 54. *Prof Exp:* Mem tech staff, Bell Tel Labs, 54-55; physicist, Stanford Res Inst, 55-70; assoc prof, 62-67, chmn dept, 67-73, PROF PHYSICS, NAVAL POSTGRAD SCH, 67- *Mem:* Am Phys Soc; Am Geophys Union. *Res:* Geomagnetic fields in the sea and on the ocean floor. *Mailing Add:* Dept Physics Naval Postgrad Sch Code 0223 Monterey CA 93940

HEINZ, RICHARD MEADE, b Toledo, Ohio, June 16, 39; div; c 2. HIGH ENERGY PHYSICS, HIGH ENERGY ASTROPHYSICS. *Educ:* Univ Toledo, BS, 61; Univ Mich, MS, 62, PhD(physics), 64. *Prof Exp:* Instr physics & res assoc, Univ Mich, 65; NSF fel, Europ Orgn Nuclear Res, Univ Geneva, 65-66; from asst prof to assoc prof, 66-72, PROF PHYSICS, IND UNIV, BLOOMINGTON, 72- *Concurrent Pos:* Prog Off Elem Particle Physics, Nat Sci Found, 80-82. *Mem:* Am Phys Soc. *Res:* Collaborator in MACRO (Monopole, Astrophysics, and Cosmic Ray Observatory) and SMILI (Superconducting Magnet Instrument for Light Isotopes). *Mailing Add:* Dept Physics Ind Univ Bloomington IN 47405

HEINZ, TONY F, b Palo Alto, Calif, Apr 30, 56. NONLINEAR OPTICS, LASER SPECTROSCOPY. *Educ:* Stanford Univ, BS, 78; Univ Calif, Berkeley, PhD(physics), 82. *Prof Exp:* NSF grad fel, 78-81; IBM fel, 82; RES STAFF MEM, IBM THOMAS J WATSON RES CTR, 83- *Concurrent Pos:* Prog comt, CLEO, 86, IQEC, 87-88, 90 & QELS, 89, 91. *Mem:* Am Phys Soc; Mats Res Soc; Optical Soc Am. *Res:* Nonlinear optics with particular interest in surface problems. *Mailing Add:* IBM Res Div TJ Watson Res Ctr PO Box 218 Yorktown Heights NY 10598-0218

HEINZ, ULRICH WALTER, b Ludwighafen, WGermany, Apr 25, 55; m 80; c 3. HEAVY-ION COLLISIONS. *Educ:* J W Goethe Univ, Frankfurt, WGer, Dipl, 78, DPhil, 80, DHabil(physics), 84. *Prof Exp:* Res & teaching asst theoret physics, Inst Theoret Physics, J W Goethe Univ, Frankfurt, WGer, 78-80; teaching fel, Yale Univ, 80-82; res assoc, Inst Theoret Physics, JW Goethe Univ, 82-84; assoc physicist nuclear theory, Brookhaven Nat Lab, 84-87; PROF PHYSICS, UNIV REGENSBURG, 87- *Concurrent Pos:* Vis asst prof, Vanderbilt Univ, Nashville, Tenn, 83-84; consult, Oak Ridge Nat Lab, 83-84; guest assoc physicist, Brookhaven Nat Lab, 87- *Honors & Awards:* Hess Prize, 88. *Mem:* Am Phys Soc; Deutsche Physikalische Ges. *Res:* Theory of heavy-ion collisions at low and high energy; positron creation in collisions between very heavy ions; quark-gluon plasma formation in relativistic nuclear collisions. *Mailing Add:* Inst Theoretishe Physik Univ Regensburg Postfach 397 Regensburg D-8400 Germany

HEINZE, WILLIAM DANIEL, b St Louis, Mo, Apr 26, 48; m 77; c 2. SEISMIC MODELLING, INVERSE. *Educ:* Tex A&M Univ, BS, 70, MS, 72, PhD(geophysics), 77; Mass Inst Technol, MS, 73. *Prof Exp:* Instr geophysics, Tex A&M Univ, 75-77, res fel, 77-78; res fel geophysics, Dept Terrestrial Magnetism, Carnegie Inst, Washington, DC, 78-79; OWNER, APPL GEOPHYS SOFTWARE, INC, 79- *Concurrent Pos:* Co-investr, Dept Geophys, Tex A&M Univ, 76-77, prin investr, 77-78. *Mem:* Am Geophys Union; Sigma Xi, Soc Explor Geophysicists. *Res:* Seismic modelling; image processing; rock deformation. *Mailing Add:* PO Box 218470 Houston TX 77218

HEIPLE, CLINTON R, b Tacoma, Wash, Nov 12, 39; div; c 2. PHYSICAL METALLURGY. *Educ:* Stanford Univ, BS, 61; Univ Sheffield, MMet, 62; Univ Ill, Urbana, PhD(metall), 67. *Prof Exp:* Fel, Sci Ctr, NAm Rockwell Corp, 67-68; res metallurgist, Rocky Flats Div, Dow Chem USA, 68-76; sr res specialist, Rocky Flats Plant, Rockwell Int, 76-79, assoc scientist, 79-89; ASSOC SCIENTIST, EG&G ROCKY FLATS, ROCKWELL INT, 90- *Concurrent Pos:* Fel, Acoust Emission Working Group, 90. *Honors & Awards:* William Spraragen Award, Am Welding Soc, 83; McKay-Helm Award, Am Welding Soc, 87. *Mem:* Am Soc Metals; Am Welding Soc; Int Inst Welding. *Res:* Effect of impurities on shape of gas tungsten arc welds; dislocation sources of acoustic emission; beryllium; stainless steel; liquid metal embrittlement; mechanical testing. *Mailing Add:* EG&G Rocky Flats Bldg 779 Box 464 Golden CO 80402-0464

HEIPLE, LOREN RAY, b Oakwood, Ill, Apr 19, 18; m 44; c 2. SANITARY ENGINEERING. *Educ:* Iowa State Univ, BS, 39, CE, 50; Harvard Univ, MS, 40; Stanford Univ, PhD, 67. *Prof Exp:* Serv & develop engr, Infilco, Inc, Ill, 40-41; jr sanit engr, Iowa Ord Plant, 41; instr, 41-42, asst prof civil eng, Iowa State Univ, 46-48; city engr, Boone, Iowa, 48-49; consult engr, Pub Admin Serv, Ill, 49-50 & Little Rock Wastewater Utility, 84-85; head civil eng, Univ Ark, Fayetteville, 50-71, dean, Col Eng, 71-79, prof civil eng, 79-84; CONSULT, 84 - *Concurrent Pos:* Consult, New Wonder World Encycl. *Mem:* Am Soc Civil Engrs; Am Soc Eng Educ; Nat Soc Prof Engrs; Water Pollution Control Fedn. *Res:* Sewage treatment processes; stream pollution and recovery; conservation of water resources, both surface and underground supplies. *Mailing Add:* Univ Ark Col Eng Fayetteville AR 72701

HEIRMAN, DONALD N, b Mishawaka, Ind, Aug 16, 40. ELECTROMAGNETIC COMPATIBILITY, MEASUREMENT TECHNIQUES. *Educ:* Purdue Univ, BSEE, 62, MSEE, 63. *Prof Exp:* Mem tech staff, AT&T Bell Labs, 63-83, Am Bell, 83-84, supvr info systs, AT&T, 84-88, SUPVR, AT&T BELL LABS, 88- *Concurrent Pos:* Past pres, vpres, current mem bd dirs & dir tech servs, Electromagnetic Compatibility Soc, Inst Elec & Electronics Engrs, 74-; lectr, Ctr Prof Advan, East Brunswick, NJ, 77-, Course dir, 87-; mem & chmn, Comt C63, subcomt One, Three & Five, Am Nat Standards Inst, 81-; tech expert, US Deleg to Int Spec Comt on Radio Interference, 85- *Honors & Awards:* Lawrence G Cumming Award, Electromagnetic Compatibility Soc, Inst Elec & Electronics Engrs, 84 & Centennial Medal, 84. *Mem:* Fel Inst Elec & Electronics Engrs; Am Nat Standards Inst; Nat Asn Radio & Telecom Engrs. *Res:* Accurate electromagnetic emission measurements; methods of measurement standards; test facility construction, calibration and improvements. *Mailing Add:* 143 Jumping Brook Rd Lincroft NJ 07738-1442

HEIRTZLER, JAMES RANSOM, b Baton Rouge, La, Sept 16, 25; div; c 2. EARTH SCIENCES. *Educ:* La State Univ, BS, 47, MS, 48; NY Univ, PhD(physics), 53. *Prof Exp:* Res assoc physics, La State Univ, 47-48; asst, NY Univ, 48-50, asst & instr, 50-53; asst prof, Am Univ, Beirut, 53-56; sr physicist, Gen Dynamics Corp, 56-60; res scientist, Lamont-Doherty Geol Observ, Columbia Univ, 60-64, sr res scientist, 64-67, dir, Hudson Lab, 67-69; chmn dept, 69-76, sr scientist geol & geophys, Woods Hole Oceanog Inst, 69-86; GEOPHYSICIST & HEAD GEOL & GEOMAGNETISM BR, NASA/GODDARD SPACE FLIGHT CTR, 86- *Concurrent Pos:* Mem, JOIDES Planning Comt, 69-79, chmn, 78-79; US chief scientist, Proj Famous, 71-75; reporter, US Geodynamics Comt, 71-83; mem, US & NAm Magnetic Anomaly Map Comt, 75-83; dir sci res & sr adv to pres, Joint Oceanog Inst Inc, 79-80; pres-elect & pres, Am Geophys Union Sect on Geomagnetism & Paleomagnetism, 80-84; ed, J Reviews Geophys, 84-88; bd gov, Am Inst Physics, 90- *Honors & Awards:* Microfossil named in honor, Pithonella heirtzleri, 72; Antartica feature named in honor, Heirtzler Ice Piedmont, 85. *Mem:* Fel AAAS; fel Geol Soc Am; fel Am Geophys Union; Am Phys Soc; Int Asn Geomag & Aeronomy; Sigma Xi. *Res:* Geophysics; author of more than 140 publications. *Mailing Add:* NASA Goddard Space Flight Ctr Code 922 Greenbelt MD 20771

HEISE, EUGENE ROYCE, b Hamlin, Kans, July 11, 32; m 62. TRANSPLANTATION. *Educ:* Wittenberg Univ, BS, 56; Univ Iowa, MS, 60; Bowman Gray Sch Med, PhD(microbiol), 66. *Prof Exp:* Sr fel microbiol, Sch Med, Univ Wash, 66-69; asst prof, 69-74, ASSOC PROF MICROBIOL & IMMUNOL, BOWMAN GRAY SCH MED, WAKE FOREST UNIV, 74- *Concurrent Pos:* Chmn, Am Bd Histocompatibility & Immunogenetics, 84-90. *Mem:* Am Asn Clin Histocompatibility Testing; Am Soc Microbiol; Reticuloendothelial Soc. *Res:* Immunogenetics; transplantation immunology; autoimmunity. *Mailing Add:* Dept Microbiol & Immunol Wake Forest Univ Bowman Gray Sch Med 300 Hawthorne Rd SW Winston-Salem NC 27103

HEISE, JOHN J, b Paoli, Ind, Feb 6, 31. BIOLOGY, CHEMISTRY. *Educ:* Earlham Col, AB, 53; Wash Univ, PhD(bot), 62. *Prof Exp:* Phys sci aide, Radioisotope Unit, Vet Admin Hosp, Indianapolis, Ind, 53-54; asst bot, Wash Univ, 55-60; res assoc biol sci, Fla State Univ, 60-62; fel, C F Kettering Res Lab, 62-63; molecular pathologist, Abbott Labs, 63-65; ASSOC PROF BIOL, GA INST TECHNOL, 65- *Mem:* AAAS; Am Chem Soc; Am Inst Biol Sci; Am Soc Photobiol; Sigma Xi. *Res:* Molecular and sub-molecular aspects of photosynthesis; enzyme processes and metabolism by the study of free radicals and transition elements; trace elements in the environment. *Mailing Add:* 1114 Westshire Pl NW Atlanta GA 30318

HEISER, ARNOLD M, b New York, NY, Feb 9, 33; m 64; c 2. ASTRONOMY. *Educ:* Ind Univ, AB, 54, AM, 57; Univ Chicago, PhD(astrophys), 61. *Prof Exp:* Asst prof, A J Dyer Observ, 61-65, actg dir, 71-72, dir, 72-86; ASSOC PROF PHYSICS & ASTRON, VANDERBILT UNIV, 65- *Concurrent Pos:* H Shapley vis lectr. *Mem:* Am Astron Soc; Int Astron Union; Royal Astron Soc; Sigma Xi. *Res:* Stellar photometry; galactic structure; variable stars. *Mailing Add:* A J Dyer Observ Box 1803-B Nashville TN 37235

HEISER, CHARLES BIXLER, JR, b Cynthiana, Ind, Oct 5, 20; m 44; c 3. BOTANY. *Educ:* Wash Univ, AB, 43, MA, 44; Univ Calif, PhD(bot), 47. *Prof Exp:* Instr bot, Wash Univ, 44-45; botanist, Herbarium, Univ Calif, 45-46, assoc bot, Exp Sta, 46-47; from asst prof to prof, 47-79, distinguished prof, 79-86, EMER DISTINGUISHED PROF BOT, IND UNIV, BLOOMINGTON, 86-, DIR HERBARIUM, 47- *Concurrent Pos:* Guggenheim fel, 53; NSF sr fel, 62; vis prof, Univ Tex, 78. *Honors & Awards:* Gleason Award, NY Bot Garden, 69; Merit Award, Bot Soc Am, 72; Distinguished Econ Botanist, Soc Econ Bot, 84; Pustovoit Award, Int Sunflower Asn, 85; Asa Gray Award, Am Soc Plant Taxonomists, 88. *Mem:* Nat Acad Sci; Bot Soc Am (pres, 80); Am Soc Plant Taxonomists (pres, 67); Soc Study Evolution (pres, 75); Soc Econ Bot (pres, 78). *Res:* Systematics; evolution; ethnobotany; cytogenetics; author of 5 books. *Mailing Add:* Dept Biol Ind Univ Bloomington IN 47405

HEISERMANN, GARY J, EXPERIMENTAL BIOLOGY. *Prof Exp:* POSTDOCTORAL FEL CANCER RES, COLD SPRING HARBOR LAB, 90- *Mailing Add:* Cold Spring Harbor Lab PO Box 100 Cold Spring NY 11746

HEISEY, LOWELL VERNON, b Ping Ting Chow, China, Oct 1, 19; m 45; c 4. ORGANIC CHEMISTRY, SYNTHETIC ORGANIC & NATURAL PRODUCTS CHEMISTRY. *Educ:* Manchester Col, AB, 41; Purdue Univ, MS, 44, PhD(org chem), 47. *Prof Exp:* Asst chem, Purdue Univ, 42-43; from asst prof to assoc prof, McPherson Col, 47-50; from assoc prof to prof chem, Bridgewater Col, 50-85; Fulbright lectr chem, Cuttington Univ Col, Liberia, 86-87; RETIRED. *Mem:* Am Chem Soc. *Res:* Organic synthesis; heterocyclic compounds; plant hormones. *Mailing Add:* Dept Chem Bridgewater Col Bridgewater VA 22812

HEISEY, S(AMUEL) RICHARD, b Elizabethtown, Pa, Oct 16, 28; m 58; c 2. MEDICAL PHYSIOLOGY. *Educ:* Elizabethtown Col, BS, 51. *Hon Degrees:* ScD, Johns Hopkins Univ, 59. *Prof Exp:* Pharmacologist, Med Res Lab, Army Chem Ctr, Md, 53-56; res fel physiol, Harvard Med Sch, 59-60, from instr to assoc, 60-67; assoc prof, 67-71, PROF PHYSIOL, MICH STATE UNIV, 71- *Concurrent Pos:* Biol asst, Army Chem Ctr, Md; hon res fel, Univ Exeter, Eng, 85-86; Nat Inst Neurol Dis & Blindness career develop award, 68-72. *Mem:* fel AAAS; Am Physiol Soc; Soc Exp Biol & Med; Soc Neurosci; Sigma Xi. *Res:* Comparative physiology of cerebrospinal fluid; respiratory control. *Mailing Add:* Dept Physiol Mich State Univ East Lansing MI 48824-1101

HEISIG, CHARLES G(LADSTONE), b Minneapolis, Minn, Feb 8, 24; m 60; c 4. CHEMICAL ENGINEERING. *Educ:* Univ Minn, BS, 44; Univ Tex, MS, 48, PhD, 51. *Prof Exp:* Develop engr, Oak Ridge Nat Lab, 50-53 & Union Carbide Chem Co Div, Union Carbide Corp, 53-56; res engr, Eng Res Inst, Univ Mich, 56-57; proj mgr res, Taylor Instrument Co, Div Sybron Corp, 57-67, res specialist, 67-72, anal systs specialist, Taylor Instruments Process Control, 72-83; sr engr, Eastman-Kodak Co, Rochester, NY, 84-90; RETIRED. *Concurrent Pos:* Consult engr, 83-84. *Mem:* Am Chem Soc; Instrument Soc Am. *Res:* Development of continuous measuring instruments and sampling systems for chemical process control systems. *Mailing Add:* Four Chapman Rd West Rush NY 14543

HEISINGER, JAMES FREDRICK, b Jefferson City, Mo, Nov 4, 35; m 61; c 2. TOXICOLOGY. *Educ:* Univ Mo, BS, 56, MA, 58, PhD(zool), 65. *Prof Exp:* Asst prof biol, physiol, Univ Mo, St Louis, 65-68; assoc prof, 68-75, PROF BIOL, UNIV SDAK, 75-, ASSOC DEAN ARTS & SCI, 89- *Mem:* Ecol Soc Am; Sigma Xi; Am Inst Biol Sci; Soc Environ Toxicol & Chem. *Res:* Physiological adaptations of vertebrates to natural and man-made environmental conditions and aquatic toxicology. *Mailing Add:* RR 1 Box 69 Vermillion SD 57069

HEISLER, CHARLES RANKIN, b Burlington, Iowa, July 1, 24; m 56; c 3. BIOCHEMISTRY. *Educ:* Monmouth Col, BS, 48; Univ Chicago, PhD(biochem), 57. *Prof Exp:* Asst prof agr chem, Ore State Univ, 57-65; from assoc prof to prof biochem, 65-91, chmn dept 70-74, EMER PROF BIOCHEM, UNIV NEV, RENO, 91- *Concurrent Pos:* Vis prof path, Univ Auckland, 75. *Mem:* AAAS; Am Chem Soc; Sigma Xi. *Res:* Mechanisms of enzyme action; comparative biochemistry in insects; mitochondrial function and activity; natural resistance to pests. *Mailing Add:* Dept Biochem Univ Nev Reno NV 89557

HEISLER, JOSEPH PATRICK, b Decatur, Ill, Aug 9, 34. MATHEMATICS. *Educ:* St Edward's Univ, BS, 56; Univ Notre Dame, MS, 59; Univ Mich, PhD(math), 65. *Prof Exp:* Teacher high schs, Ill, 56-58, Fla, 59-61 & Tex, 61-62; from asst prof to prof math, St Edward's Univ, 65-81; CONSULT, 81- *Mem:* Math Asn Am; Am Math Soc. *Res:* Diophantine analysis; finite rings. *Mailing Add:* Servants Paraclete VLM Jemez Springs NE 87025

HEISLER, RODNEY, b Tamaqua, Pa, Jan 1, 42; m 64; c 3. RADIO ENGINEERING, GEOPHYSICS. *Educ:* Walla Walla Col, BSEE, 65; Wash State Univ, MSEE, 67, PhD(eng sci), 70. *Prof Exp:* Asst prof, 70-77, PROF ENG, WALLA WALLA COL, 77- *Concurrent Pos:* Consult, Goddard Space Flight Ctr, NASA, 77-81. *Mem:* Inst Elec & Electronics Engrs; Am Soc Eng Educ. *Res:* Application of multidimensional Fourier transforms to paraboloidal reflecting antennas; multidimensional transform algorithms. *Mailing Add:* Dept Eng Walla Walla Col College Place WA 99324

HEISLER, SEYMOUR, b Montreal, Que, Sept 24, 43; m 67; c 2. ENDOCRINOLOGY. *Educ:* McGill Univ, BSc, 64, MSc, 66, PhD(pharmacol), 68. *Prof Exp:* Asst prof pharmacol, Univ Sherbrooke, 71-76; from asst prof to assoc prof, 76-82, PROF PHARMACOL, LAVAL UNIV, 82- *Concurrent Pos:* Coun mem, Pharmacol Soc Can, 82-85 & Can Soc Clin Invest, 83-86; dir, bioregulation unit, Med Ctr, Laval Univ, 84- *Mem:* Am Soc Pharmacol & Exp Therapeut; Am Soc Hypertension; Pharmacol Soc Can; Can Soc Clin Invest; Endocrine Soc. *Res:* Regulation of ACTH secretion; cellular, biochemical, molecular and genetic mechanisms. *Mailing Add:* CF Res Lab Montreal Chest Hosp McGill Univ 3650 St Urbain St Montreal PQ H2X 2P4 Can

HEISS, JOHN F, b Altoona, Pa, July 2, 20; m 51; c 5. CHEMICAL ENGINEERING. *Educ:* Univ Pittsburgh, BS, 42, MS, 44, PhD(chem eng), 50. *Prof Exp:* Asst prof chem eng, Univ Pittsburgh, 50-51; proj leader, Westvaco Chlor-Alkali Div, 51-55; engr asst to res dir, Stauffer Chem Co, 55-60; dir res & develop, Diamond Crystal Salt Co, 60-67, dir tech & opers res & develop, 67-68, dir res & develop, 68-85; RETIRED. *Mem:* Am Chem Soc; Am Inst Chem Engrs; Sigma Xi. *Res:* Fluid dynamics; metal chlorides; salt technology. *Mailing Add:* 1039 N Second St St Clair MI 48079

HEIST, HERBERT ERNEST, b Waverly, Iowa, Nov 30, 24; m 47; c 3. ZOOLOGY. *Educ:* Wartburg Col, BA, 49; Univ Iowa, MA, 51, PhD(zool, bact), 57. *Prof Exp:* Bacteriologist, Wis State Hyg Lab, 54-59; assoc tech dir, Cancer Screening Dept, Alumni Res Found, Wis, 59-61; bacteriologist, Children's Med Ctr, 61-62; prin res scientist, Honeywell Res Ctr, 62-72, mgr, Chem dept, Honeywell Corp Res Ctr, 72-77, DIR TECH ASSESSMENT & PLANNING, SCI & TECH, HONEYWELL CORP, 77- *Concurrent Pos:* Res assoc, Harvard Med Sch, 61-62. *Mem:* Am Soc Microbiol. *Res:* Medical bacteriology; cell biology. *Mailing Add:* Honeywell Corp Res Ctr 10701 Lyndale Ave South Minneapolis MN 55420

HEISTAD, DONALD DEAN, b Chicago, Ill, Apr 2, 40; m 64; c 2. CARDIOVASCULAR DISEASES. *Educ:* Univ Chicago, MD, 63. *Prof Exp:* From intern to resident, Univ Chicago, 63-66; cardiovasc trainee, Univ Iowa, 66-67; res internist, US Army Res Inst Environ Med, 67-70; from asst prof to assoc prof, 70-76, PROF, COL MED, UNIV IOWA, 76- *Concurrent Pos:* Clin investr, Vet Admin Hosp, Iowa City, 71-74; traveling fel, Royal Soc Med Found, 73; res career develop award, Nat Heart & Lung Inst, 75-; chmn, Midwest Sect, Am Fed Clin Res, 77-81; med investr, Vet Admin Hosp, Iowa City, 78-; assoc ed, Circ Res, 80-85. *Honors & Awards:* Cecile Lehman Mayer Res Award, Am Col Chest Physicians, 73; Irving S Wright Award, Am Heart Asn, 76 & Harry Goldblatt Award, 80. *Mem:* Am Physiol Soc; Am Soc Pharmacol & Exp Therapeut; Am Fed Clin Res (secy-treas, 77-81); Am Soc Clin Invest; Asn Am Physicians. *Res:* Nourishment of blood vessels; control of cerebral blood flow; reflex control of the circulation; vascular effects of atherosclerosis. *Mailing Add:* Dept Med Univ Iowa Col Med Iowa City IA 52242

HEITKAMP, NORMAN DENIS, b Houston, Tex, Jan 3, 40; m 63; c 3. RESEARCH ADMINISTRATION, APPLIED STATISTICS. *Educ:* Univ Tex, BS, 61; Southwest Tex State Col, MA, 63; Tex A&M Univ, PhD(phys chem), 65. *Prof Exp:* Sr res chemist, Southwest Res Inst, 65-68; res chemist, Brown & Williamson Tobacco Corp, 68-70, area supvr, 70-78; from res mathematician to sr res mathematician, 79-85, supvr, 84-85, supvr/staff res mathematician, 85-87, RES MGR STATIST, SHELL DEVELOP CO, 87- *Mem:* Am Chem Soc; Am Statist Asn; Opers Res Soc Am. *Res:* Experimental design, linear models and regression analysis; multivariate analysis. *Mailing Add:* Shell Oil Co PO Box 711 Martinez CA 94553-1391

HEITKEMPER, MARGARET M, b Longview, Wash, Aug 21, 51. GASTROINTESTINAL PHYSIOLOGY. *Educ:* Univ Ill, PhD(physiol & biophysics), 81. *Prof Exp:* ASSOC PROF PHYSIOL, UNIV WASH, 81- *Mem:* Am Physiol Soc; Am Geront Soc; Am Nurses Asn. *Mailing Add:* Dept Nursing Univ Wash Seattle WA 98632

HEITMAN, HUBERT, JR, b Berkeley, Calif, June 2, 17; m 41; c 2. ANIMAL NUTRITION, ENVIRONMENTAL PHYSIOLOGY. *Educ:* Univ Calif, BS, 39; Univ Mo, AM, 40, PhD(animal nutrit), 43. *Prof Exp:* Asst, Inst Animal Husb, Univ Mo, 39-43; from asst prof to assoc prof & from asst to assoc, Exp Sta, 46-61, chmn dept, 63-68, acad asst to vchancellor acad affairs, Univ, 71-78, chmn dept, 81-82, prof, 61-87, EMER PROF ANIMAL SCI & NUTRITIONIST, AGR EXP STA , UNIV CALIF, DAVIS, 87- *Concurrent Pos:* Nutrit officer captain, US Army, 43-46. *Mem:* Am Soc Animal Sci; Animal Behav Soc; Int Soc Biometeorol; Brit Soc Animal Prod; Brit Nutrit Soc. *Res:* Swine nutrition; swine behavior; environmental physiology. *Mailing Add:* Dept Animal Sci Univ Calif Davis CA 95616

HEITMAN, RICHARD EDGAR, b Bronx, NY, Mar 30, 30; m 68; c 1. MANAGEMENT, PRODUCT DEVELOPMENT. *Educ:* Mass Inst Technol, BS, 52; Princeton Univ, MS, 53, PhD(chem eng), 60. *Prof Exp:* Asst, Princeton Univ, 53-54; mem prof staff opers res, Arthur D Little, Inc, 56 & 58-65, head, London Mgt Sci Group, Arthur D Little, Ltd, 65-69, sr div staff, Mgt Sci Div, 69-73, SR VPRES, ARTHUR D LITTLE, INC, 73- *Concurrent Pos:* Mem opers res Chem Corps, US Army, 54-55; dir, Multibank Financial Corp, 77-; gov bd mem, Asn Princeton Grad Alumni, 79- *Mem:* Oper Res Soc UK; Opers Res Soc Am; Am Inst Chem Engrs. *Res:* Application of technology to meet business and scientific needs; identification and promotion of market opportunities for new technology-based products; design, evaluation and implementation of automated systems. *Mailing Add:* 117 Hosmer St Acton MA 07120

HEITMEIER, DONALD ELMER, organic chemistry; deceased, see previous edition for last biography

HEITNER, CYRIL, b Montreal, Que, July 8, 41; m 66; c 3. WOOD CHEMISTRY, PHOTOCHEMISTRY. *Educ:* Sir George Williams Univ, BSc, 63; Dalhousie Univ, MSc, 66; McGill Univ, PhD(chem), 71. *Prof Exp:* Nat Res Coun indust fel, 71-72, SCIENTIST, PULP & PAPER RES INST CAN, 72- *Honors & Awards:* I H Weldon Medal, Can Pulp & Paper Asn, 81. *Mem:* Am Chem Soc; Can Pulp & Paper Asn; Tech Asn Pulp & Paper Indust. *Res:* Modification of lignin, hemicellulose and cellulose in wood for the purpose of paper manufacture. *Mailing Add:* 4466 Glendale Pierrefonds PQ H9H 2L2 Can

HEITSCH, CHARLES WEYAND, b Pontiac, Mich, July 5, 31; m 52; c 5. INORGANIC CHEMISTRY. *Educ:* Univ Mich, BS, 56, MS, 57, PhD(chem), 60. *Prof Exp:* Asst chem, Res Inst, Univ Mich, 55-59; instr inorg chem, Iowa State Univ, 59-63; res chemist, E I du Pont de Nemours & Co, Inc, 63-67; mem staff & sr res specialist, Monsanto Co, 67-85; instr, St Louis Community Col & Bellville Area Col, 86; ASST CHMN & RES ENGR, UNIV MO, ROLLA, 86- *Mem:* Am Chem Soc. *Res:* Synthetic and physical chemistry of inorganic and organometallic compounds; industrial process chemistry of inorganic and organo-phosphorous compounds. *Mailing Add:* R&HC 01 Box 66 Bourbon MO 65441

HEITSCH, JAMES LAWRENCE, b Ypsilanti, Mich, July 15, 46; m 67; c 2. FOLIATED MANIFOLDS. *Educ:* Univ Ill, BS, 67; Univ Chicago, MS, 68, PhD(math), 71. *Prof Exp:* Lectr math, Univ Calif, Berkeley, 71-73; vis asst prof, Univ Calif, Los Angeles, 73-74; from asst prof to assoc prof, 73-85, PROF MATH, UNIV ILL, CHICAGO, 85- *Concurrent Pos:* Consult, LaSalle State Securities, 72-75; vis assoc prof, Cath Pontifical Univ, Rio de Janeiro, 75-76 & Univ Lille, France, 81; mem math, Inst Advan Study, 82; vpres res, Burrito & Burrito, 80-84, Math Sci Res Inst, 84-85; fel, Japan Soc Promotion Sci, 86-87; vis prof, Univ Lyon, France, 89. *Mem:* Am Math Soc. *Res:* Analysis on foliated manifolds. *Mailing Add:* Dept Math Univ Ill Chicago IL 60680

HEITSCHMIDT, RODNEY KEITH, b Hays, Kans, Oct 28, 44; m 65; c 2. RANGE SCIENCE, RANGE MANAGEMENT. *Educ:* Fort Hays State Univ, BS, 67, MS, 68; Colo State Univ, PhD(range sci), 77. *Prof Exp:* prof, Tex Agr Exp Sta, Tex A&M Univ, 77-90; RES LEADER, FT KEOGH LIVESTOCK & RANGE RES LAB, USDA, AGR RES STA, 90- *Honors & Awards:* Outstanding Achievement Award, Soc Range Mgt, 91. *Mem:* Soc Range Mgt; Ecol Soc Am; Am Inst Biol Sci; Am Soc Soil & Water Conserv. *Res:* Grazing management. *Mailing Add:* USDA Agr Res Sta Rte 1 Box 2021 Miles City MT 59301

HEITZ, JAMES ROBERT, b Louisville, Ky, Feb 22, 41; m 69; c 3. BIOCHEMISTRY. *Educ:* Bellarmine Col, Ky, AB, 63; Univ Tenn, Knoxville, PhD(biochem), 67. *Prof Exp:* Res assoc biochem, Johns Hopkins Univ, 68-70; from asst prof to assoc prof, 70-79, PROF BIOCHEM, MISS STATE UNIV, 79- *Mem:* Am Soc Biol Chem; Am Chem Soc. *Res:* Enzymology; active site chemistry and applications of fluorescence spectroscopy; insect biochemistry; development of improved specific pesticides. *Mailing Add:* Analytical Support Lab Miss State Univ Mississippi State MS 39762

HEIZER, KENNETH W, b Iola, Tex, Nov 21, 23; m 60; c 4. ELECTRICAL ENGINEERING. *Educ:* Southern Methodist Univ, BS, 50, MS, 51; Univ Ill, PhD(elec eng), 62. *Prof Exp:* From instr to assoc prof elec eng, Southern Methodist Univ, 51-60; instr, Univ Ill, 61-62; PROF ELEC ENG, SOUTHERN METHODIST UNIV, 62- *Concurrent Pos:* Consult, Tex Instruments Inc, 55-56, 57-58, 62-63 & 65 & Nat Data Processing Corp, 58-60. *Mem:* Inst Elec & Electronics Engrs; Am Soc Eng Educ; Sigma Xi. *Res:* Electric circuit theory. *Mailing Add:* 6119 Brandeis Dallas TX 75214

HEIZER, WILLIAM DAVID, b Rawlings, Va, Mar 23, 37; m 60; c 2. MEDICAL SCIENCES. *Educ:* King Col, BA, 58; Johns Hopkins Univ, MD, 63. *Prof Exp:* Actg instr chem, King Col, 58-59; intern med, Johns Hopkins Univ, 63-64, asst resident, 64-65; clin assoc med, NIH, 65-67; from asst prof to assoc prof, 70-78, PROF MED, UNIV NC, 78- *Concurrent Pos:* Career develop award, NIH, 72; res grant, NIH, 71, Nat Cancer Inst, 90; prof, Nutrit Dept, Sch Pub Health, 91-; mem bd dirs, Am Soc Parenteral & Enteral Nutrit. *Mem:* Am Gastroenterol Asn; Am Soc Parenteral & Enteral Nutrit; Am Soc Clin Nutrit; Am Inst Nutrit. *Res:* Absorption of nutrients and drugs from various segments of the intestinal tract. *Mailing Add:* Digestive Dis & Nutrit CB No 7080 318 Burnett-Womack Bldg Univ NC Chapel Hill NC 27599-7080

HEJHAL, DENNIS ARNOLD, b Chicago, Ill, Dec 10, 48. COMPLEX VARIABLE & NUMBER THEORY, SUPERCOMPUTING. *Educ:* Univ Chicago, BS, 70; Stanford Univ, PhD(math), 72. *Prof Exp:* Asst prof math, Harvard Univ, 72-74; Assoc prof, Columbia Univ, 74-78; PROF MATH, UNIV MINN, 78- *Concurrent Pos:* Fel, Sloan Foundation, 74-76, Minn Supercomput Inst, 85-; vis prof, Univ Calif, San Diego, 81 & Chalmers Univ Technol, Sweden, 87; mem, Inst Advan Study, Princeton, 83, 84 & 85. *Mem:* Am Math Soc; Math Asn Am. *Res:* Complex variable; analytic number theory; trace formulas; supercomputers. *Mailing Add:* Sch Math Univ Minn Minneapolis MN 55455

HEJNA, WILLIAM FRANK, b Chicago, Ill, May 13, 32; m 55; c 4. ORTHOPEDIC SURGERY. *Educ:* Grinnell Col, BA, 54, DSc, 74; Washington Univ, MD, 58; Am Bd Orthop Surg, dipl. *Hon Degrees:* DSc, Grinnell Col, 74. *Prof Exp:* Intern, Presby-St Luke's Hosp, 58-59; resident, Univ Ill Res & Educ Hosps, 59-63; dir electrolyography lab & coordr orthop clins & med student training, 63-70, asst chmn dept orthop surg, 65-70, assoc attend surgeon, 67-70, coordr hip clin & adult orthop clin, 68-70, secy staff, 69-71, assoc dean, Off Surg Sci & Serv, 70-73, dean, Rush Med Col & vpres med affairs, Med Ctr, 73-76, PROF ORTHOPEDIC SURG, RUSH MED COL, 76-; SR VPRES, RUSH PRESBY ST LUKES MED CTR, 76- *Concurrent Pos:* Examr, Am Bd Orthop Surg, 69-72; sr attend surg, Dept Orthop Surg, Presby-St Luke's Hosp, 71-; pres, Coun Med Deans, State of Ill, 74-76; pres, Bioserv Corp & Bus Consult Inc; consult, Medicus Affil Inc & Whittaker Corp; chmn bd trustees, Ancor Orgn, Health Maintenance Orgn, 81- *Mem:* Fel Am Col Surgeons; fel Am Acad Orthop Surgeons; Am Asn Med Cols; Orthop Res Soc; Clin Orthop Soc; Sigma Xi. *Res:* Total hip replacement; cartilage studies in arthritis; chemonucleolysis; osteoporosis; sports medicine. *Mailing Add:* Rush-Presbyterian St Lukes Med Ctr 1725 W Harrison St Suite 1192 Chicago IL 60612

HEJTMANCIK, KELLY ERWIN, b Galveston, Tex, Sept 10, 48; m 69; c 2. MEDICAL MICROBIOLOGY. *Educ:* Southwest Tex State Univ, BS, 70; Trinity Univ, MS, 72; Univ Tex Med Br, Galveston, PhD(med microbiol & immunol), 78. *Prof Exp:* Lab instr cellular physiol, Trinity Univ, 70-71; high school teacher biol & phys sci, San Antonio Independent Sch Dist, 71-72; res assoc, 72-74, teaching & res asst med microbiol, 74-76, res fel, 76-78, sr res assoc & fel, Univ Tex Med Br, Galveston, 78-79; chmn dept, 79-85, PROF BIOL, GALVESTON COL, 79- *Concurrent Pos:* Reviewer, WB Saunders Publ, Holt Reinhart & Winston, CV Mosby Co, 79-; Consult, Clear Lake City, NASA, 86- *Mem:* Sigma Xi; AAAS; Am Soc Microbiol. *Res:* Immunochemical techniques for research and diagnostic applications for the antigenic analysis and molecular action of microbiol toxins and other products, and the immunologic response of the host to them. *Mailing Add:* 15 Back Bay Circle E Galveston TX 77551

HEJTMANCIK, MILTON R, b Caldwell, Tex, Sept 27, 19; m 43, 76; c 3. INTERNAL MEDICINE, CARDIOLOGY. *Educ:* Univ Tex, BS, 39, MD, 43; Am Bd Internal Med, dipl, 51; Am Bd Cardiovasc Dis, dipl, 63. *Prof Exp:* Instr internal med, Univ Tex Med Br Galveston, 49-51, asst dir heart sta, 49-65, dir, Heart Sta, 49-65, dir heart clin, 49-80, asst prof internal med, 51-54, assoc prof, 54-65, prof internal med & dir heart sta, 65-80; prof internal med, Tex A&M Col Med, Temple, Tex, 81-82; cardiologist, Olin E Teague Vet Admin Hosp, Temple, Tex, 81-82; cardiologist, Beaumont Vet Admin Clin, 82-86; RETIRED. *Concurrent Pos:* Consult, St Mary's Infirmary, 51-80, Galveston County Mem Hosp, 53-80 & USPHS Hosp, 57-80; med adv, Bur Hearings & Appeals, Social Security Admin, 68-80; fel, Coun Clin Cardiol, Am Heart Asn. *Mem:* Am Heart Asn; fel Am Col Physicians; fel Am Col Chest Physicians; fel Am Col Cardiol; Am Fedn Clin Res. *Res:* Clinical cardiology, especially electrocardiography, vectorcardiography, echocardiography, arrhythmias of the heart, drug therapy of cardiac disorders. *Mailing Add:* 500 N Spruce St Hammond LA 70401

HEKAL, IHAB M, b Cairo, Egypt, Oct 27, 38; US citizen; m 64; c 1. MATERIALS CHEMISTRY, PROCESS DEVELOPMENT. *Educ:* Univ Graz, Austria, BSc, 62, MSc, 64; Univ Hannover, Ger, PhD(phys chem), 66. *Prof Exp:* Scientist, Nat Res Ctr, Cairo, Egypt, 67-68; sr scientist, Contiental Can Co, Ill, 68-72, mgr surface chem, 72-76, consult process develop, 76-78; DIR ADVAN TECHNOL, CONTINENTAL GROUP, 78- *Concurrent Pos:* Lectr chem, Roosevelt Univ, 70-71 & Hebrew Theol Col, 71-72. *Mem:* Am Chem Soc. *Res:* Coating material and process (wash coat for cans); compile recycling of aluminum treated chemicals with zeno effluent; composits (metals/minerals); oxygen Barries material for packaging; electro chemical process for oxygen removal from juices for shelf life improvement. *Mailing Add:* 121 Black Berry Dr Stamford CT 06904

HEKIMIAN, NORRIS CARROLL, electronics engineering, communications engineering, for more information see previous edition

HEKKER, ROELAND M T, b Djakarta, Indonesia, Nov 25, 53; Neth citizen; m 76; c 2. MACHINE VISION, PATTERN RECONGNITION. *Educ:* Univ Technol Delft, Neth, BD, 77, MS, 79. *Prof Exp:* Optical syst design indust, Philips-Data Systs, 79-82; dept mgr optical disk technol, OSI Laser Magnetic Syst, 82-85; mem staff advan optical rec, Hewlett & Parkard Res Labs, 85; vpres eng optical signal processing, Global Holonetics Corp, 85-89; VPRES ENG PATTERN RECOGNITION, IMPACQ TECHNOL CORP, 90- *Concurrent Pos:* Pres, ELA-Hi Tech Eng, 90- *Mem:* Int Soc Optical Eng; Soc Mfg Engrs; affil Inst Elec & Electronics Engrs. *Res:* Digital signal processing; pattern recognition; artificial neural nets and fuzzy logic as they apply to image processing; scientific visulization and machine vision. *Mailing Add:* ELA Hi Tech Eng 608 W Fillmore Ave Fairfield IA 52556

HEKMAT, HAMID MOAYED, b Tehran, Iran, Aug 24, 40; US citizen; m 69; c 1. ANXIETY & FEAR, PAIN ETIOLOGY & INTERVENTION. *Educ:* Huntingdon Col, BA, 64; Univ Southern Calif, PhD(psychol), 68. *Prof Exp:* PROF PSYCHOL, UNIV WIS-STEPHENS POINT, 68- *Mem:* Am Psychol Asn; Asn Advan Behav Ther. *Res:* Etiology of phobias; management of anxiety disorders; treatment of cold pressor pain; psychological mechanism in arthritis and other chronic pain disorders. *Mailing Add:* 1758 Pine St Stevens Point WI 54481

HEKMATPANAH, JAVAD, b Isfahan, Iran, Mar 25, 34; m 59; c 3. NEUROSURGERY, NEUROLOGY. *Educ:* Univ Tehran, MD, 56. *Prof Exp:* Intern, Mt Sinai Hosp, Chicago, 57-58; resident neurol, Univ Wis Hosp, 58-61; resident neurosurg, 61-63, instr & chief resident, 63-64, asst prof, 64-70, assoc prof, 70-75, PROF NEUROSURG, UNIV CHICAGO HOSPS, 75- *Mem:* Am Col Surgeons; Am Asn Neurologists; Am Acad Neurol; Sigma Xi. *Res:* Cerebral circulation in trauma; brain tumors; multimodal therapy in man and viral induction in rats. *Mailing Add:* 5000 East End Chicago IL 60615

HELANDER, DONALD P(ETER), b Milwaukee, Wis, May 26, 31; m 56; c 3. PETROLEUM ENGINEERING. *Educ:* Univ Tulsa, BS, 57, MS, 60; Univ Okla, PhD(petrol eng), 65. *Prof Exp:* Jr field logging engr, Schlumberger Well Surv Corp, 57-58; instr petrol eng, Univ Tulsa, 58-62, asst prof, 65-66, asst dir info serv dept, 65-66, dir, 66-68, assoc prof & head dept, 68-77; PRES, INT PETROL CONSULT, INC, 77-; VPRES & CO-OWNER, BNJ OIL PROPERTIES INC, 81- *Concurrent Pos:* Teacher formation eval & reservoir eng, 65- *Mem:* Sigma Xi; Soc Prof Well Log Analysts; Soc Petrol Engr; Can Well Logging Soc. *Res:* Formation evaluation, especially electric, acoustic and radioactive well logging; oil and gas reservoir analysis; author. *Mailing Add:* 4124 E 98th St Tulsa OK 74137-4807

HELANDER, HERBERT DICK FERDINAND, b Lund, Sweden, Dec 7, 35; m 61; c 3. CELL BIOLOGY, GASTROENTEROLOGY. *Educ:* Univ G06teborg, Sweden, BM, 56, PhD(anat), 62, MD, 63. *Prof Exp:* Asst prof anat, Univ Göteborg, Sweden, 63-65; prof anat, Univ Umeå, Sweden, 66-84; SR RES ADV, AB HÄSSLE PHARMACEUTICALS, MÖLNDAL, SWEDEN, 84- *Concurrent Pos:* vis asst prof, Dept Med, Univ Okla Med Ctr, 64-65; Fulbright scholar, 64; sr foreign scientist, NSF, 70; vis prof, Dept Physiol, Univ Ala Med Ctr, 70-82, Cardiovasc Res Inst, Univ Calif, San Francisco, 76, Dept Med, Univ Calif, Los Angeles, 83-; chmn, Dept Anat, Univ Umeå, Sweden, 74-84; adj prof, Dept Anat, Univ Göteborg, Sweden, 85- *Mem:* Am Gastroenterol Asn; Am Soc Cell Biol; Am Soc Histotechnol. *Res:* Structure and function of the gastric mucosa; morphological studies of gastric acid secretion. *Mailing Add:* AB Hassle Pharmaceuticals Molndal 431 83 Swedeni

HELANDER, MARTIN ERIK GUSTAV, b Uppsala, Sweden, May 13, 43; div; c 2. USER INTERFACE AUTOMATION. *Educ:* Chalmers Univ Technol, Gotenborg, Sweden, MS, 67, PhD(civil eng), 73; Lulea Univ, Sweden, PhD(eng psychol), 77. *Prof Exp:* Prof ergonomics, Lulea Univ, 75-83; assoc prof indust eng, Univ SFla, 83-85; ASSOC PROF INDUST ENG, STATE UNIV NY BUFFALO, 85- *Concurrent Pos:* Sr scientist human factors, Canyon Res Group, 77-82; secy, TRB comt on vehicle users characteristics, Nat Res Coun, 79-; chmn tech standards comt, Human Factors Soc, 80-; vis prof, Va Tech, 82-83; US deleg, Int Ergonomics Asn, 85. *Mem:* Fel Human Factors Soc; fel Ergonomics Soc; Inst Indust Engrs; Robotics Int. *Res:* Human factors or ergonomics research in the areas of traffic safety and driver performance; office and industrial automation; human-computer interaction. *Mailing Add:* Dept Indust Eng State Univ NY Buffalo Amherst NY 14260

HELAVA, U(UNO) V(ILHO), b Kokemaki, Finland, Mar 1, 23; m 42; c 3. PHOTOGRAMMETRY, GEODESY. *Educ:* Finnish Inst Technol, MSc, 47. *Hon Degrees:* DrEng, Helsinki Technol Univ, 78. *Prof Exp:* Geodetist, Topog Corps, Finnish Army, 46-47; photogrammetrist, Gen Surv Off, Finland, 47-53; res fel, Nat Res Coun Can, 54-55, assoc res off, 55-64, prin res off, 64-65; consult, Ottico Meccanica Italiana, Rome, 65-66; prin scientist, Bendix Res Labs, 67-69, consult scientist, 69-79; pres & chief scientist, Helava Assoc Inc, 79-91; RETIRED. *Honors & Awards:* Sr Am Scientist Award, Alexander von Humboldt Stiftung, 77. *Mem:* Am Soc Photogram; Can Inst Surv; Finnish Soc Photogram; Sigma Xi; Inst Elec & Electronics Engrs. *Res:* Aerial photogrammetry; photogrammetric instruments. *Mailing Add:* 12635 Mallet Circle West Palm Beach FL 33414-8408

HELBERT, JAMES RAYMOND, b Miles City, Mont, Aug 4, 18; m 49; c 4. BIOCHEMISTRY. *Educ:* St John's Univ, Minn, AB, 47; Marquette Univ, MS, 58; Northwestern Univ, PhD(biochem), 63. *Prof Exp:* Chief chemist, Orthmann Labs, Inc, 47-51; res chemist, Red Star Yeast & Prod Co, 51-58; res biochemist, Geriat Res Proj, Vet Admin Hosp, 58-62, actg chief, 62-63; res assoc, Dept Hemat, Michael Reese Hosp & Med Ctr, 63-67; supvr biochem res, Miller Brewing Co, 67-76, mgr biochem res & admin, 76-85; RETIRED. *Mem:* Am Chem Soc; Am Statist Asn; fel Am Inst Chem. *Res:* Physical biochemistry; statistical design of experiments; fermentation; fermentation kinetics. *Mailing Add:* PO Box 47049 Chicago IL 60647-0049

HELBERT, JOHN N, b Wichita, Kans, Aug 3, 46; m 69; c 2. RADIATION CHEMISTRY & POLYMER CHEMISTRY, INDUSTRIAL & MANUFACTURING ENGINEERING. *Educ:* Ariz State Univ, BS, 68; Wayne State Univ, PhD(chem), 72; US Army Command & Gen Staff Col, MA, 87. *Prof Exp:* Res asst chem, Wayne State Univ, 69-72; res physicist, US Army Electronics Command, 72-75, res chemist, 75-80; processing engr, 80-85, mem tech staff, 84-90, SR MEM TECH STAFF, MOTOROLA, 90- *Honors & Awards:* Res & Develop Award, US Army, 79; Motorola Silver Quill Award, 82. *Mem:* Am Chem Soc; Sigma Xi; Electrochem Soc. *Res:* Radiation chemistry of polymers to include magnetic resonance investigations of polymer radicals; photolithography and resist processing engineering investigations; semiconductor process engineering. *Mailing Add:* Motorola BTC M350 2200 W Broadway Mesa AZ 85202

HELBIG, HERBERT FREDERICK, b Bronx, NY, Jan 2, 34; m 57. ATOMIC PHYSICS. *Educ:* Yale Univ, AB, 56; Univ Conn, MS, 62, PhD(physics), 65. *Prof Exp:* From asst prof to assoc prof, 65-80, exec officer, 68-73, actg chmn, 84-85, PROF PHYSICS, CLARKSON COL TECHNOL, 80- *Concurrent Pos:* Consult, Lawrence Livermore Nat Lab, 79-81. *Mem:* Am Phys Soc; Sigma Xi; Am Vacuum Soc. *Res:* Ion-atom and ion-surface scattering experiments and simulations. *Mailing Add:* Dept Physics Clarkson Col Technol Potsdam NY 13676

HELBING, REINHARD KARL BODO, b Stettin, Ger, June 1, 35; m 60; c 1. PHYSICS. *Educ:* Univ Bonn, dipl, 61, Dr rer nat(physics), 66. *Prof Exp:* Asst physics, Inst Appl Physics, Univ Bonn, 62-68; assoc prof eng & appl sci, Univ Calif, Los Angeles, 69-72; PROF PHYSICS, UNIV WINDSOR, 72- *Concurrent Pos:* Staff scientist, Gen Dynamics/Convair, 66-69. *Mem:* Am Phys Soc; Ger Phys Soc; Europ Phys Soc; NY Acad Sci. *Res:* Properties of atoms and molecules investigated by applying molecular beams of thermal and higher energies; elastic and inelastic scattering; theoretical treatment of collision phenomena. *Mailing Add:* Dept Physics Univ Windsor Windsor ON N9B 3P4 Can

HELD, ABRAHAM ALBERT, b Vienna, Austria, Mar 27, 34; nat US; m 67. MYCOLOGY. *Educ:* Hebrew Univ, Israel, MSc, 63, Univ Calif, Berkeley, PhD(microbiol), 67. *Prof Exp:* Guest investr parasitol, Rockefeller Univ, 67-70, res assoc, 70-71; from asst prof to assoc prof, 71-82, PROF BIOL SCI, DEPT BIOL SCI, LEHMAN COL, CITY UNIV NEW YORK, 83-; MEM ADJ FAC, ROCKEFELLER UNIV, 71- *Mem:* Am Soc Microbiol. *Res:* Biology of aquatic phycomycetes; fungal nutrition and cultivation; biomedical terminology. *Mailing Add:* Dept Biol Sci Lehman Col City Univ New York Bronx NY 10468

HELD, BEREL, b Brooklyn, NY, Aug 31, 38; m 62; c 4. OBSTETRICS & GYNECOLOGY. *Educ:* Tulane Univ, MD, 62. *Prof Exp:* From asst prof to assoc prof obstet & gynec, Col Med, Univ Fla, 68-72; PROF OBSTET & GYNEC & CHMN DEPT, UNIV TEX MED SCH HOUSTON, 72- *Concurrent Pos:* Fel path, Harvard Med Sch, 66; consult gynec, M D Anderson Hosp & Tumor Inst, 72-; chief obstet & gynec, Hermann Hosp, 72- *Mem:* Am Col Obstet & Gynec. *Res:* Perinatology. *Mailing Add:* Tex Med Ctr Univ Tex Med Sch Houston TX 77024

HELD, IRENE RITA, b Milwaukee, Wis. NEUROCHEMISTRY. *Educ:* Marquette Univ, BS, 55; Northwestern Univ, MS, 62, PhD(biochem), 65. *Prof Exp:* Res biochemist, Vet Admin Hosp, Downey, Ill, 65-72; CHIEF NEUROSCI RES SECT, VET ADMIN HOSP, HINES, ILL, 72- *Concurrent Pos:* Instr biochem, Med Sch, Northwestern Univ, Chicago, 66-70, assoc, 70-74, asst prof pharmacol, Stritch Sch Med, Loyola Univ Chicago, 74-77 & assoc prof pharmacol & biochem, 77- *Mem:* Am Soc Neurochem; Soc Neurosci; AAAS; Am Chem Soc. *Res:* Biochemical mechanisms in the neurotrophic relation between nerve and muscle. *Mailing Add:* Chief Neurosci Hines Vet Hosp Hines IL 60141

HELD, ISAAC MEYER, b Germ Oct 23, 48; m 84; c 1. GEOPHYSICAL FLUID DYNAMICS, CLIMATE DYNAMICS. *Educ:* Univ Minn, BPhys, 69; State Univ Ny, Stony Brook, MA, 70; Princeton Univ, PhD, 76. *Prof Exp:* Res fel climate, Harvard Univ, 76-78; METEOROLOGIST, GEOPHYS FLUID DYNAMICS LAB, NAT OCEANIC & ATMOSPHERIC ADMIN, PRINCETON UNIV, 78- *Concurrent Pos:* Assoc ed, J Atmospheric Sci, 79-87; ed, J Atmospheric Sci, 87-90. *Honors & Awards:* Meisinger Award, Am Meteorol Soc, 87. *Mem:* Fel Am Meteorol Soc; Am Geophys Union. *Res:* Large-scale atmospheric dynamics; models of climatic sensitivity and variability; fluid dynamics. *Mailing Add:* US Dept Com Nat Oceanic & Atmospheric Admin Princeton Univ PO Box 308 Princeton NJ 08542

HELD, JOE R, b Los Angeles, Calif, June 23, 31; m 56; c 4. PRIMATOLOGY, PARASITOLOGY. *Educ:* Univ Calif, Davis, BS, 53, DVM, 55; Tulane Univ, MPH, 59. *Prof Exp:* Vet epidemiologist, Commun Dis Ctr, 55-57; pvt pract, 57-58; asst chief rabies control unit, Vet Sect, Epidemiol Br, Commun Dis Ctr, 59-60, asst to sect chief, 60-61, asst to br chief, 61-62; sr vet officer, Primate Ctr Grant Admin, Animal Resources Br, NIH, 62-64, sr vet officer malaria res, Nat Inst Allergy & Infectious Dis, 64-67; vet epidemiologist, Pan Am Zoonoses Ctr, Arg, 67-69; chief vet res br, 69-72, dir, Div Res Serv, NIH, 72-84; dir, Pan Am Zoonoses Ctr, Pan Am Health Orgn, 84-87, coordr, Vet Pub Health Prog, 87-89; VPRES PRIMATE OPERS, CHARLES RIVER LABS, 89-90. *Concurrent Pos:* Chief vet off, USPHS, 75-84. *Honors & Awards:* James H Steele Award, World Vet Epidemiol Soc, 77; KF Meyer Award, Am Vet Epidemiol Soc, 82; Charles River Prize, Am Vet Med Asn, 84; XII Int Vet Cong Prize, Am Vet Med Asn, 89; J A McCallam Award, Asn Mil Surgeons, 90. *Mem:* Am Vet Med Asn; hon mem Am Col Lab Animal Med; Am Soc Primates; Int Primate Soc; Nat Asn Fed Vet; Am Asn Lab Animal Sci; Am Col Epidemiol; Am Col Vet Prev Med; AAAS. *Res:* Primate malarias, particularly the exo-erythrocytic stages; brucellosis; infectious hepatitis; rabies; toxoplasmosis; dermatomycoses. *Mailing Add:* 1300 Crystal Dr No 505 Arlington VA 22202-3234

HELD, ROBERT PAUL, b Brooklyn, NY, Nov 18, 39; wid; c 3. PHYSICAL CHEMISTRY. *Educ:* Queens Col, NY, BS, 59; Brooklyn Col, MA, 62; Univ Vt, PhD, 65. *Prof Exp:* Jr chemist, Kings County Hosp, 59-60; chemist, Hempstead Gen Hosp, 60-62; from res chemist to sr res chemist, 65-74, res assoc, 74-76, RES FEL, PHOTO PROD DEPT, E I DU PONT DE NEMOURS & CO, INC, 76- *Mem:* Am Chem Soc. *Res:* Thermodynamics of nonaqueous solutions; entropy correlations of electrolytic solutions; photosensitive imaging systems; gelatin interactions with silver halide crystals; photopolymer imaging systems; dispersions; emulsions. *Mailing Add:* Imaging Systs Dept E I du Pont de Nemours & Co Inc Parlin NJ 08859

HELD, WILLIAM ALLEN, b West Bend, Wis, Nov 14, 40. BIOCHEMISTRY, MOLECULAR BIOLOGY. *Educ:* Marquette Univ, BS, 62, MS, 65, PhD(biol), 69. *Prof Exp:* Fel biochem, genetics, Univ Wis, 69-74; asst prof, 74-80, ASSOC PROF CELL & MOLECULAR BIOL, ROSWELL PARK MEM INST, 80-, SR CANCER RES SCIENTIST, 74- *Concurrent Pos:* NIH fel, 69-71; NIH grants, 75-78 & 79-81. *Mem:* Am Soc Microbiol; NY Acad Sci; Am Soc Biol Chemists. *Res:* Regulation of gene expression and protein synthesis. *Mailing Add:* Dept Molecular & Cellular Biol Roswell Park Mem Inst 666 Elm St Buffalo NY 14263

HELDERMAN, J HAROLD, b Newark, NJ, Feb 5, 45. RENAL TRANSPLANTS, NEPHROLOGY. *Educ:* Univ Rochester, AB, 67; State Univ NY, Downstate Med Ctr, MD, 71. *Prof Exp:* PROF MED, HEALTH SCI CTR, UNIV TEX, 77-, DIR KIDNEY TRANSPLANT PROG, 83- *Mem:* Am Soc Clin Invest; Int Transplantation Soc; Am Asn Immunologist; Am Soc Nephrology; Am Diabetes Asn; Am Fedn Clin Res. *Mailing Add:* Vanderbilt Tex Ctr Dept Int Med Vanderbilt Univ Med Ctr N S 3223 Nashville TN 37232

HELDMAN, DENNIS RAY, b Findlay, Ohio, June 12, 38; m 56, 90; c 3. FOOD ENGINEERING. *Educ:* Ohio State Univ, BS, 60, MS, 62; Mich State Univ, PhD(agr eng), 65. *Prof Exp:* From instr to assoc prof agr eng & food sci, Mich State Univ, 65-71, prof agr eng, 71-84, chmn dept, 75-79, prof food eng, 79-84; vpres, Process Res & Develop, Campbell Inst Res & Technol,

Campbell Soup Co, 84-88; exec vpres, Sci Affairs, Nat Food Processor Asn, 88-91; PRIN, WEINBERG CONSULT GROUP, INC, 91- *Concurrent Pos:* Vis assoc prof, Univ Calif, Davis, 70 & vis prof, 80; consult ed, McGraw-Hill Publ Co, 80- & ed, J Food Process Eng, 77- *Honors & Awards:* Food Engrs Award, Am Soc Agr Engrs, 81. *Mem:* Am Soc Agr Engrs; fel Inst Food Technol; Am Dairy Sci Asn; Inst Asn Milk, Food & Environ Sanit. *Res:* Thermal, rheological and thermodynamic properties of processed foods; mathematical description of processes for food; thermal processing of food; air quality in food processing facilities; characteristics of microbial particles in enclosed spaces. *Mailing Add:* Weinberg Consult Group Inc 2828 Pennsylvania Ave NW Suite 305 Washington DC 20007

HELDMAN, JULIUS DAVID, b Cleveland, Ohio, May 9, 19; m 42; c 2. PETROLEUM CHEMISTRY, ENERGY CONVERSION. *Educ:* Univ Calif, Los Angeles, AB, 39, MA, 40; Stanford Univ, PhD(phys chem), 42. *Prof Exp:* Asst chem, Univ Calif, Los Angeles, 39-40 & Stanford Univ, 40-41; Nat Res fel, Univ Calif, 42-43, instr, 43-45; technologist, Shell Oil Co, San Francisco, 45-48, Wilmington, 48-49, chief res chemist, 49, tech adv, Houston, 49-51, chief res technologist, 51-53, sr technologist, 53-57, asst mgr mfg res, 57-61, mgr prod develop, 61-65, develop mgr petrochem div, Shell Chem Co, 65-69, vpres, Shell Develop Co, 69-80; pres, Cell Sci Ltd, 80-84; CONSULT, 80- *Concurrent Pos:* Woodrow Wilson vis fel, 78-83. *Mem:* AAAS; Am Chem Soc; Am Inst Chem Eng. *Res:* Chemistry and technology of petroleum; solar energy; research administration; science policy; physical chemistry; semi-conductors. *Mailing Add:* 1002 Old Pecos Trail Santa Fe NM 87501

HELDMAN, MORRIS J, b Cleveland, Ohio, June 18, 14; m 78; c 2. COLLOID CHEMISTRY, ACADEMIC ADMINISTRATION. *Educ:* Univ Calif, Los Angeles, BA, 34, MA, 36; Univ Southern Calif, PhD(colloid chem), 47. *Prof Exp:* Asst chem, Univ Calif, Los Angeles, 35-36; city testing engr, Los Angeles, 36-37, teacher pub sch, Bd Educ, 37-42; asst, Univ Southern Calif, 42-44, lectr, 44; teacher, Los Angeles Pub Sch, Bd Educ, 44-46; instr chem & electronics, E Los Angeles Col, 46-57, asst dean, 57-62; head chemist, US Naval Ord Test Sta, Calif, 51-52; mem staff, Wichita Found Indust Res, 53; res engr, Hughes Aircraft Co, 56-57; dean, Los Angeles Metrop Col, 63-66; dean, Los Angeles Pierce Col, 66-68; pres, 68-76, EMER PRES, WEST LOS ANGELES COL, 76- *Concurrent Pos:* Educ consult, 70-83. *Mem:* AAAS; Am Chem Soc; Sigma Xi; fel AAAS. *Res:* Kinetics of inorganic reactions in solution; liquid crystals-phase studies; electrochemistry. *Mailing Add:* 5920 Hill Rd Culver City CA 90230

HELDT, LLOYD A, b Evansville, Ind, July 27, 34; m 57; c 2. METALLURGY. *Educ:* DePauw Univ, BA, 56; Ind Univ, MS, 58; Univ Pa, PhD(metall), 62. *Prof Exp:* From asst prof to assoc prof, 61-69, PROF METALL ENG, MICH TECHNOL UNIV, 69-, HEAD DEPT, 77- *Mem:* Fel Am Inst Mining, Metall & Petrol Engrs; Electrochem Soc; fel Am Soc Metals Int; Nat Asn Corrosion Engrs. *Res:* Corrosion; fracture; environment-sensitive mechanical properties; semiconducting materials; electrometallurgy. *Mailing Add:* Dept Metall & Mat Eng Mich Technol Univ Houghton MI 49931

HELDT, WALTER Z, b Danzig, Poland, July 2, 28; nat US; m 56; c 3. ORGANIC CHEMISTRY, PHYSICAL CHEMISTRY. *Educ:* Univ Heidelberg, Dipl chem, 49; Rutgers Univ, MS, 52, PhD(chem), 54; Univ Pa, MBA, 58. *Prof Exp:* res, E I Du Pont de Nemours & Co, Inc, 53-81; pres & chmn bd, Helix Assocs, Inc, 81-89; PRES, HOBSON TECHNOL CORP, 87- *Concurrent Pos:* gen partner, Helix Ltd Partnership, 81-89; chmn bd, RWJ Industs, Inc, 90- *Mem:* Am Chem Soc; Am Mkt Soc. *Res:* Reaction mechanisms; polymer intermediates; organic syntheses and catalysis; plastics composites; operations research; financial planning, licensing, acquisitions. *Mailing Add:* 106 Willow Spring Rd Wilmington DE 19807

HELE, PRISCILLA, b Cambridge, Eng, Nov 22, 20. BIOCHEMISTRY. *Educ:* Cambridge Univ, BA, 41 & 42, MA & MB, BCh, 45, PhD(biochem), 50. *Prof Exp:* Williams-Waterman fel enzym, Univ Wis, 50-51; res assoc, Exp Radiopath Res Unit, Hammersmith Hosp, London, Eng, 54-57, independent investr sr staff, 58-62 & mem sci staff, Med Res Coun Gt Brit, 54-62; mem res unit chem path ment dis, Univ Birmingham, 62-64; sr scientist, Worcester Found Exp Biol, 64-78; RETIRED. *Mem:* AAAS; Am Soc Biol Chemists; Brit Biochem Soc; Brit Med Asn. *Res:* Enzymology; nucleic acids; amino acid activation; protein biosynthesis. *Mailing Add:* 40 Dennis Circle Northborough MA 01532

HELFAND, DAVID JOHN, b New Bedford, Mass, Dec 7, 50; m 82; c 1. ASTROPHYSICS. *Educ:* Amherst Col, BA, 73; Univ Mass, MS & PhD(astron), 77. *Prof Exp:* Res asst, Five Col Radio Astron Observ, 73-77; res assoc, 77-78, asst prof astron, 78-82, assoc prof, physics, 82-87, PROF PHYSICS, COLUMBIA UNIV, 87-, CHAIR, DEPT ASTRON, 86- *Concurrent Pos:* Shapely lectr, Am Astron Soc, 80-; vpres, NY Astron Corp, 81-86, pres, 86-90; AP Sloan fel, 83-87; co-dir, Astrophys Lab, Columbia Univ, 86- *Mem:* Am Astron Soc; Sigma Xi; NY Acad Sci. *Res:* Observational high energy astrophysics; radio astronomy and x-ray astronomy concentrating on neutron stars, pulsars, supernova remnants, quasars and active stars; cosmic x-ray background. *Mailing Add:* Columbia Astrophys Lab 538 W 120th St New York NY 10027

HELFAND, EUGENE, b Brooklyn, NY, Jan 8, 34; m 57; c 3. POLYMER THEORY, STATISTICAL MECHANICS. *Educ:* Polytech Inst Brooklyn, BS, 55; Yale Univ, MS, 57, PhD(chem), 58. *Prof Exp:* Mem tech staff chem, 58-83, DISTINGUISHED MEM TECH STAFF, BELL LABS, 83- *Concurrent Pos:* Adj prof, Yeshiva Univ, 60-62 & Polytech Inst Brooklyn, 64-65; John Simon Guggenheim fel, Stanford Univ, 69-70; chmn, Div High Polymer Physics, Am Phys Soc, 87-88. *Honors & Awards:* High Polymer Physics Prize, Am Phys Soc, 89. *Mem:* Am Phys Soc; Am Chem Soc. *Res:* Polymer theory. *Mailing Add:* AT&T Bell Labs 1A-361 600 Mountain Ave Murray Hill NJ 07974

HELFER, HERMAN LAWRENCE, b New York, NY, Nov 11, 29; m 56; c 4. ASTRONOMY. *Educ:* Univ Chicago, PhB, 48, PhD(astron), 53. *Prof Exp:* Res fel radio astron, Carnegie Inst, 53-54 & 56-57; res assoc astron, Calif Inst Technol, 57-58; from asst prof to prof physics & astron, 58-77, PROF ASTRON, UNIV ROCHESTER, 77- *Concurrent Pos:* Guggenheim Mem Found fel, 65-66. *Mem:* Am Astron Soc; Royal Astron Soc; Sigma Xi. *Res:* Radio astronomy; hydromagnetics; theoretical astrophysics; astronomical spectroscopy. *Mailing Add:* Dept Physics Univ Rochester Rochester NY 14627

HELFFERICH, FRIEDRICH G, b Berlin, Ger, Aug 1, 22; m 47, 61; c 3. CHEMICAL ENGINEERING. *Educ:* Univ Hamburg, Ger, Vordipl chem, 49, dipl chem, 52; Univ Göttingen, Ger, Dr rer nat, 55. *Prof Exp:* Res asst, Max Planck Inst Phys Chem, Göttingen, Ger, 51-56, Mass Inst Technol, Cambridge, Mass, 54 & Calif Inst Technol, Pasadena, Calif, 56-58; supvr & sr staff res chemist, Shell Develop Co, Emeryville, Calif, & Houston Tex, 58-59; prof chem eng, 80-90, EMER PROF, PA STATE UNIV, 90- *Concurrent Pos:* Vis scientist, Max Planck Inst Phys Chem, Göttingen, Ger, 58; lectr, Univ Calif, Berkeley, 62-63 & Am Inst Chem Engrs Today Ser, 82-; chmn, Gordon Res Conf Ion Exchange, 67; vis prof & lectr, Univ Tex, Austin, Univ Houston & Rice Univ, Houston, Tex, 80; founder & ed-in-chief, Reactive Polymers, 81-; fac mem, Sch Artificial Membranes, Warsaw, Coun Mutual Econ Assistance, Warsaw, 83- & Pa State Environ Eng Prog, 84-90; consult, Exxon Chem, Duracell Int, Raychem, 84-; fac consult, Pa State Instrnl Develop Prog, 85-86; vis prof, East China Univ Chem Technol, Shanghai, People's Repub China, 87. *Honors & Awards:* Cert Merit, Am Chem Soc, 67, Separation Sci & Technol Award, 87; Kurt Wohl Mem Lectr, Univ Del, 75; AT&T Found Award for Excellence in Instr Eng Students, Am Soc Eng Educ, 85. *Mem:* Fel Am Inst Chemists; Am Inst Chem Engrs; Soc Petrol Engrs. *Res:* Dynamics of multivariable systems; nonlinear wave propagation; reaction kinetics; ion exchange; nonlinear chromatography. *Mailing Add:* Dept Chem Eng Pa State Univ University Park PA 16802

HELFGOTT, CECIL, b Brooklyn, NY, Nov 29, 26; m 51; c 2. PHYSICAL CHEMISTRY, POLYMER CHEMISTRY. *Educ:* Brooklyn Col, BA, 49; Rutgers Univ, New Brunswick, PhD(phys chem), 68. *Prof Exp:* Res chemist, Am Cyanamid Co, 54-62 & Colgate Palmolive Co, 62-63; res asst, Rutgers Univ, 63-66; sr res chemist, Celanese Res Co, 66-70; sr res chemist, St Regis Paper Co, 70-77, sr staff chemist, 77-79, group leader, 79-83; SR SCIENTIST, CHAMPION INT, 84- *Mem:* Sigma Xi; Am Chem Soc; Tech Asn Pulp & Paper Indust. *Res:* Desalination membranes; polyelectrolytes; fibers; films; textile finishes; converted products, paper/plastics; surfactants. *Mailing Add:* 35 Eberling Dr New City NY 10956

HELFMAN, HOWARD N, b Minneapolis, Minn, Dec 13, 20; m 45; c 3. MECHANICAL ENGINEERING. *Educ:* Univ Southern Calif, BE, 44. *Prof Exp:* Radio res engr, Signal Corps Eng Labs, US Army, 44-46; chief engr, Air Conditioning Co Inc, 46-56; exec vpres, Climate Conditioning Co, 56-58; pres, Helfman Air Conditioning Co, 58-71 & Helfman, Haloossim & Assoc Consult Engrs, 71-88; CONSULT FORENSIC ENGR, 88- *Concurrent Pos:* Sr lectr, Eng Exten, Univ Calif, Los Angeles, 52-82. *Mem:* Fel Am Soc Heating, Vent & Air Conditioning Engrs; Nat Soc Prof Engrs; Am Soc Plumbing Engrs; Am Soc Heating Refrig & Air Conditioning Engrs; Asn Energy Engrs. *Mailing Add:* 295 Bronwood Ave Los Angeles CA 90049

HELFRICH, CHARLES THOMAS, b Dallastown, Pa, Jan 25, 32; m 59; c 6. PALEONTOLOGY, STRATIGRAPHY. *Educ:* St Charles Sem, BA, 54; Villanova Univ, MSS, 67; Va Polytech Inst & State Univ, PhD(geol), 72. *Prof Exp:* Teacher, York City Sch Dist, 57-67; asst prof geol, 71-74, ASSOC PROF GEOL, EASTERN KY UNIV, 74- *Mem:* Geol Soc Am. *Res:* Mid-Paleozoic conodont biostratigraphy of North America. *Mailing Add:* Dept Geol Eastern Ky Univ Richmond KY 40475

HELFRICH, PHILIP, b San Francisco, Calif, June 22, 27; m 56; c 4. GEN MARINE SCIENCES, RESEARCH ADMINISTRATION. *Educ:* Univ Santa Clara, BS, 51; Univ Hawaii, PhD(zool), 58. *Prof Exp:* Asst marine biologist, Hawaii Inst Marine Biol, Univ Hawaii, 58-63, from asst dir to actg dir, 64-70, assoc marine biologist, 64-70, head aquacult prog, 70-72; dir int ctr for living aquatic res mgt, 74-76; assoc dean, Res Training & Spec Progs, 76-80 DIR, HAWAII INST MARINE BIOL, UNIV HAWAII, 80-, PROF ZOOL, 86- *Concurrent Pos:* Mem fac zool, 61-; consult, SPac Comn, 62; mem, French Polynesia Inst Med Res, 64-68; mem, US-Japan Coop Res Prog Marine Toxins, 66-69; bd dirs, Micronesia Found, 70-71; dir, Eniwetok Marine Biol Lab, Mid-Pac Marine Lab, Marshall Islands, 70-75, 80-86; mem, Christmas Island Artemia Res, 71; SSC, West Pac Fish Mgt Coun, 77-81, Gov's Marine Affairs Adv Coun, Hawaii, 77-, Aquacult Develop Coun, Hawaii; aquacult adv, Govt Thailand, 78-; coun chmn, Consortium Int Fisheries & Agr Develop, 81-; bd dir, Ctr Trop & Subtrop Aquaculture, 87- *Mem:* AAAS; Ecol Soc Am; World Aquaculture Soc. *Res:* Marine ecology; fisheries; aquaculture. *Mailing Add:* Hawaii Inst Marine Biol PO Box 1346 Kaneohe HI 96744

HELFRICH, WAYNE J(ON), b Milwaukee, Wis, July 14, 33; m 62; c 3. METALLURGY. *Educ:* Univ Wis, BS, 56, MS, 60, PhD(metall), 64. *Prof Exp:* Res engr metall, Allis-Chalmers Mfg Co, 58-60; SECT HEAD, KAISER ALUMINUM & CHEM CORP, 64- *Mem:* Am Soc Metals. *Res:* Alloy and process development in aluminum alloys. *Mailing Add:* 123 Redondo Way Danville CA 94526

HELGASON, SIGURDUR, b Akureyri, Iceland, Sept 30, 27; m 57; c 2. MATHEMATICS. *Educ:* Univ Iceland, BS, 46; Univ Copenhagen, MS, 52; Princeton Univ, PhD, 54. *Hon Degrees:* Dr, Univ Iceland, 86 & Univ Copenhagen, 86. *Prof Exp:* Moore instr math, Mass Inst Technol, 54-56; Louis Block lectr, Univ Chicago, 57-59; from asst prof to assoc prof, 59-65, PROF MATH, MASS INST TECHNOL, 65- *Concurrent Pos:* Vis asst prof, Columbia Univ, 59-60; mem, Inst Advan Study, 64-66, 74-75 & 83-84. *Honors & Awards:* Steele Prize, Am Math Soc, 88. *Mem:* Am Math Soc; Am Acad Arts & Sci; Royal Danish Acad Sci & Lett. *Res:* Geometric analysis on Lie groups and coset spaces. *Mailing Add:* Dept Math Mass Inst Technol Cambridge MA 02139

HELGASON, SIGURDUR BJORN, b Elfros, Sask, Mar 31, 13; m 42; c 2. PLANT BREEDING, CYTOGENETICS. *Educ:* Univ Man, BSA, 39; Univ Minn, St Paul, MSc, 42, PhD(plant genetics), 53. *Prof Exp:* Agr asst corn breeding, Can Dept Agr, 39-47; from asst prof to prof, 47-78, head dept, 77-78, EMER PROF PLANT SCI, UNIV MAN, 78- *Concurrent Pos:* Res consult, Can Hail Underwriters Assoc, 53-57; spec lectr, Macdonald Col, McGill Univ, 55-56; plant breeder, Can Int Develop Agency, Kenya, 71-72; consult, Indust Develop Res Coun, 81-82. *Mem:* AAAS; fel Agr Inst Can; Can Soc Agron; Genetics Soc Can; hon life mem Can Seed Grass Asn. *Res:* Genetics and breeding in barley and maize; cytogenetics of tetraploids and telocentrics in barley; disease resistance in barley and wheat; photoperiod and temperature reactions in maize. *Mailing Add:* 810 Oakenwald Ave Winnipeg MB R3T 1N1 Can

HELGESEN, ROBERT GORDON, b Cleveland, Ohio, Oct 9, 42; m 64. INSECT ECOLOGY. *Educ:* Univ Mich, BS, 65; NDak State Univ, MS, 67; Mich State Univ, PhD(entom), 69. *Prof Exp:* Prof entom, Cornell Univ, 69-80; dept head entom, Kans State Univ, 80-89; DEAN, FOOD & NATURAL RESOURCES, UNIV MASS, 89- *Mem:* AAAS; Entom Soc Am; Sigma Xi. *Res:* Insect ecology; insect pest management. *Mailing Add:* Univ Mass 117 Stockbridge Hall Amherst MA 01003

HELGESON, HAROLD CHARLES, b Minneapolis, Minn, Nov 13, 31; m 56, 78, 86; c 3. ORGANIC & INORGANIC REACTIONS IN GEOPHYSICAL PROCESSES. *Educ:* Mich State Univ, BSc, 53; Harvard Univ, PhD(geol), 62. *Prof Exp:* Geologist, Tech Mine Consult Ltd, Can, 53-54 & Anglo Am Corp, SAfrica, 56-59; consult, BC & Alaska, 59; chemist, Shell Develop Co, Tex, 62-63, res chemist, 63-65; from asst prof to assoc prof geol, Northwestern Univ, 65-70; PROF GEOCHEM, UNIV CALIF, BERKELEY 70- *Concurrent Pos:* Miller res prof, Univ Calif, Berkeley, 74-75; Guggenheim fel, 77-78. *Honors & Awards:* Goldschmidt Medal Distinguished Res, Geochem Soc, 88. *Mem:* Am Chem Soc; Geochem Soc; Geol Soc Am; Mineral Soc Am; Europ Asn Geochem; Int Asn Geochem & Cosmochem. *Res:* High temperature solution chemistry; chemical thermodynamics; phase equilibria; inorganic and organic geochemistry; application of the principles of physical chemistry and mathematics to geologic problems; chemical petrology; chemical kinetics and mass transfer in geochemical processes. *Mailing Add:* Dept Geol & Geophys Univ Calif Berkeley CA 94720

HELGESON, JOHN PAUL, b Barberton, Ohio, July 25, 35; m 57; c 3. PLANT PHYSIOLOGY. *Educ:* Oberlin Col, AB, 57; Univ Wis, PhD(bot), 64. *Prof Exp:* NSF fel org chem, 64-66; from asst prof to assoc prof plant, 66-75, PROF PLANT PATH & BOT, UNIV WIS-MADISON, 75-, RES PLANT PHYSIOLOGIST, PIONEERING RES LAB, AGR RES SERV, 66- *Concurrent Pos:* Prog mgr biol stress, USDA competitive grants, 82-83; vis scientist, Cellular Biol Lab, CNRA, Versailles, France, 85-86. *Mem:* AAAS; Am Phytopath; Am Soc Plant Physiol; Bot Soc Am. *Res:* Growth and development in plants; genetic modification of plants to obtain novel disease resistances; physiology of host-parasite interactions; use of plant tissue cultures for metabolic and host-parasite studies and to obtain modified crop plants. *Mailing Add:* Dept Plant Path Univ Wis Madison WI 53706

HELIN, ELEANOR KAY, b Pasadena, Calif; c 1. PLANETARY SCIENCE, GEOLOGY. *Prof Exp:* Res asst meteorite statist, 60-61 & meteorite anal, 61-68, assoc scientist asteroid res & surv, 69-76, sr scientist asteroid surv & planetary sci, 76-79, mem prof staff, 79-80, MEM TECH STAFF, JET PROPULSION LAB, CALIF INST TECHNOL, 80- *Mem:* Am Astron Soc; Int Astron Union; Meteoritical Soc. *Res:* Systematic surveys for planet-crossing asteroids; origin and history of earth-approaching asteroids; impact history of terrestrial planets; meteoritics. *Mailing Add:* Jet Propulsion Lab Calif Inst Technol 4800 Oak Grove Dr 183-501 Pasadena CA 91109

HELINSKI, DONALD RAYMOND, b Baltimore, Md, July 7, 33; m 62; c 2. BIOCHEMISTRY, GENETICS. *Educ:* Univ Md, BS, 54, MS, 56; Case Western Reserve Univ, PhD(biochem), 60. *Prof Exp:* USPHS fel biochem genetics, Stanford Univ, 60-62; asst prof, Dept Biol, Princeton Univ, 62-65; assoc prof, 65-70, chmn, 79-81, PROF, DEPT BIOL, UNIV CALIF, SAN DIEGO, 70-, DIR, CTR MOLECULAR GENETICS, 84- *Concurrent Pos:* USPHS career develop award, 68-73; mem, Recombinant DNA Adv Comt, NIH, 75-78; Guggenheim fel, 81-82; mem bd dirs, Coun Res Planning Biol Sci, 82-88; mem, Microbiol Physiol & Genetics Study Sect 2, Dept Health & Human Serv, NIH, 83-87; mem adv comt, Frederick Cancer Res Facil, 89-, Nat Cancer Inst-Frederick Cancer Res & Develop Ctr, 90-; actg dir, Systemwide Biotechnol Res & Educ Prog, Univ Calif, 90- *Mem:* Nat Acad Sci; Am Soc Biol Chem; fel AAAS; Genetics Soc Am; Am Soc Microbiol; Am Acad Arts & Sci. *Res:* Plasmids; DNA replication; biological nitrogen fixation. *Mailing Add:* Dept Biol Univ Calif 422 Bonner Hall La Jolla CA 92093

HELKE, CINDA JANE, b Waterloo, Iowa, Feb 27, 51; m 74. NEUROPHARMACOLOGY, AUTONOMIC REGULATION. *Educ:* Creighton Univ, BS, 74; Georgetown Univ, PhD(pharmacol), 78. *Prof Exp:* Staff fel neurosci, NIMH, NIH, 78-80; from asst prof, to assoc prof, 80-88, PROF, UNIFORMED SERV UNIV HEALTH SCI, 88- *Concurrent Pos:* Neurol BII Study Sect, NIH, 87-91. *Honors & Awards:* Deane N Calvert Award, Am Soc Pharmacol & Exp Therapeut, 78. *Mem:* Soc Neurosci; Asn Women Sci; AAAS; Am Soc Pharmacol & Exp Therapeut. *Res:* Using cellular and system neuropharmacologic and neuroanatomic approaches, aspects of chemical neurotransmission (transmitter localization and coexistence, plasticity in the adult animal, et cetera) in autonomic control systems. *Mailing Add:* Dept Pharmacol Uniformed Serv Univ Health Sci Bethesda MD 20814-4799

HELLACK, JENNA JO, b Ardmore, Okla, Apr 6, 45. ZOOLOGY. *Educ:* ECent Univ, BS, 68; Okla State Univ, MS, 69, PhD(zool), 75. *Prof Exp:* Vis prof zool, Univ Okla, 75-76; PROF BIOL, CENT STATE UNIV, 77- *Mem:* Genetics Soc Am. *Res:* Evolutionary genetics; Drosophila population genetics. *Mailing Add:* Dept Biol Cent State Univ 1100 N University Dr Edmond OK 73034

HELLE, JOHN HAROLD (JACK), b Williston, NDak, Apr 26, 35; m 59; c 2. FISH BIOLOGY. *Educ:* Univ Idaho, BS, 58, MS, 61; Oregon State Univ, PhD(fisheries sci), 79. *Prof Exp:* Res biologist, Bur Com Fish, US Fish & Wildlife Serv, 60-70; RES BIOLOGIST, US DEPT COM, NAT MARINE FISHERIES SERV, NAT OCEANIC & ATMOSPHERIC ADMIN, 70- *Concurrent Pos:* Hon fel, Dept Natural Hist, Univ Aberdeen, Scotland, 64; pres, Am Inst Fishery Res Biologists, 90-92. *Mem:* Am Inst Fishery Res Biologists; Am Fisheries Soc; AAAS. *Res:* Influence of environment and heredity on age and size at maturity; growth and survival of pacific salmon; genetic considerations in hatchery and wild populations; stock identification. *Mailing Add:* Auke Bay Lab PO Box 210155 Auke Bay AK 99821

HELLEBUST, JOHAN ARNVID, b Tysfjord, Norway, Mar 7, 33; Can citizen; m 58; c 3. ALGOLOGY. *Educ:* Univ Toronto, BA, 58, MA, 59, PhD(plant physiol), 62. *Prof Exp:* Ford Found fel marine biol, Woods Hole Oceanog Inst, 62-63, asst scientist, 63-64; asst prof, Harvard Univ, 64-69; assoc prof, 69-73, PROF BOT, UNIV TORONTO, 73-, CHMN, 87- *Concurrent Pos:* Ed, J Phycol, 84-87. *Mem:* Am Soc Plant Physiol; Can Soc Plant Physiol; Phycol Soc Am. *Res:* Physiology, biochemistry and ecology of algae; heterotrophy, transport systems and osmoregulation. *Mailing Add:* Dept Bot Univ Toronto Toronto ON M5S 3B2 Can

HELLEINER, CHRISTOPHER WALTER, b Vienna, Austria, Mar 21, 30; Can citizen; m 55; c 2. BIOCHEMISTRY. *Educ:* Univ Toronto, BA, 52, PhD(biochem), 55. *Prof Exp:* Nat Res Coun Can fel, Univ Oxford, 55-57; asst prof med biophys, Univ Toronto, 57-63, from asst prof to assoc prof, 63-65, head dept, 65-79, PROF BIOCHEM, DALHOUSIE UNIV, 65- *Concurrent Pos:* Asst scientist, Ont Cancer Inst, Toronto, 57-63; Med Res Coun Can vis scientist award, Univ Glasgow, 69-70. *Mem:* Can Biochem Soc. *Res:* DNA chemistry and repair. *Mailing Add:* Dept Biochem Dalhousie Univ Halifax NS B3H 4H7 Can

HELLEMS, HARPER KEITH, b Sinks Grove, WVa, Mar 16, 20; m 73; c 3. CARDIOLOGY. *Educ:* Univ Va, MD, 43; Am Bd Internal Med, dipl, 52. *Prof Exp:* Intern, Montreal Gen Hosp, 44-45; asst med, Peter Bent Brigham Hosp, 46-48, sr asst resident, 49-50; from asst prof to prof, Col Med, Wayne State Univ, 50-60; prof med & dir div cardiovasc dis, NJ Col Med & Dent, 60-65; PROF MED & CHMN DEPT, SCH MED, UNIV MISS, 65- *Concurrent Pos:* Res fel med, Harvard Med Sch, 46-48, teaching fel, 49-50; asst resident, West Roxbury Vet Admin Hosp, 48-49; attend physician & dir med outpatient dept, City Detroit Receiving Hosp, 50-; consult, Dearborn Vet Admin Hosp, Vet Admin Regional Off, Detroit, 50- & Grace Hosp, 51; mem courtesy staff, Harper Hosp, 55, dir cardiovasc serv, Children's Hosp, Mich, 58; dir, White Cardiopulmonary Inst & attend physician, Berthold S Pollak Hosp Chest Dis, 60-65; mem cardiovasc study sect, NIH, 62-66; physician-in-chief, Med Ctr, Univ Miss, 65-; fel coun clin cardiol, Am Heart Asn. *Mem:* Am Soc Clin Invest; Cent Soc Clin Res; Am Asn Physicians; Am Clin & Climatol Soc; fel Am Col Cardiol. *Res:* Cardiovascular diseases, especially hemodynamics. *Mailing Add:* Dept Med Univ Miss Sch Med Jackson MS 39216

HELLENTHAL, RONALD ALLEN, b Tucson, Ariz, June 9, 45; m 76. SYSTEMATICS, AQUATIC ECOLOGY. *Educ:* Los Angeles Valley Col, AA, 65; Calif State Univ, Northridge, BA, 67; Univ Minn, PhD(entom), 77. *Prof Exp:* Vis asst prof, 77-78, asst prof, 78-84, ASSOC PROF BIOL, DEPT BIOL SCI, UNIV NOTRE DAME, 84-, GILLEN DIR, ENVIRON RES CTR, 88- *Concurrent Pos:* Environ adv comt, Northern States Power Co, 70-71; higher educ comt, Minn Educ Comput Consortium, 75-77; res assoc, Dept Entom, Fish & Wildlife, Univ Minn, St Paul, 78-80; dir, Biocomput Facil, Univ Notre Dame, 80-; chmn, Finance Comt, NAm Benthological Soc, 85-; mem, Water Qual Standards Adv Comt, State Ind, 86- & Steering Comt, Entom Collections Network, 89- *Mem:* Sigma Xi; Entom Soc Am; NAm Benthological Soc; Am Soc Limnol & Oceanog; Soc Syst Zool; Water Pollution Control Fedn. *Res:* Host-parasite relationships; systematics and ecology of aquatic insects; assessment of environmental data; computerized systems for the automation of taxonomic and environmental data analyses. *Mailing Add:* Dept Biol Sci Univ Notre Dame Notre Dame IN 46556

HELLER, ABRAHAM, b Calremont, NH, Mar 17, 17; m 57; c 1. MENTAL HEALTH, FORENSIC PSYCHIATRY. *Educ:* Brandeis Univ, BA, 53; Boston Univ, MD, 57. *Prof Exp:* Intern, Gen Rose Mem Hosp, Denver, 57-58; resident psychiat, Med Ctr, Univ Colo, 58-61; from asst to assoc dir & chief in patient serv, Community Ment Health, Psychiat Serv, Denver Gen Hosp, 61-73; chief psychiat & dir, Community Ment Health, Newport Hosp, RI, 73-77; prof, 77-90, vchmn, 80-90, EMER PROF, DEPT PSYCHIAT & COMMUNITY MED, SCH MED, WRIGHT STATE UNIV, 90- *Concurrent Pos:* Assoc clin prof, Sch Med, Brown Univ, 73-77. *Mem:* Fel Am Psychiat Asn; fel Am Orthopsychiat Asn; fel Am Asn Social Psychiat. *Res:* Forensic psychiatry studies on the right to treatment of forensic patients in psychiatric hospitals; impact of unemployment on mental and physical health. *Mailing Add:* Dept Psychiat Sch Med Wright State Univ Dayton OH 45401

HELLER, ADAM, b Cluj, Roumania, June 25, 33; m 56; c 2. CHEMISTRY. *Educ:* PhD, Uppsala Univ, Sweden, 91. *Prof Exp:* Mem staff, Weizman Inst, 55-56; French govt scholar, Univ Paris, 57-58; sr scientist, Res Ctr, Israel Atomic Energy Comn, 59-62; postdoc assoc, Univ Calif, Berkeley, 62-63; mem tech staff, Bell Labs, 63-64; res mgr explor res, GTE Labs, 64-75; mem tech staff, AT&T Bell Labs, 75-77, head, Electronics Mat Res Dept, 77-88; ERNEST COCKRELL SR CHAIR & PROF CHEM ENG, CHEM & MAT SCI, UNIV TEX, AUSTIN, 88- *Concurrent Pos:* Mem rev comt, Chem & Mat Res Div, Lawrence Berkeley Labs; mem adv bd, Solar Energy Res Inst; Case Centennial scholar, 80; guest prof, Col France, 82; div ed, J Electrochem Soc; lectr, Univ Calif, Berkeley, 91. *Honors & Awards:* Battery Res Award, Electrochem Soc, 78, David C Grahame Award, 87; Regents Lectr, Univ, 84; Robert A Welch Lectr, 86; Raymond & Beverly Sackler Distinguished Lectr, Tel Aviv Univ, 87; Vittorio de Nora Award, 88. *Mem:* Nat Acad Eng; Am Chem Soc; Am Phys Soc; Int Soc Electrochem; Electrochem Soc; AAAS. *Res:* Transparent Metals; hydrogen-evolving solar cells; establishment of direct electrical communication betweenredox enzymes and metal electrodes, through chemical modification of the enzymes; electrical micro-engineering of enzymes; author of 1,750 publications in the field of electrochemistry of inorganic oxyhalides. *Mailing Add:* Dept Chem Eng Univ Tex Austin TX 78712-1062

HELLER, AGNES S, b Hungary; US citizen; c 2. ENGINEERING, STATISTICS. *Educ:* City Col NY, BBA, 54; Columbia Univ, MA, 63. *Prof Exp:* Res assoc eng, Columbia Univ, 57-67; instr math statist, Va Polytech Inst & State Univ, 67-74; prin engr statist & probabilistic methods, Babcock & Wilcox Co, 75-85; SR ANALYST, EVAL RES CORP, 85- *Concurrent Pos:* Comt Pressure Vessel Res Coun, 78. *Mem:* Am Nuclear Soc; Sigma Xi. *Res:* Probabilistic methods and statistical analyses in energy related systems. *Mailing Add:* 3951 Bosworth Dr Roanoke VA 24014

HELLER, ALEX, b New York, NY, July 9, 25; m 56; c 1. MATHEMATICS. *Educ:* Columbia Univ, AB, 47, PhD(math), 50. *Prof Exp:* Lectr math, Columbia Univ, 47-48, Nat Res Coun fel, 50-51; Benjamin Pierce instr, Harvard Univ, 52-54; from asst prof to prof, Univ Ill, 54-65; prof, 65-71, DISTINGUISHED PROF MATH, GRAD SCH, CITY UNIV NEW YORK, 71- *Concurrent Pos:* Sloan Found fel, Univ Ill, 55-59; mem, Div Math Sci, Nat Acad Sci-Nat Res Coun, 65-67; managing ed, J Pure & Appl Algebra, 71-; assoc prof, Univ Paris VII, 79; vis prof, Milan, Italy, 84. *Mem:* Am Math Soc. *Res:* Algebraic topology; homological algebra; category theory; recursion theory. *Mailing Add:* City Univ New York Grad Sch 33 W 42nd St New York NY 10036-8099

HELLER, ALFRED, b Chicago, Ill, July 23, 30; m 56. PHARMACOLOGY. *Educ:* Univ Ill, BS, 52; Univ Chicago, PhD(pharmacol), 56, MD, 60. *Prof Exp:* Res assoc, 56-60, from asst prof to assoc prof, 60-73, chmn dept, 73-88, PROF PHARMACOL & PHYSIOL SCI, UNIV CHICAGO, 73- *Concurrent Pos:* Mem, Nat Bd Med Examiners Pharmacol; nat adv med sci coun, NIH, 79- *Mem:* AAAS; Am Soc Pharmacol & Exp Therapeut; Am Soc Neurochem. *Res:* Neuropharmacology; neurochemical consequences of central nervous system lesions. *Mailing Add:* Dept Pharmacol & Physiol Sci Univ Chicago 947 E 58th St Chicago IL 60637

HELLER, BARBARA RUTH, b Milwaukee, Wis, May 15, 31; m 56; c 1. ESTIMATION & STATISTICAL TESTS, STEREOLOGY. *Educ:* Roosevelt Univ, BS, 53; Univ Chicago, MS, 65, PhD(statist), 79. *Prof Exp:* ASSOC PROF STATIST, PROBABILITY & MATH, ILL INST TECHNOL, 80- *Mem:* Inst Math Statist; Am Statist Asn; Soc Indust & Appl Math; Am Math Soc; NY Acad Sci; Sigma Xi. *Res:* Characterization of probability distributions, goodness-of-fit test for the negative binomial distribution; use of the computer algebra program Macsyma for problems in statistics; symbolic manipulation programs on the computer; problems in stereology, cell counting and measuring. *Mailing Add:* Dept Math Ill Inst Technol Chicago IL 60616

HELLER, CHARLES O(TA), b Prague, Czech, Jan 25, 36; US citizen; m 59; c 1. STRUCTURAL MECHANICS, COMPUTER SCIENCE. *Educ:* Okla State Univ, BS, 59, MS, 60; Cath Univ Am, PhD(struct mech), 68. *Prof Exp:* Engr, Missiles & Space Systs Div, Douglas Aircraft Co, 60-62; mem tech staff, Bell Tel Labs, 62-63; assoc prof aerospace eng, US Naval Acad, 63-68; vpres, Bay Tech Assocs, Inc, 68-69; pres, Cadcom, Inc, 69-79, exec dir, Cadcom Div, 79-81, vpres corp develop, Mantech Int Corp, 82-83; pres, Intercad Corp, 84-86; dir, Md Indust Partnerships, 86-89, DIR, DINGMAN CTR ENTREPRENEURSHIP, UNIV MD, 89- *Concurrent Pos:* Lectr, Univ Calif, Los Angeles, 61-62; consult, US Naval Ship Res & Develop Lab, 64-66, Lockheed Electronics Co, 66-67, Hydrodyn Develop Corp, 66-68 & Head Ski Co, 67-68; partic & lectr, Nat Acad Sci Exchange Prog, 70. *Mem:* Am Soc Civil Engrs; Asn Comput Mach; Am Inst Aeronaut & Astronaut; Am Mgt Asn. *Res:* Computer-aided design of aerospace and marine vehicles; structural analysis of plates and shells; finite elements techniques; use of interactive time-shared computing in engineering education; computer graphics; applications in engineering design; entrepreneurship. *Mailing Add:* 1211 Hillcrest Rd Arnold MD 21012

HELLER, DOUGLAS MAX, physics; deceased, see previous edition for last biography

HELLER, EDWARD LINCOLN, b Newark, NJ, Feb 12, 12; m 46; c 2. MECHANICAL ENGINEERING, MINING ENGINEERING. *Educ:* Lehigh Univ, BS, 34; Harvard Univ, MBA, 39. *Prof Exp:* Mining engr, Empire Zinc Co, 34-38; transp eng, Standard Oil Co, 39-41; chief test sect, QM Res & Develop, US Army, 41-42, chief qual control, QM Inspection Serv, 43-46; chief engr, McLaughlin Carr Assoc, 46-47; staff engr nuclear, Joint Cong Comt Atomic Energy, US Cong, 47-55; chief engr nuclear proj, H K Ferguson Div Morrison Knudsen, 56-59; regional mgr new prod, Gen Atomic Co, 59-75; consult, energy, Res & Develop Div Gulf Oil Corp, 75-78; ASST INSPECTOR GEN, DEPT ENERGY, US GOVT, 78- *Concurrent Pos:* Mem, Ohio State Adv Bd Atomic Energy, 57-58, Atomic Energy Panel; Soc Naval Archit & Marine Eng, 64-65. *Mem:* Am Soc Mech Eng; Am Nuclear Soc; Marine Tech Soc. *Res:* Uranium ore deposits, process heat reactors, synthetic fuels from coal and oil shale; fusion reactors for electric power production, solar heat and power production, geothermal heat sources. *Mailing Add:* 3517 Hamlet Pl Chevy Chase MD 20815

HELLER, ERIC JOHNSON, b Washington, DC Jan 10, 46; m 79; c 3. CHEMICAL PHYSICS. *Educ:* Univ Minn, Minneapolis, BA, 68; Harvard Univ, PhD(chem physics), 73. *Prof Exp:* Res assoc chem physics, James Franck Inst, Univ Chicago, 73-75; from asst prof to prof chem, Univ Calif, Los Angeles, 75-84; PROF PHYSICS, DEPT CHEM, UNIV WASH, 84- *Concurrent Pos:* Staff scientist, Los Alamos Nat Lab. *Mem:* AAAS; fel Am Phys Soc; Am Chem Soc. *Res:* Semiclassical theory of quantum mechanics with applications to molecules; intramolecular dynamics and spectroscopy; theoretical molecular spectroscopy; radiationless transitions; unimolecular decay; quantum mechanics of chaotic systems. *Mailing Add:* Dept Chem BG-10 Univ Wash Seattle WA 98195

HELLER, GERALD S, b Detroit, Mich, Sept 5, 20; m 43; c 2. SOLID STATE PHYSICS. *Educ:* Wayne Univ, ScB, 42; Brown Univ, ScM, 46, PhD(physics), 48. *Prof Exp:* Mem staff, Radiation Lab, Mass Inst Technol, 42-45; asst prof physics, Brown Univ, 48-54; group leader resonance physics, Lincoln Lab, Mass Inst Technol, 54-62, vis prof elec eng, Inst, 62-63; PROF ENG, BROWN UNIV, 63-, DIR MAT RES LAB, 68- *Concurrent Pos:* Mem staff, Radiophys Lab, Sydney Univ, 44-45. *Mem:* Am Phys Soc; fel Inst Elec & Electronics Engrs. *Res:* Ferro-ferrimagnetism; ferrite devices; millimeter spectroscopy and components; electromagnetic theory; material science; antiferromagnetism; far infrared spectroscopy. *Mailing Add:* Box M Brown Univ Providence RI 02912

HELLER, HANAN CHONON, b Lida, Poland, Aug 3, 30; US citizen; m 54; c 2. PHYSICAL CHEMISTRY. *Educ:* NY Univ, BA, 52, PhD(phys chem), 57. *Prof Exp:* Asst chem, NY Univ, 54-56; chemist, Tech Res Group, Inc, 56-57; instr chem, NY Univ, 58; A A Noyes res fel chem, Calif Inst Technol, 58-59; res chemist, 59-65, STAFF SCIENTIST, SCI LAB, FORD MOTOR CO, 65- *Mem:* Am Phys Soc; Am Chem Soc. *Res:* Electron spin resonance; molecular and free radical structure; photochemistry; radiation chemistry; supply and recycling of materials; environmental pollution; solar energy conversion; modeling of IC fabrication. *Mailing Add:* 29450 Leemoor Dr Southfield MI 48076-1693

HELLER, HORACE CRAIG, b Philadelphia, Pa, Aug 5, 43; m 70; c 1. BIOLOGY, PHYSIOLOGY. *Educ:* Ursinus Col, BS, 65; Yale Univ, MPhil, 67, PhD(biol), 70. *Prof Exp:* Fel physiol, Scripps Inst Oceanog, 70-72; asst prof, 72-77, ASSOC PROF BIOL, STANFORD UNIV, 77- *Mem:* Am Physiol Soc; Am Soc Naturalists; Am Zool Soc. *Res:* Thermoregulatory physiology; hibernation; sleep; environmental physiology. *Mailing Add:* Dept Biol Sci Stanford Univ Stanford CA 94305

HELLER, IRVING HENRY, b Montreal, PQ, Mar 26, 26; m 48; c 2. MEDICINE, NEUROLOGY. *Educ:* McGill Univ, BSc, 46, MD, CM, 50, MSc, 54, PhD(neurochem), 62. *Prof Exp:* Asst prof neurol, 60-72, assoc prof, 72-81, PROF NEUROL & NEUROSURG, MCGILL UNIV, 81- *Concurrent Pos:* Neurologist, Montreal Neurol Hosp; sr neurologist, Royal Victoria Hosp; consult neurologist, Douglas Hosp & Queen Elizabeth Hosp. *Mem:* Am Acad Neurol; Can Neurol Soc. *Res:* Migraine. *Mailing Add:* Dept Neurol-Neurosurg Montreal Neurol Hosp 3801 University St Montreal PQ H3A 2B4 Can

HELLER, JACK, b US, Sept 11, 22; m 46; c 4. COMPUTER SCIENCE. *Educ:* Polytech Inst Brooklyn, BAeE, 44, MAeE, 46, PhD(physics), 50. *Prof Exp:* Asst, Polytech Inst Brooklyn, 44-46; stress analyst, Repub Aviation Corp, 46; instr math, Newark Col Eng, 46-50; asst, Princeton Univ, 52; res scientist, Inst Math Sci, NY Univ, 53-61, assoc prof math, 61-64, prof math, dir Heights Acad Comput Facil & dir Inst Comput Res in Humanities, 65-69; PROF COMPUT SCI, STATE UNIV NY, STONY BROOK, 69- *Mem:* Asn Comput Mach. *Res:* Applied mathematics; information retrieval; data base management systems; software systems. *Mailing Add:* Dept Comput Sci State Univ NY Stony Brook NY 11794

HELLER, JOHN HERBERT, b New York, NY, Nov 28, 21; m 46, 68; c 8. MEDICAL PHYSICS, ANTHROPOLOGY. *Educ:* Yale Univ, BA, 42; Western Reserve Univ, MD, 45. *Prof Exp:* Resident internal med, Cornell Univ & Bellevue Med Ctr, 47; instr path, Yale Univ, 47-48, asst prof med, 51-52, chmn, Sect Med Physics, 52-54; exec dir, New Eng Inst Med Res, 54-66, pres, 66-76, prof med, Grad Sch, 66-81, lectr interdisciplinary studies, 69-81, chmn bd, 76-81; CONSULT MED MALPRACTICE & PROD LIABILITY, 81- *Concurrent Pos:* Brown res fel, Yale Univ, 48, Am Heart Asn res fel, 49-51; fel, Oceanic Inst Alliance, 69-; mem biophys div, Yale Univ, 49-54, proj dir, AEC, 50-54, asst prof med & physiol, 51, mem isotopes comt, 51-54, radiophys health officer, 50-54; asst physician, Yale-New Haven, Commun Hosp, 50-51, assoc physician, 51-54; estab investr, Am Heart Asn, 51-54; lectr, Univ Conn, 55-56; trustee, Found Instrumentation Educ & Res, 56-58; res assoc, Oceanog Inst, Woods Hole, 57-58 & Cape Haze Marine Lab, 57-64; lectr, US Mil Acad, 60-64; mem, Panel Atmospheric Sci, Space Sci Bd, Nat Acad Sci-Nat Res Coun, 61 & 62; Wilhelmina Key lectr, 69; chmn, Yale-New Eng Inst Task Force Biol Parameters Human Behav, 72-73; lectr biobehav, Fairfield Univ, 73; consult environmentalist, JHH Assocs, Inc, 74-; consult, var industs, govt agencies, acad insts & orgns; prof interdisciplinary studies, New Eng Inst, 75; fac med malpractice & prod liability, Am Bar Assoc, Continuing Legal Educ, 85; co-responsible investigator, long range develop & clin trials, FDA, Phase I & II, Yale-New England Inst, 73-81; res scientist, Shroud Turin Res Proj, 78-85; adj prof anthropology, Univ Maine, Orono, 87-; consult, Asia Found, 90- *Honors & Awards:* Int Gold Medal, Int Soc Res Reticuloendothelial Systs, 58; Silver Medal, Pasteur Inst. *Mem:* Biophys Soc; Reticuloendothelial Soc; NY Acad Sci; Bioelectromagnetic Soc; Royal Soc Health. *Res:* Biophysics; biogeophysics; host defense mechanism and clinical studies of host defense stimulants; human behavior; Pleistocene extinction of megafauna. *Mailing Add:* New Eng Inst 74 Horseshoe Rd Wilton CT 06897

HELLER, JOHN PHILIP, b Brooklyn, NY, Apr 15, 23; m 46; c 3. PHYSICS. *Educ:* Queens Col, NY, BS, 44; Iowa State Univ, PhD(physics), 53. *Prof Exp:* Asst microwaves, Columbia Radiation Lab, 44; instr physics, Iowa State Univ, 50-53; sr res physicist, Socony-Mobil Oil Co, 53-61, res assoc, Field Res Lab, Mobil Res & Develop Corp, 61-79; SR SCIENTIST, PETROL RECOVERY RES CTR, NMEX TECH, 79- *Honors & Awards:* Enhanced Oil Recovery Pioneer, Soc Petrol Engrs. *Mem:* Am Phys Soc; Sigma Xi; Soc Petrol Engrs; AAAS; Am Chem Soc; Am Inst Chem Engrs. *Res:* Spectroscopy; optics; low speed hydrodynamics and flow in porous media; instrumentation; dispersion and drying in porous media; computer control; petroleum production processes, especially enhanced oil recovery by carbon dioxide flooding; effects of both viscous and gravitationally; induced frontal instabilities, and of permeability variability on displacement patterns in oil reservoirs and aquifers. *Mailing Add:* Petrol Recovery Res Ctr NMex Tech Socorro NM 87801

HELLER, JOHN RODERICK, medicine; deceased, see previous edition for last biography

HELLER, JORGE, b Liberec, Czech, Aug 17, 27; m 55, 78; c 2. ORGANIC CHEMISTRY. *Educ:* Univ Calif, BS, 52; Univ Wash, PhD(org chem), 57. *Prof Exp:* Res chemist, Union Carbide Plastics Co, 57-59; sr polymer chemist, 59-67, mgr polymer chem dept, 67-70, dir phys sci, Alza Res, 70-74, DIR POLYMER SCI, SRI INT, 74- *Concurrent Pos:* Adj prof pharm, Univ Calif, San Francisco, 84- *Mem:* Biomats Soc; Am Chem Soc; Sigma Xi; Controlled Release Soc. *Res:* Synthesis of monomers and polymers; biomedical application of polymers; controlled drug release. *Mailing Add:* SRI Int 333 Ravenswood Menlo Park CA 94025-3446

HELLER, KENNETH JEFFREY, b Port of Spain, Trinidad, Nov 7, 43; m 72. EXPERIMENTAL HIGH ENERGY PHYSICS. *Educ:* Univ Calif, BA, 65; Univ Wash, PhD(physics), 73. *Prof Exp:* Res assoc & instr, Univ Mich, 72-78; from asst prof to assoc prof, 78-86, PROF PHYSICS, UNIV MINN, 86- *Concurrent Pos:* Fel, Univ Mich, 72-75; mem bd trustees, Univs Res Asn, 85-88; mem Univs Res Asn bd overseers, Fermilab, 88-; prin investr, HEP, Univ Minn, 89- *Mem:* Am Phys Soc. *Res:* High energy physics experiments; quark dynamics from strong interactions of hadrons, quark confinement from magnetic moments of baryons and their weak decay properties; muons from high energy interactions. *Mailing Add:* Sch Physics & Astron Univ Minn Minneapolis MN 55455

HELLER, LEON, b New York, NY, Dec 16, 29; m 52; c 3. NUCLEAR PHYSICS. *Educ:* Brooklyn Col, BA, 51; Cornell Univ, PhD(physics), 57. *Prof Exp:* Group leader, Medium Energy Physics Theory Group, 71-73, STAFF MEM, LOS ALAMOS NAT LAB, 56- *Concurrent Pos:* Vis asst prof, Case Inst Technol, 62-63; exchange scientist, AEC-UK Atomic Energy Authority, Harwell, 69-70; staff mem ctr theoret physics, Mass Inst Technol, 76-77; Fulbright sr scholar, Univ Adelaide, 85-86. *Mem:* Fel Am Phys Soc. *Res:* Quark structure of hadrons; magnetoencephalography. *Mailing Add:* MS B283 Los Alamos Nat Lab Los Alamos NM 87545

HELLER, LOIS JANE, b Detroit, Mich, Jan 4, 42; m 66; c 2. MEDICAL PHYSIOLOGY. *Educ:* Albion Col, BA, 64; Univ Mich, Ann Arbor, MS, 66; Univ Ill Med Ctr, Chicago, PhD(physiol), 70. *Prof Exp:* Asst prof physiol, Univ Ill Med Ctr, Chicago, 70-71; asst dean student affairs, 73-76, assoc prof, 72-89, PROF PHYSIOL, SCH MED, UNIV MINN, DULUTH, 89- *Mem:* Am Physiol Soc; Sigma Xi; Am Heart Asn; Soc Exp Biol Med. *Res:* Contractile properties and excitation processes in cardiac muscle and coronary vasculature as influenced by aging hypertension and immediate hypersensitivity reactions. *Mailing Add:* Univ Minn Sch Med Med Sci Bldg Duluth MN 55812

HELLER, MARVIN W, b Grundy Center, Iowa, Oct 4, 26; m 53; c 3. PHYSICS. *Educ:* Grinnell Col, BA, 51; Iowa State Univ, PhD(physics), 60. *Prof Exp:* Res assoc solid state physics, Iowa State Univ, 60-61; asst prof, 61-68, ASSOC PROF PHYSICS, COLO STATE UNIV, 68- *Concurrent Pos:* Consult, Naval Res Lab, 69-71. *Mem:* Am Asn Physics Teachers. *Res:* Transport properties of semiconductors. *Mailing Add:* Dept Physics Colo State Univ Ft Collins CO 80523

HELLER, MELVIN S, b Boston, Mass, Aug 11, 22; m 54; c 3. MEDICINE, PSYCHIATRY. *Educ:* Tufts Univ, BS, 44, MS & MD, 48. *Prof Exp:* Intern surg, Beth Israel Hosp, 48-49; resident, Vet Admin Hosp, Newington, Conn, 49-50, resident psychiat, 50-51; staff psychiatrist, Fed Penitentiary, Ind, 51-53; resident psychiat, Sch Med, Yale Univ, 53-54; consult, Fed Bur Prisons, 58-59; assoc prof, 63-67, CLIN PROF PSYCHIAT, SCH MED, TEMPLE UNIV, 67-, DIR DIV FORENSIC PSYCHIAT, 65- *Concurrent Pos:* Fel, Child Study Ctr, Philadelphia, 54-56; lectr, Temple Univ, 57-, co-dir, Unit Law & Psychiat, 58-; dir psychiat, State Correctional Inst Philadelphia, 59-70; dir psychiat serv, Children's Aid Soc, 62-70; mem, Govr's Comn Comprehensive Ment Health Planning, Pa; mem, Nat Coun Juvenile Court Judges. *Mem:* Am Psychiat Asn; AMA. *Res:* Forensic psychiatry. *Mailing Add:* Temple Univ Hosp Two Schiller Ave Narberth PA 19072

HELLER, MILTON DAVID, b Newark, NJ, May 18, 21; m 46; c 4. MEDICAL RESEARCH. *Educ:* Univ Mich, BS, 42, MS, 47, PhD(chem), 52. *Prof Exp:* Chemist, E I du Pont de Nemours & Co, 42; instr, Univ Mich, 50-51; chemist, Am Cyanamid Co, 51-74, ed & analyst, Lederle Labs Div, 74-77, clin study analyst, Med Res Div, 71-88, clin res assoc, 81-87; RETIRED. *Mem:* Am Chem Soc; Am Inst Chemists. *Res:* Natural product synthesis; steroid synthesis; pulmonary and allergy research; oncology research. *Mailing Add:* Seven Highview Ave New City NY 10956-2856

HELLER, MORGAN SILLIMAN, organic chemistry; deceased, see previous edition for last biography

HELLER, PAUL, b Komotau, Czech, Aug 8, 14; nat US; m 46, 88; c 2. INTERNAL MEDICINE, HEMATOLOGY. *Educ:* Ger Univ, Prague, MD, 38; Am Bd Internal Med, dipl. *Prof Exp:* Demonstr biochem, Biochem Inst, Ger Univ, Prague, 35-37; resident pulmonary dis, Montefiore Hosp, 46-47; intern, Beth Israel Hosp, 47-48; clin instr med, Sch Med, George Washington Univ, 50-51; assoc, Col Med, Univ Nebr, 52-54; clin asst prof, 54-60, assoc prof, 60-63, chief hemat sect, Dept Med, 67-79, PROF MED, COL MED, UNIV ILL MED CTR, 63-, CONSULT HEMAT, 78-; PHYSICIAN & HEMATOLOGIST, VET ADMIN WESTSIDE HOSP, 54-, SR MED INVESTR, 69- *Concurrent Pos:* Physician, Group Health Asn, 48-51 & Vet Admin, Nebr, 52-54; chief med serv, Vet Admin WestSide Hosp, 67-69; mem hemat study sect, NIH, 67-71; Ann Langer Award, Cancer Res, 80. *Honors & Awards:* Middleton Award, 75. *Mem:* Am Soc Hemat; Am Fedn Clin Res; fel Am Col Physicians; Asn Am Immunologists; Asn Am Physicians. *Res:* Hemoglobinopathies; molecular biology of fetal hemoglobin. *Mailing Add:* 1522 Dobson St Evanston IL 60202

HELLER, PAUL R, b Wooster, Ohio, May 11, 48. TURFGRASS, ORNAMENTAL HORTICULTURE. *Educ:* Malone Col, BA, 70; Ohio State Univ, MS, 72, PhD(entom), 76. *Prof Exp:* Res assoc entom, Dept Entom, Ohio Agr Res & Dev Ctr, 70-76; supvr pest mgt, Pest Mgt Consults, Inc, 76; asst prof, 76-86, ASSOC PROF ENTOM EXTEN, DEPT ENTOM, PA STATE UNIV, 76- *Concurrent Pos:* Consult, E C Geiger Supply Co, 78 & 81 & J T Baker Chem Co, 79-80; lectr, Pa Turfgrass Conf, 78-, Univ Md Turfgrass Conf, Mass Hort Cong, Univ Mass Turfgrass Conf, Pa State Christmas Tree Growers Asn & Dept Plant Sci, Univ Mass Turfgrass Winter Courge. *Mem:* Entom Soc Am; Am Registry Prof Entomologists. *Res:* Effect of nitrogen levels on piercing-sucking insects associated with turfgrass; effect of turfgrass fungicides on the latter pests; evaluation of entomgeneous nematodes to control turfgrass arthropods; evaluation of pestcides to control turfgrass arthropods. *Mailing Add:* Dept Entom 106 Patterson Bldg Pa State Univ University Park PA 16802

HELLER, RALPH, b Berlin, Ger, Feb 28, 14; nat US. GENERAL PHYSICS. *Educ:* Yale Univ, PhD(physics), 40. *Prof Exp:* From instr to prof physics, Worcester Polytech Inst, 41-79; RETIRED. *Mem:* Am Phys Soc; Sigma Xi. *Mailing Add:* Dept Physics Worcester Polytech Inst 100 Institute Rd Worcester MA 01609

HELLER, ROBERT, JR, b San Angelo, Tex, June 13, 31; m 67; c 2. MATHEMATICS. *Educ:* Univ Houston, BA, 50; Univ Tex, MA, 54, PhD(math), 58. *Prof Exp:* Instr math, Univ Tex, 53-58; asst prof, Univ Houston, 58-64; ASSOC PROF, MISS STATE UNIV, 64- *Mem:* Am Math Soc; Math Asn Am. *Res:* Analysis; continued fractions. *Mailing Add:* Dept Math Miss State Univ Mississippi State MS 39762

HELLER, ROBERT A, b Budapest, Hungary, Feb 12, 28; nat US; m 54; c 2. ENGINEERING MECHANICS, CIVIL ENGINEERING. *Educ:* Columbia Univ, BS, 51, MS, 53, PhD(eng mech), 58. *Prof Exp:* Asst, Columbia Univ, 51-57, from instr to assoc prof civil eng, 57-65, assoc dir, Inst Study of Fatigue & Reliability, 65-67; PROF ENG SCI & MECH, VA POLYTECH INST & STATE UNIV, 67- *Concurrent Pos:* Consult, USAF, 72- & US Army, 75- *Honors & Awards:* Western Elect Award, Am Soc Engr Educ. *Mem:* Assoc fel Am Inst Aeronaut & Astronaut; Soc Exp Stress Anal; Am Acad Mech; Am Soc Civil Engr. *Res:* Inelastic behavior of engineering materials and structures; fatigue and reliability; educational models and motion pictures; probabilistic mechanics. *Mailing Add:* Dept Eng Sci & Mech Va Polytech Inst & State Univ Blacksburg VA 24061-0219

HELLER, ROBERT LEO, b Dubuque, Iowa, Apr 10, 19; m 46; c 3. STRATIGRAPHY, PALEONTOLOGY. *Educ:* Iowa State Col, BS, 42; Univ Mo, MS, 43, PhD(statig & paleont), 50. *Prof Exp:* Geologist, US Geol Surv, 43-44; from asst prof to prof, 50-60, head dept, 52-69, dir earth sci curric proj, 62-65, from asst provost to assoc provost, 65-76, provost, 77-85, chancellor, 85-87, PROF GEOL, UNIV MINN, DULUTH, 60-, EMER CHANCELLOR, 87- *Concurrent Pos:* Ed, Environ Times, 76-80; vchmn bd trustees, Am Geol Inst Found, 84-86, chmn, 86-; chmn bd dirs, Wolf Ridge Environ Learning Ctr, 90-; chmn, Coun Sci Soc Pres, 82-83. *Honors & Awards:* Neil A Miner Award, 65; Ian Campbell Medal, 85. *Mem:* AAAS; Geol Soc Am; Am Asn Petrol Geologists; Nat Asn Geol Teachers (pres, 76-77); Am Geol Inst (pres, 78-79). *Res:* Lower Paleozoic stratigraphy and paleontology. *Mailing Add:* Rm 213 Heller Hall Univ Minn Duluth MN 55812

HELLER, STEPHEN RICHARD, b Manhattan, NY, Jan 5, 43; m 64; c 3. SPECTROSCOPY, COMPUTER SCIENCE. *Educ:* State Univ NY Stony Brook, BS, 63; Georgetown Univ, PhD(chem), 67. *Prof Exp:* Res chemist biophys dept, Div Nuclear Med, Walter Reed Army Inst Res, Walter Reed Army Med Ctr, 67-70; sr staff fel, NIH, 70-73; res chemist, enciron protection agency, 73-84; RES SCIENTIST, AGR RES SERV, 84- *Concurrent Pos:* Vis prof, Hebrew Univ, 81. *Mem:* AAAS; Am Chem Soc; Inst Elec & Electronics Eng; Sigma Xi; Coblentz Soc; Am Soc Mass Spectrometry. *Res:* Organic spectroscopy; mass spectrometry; chemical information systems; pesticide databases, expert systems. *Mailing Add:* Bldg 005 Rm 337 Beltsville MD 20705

HELLER, STEVEN NELSON, b Gettysburg, Pa, Nov 14, 50; m 76. FOOD SCIENCE. *Educ:* Univ Md, BS, 72; Cornell Univ, MS, 75, PhD(food sci), 77. *Prof Exp:* Food technologist prof develop, Armour-Dial Co, 77-81; RES & DEVELOP DIR, GEN MIllS, 81- *Mem:* Inst Food Technol. *Res:* Development of new food products. *Mailing Add:* Gen Mills 9000 Plymouth Ave N Minneapolis MN 55427

HELLER, WILLIAM MOHN, b Orrville, Ohio, Mar 15, 26; m 50; c 1. PHARMACY, DRUG INFORMATION & STANDARDS. *Educ:* Univ Toledo, BS, 49; Univ Md, MS, 51, PhD, 55. *Hon Degrees:* DSc, Philadelphia Col Pharm & Sci, 81, Univ Md, 87, Univ Ark, 88. *Prof Exp:* Intern, Johns Hopkins Univ Hosp, 49-51; instr pharm, Univ Md, 53-54; asst prof, Sch Pharm, Univ Ark, Little Rock, 54-60, asst prof, Sch Med, 60-62, chief pharmacist, Univ Hosp, 54-66; dir dept sci serv, Am Soc Hosp Pharmacists, 66-68; exec dir designate, US Pharmacopeia, 68-70, exec dir, 70-90; RETIRED. *Concurrent Pos:* Pharmacist, Retail Stores, 50-53; dir, Am Hosp Formulary Serv, 55-62; chief investr drug systs res, Univ Ark, Little Rock, 62-66. *Honors & Awards:* Whitney Lect Award, Am Soc Hosp Pharmacists, 72; Remington Hon Medal, Am Pharmaceut Asn, 84; Distinguished Career Award, Drug Info Asn, 90. *Mem:* AAAS; Am Soc Hosp Pharmacists; Am Pharmaceut Asn; Int Pharmaceut Fedn; Drug Info Asn; Acad Pharmaceut Sci; Am Asn Pharmaceut Scientists; Am Asn Cols Pharm. *Res:* Hospital pharmacy; pharmacy education; drug standards; operations research. *Mailing Add:* 11915 Stonewood Lane Rockville MD 20852

HELLER, WILLIAM R, b Hillside, NJ, Oct 16, 20; m 47; c 3. SOLID STATE PHYSICS, APPLIED MATHEMATICS. *Educ:* Queens Col, NY, BS, 42; Brown Univ, ScM, 45; Wash Univ, PhD(physics), 50. *Prof Exp:* Res assoc physics, Univ Ill, 50-52; asst prof, Yale Univ, 52-53; staff mem, Shell Develop Co, 53-59; staff mem physics & metall, T J Watson Res Lab, 59-64, head, Physics Dept, San Jose Res Lab, 64-67, asst to dir, 67-69, mgr design technol develop, E Fishchill, 69-77, SR PHYSICIST, POUGHKEEPSIE LAB, IBM CORP, 79- *Concurrent Pos:* Vis prof comput sci, Calif Inst of Technol, 78-79.

Mem: Fel Am Phys Soc; NY Acad Sci; fel Inst Elec & Electronics Engrs; Comput Soc. *Res:* Design techniques for digital systems; electrical and mechanical packaging analysis for digital systems. *Mailing Add:* 42 Brentwood Dr Poughkeepsie NY 12603-5438

HELLER, ZINDEL HERBERT, b Brooklyn, NY, Jan 3, 27; wid; c 2. MEDICAL PHYSICS, INSTRUMENTATION. *Educ:* Brooklyn Col, BS, 47; Purdue Univ, MS, 49; Polytech Inst Brooklyn, PhD(electrophys), 67. *Prof Exp:* Asst instr physics, Purdue Univ, 47-52; res engr, Sperry Gyroscope Co, 52-59; sr res scientist, Airborne Instrument Labs, 59-70; staff scientist, Technicon Instrument Corp, 70-71; staff physicist, Perkin-Elmer Corp, 72; INSTRUMENT RES & DEVELOP DIR, DADE DIV, BAXTER HEALTHCARE CORP, 72- *Concurrent Pos:* Adj assoc prof biomed eng, Univ Miami, Coral Gables. *Mem:* Inst Elec & Electronics Engrs; Prof Group Eng Med & Biol; AAAS; Asn Advan Med Instruments. *Res:* Development of instruments for the hospital laboratory; exploration of new instrumental techniques; laboratory automation; chemistry; biology; hematology. *Mailing Add:* Dade Div Baxter Healthcare Corp PO Box 520672 Miami FL 33152

HELLERMAN, HERBERT, b Brooklyn, NY, July 8, 27; m 48; c 3. COMPUTER SCIENCE, ELECTRICAL ENGINEERING. *Educ:* Purdue Univ, BSEE, 49; Syracuse Univ, PhD(elec eng), 55. *Prof Exp:* Electronic scientist, Wright Field, US Air Force, Ohio, 49; from instr to asst prof elec eng, Syracuse Univ, 49-57; assoc prof, Univ Del, 57-59; staff mem, Int Bus Mach Corp, 59-64, sr staff mem, systs res inst, 64-66, sr systs consult, systs develop div, 66-69; prof comput sci, Sch Advan Technol, State Univ NY, Binghamton, 69-77, chmn dept, 77; prin comput architect, 77-84, CONSULT STRATEGIC PLANNER, AMDAHL CORP, 84- *Mem:* Asn Comput Mach; Inst Elec & Electronics Engrs. *Res:* Computer system organization; education in computer systems; computer performance; software engineering. *Mailing Add:* Amdahl Corp 1250 E Arques Ave Sunnyvale CA 94086

HELLERSTEIN, HERMAN KOPEL, b Dillonvale, Ohio, June 6, 16; m 47; c 6. MEDICINE. *Educ:* Western Reserve Univ, BA, 37, MD, 41; Am Bd Internal Med, dipl, 50. *Prof Exp:* Asst physician, Univ Hosps, 50-66, clin instr, Sch Med, 50-52, sr instr, 52-54, from asst prof to assoc prof med, 54-73, PROF MED, SCH MED, CASE WESTERN RESERVE UNIV, 74-, PHYSICIAN, UNIV HOSPS, 74- *Concurrent Pos:* Fel cardiovasc res, Postgrad Sch Med, Univ Chicago & Michael Reese Hosp, 46-47; Dazian Found fel cardiol, Western Reserve Univ Hosps, 47-48, USPHS res fel, 48-49; dir work classification clin, Cleveland Area Heart Soc, 50-63, pres sci coun, 62-63; consult cardiologist, Sunny Acres Tuberc, Benjamin Rose, Babies & Children's Hosps, Cleveland, 53-; mem, Inter-Am Heart Cong; chmn cardiovasc comt, Sch Med, Case Western Reserve Univ, 60-67; assoc physician, Univ Hosp, 66-74; mem sci coun & rehab comt, Coun Arteriosclerosis, fel coun clin cardiol & mem bd dirs, Am Heart Asn, 61-64; mem adv comt, Ohio State Bur Voc Rehab, 62-; sr attend cardiol, Crile Vet Admin Hosp, 63- & WHO, Geneva, 63-; consult div chronic dis & mem adv bd heart dis control prog, USPHS, 63-; William Bunn Mem Lectr, Youngstown Hosp, Ohio, 68; partic, White House Conf on Food, Nutrit & Health, 69; mem rehab study group, Inter-Soc Comn Heart Dis Resources, 69-71, consult, 71; lectr, Univ Sask, 70; mem, President's Comt Employ of Handicapped, 70. *Mem:* Am Col Cardiol (asst secy, 63-64, secy, 64-65); Am Heart Asn (vpres, 78-); fel Soc Exp Biol & Med; Sigma Xi (secy, 62, vpres, 63, pres, 64); Am Soc Med Res. *Res:* Cardiovascular diseases; electrophysiology of the heart; rehabilitation of the cardiac; ergonomics and physiology of work; congenital heart disease. *Mailing Add:* 2182 Chatfield Rd Cleveland OH 44106

HELLERSTEIN, STANLEY, b Denver, Colo, Oct 3, 26; m 49; c 3. PEDIATRIC NEPHROLOGY, CLINICAL CHEMISTRY. *Educ:* Univ Colo, BA, 48, MD, 52. *Prof Exp:* PROF PEDIAT, SCH MED, UNIV MO-KANSAS CITY, 63- *Concurrent Pos:* Physician clin chem, Children's Mercy Hosp, Kansas City, 63-; clin prof, Sch Med, Univ Kans, 65-; chief, Sect Pediat Nephrology, Children's Mery Hosp, 80- *Res:* Fluid and electrolyte metabolism; red blood composition, alterations in disease states and with growth. *Mailing Add:* Children's Mercy Hosp 24th at Gillham Rd Kansas City MO 64108

HELLICKSON, MARTIN LEON, b Dickinson, NDak, April 6, 45; m 70; c 2. AGRICULTURAL ENGINEERING. *Educ:* NDak State Univ, BS, 68; SDak State Univ, MS, 72; Univ Minn, PhD(agr eng), 75. *Prof Exp:* Asst prof, 75-80, ASSOC PROF AGR ENG, ORE STATE UNIV, 80- *Concurrent Pos:* Consult engr, Calif Law, 79-80, ground-source heat pumps, 85. *Mem:* Am Soc Agr Engrs; Sigma Xi. *Res:* Agricultural structures and environment; applications of alternative energies to agricultural situations; energy conservation associated with agricultural situations; postharvest preservation, storage and transportation of fresh fruits and vegtables. *Mailing Add:* Agr Eng Dept Ore State Univ Corvallis OR 97331

HELLICKSON, MYLO A, b Dickinson, NDak, July 17, 42; m 64. AGRICULTURAL ENGINEERING. *Educ:* NDak State Univ, BS, 64, MS, 66; WVa Univ, PhD(eng), 69. *Prof Exp:* Res asst, WVa Univ, 66-67, instr & res asst, 67-69; assoc prof, 69-78, PROF AGR ENG, SDAK STATE UNIV, 78-, HEAD DEPT, 82- *Concurrent Pos:* Sr sabbatical leave, Los Alamos Sci Lab. *Mem:* Am Soc Agr Engrs; Am Soc Eng Educ; AAAS; Coun Agr Sci & Technol; Nat Soc Prof Engrs; Sigma Xi; Am Soc Heating, Refrig & Air Conditioning Engrs. *Res:* Structures and environment for livestock; energy use in agriculture and food production. *Mailing Add:* Box 2120 University Station Brookings SD 57007

HELLIER, THOMAS ROBERT, JR, b Ft Pierce, Fla, Dec 24, 28; m 52; c 3. AQUATIC BIOLOGY. *Educ:* Univ Fla, BA, 55, MS, 57; Univ Tex, PhD(zool, marine sci), 61. *Prof Exp:* Res scientist ecol, Inst Marine Sci, Univ Tex, 57-59; from asst prof to assoc prof, 60-71, PROF BIOL, UNIV TEX, ARLINGTON, 71- *Mem:* AAAS; Ecol Soc Am; Am Fisheries Soc; Am Soc Ichthyologists & Herpetologists; Am Soc Limnol & Oceanog. *Res:* Environmental relationships of fishes. *Mailing Add:* Dept Biol Box 19498 Univ Tex Arlington TX 76019

HELLING, CHARLES SIVER, b Madelia, Minn, Jan 9, 40; m 64; c 1. SOIL CHEMISTRY, PESTICIDE CHEMISTRY. *Educ:* St Olaf Col, BA, 61; Univ Wis, MS, 63, PhD(soil chem), 66. *Prof Exp:* Res assoc agron, Cornell Univ, 65-67; res soil scientist, Natural Resources Inst, 67-89, RES SOIL SCIENTIST, PLANT SCIENCES INST, AGR RES SERV, US DEPT AGR, 89- *Mem:* Am Soc Agron; Soil Sci Soc Am; Weed Sci Soc Am; Am Chem Soc; Coun Agr Sci Technol. *Res:* Pesticide movement in soils; pesticide-soil adsorption mechanisms; soil organic chemistry; pesticide metabolism by soil microorganisms; cation exchange in soils. *Mailing Add:* Agr Res Serv US Dept Agr Agr Res Ctr-W 10301 Baltimore Ave Beltsville MD 20705-2351

HELLING, JOHN FREDERIC, b Madelia, Minn, June 4, 33; m 62; c 2. ORGANOMETALLIC CHEMISTRY. *Educ:* St Olaf Col, BA, 55; Ohio State Univ, PhD(org chem), 60. *Prof Exp:* Res assoc chem, Mass Inst Technol, 60-61; asst prof, Mich State Univ, 61-62; from asst prof to assoc prof, 62-78, PROF CHEM, UNIV FLA, 78- *Concurrent Pos:* Nat counr, Am Chem Soc, 81- *Mem:* Am Chem Soc. *Res:* Metallocenes; bisarene complexes of transition metals; transition metal intermediates in organic syntheses. *Mailing Add:* Dept Chem Univ Fla Gainesville FL 32611-2046

HELLING, ROBERT BRUCE, b Madelia, Minn, Mar 21, 36; m 61; c 1. GENETICS. *Educ:* St Olaf Col, BA, 58; Univ Pittsburgh, MS, 50, PhD(bact), 63. *Prof Exp:* NSF fel microbial genetics, Karolinska Inst, 63-65; asst prof, 65-70, ASSOC PROF BOT, UNIV MICH, ANN ARBOR, 70- *Mem:* Genetics Soc Am; Am Soc Microbiol. *Res:* Microbial genetics; genetic control mechanisms. *Mailing Add:* Univ Mich Main Campus 4042A Natural Sci Biol Ann Arbor MI 48109-1048

HELLIWELL, R(OBERT) A(RTHUR), b Red Wing, Minn, Sept 2, 20; m 42; c 4. RADIOSCIENCE, WAVE PROPAGATION. *Educ:* Stanford Univ, AB, 42, MS, 43, EE, 44, PhD(elec eng), 48. *Prof Exp:* Actg instr elec eng, Stanford Univ, 42-43, instr, 43-44, res assoc, 44-46, actg asst prof, 46-50, from asst prof to assoc prof, 50-58, PROF ELEC ENG, STANFORD UNIV, 58- *Concurrent Pos:* Consult, Nat Bur Standards, ionospheres & radiophysics subcomt space sci steering comt, NASA; mem comns & US nat comt, Int Union Radio Sci, chmn comn, 57-63; mem, Polar Res Bd Exec Comt, Nat Res Coun, 78-81. *Honors & Awards:* Appleton Prize, Royal Soc London, 72. *Mem:* Nat Acad Sci; Am Geophys Union. *Res:* Whistlers and related ionospheric phenomena; ionospheric and magnetospheric research including plasma wave experiments in space physics; analysis of power line radiation effects; VLF wave amplification studies; HF and VLF direction-finding techniques; modification of the upper atmosphere. *Mailing Add:* Starlab Stanford Univ Stanford CA 94305

HELLIWELL, THOMAS MCCAFFREE, b Minneapolis, Minn, June 8, 36. THEORETICAL PHYSICS. *Educ:* Pomona Col, BA, 58; Calif Inst Technol, PhD(physics), 63. *Prof Exp:* Chair physics dept, 81-89, PROF PHYSICS, HARVEY MUDD COL, 62-, FAC CHAIR, 90- *Concurrent Pos:* NSF fac fel, Univ Md, 68-69& Cambridge Univ, 75-76. *Mem:* AAAS; Am Asn Physics Teachers. *Res:* General relativity, including fundamental problems and applications to astrophysics. *Mailing Add:* Harvey Mudd Col Claremont CA 91711

HELLMAN, ALFRED, b Ger, Oct 11, 31; US citizen; m 56; c 2. VIROLOGY, CELL PHYSIOLOGY. *Educ:* Univ Md, BS, 54; Univ Mich, MS, 55, PhD(microbiol), 61. *Prof Exp:* Biologist, Wright-Patterson AFB, 57-59; dep chief virol, US Air Force Sch Aerospace Med, 61-62, chief, 62-64; scientist, Nat Cancer Inst, 64, sr scientist etiology group, 64-67, chief off biohazard & environ control, 67-76, asst chief, Cellular & Molecular Lab, 76-83; sr scientist, Cong Res Serv, 84-85; consult, export control, 84-85, SR TECH ADV ASST SECY BIOTECHNOLOGY, DEPT COMMERCE, 85- *Concurrent Pos:* Lectr, San Antonio Col, 62-63 & St Mary's Univ, Tex, 63-64; consult biomed, 85- *Mem:* AAAS; Am Soc Microbiol; Tissue Cult Asn. *Res:* Hormonal influence on viral synthesis and environmental influences on cell susceptibility to infection; hormonal factors associated with oncogenesis and co-carcinogenic induction of malignancy. *Mailing Add:* PO Box 289 Clarksburg MD 20871

HELLMAN, FRANCES, b Gt Brit, Oct 28, 56; US citizen. SPECIFIC HEAT MEASUREMENT, THIN FILM GROWTH. *Educ:* Dartmouth Col, BA, 78; Stanford Univ, PhD(appl physics), 85. *Prof Exp:* Mem tech staff, AT&T Bell Labs, 85-87; ASST PROF PHYSICS, UNIV CALIF, SAN DIEGO, 87- *Concurrent Pos:* Assoc ed, J Mat Sci & Eng, 87-; prog & exec comt, Thin film div, Am Vacuum Soc. *Mem:* Am Phys Soc; Mat Res Soc. *Res:* Characterization and physical properties of novel solid materials in film form; primary experimental tool, vapor deposition chambers, anultra-sensitive wide temperature and magnetic field range specific heat experiment; amorphous magnetic materials; epitaxial sillicides and superconductors. *Mailing Add:* Dept Physics B-019 Univ Calif San Diego La Jolla CA 92093

HELLMAN, HENRY MARTIN, b Norrfors, Sweden, July 4, 20; nat US; m 51. CHEMISTRY. *Educ:* Ind Univ, BS, 43; Purdue Univ, MS, 45, PhD(org chem), 47. *Prof Exp:* Asst chem, Purdue Univ, 43-44; from instr to assoc prof, NY Univ, 47-69, from actg chmn to chmn dept, 64-73, prof chem, 69-90; RETIRED. *Mem:* Am Chem Soc. *Res:* Organic chemistry; mechanism of organic reactions; haloalkylation; organo-metallic compounds; oxidation reactions. *Mailing Add:* Dept Chem NY Univ Washington Sq New York NY 10003

HELLMAN, KENNETH P, b Baltimore, Md, July 28, 34; m 58; c 2. BIOCHEMISTRY. *Educ:* Drew Univ, AB, 56; Mich State Univ, MS, 59, PhD(chem), 62. *Prof Exp:* From asst prof to assoc prof, 61-76, PROF CHEM, SMITH COL, 76- *Concurrent Pos:* Res assoc, Inst Biol Chem, Univ Rome, 69-70; vis res assoc, Microbiol Res Estab, Porton, Eng, 75-76, 78 & 82; chmn, chem dept, 66-69, 78-81 & 85-88. *Mem:* AAAS; Am Chem Soc. *Res:* Enzyme mechanisms; protein structure and function. *Mailing Add:* Dept Chem Smith Col Northampton MA 01063

HELLMAN, LOUIS M, b St Louis, Mo, Mar 22, 08; m 34; c 2. OBSTETRICS, GYNECOLOGY. *Educ:* Yale Univ, PhB, 30; Johns Hopkins Univ, MD, 34. *Prof Exp:* Intern surg, Cornell Univ, 34-35; resident & asst path, 35-36; intern obstet, Johns Hopkins Hosp, 37-38, asst resident obstet, 38-39; asst resident gynec, 39-40, resident & instr obstet, 40-41, asst prof, 41-46, assoc prof & co-dir obstet anesthesia res prog, 46-50; prof obstet & gynec & chmn dept, State Univ NY Downstate Med Ctr, 50-70; dep asst secy pop affairs, HEW, 70-77, dir health affairs, Pop Ref Bur, 77-79; RETIRED. *Concurrent Pos:* Dir obstet & gynec, Kings County Hosp, 50-70; adminr, Health Serv Admin, 76-77. *Mem:* Am Fertil Soc; Am Gynec Soc (vpres, 60, pres, 74); Am Asn Obstet & Gynec; Am Col Surg; Am Col Obstet & Gynec. *Res:* Pathology of placenta. *Mailing Add:* 2475 Virginia Ave NW Washington DC 20037

HELLMAN, MARTIN EDWARD, b New York, NY, Oct 2, 45; m 67; c 2. ELECTRICAL ENGINEERING. *Educ:* NY Univ, BE, 66; Stanford Univ, MS, 67, PhD(elec eng), 69. *Prof Exp:* Mem staff, Thomas J Watson Res Ctr, IBM Corp, 68-69; asst prof elec eng, Mass Inst Technol, 69-71; PROF ELEC ENG, STANFORD UNIV, 71- *Concurrent Pos:* Vinton Hayes fel, 69-71. *Honors & Awards:* Donald G Fink Award, Inst Elec & Electronic Engrs, 81. *Mem:* Inst Elec & Electronic Engrs; Inst Math Statist. *Res:* Information and communication theory; cryptography and data security; international security. *Mailing Add:* Dept Elec Eng Stanford Univ Stanford CA 94305

HELLMAN, NISON NORMAN, b Milwaukee, Wis, May 19, 20; m 42; c 4. CHEMISTRY. *Educ:* Univ Wis, BS & MS, 42, PhD, 43. *Prof Exp:* Asst chemist, Metall Lab, Univ Chicago, 43-45; chemist, Northern Regional Res Lab, Bur Agr & Indust Chem, USDA, 45-50, in chg phys chem group, Northern Utilization Res Br, 50-55; res chemist, Froedtert Malt Corp, 56-59, asst tech dir, 59-60; consult chemist, 60-69; VPRES & HEAD LAB, TRANSFORMER DESIGN INC OF MILWAUKEE, 69- *Mem:* AAAS; Am Chem Soc; Am Asn Cereal Chem; Inst Food Technologists. *Res:* X-ray diffraction; physical properties of colloidal systems; molecular characterization of biological polymers; synthetic and degradative enzymatic activities; foods development. *Mailing Add:* 2135 W Edward Lane Milwaukee WI 53209-2814

HELLMAN, SAMUEL, b New York, NY, July 23, 34; m 57; c 2. RADIOTHERAPY. *Educ:* Allegheny Col, BS, 55; State Univ NY, MD, 59. *Hon Degrees:* MS, Harvard Univ, 68. *Prof Exp:* Fel radiother, Sch Med, Yale Univ, 62-64; fel, Inst Cancer Res & Royal Marsden Hosp, London, 65-66; asst prof radiol, Yale Univ, 66-68; assoc prof, Harvard Med Sch, 68-70, Fuller-Am Cancer Soc prof radiol & chmn Dept Radiation Ther, 68-83; physician & chief, Mem Sloan Kettering Cancer Ctr, New York, 83-88; DEAN, BIOL SCI DIV & PRITZKER SCH MED & VPRES MED CTR, UNIV CHICAGO, 88- *Concurrent Pos:* Prof Dept Radiation & Cellular Oncol, Univ Chicago, 88- *Mem:* Radiation Res Soc; Am Soc Therapeut Radiol; Am Col Radiologists; Asn Univ Radiologists. *Res:* Cancer; radiation biology; cell kinetics; hematopoiesis. *Mailing Add:* Biol Sci Div & Pritzker Sch Med 5841 S Maryland Ave Box 417 Chicago IL 60637

HELLMAN, WILLIAM S, b Brooklyn, NY, May 14, 31; m 54; c 1. PHYSICS. *Educ:* Brooklyn Col, BS, 53; Syracuse Univ, MS, 58, PhD(physics), 61. *Prof Exp:* Res assoc physics, US Naval Ord Lab, 62-64; asst prof, 64-69, ASSOC PROF PHYSICS, BOSTON UNIV, 69- *Concurrent Pos:* Nat Acad Sci res assoc, 62-64. *Mem:* Am Phys Soc; Sigma Xi. *Res:* Quantum field theory; elementary particle physics. *Mailing Add:* 31 Longfellow Rd Arlington MA 02174

HELLMANN, MAX, b Beroun, Czech, Nov 27, 19; nat US; m 57. SCIENCE ADMINISTRATION. *Educ:* Col Wooster, BA, 42; Univ Buffalo, MA, 47, PhD(chem), 50. *Prof Exp:* Instr chem, Univ Buffalo, 48-49; res fel radiochem, Univ NC, 49-51; res chemist, Nat Bur Standards, 51-60; proj dir, Int Sci Activ, NSF, 60-62, dep chief scientist, NSF/Rio, 62-64, prog dir, 64-68, head coop sci activ sect, Off Int Sci Activ, 68-69, head liaison staff, New Delhi, 69-71, mgr, Latin Am & Pac Sect, 71-76, dep dir, Div Int Prog, 76-81; dep dir, US-Israel Binational Sci Found, Jerusalem, 81-83; CONSULT, 83- *Honors & Awards:* Meritorious Serv Award, NSF, 81. *Mem:* Fel AAAS. *Res:* Isotope effect studies on carbon-14; aromatic fluorine compounds; international science activities. *Mailing Add:* 1235 34th St NW Washington DC 20007

HELLMERS, HENRY, b Palmerton, Pa, Sept 5, 15; m 45; c 2. PLANT PHYSIOLOGY, HYDROLOGY & WATER RESOURCES. *Educ:* Pa State Univ, BS, 37, MS, 39; Univ Calif, Berkeley, PhD(plant physiol), 50. *Prof Exp:* Mem, New Eng Emergency Proj, US Forest Serv, 39-40; forestry inspector, Pa Turnpike Comn, 40-41; plant physiologist, Pac Southwest Forest & Range Exp Sta, US Forest Serv, 49-65; dir phytotron, Duke Univ, 65-79, prof bot, 65-83, prof forestry, 68-83; RETIRED. *Concurrent Pos:* Res fel, Calif Inst Technol, 50-55, sr res fel, 55-65; lectr, Pasadena City Col, 61-65; NZ sr res fel forestry, 73-74; vis prof, Japan Soc Prom Sci. *Mem:* Fel AAAS; Soc Am Foresters; Am Soc Plant Physiologists; Ecol Soc Am; Am Inst Biol Scis. *Res:* Metabolism of pine pollen; soil fertility of wildland areas; temperature and photoperiodic effects upon growth of grasses, shrubs and trees; long term effects on plants of elevated carbon dioxide concentrations. *Mailing Add:* Dept Bot Duke Univ Durham NC 27706

HELLMUTH, WALTER WILHELM, b New York, NY, Nov 6, 38; m 61; c 3. ORGANIC CHEMISTRY. *Educ:* Iona Col, BS, 60; Univ Conn, MS, 63, PhD(org chem), 65. *Prof Exp:* Res chemist, 64-74, group leader res & technol, 74-79, GROUP LEADER ANALYSIS & TESTING, TEXACO INC, 79- *Mem:* Am Chem Soc. *Res:* Automotive lubricants and fuels, additives and new technology; bench test development; thermal analysis. *Mailing Add:* Texaco Res Ctr PO Box 509 Beacon NY 12508-0509

HELLSTROM, HAROLD RICHARD, b Marianna, Pa, Mar 22, 28. EXPERIMENTAL BIOLOGY. *Educ:* Waynesburg Col, BS, 50; Univ Pittsburgh, MD, 52; Am Bd Path, cert anat path, 57, cert clin path, 60. *Prof Exp:* Teaching fel, Dept Path, Sch Med, Univ Pittsburgh, 53-57, instr, 58, clin asst prof, 60-66, from asst prof to assoc prof path, 66-79; PROF, DEPT PATH, STATE UNIV NY, UPSTATE MED CTR, 79- *Concurrent Pos:* Staff pathologist, Vet Admin Hosp, Pittsburgh, 57-70, chief lab serv, 70-79; chief lab serv, Vet Admin Med Ctr, Syracuse, 79-; mem, Coun Basic Sci, Am Heart Asn. *Mem:* Col Am Pathologists; Am Soc Clin Path; AAAS; Int Acad Path; Am Asn Pathologists. *Res:* Color description for use in gross pathology; numerous publications. *Mailing Add:* Vet Admin Med Ctr 113 800 Irving Ave Syracuse NY 13210

HELLSTROM, INGEGERD ELISABET, b Falun, Sweden, Jan 3, 32; m 60; c 2. IMMUNOLOGY. *Educ:* Karolinska Inst Med Sch, Sweden, lic med, 64, Dr Med, 66. *Prof Exp:* Res assoc tumor biol, Karolinska Inst Med Sch, 59-66; asst prof microbiol & nursing, Univ Wash, 66-69, res assoc microbiol, Bristol-Myers Squibb PRI, 69-72, PROF MICROBIOL & IMMUNOL, UNIV WASH, 72-, VPRES, EXPLOR BIOMED RES, BRISTOL-MYERS SQUIBB PRI, 91- *Concurrent Pos:* Mem, Gen Assembly, Gen Motors Cancer Res Found; Humboldt Award, Bonn, 80. *Honors & Awards:* Lucy Wortham James Award, Ewing Soc, 72; Pap Award, Papanicolaou Cancer Res Inst, 73; Nat Award, Am Cancer Soc, 74. *Mem:* Am Asn Immunologists; Am Fedn Clin Res. *Res:* Experimental studies on the organism's immunological defense mechanisms to tumors. *Mailing Add:* Bristol-Myers Squibb Pharmaceut Res Inst 3005 First Ave Seattle WA 98121

HELLSTROM, KARL ERIK LENNART, b Stockholm, Sweden, Apr 16, 34; m 60; c 2. IMMUNOLOGY, ONCOLOGY. *Educ:* Karolinska Inst, Sweden, MD, 64, PhD, 66. *Prof Exp:* Res assoc histol, Sch Med, Karolinska Inst, 54-57, res assoc tumor biol, 58-62, asst prof, 62-66; assoc prof, 66-69, PROF PATH, SCH MED, UNIV WASH, 69- *Concurrent Pos:* mem & prog head, Fred Hutchinson Cancer Res Ctr, 75-83; lab dir, Oncogen, 83-85; vpres, Bristol-Myers Squibb Pharmaceut Res Inst, 85- *Honors & Awards:* Lucy Wortham James Award, Ewing Soc, 72; Pap Award, Papanicolaou Cancer Res Inst, 73; Nat Award, Am Cancer Soc, 74; Humboldt Award, Sr Am Scientist, Bonn, Germany, 80. *Mem:* AAAS; NY Acad Sci; Am Asn Cancer Res; Am Soc Exp Path; Fedn Am Socs Exp Biol; Sigma Xi. *Res:* Experimental studies on the organism's immunological defense mechanisms to tumors. *Mailing Add:* Bristol Myers Squibb Pharmaceut Res Inst 3005 First Ave Seattle WA 98121

HELLUMS, JESSE DAVID, b Stamford, Tex, Aug 19, 29; m 57; c 2. CHEMICAL ENGINEERING. *Educ:* Univ Tex, BS, 50, MS, 58; Univ Mich, PhD(chem eng), 61. *Prof Exp:* Process engr, Mobil Oil, Tex, 50-53; from asst prof to assoc prof chem eng, 60-68, chmn dept, 70-76, dir, 68-80, dean eng, 80-88, PROF CHEM ENG, 68-, AJ HARTSOOK PROF, RICE UNIV, 85- *Concurrent Pos:* NSF sci fac fel, Cambridge & Edinburgh Univs, 67-68; adj prof med, Baylor Col Med, 70- & Univ Tex, 77-; vis prof, Imperial Col, London, 73-74; vis scholar, Univ Calif, San Diego, 88; spec vis prof, Tokyo Inst Technol, 89. *Honors & Awards:* Res Merit Award, Nat Inst Health, 86. *Mem:* Am Inst Chem Engrs; Am Chem Soc; AAAS; Am Soc Artificial Internal Organs; Microcirculation Soc; fel Am Inst Chem Engr; Am Heart Coun on Thrombosis. *Res:* Fluid mechanics; biomedical engineering; various aspects of chemical engineering in medicine and biology including artificial organs, blood rheology, and mass transport in the microcirculation. *Mailing Add:* Dept Chem Eng Rice Univ Houston TX 77251-1892

HELLWARTH, ROBERT WILLIS, b Ann Arbor, Mich, Dec 10, 30; div; c 3. OPTICS. *Educ:* Princeton Univ, BSE, 52; Oxford Univ, DPh, 55. *Prof Exp:* Mem tech staff, Hughes Res Labs, 55-65, sr staff physicist, 65-70; GEORGE PFLEGER PROF ELEC ENG & PHYSICS, UNIV SOUTHERN CALIF, 70- *Concurrent Pos:* Lectr, Calif Inst Technol, 55-66, Hughes fel, 55-56, sr res fel, 66-70; vis assoc prof, Univ Ill, 64-65; mgr theoret studies dept, Hughes Res Labs, 68-70; sr fel, NSF, Clarendon Lab, Oxford, Eng, 70-71, vis fel, St Peter's Col, 70-71; assoc ed, IEEE J Quantum Electronics. *Honors & Awards:* Charles Hard Tounes Award, Optical Soc Am, 83; Quantum Elec Award, Inst Elec & Electronic Engrs, 85. *Mem:* Nat Acad Sci; Nat Acad Eng; fel Am Phys Soc; fel Inst Elec & Electronics Eng; fel AAAS; fel Optical Soc Am. *Res:* Nonlinear optics; solid state physics; laser devices. *Mailing Add:* Dept Elec Eng & Physics Univ Southern Calif Los Angeles CA 90089-0484

HELLWEGE, HERBERT ELMORE, b Harsefeld, Ger, Mar 3, 21; US citizen; m 53; c 2. INORGANIC CHEMISTRY, CLINICAL CHEMISTRY. *Educ:* Univ Hamburg, DSc(chem), 53. *Prof Exp:* Asst geochem, Univ Hamburg, 51-53; chemist, Food Res Labs, NY, 53-54; prof chem, Rollins Col, 54-74, Archibald Granville Bush prof sci, 74-86, chmn pre-med comt, 76-86; RETIRED. *Concurrent Pos:* Consult, Radiation Incorp, 61-63; NSF fac fel, Gothenburg Univ, 64-65; lab dir, Louis C Herring & Co, 74- *Mem:* Am Chem Soc; Am Asn Bioanal. *Res:* Solvent extraction of metal chelates; coordination chemistry; trace elements in hair; trace elements in urinary calculi. *Mailing Add:* 1141 Banbury Trail Maitland FL 32751

HELLWIG, HELMUT WILHELM, b Berlin, Ger, May 7, 38; US citizen; m 60; c 2. TIME & FREQUENCY, QUANTUM ELECTRONICS. *Educ:* Tech Univ Berlin, Dipl-Ing, 63, Dr-Ing, 66. *Hon Degrees:* Dr, Univ Besancon, 88. *Prof Exp:* Res physicist masers, Heinrich Hertz Inst, Berlin, 63-66, US Army Electronics Command, Ft Monmouth, NJ, 66-69 & Nat Bur Standards, Boulder, Colo, 69-74, sect chief atomic clocks, 74-79; pres & gen mgr atomic clocks & crystal oscillators, Frequency & Time Systs Inc, 79-86; vpres, Datum Inc, 83-86; assoc dir, Nat Bur Standards, Gaithersburg, 87-90; DIR, OFF SCI RES, USAF, WASHINGTON, DC, 90- *Concurrent Pos:* Mem, Int Radio Consult Comt, 72-; mem, US Nat Comt, Int Union Radio Sci, 74-, vchmn Comn A, 78-81, chmn, 81- *Honors & Awards:* Sci Award, Dept Defense, Secy Army, 69; E V Condon Award, Secy Com, 79. *Mem:* Fel Inst Elec & Electronic Engrs; Am Phys Soc; Sigma Xi; Int Union Radio Sci. *Res:* Atomic clocks and frequency standards; atomic and molecular beams; masers; lasers; quantum electronics; time and frequency metrology. *Mailing Add:* 13210 Colton Lane Darnestown MD 20878-2104

HELLWIG, L(ANGLEY) R(OBERTS), b Beaumont, Tex, May 21, 28; m 51; c 1. CHEMICAL ENGINEERING. *Educ:* Univ Tex, BS, 49, MS, 51, PhD(chem eng), 55. *Prof Exp:* Tech serv engr, Ethyl Corp, 54-55, develop engr, 55-56, spec coordr, 57-58; head eng sect, Res Div, Petrol Chem, Inc, 58-60; mgr planning dept, Cities Serv Res & Develop Co, 60-63; dir chem res dept, Columbian Carbon Co, 64-67; vpres & mgr res, Cities Serv Res & Develop Co, 67-69, mgr corp planning, Cities Serv Co, 69-72, vpres, Planning & Econ Div, 72-76, exec vpres & pres chem group, 76-79; PRES & CHIEF EXEC OFFICER, COLUMBIAN CHEM CO, 80- *Mem:* Am Inst Chem Engrs; Am Chem Soc; AAAS. *Res:* Process development; research planning; petroleum and petroleum-derived chemicals and polymers. *Mailing Add:* 2747 E 68th St Tulsa OK 74136

HELLY, WALTER S, b Vienna, Austria, Aug 22, 30; US citizen; m 56; c 1. OPERATIONS RESEARCH. *Educ:* Cornell Univ, BA, 50; Univ Ill, MS, 51; Mass Inst Technol, PhD(physics), 60. *Prof Exp:* Sr engr mil control systs, Sylvania Elec Co, 54-56; sr engr commun & reconnaissance, Melpar, Inc, 56-59; mem tech staff tel traffic, Bell Tel Labs, 59-62; res engr traffic & gen opers res, Port NY Authority, 62-65; proj planner tel opers res, NY Tel Co, 65-66; PROF OPERS RES, POLYTECH INST NY, 66- *Mem:* Opers Res Soc Am; Inst Indust Engrs. *Res:* Single lane, network and queueing problems of vehicular traffic; congestion in telephone switching systems and networks; applications of operations research to urban areas. *Mailing Add:* Dept Mech & Indust Eng Polytech Univ NY 333 Jay St Brooklyn NY 11201

HELM, DONALD CAIRNEY, b Yokohama, Japan, Mar 26, 37; US citizen; div; c 1. GROUNDWATER HYDROLOGY, MECHANICS OF AQUIFER MOVEMENT. *Educ:* Amherst Col, BA, 59; Hartford Sem Found, MDiv, 62; Univ Calif, Berkeley, MS, 70, PhD(geol eng), 74. *Prof Exp:* Asst rural develop India, Community Serv, Inc, 63-64; hydraul engr groundwater hydrol, 65-69, res hydrologist, Water Resources Div, US Geol Surv, 69-78; physicist, earth sci div, Lawrence Livermore Nat Lab, Univ Calif, 78-84, group leader, Geohydrology & Environ Studies Group, 81-84; prin res scientist, Groundwater Mech's Div Geo-Mech's, Commonwealth Sci & Indust Res Orgn, 84-92; RES HYDROGEOLOGIST, NEV BUR MINES & GEOL, UNIV NEV LAS VEGAS, 92- *Concurrent Pos:* Instr, UNESCO Int Sem, Land Subsidence, Mexico City, 79; mem steering comt, Geothermal Subsidence Res Prog, US Dept Energy, 76-81; found mem, Bd Dirs, Ctr Study Theol & Natural Sci, Grad Theol Union, Berkeley, 81-84; vis scientist, State Elec Comn Victoria, Australia, 82-83; exchange scientist, Chinese Acad Sci, 88; mem comt, D-34 on waste disposal, Am Soc Testing & Mat, 81- 83; vis scientist, Nev Bur Mines & Geol, 89-92, grad fac, Univ Nev, 90- *Mem:* Am Geophys Union; Am Soc Civil Engr; Asn Eng Geologists; Nat Water Well Asn; Asn Geoscientists Int Develop; Am Soc Testing & Mat; Int Asn Hydrogeologist; Int Asn Eng Geol; Int Soc Rock Mech; Int Soc Soil Mech & Found Eng; Geol Soc Am; AAAS. *Res:* Geohydrology; underground waste migration; mechanics of aquifer systems; consolidation theory; land subsidence; response of granular skeleton to transient groundwater flow; flow through fractured rock. *Mailing Add:* Nev Bur Mines & Geol Dept Geosci Univ Nev Las Vegas Las Vegas NV 89154-4010

HELM, JAMES LEROY, b Beach, NDak, Dec 28, 32; m 60; c 3. PLANT BREEDING, GENETICS. *Educ:* NDak State Univ, BS, 60, MS, 63; Univ Mo-Columbia, PhD(genetics), 68. *Prof Exp:* Dir res agron, AGSCO, Inc, 60-61; instr, Univ Mo-Columbia, 63-68; geneticist, Anheuser-Busch, Inc, 68-73; dir res, McNair Seed Co, 73-80; with Asgrow Seed Co, Subsid Upjohn Co, 80-; AT EXTEN CROP & WEED SCI, NDAK STATE UNIV. *Mem:* Genetics Soc Am; Am Genetic Asn; Crop Sci Soc Am; Am Inst Biol Sci; Am Soc Agron. *Res:* Corn genetics; identification of mutant genes; incorporation, evaluation and development of commercial hybrids with desirable genotypes. *Mailing Add:* Exten Crop & Weed Sci NDak State Univ PO Box 5051 Fargo ND 58105

HELM, RAYMOND E, b West, Tex, Sept 11, 30; m 51; c 4. DAIRY SCIENCE, NUTRITION. *Educ:* Tex A&M Univ, BS, 59, MS, 60, PhD(dairy sci), 70. *Prof Exp:* Instr dairy sci, Tex A&M Univ, 65-67; asst prof, 67-77, ASSOC PROF AGR, SOUTHWEST TEX STATE UNIV, 77- *Mem:* Am Dairy Sci Asn. *Res:* Dairy production and nutrition. *Mailing Add:* Dept Agr Southwest Tex State Univ San Marcos TX 78666

HELM, RICHARD H, b National City, Calif, Aug 2, 22; m 49; c 2. HIGH ENERGY PARTICLE ACCELERATOR PHYSICS. *Educ:* Stanford Univ, AB, 47, MS, 50, PhD(physics), 56. *Prof Exp:* Staff mem physics, Univ Calif, Los Alamos Sci Lab, 56-58; STAFF MEM PHYSICS, STANFORD LINEAR ACCELERATOR CTR, 58- *Mem:* Am Inst Physics; Sigma Xi. *Res:* Beam dynamics and beam transport design and analysis in high energy particle accelerators. *Mailing Add:* 334 Lincoln Ave Stanford CA 94301

HELM, ROBERT ALBERT, b Cincinnati, Ohio, Mar 28, 21; m 59; c 1. INTERNAL MEDICINE. *Educ:* Univ Cincinnati, BS, 42, MD, 45. *Prof Exp:* From instr to assoc prof, 50-64, PROF MED, UNIV CINCINNATI, 64- *Concurrent Pos:* Consult, Vet Admin Hosp, 58-; mem coun clin cardiol & coun circulation, Am Heart Asn. *Mem:* Fel Am Col Cardiol; fel Am Col Physicians; Sigma Xi. *Res:* Cardiovascular disease with emphasis on electrocardiography. *Mailing Add:* Christian R Holmes Hosp Cincinnati OH 45219

HELM, WILLIAM THOMAS, b Niagara Falls, NY, May 12, 23; m 48; c 2. TROUT HABITAT, AQUATIC ECOLOGY. *Educ:* Univ Wis, BS, 50, MS, 51, PhD(zool), 58. *Prof Exp:* Biologist, Wis Conserv Dept, 51; limnologist, Tenn Valley Authority, 51-53; from asst to res assoc, Univ Wis, 55-58, proj assoc, 58-59; asst prof wildlife sci, Utah State Univ, 59-69, dir Bear Lake Biol Labs, 59-73, assoc prof wildlife sci, 69-88; RETIRED. *Concurrent Pos:* Consult lake & stream projs; prin investr numerous res projs. *Mem:* AAAS; Am Fisheries Soc; Am Soc Limnol & Oceanog; Am Inst Biol Sci; Ecol Soc Am. *Res:* Stream trout microhabitat; basic ecology of freshwater fishes, distribution and movements within lakes and streams; life history of fishes. *Mailing Add:* Dept Fisheries & Wildlife Utah State Univ Logan UT 84322-5200

HELMAN, EDITH ZAK, clinical chemistry, radiochemistry, for more information see previous edition

HELMAN, SANDY I, b Winnipeg, Man, May 30, 39; nat US. PHYSIOLOGY, BIOPHYSICS. *Educ:* Univ Man, BSEE, 61; Drexel Inst Technol, MS, 62; Marquette Univ, PhD(physiol), 66. *Prof Exp:* Res assoc, Presby Hosp, Drexel Inst Technol, Pa, 61-62 & Vet Admin Hosp, Wis, 62-66; USPHS postdoctoral fel, Lab Kidney & Electrolyte Metab, NIH, Bethesda, Md, 66-69; from asst prof to assoc prof, Dept Physiol, Sch Med, Tulane Univ, La, 69-72; from asst prof to assoc prof, 72-80, PROF, DEPT PHYSIOL & BIOPHYS, UNIV ILL, URBANA-CHAMPAIGN, 80- *Concurrent Pos:* Mem, Prog Adv Comt, Am Physiol Soc, 77-79; secy-treas, Membrane Biophys Group, Biophys Soc, 77-83. *Mem:* Am Physiol Soc; Am Soc Nephrology; Biophys Soc; Int Soc Nephrology; fel AAAS; Sigma Xi; Soc Gen Physiologists. *Mailing Add:* Dept Physiol & Biophys 524 Burrill Hall Univ Ill 407 S Goodwin Ave Urbana IL 61801

HELMAN, WILLIAM PHILLIP, b Grand Junction, Colo, Nov 2, 36. PHYSICAL CHEMISTRY. *Educ:* Calif Inst Technol, BS, 58, MS, 60; Univ Minn, PhD(phys chem), 64. *Prof Exp:* Res chemist, Jackson Lab, E I du Pont de Nemours & Co, Inc, 64-66; fel chem, 66-67, jr fac fel, 67-68, asst fac fel, 68-72, ASSOC PROF SPECIALIST, RADIATION LAB, UNIV NOTRE DAME, 72- *Mem:* Am Phys Soc; Am Chem Soc. *Res:* Luminescence and reactions of excited organic molecules; pulse radiolysis. *Mailing Add:* Radiation Lab Univ Notre Dame Notre Dame IN 46556-5600

HELMBOLD, ROBERT LAWSON, mathematics, for more information see previous edition

HELMER, JOHN, b Evanston, Ill, Nov 18, 26; m 53; c 2. PHYSICS, ELECTRONICS. *Educ:* Lawrence Col, BS, 50; Calif Inst Technol, MS, 51; Stanford Univ, EE, 54, PhD(elec eng), 57. *Prof Exp:* Engr, Northrup Comput Div, Northrup Aircraft, Inc, 51-52; microwave engr, Bell Tel Labs, 53; asst, Stanford Univ, 54-57; res engr, 57-64, dir res, Vacuum Prod Div, 64-68, mgr new anal methods, Instrument Div, 68-79, SECT MGR, VARIAN RES CTR, 79- *Mem:* Am Phys Soc; Am Vacuum Soc. *Res:* Analytic instruments; x-rays; optics; electron optics; quantum electronics; microwaves; chromatography; ultrasonics; chemistry; thinfilms. *Mailing Add:* Varian Assocs 611 Hansen Way Palo Alto CA 94303

HELMER, RICHARD GUY, b Homer, Mich, Feb 19, 34; m 56; c 2. EXPERIMENTAL NUCLEAR PHYSICS. *Educ:* Univ Mich, BS, 56, MS, 57, PhD(physics), 61. *Prof Exp:* Physicist, Atomic Energy Div, Phillips Petrol Co, 61-66; physicist, Idaho Nuclear Corp, 66-71; physicist, Aerojet Nuclear Co, 71-76; PRIN SCIENTIST, EG&G IDAHO INC, 76- *Concurrent Pos:* Asst prof & collabr, Utah State Univ, 65- *Mem:* AAAS; fel Am Phys Soc; Am Nuclear Soc; NY Acad Sci. *Res:* Decay schemes of radioactive nuclides; nuclear structure studies and precise gamma-ray and electron spectroscopy. *Mailing Add:* Physics & Math Div EG&G Idaho PO Box 1625 Idaho Falls ID 83415-2219

HELMERICK, ROBERT HOWARD, b Greybull, Wyo, Feb 25, 26; m 54; c 5. PLANT BREEDING. *Educ:* Univ Wyo, BS, 50 & MS, 54; Univ Nebr, PhD(genetics), 58. *Prof Exp:* Plant breeder, Am Crystal Sugar Co, 58-64; asst dir agr res, Holly Sugar Co, 64-80, dir, Holly Seed Div, 80-82; plant breeder, Great Western Sugar Co, 82-84; SR PLANT BREEDER, HILLESHOG-MONOHY INST, 84- *Mem:* Am Soc Agron; Am Study Sugarbeet Technologists. *Res:* Development of sugar beet varieties for North America. *Mailing Add:* 11939 Sugarmill Rd Hilleshog-Mono Hy Longmont CO 80501

HELMERS, DONALD JACOB, b Quincy, Ill, Jan 8, 17; m 43; c 3. MECHANICAL ENGINEERING. *Educ:* Tex Tech Col, BS, 48; Univ Mich, MS, 50; Tex A&M Univ, PhD(mech eng), 65. *Prof Exp:* Power plant engr, Ford Motor Co, 36-41; mem teaching staff, Ford Aircraft Sch, 41-43; from instr to assoc prof, 48-64, PROF MECH ENG, TEX TECH UNIV, 65- *Concurrent Pos:* Consult, Hughes Aircraft Co. *Mem:* Am Soc Mech Engrs; Am Soc Eng Educ; Nat Soc Prof Engrs. *Res:* Fluid dynamics and heat transfer, especially the use of analog and digital computers for a model study. *Mailing Add:* Dept Mech Eng Tex Tech Univ Lubbock TX 79409

HELMHOLZ, AUGUST CARL, b Evanston, Ill, May 24, 15; m 38; c 4. HIGH ENERGY PHYSICS. *Educ:* Harvard Univ, AB, 36; Univ Calif, PhD(physics), 40. *Hon Degrees:* ScD, Univ Strathclyde, 79. *Prof Exp:* Physicist, Radiation Lab, 40-42 & 46-58, physicist, Manhattan Proj, 43-46, from instr to assoc prof, Univ, 40-51, chmn dept, 55-62, prof, 51-80, EMER PROF PHYSICS, UNIV CALIF, BERKELEY, 80-, PHYSICIST, LAWRENCE BERKELEY LAB, 58- *Concurrent Pos:* Guggenheim fel, 62-63; with Europ Orgn Nuclear Res, 62-63; mem gov bd, Am Inst Physics, 64-66. *Mem:* AAAS; Am Phys Soc; Am Asn Physics Teachers; Sigma Xi. *Res:* Nuclear reactions; nuclear physics; pion nucleon interactions. *Mailing Add:* 28 Crest Rd Lafayette CA 94549

HELMICK, LARRY SCOTT, b Traverse City, Mich, Nov 18, 41; m 62; c 2. ORGANIC CHEMISTRY, SCIENCE EDUCATION. *Educ:* Cedarville Col, BS, 63; Ohio Univ, PhD(org chem), 68. *Prof Exp:* PROF CHEM, CEDARVILLE COL, 68- *Concurrent Pos:* Fel, Univ Fla, 69-71 & 74-75, Univ Ill, 72-74; fac fel, NASA-Lewis Res Ctr, 80-87, fel, Nat Res Coun, 88-89. *Mem:* AAAS; Am Chem Soc; Creation Res Soc. *Res:* Base-catalyzed hydrogen-deuterium exchange in aromatic nitrogen heterocycles; synthesis, structure determination, reactivity and theoretical chemistry of nitrogen heterocycles; racemization of amino acids; rates of dripstone deposition; origin of optical activity; search for Noah's Ark; Meisenheimer Adducts; thermal stability of fuels and lubricants. *Mailing Add:* Dept Chem Cedarville Col Cedarville OH 45314-0601

HELMINGER, PAUL ANDREW, b Mills Springs, NC, Sept 5, 41; m 64; c 1. MOLECULAR SPECTROSCOPY. *Educ:* NC State Univ, BS, 63; Duke Univ, PhD(physics), 69. *Prof Exp:* Teaching asst physics, Duke Univ, 63-64, res asst, 64-69, res assoc & instr, 69-73; asst prof, 73-77, assoc prof, 77-81, PROF PHYSICS, UNIV SOUTH ALA, 81- *Mem:* Am Phys Soc; Sigma Xi; Am Asn Physics Teachers. *Res:* Microwave rotational spectroscopy of gases at high microwave frequencies. *Mailing Add:* Dept Physics Univ SAla Mobile AL 36688

HELMINIAK, THADDEUS EDMUND, b Chicago, Ill, Aug 11, 35; m 59; c 2. CHEMISTRY. *Educ:* John Carroll Univ, BSNS, 57, MS, 59; Univ Akron, PhD(chem), 62. *Prof Exp:* Res chemist, Inst Rubber Res, Univ Akron, 60-62; Nat Acad Sci res fel polymer chem, aeronaut systs div, Wright-Patterson AFB, 62-63; res chemist, Res Inst, Univ Dayton, 63-64; RES CHEMIST, POLYMER BR, AIR FORCE MAT LAB, WRIGHT-PATTERSON AFB, 64- *Concurrent Pos:* Adj prof, Univ Dayton, 73- *Mem:* Am Chem Soc. *Res:* Physics and physical chemistry of polymers; solution properties of macromolecules; physical property-molecular structure relationships. *Mailing Add:* 405 Orchard Dr Dayton OH 45419-1726

HELMKAMP, GEORGE KENNETH, b Nebr, Feb 24, 21; m 47; c 4. ORGANIC CHEMISTRY. *Educ:* Wartburg Col, BA, 42; Claremont Cols, MA, 50; Calif Inst Technol, PhD(chem), 53. *Prof Exp:* Control chemist, Atmospheric Nitrogen Corp, 42-43; analyst, Wilshire Oil Co, 43-45; chemist, Socal Oil & Refining Co, 47-49; instr org chem, Pomona Col, 52-53; from instr to assoc prof, 53-65, chmn dept, 70-75, assoc dean, Col Nat & Agr Sci, 75-77, PROF CHEM, UNIV CALIF, RIVERSIDE, 65- *Mem:* Am Chem Soc. *Res:* Synthesis and properties of small-ring systems; stereochemistry of the reaction of epoxides; interactions of purines and pyrimidines. *Mailing Add:* Dept Chem Univ Calif PO Box 112 Riverside CA 92521

HELMKAMP, GEORGE MERLIN, JR, b Washington, DC, Nov 29, 43; m 67; c 2. BIOCHEMISTRY. *Educ:* Rensselaer Polytech Inst, BS, 65; Harvard Univ, PhD(biochem & molecular biol), 70. *Prof Exp:* Res chemist, US Army Inst Surg Res, 70-73; fel biochem, State Univ Utrecht, 73-75; from asst prof to assoc prof, 75-86, PROF BIOCHEM, UNIV KANS MED CTR, 86- *Mem:* Sigma Xi; AAAS; Am Soc Biochem Molecular Biol; Biophys Soc. *Res:* Lipid-protein interactions, especially in biological and artificial membranes; phospholipid transfer proteins and their mechanism of action; metabolism of phosphatidylinositol. *Mailing Add:* Dept Biochem Med Ctr Univ Kans 39th & Rainbow Kansas City KS 66103-3410

HELMREICH, ERNST J M, b Munich, Germany, July 1, 22; nat US; m 49; c 2. BIOCHEMISTRY. *Educ:* Univ Munich & Erlangen, MD, 49. *Prof Exp:* Fel, German Res Coun, Munich Tech Univ, 49-53; Nat Acad Sci-Int Coop Admin foreign fel & privat docent, Med Sch, Univ Munich, 54-56; from asst prof to assoc prof biol chem, Med Sch, Wash Univ, 56-68; PROF PHYSIOL CHEM & CHMN DEPT, MED SCH, WURZBURG UNIV, 68- *Mem:* Am Soc Biol Chem; German Chem Soc; German Soc Biol Chem (pres, 73-75); Am Chem Soc; Sigma Xi; German Acad Sci. *Res:* Regulation of enzymes on the level of cellular and molecular organization; hormonal activation of adenylate cyclase; role of pyridoxal-5'-P in glycogen phosphorylase. *Mailing Add:* Dept Physiol Chem Wurzburg Univ Koellikerstr Two Wurzburg 8700 Germany

HELMS, BOYCE DEWAYNE, b Tuckerman, Ark, June 27, 37; m 59; c 2. INDUSTRIAL CHEMISTRY, ANALYTICAL CHEMISTRY. *Educ:* Harding Col, BS, 59; Univ Ark, MS, 62. *Prof Exp:* Chemist, Escambia Chem Corp, 62-67; chief chemist, Arkla Chem Corp, 67-70; plant chemist, Air Prod & Chem Inc, 70-80; CONSULT, 80- *Mem:* Am Chem Soc. *Mailing Add:* 3900 Buckner Lane Paducah KY 42001

HELMS, CARL WILBERT, b New Brighton, Pa, May 26, 33; m 70; c 5. ZOOLOGY, PHYSIOLOGICAL ECOLOGY. *Educ:* Univ Colo, BA, 55; Harvard Univ, AM, 56, PhD(biol), 60. *Prof Exp:* From asst prof to assoc prof biol, Bucknell Univ, 60-68, chmn dept, 64-68; mem Inst Ecol, NSF sci fac fel zool, Univ Ga, 66-67, assoc prof, 68-75, grad coordr, 69-75; head dept, 75-87, PROF ZOOL, CLEMSON UNIV, 75- *Concurrent Pos:* NIH grant, 62-65; health scientist adminstr & grants assoc, NIH, 70-71. *Mem:* Fel AAAS; Am Soc Zoologists; Ecol Soc Am; Animal Behavior Soc; Am Ornith Union. *Res:* Bioenergetics; annual cycles in birds; seasonal variations in fat and fatty acid reserves; seasonal variations in endocrine function; control of lipogenesis; migratory restlessness; proximate factors and migration; biometry. *Mailing Add:* Dept Biol Sci Clemson Univ Clemson SC 29634-1903

HELMS, JOHN ANDREW, b Esperance, Western Australia, Sept 29, 31; m 60. FORESTRY. *Educ:* Univ Sydney, BSc, 53; Univ Wash, MF, 60, PhD(silvicult), 63. *Prof Exp:* Forestry officer, Tasmanian Forestry Comn, 53-58; res assoc silviculture, Univ Wash, 63-64; lectr & assoc specialist, 64-65, asst prof forestry, 65-71, ASSOC PROF FORESTRY, UNIV CALIF, BERKELEY, 71-, ASST CHMN DEPT, 73-, SILVICULTURIST, 65- *Res:* Tree physiology; growth of trees in relation to environment; evaluation of photosynthesis; transpiration and microclimate under field conditions. *Mailing Add:* Sch Forestry & Conserv Univ Calif Berkeley CA 94720

HELMS, JOHN F, b Neenah, Wis, Apr 7, 19; m 46; c 2. ORGANIC CHEMISTRY. *Educ:* Lawrence Col, BA, 41; Univ Wis, PhM, 43. *Prof Exp:* Asst, Univ Wis, 42-43; chemist, Am Can Co, 43-44 & 45-47; chemist, Marathon Corp, 47-53, group leader coatings, 53-57; group leader coatings, Am Can Co, 57-65, supvr formulations, 65-67, mgr lab serv, Packaging Tech Serv, 67-73, assoc dir lab serv, packaging res & develop, 73-80; RETIRED. *Mem:* Am Chem Soc; Tech Asn Pulp & Paper Indust; Am Soc Testing & Mat. *Res:* Protective and functional coatings and laminations for the food packaging industry. *Mailing Add:* 693 Congress Pl Neenah WI 54956

HELMS, LESTER LAVERNE, b Peoria, Ill, Dec 29, 27; m 52; c 3. MATHEMATICS. *Educ:* Bradley Univ, BS, 52; Purdue Univ, MS, 53, PhD, 56. *Prof Exp:* Mathematician, Convair Div, Gen Dynamics Corp, 55-57; asst prof math, Mich State Univ, 56-58; from asst prof to assoc prof, 58-68, PROF MATH, UNIV ILL, URBANA, 68- *Mem:* Am Math Soc; Inst Math Statist. *Res:* Differential equations and probability theory and their interrelations. *Mailing Add:* Dept Math 1409 W Green St Univ Ill Urbana IL 61801

HELMS, THOMAS JOSEPH, b St Louis, Mo, Dec 20, 39; m 58; c 4. INSECT PATHOLOGY, BIOLOGICAL CONTROL. *Educ:* Ark State Univ, BS, 62, MS, 63; Iowa State Univ, PhD(entom), 67. *Prof Exp:* Instr entom, Iowa State Univ, 66-67; from asst prof to assoc prof, Univ Nebr, 67-78, asst dean agr, 73-78; prod develop mgr, Monsanto Co, 78-81; prof entom & head dept, Miss State Univ, 81-88; ASST DIR, MISS AGR & FORESTRY EXP STA, 88- *Concurrent Pos:* Chair, Sect F, Entom Soc Am, 85-86; pres, Miss Asn Regist Prof Entomologists, 87-88. *Mem:* Entom Soc Am. *Res:* Agriculture and forestry; insect pathology and biological control. *Mailing Add:* Miss Agr & Forestry Exp Sta PO Drawer ES Mississippi State MS 39762

HELMS, WARD JULIAN, b Everett, Wash, June 12, 38; m 64; c 3. ELECTRICAL ENGINEERING, GEOPHYSICS. *Educ:* Wash State Univ, BS, 60; Univ Wash, MS, 63, PhD(elec eng), 68. *Prof Exp:* Field engr, Stanford Univ, 61-63; res assoc elec eng, 63-68, asst prof elec eng & geophys & NSF grant, Byrd Sta, Antarctica, 68-75, ASSOC PROF ELEC ENG & GEOPHYS, UNIV WASH, 75- *Concurrent Pos:* NASA grant, 71- *Mem:* Inst Elec & Electronics Engrs; Am Geophys Union; Sigma Xi. *Res:* Very low frequency sounding of the D region; electronic circuit design philosophy; radioscience. *Mailing Add:* 5415 157th Dr NE Redmond WA 98052

HELMSEN, RALPH JOHN, b St Louis, Mo, May 28, 32; m 65; c 3. BIOCHEMISTRY, OPHTHALMOLOGY. *Educ:* Wash Univ, AB, 53; Univ Mo, MS, 59; St Louis Univ, PhD(biochem), 66. *Prof Exp:* Staff fel ophthal chem, Nat Inst Neurol Dis & Blindness, 65-69; res chemist, 69-76, CHIEF, ANTERIOR SEGMENT DISEASE BR NAT EYE INST, 75- *Mem:* Am Chem Soc; Asn Res Vision & Ophthal. *Res:* Isolation and characterization of corneal antigens; macromolecules of vitreous body of eye; nutritional effects on ocular diseases. *Mailing Add:* 9602 Dewmar Lane Kensington MD 20895

HELMSTETTER, CHARLES E, b Newark, NJ, Oct 18, 33; m 57; c 2. MICROBIOLOGY. *Educ:* Johns Hopkins Univ, BA, 55; Univ Mich, MS, 56; Univ Chicago, MS, 57, PhD(biophys), 61. *Prof Exp:* Scientist molecular biol, NIH, 61-63; USPHS fel, Univ Inst Microbiol, Copenhagen, 63-64; assoc cancer res scientist, Roswell Park Mem Inst, 64-69, prin cancer res scientist, 69-74, dir cancer res & head dept biol, 74-89; PROF & DIR CELL BIOL, FLA INST TECHNOL, 89- *Concurrent Pos:* Nat Cancer Inst res grant, Roswell Park Mem Inst, 65-89; res asst prof, State Univ NY Buffalo, 65-70, res prof, 70-90; res prof, Niagara Univ, 68-89 & Canisius Col, 68-89. *Honors & Awards:* Selman A Waksman Award, 70. *Mem:* AAAS; Am Soc Microbiol; Am Soc Biol Chemists. *Res:* Cell growth and division; macromolecular synthesis; cell cycle. *Mailing Add:* Dept Biol Fla Inst Technol Melbourne FL 32901-6988

HELMSWORTH, JAMES ALEXANDER, b Jamestown, NDak, Mar 31, 15; m; c 3. THORACIC SURGERY. *Educ:* Jamestown Col, BS, 35; Univ Pa, MD, 39. *Prof Exp:* Intern, Hosp Univ Pa, 39-41; surg resident, Cincinnati Gen Hosp, 41-44 & 47-49; from asst clin prof to assoc prof, 49-69, PROF SURG, UNIV CINCINNATI, 69-, DIR CARDIOVASC & THORACIC SURG, 70- *Concurrent Pos:* Mem attend staff, Cincinnati Gen Hosp, Christian R Holmes Hosp & Children's Hosp; assoc attend staff, Good Samaritan Hosp; surg consult, Vet Admin Hosp; consult thoracic surgeon, Bethesda Hosp & Shriners Burns Inst; prin investr, USPHS, Ohio State-Am Heart Asn & Southwestern Ohio Heart Asn res grants; mem, Am Bd Surg. *Mem:* Sr mem Am Asn Thoracic Surg; Am Col Cardiol; Am Col Surgeons; Am Heart Asn; Soc Vascular Surg. *Res:* Cardiac transplants; heterografts; homograft reconstruction; bronchography; counterpulsation with intra-aortic balloon in treatment of cardiogenic shock; deep hypothermia and circulatory arrest for intracardiac surgery in infants; use of aortic shunt in treatment of acute and chronic thoracic aneurysms. *Mailing Add:* Dept Surg Cincinnati Gen Hosp Cincinnati OH 45229

HELMUTH, HERMANN SIEGFRIED, b Dresden, Germany, Mar 27, 39; m 64; c 2. PHYSICAL ANTHROPOLOGY, PRIMATOLOGY. *Educ:* Univ Kiel, Dr rer nat, 64, Dr habil(phys anthrop), 75. *Prof Exp:* Sci asst, Anthrop Inst, Kiel, WGermany, 64-69; asst prof, 68-71, assoc prof, 72-79, PROF PHYS ANTHROP, TRENT UNIV, 79- *Concurrent Pos:* Vis prof, Univ Guelph, 82-83. *Mem:* Ger Anthrop Soc; Can Asn Phys Anthrop. *Res:* Human osteology, evolution; anthropometry and human factors engineering; behavior of fossil man. *Mailing Add:* Dept Anthrop Trent Univ Peterborough ON K9J 7B8 Can

HELQUIST, PAUL M, b Duluth, Minn, Mar 5, 47; m 70; c 2. ORGANOMETALLIC CHEMISTRY. *Educ:* Univ Minn, BA, 69; Cornell Univ, MS, 71, PhD(org chem), 72. *Hon Degrees:* PhD, Univ Uppsala, Sweden, 88. *Prof Exp:* Teaching asst chem, Cornell Univ, 69-71, res asst, 71-72; fel, Harvard Univ, 73-74; from asst prof to prof chem, State Univ NY, Stony Brook, 74-86; PROF CHEM, UNIV NOTRE DAME, 84-, CHMN, DEPT CHEM & BIOCHEM, 88- *Concurrent Pos:* NSF fel chem, Cornell Univ, 69-72; instr synthetic org chem, Am Chem Soc, 81-; Bergquist fel, Am Scandinavian Found, 82. *Mem:* Am Chem Soc. *Res:* New Synthetic methods; applications of organometallic compounds in synthetic organic chemistry; total synthesis of biologically active compounds. *Mailing Add:* Dept Chem Univ Notre Dame Notre Dame IN 46556

HELRICH, CARL SANFRID, JR, b Everett, Mass, Sept 12, 41; m 67; c 2. STATISTICAL MECHANICS, QUANTUM MAGNETISM. *Educ:* Case Inst Technol, BS, 63; Northwestern Univ, PhD(plasma physics), 69. *Prof Exp:* Asst prof statist mech plasma physics, Space Inst, Univ Tenn, 69-71; scientist res, Kernsforschungsanlage Jülich, Germany, 71-74; assoc prof & dept chair physics beginning & advan physics & lab, Bethel Col, Kansas, 76-85; PROF & DEPT CHAIR PHYSICS BEGINNING & ADVAN PHYSICS & LAB, GOSHEN COL, 85- *Concurrent Pos:* NASA/Am Soc Elec Eng fel res, Marshall Space Flight Ctr, 71; actg dir, Turner Lab, Goshen Col, 86-87, 88-89; consult Mat Sci, CTS Corp, 86; guest scientist, Kernforschungsanlage Jülich, 82-83. *Mem:* Am Phys Soc; Sigma Xi; Coun Undergrad Res; NY Acad Sci; Fed Am Scientists. *Res:* Condensed matter theory; statistical mechanics of quantum electronics including diluted magnetic semiconductors and heterostructures: superlattices/quantum wells. *Mailing Add:* Dept Physics Goshen Col Goshen IN 46526

HELRICH, KENNETH, pesticide chemistry, for more information see previous edition

HELRICH, MARTIN, b New York, NY, Mar 31, 22; m 50; c 2. ANESTHESIOLOGY. *Educ:* Dickinson Col, BS, 46; Univ Pa, MD, 46. *Prof Exp:* Resident anesthesiol, NY Univ, 48-51; res assoc, Sch Med, Univ Pa, 53-54, asst prof, 55-56; PROF ANESTHESIOL & HEAD DEPT, SCH MED, UNIV MD, BALTIMORE CITY, 56- *Concurrent Pos:* Consult, US Naval Med Ctr, Bethesda, Md, Walter Reed Army Med Ctr, Washington, DC, US Naval Med Ctr, Portsmouth, Va, Baltimore City Hosp; gov, Am Col Anesthesiologists, 71-74; dir, Am Bd Anesthesiol, 74-, secy-treas, 81-85, pres 85-86. *Mem:* Am Soc Pharmacol & Exp Therapeut; AMA; Asn Univ Anesthetists; Am Soc Anesthesiol; Am Col Anesthesiologists. *Res:* Physiology and pharmacology of anesthesia. *Mailing Add:* Anesthesia Educ & Res 10 S Pine St Baltimore MD 21201

HELSDON, JOHN H, JR, b Buffalo, NY, Oct 29, 48; m 77; c 2. ATMOSPHERIC ELECTRICITY, NUMERICAL CLOUD MODELING. *Educ:* Trinity Col, BS, 70; State Univ NY Albany, MS, 73, PhD(atmospheric sci), 79. *Prof Exp:* Res asst, Res Found, State Univ NY, 77-79; RES SCIENTIST, INST ATMOSPHERIC SCI, SDAK SCH MINES & TECHNOL, 79-, ASSOC PROF METEROL, 81- *Concurrent Pos:* Prin investr, NSF res grant 84-90; prin investr, NASA res grant, 84-86; mem, Comt Atmospheric Elec, Am Meteorol Soc, 86-88 & 90- *Mem:* Am Meteorol Soc; Am Geophys Union; Sigma Xi. *Res:* Numerical modeling of thunderstorm electrification mechanisms in the context of a two dimensional cloud model with emphasis on determining which charge separation mechanisms are important. *Mailing Add:* Inst Atmospheric Sci SDak Sch Mines & Technol Rapid City SD 57701-3995

HELSEL, ZANE ROGER, b East Freedom, Pa, Mar 5, 49; m 78. CROP SCIENCE, AGRONOMY. *Educ:* Penn State Univ, BS, 71, MS, 73; Iowa State Univ PhD(agron), 77. *Prof Exp:* Res assoc, 73-75, instr agron, Iowa State Univ, 75-77; asst prof crop & soil sci, Mich State Univ, 77-; ASST PROF, DEPT AGRON, UNIV MO. *Mem:* Am Soc Agron; Crop Sci Soc Am; Am Soc Agr Eng. *Res:* Soybean production; forage quality and production; energy in agriculture. *Mailing Add:* Dept Agron 214 Waters Hall Univ Mo Columbia MO 65211

HELSETH, DONALD LAWRENCE, JR, b Corvallis, Ore, Sept 23, 53; m; c 2. HEALTHCARE. *Educ:* Tex Christian Univ, Ft Worth, BS, 75; Northwestern Univ, PhD(biochem), 81. *Prof Exp:* Postdoctoral res fel, Dept Biochem, Rutgers Med Sch, Univ Med & Dent NJ, Piscataway, 81-82; postdoctoral res fel, Dent Sch, Northwestern Univ, 82-83, from instr to asst prof, 83-89; sr info specialist, 89-90, info consult, 90-91, SCI COLLAB CONSULT, INFO RESOURCE CTR, BAXTER HEALTHCARE CORP, 91- *Concurrent Pos:* Partic, Gordon Res Conf Collagen, 83; spec consult, Site Visit Team, Nat Inst Dent Res, 84; mem, Radiation Safety Comt, Northwestern Univ, 85-88 & Dent Coun, Dent Sch, 87-89; mem, Spec Grants Rev Comt, Nat Inst Dent Res, 87-91. *Mem:* AAAS; Am Chem Soc; Am Soc Biochem & Molecular Biol; Sigma Xi. *Mailing Add:* Info Resource Ctr William Graham Bldg 1-1N Baxter Healthcare Corp Rte 120 & Wilson Rd Round Lake IL 60073-0490

HELSLEY, CHARLES EVERETT, b Oceanside, Calif, June 24, 34; m 55; c 4. GEOLOGY, GEOPHYSICS. *Educ:* Calif Inst Technol, BS, 56, MS, 57; Princeton Univ, PhD(geol), 60. *Prof Exp:* Asst prof geol, Calif Inst Technol, 60-62; asst prof, Case Western Reserve Univ, 62-63; from asst prof to assoc prof, Southwest Ctr for Advan Studies, Univ Tex, Dallas, 63-69, prof geosci, 69-76, assoc head geosci div, 71-72, head geosci prog & dir inst geosci, 72-75; PROF GEOL & GEOPHYS, UNIV HAWAII, MANOA; DIR, HAWAII INST GEOPHYS,. *Concurrent Pos:* Adj prof, Southern Methodist Univ, 63-76 & Univ Tex Marine Sci Inst, Galveston, 73-76. *Mem:* Geol Soc Am; Am Geophys Union. *Res:* Rock magnetism and paleomagnetism and their implications regarding continental drift; marine geophysics; magnetostratigraphy; geothermal resource exploration. *Mailing Add:* Hawaii Inst Geophysics HIG 131 Univ Hawaii Manoa 2525 Correa Rd Honolulu HI 96822

HELSLEY, GROVER CLEVELAND, b Strasburg, Va, Sept 26, 26; m 48; c 3. ORGANIC CHEMISTRY, MEDICINAL CHEMISTRY. *Educ:* Shepherd Col, BS, 54; Univ Va, MS, 56, PhD(chem), 58. *Prof Exp:* Res chemist, E I du Pont de Nemours & Co Inc, 58-62 & A H Robins Co, 62-64, group leader, 64-68, assoc dir chem res, 68-70; dir res, 70-72, vpres, 72-87, SR VPRES PHARMACEUT RES, HOECHST-ROUSSEL PHARMACEUTICAL, INC, 87- *Mem:* Am Chem Soc; Sigma Xi. *Res:* Chemistry of heterocyclic compounds; fluoride displacement on aromatic rings. *Mailing Add:* Pharmaceut Res Dept Hoechst Roussel Pharmaceut 1041 Rte 202-206N Somerville NJ 08876

HELSON, HENRY, b Lawrence, Kans, June 2, 27; m 54; c 3. MATHEMATICS. *Educ:* Harvard Univ, AB, 47, PhD(math), 51. *Prof Exp:* Lectr math, Univ Uppsala, 50-51; instr, Yale Univ, 51-52, Jewett fel, 52-54, asst prof math, 54-55; from asst prof to assoc prof, 55-61, PROF MATH, UNIV CALIF, BERKELEY, 61- *Concurrent Pos:* Vis prof, France, Ghana, India & Sweden. *Mem:* Am Math Soc. *Res:* Harmonic analysis; function theory. *Mailing Add:* Dept Math Univ Calif Berkeley CA 94720

HELSON, LAWRENCE, b New York, NY, Mar 21, 31; c 3. ONCOLOGY, NEUROBIOLOGY. *Educ:* City Col New York, AB, 53; NY Univ, MS, 57; Univ Geneva, MD, 62; Am Bd Pediat, cert, 69. *Prof Exp:* Asst res & teaching, Dept Bact, Inst Med & Hyg, Univ Geneva, 62-63; instr path, New York City Community Col, 65-66; USPHS sr clin trainee, Mem Hosp, 67-68; from instr to assoc prof, Med Col, Cornell Univ, 68-86; RETIRED. *Concurrent Pos:* From clin asst pediatrician to assoc attend pediatrician, Mem Hosp, 68-74, attend pediatrician, 74-86; assoc, Sloan-Kettering Inst Cancer Res, 72-86. *Mem:* Harvey Soc; Am Soc Clin Oncol; fel Am Acad Pediat; Am Asn Cancer Res; Soc Pediat Res; Sigma Xi. *Res:* Clinical cancer, particularly pediatric; chemotherapy, neurobiology, specific tumors, neuroblastoma, neurosarcoma, neurofibroma; brain tumors; laboratory-nerve growth stimulating substance; tumor cell cyclic nucleotide effectors, phosphodiesterases; vitamins E, B6, C and cancer autologous bone marrow transplantation, catecholamine metabolism. *Mailing Add:* New York Med Col Valhalla NY 10595

HELSTROM, CARL WILHELM, b Easton, Pa, Feb 22, 25; m 56; c 2. APPLIED PHYSICS. *Educ:* Lehigh Univ, BS, 47; Calif Inst Technol, MS, 49, PhD(physics), 51. *Prof Exp:* Adv mathematician, Res Labs, Westinghouse Elec Corp, 51-66; PROF ELEC ENG, UNIV CALIF, SAN DIEGO, 66 - *Concurrent Pos:* Lectr, Univ Calif, Los Angeles, 63-64; ed, Inst Elec & Electronics Engrs Trans Info Theory, 67-71. *Honors & Awards:* Centennial Medal, Inst Elec & Electronics Engrs, 84. *Mem:* Fel Inst Elec & Electronics Engrs; fel Optical Soc Am. *Res:* Signal detection theory; optics; stochastic processes; communication theory. *Mailing Add:* Univ Calif R-007 La Jolla CA 92093-0407

HELTEMES, EUGENE CASMIR, b St Cloud, Minn, Sept 26, 34; m 56; c 2. PHYSICS, MAGNETISM. *Educ:* St John's Univ, Minn, BA, 56; Iowa State Univ, PhD(physics), 61. *Prof Exp:* Asst, Iowa State Univ, 56-57; asst, Ames Labs, AEC, 57-61; sr res physicist, Cent Res Labs, Minn Mining & Mfg Co, 63-67, supvr, Instrumentation & Control Systs Eng Dept, 67-71, res specialist, 71-75, Lab Mgr, 75-83, SR RES SPECIALIST, 3M CO, 83- *Mem:* Am Phys Soc. *Res:* Low temperature physics; optics; solid state physics; instrumentation; magnetic properties. *Mailing Add:* 2537 Orchard Lane White Bear Lake MN 55110

HELTNE, PAUL GREGORY, b Lake Mills, Iowa, July 4, 41; div; c 2. EVOLUTIONARY BIOLOGY, GROSS ANATOMY. *Educ:* Luther Col, Iowa, BA, 62; Univ Chicago, PhD(primatology), 70. *Prof Exp:* Instr anat & biol, Univ Chicago, 70; asst prof anat, animal med & pathobiol, Johns Hopkins Univ, 70-79; DIR, CHICAGO ACAD SCI, 82- *Concurrent Pos:* Consult, Sangamon State Univ, 70 & 71 & Dept Anat, Med Sch, Harvard Univ, 71; mem, Comt Conservation Nonhuman Primates, Ilar, Nat Res Coun, 72-75; assoc ed, Growth, 73-, assoc ed-in-chief, 78-87; planning coordr, Baltimore Zoo & Druid Hill Park, 74-76; consult, Pan Am Health Orgn, Bolivia, 75, Northern Colombia, 77 & Peru, 80-82. *Mem:* AAAS; Soc Study Evolution; Am Soc Primatologists; Int Primatological Soc; Sigma Xi. *Res:* Biology of South American primates; demography of New World primates; long term capture reproduction of Aotus, the owl monkey; physical anthropology of humans specifically Hottentots and XYY males, including their cytogenetics; evolution and ontogeny of form; morphology of genetic syndromes; population biology and ecology; sustained yield management of natural ecosystems. *Mailing Add:* 2001 N Clark St Chicago IL 60614

HELTON, AUDUS WINZLE, b Bethel, Okla, Oct 5, 22; m 45; c 5. PLANT PATHOLOGY. *Educ:* Ohio Wesleyan Univ, BA, 47, MS, 48; Ore State Univ, PhD(plant path), 51. *Prof Exp:* From asst prof & asst plant pathologist to assoc prof & assoc plant pathologist, 51-63, PROF PLANT PATH & PLANT PATHOLOGIST, UNIV IDAHO, 63- *Mem:* AAAS; Am Phytopath Soc. *Res:* Tree fruit pathology. *Mailing Add:* Dept Plant & Soil Sci Univ Idaho Moscow ID 83844

HELTON, F JOANNE, b Can, 45; m 68. APPLIED MATHEMATICS, PLASMA PHYSICS. *Educ:* Univ BC, BS, 64; Stanford Univ, MS, 66, PhD(math), 68. *Prof Exp:* From instr to asst prof appl anal, State Univ Ny, Stony Brook, 68-70; asst prof, 70-74, dept coordr, Dowling Col, 71-73; SR STAFF SCIENTIST, FUSION DIV, GEN ATOMIC, 74- *Mem:* Am Math Soc; Soc Indust & Appl Math; Am Phys Soc; Inst Elec & Electronic Engrs. *Res:* Algebraic automata theory; computational physics. *Mailing Add:* Fusion Div Gen Atomics San Diego CA 92138-5608

HELTON, JOHN W, b Jacksonville, Tex, Nov 21, 44; m 68. MATHEMATICS, ELECTRONICS ENGINEERING. *Educ:* Univ Tex, BA, 64; Stanford Univ, MS, 66, PhD(math), 68. *Prof Exp:* Asst prof math, State Univ NY Stony Brook, 68-74; assoc prof, 74-78, PROF MATH, UNIV CALIF, SAN DIEGO, 78- *Concurrent Pos:* Guggenheim Fel, 85. *Mem:* Am Math Soc; Inst Elec & Electronic Engrs; Soc Indust & Appl Math. *Res:* Generalized spectral operators; electrical networks and operator theory; integral operators; theory of electronic circuits with prescribed power gain; contol theory. *Mailing Add:* Dept Math C-012 Univ Calif San Diego Box 109 La Jolla CA 92093

HELTON, WALTER LEE, b Sloans Valley, Ky, Nov 4, 33; m 58; c 2. ECONOMIC GEOLOGY, STRATIGRAPHY. *Educ:* Univ Ky, BS, 59, MS, 64; Univ Tenn, Knoxville, PhD(geol), 67. *Prof Exp:* Petrol geologist, Ferguson & Bosworth Oil Co, 59-61; econ geologist, Ky Geol Surv, 61-64; from asst prof to assoc prof, 66-77, PROF GEOL, TENN TECHNOL UNIV, 77-, CHMN DEPT, 68- *Mem:* AAAS; Geol Soc Am; Nat Asn Geol Teachers. *Res:* Silurian and Devonian stratigraphy of Kentucky; geology of the Conasauga group of the southern Appalachians. *Mailing Add:* Box 5062 Cookeville TN 38501

HELWEG, OTTO JENNINGS, b Kalamazoo, Mich, Feb 1, 36; m 64; c 3. WATER RESOURCES ENGINEERING, GROUND WATER. *Educ:* US Naval Acad, BS, 58; Fuller Theological Seminary, MDiv, 66; Univ Calif, Los Angeles, MS, 67; Colo State Univ, PhD(civil eng), 75. *Prof Exp:* Officer, USN, 58-62; youth worker, Hollywood Presby Church, 62-66; missionary, United Presby Church, 67-73; from asst prof to assoc prof civil eng, Univ Calif, Davis, 75-83; vis prof, Tex A&M Univ, 83-87; CHAIR & PROF CIVIL ENG, MEMPHIS STATE UNIV, 87- *Concurrent Pos:* Consult, UN Develop Prog/UN Educ Sci & Cult Orgn, India, 79; pres, Helweg & assoc, 80-; ed, J Irrigation & Drainage, Am Soc Civil Engrs, 89- *Honors & Awards:* Nat Sci Award, Nat Water Well Assoc, 83. *Mem:* Fel Am Soc Civil Engrs; Am Geophys Union; Am Water Works Asn; Am Soc Agr Engrs; Am Soc Eng Educ. *Res:* Water resources planning and management; ground water hydrology; well hydraulics; computer modeling. *Mailing Add:* 3684 Briar Trail Cove Bartlett TN 38135

HELWIG, HAROLD LAVERN, b Mendon, Mich, May 20, 17; m 44; c 3. BIOCHEMISTRY. *Educ:* Mich State Univ, BS, 39; Univ Calif, PhD(biochem), 52. *Prof Exp:* Res biochemist, Donner Lab Med Physics, Univ Calif, 51-53; prin scientist, Radioisotope Unit, Vet Admin Hosp, Calif, 53-56; chief air & indust hyg lab, Calif State Dept Pub Health, 56-63, chief cancer biochem lab, 63-66; independent consult chem, 66-78; RETIRED. *Mem:* AAAS; Am Chem Soc. *Res:* Trace element nutrition and intermediary metabolism; hematopoiesis and red blood cell dyscrasias; biochemical defects in alcoholism; radiochemical tracer techniques; chemistry and biological effects of air pollution; chemical carcinogenesis. *Mailing Add:* 442 Panoramic Way Berkeley CA 94704

HELWIG, JAMES A, b Abington, Pa, Sept 25, 41; m 67; c 1. TECTONICS. *Educ:* St Louis Univ, BS, 63; Columbia Univ, PhD(geol, stratig), 67. *Prof Exp:* Asst prof geol, Case Western Reserve Univ, 67-75; res geologist, Arco Oil & Gas Co, 75-80, dir tectonics res, 81-83, sr res adv, 83-85; sr consult, Schlumberger-Doll Res, 85-88; RES SCIENTIST, MOBIL OIL RES & DEVELOP, 88- *Concurrent Pos:* Vis prof, Univ Vienna, 75; vis sr res assoc, Lamont-Doherty Geol Observ, 84. *Mem:* Am Asn Petrol Geologists; Geol Soc Am. *Res:* Tectonics and stratigraphy of mountain belts and sedimentary basins; modeling of geological environments and processes. *Mailing Add:* Mobil Res & Develop PO Box 819047 Dallas TX 75381

HELWIG, JOHN, JR, b Philadelphia, Pa, Dec 10, 27; m; c 4. INTERNAL MEDICINE, CARDIOLOGY. *Educ:* La Salle Col, AB, 50; Univ Pa, MD, 54; Am Bd Internal Med & Am Bd Cardiovasc Dis, dipl. *Prof Exp:* Intern, Univ Hosp, Univ Pa, 54-55, asst resident & asst instr surg, 55-56, resident & asst instr med, 56-57, fel cardiol & instr med, 57-59, asst prof med, Sch Med & assoc dir cardiovasc clin res ctr, Univ Hosp, 59-65; clin assoc prof, 69-74, CLIN PROF MED, TEMPLE UNIV, 74-; CHIEF CARDIOL, GERMANTOWN HOSP, 65- *Mem:* AAAS; AMA; fel Am Col Physicians; Am Col Cardiol. *Res:* Cardiovascular disease. *Mailing Add:* Germantown Hosp One Penn Blvd Philadelphia PA 19144

HELZ, GEORGE RUDOLPH, b Silver Spring, Md, Mar 4, 42; m 70; c 1. ENVIRONMENTAL GEOCHEMISTRY. *Educ:* Princeton Univ, AB, 64; Pa State Univ, PhD(geochem), 71. *Prof Exp:* From asst prof to assoc prof, 70-84, PROF CHEM, UNIV MD, COLLEGE PARK, 84-; DIR, MD WATER RESOURCES RES CTR, 90- *Concurrent Pos:* Mem disinfectants chem subcomt, Nat Acad Sci-Nat Res Coun, 78; vis prof, Stanford Univ, 83-84; AAAS environ fel, 88; chmn, geochem div, Am Chem Soc, 85; sr vis fel, Manchester Univ, UK, 89-90. *Mem:* Am Geophys Union; Geochem Soc (treas, 75-78); Geol Soc Am; Am Mineral Soc; Am Chem Soc. *Res:* Aqueous geochemistry; geochemistry of mineral deposits; environmental chemistry; fate of pollutants in estuaries. *Mailing Add:* Dept Chem & Biochem Univ Md College Park MD 20742

HELZ, ROSALIND TUTHILL, b Ames Iowa, July 6, 44; m 70; c 1. PETROLOGY. *Educ:* Stanford Univ, BS, 65; Pa State Univ, MS, 68, PhD(geochem/mineral), 78. *Prof Exp:* geologist, US Geol Surv Br of Exp Geochem & Mineral, 68-82, BR OF IGNEOUS & GEOTHERMAL PROCESSES, 82- *Mem:* Mineral Soc Am; Geol Soc Am; Am Geophys Union. *Res:* Petrogenesis of basalts; petrology of the earth's mantle; chemistry of igneous rocks; mineralogy of igneous rocks; experimental determination of phase relations of igneous rocks. *Mailing Add:* US Geol Surv Nat Ctr 959 Reston VA 22092

HELZER, GERRY A, b Portland, Ore, Dec 7, 37; m 65. ALGEBRA. *Educ:* Portland State Col, BA, 59; Northwestern Univ, MA, 61, PhD(math, algebra), 64. *Prof Exp:* Asst prof, 64-74, ASSOC PROF MATH, UNIV MD, COLLEGE PARK, 74- *Mem:* Am Math Soc. *Res:* Linear algebra; numerical analysis. *Mailing Add:* Dept Math Univ Md College Park MD 20742

HEM, JOHN DAVID, b Starkweather, NDak, May 14, 16; wid; c 2. AQUEOUS GEOCHEMISTRY, NON-EQUILIBRIUM THERMODYNAMICS. *Educ:* George Washington Univ, BS, 40. *Prof Exp:* Sci aide chem, Washington, DC, 39-40, jr chemist, Ariz, 40-42, assoc chemist, 43-45, assoc chemist, NMex, 45-48, dist chemist, NMex & Ariz, 48-53, staff chemist, Colo, 53-57, res chemist, 58-63, RES CHEMIST, WATER RES DIV, US GEOL SURV, CALIF, 63- *Concurrent Pos:* Comt on Effects Environ Pollutants, Nat Acad Sci, 76-78. *Honors & Awards:* Sci Award, Nat Water Well Asn, 86; O E Meinzer Award, Geol Soc Am, 90. *Mem:* AAAS; Am Chem Soc; Geochem Soc; Am Water Works Asn; Am Geophys Union; Int Asn Geochem & Cosmochem; Soc Environ Geochem & Health. *Res:* Chemistry of natural waters; geochemical relations of waters to rocks and minerals; chemistry of iron, manganese and aluminum; non-equilibrium controls of aqueous solubility for lead, zinc, cadmium and transition metals. *Mailing Add:* Water Resources Div US Geol Surv MS 427 345 Middlefield Rd Menlo Park CA 94025

HEM, STANLEY L, b New York, NY, Oct 5, 39; m 63; c 2. PHYSICAL PHARMACY. *Educ:* Rutgers Univ, BS, 61; Univ Conn, PhD(phys pharm), 66. *Prof Exp:* Sr res pharmacist, Wyeth Labs, Inc, 65-69; from asst prof to assoc prof, 69-76, assoc dean grad sch, 85-87, PROF PHYS PHARM, SCH PHARM, 76-, ASST VPRES RES, PURDUE UNIV, 87- *Mem:* AAAS; fel Acad Pharmaceut Sci; Am Chem Soc; fel Am Asn Pharmaceut Scientists. *Res:* Suspensions; penicillin kinetics; aluminum hydroxide chemistry. *Mailing Add:* Purdue Univ Sch Pharm West Lafayette IN 47907

HEMAMI, HOOSHANG, b Esfahan, Iran, Apr 9, 36; m 62; c 1. ELECTRICAL ENGINEERING. *Educ:* Univ Tehran, BS, 58; Mass Inst Technol, MS & EE, 62; Ohio State Univ, PhD(elec eng), 66. *Prof Exp:* Res asst elec eng, Mass Inst Technol, 62-63; sr res engr, Nat Cash Register Co, 63-64; res assoc 65-66, asst prof, 66-70, assoc prof, 70-80, PROF ELEC ENG, OHIO STATE UNIV, 80- *Mem:* Inst Elec & Electronics Engrs. *Res:* Pattern recognition; human locomotion; control and nonlinear systems. *Mailing Add:* Dept Elec Eng 2015 Neil Ave Columbus OH 43210-1272

HEMAN-ACKAH, SAMUEL MONIE, b Dunkwa, Ghana, Nov 16, 26; m; c 3. CHEMOTHERAPY, BIOPHARMACEUTICS. *Educ:* Nottingham Univ, BPharm, 54; Pharm Soc Gt Brit, MPS, 55; Univ London, PhD(pharmaceut), 65. *Prof Exp:* From lectr to sr lectr pharmaceut, Univ Sci & Technol, Kumasi, Ghana, 56-61, assoc prof & actg chmn dept, 65-69; res, Univ Fla, 69-71; assoc prof pharmaceut & asst dean, Fla A&M Univ, 72-73; asst dean, 73-76, PROF PHARM & CHMN DEPT, HOWARD UNIV, 73- *Mem:* Brit Pharmaceut Soc; Am Pharmaceut Asn; Acad Pharmaceut Sci; Am Soc Microbiol; Am Chem Soc; Sigma Xi. *Res:* Kinetics and mechanisms of action of drugs on microorganisms; cancer chemotherapy; bioavailability of drugs from multiphase dosage formulations; combined action of mutagenic carcinogens. *Mailing Add:* 1207 Gersham Rd Silver Spring MD 20904

HEMBREE, GEORGE HUNT, b Richmond, Ky, Sept 2, 30; m 52; c 3. PHYSICAL CHEMISTRY. *Educ:* Eastern Ky State Col, BS, 52; Ohio State Univ, PhD(chem), 58. *Prof Exp:* Asst chem, Ohio State Univ, 52-53, asst, Res Found, 55-57; from res chemist to sr res chemist, 58-64, res supvr, 65-68, res mgr, Parlin, NJ, 68-69 & 71-73 & Rochester, NY, 69-71, mkt mgr, Wilmington, Del, 73-74, lab dir, Photo Prod Dept, E I Du Pont DE Nemours & Co, Inc, 74-88; RETIRED. *Mem:* Am Chem Soc; Soc Photog Scientists & Engrs. *Res:* Hydrocarbon oxidation; photographic systems. *Mailing Add:* 207 Newport Rd Hendersonville NC 28739-4395

HEMBROUGH, FREDERICK B, b Jacksonville, Ill, May 1, 24; m 47; c 3. CARDIOVASCULAR PHYSIOLOGY. *Educ:* Univ Ill, BS, 52, DVM, 54, MS, 63, PhD(physiol), 66. *Prof Exp:* Pvt pract vet med, 54-60; instr physiol, Univ Ill, 60-65; from asst prof to assoc prof, 66-73, PROF PHYSIOL, IOWA STATE UNIV, 73- *Concurrent Pos:* NIH fel, 64-65; Iowa Heart Asn grant, 66-67; NIH grant geront studies, 71-73. *Mem:* Am Vet Med Soc; Am Physiol Soc; Am Soc Vet Physiol & Pharmacol; Sigma Xi. *Res:* Cardiovascular research, especially effects of starvation-refeeding and aging on the distensibility of the arterial system; effect of age on isolated canine papillary muscle function; effect of cardiac lesions on heart function. *Mailing Add:* Physiol Dept 2018 Vet Med Iowa State Univ Ames IA 50011

HEMBRY, FOSTER GLEN, b Decatur, Iowa, Jan 3, 41; m 64; c 2. ANIMAL NUTRITION. *Educ:* Iowa State Univ, BS, 63; Univ Tenn, MS, 66; Univ Mo, PhD(animal nutrit), 69. *Prof Exp:* From asst prof to prof animal nutrit, La State Univ, 69-80, prof animal sci & assoc dean grad sch, 80-90; PROF & CHMN ANIMAL SCI, UNIV FLA, 90- *Mem:* Am Soc Animal Sci. *Res:* Forage evaluation; protein metabolism; climatic-nutritional interrelationships; growth stimulants. *Mailing Add:* Dept Animal Sci Univ Fla Gainesville FL 32611

HEMDAL, JOHN FREDERICK, b Peru, Ind, July 29, 34; m 58; c 3. ACOUSTICS. *Educ:* Purdue Univ, BSEE, 57, MSEE, 59, PhD(elec eng), 64. *Prof Exp:* Assoc engr, Appl Physics Lab, Johns Hopkins Univ, 57-58; instr elec eng, Purdue Univ, 58-64; res asst acoust, Univ Mich, Ann Arbor, 64-68, assoc res acoustician, 68-72; res acoustician, Environ Res Inst Mich, 72-74, head acoust & seismics, 74-77; consult, Hearing & Noise Assocs, 77-82; CONSULT, UNIV TOLEDO, 82- *Concurrent Pos:* Consult, Purdue Univ, 67-68,; noise control consult, 75- *Mem:* Inst Elec & Electronics Eng; Acoust Soc Am; Pattern Recognition Soc; Inst Noise Control Eng; Nat Soc Prof Eng. *Res:* Pattern recognition; automatic recognition of speech; acoustic phonetics; atmospheric acoustics; noise control; vibrations; signal processing. *Mailing Add:* Hearing & Noise Assocs 1211 Meanwell Rd Dundee MI 48131-9716

HEMENWAY, MARY-KAY MEACHAM, b Akron, Ohio, Nov 20, 43; m 68; c 2. ASTRONOMY. *Educ:* Notre Dame Ohio, BS, 65; Univ Va, MA, 67, PhD(astron), 71. *Prof Exp:* Lectr physics, Mary Baldwin Col, 70-71 & astron, Univ Va Exten, 70-73; lectr astron, 74-87, SR LECTR ASTRON, UNIV TEX, AUSTIN, 87- *Concurrent Pos:* Consult var sch dist & reg serv ctrs, 81- & Chautauqua Prog Col Sci Teachers, 85-89; reviewer, Choice, 82-; prin investr, NSF, 90-93; educ officer, Am Astron Soc, 91-94. *Mem:* Am Astron Soc; Sigma Xi; Int Astron Union. *Res:* Astronomy education; astronomy laboratory activities; author of various publications. *Mailing Add:* 2508 Rollingwood Dr Austin TX 78746-5648

HEMENWAY, WILLIAM GARTH, otolaryngology, for more information see previous edition

HEMILY, PHILIP WRIGHT, b Newaygo, Mich, June 2, 22; c 3. CRYSTALLOGRAPHY. *Educ:* Univ Mich, BS, 47; Univ Paris, Dr, 53. *Prof Exp:* Instr math, Auburn Univ, 47-49; res assoc crystallog res, Nat Sci Res Ctr, France, 55-57; prog dir, NSF, 57-61, dep head, Off Int Sci Activ, 61-65, sci attache, US Mission to Orgn Econ Coop & Develop, Paris, State Dept, 65-74; dep off dir, Bur Oceans & Int Environ & Sci Affairs, State Dept, 74-76; dep asst sect gen sci affairs, NATO, Brussels, Belg, 76-83; SR PROG OFFICER, OFF INT AFFAIRS, NAT ACAD SCI, 84- *Concurrent Pos:* Consult, Int Sci & Technol, State Dept, NATO, Stanford Res Inst, 83-, Carnegie Comn Sci, Technol & Govt. *Mem:* Fel AAAS; Am Crystallog Asn. *Res:* Inorganic crystal structure determinations at low temperatures by x-ray diffraction methods. *Mailing Add:* Off Int Affairs NAS Washington DC 20418

HEMING, ARTHUR EDWARD, b Detroit, Mich, June 3, 13; m; c 2. BIOCHEMISTRY. *Educ:* Kalamazoo Col, AB, 37, MS, 38; Univ Wis, PhD(physiol chem), 41. *Prof Exp:* Chief chemist, Johnson & Johnson, Brazil, 41-48; res biochemist, Smith, Kline & French Labs, 48-49, head biochem sect, 49-57, assoc dir res, 57-60, assoc dir res & develop, 60-67; chief pharmacol-toxicol sect, 68, asst chief res grants br, 68-70, actg chief, 70-71, ASSOC DIR PROG ACTIVITIES, NAT INST GEN MED SCI, 71- *Mem:* Am Chem Soc; Endocrine Soc; Am Soc Pharmacol & Exp Therapeut; Royal Soc Med; NY Acad Sci; Sigma Xi. *Res:* Pharmacology; toxicology; endocrinology; enzymology; digitalis glycosides; thyroid physiology; metabolism. *Mailing Add:* 12604 St James Rd Rockville MD 20850

HEMING, BRUCE SWORD, b Ithaca, NY, Dec 1, 39; Can citizen; m 76; c 2. ENTOMOLOGY. *Educ:* Ont Agr Col, Univ Toronto, BSA, 63; NC State Univ, PhD(entom), 68. *Prof Exp:* From asst prof to assoc prof, 68-80, PROF ENTOM, UNIV ALTA, 80- *Honors & Awards:* G Gordon Hewitt Award, Entom Soc Can, 76. *Mem:* AAAS; Can Soc Zoologists; Entom Soc Can; Entom Soc Am. *Res:* Insect morphology; embryonic and postembryonic development; taxonomy and morphology of Thysanoptera. *Mailing Add:* Dept Entom Univ Alta Edmonton AB T6G 2E3 Can

HEMINGHOUS, WILLIAM WAYNE, SR, b Mattoon, Ill, Nov 11, 45; m 70; c 3. MATERIALS & PROCESS PRODUCT DEVELOPMENT, FAILURE ANALYSIS AND PRODUCT IMPROVEMENT. *Educ:* Univ Ill, BS, 68; Purdue Univ, BS, 70; Indiana Univ, BS, 83. *Prof Exp:* Exp metallurgist turbine engine metall, Allison Div, GMC, 68-79; pipe line metallurgist gas transmission indust, Tex Gas Transmission, 71-72; group leader reliability, Cummins Engine Co, 72-83; tech serv mgr metall qual assurance, Teledyne Continental Motors, Mobile, Ala, 83-86; sci lab mgr phy sci, Sverdrup Technol, Stennis Space Ctr, NASA, 86-88; DIR MAT ENG MAT SCI, THOMSON SAGINAW BALL SCREW CO, THOMSON INDUST, 88- *Concurrent Pos:* Sec adv, Kings Serv, Indianapolis, Ind, 78-81; Customer qual assurance mgr, metall qual assurance, Consolidated Disel Co, Cummins Engine Co, 81-83; inventor, Teledyne Continental Motors, Mobile, Ala, 83-86; prin investr plume diag, Sverdrup Technol, Stennis Space Ctr, NASA, 86-88; level III inspector, NDE Thomson Industs, 89-91, NDE instr, Thomson Sagmaw Ball Screw Co, 90-91; prog comt, Am Soc Metals, 90-91. *Mem:* Am Soc Metals; Soc Automotive Engrs; Am Soc Nondestructive Testing; Am Soc Testing & Mat. *Res:* Advanced aerospace lubricant research and development; exploratory research in plume diagnostics of liquid fueled space shuttle engines; development and patent for nitrocarburization heat treatment of aircraft components; concept, development, and international patents on advanced turbo-compounded rotary aircraft engines. *Mailing Add:* Thomson Saginaw Ballscrew Co 628 N Hamilton St Saginaw MI 48603

HEMINGWAY, BRUCE SHERMAN, b Chicago, Ill, Nov 19, 39; m 71; c 1. THERMODYNAMICS, CALORIMETRY. *Educ:* Macalester Col, AB, 62; Univ Minn, MS, 65, PhD(geol & geophys), 71. *Prof Exp:* GEOLOGIST, US GEOL SURV, 69-, ASST CHIEF, BR IGNEOUS & GEOTHERMAL PROCESSES, 86-, CHIEF, NAT CTR THERMODYN DATA MINERALS, 88- *Concurrent Pos:* Mem, Conf Thermodyn & Nat Energy Probs, Nat Acad Sci, 74; mem, Panel Rock-Mech Res Requirements, Nat Res Coun & Nat Acad Sci, 79-81; vchmn & US nat rep, Working Group Thermodyn Natural Processes, IAGC, 83-; chair, financial adv comt, Mineral Soc Am. *Mem:* Geochem Soc; fel Mineral Soc Am. *Res:* Thermodynamic data measured and evaluated, used to test models in geologic and industrial processes and are provided in the form of a data base for use by others. *Mailing Add:* Nat Ctr Thermodyn Data Minerals US Geol Surv 959 Nat Ctr Reston VA 22092

HEMINGWAY, GEORGE THOMSON, b Corvallis, Ore, Aug 23, 40; m; c 1. THEOLOGY. *Educ:* San Diego State Univ, BS, 66, MS, 73. *Prof Exp:* Lectr practical oceanog, San Diego State Univ, 72-73; prof & chair biol, fac marine sci, Autonomous Univ, Baja, Calif, 73-75; ASST DIR, MARINE LIFE RES GROUP, SCRIPPS INST OCEANOG, UNIV CALIF, SAN DIEGO, 78- *Concurrent Pos:* Adj prof biol, fac marine sci, Autonomous Univ, Baja, Calif, 75-; coord, Calif Coop Ocean Fisheries Invest, 79-81, 85-88 & 91-, Interamericas Prog, Scripps Inst Oceanog, 79-, Tinker Found, Latin Am Prog, Scripps Inst, 82-85. *Mem:* Am Soc Zool; Am Inst Biol Sci; AAAS; Western Soc Naturalists; Ctr Theol & Natural Sci. *Res:* Longterm, largescale study of the California current ecosystem, integrating its biology, chemistry and physics. *Mailing Add:* Scripps Inst Oceanog Univ Calif San Diego 9500 Gilman Dr La Jolla CA 92093-0227

HEMKEN, ROGER WAYNE, b Pontiac, Ill, Dec 9, 28; m 55; c 3. DAIRY NUTRITION. *Educ:* Univ Ill, BS, 50, MS, 54; Cornell Univ, PhD(dairy husb), 57. *Prof Exp:* From asst prof to prof dairy husb, Univ Md, 57-70; assoc prof, 70-72, PROF DAIRY HUSB, UNIV KY, 72-, DAIRY COMMODITY CHMN, 70- *Concurrent Pos:* Moormans Travel fel, Animal Nutrit Res, Nat Feed Ingredients Asn, 84. *Honors & Awards:* Am Feed Mfrs Award, Am Dairy Sci Asn, 74. *Mem:* AAAS; Am Soc Animal Sci; Am Dairy Sci Asn; Am Inst Nutrit. *Res:* Ruminant nutrition; nutritive evaluation of forages; mineral nutrition. *Mailing Add:* Dept Animal Sci Univ Ky Lexington KY 40546

HEMLEY, JOHN JULIAN, b El Paso, Tex, Nov 8, 26; m 52; c 3. GEOLOGY, CHEMISTRY. *Educ:* Tex Col Mines & Metall, BS, 49; Northwestern Univ, MS, 52; Univ Calif, Berkeley, PhD(geol), 58. *Prof Exp:* Geologist, Calif Res Corp, Standard Oil Co Calif, 52-54; geologist, US Geol Surv, 58-68; chief geochemist, Anaconda Co, 68-72; GEOLOGIST, US GEOL SURV, 72- *Honors & Awards:* Silver Medal, Soc Econ Geologist, 87. *Mem:* Soc Econ Geologists; Geol Soc Am; Geochem Soc; Am Mineral Soc; Am Geophys Union; AAAS. *Res:* Geochemistry; petrology; mineral genesis; aqueous chemistry and fields of stability of mineral phases characteristic of hydrothermal alteration and ore deposition. *Mailing Add:* US Geol Surv Reston VA 22092

HEMM, ROBERT VIRGIL, b St Louis, Mo, June 20, 21; m 42; c 3. CHEMICAL ENGINEERING. *Educ:* Wash Univ, BS, 42, DSc, 53; Auburn Univ, MBA, 72. *Prof Exp:* Sect chief packaging mat, US Air Force, 43-46, head packaging res, Mallinckrodt Chem Works, 46-51, proj engr, Power Plant Lab, 51-52, chief decontamination & cleaning sect, Mat Lab, 52-54, protective processes br, 54-56, staff officer mat res & develop, 56-57, tech liaison officer, Foreign Technol Div, 57-61, res & develop dir hq res & technol div Air Force Systs Command, 62-66, res & develop staff engr, 66-68, chief technol br, Sci & Technol Div, 68-70, chief sci & technol studies, Air War Col, 70-73, Commander European Off Aerospace Res & Develop Air Force Systs Command, 73-77; exec secy, Nat Mat Adv Bd, Nat Acad Sci, 77-82; TEACHER, FAIRFAX COUNTY, VA, PUBLIC SCHS. *Mem:* Am Chem Soc; Sigma Xi. *Res:* Materials and techniques for packaging military equipment; packaging of chemicals; research and development of aerospace materials; management of defense research and development; technical information management; materials policy studies. *Mailing Add:* 8218 Dabney Ave Springfield VA 22152

HEMMENDINGER, ARTHUR, b Bernardsville, NJ, July 11, 12; m 45; c 3. NUCLEAR PHYSICS. *Educ:* Cornell Univ, AB, 33; Calif Inst Technol, PhD(physics), 37. *Prof Exp:* From instr to asst prof physics, Univ Okla, 37-41; physicist, Naval Ord Lab, Washington, DC, 41-44 & Metall Lab, Univ Chicago, 45; physicist, Los Alamos Sci Lab, 45-77; RETIRED. *Mem:* Fel Am Phys Soc; AAAS. *Res:* Mass spectroscopy; underwater sound; instrumentation for nuclear physics; separation of the isotopes of potassium and rubidium and determination of naturally radioactive isotope; light particle reactions, especially those involving tritium and helium three; neutron and radiation physics; cross section measurements. *Mailing Add:* 1117 Sangre de Cristo Dr Santa Fe NM 87501

HEMMENDINGER, HENRY, b Bernardsville, NJ, Apr 1, 15; m 40; c 3. PHYSICS. *Educ:* Harvard Univ, AB, 35, AM, 37; Princeton Univ, PhD(astron), 39. *Prof Exp:* Bausch & Lomb fel physics, Mass Inst Technol, 39-41; physicist, Inst Optics, Univ Rochester, 42-44; consult opers res group, US Navy, 44-45; physicist & group leader spectros, Cent Res Lab, Gen Aniline & Film Corp, Pa, 46-52; partner, Davidson & Hemmendinger, 52-63, vpres, treas & dir color ctr, 63-70, DIR, HEMMENDINGER COLOR LAB, 70- *Concurrent Pos:* Mem, Harvard-Mass Inst Technol Eclipse Exped, USSR, 36. *Honors & Awards:* Armin J Bruning Award, Fedn Socs Paint Technol, 66; chmn, US Nat Comt 1-3 Colorimetry, Int Comn Illumination, 77-79. *Mem:* Optical Soc Am; Inter-Soc Color Coun. *Res:* Spectroscopy; solar physics; excitation conditions in light sources; operations research; spectrophotometry; luminescence; color and constitution; colorimetry; color standards. *Mailing Add:* Hemmendinger Color Lab 438 Wendover Dr Princeton NJ 08540

HEMMERLE, WILLIAM J, b Des Plaines, Ill, May 8, 27; m 48; c 4. STATISTICS, COMPUTER SCIENCE. *Educ:* Univ Colo, BS, 50; Univ Wis, MS, 51; Iowa State Univ, PhD(math statist), 63. *Prof Exp:* Analyst, Nat Security Agency, DC, 51-55; asst mgr data processing ctr, mgr comput ctr & sr appl sci rep to fed govt, Int Bus Mach Corp, DC & NY, 55-60; from asst prof to assoc prof & head numerical anal prog group, Iowa State Univ, 60-65; dir comput lab, Univ RI, 65-73, prof comput sci, 65-82, chmn dept comput sci & exp statist, 67-82, EMER PROF COMPUT SCI, UNIV RI, 82- *Concurrent Pos:* Consult, Nat Sch Agr Mex, 64; NSF grants, 64-; vis scientist, Inst Statist, Tex A&M Univ, 71-72; vis res prof, Dept Statist, NC State Univ, 78-79. *Mem:* Am Statist Asn. *Res:* Statistical computations. *Mailing Add:* 350 Ministerial Rd Wakefield RI 02882

HEMMES, DON E, b Hampton, Iowa, Dec 28, 42; m 67; c 2. CELL BIOLOGY, MYCOLOGY. *Educ:* Cent Col, Iowa, BA, 65; Univ Hawaii, Honolulu, MS, 67, PhD(microbiol), 70. *Prof Exp:* Fel, Univ Zurich, Switz, 70-72 & Univ Calif, Riverside, 72-73; PROF BIOL, UNIV HAWAII, HILO, 73-, CHMN NATURAL SCI DIV, 85- *Concurrent Pos:* Pac Island Minority Schs Biomed Res grant, 75. *Mem:* Sigma Xi; Bot Soc Am; Mycol Soc Am. *Res:* Ultrastructural investigations of morphogenesis in fungi. *Mailing Add:* 523 W Lanikaula St Hilo HI 96720

HEMMES, PAUL RICHARD, b Staten Island, NY, June 14, 44; c 2. CHEMICAL BATCH PROCESS DEVELOPMENT, TECHNOLOGY TRANSFER. *Educ:* Clarkson Col, BS, 66; Polytechnic Inst Brooklyn, PhD(phys chem), 70. *Prof Exp:* Postdoctoral, Chem Dept, Univ Utah, 69-70; from asst prof to prof, Rutgers Univ, Newark, NJ, 70-81; supvr res & develop, Diag Div, Miles Labs, 81-85, dir process develop, 85-88; vpres mfg, Angenics Inc, 88-89; STAFF SCIENTIST, ENVIRON TEST SYSTS, 89- *Concurrent Pos:* Nat Acad Sci exchange visitor, USSR, 78 & 81; assoc ed, Microchem J, 85-88; consult, 89- *Res:* Application of medical diagnostic test technology to environmental tests; improving technology transfer between research and manufacturing, including batch chemical process development. *Mailing Add:* Environ Test Systs PO Box 4659 Elkhart IN 46514

HEMMINGER, JOHN CHARLES, b Lyons, NY, Apr 2, 49; m 78. SURFACE CHEMISTRY. *Educ:* Univ Calif, Irvine, AB, 71; Harvard Univ, PhD(chem physics), 76. *Prof Exp:* NSF fel, Univ Calif, Berkeley, 76-77, fel chem, 77-78; from asst prof to assoc prof, 78-87, PROF CHEM, UNIV CALIF, IRVINE, 87- *Honors & Awards:* Alfred P Sloan Res Award. *Mem:* Am Chem Soc; Am Vacuum Soc. *Res:* Chemistry and physics of adsorbates on well characterized surfaces; molecular spectroscopy of chemisorbed species. *Mailing Add:* Dept Chem Univ Calif Irvine CA 92717

HEMMINGS, RAYMOND THOMAS, b London, Eng, Apr 8, 48; Can citizen; m 78; c 2. MATERIALS SCIENCE & TECHNOLOGY, MATERIALS CHEMISTRY. *Educ:* Univ Southampton, Eng, BSc, 69; Univ Windsor, Ont, PhD(chem), 73. *Prof Exp:* Fel chem, Univ Windsor, 73-75, res assoc, 75-76; lectr, Univ W Indies, 76-78; asst prof chem, Scarborough Col, Univ Toronto, 78-81; res scientist, Ont Res Found, 81-88; PRIN & VPRES, MATEX CONSULT, INC, 88- *Mem:* Chem Inst Can; Royal Soc Chem; Mat Res Soc; Mineralogical Soc Am; Can Inst Mining & Metall; Am Concrete Inst. *Res:* Materials chemistry; cement and concrete science and technology; waste materials technology; industrial minerals; construction materials; technical-economic assessments. *Mailing Add:* 2575 Chalkwell Close Mississauga ON L5J 2C1 Can

HEMMINGSEN, BARBARA BRUFF, b Whittier, Calif, Mar 25, 41; m 67; c 1. ENVIRONMENTAL SCIENCES. *Educ:* Univ Calif, Berkeley, BA, 62, MA, 64; Univ Calif, San Diego, PhD(marine biol), 71. *Prof Exp:* Res microbiologist, Ames Res Ctr, NASA, 64-65; vis asst prof microbial ecol, Aarhus Univ, Denmark, 71-72; lectr, 73-77, from asst prof to assoc prof, 73-88, PROF MICROBIOL, SAN DIEGO STATE UNIV, 88- *Concurrent Pos:* Consult, hydrocarbon-degrading bacteria, 88- *Mem:* AAAS; Am Soc Microbiol; Sigma Xi; Soc Gen Microbiol; Soc Protozoologists. *Res:* Physiology and taxonomy of marine microorganisms; intracellular gas supersaturation tolerances; bioremediation of petroleum spills; methods for enumerating and stimulating the in situ growth of hydrocarbon-degrading bacteria. *Mailing Add:* Dept Biology San Diego State Univ San Diego CA 92182-0057

HEMMINGSEN, EDVARD A(LFRED), b Tromso, Norway, July 15, 32; m 67; c 1. ANIMAL PHYSIOLOGY. *Educ:* Univ Oslo, Norway, MagSci, 58, DrPhil, 65. *Prof Exp:* Res fel physiol, Norweg Res Coun, 58-59; vis fel glaciol, Arctic Inst NAm, 59-60; from jr res physiologist to assoc res physiologist, 60-

77, RES PHYSIOLOGIST & LECTR PHYSIOL, UNIV CALIF, SAN DIEGO, 77- *Concurrent Pos:* Vis prof physiol, Univ Aarhus, Denmark, 71-72. *Mem:* Am Physiol Soc; AAAS. *Res:* Diffusive transport of oxygen in animals; functions and properties of myoglobin; respiratory and cardiovascular functions in fishes; gas-water interactions at high pressures; gas bubble formation in animals. *Mailing Add:* Scripps Inst Oceanog A-004 La Jolla CA 92093-0204

HEMMINGSEN, ERIK, b Espergaerde, Denmark, Sept 16, 17; nat US; m 46; c 2. MATHEMATICS. *Educ:* Temple Univ, BS, 38; Univ Notre Dame, MS, 40; Univ Pa, PhD(math), 46. *Prof Exp:* Asst instr math, Univ Notre Dame, 39-41; asst instr, Univ Pa, 41-42; physicist, Naval Air Exp Sta, Philadelphia, 42-45; asst prof math, Univ Ga, 46-47; from instr to prof, 47-87, chmn dept, 71-79, EMER PROF MATH, SYRACUSE UNIV, 88- *Concurrent Pos:* Vis prof, Univ Trondheim, 65-66; Chmn Dept Math, Vanderbilt Univ, 69-70. *Mem:* Am Math Soc; Math Asn Am. *Res:* Topology; structure and design of gyroscopic aircraft flight instruments; theorems in dimension theory for normal Hausdorff spaces; plane homeomorphisms with equicontinuous iterates; light interior maps of n-manifolds on n-manifolds. *Mailing Add:* Dept Math Syracuse Univ Syracuse NY 13210

HEMOND, CONRAD J(OSEPH), JR, b Holyoke, Mass, June 8, 16; m 52; c 3. CIVIL ENGINEERING, ACOUSTICS. *Educ:* Univ Mass, BS, 38, MS, 46. *Prof Exp:* Asst city engr, Holyoke, Mass, 38-40; field engr, US Eng Dept, Westover Field, 40-43; instr physics, Rensselaer Polytech Inst, 43-45; asst prof, Amherst Col, 45-50; proj engr, Bolt Beranek & Newman, 50-52 & Indust Sound Control, Inc, 52-55; chmn dept, 55-77, dir eng res, 68-71, PROF MECH ENG, UNIV HARTFORD, 55-, DEAN, 81- *Concurrent Pos:* Mem, noise in bldg comt, Bldg Res Inst, 59, membership comt, 59-60; Nat Coun Acoust consult. *Mem:* Acoust Soc Am; Sigma Xi; Am Soc Mech Engrs; Am Soc Eng Educ. *Res:* Fluid flow; underwater sound; architectural acoustics; industrial noise control. *Mailing Add:* Three Cricket Lane East Granby CT 06026

HEMP, GENE WILLARD, b Minneapolis, Minn, Dec 6, 38; m 60; c 2. APPLIED MECHANICS. *Educ:* Univ Minn, BS, 61, BSB, 62, MS, 63, PhD(eng mech), 67. *Prof Exp:* Res technol asst, Univ Minn, 59-63, instr, 63-64, vis lectr mech, 66-67; from asst prof to assoc prof, Univ Fla, 67-76, asst vpres, 74-76, assoc vpres, 76-89, actg provost, 89-90, PROF ENG SCI & MECH, UNIV FLA, 76-, VPROVOST, 90- *Mem:* Am Soc Mech Engrs; Am Inst Aeronaut & Astronaut; Soc Eng Sci; Am Soc Eng Educ. *Res:* Nonlinear oscillations; biomechanics; dynamic material properties; analog simulation; vibrations of discrete and continuous media; rigid body dynamics. *Mailing Add:* 9909 NW 59th Pl Gainesville FL 32606

HEMPEL, FRANKLIN GLENN, b Gillett, Tex, Sept 22, 39; m 63; c 1. RESPIRATION PHYSIOLOGY, SCIENCE ADMINISTRATION. *Educ:* Univ Tex, Austin, BA, 64, MA, 66, PhD(zool), 69. *Prof Exp:* NIH fel, Med Ctr, Duke Univ, 70-72; Stapells fel, Univ Toronto, 72-73; asst prof physiol, Med Ctr, Duke Univ, 73-79; sci officer physiol prog, Off Naval Res, Arlington, Va, 81-88; DEP DIR, BASIC NEUROSCI, NIH, 88- *Concurrent Pos:* Actg chief, Lab Molecular Biol, Nat Inst Neurol Dist Blindness. *Mem:* Am Physiol Soc. *Res:* Metabolism of retina; visual electrophysiology; oxygen toxicity; bioenergetics of central nervous system. *Mailing Add:* Basic Neurosci Labs NINDS NIH Bldg 36 Rm 5A05 Bethesda MD 20892

HEMPEL, JOHN PAUL, b Salt Lake City, Utah, Oct 14, 35; m; c 1. TOPOLOGY. *Educ:* Univ Utah, BS, 57; Univ Wis, MS, 59, PhD(math), 62. *Prof Exp:* Asst prof math, Fla State Univ, 62-63; mem, Inst Advan Study, 63-64; from asst prof to assoc prof, 64-75, PROF MATH, RICE UNIV, 75- *Concurrent Pos:* Mem, Inst Advan Study, 71-72. *Mem:* Am Math Soc. *Res:* Topology of manifolds, positional properties of submanifolds; group theory; geometry; three-dimensional manifolds. *Mailing Add:* Dept Math Rice Univ Houston TX 77251

HEMPEL, JUDITH CATO, b Stephenville, Tex, Feb 16, 41; m 63; c 1. INORGANIC CHEMISTRY, THEORETICAL CHEMISTRY. *Educ:* Univ Tex, Austin, BS, 62, MA, 65, PhD(chem), 67. *Prof Exp:* Fel chem, Univ Tex, Austin, 67-70; res assoc chem, Univ NC, Chapel Hill, 70-72, York Univ, 72-73 & Duke Univ, 73-75; vis asst prof chem, Wake Forest Univ, 75-76; asst prof, Univ NC, Greensboro, 76-77; asst prof chem, Swarthmore Col, 77-80; sr investr, Smith, Kline & French Labs, 80-89; SR SCI PROJ LEADER, BIO SYM TECH, 89- *Mem:* Sigma Xi; Am Chem Soc. *Res:* Drug design. *Mailing Add:* c/o Bio Sym Tech 10065 Barnes Cnyn Rd San Diego CA 92121

HEMPERLY, JOHN JACOB, b Ft Belvoir, Va, Dec 22, 51; m 74; c 4. NEUROBIOLOGY. *Educ:* Ind Univ, BS, 73; Rockefeller Univ, PhD, 79. *Prof Exp:* Asst prof, Rockefeller Univ, 80-87; GROUP LEADER NEUROBIOL, BECTON DICKINSON RES CTR, 87- *Mem:* Soc Neurosci; Am Soc Biochem & Molecular Biol; Protein Soc; AAAS. *Res:* Structures and regulations of biomolecules involved in cell-cell adhesion in vertebrate tissues; human cell adhesion molecules, particularly their involvement in disease. *Mailing Add:* Becton Dickinson & Co Res Ctr PO Box 12016 Research Triangle Park NC 27709

HEMPFLING, WALTER PAHL, b Cincinnati, Ohio, Mar 21, 38; m 60; c 2. MICROBIOLOGY. *Educ:* Univ Cincinnati, BS, 59; Yale Univ, PhD(microbiol), 64. *Prof Exp:* NIH trainee phys biochem, Johnson Found, Univ Pa, 63-65, res assoc & Pa Plan scholar, 65-66; asst prof microbial biochem, Temple Univ, 66-68; assoc prof, 68-81, actg chmn biol, 79-80, prof biol, Univ Rochester, 81-, assoc chmn biol, 81-; AT PHILLIP MORRIS RES CTR, RICHMOND, VA. *Concurrent Pos:* Vis investr, Scripps Clin Res Inst, 78-79 & Gastprof, Univ Bonn, 79. *Mem:* AAAS; Am Soc Biol Chemists. *Res:* Microbial biochemistry and biotechnology control of electron transport and oxidative phosphorylation, techniques of continuous culture and maintenance metabolism; solvent effects on microbial membranes; physiology of immobilized cells. *Mailing Add:* Phillip Morris Res Ctr PO Box 26583 Richmond VA 23261

HEMPHILL, ADLEY W(ALTON), b Pittsburgh, Pa, Dec 19, 25; m 50; c 3. CHEMICAL ENGINEERING. *Educ:* Grove City Col, BS, 47; Univ Pittsburgh, MS, 50. *Prof Exp:* Mem staff, Pittsburgh Coke & Chem Co, 49-50; process engr, W R Grace & Co, 51-54, mgr, 55-59, res coordr, 60-62, div tech serv, 62-64, mgr desiccants, 65-69, vpres, 69-76, pres, Davison Chem Div, 76-87; RETIRED. *Mem:* Am Chem Soc. *Res:* Chemical engineering and design of parathion and phthalate plasticizer units; engineering, design and operation of specialty catalyst facility and silica gel unit; natural gas treating, petroleum catalysis, dehydration and molecular sieve development. *Mailing Add:* 60 Woodenbridge Lane Pinehurst NC 28374

HEMPHILL, ANDREW FREDERICK, b Louisville, Miss, Aug 6, 27; m 47; c 2. ZOOLOGY. *Educ:* Univ Ala, BS, 49, MS, 54, PhD, 60. *Prof Exp:* PROF BIOL & DIR DEPT, SPRING HILL COL, 70- *Res:* Fresh-water ichthyology. *Mailing Add:* Dept Biol Spring Hill Col 4000 Dauphin St Mobile AL 36608

HEMPHILL, DELBERT DEAN, b Crane, Mo, Nov 8, 18; m 43; c 4. HORTICULTURE, ENVIRONMENTAL HEALTH. *Educ:* Univ Mo-Columbia, BS, 40, PhD(hort), 48. *Prof Exp:* Asst hort, 40-42, 46-48, from asst prof to assoc prof, 48-58, PROF HORT, UNIV MO-COLUMBIA, 58- *Concurrent Pos:* Res Assoc, Environ Health, 64-; consult, Midwest Res Inst, 73- *Mem:* Fel AAAS; Am Soc Hort Sci; Soc Environ Geochem & Health (pres, 71-73); fel Weed Sci Soc Am; Am Pub Health Asn. *Res:* Phytohormones; effects of plant growth-regulating substances on plant growth and development; herbicides; effects of pesticides and trace metals on human health; pollution of environment by pesticides and toxic metals. *Mailing Add:* Dept Hort Univ Mo Columbia MO 65211

HEMPHILL, LOUIS, b Gainesville, Tex, Jan 25, 27; m 51; c 2. SANITARY ENGINEERING, CHEMISTRY. *Educ:* NTex State Univ, BA, 49, MS, 51; Univ Mich, MPH, 51; Univ Mo, PhD(civil eng), 67. *Prof Exp:* Instr chem & biol, NTex State Univ, 49-51; instr chem, Sch Pub Health, Univ Mich, 51-53, res assoc environ health, 53-55; eng leader, Oak Ridge Nat Lab, 56-61; asst prof sanit eng, Okla State Univ, 61-64; instr, Univ Mo-Columbia, 64-67, assoc prof civil eng, 67-; RETIRED. *Concurrent Pos:* Univ Mo Water Resources grant, 67-68. *Mem:* Am Water Works Asn; Water Pollution Control Fedn; Health Physics Soc; Am Chem Soc. *Res:* Instrumentation; nuclear wastes. *Mailing Add:* Dept Civil Eng 0066 Eng Univ Mo Columbia MO 65211

HEMPLING, HAROLD GEORGE, b Brooklyn, NY, May 21, 26; m 47; c 3. PHYSIOLOGY. *Educ:* NY Univ, AB, 48; Oberlin Col, MA, 50; Princeton Univ, PhD(biol), 53. *Prof Exp:* Instr physiol, Sch Med, Univ Pa, 53-55, assoc, 55-57; from asst prof to assoc prof, Med Col, Cornell Univ, 57-71; chmn dept, 71-87, PROF PHYSIOL, MED UNIV SC, 71- *Mem:* Am Physiol Soc; Biophys Soc; Soc Gen Physiol. *Res:* Permeability of cell membranes; ion transport; electrolyte movements; volume regulation; computer modelling; computer assisted instruction. *Mailing Add:* Dept Physiol Med Univ SC Charleston SC 29425

HEMPSTEAD, CHARLES FRANCIS, b Gloucester, Mass, May 15, 25; m 50; c 3. EXPERIMENTAL PHYSICS, SOLID STATE PHYSICS. *Educ:* Northwestern Univ, BS, 49; Cornell Univ, PhD(physics), 55. *Prof Exp:* Mem staff, Electronics & Radio Res Dept, Bell Labs, 54-60, Quantum Electronics Res Dept, 60-65, Visual Systs Res Dept, 65-66 & Lab Measurements Dept, 66-69, supvr, Comput Aided Anal & Characterization Dept, 69-82 & Magnetics & Characterization Dept, 82-88, mgr, Built-In-Self-Test & Thermal Characterization Dept, 88-90; RETIRED. *Mem:* Inst Elec & Electronics Eng. *Res:* Microwave spectroscopy and solid state masers; millimeter wave length backward wave oscillators; x-ray spectroscopy; superconductivity research; visual processes research. *Mailing Add:* Eight Sagamore Dr Andover MA 01810

HEMPSTEAD, J(EAN) C(HARLES), b Woden, Iowa, Aug 6, 04; m 28. INDUSTRIAL ENGINEERING. *Educ:* Iowa State Univ, BS, 26; Univ Pa, MA, 30; Univ Iowa, CE, 42. *Prof Exp:* Instr math, Drexel Inst Technol, 27-30; instr gen eng, Iowa State Univ, 30-42 & 46-51; instr marine eng, US Naval Acad, 52-55; assoc prof math, US Air ForceAcad, 55-64, fac exec officer, 60-63, dep head math, 63-64; prof, 64-76, EMER PROF INDUST ENG, IOWA STATE UNIV, 76- *Mem:* Am Soc Eng Educ. *Res:* Mathematics of industrial property retirements; engineering valuation and depreciation. *Mailing Add:* 126-24 St Ames IA 50010

HEMPSTEAD, ROBERT DOUGLAS, b Cincinnati, Ohio, Aug 20, 43; m 69; c 1. THIN FILM MAGNETIC DEVICES. *Educ:* Mass Inst Technol, SB & SM, 65; Univ Ill, Urbana-Champaign, PhD(physics), 70. *Prof Exp:* Adv engr magnetic rec, Gen Prod Div, IBM Corp, 70-81; sr vpres eng & develop, Cybernex, 81-84, vpres res & develop, 84-86; vpres develop, Domain, 86-89; CONSULT, 89- *Mem:* Inst Elec & Electronics Engrs. *Res:* Materials research, design, analysis and process development for thin film magnetic recording heads, magnetic recording media and magnetic bubbles. *Mailing Add:* 342 Bean Ave Los Gatos CA 95030

HEMSKY, JOSEPH WILLIAM, b Cedar Rapids, Iowa, Oct 26, 36; m 64; c 1. NUCLEAR PHYSICS. *Educ:* Mo Sch Mines, BS, 58; Purdue Univ, PhD(physics), 67. *Prof Exp:* Asst prof, 66-70, ASSOC PROF PHYSICS, WRIGHT STATE UNIV, 70- *Mem:* Am Asn Physics Teachers. *Res:* Nuclear reactions. *Mailing Add:* Dept Physics Wright State Univ Colonel Glenn Hghwy Dayton OH 45435

HEMSWORTH, MARTIN C, Waterloo, Iowa, June 3, 18. ENGINEERING. *Educ:* Univ Nebr, BSME, 40. *Prof Exp:* Engr, Test Facil Aircraft Engine Bus Group, Gen Elec Co, 40-48, mgr test facil, Evandale Plant, 48-52, engr, Mgt Engine Develop Dept, 52-80, chief engr, 80-85, sr consult engine design, Cincinnati, 85-87; RETIRED. *Mem:* Nat Acad Eng; fel Am Soc Mech Engrs; Soc Automotive Eng. *Mailing Add:* 8040 S Clippinger Dr Cincinnati OH 45243

HEMWALL, EDWIN LYMAN, CARDIOVASCULAR, ALLERGY. *Educ:* Hahnemann Univ, PhD(pharmacol), 81. *Prof Exp:* SR INVESTR, SMITHKLINE & FRENCH LABS, 83- *Mailing Add:* PO Box 1539 King of Prussia PA 19406

HEN, JOHN, b Manila, Philippines, June 14, 41; m 65; c 1. SURFACE CHEMISTRY. *Educ:* Mapua Inst Technol, BS, 63; Utah State Univ, MS, 65; Rutgers Univ, PhD(phys chem), 69. *Prof Exp:* Anal chemist, Johnson & Johnson Res Ctr, 65-66; res chemist, Uniroyal Res Ctr, Wayne, NJ, 69-77; SR RES CHEMIST & ASSOC MGR RES & DEVELOP PLANNING, MOBIL CHEM CORP, 77- *Mem:* Am Chem Soc. *Res:* Surface and colloid chemistry; polyelectrolytes; kinetics; photochemistry; analytical chemistry; biochemistry; polymer science, process development. *Mailing Add:* Mobil Tech Ctr Cent PO Box 1025 Princeton NJ 08540

HENCH, DAVID LE ROY, b Cherokee, Iowa, May 19, 41; m 75; c 1. ELECTRO-OPTICS DIGITAL SIGNAL PROCESSING. *Educ:* Iowa State Univ, BS, 63, MS, 65; Rice Univ, PhD(elec eng), 72. *Prof Exp:* Elec eng bio-instrumentation, NASA, Manned Space Craft Ctr, Houston, 65-66; res engr clin automation, Pulmonary Physiol Lab, Methodist Hosp, Houston, 71-75; mem tech staff, Image Res, Electron Res Ctr. Rockwell Int, Anaheim, 75-83; SR SCIENTIST, OPTICAL SCIS CO, PLACENTIA, CALIF, 83- *Concurrent Pos:* NDEA fel, 67-69, PHS trainee, Rice Univ, Houston, 69-71. *Mem:* Inst Elec & Electronics Engrs. *Res:* Analysis and systems design of infrared systems used for target detection and of visible systems used for space object imaging. *Mailing Add:* Optical Scis Co PO Box 1329 Placentia CA 92670

HENCH, LARRY LEROY, b Shelby, Ohio, Nov 21, 38; m 62, 80; c 2. MATERIALS SCIENCE ENGINEERING. *Educ:* Ohio State Univ, BCE, 61, PhD(ceramic eng), 64. *Prof Exp:* Res engr, Battelle Mem Inst, 64; asst prof ceramic & mat eng, Univ Fla, 64-68, assoc prof, Dept Metall & Mat Eng, 68-70, head prof ceramics, Dept Metall & Mat Eng, 70-72, prof & head ceramics div, Dept Mat Sci & Eng, 72- 83, dir Biomed Eng Prog, 74-79, prof, Dept Mat Sci & Eng, 83-86, GRAD RES PROF, DEPT MAT SCI & ENG, UNIV FLA, 86-, DIR BIOGLASS RES CTR, 88- *Concurrent Pos:* Consult, Holmes Co, Inc, 65-66, E I du Pont de Nemours & Co & Knox Glass, Inc, 66- & Phys Sci Br, Picatinny Arsenal, US Army, 69-; co-chmn, workshop on urolithiasis, Nat Res Coun; chmn, 7th Univ Conf Ceramic Res. *Honors & Awards:* Clemson Award, Soc Biomat; George Morey Award, Am Ceramic Soc, Whitewares Award, PACE Award; Scholes Lectr, Alfred Univ; Wedgewood Lectr, Wedgewood Found. *Mem:* Fel Am Ceramic Soc; Nat Inst Ceramic Engrs; Soc Biomat (secy-treas, 77-79, pres, 79); fel Brit Soc Glass Technol; Acad Ceramics; Mat Res Soc; Sigma Xi. *Res:* Materials engineering; structure and properties of glasses, bio-materials, radiation damage and glass-ceramics; electrical behavior of semiconducting glasses and oxides; dielectric and materials; nuclear waste disposal; sol-gel processing; optical materials; technology transfer theory; ethics of science, engineering & technology. *Mailing Add:* Dept Mat Sci Univ Fla Gainesville FL 32611

HENCH, MILES ELLSWORTH, b Minneapolis, Minn, Oct 28, 19; m 45; c 3. MEDICAL MICROBIOLOGY. *Educ:* Lawrence Col, BS, 41; Univ Mich, MS, 49, PhD(bact), 52; Am Bd Med Microbiol, dipl. *Prof Exp:* Asst, Univ Mich, 49-52; from asst prof to prof clin path & microbiol, 52-82, asst dean admis, 61-67, assoc dean admis, 67-82, EMER PROF CLIN PATH MED COL, VA COMMONWEALTH UNIV, 82-, DIR ALUMNI RES, SCH MED, 83- *Mem:* AAAS; Am Asn Med Clins; Am Soc Microbiol. *Res:* Diagnostic microbiology; environmental sanitation. *Mailing Add:* 3115 Patterson Ave No 11 Richmond VA 23221

HENCHMAN, MICHAEL J, b London, Eng, Feb 28, 35; m 66. PHYSICAL CHEMISTRY. *Educ:* Univ Cambridge, BA, 56, MA, 60; Yale Univ, MS, 58, PhD(chem), 61. *Prof Exp:* Asst lectr phys chem, Univ Leeds, 61-63, lectr, 63-67; ASSOC PROF CHEM, BRANDEIS UNIV, 67- *Mem:* Royal Soc Chem; Am Chem Soc. *Res:* Chemical dynamics of ion-molecule collision processes in the gas phase; charge transfer; mass spectrometry. *Mailing Add:* Dept Chem Brandeis Univ Waltham MA 02254

HENDEE, JOHN CLARE, b Duluth, Minn, Nov 12, 38; c 6. FORESTRY. *Educ:* Mich State Univ, BS, 60; Ore State Univ, MF, 62; Univ Wash, PhD(forestry), 67. *Prof Exp:* Forester, Sinslaw Nat Forest, US Forest Serv, 61-64, res scientist, Pac Southwest Forest Exp Sta, 64-65, Pac Northwest Forest Exp Sta, 65-76, legis coord forestry, Wash, 76-78, asst dir, Southeast Forest Exp Sta, 78-85; PROF & DEAN, WILDLIFE & RANGE SCI, COL FORESTRY, UNIV IDAHO, 85- *Concurrent Pos:* Adj assoc prof, Univ Wash, 68-76; Cong fel, Am Polit Sci Asn, 76-77. *Honors & Awards:* Nat Conserv Award, Am Motors, 74; Lifetime Achievement Award, Am Soc Pub Admin, 88. *Mem:* Soc Am Foresters; Soc Range Mgt; Wildlife Soc. *Res:* Author or co-author of one hundred articles reporting research results on human behavior aspects of natural resources; textbook written on wilderness management; forestry; wildlife in wilderness. *Mailing Add:* 497 Ridge Rd Moscow ID 83843

HENDEE, WILLIAM RICHARD, b Owosso, Mich, Jan 1, 38; m 60; c 7. MEDICAL PHYSICS. *Educ:* Millsaps Col, BS, 59; Univ Tex, PhD(radiation physics), 62. *Prof Exp:* From asst prof to assoc prof physics, Millsaps Col, 62-64, assoc prof astron & physics & chmn dept, 64-65; from asst prof to assoc prof radiol, Univ Colo Med Ctr, Denver, 65-73, prof, 74-85, chmn dept, 78-85; VPRES, AM MED ASN, 85- *Concurrent Pos:* Dir, Nat Cancer Inst Cancer Control Projs, Southwestern Med Physics Ctr; consult ed, J Nuclear Med Technol; assoc ed, J Optical Eng & Int J Radiation Oncol, Biol & Physics. *Mem:* Health Physics Soc; Am Asn Physicists in Med; Soc Nuclear Med; Am Col Radiol; Soc Photo-Optical Instrument Eng. *Res:* Visual perception and cognitive recognition of visual information; trend analysis in technology development and diffusion. *Mailing Add:* Am Med Asn 535 N Dearborn St Chicago IL 60610

HENDEL, ALFRED Z, b Vienna, Austria, Oct 19, 16. PHYSICS. *Educ:* Univ Vienna, dipl, 48; Univ Paris, DUniv, 55. *Prof Exp:* Prof physics, La Paz Univ, 45-56; res assoc, Princeton Univ, 57-58; vis asst prof, 59-64, assoc prof, 64-70, PROF PHYSICS, UNIV MICH, ANN ARBOR, 70- *Concurrent Pos:* Assoc prof, Brazilian Ctr Physics Res, 52-56; res physicist, Polytech Sch, Paris, 54-55. *Res:* Cosmic rays; strange particles; air showers; electromagnetic interactions. *Mailing Add:* Dept Physics Univ Mich 738 Bennison Ann Arbor MI 48109

HENDEL, HANS WILLIAM, b Woldenberg, Ger, Nov 27, 22; US citizen; m 53; c 3. MEASUREMENT OF FUSION REACTION RATES. *Educ:* Munich Tech Univ, BS, 46, MS, 49, PhD(physics), 53. *Prof Exp:* Physicist, AGFA Camera Works, 53-57; res physicist, US Army Eng Res & Develop Labs, Ft Monmouth, NJ, 57-61; mem tech staff, David Sarnoff Res Lab, RCA Corp, 61-89; vis prin physicist, Plasma Physics Lab, Princeton, Univ, 65-89; RETIRED. *Concurrent Pos:* Leader plasma & space physics, Astro-electronics Div, RCA Corp, 64-68. *Mem:* Fel Am Phys Soc; Sigma Xi. *Res:* Plasma waves and instabilities; fusion reaction products. *Mailing Add:* 214 Riverside Dr Princeton NJ 08540

HENDERLONG, PAUL ROBERT, b Marshallville, Ohio, Sept 9, 37; m 62; c 3. FORAGE CROPS, CROPPING SYSTEMS. *Educ:* Ohio State Univ, BSc, 59, MSc, 61; Va Polytech Inst, PhD(physiol & biochem), 64. *Prof Exp:* Asst prof soil fertil, WVa Univ, 64-65, pasture & turfgrass res, 65-68; assoc prof agron, 68-71, PROF AGRON, OHIO STATE UNIV, 71- *Concurrent Pos:* Vis prof pasture agron, Makerere Univ, Uganda, 72-73; assoc ed, agron jour, 76-79; chmn, CSSA Div, C-3, 81; agron ed adv, Longman Publ Inc, 83-85; campus coordr, Int Progs Agr, 84-88. *Honors & Awards:* Alfred J Wright Award, 77. *Mem:* Am Soc Agron; Crop Sci Soc Am; Am Soc Plant Physiol; Weed Sci Soc Am; Am Forage Grassland Coun. *Res:* Legume and grass physiology; biochemistry and nutrition as related to crop production and utilization; cropping systems. *Mailing Add:* Dept Agron Ohio State Univ 2021 Coffey Rd Columbus OH 43210-1086

HENDERSHOT, WILLIAM FRED, b Dalton, Ga, Feb 21, 30; m 49; c 4. BIOCHEMISTRY. *Educ:* Ind Univ, BS, 54; Univ Wis, PhD(biochem), 59. *Prof Exp:* Asst chemist, Forest Prod Lab, USDA, 54-58, assoc biochemist, Northern Utilization Res & Develop Div, 58-62; sr res biochemist & sect head, Ames Res Labs Div, 62-67, mgr qual control develop, 67-68, dir qual control, Oper Serv Div, 68-70, DIR CORP QUAL ASSURANCE, MILES LABS, INC, 70- *Mem:* AAAS; Pharmaceut Mfrs Asn; Am Chem Soc; Sigma Xi. *Res:* Diagnostic research; clinical enzymology; microbiological investigations; fermentation products; intermediary metabolism and enzymology; control systems; automation of analytical procedures; computerization of data acquisition. *Mailing Add:* 58386 Co Rd 9 Elkhart IN 46517

HENDERSHOTT, CHARLES HENRY, JR, b Marked Tree, Ark, Oct 13, 23; m 44; c 5. HORTICULTURE. *Educ:* Univ Ark, MS, 52; NC State Col, PhD(plant physiol), 59. *Prof Exp:* Jr horticulturist & instr hort, Univ Ark, 52-54, horticulturist, Mission to Panama, 54-57; from asst plant physiologist to assoc plant physiologist, Citrus Exp Sta, Fla Citrus Comn, 59-64 & Dept Fruit Crops, Univ Fla, 64-67; head dept & chmn div, Univ Ga, 67-74, prof hort, 74-79, head dept & chmn div, 79-; RETIRED. *Mem:* Fel Am Soc Hort Sci; Am Soc Plant Physiol. *Res:* Horticultural physiology. *Mailing Add:* 155 Chinguapin Way Athens GA 30605

HENDERSON, ALEX, b Bicknell, Ind, July 2, 24; m 48; c 1. VERTEBRATE MORPHOLOGY, BIOLOGY. *Educ:* Pa State Univ, AB, 47, DEd, 57; Univ Wichita, MS, 49. *Prof Exp:* Operator, Rohm and Haas Chem Co, 43-44; pub sch teacher, Kans, 50-53; instr, Pa State Univ, 54-55; assoc prof sci, 55-57, chmn dept, 64-73, PROF BIOL, MILLERSVILLE STATE COL, 57- *Concurrent Pos:* Sci consult, Pa Dept Educ, 63-; mem, Pa Coun Educ Dept Pub Instr Col Eval Team, 63-; vis res prof, Glasgow, 65; pres, Environ Sci Res Assocs, Inc; biologist in residence, Marine Sci Consortium, 77-80; mem bd, Commonwealth Pa, Univ Biologists. *Mem:* Sigma Xi. *Res:* Morphology of Terrapene ornata; biology of Coenobita clypeatus; biology of Coregonus clupeoides; atlas of fisheries of Virginia eastern shore; scientific illustration. *Mailing Add:* Dept Biol Millersville univ Millersville PA 17551

HENDERSON, ARNOLD RICHARD, b State Line, Miss, Oct 28, 32; m 67. GEOLOGY. *Educ:* Miss Southern Col, BS, 53; Univ Tenn, Knoxville, MS, 61; Miss State Univ, EdD(geol, geog), 70. *Prof Exp:* From asst computer to computer, Western Geophys Co Am, 53-54, computer & asst consult, 56-58; asst prof geol, Ga Southwestern Col, 61-80; southeastern dist geologist, Clay Div, J M Huber Corp, 80-82; SUPVR, STATE OIL & GAS BD, 82- *Concurrent Pos:* Consult, Perry County Pub Schs, Miss, 70 & Coahoma Jr Col, 70-71; indust consult geol, 71-; adminr & coordr consult serv for teachers sci staff develop, Griffin Coop Educ Serv Agency, NSF, 77-78 & 78-79. *Mem:* Geol Soc Am. *Res:* Earth science education. *Mailing Add:* State Oil & Gas Bd 500 Graymont Ave Suite E Jackson MS 39205

HENDERSON, BEAUFORD EARL, b Dothan, Ala, June 26, 39; m 65; c 3. COMMUNICATION ENGINEERING. *Educ:* Newark Col Eng, BSEE, 61; Univ Southern Calif, MSEE, 68; Univ Calif, Los Angeles, MSE, 72. *Prof Exp:* Electronic officer radar, US Air Force, 61-64; engr elec, Bunker Ramo Corp, 64-65 & Hughes Aircraft Co, 65-68; staff engr, Magnavox Res Labs, 68-72; br chief eng, 72-83, DEP DIR, LISTER HILL NAT CTR BIOMED COMMUN, 83- *Mem:* Sr mem Inst Elec & Electronics Engrs. *Res:* Broadband communication systems and techniques related to the storage, transmission and display of video signals; image processing. *Mailing Add:* Lister Hill Nat Ctr 8600 Rockville Pike Bethesda MD 20014

HENDERSON, BILLY JOE, b Texarkana, Tex, Aug 10, 37; m 67. ELECTROMAGNETISM. *Educ:* NC State Col, BS, 59, MS, 62, MBA, 83; Univ Ga, PhD(physics), 67. *Prof Exp:* Nuclear engr, Aircraft Nuclear Propulsion Dept, Gen Elec Co, 59-60; asst prof physics, Univ Ga, 67-68; asst prof physics, Fla Technol Univ, 68-77 & BoeingCo, 77-88; CONSULT, 88- *Mem:* Sigma Xi. *Res:* Theoryof direct nuclear reactions; analysis of nuclear reactions which proceed via compound nucleus formation; molecular vibrational energy levels; nuclear reactor analysis; bioelectric effects. *Mailing Add:* 1057 Summit E Seattle WA 98102

HENDERSON, CHARLES B(ROOKE), b Washington, DC, Mar 13, 29; m 54; c 3. CHEMICAL ENGINEERING. *Educ:* Purdue Univ, BS, 50; Mass Inst Technol, SM, 52. *Prof Exp:* Chem engr, Altantic Res Corp, 54-57, head chem eng group, 57-59, dir, Chem Eng Div, 59-66, asst gen mgr, Propulsion

Div, 66-71, dir res & technol, 71-75, vpres, 75-80, sr vpres, 80-88, corp dir, 77-90; DIR, ARCTECH INC, 87- *Concurrent Pos:* Res adv coun, Va Ctr Innovative Technol. *Mem:* Sigma Xi; Am Inst Aeronaut & Astronaut; Am Chem Soc. *Res:* Rocket propulsion; alternate fuels; thermodynamics; fluid dynamics; heat transfer; combustion. *Mailing Add:* RR 3 Box 141 Leesburg VA 22075

HENDERSON, CHARLES R, animal breeding, animal genetics; deceased, see previous edition for last biography

HENDERSON, COURTLAND M, b Sullivan, Ind, Sept 2, 15; m 40; c 3. CHEMICAL ENGINEERING. *Educ:* Purdue Univ, BChE, 40. *Prof Exp:* Chem engr, Columbian Enameling & Stamping Co, Ind, 40-44; res engr, Linde Co, div Union Carbide Corp, 44-52; res develop engr, Mallinckrodt Chem Works, 52-55; res engr, uranium plant, 55-56, tech supt uranium metal plant, 56-57; res group leader, Res & Eng Div, Monsanto Chem Co, 57-61, res mgr, 61-66, supvr, Mound Facil, 66-68, mgr prog planning & tech coordr nuclear opers, 68-76, mgr liaison-coal conversion progs, Mound Facil, 76-79; pres, Site-Tac, Inc, 79-83; CONSULT, 83. *Concurrent Pos:* Pres, Site-Tac Inc, 79- *Mem:* Am Chem Soc; Am Inst Chem Engrs. *Res:* Ultrahigh temperature chemical reactors; dispersion-strengthened metals; flame and arc-plasma research; thermoelectric materials and devices; radioisotopes; advanced coal conversion research and development; technical monitoring of engineering and plant design and construction of coal conversion demonstration plants; pressure sensitive adhesives; gaseous diffusion equipment design. *Mailing Add:* 3560 Jonathon Dr Beavercreek OH 45385-5914

HENDERSON, D(ELBERT) W, b Colo, Apr 8, 19; m 45. IRRIGATION. *Educ:* Univ Ariz, BS, 46; Univ Calif, PhD(soil sci), 50. *Prof Exp:* Asst irrig, 46-50, from instr & jr irrigationist to assoc prof & assoc irrigationist, 50-65, PROF WATER SCI & IRRIGATIONIST, UNIV CALIF, DAVIS, 65- *Res:* Soil physics; plant-soil-water relations; irrigation management. *Mailing Add:* Dept Land, Air & Water Resources Univ Calif Davis CA 95616

HENDERSON, DALE BARLOW, b Tulsa, Okla, June 6, 41; m 64; c 4. PLASMA PHYSICS. *Educ:* Cornell Univ, BEngr Physics, 63, PhD(appl physics), 67. *Prof Exp:* Mem staff controlled fusion, Thermonuclear Res, Los Alamos Nat Lab, 66-71, alt group leader, 72-75, group leader, 75-80, mem staff laser-fusion, 71-80, assoc div leader, nuclear explosive physics, Appl Theoret Physics Div, 80-84, dep assoc dir, Theoret & Computative Physics, 84-85, chief scientist, Studies & Anal Div, 86-88, CHIEF SCIENTIST, COMPUT DIV, LOS ALAMOS NAT LAB, 88- *Concurrent Pos:* Chief scientist SDI, Nat Test Bd, FY, 89. *Mem:* AAAS; Am Inst Astronaut & Aeronaut; Am Phys Soc. *Res:* Computer simulation; physics of nuclear weapons and related computer codes; laser-fusion; hydrodynamics; controlled fusion. *Mailing Add:* C-DO MS B260 Los Alamos Nat Lab Los Alamos NM 87545

HENDERSON, DAVID ANDREW, b Alyth, Scotland, Apr 9, 48; UK citizen; m 75; c 3. TUMOUR THEORY. *Educ:* Univ Edinburgh, Scotland, BSc, 70; Vanderbilt Univ, PhD(molecular biol), 74. *Prof Exp:* Postdoctoral fel cell biol, Max Planck Inst Biophys Chem, Göttingen, Ger, 76-77, staff scientist, 77-82; staff scientist endocrinol, Dept Biochem Pharmacol, 82-89, HEAD, DEPT EXP ONCOL, SCHERING AGR, 89- *Mem:* Sigma Xi; NY Acad Sci; Am Soc Cell Biol. *Res:* Growth control processes in normal and malignant cells, especially in endocrine dependent tumours; drug discovery and development of therapeutic agents for tumour therapy. *Mailing Add:* Exp Oncol Schering Agr Postfach 650311 W-1000 Berlin 65 Germany

HENDERSON, DAVID EDWARD, b Richmond, Va, Sept 22, 46; m 72; c 2. CHROMATOGRAPHY, MASS SPECTROMETRY. *Educ:* St Andrews Presby Col, BA, 68; Univ Mass, PhD(chem), 75. *Prof Exp:* Res assoc chem, Univ Mass, 74-77; from asst prof to assoc prof, 77-84, PROF CHEM, DEPT CHEM, TRINITY COL, 88- *Concurrent Pos:* Teaching assoc, Dept Chem, Mt Holyoke Col, 74-77; lectr consult, Ctr Prof Adv, 75- & Univ Mass, 77; consult examr, Charter Oak Col. *Mem:* Am Soc Mass Spectrometry; Sci Res Soc NAm; Sigma Xi; AAAS. *Res:* Low temperature high performance liquid chromatography of labile metal complexes and peptides; microbore high performance liquid chromatography of proteins; mass spectrometry of peptides and metal complexes; air pollution; lipid-carbohydrate thermal interactions. *Mailing Add:* Dept Chem Trinity Col Hartford CT 06106

HENDERSON, DAVID MICHAEL, b Zanesville, Ohio, May 14, 40; m 68; c 2. ELECTROOPTICS. *Educ:* Miami Univ, AB, 62, MS, 64; Yale Univ, PhD(appl sci), 69. *Prof Exp:* Mem tech staff laser physics, Bell Lab, 69-74; sr proj eng, 74-76, sect head electrooptics, 76-81, dept mgr, Hughes Aircraft Co, 81-84, LAB MGR, HUGHES RES LAB, 84- *Concurrent Pos:* Yale Univ fel. *Mem:* Am Phys Soc; Inst Elec & Electron Eng. *Res:* Interaction of infrared radiation with materials for purposes of detection, imaging and modulation. *Mailing Add:* Hughes Aircraft Co E1-B131 PO Box 902 El Segundo CA 90245

HENDERSON, DAVID RIPPEY, organic chemistry, for more information see previous edition

HENDERSON, DAVID WILSON, b Walla Walla, Wash, Feb 23, 39; m 83; c 2. MATHEMATICS. *Educ:* Swarthmore Col, BA, 61; Univ Wis, MS, 62, PhD(math), 64. *Prof Exp:* Mem math, Inst Advan Study, 64-66; asst prof, 66-68, assoc prof, 68-82, PROF, CORNELL UNIV, 82-, MEM FAC MATH, 66- *Concurrent Pos:* Alfred P Sloan res fel, 68-72; vis prof, Birzeit Univ, Jordan, 81; exchange scientist, Soviet Union & Poland, 70. *Mem:* Am Math Soc; Math Asn Am; Asn Women Math. *Res:* Basic geometry; geometric topology; meaning in mathematics; mathematics education. *Mailing Add:* Dept Math Cornell Univ Ithaca NY 14853-7901

HENDERSON, DONALD, b Hamilton, Ont, Oct 3, 38; m 63. SENSORY PSYCHOLOGY, SENSORY PHYSIOLOGY. *Educ:* Western Wash State Col, BA, 62; Univ Tex, Austin, PhD(psychol), 66. *Prof Exp:* Res assoc psychol, Radiobiol Labs, Univ Tex, 62-64 & Tracor, Inc, 64-66; res assoc physiol, Cent Inst Deaf, 66-68; asst prof otolaryngol, Upstate Med Ctr, State Univ NY, 68-80; mem staff, Callier Ctr Commun Disorders, 80-87; MEM STAFF, STATE UNIV NEW YORK, BUFFALO, 87- *Concurrent Pos:* Lectr, Univ Tex, 65-66, Wash Univ, 66-68 & Syracuse Univ, 69- *Mem:* AAAS; Acoust Soc Am; Am Speech & Hearing Asn. *Res:* Relation between sensory phenomena and the underlying physiological mechanisms, specifically how the auditory system reacts to various parameters of high intensity noise. *Mailing Add:* State Univ NY 109 Park Hall Buffalo NY 14200

HENDERSON, DONALD AINSLIE, b Cleveland, Ohio, Sept 7, 28; m 51; c 3. MEDICINE, EPIDEMIOLOGY. *Educ:* Oberlin Col, BA, 50; Univ Rochester, MD, 54; Johns Hopkins Univ, MPH, 60. *Hon Degrees:* ScD, Univ Rochester, Oberlin Col, Univ Ill, Yale Univ, Albany Med Col & Univ Maryland; LLD, Marietta Col; MD, Universite de Geneve; LHD, State Univ New York. *Prof Exp:* Intern, Mary Imogene Bassett Hosp, 54-55, resident, 57-59; chief epidemic intel serv, Commun Dis Ctr, USPHS, 55-57, asst chief epidemiol br, 60-61, chief surveillance sect, 61-67; chief smallpox eradication, WHO, Geneva, 67-77; dean, 77-90, EDGAR BERMAN PROF EPIDEMIOL & INT HEALTH, SCH HYG & PUB HEALTH, JOHNS HOPKINS UNIV, 77- *Concurrent Pos:* Mem, US Dept Health & Human Serv, Adv Comt Oceans, Environ & Sci Affairs, US State Dept & Expert Adv Panel Virus Dis, WHO; chmn, Nat Vaccine Adv Comt, Tech Adv Group Immunization-Pan Am Health Orgn & bd overseers, Am J Epidemiol. *Honors & Awards:* Nat Medal Sci, 86; Medal for Health, Govt India, 78; Indian Soc Malaria & Other Commun Dis Award, 75; Award, Govt India, 75; Ernst-Jung Preis fur Medizin, 76; Public Welfare Medal, Nat Acad Sci, 78; Joseph C Wilson Award Int Affairs, 78; George MacDonald Prize & Medal, London Sch Hygiene & Trop Med & Royal Soc Trop Med & Hygiene; James Bruce Award, Am Col Physicians, 78; Gairdner Found Int Award Merit, 83; Albert Schweitzer Int Prize Med, 85; Charles S Dana Found Award, 86; Richard T Hewitt Award, Royal Soc Med, 86; Japan Prize, 88; Health for All Medal, WHO, 90. *Mem:* Inst Med-Nat Acad Sci; Am Epidemiol Soc; Infectious Dis Soc Am; hon fel Am Acad Pediat; hon fel Royal Col Physicians; fel AAAS. *Res:* Infectious diseases; public health administration; educational administration. *Mailing Add:* Johns Hopkins Sch Hyg & Pub Health 615 N Wolfe St Baltimore MD 21205

HENDERSON, DONALD LEE, b Holstein, Iowa, July 6, 33; c 2. COMPUTER SCIENCES, DATA COMMUNICATIONS. *Educ:* Wayne State Col, BA, 58; Univ SDak, MA, 59; Okla State Univ, EdD(comput sci), 67. *Prof Exp:* Instr math, Mankato State Col, 59-62, asst registr & dir comput serv, 62-68, dir comput serv & chmn comput sci, 68-73; exec dir Minn Educ Comput Consortium, 73-75; chmn comput sci, 75-83, PROF COMPUT SCI, MANKATO STATE UNIV, 84- *Concurrent Pos:* Comt mem, Minn Governor's Comput Adv Comt, 75-83; chmn comput sci, Western Ky, 83-84. *Mem:* Asn Comput Mach; Asn Educ Data Systs. *Res:* Data communication curriculum development. *Mailing Add:* Dept Comput Sci Mankato State Univ Ellis Ave Mankato MN 56001

HENDERSON, DONALD MUNRO, b Boston, Mass, Nov 8, 20; m 48; c 5. MINERALOGY. *Educ:* Brown Univ, AB, 43; Harvard Univ, AM, 46, PhD(geol), 50. *Prof Exp:* Geologist strategic minerals invests, US Geol Surv, 44-45; from instr to prof, 48-89, EMER PROF GEOL, UNIV ILL, URBANA, 89- *Concurrent Pos:* Guggenheim fel, 58-59. *Mem:* Mineral Soc Am; Mineral Soc Gt Brit & Ireland; Norweg Geol Soc; Nat Asn Geol Teachers; Am Geophys Union. *Res:* Mineralogy; crystallography. *Mailing Add:* 245 NHB-Geol 1301 W Green UIUC Urbana IL 61801

HENDERSON, DOUGLAS J, b Calgary, Alta, July 28, 34; m 60; c 3. THEORETICAL PHYSICS, THEORETICAL CHEMISTRY. *Educ:* Univ BC, BA, 56; Univ Utah, PhD(physics), 61. *Prof Exp:* Asst prof theoret physics, Univ Idaho, 61-62 & Ariz State Univ, 62-64; assoc prof, Univ Waterloo, 64-67, prof theoret physics & appl math, 67-69; res scientist, res lab, IBM Corp, 69-90; RES SCIENTIST, IBM/UTAH SUPERCOMPUT INST PARTNERSHIP, 91- *Concurrent Pos:* Sloan Found fel, 64-69; Ian Potter Found fel & vis scientist, Chem Res Labs, Commonwealth Sci & Indust Res Orgn, Australia, 66-67; adj prof, Univ Waterloo, 69-; vis prof physics, Nat Univ La Plata, Arg, 73, Univ Guelph, 91; vis prof chem, Korea Advan Inst Sci, Seoul, 74; assoc ed, J Chem Physics, 74-76; Manuel Sandoval Vallarta, distinguished vis prof physics, Universidad Autonoma Metropolitana Mex, 88. *Honors & Awards:* IBM Outstanding Res Contrib Award, 73. *Mem:* Fel Am Phys Soc; Am Chem Soc; Math Asn Am; fel Brit Inst Physics; Can Asn Physicists; Sigma Xi. *Res:* Investigation of intermolecular forces and development of theory to use these potentials to calculate properties of matter, particularly in liquid state. *Mailing Add:* Utah Supercomput Inst 85 SSB Univ Utah Salt Lake City UT 84112

HENDERSON, DOUGLASS MILES, b Long Beach, Calif, July 9, 38; m 71; c 3. BIOSYSTEMATICS, BOTANY. *Educ:* Fresno State Col, BA, 65; Univ Wash, PhD(bot), 72. *Prof Exp:* Inspector food prod, Consumer & Mkt Serv, USDA, 65-67; asst prof, 72-78, ASSOC PROF BOT, UNIV IDAHO, 78-, DIR HERBARIUM, 72- *Concurrent Pos:* Managing ed, Syst Bot, Am Soc Plant Taxonomists, 84- *Mem:* Am Soc Plant Taxonomists; Int Asn Plant Taxonomists; Am Bot Soc. *Res:* Biosystematic studies in the genus Sisyrinchium and Idaho floristics; biology of rare and endangered species. *Mailing Add:* Dept Biol Sci UI Herbarium Univ Idaho Moscow ID 83843

HENDERSON, EARL ERWIN, b Detroit, Mich, Aug 2, 46; c 2. MICROBIOLOGY, VIROLOGY. *Educ:* Va Polytechnic Inst & State Univ, BS, 69; Univ Chicago, PhD(microbiol), 75. *Prof Exp:* Res specialist, WalterReed Army Med Ctr, 70-71; fel pediat & epidemiol, Yale Univ, 75-77; ASST PROF MICROBIOL, TEMPLE UNIV, 77- *Concurrent Pos:* Special fel, Leukemia Soc Am, 78-80. *Mem:* Am Soc Microbiol; Sigma Xi. *Res:* Development and characterization of cell lines transformed by Epstein-Barr Virus. *Mailing Add:* Dept Microbiol & Immunol Temple Univ Sch Med 3400 N Broad Philadelphia PA 19140

HENDERSON, EDWARD GEORGE, b Bridgeport, Conn, Apr 6, 35; m 58; c 2. BIOPHYSICS, PHARMACOLOGY. *Educ:* Univ Conn, BA, 61; Univ Md, PhD(biophys), 66. *Prof Exp:* Group leader radiochem, Wyeth Labs, 65-68; asst prof, 68-73, ASSOC PROF PHARMACOL, UNIV CONN HEALTH CTR, 73- *Mem:* Biophys Soc; Am Soc Pharmacol & Exp Therapeut; Neurosci Soc. *Res:* Ion transport; electrophysiology; neurophysiology; membrane pharmacology. *Mailing Add:* Dept Pharmacol Univ Conn Health Ctr Farmington CT 06032

HENDERSON, EDWARD S, b Ventura, Calif, July 19, 32; m 54, 70; c 5. INTERNAL MEDICINE, ONCOLOGY. *Educ:* Stanford Univ, BA, 53, MD, 56; Am Bd Internal Med, dipl, 64, cert med oncol, 74. *Prof Exp:* Intern, Los Angeles County Gen Hosp, 56-57, resident med, 57-59; instr internal med & fel hemat, Sch Med, Univ Southern Calif, 59-61; clin assoc med oncol, Nat Cancer Inst, 61-64, sr investr pharm, 64-65, head leukemia serv, 65-71, actg chief hemat & supportive care br, 71-73; chief dept med, Roswell Park Mem Inst, 73-89; res prof med, 73-81, chief med oncol, 79-89, PROF MED, STATE UNIV, NY, BUFFALO, 81- *Mem:* Am Fedn Clin Res; Am Soc Clin Oncol; Am Soc Hemat; Am Asn Cancer Res; Transplantation Soc; fel Am Col Physician. *Res:* Clinical cancer chemotherapy; clinical oncology; pharmacology of cancer chemotherapeutic agents; complications of cancer therapy. *Mailing Add:* Dept Med Roswell Park Mem Inst Buffalo NY 14263

HENDERSON, ELLEN JANE, b Little Rock, Ark, Feb 16, 40. CELL BIOLOGY, DEVELOPMENTAL BIOLOGY. *Educ:* Purdue Univ, BS, 66, PhD(biochem), 71. *Prof Exp:* Jane Coffin Childs Mem Fund fel, Univ Edinburgh, 71-73, res staff molecular biol, 73-74; asst prof biochem, Mass Inst Technol, 74-80; asst prof biol, 80-83, ASSOC PROF BIOL, GEORGETOWN UNIV, 83- *Concurrent Pos:* Mem adv subcomt develop biol, NSF, 80-84; vis scientist, Imp Cancer Res Fund, London/Clare Hall Lab, 85 & 86; nat lectr, Sigma Xi, 87 & 88. *Mem:* Soc Complex Carbohydrates; Am Soc Cell Biol; Soc Develop Biol; Am Soc Micros; AAAS; Sigma Xi; fel AAAS. *Res:* Glycoprotein synthesis in Dictyostelium discoideum; structure-function relationships of glycoconjugates intercellular cohesion. *Mailing Add:* Dept Biol Georgetown Univ Washington DC 20057

HENDERSON, FLOYD M, b Denison, Iowa, Feb 23, 46; m 68. REMOTE SENSING, LAND USE ANALYSIS. *Educ:* Nebr Wesleyan Univ, BA, 68; Univ Kans, MA, 70, PhD(geog), 73. *Prof Exp:* Asst prof geog, San Diego State Univ, 72-73; from asst prof to assoc prof, 73-85, PROF GEOG, STATE UNIV NY, ALBANY, 85-, ASSOC DEAN, COL SOCIAL & BEHAV SCI, 87- *Concurrent Pos:* Co-dir, Geog Info Systs & Remote Sensing Lab, State Univ NY, Albany, 75-; vis asst prof, La State Univ, 77; vis scientist, Inst Bodenkultar, Vienna, Austria, 85. *Mem:* Am Soc Photogram & Remote Sensing; Int Soc Photogram & Remote Sensing; Inst Elec & Electronics Engrs; Soc NAm Cult Surv. *Res:* Effects of environmental modulation on imaging radak signal response; digital and contextual analysis of remote sensing imagery for terrain features. *Mailing Add:* Dept Geog & Planning State Univ NY Albany NY 12222

HENDERSON, FREDERICK BRADLEY, III, b Oakland, Calif, Dec 5, 35; m 62, 86; c 3. GEOLOGY. *Educ:* Stanford Univ, BS, 57, MS, 60; Harvard Univ, PhD(econ geol), 66. *Prof Exp:* Mining geologist, St Joseph Lead Co, 65-69; explor geologist, Kaiser Aluminum, 69-71; consult geologist, Hendco, 71-74; res geologist, Lawrence Berkeley Lab, Univ Calif, Berkeley, 74-76; PRES, GEOSAT COMT, 76- *Concurrent Pos:* Consult, Nat Security Coun, 78-80; mem, Civil Oper Remote Sensing Adv Comn, 82-85, technol transfer, 85-88. *Mem:* Assoc mem AAAS; Sigma Xi. *Res:* Application of sattelite remote sensing systems to geological, agricultural and oceanographic problems; deposition of base and precious metal deposits. *Mailing Add:* Geosat Comt 601 Elm St Rm 438C Norman OK 73019

HENDERSON, GARY BORGAR, b Seattle, Wash, Feb 8, 46; m 70; c 2. BIOCHEMISTRY. *Educ:* Univ Wash, BS, 68; Univ Calif, PhD(biochem), 73. *Prof Exp:* Asst mem, 76-83, ASSOC MEM, SCRIPPS CLIN & RES FOUND, 83- *Mem:* Am Soc Biol Chemists; Am Asn Cancer Res; Am Soc Microbiol. *Res:* Mechanisms and components of transport systems for folate compounds, vitamins, and anions; antineoplastic drugs; drug uptake and delivery systems. *Mailing Add:* Dept Biochem Scripps Clin & Res Found La Jolla CA 92037

HENDERSON, GARY LEE, b Springfield, Mo, Dec 13, 38; m 60. PHARMACOLOGY. *Educ:* Univ Calif, Berkeley, AB, 65; Univ Calif, Davis, PhD(comp pharmacol & toxicol), 69. *Prof Exp:* Asst res pharmacologist, Sch Med, Univ Calif, Los Angeles, 70-71; ASSOC PROF PHARMACOL, SCH MED, UNIV CALIF, DAVIS, 71- *Concurrent Pos:* USPHS scholar, Brain Res Inst, Ctr Health Sci, Sch Med, Univ Calif, Los Angeles, 69-70. *Mem:* AAAS; Am Chem Soc. *Res:* Drug metabolism and pharmacokinetics; development of methodology for micro-determination of drugs and their metabolites in biological fluids and tissues. *Mailing Add:* Dept Pharmacol Univ Calif Sch Med Davis CA 95616-5224

HENDERSON, GEORGE ASA, b Charleston, WVa, Aug 22, 40; m 63; c 3. THEORETICAL PHYSICS, QUANTUM CHEMISTRY. *Educ:* Georgetown Univ, BS, 62, MS, 67, PhD(physics), 70. *Prof Exp:* Res assoc quantum chem, Johns Hopkins Univ, 69-70; from asst prof to assoc prof, 70-78, PROF PHYSICS, SOUTHERN ILL UNIV, EDWARDSVILLE, 78- *Concurrent Pos:* Fac res partic, Argonne Nat Lab, 72-75; res assoc Univ NC, 75-76; fac fel NASA-Langley Res Ctr, 81; res, McDonnell Douglas Corp, 82- *Mem:* Am Phys Soc; Am Asn Physics Teachers; AAAS; Sigma Xi. *Res:* Computational models for lasers; density matrix theory; electronic properties of atoms and molecules; density fuctional theory. *Mailing Add:* Dept Physics Southern Ill Univ Edwardsville IL 62026-1654

HENDERSON, GEORGE EDWIN, b Kenton, Ohio, Sept 10, 06; m 33; c 2. AGRICULTURAL ENGINEERING. *Educ:* Ohio State Univ, BSAgr, 29. *Prof Exp:* Rural engr, Ohio Edison Power Co, 29-32 & Dayton Power & Light Co, 32-37; rural elec eng ed, Tenn Valley Authority, 37-39, head ed sect, 39-41, asst chief agr eng div, 41-45, chief, 45-47, agr eng & food processing div, 47-48; prof 49-73, EMER PROF AGR ENG, UNIV GA, 73- *Concurrent Pos:* Exec dir, Am Asn Voc Instructional Mat. *Mem:* Fel Am Soc Agr Engrs. *Res:* Rural electrification; structures and farm power. *Mailing Add:* 150 Valley Rd Athens GA 30606

HENDERSON, GEORGE I, b Indianapolis, Ind, July 2, 43. TERATOLOGY, AGING. *Educ:* Vanderbilt Univ, PhD(pharmacol), 72. *Prof Exp:* Asst prof pharmacol, Vanderbilt Univ, 76-82; ASSOC PROF MED & PHARMACOL, UNIV TEX HEALTH SCI CTR, SAN ANTONIO, 82- *Mem:* Teratology Soc; Am Soc Pharmacol & Exp Therapeut; Am Inst Nutrit; Sigma Xi; Res Soc Alcoholism. *Mailing Add:* Dept Med Univ Tex Health Sci Ctr 7703 Floyd Curl Dr San Antonio TX 78284

HENDERSON, GEORGE RICHARD, b New York, NY, Oct 28, 45; m 77; c 1. ENZYMOLOGY, TRANSPORT. *Educ:* St Lawrence Univ, BS, 67; Cornell Univ, PhD(pharmacol), 76. *Prof Exp:* Instr, 76-77, ASST PROF PHARMACOL, MED COL OH, 77- *Mem:* Sigma Xi. *Res:* Properties and functions of sodium. *Mailing Add:* Wyeth-Ayerst 145 King of Prussia Rd Radnor PA 19087

HENDERSON, GERALD GORDON LEWIS, b Vernon, BC, June 10, 26; m 55; c 3. GEOLOGY. *Educ:* McGill Univ, BSc, 48, MSc, 50; Princeton Univ, PhD(geol), 53. *Prof Exp:* Asst geologist, BC Dept Mines, 50-53; geologist, Calif Standard Co, 53-62, chief geologist, 62-68, vpres explor, Standard Oil Calif, 68-; at Chevron Standard Ltd, Calgary, Alta; RETIRED. *Mem:* Am Asn Petrol Geologists; Geol Asn Am. *Res:* Structural geology. *Mailing Add:* 3240 66th Ave SW Suite 915 Calgary AB T3E 6M5 Can

HENDERSON, GERALD VERNON, geology; deceased, see previous edition for last biography

HENDERSON, GILES LEE, b Aberdeen, SDak, July 19, 43; m 62; c 2. MOLECULAR SPECTROSCOPY, PHYSICAL CHEMISTRY. *Educ:* Mont State Col, BS, 65, MS, 66; Ind Univ, PhD(chem physics), 73. *Prof Exp:* Assoc prof, 67-80, PROF CHEM, EASTERN ILL UNIV, 80- *Res:* Spectroscopic studies of the properties and structures of van der Waals molecules. *Mailing Add:* Dept Chem Eastern Ill Univ Charleston IL 61920

HENDERSON, GRAY STIRLING, b St Paul, Minn, Dec 23, 41; m 68. FOREST SOILS, WATER QUALITY. *Educ:* Iowa State Univ, BS, 63; Cornell Univ, MS, 66, PhD(forest soils), 68. *Prof Exp:* Soil scientist, Cold Regions Res & Eng Lab, US Army, 68-69, agr adv, Vietnam, 69-70; res ecologist, Oak Ridge Nat Lab, US Energy Res & Develop Admin, 70-78; ASSOC PROF FORESTRY, SCH FORESTRY, FISHERIES & WILDLIFE, UNIV MO-COLUMBIA, 78- *Concurrent Pos:* Nutrient cycling process coordr, Eastern Deciduous Biome, US Int Biol Prog, 71-74. *Mem:* Soil Sci Soc Am; Am Soc Agron; Am Foresters Asn; Sigma Xi. *Res:* Investigation of processes controlling nutrient cycling in forested watersheds, including input-output relationships; effects of acid precipitation on nutrient cycling processes; relationships between soil fertility and tree nutrition. *Mailing Add:* Forestry/1-30 Agr Bldg Univ Mo Columbia MO 65211

HENDERSON, H(OMAN) THURMAN, b Berea, Ky, Dec 28, 32; m 54; c 3. SOLID STATE ELECTRONICS & PHYSICS. *Educ:* Ind Inst Technol, BSEE, 58; Southern Methodist Univ, MSEE, 61, PhD(elec eng), 68; Harvard Univ, SM, 67. *Prof Exp:* Electronics engr & sr tech writer, Chance Vought Corp, 58-60; from instr to asst prof elec eng, Univ Tex, Arlington, 60-68; ASSOC PROF ELEC ENG, UNIV CINCINNATI, 68- *Concurrent Pos:* Prin investr, NSF grant, 69-70, Inst Space Sci, Univ Cincinnati-NASA res grant, 70 & NASA res grant, 71-; mem bd sci adv, KDI Corp, 70. *Mem:* Am Soc Eng Educ; Am Phys Soc; Inst Elec & Electronics Engrs; Cryogenic Soc Am; Sigma Xi. *Res:* Electrical materials science, particularly the effects of radiation on the electrical properties of materials; high level injection in semi-insulators and the role of deep impurities in achieving unique volt-ampere characteristics and instabilities; microelectronics. *Mailing Add:* Dept Elec Engr M L 30 Rhodes Hall Univ Cincinnati Cincinnati OH 45221

HENDERSON, HALA ZAWAWI, b Karachi, Pakistan, July 11, 31; US citizen; m 61; c 1. PEDODONTICS. *Educ:* Bombay Univ, BDS, 56; Ind Univ, MSD, 59, DDS, 70; Am Bd Pedodontics, dipl, 79. *Prof Exp:* Jr lectr dent, Govt Dent Col, Bombay Univ, 59-60; head, Dent Div, Sch Health Serv, Ministry Health, Kuwait, Arabia, 60-68; from asst prof to assoc prof, 70-82, PROF PEDODONTICS, SCH DENT, IND UNIV, 82- *Concurrent Pos:* Dent consult, Ministry Health, Sultanate, Oman, 81. *Mem:* Am Soc Dent Children; Am Dent Asn; Am Acad Pedodontics; Am Asn Dent Sch. *Res:* Clinical dentistry, including correlation between clinical, radiographic and histologic findings in ankylosed teeth; use of chrome crowns on primary molars; diet counseling for the child patient. *Mailing Add:* Div Pedodontics Sch Dent Ind Univ Indianapolis IN 46202

HENDERSON, JAMES B(ROOKE), b Washington, DC, Feb 26, 26; m 48; c 2. CHEMICAL ENGINEERING. *Educ:* Purdue Univ, BS, 46, PhD(chem eng), 49. *Prof Exp:* Asst instr, Purdue Univ, 46-47, res assoc, 47-49; from technologist to gen mgr synthetic rubber div, Shell Chem Co, Houston, 49-66, indust chem div, 66-70, vpres, 70-72, vpres, Shell Oil Co, 72-77, mkt coordr, Shell Int Petrol Co, 77-79, PRES, SHELL CHEM CO & VPRES & DIR, SHELL OIL CO, 79- *Mem:* Am Inst Chem Engrs; Am Chem Soc; Sigma Xi. *Mailing Add:* 72 Fairway Ridge Lake Wylie SC 29710

HENDERSON, JAMES HENRY MERIWETHER, b Falls Church, Va, Aug 10, 17; m 48; c 4. PLANT PHYSIOLOGY. *Educ:* Howard Univ, BS, 39; Univ Wis, MPh, 40, PhD(plant physiol), 43. *Prof Exp:* Jr chemist, Badger Ord Works, Wis, 42-43; asst pharmacol, Nat Defense Res Comt, Calif, 43-45; asst prof & res assoc plant physiol, Carver Res Found, 45-50, res assoc prof biol, 50-68, head dept, 57-68, dir res found, 68-75, SR RES PROF BIOL, CARVER RES FOUND, TUSKEGEE INST, 75-, CHMN NATURAL SCI DIV, 68-, DIR MINORITY BIOMED RES PROG, 73- *Concurrent Pos:* Res fel, Kerckhoff Biol Labs, Calif Inst Technol, 48-50; NSF sr fac fel, Nat Ctr Sci Res, France, 61-62; comnr, Coun Undergrad Educ Biol Sci, 67-70; res biochem & cell cult, Hawaii Sugar Planters Ans, 74. *Mem:* Fel AAAS; Am Soc Plant Physiol; Nat Inst Sci; Sigma Xi. *Res:* Tissue culture; auxin and plant growth regulator metabolism and physiology, including the mechanism of action; physiology of normal and abnormal plant tissues; gibberellins. *Mailing Add:* Dept Nat Sci/Biol Tuskegee Institute AL 36088

HENDERSON, JAMES MONROE, b Brazil, Ind, Feb 23, 21; m 45, 68; c 2. CHEMICAL ENGINEERING, PROCESS METALLURGY. *Educ:* Purdue Univ, BS, 42. *Prof Exp:* Res engr chem eng, US Rubber Co, 46-48; dept supt, Am Smelting & Refining Co, 48-55, res engr, 55-58, supt western res sect, 59-64, res supt cent res dept, 64-76; gen supt process res, Asarco Inc, 77-82; RETIRED. *Mem:* Am Inst Mining, Metall & Petrol Engrs. *Res:* Process metallurgy of non-ferrous metals; recovery of sulfur dioxide; environmental control pertinent to recovery of non-ferrous metals. *Mailing Add:* 2571 Avenida Loma Linda No 17208 Green Valley AZ 85614-1273

HENDERSON, JAMES STUART, b Dundee, Scotland, May 26, 28; m 56; c 2. PATHOLOGY, ONCOLOGY. *Educ:* Univ St Andrews, MB, ChB, 51. *Prof Exp:* Intern med, Muhlenberg Hosp, Plainfield, 51-52; secondment, Atomic Bomb Casualty Comn, Hiroshima, 54; instr path, Duke Univ, 55-57; res assoc, Rockefeller Inst Med Res, 57-60; asst prof, Rockefeller Univ, 60-70; PROF PATH, UNIV MANITOBA, 70- *Concurrent Pos:* Consult pathologist & dir exp path, Deer Lodge Vet Admin Hosp, Winnipeg, 70-83; consult pathologist, Health Sci Ctr, Winnipeg, 80 & St Boniface Hosp, 83. *Mem:* NY Acad Sci; Reticuloendothelial Soc; Am Asn Cancer Res; Am Asn Path & Bact; fel Royal Soc Med; Group Res Path Educ. *Res:* Cancer, especially co-factors which cause progression in certain mouse tumors; medical education, especially the use of interactive computer programs. *Mailing Add:* Dept Path Univ Manitoba Fac Med Winnipeg MB R3E 0W3 Can

HENDERSON, JOHN FREDERICK, b Montreal, Que, Apr 26, 33; m 61; c 2. PHYSICAL CHEMISTRY, POLYMER CHEMISTRY. *Educ:* McGill Univ, BSc, 54, PhD(phys chem), 58. *Prof Exp:* Postdoctorate fel, Nat Res Coun Can, 57-59; sr res chemist, Polymer Corp Ltd, 59-62, proj supvr exp res, 62-66, proj supvr resins, 67-69, mgr current prod res, 69-71; tech mgr emulsion polymers, Polysar Ltd, 71-73, tech develop mgr, 73-79, global prod mgr emulsion rubbers, 79-82, mgr technol scanning & assessment, 82-86, Polysar fel, 86-90; MGR TECHNOL SCANNING & ASSESSMENT, NOVA CORP, ALTA, 90- *Concurrent Pos:* Mem assoc comt high polymer res, Nat Res Coun Can, 68-69. *Mem:* Fel Chem Inst Can. *Res:* Chemical kinetics in gas and solution phases; sterospecific, anionic and free radical polymerization; polymer characterization. *Mailing Add:* Technol Mgt Off Nova Corp Calgary AB T2E 7K7 Can

HENDERSON, JOHN WARREN, b Nebr, Sept 11, 12; m 33; c 2. OPHTHALMOLOGY. *Educ:* Univ Nebr, MA, 36, MD, 37; Univ Minn, MS, 43. *Prof Exp:* From asst prof to assoc prof, Mayo Grad Sch Med, 47-67, chmn sect ophthal, Mayo Clin, 67-72, prof, 67-77, EMER PROF OPHTHAL, MAYO GRAD SCH MED, 77- *Mailing Add:* 1227 19th Ave NE La State Univ Med 1501 Kings Hwy Rochester MN 55904

HENDERSON, JOHN WOODWORTH, b Clarinda, Iowa, Mar 8, 16; m 42; c 2. OPHTHALMOLOGY. *Educ:* Occidental Col, BA, 37; Northwestern Univ, MD & MS, 42; Am Bd Ophthal, dipl, 46; Univ Mich, PhD(neuro-ophthal), 48. *Prof Exp:* From intern to resident, Univ Hosp, Univ Mich, Ann Arbor, 42-45, from instr to assoc prof, Med Sch, 45-62, chmn dept, 68-78, PROF OPHTHAL, MED SCH, UNIV MICH, ANN ARBOR, 62- *Mem:* Am Ophtal Soc (pres, 80); fel Am Acad Opthal & Otolaryngol. *Res:* Neuro-ophthalmology; anatomical research in Macaca mulatta; human embryology; anatomical basis for certain reflex and automatic eye movements; ocular signs in brain tumor and in aneurysms; corneal disease and surgery. *Mailing Add:* 261 Memory Lane Naples FL 33962

HENDERSON, JOSEPH FRANKLIN, biochemistry, for more information see previous edition

HENDERSON, LAVELL MERL, b Swan Lake, Idaho, Sept 9, 17; m 39; c 3. BIOCHEMISTRY. *Educ:* Utah State Univ, BS, 39; Univ Wis, MS, 41, PhD(biochem), 47. *Hon Degrees:* DSc, Utah State Univ, 74. *Prof Exp:* Instr biochem, Univ Wis, 47-48; asst prof, Univ Ill, 48-57; prof & head dept, Okla State Univ, 57-63; head dept biochem, Univ Minn, St Paul, 63-74, prof, 63-84, assoc dean, Col Biol Sci, 78-84; RETIRED. *Honors & Awards:* Borden Award in Nutrit, 70. *Mem:* AAAS; Am Chem Soc; Am Soc Biol Chemists; Am Inst Nutrit; Am Soc Microbiol. *Res:* Metabolism of amino acids; amino acid assay; vitamin B complex in animal nutrition; tryptophan metabolism; niacin, carnitine and alkaloid biosynthesis; hydroxylysine metabolism. *Mailing Add:* 8612 Mt Majestic Rd Sandy UT 84093

HENDERSON, LAWRENCE J, b Oceanside, Calif, Aug 10, 52; m 73; c 5. SOIL FERTILITY, SOIL CHEMISTRY. *Educ:* Brigham Young Univ, BS, 77; Okla State Univ, MS, 80, PhD(soil sci), 83. *Prof Exp:* SOIL TECHNOLOGIST, US SUGAR CORP, 82- *Mem:* Am Soc Agronomy; Soil Sci Soc Am; Am Soc Sugarcane Technologists. *Res:* Conducting and evaluating fertility, irrigation and drainage and cultural studies for sugarcane on sand soils and histisols; supervision of routine soil and tissue analyses for sugarcane, pasture, citrus and corn production. *Mailing Add:* Res Dept US Sugar Corp PO Drawer 1207 Clewiston FL 33440

HENDERSON, LINDA SHLATZ, b Johnson City, NY, Feb 21, 46; m 77. BIOCHEMISTRY, ENDOCRINOLOGY. *Educ:* St Lawrence Univ, BS, 67; Univ Rochester, PhD(biochem), 72. *Prof Exp:* Fel physiol, Mt Sinai Sch Med, 71-73, res assoc, 73-74; res assoc biochem, Med Sch, Univ Mass, 74-75; asst prof med & biochem, Sch Med, Univ Rochester, 75-77; asst prof biochem, Med Col Ohio, 77-; AT SMITH KLINE/FRENCH LABS. *Concurrent Pos:* USPHS fel, 71-74; res assoc, Max Planck Inst Biophys, Frankfurt, 72-73; USPHS grants, 75-77 & 77- *Mem:* AAAS; Sigma Xi. *Res:* Effects of hormones and calcium on membrane structure and function; importance of membrane polarity in the kidney and small intestine with respect to hormone action; regulation of gastric secretion. *Mailing Add:* Smith Kline/French Labs l-217 PO Box 1539 1500 Spring Garden St King of Prussia PA 19406-0939

HENDERSON, LOUIS E, b Rulo, Nebr, Jan 1, 34; m 71; c 1. BIOCHEMISTRY. *Educ:* Univ Omaha, BA, 57; Univ Colo, PhD(biochem), 65. *Prof Exp:* NIH fel biochem, Harvard Univ, 65-68, mem tutorial bd, 66-68; Nobel vis scientist, Chalmers Univ Technol, Sweden, 69; mem staff, Sweden Indust Res Coun, 70-73; res assoc biochem & biophys, Yale Univ, 73-76; res scientist II, Frederick Cancer Res Ctr, 76-89; SECT SUPVR, PRI FT DETRICK, FREDERICK, MD, 89- *Mem:* AAAS; Am Chem Soc; NY Acad Sci. *Res:* Protein chemistry; radiation research; protein structure; vetrovirology; AIDS. *Mailing Add:* PRI AIDS Vaccine Prog Frederick Cancer Res & Develop Ctr Frederick MD 21701

HENDERSON, MADELINE M BERRY, b Merrimac, Mass, Sept 3, 22; m 57; c 4. INFORMATION RESOURCES MANAGEMENT, APPLICATION STANDARDS INFORMATION SYSTEMS. *Educ:* Emmanuel Col, AB, 44; Am Univ, MPA, 77. *Prof Exp:* Chemist, Anal Res, E I du Pont de Nemours & Co, 44-45 & Org Res, C S Batchelder Co, 45-46; res assoc chem eng, Mass Inst Technol, 46-50, document, 50-52; consult, US Govt, 53; res engr, Info Systs, Battelle Mem Inst, 53-56; prof asst, Sci Info Serv, NSF, 56-58, consult, 58-62; data processing applns analyst, Ctr Comput Sci & Technol, Nat Bur Standards, 64-69, consult to dir, 69-71, staff asst to dir, 72-75, chief comput info sect, 75-78, chief automatic data processing standards anal, 78-79; consult info mgt, anal & assessment, 79-91; RETIRED. *Concurrent Pos:* US Dept Com Sci & Technol fel, 71-72. *Honors & Awards:* Watson Davis Award, Am Soc Info Sci, 89. *Mem:* AAAS; Am Chem Soc; Am Soc Info Sci. *Res:* Development of information systems; use of electronic aids to literature searching; automation of library operations and services; role of standardization in systems design and operation. *Mailing Add:* 5021 Alta Vista Rd Bethesda MD 20814

HENDERSON, MAUREEN MCGRATH, b Tynemouth, Eng, May 11, 26; nat US. EPIDEMIOLOGY. *Educ:* Univ Durham, MB, BS, 49, DPH, 56. *Prof Exp:* Fel cancer epidemiol, Dept Path, St Bartholomew's Hosp & Med Sch, London, Eng, 56-58, clin epidemiologist, Med Res Coun Group Res on Atmospheric Pollution, 58-60; from instr to prof prev med, Sch Med, Univ Md, 60-75, chmn, Dept Social & Prev Med, 71-75; actg dean, Sch Dent, Univ Wash, 77, assoc vpres health sci, 79-81, dir, Pac Northwest Long-Term Care Ctr, 81-83, PROF EPIDEMIOL & MED, UNIV WASH, 75-; HEAD, CANCER PREV RES PROG, FRED HUTCHINSON CANCER RES CTR, 83- *Concurrent Pos:* Vis prof chronic dis, Sch Hyg & Pub Health, Johns Hopkins Univ, 68-69, assoc in epidemiol, 70-75, assoc dir regional med prog, Epidemiol & Statist Ctr, 68-73; deleg in US-USSR Joint Working Group in Epidemiol of Cancer, 74-; mem steering comt, Social Security Res Studies, Inst of Med, Nat Acad Sci, 74-76; mem panel health serv res, Nat Acad Sci-Nat Res Coun Comt on Study Nat Needs for Biomed & Behav Res Personnel, 74-76; chmn expert panel on reserpine & breast cancer, Food & Drug Admin, 74-76; mem, President's Biomed Res Panel with Interdisciplinary Cluster on Epidemiol, Biostatics & Bioeng, 75; mem task force theory & appln prev med in health care delivery, Fogarty Nat Ctr & Am Col Prev Med, 75; mem expert comt for review of data on carcinogenicity of cyclamate, Nat Cancer Inst, 75-76; mem comt on impacts of stratospheric change, Nat Res Coun, 75-; mem adverse drug effects adv panel, Off Technol Assessment, US Cong, 76. *Honors & Awards:* John A Snow Award, Am Pub Health Asn, 90. *Mem:* Inst Med-Nat Acad Sci; Am Epidemiol Soc (pres-elect, 89-90, pres, 90); Int Epidemiol Soc; Soc Epidemiol Res (pres, 69-70); Asn Teachers Prev Med (vpres, 70-72, pres, 72-73); Am Soc Prev Oncol; Am Asn Cancer Res; Am Col Epidemiol; Am Pub Health Asn; Royal Soc Med. *Res:* Health services research with emphasis on health manpower; adverse drug reaction; preventive services, clinical trials. *Mailing Add:* Dept Epidemiol Univ Wash SC-36 Seattle WA 98195

HENDERSON, MERLIN THEODORE, b Oakdale, La, Dec 5, 14; m 41; c 3. AGRONOMY. *Educ:* Southwestern La Inst, BS, 39; La State Univ, MA, 41; Univ Minn, PhD(plant genetics), 45. *Prof Exp:* Asst potato breeding, USDA, 39-41; instr & asst prof plant breeding, Univ Minn, 41-45; assoc prof agron, Pa State Col, 45-48; prof, 48-77, alumni prof agron, La State Univ, Baton Rouge, 77-80; RETIRED. *Concurrent Pos:* Agt, Regional Pasture Res Lab, Bur Plant Indust, Soils & Agr, USDA, Pa, 47-48. *Mem:* Am Soc Agron; Am Genetic Asn. *Res:* Potato breeding; barley breeding and genetics; breeding of forage plants; inheritance of reaction to leaf rust in barley; breeding and genetics of cotton; cytogenetics of rice; genetics of sugarcane. *Mailing Add:* 5538 Riverstone Dr Baton Rouge LA 70820

HENDERSON, NANNETTE SMITH, b Washington, DC, June 9, 46; m 69; c 2. PLANT PATHOLOGY. *Educ:* Howard Univ, BS, 67, MS, 69; NC State Univ, PhD(plant path), 73. *Prof Exp:* Lab asst virol, Plant Indust Sta, 67; microbiologist, Walter Reed Army Hosp Med Res, 67-68; teaching asst biol, Howard Univ, 68-69; lab asst mycol, Forestry Res Sta, 69; asst prof plant path, NC State Univ, 73-76; dir, Col Transfer Prog, 76-78, HEAD DIV MATH-SCI PUB SERV PROGS, VANCE-GRANVILLE COMMUNITY COL, 78- *Mem:* Nat Educ Asn; Nat Sci Teachers Am. *Mailing Add:* Dept Sci Vance-Granville Communtiy Col PO Box 917 Henderson NC 27536

HENDERSON, NORMAN LEO, b Jersey City, NJ, Feb 18, 32; m 56; c 3. PHYSICAL PHARMACY. *Educ:* Rutgers Univ, BSc, 53; Temple Univ, MSc, 58, PhD(pharm), 62. *Prof Exp:* Proj leader pharmaceut res & develop, Lederle Labs Div, Am Cyanamid Co, 61-67, group leader tablet & powder technol, 67-72; dir pharmaceut develop, 72-84, DIR PHARMACEUT RES, HOECHST-ROUSSEL PHARMACEUTICALS INC, 84- *Concurrent Pos:* Mem adv comt pharm, Rutgers Univ, 80-92; mem pharmaceut develop comt, Pharmaceut Mfrs Asn, 82-85. *Mem:* Am Pharmaceut Asn; Acad Pharmaceut Sci; AAAS; Sigma Xi. *Res:* Pharmaceutical dosage forms; research and development of new drug entities from pre-formulation studies through production; evaluation of formulation factors upon clinical efficacy of dosage forms; novel drug delivery systems. *Mailing Add:* Patriot Rd Gladstone NJ 07934

HENDERSON, RALPH JOSEPH, JR, b Galveston, Tex, Sept 21, 40; m 64; c 2. BIOCHEMISTRY. *Educ:* Univ Tex, Austin, BS, 63; Univ Tex Med Br Galveston, PhD(biochem), 70. *Prof Exp:* Chemist, Jefferson Chem Co, 64-66; trainee biochem, Univ Tex Southwestern Med Sch Dallas & Vet Admin Hosp, Dallas, 70-71; asst prof, 71-76, ASSOC PROF BIOCHEM, SCH MED, LA STATE UNIV, SHREVEPORT, 76- *Mem:* Am Chem Soc; AAAS; Sigma Xi. *Res:* Nucleotide and nucleoside metabolism; metabolic regulation; effects of extracellular matrix on cells; cell biology. *Mailing Add:* Dept Biochem La State Univ Sch Med Shreveport LA 71130

HENDERSON, RICHARD ELLIOTT LEE, b Lincolnton, NC, Feb 17, 45. MEDICINAL CHEMISTRY. *Educ:* Univ NC, Chapel Hill, BS, 67; Univ Ill, Urbana-Champaign, PhD(org chem), 73; Ill Inst Technol, JD, 86. *Prof Exp:* Res investr chem & coord res, Develop Div, Corp Patent Dept & Patent Dept, G D Searle & Co, 73-87; Patent Dept, Merck & Co, 87; PATENT DEPT, MOBAY CORP, 88- *Mem:* Am Chem Soc; Am Intellectual Property Law Asn. *Mailing Add:* 316 Dickson Ave Pittsburgh PA 15202

HENDERSON, RICHARD WAYNE, b Baton Rouge, La. ORGANIC CHEMISTRY, PHYSICAL CHEMISTRY. *Educ:* La State Univ, BS, 66, PhD(org chem), 71. *Prof Exp:* ASSOC PROF CHEM, FRANCIS MARION COL, 72- *Concurrent Pos:* Fel, Memphis State Univ, 71-72. *Mem:* Am Chem Soc; Am Acad Forensic Sci; Am Soc Testing & Mat; Nat Asn Fire Invest; Int Asn Arson Invest. *Res:* Polar effects in radical reactions; heavy metal chemistry; arson research; analysis for accelerant residues. *Mailing Add:* Southeastern Res Lab PO Box 15010 Quinby SC 29506

HENDERSON, ROBERT BURR, organic chemistry; deceased, see previous edition for last biography

HENDERSON, ROBERT E, b Olean, NY, Nov 1, 35; m 57; c 3. TURBOMACHINERY. *Educ:* Penn State Univ, BS, 58, MS, 62; Cambridge Univ, PhD(mech eng), 73. *Prof Exp:* Res asst, Appl Res Lab, 59-72, from assoc prof to prof, Mech Eng & Appl Res Lab, 72-91, EMER PROF, MECH ENG & APPL RES LAB, PA STATE UNIV, 91- *Concurrent Pos:* Underwater Comt, Am Inst Aeronaut & Astronaut, 66-69; Turbomach Comt, Am Soc Mech Engrs, 79-91. *Mem:* Fel Am Soc Mech Engrs; Am Soc Naval Engrs; sr mem Am Inst Aeronaut & Astronaut. *Res:* Fluid dynamics of turbomachinery. *Mailing Add:* 204 Fairfield Dr State College PA 16801

HENDERSON, ROBERT EDWARD, b Kokomo, Ind, Feb 28, 25; m 52; c 6. PHYSICS, ENGINEERING. *Educ:* Carleton Col, BA, 49; Univ Mo, MA, 51, PhD(physics), 53. *Prof Exp:* Instr physics, Univ Mo, 51-53; exp engr, Allison Div, Gen Motors Co, 53-56, supvr appliedphysics, 56-58, res mgr applied sci, 58-66, mgr exp res, 66-69, dir res, 69-73; chmn sci adv bd, Ind Ctr Adv Res Inc, 70-83, exec dir, 73-78, pres, Ind Ctr Adv Res Inc, 78-82; pres, Technol Develop Corp, 82-83; DIR, SC RES AUTHORITY, 83- *Concurrent Pos:* Mem comt on solar energy for heating & cooling of bldgs, Bldgs Res Adv Bd, Nat Acad Sci; adv, Aeronaut Systs Div, Div Adv Group, Dept Air Force; prof, Purdue Sch Engr & Technol, Ind; prof, dept radiol, Ind Univ Sch Med; adj prof, Ind Univ Sch Pub & Environ Affairs; pres, Ind Sci & Eng Found, 82. *Mem:* Fel Am Inst Aeronaut & Astronaut; Am Phys Soc; Solar Energy Soc; Sigma Xi. *Res:* Energy conversion research. *Mailing Add:* 225 Pebble Creek Rd Columbia SC 29223

HENDERSON, ROBERT WESLEY, b Omaha, Nebr, Mar 31, 14; m 38; c 4. NUMISMATICS, PHOTOGRAPHY. *Educ:* Ore State Univ, BS, 38; Cornell Univ & Univ Minn, PhD(plant genetics), 50. *Prof Exp:* Asst, Sherman Exp Sta, Moro, Ore, 38-41; asst plant breeding, Cornell Univ, 41-42; assoc geneticist, USDA, Univ Minn, 42-46; assoc prof & assoc agronomist, Dept Farm Crops, 46-48, asst to dir agr exp sta, 47-53, asst dir agr exp sta, 53-76, EMER PROF AGRON CROP SCI, ORE STATE UNIV, 77- *Concurrent Pos:* Geneticist, USDA, 46-52; chief-of-party, Ore State Univ Adv Staff to Kasetsart Univ, Bangkok, 59-61; proprietor, Henderson Enterprises, 75- *Mem:* Fel AAAS; Sigma Xi. *Res:* Plant breeding, rubber plants, sugar beets, red clover and wheat. *Mailing Add:* 3732 NW Van Buren Corvallis OR 97330

HENDERSON, ROGENE FAULKNER, b Breckenridge, Tex, July 13, 33; m 57; c 3. BIOCHEMISTRY. *Educ:* Tex Christian Univ, BA, 55; Univ Tex, Austin, PhD(chem), 60; Am Bd Toxicol, dipl, 81. *Prof Exp:* NIH fel, Sch Med, Univ Ark, 60-62, res assoc biochem, 62-66; BIOCHEMIST & TOXICOLOGIST, LOVELACE ENVIRON & BIOMED RES INST, INC, 67- *Concurrent Pos:* Mem, NIH Toxicol Study Sect, 82-86; mem ed bd Toxicol, 84-88, J Biochem Toxicol, 86- & Inhalation Toxicol, 88-; mem, Nat Res Coun/Nat Acad Sci Comts, Toxicol, 85-91, Biol Markers, 86-88, Epidemiol Air Pollution, 84-85, Common Risk Assessment Methodology, 90-; mem, Burroughs Wellcome Adv Comt Toxicol, Scholar Award, 87-89; adj prof, Dept Vet Microbiol, Path & Public Health, Sch Vet Med, Purdue Univ, 87-89; clin prof, Col Pharm, Univ NMex, 89-; mem, Assoc Western Univs Lab Adv Bd, 88-89; assoc ed, Toxicol Appl Pharmacol, 90- *Honors & Awards:* Frank Blood Award, Soc Toxicol, 89. *Mem:* NY Acad Sci; AAAS; Am Chem Soc; Soc Toxicol. *Res:* Early indicators of pulmonary toxicity; toxicokinetics of inhaled organic compounds; pathogenesis of noncancer diseases of respiratory tract. *Mailing Add:* Lovelace Environ & Biomed Res Inst PO Box 5890 Albuquerque NM 87115

HENDERSON, RUTH MCCLINTOCK, pharmacology, physiology, for more information see previous edition

HLNDERSON, SHERI DAWN, b Anoka, Minn, Jan 17, 53. RADIATION BIOPHYSICS, RADIATION BIOLOGY. *Educ:* Univ Kans, Lawrence, BA, 75, BA, 75, PhD(radiation biophysics), 80. *Prof Exp:* Teaching asst & res asst, Univ Kans Med Ctr, 79, res assoc, radiation biol, 80-, instr, 81-; AT LAWRENCE BERKELEY LAB. *Concurrent Pos:* Mem grad fac, Univ Kans, Lawrence, 81-; consult, Northeastern Univ, 82- *Mem:* Radiation Res Soc; Cell Kinetic Soc; Am Asn Cancer Res; Sigma Xi. *Res:* Interaction of radiation with other cytotoxic agents (hypoxic cell sensitizers, chemotherapy agents, hyperthermia) in in-vitro cell culture systems and in-vivo animal systems (both normal and tumor tissues). *Mailing Add:* Lawrence Berkeley Lab Bldg 55 1 Cyclotron Rd Berkeley CA 94720

HENDERSON, THOMAS E, b Middletown, NY, Aug 17, 34; m 81; c 3. ADULT EDUCATION PROGRAMS. *Educ:* State Univ NY, New Paltz, BS, 58; Univ Albany, MS, 79. *Prof Exp:* Teacher IV, Dept Correctional Serv, NY State, 62-89; adj prof, Mercy Col, 69-89; RETIRED. *Concurrent Pos:* Founder & coordr, Solar Energy Proj, NY State Dept Corrections, 78-80. *Honors & Awards:* One Idea Ahead of Its Time Award, Atlantic Richfield, 74; Technol Today Award, 80. *Res:* Established first solar energy program in NY State Prison System. *Mailing Add:* 130 W Main St Middletown NY 10940

HENDERSON, THOMAS OTIS, b Hickory, NC, Nov 17, 37; m 61, 70; c 5. BACTERIOLOGY, BIOCHEMISTRY. *Educ:* Lenoir-Rhyne Col, BS, 61; NC State Univ, MS, 64, PhD(microbiol, nutrit biochem), 67. *Prof Exp:* Res technician chem, Tape Div, Shuford Mills, NC, 60-61; asst prof, 68-74, assoc prof, 74-80, PROF BIOCHEM, UNIV ILL COL MED, 80-, ASSOC DEAN ACAD PROG, 77- *Concurrent Pos:* NIH res fel, Univ Chicago, 66-68; NIH res grants, 70-76; Am Heart Asn grant, 74-77. *Mem:* AAAS; Am Soc Biol Chem; Am Soc Microbiol. *Res:* Regulation of fatty acid synthesis in bacteria; bacterial phospholipid metabolism; phosphonic acid chemistry and metabolism. *Mailing Add:* Dept Biol Chem A-312 CMWT Univ Ill 1853 W Polk St Chicago IL 60612

HENDERSON, THOMAS RANNEY, b Orange, NJ, Apr 11, 41; m 63, 90; c 2. ORGANIC CHEMISTRY. *Educ:* Drew Univ, BA, 63; Rutgers Univ, PhD(org chem), 73. *Prof Exp:* Jr chemist, Hoffmann-La Roche Inc, 64-68; develop scientist III, 74-77, develop scientist IV, 77-87, ADMIN & TECH SERV MGR, BURROUGHS WELLCOME CO, 88- *Concurrent Pos:* Res assoc, Univ Ill, Urbana, 72-74. *Mem:* Am Chem Soc. *Res:* Development of new routes to pharmaceutically important compounds. *Mailing Add:* Burroughs Wellcome Co 3030 Cornwallis Rd Research Triangle Park NC 27709

HENDERSON, THOMAS RICHARD, b Edinburg, Tex, Aug 24, 32; m 57; c 3. BIOCHEMISTRY. *Educ:* Pan Am Col, BA, 53; Univ Tex, PhD(biochem), 60. *Prof Exp:* Instr biochem, Sch Med, Univ Ark, 60-63, asst prof, 63-66; BIOCHEMIST, LOVELACE ENVIRON & BIOMED RES INST, INC, 66- *Concurrent Pos:* Asst prof chem, Univ Albuquerque, 84- *Mem:* AAAS; Am Chem Soc; Am Soc Biol Chemists. *Res:* Geochemistry; environmental chemistry; inhalation toxicology; mass spectrometry; chemistry and toxicology of combustion effluents; diesel exhaust analysis. *Mailing Add:* 5609 Don Felipe Ct SW Albuquerque NM 87105

HENDERSON, ULYSSES VIRGIL, JR, b Miami, Fla, July 15, 26; m 51; c 2. ORGANIC CHEMISTRY. *Educ:* US Merchant Marine Acad, BS, 47; Ga Inst Technol, BS, 51, PhD(org chem), 54. *Prof Exp:* Res chemist, Stand Oil Co, Ind, 54; sr scientist, Exp Inc, 56-61; res chemist, Texaco Inc, 61-62, group leader, 62-65, asst supvr, 65-68, supvr fundamental & explor res, 68-71, supvr chem res, 71-72, supvr air & water conserv, 72-82, assoc dir, 82-87, dir, Environ Affairs, 88-91, gen mgr, Environ & Prod Safety, 91; RETIRED. *Mem:* AAAS; Am Chem Soc. *Res:* Reaction mechanisms; decomposition of monopropellants; high speed gas phase reactions; heterogeneous combustion; chemical vapor plating; petrochemical processes and products; solid and aqueous waste management for petroleum industry. *Mailing Add:* Texaco Inc PO Box 509 Beacon NY 12508

HENDERSON, WARREN ROBERT, b Boston, Mass, May 5, 27. HORTICULTURE. *Educ:* Univ NH, BS, 49; Harvard Univ, AM, 51; Ohio State Univ, PhD(hort), 59. *Prof Exp:* Asst hort, Ohio State Univ, 54-59; asst prof, 59-67, ASSOC PROF HORT, NC STATE UNIV, 67- *Honors & Awards:* L M Ware Award, 71. *Mem:* Am Soc Hort Sci; Sigma Xi. *Res:* Tomato breeding and genetics; disease resistance breeding of cabbage, squash, and watermelon. *Mailing Add:* Dept Hort Sci NC State Univ Raleigh NC 27650

HENDERSON, WILLIAM ARTHUR, JR, b Winthrop, Mass, Oct 5, 32; m 60; c 2. ORGANIC CHEMISTRY. *Educ:* Harvard Univ, BA, 54; Yale Univ, MS, 56, PhD(org chem), 60. *Prof Exp:* From res chemist to sr res chemist, Am Cyanamid Co, 60-84, assoc res fel org chem, 84-; RETIRED. *Mem:* Am Chem Soc; Geol Soc Am. *Res:* Chemistry of phosphines, cyclopolyphosphines, photochromic compounds and photochemistry; singlet oxygen; photostabilization of polymers; thermal and photostabilization of agricultural chemicals; photochemical methods of immunassay, chiral synthesis of pharmaceuticals; synthesis of polyisocyanates. *Mailing Add:* 47 Robin Ridge Dr Madison CT 06443

HENDLER, EDWIN, b Philadelphia, Pa, Aug 29, 22; m 45; c 2. HUMAN PHYSIOLOGY. *Educ:* Pa State Univ, BS, 43; Univ Pa, MS, 56, PhD(physiol), 59. *Prof Exp:* Physicist, Aerospace Crew Equip Lab, Naval Air Eng Ctr, 46-49, head physiol sect, Aircraft & Crew Systs Directorate, naval Air Develop Ctr, 50-52, head acceleration br,52-55, supt med sci res div, 55-63, mgr life sci res group, 63-71, head appl physiol lab, Crew Systs Dept, 71-74, head res group, Life Sci Div, Aircraft & Crew Systs Directorate, Naval Air Develop Ctr, 75-81; CONSULT, LIFE SCI, 81- *Concurrent Pos:* Assoc, Sch Med, Univ Pa, 60-63. *Honors & Awards:* Paul Bert Award for Physiol Res, Aerospace Med Asn, 73; Prof Excellence Award, Life Scis & Biomed Engr Br, Aerospace Med Asn, 88. *Mem:* Biophys Soc; fel Aerospace Med Asn; Am Physiol Soc; Sigma Xi. *Res:* Aerospace physiology; human engineering; acceleration effects; thermal physiology; temperature sensation; pulmonary physiology. *Mailing Add:* Eight Sandringham Pl Cherry Hill NJ 08003-1531

HENDLER, ERNESTO DANILO, b San Martin, Mendoza, Arg, Oct 24, 35; m 61; c 2. INTERNAL MEDICINE, NETHROLOGY. *Educ:* San Martin Nat Col, Arg, BS, 52; Univ Buenos Aires, MD, 60. *Hon Degrees:* Hon dipl, Univ Buenos Aires, 61. *Prof Exp:* Clin fel nephrology, Inst Med Res, Univ Buenos Aires, 64-66; intern med, Rivadavia Hosp, Buenos Aires, 61; resident internal med, Vet Admin Hosp, DC, 66-67; instr, Vet Admin Div, Sch Med, George Washington Univ, 67-68; from instr to asst prof, 68-76, ASSOC

PROF MED, SCH MED, YALE UNIV, 76- *Concurrent Pos:* NIH res fel, 67-68, spec res fel, 68-70; staff physician, Med Serv, West Haven Vet Admin Hosp, Conn, 70-, assoc dir dialysis unit, 70-73, dir, 73- *Mem:* Am Soc Artificial Internal Organs; Int Soc Nephrol; Am Fedn Clin Res; Am Soc Nephrology. *Res:* Metabolic aspects of kidney function, especially regarding the active transport of sodium; treatment of acute and chronic renal failure. *Mailing Add:* Vet Admin Hosp Spring St West Haven CT 06516

HENDLER, GORDON LEE, b New York, NY, Dec 11, 46. SYSTEMATICS, INVERTEBRATE ZOOLOGY. *Educ:* Rutgers Univ, BA, 68; Univ Conn, PhD(zool), 73. *Prof Exp:* Fel, Woods Hole Oceanog Inst, 73-74; Walter Rathbone Bacon fel, Smithsonian Inst, 74-75, res marine biologist, Smithsonian Trop Res Inst, 75-78, marine biologist, Benthic Sect, Smithsonian Oceanog Sorting Ctr, 78-85; assoc cur, 85-90, CURATOR NATURAL HIST MUS, LOS ANGELES COUNT, 90-, HEAD INVERT ZOOL SECT, 87- *Concurrent Pos:* Expeds, Smithsonian Philippines, 78 & research vessel Alpha Helix Moro, 79; field work in Cent Am & Antarctica, 81-85. *Mem:* Sigma Xi; Am Soc Zoologists; Soc Syst Zool; AAAS; Ecol Soc Am. *Res:* Systematics, ecology and behavior of echinoderms, primarily ophiuroids; curation of LACM echinoderm collection; study of bathyal Caribbean brittlestars with submersibles; brittlestar photoreception, reproduction, skeletal ontogeny. *Mailing Add:* Natural Hist Mus 900 Exposition Blvd Los Angeles CA 90007

HENDLER, NELSON HOWARD, b New York, NY, Aug 15, 44; m 74; c 4. PSYCHIATRY, NEUROPHYSIOLOGY. *Educ:* Princeton Univ, BA, 66; Univ Md Sch Med, MD, 72, MS, 74. *Prof Exp:* From instr to asst prof psychiat, 75-82, psychiat consult, Chronic Pain Treat Ctr, 75-82, ASST PROF NEUROSURG IN PSYCHIAT, JOHNS HOPKINS HOSP, 75-; CLIN DIR, MENSANA CLIN, 78- *Concurrent Pos:* Dir med res, Reflex Sympathetic Dystrophy Asn, 85-; Falk fel, Comt Res & Develop, Am Psychiat Asn, 74-75. *Honors & Awards:* William Meninger Award. *Mem:* Am Asn Study Headache; AMA; Hans Selye Am Inst Stress (vpres, 83-); Pavlovian Soc; Am Pain Soc; Biofeedback Soc Am. *Res:* Research into chronic pain and testing to research the validity of chronic pain and the Hendler back pain test; author of 3 books, 18 chapters, and 35 articles. *Mailing Add:* Mensana Clin 1718 Greenspring Valley Rd Stevenson MD 21153

HENDLER, RICHARD WALLACE, b Philadelphia, Pa, Mar 3, 27; m 48; c 2. BIOCHEMISTRY, BIOPHYSICS. *Educ:* Pa State Univ, BS, 48; Univ Pa, MS, 49; Univ Calif, PhD(biochem), 52. *Prof Exp:* Asst biochem, Univ Calif, 50-52; BIOCHEMIST, NAT HEART & LUNG INST, 55-, HEAD SECT MEMBRANE ENZYMOL, LAB CELL BIOL, 75- *Concurrent Pos:* Fel, Nat Found Infantile Paralysis, Nat Heart Inst & NIH, 52-54; USPHS fel, Free Univ Brussels, 54-55. *Mem:* Am Soc Biol Chemists; Biophys Soc. *Res:* Role of cellular membranes in biosynthetic and energy producing activities of cells. *Mailing Add:* Lab Cell Biol Nat Heart & Lung Inst Bethesda MD 20892

HENDLEY, EDITH DI PASQUALE, b New York, NY, Sept 5, 27; m 52; c 3. PHYSIOLOGY. *Educ:* Hunter Col, AB, 48; Ohio State Univ, MS, 50; Univ Ill, PhD(physiol), 54. *Prof Exp:* Instr physiol, Univ Chicago, 54-56; asst lectr, Univ Sheffield, 57-58; instr ophthal, Johns Hopkins Univ, 63-66, res assoc pharmacol & exp therapeut, 66-71; sr investr, Friends Med Sci Res Ctr, Inc, 72-73; assoc prof, 73-82, PROF PHYSIOL, MED COL, UNIV VT, 83- *Mem:* Am Physiol Soc; Am Soc Pharmacol & Exp Therapeut; Soc Neurosci; Asn Women in Sci; Sigma Xi; Am Asn Advan Sci. *Res:* Neurochemical basis of behavior; biogenic amines in brain. *Mailing Add:* Dept Physiol & Biophys Univ Vt Col Med Burlington VT 05405

HENDLEY, JOSEPH OWEN, b Chattanooga, Tenn, Aug 18, 37; c 4. PEDIATRICS. *Educ:* Vanderbilt Univ, BA, 59; Univ Pa, MD, 63. *Prof Exp:* Intern pediat, Med Ctr, Duke Univ, 63-64, resident, 64-65; res fel pediat & prev med, Sch Med, Univ Va, 65-67; resident pediat, Med Ctr, Duke Univ, 64-65; res fel infectious dis, Harvard Sch Pub Health, 68-70; from asst prof to assoc prof, 70-82, PROF PEDIAT, SCH MED, UNIV VA, 82- *Concurrent Pos:* Epidemic Intel Off, Dept Preventative Med, USPHS, Univ Va, 65-68. *Mem:* Am Soc Microbiol; Infectious Dis Soc Am; Am Acad Pediat; Am Epidemiol Soc; Soc Pediat Res. *Res:* Pediatric infectious disease; interruption of rhinovirus transmission; characterization of gonococcal capsule; mechanisms of human immunity to bordetella pertussis. *Mailing Add:* Dept Pediat Box 386 Univ Va Hosp Charlottesville VA 22908

HENDLIN, DAVID, b Harbin, China, Mar 8, 20; nat US; m 44; c 2. MICROBIOLOGY. *Educ:* City Univ NY, BA, 41; Iowa State Univ, MSc, 43; Rutgers Univ, PhD(microbiol), 49. *Prof Exp:* Res microbiologist, Merck Sharp & Dohme Res Lab, 43-48, head nutrit sect, 48-53, mgr microbiol dept, 53-56, asst dir, 56-67, dir bact labs, Merck Inst Therapeut Res, 67-69, dir basic microbiol, 69-74, sr dir develop microbiol, 74-81, exec adminr, Japan, 81-85, sr dir & sci liaison, 85-90; RETIRED. *Mem:* AAAS; Am Chem Soc; Am Soc Microbiol; fel NY Acad Sci; fel Am Acad Microbiol; Soc Gen Microbiol; Soc Ind Microbiol. *Res:* Antibiotics; production of chemicals by microorganisms; microbial nutrition; chemotherapy of bacterial infections. *Mailing Add:* Five Laurel Dr Springfield NJ 07081

HENDON, JOSEPH C, b Mayfield, Ky, July 19, 38; m 62; c 2. ORGANIC CHEMISTRY. *Educ:* Murray State Univ, BS, 61; Univ Ky, PhD(org chem), 68. *Prof Exp:* Res chemist, Tenn Eastman Co, 67-69; from asst prof to assoc prof org chem, Murray State Univ, 69-74; CHMN, DEPT SCI & MATH, PIKEVILLE COL, 78-; CHEMIST, PB&S CHEM CO, 81- *Mem:* Am Chem Soc. *Res:* Synthesis and determination of carcinogenic activity of isoquinoline azo dyes related to butter yellow-synthesis of cationic dyes for use in dying fibers; environmental contaminents (air and water) monitoring associated with the development of a coal gasification plant. *Mailing Add:* 331 Old Orchard Rd Henderson KY 42420-4756

HENDREN, RICHARD WAYNE, b Salisbury, NC, June 24, 48; m 69; c 2. BIOCHEMISTRY. *Educ:* Univ NC, Chapel Hill, BS, 70; Harvard Univ, PhD(chem), 77. *Prof Exp:* Res scientist biochem, Corp Res Lab, Union Carbide Corp, 77-79; SR RES BIOCHEMIST, CHEM & LIFE SCI DIV, RES TRIANGLE INST, 79- *Mem:* Am Chem Soc; AAAS; Am Asn Clin Chem. *Res:* Biochemical genetics of environmental mutagenesis; enzymology of lipid biosynthesis; chemistry of peptide hormones; applications of clinical chemistry to toxicology. *Mailing Add:* 3462 Robin Hill St Thousand Oaks CA 91360

HENDREY, GEORGE RUMMENS, b Seattle, Wash, Dec 9, 40; m 65; c 2. LIMNOLOGY, ECOLOGY. *Educ:* Univ Wash, BA, 66, MS, 70, PhD(aquatic biol), 73. *Prof Exp:* Assoc res engr chem, Boeing Co, 66-68; Pub Health Serv trainee, 68-73; res assoc aquatic biol, Univ Wash, 70-73; res scientist limnol, Norweg Inst Water Res, Oslo, 73-76, tenure, 75; vis lectr & sr res assoc, Sect Ecol & Systs, Cornell Univ, 76-77; ECOL SCIENTIST & DIV HEAD, TERRESTRIAL & AQUATIC ECOL, BROOKHAVEN NAT LAB, 77- *Concurrent Pos:* Vis fel, Ctr Environ Res, Cornell Univ, 76-; adv, Int Joint Comn, 79-81; prin investr, Effects Acid Precipitation Crop Growth & Yield, US Environ Protection Agency & Northeastern Nat Environ Res Park, Dept Energy, 77-; consult, Environ Criteria Assessment Off, Environ Protection Agency, 79-81; mem, Biol Effects Gov Bd, NADP, 79-81, EPA AdminrRound Table, 83; expert witness, Environ Defense Fund, Scenic Hudson Inc, Hudson River Fisherman's Asn, 81, City of New York, Hudson River Found, State of Fla, 83-85; DOE Intragency Personnel Assignment, 84-88. *Mem:* Am Asn Limnol & Oceanog; Int Soc Limnol; Int Asn Theoret & Appl Limnol; Am Chem Soc; NY Acad Sci. *Res:* Aquatic biology; limnology; bioassay; acid and precipitation effects; geochemistry. *Mailing Add:* Terrestrial & Aquatic Ecol Div Brookhaven Nat Lab Upton NY 11973

HENDRICH, CHESTER EUGENE, b Clinton, Mo, Jan 29, 35; m 59; c 3. PHYSIOLOGY, ENDOCRINOLOGY. *Educ:* Univ Mo-Columbia, AB, 59, MS, 61, PhD(endocrinol), 65. *Prof Exp:* Fel endocrinol, Emory Univ, 64-65 & Dartmouth Med Sch, 65-67; asst prof physiol & endocrinol, Col Med, Ohio State Univ, 67-73; assoc prof physiol & endocrinol, 73-77, PROF PHYSIOL, MED COL GA, 77- *Mem:* Am Physiol Soc; Endocrine Soc; Soc Exp Biol & Med. *Res:* Effects of thyroid and parathyroid hormone on other maternal hormone secretions and fetal metabolism and development; effects of artificial atmospheres on the endocrine system. *Mailing Add:* Dept Physiol Med Col Ga Augusta GA 30912

HENDRICK, LYNN DENSON, b Panama City, Fla, July 23, 37; m 62; c 2. NUCLEAR PHYSICS, SCIENCE EDUCATION. *Educ:* Auburn Univ, BEP, 59, MS, 60; Univ SC, PhD(physics), 66. *Prof Exp:* Asst prof physics, Maryville Col, 61-62; res assoc, Univ SC, 66-67; assoc prof & head dept phys sci, Columbia Col, 67-70; PROF PHYSICS, FRANCIS MARION COL, 70- *Concurrent Pos:* Chmn manpower & prof educ comt, Health Physics Soc, 86-88. *Mem:* Am Asn Physics Teachers; Am Phys Soc; Sigma Xi; Health Physics Soc. *Res:* Nuclear spectroscopy; structure and decay schemes. *Mailing Add:* Dept Physics Francis Marion Col Florence SC 29501

HENDRICK, MICHAEL EZELL, b Memphis, Tenn, July 31, 45; m 69; c 1. ORGANIC CHEMISTRY. *Educ:* Southwestern at Memphis, BS, 67; Princeton Univ, AM, 69, PhD(chem), 71. *Prof Exp:* Res fel org chem, Ruhr Univ, Bochum, Ger, 71-73 & Stanford Univ, 73-74; SR RES SCIENTIST ORG CHEM, PFIZER INC-CENT RES, 74- *Concurrent Pos:* NIH fel, Stanford Univ, 74. *Mem:* Am Chem Soc. *Res:* Organic chemistry of odor and taste, especially sweeteners, odorants, and flavors. *Mailing Add:* 42 Pine Island Rd Groton CT 06340

HENDRICKER, DAVID GEORGE, b Aurora, Ill, Oct 14, 38; m 65. INORGANIC CHEMISTRY. *Educ:* Northern Ill Univ, BS, 60; Iowa State Univ, MS, 63, PhD(inorg chem), 65. *Prof Exp:* From asst prof to assoc prof, 65-74, PROF CHEM, OHIO UNIV, 74- *Mem:* Am Chem Soc; Sigma Xi. *Res:* Metal carbonyl, phosphorus and aluminum hydride chemistry; coordination compounds; application of nuclear magnetic resonance and infrared spectroscopy to inorganic chemistry. *Mailing Add:* Dept Chem Ohio Univ Athens OH 45701

HENDRICKS, ALBERT CATES, b Beaumont, Tex, Dec 28, 37; m 67; c 3. AQUATIC BIOLOGY, AQUATIC TOXICOLOGY. *Educ:* Lamar Univ, BA, 63; NTex State Univ, MA, 66, PhD(biol), 70. *Prof Exp:* Hydrologist ecol, Tex Water Develop Bd, 70; prof biol, 71-80, ASSOC PROF ZOOL, VA POLYTECH INST & STATE UNIV, 80- *Mem:* Sigma Xi; Am Soc Limnol & Oceanol; Ecol Soc Am; NAm Benthol Soc. *Res:* Fate and effect of toxic chemicals in the aquatic environment; use of biomonitoring as early warning devices. *Mailing Add:* Biol Dept Va Polytech Inst Blacksburg VA 24061

HENDRICKS, CHARLES D(URRELL), JR, b Lewiston, Utah, Dec 5, 26; m 48; c 2. INERTIAL CONFINEMENT FUSION. *Educ:* Utah State Univ, BS, 49; Univ Wis, MS, 51; Univ Utah, PhD(physics), 55. *Prof Exp:* Physicist, US Naval Ord Test Sta, 52 & Off Ord Res, 54; mem staff, Lincoln Lab, Mass Inst Technol, 55-56; from asst prof to prof, 56-80, EMER PROF ELEC ENG, UNIV ILL, URBANA, 80-; DEP ASSOC PROG LEADER, AIT FLIGHT PROG, LAWRENCE LIVERMORE NAT LAB, UNIV CALIF, 88- *Concurrent Pos:* Consult, Lincoln Lab, Mass Inst Technol, 56-57, Ramo-Wooldridge Corp, 57-, Space Tech Labs, 62- & Xerox Corp, 63-, Rand Corp, 72-78, Exxon Res & Eng Ctr, 74-, Calif Inst Technol, 78-, Nuclear Eng Dept, Univ Wis, 79; assoc prog leader, laser fusion target fabrication, Lawrence Livermore Nat Lab, Univ Calif, 74-80, sr scientist, 80-88; pres, Int Microprod, Inc, 87- *Mem:* Fel Am Phys Soc; foreign fel Electrostatics Soc Japan; fel Inst Elec & Electronic Engrs; assoc fel Am Inst Aeronaut & Astronaut; Electrostatics Soc Am. *Res:* Research in inertial confinement fusion primarily in target fabrication for fusion experiments; plasma and small particle physics and electrostatic phenomena and processes. *Mailing Add:* Lawrence Livermore Nat Lab PO Box 5508 L-482 Livermore CA 94550

HENDRICKS, CHARLES HENNING, b Traverse City, Mich, Oct 26, 17; m 42; c 5. OBSTETRICS & GYNECOLOGY. *Educ:* Univ Mich, AB, 41, MD, 43; Am Bd Obstet & Gynec, dipl. *Prof Exp:* Instr obstet & gynec, Ohio State Univ, 48-51, asst prof, 51-54; from assoc prof to prof, Case Western Reserve Univ, 54-68; ROBERT A ROSS DISTINGUISHED PROF OBSTET & GYNEC & CHMN DEPT, SCH MED, UNIV NC, CHAPEL HILL, 68- *Honors & Awards:* Am Asn Obstet & Gynec Found Prize, 56. *Mem:* AAAS; AMA; Am Gynec Soc; Am Asn Obstet & Gynec; Am Col Obstet & Gynec. *Res:* Physiology of pregnancy. *Mailing Add:* Dept Obstet & Gyn NC Mem Hosp Chapel Hill NC 27514

HENDRICKS, DAVID WARREN, b Springfield, Mo, Sept 10, 31; m 59; c 3. SANITARY ENGINEERING. *Educ:* Univ Calif, Berkeley, BS, 54; Utah State Univ, MS, 60; Univ Iowa, PhD(sanit eng), 65. *Prof Exp:* Jr civil engr, State Dept Water Res, Calif, 54; instr civil eng, Univ Idaho, 58-61, asst res prof, 61-62; from asst prof to assoc prof, Utah State Univ, 65-70; ASSOC PROF CIVIL ENG, COLO STATE UNIV, 70- *Concurrent Pos:* Consult, Eimco Corp, Salt Lake City, 66-67, Pan Am Health Asn, 78, UNESCO, 78 & Yemen Gov, 80. *Mem:* Am Soc Civil Engrs; Water Pollution Control Fedn; Am Water Works Asn; Am Chem Soc; Asn Environ Eng Prof. *Res:* Water resources; physical chemistry; hydraulic engineering; waste management systems; environmental interdisciplinary problems. *Mailing Add:* Engin Res Ctr Foothell Campus Colo State Univ Ft Collins CO 80523

HENDRICKS, DELOY G, b Pocatello, Idaho, Dec 18, 38; m 62; c 7. MINERAL METABOLISM, NUTRIENT BIOAVAILABILITY. *Educ:* Univ Idaho, BS, 61; Mich State Univ, PhD(animal nutrit), 67. *Prof Exp:* Asst prof nutrit, dept nutrit, 67-72, Assoc prof, 72-77, PROF NUTRIT & FOOD SCI, UTAH STATE UNIV, 77- *Concurrent Pos:* Vis scientist, Vet Admin Hosp, Baltimore, Md, 71; vis prof, USDA Human Nutrit Lab, Beltsville, Md, 80-81. *Mem:* Am Inst Nutrit. *Res:* Mineral metabolism and bioavailabiltiy; food storage and safety; nutritional status and behavior. *Mailing Add:* Dept Nutrit & Food Sci Utah State Univ Logan UT 84322-8700

HENDRICKS, DONOVAN EDWARD, b Apr 8, 40; US citizen. ENTOMOLOGY, ECOLOGY. *Educ:* Purdue Univ, Ind, BSc, 63, MSc, 65. *Prof Exp:* RES ENTOMOLOGIST ECOL & ENTOM, AGR RES SERV, USDA, 66- *Mem:* Entom Soc Am; Am Registry Prof Entomologists. *Res:* Insect ecology and behavior in relation to field mating and migration of Lepidoptera; biophysics; insect detection systems; electronic insect detection; US patent for SODAR insect detection device. *Mailing Add:* Agr Res Serv USDA Box 346 Stoneville MS 38776-0346

HENDRICKS, GRANT WALSTEIN, b Richmond, Utah, Aug 9, 16; m 39; c 3. PETROLEUM CHEMISTRY. *Educ:* Univ Utah, BA, 38, MA, 40. *Prof Exp:* Res chemist & sr sect leader, Union Oil Co Calif, 43-59, supvr process res, 59-64, supvr reservoir mech, 64-65, mgr prod res, 65-80; CONSULT, 80- *Mem:* Am Chem Soc; Soc Petrol Eng. *Res:* Catalyst development and catalytic process development; petroleum refining and production. *Mailing Add:* 817 Larchwood Dr Brea CA 92621

HENDRICKS, JAMES OWEN, b Waynetown, Ind, Sept 3, 09; m 36; c 3. ELASTOMERS, ADHESIVES. *Educ:* Wabash Col, AB, 31; Univ Ill, PhD(phys chem), 36. *Prof Exp:* Teaching fel, Univ Calif, 31-32; chemist, Kendall Ref Co, Pa, 33-34; asst & fel, Univ Ill, 34-36; res chemist, Minn Mining & Mfg Co, 36-42, head colloid sect, Cent Res Dept, 42-47, from asst dir to assoc dir, 47-53, mgr res tape div, 53-66, sci liaison, 66-74; RETIRED. *Concurrent Pos:* Campbell fel award, Wabash Col. *Honors & Awards:* Carlton Award, Minn Mining & Mfg Co, 73. *Mem:* Am Chem Soc. *Res:* Elasticity of permanently tacky adhesives; specific adhesion of rubber adhesives; water dispersed adhesives; low adhesion surfaces. *Mailing Add:* 2292 6th Street White Bear Lake Sta St Paul MN 55110

HENDRICKS, JAMES RICHARD, b Guilford Co, NC, Oct 23, 20. PARASITOLOGY. *Educ:* Guilford Col, BS, 40; Univ NC, MS, 48, PhD(parasitol), 51. *Prof Exp:* Instr biol, Guilford Col, 46; from instr to asst prof parasitol, 48-54, ASSOC PROF PARASITOL, UNIV NC, CHAPEL HILL, 54- *Mem:* AAAS; Am Soc Parasitol; Am Soc Trop Med & Hyg; Am Soc Prof Biol; Am Pub Health Asn; Sigma Xi. *Res:* Immunological and serological studies related to animal parasites. *Mailing Add:* 107 Simpson St Carrboro NC 27510

HENDRICKS, JERRY DEAN, b Atkinson, Nebr, July 7, 44; m 68; c 3. FISH BIOLOGY, FOOD TOXICOLOGY. *Educ:* Colo State Univ, BS, 66, PhD(fishery biol), 71. *Prof Exp:* Fishery biologist histopathol, Nat Oceanic & Atmospheric Admin, Nat Marine Fisheries Serv, 71-72 & US Fish & Wildlife Serv, 73-74; res assoc, 75-78, from asst prof to assoc prof, 78-85, PROF HISTOPATHOL, ORE STATE UNIV, 85-, PROF, DEPT FOOD SCI, 80- *Mem:* Am Fisheries Soc; Sigma Xi. *Res:* Fish histology and histopathology; fish nutrition; carcinogenesis; food toxicology. *Mailing Add:* Dept Food Sci Ore State Univ Corvallis OR 97331

HENDRICKS, LAWRENCE JOSEPH, b Springfield, Ill, Aug 29, 19; m 51; c 2. LIMNOLOGY, FISHERIES. *Educ:* Western State Col Colo, AB, 47; Univ Colo, MA, 51; Univ Mo, PhD(zool), 55. *Prof Exp:* Jr res biologist, Univ Calif, Los Angeles, 55-56; asst prof biol, NDak State Teachers Col, Minot, 56-57; from asst prof to assoc prof, 57-68, PROF BIOL, SAN JOSE STATE UNIV, 68- *Mem:* Am Soc Limnol & Oceanog; Am Fisheries Soc; Wildlife Soc. *Res:* Limnology of ponds and reservoirs; culture of fish in ponds; conservation of natural resources; life histories and ecology of anadromous fishes. *Mailing Add:* Dept Biol Sci San Jose Univ San Jose CA 95192

HENDRICKS, LEWIS T, b Rome, NY, July 3, 40; m 62; c 2. FOREST PRODUCTS, MARKETING. *Educ:* State Univ NY, BS, 61, MS, 63; Mich State Univ, PhD(econ, mkt), 67. *Prof Exp:* Teaching asst forest prod, Mich State Univ, 62-64; technologist, USDA, 64-67; assoc prof, 67-76, EXTEN SPECIALIST, UNIV MINN, 67-, PROF FOREST PROD, 76- *Concurrent Pos:* Consult, Wood Prods Indust; vis prof, Univ Wis-Madison, 85-86; spec assignment, Forest Prod Lab, USDA Forest Serv; coordr, Cold Climate Housing Info Ctr, Univ Minn, 87-89, dir, 89-90. *Mem:* Sigma Xi; Forest Prod Res Soc. *Res:* Veneer and plywood industry; wood finishing; wood preservation; moisture relationships; economics of woodburning; ice-dam prevention in residential housing; performance of wood in buildings. *Mailing Add:* 220 Kaufert Lab Univ Minn 2004 Folwell Ave St Paul MN 55108

HENDRICKS, ROBERT WAYNE, b Trail, BC, Apr 30, 37; US citizen; m 58; c 2. MATERIALS SCIENCE, DIFFRACTION PHYSICS. *Educ:* Cornell Univ, BMetE, 59, PhD(phys metall), 64; Univ Tenn, MBA, 85. *Prof Exp:* Res metallurgist & mem res staff, Oak Ridge Nat Lab, 64-81, assoc dir, Nat Ctr Small-Angle Scattering Res, 78-81; mgr, specialty prods, Technol for Energy Corp, 81-85, chief scientist, 85-86; PROF MAT ENG, VA POLYTECH INST & STATE UNIV, 86- *Concurrent Pos:* Res assignment, Inst fur Festkorperforschung der Kernforschungsanlage, WGer, 72-73. *Honors & Awards:* IR-100 Award, 77. *Mem:* Am Inst Mining, Metall & Petrol Engrs; fel AAAS; Am Crystallographic Asn; fel Am Phys Soc; Am Asn Artificial Intel; Sigma Xi; Am Soc Metals Int; Soc Exp Mech; Mat Res Soc; Am Asn Physics Teachers. *Res:* Solid state physics; physical chemistry; x-ray diffraction; application of small-angle and high-angle diffuse x-ray and neutron scattering techniques to radiation damage and phase transformation problems in physical metallurgy and in structure of polymeric materials; determination of residual stresses in metals and ceramics and electronic materials by x-ray diffraction; research instrumentation; expert systems for materials research. *Mailing Add:* 2904 Wakefield Dr Blacksburg VA 24060

HENDRICKS, ROBERT WILLIAM, b Mt Home, Idaho, June 2, 29; m 51; c 3. PHYSICAL CHEMISTRY. *Educ:* Univ Rochester, BS, 51; Brown Univ, PhD(chem), 56. *Prof Exp:* Res chemist, Exp Sta, Film Res Lab, E I du Pont de Nemours & Co, Inc, 55-60, res supvr, Circleville Res & Develop Lab, 60-64, develop supvr, Circleville Mylar plant, 64-65, mgr develop dept, 65-66, res supvr, Photo Prod Detp, 66-74, res mgr, 74-77, lab dir, Photo Prod Dept, 77-90; RETIRED. *Res:* Photographic products; magnetic recording; polymer chemistry; polymeric films and coatings. *Mailing Add:* 403 French Rd Rochester NY 14618

HENDRICKS, WALTER JAMES, b Perkasie, Pa, Jan 23, 35; m 61; c 2. MATHEMATICS. *Educ:* Pa State Univ, BS, 56; Mass Inst Technol, MS, 58; Univ Minn, PhD(math), 69. *Prof Exp:* Asst prof, Case Western Reserve Univ, 69-76, assoc prof math & statist, 76-83; sr systs consult, INCO, 83-85 & Gen Tel Electronics, 85-88; MEM TECH STAFF, MITRE CORP, 88- *Concurrent Pos:* Vis lectr & sr res fel, Westfield Col, Univ London, 74-75; vis assoc prof math, Univ Va, 81-82. *Mem:* Am Math Soc; Math Asn Am; Inst Math Statist. *Res:* Markov chains; Markov processes; processes with stationary and independent increments. *Mailing Add:* Mitre Corp 7525 Colshire Dr MS Z426 McLean VA 22102

HENDRICKSON, ADOLPH C(ARL), b Kearsarge, Mich, June 19, 27; m 58; c 2. COMPUTER SCIENCE, INTELLIGENCE SYSTEMS. *Educ:* Mass Inst Technol, BS & MS, 52. *Prof Exp:* Res engr, analog missile simulation, Mass Inst Technol, 52-53; mathematician, digital comput prog, Remington Rand, Inc, 53-56; staff analyst, Gen Kinetics, Inc, 56-59; opers analyst, Res Anal Corp, 59-63; sr assoc, Planning Res Corp, 64-65; mem tech staff, Lambda Corp, 66-68; pres, Hendrickson Corp, 68-81; COMPUT SCIENTIST, COMPUT SCI CORP, 81- *Mem:* Asn Comput Mach; Sigma Xi; Audubon Naturalist Soc. *Res:* Computer application to operations research; micro-analytic simulation and gaming; software systems; management information systems; reliability, testing and fault tolerance. *Mailing Add:* 5014 Newport Ave Bethesda MD 20816

HENDRICKSON, ALFRED A, b Lake Linden, Mich, May 18, 29; m 54; c 2. METALLURGY. *Educ:* Mich Tech Univ, BS, 51; Columbia Univ, MS, 54; Northwestern Univ, PhD(metall), 60. *Prof Exp:* Metallurgist, FWD Corp, 51-52 & Ampco Metal Inc, 54-56; from asst prof to prof, Mich State Univ, 56-70, forging indust prof metall eng, 70-78; RETIRED. *Concurrent Pos:* Lectr, Northwestern Univ, 54-58; hon res fel, Univ Birmingham, 70-71; Nat Sci Coun Prof, Nat Sun-Yat Sen Univ, 82. *Mem:* Am Soc Metals; Am Inst Mining, Metall & Petrol Engrs; Soc Mfg Engrs. *Res:* Strength of solids; metal processing; forging processes. *Mailing Add:* 1762 SW Manitou Trail Lake Leelanu MI 49653

HENDRICKSON, ANITA ELIZABETH, b LaCrosse, Wis, Feb 20, 36; m 57; c 3. DEVELOPMENTAL NEUROBIOLOGY. *Educ:* Pac Luthern Col, BA, 57; Univ Wash, PhD(anat), 64. *Prof Exp:* Instr anat, Northwestern Med Sch, 64-65; res instr biol structr, 65-67, from instr to assoc prof 67-81, PROF OPHTHAL, 81-, PROF BIOL STRUCT, UNIV WASH, 84- *Concurrent Pos:* Vis assoc prof, Dept Neuropath, Harvard Med Sch, 75-76; vis B study sect, Div Res Grants, NIH, 76-80; Wellcome fel, Neurochem Pharmacol Unit, Med Res Coun, Eng, 79; sci counr, Nat Inst Ment Health, 81-84; invest ophthalmologist, vis scientist, J Neurosci; mem, nat coun, Soc Neurosci, 82-86. *Honors & Awards:* Dolly Green Scholar Res, Prevent Blindness, Inc, 81-82. *Mem:* Soc Neurosci; Asn Res Vision & Ophthal; Am Asn Anatomists; Int Soc Eye Res. *Res:* Development of the visual system of primates with emphasis on the synaptic organization and identification of neurotransmitters in retina, thalamus and visual cortex; development of new neuroanatomical techniques for synaptic organization. *Mailing Add:* Dept Biostruct SM20 Univ Wash Seattle WA 98195

HENDRICKSON, CHRIS THOMPSON, b Oakland, Calif, Mar 31, 50; m 77; c 3. TRANSPORTATION ENGINEERING. *Educ:* Stanford Univ, BS, MS, 73; Oxford Univ, BPhil, 75; Mass Inst Technol, PhD(civil eng), 78. *Prof Exp:* from asst prof to assoc prof, 78-87, PROF CIVIL ENG, CARNEGIE-MELLON UNIV, 87- *Honors & Awards:* Huber Res Award, Am Soc Civil Eng, 88. *Mem:* Opers Res Soc Am; Am Econ Asn; Am Soc Civil Eng; Trans Res Bd. *Res:* Transportation systems analysis, project management and finance, computer applications (including knowledge based expert systems); engineering economics; microeconomics. *Mailing Add:* Dept Civil Eng Schenley Park Pittsburgh PA 15213

HENDRICKSON, CONSTANCE MCRIGHT, b Baton Rouge, La, June 7, 49; m 71; c 2. BIOPHYSICAL CHEMISTRY, SURFACTANT CHEMISTRY. *Educ:* La Polytech Univ, BS, 71; La State Univ, PhD(biochem), 75; Univ N Tex, Med, 84. *Prof Exp:* Grad asst biochem, La State Univ, Baton Rouge, 71-75; NIH fel biophysics, Johns Hopkins Univ, 75-78; clin chem fel, Univ Ala Med Ctr, 78-79; asst prof, Tex Wesleyan Col, 80-81; chief chemist, Rockwood Systs Corp, 82-83; OWNER, ARKON CONSULTS, 83- *Mem:* Sigma Xi; Am Chem Soc; fel Am Inst Chemists. *Res:* Liposome and micelle formation; aqueous foam structure and formation; biological and artificial membrane and transport phenomena. *Mailing Add:* 2915 LBJ Freeway Suite 161 Dallas TX 75234

HENDRICKSON, DAVID NORMAN, b Minneapolis, Minn, Jan 1, 43; m 65; c 2. INORGANIC CHEMISTRY. *Educ:* Univ Calif, Los Angeles, BS, 66, Berkeley, PhD(chem), 69. *Prof Exp:* From asst prof to prof chem, Univ Ill, Urbana, 70-88; PROF CHEM, UNIV CALIF, SAN DIEGO, 88- *Concurrent Pos:* Fel, Calif Inst Technol, 69-70, Sloan Found & Jap Soc Promotion Sci; Dreyfus teacher scholar. *Mem:* Am Chem Soc; Sigma Xi; AAAS. *Res:* Bioinorganic chemistry; mixed valence chemistry; electron transfer; superparamagnetism; ferrofluids; clay and zeolite supported catalysts; magnetic exchange interactions. *Mailing Add:* Dept Chem Univ Calif San Diego La Jolla CA 92093-0506

HENDRICKSON, DONALD ALLEN, b New Rockford, NDak, Feb 9, 42; m 66; c 2. MEDICAL MICROBIOLOGY, MICROBIAL ECOLOGY. *Educ:* Minot State Col, BS, 64; Colo State Univ, MS, 69, PhD(microbiol), 70. *Prof Exp:* Pub sch teacher sci & math, NDak, 64-66; asst microbiol, Colo State Univ, 69-70; PROF BIOL, BALL STATE UNIV, 70- *Concurrent Pos:* Adj prof microbiol, Sch Med, Ind Univ, 71-72; Am Soc Microbiol area adv & consult commun media. *Mem:* Am Soc Microbiol. *Res:* Alflatoxin and mycotoxin production on solid wastes; factors affecting slime formation by bacteria; biological indicators of water quality. *Mailing Add:* Dept Biol Ball State Univ Muncie IN 47306

HENDRICKSON, FRANK R, b Springfield, Pa, Aug 3, 26; c 4. RADIOLOGY. *Educ:* Swarthmore Col, BA, 47; Jefferson Med Col, MD, 50; Am Bd Radiol, cert radiol & nuclear med, 55. *Prof Exp:* Intern, Jefferson Med Col Hosp, 50-51, resident radiol, 51-54; Am Cancer Soc fel, Univ Pa, 54-55 & Eng & Scand Therapeut Ctrs, 55-56; asst prof, Univ Ill, Chicago Circle, 56-64, assoc prof, 64-71; PROF THERAPEUT RADIOL, RUSH MED COL, 71-; PROF RADIOL, UNIV ILL, 68- *Concurrent Pos:* From asst radiologist to attend radiologist & dir sect radiation ther, Presby-St Luke's Hosp, Chicago, 56-, safety radiol officer, Radioisotope Comt, 56-71, mem tumor comt, 56-; pres, Nuclear Oncol Serv Corp, 71- *Mem:* Am Col Radiol; Am Soc Therapeut Radiol (pres, 77); Radiation Res Soc; Radiol Soc NAm; Radium Soc; Sigma Xi. *Mailing Add:* 870 Seminary Circle Glen Ellyn IL 60302-6158

HENDRICKSON, HERBERT T, b Jersey City, NJ, Oct 13, 40; m 62; c 2. VERTEBRATE ZOOLOGY. *Educ:* Cornell Univ, BS, 62, PhD(vert zool), 66. *Prof Exp:* Instr biol, Yale Univ, 66-68; asst prof, 68-74, ASSOC PROF BIOL, UNIV NC, GREENSBORO, 75- *Mem:* AAAS; Am Ornith Union; Soc Syst Zool; Soc Study Evolution; Wilson Ornith Soc; Cooper Ornith Soc. *Res:* Population trends of wintering birds in North Carolina piedmont. *Mailing Add:* Dept Biol 419 Eberhart Bldg Univ NC 1000 Spring Garden St Greensboro NC 27412

HENDRICKSON, HERMAN STEWART, II, b Los Angeles, Calif, May 14, 37; m 60; c 3. BIOCHEMISTRY. *Educ:* Pomona Col, BA, 59; Univ Ill, PhD(chem), 62. *Prof Exp:* Res chemist, Univ Calif, Berkeley, 62-63 & Western Regional Res Lab, USDA, 63-65; asst prof biochem, Southwestern Med Sch, Univ Tex, 65-68; asst prof, 68-73, assoc prof, 73-80, PROF CHEM, ST OLAF COL, 80- *Concurrent Pos:* Res assoc, State Univ Utrecht, 74-75; vis scholar, Univ Calif, San Diego, 82-83; res assoc, Univ Ore, 89-90. *Mem:* AAAS; Am Chem Soc; Am Soc Biochem & Molecular Biol. *Res:* Phospholipases; surface chemistry of lipids; biomembranes; lipid-protein interactions. *Mailing Add:* St Olaf Col Dept Chem 1520 St Olaf Ave Northfield MN 55057-1098

HENDRICKSON, JAMES BRIGGS, b Toledo, Ohio, Jan 3, 28; m 53; c 2. INTELLIGENT SYSTEMS, SOFTWARE SYSTEMS. *Educ:* Calif Inst Technol, BS, 50; Harvard Univ, MA, 51, PhD, 55. *Prof Exp:* Nat Res Coun res fel, Univ London, 54-55; NIH res fel, Harvard Univ, 55-57; asst prof chem, Univ Calif, Los Angeles, 57-62; assoc prof, 63-66, PROF CHEM, BRANDEIS UNIV, 66- *Concurrent Pos:* Consult, Sandoz Pharmaceut, Inc; Fulbright prof, Univ Cape Coast, Ghana, 74-76. *Mem:* Am Chem Soc; Royal Chem Soc; Ghana Sci Asn. *Res:* Computerized design of organic synthesis; design of new synthetic reactions; synthesis of natural products. *Mailing Add:* Dept Chem Brandeis Univ Waltham MA 02254-9110

HENDRICKSON, JOHN ALFRED, JR, b San Francisco, Calif, June 2, 41; m 68; c 1. BIOLOGICAL STATISTICS, STATISTICAL ECOLOGY. *Educ:* Univ San Francisco, BS, 63; Univ Kans, PhD(entom), 67. *Prof Exp:* Instr biol, Stanford Univ, 68-69; asst cur limnol, 69-77, assoc cur limnol & ecol, 77-82, RES ASSOC, ACAD NATURAL SCI, PHILADELPHIA, 83-, DIR COMP, 84- *Concurrent Pos:* Adj asst prof biol, Univ Pa, 70-76; adj lectr statist, Temple Univ, 83-85 & 88- *Mem:* Ecol Soc Am; Biomet Soc; Am Statist Asn; Soc Study Evolution; Soc Syst Zool; Sigma Xi. *Res:* Biological statistics, especially theoretical development and application of randomization tests for current questions in ecology/evolutionary biology. *Mailing Add:* Acad Natural Sci 19th St & The Parkway Philadelphia PA 19103-1156

HENDRICKSON, JOHN R(EESE), SR, chemical engineering, physical chemistry, for more information see previous edition

HENDRICKSON, JOHN ROBERT, b Erie, Pa, Feb 10, 44. SOLID STATE PHYSICS. *Educ:* Case Inst Technol, BS, 66; Brown Univ, PhD(physics), 73. *Prof Exp:* Res assoc, US Naval Lab, 72-74; asst prof physics, Univ Toledo, 74-79; asst prof, 79-81, ASSOC PROF ENG PHYSICS, UNIV TULSA, 81- *Concurrent Pos:* Physicist, US Naval Ordnance Lab, 66-67; prin investr, Dept Energy, Amoco, 80-82; consult, Sandia Lab, 81; founder & dir, Tulsa Univ Corrosion Fatigue Projs, 85- *Mem:* Am Phys Soc; Sigma Xi; Catgut Acoust Soc. *Res:* Corrosion fatigue studies of drill pipe; electromagnetic studies of black loaded elastomer; nucelar magnetic resonance; far infrared studies of oxide glasses and amorphous semiconductors. *Mailing Add:* Univ Tulsa 600 S College Tulsa OK 74104

HENDRICKSON, JOHN ROSCOE, b Tipton, Iowa, Aug 26, 21; m 46; c 4. VERTEBRATE ZOOLOGY, ECOLOGY. *Educ:* Univ Ariz, BSc, 44; Univ Calif, Berkeley, MA, 49, PhD(zool), 51. *Prof Exp:* Lectr zool, Univ Malaya, Singapore, 51-59; prof & head dept, Univ Malaya, Kuala Lumpur, 59-63; vchancellor, Inst Student Interchange, East-West Ctr, Univ Hawaii, 63-67; dir oceanic inst, Oceanic Found, Waimanalo, Hawaii, 67-69; prof biol, Univ Ariz, 69-76, prof ecol & evol biol, 76-; RETIRED. *Mem:* AAAS; Am Inst Biol Sci; Am Soc Ichthyol & Herpet; Zool Soc London; World Maricult Soc. *Res:* General tropical ecology; sea turtles; biological oceanography and ecology of Gulf of California; aquaculture, conservation and other activities applying basic biological knowledge to human impacts on ecosystems. *Mailing Add:* PO Box 3375 Cottonwood AZ 86326

HENDRICKSON, LESTER ELLSWORTH, b Republic, Mich, Nov 23, 41; m 67; c 2. METALLURGY, PHYSICS. *Educ:* Mich Technol Univ, BS, 63, MS, 64; Univ Ill, PhD(metall), 69. *Prof Exp:* Res asst metall, Mich Technol Univ, 63-64 & Univ Ill, 64-68; asst prof 68-73, fac res grant, 69-71, ASSOC PROF ENG SCI, ARIZ STATE UNIV, 73- *Concurrent Pos:* Consult to legal profession; fac res grant, Ariz State Univ, 69-71. *Mem:* Am Soc Metals; Metall Soc. *Res:* Impurity; lattice defect interactions in high purity dilute alloys of aluminum; mechanical properties of semiconducting solids; failure analysis. *Mailing Add:* Dept Chem & Mat Engr Tempe AZ 85287

HENDRICKSON, RICHARD A(LLAN), b Omaha, Nebr, Feb 1, 33; m 56; c 2. NUCLEAR ENGINEERING. *Educ:* Iowa State Univ, BS, 55, MS, 62, PhD(nuclear eng), 66. *Prof Exp:* Grad asst nuclear eng, Ames Lab, Atomic Energy Comn, 60-62, jr engr, 62-64, from instr to assoc prof, 64-85, PROF NUCLEAR ENG & MGR, REACTOR OPERS & MAINTENANCE, IOWA STATE UNIV, 85- *Honors & Awards:* Sigma Xi. *Mem:* Am Nuclear Soc; Am Soc Eng Educ. *Res:* Nuclear reactor engineering; reactor kinetics and noise analysis; radiation detection and measurement. *Mailing Add:* Dept Mech Eng Iowa State Univ Ames IA 50011

HENDRICKSON, ROBERT MARK, JR, b Corona, Calif, Jan 19, 38. ENTOMOLOGY. *Educ:* Calif State Polytech Univ, Kellogg-Voorhis Campus, BS, 65; Univ Calif, Riverside, PhD(entom), 74. *Prof Exp:* ENTOMOLOGIST FORAGE CROPS, SCI & EDUC ADMIN-AGR RES, USDA, 74- *Concurrent Pos:* Adj asst prof, Univ Del, 75- *Mem:* Entom Soc Am; Int Orgn Biol Control. *Res:* Biological control of agromyzids on alfalfa. *Mailing Add:* PO Box 1035 Livingston MT 59047

HENDRICKSON, THOMAS JAMES, b Detroit, Mich, Mar 26, 26; m 61. PHYSICS. *Educ:* Univ Mich, BS, 47, MS, 49; Iowa State Univ, PhD(physics), 56. *Prof Exp:* Instr physics, Mich Col Mining & Technol, 48-50; from instr to asst prof, Tufts Univ, 56-60; asst prof, 60-63, assoc prof, 63-80, PROF PHYSICS, GETTYSBURG COL, 80- *Mem:* Sigma Xi. *Res:* Solid state physics; ferromagnetism; antiferromagnetism; order-disorder transformations in alloys; optic properties of solids. *Mailing Add:* 65 Confederate Dr Gettysburg PA 17325

HENDRICKSON, TOM A(LLEN), b Los Angeles, Calif, Dec 25, 35; m 81; c 3. NUCLEAR ENGINEERING. *Educ:* Harvard Univ, AB, 57; Georgetown Univ, MS, 62. *Prof Exp:* Nuclear engr, Nuclear Power Div, US Navy Bur Ships, 57-64, chief submarine fluid systs br, Naval Ship Systs Command, 64-72; from chief nuclear engr to dep dir eng, 72-78, asst to pres, 78-80, vpres, Burns & Roe, Inc, 80-85; pres & chief exec officer, Proto-Technol Corp, A Kollmorgen Co, 85-87; pres, Magnetic Bearings Inc, 87-90; PRIN DEP ASST SECY NUCLEAR ENERGY, US DEPT ENERGY, 90- *Concurrent Pos:* Dir, Gen Physics Corp, 77-85, Separative Work Unit Corp (SWUCO), 85-88. *Mem:* Am Nuclear Soc; Am Soc Mech Engrs; Am Phys Soc. *Mailing Add:* 526 Hillcrest Rd Ridgewood NJ 07450-1525

HENDRICKSON, WALDEMAR FORRSEL, b Tobias, Nebr, June 7, 34; m 58, 84; c 2. HIGH LEVEL WASTE MANAGEMENT. *Educ:* Univ Idaho, BS, 56, MS, 58; Wash State Univ, MS, 64, PhD(eng sci), 71. *Prof Exp:* Chem engr, Hanford Atomic Prod Opers, Gen Elec Co, 57-58; asst nuclear engr, Wash State Univ, 60-71; scientist, Northwest Orgn Cols & Univs, 71; res physicist, US Navy Ord Lab, 72-74; CHEM ENGR, AEC-ERDA-DEPT ENERGY, 74- *Concurrent Pos:* Asst, Nat Res Coun, 71-74. *Mem:* Am Nuclear Soc; Health Physics Soc. *Res:* Electrochemical kinetics (oxidation of chloride ion) at electrode (carbon) with radiochemical absorbtion techniques; automated analysis of trace elements in sea water by neutron activation analysis using a large Californium source; nuclear fuel cycle (mixed oxide fuel fabrication and reprocessing). *Mailing Add:* Dept Energy Box 550 Richland WA 99352

HENDRICKSON, WAYNE ARTHUR, b Spring Valley, Wis, Apr 25, 41; m 69; c 2. MOLECULAR BIOPHYSICS, STRUCTURAL BIOLOGY. *Educ:* Univ Wis-River Falls, BA, 63; Johns Hopkins Univ, PhD(biophys), 68. *Prof Exp:* NIH traineee, Johns Hopkins Univ, 68-69; Nat Res Coun res assoc, 69-71, res biophysicist, US Naval Res Lab, 71-84; PROF BIOCHEM & MOLECULAR BIOPHYS, COLUMBIA UNIV, 84-; INVESTR, HOWARD HUGHES MED INST, 86- *Concurrent Pos:* Mem, NSF Molecular Biol Adv Panel, 80-83 & Review Panel, Stanford Synchrotron Radiation Lab, 80-89; Biophys Chem Study Sect, NIH, 86-89; assoc ed, J Molecular Biol, 87-; consult, Molecular Structure Corp, 85-, sci adv bd, Progenics Pharmaceut, 87-, sr adv, Molecular Biophys Tech, Inc, 83-87; proposal rev panel, Cornell High Energy Synchrotron Source, 87-; biomed adv comt, Pittsburgh Supercomputing Ctr, 87-; proposal eval bd, Advan Photon Source, 89-; sci policy bd, Stanford Synchrotron Radiation Lab, 91-

Honors & Awards: Navy Meritorious Civilian Serv Award, 78; Arthur S Flemming Award, 79; A L Patterson Award, Am Crystallog Asn, 81; Fritz Lipmann Award, Am Soc Biochem & Molecular Biol, 91. *Mem:* Fel AAAS; Biophys Soc; Am Crystallog Asn; Am Soc Biochem & Molecular Biol. *Res:* Structure and function of biological macromolecules; diffraction methods, synchrotron radiation. *Mailing Add:* Dept Biochem & Molecular Biophys Columbia Univ 630 W 168th St New York NY 10032

HENDRICKSON, WILLARD JAMES, b Battle Creek, Mich, Aug 23, 16; m 46; c 2. PSYCHIATRY. *Educ:* Calif Inst Technol, BS, 42; Univ Mich, MD, 45, MS, 53. *Prof Exp:* Intern, Univ Hosp, 45-46, resident psychiat, Neuropsychiat Inst, 48-50, from instr to assoc prof, Univ, 50-72, PROF PSYCHIAT, UNIV MICH, ANN ARBOR, 72- *Concurrent Pos:* Consult Vet Admin Hosp, 53- *Mem:* Am Psychiat Asn. *Res:* Clinical investigation of methods of psychiatric treatment of adolescent patients. *Mailing Add:* Dept Psychiat Univ Mich Ann Arbor MI 48109

HENDRICKSON, YNGVE GUST, b Virginia, Minn, Sept 3, 27; m 64; c 2. ORGANIC CHEMISTRY. *Educ:* Mass Inst Technol, SB, 51; Univ Ill, PhD, 55. *Prof Exp:* Res chemist, Chevron Res Co, Chevron Corp, 55-64, sr res chemist, 64-69, sr res assoc, 69-86; RETIRED. *Mem:* Am Chem Soc; The Chem Soc. *Res:* Physical organic chemistry; study of mechanisms of organic reactions; composition of petroleum and its fractions; stability of distillate fuels; lubricating oil additives; lubricating oils and greases. *Mailing Add:* 8364 Kent Dr El Cerrito CA 94530

HENDRICKX, ANDREW GEORGE, b Butler, Minn, July 14, 33; m 57; c 5. EMBRYOLOGY. *Educ:* Concordia Col, BS, 59; Kans State Univ, MS, 61, PhD(zool), 63. *Prof Exp:* Asst prof zool, Southern Ill Univ, 63-64; embryologist, Southwest Found Res & Educ, Tex, 64-69; assoc res physiologist, Calif Primate Res Ctr, 69-73, prof reproduction, Sch Vet Med, 73-78, RES PHYSIOLOGIST & PROF HUMAN ANAT, SCH MED, UNIV CALIF, DAVIS, 73-, DIR, CALIF PRIMATE RES CTR, 87- *Mem:* Am Asn Anat; Soc Study Reprod; Teratol Soc; Am Soc Primatologists. *Res:* Reproduction, embryology and teratology of primates, with emphasis on domestic breeding, developmental staging and effects of environmental agents on prenatal and postnatal development. *Mailing Add:* Calif Primate Res Ctr Univ of Calif Davis CA 95616

HENDRIE, DAVID LOWERY, b St Louis, Mo, Sept 22, 32; m 56; c 2. PHYSICS. *Educ:* Univ Wash, PhD(physics), 64. *Prof Exp:* Physicist, Lawrence Berkeley Lab, 64-73 & 80-85, dir cyclotron, 73-78; prof, Univ Md, 78-80; WITH DIV NUCLEAR PHYSICS, US DEPT ENERGY, 85- *Concurrent Pos:* Guggenheim fel, Saclay Nuclear Res Ctr, France, 71-72. *Mem:* Fel Am Phys Soc. *Res:* Nuclear reactions and structure. *Mailing Add:* Div Nuclear Physics Off High Energy & Nuclear Physics er-23 GTN US Dept Energy Washington DC 20585

HENDRIE, JOSEPH MALLAM, b Janesville, Wis, Mar 18, 25; m 49; c 2. APPLIED PHYSICS, NUCLEAR ENGINEERING. *Educ:* Case Inst Technol, BS, 50; Columbia Univ, PhD(physics), 57. *Prof Exp:* Asst, Radiation Lab, Columbia Univ, 50-53, asst physics, 53-55; asst physicist reactor physics, Brookhaven Nat Lab, 55-57, assoc physicist, 57-60, proj chief engr high flux beam reactor, 58-65, actg head exp reactor physics div, 65-66, assoc head eng div, Dept Appl Sci, 66-70, proj mgr pulsed fast reactor proj, 67-70, sr physicist & head eng div, Dept Appl Sci, 71-72; dep dir licensing for tech rev, AEC, 72-74; chmn dept appl sci, Brookhaven Nat Lab, 75-77; chmn, US Nuclear Regulatory Comn, 77-81; SR SCIENTIST & CONSULT ENGR, BROOKHAVEN NAT LAB, 81- *Concurrent Pos:* Consult, Columbia Radiation Safety Comt, 64-72, US Atomic Energy Comn & Nuclear Regulatory Comn, 74-75, US Gen Acct Off, 75-77 , Elec Power Res Inst, 82 & various nuclear utilities, 81-; mem adv comt reactor safeguards, AEC, 66-72, chmn 70; US rep, Sr Adv Group Reactor Safety Standards, Int Atomic Energy Agency, 74-79; dir, Am Nuclear Soc, 76-77; mem, Appl Physics Div Vis Comt, Argonne Nat Lab, 76-77, Comt Mem, Nat Acad Eng, 80-83, Comt Int Coop Magnetic Fusion Nat Res Coun, 83-85, Adv Coun, Inst Nuclear Power Opers, 84-90, Adv Comt Enforcement Policy, US Nuclear Regulatory Comn, 84-85, Comt Res & Tech Planning, Am Soc Testing & Mat, 85-88, Reactor Safety & Anal Div Rev Comt, Argonne Nat Lab, 85-87; dir, Houston Indust Inc, 85-, Houston Lighting & Power Co, 85-, Syst Energy Resources Inc, 87-90 & Entergy Opers Inc, 90-; mem, Spec Comt Integrated Fast Reactor, Univ Chicago, 87- *Honors & Awards:* E O Lawrence Award, Atomic Energy Comn. *Mem:* Nat Acad Eng; fel Am Nuclear Soc (pres, 84-85); Nat Soc Prof Engrs; Am Concrete Inst; Inst Elec & Electronics Engrs; fel Am Soc Mech Engrs; Am Phys Soc. *Res:* Physics and engineering design of research reactors; safety of nuclear power reactors; structural analysis and design of reinforced and prestressed high-strength concrete; electrical power transmission. *Mailing Add:* Brookhaven Nat Lab Upton NY 11973

HENDRIKS, HERBERT EDWARD, b West Liberty, Iowa, Jan 23, 18; m 41; c 1. GEOLOGY. *Educ:* Cornell Col, BA, 40; Univ Iowa, MS, 42, PhD(geol), 49. *Prof Exp:* Field geologist, Lone Star Steel Co, Tex, 42-44; geologist, Sun Oil Co, Miss, 44-45; asst geol, Univ Iowa, 46-47; from instr to prof, 47-83, WILLIAM HARMON NORTON EMER PROF GEOL, CORNELL COL, 83- *Concurrent Pos:* Dir, Camp Norton Geol Field Camp, 50-52; field geologist, Ashland Oil & Refining Co, 52-56; consult, Shell Oil Co, 57-58. *Mem:* AAAS; Geol Soc Am; Paleont Soc; Nat Asn Geol Teachers; Sigma Xi. *Res:* Cambro-Ordovician stratigraphy; stratigraphy and structure of the Ozark Mountains and the Wind River Range; stratigraphy and structure of Big Snowy Mountains, Montana; geology of the Crooked Creek Area, Missouri. *Mailing Add:* 616 Eighth Ave N Mt Vernon IA 52314

HENDRIX, C(HARLES) E(DMUND), b Moberly, Mo, Mar 14, 24; m 46; c 3. SYSTEMS ENGINEERING. *Educ:* Boston Univ, AB, 48; Univ Calif, Los Angeles, MS, 55, PhD(eng), 64. *Prof Exp:* Technician, Physics Lab, Tenn Eastman Corp, 46-47; mech designer, Polaroid Corp, 47; jr engr, Lab for Electronics, Inc, 48-49 & Mo Res Lab, Inc, 49-51; electronic scientist, Electronic Systs Br, US Naval Ord Test Sta, Calif, 51-56, head photophys br, 56-57, staff consult, Instrument Develop Div, 59-62; biomed engr, Chem & Biol Div, Space-Gen Corp, 62-66, sr scientist, Ctr Res & Educ, 66-69; sr scientist, Telluron, 69-71; tech staff mgr, Hughes Aircraft Co, 71-88, LAB CHIEF SCIENTIST, 88- *Concurrent Pos:* Lectr, Univ Calif, 56. *Mem:* Sigma Xi. *Res:* Physical and biological instrumentation; simulation of neuron networks; analysis, simulation and optimization of complex physical systems; communication satellites design; millimeter wave propagation. *Mailing Add:* 833 Bienvenida Ave Pacific Palisades CA 90272

HENDRIX, DONALD LOUIS, b Snohomish, Wash, Aug 21, 42; m 65; c 3. PLANT PHYSIOLOGY. *Educ:* Cent Wash Univ, BA, 65; Wash State Univ, PhD(bot), 73. *Prof Exp:* Res assoc plant physiol, Purdue Univ, 74; asst prof biol, Univ Houston, 74-81; PLANT PHYSIOLOGIST, WESTERN COTTON RES LAB, AGR RES SERV, USDA, 81- *Concurrent Pos:* Adj assoc prof bot, Ariz State Univ, 81-; fel, Orgn Econ Coop & Develop, 87. *Mem:* Am Soc Plant Physiologists; Scand Soc Plant Physiologists; Sigma Xi; Crop Sci Soc Am. *Res:* Biophysics and biochemistry of metabolite transport; plant biochemistry; photosynthate translocation; natural product analysis. *Mailing Add:* Western Cotton Res Lab USDA Agr Res Serv 4135 E Broadway Rd Phoenix AZ 85040

HENDRIX, FLOYD FULLER, JR, b Columbia, NC, Apr 18, 33; m 55; c 3. PLANT PATHOLOGY, SOIL MICROBIOLOGY. *Educ:* NC State Univ, BS, 55, MS, 57; Univ Calif, Berkeley, PhD(plant path), 61. *Prof Exp:* Plant pathologist, Southeastern Forest Exp Sta, US Forest Serv, 61-64; asst plant pathologist, 65-68, assoc prof plant path, 68-73, PROF PLANT PATH, UNIV GA, 73- *Honors & Awards:* Campbell Award, Am Inst Biol Sci, 64; Cert Merit, US Forest Serv. *Mem:* Am Phytopath Soc; Mycol Soc Am. *Res:* Ecology of fungi in soil with particular emphasis on Fusarium Phytophthora and Pythium species, fomes annosus, taxonomy of the genus Pythium; diseases of apples and peaches. *Mailing Add:* Dept Plant Path & Genetics Univ Ga Athens GA 30602

HENDRIX, JAMES EASTON, b Pensacola, Fla, Oct 31, 41; m 66; c 3. POLYMER CHEMISTRY, ORGANIC CHEMISTRY. *Educ:* Auburn Univ, BS, 66; Clemson Univ, MS, 68, PhD(chem), 70. *Prof Exp:* Res chemist, FMC Corp, 69-70; res fel, Nat Acad Sci-Nat Res Coun Southern Regional Labs, New Orleans, 70-72; res mgr, Milliken Res Corp, 72-75, dir develop, Milliken & Co, Inc, 75-78; dir res, 78-85, VPRES RES & DEVELOP, SPRINGS INDUST INC, 85- *Honors & Awards:* Super Serv Award Res, USDA, 72. *Mem:* Fiber Soc; Am Chem Soc; Am Asn Textile Chemists & Colorists. *Res:* Flame inhibition mechanisms; pyrolytic degradation of organic high polymers; high temperature reactions of organic phosphates; instrumental techniques for determining textile flammabilities; synthetic coatings for textile substrates; management. *Mailing Add:* Spring Industs Inc PO Box 70 Ft Mill SC 29715

HENDRIX, JAMES HARVEY, JR, b Newbern, Tenn, Jan 22, 20; c 3. PLASTIC SURGERY. *Educ:* Univ Tenn, Knoxville, BS, 42; Univ Tenn, Memphis, MD, 43; Am Bd Plastic Surg, dipl. *Prof Exp:* Intern, Methodist Hosp, Memphis, Tenn, 44; resident surgeon, Univ Tex Med Br, Galveston, 47-50, instr surg, 50, resident gen surg & plastic surg, Univ Tex Med Br, Galveston & Baptist Mem Hosp, Memphis, 51; asst, Dept Surg, Univ Tenn, Memphis, 51-52; chief plastic surg, Univ Miss, Jackson, 55-72; chief plastic surg, Univ Tenn, Memphis, 72-78, prof surg, 72-91; RETIRED. *Concurrent Pos:* Pvt pract, Miss & Tenn, 51-; mem, Adv Coun Plastic & Maxillofacial Surg, Am Col Surgeons, 73-77, chmn, 75; mem bd, Am Bd Plastic Surg; mem, Plastic Surg Res Coun; consult plastic surg, Miss Crippled Children's Serv, Tenn Crippled Children's Serv, St Jude Children's Res Hosp & Vet Admin Hosp. *Mem:* Am Soc Plastic & Reconstruct Surgeons (pres, 73-74); Am Asn Plastic Surgeons; fel Am Col Surgeons; Am Cleft Palate Asn; Am Trauma Soc. *Res:* Healing of tendons in hands of monkeys; treatment of fractures of mandibular condyles with growth studies using piglets. *Mailing Add:* Suite 525 910 Madison Ave Memphis TN 38103

HENDRIX, JAMES LAURIS, b Omaha, Nebr, July 19, 43; m 65. CHEMICAL ENGINEERING, BIOENGINEERING. *Educ:* Univ Nebr, BS, 66, MS, 68, PhD(chem eng), 69. *Prof Exp:* Asst prof, 69-77, assoc prof, 77-80, PROF CHEM ENG, UNIV NEV, RENO, 80- *Mem:* Am Soc Chem. *Res:* Biological fuel cells. *Mailing Add:* Dept Chem Eng Univ Nev Reno NV 89557

HENDRIX, JAMES WILLIAM, b Kinston, NC, Apr 18, 37; m 59; c 2. PLANT PATHOLOGY, MYCOLOGY. *Educ:* NC State Univ, BS, 59, PhD(plant path), 63; Univ Ark, MS, 60. *Prof Exp:* NSF fel plant path & biochem, Univ Nebr, 63-64; res plant pathologist, Crops Res Div, Agr Res Serv, USDA, 64-67; assoc prof, 67-73, PROF PLANT PATH, UNIV KY, 73- *Concurrent Pos:* Vis assoc & NIH spec fel biol, Calif Inst Technol & City of Hope Med Ctr, 68-69. *Mem:* Am Phytopath Soc; Mycol Soc Am. *Res:* Ecology of mycorrhizal fungi; ecology of soil-borne phytopathogenic fungi. *Mailing Add:* Dept Plant Path Univ Ky Lexington KY 40506

HENDRIX, JOHN EDWIN, b Van Nuys, Calif, Aug 30, 30; m 54; c 2. PLANT PHYSIOLOGY. *Educ:* Fresno State Col, BS, 56, AB, 60; Ohio State Univ, MS, 63, PhD(bot), 67. *Prof Exp:* Orchard foreman, Fresno State Col, 59-60; instr bot, Ohio State Univ, 65-67; from asst prof to assoc prof bot, 67-89, PROF PLANT PATH, COLO STATE UNIV, 89- *Mem:* AAAS; Am Soc Plant Physiol; Am Inst Biol Sci; Sigma Xi. *Res:* Phosphate uptake and transport in plants; stachyose synthesis and transport in plants; carbohydrate transport in plants; carbohydrate accumulation in developing grain; fructan metabolism. *Mailing Add:* Dept Biol & Plant Path Colo State Univ Ft Collins CO 80523

HENDRIX, JOHN WALTER, b Quincy, Ill, Nov 20, 15; m 45; c 2. PLANT PATHOLOGY. *Educ:* NC State Col, BS, 37; Yale Univ, MF, 40; Univ Minn, PhD(plant path), 48. *Prof Exp:* Asst plant pathologist, Agr Exp Sta, Univ Hawaii, 43-48, assoc plant pathologist, 48-52, head dept, 43-52; assoc prof & assoc plant pathologist, 52-58, actg head dept, 53-54 & 58-69, PROF PLANT

PATH & PLANT PATHOLOGIST, AGR EXP STA, WASH STATE UNIV, 58- *Mem:* Am Phytopath Soc; Am Soc Hort Sci. *Res:* Phytopathology; breeding and genetics of tomatoes; plant diseases caused by atmospheric pollutants; cereal rusts. *Mailing Add:* Dept Plant Path Wash State Univ Pullman WA 99164

HENDRIX, MARY J C, b La Jolla, Calif, Sept 19, 53. CELL BIOLOGY, CANCER BIOLOGY. *Educ:* Shepherd Col, BS, 74; George Washington Univ, PhD(anat), 77. *Prof Exp:* Teaching asst gross anat, Harvard Univ, 78-80; assoc prof histol, 80-85, ASSOC PROF ANAT, UNIV ARIZ, 86- *Concurrent Pos:* Comnr, Ariz Dis Control Res Comn, 86-; adj researcher path, Univ Calif, San Francisco, 81- *Mem:* Electron Micros Soc Am; AAAS; Am Soc Cell Biol; Soc Develop Biol; Am Asn Anatomists; Int Pigment Cell Soc. *Res:* The normal and abnormal interactions of cells with extracellular matrices; the remodeling of the extracellular matrices by cellular interactions during normal developmental events; the study of tumor cell degradation and modification of the extracellular matrices during transformation and metastasis. *Mailing Add:* Dept Anat Col Med Univ Ariz Tucson AZ 85724

HENDRIX, ROGER WALDEN, b San Francisco, Calif, July 7, 43; m 71. MOLECULAR BIOLOGY. *Educ:* Calif Inst Technol, BS, 65; Harvard Univ, PhD(biochem & molecular biol), 70. *Prof Exp:* Fel biochem, Stanford Univ, 71-73; asst prof microbiol, 73-78, assoc prof, 78-86, PROF, BIOL SCI, UNIV PITTSBURGH, 86- *Concurrent Pos:* Res Career Develop Award, NIH, 75. *Res:* Biochemistry and genetics of virus assembly, using bacteriophage lambda as a model system; role of protein processing, host factors, and catalytically acting proteins in assembly processes. *Mailing Add:* Dept Biol Sci Langley 340 Univ Pittsburgh 4200 5th Ave Pittsburgh PA 15260

HENDRIX, SHERMAN SAMUEL, b Bridgeport, Conn, June 1, 39; m 61; c 2. MARINE PARASITOLOGY, HELMINTHOLOGY. *Educ:* Gettysburg Col, BA, 61; Fla State Univ, MS, 64; Univ Md, PhD(parasitol), 72. *Prof Exp:* Instr, Gettysburg Col, 64-70, from asst prof to assoc prof, 70-90, chmn dept, 85-90, PROF BIOL, GETTYSBURG COL, 90- *Concurrent Pos:* Fel, InterAm Prog Trop Med, La State Univ, 73- *Mem:* Am Soc Parsitologists; Am Malacol Union; Wildlife Dis Asn; Am Fisheries Soc; Nat Sci Teachers Asn. *Res:* Systematics, zoogeography and ecology of aspidogastrid trematodes and monogenea of fishes and mollusks. *Mailing Add:* Dept Biol Gettysburg Col Gettysburg PA 17325-1486

HENDRIX, THOMAS EUGENE, b Lancaster, Pa, Feb 6, 33; m 55; c 3. GEOLOGY. *Educ:* Franklin & Marshall Col, BS, 55; Univ Wis, MS, 57, PhD(geol), 60. *Prof Exp:* From instr to assoc prof geol, Ind Univ, Bloomington, 59-78, assoc dir geol field sta, 62-68; assoc prof, 78-80, PROF GEOL, GRAND VALLEY STATE COLS, 80- *Concurrent Pos:* Ed, J Geol Educ, 69-71. *Mem:* Geol Soc Am; Nat Asn Geol Teachers (pres, 73-74). *Res:* Structural geology of Precambrian metamorphic tectonites; laramide structures of southwestern Montana; Precambrian tectonites. *Mailing Add:* Dept Geol Grand Valley State Univ One Campus Dr Allendale MI 49401

HENDRIX, THOMAS RUSSELL, b Ft Ancient, Ohio, Oct 17, 20. GASTROENTEROLOGY. *Educ:* Johns Hopkins Univ, MD, 51. *Prof Exp:* PROF MED & GASTROENTEROL, JOHNS HOPKINS UNIV, 68- *Mailing Add:* Dept Gastroenterol Johns Hopkins Univ 720 Rutland Ave Baltimore MD 21205

HENDRON, ALFRED J, JR, b Clifton, Ill, Oct 4, 37. GEOTECHNICAL & CIVIL ENGINEERING. *Educ:* Univ Ill, BS, 59, MS, 60, PhD(civil eng), 63. *Prof Exp:* From asst prof to assoc prof, 65-70, PROF CIVIL ENG, UNIV ILL, 70- *Mem:* Nat Acad Eng; Am Soc Civil Eng. *Mailing Add:* 2230 C Newmark Civil Eng Lab Univ Ill 205 N Mathews Urbana IL 61801

HENDRY, ANNE TERESA, b Hearst, Ont, Feb 29, 36; wid; c 1. MEDICAL MYCOLOGY. *Educ:* Univ Toronto, BSA, 59; Univ Guelph, MSc, 66; Univ Western Ont, PhD(microbiol), 71. *Prof Exp:* Lab scientist, Ont Ministry Health, 59-64; teacher biol, Halton County Sch Bd, 70-71; res assoc biochem, Med Fac, Univ Western Ont, 72-74; assoc prof microbiol, Med Sch, McMaster Univ, 75-; AT MICROBIOL DEPT, HAMILTON DIV, HAMILTON CIVIL HOSP. *Concurrent Pos:* Clin microbiologist, Hamilton Gen Hosp, 75- *Mem:* Int Soc Human & Animal Mycol; Am Soc Microbiol; Can Soc Microbiol; Can Asn Clin Microbiol Infectious Dis; Am Venereal Dis Asn. *Res:* Auxotyping and physiological studies of Neisseria gonorrhoeae; aromatic amino acid biosynthesis and regulation in N gonorrhoeae. *Mailing Add:* Microbiol Dept Hamilton Gen Div Hamilton Civic Hosp 237 Barton St E Hamilton ON L8L 2X2 Can

HENDRY, ARCHIBALD WAGSTAFF, b Scotland, Nov 18, 36; m 64; c 3. ELEMENTARY PARTICLE PHYSICS. *Educ:* Glasgow Univ, BSc, 58, PhD(physics), 62. *Prof Exp:* Res assoc theoret physics, Univ Calif, San Diego, 62-64; sr sci officer, Rutherford High Energy Lab & Oxford Univ, 64-67; res asst prof, Univ Ill, Urbana, 67-69; from asst prof to assoc prof, 69-77, assoc dean, 85-88, PROF THEORET PHYSICS, IND UNIV, 77- *Concurrent Pos:* Guest prof, Univ Heidelberg, 67. *Res:* Theoretical high energy physics. *Mailing Add:* Dept Physics Ind Univ Bloomington IN 47405

HENDRY, GEORGE ORR, b Oakland, Calif, June 23, 37. RADIO FREQUENCY SYSTEMS, MAGNET DESIGN. *Educ:* Univ Calif, Berkeley, BS, 60, MS, 63. *Prof Exp:* Proj engr, WM Brobeck & Assoc, 63-65; proj engr, the Cyclotron Corp, 65-69, vpres res & eng, 69-83; dir res & develop, CTI, 84-86; VPRES RES & DEVELOP, RDS, 86- *Res:* Development of positive and negative ion cyclotron particle accelerators and their associated targets and shielding. *Mailing Add:* 3104 Redwood Rd Napa CA 94558

HENDRY, HUGH EDWARD, b Edinburgh, Scotland, May 7, 44; m 67; c 3. GEOLOGY. *Educ:* Univ Edinburgh, BSc, 66, PhD(geol), 70. *Prof Exp:* Post-doctoral fel, McMaster Univ, 69-71; from asst prof to assoc prof, 71-82, PROF GEOL SCI, UNIV SASK, 82- *Mem:* Soc Econ Paleont & Mineral; Am Assoc Petrol Geol; Can Assoc Petrol Geol; Geol Assoc Can; Int Assoc Sedimentology. *Res:* Sedimentology of conglomerates; fluvial sedimentology. *Mailing Add:* Dept Geol Sci Univ Sask Saskatoon SK S7N 0W0 Can

HENDRY, RICHARD ALLAN, b Santa Barbara, Calif, Oct 13, 29; m 63; c 2. BIOCHEMISTRY, ORGANIC BIOCHEMISTRY. *Educ:* Univ Calif, Santa Barbara, BA, 51; Univ of the Pac, MA, 52; Baylor Univ, PhD(biochem), 56. *Prof Exp:* Res assoc chem, Univ Ill, 56-57; asst prof, Tex Tech Col, 57-59; assoc prof, 59-63, PROF CHEM, WESTMINSTER COL, PA, 63- *Concurrent Pos:* Vis prof, Mich State Univ, 68-69 & 90-91; res assoc chem, Univ Ill, 80-81. *Mem:* Am Chem Soc; Am Sci Affil; Sigma Xi. *Res:* Chromatographic methods of analysis, amino acid chemistry; chemistry of antibiotics; history of chemistry; chemical education; chemistry of glycosphingolipids. *Mailing Add:* Dept Chem Westminster Col New Wilmington PA 16172-0001

HENEGHAN, JAMES BEYER, b La Porte, Ind, Feb 5, 35; m 59; c 2. PHYSIOLOGY. *Educ:* Univ Notre Dame, BS, 57, PhD(physiol), 62. *Prof Exp:* From instr to assoc prof, 62-74, PROF PHYSIOL & SURG RES, LA STATE UNIV MED CTR, NEW ORLEANS, 74-, DIR GERMFREE LAB, 62- *Concurrent Pos:* Mem, NIH Proj Site Visit Comt, 67; chmn & ed, IV Int Symp Germfree Res, 72-73. *Mem:* AAAS; Am Physiol Soc; Am Soc Microbiol; Asn Gnotobiotics (vpres, 70-71, pres, 71-72); Soc Nuclear Med. *Res:* Gastrointestinal physiology and oncology; hemorrhagic shock; neonatal growth; application of gnotobiotic technology to hospital patient care; germfree animals. *Mailing Add:* Dept Surgery/Physiol LA State Univ Med Ctr 1542 Tulane Ave New Orleans LA 70112

HENEIN, NAEIM A, b Egypt; US citizen; m; c 1. MECHANICAL ENGINEERING. *Educ:* Cairo Univ, BSc, 49; Alexandria Univ, MSc, 52; Univ Mich, PhD(mech eng), 57. *Prof Exp:* From lectr to asst prof mech eng, Alexandria Univ, 57-65; vis assoc prof, Univ Mich, 65-70; assoc prof, 70-76, PROF MECH ENG, WAYNE STATE UNIV, 76-, DIR, CTR AUTOMOTIVE RES, 80- *Concurrent Pos:* Consult, Nat Res Coun, Nat Acad Sci, 74, Environ Protection Agency, Washington, DC, 83-87, US Tank Automotive Command, Warren, Mich, 79-88, Dept Transp, Transp Systs Ctr, Cambridge, Mass, 78- 80, Gen Motors, Ford Motor Co, 69. *Mem:* Fel Soc Automotive Engrs; Combustion Inst; Am Soc Mech Engrs; Sigma Xi. *Res:* Thermal sciences, combustion & friction particularly as applied to combustion engines. *Mailing Add:* Col Eng Wayne State Univ Detroit MI 48202

HENERY, JAMES DANIEL, b Forney, Tex, Sept 23, 40; m 71. INDUSTRIAL CHEMISTRY. *Educ:* Univ Tex, Austin, BS, 66, PhD(chem), 70. *Prof Exp:* Res chemist, Petro-Tex Chem Corp, 69-77, sr res chemist, 77-80; MEM STAFF, EL PASO PROD CO, 80- *Mem:* Am Chem Soc. *Res:* Synthesis of intermediates using inorganic and organometallic catalysts; preparation of petrochemical by gas phase catalysis. *Mailing Add:* Perry Equip Co PO Box 640 Mineral Wells TX 76067

HENERY-LOGAN, KENNETH ROBERT, b Montreal, Que, July 7, 21; m 61. ORGANIC CHEMISTRY. *Educ:* McGill Univ, BSc, 42, PhD(org chem), 46. *Prof Exp:* Res assoc, Mass Inst Technol, 51-60; from asst prof to assoc prof, 60-69, prof, 69-89, EMER PROF, CHEM, UNIV MD, COL PARK, 89- *Concurrent Pos:* Vis lectr, Boston Univ, 57-58; vis prof, Northeastern Univ, 80; fel org chem, Univ Chicago, 48-50. *Mem:* Am Chem Soc. *Res:* Synthesis and chemistry of compounds with potential anticancer activity. *Mailing Add:* Dept of Chem Univ of Md College Park MD 20742-2021

HENEY, LYSLE JOSEPH, JR, b Minneapolis, Minn, Oct 31, 23; m 46; c 2. CHEMICAL ENGINEERING, PHYSICAL CHEMISTRY. *Educ:* Univ Minn, Minneapolis, BChE, 45, PhD(chem eng, phys chem), 51. *Prof Exp:* Chem engr, Am Cyanamid Co, 51-85; RETIRED. *Mem:* Am Chem Soc; Am Inst Chem Engrs; Sigma Xi. *Res:* Measurement of surface area of solids; measurement of energy absorbed by solids during comminution; investigation of relationship between new surface and absorbed energy for comminution of solids; economic analysis and process designs of first-of-a-kind chemical plants, reactors, separations, recoveries and environmental controls. *Mailing Add:* 11 Mead Ave Middlesex NJ 08846

HENGEHOLD, ROBERT LEO, b Cincinnati, Ohio, June 18, 36. SOLID STATE & OPTICAL PHYSICS. *Educ:* Thomas More Col, AB, 56; Univ Cincinnati, MS, 60, PhD(physics), 65. *Prof Exp:* From instr to assoc prof physics, 61-73, PROF PHYSICS, USAF INST TECHNOL, 73- *Mem:* Am Phys Soc; Am Asn Physics Teachers; Am Soc Eng Educ; Sigma Xi. *Res:* Optical properties of solids; electron interaction with solids and surfaces; laser diagnostics. *Mailing Add:* Dept Eng Physics Afit Engr Air Force Inst Technol Wright-Patterson AFB OH 45433

HENIKA, RICHARD GRANT, b Wauwatosa, Wis, Sept 19, 21; m 41; c 4. FOOD CHEMISTRY, FOOD SCIENCE. *Educ:* Lawrence Col, BA, 43. *Prof Exp:* Chemist, Western Condensing Co, 43-44 & 46-50, proj leader res & develop, 50-53, foods sect head, 53-55, proj leader, Foremost Dairies, Inc, 55-64, group leader, 64-67, assoc dir res & develop, 67-74, MGR TECH SERV, TECH MGR DAIRY DIV & PATENT COORDR, FOREMOST FOODS CO, FOREMOST-MCKESSON, INC, 74- *Honors & Awards:* Co-recipient Indust Achievement Award, Inst Food Technol, 67. *Mem:* Inst Food Technol; Am Asn Cereal Chem; Am Chem Soc; Am Soc Bakery Eng; fel Am Inst Chem. *Res:* Development of new products and methods in bread baking, whipping agents and food proteins; research management. *Mailing Add:* 90 Stephanie Lane Alamo CA 94507-1994

HENINGER, GEORGE ROBERT, b Nov 15, 34; US citizen; m 57; c 4. PSYCHIATRY. *Educ:* Univ Utah, BS, 57, MD, 60; Am Bd Psychiat & Neurol, dipl. *Prof Exp:* Intern med, Boston City Hosp, 60-61; resident psychiat, Mass Ment Health Ctr, 61-64; clin assoc, NIMH, 64-65; prog specialist, Nat Clearinghouse Ment Health Info, 65-66; from asst prof to assoc prof, 66-76, PROF PSYCHIAT, SCH MED, YALE UNIV, 76- *Concurrent Pos:* Teaching fel psychiat, Harvard Univ, 61-64; consult, Conn Valley Hosp, 66-; prin investr, NIMH grants, 67, 70, 73 & 76. *Mem:* AAAS; Fel Am Psychiat Asn; Soc Biol Psychiat; Psychiat Res Soc; Am Psychopath Soc. *Res:* Clinical neurophysiologic correlates of psychiatric disorders; neuro pharmacology of antidepressant treatment. *Mailing Add:* Dept Psychiat 358 CMHC Yale Univ 34 Park St New Haven CT 06508

HENINGER, RICHARD WILFORD, b Raymond, Alta, Sept 28, 31; m 55; c 2. PHYSIOLOGY, BIOCHEMISTRY. *Educ:* Brigham Young Univ, BSc, 57; Okla State Univ, MSc, 59, PhD(physiol), 61. *Prof Exp:* Trainee endocrinol, Med Sch, Univ Wis-Madison, 61-62, asst prof physiol, 62-66; assoc prof zool, 66-71, PROF ZOOL, BRIGHAM YOUNG UNIV, 71- *Mem:* Endocrine Soc; Brit Soc Endocrinol; Am Physiol Soc. *Res:* Peripheral metabolism of thyroid hormones. *Mailing Add:* Dept Zool/Widb 575 Brigham Young Univ Provo UT 84601

HENINGER, RONALD LEE, b Waukegan, Ill, Dec 28, 44; m 68; c 2. FOREST MANAGEMENT. *Educ:* Mich Technol Univ, BSF, 68, MSF, 69; Mich State Univ, PhD(forest soils), 73. *Prof Exp:* Instr silvicult, Mich State Univ, 73-74; forester, Dept State Hwys & Transp, Mich, 74; scientist III, Weyerhaeuser Co, 74-76, scientist IV silvicult, 76-81, actg mgr forestry res field sta, 81-83, MGR, ORE FORESTRY RES FIELD STA, WEYERHAESER CO, 83- *Mem:* Soc Am Foresters; Soil Sci Soc Am; Sigma Xi; Am Soc Agron. *Res:* Development of silvicultural prescriptions to produce maximum fiber growth; studies in plantation establishment, spacing, weed control, stocking control, fertilization, forest soils, irrigation, commercial thinning and marking guides. *Mailing Add:* 572 N 71st St Springfield OR 97478

HENINS, IVARS, b Pampali, Latvia, June 2, 33; US citizen; m 60; c 3. PHYSICS. *Educ:* Friends Univ, BA, 55; Johns Hopkins Univ, PhD(physics), 61. *Prof Exp:* Instr physics, Johns Hopkins Univ, 61-62; staff mem, Los Alamos Sci Lab, Univ Calif, 62-90; CONSULT, 90- *Mem:* Am Phys Soc; Am Vacuum Soc. *Res:* Plasma physics; controlled thermonuclear research; high altitude plasma phenomena; optical instrumentation. *Mailing Add:* 121 Monte Rey Dr Los Alamos NM 87544

HENION, RICHARD S, b Brockport, NY, Dec 14, 39; m 59; c 2. ORGANIC CHEMISTRY. *Educ:* Alfred Univ, BA, 62; Syracuse Univ, PhD(org chem), 67. *Prof Exp:* Res assoc, Mass Inst Technol, 67-68; sr develop chemist, 68-80, TECH ASSOC, SYNTHETIC CHEM DIV, EASTMAN KODAK CO, 80- *Mem:* Am Chem Soc. *Res:* Organic syntheses; chemistry of heterocyclic compounds; reaction mechanisms; sulfur compounds. *Mailing Add:* 312 Gallup Rd Spencerport NY 14559

HENIS, JAY MYLS STUART, b New York, NY, July 9, 38; m 63; c 2. PHYSICAL CHEMISTRY. *Educ:* Alfred Univ, BA, 59; Syracuse Univ, PhD(phys chem), 64. *Prof Exp:* Res assoc molecular beams, Brown Univ, 64-66; res specialist, Cent Res Dept, 66-75, sr fel, 78, res dir, 84, SCI FEL, MONSANTO CO, ST LOUIS, 75- *Concurrent Pos:* Adj assoc prof chem, Wash Univ, St Louis, 73-84. *Honors & Awards:* Kirkpatric Award, 81; Soc Plastics Engr Award, 83. *Mem:* Am Chem Soc; Am Phys Soc. *Res:* Ion molecule reactions; mass and ion cyclotron resonance spectroscopy; molecular beams; discharges; reactions of exited intermediates; membrane separations; hollow fibers; diffusion; pollution controls; energy utilization; hydrogen production and recovery; controlled delivery of proteins and peptides; cell culture devices and systems; ion exchange and affinity separations. *Mailing Add:* Monsanto Co 800 N Lindbergh St Louis MO 63116

HENISCH, HEINZ KURT, b Neudek, Apr 21, 22; m 60. SOLID STATE PHYSICS, HISTORY OF PHOTOGRAPHY. *Educ:* Reading Univ, BSc, 42, PhD(physics), 49. *Hon Degrees:* DSc, Reading Univ, 78. *Prof Exp:* Jr sci officer physics, Royal Aircraft Estab, Eng, 42-46; lectr, Univ Reading, 48-63; prof appl physics, 63-70, assoc dir mat res lab, 69-75, PROF PHYSICS, PA STATE UNIV, 63-, PROF HIST OF PHOTOG, 73- *Concurrent Pos:* Vis scientist, Sylvania Elec Prod, Inc, NY, 55-56; consult, Mining & Chem Prod, UK, 57-63; ed, Int Ser of Monogr on Semiconductors, 57-68; consult, Eng Elec Valve Co, 58-63, Polaroid Corp, Mass, 63-67, Carborundum Corp, 67-75 & Energy Conversion Devices, 68-; ed-in-chief, Mat Res Bull, 85-; founder & ed, Hist of Photog, 87-90; guest cur exhibition, Photographic Experience, Palmer Mus Art, Pa State Univ, 88. *Mem:* Fel Am Phys Soc; fel Royal Photog Soc; fel Brit Inst Physics; fel Inst Arts & Humanistic Studies. *Res:* Semiconductors; phosphors; electroluminescence; contact and surface phenomena; crystal growth; amorphous materials; history of photography; computational physics; author, ten books. *Mailing Add:* 346 W Hillcrest Ave State College PA 16801

HENISZ, JERZY EMIL, b Warsaw, Poland, July 2, 37; m 62; c 1. PSYCHIATRY. *Educ:* Acad Med, Warsaw, MD, 61, DrMedSci, 65. *Prof Exp:* Resident psychiat, Acad Med, Warsaw, 62-67, asst, 65-69; res assoc, 70-71, from asst prof to assoc prof, 71-85, CLIN ASSOC PROF PSYCHIAT, YALE UNIV, 85- *Mem:* Polish Psychiat Asn; Am Psychiat Asn. *Res:* Social psychiatry; utilization and delivery of mental health services; evaluation of clinical services. *Mailing Add:* PO Box 1089 Sharon CT 06069

HENIZE, KARL GORDON, b Cincinnati, Ohio, Oct 17, 26; m 53; c 4. SPACE DEBRIS, PLANETARY NEBULAE. *Educ:* Univ Va, BA, 47, MA, 48; Univ Mich, PhD(astron), 54. *Prof Exp:* Observer astron, Lamont-Hussey Observ, Univ Mich, 48-51; Carnegie fel, Mt Wilson Observ, 54-56; sr astronr, Smithsonian Astrophys Observ, 56-59; actg dir, Dearborn Observ, 59-60, assoc prof astron, Northwestern Univ, 59-64, prof, 64-72; scientist & astronaut, 67-86, SR SCIENTIST, JOHNSON SPACE CTR, NASA, 86- *Concurrent Pos:* Mission specialist, Assess II Shuttle & Spacelab simulation, 77; mem NASA astron adv subcomt, Space Sci Steering Comt, 66-68; prin investr Skylab Exp S-019, 66-78, mem astronaut support crew, Apollo 15, 70-71, Skylab 2 3 & 4 missions, NASA, 72-74; adj prof astron, Univ Tex, Austin, 72-86; mission specialist, space shuttle flight 51-F (Spacelab 2), 85. *Honors & Awards:* Robert Gordon Mem Award, 68; NASA Medal for Except Sci Achievement, 74; NASA Spaceflight Medal, 85; Flight Achievement Award, Am Astronaut Soc, 85. *Mem:* Int Astron Union; Am Astron Soc; Royal Astron Soc; Astron Soc of Pac. *Res:* Emission-line stars; planetary nebulae; objective-prism spectroscopy; ultraviolet stellar spectroscopy; space debris; Magellanic Clouds. *Mailing Add:* Space Sci Br Code SN3 Johnson Space Ctr Houston TX 77058

HENKART, PIERRE, b New York, NY, Aug 28, 41. IMMUNOLOGY, BIOCHEMISTRY. *Educ:* Rensselaer Polytech Inst, BS, 63; Harvard Univ, PhD(biochem & molecular biol), 68. *Prof Exp:* NIH fel biol, Univ Calif, San Diego, 68-70, res biologist, 70-71; staff fel, 71-75, staff chemist, 75-80, SR INVESTR, NAT CANCER INST, 80- *Mem:* Am Asn Immunologists. *Res:* Mechanism of lymphocyte cytotoxicity. *Mailing Add:* Exp Immunol Br Nat Cancer Inst Bethesda MD 20892

HENKE, BURTON LEHMAN, b Ohio, Aug 27, 22; m 48; c 2. PHYSICS. *Educ:* Miami Univ, BA, 44; Calif Inst Technol, MS, 46, PhD(physics), 53. *Prof Exp:* From instr to prof physics, Pomona Col, 48-67; prof physics & astron, Univ Hawaii, 67-84; sr scientist, Lawrence Berkeley Lab, Univ Calif, Berkeley, 85-89, EMER SR STAFF SCIENTIST, 89- *Concurrent Pos:* Prin investr, Off Sci Res Grants, USAF, 54- & US Dept Energy grants, 76-; Guggenheim fel, 56; Dept Energy grant, 76-; consult X-Ray physics, gobt & indust. *Mem:* Fel Am Phys Soc; Am Asn Physics Teachers; Am Asn Univ Professors; Sigma Xi; Am Astron Soc. *Res:* Low energy x-ray and electron physics. *Mailing Add:* 1200 Mira Mar No 1324 Medford OR 97504

HENKE, MITCHELL C, b Wausau, WI, Sept 19, 49; m 71; c 2. HEAT TRANSFER-FOOD, AIR IMPINGEMENT. *Educ:* Valparaiso Univ, BSEE, 71; Purdue Univ, MSE, 78. *Prof Exp:* Elec engr, 71-80, supvr elec eng, 80-81, RES & DEVELOP MGR, LINCOLN FOOD SERV PROD, INC, 81- *Honors & Awards:* Doctorate Food Serv hon medallion, Nat Asn Food Equip Mfrs, 87. *Mem:* Soc Advan Food Serv Res (pres, 88-89); Inst Elec & Electronics Engrs; Inst Food Technologists; Nat Soc Prof Engrs; Int Microwave Power Inst. *Res:* Assess, develop and implement strategic, technology based new products; air impingement cooking technology to reduce cooking time and yet improving or maintaining cooking quality; heat transfer, air movement and control related to air impingement cooking; granted several patents; author of various publications. *Mailing Add:* 5208 Oak Chase Run Ft Wayne IN 46845

HENKE, RANDOLPH RAY, b Spokane, Wash, Mar 5, 48; m 66; c 3. BIOCHEMICAL GENETICS. *Educ:* Rochester Inst Technol, BS, 70; Miami Univ, PhD(bot), 74. *Prof Exp:* NDEA Title IV fel bot, Miami Univ, 71-74; asst prof, 75-78, assoc prof plant genetics, 78-81, PROF, DEPT BOT, US DEPT ENERGY, COMP ANIMAL RES LAB, UNIV TENN, 81- *Mem:* AAAS; Bot Soc Am; Am Soc Plant Physiologists; Crop Sci Soc Am; Tissue Cult Asn. *Res:* Plant biochemical genetics; methods of mutation induction and selection of specific biochemical mutations involving amino acid and protein metabolism in whole plants and cultured plant cells and tissues. *Mailing Add:* Box 117 Knoxville TN 37830

HENKE, RUSSELL W, b Milwaukee, Wis, Apr 28, 24; m 47; c 2. MECHANICAL ENGINEERING, FLUID POWER. *Educ:* Univ Wis, BS, 49, MS, 53, PME, 60. *Prof Exp:* Proj engr, Heil Co, Wis, 50-52; chief res & develop engr, Badger Meter Mfg Co, 52-55; dir eng, construct equip div, Am Marietta Corp, 55-57; consult engr, 57-60; dir fluid power res, Racine Hydraul & Mach, 60-62; prof mech eng & dir fluid power inst, Milwaukee Sch Eng, 62-68; exec vpres, Fluid Power Soc, 68-71; CONSULT, FLUID POWER CONSULTS INT & LAB FLUID POWER, 71- *Concurrent Pos:* Mem, US adv group to Int Standardization Orgn Tech Comt, 131; mem, Hydraul & Pneumatic Comt, Soc Automotive Engrs; mem, Fluid Power Standards Comt, Am Nat Standards Inst. *Honors & Awards:* Prod Eng-Master Designer Award, 67 & Pascal Medal, 70, Fluid Power Soc; Distinguished Achievement Award, Fluid Power Soc, 89. *Mem:* Nat Asn Corrosion Engrs; Am Soc Eng Educ; Am Soc Mech Engrs; Am Soc Testing & Mat; fel Fluid Power Soc (vpres, 65-66, pres, 66-67); Soc Automotive Engrs. *Res:* Fluid power devices and systems; related instrumentation. *Mailing Add:* PO Box 106 Elm Grove WI 53122-0106

HENKEL, ELMER THOMAS, b Toledo, Ohio, June 28, 36; m 62; c 3. PHYSICS, SCIENCE EDUCATION. *Educ:* Columbia Col, AB, 58; Univ Toledo, MEd, 63, MS & PhD(sci educ, physics), 65. *Prof Exp:* Asst prof physics, Southampton Col, Long Island Univ, 65-67; assoc prof, 67-77, chmn dept, 69-73, PROF PHYSICS, WAGNER COL, 78-, DIR, WAGNER SOLAR ENERGY DEMONSTRATION PROJ, 76- *Concurrent Pos:* Energy conserv consult, 87- *Mem:* Am Asn Physics Teachers; Int Solar Energy Soc. *Res:* Metallic friction, changes in the coefficient of friction due to variation in normal loads and relative velocity of surfaces; relationships between critical thinking abilities of students and instruction in physics; low energy nuclear physics; application of solar energy for the heating and cooling of commercial size buildings. *Mailing Add:* Dept Physics Wagner Col 631 Howard Ave Staten Island NY 10301

HENKEL, JAMES GREGORY, b Santa Monica, Calif, Nov 22, 45; m 70; c 2. MEDICINAL CHEMISTRY, ORGANIC CHEMISTRY. *Educ:* Univ Calif, Los Angeles, BS, 67; Brown Univ, PhD(chem), 73. *Prof Exp:* Res specialist, Univ Minn, 72-74, asst prof med chem, 74-77; asst prof, 77-82, ASSOC PROF MED CHEM, UNIV CONN, 82-, ASSOC DEAN, GRAD SCH, 86- *Concurrent Pos:* Secy, Sect Teachers Chem, Am Asn Col Pharm, 83-86, chair-elect, 87-88, chair, 88-89. *Mem:* Am Chem Soc; Am Asn Col Pharm; Asn Develop Comput-based Instruct. *Res:* Biological alkylations; synthesis of potential central nervous system active compounds; biological applications of spectroscopic methods; bridged polycyclic systems; electronic data processing research applications. *Mailing Add:* Dept Med Chem Univ Conn Sch Pharm Box U-92 Storrs CT 06268

HENKEL, JOHN HARMON, b Kentwood, La, Aug 14, 24; m 48; c 5. THEORETICAL SOLID STATE PHYSICS. *Educ:* Tulane Univ, BS, 47, MS, 48; Brown Univ. PhD(physics), 54. *Prof Exp:* Jr res technologist, Magnolia Petrol Co, 48-51, res res technologist, Field Res Labs, 54-55; asst, Brown Univ, 51-54; from asst prof to assoc prof, 55-64, actg head, Dept Physics & Astron, 75-77, prof physics, 64-90, EMER PROF PHYSICS, UNIV GA, 90- *Concurrent Pos:* NSF fel, 59-60; Nat Res Coun sr res assoc, Aerospace Res Labs, Wright Patterson AFB, Ohio, prog mgr, Directorate Physics, Air Force Off Sci Res, 77-78. *Mem:* Fel AAAS; Am Phys Soc; Sigma Xi. *Res:* Theoretical solid state physics; electrical geophysical exploration; induced polarization; lattice dynamics; ferroelectricity; band calculations. *Mailing Add:* Dept Physics & Astron Univ Ga Athens GA 30602

HENKEL, RICHARD LUTHER, b Toledo, Ohio, Mar 24, 21; m 47; c 3. SCIENCE EDUCATION, ELECTRICAL ENGINEERING. *Educ:* Univ Toledo, BE, 43; Univ Wis, PhD(physics), 50. *Prof Exp:* Teaching asst, Univ Toledo, 40-43; res assoc radar res, Harvard Univ, 43-46; res asst nuclear physics, Univ Wis, 46-50; nuclear physics staff, Los Alamos Nat Lab, 50-55, group leader, 55-76; RETIRED. *Mem:* Fel Am Phys Soc. *Res:* Experimental research in neutron physics, primarily fission measurements. *Mailing Add:* PO Box 128 Pitkin CO 81241

HENKELMAN, JAMES HENRY, mathematics, for more information see previous edition

HENKELS, WALTER HARVEY, b Philadelphia, Pa, Oct 10, 44; m 66; c 2. SOLID STATE PHYSICS. *Educ:* Lehigh Univ, BS, 66; Cornell Univ, PhD(appl physics), 74. *Prof Exp:* RES STAFF MEM, RES DIV, IBM CORP, 71- *Mem:* Am Phys Soc; Inst Elec & Electronics Engrs; AAAS. *Res:* Feasibility of and application of Josephson junctions as logic and memory devices for use in computer hardware. *Mailing Add:* Thomas J Watson Res Ctr IBM Corp PO Box 218 Yorktown Heights NY 10598

HENKENS, ROBERT WILLIAM, b Chicago, Ill. BIOPHYSICAL CHEMISTRY, ENZYME TECHNOLOGY. *Educ:* Univ Wash, BA, 58; Yale Univ, MS, 64, PhD(chem), 67. *Prof Exp:* Chemist, Gen Elec Co, 58-62; res fel, Harvard Univ, 68-69; asst prof, 69-73, ASSOC PROF CHEM, DUKE UNIV, 73- *Concurrent Pos:* Chmn, Enzyme Technol Res Group, Inc, 84-; mem, NC Biomolecular Eng & Mat Applns Ctr, 84- *Mem:* Am Chem Soc; Sigma Xi. *Res:* Spectroscopic, kinetic and thermodynamic studies of proteins, especially metalloenzymes and enzyme catalysis. *Mailing Add:* Dept Chem Duke Univ Durham NC 27706

HENKER, FRED OSWALD, III, b Little Rock, Ark, Sept 20, 22; m 45; c 2. PSYCHIATRY. *Educ:* Univ Ark, BS, 44, MD, 45. *Prof Exp:* Chief psychiat, Vet Admin Hosp, Jackson, Miss, 57-58; from instr to assoc prof, 58-74, PROF PSYCHIAT, MED CTR, UNIV ARK, LITTLE ROCK, 75-, CHIEF CONSULT SERV, 64- *Concurrent Pos:* Clin instr, Univ Miss, 57-58. *Mem:* AMA; Am Psychiat Asn; Acad Psychosom Med (pres, 83-84); Int Col Psychosomatic Med. *Res:* Psychosomatic medicine. *Mailing Add:* Nine Belmont Dr Little Rock AR 72204

HENKIN, HYMAN, b New York, NY, Feb 2, 15; m 40; c 3. PHYSICAL CHEMISTRY. *Educ:* City Col New York, BS, 36; NY Univ, PhD(phys chem), 40. *Prof Exp:* Assoc chemist, US Bur Mines, 44; res mgr & sect head, Colgate-Palmolive Co, 44-63; dir res, Helene Curtis Industs, Inc, 63-80, vpres res & develop, 65-80; RETIRED. *Mem:* Am Chem Soc; Soc Cosmetic Chem. *Res:* Administration; cosmetics and toiletries. *Mailing Add:* Exeter E 2073 Boca Raton FL 33434-3699

HENKIN, JACK, b Ger, Dec 23, 47; US citizen. ENZYMOLOGY, PROTEIN CHEMISTRY. *Educ:* City Col New York, BS, 69; Brandeis Univ, PhD(biochem), 75. *Prof Exp:* Res fel bio-org, dept chem, Harvard Univ, 74-77; asst prof biochem, dept biochem & molecular biol, Med Sch, Univ Tex, Houston, 77-84; SR BIOCHEMIST, PHYS BIOCHEM LAB, PHARMACEUT PROD DIV, ABBOTT LABS, 84- *Concurrent Pos:* Adj asst prof, Chicago Med Sch, 85- *Mem:* Am Chem Soc; Am Soc Biol Chemists. *Res:* Protein structure and function applied to enzymes, hormones and receptores; design of metabolic inhibitors and peptinomimetics as drugs; photochemistry and drug delivery. *Mailing Add:* 635 Gray Ave Highland Park IL 60035

HENKIN, LEON (ALBERT), b Brooklyn, NY, Apr 19, 21; m 50; c 2. MATHEMATICAL LOGIC. *Educ:* Columbia Univ, AB, 41; Princeton Univ, MA, 42, PhD(math), 47. *Prof Exp:* Mathematician, Kellex Corp, NY, 43-45 & Carbide & Carbon Chem Co, Tenn, 45-46; Fine Instr math, Princeton Univ, 47-48, Jewett fel, 48-49; from asst prof to assoc prof math, Univ Southern Calif, 49-53; from asst prof to assoc prof math, 53-58, from vchmn to chmn dept, 59-68, chmn, 83-85, PROF MATH, UNIV CALIF, BERKELEY, 58-, PROF EDUC, 83- *Concurrent Pos:* NSF grants, 52-; Fulbright res scholar, Neth, 54-55 & Israel, 79; vis prof, Dartmouth Col, 60-61; Guggenheim fel & mem, Inst Advan Study, 61-62; vis fel, All Souls Col, Oxford Univ, 68-69; vis scholar, Univ Colo, 75-76. Mem, US Nat Comt Hist & Philos Sci, 62-65, chmn comt logic & methodology sci, 62-64; mem coun, Conf Bd Math Sci, 63-64; prin investr, Community Teaching Fel Prog, Univ Calif, 69-75. *Honors & Awards:* Chauvenet Prize, Math Asn Am, 64. *Mem:* Fel AAAS; Am Math Soc; Math Asn Am; Asn Symbolic Logic (vpres, 53-55, pres, 62-64). *Res:* Mathematical logic; algebra; mathematics education. *Mailing Add:* Dept Math Univ Calif Berkeley CA 94720

HENKIN, ROBERT I, b Los Angeles, Calif, Oct 5, 30; c 6. TASTE & SMELL DISORDERS, NEUROBIOLOGY. *Educ:* Univ Southern Calif, AB, 51; Univ Calif, Los Angeles, MA, 53, PhD(music, psychol), 56, MD, 59. *Prof Exp:* Asst music, Univ Calif, Los Angeles, 52-54, intern med, Univ Hosp, 59-60, resident, Jackson Mem Hosp, Miami, Fla, 60-61; res assoc, NIMH, 61-63; sr investr, Nat Heart & Lung Inst, 63-69, chief sect neuroendocrinol, 69-75; DIR, CTR MOLECULAR NUTRIT & SENSORY DIS, TASTE & SMELL CLIN, WASHINGTON, DC, 75-; PROF NEUROL, MED CTR, GEORGETOWN UNIV, WASHINGTON, DC, 75- *Concurrent Pos:* Nat Kidney Dis Found traveling fel, 63; clin instr med, Hosp, Georgetown Univ, 64-67, asst clin prof, Sch Med, 67-75; consult, Gynec Endocrinol Clin, Lenox Hill Hosp, New York, 67-70, US Postal Serv, 74-, NIH, 75-, Hooker Chem Co, 76-77, ITT-Continental Baking Co, 76-80 & USDA, 78-, Lewis Howe Co, 79, Nat Geog Soc, 81-82, Bristol Myers Squibb, 85-86, Flornsynth, 86-, Wash Conf Zinc, 87- & Westport Pharmacol, 87-; mem sect renal dis, Coun Circulation, Am Heart Asn; chmn panel zinc, BEEP Comt, Nat Res Coun-Nat Acad Sci, 72-78; dir, Taste & Smell Clin, 75-; Sigma Xi nat lectr, 84-86; pres & chief exec officer, Sialon Corp, 87- *Honors & Awards:* Gill Mem lectr, 81. *Mem:* Biophys Soc; Am Fedn Clin Res; Endocrine Soc; Am Soc Clin Invest; Am Inst Nutrit; Sigma Xi. *Res:* Taste and olfaction, basic science and clinical disorders; nutrition; trace metal metabolism; steroid and sensory physiology; neuroendocrinology; psychology of music; United States and foreign patents. *Mailing Add:* 6601 Broxburn Dr Bethesda MD 20817

HENKIND, PAUL, b New York, NY, Dec 12, 32; m 56, 77; c 4. OPHTHALMOLOGY. *Educ:* Columbia Univ, AB, 55; NY Univ, MD, 59, MSc, 64; Univ London, PhD(path), 65. *Prof Exp:* Intern med, Henry Ford Hosp, Detroit, 59-60; resident ophthal, NY Univ-Bellevue Med Ctr, 60-63; from asst prof to assoc prof, Sch Med, NY Univ, 65-70; PROF OPHTHAL & CHMN DEPT, ALBERT EINSTEIN COL MED & DIR OPHTHAL SERV, MONTEFIORE HOSP MED CTR, 70- *Concurrent Pos:* Consult, Nat Eye Inst, 71-; mem vision & res prog comt, NIH; vis prof, Royal Soc Med, London, 81. *Mem:* Asn Res Vision & Ophthal (secy-treas, 77-81); Ophthal Soc UK; fel NY Acad Sci; Am Acad Ophthal. *Res:* Retinal vascular disease; ocular pathology; systemic ophthalmology; ocular tumors. *Mailing Add:* 111 E 210th St Bronx NY 10467

HENLE, GERTRUDE, virology, for more information see previous edition

HENLE, JAMES MARSTON, Washington, DC, Nov 13, 46; m 69; c 1. COMBINATORIAL SET THEORY. *Educ:* Dartmouth Col, AB, 68; Mass Inst Technol, PhD(math), 76. *Prof Exp:* Vis instr math, Peace Corps, Col Baguio, Univ Philippines, 68-70; teacher, Burgundy Farm Country Day Sch, 71-73; from asst prof to assoc prof, 76-88, PROF MATH, SMITH COL, 88- *Concurrent Pos:* Fulbright lectr, Univ Philippines, Quezon City, 80; co-prin investr, NSF grant, 84- *Mem:* Math Asn Am; Asn Symbolic Logic. *Res:* Models of set theory without the Axiom of Choice; models of the Axiom of Determinateness and infinite-exponent partition relations. *Mailing Add:* Dept Math Smith Col Northampton MA 01063

HENLE, ROBERT A, solid-state computer circuits; deceased, see previous edition for last biography

HENLEY, ERNEST J(USTUS), b Ger, Sept 30, 26; nat US; m 54; c 2. CHEMICAL ENGINEERING. *Educ:* Univ Del, BS, 50; Columbia Univ, MS, 51, EngScD, 53. *Prof Exp:* Asst prof chem eng, Columbia Univ, 54-58; from assoc prof to prof, Stevens Inst Technol, 58-67; PROF CHEM ENG, COL ENG, UNIV HOUSTON, 67- *Concurrent Pos:* Consult, Knolls Atomic Power Labs, 52-54, Am Cyanamid Co, 54-58 & E I du Pont de Nemours & Co, 78; dir, RAI res, 54-88, Procedyne Corp, 61-, Houston Glass Fabrication Corp, 71-81, Cache Corp, 75-86, Henley Int Inc, 76- & Continuous Learning Corp, 80-86; US deleg, Warsaw Conf, Int Atomic Energy Agency, 59; dir, US-Japanese Coop Sci Sem, Kyoto, 74; dir, NATO Advan Study Inst, Liverpool, 73, Sogesta, Italy, 78. *Mem:* Am Chem Soc; Am Nuclear Soc; Radiation Res Soc; Am Inst Chem Engrs. *Res:* Nuclear engineering; safety and reliability. *Mailing Add:* Dept Chem Eng Univ Houston Univ Park 4800 Calhoun Rd Houston TX 77004

HENLEY, ERNEST MARK, b Ger, June 10, 24; nat US; m 48; c 2. THEORETICAL NUCLEAR PHYSICS. *Educ:* City Col New York, BEE, 44; Univ Calif, PHD(physics), 52. *Prof Exp:* Elec engr, Airborne Instruments Lab, 46-48; elec engr, Microwave Lab, Univ Calif, 48-50, physicist, Radiation Lab, 50-51; assoc physics, Stanford Univ, 51-52; lectr, Columbia Univ, 52-54; from asst prof to assoc prof, 54-61, chmn, Univ Senate, 71-72, chmn dept, 74-77, dean, Col Arts Sci, 79-87, PROF PHYSICS, UNIV WASH, 61- *Concurrent Pos:* Jewett fel, Columbia Univ, 52-53; NSF sr fel, 58-59; consult, Los Alamos Sci Lab, 62-; Guggenheim Fel, 67-68; NATO sr fel, 76-77; exchange scientist to USSR, Nat Acad Sci, 77; chmn, Div Nuclear Physics, Am Phys Soc, 78-79, counr, 82-86; distinguished scholar to People's Rep China, 83; bd dir, Wash Technol Ctr, 83-87, & Pac Sci Ctr, 84-87; chair, Nuclear Sci Adv Comn, 86-89; bd trustees, Assoc Univs, 89- *Honors & Awards:* Sr Alexander Von Humboldt Award, 84; T W Bonner Award, Am Phys Soc, 89. *Mem:* Nat Acad Sci; fel Am Phys Soc (pres-elect, 91); fel AAAS. *Res:* Theoretical nuclear and particle physics; symmetries. *Mailing Add:* Dept Physics FM-15 Univ Wash Seattle WA 98195

HENLEY, KEITH STUART, b Hamburg, Ger, Feb 18, 24; nat US; m 59; c 1. GASTROENTEROLOGY. *Educ:* Univ Durham, MB & BS, 48, MD, 57. *Prof Exp:* Physician, Newcastle Gen Hosp, 48; surgeon, Royal Victoria Infirmary, 49; physician, Postgrad Med Sch, Univ London, 50, registr internal med, 52-54; from instr to assoc prof, Med Sch, 54-68, physician-in-chg, sect gastroenterol, 73-81, PROF INTERNAL MED, MED CTR, UNIV MICH, ANN ARBOR, 68- *Concurrent Pos:* Res career develop award, USPHS, 61-66. *Mem:* Am Gastroenterol Asn; Am Physiol Soc; fel Am Col Physicians; Am Fedn Clin Res; Brit Med Asn. *Res:* Biochemical properties of liver gastroenterology. *Mailing Add:* Dept Internal Med Univ Mich Ann Arbor MI 48109

HENLEY, MELVIN BRENT, JR, b Hickory Valley, Tenn, Aug 25, 35; m 54; c 5. PHYSICAL CHEMISTRY. *Educ:* Murray State Col, BS(physics, math, chem), 61; Univ Miss, Phd(chem), 64; Murray State Univ, MBA, 90. *Prof Exp:* Asst prof, 67-70, ASSOC PROF CHEM, MURRAY STATE UNIV, 70- *Concurrent Pos:* vchmn, Bd of Regents, Murray State Univ, 83-86. *Mem:* Am Chem Soc. *Res:* Catalytic effects on gas-phase dehydrohalogenations. *Mailing Add:* Dept Chem Murray State Univ Murray KY 42071

HENLEY, WALTER L, US citizen. PEDIATRICS, IMMUNOLOGY. *Educ:* Univ Calif, Los Angeles, BA, 47; Univ Pa, MD, 51. *Prof Exp:* Rotating intern, Univ Pa Hosp, 51-52; resident pediat, Bronx Hosp, NY, 52-53; res fel pediat & virol, Mt Sinai Hosp, 53-56; chief resident, 56-57, from asst attend physician to assoc attend physician, 57-74, assoc prof, Sch Med, 66-74, ATTEND PHYSICIAN PEDIAT, MT SINAI HOSP, 74-, PROF, MT SINAI SCH MED, 74- *Concurrent Pos:* Instr, Col Physicians & Surgeons, Columbia Univ, 56-65; Am Cancer Soc fel immunochem, Rockefeller Univ, 66-68; Nat Soc Prev Blindness grants & Fight for Sight Inc grants, 69 & 77; USPHS grant, Mt Sinai Sch Med, 70-79, chief pediat outpatient dept, 68-73, res assoc prof ophthal, 77-; attend pediat, Beth Israel Med Ctr, 76- *Mem:* Am Acad Pediat; Am Soc Pediat; Harvey Soc; Am Asn Immunologists; Asn Res Vision & Ophthal. *Res:* Humoral and cellular immunology in pediatrics and ophthalmology; infectious diseases. *Mailing Add:* Dept Pediat Beth Israel Med Ctr First Ave & 16th St New York NY 10003

HENLY, ROBERT STUART, b Chicago, Ill, Apr 3, 31; m 55. CHEMICAL ENGINEERING. *Educ:* Univ Ill, BS, 54; Pa State Univ, MS, 55, PhD(chem eng), 61. *Prof Exp:* Chem engr, 57-63, prod supvr gas chromatography supplies, 63-64, sr res engr, 64-66, mgr gas chromatography dept, 66-70, MGR CHROMATOGRAPHY RES & DEVELOP DEPT, APPL SCI LABS, INC, 70-, 2ND VPRES, 66-, VPRES PROM & PLANNING, 76- *Mem:* Am Chem Soc; Am Inst Chem Engrs; Sigma Xi. *Res:* Gas, preparative gas and liquid chromatography. *Mailing Add:* 111 Harvard Rd No 1 Port Matilda PA 16870-9430

HENN, FRITZ ALBERT, b Aldan, Pa, Mar 26, 41; m 64; c 2. NEUROPSYCHIATRY, NEUROCHEMISTRY. *Educ:* Wesleyan Univ, Conn, BA, 63; Johns Hopkins Univ, Baltimore, PhD(biochem), 67; Univ Va, MD, 71. *Prof Exp:* Res scientist neurochem, Inst Neurobiol, Univ Goteborg, Sweden, 70-71; res psychiatrist, Wash Univ, St Louis, Mo, 71-74; from asst prof to prof psychiat, Sch Med, Univ Iowa, Iowa City, 74-82; PROF PSYCHIAT & CHMN DEPT, STATE UNIV NY, STONY BROOK, 82- *Concurrent Pos:* Falk fel, Am Psychiat Asn, 73; consult, Drug Enforcement Agency, 73-74; mem coun res & develop, Am Psychiat Asn, 73-74; dir, Inst Ment Health Res, State Univ NY, Stony Brook, 83- *Mem:* Int Soc Neurochem; Am Soc Neurochem; Am Psychiat Asn; Psychiat Res Soc; Am Psychopath Soc. *Res:* Study of neuronal modulation by glial cells through removing and metabolizing neurotransmitters and controlling ion movements; forensic psychiatry study of relationship between psychiatric illness and criminal activity; biology markers of schizophrenia. *Mailing Add:* Dept Psychiat Health Sci Ctr State Univ NY Stony Brook NY 11794-8101

HENNEBERGER, WALTER CARL, b Bradley, Ill, Jan 17, 30; m 67; c 2. ELECTRODYNAMICS. *Educ:* Purdue Univ, BS, 52, MS, 56; Univ Gottingen, Dr nat sci(physics), 59. *Prof Exp:* Res assoc physics, Univ Notre Dame, 59-60; asst prof, Fordham Univ, 60-62; scholar, Dublin Inst Advan Studies, 62-63; from asst prof to assoc prof, 63-72, chmn dept, 74-75, PROF PHYSICS, SOUTHERN ILL UNIV, CARBONDALE, 72- *Concurrent Pos:* Consult, US Army Missile Command, Redstone Arsenal, Ala, 75-76. *Mem:* Am Phys Soc; Am Asn Physics Teachers. *Res:* Quantum electrodynamics; interaction of laser beams with matter, Aharonov-Bohm effect. *Mailing Add:* Dept Physics/Neckers C 0477 Southern Ill Univ Carbondale IL 62901

HENNEBERRY, RICHARD CHRISTOPHER, b Yonkers, NY, Nov 15, 37; m 59; c 4. CELL PHYSIOLOGY IN NERVOUS SYSTEM. *Educ:* St Michael's Col, AB, 59; Univ Mass, Amherst, MS, 66, PhD(microbiol), 69. *Prof Exp:* Fel molecular biol, Yale Univ, 69-71; chief, Molecular Neurobiol Sect, Lab Molecular Biol, Nat Inst Neurol Dis & Stroke, 71-90; VPRES & DIR NEUROSCI PROG, BERKSHIRE MED CTR, CONTE INST ENVIRON HEALTH, PITTSFIELD, MASS, 90- *Concurrent Pos:* Res consult, Dept Biol, Georgetown Univ, 78-80, Dept Physiol, George Washington Univ Sch Med, 79-90. *Honors & Awards:* Outstanding Serv Medal, USPHS, 89. *Mem:* AAAS; Sigma Xi; NY Acad Sci; Soc Neurosci. *Res:* Mechanisms of action of neurotransmitters in mammalian brain, with special emphasis on receptor-mediated neurotoxicity and role of perturbations of energy metabolism in excitatory amino acid neurotoxicity in the etiology of the neurodegenerative disorders. *Mailing Add:* Conte Inst Environ Health 742 North St Pittsfield MA 01201

HENNEBERRY, THOMAS JAMES, b Milford, Mass, Feb 26, 29; m 51; c 4. ENTOMOLOGY. *Educ:* Univ Mass, BS, 51; Univ Md, MS, 56, PhD, 60. *Prof Exp:* Res entomologist & res leader, Western Region, 51-52 & 55-74, LAB DIR, WESTERN COTTON RES LAB, SCI & EDUC ADMIN-AGR RES, USDA, 74- *Mem:* Entom Soc Am; Sigma Xi. *Res:* Insect and mite biology; radiation and chemosterilant research; cotton insect research. *Mailing Add:* 1409 E Northshore Dr Tempe AZ 85281

HENNEIKE, HENRY FRED, b Kankakee, Ill, Jan 27, 39; m 58; c 3. INORGANIC CHEMISTRY. *Educ:* Univ Ill, BS, 63, PhD(chem), 67. *Prof Exp:* Asst prof inorg chem, Univ Minn, Minneapolis, 66-70; ASSOC PROF INORG CHEM, GA STATE UNIV, 70- *Mem:* Am Chem Soc. *Res:* Lewis acid-base interactions in solution, hydrogen bonding and semi-empirical molecular orbital methods applied to inorganic systems. *Mailing Add:* 5815 Silver Ridge Dr Stone Mountain GA 30087-5799

HENNEKE, EDMUND GEORGE, II, b Baltimore, Md, May 30, 41; m 63; c 2. MATERIALS SCIENCE, SOLID MECHANICS. *Educ:* Johns Hopkins Univ, BES, 63, MSE, 66, PhD(mech), 68. *Prof Exp:* Asst prof eng sci, Fla State Univ, 68-71; from asst prof to assoc prof eng mech, 71-78, PROF ENG SCI & MECH, VA POLYTECH INST & STATE UNIV, 78-, HEAD, ENG SCI & MECH DEPT, 89- *Concurrent Pos:* Assoc tech ed, Mat Eval; bd dirs, Am Soc Nondestructive Testing, 87-89, tech coun chmn, 88-89. *Mem:* Acoust Soc Am; Am Soc Testing & Mat; Am Soc Nondestructive Testing. *Res:* Composite materials; wave propagation in anisotropic media; mechanical properties; non-destructive testing. *Mailing Add:* Dept Eng Sci & Mech Va Polytech Inst & State Univ Blacksburg VA 24061

HENNELLY, EDWARD JOSEPH, b Schenectady, NY, Mar 29, 23; m 46; c 5. PHYSICAL CHEMISTRY. *Educ:* Union Col, BS, 43; Princeton Univ, MA, 48, PhD(chem), 49. *Prof Exp:* Chemist, E I du Pont de Nemours & Co Inc, 49-51, reactor physicist, 52-55, sr res supvr, 55-67, res assoc, 67-79, prog coordr, 79-86; RETIRED. *Concurrent Pos:* Vchmn, SC Nuclear Adv Coun, 75-76, chmn, 76-79; mem, Coun Sci Soc Pres, 80-81. *Mem:* Am Chem Soc; fel Am Nuclear Soc (vpres/pres elec, 78-79 & pres, 79-80); Sigma Xi. *Res:* Reactor physics; radioisotope production and applications; nuclear fusion technology; nuclear cross sections; radioactive waste technology and disposal. *Mailing Add:* 638 Magnolia St SE Aiken SC 29801

HENNEMAN, ELWOOD, b Washington, DC, Dec 22, 15; c 2. NEUROPHYSIOLOGY. *Educ:* Harvard Univ, AB, 37; McGill Univ, MD, 43. *Prof Exp:* NEUROPHYSIOLOGIST, MASS GEN HOSP, 60-; chmn dept, 71-79, PROF PHYSIOL, SCH MED, HARVARD UNIV, 69- *Concurrent Pos:* Guggenheim fel, 49-50; mem, radiobiology panel of President's Space Sci Bd. *Mem:* Am Physiol Soc; Int Brain Res Orgn; Soc Neurosci; AAAS; Belgian Soc Electromyography & Clin Neurophysiol; Sigma Xi. *Res:* Spinal cord electrophysiology; control of movement. *Mailing Add:* Harvard Med Sch 25 Shattuck St Boston MA 02115

HENNEMAN, HAROLD ALBERT, b Belmont, Wis, Nov 20, 18; m 46; c 1. ANIMAL HUSBANDRY. *Educ:* Univ Wis, BS, 40; Mich State Col, MS, 42, PhD(animal husb), 53. *Prof Exp:* From asst prof to prof animal husb, Mich State Univ, 51-59, dir, Dept Short Courses, 59-66, prof, 66-85, EMER PROF ANIMAL HUSB, MICH STATE UNIV, 85- *Mem:* Fel Am Soc Animal Sci. *Res:* Animal physiology. *Mailing Add:* Dept Animal Sci Mich State Univ East Lansing MI 48823

HENNEN, JOE FLEETWOOD, b Sherman, Tex, Jan 6, 28; m 54; c 1. BOTANY, MYCOLOGY. *Educ:* Southern Methodist Univ, BS, 50; Purdue Univ, MS, 52, PhD, 54. *Prof Exp:* Asst bot, Purdue Univ, 50-54; asst plant pathologist, SDak State Col, 54-58; from assoc prof to prof bot, Ind State Univ, 58-68; PROF BOT & CUR ARTHUR HERBARIUM, PURDUE UNIV, 68- *Concurrent Pos:* Vis prof, Biol Inst Sao Paulo, Brazil, 75-76. *Mem:* Mycol Soc Am; Am Phytopath Soc; Soc Plant Taxon. *Res:* Taxonomy of rust fungi; biosystematics of Uredinales of Neotropica. *Mailing Add:* Dept Bot & Plant Path Purdue Univ West Lafayette IN 47907

HENNEN, SALLY, b New Rochelle, NY, Apr 18, 35. DEVELOPMENTAL BIOLOGY. *Educ:* Wellesley Col, BA, 57; Ind Univ, PhD(exp embryol), 62. *Prof Exp:* USPHS res fel, Inst Cancer Res, Pa, 62-63; USPHS res fel, Ind Univ, 63-64, res assoc zool, 64-66; NATO fel, Univ St Andrews, 66-67; res assoc zool, Ind Univ, 67-69; asst prof, 69-72, ASSOC PROF BIOL, MARQUETTE UNIV, 72- *Mem:* Soc Develop Biol; Am Soc Cell Biol; Int Soc Develop Biologists. *Res:* Nucleocytoplasmic interactions during development; developmental genetics and cytology. *Mailing Add:* Dept Biol Marquette Univ Milwaukee WI 53233

HENNES, ROBERT G(RAHAM), b Lake Linden, Mich, July 7, 05; m; c 4. CIVIL ENGINEERING. *Educ:* Univ Notre Dame, BS, 27; Mass Inst Technol, MS, 28. *Prof Exp:* Sr eng aide, City Engrs Off, Detroit, 27-34; from instr to prof, 34-73, chmn transp res group, 60-69, chmn dept, 66-71, EMER PROF CIVIL ENG, UNIV WASH, 73- *Concurrent Pos:* Mem, 1st & 2nd Int Cong Soil Mech & Triaxial Inst; consult, landside, earthwork & found problems, 34-73, US Bureau Pub Roads, Econ Anal, 64-69; dir, ed, J author, several books, 52-65; chmn, Wash State Coun Hwy Res, 54-64, Transp Res Group, Univ Wash, 61-69, Div Econ Studies, 59-67, comt econ anal, 57-59, Div Econ Studies, Hwy Res Bd, NAS, 59-67, Urban Transp Person, 63- 67, Dept Civil Eng, 65-69; hwy res bd, Nat Res Coun, Econ Data Panel, Am Asn State Hwy Officials Road Test, 56-62; guest prof, Bengal Eng Col, Calcutta, 60-61; Univ Wash fac, Dept Humanistic Social Studies, 69-73. *Mem:* Am Soc Eng Educ; fel Am Soc Civil Engrs; Am Asn Univ Professors; Am Soc Testing & Mat. *Res:* Soil mechanics; transportation engineering and engineering economics; transportation economics; social and economic consequences of transportation development; highway engineering; soil mechanics and coastal engineering. *Mailing Add:* Dept Civil Eng Univ Wash Seattle WA 98195

HENNESSEY, AUDREY KATHLEEN, b Fairbanks, Alaska, Apr 4, 36; m 63; c 2. MEDICAL INFORMATION SYSTEMS DESIGN, INDUSTRIAL SYSTEMS DESIGN. *Educ:* Stanford Univ, BA, 57; Univ Toronto, HSA(B), 68; Univ Lancaster, Eng, PhD(systs), 82. *Prof Exp:* Systs anal, Northern Telephone Co, Can, 62-63; pension admin, Manufacturers Life Ins, Toronto, 64-65; instr off systs, Adult Educ Ctr, Toronto, 65-68; lectr, off systs, Salford Col, UK, 68-70; sr lectr data processing, Manchester Polytech, UK, 70-80; lectr comput, Univ Manchester, UK, 80-82; assoc prof comput sci, 82-86, ASSOC PROF INFO SYSTS, TEX TECH UNIV, 87- *Concurrent Pos:* Prin invest, consult, Xerox Corp, 85-86, Tex Instruments, 82-, Tex Advan Technol, Prog, 88-; vis lectr, Fed law Enforcement Training Ctr, 85-; prin investr, Systs Exploration, 86-; Am Nat Standards Inst X3VI Comt, Doc Interchange Standards. *Honors & Awards:* Halliburton Found Award, for teaching & Res, 86. *Mem:* Inst Elec & Electronics Engrs; Data Processing Mgt Asn; Asn Comput Mach; Soc Mfg Engrs. *Res:* Automated visual inspection; automated document interchange; data interchange and process protocols; automated indexing and representation of knowledge from large textual data bases. *Mailing Add:* Inst Studies Orgn Automation Col Bus Admin Tex Tech Univ Lubbock TX 79409-2101

HENNESSY, JOHN LEROY, b New York, NY, Sept 22, 52; m 74; c 2. COMPUTER SCIENCE. *Educ:* Villanova Univ, BE, 73; State Univ NY, MS, 75, PhD(comput sci), 77. *Prof Exp:* Prog consult, Computation Ctr, Villanova Univ, 71-73, instr intro prog, 72-73; res asst, instr & teaching asst, Comput Sci Dept, State Univ NY, 74-77; from asst prof to assoc prof, 77-86, PROF ELEC ENG & COMPUT SCI, STANFORD UNIV, 86- *Concurrent Pos:* Prin investr, NSF, 78-80, co-prin investr, TRW, 80-81, Off Naval Res, Lawrence Livermore Lab, 80-81, Defense Advan Res Proj Agency, 81-83 & 82-85; assoc ed, Inst Elec & Electronics Engrs Micro, 81-; chief scientist, MIPS Comput Systs, 84- *Mem:* Inst Elec & Electronics Engrs; Asn Comput Mach. *Res:* Programming languages and implementation: compilers for novel architectures and the design of protable, optimizing compilers; very-large-scale integration design systems and very-large-scale integration architectures, especially processor design. *Mailing Add:* Comput Systs Lab CIS 208 Stanford Univ Stanford CA 94305

HENNEY, CHRISTOPHER SCOT, b Sutton Coldfield, Eng, Feb 4, 41; m 64; c 2. IMMUNOLOGY. *Educ:* Univ Birmingham, BSc, 62, PhD(exp path), 65, DSc, 73; ARIC, 64. *Prof Exp:* Fel, Dept Exp Path, Univ Birmingham, 65-66; fel immunol, Children's Asthma Res Inst, Denver, 66-68; immunologist, WHO, Geneva, 68-70; asst prof med, Med Sch, Johns Hopkins Univ, 70-73, assoc prof med & microbiol, 73-78; prof microbiol & immunol, Med Sch, Univ Wash, 78-82; sci dir, Immunex Corp, Seattle, 82-89; EXEC VPRES & SCI DIR, ICOS CORP, BOTHELL, WASH, 89- *Concurrent Pos:* WHO consult, Ibadan, Nigeria, 70; head prog basic immunol, Fred Hutchinson Cancer Res Ctr, Seattle, 78-; sci med lectr, Case Western Reserve, 80 & Univ Wash, 80;

chmn, Immunol Adv Comt, Am Cancer Soc, 83; Res Career Develop Award, NIH, 72-77. *Honors & Awards:* Pinkham lectr, 90. *Mem:* Brit Soc Immunologists; Am Asn Immunologists; Transplantation Soc; Am Soc Exp Pathologists; Reticuloendothelial Soc. *Res:* Quantitation of the cell-mediated immune response; correlations between humoral and cellular immunity; lymphokines; immune regulation; inflammatory diseases. *Mailing Add:* ICOS Corp 22021 20th Ave SE Bothell WA 98021

HENNEY, DAGMAR RENATE, b Berlin, Ger; US citizen; c 1. MATHEMATICS. *Educ:* Univ Miami, BS, 54, MS, 56; Univ Md, PhD(math), 65. *Prof Exp:* Asst, Univ Miami, 54-56; instr math, Univ Md, 56-65; asst prof, 65-70, ASSOC PROF MATH, GEORGE WASHINGTON UNIV, 70- *Concurrent Pos:* B'nai B'rith & Delta Delta Delta fel. *Mem:* Hon mem Am Math Soc; Math Asn Am; Sigma Xi; Nat Asn Sci Writers; Austrian Math Asn. *Res:* Funcional analysis, especially functional equations; author of numerous publications. *Mailing Add:* 6912 Prince Georges Ave Takoma Park MD 20912-5414

HENNEY, HENRY RUSSELL, JR, b Carnegie, Pa, Apr 3, 34; m 60; c 2. MICROBIOLOGY. *Educ:* Univ Ariz, BS, 60, MS, 61; Univ Tex, PhD(microbiol), 65. *Prof Exp:* Res assoc, Clayton Found Biochem Inst, Univ Tex, 64-66; from asst prof to assoc prof, 66-77, PROF MICROBIAL PHYSIOL, UNIV HOUSTON, 77- *Mem:* AAAS; Mycol Soc Am; Am Soc Microbiol. *Res:* Nutrition, physiology, molecular biology, and differentiation of eukaryotic microbes. *Mailing Add:* Dept Biol Univ Houston 4800 Calhoun Rd Houston TX 77204-5513

HENNEY, ROBERT CHARLES, b Dunlap, Iowa, June 3, 25; m 53; c 4. ANALYTICAL CHEMISTRY. *Educ:* Colo State Col Educ, BA, 52; SDak State Univ, MNS, 59; Univ Nebr, MS, 62, PhD(chem), 66. *Prof Exp:* High sch teacher, Iowa, 54-58; high sch prin, Iowa, 59-60; instr chem, Northwest Mo State Col, 62-64; assoc prof, 64-70, PROF CHEM, MANKATO STATE UNIV, 70- *Mem:* Am Chem Soc; Sigma Xi. *Res:* Polarographic behavior of complex ions. *Mailing Add:* Dept Chem Mankato State Col Mankato MN 56001

HENNIG, HARVEY, b Brazil, SAm, July 11, 19; nat US; m 42; c 1. CHEMICAL ENGINEERING. *Educ:* Stevens Inst Technol, ME, 41. *Prof Exp:* Jr res engr, Res Ctr, Pure Oil Co, Ill, 42, sr res engr, 43-59, sect supvr, 59-61, div dir, 61-65; sr eng assoc, Res Ctr, Unocal Corp, 65-67, supvr process eval, 67-84; CONSULT, 84- *Mem:* Am Inst Chem Engrs. *Res:* Petrochemical and petroleum refining processes; butadiene manufacturing, carbon disulfide ex methane and sulfur; synthetic liquid fuels; process economics and design; alternate energy. *Mailing Add:* 1528 N Shadow Lane Fullerton CA 92631

HENNIGAN, ROBERT DWYER, b Syracuse, NY, Sept 21, 25; m 50; c 8. ENVIRONMENTAL MANAGEMENT. *Educ:* Manhattan Col, BCE, 49; Syracuse Univ, MA, 64; Am Acad Environ Eng, dipl. *Prof Exp:* Regional water pollution engr, NY State Dept Health, Syracuse Univ, 49-58, chief water pollution control sect, Albany, 58-60; prin engr, Div Munic Serv, NY State Off Local Govt, 60-65; asst comnr, Div Pure Waters, Dept Health & dir, Bur Water Resources, 65-67; dir grad prog environ sci, 75-81, PROF POLICY & PROG AFFAIRS, SCH ENVIRON & RESOURCES MGT, STATE UNIV NY COL ENVIRON SCI & FORESTRY, 68-, PROF & CHMN ENVIRON SCI FAC, 86- *Concurrent Pos:* Lectr, Rensselaer Polytech Inst, 58-59; mem adv bd, Int Joint Comn US & Can, 65-67; alt mem, NY State Water Resources Comn, 65-67; dir, State Univ Water Resources Ctr, State Univ NY Col Forestry, Syracuse Univ, 67-71; consult, Fed Water Pollution Control Admin, 68, New York City Bd Water Supply, 68-69, Monroe County Pure Waters Agency, 69-70, US Army CEngrs, 70 & Dept Natural Resources, Wis, 70; exec dir, Temp State Comn Water Supply Needs Southeastern NY, 70-75; dir cent NY water qual mgt proj, 75-80; dir, Water Supply Source Protection Proj, NY State Dept Health, 79-81; prof environ mgt, Sch Environ & Resources Eng, State Univ NY Col Environ Sci & Forestry, 81- *Honors & Awards:* Bedell Award, Water Pollution Control Fedn, 80. *Mem:* Fel Am Soc Civil Engrs; Am Pub Health Asn; Am Water Resources Asn; Am Acad Environ Eng; Am Water Works Asn. *Res:* Water supply and water quality management; urban water resources; land use management and control; environmental management and control; water resources management. *Mailing Add:* State Univ NY Col Environ Sci & Forestry Syracuse Univ Syracuse NY 13210

HENNIGAR, GORDON ROSS, JR, b Halifax, NC, Dec 16, 19; m 47; c 5. PATHOLOGY. *Educ:* Dalhousie Univ, MD, 45. *Prof Exp:* Intern, Victoria Gen Hosp, NS, 44-45; asst path, Banting Inst & Dalhousie Univ, 45-46; intern, Union Mem Hosp, MD, 46-47; pathologist, South Baltimore Gen Hosp, 47-48; instr path, Sch Med, Johns Hopkins Univ, 48-50; assoc prof, Med Col Va, 50-57; prof, State Univ NY Downstate Med Ctr & dir labs, Kings County Hosp, 57-65; PROF PATH & CHMN DEPT, MED UNIV SC, 65- *Concurrent Pos:* Consult, Off Chief Med Examr, 52-57; consult, Vet Admin Hosp, Charleston, SC & US Naval Hosp. *Mem:* Soc Exp Path; fel Am Soc Clin Path; Am Col Physicians; fel Acad Forensic Sci; fel Int Acad Pathologists. *Res:* Chemical and drug reactions. *Mailing Add:* Dept Path Med Univ SC 171 Ashley Ave Charleston SC 29425

HENNIGER, BERNARD ROBERT, b Zanesville, Ohio, Sept 8, 34; m 82; c 1. SEDIMENTOLOGY, STRATIGRAPHY. *Educ:* Marietta Col, BS, 56; Miami Univ, MS, 64; WVa Univ, PhD(geol), 72. *Prof Exp:* Petrol engr, Halliburton Oil Well Cementing Co, 56; reconnaissance geologist, Ohio State Testing Lab, 59-60; explor geologist, Texaco Inc, 63-65; asst prof, WVa Univ, 68-70; from asst prof to assoc prof geol, Ashland Col, 70-82, chmn dept, 74-82; vpres, 82-84, CONSULT, EXPLOR PETROL DEVELOP CORP, 82- *Concurrent Pos:* Geologist, Petrol Res Lab, US Bur Mines, 66-, consult geologist, 75- *Mem:* Geol Soc Am; Am Asn Petrol Geologists; Sigma Xi; Nat Geog Soc. *Res:* Mather sandstone of southwestern Pennsylvania and northern West Virginia; disaggregation of sandstones by ultrasonic energy; the Clinton sandstone of Hocking and Perry Counties and remote sensing interpretation in Ashland County, Ohio; greenbrier limestone of Wayne and Lincoln Counties, West Virginia; remote sensing linement analysis of Devonian shales in Ohio and surrounding areas. *Mailing Add:* 701 Chestnut St Ashland OH 44805-3106

HENNIGER, JAMES PERRY, b Smiths Falls, Ont, May 30, 38; m 63; c 3. MATHEMATICS. *Educ:* McGill Univ, BSc, 60, MSc, 62, PhD(math), 65. *Prof Exp:* Nat Res Coun Can fel, 65-67; asst prof, 67-74, ASSOC PROF MATH, TRENT UNIV, 74- *Mem:* Can Math Cong; Am Math Soc; Math Asn Am. *Res:* Fourier analysis; almost periodic functions; ergodic theory. *Mailing Add:* Dept Math Trent Univ Peterborough ON K9J 7B8 Can

HENNING, CARL DOUGLAS, b Cleveland, Ohio, Feb 28, 39; m 62; c 2. MECHANICAL ENGINEERING. *Educ:* Ohio Univ, BS, 61; Univ Mich, MS, 63, PhD(mech eng), 65. *Prof Exp:* Res engr, Lawrence Livermore Lab, 65-71, proj engr, Controlled Thermonuclear Fusion Res, 71-73; vpres, Intermagnetics Gen Corp, 73-75; chief magnetic systs, Magnetic Fusion Dept, US Dept Energy, 75-78; dep proj mgr, 78-80, head mirror fusion prog off, 81-86, PROG LEADER, FUSION TECHNOL, LAWRENCE LIVERMORE NAT LAB, 86- *Concurrent Pos:* Dep managing dir, Int Thermonuclear Exp Reactor. *Honors & Awards:* Lincoln Arc Welding Found Award, 71 & 80; Outstanding Tech Achievement Award, Am Nuclear Soc, 82. *Mem:* Am Nuclear Soc; Fusion Power Assocs. *Res:* Heat transfer; cryogenics; superconducting magnets; magnetic field theory; magnetic fusion energy. *Mailing Add:* Lawrence Livermore Nat Lab PO Box 808 Livermore CA 94550

HENNING, EMILIE D, b Scotrun, Pa, Dec 4, 30. NURSING EDUCATION, MATERNAL-CHILD NURSING. *Educ:* Seton Hall Univ, BS, 62; Columbia Univ, MEd, 65, EdD, 74. *Prof Exp:* Instr nursing arts, Sch Nursing, Methodist Hosp, NY, 51-52; staff nurse, East Stroudsburg Gen Hosp, Pa, 52-54; off nurse, Pa, 54-56 & NJ, 59-62; instr maternity nursing, Sch Nursing, Holy Name Hosp, Teaneck, NJ, 62-63; nurse consult, Maternal Infant Care Proj, Newark, NJ, 65-66; from instr to asst prof maternal & child nursing, Rutgers Univ, NJ, 66-71, actg chmn, 71-72, adj asst prof, 72-73, chmn & assoc prof, 73-76; dean & prof, Fla State Univ, 76-82; DEAN & PROF, ECAROLINA UNIV, 82- *Concurrent Pos:* Consult to five cols & Dept Pub Health Nursing; adj asst prof, Seton Hall Univ, 72-73; mem, Bd Rev Baccalaureate Progs, Nat League Nursing, 80-86. *Mem:* Nat League Nursing; Am Asn Higher Educ; Nat Asn Women Deans, Adminr & Counr. *Res:* Use of the clinical laboratory in nursing education; employment opportunities for baccalaureate graduates of selected schools; accreditation of nursing programs. *Mailing Add:* Sch Nursing ECarolina Univ Greenville NC 27858-4353

HENNING, HARLEY BARRY, b New York, NY, May 12, 31; m 61; c 3. ELECTROOPTICS, MATHEMATICAL PHYSICS. *Educ:* Univ Mich, BS(elec eng) & BS(math), 53; Yale Univ, ME, 54; Harvard Univ, PhD(appl physics), 59. *Prof Exp:* Sr engr math physics, LFE Corp, 59-64; prin engr electrooptics, Raytheon Corp, 64-71; prin mem eng staff electronics & optics, RCA Corp, 71-72; res staff phys sci, Rand Corp, 72-76; RES ENGR, NORTHROP CORP, 76- *Concurrent Pos:* Lectr mech eng, Northeastern Univ, 67-68; ed reviewer, Inst Elec & Electronics Engrs, 72- *Mem:* Inst Elec & Electronics Engrs. *Res:* Electronics; lasers and holography; lidar, radar and sonar theory; probability theory and mathematical physics; systems analysis, signal processing and information theory; management systems and resource allocation. *Mailing Add:* 18229 Verano Dr Rancho Bernardo CA 92128

HENNING, LESTER ALLAN, b Oak Park, Ill, Apr 3, 39; m 73; c 4. CHEMICAL COATINGS. *Educ:* Wayne State Univ, BS, 64; Northwestern Univ, MBA, 70. *Prof Exp:* Sr chemist, W P Fuller Co, 65-66; group leader, Desoto, Inc, 66-68, chief chemist, 68-69, prod mgr, 70-72, tech mgr, 73-76, mgr, 77-78, dir res, 80-85, dir, Indust Res & Prod Develop, 79-90, dir qual excellence, 86-90. *Concurrent Pos:* Dir, Soc Mfr Engrs/Asn Finishing Processes, 78-82; lectr, Univ Wis, 82-83. *Mem:* Indust Res Inst; Soc Mfr Engrs/Asn Finishing Processes (vpres, 80-81); Nat Paint & Coatings Asn; Am Soc Qual Control. *Res:* Research and development of coatings and polymers; research computer operations. *Mailing Add:* 346 Sheridon Rd Winthrop Harbor IL 60096

HENNING, RUDOLF E, b Hamburg, Ger, Aug 3, 23; US citizen; m 61; c 1. MICROWAVE ENGINEERING. *Educ:* Columbia Univ, BS, 43, MS, 47, DEngSc(elec eng), 54. *Prof Exp:* Jr engr, Radio Receptor Co, 46; proj engr, Sperry Gyroscope Co Div, Sperry Rand Corp, 47-54, eng sect head, 54-58, chief engr, Sperry Microwave Electronics Div, 58-70; lectr elec eng & asst dean col eng, Univ SFla, 70-71; dept head eng sci, Naval Electronics Lab Ctr, San Diego, 71; prof & assoc dean col eng, 71-82, actg chmn elec eng, 86-87, PROF ELEC ENG, UNIV SFLA, 82- *Concurrent Pos:* Consult, ANRO Eng Inc, 89-; mem, Microwave Theory & Tech Soc Admin Comt, Inst Elec & Electronics Engrs, 66-71, & chmn, 68; evaluator, Inst Elec & Electronic Engrs/ABET Prog, 88- *Honors & Awards:* Centennial Medal, Inst Elec & Electronics Engrs, 84. *Mem:* Fel Inst Elec & Electronics Engrs; Am Soc Eng Educ; Nat Soc Prof Engrs; Sigma Xi; Soc Photo-Optical Instrumentation Engrs. *Res:* Microwave theory and techniques; high frequency instrumentation; electronic measurement techniques; interdisciplinary electronics. *Mailing Add:* Col Eng Univ SFla Tampa FL 33620

HENNING, SUSAN JUNE, b Griffith, Australia, June 6, 46; m 74; c 3. BIOCHEMISTRY. *Educ:* Univ Melbourne, BS, 67, PhD(biochem), 71. *Prof Exp:* Fel develop biochem, Stanford Univ, 71-74 & Fels Res Inst, 74-75; asst prof biol, Temple Univ, 75-79; from assoc prof to prof biol, Univ Houston, 79-89; PROF, BAYLOR COL MED, 89- *Honors & Awards:* Merit Award, NIH, 89. *Mem:* Endocrine Soc; Am Gastroenterol Asn; Soc Pediat Res; Soc Develop Biol; Soc Exp Biol Med; Sigma Xi. *Res:* Development of intestinal function in mammals, with emphasis on the changes that occur at weaning; hormonal and dietary regulation of these changes on a molecular level; development of hunger and satiety mechanisms. *Mailing Add:* Dept Pediat Baylor Col Med Houston TX 77030

HENNING, WALTER FRIEDRICH, b Bad Hersfeld, Ger, Jan 13, 39; m 60; c 2. NUCLEAR PHYSICS. *Educ:* Darmstadt Tech Univ, dipl, 66; Tech Univ Munich, PhD(physics), 68. *Prof Exp:* Asst prof physics, Tech Univ Munich, 68-76; physicist, 76-81, SR PHYSICIST, ARGONNE NAT LAB, 81- *Mem:* Am Phys Soc. *Res:* Nuclear physics research with heavy and light ion induced reactions; nuclear instrumentation in heavy ion research. *Mailing Add:* Argonne Nat Lab 9700 S Cass Ave Bldg 203 Rm G-149 Argonne IL 60439

HENNINGER, ANN LOUISE, b Chambersburg, Pa, Apr 13, 46; m 73. PHYSIOLOGY. *Educ:* Wilson Col, AB, 68; Univ Mich, PhD(physiol), 73. *Prof Exp:* Asst prof, Lebanon Valley Col, 73-81, assoc prof biol & dir continuing educ, 80-83, registrar & dir special prog, 83-85; assoc prof biol & registr, 85-89, ASSOC PROF BIOL, WARTBURG COL, 89- *Mem:* Am Physiol Soc; Nat Soc Teachers Am; Sigma Xi; Soc Col Sci Teachers. *Res:* Baroreceptor control of heart rate; effect of diet on serum cholesterol and triglyceride levels. *Mailing Add:* Wartburg Col Waverly IA 50677

HENNINGER, ERNEST HERMAN, b Indianapolis, Ind, Dec 1, 32; m 59; c 2. SCIENCE, TECHNOLOGY & SOCIETY. *Educ:* Wabash Col, BA, 55; Harvard Univ, MAT, 56; Purdue Univ, MS, 62, PhD(physics), 66. *Prof Exp:* Instr physics, Purdue Univ, 56-59, asst, 59-66; res fel chem physics, Calif Inst Technol, 66-68; asst prof, 68-74, PROF PHYSICS, DEPAUW UNIV, 68-, PRE ENG ADV, 76- *Concurrent Pos:* Res metallic glasses, Oak Ridge Nat Lab, 81-82; pres, Ind Sect, Am Asn Physics Teachers, 85-86. *Mem:* Am Phys Soc; Am Soc Eng Educ; Am Asn Physics Teachers. *Mailing Add:* Dept Physics DePauw Univ Greencastle IN 46135

HENNINGFIELD, MARY FRANCES, b Burlington, Wis, Aug 15, 61. ENTERAL NUTRITION, IMMUNOLOGY. *Educ:* Univ Wis-Madison, BS, 83, MS, 86, PhD(nutrit sci), 88. *Prof Exp:* Postdoctorate, Temple Univ, 89; CLIN RES ASSOC, ROSS LABS, 89- *Mem:* Am Inst Nutrit; Clin Immunol Soc. *Res:* Research of effects of enteral nutrition for trauma/intensive care unit patients. *Mailing Add:* 625 Cleveland Ave Columbus OH 43215

HENNINGS, GEORGE, b Mt Vernon, NY, June 16, 22; m 68. ENVIRONMENTAL BIOLOGY. *Educ:* Montclair State Col, BA, 47, MA, 48; Columbia Univ, EdD(sci educ), 56; Rutgers Univ, MS, 68. *Prof Exp:* Pub sch teacher, NJ, 47-48 & NY, 48-52; asst prof sci, Ball State Teachers Col, 52-55; chmn dept, NJ High Sch, 55-60; assoc prof, 60-63, PROF BIOL, KEAN COL NJ, 63- *Mem:* AAAS; Nat Sci Teachers Asn. *Res:* Science education; writer of teacher preparation books in science and social studies; freelance editor and writer of student study and test materials for textbooks. *Mailing Add:* 21 Flintlock Dr Warren NJ 07060

HENNINGS, HENRY, b Elyria, Ohio, June 30, 41; m 81; c 1. ONCOLOGY. *Educ:* Ohio State Univ, BS, 63; Univ Wis, PhD(exp oncol), 68. *Prof Exp:* Wis Div Res, Am Cancer Soc res fel oncol, Univ Oslo, 68-70; res assoc, McArdle Lab, Univ Wis, 70; staff fel, 71-74, RES CHEMIST, NAT CANCER INST, 74- *Mem:* AAAS; Am Asn Cancer Res; Tissue Cult Asn. *Res:* Chemical carcinogenesis in mouse skin; epidermal cell culture; control of epidermal proliferation and differentiation. *Mailing Add:* Rm 3B26 Bldg 37 Nat Cancer Inst Bethesda MD 20892

HENNION, GEORGE FELIX, b South Bend, Ind, Aug 23, 10; m 33; c 4. ORGANIC CHEMISTRY. *Educ:* Univ Notre Dame, BS, 32, MSc, 33, PhD(org chem), 35. *Prof Exp:* Asst chem, Univ Notre Dame, 32-35, from instr to assoc prof, 35-45, dir res chem, 42-45, Nieuwland prof chem, 45-76; RETIRED. *Mem:* AAAS; Am Chem Soc. *Res:* Acetylene chemistry; medicinal agents; organoboron compounds; reaction mechanisms. *Mailing Add:* 1441 E LaSalle Ave South Bend IN 46617-3327

HENNIS, HENRY EMIL, b Madelia, Minn, Oct 23, 25; m 51. ORGANIC CHEMISTRY. *Educ:* Univ Minn, BCh, 50; Univ Mo, PhD(chem), 56. *Prof Exp:* Org chemist, Dow Chem Co, 56-57, proj leader, 57-63, group leader, 63-71, assoc scientist, 71-86; RETIRED. *Concurrent Pos:* Instr org chem, Cent Mich Univ, 64-69. *Mem:* AAAS; Am Chem Soc; Sigma Xi; NY Acad Sci. *Res:* Organic chemical syntheses; chemical reaction mechanisms; chemical process development. *Mailing Add:* 4959 W Baker Rd RFD 3 Coleman MI 48618

HENNON, DAVID KENT, b Midland, Ind, Oct 20, 33; m 57; c 2. PEDIATRIC DENTISTRY, ORTHODONTICS. *Educ:* Ind Univ, AB, 57, DDS, 60, MSD, 75; Am Bd Pediat Dent, dipl, 80, cert orthod, 83. *Prof Exp:* USPHS fel, 60-62, from asst prof to assoc prof prev dent, 62-74, assoc prof, 74-79, PROF PEDIAT DENT, SCH DENT, IND UNIV-PURDUE UNIV, 79-, PROF ORTHOD, 88- *Concurrent Pos:* Pediat dentist & orthodontist, Craniofacial Anomales Team. *Mem:* Am Dent Asn; Am Asn Orthodontists; Am Acad Pediat Dent; Am Soc Clin Hypn; Am Soc Dent Children. *Res:* Orthodontic treatment and management of oral hygience in these patients; general area of craniofacial growth in cleft lip and palate patients; interrelations; preventive dentistry, pediatric dentistry, orthodontics. *Mailing Add:* Sch Dent Ind Univ-Purdue Univ Indianapolis IN 46202-5186

HENNY, CHARLES JOSEPH, b Salem, Ore, Mar 20, 43; m 67; c 2. FIELD RISK ASSESSMENT, POPULATION STUDIES. *Educ:* Ore State Univ, BS, 65, MS, 67, PhD(wildlife ecol), 71. *Prof Exp:* Res biologist, US Fish & Wildlife Serv, Laurel, Md, 70-74; leader forest environ studies, US Fish & Wildlife Serv, Denver, Colo, 74-76; LEADER, PAC NORTHWEST RES STA, US FISH & WILDLIFE SERV, CORVALLIS, ORE, 76- *Concurrent Pos:* Vis prof, Ohio State Univ, 72; assoc prof, Ore State Univ, 76-87, prof, 87-; assoc ed, J Raptor Res, 90- *Mem:* Wildlife Soc; Am Ornithologists Union; Sigma Xi; Am Soc Mammalogists; Raptor Res Found. *Res:* Study environmental contaminants, pesticides, trace elements, industrial pollutants and their effects on natural populations of wildlife-primarily birds in the Pacific Northwest with special interest in birds of prey. *Mailing Add:* Pac Northwest Res Sta Patuxent Wildlife Res Ctr 480 SW Airport Rd Corvallis OR 97333

HENRICH, CHRISTOPHER JOHN, b Brooklyn, NY, Feb 16, 42. PURE MATHEMATICS. *Educ:* Princeton Univ, AB, 62; Harvard Univ, MA, 65, PhD(math), 68. *Prof Exp:* Instr math, Stevens Inst Technol, 66-67 & Fordham Univ, 67-68; Joseph Fells Ritt instr, Columbia Univ, 68-71; asst prof, Fordham Univ, 71-75; systs programmer, Perkin-Elmer Corp, 75-76, sr mem tech staff, 76-78, prin mem tech staff, 78-88, CONSULT MEM TECH STAFF, PERKIN-ELMER CORP, 88- *Concurrent Pos:* Consult, Cornell Aeronaut Lab, 66-67. *Mem:* Am Math Soc; Math Asn Am; Asn Comput Mach. *Res:* Abstract analysis group representations; several complex variables; harmonic analysis; partial differential equations. *Mailing Add:* Concurrent Comput Corp 106 Apple St Tinton Falls NJ 07724

HENRICH, CURTIS J, CELL BIOLOGY. *Prof Exp:* SR SCIENTIST CELL BIOL, LIFE TECHNOL INC, 89- *Mailing Add:* Dept Cell Biol Life Technol Inc 8717 Grovemont Circle PO Box 9418 Gaithersburg MD 20877

HENRICH, VICTOR E, b Detroit, Mich, Oct 1, 39; m 67; c 2. SURFACE PHYSICS. *Educ:* Univ Mich, BSE, 61, MS, 62, PhD(physics), 67. *Hon Degrees:* MA, Yale Univ, 83. *Prof Exp:* Staff mem solid state physics, Lincoln Lab, Mass Inst Technol, 67-78; PROF, DEPT APPL PHYSICS, YALE UNIV, 78- *Mem:* Fel Am Phys Soc; Am Vacuum Soc; Catalysis Soc; AAAS. *Res:* Electron spectroscopy; surface structure; surface physics; chemisorption; catalysis. *Mailing Add:* Dept Appl Physics Yale Univ New Haven CT 06520-2157

HENRICH, WILLIAMS LLOYD, NEPHROLOGY. *Educ:* Baylor Univ, MD, 72. *Prof Exp:* ASSOC PROF MED, HEALTH SCI CTR, UNIV TEX, 83-; CHIEF, RENEL SECT, VET ADMIN MED CTR, DALLAS, TEX, 83- *Res:* Renin-angiotensin system; hemodynamics of dialysis patients. *Mailing Add:* Dallas Vet Admin Med Ctr Univ Tex Southwestern Med Ctr 4500 S Lancaster Rd Dallas TX 75205

HENRICHS, PAUL MARK, b Roswell, NMex, June 18, 43; m 68; c 2. PHYSICAL ORGANIC CHEMISTRY. *Educ:* Rice Univ, BA, 65; Univ Calif, Los Angeles, PhD(org chem), 70. *Prof Exp:* Fel, Univ Basel, Switz, 70-71; fel, Univ SC, 71-73; res assoc, Eastman Kodak Co, 73-90; RES ASSOC, EXXON CHEM CO, 90- *Concurrent Pos:* Chair, Exp NMR Conf, 84. *Mem:* Am Chem Soc; Royal Chem Soc; Sigma Xi. *Res:* Determination of structure and dynamics of dissolved and solid organic compounds and polymers, primarily with nuclear magnetic resonance. *Mailing Add:* Exxon Chem Co Baytown Polymer Ctr 5200 Bayway Dr Baytown TX 77522

HENRICI, CARL RESSLER, b Minneapolis, Minn, Sept 11, 15; m 40. ELECTRICAL ENGINEERING. *Educ:* Univ Minn, BEE, 37. *Prof Exp:* Engr, Collins Radio Co, 40-55, dir res & develop, Staff Div, 56-57, dir mfg, Tex Mfg Div, 57-59, proj dir, Commun Systs, Alpha Corp, 59-62, asst dir, Spacecraft Systs Div, 62-71, eng admin staff, 72-73, prog mgr, Tactical Air Navig, Rockwell Int Corp, Collins Govt Avionics Div, 73-77, prog mgr, very high frequency commun, 77-81; RETIRED. *Res:* Electronic systems. *Mailing Add:* 304 Red Fox Rd SE Cedar Rapids IA 52403

HENRICK, CLIVE ARTHUR, b Big Bell, Australia, Jan 13, 41; div; c 3. ORGANIC CHEMISTRY, SELECTIVE PESTICIDES. *Educ:* Univ Western Australia, BSc, 62, PhD(org chem), 65. *Prof Exp:* Queen Elizabeth II fel org chem, Univ Sydney, 65-67; fel org chem, Syntex Res, 67-68, sr chemist, 68-69; chief chemist, Zoecon Corp, 69-71, dir chem res, 71-87, DIR CHEM & ANALYTICAL RES, SANDOZ CROP PROTECTION, ZOECON RES INST, 88- *Mem:* Royal Soc Chem; Am Chem Soc; AAAS. *Res:* Organic chemistry with emphasis on the synthesis of biologically active compounds useful for the selective control of insect and arachnid pests and of weeds; insect hormones and pheromones; herbicides. *Mailing Add:* 975 California Ave Palo Alto CA 94304-1104

HENRICKS, DONALD MAURICE, b Quincy, Ill, Mar 13, 36; m 57; c 4. BIOCHEMISTRY, ENDOCRINOLOGY. *Educ:* Univ Mo, BS, 57, PhD(biochem), 65; Purdue Univ, MS, 61. *Prof Exp:* Asst prof animal sci, Iowa State Univ, 65-68; assoc prof, 68-75, PROF BIOCHEM, CLEMSON UNIV, 75- *Concurrent Pos:* Consult, Syntex Res Corp, 73- *Mem:* Am Chem Soc; Am Physiol Soc; Am Soc Animal Sci; Endocrine Soc; Soc Study Reproduction. *Res:* Occurance and role of steroid hormones and gonadotropins in reproductive processes; determination of steroid residues by radioimmunoassay in mammalian tissues; bioassay and radioimmunoassay of gonadotropins. *Mailing Add:* Dept Food Sci & Biochem Clemson SC 29631

HENRICKSEN, THOMAS ALVA, b Marinette, Wis. GEOLOGY. *Educ:* Univ Wis-Oshkosh, BS, 69; Ore State Univ, PhD(geol), 74. *Prof Exp:* Geologist, Asarco, Inc, 74-77; SR GEOLOGIST, US BORAX & CHEM CORP, 77- *Mem:* Geol Soc Am; Am Inst Mining Metall & Petrol Engrs; Soc Econ Geol. *Res:* Planning and carrying out base and precious metal exploration programs in the Pacific Northwest. *Mailing Add:* E 5812 25th Spokane WA 99223

HENRICKSON, CHARLES HENRY, b Cornell, Wis, Nov 27, 38; m 81; c 2. COLLEGE-HIGH SCHOOL CHEMISTRY INTERFACE. *Educ:* Univ Wis-Eau Claire, BS, 62; Univ Iowa, PhD(chem), 68. *Prof Exp:* Chemist, 3M Co, 62-64; PROF CHEM, WESTERN KY UNIV, 68- *Concurrent Pos:* Lectr gen chem, Univ Ill, 76-77; mem, Comt Chem Health Professions, Am Chem Soc, 81-; res, Univ Calif, Berkeley, 86. *Mem:* Am Chem Soc; Sigma Xi. *Res:* Textbooks and laboratory manuals in general chemistry and health-related areas; computer-based instructional materials in general chemistry. *Mailing Add:* Dept Chem Western Ky Univ Bowling Green KY 42101

HENRICKSON, EILER LEONARD, b Crosby, Minn, Apr 23, 20; m 78; c 4. ECONOMIC GEOLOGY. *Educ:* Carleton Col, BA, 43; Univ Minn, PhD, 56. *Prof Exp:* Geologist, US Geol Surv, Calif, 43-44; instr geol, Carleton Col, 46-47 & Univ Minn, 47-48; from instr to asst prof, Carleton Col, 48-53; instr, Univ Minn, 53-54; from asst prof to prof, 54-70, Charles L Denison Prof Geol & Chmn Dept, Carleton Col, 70-87; PROF & CHMN, DEPT GEOL, COLO COL, 87- *Concurrent Pos:* Consult, Jones & Laughlin Steel Co, 46-58, Fremont Mining Co, 58-61, C T Schjeldahl Co, 61-62, Bear Creek Mining Co, 65-66, Argonne Nat Lab, 66-75, Minn Messenia Exped, 66- & Exxon-Esso Eastern, 77-78; Fulbright res scholar, Greece, 69-70. *Mem:* AAAS; Mineral Soc Am; Soc Econ Geologists; Nat Asn Geol Teachers; Geol Soc Am; Sigma Xi. *Res:* Economic geology of metals; metamorphic geology; mineralogy and petrology; structural and field geology; trace element analysis of copper and bronze artifacts and copper and tin ores for correlation studies; adaptation of geochemistry and geophysics to archaeological studies; geology of Greece and Egypt. *Mailing Add:* Dept Geol Colo Col Colorado Springs CO 80903

HENRICKSON, JAMES SOLBERG, b Eau Claire, Wis, Oct 15, 40; c 1. PLANT TAXONOMY. *Educ:* Wis State Univ, Eau Claire, BS, 62; Claremont Grad Sch & Univ Ctr, MA, 64, PhD(bot), 68. *Prof Exp:* From instr to assoc prof, 66-77, PROF BOT, CALIF STATE UNIV, LOS ANGELES, 77- *Concurrent Pos:* Dir, Independent Environ Consults. *Mem:* Int Asn Plant Taxon; Bot Soc Am; Am Asn Plant Taxon. *Res:* Flora of Chihuahuan-Mojavean deserts; plant anatomy. *Mailing Add:* Dept Biol Calif State Univ Los Angeles CA 90032

HENRICKSON, ROBERT LEE, b Hays, Kans, Jan 31, 20; m 42; c 3. ANIMAL SCIENCE. *Educ:* Kans State Col, BS, 47, MS, 49; Univ Mo, PhD(meat technol), 54. *Prof Exp:* Asst prof animal sci, Kans State Col, 47-49; assoc prof, WVa Univ, 49-51; asst prof, Univ Mo, 51-58; PROF ANIMAL SCI, OKLA STATE UNIV, 59- *Mem:* Inst Food Technologists; Am Meat Sci Asn; Am Soc Animal Sci. *Res:* Histochemical characteristics of muscle as a food; removal of glucose from pork; microbial flora in fresh pork sausage; characteristics of the bovine muscle fiber as influenced by prerigor boning, precooking, degeneration; development as an instrument for measuring the physical properties of a muscle fiber. *Mailing Add:* 704 W Hillcrest Ave Stillwater OK 74075

HENRIKSEN, MELVIN, b New York, NY, Feb 23, 27; m 46, 64; c 3. MATHEMATICS. *Educ:* City Col New York, BS, 48; Univ Wis, MS, 49, PhD(math), 51. *Prof Exp:* Asst math, Univ Wis, 48-50, instr, Exten Div, 50-51; asst prof, Univ Ala, 51-52; from instr to prof, Purdue Univ, 52-65; prof & head dept, Case Inst Technol, 65-68; res assoc, Univ Calif, Berkeley, 68-69; chmn dept, 69-72, PROF MATH, HARVEY MUDD COL, 69- *Concurrent Pos:* Sloan fel, 56-58; mem, Inst Advan Study, Princeton Univ, 56-57 & 63-64; vis prof, Wayne State Univ, 60-61; res assoc math, Univ Man, 75-76; vis prof, Wesleyan Univ, 78-79, 82-83 & 86-87. *Mem:* Am Math Soc; Math Asn Am. *Res:* Algebra; ring theory; lattice-ordered rings; rings of continuous functions; general topology. *Mailing Add:* Dept Math Harvey Mudd Col Claremont CA 91711

HENRIKSEN, RICHARD NORMAN, b St John, NB, Dec 14, 40; m 62; c 3. ASTROPHYSICS. *Educ:* McGill Univ, BSc, 62; Univ Manchester, PhD(astron), 65. *Prof Exp:* From asst prof to assoc prof physics, Queen's Univ, Kingston, Ont, 66-72; sr res fel astron, Astron Ctr, Univ Sussex, Eng, 72-73; PROF PHYSICS, QUEEN'S UNIV, KINGSTON, ONT, 78- *Concurrent Pos:* Nat Res Coun grant, 66-85; Alexander von Humboldt Found fel, 72-73 & 80-81; sr res assoc, Stanford, 80-81; chmn coun, Can Inst Theoret Physics, Univ Toronto, 83-85 & fel, 84-85, coun, 88-92; vis scientist Saclay-Meudan (France) 88-89. *Mem:* Can Astron Soc; Am Astron Soc; Int Astron Union. *Res:* Galactic structure and evolution; radio galaxies; gravitational collapse; x-ray sources and pulsars; relativistic astrophysics; cosmology. *Mailing Add:* Astron Group Dept Physics Queen's Univ Kingston ON K7L 3N6 Can

HENRIKSON, ERNEST HILMER, b Portland, Ore, Sept 16, 03; m 33; c 2. SPEECH PATHOLOGY. *Educ:* Univ Ore, BA, 25; Univ Iowa, MA, 29, PhD(speech path & psychol), 32. *Prof Exp:* Prof speech, Gustavus Adolphus Col, 25-33; prof, Univ Mont, 33-36; prof & dir speech correction, Iowa State Teachers Col, 36-43, Univ Denver, 43-44 & Univ Colo, 44-47; prof & dir speech & hearing clin, 47-60, chmn div speech sci, path & audiol & dir grad training speech path, 60-67, prof speech sci, path & audiol & dir grad training speech path, 60-67, prof speech sci, path & audiol, 67-73, prof otolaryngol, 71-73, EMER PROF SPEECH SCI, PATH & AUDIOL, UNIV MINN, MINNEAPOLIS, 73- *Concurrent Pos:* Res consult, US Vet Admin, 48-; spec consult, Staff Col, Off Civilian Defense, 51. *Mem:* Fel Am Speech & Hearing Asn (exec vpres, 50-54, pres, 69-70). *Mailing Add:* Speech Clin 110 Shevlin Hall Univ Minn Minneapolis MN 55455

HENRIKSON, KATHERINE POINTER, b Erie, Pa, Oct 4, 39; m 66; c 2. ENZYMOLOGY, STEROIDS. *Educ:* Univ Rochester, BA, 61; Harvard Univ, MA, 63, PhD(biol chem), 68. *Prof Exp:* Res scientist, Commonwealth Sci & Indust Res Orgn, 67-69; res assoc, Col Physicians & Surgeons, Columbia Univ, 70-71; adj asst prof biochem, Sch Dent, Fairleigh Dickinson Univ, 73-74; res assoc, 76-79, ASST PROF BIOCHEM, ALBANY MED COL, 79-; RES SCIENTIST, DIV LABS & RES, NY STATE HEALTH DEPT, 79-; ASST PROF, GRAD SCH PUB HEALTH, NY STATE UNIV, 86- *Concurrent Pos:* Lectr biol, State Univ NY, Albany, 85. *Mem:* AAAS; Microbiol Soc; Sigma Xi; Endocrine Soc; Am Soc Biochem & Molecular Biol. *Res:* Serine proteinases; steroid-induced enzymes. *Mailing Add:* Wadworth Ctr Lab & Res NY State Health Dept Empire State Plaza PO Box 509 Albany NY 12201-0509

HENRIKSON, RAY CHARLES, b Worcester, Mass, May 22, 37; m 66; c 2. CELL BIOLOGY. *Educ:* Univ Mass, BSc, 59; Brown Univ, MSc, 61; Boston Univ, PhD(biol), 66. *Prof Exp:* Instr dermat, Sch Med, Boston Univ, 66-67; res scientist animal physiol, Commonwealth Sci & Indust Res Orgn, Sydney, Australia, 67-69; asst prof anat, Col Physicians & Surgeons, Columbia Univ, 69-76; assoc prof anat, 76-90, PROF ANAT, ALBANY MED COL, 90- *Mem:* AAAS; Am Soc Cell Biol; Am Asn Anatomists; Sigma Xi. *Res:* Structure and function of skin; comparative ultrastructure of cutaneous cells and appendages; transport across epithelia; computer assisted instruction. *Mailing Add:* Dept Anat Albany Med Col Albany NY 12208

HENRIQUEZ, THEODORE AURELIO, b Tampa, Fla, July 20, 34; m 56; c 6. ACOUSTICS. *Educ:* Univ Tampa, BS, 55. *Prof Exp:* Physicist, Naval Underwater Explosion Res Div, 56-57; mathematician, Gen Elec Corp, 57-59; physicist, Underwater Sound Ref Lab, 59-71; SR CONSULT, NAVAL RES LAB, UNDERWATER SOUND REF DIV, 71- *Mem:* Fel Acoust Soc Am. *Res:* Development of underwater electro-acoustic transducers. *Mailing Add:* PO Box 8337 Underwater Sound Ref Detachment Orlando FL 32856

HENRY, ALLAN FRANCIS, b Philadelphia, Pa, Jan 12, 25. PHYSICS. *Educ:* Yale Univ, BS, 45, MS, 47, PhD(physics), 50. *Prof Exp:* Mgr theoret physics sect, Bettis Atomic Power Lab, Westinghouse Elec Corp, 50-59, mgr reactor theory sect, 59-69; PROF NUCLEAR ENG, MASS INST TECHNOL, 69- *Mem:* Nat Acad Eng; Am Phys Soc; Am Nuclear Soc. *Res:* Molecular spectroscopy reactor theory. *Mailing Add:* Dept Nuclear Eng Mass Inst Technol Cambridge MA 02139

HENRY, ARNOLD WILLIAM, b Williamsport, Pa, June 16, 38; m 61; c 2. POLYMER SCIENCE. *Educ:* Cornell Univ, BChemE, 61; Princeton Univ, MScE, 62; Univ Akron, PhD(polymer sci), 67. *Prof Exp:* Assoc scientist polymer res, Jet Propulsion Labs, 62-64; res scientist silicone rubber, Union Carbide Corp, 66-70; mgr res & develop elec insulation, Bishop Elec Div, Sola Basic Industs, 70-76; TECH SPECIALIST & PROJ MGR, XEROX CORP, 76- *Mem:* Am Chem Soc; NY Acad Sci; Sigma Xi. *Res:* Optimization of electrical and physical properties of polymers; high temperature polymers. *Mailing Add:* 43 Deer Creek Rd Pittsford NY 14534

HENRY, ARTHUR CHARLES, polymer chemistry, polymer physics, for more information see previous edition

HENRY, BILLY WENDELL, psychiatry; deceased, see previous edition for last biography

HENRY, BOYD (HERBERT), b Bronson, Iowa, Feb 19, 25; m 48; c 5. MATHEMATICS. *Educ:* Morningside Col, BS, 46; Univ Iowa, MS, 48. *Prof Exp:* Pub sch teacher, Iowa, 46-47; asst math, Univ Iowa, 47-48; from instr to asst prof, Parsons Col, 48-55; mem actuarial dept, Bankers Life Co, 55-57; from assoc prof to prof math, 77-91, DIR MATH & SCI RESOURCE CTR, COL IDAHO, 91- *Concurrent Pos:* Dir, NSF In-Serv Inst, Col Idaho, 62-72, 77-80. *Mem:* Nat Coun Teachers Math. *Res:* Factoring of integers; audiovisual aids in mathematics; effective means of appealing to child's intuition in teaching mathematics. *Mailing Add:* Dept Math Col Idaho Caldwell ID 83605

HENRY, BRYAN ROGER, b Vancouver, BC, Nov 14, 41; m 63; c 2. PHYSICAL CHEMISTRY. *Educ:* Univ BC, BSc, 63; Fla State Univ, PhD(phys chem), 67. *Prof Exp:* Res asst, Fla State Univ, 64-66; Nat Res Coun Can fel, 68-69; from asst prof to assoc prof, Univ Man, 69-78, prof phys chem, 78-86, head dept, 80-86; PROF PHYS CHEM, UNIV GUELPH, 87- & CHMN DEPT, 88- *Concurrent Pos:* Vis prof, Australian Nat Univ, 75-76, & Univ BC, 84-85. *Mem:* Fel Chem Inst Can. *Res:* Electronic molecular spectroscopy; theory of radiationless transitions; overtone spectroscopy; intracavity dye laser photoacoustic spectroscopy; local mode description of highly vibrationally excited molecules; overtone spectroscopy, intracavity dye laser photoacoustic spectroscopy. *Mailing Add:* Dept Chem & Biochem Univ Guelph Guelph ON N1G 2W1 Can

HENRY, CHARLES H, b Chicago, Ill, May 6, 37; m 58; c 3. SEMICONDUCTOR PHYSICS, SEMICONDUCTOR LASER. *Educ:* Univ Chicago, MS, 59; Univ Ill, Urbana, PhD(physics), 65. *Prof Exp:* Dept head, 70-75, MEM TECH STAFF, SEMICONDUCTOR RES, BELL LABS, 65- *Mem:* Am Phys Soc; fel, AAAS. *Res:* Semiconductor lasers and integrated optics and its application to optical communications. *Mailing Add:* AT&T Bell Labs 600 Mountain Ave Murray Hill NJ 07974

HENRY, CHARLES STUART, b Pittsfield, Mass, Mar 16, 46; m 70, 86; c 6. EVOLUTIONARY BIOLOGY, ENTOMOLOGY. *Educ:* Harvard Univ, BA, 68, PhD(biol), 73. *Prof Exp:* Asst prof biol sci, George Washington Univ, 73-75; from asst prof to assoc prof, 75-85, PROF BIOL, UNIV CONN, 85- *Mem:* Sigma Xi; Ecol Soc Am; Soc Study Evolution; Entom Soc Am. *Res:* Acoustical communication in lacewings of the family Chrysopidae (Neuroptera); mechanisms of speciation. *Mailing Add:* 1620 Storrs Rd Storrs CT 06268

HENRY, CHRISTOPHER DUVAL, b Kansas City, Mo, Aug 21, 46; m 84; c 1. GEOCHEMISTRY, GEOCHRONOLOGY. *Educ:* Calif Inst Technol, BS, 69; Univ Tex, Austin, MA, 72, PhD(geol), 75. *Prof Exp:* Res assoc, 74-79, res scientist, 79-89, SR RES SCIENTIST, BUR ECON GEOL, UNIV TEX, AUSTIN, 90- *Concurrent Pos:* Dir, Tex Mining & Mineral Resources Res Inst, 89- *Mem:* Geol Soc Am; Geochem Soc; Am Geophys Union. *Res:* Coordination of field and geochemical studies of complex problems of petrology, tectonics, and economic geology; application of geochemistry to environmental problems resulting from mining. *Mailing Add:* Bur Econ Geol Univ Tex Austin TX 78713-7508

HENRY, CLAUDIA, b Townsville, Australia, May 19, 31. VIROLOGY, IMMUNOLOGY. *Educ:* Univ Queensland, BApplSci, 53, Hons, 54; Univ Pittsburgh, PhD, 62. *Prof Exp:* Demonstr microbiol, Univ Queensland, 53, instr biochem, 54; Am Asn Univ Women res fel virol, Univ Pittsburgh, 55-56, USPHS fel immunol, 63-64, asst prof, 65-66; res fel, Wellcome Res Labs, Eng, 56-57; Wellcome fel, Max Planck Inst Virus Res, 58; sci mem, Paul Ehrlich Inst, Ger, 66-67; asst res immunologist, Med Ctr, 68-72. *Concurrent Pos:* Adj assoc prof bact & immunol, Univ Calif, Berkeley, 72-82, adj prof microbiol & immunol, 82-89. *Mem:* Basel Inst Immunol, Switz. *Res:* Antibody formation and its control; expression and function of lymphocyte receptors, regulation of idiol type expression. *Mailing Add:* Box 2951 State Line NV 89449

HENRY, CLAY ALLEN, b Denver, Colo, Jan 13, 32; m 54; c 3. DENTISTRY, MICROBIOLOGY. *Educ:* Colo State Univ, BS, 54; Baylor Col Dent, DDS, 60; Mass Inst Technol, PhD(nutrit), 68. *Prof Exp:* Student res asst biochem, Baylor Col Dent, 57-60; pvt pract dent, 60-64; asst lab instr, Mass Inst Technol, 66-67; from asst prof to assoc prof, 68-72, PROF MICROBIOL & CHMN DEPT, BAYLOR COL DENT, 72- *Concurrent Pos:* Instr oral surg & dent anat, Baylor Col Dent, 60-64. *Mem:* AAAS; Am Dent Asn; Am Soc Microbiol; Am Asn Dent Schs; Int Asn Dent Res. *Mailing Add:* 3302 Gaston Baylor Col Dent Dallas TX 75246

HENRY, DAVID P, II, b Durham, NC, Aug 23, 42; m 75; c 3. PHARMACY, ENDOCRINOLOGY. *Educ:* Univ NC, BA, 64; Duke Univ, MD, 68. *Prof Exp:* Intern & resident med, Sch Med, Duke Univ, 68-70; res fel neuropharmacol, NIMH, 70-72; sr res fel med & endocrinol, Univ Wash, 72-74; clin pharmacologist res, 74-88, SR CLIN PHARMACOLOGIST RES, ELI LILLY & CO, 88- *Concurrent Pos:* From asst prof to assoc prof, Sch Med, Ind Univ, 75-81, prof med and pharmacol, 81- *Mem:* Am Soc Clin Invest; Endocrine Soc; Am Soc Pharmacol & Exp Therapeut; Am Soc Clin Pharmacol & Therapeut; Cent Soc Clin Res; fel Am Col Physicians. *Mailing Add:* Lilly Lab Clin Res Wishard Mem Hosp Indianapolis IN 46202

HENRY, DAVID WESTON, b Los Angeles, Calif, Oct 13, 32; m 55; c 3. MEDICINAL CHEMISTRY. *Educ:* Univ Minn, BS, 54; Univ Calif, Los Angeles, PhD(org chem), 58. *Prof Exp:* NSF fel, Imp Col, Univ London, 59-60; sr org chemist, Merck Sharp & Dohme Res Labs, 60-64; sr org chemist, SRI Int, 64-70, dir dept bio-org chem, Life Sci Res Div, 70-77; head, dept org chem, Wellcome Res Lab, Burroughs Wellcome Co, 77-88; vpres & dir res, Agouron Pharmaceut Inc, 88-90; ASSOC DIR DRUG DISCOVERY, GREEN CANCER CTR, 90- *Concurrent Pos:* Mem, Med Chem Adv Comt, Walter Reed Army Inst Res, 79-, chmn, 81-83; mem steering comt, WHO Sci Working Group on Filariasis, 77-79; mem, NIH Med Chem Study Sect A, 81-83. *Mem:* AAAS; Am Asn Cancer Res; Am Chem Soc; NY Acad Sci. *Res:* Synthetic organic chemistry; structure-activity relationships of chemotherapeutic agents; anti-parasitic drugs; anti-tumor drugs; computer assisted drug design. *Mailing Add:* Green Cancer Ctr MS 217 10666 N Torrey Pines Rd La Jolla CA 92307

HENRY, DONALD J, b Columbus, Ohio, 1910. METALLURGY. *Educ:* Otterbein Col, BS, 33; Ohio State Univ, MS, 37. *Prof Exp:* Dir mat sci, Gen Motors Res Lab, 69-73; RETIRED. *Mem:* Am Soc Metals; Sigma Xi; Am Foundrymen's Soc. *Mailing Add:* 223 Franklin Ave Worthington OH 43085

HENRY, DONALD LEE, b Akron, Ohio, Apr 24, 42. PHYSICS. *Educ:* Ohio Univ, BS, 64; Johns Hopkins Univ, PhD(physics), 70. *Prof Exp:* Fel chem eng, Rice Univ, 70-74; asst prof & res assoc physics, McMaster Univ, 74-76; ASST PROF PHYSICS, WILSON COL, 76-, COORD DIV NATURAL SCI & MATH, 79- *Mem:* Am Phys Soc; Sigma Xi. *Res:* Studies of transport properties of fluids near the critical point. *Mailing Add:* Div Sci & Math Shepherd Col Shepherdstown WV 25443

HENRY, DORA PRIAULX, b Maquoketa, Iowa, May 24, 04; m 26. SYSTEMATIC ZOOLOGY. *Educ:* Univ Calif, AB, 25, MA, 26, PhD(zool), 31. *Prof Exp:* Asst zool, Univ Calif, 25-26; res assoc oceanog & zool, Univ Wash, 32-42; asst oceanog, Hqs, Air Forces, US Dept Army, 42-43; assoc oceanogr, Hydrographic Off, US Dept Navy, 43-45, oceanogr, 45; res assoc oceanog & zool, 45-60, res assoc prof oceanog, 60-73, RES PROF OCEANOG, UNIV WASH, 73- *Mem:* Soc Syst Zool; Crustacean Soc. *Res:* Protozoan parasites of birds and mammals; amoebae of man; coccidiosis of the guinea pig; protozoan parasites of invertebrates; taxonomic study of barnacles. *Mailing Add:* Dept Oceanog Univ Wash Seattle WA 98195

HENRY, EGBERT WINSTON, b New York, NY, Apr 28, 31; m 64. PLANT PHYSIOLOGY. *Educ:* Queens Col, NY, BS, 53; Brooklyn Col, MA, 59; City Univ New York, PhD(biol), 72. *Prof Exp:* From instr to asst prof biol, Herbert L Lehman Col, City Univ New York, 71-74; from asst prof to assoc prof, 74-85, PROF BIOL, OAKLAND UNIV, 85-, CHAIRMAN, 88- *Concurrent Pos:* Grant-in-aid res, Sigma Xi. *Mem:* Am Soc Plant Physiologists; Scand Soc Plant Physiol; Japanese Soc Plant Physiologists; Sigma Xi. *Res:* Biochemical and histochemical localization of peroxidase enzymes in plant tissue, particularly abscission zone tissue; the effect of ethylene in inducing peroxidase enzymes in plant tissue; polyphenol oxidase activity and localization in chloroplasts. *Mailing Add:* Dept Biol Sci Oakland Univ Rochester MI 48309-4401

HENRY, EUGENE W, b Wichita Falls, Tex, Dec 31, 32; m 55; c 6. ELECTRICAL ENGINEERING, COMPUTER ENGINEERING. *Educ:* Univ Notre Dame, BSEE, 54, MSEE, 55; Stanford Univ, PhD(elec eng), 60. *Prof Exp:* Proj engr, Wright Air Develop Ctr, Ohio, 55-57; develop engr, Sylvania Electronic Defense Lab, Calif, 57-60; assoc prof elec eng, 60-70, dir Eng Comput Lab, 68-81, scientist, Radiation Labs, 76-86, PROF ELEC ENG, UNIV NOTRE DAME, 70- *Concurrent Pos:* Consult, Keltec Industs, Va, 61-66; mem bd dir, Nat Electronics Conf, 62-63, trustee, 66-73; mem, Simulation Coun; NASA-Am Soc Eng Educ summer fac fel, Goddard Space Flight Ctr, 70-71. *Mem:* Asn Comput Mach; Inst Elec & Electronics Engrs; Am Soc Eng Educ; Soc Comput Simulation. *Res:* Electronic computers and automatic control systems. *Mailing Add:* 537 River Ave South Bend IN 46601

HENRY, HAROLD ROBERT, b Stockbridge, Ga, June 2, 28; m 50; c 5. CIVIL ENGINEERING, HYDROLOGY. *Educ:* Ga Inst Technol, BCE, 48; Univ Iowa, MS, 50; Columbia Univ, PhD(civil eng), 60. *Prof Exp:* Instr math, Ga Inst Technol, 47-48, instr civil eng, 49-50; asst fluid mech, Univ Iowa, 48-49; lectr & asst, Columbia Univ, 50-53; civil engr, Ebasco Serv, NY, 53-54; from asst prof to assoc prof civil eng, Mich State Univ, 54-64; prof eng mech, Univ Ala, Tuscaloosa, 64-69, head dept, 69-84, prof civil eng, 69-90; CONSULT HYDROLOGIST, 90- *Concurrent Pos:* Consult for numerous pvt & govt agencies; mem, Nat Adv Environ Health Sci Coun, 74-78; assoc ed, J Hydraul Eng, 85-91. *Honors & Awards:* Stevens Award, Am Soc Civil Engrs. *Mem:* Am Soc Civil Engrs; Am Soc Eng Educ; Am Geophys Union; Nat Soc Prof Engrs; Am Water Resources Asn; Am Soc Photogammetry & Remote Sensing; Am Meteorol Soc; Am Water Resources Asn. *Res:* flow in porous media; hydrology; waste migration in groundwater flow; coastal engineering; urban hydrology; hydrologic simulation; stochastic hydrology; real-time flood forecasting; use of radar in hydrology; use of satellite derived data in hydroligic forecasting. *Mailing Add:* 36 Glenn Springs Rd Travelers Rest SC 29690

HENRY, HARRY JAMES, b Kokomo, Ind, June 28, 17; m 40; c 3. BIOCHEMISTRY. *Educ:* Hanover Col, AB, 39. *Prof Exp:* Biochemist, Eli Lilly & Co, 39-48, head dept biochem preparations, 48-62, mgr assay & insulin control, 62-67, mgr biochem prod, 67-77, tech adv, 77-81; RETIRED. *Mem:* AAAS; Am Chem Soc; Asn Contamination Control. *Res:* Hormones; antibiotics; growth factors; glandular products. *Mailing Add:* 6060 Shawnee Trail Indianapolis IN 46220-5072

HENRY, HELEN L, b Fairborn, Ohio, Sept 21, 44; m 76; c 1. ENDOCRINOLOGY. *Educ:* Wash Univ, PhD(biol), 70. *Prof Exp:* PROF BIOCHEM, UNIV CALIF, RIVERSIDE. *Honors & Awards:* Fuller Albright Award, Am Soc Bone Mineral Res, 84. *Mem:* Am Soc Bone Mineral Res; Endocrine Soc; Am Inst Nutrit; Am Soc Biochem & Molecular Biol; Am Women in Sci. *Res:* Hormonal regulation of calcium metabolism; cytochrome P-450 and steroid metabolism; gene structure and function. *Mailing Add:* Dept Biochem Univ Calif Riverside Riverside CA 92521

HENRY, HERMAN LUTHER, JR, b Arcadia, La, Mar 6, 18; m 41; c 2. INDUSTRIAL ENGINEERING, COMPUTER SCIENCE. *Educ:* La Polytech Inst, BS, 40; Ill Inst Technol, MS, 46. *Prof Exp:* Instr tech drawing, Ill Inst Technol, 42-46; from asst prof to assoc prof mech eng, La Polytech Inst, 46-51; proj engr power plants & elec distribution, Tex Div, Dow Chem Co, 51-55; assoc prof mech eng, La Tech Univ, 55-56, prof mech eng & head gen eng dept, 56-62, prof indust eng & head dept, 62-76, assoc dean eng, 76-83, prof indust, 83-85; RETIRED. *Mem:* Am Soc Eng Educ; Nat Coun Eng Examrs; Nat Soc Prof Engrs. *Res:* Engineering economy; guidance of freshman engineering students. *Mailing Add:* 2300 Hillside Rd Ruston LA 71270

HENRY, HUGH FORT, b Emory, Va, Apr 25, 16; m 42; c 4. RADIATION PROTECTION, NUCLEAR STRUCTURE. *Educ:* Emory & Henry Col, BS & BA, 36; Univ Va, MS, 38, PhD(physics), 40. *Prof Exp:* Prof, Emory & Henry Col, 38; prof physics & math, Col Ozarks, 40-41; from instr to assoc prof physics, Univ Ga, 41-49; head radiation hazards & safety dept, Carbide & Carbon Chems Co, 49-54, head safety fire & radiation control dept, Union Carbide Corp, 54-61; prof & chmn dept physics, 61-81, EMER PROF PHYSICS, DEPAUW UNIV, 81- *Concurrent Pos:* US rep, Tech Comt 85, Int Standards Orgn, 58, 60 & 65; consult, US AEC, Union Carbide Nuclear Co, Goodyear Atomic Corp & Oolithic Limestone Corp; pres bd dirs, Cent States Univs, Inc, 67; vchmn radiation protection comt N-13, Am Nat Standards Inst. *Mem:* AAAS; Optical Soc Am; Am Phys Soc; Sigma Xi; Am Nuclear Soc. *Res:* Health physics; nuclear safety; electrical discharge in liquids. *Mailing Add:* 404 Linwood Dr Greencastle IN 46135-1137

HENRY, JACK LELAND, chemistry; deceased, see previous edition for last biography

HENRY, JAMES H(ERBERT), b Toronto, Ont, Oct 29, 22; nat US; m 49; c 3. ELECTRICAL ENGINEERING. *Educ:* Univ Toronto, BASc, 43; Mass Inst Technol, SM, 48. *Prof Exp:* Asst electronics, Mass Inst Technol, 46-47, instr elec eng, 47-48; opers analyst & div chief, Opers Res Off, Johns Hopkins Univ, 48-62; sr staff mem, Sci & Technol Div, 62-79, mem staff, Strategy, Forces & Resources Div, 79-87, MEM STAFF, SCI & TECHNOL DIV, INST DEFENSE ANALYSIS, 87- *Concurrent Pos:* Consult, Off Civil Defense, 61, US Army Aviation, 65; mem explosive tagging tech comt, Bur Alcohol, Tobacco & Firearms, Dept of Treas, 72-; mem, Wash Opers Res/Mgt Sci Coun. *Mem:* Fel Opers Res Soc Am; Sigma Xi. *Res:* Operations research; air defense; tactical aviation; civil defense; protection technology; explosive detection technology; surveillance; air traffic controller training; satellite applications; enology and viticulture; technology assessment; military comparisons. *Mailing Add:* Sci & Technol Div/Inst Defense Analysis 1801 N Beauregard St Alexandria VA 22311-1772

HENRY, JAMES PAGET, b Leipzig, Ger, July 12, 14; US citizen; m 38; c 3. MEDICINE. *Educ:* Cambridge Univ, BA, 35, MA, 38, MD, 52; McGill Univ, MS, 42; Calif Inst Technol, PhD(exp med), 55. *Prof Exp:* From asst prof to assoc prof physiol, Univ Southern Calif, 43-47; chief sect, Aeromed Lab, Wright Air Develop Ctr, Wright-Patterson AFB, 47-56; flight surgeon, USAF Med Corps, 56-63; prof physiol, 63-80, PROF PHYSIOL & BIOPHYS, SCH MED, UNIV SOUTHERN CALIF, 80- *Concurrent Pos:* Panel mem, Res & Develop Bd, Environ Physiol, 49-53; mem, Behav Biol Panel, NASA, Am Inst Biol Sci, 62-65; nat consult, Surgeon Gen, USAF, 70-75; ed, Am Psychosom Soc, 72- *Honors & Awards:* Arnold Tuttle Award, Aerospace Med Asn, 53; John Jeffries Award, Am Inst Aeronaut & Astronaut, 55; Carl Ludwig Medal, Ger Heart Asn, 76. *Mem:* Am Physiol Soc; fel Aerospace Med Asn; Am Inst Int Soc Hypertension; Am Psychosom Soc; fel Int Col Psychosom Med. *Res:* Aerospace physiology, acceleration and altitude; cardiovascular physiology, blood volume regulation; psychosocial stress, hypertension and pathophysiological changes in animals and man; role of the autonomic and endocrine systems in stress response. *Mailing Add:* Physiol & Biophys Univ Southern Calif 1333 San Pablo St Los Angeles CA 90033

HENRY, JOHN BERNARD, b Elmira, NY, Apr 26, 28; m 53; c 6. MEDICINE. *Educ:* Cornell Univ, AB, 51; Univ Rochester, MD, 55. *Prof Exp:* Asst med, Wash Univ, 55-56; asst path, Col Physicians & Surgeons, Columbia Univ, 56-57, instr, 57-58; Nat Cancer Inst trainee, 58-60; asst prof path, Col Med, Univ Fla, 60-63, assoc prof, 63-64; prof path, State Univ NY Upstate Med Ctr, 64-79, dean, Col Health Related Prof, 70-77; dean, Sch Med, Georgetown Univ, 79-; OFC PRES UPSTATE MED CTR, SYRACUSE, NY. *Concurrent Pos:* Intern, Barnes Hosp, St Louis, 55-56; resident, Presby Hosp, NY, 56-58 & New Eng Deaconess Hosp, Boston, 58-60; teaching fel, Harvard Med Sch, 59-60; nat consult clin path, Surgeon Gen, USAF, 70-; pres, Am Asn Blood Banks, 70-71; pres, Am Bd Path, 76-77; pres, Am Blood Comn, 78-80. *Mem:* Am Chem Soc; Am Soc Exp Pathologists; Am Soc Clin Pathologists (vpres, 78-79); Col Am Pathologists (pres, 80-81); Int Soc Blood Transfusion. *Res:* Enzymology, especially serum enzymes of diagnostic value and localization at the tissue level; blood groups and immunohematology; clinical chemistry automation; medical education; health care delivery; HLA antigens and disease; transplantation immunopathology. *Mailing Add:* Pres State Univ NY Upstate Med Ctr 750 E Adams St Syracuse NY 13210

HENRY, JOHN PATRICK, JR, b Burley, Idaho, Dec 24, 35. CHEMICAL ENGINEERING. *Educ:* Gonzaga Univ, BS, 57; Northwestern Univ, MS, 59; Ohio State Univ, PhD(mass transfer), 63. *Prof Exp:* Res engr, Calif Res Corp, Standard Oil Co, Calif, 63-65; chem engr, 65-67, indust economist, 67-71, mgr, Chem Process Indust Econ, 71-77, DIR ENERGY CTR, STANFORD RES INST, 77-; VPRES, ENERGY & ENVIRON DIV, BOOZ-ALLEN & HAMILTON, INC, 78- *Mem:* Am Inst Chem Engrs. *Res:* Mass transfer; gas diffusion and flow in porous solids; fluid flow-atomization phenomena; industrial economic analysis; synthetic fuel production; technoeconomics. *Mailing Add:* Dept Sociol Eckerd Col PO Box 12560 St Petersburg FL 33733

HENRY, JONATHAN FLAKE, b New York, NY, Oct 25, 51; m 83; c 3. WATER POLLUTION CONTROL, ENVIRONMENTAL ENGINEERING. *Educ:* Univ Ala, BS, 74, MS, 77; Univ Ky, PhD(chem eng), 82. *Prof Exp:* Researcher Univac software, Univ Ala, 74, instr chem, 74, res asst chem eng, 75-76; process engr, Conoco, 76-78; instr chem eng, Univ Ky, 80-81; asst prof, 81-84, ASSOC PROF CHEM & PHYSICS, TENN TEMPLE UNIV, 84- *Concurrent Pos:* Grad coun fel, Univ Ala, 74-75, grad coun res fel, 75-76; res found fel, Univ Ky, 78-81; chair math & sci, Tenn Temple Univ, 84-; tech adv, Citizen's Sci Integrity, 88-; eng consult, Robert Ray & Assocs, 90- *Mem:* Am Inst Chem Engrs; Creation Res Soc. *Res:* Chemistry as applied to water pollution assessment and control; mathematical modeling of system degradation applied to water and atmospheric systems on earth, and to the systems of other planets; decay processes in the solar system and the universe. *Mailing Add:* 9108 Westminster Circle Chattanooga TN 37416

HENRY, JOSEPH L, b New Orleans, La, May 2, 24; m 43; c 5. ANATOMY, DENTISTRY. *Educ:* Xavier Univ, BS, 48; Univ Ill, MS, 49, PhD(anat), 51, Harvard Univ, MA, 79. *Hon Degrees:* DDS, Howard Univ, 46; DHL, Ill COl Optom, 73; ScD, Xavier Univ, 75. *Prof Exp:* Instr oral med, Howard Univ, 46-48; extern, Res & Educ Hosps, Univ Ill, 49-51; assoc prof, Howard Univ, 51-53, supt clins, 53-65, prof oral med, 58-76, dir clins, 65-76, dean col dent, 66-76; assoc dean, 78-90, PROF & CHMN ORAL DIAG & RADIOL, SCH DENT MED, HARVARD UNIV, 76-, DEAN, 90- *Concurrent Pos:* Consult, Freedmen's Hosp, 51 & Crownsville State Hosp, 60; consult path, Vet Admin Hosp, 62; mem, Nat Comt Advan Educ, Am Asn Dent Schs, 65, White House Conf Int Coop, 65 & Conf Aid to Handicapped, 67; consult, St Elizabeth's Hosp, Washington, DC, 68; trustee, Community Group Health Found, Inc, Neighborhood Health Ctr, 68; mem, Nat Adv Comt Policy & Award of Scholar for Negroes to Study Dent, Am Fund Dent Educ, 68-69; consult, Nat Adv Comt Health Affairs, NIH, 69-72; dir, Pub Health & chmn, DC Task Force Dent Affairs, 70; mem, Bd External Visitors, Sch Dent Med, Harvard Univ, 70; mem, Nat Study Comn Optom Educ, 71, Dean's Comt, Mt Alto Vet Admin Hosp & DC Gen Hosp & Nat Steering Comt, Ann Am Conf Teachers Pract Admin; mem attend staff, Brigham & Women's Hosp, 80; chmn bd trustees, Illuni Col Optom, 81-86. *Mem:* Inst Med-Nat Acad Sci; fel Am Col Dent; Am Acad Periodont; Am Dent Asn; Int Asn Dent Res. *Res:* Colchicine; root resorption; periodontal diseases; cancer; hereditary dental diseases; dental education. *Mailing Add:* Dept Oral Diag & Radiol Harvard Sch Dent Med Boston MA 02115

HENRY, JOSEPH PATRICK, b Mojave, Calif, Apr 16, 47. ASTROPHYSICS. *Educ:* La Salle Col, Pa, BA, 69; Univ Calif, Berkeley, PhD(physics), 74. *Prof Exp:* Physicist astrophysics, Astrophys Observ, Smithsonian Inst, 74-81; dir 2.2m telescope, 84-87, ASSOC PROF, DEPT PHYSICS & ASTRON, UNIV HAWAII, 87- *Concurrent Pos:* Prin investr, NASA & NSF grants, 82-; consult, Patrick Henry & Assocs, 84-; Woodrow Wilson Fel, 69. *Mem:* Am Astron Soc; Soc Photo-Optical Instrumentation Engrs. *Res:* Observations of clusters of galaxies and quasars at optical, infrared and x-ray wavelengths; imaging detector development. *Mailing Add:* Inst Astron 2680 Woodlawn Dr Honolulu HI 96822

HENRY, JOSEPH PETER, organic chemistry; deceased, see previous edition for last biography

HENRY, KENNETH ALBIN, b Everett, Wash, July 10, 23; m 56; c 3. FISH BIOLOGY, STATISTICS. *Educ:* Univ Wash, BS, 49, PhD, 61; Iowa State Univ, MS, 52. *Prof Exp:* Fishery aide, Fish Comn Ore, 48, fishery biologist, 49-52, chg coastal rivers invest, 52-56 & Columbia River invest, 56; fishery biologist, Int Pac Salmon Fishery Comn, 56-62, chief biologist, 62-63; fishery biologist, Wash, 63, lab dir, NC, 63-69, FISHERY BIOLOGIST, BIOL LAB, NAT MARINE FISHERIES SERV, 69- *Concurrent Pos:* Chmn, Pac Fish Mgt Coun Salmon Tech Team; mem, US/Can Salmon Treaty Tech Comts. *Honors & Awards:* Silver Medal, Dept Com, 83. *Mem:* Am Inst Fisher Res Biol. *Res:* Population dynamics; life history studies; migrations; scale patterns as a basis for racial identification of sockeye salmon; effect of various environmental factors on salmon production; fishery management. *Mailing Add:* Alaska Fisheries Ctr Bldg 4 7600 Sandpoint Way NE Seattle WA 98115

HENRY, LEONARD FRANCIS, III, b Baltimore, Md, Feb 14, 42; m 64; c 2. MATERIALS SCIENCE, POLYMER SCIENCE. *Educ:* Fla Presby Col, BS, 64; Univ Va, MS, 66, PhD(physics), 69. *Prof Exp:* Res scientist, Uniroyal Corp Res Ctr, 69-74, sr res scientist, 74-76, mgr passenger tire mat, 76-80, mgr mat develop, 80-81, dir tire res, 81-84, dir res & develop, 84-87; VPRES MKT & PLANNING, BRIDGESTONE-FIRESTONE INC, 87- *Mem:* Am Phys Soc; Soc Rheology; Am Chem Soc; Soc Automotive Engrs. *Res:* Vinyl dispersion rheology; deformation mechanisms in metallic alloys; effect of environment on the yielding of glassy plastics; fracture characteristics of plastics; polymer processing; deformation mechanisms in plastics and metals; tire compound development; textile reinforcement of tires; tire mechanics and design. *Mailing Add:* Bridgestone-Firestone Inc 1200 Firestone Pkwy Akron OH 44317-0001

HENRY, MARY GERARD, b Detroit, Mich, Aug 11, 54. AQUATIC TOXICOLOGY, FISH BEHAVIORAL ECOLOGY. *Educ:* Univ Detroit, BS, 75; Purdue Univ, MS, 78; Iowa State Univ, PhD(animal ecol), 84. *Prof Exp:* Fishery biologist, Columbia Nat Fishery Res Lab, 80-82, sect chief, 82-84, proj leader physiol & behav, Great Lakes Fishery Res Lab, US Fish & Wildlife Serv, 84-; MINN COOP FISH & WILDLIFE RES UNIT. *Concurrent Pos:* Res assoc, Univ Mo, 82-; proj coordr, Energy Prod, Columbia Nat Fishery Res Lab, 83-84, Acid Rain, 84-85; vis scientist, US-USSR Sci Exchange, 84; adj prof, Mich State Univ, 85- *Mem:* Soc Environ Toxicol & Chem; Am Fisheries Soc; Sigma Xi; Int Asn Great Lakes Res; Am Soc Testing & Mat. *Res:* Impact of aquatic contaminants on fish; behavioral, physiological and toxicological assessments made to determine the ecological significance of metals, organochlorines, organophosphates, PCB's and fossil fuels on various species of warm water and cold water fish and invertebrates. *Mailing Add:* Minn Coop Fish & Wildlife Res Unit 141 Hodson Hall 1980 Fowell Ave St Paul MN 55108

HENRY, MYRON S, b Logansport, Ind, Nov 13, 41; m 62; c 2. APPLIED MATHEMATICS. *Educ:* Ball State Univ, BS, 62; Colo State Univ, MS, 65, PhD(math), 68. *Prof Exp:* High sch teacher, Ind, 62-63; prof math, Mont State Univ, 68-80; prof math & dean, Col Arts & Sci, Cent Mich Univ, 80-87; VPRES ACAD AFFAIRS, OLD DOMINION UNIV, 87- *Concurrent Pos:* Vis assoc prof math, NC State Univ, 75-76; actg dean, Col Lett & Sci, Mont State Univ, 78-79; vis prof, Old Dominion Univ, 79-80; dean, Col Arts & Sci, Cent Mich Univ, 80-87. *Mem:* Am Math Soc; Math Asn Am; Soc Indust & Appl Math. *Res:* Approximation theory; numerical analysis; differential equations. *Mailing Add:* Provost Old Dominion Univ Norfolk VA 23529-0011

HENRY, NEIL WYLIE, b Winchester, Mass, Nov 30, 37; m 64; c 1. MULTIVARIATE STATISTICS, DATA ANALYSIS. *Educ:* Wesleyan Univ, BA, 58; Dartmouth Col, MA, 60; Columbia Univ, PhD(math statist), 70. *Prof Exp:* Lectr & res assoc, Columbia Univ, 62-67; asst prof sociol, Cornell Univ, 67-73; sr staff statistician, Gary Income Maintenance Exp, Ind Univ, 73-75; ASSOC PROF MATH & SOCIOL, VA COMMONWEALTH UNIV, 75- *Concurrent Pos:* Assoc ed, Sociol Methods & Res, 71-, Am Sociol Rev, 72-75, J Am Statist Asn, 75-76 & J Educ Statist, 81-92. *Mem:* Am Statist Asn; Am Sociol Asn; Math Asn Am; Am Educ Res Asn. *Res:* Latent structure analysis and measurement theory; statistical design and analysis of evaluation research studies; statistical computing. *Mailing Add:* Dept Math Sci Va Commonwealth Univ Richmond VA 23284-2014

HENRY, PATRICK M, b Joliet, Ill, Sept 29, 28; m 56; c 3. ORGANOMETALLIC CHEMISTRY, CATALYSIS. *Educ:* De Paul Univ, BS, 51, MS, 53; Northwestern Univ, PhD(chem), 56. *Prof Exp:* Res chemist, Hercules Inc, 56-71; from assoc prof to prof chem, Univ Guelph, 71-81; prof & chmn chem, Loyola Univ, Chicago, 81-86; CONSULT, 86. *Mem:* Am Chem Soc; Sigma Xi; fel Chem Inst Can. *Res:* Heterogeneous catalysis; physical chemistry of polymers; mechanism of metal ion catalysis, especially noble metals; environmental chemistry energy. *Mailing Add:* 3506 Greenwood Ave Wilmette IL 60091

HENRY, RAYMOND LEO, b Shabbona, Ill, July 14, 30; m 53; c 4. ANATOMY, PHYSIOLOGY. *Educ:* Univ Ill, BS, 53, MS, 57; Med Col SC, PhD(anat), 61. *Prof Exp:* From instr to assoc prof, 61-73, actg chmn, 81-82, PROF PHYSIOL, SCH MED, WAYNE STATE UNIV, 73- *Mem:* Am Asn Anatomists; Microcirculatory Soc. *Res:* Rheology of blood in the microcirculatory system, especially mechanisms of genesis and lysis of intravascular thrombosis; erythrocyte and platelet function in the spleen. *Mailing Add:* Dept Physiol 5374 Scott Hall Wayne State Univ Sch Med 540 E Canfield Detroit MI 48201

HENRY, RICHARD CONN, b Toronto, Ont, Mar 7, 40; nat US; m 75; c 2. SPACE ASTRONOMY, COSMOLOGY. *Educ:* Univ Toronto, BSc, 61, MA, 62; Princeton Univ, PhD(astron), 67. *Prof Exp:* Res assoc, Inst Advan Study, Princeton, 67; res appointee astrophys, Naval Res Lab, E O Hulburt Ctr Space Res, 67-69; from asst prof to assoc prof, 68-77, PROF PHYSICS & ASTRON, JOHNS HOPKINS UNIV, 77- *Concurrent Pos:* Lectr, Int Sch Space Sci, Cordoba, Arg, 69; res physicist, E O Hulburt Ctr Space Res, Naval Res Lab, 69-76; Alfred P Sloan res fel, 71-75; managing ed, Astrophys Lett, 75-77, ed-in-chief, 77-86; dep dir astrophys div, NASA Hq, 76-78. *Honors & Awards:* Gold Medal, Royal Astron Soc Can, 61. *Mem:* fel AAAS; Am Astron Soc; fel Royal Astron Soc; Royal Astron Soc Can; Am Physical Soc. *Res:* Ultraviolet astronomy; abundances of the elements; cosmology. *Mailing Add:* Dept Physics & Astron Johns Hopkins Univ Baltimore MD 21218

HENRY, RICHARD JOSEPH, b Harrisburg, Pa, Mar 4, 18; m 43; c 2. BIOCHEMISTRY. *Educ:* Gettysburg Col, AB, 40; Univ Pa, MD, 43. *Prof Exp:* Investr bact, Sch Med, Univ Pa, 42-46; biochemist, Civil Serv, US Govt, 46-48; dir, Bio-Sci Labs, Bio-Sci Enterprises, 48-71, vpres in chg clin labs, 71-75; RETIRED. *Concurrent Pos:* Lit researcher, Smith, Kline & French Labs, Pa, 43; lectr, Hood Col, 46-47; instr, Sch Med, Univ Southern Calif, 49-53, asst clin prof, 51-58, adj prof, 58-75; photog res, 75- *Honors & Awards:* Ames Award, 67; Reinhold Award, 73; Vanslyke Award, 73; Chaney Award, 73. *Mem:* Am Chem Soc; Soc Exp Biol & Med; Am Asn Clin Chemists (pres, 63-64); Asn Clin Sci; Soc Photo Scientists & Engrs. *Res:* Antibiotics and other chemotherapeutic drugs; analytical biochemistry; bacterial biochemistry and physiology; clinical chemistry; photographic research. *Mailing Add:* 2595 Ribera Rd Carmel CA 93923-9707

HENRY, RICHARD LYNN, b Lewistown, Pa, May 30, 48; m 71. INORGANIC CHEMISTRY, SOLID STATE CHEMISTRY. *Educ:* Millersville State Col, BA, 72; Brown Univ, PhD(inorg chem), 76. *Prof Exp:* Res chemist, Code 6821, 75-90, RES CHEMIST SYNTHESIS & CRYSTAL GROWTH ELECTRON MAT, NAVAL RES LAB CODE 6870, 90- *Mem:* Am Chem Soc. *Res:* Synthesis, bulk crystal growth, epitaxial growth and solid state chemistry of semi-conductor compounds. *Mailing Add:* Naval Res Lab Code 6870 4555 Overlook Ave SW Washington DC 20375-5000

HENRY, RICHARD WARFIELD, b New York, NY, Nov 23, 32; m 54; c 2. PHYSICS. *Educ:* Union Univ, NY, BA, 54; Univ Ill, MS, 56, PhD, 58. *Prof Exp:* Asst physics, Univ Ill, 54-56, asst prof, 58-59; from asst prof to assoc prof, Union Univ, NY, 59-70; chmn dept, 70-80, pres & head dept, 80-84, PROF PHYSICS, BUCKNELL UNIV, 70- *Concurrent Pos:* NSF fel, Calif Inst Technol, 63-64; vis assoc prof, Mass Inst Technol, 67-69. *Mem:* Am Phys Soc. *Res:* Optics; quantum electronics. *Mailing Add:* Dept Physics Bucknell Univ Lewisburg PA 17837

HENRY, ROBERT DAVID, b Columbus, Ohio, Aug 25, 27; m 75. PLANT MORPHOLOGY, FLORISTICS. *Educ:* Ohio State Univ, BS, 51, PhD(bot), 58; Univ Ill, MS, 53. *Prof Exp:* Asst bot, Univ Ill, 51-53; asst & asst instr, Ohio State Univ, 55-58; vis instr, Ohio Wesleyan, 58-59, asst prof, 59-60; asst prof, 60-64, ASSOC PROF BIOL SCI, WESTERN ILL UNIV, 65- *Mem:* AAAS; Bot Soc Am; Int Soc Plant Morphol; Am Soc Plant Taxon; Sigma Xi. *Res:* Vascular plant floristics of West Central Illinois; vascular plant morphology and anatomy. *Mailing Add:* Dept Biol Sci Western Ill Univ Macomb IL 61455

HENRY, ROBERT EDWIN, b Princeville, Ill, Jan 11, 11; m 38; c 2. CHEMISTRY. *Educ:* Monmouth Col, BS, 32; Univ Ill, MS, 34, PhD(anal chem), 38. *Prof Exp:* Asst, Univ Ill, 34-37; jr fel, Mellon Inst, 38-41; chief chem res sect, Chem & Phys Labs, Continental Can Co, 41-52, mgr, Chem Labs, 52-58, mgr, Chem & Phys Labs, 58-64, dir, 64-75; RETIRED. *Mem:* Am Inst Chemists; Am Chem Soc; Inst Food Technologists. *Res:* Container research and development. *Mailing Add:* 823 S Quincy St Hinsdale IL 60521-4385

HENRY, ROBERT LEDYARD, b New Ulm, Minn, July 4, 20; m 42; c 6. PHYSICS. *Educ:* Carleton Col, BA, 41; Johns Hopkins Univ, PhD(physics), 47. *Prof Exp:* Instr physics, Johns Hopkins Univ, 43-46; instr, Carleton Col, 46-48, asst prof, 48-52; assoc prof, Ripon Col, 52-55, prof, 55-56, chmn dept, 52-56; chmn dept, 56-86, prof, 56-89, EMER PROF PHYSICS, WABASH COL, 89- *Concurrent Pos:* Mem, exec bd, Am Asn Physics Teachers, 74-77. *Mem:* Am Phys Soc; Am Asn Physics Teachers. *Res:* Nuclear magnetic resonance; x-ray diffraction. *Mailing Add:* Dept Physics Wabash Col Crawfordsville IN 47933

HENRY, ROBERT R, b Pasadena, Calif, Oct 31, 55. CODE GENERATION FOR COMPILERS. *Educ:* Univ Calif, Davis, BS, 77, Berkeley, MS, 81, PhD(comput sci), 84. *Prof Exp:* Eng aid, Naval Weapon Ctr, 73-77; teaching assoc comput sci, Univ Calif, Berkeley, 77-78, res assoc, 78-84; ASST PROF COMPUT SCI, UNIV WASH, 84- *Concurrent Pos:* Prin investr comput sci, NSF, 85- *Mem:* Asn Comput Mach. *Res:* Compiler construction; retargetable compilers; machine dependent Graham-Glanville code generators; table driven tree pattern matchers; algorithm animation. *Mailing Add:* Comput Sci Dept Univ Wash FR-35 Seattle WA 98195

HENRY, ROGER P, b Toronto, Ont, Apr 2, 38; m 63; c 2. MECHANICAL ENGINEERING. *Educ:* Univ Toronto, BASc, 61, MASc, 64, PhD(mech eng), 68. *Prof Exp:* From asst prof to assoc prof, Univ Ottawa, 68-75; specialist eng training, Technol Res & Develop Br, 75-79, specialist solar design, 79-81, chief, Tech Systs Secretariat, 81-86, CAD SPECIALIST, DEPT PUB WORKS CAN, 86- *Concurrent Pos:* Ont Dept Univ Affairs grant, 68-69; Nat Res Coun operating grants, 68-75; Am Soc Mech Eng diesel & gas power grant, 69. *Honors & Awards:* Teetor Award, Soc Automotive Engrs, 70. *Mem:* Am Soc Mech Engrs; Soc Automotive Engrs. *Res:* Applied thermodynamics; pollution control; engine environmental problems; computer-aided-design in buildings; lighting design; building energy systems analysis; solar energy. *Mailing Add:* Dept Pub Works Rm No E322 Sir Charles Tupper Bldg Riverside & Heron Ottawa ON K1A 0M2 Can

HENRY, RONALD ALTON, b La Crosse, Wis, Oct 30, 40; c 2. MICROBIOLOGY, BIOCHEMISTRY. *Educ:* Univ Wis, La Crosse, BS, 65; Univ Minn, Minneapolis, PhD(microbiol), 72. *Prof Exp:* Res asst, Dept Microbiol, Univ Minn, 65-71; res assoc, Diamond Lab Inc, 71-73; asst prof, 73-80, ASSOC PROF BIOL, ST JOHN'S UNIV, 80- *Concurrent Pos:* Consult, Fromm Lab Inc, 74-78; res grant, Dr J E Salsbury Found, 74-78 & MacPherson Found, 78-79; sec chmn, Am Leptospirosis Res Conf Inc, 75 & 78. *Mem:* Am Soc Microbiol; AAAS. *Res:* Survival and maintenance of virulence of pathogenic bacteria outside the normal host; physiological differences between avirulent and virulent microorganisms; isoenzyme patterns and activity in aquatic animals at different temperatures. *Mailing Add:* Dept Biol St John's Univ Collegeville MN 56321

HENRY, RONALD ANDREW, b Yakima, Wash, Jan 26, 16; m 42; c 3. CHEMISTRY. *Educ:* Univ Wash, BS, 38, PhD(chem), 42. *Prof Exp:* Res chemist, Procter & Gamble Co, 42-47; sr chemist, US Naval Ord Test Sta, 47-53, head, Org Chem Br, 53-74, SR RES SCIENTIST, US NAVAL WEAPONS CTR, 74- *Mem:* Am Chem Soc; Sigma Xi. *Res:* Chemistry of nitrogen compounds; chemiluminescence; tetrazole complexes; chemistry of explosives and propellants; dyes for lasers. *Mailing Add:* Michelson Lab Code 38505 US Naval Weapons Ctr China Lake CA 93555-6001

HENRY, RONALD JAMES WHYTE, b Belfast, Ireland, Feb 5, 40; m 65; c 2. ATOMIC PHYSICS. *Educ:* Queen's Univ, Belfast, BSc, 61, PhD(appl math), 64. *Prof Exp:* Asst lectr appl math, Queen's Univ, Belfast, 64-65; Nat Acad Sci-Nat Res Coun resident res fel physics, Goddard Space Flight Ctr, 65-66; asst physicist, Kitt Peak Nat Observ, Ariz, 66-69; assoc prof, La State Univ, Baton Rouge, 69-74, chmn dept physics & astron, 76-82, prof physics, 74-89, dean Col Basic Sci, 82-89; VPRES ACAD AFFAIRS & PROF PHYSICS, AUBURN UNIV, 89- *Mem:* Fel Am Phys Soc. *Res:* Atomic collision theory. *Mailing Add:* Vpres Acad Affairs Auburn Univ 208 Samford Hall Auburn AL 36849

HENRY, SUSAN ARMSTRONG, b Alexandria, Va, June 27, 46; m 68; c 2. MOLECULAR GENETICS OF YEAST, LIPID BIOCHEMISTRY. *Educ:* Univ Md, BS, 68; Univ Calif, PhD(genetics), 71. *Prof Exp:* Postdoctoral fel molecular biol & genetics, Brandeis Univ, 71-72; from asst prof to assoc prof genetics & molecular biol, Albert Einstein Col Md, 72-82, dir, Grad Div, 83-87; PROF & DEPT HEAD MOLECULAR BIOL, CARNEGIE MELLON UNIV, 87- *Concurrent Pos:* Prin investr, 72-; vis prof, Harvard Univ, 82-83; mem, Genetics Study Sect, NIH, 82-86 & Genetics Basis Dis Rev Comt, 89-; ed, Molecular & Cellular Biol, 82-86, Yeast, 86-; dir, Undergrad Biol Sci Educ Initiative, Howard Hughes Med Inst, 89- *Mem:* Genetics Soc Am; Am Soc Biochem & Molecular Biol; Am Soc Microbiol. *Res:* Genetic and molecular studies on the mechanism of regulation of synthesis of membrane phospholipids in bakers yeast, Saccharomyces cerevisiae; emphasis is on the role of inositol and inositol-containing lipids in regulation and cell physiology. *Mailing Add:* Dept Biol Sci Carnegie Mellon Univ 4400 Fifth Ave Pittsburgh PA 15213

HENRY, SYDNEY MARK, b Brooklyn, NY, Dec 27, 30; m 53; c 5. MICROBIOLOGY, PHYSIOLOGY. *Educ:* St John's Univ, BS, 53, MS, 55; NY Univ, PhD(physiol), 61. *Prof Exp:* Res scientist, Boyce Thompson Inst Plant Res, NY, 55-64; head dept microbiol, 64-73, mgr biol res, 73-84, MGR CLIN RES PERSONAL CARE, BRISTOL-MYERS CO, 84- *Concurrent Pos:* NIH grants, 62, 63 & 64; lectr, Col Pharmaceut Sci, Columbia Univ, 70-74, adj asst prof, 74-76. *Honors & Awards:* Perry Bros Award, Soc Cosmetic Chemists, 77. *Mem:* Am Soc Microbiol; Soc Indust Microbiol; Am Chem Soc; NY Acad Sci; fel Soc Cosmetic Chemist; fel Royal Soc Med. *Res:* Biochemistry and physiology of symbiotic relationships; sulfur metabolism; amino acid metabolism; insect nutrition; microbiology of scalp, skin, mouth; intermicrobial influences; chromatographic techniques; microbial spoilage and preservation; efficacy testing of deodorants and antiperspirants; research administration. *Mailing Add:* 132 Hardwick Ave Westfield NJ 07090

HENRY, TIMOTHY JAMES, b Doylestown, Ohio, Sept 21, 39; c 2. HEALTH SCIENCE ADMINISTRATION. *Educ:* Kent State Univ, BS, 61; Univ Calif, Berkeley, MS, 64; Univ Wis-Madison, PhD(molecular biol), 69. *Prof Exp:* Fel molecular biol, Univ Pittsburgh, 68-70; Helen Hay Whitney fel, Freidrich-Miescher-Lab, Tubingen, WGer, 71-73; fel, Princeton Univ, 73; res assoc, Sloan-Kettering Inst, 74-76; cancer expert, Lab Molecular Biol, Nat Cancer Inst, 77-79; staff sci adv, molecular biol, 79-80, reviewing toxicologist, Bur Foods, Food & Drug Admin, 80-82; dir biol sci, Health Ind Mfg Asn, 82-85; assoc dir, Biol Instr Prog, NSF, 86; HEALTH SCI ADMINR, BACT STUDY SECT, NIH, 87- *Mem:* NY Acad Sci; Genetic Toxicol Asn; Fedn Am Scientists; AAAS; Am Soc Microbiol; Sigma Xi. *Res:* Identification and isolation of virus directed proteins made in vivo; electron microscopy of virus maturation and release; molecular biology of RNA tumor viruses; nucleic acid hybridization; recombinant DNA technology; short-term carcinogenesis/mutagenesis tests; genetic toxicology; risk assessment. *Mailing Add:* 6905 Wilson Lane Bethesda MD 20817

HENRY, VERNON JAMES, JR, b Port Arthur, Tex, Oct 13, 31; m 54; c 3. OCEANOGRAPHY, GEOLOGY. *Educ:* Lamar State Col, BS, 53; Tex A&M Univ, MS, 55, PhD(oceanog), 61. *Prof Exp:* Res assoc, Marine Inst & asst prof geol, Univ, 61-64, dir inst & assoc prof, 64-71, PROF GEOL, UNIV GA & MARINE INST, 71-; RES SCIENTIST, SKIDAWAY INST OCEANOG, 71- *Concurrent Pos:* Prog dir marine geol & geophys, NSF, 68-69. *Mem:* Geol Soc Am; Am Asn Petrol Geologists; Soc Econ Paleont & Mineral. *Res:* Coastal geology; shallow water oceanography. *Mailing Add:* Dept Geol Ga State Univ Univ Plaza Atlanta GA 30303

HENRY, WALTER LESTER, JR, b Philadelphia, Pa, Nov 19, 15; m 42. MEDICINE, ENDOCRINOLOGY. *Educ:* Temple Univ, AB, 36; Howard Univ, MD, 41. *Prof Exp:* Michael Reese fel, Am Cancer Soc, 51-53; from asst prof to prof, 53-71, chief dept, 63-73, JOHN B JOHNSON PROF MED, COL MED, HOWARD UNIV, 71- *Concurrent Pos:* Markle scholar, 53-58; consult, Vet Admin Hosp, Washington, DC, 53- *Mem:* AMA; Am Col Physicians; Am Fedn Clin Res; Asn Am Physicians. *Res:* Etiology of diabetes mellitus; mechanism of action of insulin. *Mailing Add:* 1780 Redwood Terr NW Washington DC 20012

HENRY, WILLIAM MELLINGER, b Westerville, Ohio, Apr 20, 18; m 49; c 3. ENVIRONMENTAL CHEMISTRY. *Educ:* Otterbein Col, AB, 40. *Prof Exp:* Lab technician, Repub Steel Corp, Ohio, 40-42; engr, Chrysler Corp, Ill, 42-45; from asst div chief anal dept to chief anal spectros, 45-63, chief anal chem div, 63-71, chief environ & mat characterization div, 71-73, PROJ MGR, BATTELLE-COLUMBUS LABS, 73- *Mem:* Am Soc Testing & Mat; Am Chem Soc; Am Acad Forensic Sci; Air Pollution Control Asn. *Res:* Technical direction and management of research and development for sampling and analysis of environmental pollutants; spectroscopy; metallurgy. *Mailing Add:* 2335 Arlington Columbus OH 43221

HENRY, ZACHARY ADOLPHUS, b Stockbridge, Ga, Apr 25, 30; m 53; c 5. AGRICULTURAL ENGINEERING, PHYSICS. *Educ:* Univ Ga, BSAE, 51; Clemson Col, MSAE, 59; NC State Univ, PhD(agr eng), 62. *Prof Exp:* Construct engr, E I du Pont de Nemours & Co, Ga, 51-52; sales engr, Tri-State Culvert & Mfg Div, Fla Steel Corp, Ga, 55-57; from asst prof to assoc prof, 61-82, PROF AGR ENG, UNIV TENN, KNOXVILLE, 82- *Mem:* AAAS; Am Phys Soc; Am Soc Eng Educ; Am Soc Naval Engrs; Am Soc Testing & Mat; Am Soc Agr Engrs. *Res:* Development of instruments and techniques for research on the physical properties of agricultural products; curing and drying agricultural products; materials handling; pneumatic conveying and systems engineering; waste management systems; mechanization for tobacco production. *Mailing Add:* Dept Agr Eng Univ Tenn Knoxville TN 37996-4500

HENSCHEL, AUSTIN, Princeton, Minn, Mar 13, 06. ERGONOMICS. *Educ:* Univ Minn, PhD(physiol), 38. *Prof Exp:* Chief, physiol & ergonomics br, 61-73, CONSULT, NAT INST OCCUP SAFETY & HEALTH, USPHS, 73- *Mem:* Am Physiol Soc; Sigma Xi; Soc Exp Biol & Med; Am Conf Am Indust Hygenists; AAAS. *Mailing Add:* 3580 Shaw Ave Apt 429 Cincinnati OH 45208

HENSCHEN, LAWRENCE JOSEPH, b Joliet, Ill, Oct 11, 44. COMPUTER SCIENCE. *Educ:* Univ Ill, Urbana, BA, 66, MA, 68, PhD(math), 71. *Prof Exp:* PROF COMPUT SCI, NORTHWESTERN UNIV, 71- *Concurrent Pos:* Vis scientist, Argonne Nat Lab, 72-; vpres micro-comput appln, Hank's Racing Bodies, 74-; vis scientist, Bell Labs, 80- *Mem:* Asn Comput Mach; Am Asn Artificial Intel. *Res:* Develop programs for digital computers that display intelligence; mathematical logic as a vehicle for implementing automated deduction in mathematics and other areas; logic and data bases. *Mailing Add:* Dept Comput Sci Northwestern Univ 633 Clark St Evanston IL 60208

HENSEL, DALE ROBERT, b Carmel, Ind, Sept 8, 31; m 54; c 4. SOIL FERTILITY. *Educ:* Purdue Univ, BS, 53, MS, 58; Rutgers Univ, PhD(soil fertil), 60. *Prof Exp:* Asst prof & asst soils chemist, 60-63, DIR, AGR RES CTR, UNIV FLA, 64 - *Concurrent Pos:* Dir, Potato Asn Am, 84-87. *Mem:* Am Soc Agron; Soil Sci Soc Am; Potato Asn Am; Am Inst Chemists; Sigma Xi. *Res:* Pest management programs; soil fertility and soil management problems associated with sandy flatwood soils which includes irrigation and drainage of potatoes and cabbage. *Mailing Add:* Agr Res Ctr Univ Fla PO Box 728 Hastings FL 32145-0728

HENSEL, GUSTAV, b Sheboygan, Wis, Nov 11, 34; m 63; c 2. MATHEMATICS. *Educ:* Cath Univ Am, AB, 57; Princeton Univ, PhD(math), 63. *Prof Exp:* From instr to asst prof, 61-68, ASSOC PROF MATH, CATH UNIV AM, 69- *Mem:* Am Math Soc; Math Asn Am; Asn Symbolic Logic. *Res:* Mathematical logic. *Mailing Add:* Dept Math Cath Univ Am 620 Mich Ave NE Washington DC 20064

HENSEL, JOHN CHARLES, b Pontiac, Mich, Dec 5, 30; m 58; c 4. SOLID STATE PHYSICS. *Educ:* Univ Mich, BSE, 52, MS, 53, PhD(physics), 58. *Prof Exp:* Mem Tech Staff, Bell Labs, 58-90; DISTINGUISHED RES PROF, NJ INST TECHNOL, NEWARK, 90- *Mem:* Fel Am Phys Soc; Sigma Xi. *Mailing Add:* Six Hillcrest Ave Summit NJ 07901

HENSENS, OTTO DERK, b Zuidlaren, Holland, Nov 27, 40; Australian citizen; m 70. PHYSICAL BIOCHEMISTRY. *Educ:* Univ NEng, Australia, BSc, 62, Hons, 63, PhD(org chem), 69. *Prof Exp:* Fel, Univ Hawaii, 68-70; res fel, Calif Inst Technol, 70-72; sr res fel, Oxford Univ, 72-74; SR INVESTR, NATURAL PROD CHEM, MERCK SHARPE & DOHME RES LABS, 74- *Mem:* Am Chem Soc. *Res:* Chemistry of natural products and elucidation of biosynthetic pathways, especially by Fourier transform nuclear magnetic resonance techniques. *Mailing Add:* Merck Inst Therapeut Res Labs PO Box 2000 RY8Y-345 Rahway NJ 07065

HENSHAW, CLEMENT LONG, b Pittsfield, Mass, Dec 18, 06; m 38; c 3. PHYSICS, SCIENCE EDUCATION. *Educ:* Union Univ, NY, BS, 28; Univ Mich, MA, 29; Yale Univ, PhD(physics), 36. *Prof Exp:* Instr physics, Lehigh Univ, 29-31; instr, Yale Univ, 31-33, asst, 33-34, instr, 35-36; from instr to prof physics & chmn dept physics & astron, 36-73, EMER PROF PHYSICS, COLGATE UNIV, 73- *Concurrent Pos:* Consult, Col Entrance Exam Bd, 49-61; fac fel, Ford Fund Advan Educ, 52-53; vis prof, Univ Chicago, 56-57; consult, Educ Testing Serv, 63-72. *Mem:* Am Phys Soc; Am Asn Physics Teachers. *Res:* Evaluation in science education. *Mailing Add:* 31 University Ave Hamilton NY 13346

HENSHAW, EDGAR CUMMINGS, b Cincinnati, Ohio, Dec 14, 29; m 58; c 1. PROTEIN SYNTHESIS INITIATION. *Educ:* Harvard Univ, AB, 52, MD, 56. *Prof Exp:* Intern, Harvard Med Serv, Boston City Hosp, 56-57, asst resident med, 59-60; res fel bact, Harvard Med Sch, 60-62, assoc med, 64-69, from asst prof to assoc prof, 69-76; PROF ONCOL-BIOCHEM, UNIV ROCHESTER, 76-, ASSOC DIR BASIC SCI, CANCER CTR, 84- *Concurrent Pos:* Lieutenant, USN Med Corps, 57-59; prin investr, Nat Cancer Inst grant, 68-; vis lectr, Nat Inst Med Res, Mill Hill, Eng, 74-75. *Mem:* Am Soc Biochem & Molecular Biol; Am Asn Cancer Res; Am Chem Soc; Biochem Soc; NY Acad Sci; Am Asn Cancer Educ. *Res:* Regulation of protein synthesis and growth in mammalian cells; response to growth factors, hormones, nutrient availability and physical stresses (heat shock) in normal and malignant cells; metabolic regulation. *Mailing Add:* Univ Rochester Cancer Ctr Box 704 601 Elmwood Ave Rochester NY 14642

HENSHAW, PAUL CARRINGTON, b Rye, NY, Nov 15, 13; m 39; c 3. ECONOMIC GEOLOGY. *Educ:* Harvard Univ, AB, 36; Calif Inst Technol, MS, 38, PhD(econ geol), 40. *Prof Exp:* Geologist, Cerro de Pasco Copper Corp, Peru, 40-43, Consorcio Mining Co Peru, Lima, 43-45, Peru Portland Cement Co, 45 & Day Mines, Inc, Idaho, 45; assoc prof geol & actg head dept, Sch Mines, Univ Idaho, 46-47; geologist, San Luis Mining Co, 47-53; geologist, 53-60, vpres, 61-70, pres & dir, 70-76, chief exec officer, 70-80, CHMN BD, HOMESTAKE MINING CO, SAN FRANCISCO, 77- *Mem:* AAAS; Mining & Metall Soc Am; Soc Econ Geologists; Geol Soc Am; Am Inst Mining, Metall & Petrol Engrs. *Mailing Add:* 3258 Ptarmigan Dr No 2B Walnut Creek CA 94595

HENSHAW, PAUL STEWART, b Salt Fork, Okla, Aug 23, 02; m 30; c 2. RADIOBIOLOGY, INFORMATION BIOLOGY. *Educ:* Southwestern Col, Kans, AB, 25; Univ Wis, MS, 27, PhD(zool), 30. *Hon Degrees:* LLD, Southwestern Col, Kans, 47. *Prof Exp:* Asst instr zool, Univ Wis, 25-29; biophysicist, Mem Hosp, NY, 29-38; fel, Nat Cancer Inst, 38-39, sr radiobiologist, 40-44; prin biologist, Manhattan Proj, Univ Chicago, 44-45 & Oak Ridge, Tenn, 45-47; chief fundamental res br, Rehab Japanese Sci, Gen Hq, Supreme Command Allied Powers, Tokyo, 47-49; asst chief div int health, USPHS, 49-52; spec res assoc, Conserv Found, 52; dir res, Planned Parenthood Fedn Am, Inc, 52-54; exec dir, Nat Comt Maternal Health, NY, 54-55; biophysicist, AEC, 56-66; dir, Avery Postgrad Inst Achievement of Human Potential, 66-69; vis prof, Sch Govt, 69-70, VIS SCHOLAR GEOSCI, UNIV ARIZ, 70- *Concurrent Pos:* Co-leader, Atomic Bomb Casualty Comn, Japan, 46-47. *Mem:* AAAS; Radiation Res Soc; Am Soc Cybernet; Pop Asn Am. *Res:* Neurological organization; science of information use; radiation carcinogenesis; experimental embryology; senescence; human fertility; population behavior; neural function; human achievement; regulatory biology; information dynamics; cybernetics; simulation; information dynamics. *Mailing Add:* 10045 Royal Oak Rd Sun City AZ 85351

HENSHAW, ROBERT EUGENE, b New York, NY, July 27, 34; div; c 3. ENVIRONMENTAL ANALYSIS & PHYSIOLOGY. *Educ:* Ohio Wesleyan Univ, BA, 56; Univ Mich, MS, 58; Univ Iowa, PhD(physiol), 65. *Prof Exp:* Instr physiol, Univ Iowa, 63-65; asst prof biol, Carnegie-Mellon Univ, 65-68 & Pa State Univ, University Park, 68-74; assoc environ anal, 74-82, ENVIRON MGT SPECIALIST, NY STATE DEPT ENVIRON CONSERV, 82- *Concurrent Pos:* Adj prof, Rensselaer Polytech Inst, 85-86, State Univ NY, Albany, 86- *Mem:* AAAS; Am Soc Zoologists; Am Soc Mammalogists; Wildlife Soc; Arctic Inst NAm. *Res:* Methods of environmental decision making; physiology of acclimatization; energy metabolism; water balance; comparative environmental physiology; energetic and vascular adaptations of homeothermic and hibernating mammals. *Mailing Add:* NY State Dept Environ Conserv 50 Wolf Rd Albany NY 12233

HENSLER, DONALD H, b Baltimore, Md, Mar 2, 33; m 54; c 3. SOLID STATE ELECTRONICS. *Educ:* Western Md Col, BA, 55; Dartmouth Col, MA, 57; Univ Wis, PhD(physics), 62. *Prof Exp:* MEM TECH STAFF, BELL LABS, 62- *Mem:* Am Phys Soc; Optical Soc Am; Sigma Xi. *Res:* Galvanomagnetic effects in thin vanadium dioxide films; deposition and structural effects on the metal-semiconductor transition in vanadium dioxide films; optical propagation in thin films of Ta-2O-5; x-ray diffraction in thin tantalum films; light scattering from fused aluminum oxide surfaces; metal oxide semiconductor, large scale memory design; reliability physics of semiconductor integrated circuits. *Mailing Add:* 417 Iroquois St Emmaus PA 18049

HENSLER, GARY LEE, mathematical statistics, biometrics; deceased, see previous edition for last biography

HENSLER, JOSEPH RAYMOND, b Pittsburgh, Pa, May 1, 24; m 56; c 3. CHEMISTRY. *Educ:* Pa State Col, BS, 46, PhD(ceramics), 51. *Prof Exp:* Sect head glass chem lab, 54-60, head ceramics res & develop labs, 60-65, dir mat res & develop, 65-68, dir cent res lab, 68-71, mgr, glass & mat div, 71-77, MEM STAFF, MAT RES & DEVELOP, BAUSCH & LOMB, 77- *Mem:* Am Chem Soc; fel Am Ceramic Soc; Am Crystallog Asn; Sigma Xi. *Res:* Ultraviolet transmission in glass; glass luminescence. *Mailing Add:* 415 Yarmouth Rd Rochester NY 14610

HENSLER, RONALD FRED, b Watertown, Wis, Aug 15, 44; m 71. SOIL FERTILITY, SOIL CHEMISTRY. *Educ:* Univ Wis-Madison, BS, 66, MS, 68, PhD(soils), 70. *Prof Exp:* Proj assoc soils fertil, Univ Wis-Madison, 70-71; asst prof soil chem, 71-77, ASSOC PROF SOIL SCI, UNIV WIS-STEVENS POINT, 77- *Mem:* Am Soc Agron; Soil Sci Soc Am; Sigma Xi. *Res:* Effect of manure handling on crop yields, nutrient recovery, and runoff losses; movement of nitrogen and sulfur through soil profiles; soil and water contamination from sewage sludge disposal. *Mailing Add:* 1619 County One Custer WI 54423

HENSLEY, DOUGLAS AUSTIN, b New York, NY, Mar 25, 49; m 70; c 5. NUMBER THEORY. *Educ:* Univ Kans, BA, 70; Univ Minn, PhD(math), 74. *Prof Exp:* Vis lectr math, Univ Ill, Urbana, 74-76; mem, Inst Advan Study, Princeton Univ, 76; from asst prof to assoc prof, 77-89, PROF MATH, TEX A&M UNIV, 90- *Mem:* Am Math Soc. *Res:* Analytic number theory, inequalities; probabilistic number theory and related areas of analysis and probability which can be used to give asymptotic estimates of the density of numbers with a given multiplicative structure; fractal structure of continued fractions. *Mailing Add:* Dept Math Tex A&M Univ College Station TX 77843

HENSLEY, EUGENE BENJAMIN, b Augusta, WVa, Jan 6, 18; m 54. SOLID STATE PHYSICS. *Educ:* Cent Col, Mo, AB, 47; Univ Mo, Columbia, MA, 48, PhD(physics), 51. *Prof Exp:* Mem staff, Res Lab Electronics, Mass Inst Technol, 51-53; from asst prof to prof, 53-85, EMER PROF PHYSICS, UNIV MO, COLUMBIA, 85- *Mem:* Am Phys Soc. *Res:* Color centers and electronic structure of alkaline earth chalogenides; thermionic and photoelectric emission; oxide coated cathodes. *Mailing Add:* 802 Greenwood Ct Columbia MO 65203

HENSLEY, JOHN COLEMAN, II, b Jasper, Tex, May 31, 33; m 53; c 4. EXPERIMENTAL PATHOLOGY. *Educ:* Southwestern La Univ, BS, 58; Tex A&M Univ, DVM, 62. *Prof Exp:* Instr diag path, Tex A&M Univ, 62-63, fel vet path, 65-66; staff vet health res, Los Alamos Sci Lab, Univ Calif, 63-65; dir, Dept Animal & Vet Sci, Gulf Southern Res Inst, 66-68, dir, Div Environ Eng, 68-69, prog dir defense & related progs, 69-71; chief, Biophys Lab & sci liaison officer, Animal & Plant Health Inspection Serv, USDA, 71-76; pres, Identronix, Inc, Santa Cruz, 76-80; chmn & chief exec officer, Teledata, Inc, 80-82; chmn, Exec Consults, 82-89; SUPVRY VET MED OFFICER FOOD SAFETY, USDA, 89- *Concurrent Pos:* Consult, Int Animal Exchange, 64- & Tex Agr Exp Sta, 65-66; NIH training grant, 65-66. *Mem:* Am Vet Med Asn; assoc Am Col Lab Animal Med; Am Asn Lab Animal Sci; US Animal Health Asn. *Res:* Evaluation of the effects of adverse environmental stresses on the biological systems, especially the effects as seen in pathologic physiology and behavioral reaction; single cell diagnostic pathology; remote electronic temperature monitoring and animal identification; serologic and hematologic diagnostic test automation; image enhancement. *Mailing Add:* 332 Meadowbrook Nacogdoches TX 75961

HENSLEY, MARBLE JOHN, SR, b Ball Ground, Ga, Nov 6, 22; m 48; c 4. TRANSPORTATION & TRAFFIC ENGINEERING. *Educ:* Ga Inst Technol, BCE, 49. *Prof Exp:* Jr engr, Ga Hwy Dept, 49-50; asst traffic engr, City Atlanta, Ga, 50-54; traffic engr, City Chattanooga, 54-58, city coordr, 58-63; pres, 63-81, chief exec officer, 63-85, CHMN, HENSLEY-SCHMIDT INC, 72- *Concurrent Pos:* Pres, Hensley & Assocs Inc, 57-63; mem, Nat Comt Uniform Traffic Control Devices, 63-74. *Mem:* Fel Am Soc Civil Engrs; fel Am Consult Eng Coun; Inst Transp Engrs (secy, tres, vpres, 66-69, pres, 69-70). *Mailing Add:* 728 Bacon Trail 57 Chattanooga TN 37412

HENSLEY, MARVIN MAX, herpetology, for more information see previous edition

HENSLEY, SESS D, b Smithville, Okla, Jan 18, 20; m 50; c 3. SUGARCANE PESTS INSECTS WEEDS & DISEASES. *Educ:* Okla State Univ, BS, 54, MS, 55, PhD, 60. *Prof Exp:* Entomologist, Sugar Cane Field Lab, Agr Res Serv, USDA, 57-61; assoc prof, 61-67, prof entom, 67-76, AFFIL PROF ENTOM, LA STATE UNIV, BATON ROUGE, 67-; CONSULT, SUGARCANE CROP PROTECTION & PRODUCTION, SUGARCANE FIELD LAB, HOUMA, 85- *Concurrent Pos:* Res leader crop protection, Sugarcane Field Lab, Houma, 76-85. *Mem:* Entom Soc Am; Int Soc Sugar Cane Technol; Sigma Xi. *Res:* Pest management of insects attacking sugar cane. *Mailing Add:* 204 Pendleton Dr Houma LA 70360

HENSON, ANNA MIRIAM (MORGAN), b Springfield, Mo, Nov 7, 35; m 64; c 1. NEUROBIOLOGY. *Educ:* Park Col, AB, 57; Smith Col, MA, 59; Yale Univ, PhD(biol), 67. *Prof Exp:* Instr zool, Smith Col, 60-61; res assoc, Yale Univ, 67-74; instr, 75-78, from res asst prof to res assoc prof, 78-87, RES PROF DEPT SURG, DIV OTOLARYNGOL, SCH MED, UNIV NC, 87- *Concurrent Pos:* Fulbright scholar, Australia, 59-60. *Honors & Awards:* Claude Pepper Award. *Mem:* Sigma Xi; Assoc Res Otolaryngol. *Res:* Structural and functional correlates of hearing and biosonar; normal and pathological conditions in the inner ear; quantitative morphometry and three dimensional reconstruction (computer assisted) at light and electron microscopic levels of the inner ear; distribution of cells with contractile properties, using immunofluorescence techniques, and their role in hearing; efferent auditory system. *Mailing Add:* Dept Surg Div Otolaryngol Univ NC Chapel Hill NC 27599

HENSON, BOB LONDES, b Pierce City, Mo, July 28, 35; m 60. PHYSICS, APPLIED MATHEMATICS. *Educ:* Univ Mo, BS, 57, MS, 60; Wash Univ, PhD(physics), 64. *Prof Exp:* Res assoc physics, Wash Univ, 64; asst prof, Univ NDak, 64-66; FROM ASST PROF TO PROF PHYSICS, UNIV MO-ST LOUIS, 66- *Mem:* AAAS; Am Phys Soc; Soc Indust & Appl Math; Am Math Soc; Sigma Xi. *Res:* Weakly ionized gases; ion transport phenomena; electrical corona discharges; fluid dynamics; mathematical modeling in physics; hydrology and biophysics. *Mailing Add:* Dept Physics Univ Mo St Louis MO 63121

HENSON, C WARD, b Worcester, Mass, Sept 25, 40; m 63; c 3. MATHEMATICAL LOGIC. *Educ:* Harvard Univ, AB, 62; Mass Inst Technol, PhD(math), 67. *Prof Exp:* Asst prof, Duke Univ, 67-74; assoc prof, NMex State Univ, 74-75; assoc prof, 75-81, PROF MATH, UNIV ILL, 81- *Concurrent Pos:* Dept Chair, Univ Ill, 88-92. *Mem:* Am Math Soc; Asn Symbolic Logic; Math Asn Am; Asn Comput Mach. *Mailing Add:* Dept Math Univ Ill 1409 W Green St Urbana IL 61801

HENSON, CARL P, b Buffalo, NY, Sept 14, 38. CLINICAL CHEMISTRY, HEMATOLOGY. *Educ:* Bethany Col, WVa, BS, 60; Univ Wis-Madison, MS, 63, PhD(biochem), 66. *Prof Exp:* NSF fel, Ind Univ, 66-68; chemist, Xerox Med Diag Opers, 68, sect mgr, 68-69; dir & mgr automated reagents, Biodiagnostics, Inc, 70-71; mgr prod develop, Abbott Sci Prod Div, 70-73, mgr clin chem, Abbott Diag Div, Abbott Labs, 73-76; sr mgr lab systs, 76-80, staff scientist, Systs Support, Coulter Diag Div, 80-84, MGR MKT RES, COULTER ELECTRONICS, INC, 84- *Mem:* Am Chem Soc; Am Asn Clin Chem. *Res:* Enzyme kinetics, mechanisms and protein structure; formation of mitochondria; development of high quality analytical methodology and instruments for clinical chemistry; analytical instrument development. *Mailing Add:* Coulter Electronics Inc 601 W 20th St Hialeah FL 33010

HENSON, DONALD E, b St Louis, Mo, Mar 12, 35. EARLY CANCERS. *Educ:* St Louis Univ, MD, 62. *Prof Exp:* PROG DIR, NAT CANCER INST, NIH, 81- *Mem:* Int Acad Path; Col Am Pathologists; Asn Pathologists. *Mailing Add:* Early Detection Br Nat Cancer Inst-NIH 9000 Rockville Pike Bethesda MD 20892

HENSON, EARL BENNETTE, b Charleston, WVa, June 13, 25; m 51; c 2. LIMNOLOGY. *Educ:* Marshall Univ, BS, 49; WVa Univ, MS, 50; Cornell Univ, PhD(limnol), 54. *Prof Exp:* Instr biol & geol, Baldwin-Wallace Col, 54-56; asst prof zool, Univ Md, 56-58; tutor, St John's Col, Md, 58-62; in-chg benthos studies, Basic Data Br, USPHS, Ohio, 62-65; assoc prof, 65-69, PROF ZOOL, UNIV VT, 69-, DIR LAKE CHAMPLAIN STUDIES CTR, 65- *Mem:* Am Soc Limnol & Oceanog; Am Micros Soc; Ecol Soc Am; Brit Ecol Soc; Int Soc Limnol; Sigma Xi. *Res:* Great Lakes drainage limnology; aquatic fauna; aquatic invertebrate zoology and entomology. *Mailing Add:* 102 Adams St Burlington VT 05401

HENSON, JAMES BOND, b Colorado City, Tex, Nov 13, 33; m 56; c 3. VETERINARY PATHOLOGY, INTERNATIONAL DEVELOPMENT. *Educ:* Tex A&M Univ, BS, 56, DVM, 58, MS, 59; Wash State Univ, PhD(vet path), 64. *Prof Exp:* Asst vet, Tex Agr Exp Sta, 58-60; NIH fel vet path, Wash State Univ, 60-62; assoc prof, Tex A&M Univ, 62-65; assoc prof exp path, Wash State Univ, 65-68, chmn dept vet path, 68-74, dir res & grad educ, Col Vet Med, 73-74, dir gen, Int Lab Res on Animal Dis, 74-78, proj dir, Western Sudan Agr Res Proj, 79-84, PROF VET PATH, WASH STATE UNIV, 68-, DIR INT PROG DEVELOP, 78- *Concurrent Pos:* NIH res career develop award, 65-; prof exp animal med, Univ Wash Col Med, Seattle, 71-74; dir gen, Int Lab Res Animal Dis, Nairobi, Kenya, 74-78; mem, Prog Res Training Trop Dis, WHO, 78-82; Int Educ & Develop, 78-; Strategic Planning Res Technol Transfer & Econ Develop, 78- *Honors & Awards:* Nancy Kay Dunkee Mem Award, 67. *Mem:* Am Vet Med Asn; Am Soc Exp Pathologists; Am Soc Immunologists; Conf Res Workers Animal Dis; Am Col Vet Path. *Mailing Add:* Int Prog Develop Off Wash State Univ Pullman WA 99164-1034

HENSON, JAMES WESLEY, b Pierce City, Mo, Dec 30, 29; m 53; c 3. PLANT PHYSIOLOGY, CELL BIOLOGY. *Educ:* Univ Mo-Columbia, BS, 51, MST, 63, PhD(bot), 68; Wash Univ, MA, 62. *Prof Exp:* Instr pub schs, Mo, 51-59, teacher, 59-65; from asst prof to assoc prof biol, 68-77, PROF BIOL, UNIV TENN, MARTIN, 77- *Concurrent Pos:* Res asst, Univ Mo-Columbia, 65-68. *Mem:* AAAS. *Res:* Hormonal control of plant growth and development; productivity of aquatic environments. *Mailing Add:* Dept Biol Univ Tenn Martin TN 38238

HENSON, JOSEPH LAWRENCE, b Hollis, Okla, Aug 13, 32; m 53; c 2. ENTOMOLOGY, MICROBIOLOGY. *Educ:* Bob Jones Univ, BSc, 53; Clemson Univ, MSc, 63, PhD(entom), 65. *Prof Exp:* Chief chemist, Vanadium Corp Am, 53, chemist, 56-57; from instr to prof biol, 57-65, chmn dept biol, 65-73, CHMN DIV NATURAL SCI, BOB JONES UNIV, 74- *Concurrent Pos:* Lectr, Univ SC, 64-65 & Clemson Univ, 65-67. *Mem:* AAAS; Am Sci Affil; Entom Soc Am. *Res:* Phytoplankton and zooplankton in sewage lagoons; the effect of apholate on the reproductive organs of face flies; plant pathology; medical entomology; parasitology. *Mailing Add:* Box 34621 Bob Jones Univ Greenville SC 29614

HENSON, O'DELL WILLIAMS, JR, b Kansas City, Mo, Jan 11, 34; m 64; c 1. ANATOMY. *Educ:* Univ Kans, BA, 57, MA, 60; Yale Univ, PhD(anat), 64. *Prof Exp:* From instr to assoc prof anat, Sch Med, Yale Univ, 63-74; PROF ANAT, SCH MED, UNIV NC, CHAPEL HILL, 74- *Concurrent Pos:* Chmn, Comn Anat, State NC. *Honors & Awards:* Alexander von Humboldt Award; Claude Pepper Award. *Mem:* Am Asn Anatomists; Asn Res Otolaryngol; fel AAAS. *Res:* Bio-sonar, particularly neurophysiological correlates of echolocation in bats; comparative anatomy; sensory physiology. *Mailing Add:* Dept Cell Biol & Anat Univ NC Sch Med Swing Blvd 217H CB7090 Chapel Hill NC 27599

HENSON, PAUL D, b Elliston, Va, Dec 25, 37; m 65; c 1. ORGANIC CHEMISTY, ENVIRONMENTAL SCIENCE & HEALTH. *Educ:* Roanoke Col, BS, 60; Va Polytech Inst & State Univ, MS, 62, PhD(org chem), 65. *Prof Exp:* From asst prof to assoc prof chem, Roanoke Col, 64-75, prof & chmn dept, 75-80; dir environ protection, Norfolk & Western Rwy Co, 80-84; dir environ protection, 84-88, mgr hazardous mat protection, 88-91, DIR SAFETY & HAZARDOUS MAT, NORFOLK SOUTHERN CORP, 91- *Concurrent Pos:* Pub Health Inst spec res fel, Princeton Univ, 68-69; res fel, Med Col Va, Va Commonwealth Univ, 75-76. *Mem:* Am Chem Soc; Sigma Xi; Water Pollution Control Fedn; AAAS; Am Rwy Eng Soc; Am Bd Indust Hyg. *Res:* Organophosphorus chemistry; cancer chemotherapy. *Mailing Add:* 6836 Trevilian Rd NE Roanoke VA 24019

HENSON, PETER MITCHELL, b Eng, Aug 25, 40; m 62; c 2. IMMUNOLOGY, CELL BIOLOGY. *Educ:* Univ Edinburgh, BVM & S, 63, BSc, 64; Cambridge Univ, PhD(immunol), 67. *Prof Exp:* USPHS training grant, Scripps Clin & Res Found, 67-69, assoc, 69-72, assoc mem immunopath, 72-77; dir res, Dept Pediat, 77-82, EXEC VPRES BIOMED AFFAIRS, NAT JEWISH CTR, 82-; PROF PATH & MED, UNIV COLO MED CTR, 77- *Mem:* AAAS; Royal Col Vet Surg; Am Asn Immunologists; Am Soc Exp Pathologists; Reticuloendothelial Soc. *Res:* Mediation of tissue injury; activation of mediator cells and secretion of mediators; inflammation; complement. *Mailing Add:* Nat Jewish Ctr Immunol & Respiratory Med 1400 Jackson St Rm F305 Denver CO 80206

HENSON, RICHARD NELSON, b Asheville, NC, Mar 13, 41; m 66; c 2. PARASITOLOGY. *Educ:* Lamar State Univ, BS, 63; Tex A&M Univ, MS, 66, PhD(zool), 70. *Prof Exp:* Instr invert zool, Tex A&M Univ, 69-70; from asst prof to assoc prof, 70-79, PROF INVERT ZOOL & PARASITOL, APPALACHIAN STATE UNIV, 79- *Mem:* Am Soc Parasitol; Am Soc Zool. *Res:* Cestodes of elasmobranch fishes; veterinary and human parasites as they relate to various economic problems. *Mailing Add:* Dept Biol Appalachian State Univ Boone NC 28608

HENSON, W(ILEY) H(IX), JR, b Cohutta, Ga, June 7, 28; m 54; c 3. AGRICULTURAL ENGINEERING. *Educ:* Univ Ga, BS, 52; NC State Univ, MS, 56, PhD(agr eng), 62. *Prof Exp:* Agr engr, Agr Eng Res Div, 56-61, res invest leader, 61-72, RES LEADER, SOUTHERN REGION, AGR RES SERV, USDA, 72-; ASSOC PROF AGR ENG, UNIV KY, 62-, GRAD FAC, 66- *Concurrent Pos:* Consult & prin investr, US Dept Energy, 75- *Mem:* Sigma Xi; Am Soc Agr Engrs. *Res:* Tobacco curing, including electronic methods of moisture measurement and control; investigation of mathematical models which relate the environmental factors to drying of foliar materials, such as tobacco leaves. *Mailing Add:* 1130 Spendthrift Rd No D Lexington KY 40517

HENSON, WALTER ROBERT, entomology, for more information see previous edition

HENSZ, RICHARD ALBERT, b Evansville, Ind, July 17, 29; m 54; c 2. HORTICULTURE. *Educ:* Tex A&M Univ, BS, 54, MS, 58; Univ Fla, PhD(fruit crops), 64. *Prof Exp:* Instr citricult, 58-62, PROF HORT & DIR CITRUS CTR, TEX A&I UNIV, 64- *Concurrent Pos:* Coordr citrus res, Int Bank Reconstruct & Develop-Nat Inst Investigative Agr, Valencia, Spain, 75-76. *Mem:* Sigma Xi; Int Soc Citricult; Int Org Citrus Virologists; Am Soc Hort Sci. *Res:* Citrus variety improvement, and cultural and production practices. *Mailing Add:* 306 N International Ave Weslaco TX 78596

HENTGES, DAVID JOHN, b Le Mars, Iowa, Sept 18, 28; m 57; c 3. MICROBIOLOGY, BIOCHEMISTRY. *Educ:* Univ Notre Dame, BS, 53; Loyola Univ, Ill, MS, 58, PhD(microbiol), 61. *Prof Exp:* Res assoc, La State Int Ctr Med Res & Training, Costa Rica, 62-64; from asst prof to assoc prof microbiol, Sch Med, Creighton Univ, 64-68; from assoc prof to prof microbiol, Sch Med, Univ Mo-Columbia, 68-81; PROF & CHMN MICROBIOL, SCH MED, HEALTH SCI CTR, TEX TECH UNIV, 81- *Concurrent Pos:* Contract, US Army Med Res & Command, 66-74 & Nat Cancer Inst, 73-76; Thrasher res fund, 86-89; fel, Am Acad Microbiol, 87-; regional ed, Microbiol Ecol Health & Dis, 87- *Mem:* Am Soc Microbiol; Asn Gnotobiotics; Sigma Xi; Soc Microbiol Ecol & Dis (pres, 87-89). *Res:* Enteric bacteriology; anaerobes. *Mailing Add:* Sch Med Health Sci Ctr Tex Tech Univ Lubbock TX 79430

HENTGES, JAMES FRANKLIN, JR, b Perry, Okla, Feb 6, 25; m 46; c 3. RUMINANT NUTRITION. *Educ:* Okla Agr & Mech Col, BS, 48; Univ Wis, MS, 50, PhD(animal husb & biochem), 52. *Prof Exp:* Asst biochem, Univ Wis, 48-51, instr animal husb, 51-52; asst prof & asst animal husbandman, Univ

Fla, 52-56, assoc prof, 56-66, prof animal husb, 66-; RETIRED. *Concurrent Pos:* Consult to Cuban Mission, Int Coop Admin, 57, US Feed Grains Coun, Venezuela, 84. *Mem:* Am Soc Animal Sci; Am Dairy Sci Asn; Am Inst Nutrit; Soc Range Mgt; Nat Cattlemen's Asn. *Mailing Add:* 550 NW 55th St Gainesville FL 32607

HENTGES, LYNNETTE S W, NUTRITION. *Educ:* Cornell Univ, BS, 79; Iowa State Univ, MS, 81, PhD(nutrit physiol), 84. *Prof Exp:* Res data analyst, Dept Animal Sci, Cornell Univ, 77-79, grad res asst, Nutrit Physiol Sect, 79-84; res assoc, Dept Foods & Nutrit, Univ Ga, 84-87; res scientist, Dept Nutrit, Health & Toxicol, Kraft, Inc, 87-88; nutrit res assoc, Nutrit Res & Tech Serv, Nat Dairy Coun, 88-91; NUTRIT-RES CONSULT, 91- *Mem:* Am Inst Nutrit; AAAS; Fedn Am Socs Exp Biol; Sigma Xi. *Res:* Dietary influences on cholesterol and lipoprotein metabolism; diet and chronic disease relationships; endocrine and dietary control of maternal and fetal lipid metabolism during gestation; effects of genetically-induced obesity on serum lipid metabolism. *Mailing Add:* Nutrit-Res Consult 205 E Clarendon St Prospect Heights IL 60070

HENTSCHEL, ROBERT A(DOLF) A(NDREW), b New York, NY, Oct 1, 10; m 38; c 3. METALLURGY. *Educ:* Mass Inst Technol, ScB, 33, ScD, 36. *Prof Exp:* Metallurgist, Remington Arms Co, Conn, 36-37, res supvr, 37-40, asst mgr res div, NY, 40-42, asst supt process eng & control, 42-44, eng supt tech div, 44-46; res supvr, E I du Pont de Nemours & Co, Inc, 46-47, res mgr, 47-59, prod develop mgr, 59-64, res mgr tech div, 64-75; RETIRED. *Res:* Metallurgy of arms, ammunition and cutlery; design of fire arms; manufacturing processes for fire arms, and rayon and plastic films; properties of and manufacturing processes for synthetic fibers and related products. *Mailing Add:* 3204 Swarthmore Rd Wilmington DE 19807

HENTZ, FORREST CLYDE, JR, b Mullins, SC, Aug 20, 33; m 58; c 3. INORGANIC CHEMISTRY, PHYSICAL CHEMISTRY. *Educ:* Newberry Col, BS, 55; Univ NC, MA, 58, PhD(chem), 63. *Prof Exp:* From instr to asst prof chem, Randolph-Macon Woman's Col, 58-64; asst prof, 64-70, ASSOC PROF CHEM, NC STATE UNIV, 70-, DIR GEN CHEM, 68- *Concurrent Pos:* Res grant, Water Resources Res Inst NC, 65-68; consult, Am Enka Corp, 64-65. *Mem:* AAAS; Am Chem Soc. *Res:* Hydrolysis and aggregation of metal and metalate ions; chemistry of basic salts, coordination chemistry of second and third-row transition metal ions. *Mailing Add:* Dept Chem Box 8204 NC State Univ Raleigh NC 27695-8204

HENTZEL, IRVIN ROY, b Burlington, Iowa, July 4, 43; m 70; c 4. MATHEMATICS. *Educ:* Univ Iowa, BA, 64, MA, 66, PhD(math), 68. *Prof Exp:* From asst prof to assoc prof, 68-79, PROF MATH, IOWA STATE UNIV, 79- *Concurrent Pos:* Res appointment, Iowa State Univ Sci & Humanities Res Inst, 74- *Mem:* Am Math Soc: Math Asn Am. *Res:* Algebra and ring theory; representation theory; computers in non-associative algebra. *Mailing Add:* Dept Math/432 Carver Iowa State Univ Ames IA 50011

HENZEL, RICHARD PAUL, b Philadelphia, Pa, Jan 15, 45; m 68; c 2. ORGANIC CHEMISTRY. *Educ:* Lebanon Valley Col, BS, 66; Bucknell Univ, MS, 68; Ohio State Univ, PhD(chem), 72. *Prof Exp:* RES ASSOC, EASTMAN KODAK CO RES LABS, 72- *Mem:* Am Chem Soc. *Res:* Quinone-hydroquinone chemistry; aromatic chemistry; intramolecular reactions; photographic chemistry. *Mailing Add:* 699 Ashdon Circle Webster NY 14580-9166

HENZL, MILAN RASTISLAV, b Bratislava, Czech, Feb 2, 28; m 54; c 3. MEDICINE, ENDOCRINOLOGY. *Educ:* Palacky Univ, Czech, MD, 51; Charles Univ, Prague, CSc, 61. *Prof Exp:* Gynecologist & obstetrician, County Hosp, Ostrava, Czech, 51-55; sr clin res assoc gynec endocrinol, Inst for Care of Mother & Child, Prague, 55-66; res assoc reproductive endocrinol, Sch Med, Stanford Univ, 67-68; assoc med dir, Syntex Res Div, Syntex Corp, 68-76, SECT HEAD REPRODUCTIVE PHYSIOL, SYNTEX RES, 76- *Concurrent Pos:* Clin asst prof obstet/gynec & reproductive med, Sch Med, Univ Calif, San Francisco, 74. *Mem:* NY Acad Sci; Endocrine Soc. *Res:* Human reproductive physiology and pathology; diagnosis of ovarian disturbances; prostaglandin inhibitors in gynecology and obstectrics; effect of contraceptive steroids on subcellular structures of endometrium; gonsdotropin releasing hormones and their analogs. *Mailing Add:* Syntex Res Inst Clin Med A4 ICM Res 3401 Hillview Ave Palo Alto CA 94305

HENZLIK, RAYMOND EUGENE, b Casper, Wyo, Dec 26, 26; m 50; c 2. PHYSIOLOGY, HEALTH SCIENCES. *Educ:* Univ Nebr, BS, 48, MS, 52, PhD(zool), 60. *Prof Exp:* Teacher high schs, Nebr, 48-50; supvr biol, Univ Nebr, 52-53; teacher high sch, Nebr, 53-56; instr biol, Nebr Wesleyan Univ, 57-58; from instr to asst prof zool, Univ Nebr, 59-61; asst prof biol physiol, 62-63, assoc prof physiol, 63-68, radiol health officer, Ctr Med Educ, 67-68, 70-73, PROF PHYSIOL, BALL STATE UNIV, 68- *Concurrent Pos:* NSF grants, Acad Year Inst, Cornell Univ, 61-62; consult, Nat Prescription Footwear Applicator's Asn, 62-, lectr pedorthosis mgt, 75-; res grants, AEC, PR Nuclear Ctr, 67; res grants, Argonne Nat Lab, Ill, 69, consult, Ctr Educ Affairs; fel physiol, Baylor Col Med, 70-71; mem staff, Ball State Europ Ctr Educ, 77-78; vis prof physiol & pharmacol, Col Vet Med, Tex A&M Univ, 84-85. *Mem:* Sigma Xi; Nutrit Today Soc; AAAS; Ecol Soc Am. *Res:* Science education; animal ecology and zoogeography; radiation effects and radioisotopic tracer methodology; physiology-anatomy instructional material. *Mailing Add:* Physiol & Health Sci Dept Ball State Univ Muncie IN 47306-0510

HEPBURN, FRANK, b Peoria, Ill, June 23, 22. NUTRITION. *Educ:* Cornell Univ, MA, 48. *Prof Exp:* chief, Nutrit Data Res Br, Human Nutrit Info Serv, USDA, 77-87; RETIRED. *Mailing Add:* 923 Glyndon St SE Vienna VA 22180

HEPBURN, JOHN CHRISTOPHER, b Glen Ridge, NJ, Nov 10, 42; m 71; c 1. GEOCHEMISTRY OF METAMORPHIC ROCKS, TECTONICS. *Educ:* Colgate Univ, AB, 64; Harvard Univ, AM, 66, PhD(geol), 72. *Prof Exp:* From instr to assoc prof geol, 71-89, chmn, Dept Geol & Geophys, 80-89, PROF GEOL, BOSTON COL, 89- *Mem:* Fel Geol Soc Am; Geochem Soc; fel AAAS; Am Geophys Union. *Res:* Petrology of igneous and metamorphic rocks; geochemistry; structure and tectonics of the New England Appalachians. *Mailing Add:* 132 Stanley Rd Newton MA 02168

HEPBURN, JOHN DUNCAN, b Ganges, Can, Sept 29, 44; m 77; c 3. ACCELERATOR PHYSICS, COMPUTER CONTROL. *Educ:* Univ BC, BSc, 65, MSc, 67; Univ Alta, PhD(elec eng), 73. *Prof Exp:* Asst res officer, Atomic Energy Can Ltd, 67-69, assoc res officer, 73-83, res officer, 83-85, sci admin officer, Res Group, 85-86, admin asst to vpres physics & health sci, 86-90, interim proj mgr, SNO Proj, 89-90, DETECTOR PROJ MGR, SNO PROJ, ATOMIC ENERGY CAN LTD, 90- *Mem:* Can Asn Physicists. *Res:* High voltage techniques; direct current ion accelerators, electron guns, neutron generators, microprocessors and smart instruments; computer control of accelerator components; project management. *Mailing Add:* Atomic Energy Can Chalk River Nuclear Labs Chalk River ON K0J 1J0 Can

HEPFINGER, NORBERT FRANCIS, b Erie, Pa, Oct 4, 32; m 56; c 4. ORGANIC CHEMISTRY, POLYMER CHEMISTRY. *Educ:* Gannon Col, BA, 54; Univ Pittsburgh, PhD(chem), 63. *Prof Exp:* Res chemist, Celanese Corp Am, 63-65; Nat Inst Gen Med Sci spec fel chem, Ill Inst Technol, 65-66; asst prof org chem, 66-70, ASSOC PROF ORG CHEM, RENSSELAER POLYTECH INST, 70-, ASST CHMN DEPT CHEM, 76- *Mem:* Am Chem Soc. *Res:* Additions of aromatic nitroso compounds to olefins; cationic polymerization of cyclic ethers; nuclear magnetic resonance spectroscopy. *Mailing Add:* 25 24th St Troy NY 12180-1914

HEPLER, CHARLES DOUGLAS, b Newton, Mass, Dec 18, 38; m 61; c 3. PHARMACY. *Educ:* Univ Conn, BS, 60; Univ Iowa, MS, 65, PhD(pharm & socio-econ), 73. *Prof Exp:* Asst chief pharm serv, USPHS Indian Hosp, 61-63; clin res pharm, Univ Iowa, 64-65; assoc dir pharm serv, Duke Univ Med Ctr, 65-66, instr, 66-73, asst prof hosp pharm, Col Pharm, Univ Iowa, 73-80; assoc prof, Dept Pharm & Pharmaceut, Va Commonwealth Univ, 79-88; PROF & CHMN, DEPT PHARM HEALTH CARE ADMIN, COL PHARM, UNIV FLA, 88- *Concurrent Pos:* Pharm educ consult, Vet Admin Hosp, Iowa City, 68-; ed bd Drugs in Health Care, Am Soc Hosp Pharm Res & Educ Found, 74-75; bd dir, Iowa Soc Hosp Pharm, 73-76; ed, Pharm Mgt, 85-; chmn, Am Soc Hosp Pharm Coun Educ & Manpower, 83-84; res award, Am Soc Hosp Pharm, 89. *Honors & Awards:* Parke-Davis Distinguished lectr, Ferris State Univ Col Pharm, 88. *Mem:* Am Soc Hosp Pharm; Am Asn Col Pharm; Am Pub Health Asn; Sigma Xi; AAAS. *Res:* Professional service subsystems of health service delivery systems; drug prescribed behavior and drug utilization; prescribing and drug use process; sociology of pharmacy. *Mailing Add:* Dept Pharm Health Care Admin Univ Fla Box J496 Health Sci Ctr Gainesville FL 32610

HEPLER, LOREN GEORGE, b Ottawa, Kans, Aug 26, 28; m 54; c 4. PHYSICAL CHEMISTRY, WATER CHEMISTRY. *Educ:* Univ Kans, BS, 50; Univ Calif, PhD(chem), 53. *Hon Degrees:* DSc, Univ Lethbridge, 89. *Prof Exp:* Asst chem, Univ Calif, 50-51, res chemist, Radiation Lab, 51-53; Du Pont res fel, Univ Minn, 53-54; from asst prof to assoc prof chem, Univ Va, 54-61; from assoc prof to prof, Carnegie Inst Technol, 61-67; prof & head dept, Univ Louisville, 67-68; prof chem, Univ Lethbridge, 68-83; PROF CHEM & CHEM ENG, UNIV ALTA, 83- *Concurrent Pos:* Sloan Found res fel, 57-61; NSF sr res fel, 63-64; res prof, Alta Oil Sands Technol & Res Authority, 76- *Honors & Awards:* Huffman Award, Calorimetry Conf, 80. *Mem:* Fel Chem Inst Can. *Res:* Thermodynamics; calorimetric and electromotive force investigations; aqueous and nonaqueous solutions; acids; transition metal chemistry, oil sands tailings, surfactants, clay chemistry, water treatment, production and uprading heavy oil and bitumen. *Mailing Add:* Dept Chem Univ Alta Edmonton AB T6G 2G2 Can

HEPLER, OPAL ELSIE, b Streator, Ill, Oct 26, 99. PATHOLOGY. *Educ:* Northwestern Univ, BS, 21, MS, 29, PhD(path), 34, MD, 41; Am Bd Path, dipl, 44. *Prof Exp:* From asst to assoc prof path, 24-67, dir, Sch Med Technol & Clin Labs, 34-78, PROF CLIN PATH, MED SCH, NORTHWESTERN UNIV, CHICAGO, 67- *Concurrent Pos:* Dir clin labs, Passavant Mem Hosp, 39-74; consult, Alexian Bros Hosp, 48-63. *Mem:* AAAS; Soc Exp Biol & Med; Am Soc Exp Path; hon fel Am Soc Clin Path; fel AMA. *Res:* Allergic inflammation; guanidine in blood fat metabolism in hyperthyroidism; spontaneous nephritis in rabbits; blood volume during mechanical constriction of hepatic veins; phosphatase; experimental nephropathies; clinical laboratory methods. *Mailing Add:* 4431 Cameron Rd Cameron Park CA 95682

HEPLER, PAUL RAYMOND, b Dover, NH, Nov 27, 25; m 59; c 4. HORTICULTURE. *Educ:* Mich State Univ, BS, 48; Univ Ill, MS, 50, PhD(hort), 56. *Prof Exp:* Asst prof, 56-60, ASSOC PROF HORT, EXP STA, UNIV MAINE, ORONO, 60- *Concurrent Pos:* Res assoc, Mass Inst Technol, 59-60. *Mem:* Am Soc Sugar Beet Technol; Am Soc Hort Sci; Int Inst Sugar Beet Res. *Res:* Physiology and plant breeding of sugar beets, small fruits and vegetables. *Mailing Add:* Six Chapel St Orono ME 04473

HEPLER, PETER KLOCK, b Dover, NH, Oct 29, 36; m 64; c 3. CELL MOTILITY, CYTOMORPHOGENESIS. *Educ:* Univ NH, BA, 58; Univ Wis, PhD(bot), 64. *Prof Exp:* Res parasitologist malaria res, Walter Reed Army Inst Res, 64-66; res fel, Harvard Univ, 66-67, lectr cell biol, 67-68, asst prof, 68-70; asst prof, Stanford Univ, 70-77; from assoc prof to prof, 77-89, RAY ETHAN TORREY PROF BOT, UNIV MASS, AMHERST, 89- *Concurrent Pos:* Mem, Develop Biol Panel, NSF, 72-75; investr, Marine Biol Lab, Woods Hole, Mass, 78-; vis prof, Australian Nat Univ, Canberra, 81-82, Univ Siena, Siena, Italy, 85; dir, Plant Growth & Develop Panel, USDA, 86-87. *Honors & Awards:* Pelton Award, Bot Soc Am, 75. *Mem:* Am Soc Cell Biol; Am Soc Plant Physiol; AAAS. *Res:* Formation, structure and function of the mitotic apparatus in dividing cells, with special reference to the role of calcium ions in the regulation of mitosis; the role of calcium ions as mediators of development in plant cells. *Mailing Add:* Bot Dept Morrill Sci Ctr Univ Mass Amherst MA 01003

HEPLER, ROBERT S, b Indianapolis, Ind, Dec 27, 34; m 57; c 4. OPHTHALMOLOGY. *Educ:* Occidental Col, BA, 57; Univ Calif, Los Angeles, MD, 61. *Prof Exp:* From asst prof to assoc prof, 67-75, PROF OPHTHAL, UNIV CALIF, LOS ANGELES, 75- *Concurrent Pos:* NIH spec fel neuro-ophthal, Mass Eye & Ear Infirmary, Boston, 65-66 & NIH spec fel, Med Ctr, Univ Calif, San Francisco, 66-67. *Res:* Neuro-ophthalmology and pupillography. *Mailing Add:* Jules Stein Eye Inst 2-142 Univ Calif Sch Med Los Angeles CA 90024

HEPPA, DOUGLAS VAN, b Brooklyn, NY, May 26, 45; m 75. NUMERICAL ANALYSIS, NUMERICAL INTEGRATION TECHNIQUES. *Educ:* Polytech Inst, BS, 68, BS, 71, MS, 73. *Prof Exp:* Assoc engr, Submarine Signal Div, Raytheon Co, 68-70; systs engr, PRD Electronics, 70-71; mathematician, Underwater Sound Lab, USN, 71; asst computer engr, George Sharp-Marine Design, 72-73; programmer, New York Dept Social Servs, 75; quant analyst, 76-80, assoc staff analyst, 80-81, COMPUTER SPECIALIST, NY FIRE DEPT, 81- *Concurrent Pos:* Pres, Algorithm Develop Co Inc, 85- *Mem:* Math Asn Am; Am Mgt Asn; Soc Indust & Appl Math; Asn Comput Mach; Inst Elec & Electronics Engrs; Am Math Soc. *Res:* Investigating new numerical techniques in both two and three dimensional numerical integration and modeling. *Mailing Add:* 64-08 60 Rd Maspeth NY 11378

HEPPE, R RICHARD, b Kansas City, Mo, Mar 4, 23; m 47; c 3. AERONAUTICAL ENGINEERING. *Educ:* Stanford Univ, BS, 44, MS, 45; Calif Inst Technol, AE, 47. *Prof Exp:* Vpres & gen mgr govt projs, Lockheed Corp, 74-79, vpres opers, 79-81, vpres & asst gen mgr adv develop projs, 81-84, vpres, 74-88, pres, Lockheed-Calif Co, 84-88; RETIRED. *Concurrent Pos:* Sr consult, Rand Corp. *Honors & Awards:* Slvanus Albert Reed Aeronaut Award, 87. *Mem:* Nat Acad Eng; fel Am Inst Aeronaut & Astronaut; Sigma Xi. *Res:* Aerodynamics; dynamics and aircraft design. *Mailing Add:* 1341 Fairlawn Way Pasadena CA 91105

HEPPEL, LEON ALMA, b Granger, Utah, Oct 20, 12; m 44; c 2. CELL BIOLOGY. *Educ:* Univ Calif, BS, 33, PhD(biochem), 37, MD, 41. *Prof Exp:* From asst surgeon to surgeon, NIH, 42-51, sr surgeon, 51-57, med dir, 57-58, chief lab biochem & metab, Nat Inst Arthritis & Metab Dis, 58-67; PROF BIOCHEM, CORNELL UNIV, 67- *Concurrent Pos:* Guggenheim fel, 53 & 75; Fogarty fel, NIH, 80-82. *Honors & Awards:* 3M Award Life Sci, Fedn Am Socs Exp Biol, 78. *Mem:* Nat Acad Sci; Am Soc Biol Chemists; Am Chem Soc. *Res:* Study of growth factors and permeable agents in cultured mammalian cells. *Mailing Add:* Dept Biochem Wing Hall Cornell Univ Ithaca NY 14850

HEPPNER, FRANK HENRY, b San Francisco, Calif, Oct 21, 40; m; c 1. ORNITHOLOGY. *Educ:* Univ Calif, Berkeley, AB, 62; San Francisco State Col, MA, 64; Univ Calif, Davis, PhD(zool), 67. *Prof Exp:* Res assoc zool, Univ Wash, 67-68, actg asst prof, 68-69; from asst prof to prof, 69-85, prof, 80-85, HONORS PROF ZOOL, UNIV RI, 85- *Concurrent Pos:* Partic vis biologists prog, Am Inst Biol Sci; consult, Am Petrol Inst, Bur Land Mgt, 73-; Lilly Found fel, 79-80, Fulbright fel, 81-82 & Univ RI honors fel, 85-; vis prof biol, Xavier Univ, 81-82; Am Assoc Col Nat Comt Biol Maj, 89-90. *Mem:* Am Soc Zool; Am Ornith Union; Cooper Ornith Soc; Wilson Ornith Soc; Ecol Soc Am; fel Explorer's Club. *Res:* Mechanisms of flocking behavior in birds; population studies of coastal birds; bird aerodynamics. *Mailing Add:* Dept Zool Univ RI Kingston RI 02881

HEPPNER, GLORIA HILL, b Great Falls, Mont, May 30, 40; div; c 1. IMMUNOLOGY. *Educ:* Univ Calif, Berkeley, BA, 62, MA, 64, PhD(bact, immunol), 67. *Prof Exp:* Asst prof biomed sci, Brown Univ, 69-75, assoc prof path, 75-79; CHMN, DEPT IMMUNOL, MICH CANCER FOUND, 79-, SR VPRES, 85- *Concurrent Pos:* Damon Runyon fel path, Univ Wash, 67-69; chmn, Nat Bladder Cancer Working Group, 84-89; bd dir, Am Asn Cancer Res, 83-86; mem, NIH Study Sect, exp therapeut, 76-82, path B, 86-88; sr vpres, Mich Cancer Found, 85-; pres, Women Cancer Res, 89-90; vpres, Int Soc Differentiation, 90- *Mem:* Am Asn Cancer Res; Am Asn Immunol; AAAS; Int Soc Differentiation. *Res:* Immune responses to tumors; in vitro tests of cellular immunity; mouse mammary tumors; tumor heterogeneity; tumor progression; breast cancer. *Mailing Add:* Mich Cancer Found 110 E Warren Ave Detroit MI 48201

HEPPNER, JAMES P, b Winona, Minn, Aug 9, 27; m 55; c 3. SPACE PHYSICS, GEOPHYSICS. *Educ:* Univ Minn, BS, 48; Calif Inst Technol, MS, 50, PhD(geophys), 54. *Prof Exp:* Res proj leader, Geophys Inst, Univ Alaska, 50-52; physicist, Naval Res Lab, 54-58, head electromagnetic fields sect, 58; head magnetic fields sect, Goddard Space Flight Ctr, NASA, Greenbelt, 59-60, asst head fields & particles br, 61-63, head fields & plasmas br, 63-69, assoc chief lab space physics, 69-74, head, Electrodynamics Br, 74-89; mgr space sci, BDM Int, 89-90; SR STAFF ADV, S T SYSTS CORP, 90- *Concurrent Pos:* Tech adv, Bilateral US-USSR Coop in Space, 62-; mem panel world magnetic surv, Geophys Res Bd, Nat Acad Sci, 62-; mem, Int Sci Radio Union; proj scientist, Orbiting Geophys Observ, 65-; discipline rep, Inter-Union Comn Solar Terrestrial Physics, 69- *Honors & Awards:* Except Sci Achievement Medal, NASA, John C Lindsay Mem Award, 83. *Mem:* Fel Am Geophys Union. *Res:* Measurement of magnetic and electric fields in space; chemical release measurements of ionospheric winds; auroral physics. *Mailing Add:* 6201 Green Valley Rd Mt Airy MD 21771

HEPPNER, JOHN BERNHARD, b Timmendorfer Strand, WGer, Nov 18, 47; US citizen; m 80; c 1. SYSTEMATIC ENTOMOLOGY. *Educ:* Univ Calif, Berkeley, BA, 70, BS, 72; Univ Fla, PhD(entom), 78. *Prof Exp:* Asst cur entom, Smithsonian Inst, 78-82; CUR LEPIDOPTERA, FLA STATE COLLECTION ARTHROPODS, 83- *Concurrent Pos:* NSF grants, 76-77, 82-85 & 87-89; exec dir, Asn Trop Lepidop, 89- *Mem:* Entom Soc Am; Lepidop Soc; Royal Entom Soc London; Soc Europaea Lepidop; Asn Trop Lepidop; Am Inst Biol Sci. *Res:* Systematics of microlepidoptera, especially Glyphipterigidae, Choreutidae, Brachodidae, Immidae, Tortricidae and families in Copromorphoidea and Yponomeutoidea. *Mailing Add:* Fla State Collection Arthropods DPI FDACS PO Box 1269 Gainesville FL 32602

HEPTING, GEORGE HENRY, forest pathology; deceased, see previous edition for last biography

HEPTINSTALL, ROBERT HODGSON, b Keswick, Eng, July 22, 20; m 50; c 6. MEDICINE, PATHOLOGY. *Educ:* Univ London, MB & BS, 44, MD, 48. *Prof Exp:* Registr bact, Wright-Fleming Inst, Eng, 47; registr path, St Mary's Hosp, 47-49 & from jr lectr to sr lectr, 49-60; vis prof, Wash Univ, 60-62; from assoc prof to prof, 62-69, BAXLEY PROF PATH & DIR DEPT, SCH MED, JOHNS HOPKINS UNIV, 69- *Concurrent Pos:* Med Res Coun Eli Lilly fel, Johns Hopkins Hosp, 54-55; pathologist, Johns Hopkins Hosp, 62-69, pathologist-in-chief, 69-; consult, Path Study Sect, NIH, 63-71; ed, Lab Invest, 76-81. *Mem:* Am Asn Path & Bact; Path Soc Gt Brit & Ireland. *Res:* Human and experimental arteriosclerosis; hypertension and renal disease in general. *Mailing Add:* Dept Path Johns Hopkins Hosp 600 N Wolfe St Baltimore MD 21205

HEPTON, ANTHONY, b Matlock, Eng, Apr 23, 36; m 60. VEGETABLE CROPS. *Educ:* Univ Nottingham, BSc, 57; Cornell Univ, PhD(veg crops), 63. *Prof Exp:* Instr bot, Cornell Univ, 62-63; plant physiologist, Dole Co Div, 63-70, tech dir, Dole Philippines, Inc, 70-75, supt res Hawaii pineapple, Castle & Cooke Foods, Inc, 75-85, DIR RES, CASTLE & COOKE PROCESSED PROD, 85- *Res:* Physiology and anatomy of pineapple; post harvest physiology of potato and temperature effects on germination; post harvest physiology of pineapple; agricultural chemicals used in pineapple culture. *Mailing Add:* 46403 Holoanai Way Kaneohe HI 96744

HEPWORTH, H(ARRY) KENT, b Phoenix, Ariz, Aug 14, 42; m 74; c 2. MECHANICAL ENGINEERING. *Educ:* Okla State Univ, BS, 64; Ariz State Univ, MS, 66, PhD(eng), 70. *Prof Exp:* Res assoc fluid mech, Ariz State Univ, 69-70; asst prof mech eng, 70-75, assoc prof mech eng, 75-81, PROF AREA COORDR MECH ENG, NORTHERN ARIZ UNIV, 76-, PROF ENG & TECHNOL, 81- *Concurrent Pos:* Consult, Aires Mfrs Co Calif, 79-84, Gen Dynamics Convair & Space Systs Div, 85- *Mem:* Am Soc Mech Engrs; assoc fel Am Inst Aeronaut & Astronaut; Am Soc Eng Educ; Sigma Xi. *Res:* Fluid mechanics with special emphasis in non-steady internal flows, thin film lubrication and vortex phenomenon; gas turbine engine operation; thermodynamics with specialty in cryogenics. *Mailing Add:* CU Box 15600 Col Eng Northern Ariz Univ Flagstaff AZ 86011-1560

HEPWORTH, MALCOLM T(HOMAS), b Singapore, Malaya, Oct 1, 32; nat US; m 57; c 5. METALLURGY. *Educ:* Mass Inst Technol, BS, 54; Purdue Univ, PhD(metall), 58. *Prof Exp:* Asst, Purdue Univ, 54-58; asst prof metall eng, Colo Sch Mines, 58-61; assoc prof, Purdue Univ, 61-68; from assoc prof to prof metall & chmn, Univ Denver, 68-75; sr res metallurgist, 75-79, SECT SUPVR, AMAX EXTRACTIVE RES & DEVELOP INC, 79- *Concurrent Pos:* Metallurgist, US Bur Mines, Nev, 68. *Mem:* Am Inst Mining, Metall & Petrol Engrs; Am Inst Chem Engrs; Sigma Xi. *Res:* Extractive metallurgy. *Mailing Add:* Dept Mineral Resources Res 122 Civil & Mineral Eng 500 Hillsbury Dr SE Minneapolis MN 55455

HERAKOVICH, CARL THOMAS, b East Chicago, Ind, Aug 6, 37; m 60; c 4. ENGINEERING MECHANICS. *Educ:* Rose-Hulman Inst Technol, BS, 59; Univ Kans, MS, 62; Ill Inst Technol, PhD(mech), 68. *Prof Exp:* Asst prof civil eng, Rose-Hulman Inst Technol, 62-64; instr mech, Ill Inst Technol, 64-67; from asst prof to assoc prof eng mech, Va Polytech Inst & State Univ, 67-76, asst head dept, 71- 72, prof eng sci & mech, 76-87; PROF APPL MECH & DIR APPL MECH PROG, UNIV VA, 87- *Mem:* Fel Am Soc Civil Engrs; Am Soc Eng Educ; Am Soc Mech Engrs; Soc Exp Stress Anal; Am Soc Testing & Mat; Soc Eng Sci; Soc Advan Mat Process Eng. *Res:* Composite materials. *Mailing Add:* Dept Civil Eng Univ Va Charlottesville VA 22903

HERALD, CHERRY LOU, b Beeville, Tex, Dec 23, 40; m 64; c 2. ORGANIC CHEMISTRY. *Educ:* Ariz State Univ, BS, 62, MS, 65, PhD(org chem), 68. *Prof Exp:* SR RES CHEMIST NATURAL PROD ISOLATION, CANCER RES INST, ARIZ STATE UNIV, 73- *Mem:* Am Chem Soc; Am Soc Pharmacog. *Res:* Isolation of potential antineoplastic agents from natural sources of plants, insects, marine animals. *Mailing Add:* 1324 W Seventh St Tempe AZ 85281

HERALD, DELBERT LEON, JR, b Fondis, Colo, Oct 9, 41; m 64; c 2. NATURAL PRODUCTS CHEMISTRY. *Educ:* Univ Colo, BA, 63; Ariz State Univ, PhD(org chem), 70. *Prof Exp:* Ga Forest Res Coun-NSF fel, 69-70, univ gen res grant, Univ Ga, 70-71; sea grant chem, Univ Okla, 71-73; ASSOC RES PROF, CANCER RES INST, ARIZ STATE UNIV, 73- *Mem:* Am Chem Soc. *Res:* Chemistry and reactions of epoxides; x-ray crystallography; isolation, synthesis and structure determination of antineoplastic drugs, particularly of plant and marine origin; computer assisted synthetic and structure analysis. *Mailing Add:* 1324 W Seventh St Tempe AZ 85281

HERB, JOHN A, b Wauseon, Ohio, Dec 1, 46. SEMICONDUCTOR PHYSICS, LOW TEMPERATURE PHYSICS. *Educ:* Miami Univ, BS, 68; Univ Wash, MS, 69, PhD(physics), 74. *Prof Exp:* Teaching asst physics, Univ Wash, 68-70; physicist, Lawrence Livermore Lab, 69; res asst, Univ Wash, 70-74; res fel physics, Calif Inst Technol, 74-77; physicist, optoelectronic div, Hewlett Packard, 77-83; engr & mgr, Mat Develop, Gould Elect, 83-; MEM STAFF, CRYSTALLUME. *Mem:* Am Inst Physics. *Res:* Kaptiza resistance and other properties of noble gases adsorbed on surfaces; effects of dimensionality and impurities on superfluid properties; device research and epitaxial growth of compound semiconductors. *Mailing Add:* Crystallume 125 Constitution Dr Menlo Park CA 94025

HERB, RAYMOND GEORGE, b Navarino, Wis, Jan 22, 08; m 45; c 5. PHYSICS. *Educ:* Univ Wis, BA, 31, PhD(physics), 35. *Hon Degrees:* Dr, Univ Basel & Univ Sao Paulo; DSc, Univ Wis, 88. *Prof Exp:* Res assoc physics, Univ Wis-Madison, 35-39, res assoc & asst prof, 39-40, from assoc prof to prof, 40-61, Charles Mendenhall prof physics, 61-72; PRES & CHMN BD, NAT ELECTROSTATICS CORP, 65- *Concurrent Pos:* Staff mem radiation lab, Univ Wis-Madison, 41-46. *Honors & Awards:* Tom W Bonner Award, 68. *Mem:* Nat Acad Sci; fel Am Phys Soc. *Res:* Design of high voltage accelerators and study of light nuclei. *Mailing Add:* Nat Electrostatics Corp Box 310 Graber Rd Middleton WI 53562

HERBACH, LEON HOWARD, b Brooklyn, NY, Jan 9, 23; m 48; c 2. STATISTICS, MATH. *Educ:* Brooklyn Col, AB, 43; Columbia Univ, AM, 47, PhD(math statist), 57. *Prof Exp:* Instr math, Brooklyn Col, 46-50 & 51-53; lectr math & statist, Hunter Col, 50; from res assoc to sr res scientist, Res Div, Sch Eng, NY Univ, 53-62, lectr & adj asst prof indust eng, 56-57, adj asst prof math, 57-58, adj assoc prof, 58-61, adj assoc prof opers res, 61, from assoc prof to prof, 62-73; admin officer, 73-74, head dept, 74-78, prof opers res & syst anal, 73-77, PROF MATH & STATIST, POLYTECH UNIV, 78- *Concurrent Pos:* Consult, Mus Natural Hist, 49-50; statist consult serv, Columbia Univ, 50-51, Radio Corp Am, 58, Am Power Jet Co, 62-, Naval Appl Sci Lab, 63-70, US Army, 76-77 & Mobil, 79-81; Inst Math Statist rep, Int Adv Comt Statist Nomenclature, Int Stand Orgn, 67-72; vis prof, Univ Grenoble, 68-69, Technion, 79 & Univ Lisbon, 80. *Mem:* Am Statist Asn; Inst Math Statist; Opers Res Soc Am; Int Statist Inst; Sigma Xi. *Res:* Reliability; probabilistic models and statistical inference as applied in engineering and physical science; analysis of variance; teaching statistics to engineers. *Mailing Add:* Dept Math & Statist Polytech Univ Brooklyn NY 11201-2990

HERBEIN, JOSEPH HENRY, JR, b Dec 16, 43; m 66; c 2. RUMINANT NUTRITION, HEPATIC METABOLISM. *Educ:* Pa State Univ, PhD(animal nutrit), 76. *Prof Exp:* ASSOC PROF ANIMAL NUTRIT, VA POLYTECH & STATE UNIV, 78- *Mem:* Am Inst Nutrit; Am Dairy Sci Asn. *Res:* Nutritional physiology. *Mailing Add:* Dept Dairy Sci Va Polytech & State Univ Blacksburg VA 24061-0315

HERBEL, CARLTON HOMER, b San Antonio, Tex, June 2, 27; m 52; c 2. RANGE SCIENCE. *Educ:* Tex Col Arts & Indust, BS, 49; Kans State Univ, MS, 54, PhD(agron), 56. *Prof Exp:* Asst agronomist, Southwest Found Res, 49-50; asst, Kans State Univ, 53-56; agronomist, Southwest Found Res, 56; res range scientist, Jornada Exp Sta, 56-72, RES LEADER JORNADA EXP RANGE & TECH ADV RANGE MGT, WESTERN REGION, SCI & EDUC ADMIN-AGR RES, USDA, 72- *Concurrent Pos:* Mem grad fac, NMex State Univ; mem range mgt task force, Great Plains Agr Coun, 74-78; mem directorate, US Man & Biosphere Prog Grazing Lands, 75-88. *Mem:* Soc Range Mgt; Soil Sci Soc Am; Am Soc Agron; Crop Sci Soc Am; Ecol Soc Am; Sigma Xi; fel AAAS. *Res:* Physiological effects of environment on range plants; microclimate; reseeding; brush control; grazing management; soil-plant-environment-animal relationships. *Mailing Add:* USDA Agr Res Serv Jornada Exp Range NMex State Univ PO Box 30003 Dept 3JER Las Cruces NM 88003-0003

HERBENER, GEORGE HENRY, b Watertown, Wis, Oct 19, 29. QUANTITATIVE ELECTRON MICROSCOPY, IMMUNOCYTOCHEMISTRY. *Educ:* Wartburg Col, BA, 60; Univ Louisville, MS, 60, PhD(anat), 68. *Prof Exp:* Res biologist path, US Army Med Res Lab, Ft Knox, Ky, 63-67; from instr to assoc prof, 68-89, PROF, SCH MED, UNIV LOUISVILLE, 89- *Mem:* Am Asn Anat; Electron Micros Soc Am; Am Soc Cell Biol. *Res:* Immunocytochemical, cytochemical and stereological studies on protein secretion; cell biology. *Mailing Add:* Dept Anat Sci & Neurobiol Health Sci Ctr Univ Louisville Louisville KY 40292

HERBENER, ROLAND EUGENE, b Wilmington, Del, Mar 5, 23; m 45; c 4. ORGANIC CHEMISTRY. *Educ:* Western Mich Univ, AB, 48; Univ Del, MS, 52; Univ NH, MS, 62. *Prof Exp:* Teacher high sch, Mich, 48-61; PROF CHEM & HEAD DEPT, HILLSDALE COL, 61- *Concurrent Pos:* Mem, NSF Res Partic & Acad Year Exten for Col Chem Teachers, Ohio, 64-65, NSF sci fac fel, 66-67. *Mem:* Am Chem Soc. *Res:* Derivatives of 1, 2, 4-Triazine, its reactions; computer applications in chemistry and chemical education. *Mailing Add:* Dept Chem Hillsdale Col Hillsdale MI 49242-1298

HERBER, JOHN FREDERICK, b Ft Wayne, Ind, Mar 27, 33; m 55; c 4. ORGANIC CHEMISTRY. *Educ:* Univ Notre Dame, BS, 55; Univ Pa, PhD(org chem), 60. *Prof Exp:* Res chemist, 60-65, res specialist, 65-66, group leader res & develop, 66-76, mgr res & develop, 76-79, technol mgr, process chem, Monsanto Flavor & Essence, Monsanto Chem Intermediates Co, 79-82, MGR RES & DEVELOP, MONSANTO INDUST CHEM CO, 83- *Mem:* Am Chem Soc. *Res:* Synthesis, process development and manufacture of chemical intermediates. *Mailing Add:* Monsanto Co 800 N Lindbergh Blvd St Louis MO 63166-0001

HERBER, LAWRENCE JUSTIN, b Ft Wayne, Ind, Sept 5, 37; m 61; c 6. MINERALOGY, ENGINEERING GEOLOGY. *Educ:* St Joseph's Col, Ind, BS, 59; NMex Inst Mining & Technol, MS, 62; Univ Nev, PhD(geol), 68. *Prof Exp:* Asst prof geol, 66-70, assoc prof physics & earth sci, 70-74, PROF GEOL, CALIF STATE POLYTECH UNIV, POMONA, 74- *Concurrent Pos:* Consult geologist. *Mem:* AAAS; Nat Asn Geol Teachers; Am Geol Inst; Inland Geol Soc. *Res:* Order and disorder in co-existing feldspars; Cucamonga fault; floatation-separation of feldspars from quartz; geology in land use and urban planning; erosion in granite monuments. *Mailing Add:* Dept Geol Sci Calif State Polytech Univ 3801 W Temple Ave Pomona CA 91768

HERBER, RAYMOND, b Sattuck, Okla, Mar 1, 32; m 54; c 2. GASTROENTEROLOGY, BIOCHEMISTRY. *Educ:* Union Col, Nebr, BA, 53; Loma Linda Univ, MD, 57; Am Bd Internal Med, dipl, 65. *Prof Exp:* Instr, Sch Med, Loma Linda Univ, 63-64; fel gastroenterol, Sch Med, Washington Univ, 64-66; asst prof, 66-72, assoc prof, 72-81, PROF INTERNAL MED, SCH MED, LOMA LINDA UNIV, 81- *Mem:* Am Fedn Clin Res; Am Asn Study Liver Dis; fel Am Col Physicians; Am Gastroenterol Asn; Sigma Xi. *Res:* Pathogenesis and genetic considerations in disaccharidase deficiency in the human intestinal mucosa; bilirubin metabolism in the isolated rat liver. *Mailing Add:* 11561 Richardson St Loma Linda CA 92354

HERBER, ROLFE H, b Ger, Mar 10, 27; nat US; m 54; c 3. CHEMICAL PHYSICS, SOLID STATE PHYSICS. *Educ:* Univ Calif, Los Angeles, BS, 49; Ore State Univ, PhD(phys chem), 52. *Prof Exp:* Res assoc chem, Mass Inst Technol, 52-55; asst prof, Univ Ill, 55-59; from asst prof to assoc prof, 59-77, DISTINGUISHED PROF CHEM, RUTGERS UNIV, NEW BRUNSWICK, 77. *Concurrent Pos:* NSF sr fel, 65-66; vis sr scientist, Ctr Nuclear Studies, Grenoble, France, 74; mem exec bd, Int Comt Application of Mossbauer Effect, 75-; Karl Taylor Compton prof physics, Technion, 81, 82; JSPS lectr, Japan, 78. *Mem:* Fel Am Phys Soc; Fel AAAS. *Res:* Nuclear chemistry; radiochemistry; Mossbauer effect; hot atom chemistry; FTIR low temp vibrational spectroscopy. *Mailing Add:* Dept Chem Rutgers Univ Busch Campus Piscataway NJ 08855-0939

HERBERMAN, RONALD BRUCE, b Brooklyn, NY, Dec 26, 40; m 63; c 2. CANCER CENTER ADMINISTRATION, CANCER IMMUNOLOGY. *Educ:* NY Univ, BA, 60, MD, 64. *Hon Degrees:* MD, Univ Rome, 86. *Prof Exp:* Intern & asst resident med, Mass Gen Hosp, 64-66; clin assoc immunol, USPHS, 66-68; sr investr immunol br, Nat Cancer Inst, NIH, 68-71, head cellular & tumor immunol sect, Lab Cell Biol, 71-74, chief, Lab Immunodiag, 75-81 & Biol Therapeut Br, 81-85; PROF MED & PATH, UNIV PITTSBURGH, 85-; DIR, PITTSBURGH CANCER INST, 85- *Concurrent Pos:* Sect ed, J Immunol, 74-77; assoc ed, Cancer Res, 75-, Clin Immunol & Immunopath, 78-, J Immunol Methods & Clin Immunol Ther, 80-, J Clin Immunol, 81- & J Nat Cancer Inst, 72-; actg assoc dir biol response prog, Nat Cancer Inst, NIH, 81-85, dir immunodiag contract prog, 72-76; mem, Immunol & Immunother Study Sect, Am Cancer Soc, 84-; mem, FDA rev panel diag tests, 79-83; mem, AIDS Clin Drug Develop Comt, Nat Inst Allergy & Infectious Dis, 86-; bd dirs, Am Asn Cancer Res, 87-90. *Mem:* Reticuloendothelial Soc (pres, 84); Am Soc Clin Invest; Am Asn Immunologists; Am Asn Cancer Res; Int Soc Interferon Res; fel Am Acad Microbiol; Clin Immunol Soc; Soc Biol Ther. *Res:* Cancer immunology and immunotherapy; immunodiagnostic tests for cancer; natural killer cells characterization and in vivo role in resistance to cancer and AIDS. *Mailing Add:* Pittsburgh Cancer Inst 200 Meyran Ave Pittsburgh PA 15213

HERBERT, DAMON CHARLES, b New York, NY, Apr 15, 45; m 67; c 2. ANATOMY. *Educ:* Calif State Univ, Chico, AB, 67; Univ Calif, San Francisco, PhD(anat), 73. *Prof Exp:* Instr, 73-74, asst prof, 74-78, ASSOC PROF CELLULAR & STRUCT BIOL, UNIV TEX HEALTH SCI CTR, SAN ANTONIO, 78- *Concurrent Pos:* Edith Claypole fel, Univ Calif, San Francisco, 70-72; res grants, NIH, 71-81, NSF, 82-85, March of Dimes, 86-91; assoc ed, The Anat Record, 78- *Mem:* Endocrine Soc; Am Asn Anat; Soc Study Reproduction; Histochem Soc. *Res:* Reproductive endocrinology; pituitary histophysiology. *Mailing Add:* Dept Cellular & Struct Biol Univ Tex Health Sci Ctr 7703 Floyd Curl Dr San Antonio TX 78284-7762

HERBERT, ELTON WARREN, JR, b Sykesville, Md, Nov 5, 43. ENTOMOLOGY, BIOCHEMISTRY. *Educ:* Univ Md, BS, 66, MS, 69, PhD(entomol biochem), 75. *Prof Exp:* Res entomologist, Beltsville Agr Res Ctr, 66-69; cpt med serv med entomol, US Army, 69-72; RES ENTOMOLOGIST, BELTSVILLE AGR RES CTR, 72- *Honors & Awards:* James I Hambleton Award, 81. *Mem:* Eastern Apicult Soc; Entom Soc Am; Sigma Xi. *Res:* Studying nutritional requirements of honeybees to develop substitutes for pollen; mineral and fat soluble vitamin requirements. *Mailing Add:* Bioenviron Bee Lab Bldg 476 Beltsville Agr Ctr Beltsville MD 20705

HERBERT, FLOYD LEIGH, b Orange, Calif, Apr 16, 42. ASTROPHYSICS, GEOPHYSICS. *Educ:* Calif Inst Technol, BS, 64; Univ Ariz, PhD(physics), 75. *Prof Exp:* Res assoc, Solar Syst Astrophys, 75-79, SR RES ASSOC, LUNAR & PLANETARY LAB, UNIV ARIZ, 79- *Mem:* Am Astron Soc; Am Geophys Union; AAAS. *Res:* Origin and evolution of solar system bodies; planetary thermospheres, exospheres, magnetospheres; physics of the plasma interactions with planetary atmospheres, fields and surfaces. *Mailing Add:* Lunar & Planetary Lab W Gould-Simpson Bldg Univ Ariz Tucson AZ 85721

HERBERT, JACK DURNIN, b Hammond, La, Aug 2, 40; m 68; c 2. BIOCHEMISTRY, NUTRITION. *Educ:* Southwestern Univ, Memphis, BA, 62; La State Univ, New Orleans, MS, 65, PhD(biochem), 67. *Prof Exp:* Instr pharmacol, Med Ctr, La State Univ, New Orleans, 67-68; tech dir clin chem, Cent Path Lab, Path Assocs New Orleans, 68-70; from instr to asst prof, 70-76, ASSOC PROF BIOCHEM, MED CTR, LA STATE UNIV, NEW ORLEANS, 76- *Concurrent Pos:* Chief chemist, La Community Pesticide Study, 67-68; consult, Cent Path Lab, Path Assocs New Orleans, 70-72. *Mem:* Am Inst Nutrit; Am Soc Biol Chemists; NY Acad Sci. *Res:* Intermediary metabolism of amino acids; protein nutrition; role of carbonic anhydrase in intermediary metabolism. *Mailing Add:* Dept Biochem/Molecular Biol La State Univ Med Ctr New Orleans LA 70112

HERBERT, MARC L, b Kew Gardens, NY, Sept 2, 48; m 82; c 2. SOFTWARE ENGINEERING. *Educ:* State Univ NY, BA, 69; Purdue Univ, MS, 71; Univ Pittsburgh, PhD(physics), 78. *Prof Exp:* Sr engr, Singer-Kearfott, 78-82; SR ENGR, UNISYS CORP, 82- *Concurrent Pos:* Adj asst prof, William Paterson State Col, 80-81, Hofstra Univ, 82-85. *Mem:* Am Phys Soc; Inst Elec & Electronics Engrs. *Res:* Software engineering; computer aided software engineering; software testing; automated documentation; real-time systems. *Mailing Add:* 3450 Manchester Rd Wantagh NY 11793

HERBERT, MICHAEL, b Lansford, Pa, May 29, 28. BACTERIOLOGY, BIOCHEMISTRY. *Educ:* Univ Md, BS, 53; Lehigh Univ, MS, 55, PhD(bact), 60. *Prof Exp:* Chemist, Lehigh Valley Coop Dairy, Pa, 53-55; instr bact, Univ Mass, 60-61; res assoc sanit eng, Johns Hopkins Univ, 61-63; assoc prof, 63-

67, PROF BIOL SCI, BLOOMSBURG STATE COL, 67- *Mem:* AAAS; Am Soc Microbiol; Am Pub Health Asn. *Res:* Interaction of nitrated and halogenated phenols with amino acids; effect of phenols on bacteria and their reversal by amino acids; water pollution; environmental health; aquatic microbiology; dairy bacteriology. *Mailing Add:* Dept Biol Sci Bloomsburg Univ Bloomsburg PA 17815

HERBERT, MORLEY ALLEN, b Toronto, Ont, July 7, 44; m 68; c 3. FUNCTIONAL ELECTRICAL STIMULATION. *Educ:* Univ Toronto, BSc, 66, MSc, 68, PhD(med biophys), 72. *Prof Exp:* Res assoc, Hosp Sick Children, 72-77, scientist, 77-81, from asst prof to assoc prof surg, 81-88; res assoc, Univ Toronto, 79-82, asst prof surg, 82-88, adj asst prof biomed eng, 83-88; DIR RES, HUMANA ADVAN SURG INST, DALLAS, TEX, 88- *Concurrent Pos:* Consult, dept orthod, Univ Toronto, 83-88, Ont Ctr Crippled Children, 80-88, Bur Med Devices, Govt Can, 84-85. *Mem:* Scoliosis Res Soc; NY Acad Sci; Can Med & Biol Eng Soc; Biomat Soc Can; Can Standards Asn. *Res:* Investigations of effect of electrical muscle and nerve stimulation on the body and its clinical application in areas of rehabilitation; electro-magnetic fields and growth. *Mailing Add:* Humana Advan Surg Inst 7777 Forest Lane Dallas TX 75230

HERBERT, STEPHEN AVEN, b Houston, Tex, July 1, 23; m 46; c 3. PETROLEUM CHEMISTRY, RESEARCH ADMINISTRATION. *Educ:* Pa State Univ, BS, 44, MS, 48; Purdue Univ, PhD(chem), 52. *Prof Exp:* Res assoc fuel technol, Mineral Indust Exp Sta, Pa State Univ, 44-45, 47-48; asst chem, Purdue Univ, 48-49; res chemist, Res Lab, Shell Oil Co, 52-54, technologist, Mfg Res Dept, NY, 54-57, group leader res lab, 57-60, spec engr prod appln dept, 61-64, mgr eastern region prod appl dept, 65-67, gen supt int lubricant corp, 67-69, mgr sales & tech develop lubricants, 69-70, mgr prod dept, MTM Res Lab, Ill, 70-71, mgr indust prod, Oil Prod Res & Develop Lab, Shell Develop Co, 71-74, mgr lubricants, 75-79, MGR FUELS & LUBRICANTS DEPT, OIL PROD RES & DEVELOP LAB, SHELL DEVELOP CO, 80- *Mem:* Am Chem Soc; fel Am Inst Chem; NY Acad Sci; Soc Automotive Engrs. *Res:* High pressure hydrogenation of coal; reaction of alkyl halides with silver salts free radical chemistry; polymers; lubricant research; rubber chemistry; all types of lubrication; electrochemical studies related to fuel cells and super batteries. *Mailing Add:* 3307 El Dorado Blvd Missouri City TX 77459

HERBERT, THOMAS JAMES, b Seattle, Wash, Nov 3, 42; m 72; c 2. BIOPHYSICS. *Educ:* Mass Inst Technol, BS, 64; Johns Hopkins Univ, PhD(biophys), 70. *Prof Exp:* Fel biophys, Johns Hopkins Univ, 70-71 & Brown Univ, 71-72; res assoc biophys, Cornell Univ, 72-74; asst prof, 74-79, ASSOC PROF BIOL, UNIV MIAMI, 80- *Mem:* Biophys Soc; Am Chem Soc; Am Optical Soc; Sigma Xi. *Res:* Laser light scattering from muscle proteins cytoplasmic contractile systems; electro-optical methods applied to biological systems. *Mailing Add:* 7861 SW 67th Ct Miami FL 33143

HERBERT, THORWALD, b Horba, Ger, July 13, 37; c 3. NONLINEAR PHENOMENA, COMPUTATIONAL SCIENCE. *Educ:* Univ Gottingen, Ger, dipl, 63; Univ Karlsruhe, Dr Eng, 74; Univ Stuttgart, Dr Eng, 78. *Prof Exp:* Res asst astrophys, Max-Planck Inst Physics, 58-59; res asst comput physics, AVA Gottingen, 59-63; res scientist, Max-Planck Inst Aeronomy, 63-66, Max-Planck Inst Fluid Mech, 66-67 & DFVLR-AVA Gottingen, 67-71; asst appl math, Univ Freiburg, 71-77; pvt docent aerospace eng, Univ Stuttgart, 77-80; prof eng sci, Va Polytech Inst & State Univ, 80-87; PROF MECH & AEROSPACE ENG, OHIO STATE UNIV, 87- *Concurrent Pos:* Prin investr, NSWC, White Oak, Md, 81-82, Off Naval Res, 83-85, US Army, Aberdeen Proving Ground, 83- & Off Sci Res, Bolling AFB, 84-; consult, Batelle Res, Triangle Park, NC, 82 & ICASE, Hampton, Va, 87-90; pres, Dyna Flow, Inc, 90- *Mem:* Fel Am Phys Soc; Am Inst Aeronaut & Astronaut; Am Soc Mech Eng; Sigma Xi; Soc Eng Sci. *Res:* Fluid dynamics: hydrodynamic stability, laminar-turbulent transition, wave phenomena, steady and unsteady viscous flows, stratified, rotating, compressible and real gas flows; methods: analytical modeling and numerical simulation, diagnostic tools and visual methods for analysis of computational data. *Mailing Add:* Dept Mech Eng 206 W 18th Ave Columbus OH 43210-1107

HERBERT, VICTOR, b New York, NY, Feb 22, 27; m 53, 68; c 5. MEDICINE. *Educ:* Columbia Univ, BS, 48, MD, 52, JD, 74. *Prof Exp:* Intern, Walter Reed Army Med Ctr, 52-53; jr asst resident med, Montefiore Hosp, NY, 54-55; asst hemat, Mt Sinai Hosp, 58-59; from instr med to asst prof, Harvard Med Sch, 59-64; assoc clin prof, Col Physicians & Surgeons, Columbia Univ, 64-70, clin prof med, 74-76; vchmn, Downstate Med Ctr, 76-80, prof med, State Univ NY Downstate Med Ctr, 76-84, attend physician, Univ Hosp, 76-84; chair med, Hahnemann Univ, 84-85; PROF MED, MT SINAI MED CTR, 85- *Concurrent Pos:* Sr res fel hemat, Montefiore Hosp, NY, 57-58; res fel med, Albert Einstein Col Med, Yeshiva Univ, 55-57; res assoc, Thorndike Mem Lab, Boston City Hosp, 59-64; med consult, WHO, 62-; assoc dir hemat, Mt Sinai Hosp, 64-69, attend hematologist, 65-69, attend physician, 69-70; prof, Mt Sinai Sch Med, 66-70; chief hemat & nutrit lab, Vet Admin Bronx Hosp, 70-; chief med serv, Vet Admin Brooklyn Hosp, 76-77; clin prof path, Columbia Univ, 70-76; mem, Food & Nutrit Bd, Nat Acad Sci, 79-85; Vischer prof & chmn med, Hahnemann Univ, 84-85. *Honors & Awards:* McCollum Award, Am Soc Clin Nutrit, 72; Middleton Award, US Vet Admin, 78; Herman Award, Am Soc Clin Nutrit, 86; Dupont Award, Clin Ligand Assay Soc, 89; Van Slyke Award, Am Asn Clin Chem, 90. *Mem:* Am Soc Clin Nutrit (pres, 80-81); Am Soc Clin Invest; Asn Am Physicians; Am Soc Hemat; Asn Prof Med. *Res:* Nature, diagnosis and treatment of nutritional anemias; mechanisms of absorption, transport, delivery, metabolism and excretion of nutrients; development of assays for vitamins, minerals and hormones. *Mailing Add:* 130 W Kingsbridge Rd Bronx NY 10468

HERBETTE, LEO G, b Derby, Conn, May 23, 53; m 76; c 3. DRUG-MEMBRANE INTERACTIONS, MEMBRANE STRUCTURE. *Educ:* Univ Conn, Storrs, BS, 75; Univ Pa, Philadelphia, PhD(biophys), 80. *Prof Exp:* Instr med, Univ Conn Health Ctr, 80-81, asst prof med, 81-86, asst prof biochem, 83-86, asst prof radiol, 86, DIR BIOMOLECULAR STRUCT ANALYSIS CTR, UNIV CONN HEALTH CTR, 84-, ASSOC PROF RADIOL, MED & BIOCHEM, 87-, HEAD BASIC RES, DEPT RADIOL, 87- *Concurrent Pos:* Asst scientist, Brookhaven Nat Lab, Upton, NY, 80-86, sr scientist, 86-, chmn, Adv Comt, Partic Res Team, High Flux Beam Reactor, 89-; consult to major pharmaceut co, 81-; mem, High Flux Beam Reactor Prog Adv Comt, 90- *Mem:* Biophys Soc; Int Soc Heart Res; Am Soc Biol Chemists; Protein Soc; Am Crystallog Soc. *Res:* Structure of model and native biological membranes; understanding molecular mechanisms for the way drugs and toxins interact with cell membranes. *Mailing Add:* Biomolecular Struct Analysis Ctr Univ Conn Health Ctr Farmington CT 06030

HERBICH, JOHN BRONISLAW, b Warsaw, Poland, Sept 1, 22; US citizen; m 51; c 3. CIVIL ENGINEERING, PHYSICAL OCEANOGRAPHY. *Educ:* Univ Edinburgh, BSc, 49; Univ Minn, Minneapolis, MS, 53; Pa State Univ, PhD(civil eng), 63. *Prof Exp:* Field engr, John Laing & Son, Eng, 48; res engr coastal eng, Tech Univ Delft, 49-50; intermediate engr, Aluminum Co Can Ltd, 50-53; res fel hydromech, Univ Minn, 53-57; from asst prof to prof fluid mech & hydraul eng, Lehigh Univ, 57-67; PROF COASTAL & OCEAN ENG, HYDRAUL ENG & FLUID MECH, TEX A&M UNIV, 67-, HEAD FLUID MECH DIV & DIR CTR DREDGING STUDIES, 70- *Concurrent Pos:* Consult hydraul engr, Bethlehem Steel Corp & Dragon Cement Co, 57-67 & Ellicott Mach Corp, Hale Fire Pump Co, Pekor Iron Works Inc & Pa Dept Hwys, 57-; Ford Found fel civil eng, Hydraul Lab, Pa State Univ, 63, NSF fac fel, 63; fel water resources, Utah State Univ, 66; consult ocean engr, Timewealth Corp, Chevron Oil Co & Ocean Pollution Control, Inc, 67- *Honors & Awards:* Karl Emil Hilgard Hydraul Prize, Am Soc Civil Engrs, 66. *Mem:* Am Soc Eng Educ; Am Soc Civil Engrs; Am Soc Oceanog; Int Asn Hydraul Res; World Dredging Asn; Sigma Xi. *Res:* Coastal engineering; dredging; wave mechanics; hydromechanics; physical oceanography. *Mailing Add:* Civil Eng Dept Tex A&M Univ College Station TX 77843

HERBIG, GEORGE HOWARD, b Wheeling, WVa, Jan 2, 20; m 43, 68; c 4. ASTRONOMY. *Educ:* Univ Calif, Los Angeles, AB, 43; Univ Calif, PhD(astron), 48. *Prof Exp:* Asst, 44-46, jr astronr, 48-50, from asst astronr to astronr, 50-66, asst dir, 60-63, actg dir, 70-71, PROF ASTRON, LICK OBSERV, UNIV CALIF, SANTA CRUZ, 66-; ASTRONR INST ASTRON, UNIV HAWAII, 88- *Concurrent Pos:* Nat Res Coun fel, Mt Wilson Observ, Carnegie Inst, Palomar Observ, Yerkes Observ, Chicago & McDonald Observ, Chicago-Texas, 48-49; vis prof, Yerkes Observ, Chicago, 59, Nat Univ Mex, 61 & Observ Paris, 65; NSF sr fel, 65; vis prof, Max Planck Inst Astron, 69, foreign sci mem, 71; nat lectr, Sigma Xi, 72-73. *Honors & Awards:* Warner prize, Am Astron Soc, 55; Catherine Wolfe Bruce Gold Medal, Astron Soc Pac, 80; H N Russell lectr, Am Astron Soc, 75. *Mem:* Nat Acad Sci; Am Acad Arts & Sci; corresp mem Royal Belg Soc Sci; Int Astron Union; Am Astron Soc. *Res:* Stellar and interstellar spectroscopy. *Mailing Add:* Inst Astron Univ Hawaii 2680 Woodlawn Dr Honolulu HI 96822

HERBISON, GERALD J, b Cleveland, Ohio, Aug 28, 37. EXERCISE PHYSIOLOGY, MUSCLE DISEASE. *Educ:* Loyola Univ, MD, 62. *Prof Exp:* PROD MED, THOMAS JEFFERSON COL, 76- *Mem:* Fel Am Col Physicians; fel Am Acad Phys Med & Rehab. *Mailing Add:* Dept Rehab Med Thomas Jefferson Univ Med Col 1025 Walnut St Philadelphia PA 19107

HERBLIN, WILLIAM FITTS, b Birmingham, Ala, Jan 21, 37; m 59; c 2. BIOCHEMISTRY, PHARMACOLOGY. *Educ:* Rollins Col, BS, 58; Am Univ, MS, 63; Cornell Univ, PhD(physiol), 65. *Prof Exp:* Biochemist, E I du Pont de Nemours & Co, Inc, 65-90, BIOCHEMIST, DU PONT MERCK PHARMACEUT CO, 91- *Mem:* Soc Neurosci. *Res:* Biochemistry and pharmacology of central nervous system. *Mailing Add:* Dept Biomed Bldg 400 Du Pont Merck Pharmaceut Co Wilmington DE 19898

HERBOLSHEIMER, GLENN, b East Moline, Ill, Feb 13, 12; m 40; c 2. CHEMICAL ENGINEERING. *Educ:* Univ Ill, BS, 37; Pa State Col, MS, 39, PhD(chem eng), 42. *Prof Exp:* Asst, Pa State Col, 37-43; process engr, Phillips Petrol Co, 43-49, sr develop engr, 49-76; RETIRED. *Mem:* Am Chem Soc; Am Inst Chem Engrs; Am Inst Chem. *Res:* Hydrocarbon oxidation; liquid-liquid solvent extraction of hydrocarbons; conversion of hydrocarbons to chemicals. *Mailing Add:* 1425 Shawnee Ave Bartlesville OK 74003

HERBRANDSON, HARRY FRED, b Watertown, SDak, July 25, 21; m 46; c 3. ORGANIC CHEMISTRY. *Educ:* Univ Minn, BChem, 42; Univ Ill, PhD(org chem), 45. *Prof Exp:* Asst org chem, Univ Ill, 42-44, asst antimalarial res, Comt Med Res & Off Sci Res & Develop, 44-45; res chemist, Nat Aniline Div, Allied Chem & Dye Corp, 45-46; Pittsburgh Plate Glass Co fel, Harvard Univ, 46-47; asst prof chem, Union Univ, NY, 47-49; from asst prof to prof, 49-86, EMER PROF CHEM, RENSSELAER POLYTECH INST, 86- *Concurrent Pos:* Consult, Robert G Allen Co, 53-55, IBM Corp, 62-66, J De Beer & Son, Inc, 81-82, NJ Dept Higher Educ, 82-83, Dionex Corp, 83-85; vis lectr, Albany Med Col, 69-73. *Mem:* Fel AAAS; Am Chem Soc; Sigma Xi; Royal Soc Chem; NY Acad Sci. *Res:* Physical organic chemistry & organic reaction mechanisms; heterocyclic chemistry; organic-sulfur compounds. *Mailing Add:* 214 Forts Ferry Rd Latham NY 12110

HERBST, EDWARD JOHN, b Jacksonport, Wis, Dec 14, 18; m 51; c 3. BIOCHEMISTRY. *Educ:* Univ Wis, BS, 43, MS, 44, PhD(biochem), 49. *Prof Exp:* From asst prof to assoc prof biochem, Med Sch, Univ Md, 49-62; chmn dept, Univ NH, 62-74, prof biochem, 74-; RETIRED. *Mem:* Am Soc Biol Chem; Sigma Xi. *Res:* Identification of putrescine as a growth factor and continued investigations of the role of the new growth factor in the metabolism of microorganisms and other forms of life. *Mailing Add:* 183 Reservoir Rd Perryville MD 21903

HERBST, ERIC, b Brooklyn, NY, Jan 15, 46; m 72; c 3. MOLECULAR ASTRONOMY, MOLECULAR SPECTROSCOPY. *Educ:* Univ Rochester, AB, 66; Harvard Univ, MA, 69, PhD(chem), 72. *Prof Exp:* Res assoc chem, Harvard Univ, 72-73; res assoc lab astrophysics, Joint Inst Lab Astrophys,

Univ Colo, 73-74; from asst prof to assoc prof chem, Col William & Mary, 74-80; assoc prof, 80-85, PROF PHYSICS, DUKE UNIV, 85- *Concurrent Pos:* Prin investr, NSF & NASA grants, 79-; consult, NASA, 82-; mem, Comt Planetary Biol & Chem Evolution, Nat Acad Sci, 85-; Humboldt Sr Fel, 88-89; mem, astron adv com, NSF, 89-92. *Mem:* Am Astron Soc; Am Chem Soc; Am Phys Soc; Int Astron Union; Sigma Xi. *Res:* Molecular astrophysics: the development of molecular complexity throughout the universe; molecular spectroscopy, particularly gas phase molecular ions; chemical dynamics. *Mailing Add:* Dept Physics Duke Univ Durham NC 27706

HERBST, JAN FRANCIS, b Tucson, Ariz, May 1, 47; m 82; c 3. SOLID STATE PHYSICS. *Educ:* Univ Pa, BA & MS, 68; Cornell Univ, PhD(physics), 74. *Prof Exp:* Res assoc, Nat Bur Standards, 74-76; asst physicist, Brookhaven Nat Lab, 76-77; assoc sr res physicist, 77-80, staff res physicist, 80-84, SR STAFF RES SCIENTIST SOLID STATE PHYSICS, GEN MOTORS RES LABS, 85- *Concurrent Pos:* Secy-treas, Div Condensed Matter Physics, Am Phys Soc, 85- *Honors & Awards:* John M Campbell Award, Gen Motors Res Labs, 83, Charles L McCuen Award, 87 & Kettering Award, 87. *Mem:* Am Phys Soc; Sigma Xi. *Res:* Physics of rare earth materials, especially magnetic and electronic properties; superconductivity. *Mailing Add:* Dept Physics Gen Motors Res Labs Warren MI 48090-9055

HERBST, JOHN A, RESEARCH ADMINISTRATION. *Prof Exp:* MEM STAFF, CALLAWAY CHEM CO, EXXON CORP. *Mailing Add:* Callaway Chem Co Exxon Corp, PO Box 2335 Columbus GA 31993

HERBST, JOHN J, b Cincinnati, Ohio, July 18, 35; m 61; c 2. PEDIATRICS, GASTROENTEROLOGY. *Educ:* Xavier Univ, BS, 57; St Louis Univ, MD, 61. *Prof Exp:* Intern pediat, Cincinnati Gen Hosp, 61-62, resident, 62-64; lt comdr, USPHS Indian Health Serv, Gallup, NMex, 64-66; fel, Stanford Med Ctr, 66-69; asst prof, 69-74, assoc prof, 74-79, PROF PEDIAT, MED CTR, UNIV UTAH, 79- *Concurrent Pos:* Co-chmn, Nat Digestive Dis Comn, NIH, 77-78. *Mem:* Am Soc Pediat Gastroenterol (pres, 77-78); NAm Soc Pediat Gastroenterol (pres, 78-79); Am Gastroenterol Asn; Soc Pediat Res. *Res:* Development relationship to intestine and esophagus; esophageal function in children. *Mailing Add:* Dept Pediat Utah Med Ctr Salt Lake City UT 84132

HERBST, MARK JOSEPH, US citizen. LASER-PLASMA INTERACTIONS, PLASMA INSTABILITIES. *Educ:* Mass Inst Technol, BS, 73; Univ Calif, Los Angeles, PhD(elec eng), 79. *Prof Exp:* Asst res engr, Univ Calif, Los Angeles, 79; res scientist, Mission Res Corp, 79-80; RES PHYSICIST, US NAVAL RES LAB, 80- *Concurrent Pos:* Cong fel, 85. *Mem:* Am Phys Soc; AAAS. *Res:* Laser-matter interaction and basic plasma physics relevant to laser-induced thermonuclear fusion. *Mailing Add:* 7709 Elgar St Springfield VA 22151

HERBST, NOEL MARTIN, b New York, NY, Sept 10, 37; m 65. ELECTRICAL ENGINEERING. *Educ:* Cornell Univ, BSEE, 59, MSEE, 61, PhD(elec eng), 63. *Prof Exp:* Res staff mem, comput sci dept, IBM Res Ctr, 63-82; SOFTWARE DESIGNER, TANDEM COMPUTERS, 82- *Mem:* Asn Comput Mach. *Res:* Applications of computers to non-numeric data; pattern recognition; character recognition; artificial intelligence; large programming systems. *Mailing Add:* 20 Spiros Way Menlo Park CA 94025

HERBST, RICHARD PETER, b Milwaukee, Wis, Nov 16, 40; m 65; c 1. ALGOLOGY, FRESHWATER ECOLOGY. *Educ:* Univ Wis-Milwaukee, BS, 64, MS, 66; Univ Pittsburgh, PhD(aquatic ecol), 69. *Prof Exp:* Asst bot, Univ Wis-Milwaukee, 64-66; asst prof biol, Univ Wis-Whitewater, 69-71; asst prof bot, Univ Wis-Waukesha, 71-78; MGR, ENVIRON MEASUREMENTS DIV, ENVIRON RES & TECH, INC, 78-; ENG ASSOC, EXXON CHEM CO. *Concurrent Pos:* Consult, Limnetics, Inc, 71-, assoc dir, Environ Res Div, 72-74; consult, Wis Elec Power Co, 72- & Rainbow Springs Trout Farm, Inc, 74- *Mem:* AAAS; Am Mgt Asn; Ecol Soc Am; Int Soc Petrol Indust Biologists; Am Soc Limnol & Oceanog. *Res:* Toxicology; ecology of aquatic systems; hazardous waste disposal and ground water problems; ecosystem planning and siting; wetlands. *Mailing Add:* Exxon Chem Co PO Box 400 Baytown TX 77522-0400

HERBST, ROBERT MAX, b Mt Vernon, NY, Sept 29, 04; m 38; c 4. ORGANIC CHEMISTRY. *Educ:* Cornell Univ, BChem, 26; Yale Univ, PhD(org chem), 30. *Prof Exp:* Instr biochem, Col Physicians & Surgeons, Columbia Univ, 31-39; asst prof org chem, NY Univ, 39-43; dir res, E Bilhuber, Inc, NJ, 43-47; from assoc prof to prof, 47-70, EMER PROF CHEM, MICH STATE UNIV, 70- *Mem:* Fel AAAS; Am Chem Soc; Am Soc Biol Chem; fel NY Acad Sci. *Res:* Chemistry of amino acids, peptides, tetrazole derivatives and cyclohexane derivatives; condensation reactions of amides; organic medicinal chemicals. *Mailing Add:* PO Box 2001 Southern Pines NC 28388

HERBST, ROBERT TAYLOR, b Wilmington, NC, Mar 5, 26; m 48; c 2. MATHEMATICS. *Educ:* Duke Univ, AB, 45, PhD(math), 51; NC State Col, MS, 49. *Prof Exp:* Asst math, Duke Univ, 48-49, instr, 50-51; mem sr staff, Appl Physics Lab, Johns Hopkins Univ, 53-54; chief mech br, Math Sci Div, Off Ord Res, US Dept Army, 54-57; mem tech staff, Bell Tel Labs, 57-66, dir Math Anal & Mil Apparatus Lab, 65-66, dir, Data Switching Lab, 66-68, dir, Safe Guard Tactical Design & Prog Ctr, 68-71, Tactical Syst Design Ctr, 71-73, dir tech employ & univ rels, 73-76, dir, Loop Plant Eng & Assignment Software Lab, 76-81; PROF, DEPT COMPUT SCI, UNIV NC, 81- *Mem:* Am Math Soc; Math Asn Am; Asn Comput Mach; Inst Elec & Electronics Eng. *Res:* Electrical circuit theory; partial differential equations; numerical analysis; computer programming languages. *Mailing Add:* 6236 Towles Rd Wilmington NC 28409

HERBST, WILLIAM, b Doylestown, Pa, Nov 12, 47; m 70; c 2. ASTRONOMY. *Educ:* Princeton Univ, AB, 70; Univ Toronto, MSc, 72, PhD(astron), 74. *Prof Exp:* Lectr physics, York Univ, 74-76; fel astron, Carnegie Inst Washington, 76-78; from asst prof to assoc prof astron, 78-90, PROF ASTRON, VAN VLECK OBSERV, WESLEYAN UNIV, 90- *Mem:* Am Astron Soc; Int Astron Union; Sigma Xi. *Res:* Observational stellar astronomy; star associations and clusters; star formation; galactic structure. *Mailing Add:* Astron Dept Wesleyan Univ Middletown CT 06457

HERBSTMAN, SHELDON, b New York, NY, June 15, 31; m 60; c 2. ORGANIC CHEMISTRY. *Educ:* Brooklyn Col, BS, 54, MA, 59; NY Univ, PhD(chem), 63. *Prof Exp:* Sr res chemist, Eastern Res Ctr, Stauffer Chem Co, 61-65; sr chemist, 65-68, res chemist, 68-76, res assoc, 88, SR RES CHEMIST, BEACON RES LABS, TEXACO INC, 76- *Mem:* Am Chem Soc. *Res:* Organometallic chemistry, especially Grignard reaction mechanisms; sulfur mustards and episulfides; petroleum chemistry; petroleum chemistry; chemistry of lubricant and fuel additives; investigation of petroleum process and catalyst systems; alcohol and ether chemistry and processing related to use in gasoline. *Mailing Add:* Texaco Inc PO Box 509 Beacon NY 12508

HERCULES, DAVID MICHAEL, b Somerset, Pa, Aug 10, 32; m 70; c 3. ANALYTICAL CHEMISTRY. *Educ:* Juniata Col, BS, 54; Mass Inst Technol, PhD(anal chem), 57. *Prof Exp:* Asst chem, Mass Inst Technol, 54-57; asst prof, Lehigh Univ, 57-60; assoc prof, Juniata Col, 60-63; from asst prof to assoc prof, Mass Inst Technol, 63-69; from assoc prof to prof, Univ Ga, 69-76; chmn dept, 80-89, PROF CHEM, UNIV PITTSBURGH, 76- *Concurrent Pos:* Mem, Vis Scientist Prog, NSF, 64-76; chmn, Gordon Conf Anal Chem, 66 & Gordon Conf Electron Spectros, 74; Guggenheim fel, 73-74; chmn, Div Anal Chem, Am Chem Soc, 77-78, Petrol Res Fund adv bd, 78-80; res chemist, Geol Div, US Dept Interior, 84-; mem, Appl Res Ctr, Univ Pittsburgh, 88-; adv bd, Marietta Col Chem, 88-; task force, anal chem, Am Chem Soc, 89. *Honors & Awards:* Lester W Strock Medal, Soc Appl Spectros, 81; Alexander von Humboldt Prize, 83; Fisher Nat Award Anal Chem, Am Chem Soc, 86; Benedetti-Pichler Award, Am Microchem Soc, 87. *Mem:* AAAS; Am Vacuum Soc; Am Chem Soc; Soc Appl Spectros; Mat Res Soc; Sigma Xi. *Res:* Analytical chemistry of surfaces; use of techniques like ESCA, Auger spectroscopy, ion scattering, FT infrared and Raman spectroscopy; types of investigations involve catalysts, polymers, biomedical materials, quantitative analysis, trace analysis and chemically modified surfaces; mass spectrometry of solids; use of pulsed layers and pulsed ion beams to study mass spectra of solids; emphasis on understanding volatilization/ionization mechanisms with main emphasis on mass spectrometry polymers; other types of materials include polymers, silicates, coordination compounds, carbohydrates and large biomolecules. *Mailing Add:* Dept Chem Univ Pittsburgh Pittsburgh PA 15260

HERCZEG, JOHN W, b Cleveland, Ohio, Sept 6, 42; m 77. REACTOR PHYSICS, SOLID STATE PHYSICS. *Educ:* Bowling Green State Univ, BS, 64, MA, 70; Purdue Univ, MS, 74, PhD(nuclear eng), 77. *Prof Exp:* Res assoc, 75-76, ASSOC NUCLEAR ENGR REACTOR PHYSICS, BROOKHAVEN NAT LAB, 76- *Mem:* Am Nuclear Soc. *Res:* Nuclear reactor physics and safety research in fuel management and economics of nuclear fuel cycle. *Mailing Add:* 9050 N Capital Tex Hwy No 180 Austin TX 78759

HERCZFELD, PETER ROBERT, b Budapest, Hungary, Aug 9, 36; US citizen; m 66; c 2. PHYSICS, ELECTRICAL ENGINEERING. *Educ:* Colo State Univ, BS, 61; Univ Minn, MS, 63, PhD(elec eng), 67. *Prof Exp:* Teaching asst physics, Univ Minn, 61-63, res assoc & instr elec eng, 63-67; from asst prof to assoc prof, 67-79, PROF ELEC ENG, DREXEL UNIV, 79- *Concurrent Pos:* Guest ed, joint issue, Trans on MTT & J Lightwave Technol. *Honors & Awards:* European Microwave Prize. *Mem:* Am Phys Soc; fel Inst Elec & Electronics Engrs; Int Soc Optical Eng; Int Solar Energy Soc. *Res:* Fluctuation phenomena in solids; microwave studies in photoconductors; solid state properties of photodetectors; bioelectronics. *Mailing Add:* Dept Elec Eng Drexel Univ 32nd Chestnut St Philadelphia PA 19104

HERCZOG, ANDREW, b Salgotarjan, Hungary, Dec 21, 17; m 49; c 1. PHYSICAL CHEMISTRY, MATERIALS SCIENCE. *Educ:* Milan Polytech Inst, EngD, 41; Univ Zurich, PhD(phys chem), 49. *Prof Exp:* Res asst catalysis, Inst Indust Chem, Milan Polytech Inst, 41-43; res assoc, Phys Chem Inst, Univ Zurich, 49-51; chemist, Defensor AG, Switz, 51-53; res assoc thermochem, Dept Metall, Univ Toronto, 53-55; sect head semiconductors, P R Mallory & Co, Ind, 55-57; sr chemist, Corning Glass Works, 57-61, res assoc, 61-66, mgr solid state res, 66-77, sr res assoc, 77-83; CONSULT, 83- *Concurrent Pos:* Lectr, Univ Milan, 41-42; consult, Katadyn AG, Switz, 47-53. *Honors & Awards:* E C Sullivan Award, Am Chem Soc, 83. *Mem:* Electrochem Soc; Am Chem Soc; fel Am Ceramic Soc. *Res:* Vapor phase reactions; thin film research; optical and electrical properties of glasses and glass-crystallized ferroelectrics and ferromagnetic materials; electroceramics; ionic conductors applications; catalysis-automobile emission control. *Mailing Add:* PO Box 96 Hammondsport NY 14840

HERCZYNSKI, ANDRZEJ, b Warsaw, Poland, Apr 29, 56. THERMO-MECHANICAL PHENOMENA GASES, KINETIC THEORY. *Educ:* Warsaw Univ, MS, 80; Lehigh Univ, MS, 83, PhD(physics), 87. *Prof Exp:* Teaching asst physics, Lehigh Univ, 80-85; res assoc fluid dynamics, Univ Colo, Boulder, 87-90; ASST TO EXEC DIR, AM INST PHYSICS, 90- *Concurrent Pos:* Lectr math, Lehigh Univ, 87. *Mem:* Am Phys Soc. *Res:* Mathematical modelling of continuum mechanical systems; eigenfunction expansions for elasto-dynamic problems; developing analytical models for response of compressible fluids to heating; application to thermo-mechanical processes in gases, subject to boundary and/or volumetric heating. *Mailing Add:* 65 N Woodhull Rd Huntington NY 11743

HERD, DARRELL GILBERT, b Logansport, Ind, Jan 13, 49. ACTIVE FAULTS, GEOMORPHOLOGY. *Educ:* Ind Univ, AB, 71; Univ Wash, MS, 72, PhD(geol), 74. *Prof Exp:* RES GEOLOGIST, US GEOL SURV, 74- *Concurrent Pos:* Govt consult, Geosci Br, US Nuclear Regulatory Comn, 77- & US Army CEngrs, San Francisco District, 77-78. *Mem:* AAAS. *Res:* Neotectonics of the San Francisco Bay area; study of the active faults of central and northern coastal California; Sonoran earthquake of 1887; investigation of the earthquake tectonics of northern Mexico and southern Arizona. *Mailing Add:* Nat Ctr 905 US Geol Surv 12201 Sunrise Valley Dr Reston VA 22092

HERD, G RONALD, mathematical statistics, operations research, for more information see previous edition

HERD, JAMES ALAN, b Vancouver, BC, Feb 7, 32; US citizen; m 55; c 3. PHYSIOLOGY. *Educ:* Univ BC, MD, 56. *Prof Exp:* Intern, Vancouver Gen Hosp, BC, 56-57, asst resident med, 57-58, assoc resident path, 58-59; asst resident med, Peter Bent Brigham Hosp, Boston, 62-63; from instr to assoc prof physiol, Harvard Med Sch, 63-74, assoc prof psychobiol, 74-81; PROF MED, BAYLOR COL MED, 81- *Concurrent Pos:* Life Ins Med Res Fund fel med, Harvard Med Sch, 59-60 & fel physiol, 60-61; Am Heart Asn Adv Res Fund fel, 61-62; Webster Underhill fel, 62-63; NIMH res scientist develop grant, 69-; estab investr, Am Heart Asn, 63. *Mem:* AAAS; Am Physiol Soc; Sigma Xi. *Res:* Mammalian cardiovascular physiology and pathophysiology; renal blood flow and function; physiology of behavior; preventive cardiology. *Mailing Add:* 3635 Merrick St Houston TX 77025

HERDA, HANS-HEINRICH WOLFGANG, b Berlin, Ger, July 6, 38; US citizen; m 62; c 3. ROCK FRACTURE GEOMETRY, FEVER CONCEPTS HISTORY. *Educ:* Wayne State Univ, BA, 60, MA, 66, PhD(math), 68. *Prof Exp:* Math ed, Info Process J, Cambridge Commun, Mass, 65; asst prof math, Salem State Col, 66-68 & Tufts Univ, 68-70; from asst prof to assoc prof, 70-80, prof math, Boston State Col, 80-82; PROF MATH, UNIV MASS, BOSTON, 82- *Concurrent Pos:* Vis scientist, Weizman Inst Sci, 78-79, 85-86; vis scientist, Dept Civil Eng, Mass Inst Technol, 82-87, res affil, 87- *Mem:* Am Math Soc; Math Asn Am. *Res:* Rock fracture geometry; history of concepts of fever. *Mailing Add:* Dept Math Univ Mass Harbor Campus Boston MA 02125-3393

HERDENDORF, CHARLES EDWARD, III, b Sheffield Lake, Ohio, Oct 2, 39; m 81; c 1. GEOLOGY, LIMNOLOGY. *Educ:* Ohio Univ, BS, 61, MS, 63; Ohio State Univ, PhD(zool), 70. *Prof Exp:* Geologist, Div Shore Erosion, Ohio Dept Natural Resources, 60-61 & Div Geol Surv, 61-64, sect head, 64-71; dir res, Ctr Lake Erie Area Res, 71-87, Ohio Sea Grant Prog, 78-87; assoc prof zool, 71-80, assoc prof geol, 74-80, PROF ZOOL & GEOL, OHIO STATE UNIV, 80- *Concurrent Pos:* Consult, Erie Regional Planning Comn, 67-68; Mem, Nat Coastal Zone Mgt Adv Comt, US Dept Com, 74-76; Consult, Columbus-Am Discovery Group, 88-91. *Mem:* Am Inst Prof Geol; Am Soc Limnol & Oceanog; fel Geol Soc Am; Int Asn Gt Lakes Res; Oceanog Soc. *Res:* Great Lakes research in physical limnology; water quality, shore erosion, sedimentology, paleolimnology, mineral resources and aquatic ecology; environmental impact studies on Lake Erie; physical oceanography, New Zealand estuaries; deep-sea benthic fauna; North Atlantic Ocean. *Mailing Add:* Dept Zool Ohio State Univ 1507 Cleveland Rd E Suite 421 Huron OH 44839

HERDKLOTZ, JOHN KEY, b Rockford, Ill, Apr 23, 43; m 64; c 2. CHEMISTRY. *Educ:* Rockford Col, BA, 65; Rice Univ, PhD(phys chem), 69. *Prof Exp:* Res assoc & fel phys chem, Rice Univ, 68-69; NIH fel lipid res, Baylor Col Med, 69-70; res chemist, Fibers Indust Inc, 70-74, res & develop mgr, 79-84; group leader, Eval Labs, Celanese Fibers Mkt Co, 74-90, EXEC VPRES RES MFG, HOECHST ROUSSEL PHARMACEUT INC, HOECHST CELANESE CORP, 90-; PROD MGR CELANESE FIBERS, FIBERS INDUST INC, 85- *Mem:* Am Chem Soc; Am Asn Textile Chemists & Colorists. *Res:* Crystal structure analysis of interesting organic molecules; solid state conformation of molecules of biological importance; structural characterization of man-made polymers; textile test methods. *Mailing Add:* 28 Olden Dr Flemington NJ 08822

HERDKLOTZ, RICHARD JAMES, b Rockford, Ill, Dec 26, 40; m 63; c 4. PHYSICAL CHEMISTRY. *Educ:* Bob Jones Univ, BS, 63; Univ Tenn, PhD(phys chem), 70. *Prof Exp:* Instr phys sci & chem, 63-67, prof chem, Bob Jones Univ, 70-79, chmn dept, 72-79; proj engr, Universal Serv SC, 78-84; chem engr, Arce & Hendrix Eng, 85; ENG PROJ MGR, FLUOR DANIEL ENG, 85- *Concurrent Pos:* Mech engr, Piedmont Engrs, Architects & Planers, 73-78; mem, Greenville County Coun, 85- *Mem:* Sigma Xi; Am Chem Soc. *Res:* Thermodynamics of aqueous electrolyte solutions over wide ranges of temperature and concentration; mathematical treatment of experimental data; spontaneous precipitation from sea salt solutions; thermodynamic properties of sea water. *Mailing Add:* 424 Leyswood Dr Greenville SC 29615

HERDLE, LLOYD EMERSON, b Homerville, Ohio, Oct 6, 13; m 46; c 6. CELLULOSE CHEMISTRY. *Educ:* Mt Union Col, BS, 34; Ohio State Univ, PhD(org chem), 37. *Prof Exp:* Res chemist, Procter & Gamble Co, Ohio, 37-41; cellulose res supvr, Eastman Kodak Co, 41-78; RETIRED. *Mem:* Am Chem Soc. *Res:* Cellulose; cellulose esters; vapor and sulfuric acid sorption by cellulose; cellulose crystallinity and swelling; pulping of wood; photographic paper; paper sizing; polymer solubility and plasticizers. *Mailing Add:* 315 Colebrook Dr Rochester NY 14617

HERDMAN, TERRY LEE, b LaCrosse, Kans, Dec 2, 45; m 63; c 3. APPLIED MATHEMATICS. *Educ:* Fort Hays Kans State Col, BS, 67; Univ Okla, MA, 70, PhD(math), 74. *Prof Exp:* Asst math, Univ Okla, 67-72, spec instr, 72-74; vis asst prof, 74-75, from asst prof to assoc prof, 75-88, PROF MATH, VA POLYTECH INST & STATE UNIV, 88- *Concurrent Pos:* Dir, Interdisciplinary Ctr Appl Math, Va Polytech Inst & State Univ, 87- *Mem:* Sigma Xi; Am Math Soc; Math Asn Am. *Res:* Functional differential equations; Volterra integral equations; differential equations. *Mailing Add:* Dept Math Va Polytech Inst Blacksburg VA 24060

HERDT, ROBERT WILLIAM, b New York, NY, 39. AGRICULTURAL ECONOMICS. *Educ:* Cornell Univ, BS, 61, MS, 63; Univ Minn, PhD(agr econ), 69. *Prof Exp:* Asst agr econ, Cornell Univ, 61-62; training assoc farm mgt, Ford Found, New Delhi, India, 62-64; res asst agr econ, Univ Minn, 64-66; res asst, Rockefeller Found, India, 67-68; from asst prof to assoc prof agr econ, Univ Ill, Urbana, 69-75; agr economist & head agr econ dept, Int Rice Res Inst, 78-83; sci adv, Consultative Group Int Agr Res, World Bank, 83-86; sr scientist, 86-87, DIR AGR SCI, ROCKEFELLER FOUND, 87- *Concurrent Pos:* Agr economist, Int Rice Res Inst, Los Banos, Philippines, 73-78; adj prof, Agr Econ Dept, Cornell Univ, 81- *Mem:* Am Agr Econ Asn; Am Econ Asn; Int Asn Agr Econ; Indian Soc Agr Econ. *Res:* Economics of agricultural development, with special interest in South and South East Asia and the role of technological change in development. *Mailing Add:* 226 Hunter Ave North Tarrytown NY 10591

HEREFORD, FRANK LOUCKS, JR, b Lake Charles, La, July 18, 23; m 48; c 3. PHYSICS. *Educ:* Univ Va, BA, 43, PhD(physics), 47. *Hon Degrees:* DSc, Fla Inst Technol, 74; LLD, Hampden-Sydney Col, 74. *Prof Exp:* Res assoc, Off Sci Res & Develop, 43-45; physicist, Bartol Res Found, Franklin Inst, 47-49; from assoc prof to prof physics, 49-66, dean grad sch arts & sci, 62-65, univ vpres & provost, 65-71, pres, 74-85, ROBERT C TAYLOR PROF, UNIV VA, 66- *Concurrent Pos:* Fulbright scholar, Univ Birmingham, 57-58; vis prof, Univ St Andrews, 71. *Mem:* AAAS; fel Am Phys Soc. *Res:* Luminescence of liquid helium; nuclear reactions; interactions of polarized neutrons; color vision. *Mailing Add:* Dept Physics McCormick Rd Charlottesville VA 22901

HEREMANS, JOSEPH P, b Leuven, Belg, Jan 8, 53; m 78; c 2. THERMAL CONDUCTIVITY, NARROW-GAP SEMICONDUCTORS. *Educ:* Cath Univ Louvain, Belg, MS, 75, PhD(appl physics), 78. *Prof Exp:* Researcher appl physics, Fonds Nat Res Sci, Belg, 78-83 & Cath Univ Louvain, 83-84; sr res scientist, 84-85, staff res scientist & group leader, 85-87, MGR ATOMICALLY ENG MAT & SR STAFF RES SCIENTIST, GEN MOTORS RES LABS, 87- *Concurrent Pos:* Vis scientist, Francis Bitter Nat Magnet Lab, Mass Inst Technol, 80 & Ctr Mat Sci & Eng, 81, Inst Solid State Physics, Univ Tokyo, 82 & H C Ørsted Inst, Univ Copenhagen, 83; mem, Panel Diluted Magnetic Semiconductors, Nat Res Coun, 89; vis prof, Cath Univ Louvain, 89- *Mem:* Fel Am Phys Soc; AAAS; Mat Res Soc. *Res:* Thermal and electrical transport in narrow-gap semiconductors IV-VI and III-V compounds, two dimensional systems; semimetals, graphite, intercalation compounds and group V semimetals, and superconductors; magnetic field sensors. *Mailing Add:* Physics Dept Gen Motors Res Lab 30500 Mound Rd Warren MI 48090-9055

HERENDEEN, ROBERT ALBERT, b Freeport, NY, Oct 18, 40; div; c 2. ENERGY ANALYSIS. *Educ:* Rensselaer Polytech Inst, BS, 62; Cornell Univ, PhD(physics), 70. *Prof Exp:* Fel, Oak Ridge Nat Lab, 71-72; res asst prof, Ctr Adv Computation, Univ Ill, 72-75; fel, Tech Univ Norway, 75-77; res asst prof, Energy Res Group, 77-81, asst prof, dept forestry, Univ Ill, 81-85; ASSOC PROF SCIENTIST, ILL NATURAL HIST SURV, 85- *Mem:* Int Soc Ecol Modelling; Int Soc Ecol Econ. *Res:* Energy analysis; efficiency of energy and other resource use, environmental impacts of resource use; ecological modelling. *Mailing Add:* Ill Natural Hist Surv Champaign IL 61820

HERFORTH, ROBERT S, b Wahoo, Nebr, July 13, 39; m 70. GENETICS. *Educ:* Wartburg Col, BA, 60; Univ Nebr, MS, 63, PhD(genetics), 68. *Prof Exp:* From asst prof to assoc prof, 67-84, PROF BIOL, AUGSBURG COL, 84- *Honors & Awards:* Harold Winfred Manter Prize, 68. *Mem:* AAAS; Sigma Xi. *Res:* Studies on hereditary viruses of Drosophila. *Mailing Add:* Dept Biol Augsburg Col Minneapolis MN 55454

HERGENRADER, GARY LEE, b Scottsbluff, Nebr, Jan 18, 39; m 59; c 2. LIMNOLOGY, FISH BIOLOGY. *Educ:* Univ Nebr, BS, 61; Univ Wis, MS, 63, PhD(zool), 67. *Prof Exp:* From asst prof to assoc prof zool, 67-76, vchmn dept, 71-73, chmn sect organismic biol, 73-74, interim dir sch life sci, 74-75, prof zool, 76-81, HEAD & STATE FORESTER, DEPT FORESTRY FISHERIES & WILDLIFE, UNIV NEBR-LINCOLN, 81- *Mem:* AAAS; Am Fisheries Soc; Am Soc Limnol & Oceanog; Sigma Xi; Am Inst Fishery Res Biologists; Soc Am Foresters; Am Forestry Asn. *Res:* Biology of freshwater fishes; experimental and comparative limnology. *Mailing Add:* Inst Agr Natural Resources Univ Nebr Dept Forestry Fisheries & Wildlife Lincoln NE 68588

HERGENROTHER, WILLIAM LEE, b Springfield, Ill, Oct 4, 38; m 64; c 4. ORGANIC POLYMER CHEMISTRY. *Educ:* Univ Notre Dame, BS, 60, PhD(chem), 63. *Prof Exp:* Res chemist, Firestone Tire & Rubber Co, 63-68, group leader, 68-73, res assoc, 73-79, sr res assoc, 79-88, SR RES ASSOC, BRIDGESTONE-FIRESTONE RES, 88- *Mem:* Am Chem Soc; Sigma Xi; AAAS. *Res:* Dehydration of alcohols in dimethyl sulfoxide; synthesis, characterization and evaluation of polar-nonpolar block copolymers; mechanism and characterization cure of rubbers and thermosetting resins; synthesis, stabilization and characterization of polyethylene terephthalate; synthesis of substituted polyphosphazenes; synthesis of rubber impact modifiers for use in nylon or polyethylene terephthalate. *Mailing Add:* 195 Dorchester Rd Akron OH 44313

HERGERT, HERBERT L, b Portland, Ore, Feb 20, 27; m 49; c 4. ORGANIC CHEMISTRY. *Educ:* Reed Col, BA, 48; Ore State Col, MS, 51, PhD(chem), 54. *Prof Exp:* Res chemist, Ore Forest Prod Lab, 52-54; res chemist, Rayonier, Inc, 54-60, group leader, Res Div, 60-64, sect head, 64-69, res supvr, ITT Rayonier Inc, 69-70, asst dir res, 70-72, dir res & develop, 72-74, vpres & dir res & develop, 74-80, VPRES & DIR TECH MKT, ITT RAYONIER INC, 80- *Mem:* Am Chem Soc; Am Sci Affil; fel Tech Asn Pulp & Paper Indust; Phytochem Soc NAm; AAAS; Int Org Paleobot; fel Int Acad Wood Sci. *Res:* Wood chemicals, extractives, lignin; infrared spectroscopy; paleobotany; cellulose derivatives; pulp and paper technology; textiles; plant taxonomy. *Mailing Add:* Repap Tech Inc 2650 Eisenhower Ave Box 766 Valley Forge PA 19482-0766

HERGET, CHARLES JOHN, b Los Angeles, Calif, Dec 2, 37; m 64; c 2. ELECTRICAL ENGINEERING. *Educ:* Univ Calif, Los Angeles, BS, 59, MS, 64, PhD(eng), 67. *Prof Exp:* Mem tech staff, Hughes Aircraft Co, 59-64; res asst eng, Univ Calif, Los Angeles, 64-67; group head, Hughes Aircraft Co, 67-69; from asst prof to assoc prof elec eng, Iowa State Univ, 69-78; ENGR, LAWRENCE LIVERMORE LAB, UNIV CALIF, 78- *Mem:* Inst Elec & Electronics Engrs. *Res:* Control systems theory; applied mathematics; system identification; process control. *Mailing Add:* 2714 Farnsworth Dr Livermore CA 94550

HERGET, WILLIAM F, b Wheeling, WVa, Sept 29, 31; div; c 4. GAS ANALYSIS BY FTIR SPECTROSCOPY. *Educ:* Univ Richmond, BS, 52; Vanderbilt Univ, MS, 55; Univ Tenn, PhD(physics), 62. *Prof Exp:* From res asst to res assoc physics, Univ Tenn, 60-62; sr res physicist, Rocketdyne Div, Rockwell Int, 62-70; res physicist, Environ Res Ctr, Environ Protection Agency, 70-75, head spec tech group, 75-81; proj scientist, 81-83, SR SCIENTIST, NICOLET INSTRUMENT CORP, 83- *Concurrent Pos:* Nicolet fel award, 83. *Honors & Awards:* Bronze Medal Award, Environ Protection Agency, 78. *Mem:* Optical Soc Am; Air & Waste Mgt Asn; Coblentz Soc; Soc Automotive Engrs; Soc Appl Spectros. *Res:* Pressure induced spectral line broadening and shifts; infrared and ultraviolet characteristics of rocket exhaust radiation; remote sensing of air pollutants by spectroscopic methods; applications of Fourier transform infrared spectroscopy to gas analysis. *Mailing Add:* 5146 Anton Dr 208 Madison WI 53719

HERGLOTZ, HERIBERT KARL JOSEF, b Dux, Czech, Dec 26, 19; m 42; c 4. PHYSICAL CHEMISTRY, X-RAY & POLYMER PHYSICS. *Educ:* Tech Univ Prague, MS, 44; Mining Acad, Freiberg, PhD(phys chem), 48; Vienna Tech Univ, x-ray physics, 55. *Prof Exp:* Asst phys chem, Mining Acad, Freiberg, 46-47, chief asst, 47-49; asst exp physics, Vienna Tech Univ, 49-55; physicist, Pigments Dept, E I du Pont de Nemours & Co, 56-59; sect head physics res, Chemie Linz AG, 59-61; sr res physicist, Eng Physics Lab, E I du Pont de Nemours & Co, Inc, 61-62, res assoc, 62-71, res fel, 71-84; RETIRED. *Concurrent Pos:* Asst, Charles Univ Prague, 44-55; consult, US Dept Army, 55-56; privat-dozent, Vienna Tech Univ, 55-62, extraordinary prof, 62-; trustee, Mt Cuba Astron Observ; consult, Condux. *Mem:* Am Phys Soc; Austrian Phys Soc; Sigma Xi. *Res:* Polymer physics; x-ray diffraction and spectroscopy; astronomy. *Mailing Add:* 2409 Hartley Pl Wilmington DE 19808-4258

HERIC, EUGENE LEROY, b Bellingham, Wash, Aug 30, 24; m 52; c 3. PHYSICAL CHEMISTRY. *Educ:* Univ Wash, BSChem, 48; Western Reserve Univ, MS, 50, PhD(chem), 52. *Prof Exp:* Res chemist, Shell Oil Co, 52-53; instr chem, Williams Col, 53-54; asst prof, Va Mil Inst, 54-56; asst prof, 56-60, ASSOC PROF CHEM, UNIV GA, 60-, ASST DEAN GRAD SCH, 72- *Mem:* Am Chem Soc. *Res:* Thermodynamics; thermochemistry; physical properties and heterogeneous equilibrium in binary and multicomponent systems; phase equilibrium. *Mailing Add:* 186 Catawba Ave Athens GA 30606-4304

HERIN, REGINALD AUGUSTUS, b Jackson, Miss, June 17, 24; m 46; c 1. TOXICOLOGY, PHARMACOLOGY. *Educ:* Colo State Univ, BS, 55, DVM, 58, MS, 59; Univ Minn, PhD, 66. *Prof Exp:* From instr to prof physiol, 59-68, PROF PHARMACOL, COLO STATE UNIV, 68- *Concurrent Pos:* NSF fel, Colo State Univ, 58-59; NIH spec fel, 65-66. *Mem:* AAAS; Am Vet Med Asn. *Res:* Toxicology, electroencephalography and neuropharmacology. *Mailing Add:* Dept Anat Colo State Univ Ft Collins CO 80523

HERING, ROBERT GUSTAVE, b Chicago, Ill, Feb 18, 34; m 56; c 4. HEAT TRANSFER, FLUID MECHANICS. *Educ:* Univ Ill, BS, 56; Univ Southern Calif, MS, 58; Purdue Univ, PhD(mech eng), 61. *Prof Exp:* Res asst engr, Armour Res Found, 56; mem tech staff, Hughes Res & Develop Lab, 56-59; supvr, Allison Div Gen Motors, 60-61; from asst prof to prof mech eng, Univ Ill, 61-71; prof & chmn, 71-72, actg dean, 72-73, DEAN, COL ENG, UNIV IOWA, 73- *Concurrent Pos:* Consult, Hughes Res & Develop Lab, 58-60 & Jet Propulsion Lab, 62-65; res adv, Argonne Nat Lab, 62; tech staff, Bell Tel Lab, 63; vis prof, Hokkaido Univ, Japan Soc Promotion Sci, 78; mem, Iowa Hwy Res Bd, Dept Transp, 71-; Fulbright res scholar, Univ Stutgart, WGer, 80-81. *Mem:* Am Soc Mech Eng; Am Inst Aeronaut & Astronaut; Am Soc Eng Educ. *Mailing Add:* 918 Bluffwood Dr Iowa City IA 52245-3516

HERING, THOMAS M, b Mar 22, 52. ORTHOPEDICS. *Educ:* Kent State Univ, BS, 74; Cleveland State Univ, MS, 77; Case Western Reserve Univ, PhD(path), 85. *Prof Exp:* Teaching asst, Dept Biol, Cleveland State Univ, 74-75; res asst, Dept Pediat, Case Western Reserve Univ, 75-78; instr, Dept Biochem & Orthop, Rush Univ, 85-87; postdoctoral res assoc, Dept Orthop, Univ Wash, 87-88; ASST PROF, DEPT MED, DIV RHEUMATIC DIS, CASE WESTERN RESERVE UNIV, 89-, ASST PROF, DEPT ANAT, 91- *Concurrent Pos:* USPHS postdoctoral trainee orthop, 85; postdoctoral fel, Dept Biochem & Orthop, Rush Univ, 85-87; NIH postdoctoral fel, 87-88, first independent res & support award, 91. *Mem:* AAAS; Am Soc Cell Biol; Orthop Res Soc. *Res:* Author of 17 technical publications. *Mailing Add:* Dept Med Lakeside Hosp Case Western Reserve Univ Cleveland OH 44106

HERION, JOHN CARROLL, b Salisbury, NC, Sept 5, 27; m 53; c 5. MEDICINE. *Educ:* Davidson Col, BS, 49; Harvard Univ, MD, 53. *Prof Exp:* Intern, NC Mem Hosp, 53-54 & from jr asst resident to sr asst resident med, 54-56; from instr to assoc prof, 57-70, PROF MED, SCH MED, UNIV NC, CHAPEL HILL, 70- *Concurrent Pos:* Nat Heart Inst res fel, 56-57; USPHS grant & career res develop award, 62-72; fel, Am Col Phys, 73. *Mem:* AAAS; Am Fedn Clin Res; Am Soc Hemat; Am Soc Int Med; Am Soc Law & Med; Royal Soc Med. *Res:* Physiology of blood cells, especially kinetics of leukocytes, their role in production of fever and in hemostasis; control of production and differentiation of marrow progenitor cells. *Mailing Add:* Dept Med Rm 024-030 Glaxo Bldg Univ NC Chapel Hill NC 27599-7037

HERK, LEONARD FRANK, b Milwaukee, Wis, Apr 16, 31; m 53; c 2. TECHNICAL CONTRACT PROMOTION. *Educ:* Marquette Univ, BS, 53; State Univ NY, PhD(phys chem), 62. *Prof Exp:* Res chemist, E I du Pont de Nemours & Co, 53-55; phys sci specialist, US Army, Quartermaster R&E Ctr, 55-57; res specialist, 3M Co, 61-66, photo prods lab supvr, 66-71, pavement mark lab mgr, 71-74, prog mgr carbide prod, 74-79, bus opportunities mgr, New Ventures Div, 79-82; dir, bus & indust develop inst, Saginaw Valley State Univ, 82-88; DIR, DETROIT OFF, SOUTHWEST RES INST, 88- *Mem:* Soc Automotive Engrs; Soc Plastic Engrs. *Res:* Corporate contacts and team formation for major research and development contracts with Southwest Research Institute in the midwest. *Mailing Add:* Southwest Res Inst 26100 American Dr Suite 603 Southfield MI 48034

HERKES, FRANK EDWARD, b Chicago, Ill, July 2, 39; m 66; c 3. INDUSTRIAL ORGANIC CHEMISTRY. *Educ:* DePaul Univ, BS, 62; Univ Iowa, MS, 64, PhD(org chem), 66. *Prof Exp:* NSF fel, Harvard Univ, 66-68; sr res chemist, Exp Sta, 68-80, res assoc, Laplace La, 80-88, SR RES ASSOC, EXP STA, E I DU PONT DE NEMOURS & CO, INC, WILMINGTON, DE, 88- *Mem:* Am Chem Soc; Royal Soc Chem; Sigma Xi. *Res:* Aromatic free radical chemistry; aromatic and diazonium chemistry; heterogeneous and hydrogenation catalysts. *Mailing Add:* 223 Charleston Dr Wilmington DE 19808

HERKSTROETER, WILLIAM G, b St Louis, Mo, Nov 30, 38; m 67; c 2. PHOTOCHEMISTRY. *Educ:* Wesleyan Univ, BA, 60; Calif Inst Technol, PhD(chem), 66. *Prof Exp:* NSF fel, Univ Chicago, 65-66; sr res chemist, 66-74, RES ASSOC, EASTMAN KODAK CO, 75- *Mem:* Inter-Am Photochem Soc; European Photochem Asn; Am Soc Photobiol. *Res:* Mechanisms of photochemical reactions; flash photolysis and transient absorption spectroscopy; luminescence spectroscopy. *Mailing Add:* Corp Res Labs, B-82 Eastman Kodak Co Rochester NY 14650-2109

HERLANDS, CHARLES WILLIAM, US citizen. CATEGORY THEORY. *Educ:* Stanford Univ, BS, 68; Univ Calif, San Diego, MA, 69; Univ Calif, Irvine, PhD(math), 73. *Prof Exp:* Asst prof, 75-82, ASSOC PROF MATH, STOCKTON STATE COL, 82- *Mem:* Am Math Soc; Math Asn Am. *Res:* Category theory; homological algebra; triple cohomology; maps of triples. *Mailing Add:* Div Natural Sci & Math Stockton State Col Pomona NJ 08240

HERLEY, PATRICK JAMES, b Johannesburg, SAfrica, July 14, 34; m 60. SOLID STATE CHEMISTRY. *Educ:* Rhodes Univ, SAfrica, BSc, 56, Hons, 57, MSc, 58, PhD(chem), 60; Univ London, PhD(chem) & DIC, 64. *Hon Degrees:* DSc, Univ London, 82. *Prof Exp:* Fel, Univ Notre Dame, 64; res assoc physics, 65, GUEST SCIENTIST, BROOKHAVEN NAT LAB, 65-; PROF MAT SCI, STATE UNIV NY, STONY BROOK, 78- *Concurrent Pos:* From asst prof to assoc prof mat sci, State Univ NY, Stony Brook, 71-78; sr vis fel chem dept, UK Sci & Eng Res Coun, Univ Cambridge, 87-88. *Mem:* The Chem Soc; Am Chem Soc. *Res:* Thermal and photolytic decomposition processes in pseudo-stable solids; x-ray topography of structural defects in inorganic single crystals; electron microscopy of metal hydrides. *Mailing Add:* Dept Mat Sci State Univ NY Stony Brook NY 11794-2275

HERLICZEK, SIEGFRIED H, b Breslau, Ger, June 17, 40; US citizen; m 67. ORGANIC POLYMER CHEMISTRY. *Educ:* Univ Mass, Amherst, BS, 63; Northeastern Univ, PhD(chem), 70. *Prof Exp:* Res chemist, Consumer Prod Corp, Union Carbide Corp, 63-64; asst, Northeastern Univ, 65-68; sr res scientist, 70-78, DIR, PROD TECHNOL LAB, LIBBEY-OWENS-FORD CO, 78- *Mem:* Am Chem Soc; Soc Plastic Eng; Soc Automotive Engrs. *Res:* Synthetic organic chemistry; studies on the mechanism of separation of diastereoisomeric compounds by gas-liquid chromatography; synthesis of polyurethane polymers; research and development of plastic film-glass laminates. *Mailing Add:* Libbey-Owens-Ford Co Tech Ctr 1701 E Broadway Toledo OH 43605

HERLIHY, JEREMIAH TIMOTHY, CARDIOVASCULAR, AGING. *Educ:* Univ Va, PhD(physiol), 72. *Prof Exp:* ASST PROF PHYSIOL, HEALTH & SCI CTR, UNIV TEX, SAN ANTONIO, 75- *Mailing Add:* Dept Physiol Health & Sci Ctr Univ Tex 7703 Floyd Curl Dr San Antonio TX 78284

HERLIN, MELVIN ARNOLD, b Salt Lake City, Utah, Apr 25, 23; m 43; c 2. PHYSICS. *Educ:* Univ Utah, BS, 43; Mass Inst Technol, PhD(physics), 48. *Prof Exp:* Mem staff magnetron res, Radiation Lab, Mass Inst Technol, 43-45, res assoc microwave gaseous electronics, Res Lab Electronics & Dept Physics, 45-48, mem staff, 48-49, asst prof low temperature physics, 49-55, group leader, Lincoln Lab, Mass Inst Technol, 55-63, assoc div head, 63-70 & 78-79, div head, 71-78, asst to dir, 79-87; RETIRED. *Mem:* Am Inst Aeronaut & Astronaut; Am Phys Soc. *Res:* Microwave magnetrons; microwave gaseous electronics; low temperature physics; radar; reentry physics; data systems. *Mailing Add:* 21 Stonehenge Lincoln MA 01773

HERLINGER, ALBERT WILLIAM, b Philadelphia, Pa, Nov 9, 43; m; c 2. INORGANIC CHEMISTRY. *Educ:* Hobart Col, BS, 65; Pa State Univ, PhD(chem), 70. *Prof Exp:* Res assoc, Univ Ill, Urbana, 70-71; vis asst prof chem, Bucknell Univ, 71-72; asst prof, 72-76, ASSOC PROF CHEM, LOYOLA UNIV, CHICAGO, 76- *Concurrent Pos:* Vis scholar, Northwestern Univ, 72-75. *Mem:* Am Chem Soc; Sigma Xi. *Res:* Raman spectroscopy; inorganic and organometallic chemistry; transition metal cluster compounds. *Mailing Add:* Dept Chem Loyola Univ 6525 N Sheridan Rd Chicago IL 60626

HERLYN, DOROTHEE MARIA, b Seltmans, WGer, Apr 17, 45; m 70; c 2. TUMOR IMMUNOLOGY, TUMOR BIOLOGY. *Educ:* Univ Munich, DVM, 70. *Prof Exp:* Res asst endocrinol, Inst Physiol, Freising Univ, Ger, 69-70; res asst immunol, Inst Physiol, Univ Munich, Ger, 71-76; assoc scientist immunol, Wistar Inst, 76-79, assoc scientist tumor immunol, 79-81, res assoc, 81-85, asst prof, 85-91, ASSOC PROF TUMOR IMMUNOL, WISTAR INST, PHILADELPHIA, 91- *Mem:* Am Asn Immunologists; Am Asn Cancer Res. *Res:* Use of monoclonal anti-human tumor antibodies in localization and therapy of human tumors; approaches to immunotherapy of human cancer with anti-idiotypic antibodies. *Mailing Add:* Wistar Inst Anat & Biol 36th & Spruce St Philadelphia PA 19104

HERM, RONALD RICHARD, b Louisville, Ky, Jan 18, 40; m 63; c 2. CHEMICAL PHYSICS. *Educ:* Univ Notre Dame, BS, 61; Harvard Univ, PhD(chem physics), 65. *Prof Exp:* Nat Acad Sci-Nat Res Coun & Air Force Off Sci Res fels photoionization, Univ Chicago, 65-66; asst prof chem, Univ Calif, Berkeley, 66-74; assoc prof chem, Iowa State Univ, 74-76; dept head, Iran A Getting Labs, 76-89, SYSTS DIR, ADVAN ORBITAL SYSTS OPERS, AEROSPACE CORP, 89- *Concurrent Pos:* Alfred P Sloan Found fel, 70-72. *Mem:* Am Chem Soc; Am Phys Soc. *Res:* Chemical kinetics; atomic and molecular collisions. *Mailing Add:* Aerospace Corp PO Box 92957 Los Angeles CA 90009-2957

HERMACH, FRANCIS L, b Bridgeport, Conn, Jan 8, 17; m 40; c 2. ELECTRICAL MEASUREMENTS. *Educ:* George Washington Univ, BEE, 43. *Prof Exp:* Elec engr, Nat Bur Standards, 39-63, chief elec instruments sect, 63-72; INDEPENDENT CONSULT ELEC MEASUREMENTS, 72- *Concurrent Pos:* Assoc dir, Metrol Div, Instrument Soc Am, 67-69. *Honors & Awards:* Silver Medal, US Dept Com, 54; Morris E Leeds & Centennial Awards, Inst Elec & Electronics Engrs, 76 & 84. *Mem:* Fel Inst Elec & Electronics Engrs Instrumentation & Measurement Soc (pres, 68-69); fel Instrument Soc Am; Precision Instruments Asn. *Res:* Electronic instruments and measurements fundamental standards for ac voltage and current. *Mailing Add:* 2201 Colston Dr Apt 311 Silver Spring MD 20910-2546

HERMAN, BARBARA HELEN, b New York, NY, June 4, 50. NEUROPEPTIDES, PEDIATRIC PSYCHIATRIC DISORDERS. *Educ:* State Univ NY, Binghamton, BA, 72; Bowling Green State Univ, MA, 74, PhD(biopsychol), 79. *Prof Exp:* Fel pharmacol, Addiction Res Found, Palo Alto, Calif, 78-82; assoc pharmacol, Dept Pharmacol, Emory Univ Sch Med, 82-83; chief immunocytochem, Brain Res Ctr, Children's Hosp Nat Med Ctr, 83-86, chief lab opers, 86-88; asst res prof psychiat & behav sci, 83-89, asst res prof child health & develop, 84-89, ASSOC RES PROF PEDIAT, PSYCHIAT & BEHAV SCI, SCH MED, GEORGE WASHINGTON UNIV, 89-; CHIEF, BRAIN RES CTR, CHILDREN'S HOSP NAT MED CTR, 88- *Concurrent Pos:* Mem, Strategic Planning Steering Comt, Children's Hosp Nat Med Ctr, 86, chair, Dept Psychiat, Children's Res Inst Comt, 90-; mem, Senate Comt on Res, George Washington Univ, 88-89; mem comt on autism, Off Spec Serv & State Affairs, Div Spec Educ & Pupil Personnel Serv, DC Pub Schs, Washington, 88-; consult, Calif State Develop Res Insts, 88-, Disneyland Proj on Frontiers in Med Neurosci, 89-; res training dir, Dept Psychiat, Children's Nat Med Ctr, Washington, DC, 90-, dir grand rounds, 90- *Mem:* Sigma Xi; Soc Neurosci; AAAS; NY Acad Sci; Women Neurosci. *Res:* Conduct preclinical and clinical studies designed to determine the role of neuropeptides, particularly opioid peptides, in the development of the central nervous system and the expression of the functions of the central nervous system. *Mailing Add:* Brain Res Ctr Children's Hosp 111 Michigan Ave NW Washington DC 20010

HERMAN, BENJAMIN MORRIS, b Port Chester, NY, Aug 13, 29; m 54; c 3. METEOROLOGY. *Educ:* NY Univ, BS, 51, MS, 54; Univ Ariz, PhD(meteorol), 64. *Prof Exp:* Weather officer, USAF, 51-57; res meteorologist severe storms, US Weather Bur, 57; assoc scientist physics, Res & Adv Develop Div, Avco, 57-59; res assoc, 59-64, from asst prof to assoc prof meteorol, 64-68, PROF ATMOSPHERIC SCI, INST ATMOSPHERIC PHYSICS, UNIV ARIZ, 68- *Mem:* Am Meteorol Soc; Am Geophys Union. *Res:* Radiative transfer through atmosphere; single and multiple scattering and polarization; remote sensing of atmosphere; influence of aerosols on radiative transfer. *Mailing Add:* Inst Atmospheric Physics Univ Ariz Tucson AZ 85721

HERMAN, BERTRAM, b Brooklyn, NY, Oct 24, 35; m 67; c 1. EPIDEMIOLOGY. *Educ:* Hunter Col, BA, 57; Pa State Univ, MS, 59; Univ Pittsburgh, MPH, 67, ScD(epidemiol), 69. *Prof Exp:* Res asst, Sloan-Kettering Inst Cancer Res, 64-65; res coordr, Environ Health Found, Inc, New York, 65-66; asst prof epidemiol & statist, Univ Tex Med Br Galveston, 69-71; chief epidemiologist & lectr, Dept Training, Neth Inst Prev Med, 71-73; chief sci collabr, Dept Epidemiol, Med Fac, Erasmus Univ, 73-77; CHIEF SCI COLLABR, DEPT EPIDEMIOL, MED FAC, UNIV UTRECHT, 77- *Concurrent Pos:* NIMH res grant, 70-71; proj dir, Tilburg Epidemiol Study of Stroke, 77- *Mem:* Am Pub Health Asn; Am Statist Asn; NY Acad Sci; Dutch Nat Soc Social Med; Sigma Xi. *Res:* Chronic disease; preventive medicine, medical care and health education, specifically cancer, cardiovascular and mental diseases, air and water pollution, smoking and health, ethnic effects, heavy metals, social epidemiology, education of preventive medicine personnel. *Mailing Add:* Grindweg 54 Rotterdam Holland Netherlands

HERMAN, BRIAN, CELL BIOLOGY, QUANTITATIVE FLUORESCENCE. *Educ:* Univ Conn, Farmington, PhD(cell biol & biophys), 80. *Prof Exp:* ASST PROF ANAT, SCH MED, UNIV NC, 83- *Res:* Microscopy. *Mailing Add:* Dept Anat Sch Med Univ NC 232 Swing Bldg Chapel Hill NC 27599

HERMAN, CEIL ANN, b New York, NY, Nov 12, 46; m 68. ENDOCRINOLOGY, COMPARATIVE PHYSIOLOGY. *Educ:* Carnegie-Mellon Univ, BS, 68; Duquesne Univ, MS, 69; Univ Mo, PhD(zool), 74. *Prof Exp:* Teaching asst zool, Duquesne Univ, 68-69; res asst, Univ Mo, 69-72, teach asst, 72-74, res assoc biochem, 74-76; res chemist physiology, Vet Admin Hosp, St Louis, 76-78; res assoc, Washington Univ, 78-79; from asst prof to assoc prof, 79-89, PROF, DEPT BIOL, NMEX STATE UNIV, 89- *Mem:* Am Physiol Soc; Sigma Xi; Am Fedn Clin Res; fel AAAS. *Res:* Evolution of hormonal systems in non-mammalian vertebrates, with particular emphasis on synthesis and physiological actions of prostaglandins; interactions of hormones, with adenylate cyclase in plasma membranes of amphibians. *Mailing Add:* 1440 Carol Ann Ct Las Cruces NM 88005

HERMAN, CHESTER JOSEPH, b Cincinnati, Ohio, Nov 24, 41; m 76. LABORATORY INSTRUMENTS. *Educ:* Univ Rochester, PhD(path), 70. *Prof Exp:* Chief, Quant Cytol, Nat Cancer Inst, Bethesda, Md, 75-79; assoc prof path, Cath Univ, Nijmegen, Holland, 79-83; pathologist, Lab Delft Hosps, Holland, 83-86; prof & chair path, Loyola Univ Chicago, 86-90; PROF PATH, EMORY UNIV, 91- *Concurrent Pos:* Clin assoc, NIH, Bethesda, Md, 70-72, resident path, 72-76; vis scientist, Nat Cancer Ctr, Tokyo, Japan, 78. *Honors & Awards:* Commendation Medal, USPHS, 78. *Mem:* Int Acad Pathologists; fel Col Am Pathologists; Int Acad Cytol; Am Soc Cytol; Int Soc Anal Cytol. *Res:* Application of new technology in laboratory and clinical medicine; new medical technology evaluation. *Mailing Add:* Dept Path Grady Mem Hosp Atlanta GA 30335

HERMAN, DANIEL FRANCIS, b Brooklyn, NY, Feb 19, 19; m 47; c 1. ORGANIC CHEMISTRY. *Educ:* Purdue Univ, BS, 39, MS, 40; Pa State Univ, PhD(chem), 43. *Prof Exp:* Res chemist, Grosvenor Lab, 43-44; group leader, Publicker Industs, Inc, 44-47; head, Org Res Dept, NL Industs, Inc, 47-61, asst tech dir, Hightstown Cent Res Lab, 61-69, assoc dir res, Hightstown Cent Res Lab, 69-82; CONSULT, POLYMER CHEM, 82- *Concurrent Pos:* Mem adv coun, plastics dept, Princeton Univ; Plenary lect, 5th Int Cellulose Cong. *Mem:* AAAS; Am Inst Chem; Am Chem Soc. *Res:* Locus controlled polymerization process for encapsulation of cellulose; glass fibers with polyolefins; organometallics; first synthesis of organotitanium compound; catalysts; paper technology; vehicles; polyelectrolytes; super absorbing polymers surface chemistry; gellants; paints; microviod coatings. *Mailing Add:* 39 Hemlock Circle Princeton NJ 08540

HERMAN, DAVID S, solid state physics; deceased, see previous edition for last biography

HERMAN, E(LVIN) E(UGENE), b Sigourney, Iowa, Mar 17, 21; m 45; c 1. ELECTRONICS ENGINEERING. *Educ:* Univ Iowa, BS, 42. *Prof Exp:* Engr, Naval Res Lab, Washington, DC, 42-51; engr circuits group, Nat Bur Standards, Calif, 51-53; sr staff engr, Hughes Aircraft Co, 53-55, head radar systs dept, 55-58, mgr signal processing & display lab, 58-68, asst mgr res & develop div, 63-70, tech dir radar systs group, 70-84. *Concurrent Pos:* Eng consult, 84- *Honors & Awards:* Lawrence A Hyland Patent Award, Hughes Aircraft Co, 71. *Mem:* Sigma Xi; fel Inst Elec & Electronics Engrs. *Res:* Design and development of counter-countermeasures; display; signal processing; air-to-air and high resolution radar; twenty-two patents. *Mailing Add:* 1200 Lachman Lane Pacific Palisades CA 90272

HERMAN, ELIOT MARK, b Boston, Mass, Sept 26, 52; m 80; c 2. PLANT PHYSIOLOGY. *Educ:* Univ Calif, Santa Barbara, BA, 73, MA, 75; Univ Calif, San Diego, PhD(biol), 80. *Prof Exp:* Res assoc, Calif Inst Technol, 80-81; postdoctoral assoc biochem, Univ Calif, Riverside, 81-83, asst res biochemist, 84; PLANT PHYSIOLOGIST, AGR RES SERV, USDA, BELTSVILLE, MD, 85- *Concurrent Pos:* Adj assoc prof bot, Univ Md, College Park, 90- *Mem:* AAAS; Am Soc Plant Physiologists; Am Soc Biochem & Molecular Biol; Protein Soc; Sigma Xi. *Res:* Ontogeny of seed oil and protein storage organelles using soybean and tobacco as model systems. *Mailing Add:* Plant Molecular Biol Lab USDA Agr Res Serv Bldg 006 Beltsville MD 20705

HERMAN, EUGENE ALEXANDER, b Washington, DC, Apr 25, 37; m 63; c 2. MATHEMATICS. *Educ:* Univ Chicago, BS, 59; Univ Calif, Berkeley, MA, 61, PhD(math), 64. *Prof Exp:* Instr math, Univ Calif, Berkeley, 64-65; from asst prof to assoc prof, 65-74, PROF MATH, GRINNELL COL, 74- *Concurrent Pos:* Alexander von Humboldt Found fel, Univ Hamburg, 67-68. *Mem:* Am Math Soc; Math Asn Am. *Res:* Educational computing; functional analysis. *Mailing Add:* Dept Math & Computer Sci Grinnell Col Grinnell IA 50112

HERMAN, EUGENE H, b Minneapolis, Minn, Mar 26, 36; m 58; c 4. CARDIOVASCULAR PHARMACOLOGY, CARDIOVASCULAR TOXICOLOGY. *Educ:* Univ Calif, Berkeley, AB, 59; Univ Calif, San Francisco, MS, 63, PhD(pharmacol), 65. *Prof Exp:* Res pharmacologist, Med Ctr, Univ Calif, San Francisco, 65; pharmacologist, Walter Reed Army Inst Res, 65-67; sr investr preclin pharmacol div, Cancer Chemother Dept, Microbiol Assocs, Inc, 67-73; PHARMACOLOGIST RES TESTING, DIV DRUG BIOL, DRUG DIV RES & TESTING, FOOD & DRUG ADMIN, 73- *Honors & Awards:* Accommendation Medal, Pub Health Serv, 81, Outstanding Serv Medal, 89. *Mem:* Am Soc Pharmacol & Exp Therapeut; Soc Toxicol; Soc Exp Biol & Med; Am Asn Cancer Res. *Res:* Drugs acting on the cardiovascular system; neuropharmacology of the spinal cord; adrenergic blocking agents; adverse cardiovascular effects of anticancer drugs, antidiabetic drugs and antihypertensive drugs. *Mailing Add:* HFD-472 Food & Drug Admin 200 C St SW Washington DC 20204

HERMAN, FRANK, b Brooklyn, NY, Mar 21, 26; m 53; c 3. SOLID STATE PHYSICS, THEORETICAL PHYSICS. *Educ:* Columbia Univ, BS, 45, MS, 49, PhD(physics), 53. *Prof Exp:* Instr elec eng, Cooper Union, 47-49; res physicist, David Sarnoff Res Ctr, Radio Corp Am, 49-61; sr consult scientist & head theoret chem, Lockheed Palo Alto Res Labs, 62-69; mgr, Large-Scale Sci Computations Dept, 69-72, RES PHYSICIST, IBM RES LAB, SAN JOSE, 72- *Concurrent Pos:* RCA Labs award, 50,52 & 55; sci consult, Lawrence Radiation Lab, Livermore, 65-69; consult, NSF, 71; Alexander von Humboldt sr US scientist award, 85-88; vis scientist, Max Planck Inst Solid State Res, Stuttgart, WGer, 86; Nordita guest prof, Chalmers Univ, Gothenberg, Sweden, 77, 85. *Mem:* Fel Am Phys Soc; fel Inst Elec & Electronics Engrs; AAAS; Sigma Xi. *Res:* Theoretical solid state physics, especially band structure, space groups, lattice dynamics, alloys and disordered solids; atomic structure; exchange, correlation, relativistic effects; computational physics; theoretical chemistry; surface physics; organic solids; quantum chemistry; atomic and molecular physics; theory of superconductivity magnetism. *Mailing Add:* 15 Anderson Way Menlo Park CA 94025

HERMAN, FREDERICK LOUIS, b Brooklyn, NY, Jan 8, 51. MEMBRANE SEPARATION PROCESSES, ORGANIC CHEMISTRY. *Educ:* Brooklyn Col, BS, 71; Columbia Univ, PhD(chem), 75; Lehigh Univ, MBA, 81. *Prof Exp:* Res chemist, 75-79, contract develop mgr, 79-81, GROUP LEADER, AIR PROD & CHEM, 81- *Mem:* Am Chem Soc. *Res:* Functional monomer-polymer chemistry; organofluorine chemistry; catalysis; nitration chemistry; separation science. *Mailing Add:* Air Prod & Chem Inc 7201 Hamilton Blvd Allentown PA 18195

HERMAN, GEORGE, b New York, NY, Apr 19, 22; m 46; c 2. SPEECH PATHOLOGY. *Educ:* Brooklyn Col, AB, 41; Univ Mich, MS, 43, PhD(speech), 52. *Prof Exp:* Asst prof speech, Wayne Univ, 46-52 & Univ Mich, 52-58; from asst prof to assoc prof speech, 58-71, vprovost, 71-75, prof speech, Bowling Green State Univ, 71-81; dir bus commun, Dacor Comput Systs, 81-83; col prof spec ed, NMex State Univ, 85-88; RETIRED. *Mem:* Acoust Soc Am. *Res:* Experimental phonetics. *Mailing Add:* 2861 Buena Vida Ct Las Cruces NM 88001

HERMAN, HARVEY BRUCE, b Brooklyn, NY, Oct 15, 36; m 60; c 4. ANALYTICAL CHEMISTRY. *Educ:* Polytech Inst Brooklyn, BS, 58. Syracuse Univ, PhD(anal chem), 64. *Prof Exp:* Asst prof chem, Univ Ga, 64-69; assoc prof, 69-73, actg head, 81-82, PROF CHEM, UNIV NC, GREENSBORO, 73-, HEAD, 82- *Concurrent Pos:* Assoc ed, Compute Mag. *Mem:* AAAS; Am Chem Soc; Electrochem Soc; Sigma Xi. *Res:* Electroanalytical chemistry including chronopotentiometry polarography and coulometry; ion selective electrodes; adsorption and kinetics; instrumentation and computers. *Mailing Add:* Dept Chem Univ NC Greensboro NC 27412

HERMAN, HERBERT, b Brooklyn, NY, June 15, 34; m 63; c 1. MATERIALS SCIENCE ENGINEERING, OCEAN ENGINEERING. *Educ:* De Paul Univ, BS, 56; Northwestern Univ, MS, 58, PhD(metall), 61. *Prof Exp:* Fulbright fel metall, Univ Paris, 61-62; fel, Argonne Nat Lab, 62-63; asst prof, Univ Pa, 63-68; assoc prof mat sci, 68-72, chmn dept mat sci, 74-80, PROF MAT SCI, STATE UNIV NY, STONY BROOK, 72-, PROF, MARINE SCI RES CTR, 80- *Concurrent Pos:* Ford Found prof indust, Western Elec Eng Res Ctr, 67-68; mem undersea vehicle comt, Marine Technol Soc, 68; liaison scientist, London Br, US Off Naval Res, 75-76; guest scientist physics, Brookhaven Nat Lab, 76-, consult mat div, 78-; ed-in-chief, Mat Sci & Eng J & ed, Treatise Mat Sci & Eng Series; chair, Thermal Spray Div, Am Soc Metals, 86-89. *Mem:* AAAS; fel Am Soc Metals; Am Welding Soc; Marine Technol Soc; Am Inst Mining, Metall & Petrol Engrs; fel Am Ceramic Soc; Am Soc Eng educ; Mat Res Soc. *Res:* Thermal sprayed protective coatings and free standing structural components; intermetallic compounds; metal and ceramic matrix composites; ceramics; fuel cell processing; powder metallurgy; powder processing; corrosion protection of transportation and marine structures. *Mailing Add:* Dept Mat Sci & Eng State Univ NY Stony Brook NY 11794-2275

HERMAN, IRA MARC, b New York, NY, June 9, 52. CELL MOTILITY, MITOSIS & CYTOKINESIS. *Educ:* State Univ NY, Buffalo, BA, 74; Tulane Univ, PhD(cell biol), 78. *Prof Exp:* Teaching asst biol sci, Tulane Univ, 75-78; Muscular Dystrophy Asn fel, Med Sch, Johns Hopkins Univ, 78-81, lab instr histol, 79-81; ASST PROF CELL BIOL & ANAT, SCH MED, TUFTS UNIV, 81- *Concurrent Pos:* Estab Investigatorship Award, Am Heart Asn, 83-88. *Mem:* Am Soc Cell Biol; Sigma Xi. *Res:* Molecular mechanism utilized by non-muscle cells for powering cell locomotion, division and chromosome movement. *Mailing Add:* Dept Anat & Cell Biol Sch Med 136 Harrison Ave Tufts Univ Boston MA 02111

HERMAN, IRVING PHILIP, b Brooklyn, NY, Oct 18, 51; m 77. LASER PHYSICS, CHEMICAL PHYSICS. *Educ:* Mass Inst Technol, SB, 72, PhD(physics), 77. *Prof Exp:* STAFF PHYSICIST, LAWRENCE LIVERMORE LAB, 77- *Mem:* Am Phys Soc; Am Chem Soc. *Res:* Laser-initiated chemical reactions; laser spectroscopy; deuterium isotope separation; laser development. *Mailing Add:* Dept Appl Physics Columbia Univ 202 Seeley W Mudd New York NY 10027

HERMAN, JAN ALEKSANDER, b Warszawa, Poland, Oct 4, 23; m 51. PHYSICAL CHEMISTRY. *Educ:* Cath Univ Louvain, Dr en Sc, 51. *Prof Exp:* With lab phys chem, Queen Elizabeth Med Found, Belg, 51-57; from asst prof to assoc prof, 57-66, PROF CHEM, LAVAL UNIV, 66- *Mem:* Am Chem Soc; Chem Inst Can; Faraday Soc. *Res:* Radiation and photochemistry; mass spectrometry. *Mailing Add:* Dept Chem 2214B Univ Laval Pavillon Vachon Quebec PQ G1K 7P4 Can

HERMAN, JEROME HERBERT, b Baltimore, Md, Dec 1, 34; m 62; c 3. RHEUMATOLOGY, IMMUNOLOGY. *Educ:* Univ Md, BS, 58, MD, 60. *Prof Exp:* Intern/residency, Sinai Hosp, Baltimore, 60-62; residency internal med, Univ Colo Med Ctr, Denver, 62-64; from asst prof to assoc prof, 69-77, PROF MED IMMUNOL/CONNECTIVE TISSUE DIS, UNIV CINCINNATI MED CTR, 77-; ATTEND PHYSICIAN INTERNAL MED, UNIV HOSPS, 69- *Concurrent Pos:* Fel rheumatol, Univ Colo Med Ctr, Denver, 64-66; fel immunol & rheumatol, Univ Tex Southwestern Med Sch, Dallas, 66-68; Arthritis Found fel, 70-73; consult physician, Vet Admin Hosp, Cincinnati, 69- *Mem:* Am Col Rheumatol; Am Asn Immunol; Am Fedn Clin Res; Cent Soc Clin Res; NY Acad Sci. *Res:* Biochemistry/immunology of connective tissue diseases; immunobiology of cartilage; laser modulation of connective tissue metabolism. *Mailing Add:* Med Col ML No 563 Univ Cincinnati 231 Bethesda Ave Cincinnati OH 45267

HERMAN, JOHN EDWARD, b Port Huron, Mich, June 6, 38; m 61; c 2. SOFTWARE SYSTEMS. *Educ:* Univ Mich, BSE, 60, MSE, 61, MS, 63, PhD(physics, elec eng), 66. *Prof Exp:* Asst res engr, Cooley Res Lab, Univ Mich, 62-64, res asst electron magnetic resonance, Univ, 64-66; asst prof, 66-74, ASSOC PROF COMPUT SCI, WESTERN MICH UNIV, 74- *Mem:* Asn Comput Mach; AAAS. *Res:* Use of computers in instruction; graphics. *Mailing Add:* Upjohn Co 7000 Portage Rd Kalamazoo MI 49001

HERMAN, LAWRENCE, b New York, NY, May 22, 24; m 61; c 2. CELL BIOLOGY, ELECTRON MICROSCOPY. *Educ:* NY Univ, BA, 47; Columbia Univ, MA, 48; Univ Chicago, PhD(zool), 56. *Prof Exp:* Instr biol, Olivet Col, 48-49; instr, Allegheny Col, 49-51; teaching asst, Univ Chicago, 55-56; res assoc path, Sch Med, Cornell Univ, 56-57; from asst prof to prof path, State Univ NY Downstate Med Ctr, 57-76; prof & chmn, 76-85, PROF, DEPT ANAT, NY MED COL, VALHALLA, 86- *Concurrent Pos:* NIH spec res fel, 68-69; Fulbright res scholar biophys, All-India Inst Med Sch, New Delhi, India, 75-76; Fogarty scholar, dept med biochem, Cambridge Univ Sch Med, Eng, 78. *Mem:* Am Soc Cell Biol; Am Soc Exp Path; Am Soc Zool; Electron Micros Soc Am; Biophys Soc. *Res:* Mechanisms of secretion; comparative morphology of endocrine organs and muscle. *Mailing Add:* Dept Anat NY Med Col Valhalla NY 10595

HERMAN, MARTIN, b Tel-Aviv, Israel; US citizen; m 79. COMPUTER VISION. *Educ:* Cooper Union, BS, 72; Univ Md, MS, 77, PhD(comput sci), 79. *Prof Exp:* Tech info specialist, Prog Methods, Inc, 73; data syst analyst, NASA, 74-75; RES ASSOC, CARNEGIE-MELLON UNIV, 79- *Mem:* Asn Comput Mach; Am Asn Artificial Intel. *Res:* Artificial intelligence; computer vision and knowledge acquisition; automated computer acquisition of three-dimensional geometric descriptions from two-dimensional images of objects and scenes. *Mailing Add:* Dept Comput Sci Carnegie-Mellon Univ 5000 Forbes Ave Pittsburgh PA 15213

HERMAN, MARVIN, b New York, NY, Mar 2, 27; m 51; c 4. METALLURGY. *Educ:* Drexel Inst, BS, 51; Univ Pa, MS, 53, PhD(metall), 65. *Prof Exp:* Develop engr metall, Westinghouse Elec Corp, 50-51; asst, Univ Pa, 51-55; lab mgr, Franklin Inst, 55-65; SECT CHIEF MAT SCI, ALLISON DIV, GEN MOTORS CORP, 65- *Mem:* Sigma Xi; Am Inst Mining, Metall & Petrol Engrs; Am Soc Metals. *Res:* Sintering and mechanical behavior of metals; substructure in metals; purification of reactive metals; beryllium metallurgy; composite materials; ceramics; materials selection. *Mailing Add:* 1103 Fairway Dr Washington Township Indianapolis IN 46260

HERMAN, MARY M, b Plymouth, Wis, July 26, 35; wid. PATHOLOGY, NEUROPATHOLOGY. *Educ:* Univ Wis, BS, 57, MD, 60; Am Bd Path, dipl, 67, cert neuropath, 68. *Prof Exp:* Intern, Mary Hitchcock Mem Hosp, Hanover, NH, 60-61; resident neurol, Wis Univ Hosps, 61-62; intern path, Yale Univ, 62-63, asst resident, 63-64; actg instr path, Stanford Univ, 66-67, from asst prof to assoc prof, 67-81; PROF PATH & CO DIR, DIV NEUROPATH, UNIV VA, CHARLOTTESVILLE, 81- *Concurrent Pos:* USPHS fel, Wis Univ Hosps, 61-62 & fel path, Yale Univ, 63-64; Nat Inst Neurol Dis & Blindness spec fel neuropath, 64-65 & Stanford Univ, 65-66, spec fel, 66-67; Regents, Hazel Duling & Johnson Found scholar, Univ Wis; Mary Putnam Jacobi fel, Yale Univ; Nat Inst Neurol Dis & Blindness res develop award, Stanford Univ, 67-72, Merck fac develop award, 69, Weil Award, 74; Nat Inst Neurol Dis & Stroke res grant, 69-72; vis asst prof, Albert Einstein Col Med, Bronx, NY, 71-72; co-prin investr, Nat Cancer Inst res grant, 71- & Nat Inst Neurol Dis & Stroke Training grant, 72-91; mem spec subcomt, Neurol Dis Prog Proj Rev Comt, Spinal Cord Injury Ctr, Nat Inst Neurol Dis & Stroke, 72-73, mem, Neurol Dis Prog Proj Rev Comt, 73-77; Nat Inst Alcohol Abuse & Alcoholism grant, 74-77; ad hoc mem, Path A Study Sect, NIH, 86-; consult, Lab Serv, VA Hosp, Salem, Va & Cent Va Training Ctr, Lynchburg, Va, 81-; mem med staff, Univ Va Hosps, 81- *Honors & Awards:* Weil Award, Am Asn Neuropathologists, 74. *Mem:* AAAS; Am Soc Cell Biol; Am Tissue Cult Asn; Soc Develop Biol; Am Asn Neuropathologists; Am Asn Pathologists; Int Acad Path. *Res:* Neoplastic differentiation in central nervous system tumors; aluminum neurotoxicity. *Mailing Add:* Dept Path Div Neuropath Univ Va Sch Med Charlottesville VA 22908

HERMAN, PAUL THEODORE, b Harrisburg, Pa, July 25, 39; m 57; c 1. PHYSICS. *Educ:* Muhlenberg Col, BS, 60; Lehigh Univ, MS, 62, PhD(physics), 66. *Prof Exp:* Res fel physics, Univ Md, College Park, 66-68; physicist, Lawrence Radiation Lab, 68-71, PHYSICIST, LAWRENCE LIVERMORE LAB, 71- *Concurrent Pos:* NSF int travel grant, Neth, 67. *Mem:* Am Phys Soc. *Res:* Equilibrium and nonequilibrium statistical mechanics; nuclear weapons research. *Mailing Add:* 2080 Pulsar Ave Livermore CA 94550

HERMAN, RAY MICHAEL, b Chicago, Ill, Oct 19, 49; m 74; c 1. SOLID STATE PHYSICS. *Educ:* Univ Ill, BS, 71,MS, 72, PhD(physics), 76. *Prof Exp:* Mem staff, Hughes Aircraft, 76-; MAGNAVOX, TORRANCE, CALIF. *Mem:* Inst Elec & Electronics Engrs. *Mailing Add:* Magnavox 2829 Maricopa St MS 15 Torrance CA 90503

HERMAN, RICHARD GERALD, b Springville, NY, Mar 11, 44; m; c 3. CATALYSIS, ENERGY CHEMICALS. *Educ:* State Univ NY Fredonia, BS, 66; Ohio Univ, PhD(chem), 72. *Prof Exp:* Postdoctoral, Dept Chem, Univ Lund, Sweden, 72-73 & Tex A&M Univ, Col Sta, 73-75; res scientist I, 75-82, res scientist II, 82-89, PRIN RES SCIENTIST, ZETTLEMOYER CTR SURFACE STUDIES, LEHIGH UNIV, BETHLEHEM, PA, 89- *Concurrent Pos:* Adj assoc prof, Dept Chem, Lehigh Univ, 81; ed, Catalytic Conversions Synthesis Gas & Alcohols to Chem, 84; interim dir Zettlemoyer Ctr Surface Studies, Lehigh Univ, 89; chmn, Lehigh Valley Sect, Am Chem Soc, 89. *Mem:* Am Chem Soc; Catalysis Soc NAm; Sigma Xi. *Res:* Synthesis of high octane alcohols and ethers from synthesis gas; preparation and characterization of inorganic catalysts; ion exchange; catalytic testing; selective methane oxidation; granted four patents. *Mailing Add:* Sinclair Lab No 7 Lehigh Univ Bethlehem PA 18015

HERMAN, RICHARD HOWARD, b Huntington, NY, Sept 4, 41; m 64; c 3. MATHEMATICAL PHYSICS, OPERATOR ALGEBRAS. *Educ:* Stevens Inst Technol, BS, 63; Univ Md, College Park, PhD(math), 67. *Prof Exp:* Res assoc math physics, Univ Rochester, 66-68; asst prof math, Univ Calif, Los Angeles, 68-72; from assoc prof to prof math, Pa State Univ, University Park, 78-90; PROF MATH & DEAN COMPUTER SCI, UNIV MD, 90- *Mem:* Am Math Soc. *Res:* Operator algebras. *Mailing Add:* Dean's Off 2300 Math Bldg Univ Md College Park MD 20742

HERMAN, ROBERT, b New York, NY, Aug 29, 14; m 39; c 3. CIVIL ENGINEERING. *Educ:* City Col New York, BS, 35; Princeton Univ, MS & PhD(physics), 40. *Hon Degrees:* DCEng, Univ Karlsruhe, 84. *Prof Exp:* Fel physics dept, City Col New York, 35-36; res assoc differential analyzer, Moore Sch Elec Eng, Univ Pa, 40-41; instr physics, City Col New York, 41-42; supvr chem physics group, physicist & asst to dir, Appl Physics Lab, Johns Hopkins Univ, 42-55; vis prof, Univ Md, 55-56; head theoret physics dept, Traffic Sci Dept, Res Labs, Gen Motors Corp, 59-79; L P Gilvin prof 79-88, EMER L P GLIVIN CENTENNIAL PROF CIVIL ENG, UNIV TEX, 88- *Concurrent Pos:* Assoc ed, Rev Mod Physics, 53-55 & Opers Res Soc Am, 60-74; consult physicist, Gen Motors Res Labs, 56; asst chmn, Basic Sci Group, Gen Motors, 56-59; ed, Transp Sci, 66-73; prof physics, Univ Tex,

79-; Philip McCord Morse lectureship, Opers Res Soc Am, 89-91; Nat Sci Coun Lectr, Repub China, 90. *Honors & Awards:* Lanchester Prize, Johns Hopkins Univ & Opers Res Soc Am, 59; Magellanic Premium, Am Philos Soc, 75; Georges Vanderlinden Prize, Belg Royal Acad, 75; Regents Lectr, Univ Calif, Santa Barbara, 75; George E Kimball Medal, Opers Res Soc Am, 76, Transp Sci Sect Lifetime Achievement Award, 90; John Price Wetherill Gold Medal, Franklin Inst, 80; NY Acad Sci Award in Phys & Math Sci, 81; Rebuen Sneed Mem lectr, Univ Col, London, 83. *Mem:* Nat Acad Eng; Sigma Xi; Fel Am Phys Soc; Opers Res Soc Am; fel Acad Arts & Sci. *Res:* Vibration-rotation spectra and molecular structure; infrared spectroscopy; solid state physics; astrophysics and cosmology; theory of traffic flow; high energy electron scattering; infrastructure, technology and the environment. *Mailing Add:* Dept Civil Eng Univ Tex Austin TX 78712-1076

HERMAN, ROGER MYERS, b Torrington, Conn, Dec 10, 34; m 58; c 3. ATOMIC PHYSICS, MOLECULAR PHYSICS. *Educ:* Lehigh Univ, BS, 57; Yale Univ, MS, 59, PhD(physics), 63. *Prof Exp:* Mem tech staff, TRW Space Tech Labs, 62-64; from asst prof to assoc prof, 64-68, PROF THEORET PHYSICS, PA STATE UNIV, UNIVERSITY PARK, 68- *Concurrent Pos:* Consult, Phys Studies, Inc, 65-70; vis prof, Mem Univ St Johns, 73, Univ Guelph, Ont, 79, Univ Heidelberg, Ger, 80-81; US Sr Scientist Award, Alexander von Humboldt Found, Ger, 80-81; proj specialist, Chinese Ministry Educ, 85; distinguished vis prof, Nankai Univ, Peoples Repub China, 85; vis scholar, Univ Toronto, 87; vis scientist, Joint Inst Lab Astrophys, Boulder, Colo, 88; consult, AMPAC Inc, 87- *Mem:* Fel Am Phys Soc. *Res:* Theoretical aspects of atomic and molecular physics; level shifts and spin relaxation in gases; molecular collisions and coherence relaxation; intramolecular physics; nonlinear optics; atom-surface interactions; liquid crystal studies; molecular hydrogen properties; classical and quantum optics. *Mailing Add:* 104 Davey Lab Pa State Univ University Park PA 16802

HERMAN, SAMUEL SIDNEY, b Boston, Mass, Oct 29, 17; m 54; c 2. RESEARCH ADMINISTRATION. *Educ:* Harvard Univ, AB, 40; Loyola Univ, DDS, 44; Yale Univ, MPH, 48, PhD(pub health), 50. *Prof Exp:* Dent officer, US Dept Interior, 46-47; staff officer div pub health methods, USPHS, 49-50, chief dent resources planning br, 50-55, asst to chief dent off, 54-55; asst chief div med serv, Off Voc Rehab, US Dept HEW, 56-57, chief, 56-59, exec secy radiation study sect & dir Russian Sci Trans Prog, NIH, 59-61, prog coordr & head foreign grants & awards, Off Int Res, 61-63, dep asst comnr health & med affairs, & exec secy med study sect, Voc Rehab Admin, 63-64, spec asst to assoc dir, Nat Cancer Inst, 64-65, dep assoc dir grants & training, 65-67, assoc dir for extramural res, Nat Inst Environ Health Sci, 67-69, assoc dir for extramural progs, Nat Eye Inst, 69-71; assoc dean, Grad Sch, Temple Univ, 71-79, assoc vpres, Health Sci Campus, 71-84, assoc dean, Sch Med, 73-84; CONSULT, 84- *Concurrent Pos:* Co-ed, Child Develop Abstr & Bibliog, Am Soc Res & Child Develop, 52-54; lectr, Sch Med, Howard Univ, 54-58; mem bd dirs, Nat Health Coun, 57-58; comt prosthetics educ & info, Nat Acad Sci-Nat Res Coun, 58-59, consult comn human resources, 74-; consult, Comt Rehab, Am Heart Asn, 58-59; tech adv, Alexander & Margaret Stewart Trust, 66-71; consult, Nat Eye Inst, 71-73, comt nat needs biomed & behav res personnel, Nat Acad Sci-Nat Res Coun, 74-85 & Lucille P Markey Charitable Trust, 84-; mem bd dirs, Federated Med Resources, 75-84, pres, 77-80, vpres, 80-81; adj prof, Sch Dent, Temple Univ, 71-84; mem bd dirs, Philadelphia Asn Clin Trials, 86-88, chmn human res bd, 81- *Mem:* AAAS; Am Pub Health Asn; Am Inst Biol Sci; NY Acad Sci. *Res:* Science and public policy; biomedical research manpower analyses; research administration; ethics of human experimentation. *Mailing Add:* 4705 Chevy Chase Blvd Chevy Chase MD 20815-5341

HERMAN, SIDNEY SAMUEL, b Pittsfield, Mass, Sept 8, 30; m 54; c 2. BIOLOGICAL OCEANOGRAPHY. *Educ:* Georgetown Univ, BS, 53; Univ RI, MS, 58, PhD(biol oceanog), 62. *Prof Exp:* Fishery biologist aide, US Fish & Wildlife Serv, 57-58; asst biol oceanog, Narragansett Marine Lab, Univ RI, 58-62; from asst prof to assoc prof, 62-71, dir Wetlands Inst, 72-79, chmn dept biol, 80-82, PROF BIOL & BIOL OCEANOG, LEHIGH UNIV, 71- *Concurrent Pos:* NSF grant, 62-74; US Off Naval Res grant, 64-65; Nat Oceanic & Atmospheric Admin sea grants, 79, 80 & 81. *Mem:* AAAS; Am Soc Limnol & Oceanog; Crustacean Biol Soc. *Res:* Ecology of planktonic organisms and specifically the diurnal vertical migration of the animal forms; wetlands ecology. *Mailing Add:* Dept Biol Lehigh Univ Bethlehem PA 18015

HERMAN, WARREN NEVIN, b Danville, Pa, Mar 10, 48; m 70; c 2. PHYSICS. *Educ:* Bloomsburg State Col, BA, 70; Franklin & Marshall Col, MS, 72; Temple Univ, PhD(physics), 77. *Prof Exp:* ASSOC PROF PHYSICS, SPRING GARDEN COL, 76- *Mem:* Am Phys Soc. *Res:* Classical relativistic theories of interacting particles; general relativity. *Mailing Add:* 3410 Larch Rd Huntingdon Valley PA 19006

HERMAN, WILLIAM S, b Seattle, Wash, Oct 12, 31; m 62; c 3. ZOOLOGY. *Educ:* Portland State Univ, BS, 58; Northwestern Univ, MS, 60, PhD(zool), 64. *Prof Exp:* USPHS fel, 64-66; from asst prof to prof, Univ Minn, Minneapolis, 66-75, head, dept genetics & cell biol, 80-89, PROF CELL BIOL, UNIV MINN, ST PAUL, 75- *Mem:* Am Soc Zool; AAAS. *Res:* Anthropod neuroendocrinology. *Mailing Add:* Dept Genetics & Cell Biol Univ Minn St Paul MN 55108

HERMAN, ZELEK SEYMOUR, b Denver, Colo, July 10, 45. THEORETICAL CHEMISTRY & PHYSICS. *Educ:* Case Inst Technol, BS, 67; Univ Uppsala, Sweden, MS, 68, PhD(quantum chem), 75. *Prof Exp:* Res assoc, dept chem, Univ Denver, 77-78; scholar, Molecular Theory Lab, dept genetics, Med Ctr, Stanford Univ, 78-79 & Molecular Theory Lab, Rockefeller Univ, 79-80; res fel, 80-82, RES ASSOC CHEM & PHYSICS, LINUS PAULING INST, 82- *Concurrent Pos:* Fulbright Distinguished Prof, Rudjer Bošković Inst, Zagreb, Yugoslavia, 85. *Mem:* Sigma Xi; Int Soc Quantum Biol. *Res:* Inorganic chemistry; spectroscopy; mathematical analysis; analytical chemistry; computer science. *Mailing Add:* Linus Pauling Inst Sci & Med 440 Page Mill Rd Palo Alto CA 94306-9979

HERMANCE, C(LARKE) E(DSON), b Windsor, Ont, Oct 6, 36; US citizen; m 63; c 3. AERONAUTICAL ENGINEERING. *Educ:* Yale Univ, BEng, 58; Princeton Univ, PhD(aeronaut eng), 63. *Prof Exp:* Res assoc aeronaut eng, Princeton Univ, 63-64; res scientist, Aeronaut Res Inst Sweden, 64-65; from asst prof to prof mech eng, Univ Waterloo, 65-82; prof mech eng & chmn civil & mech eng dept, 82-91, PROF MECH ENG, UNIV VT, 91- *Concurrent Pos:* Vis prof, Royal Inst Technol, Sweden, 71-72. *Mem:* Am Inst Aeronaut & Astronaut; Sigma Xi; Am Soc Mech Engrs. *Res:* Combustion and propulsions; ignition processes; heat transfer; gaseous fire hazards; mathematical modeling; aerothermochemistry; welding heat transfer; pyrotechnics. *Mailing Add:* CEME Dept Univ Vt Burlington VT 05405-0516

HERMANCE, JOHN FRANCIS, b Kingston, NY, Jan 9, 39; m 83; c 2. ELECTROMAGNETISM. *Educ:* State Univ NY Col New Paltz, BSc, 61; Syracuse Univ, MSc, 64; Univ Toronto, PhD(physics), 67. *Prof Exp:* Res assoc geophys, Mass Inst Technol, 67-68; from asst prof to assoc prof, 68-82, PROF GEOL SCI, BROWN UNIV, 82- *Concurrent Pos:* Mem sci adv comt, Magma Energy Prog, Sandia Nat Labs, 74-81, 85-; sr vis res assoc, Lamont-Doherty Geol Observ, 75-76; mem thermal regimes panel, Continental Sci Drilling Comt, Nat Acad Sci, 80-85, chmn, 85-86; exec comt & bd dir, Deep Observation and Sample Earth's Crust; assoc ed, Tectonophysics, 87-; vis fac fel, Res Div, Phillips Petrol Co, 74; chmn, US Geodynamics Comt Task Group Mobilizing Nat Geomagnetic Initiative, Nat Res Coun/Nat Acad Sci, 90- *Mem:* Am Geophys Union; Soc Explor Geophys; Europ Asn Explor Geophys; Soc Terrestrial Magnetism & Elec Japan. *Res:* Physical processes in the earth's crust and upper mantle associated with its dynamical evolution; characterization of fluids in the crust; regional and global studies of the earth's thermal regime using satellite and ground-based observations; geophysical and far-field sensing experiments associated with continental scientific drilling. *Mailing Add:* Dept Geol Sci Brown Univ Providence RI 02912

HERMANN, ALLEN MAX, b New Orleans, La, July 17, 38; m 79; c 4. SUPERCONDUCTIVITY, PHOTO-VOLTAICS. *Educ:* Loyola Univ, BS, 60; Univ Notre Dame, MS, 62; Tex A&M Univ, PhD(physics), 65. *Prof Exp:* Sr scientist, Jet Propulsion Lab, Calif Inst Technol, 65-67; from asst prof to prof physics, Tulane Univ, 67-81; task mgr, Solar Energy Res Inst, 80-85; tech mgr, Jet Propulsion Lab, Calif Inst Technol, 85-86; prof physics, Univ Ark, 86-88, distinguished prof, 88-90; PROF PHYSICS, UNIV COLO, 90- *Concurrent Pos:* Consult, Meditronics Corp, 76-78, Cardiac Pacemakers Inc, 77-80, Lewis Res Ctr, NASA, 80, USAF, 80, Jet Propulsion Lab, 87, Ethyl Corp, 88, Radiation Monitoring Devices, 89-90, Superconducting Core Technologies Inc, 90. *Honors & Awards:* Distinguished Scientist Award, Am Asn Physics Teachers, 86. *Mem:* Fel Am Phys Soc; sr mem Inst Elec & Electronics Engrs. *Res:* High temperature superconductors; new materials synthesis; characterization; thin-films; devices; thin-film photovoltaic materials and devices. *Mailing Add:* Dept Physics Univ Colo Campus Box 390 Boulder CO 80309-0390

HERMANN, EDWARD ROBERT, b Newport, Ky, Oct 9, 20; m 46; c 7. ENVIRONMENTAL ENGINEERING, HEALTH SCIENCES. *Educ:* Univ Ky, BS, 42, CE, 53; Mass Inst Technol, SM, 49; Univ Tex, PhD(eng/health sci), 57; Am Acad Environ Eng, dipl, 57; Am Acad Indust Hyg, dipl, 62. *Prof Exp:* Jr engr, Tenn Valley Authority, 42; instr eng & physics, Univ Ky, 43-44; partner, Hermann Consult Engrs, 46-47; teaching asst, Mass Inst Technol, 47-48; asst sanit engr, US Atomic Energy Comn, Los Alamos, NMex, 49, pub health engr, 50-51, chief Health & Sanit Sect, 51-52, sanit engr, 52-53, chief sanit engr, 53-54; res engr, Univ Tex, 54-57; indust health engr, Humble Oil & Refining Co, 57-62; prof & assoc prof, Northwestern Univ, 63-75; prof occup & environ med, Med Ctr, 75-80, prof environ & occup health sci, 81-88, dir, Indust Hygiene Progs, 76-88, EMER PROF, SCH PUB HEALTH, UNIV ILL, CHICAGO, 88-; CONSULT, HERMANN ASSOCS, 63- *Concurrent Pos:* Pres, Bayshore Munic Utility Dist, 61-62; pvt pract consult engr, 63-, pres sci adv comt, Energy Study Group, 63-64 & Space Sci Bd, Panel Waste in Space, Nat Acad Sci, 65-67, Panel Spacecraft Solid & Liquid Wastes, 69-70. *Honors & Awards:* Harrison Prescott Eddy Medal, Water Pollution Control Fedn, 59; Resources Div Award, Am Water Works Asn, 60; Radebaugh Award, Cent States Water Pollution Control Asn, 76; Borden Found Award, Am Indust Hyg Asn, 88. *Mem:* Fel AAAS; Am Indust Hyg Asn; life mem Water Pollution Control Fedn; fel Am Soc Civil Engrs; fel Am Pub Health Asn; life mem Am Acad Environ Eng. *Res:* Establishment of thresholds, of health effects, development of scientific approaches to environmental standard setting; effects of noise on humans; measurement, evaluation and control of substances and energy releases of health significance; toxicity of cyanides sulfides and heavy metals; heat stress. *Mailing Add:* 117 Church Rd Winnetka IL 60093

HERMANN, HENRY REMLEY, JR, b New Orleans, La, Sept 6, 35; m 70. INSECT MORPHOLOGY & BEHAVIOR. *Educ:* La State Univ, New Orleans, BS, 63; La State Univ, Baton Rouge, MS, 65, PhD(entom), 67. *Prof Exp:* Res assoc entom, La State Univ, Baton Rouge, 67; asst prof, 67-71, assoc prof, 71-77, MEM STAFF ENTOM, UNIV GA, 77- *Concurrent Pos:* Ed, Social Insects, Vols I-IV. *Mem:* Entom Soc Am; Royal Entom Soc London. *Res:* Insect morphology, evolution and behavior, primarily involving the hymenopterous venom apparatus, defensive behavior and sociality. *Mailing Add:* Dept Entom Univ Ga Athens GA 30601

HERMANN, HOWARD T, b Newark, NJ, July 20, 25; m; c 5. INTERHEMISPHERIC COORDINATION, VISUAL PHYSIOLOGY. *Educ:* Albany Med Col, MD, 48. *Prof Exp:* PROF PSYCHIAT, SCH MED, BOSTON UNIV, 80- *Concurrent Pos:* Res affil, dept aeronaut & astronaut, Mass Inst Technol, 82-; lectr, Harvard Univ, 86- *Mem:* Int Soc Pure & Appl Biophys; Soc Neurosci; Physiol Soc. *Mailing Add:* 7-32 Blackstone Dr Nashua NH 03060

HERMANN, JOHN ALEXANDER, b San Francisco, Calif, Nov 22, 21; m 49; c 3. APPLIED CHEMISTRY. *Educ:* NJ Inst Technol, BS, 42; Univ NMex, PhD(chem), 56. *Prof Exp:* Line supvr explosives, Hercules Powder Co, 42-43; res engr, Manhattan Proj, 44; staff mem, Los Alamos Sci Lab,

44-56; sr res engr, 56-63, chief spec proj, 63, sr res group leader chem, 64-67, asst mgr, Hobbs Potash Facil, 67-69, PROJ MGR, KERR MCGEE CORP, 69- *Mem:* Am Chem Soc; Metall Soc. *Res:* Separation processes; precipitation and crystallization; coprecipitation, solvent extraction; extractive metallurgy of uranium, vanadium, lithium, molybdenum; potash, phosphate and fertilizer chemistry. *Mailing Add:* 320 NW 21st St Oklahoma City OK 73103

HERMANN, R(UDOLF), b Leipzig, Ger, Dec 15, 04; nat US; m 30; c 4. AEROSPACE SCIENCE, ENGINEERING. *Educ:* Univ Leipzig, PhD(physics), 29; Aachen Inst Technol, Dr habil(aerodyn), 35. *Prof Exp:* Asst, Dept Appl Mech & Thermodyn, Univ Leipzig, 29-33; head supersonic wind tunnel div, Aachen Inst Technol, 34-37; dir, Supersonic Aerodyn Inst, Rocket Exp Sta, Peenemuende, 37-43; dir, Res Inst Supersonic Aerodyn & Ballistics, Kochel, Bavaria, 44-45; tech consult, Air Eng Develop Div, Wright-Patterson AFB, 45-50; prof aeronaut eng, Univ Minn, 50-62, tech dir, Hypersonic Facil, Rosemount Aeronaut Labs, 59-62; dir, Res Inst, 62-70, prof aerospace sci & eng, 62-74, EMER PROF AEROSPACE SCI & ENG, UNIV ALA, HUNTSVILLE, 74- *Concurrent Pos:* Lectr, Aachen Inst Technol, 36-37 & Berlin Inst Technol, 37-45; consult, Aero Div, Minneapolis-Honeywell Regulator Co, 54-62, Gen Mills, 56-58, Winzen Res, Inc, 57-60, Rensselaer Polytech Inst, 57-60, Crosley Div, Avco Corp, 58-59 & NSF, 62-64. *Honors & Awards:* Hermann Oberth Medal, 69. *Mem:* Fel Am Inst Aeronaut & Astronaut; Ger Soc Aeronaut & Astronaut; Sigma Xi. *Res:* Pipe friction; free convection heat transfer; viscosity of non-Newtonian fluids; fin stabilized projectiles; guided missiles; supersonic wind tunnels and diffusers; supersonic inlet diffusers; hypersonic nozzles; film cooling; re-entry of satellite vehicles; hypersonic nonequilibrium gasdynamics and re-entry physics. *Mailing Add:* 1810 Inspiration Lane SE Huntsville AL 35801

HERMANN, RICHARD KARL, b Munich, Ger, Feb 16, 24; US citizen; m 48; c 1. PLANT ECOLOGY. *Educ:* Univ Munich, BS, 51; Yale Univ, MF, 56; Ore State Univ, PhD(plant ecol), 60. *Hon Degrees:* Dr forest, Georg-August Univ, Gottingen, 79. *Prof Exp:* Instr bot, Ore State Univ, 58-59; res assoc forest ecol, Ore Forest Res Ctr, Forest Protection & Conserv Comt, 59-61; from asst prof to prof forest ecol, Ore State Univ, 61-90; RETIRED. *Concurrent Pos:* Vis prof, Georg-August Univ, Goettingen, 82 & 89; chmn, div I, Silvicult & Forest Environ, Int Union Forest Res Orgn, 82-86. *Mem:* Ecol Soc Am; Soc Am Foresters; Am Inst Biol Sci; corresp mem Ital Acad Forest Sci. *Res:* Ecology and physiology of tree roots. *Mailing Add:* Col Forestry Ore State Univ Corvallis OR 97331

HERMANN, ROBERT J, SCIENCE & TECHNOLOGY. *Educ:* Iowa State Univ, BS, MS & PhD. *Prof Exp:* Prin dep asst secy defense, Nat Security Agency, 77-79, asst secy, USAF, 79-81, spec asst intel, 81-82; vpres systs technol, 82-84, vpres advan systs, 84-87, VPRES SCI & TECHNOL, UNITED TECHNOL CORP, 87- *Concurrent Pos:* Mem, Defense Sci Bd; chmn, Naval Studies Bd; mem, Indust Adv Group, Nat Soc Prof Engrs. *Mem:* Nat Acad Eng. *Res:* Development of technical resources. *Mailing Add:* United Technol Corp One Financial Plaza Hartford CT 06101

HERMANS, COLIN OLMSTED, b Seattle, Wash, May 4, 36; m 58; c 3. INVERTEBRATE ZOOLOGY. *Educ:* Pomona Col, BA, 58; Univ Wash, MS, 64, PhD(zool), 66. *Prof Exp:* NATO fel zool, Univ Newcastle, 66-67; NIH fel, Univ Calif, Berkeley, 67-68, asst res zoologist, 68-69; from asst prof to assoc prof, 69-77, PROF BIOL, SONOMA STATE UNIV, 77- *Concurrent Pos:* Alexander von Humboldt fel, Univ Göttingen, WGer, 75-76; vis prof, Univ Osnabrück, 85, Ariz State Univ, 88-89; treas, Western Soc Naturalists, 80-81. *Mem:* Fel AAAS; Am Soc Zool; Electron Micros Soc Am; Int Asn Meiobenthologists. *Res:* Polychaetous annelids; invertebrate photoreceptors; invertebrate adhesion; electron microscopy. *Mailing Add:* Dept Biol Sonoma State Univ Rohnert Park CA 94928

HERMANS, HANS J, b The Hague, Neth, Mar 21, 38; m 62; c 5. COMPUTER SCIENCE. *Educ:* Boston Col, BS, 60; Univ Notre Dame, PhD(chem), 67. *Prof Exp:* Assoc res scientist, Radiation Lab, Univ Notre Dame, 67-68; staff consult, Univac Div, Sperry Rand Corp, 68-85, sr res scientist, 89-91, DIR, ADVAN CASE DEVELOP, UNISYS CORP, 91- *Mem:* Am Chem Soc; Asn Comput Mach. *Res:* Microwave spectroscopy. *Mailing Add:* 1611 Stephens Dr Wayne PA 19087

HERMANS, JAN JOSEPH, b Leiden, Neth, Nov 1, 09; nat US; m; c 1. PHYSICAL CHEMISTRY. *Educ:* Univ Leiden, PhD(chem), 37. *Prof Exp:* Instr chem, Univ Leiden, 36-37, res assoc, Sch Pediat, 41-42, prof phys chem, 53-58; Ramsay Mem fel, Univ London, 38-39; res assoc, Wageningen, 40-41; res chemist, Inst Cellulose Res, Utrecht, Neth, 42-46; prof phys chem, Groningen Univ, 46-53; dir cellulose res inst, Col Forestry, State Univ NY, 58-61; sr staff scientist, Chemstrand Res Ctr, 61-68; distinguished prof, 68-80, EMER PROF CHEM, UNIV NC, CHAPEL HILL, 80- *Concurrent Pos:* Rockefeller fel, Polytech Inst Brooklyn, 50-51; vis prof, Univ Toronto, 56 & Univ NC, Chapel Hill, 65-66. *Mem:* Am Chem Soc; NY Acad Sci; Royal Neth Chem Soc; Royal Neth Acad Sci. *Res:* Polymers. *Mailing Add:* Dept Biochem/Nutrit Univ NC Chapel Hill Sch Med Chapel Hill NC 27514

HERMANS, PAUL E, b Rotterdam, Neth, Oct 17, 27; US citizen; m 54; c 3. INTERNAL MEDICINE, INFECTIOUS DISEASES. *Educ:* Univ Utrecht, MD, 57; Univ Minn, MS, 62. *Prof Exp:* From instr internal med to assoc prof med, 63-76, chmn Div Infectious Dis, 73-83, PROF MED, MAYO SCH MED, UNIV MINN, 76- STAFF MEM, MAYO CLIN, 62- *Concurrent Pos:* Fel internal med, Mayo Clin, 58-61, asst to staff, 61-62, consult infectious dis, 62- *Mem:* Infectious Dis Soc Am; Med Mycol Soc Americas; Am Soc Microbiol; NY Acad Sci; Am Col Physicians. *Res:* Clinical immunology, especially immunologic deficiency diseases. *Mailing Add:* 200 First St SW Rochester MN 55902

HERMANSON, HARVEY PHILIP, b Spring Valley, Minn, Oct 7, 32; m 55; c 1. AGRICULTURAL & FOOD CHEMISTRY. *Educ:* Univ Minn, BS, 54, PhD(soils), 61; Cornell Univ, MS, 57. *Prof Exp:* Asst prof soil chem & asst agr chemist, Univ Idaho, 61-64; NIH trainee soil chem pesticide residues, Univ Calif, Riverside, 64-67, univ fel, 65-67; asst res prof agr chem, Rutgers Univ, 67-70; asst res prof chem, NC A&T State Univ, 70-71, assoc prof plant sci, 73-78. *Concurrent Pos:* Consult, Chevron Chem Co, 64-67. *Mem:* Am Soc Agron; Int Soc Soil Sci; Sigma Xi. *Res:* Soil chemistry to pesticides, especially insecticides; special techniques of regression statistics applied to soil fertility and chemistry; problems; soil fertility problems of peat soils; ecology of birdsfoot trefoil; disposal of effluent from septic tanks in problem soils; tobacco and pesticide research. *Mailing Add:* 1401 B Whilden Pl Greensboro NC 27408

HERMANSON, JOHN CARL, b Stevens Point, Wis, June 2, 40; m 65; c 3. SOLID STATE PHYSICS. *Educ:* Mass Inst Technol, BS, 62; Univ Chicago, MS, 64, PhD(physics), 66. *Prof Exp:* Res assoc physics, Univ Ill, Urbana, 66-68; res fel, Miller Inst Basic Res in Sci & Univ Calif, Berkeley, 68-69, res assoc, 69-70; from asst prof to assoc prof , 73-80, PROF PHYSICS, MONT STATE UNIV, 80- *Concurrent Pos:* Theoret consult, Michelson Labs, China Lake, Calif, 73-77; vis prof, Univ Munich, Fed Repub Ger, 78-79. *Mem:* Am Phys Soc; Am Inst Physics; Sigma Xi. *Res:* Optical properties of solids; theory of excitons; electronic structure of solid surfaces; photoemission spectroscopy. *Mailing Add:* Dept Physics Mont State Univ Bozeman MT 59717

HERMANSON, RONALD ELDON, b Kiester, Minn, Mar 31, 33; m 55; c 3. WATER RESOURCES, ANIMAL WASTE MANAGEMENT. *Educ:* Iowa State Univ, BS, 58, MS, 64, PhD(water resources), 67. *Prof Exp:* Staff engr water resources, Iowa Natural Resources Coun, 58-62; res assoc agr eng, Iowa State Univ, 62-64, fel water resources, 64-66; asst prof agr eng, Auburn Univ, 66-72; EXTEN AGR ENGR WATER QUALITY & SOIL & WATER ENG, WASH STATE UNIV, 73- *Concurrent Pos:* Off Water Resources Res grant, 66-72; mem, USDA task force animal waste mgt, 68 & Environ Protection Agency task force fruit & veg waste guidelines, 74. *Mem:* Sigma Xi; Am Soc Agr Engrs. *Res:* Animal waste management especially to develop economical methods of handling animal waste that will avoid air, water and land pollution; control of groundwater contamination from agrichemicals. *Mailing Add:* Smith Agr Eng Bldg Rm 220 Wash State Univ Pullman WA 99164-6120

HERMENS, RICHARD ANTHONY, b Forest Grove, Ore, Nov 17, 35; m 61; c 4. PHYSICAL CHEMISTRY. *Educ:* Pac Univ, BS, 57; Ore State Univ, MS, 60; Univ Idaho, PhD(phys chem), 63. *Prof Exp:* Instr chem, Univ Idaho, 62-63; asst prof, Millikin Univ, 63-66; from asst prof to assoc prof, 66-70, PROF CHEM & CHMN DEPT, EASTERN ORE STATE COL, 70- *Concurrent Pos:* Consult, 80-86. *Mem:* Am Chem Soc; fel Am Inst Chemists; AAAS. *Res:* Instrumental, polarographic and spectrophotometric analyses. *Mailing Add:* Dept Chem Eastern Ore State Col La Grande OR 97850

HERMES, HENRY, b Jersey City, NJ, July 24, 33; m 58; c 2. MATHEMATICS. *Educ:* Montclair State Col, BA, 54; Univ NMex, MS, 59, PhD(math), 62. *Prof Exp:* Assoc res scientist, Martin Co, Colo, 61-64; asst prof appl math, Brown Univ, 64-66; chmn dept math, 74-76, PROF MATH, UNIV COLO, BOULDER, 67- *Concurrent Pos:* Assoc ed, J on Control, Soc Indust & Appl Math, 67- *Mem:* Am Math Soc; Soc Indust & Appl Math. *Res:* Ordinary differential equations and control theory. *Mailing Add:* Dept Math Box 426 Univ Colo Ecot 3-42 Boulder CO 80309

HERMES, O DON, b Brighton, Ill, Mar 13, 39; m 62; c 2. PETROLOGY, GEOCHEMISTRY. *Educ:* Washington Univ, AB, 61; Univ NC, Chapel Hill, MS, 63, PhD(geol), 66. *Prof Exp:* Fel, Univ Calif, Los Angeles, 66-68; from asst prof to assoc prof, 68-78, PROF GEOL, UNIV RI, 78- *Mem:* Am Geophys Union; Geol Soc Am. *Res:* Igneous and metamorphic petrology; mineralogy; analytical chemistry. *Mailing Add:* Dept Geol Univ RI Kingston RI 02881

HERMODSON, MARK ALLEN, b Crookston, Minn, Jan 30, 42; m 68; c 2. BIOCHEMISTRY. *Educ:* St Olaf Col, BA, 64, Univ Wis, PhD(biochem), 68. *Prof Exp:* Fel biochem, Univ Wis, 68-69 & Univ Wash, 69-72; res asst prof med genetics, Univ Wash, 72-77; assoc prof biochem, 77-80, PROF BIOCHEM, PURDUE UNIV, 80-, HEAD DEPT BIOCHEM, 81- *Mem:* Am Soc Biol Chem; Protein Soc. *Res:* Protein structure; amino acid sequencing; enzymology; membrane transport. *Mailing Add:* Dept Biochem Purdue Univ West Lafayette IN 47907

HERMRECK, ARLO SCOTT, VASCULAR & GENERAL SURGERIES. *Educ:* Univ Kans, PhD(physiol), 70. *Prof Exp:* PROF SURG, UNIV KANS MED CTR, 71- *Mailing Add:* 511 Terrace Trail E Lake Quivira Kansas City KS 66106

HERMSEN, RICHARD J, b Little Chute, Wis, Oct 20, 28; m 53. ELECTRICAL ENGINEERING. *Educ:* Univ Wis, BS, 59, MS, 60, PhD(nuclear eng), 63. *Prof Exp:* Design specialist, Gen Dynamics Corp, Calif, 63-65; mem tech staff, Aerospace Corp, 65-67; assoc prof, 67-72, chmn dept, 73-81, PROF ELECTRONICS, CALIF STATE POLYTECH UNIV, POMONA, 72- *Concurrent Pos:* Vis asst prof, Univ Redlands, 66-67. *Mem:* Inst Elec & Electronics Engrs. *Res:* Analytical aspects of automatic control systems; analog and digital communication systems; digital signal processing. *Mailing Add:* Calif State Polytech Univ ECE Dept Pomona CA 91768

HERMSEN, ROBERT W, b Baker, Ore, Apr, 25, 34; m 57; c 4. COMBUSTION RESEARCH. *Educ:* Ore State Univ, BS, 56; Univ Calif, Berkeley, PhD(chem eng), 62. *Prof Exp:* Res engr, Titanium Metals Corp Am, 56-57; staff scientist res combustion kinetics, United Technologies Chem Systs, 61-70, sr staff scientist Phys Sci Lab, 70-71, mgr, 71-75, chief, kinetics & combustion res chem Systs Div 75-85, MGR COMBUSTION RES & DEVELOP, UNITED TECHNOLOGIES CHEM SYSTS DIV, 85- *Mem:* Combustion Inst; Am Inst Chem Engrs; Am Inst Aeronaut & Astronaut. *Res:* High temperature thermodynamics; chemical kinetics; combustion of metals; chemical propulsion; solid propellant combustion; production of sound from flames; coal gasification; two-phase flow in combustion systems; rocket motor performance; detonation hazards of rocket motors. *Mailing Add:* 3563 Evergreen Dr Palo Alto CA 94303

HERMSMEIER, LEE F, b Quincy, Ill, Aug 13, 17; m 49; c 3. AGRICULTURAL ENGINEERING. *Educ:* Univ Minn, BS, 49; Iowa State Univ, MS, 50. *Prof Exp:* Area engr, Soil Conserv Serv, USDA, 50-56, res agr engr, Soil & Water Conserv Div, Agr Res Serv, Minn, 57-68, res agr engr, Agr Res Serv, Imp Vally Conserv Res Ctr, 68-83; RETIRED. *Concurrent Pos:* Mem, US Comt Irrig, Drainage & Flood Control; drainage consult, several foreign countries, Algeria, Cuba & Egypt, 73-81, Pakistan, 85; consult, 83- *Res:* Irrigation efficiencies and irrigation scheduling for arid lands; development of improved drainage methods and procedures; use of alternate materials for drainage of arid lands. *Mailing Add:* 352 J St Brawley CA 92227

HERMSMEYER, RALPH KENT, b Litchfield, Ill, Oct 13, 42; m 64. PHYSIOLOGY. *Educ:* Univ Ill, AB, 64, MS, 66, PhD(physiol), 69. *Prof Exp:* Asst prof physiol & biophys, Univ Nebr Med Ctr, Omaha, 70-73; from asst prof to assoc prof, 73-80, PROF PHARMACOL, COL MED UNIV IOWA, 80- *Concurrent Pos:* Nat Heart Inst fel, Sch Med, Univ Va, 68-70; Nebr Heart Asn res grant, Univ Nebr Med Ctr, Omaha, 71-73, USPHS res grant, 71-74; Iowa Heart Asn res grant, Univ Iowa, 73-75, USPHS res grant, 74-81; res cancer develop award, NIH, 75-80; mem circulation & high blood pressure coun, Am Heart Asn. *Honors & Awards:* Katz Prize, Am Heart Asn, 75. *Mem:* Am Physiol Soc; Biophys Soc; Soc Gen Physiol; Am Soc Zool; Sigma Xi. *Res:* Electrophysiology of vascular muscle and heart. *Mailing Add:* Dept Physiol Sch Med Univ Nev Reno NV 89557

HERN, THOMAS ALBERT, b Cincinnati, Ohio, Dec 23, 41. MATHEMATICS, COMPUTER GRAPHICS. *Educ:* Univ Cincinnati, AB, 64; Ohio State Univ, MS, 66, PhD(math), 69. *Prof Exp:* From instr to asst prof, 69-74, ASSOC PROF MATH, BOWLING GREEN STATE UNIV, 74- *Concurrent Pos:* Vis assoc prof, comput sci, Univ NC, Chapel Hill, 82-84; mem, Spec Interest Group Comput Graphics & Interactive Tech of Asn Comput Mach. *Mem:* Math Asn Am; Asn Comput Mach; Soc Indust & Appl Math. *Mailing Add:* Dept Math & Statist Bowling Green State Univ Bowling Green OH 43403

HERNANDEZ, GONZALO J, b San Jose, Costa Rica. OPTICS, ATMOSPHERIC PHYSICS. *Educ:* Univ Notre Dame, BS, 58; Univ Rochester, PhD(phys chem), 62. *Prof Exp:* Physicist, Air Force Cambridge Res Lab, 61-69; res scientist, Space Physics Res Lab, Univ Mich, 85-87; dir, Fritz Peak Observ, 70-85, PHYSICIST, ENVIRON RES LABS, NAT OCEANIC & ATMOSPHERIC ADMIN, 69-; PRIN RES SCIENTIST, GRAD PROG GEOPHYS, UNIV WASH, 88- *Concurrent Pos:* Affil prof geophys, Univ Alaska. *Mem:* Am Geophys Union; Am Phys Soc; Optical Soc Am; Sigma Xi. *Res:* Upper atmosphere physics; high resolution spectroscopy; optical aeronomy. *Mailing Add:* 15545 65th Pl NE Bothell WA 98011

HERNANDEZ, JOHN PETER, b Madrid, Spain, Sept 6, 40; nat US; m 66, 84; c 4. CONDENSED MATTER PHYSICS. *Educ:* Manhattan Col, BEE, 62; Stanford Univ, MS, 63; Univ Rochester, PhD(optics), 67. *Prof Exp:* From instr to assoc prof, 66-77, PROF PHYSICS, UNIV NC, CHAPEL HILL, 77- *Concurrent Pos:* Sr Fulbright lectr, Univ Autonoma Madrid, 72-73; Kenan Res prof, Univ Oxford, 82-83. *Mem:* Am Phys Soc; Sigma Xi. *Res:* Electronic properties of condensed matter; density of states of disordered materials; chemical physics; vapor-liquid and insulator metal transitions. *Mailing Add:* Dept Physics Univ NC Chapel Hill NC 27599-3255

HERNANDEZ, JOHN W(HITLOCK), b Albuquerque, NMex, Aug 17, 29; m 51. SANITARY ENGINEERING. *Educ:* Univ NMex, BS, 51; Purdue Univ, MS, 59; Harvard Univ, MS, 61, PhD(water resources), 65. *Prof Exp:* Asst engr dam design, State Game & Fish Dept, NMex, 54-55, engr waterworks design, State Engrs Off, 55-57, assoc engr, State Health Dept, 57-63; assoc prof sanit eng, 65-68, dean eng, 75-80, PROF CIVIL ENG, NMEX STATE UNIV, 68- *Concurrent Pos:* Dep adminr, US Environ Protection Agency, 81-83. *Mem:* Am Soc Civil Engrs; Nat Soc Prof Engrs; Am Water Works Asn; Water Pollution Control Fedn; Am Acad Environ Eng; Sigma Xi. *Res:* Biological treatment of sanitary area waste waters; application of systems analysis technique to water resources development. *Mailing Add:* Box 3196 Las Cruces NM 88003

HERNANDEZ, JUAN ANTONIO, b Holguin, Cuba, Sept 4, 29; US citizen; m 51; c 3. PATHOLOGY. *Educ:* Inst Sec Teachings, Holguin, Cuba, BS, 48; Univ Havana, MD, 56; Am Bd Path, dipl, 64. *Prof Exp:* Intern, Ga Baptist Hosp, Atlanta, Ga, 58-59, trainee path, 59-61; res fel pulmonary path, Baptist Mem Hosp, Jacksonville, 63-76; CLIN ASST PROF PATH, COL MED & DENT NJ, NEWARK, 76-; ASSOC PATHOLOGIST, ST ELIZABETH HOSP, 76- *Mem:* Col Am Path; Am Soc Clin Path. *Res:* Pulmonary pathology; gross and microscopic morphology; pathogenesis of emphysema. *Mailing Add:* Seven Crest Circle South Orange NJ 07209

HERNANDEZ, NORMA, b El Paso, Tex, May 19, 34; m 54; c 4. SCIENCE ADMINISTRATION, HISTORY & PHILOSOPHY OF SCIENCE. *Educ:* Tex Western Col, BS, 54; Univ Tex, MS, 60, PhD(math educ), 70. *Prof Exp:* Instr math educ, Univ Tex, Austin, 66-69; full prof & dean, Col Educ, 74-80, ASST PROF MATH EDUC, UNIV TEX, EL PASO, 69-, PROF, 80- *Concurrent Pos:* Founding pres, El Paso Coun, Teachers Math, 65-66; pres, Tex Coun Teachers Math, 68-69; mem, panel Labs & Ctrs, NIE, 78-80 & bd dir, Southwest Educ Develop Lab, Austin, Tex, 83- *Mem:* Nat Coun Teachers Math; Sch Sci & Math Asn; Math Asn Am. *Res:* Identification of teaching/learning strategies that will increase mathematical abilities for all children; factors affecting achievement of Mexican Americans. *Mailing Add:* Col Educ Univ Tex El Paso TX 79968

HERNANDEZ, TEME P, b Lafayette, La, May 15, 19; m 56; c 3. HORTICULTURE, PLANT BREEDING. *Educ:* La State Univ, BS, 40, MS, 42; Univ Wis, PhD(hort), 49. *Prof Exp:* Supt sweet potato res ctr, 49-57, PROF HORT, LA STATE UNIV, BATON ROUGE, 58- *Mem:* AAAS; Am Soc Hort Sci; Genetics Soc Am. *Res:* Breeding and testing of horticultural crops, including sweet potatoes and tomatoes. *Mailing Add:* Hort Dept La State Univ Baton Rouge LA 70803

HERNANDEZ, THOMAS, b Lafayette, La, May 17, 14; m 43; c 4. PHARMACOLOGY. *Educ:* La State Univ, BA, 36, MS, 38, MD, 47; Univ Iowa, PhD(zool), 42. *Prof Exp:* Asst surgeon, Marine Hosp, USPHS, New Orleans, 47-48; instr biochem, Sch Med, La State Univ, 48-49, asst prof, 49-51; sr asst surgeon, NIH, 51-53; from asst prof to prof biochem, La State Univ Med Ctr, New Orleans, 53-60, chmn dept, 60-76, prof pharmacol, 60-80, prof exp therapeut, 76-80; RETIRED. *Concurrent Pos:* Assoc mem, Armed Forces Epidemiol Bd Comn Malaria, 65. *Mem:* AAAS; Soc Exp Biol & Med; Am Physiol Soc; Am Soc Pharmacol & Exp Therapeut. *Res:* Endocrinology; antimalarial drugs; biochemical studies of cold blooded animals; fluid and electrolytes; carbohydrate and nitrogen metabolism. *Mailing Add:* 145 Saul Dr Duson LA 70529

HERNANDEZ AVILA, MANUEL LUIS, b Quebradillas, PR, Apr 15, 35; m 56. PHYSICAL OCEANOGRAPHY. *Educ:* Univ PR, BS, 67, MS, 70; La State Univ, PhD(marine sci, phys oceanog), 74. *Prof Exp:* Res asst marine geol, Univ PR, Mayaguez, 67-69; res asst phys oceanog, Coastal Studies Inst, La State Univ, Baton Rouge, 70-73; asst prof, 74-77, PROF PHYS OCEANOG, UNIV PR, MAYAGUEZ, 78-, DIR, DEPT MARINE SCI, 78-, DIR, SEA GRANT PROG, 80- *Concurrent Pos:* Dir, MBRS-NIH Prog, Makagüez Campus, 86. *Mem:* Am Geophys Union; Sigma Xi; Marine Technol Soc; Am Meteorol Soc. *Res:* Coastal physical processes; estuarine dynamics; coral reefs dynamics. *Mailing Add:* PO Box 5000 Mayaquez PR 00709

HERNDON, CHARLES HARBISON, b Dublin, Tex, Dec 12, 15; m 44; c 2. ORTHOPEDIC SURGERY. *Educ:* Univ Tex, BA; Harvard Univ, MD, 40; Am Bd Orthop Surg, dipl, 49. *Prof Exp:* Surg intern, Univ Hosps, Cleveland, 40-41; jr orthop surgeon, Am Hosp Brit, Oxford, Eng, 42; orthop resident, Hosp Spec Surg, New York, 46-47; instr, 47-49, asst prof & demonstr anat, 49-53, assoc prof, 53-60, prof clin orthop, 60-61, Rainbow prof, 61-82, EMER RAINBOW PROF ORTHOP SURG, SCH MED, CASE WESTERN RESERVE UNIV, 85- *Concurrent Pos:* Beekmen fel orthop surg, Columbia Univ, 45-46; Bunts, Crile & Lower fel, Case Western Reserve Univ, 47-53; assoc orthop surgeon, Rainbow Hosp, 47-52, dir orthop surg, 52-82; asst orthop surgeon, Univ Hosps, Cleveland, 47-53, dir dept orthop, 53-82; assoc orthopedist, Highland View Hosp, 53-82; consult orthopedist, Elyria Mem & Gates Mem Hosps, Elyria, Ohio, 48-59 & Vet Admin Hosp, 56-82; mem, Skeletal Syst Comt, Nat Res Coun-Nat Acad Sci, 58-67, chmn, 62-67; mem, Am Bd Orthop Surg, 60-66, chmn, Exam Comt, 61-64, pres, 64-66; mem, Bd Trustees, J Bone & Joint Surg, 69-75; mem, Adv Bd, Ohio State Serv Crippled Children; assoc mem, Orthop Res & Educ Found. *Mem:* Clin Orthop Soc; Am Acad Orthop Surg (pres, 68-69); Am Col Surgeons (2nd vpres, 74); Orthop Res Soc (secy-treas, 54-55, pres, 57); Am Orthop Asn. *Res:* Problems related to bone transplantation and bone induction. *Mailing Add:* 13600 Shaker Blvd Suite 802 Cleveland OH 44120

HERNDON, CHARLES L(EE), b Grant City, Mo, Apr 19, 21; m 64; c 2. FORENSIC ENGINEERING. *Educ:* Univ Tex, BSME, 43, MSME, 57. *Prof Exp:* Liaison engr, Douglas Aircraft Co, Okla, 43-47; sales & design engr, Frigidaire Dept, Straus-Frank Co, Tex, 47-49, Air Conditioning Div, Alamo Lumber Co, 49-54 & Deansteel Prod, 54-56; asst prof eng & physics, San Antonio Col, 56-58; assoc prof eng, Evansville Col, 58-62; asst prof, Calif State Col Los Angeles, 62-65; from asst prof to prof, 65-84, chmn dept, 76-80, EMER PROF ENG SCI, MONT COL MINERAL SCI & TECHNOL, 84- *Concurrent Pos:* Instr, Univ Tex, 46-47; sr proj engr, US Naval Civil Eng Lab, Calif, 62-65; forensic engr, Herndon Assocs, 66-84, forensic consult, 84- *Honors & Awards:* Teetor Award, Soc Automotive Engrs, 75. *Mem:* Am Soc Eng Educ; Am Soc Heating, Refrig & Air-Conditioning Engrs; Nat Soc Prof Engrs; Soc Automotive Engrs. *Res:* Air conditioning and refrigeration; solar energy; machine design. *Mailing Add:* Herndon & Assoc 1321 W Platinum Butte MT 59701-2125

HERNDON, DAVID N, SURGERY. *Prof Exp:* CHIEF STAFF, SHRINERS BURN INST, 81- *Mailing Add:* Shriners Burns Inst 610 Texas Ave Galveston TX 77550

HERNDON, JAMES HENRY, JR, b Dallas, Tex, June 2, 39; m 60; c 3. DERMATOLOGY. *Educ:* Univ Tex Southwestern Med Sch, MD, 63. *Prof Exp:* Assoc prof internal med, 74-79, ASSOC CLIN PROF DERMATOL, UNIV TEX HEALTH SCI CTR, DALLAS, 79- *Concurrent Pos:* Consult dermatol, Vet Admin Hosp, Dallas & Children's Med Ctr, 70-79, Presby Hosp, Dallas, 79-; USPHS res career develop award, 70-75. *Mem:* Am Fedn Clin Res; Soc Invest Dermat; fel Am Col Physicians. *Res:* Pathogenesis of psoriasis, regulation of epidemial proliferation mechanisms of pruritus. *Mailing Add:* 8220 Walnut Hill Lane Suite 408 Dallas TX 75231

HERNDON, JAMES W, b Greensboro, NC, Aug 3, 57; m 87. ORGANIC SYNTHESIS, ORGANOMETALLIC CHEMISTRY. *Educ:* Univ NC, Greensboro, BS, 79; Princeton Univ, MA, 80, PhD(chem), 83. *Prof Exp:* NIH postdoctoral fel chem, Univ Wis-Madison, 83-85; ASST PROF, DEPT CHEM, UNIV MD, COLLEGE PARK, 85- *Mem:* Am Chem Soc. *Res:* Design of new reagents for use in organic synthesis and organometallic chemistry. *Mailing Add:* Dept Chem Univ Md College Park MD 20742-2021

HERNDON, ROBERT MCCULLOCH, b Richmond, Va, May 29, 35; m 55; c 3. NEUROLOGY, NEUROPATHOLOGY. *Educ:* Univ Chicago, BA, 55; Univ Tenn, MD, 58. *Prof Exp:* From intern to chief resident neurol, Detroit Receiving Hosp, 59-61; neurologist, USAF Hosp, Travis AFB, Calif, 63-64, chief neurol, 64-65; asst prof neurol, Med Sch, Stanford Univ, 66-69; assoc prof neurol, Med Sch, Johns Hopkins Univ, 69-77, prof & chmn, Ctr Brain Res & Neurol, Med Sch, Univ Rochester, 77-88; CHIEF NEUROL, GOOD SAMARITAN HOSP, 88-; PROF NEUROL, ORE HEALTH SCI UNIV, 88- *Concurrent Pos:* Fel neuropath, Montreal Neurol Inst, 62-63; fel anat, Harvard Med Sch, 65-66; chief neurol, Palo Alto Vet Admin Hosp, 68-69. *Honors & Awards:* Arthur Weil Award, Am Asn Neuropath, 68 & 72 & Moore Award, Am Asn Neuropath, 83. *Mem:* Am Asn Neuropath; Am Neurol Asn; Am Acad Neurol. *Res:* Electron microscopic studies of the normal cerebellum and of chemical and virus induced abnormalities; viral disease of the nervous system and multiple sclerosis; experimental neuropathology. *Mailing Add:* 1022 SW 22nd Ave Suite 460 Portland OR 97210

HERNDON, ROY C, b Washington, DC, Sept 25, 34. NUCLEAR PHYSICS, HAZARDOUS WASTE MANAGEMENT. *Educ:* Washington & Lee Univ, BS, 55; Univ Ill, MS, 58; Fla State Univ, PhD(nuclear physics), 62. *Prof Exp:* Physicist, Lawrence Livermore Lab, Calif, 62-67; assoc prof physics, Phys Oceanog Lab, Nova Univ, 68-71, prof physics & comput sci, 71-74; DIR, CTR BIOMED & TOXICOL RES & WASTE MGT, FLA STATE UNIV, TALLAHASSEE, FLA, 80- *Concurrent Pos:* Exec dir, Gov Hazardous Waste Policy Adv Coun; dir, Fla Waste Info Exchange; vis prof physics, Fla State Univ, Tallahassee, 74-77, Univ Lausanne, Switz, 77, 79, 81, 84 & 90. *Mem:* Am Phys Soc; Sigma Xi; AAAS; NY Acad Sci; Am Inst Biol Sci. *Res:* Environmental research. *Mailing Add:* 232 Westminster Dr Tallahassee FL 32304

HERNDON, WALTER ROGER, b Birmingham, Ala, Sept 7, 26; m 49; c 4. BOTANY. *Educ:* Univ Ala, BS, 47, MS, 48; Vanderbilt Univ, PhD(biol), 54. *Prof Exp:* Asst biol, Univ Ala, 47-48; asst, Vanderbilt Univ, 48-50; instr, Mid Tenn State Univ, 50-54, asst prof, 54-55; from asst prof to assoc prof, Univ Ala, 55-61; head dept bot, 61-64, assoc dean col lib arts, 64-67, asst vpres acad affairs, 67-68, assoc vchancellor acad affairs, 68-70, PROF BOT, UNIV TENN, KNOXVILLE, 61-, VCHANCELLOR ACAD AFFAIRS, 70- *Concurrent Pos:* Asst biol, Vanderbilt Univ, 53-54. *Mem:* AAAS; Am Bot Soc; Phycol Soc Am (secy-treas, 61-64, vpres, 64-65, pres, 66-); Int Phycol Soc; Sigma Xi. *Res:* Phycology; marine botany. *Mailing Add:* Dept Bot Univ Tenn Knoxville TN 37996-1100

HERNDON, WILLIAM CECIL, b El Paso, Tex, Aug 12, 32; m 56; c 2. PHYSICAL ORGANIC CHEMISTRY. *Educ:* Tex Western Col, BS, 54; Rice Univ, PhD(chem), 59. *Prof Exp:* Res chemist, Cent Res Div, Am Cyanamid Co, 58-61; asst prof chem, Univ Miss, 61-64; assoc prof, Fla Atlantic Univ, 64-66; from asst prof to prof, Tex Tech Univ, 66-72; chmn dept chem, 72-82, dean Col Sci, 82-87, DUDLEY MEM PROF CHEM, UNIV TEX, EL PASO, 87-, CHMN DEPT CHEM, 89- *Concurrent Pos:* Consult, Bell Tel Labs, 65; prog dir chem dynamics, NSF, 75-76. *Mem:* Fel NY Acad Sci; Sigma Xi; AAAS; Am Chem Soc. *Res:* Theoretical organic chemistry; organic photochemistry; benzonoid hydrocarbon chemistry. *Mailing Add:* Dept Chem Univ Tex El Paso TX 79912

HERNER, ALBERT ERWIN, b Brooklyn, NY, Jan 24, 31; div; c 3. ENVIRONMENTAL CHEMISTRY, DATABASE DEVELOPMENT. *Educ:* Union Col, Schenectady, BS, 52; NY Univ, MS, 54; Fla State Univ, PhD(biochem), 60. *Prof Exp:* Postdoctoral biochem, USPHS, 60-62; assoc chemist, Midwest Res Inst, 62-64; chemist, New Eng Nuclear Corp, 64-67; res assoc molecular biol, Beth Israel Hosp, Boston, 67-69; chief radiochem, Clin-Chem Labs, Boston, 67-69; dir, Herner Anal, Inc, 72-84; assoc dir, Bionetics-Metpath, 84-89; RES CHEMIST, ENVIRON CHEM LAB, USDA, BELTSVILLE, 89- *Concurrent Pos:* Lectr chem, George Washington Univ, 87-90. *Mem:* Sigma Xi; NY Acad Sci; Am Asn Clin Chem. *Res:* Analysis of soil, water, and agricultural products for pesticides and other agricultural chemical and methodology development; organized pesticide properties database. *Mailing Add:* ARS NRI Environ Chem Lab USDA Beltsville MD 20705

HERNQVIST, KARL GERHARD, b Boras, Sweden, Sept 19, 22; US citizen; m 48; c 3. QUANTUM ELECTRONICS, LASERS. *Educ:* Royal Inst Technol Sweden, EE, 45, PhD(electronics), 59. *Prof Exp:* Mem tech staff electronics, Res Inst Nat Defense Sweden, 46-52; FEL ELECTRONICS, RCA LABS, 52- *Concurrent Pos:* Fel RCA Labs, 70. *Honors & Awards:* Indust Res-100 Award, 67; David Sarnoff Award, RCA Corp, 74 & 82. *Mem:* Sigma Xi; Am Phys Soc; fel Inst Elec & Electronics Engrs. *Res:* Gas lasers; optics; gaseous discharges; energy conversion; electron physics; display tubes. *Mailing Add:* 667 Lake Dr Princeton NJ 08540

HEROD, JAMES VICTOR, b Selma, Ala, May 4, 37; m 60; c 3. MATHEMATICAL ANALYSIS. *Educ:* Univ Ala, BS, 59, MS, 60; Univ NC, Chapel Hill, PhD(math), 64. *Prof Exp:* From asst prof to assoc prof, 66-79, PROF MATH, SCH MATH, GA INST TECHNOL, 79- *Concurrent Pos:* Vis prof, US Mil Acad, 81-82. *Mem:* Am Math Soc; Math Asn Am; Am Asn Univ Professors. *Res:* Connections between operator equations, evolution systems and product integrals; product integral representation for generalized inverses of operators that are unbounded; nonlinear Boltzmann equations. *Mailing Add:* Sch Math Ga Inst Technol 225 North Ave Atlanta GA 30332

HEROLD, E(DWARD) W(ILLIAM), b New York, NY, Oct 15, 07; m 31; c 1. ELECTRONICS. *Educ:* Univ Va, BS, 30; Polytech Univ, MSc, 42. *Hon Degrees:* DSc, Polytech Univ, 61. *Prof Exp:* Tech asst, Bell Tel Labs, NY, 24-26; radio engr, E T Cunningham, Inc, 27-29; develop engr, Radio Corp Am, 30-38, res engr, 38-42, res engr labs, 42-46, sect head, 46-51, dir radio tube lab, 51-54, dir electronic res lab, 54-59; vpres res, Varian Assocs, 59-64; corp staff, RCA Corp, 65-69, dir technol, 70-72; chmn bd trustees, Palisades Inst Res Servs, 69-84; RES MGT CONSULT, 72- *Concurrent Pos:* Consult, US Dept Defense, 50-76; mem adv coun, Elec Eng Dept, Princeton Univ, 57-71. *Honors & Awards:* Founders Medal, Inst Elec & Electronics Engrs, 76. *Mem:* Fel Inst Elec & Electronics Engrs; Sigma Xi. *Res:* Electron tubes; transistors and solid-state devices; microwave electronics; television; controlled fusion; semiconductor devices; development of the first color television picture tube. *Mailing Add:* 332 Riverside Dr E Princeton NJ 08540-5414

HEROLD, RICHARD CARL, b Butler, Pa, June 13, 27. DEVELOPMENTAL BIOLOGY. *Educ:* Pa State Univ, BS, 50; Univ Pa, PhD(zool), 61. *Prof Exp:* Chemist, Aeroprojs, Inc, Pa, 52-55; assoc histol & embryol, 61-62, asst prof, 62-66, ASSOC PROF HISTOL & EMBRYOL, UNIV PA, 66- *Concurrent Pos:* Trustee & secy-treas, Swans Island Marine Sta, Maine, 65-; hon res fel, Univ Col London, 83-84. *Mem:* AAAS; Electron Micros Soc Am; Sigma Xi. *Res:* Hard tissue, development, ultrastructure, chemistry; development of face and palate; immunogold cytochemistry. *Mailing Add:* 200 Gray Line Ave Philadelphia PA 19072-1904

HEROLD, ROBERT JOHNSTON, b Hasbrouck Heights, NJ, Oct 7, 20; m 46; c 5. ORGANIC CHEMISTRY. *Educ:* Temple Univ, AB, 42, MA, 50. *Prof Exp:* Analyst explosives, Gen Chem Co, 42-43; asst hydrocarbon chem, Magnolia Petrol Co, 43-46; instr anal chem, Southern Methodist Univ, 46; res chemist polymer chem, Socony Mobil Oil Co, 46-56; sr res chemist, Gen Tire & Rubber Co, 57-63, group leader rubber res, 63-71, res scientist, 71-84; RETIRED. *Mem:* Am Chem Soc; Sigma Xi. *Res:* Hydrocarbon reactions; hydrocarbon polymers; unsaturated polyesters; polyethers; polyurethanes; catalysis; development of catalyst systems for polymerization; synthesis of oligomeric polythers and polyesters; investigations of curing conditions for polyurethane and for unsaturated polyesters. *Mailing Add:* 362 Keith Ave Akron OH 44313-5304

HERON, S(TEPHEN) DUNCAN, (JR), b Jackson, Miss, Sept 18, 26; m 48; c 2. GEOLOGY. *Educ:* Univ SC, BS, 48, MS, 50; Univ NC, PhD, 58. *Prof Exp:* From instr to assoc prof, 50-71, chmn dept, 68-78, PROF GEOL, DUKE UNIV, 71- *Concurrent Pos:* Geologist, SC State Develop Bd, 58-70; ed-in-chief, Southeastern Geol; assoc ed, Bull Geol Soc Am, 89-91. *Mem:* Geol Soc Am; Am Asn Petrol Geol; Sigma Xi; Soc Sedimentary Geol; Asn Earth Sci Ed (pres, 90). *Res:* Cretaceous and Tertiary stratigraphy of the Atlantic Coastal Plain; Pleistocene and Holocene barrier island development; clay mineralogy of Atlantic Coastal Plain sediments; nonmetallic economic and general geology. *Mailing Add:* Dept Geol Duke Univ Durham NC 27706

HEROUX, LEON J, b Haverhill, Mass, Aug 26, 27; m 47; c 2. PHYSICS. *Educ:* Harvard Univ, AB, 49; Johns Hopkins Univ, PhD(physics), 55. *Prof Exp:* Mem staff, Baird Assocs, Mass, 49-51; res physicist, Corning Glass Works, NY, 55-57; RES PHYSICIST, USAF GEOPHYS LAB, 57-, BR CHIEF, 85- *Mem:* Am Phys Soc; fel Optical Soc Am. *Res:* Electrical properties of glass; Zeeman effects and absorption spectra of the rare earths; photoelectric detectors in the extreme ultraviolet, solar spectroscopy below 1200 angstroms; plasma spectroscopy; atomic lifetimes; photoelectron emission processes in the extreme ultraviolet; upper atmospheric physics; measuring ultraviolet missile plume radiation for detection of rockets in space. *Mailing Add:* 11 Concord Greene St Concord MA 01742

HEROY, WILLIAM BAYARD, JR, b Washington, DC, Aug 13, 15; m 37; c 4. GEOLOGY, GEOPHYSICS. *Educ:* Dartmouth Col, AB, 37; Princeton Univ, PhD(geol), 41; Harvard Univ, Advan Mgt Prog, 61. *Prof Exp:* Asst geol, Princeton Univ, 39-41; from asst geologist to geologist field party, Tex Co, 41-45; geologist, Geotech Corp, 45-46, supvr, 46-50, vpres & dir, 50-59, exec vpres, 59-61, pres, 61-65, exec vpres, Teledyne Inc, 65-68, pres, Geotech Div, 65-67, group mgr, 67-68, asst to pres, Teledyne, Inc, 68-70; vpres-treas, 70-76, prof, 76-81, EMER PROF GEOL SCI, SOUTHERN METHODIST UNIV, 81- *Concurrent Pos:* Pres, Inst Study Earth & Man, treas, 76-81, sr scientist, 81-; mem bd, Ft Burgwin Res Ctr. *Honors & Awards:* Campbell Medal, Am Geol Inst, 86; Distinguished Serv Award, Geol Soc Am, 90. *Mem:* Fel Geol Soc Am (treas, 76-); Soc Explor Geophys; Am Inst Prof Geologists; Am Asn Petrol Geol (treas, 70-72); Am Geol Inst (pres, 69). *Res:* Geology of the Shell Canyon area; economic geological petroleum; geophysical prospecting; surface and structural geology; stratigraphy, sedimentation and seismology. *Mailing Add:* 3901 Montecito Apt 610 Denton TX 76205

HERPEL, COLEMAN, b McKeesport, Pa, Nov 21, 11; m 39; c 4. MATHEMATICS. *Educ:* Pa State Col, BA, 32; Harvard Univ, MA, 33. *Hon Degrees:* LLD, Elizabethtown Col, 77. *Prof Exp:* Instr math & Ger, Pa State Univ, Hazleton, 36-37, instr math & physics, 37-39, asst prof math & admin head, 39-43; asst prof & asst in admin, Pa State Univ, Altoona, 46-47, assoc prof & actg dean fac, 47-55; assoc prof, Pa State Univ, Ogontz Campus, 55-66, admin head, 55-59, dir, 59-66; dir, 66-72, EMER DIR, PA STATE UNIV, CAPITOL CAMPUS, 72- *Concurrent Pos:* Exec dir, Univ Ctr Harrisburg, 76-78. *Mem:* Math Asn Am. *Res:* University administration. *Mailing Add:* Eleven Gale Circle Camp Hill PA 17011

HERR, DAVID GUY, b Abbington, Pa, Mar 12, 37; m 61; c 2. STATISTICS, MATHEMATICS. *Educ:* Ga Inst Technol, BEE, 59, MS, 61; Univ NC, Chapel Hill, PhD(statist), 67. *Prof Exp:* Instr math, Ga Inst Technol, 61-62; instr, Univ NC, Charlotte, 65-66, asst prof, 66-67; asst prof, Duke Univ, 67-73; asst prof, 73-78, ASSOC PROF MATH, UNIV NC, GREENSBORO, 78- *Concurrent Pos:* Statist consult, Western Elec, 79- *Mem:* Am Math Asn; Am Statist Asn; Math Asn Am; Inst Math Statist. *Res:* Linear models; unbalanced designs; coordinate-free or geometric approach. *Mailing Add:* Dept Math/383 Bus ECO Bldg Univ NC 1000 Spring Garden St Greensboro NC 27412

HERR, DONALD EDWARD, b Dec 23, 26; US citizen; m 54; c 2. AGRONOMY. *Educ:* Ohio State Univ, BS, 51, PhD(weed control, agron), 65; Mich State Univ, MS, 62. *Prof Exp:* Assoc exten agent, Ohio State Univ, 52-54, mgr res farm, Ohio Agr Res & Develop Ctr, 54-62, res asst weed control, 62-64; from instr to prof agron, Ohio State Univ, 64-90; RETIRED. *Concurrent Pos:* US AID adv, India, 69-70 & Somalia, 78. *Mem:* Am Soc Agron; Weed Sci Soc Am. *Res:* Movement and persistence of herbicides in soil; agronomic crop production. *Mailing Add:* 4454 Sussex Dr Columbus OH 43220

HERR, EARL BINKLEY, JR, b Lancaster, Pa, Apr 14, 28; m 50; c 2. BIOCHEMISTRY. *Educ:* Franklin & Marshall Col, BS, 48; Univ Del, MS, 50, PhD(biochem), 53. *Hon Degrees:* DSc, Ind Univ. *Prof Exp:* AEC asst, Univ Del, 50-53; res fel, Cornell Univ, 53-55; res fel, Brookhaven Nat Lab, 55-57; sr biochemist, Eli Lilly & Co, 57-62, mgr antibiotic purification develop dept, 62-64, head pharmaceut res, 64-65, asst dir prod develop, 65, div dir antibiotics mfg & develop, 65-68, exec dir biochem & biol opers, 68-69, vpres, biochem & biol opers, 69-70, vpres res, develop & control, 70-74, pres,

Lilly Res Labs, 74-86, EXEC VPRES, ELI LILLY & CO, 86- *Concurrent Pos:* Bd dirs exec comt, Eli Lilly & Co, 71- *Mem:* AAAS; Chem Soc; Am Chem Soc; Sigma Xi. *Res:* Enzyme mechanisms and purification; antibiotics; pharmaceutical research and development. *Mailing Add:* 12011 Eden Glen Dr Carmel IN 46032

HERR, FRANK LEAMAN, JR, b Schenectady, NY, Feb 17, 48; m 70; c 2. GEOCHEMISTRY, PHYSICAL CHEMISTRY. *Educ:* Hamilton Col, BA, 70; Univ Md, PhD(chem), 75. *Prof Exp:* Nat Res Coun-Nat Acad Sci Res Assoc, 75-77, chemist, Ocean Sci Div, 77-80, CHEMIST, ENVIRON SCI DIV, NAVAL RES LAB, WASHINGTON, DC, 80- *Mem:* Am Geophys Union. *Res:* Aqueous solution thermodynamics of the electrolyte and inorganic non-electrolyte constituents of seawater and natural waters; analysis of trace constituents in natural waters (gases, oxidants, organics) resulting from photochemical or interfacial phenomena. *Mailing Add:* Naval Res Lab Code 1221 4555 Overlook Ave Sw Washington DC 20375-5000

HERR, HARRY WALLACE, b St Louis, Mo, Oct 1, 43; m 65; c 3. UROLOGY, ONCOLOGY. *Educ:* Univ Calif, Davis, AB, 65; Univ Calif, Irvine, MD, 69; Am Bd Urol, dipl, 76. *Prof Exp:* Intern, Los Angeles County-Univ Southern Calif Med Ctr, 69-70; resident gen surg, Univ Calif, Irvine, 70-71 & urol, 71-74; res fel tumor immunol, Sloan-Kettering Inst Cancer Res, 74-76; clin fel urol & oncol, 75-80, ASSOC ATTEND SURG, MEM SLOAN-KETTERING CANCER CTR, 80-; ASSOC PROF SURG & UROL, MED COL, CORNELL UNIV, 80- *Concurrent Pos:* Fel surg, Cornell Univ Med Col, 75-76; clin fel, Am Cancer Soc, 75-76; assoc attend surgeon, New York Hosp, 80- *Honors & Awards:* F C Valentine Award, NY Acad Med, 75; Meller Award, Mem Sloan-Kettering Cancer Ctr, 75. *Mem:* AAAS; NY Acad Sci; Am Col Surgeons; Am Asn Cancer Res; Acad Appl Sci. *Res:* Clinical and laboratory investigation on genitourinary cancers. *Mailing Add:* Dept Surg Cornell Univ Med Col 1275 York Ave New York NY 10021

HERR, JOHN CHRISTIAN, b Dubuque, Iowa, June 29, 48. PROTEIN SECRETION, REPRODUCTIVE IMMUNOLOGY. *Educ:* Univ Iowa, PhD(anat & cell biol), 78. *Prof Exp:* Asst prof, 81-87, ASSOC PROF ANAT & CELL BIOL, UNIV VA, 87-, DIR, LYMPHOCITE CULTURAL CTR, 82- *Mem:* Am Soc Cell Biol; Am Asn Anatomists; AAAS; Am Soc Reproductive Immunol; Soc Study Reproduction. *Res:* Molecular biology of human spermatogenesis; contraceptive vaccine development. *Mailing Add:* Anat Dept Univ Va Charlottesville VA 22908

HERR, JOHN MERVIN, JR, b Charlottesville, Va, July 26, 30; m 52, 74; c 3. PLANT EMBRYOLOGY. *Educ:* Univ Va, BA, 51, MA, 52; Univ NC, PhD(bot), 57. *Prof Exp:* Instr biol, Washington & Lee Univ, 52-54; Fulbright fel, India, 57-58; asst prof biol, Pfeiffer Col, 58-59; from asst prof to assoc prof, 59-69, PROF BIOL, UNIV SC, 69- *Mem:* Bot Soc Am; Int Soc Plant Morphol. *Res:* Angiosperm embryology; progress in both descriptive experimental phases of angiosperm embryology. *Mailing Add:* Dept Biol Univ SC Columbia SC 29208

HERR, LEONARD JAY, b Orville, Ohio, Dec 21, 28; m 54; c 3. PLANT PATHOLOGY. *Educ:* Ohio State Univ, BSc, 52, MSc, 53, PhD(plant path), 56. *Prof Exp:* From instr to assoc prof, 56-77, PROF PLANT PATH, OHIO AGR RES & DEVELOP CTR, OHIO STATE UNIV, 77- *Mem:* AAAS; NY Acad Sci; Sigma Xi; Am Inst Biol Scientists; Am Phytopath Soc; Asn Appl Biologists. *Res:* Soil microbiology in relation of soil-borne plant pathogens; diseases of sugar beets and tobacco; biocontrol of rhizoctonia. *Mailing Add:* Dept Plant Path Ohio State Univ Ohio Agr Res & Develop Ctr Wooster OH 44691-4096

HERR, RICHARD BAESSLER, b Philadelphia, Pa, Mar 3, 36; m 58; c 1. OBSERVATIONAL ASTRONOMY, VARIABLE STARS. *Educ:* Franklin & Marshall Col, BS, 57; Univ Del, MS, 60; Case Inst Technol, PhD(astron), 65. *Prof Exp:* Asst prof, 64-70, ASSOC PROF ASTRON, MT CUBA ASTRON OBSERV, UNIV DEL, 70- *Mem:* Royal Astron Soc Can; Int Astron Union; Am Astron Soc. *Res:* Observational astrophysics; photoelectric photometry. *Mailing Add:* Dept Physics & Astron Univ Del Newark DE 19716

HERR, ROSS ROBERT, b Chicago, Ill, Aug 11, 26; m 48; c 4. ORGANIC CHEMISTRY, REGULATORY AFFAIRS. *Educ:* Univ Ill, BS, 47; Northwestern Univ, PhD(chem), 51. *Prof Exp:* Res chemist, Armour Labs, 51-56; res chemist, 56-66, mgr pesticide regulatory affairs, 66-77, EXEC DIR PHARMACEUT REGULATORY PRACT, UPJOHN CO, 77- *Mem:* Am Chem Soc. *Res:* Isolation and chemistry of antibiotics; natural products. *Mailing Add:* 8318 Brookwood Dr Portage MI 49002-5208

HERREID, CLYDE F, II, b Grand Forks, NDak, Apr 8, 34; m 57; c 2. COMPARATIVE & ECOLOGICAL PHYSIOLOGY. *Educ:* Colo Col, BA, 56; Johns Hopkins Univ, MS, 59; Pa State Univ, PhD(zool), 61. *Prof Exp:* Asst ecol, Johns Hopkins Univ, 56-59; asst zool, Pa State Univ, 59-61; NIH fel marine biol, Inst Marine Sci, Univ Miami, 61; asst prof zool, Univ Alaska, 62-65 & Duke Univ, 65-67; assoc prof, 68-82, PROF BIOL, STATE UNIV NY, BUFFALO, 82-, ACAD DIR HONORS PROF & DISTINGUISHED TEACHING PROF, 88- *Concurrent Pos:* Vis prof biol, Univ Nairobi, 88-89. *Mem:* AAAS; (secy, 84-86); Am Physiol Soc. *Res:* Temperature regulation, metabolism and water balance in vertebrates; behavior and ecology of bats, crabs, frogs, sea cucumbers and snails; locomotion and energetics of invertebrates. *Mailing Add:* Dept Biol Sci State Univ NY Buffalo NY 14260

HERRELL, ASTOR Y, b Fork Ridge, Tenn, Feb 13, 35; m 60; c 1. SCIENCE EDUCATION & ADMINISTRATION. *Educ:* Berea Col, Ky, BA, 57; Tuskegee Inst, MSEd, 61; Wayne State Univ, PhD(chem), 73. *Prof Exp:* Instr chem, Morristown Col, Tenn, 59-60 & St Augustine's Col, NC, 61-63; from asst prof to prof chem, Knoxville Col, Tenn, 63-79, chmn dept, 72-79; PROF CHEM & CHMN DEPT, WINSTON-SALEM STATE UNIV, NC, 79- *Concurrent Pos:* At Sandia Labs, NMex, 74; mem fac, Lawrence Livermore Nat Lab, Univ Calif, Berkeley, 75; consult, Fisk Univ, Tenn, 77-83; fac res, Arnold Eng Ctr, Tenn, 85. *Mem:* Sigma Xi; Am Chem Soc. *Res:* Synthesis and energy; thermodynamic properties of materials. *Mailing Add:* 1843 Pope Rd Winston-Salem NC 27127

HERRERO, FEDERICO ANTONIO, b Santo Domingo, Dominican Repub, July 17, 41; US citizen; m 64; c 4. GEOPHYSICS, ATMOSPHERIC PHYSICS. *Educ:* Spring Hill Col, BS, 63; Univ Fla, Gainesville, MS, 65, PhD(physics), 68. *Prof Exp:* Res assoc excitation collisions, Johns Hopkins Univ, 70-72; asst prof physics, Univ PR-Mayagüez, 72-75; assoc prof physics, Univ PR, Rio Piedras, 75-81, chmn dept & lab collisions, 77-81; ASTROPHYSICIST, GODDARD SPACE FLIGHT CTR, NASA, 81- *Concurrent Pos:* Scientist I, PR Nuclear Ctr, Mayagüez, 73-75; investr, NIH grant, 74-75; proj dir, NSF grant, 74-75; guest worker, Nat Astron & Ionosphere Ctr, PR, 75-; proj dir, Energy Domonst Lab, Dept Energy grant, 77-; prin investr, NSF grant, 78-80. *Mem:* Am Phys Soc; Am Geophys Union; AAAS. *Res:* Study of the dynamics of the earth's ionosphere with measurements of airglow and radar backscatter intensities; theoretical and experimental studies of vibrational excitation of diatomic molecules by ion impact. *Mailing Add:* NASA Goddard Space Flight Ctr Code 692 Greenbelt MD 20771

HERRETT, RICHARD ALLISON, b Buffalo, NY, Aug 4, 32; m 58; c 4. BIOCHEMISTRY, PLANT PHYSIOLOGY. *Educ:* Rutgers Univ, BS, 54; Univ Minn, MS, 56, PhD(agr bot), 59. *Prof Exp:* Asst agron & plant genetics, Univ Minn, 54-56, asst plant path & bot, 56-58; Union Carbide Corp fel plant physiol, Boyce Thompson Inst Plant Res, Inc, 59-60; plant physiologist, Olefins Div, Union Carbide Corp, 60-67, dir biol res, 67-70; tech mgr, agr chem div, 70-75, dir res & develop, 75-87, GOVT RELS-SCI LIAISON, ICI AMERICAS INC, 87- *Concurrent Pos:* Treas & mem exec bd, NC Biotechnol Ctr; pres, C V Riley Mem Found. *Mem:* Am Soc Plant Physiol; Am Chem Soc; Weed Sci Soc Am; AAAS; Agr Res Inst. *Res:* Chemical control of plant growth; differentiation and the mechanism of action of plant growth regulators; pesticide metabolism in plants. *Mailing Add:* 16708 Shea Lane Gaithersburg MD 20877-1230

HERRICK, CLAUDE CUMMINGS, b Montello, Wis, Jan 27, 27; m 51; c 4. PHYSICAL CHEMISTRY. *Educ:* Ill Inst Technol, BS, 50, MS, 54, PhD(chem), 58. *Prof Exp:* Staff mem, Armour Res Found, 53-55; STAFF MEM, LOS ALAMOS SCI LAB, 56- *Concurrent Pos:* Vis lectr, Ariz State Univ, 69-70. *Mem:* Am Chem Soc; Am Soc Metals. *Res:* Organic chemistry; high nitrogen compounds; kinetics-acid-base catalysis; thermodynamics of metals and alloys. *Mailing Add:* Ten Goshen Court Gaithersburg MD 20882-1016

HERRICK, DAVID RAWLS, b Lafayette, Ind, May 9, 47; m 68; c 1. THEORETICAL CHEMISTRY. *Educ:* Univ Rochester, BS, 69; Yale Univ, PhD(chem), 73. *Prof Exp:* Fel chem physics, Bell Labs, 73-75; asst prof chem & assoc, 75-80, ASSOC PROF CHEM, INST THEORET SCI, UNIV ORE, 81- *Mem:* Am Phys Soc; Am Chem Soc. *Res:* Theoretical chemical physics; Lie algebraic descriptions of electron correlation; collisional and field induced Rydberg angular momentum transfer; decaying state resonances, multiply-excited atoms. *Mailing Add:* Dept Chem Univ Ore Eugene OR 97403

HERRICK, ELBERT CHARLES, b Joliet, Mont, Oct 16, 19; m 43, 62; c 4. HETEROGENEOUS CATALYSIS. *Educ:* Mont State Col, BS, 41; Princeton Univ, ChemE, 42; Mass Inst Technol, PhD(org chem), 49. *Prof Exp:* Res chem engr, Plastics Div, Monsanto Chem Co, 42; asst, Mass Inst Technol, 48-49; res chemist, E I du Pont de Nemours & Co, 49-54, Am Viscose Corp, 54-55 & Houdry Process Corp, 55-58; supvr chem res, Climax Molybdenum Co, Mich, 58-59; res chemist, Basic Res Div, Res & Eng Dept, Sun Oil Co, 59-61; sr res chemist, Dow Chem Co, 62-63; consult chemist, 63-64; sect head org res, Great Lakes Res Corp, 65-67; dir chem res, Escambia Chem Corp, 67-71; sr org chemist, Houdry Labs, Air Prod & Chem Inc, Marcus Hook, 71-77; environ systs scientist, Mitre Corp, McLean, Va, 77-88; sr scientist, Dynamac Corp, 88-89; CONSULT CHEMIST & ENGR, 84- *Mem:* AAAS; Am Chem Soc; Sigma Xi; Am Inst Chem Eng; NY Acad Sci. *Res:* Cyclooctatetraene derivatives; high pressure hydrogenation of carbon monoxide to polymethylene; polymerization of acetylene to monovinylacetylene; synthesis of peptides; catalytic synthesis of nitrogen heterocycles including the polyurethane catalyst, DABCO; carbene reaction; olefin polymerization by Ziegler type catalysts; preparation of dyeable polyproplene fibers; catalysts for toluenediamine synthesis; thermal decomposition of methyl carbamates to toluene dilsocyanates; decontamination of nitroaromatic waste streams. *Mailing Add:* 2740 Florence Rd Woodbine MD 21797-7841

HERRICK, FRANKLIN WILLARD, b Seattle, Wash, Dec 14, 22; m 45; c 2. ORGANIC CHEMISTRY. *Educ:* Willamette Univ, AB, 46; Mich State Univ, PhD(org chem), 50. *Prof Exp:* Asst anal chem, Mich State Univ, 46-47 & org chem, 47-49; chemist, Res Div, ITT Rayonier Res Ctr, 50-58, group leader, 58-63, sect leader, 63-69, res supvr, Olympic Res Div, 69-75, res assoc, 75-83, sr res assoc, 83-86; RETIRED. *Mem:* Am Chem Soc; Forest Prod Res Soc. *Res:* Synthesis of organic chemicals; chemistry of cellulose, lignin and bark; resins and adhesives; chemical grouting, soil stabilization; soil mechanics; forest genetics; fermentation of wood carbohydrates or components; enzymatic hydrolysis of wood carbohydrates; cellulose derivatives; cellulose acetate; regenerated cellulose; microfibrillated cellulose; vanilla and derivatives. *Mailing Add:* 705 Holly Lane Shelton WA 98584

HERRICK, GLENN ARTHUR, b Mar 10, 45; US citizen. MOLECULAR BIOLOGY, BIOCHEMISTRY. *Educ:* Col Wooster, BA, 67; Princeton Univ, MA, 70, PhD(biochem), 73. *Prof Exp:* ASSOC PROF CELLULAR VIRAL MOLECULAR BIOL, UNIV UTAH MED CTR, 77- *Concurrent Pos:* Jane Coffin Childs fel, Boulder, Colo, 73-75, Nat Inst GMS fel, 75-77. *Res:* Eukaryotic genome; organization and function; somatic genome reorganization in ciliated protozoa, telomeres. *Mailing Add:* Dept Microbiol Univ Utah Med Ctr 50 N Medical Dr Salt Lake City UT 84132

HERRICK, JOHN BERNE, b Sheffield, Iowa, Dec 18, 19; m 41; c 10. VETERINARY MEDICINE. *Educ:* Iowa State Univ, BS, 41, DVM, 46, MS, 50. *Prof Exp:* With Voc Agr Inst, Iowa, 41-42; asst vet physiol, Iowa State Univ, 43-46; pvt pract, 46-47; in-chg artificial insemination, Iowa State Univ, 48-51, prof exten vet in-chg cattle, 51-80, prof vet clin serv, 80-83; RETIRED. *Concurrent Pos:* Consult, Nat Asn Artificial Breeding Indust; vpres, Pan Am Vet Cong, 68-70; animal health consult, 83- *Honors & Awards:* Distinguished Serv Award, USDA, 67; Borden Award, Am Vet Med Asn, 70; Am Asn Bovine Practr Award, 86; Am Vet Med Asn Award, 87. *Mem:* Am Vet Med Asn (pres, 69-70). *Res:* Reproductive disorders and diseases of cattle; organized educational programs on disease control. *Mailing Add:* 7807 N Calle Caballeros Iowa State Univ Paradise Valley AZ 85253

HERRICK, JOSEPH RAYMOND, b McKees Rocks, Pa, Nov 2, 32; m 60; c 2. MECHANICAL ENGINEERING. *Educ:* Univ Pittsburgh, BS, 60. *Prof Exp:* Res engr aluminum, Alcoa Res Labs, 60-62, continuous casting, 62-67, sr res engr ingot casting & melting, 68-75, staff engr ingot casting prod equip, Alcoa Tech Ctr, 75-77, staff engr smelting magnesium, Alcoa-Pittsburgh Off, 77-79, staff engr, Equip Develop Div, 79-81, TECH SPECIALIST CHEM PROCESSES, EQUIP DEVELOP DIV, ALCOA TECH CTR, 81- *Mem:* Sigma Xi. *Res:* Aluminum ingot casting; melting; aluminum scrap reclamation and furnace energy efficiency. *Mailing Add:* Alcoa Labs Alcoa Tech Ctr Alcoa Center PA 15069

HERRICK, THOMAS J(EFFERSON), b Chicago, Ill, Aug 6, 13; m 40, 75; c 3. AERONAUTICAL ENGINEERING. *Educ:* Univ Ill, BSME, 36; Univ Mich, MS, 40. *Prof Exp:* Asst appl mech, Purdue Univ, 36-38, instr, 38-42; res engr, McDonnell Aircraft Corp, 42-45; asst prof aeronaut eng, 45-49, exec asst, Sch Aeronaut & Astronaut, 67-70, ASSOC PROF AERONAUT ENG, PURDUE UNIV, 49-, HEAD, SCH AERONAUT & ASTRONAUT, 70- *Mem:* Asn Comput Mach; Am Inst Aeronaut & Astronaut; Soc Eng Scientists; Sigma Xi. *Res:* Engineering mechanics; aircraft structure. *Mailing Add:* 1560 Marilyn Ave West Lafayette IN 47906

HERRICKS, EDWIN E, b Axtell, Kans, Dec 24, 46; m 68; c 2. ENVIRONMENTAL BIOMONITORING, AQUATIC ECOLOGY. *Educ:* Univ Kans, BA, 68; Johns Hopkins Univ, MS, 70; Va Polytech Inst & State Univ, PhD(biol), 74. *Prof Exp:* Res assoc civil eng & instr agr eng, Va Polytech Inst & State Univ, 73; aquatic res biologist & field edologist, Union Carbide Corp, 73-75; from asst prof to assoc prof, 75-91, PROF ENVIRON BIOL, UNIV ILL, URBANA-CHAMPAIGN, 91- *Concurrent Pos:* Res scientist, Rocky Mountain Biol Lab, Colo, 77; vis asst prof & assoc prof evolution & environ biol, Univ Pa, 79-82; vis fel, Johns Hopkins Univ, 83; Fulbright fel & distinguished prof, 89. *Mem:* Affil mem Am Soc Civil Engrs; Am Water Res Asn; Int Asn Water Pollution Res & Control; NAm Benthological Soc; Soc Environ Toxicol & Chem; Water Pollution Control Fedn. *Res:* Effects of contaminants and other environmental alterations on communities of organisms, including aquatic and terrestrial ecosystems; analysis of habitat factors which affect the ecological interactions; connections between engineering practice and environmental consequences; biological monitoring procedures for environmental decision- making. *Mailing Add:* Dept Civil Eng 3215 NCEL MC-250 Univ Ill 205 N Mathews Urbana IL 61801

HERRIN, CHARLES SELBY, b Mechanicsburg, Miss, May 16, 39; m 61; c 9. FORTRAN COBOL & C PROGRAMMING, DATABASE MANAGEMENT. *Educ:* Brigham Young Univ, BS, 63, MS, 66; Ohio State Univ, PhD(entom), 69. *Prof Exp:* Res assoc entom, Univ Ga, 69-70; NSF res assoc biol, Ga Southern Col, 70-72; consult biostatist & comput data-base mgr, Res Div, Brigham Young Univ, 75-78, mus comput mgr & programmer, 78-83; systs programmer & cust support serv supvr, Code 3 Health Info systs, 3M Corp, 83-86; SOFTWARE TESTING MGR, NOVELL INC, 86- *Concurrent Pos:* Proj dir ectoparasites of African mammals, US Army (MR & D), Ctr Health & Environ Studies, Brigham Young Univ, 72-76. *Res:* Software testing. *Mailing Add:* 920 W 1020 S Provo UT 84601

HERRIN, EUGENE THORNTON, JR, b Dallas, Tex, Nov 19, 29; m 53; c 3. GEOPHYSICS. *Educ:* Southern Methodist Univ, BS, 51, MS, 53; Harvard Univ, PhD(geol, geophys), 58. *Prof Exp:* Asst geophys, Harvard Univ, 53-55; instr geol, Southern Methodist Univ, 55-57; asst geophys, Harvard Univ, 57-58; from asst prof to prof geol & geophys, 58-74, chmn dept, 71-74, Shuler-Foscue prof geol sci, 74-80, SHULER-FOSCUE EARTH SCI, SOUTHERN METHODIST UNIV, 80- *Concurrent Pos:* Fel, Carnegie Inst Washington, 64 c. *Honors & Awards:* Grove Karl Gilbert Award, 64. *Mem:* Seismol Soc Am; fel Geol Soc Am; fel Am Geophys Union. *Res:* Experimental and theoretical seismology; problems of terrestrial heat flow and heat production; structural geology; experimental tectonics. *Mailing Add:* Dept Geol Sci Southern Methodist Univ Dallas TX 75275

HERRIN, MORELAND, b Morris, Okla, Nov 14, 22; m 46; c 3. CIVIL ENGINEERING. *Educ:* Okla State Univ, BSCE, 47, MSCE, 49; Purdue Univ, PhD(hwy mat), 54. *Prof Exp:* Instr struct, Okla State Univ, 47-50; design engr, Hudgins, Thompson & Ball, Engrs, 50-51; from asst prof to assoc prof hwy eng, Okla State Univ, 54-58; assoc prof mat, 58-63, PROF MAT, UNIV ILL, URBANA, 63- *Concurrent Pos:* Comt chmn, Hwy Res Bd, 63-; pvt consult hwy mat. *Honors & Awards:* Epstein Mem Award, 63. *Mem:* Asn Asphalt Paving Technol (pres, 80); Am Soc Eng Educ; Am Soc Civil Engrs. *Res:* Rheological behavior and failure mechanics of bituminous materials; durability and strength of stabilized materials. *Mailing Add:* 1414 W William St Champaign IL 61821

HERRING, CAREY REUBEN, b Tifton, Ga, Oct 4, 43; m 65; c 2. MATHEMATICS EDUCATION. *Educ:* Carson-Newman Col, BS, 65; Samford Univ, MA, 67; Univ Tenn, EdD, 75. *Prof Exp:* PROF MATH, CARSON-NEWMAN COL, 67- *Mem:* Nat Coun Teachers Math; Math Asn Am. *Mailing Add:* Dept Nat Sci Carson-Newman Col Russell Ave Jefferson City TN 37760

HERRING, H(UGH) JAMES, b Boston, Mass, Aug 3, 39; m 60; c 3. AERONAUTICAL ENGINEERING. *Educ:* Harvard Col, BA, 61, BS, 62; Princeton Univ, MA & MS, 66, PhD(aeronaut eng), 67. *Prof Exp:* Mem res staff, Dept Aerospace & Mech Sci, Princeton Univ, 67-76; PRES, DYNALYSIS OF PRINCETON, 70- *Concurrent Pos:* Assoc ed, J Fluids Eng, 78-81. *Mem:* AAAS; Sigma Xi; Am Inst Aeronaut & Astronaut (treas, 76-79); Am Soc Mech Engr. *Res:* Fluid mechanics and numerical simulation. *Mailing Add:* Eighteen Winfield Rd Princeton NJ 08540

HERRING, HAROLD KEITH, b Woodward, Okla, May 16, 39; m 65; c 2. ANIMAL SCIENCE, BIOCHEMISTRY. *Educ:* Okla State Univ, BS, 61; Univ Wis, MS, 66, PhD(meat & animal sci), 68. *Prof Exp:* Res asst meat sci, Univ Wis, 63-67; scientist, Food Res Div, 68-70, sect head, 70-73, asst dir res, Food Res Div, Armour & Co, 73-77; res div mgr, Wilson Foods Corp, 77-84; CAMPBELL INST RES TECHNOL, CAMDEN, NJ, 84- *Mem:* Am Chem Soc; Am Soc Animal Sci; Inst Food Technologists; Am Meat Sci Asn. *Res:* Meat science and biochemistry; product development, fresh, processed and sausage products. *Mailing Add:* Campbell Inst Res Technol Campbell Pl Camden NJ 08103

HERRING, JACKSON REA, b Ashland, Ky, Oct 2, 31; m 59; c 2. THEORETICAL PHYSICS, FLUIDS. *Educ:* Wake Forest Col, BS, 53; Univ NC, MA, 56, PhD(physics), 59. *Prof Exp:* Theoret physicist, Goddard Space Flight Ctr, NASA, 59-72; SR SCIENTIST, NAT CTR ATMOSPHERIC RES, 72- *Concurrent Pos:* Mem, ed adv comt, World Sci Publ Co, 85-; assoc ed, Physics Fluids, 86- *Mem:* Am Phys Soc. *Res:* Nuclear structure; statistical physics; fluid turbulence theory. *Mailing Add:* 2581 Briarwood Dr Boulder CO 80303

HERRING, JOHN WESLEY, JR, b Brookhaven, Miss, Aug 8, 27; m 59; c 3. ELECTRIC POWER SYSTEMS & CONTROLS. *Educ:* Miss State Univ, BSEE, 50; Univ Ariz, MSEE, 60, PhD(elec eng), 67. *Prof Exp:* Chief comput, Western Geophys Co Am, 50-52; engr, Miss Power Co, 52-57; instr elec eng, Miss State Univ, 57-58, asst prof, 59-63; instr, Univ Ariz, 66-67; assoc prof, 67-80, PROF ELEC ENG, MISS STATE UNIV, 80- *Mem:* Inst Elec & Electronics Engrs; Am Soc Eng Educ. *Res:* Power system stability; nonlinear stability theory; design and analysis of control systems. *Mailing Add:* Dept Elec Eng Miss State Univ PO Drawer EE Mississippi State MS 39762

HERRING, RICHARD NORMAN, b Reading, Pa, Dec 25, 38; m 61; c 4. CHEMICAL ENGINEERING, CRYOGENICS. *Educ:* Lehigh Univ, BS, 60; Univ Colo, MS, 62, PhD(chem eng), 64. *Prof Exp:* Chem engr, Cryogenic Div, Nat Bur Standards, 60-64; res dir thermodyn, P-V-T Inc, 64-65; adv design engr cryogenics, Beech Aircraft Corp, 65-68; staff scientist, Ball Bros Res Corp, 68-72, mgr advan progs, 72-78, dir sci & applns exp, 78-79, exec vpres, 79-84, PRES, BALL AEROSPACE SYSTS DIV, 85- *Res:* Management of space experiments and spacecraft for astrophysics, earth applications and Department of Energy requirements. *Mailing Add:* Ball Aerospace Systs Div PO Box 1062 Boulder CO 80306

HERRING, SUSAN WELLER, b Pittsburgh, Pa, Mar 25, 47; div. ANATOMY. *Educ:* Univ Chicago, BS, 67, PhD(anat), 71. *Prof Exp:* From asst prof to assoc prof oral anat, Univ Ill, 72-86, prof oral anat & anat, 86-90; PROF ORTHODS & ADJ PROF ZOOL, UNIV WASH, SEATTLE, 90- *Concurrent Pos:* NIH fel oral anat, Univ Ill Med Ctr, 71-72, Nat Inst Dent Res grants, 74-77, 82-; vis assoc prof biol sci, Univ Mich, 82; chmn, Div Vert Morphol, Am Soc Zool, 81-84; Muscular Dystrophy Asn grant, 82-83; Study Sect (Oral Biol & Med), NIH, 86-89; NSF grant, 90-92. *Mem:* AAAS; Am Soc Mammal; Am Soc Zool; Soc Vert Paleont; Am Asn Anat; Int Asn Dent Res. *Res:* Functional mammalian anatomy; evolutionary biology. *Mailing Add:* Dept Orthods Univ Wash SM-46 Seattle WA 98195

HERRING, WILLIAM BENJAMIN, b Pender County, NC, July 16, 28; m 50; c 2. INTERNAL MEDICINE. *Educ:* Wake Forest Col, BS, 49, MD, 53; Am Bd Internal Med, dipl, 63. *Prof Exp:* Asst resident internal med, Univ Va Hosp, 56-57, fel cardiol, 57-58, fel hemat, 58-59; res fel hemat, 61-63, from asst prof to assoc prof internal med, 63-78, chief med teaching serv, Moses H Cone Mem Hosp, 67-81, PROF MED, SCH MED, UNIV NC, CHAPEL HILL, 78- *Mem:* Am Fedn Clin Res; Am Soc Hemat; fel Am Col Physicians. *Res:* Hematology. *Mailing Add:* Moses H Cone Mem Hosp 1200 N Elm St Greensboro NC 27402

HERRING, WILLIAM CONYERS, b Scotia, NY, Nov 15, 14; m 46; c 4. THEORETICAL SOLID STATE PHYSICS. *Educ:* Univ Kans, AB, 33; Princeton Univ, PhD(math physics), 37. *Prof Exp:* Nat Res Coun fel, Mass Inst Technol, 37-39; instr math & res assoc math physics, Princeton Univ, 39-40; instr physics, Univ Mo, 40-41; mem sci staff, Div War Res, Columbia Univ, 41-45; prof, Univ Tex, 46; mem tech staff, Bell Labs, 46-78; prof, 78-81, EMER PROF APPL PHYSICS, STANFORD UNIV, 81- *Concurrent Pos:* Mem, Inst Advan Study, 52-53. *Honors & Awards:* Buckley Prize, Am Phys Soc, 59; Luck Award, Nat Acad Sci, 80; Von Hippel Award, Mat Res Soc, 80; Wolf Prize physics, 85. *Mem:* Nat Acad Sci; AAAS; fel Am Phys Soc; fel Am Acad Arts & Sci; Am Soc Info Sci; Mat Res Soc. *Res:* Theory of solids; scientific communication. *Mailing Add:* Dept Appl Physics Stanford Univ Stanford CA 94305

HERRING, WILLIAM M(AYO), chemical engineering, systems engineering, for more information see previous edition

HERRINGTON, KERMIT (DALE), b Sioux Falls, SDak, Apr 19, 23; m 47; c 3. CHEMISTRY. *Educ:* Augustana Col, AB, 44; Wayne State Univ, MS, 46, PhD(chem), 50. *Prof Exp:* Res chemist, Titanium Div, Nat Lead Co, 50-57, supvr basic sci dept, 57-70, mgr pigments res, NL Industs, Inc, 70-75. *Mem:* Am Chem Soc. *Res:* Physical chemistry of titanium dioxide pigments; research and development on fine particle technology. *Mailing Add:* Box 338 Peapack NJ 07977

HERRINGTON, LEE PIERCE, b New Haven, Conn, June 11, 33; m 55; c 4. URBAN FORESTRY, FOREST METEOROLOGY. *Educ:* Univ Maine, BS, 59; Yale Univ, MF, 60, PhD(forest meteorol), 64. *Prof Exp:* Sr meteorologist, Melpar, Inc, Westinghouse Air Brake Co, 64-65; from asst prof to prof meteorol, State Univ NY Col Environ Sci & Forestry, 65-78; urban forest res coordr, Northeast Forest Exp Sta, US Forest Serv, 78-79; exec secy, Consortium Environ Forestry Studies & Coordr, Res & Grad Studies, Sch Forestry, Col Environ Sci & Forestry, State Univ NY, 79-83; vpres, 81-83, PRES, FORESTRY SOFTWARE ASSOCS, SYRACUSE, NY, 84- *Mem:* Am Meteorol Soc; Am Forestry Asn; Sigma Xi; Soc Am Foresters. *Res:* Computer applications to forest management; expert systems for resources management; urban meteorology; energy transfer within plant parts and between plants and the environment. *Mailing Add:* 113 Janet Dr Syracuse NY 13224

HERRINTON, PAUL MATTHEW, b Detroit, Mich, Sept 21, 57. PROCESS RESEARCH & DEVELOPMENT, ORGANIC SYNTHETIC METHODS. *Educ:* Wayne State Univ, BS, 79; Mich State Univ, MS, 82, PhD(chem), 84. *Prof Exp:* Postdoctoral fel, Univ Calif, Irvine, 85-86; asst prof, Nazareth Col, 90-91; RES CHEMIST, UPJOHN CO, 86- *Mem:* Am Chem Soc; affil mem Int Union Pure & Appl Chem. *Res:* Identify and develop new reactions and processes used in manufacturing pharmaceutical and fine chemicals; steroids; antibiotics. *Mailing Add:* Upjohn Co 1500-91-2 Kalamazoo MI 49002

HERRIOT, JOHN GEORGE, b Winnipeg, Man, Mar 7, 16; nat US; m 41; c 4. COMPUTER SCIENCE, NUMERICAL ANALYSIS. *Educ:* Univ Man, BSc, 37; Brown Univ, ScM, 38, PhD(math), 41. *Prof Exp:* Instr math, Brown Univ, 39-41 & Yale Univ, 41-42; from instr to prof math, 42-61, instr eng, sci & mgt war training, 45, prof, 61-82, EMER PROF COMPUT SCI, STANFORD UNIV, 82- *Concurrent Pos:* Physicist, Ames Aeronaut Lab, Nat Adv Comt Aeronaut, Moffett Field, Calif, 44-46; Fulbright lectr, Univ Grenoble, France, 62-63; guest prof, Math Inst, Tech Univ, Munich, Ger, 69-70; guest mem, Math Res Inst, Fed Inst Technol, Zurich, Switz, 76-77; guest prof math, Inst Tech Univ, Munich, Ger, 80. *Mem:* Am Math Soc; Math Asn Am; Asn Comput Mach; Soc Indust & Appl Math. *Res:* Multiple Fourier series; potential theory; numerical analysis; computing machines; spline fuctions; partial differential equations. *Mailing Add:* Dept Comput Sci Stanford Univ Stanford CA 94305

HERRIOTT, ARTHUR W, b New Brighton, Pa, June 17, 41; m 64; c 2. ORGANIC CHEMISTRY. *Educ:* Col Wooster, BA, 63; Univ Fla, PhD(chem), 67. *Prof Exp:* NSF fel chem, Princeton Univ, 67-68; asst prof, State Univ NY Albany, 68-73; assoc prof, 73-80, ASSOC DEAN COL ARTS & SCI, FLA INT UNIV, 76-, PROF, 80- *Mem:* Am Chem Soc; AAAS. *Res:* Physical organic chemistry; catalysis mechanisms; organophosphorus chemistry. *Mailing Add:* Dept Chem Fla Int Univ Miami FL 33199

HERRIOTT, DONALD R, b Rochester, NY, Feb 2, 28; c 4. OPTICS, ELECTRICAL ENGINEERING. *Educ:* Duke Univ, BA, 49; Univ Rochester, MA, 51. *Prof Exp:* Mem sci bur, Bausch & Lomb Co, 49-56; mem staff, res dept, Bell Lab, 56-68, head dept, lithographic syst, 68-81; sr sci adv, Perkin-Elmer Corp, 81-90; RETIRED. *Honors & Awards:* Cledo Brunetti Award, Inst Elec & Electronics Engrs; Franhofer Medal, Optical Soc Am; Res Coun Patent Year, 79; Edison Award, 86. *Mem:* Nat Acad Eng; fel Optical Soc Am (pres, 84); sr mem Inst Elec & Electronics Engrs. *Res:* Lasers-optics-interferometry-electron beams. *Mailing Add:* 1237 Isabel Dr Sanibel FL 33957

HERRIOTT, JON R, b New York, NY, Apr 8, 37; m 61; c 3. BIOCHEMISTRY, BIOPHYSICS. *Educ:* Dartmouth Col, BA, 59; Johns Hopkins Univ, PhD(biophys), 67. *Prof Exp:* Sr fel, Dept Biol Struct, 67-69, asst prof biochem, 69-77, ASSOC PROF BIOCHEM, UNIV WASH, 77- *Res:* Protein structure; x-ray crystallography; structure-function relationships in macromolecules; proton nuclear magnetic resonance. *Mailing Add:* Dept Biochem Univ Wash Seattle WA 98195

HERRIOTT, ROGER MOSS, b Des Moines, Iowa, Mar 13, 08; m 32; c 2. BIOCHEMISTRY. *Educ:* Drake Univ, AB, 28; Columbia Univ, AM, 29, PhD(chem), 32. *Prof Exp:* Asst gen physiol, Rockefeller Inst, 32-37, assoc, 37-48; head dept, 48-75, PROF BIOCHEM, SCH HYG & PUB HEALTH, JOHNS HOPKINS UNIV, 48- *Mem:* Am Chem Soc; Am Soc Biol Chemists; Am Soc Microbiol; Biophys Soc. *Res:* Chemical reactions of enzymes, viruses and proteins; solubility analysis as a criterion of purity; chemical structure of pepsin; action of bacteriophage; mechanism of bacterial genetic transformation; in vitro photoreactivation; hybrid DNA formation; heat labile precursors of mutagens in fresh beef. *Mailing Add:* 504 Highland Ave Towson MD 21204

HERRMANN, CHRISTIAN, JR, b Lansing, Mich, Jan 25, 21. CLINICAL NEUROLOGY. *Educ:* Univ Mich, AB, 42, MD, 44. *Prof Exp:* Asst neurol, Columbia Univ, 50-52, instr, 52-53; from asst prof to assoc prof, 54-69, PROF NEUROL MED, SCH MED, UNIV CALIF, LOS ANGELES, 69- *Concurrent Pos:* USPHS res fel, Columbia Univ, 52-54, vis fel, 53-54. *Mem:* Asn Res Nervous & Ment Dis; Am Neurol Asn; Am Acad Neurol; Sigma Xi. *Res:* Neurological and neuromuscular disorders, especially myasthenia gravis. *Mailing Add:* Neurol/Reed Res Ctr Univ Calif 405 Milgard Ave Los Angeles CA 90024

HERRMANN, ERNEST CARL, JR, b Elkhart, Ind, Jan 1, 25; m 70; c 2. VIROLOGY. *Educ:* Univ Md, BS, 50, MS, 51, PhD(bact), 53. *Prof Exp:* Res assoc virol, Inst Microbiol, Rutgers Univ, 53-54; head virol sect, Chemother Dept, E R Squibb & Sons, 54-56; head virol lab, Biochem Dept, Res Div, Schering Corp, 56-61; assoc prof virol, Mayo Grad Sch Med, Univ Minn & head virol lab, Mayo Clin, 61-71; assoc prof microbiol, Univ Ill Col Med, Peoria, 71-87; VPRES, MOBILAB INC, 75-, LAB DIR, 81- *Concurrent Pos:* Consult, Am Med Asn, NIH, Am Bd Bioanal & various firms & insts. *Mem:* Int Soc Antiviral Res; Am Soc Microbiol; Pan-Am Group Rapid Viral Diag. *Res:* Antiviral agents; relation of viruses to human diseases; diagnostic virology. *Mailing Add:* Mobilab Inc 5116 Big Hollow Rd Peoria IL 61615

HERRMANN, GEORGE, b USSR, Apr 19, 21; US citizen. STRUCTURAL MECHANICS, STRUCTURAL ACOUSTICS. *Educ:* Swiss Fed Inst Technol, Zurich, CE, 45, PhD(mech), 49. *Prof Exp:* Asst prof mech, Ecole Polytech, Montreal, 49-50; from asst prof to assoc prof, Columbia Univ, 50-62; prof, Northwestern Univ, 62-70; chmn, Div Appl Mech, 75-84, PROF APPL MECH & CIVIL ENG, STANFORD UNIV, 70-, CHMN, DEPT APPL MECH, 70- *Concurrent Pos:* Sci liaison officer, Dept Navy, Off Naval Res Br Off, London, 60-61; consult, Off Sci Res, USAF, 57-71, 64-69, Eng Sci Div, US Army, Stanford Res Inst, 70-80, Aerospace Corp, 72-76; adv eval panels, Grad Traineeship Progs, NSF, 64 & 65; res assoc, Lab Exp Surg, Swiss Res Inst, Davos, Switz, 69-70; co-organizer, Conf Continuum Mech Solids, Oberwolfach, Ger, 81, 83, 86 & 90; mem, Comt Recommendations, US Army Basic Sci Res, Nat Res Coun, 81-84. *Honors & Awards:* Centennial Medal, Am Soc Mech Engrs, 80; Theodore von Karman Medal, Am Soc Civil Engrs, 81; Silver Anniversary, Soc Eng Sci, 88; Distinguished Serv Award, Am Acad Mech, 89. *Mem:* Nat Acad Eng; Am Soc Civil Engrs; hon mem Am Soc Mech Engrs; assoc fel Am Inst Aeronaut & Astronaut; Am Soc Eng Educ; Am Acad Mech; Acoust Soc Am; Biomed Eng Soc; Soc Appl Math & Mech; Int Soc Interaction Mech & Math. *Res:* Wave propagation; acoustoelasticity; composite materials dynamics; dynamic stability; dynamic fracture mechanics; conservation laws in continuum mechanics. *Mailing Add:* Dept Mech Eng Div Appl Mech Durand Bldg Stanford Univ Stanford CA 94305-4040

HERRMANN, HEINZ, b Vienna, Austria, Oct 17, 11; US citizen; m 47; c 1. CHEMICAL EMBRYOLOGY. *Educ:* Univ Vienna, MD, 36. *Prof Exp:* Res asst biochem, Univ Vienna, 32-36; Rask-Oerstedt fel tissue culture & protein chem, Carlsberg Labs, Carlsberg Found, 36-39; instr ophthal, Sch Med, Johns Hopkins Univ, 39-46; asst prof embryol, Yale Univ, 46-49; assoc prof pediat & head lab chem embryol, Sch Med, Univ Colo, 49-59; prof, Univ Conn, 59-79, emer prof biol, 79-; RETIRED. *Concurrent Pos:* Corresp comt develop biol, Nat Acad Sci, 53-55; mem study sect morphogenesis & genetics, NIH, 55-58 & cell biol, 60-64; res prof, Am Cancer Soc, 60-80; vis prof NATO, Univ Milan, 71. *Mem:* Am Soc Cell Biol; Am Soc Biol Chem; Soc Develop Biol. *Res:* Molecular mechanisms of embryonic development, particularly control of protein synthesis; relationship of cell structure and function on the molecular level; author of several books and papers about cell biology, philosophy of biology and philosophy of science. *Mailing Add:* 82 Willowbrook Rd Storrs CT 06268

HERRMANN, JOHN BELLOWS, b Cincinnati, Ohio, Nov 21, 32; m 60; c 3. MEDICINE, SURGERY. *Educ:* Dartmouth Col, AB, 54, cert med, 55; Harvard Med Sch, MD, 57; Am Bd Surg, dipl, 63. *Prof Exp:* Intern surg, Mass Gen Hosp, Boston, 57-58, resident, 58-62, clin asst, 64-65; resident surgeon, Walter Reed Army Inst Res, 62-64; from asst prof to assoc prof surg, Georgetown Univ, 65-72; chief of surg, Worcester City Hosp, 72-83; PROF SURG, UNIV MASS MED SCH, WORCESTER, 72-, VCHMN DEPT SURG, 83- *Concurrent Pos:* Asst chief of surg, Vet Admin Hosp, Washington, DC, 65-72. *Mem:* AMA; Fel Am Col Surg; Asn Acad Surg. *Res:* Experimental surgery; vascular physiology; wound healing; experimental pathology; surgical education. *Mailing Add:* Dept Surg Univ Mass Med Sch 55 Lake Ave N Worcester MA 01656

HERRMANN, KENNETH L, b Cincinnati, Ohio, Nov 16, 34; m 59; c 2. MEDICINE, VIROLOGY. *Educ:* Dartmouth Col, AB, 56; Harvard Med Sch, MD, 59. *Prof Exp:* Rotating intern, Strong Mem Hosp, Med Ctr, Univ Rochester, 59-61, resident pediat, 61-62; med officer, Commun Dis Ctr, 62-64, chief virus reference lab, 64-66, chief viral exanthems unit, 67-70, chief perinatal virol br, 70-81, chief viral exanthems and herpes virus br, 81-85, ASST DIR, DIR VIRAL DIS, CTR DIS CONTROL, USPHS, 85- *Concurrent Pos:* Asst chief resident pediat, Buffalo Children's Hosp, NY, 66-67. *Mem:* AMA; Sigma Xi. *Res:* Methods of diagnostic virology; rubella virus infection. *Mailing Add:* CDC Virol Div 1600 Clifton Rd NE Atlanta GA 30333

HERRMANN, KENNETH WALTER, b Chicago, Ill, Dec 13, 29; m 51; c 5. PHYSICAL CHEMISTRY, SURFACE CHEMISTRY. *Educ:* Valparaiso Univ, BA, 51; Iowa State Univ, PhD(rare earth phys properties), 55. *Prof Exp:* Res chemist, 55-62, head surface chem sect, 62-66 & oral prods res sect, Winton Hill Tech Ctr, 66-68, assoc dir, Toilet Goods Prod Res, 68-72, assoc dir, Corp Res & Develop, Int Toilet Goods, 72-74, assoc dir, tech serv dept, Bar Soap & Household Cleaning Prod Div, 74-81, MGR, PROD SAFETY & REGULATORY SERV, SHARON WOODS TECH CTR, PROCTER & GAMBLE CO, 81- *Mem:* Am Chem Soc; Int Asn Dent Res. *Res:* Physical-metallurgical properties of rare earth elements; modification of chemical, surface and mechanical properties of fibrous proteins; micellar, surface, adsorption and phase properties of surfactants; dental research; chemistry of biological systems; toxicology. *Mailing Add:* 830 Carini Lane Cincinnati OH 45218

HERRMANN, KLAUS MANFRED, b Lingen, Ger, June 27, 37; m 70; c 2. BIOCHEMISTRY, MOLECULAR BIOLOGY. *Educ:* Univ Munster, Diplomchem Vorexamen, 60; Univ Freiburg, Diplomchem Hauptexamen, 64, Dr rer nat(chem), 66. *Prof Exp:* German Res Found grants, Univ Freiburg, 66-67 & Stanford Univ, 67-68, res assoc, 68-69; from asst prof to prof, 69-83, PROF BIOCHEM, PURDUE UNIV, 83- *Concurrent Pos:* Chmn, Univ Senak, 85-86; asst dean, Grad Sch, 86-87. *Mem:* Am Soc Biol Chemists; Am Soc Plant Physiol. *Res:* Molecular genetics; allosteric enzymes. *Mailing Add:* Dept Biochem Purdue Univ West Lafayette IN 47907-6799

HERRMANN, LEO ANTHONY, b San Antonio, Fla, Oct 14, 25; m 47; c 4. GEOLOGY. *Educ:* Miami Univ, Ohio, BS, 47; Johns Hopkins Univ, PhD, 51. *Prof Exp:* Party chief geol & geophys, NJ Zinc Co, 51-53, resident geologist, 53-56; explor geologist, Magnolia Petrol Co Div, Socony Mobil Oil Co, Inc, 56-59; sr explor geologist, Mobil Oil Co Div, 59-66; assoc prof geol, 66-70, PROF GEOL, LA TECH UNIV, 70-, HEAD DEPT GEOSCI, 72- *Concurrent Pos:* Asst geologist, Ga Geol Surv, 49-50; consult various industs, 66- *Mem:* Fel Geol Soc Am; Am Asn Petrol Geologists; Sigma Xi. *Res:* Exploration for petroleum; computer applications in exploration; subsurface stratigraphy. *Mailing Add:* Geosci Dept La Tech Univ Ruston LA 71270

HERRMANN, LEONARD R(ALPH), b Quinn, SDak, Mar 21, 36; m 57, 88; c 2. CIVIL ENGINEERING, APPLIED MECHANICS. *Educ:* Univ Calif, Berkeley, BS, 58, MS, 60, PhD(civil eng), 62. *Prof Exp:* Sr res engr, Aerojet-Gen Corp, 62-65; chmn civil eng dept, 72-76, PROF CIVIL ENG, UNIV CALIF, DAVIS, 65- *Concurrent Pos:* Consult, Aerojet-Gen Corp, 65-, Thiokol Chem Corp, 67-69, Calif Dept Water Resources, 68-69 & Div Hwys, 69-, Lockheed Corp, 70, Sandia Corp, 71 & United Technol, 77. *Mem:* Am Soc Civil Engrs. *Res:* Structural mechanics; development and application of finite element procedure; material characterization. *Mailing Add:* Dept Civil Eng Univ Calif Davis CA 95616

HERRMANN, RAYMOND, b Chicago, Ill, July 16, 41; m 65; c 2. PARK RESOURCES MANAGEMENT. *Educ:* Columbia Univ, BS, 68; Univ Wyo, MS, 72, PhD(geol), 72. *Prof Exp:* Hydrologist/geologist, Southeast Region, Nat Park Serv, 73-74, chief, Natural Sci Res Div, 74-79, chief, Air & Water Resources Div, 79-81, dir, Water Resources, Field Support Lab, 81-85, chief appl res br, Water Resources Div, 85-90, LEADER, WATER RESOURCES COOP PARK STUDIES UNIT, NAT PARK SERV, 90- *Concurrent Pos:* Assoc fac, Colo State Univ, 81-; adj assoc prof, Univ Tenn, Knoxville, 80-82, fac assoc, 82- *Mem:* Am Water Resources Asn (vpres, 86, pres, 88); fel Geol Soc Am; AAAS; Sigma Xi. *Res:* Study of the effects of low level acid inputs on park streams and aquatic resources; technology assessment research as related to development of remote natural area monitoring and research in parks and equivalent reserves; geologic and water resources research and applications directed towards solution of park resources management problems; long term ecological investigations: US/USSR bilateral research. *Mailing Add:* 927 Greenfields Ft Collins CO 80524

HERRMANN, ROBERT ARTHUR, b Baltimore, Md, Apr 29, 34; m 69; c 3. NONSTANDARD LOGIC. *Educ:* Johns Hopkins Univ, BA, 63; Am Univ, MA, 68, PhD(math), 73. *Prof Exp:* PROF MATH, US NAVAL ACAD, ANNAPOLIS, MD, 68- *Concurrent Pos:* Prin investr, US Naval Acad Res Coun, 76-88; adv, US Cong, 68-; dir, Inst Math Philos, 80- *Mem:* Sigma Xi; Am Math Soc; Math Asn Am; Am Sci Affil. *Res:* Mathematical philosophy; mathematical logic; nonstandard analysis; model theory; applied mathematics; general topology; philosophical and natural system models; author and coauthor of over 130 articles and reports. *Mailing Add:* Math Dept US Naval Acad Annapolis MD 21402-5002

HERRMANN, ROBERT BERNARD, b Cincinnati, Ohio, Dec 22, 44; m; c 2. SEISMOLOGY. *Educ:* Xavier Univ, Ohio, BS, 67; St Louis Univ, PhD(geophys), 74. *Prof Exp:* Res assoc geophys, Coop Inst Res Environ Sci, Univ Colo, 74-75; from asst to assoc prof, 75-83, PROF GEOPHYS, ST LOUIS UNIV, 83- *Mem:* Seismol Soc Am; Am Geophys Union; Soc Explor Geophysicists. *Res:* Transmission of elastic waves; prediction of ground motion due to local earthquakes. *Mailing Add:* Dept Earth Sci St Louis Univ 221 N Grand Blvd St Louis MO 63103

HERRMANN, ROBERT LAWRENCE, b New York, NY, July 17, 28; m 50; c 4. BIOCHEMISTRY, BIOETHICS. *Educ:* Purdue Univ, BS, 51; Mich State Univ, PhD(biochem), 56. *Prof Exp:* Asst biochem, Mich State Univ, 52-56; res assoc, Mass Inst Technol, 58-59; from asst prof to assoc prof biochem, Sch Med, Boston Univ, 59-76; prof biochem & chmn dept, Schs Med & Dent, Oral Roberts Univ, 76-81, assoc dean biomed sci, 77-79; lectr, 81-82, ADJ PROF CHEM, GORDON COL, 82-, EXEC DIR, AM SCI AFFIL, 81- *Concurrent Pos:* Damon Runyon fel, Mass Inst Technol, 56-58; trustee, Barrington Col, 76-79, Templeton Found, 86-, Southeastern Mass Univ, 88-; chmn med ethic comn, Christian Med Soc, 76-83. *Honors & Awards:* Staley Lectr, Gordon Col, 74, Stony Brook, 84, Eastern Col, 85, Trinity Col, 87. *Mem:* Fel AAAS; Soc Health Human Values; fel Geront Soc; Am Soc Biol Chem; fel Am Sci Affil; Victoria Inst, Philos Soc Gt Brit; Sigma Xi. *Res:* Biochemical genetics; enzyme structure and mechanism; viral etiology of cancer; genetic mechanisms of aging. *Mailing Add:* Dept Chem Gordon Col Wenham MA 01984

HERRMANN, ROY G, pharmacology; deceased, see previous edition for last biography

HERRMANN, SCOTT JOSEPH, b Chicago, Ill, Apr 6, 42; m 67; c 3. TRICHOPTERA & CHIRONOMIDAE, SYSTEMATICS & ECOLOGY. *Educ:* Northern Ill Univ, BS, 64; Univ Colo, PhD(zool), 68. *Prof Exp:* From asst prof to assoc prof, 68-77, PROF BIOL, UNIV SOUTHERN COLO, 77- *Mem:* Am Entom Soc; Sigma Xi; NAm Benthol Soc. *Res:* Chemical and physical limnology as applied to trace metal pollution; systematics and ecology of Trichoptera, Plecoptera and Chironomidae. *Mailing Add:* Dept Biol Univ Southern Colo Pueblo CO 81001-4901

HERRMANN, STEVEN H, BIOCHEMISTRY, CELL BIOLOGY. *Educ:* Wash Univ, PhD(biochem), 78. *Prof Exp:* Assoc prof path, Med Sch, Harvard Univ, 84-; AT DEPT CELL IMMUNOL, GENETICS INST. *Mailing Add:* Dept Cell Immunol Genetics Inst 87 Cambridge Park Dr Cambridge MA 02140-2387

HERRMANN, ULRICH OTTO, b Schneidemühl, Ger, Mar 27, 25; m 53; c 2. PHYSICS. *Educ:* Univ Giessen, BS, 51, MS, 53, PhD(physics), 56. *Prof Exp:* Res physicist, Well Surv, Inc Div, Dresser Industs, 56-58; staff physicist, Jet Res Ctr, Inc, 58-61; from asst prof to assoc prof, 61-75, PROF PHYSICS, UNIV TEX, ARLINGTON, 75- *Mem:* AAAS; Am Asn Physics Teachers; Am Astronaut Soc. *Res:* Radioactive well logging; explosive shaped charges; crystal optics and colorimetry. *Mailing Add:* Dept Physics Univ Tex Box 19088 UTA Sta Arlington TX 76019

HERRMANN, WALTER, b Johannesburg, SAfrica, May 2, 30; US citizen; m 54; c 2. CONSTITUTIVE EQUATIONS, WAVE PROPAGATION IN SOLIDS. *Educ:* Univ Witwatersrand, BSc, 51, PhD(eng), 55. *Prof Exp:* Res engr, Aeroelastic & Struct Res Lab, Mass Inst Technol, 53-55, sr res engr, 57-59, aeronaut res engr, 59-64; lectr mech eng, Univ Cape Town, 55-57; div supvr, Sandia Corp, 64-67, dept mgr, 67-82, DIR ENG SCI, SANDIA NAT LABS, 82- *Mem:* Am Phys Soc; Am Soc Mech Engrs; Am Acad Mech; Soc Natural Philos. *Res:* Transient compressible aerodynamics; shock tubes; hypervelocity impact; stress waves in solids; elastic-plastic shell response; constitutive equations. *Mailing Add:* Sandia Nat Labs Org 1600 PO Box 5800 Albuquerque NM 87185

HERRMANN, WALTER L, b Berlin, Ger, Mar 28, 23; nat US; m 52; c 3. OBSTETRICS & GYNECOLOGY. *Educ:* Univ Geneva, Switz, BMedSc, 45, MD, 49; Am Bd Obstet & Gynec, dipl, 59. *Prof Exp:* Assoc obstetrician & gynecologist, Grace New Haven Community Hosp, 55; from instr to assoc prof obstet & gynec, Sch Med, Yale Univ, 55-61; prof obstet & gynec, Sch Med, Univ Wash, 61-76, chmn dept, 68-76; prof & chmn dept, 76-88, HON PROF OBSTET & GYNEC, SCH MED, UNIV GENEVA, 88- *Concurrent Pos:* Vis fel, Harvard Univ, 51-52; assoc examr, Am Bd Obstet & Gynec, 65-; consult, Ford Found, Study Sect NIH, Swiss Acad Med Sci. *Honors & Awards:* Purdue Frederick Award, Am Col Obstet & Gynec, 67, President's First Award, 70. *Mem:* AAAS; Am Gynec Obstet Soc; Endocrine Soc; Am Col Obstet & Gynec; Soc Gynec Invest. *Res:* Endocrinology; infertility; steroid chemistry. *Mailing Add:* Dept Obstet & Gynec 9 rue John-Grasset Geneva 1205 Switzerland

HERRMANNSFELDT, WILLIAM BERNARD, b Chicago, Ill, Apr 22, 31; m 54; c 2. ELECTRON OPTICS. *Educ:* Univ Educ: Miami Univ, Ohio, AB, 53; Univ Ill, MS, 55, PhD(physics), 58. *Prof Exp:* Mem staff, Los Alamos Sci Lab, 58-62; MEM STAFF, STANFORD LINEAR ACCELERATOR CTR, 62- *Concurrent Pos:* Consult work in electron gun design, 65-; mem fusion policy adv comt, DOE, 90. *Mem:* Fel Am Phys Soc. *Res:* Design and planning for high energy accelerators and storage rings; heavy ion accelerators for inertially confined fusions; computer modeling of charged particle beams. *Mailing Add:* Stanford Linear Accelerator Ctr Stanford CA 94309

HERRNKIND, WILLIAM FRANK, b Bayshore, NY, Oct 15, 40; m 62; c 2. ANIMAL BEHAVIOR, MARINE BIOLOGY. *Educ:* State Univ NY Albany, BS, 61; Univ Miami, MS, 65, PhD(marine biol), 68. *Prof Exp:* From asst prof to assoc prof, 67-80, PROF BIOL SCI, FLA STATE UNIV, 80- *Mem:* AAAS; Am Inst Biol Sci; Am Soc Zoologists; Animal Behav Soc; Sigma Xi. *Res:* Orientation of marine animals, especially crustaceans, the adaptive significance and functional mechanisms; ontogenetic development of orientational and related behavior. *Mailing Add:* Dept Biol Sci Fla State Univ Tallahassee FL 32306

HERROD, HENRY GRADY, IMMUNOLOGY, PEDIATRICS. *Educ:* Univ Ala, MD, 72. *Prof Exp:* PROF PEDIAT, UNIV MEMPHIS, 78- *Mem:* Soc Pediat Res; Am Asn Immunologists; Am Acad Allergy Immunol. *Res:* Allergy immunology. *Mailing Add:* Univ Tenn Coleman Bldg Rm B310 Memphis TN 38163

HERRON, DAVID KENT, b Toledo, Ohio, Apr 30, 42; m 66; c 2. MEDICINAL CHEMISTRY, LEUKOTRIENE RESEARCH. *Educ:* Carleton Col, BA, 64; Rockefeller Univ, PhD(life sci), 69. *Prof Exp:* Fel chem, Harvard Univ, 69-71; sr chemist, 71-77, res scientist, 77-84, RES ASSOC, ELI LILLY & CO, 84- *Mem:* Am Chem Soc; AAAS; NY Acad Sci; Shock Soc. *Res:* Design and synthesis of new agents for the treatment of diseases involving leukotrienes; scientific applications of supercomputers. *Mailing Add:* 5945 Andover Rd Indianapolis IN 46220

HERRON, GEORGE M, b Caddo, Okla, Feb 22, 25; m 53; c 3. SOIL SCIENCE. *Educ:* Okla State Univ, BS, 49, MS, 50; Univ Nebr, Lincoln, PhD(soils), 68. *Prof Exp:* High sch teacher, Okla, 50-51; asst supvr agr credit, Farmers Home Admin, USDA, Okla, 51-53; instr agron, Murray State Agr Col, 53-56; assoc prof soils, Exp Sta, Kans State Univ, Garden City, 56-90; RETIRED. *Concurrent Pos:* Res assoc agron, Univ Nebr, 63-67. *Mem:* Am Soc Agron; Soil Sci Soc Am. *Res:* Crop production as related to soil fertility and moisture use; accumulation and distribution of nitrogen in soil profiles used for grain crop production in the midwest states; use of chemical test to predict the need for chemical fertilizers in crop production. *Mailing Add:* 910 Fitz Garden City KS 67846

HERRON, ISOM H, b St Louis, Mo, Sept 8, 46; m 81; c 1. FLUID MECHANICS, STABILITY THEORY. *Educ:* Mass Inst Technol, BS, 67; Johns Hopkins Univ, PhD(mechanics), 73. *Prof Exp:* Res fel appl math, Calif Inst Technol, 72-73; from asst to assoc prof, 74-84, PROF MATH, HOWARD UNIV, 84- *Concurrent Pos:* Vis scholar & vis assoc prof, Northwestern Univ, 81-82; Ford Found postdoctoral fel, Nat Res Coun, 88-; vis res assoc, Univ Md, Inst Phys Sci & Technol, 88-89. *Mem:* Am Math Soc; Math Asn Am; Am Phys Soc; Soc Indust & Appl Math. *Res:* Applied mathematics; problems in fluids mechanics; stability of flows; ORR-Sommerfeld equation; spectral theory. *Mailing Add:* Dept Math Howard Univ Washington DC 20059

HERRON, JAMES DUDLEY, b Providence, Ky, June 15, 36; m 56; c 2. SCIENCE EDUCATION, CHEMISTRY. *Educ:* Univ Ky, AB, 58; Univ NC, MS, 60; Fla State Univ, PhD(sci educ), 65. *Prof Exp:* Head, Div Chem & Educ, 83-86, dir, Sch Math & Sci Ctr, 87-89, PROF CHEM & EDUC, PURDUE UNIV, 77-, HEAD, DEPT CURRIC & INSTR, 89- *Concurrent Pos:* Consult & ctr coordr intermediate sci curric study, 66-72; training adv, Regional Educ Ctr for Sci & Math, Penang, Malaysia, 72-73. *Honors & Awards:* Catalyst of the Year, Chem Mfrg Asn, 83. *Mem:* fel AAAS; Am Chem Soc; Nat Asn Res Sci Teaching; Nat Sci Teachers Asn; Asn Educ Teachers Sci. *Res:* Science curriculum development; techniques of evaluating science achievement; application of theories of cognitive psychology to teaching chemistry; problem solving; curriculum development. *Mailing Add:* Dept Curric & Instr Purdue Univ West Lafayette IN 47907-1442

HERRON, JAMES WATT, b Beaver, Pa, May 27, 20; m 43; c 2. BOTANY. *Educ:* Pa State Teachers Col, BS, 41; Cornell Univ, MS, 48, PhD(econ bot), 51. *Prof Exp:* Pub sch teacher, Pa, 41-44 & 45-47; instr econ bot, Cornell Univ, 51-52; from asst botanist to assoc botanist, 52-57, assoc prof hort &

assoc horticulturist, 57-67, EXTEN PROF AGRON, AGR EXP STA, UNIV KY, 67- *Mem:* Weed Sci Soc Am; Am Soc Hort Sci. *Res:* Weed control; agronomic crops, horticultural crops, turf and aquatic weeds; economic botany. *Mailing Add:* Dept Agron Univ Ky Lexington KY 40506

HERRON, JOHN THOMAS, b Winnipeg, Man, Jan 19, 31; US citizen; m 59; c 3. PHYSICAL CHEMISTRY. *Educ:* Univ Man, BSc, 53, MSc, 55; McGill Univ, PhD(chem), 57. *Prof Exp:* res chemist, 57-80, chief, Chem Kinetics Div, 80-84, DIR, CHEM KINETICS DATA CTR, NAT BUR STANDARDS, 84- *Mem:* AAAS; Am Chem Soc; Combustion Inst. *Res:* Application of mass spectrometry to the study of free radicals and gas kinetics; air pollution and combustion chemistry; data evaluation. *Mailing Add:* Chem 260 Nat Bur Standards Gaithersburg MD 20899-0001

HERRON, MICHAEL MYRL, b Chula Vista, Calif, Oct 12, 49; m 77; c 2. SEDIMENTARY GEOCHEMISTRY, GEOCHEMICAL WELL LOGGING. *Educ:* Univ Calif, San Diego, BA, 70; State Univ NY, PhD(geol sci), 80. *Prof Exp:* Res asst prof geol, State Univ NY, 80-82; RES SCIENTIST, SCHLUMBERGER-DOLL RES, 82- *Mem:* Soc Prof Well Log Analysts; Soc Petrol Engrs; Am Asn Petrol Geologists. *Res:* Sedimentary geochemistry and geochemical well logging. *Mailing Add:* Schlumberger-Doll Res Old Quarry Rd Ridgefield CT 06877

HERRON, NORMAN, b Newcastle-upon-Tyne, UK, July 14, 54; m 77; c 2. ZEOLITE CHEMISTRY, CATALYSIS. *Educ:* Univ Warwick, Coventry, UK, BSc, 75, PhD(chem), 78. *Prof Exp:* Teaching fel inorg chem, Ohio State Univ, Columbus, 79-81, sr res asst, 81-83; RES CHEMIST INORG CHEM, E I DU PONT, CR&D, WILMINGTON, DEL, 83- *Mem:* Royal Soc Chem. *Res:* Inorganic coordination chemistry; synthesis and characterization of novel homogeneous and solid state materials. *Mailing Add:* E 328-334 Exp Sta E I DuPont de Nemours Wilmington DE 19880-0328

HERRUP, KARL FRANKLIN, b Pittsburgh, Pa, July 16, 48; c 3. GENETICS. *Educ:* Brandeis Univ, BA, 70; Stanford Univ, PhD(neurosci, biobehav sci), 75. *Prof Exp:* Res assoc neuropath, Harvard Med Sch, 74-78, res asst neurosci, Children's Hosp Med Ctr, 74-78; from asst prof to assoc prof human genetics, Med Sch, Yale Univ, 78-88; DIR DIV DEVELOP NEUROBIOL, E K SHRIVER CTR, 88-; ASSOC PROF NEUROL, MASS GEN HOSP, 88-; ASSOC PROF NEUROSCI PROG, HARVARD MED SCH, 88- *Concurrent Pos:* Jane Coffin Childs Mem Fund Med Res fel, 74-76; Med Found fel, 76-78; fel, Dept Pharmacol, Bioctr, Univ Basel, 78; Andrew Mellon Found fel, 84; Cornelius Wiersma vis prof neurosci, Div Biol, Calif Tech, 86. *Mem:* Soc Neurosci; Am Soc Microbiol; Soc Develop Biol. *Res:* Developmental genetics of the nervous system from the perspective of quantitative cell biology with emphasis on neurological mutants of cerebellar development, mutant-wild type chimeric mice and transgenic mice. *Mailing Add:* Div Develop Neurobiol E K Shriver Ctr 200 Trapelo Rd Waltham MA 02254

HERSCHBACH, DUDLEY ROBERT, b San Jose, Calif, June 18, 32; m 64; c 2. CHEMICAL PHYSICS. *Educ:* Stanford Univ, BS, 54, MS, 55; Harvard Univ, AM, 56, PhD(chem physics), 58. *Hon Degrees:* DSc, Univ Toronto, 77, Cornell Col, 88, Framingham State, 89, Adelphi Univ, 90. *Prof Exp:* Jr fel chem, Harvard Univ, 57-59; from asst prof to assoc prof, Univ Calif, Berkeley, 59-63; prof chem, 63-76, chmn, Dept Chem, 77-80, BAIRD PROF SCI, HARVARD UNIV, 76- *Concurrent Pos:* Alfred P Sloan res fel, 59-63; Falk-Plaut lectr, Columbia Univ, 63; vis prof, Univ Gottingen, 63; Harvard lectr, Yale Univ, 64; vis fel, Joint Inst Lab Astrophys, Colo, 68-69; Fairchild scholar, Calif Inst Technol, 76. *Honors & Awards:* Nobel Prize, 86; Nat Medal of Sci, 91; Pure Chem Prize, 65; Spiers Medal, 76; Centenary Medal, 77; Pauling Medal, 78; Phillips Lectr, Haverford Col, 62; Rollefson lectr, Univ Calif, Berkeley, 69; Reilly lectr, Univ Notre Dame, 69; Phillips lectr, Univ Pittsburgh, 71; Gordon lectr, Univ Toronto, 71; Clark Mem lectr, San Jose State Univ, 79; Clark lectr, WVa Univ, 81, Flygare lectr, Univ Ill, 88, Kaufman lectr, Univ Pittsburgh, 89, Priestly lectr, Pa State Univ, 90; Polanyi Medal, 82; Langmuir Prize, 83. *Mem:* Nat Acad Sci; fel AAAS; fel Am Acad Arts & Sci; fel Am Phys Soc; Am Chem Soc; Royal Soc Chem; Am Philosophical Soc. *Res:* Theory of molecular spectra, collision processes and electronic structure; molecular beam scattering; dynamics of chemical reactions. *Mailing Add:* Dept Chem Harvard Univ Cambridge MA 02138

HERSCHLER, MICHAEL SAUL, b New York, NY, July 22, 36; m 57; c 3. EMBRYOLOGY, CYTOGENETICS. *Educ:* Cornell Univ, BS, 58; Ohio State Univ, MSc, 61, PhD(cytogenetics), 64. *Prof Exp:* From asst prof to assoc prof biol, 64-80, PROF LIFE SCI, OTTERBEIN COL, 80- *Concurrent Pos:* NIH res grant, 66-69. *Res:* Cytogenetics of the genus anthurium. *Mailing Add:* Dept Life Sci Otterbein Col Westerville OH 43081

HERSCHMAN, ARTHUR, theoretical physics; deceased, see previous edition for last biography

HERSCHMAN, HARVEY R, b Cleveland, Ohio, June 21, 40; m 69. BIOLOGICAL CHEMISTRY, CELL BIOLOGY. *Educ:* Rice Univ, BA, 62; Univ Calif, San Diego, PhD(cell biol), 67. *Prof Exp:* from asst prof to assoc prof, 69-79, PROF BIOL CHEM, SCH MED, UNIV CALIF, LOS ANGELES, 79- *Concurrent Pos:* USPHS trainee, Brandeis Univ, 67-69; mem, Molecular Biol Inst, Univ Calif, Los Angeles, 69-, mem, Brain Res Inst, 70-; assoc dir, Lab Biomed & Environ Sci & dir basic res progs, Univ Calif, Los Angeles-Jonsson Comprehensive Cancer Ctr, 81- *Res:* Control of gene expression; regulation of cell division and cell cycle; mechanisms of carcinogenesis. *Mailing Add:* 33-257 Dept Biol Chem Univ Calif Sch Med Los Angeles CA 90024

HERSCHORN, MICHAEL, b Montreal, Que, Apr 21, 33; m 54, 89; c 2. ANALYSIS, FUNCTIONAL ANALYSIS. *Educ:* McGill Univ, BA, 53, MA, 56, PhD(math), 58. *Prof Exp:* Lectr, McGill Univ, 54-59, from asst prof to assoc prof, 59-79, assoc dean fac sci, 69-78, dean students, 78-82, chmn, dept math & statist, 82-88, PROF MATH, MCGILL UNIV, 79- *Concurrent Pos:* Vis mem, Courant Inst Math Sci, NY Univ, 60-61. *Mem:* AAAS; Am Math Soc; Math Asn Am; Can Math Soc; Am Schs Oriental Res. *Res:* Uniqueness, stability and oscillation theorems for ordinary differential equations. *Mailing Add:* Dept Math & Statist McGill Univ 805 Sherbrooke St W Montreal PQ H3A 2K6 Can

HERSCOVICS, ANNETTE ANTOINETTE, b Paris, France, June 29, 38; c 1. BIOCHEMISTRY. *Educ:* McGill Univ, Montreal, BSc, 59, PhD(biochem), 63. *Prof Exp:* Res fel biochem, McGill-Montreal Gen Hosp Res Inst, 62-63; res fel biochem, Dept Anat, McGill Univ, 63-67, lectr, 67-69, asst prof, 69-71; res assoc biol chem, Harvard Univ Med Sch, 71-75, prin res assoc, 75-81; assoc prof, Fac Med, 81-87, PROF, MCGILL CANCER CTR, DEPT MED & BIOCHEM, MCGILL UNIV, 87- *Concurrent Pos:* Damon Runyon Mem Fund res fel, 62-63; vis fel biochem, Mass Gen Hosp, 71-74, asst biochemist, 74-81. *Mem:* Biochem Soc; Can Biochem Soc; Soc Complex Carbohydrates; Am Soc Biol Chemists; Am Soc Cell Biol. *Res:* Structure function and biosynthesis of complex carbohydrates in animal cells. *Mailing Add:* Dept Biochem McGill Univ Sherbrooke St W Montreal PQ H3A 2M5 Can

HERSCOWITZ, HERBERT BERNARD, b Brooklyn, NY, June 19, 39; m 61; c 3. IMMUNOLOGY, MICROBIOLOGY. *Educ:* Brooklyn Col, BS, 61; Long Island Univ, MS, 63; Hahnemann Med Col, PhD(microbiol), 68. *Prof Exp:* Asst instr microbiol, Sch Dent Med, Univ Pa, 62-65; chief bacteriologist, Med Arts Labs, 65-66; from asst prof to assoc prof, Sch Med & Dent, 70-81; PROF MICROBIOL, SCH MED, GEORGETOWN UNIV, 81- *Concurrent Pos:* USPHS fel immunol, Case Western Reserve Univ, 68-70; consult, Dynatech Labs-Cooke Eng, Coleman Assocs, Nat Geog Soc, Prentice-Hall Publishers, Westinghouse Elec Corp, Off Extramural Res & Training, NIH, Off Protection from Res Risks, NIH & Hybritech, Inc; guest worker, Lab Immunol, Nat Inst Allergy & Infectious Dis, 78-79; mem, Vincent T Lombardi Cancer Res Ctr, Georgetown Univ Med Ctr, 89-, Exp Immunol Study Sect, NIH, DRG, 89-93; chmn, Exp Immunol Study Sect, NIH, DRG, 90-93. *Honors & Awards:* Reteculoendothelial Soc Award, 85. *Mem:* AAAS; Reticuloendothelial Soc; Am Asn Immunol; Am Soc Microbiol; Sigma Xi. *Res:* Mechanisms of antibody formation, cellular interactions, alveolar macrophage function, control of immune response by macrophases. *Mailing Add:* Dept Microbiol Georgetown Univ Sch Med Washington DC 20007

HERSEY, JOHN B, b Wolfeboro, NH, Aug 20, 13. UNDERWATER ACOUSTICS. *Educ:* Princeton Univ, AB, 34, MA, 35; Lehigh Univ, PhD(physics), 43. *Prof Exp:* Recorder, gravity party, US Coast & Geodesic Survey, 35-36; recorder, seismic exploration crew, Phillips Petrol Co, 36- 39; physicist, Naval Ordnance Lab, Naval Gun Factory, Washington DC, 41-46; sr scientist, Woods Hole Oceanog Inst, 46-66; dep to asst oceanographer for ocean sci, chief of naval res, USN, 66-79; consult, ocean sci, Sci Appl Inter Corp, 79-86; CONSULT, OCEAN SCI, 86- *Honors & Awards:* Award for Excellence, Inst Elec & Electronics Engrs; Civilian Award for Excellence, USN; Flemming Award. *Mem:* Fel Geol Soc Am; Am Geophys Union; AAAS; Soc Explor Geophysics. *Mailing Add:* 923 Harriman St Great Falls VA 22066

HERSEY, STEPHEN J, b Buffalo, NY, Jan 17, 43; m 63; c 3. PHYSIOLOGY. *Educ:* Cornell Univ, AB, 64; Duke Univ, PhD(physiol), 68. *Prof Exp:* Asst prof physiol, Duke Univ, 69-70; from asst prof to assoc prof, 70-79, PROF PHYSIOL, EMORY UNIV, 79-, DIR GRAD STUDIES, 74- *Concurrent Pos:* NIH trainee, Duke Univ, 68-69, career develop award, 71-76. *Mem:* Am Physiol Soc. *Res:* Membrane transport; ion secretion; gastric acid secretion. *Mailing Add:* Dept Physiol Woodruff Mc Adm Bldg Emory Univ Atlanta GA 30322

HERSH, CHARLES K(ARRER), b Cleveland, Ohio, Sept 19, 24; m 49; c 4. CHEMICAL ENGINEERING. *Educ:* Univ Ill, AB, 46, BS, 50; Ill Inst Technol, MS, 55, PhD(chem eng), 60. *Prof Exp:* Res engr plastics, Union Carbide Chem Co, 50-52; assoc engr, Armour Res Found, Ill Inst Technol, 52-55, res engr, 55-62, mgr Res Inst, 62-68, asst dir, 68-72; CHEM ENGR, WITCO CORP, 72- *Mem:* Am Chem Soc; Am Inst Chem Eng. *Res:* Ozone, heat and mass transfer; thermodynamics, molecular sieves, adsorption, process design; high energy materials. *Mailing Add:* Witco Corp 520 Madison Ave New York NY 10022-4236

HERSH, EVAN MANUEL, b New York, NY, Mar 18, 35; m 56; c 3. IMMUNOLOGY, ONCOLOGY. *Educ:* City Col New York, BS, 56; Columbia Univ, MD, 60. *Prof Exp:* Intern med, St Luke's Hosp, NY, 60-61; resident, Stanford Univ, 61-62; clin assoc, Nat Cancer Inst, 62-65; from asst prof to assoc prof, 66-75, PROF MED, UNIV TEX M D ANDERSON HOSP & TUMOR INST, HOUSTON, 75- *Concurrent Pos:* Am Cancer Soc fel, Stanford Univ, 65-66. *Honors & Awards:* Ward Medal, City Col New York. *Mem:* Am Asn Cancer Res; AMA; Am Soc Clin Oncol; Am Soc Hemat; Am Asn Immunol. *Res:* Host defense mechanisms; cancer immunology and chemotherapy; immunosuppression; immunotherapy of cancer. *Mailing Add:* Dept Med Hemat Univ Ariz Col Med 1501 N Campbell Tucson AZ 85724

HERSH, HERBERT N, b New York, NY, Feb 6, 23; m 48; c 3. MATERIALS SCIENCE, TECHNICAL MANAGEMENT. *Educ:* Ohio State Univ, PhD(chem), 50. *Prof Exp:* Res scientist, Nat Adv Comt Aeronaut, 50-53; phys chemist, Zenith Radio Corp, 54-60, div chief, res dept, 60-78; prog mgr, Argonne Nat Lab, 78-85; consult, 86-87; Coodr, US ENVIRON PROTECTION AGENCY, 88- *Mem:* Am Chem Soc; fel Am Phys Soc; Electrochem Soc; Tech Asn Pulp & Paper Indust. *Res:* Materials science; low temperature phenomena; coloration of ionic crystals; solid state diffusion; energy analysis; spectroscopy of solids; photochemistry; radiation damage; luminescence; image display; electrochemistry. *Mailing Add:* 706 W Buena Ave No 1W Chicago IL 60613

HERSH, JOHN FRANKLIN, b Allentown, Pa, Mar 27, 20. METROLOGY. *Educ:* Oberlin Col, AB, 41; Harvard Univ, MA, 42, PhD(appl physics), 57. *Prof Exp:* Res assoc underwater sound lab, Harvard Univ, 42-45, asst acoust res lab, 53-57; from instr to asst prof physics, Wellesley Col, 47-53; engr, Gen Radio Co, 57-80, mgr Eng Standards Lab, 80-85; RETIRED. *Res:* Applied physics; electrical impedance standards and measurements; electromechanical transducer analysis and measurement. *Mailing Add:* 225 Prairie St Concord MA 01742-2927

HERSH, LEROY S, b New York, NY, Aug 15, 31; m 68; c 2. SURFACE CHEMISTRY. *Educ:* Polytech Inst Brooklyn, BChE, 53; Univ Chicago, PhD(chem), 64. *Prof Exp:* Tech specialist nuclear eng, Brookhaven Nat Lab, 53-54; instr chem, Barat Col, 58-59; USPHS fel, 64-66; res chemist, Corning Glass Works, 66-69, sr res chemist, 69-78, RES ASSOC CHEM, CORNING INC, 78- *Mem:* AAAS; Sigma Xi; Am Chem Soc. *Res:* Surface chemistry; cell culture surface studies. *Mailing Add:* Res & Develop Labs Corning Inc Corning NY 14831

HERSH, LOUIS BARRY, b Baltimore, Md, Mar 28, 40; m 67; c 2. BIOCHEMISTRY. *Educ:* Drexel Inst, BSc, 62; Brandeis Univ, PhD(biochem), 67. *Prof Exp:* Fel biochem, Nat Heart Inst, 66-68; asst prof, 68-72, ASSOC PROF BIOCHEM, UNIV TEX HEALTH SCI CTR DALLAS, 72- *Concurrent Pos:* NIH res grant, Univ Tex Health Sci Ctr Dallas, 69- *Mem:* Am Soc Biol Chem; Am Chem Soc. *Res:* Metabolism; flavin chemistry; enzymology. *Mailing Add:* Dept Biochem L3 122 Univ Tex Health Sci Ctr Dallas TX 75235-7200

HERSH, REUBEN, b New York, NY, Dec 9, 27; m 49; c 2. MATHEMATICAL ANALYSIS. *Educ:* Harvard Univ, BA, 46; NY Univ, MS, 60, PhD(math), 62. *Prof Exp:* Asst ed, Sci Am Mag, 48-52; machinist, Varityper Corp, 55-57; asst prof math, Fairleigh Dickinson Univ, 62; instr, Stanford Univ, 62-64; from asst prof to assoc prof, 64-70, PROF MATH, UNIV NMEX, 70- *Concurrent Pos:* Vis mem, Courant Inst, NY Univ, 70-71; vis prof, Brown Univ, 79, Univ Calif, Berkeley, 79, Ctr Invest & Studies, Avanzadus, Mex, 73. *Honors & Awards:* Chauvenet Prize, Math Asn Am, 75; Am Book Award, 82. *Mem:* Am Math Soc; Math Asn Am. *Res:* Differential equations; stochastic processes; operator theory; applied mathematics; philosophy of mathematics. *Mailing Add:* Dept Math Univ NMex Albuquerque NM 87131

HERSH, ROBERT TWEED, b Cleveland, Ohio, Oct 22, 27; m 59; c 2. BIOPHYSICS. *Educ:* Columbia Univ, AB, 47, MA, 51; Univ Calif, PhD(biophys), 55. *Prof Exp:* USPHS fel, Univ Calif, 56; physicist cent res, E I du Pont de Nemours & Co, 57-58; chmn dept, 71-78, from asst prof to assoc prof, 58-67, PROF BIOCHEM, UNIV KANS, 67-, DIR, HUMAN BIOL, 85- *Concurrent Pos:* NIH career develop award, 60-70. *Mem:* AAAS; Biophys Soc; NY Acad Sci. *Res:* Physical biochemistry; properties of biological macromolecules; cell division; mathematical biology. *Mailing Add:* 3030 W Ninth St Lawrence KS 66049

HERSH, SOLOMON PHILIP, b Winston-Salem, NC, Jan 14, 29; m 54; c 2. FIBER & TEXTILE SCIENCE, STATIC ELECTRICITY. *Educ:* NC State Col, BS, 49; Inst Textile Technol, MS, 51; Princeton Univ, MA, 53, PhD(phys chem), 54. *Prof Exp:* Res fel, Textile Res Inst, 51-54; res chemist, Res Dept, Union Carbide Chem Co, 54-62; sr res chemist, Chemstrand Res Ctr, Inc, 62-66; from assoc prof to prof textile technol, 66-73, actg head, Dept Textile Mat & Mgt, 83-85, head, Dept Textile Eng & Sci, 85-88, CHARLES A CANNON PROF TEXTILES, NC STATE UNIV, 73-, DIR GRAD PROGS, 88- *Honors & Awards:* O Max Gardner Award, Univ NC Syst, 79. *Mem:* Am Chem Soc; Fiber Soc; Electrostatics Soc Am; Am Asn Textile Chem & Colorists; Int Inst Conserv Hist & Artistic Works; Air & Waste Mgt Asn. *Res:* Cellulose chemistry; electrical properties of filaments; synthetic fibers; viscoelastic properties of polymers; cotton dust formation and analysis; preservation of historic and artistic textiles; byssinosis; electrostatic phenomena; expert systems for cad; fiber fatigue. *Mailing Add:* Sch Textiles NC State Univ Campus Box 8301 Raleigh NC 27695-8301

HERSH, SYLVAN DAVID, b Philadelphia, Pa, July 18, 40; m 61; c 2. CHEMISTRY. *Educ:* Drexel Univ, BS, 62; Univ Del, PhD(chem), 69. *Prof Exp:* Res chemist, Am Viscose Div, FMC Corp, 69-76; mem staff, AM McGraw Lab, 76-82; SR PROJ SCIENTIST, ITT CORP, 82- *Mem:* Am Chem Soc. *Res:* Infrared and nuclear magnetic resonance spectroscopy to investigate composition of industrial unknowns; thin layer chromatographic separations and identifications; pharmaceutical analyses. *Mailing Add:* Rte 9 Box 59E Santa Fe NM 87505

HERSH, THEODORE, INTESTINAL ABSORPTION, GALLSTONES. *Educ:* Columbia Univ, MD, 59. *Prof Exp:* PROF MED, DEPT RES CLIN CARE, SCH MED, EMORY UNIV, 73- *Mailing Add:* Sch Med Emory Univ 1365 Clifton Rd NE Atlanta GA 30322

HERSHBERG, PHILIP I, b Albany, NY, Aug 5, 35; m 58; c 3. MEDICAL ENGINEERING. *Educ:* Rensselaer Polytech Inst, BEE, 57; Yale Univ, MEE, 58; State Univ NY, MD, 66. *Prof Exp:* Res fel med, Harvard Sch Pub Health, 68-70, asst prof, 70-73; asst prof med, Med Sch, Boston Univ, 73-81; dir res, Neico Co, 81-83; vpres, Invocom, Inc, 83-85; staff physician, Richmond Mem Hosp, 87-89; ASSOC MED DIR, WORCESTER COUNTY HOSP, 90- *Concurrent Pos:* Consult indust, 66-; asst med, Brigham Hosp, 68-73; investment adv, Deermont Market Ctr, 85- *Mem:* Inst Elec & Electronics Engrs; Am Col Prev Med; Am Soc Internal Med. *Res:* Electronics in medicine. *Mailing Add:* PO Box 332 Needham MA 02192

HERSHBERGER, CHARLES LEE, b Louisville, Ill, May 1, 42; m 62; c 3. MOLECULAR BIOLOGY. *Educ:* Eureka Col, BS, 64; Univ Ill Col Med, PhD(biochem), 67. *Prof Exp:* Res assoc molecular biol, Univ Wis-Madison, 67-69; asst prof microbiol, Univ Ill, Urbana, 69-76; sr microbiologist, 76-80, res scientist, 81-86, SR RES SCIENTIST, ELI LILY & CO, 87-, GROUP LEADER, 88- *Concurrent Pos:* USPHS-NIH trainee, Univ Wis-Madison, 67-68, Am Cancer Soc fel, 68-69. *Honors & Awards:* Sci Achievement Award, Am Soc Microbiol, 87; Charles Thomas Award, Soc Indust Microbiol, 90. *Mem:* AAAS; Am Soc Biochem & Molecular Biol; Soc Indust Microbiol; Sigma Xi. *Res:* Synthesis of nucleic acids; recombinant DNA; cellular regulatory mechanisms; cloned genes for peptide hormones and biosynthesis; cloned genes for industrial process enzymes. *Mailing Add:* Lilly Corp Ctr Eli Lilly & Co Indianapolis IN 46285

HERSHBERGER, TRUMAN VERNE, b Walnut Creek, Ohio, Oct 10, 27; m 49; c 3. ANIMAL NUTRITION. *Educ:* Goshen Col, BA, 49; Ohio State Univ, MSc, 51, PhD(agr biochem), 55. *Prof Exp:* Asst natural sci, Goshen Col, 47-49; asst agr biochem, Ohio State Univ, 49-51; instr animal sci, Ohio Agr Exp Sta, 51-55, asst prof animal nutrit, 55-60, res fel, 65-66, ASSOC PROF ANIMAL NUTRIT, PA STATE UNIV, 60- *Mem:* Fel AAAS; Am Inst Nutrit; Am Chem Soc; Am Soc Animal Sci; Am Sci Affil; NY Acad Sci. *Res:* Rumen physiology; energy metabolism; protein requirements; equine nutrition. *Mailing Add:* Dept Dairy & Animal Sci Pa State Univ 328 W L Henning Bldg University Park PA 16802

HERSHENOV, B(ERNARD), b New York, NY, Sept 22, 27; m 50; c 1. ELECTRICAL ENGINEERING, PHYSICS. *Educ:* Univ Mich, BS, 50, MS, 52, PhD(elec eng), 59. *Prof Exp:* Dir res labs, RCA, Tokyo, 72-75, staff adv res labs, Princeton, NJ, 75-77, head, Energy Systs Group, 77-79, dir optoelectronics res lab, RCA Res Labs, 79-87, dir, Mkt Coord, 87-89, DIR, E ASIA BUS DEVELOP, DAVID SARNOFF RES CTR, PRINCETON, 89- *Concurrent Pos:* Co-adj staff mem math dept, Rutgers Univ, 64-66. *Mem:* fel Inst Elec & Electronics Engrs. *Res:* Microwave ferrites and integrated circuits; millimeter wave devices; energy management; surface state physics; optoelectronics. *Mailing Add:* 22 Raleigh Rd Kendall Park NJ 08824

HERSHENOV, JOSEPH, b Brooklyn, NY, Mar 2, 35. MATHEMATICS. *Educ:* Yeshiva Univ, BA, 55; Mass Inst Technol, SM, 57, PhD(math), 61. *Prof Exp:* From res asst to res assoc math, Mass Inst Technol, 55-62; from asst prof to assoc prof, 62-76, chmn dept, 69-80, PROF MATH, QUEEN'S COL, NY, 77- *Mem:* Math Asn Am; Am Math Soc; Soc Indust & Appl Math. *Res:* Hydrodynamics stability theory; complex variables; numerical analysis. *Mailing Add:* Dept Math Queen's Col Flushing NY 11367

HERSHENSON, BENJAMIN R, b Boston, Mass, July 13, 40. PHARMACOGNOSY, ACADEMIC ADMINISTRATION. *Educ:* Mass Col Pharm, BS, 62, MS, 64, PhD(pharm), 68. *Prof Exp:* From instr to asst prof pharmacog & bot, Mass Col Pharm, 64-71, assoc prof pharmacog & bot, 71-80, dir div spec prog, 74-78, asst dean, 78-81, dean admis, 81-83, dean admin, 83-88, PROF BIOL, MASS COL PHARM, 81-, DEAN COL & VPRES ACAD AFFAIRS, 88- *Concurrent Pos:* Consult, Sch Nursing, Mass Gen Hosp, 68-, Pharmaceut Res Assocs, Inc, 68- & Mass Dept Educ; chmn, Coun Deans, Am Asn Col Pharm, 88-90. *Mem:* Am Asn Col Pharm; Am Pharmaceut Asn; Am Soc Allied Health Profession. *Res:* Natural products research, primarily phytochemical screening of West Indian plants, including extraction, isolation and characterization of active constituents. *Mailing Add:* Mass Col of Pharm 179 Longwood Ave Boston MA 02115

HERSHENSON, HERBERT MALCOLM, b Brooklyn, NY, Feb 17, 29; m 53; c 3. CHEMISTRY. *Educ:* Mass Inst Technol, BS, 49, PhD(anal chem), 52. *Prof Exp:* Asst, Mass Inst Technol, 49-51; asst prof chem, Univ Kans, 52 & Wesleyan Univ, 52-55; supvr anal chem, Aircraft Nuclear Engine Lab, Pratt & Whitney Aircraft Div, United Aircraft Corp, 55-57; group leader, Olin Mathieson Chem Corp, 58; asst tech dir, Baird-Atomic, Inc, 59-60; asst dir, Advan Mat Res & Develop Lab, 60-70, asst mgr, Mat Eng & Res Lab, Pratt & Whitney Aircraft Group, United Technologies Corp, 71-77; vpres & gen mgr, Coatings Technol Corp, 77-83; VPRES, TURBINE COMPONENTS CORP, 86- *Concurrent Pos:* Consult engr, Pratt & Whitney Aircraft Div, United Aircraft; adj assoc prof, Hartford Grad Ctr, Rensselaer Polytech Inst, 56-58; pres, Amity Group Investment Adv, 84- *Mem:* Am Chem Soc. *Res:* Chemical problems related to materials development; diffusion and plasma spray coatings. *Mailing Add:* PO Box 31 Branford CT 06405

HERSHEY, ALFRED D, b Owosso, Mich, Dec 4, 08; m 45; c 1. CHEMISTRY. *Educ:* Mich State Univ, BS, 30, PhD(chem), 34. *Hon Degrees:* DSc, Univ Chicago, 67. *Prof Exp:* Asst bacteriologist, Sch Med, Wash Univ, 34-36, from instr to assoc prof, 36-50; mem staff, genetic res unit, Carnegie Inst Wash, 50-62, dir, 62-74; RETIRED. *Honors & Awards:* Nobel Prize Med, 69. *Mem:* Nat Acad Sci. *Mailing Add:* Moores Hill Rd RD 1640 Syosset NY 11791

HERSHEY, ALLEN VINCENT, b Kellogg, Idaho, Aug 15, 10; m 49; c 1. MATHEMATICAL PHYSICS, COMPUTER TYPOGRAPHY. *Educ:* Univ Calif, BS, 32, MA, 36, PhD(physics), 38. *Prof Exp:* Res engr, Gen Elec Co, NY, 38-39; contract employee, 39-43, physicist, 46-61, scientist, Naval Surface Weapons Ctr, Dahlgren, Va, 61-79, RES AFFIL, NAVAL POSTGRAD SCH, 79- *Honors & Awards:* Ed Rosse Award, Comput Micrographic Technol Users Group, 78. *Mem:* AAAS; Am Phys Soc. *Res:* Chemical kinetics; gaseous ionics; terminal ballistics; polycrystal plasticity; hydroballistics; visual perception; ellipsoidal potentials; rational approximations; cartography and typography; computers and cathode ray printers; fluid dynamics. *Mailing Add:* 722 Redwood Lane Pacific Grove CA 93950

HERSHEY, DANIEL, b New York, NY, Feb 12, 31; m 65; c 2. CHEMICAL ENGINEERING, SYSTEMS SCIENCE. *Educ:* Cooper Union, BS, 53; Univ Tenn, MS, 59, PhD(chem eng), 61. *Prof Exp:* Chem engr, Pall Filtration Co, 53-54 & Merck, Sharpe & Dohme, 61-62; from asst prof to assoc prof, 62-68, asst to pres, 73-76, PROF CHEM ENG, UNIV CINCINNATI, 68- *Concurrent Pos:* NIH res grant, Univ Cincinnati, 64-69; Fulbright fel, 75; vpres, Basal Tech, 82- *Honors & Awards:* Clin Res Award, Bariatric Physicians, 89. *Mem:* NY Acad Sci; Soc Gen Systs Res. *Res:* Entropy analysis; living systems; corporations. *Mailing Add:* Dept Chem Eng Univ Cincinnati Cincinnati OH 45221

HERSHEY, FALLS BACON, b Chicago, Ill, Aug 16, 18; m 45; c 4. SURGERY. *Educ:* Univ Ill, BS, 39; Harvard Med Sch, MD, 43. *Prof Exp:* Intern, Peter Bent Brigham Hosp, Mass, 43-44; intern, Mass Gen Hosp, 46-47, asst resident, 47-48 & 50-51, chief resident surg, East Surg Serv, 52; res assoc biol, Mass Inst Technol, 49-50; from instr to assoc prof surg, Sch Med, Wash Univ, 53-59, assoc prof clin surg, 59-64; prof surg, Chicago Med Sch, 64-66; chmn div surg & dir surg res, Michael Reese Hosp & Med Ctr, Univ Chicago, 64-66; assoc prof clin surg, Sch Med, Wash Univ, 66-77; DIR VASCULAR SURG & DIR BLOOD FLOW LAB, ST JOHN'S MED CTR, ST LOUIS, 78- *Concurrent Pos:* Fel, Nat Cancer Inst, 49-50; teaching fel, Harvard Med Sch, 52. *Mem:* Soc Univ Surg; Int Cardiovasc Soc; Am Col Surg; Sigma Xi. *Res:* Vascular surgery; cancer surgery; biochemical research in the enzymes of skin and cancer. *Mailing Add:* Eleven Wydown Terr Clayton MO 63105

HERSHEY, H GARLAND, b Quarryville, Pa, Oct 1, 05. GEOLOGY, HYDROLOGY. *Educ:* Johns Hopkins Univ, BS, 29, PhD(geol), 36. *Prof Exp:* From geologist to state geologist, 36-69, EMER STATE GEOLOGIST, IOWA GEOL SURVEY, 69- *Concurrent Pos:* Chmn, Iowa Nat Resources Coun, 48-69; oil & gas adminr, State of Iowa, 55-69; chmn, US Comt Int Hydrol Decade, 68-72; dir, Off Water Resource Res, US Dept Interior, 69-72. *Mem:* Fel Geol Soc Am. *Mailing Add:* 708 McLean Street Iowa City IA 52246

HERSHEY, HARRY CHENAULT, b Baton Rouge, La, Nov 10, 38; m 61; c 3. CHEMICAL ENGINEERING. *Educ:* Univ Mo, Rolla, BS, 60, MS, 63, PhD(rheol), 65. *Prof Exp:* Assoc engr, Union Carbide Nuclear Co, 60-62; asst prof chem eng, Univ Mo, Rolla, 65-66; res assoc, 66-67, from asst prof to assoc prof, 67-78, PROF CHEM ENG, OHIO STATE UNIV, 78- *Concurrent Pos:* Consult, Battelle Mem Inst, Sohio. *Mem:* Am Chem Soc; Am Inst Chem Engrs; Asn Comput Mach; Soc Rheol; Sigma Xi. *Res:* Fluid mechanics of viscoelastic fluids; dilute polymer solutions; mathematical methods in chemical engineering, especially statistics and numerical analysis; thermodynamics. *Mailing Add:* Dept Chem Eng 140 W 19th Ave Columbus OH 43210

HERSHEY, JOHN LANDIS, b Lancaster, Pa, Sept 6, 35; m 60; c 3. ASTRONOMY. *Educ:* Eastern Mennonite Col, AB, 58; Univ Va, MA, 66, PhD(astron), 69. *Prof Exp:* Res assoc astron, 68-72, from asst prof to assoc prof, 72-80, actg dir, 80-81, astron, Sproul Observ, Swarthmore Col, 83-85; ASTRON, US NAVAL OBSERV, 82- *Concurrent Pos:* Vis assoc prof, Univ Md, 80-82. *Mem:* Am Astron Soc; Sigma Xi. *Res:* Long focus astronomy for determining distances, orbital motions and masses of stars; astrometric applications of CCD arrays; high precision measuring machines; long base line optical interferometry. *Mailing Add:* 9006 Sudbury Rd Silver Spring MD 20901

HERSHEY, JOHN WILLIAM BAKER, b Lancaster, Pa, Jan 27, 34; m 60; c 2. BIOCHEMISTRY, ORGANIC CHEMISTRY. *Educ:* Haverford Col, BA, 58; Rockefeller Inst, PhD(biochem), 63. *Prof Exp:* Jane Coffin Childs Mem Fund fel org chem, Cambridge Univ, 63-65; res biochemist, Huntington Labs, Mass Gen Hosp, 65-67; res fel, tutor & lectr, Harvard Univ, 67-70; asst prof, 70-74, assoc prof, 74-78, PROF BIOL CHEM, UNIV CALIF, DAVIS, 78- *Mem:* Am Soc Biol Chemists; Sigma Xi. *Res:* Initiation of protein synthesis; translational control mechanisms; ribosome structure and function. *Mailing Add:* Dept Biol Chem Univ Calif Sch Med Davis CA 95616

HERSHEY, LINDA ANN, b Marion, Ind, Jan 15, 47; m 76; c 3. NEUROPHARMACOLOGY, CLINICAL PHARMACOLOGY. *Educ:* Purdue Univ, BS, 68; Washington Univ, PhD(neurobiol), 73, MD, 75. *Prof Exp:* Intern med, St John's Mercy Hosp, St Louis, 75-76; resident neurol, Barnes Hosp, 76-78; fel & instr, Strong Mem Hosp, Rochester, NY, 78-80; ASST PROF NEUROL & CLIN PHARMACOL, UNIV HOSP, CASE WESTERN RESERVE UNIV, CLEVELAND, 80- *Concurrent Pos:* Med adv, Comt Combat Huntington's Dis, 80-; chief, Div Neuropharmacol, Univ Hosp, Case Western Reserve Univ, Cleveland, 80-, chmn, Pharm & Therapeut Comt, 81-; prin investr, Tidopidine-Aspirin Stroke Study, 81- & Pergolide Study in Parkinson's Dis, 83-; Andrew W Mellon Found scholar, 80-81. *Mem:* Soc Neurosci; Am Acad Neurol; Am Soc Clin Pharmacol & Therapeut; Cent Soc Neurol Res. *Res:* Controlled trials of new drugs that have therapeutic potential for neurologic diseases; pharmacokinetics of antiparkinsonian drugs; adverse neurologic and psychiatric drug reactions. *Mailing Add:* 66 North Dr Buffalo NY 14226

HERSHEY, NATHAN, b New York, NY, Apr 28, 30; m 58; c 2. PUBLIC HEALTH EDUCATION. *Educ:* NY Univ, AB, 50; Harvard Univ, LLB, 53. *Prof Exp:* Pvt pract law, New York, 55-56; res assoc, 56-58, from asst prof to assoc prof, 58-68, PROF HEALTH LAW, SCH PUB HEALTH, UNIV PITTSBURGH, 68- *Concurrent Pos:* Mem, Secy's Comn Med Malpract, Legal Issues Adv Panel, HEW, 72, Pa Bd Med Educ, 74-80, Vision Serv Assocs, 76, bd dirs, Women's Health Serv, 76- & Comt Utilization Mgt Third Parties, Inst Med, 88-89; consult, Pa State Comt Pub Health & Welfare, 73-80; counr, Markel, Schafer & Means, Pittsburgh & Post & Schell, Philadelphia; pres, Soc Hosp Attorneys Western Pa, 79. *Mem:* Inst Med Nat Acad Sci; Am Pub Health Asn; Am Acad Hosp Attorneys (pres, 72). *Res:* Law affecting health services delivery; author of 150 technical publications. *Mailing Add:* Sch Pub Health Univ Pittsburgh Pittsburgh PA 15261

HERSHEY, ROBERT LEWIS, b Chicago, Ill, Dec 18, 41. AIR POLLUTION CONTROL TECHNOLOGIES, RESEARCH PROGRAM APPRAISAL METHODOLOGIES. *Educ:* Tufts Univ, BS, 63; Mass Inst Technol, MS, 64; Cath Univ Am, PhD(eng), 73. *Prof Exp:* Mem tech staff, Bell Tel Labs, 63-67; mgr acoust, Weston Instruments Inc, 67-68; sr scientist, Bolt Beranek & Newman, 68-71; mgr acoust progs, Booz Allen & Hamilton, 71-79; prog vpres, 79-80, div vpres, 80-88, EXEC ENGR, SCI MGT CORP, 88- *Concurrent Pos:* Mem, Joint Bd Sci & Eng Ed, 72-78; nat secy, Prof Engrs in Indust Div, Nat Soc Prof Engrs, 73-75; secy, DC Prof Coun, 74, Dc Prof Eng Regist Bd, 87-; mem, Coord Comt Productivity & Innovation, Am Asn Eng Soc, 84-88; sci policy analyst, George Bush Presidential Campaign, 88; consult, Energetics Inc, 88-89 & Franklin Assocs, 89-90. *Mem:* Nat Soc Prof Engrs; Am Soc Mech Engrs; Acoust Soc Am; Soc Mfg Engrs. *Res:* Technologies for control of air toxics, flue gas desulfurization, and reduction of NOx; increased energy efficiency for industrial processes; automotive engine combustion; noise abatement; sonar; statistical modeling of glass fractures under dynamic loading. *Mailing Add:* 1255 New Hampshire Ave NW Washington DC 20036

HERSHEY, SOLOMON GEORGE, b New York, NY, June 23, 14; m 41; c 1. ANESTHESIOLOGY. *Educ:* City Col New York, BS, 34; NY Univ, MD, 39; Am Bd Anesthesiol, dipl. *Prof Exp:* Intern, Beth Israel Hosp, 39-41; resident anesthesiol, Bellevue Hosp, 42-44; instr anesthesia, Col Med, NY Univ, 45, asst clin prof anesthesiol, 46-50, clin prof, Post-Grad Med Sch, 50-63, prof, Med Ctr, 64-67, attend physician & assoc dir, Univ Hosp, 64-67; PROF ANESTHESIOL, ALBERT EINSTEIN COL MED, 67-, DIR OFF CONTINUING MED EDUC, 71- *Concurrent Pos:* Fel anesthesia, Col Med, NY Univ, 44; from asst vis anesthesiologist to assoc vis anesthesiologist, 45-59, vis anesthetist, 59-; attend anesthesiologist & chief serv, Beth Israel Hosp, 46-59, dir anesthesiol, 59-63, consult, 64-; res assoc, Washington Sq Col, 52-55; lectr, Univ London, 59; vis prof, Univ Calif, San Francisco, 63; mem comt shock, Nat Res Coun-Nat Acad Sci, 65-71; attend anesthesiologist, Bronx Munic Hosp Ctr, NY, 67-, Hosp Albert Einstein Col Med, 67- & Lincoln Hosp Med Ctr, 67-76; ed, Anesthesiol, Am Soc Anesthesiol, 63-72; mem consult panel, Nat Heart, Blood & Lung Inst, 77- *Mem:* Fel Am Col Anesthesiol; Am Asn Med Col; Asn Univ Anesthetists; fel Royal Soc Med; Soc Exp Biol & Med. *Res:* Physiology of circulation. *Mailing Add:* 750 Ladd Rd Riverdale Bronx NY 10471

HERSHFIELD, EARL S, b Winnipeg, Man, May 17, 34; m 57; c 3. INTERNAL MEDICINE. *Educ:* Univ Man, MD & BSc, 58; FRCP(C). *Prof Exp:* Asst physician, Cent Tuberc Clin, 64-67; from assoc med dir to med dir, D A Stewart Ctr, 68-73; from asst prof to assoc prof, 67-82, PROF MED, UNIV MAN, 82-; DIR, RESPIRATORY HOSP, WINNIPEG, 75- *Concurrent Pos:* Asst physician, Health Sci Ctr, Manitoba, 66; chmn, Med Adv Comt, Man Lung Asn Div, Sanatorium Bd Man, 73-; med dir, Can Tuberc & Respiratory Dis Asn, 76-82. *Mem:* Am Thoracic Soc; assoc Am Col Physicians; fel Am Col Chest Physicians; Can Thoracic Soc. *Res:* Chest diseases. *Mailing Add:* Dept Med Univ Manitoba Winnipeg MB R3T 2N2 Can

HERSHFIELD, MICHAEL STEVEN, b Wilkes Barre, Pa, June 28, 42. BIOCHEMICAL GENETICS, RHEUMATOLOGY. *Educ:* Franklin & Marshall Col, AB, 63; Univ Pa, MD, 67. *Prof Exp:* Intern Med, Philadelphia Gen Hosp, 67-68; staff fel & sr staff fel biochem, NIH, 68-72; res fel genetics, Sch of Med, Univ Calif, San Diego, 72-74, resident internal med, 74-75, fel rheumatology, 75-76; asst prof med, 76-80, ASST PROF BIOCHEM, DIV RHEUMATIC & GENETIC DIS, DUKE UNIV, 76-, ASSOC PROF MED, 80- *Concurrent Pos:* NIH res fel, 75-76; res grant, Nat Inst Arthritis & Metab Dis, NIH, 78-; Basil O'Connor res grant, Nat Found March Dimes, 77-; NIH career develop award, 78- *Mem:* Am Fedn Clin Res; Am Soc Clin Investr; Am Soc Biol Chem. *Res:* Human biochemical genetics; diseases of purine metabolism; immunodeficiency diseases; somatic cell genetics; nucleic acid metabolism; biological transmethylation. *Mailing Add:* Dept Med Duke Univ Sch Med Box 3049-M Durham NC 27710

HERSHKOWITZ, NOAH, b Brooklyn, NY, Aug 16, 41; m 62; c 2. PHYSICS. *Educ:* Union Col, NY, BS, 62; Johns Hopkins Univ, PhD(physics), 66. *Prof Exp:* Instr physics, Johns Hopkins Univ, 66-67; asst prof, 67-71, assoc prof, 71-76, prof physics, 76-80, PROF PHYSICS & ASTRON, UNIV IOWA, 80- *Concurrent Pos:* Irving Langmuir prof, Nuclear Eng & Eng Physics Dept, Inst Elec & Electronics Engrs; Dept Nuclear Eng, Univ Wis. *Honors & Awards:* Merit Award, Inst Elec & Electronics Engrs, 87. *Mem:* Fel Am Phys Soc; Inst Elec & Electronics Engrs; Am Geophys Union; Am Vacuum Soc. *Res:* Experimental plasma physics; plasma etching; ICRF; tandem mirror physics; tokamak physics; basic plasma physics. *Mailing Add:* Dept Nuclear Eng Univ Wis 1500 Johnson Dr Madison WI 53706-1687

HERSHKOWITZ, ROBERT L, b New York, NY, Feb 11, 38; m 59; c 4. TEXTILE CHEMISTRY. *Educ:* Brooklyn Col, AB, 57; Syracuse Univ, PhD(org chem), 64. *Prof Exp:* RES CHEMIST, TEXTILE FIBERS DEPT, E I DU PONT DE NEMOURS & CO, INC, 64- *Mem:* Am Chem Soc. *Res:* Organophosphorus chemistry; substitution reactions; polyamidation. *Mailing Add:* 2504 Bona Rd Wilmington DE 19810

HERSHMAN, JEROME MARSHALL, b Chicago, Ill, July 20, 32; m 67; c 3. INTERNAL MEDICINE, ENDOCRINOLOGY. *Educ:* Northwestern Univ, BS, 52; Calif Inst Technol, MS, 53; Univ Ill, Chicago, MD, 57. *Prof Exp:* Clin investr, Vet Admin Res Hosp, 64-67; from asst prof to prof med, Univ Ala, Birmingham, 67-72; PROF MED, SCH MED, UNIV CALIF, LOS ANGELES, 72-; CHIEF ENDOCRINE SECT, WADSWORTH VET ADMIN HOSP, 72- *Concurrent Pos:* Fel endocrinol, Tufts Univ-New Eng Med Ctr, 61-63; NIH & Vet Admin res grants, 64- *Mem:* Am Soc Clin Invest; Endocrine Soc; Am Thyroid Asn; Am Fedn Clin Res; Am Col Physicians; Asn Am Physicians. *Res:* Control of thyroid function; thyroid physiology and disease; pituitary hormones. *Mailing Add:* Dept Med Univ Calif Los Angeles CA 90024

HERSHNER, IVAN RAYMOND, JR, b Lincoln, Nebr, Aug 24, 16; m 48; c 4. MATHEMATICS. *Educ:* Univ Nebr, BS, 38, MA, 40; Harvard Univ, MA, 41, PhD(math), 47. *Prof Exp:* Asst math, Univ Nebr, 38-40; instr, Univ Chicago, 47-48; asst prof, Univ NC, 48-51; prof & chmn dept, Univ Vt, 53-56; consult, US Dept Army, 53-56, math adv, Chief Res & Develop, 56-58, chief phys sci div, Off Chief Res & Develop, 58-72, sci dir army res, 72-75, asst dir res progs, Off Dep Chief of Staff Res, Develop & Acquisition, 75-80; prof math, George Mason Univ, 80-87; RETIRED. *Concurrent Pos:* Mem appl math adv coun, Nat Bur Standards, 52-54. *Mem:* Am Math Soc; Math Asn Am. *Res:* Conformal mapping; complex variable analysis. *Mailing Add:* 5427 37th St N Arlington VA 22207

HERSKOVITS, LILY EVA, petroleum chemistry, for more information see previous edition

HERSKOVITS, THEODORE TIBOR, b Kosice, Czech, June 11, 28; US citizen; m; c 2. PHYSICAL CHEMISTRY. *Educ:* City Col New York, BS, 55; Yale Univ, MS, 57, PhD, 60. *Prof Exp:* Fel chem, Purdue Univ, 59-61; res assoc biophys, Univ Chicago, 61-62; phys chemist, Agr Res Serv, USDA, 62-65; assoc prof, 65-74, PROF CHEM, FORDHAM UNIV, 74- *Mem:* Am Chem Soc; Biophys Soc; Fedn Am Socs Exp Biol; Sigma Xi. *Res:* Macromolecular chemistry; proteins; nucleic acids. *Mailing Add:* 100 Bennet Ave No 6J New York NY 10033

HERSKOVITZ, THOMAS, b Budapest, Hungary, Apr 15, 47; US citizen; m 72. ORGANOMETALLIC CHEMISTRY, BIOINORGANIC CHEMISTRY. *Educ:* Case Western Reserve Univ, BS, 68; Mass Inst Technol, PhD(inorg chem), 73. *Prof Exp:* RES CHEMIST, E I DU PONT DE NEMOURS & CO, INC, 73- *Mem:* Am Chem Soc. *Res:* Homogeneous catalysis; coordination chemistry of carbon dioxide. *Mailing Add:* 52 Rockglen Rd Winwood PA 19096

HERSKOWITZ, GERALD JOSEPH, b Brooklyn, NY, Feb 20, 36; m 59; c 3. ELECTRICAL ENGINEERING. *Educ:* Polytech Inst Brooklyn, BSEE, 57; Rutgers Univ, MSEE, 59; NY Univ, Eng ScD(elec eng), 63. *Prof Exp:* Mem res staff, RCA Labs, NJ, 57-59; mem tech staff, Bell Tel Labs, 63-65; prof elec eng, Stevens Inst Technol, 65-77; DIR, STEVENS ASSOCS, 75- *Concurrent Pos:* Chmn arrangements comt, Solid State Circuits Conf, 65-67; consult, Bell Tel Labs, 65-, US Army, 65 & Lockheed Electronics Co, 65-; lectr, Univ Wis, 67 & Univ Mo, 67; ed, Design Automation Workshop, 69-71; vprog chmn, Int Comn Commun, 76. *Mem:* Inst Elec & Electronics Engrs. *Res:* Solid-state device and integrated circuit research; lasers; magnetic memory systems; microwaves; computers; optical communications; microprocessors. *Mailing Add:* Seven Clover St Tenafly NJ 07670

HERSKOWITZ, IRA, b Brooklyn, NY, July 14, 46. MOLECULAR GENETICS. *Educ:* Calif Inst Technol, BS, 67; Mass Inst Technol, PhD(microbiol), 71. *Prof Exp:* Instr biol, Mass Inst Technol, 72; asst prof to prof biol, Univ Ore, 72-81, assoc, Inst Molecular Biol, 72-81; PROF GENETICS & HEAD DEPT BIOCHEM & BIOOPHYS, DIV GENETICS, UNIV CALIF, SAN FRANCISCO, 81-, CHMN DEPT, 90- *Concurrent Pos:* MacArthur Found fel, 88- *Honors & Awards:* Eli Lilly Award, Microbiol & Immunol, 83; Nat Acad Sci Award, Sci Reviewing, 85; Streisinger Lectr, 85; Genetics Soc Am Medal, 88. *Mem:* Nat Acad Sci; Genetics Soc Am (pres, 85). *Res:* Control of gene expression in yeast, bacteria, and bacterial viruses; pathogen-host interactions; control of cell growth. *Mailing Add:* Dept Biochem & Biophys Univ Calif San Francisco CA 94143-0448

HERSKOWITZ, IRWIN HERMAN, b New York, NY, Mar 18, 20; m 45; c 4. GENETICS. *Educ:* Brooklyn Col, AB, 40; Univ Mo, MA, 42; Columbia Univ, PhD(zool), 49. *Prof Exp:* Res assoc genetics & cancer, Sch Med, La State Univ, 48-49, instr anat, 49-51; res exec zool, Ind Univ, 51-57; from assoc prof to prof biol, St Louis Univ, 57-63; prof biol, Hunter Col, 63-85; RETIRED. *Mem:* Am Soc Naturalists; Am Soc Zoologists; Soc Human Genetics. *Res:* Point mutation; chromosomal breakage and rearrangement induced by chemicals and radiations; chromosome arrangement in sperm; salivary gland chromosome electron microscopy and histochemistry; homoeosis; tumors; Drosophila bibliography. *Mailing Add:* 3326 Arcara Way Lake Worth FL 33467

HERSON, DIANE S, b New York, NY, Apr 23, 44; m 73; c 2. MICROBIOLOGY. *Educ:* Cornell Univ, BS, 64; Rutgers Univ, MS, 66, PhD(bact), 68. *Prof Exp:* Res asst, Rutgers Univ, 64-66; from asst prof to assoc prof biol sci, 68-74, ASSOC PROF, SCH LIFE & HEALTH SCI, UNIV DEL, 74- *Concurrent Pos:* Consult. *Mem:* NY Acad Sci; Sigma Xi; AAAS; Am Soc Microbiol; Soc Indust Microbiol. *Res:* Microbial of drinking water; assessment of the viability and activity of microbes in the environment; biodegradation of toxic chemicals. *Mailing Add:* Sch Life & Health Sci Univ Del Newark DE 19716

HERTEL, GEORGE ROBERT, b Detroit, Mich, Apr 26, 34; m 63; c 3. COMPUTER APPLICATIONS. *Educ:* Mich Technol Univ, BS, 56; Rensselaer Polytech Inst, MS, 58; Johns Hopkins Univ, PhD(phys chem), 64. *Prof Exp:* Honor prog trainee res & develop, Gen Elec Co, NY, 56-58, mass spectroscopist, Knolls Atomic Power Lab, 59-60; res chemist, Oak Ridge Nat Lab, 64-68; from asst prof to assoc prof chem, Fla Technol Univ, 68-76; PROF CHEM, UNIV CENT FLA, 76- *Mem:* Am Soc Test Mat; Am Chem Soc. *Res:* Computer applications in chemistry lab; estimation methods in thermodynamics. *Mailing Add:* Dept Chem Univ Cent Fla Box 25000 Orlando FL 32816

HERTELENDY, FRANK, b Velikigaj, Yugoslavia, June 19, 31; m 62. REPRODUCTIVE PHYSIOLOGY. *Educ:* Univ Budapest, BSc, 55; Univ Reading, PhD(physiol chem), 62. *Prof Exp:* Nat Res Coun Can fel, McGill Univ, 62-64; res scientist, Dept Life Sci, Monsanto Co, Mo, 64-71; from asst prof to prof med, 71-80, assoc prof physiol, 71-76, PROF PHYSIOL, SCH MED, ST LOUIS UNIV, 76-, PROF OBSTET & GYNEC, 80- *Concurrent Pos:* Asst prof, Univ Mo-St Louis, 70-71. *Mem:* Brit Biochem Soc; Endocrine Soc; Soc Study Reproduction; Soc Gynec Invest. *Res:* Reproductive physiology; mode of action of prostaglandins; regulation of ovarian steroidogenesis. *Mailing Add:* Dept Obstet-Gynec 1402 S Grand Blvd St Louis Univ Sch Med St Louis MO 63104

HERTER, FREDERIC P, b Brooklyn, NY, Nov 12, 20; m 42; c 3. SURGERY. *Educ:* Harvard Univ, MD, 44; Am Bd Surg, dipl, 53. *Prof Exp:* From instr to assoc prof surg, 54-69, asst prof clin surg, 60-61, actg chmn dept, 69-71, prof, 69-72, AUCHINCLOSS PROF SURG, COLUMBIA UNIV, 72-, VCHMN DEPT, 71- *Concurrent Pos:* Health Res Coun New York City investr, 61-; dir surg, Francis Delafield Hosp, NY, 66-69; assoc attend surgeon, Presby Hosp & attend surgeon, 69-, actg dir surg serv, 69-71, dep dir surg serv, 71-; consult, St Luke's Hosp, 74- & Goldwater Mem Hosp. *Mem:* Am Col Surg; Am Surg Asn; Am Soc Clin Oncol; Int Surg Soc; AMA. *Res:* Surgical aspects of gastrointestinal tract cancer; cancer chemotherapy; surgical application. *Mailing Add:* c/o American Univ Beirut 850 Third Ave New York NY 10022

HERTIG, ARTHUR TREMAIN, b Minneapolis, Minn, May 12, 04; m 32; c 2. EMBRYOLOGY, PATHOLOGY. *Educ:* Univ Minn, BS, 28; Harvard Univ, MD, 30; Am Bd Obstet & Gynec & Am Bd Path, dipl. *Prof Exp:* Kala Azar field studies, Rockefeller Found Peking, China, 25-27; from asst to prof path, 31-52, Shattuck prof, 52-70, prof, 70-74, sr scientist & chief div pathobiol, New Eng Regional primate Res Ctr, 68-74, assoc dir res, 72-74, EMER PROF PATH ANAT, HARVARD MED SCH, 74- *Concurrent Pos:* Fel, Nat Res Coun, Carnegie Inst, 33-34; path trainee, Peter Bent Brigham Hosp, 30-31; res pathologist, Boston Lying-in Hosp, 31-32, from asst pathologist to pathologist, 34-52, consult pathologist, 52-80, obstet training, 36-38; path training, Children's Hosp, 32-33; pathologist, Free Hosp for Women, Brookline, 38-52, consult pathologist, 52-80; trustee, Boston Med Libr, 45-48; consult, Armed Forces Inst Path, Washington, DC, 48-78, sci adv bd consult, 70-74; consult pathologist, USN, Chelsea, Mass, 57-68; consult, Peter Bent Brigham & Children's Hosps, Boston, 54-74, Beth Israel Hosp, 60-74 & Cambridge City Hosp, 65-68; trustee, Am Bd Path, 59-70, life trustee, 71- *Honors & Awards:* Am Gynec Soc Award, 49; Outstanding Achievement Award, Univ Minn, 51; Ward Burdick Award, Am Soc Clin Path, 67; Distinguished Serv Award, Col Am Pathologists & Am Soc Clin Path, 72; Distinguished Serv Award, Am Col Obstet & Gynec, 75. *Mem:* Am Gynec & Obstet Soc; AMA; fel Royal Col Obstet & Gynec, London. *Res:* Primatology; anatomy and pathology of female reproduction in subhuman primates. *Mailing Add:* Brookhaven at Lexington 1010 Waltham St Apt 442B Lexington MA 02173-7940

HERTIG, BRUCE ALLERTON, environmental physiology, for more information see previous edition

HERTING, DAVID CLAIR, b Pottstown, Pa, Sept 29, 28; m 52; c 4. NUTRITIONAL BIOCHEMISTRY, ANALYTICAL CHEMISTRY. *Educ:* Pa State Univ, BS, 50; Univ Wis, MS, 52, PhD(biochem), 54. *Prof Exp:* Res biochemist, Distillation Prod Industs, 54-58, sr res biochemist, 58-63, res assoc, 63-71; res assoc, Health & Nutrit Div, Tenn Eastman Co, 71-76, sr res assoc, 76-88; ADJ PROF, DEPT SURG & TENN STATE UNIV, 87- *Concurrent Pos:* Mem bd vis, Sch Appl Sci & Technol, Tenn State Univ, 86-89. *Honors & Awards:* Glycerine Producers Asn Basic Res Award, 60. *Mem:* Fel AAAS; Am Chem Soc; Am Inst Nutrit; fel Am Inst Chemists; NY Acad Sci. *Res:* Nutrition, metabolism, toxicology and analysis of fats, fat-soluble vitamins, sterols and carotenoids; physiological disposition of antibiotics, coating agents, fungistats and other materials useful in animal science and agriculture; endocrinology of animal production; growth promoters for monogastric animals. *Mailing Add:* Rte 8 Box 580 Johnson City TN 37601

HERTING, ROBERT LESLIE, b Aurora, Ill, Jan 26, 29; m 54; c 1. INTERNAL MEDICINE, CLINICAL PHARMACOLOGY. *Educ:* Univ Ill, BS, 50, MD, 54; Ill Inst Technol, MS, 61, PhD(biol), 70; Am Bd Internal Med, dipl, 69. *Prof Exp:* Clin investr, 57-63, clin pharmacologist, 63-68, med dir pharmaceut prod div, 68-69, dir exp med, 69-71, vpres exp med, 71-73, vpres clin res & develop, Abbott Labs, 73-77, vpres corp med res, Schering-Plough, 77-80; VPRES, CLIN RES, G D SEARLE, 80- *Concurrent Pos:* From clin asst to clin asst prof, Univ Ill Col Med, 57-71, clin assoc prof, 71-77; from assoc attend physician to attend physician, Cook County Hosp, Chicago, 63-73; assoc prof, Cook County Grad Sch Med, 66-69, prof, 69-77; consult, Ill Cent Hosp, Chicago, 67-77. *Mem:* AAAS; Am Fedn Clin Res; AMA; Am Soc Microbiol; fel Am Col Physicians. *Res:* Clinical pharmacology of the cardiovascular system; central nervous system drugs; antimicrobial agents; appetite suppressants; enzymology of quanine deaminase. *Mailing Add:* G D Searle 4901 Searle Pkwy Skokie IL 60077

HERTL, WILLIAM, b Philadelphia, Pa, July 2, 32; m 64; c 2. PHYSICAL CHEMISTRY. *Educ:* Univ Pa, BS, 54; Cambridge Univ, PhD(chem), 62. *Prof Exp:* Res chemist, Ciba, Ag, Switz, 62-63; SR RES ASSOC, CORNING, INC, 63- *Concurrent Pos:* Prof, Univ of the Andes, Venezuela, 70-71; vis scientist, Cornell Univ, 87. *Mem:* Am Chem Soc; Royal Soc Chem; Mat Res Soc; Am Phys Soc. *Res:* Kinetics of heterogeneous reactions and infrared spectroscopy; clinical chemistry; bio-chemistry; surface chemistry; rheology; electrochemistry; ceramics; heterogeneous catalysis. *Mailing Add:* Res & Develop Lab Corning Inc Corning NY 14831

HERTLEIN, FRED, III, b San Francisco, Calif, Oct 17, 33; div; c 4. INDUSTRIAL HYGIENE, ANALYTICAL CHEMISTRY. *Educ:* Univ Nev, Reno, BS, 56; Am Bd Indust Hyg, cert, 72; Bd Cert Safety Prof, cert, 74; Am Inst Chemists, cert, 75; Hazard Control Mgr Bd, cert, 77. *Prof Exp:* Grad asst chem, Dept Chem, Univ Hawaii, 56-58; oceanog chemist, US Dept Interior, Fish & Wildlife Serv, Pac Oceanic Fishery Invest, 57-59; radiochemist, Pearl Harbor Naval Shipyard Med Dept, Indust Hyg Div, 59-62, indust hygienist, 62-72; indust hyg prog mgr, US Naval Regional Med Clin, Indust Hyg Br, Hawaii, 72-78; PRES, INDUST ANALYSIS LAB, INC, 78- *Concurrent Pos:* Pres, Fred Hertlein & Assoc Environ Consult, 70-78; lab dir & indust hygiene dir, Indust Anal Lab, 78-; tech book reviewer, Am Indust Hyg Asn, 76-80; chmn hyperbarics comt, Am Conf Govt Indust Hyg, 64-71; mem, Occup Safety & Health Comt, Am Inst Chemists; asst clin prof, Univ Hawaii, Sch Pub Health, 74-; licensed lab dir, Hawaii State Dept Health, 73-84; cert Ahera inspector, mgt planner & asbestos abatement supvr, 87. *Mem:* Am Indust Hyg Asn; Am Acad Indust Hyg; Am Chem Soc; AAAS; Am Inst Chem; Ger Chem Soc; Asn Aerosol Res. *Res:* Atmospheric sampling and analysis of a wide variety of gases, vapors, dusts, mists and fumes in the occupational environment; evaluation of physical, chemical and biological stressors in occupational environments, noise, ionizing and nonionizing radiation, heat, humidity and pressure. *Mailing Add:* Indust Analysis Lab Inc 3615 Harding Ave Suite 304 Honolulu HI 96816

HERTLER, WALTER RAYMOND, b Philadelphia, Pa, Apr 10, 33; m 61; c 2. ORGANIC CHEMISTRY. *Educ:* Univ Pa, AB, 55; Univ Ill, PhD, 58. *Prof Exp:* RES CHEMIST, E I DU PONT DE NEMOURS & CO, INC, 58- *Mem:* Am Chem Soc; Sigma Xi. *Res:* Polymer chemistry. *Mailing Add:* 1375 Parkersville Rd Kennett Square PA 19348

HERTSGAARD, DORIS M FISHER, b Mandan, NDak, Feb 14, 39; c 2. MATHEMATICAL STATISTICS. *Educ:* Univ NDak, BA, 60; NDak State Univ, MS, 64; Univ Iowa, PhD(statist), 72. *Prof Exp:* From instr to assoc prof, 64-81, PROF MATH, NDAK STATE UNIV, 81- *Concurrent Pos:* Bd dir, Asn Women Math Educ. *Mem:* Am Statist Asn; Sigma Xi; Nat Coun Teachers Math; Asn Women Math. *Res:* Testing procedures for asymmetric distributions, specifically with adaptive procedures. *Mailing Add:* RR 2 Box 53 Fargo ND 58102

HERTWECK, GERALD, b Evansville, Ind, Sept 19, 34; m 56; c 3. SYSTEMS ENGINEERING, TECHNICAL MANAGEMENT. *Educ:* Southern Ill Univ, BA, 59; Northeastern Univ, MS, 70; Univ Mass, PhD, 76. *Prof Exp:* Mathematician & physicist, US Naval Weapons Lab, 59-62, supvry opers res analyst, 62-65; fel scientist, Advan Studies Group, Westinghouse Elec Corp, Waltham, 66-71; opers res analyst, Opers Res & Systs Anal Off, 71-79, asst Army Combat Food Serv Syst, 79-80, spec asst Dept Defense Food Prog, US Army Natick Res & Develop Labs, 80-83, CHIEF, FOOD EQUIP & SYSTS DIV, FOOD ENG DIR, US ARMY NATICK RES, DEVELOP & ENG CTR, 83- *Concurrent Pos:* Naval res assoc, Inst Naval Studies, Ctr Naval Anal, 65-66; lectr sch mgt, Boston, 77-79. *Mem:* Opers Res Soc Am; Sigma Xi. *Res:* Military food service systems and operations. *Mailing Add:* 907C Ridgefield Circle Clinton MA 01510

HERTWIG, WALDEMAR R, b San Francisco, Calif, Sept 28, 20; m 46; c 2. CHEMICAL ENGINEERING. *Educ:* Columbia Univ, BS, 42, ChE, 43. *Prof Exp:* Chem engr, Standard Oil Co Ind, Am Oil Co, 43-51, sect leader, Pilot Plant Div, Res Dept, 51-61, res assoc, process econ div, 61-80; RETIRED. *Mem:* Am Chem Soc; Am Inst Chem Engrs. *Res:* Development and economics of petroleum process and synthetic fuels. *Mailing Add:* 1121 Jugador Ct San Marcos CA 92069-4834

HERTZ, DAVID BENDEL, b Yoakum, Tex, Mar 25, 19; m 41; c 2. OPERATIONS ANALYSIS. *Educ:* Columbia Univ, BA, 39, BS, 40, PhD(indust eng), 49. *Prof Exp:* Prod engr, Radio Corp Am, 40-41; asst dir eng res, Celanese Corp, 45-49; from asst prof to assoc prof indust eng, Columbia Univ, 49-54; mgr opers res, Popular Merchandise Co, 54-56; prin, opers res, Arthur Andersen & Co, 57-61; dir, McKinsey & Co, Inc, 62-83; DISTINGUISHED PROF, ART INTEL, UNIV MIAMI, 83- *Concurrent Pos:* Lectr, Columbia Univ, 48-49, 55-83; grant, Columbia Coun, 70-71; mem sci adv coun, Picatinny Arsenal, 50-55; consult, Nat Res Coun, 51-52; dir proj team res, Off Naval Res, 51-53; dir eng, Plastics Div, Celanese Corp, 53-54; trustee, Columbia Univ, 77-83 & Columbia Univ Press, 74-; tech anal dir, Adv Panel, Nat Bur Standards, chmn, 73-77; chmn bd gov, 77-83, Opportunity Fund Corp, 67-; vis prof, Grad Sch Bus Studies, Univ London, 77-82; adj prof, Grad Sch Bus, Columbia Univ, 77-83; mem exec comt minority enterprise, President's Adv Coun; publs ed, Opers Res Soc, pres, 74-75; Nat Acad Sci liaison subcomt, Int Inst Systs Anal, 84-85; vis prof, Univ Vienna, 86-87. *Honors & Awards:* George S Kimmbel Award, Opers Res Soc, 81. *Mem:* Fel AAAS; Int Fedn Opers Res Soc (pres, 76-); NY Acad Sci (secy, 52-54); Economet Soc (pres, 73); Inst Mgt Sci (vpres, 56-63, pres, 64-65). *Res:* Applications of management science, systems analysis and operations research; systems engineering; computer systems. *Mailing Add:* 441 Valencia Ave Coral Gables FL 33134

HERTZ, HARRY STEVEN, b New York, NY, Feb 25, 47; m 69; c 2. ANALYTICAL CHEMISTRY, TECHNICAL MANAGEMENT & RESEARCH PLANNING. *Educ:* Polytech Inst Brooklyn, BS, 67; Mass Inst Technol, PhD(org chem), 71. *Prof Exp:* Alexander von Humboldt fel biochem, Univ Munich, 71-73; res chemist, Nat Bur Standards, 73-78, chief org anal res div anal chem, 78-83; DIR, CTR ANALYTICAL CHEM, NAT INST STANDARDS & TECHNOL, 83- *Concurrent Pos:* Secy, Nat Comt Clin Lab Standards, 83-84, pres-elect, 84-86, pres, 86-88; dept energy, Health & Environ Res Adv Comm, 84-88; adv comt, Health & Human Serv Good Mfg Practices, 88-89. *Honors & Awards:* Bronze Medal, US Dept Com, 81, Silver Medal, 86; Arthur S Flemming Award Outstanding Fed Serv, 86. *Mem:* Am Chem Soc; Sigma Xi; Am Soc Mass Spectrometry (secy, 83-85); The Chem Soc. *Res:* Roles of quality assurance and accuracy in chemical measurement programs and systems; organic mass spectrometry; trace organic analyses in complex matrices; health and environmental measurements. *Mailing Add:* Ctr Analytical Chem Nat Inst Standards & Technol Gaithersburg MD 20899

HERTZ, JOHN ATLEE, b Bethlehem, Pa, Jan 10, 45; m 66; c 2. STATISTICAL PHYSICS. *Educ:* Harvard Univ, BA, 66; Univ Pa, MS, 67, PhD(physics), 70. *Prof Exp:* Res assoc physics, Univ Pa, 71 & Univ Cambridge, 71-73; asst prof physics, Univ Chicago, 73-80; PROF, NORDITA, 80- *Honors & Awards:* Sloan Fel, 74-77. *Mem:* Am Phys Soc; Danish Phys Soc; Europ Phys Soc. *Res:* Computational networks; spin glasses. *Mailing Add:* Nordita Blegdamsvej 17 Copenhagen 2100 Denmark

HERTZ, LEONARD B, b South Milwaukee, Wis, Nov 26, 24; m 50; c 4. HORTICULTURE. *Educ:* Univ Wis, BS, 49, MS, 50, PhD(agron), 56. *Prof Exp:* Conservationist, Soil Conserv Serv, USDA, 50-53; agronomist, Kans State Univ, 56-61; res biologist, Niagara Chem Div, FMC Corp, 61-67; horticulturist, 67-80, PROF HORT, UNIV MINN, ST PAUL, 80- *Mem:* Weed Sci Soc Am; Am Soc Hort Sci. *Res:* Weed control in fruit crops; culture of fruit crops. *Mailing Add:* Dept Hort Univ Minn St Paul MN 55108

HERTZ, PAUL ERIC, b Brooklyn, NY, July 8, 51. EVOLUTIONARY BIOLOGY, ECOLOGY. *Educ:* Stanford Univ, BS, 72; Harvard Univ, AM, 73, PhD(biol), 77. *Prof Exp:* Fel biol, Dalhousie Univ, 77-79; asst prof, 79-85, ASSOC PROF BIOL, BARNARD COL, COLUMBIA UNIV, 85- *Concurrent Pos:* Res assoc zool, Univ Wash, 80-81; res assoc, Inst Evolution, Univ Haifa, Israel, 80. *Mem:* Am Soc Zoologists; Soc Study Evolution; Sigma Xi; Ecol Soc Am; Am Soc Naturalists. *Res:* Strategies of adaptation in reptiles; thermal biology and physiological ecology of reptiles; structure of reptile communities; physiological ecology of arthropod predator-prey systems. *Mailing Add:* Dept Biol Sci Barnard Col 3009 Broadway New York NY 10027

HERTZ, ROY, b Cleveland, Ohio, June 19, 09; m 34, 62; c 2. ENDOCRINOLOGY, REPRODUCTION. *Educ:* Univ Wis, AB, 30, PhD(physiol), 33, MD, 39; Johns Hopkins Univ, MPH, 41. *Hon Degrees:* DSc, Univ Wis, 86. *Prof Exp:* Asst zool, Univ Wis, 30-34; instr pharmacol, Med Sch, Howard Univ, 34-35; intern, Wis Gen Hosp, 39-40; res officer, USPHS, 41-47, endocrinologist, chmn, Endocrinol Sect & mem, Study Sect Endocrinol & Metab, Nat Cancer Inst, 47-65; sci dir, Nat Inst Child Health & Develop, 65-66; prof obstet & gynec, Sch Med, George Wash Univ, 66-67; dir, Reproduction Res Br, Nat Inst Child Health & Develop, 67-69; sr physician & assoc dir, Biomed Div, Pop Coun, Rockefeller Univ, 69-72; prof obstet & gynec & med, NY Med Col, 72-73; res prof, 73-84, EMER PROF PHARMACOL, OBSTET & GYNEC, MED CTR, GEORGE WASHINGTON UNIV, WASHINGTON, DC, 84-; EMER SCIENTIST, NAT INST CHILD HEALTH & HUMAN DEVELOP, NIH, BETHESDA, MD, 89- *Concurrent Pos:* Mem, Endocrinol & Metab Study Sect, NIH, 47-55, lectr, 61, adj scientist, Nat Inst Child Health & Human Develop, Bethesda, Md, 87-89; chmn, Panel Endocrinol, Nat Res Coun, 53-56; chmn, Adv Comt Ther, Am Cancer Soc, 56-59, Res Adv Coun, 59-62; Anne Frankel Rosenthal Mem award for cancer res, AAAS, 57; mem coun, Endocrine Soc, 59-62; cancer res award, Int Col Surgeons, 69; mem, Int Comt Contraceptive Res, 71-; med res award, Lasker Found, 72. *Honors & Awards:* Distinguished Serv Award, Am Col Obstetricians & Gynecologists, 75; Twenty-fifth Medal, Barren Found, 85; Axel Munthe Found Medallion, 85. *Mem:* Nat Acad Sci; fel Am Col Physicians; hon fel Am Asn Obstet & Gynec; Endocrine Soc (vpres, 60-61); Soc Exp Biol & Med; Sigma Xi; AAAS; Am Physiol Soc; AMA; Fedn Clin Invest. *Res:* Endocrinology; reproduction; cancer; metabolism; nutrition; author or co-author of over 130 publications. *Mailing Add:* Rte Three Box 582 Hollywood MD 20636

HERTZBERG, ABRAHAM, b New York, NY, July 8, 22; m 50; c 3. AERODYNAMICS, HEAT TRANSFER. *Educ:* Va Polytech Inst, BS, 43; Cornell Univ, MS, 49. *Prof Exp:* Engr, Cornell Aeronaut Lab, 49-57, asst head, Aerodyn Res Dept, 57-59, head, 59-65; PROF, DEPT AERONAUT & ASTRONAUT & DIR AEROSPACE & ENERGETICS RES PROG, UNIV WASH, 65- *Concurrent Pos:* Vis Lectr, Chinese Acad Sci, 83. *Honors & Awards:* Dryden Medal Award, Am Inst Aeronaut & Astronaut; Minta-Martin lectr, Univ Md, 75; Paul Vieille Lectr. *Mem:* Nat Acad Eng; fel Am Inst Aeronaut & Astronaut; Am Phys Soc; Am Asn Univ Profs; Int Acad Astronaut. *Res:* Energy conversion; propulsion; high powered lasers; fusion; advanced energy conversion techniques and concepts; laser applications; space laser concepts; high energy gasdynamics, gas physics as related to lasers phenomenon; reenty phenomena. *Mailing Add:* Aerospace & Energetics Res Prog FL-10 Univ Wash Seattle WA 98195

HERTZBERG, MARTIN, b New York, NY, Jan 23, 30; m 51; c 5. PHYSICAL CHEMISTRY, THERMAL PHYSICS. *Educ:* NY Univ, BA, 49; Stanford Univ, PhD(chem), 59. *Prof Exp:* Phys chemist, US Naval Res Lab, DC, 51-52; res scientist, Burndy Eng Co, Conn, 52-53; res scientist, Lockheed Aircraft Corp, 56-61; phys chemist, Repub Aviation Corp, NY, 61-64; proj mgr, Atlantic Res Corp, 64-70; SUPVRY RES CHEMIST, PITTSBURGH RES CTR, US BUR MINES, 70- *Concurrent Pos:* Adj asst prof chem, C W Post Col Long Island, 62-65; mem teaching staff, USDA Grad Sch, 66-70; foreign vis scholar, CNRS Orleans, 88. *Mem:* Combustion Inst; Am Chem Soc. *Res:* Kinetics; atmospheric sciences; optical and mass spectrometry; optical masers; atomic and molecular physics; combustion research; safety research; mining research; fire safety; flammability; fire and explosion prevention; mining engineering; other engineering (combustion and safety). *Mailing Add:* 39 Jaycee Dr Scot Twp Pittsburgh PA 15243

HERTZBERG, RICHARD WARREN, b New York, NY, Aug 17, 37; m 61; c 2. MECHANICAL METALLURGY. *Educ:* City Col New York, BME, 60; Mass Inst Technol, MMet, 61; Lehigh Univ, PhD(metall eng), 65. *Prof Exp:* Res asst metall, Mass Inst Technol, 60-61; assoc res scientist, Res Labs, United Aircraft Corp, 61-64; res assoc metall, 64-65, prof, 65-78, NJ ZINC PROF METALL & MAT SCI, LEHIGH UNIV, 78-, DIR MECH BEHAV LAB, MAT RES CTR, 65-, CHAIRPERSON, 87- *Concurrent Pos:* Vpres, Del Res Corp, 70-73; consult to numerous corp; Libsch Res Award, 84. *Honors & Awards:* William Woodside Mem lectr, Am Soc Metals, 78. *Mem:* fel Am Soc Metals; Am Inst Mining, Metall & Petrol Engrs; Am Soc Testing & Mat. *Res:* Deformation and fracture of materials; special efforts related to fatigue crack propagation of metals and polymers; mechanical response of unidirectionally solidified eutectic composites; fractography and failure analysis. *Mailing Add:* Dept Metall & Mat Eng Lehigh Univ Bethlehem PA 18015

HERTZENBERG, ELLIOT PAUL, b Brooklyn, NY, Oct 4, 38; m 68; c 2. INORGANIC CHEMISTRY. *Educ:* Hofstra Univ, BS, 60; Adelphi Univ, MS, 65; Univ Ill, Urbana, PhD(inorg chem), 69. *Prof Exp:* Chemist, Fairchild Camera & Instrument Corp, 60-65; asst chem, Univ Ill, 65-69; res chemist, Pigment Dept, E I du Pont de Nemours & Co, Inc, 69-72; RES & DEVELOP ASSOC, PQ CORP, 72- *Mem:* Am Chem Soc; Sigma Xi; AAAS. *Res:* Coordination chemistry; solid phase reactions of coordination compounds; industrial inorganic chemistry; silica gels, zeolites. *Mailing Add:* 2420 Graydon Rd Chatham Wilmington DE 19803

HERTZIG, DAVID, b Brooklyn, NY, Dec 15, 32; div; c 4. ALGEBRA. *Educ:* Cornell Univ, BA, 53; Univ Chicago, PhD(math), 57; Univ Miami, JD, 78. *Prof Exp:* Vis mem, Sch Math, Inst Advan Study, 57; asst prof math, Cornell Univ, 58-61; vis asst prof, Princeton Univ, 61-62; assoc prof, Purdue Univ, 62-70; chmn dept math, 70-76, PROF MATH, UNIV MIAMI, 70- *Concurrent Pos:* Consult, Inst Defense Anal, 62-71. *Mem:* Math Asn Am; Am Math Soc; Math Soc France. *Res:* Algebraic geometry; algebraic groups. *Mailing Add:* Dept Math/Comput Sci Univ Miami Univ Sta Coral Gables FL 33124

HERTZLER, BARRY LEE, b Elizabethtown, Pa, May 5, 47; div. PLASMA CHEMISTRY, PHYSICAL CHEMISTRY. *Educ:* Elizabethtown Col, BS, 68; Princeton Univ, MS, 70, PhD(phys chem), 74. *Prof Exp:* Grad res asst reaction mech, Dept Chem, Princeton Univ, 69-74; res assoc, Emory Univ, Atlanta, 74-75; res scientist plasma chem, Palo Alto Res Lab, Lockheed Missiles & Space Co, Inc, 75-81, mat scientist, 81-91. *Mem:* Am Chem Soc. *Res:* Aging of materials--polymers. *Mailing Add:* Lockheed Palo Alto Res Lab 3251 Hanover St Orgn 93-50 Bldg 204 Palo Alto CA 94304

HERTZLER, DONALD VINCENT, b Cherokee, Okla, June 7, 38; m 63; c 2. POLYMER CHEMISTRY. *Educ:* Northwestern State Col, Okla, BS, 60; Okla State Univ, MS, 65, PhD(chem), 69. *Prof Exp:* From instr to asst prof chem, Panhandle State Col, 63-66; from asst prof to assoc prof, 69-84, PROF CHEM, SOUTHWESTERN OKLA STATE UNIV, 84- *Concurrent Pos:* NSF fac sci fel, 65-67 & 81-82. *Mem:* Am Chem Soc; Am Asn Univ Professors. *Res:* Synthetic organic chemistry; polymer chemistry. *Mailing Add:* Dept Chem Southwestern Okla State Univ Weatherford OK 73096

HERTZLER, EMANUEL CASSEL, b Norristown, Pa, Jan 7, 17; m 42; c 3. PHYSIOLOGY. *Educ:* Goshen Col, BA, 38; Univ Mich, MA, 40, PhD(zool), 51. *Prof Exp:* From asst prof to prof biol, Kent State Univ, 46-60; spec fel neurophysiol, NIH, Wash Univ, 60-61; assoc prof, Univ Mich, Dearborn, 61-62, chmn math & sci div, 63-67, chmn dept natural sci, 71-73, prof, 62-82, assoc dean acad affairs, 73-82, emer prof biol, 83-; RETIRED. *Concurrent Pos:* Prof zool, Univ Mich, Ann Arbor, 62-82. *Mem:* AAAS; Am Soc Zool. *Res:* Pigmentation; bioelectric potentials; neurophysiology; sympathetic nervous system; serotonin and blocking agents; bioengineering; biotelemetry. *Mailing Add:* 1539 Spring Book Dr Greenleaf Manor Elkhart IN 46514

HERUM, FLOYD L(YLE), b Dolliver, Iowa, Mar 2, 28; m 52; c 3. AGRICULTURAL ENGINEERING, GRAIN PROCESSING & STORAGE. *Educ:* Iowa State Univ, BS, 53, MS, 54; Purdue Univ, PhD(agr eng), 64. *Prof Exp:* Res assoc agr eng, Iowa State Univ, 54-55; Int Coop Admin contract specialist for Univ Ariz in Iraq, 55-57; asst prof, Univ Ill, 57-63; prof, 64-87, EMER PROF AGR ENG, OHIO STATE UNIV, 87- *Mem:* Am Soc Agr Engrs. *Res:* Measurement of physical properties of agricultural products for evaluating natural quality and effects of harvesting, handling and storage parameters, design of grain processing and storage systems. *Mailing Add:* 1580 Pemberton Dr Columbus OH 43221

HERWALD, SEYMOUR W(ILLIS), b Cleveland, Ohio, Jan 17, 17; m 41; c 3. AUTOMATIC CONTROL SYSTEMS, ELECTRONICS. *Educ:* Case Inst Technol, BS, 38; Univ Pittsburgh, MS, 40, PhD(math), 44. *Prof Exp:* Cent engr, Westinghouse Elec Corp, 39-46, spec prod engr, 46-47, mgr develop sect, 47-51, mgr eng dept, Air Arm Div, 51-56, mgr, Air Arm Div, 56-59, vpres res, 59-62, vpres electronic components & specialty prod group, 62-68, vpres eng, 68-70, vpres eng & develop, 70-77, vpres serv, 77-79; RETIRED. *Concurrent Pos:* Mem air force sci adv bd, 56-71; consult, NASA, 60-64, NSF, 74-76; consult, 79-85. *Mem:* Nat Acad Eng; Sigma Xi; Am Soc Mech Engrs; Inst Elec & Electronics Engrs (pres, 68); Am Inst Aeronaut & Astronaut. *Res:* Design servomechanisms; feedback control systems design; autopilots; airborne radar directed fire control systems. *Mailing Add:* Two Summer Sea Rd Mashpee MA 02649

HERWIG, LLOYD OTTO, b West Union, Iowa, Nov 4, 21; m 49; c 5. ENGINEERING PHYSICS, MATERIALS SCIENCE ENGINEERING. *Educ:* Luther Col, Iowa, BA, 43; Univ Iowa, MS, 47; Iowa State Univ, PhD, 53. *Prof Exp:* Asst physics, Univ Iowa, 43-44 & 46-47, jr physicist, Weapon Res & Develop, 44-46; instr physics, Iowa State Univ, 47-50, asst physics, 50-53, res assoc, 53-54; sr scientist & supvr, Naval Reactors Proj, Bettis Lab, Westinghouse Elec Corp, 54-58, mgr, High Temperature Test Reactor Facil, 58-61; sr res scientist, Res Labs, United Technol Corp, Conn, 61-64; staff assoc, Div Inst Progs, NSF, 64-70, prog mgr, Advan Technol Appln Div, 71-73, dir, Advan Solar Energy Res & Technol, 73-75; sci adv & actg dep dir, Solar Energy Div, Energy Res & Develop Admin, 75-77; solar div sci adv & sr int specialist, 77-80, chief scientist, 81-83, PROG MGR, PHOTOVOLTAIC DIV, DEPT ENERGY, 83- *Concurrent Pos:* US deleg to UN Conf New & Renewable Energy Resources, 81. *Mem:* Int Solar Energy Soc; AAAS; Am Phys Soc; Sigma Xi; Inst Elec & Electronics Engrs. *Res:* Optical stimulated emission; solar energy research and development in solar heating and cooling, biomass for fuels, solar thermal, photovoltaic, wind and ocean thermal systems; nuclear and atomic physics; particle ionization phenomena; reactor neutron physics, nuclear reactor analysis, accelerators and electronic instrumentation; solar energy research and development; applied mathematics. *Mailing Add:* 2309 N Stafford St Arlington VA 22207

HERWITZ, PAUL STANLEY, b Cincinnati, Ohio, June 10, 23; m 49; c 2. MATHEMATICS, STATISTICS. *Educ:* Univ Cincinnati, AB, 48, MA, 50; Univ NC, PhD(math), 54. *Prof Exp:* Res assoc, Inst Coop Res, Johns Hopkins Univ, 53-55; mathematician, IBM Corp, Armonk, 55-59, mgr mach oriented prog, 59-63, adv prog develop, 63, sr mem tech staff to dir res, 63-64, tech asst to dir exp comput & prog, 64-65, prog dir prog resources, 65-74, mem develop staff, IBM Corp, Poughkeepsie, 74-77, MATHEMATICIAN, CORP STAFF, IBM CORP, ARMONK, 77- *Mem:* Am Statist Asn; Am Math Soc; Math Asn Am; Sigma Xi. *Res:* Computer systems planning, organization, systems programming; problems inherent in programming as a profession; problems of the programming development process. *Mailing Add:* Four Lakeview Ave W Peekskill NY 10566

HERZ, ARTHUR H, b Ger, Feb 11, 21; nat US; m 47; c 4. SURFACE CHEMISTRY. *Educ:* Univ Rochester, BS, 50, PhD(chem), 53. *Prof Exp:* Res Chemist, 53-89, CONSULT, RES LAB, EASTMAN KODAK CO, 90- *Mem:* Am Chem Soc; Soc Imaging Sci & Technol; Royal Photog Soc. *Res:* Correlation of structure of compounds and their effect on properties of silver halides. *Mailing Add:* 1020 Park Ave Rochester NY 14610

HERZ, CARL SAMUEL, b New York, NY, Apr 10, 30; m 60; c 2. MATHEMATICS. *Educ:* Cornell Univ, AB, 50; Princeton Univ, PhD, 53. *Prof Exp:* From instr to prof math, Cornell Univ, 53-70; PROF MATH, MCGILL UNIV, 70- *Concurrent Pos:* Mem Inst Advan Study, 57-58; Alfred P Sloan fel, 62-63; vis prof, Paris-Orsay, 65 & 68 & Brandeis Univ, 69-70. *Mem:* Am Math Soc; Royal Soc Can; Can Math Soc (pres, 87-89). *Res:* Fourier analysis; functional analysis; probability. *Mailing Add:* Dept Math 805 Rue Sherbrooke Ouest Montreal PQ H3A 2K6 Can

HERZ, FRITZ, b Heilbronn, Ger, July 16, 30; US citizen; m 66; c 2. BIOCHEMISTRY, CELL BIOLOGY. *Educ:* Univ Guayaquil, Chemist, 54, PhD(biochem), 55. *Prof Exp:* Assoc head dept tuberc, Nat Inst Hyg, Ecuador, 55-57; from res asst to res assoc, Sinai Hosp, 62-67; assoc dir pediat res, 67-73; asst prof pediat, Sch Med, Johns Hopkins Univ, 70-73; ASSOC PROF PATH, ALBERT EINSTEIN COL MED, YESHIVA UNIV, 73-; HEAD DIV TISSUE CULT, DEPT PATH, MONTEFIORE MED CTR, 73- *Concurrent Pos:* Humboldt res fel org chem, Tech Univ Berlin & Humboldt res fel microbiol, Free Univ Berlin, 57-59; USPHS res fel pediat res, Sinai Hosp, 60-62. *Mem:* Am Chem Soc; Am Soc Biochem & Molecular Biol; Tissue Cult Asn. *Res:* Biochemistry of cultured cells; regulation of enzyme synthesis; tumor marker analysis. *Mailing Add:* Dept Path Montefiore Med Ctr Bronx NY 10467-2490

HERZ, JACK L, b Frankfurt, Ger, July 10, 38; US citizen; m 64; c 2. PHYSICAL ORGANIC CHEMISTRY. *Educ:* City Col New York, BS, 60; Cornell Univ, PhD(org chem), 64. *Prof Exp:* Sr res chemist, Cent Res Labs, Allied Chem Corp, 64-67; group leader, Cowles Chem Co, 67-68; sr res chemist, 68-72, asst to dir, Eastern Res Ctr, 72-74, sr bus analyst corp planning & develop, 74-76, mgr corp develop, 76-77, bus dir, Whey Proteins, 77-81, dir formulated food systs, 81-83, corp prod dir, Stauffer Chem Co, 83-85; PRES & CHIEF EXEC OFFICER, DELTOWN CHEMURGIC CORP, 86- *Concurrent Pos:* NIH fel, 60-64; adj assoc prof chem, Hunter Col, 74-76. *Honors & Awards:* Medal, Am Inst Chemists, 60. *Mem:* Am Chem Soc; Am Soc Lubrication Engrs; Inst Food Technologists. *Res:* Structure-properties relationships; flamability properties of organic compounds; organic reactions and polymerization kinetics and mechanisms; lubrication technology; food technology; whey processing; whey derived proteins; food science and technology; chemical marketing; research management; general management. *Mailing Add:* 191 Mason Ct Greenwich CT 06830

HERZ, JOSEF EDWARD, steroid chemistry, radio pharmaceuticals, for more information see previous edition

HERZ, MARVIN IRA, b New York, NY, Dec 24, 27; m 52; c 3. PSYCHIATRY. *Educ:* Univ Mich, BA, 49; Yale Univ, MS, 51, Chicago Med Sch, MD, 55; Columbia Univ, cert psychoanal, 68; Am Bd Psychiat & Neurol, dipl. *Prof Exp:* Intern, Univ Ill Res & Educ Hosps, 55-56; residency, Michael Reese Hosp, Chicago, 56-59; Dir inpatient serv, Div Psychiat, Montefiore Hosp, NY, 61-63; dir day hosp & asst prof psychiat, Albert Einstein Col Med, 63-65; assoc in psychiat, Col Physicians & Surgeons, Columbia Univ, 65-72, assoc prof clin psychiat, 72-77; prof psychiat, Emory Univ, 77-78; PROF PSYCHIAT & CHMN DEPT, STATE UNIV NY, BUFFALO, 78-; DIR PSYCHIAT, ERIE COUNTY MED CTR, 78-; DIR, PSYCHIAT DEPT, BUFFALO GEN HOSP, 78- *Concurrent Pos:* Ward admin to dir, Community Serv, NY State Psychiat Inst, 65-77, dir, 68-72, actg clin dir, 75-76; educ consult, Nat Inst heart & Lung Dis, 75-78; consult, Task Panel of President's Comn Ment Illness, 77; med dir & dir res, Ga Ment Health Inst, 77-78; consult, Psychiat Educ Br, Nat Inst Ment Health, 78-; consult psychiat, Vet Admin Med Ctr, 78-; chmn, Psychiat Adv Comt, NY Off Ment Health, 80-87; mem, Spec Rev Comn, NIMH, 81, Psycho Social Res Rev Comt, 81-85; chmn, comt educ, Am Col Psychiatrists, 87-90; sr sci adv/dir, NIMH, 89-; bd regents, Am Col Psychol, 90-93. *Honors & Awards:* PIA Found Prize, Am Psychiat Asn, 88. *Mem:* AAAS; fel Am Psychiat Asn; World Fedn Ment Health; AMA; fel Am Psychopath Asn; fel Am Col Psychiatrists; fel Am Col Psychoanal; fel Am Col Ment Health Admin. *Res:* Efficacy of psychiatric treatments; clinical research in schizophrenia; social and community psychiatry; delivery of mental health services. *Mailing Add:* Dept Psychiat State Univ NY 462 Grider St Buffalo NY 14215-3098

HERZ, MATTHEW LAWRENCE, b Newport, RI, Dec 18, 41; m 65; c 2. PHYSICAL ORGANIC CHEMISTRY. *Educ:* Tufts Univ, BS, 63; Univ RI, PhD(phys org chem), 69; Boston Col, MBA, 78. *Prof Exp:* res chemist, 69-77, oper res analyst, 77-81, supvry opers res analyst, Res & Develop Command, 81-84, chief, Phys Sci Div, 84-88, DIR TECHNOL, RES & DEVELOP CTR, US ARMY NATICK LABS, 88- *Concurrent Pos:* Instr, Metropolitan Col, Boston Univ, 69-70 & Boston Col, 78-81. *Mem:* Am Chem Soc; Sigma Xi. *Res:* Mechanistic organic chemistry; laser photochemistry, photophysical studies, polymers; mathematical modeling of chemistry. *Mailing Add:* 20 Swanson Rd Framingham MA 01701

HERZ, NORMAN, b New York, NY, Apr 12, 23; div; c 3. ARCHAEOLOGICAL GEOLOGY. *Educ:* City Col New York, BS, 43; Johns Hopkins Univ, PhD(geol), 50. *Prof Exp:* Instr geol, Univ Ill, 46-47 & Wesleyan Univ, 50-51; Fulbright Act res scholar, Am Sch Classical Studies, Greece, 51-52; head dept geol, Univ Ga, 70-78; geologist, US Geol Surv, 52-85; PROF GEOL, UNIV GA, 78-, DIR, CTR ARCHAEOL SCI, 84- *Concurrent Pos:* Geologist, Conn Geol & Natural Hist Surv, 50-51 & AID, US Geol Surv, 56-64; vis prof, Univ Sao Paulo, 62-64, George Washington Univ, 69, Univ Orléans, France, 83. *Mem:* AAAS; Geol Soc Am; Sigma Xi; Geochem Soc; Geol Soc Brazil; Soc Econ Geologists. *Res:* Igneous petrology; economic geology; plate tectonic models; archaeological geology; classical Greek and Roman marble provenance signatures. *Mailing Add:* Dept Geol Univ Ga Athens GA 30602

HERZ, RICHARD KIMMEL, chemical engineering, catalysis, for more information see previous edition

HERZ, WERNER, b Stuttgart, Ger, Feb 12, 21; nat US; m 45; c 4. ORGANIC CHEMISTRY. *Educ:* Univ Colo, BA, 43, MA, 45, PhD(org chem), 47. *Prof Exp:* Instr math, Univ Colo, 45-47; Am Cyanamid fel, Univ Ill, 47-49; from asst prof to prof chem, 49-87, R O LAWTON DISTINGUISHED PROF CHEM, FLA STATE UNIV, 87- *Concurrent Pos:* Mem chem panel, Cancer Chemother Nat Serv Ctr, Nat Cancer Inst, 59-62, consult, 62-65, mem chemother study sect, NIH, 62-66, mem med chem study sect, 70-74; adv panel chem, NSF, 61-64; sr ed, J Org Chem, Am Chem Soc, 63-89; ed, Fortschritte der Chemie Organischer Naturstoffe, 69-; vis prof, Univ Munster, 70. *Honors & Awards:* Fla Award, Am Chem Soc, 65. *Mem:* Am Chem Soc; Royal Soc Chem. *Res:* Isolation and structure of natural products; terpene chemistry; molecular rearrangements; chemotaxonomy of Compositae; antitumor substances. *Mailing Add:* Dept Chem Fla State Univ Tallahassee FL 32306

HERZBERG, GERHARD, b Hamburg, Ger, Dec 25, 04; Can citizen; m 29, 72; c 2. MOLECULAR SPECTROSCOPY. *Educ:* Darmstadt Tech Univ, Dipl Ing, 27, Dr Ing(physics), 28. *Hon Degrees:* Various from US & foreign univs. *Prof Exp:* Privatdozent & asst, Darmstadt Tech Univ, 30-35; res prof physics, Univ Sask, 35-45; prof spectros, Yerkes Observ, Chicago, 45-49; dir div physics, 49-55, div pure physics, 55-69, DISTINGUISHED RES SCIENTIST, NAT RES COUN CAN, 69- *Honors & Awards:* Nobel Prize Chem, 71; Tory Medal, Royal Soc Can, 53; Gold Medal, Can Asn Physicists, 57; Companion, Order of Can, 68; Willard Gibbs Medal, Am Chem Soc, 69; Faraday Medal, The Chem Soc, 70; Linus Pauling Medal, Am Chem Soc, 71; Royal Medal, Royal Soc, 71. *Mem:* Foreign assoc Nat Acad Sci; hon foreign mem Am Acad Arts & Sci; foreign mem Am Philos Soc; centennial foreign fel Am Chem Soc; fel Am Phys Soc; fel Royal Soc London. *Res:* Atomic and molecular spectroscopy; structure of atoms and molecules; astrophysics. *Mailing Add:* Herzberg Inst Astrophys Nat Res Coun Can Ottawa ON K1A 0R6 Can

HERZBERGER, MAXIMILIAN JACOB, optics, mathematical physics, for more information see previous edition

HERZENBERG, ARVID, b Vienna, Austria, Apr 16, 25; m 49; c 3. THEORETICAL PHYSICS. *Educ:* Victoria Univ Manchester, BSc, 49, PhD(physics), 52, DSc(physics), 64. *Prof Exp:* Asst lectr physics, Victoria Univ Manchester, 52-55, lectr, 55-58, sr lectr, 58-64, reader, 64-69; prof eng & appl sci, 69-80, PROF APPL PHYSICS, YALE UNIV, 80-, PROF PHYSICS, 90- *Mem:* Am Phys Soc; Inst Physics London. *Res:* Theoretical physics of atoms and molecules; theory of electron collisions with atoms, molecules, surfaces and in solids. *Mailing Add:* Appl Physics Yale Univ New Haven CT 06520

HERZENBERG, CAROLINE STUART LITTLEJOHN, b East Orange, NJ, Mar 25, 32; m 61; c 2. APPLIED PHYSICS. *Educ:* Mass Inst Technol, SB, 53; Univ Chicago, MS, 55, PhD(physics), 58. *Hon Degrees:* DSc, State Univ NY, 91. *Prof Exp:* Res assoc nuclear physics, Univ Chicago, 58-59 & Argonne Nat Lab, 59-61; asst prof physics, Ill Inst Technol, 61-67, Res Corp grant, 63-64; res physicist, IIT Res Inst, 67-70, sr physicist, 70-71; consult, 71-72; vis assoc prof physics, Univ Ill Med Ctr, 72-74; consult, 74-75; lectr physics, Calif State Univ, Fresno, 75-76; PHYSICIST, ARGONNE NAT LAB, 77- *Concurrent Pos:* Consult, Argonne Nat Lab, 61-62, delegate-at-large, Fedn Am Scientists, 64-65; prin investr, NASA Apollo Lunar Sample Anal Prog, 67-71; TV producer-host, Freeport Cablevision Inc, 74-75; past pres & mem exec bd, Asn Women Sci, 90-92. *Mem:* fel AAAS; fel Am Phys Soc; Asn Women Sci (secy, 82-84, pres elect, 86-88, pres, 88-90); Fedn Am Scientists; Sigma Xi. *Res:* Low energy experimental nuclear physics; Mossbauer effect studies; lunar sample analysis; nuclear methods for the analysis of coal; instrumentation development; fossil energy utilization technology; radioactive isotope applications; nuclear radiation physics; history of science. *Mailing Add:* EAIS Div Bldg 331 Argonne Nat Lab Argonne IL 60439-4817

HERZENBERG, LEONARD ARTHUR, b Brooklyn, NY, Nov 5, 31; m 53; c 4. IMMUNOLOGY, GENETICS. *Educ:* Brooklyn Col, BA, 52; Calif Inst Technol, PhD(biochem), 56. *Prof Exp:* From asst prof to assoc prof, 59-69, PROF GENETICS, STANFORD UNIV, 69- *Concurrent Pos:* Am Cancer Soc fel, Pasteur Inst, Paris, 55-57; mem, Genetics Study Sect, NIH. *Mem:* Nat Acad Sci; AAAS; Genetics Soc Am; Am Asn Immunologists; fel Soc Anal Cytol; Soc Develop Biol. *Res:* Genetics of mammalian cells; immunogenetics; genetics of immunoglobulins; membrane proteins; cellular immunology; mechanism of antibody formation; cell separation; genetics of immune response; molecular biology; flourescence activated cell sorting. *Mailing Add:* Dept Genetics Stanford Univ Med Ctr Stanford CA 94305-5120

HERZFELD, CHARLES MARIA, b Vienna, Austria, June 29, 25; nat US; c 3. PHYSICS ENGINEERING. *Educ:* Cath Univ, BChE, 45; Univ Chicago, PhD(phys chem), 51. *Prof Exp:* Lectr, Univ Chicago, 46-47 & DePaul Univ, 48-50; physicist, Ballistic Res Lab, Aberdeen Proving Ground, 51-53 & US Naval Res Lab, 53-55; consult to chief, Heat & Power Div, Nat Bur Standards, 55-56, actg asst chief, 56-57, chief heat div, 57-61, assoc dir, 61; dir ballistic missile defense, Adv Res Projs Agency, US Dept Defense, 61-63, dep dir, 63-65, dir, 65-67; tech dir, ITT Corp Defense Space Group, ITT Corp, 67-75, tech dir, ITT Aerospace, Electronics, Components & Energy Group, 75-77, tech dir, ITT Telecommun & Electronics Group-NA, 78-79, dir res, 79-85, vpres, 81-85; VCHMN, AETNA, JACOBS RAMO TECHNOL VENTURES, 85- *Concurrent Pos:* Lectr, Univ Md, 53-57, prof, 57-61; mem, Brookings Inst Fifth Conf Career Execs Fed Govt, 58, Defense Sci Bd, 68-82; consult, USN & Nat Security Coun; chief, Naval Opers Exec Panel, defense policy bd, 83-; bd dirs, Westronix, 85-, chmn, 85-88; bd dirs, T Cell Sci, Coordn Technol, Inc, Memorexi, 87- *Honors & Awards:* Flemming Award, 63; Civilian Meritorious Serv Medal, Dept Defense, 67. *Mem:* Am Inst Astronaut & Aeronaut; fel Am Phys Soc; Int Inst Strategic Studies (London); Coun Foreign Rels; fel AAAS. *Res:* Heterogeneous catalysis; theory of interior ballistics; magnetism and spectra of solids; trapped radicals; group theory; missile defense; strategy; arms control; technology management. *Mailing Add:* 1531 Live Oak Dr Silver Spring MD 20910

HERZFELD, JUDITH, b Guayaquil, Ecuador, Jan 12, 48; US citizen; m 74; c 2. PROTEINS. *Educ:* Barnard Col, AB, 67; Mass Inst Technol, PhD(chem physics), 72; Harvard Univ, MPP, 73. *Prof Exp:* Asst prof chem, Amherst Col, 73-74; res assoc, Sch Med, Harvard Univ, 74-75, lectr biophys, 75-76, asst prof, 76-83, assoc prof physiol & biophys, 83-85; assoc prof, 85-90, PROF BIOPHYS CHEM, BRANDEIS UNIV, 90- *Concurrent Pos:* Exec comt mem-at-large, Div Biol Physics, Am Phys Soc, 88-91; coun mem, Biophys Soc, 89-93 & mem exec bd, 90-92; mem, Biophys Chem Study Sect, NIH, 89-93. *Mem:* Am Chem Soc; Biophys Soc; Am Phys Soc; Am Soc Biochem & Molecular Biol; NY Acad Sci; AAAS. *Res:* Phase behavior of self-assembling systems; structure and function of membrane transport systems; cooperativity in oligomeric proteins. *Mailing Add:* Dept Chem Brandeis Univ Waltham MA 02254-9110

HERZFELD, VALERIUS E, b Weyauwega, Wis, Mar 1, 21; m 49; c 2. ELECTRICAL ENGINEERING, EDUCATION ADMINISTRATION. *Educ:* Univ Wis, BS, 49, MS, 51, PhD(elec eng), 53. *Prof Exp:* Elec engr, Broadcast Eng, Wis, 48; field engr, Ill Northern Utilities Co, 49-50; res asst, Univ Wis, 50-53; elec engr, Remington Rand, Univac, 53-54, proj engr, Sperry Rand Corp, 54-56, supvr engr, 56-58, mgr spec prod develop, 59-60, mgr tele-control dept, 60-61, mgr eng & tele-systs, 61, mgr indust control comput, 61-62, dir com eng, Univac, 62-64, mgr com opers, 64, vpres res & eng, Data Processing Div, 64-68, vpres prod develop, 68-71, vpres bus planning & develop, 71-81; asst dean, grad bus & prof admin sci, St Joseph's Univ, 81-89; RETIRED. *Mem:* Inst Elec & Electronics Engrs; Sigma Xi. *Res:* Computer peripherals and systems; textbook and management science. *Mailing Add:* 1749 Hamilton Dr Box 175 Valley Forge PA 19481

HERZIG, DAVID JACOB, b Cleveland, Ohio, Dec 13, 36; m 62; c 4. PHARMACOLOGY. *Educ:* Oberlin Col, BA, 58; Univ Cincinnati, PhD(chem), 63. *Prof Exp:* Vis scientist protein chem, Lab Chem Biol, Nat Inst Arthritis & Metab Dis, NIH, 63-65, staff fel, Lab Gen & Comp Biochem, NIMH, 65-67; sr res assoc neurochem, Dept Psychiat, Sch Med, NY Univ, 67-68; sr scientist immunopharmacol, Warner Lambert Res Inst, 68-75, sr res assoc, 75- 77, dir immunopharmacol, 77-81, SR DIR SCI DEVELOP LICENSING, WARNER LAMBERT-PARKE DAVIS, 81- *Concurrent Pos:* Damon Runyon fel, Lab Chem Biol, Nat Inst Arthritis & Metab Dis, NIH, 63-65; instr, NIH Grad Prog, 65-67. *Mem:* Am Acad Allergy; Am Soc Pharmacol & Exp Therapeut; AAAS; Sigma Xi. *Res:* Pharmacological control of the immune response; new therapies for allergies and immune diseases; control of bronchoconstriction; pulmonary diseases. *Mailing Add:* Parke-Davis Pharmaceut Res 2800 Plymouth Rd Ann Arbor MI 48105

HERZIG, GEOFFREY PETER, b Cleveland, Ohio, Dec 6, 41; c 1. HEMATOLOGY, ONCOLOGY. *Educ:* Univ Cincinnati, BS, 63; Western Reserve Univ, MD, 67. *Prof Exp:* Intern-resident, Bronx Munic Hosps, NY, 67-69; clin assoc, Nat Cancer Inst, 69-72, sr investr, 73-75; fel hemat, 72-73, from asst prof to assoc prof, 75-88, PROF MED, SCH MED, WASH UNIV, 88-, DIR BONE MARROW TRANSPLANT PROG, 75- *Mem:* Am Soc Hemat; Am Soc Clin Oncol; Am Asn Cancer Res. *Res:* Clinical studies of bone marrow transplantation and experimental chemotherapy for hematologic disorders and cancer. *Mailing Add:* Med Sch Box 8125 Univ Wash 660 S Euclid Ave St Louis MO 63110

HERZLICH, HAROLD JOEL, b Brooklyn, NY, Aug 1, 34; m 57; c 2. TIRE SAFETY & RELIABILITY, OPTIMIZATION OF CHEMICAL & STRUCTURAL SYSTEMS USED IN TIRES. *Educ:* NY Univ, BChE, 56. *Prof Exp:* engr, Tires Mfg, Goodyear Tire & Rubber, 56-57; chemist, tires res & develop, Armstrong Rubber Co, 58-62, mgr, 62-85; dir, Tire Safety, Pirelli Armstrong Tire, 85-89; CONSULT ENGR TIRE SAFETY, HERZLICH CONSULT, 90- *Concurrent Pos:* Dir, Connective Rubber Group, 58-62, treas, 62-64, vchmn, 65 & chmn, 66; treas, Rubber Div, Am Chem Soc, 78-81, chmn-elect, 81 & chmn, 82; chmn, Elastomed, 83. *Mem:* Am Chem Soc; Am Soc Testing & Mat; Am Acad Forensic Sci; Soc Automotive Engrs. *Res:* Pioneered usage of silanes, antiozonants, adhesion promoters, polybutadiene rubber, ethylene propylene rubber, halobutyl rubber and hypalon rubber in tires; aging, adhesion and fracture resistance of chemical/polymer systems in tires; optimization of tire manufacturing processes. *Mailing Add:* 8908 Desert Mound Dr Las Vegas NV 89134-8801

HERZLINGER, GEORGE ARTHUR, b Newark, NJ, June 16, 43; m 66; c 2. BIOPHYSICS, BIOENGINEERING. *Educ:* Mass Inst Technol, SB, 65, PhD(physics), 71. *Prof Exp:* Instr, Dept Physics, Mass Inst Technol, 71-72; sr scientist med group, Avco-Everett Res Lab Inc, Everett, Mass, 72-80; PRES, BELMONT INSTRUMENT CORP, 80- *Mem:* Am Heart Asn. *Res:* New methods for monitoring cardiovascular function; cardiac assist devices; cardiac simulation; applications of fluid control systems to medicine. *Mailing Add:* Belmont Instrument Corp Eight Cook St Billerica MA 01821

HERZOG, BERTRAM, b Offenburg, Ger, Feb 28, 29; US citizen; m 50, 71; c 3. ENGINEERING MECHANICS. *Educ:* Case Inst Technol, BS, 49, MS, 55; Univ Mich, PhD(eng mech), 61. *Prof Exp:* Struct engr, Dalton-Dalton Assocs, 52-53; instr eng mech, Case Inst Technol, 53-54, res assoc Proj Doanbrook, 54-55; from instr to assoc prof, Univ Mich, 55-63; mgr eng methods, Ford Motor Co, 63-65, adv comput systs planning, 65; from assoc prof to prof indust eng, Univ Mich, Ann Arbor, 65-75; dir, Univ Comput Ctr, Univ Colo, Boulder, 76-79, prof elec eng & computer sci, 76-81; pres, Herzog Assocs, Inc, 79-87; DIR, CTR INFO TECHNOL INTEGRATION, UNIV MICH, ANN ARBOR, 87-, DIR, RES SYSTS, 90- *Concurrent Pos:* Consult, Gen Elec Co, 47 & Builders Struct Steel Co, 61; consult res tire proj, Univ Mich, 61-63; consult, Bendix Systs Div, 62 & Boeing Co, 62; dir, Mich Educ Res & Info Triad, Inc, 68-74; vis staff mem, Los Alamos Nat Lab, 76-83; vis fac mem, Boeing Aircraft Co, Seattle, Wash, 62; nat lectr, Asn Comput Mach, 65-74; assoc ed, Computer Graphics & Image Processing, Res J, 72-78 & Computer & Graphics, 74- *Mem:* Am Soc Mech Engrs; Am Soc Civil Engrs; Asn Comput Mach; Sigma Xi. *Res:* Computer networks, especially as applied to needs of educational institutions; computer graphics as an interactive man-computer communication mechanism for computer-aided design and other activities. *Mailing Add:* Info Technol Div Argus Bldg Univ Mich 535 W William St Ann Arbor MI 48103-4943

HERZOG, EMIL RUDOLPH, b Basel, Switz, Mar 18, 17; nat US; m 46; c 1. ASTRONOMY. *Educ:* Univ Basel, PhD, 46. *Prof Exp:* Teacher math, physics & astron, Athenaeum, Basel, Switz, 46-49, 51-56; res scientist, Extra Galactic, Calif Inst Technol, 49-51, 56-68; prof math, Calif State Polytech Univ, Pomona, 68-85, chmn dept, 73-77; RETIRED. *Concurrent Pos:* Vis assoc prof, Univ Southern Calif, 63-71. *Mem:* AAAS; Am Astron Soc; Swiss Math Soc; Am Math Soc; Math Asn Am. *Res:* Galaxies; novae and supernovae; celestial mechanics; theory of numbers; complex variables. *Mailing Add:* 580 S Rancho Simi Dr Covina CA 91724

HERZOG, FRITZ, b Posen, Ger, Dec 6, 02; nat US; m 37. MATHEMATICS. *Educ:* Columbia Univ, PhD(math), 35. *Prof Exp:* Res assoc elec eng, Cornell Univ, 37-39, instr math, 38-43; from asst prof to prof, 43-73, EMER PROF MATH, MICH STATE UNIV, 73- *Concurrent Pos:* Vis assoc prof, Wash Univ, 48; vis assoc prof, Univ Mich, 49-50, vis prof, 56 & 62; NSF res grant, 63-64; Sigma Xi sr res award, Mich State Univ, 69. *Mem:* Am Math Soc; Math Asn Am. *Res:* Functions of a complex variable; power series; boundary behavior; number theory. *Mailing Add:* Dept Math Mich State Univ East Lansing MI 48824

HERZOG, GERALD B(ERNARD), b Minneapolis, Minn, Aug 19, 27; m 47; c 4. ELECTRICAL ENGINEERING. *Educ:* Univ Minn, BSEE, 50, MSEE, 51. *Prof Exp:* Res engr transistor TV circuits, RCA Corp, 51-54, res engr color TV reproducers, 54-57, res engr & group leader high speed comput, 57-67, dir process res & develop lab, 67-71, dir digital res lab, 71-72, sr vpres technol ctrs, David Sarnoff Res Ctr, 72-79; dir eng, Consumer Prod Group, Tex Instruments, 79-; GEN MGR, DATA GENERAL, SUNNYVALE, CALIF. *Mem:* Fel Inst Elec & Electronics Engrs. *Res:* Application of new semiconductor devices to the fields of digital logic and memory, including semiconductor memories; calculators, speaking products, home computers. *Mailing Add:* 1452 Graywood Dr San Jose CA 95129

HERZOG, GREGORY F, b New York, NY, Apr 14, 44. NUCLEAR CHEMISTRY. *Educ:* Cornell Univ, BA, 64; Columbia Univ, MA, 65, PhD(chem), 69. *Prof Exp:* Res assoc chem, Univ Chicago, 69-71; from asst to assoc prof, 71-85, PROF CHEM, RUTGERS UNIV, 85- *Concurrent Pos:* NASA prin investr, 80-90, mem, Meteorite Working Group, 88-89. *Mem:* AAAS; Am Geophys Union; Meteoritical Soc. *Res:* Heavy-ion-induced nuclear reactions; cosmic ray interactions in meteorites; accelerator mass spectrometry. *Mailing Add:* Dept Chem Rutgers Univ New Brunswick NJ 08903

HERZOG, HERSHEL LEON, b Brooklyn, NY, May 4, 24; m 48; c 3. CHEMISTRY. *Educ:* Univ Ill, BS, 44; Univ Southern Calif, MS, 47, PhD(org chem), 50. *Prof Exp:* Org chemist, E R Squibb & Sons, 44-45; chemist nat prod, Schering Corp, 50-58, mgr process res, 58-64, mgr natural prod res, 64-67, dir chem develop, 67-70, dir chem & microbiol develop, 70-73, dir chem res & develop, 73-75, vpres develop, 75-77, sr vpres, develop opers, 77-82; PRES, L & H HERZOG ASSOCS, INC, 83- *Concurrent Pos:* Adj prof chem, Stevens Inst Technol, 73-; fel, Drew Univ, 83-; consult, Drexel Burnham Lambert, 83-85, E I du Pont de Nemours, 83-, Merrill Lynch, 85- & People Repub China, 86-91. *Mem:* Am Chem Soc; Am Soc Microbiol. *Res:* Steroid chemistry; antibiotics; natural products; microbiological transformations. *Mailing Add:* 16 Evergreen Ct Glen Ridge NJ 07028

HERZOG, JAMES HERMAN, b Chicago, Ill, Nov 28, 39; m 62. ELECTRICAL ENGINEERING, COMPUTER SCIENCE. *Educ:* Northwestern Univ, BS, 62; Univ Mich, MS, 63, PhD(elec eng), 67. *Prof Exp:* Res asst man-mach systs, Univ Mich, 65-67; assoc prof elec eng, 67-70, ASSOC PROF ELEC & COMPUT ENG, ORE STATE UNIV, 70- *Mem:* Inst Elec & Electronics Engrs. *Res:* Real-time systems; digital systems; robotics. *Mailing Add:* Dept Elec Eng Ore State Univ Corvallis OR 97331

HERZOG, JOHN ORLANDO, b Ulen, Minn, Apr 6, 35; c 5. MATHEMATICS. *Educ:* Concordia Col, Moorhead, Minn, BA, 57; Univ Nebr, MA, 59, PhD(math), 63. *Prof Exp:* Instr math, Univ Nebr, 59-61; asst prof, Idaho State Univ, 63-67; assoc prof, Pac Lutheran Univ, 67-72, chmn, Div Natural Sci, 75-81, dean, 84-90, PROF MATH, PAC LUTHERAN UNIV, 72-, CHMN DEPT, 68-74, 83-84 & 91- *Concurrent Pos:* NSF In-serv Inst grant, 66-67 & 68-72; energy educ grants, 79, 82 & 85. *Mem:* Am Math Soc; Math Asn Am; Nat Coun Teachers Math. *Res:* Mathematical analysis. *Mailing Add:* Dept Math Pac Lutheran Univ Tacoma WA 98447

HERZOG, KARL A, b Philadelphia, Pa, May 2, 40; m 73; c 3. PHARMACEUTICAL RESEARCH & DEVELOPMENT, QUALITY ASSURANCE CONTROL. *Educ:* Gettysburg Col, BA, 62; Philadelphia Col Pharm, BSc, 67; Univ Conn, PhD(pharmaceut), 71. *Prof Exp:* Sr pharmaceutical chemist, Smith, Kline & French Labs, 71-77; qual control mgr, Merck, Sharp & Dohme, 77-83; MGR & DIR CORP QUAL ASSURANCE, SMITHKLINE BECKMAN CORP, 83- *Mem:* Am Pharmaceut Asn; Acad Pharmaceut Sci; Am Chem Soc. *Res:* Drug transfer through polymeric membranes; application of physical chemical principles to pharmaceutical systems during preformulation, product development and pilot scale-up; quality control service to computer-controlled, continuous-process film coated tablet and parenteral antibiotic manufacturing and packaging operations; quality control; corporate quality assurance in the health care industry. *Mailing Add:* PO Box 352 Kulpsville PA 19443

HERZOG, LEONARD FREDERICK, II, b Syracuse, NY, June 17, 26; m 55; c 3. GEO-COSMO CHEMISTRY, MASS SPECTROMETRY. *Educ:* Calif Inst Technol, BS, 48; Mass Inst Technol, PhD(geol), 52. *Prof Exp:* Asst metall, Mass Inst Technol, 48-49, asst geol, 51-52, res assoc geol & geophysics & dir mass spectrometry labs, 52-56, prof geophys & geochem, Pa State Univ, 56-67; chmn & pres, Nuclide Corp, 54-87; SCI CENTRE INC, 88-; MGR TECH OPERS, MEASUREMENT & ANALYTIC SYSTS, INC, 87- *Concurrent Pos:* Vis scientist, Dept Terrestrial Magnetism, Carnegie Inst, 51-52; mem comt isotopic measurements standards, US AEC. *Honors & Awards:* Free Enterprise Assoc Award, 64. *Mem:* Am Soc Mass Spectros; Geol Soc Am; Am Soc Testing & Mat; Am Phys Soc; Meteoritical Soc. *Res:* Development of mass spectrometers, especially for compositional analysis of solids, liquids and gases; age determination by radioactivity; isotope abundance variations in nature; stable isotope dilution analysis, especially isotopic label applications in medicine, physiology and biochemistry; origin of universe and earth; spectroscopy; design of systems for implanting and depositing ions and atoms; cathodoluminescence instrumentation. *Mailing Add:* 1301 Boal Ave Boalsburg PA 16827

HERZOG, RICHARD (FRANZ KARL), b Vienna, Austria, Mar 13, 11; nat US. PHYSICS. *Educ:* Univ Vienna, PhD, 33. *Prof Exp:* From asst prof to assoc prof physics, Univ Vienna, 39-53; br chief, Air Force Cambridge Res Ctr, 53-58; sci dep dir, Space Sci Lab, GCA Corp, 58-66, chief scientist space sci opers, 66-73; prof physics & astron, Univ Southern Miss, 73-77; RETIRED. *Res:* Mass spectroscopy; electron optics; vacuum technique. *Mailing Add:* 9221 W Broward Blvd No 2304 Plantation FL 33324

HESKETH, HOWARD E, b Erie, Pa, Feb 24, 31; m 55; c 5. CHEMICAL ENGINEERING. *Educ:* Pa State Univ, BS & MS, 53, PhD(chem eng), 68; Environ Eng Intersoc, dipl. *Prof Exp:* Sr chem engr, E I du Pont de Nemours & Co, Inc, 53-60; supvr high purity chems, Beryllium Corp, 60-63; proj coordr, Western Elec Corp, 63; asst prof chem eng, Pa State Univ, 65; scientist, Kutztown State Col, 63-66; PROF ENG, SOUTHERN ILL UNIV, 66-; CONSULT AIR POLLUTION CONTROL, 66- *Concurrent Pos:* External examr, chem eng grad studies, Univ Windsor & Univ Waterloo, Can. *Honors & Awards:* L A Ripperton Award Outstanding Educr, Air & Waste Mgt Asn, 86; Stanley Kappe Award Outstanding Serv, Am Acad Environ Engrs, 88. *Mem:* Nat Soc Prof Engrs; Am Inst Chem Engrs; Am Chem Soc; Am Soc Mech Engrs; Am Acad Environ Engrs. *Res:* Air pollution control; atomization devices, especially Venturi wet scrubber for both particulate and gaseous air pollution removal; fluidization related to material handling and combustion related to air pollution control and hazardous waste management. *Mailing Add:* Col Eng & Technol Southern Ill Univ Carbondale IL 62901

HESKETH, J D, b Sebec, Maine, Mar 12, 35; m 63; c 3. PLANT PHYSIOLOGY, PLANT ECOLOGY. *Educ:* Univ Maine, BS, 56; Cornell Univ, MS, 58, PhD(crop ecol), 61. *Prof Exp:* Fel crop physiol, Conn Agr Exp Sta, 61-63; asst plant breeder, Univ Ariz, 63-65; plant physiologist, Div Plant Indust, Commonwealth Sci & Indust Res Orgn, Australia, 65-68; adj prof & plant physiologist, Miss State Univ, 68-71, NC State Univ, 71-73 & Miss State Univ, 73-78, PROF PLANT PHYSIOL & PLANT PHYSIOLOGIST, UNIV ILL, USDA, 78- *Mem:* Am Soc Agron. *Res:* Leaf and canopy photosynthetic rates; carbon budgets; crop productivity; crop yield prediction. *Mailing Add:* Physiol S-212 Turner Hall W Gregory St Urbana IL 61801

HESLEP, JOHN MCKAY, health physics, for more infonnation see previous edition

HESPENHEIDE, HENRY AUGUST, III, b Norfolk, Va, Dec 3, 42; m 76; c 2. TROPICAL ECOLOGY, TAXONOMY. *Educ:* Duke Univ, BSc, 64; Univ Pa, PhD(pop biol), 69. *Prof Exp:* Smithsonian fel, Smithsonian Trop Res Inst, 69-70; asst prof ecol, Univ Conn, 70-73; asst prof, 73-79, ASSOC PROF BIOL, UNIV CALIF, LOS ANGELES, 79- *Concurrent Pos:* Bd dirs, Orgn Trop Studies; Woodrow Wilson fel. *Mem:* Asn Trop Biol; Am Ornith Union; Entom Soc Am; Ecol Soc Am; Am Soc Naturalists. *Res:* Ecology of birds and insects, especially coexistence, resource use and predation; taxonomy and ecology of leaf-mining beetles (Buprestidae and Curculonidae); taxonomy of Lepanthes (Orchidaceae); mimicry; ecology of extrafloral nectaries. *Mailing Add:* Dept Biol Univ Calif Los Angeles CA 90024-1606

HESS, ADRIEN LEROY, b Aullville, Mo, Jan 18, 08; wid; c 2. MATHEMATICS EDUCATION. *Educ:* Mo Valley Col, BS, 30; Cent Mo State Col, BS, 31; Univ Mont, MA, 41; Univ Wis, PhD, 54. *Prof Exp:* Teacher pub schs, Mont, 31-45; from instr to prof math, 45-74, EMER PROF MATH, MONT STATE UNIV, 74- *Concurrent Pos:* Partic, Adrien L Hess Conf, Mont Coun Teachers Math, 71; mem, Nat Coun Teachers Math. *Mem:* Sch Sci & Math Asn; Math Asn Am; Sigma Xi. *Res:* Problems in teaching of mathematics at elementary and high school levels. *Mailing Add:* Dept Math Mont State Univ Bozeman MT 59715

HESS, ALLAN DUANE, CELLULAR IMMUNOLOGY. *Educ:* Univ Ill, PhD(path & immunol), 76. *Prof Exp:* ASST PROF ONCOL & IMMUNOL, ONCOL CTR SCH MED, JOHNS HOPKINS UNIV, 78- *Res:* Bone marrow transplantation; immunosuppression. *Mailing Add:* Oncol Ctr Sch Med Johns Hopkins Univ 600 N Wolfe St Rm 3-127 Baltimore MD 21205

HESS, ARTHUR, b New York, NY, Feb 19, 27; m 53; c 2. NEUROSCIENCES. *Educ:* Univ Ark, BS, 46, MS, 47; Univ London, PhD(anat), 49, DSc, 59. *Prof Exp:* From instr to asst prof anat, Sch Med, Wash Univ, 51-61; assoc prof physiol, Col Med, Univ Utah, 61-67; prof & chmn, Anat Dept, Rutgers Med Sch, 67-84; PROF NEUROSCI & CELL BIOL, UMDNJ RW JOHNSON MED SCH, 84- *Mem:* Soc Neurosci; Am Asn Anat. *Res:* Neuroanatomy; neurohistology; neurophysiology. *Mailing Add:* Dept Neurosci & Cell Biol UMDNJ RW Johnson Med Sch Piscataway NJ 08854-5635

HESS, BERNARD ANDES, JR, b Wilmington, Del, Apr 20, 40; m 63; c 2. ORGANIC CHEMISTRY. *Educ:* Williams Col, BA, 62; Yale Univ, MS, 63, PhD(org chem), 66. *Prof Exp:* NIH fel org chem, Univ Ore, 66-68; from asst prof to assoc prof, 68-80, PROF CHEM, VANDERBILT UNIV, 80-, CHMN DEPT, 82- *Concurrent Pos:* Am Chem Soc Petrol Res Fund starter grant, 68-70; Nat Acad Sci Exchange fel, Czech Acad Sci, Prague, 73-74. *Mem:* Am Chem Soc. *Res:* Physical organic chemistry. *Mailing Add:* Box 6220 Vanderbilt Univ Nashville TN 37235

HESS, CARROLL V, b Waynesboro, Pa, Jan 14, 23; m 49; c 4. AGRICULTURAL ECONOMICS. *Educ:* Pa State Univ, BS, 47; Iowa State Univ, MS, 48, PhD(agr econ), 53. *Prof Exp:* Instr agr econ, Pa State Univ, 47-48, asst prof, 50-54; assoc prof, Southern Ill Univ, 54-56; agr economist, Prod Econ Res Div, Econ Res Serv, USDA, Cornell Univ, 56-59; prof agr econ, Univ Minn, 59-66; dean, Col Agr & assoc dir, Agr Exp Sta, 66-80, PROF AGR ECON, KANS STATE UNIV, 81- *Mem:* Am Agr Econ Asn; Sigma Xi. *Res:* Factors influencing the decision-making processes of farmers; agricultural adjustment analysis of representative farm situations; marketing efficiency studies. *Mailing Add:* 2104 Blue Hills Rd Manhattan KS 66506

HESS, CHARLES, b Paterson, NJ, Dec 20, 31; m 53; c 4. HORTICULTURE, PLANT PHYSIOLOGY. *Educ:* Rutgers Univ, BS, 53; Cornell Univ, MS, 54, PhD(plant physiol, hort), 57. *Hon Degrees:* Dr Agr, Purdue Univ, 83. *Prof Exp:* From asst prof to prof hort, Purdue Univ, 58-66; res prof & chmn dept hort & forestry, Rutgers Univ, 66-70, dir, NJ Agr Exp Sta & assoc dean col agr & environ sci, 70, actg dean, 71-72, dean, Cook Col, 72-75; DEAN COL AGR & ENVIRON SCI, UNIV CALIF, DAVIS, 75- *Concurrent Pos:* Int ed, Int Plant Propagators Soc, 62-72; pres, West Lafayette Community Sch Bd, 64-65; consult, AID, 65 & Off Technol Assessment, 77-; mem, Nat Sci Bd, 82-, vchmn, 84- *Honors & Awards:* Int Plant Propagators Soc Award, 63; Norman Jay Coleman Award, 67; Jackson Dawson Mem Medal, 71. *Mem:* Am Soc Plant Physiologists; fel Am Soc Hort Sci (pres elect, 71, pres, 72); Int Plant Propagators Soc (pres, 70); fel AAAS; Sigma Xi. *Res:* Physiology of the initiation of roots in stem cuttings. *Mailing Add:* Col Agr & Environ Sci Univ Calif Davis CA 95616

HESS, CHARLES THOMAS, b Cincinnati, Ohio, Nov 21, 40; m 65. NUCLEAR PHYSICS. *Educ:* Wabash Col, BA, 62; Ohio Univ, PhD(nuclear physics), 67. *Prof Exp:* Fels, Ohio Univ, 67-68 & Fla State Univ, 68-69; assoc prof, 69-83, PROF PHYSICS, UNIV MAINE, 83- *Concurrent Pos:* Fels, Univ Tex, Sch Pub Health, Houston, 81. *Mem:* AAAS; Am Asn Physics Teachers; Am Phys Soc; Am Nuclear Soc; Health Physics Soc. *Res:* Theoretical models of odd-odd deformed nuclei; environmental radioactivity; radon in water and air; lead 210 dating of sediment. *Mailing Add:* Dept Physics Univ Maine Orono ME 04469

HESS, DANIEL NICHOLAS, b Milwaukee, Wis, May 3, 20; m 48; c 5. ENGINEERING SCIENCE WRITING. *Educ:* Univ Wis, BSc, 44. *Prof Exp:* Control chemist brewing, Kurth Malting & Brewing Co, 40-42; chemist petrol, Shell Develop Co, 44-46; scientist & engr, Oak Ridge Nat Lab, 46-73; sr engr & tech writer, Carolina Power & Light Co, 73-85; RETIRED. *Concurrent Pos:* Univ Wis Alumni Res Found fel, 43-44. *Mem:* Health Physics Soc; Soc Tech Commun; Am Nuclear Soc. *Res:* Advances in state-of-the-art in all disciplines affecting the electric power industry. *Mailing Add:* 4705 Yadkin Dr Raleigh NC 27609

HESS, DAVID CLARENCE, b USA, Oct 30, 16; m 41; c 3. PHYSICS. *Educ:* Univ Denver, BS, 37; Univ Chicago, PhD(physics), 49. *Prof Exp:* Asst, Univ Chicago, 40-42; physicist, Ballistic Res Lab, Aberdeen Proving Grounds, 42-46; physicist, Argonne Nat Lab, 46-77; RETIRED. *Concurrent Pos:* Guest investr, Max Planck Inst Chem, Mainz, Germany, 62-63; assoc ed, Appl Physics Lett & J Appl Physics, 70-83. *Mem:* Inst Elec & Electronics Engrs. *Res:* Mass spectroscopy; cosmology; ion optics applied to surface studies. *Mailing Add:* 4425 Oakwood Downers Grove IL 60515-2712

HESS, DAVID FILBERT, b Lebanon, Pa, Aug 17, 40; m; c 2. GEOLOGY, PETROLOGY. *Educ:* Franklin & Marshall Col, AB, 62; Ind Univ, MA, 64, PhD, 67. *Prof Exp:* Asst prof to assoc prof geol, Western Ill Univ, 66-91; INDUST MINERAL & PETROL CONSULT & ED, 91- *Mem:* Geol Soc Am; Friends of Mineral; Mineral Soc Am; Nat Asn Geol Teachers; Nat Earth Sci Teachers Asn. *Res:* Metamorphic petrology; mineralogy and geochemistry; Precambrian geology; hydrogeology and water resources; earth science education; geologic and biogeographic factors in the distribution of Rhopalocera (butterflies); groundwater geology; Caribbean mollusks. *Mailing Add:* Dept Geol Western Ill Univ Macomb IL 61455

HESS, DELBERT COY, b Sweetwater, Tex, May 5, 36; m 58; c 3. AGRONOMY. *Educ:* Tex Tech Univ, BS, 58; Univ Wis, MS, 60, PhD(agron), 65. *Prof Exp:* Res asst agron, Univ Wis, 58-60 & 63-65; res agronomist, Paymaster Seeds, 65-68; dir cotton res, Acco Seed Div, Cargill Seed Co, 68-80, cotton res mgr, Cargill Seed Div, 80-84, Southwest Regional Mgr, 84-86, vpres, Int Seed, Cargill Inc, 86-91. *Concurrent Pos:* Pres, Tex Seed Trade Asn, 84; mem adv bd, Plant Variety Protection Off, 84-88. *Mem:* Am Soc Agron; Crop Sci Soc Am. *Res:* Plant breeding with cotton; plant genetics as related to breeding problems; hybrid cotton development; research management. *Mailing Add:* PO Box 5645 Minneapolis MN 55440

HESS, DENNIS WILLIAM, b Reading, Pa, Mar 1, 47; m 68; c 2. ELECTRONIC & PHOTONIC MATERIALS, PLASMA CHEMISTRY. *Educ:* Albright Col, BS, 68; Lehigh Univ, MS, 70, PhD(phys chem), 73. *Prof Exp:* Mem res staff, Fairchild Semiconductor, 73-77; from asst prof to prof chem eng, Univ Calif, Berkeley, 77-91; PROF CHEM ENG & CHMN DEPT, LEHIGH UNIV, 91- *Concurrent Pos:* Div ed, J Electrochem Soc, 78-90; consult, var integrated circuit & semiconductor equip mfg cos, 78-; asst dean, Col Chem, Univ Calif, Berkeley, 82-87, actg vchmn, Dept Chem Eng, 88, vchmn, 89-91; assoc ed, Chem Mat, 88-; chmn, Gordon Conf Chem Electronic Mat, 88- *Mem:* AAAS; Am Chem Soc; Am Inst Chem Engrs; Electrochem Soc; Sigma Xi. *Res:* Science and technology of thin film formation and etching techniques; design of thin film materials for specific applications; chemistry and physics of glow discharges, plasmas, used in thin film processing and surface modification. *Mailing Add:* Dept Chem Eng Bldg A Lehigh Univ 111 Research Dr Bethlehem PA 18015

HESS, DEXTER WINFIELD, b Rome, Ga, Aug 1, 27; m 56; c 3. PLANT TAXONOMY & HISTORICAL PLANT RESEARCH, FOOD PRODUCT FORMULATIONS. *Educ:* Duke Univ, AB, 52, MA, 53; Univ Colo, PhD, 59. *Prof Exp:* Asst biol, Univ Colo, 53-56; researcher ecol, Inst Arctic Alpine Res, Univ Colo, 56-59; teacher bot & bact, Otero Jr Col, 59-83, chmn dept sci & math, 71-77, dean fac & acad instruct, 77-83; mgr Qual Assurance Labs, Western Food Prod Co, Inc, 83-87; TEACHER BIO & NUTRIT, OTERO JR COL. *Concurrent Pos:* Partic, Inst Desert Biol, Ariz State Univ, 69 & Inst Radiation Biol, Tex Woman's Univ, 70; guest lectr, Univ Denver, 71 & Univ Colo & Colo State Univ, 84 & 85; dir, Winfield Biol, 88- *Mem:* AAAS. *Res:* Compiling local flora of Southeastern Colorado; historical changes local floro Southeastern Colorado. *Mailing Add:* Otero Jr Col La Junta CO 81050

HESS, EARL HOLLINGER, b Lancaster, Pa, June 16, 28; m 51; c 3. AGRICULTURAL CHEMISTRY, ENVIRONMENTAL CHEMISTRY. *Educ:* Franklin & Marshall Col, BS, 52; Univ Ill, PhD(org chem), 55. *Prof Exp:* Asst chem, Univ Ill, 52-54; asst prof, Franklin & Marshall Col, 55-57; proj leader & org res chemist, Res Lab, Gen Cigar Co, Inc, 57-61; PRES & CHIEF EXEC OFFICER, LANCASTER LABS, INC, 61- *Concurrent Pos:* Pres, Am Coun Independent Labs, 84-86, co-chmn, Pub Relations Comt, 89-90; treas & mem bd dirs, Commonwealth Found; trustee, Franklin & Marshall Col; mem bd dirs, Conestoga Valley Educ Found, Meridian Bank, Susquehanna Valley Div & Pa Coun Econ Educ; chmn, Accreditation Coun, Am Asn Lab Accreditation, mem bd dirs; chmn, Environ Comt, US Chamber Com, mem, Accrediting Bd, Coun Small Bus & Econ Policy Comt. *Honors & Awards:* Spec Serv Citation, Am Coun Independent Labs, 79, Roger W Truesdail Award Outstanding Serv, 83. *Mem:* Am Asn Lab Accreditation; AAAS; Am Chem Soc; Am Soc Testing & Mat; NY Acad Sci; Sigma Xi; Am Coun Independent Labs. *Res:* Agricultural and environmental chemistry and biology, including structure-biological activity relationships; agricultural and industrial by-product recovery and utilization; industrial waste treatability. *Mailing Add:* Lancaster Labs Inc 2425 New Holland Pike Lancaster PA 17601

HESS, EUGENE LYLE, b Superior, Wis, May 14, 14; m 41; c 2. BIOCHEMISTRY. *Educ:* Univ Minn, BCh, 38; Univ Wis, MS, 42, PhD(chem), 47. *Prof Exp:* Asst prof mil sci & tactics, Univ Minn, 40-42; res assoc, Univ Wis, 47-48; res assoc rheumatic fever res inst, Northwestern Univ, 48-57; sr scientist, Worcester Found Exp Biol, 57-65; prog dir metab biol, NSF, 65-66, head molecular biol sect, 66-70, sr staff assoc, Off Asst Dir Res, 70-71; exec dir, Fedn Am Socs Exp Biol, 71-79; RETIRED. *Concurrent Pos:* Affil prof, Clark Univ, 61-65; adv comt, NSF, 76-80 & Nat Res Coun, 71-79. *Mem:* AAAS; Am Chem Soc; Am Soc Biol Chem; Biophys Soc. *Res:* Ultracentrifugation; moving boundary electrophoresis; light scattering; proteins; nucleic acids. *Mailing Add:* 717 Laurel Lane Wayne PA 19087

HESS, EVELYN V, b Dublin, Ireland, Nov 8, 26; US citizen; m 54. INTERNAL MEDICINE, IMMUNOLOGY. *Educ:* Univ Dublin, MB, BCh & BAO, 49, MD, 80. *Hon Degrees:* MD, Univ Del Nord, Columbia, 87. *Prof Exp:* Registr med, Rheumatic Dis Unit, Royal Free Hosp & Med Sch, 56-57; from instr to asst prof internal med, Univ Tex Southwestern Med Sch, 60-64; assoc prof, Col Med & dir div immunol & connective tissue dis, 64-69, McDONALD PROF MED & DIR DIV IMMUNOL, MED CTR, UNIV CINCINNATI, 69- *Concurrent Pos:* Empire Rheumatism Coun traveling fel, 58-59; attend physician, Vet Admin Hosp & Parkland Mem Hosp, Dallas, Tex, 61-64; sr investr, Arthritis & Rheumatism Found, 63-68; mem bd gov, Arthritis Found, 67-70; attend physician, Univ Hosps & Vet Admin Hosp; consult physician, Children's Hosp & Convalescent Hosp. *Honors & Awards:* Oscar Schmidt Award, Univ Cincinnati; Award for Excellence, Arthritis Found. *Mem:* Fel Am Acad Allergy & Immunol; fel Am Col Physicians; Transplantation Soc; Am Soc Nephrology; Cent Soc Clin Res; Am Asn Immunol; hon mem Italian Soc Rheumatism; hon mem Peruvian Soc Rheumatol; hon mem Japanese Soc Clin Immunol; Am Rheumatism Soc (vpres, 82-83); fel Am Col Rheumatol. *Res:* Immunologic drug reactions; immunology of rheumatic diseases; AIDS and related diseases; tolerance mechanisms; computers in medicine. *Mailing Add:* Univ Cincinnati Med Ctr Rm 7562 Cincinnati OH 45267

HESS, FREDERICK DAN, b Tacoma, Wash, Mar 22, 46. CELL BIOLOGY, ELECTRON MICROSCOPY. *Educ:* Mich State Univ, BS, 69; Univ Calif, Davis, MS, 73, PhD(plant physiol), 75. *Prof Exp:* Asst prof bot & weed sci, Colo State Univ, 75-76; from asst prof to assoc prof plant physiol & weed sci, Purdue Univ, 76-86; DIR BIOL & BIOCHEM RES, SANDOZ CROP PROTECTION, 86- *Concurrent Pos:* Assoc ed, Weed Sci, 85-88, abstract ed, 90-92. *Honors & Awards:* Outstanding Young Researcher Award, Weed Sci Soc Am. *Mem:* Am Soc Plant Physiologists; Electron Microscope Soc Am; Weed Sci Soc Am. *Res:* Mechanism of action of herbicides; physiological and cell biological effects of herbicides; analytical electron microscopy of plant tissue; absorption and translocation of postemergence herbicides; plant cuticle development. *Mailing Add:* Sandoz Crop Protection Res Div 975 California Ave Palo Alto CA 94304-1104

HESS, GEORGE BURNS, b Princeton, NJ, Sept 17, 36; m 65; c 2. LOW TEMPERATURE PHYSICS. *Educ:* Princeton Univ, BA, 58; Stanford Univ, PhD(physics), 67. *Prof Exp:* Res assoc physics, Stanford Univ, 67-68; from asst prof to assoc prof, 68-88, PROF PHYSICS, UNIV VA, 88- *Concurrent Pos:* Vis assoc prof, Univ Wash, 79-80. *Mem:* Am Phys Soc. *Res:* Surface physics; superfluid helium; fluid dynamics; study of multilayer physisorption by ellipsometry and other techniques. *Mailing Add:* Dept Physics Univ Va Charlottesville VA 22901

HESS, GEORGE G, b Collingswood, NJ, Jan 4, 38; m 62; c 2. ORGANIC CHEMISTRY OF GROUNDWATER. *Educ:* Juniata Col, BS, 59; Pa State Univ, PhD(chem), 64. *Prof Exp:* Res assoc chem, Pa State Univ, 64-65; asst prof, 65-70, ASSOC PROF CHEM, WRIGHT STATE UNIV, 70-, chmn, chem dept, 84-88. *Concurrent Pos:* Consult, gas chromatography & mass spectrometry technol, Monsanto Res Corp, Dayton, 80-85; vis scientist, USEPA-ERL, Athens, Ga, 90-91. *Mem:* Am Chem Soc; Am Asn Univ Prof. *Res:* Analyses of organics in water and sediments in multidisciplinary projects. *Mailing Add:* Dept Chem Wright State Univ Dayton OH 45435

HESS, GEORGE PAUL, b Vienna, Austria, Nov 18, 26; nat US; m 53, 80; c 4. BIOCHEMISTRY. *Educ:* Univ Calif, AB, 49, PhD(biochem), 53. *Prof Exp:* From instr to assoc prof biochem, 55-64, prof chem, 64-66, PROF BIOCHEM & MOLECULAR BIOL, CORNELL UNIV, 66- *Concurrent Pos:* Nat Found Infantile Paralysis fel chem, Mass Inst Technol, 53-55; Guggenheim fel & sr Fulbright grantee, Max Planck Inst Phys Chem, Ger, 62-63; NIH spec fel, Med Res Coun Lab Molecular Biol & fel, Churchill Col, 69-70; mem numerous adv panels, NIH; Fulbright selectee, 66-67; mem ed bd, Fedn Am Socs Exp Biol, 72-76; vis prof, Johnson Found, Univ Pa, 64-65; adv panel mem, NSF, 77- *Honors & Awards:* Alexander von Humboldt Award, 82. *Mem:* Nat Acad Sci; Am Biophys Soc; Am Soc Biol Chem; Soc Neurosci; fel, Am Acad Arts & Sci; Am Chem Soc. *Res:* Structural and functional interrelationships in membrane-bound proteins of nerve and muscle cells; structure and function relationships of receptor proteins involved in cellular communication; laser-flash photolysis, optical methods; molecular biological approaches. *Mailing Add:* Sect Biochem Molecular & Cell Biol Cornell Univ 216 Biotechnology Bldg Ithaca NY 14853-2703

HESS, HANS-JURGEN ERNST, b Helmstedt, Ger, June 20, 30; m 64; c 2. ORGANIC CHEMISTRY. *Educ:* Brunswick Tech Univ, dipl, 55, Dr rer nat(org chem), 57. *Prof Exp:* Fel org chem, Univ Va, 57-58 & Univ Ill, 58-59; res chemist, Pfizer Inc, 59-64, res supvr, 64-67, sect mgr, 67-71, asst dir, 71-72, dir, 72-80, exec dir, 80-84, SR EXEC DIR, MED CHEM RES, CENT RES, PFIZER, INC, 84- *Concurrent Pos:* Ed-in-chief, annual reports in Med Chem, 79-84. *Mem:* Am Chem Soc; Soc Ger Chem; NY Acad Sci. *Res:* Peptides; cardiovascular-renal-pulmonary-gastrointestinal drugs; prostaglandins, antibodies, anti-inflammatory drugs, natural products, chemistry. *Mailing Add:* Pfizer Inc Groton CT 06340

HESS, HELEN HOPE, b Clarksburg, WVa, Aug 26, 23. BIOCHEMISTRY, NEUROPATHOLOGY. *Educ:* WVa Univ, BA, 46, BS, 48; Harvard Med Sch, MD, 51. *Prof Exp:* Instr anat, WVa Univ, 49-50; asst in neuropath, McLean Hosp, Mass, 53-56, from asst neuropathologist to assoc neuropathologist, 56-71; asst in neuropath, Harvard Med Sch, 53-56, instr, 56-59, res assoc, 59-69, prin res assoc, 69-70, assoc prof neuropath & neurochem, 70-71; prof neurol & biochem, Med Sch, George Washington Univ, 71-75; RES MED OFFICER, BIOCHEM SECT, LAB VISION RES, NAT EYE INST, 71- *Concurrent Pos:* USPHS fel neurochem, Res Lab, McLean Hosp, Mass, 51-53; mem, Neurol Study Sect, NIH, 73-75. *Mem:* Fel AAAS; Asn Res Vision Ophthal; Am Soc Neurochem; Am Chem Soc; Int Soc Neurochem. *Res:* Neurochemistry and neuropathology of retina and brain; cerebral cortex; photoreceptor, neuronal and glial membranes; correlation of histological structure, biochemical composition and metabolic activity in nervous tissues, normal and diseased. *Mailing Add:* Off Sci Dir Bldg Nine Rm 1E-118 NIH-Nat Eye Inst Bethesda MD 20892

HESS, HOWARD M, b Akron, Mich, Sept 11, 08. POWER GENERATION & DISTRIBUTION. *Educ:* Wayne State Univ, BS, 34; Univ Mich, MS, 37. *Prof Exp:* Installer Western Electric, 27-30; prof elec eng, Wayne State Univ, 34-76; RETIRED. *Mem:* Fel Inst Elec & Electronics Engrs. *Mailing Add:* 2609 Elmhurst Cir Longmont CO 80503

HESS, JOHN BERGER, b Dayton, Ohio, Apr 4, 42; c 1. ECOLOGY, GENETICS. *Educ:* Wheaton Col, BS, 64; Southern Ill Univ, MS, 66; Colo State Univ, PhD(zool), 70. *Prof Exp:* PROF BIOL, CENT MO STATE UNIV, 69- *Mem:* Sigma Xi; AAAS; Soc Study Amphibians & Reptiles; Biol Photog Asn. *Res:* Genetics of natural populations, particularly the influence of environmental conditions on composition of the gene pool. *Mailing Add:* Dept Biol Cent Mo State Univ Warrensburg MO 64093

HESS, JOHN LLOYD, b Landisville, Pa, Nov 25, 39; m 66; c 4. METABOLIC REGULATION, OXIDATIVE STRESS. *Educ:* Franklin & Marshall Col, BA, 61; Univ Del, MS, 63; Mich State Univ, PhD(biochem), 66. *Prof Exp:* Res assoc biochem, Mich State Univ, 66; res biochemist, Scripps Inst, Univ Calif, 66-67; asst prof biochem & nutrit, 67-75, ASSOC PROF BIOCHEM & NUTRIT, VA POLYTECH INST & STATE UNIV, 75- *Concurrent Pos:* Res Corp NY res grant, 68-69; NIH res grant, 72-90, USDA Coop State Res Serv, 73-76, Environ Protection Agency, 85-90 & USDA Forest Serv, 90; dir, Va Jr Acad Sci, 75-78; fel, Dept Plant Biol, Carnegie Inst Washington, 78-79. *Mem:* Am Chem Soc; Am Soc Plant Physiol; Am Soc Biochem & Molecular Biol; Sigma Xi; Asn Res Vision & Ophthal. *Res:* Investigations in plant metabolism in response to ozone, sulfur dioxide and enzyme regulation; mechanisms of cataract formation in lens; intercellular transport and cellular compartmentation. *Mailing Add:* Dept Biochem & Nutrit Va Polytech Inst & State Univ Blacksburg VA 24061-0308

HESS, JOHN MONROE CONVERSE, b Sunbury, Pa, Nov 22, 31; m 55, 75; c 2. PHYSICAL CHEMISTRY. *Educ:* Pa State Univ, BS, 53; Univ Maine, MS, 55, PhD(phys chem), 61. *Prof Exp:* Asst chemist, Maine Tech Exp Sta, Univ Maine, 55-56, instr chem, 56-62; sr chemist, Sprague Res Ctr, Sprague Elec Co, Mass, 62-63; assoc prof, 63-65, chmn sci div, 65-69, chmn dept chem, 69-85, PROF CHEM, NORTH ADAMS STATE COL,65- *Mem:* Sigma Xi; Am Chem Soc. *Res:* Thermodynamics and determination of formation constants for complex association in electrolytes; molten salts and concentrated aqueous solutions of electrolytes. *Mailing Add:* Dept Chem North Adams State Col North Adams MA 01247

HESS, JOHN WARREN, b Lancaster, Pa, May 6, 47; m 71; c 2. HYDROGEOCHEMISTRY, ISOTOPE HYDROLOGY. *Educ:* Pa State Univ, BS, 69, PhD(geol, hydrogeol), 74. *Prof Exp:* Asst res prof, Water Resources Ctr, Univ Nev Syst, 74-78, assoc res prof, hydrogeochem, 78-86, dir, Isotope Lab, 81-87, dep exec dir, 87-89, RES PROF, DESERT RES INST, WATER RESOURCES CTR, UNIV NEV SYST, 86-, EXEC DIR, 89- *Concurrent Pos:* Chmn, Res Adv Comt, Nat Speleol Soc, 77-84; res fel, Scottish Univ Res & Reactor Ctr, 80-81; prof, Univ Nev-Las Vegas, 81-; mem, IAH Karst Comn, 90- *Mem:* Am Geophys Union; Geol Soc Am; Nat Speleol Soc; Sigma Xi; Geochem Soc; Int Asn Hydrogeologists. *Res:* Hydrogeologic, isotopic, hydrogeochemical, and geophysical research pertaining to carbonate rock terrains, unsaturated zones and groundwater monitoring. *Mailing Add:* Desert Res Inst 2505 Chandler Ave No 1 Las Vegas NV 89120-4004

HESS, JOSEPH W, JR, b Farmington, Utah, June 7, 26; m 53; c 6. MEDICAL EDUCATION. *Educ:* Utah State Univ, BS, 53; Univ Utah, MD, 56; Univ Ill, Urbana, MEd(med), 68. *Prof Exp:* From instr to prof internal med, Sch Med, Wayne State Univ, 60-72, chief sect rheumatol, 65-68, dir educ serv & res, 68-74, prof family med & chmn dept, 74-88; DIV CHIEF, FAMILY MED, UNIV UTAH, 88- *Concurrent Pos:* NSF fel, 60-61; guest worker, Nat Inst Arthritis & Metab Dis, 65; mem med adv comt, Nat Myasthenia Gravis Found, 66-67; consult, WHO, 70-; mem nat rev comt, Regional Med Progs, 70-73. *Mem:* Am Col Physicians; Am Educ Res Asn; Asn Am Med Cols; Am Acad Family Physicians; Soc Teachers Family Med. *Res:* Muscles and connective tissue metabolism; methods for improving the assessment of student and physician performance and for improving the effectiveness of the learning environment in undergraduate medical education. *Mailing Add:* Dept Med Univ Utah Salt Lake City UT 48202

HESS, KARL, b Trumau, Austria, June 20, 45; US citizen; m 67; c 2. PHYSICS. *Educ:* Univ Vienna, Austria, PhD(energy relaxation, semiconductors), 70. *Prof Exp:* Dozent, Univ Vienna, 77-80; assoc dir, Beckman Inst, 87-90, PROF ELEC ENG, UNIV ILL, URBANA-CHAMPAIGN, 80- *Concurrent Pos:* Vis assoc prof, Coord Sci Lab, Univ Ill, Urbana-Champaign, 77-80; Beckman assoc, Ctr for Advan Study, Univ Ill; adj prof, Supercomput Appln, Univ Ill, 88-; consult, US Army, Ft Momouth, Naval Res Lab, Siemens,Ger, Gould, Chicago; Fulbright fel, 73-74. *Mem:* Fel Inst Elec & Electronics Engrs. *Res:* Semiconductor physics. *Mailing Add:* Univ Ill Coord Sci Lab 1101 W Springfield Ave Urbana IL 61801

HESS, LAVERNE DERRYL, b Stockton, Ill, Oct 28, 33; m 55; c 6. PHYSICAL CHEMISTRY. *Educ:* Univ Calif, Riverside, BA, 61, PhD(phys chem), 65. *Prof Exp:* Res chemist, Hughes Res Labs, 65-89; RETIRED. *Res:* Photochemistry; chemical lasers; chemical kinetics--relaxation phenomena; nonlinear optics; excimer lasers; laser annealing; laser-induced epitaxy; liquid crystal light valves; superconductivity. *Mailing Add:* 505 N Roosevelt Blvd Apt C 714 Falls Church VA 22044

HESS, LAWRENCE GEORGE, b Humboldt, Kans, Feb 25, 16; m 43, 86; c 3. SYNTHETIC ORGANIC CHEMISTRY. *Educ:* Univ Notre Dame, BS, 38, MS, 39, PhD(org chem), 41. *Prof Exp:* Asst, Univ Notre Dame, 38-41; chemist, Res & Develop Dept, Union Carbide Corp, 41-71, develop scientist chem & plastics, 71-78; RETIRED. *Mem:* Am Chem Soc. *Res:* Process development. *Mailing Add:* 868 Alta Rd Charleston WV 25314

HESS, LINDSAY LAROY, b Great Falls, MT, Apr 30, 40; m 77; c 6. MATHEMATICAL MODELING, COMPUTER SIMULATION. *Educ:* Mont State Univ, BS, 62, MS, 65; Univ Ill, MA, 67; Ohio Univ, PhD(physics), 71. *Prof Exp:* Res physicist, ERDA, 74-76; asst prof elec eng, Univ NDak, 76-77; assoc prof math & physics, Carroll Col, Mont, 77-83; VPRES ACAD AFFAIRS, MONT TECH, 83- *Concurrent Pos:* NSF fel, Univ Ill, 64-67; NDEA fel, Ohio Univ, 68-71; educ consult, Indians Into Med, 73-74; dir, Mont Sci Talent Search, 86-88; pres, Mont Sect, Am Asn Physics Teachers, 86-87 & Mont Acad Sci, 87-88. *Mem:* Sigma Xi; Am Asn Physics Teachers. *Res:* Applied modeling of systems; math and science education. *Mailing Add:* 3410 Parkway Butte MT 59701

HESS, MARILYN E, b Erie, Pa, Dec 31, 24. PHARMACOLOGY. *Educ:* Villa Maria Col, BS, 46; Univ Pa, MS, 49, PhD(pharmacol), 57. *Prof Exp:* Asst pharmacol, 49-50, asst instr physiol & from asst instr pharmacol to asst prof, 52-68, assoc prof, 68-76, PROF PHARMACOL, SCH MED, UNIV PA, 76- *Concurrent Pos:* Am Heart Asn res fel, 60-62; NIH career develop award, 67-72; estab investr, Am Heart Asn, 62-67, mem coun basic sci. *Honors & Awards:* Lindback Award. *Mem:* AAAS; Am Soc Pharmacol & Exp Therapeut; NY Acad Sci. *Res:* Circulatory and respiratory reflexes; effect of antiarrhythmic drugs on carbohydrate metabolism of heart muscle; toxic actions of oxygen; effect of sympathomimetic amines on phosphorylase activity in heart muscle; influence of hormones on cardiac function and metabolism. *Mailing Add:* Dept Pharmacol 6084 Univ Pa Sch Med Philadelphia PA 19104-6084

HESS, MELVIN, endocrinology; deceased, see previous edition for last biography

HESS, MICHAEL L, MEDICINE. *Prof Exp:* PROF & CHMN, DIV CARDIOPULMONARY LAB & RES, VA MED COL, 91- *Mailing Add:* Dept Med Va Med Col Box 281 Richmond VA 23298

HESS, PATRICK HENRY, b Albia, Iowa, Aug 6, 31; m 59; c 5. RESEARCH MANAGEMENT, ENHANCED OIL RECOVERY. *Educ:* Univ Iowa, BS, 53; Univ Nebr, MS, 58, PhD(org chem), 60. *Prof Exp:* Chemist, State Hyg Labs, Iowa, 53-54; asst, Univ Nebr, 56-58; res chemist, Chevron Res Co, 60-65, sr res chemist, 65-69, SR RES ASSOC, CHEVRON OIL FIELD RES CO, 69- *Mem:* Am Chem Soc; Soc Petrol Engrs. *Res:* Chemical flooding for enhanced oil recovery; control of sand and water associated with petroleum production; addition and condensation polymers; furans, polyesters, epoxys, acrylics; mechanisms of organic elimination reactions. *Mailing Add:* Chevron Oil Field Res Co PO Box 446 La Habra CA 90633-0446

HESS, PAUL C, b Sanremo, Italy, July 2, 40; US citizen; m 63; c 3. PETROLOGY, GEOCHEMISTRY. *Educ:* Tufts Univ, AB, 63; Harvard Univ, MS, 65, PhD(geol), 69. *Prof Exp:* From asst prof to assoc prof, 68-80, chmn dept, 80-84, PROF GEOL, BROWN UNIV, 80- *Concurrent Pos:* Assoc ed, Geochem Cosmochem Act, 83-; series ed, Cambridge Press, 86- *Mem:* Mineral Soc Am; Geochem Soc. *Res:* Metamorphic petrology; solution theory of silicate melts and glasses; phase equilibria of solids and fluids; planetary petrology and geochemistry. *Mailing Add:* Dept Geol Sci Brown Univ Brown Sta Providence RI 02912

HESS, R(ONALD) A(NDREW), b Norwalk, Ohio, Mar 12, 42; m 67; c 2. AEROSPACE ENGINEERING. *Educ:* Univ Cincinnati, BS, 65, MS, 67, PhD(aerospace eng), 70. *Prof Exp:* Asst prof aeronaut, Naval Postgrad Sch, 70-76; res scientist, Ames Res Ctr, NASA, 76-82; assoc prof, 82-84, PROF MECH ENG, UNIV CALIF, DAVIS, 84- *Concurrent Pos:* Assoc ed, Journal Aircraft, 77-, Inst Elec Electronics Engrs, Trans on Syst, Man & Cybernet, 80- *Mem:* Assoc fel Am Inst Aeronaut & Astronaut; Sigma Xi; Inst Elec Electronics Engrs. *Res:* Manual control theory; man-machine systems; automatic control systems; vehicle handling qualities. *Mailing Add:* Dept Mech Aeron & Mat Eng Univ Calif Davis CA 95616

HESS, RICHARD WILLIAM, b Rochester, NY, Nov 18, 44; m 67. INORGANIC PIGMENT CHEMISTRY. *Educ:* Kalamazoo Col, BA, 66; Univ Wis-Madison, PhD(inorg chem), 71. *Prof Exp:* res chemist, 71-73, tech supvr res & develop, Pigments Dept, 74-76, prod mgr ferric chloride, 76-79, RES MGR WHITE PIGMENT & MINERAL PRODS, C&P DEPT, E I DU PONT DE NEMOURS & CO, INC, 79- *Mem:* Am Chem Soc. *Res:* Inorganic pigments. *Mailing Add:* 103 N Rd Wilmington DE 19809

HESS, ROBERT L(AWRENCE), b Orange, NJ, Sept 29, 24; m 45; c 3. ENGINEERING MECHANICS. *Educ:* Univ Mich, BSE(eng math) & BSE(eng mech), 45, MS, 48, PhD(eng mech), 50. *Prof Exp:* Instr eng mech, Univ Mich, 46-49; mem tech staff, Bell Tel Labs, 49-52; asst prof eng mech, chem & metall eng, 52-54, assoc prof eng mech, 54-58, asst dir, Willow Run Labs, 58-60, prof eng mech, 58-80, PROF APPL MECH, UNIV MICH-ANN ARBOR, 80-, ASSOC DIR INST SCI & TECHNOL, 60-, DIR HWY SAFETY RES INST, 67- *Concurrent Pos:* Consult, US Army Sci Adv Bd, 75- *Mem:* Sigma Xi. *Res:* Electron tube envelopes; glass structure; stress analysis; design of magnetrons and transistors; computer modeling of automobile braking; handling and crash reconstruction. *Mailing Add:* 2629 Danbury Lane Ann Arbor MI 48103

HESS, ROBERT WILLIAM, cardiovascular physiology, temperature regulation, for more information see previous edition

HESS, RONALD EUGENE, b Lock Haven, Pa, Nov 22, 38; m 61; c 2. ORGANIC CHEMISTRY. *Educ:* Lock Haven State Col, BS, 60; Cornell Univ, PhD(org chem), 67. *Prof Exp:* Teacher high sch, Pa, 60-62; from asst prof to assoc prof, 66-83, PROF CHEM, URSINUS COL, 83- *Concurrent Pos:* Premed adv, Ursinus Col, 81- *Mem:* Am Chem Soc; Sigma Xi; Nat Asn Adv to Health Prof. *Res:* Dimerization and trimerization of aldoketenes; preparation and chemistry of small-ring carbocyclic ketones, especially cyclobutanones; nuclear magnetic resonance spectroscopy of organometallic compounds. *Mailing Add:* Dept Chem Ursinus Col Collegeville PA 19426

HESS, WENDELL WAYNE, b Sweetwater,Tex, Jan 1, 35; m 55; c 2. INORGANIC CHEMISTRY. *Educ:* McMurry Col, BA, 57; Univ Kans, PhD(inorg chem), 63. *Prof Exp:* Chemist rocket fuel, Phillips Petrol Co, 57-58, reactor engr atomic energy, 58-59; from asst prof to assoc prof, 63-71, PROF CHEM, ILL WESLEYAN UNIV, 71-, DEAN, 80- *Mem:* AAAS; Am Chem Soc; Sigma Xi. *Res:* Reactions of inorganic free radicals, particularly those concerning the halogen and pseudohalogen compounds; coordination compounds. *Mailing Add:* Dept Chem Ill Weslyan Univ 1317 Fell Ave Bloomington IL 61701

HESS, WILFORD MOSER BILL, b Clifton, Idaho, Feb 18, 34; m 54; c 2. BOTANY, PLANT PATHOLOGY. *Educ:* Brigham Young Univ, BS, 57; Ore State Univ, MS, 60, PhD(plant path), 62. *Prof Exp:* From asst prof to assoc prof bot, 62-71, PROF BOT, 71-, DIR, ELECTRON OPTICS LAB, 78-, CHMN DEPT BOT & RANGE SCI, BRIGHAM YOUNG UNIV, 87- *Concurrent Pos:* NSF fel cell res inst, Univ Tex, 64-65; NIH fel, Swiss Fed Inst Technol, 66-67; NSF grants, 67-79, 87-; NIH career develop award, 69-74; USDA grant, 72-74; Nat Sci Found grant, 87- *Mem:* AAAS; Am Phytopath Soc; Am Bot Soc; Mycol Soc Am; Royal Microscopical Soc; Electron Micros Soc Am; Am Soc Econ Bot; Sigma Xi. *Res:* Fungus and host ultrastructure and interaction; ultrastructure and biochemistry of fungus spores; x-ray microanalysis of fungal spores and interaction; morphological and chemical studies of hair. *Mailing Add:* 129 W1DB Dept Bot & Range Sci Brigham Young Univ Provo UT 84602

HESS, WILMOT NORTON, b Oberlin, Ohio, Oct 16, 26; m 50; c 3. ATMOSPHERIC PHYSICS, OCEANOGRAPHY. *Educ:* Columbia Univ, BE, 46; Oberlin Col, MA, 49; Univ Calif, PhD(physics), 54. *Hon Degrees:* DSc, Oberlin Col, 70. *Prof Exp:* Instr physics, Mohawk Valley Community Col, 46-47 & Oberlin Col, 49-50; physicist, Lawrence Radiation Lab, Univ Calif, 54-61; chief theoret div, Goddard Space Flight Ctr, 61-67; dir sci & appln directorate, Manned Spacecraft Ctr, NASA, Tex, 67-69; dir environ res labs, Environ Sci Serv Admin, Colo, 69-71 & Nat Oceanic & Atmospheric Admin, 71-80; dir, Nat Ctr Atmospheric Res, NSF, 80-86; ASSOC DIR, HIGH ENERGY & NUCLEAR PHYSICS RES, DEPT ENERGY, WASHINGTON, DC, 86- *Mem:* Nat Acad Eng; Am Meteorol Soc; Am Geophys Union; Am Phys Soc. *Res:* High energy nuclear physics; neutron scattering; cosmic ray neutrons; production of Van Allen radiation belt. *Mailing Add:* 14508 Pebble Hill Lane Gaithersburg MD 20878-2473

HESSE, CHRISTIAN AUGUST, b Chemnitz, Ger; Can citizen; m 64; c 2. MINERAL PROJECT EVALUATION & CONCEPTURAL DESIGN, SHAFT CONSULTATION SERVICES. *Educ:* Univ Toronto, BASc, 48. *Prof Exp:* Asst layout engr, Inco Ltd, Sudbury, Can, 49-52; construct engr & proj engr, Perini-Walsh, Niagara Falls, Ont & Perini Lt, 52-55; field engr, Pluton Uranium Mines & Aries Copper Mines, 55-56; instr mining eng, Univ Toronto, Can 56-57; chief engr & planning engr, Stanleigh Uranium Mines, Elliot Lake, Ont, 57-60; field construct engr, Johnson-Perini-Kiewit, Toronto, Can, 60-61; sr mining engr, mine engr & staff engr, US Borax, Carlsbad, NM, 61-64, prin engr, US Borax & Chem Corp, Los Angeles, 69-70, mgr mining develop, 72-74, chief engr & proj dir major plant expansion, 74-77, proj dir, Boric Acid Plant, 77-81, vpres eng, 77-90, vpres mining devel, 84-90; opers mgr & shaft & mine supt, Allan Potash Mines, Sask, 64-69; managing dir & proj mgr, Yorkshire Potash Ltd, London, Eng, 70-71; pres & gen mgr, APM Opers, Ltd, 74; CONSULT, 74- *Concurrent Pos:* Deleg, 7th Commonwealth Mining Long, Southern Africa, 61; vpres & proj mgr, Quartz Hill Proj, Alaska, 81-90; proj dir & chmn Mgt Comt, Trinity Silver Mine, Nev, 87-90. *Mem:* Am Inst Mining Engrs; Can Inst Mining & Metall. *Mailing Add:* 2701 Lake Hollywood Dr Los Angeles CA 90068-1629

HESSE, M(AX) H(ARRY), b Milwaukee, Wis, Mar 20, 27; m 58; c 2. ELECTRICAL ENGINEERING. *Educ:* Marquette Univ, BEE, 51; Ill Inst Technol, MSEE, 53; Aachen Tech Univ, Dr Ing, 55. *Prof Exp:* Design engr, Gen Elec Co, Mass, 56-58, anal engr, Aircraft Power Syst Anal, NY, 58-63, Transmission Line Anal, 63-65 & Excitation Syst Anal, 65-67, sr anal engr, 67-70; PROF ELEC POWER ENG, RENSSELAER POLYTECH INST, 70- *Concurrent Pos:* Consult engr, 70- *Honors & Awards:* Power Eng Educ Award, Edison Electric Inst, 80. *Mem:* Fel Inst Elec & Electronics Engrs; fel Brit Inst Elec Engrs; Am Soc Eng Educ. *Res:* Analytical techniques for solution of electric power transmission and component design problems; transformer noise and vibrations; radio noise on transmission lines; surge phenomena and other electromagnetic phenomena on lines. *Mailing Add:* Elec Power Eng JEC 5008 Rensselaer Polytech Inst Troy NY 12181

HESSE, REINHARD, b Halle, Ger, Mar 9, 36; wid. SEDIMENTOLOGY, MARINE GEOLOGY. *Educ:* Univ Gottingen, Vordipl, 57; Munich Tech Univ, diplom, 61, Dr rer nat, 64, Habilitation, 69. *Prof Exp:* prof res assoc sedimentol & marine geol, Munich Tech Univ, 64-68, asst prof, 69; from asst prof to assoc prof, 69-80, PROF GEOL, McGILL UNIV, 80- *Concurrent Pos:* Pvt docent, Munich Tech Univ, 69- *Honors & Awards:* Hermann Credner Award, Ger Geol Soc, 71. *Mem:* Geol Soc Am; Swiss Geol Soc; Austrian Geol Soc; Ger Geol Soc; Soc Econ Paleontologists & Mineralogists; Int Asn Geol Chem & Cosmochem. *Res:* Clastic sediments, diagenesis, processes and environments of deposition, turbidites, pelagic sediments; continental margin evolution; alpine-carpathian arc; marine geology-Labrador Sea, West Pacific margin of Middle America, Ocean Drilling Project; sedimentary petrography and geochemistry. *Mailing Add:* Dept Geol Sci McGill Univ 3450 Univ St Montreal PQ H3A 2A7 Can

HESSE, WALTER HERMAN, b New York, NY, Dec 17, 20; m 42; c 1. AGRONOMY. *Educ:* Calif State Polytech Col, BS, 52; Cornell Univ, MS, 54, PhD(agron), 55. *Prof Exp:* Asst agron, Cornell Univ, 52-55; asst prof, Univ Nev, 55-56; from asst prof to assoc prof phys sci, 56-70, prof physics & earth sci, 70-80, PROF EARTH SCI, CALIF STATE POLYTECH UNIV, POMONA, 80- *Concurrent Pos:* Asst, Calif Inst Technol, 57- *Mem:* Am Soc Agron; Soc Range Mgt. *Res:* Genetic aspects of nutrient uptake by corn; soil and plant nutrient relationships. *Mailing Add:* Dept Earth Sci Calif State Polytech Univ 3801 WTemple Ave Pomona CA 91768

HESSE, WALTER J(OHN), b St Louis, Mo, Apr 4, 23; m 47, 75; c 5. SOLAR ENERGY, TRANSPORTATION SYSTEMS. *Educ:* Purdue Univ, BSME, 44, MSME, 49, PhD, 51. *Prof Exp:* Instr, Purdue Univ, 46-49; chief eng, Test Pilot Training Div, Naval Air Test Ctr, Md, 49-56; mgr adv systs eng, Chance Vought Aircraft Corp, 56-60; prog dir nucleonic systs, LTV Aerospace Corp, 60-64, adv missile systs, 64-65, vpres & dir vertical short takeoff & landing progs, 65-69, vpres plans & requirements, Vought Aeronaut Div, 69-71, vpres transp progs, 71-73; vpres, Advan Transp Syst Div, Rohr Industs, 73-77; vpres & gen mgr, Energy Technol Ctr, E Systs, Inc, 77-83; PRES & CEO, ENTECH, INC, 83- *Concurrent Pos:* Vis prof, Univ Md, 49-56; lectr, Southern Methodist Univ, 56-57; consult, Adv Panel Aeronaut, US Dept Defense; mem panel sci & technol, Comt Sci & Astronaut, US House of Rep, 63- *Mem:* Solar Energy Indust Asn; Sigma Xi. *Res:* Aeronautics; solar energy. *Mailing Add:* Wolf Creek Ranch 847 Wolf Creek Rd Valley View TX 76272

HESSEL, ALEXANDER, b Vienna, Austria, Oct 19, 16; US citizen; m 49; c 2. PHYSICS. *Educ:* Hebrew Univ, Israel, MSc, 44; Polytech Inst Brooklyn, DEE, 60. *Prof Exp:* Engr, Broadcasting Sta, Israel, 45-48; res scientist, Israeli Ministry Defense, 48-51, head microwave group, 53-56; res assoc, Microwave Res Inst, 57-60, from asst prof to assoc prof, 60-67, prof, 67, EMER PROF ELEC ENG, POLYTECH UNIV, 88- *Concurrent Pos:* Mem US comn B, URSI, Int Sci Radio Union. *Mem:* Fel Inst Elec & Electronics Engrs; Sigma Xi. *Res:* Electromagnetics; antennas; phased arrays; conformal arrays. *Mailing Add:* 2128 E Fourteenth St Brooklyn NY 11229

HESSEL, DONALD WESLEY, b Menomonie, Wis, May 12, 22; m 52; c 2. CLINICAL CHEMISTRY, TOXICOLOGY. *Educ:* Union Col, Nebr, BA, 49; Univ Wis, MS, 51; Univ Calif, Los Angeles, PhD, 57. *Prof Exp:* Res chemist cancer, White Mem Hosp, 51-52; asst gen chem, Qual Anal & Org Chem, Univ Calif, Los Angeles, 52-56; res assoc chem marine biotoxins, Loma Linda Univ, 56-59; res chemist, World Life Res Inst, 59-61; mem staff, Bio Labs, 61-80, MEM STAFF, LOMA LINDA UNIV MED CTR & FAC MED LAB, 80- *Mem:* AAAS; Am Chem Soc; Int Asn Forensic Toxicologists. *Res:* Development of new analytical methods and the application of new instrumental techniques to clinical chemistry and forensic toxicology. *Mailing Add:* 22854 Minona Dr Grand Terrace CA 92324-5162

HESSEL, KENNETH RAY, b West, Tex, Dec 16, 39. OPTICS. *Educ:* Univ Tex, BS, 65, MS, 66, PhD(elec eng), 69. *Prof Exp:* SUPVR LASERS & OPTICS, SANDIA LABS, 69- *Mem:* Inst Electronics & Elec Engrs; Optical Soc Am. *Res:* Laser applications; noise limitations of optical systems; coherent and incoherent optical processing; fiber optics; electro-optics. *Mailing Add:* PO Box 13861 Albuquerque NM 87192

HESSEL, MERRILL, b Brooklyn, NY, Nov 25, 33; div; c 2. MOLECULAR PHYSICS, LASERS. *Educ:* Cornell Univ, BChE, 56; Columbia Univ, MA, 62, PhD(physics), 65. *Prof Exp:* Engr, Savannah River Plant, E I du Pont de Nemours & Co, 56-59; res physicist, Columbia Univ, 65-68; from asst prof to assoc prof physics, Fordham Univ, 68-74; sect chief, Nat Bur Standards, Colo, 74-78, chief laser spectros sect, 76-78, chief molecular spectros div, 78-82, dep dir, Ctr Chem Eng, 82-87, info technol advr to NBS dir, 88-89, ASST TO DIR, NAT BUR STANDARDS & TECHNOL, WASHINGTON, DC, 89-

Concurrent Pos: Vis scientist, Quantum Electronics Div, Nat Bur Standards, 73-74; pres exchange exec, The Boeing Co, 87-88. *Mem:* AAAS; Am Phys Soc; Am Chem Soc; Sigma Xi; Am Inst Chem Engrs. *Res:* Molecular beams and spectroscopy; molecular and heat pipe spectroscopy as applied to lasers and laser processes; computing and computers. *Mailing Add:* Rm A1019 Admin Nat Inst Standards & Technol Washington DC 20899

HESSELINK, LAMBERTUS, b Enschede, Neth, Dec, 48. NON-LINEAR OPTICS, IMAGE PROCESSING. *Educ:* Twente Inst Technol, BS(med eng), 70, BS(appl physics), 71; Calif Inst Technol, MS, 72, PhD(appl mech & appl physics), 77. *Prof Exp:* Res fel fluid mech, Grad Aeronaut Lab, Calif Inst Technol, 77-79, instr appl physics, 78-80; from asst prof to assoc prof, 80-90, PROF, ELEC ENG DEPT & AERONAUT/ASTRONAUT DEPT, STANFORD UNIV, 90- *Concurrent Pos:* Consult, Space & Commun Group, Hughes Aircraft Co, 78-88, Microelectronics Corp, Visulux Corp, NSF, VSL Corp, Phys Optics Corp, Air Force, 80-88; invited lectr, Von Karman Inst, Brussels, Belg, 84 & 86, Int Sch Quantum Electronics, Erice, Italy, 88; mem adv panel, NSF, 87-88, mem sci adv bd, 88-90; assoc ed, J Appl Sci Res, 87-90. *Mem:* Sigma Xi; Am Phys Soc; fel Optical Soc Am; Int Soc Optical Eng; Am Inst Aeronaut & Astronaut. *Res:* Non-linear optics, photorefractives and optical processors; application of new image processing techniques to optical diagnostics and visualization of scientific data; author of over 60 articles and papers. *Mailing Add:* Dept Aeronaut/Astronaut & Elec Eng Stanford Univ Durand Bldg Rm 370 Stanford CA 94305

HESSELTINE, CLIFFORD WILLIAM, b Brighton, Iowa, Apr 4, 17; m 41; c 4. MYCOLOGY. *Educ:* Univ Iowa, BA, 40; Univ Wis, PhD, 50. *Prof Exp:* Res mycologist, Lederle Labs, Am Cyanamid Co, 48-53; prin microbiologist in-chg cult collection group, Northern Utilization Res Br, USDA, 53-67, chief fermentation lab, Northern Regional Res Lab, Agr Res Serv, 67-86; RETIRED. *Concurrent Pos:* Mycol Soc Am rep, Nat Res Coun; chmn, US Toxic Microorganisms Panel, Joint US-Japan Coop Develop & Utilization of Natural Resources. *Honors & Awards:* Ciba-Geigy Microbiol Award, 76; Pasteur Award, 78; Thom Award, 80. *Mem:* AAAS; assoc Mycol Soc Am (pres, 63-64); assoc Int Soc Plant Path; assoc Torrey Bot Club; Soc Indust Microbiol (pres, 58); Brit Mycol Soc; Soc Appl Bacteriologists; Japanese Mycol Soc. *Res:* Soil fungi; taxonomy of mucorales and actinomycetes; maintenance of culture collections; mycotoxins; fungus food and industrial fermentations. *Mailing Add:* Northern Regional Res Lab USDA 1815 N University St Peoria IL 61604

HESSELTINE, WILBUR R, b Malone, NY, Sept 14, 19; m 47; c 4. DAIRY HUSBANDRY. *Educ:* Cornell Univ, BS, 47; Univ Conn, MS, 52; Univ Wis, PhD(dairy), 59. *Prof Exp:* Asst county agr agent dairy, Allegany County Exten Serv. 47-49; asst prof, Univ Conn, 49-59; from assoc prof to prof, 59-80, EMER PROF DAIRY, UNIV DEL, 80- *Mem:* Am Dairy Sci Asn. *Res:* Dairy production. *Mailing Add:* 122 Oliver Guessford Rd Townsend DE 19734

HESSEMER, ROBERT A(NDREW), JR, b Montesano, Wash, June 20, 23; m 51; c 2. ELECTRICAL ENGINEERING. *Educ:* Univ Wash, BS, 47; Stanford Univ, MS, 48, PhD(elec eng), 53. *Prof Exp:* From instr to asst prof elec eng, Univ NMex, 48-55; PROF ELEC ENG, UNIV ARIZ, 55- *Concurrent Pos:* Res assoc, Stanford Univ, 50-51; consult, Sandia Corp, 53-55, RCA Corp, 57-59, Bell Aerosysts Co, 62-65 & Electro Tech Anal Corp, 66-70. *Mem:* Am Soc Eng Educ; Inst Elec & Electronics Engrs. *Res:* Passive non-invasive temperature measurement of the interior of a body using acoustic or electromagnetic thermal noise spectra of the body. *Mailing Add:* Dept Elec Eng Univ Ariz Tucson AZ 85614

HESSER, JAMES EDWARD, b Wichita, Kans, June 23, 41; m 63; c 3. ASTROPHYSICS, MOLECULAR SPECTROSCOPY. *Educ:* Univ Kans, BA, 63; Princeton Univ, MA, 65, PhD(molecular & atomic physics), 66. *Prof Exp:* Res asst atomic & molecular physics, Princeton Univ, 63-66, res assoc observational astrophys & lab spectros, 66-68; jr astronr, Cerro Tololo Interam Observ, 68-69, asst astronr, 69-71, asst dir, 72-74, assoc astronr, 71-77; assoc res officer, 77-80, assoc dir, 84-86, SR RES OFFICER, DOMINION ASTROPHYS OBSERV, NAT RES COUN CAN, 80-, DIR, 86- *Concurrent Pos:* Vis prof astron, Univ Chile, Santiago, 74; mem, bd dirs, Astron Soc Pac, 81-88, assoc comt astron, 83-89, assoc comt space res, 84-89, chmn Joint Subcomt Space Astron, 84-89; counr, Am Astron Soc, 84-87; adj prof physics & astron, Univ Victoria, 87- *Honors & Awards:* Northcutt lectr, Royal Astron Soc Can, 81. *Mem:* Am Astron Soc; Int Astron Union; Can Astron Soc; Astron Soc Pac (vpres, pres, 85-88). *Res:* Determination of atomic and molecular oscillator strengths in the ultraviolet; power spectrum analysis of ultrashort period variable stars; precision photometry of globular and galactic clusters; K-line photometry of A stars; spectroscopy of galactic halo stars; analysis of globular clusters in extragalactic systems; interstellar lines in the carina nebula; development of instrumentation for space astronomy. *Mailing Add:* Dominion Astrophys Observ 5071 W Saanich Rd Victoria BC V8X 4M6 Can

HESSINGER, DAVID ALWYN, b Niagara Falls, NY, May 28, 42; m 69; c 3. CELL BIOLOGY, TOXINOLOGY. *Educ:* Kenyon Col, BA, 64; Univ Miami, PhD(cell biol), 70. *Prof Exp:* Fel biophys, Sch Med, Univ Miami, 70-71 & immunobiol, Univ Calif, Irvine, 71-73; asst prof biol, Univ SFla, 73-78; ASSOC PROF PHYSIOL & PHARMACOL, SCH MED, LOMA LINDA UNIV, CALIF, 78- *Mem:* AAAS; Sigma Xi; Am Soc Cell Biol; Int Soc Toxicol. *Res:* Mechanism of action of protein toxins from nematocysts on the structure and function of biological membranes; nematocyst structure, function and cnidocyte control of nematocyst discharge. *Mailing Add:* Dept Physiol Pharmacol Sch Med Loma Linda Univ Loma Linda CA 92350

HESSLER, JACK RONALD, b St Louis, Mo, Oct 31, 39; m 62; c 3. LABORATORY ANIMAL MEDICINE, PHYSIOLOGY. *Educ:* Univ Mo, Columbia, BS & DVM, 63; Univ Fla, MS, 68; Am Col Lab Animal Med, dipl, 69. *Prof Exp:* Captain Vet Corp, US Army, 63-65; USPHS fel comp med, Univ Fla, 65-68, from asst prof to assoc prof, 68-75; CHIEF VET MED UNIT, VET ADMIN MED CTR, 75-; ASSOC PROF PHYSIOL & DIR ANIMAL RESOURCE DIV, CTR HEALTH SCI, UNIV TENN, 75- *Concurrent Pos:* Consult, Am Asn Accreditation Lab Animal Care, 70-81, mem coun accreditation, 81-; pres, Asn Vet Admin Veterinary Med Officers, 85- *Mem:* Am Asn Lab Animal Med; Am Vet Med Asn; Am Physiol Soc; Asn Am Vet Med Col. *Res:* Perinatal physiology; respiration physiology; pulmonary circulation; diseases of laboratory animals and animal models of human disease; control of the research animal's environment. *Mailing Add:* Comp Med Div Wash Univ 660 S Euclid Ave Box 8061 St Louis MO 63110

HESSLER, JAN PAUL, b Detroit, Mich, Jan 13, 44; m 68. CHEMICAL PHYSICS, PHYSICAL CHEMISTRY. *Educ:* Kalamazoo Col, BA, 65; Mich State Univ, PhD(physics), 71. *Prof Exp:* Res assoc chem phys, Enrico Fermi Inst, Dept Chem, Univ Chicago, 71-73; instr physics, Queen's Univ, 73-75; vis scientist chem physics, 75-78, asst scientist chem physics, 78-81, PHYSICIST, CHEM DIV, ARGONNE NAT LAB, 81- *Mem:* Am Phys Soc; Optical Soc Am; Inst Elec & Electronics Engrs; Sigma Xi; AAAS. *Res:* Magnetic and electric phase transitions of rare earth systems; nuclear and electronic spin resonance studies of rare earth ions; laser studies of static and dynamic optical properties of actinide systems. *Mailing Add:* 4728 Middaugh Downers Grove IL 60515

HESSLER, ROBERT RAYMOND, b Chicago, Ill, Nov 22, 32; m 54. ZOOLOGY. *Educ:* Univ Chicago, AB, 53, BS, 55, PhD(invert paleont), 60. *Prof Exp:* Assoc scientist, Dept Biol, Woods Hole Oceanog Inst, 60-69; from assoc prof to prof zool, 69-80, PROF BIOL OCEANOG, SCRIPPS INST OCEANOG, UNIV CALIF, SAN DIEGO, 80- *Res:* Arthropod morphology and evolution, especially in Crustacea; taxonomy of Isopod; deep-sea benthic ecology. *Mailing Add:* Dept Oceanog A-002 Univ Calif San Diego Box 109 La Jolla CA 92093

HESSLEY, RITA KATHLEEN, b Warren, Pa, Dec 6, 46. ORGANIC CHEMISTRY, ANALYTICAL CHEMISTRY. *Educ:* Univ Mo-Rolla, PhD(chem), 74. *Prof Exp:* Asst prof chem, Tenn Tech Univ, 74-77; res assoc, Univ Tex, Arlington, 77-79 & Oak Ridge Nat Lab, 79-81; assoc prof, 81-87, PROF CHEM, WESTERN KY UNIV, 87- *Mem:* Am Chem Soc; Sigma Xi. *Res:* Redox reaction kinetics; structural studies and depolymerization processes for coal. *Mailing Add:* 1034 Nutwood Ave Bowling Green KY 42103

HESTENES, DAVID, b Chicago, Ill, May 21, 33; m 54; c 5. THEORETICAL PHYSICS. *Educ:* Pac Lutheran Univ, BA, 54; Univ Calif, Los Angeles, MA, 58, PhD(physics), 63. *Prof Exp:* Res assoc physics, Univ Calif, Los Angeles, 63-64; NSF fel, Palmer Phys Lab, Princeton Univ, 64-66; asst prof, 66-69, assoc prof, 69-76, PROF PHYSICS, ARIZ STATE UNIV, 76- *Honors & Awards:* Fulbright, England, 88. *Mem:* Am Asn Physics Teachers; AAAS; Int Neural Network Soc; Soc Neurosci. *Res:* Relativity; quantum electrodynamics; neural networks; physics education. *Mailing Add:* Dept Physics Ariz State Univ Tempe AZ 85287

HESTER, DONALD L, b Stanley, NC, Mar 28, 34; m 58; c 2. ELECTRICAL ENGINEERING. *Educ:* NC State Col, BSEE, 61; Univ NMex, MSEE, 63; Duke Univ, PhD(nonlinear oscillator theory), 67. *Prof Exp:* Staff mem, Sandia Corp, NMex, 61-64; asst prof elec eng, Va Polytech Inst, 66-68; mem tech staff, Bell Tel Labs, 68-77; pres, Micro Comput Systs, Inc, 77-; RETIRED. *Res:* Microcomputers; telephone loop plant testing. *Mailing Add:* 1295 Old Salem Rd Kernersville NC 27284

HESTER, JACKSON BOLING, JR, b Norfolk, Va, Sept 28, 33; m 60; c 2. MEDICINAL CHEMISTRY, ORGANIC SYNTHESIS. *Educ:* Mass Inst Technol, SB, 55; Univ Wis, PhD(org chem), 60. *Prof Exp:* RES SCIENTIST CHEM, UPJOHN CO, 60- *Honors & Awards:* UpJohn Award, 72. *Mem:* Am Chem Soc; AAAS. *Res:* Heterocyclic organic chemistry; pharmaceutical chemistry, including indoles and alkaloids; chemistry of benzodiazepines which have antianxiety activity; chemistry of medicinal agents in central nervous system and cardiovascular areas. *Mailing Add:* 9219 E ML Ave Galesburg MI 49053

HESTER, JARRETT CHARLES, b Mt Vernon, Tex, Dec 14, 38; m 61; c 3. MECHANICAL ENGINEERING. *Educ:* Univ Tex, Arlington, BS, 62; Okla State Univ, MS, 64, PhD(mech eng), 66. *Prof Exp:* Eng specialist, LTV Aerospace Corp, Tex, 65-66; asst prof mech eng, Univ Tex, 66-67; eng specialist, LTV Aerospace Corp, 67-70; assoc prof mech eng, Clemson Univ, 70-74, dept head mech eng, 71-74, assoc dean eng, 74-77, prof mech eng, 74-86; VPRES ENG, CRSS CAPTIAL INC, GREENVILLE, 86- *Mem:* Am Soc Mech Engrs. *Res:* Energy conversion; solar energy; industrial energy processes; power generation. *Mailing Add:* CRSS Capital Inc Two Patewood Dr Greenville SC 29615

HESTER, JOHN NELSON, b Washington, DC, Dec 1, 30. CHEMICAL ENGINEERING. *Educ:* Tri-State Col, BS, 58; Mich State Univ, MS, 59; Univ Calif, Davis, PhD(appl math), 71. *Prof Exp:* Eng supvr fluid dynamics, Aerojet-Gen Corp, 60-69; assoc prof mech eng, 69-74, PROF MECH ENG, CALIF STATE UNIV, SACRAMENTO, 74- *Concurrent Pos:* Consult, Alternative Energy Co, 79-, Calif Energy Comn, 79-, Agroset Tech Systs Co, 84- *Mem:* Am Chem Soc; Am Inst Chem Engrs; Am Soc Eng Educ. *Res:* Thermodynamics; chemical kinetics; waste product energy conversion and thermal destruction of hazardous wastes. *Mailing Add:* Dept Mech Engr Calif State Univ 6000 J St Sacramento CA 95819

HESTER, LAWRENCE LAMAR, JR, b Anderson, SC, May 23, 20; m 47; c 4. OBSTETRICS & GYNECOLOGY. *Educ:* The Citadel, BS, 41; Med Col SC, MD, 44. *Hon Degrees:* Dsc, The Citadel, 80. *Prof Exp:* prof obstet & gynec & chmn dept, Med Univ SC, 56-84; RETIRED. *Mem:* Am Col Obstet & Gynec; Am Asn Obstet & Gynec; Am Gynec Club. *Res:* Chemotherapy of gynecological cancer and the diagnosis and treatment of fetal distress. *Mailing Add:* Dept Obstet & Gynec Med Univ SC 171 Ashley Ave Charleston SC 29425

HESTER, NORMAN ERIC, b Niangua, Mo, Dec 16, 46; m 73; c 3. ENVIRONMENTAL CHEMISTRY, ORGANIC CHEMISTRY. *Educ:* Calif State Univ, BS, 68; Univ Calif, Riverside, MS, 71, PhD(org chem), 72. *Prof Exp:* Chemist air pollution, Statewide Air Pollution Res Ctr, Univ Calif, Riverside, 72-74; air qual chemist, US Environ Protection Agency, 74-77; mem tech staff environ chem, Environ Monitoring & Serv Ctr, Rockwell Int Corp, 77-80; group head, Environ Res, Occidental Res Corp, 80-83; TECH DIR, TRUESDAIL LABS, 83- *Mem:* Am Chem Soc. *Res:* Development of procedures for the collection and analysis of trace levels of environmental pollutants. *Mailing Add:* Truesdail Labs Inc 14201 Franklin Ave Tustin CA 92680

HESTER, RICHARD KELLY, b Austin, Tex, July 30, 47; m 79; c 2. PHARMACOLOGY, PHYSIOLOGY. *Educ:* Austin Col, BA, 69; Health Sci Ctr, Univ Tex, San Antonio, PhD(pharmacol), 75. *Prof Exp:* From instr to asst prof, Dept Pharmacol, Health Sci Ctr, Univ Tex, Dallas, 77-79; asst prof, 79-85, ASSOC PROF, DEPT MED PHARMACOL, COL MED, TEX A&M UNIV, 85-, INVESTR, MICROCIRCULATION RES INST, 81- *Concurrent Pos:* USPHS fel & Am Heart Asn fel, Sch Med, Univ Miami, 75-76; res fel, Health Sci Ctr, Univ Tex, Dallas, 76-77; vis res assoc, Dept Pharmacol, John A Burns Sch Med, Univ Hawaii, Honolulu, 78-79; young investr res grant, Nat Heart, Lung & Blood Inst, 80-84; teaching consult, Dept Pharmacol, Sch Med, Tex Tech Univ Health Sci Ctr, Lubbock, Tex, 85-86. *Mem:* AAAS; Am Heart Asn; Sigma Xi; Western Pharmacol Soc; Am Soc Pharmacol & Exp Therapeut; Microcirculatory Soc. *Res:* Excitation/contraction (relaxation) coupling in vascular smooth muscle; correlation of calcium kinetics with tension development and-or relaxation; receptor-mediated calcium ion entry; microvascular physiology and pharmacology. *Mailing Add:* 4595 Barnstable Harbor Box 24 College Station TX 77845

HESTER, ROBERT LESLIE, b Sept 12, 53; m; c 1. MICROCIRCULATION, OXYGEN DELIVERY. *Educ:* Univ Miss, PhD(physiol), 82. *Prof Exp:* ASST PROF CARDIOVASC PHYSIOL, UNIV MISS, 85- *Mem:* Am Phys Soc; Microcirculatory Soc; Biomed Eng Soc. *Mailing Add:* Dept Physiol & Biophysics Univ Miss Med Ctr 2500 N State St Jackson MS 39216

HESTERBERG, GENE ARTHUR, b Cincinnati, Ohio, Aug 30, 18; m 41; c 2. FOREST PATHOLOGY. *Educ:* Purdue Univ, BS, 41; Univ Mich, MS, 47, PhD, 56. *Prof Exp:* Biologist, State Dept Conserv, Mich, 47-48; prof forestry, Mich Technol Univ, 48-81, head dept, 67-81; RETIRED. *Concurrent Pos:* Collabr, Lake States Forest Exp Sta, US Forest Serv. *Mem:* Soc Am Foresters; Wilson Ornith Soc. *Res:* Pathology in forest practice. *Mailing Add:* Nine N Royce Rd Hancock MI 49930

HESTERBERG, THOMAS WILLIAM, b Tucson, Ariz, Mar 16, 50; m 71; c 2. CARCINOGENESIS. *Educ:* Univ Calif, Los Angeles, BA & MA, Univ Calif, Davis, PhD(pharmacol/toxicol), 81. *Prof Exp:* fel cancer res, Nat Inst Environ Health, Sci, 81-84; scientist, Chem Indust Inst Toxicol, 84-88; CONSULT, 88- *Mem:* Am Asn Cancer Res; Soc Toxicol. *Res:* Direct the testing of man-made fibers and other particulates using animal inhalation models; study the mechanisms of chemical carcinogenesis using cell culture models. *Mailing Add:* Health Safety & Environ Dept Manville Corp PO Box 5108 Denver CO 80217-5108

HESTON, LEONARD L, b Burns, Ore, Dec 16, 30; m 66; c 6. PSYCHIATRY, GENETICS. *Educ:* Univ Ore, BS, 55, MD, 61. *Prof Exp:* Intern med, Bernalillo County-Indian Hosp, 61-62; resident psychiat, Med Sch, Univ Ore, 62-65; guest worker, Psychiat Genetics Res Unit, Med Res Coun, London, 65-66; asst prof psychiat, Univ Iowa, 66-70; assoc prof, 70-74, dir psychiat res unit, 77-80, PROF PSYCIAT, UNIV WASH, SEATTLE, 74-; DIR, WASH INST MENTAL ILLNESS RES. *Concurrent Pos:* Spec fel, NIMH, 65-66; mem epidemiol studies rev comt, NIMH, Nat Adv Coun, Nat Inst Drug Abuse, Nat Coun, Alzheimer's Dis & Related Dis Asn. *Honors & Awards:* Dobzhansky Award, Behav Genetics Asn; Paul Hoch Award, Am Psychopath Asn. *Mem:* AMA; Am Eugenics Soc; Am Psychopath Asn; Am Soc Human Genetics. *Res:* Genetics of psychiatric and neurological disorders; genetics of normal behavior; medical history. *Mailing Add:* Dept Psychiat Univ Wash Seattle WA 98195

HESTON, WILLIAM MAY, JR, b Toledo, Ohio, Nov 2, 22; m 50; c 4. PHYSICAL CHEMISTRY. *Educ:* Ohio State Univ, BSc, 43; Princeton Univ, MA, 48, PhD(phys chem), 49. *Prof Exp:* Asst, Princeton Univ, 46-48; res chemist, E I du Pont de Nemours & Co, 49-51; res physicist, Argonne Nat Lab, 51-52, staff scientist, Del, 52-53; res supvr, Radiochem Eng, Savannah River Lab, 53-54, personnel supvr tech placement, 54-55; tech asst to lab dir, Employee Rels Dept, 55-58, placement rep, 58-59; dir, Univ Off Res & Assoc Chem, Case Western Reserve Univ, 59-63, vpres res, 63-64, vpres student serv, 64-66, vprovost & assoc dean arts & sci, 66-67, vpres, Univ Plans & Progs, 67-69; vpres, Hofstra Univ, 69-73; assoc provost, NY Inst Technol, 77-78, dean, 83-88, dir, Ctr Natural Sci, 78-88, Pro Life Sci, 81-90; CONSULT, 90- *Concurrent Pos:* Mem prog-proj comt, Nat Inst Dent Res, 64-68; AID consult, Govt India, 65; vis prof, Univ Mysore, 65; consult grad chem prog develop, NSF, 65-68, instnl sci develop prog, 68-70; trustee, Laurel Sch Girls, Shaker Heights, 67-69, AMD Res Found, 77-78; consult sci develop prog, Cleveland Pub Schs, 67-69; spec consult to dir, Nat Inst Dent Res, 68-70, mem dent res insts & spec progs adv comt, 70-74; mem bd adv, Heald, Hobson & Co, 69-70; mem sci adv comt, Nassau County Police Dept, 70-72; exec dir, Nassau Higher Educ Consortium, 73-75 & Long Island Regional; Adv Coun Higher Educ, 75-76; consult, Nat Inst Gen Med Sci, 76-80, Standford Res Inst, 81-84 & Nat Can Inst, 81-90. *Mem:* Fel AAAS; NY Acad Sci. *Res:* Microwave adsorption; molecular structure; inorganic polymers; ion exchange; chemical problems associated with nuclear reactor operation; colloids; comprehensive postsecondary continuing education plan. *Mailing Add:* 47 Hilton Ave Garden City NY 11530-4427

HETENYI, GEZA JOSEPH, b Budapest, Hungary, Sept 26, 23; m 47. MEDICAL PHYSIOLOGY. *Educ:* Univ Budapest, MD, 47; Univ Toronto, PhD, 60; FRCP(C), 87. *Prof Exp:* Lectr med sch, Univ Szeged, 47-51, asst prof, 51-56; res assoc physiol, Charles H Best Inst, Univ Toronto, 57-60, from asst prof to prof, 70-79; prof physiol & head dept, Univ Ottawa, 70-79, vdean, Fac Health Sci, 79-86; SCI ADV TO PRES, MED RES COUN CAN, 86- *Mem:* Am Physiol Soc; Can Physiol Soc (vpres, 75-76, pres, 76-77); Can Diabetic Asn. *Res:* Physiology of carbohydrate metabolism; glucose; control of blood sugar level; tracer methods; mathematical modeling. *Mailing Add:* Dept Physiol Univ Ottawa Ottawa ON K1H 8M5 Can

HETERICK, ROBERT CARY, JR, b Washington, DC, Apr 9, 36; m 61; c 4. COMPUTER SCIENCE, STRUCTURAL ENGINEERING. *Educ:* Va Polytech Inst, BS, 59, MS, 61, PhD(eng), 68. *Prof Exp:* From instr to assoc prof civil eng, 59-68, dir comput ctr, 68-74, PROF BLDG CONSTRUCTION & CHMN DEPT, COL ARCHIT, VA POLYTECH INST & STATE UNIV, 75-, DIR DESIGN AUTOMATION LAB, 75- *Concurrent Pos:* Consult, Hayes, Seay, Mattern & Mattern, 61-68; mem, Va Adv Coun Educ Data Processing, 68-; pres, Commonwealth Comput Consults, Inc, 70-75; consult, Gov Va Mgt Study, 70-71; lectr, IBM Corp, 71-; consult, Penton Learning Systs Inc, 77- *Mem:* Nat Soc Prof Engrs; Am Soc Civil Engrs; Am Soc Eng Educ; Am Inst Constructors. *Res:* Effect of management information systems on organizational structure; computer-aided design. *Mailing Add:* 201 Burruss Hall Va Tech Blacksburg VA 24061-0152

HETHCOTE, HERBERT WAYNE, b Villisca, Iowa, Nov 18, 41; div; c 2. MATHEMATICAL EPIDEMIOLOGY. *Educ:* Univ Colo, BS, 64; Univ Mich, MS, 65, PhD(math), 68. *Prof Exp:* From asst prof to assoc prof, 69-79, PROF MATH, UNIV IOWA, 79- *Concurrent Pos:* Boettcher Found scholar, 59-64; NSF fel, 64-66; grantee, NSF, 68, NIH, 77-80, Ctrs Dis Control, 76, 80, 87-91; vis mathematician, M D Anderson Hosp & Tumor Inst, 74-75 & NIH, 80-81; vis assoc prof, Ore State Univ, 77-78. *Mem:* Am Math Soc; Math Asn Am; Soc Indust & Appl Math; Soc Math Biol. *Res:* Mathematical analysis of models for the spread of infectious diseases; modeling the transmission and control of specific diseases such as measles, rubella, influenza and gonorrhea; using models to study HIV transmission and AIDS, for the Centers for Disease Control. *Mailing Add:* Univ Iowa City Iowa City IA 52242

HETHERINGTON, DONALD WORDSWORTH, b Montreal, Can, Feb 27, 45; m 68; c 2. BETA SPECTROSCOPY, NEUTRINO MASS. *Educ:* Univ Toronto, BASc, 67, MSc, 68; McGill Univ, PhD(nuclear physics), 85. *Prof Exp:* Lectr physics, Ahmach Bello Univ, Nigeria, 68-70; engr, Canatom Ltd, Montreal, Can, 70-71; teacher physics, Vanier Col, 71-82; res fel, Atomic Energy Can, Ltd, 84-86; RES ASST, APPL PHYSICS, UNIV MONTREAL, 87- *Mem:* Can Asn Physicists. *Res:* Low energy nuclear physics; beta spectroscopy; measurement of spectral shapes to study weak magnetism and neutrino mass. *Mailing Add:* 4019 Melrose Montreal PQ H4A 2S5 Can

HETNARSKI, RICHARD BOZYSLAW, b Stopnica, Poland, May 31, 28; nat US; m 60; c 2. MECHANICS, APPLIED MATHEMATICS. *Educ:* Gdansk Tech Univ, MSc, 52; Warsaw Univ, MSc, 60; Polish Acad Sci, Dr Tech Sci, 64. *Prof Exp:* Teaching asst math, Gdansk Tech Univ, 50-51; designer, Design Bur Diesel Engines, Warsaw, 52-54; res scientist, Inst Aircraft Res, 55-59 & Polish Acad Sci, 59-69; vis assoc prof theoret & appl mech, Cornell Univ, 69-70; distinguished vis prof mech eng, 70-71, PROF MECH ENG, ROCHESTER INST TECHNOL, 71- *Concurrent Pos:* Polish Acad Sci fel, Columbia Univ, 64-65; res fel, Northwestern Univ, 65; ed-in-chief, J Thermal Stresses, An Int Quart, 78-; NASA-ASEE fel, NASA Lewis Res Ctr, 79; lectr, Int Ctr Mech Sci, Udine, Italy, 79; aeronaut engr, NASA Lewis Res Ctr, 79-80; ed, Thermal Stress Handbook; vis prof, Univ Paderborn, Germany. *Mem:* Am Acad Mechanics; Am Soc Eng Educ; fel Am Soc Mech Engrs. *Res:* Theory of elasticity, thermoelasticity; vibrations, stress analysis; integral transforms; mechanical design. *Mailing Add:* Dept Mech Eng Rochester Inst Technol Rochester NY 14623

HETRICK, BARBARA ANN, b San Francisco, Calif, Sept 10, 51; m 83; c 1. MYCORRHIZAL FUNGI, SOIL MICROBIOLOGY. *Educ:* Ohio Wesleyan Univ, BA, 73; Wash State Univ, MS, 75; Ore State Univ, PhD(plant path), 78. *Prof Exp:* Fel, Univ Calif, Riverside, 78-80; ASSOC PROF PLANT PATH, KANS STATE UNIV, 80- *Concurrent Pos:* Consult, Lehigh Univ, 82-83. *Mem:* Am Phytopath Soc; Sigma Xi; Mycol Soc Am. *Res:* Ecology of vesicular-arbuscular mycorrhizal fungi and their potential for commercialization; use of these fungi to retard drought stress or to stimulate plant growth in infertile soils. *Mailing Add:* Dept Plant Path Kans State Univ Manhattan KS 66506

HETRICK, DAVID LEROY, b Scranton, Pa, Jan 26, 27; m 48; c 3. NUCLEAR SAFETY. *Educ:* Rensselaer Polytech Inst, BS, 47, MS, 50; Univ Calif, Los Angeles, PhD(physics), 54. *Prof Exp:* Instr physics, Rensselaer Polytech Inst, 47-50; res engr, Atomic Energy Res Dept, NAm Aviation, Inc, 50-53 & 54-55, supvr reactor theory, 55-56, res specialist, 56-59; pvt consult, 59; mem tech staff, Systs Labs Div, Electronic Specialty Co, 59-60; assoc prof physics, Calif State Univ Northridge, 60-63; PROF NUCLEAR & ENERGY ENG, UNIV ARIZ, 63- *Concurrent Pos:* Admin judge, US Nuclear Regulatory Comm, 72- *Mem:* AAAS; Am Nuclear Soc; Am Phys Soc; Am Soc Eng Educ; Soc Computer Simulation; Soc Indust & Applied Math. *Res:* Dynamics and safety of nuclear power plants; applied mathematics. *Mailing Add:* Dept Nuclear & Energy Eng Univ Ariz 8740 E Dexter Dr Tucson AZ 85715

HETRICK, FRANK M, b York, Pa, Aug 28, 32; m 54; c 5. VIROLOGY. *Educ:* Mich State Univ, BS, 54; Univ Md, MS, 60, PhD(microbiol), 62. *Prof Exp:* Microbiologist, Mich Dept Health, 56-58; from asst microbiol to assoc prof virol, 58-68, actg chmn, 68-69 & 75-76, PROF VIROL, UNIV MD, 68-, CHMN, 89- *Concurrent Pos:* Int work, Egypt, Brazil, Japan & China; Sea Grant Prog, 77-; Am Soc Microbiol Found lectr, 84-85. *Mem:* AAAS; Am

Soc Microbiol; Tissue Cult Asn; fel Am Acad Microbiol; Am Fisheries Soc; Am Inst Biol Sci. *Res:* Viral and bacterial diseases of marine and freshwater fish; developmental and comparative immunology; oncogene activation and tumor development in fish. *Mailing Add:* Dept Microbiol Univ Md College Park MD 20742

HETRICK, JOHN HENRY, b Beavertown, Pa, Sept 16, 16; m 40; c 2. DAIRY TECHNOLOGY. *Educ:* Pa State Col, BS, 38; Ohio State Univ, MS, 39; Univ Ill, PhD(dairy technol), 47. *Prof Exp:* Lab technician & plant mgr, Supplee-Wills-Jones Milk Co, Pa, 39-41; instr dairy mfg, Okla Agr & Mech Col, 41-43; assoc, Univ Ill, 43-45; dir res, Dean Milk Co, Ill, 45-67; prof food sci, Univ Ill, Urbana, 67-74; DIR ENVIRON CONTROL, DEAN FOODS CO ILL, 74- *Mem:* Am Chem Soc; Am Dairy Sci Asn; Inst Food Technol; Sigma Xi. *Res:* Dehydrated milk products; frozen dairy products; high temperature short time heat treatment of milk; dairy chemistry; analytical determination of copper and nitrogen in milk. *Mailing Add:* 5851 Page Pl Rockford IL 61101

HETRICK, LAWRENCE ANDREW, b Harrisburg, Pa, Feb 9, 10; m 35; c 4. ENTOMOLOGY, FORESTRY. *Educ:* Am Univ, AB, 31; La State Univ, MS, 32; Ohio State Univ, PhD(entom), 51. *Prof Exp:* Lab asst biol, Am Univ, 30-31; inspector & technician entom, State Dept Agr, La, 34-38; asst entomologist, Va Agr Exp Sta, 38-46; asst entom, Res Found, Ohio State Univ, 46-47; from assoc prof to prof, 47-72, EMER PROF ENTOM, UNIV FLA, 72- *Concurrent Pos:* Sanitarian, USPHS. *Mem:* Entom Soc Am. *Res:* Life history and control studies of various forest insects; biology of termites and their control. *Mailing Add:* 1614 N W Twelfth Rd Gainesville FL 32605

HETT, JOHN HENRY, b New York, NY, July 23, 09; m 37 & 62. OPTICS, ASTRONOMY. *Educ:* Manhattan Col, BA, 32; Columbia Univ, MS, 37, PhD(astron), 42. *Prof Exp:* Instr physics, Manhattan Col, 32-34, from asst prof to assoc prof, 34-42; tech consult, US Mgt & Eng Co, 44-45; sr res scientist, NY Univ, 46-61, assoc tech dir, Proj Squid, 48-50, dir jet & flame lab, 50-52; SR CONSULT, AM CYSTOSCOPE MAKERS INC, 42- *Concurrent Pos:* Consult, Merganthaler Linotype Co, 41-44, NY Ord Dist, US Army, 42-45, Raytheon Mfg Co, NY, 46-47 & Bell Tel Labs, NJ, 61-64. *Mem:* Am Astron Soc; Am Phys Soc; Optical Soc Am; Soc Motion Picture & TV Eng. *Res:* Medical optical instrumentation. *Mailing Add:* 6612 Case Ave Bradenton FL 33507

HETTCHE, LEROY RAYMOND, b Baltimore, Md, Mar 24, 38; m 65; c 3. MATERIALS SCIENCE, SOLID MECHANICS. *Educ:* Bucknell Univ, BS, 61; Carnegie-Mellon Univ, MS, 63, PhD(civil eng), 65. *Prof Exp:* Asst prof civil eng, Rutgers Univ, 64-66; Nat Res Coun res assoc, Nat Bur Standards, 66-68; struct engr, Naval Res Lab, 68-71, phys scientist, 71-74, supt, Mat Sci & Tech Div, 74-81; DIR, APPL RES LAB & PROF ENG RES, PA STATE UNIV, 81- *Mem:* Acoust Soc Am; Am Soc Test & Mat; Am Soc Mech Engrs. *Res:* Response of materials to pulsed radiation heating; dynamic plasticity; stress wave analysis; fracture mechanics; impact testing; numerical analysis. *Mailing Add:* Appl Res Lab Pa State Univ PO Box 30 State Col PA 16804

HETTINGER, DEBORAH D R, TEMPERATURE REGULATION BIOENERGETICS. *Educ:* Univ Calif, PhD(physiol & biochem), 80. *Prof Exp:* ASST PROF BIOL & CHEM, TEX LUTHERAN COL, 80- *Mailing Add:* 241 Ridge Crest Dr Seguin TX 78155

HETTINGER, WILLIAM PETER, JR, b Aurora, Ill, Sept 13, 22; m 44; c 4. PHYSICAL CHEMISTRY, CATALYSIS. *Educ:* Purdue Univ, BS, 47; Northwestern Univ, PhD(phys chem), 51; Univ Chicago, cert adv mgt, 64. *Prof Exp:* Res chemist catalytic res, Sinclair Res Lab, 50-56, dir catalytic res new catalysts develop, 56-57; gen mgr res, Nalco Chem Co, 57-65; vpres & develop, Davison Chem Div W R Grace & Co, 65-68; NIH res fel, Gerontology Res Ctr, 68-71; dir corp develop, Ga Koalin Co, 71-72; dir res & develop, Engelhard Indust Engelhard Minerals & Chem Corp, 72-74; corp vpres res & develop, N L Industs Inc, 74-76; spec staff, Arthur D Little Inc, 76-77; dir res & develop, Ashland Petrol Co, Div Ashland Oil Inc, 77-79, vpres res & develop, Automotive & Prod Appl Lab, 79-84, pres, Ashland Carbon Fibers Div, 84-87, vpres & tech adv to sr mgt, 87-91; RETIRED. *Concurrent Pos:* Res fel, NIH, 68-71; pres, Ky Acad Sci, 87-88. *Mem:* Asn Res Dirs (pres, 76-77); Am Chem Soc; Am Inst Chem Engrs; hon mem NY Acad Sci; Sigma Xi; Indust Res Inst; Catalysis Soc; Gerontol Soc; fel AAAS. *Res:* Heat capacities high polymers; heterogeneous catalysis; petroleum processing; scale and corrosion prevention; colloid chemistry; industrial water problems; coagulation; petrochemicals; research administration; paper chemistry; herbicides; biocides; molecular and cellular biology; medicine. *Mailing Add:* 203 Meadowlark Rd Russell KY 41169

HETTMANSPERGER, THOMAS PHILIP, b Wabash, Ind, Aug 30, 39; m 61; c 3. STATISTICS. *Educ:* Ind Univ, BA, 61; Univ Iowa, MS, 65, PhD(statist), 67. *Prof Exp:* Assoc prof, 67-76, head dept, 88-90, PROF STATIST, PA STATE UNIV, UNIVERSITY PARK, 76-,. *Concurrent Pos:* Assoc ed, Am Statistician, 72-73 & J Am Statist Asn, 78-80, 88-91, J Nonparametric Statist, 90-, J Educ Statist, 89-, ed bd comn statist, 87-; NSF fac fel, 74-75; res assoc, dept statist, Univ Calif, Berkeley, 74-75; vis assoc prof, dept statist, Princeton Univ, 76-77; bd dir, Am Statist Asn, 83; vis fel, dept statist, Univ Melbourne & Univ Latrobe, 84-85; vis prof, Univ Bern, 90; Walter Scott vis prof, Univ New SWales, 90. *Mem:* Fel Inst Math Statist; fel Am Statist Asn; Psychomet Soc. *Res:* Nonparametric statistics and robustness. *Mailing Add:* Dept Statist Pa State Univ University Park PA 16802

HETZEL, DONALD STANFORD, b Philadelphia, Pa, July 1, 41; m 64. ORGANIC CHEMISTRY. *Educ:* Ohio Wesleyan Univ, BA, 63; Univ Ill, MS, 65, PhD(org chem), 68. *Prof Exp:* Res chemist, Chem Div, Pfizer, Inc, 67-72, dir licensing health care prods, 72-74; dir corp res, Howmedia, Inc, 74-76, vpres corp res, 76-81; VPRES CORP RES, BECTON, DICKINSON & CO, 81- *Mem:* Am Chem Soc; Soc Biomat Res; NY Acad Sci. *Res:* Organic synthesis; biomedical polymers; new technology aquisition; research management. *Mailing Add:* Becton Dickinson & Co One Becton Dr Franklin Lakes NJ 07417-1815

HETZEL, HOWARD ROY, b Rochester, NY, July 17, 31. INVERTEBRATE ZOOLOGY. *Educ:* Univ Buffalo, BA, 53, MA, 55; Univ Wash, PhD(zool), 60. *Prof Exp:* From instr to asst prof biol, Whitman Col, 60-62; assoc prof, 62-67, chmn, Dept Biol Sci, 74-79, PROF ZOOL, ILL STATE UNIV, 67- *Mem:* AAAS; Am Soc Zool; Marine Biol Asn; Am Micros Soc; Sigma Xi. *Res:* Biology of the Holothuroidea; life cycles and behavior of fresh water Isopods. *Mailing Add:* Dept Biol Sci PSA 322 Ill State Univ Normal IL 61761

HETZEL, RICHARD ERNEST, b New York, NY, 1935. CONTROL SYSTEM & INSTRUMENT ENGINEERING. *Educ:* Stevens Inst Technol, BS, 56. *Prof Exp:* Serv engr, Repub Flow Meters Co, 56-59; mem qual control dept, Tech Mat Corp, 61-65; sr instrument engr, Crawford & Russell Inc, 65-69; sr instrument engr, Stauffer Chem Co, 69-88; CONSULT, CONTROL SYST ENG, 88- *Mem:* Sr mem, Instrument Soc Am. *Mailing Add:* 200 G High Point Dr Hartsdale NY 10530-1139

HETZEL, THEODORE B(RINTON), mechanical engineering; deceased, see previous edition for last biography

HETZLER, BRUCE EDWARD, b St Louis, Mo, Oct 18, 48; m 72; c 2. PHYSIOLOGICAL PSYCHOLOGY, NEUROPHARMACOLOGY. *Educ:* DePauw Univ, BA, 70; Northwestern Univ, MA, 73, PhD(psychol), 78. *Prof Exp:* Instr, 76-77, asst prof, 77-83, ASSOC PROF PSYCHOL, LAWRENCE UNIV, 83- *Concurrent Pos:* Res psychologist, Neurotoxicol Div, US Environ Protection Agency, 83-84; dir, Lawrence Univ London Centre, London, 87-88. *Mem:* AAAS; Soc Neurosci; Int Soc Biomed Res Alcoholism. *Res:* Behavioral electrophysiology; operant control of neural events; pharmacological effects on evoked brain acitivity; hypothermia. *Mailing Add:* Dept Psychol Lawrence Univ Box 599 Appleton WI 54911

HETZLER, MORRIS CLIFFORD, JR, b Chattanooga, Tenn, Nov 1, 37; m 65; c 2. SOLID STATE PHYSICS. *Educ:* Univ Chattanooga, AB, 59; Vanderbilt Univ, PhD(physics), 70. *Prof Exp:* Asst prof, 64-73, ASSOC PROF PHYSICS, UNIV TENN, CHATTANOOGA, 73- *Mem:* Am Phys Soc; Am Asn Physics Teachers; Sigma Xi. *Res:* Experimental investigation of dielectric properties of insulators as functions of temperature and of externally applied fields. *Mailing Add:* 3359 Northbrook Dr Atlanta GA 30340

HEUBERGER, GLEN (LOUIS), animal science, animal nutrition, for more information see previous edition

HEUBERGER, OSCAR, b Berne, Switz, May 30, 24; nat US; m 55; c 2. ORGANIC CHEMISTRY. *Educ:* Swiss Fed Inst Technol, PhD(chem), 49. *Prof Exp:* Instr, Swiss Fed Inst Technol, 48-50; fel, Imp Col, London, 50-51; res chemist, E I du Pont de Nemours & Co, Inc, 51-74, lab dir textile fibers dept, 74-83, mgr res & develop, 83-85; RETIRED. *Honors & Awards:* Silver Medal, Swiss Fed Inst Technol, 50. *Mem:* Am Chem Soc. *Res:* Polymer chemistry; textile fibers; textile technology. *Mailing Add:* E I du Pont de Nemours & Co Inc 830 Fairway Dr VA 22980

HEUCHLING, THEODORE P, b Chicago, Ill, June 24, 25; m 49; c 3. ELECTRICAL ENGINEERING. *Educ:* Mass Inst Technol, SB, 46, SM, 48. *Prof Exp:* Res asst elec eng, Servomech Lab, Mass Inst Technol, 46-48, staff mem, 48-51; proj engr & sect head comput controls, Ultrasonic Corp, 51-55; vpres & chief eng, Feedback Controls, Inc, 55-59; sect head eng sci, Arthur D Little, Inc, 59-68, vpres, 65-85, head eng div, 68-74, corp tech staff, 74-85, pres, Arthur D Little Enterprises, Inc, 86-91, SR VPRES, ARTHUR D LITTLE, INC, 85- *Concurrent Pos:* Consult environ mgt & energy resources. *Mem:* Sr mem Inst Elec & Electronics Engrs; Am Inst Aeronaut & Astronaut. *Res:* Feedback controls; analog computers; instrumentation. *Mailing Add:* Arthur D Little Inc Twenty Acorn Park Cambridge MA 01742

HEUER, ANN ELIZABETH, b Irvington, NJ, Oct 11, 30. MICROBIOLOGY, IMMUNOLOGY. *Educ:* Rutgers Univ, BS, 52, MS, 54, PhD(bact), 62. *Prof Exp:* Instr bact, Douglass Col, Rutgers Univ, 54-57, res asst microbiol, Inst Microbiol, 57-59, instr bact, Douglass Col, 58-59; instr bact, Vassar Col, 59-60; res assoc, Inst Microbiol, Douglass Col, Rutgers Univ, 60-62; USPHS fel immunol, Univ Milan, 62-63; res assoc, Inst Microbiol, Douglass Col, Rutgers Univ, 64; asst prof microbiol, Carnegie Inst Technol, 64-68; assoc prof, 68-74, prof biol sci, 74-89, EMER PROF BIOL SCI, CALIF STATE UNIV, HAYWARD, 89- *Mem:* Am Soc Microbiol. *Res:* Bacterial population dynamics; genetic control of pigment synthesis in Serratia; immune response and mechanism of enhancement. *Mailing Add:* Dept Biol Sci Calif State Univ Hayward CA 94542

HEUER, ARTHUR HAROLD, b New York, NY, Apr 29, 36; m 56; c 3. CERAMICS. *Educ:* City Col New York, BS, 56; Univ Leeds, PhD(ceramics), 66, DSc, 77. *Prof Exp:* Res chemist, Electronics Div, Ind Gen Corp, 56-60; res engr, Electron Tube Div, Bendix Corp, 60-61; res asst ceramics, Univ Leeds, 61-65; staff scientist, Space Systs Div, Avco Corp, 65-67; asst prof metall, 67-70, from assoc prof to prof ceramics, 70-85, dir, Mats Res Lab, 77-80, KYOCERA PROF CERAMICS, CASE WESTERN RESERVE UNIV, 85- *Concurrent Pos:* Chmn, Ceramics Gordon Conf, 72; ed J Am Ceramic Soc, 88-90. *Honors & Awards:* Ross Coffin Purdy Award, Am Ceramic Soc, 81. *Mem:* Nat Acad Eng; Am Inst Mining, Metall & Petrol Engrs; fel Am Ceramic Soc; fel Inst Physics UK; AAAS; Am Ceramic Soc. *Res:* Physical ceramics; mechanical properties; study of phase transformation and diffusion-controlled processes; transmission electron microscopy of ceramics; magnetic properties of ceramics; fracture mechanics; properties of ceramics; electron microscopy of rock and minerals; displacive phase transformations in oxides and silicates; solid-state precipitation in ceramics and minerals; fracture mechanics and deformation behavior in single crystal and poly-crystalline ceramics; oxidation of non-oxide ceramics; transformation toughening; author of over 200 publications. *Mailing Add:* Mat Sci & Eng Case Western Reserve Univ 10900 Euclid Ave Cleveland OH 44106

HEUER, CHARLES VERNON, b Bertha, Minn, Apr 27, 37; m 57; c 4. MATHEMATICS. *Educ:* Concordia Col, Moorhead, Minn, BA, 58; Univ Nebr, MA, 60, PhD(math), 63. *Prof Exp:* Asst prof math, Univ Mo-Columbia, 63-66; assoc prof & chmn dept, 66-73, PROF MATH, CONCORDIA COL, MOORHEAD, MINN, 73- *Concurrent Pos:* Hon fel, Univ Wis-Madison, 74. *Mem:* Am Math Soc; Math Asn Am. *Res:* Group theory; algebraic theory of semigroups. *Mailing Add:* Dept Math Concordia Col Moorhead MN 56560

HEUER, GERALD ARTHUR, b Bertha, Minn, Aug 31, 30; m 54; c 4. MATHEMATICS, ANALYSIS & FUNCTIONAL ANALYSIS. *Educ:* Concordia Col, Moorhead, Minn, BA, 51; Univ Nebr, MA, 53; Univ Minn, PhD(math), 58. *Prof Exp:* Asst, Univ Nebr, 51-53 & Univ Minn, 53-55; instr math, Hamline Univ, 55-56; from instr to assoc prof, 56-61, PROF MATH, CONCORDIA COL, MOORHEAD, MINN, 62- *Concurrent Pos:* Vis asst prof, Univ Nebr, 60-61; consult, Control Data Corp, 60-63; res assoc math, Univ Calif, Berkeley, 66-67; vis mathematician, Math Inst, Cologne Univ, Ger, 73-74; vis prof math, Wash State Univ, 80-81; vis prof, Institut für Statistik und Opers Res, Graz Univ, Austria, 87-88, 90; leader Am team, Int Math Olympiad, 88-90. *Mem:* Am Math Soc; Math Asn Am; Soc Indust & Appl Math; Deut Math Ver Ger; Austrian Math Soc. *Res:* Theory of 2-person games; probability. *Mailing Add:* Dept Math Concordia Col Moorhead MN 56562

HEUMANN, KARL FREDRICH, b Chicago, Ill, Mar 3, 21; m 47; c 3. ORGANIC CHEMISTRY. *Educ:* Iowa State Col, BS, 42, MS, 43; Univ Ill, PhD(chem), 51. *Prof Exp:* Chemist, Tech Info Sect, Minn Mining & Mfg Co, 50-52; dir chem-biol coord ctr, Nat Res Coun, 52-55; res, Chem Abstracts, Am Chem Soc, 55-59 & Off Doc, Nat Acad Sci, 59-66; asst exec ed, Fedn Am Socs Exp Biol, 66-67, exec ed, 67-68, dir ed & info serv, 68-71, dir off publ, 71-85; RETIRED. *Concurrent Pos:* Vpres, Int Fedn Doc, 61-64. *Mem:* Am Chem Soc; Am Soc Info Sci (pres, Am Doc Inst, 59); Coun Biol Ed (secy, 69-74, chmn, 76). *Res:* Catalytic hydrogenation; furans; quinone diamines; scientific literature; chemical structure and biological activity; punched cards; mechanical aids in literature searching; international documentation; scientific editing. *Mailing Add:* 6410 Earlham Dr Bethesda MD 20817

HEUSCH, CLEMENS AUGUST, b Aachen, Ger, Apr 19, 32; m 68; c 2. ELEMENTARY PARTICLE PHYSICS. *Educ:* Aachen Tech Univ, dipl physics, 55; Munich Tech Univ, Dr rer nat, 59. *Prof Exp:* Res asst neutron physics, Munich Tech Univ, 56-59; proj supvr semiconductor physics, AEG Res Labs, Frankfurt, Ger, 60-61; res assoc high energy physics, DESY, Hamburg, 61-63; res fel, Calif Inst Technol, 63-65, sr res fel, 65-67, assoc prof physics, 67-69; PROF PHYSICS, UNIV CALIF, SANTA CRUZ, 69- *Concurrent Pos:* Consult, AEG, Frankfurt, 56-62; vis prof & vis assoc, Calif Inst Technol, 70-72; vis assoc, Max Planck Inst Physics, Ger, 74; vis prof, Univ Munich, 74-75 & 90-91 Univ Aachen, 80 & Academia Sinica, Beijing, 87; vis scientist, European Ctr Nuclear Res, Switzerland, 74-75 & 83- *Honors & Awards:* Fulbright Award, Bowdoin Col, 51-52; Humboldt Prize, 91. *Mem:* Am Phys Soc. *Res:* Electron-positron annihilation studies; nuclear instrumentation; interactions of leptons and photons with nucleons at high energies; flavor conservation in elementary particle interactions. *Mailing Add:* Inst Particle Physics Univ Calif Santa Cruz CA 95064

HEUSCHELE, ANN, b Cadillac, Mich, Sept 5, 38; m 64; c 2. AQUATIC BIOLOGY. *Educ:* Univ Mich, BS & MS, 60; Univ Minn, PhD(zool), 68. *Prof Exp:* Instr biol, Canisius Col, 61-64; asst prof, Carleton Col, 68-69; MEM BIOL FAC, NORMANDALE COMMUNITY COL, 70- *Mem:* Asn Women Sci; NAm Benthological Soc; Ecol Soc Am. *Res:* Lake Superior benthos; invertebrate biology; aquatic ecology; biology of women. *Mailing Add:* Dept Biol Normandale Community Col Bloomington MN 55431

HEUSCHELE, WERNER PAUL, b Ludwigsburg, Ger, Aug 28, 29; US citizen; m 53; c 6. VETERINARY VIROLOGY, PREVENTIVE MEDICINE. *Educ:* Univ Calif, Davis, AB, 52, DVM, 56; Univ Wis-Madison, PhD(vet sci), 69. *Prof Exp:* Vet & mgr hosp & lab, Zool Soc San Diego, 56-61; vet res officer, USDA, Plum Island Animal Dis Lab, 61-70; assoc prof microbiol, Col Vet Med, Kans State Univ, 70-71; head virol, Jensen-Salsbery Labs, 71-76; assoc prof, 76-78, PROF, DEPT VET PREV MED, COL VET MED, OHIO STATE UNIV, 78- *Concurrent Pos:* Resident trainee vet path, Armed Forces Inst Path, 65-66. *Mem:* Am Vet Med Asn; US Animal Health Asn; Conf Res Workers Animal Dis; Wildlife Dis Asn. *Res:* Foreign animal diseases, especially African swine fever; bovine and equine viral diseases. *Mailing Add:* Dept Vet Prev Med 188 Montrose Way Columbus OH 43214

HEUSER, EVA T, b Warsaw, Poland, Nov 8, 32; US citizen; c 2. PEDIATRICS, PATHOLOGY. *Educ:* Queen's Univ, Ont, MD, CM, 56. *Prof Exp:* Assoc pathologist, Children's Hosp, Los Angeles, 64-80; asst prof med, 64-76, ASST PROF PATH, UNIV SOUTHERN CALIF, 76-; DEP MED EXAMR, LOS ANGELES COUNTY, 80- *Mem:* Int Acad Path. *Res:* Pediatric forensic pathology. *Mailing Add:* Coroners Off 1104 N Mission Rd Los Angeles CA 90033

HEUSER, GUNNAR, b Hamburg, Germany, July 17, 27. HEADACHE & PAIN MEDICINE. *Educ:* Cologne Univ, Germany, MD, 52; Univ Montreal, PhD(exp med & surg), 57. *Prof Exp:* ASST CLIN PROF MED, UNIV CALIF, LOS ANGELES, 70-; MED DIR, BEVERLY HILLS HEADACHE & PAIN MED GROUP, 75- *Concurrent Pos:* Med dir, Environ Med Res & Info Ctr. *Mem:* Fel Am Col Physicians. *Mailing Add:* Dept Med Univ Calif Los Angeles CA 90024

HEUSINKVELD, MYRON ELLIS, b Hull, Iowa, Aug 3, 21; m 53; c 3. PHYSICS. *Educ:* Iowa State Col, BS, 43; Univ Minn, PhD(physics), 51. *Prof Exp:* Elec engr, US Naval Res lab, Washington, DC, 44-47; PHYSICIST, LAWRENCE LIVERMORE LAB, UNIV CALIF, 51- *Mem:* AAAS; Am Phys Soc; Soc Eng Sci; Am Geophys Union. *Res:* Nuclear physics; particle accelerator development; shock and other transient stress experimentation. *Mailing Add:* 3875 Pestana Way Livermore CA 94550

HEUSNER, ALFRED AUGUST, BASAL METABOLISM & BODY SIZE, CIRCADIAN & SEASONAL RHYTHMS. *Educ:* Univ Strasbourg, France, ScD, 63. *Prof Exp:* PROF PHYSIOL, SCH VET MED, UNIV CALIF, DAVIS, 67- *Mailing Add:* Physiol Sci Dept Sch Vet Med Univ Calif Davis CA 95616

HEUSSER, CALVIN JOHN, b North Bergen, NJ, Sept 10, 24; m 47; c 2. BOTANY. *Educ:* Rutgers Univ, BS, 47, MS, 49; Ore State Col, PhD, 52. *Prof Exp:* Seessel fel, Yale Univ, 52-53; res assoc, Am Geog Soc, 53-67, adminr, Juneau Ice Field Proj, 53-59; assoc prof dept biol & geol, 67-71, prof biol, 71-91, EMER PROF BIOL, NY UNIV, 91- *Concurrent Pos:* Guggenheim & Fulbright fel, Chile, 62-63; assoc ed, Torrey Bot Club, 71-77, pres, 75-76; fel, Clare Hall, Univ Cambridge, 86- *Honors & Awards:* David Livingstone Centenary Medal, Am Geog Soc, 87. *Mem:* AAAS; Ecol Soc Am; Torrey Bot Club; Am Asn Stratig Palynologists; Sigma Xi. *Res:* Quaternary palynology of western North America and southern Chile and Argentina; paleoecological employing fossil pollen and spores and radiometric dating techniques in stratigraphic settings to trace vegetation and climate during the quaternary as a basis for paleoclimatic theory. *Mailing Add:* Clinton Woods Tuxedo NY 10987

HEUSSER, LINDA OLGA, b Sharon, Pa, Apr 12, 32; m 70; c 5. QUARTERNARY PALEOCLIMATOLOGY. *Educ:* Wellesley Col, BA, 54; Columbia Univ, MA, 66; NY Univ, PhD(geol), 71. *Prof Exp:* Lectr geol, Iona Col, 65; asst prof, Orange County City Col, 75; geologist, US Geol Surv, 78-82; sr res scientist, NY Univ, 87-90; RES SCIENTIST, LAMONT-DOHERTY GEOL OBSERV, COLUMBIA UNIV, 78- *Concurrent Pos:* Sr res scientist, NY Univ, 70-81. *Mem:* Geol Soc Am; AAAS; Am Geophys Union. *Res:* Marine palynology; analysis of pollen from North Pacific, North and South Atlantic to reconstruct vegetational and climate history of Northeast Asia. *Mailing Add:* Clinton Rd Tuxedo NY 10987

HEUSTIS, ALBERT EDWARD, b Fitchburg, Mass, Apr 22, 13; m 36; c 5. PUBLIC HEALTH ADMINISTRATION. *Educ:* Univ Mich, AB, 33, MD, 36; Johns Hopkins Univ, PM MPH, 42. *Hon Degrees:* LLD, Mich State Univ, 55. *Prof Exp:* From asst dir to dir, Monroe County Health Dept, 42-45; dir, Branch County Health Dept, 45-48 & Branch County Community Hosp, 46-48; comnr, State Dept Health, Mich, 48-67; prog coordr, Mich Asn Regional Med Progs, 67-71; HEALTH PROGS CONSULT, 71- *Honors & Awards:* Cert Appreciation, Nat Found Infantile Paralysis, 54; Citation Merit Serv, US Civil Defense Coun, 63; McCormack Award, Asn State & Territorial Health Officers, 63. *Mem:* Am Pub Health Asn; AMA; Asn State & Territorial Health Officers; hon fel Am Col Chest Physicians. *Res:* Most effective and least expensive application of knowledge to build better health for the individual, the family and the community. *Mailing Add:* Rte One 13160 Spence Rd Three Rivers MI 49093

HEUVERS, KONRAD JOHN, b Stockton, Calif, Mar 15, 40; m 67; c 2. MATHEMATICS. *Educ:* Stanford Univ, BS, 62, MS, 64; Ohio State Univ, PhD(math), 69. *Prof Exp:* From asst prof to assoc prof, 69-82, PROF MATH, MICH TECHNOL UNIV, 82- *Mem:* Am Math Soc; Math Asn Am; Sigma Xi; German Mathematicians' Union; Austrian Math Soc; Soc Indust & Appl Math. *Res:* Functional equations; linear algebra; combinatorics and finite mathematics. *Mailing Add:* Dept Math Mich Technol Univ Houghton MI 49931

HEUZE, FRANCOIS E, b Oran, Algeria, Dec 3, 41; US citizen; m 72; c 2. GEOLOGICAL ENGINEERING, ROCK MECHANICS. *Educ:* Nat Sch Mines, France, MSc, 64; Univ Calif, Berkeley, MSc(civil eng), 67, D Eng(geol eng), 70. *Prof Exp:* Asst res specialist, Univ Calif, Berkeley, 67-70, asst res engr, 70-75, lectr, 70-75; assoc prof civil eng, Univ Colo, 75-79; HEAD, GEOTECH GROUP & MGR, UNCONVENTIONAL GAS PROG, LAWRENCE LIVERMORE NAT LAB, 79- *Concurrent Pos:* Consult geol eng & rock mech, 72-; app mem, Geotech Bd Nat Res Coun/Nat Acad Sci, 87. *Honors & Awards:* Publs Bd Award, Soc Mining Eng of AIME, 82; Case Histories Award, US Nat Comt Rock Mech/Nat Res Coun, 86. *Mem:* Fel Am Soc Civil Eng; Soc Mining Engrs; Int Soc Rock Mech; Eng & Sci Soc France; Am Soc Testing Mat. *Res:* Analysis and design of rock structures; foundations; tunnels; slopes; mines; gas recovery; hydraulic fracturing; dynamic ground motion; cratering; penetration; subsidence. *Mailing Add:* Code L-200 Lawrence Livermore Nat Lab Livermore CA 94551

HEVERAN, JOHN EDWARD, b New York, NY, Aug 7, 38; m 59; c 2. ANALYTICAL CHEMISTRY. *Educ:* Manhattan Col, BS, 60; Purdue Univ, MS, 62, PhD(chem), 65; Am Bd Clin Chem, dipl, 77. *Prof Exp:* Proj chemist, Am Cyanamid Co, 65-66; sr chemist, Hoffmann-La Roche Inc, 66-67, group leader, 67-68, mgr anal res, 68-73, res group chief, 73-78, sr res group chief, 78-80; dir res & develop, gen diag div, Warner Lambert Co, 80-82; managing dir, Heveran & Assocs, Inc, 82-88; tech dir, Lab Serv Inc, 83-88; DIR, METPATH, 88- *Concurrent Pos:* Adj assoc prof, Univ Med & Dent, NJ. *Honors & Awards:* Cert Honor, Am Asn Clin Chem, 87. *Mem:* Am Chem Soc; Am Asn Clin Chem; NY Acad Sci; Sigma Xi. *Res:* Pharmaceutical analysis; clinical chemistry; immunochemistry; diagnostic immunology; toxicology. *Mailing Add:* Six Caryn Pl Fairfield NJ 07006

HEVNER, ALAN RAYMOND, b Marion, Ind, Dec 9, 50; m 85; c 1. DATABASE SYSTEMS. *Educ:* Purdue Univ, BS, 73, MS, 76, PhD(comput sci), 79. *Prof Exp:* Prof comput sci, Univ Minn, 79-81; PROF INFO SYSTS, UNIV MD, 81- *Concurrent Pos:* Consult, Honeywell, Inc, 80-, IBM, 85- & MCI, 90- *Mem:* Asn Comput Mach; Inst Elec & Electronic Engrs Comput Soc; Opers Res Soc Am. *Res:* Database systems design; distributed database systems; systems engineering; information systems analysis and design. *Mailing Add:* Col Bus & Mgt College Park MD 20742

HEW, CHOY-LEONG, b Ipoh, Malaysia, Mar 10, 42; m 68. BIOCHEMISTRY. *Educ:* Nanyang Univ, BSc, 63; Simon Fraser Univ, MSc, 66; Univ BC, PhD(biochem), 70. *Prof Exp:* Res assoc biochem, Yale Univ, 70-72; C H Best fel, Univ Toronto, 73-74; from asst prof to prof biochem,

Mem Univ Nfld, 74-82; PROF BIOCHEM, UNIV TORONTO, 83-; SR SCIENTIST, HOSP SICK CHILDREN, 83- *Honors & Awards:* Fraser Award, Atlantic Prov Inter Univs Coun Soc, Can. *Mem:* Can Biochem Soc; Am Soc Biol Chemists; NY Acad Sci; AAAS. *Res:* Proteins structure and function; biosynthesis and regulation of peptide hormones in marine organisms; structure and biosynthesis of antifreeze proteins; hydroxylation of collagen. *Mailing Add:* Banting Inst Rm 351B Univ Toronto 100 College St Toronto ON M5G 1L5

HEWER, GARY ARTHUR, b Mitchell, SDak, Aug 11, 40; m 69. APPLIED MATHEMATICS, SYSTEMS SCIENCE. *Educ:* Yankton Col, BA, 62; Wash State Univ, MA, 64, PhD(math), 68. *Prof Exp:* MATHEMATICIAN, NAVAL WEAPONS CTR, CHINA LAKE, 68- *Mem:* Soc Indust & Appl Math; Sigma Xi. *Res:* Controllability and stabilizability of control systems; general theory of Riccati equations; numerical solutions of linear and quadratic matrix equations; applied stochastic process theory; robust kalman filtering; radar signature analysis. *Mailing Add:* 636 Las Flores Ridgecrest CA 93555

HEWES, RALPH ALLAN, b Columbus, Ohio, July 4, 39; m 62; c 2. SOLID STATE PHYSICS. *Educ:* Case Inst Technol, BS, 61; Univ Ill, MS, 63, PhD(physics), 66. *Prof Exp:* Res assoc physics, Univ Ill, 66; RES PHYSICIST, GEN ELEC CO, 66- *Mem:* Am Phys Soc; Sigma Xi. *Res:* Radiation damage in semiconductors; energy transfer; luminescence of solids. *Mailing Add:* 39 Velina Dr Burnt Hills NY 12027

HEWETSON, JOHN FRANCIS, b Chicago, Ill, July 3, 39; m 64, 83; c 2. VIROLOGY, IMMUNOLOGY. *Educ:* Pa State Univ, BS, 61; Fla State Univ, MS, 65; Rutgers Univ, PhD(virol & microbiol), 69. *Prof Exp:* Asst prof biol, Newark State Col, 65-66; res assoc, Univ Pa, 71-72, asst prof virol, 72-76; assoc prof microbiol, Med Col Pa, 77-81; CHIEF, DEPT IMMUNOTHER & DIAG, US ARMY MED RES INST INFECTIOUS DIS, FREDERICK, MD, 81- *Concurrent Pos:* Damon Runyon Fund fel tumor biol, Karolinska Inst, Sweden, 69-71; res assoc, Children's Hosp, Philadelphia, 71-72; scholar, Leukemia Soc Am, 75-80. *Mem:* AAAS; Am Soc Microbiol; Int Soc Toxinol. *Res:* Virology and immunology of tumors; cellular immune response associated with neoplasia; immunological response to low molecular weight toxins. *Mailing Add:* 126 Kline Blvd Frederick MD 21701

HEWETT, DENNIS W, b Girard, Kans, Sept 17, 47; m; c 2. NUMERICAL SIMULATION PLASMA & PARTICLE ACCELERATORS, MAGNETIC & INERTIAL CONFINEMENT FUSION. *Educ:* Pittsburgh State Univ, BS, 69; Univ Kans, PhD(plasma physics), 73. *Prof Exp:* Staff physicist, Los Alamos Nat Lab, 73-78, assoc group leader, 78-82; assoc group leader, Plasma Fusion Ctr, Mass Inst Technol, 82-83; STAFF MEM, LAWRENCE LIVERMORE LABS, 83- *Mem:* Am Physiol Soc; Am Geophys Union. *Res:* Investigating microscopic properties of magnetic reconnection by numerical simulation and numerical investigation of particle acceleration, especially physics of high current heavy ion beams. *Mailing Add:* L-472 Lawrence Livermore Nat Lab PO Box 5508 Livermore CA 94550

HEWETT, JAMES VEITH, b Broken Bow, Nebr, July 25, 21; m 50; c 2. ORGANIC CHEMISTRY. *Educ:* Univ Nebr, BS, 43; Univ Purdue, MS, 48, PhD(chem), 50. *Prof Exp:* Chemist, Phillips Petrol Co, 43-46; asst chem, 46-48, fel, Purdue Univ, 48-50; chemist, E I du Pont de Nemours & Co, Inc, 50-53, group supvr, 53-55, from supvr to res assoc, 55-83; PRIVATE INVESTR, 83-; CONSULT, CONDUX, 90- *Mem:* Am Chem Soc. *Res:* Acrylic, spandex and high temperature fiber technology; process economics; process and product modeling; open end spinning of staple fibers; time series analysis and modeling. *Mailing Add:* 1211 Keesling Ave Waynesboro VA 22980-5217

HEWETT, JOANNE LEA, b Boulder, Colo, Mar 15, 60; m 85. ELECTROWEAK PHENOMENOLOGY. *Educ:* Iowa State Univ, BS, 82; Univ Calif, Irvine, MS, 84, Iowa State Univ PhD, 88. *Prof Exp:* Teaching asst physics, Iowa State Univ, 82-83 & Univ Calif, Irvine, 83-84; res asst physics, Ames Lab, Iowa State Univ, 84-88; ASST PROF PHYSICS, UNIV WIS-MADISON, 88- *Mem:* Am Women Sci. *Res:* Theoretical high energy physics: composite models, fermion masses, extended electroweak theories, grand unified theories and superstring theories. *Mailing Add:* Physics Dept Univ Wis-Madison 1150 Univ Ave Madison WI 53706

HEWETT, JOHN EARL, b Fairfield, Iowa, Feb 20, 37; m 60; c 3. STATISTICS. *Educ:* Parsons Col, BS, 58; Univ Iowa, MS, 62, PhD(math), 65. *Prof Exp:* Teacher math, Monroe Community High Sch, 58-60; lectr, Univ Iowa, 65; from asst prof to assoc prof, 65-78, PROF STATIST, UNIV MO, COLUMBIA, 78- *Concurrent Pos:* Consult, Harry S Truman Mem Vet Hosp, 73- *Mem:* Fel Am Statist Asn; Inst Math Statist; Biomet Soc. *Res:* Statistical distribution theory; regression; hypothesis testing; prediction. *Mailing Add:* Dept Statist Univ Mon Columbia MO 65211

HEWETT, LIONEL DONNELL, b Cleburne, Tex, July 20, 38; m 70; c 2. SOLAR PHYSICS. *Educ:* Tex A&I Univ, BS, 60; Univ Mo-Rolla, PhD(eng physics), 65. *Prof Exp:* From asst prof to assoc prof, 64-77, PROF PHYSICS, TEX A&I UNIV, 77-, CHMN, PHYSICS DEPT, 87- *Mem:* Am Asn Physics Teachers. *Res:* Physics teaching techniques. *Mailing Add:* Dept Physics Tex A&I Univ Kingsville TX 78363

HEWETT'EMMETT, DAVID, b Bromley, Eng, June 2, 46; m 87. HUMAN MOLECULAR GENETICS. *Educ:* Univ London, Eng, PhD(biochem), 73. *Prof Exp:* ASSOC PROF GENETICS, UNIV TEX HEALTH SCI CTR, 84- *Concurrent Pos:* Asst res scientist human genetics, Univ Mich, 79-84. *Mem:* Am Soc Humam Genetics; Am Soc Biochem & Molecular Biol; NY Acad Sci. *Res:* Human genetics; molecular evolution; polygenic disease. *Mailing Add:* Univ Tex Health Grad Sch Biomed Sci PO Box 20334 Houston TX 77225

HEWGILL, DENTON ELWOOD, b Collingwood, Ont, Feb 11, 40; m 64; c 2. MATHEMATICS. *Educ:* Univ BC, BSc, 63, PhD(math), 66. *Prof Exp:* Res assoc & fel math, Univ BC, 66-67; Nat Res Coun Can fel, NY Univ, 67-69; asst prof, 69-80, ASSOC PROF MATH, UNIV VICTORIA, 80- *Res:* Singular elliptic partial differential equations; mathematical analysis; fluid dynamics; numerical analysis of partial differential equations. *Mailing Add:* Dept Math Univ Victoria Victoria BC V8W 2Y2 Can

HEWINS, ROGER HERBERT, b Farnham, Eng, Nov 29, 40; US citizen; m 63; c 2. MINERALOGY, PETROLOGY. *Educ:* Aberdeen Univ, BSc, 62; Univ Toronto, PhD(geol), 71. *Prof Exp:* Geologist, Geol Surv Brit Guiana, 63-65 & Falconbridge Nickel Mines Ltd, 65-67; teaching asst mineral, Univ Toronto, 67-71, lectr, 71-72; res assoc lunar petrol, Lehigh Univ, 72-75; from asst prof to assoc prof, 75-85, PROF MINERAL & PETROL, RUTGERS UNIV, 85- *Concurrent Pos:* Res assoc, Am Mus Natural Hist, NY, 76-; abstractor mineral abstracts, London, 80-; prin investr, grants on metal & silicates in meteorites, NASA. *Mem:* Meteoritical Soc; Geol Asn Can; Mineral Asn Can; Mineral Soc Am. *Res:* Thermal and collisional history of parent bodies of igneous meteorites; dynamic crystallization experiments on chondrule compositions; microprobe analysis of minerals; petrology of mafic igneous rocks; geothermometry. *Mailing Add:* Dept Geol Sci Rutgers Univ New Brunswick NJ 08903

HEWISH, ANTONY, b Fowey, Cornwall, Eng, May 11, 24. RADIOASTRONOMY. *Educ:* Kings Col, Caius Col. *Prof Exp:* Asst dir res, Gonville & Caius Col, 53-61; reader radio astron, Univ Cambridge, 69-71; DIR, MULLARD RADIO ASTRON OBSERV, 82- *Honors & Awards:* Nobel Prize in Physics, 74; Hamilton Prize, 51; Eddington Medal, Royal Astron Soc, 68. *Mem:* Am Acad Arts & Sci; Indian Nat Sci Acad. *Mailing Add:* Pryor's Cottage Kingston Cambridge CB3 0HE England

HEWITSON, WALTER MILTON, b Neptune, NJ, Dec 21, 33; m 71. PLANT MORPHOLOGY. *Educ:* Miami Univ, Ohio, AB, 56; Cornell Univ, MS, 59; Wash Univ, St Louis, PhD(plant morphol), 62. *Prof Exp:* Instr bot, Miami Univ, Ohio, 58-59 & Parsons Col, 62-63; asst prof, Southern Ill Univ, 63-65 & Univ Pac, 65-70; from asst prof to assoc prof, 70-77, PROF BIOL, BRIDGEWATER STATE COL, 77- *Concurrent Pos:* Res fel, Highlands Biol Sta, NC, 63; consulting work, wetlands delineatious & replicatious. *Mem:* Bot Soc Am; Torrey Bot Club. *Res:* Comparative morphology of lower vascular plants. *Mailing Add:* Dept Bot Bridgewater State Col Jct Rte 18-28-104 Bridgewater MA 02324

HEWITT, ALLAN A, b Sacramento, Calif, Mar 1, 34; m 55; c 2. PLANT PHYSIOLOGY. *Educ:* Univ Calif, BS, 55, MS, 57; Univ Md, PhD(hort), 62. *Prof Exp:* Res horticulturist, Crops Res Div, USDA, 62-64; asst res pomologist, Univ Calif, Davis, 64-68; assoc prof, 68-70, PROF POMOL, CALIF STATE UNIV, 70- *Honors & Awards:* Joseph H Gourley Award, 60. *Mem:* Am Soc Hort Sci; Am Soc Plant Physiol. *Res:* Nitrogen uptake and nitrogen-carbohydrate relationships in deciduous fruits; pistachio research. *Mailing Add:* Dept Pomol Calif State Univ 6241 N Maple Ave Fresno CA 93740

HEWITT, ANTHONY VICTOR, b Witney, Eng, Feb 2, 43; m 65. ASTRONOMY. *Educ:* Oxford Univ, BA, 64, DPhil(astron), 67. *Prof Exp:* Astron, Flagstaff Sta, US Naval Observ, 67-87; LORAL ELECTRONIC SYSTS, 87- *Mem:* Am Astron Soc; Int Astron Union. *Res:* Application of image tubes to astronomy; development and use of the navy electronic camera; real-time data processing; image analysis. *Mailing Add:* PO Box 509 Ft Ashby WV 26719-0509

HEWITT, ARTHUR TYL, SURFACE PROTEOGLYCANS, GLYCOPROTEINS. *Educ:* Emory Univ, PhD(develop biol), 78. *Prof Exp:* ASST PROF OPHTHAL, WYNN CTR RETINAL DEGENERATION, WILMER INST, JOHNS HOPKINS MED SCH, 82- *Res:* Cell surface properties. *Mailing Add:* Wilmer Eye Inst John Hopkins Hosp Maumenee Bldg 519 Baltimore MD 21205

HEWITT, CHARLES HAYDEN, b Butte, Mont, May 18, 29; m 54; c 2. GEOLOGY, MINERALOGY. *Educ:* Mont Sch Mines, BS, 51; Univ Mich, MS, 53, PhD(mineral), 57. *Prof Exp:* Asst, Eng Res Inst, Univ Mich, 51-53; sr res geologist, Denver Res Ctr, Marathon Oil Co, Ohio, 56-57, assoc res dir, 67-74, coord mgr prod explor, Ohio, 74-76, vpres minerals, 77-80; MGR MINING SYNTHETIC FUELS DIV, EXXON PROD RES CO, HOUSTON, TEX, 80- *Mem:* Fel Geol Soc Am; Soc Econ Paleontologists & Mineralogists; Am Asn Petrol Geol; Soc Petrol Eng; Sigma Xi. *Res:* Petrology and petrography of sandstones; geologic characteristics of petroleum reservoirs; petroleum exploration and production; mineral exploration and production. *Mailing Add:* 13903 Calmont Dr Houston TX 77070

HEWITT, DAVID, b Leeds, Eng, Mar 15, 27; Can citizen; m 62; c 4. MEDICAL STATISTICS. *Educ:* Oxford Univ, BA, 47, dipl, 50, MA, 51. *Prof Exp:* Res asst soc med, Oxford Univ, 50-66; assoc prof, 66-76, PROF MED STATIST & PREV MED, UNIV TORONTO, 76-, COORDR GRAD STUDIES, DEPT COMMUNITY HEALTH, 75- *Concurrent Pos:* Vis prof, State Univ NY Downstate Med Ctr, 60 & Sch Med, Univ Wash, 61-62; sr res scholar, Corpus Christi Col, Cambridge, Eng, 86-87; chair, Univ Toronto, 89- *Honors & Awards:* Frances Wood Mem Prize, Royal Statist Soc, 58. *Mem:* Biomet Soc; Int Epidemiol Asn; Soc Epidemiol Res; fel Am Col Epidemiol. *Res:* Epidemiology; child development. *Mailing Add:* Prev Med Univ Toronto Toronto ON M5S 1A1 Can

HEWITT, EDWIN, b Everett, Wash, Jan 20, 20; m 44, 64; c 2. MATHEMATICS. *Educ:* Harvard Univ, AB, 40, MA, 41, PhD(math), 42. *Prof Exp:* Instr, Harvard Univ, 42-43; opers analyst, US Air Force, 43-45; Guggenheim fel, Inst Adv Study & Princeton Univ, 45-46; asst prof math, Bryn Mawr Col, 46-47; lectr, Univ Chicago, 47-48; from asst prof to assoc prof, 48-54, PROF MATH, UNIV WASH, 54- *Concurrent Pos:* Vis prof, Univ Uppsala, 51-52, Univ Hokkaido, 82 & Univ Alaska, 83; Guggenheim fel,

Inst Adv Study, 55-56; mem, Div Math, Nat Res Coun, 57-65; vis res assoc, Yale Univ, 59; vis prof, Australian Nat Univ, 63, vis res assoc, 70; vis scholar, Steklov Inst, Moscow, 69-70 & Univ New S Wales, 80, 82, & 84; ed, Pac J Math; vis scholar, Steklov Inst, Moscow, 73 & 76; comt mem, US Nat Comt Math, 74-77; Alexander Von Humboldt Found fel, Ger, 75-76; fac lectr, Univ Wash, 81. *Honors & Awards:* Sr US Sci Award, Alexander Von Humboldt Found. *Mem:* Am Math Soc; Math Asn Am. *Res:* Harmonic analysis on groups, measure theory, number theory. *Mailing Add:* Dept Math GN-50 Univ Wash Seattle WA 98195

HEWITT, FREDERICK GEORGE, b Aurora, Ill, Apr 4, 29; m 54; c 5. PHYSICS. *Educ:* St Procopius Col, BS, 51; Univ Notre Dame, PhD(physics), 58. *Prof Exp:* Asst prof physics, Col St Thomas, 57-61; sr staff scientist, Unisys Corp, 61-88; RES DIR, ORFIELD ASSOC, 88- *Concurrent Pos:* Asst, Univ Notre Dame, 58. *Res:* Optics; polymer physics; acoustics; magnetics; visibility; lighting. *Mailing Add:* 545 Chapel Lane Eagan MN 55121

HEWITT, GEORGE BERLYN, b Exeland, Wis, May 10, 30; m 57; c 4. ENTOMOLOGY. *Educ:* Mont State Univ, BS, 60; Univ Idaho, MS, 64. *Prof Exp:* Res technician, USDA, 61-62, res entomologist, Sci & Educ Admin, Fed Res, USDA, 64-; RETIRED. *Mem:* Entom Soc Am; Soc Range Mgt; Sigma Xi. *Res:* Insect and plant ecology; taxonomy of Acrididae; forage plants resistant to grasshopper feeding; determination of forage losses caused by rangeland insects. *Mailing Add:* PO Box 3556 Flagstaff AZ 86003

HEWITT, HUDY C, JR, b Mead, Okla, Apr 9, 37; m 63; c 3. MEASUREMENTS, THERMAL SCIENCE. *Educ:* Okla Univ, BS, 60; Ohio State Univ, MS, 61; Okla State Univ, PhD(mech eng), 66. *Prof Exp:* From asst prof to prof mech eng, Tenn Technol Univ, 66-90, chmn dept, 80-90; PROF & CHMN MECH ENG DEPT, UNIV NEW ORLEANS, 90- *Concurrent Pos:* Pres, Solar Sales & Serv, 79-90; chmn, MEDHC, Am Soc Mech Engrs, 87-88. *Mem:* Fel Am Soc Mech Engrs; Am Soc Eng Educ. *Res:* Solar energy; measurements; heat transfer; energy conservation; fluid flow; dynamic response. *Mailing Add:* Dept Mech Eng Univ New Orleans New Orleans LA 70148

HEWITT, PHILIP COOPER, b Boston, Mass, Sept 28, 25; div; c 4. GEOLOGY, PALEONTOLOGY. *Educ:* Harvard Univ, AB, 49; Univ Tenn, MS, 53; Cornell Univ, PhD(paleont, stratig), 58. *Prof Exp:* Assoc prof geol, Union Col, NY, 56-67, chmn dept, 60-67; chmn dept geol & earth sci, 67-74, coordr, Fac Sci, 68-70, PROF GEOL & EARTH SCI, STATE UNIV NY COL BROCKPORT, 67- *Concurrent Pos:* NSF grant, Int Field Inst, Am Geol Inst, Paris Basin, 65. *Mem:* Geol Soc Am; Paleont Soc; Am Asn Geol Teachers (vpres, 64-65, pres, 65-66). *Res:* Foraminifera; stratigraphy and sedimentology; Taconic geology; sedimentologic and paleontologic environmental studies. *Mailing Add:* Dept Geol & Earth Sci State Univ NY Col Brockport NY 14420

HEWITT, ROBERT LEE, b Paducah, Ky, Nov 2, 34; m 59; c 4. THORACIC SURGERY, CARDIOVASCULAR SURGERY. *Educ:* Tulane Univ, MD, 59; Am Bd Surg, cert; Am Bd Thoracic Surg, cert. *Prof Exp:* Intern surg, Charity Hosp of La, 59-60, resident, 60-63, instr surg, 64-65, instr thoracic surg, 65-66, from asst prof to prof surg, 68-76, chief cardiac surg, 70-76, CLIN PROF SURG, SCH MED, TULANE UNIV, 76- *Concurrent Pos:* Fel cardiovasc surg, Sch Med, Tulane Univ, 63-64; chief resident, Charity Hosp of La, 64-66, mem vis staff, 68-; consult, Keesler Air Force Base Hosp, 68-; mem staff, Southern Baptist Hosp, Touro Infirmary & Tulane Univ Hosp, New Orleans. *Mem:* Am Col Surg; Int Cardiovasc Soc; Am Asn Thoracic Surg; Soc Thoracic Surg; Soc Vascular Surg. *Res:* Cardiovascular research and surgery. *Mailing Add:* Suite 300 4400 Magnolia St New Orleans LA 70115

HEWITT, ROBERT T, b Roth, NDak, June 28, 09; m 37; c 3. PSYCHIATRY. *Educ:* Univ Minn, BM & MD, 38; Am Bd Neurol & Psychiat, dipl, 47; Johns Hopkins Univ, MPH, 48. *Prof Exp:* Intern, USPHS Hosp, Stapleton, NY, 37-38; resident psychiat, Ellis Island, NY & Lexington, Ky, 38-41, from asst clin dir to clin dir, Lexington, KY, 41-45, clin dir, Ft Worth, Tex, 45-49; dir, Phoenix Ment Health Ctr, Ariz, 49-52; asst dir clin ctr, NIH, 52-55, chief hosp consult serv, NIMH, 55-61; dir ment health prog, Western Interstate Comn Higher Educ, Univ Colo, 61-64; chief dep dir, Calif State Dept Ment Hyg, 64-73; field rep, Accreditation Coun Psychiat Facil, Joint Comn Accreditation Hosps, 73-77; CONSULT NIMH, 75- *Mem:* Fel Am Psychiat Asn; AMA. *Res:* Administration of mental health programs. *Mailing Add:* 1617 Gary Way Carmichael CA 95608

HEWITT, ROGER R, b Portland, Ore, Nov 27, 37; div; c 4. BIOCHEMISTRY. *Educ:* Willamette Univ, BA, 59; Univ Rochester, MS, 60, PhD(radiation biol), 63. *Prof Exp:* USPHS res trainee radiobiol, Univ Tex M D Anderson Hosp & Tumor Inst, Houston, 63-65, Am Cancer Soc fel, 65-67, from asst prof to assoc prof, 67-77; actg dean, 78-79, MEM GRAD FAC, UNIV TEX GRAD SCH BIOMED SCI HOUSTON, 67-, ASSOC DEAN, 79-; PROF BIOL, UNIV TEX M D ANDERSON HOSP & TUMOR INST HOUSTON, 77- *Concurrent Pos:* Mem, NIH Chem Path Study Sect, 78-81. *Mem:* Radiation Res Soc; Biophys Soc; Am Soc Cell Biol; Am Soc Photobiol; Am Soc Biol Chem. *Res:* Ultraviolet light effects on DNA replication; roles of dexyribonucleases in DNA metabolism; identification of chemical carcinogens. *Mailing Add:* Dept Biol Univ Tex Cancer Ctr 6723 Berther Houston TX 77030

HEWITT, WILLIAM BORIGHT, b Fayetteville, Ark, July 17, 08; m 34. PHYTOPATHOLOGY. *Educ:* Univ Calif, BS, 33, MS, 34, PhD(plant path), 36. *Prof Exp:* Plant pathologist, Agr Exp Sta, 37-68, chmn dept plant path, 68-69, from instr plant path & jr plant pathologist to prof, 73-74, emer prof plant path, Univ Calif, Davis, 74-; RETIRED. *Concurrent Pos:* Pres, Int Coun Study Viruses & Virus Dis Grapevine. *Honors & Awards:* Ruth Allen Award, Am Phytopath Soc, 74. *Mem:* Fel Am Phytopath Soc (pres, 62); Italian Acad Vines & Wines; Italian Soc Phytopath. *Res:* Disease of grapevines; ecology and epidemiology of fungi associated with post harvest rots of grapes; viruses that infect grapevines; transmission and survival of soil-borne viruses. *Mailing Add:* 102 Sunnyside Pl Sequim WA 98382

HEWITT, WILLIAM LANE, b Hebron, Nebr, Nov 25, 16; m 42; c 1. MEDICINE. *Educ:* Univ Calif, AB, 38, MD, 42. *Prof Exp:* Intern, San Francisco Hosp, Calif, 41-42; ward officer, San Francisco Marine Hosp, 42; chief med officer, USS Albireo, USPHS, 43; instr med, Sch Med, Boston Univ, 46-53; assoc prof, 53-58, PROF MED & PHARMACOL, SCH MED, UNIV CALIF, LOS ANGELES, 58- *Concurrent Pos:* NIH fel, Evans Mem Hosp, 46-48; mem staff, Div Infectious Dis, NIH, 46-50; asst resident, Mass Mem Hosps, Boston, 46-47, vis physician, 47-48; mem training grant comt, Nat Inst Allergy & Infectious Dis. *Mem:* Am Col Physicians; Infectious Dis Soc Am. *Res:* Infectious diseases; chemotherapy; cardiovascular diseases; pharmacology; antibiotic agents. *Mailing Add:* Dept Med 3-7-121 Ctr Health Sci Univ Calif 405 Hilgard Ave Los Angeles CA 90024

HEWLETT, JOHN DAVID, b Philadelphia, Pa, Mar 29, 22; m 50; c 2. HYDROLOGY, ECOLOGY. *Educ:* State Univ NY Col Forestry, Syracuse Univ, BS, 49, MS, 56; Duke Univ, PhD(plant physiol), 62. *Prof Exp:* Res forester, Southeastern Forest Exp Sta, Forest Serv, USDA, 56-59, res ctr leader, Coweeta Hydrol Lab, 59-64; assoc prof, 64-70, prof forest hydrol, Sch Forest Resources, Univ GA, 70-84; RETIRED. *Concurrent Pos:* Panel mem comt water resources res, Off Sci & Technol, 65; Univ Ga del, Univs Coun Water Resources, 65-; mem work group on rep & exp basins, US Nat Comt, Int Hydrol Decade, Nat Acad Sci-Nat Res Coun, 69-75; consult to SAfrican countries, 70; Food & Agr Orgn specialist, USSR, 70; consult, Corps Engrs, 74- *Mem:* AAAS; Am Geophys Union; Sigma Xi; Int Asn Sci Hydrol; Am Water Resources Asn. *Res:* Runoff and evapo-transpiration from forests, clarifying the role of plant and soil in the water cycle; originator of variable source area concept in hydrology; authority on small water shed experimentation. *Mailing Add:* 360 Ashton Dr Athens GA 30606

HEWLETT, MARTINEZ JOSEPH, b Los Angeles, Calif, Dec 6, 42; m 66; c 1. MOLECULAR BIOLOGY, VIROLOGY. *Educ:* Univ Southern Calif, BA, 64; Univ Ariz, PhD(biochem), 73. *Prof Exp:* Res biochemist, Vet Admin Hosp, Sepulveda, Calif, 64-69; fel virol, Ctr Cancer Res, Mass Inst Technol, 73-76; asst prof cellular & develop biol, ASSOC PROF MOLECULAR & CELLULAR BIOL & BIOCHEM, UNIV ARIZ, TUCSON, 82- *Concurrent Pos:* mem, Virol Study Sect, NIH, 80-84; fac res award, 82-86, Fogarty Sr Int fel, Am Cancer Soc, 86-87; mem, Neurol Dis Problem Proj Review B Comt, NIH. *Mem:* Am Soc Microbiol; AAAS; Am Soc Virol. *Res:* Structure and function of viral nucleic acids with emphasis on the RNA of bunyaviruses; structure of enveloped viruses. *Mailing Add:* Dept Molecular & Cellular Biol Univ Ariz Tucson AZ 85721

HEWLETT, W(ILLIAM) R(EDINGTON), b Ann Arbor, Mich, May 20, 13; m 39, 78; c 5. ELECTRONICS. *Educ:* Stanford Univ, BA, 34, EE, 39; Mass Inst Technol, MS, 36. *Hon Degrees:* LLD, Univ Calif, Berkeley, 66, Yale Univ, 76; Mills Col, 83; DSc, Kenyon Col, 78, Polytechnic Inst NY, 78; DEng, Univ Notre Dame, 80, Utah State Univ, 80, Dartmouth Col, 83; LHD, Johns Hopkins Univ, 85; PhD, Rand Grad Inst, 85; Dr Electronic Sci, Univ Bologna, Italy, 89. *Prof Exp:* Electro-med res, San Francisco, 36-38; co-founder & partner, Hewlett-Packard Co, 39-47, vpres & dir, 47-57, exec vpres & dir, 57-64, pres & dir, 64-68, pres-chief exec officer & dir, 69-77, chmn exec comt, chief exec officer & dir, 77-78, chmn exec comt & dir, 78-83, vchmn, bd dir, 83-87, EMER DIR, BD DIRS, HEWLETT-PACKARD CO, 87- *Concurrent Pos:* Mem, President's Sci Adv Comt, 66-69, bd trustees, Stanford Univ, 63-74 & Nat Acad Sci, Inst Med, 71-72; dir, Inst Radio Engrs, 50-57, Chrysler Corp, 66-83, FMC Corp, 65-74, Overseas Develop Coun, 69-77, Chase Manhattan Bank, 69-80, Utah Int Inc, 74-85, Nat Acad Corp, 86-, Univ Corp Atmospheric Res, 86-88, Monterey Bay Aquarium Res Inst, 87-, Drug Abuse Coun, Wash DC, 72-74, Kaiser Found Hosp & Health Plan Bd, 72-78; pres bd dir, Palo Alto Stanford Hosp Ctr, 56-58 & dir, 58-62; coordr, Res Indust, Nat Acad Sci Five-Year Outlook Rept, 80- 81; mem, Pres's Gen Adv Comt Foreign Assistance Progs, Wash DC, 65-68, San Francisco Bay Area Coun, 69-81, San Francisco Regional Panel Comn White House Fels, 69-70, chmn, 70, asn Quadrato della Radio, Bologna, Italy, 78, panel on advan technol competition, Nat Acad Sci, 82-83, adv coun Educ & New Technol, Technol Ctr, Silicon Valley, 87-88, int adv coun, Wells Fargo Bank, 86-, President's Circle, Nat Acad Sci, 89-; trustee, Mills Col, Oakland, Calif, 58-68, Rand Corp, 62-72, Stanford Univ, Calif, 63-74, Calif Acad Sci, 63-68; pres bd, Palo Alto Stanford Hosp Ctr, 56-58, dir, 58-62; consult, Rand Corp, 72-74; coordr, Chapter Res Indust, Five-Year Outlook Report, Nat Acad Sci, 80-81; chmn, bd trustees, Carnegie Inst Wash, 80-86, trustee, 71-90, emer trustee, 90-; hon trustee, Calif Acad Sci, 68-90. *Honors & Awards:* Co-recipient, Inst Elec & Electronics Engrs Founders Medal, 73; Co-recipient, Vermilye Medal, Franklin Inst, 76; Corp Leadership Award, Mass Inst Technol, 76; Nat Medal Sci, US Nat Sci Comt, 85; Degree Uncommon Man, Stanford Univ, 87; Comdr's Cross Order of Merit, Fed Rep Ger, 87; John M Fluke Sr Mem Pioneer Award, 90. *Mem:* Nat Acad Sci; Nat Acad Eng; fel Inst Elec & Electronics Engrs (pres, 54); fel Am Acad Arts & Sci; hon life mem Am Instrument Soc; Am Philos Soc. *Res:* Electroencephalography; electronic measurements; a practical resistance capacity audio oscillator; circuitry; instrumentation; microwaves; patents on resistance-capacitance oscillators and other electronic devices. *Mailing Add:* Hewlett-Packard Co 1501 Page Mill Rd Palo Alto CA 94304

HEWSON, EDGAR WENDELL, b Amherst, NS, July 12, 10; nat US; m 35; c 2. METEOROLOGY. *Educ:* Mt Allison Univ, BA, 32; Dalhousie Univ, MA, 33; Univ Toronto, MA, 35; Univ London, PhD(meteorol), 37, DIC, 37. *Prof Exp:* Res meteorologist, Meteorol Serv Can, 38-47, asst controller training & res serv, 47-48; proj dir, Mass Inst Technol, 48-53; lectr meteorol, Univ Mich, 53-54, prof, 54-68, res physicist, Eng Res Inst, 53-54; prof physics, 68-69, chmn dept atmospheric sci, Sch Sci, 69-76, prof atmospheric sci, 76-81, EMER PROF ATMOSPHERIC SCI, ORE STATE UNIV, 81- *Concurrent Pos:* Hon spec lectr, Univ Toronto, 38-41 & 48; consult meteorologist, Consol Mining & Smelting Co Can, Ltd, 39-40, US Bur Mines, 39-40 & 45-46 & Bendix Aviation Corp, 58-59; ed, Monogr, Am Meteorol Soc, 47-57; mem comt climat, Nat Acad Sci, 57-61; trustee, Univ Corp Atmospheric Res, 59-68; ed, Environ Sci Monogr, Acad Press. *Honors & Awards:* Am Meteorol Soc Award, 69; Buchan Prize, Royal Meteorol Soc,

39; Am Wind Power Asn Award, 83. *Mem:* Am Geophys Union; fel Am Meteorol Soc; fel Royal Meteorol Soc. *Res:* Atmospheric thermodynamics and radiation; meteorological aspects of atmospheric pollution; atmospheric diffusion in transitional states; mountain and valley winds; industrial and engineering meteorology; meteorological aspects of wind and solar power. *Mailing Add:* 1770 Avenida del Mundo No 1604 Coronado CA 92118-3060

HEWSTON, JOHN G, b Roy, Wash, Aug 21, 23; m 55; c 3. WILDLIFE CONSERVATION, FISH MANAGEMENT. *Educ:* Pac Lutheran Col, BA, 50; Ore State Univ, MS, 55; Utah State Univ, PhD(fisheries), 66. *Prof Exp:* Mem staff game farm, Ore & Wash Game Depts, 48-52; res biologist, Ore Fish Comn, 52 & Wash Dept Fish, 53; dist biologist, NDak Game & Fish Dept, 53-55, div chief info & ed, 55-62; res biologist, Coop Fish Unit, Utah State Univ, 62-66; asst prof wildlife, 66-70, assoc prof natural resources, 70-77, prof, 77-88, EMER PROF RESOURCE PLANNING & INTERPRETATION, HUMBOLDT STATE UNIV, 88- *Mem:* Wildlife Soc; Conserv Educ Asn (pres, 81-85); Western Interpreters Asn. *Res:* Artificial propagation of Northern Pike eggs; role of fishery in establishment of recreational-use patterns in a new reservoir; natural resource management, interpretation and conservation education. *Mailing Add:* Col Natural Resources Humboldt State Univ Arcata CA 95521

HEXTER, ROBERT MAURICE, b Atlanta, Ga, Oct 15, 25; m 48; c 3. SPECTROSCOPY. *Educ:* Univ Minn, AB, 48; Columbia Univ, AM, 50, PhD(chem), 52. *Prof Exp:* Asst, Univ Minn, 48 & chem, Columbia Univ, 48-50, lectr, 51-52; from instr to asst prof chem, Cornell Univ, 52-57; sr fel fundamental res, Mellon Inst, 57-69, prof chem, Carnegie-Mellon Univ, 67-69; chmn dept, 69-75, actg dir, Ctr Microelectronic & Info Sci, 81-83, PROF CHEM, UNIV MINN, MINNEAPOLIS, 69- *Concurrent Pos:* Vis prof, Fla State Univ, 60; Guggenheim fel, 61-62; Fulbright res scholar, 61-62; vis prof Israel Inst Technol, 61-62; adj prof, Carnegie Inst Technol, 65-67; consult, 3M Co, 75-77 & IBM Co, 77-; mem bd, Control Data Corp & North Star Res Inst, 81-; co-dir, NSF Regional Instrumentation Facil Surface Anal, 79- *Mem:* Am Chem Soc; Am Phys Soc. *Res:* Infrared spectroscopy at very low temperatures; theoretical and experimental studies of intermolecular forces in molecular crystals; very rapid chemical reactions by rapid scanning infrared spectroscopy; critical point analysis in molecular crystals by use of modulation spectroscopy; fluorescence and raman probes of metal surface-adsorbed molecule interactions. *Mailing Add:* Dept Chem 241A Smith Hall 207 Pleasant St SE Minneapolis MN 55455

HEXTER, WILLIAM MICHAEL, b Canton, Ohio, Aug 10, 27; m 50; c 3. BIOLOGY, GENETICS. *Educ:* Univ Calif, AB, 49, MA, 51, PhD(zool), 53. *Prof Exp:* Asst zool, Univ Calif, 50-53; from instr to prof, 53-77, EDWARD SHARKUESS PROF BIOL, AMHERST COL, 77- *Concurrent Pos:* NSF sci fac fel, 63-64, lectr, 71-72; lectr, AAAS, 71-72. *Mem:* AAAS; Genetics Soc Am; Am Soc Zoologists; Sigma Xi. *Res:* Genetics of Drosophila. *Mailing Add:* Dept Biol Amherst Col Amherst MA 01002

HEXUM, TERRY DONALD, b Appleton, Wis, July 20, 41; m 66; c 3. PHARMACOLOGY. *Educ:* Univ Wis, River Falls, BS, 63; Univ Kans, PhD(biochem), 68. *Prof Exp:* Teaching asst biochem, Univ Kans, 63-65, res asst, 65-68; NIH fel pharmacol, Univ Wis-Madison, 68-71; asst prof, 71-75, ASSOC PROF PHARMACOL, UNIV NEBR MED CTR, 75- *Concurrent Pos:* Pharmacologist, Ctr Ment Retardation, Univ Nebr Med Ctr, 71-74; vis assoc, Nat Inst Mental Health, 79-80. *Honors & Awards:* Int Travel Award, Am Soc Pharmacol & Exp Therapeut; Golden Apple Award, Am Med Students Asn, 77 & 80. *Mem:* Sigma Xi; Soc Neurosci; AAAS; Am Soc Pharmacol & Exp Therapeut; Am Soc Neurochem. *Res:* Opiate peptides in adrenal junction, hypertension; role of catechalameres in membrane function; adrenergic receptor mechanisms; autonomic pharmacology. *Mailing Add:* Pharmacol Med Sch Univ Nebr Omaha NE 68198-6260

HEY, JOHN ANTHONY, b New York, NY, Sept 7, 59. PULMONARY PHARMACOLOGY & PHYSIOLOGY, NEUROPHARMACOLOGY. *Educ:* Univ Okla, PhD(pharmacol), 86. *Prof Exp:* Postdoctoral fel ocular pharmacol, Dean McGee Eye Inst, 86-88; sr scientist, 88-90, ASSOC PRIN SCIENTIST ALLERGY & PULMONARY PHARMACOL, SCHERING-PLOUGH RES, 90- *Concurrent Pos:* Adj asst prof pharmacol, Rutgers Univ, Newark, 88- *Mem:* Am Soc Pharmacol & Exp Therapeut; Soc Neurosci; NY Acad Sci; Sigma Xi. *Res:* Studies of mechanisms of allergic disease, autonomic control of pulmonary function and its role in airway disease; Histamine-3 receptor mechanisms in vivo; models of asthma and pulmonary pharmacology; autonomic pharmacology. *Mailing Add:* Dept Allergy Schering Plough Res 60 Orange St Bloomfield NJ 07003

HEY, RICHARD N, b Lebanon, Tenn, June 2, 47. PLATE TECTONICS, COMPUTER GRAPHICS. *Educ:* Calif Inst Technol, BS, 69; Princeton Univ, PhD(geophysics), 75. *Prof Exp:* Res assoc, Marine Sci Inst, Univ Tex, 74-75; res geophysicist, Hawaii Inst Geophysics, 75-81; ASST RES GEOPHYSICIST, SCRIPPS INST OCEANOG, UNIV CALIF, SAN DIEGO, 81- *Concurrent Pos:* Mem, Oceanog Review Panel, NSF, 79, Crustal Geodynamics Review Panel, NASA, 80- & Int Lithosphere Proj, Working Group 6, 81- *Mem:* Am Geophys Union; Geol Soc Am; AAAS. *Res:* Investigation of the plate tectonic history of the Earth and development of the propagating rift hypothesis. *Mailing Add:* 401 Harriett Ave Minneapolis MN 55409

HEYBACH, JOHN PETER, b Chicago, Ill, Sept 25, 50; m 71; c 2. ENDOCRINE PHYSIOLOGY, BEHAVIORAL PHYSIOLOGY. *Educ:* Northern Ill Univ, BA, 72, MA, 74, PhD(psychol), 76; Kellogg Grad Sch Bus Admin, Northwestern Univ, 88. *Prof Exp:* Nat Acad Sci/Nat Res Coun fel, Biomed Res Div, Ames Res Ctr, NASA, 76-78; res assoc, Dept Physiol & Pharmacol, Bowman Gray Med Sch, 78-79; res specialist nutrit & physiol, Cent Res Div, Gen Foods Corp, White Plains, NY, 79-85; dir sci affairs, G D Searle & Co, Nutrasweet Group, 85-87; dir, Nutrasweet Co, 87-89; ASST GEN MGR, NUTRASWEET AG, ZUG, SWITZ, 89- *Concurrent Pos:* Partic, workshop methods assessing food & nutrient intake, Nat Acad Sci, 84. *Mem:* Endocrine Soc; Soc Neurosci; Inst Food Technologists; Am Inst Nutrit. *Res:* Neuroendocrine physiology and regulation of endocrine stress responses; central nervous system and behavioral; regulation of body weight and food intake; regulation of sensory systems; human food intake and selection; human nutrition status; body weight regulation. *Mailing Add:* The NutraSweet Co 1751 Lake Cook Rd Deerfield IL 60015

HEYBEY, OTFRIED WILLIBALD GEORG, b Swinemuende, Ger, Nov 9, 38; US citizen. POLYMER PHYSICS, MATERIALS SCIENCE. *Educ:* George Washington Univ, BS, 59; Cornell Univ, MS, 62, PhD(physics), 69. *Prof Exp:* Instr physics, Hamilton Col, 62-64; res assoc, Rice Univ, 69-70; res assoc polymer physics, Marburg Univ, Ger, 71-72 & Univ Mass, 72-73; supvr mat technol, Batavia Lab, Am Can Co, 73-75; res scientist mat technol, Corp Res, Whirlpool Corp, 76-80; MEM STAFF, MAT DEVELOP LAB, AMP INC, 80- *Mem:* Am Phys Soc; Soc Plastics Engrs; Sigma Xi. *Res:* Polymer processing; mechanical properties of polymers. *Mailing Add:* 2121 E Philadelphia St York PA 17402

HEYBORNE, ROBERT L(INFORD), b McCornick, Utah, Apr 17, 23; m 42; c 2. ELECTRICAL ENGINEERING, ENGINEERING EDUCATION. *Educ:* Utah State Univ, BS, 49, MS, 60; Stanford Univ, PhD(elec eng), 67. *Prof Exp:* Prog mgr, Radio Broadcasting Co, KSUB, Utah, 49-52, asst mgr, 53-57; from instr to assoc prof elec eng, Utah State Univ, 57-69; DEAN, SCH ENG, UNIV OF THE PAC, 69- *Concurrent Pos:* Consult, var com broadcasting & pvt tel companies, 57-67; fac fel, Sch Indust Prog, Hughes Aircraft Co, 59; consult, Sperry Utah Eng Lab, 60 & 62; lectr, Vis Scientist Prog, NSF, 66; nat chmn, Coop Educ Div, Am Soc Eng Educ, 77-78. *Mem:* Am Geophys Union; Inst Elec & Electronics Engrs; Soc Eng Educ; Int Union Radio Sci; Am Soc Eng Educ; Sigma Xi. *Res:* Very low frequency whistler-mode propagation and ionospheric physics. *Mailing Add:* Col Eng Univ of the Pac Stockton CA 95211

HEYD, ALLEN, b Leola, SDak, May 27, 35; m 58; c 1. PHARMACEUTICS. *Educ:* Columbia Univ, BS, 62, MS, 65; Purdue Univ, PhD(indust & phys pharm), Purdue 68. *Prof Exp:* Asst prof pharm, Sch Pharm, Univ Conn, 68-70; sr pharmaceut scientist, Vick Div Res & Develop, 70-72, group leader, 72-78, sect head oral hyg, 78-80; ASSOC DIR MED RES, MILES PHARMACEUT, 80- *Mem:* Am Pharmaceut Asn; Acad Pharmaceut Sci; Assocs Clin Pharmacol. *Res:* Dissolution of macromolecules and their application in pharmaceutical dosage forms; study of formulation and manufacturing parameters affecting oral hygiene dosage forms. *Mailing Add:* Eight Three Seasons Ct Norwalk CT 06851

HEYD, CHARLES E, b Detroit, Mich, Mar 3, 28; m 50; c 2. ORGANIC CHEMISTRY. *Educ:* Univ Detroit, BS, 50, MS, 52; Mich State Univ, PhD(org chem), 56. *Prof Exp:* Res chemist, E I du Pont de Nemours & Co, Inc, 56-60; res supvr, Avisun Corp, 60-69; coordr packaging & fabricated prods depts, Res & Develop Dept, Amoco Chem Corp, 69-73, mgr com develop, Plastic Prods Div, Chicago, 73-75, sr proj mgr, Packaging Div, 75-78; MGR MKT & MKT RES, AMOCO CONTAINER CO, 78- *Mem:* Am Chem Soc; Soc Plastics Eng; Brit Plastics Inst. *Res:* Development of polyolefin resins for fiber and film applications; development of polyolefin stabilizer systems and formulations; industrial applications for new synthetic fibers produced by unique methods. *Mailing Add:* 520 Brook Hollow Circle Marietta GA 30067

HEYD, WILLIAM ERNST, b Cleveland, Ohio, Oct 2, 45; m 68; c 2. MEDICINAL CHEMISTRY, ORGANIC CHEMISTRY. *Educ:* Princeton Univ, AB, 67; Case Western Reserve Univ, PhD(org chem), 72. *Prof Exp:* Fel, Univ Saarland, 71-72 & Ohio State Univ, 72-73; res chemist org med chem, Upjohn Co, 73-78, sr staff scientist, 78-86, contracts assoc dir, 86-89, SR ACQUISITIONS REV MGR, UPJOHN CO, 89- *Mem:* Am Chem Soc; AAAS; Licensing Execs Soc. *Res:* Patent liaison specialist; interest in a wide variety of intellectual property agreements and licensing. *Mailing Add:* Upjohn Co Kalamazoo MI 49001

HEYDA, DONALD WILLIAM, b Chicago, Ill, Apr 11, 46; m 72. PHYSICS, NUCLEAR MEDICINE. *Educ:* Univ Ill, Urbana, BS, 68; Harvard Univ, AM, 70, PhD(physics), 76. *Prof Exp:* Sr scientist, 74-76, MGR ENG NUCLEAR MED, NUCLEAR DIV, BAIRD CORP, 76- *Mem:* Am Phys Soc; AAAS; Soc Nuclear Med. *Res:* Nuclear medicine imaging devices; computer processing in cardiology and nuclear medicine. *Mailing Add:* 378 Davis Rd Bedford MA 01730

HEYDANEK, MENARD GEORGE, JR, b Chicago, Ill, Sept 19, 42; m 64; c 2. FOOD CHEMISTRY. *Educ:* Northern Ill Univ, BS, 64; Utah State Univ, MS, 66; Northwestern Univ, PhD(chem), 68. *Prof Exp:* Res chemist, Kraftco Corp, 68-71; mgr flavor technol, 71-84, sr mgr tech & new bus develop, 84-86, DIR, GOLDEN GRAIN RES & DEVELOP, QUAKER OATS CO, 86- *Concurrent Pos:* Instr, Ill Inst Technol, 75. *Mem:* Am Chem Soc; Inst Food Technologists; Am Asn Cereal Chem. *Res:* Application of chemical instrumentation to flavor studies and correlation of flavor chemistry with organoleptic properties of foods. *Mailing Add:* Four Jane Ct Hawthorn Woods Lake Zurich IL 60047-9292

HEYDEGGER, HELMUT ROLAND, b Philadelphia, Pa, Dec 3, 35. PHYSICAL CHEMISTRY, ANALYTICAL CHEMISTRY. *Educ:* Queens Col, NY, BS, 56; Univ Ark, Fayetteville, MS, 58; Univ Chicago, PhD(chem), 68. *Prof Exp:* Phys chemist, Petrol Res Ctr, US Bur Mines Okla, 58; instr, chem, Prairie State Col, 61-62; res assoc, Enrico Fermi Inst, Univ Chicago, 68-77; Sr Res Assoc, Enrico Fermi Inst, Univ Chicago, 78-88; from asst prof to assoc prof, 70-81, actg dean, 83-84, PROF CHEM, PURDUE UNIV, CALUMET, 81-, DEPT HEAD, 79- *Concurrent Pos:* Consult, Argonne Nat Lab, 73-74; vis fel, Res Sch Earth Sci, Australian Nat Univ, 76-77, 84 & 90; vis staff mem, Los Alamos Sci Lab, Univ Calif, 78-85. *Mem:* Am Chem Soc; Am Phys Soc; Am Geophys Union; Geochem Soc; Meteoritical Soc; Int Asn Geochem. *Res:* Applications of nuclear and radiochemistry in geochemistry, cosmochemistry, physical and analytical chemistry; cosmic ray and solar particle induced nuclear reactions; activation analysis; isotopic anomalies in the early solar system. *Mailing Add:* Dept Chem Physics Purdue Univ Calumet Hammond IN 46323-2094

HEYDEMANN, PETER LUDWIG MARTIN, b Gottingen, Ger, Nov 10, 28; US citizen; m 58; c 2. PHYSICS, NEUROPHYSIOLOGY. *Educ:* Univ Gottingen, PhD(physics), 58. *Prof Exp:* Asst prof physics, Univ Gottingen, 57-61, asst to dir, 61-64; physicist, Nat Bur Standards 64-70, chief pressure & vacuum sect, 70-78, prog analyst, 78-80, dir, Ctr Chem Physics, 80-81, assoc dir, 80-86; dir, Ctr Basic Standards, 86-87, DEP DIR, INDUS TECHNOL SER, 87- *Concurrent Pos:* Assoc mem, Comn Thermodyn, Int Union Pure & Appl Chem, 70-; mem, Indo-US Subcomn, Sci & Technol, 75- *Honors & Awards:* IR-100 Award. *Mem:* Am Soc Mech Engrs; Am Phys Soc; Sigma Xi. *Res:* Physics of polymer, mechanical and dielectric properties; high pressure technology, ultrasonics, pressure and vacuum measurements, solid and liquid states at very high pressures; manometry; neurology of the auditory system; chemistry; science administration; program analysis; budget preparation; chemical physics; interferometry. *Mailing Add:* Dept State US Embassy New Delhi Washington DC 20520

HEYDEN, FRANCIS JOSEPH, astronomy; deceased, see previous edition for last biography

HEYDING, ROBERT DONALD, b Regina, Sask, July 25, 25; m 47; c 3. SOLID STATE CHEMISTRY. *Educ:* Univ Sask, BE, 47, MSc, 49; McGill Univ, PhD(chem), 51. *Prof Exp:* Nat Res Coun Can fel, 51-53; Nat Res Coun fel, State Univ Leiden, 53-54; res officer inorg chem, Div Appl chem, Nat Res Coun Can, 54-61; PROF INORG CHEM, QUEEN'S UNIV, ONT, 62-, HEAD DEPT, 76- *Mem:* Chem Inst Can. *Res:* Solid state inorganic chemistry; properties of compounds formed by transition metals and the elements of groups VA and VIA. *Mailing Add:* Dept Chem Queen's Univ Kingston ON K7L 3N6 Can

HEYDRICK, FRED PAINTER, b Clearfield, Pa, Jan 30, 34; m 60; c 4. MICROBIOLOGY, SCIENCE ADMINISTRATION. *Educ:* Juniata Col, BS, 55; Univ NH, MS, 61; Pa State Univ, PhD(microbiol), 67. *Prof Exp:* Microbiologist, US Army Biol Labs, Ft Detrick, 61-71; health sci adminr, Nat Heart, Lung & Blood Inst, 71-89; PRES, BIO REV INC, 89- *Mem:* Sigma Xi. *Res:* Isolation and characterization of viruses; lipids of host cells and virus progeny; management of research grant and contract review groups; management of evaluation and advisory panels in biotechnology research. *Mailing Add:* 6813 Sunnybrook Dr Frederick MD 21702

HEYDT, GERALD THOMAS, b New York, NY, Oct 1, 43. ELECTRICAL ENGINEERING. *Educ:* Cooper Union, BS, 65; Purdue Univ, MS, 67, PhD(elec eng), 70. *Prof Exp:* Elec engr, US Atomic Energy Comn, 65-66 & E G & G Inc, 66-67; ASSOC PROF ELEC ENG, PURDUE UNIV, 67- *Concurrent Pos:* Consult, US Army, 74-78, Peabody Coal Co, 74-76, Westmoreland Coal Co, 74-76; NSF grant, 73-75; US Dept Energy grant, 76-80; instr, Am Elec Power Co, 70-72 & Politecnica Nacional, Quito, 77. *Mem:* Sigma Xi; Am Soc Eng Educrs. *Res:* Electric power engineering; reliability of power systems. *Mailing Add:* Dept Elec Eng Purdue Univ West Lafayette IN 47907

HEYERDAHL, EUGENE GERHARDT, b Williston, NDak, Feb 11, 40; m 60; c 3. FISHERIES BIOLOGY, DATA MANAGEMENT. *Educ:* Luther Col, Iowa, BA, 62; Univ Minn, St Paul, PhD(fisheries biol), 68. *Prof Exp:* Res assoc fisheries, Univ Minn, St Paul, 68-71; fishery biologist, 71-76, NE REGIONAL DATA BASE ADMINR, NAT MARINE FISHERIES SERV, NAT OCEANIC & ATMOSPHERIC ADMIN, US DEPT COM, 77- *Mem:* Sigma Xi; Am Inst Fish Res Biol. *Res:* Regional data management coordination, including technical supervision of computer systems development and implementation, and equipment procurement and installation. *Mailing Add:* US Dept Comm-NOAA NMFS- Northeast Fisheries Ctr Woods Hole MA 02543

HEYING, THEODORE LOUIS, b Baltimore, MD, Oct 19, 27; m 52. ORGANIC CHEMISTRY, RESEARCH ADMINISTRATION. *Educ:* Loyola Col, Md, BS, 48; Col of the Holy Cross, MS, 49; Univ Md, PhD(chem), 54. *Prof Exp:* Res chemist, Olin Corp, 53-58, sr res chemist, 58-59, chem proj leader, 59-60; chief synthesis sect, United Tech Corp, 60-63; proj mgr, 60-63, sect mgr, 63-64, tech dir, 65-70, mgr spec prod, 70-72, tech dir, 72-74, dir res, 74-82, DIR INT TECHNOL, OLIN CORP, 82- *Mem:* Am Chem Soc; Sigma Xi. *Res:* Inorganic, organometallic and polymer chemistry. *Mailing Add:* 1290 Gulf Blvd Apt 308 Clearwater FL 33515-8718

HEYL, ALLEN VAN, JR, b Allentown, Pa, Apr 4, 18; m 45; c 2. GEOLOGY, GEOPHYSICS. *Educ:* Pa State Univ, BS, 41; Princeton Univ, PhD(geol), 50. *Prof Exp:* Field geologist, Nfld Geol Surv, Field Seasons, 37-40, proj chief, 42; asst mineral, Princeton Univ, 42-43; geologist, Md, Wisc, DC, US Geol Surv, staff geologist, Denver, 68-90; GEOL CONSULT, 90- *Concurrent Pos:* Chmn, Int Asn Genesis Ore Deposits; silver, lead, zinc resource specialist, US Geol Surv; Thayer-Lindley distinguished lectr, Soc Econ Geologists, 85-86; chmn, Soc of Econ Geologists, Int Exchange Lectr Comt; Invited lectr to grad col, Beijing Univ Geol Sci. *Honors & Awards:* Interior Dept Meritorious Award; Hon Life Mem, Friends of Mineral. *Mem:* Fel Mineral Soc Am; fel Geol Soc Am; Soc Econ Geologists; fel Brit Inst Mining & Metall. *Res:* Geochemistry and geology of Mississippi Valley mineral deposits; geology of eastern states mineral deposits; oxidized sulfide deposits and lead, zinc, silver resources of United States; geology and ore deposits of Southern NMex and West; basement structure and earthquake geology of USA. *Mailing Add:* PO Box 1052 Evergreen CO 80439

HEYL, GEORGE RICHARD, b Allentown, Pa; m 68; c 4. GEOLOGY, ECONOMIC GEOLOGY & GEOPHYSICS. *Educ:* Pa State Univ, BSc, 32; Princeton Univ, AM, 34, PhD(econ geol, petrol), 35. *Prof Exp:* Instr geol, Rutgers Univ, 35-37; dist geologist, Standard Oil Co, Calif, 37-40; asst state geologist, Ark, 40-41; consult geologist, 41-42; geologist, US Geol Surv, 42-48; res geologist, Creole Petrol Corp, 48-49, chief surface geologist, 49-55; mgr explor geol, Frome-Broken Hill Co, 55-59; prof geol & geophys, 59-66, PROF GEOL, STATE UNIV NY COL NEW PALTZ, 66- *Concurrent Pos:* Mem, Princeton Exped, Nfld, 33-34; mem, Nfld Geol Surv, 36-37; consult, Indust & Govt Orgns, 34-; trustee, Inst Advan Studies & Res, NY; J W White fel, Penn State; C E Procter fel, Princeton Univ, mem, coun dean Grad Sch. *Mem:* Fel AAAS; fel Geol Soc Am; fel Mineral Soc Am; fel Am Geog Soc; Int Asn Genesis Ore Deposits; Sigma Xi; Soc Econ Geologists; Am Asn Petrol Geologists. *Res:* Metallic and non-metallic mineral deposits; geological occurrence of petroleum; structural relationships in ore deposits; stratigraphic and structural geological problems; geothermal gradients and related phenomena. *Mailing Add:* PO Box 582 New Paltz NY 12561

HEYMAN, ALBERT, b Baltimore, Md, May 30, 16; m 42; c 2. INTERNAL MEDICINE, NEUROLOGY. *Educ:* Univ Md, BS, 36, MD, 40; Am Bd Internal Med, dipl; Am Bd Psychiat & Neurol, dipl, 61. *Prof Exp:* Intern, Baltimore City Hosps, 40-41; med intern, Grady Mem Hosp, Emory Univ, 41-42, asst resident med, 42-43, from asst to asst prof med, Sch Med, 43-53; from assoc prof to prof med, 53-77, PROF NEUROL, SCH MED, DUKE UNIV, 77- *Concurrent Pos:* Fel, WHO, 51; travel award, NSF, 57; pub health physician, State Dept Pub Health, Ga, 43-53; chief neurol sect, Vet Admin Hosp, Durham, NC, 54-69; mem subcomt venereal dis, Nat Res Coun, 50-53; mem comt cerebral vascular dis, USPHS, 55. *Mem:* Am Soc Clin Invest; Am Fedn Clin Res. *Res:* Cerebrovascular disease; cerebral circulation. *Mailing Add:* Duke Hosp Durham NC 27710

HEYMAN, DUANE ALLAN, b Toledo, Ohio, June 11, 41; m 68; c 1. ORGANIC CHEMISTRY, POLYMER CHEMISTRY. *Educ:* Case Inst Technol, BS, 63; Univ Calif, Berkeley, PhD(org chem), 68. *Prof Exp:* Asst prof chem, Whitman Col, 68-69; chemist org chem, Owens-Ill, Toledo, 69-74; sr chemist, Sherwin-Williams Chem, 74-78; SR CHEMIST POLYMER CHEM, BASF WYANDOTTE CORP, 78- *Concurrent Pos:* Chemist org chem, Battelle-Northwest, 69. *Mem:* Am Chem Soc; Am Inst Chem; Sigma Xi. *Res:* Synthetic aspects of urethane chemicals, block and graft copolymers, aromatic nitrogen heterocycles, silicon-carbon multiple bonded systems and organothallium chemistry. *Mailing Add:* 1902 S Raisinville Rd Monroe MI 48161

HEYMAN, JOSEPH SAUL, b New Bedford, Mass, Nov 4, 43; m 68; c 1. ULTRASONICS, PHYSICS. *Educ:* Northeastern Univ, BA, 68; Wash Univ, MA, 71, PhD(physics), 75. *Prof Exp:* Coop student physics, NASA Langley Res Ctr, 64-68, aerospace technologist, 68-69; teaching asst, Wash Univ, 69-70, res asst, 70-71; res leader ultrasonics, Lab Ultrasonics, 71-81, head, Mats Characterization Instrumentation Group, Instrument Res Div, 81- 87, HEAD, NONDESTRUCTIVE EVAL SCI BR, NASA LANGLEY RES CTR, 87-, MGR, NASA NONDESTRUCTIVE EVAL RES PROG. *Concurrent Pos:* Adj prof physics, Col William & Mary, Williamsburg, Va, 79-; ad com comt, Inst Elec & Electronics Engrs Ultrasonics, Ferroelectrics & Frequency Control Soc, 83-87; distinguished int lectr, Insts Elec & Electronics Engrs, 89. *Honors & Awards:* Arthur S Fleming Award, 81. *Mem:* AAAS; Am Phys Soc; Inst Elec & Electronics Engrs; Sigma Xi; Soc Exp Mech. *Res:* Physics of ultrasonic propagation in materials and application to the development of a better understanding and measurement of materials; author and coauthor of over 80 publications; granted over 17 patents. *Mailing Add:* 130 Indian Springs Rd Williamsburg VA 23185

HEYMAN, KARL, b Elberfeld, Ger, July 30, 04; nat US; m 41. ORGANIC CHEMISTRY. *Educ:* Univ Freiburg, BA, 25; Univ Munich, MA, 27, PhD, 29. *Prof Exp:* Asst, Rockefeller Inst Med Res, 29-30; asst patent law, Eng, 31; res chemist dyestuffs, I G Farbenindust, Ger, 33-36; res chemist rubber thread, Filatex Corp, NJ & Va, 37-38; group leader vinyon synthetic fibers, Am Viscose Corp, Pa, 39-43; chief chemist textile chem specialties, Kearny Mfg Co, NJ, 43-51; pres, 51-68, CHMN BD, MONA INDUST, INC, 68- *Mem:* AAAS; Am Chem Soc; Am Asn Textile Chemists & Colorists; Am Inst Chemists. *Res:* Surface active agents; dyestuffs; synthetic fibers; rubber and elastomers; corrosion inhibitors. *Mailing Add:* MONA Indust Inc 65 E 23rd St Patterson NJ 07524

HEYMAN, LAUREL ELAINE, b West Elizabeth, Pa, Jan 2, 41; m 68; c 1. INORGANIC CHEMISTRY, TECHNICAL MATHEMATICS. *Educ:* Mt Union Col, BS, 62; Case Inst Technol, PhD(inorg chem), 66. *Prof Exp:* Asst prof chem, Eastern Wash State Col, 66-69; vis lectr, Bowling Green State Univ, 70; vis prof, Defiance Col, 70-72; assoc prof, 72-80, PROF CHEM, OWENS TECH COL, 80- *Mem:* Am Chem Soc. *Res:* Coordination compounds; synthesis and physical properties, particularly Schiff base complexes of multidentate ligands and their characterization by spectroscopic methods. *Mailing Add:* 1902 S Raisinville Rd Monroe MI 48161-9704

HEYMAN, LOUIS, b New York, NY, Feb 25, 23; m 57; c 3. GEOLOGY. *Educ:* Brooklyn Col, BS, 47; Univ Mich, MS, 49; Va Polytech Inst & State Univ, PhD(geol), 70. *Prof Exp:* Geologist petrol geol, Kerr-McGee Oil Indust, Inc, 50-56; asst regional res geologist, Continental Oil Co, 56-63; teaching asst, Dept Geol Sci, Va Polytech Inst & State Univ, 63-66; staff geologist energy & gen geol, Pa Geol Surv, 66-79; sr staff geologist petrol geol, Coastal Corp, 79-81; dist exp geologist, Coastal Oil & Gas Corp, 81-84; CONSULT PETROL GEOLOGIST, 84- *Concurrent Pos:* Consult geologist, 63; chmn, District 6, Geothermal Map of NAm, 70-71; instr & lectr geol, Dept Phys Sci, Amarillo Col, 81-83; chmn awards comt, Middle Continental Sect, Am Asn Petrol Geologists, 85. *Mem:* Fel AAAS; fel Geol Soc Am; Am Asn Petrol Geologists; Soc Econ Paleontologists & Mineralogists. *Res:* Analysis of petroleum reservoir rocks in Texas, Oklahoma and Kansas; subsurface geology of Central Appalachian Basin; environmental geology of Western Pennsylvania; sedimentary petrology of Ordovician rocks of Virginia; ancient soils unconformities. *Mailing Add:* 3318 Edenburg Dr Amarillo TX 79106

HEYMANN, DIETER, b Mannheim, Ger, Aug 4, 27; m 58; c 3. ASTRONOMY, GEOLOGY. *Educ:* Univ Amsterdam, PhD(chem), 58. *Prof Exp:* Jr scientist, Found Fundamental Res Matter Lab, Neth, 54-59, sr scientist, 61-63; res assoc chem, Brookhaven Nat Lab, 59-61; res asst meteorites, Enrico Fermi Inst Nuclear Studies, Univ Chicago, 63-66; assoc prof, 66-77, PROF GEOL & SPACE SCI, RICE UNIV, 77- *Res:* Isotope separation; thermal diffusion; meteorites and planets, particularly mass spectrometry; lunar samples. *Mailing Add:* Rice Univ PO Box 1892 Houston TX 77251-1892

HEYMANN, MICHAEL ALEXANDER, b Johannesburg, SAfrica, Feb 24, 37; m 63; c 1. PEDIATRICS, CARDIOLOGY. *Educ:* Univ Witwatersrand, Johannesburg, MB & BCh, 59. *Prof Exp:* House surgeon & physician, Johannesburg Gen Hosp, 60; sr house physician pediat, Transvaal Mem Hosp Children, 61; resident & chief resident pediat, Albert Einstein Col Med, 62-64; asst clin prof, 67-69, asst prof in residence, 69-73, assoc prof in residence pediat, 73-74, assoc prof 74-78, PROF PEDIAT & OBSTET, GYNEC & REPRODUCTIVE SCI, UNIV CALIF, SAN FRANCISCO, 78- *Concurrent Pos:* Res fel, Albert Einstein Col Med, 64-66; res fel, Cardiovasc Res Inst, Univ Calif, San Francisco, 66-67, assoc staff mem, 69-76, sr staff mem, 76-; mem exam comt, Am Bd Neonatal-perinatal Med, 76- *Honors & Awards:* Young Investr Award, Am Acad Pediat, 68; Res Career Develop Award, Nat Inst Child Health & Human Develop, 69. *Mem:* Am Acad Pediat; Am Heart Asn; Am Pediat Soc; Soc Gynec Invest; Soc Pediat Res. *Res:* Fetal and perinatal cardio-pulmonary physiology. *Mailing Add:* Pediat/Obstet-Gynec Univ Calif San Francisco CA 94143-0544

HEYMSFIELD, ANDREW JOEL, b Brooklyn, NY, June 12, 47; m 75. ATMOSPHERIC SCIENCES, METEOROLOGY. *Educ:* State Univ NY, Fredonia, BA, 69; Univ Chicago, MA, 70, PhD(meteorol), 73. *Prof Exp:* Scientist cloud physics, Meteorol Res Inc, 73-75; SCIENTIST CLOUD PHYSICS, NAT CTR ATMOSPHERIC RES, 75- *Concurrent Pos:* Adv Comt Field Observ Facil, Nat Ctr Atmospheric Sci, 78-81; pres, Cloud Physics Comt, Am Meteorol Soc, 86-88; mem, Int Cloud Physics Comt, 89. *Mem:* Am Meteorol Soc. *Res:* Area of cloud physics, specifically cirrus clouds, climate and hail research; precipitation initiation processes; cloud dynamics. *Mailing Add:* Nat Ctr Atmospheric Res PO Box 3000 Boulder CO 80303

HEYMSFIELD, GERALD M, b Queens, NY, Nov 15, 49; m 82; c 2. RADAR METEOROLOGY, MESOMETEOROLOGY. *Educ:* State Univ NY, Fredonia, BA, 71; Univ Chicago, MA, 72; Univ Okla, PhD(meteorol), 76. *Prof Exp:* res assoc radar meteorol, Univ Chicago, 76-; METEOROLOGIST, NASA-GODDARD SPACE FLIGHT CTR, 79- *Mem:* Am Meteorol Soc. *Res:* Severe storm dynamics, winter cyclonic storm wind and precipitation structure; satellite and radar meteorology. *Mailing Add:* NASA Goddard Space Flight Ctr Code 912 Greenbelt MD 20771

HEYMSFIELD, STEVEN B, b New York, July 15, 44. INTERNAL MEDICINE. *Educ:* Mt Sinai Sch Med, MD, 71. *Prof Exp:* ASSOC PROF MED, EMORY UNIV HOSP, 79- *Mem:* Am Fedn Clin Res; Am Soc Clin Nutrit; Am Heart Asn. *Res:* Obesity; nutrition support. *Mailing Add:* St Luke's Roosevelt Hosp Columbia Univ Sch Med 411 W 114th St New York NY 10025

HEYN, ANTON NICOLAAS JOHANNES, b Delft, Holland, Jan 25, 06; nat US; m 52. BIOLOGY, BIOPHYSICS. *Educ:* State Univ Utrecht, BSc, 26, MSc, 29, PhD(biol), 31. *Prof Exp:* Rockefeller Found fel, Col France, Leeds, 31-32; chief asst, State Univ Utrecht, 32-35; Treub fel, Dutch East Indies Sci Inst & Exp Sta, 36; indust adv, Netherland East Indies Govt, Batavia, 36-46; prof natural & synthetic fibres, Clemson Univ, 47-60 & biophysics, Auburn Univ, 60-63; prof biol, 63-76, EMER PROF BIOL, LA STATE UNIV, NEW ORLEANS, 76- *Mem:* Fel AAAS; fel Am Inst Chem; Biophys Soc; fel Royal Micros Soc; NY Acad Sci. *Res:* Molecular structural physical chemical properties of biopolymers and cell wall; cell growth and hormones; cytology of cultivated plants; mutations; research of fibers; high polymers; x-ray and electron microscope research; molecular biology; molecular biophysics. *Mailing Add:* 2363 Killdeer New Orleans LA 70122

HEYN, ARNO HARRY ALBERT, b Breslau, Ger, Oct 6, 18; US citizen; m 42; c 3. ANALYTICAL CHEMISTRY. *Educ:* Univ Mich, BS, 40, MS, 41, PhD(anal chem), 44. *Prof Exp:* Exp chemist, Sun Oil Co, Pa, 44-47; from instr to prof, 47-84, EMER PROF CHEM, BOSTON UNIV, 84- *Concurrent Pos:* Sci Adv, US Food & Drug Admin, Boston Dist, 67-72; vis scientist, Kernforschungszentrum Karlsruhe, 73, 80-82; chmn, Const & Bylaws Comt, Am Chem Soc, 83-85 & Coun Policy Comt, 86-; ed, Nucleus, Am Chem Soc, 89- *Honors & Awards:* Henry A Hill Award, Am Chem Soc, 86. *Mem:* Fel AAAS; Am Chem Soc; Am Asn Univ Prof. *Res:* Physical methods of analysis; photochemical methods in analysis; environmental analysis. *Mailing Add:* 21 Alexander Rd Newton MA 02161

HEYNE, ELMER GEORGE, b Wisner, Nebr, Apr 4, 12; m 38; c 4. AGRONOMY. *Educ:* Univ Nebr, BSc, 35; Kans State Col, MSc, 38; Univ Minn, PhD, 52. *Prof Exp:* Jr agronomist, Soil Conserv Serv, USDA, Tex, 35-36; agron agent, Div Cereal Crops & Dis, Exp Sta, 36-38, from jr agronomist to agronomist, 38-56, prof agron, 47-82, prof plant breeding, 57-82, EMER PROF, KANS STATE UNIV, 82-,. *Honors & Awards:* Dekalb-Pfizer Crop Sci Distinguished Career Award, Crop Sci Soc Am; Agron Achievement Award, Am Soc Agron. *Mem:* Fel AAAS; fel Am Soc Agron; Genetics Soc Am; Am Phytopath Soc; Am Genetic Asn; fel Crop Sci Soc Am. *Res:* Plant breeding and genetics of wheat, oats and barley; disease and insect resistance; quality and yield. *Mailing Add:* Dept Agron Throckmorton Hall Kans State Univ Manhattan KS 66506

HEYNEMAN, DONALD, b San Francisco, Calif, Feb 18, 25; m 71; c 5. PARASITOLOGY, EDUCATION ADMINISTRATION. *Educ:* Harvard Univ, AB, 50; Rice Inst, MA, 52, PhD, 54. *Prof Exp:* Asst biol, Rice Inst, 52-54; from instr to asst prof zool, Univ Calif, Los Angeles, 54-59; head dept parasitol, US Naval Med Res Unit, Cairo, 60-62; assoc res parasitologist & assoc prof parasitol, Univ Calif, San Francisco, 62-68, res parasitologist, Hooper Found, 68-77, asst dir, 70-77, actg chmn, Dept Epidemiol & Int Health, 75-78, PROF PARASITOL, MED CTR, UNIV CALIF, SAN FRANCISCO, 68- *Concurrent Pos:* Resident coordr, Int Ctr Med Res & Training, Inst Med Res, Kuala Lumpur, Malaysia, 64-66; assoc prof, Med Ctr, Univ Calif, San Francisco, 65-68; consult-rapporteur, WHO Traveling Sem Leishmaniasis USSR, 67, consult, US Navy Med Res Unit 3, Cairo, Egypt, 67-88; mem, Adv Sci Bd, Gorgas Mem Inst, 67-91; vis prof parasitol, Univ Toronto, 71; vis biologist, Am Inst Biol Sci, 71-74; mem, Trop Med & Parasitol Study Sect, Nat Inst Allergy & Infectious Dis, NIH, 73-76; external examr trop med & parasitol, State La Bd Regents, La State Univ, 78; mem, UN Develop Prog/WHO Joint Schistosomiasis Control Proj Rev Mission, Ghana, 78; mem, Addis Ababa Univ,Univ Calif, San Francisco, Trop Dis Surv, Southwest Ethiopia, 81; consult, WHO, UN Develop Prog & Agency Int Develop; external examr, Sch Biol Sci, Univ Nebr, Lincoln, 86; chair, Joint Med Prog, dir, Health & Med Sci & assoc dean, Sch Pub Health, Univ Calif, Berkeley, 87-91. *Mem:* Sigma Xi; Am Soc Parasitol (pres, 82-83); Soc Protozool; Am Micros Soc; Am Soc Trop Med & Hyg. *Res:* Parasite immunology and host-parasite studies; epidemiology; evolution and systematics of helminths; biological control of trematode infections in snail hosts; disease impact of development in Third World. *Mailing Add:* Joint Med Prog Bldg T-7 Univ Calif Berkeley CA 94720

HEYNER, SUSAN, b Hemel Hempstead, UK, Jan 21, 36; m 63; c 2. CELLULAR PROLIFERATION & DEVELOPMENT. *Educ:* Univ Southampton, BSc, 57; Univ London, PhD(physiol), 60. *Prof Exp:* Res fel, Royal Col Surgeons Eng, 61-63; res assoc, Univ Pa, 64-69; from asst prof to prof, Philadelphia Col Pharm & Sci, 69-85; prof, Sch Med, Temple Univ, 85-90; PROF, UNIV PA, 90-, DIR IN VITRO FERTIL & ANDROLOGY LAB, MED CTR, 90- *Concurrent Pos:* Mem, HED Study Sect, 75-79, REB Study Sect, NIH, 81-85; assoc ed, Contraception, 82-; bd mem, Soc Develop Biol, 83-85; assoc ed, Molecular Reprod & Develop, 91- *Honors & Awards:* Res Career Develop Award, NIH, 78. *Mem:* Am Soc Cell Biol; AAAS; Soc Develop Biol; Soc Study Reproduction. *Res:* Underlying early mammalian development; intrinsic and extrinsic factors that govern cellular proliferation and development. *Mailing Add:* Dept Obstet & Gynec Univ Pa Med Ctr Philadelphia PA 19104-6080

HEYNICK, LOUIS NORMAN, b Brooklyn, NY, Mar 17, 19; m 41; c 2. PHYSICS, MATHEMATICS. *Educ:* Brooklyn Col, BS, 40; Columbia Univ, MA, 48. *Prof Exp:* Asst chief electron physics sect, US Naval Appl Sci Lab, 47-56; chief mat & processing, US Army Electronics Command, 56-59, res team leader, 59-63; mgr phys electronics, SRI Int, 63-73, staff scientist, 73-84. *Concurrent Pos:* Assoc ed, Trans Electron Devices, Inst Elec & Electronics Engrs, 69-77, comt man & radiation, 78-; consult, 85- *Mem:* Am Phys Soc; Sigma Xi; Inst Elec & Electronics Engrs; Int Microwave Power Inst; Bioelectromagnetics Soc. *Res:* Electron device research and development; imaging and display devices; biological effects of radiation; electron optics and lithography; microwave devices; sensors and instrumentation. *Mailing Add:* 833 Richardson Ct Palo Alto CA 94303

HEYS, JOHN RICHARD, b Dayton, Ohio, Oct 15, 47; m 75; c 1. ORGANIC RADIOCHEMICAL SYNTHESIS, RADIOCHEMISTRY. *Educ:* DePauw Univ, BA, 69; Stanford Univ, PhD(org chem), 76. *Prof Exp:* Assoc org chem, Dept Chem, Yale Univ, 76; assoc chemist, Midwest Res Inst, Kansas City, 77-78, sr radiochemist & prog mgr, org & radiosynthesis, 78-83; sr investr, 83-87, ASST DIR, SYNTHETIC CHEM DEPT, SMITH KLINE & FRENCH LABS, 87- *Mem:* Am Chem Soc; AAAS; Int Union Pure & Appl Chem; Int Isotope Soc. *Res:* Organic synthesis of isotopically labeled compounds; development of synthetic methods for isotopically labeled compounds of medicinal and biological interest; peptide labeling methods; metabolism of xenobiotics. *Mailing Add:* Smith Kline Beecham Pharmaceut 709 Swedeland Rd PO Box 1539 Mail Stop L-830 King Prussia PA 19406

HEYSE, STEPHEN P, b New York, NY, May 1, 48. PREVENTATIVE MEDICINE. *Educ:* Univ NC, BA, 70; State Univ NY, MD, 74; Johns Hopkins Univ, MA, 78. *Prof Exp:* DIR, OFF DIS PREV, EPIDEMIOL & CLIN APPLN, NAT INST ARTHRITIS, MUSCULOSKELETAL & SKIN DIS, NIH, 87- *Mem:* Am Col Rheumatology. *Mailing Add:* Off Dis Prev Epidemiol & Clin Appln Nat Inst Arthritis Musculoskeletal & Skin Dis NIH Bldg 31 Rm 4C13 Bethesda MD 20892

HEYSSEL, ROBERT M, b Jamestown, Mo, June 19, 28; m 55; c 5. HEMATOLOGY, NUCLEAR MEDICINE. *Educ:* Univ Mo, BS, 51; St Louis Univ, MD, 53. *Prof Exp:* Intern, St Louis Univ, 53-54; asst resident, Vet Admin Hosp, St Louis, 54-55; asst resident, Barnes Hosp, 55-56; sr asst, Surgeon Sta Atomic Bomb Casualty Comm, United States Pub Health Serv, 56-58; fel hemat, Wash Univ Sch Med, 58-59; from instr to assoc prof med, Vanderbilt Univ, 59-68; assoc prof, 68-71, PROF MED, JOHNS HOPKINS UNIV SCH MED, 71-; PRES, JOHNS HOPKINS HEALTH SYST, 86- & JOHNS HOPKINS HOSP, 88- *Concurrent Pos:* Consult radiation, Tenn Dept Pub Health, 54; asst dir, Radioisotope Ctr, Vanderbilt Univ Hosp, 59-61; dir, Div Nuclear Med & Biophys, dept med, Vanderbilt Sch Med & Radioisotope Ctr, Vanderbilt Univ Hosp, 62-68; assoc dean, Johns Hopkins Univ, Sch Med & physician & dir, Outpatient Serv & Off Health Care Prog, Johns Hopkins Hosp, 68-72; mem expert panel, Int Atomic Energy Agency, Vienna, 65; res consult, Oak Ridge Assoc Univ, 67, chmn, med prog review comt, 70-71; chmn, health serv adv comt, Asn Am Med Col, 71-74, comt emergency med serv, NAS, 73-76 & controlling supply short term gen hosp beds study, 75-76, Asn Am Med Col, 83-84, Commonwealth Fund Task Force Acad Health Ctr, 83-; mem, liaison comt grad med educ, 74-77, Gen Assembly, Asn Am Med Col, 74-80, adv comt, subcomt health ways & means, US House Rep, 74-78, Coun Teaching Hosp Admin Bd, As Am Med Col, 75-79. *Mem:* Nat Acad Sci; Inst Med NAS; AAAS; Am Soc Hemat; Am Fedn Clin Res; fel Am Col Physicians; Asn Am Physicians; Sigma Xi; Reticuloendothelial Soc; fel Int Soc Hemat. *Res:* Radiation induced leukemia; epidemiology; thrombocyte kinetics; serotonin metabolism; vitamin B-12 absorption, turnover and requirements; human whole body counting; iron metabolism; health services research; author of over 50 publications. *Mailing Add:* Off Pres Johns Hopkins Health Syst Baltimore MD 21205

HEYSTEK, HENDRIK, b Feb 1, 24; US citizen; m 46; c 2. CERAMICS ENGINEERING. *Educ:* Univ Pretoria, BSc, 43, DSc(mineral), 55; State Univ NY Col Ceramics, Alfred Univ, MSc, 47; Pa State Univ, MSc, 49. *Prof Exp:* Sr res officer ceramics, SAfrican Coun Sci & Indust Res, 48-60; mgr res, Int Pipe & Ceramics Corp, 60-66, dir res, 66-69; dir res, Brick Corp SAfrica, Ltd, 69-73; RES SUPVR, TUSCALOOSA RES CTR, US BUR MINES, 73-

Concurrent Pos: Nat Res Coun Can fel, 55-56. *Mem:* Fel Am Ceramic Soc; Clay Minerals Soc; Mineral Soc Gt Brit & Ireland; Am Soc Testing & Mat (secy, 77-). *Res:* Clay mineralogy; ceramic research; refractories. *Mailing Add:* Bur Mines PO Box L University AL 35486

HEYTLER, PETER GEORGE, b Prague, Czech, Apr 24, 30; nat US; m 50; c 3. BIOCHEMISTRY. *Educ:* Cornell Univ, BA, 50, MNS, 52, PhD(biochem), 56. *Prof Exp:* Chemist, Lederle Labs, Am Cyanamid Corp, 52-54; asst, Cornell Univ, 54-56; biochemist, Cent Res Dept, E I du Pont de Nemours & Co, Inc, 56-61, res supvr, 61-86, med prod dept, 86-90; RES ASSOC, DUPONT MERCK PHARMACEUT CO, 90- *Mem:* Am Soc Biol Chem; AAAS. *Res:* Bioenergetics; membrane biochemistry; endocytosis. *Mailing Add:* Neurosci Div Du Pont Merck Pharmaceut Co Wilmington DE 19898

HEYWOOD, JOHN BENJAMIN, b Sidcup, Eng, Jan 11, 38; m 61; c 3. MECHANICAL ENGINEERING. *Educ:* Cambridge Univ, BA, 60, ScD, 84; Mass Inst Technol, SM, 62, PhD(mech eng), 65. *Prof Exp:* Res assoc mech eng, Mass Inst Technol, 64-65; res officer, Cent Elec Generating Bd, UK, 65-67, group leader, Leatherhead, 67-68; from asst prof to prof mech eng, 76-89, LEADERS FOR MFG PROF, MASS INST TECHNOL, 89-, DIR, SLOAN AUTOMOTIVE LAB, 72- *Concurrent Pos:* Lectr, Northeastern Univ, 63-65; consult, Avco Systs Div, Bendix, Ford Motor Co, Jaguar Cars, Nat Acad Sci, Mobil Res & Develop Corp and var others. *Honors & Awards:* Ayreton Premium Award, Brit Inst Elec Engrs, 69; Ralph R Teetor Award, Soc Automotive Engrs, 71, Arch T Colwell Merit Award, 73, 81 & 89, Fel, 82, Horning Mem Award, 84; Freeman Scholar, Am Soc Mech Engrs, 86, Soichiro Honda Lectr, 90. *Mem:* Assoc fel Am Inst Aeronaut & Astronaut; fel Brit Inst Mech Engrs; Combustion Inst; Am Soc Mech Engrs; fel Soc Automotive Engrs. *Res:* Internal combustion engines; combustion; power generation; thermodynamics. *Mailing Add:* Dept Mech Eng Rm 3-340 77 Massachusetts Ave Cambridge MA 02139

HEYWOOD, PETER, b Manchester, UK, Apr 1, 43. CELL BIOLOGY, PHYCOLOGY. *Educ:* London Univ, BSc, 64, PhD(bot), 68. *Prof Exp:* Maria Moors Cabot res fel bot, Harvard Univ, 68-70; asst prof microbiol, Sch Med, Yale Univ, 70-74; asst prof, 74-78, assoc prof, 78-88, assoc dean, 80-81, PROF BIOL, BROWN UNIV, 88- *Honors & Awards:* Nat Sci Teacher Asn. *Mem:* Nat Sci Teachers Asn; Bot Soc Am; Brit Phycol Soc; Sigma Xi; Phycol Soc Am. *Res:* Botany; nuclear cytology of eukaryotic microorganisms; evolution of eukaryotic microorganisms; development of the inner ear. *Mailing Add:* Div Biol & Med Sci Brown Univ Providence RI 02912

HEYWOOD, STUART MACKENZIE, b Concord, Mass, Oct 15, 34; m 58; c 2. MOLECULAR BIOLOGY, BIOCHEMISTRY. *Educ:* Univ Mass, BS, 57; Syracuse Univ, PhD(biochem), 65. *Prof Exp:* Am Heart Asn fel molecular biol, Mass Inst Technol, 65-66, NIH fel, 66-67; from asst prof to assoc prof, 67-73, PROF CELL BIOL, UNIV CONN, 73-, HEAD SECT CELL BIOL & GENETICS, 74-, HEAD DEPT MOLECULAR & CELL BIOL, 85- *Concurrent Pos:* NIH res grant, 67- & career develop award, 70- *Mem:* Biophys Soc; Am Soc Cell Biol; Am Soc Biol Chem. *Res:* Protein synthesis in higher organisms; transcription and translational controls during differentiation; ribonucleic acid metabolism. *Mailing Add:* 367 Wormwood Hill Rd Mansfield Center CT 06250

HIATT, ANDREW JACKSON, b Wildie, Ky, Apr 14, 32; m 54; c 2. PLANT PHYSIOLOGY, AGRONOMY. *Educ:* Univ Ky, BS, 53, MS, 57; NC State Univ, PhD(plant physiol), 60. *Prof Exp:* From asst prof to assoc prof, 60-67, chmn dept agron, 69-88, PROF AGRON, UNIV KY, 67-, ASSOC DEAN ADMIN, 88- *Concurrent Pos:* Mem coun, Agr Sci & Technol; res award, Univ Ky, 65, fac-alumni res award, 66. *Mem:* Am Soc Plant Physiol; fel Am Soc Agron; Soil Sci Soc Am; Crop Sci Soc Am. *Res:* Investigations of role of ions in metabolism of plants, mechanism of absorption of salts by cells and translocation of ions by plants. *Mailing Add:* Assoc Dean Admin S-129 Agr Sci Ctr N Univ Ky Lexington KY 40546-0915

HIATT, CASPAR WISTAR, III, b Lakewood, Ohio, Sept 23, 19; m 52; c 4. BIOPHYSICAL CHEMISTRY. *Educ:* Case Western Reserve Univ, BS, 42, PhD(immunochem), 48. *Prof Exp:* Res chemist, Standard Oil Co, Ohio, 42-43; asst, Inst Path, Cleveland, 46-48; Merck fel natural sci, Rockefeller Inst, 48-50; chief vet chem, Army Med Serv Grad Sch, Walter Reed Med Ctr, 50-56; biochemist, Viral Prod Lab, Div Biol Standards, NIH, 56-60, chief lab biochem & biophys, 60-67; prof chem & chmn dept, Fla Atlantic Univ, 67-68; chmn dept, Univ Tex Health Sci Ctr, San Antonio, 68-77, prof, 77-79; DIR, TECHSCAN LABS, 77- *Mem:* AAAS; Am Chem Soc; NY Acad Sci. *Res:* Chemical factors in immunity; physical properties of viruses; ultracentrifugation; extraction of spirochetal antigens; kinetics of inactivation of viruses. *Mailing Add:* 861 Cumberstone Rd Harwood MD 20776

HIATT, HAROLD, b Wilmington, Ohio, Oct 15, 21; m 62; c 3. PSYCHIATRY. *Educ:* Wilmington Col, BS, 43; Univ Cincinnati, MD, 46. *Prof Exp:* Rotating internship, Cincinnati Gen Hosp, 46-47; from jr resident to chief resident, Dept Psychiat, Univ Cincinnati, 49-52, from instr to asst prof psychiat, 52-64, assoc clin prof, 64-68; assoc prof, 68-74, PROF PSYCHIAT, UNIV CINCINNATI, 74-, CHIEF PSYCHIAT SERV, VET ADMIN HOSP, CINCINNATI, 54- *Concurrent Pos:* Attend psychiatrist & clinician, Cincinnati Gen Hosp Dept Psychiat; mem admis comt, Cincinnati Col Med, 64-71; mem bd, Cincinnati Ment Health Asn, 66-72; coordr psychol curric, Dept Psychiat, Univ Cincinnati, 66-68; pres, Cincinnati Soc Neurol & Psychiat, 67-68; mem bd, Marjorie P Lee Home Aged, 67-70; mem prog comt, Cincinnati Acad Med, 68; mem bd cancer control coun, Pub Health Fedn, 68-74; asst ed, Cincinnati Col Med Alumni Bull, 71. *Mem:* AMA; Am Psychiat Asn; Am Col Psychiat. *Res:* Geriatrics; psychotherapy. *Mailing Add:* Dept Psychiat Med Sci Bldg 7211 231 Bethesda Cincinnati OH 45267

HIATT, HOWARD HAYM, b Patchogue, NY, July 22, 25; m 47; c 3. INTERNAL MEDICINE, PUBLIC HEALTH. *Educ:* Harvard Univ, MD, 48. *Hon Degrees:* DSc, Northeastern Univ & Mass Col Pharm. *Prof Exp:* Investr, Nat Inst Arthritis & Metab Dis, 53-55; instr & assoc to asst prof med, 55-63, Herman L Blumgart prof, 63-72, dean Sch Pub Health, 72-84, PROF MED, HARVARD UNIV, 72- *Concurrent Pos:* Res fel med, Med Col, Cornell Univ & NY Hosp, 51-53; Am Cancer Soc res scholar, 58-59; Lederle med fac award, 59-62; Commonwealth Fund travel fel, Pasteur Inst, Paris, 60-61; vis scientist, Imp Cancer Res Fund Lab, London, 69-70; assoc med, Beth Israel Hosp, 55-63, physician-in-chief, 63-72. *Mem:* Inst Med - Nat Acad Sci; Am Soc Biol Chem; Am Soc Clin Invest; AAAS; Asn Am Physicians. *Res:* Research in social medicine. *Mailing Add:* Dept Med Brigham & Woman's Hosp Boston MA 02115

HIATT, JAMES LEE, b Lebanon, Ind, July 25, 34; m 54; c 3. ADMINISTRATION, ANATOMICAL SCIENCES & EDUCATION. *Educ:* Ball State Univ, BS, 59, MS, 68; Univ Md, PhD(anat), 73. *Prof Exp:* Teacher biol, Wendell Wilkie High Sch, Elwood, Ind, 59-67; res assoc, 67-72, from instr to asst prof, 72-76, ASSOC PROF ANAT, DENT SCH, UNIV MD, BALTIMORE, 76- *Concurrent Pos:* Consult, US Army Inst Dent Res, Walter Reed Army Med Ctr, Washington, DC, 82-; Oral Surg, Johns Hopkins Hosp; Dent Corp, Ft Meade Army Med Ctr. *Mem:* Int Asn Dent Res; Am Asn Dent Res; Am Asn Anatomists. *Res:* Craniofacial development in rodents related to teratogenic effects on development; tooth development in mammals; craniofacial development in trisomic mice; CAI development; 3-D computer reconstruction. *Mailing Add:* Dept Anat Dent Sch Univ Md Baltimore MD 21201

HIATT, NORMAN ARTHUR, b Worcester, Mass, Nov 22, 37; m 65; c 2. POLYMER CHEMISTRY, ORGANIC CHEMISTRY. *Educ:* Worcester Polytech Inst, BS, 59; Lowell Technol Inst, PhD(org polymer chem), 68. *Prof Exp:* Engr plastics develop, Gen Dynamics Corp, 59-60 & AVCO Corp, 60-62; res scientist polymer res & develop, Uniroyal Inc, 69-73, sr group leader rubber res & develop, 73-74; sr res engr develop high temperature plastics, Norton Co, 74-76; res assoc heat transfer labels, Therimage Prod Div, Dennison Mfg Co, 76-78, res sect head heat transfer labels & indust crepe, 78-81, mgr tech develop heat transfer labels, 81-87; CONSULT, CHEM, PLASTICS & PACKAGING FIELDS, 87- *Concurrent Pos:* Res fel, Univ Louvain, Belgium, 68-69. *Mem:* Am Chem Soc; Soc Plastics Engrs. *Res:* Photochromic polymers; organo-sulfur polymers; polyurethanes; high-temperature polymers; adhesives; coating; printing technology. *Mailing Add:* Three Barry Dr Framingham MA 01701-6101

HIATT, RALPH ELLIS, b Portland, Ind, Apr 12, 10; m 40; c 3. ELECTRICAL ENGINEERING. *Educ:* Ind Cent Univ, AB, 32; Ind Univ, MA, 39. *Prof Exp:* Br head Microwave Physics Lab, Air Force Cambridge Res Ctr, 45-55, head, 55-58; assoc dir Radiation Lab, Univ Mich, 58-61, dir, 61-75, prof elec eng, 66-80, emer prof elec & comput eng, 80-; RETIRED. *Concurrent Pos:* Chief, Ipswich Antenna Field Sta, Mass Inst Technol, 43-45. *Mem:* Fel Inst Elec & Electronics Engrs; fel AAAS; Sigma Xi. *Res:* Research, design and development of microwave antennas; radar scattering characteristics and radar absorbing materials. *Mailing Add:* Bldg Elec Eng & Comput Sci Univ Mich Ann Arbor MI 48109

HIATT, ROBERT BURRITT, b Wilmington, Ohio, Apr 30, 17. SURGERY, PHYSIOLOGY. *Educ:* Wilmington Col, Ohio, BA, 38; Univ Cincinnati, MD, 42; Am Bd Surg, dipl. *Prof Exp:* From intern to resident, Columbia-Presby Med Ctr, 42-46; asst surg, Col Physicians & Surgeons, Columbia Univ, 46-50, instr, 50-56, asst prof clin surg, 56-61, from assoc prof to prof surg, 61-83, dir undergrad surg teaching, 62-68. *Concurrent Pos:* NIH career develop award, 61-66, grant, 64-66; New York City Health Res Coun grant, 64-66; asst attend surgeon, Babies & Bellevue Hosps, New York, 46-50; asst attend surgeon, Presby Hosp, New York, 50-56, assoc attend, 56-62, chief West Surg Serv, 62-68; dir surg, Shiraz Med Ctr, Iran, 54-56; pres, Coherin Res Found, 80- *Mem:* Fel Am Col Surgeons; AMA; Pan-Am Med Asn; Am Surg Asn. *Res:* Intestinal physiology; abdominal surgery. *Mailing Add:* Box 222 Dresden ME 04342

HIBBARD, AUBREY D, horticulture; deceased, see previous edition for last biography

HIBBARD, MALCOLM JACKMAN, b Rockport, Mass, Feb 20, 36; m 65. GEOLOGY. *Educ:* Dartmouth Col, BA, 58; Univ Wash, MS, 60, PhD(geol), 62. *Prof Exp:* From asst prof to assoc prof, 62-73, PROF GEOL, MACKAY SCH MINES, UNIV NEV, RENO, 73- *Mem:* Geol Soc Am. *Res:* Igneous and metamorphic petrology. *Mailing Add:* Dept Geol Univ Nev Reno NV 89557

HIBBARD, WALTER ROLLO, JR, b Bridgeport, Conn, Jan 20, 18; m 42, 72; c 3. MATERIALS SCIENCE, ENERGY. *Educ:* Wesleyan Univ, AB, 39; Yale Univ, DEng(metall), 42. *Hon Degrees:* LLD, Mich Technol Univ, 66; DEng, Mont Col Mineral Technol, 68. *Prof Exp:* From asst to assoc prof metall, Yale Univ, 46-51; res assoc & mgr metals & ceramics, Gen Elec Res Lab, 51-65; dir, Bur of Mines, US Dept of Interior, 65-68; vpres res & develop tech serv, Owens Corning Fiberglas, 68-74; vpres res & develop tech serv, Energy Res & Develop Off, Fed Energy Admin, 74; emer prof eng, Va Polytech Inst & State Univ, 74-88; RETIRED. *Concurrent Pos:* Chmn Mat Adv Bd, Nat Res Coun, 63-65 & Bldg Adv Bd, 75-76; bd dir, Norton Co, Mass, 72-88; ed, Mat & Soc, 75-88; emer dir, Va Ctr for Coal & Energy Res, 77-88. *Honors & Awards:* Douglas Medal, Am Inst Mining, Metall & Petrol Engrs, 67; Raymond Award, 50; Mineral Econ Award, 83. *Mem:* Nat Acad Eng; fel Am Ceramic Soc; fel Am Soc Metals; fel Am Acad Arts & Sci; Am Inst Mining, Metall & Petrol Engrs (pres, 67); fel Metall Soc; distinguished mem Soc Mining Engrs. *Res:* Materials, energy, environment, mineral processing, mineral economics. *Mailing Add:* 1403 Highland Cir Blacksburg VA 24060-5624

HIBBELER, RUSSELL CHARLES, b Evanston, Ill, Jan 18, 44; m 74; c 1. ENGINEERING MECHANICS, NUCLEAR ENGINEERING. *Educ:* Univ Ill, Urbana, BS, 65, MS, 66; Northwestern Univ, PhD(theoret & appl mech), 68. *Prof Exp:* Instr, Ill Inst Technol, 72-73; asst prof, Union Col, 73-76; PROF, UNIV SOUTHWESTERN LA, 77- *Concurrent Pos:* Pres, March

Eng, Inc, 81- *Honors & Awards:* AMCO Teaching Excellence Award. *Mem:* Am Soc Eng Educ; Am Soc Civil Engrs; Am Soc Mech Engrs. *Res:* Thermal and creep problems in nuclear reactor components; problems in the theory of vibrations and theory of elasticity. *Mailing Add:* 149 Shannon Rd Lafayette LA 70504

HIBBEN, CRAIG RITTENHOUSE, b Montclair, NJ, May 25, 30; m 58; c 2. PLANT PATHOLOGY. *Educ:* Pa State Univ, BS, 53; Cornell Univ, MS, 59, PhD(plant path), 62. *Prof Exp:* RESEARCHER PLANT PATH, KITCHAWAN RES LAB, BROOKLYN BOT GARDEN, 62- *Concurrent Pos:* Mem, Int Shade Tree Conf. *Mem:* Am Phytopath Soc; Sigma Xi. *Res:* Forest and ornamental tree diseases, especially virus infection and air pollution injury; diebacks of forest tree species. *Mailing Add:* 712 Kitchawan Rd Ossining NY 10562

HIBBERT, LARRY EUGENE, b LaGrande, Ore, May 17, 37; m 59; c 7. PARASITOLOGY, PROTOZOOLOGY. *Educ:* Eastern Ore Col, BS, 62; Utah State Univ, MS, 67, PhD(zool), 69. *Prof Exp:* PROF BIOL, RICKS COL, 69-, DEPT HEAD, 72- *Concurrent Pos:* Vis asst prof, Univ Utah, 68-69. *Mem:* Sigma Xi; Soc Protozoologists; Am Inst Biol Sci; Human Anat & Physiol Soc. *Res:* Excystation of bovine coocidia, both invitro, and comparing their stimuli for excystation with other mammalian coocidia. *Mailing Add:* Dept Biol Ricks Col Rexburg ID 83460-1100

HIBBITS, JAMES OLIVER, JR, b St Louis, Mo, Oct 27, 24; m 47; c 3. ANALYTICAL CHEMISTRY. *Educ:* St Louis Univ, BS, 50, MS, 52. *Prof Exp:* Assoc chemist, Union Carbide Nuclear Corp, 51-55; chemist, Mallinckrodt Chem Works, 55-56; sr chemist aircraft nuclear propulsion, Gen Elec Co, 56-60, prin chemist, 60-64, mgr anal chem, Nuclear Mat & Propulsion Oper, Ohio, 64-70; mgr anal chem, Owens-Ill Inc, Toledo, 70-73, mgr chem & phys testing dept, 73-87; pres, Monarch Anal Labs, 87-90; CONSULT, 90- *Concurrent Pos:* Bd dirs, Eitel Inst Silicate Res, 87- *Honors & Awards:* Tech Excellence, Gen Electric, 65. *Mem:* Am Chem Soc. *Res:* Analytical inorganic chemistry; analytical chemistry of the rare earths, uranium and less familiar elements; specific reactions obtained by liquid-liquid extraction and semi-specific reagents; potentiometry and other phases of electro-analytical chemistry; gas chromatography. *Mailing Add:* 4555 Crestview Dr Sylvania OH 43560

HIBBS, ANDREW D, LOW-TEMPERATURE PHYSICS. *Educ:* Cambridge Univ, Eng, MA, 85, PhD(physics), 89. *Prof Exp:* Res assoc, Nat Ctr Superconductinity, 87-89; RES PHYSICIST, QUANTUM MAGNETICS, 89- *Concurrent Pos:* Prin investr, Army Elec Tech Lab, 91. *Mem:* Cryog Soc Am. *Res:* Designed and developed custom built cryogenic instrumentation for magnetic imaging. *Mailing Add:* Quantum Magnetics Inc 11578 Sorrento Valley Rd No 30 San Diego CA 92121

HIBBS, CLAIR MAURICE, b Lucerne, Mo, Oct 10, 23; m 46; c 2. VETERINARY PATHOLOGY, MICROBIOLOGY. *Educ:* Univ Mo, BS, 49, DVM, 53; Kans State Univ, MS, 62, PhD(path), 65. *Prof Exp:* Technician, Univ Mo, 50-52; technician pvt pract, 53-60; res assoc, Kans State Univ, 60-62, instr parasitol, 62-63 & diag lab, 63-65, from asst prof to assoc prof path, 65-69. assoc prof, 69-73, PROF PATH, N PLATTE AGR EXP STA, UNIV NEBR, LINCOLN, 73-; DIR, NMEX VET DIAG SERV, 79-; AT DEPT PATH, UNIV NMEX. *Concurrent Pos:* Consult pathologist, Chemagro Corp, 67-69; Agr Res Serv, USDA coop agreement, Kans State Univ, 69-; Mark Morris Animal Found fel. *Honors & Awards:* Rotary Paul Harris Award, 88. *Mem:* Conf Res Workers Animal Dis; Am Asn Vet Lab Diagnosticians; Am Asn Bovine Practitioners. *Res:* Infectious diseases, anomalies and related pathology. *Mailing Add:* 3401 Treesmill Manhattan KS 66502

HIBBS, EDWIN THOMPSON, entomology, for more information see previous edition

HIBBS, JOHN BURNHAM, MACROPHAGE PHYSIOLOGY *Educ:* Univ Pittsburgh, MD, 62. *Prof Exp:* PROF MED, COL MED, UNIV UTAH, 80- *Mailing Add:* Med Sect 151G Vet Admin Med Ctr 500 Foothill Dr Salt Lake City UT 84148

HIBBS, JOHN WILLIAM, animal nutrition; deceased, see previous edition for last biography

HIBBS, LEON, mathematics, academic administration, for more information see previous edition

HIBBS, RICHARD GUYTHAL, b Winner, SDak, Feb 17, 22; m 46; c 3. HISTOLOGY. *Educ:* Univ SDak, BA, 50; Univ Minn, PhD(anat), 55. *Prof Exp:* From asst to instr anat, Univ Minn, 51-55; from instr to prof, Tulane Univ, 55-75; prof anat, Sch Med, La State Univ, Shreveport, 75-87, head dept, 77-87; RETIRED. *Honors & Awards:* Gold Award, Am Acad Dermat, 59. *Mem:* Am Asn Anat; Am Soc Cell Biol. *Res:* Electron microscopy and histochemistry of the cardiovascular system. *Mailing Add:* 6231 S Lakeshore Dr Shreveport LA 71119

HIBBS, ROBERT A, b Cocoa, Fla, Sept 9, 23; m 52; c 6. CHEMISTRY, BACTERIOLOGY. *Educ:* Univ Fla, BSA, 47, MS, 48; Wash State Univ, PhD, 51. *Prof Exp:* Asst, Univ Fla, 47-48 & Wash State Univ, 48-51; qual control supvr, Darigold Farms, Wash, 51-54; asst prof dairy sci, Univ Idaho, 54-61; DIR, HIBBS LABS, 61-; PROF CHEM, BOISE STATE UNIV, 71- *Concurrent Pos:* From asst prof to assoc prof phys sci, Boise State Col, 65-71. *Res:* Food technology; quality improvement in processing food products; new product development. *Mailing Add:* Dept Chem Boise State Univ 1910 Univ Dr Boise ID 83725

HIBBS, ROGER FRANKLIN, b St Louis, Mo, Jan 28, 21; m 45; c 8. CHEMISTRY. *Educ:* Eastern Ill State Teachers Col, Charleston, BEd, 43. *Prof Exp:* Jr chemist anal chem, Tenn Eastman Corp, 43-44, lab supvr mass spectrometry, 44-47, lab dept supt, 47; lab dept supt, Carbide & Carbon Chem Co, 47-52, lab div head chem & phys measurements, Union Carbide Corp, 52-54, supt, Process Div, 54-58, supt, Tech Div, Union Carbide Nuclear Co Div, 58-62, from plant mgr to vpres, Nuclear Div, 62-70, pres, Nuclear Div, 70-84; RETIRED. *Mem:* AAAS; Am Chem Soc; Am Nuclear Soc. *Res:* Mass spectrometry. *Mailing Add:* RR 2 Harriman TN 37748

HIBLER, CHARLES PHILLIP, b Austin, Tex, Aug 19, 30; m 53; c 4. ZOOLOGY, BIOCHEMISTRY. *Educ:* NMex State Univ, BS, 56; Utah State Univ, MS, 59; Colo State Univ, PhD(zool), 63. *Prof Exp:* Res scientist, Animal Dis & Parasite Res Div, USDA, 62-65; prof parasitol, Col Vet Med, Colo State Univ, 65-87, dir, Wild Animal Dis Ctr, 71-87; RETIRED. *Mem:* Am Soc Parasitologists; Wildlife Dis Asn. *Res:* Biology of parasitic nematodes, especially the Filarioidea. *Mailing Add:* 2012 Derby Ct Ft Collins CO 80526

HIBLER, WILLIAM DAVID, III, b Brunswick, Mo, Feb 8, 43; m 65; c 2. SEA ICE DYNAMICS, GLACIOLOGY. *Educ:* Univ Mo Columbia, BS, 65; Cornell Univ, PhD(physics), 69. *Prof Exp:* Vis asst prof physics, Univ Cincinnati, 69-70; res physicist glaciol, US Army Cold Regions Res & Eng Lab, 70-86; CONSULT. *Concurrent Pos:* Vis fel, Geophys Fluid Dynamics Prog, Princeton Univ, 76-78 & 81-82; mem Global Atmospheric Res Prog Polar Subprog Panel, Nat Acad Sci, 77-79, glaciol Subcomt, Polar Res Bd, 85-88; vis scientist, Max Planck Inst Meteorol, Hamburg, Fed Repub Ger, 84-85. *Mem:* Am Geophys Union; Int Glaciol Soc; Am Meteorol Soc; Sigma Xi. *Res:* Large scale numerical modeling of sea ice and ice covered oceans; geophysical scale ice mechanics. *Mailing Add:* Thayer Sch Eng Dartmouth Col Hanover NH 03755

HICHAR, JOSEPH KENNETH, b Allentown, Pa, Aug 5, 28; m 55; c 2. PHYSIOLOGY, ZOOLOGY. *Educ:* Univ Pittsburgh, BS, 50; Pa State Univ, MS, 52; Harvard Univ, PhD(biol), 58. *Prof Exp:* Mgr sales promotion, Everson Elec Co, 50; investr, Liberty Mutual Ins Co, 50-51; asst, Pa State Univ, 51-52; asst prof biol, Moravian Col Women, 52-53; from asst prof to assoc prof, State Univ NY Col, Brockport, 53-55; teaching res fel, Harvard Univ, 55-58; asst prof zool, Ohio Wesleyan Univ, 58-60; from assoc prof to prof biol, Parsons Col, 60-65, dean grad sch, 61, dean col, 61-62, dean fac sci, 62-64; prof biol, dean col & vpres acad affairs, Hiram Scott Col, 65-70; dean col arts & sci, 70-72, dean fac natural sci, 72-75, prof, 70-85, EMER PROF BIOL, STATE UNIV NY COL BUFFALO, 85- *Concurrent Pos:* NIH res grants, 60-65; Fulbright prof, Univ Ceylon, 63-64; admin dean, Hiram Scott Col, 65-66, vpres acad affairs, 66-70; NSF grants, Int Conf Med Electronics, Paris, 59 & biomed eng, Univ Vt, 60; Nat Acad Sci-Nat Res Coun grant, Int Cong biophysics, Stockholm & Moscow, 61; NIH grant, Int Conf Med Electronics & Biol Eng, Tokyo, 65; US Off Educ grant newer mediated learning systs, Univ Nebr, 68-69; mem ad comt, Erie Community Col, 72- *Mem:* Fel AAAS; Am Soc Zool; Biophys Soc; Sigma Xi; fel Acad Zool India; NY Acad Sci; Am Inst Biol Sci. *Res:* Comparative and general physiology; invertebrate general zoology; neuro-electrophysiology; neuropharmacology; neurohormones; central nervous system spontaneous electrical activity and the effects of drugs on this activity; behavior; peripheral nervous system; isolated nerves. *Mailing Add:* 219 Cotuit Rd Sandwich MA 02563

HICKENBOTTOM, JOHN POWELL, b Burns, Ore, Apr 12, 38; m 68; c 2. BIOCHEMISTRY. *Educ:* Idaho State Univ, BS, 60; Univ Wash, PhD(biochem), 67. *Prof Exp:* From asst prof to assoc prof pharmacol & biochem, Sch Pharm, Univ Miss, 69-80; ASSOC PROF BIOL, D'YOUVILLE COL, BUFFALO, NY, 80- *Concurrent Pos:* Nat Heart Inst fel, Dept Pharm, Sch Med, Emory Univ, 67-69. *Mem:* NY Acad Sci; Am Chem Soc; Sigma Xi. *Res:* Mechanisms of hormonal control of glycogen metabolism in muscle and liver; effects of chlorinated hydrocarbons on hepatic glycogen metabolism in mammals and marine animals. *Mailing Add:* 17 Townsend Pl Athens OH 45701

HICKERNELL, FRED SLOCUM, b Phoenix, Ariz, Jan 16, 32; m 54; c 4. ACOUSTICAL & OPTICAL MICROELECTRONICS. *Educ:* Ariz State Univ, BA, 53, MS, 59, PhD(physics), 66. *Prof Exp:* Instr physics, Ariz State Univ, 57-58; engr math, Goodyear Aerospace, 58-60; MEM TECH STAFF, MOTOROLA GOVT ELECTRONICS GROUP, 60- *Concurrent Pos:* Fac assoc, Dept Math, Ariz State Univ, 81-83; vis prof, Univ Ariz, 85-87; adj prof, Univ Ariz, 87-; Dan Noble Fel, Motorola Inc, 87. *Honors & Awards:* Eng Achievement, Inst Elec & Electronic Engrs, 69; Tech Achievement, NASA, 74. *Mem:* Am Phys Soc; Am Meteorol Soc; fel Inst Elec & Electronics Engrs; fel Am Sci Affil; Sigma Xi. *Res:* Microwave acoustics and integrated optics; bulk and surface acoustic wave materials and devices; integrated acousto-optic devices; microwave ferrites; vacuum thin films; silicon microelectronic circuits. *Mailing Add:* Motorola Govt Electronics Group 8201 E McDowell Rd Scottsdale AZ 85252

HICKERNELL, GARY L, b Meadville, Pa, Sept 30, 42; m 65; c 2. FOOD CHEMISTRY, AGRICULTURAL & FOOD CHEMISTRY. *Educ:* Allegheny Col, BS, 64; Univ Wash, PhD(org chem), 68. *Prof Exp:* Sr chemist coffee res, Gen Foods Corp, 68-88; vis asst prof, Vassar Col, 88-90; ASST PROF, KEUKA COL, 90- *Mem:* AAAS; Am Chem Soc; Inst Food Technologists. *Res:* Natural products structural elucidation, especially papuanic acid; chemistry of coffee, particularly with regards to its effect on flavor; QSAR re sweetness, bitterness; cereal chemistry and flavor. *Mailing Add:* 12 Terrich Ct Ossining NY 10562

HICKEY, BARBARA MARY, b Toronto, Ont, June 7, 45. PHYSICAL OCEANOGRAPHY. *Educ:* Univ Toronto, BSc, 67; Univ Calif, San Diego, MS, 69, PhD(phys oceanog), 75. *Prof Exp:* Res asst prof, 73-80, RES ASSOC PROF PHYS OCEANOG, UNIV WASH, 80- *Mem:* Am Geophys Union. *Res:* Physical oceanography of the Washington-Oregon continental shelf and slope regions; dynamics of Eastern Boundary Current Systems; dynamics of topographic effects in coastal areas; sediment transport. *Mailing Add:* Dept Oceanog Univ Wash Seattle WA 98195

HICKEY, DONAL ALOYSIUS, b Co Kerry, Ireland, July 13, 48. POPULATION GENETICS, EVOLUTIONARY THEORY. *Educ:* Nat Univ Ireland, BSc, 70; Harvard Univ, PhD(biol), 77. *Prof Exp:* Asst prof genetics, Brock Univ, 78-81; from asst prof to assoc prof, 81-88, PROF GENETICS & EVOLUTION, UNIV OTTAWA, 88- *Honors & Awards:* Assoc Fel, Can Inst Advan Res. *Mem:* Genetics Soc Am; Genetics Soc Can; Soc Study Evolution; Am Soc Naturalists; AAAS. *Res:* Gene regulation in eukaryotes; biochemical genetics of drosophila populations; population genetics and evolutionary theory; importance of regulatory gene polymorphisms in adaptive evolution; evolution of eukaryotic genome structure. *Mailing Add:* Dept Biol Univ Ottawa Ottawa ON K1N 6N5 Can

HICKEY, JIMMY RAY, b Hallsville, Tex, Feb 12, 29; m; c 2. LAPLACE TRANSFORM, DIFFERENTIAL EQUATION. *Educ:* ETex State Univ, BS, 53; Baylor Univ, MA, 59; Univ Tex at Austin, PhD(math), 71. *Prof Exp:* Instr math, Talco High Sch, 53-55; seismologist, Shell Oil Co, 55-57; PROF MATH, BAYLOR UNIV, 59- *Mem:* Math Asn Am; Soc Indust & Appl Math. *Res:* Operational calculus for generalized functions; the theory of distributions, generalized transforms and solution of differential equations using generalized functions. *Mailing Add:* Box 7328 Baylor Univ Waco TX 76798

HICKEY, JOSEPH JAMES, b New York, NY, Apr 16, 07; m 42, 76; c 1. WILDLIFE ECOLOGY. *Educ:* NY Univ, BS, 30; Univ Wis, MS, 43; Univ Mich, PhD, 49. *Prof Exp:* Asst, State Soil Conserv Comn, Wis, 41-43; ed chem warfare, Toxicity Lab, Univ Chicago, 43-44; asst cur mus zool, Univ Mich, 44-46; from asst prof to prof, 48-77, EMER PROF WILDLIFE ECOL, UNIV WIS-MADISON, 77- *Concurrent Pos:* Ed, J Wildlife Mgt, 56-59; mem bd dirs, Western Fedn Vert Zool, Bodega Bay Inst, World Pheasant Asn-US, Am Ornith Union & Dane Co Natural Heritage Found. *Honors & Awards:* Leopold Medal, Wildlife Soc, 72; Allen Medal, Cornell Lab Ornith, 76; Coues Award, Am Ornith Union, 78; Eisenmann Medal, Linnaean Soc, NY, 84; Audubon Medal, Nat Audubon Soc, 84. *Mem:* Ecol Soc Am; Wildlife Soc; Wilson Ornith Soc; Cooper Ornith Soc; Am Ornith Union (vpres, 70-72, pres, 72-73). *Res:* Population of birds; ecological effects surface coal mining; insecticide-wildlife relationships. *Mailing Add:* 1520 Wood Lane Madison WI 53705

HICKEY, KENNETH DYER, b Sparta, Tenn, Mar 4, 33; m 60; c 3. PLANT PATHOLOGY. *Educ:* Tenn Polytech Univ, BS, 55; Pa State Univ, MS, 58, PhD(plant path), 60. *Prof Exp:* Asst, Pa State Univ, 55-60; asst prof plant path & exten plant pathologist, Cornell Univ, 60-66; assoc prof plant path & res fruit pathologist, Va Polytech Inst & State Univ, 66-76; PROF PLANT PATH & SCIENTIST-IN-CHG, FRUIT RES LAB, PA STATE UNIV, 76- *Mem:* Am Phytopath Soc. *Res:* Development of disease control programs including evaluation of fungicides for control of fruit diseases. *Mailing Add:* Fruit Res Lab Box 309 Biglerville PA 17307

HICKEY, LEO JOSEPH, b Philadelphia, Pa, Apr 26, 40; m 68; c 3. GEOLOGY, PALEOBOTANY. *Educ:* Villanova Univ, BS, 62; Princeton Univ, MA, 64, PhD(geol), 67. *Prof Exp:* Nat Acad Sci res assoc, Smithsonian Inst, 66-69, assoc cur, 69-80, curator, 80-82; PROF DEPT GEOL & BIOL, YALE UNIV, 82-, dir Yale Peabody Mus, 82-87. *Concurrent Pos:* counr, Yellowstone Bighorn Res Asn, 73-79 & 85-, vpres, 79-81, pres, 81-83 & past pres, 83-85; chmn, Exhibits Comt, Nat Mus Nat Hist, 73-75, chmn-elect & treas, Senate Scientists; mem, Smithsonian Inst Trop Biol Steering Comt, 77-79; assoc ed, Paleobiol, 80-83, 87-90; trustee, Sheffield Sci Sch, Yale Univ, 82-87; sci adv bd, Sci Mus Conn, 84-87; sci adv comt, Yale Press, 85-86, publ comt, 88-91. *Honors & Awards:* All-Cong Lectr, Thirteenth Int Bot Cong, Sydney, 81; H A Gleason Award, NY Bot Garden, 78. *Mem:* AAAS; Geol Soc Am; Bot Soc Am; Paleont Soc; Am Geol Inst. *Res:* Early flowering plant evolution; Late Cretaceous and Early Tertiary paleobotanical and stratigraphic studies of the Rocky Mountains and Arctic; application of leaf architecture to angiosperm phylogeny and systematics. *Mailing Add:* Peabody Mus Natural Hist Yale Univ P O Box 6666 New Haven CT 06511

HICKEY, RICHARD JAMES, b Rock Island, Ill, Sept 18, 13; m 41; c 4. BIOCHEMISTRY, HUMAN ECOLOGY. *Educ:* Univ Ill, BS, 35; Iowa State Univ, PhD(biophys chem), 41. *Prof Exp:* Asst chem, Univ Buffalo, 35-36 & Iowa State Univ, 36-41; res microbiol chemist, Com Solvents Corp, 41-53; sr res biochemist, Inst Coop Res, Univ Pa, 53-58; res & develop dept, Com Solvents Corp, 58-63; sr res investr, Inst Environ Studies, Wharton Sch, Univ Pa, 68-71, sr res investr, Mgt & Behav Sci Ctr, 71-74, sr res investr, dept statist, 75-86. *Mem:* AAAS; Am Chem Soc; NY Acad Sci. *Res:* Air pollution and health; theoretical biology; evolutionary biology; psychobiology; complex biological systems analysis; genetics; ethology; population biology; environmental mutagens and mutagenesis; biological senescence; etiologies of chronic diseases; cancer; cardiovascular diseases; birth defects; statistics; radiation biology; radiation hormesis; human epidemiology; scientific malpractices. *Mailing Add:* 43 E Clearfield Rd Upper Darby PA 19083-1401

HICKEY, ROBERT CORNELIUS, b Hallstead, Pa, Dec 9, 17; m 42; c 5. MEDICINE. *Educ:* Cornell Univ, BS, 38, MD, 42; Am Bd Surg, dipl. *Prof Exp:* Asst, Med Col, Cornell Univ, 40-41; intern, Univ Hosps, Univ Iowa, 42, asst resident surg, Col Med, 46-47; asst resident, Mem Hosp Cancer & Allied Dis, NY, 47-48; assoc surg, Col Med, Univ Iowa, 51, clin asst prof, 51-53, from assoc prof to prof, 51-62, asst & assoc dean med res, 55-62; assoc dir res, Univ Tex M D Anderson Hosp & Tumor Inst, Houston, 62-63; prof surg & chmn dept, Sch Med, Univ Wis, 63-68; prof surg & dep dir, M D Anderson Hosp & Tumor Inst, Univ Tex, Houston, 68-69, dir, 69-80, exec vpres, 69-84, M G & Lillie Johnson chmn, 84-88, PROF SURG, M D ANDERSON HOSP & TUMOR INST, UNIV TEX, HOUSTON, 84- *Concurrent Pos:* Nat Cancer Inst trainee, Univ Iowa, 48-51; consult, Surgeon Gen, USPHS, 59-68; mem, Nat Cancer Adv Bd, 80-86. *Mem:* AAAS; Am Radium Soc (vpres, 64-65); Am Col Surg; AMA; James Ewing Soc (vpres, 64-65); Sigma Xi; Am Surg Asn; Am Asn Endocrine Surgeons (vpres, 87-88). *Res:* Surgical physiology; cancer and allied diseases. *Mailing Add:* 435 Tallowood Dr Houston TX 77024

HICKEY, ROBERT JOSEPH, b Jamaica, NY; m 81. DNA REPLICATION, MOLECULAR CLONING. *Educ:* Queens Col, BA, 72; City Univ New York, PhM, 79, PhD(biochem), 79. *Prof Exp:* Fel molecular biol, Albert Einstein Col Med, 80-85; res assoc molecular biol, Worcester Found, 85-86, sr res assoc biochem, 86-89; ASST PROF PHARMACOL, SCH MED, UNIV MD, 89-, ADJ ASST PROF MOLECULAR & CELL BIOL, 89-, ONCOL, 90- & TOXICOL, 91- *Concurrent Pos:* Res initiation award, Md Cancer Prog/Am Cancer Soc, 89-90; Bressler res award, Univ Md, 91-92. *Mem:* AAAS; Am Asn Cancer Res. *Res:* Activity, function and assembly of a human cell multiprotein DNA replication complex in normal and neoplastic cells; molecular cloning and biochemical analysis of transporters for neurotransmitters; effects of heavy metals and organic toxins on DNA replication in human cells. *Mailing Add:* Dept Pharmacol Sch Med Univ Md 655 W Baltimore Baltimore MD 21201

HICKEY, ROGER, b Troy, NY, June 8, 42; m 64; c 2. EXPERIMENTAL PHYSICS. *Educ:* Siena Col, NY, BS, 64; Clarkson Col Technol, MS, 67, PhD(physics), 70. *Prof Exp:* From asst prof to assoc prof, 69-82, chmn sci div, 77-79, 84-86, chmn dept, 75-91, PROF PHYSICS, HARTWICK COL, 82- *Concurrent Pos:* Col Ctr Finger Lakes res grant, 71. *Mem:* Am Phys Soc; Am Asn Physics Teachers. *Res:* Computers in physics education, chaos. *Mailing Add:* Dept Physics Hartwick Col Oneonta NY 13820

HICKEY, WILLIAM AUGUST, b Stroudsburg, Pa, June 24, 36; m 57; c 4. ENTOMOLOGY, GENETICS. *Educ:* King's Col, BS, 57; Univ Notre Dame, MS, 59, PhD(biol), 65. *Prof Exp:* Asst biol, Univ Notre Dame, 57-59; from instr to prof biol & chmn dept, St Mary's Col, 59-72, vpres acad affairs, 72-74, actg pres, 74-75 & 85-86, vpres & dean fac, 75-85, PRES, ST MARY'S COL, IND, 86- *Concurrent Pos:* Res assoc, Mosquito Genetics Proj, Univ Notre Dame, 59-62; Entom Soc Am travel grant, Int Cong Entom, London, 64. *Honors & Awards:* President's Medal, Saint Mary's Col, 85. *Mem:* AAAS; Am Asn Higher Educ; Entom Soc Am; Genetics Soc Am; Am Inst Biol Sci; Sigma Xi; Am Coun Educ; Asn Am Cols. *Res:* Culicidae; transmission genetics, population genetics and genetics of sex ratios in Aedes aegypti. *Mailing Add:* 51263 Lake Pointe Ct Granger IN 46530

HICKIE, ROBERT ALLAN, b Melville, Sask, Aug 29, 36; m 59; c 4. PHARMACOLOGY. *Educ:* Univ Sask, BSc, 58, MSc, 60; Univ Toronto, PhD(pharmacol), 65. *Prof Exp:* Lectr pharmacol, Univ Toronto, 65-66; from asst prof to assoc prof, 66-76, PROF PHARMACOL, FAC MED, UNIV SASK, 76- *Concurrent Pos:* Nat Cancer Inst fel, Univ Toronto & Univ Sask, 65-69; Med Res Coun Can fel, Univ Sask, 69- *Mem:* Am Asn Cancer Res; Pharmacol Soc Can; AAAS; NY Acad Sci. *Res:* Cell membranes; cyclic nucleotides; cancer chemotherapy. *Mailing Add:* Dept Pharmacol Univ Sask Health Sci Bldg Saskatoon SK S7N 0W0 Can

HICKLING, ROBERT, acoustics, combustion, for more information see previous edition

HICKMAN, CAROLE STENTZ, b LaSalle, Ill, Jan 5, 42; m 64. PALEOBIOLOGY, MALACOLOGY. *Educ:* Oberlin Col, BA, 64; Univ Ore, MS, 68; Stanford Univ, PhD(geol), 75. *Prof Exp:* Adj res assoc biol, Swarthmore Col, 70-77; from asst prof to assoc prof, 77-85, PROF PALEONT, UNIV CALIF, BERKELEY, 85- *Concurrent Pos:* Consult, US Geol Surv, 73- & NSF, 81-84; vis invest, Dept Paleobiol, Smithsonian Inst, 75-76; res assoc, Dept Malacol, Acad Natural Sci Philadelphia, 76-77; NSF grants, 77- *Mem:* Paleont Soc; Paleont Res Inst; Am Malacol Union; AAAS; Soc Econ Paleontologists & Mineralogists. *Res:* Systematics, evolution and historical zoogeography of Cenozoic marine mollusks; age, origin and evolution deep-water mollusk faunas; functional morphology of gastropod radulae; systematics and ecology of deep-water limpets. *Mailing Add:* Dept Paleont Univ Calif Berkeley CA 94720

HICKMAN, CHARLES GARNER, animal genetics, animal production, for more information see previous edition

HICKMAN, CLEVELAND PENDLETON, JR, b Greencastle, Ind, Oct 29, 28; m 50; c 2. ZOOLOGY. *Educ:* DePauw Univ, AB, 50; Univ NH, MS, 53; Univ BC, PhD(zool), 58. *Prof Exp:* Fishery res, Univ Wash, Seattle, 54-55; asst zool, Univ BC, 55-58; from asst prof to assoc prof, Univ Alta, 58-67; assoc prof, 67-69, PROF ZOOL, WASHINGTON & LEE UNIV, 69- *Concurrent Pos:* Vis researcher, Duke Univ Marine Lab, 65-66, Univ Uppsala, Sweden, 72-73; vis prof, Univ Oxford, 78. *Mem:* AAAS; Am Soc Zool. *Res:* Physiological adaptation in fish; electrolyte balance; kidney function in fish with emphasis on glomerular permeability. *Mailing Add:* Dept Biol Washington & Lee Univ Lexington VA 24450

HICKMAN, EUGENE, SR, b DeRidder, La, Nov 28, 28; m 52; c 3. PHARMACEUTICS. *Educ:* Tex Southern Univ, BS, 52; Univ Tex, MS, 55; Univ Iowa, PhD(pharm), 59. *Prof Exp:* Asst, Univ Iowa, 56-59; assoc prof, 59-74, PROF PHARMACEUT, TEX SOUTHERN UNIV, 74- *Mem:* Am Pharmaceut Asn. *Res:* Synthesis and testing of organic compounds for their possible application as anti-fungal agents. *Mailing Add:* Dept Pharmaceut Tex Southern Univ 3201 Wheever St Houston TX 77004

HICKMAN, HOWARD MINOR, b Sept 6, 25; US citizen; m 52; c 3. ORGANIC CHEMISTRY. *Educ:* Southwestern Col, AB, 48; Kans State Univ, MS, 49; Univ Ill, PhD(food technol), 53. *Prof Exp:* Res chemist, Archer-Daniels-Midland Co, 53-60; res chemist, Gustin-Bacon Mfg Co, 60-62, mgr reinforced plastics res, 62-68; dir res & develop, Clark-Schwebel Fiberglass, 68-69; group leader chem spec, Ashland Chem Co, 69-79; sect mgr, Sherex Chem Co, 79-80, res admin, 81-87; PRES, HICKMAN ASSOCS, 88- *Mem:* Am Chem Soc; Sigma Xi; AAAS. *Res:* Glass fiber reinforced plastics; fatty chemicals; antioxidants; chemical specialties; fuel, lube oil additives, environmental. *Mailing Add:* 596 Whitney Ave Worthington OH 43085

HICKMAN, JACK WALTER, internal medicine; deceased, see previous edition for last biography

HICKMAN, JAMES BLAKE, b Charleston, WVa, Nov 29, 21; m 48. PHYSICAL CHEMISTRY. *Educ:* WVa Univ, BS, 42, MS, 43; Pa State Univ, PhD(chem), 50. *Prof Exp:* Asst chem, WVa Univ, 41-43; chemist, Carbide & Carbon Chem Div, Union Carbide Corp, 43-45; asst, Pa State Univ, 45-46 & 49-50; instr chem, WVa Univ, 46-49; from instr to prof, 50-82, EMER PROF CHEM, W VA UNIV, 83- *Concurrent Pos:* Consult bibliog, 83- *Mem:* AAAS; Am Chem Soc; Royal Soc Chem; Am Inst Chemists. *Res:* Measurement and interpretation of data for mixtures of non-electrolytes; digital computer analysis of data; strategy of minimum cost maximum coverage literature searches. *Mailing Add:* Dept Chem WVa Univ PO Box 924 Morgantown WV 26507-0924

HICKMAN, JAMES CHARLES, b Indianola, Iowa, Aug 27, 27; m 50; c 3. ACTUARIAL MATHEMATICS, STATISTICS. *Educ:* Simpson Col, BA, 50; Univ Iowa, MS, 52, PhD(math), 61. *Prof Exp:* Actuarial asst, Bankers Life Co, Des Moines, 52-57; from instr math to prof statist, Univ Iowa, 57-72; dean, Sch Bus, 85-90, PROF BUS-STATIST, UNIV WIS-MADISON, 72- *Concurrent Pos:* Mem, Adv Coun, Iowa Comprehensive Health Planning, 70-72; consult, Panel Consults, Financial Probs of Soc Security, Cong Res Serv, 75-76; mem, Med Malpract Comt, Wis Legis, 76. *Honors & Awards:* Halmstad Prize for actuarial res, 81 & 84. *Mem:* Fel Soc Actuaries(vpres, 75-77); assoc Casualty Actuarial Soc; Am Acad Actuaries; Am Statist Asn; Math Asn Am. *Res:* Mortality estimation and projection; pension funding models under dynamic economic and demographic conditions. *Mailing Add:* Sch Bus Univ Wis Madison WI 53706

HICKMAN, JAMES JOSEPH, b Medford, Mass, Aug 30, 42; m 64; c 2. PHOTOCHEMISTRY. *Educ:* Tufts Univ, BS, 64; Univ Notre Dame, PhD(phys org chem), 68. *Prof Exp:* Res chemist, 67-70, res chem & tech rep, 70-73, supvr customer serv, 73-76, prog mgr new prod, 76-78, eastern sales mgr, Riston Div, Photoprod Dept, 78-80, EUROP SALES MGR, RISTON DIV, E I DU PONT DE NEMOURS & CO, INC, 80- *Mem:* Am Chem Soc; Sigma Xi; Am Electroplaters Soc; Inst Printed Circuits. *Res:* New photo polymers for extrusion into film form. *Mailing Add:* 45 Concord Ave Somerville MA 02143

HICKMAN, JOHN ROY, b Wellington, Eng, Apr 19, 34; m 64; c 2. ENVIRONMENTAL HEALTH. *Educ:* Univ London, BPharm, 53; Univ Aston, ACT, 55; Univ Birmingham, MSc, 56. *Prof Exp:* Asst pharm, Queen Elizabeth Hosp, Birmingham, Eng, 53-54; asst toxicol, Univ Birmingham, 54-57; sci officer, Wantage Res Lab, Atomic Energy Auth, Harwell, 57-60, sr sci officer, 60-67; sci adv, Div Toxicol, Food Adv Bur, Food & Drug Directorate, Ottawa, Can, 67-74, chief planning & eval, Health Protection Br, 74-75, dir, Bur Chem Hazards, Environ Health Directorate, 75-79, DIR GEN, ENVIRON HEALTH DIRECTORATE, HEALTH & WELFARE, OTTAWA, CAN, 79- *Concurrent Pos:* Consult, Food & Agr Orgn, UN, Rome, 62-64; guest lectr, Food & Agr Orgn-Int Atomic Energy Agency, Sofia, Bulgaria, 68, Bangkok, Thailand & Bombay, India, 71; consult, WHO, Geneva, 68-69, Int Atomic Energy Agency, Vienna, 70 & Europ Nuclear Energy Agency, Orgn Econ Coop & Develop, 70; proj leader, Int Food Irradiation Proj, WGer, 71-74; consult, WHO, 74, Int Atomic Energy Agency, Rio de Janeiro, 75 & WHO, Copenhagen, 78; temp adv, UN Environ Prog, 79- *Res:* Health effects of environmental contaminants; industrial health; product safety. *Mailing Add:* Environ Health Directorate Health Protection Br Ottawa ON K1A 0L2 Can

HICKMAN, MICHAEL, b Evesham, Eng, May 10, 43; m 68; c 2. PHYCOLOGY, PALEOLIMNOLOGY. *Educ:* Univ Western Ont, BSc, 66; Bristol Univ, PhD(bot), 70. *Prof Exp:* Res asst bot, Bristol Univ, 66-70; asst prof, 70-75, assoc prof, 75-81, PROF BOT, UNIV ALTA, 81- *Mem:* Can Bot Asn; Brit Phycol Soc; Brit Freshwater Biol Asn; Brit Ecol Soc; Phycol Soc Am. *Res:* Palaeolimnology, effects of holoceneclimate change upon Western Canadian lakes, particularly those of central Alberta and alpine and subalpine lakes of British Columbia; interpretations and reconstructions based upon pollen, diatom, chrysophyte stomato cyst, pigment and inoreanic geochemical analyses. *Mailing Add:* Dept Bot Univ Alta Edmonton AB T6G 2E9 Can

HICKMAN, ROY D, b Kress, Tex, Sept 12, 32; m 55; c 2. STATISTICS. *Educ:* Tex A&M Univ, BS, 54, MEd, 60; Iowa State Univ, PhD(educ), 67. *Prof Exp:* Foreign area training adv, Tex A&M Univ, 58-62, asst registr, 62-63; instr educ, 64-66, assoc statist, 66-67, from asst prof to assoc prof, 67-76, PROF STATIST, IOWA STATE UNIV, 76- *Mem:* Am Statist Asn; Am Agr Econ Asn; Biomet Soc. *Res:* Research design and analysis of research projects in the social sciences; operational aspects of surveys; statistical methods. *Mailing Add:* Dept Statist Iowa State Univ 102C Snedecor Hall Ames IA 50011

HICKMAN, ROY SCOTT, b Ponca City, Okla, Apr 13, 34; m 55; c 4. AEROSPACE & MECHANICAL ENGINEERING. *Educ:* Univ Calif, BS, 57, PhD(mech eng), 62. *Prof Exp:* Mech engr, Lawrence Radiation Lab, Univ Calif, 57-60; sr res engr, jet propulsion lab, Calif Inst Technol, 61-62; sr res scientist, Heliodyne Corp, 62-64; from asst prof to assoc prof aerospace eng, Univ Southern Calif, 64-68; assoc prof mech eng, 68-77, PROF MECH & ENVIRON ENG, UNIV CALIF, SANTA BARBARA, 77- *Concurrent Pos:* Lectr, Univ Calif, Los Angeles, 62-63; consult, aerophys lab, Douglas Aircraft Co, 65- *Mem:* Nat Soc Prof Engrs; Am Soc Mech Engrs; Nat Acad Forensic Engrs. *Res:* Rarefied gas dynamics; heat transfer; spacecraft temperature control; cryogenics; Raman spectroscopy; electron beam; shock tubes. *Mailing Add:* Dept Mech Eng Univ Calif Santa Barbara CA 93106

HICKMAN, WARREN DAVID, b Pittsburgh, Pa, Aug 20, 41; m 65; c 3. MATHEMATICS EDUCATION. *Educ:* Capital Univ, Columbus, BS, 63; Case Western Univ, Cleveland, PhD(math), 74. *Prof Exp:* PROF MATH & CHAIR DEPT, WESTMINSTER COL, 68- *Mem:* Am Math Soc; Math Asn Am; Nat Coun Teachers Math. *Res:* Mathematics education; why past mathematics reforms have not succeeded; geometry; using the computer in teaching calculus. *Mailing Add:* Dept Math Westminster Col New Wilmington PA 16172

HICKMOTT, THOMAS WARD, b Kalamazoo, Mich, May 1, 29; m 59; c 4. PHYSICAL CHEMISTRY. *Educ:* Yale Univ, BS, 50; Northwestern Univ, PhD(chem physics), 54. *Prof Exp:* Phys chemist, Res & Develop Ctr, Gen Elec Co, 54-66; phys chemist, IBM Components Div, 66-70, PHYS CHEMIST, THOMAS J WATSON RES CTR, IBM CORP, 70- *Mem:* Am Phys Soc; AAAS; Electrochem Soc; Am Vacuum Soc. *Res:* Surface chemistry; gas-metal interactions; semiconductor surfaces; conduction in oxide films; properties of sputtered insulator films. *Mailing Add:* IBM Corp Thomas J Watson Res Ctr PO Box 218 Yorktown Heights NY 10598

HICKNER, RICHARD ALLAN, b Baudette, Minn, June 29, 32; m 56; c 7. ORGANIC POLYMER CHEMISTRY. *Educ:* St John's Univ, Minn, BA, 54; Purdue Univ, MS, 56, PhD(org chem), 58. *Prof Exp:* From res chemist to sr res chemist, 58-67, group leader designed prod dept, 67-72, develop specialist, 72-81, RES ASSOC, DOW CHEM CO USA, 81- *Mem:* Am Chem Soc. *Res:* Heterocyclic vinyl compounds and their reactions; reactions of urea with polyamines and alkanolamines; urethane coatings; thermoset adhesives; organosulfur chemistry; radiation cure of coatings, printing inks and adhesives. *Mailing Add:* 119 Bougainvillea Lake Jackson TX 77566

HICKOK, LESLIE GEORGE, b Schenectady, NY, July 15, 46; m 67; c 1. MUTATION SELECTION, CROP IMPROVEMENT. *Educ:* Murray State Univ, BA, 69; Ohio Univ, MS, 71; Univ Mass, PhD(bot), 75. *Prof Exp:* Asst prof bot, Miss State Univ, 74-78; from asst prof to assoc prof, 79-86, PROF BOT, UNIV TENN, 86- *Mem:* Bot Soc Am; AAAS; Sigma Xi; Am Fern Soc. *Res:* Genetics and cytogenetics; mutation induction and selection; genetic improvement of crops. *Mailing Add:* Dept Bot Univ Tenn Knoxville TN 37916

HICKOK, ROBERT BAKER, b Ypsilanti, Mich, Apr 18, 10; m 32; c 2. HYDROLOGY & WATER RESOURCES. *Educ:* Mich State Univ, BS, 32. *Prof Exp:* Jr engr soil erosion control, civilian conserv corps, State Dept Agr, Wis, 33-34; asst engr, soil conserv serv, USDA, 34-35, assoc res engr, 35-37, supvr, soil & water res, coop with Purdue Univ, 37-52, watershed hydrol res, southwestern region, 52-56; proj engr, Holmes & Narver, Inc, 56-60; air res & develop command, Ballistic Missiles Div, US Dept Air Force, 60-61; dir watershed eng res, southwest br, soil & water conserv res div, Agr Res Serv, USDA, 62-68; CONSULT LAND & WATER RESOURCES DEVELOP & MGT, 68- *Concurrent Pos:* US Comt, Int Comm Irrig, Drainage & Flood Control; mem, Southwest Inter-Agency Comn River Basin Planning. *Mem:* AAAS; Am Soc Agr Engrs; Am Soc Civil Engrs; Soil & Water Conserv Soc; Sigma Xi. *Res:* Soil and water resources; development planning research; facilities site selection; management. *Mailing Add:* 8290 Coleman St Riverside CA 92504

HICKOK, ROBERT L, JR, b Schenectady, NY, Feb 25, 29; m 49; c 3. PLASMA DIAGNOSTICS. *Educ:* Rensselaer Polytech Inst, BS, 51, PhD(physics), 56; Dartmouth Col, MA, 53. *Prof Exp:* Fel, Yale Univ, 56-58; res assoc, Mobil Res & Develop Corp, 58-71; FROM ASSOC PROF TO PROF ELECTROPHYS, RENSSELAER POLYTECH INST, 71- *Honors & Awards:* Fel, Am Phys Soc; Fel, Inst Elec & Electronics Engrs. *Mem:* Am Phys Soc; Inst Elec & Electronics Engrs; Nuclear & Plasma Sci Soc; Sigma Xi. *Res:* Plasma and nuclear physics; development of controlled thermonuclear fusion energy source. *Mailing Add:* ECSE Dept Rensselaer Polytech Inst Troy NY 12180-3590

HICKOK, ROBERT LEE, b Hillsdale, Mich, June 23, 29; m 61; c 3. PHYSICAL CHEMISTRY, MATERIALS SCIENCE. *Educ:* Antioch Col, BS, 52; Univ Colo, PhD(chem), 57. *Prof Exp:* Res chemist, Atomic Energy Div, Phillips Petrol Co, 56-60, res phys chemist, Lamp Div, 60-69, MGR LAMP MAT RES LAB, LAMP BUS DIV, GEN ELEC CO, 69- *Mem:* AAAS; Sigma Xi; Am Mgt Asn. *Mailing Add:* 1360 Lamp MTLS Res Gen Elec Co Neid Park Cleveland OH 44112

HICKS, ARTHUR M, b Atlanta, Ga, Nov 29, 17; m 43; c 4. ORGANIC CHEMISTRY. *Educ:* Emory Univ, AB, 40, MS, 41; Auburn Univ, PhD(org chem), 65. *Prof Exp:* Asst, Emory Univ, 39-41 & Rutgers Univ, 41-43; chemist, Johnson & Johnson, 43-45 & Graton & Knight Co, 45-50; from assoc prof to prof chem, 50-86, chmn, Div Sci & Math, 66-86, EMER PROF CHEM, LA GRANGE COL, 86- *Mem:* Am Chem Soc. *Res:* Organic synthesis. *Mailing Add:* Dept Math & Sci La Grange Col La Grange GA 30240

HICKS, BRUCE BOUNDY, b Melbourne, Australia, June 24, 40; m 66; c 2. ATMOSPHERIC PHYSICS. *Educ:* Univ Tasmania, BS, 62; Melbourne Univ, MS, 68. *Prof Exp:* Exp officer atmospheric physics, Div Meteorol Physics, Commonwealth Sci & Indust Res Orgn, 62-68, from res scientist to sr res scientist, 68-73; meteorologist, Argonne Nat Lab, 73-76, sect head atmospheric physics, Radiol & Environ Res Div, 76-81; dir, Atmospheric Turbulence & Diffusion Div, Oak Ridge, Tenn, 81-90, DIR AIR RESOURCES LAB, NAT OCEANIC & ATMOSPHERIC ADMIN, SILVER SPRING, MD, 90- *Concurrent Pos:* Vis res assoc, Biol & Med Res Div, 65-66 & Radiol Physics Div, Argonne Nat Lab, 70-71; chmn, Task Group Atmospheric Chem, Nat Acid Precipitation Assessment Prog, 85- *Honors & Awards:* David Syme Res Award, Univ Melbourne, 72. *Mem:* Royal Meteorol Soc; Am Meteorol Soc; Am Geophys Union. *Res:* Micrometeorology; air-surface exchange; planetary boundary layer studies. *Mailing Add:* Admin/Air Resources Lab Nat Oceanic & Atmospheric Admin 20910 East West Hwy Silver Spring MD 20910

HICKS, BRUCE LATHAN, computer graphics, computer-based instruction, for more information see previous edition

HICKS, CHARLES ROBERT, b Syracuse, NY, Apr 7, 20; m 42; c 3. APPLIED STATISTICS. *Educ:* Syracuse Univ, AB, 42, MS, 44, PhD(educ), 53. *Prof Exp:* Asst physics, Syracuse Univ, 42-44; qual control engr, Eastman Kodak, 44-46; instr math & educ, Syracuse Univ, 46-53; asst prof math, 53-57, assoc prof math & statist, 57-60, asst dean, Sch Sci Educ & Humanities, 60-64, head educ dept, 64-74, dir teacher educ, 70-74, PROF EDUC, PURDUE UNIV, WEST LAFAYETTE, 70-, PROF STATIST, 74- *Honors & Awards:* Brumbaugh Award, Am Soc Qual Control, 57. *Mem:* Fel Am Soc Qual Control (vpres, 69-71); Am Statist Asn; Comp & Int Educ Soc; Biomet Soc. *Res:* Quality control. *Mailing Add:* 1016 S 22nd West Lafayette IN 47907

HICKS, DALE R, b Odin, Ill, Oct 10, 38; m 60; c 3. PLANT BREEDING, STATISTICS. *Educ:* Univ Ill, BS, 60, MS, 66, PhD(agron), 68. *Prof Exp:* Asst farm adv, Univ Ill, 64-65, teaching asst statist, 66-68; from asst prof to assoc prof, 68-76, PROF AGRON, UNIV MINN, ST PAUL, 76- EXTEN AGRONOMIST, 68- *Mem:* Fel Am Soc Agron; Crop Sci Soc Am. *Res:* Growth regulator effects on yield; genotypes and yield of both corn and soybeans; production practices and maximum yield. *Mailing Add:* Dept Agron Univ Minn St Paul MN 55108

HICKS, DARRELL LEE, b Clovis, NMex, July 3, 37; m 79, 85; c 3. APPLIED MATHEMATICS, COMPUTER SCIENCE. *Educ:* Univ NMex, BS, 61, PhD(math), 69. *Prof Exp:* Mathematician, Air Force Weapons Labs, 62-69; mathematician, Sandia Labs, 69-81; prof math, Univ Colo, Denver, 81-83; dir, Ctr Exp Computation, 87-88, PROF MATH SCI, MICH TECH UNIV, 83- *Concurrent Pos:* consult, Idaho Nat Eng Labs, 82-87, KMS Fusion Inc, Ann Arbor, 83-88; vis scholar, NSF, Res Parallel Computation, Rice Univ, Houston, Tex, 89-90. *Mem:* Math Asn Am; Am Math Soc; Soc Indust & Appl Math; Am Phys Soc; AAAS; Sigma Xi; Am Acad Mech; Am Nuclear Soc; Asn Comput Mach; Inst Elec & Electronics Engrs Comput Soc; Int Assoc Math & Comput Modelling; Int Assoc Math & Comput Simulation; Int Soc Sci & Tech Develop; Soc Comput Simulation. *Res:* Analysis of the equations of material dynamics; applied mathematical and computer sciences; mathematical modelling. *Mailing Add:* Dept Math Mich Tech Univ Houghton MI 49931

HICKS, DAVID L, b Alliance, Ohio, July 20, 35; m 56; c 3. PHYSIOLOGICAL ECOLOGY, ORNITHOLOGY. *Educ:* Cascade Col, BA, 56; Univ Ga, MS, 65, PhD(zool), 67. *Prof Exp:* PROF BIOL, WHITWORTH COL, WASH, 67- *Mem:* Ecol Soc Am; Am Ornithologists Union; Cooper Ornith Soc. *Res:* Physiological effects and ecological significance of depletion and utilization of adipose tissue in migrating passerine birds. *Mailing Add:* Dept Biol Whitworth Col Spokane WA 99251-0002

HICKS, DONALD GAIL, b Quanah, Tex, Feb 11, 34; m 55; c 2. ANALYTICAL CHEMISTRY, INORGANIC CHEMISTRY. *Educ:* Murray State Univ, BS, 55; Univ Ky, MS, 58; Univ Tenn, PhD(anal chem), 65. *Prof Exp:* Asst prof chem, The Citadel, 58-59 & Murray State Univ, 59-65; ASSOC PROF CHEM, GA STATE UNIV, 65- *Mem:* AAAS; Am Chem Soc; fel Am Inst Chemists; Nat Sci Teachers Asn; Soc Appl Spectros. *Res:* Atomic absorption and emission flame spectrophotometry; trisubstituted group V element coordination compounds; solvent extraction; cyclic group V element ligands; mechanism of decomposition in flames. *Mailing Add:* Ga State Univ Univ Plaza Box 240 Atlanta GA 30303-3044

HICKS, EDWARD JAMES, b Gainesville, Fla, Oct 18, 36; m 67. IMMUNOLOGY, PREVENTIVE MEDICINE. *Educ:* Clark Col, BS, 59; Univ Iowa, MS, 66, PhD, 68. *Prof Exp:* Clin chemist, Univ Fla, 59-60; microbiologist, Dept Prev Med, Univ Iowa, 62-65, clin immunologist, Dept Path, 65-69; ASSOC PROF PATH, MED CTR, IND UNIV, INDIANAPOLIS, 69- *Concurrent Pos:* Co-investr grant sickle cell dis, Nat Heart & Lung Inst, 72-77; co-investr grant ion-affinity electrophoresis, Nat Inst Arthritis, Metab & Digestive Dis, 74-76. *Honors & Awards:* Cert, Nat Registry Clin Chem. *Mem:* Am Soc Clin Path. *Res:* Myelomatosis; macrogammaglobulinemia; hemoglobinopathies detection methods; developmental work using radioimmunoassay and other immunochemical methods for identification of hormones, peptides, steroid and proteins and immunochemical studies on penicillin antibody activity. *Mailing Add:* Dept Clin Path FH-416 Ind Univ Med Ctr 1120 South Dr Indianapolis IN 46223

HICKS, ELIJA MAXIE, JR, b Florence, SC, Aug 26, 20; m 44; c 2. ORGANIC POLYMER CHEMISTRY. *Educ:* Furman Univ, BS, 41; Princeton Univ, MS, 43, PhD(chem), 44. *Hon Degrees:* LLD, Furman Univ, 77. *Prof Exp:* Res chemist, E I du Pont De Nemours & Co, Inc, Va, 44-47, res supvr, 47-50, tech supt, NY, 50-53, mfg supt, Va, 53-57, tech serv mgr, 57-59, orlon tech dir, 59-72, dir, Nylon Mfg Div, Textile Fibers Dept, 72-73, dir, Int Develop & Opers Div, 73-76, dir manufacture, Textile Fibers Dept, 76-81; RETIRED. *Concurrent Pos:* Consult, 81-90. *Mem:* Fel AAAS; Am Chem Soc; NY Acad Sci; fel Textile Inst London. *Res:* Steroid synthesis; cellulose chemistry; synthetic fiber technology. *Mailing Add:* 4777 Ringwood Meadow The Meadows Sarasota FL 34235

HICKS, ELLIS ARDEN, b Lamoni, Iowa, Feb 12, 15; m 46; c 1. ZOOLOGY. *Educ:* Iowa State Col, BS, 38, MS, 40, PhD(econ zool), 47. *Prof Exp:* Exten asst wildlife conserv, 38-41, exten assoc, 46, from instr to prof zool & entom, 47-76, PROF ZOOL, IOWA STATE UNIV, 76- *Res:* Arthropod fauna of bird nests; insect-mite symbioses, acarology. *Mailing Add:* 3012 Woodland St Ames IA 50010

HICKS, GARLAND FISHER, JR, b Southampton, NY, Nov 19, 45; m 67. MICROBIOLOGY. *Educ:* St Lawrence Univ, BS, 67; Mich State Univ, PhD(microbiol), 75. *Prof Exp:* ASSOC PROF BIOL, VALPARAISO UNIV, 74- *Mem:* Am Soc Microbiol. *Res:* Chestnut canker disease; natural areas of west Tennessee. *Mailing Add:* Dept Biol Valparaiso Univ Valparaiso IN 46383

HICKS, HAROLD E(UGENE), b Minneapolis, Minn, Jan 20, 19; m 41, 90; c 4. CHEMICAL ENGINEERING. *Educ:* Univ Minn, BChE, 41. *Prof Exp:* Res chemist, Hercules Powder Co, 41-54, oper supvr, Hercules, Inc, 54-57, safety supvr, 57-62, resin supvr, 62-64, plant supt, 64-66, plant mgr, 66-81; RETIRED. *Concurrent Pos:* Tech adv, Dawood Hercules Chem Ltd, Pakistan, 76-78. *Mem:* Am Chem Soc; Am Inst Chem Engrs. *Res:* Ammoniation; chlorination; hydrogenation; polymerization; hydrogen manufacture; industrial health and safety; labor relations; plant management. *Mailing Add:* 133 Shore Rush Dr St Simons Island GA 31522

HICKS, HARRY GROSS, b Reno, Nev, Dec 13, 22; m 48, 69; c 4. NUCLEAR CHEMISTRY. *Educ:* Univ Calif, BS, 43, PhD(chem), 49. *Prof Exp:* Asst, Univ Calif, 46-47, asst, Radiation Lab, 47-49; res chemist, Hanford Works, Gen Elec Co, 49-51 & Calif Res & Develop Corp, 51-53; res chemist, 53-63, sect leader, Radiochem Div, 63-69, scientist, Spec Projs Div, 69-72, chemist, nuclear chem div, 72-84, CONSULT, LAWRENCE LIVERMORE NAT LAB, UNIV CALIF, 84-; CONSULT, UNIV UTAH, 87- *Mem:* AAAS. *Res:* Radiochemical separations, principally fission products; nuclear fission; nuclear device debris diagnostics; very rapid automated radiochemical separations of fission products; fractionation of nuclear detonation debris. *Mailing Add:* 727 Canterbury Ave Livermore CA 94550

HICKS, HERALINE ELAINE, b Beaufort, SC, Sept 27, 51. CRANIOFACIAL BIOLOGY, TERATOLOGY. *Educ:* Ohio Wesleyan Univ, BA, 73; Atlanta Univ, MS, 78, PhD(develop biol), 80. *Prof Exp:* Teacher & counr biol sci, Upward Bound Prog, Ohio Wesleyan Univ, 73; teacher, Ohio Bd Educ, Dayton, 73-76; teaching asst, Atlanta Univ, 76-77; instr, Morris Brown Col, Atlanta, 78-80; res assoc dent, 80-81, NIH teaching fel, 81-84, PROF DENT, UNIV NC, CHAPEL HILL, 85- *Concurrent Pos:* Vis prof, Naval Med Res Inst, Bethesda, Md, 85- *Mem:* Sigma Xi; NY Acad Sci; Teratology Soc; Am Soc Cell Biol; Am Soc Dent Res. *Res:* Effects of various anti-convulsant drugs on craniofacial development; effects of hormones on bone in vitro. *Mailing Add:* Dept Biol Morris Brown Col 645 Martin Luther King Jr Dr Atlanta GA 30314

HICKS, JACKSON EARL, b Fruitland Park, Fla, Dec 12, 34; m 58; c 1. WASTEWATER TREATMENT, INCINERATION. *Educ:* Erskine Col, BA, 58; Univ Va, MS, 60; Ga Inst Technol, PhD(anal chem), 67. *Prof Exp:* Chemist, WVa Pulp & Paper Co, Va, 60-63; chemist, 67-69, sr chemist, Acid Div, Anal Develop Lab, 69-70, sr chemist & dir serv anal lab, 70-76, DEPT SUPT, WATER & WASTE TREATMENT DEPT, TENN EASTMAN CO, 76- *Mem:* Am Soc Testing & Mat; Am Chem Soc; Sigma Xi. *Res:* Application of organic complexing reagents to analytical chemistry, especially photometric titrations, atomic absorption spectrophotometry, gas chromatography and spectrophotometric methods; analytical application of infrared spectroscopy; analytical aspects of air and water pollution. *Mailing Add:* 4529 Chickasaw Rd Kingsport TN 37664

HICKS, JAMES THOMAS, b Brownsville, Pa, June 5, 24; m 50; c 2. FORENSIC MEDICINE. *Educ:* Univ Pittsburgh, BS, 45, AB & MS, 46; George Washington Univ, PhD(physiol), 50; Univ Ark, MD, 56; DePaul Univ, JD, 75. *Prof Exp:* Asst physiol, Univ Pittsburgh, 43-46; lectr biol, Harcum Jr Col, 47; asst prof anat, St Lawrence Univ, 47-48; instr comp anat, Montgomery Jr Col, 48-49; assoc physiol, Sch Med, George Washington Univ, 49-50; instr anat, Med Col Va, 50; asst prof path, Sch Med, WVa Univ, 51-53; asst prof path & oncol, Sch Med, Univ Ark, 53-56; mem staff, USPHS, 56-57; physician, Health Serv, WVa, 57-58; pathologist, Sch Med, Univ Pittsburgh, 58-60; pathologist, Charleroi Monessen Hosp, 60-64; CLIN PATHOLOGIST & DIR LABS, OAK PARK HOSP, 65- *Concurrent Pos:* Pathologist, Vet Admin Hosp, 58- & Brownsville, Pa Hosp, 60-64. *Mem:* AAAS; Soc Exp Biol & Med; AMA; fel Col Am Path; fel Am Soc Clin Path. *Res:* Effect of blood pressure on muscle tone; morphologic changes in adrenal nodules. *Mailing Add:* 7980 W Chicago Ave River Forest IL 60305

HICKS, JIMMIE LEE, b 1936; m; c 2. ADMINISTRATION. *Educ:* Colo State Univ, PhD(physiol-biophys), 71. *Prof Exp:* From asst prof to assoc prof physiol, Univ Health Sci, Col Osteop Med, Kansas City, 71-83; ASST DEAN CLIN ROTATIONS & PROF PHYSIOL, COL OSTEOP MED, PAC, 83- *Concurrent Pos:* Instr Biol, Col St Thomas, St Paul, Minn, 63-67. *Mem:* Am Physiol Soc. *Mailing Add:* Col Osteop Med Pac Col Plaza Pomona CA 91766-1889

HICKS, JOHN T, ARTHRITIS. *Prof Exp:* PVT PRACT MED DOCTOR, 84- *Mailing Add:* 2465 Army-Navy Dr Arlington VA 22206

HICKS, JOHN W, III, b Sydney, Australia, Dec 2, 21; US citizen; m 47; c 8. AGRICULTURAL ECONOMICS. *Educ:* Mass State Col, BS, 46; Purdue Univ, MS, 48, PhD(agr econ), 50. *Hon Degrees:* DEd, Vincennes Univ, 84. *Prof Exp:* Instr agr econ, Univ Mass, 46-47; from asst prof to assoc prof, 50-60, exec asst to pres, 55-82, actg pres, 82-83, sr pres, 83-87, PROF AGR ECON, PURDUE UNIV, WEST LAFAYETTE, 60- *Concurrent Pos:* Dir, Ind Comn State Tax & Financing Policy, 53-55; dir, Comt Inst Coop, Coun of 10 & Univ Chicago, 60-62; chmn, Ind Post High Sch Study Comn, 61-63; secy, Bd Trustees, Purdue Univ, Lafayette, 72-73; actg comnr voc & tech educ, State of Ind, 87. *Mem:* Am Agr Econ Asn; Sigma Xi. *Res:* Political economy; higher education. *Mailing Add:* Exec Bldg Purdue Univ West Lafayette IN 47907

HICKS, KENNETH WARD, b Cleveland, Ohio, Sept 5, 40; m 71; c 2. PHYSICAL INORGANIC CHEMISTRY. *Educ:* Miami Univ, BA, 62, MS, 64; Howard Univ, PhD(phys chem), 69. *Prof Exp:* Res assoc chem, Univ Calif, San Diego, 68-70; asst prof, 70-76, assoc prof chem, 76-79, prof chem, Eastern Mich Univ, 79-86; PROF CHEM, NC A&T STATE UNIV, 86- *Concurrent Pos:* Consult, NIH, NIGMS, MARC; vis prof, Univ Utah, 77-78, Battelle Nat Lab, 89. *Mem:* Am Chem Soc; Nat Asn Black Chemists & Chem Engrs; Sigma Xi. *Res:* Oxidation of dimeric Mo(V) complexes by MnO4, VO2 and Ce(IV); prep and kinetics of bioinorganic DiMeric Mo(V) model complexes. *Mailing Add:* Dept Chem NC A&T State Univ Greensboro NC 27411

HICKS, KEVIN B, b Ironton, Mo, Apr 3, 52; m 78. CARBOHYDRATE CHEMISTRY, CARBOHYDRATE BIOCHEMISTRY. *Educ:* Univ Mo, BS, 74, MS, 76, PhD(biochem), 79. *Prof Exp:* Assoc res fel, Dairy Res Found, 79-80; res chemist, 80-85, LEAD SCIENTIST, PLANT SCI RES, EASTERN REGIONAL RES CTR, USDA, 85- *Concurrent Pos:* Mem exec comt, Div Carbohydrate Chem, Am Chem Soc, 83, exec secy, 85- *Mem:* Am Chem Soc (secy, 84-85); AAAS. *Res:* Chemistry and biochemistry of carbohydrates; conversion of known carbohydrates into new and potentially useful forms; new analytical methods especially helpful for evaluation, separation and identification of sugars and related bio-molecules; preparative of biomolecules. *Mailing Add:* Eastern Regional Res Ctr US Dept Agr 600 E Mermaid Lane Philadelphia PA 19118-2551

HICKS, PATRICIA FAIN, b Brownwood, Tex, Dec 16, 27; m 66. BIOLOGICAL CHEMISTRY, COMMUNICATION SCIENCE. *Educ:* Univ Tex, BA, 47; Tex Tech Col, MS, 51, PhD(biol chem), 53. *Prof Exp:* Anal chemist, Mallinckrodt, Inc, 48-50; mem staff mass spectrometry, Los Alamos Sci Lab, 53-55; asst prof chem, Tex Tech Col, 55-57; tech ed, 57-70, sr res assoc, 70-71, ASST TO DIR PUB RELS, MALLINCKRODT, INC, 71- *Concurrent Pos:* Robert Welch Found grant, Houston, Tex, 56-57. *Mem:* Am Chem Soc; Sigma Xi. *Res:* Fine chemicals; technical writing. *Mailing Add:* 12930 Ferntop Lane St Louis MO 63141

HICKS, ROBERT EUGENE, b McAllen, Tex, Oct 9, 43; m 65; c 2. NEUROPSYCHOLOGY, PSYCHOPHARMACOLOGY. *Educ:* NMex Highlands Univ, BA, 68; Tulane Univ, MS, 70; Univ Tex, Austin, PhD(psychol), 73. *Prof Exp:* From asst prof to assoc prof psychol, State Univ NY, Albany, 73-78; assoc prof psychol, Univ Lethbridge, 78-81; RES SCIENTIST BIOL SCI & ASSOC PROF PSYCHIAT, SCH MED, UNIV NC, 81-, RES SCIENTIST ALCOHOL STUDIES, 85- *Concurrent Pos:* Vis scientist, neuropsychol unit, Host for Sick Children, Toronto, Ont, 75-76; vis prof, Univ Ala, 84-85. *Mem:* Psychonomic Soc; Sigma Xi; Int Neuropsychol Soc; AAAS. *Res:* Neuropsychobiological analysis of psychotropic drug effects; interactions of genotype and drug effects; neuropsychological analysis of neurological and psychiatric syndromes. *Mailing Add:* Dept Psychiat C No 7160 Med Sch Wing B Univ NC Chapel Hill NC 27599

HICKS, SAMUEL PENDLETON, b Bryn Athyn, Pa, Nov 21, 13; m 41; c 2. NEUROPATHOLOGY. *Educ:* Univ Pa, AB, 36, MD, 40. *Prof Exp:* Assoc prof path, Sch Med, Georgetown Univ, 47-48; pathologist, New Eng Deaconess Hosp, 48-62; assoc prof path, Harvard Med Sch, 52-62; PROF PATH, SCH MED, UNIV MICH, ANN ARBOR, 62- *Concurrent Pos:* Head dept path & consult neuropathologist, Nat Naval Med Ctr, 47-48; pathologist-in-chief, Peter Bent Brigham Hosp, Boston, 50-51. *Honors & Awards:* Max Weinstein Award, Cerebral Palsy, 51. *Mem:* Am Asn Neuropathologists; AMA; Am Acad Neurol; Soc Neurosci; Am Soc Human Genetics. *Res:* Environmental and genetic factors in morphologic and functional development of nervous system; neuropathology; radiobiology. *Mailing Add:* Dept Path Box 0602 Univ Mich Med Ctr Ann Arbor MI 48109-0602

HICKS, SONJA ELAINE, b Machiasport, Maine, May 25, 40. BIOCHEMISTRY. *Educ:* Univ Maine, BS, 62; Ind Univ, PhD(biochem), 66. *Prof Exp:* Fel biochem, Woods Hole Oceanog Inst, 66-67; res assoc, Mass Inst Technol, 67-68; asst prof, 68-74, dir molecular biol prog, 70-74, chmn dept, 74-77, ASSOC PROF CHEM, WELLESLEY COL, 74- *Concurrent Pos:* Asst prof pediat, Harvard Med Sch, 71-73; res scientist biochem, Mass Gen Hosp, 71-73; chmn, Premed Adv Comt, Wellesley Col, 73- *Mem:* Sigma Xi. *Res:* Studies of biochemical effects of prostaglandins on membrane structure and function; development of new biochemical clinical tests. *Mailing Add:* Dept Chem Wellesley Col Wellesley MA 02181

HICKS, STEACY DOPP, b Detroit, Mich, Apr 19, 25; m 48; c 3. PHYSICAL OCEANOGRAPHY. *Educ:* Univ Calif, Los Angeles, BA, 50; Scripps Inst Oceanog, MS, 52. *Prof Exp:* From instr to asst prof phys oceanog, Univ RI, 52-62; phys oceanogr, Nat Ocean Surv, Nat Oceanic & Atmospheric Admin, 62-65 & 71-90, chief oceanog div, 65-70, chief oceanog res group, 70-71; RETIRED. *Concurrent Pos:* Sci fel, US Dept Com, 64-65; US Naval War Col, 69-70; assoc prof lectr oceanog, George Washington Univ, 66- *Honors & Awards:* Silver Medal Res Phys Oceanog, US Dept Com. *Mem:* Am Shore & Beach Preserv Asn. *Res:* Physical oceanography; tides and sea level. *Mailing Add:* 6857 Grande Lane Falls Church VA 22043

HICKS, T PHILIP, b Fredericton, NB, Dec 30, 52; m 75; c 1. SENSORY NEUROPHARMACOLOGY, NEUROPHYSIOLOGY. *Educ:* Carleton Univ, BA, 73; Dalhousie Univ, BSc, 76; Univ BC, PhD(physiol), 79. *Prof Exp:* Fel, neurobiol, Max Planck Inst Biophys Chem, 79-81; fel, vision res, Inst Equilibrium Res, Gifu Univ, Japan, 81; from asst prof to assoc prof med physiol, Sch Med, 81-88, ASSOC PROF FAC ARTS & SCI, UNIV NC, GREENSBORO, 88- *Mem:* Can Physiol Soc; Can Fedn Biol Soc; AAAS; Int Brain Res Orgn; Soc Neurosci; Brit Physiol Soc. *Res:* Characterization by pharmacological techniques and neurochemical methods of receptors for endogenous synaptic transmitters in ascending sensory pathways, and studies of their physiological roles using electrophysiological procedures. *Mailing Add:* Univ NC 296 Everhart Bldg Greensboro NC 27412

HICKS, TROY L, b Pampa, Tex, Sept 30, 32; m 55; c 2. MATHEMATICS. *Educ:* Southwest Mo State Col, BS, 57; Univ Mo, Columbia, MEd, 60; Univ Kans, MA, 61; Univ Cincinnati, PhD(math), 65. *Prof Exp:* Teacher, Rural Sch, Mo, 49-53; asst prof math, Col Sch of the Ozarks, 57-60; instr, Univ Cincinnati, 61-65; assoc prof, Ill State Univ, 65-67; from asst prof to assoc prof, 67-78, PROF MATH, UNIV MO, ROLLA, 78- *Honors & Awards:* Cert Meritorious Serv, Math Asn Am, 88. *Mem:* Am Math Soc; Math Asn Am; Am Asn Univ Prof; Sigma Xi. *Res:* Topology and functional analysis. *Mailing Add:* Dept Math Univ Mo Rolla MO 65401

HICKS, WILLIAM B(RUCE), b Washington Court House, Ohio, May 9, 25; m 47; c 2. CHEMICAL ENGINEERING. *Educ:* Univ Ill, BS, 45. *Prof Exp:* Res chem engr, Cent Res Dept, Ohio, 46-51, proj specialist, Org Div, Develop Dept, Mo, 51-56, develop mgr, 56-65, dir, Int Org Chem, Europe, 69-71, dir growth, Int Div, 70-73, dir, Int Planning & Admin, 73-79, DIR, PROCESS CHEM & PHOSPHORUS, MONSANTO CO, 79- *Mem:* Am Chem Soc; Soc Plastics Indust; Com Develop Asn. *Res:* Polymers; plastics; surface coatings; adhesives; paper chemistry; commercial chemical development. *Mailing Add:* 114 Augusta St Simons Island GA 31522-2437

HICKSON, DONALD ANDREW, b Windsor, Ont, July 28, 31; m 53; c 3. SURFACE CHEMISTRY, PETROLEUM CHEMISTRY. *Educ:* Wayne State Univ, BS, 53; Iowa State Univ, PhD(phys chem), 58. *Prof Exp:* Chemist, Ames Lab, AEC, 52-58, res chemist catalysis, Calif Res Corp, 58-65, Chevron Res Co, 65-70, sr res assoc, Chevron Oil Field Res Co, 72-75, SR RES ASSOC, CHEVRON RES CO, 75- *Mem:* Am Chem Soc; Clays & Clay Mineral Soc. *Res:* Electrochemical phenomena at interphases; electrophillic properties and structure of mixed-oxide catalysts; infrared spectroscopy of solids and gas-solid interactions; petroleum processing catalysis; in-situ processing of carbonaceous materials and assisted recovery methods. *Mailing Add:* 189 Carlisle Way Benicia CA 94510-1615

HICKSON, JAMES FORBES, b May 10, 54. PROTEIN METABOLISM, SPORTS NUTRITION. *Educ:* Va Polytech Inst, BS, 76, PhD(human nutrit), 80. *Prof Exp:* Asst prof nutrit, Indiana Univ, Bloomington, IN, 80-82; ASSOC PROF NUTRIT, PROG NUTRIT & DIETETICS, UNIV TEX, 82- *Mem:* Am Dietetic Asn; Am Inst Nutrit; Inst Food Technologist. *Res:* Emphasizing bodybuilding exercise as it relates to the efficiency of protein utilization in humans; characterization and explaination of the dietary behavior of athletes. *Mailing Add:* Sch Allied Health Sci Univ Tex Health Sci Ctr PO Box 20708 Houston TX 77225

HICKSON, JOHN LEFEVER, b Milford, NH, June 17, 16; m 38, 71; c 3. SUGAR CHEMISTRY. *Educ:* Ohio Wesleyan Univ, BA, 37; Purdue Univ, MS, 52, PhD(biochem), 53. *Prof Exp:* Chemist, Nat Aniline Div, Allied Chem & Dye Corp, 38-46; asst prof chem, Kans Wesleyan Univ, 46-48; chemist, Ind State Chemist's Labs, 48-50; Corn Industs fel, 50-53; asst to pres, Int Sugar Res Found, Inc, 53-60, vpres & sci dir, 60-68, vpres & dir res, 68-71; sr assoc, Sidney M Cantor Assocs, Inc, 71; sci dir, Cigar Res Coun, 71-75, consult, Int Sugar Res Found, Inc, 74-77; RETIRED. *Mem:* Am Chem Soc; hon mem Am Inst Chem (pres, 66-67). *Res:* Chemical, physical, physiological and food technological properties of sugars and oligosaccharides as well as pleasure and health aspects of cigar smoking. *Mailing Add:* 1045 Linn-Hipsher Rd Marion OH 43302

HIDALGO, HENRY, b Ecuador, Sept 1, 22; US citizen; m 47; c 4. FLUID MECHANICS, AEROSPACE SCIENCE. *Educ:* Tri-State Col, BS, 49; Mass Inst Technol, MS, 51. *Prof Exp:* Res engr, United Aircraft Res Lab, United Aircraft Corp, 51-54; tech engr, Small Engine Dept, Gen Elec Co, 54-56; sr engr, Avco-Everett Res Lab, Avco Corp, 56-60, prin engr, 60-62; prin engr, Heliodyne Corp, 62-64; staff mem, Inst Defense Anal, 64-81; mgr advan technol & mission anal, Advan Systs Div, Gould Advan Technol Ctr, 81-84; MEM TECH STAFF, ROCKWELL INT CORP, 84- *Concurrent Pos:* Consult, Univ Mich & Stanford Res Inst, 62 & Atomics Int Div, NAm Rockwell Corp, 63-64. *Mem:* Assoc fel Am Inst Aeronaut & Astronaut. *Res:* Ballistic missile defense technology and systems; air to air warfare technology and systems; naval warfare technology and system; directed energy technology and advanced space systems; aeronautical and astronautical engineering. *Mailing Add:* 3750 N Woodrow Ave Arlington VA 22207-4322

HIDALGO, JOHN, b Caracas, Venezuela, Mar 11, 16; m 41; c 2. PHARMACOLOGY. *Educ:* Columbia Univ, BS, 42; Univ Colo, PhD(pharmacol), 51. *Prof Exp:* Asst toxicologist, Walter Reed Hosp, 45-46 & Mt Alto Hosp, Washington, DC, 46-48; sect head pharmaceut & pharmacol res, 51-54, sect leader pharmacol res, 54-72, DIR EXP THER, CUTTER LABS, 72- *Mem:* Am Soc Pharmacol & Exp Therapeut. *Res:* Fibrinolysis; central nervous system pharmacology. *Mailing Add:* 2929 Driftwood Dr San Jose CA 95128

HIDORE, JOHN J, b Cedar Falls, Iowa, July 6, 32; m 84; c 2. PHYSICAL GEOGRAPHY. *Educ:* Univ Northern Iowa, BA, 54; Univ Iowa, MA, 58, PhD(geog), 60. *Prof Exp:* Instr geog, Univ Wis, 60-62; assoc prof, Okla State Univ, 62-66; from assoc prof to prof geog, Ind Univ, Bloomington, 66-80, chmn dept, 68-71; dept head, 80-87, PROF GEOG, UNIV NC, GREENSBORO, 80- *Concurrent Pos:* Vis prof, Cent Wash State Col, 65, 69 & 73, Univ Ife, Nigeria, 71-72, Univ Khartoum, Sudan, 74-75 & Ben Gurion Univ Negev, 78. *Mem:* Asn Am Geog; Nat Coun Geog Educ; Am Water Resources Asn; Int Water Resources Asn; Am Geog Soc. *Res:* Hydroclimatology; climatology. *Mailing Add:* Dept Geog Univ NC Greensboro NC 27412

HIDU, HERBERT, b Bridgeport, Conn, Dec 18, 31; m 57; c 3. MARINE ECOLOGY. *Educ:* Univ Conn, BS, 58; Pa State Univ, MS, 60; Rutgers Univ, PhD(zool), 67. *Prof Exp:* Fisheries res biologist, Biol Lab, US Bur Com Fisheries, Conn, 60-63; asst res prof, Chesapeake Biol Lab, Univ Md, Solomons, 67-70; asst prof oceanog & zool, 70-74, ASSOC PROF OCEANOG, IRA C DARLING CTR, UNIV MAINE, WALPOLE, 74- *Concurrent Pos:* Proj coordr, Nat Park Serv grant, Univ Md, 69-70; prin investr, NSF grant, Univ Maine, 71-72 & 80-82, proj coordr, Sea Grant Coherent Areas Award, 71-78. *Mem:* Nat Shellfisheries Asn (pres, 80-81); Atlantic Estuarine Res Soc; New Eng Estuarine Res Soc. *Res:* Marine shellfish biology and ecology, especially the ecology and physiology of recruitment in bivalve mollusks; shellfish mariculture. *Mailing Add:* Dept Animal/Vet Sci Univ Maine Hitchner Hall Orono ME 04469

HIDY, GEORGE MARTEL, b Kingman, Ariz, Jan 5, 35; m 58; c 3. PHYSICAL CHEMISTRY, CHEMICAL ENGINEERING. *Educ:* Columbia Univ, AB, 56, BS, 57; Princeton Univ, MSE, 58; Johns Hopkins Univ, DEng, 62. *Prof Exp:* Res engr, Miami Valley Labs, Procter & Gamble Co, 58-59; prog head aerosol physics & fluid dynamics, Nat Ctr Atmospheric Res, 62-67, asst head dept chem & microphys, Lab Atmospheric Sci, 67-68; staff scientist, Sci Ctr, NAm Rockwell Corp, 68-70, group leader, Atmospheric Sci, 70-73, dir air qual res & monitoring, 73-74; gen mgr, Environ Chem Ctr, 74-81, chief scientist, Environ Res & Technol, Inc, 81-84, vpres, 74-84; pres, Desert Res Inst, 84-87; VPRES, ENVIRON DIV, ELEC POWER RES INST, 87- *Concurrent Pos:* Sr res fel environ eng, Calif Inst Technol, 69-73; sci adv bd, Welded Electronic Packaging Asn, 81-; prof eng, Univ Nev Las Vegas, 84-87. *Mem:* AAAS; Am Phys Soc; Am Chem Soc; Am

Meteorol Soc; Am Geophys Soc; Air Pollution Control Asn; Am Asn Aerosol Res; Am Inst Chem Eng. *Res:* Aerosol physics and chemistry; atmospheric chemistry; air-sea interaction; fluid dynamics; environmental science. *Mailing Add:* Elec Power Res Inst PO Box 10412 Palo Alto CA 94303

HIDY, PHIL HARTER, b North Manchester, Ind, Feb 11, 15; m 39; c 2. MICROBIAL BIOCHEMISTRY. *Educ:* Ind Univ, BS, 38, AM, 42, PhD(chem), 44. *Prof Exp:* Chemist, Inland Steel Corp, 39-40; asst prof biochem, Col Med, Baylor Univ, 43-47; res biochemist, Com Solvents Corp, 47-53, group leader biochem, 53-76, mgr microbiol res, Int Minerals & Chem Corp, 76-79; RETIRED. *Mem:* Am Chem Soc. *Res:* Carbohydrate metabolism; antibiotics; organic chemistry; physiology; effect of different starches and starch degradation products on the activity of potato phosphorylase; fermentation biochemistry. *Mailing Add:* 307 Gardendale Rd Terre Haute IN 47803

HIEBER, THOMAS EUGENE, b Dayton, Ohio, Aug 6, 36; m 64; c 2. BIOCHEMISTRY. *Educ:* Univ Dayton, BS, 59; Univ Cincinnati, MS, 63; St Thomas Inst, PhD(biochem), 72. *Prof Exp:* Chemist cancer & heart res, Miami Valley Hosp, 59-60; asst org chem, Univ Cincinnati, 61-63; chemist res & develop, MC/B Mfg Chemists, 63-74; tech dir drug prod res & develop, Sperti Drug Co, 75-; VPRES, MDH LAB. *Mem:* Am Chem Soc. *Res:* Antihemolytic drugs, such as adrenochrome semicarbazone and over the counter drugs containing salicylic acid and methyl salicylate. *Mailing Add:* 4764 Clevesdale Dr Cincinnati OH 45238

HIEBERT, ALLEN G, b Goessel, Kans, Oct 17, 41; m 63; c 2. ANALYTICAL CHEMISTRY. *Educ:* Tabor Col, BA, 63; Univ Ill, Urbana, MS, 65, PhD(chem), 67. *Prof Exp:* Postdoctoral res chem, Mich State Univ, 67-68; prof chem, Knox Col, 68-84; PROF CHEM, TABOR COL, 84- *Concurrent Pos:* Vis prof, Univ Ill, Urbana, 72, Univ Ariz, Tucson, 77-78. *Mem:* Am Chem Soc. *Res:* Electroanalytical methods applied to trace analysis, primarily metal ions in the environment. *Mailing Add:* Tabor Col Hillsboro KS 67063

HIEBERT, ERNEST, b Rosenfeld, Man, May 28, 41; m 64; c 1. VIROLOGY, PLANT PATHOLOGY. *Educ:* Univ Man, BSA, 64; Purdue Univ, MS, 67, PhD(plant virol), 69. *Prof Exp:* From asst prof to assoc prof, 69-81, PROF VIROL, UNIV FLA, 81- *Concurrent Pos:* NSF grants, 72-74, 75-78, 78-79 & 84-87, USDA, 82-85 & 84-87, BARD, 85-88; vis scientist, Agr Can, Vancouver, 77-78. *Mem:* Am Phytopath Soc; Can Phytopath Soc; Am Soc Virol; Am Soc Microbiol. *Res:* Molecular biology of plant viruses and their nonstructural proteins; cloning of viral genomes. *Mailing Add:* 2201 NW 36th Terr Univ Fla 1435 Fitfield Hall Gainesville FL 32605

HIEBERT, ERWIN NICK, b Waldheim, Sask, May 27, 19; US citizen; m 43; c 3. HISTORY OF SCIENCE. *Educ:* Bethel Col, Kans, AB, 41; Univ Kans, AM, 43; Univ Chicago, MS, 49; Univ Wis, PhD(chem & hist sci), 54. *Prof Exp:* Res chemist, Manhattan Proj, Stand Oil Co, 43-46; res chemist, Inst Metals, Univ Chicago, 47-50; asst prof chem, San Francisco State Col, 52-54; instr hist sci, Harvard Univ, 55-57; from asst prof to prof, Univ Wis-Madison, 57-70; chmn dept, 77-84, prof, 70-90, EMER PROF HIST SCI, HARVARD UNIV, 90- *Concurrent Pos:* Fulbright lectr, Max Planck Inst Physics, Gottingen, 54-55; vis prof, Univ Tubingen, 65; vis scholar, Sch Hist Studies, Inst Advan Study, Princeton, 61-62 & 68-69; fel Churchill Col, Cambridge Univ, 84-85. *Mem:* Fel AAAS; Sigma Xi; Hist Sci Soc; fel Am Acad Arts & Sci; Am Phys Soc; fel Acad Int d Hist des Sci; Czech Soc Hist Sci & Technol. *Res:* History of physical sciences since 1800. *Mailing Add:* 40 Payson Rd Belmont MA 02178

HIEBERT, GORDON LEE, b Boston, Mass, Sept 15, 27; m 73; c 1. PHYSICAL CHEMISTRY. *Educ:* Bates Col, BS, 49; Brown Univ, PhD(chem), 54. *Prof Exp:* Res asst biochem, NY Univ-Bellevue Med Ctr, 48-50; from instr to assoc prof chem & chmn dept, Bowdoin Col, 54-65; adv to Govt India & US Agency Int Develop Mission to India on educ in chem, 66; dept head, Sci Liaison Staff, NSF, New Delhi, 67-68, prog dir, Off Int Progs, 69-71, head sci liaison staff, New Delhi, 71-73, prog mgr, Div Int Progs, 73-86; sci adv, Mission Bangkok, USAID, 86-90; SCI ADV, OFF PRES, CHIANG MAI UNIV, CHIANG MAI THAILAND, 91- *Honors & Awards:* Potter Prize, Brown Univ, 54. *Mem:* AAAS; Am Chem Soc; Sigma Xi. *Res:* Infrared spectra of crystals; education in chemistry; science and developing countries. *Mailing Add:* 4000 Gibbs St Alexandria VA 22309

HIEBERT, JOHN COVELL, b Mt Lake, Minn, Aug 18, 34; m 62; c 2. NUCLEAR PHYSICS. *Educ:* Harvard Univ, AB, 56; Yale Univ, MS, 60, PhD(physics), 64. *Prof Exp:* AEC spec fel, Oak Ridge Nat Lab, 63-65; from asst prof to assoc prof, 65-74, PROF PHYSICS, TEX A&M UNIV, 74- *Concurrent Pos:* Guest scientist, Kernfoschungszentrum, Karlsruhe, Fed Rep Germany, 81-82. *Mem:* Am Phys Soc; Am Asn Physics Teachers. *Res:* Few particle problems in nuclear physics using neutron beams. *Mailing Add:* Cyclotron Inst Tex A&M Univ College Station TX 77843-3366

HIEBLE, J PAUL, b Lansing, Mich, Jan 13, 48; m 85; c 1. CARDIOVASCULAR PHARMACOLOGY. *Educ:* N Tex State Univ, PhD(org chem), 71; Univ Tex Med Br, PhD(pharmacol), 77. *Prof Exp:* Asst dir, 82-84, res fel pharmacol, 84-90, SR RES FEL, SMITHKLINE BEECHAM PHARMACEUT, 91- *Mem:* Am Chem Soc; Sigma Xi; Am Soc Pharmacol & Exp Therapeut. *Res:* Adrenergic pharmacology, specifically on the role of adrenergic and dopaminergic receptors in neurotransmission and in maintenance of smooth muscle tone; neuromodulators and co-transmitters. *Mailing Add:* Dept Pharmacol Smithkline Beecham Pharmaceut 709 Swedeland Rd King of Prussia PA 19406-2799

HIEFTJE, GARY MARTIN, b Zeeland, Mich, Oct 1, 42. ANALYTICAL CHEMISTRY. *Educ:* Hope Col, AB, 64; Univ Ill, PhD(anal chem), 69. *Prof Exp:* Res asst phys chem, Ill State Geol Surv, 64-65; from asst prof to prof anal chem 69-85, assoc chmn dept chem, 78-80, DISTINGUISHED PROF ANAL CHEM, IND UNIV, BLOOMINGTON, 85- *Concurrent Pos:* Consult, Lawrence Livermore Lab, 70-, Upjohn Co, 79-, Los Alamos Nat Lab, 83-, Am Cyanamid, 84-, LECO, 87-; sr fel, Sci & Eng Res Coun, Eng, 83. *Honors & Awards:* Can Test Award, Chem Ins Can, 79; Anachem Award, 84; Meggars Award, Soc Appl Spectros, 84; Lester W Strock Medal, 84; Chem Instrumentation Award, Am Chem Soc, 85; Pittsburgh Anal Chem Award, 86; Theophilus Redwood Award, Royal Soc Chem, 86; Anal Chem Award, Am Chem Soc, 87; Tracy M Sonneborn Award, Indiana Univ, 87. *Mem:* Am Chem Soc; Soc Appl Spectros; Optical Soc Am; Sigma Xi; fel AAAS. *Res:* Basic mechanisms in atomic emission, absorption and fluorescence flame spectrometric analysis; development of flame methods of analysis; computer interfacing and control in analysis; chemical instrumentation; correlation spectroscopy. *Mailing Add:* Dept of Chem A169 Chem Bldg Ind Univ Bloomington IN 47405-4001

HIELSCHER, FRANK HENNING, b Jeserig, Ger, Dec 8, 38; US citizen. ELECTRICAL ENGINEERING. *Educ:* Drexel Univ, BSEE, 61; Univ Denver, MSEE, 63; Univ Ill, Urbana, PhD(elec eng), 66. *Prof Exp:* Res assoc, Univ Ill, Urbana, 66-67; mem tech staff, Sprague Elec Co, 67-70; from asst prof to assoc prof elec eng, 71-84, PROF COMPUT SCI & ELEC ENG, LEHIGH UNIV, 84- *Mem:* Sr mem Inst Elec & Electronic Engrs. *Res:* Semiconductor device modeling; design automation and testing; design of analog and digital integrated circuits. *Mailing Add:* Dept Comput Sci & Elec Eng Packard Lab No 19 Lehigh Univ Bethlehem PA 18015

HIEMENZ, PAUL C, b Los Angeles, Calif, June 23, 36. PHYSICAL CHEMISTRY, COLLOID CHEMISTRY. *Educ:* Loyola Univ, BS, 58; Univ Southern Calif, PhD(chem), 64. *Prof Exp:* Res chemist, Dow Chem Co, 64-65; from asst prof to assoc prof, 65-73, PROF CHEM, CALIF STATE POLYTECH UNIV, POMONA, 73- *Concurrent Pos:* Developer & admin, Minority Sci Prog; outstanding adv, Nat Acad Advan Asn. *Mem:* AAAS; Am Chem Soc; Sigma Xi; Nat Sci Teachers Asn. *Res:* Colloid, surface and polymer chemistry; chemical education; minority education. *Mailing Add:* 121 S Hollenbeck Ave Covina CA 91723

HIENZ, ROBERT DOUGLAS, b Detroit, Mich, Dec 21, 44; m 67; c 2. SENSORY PSYCHOLOGY, BEHAVIORAL PHARMACOLOGY. *Educ:* Univ Mich, BA, 67; Western Mich Univ, MA, 68; Univ NC, Chapel Hill, PhD(exp psychol), 71. *Prof Exp:* Fel psychol, Univ NC, Chapel Hill, 68-69, res asst, 69-70, teaching fel, 70-71; fel neurophysiol, Univ Wash Med Sch, 71-73; asst prof behav biol, 75-82, RES ASSOC BIOMED ENGR, 73-, ASSOC PROF BEHAV BIOL, JOHNS HOPKINS MED SCH, 82- *Mem:* AAAS; Acoust Soc Am; Asn Res Otolayngol; Behav Pharmacol Soc; Behav Toxicol Soc. *Res:* Animal psychophysics; audition; neurophysiology of audition; behavioral pharmacology. *Mailing Add:* Behav Biol 619 Traylor Bldg Johns Hopkins Med Sch 720 Rutland Ave Baltimore MD 21205

HIERGEIST, FRANZ XAVIER, b Pittsburgh, Pa, July 31, 38; m 61; c 3. MATHEMATICAL ANALYSIS. *Educ:* Univ Pittsburgh, BS, 59, PhD(math), 64. *Prof Exp:* From asst prof to prof math, 64-82, assoc chmn dept, 75-82, PROF COMPUTER SCI, WVA UNIV, 82- *Mem:* Math Asn Am; Asn Comput Mach. *Res:* Compiler design and programming languages; mathematics of computer science. *Mailing Add:* Dept Statist & Computer Sci WVa Univ Morgantown WV 26506

HIERHOLZER, JOHN CHARLES, b Gravenhurst, Ont, July 1, 38; US citizen; m 67; c 3. MEDICAL MICROBIOLOGY, BIOCHEMISTRY. *Educ:* Spring Hill Col, BS, 60; Univ Fla, MS, 62; Univ Md, PhD(microbiol), 66. *Prof Exp:* Microbiologist, Blood Antigen Pioneering Res Lab, Agr Res Ctr, USDA, 63-66; SUPVRY RES MICROBIOLOGIST, VIROL DIV, CTR DIS CONTROL, USPHS, 66- *Concurrent Pos:* Consult, Fernbank Sci Ctr, Atlanta, 75-; vis fel, Fac Med, Univ Newcastle, Australia, 83-84. *Mem:* AAAS; Am Soc Microbiol; Am Chem Soc; NY Acad Sci; Sigma Xi; Am Soc Virol. *Res:* Anaerobic cellulolytic bacteria; erthrocyte antigens; respiratory viruses; purification of viral antigens; laboratory diagnosis of respiratory virus infection; automated serological techniques; immunogenetics; immunology of viral infections; viral serology; viral conjunctivitis. *Mailing Add:* Resp Virol Br Ctr Dis Control 1600 Clifton Rd NE Atlanta GA 30333

HIERL, PETER MARSTON, b Brooklyn, NY, May 17, 41; m 65; c 5. CHEMICAL KINETICS. *Educ:* Mass Inst Technol, BS, 63; Rice Univ, PhD(phys chem), 67. *Prof Exp:* Res assoc chem, Yale Univ, 67, Univ Colo, 67-69; from asst prof to assoc prof, 69-81, PROF CHEM, UNIV KANS, 81-, ASSOC CHMN, CHEM DEPT, 87- *Mem:* AAAS; Am Chem Soc; Am Phys Soc; Am Soc Mass Spectrometry. *Res:* Mass spectrometry; electron impact phenomena; ion-molecule reactions; molecular beam studies of reactive scattering. *Mailing Add:* Dept Chem Univ Kans Lawrence KS 66045

HIESERMAN, CLARENCE EDWARD, b Iowa Park, Tex, Jan 15, 17; m 42; c 3. CHEMISTRY. *Educ:* Tex Tech Col, BS, 37; Mich State Univ, MS, 39. *Prof Exp:* Res chemist, Celanese Corp Am, 42-47, mfg supt, 47-49, sect head res & develop, 49-51, supt res, Amcelle Plant, 51-52; sr res chemist, Chemstrand Co, 52-54, group leader res & develop, 54-55, sect head pilot plants, 55-57, sr sect head, 57-58, mgr acrilan develop, 58-61 & mgr acrilan mfg, 61-66; plant mgr textiles div, Monsanto Co, Ala, 66-72, PLANT MGR, MONSANTO TEXTILES CO, PENSACOLA, FLA, 72- *Mem:* Am Chem Soc. *Res:* Dyestuff and other organic chemical syntheses and identifications; polymer synthesis and spinning of chemical fibers; manufacturing processes in the field of cellulose acetate, polyamide, polyester and acrylic chemical fibers. *Mailing Add:* 212 N Cliff Dr Gulf Breeze FL 32561-4440

HIESTAND, EVERETT NELSON, b Hancock Co, Ohio, Aug 17, 20; m 53; c 2. PHYSICAL PHARMACY, PHARMACEUTICS. *Educ:* Bluffton Col, AB, 42; Ohio State Univ, PhD(phys chem), 50. *Prof Exp:* Asst prof chem, SDak State Col, 49-51; prin chemist, Battelle Mem Inst, 51-55; sr res scientist, Upjohn Co, 55-90; RETIRED. *Concurrent Pos:* Lectr, Univ Mich, 65-66, adj prof, 67- *Honors & Awards:* Ebert Prize, Am Pharmaceut Asn, 78; Indust Pharmaceut Technol Award, Am Pharmaceut Asn, 82. *Mem:* Am Chem Soc; Am Asn Pharmaceut Scientists; Soc Rheol; fel AAAS; fel Acad Pharmaceut Sci. *Res:* Suspensions; solid dosage forms; compaction physics; micromeretics; rheology; powder technology. *Mailing Add:* 11 378 East G Ave Galesburg MI 49053

HIETANEN-MAKELA, ANNA (MARTTA), b Isokyro, Finland, Apr 10, 09; US citizen; m 46; c 1. GEOLOGY. *Educ:* Univ Helsinki, MS, 33, PhD(geol), 38. *Prof Exp:* Instr geol & mineral, Univ Helsinki, 36-46, docent, 44-46; instr mineral, Stanford Univ, 46-47; asst prof geol, Univ Ore, 48-49; GEOLOGIST, US GEOL SURV, 49- *Concurrent Pos:* Collins fel, Bryn Mawr Col, 38-39; Finnish Govt fel, Johns Hopkins Univ, 39-40. *Mem:* Fel AAAS; fel Geol Soc Am; fel Mineral Soc Am; Am Geophys Union; Geochem Soc. *Res:* Petrology; structural geology; petrofabrics; rock forming minerals; metamorphism and metasomatism. *Mailing Add:* 1134 Palo Alto Ave Palo Alto CA 94301

HIETBRINK, BERNARD EDWARD, b Strasburg, NDak, Nov 23, 30; m 51; c 4. DRUG METABOLISM, ENZYME ACTIVITY. *Educ:* SDak State Univ, BS, 58; Univ Chicago, PhD(pharmacol), 61. *Prof Exp:* Res assoc toxicol, Univ Chicago, 61-64; from asst prof to prof, dept pharmacol, 64-83, head dept, 66-83, prof pharmacol & toxicol & head, Dept Pharm Sci, 83-86, DEAN, COL PHARM, SDAK STATE UNIV, 87- *Mem:* Soc Toxicol; Am Soc Pharmacol & Exp Therapeut; Am Asn Cols Pharm; Sigma Xi. *Res:* Effect of chemical agents on the activity of microsomal enzymes; influence of pentoxifylline on permeability of the red cell. *Mailing Add:* Dept Pharm Sci Col Pharm SDak State Univ Brookings SD 57007

HIETBRINK, EARL HENRY, b Firth, Nebr, Apr 28, 30; m 52; c 3. APPLIED MATHEMATICS. *Educ:* Cent Col, Iowa, BA, 51; Univ Nebr, MA, 56. *Prof Exp:* Sr mathematician, Res Dept, Allison Div, Gen Motors Corp, Ind, 56-67, survry res engr, Gen Motors Res Labs, 67-87; RETIRED. *Concurrent Pos:* Adj prof math, 87-90. *Mem:* Soc Automotive Engrs; Math Asn Am; Electrochem Soc. *Res:* Advanced battery and fuel cell systems for vehicular applications; direct conversion of energy to electricity by electrochemical power systems. *Mailing Add:* Rte 3 Box 228H Fairfield Bay AR 72088

HIGA, HARRY HIROSHI, US citizen. MEDICAL MICROBIOLOGY. *Educ:* Univ Hawaii, AB, 40; Univ Southern Calif, MS, 46; Syracuse Univ, PhD(microbiol), 54. *Prof Exp:* Fel microbiol chem & res assoc microbiol, Purdue Univ, 54-56; microbiologist, USDA, 56-59, biochemist, 59-61; res assoc biochem, Univ Hawaii, 61-64, asst agr biochemist, 64-65; tech dir, Pac Labs, Inc, 66-67; MICROBIOLOGIST, HAWAII STATE DEPT HEALTH, 67- *Concurrent Pos:* Res assoc, Chem Corps, Dept Army, 54-56. *Mem:* AAAS; Am Chem Soc; Am Soc Microbiol. *Res:* Zoonoses, especially rodent plague, typhus and leptospirosis surveillances, detection, isolation and identification of causative organisms; biochemistry, especially enzymology and nucleic acids. *Mailing Add:* Hawaii State Dept Health 98-346 Puahoku Pl Aiea HI 96701

HIGA, LESLIE HIDEYASU, b Hawi, Hawaii, Oct 8, 25; m 61; c 3. ORAL PATHOLOGY. *Educ:* Grinnell Col, BA, 53; Univ Mo, Kansas City, DDS, 60; Univ Chicago, MS, 67; Am Bd Oral Path, dipl, 72. *Prof Exp:* Fel dent, Zoller Dent Clin, Univ Chicago, 60-65; PROF ORAL SURG & ORAL PATH & DIAG, COL DENT, UNIV IOWA, 65- *Concurrent Pos:* Fel Am Acad Oral Path, 65; consult, Vet Admin Hosp, Iowa City, Iowa, 73. *Mem:* Am Acad Oral Path; Int Asn Dent Res; Am Dent Asn; AAAS. *Res:* Pain control. *Mailing Add:* Col Dent Univ Iowa Iowa City IA 52242

HIGA, WALTER HIROICHI, b Maui, Hawaii, Sept 21, 19; m 46; c 3. APPLIED PHYSICS, ELECTRICAL ENGINEERING. *Educ:* Tri-State Col, BS, 42; Univ Cincinnati, MS, 47, PhD(physics), 49. *Prof Exp:* Instr physics, Seattle Univ, 49-50; res engr electronics, NAm Aviation, 50-54; res scientist appl physics, Jet Propulsion Lab, 54-80; RETIRED. *Mem:* Sr mem Inst Elec & Electronics Engrs; Cryogenic Soc Am; Am Phys Soc; fel Inst Advan Eng. *Res:* Quantum electronics and cryogenic refrigeration technology. *Mailing Add:* 600 Tamarac Dr Pasadena CA 91105

HIGASHI, GENE ISAO, vaccines, epidemiology; deceased, see previous edition for last biography

HIGASHIYAMA, TADAYOSHI, b Osaka, Japan, Mar 24, 33; m 59; c 1. BIOCHEMISTRY, ENDOCRINOLOGY. *Educ:* Osaka Univ, Japan, BS, 58, MS, 60, PhD(biochem), 65. *Prof Exp:* Asst prof inorg biochem, Osaka Univ, Japan, 66-75; res asst prof biochem, State Univ NY, Buffalo, 73-77; RES SCIENTIST ENDOCRINE BIOCHEM, MED FOUND BUFFALO, 77- *Res:* Isolation and purification of cytochrome P-450-containing enzymes related to steroid hormone biosynthesis; study of topography of hormone action and biosynthesis. *Mailing Add:* Med Found Buffalo 73 High St Buffalo NY 14203-1196

HIGDEM, ROGER LEON, b Fosston, Minn, Nov 20, 33; m 58; c 1. MATHEMATICS. *Educ:* Univ NDak, BS, 55, MS, 59; Ore State Univ, PhD(math), 70. *Prof Exp:* From asst prof to assoc prof, 59-71, chmn dept, 71-73, PROF MATH, COL IDAHO, 71-, CHMN DIV NATURAL SCI, 73- *Concurrent Pos:* Acad vpres, Col of Idaho, 81-83. *Mem:* Math Asn Am; Am Math Soc; Nat Coun Teachers Math. *Res:* Additive number theory; cardinality theorems in set theory; generalized number density. *Mailing Add:* Dept Math Col Idaho Caldwell ID 83605

HIGDON, ARCHIE, engineering mechanics, mathematics; deceased, see previous edition for last biography

HIGERD, THOMAS BRADEN, b Pittsburgh, Pa, July 1, 42; m 64; c 2. MICROBIAL PHYSIOLOGY, IMMUNOLOGY. *Educ:* Washington & Jefferson Col, BA, 64; Wayne State Univ, PhD(med microbiol), 69. *Prof Exp:* Am Cancer Soc Dernham jr fel oncol, Scripps Clin & Res Found, 69-72; asst prof, 72-79, ASSOC PROF MICROBIOL & IMMUNOL, MED UNIV SC, 79-, ASSOC DEAN, RESOURCE PLANNING, COL MED, 90- *Concurrent Pos:* Intermittent fac appointee, Nat Marine Fisheries Serv & Nat Oceanic & Atmospheric Admin, 80-86. *Mem:* Am Soc Microbiol; Int Asn Dent Res. *Res:* Microbial physiology, metabolism and genetics, characteristics of pyocin, a Pseudomonas bacteriocin; role of proteolytic enzymes in sporulation and genetics of Bacillus subtilis; bacterial modulators of the immune response; dental research; ciguatera seafood poisoning. *Mailing Add:* Off Dean Col Med Med Univ SC Charleston SC 29425

HIGGINBOTHAM, ROBERT DAVID, b Salt Lake City, Utah, Mar 15, 21; m 46; c 2. MICROBIOLOGY. *Educ:* Univ Utah, BA, 49, MS, 50, PhD(bact), 55. *Prof Exp:* Asst bact, Univ Utah, 50-53, res assoc anat, 53-55, res instr, 55-57, instr, 57-58; from asst prof to prof microbiol, Med Br, Univ Tex, 58-67; prof microbiol, Med Sch, Univ Louisville, 67-80, actg chmn dept, 70-72, prof microbiol & immunol, 80-91, EMER PROF MICROBIOL & IMMUNOL, MED SCH, UNIV LOUISVILLE, 91- *Concurrent Pos:* Lederle med fac award, 57-60. *Mem:* AAAS; Am Soc Microbiol; Am Asn Immunologists; Am Soc Exp Path; Reticuloendothelial Soc. *Res:* Mast cells; bacterial endotoxins; viruses; immunology; virology; connective tissue function and resistance with regard to cells and endogenous secretions. *Mailing Add:* 4017 Ashridge Dr Louisville KY 40241

HIGGINS, BRIAN GAVIN, b Umtata, Transkei, SAfrica, July 24, 48; div; c 1. FLUID MECHANICS OF COATING FLOWS, LANGMIER-BLODGETT FILMS. *Educ:* Univ Witwatersrand, Johannesburg, SAfrica, BSc, 72, MSc, 75; Univ Minn, PhD(chem), 80. *Prof Exp:* Res asst, Inst Paper Chem, 80-82, res assoc, 82-83; from asst prof to assoc prof, 83-90, PROF CHEM ENG, UNIV CALIF, DAVIS, 90- *Concurrent Pos:* Actg chmn, Dept Chem Eng, 88-90, chmn, 90- *Mem:* Am Inst Chem Engrs; Am Phys Soc. *Res:* Fluid mechanics of coating flows; interfacial mechanics; hydrodynamic stability of thin films; Langmier-Blodgett films for nonlinear optics. *Mailing Add:* Dept Chem Eng Univ Calif Davis CA 95616

HIGGINS, CHARLES GRAHAM, b Oak Park, Ill, Nov 18, 25; m 74; c 2. PHYSICAL GEOLOGY. *Educ:* Univ Chicago, SB, 46, SM, 47; Univ Calif, Berkeley, PhD(geol), 50. *Prof Exp:* Asst dept geol sci, Univ Calif, Berkeley, 48-49; instr geol, Univ Mich, 50-51; from asst prof to prof, 51-90, EMER PROF GEOL, UNIV CALIF, DAVIS, 90- *Concurrent Pos:* Mem, Brit Geomorphol Res Group. *Mem:* Geol Soc Am; Sigma Xi. *Res:* Geomorphology; coastal geology; geomorphic theory; origin and significance of tiered slopes and other microrelief features; development of drainage networks and landforms by subsurface outflow and sapping; subsurface physical environments in beaches. *Mailing Add:* Dept Geol Univ Calif Davis CA 95616

HIGGINS, DOROTHY, b Lawrence, Mass, May 1, 30. INORGANIC CHEMISTRY. *Educ:* Emmanuel Col, BA, 51; Catholic Univ, MS, 61; Boston Col, PhD(inorg & org chem), 66. *Prof Exp:* Teacher high sch, Mass, 54-58; instr anal org chem, Emmanuel Col, Mass, 66, assoc prof inorg phys chem, 66-87, chmn dept chem, 69-77, res assoc chem, 75-87; fac consult, Zymark Corp, 83; DIV CHMN MATH, SCI & TECHNOL, ROXBURY COMMUNITY COL, 88- *Concurrent Pos:* Partic, Chautague Course interfacing microcomput & lab instrumentation, NSF, 81-82; mem, Extramural Assoc Prog, NIH, 84; educ consult, high schs & cols, Japan, 86. *Mem:* Am Chem Soc; Nat Sci Teachers Asn; Sigma Xi; Asn Am Med Cols; Soc Col Sci Teaching. *Res:* Investigation of equilibria involved in the formation of mixed ligand chelates of indium (III); transition metal complexes of tameta 3HBr. *Mailing Add:* 400 Fenway Emmanuel Col Boston MA 02115

HIGGINS, E ARNOLD, Norman, Okla, May 5, 30. RESPIRATORY PHYSIOLOGY. *Educ:* Univ Okla, PhD(physiol), 65. *Prof Exp:* SUPVR, SURVIVAL RES UNIT, CIVIL AEROMED INST, 82- *Mem:* Fel Aerospace Med Asn; Am Physiol Soc. *Mailing Add:* Civil Aeromed Inst FAA AAM-600 PO Box 25082 Oklahoma City OK 73125

HIGGINS, EDWIN STANLEY, b New York, NY, Mar 12, 25; m 58; c 3. BIOCHEMISTRY. *Educ:* Alfred Univ, BA, 52; Syracuse Univ, PhD(biochem), 56. *Prof Exp:* Asst physiol, anat & biol, Alfred Univ, 51-52; from asst prof to assoc prof, 56-68, PROF BIOCHEM & MOLECULAR BIOPHYS, MED COL VA, 68-, FROM ASSIST TO DEAN, SCH BASIC HEALTH SCIENCES, 85- *Concurrent Pos:* Dir biochem res, Bur Alcohol Studies & Rehab, Va State Health Dept, 59-73; vis lectr, Univ Ctr Va, Inc, 60-67; mem, Int Cong Alcohol & Alcoholism, 68. *Mem:* AAAS; Am Chem Soc; Soc Exp Biol & Med; Am Soc Cell Biol; Am Inst Nutrit. *Res:* Biochemistry of mitochondrial functions and changes accompanying hepatotoxicities and tissue regeneration; brain metabolism; alcohol metabolism and toxic mechanisms; fungal metabolism. *Mailing Add:* Dept Biochem & Molecular Biophysics Med Col Va Commonwealth Univ Richmond VA 23298-0614

HIGGINS, FREDERICK B(ENJAMIN), JR, b Portsmouth, Va, Sept 28, 36; m 58; c 2. SANITARY ENGINEERING. *Educ:* Ga Inst Technol, BCE, 58, MSSE, 60, PhD(sanit eng), 64. *Prof Exp:* Sanit engr water pollution control, Environ Hyg Agency, US Army, 63-65, asst for air pollution eng serv directorate, 65-66, chief, Air Pollution Eng Div, 66-67; proj engr, Roy F Weston, Inc, Pa, 67-68, proj mgr, 68-69; asst prof environ eng, Drexel Univ, 69-73; assoc prof civil & environ eng technol & chmn dept, Temple Univ, 73-78, prof environ eng, 73-78, dean, Col Eng & Archit, 81-89, DIR, CTR ENVIRON STUDIES, TEMPLE UNIV, 89- *Mem:* AAAS; Am Soc Civil Engrs; Am Soc Eng Educ. *Res:* Radiological hygiene; air and water pollution; solid wastes. *Mailing Add:* 515 Welsh Rd Apt D-1 Huntingdon Valley PA 19006

HIGGINS, GEORGE A, JR, b Wichita, Kans, Mar 30, 17; m 45; c 5. SURGERY. *Educ:* Univ NMex, BS, 38; Harvard Univ, MD, 42; Am Bd Surg, dipl; Am Bd Thoracic Surg, dipl. *Prof Exp:* Intern surg, Boston City Hosp, 42-43; resident, Mt Alto Hosp, 46-49; chief surg serv, Vet Admin Hosp, Wichita, 49-52; chief surg serv, Vet Admin Hosp, Kansas City, Mo, 52-60; CHIEF SURG SERV, VET ADMIN HOSP, WASHINGTON, DC, 60- *Concurrent Pos:* Instr, George Washington Univ, 48-50, clin prof, Sch Med, 60-; instr, Georgetown Univ, 49-60, clin prof, Sch Med, 60-68, prof, 68-; from asst prof to prof, Med Sch, Univ Kans, 52-60. *Mem:* Western Surg Asn; Am Col Surg; Am Surg Asn. *Res:* Clinical surgery. *Mailing Add:* Santa Barbara Cottage Hosp Santa Barbara CA 93102

HIGGINS, IAN T, b Edinburgh, Scotland, Mar 12, 19; m 64. EPIDEMIOLOGY. *Educ:* Univ London, MB, BS, 46, MD, 51. *Prof Exp:* Resident med officer, Royal Nat Hosp, Ventnor, Eng, 43-44; house physician & receiving officer, London Hosp, 44-45; house physician, Hosp Sick Children, 45-46; house surgeon, Norfolk & Norwich Hosp, 46-47, resident med officer, 47-48; sr registr chest clin, Grimsby County Borough, 48-49; resident physician, Mont Hall, Switz, 50-53; sr registr, Cambridge Chest Clin, 53; mem, Sci Staff, Pneumoconiosis Res Unit, Med Res Coun, 53-56, asst dir, Epidemiol Res Unit, 62-63; prof chronic dis epidemiol, Grad Sch Pub Health, Univ Pittsburgh, 63-67; prof community health serv & adult health & aging unit, 67-70, prof epidemiol, 67-85, prof environ & indust health, 70-85, EMER PROF EPIDEMIOL, ENVIRON & INDUST HEALTH, UNIV MICH, ANN ARBOR, 85- *Concurrent Pos:* Wander scholar & registr, Children's Dept, Westminster Hosp, Eng, 45-46; actg physician in chg, Cotswold Sanatorium, Cranham, 62; consult, Div Occup Health, USPHS & Epidemiol Br, Commun Dis Ctr, Ga, 65-66; fel coun arteriosclerosis & fel coun epidemiol, Am Heart Asn. *Mem:* Am Epidemiol Soc; fel Brit Thoracic Soc. *Res:* Epidemiology of chronic diseases, especially pneumoconiosis and chronic non-specific respiratory disease; effects of occupation and tobacco smoking on heart and lung conditions. *Mailing Add:* 252 Indian River Pl Ann Arbor MI 48104

HIGGINS, IRWIN RAYMOND, b Mapleton, Maine, Feb 15, 19; m 43; c 3. ION EXCHANGE, ELECTROCHEMISTRY. *Educ:* Univ Maine, BS, 42. *Prof Exp:* Process engr, TNT Plant, Manhattan Proj, DDT-2-4-D, Ala, Chicago, Tenn, Wash & NJ, E I du Pont de Nemours & Co, 42-46; chemist, Oak Ridge Nat Lab, 46-58; tech dir process develop, Chem Sep Corp, 58-82; TECH DIR PROCESS DEVELOP, CSA, LAKE IND, 82- *Mem:* Fel Am Inst Chem Engrs; Am Nuclear Soc; AAAS; Assoc Inst Mining Eng. *Res:* Continuous ion exchange process development. *Mailing Add:* 113 Clarion Rd Oak Ridge TN 37830

HIGGINS, JAMES JACOB, b Canton, Ill, Oct 31, 43; m 67; c 2. STATISTICS. *Educ:* Univ Ill, BS, 65; Ill State Univ, MS, 67; Univ Mo, Columbia, PhD(statist), 70. *Prof Exp:* Asst prof math, Univ Mo, Rolla, 70-74; from asst prof to assoc prof math, Univ SFla, 74-80; PROF STATIST, KANS STATE UNIV, 80- *Mem:* Am Statist Asn; Inst Math Statist. *Res:* Reliability theory; classical and Bayesian estimation theory; statistical modelling; experimental design. *Mailing Add:* Dept Statist Kans State Univ Manhattan KS 66506

HIGGINS, JAMES THOMAS, JR, b Columbia, SC, July 13, 34; m 56; c 4. INTERNAL MEDICINE, PHYSIOLOGY. *Educ:* Duke Univ, Durham, BS, 59, MD, 59. *Prof Exp:* Intern, Duke Univ Hosp, 59-60; fel, Dept Med, Duke Univ, 60-61; investr, Nat Heart Inst, Bethesda, 61-63; resident, Yale-New Haven Hosp, 63-65; clin investr, Vet Admin Hosp, Conn, 65-67; instr med, Yale Univ, 65-66, asst prof, 66-67; from asst prof to prof, Ind Univ, Indianapolis, 67-76, prof med & chief nephrol, Med Col Ohio, 76-87; CHIEF MED SERV, ALBANY VET ADMIN MED CTR & PROF MED, ALBANY MED COL, 87- *Concurrent Pos:* Nat Heart Inst career develop award, 69-74; dir, Specialized Ctr Res in Hypertension, Sch Med, Ind Univ, Indianapolis, 71-76. *Mem:* Am Physiol Soc; Am Soc Nephrology; Cent Soc Clin Res; Endocrine Soc; fel Am Col Physicians. *Res:* Renal physiology; control of water and electrolyte balance; membrane physiology; causes and treatment of hypertension. *Mailing Add:* Med Service III Vet Admin Med Ctr 113 Holland Ave Albany NY 12208

HIGGINS, JAMES VICTOR, b Sandusky, Ohio, Jan 24, 33; m 53; c 3. HUMAN GENETICS. *Educ:* Mich State Univ, BS, 54; Univ Minn, MS, 58, PhD(zool), 61. *Prof Exp:* Asst prof zool, 61-70, PROF ZOOL & HUMAN DEVELOP, MICH STATE UNIV, 70- *Mem:* AAAS; Am Soc Human Genetics; Am Eugenics Soc; Am Genetic Asn; Am Asn Ment Deficiency; Sigma Xi. *Res:* Gene action in relation to mental retardation; population frequency and biochemical variation. *Mailing Add:* Dept Zool Mich State Univ East Lansing MI 48823

HIGGINS, JAMES WOODROW, b Seaman, Ohio, Feb 5, 21; m 44; c 2. GEOLOGY. *Educ:* Miami Univ, AB, 43; Univ Chicago, PhD(geol), 47. *Prof Exp:* Asst geol, Miami Univ, 40-42 & Univ Chicago, 45-47; geologist & geophysicist, Standard Oil Co, Calif, 47-53, area geologist, 53-55, dist geologist, 55-58, geologist, Staff of Vpres Oil Field Res, Calif Res Corp, 58-61, dist geologist, Alaskan Dist, Standard Oil Co Calif, 61-65, financial analyst, Comptroller's Dept, 65-68, sr geologist, 68-77; RETIRED. *Concurrent Pos:* Instr, Exten Sch, Univ Calif, 49. *Mem:* Geol Soc Am; Am Asn Petrol Geol. *Res:* Petroleum exploration; investment analysis. *Mailing Add:* 12225 Lakeshore S Lake of the Pines Auburn CA 95603

HIGGINS, JERRY MITCHELL, b St Louis, Mo, Oct 28, 30; m 56; c 2. SPEECH PATHOLOGY. *Educ:* Asbury Col, BA, 56; Univ Iowa, MA, 65; Mich State Univ, PhD(speech path), 71. *Prof Exp:* Asst prof speech path & dir speech & lang clin, Med Ctr, Univ Ala, Birmingham, 71-74; assoc prof speech path, Asbury Col, 74-78; asst prof speech path, Mich State Univ, 78-82; EXEC VPRES, PRENTKE RUMICH CO, WOOSTER, OH, 86- *Mem:* Am Speech & Hearing Asn. *Res:* Development of a more versatile electro-larynx; modification of nasality in the deaf. *Mailing Add:* Prentke Romich Co 1022 Heyl Rd Wooster OH 44691

HIGGINS, JOHN CLAYBORN, b Logan, Utah, June 14, 34; m 61; c 3. MATHEMATICS, COMPUTER SCIENCE. *Educ:* Brigham Young Univ, BA, 58, MA, 60; Univ Calif, Davis, PhD(math), 66. *Prof Exp:* Instr math, Church Col Hawaii, 60-62; from instr to assoc prof, 62-74, PROF MATH, BRIGHAM YOUNG UNIV, 74- *Concurrent Pos:* NSF sci fac fel, 65-66; vis prof, Shimane Univ Matsue, Japan, 76. *Mem:* Am Math Soc. *Res:* Semigroups; topological semigroups; category theory; recursive functions; automata; formal languages. *Mailing Add:* 3374 TMCB Brigham Young Univ Provo UT 84602

HIGGINS, JOHN J, b Norfolk, Va, Nv 20, 43. SEMI-CONDUCTOR DEVICES. *Educ:* NC State Univ, BS, 66; Mich State Univ, PhD(physics), 72, MS, 82. *Prof Exp:* Res assoc, Mich State Univ, 77-82; ENGR SCIENTIST, IBM CORP, 82- *Mem:* Am Phys Soc; Inst Elec & Electronics Engrs. *Mailing Add:* IBM M65/972-1 River Rd Essex Junction VT 05452

HIGGINS, JOSEPH JOHN, b Philadelphia, Pa, Jan 26, 32; m 56; c 5. BIOPHYSICS. *Educ:* Univ Pa, BA, 54, PhD(physics), 59; Harvard Univ, MA, 55. *Prof Exp:* Asst analogue comput, Johnson Found, Univ Pa, 49-59, asst digital comput, 56-59, lectr reaction kinetics & med math, Univ, 56; USPHS fel, Free Univ Brussels, Inst Phys Chem, Univ Copenhagen & Quantum Chem Group, Univ Uppsala, 59-61; res assoc, Johnson Found, 61-63, asst prof biophys, 63-67, chmn biophys grad group, 67-77, assoc prof biophys, 67-80, ASSOC PROF BIOPHYS & BIOCHEM, UNIV PA, 80- *Concurrent Pos:* NIH career develop award, 70- *Mem:* AAAS; Am Phys Soc; Biophys Soc; NY Acad Sci. *Res:* Chemical and cellular kinetics; irreversible thermodynamics; electronic analogue and digital computers. *Mailing Add:* 410 S 45th St Philadelphia PA 19104

HIGGINS, LARRY CHARLES, b St George, Utah, Oct 21, 36; m 60; c 4. PLANT TAXONOMY, PLANT MORPHOLOGY. *Educ:* Utah State Univ, BS, 64; Brigham Young Univ, MS, 67, PhD(plant taxon), 69. *Prof Exp:* Asst prof, 69-75, ASSOC PROF BIOL, W TEX STATE UNIV, 75- *Mem:* Am Soc Plant Taxon; Int Asn Plant Taxon. *Res:* Family Boraginaceae, specifically genus Cryptantha. *Mailing Add:* Dept Biol WTex State Univ 500 12th Ave Canyon TX 79016

HIGGINS, MICHAEL LEE, b Ft Worth, Tex, Sept 2, 40. MICROBIOLOGY, BIOCHEMISTRY. *Educ:* Univ NTex, BA, 63, MS, 64; Rutgers Univ, PhD(microbiol), 68. *Prof Exp:* Vis investr mycol, Rutgers Univ, 64-65; from asst prof to assoc prof, 68-78, PROF MICROBIOL, SCH MED, TEMPLE UNIV, 78- *Concurrent Pos:* Chmn, Ultrastruct Div, Am Soc Microbiol, 86. *Honors & Awards:* Rutgers Microbiol Res Award, 68. *Mem:* Am Soc Microbiol. *Res:* Physiology and ultrastructure of procaryote organisms. *Mailing Add:* Dept Microbiol & Immunol Temple Univ Sch Med 3400 N Broad St Philadelphia PA 19140

HIGGINS, MILLICENT WILLIAMS PAYNE, b Halifax, Eng, Mar 5, 28; m 64; c 2. MEDICINE, EPIDEMIOLOGY. *Educ:* Univ Durham, MB & BS, 51, DPH, 56, MD, 59. *Prof Exp:* Trainee asst to Dr A B Follows, York, Eng, 51-52; casualty officer, Nottingham Children's Hosp, 52-53; intern med, Royal Victoria Hosp, Montreal, Que, 53-54; sr house officer, Newcastle Gen Hosp, Eng, 54-56; res asst pediat, Durham Univ, 56-58; med officer, Gateshead Health Dept, Eng, 58-59; res assoc epidemiol, Univ Mich, 59-62, asst prof, 63-64; asst prof, Univ Pittsburgh, 64-67; from assoc prof to prof epidemiol, Sch Pub Health, Univ Mich, Ann Arbor, 67-85, dir, prog gen epidemiol 73-85, prof internal med, 82-85; ASSOC DIR EPIDEMIOL & BIOMETRY, NAT HEART, LUNG & BLOOD INST, NIH, 84- *Concurrent Pos:* Fel coun epidemiol, Am Heart Asn; mem adv coun, Nat Heart & Lung Inst. *Mem:* Am Thoracic Soc; Soc Epidemiol Res; Brit Med Asn; hon fel Am Col Chest Physicians; fel Am Col Epidemiol. *Res:* Epidemiological studies of chronic diseases, especially chronic respiratory diseases; diabetes, hypertension and coronary disease; relationships between pregnancy and childbirth and health and disease. *Mailing Add:* NIH 2C08 Fed Bldg 7550 Wisconsin Ave Bethesda MD 20892

HIGGINS, N PATRICK, b Wichita, Kans, Mar 17, 46; m 69; c 3. DNA ENZYMOLOGY, GENE TRANSPOSITION. *Educ:* Wichita State Univ, BS, 68, MS, 69; Univ Chicago, PhD(microbiol), 76. *Prof Exp:* Asst prof biochem, Univ Wyo, 79-83; ASSOC PROF BIOCHEM, UNIV ALA, BIRMINGHAM, 83- *Mem:* Am Soc Microbiol; AAAS; Am Soc Biol Chem. *Res:* Genetic recombination and DNA transposition in bacteria; understanding the regulatory mechanisms by analyzing how purified proteins interact with their DNA substrates. *Mailing Add:* Dept Biochem Univ Ala Birmingham AL 35294

HIGGINS, PAUL DANIEL, b Lexington, Ky, May 11, 46; m 71; c 3. PHYSIOLOGICAL ECOLOGY, BOTANY. *Educ:* Georgetown Col, BA, 68; Univ Louisville, MS, 70; Univ Idaho, PhD(bot), 75. *Prof Exp:* PRIN ENGR, ENVIRON SERV, PAC POWER & LIGHT CO, 75- *Concurrent Pos:* Mgr, dist & trans, MTC & Environ Systs Planning. *Mem:* Ecol Soc Am; Am Soc Plant Physiologists; AAAS. *Res:* Physiological adaptations of alpine and subalpine plants to low soil temperatures; role of soil moisture and heat in reclamation of arid lands; methods for coal strip mine reclamation. *Mailing Add:* Pac Power & Light Co 920 SW Sixth Ave Portland OR 97204

HIGGINS, RICHARD J, b Winchester, Mass, May 15, 39; m 65; c 2. SOLID STATE PHYSICS. *Educ:* Mass Inst Technol, BS, 60; Northwestern Univ, PhD(mat sci), 65. *Prof Exp:* Mem staff phys chem, Fundamental Res Lab, US Steel Corp, 60; asst prof, 65-70, assoc prof, 70-75, PROF PHYSICS, UNIV ORE, 75- *Concurrent Pos:* Vis prof, Univ Islamabad & consult, Ford Found, Pakistan, 71; consult & resident vis, Bell Labs, Murray Hill, NJ, 72-73; prog chmn, Int Conf on Electron Lifetimes in Metals, 74. *Mem:* Am Phys Soc; Fedn Am Sci; Inst Elec & Electronics Engrs; Am Asn Univ Profs; Sigma Xi. *Res:* Experimental study of the electronic structure of metals, alloys and semiconductors; Fermi surface measurements; electronic instrumentation; application of mini-computers and microcomputers in scientific research. *Mailing Add:* 1079 Commonwealth Ave Apt 437 Boston MA 02215

HIGGINS, ROBERT ARTHUR, b Watertown, SDak, Sept 5, 24; m 58; c 3. ELECTRONICS ENGINEERING, SOFTWARE SYSTEMS. *Educ:* Univ Minn, Minneapolis, BEE, 48; Univ Wis, Madison, MS, 64; Univ Mo, Columbia, PhD(elec eng), 69. *Prof Exp:* Engr, Schlumberger Well Surv Corp, Tex, 48-57; sr res technologist, Mobil Res & Develop Corp, 58-61; sr res engr, United Aircraft Res Labs, Conn, 65; staff specialist, Remote Sensing Inst, 69-71, asst prof elec eng, SDak State Univ, 69-74, assoc dir, eng exp sta, 73-77, prof elec eng, 74-77; consult, Control Data Corp, 77-80; prin engr, Sperry Univac, 81-86; PROF ELEC ENGR, ST CLOUD STATE UNIV, 85-

Concurrent Pos: US Dept Interior Off Water Resources res grant, SDak State Univ, 71-74; consult, Lawrence Livermore Lab, 71-73 & USAF Off Sci Res, 76; proj dir, NSF, 73-76 & 88-89; prof elec eng, SDak State Univ, 77-79; consult, NCR Comten, 88. *Mem:* Inst Elec & Electronics Engrs; Sigma Xi. *Res:* Adaptive system theory; data communications; nonlinear system identification; instrumentation; hybrid control; power system stability; memory systems; computer architecture. *Mailing Add:* Dept Elec Eng St Cloud State Univ St Cloud MN 56301-4498

HIGGINS, ROBERT H, b Warren, Ohio, Dec 9, 42; div; c 4. STEREOCHEMISTRY, SPECTROSCOPY. *Educ:* Ohio Univ, BS, 65, MS 67; Univ Nebr, PhD(org chem), 71. *Prof Exp:* Instr chem, Doane Col, 68-70; res fel, Univ Nebr, 71-72; res chemist, Dorsey Labs, 72-74; res fel, Univ Colo, 74-75; asst prof chem, Metrop State Col, 75; asst prof, 76-79, coordr, 79-85, ASSOC PROF CHEM, FAYETTEVILLE STATE UNIV, 82- *Concurrent Pos:* Dow fel, Univ Nebr, 68; consult, Am Inst Res, 84-87; prin investr, NIH, 85- *Mem:* Sigma Xi; Am Chem Soc; fel Am Inst Chemists. *Res:* Synthesis, sterochemistry and reactions of l-alkylazetidinols, particularly as applicable to a new route of beta-adrenolytics and anti-tumor drugs. *Mailing Add:* Rte 20 Box 1254 Fayetteville NC 28306

HIGGINS, ROBERT PRICE, b Denver, Colo, Oct 8, 32; m 54; c 3. INVERTEBRATE ZOOLOGY. *Educ:* Univ Colo, BA, 56, MA, 58; Duke Univ, PhD(zool), 61. *Prof Exp:* From asst prof to assoc prof biol, Wake Forest Univ, 61-68; resident systematist, Systs-Ecol Prog, Marine Biol Lab, Woods Hole, 68; assoc prof & dir marine prog, Boston Univ, 68; oceanogr, 68-69, dir, Mediter Marine Sorting Ctr, Tunisia, 69-71, dir oceanog & limnol prog, 71-74, actg dir int environ sci prog, 74, sr zoologist, 74-78, CUR, DEPT INVERT ZOOL, SMITHSONIAN INST, 78- *Concurrent Pos:* Mem, Int Indian Ocean Exped, 64; NSF grants syst biol, 64, 68; mem adv comt, Smithsonian Oceanog Sorting Ctr, 64-69; NASA grant, 65; mem, Southeast Pac Biol Oceanog Prog, 66; consult zoologist, Smithsonian Inst, 66-68; assoc ed trans, Am Micros Soc, 66-; res assoc, Baruch Inst Marine Biol & Coastal Res, 75-; adj prof biol, W Carolina Univ, Cullowhee, 90- *Mem:* Am Asn Zool Nomenclature; Am Soc Zoologists; Soc Syst Zool; Am Micros Soc (vpres, 76, pres, 79); Int Asn Meiobenthologists (chmn, 80-81); Biol Soc Wash. *Res:* Systematics, life history and ecology of meiobenthic and bryophilic invertebrates, especially Kinorhyncha Loricifera Priapulida and Tardigrada. *Mailing Add:* Smithsonian Inst Washington DC 20560

HIGGINS, TERRY JAY, b Napa, Calif, July 22, 47; m 71; c 4. T-CELL ACTIVATION, CARBOHYDRATE ANTIGENS-LIGANDS. *Educ:* Univ Calif, Davis, BS, 70; Harvard Univ, PhD(immunol), 76. *Prof Exp:* Res asst biochem, Univ Calif, Davis, 70-71; postdoctoral immunol, Australian Nat Univ, 77-79, res fel, 79-80; asst prof microbiol, Sch Med, Univ Pa, 81-87; sr res investr, 88-90, PRIN RES INVESTR IMMUNOPHARMACOL, STERLING DRUG, INC, 90- *Mem:* NY Acad Sci; Am Asn Immunologists; Complex Carbohydrate Soc; AAAS. *Res:* Basic molecular mechanisms of T cell, B cell and macrophage activation and effector function to discover ways of immunomodulating the lost response to self-antigens (autoimmunity/inflammation) and transplants. *Mailing Add:* Sterling Drug Inc 25 Great Valley Pkwy Malvern PA 19355

HIGGINS, THEODORE PARKER, b Houston, Tex, Jan 18, 20; m 48; c 1. MATHEMATICS. *Educ:* Agr & Mech Col, Tex, BS, 40; Univ Tex, MS, 53, PhD(math), 55. *Prof Exp:* Geophysicist, Carter Oil Co, Okla, 40-50; instr math, Univ Tex, 52-55; res specialist, Boeing Airplane Co, Wash, 55-58; sr scientist, Dalmo Victor Co, Calif, 58; MEM STAFF MATH, SCI RES LABS, BOEING AIRPLANE CO, 58- *Mem:* Am Math Soc; Soc Indust & Appl Math; Math Asn Am; Sigma Xi. *Res:* Applications of techniques of applied mathematics to problems in electromagnetic theory. *Mailing Add:* 4222 Mercerwood Dr Mercer Island WA 98040

HIGGINS, THOMAS ERNEST, b Plattsburgh, NY, Aug 24, 48; m 68; c 1. ENVIRONMENTAL ENGINEERING, CIVIL ENGINEERING. *Educ:* Univ Notre Dame, BS, 70, MS, 73, PhD(environ health eng), 75. *Prof Exp:* Proj engr environ eng, Harza Eng Co, 75-77; ASST PROF ENVIRON ENG, DEPT CIVIL ENG, ARIZ STATE UNIV, 77- *Concurrent Pos:* Consult & environ engr, Ten-Ech Environ Consult, 70-74; partner, Higgins Eng, 74-75; teaching asst, Dept Civil Eng, Notre Dame, 73-75; res grant, Eng Found, New York, 78-79; USAF, 80-82; res fel, USAF, 80 & US Dept Eng, 81. *Mem:* Am Chem Soc; Am Soc Civil Eng; Am Water Works Asn; Water Pollution Control Fedn; Am Acad Environ Eng. *Res:* Physiochemical treatment of water and wastewater; water chemistry; design of water and wastewater treatment systems; wastewater reuse; movement of heavy metals in soil systems; treatment of metal plating wastewaters; treatment of shale oil process waste waters. *Mailing Add:* 2112 Lirio Ct Reston VA 22091

HIGGINS, THOMAS JAMES, b Charlottesville, Va, July 4, 11; m 42, 76; c 2. ELECTRICAL ENGINEERING. *Educ:* Cornell Univ, EE, 32, MA, 37; Purdue Univ, PhD(elec eng), 41. *Prof Exp:* Instr math, Auburn Collegiate Ctr, NY, 33-34; plant engr, Agfa Ansco Corp, 34-35; instr elec eng, Wyomissing Polytech Inst, 35-37; instr, Purdue Univ, 37-41; asst prof, Tulane Univ, 41-42; from assoc prof to prof, Ill Inst Technol, 42-48; prof, 48-82, EMER PROF, ELEC ENG, UNIV WIS-MADISON, 82- *Concurrent Pos:* Consult elec engr. *Honors & Awards:* Westinghouse Award, Am Soc Eng Educ, 54; Eckman Award, Instrument Soc Am, 64. *Mem:* Nat Soc Prof Engrs; fel Am Soc Eng Educ; fel Instrument Soc Am; Inst Elec & Electronics Engrs; Tensor Soc (vpres, 64-86); Sigma Xi; fel AAAS. *Res:* Electric circuit theory; applied elasticity; electromagnetic theory; servomechanism and automatic control theory; power systems engineering; electric machine theory; nuclear reactor control theory. *Mailing Add:* 12 Pin Oak Trail Madison WI 53717

HIGGINS, VERNA JESSIE, b Brookvale, NS, Feb 11, 43. PLANT PATHOLOGY. *Educ:* Acadia Univ, BS, 64; Cornell Univ, MS, 66, PhD(plant path), 69. *Prof Exp:* From asst prof to assoc prof, 69-81, PROF PLANT PATH, UNIV TORONTO, 81- *Mem:* Am Phytopath Soc; Can Phytopath Soc. *Res:* Physiological plant pathology; physiology of host-parasite interactions; mechanisms by which plants actively resist infection and response of fungi to such resistance mechanisms. *Mailing Add:* Dept Bot Univ Toronto 25 Willcocks St Toronto ON M5S 3B2 Can

HIGGINS, W(ILLIAM) F(RANCIS), b Boston, Mass, Sept 2, 20; m 48; c 2. ELECTRICAL ENGINEERING. *Educ:* Univ Mass, BS, 50, MS, 53. *Prof Exp:* Engr, Vitro Corp Am, 50-52; instr elec eng, Univ Mass, 52-53 & Brown Univ, 53-55; engr, Sylvania Elec Prod, Inc, Gen Tel & Electronics Corp, 55-60; mem staff, Mass Inst Technol, 60-64 & Sylvania Elec Prod, Inc, Gen Tel & Electronics Corp, 64-68; staff engr, 68-77, GROUP LEADER, MITRE CORP, BEDFORD, 77- *Mem:* Inst Elec & Electronics Engrs. *Res:* Statistical communications and systems engineering. *Mailing Add:* Mitre Corp Burlington Rd Box 208 Bedford MA 01778

HIGGINS, WILLIAM JOSEPH, b Lakewood, Ohio, May 13, 47; m; c 3. COMPARATIVE PHYSIOLOGY. *Educ:* Boston Col, BS, 69; Fla State Univ, PhD(physiol), 73. *Prof Exp:* Asst prof, 73-77, ASSOC PROF ZOOL, UNIV MD, COLLEGE PARK, 77-, ASSOC DEAN, COLS AGR & LIFE SCI, 90- *Mem:* Am Soc Zoologists; Sigma Xi. *Res:* Cyclic nucleotide control of contractility and membrane permeability of molluscan muscles; cyclic nucleotide regulation of cell growth and division. *Mailing Add:* Dept Zool Univ MD College Park MD 20742

HIGGINSON, GEORGE W, JR, b Philadelphia, Pa, Oct 6, 23; m 47; c 3. CHEMICAL ENGINEERING. *Educ:* Univ Pa, BS, 43, MS, 48. *Prof Exp:* Chem engr, Bakelite Corp, NY, 43-44; asst instr chem, Univ Pa, 46-47; chem engr, Firestone Plastics Co, Pa, 47-51; develop supvr, Attapulgus Clay Co, NJ, 51-52; process develop engr, Houdry Process & Chem Co Div, Pa, 52-54, asst plant mgr, Paulsboro, NJ, 54-56, plant mgr, 56-68, mgr mfg & mfg serv, 68-69, dir catalyst sales, 69-71, mgr catalyst sales, 71-74, proj mgr, 74-76, sales mgr, Catalytic, Inc, Subsidiary, Air Prods & Chem, Inc, 76-81; RETIRED. *Mem:* Am Chem Soc; Am Inst Chem Engrs. *Res:* Chemical production and sales management. *Mailing Add:* 411 42nd St Ocean City NJ 08226

HIGGINSON, JOHN, b Belfast, Northern Ireland, Oct 16, 22; m 49; c 2. PATHOLOGY. *Educ:* Univ Dublin, BA, 45, MB, BCh & BAO, 46, MD, 61; FRCP, 71. *Prof Exp:* Asst path & bact, Glasgow Univ, 47-49; pathologist, SAfrican Inst Med Res, Johannesburg, 50-58, head geog path unit, 54-58; assoc prof path, Med Ctr, Univ Kans, 58-62, prof & Am Cancer Soc prof, 62-66, Lewis & Hubbard traveling prof geog path, 60-66; dir, Int Agency Res Cancer, WHO, 66-81; sr consult scientist, Univ Assoc Res & Educ Path, Inc, 82-84; SR FEL & PROF, GEORGETOWN UNIV MED CTR, 85- *Concurrent Pos:* Vis prof, Univ Pittsburgh, 54; Sommer mem lectr, 54; consult, Armed Forces Inst Path, 60; mem path comt, Nat Acad Sci. *Mem:* AAAS; Am Asn Path & Bact; Am Asn Pathologists; Am Asn Cancer Res; Sigma Xi. *Res:* Geographical pathology, environmental cancer. *Mailing Add:* Kober-Cogan Hall Georgetown Univ Med Ctr Washington DC 20007

HIGGS, DONALD VAL, geology, for more information see previous edition

HIGGS, LLOYD ALBERT, b Moncton, NB, June 21, 37; m 66; c 3. RADIO ASTRONOMY. *Educ:* Univ NB, BSc, 58; Oxford Univ, DPhil, 61. *Prof Exp:* Asst res officer, Radio Astron Sect, Nat Res Coun Can, 61-67, assoc res officer, Astrophys, 67-74, sr res officer, Herzberg Inst Astrophys, 75-81; DIR, DOMINION RADIO ASTROPHYS OBSERV, BC, 81- *Concurrent Pos:* Sci civil servant, Observ, State Univ Leiden, 64-65; ed, J Royal Astron Soc Can, 76-80. *Mem:* Am Astron Soc; Royal Astron Soc Can (pres, 88-); Royal Astron Soc; Can Astron Soc (vpres, 88-). *Res:* Physics of interstellar medium, diffuse nebulae; astronomical automation and data processing. *Mailing Add:* Dominion Radio Astrophys Observ Nat Res Coun Can Box 248 Penticton BC V2A 6K3 Can

HIGGS, ROBERT HUGHES, b Morgantown, WVa, July 29, 32; m 57; c 3. GEOMAGNETICS. *Educ:* Pa State Univ, BS, 54; Univ Ariz, 58. *Prof Exp:* Geophysicist, 58-63, head magnetics div, Marine Br, 63-74, dir magnetics div, 74-76, sci & tech dir, 85-86, dir Geophys Dept, 86-87, geophys & mgt comn, 87-90, PHYS SCI ADMINR & DIR HYDROGRAPHIC DEPT, US NAVAL OCEANOG OFF, 76- *Concurrent Pos:* Mem, Soc Explor Geophysicists Comt for Nat Magnetic Anomaly Map, 75-81, NAm Magnetic Anomaly Map Comt, Geol Soc Am. *Honors & Awards:* Navy Civilian Meritorious Serv Award & Metal, 77. *Mem:* Am Geophys Union; Soc Explor Geophysicists; Geol Soc Am; Sigma Xi; Archaeol Inst Am; Nat Speleol Soc. *Res:* Application of geomagnetic survey data to navigation and antisubmarine warfare problems; geological interpretation of geomagnetic data; geomagnetic temporal variation studies; geophysics and hydrographic surveying. *Mailing Add:* 112 Fischer Dr Pearl River LA 70452-9327

HIGGS, ROGER L, b Trivoli, Ill, Apr 10, 38; m 62; c 2. CROP BREEDING, GENETICS. *Educ:* Univ Ill, BS, 60, MS, 61; Iowa State Univ, PhD(crop breeding), 66. *Prof Exp:* Assoc prof, 66-77, PROF AGRON, UNIV WIS-PLATTEVILLE, 77- *Mem:* Am Forage & Grassland Coun; Am Soc Agron; Crop Sci Soc Am; Soil Conserv Soc Am. *Res:* Corn breeding and hybrid evaluation; crop sequence; alfalfa fertilization; corn cultural experiments; agricultural mathematics. *Mailing Add:* Col Agr Univ Wis Platteville WI 53818

HIGH, EDWARD GARFIELD, biochemistry, nutrition; deceased, see previous edition for last biography

HIGH, LEE RAWDON, JR, b Pine Bluff, Ark, Feb 6, 41; c 3. GEOLOGY. *Educ:* Princeton Univ, AB, 63; Rice Univ, PhD(geol), 67. *Prof Exp:* Instr geol, Univ Nebr, 66-67; asst prof, 67-72, assoc prof geol, Oberlin Col, 72-78; STAFF GEOLOGIST, MOBIL EXPLOR & PROD SERV INC, 78- *Concurrent Pos:* Petrol Res Fund res grants, 67-69 & 71-; Res Corp res grant, 70-71. *Mem:* Am Asn Petrol Geol; Geol Soc Am; Soc Econ Paleont & Mineral. *Res:* Environmental interpretation of sedimentary rocks; lacustrine rocks; recent coastal sediments; sedimentary structures; Triassic stratigraphy; environmental and archaeological geology; regional geology. *Mailing Add:* 7826 Meadow Park Dr Dallas TX 75230

HIGH, LEROY BERTOLET, b Menomonie, Wis, May 20, 14; m 39; c 3. ORGANIC CHEMISTRY. *Educ:* Univ Mich, BS, 36, MS, 37; Wayne State Univ, PhD(chem), 56. *Prof Exp:* Res chemist, Udylite Corp, 37-56; sr res chemist, Squibb Inst Med Res, 57-80, res fel, 80-81; consult, McNeil, 82-88; RETIRED. *Mem:* Am Chem Soc; NY Acad Sci. *Res:* Steroids; alkaloids; antibiotics; peptides; heterocyclics; carbohydrates. *Mailing Add:* Five Piedmont Dr Cranbury NJ 08512-9728

HIGHET, ROBERT JOHN, b Springfield, Ill, Oct 6, 25; m 55; c 3. ORGANIC CHEMISTRY. *Educ:* Univ Ill, BS, 50; Univ Wis, PhD(chem), 54. *Prof Exp:* Chemist, E I du Pont de Nemours & Co, Inc, 52; CHEMIST, NAT HEART, LUNG & BLOOD INST, 54- *Concurrent Pos:* Am Cancer Soc res fel, Swiss Fed Inst Technol, 57-58. *Mem:* Am Chem Soc; The Chem Soc; NY Acad Sci. *Res:* Alkaloids of the Amaryllidaceae and Cassia species; nuclear magnetic resonance and mass spectrometry. *Mailing Add:* Bldg 10-7N320 NIH Bethesda MD 20892

HIGHLAND, HENRY ARTHUR, b Patchogue, NY, July 17, 24; m 51; c 4. ECONOMIC ENTOMOLOGY. *Educ:* Wash Col, BS, 50; Univ Md, MS, 55, PhD, 57. *Prof Exp:* Biol aide, NIH, 50-51 & USDA, 51-53; asst, Univ Md, 53-57; ENTOMOLOGIST, USDA, 57- *Mem:* Am Asn Cereal Chem; Entom Soc Am. *Res:* House fly attractants and baits; biology and control of insects affecting ornamental plants; stored-product insects; insect resistant food packaging. *Mailing Add:* 424 E Macon St No B Savannah GA 31401

HIGHLAND, VIRGIL LEE, b Clarksburg, WVa, Oct 12, 35; m 63; c 3. PHYSICS. *Educ:* Yale Univ, BS, 57; Cornell Univ, PhD(physics), 63. *Prof Exp:* Res assoc physics, Univ Pa, 63-65, asst prof, 65-67; assoc prof, 67-79, PROF PHYSICS, TEMPLE UNIV, 79- *Mem:* Am Phys Soc; Am Asn Physics Teachers; AAAS; Sigma Xi. *Res:* Experimental elementary particle physics; weak and strong interactions. *Mailing Add:* Dept Physics Temple Univ Philadelphia PA 19122

HIGHLANDS, MATTHEW EDWARD, b Huntington, Ind, June 19, 05; m 31. BACTERIOLOGY, FOOD SCIENCE. *Educ:* Univ Maine, AB, 28; Mass Inst Technol, SM, 34; Univ Mass, PhD, 51. *Prof Exp:* Asst microbiol, Mass Inst Technol, 28-29, 32-34, res assoc, 35; assoc bacteriologist, Frigidaire Corp, Div, Gen Motors Corp, Ohio, 29-32; bacteriologist & food technologist, Palo Prod Co, PR, 34-35; from instr to asst prof bact, Univ Maine, 35-42; mgr res & processing, Friend Bro, Inc, Mass, 46; prod mgr, Lange Canning Corp, Wis, 47; assoc food technologist, Agr Exp Sta, Maine, 47-68, head dept food sci, 55-68, prof food technol, 55-70, EMER PROF FOOD SCI, UNIV MAINE, ORONO, 70- *Concurrent Pos:* Dir, Food & Container Assocs, 58-60. *Mem:* Am Soc Microbiol; Inst Food Tech; Am Inst Chem Eng. *Res:* Freezing of fruits; potato product development; canning technology; dehydration of vegetables; technology of processed fisheries products. *Mailing Add:* 111 Forest Ave Orono ME 04473

HIGHLEY, TERRY L, b Anamosa, Iowa, July 3, 40; m 67; c 2. FOREST PATHOLOGY. *Educ:* Iowa State Univ, BS, 62; Ore State Univ, MS, 64, PhD(plant path), 68. *Prof Exp:* PLANT PATHOLOGIST, US FOREST PROD LAB, 67- *Honors & Awards:* USDA Super Serv Award. *Mem:* Int Res Group Wood Preserv; Am Phytopath Soc. *Res:* Forest products pathology; fungal hydrolytic enzymes; natural durability of wood; nontoxic methods of improving durability of wood; deterioration of wood in the marine environment. *Mailing Add:* US Forest Prod Lab Madison WI 53705

HIGHMAN, BENJAMIN, pathology; deceased, see previous edition for last biography

HIGHSMITH, PHILLIP E, b Glenmary, Tenn, Dec 2, 25; m 52; c 3. PHYSICS. *Educ:* Tenn State Univ, BS, 51; Univ Va, MEd, 55; Ohio State Univ, PhD(physics), 66. *Prof Exp:* Instr physics & math, Greenbrier Mil Sch, 52-54 & Fork Union Mil Acad, 54-60; asst prof physics, Transylvania Col, 60-66; assoc prof, 66-73, PROF PHYSICS, CONVERSE COL, 73- *Mem:* Am Asn Physics Teachers; Nat Sci Teachers Asn. *Res:* Acceleration of macroscopic ferromagnetic materials in varying magnetic fields. *Mailing Add:* Dept Physics Converse Col Spartanburg SC 29301

HIGHSMITH, ROBERT F, b June 22, 45; m 67; c 2. BASIC SCIENCE RESEARCH, CELL CULTURE. *Educ:* Univ Cincinnati, PhD(physiol), 72. *Prof Exp:* Fel biophys, Harvard Med Sch, 72-74; PROF PHYSIOL & BIOPHYS, COL MED, UNIV CINCINNATI, 75- *Concurrent Pos:* Consult & prin investr, NIH; mem comt, Soc Exp Biol & Med. *Mem:* Am Physiol Soc; Int Soc Thrombosis/Hemostasis; Soc Exp Biol & Med; AAAS. *Res:* Intercellular communication in the blood vessel wall; endothelial cell-vascular smooth muscle coupling; ion channels. *Mailing Add:* Physiol & Biophys Univ Cincinnati 576 4108 MSB Cincinnati OH 45267-0576

HIGHSMITH, RONALD EARL, b Olney, Ill, Feb 2 39; m 63; c 2. POLYMER FLOCCULANTS, POLYELECTROLYTES. *Educ:* Eastern Ill Univ, BS, 63; Univ Fla, PhD(chem), 68. *Prof Exp:* Fel & lectr, chem dept, Univ Pa, 68-69; chemist, Carborundum Co, Niagara Falls, NY, 69-72, proj mgr, 72-77, new prod mgr, Knoxville Tenn, 77-78; gen mgr, Micron Chem Co, Knoxville, Tenn, 78-80; dir res & develop water treat polymers, Allied-Signal Inc, Syracuse, NY, 80-88; DIR RES & DEVELOP, POLYPURE, INC, SUBSID RHÔNE-POULENC, 88- *Res:* Water purification; water pollution control; sludge dewatering equipment; polymer flocculants; polymer hydrogels; microbial polysaccharides; mechanisms of polymer flocculation; synthesis of polymer flocculants; colloid chemistry. *Mailing Add:* 28 E Austin St Skaneateles NY 13152

HIGHSTEIN, STEPHEN MORRIS, b Baltimore, Md, Aug 23, 39; m 69; c 2. ANIMAL PHYSIOLOGY. *Educ:* Rensselaer Polytech Inst, BS, 61; Med Sch, Univ Md, MD, 65, Univ Tokoyo, PhD(physiol), 76. *Prof Exp:* Intern med, Maimonides Hosp, NY, 65-66; resident med, Mt Sinai Hosp, NY, 66-69; fel physiol, Fac Med, Univ Tokyo, 69-72; asst prof neurol, Mt Sinai Hosp, NY, 73-74; from asst prof to prof neurosci, Alber Einstein Col Med, NY, 74-83; PROF OTOLARYNGOL, ANAT & NEUROBIOL, WASH UNIV, ST LOUIS, 83- *Concurrent Pos:* Fel, Nat Inst Neruol, Commun Disorders & Stroke; scholar, New York Health Coun Career Scientist, 73-75, NIH Res Career Develop Award, 75-80; chair, Nat Acad Res Study Sect; mem, Soc Neurosci Prog Comt & Panel NIH/NINDS Nat Strategic Res Plan. *Mem:* Soc Neurosci; Am Physiol Soc; Int Brain Res Orgn. *Res:* Structure function studies of the central vestibular and oculomotor systems using intracellular recording and staining; cellular basis of responses of the peripheral vestibular apparatus including efferent control, using intracellular recording and staining, molecular techniques. *Mailing Add:* Dept Anat & Otol & Neurobiol Wash Univ Med 660 S Euclid Ave Box 8115 St Louis MO 63110

HIGHT, DONALD WAYNE, b Neodesha, Kans, Aug 23, 31; m 51; c 2. MATHEMATICS. *Educ:* Kans State Col Pittsburg, BS, 53; Okla State Univ, MS, 58, EdD(math educ), 61. *Prof Exp:* Teacher high schs, Kans, 53, 55-57, 58-60; asst prof math, Ark State Col, 61-62; assoc prof, 62-67, PROF MATH, PITTSBURG STATE UNIV, 67- *Concurrent Pos:* NSF sci fac fel, Univ Mass, 67-68; dir, Am Asn Col Teachers Educ, 71-74; mem adv coun, Assoc Orgns for Teacher Educ, 70. *Mem:* Math Asn Am. *Res:* Mathematics education and participation in accreditation of teacher education by learned societies and academic organizations. *Mailing Add:* Dept Math Pittsburg State Univ Pittsburg KS 66762

HIGHT, RALPH DALE, b San Antonio, Tex, Mar 6, 45; m 67; c 1. ATOMIC PHYSICS. *Educ:* NTex State Univ, BS, 67, MS, 69; Mont State Univ, PhD(atomic physics), 75. *Prof Exp:* Fel atomic physics, Univ Toledo, 75-77; vis asst prof physics, Bowling Green State Univ, 77; vis asst prof atomic res, Univ Nebr, Lincol, 77-79; res scientist, Dale Electronics, 79-85; sr assoc prin engr, Harris Corp, 85-89; PROJ MGR, BURR BROWN CORP, 89- *Concurrent Pos:* Nat Defense Educ Act fel, Mont State Univ, 69-72; NSF fel, Univ Toledo, 75-77. *Mem:* Am Inst Physics; Int Soc Hybrid Microelectronics. *Res:* Experimental atomic and molecular physics, specifically beam-foil spectroscopy and optical excitation functions of diatomic and triatomic molecules; sputtering; thin films; hybrid technol. *Mailing Add:* 7151 E Pintail Dr Tucson AZ 85715-6158

HIGHT, ROBERT, b Springfield, Mo, Mar 16, 30; m 62. SOLID STATE PHYSICS. *Educ:* Southwest Mo State Col, BS, 55; Univ Mo, MS, 59, PhD(physics), 62. *Prof Exp:* Assoc physicist, Midwest Res Inst, 61-62; asst prof physics, Kans State Col Pittsburg, 62-65; asst prof, 65-69, ASSOC PROF PHYSICS, UNIV MO, ST LOUIS, 69- *Mem:* Am Phys Soc; Am Asn Physics Teachers. *Res:* Debye-Waller characteristics of II-VI and III-V compounds. *Mailing Add:* Dept Physics Univ Mo 8001 Nat Bridge St Louis MO 63121

HIGHTON, RICHARD, b Chicago, Ill, Dec 24, 27; m 50; c 4. EVOLUTIONARY BIOLOGY, HERPETOLOGY. *Educ:* NY Univ, AB, 50; Univ Fla, MS, 53, PhD(biol), 56. *Prof Exp:* From asst prof to assoc prof, 56-73, PROF ZOOL, UNIV MD, COLLEGE PARK, 73- *Mem:* Fel AAAS; Am Soc Naturalists; Soc Study Evolution; Genetics Soc Am; Am Soc Ichthyologists & Herpetologists (pres, 76); Soc Syst Zool; Ecol Soc Am. *Res:* Evolution; population genetics; systematics; life history of salamanders. *Mailing Add:* Dept Zool Univ Md College Park MD 20742

HIGHTOWER, COLLIN JAMES, b Liverpool, Eng, Aug 12, 36; US citizen; m 65; c 2. NUMBER THEORY. *Educ:* Univ Ark, BS, 58; Tulane Univ, PhD(math), 63. *Prof Exp:* Asst prof, Boulder, 65-72, vchmn in charge Denver Ctr, 67-69, math chmn, 72-76, PROF MATH, UNIV COLO, DENVER, 72- *Mem:* Math Asn Am; Am Math Soc. *Res:* Geometry of numbers. *Mailing Add:* Dept Math Univ Colo 1200 Larimer Denver CO 80204

HIGHTOWER, DAN, b Eastland, Tex, Oct 15, 25; m 51. VETERINARY NUCLEAR MEDICINE. *Educ:* Tex A&M Univ, DVM, 46; NC State Univ, MS, 61. *Prof Exp:* Vet officer, Vet Corps, US Army, 46-56, radiation planning officer, Qm Radiation Planning Agency, 57-59, chief reactor sect, Walter Reed Army Inst Res, 61-63 & Dept Biophys, 63-66; assoc prof, 66-72, PROF VET PHYSIOL & PHARMACOL, TEX A&M UNIV, 72- *Mem:* AAAS; Am Vet Med Asn; Radiation Res Soc; Soc Nuclear Med. *Res:* Biological effects of particulate radiations; neutrons, use of radioisotopes in biology and medicine and applications of activation analysis to biology and medicine. *Mailing Add:* Dept Vet Physiol & Pharmacol Tex A&M Univ College Station TX 77843-4466

HIGHTOWER, FELDA, b Wadesboro, NC, Dec 21, 09; m 35; c 3. SURGERY. *Educ:* Wake Forest Col, BS, 31; Univ Pa, MD, 33; Am Bd Surg & Am Bd Thoracic Surg, dipl. *Prof Exp:* Med dir & surgeon, NC State Prison Hosp, 38-40; from instr anat to assoc prof surg, 40-68, PROF SURG, BOWMAN GRAY SCH MED, 68- *Mem:* Fel Am Col Surg; Soc Surg Alimentary Tract; Int Col Digestive Surg. *Res:* General and thoracic surgery; gastrointestinal tract; peripheral vascular diseases. *Mailing Add:* Dept Surg Bowman Gray Sch Med Winston-Salem NC 27103

HIGHTOWER, JAMES K, b Kalamazoo, Mich, Mar 22, 37; m 57; c 2. ARTIFICIAL INTELLIGENCE. *Educ:* Kalamazoo Col, AB, 58; Claremont Grad Sch, MA, 67, PhD(econ), 70. *Prof Exp:* Nat Defense Emergency Authorization fel, 60-63; asst prof math & econ, Univ Richmond, 64-67; asst prof econ, Calif State Polytech Univ, Pomona, 67-68; assoc dean grad prog, Calif State Univ, Fullerton, 70-73, assoc prof quant methods, 69-76; assoc dir, Comput & Commun Resources, 87-90, DIR ACAD COMPUT, CALIF STATE UNIV, 90- *Concurrent Pos:* Lectr, Claremont Grad Sch, 67-69 & 73-; adj prof computer sci, Calif State Univ, Dominguez Hills, 83-87, Fullerton, 88- *Mem:* Asn Comput Mach; Opers Res Soc Am; Inst Elec & Electronics Engrs; NY Acad Sci; Am Asn Artificial Intel. *Res:* Application of artificial intelligence techniques to educational systems (computer aided instruction); ethical problems in application of computer technology. *Mailing Add:* Off of the Chancellor Calif State Univ PO Box 3842 Seal Beach CA 90740-7842

HIGHTOWER, JESSE ROBERT, b Cleveland, Tenn, Apr 9, 39; m 61; c 1. CHEMICAL ENGINEERING. *Educ:* Univ Miss, BS, 61; Tulane Univ, MS, 63, PhD(chem eng), 64. *Prof Exp:* Staff mem process develop, Oak Ridge Nat Lab, 64-74, prog mgr, 74-76, sect head process res & develop, Martin Marietta Energy Systs, Inc, 76-88, ASSOC DIV DIR, RADIO CHEM PROC, OAK RIDGE NAT LAB, 88- *Mem:* Am Inst Chem Engrs; AAAS. *Res:* Nuclear fuel processing development; development of fossil fuel conversion technology; chemical engineering research and development. *Mailing Add:* Oak Ridge Nat Lab PO Box 2008 Oak Ridge TN 37831-6268

HIGHTOWER, JOE W(ALTER), b Morrilton, Ark, Sept 14, 36; m 80; c 1. PHYSICAL CHEMISTRY, CHEMICAL ENGINEERING. *Educ:* Harding Col, BS, 59; Johns Hopkins Univ, MS, 61, PhD(catalysis), 63. *Prof Exp:* NSF fel, catalysis, Queen's Univ, Belfast, 63-64; Gulf fel, Mellon Inst, 64-67; assoc prof chem eng, 67-71, PROF CHEM ENG, RICE UNIV, 71- *Concurrent Pos:* Res grants, Petrol Res Fund & NASA, 68-70, NSF, 69-71, Welch Found, 69-72 & Baroid Div grant, NL Industs, 71-75; chmn, Gordon Res Conf Catalysis, 73; consult, Exxon Res & Eng, 68-71, Chevron Oil Field Res Co, 72-76, several Nat Res Coun panels on auto emission control catalysts, 72-77, Monsanto Co, 74- & Catalytica Assocs, 74-; chmn adv bd, Petrol Res Found, 82-; founder & pres, Human Resources Develop Found, 68-; pres, Dow Chem, 84- *Honors & Awards:* Nat Petrol Chem Award, Am Chem Soc, 73; Jefferson Prize, 83. *Mem:* Am Chem Soc; Catalysis Soc; Am Inst Chem Engrs; Southwest Catalysis Soc (pres, 68-70). *Res:* Heterogeneous catalysis, mainly with the use of isotopic tracers to study reaction mechanisms. *Mailing Add:* Dept Chem Eng Rice Univ Box 1892 Houston TX 77251-1892

HIGHTOWER, KENNETH RALPH, b Honolulu, Hawaii, Mar 20, 47; m 67; c 2. BIOPHYSICS. *Educ:* Southern Ill Univ, BA, 69, MA, 70, PhD(molecular sci), 74. *Prof Exp:* Teaching asst physics, Southern Ill Univ, 69-71, preceptor, 71-74; RES ASSOC, INST BIOL SCI, 74- *Concurrent Pos:* NIH fel, Inst Biol Sci, Oakland Univ, 74-77, res assoc & asst prof health sci electrophysiol, 78- *Mem:* Sigma Xi; Asn Res Vision & Opthal. *Res:* Lens biophysics; kinetics of transport and electrophysiology of cataractous lenses; pigment epithelium of retina and associated transport and electrical properties. *Mailing Add:* 3373 Innsbrook Dr Rochester MI 48063

HIGHTOWER, LAWRENCE EDWARD, b Fresno, Calif, June 2, 46; m 70; c 1. CELL BIOLOGY, VIROLOGY. *Educ:* Hampden-Sydney Col, BS, 68; Harvard Univ, PhD(biochem), 74. *Prof Exp:* Instr microbiol, Med Sch, Univ Mass, 74-75; from asst prof to assoc prof, 75-85, PROF BIOL, UNIV CONN, 85- *Concurrent Pos:* Cell biol adv panel, NSF, 81-84. *Mem:* Am Soc Microbiol; Am Soc Cell Biol; AAAS. *Res:* Molecular and cell biology of environmental stress responses; the heat shock response in mammalian, avian, and fish cells. *Mailing Add:* Dept Molecular & Cell Biol Univ of Conn Storrs CT 06269-3044

HIGHTOWER, NICHOLAS CARR, JR, b Nashville, Tenn, Sept 26, 18; m 41; c 4. PHYSIOLOGY. *Educ:* NTex State Univ, BS, 41; Univ Tex, MD, 44; Univ Minn, MS, 49, PhD, 52. *Prof Exp:* Instr physiol, Univ Minn, 50-51, asst to staff, Sect Physiol, Mayo Found, 51-52; consult internal med, 52-54, dir dept clin res, 54-69, dir, Div Res & Educ, 69-79, sr consult, res & gastroenterol, Scott & White Clin, 79-86; EMER PROF MED, TEX A&M UNIV COL MED, 86- *Concurrent Pos:* Secy-treas, Gastroenterol Res Group, 60-62, chmn, 63; consult gastroenterol, Vet Admin Hosp, Temple, Tex. *Honors & Awards:* Seale Harris Award, Southern Med Asn, 60; Outstanding Achievement Awards, Univ Minn & Mayo Found, 64 & NTex State Univ, 65; Ashbel Smith Award, Galveston, 74. *Mem:* AAAS; Am Physiol Soc; AMA; Am Gastroenterol Asn (pres, 71-72); Am Fedn Clin Res. *Res:* Gastrointestinal physiology. *Mailing Add:* Scott & White Clin Temple TX 76508

HIGINBOTHAM, WILLIAM ALFRED, b Bridgeport, Conn, Oct 25, 10; m 49, 76, 83; c 3. PHYSICS. *Educ:* Williams Col, AB, 32. *Hon Degrees:* DSc, Williams Col, 63. *Prof Exp:* Mem staff, Radiation Lab, Mass Inst Technol, 41-43; mem staff, Manhattan Proj, Los Alamos Sci Lab, Univ Calif, 43-44, head electron group, 44-45; exec secy, Fedn Am Scientists, Washington, DC, 46-47; assoc head, Instrumentation Div, Brookhaven Nat Lab, 48-51, head, 52-68, sr physicist, Dept Appl Sci, 68-84; RETIRED. *Concurrent Pos:* Tech ed, J Inst Nuclear mat mgt, 73- *Honors & Awards:* Nuclear Sci Contrib Award, Inst Elec & Electronics Engrs, 72; Distinguished Serv Award, Inst Nuclear Mat Mgt, 78. *Mem:* Fel AAAS; fel Am Phys Soc; fel Am Nuclear Soc; fel Inst Nuclear Mat Mgt; fel Inst Elec & Electronics Engrs. *Res:* Radar; instrumentation for nuclear research; nuclear materials safeguards. *Mailing Add:* 11 N Howell's Pt Rd Bellport NY 11713

HIGLEY, LEON GEORGE, b Sacramento, Calif, Mar 7, 58; m 81; c 1. ARTHROPOD & PLANT STRESS RELATIONSHIPS, PEST MANAGEMENT THEORY. *Educ:* Cornell Univ, BA, 80; Iowa State Univ, MS, 84, PhD(entom & crop physiol), 88. *Prof Exp:* Temp asst prof, Iowa State Univ, 88-89; ASST PROF, UNIV NEBR, 89- *Honors & Awards:* J H Comstock Award, Entom Soc Am. *Mem:* Entom Soc Am; Crop Sci Soc; Am Agron Soc. *Res:* Plant physiological responses to biotic stress; arthropod and plant stress interactions; photosynthetic responses to injury; pest management theory; soybean entomology; insect ecology. *Mailing Add:* 202 Plant Indust Bldg Univ Nebr Lincoln NE 68583-0816

HIGMAN, DONALD GORDON, b Vancouver, BC, Sept 20, 28; m 49; c 3. MATHEMATICS. *Educ:* Univ BC, BA, 49; Univ Ill, MA, 50, PhD(math), 52. *Prof Exp:* Nat Res Coun fel, 52-54; assoc prof math, Univ Mont, 54-57; from asst prof to assoc prof, 57-63, PROF MATH, UNIV MICH, ANN ARBOR, 63- *Concurrent Pos:* Vis prof, Univ Frankfurt, 62-63. *Res:* Group theory. *Mailing Add:* Dept Math Univ Mich Ann Arbor MI 48104

HIGMAN, HENRY BOOTH, b Millington, Md, Mar 17, 27; m 52; c 4. NEUROCHEMISTRY, PHARMACOLOGY. *Educ:* St John's Col, BA, 50; Univ Md, MD, 55. *Prof Exp:* Intern, Del Hosp, Wilmington, 55-56; resident, Dept Neurol, La State Univ, 56-59, instr neurol, Univ, 62-64; from asst prof to assoc prof, Univ Ill, 64-67; PROF NEUROL & CHMN DEPT, SCH MED, UNIV PITTSBURGH, 68- *Concurrent Pos:* NIH spec training fel neurochem, Col Physicians & Surgeons, Columbia Univ, 59-62; NIH grant, 64-68, res career develop award, 66-68; mem, Neurol Sci Res Training Comt, Nat Inst Neurol Dis & Blindness, 67-71. *Mem:* Am Acad Neurol; Am Neurol Asn; assoc Am Asn Neuropath; Int Soc Neurochem. *Res:* Molecular mechanism of cholinergic transmission; ion flux in excitable membranes. *Mailing Add:* Dept Neurol Scaife Hall 0322 Sch Med Univ Pittsburgh 4200 Fifth Ave Pittsburgh PA 15213

HIGMAN, JAMES B, b Millington, Md, Feb 22, 22; m 49; c 4. FISHERIES. *Educ:* Western Md Col, AB, 43; Univ Miami, MS, 55. *Prof Exp:* Fishery biologist, Marine Lab, Univ Miami, 50-56; fisheries methods & equip specialist, US Fish & Wildlife Serv, Fla & Washington, DC, 56-59; fishery biologist, Inst Marine Sci, 59-67, RES SCIENTIST, ROSENSTIEL SCH MARINE & ATMOSPHERIC SCI, UNIV MIAMI, 67-, EXEC DIR, GULF & CARIBBEAN FISHERIES INST, 59- *Mem:* Am Fisheries Soc; Am Inst Fishery Res Biol. *Res:* Fisheries. *Mailing Add:* Bullentine Marine Sci Rosensteil Sch Marine & Atmospheric Sci 4600 Rickenbacker Causeway Miami FL 33149

HIGNETT, TRAVIS P(ORTER), b Maxwell, Iowa, Dec 30, 07; m 33. CHEMICAL ENGINEERING. *Educ:* Drake Univ, AB, 29. *Prof Exp:* Chemist, fixed nitrogen res lab, USDA, 29-33; chem engr, Res Assocs, Inc, 33-38; chem engr, Wilson Dam, 38-42, proj leader, 42-47, chief develop br, 47-59, appl res br, 59-62, dir chem develop, 62-73, SPEC CONSULT TO MANAGING DIR, INT FERTILIZER DEVELOP CTR, TENN VALLEY AUTHORITY, 74- *Mem:* Am Chem Soc; Am Inst Chem Engrs; Brit Fertilizer Soc. *Res:* Fertilizer technology. *Mailing Add:* 411 Oak Hill Ave Sheffield AL 35660

HIGUCHI, HIROSHI, b Japan. FLUID MECHANICS. *Educ:* Univ Tokyo, BS, 70; Calif Inst Technol, MS, 71, PhD(aeronaut), 77. *Prof Exp:* Res assoc, Nat Res Coun, NASA Ames, 76-77; res scientist, Dynamics Technol, 77-80; sr res assoc, Univ Santa Clara, 80-81; res assoc, Univ Minn, 81-89; ASSOC PROF, DEPT MECH & AEROSPACE ENG, SYRACUSE UNIV, 89- *Concurrent Pos:* Prin investr, NASA Ames Res Ctr, 76-81; mem, Nat Parachute Technol Coun, 90- *Honors & Awards:* Space Act Award, NASA, 88. *Mem:* Assoc fel Am Inst Aeronaut & Astronaut; Am Soc Mech Engrs; Am Phys Soc. *Res:* Bluff body aerodynamics; vortex interactions; turbulent boundary layers and wakes; hydroacoustics. *Mailing Add:* Dept Mech & Aerospace Eng Syracuse Univ Syracuse NY 13244

HIGUCHI, WILLIAM IYEO, b San Jose, Calif, Mar 16, 31; m 56; c 4. PHARMACEUTICAL CHEMISTRY. *Educ:* San Jose State Col, AB, 52; Univ Calif, PhD(chem), 57. *Prof Exp:* Proj assoc phys pharm, Univ Wis, 56-58; res chemist phys chem, Calif Res Corp, 58-59; asst prof phys pharm, Univ Wis, 59-62; from assoc prof to prof pharm, 62-76, PROF DENT, UNIV MICH, ANN ARBOR, 66-, ALBERT B PRESCOTT PROF PHARM, 76-; AT PHARMACEUT DEPT, COL PHARM, UNIV UTAH. *Concurrent Pos:* Mem dent training comt, Nat Inst Dent Res. *Honors & Awards:* Ebert Prize, Am Pharmaceut Asn, 68 & 70, Res Achievement Award Phys Pharm, 70. *Mem:* AAAS; Am Chem Soc; Am Pharmaceut Asn; fel Acad Pharmaceut Sci; Am Asn Dent Res; Sigma Xi. *Res:* Physical pharmaceutical chemistry; thermodynamic and kinetic behavior of polyphase systems; nucleation kinetics; dissolution phenomena; adsorption and surface properties; colloids, emulsions and suspensions; solubilization phenomena in aqueous and nonaqueous systems; enamel demineralization mechanisms. *Mailing Add:* Pharmaceut Dept Col Pharm Univ Utah Salt Lake City UT 84112

HIHARA, LLOYD HIROMI, b Honolulu, Hawaii, Oct 29, 61. CORROSION, MECHANICAL BEHAVIOR OF MATERIALS. *Educ:* Univ Hawaii, Manoa, BS, 83; Mass Inst Technol, SM, 85, PhD(mat sci & eng), 89. *Prof Exp:* ASST PROF MECH ENG, UNIV HAWAII, MANOA, 89- *Concurrent Pos:* Prin investr, NSF, 90-; Am Soc Mech Engrs outstanding fac award, Univ Hawaii, 90. *Mem:* Nat Asn Corrosion Engrs; Electrochem Soc; Am Inst Mining Metall & Petrol Engrs, Metall Soc; Am Soc Mech Engrs; Sigma Xi. *Res:* Corrosion studies of metal-matrix composites. *Mailing Add:* Dept Mech Eng Holmes Hall 302 Univ Hawaii Manoa Honolulu HI 96822

HIIEMAE, KAREN MARION, anatomy, evolutionary biology, for more information see previous edition

HIKIDA, ROBERT SEIICHI, b Long Beach, Calif, June 3, 41; m 64; c 1. EXPERIMENTAL MORPHOLOGY. *Educ:* Univ Ill, BS, 63, MS, 65, PhD(zool), 67. *Prof Exp:* Nat Inst Neurol Dis & Blindness res assoc biol, Columbia Univ, 67-69; asst prof, 69-73, assoc prof, 73-77, PROF ZOOL, OHIO UNIV, 77- *Concurrent Pos:* Assoc ed, Anatomical Record, 80- *Mem:* Am Soc Zool; Am Asn Anat; NY Acad Sci; Am Soc Cell Biol. *Res:* Comparative vertebrate muscle structure and activity; maintenance of skeletal muscle integrity; regeneration, neurotrophism, development; ultrastructure, histochemistry and histology; adaptations of human muscle to activity. *Mailing Add:* Dept Zool & Biomed Sci Ohio Univ Athens OH 45701-2979

HILAL, SADEK K, b Cairo, Egypt, Mar 11, 30; US citizen; m 64; c 4. RADIOLOGY. *Educ:* Univ Cairo, MD, 55; Univ Minn, PhD(radiol), 66. *Prof Exp:* Resident radiation ther, Mem Ctr Cancer & Allied Dis, 57-58; resident radiodiag, Univ Minn, 59-61; from asst prof to assoc prof, 63-69, PROF RADIOL, COL PHYSICIANS & SURGEONS, COLUMBIA UNIV, 69-, DIR NEURORADIOL, 75- *Concurrent Pos:* Res grants diag & therapeut radiol & radiologist training grant; assoc ed, Invest Radiol & Child's Brain; chmn, Int Symp Small Vessel Angiography; consult, Study Sect, NIH & Am Bd Radiol. *Honors & Awards:* Gold Medal, Asn Univ Radiologists, 62. *Mem:* Am Roentgen Ray Soc; Am Col Radiol; Asn Univ Radiologists; Radiol Soc NAm; Am Soc Neuroradiol (pres, 77-78). *Res:* Neuroradiology radiodiagnosis; cerebral blood flow and angiography; methodology; radiographic image improvement; exploration of the small intracranial vessels in man with a magnetic catheter and the conduction of interventional procedures; toxicity of radiographic contrast media; developed the fourth generation computed tomography scanner. *Mailing Add:* One Loretta Ct Englewood Cliffs NJ 07632

HILBERG, ALBERT WILLIAM, b Michigan City, Ind, Apr 5, 22; m 43; c 5. PATHOLOGY. *Educ:* Elmhurst Col, BS, 44; Ind Univ, MD, 46. *Hon Degrees:* DSc, Elmhurst Col, 61. *Prof Exp:* Intern, Evangel Hosp, Chicago, 46-47; from asst to instr path, Sch Med, Univ Iowa, 49-51; pathologist, Lab Path, Nat Cancer Inst, 52-57, head cytodiag serv, Dept Path Anat, 53-57, path consult, Field Invests & Demonstrations Br & head cytol sect, 57-62; chief radiopath div, Armed Forces Inst Path, 62-65; dep chief res br, Div Radiol Health, USPHS, Md, 65-67, chief pop studies prog, Nat Ctr Radiol Health, 67-68; prof assoc, Div Med Sci, Nat Acad Sci-Nat Res Coun, 68-79; RETIRED. *Concurrent Pos:* Fel, Hosp for Joint Dis, NY, 51-52. *Mem:* Am Asn Path & Bact; emer fel Col Am Path; emer fel Am Soc Clin Path; emer mem Soc Nuclear Med; NY Acad Sci; emer mem Int Acad Path. *Res:* Long-term effects of ionizing radiation induction; histogenesis and morphology of bone tumors; exfoliative cytology. *Mailing Add:* 12512 Davan Dr Silver Spring MD 20904

HILBERT, MORTON S, b Pasadena, Calif, Jan 3, 17; m 72; c 3. ENVIRONMENTAL HEALTH, PUBLIC HEALTH. *Educ:* Univ Calif, Berkeley, BSCE; Univ Mich, MPH, 46; Am Acad Environ Engr, dipl. *Prof Exp:* Sanit engr, Barry County Dept Health, Mich, 40-42, Chippewa County, Mich, 42-43 & Wayne County Health Dept, Mich, 44-58, asst dir, 58-62; assoc prof, 61-66, assoc dir, Inst Environ & Indust Health, 70-78, PROF ENVIRON HEALTH, SCH PUB HEALTH, UNIV MICH, ANN ARBOR, 66-, CHMN DEPT ENVIRON & INDUST HEALTH, 70-, DIR INST ENVIRON & INDUST HEALTH, 78- *Concurrent Pos:* Field engr construct, Great Lakes Dredge & Dock Co, 42-43; instr, Wayne State Univ, 51-63; lectr pub health, Mercy Col, 58-65; consult eng, Vietnamese govt, 54; coun consult, Nat Sanit Found, 56- consult environ health, Virgin Islands Govt, 62-71, aide to gov, 70; consult, Pan-Am Health Orgn, 71; environ health consult, Egyptian Acad Sci Res & Technol, 74-76, Yugoslavia, 76, WHO, WPac Region, 67, Geneva, Switz, 68, Thailand, 71 & Copenhagen, 82; consult cancer control, Stanford Res Inst, 75-76; mem, Cancer Control & Rehab Comt, NIH, 75-79; chmn, Univ Senate, Univ Mich, 84-85; chmn, Water Qual Rev Panel, Water Qual Asn, 85- *Mem:* Fel Am Pub Health Asn (pres-elect, 74-75, pres, 75-76); Am Acad Health Admin (pres, 71); Am Acad Environ Engrs; Water Pollution Control Fedn; Conf Local Environ Health Adminr (pres, 52). *Mailing Add:* Ave Grand Champe 148 Brussels 1150 Belgium

HILBERT, ROBERT S, b Washington, DC, Apr 29, 41; m 66; c 2. OPTICS. *Educ:* Univ Rochester, BS, 62, MS, 64. *Prof Exp:* Fel, Am Optical Co, Univ Rochester, 62; engr, Itek Corp, 63-65, supvr, 65-67, asst mgr, 67-69, mgr, 69-74, dir, 74-75; SR VPRES ENG, OPTICAL RES ASSOCS, INC, 75- *Concurrent Pos:* Lectr, Grad Sch Eng, Northeastern Univ, 67-69. *Mem:* Optical Soc Am; Soc Photo-Optical Instrumentation Engrs. *Res:* Development of new optical products or systems from conception through early production; development of new concepts and designs for optical systems; optical design for a broad variety of fabricated optics. *Mailing Add:* Optical Res Assocs Inc 550 N Rosemead Blvd Pasadena CA 91107

HILBERT, STEPHEN RUSSELL, b Brooklyn, NY, Dec 27, 42. NUMERICAL ANALYSIS, MATHEMATICS. *Educ:* Univ Notre Dame, BS, 64; Univ Md, College Park, PhD(math), 69. *Prof Exp:* Asst prof, 69-74, ASSOC PROF MATH, ITHACA COL, 74- *Mem:* Am Math Soc; Math Asn Am; Soc Indust & Appl Math. *Res:* Galerkin methods; approximation theory. *Mailing Add:* Dept Math Ithaca Col Ithaca NY 14850

HILBORN, DAVID ALAN, b Norristown, Pa, Apr 14, 45; m 68; c 2. BIOCHEMISTRY. *Educ:* Lafayette Col, BS, 67; Cornell Univ, PhD(chem), 73. *Prof Exp:* NIH fel cell biol, Salk Inst Biol Studies, 72-74; asst prof chem, Rochester Inst Technol, 74-79, assoc prof, 79-80; RES CHEMIST, EASTMAN KODAK CO, 81- *Mem:* Sigma Xi; Am Chem Soc. *Res:* Isolation and characterization of plant lectins; molecular basis of growth regulatory mechanisms in mammalian cells. *Mailing Add:* Ten Southern Hills Circle Henrietta NY 14467

HILBORN, ROBERT CLARENCE, b Norristown, Pa, June 24, 43; m 70; c 2. ATOMIC PHYSICS, LASERS. *Educ:* Lehigh Univ, BA, 66; Harvard Univ, MA, 67, PhD(physics), 71. *Hon Degrees:* MA, Amherst Col, 87. *Prof Exp:* Res assoc physics, State Univ NY, Stony Brook, 71-72, lectr, 72-73; from asst prof to prof physics, Oberlin Col, 73-86; PROF PHYSICS, AMHERST COL, 86- *Concurrent Pos:* Vis researcher, Univ Calif, Santa Barbara, 79-80; consult, Gilford Instrument Labs, 81-83; lectr, Taiyuan Univ Tech, Shanxi, People's Repub China, 84. *Mem:* Sigma Xi; Am Asn Physics Teachers; Am Phys Soc; Optical Soc Am. *Res:* Tunable dye lasers; atomic and molecular spectroscopy; Hanle effect and level-crossing experiments; molecular beam experiments; non-linear dynamics. *Mailing Add:* Dept Physics Amherst Col Amherst MA 01002

HILBURN, JOHN L, b Mobile, Ala, June 7, 38; m 61; c 2. ELECTROMAGNETICS. *Educ:* Auburn Univ, BEP, 62, MSEE, 64, PhD(elec eng), 67. *Prof Exp:* Engr, Irby & Rester Eng Co, Ala, 56-62; asst elec eng, Auburn Univ, 62-64, instr, 64-67; from asst prof to prof elec eng, La State Univ, Baton Rouge, 67-80; PRES, MICROCOMPUT SYSTS, INC, 76- *Mem:* Inst Elec & Electronics Engrs. *Res:* Microcomputers and microcomputer-based industrial instrumentation; antenna and microwave engineering; electromagnetic fields. *Mailing Add:* 1212 Baird Dr Baton Rouge LA 70808

HILCHIE, DOUGLAS WALTER, b Calgary, Alta, Aug 24, 30; US citizen; m 54; c 2. ENGINEERING SCIENCE, GEOPHYSICS. *Educ:* Univ Okla, BS, 54, PhD(eng sci), 64; Univ Tex, MS, 60. *Prof Exp:* Field engr, Schlumberger Well Surv Corp, 54-58; engr, Socony Mobil Oil Corp, Inc, 62-68; mem fac petrol eng, Mont Col Mineral Sci & Technol, 68-69; tech dir, Dresser Atlas Div, Dresser Industs, Inc, 69-74; pres petrol eng, Colo Sch Mines, 75-81; PRES, DOUGLAS W HILCHIE, INC, 76- *Mem:* Soc Explor Geophys; Sigma Xi; Soc Prof Well Log Analysts; Soc Petrol Eng. *Res:* Measurement and interpretation of physical and chemical properties of rocks in laboratory and in situ as measured by geophysical probes. *Mailing Add:* 927 Pine St Boulder CO 80302-4020

HILCKEN, JOHN ALLEN, b Brooklyn, NY, Mar 5, 16; m 41; c 1. PUBLIC HEALTH. *Educ:* Mass Inst Technol, SB & SM, 39; Univ Rochester, PhD(pharmacol & toxicol), 54. *Prof Exp:* Asst biol, Mass Inst Technol, 38-39; jr document examr, Tech Lab, Fed Bur Invest, 39-41; instr forensic lab, Provost Marshal General's Sch, US Army, 44-46, chief health physics group, Army Chem Ctr, Md, 48-50, chief spec projs & biophys res br, Res & Develop Div, Off Surgeon Gen, 54-57, chief fallout br & exec officer radiation div, Defense Atomic Support Agency, 57-61, chief weapons effects br, Nuclear Energy Div, Hqs, Army Med Res & Develop Command, 61-64, chief radiol hyg div & dir radiation serv, Army Environ Hyg Agency, 64-67, Surgeon Gen rep to Nuclear Systs Health & Safety Rev Comt, 67-68, chief med appln div, Defense Atomic Support Agency, 68-71; chief eng sect, Arlington County Environ Health Bur, Arlington, chief environ eng div, 71-76; dir, Toxic Substances Info Bur, Va State Health Dept, 77-83. *Mem:* Am Indust Hyg Asn. *Res:* Research and development management; industrial hygiene; toxicology. *Mailing Add:* 5035 Edgemere Blvd Richmond VA 23234

HILD, WALTER J, b Wesel, Ger, Nov 3, 19; nat US; m 55; c 2. ANATOMY. *Educ:* Univ Kiel, MD, 49. *Prof Exp:* Asst gen anat, Univ Kiel, 49-54; res assoc, Tissue Cult Lab, 54-59, from asst prof to assoc prof, 59-66, chmn dept, 68-86, PROF ANAT, UNIV TEX MED BR GALVESTON, 66- *Concurrent Pos:* Rockefeller Found fel, 52-53; mem neurol B study sect, NIH, 68-72; ed, Advan in Anat, Embryol & Cell Biol, 65- *Mem:* Am Asn Anat; Am Acad Neurol; Ger Anat Soc; Ger Endocrinol Soc. *Res:* Neuroanatomy; neurosecretion; tissue culture of nervous system; cytophysiology of neurons and neuroglia. *Mailing Add:* Dept Anat Univ Tex Med Sch 301 University Blvd Galveston TX 77550

HILDE, THOMAS WAYNE CLARK, b Stanley, NDak, May 4, 38; m 59; c 2. MARINE GEOPHYSICS, PLATE TECTONICS. *Educ:* San Diego State Univ, BA, 63; Univ Tokyo, DSc, 73. *Prof Exp:* Res asst marine geol, Scripps Inst Oceanog, Univ Calif, San Diego, 59-67; oceanographer marine geol & geophysics, US Naval Oceanog Off, 67-70; adv oceanog marine sci, Govt Taiwan, Repub China, 70-73; vis marine geophysicist, Earthquake Res Inst, Univ Tokyo, 73; sr marine geologist & geophysicist, UN, Bangkok, Thailand, 74-76; DIR GEODYNAMICS RES PROG & PROF GEOPHYSICS MARINE GEOPHYSICS, TEX A&M UNIV, 77- *Concurrent Pos:* Mem, Western Pac Working Group Inter-Union Comn Geodynamics, 74-80, Ocean Crustal Dynamics Comt, JOI, Inc, 78-81, Gulf Coast Working Group, US Geodynamics Comn, 79-; ed, Western Pac Final Report, Inter-Union Comn Geodynamics, 79-; chmn, oceanic lithosphere working group 4, Int Comn Lithosphere. *Mem:* AAAS; Am Geophys Union; Geol Soc Am. *Res:* Marine geophysics and tectonic evolution of the western Pacific marginal seas; origin and evolution of the Pacific Ocean basin; studies of convergence, subduction and the geodynamics of trench-arc-back arc systems; plate tectonics. *Mailing Add:* Dept Oceanog Tex A&M Univ College Station TX 77843

HILDEBOLT, WILLIAM MORTON, b Richmond, Ind, Dec 7, 43; m 64; c 2. FOOD TECHNOLOGY. *Educ:* Ohio State Univ, BS, 66, MS, 67, PhD(food technol), 69. *Prof Exp:* Teaching assoc food technol, Ohio State Univ, 66-69; from res technologist to sr res tecyhnologist, Campbell Soup Co, 69-72, sr res scientist, 72-74, mgr prod develop, 74-77, dir, 77-79, vpres prod technol, 79-83, vpres prod develop, 83-84, vpres int res & develop, 84-85; VPRES RES & DEVELOP, R J REYNOLDS TOBACCO CO, 85- *Concurrent Pos:* Mem bd dirs, Ohio State Univ Res Found, 86-; vchmn & mem bd dirs, Res & Develop Assoc for Mil Food & Packaging Systs, Inc; Nat Acad Sci & Eng partic, Seminar Post-Harvest Food Losses Fruits & Veg, China State Sci & Technol Comn, 84. *Mem:* Sigma Xi; Inst food Technologists. *Mailing Add:* R J Reynolds Tobacco Co Bowman Gray Tech Ctr Reynolds Blvd Winston Salem NC 27102

HILDEBRAND, ADOLF J, b Leutkirch, WGer, Nov 25, 56. ANALYTIC NUMBER THEORY, PROBABILISTIC NUMBER THEORY. *Educ:* Univ Freiburg, PhD(math), 83. *Prof Exp:* From asst prof to assoc prof, 86-91, PROF MATH, UNIV ILL, URBANA, 91- *Concurrent Pos:* Vis mem math, Inst Advan Study, Princeton, NJ, 90-91; assoc ed, J Number Theory, 90-, Ill J Math, 90- *Mem:* Am Math Soc. *Res:* Problems in number theory such as the distribution of prime numbers or other special sets of integers; questions at the interface between number theory and other areas of mathematics such as analysis, combinatorics and probability theory. *Mailing Add:* Dept Math Univ Ill Urbana IL 61801

HILDEBRAND, BERNARD, b New York, NY, Jan 23, 24; m 50; c 2. FUNDING UNIVERSITY RESEARCH, INTERNATIONAL COOPERATIVE RESEARCH. *Educ:* Brooklyn Col, BA, 44; Univ Ill, MS, 46; Rensselaer Polytech Inst, PhD(physics), 53. *Prof Exp:* Res asst radar res & develop, Columbia Radiation Lab, Columbia Univ, 44-45; instr & res asst physics, Rensselaer Polytech Inst, 46-52; radiation det engr res & develop, Westinghouse Elec Corp, 52; consult cosmic rays, US Naval Res Lab, 52-60; from asst to assoc prof physics, Worcester Polytech Inst, 53-60; res physicist, US Naval Res Lab, 60-65; physicist high energy physics, AEC, Dept Energy, 65-68, chief physics res, Div High Energy Physics, Energy Res & Develop Admin, 69-88; CONSULT, SOUTHEASTERN UNIVS RES ASN, 89- *Concurrent Pos:* Exec secy, High Energy Physics Adv Panel, 69-76, US-USSR Joint Coord Comt on Fundamental Properties of Matter, 76-88, US-Japan Comt on High Energy Physics, 79-88. *Mem:* Fel Am Phys Soc; Hist Sci Soc; Fedn Am Scientists; Sigma Xi. *Res:* Intergalactic matter path from cosmic ray nuclei transformations; international cooperative emulsion flight studies of ultra high energy jets; funding high energy physics research; high energy physics achievements; continuous electron beam accelerator facility and SSC particle physics in southeastern United States. *Mailing Add:* 11801 Rockville Pike No 1504 Rockville MD 20852

HILDEBRAND, BERNARD PERCY, b Emerson, Man, May 5, 30; US citizen; m 52; c 2. OPTICS, ULTRASOUND. *Educ:* Univ BC, BASc, 54, MASc, 56; Univ Mich, PhD(elec eng), 67. *Prof Exp:* Electronics engr, Canadair, Ltd, 55-59; sr electronics engr, Hallamore Electronics Co, 59-60; mem tech staff undersea warfare, Hughes Aircraft Co, 60-61; res assoc radar & optics, Inst Sci & Technol, Univ Mich, 61-63, assoc res engr, 63-67; sr res scientist, Pac Northwest Labs, Battelle Mem Inst, 67-69, mgr, Phys Measurements Sect, 69-71, sr staff engr, 71-89, LEAD SCIENTIST, PAC NORTHWEST LABS, BATTELLE MEM INST, 89- *Concurrent Pos:* Affil prof, Univ Wash, Wash State Univ & Ore State Univ, 69-; Battelle res fel, Battelle Seattle Res Ctr, Wash, 70. *Mem:* Optical Soc Am; Inst Elec & Electronics Engrs; Am Inst Ultrasound Med; Soc Explor Geophysicists. *Res:* Random processes; application of communications and systems technology to the physical sciences such as optics, ultrasonic imaging and seismology; biological effects of ultrasound; digital processing of ultrasonic and television images. *Mailing Add:* Battelle Mem Inst Pac Northwest Labs MS K2-28 Battelle Blvd Richland WA 99352

HILDEBRAND, CARL EDGAR, b Gettysburg, Pa, June 3, 44; m 66; c 1. BIOCHEMISTRY, BIOPHYSICS. *Educ:* Gettysburg Col, BA, 66; Pa State Univ, MS, 68, PhD(biophys), 70. *Prof Exp:* Appointee res, 71-72, actg dep group leader, 81-82, mem staff res, 72-84, DEP GROUP LEADER, LOS ALAMOS NAT LAB, 84- *Concurrent Pos:* Mem Nat Inst Child Health & Human Develop, NIH, 83-84. *Mem:* Am Soc Biochem & Molecular Biol; Am Soc Human Genetics; Am Soc Cell Biol; AAAS. *Res:* Regulation of gene structure and activity; chromosome organization; human genome physical mapping; US patent, on nucleic acid isolation. *Mailing Add:* Genetics Group Los Alamos Nat Lab MS886 Los Alamos NM 87545

HILDEBRAND, DAVID KENT, b Minneapolis, Minn, June 24, 40; m 64; c 2. STATISTICS, OPERATIONS RESEARCH. *Educ:* Carleton Col, BA, 62; Carnegie Inst Technol, MS, 65, PhD(statist), 67. *Hon Degrees:* MA, Univ Pa, 71. *Prof Exp:* Lectr, 65-66, from asst prof to assoc prof, 66-77, PROF STATIST, UNIV PA, 77- *Concurrent Pos:* Vis asst prof statist, Carnegie-Mellon Univ, 70-71. *Mem:* Inst Math Statist; Am Statist Asn. *Res:* Methods for analyzing qualitative data, arising in social science problems. *Mailing Add:* Dept Statist Univ Pa Philadelphia PA 19104

HILDEBRAND, DONALD CLAIR, b Astoria, Ore, Feb 27, 32; m 76; c 2. PHYTOBACTERIOLOGY. *Educ:* Wash State Univ, BS, 53; Univ Calif, PhD(plant path), 62. *Prof Exp:* Asst specialist plant path, 62-63, asst res plant pathologist, 64-68, ASSOC RES PLANT PATHOLOGIST, UNIV CALIF, BERKELEY, 68- *Concurrent Pos:* NIH res career develop award, 67-71. *Mem:* Am Soc Microbiol; Am Phytopath Soc; Brit Soc Appl Bacteriologists. *Res:* Bacterial pathogens and diseases of plants; bacterial taxonomy. *Mailing Add:* Dept Plant Path Univ Calif Berkeley CA 94720

HILDEBRAND, FRANCIS BEGNAUD, b Washington, Pa, Sept 1, 15; m 43; c 3. MATHEMATICS. *Educ:* Washington & Jefferson Col, BS, 36, MA, 38; Mass Inst Technol, PhD(math), 40. *Hon Degrees:* ScD, Washington & Jefferson Col, 69. *Prof Exp:* From instr to prof math, Mass Inst Technol, 40-84, staff mem, Radiation Lab, 42-47; RETIRED. *Mem:* Am Math Soc; Math Asn Am. *Res:* Integral equations; aerodynamics; theory of elasticity; numerical analysis. *Mailing Add:* Seven Bucknell Rd Wellesley MA 02181

HILDEBRAND, HENRY H, b Greensburg, Kans, Aug 19, 22. MARINE BIOLOGY. *Educ:* Univ Kans, BA, 46; McGill Univ, MS, 48; Univ Tex, PhD(zool), 54. *Prof Exp:* Asst eastern Arctic investr, Can Fish Res Bd, 47-48; aquatic biologist, US Fish & Wildlife Serv, Bering Sea & NPac, 49; res scientist, Inst Marine Sci, Tex, 53-56; prof marine biol, Univ Corpus Christi, 57-73; ASSOC PROF BIOL, TEX A&I UNIV, 73- *Mem:* Am Soc Ichthyologists & Herpetologists; Am Soc Mammal; Soc Vert Paleont; Arctic Inst NAm; Sigma Xi. *Res:* Arctic fishes; natural history of Gulf of Mexico fishes and shrimp; pesticides; fish-eating birds. *Mailing Add:* 413 Millbrook Corpus Christi TX 78418

HILDEBRAND, JOHN GRANT, III, b Boston, Mass, Mar 26, 42; m 85. NEUROBIOLOGY, BIOCHEMISTRY. *Educ:* Harvard Univ, AB, 64; Rockefeller Univ, PhD(biochem), 69. *Prof Exp:* From instr to asst prof, Harvard Med Sch, 70-77, assoc prof neurobiol, 77-80, tutor biochem sci, 70-80, prof biol sci, Columbia Univ, 80-85; ASSOC BEHAV BIOL, MUS COMP ZOOL, HARVARD UNIV, 80-; PROF NEUROBIOL, BIOCHEM, ENTOM, MOLECULAR & CELLULAR BIOL, UNIV ARIZONA, 85-, DIR, ARIZONA RES LABS, DIV NEUROBIOL, 85- *Concurrent Pos:* Teaching fel neurobiol, Harvard Med Sch, 69-70; Helen Hay Whitney Found res fel, 69-72; A P Sloan res fel, 73-75; trustee, Rockefeller Univ, 70-73; estab investr, Am Heart Asn, 72-75; mem, NSF adv panel neurobiol, 74-77; prog comt, Soc Neurosci, 78-83; dir, Neurobiol course, 80-84, trustee Marine Biol Lab, Woods Hole, 81-89; regents prof, Univ Ariz, 89- *Honors & Awards:* Givaudan lectr, Asn Chemoreception Sci, 85; Lang lectr, Marine Biol Lab, Woods Hole, 85; Javits Award, NIH, 86; Merit Award, NIH, 86; Spencer Mem lectr, Univ BC, 90; RH Award Olfactory Res, 90; Max Planck Res Award, Alexander von Humboldt-Stiftung, 90. *Mem:* Am Soc Neurochem; Animal Behav Soc; Am Soc Biochem & Molecular Biol; Soc Exp Biol; Soc Neurosci; fel Royal Entom Soc (UK). *Res:* Chemical aspects of synaptic transmission; physiology and anatomy of chemical senses; developmental neurobiology; behavior and biology of arthropods. *Mailing Add:* Ariz Res Labs Div Neurobiol 603 Gould-Simpson Bldg Univ Ariz Tucson AZ 85721

HILDEBRAND, MILTON, b Philadelphia, Pa, June 15, 18; m 43; c 3. VERTEBRATE MORPHOLOGY. *Educ:* Univ Calif, AB, 40, MA, 48, PhD(zool), 51. *Prof Exp:* Lectr zool, 48-52, from asst prof to assoc prof, 52-61, chmn col lett & sci, 67-69, emer prof zool, Univ Calif, Davis, 62-86; RETIRED. *Mem:* Am Soc Zool; Am Soc Mammal. *Res:* Analysis of vertebrate locomotion; bone-muscle systems; effective university teaching; morphology of fossorial mammals. *Mailing Add:* 30 Parkside Dr Davis CA 95616

HILDEBRAND, ROGER HENRY, b Berkeley, Calif, May 1, 22; m 44; c 4. ASTRONOMY, ELEMENTARY PARTICLE PHYSICS. *Educ:* Univ Calif, Berkeley, AB, 47, PhD(physics), 51. *Prof Exp:* Physicist, Radiation Lab, Univ Calif, 42-51; from asst prof to assoc prof physics, 52-60, dir inst, 65-68, dean col, 69-73, chmn dept astron & astrophys, 84-88, PROF PHYSICS, ENRICO FERMI INST & DEPT PHYSICS, UNIV CHICAGO, 60-, PROF ASTRON & ASTROPHYS, 78-, SAMUEL K ALLISON DISTINGUISHED SERV PROF, 85- *Concurrent Pos:* Assoc lab dir high energy physics, Argonne Nat Lab, 58-64, actg dir high energy physics div, 59-64; chmn sci policy comt, Stanford Linear Accelerator Ctr, 62-66; mem policy adv comt, Nat Accelerator Lab, 66-69; Guggenheim fel, Univ Calif, Berkeley, 68-69; mem sci & educ adv comt, Lawrence Berkeley Lab, 71-79; chmn, AEC-NSF Rev Comt on US Medium Energy Sci, 74, Airbone Astron Users Group, 83-84, consult, Group Strotospheric Observ Infrared Astron, Ames Res Ctr, NASA, 85-89; mem, Comt Space Astron & Astrophys Space Sci Bd, 87-90; mem, Astron & Astrophys Surv Comt, Nat Acad Sci, Panel Infrared Astron, 89-90; mem, Sci & Tech Adv Comt Submillimeter Array, Smithsonian Astrophys Observ, 89-; chmn, Donnie Heineman Prize Comt, Am Inst Physics, 90. *Mem:* Fel Am Phys Soc; Am Astron Soc; Int Aston Union. *Res:* Neutron scattering; conservation of isospin in particle production; bubble chamber development; muon capture and decay; electronic component of primary cosmic radiation; kaon decay; infrared and submillimeter astronomy photometry and polarimetry; properties of interstellar material; magnetic fields in interstellar clouds. *Mailing Add:* Enrico Fermi Inst Univ Chicago Chicago IL 60637

HILDEBRAND, STEPHEN GEORGE, b Reedsburg, Wis, Apr 14, 44; m 69; c 2. AQUATIC ECOLOGY. *Educ:* Wabash Col, BA, 66; Univ Mich, MS, 69, PhD(fisheries), 73. *Prof Exp:* Res assoc environ impacts nuclear plants, Oak Ridge Nat Lab, 73-74, res assoc mercury and trace elements in aquatic systs, 74-76, res staff mem & group leader environ impacts energy technol, 76-78, res staff mem & group leader environ res, 78-80, mgr environ impacts prog, 81-85, head Environ Anal Sect, 85-89, HEAD ECOSYST STUDIES SECT, OAK RIDGE NAT LAB, 90- *Concurrent Pos:* Adv, Inst Ecol & Nat Comn on Water Qual, 73-74; adj fac, Univ Tenn, 84-; mem, Unesco Int Hydrol prog, 86-, CEQ/NSF panel ecol res, 85, assessment & policy task group, Nat Acid Precipitation Assessment Prog, 84-86. *Mem:* Ecol Soc Am; Am Fisheries Soc; fel AAAS. *Res:* Aquatic ecology; structure and function of aquatic ecosystems; stream invertebrates; environmental assessment of stress on aquatic systems; environmental risk assessment. *Mailing Add:* Environ Sci Div PO Box 2008 Oak Ridge TN 37831-6038

HILDEBRANDT, ALVIN FRANK, b Spring, Tex, Dec 31, 25; m 50; c 2. SOLAR ENERGY, LOW TEMPERATURE PHYSICS. *Educ:* Univ Houston, BS, 49; Tex A&M Univ, PhD(physics), 56. *Prof Exp:* Instr math, Tex A&M Univ, 54-55, res assoc, 55-56; staff scientist, Jet Propulsion Lab, Calif Inst Technol, 56-65, sr res fel chem, 60-63 & physics, 63-64; assoc prof, Univ Houston, 65-69, chmn dept, 69-74, dir solar energy lab, 74-78, PROF PHYSICS, UNIV HOUSTON, 69-, DIR ENERGY LAB, 78- *Concurrent Pos:* Pres, Energy Found Tex, 75-80; chmn, Solar Thermal Test Facil Users Asn, 77-79; dir, Houston Petroleum Res Ctr, 87- *Honors & Awards:* Outstanding Contribution Solar Thermal Technol, Dept Energy, 81. *Mem:* AAAS; Am Phys Soc. *Res:* Low temperature physics; solar energy; solar thermal tower plants with electrical applications and development of chemical conversion cycles; exploration geophysics. *Mailing Add:* Energy Lab Univ Houston Houston TX 77004

HILDEBRANDT, JACOB, b Ernfold, Sask, Sept 21, 30; div; c 2. RESPIRATORY PHYSIOLOGY. *Educ:* Univ BC, BA, 57, MSc, 60; Univ Wash, PhD(physiol, biophys), 66. *Prof Exp:* Sr investr, Virginia Mason Res Ctr, 68-86; instr, 66, from asst prof to assoc prof, 69-84, PROF PHYSIOL BIOPHYS, UNIV WASH, 84- *Concurrent Pos:* NRC fel physics, 58-60; NIH fel, 61-65, res career develop award, 71-76; mem, Cardiovasc & Pulmonary Study Sect, Nat Heart & Lung Inst, 74-78. *Mem:* Am Physiol Soc. *Res:* Mechanics of respiration; control of breathing; lung fluid balance and solute exchange; acute lung injury and respiratory failure; lung circulation and gas exchange. *Mailing Add:* Dept Physiol SJ-40 Univ Wash Seattle WA 98195

HILDEBRANDT, PAUL KNUD, b Lamar, Colo. PATHOLOGY. *Educ:* Colo A&M Col, BS, 55; Colo State Univ, DVM, 59; Am Col Vet Pathologists, dipl, 68. *Prof Exp:* Vet clinician, Ft Detrick, 59-62; vet path resident, Armed Forces Inst Path, 62-64, asst chief, Dept Vet Path, 64-65; vet pathologist, Navy Med Res Unit No 3, Cairo, 65-67; chief, Dept Vet Path, Walter Reed Army Inst Res, 67-71, dir, Div Path, 71-78; staff vet pathologist, US Army Bioeng Res & Develop Lab, Ft Detrick, Md, 78-80; vet pathologist, Tracor Jitco, Inc, 80-85; VPRES & SR PATHOLOGIST, PATHCO, INC, 85- *Concurrent Pos:* Chmn, Case of Month Prog, Am Col Vet Pathologists, 72-78; consult vet path, Surgeon Gen, 72-80, Dept Path, Litton Bionetics Inc, 74-78, USDA, 78-80. *Mem:* Am Vet Med Asn; Am Col Vet Pathologists; Fedn Am Soc Exp Biol; Soc Trop Med & Hyg; Soc Trop Vet Med. *Res:* Veterinary pathology; radiation injury; pathophysiology of host-parasite relationships. *Mailing Add:* Pathco Inc Hyatt Park II 10075 Tyler Pl No 6 Ijamsville MD 21754

HILDEBRANDT, THEODORE WARE, b Ann Arbor, Mich, Dec 8, 22; m 53, 62; c 5. MATHEMATICS, COMPUTER SCIENCE. *Educ:* Univ Mich, AB, 42, AM, 47, PhD(math), 56; Mass Inst Technol, SM, 51. *Prof Exp:* Asst physics, Univ Mich, 43-44; design engr, Electronic Comput Proj, Inst Advan Study, Princeton, NJ, 47-48; asst math, Mass Inst Technol, 48-51; assoc scientist, Oak Ridge Inst Nuclear Studies, 56-57; from asst prof to prof math, Ohio State Univ, 57-68, assoc chmn comput & info sci, 67-68; prof comput sci & dir, Comput Ctr, Kans State Univ, 68-69; head comput facil, Nat Ctr Atmospheric Res, Colo, 69-73, consult, 73-74; expert consult, Inst Telecommun Sci, US Dept Com, 74-75; dir, Acad Comput Ctr, 76-86, PROF MATH, UNIV NC, GREENSBORO, 76- *Concurrent Pos:* Asst dir, Comput Ctr, Ohio State Univ, 57-66, actg chmn, div comput sci, 66-67. *Mem:* Am Math Soc; Math Asn Am; Asn Comput Mach. *Res:* Numerical analysis; computing machines; mathematical programming; data communication; micro processors. *Mailing Add:* Dept Math Univ NC Greensboro NC 27412-5001

HILDEBRANDT, WAYNE ARTHUR, b Tacoma, Wash, June 11, 47; div. ELECTROANALYTICAL CHEMISTRY. *Educ:* Univ Wash, BS, 69; Wash State Univ, PhD(chem), 74. *Prof Exp:* Asst prof chem, Univ Mich, Flint, 74-75; asst prof, 75-83, ASSOC PROF CHEM, NORTHERN ARIZ UNIV, 83-, ASST TO THE DEAN, 88- *Mem:* Am Chem Soc Soc. *Res:* Development and characterization of new ion selective electrodes and new data manipulation techniques for their use; computer automation and interfacing techniques for use with ISE. *Mailing Add:* 501 W Santa Fe Ave No 18 Flagstaff AZ 86001

HILDEBRANT, JOHN A, b Aug 8, 35; US citizen; m 64; c 1. ALGEBRA, TOPOLOGY. *Educ:* Univ Okla, BS, 57; Univ Tenn, MA, 64, PhD(topological algebra), 65. *Prof Exp:* Asst prof math, Univ Tenn, 66; asst prof, 66-71, ASSOC PROF MATH, LA STATE UNIV, BATON ROUGE, 71-, PROF. *Mem:* Am Math Soc. *Res:* Topological algebra; compact divisible topological semigroups. *Mailing Add:* Dept Math 301 Lockett Hall La State Univ Baton Rouge LA 70803

HILDEMAN, GREGORY JOHN, b Milwaukee, Wis, May 16, 47; m 70; c 2. MATERIALS SCIENCE, METALLURGY. *Educ:* Univ Wis, Madison, BS, 70; Univ Wis, Milwaukee, MS, 74; Mass Inst Technol, DSc(metall), 78. *Prof Exp:* Sr develop engr process develop, Allen-Bradley Co, 70-74; sr scientist alloy res, 77-81, staff scientist, 81-85, SECT HEAD MGR, ALUMINUM CO AM, 85- *Mem:* Am Soc Metals; Am Inst Metall Engrs; Sigma Xi. *Res:* Development of new aluminum powder metallurgy alloys by rapid solidification processes. *Mailing Add:* 3108 Treeline Dr Murrysville PA 15668-1520

HILDEN, SHIRLEY ANN, b Montevideo, Minn, July 2, 40. CELL PHYSIOLOGY. *Educ:* St Olaf Col, BA, 62; Stanford Univ, MA & PhD(biol), 69. *Prof Exp:* Asst prof biol, Univ Idaho, 70-71; res assoc physiol, Med Sch, Univ Wis-Madison, 72-76; MEM STAFF, NAT INST AGING, 76- *Concurrent Pos:* Am Asn Univ Women fel, Hokkaido Univ, 69-70. *Mem:* AAAS; Biophys Soc; Am Soc Cell Biologists. *Res:* Ionic regulation in protozoa; membrane physiology. *Mailing Add:* 127 Brooks St East Boston MA 02128

HILDERBRAND, DAVID CURTIS, b California, Mo, Apr 1, 46; m 69; c 2. CHEMISTRY, ATOMIC SPECTROSCOPY. *Educ:* Southwest Baptist Univ, BA, 67; Univ Mo, Columbia, MA, 69, PhD(chem), 71. *Prof Exp:* Instr res, dept obstet & gynec, Sch Med, Univ Mo, 71-74; from asst prof to assoc prof, 74-83, PROF CHEM, SDAK STATE UNIV, 83-, HEAD DEPT, 80-, DIR INT PROGS, 88- *Concurrent Pos:* Bd cur, Southwest Baptist Univ, 81-84; mem bd regents, Bethel Col, 89-94; mem bd dirs, Arid Land Consortium, Asn Int Educ Adminr. *Mem:* Am Chem Soc; Sigma Xi; Asn Int Educ Adminr. *Res:* Method development for analyses using flame and flameless atomic spectroscopy including sample preparation and interpretation of chemical processes occuring in high temperature flame and flameless systems. *Mailing Add:* Box 2202 University Station Brookings SD 57007

HILDING, STEPHEN R, b Duluth, Minn, Aug 5, 36; m 58; c 4. MATHEMATICS. *Educ:* Gustavus Adolphus Col, BA, 58; Kans State Univ, MS, 60; Univ Mich, DEd, 65. *Prof Exp:* Instr math, NDak State Univ, 60-61, Mankato State Col, 61-62 & Univ Mich, 64-65; assoc prof, 65-68, chmn dept, 68-72, PROF MATH, GUSTAVUS ADOLPHUS COL, 68- *Concurrent Pos:* Assoc math, Univ Minn, 74-75. *Mem:* Am Math Soc; Math Asn Am. *Res:* Functional analysis; Banach algebras. *Mailing Add:* Dept Math Gustavus Adolphus Col St Peter MN 56082

HILDNER, ERNEST GOTTHOLD, III, b Jacksonville, Ill, Jan 23, 40; m 68; c 2. SOLAR PHYSICS. *Educ:* Wesleyan Univ, BA, 61; Univ Colo, MA, 64, PhD(physics), 71. *Prof Exp:* Res physicist, Nat Oceanog & Atmospheric Admin, 61-67; fel, Nat Ctr Atmospheric Res, 71-72, scientist, 72-80; br chief, Marshall Space Flight Ctr, NASA, 80-85; vis scientist, Nat Ctr Atmospheric Res, 85-86, DIR, SPACE ENVIRON LAB, NAT OCEANIC & ATMOSPHERIC ADMIN, 86- *Mem:* Int Astron Union; Am Astron Soc; Am Geophys Union; AAAS; Sigma Xi. *Res:* Solar coronal physics; magnetohydrodynamics; interplanetary medium; solar wind; solar cosmic rays; solar radio bursts. *Mailing Add:* Nat Oceanic & Atmospheric Admin-Space Environ Lab 325 Broadway Boulder CO 80303

HILDRETH, EUGENE AUGUSTUS, b St Paul, Minn, Mar 11, 24; m 46; c 4. INTERNAL MEDICINE. *Educ:* Washington & Jefferson Col, BS, 43; Univ Va, MD, 47; Am Bd Internal Med, dipl, 54; Am Bd Allergy & Immunol, dipl, 64. *Prof Exp:* Intern, Johns Hopkins Hosp, 47-48; dir, Dept Med, Reading Hosp & Med Ctr, 69-89; asst instr med, Hosp Univ Pa, 48-53, instr med, 53-54, assoc prof med, 54-55, from asst to assoc prof, 55-69, assoc dean, Sch Med, 64-67, chief allergy & immunol, 54-70, PROF CLIN MED, HOSP UNIV PA, 72- *Concurrent Pos:* Res fel, USPHS, Univ Pa, 49-51; chief resident med, Hosp Univ Pa, 53-54; consult, Vet Admin, Philadelphia, 54-76; Markle scholar, 58-63; consult med, Wernersville State Hosp; mem adv comt on reactions to drugs, Nat Acad Sci, 67-69; mem nat adv comt, Int Medic Alert Found, 67-; mem, Am Bd Internal Med, 68-80, chmn, 81-82, chmn specialty bd allergy & immunol, 69-71; founding comt, Am Bd Allergy & Immunol, 70-71, chmn med, 71-72; consult med, Wernersville State Hosp, 72-80; consult, Annals Internal Med; chmn, Federated Coun Internal Med, 81-82; lect bioethics, Hosp Asn Pa, 84-85, Darmouth-Hitchcock Med Ctr, 84, Univ NMex, 86; chmn bd, Am Col Physicians, 89-91. *Mem:* Inst Med-Nat Acad Sci; Am Col Physicians; Am Acad Allergy; NY Acad Sci; AAAS; Fedn Am Soc Exp Biol; Am Bd Internal Med; Am Bd Allergy & Immunol. *Res:* Diseases of antigen-antibody interaction; lipid metabolism; nephrology; biothetics. *Mailing Add:* Reading Hosp & Med Ctr PO Box 16052 Reading PA 19612-6052

HILDRETH, MICHAEL B, b Bisbee, Ariz, May 4, 55; m 79; c 2. PARASITOLOGY, CELL BIOLOGY. *Educ:* Westmar Col, BA, 77; Tulane Univ, PhD(biol), 83. *Prof Exp:* Teaching & res assoc hist, Sch Vet Med, Univ Wis, 83-86, lectr histol, 86-87; asst prof, 87-91, ASSOC PROF PARASITOL, SDAK STATE UNIV, 91-; VIS PROF, SCH VET MED, UNIV WIS, 86- *Concurrent Pos:* Asst prof, Dept Biol & Microbiol, SDak State Univ, 87-91, assoc prof, 91-; prin investr, NIH Area grant, 90-92. *Mem:* Am Soc Parasitologists; Am Soc Trop Med & Hygiene. *Res:* Molecular and immunological methods for cestode diagnosis; resistance of membranes to digestion; biocontrol of grasshoppers; stress on parasite loads. *Mailing Add:* Dept Biol & Microbiol SDak State Univ Brookings SD 57007

HILDRETH, PHILIP ELWIN, b Marlboro, NH, Jan 14, 23; m 52; c 3. GENETICS. *Educ:* Dartmouth Col, AB, 47; Univ Calif, MA, 50, PhD(zool), 55. *Prof Exp:* Biologist, Radiation Lab, Univ Calif, 51-56; asst prof biol, Long Beach State Col, 56-59; biologist, Radiation Lab, Univ Calif, Berkeley, 59-67, consult, 58-59; chmn dept biol, 67-74, chmn div math & natural sci, 68-70, dean col sci & math, 70-74, DISTINGUISHED PROF BIOL, UNIV NC, CHARLOTTE, 67-, VCHANCELLOR ACAD AFFAIRS, 74- *Mem:* AAAS; Genetics Soc Am; Am Inst Biol Sci; Sigma Xi. *Res:* Lethal production in Drosophila melanogaster; spontaneous and induced lethal mutation rates; developmental genetics. *Mailing Add:* Biol Dept Rural Sta Univ NC Charlotte NC 28223

HILDRETH, ROBERT CLAIRE, b Minneapolis, Minn, Dec 23, 24; m 46; c 2. PLANT PATHOLOGY. *Educ:* Univ Wyo, BS, 49, MS, 50; Univ Minn, PhD(plant path), 57. *Prof Exp:* Sci aide, USDA, Wyo, 46; instr bot, Western Col, 50-51; asst, univ & plant pathologist, Exp Sta, USDA, Univ Minn, 52-55; plant pathologist, Div Trop Res, United Fruit Co, Honduras, 55-61; plant pathologist, Rohm & Haas Co, 61-87; RETIRED. *Concurrent Pos:* Agr & chem res & develop, Asia, Oceania, Cent & SAm; agr consult, 87- *Mem:* Fel AAAS; Sigma Xi; Western Weed Sci Soc. *Res:* Agricultural chemical development; administration; tropical agriculture; foreign agriculture; agricultural research consulting and contract work. *Mailing Add:* Agr Consult 1031 White Gate Rd Danville CA 94526

HILE, MAHLON MALCOLM SCHALLIG, b Sacramento, Calif, Feb 17, 46. CROP PHYSIOLOGY, VEGETABLE CROPS. *Educ:* Chico State Col, BA, 68; Calif State Univ, Chico, MS, 73; Ore State Univ, PhD(crop physiol), 76. *Prof Exp:* Crop physiologist, Cel Pril Indust Inc, 76-77; PROF VEG CROPS, CALIF STATE UNIV, FRESNO, 77- *Mem:* Am Soc Agron; Crop Sci Soc Am; Sigma Xi. *Res:* Physiological factors affecting adaptability, production, quality and cultural management of crop plants. *Mailing Add:* 6309 N Nineth Fresno CA 93710

HILEMAN, ANDREW R, b New Bethlehem, Pa, Aug 28, 26. POWER SYSTEMS, TRANSIENT LIGHTING. *Educ:* Lehigh Univ, BSEE, 51; Univ Pittsburgh, MSEE, 54. *Prof Exp:* Staff, 51-68, ADV ENGR, WESTINGHOUSE CORP, 68- *Concurrent Pos:* Lamme fel, Westinghouse Elec Corp, 66-67; chmn, Inst Elec & Electronics Engrs Surge Protective Devices, 64-66, Am Nat Standards Inst, 70-86. *Mem:* Fel Inst Elec & Electronics Engrs. *Mailing Add:* 109 Overlook Circle Monroeville PA 15146

HILEMAN, ORVILLE EDWIN, JR, b Celina, Ohio, Dec 23, 36; div; c 2. ANALYTICAL CHEMISTRY. *Educ:* Bowling Green State Univ, BScEd, 58; Case Inst Technol, PhD(chem), 64. *Prof Exp:* Teacher high sch, Ohio, 58-60; from asst prof to assoc prof, 64-81, PROF CHEM, MCMASTER UNIV, 81- *Concurrent Pos:* Consult, Peabody Holmes Ltd, 74-82, Radian Corp & Gas Res Inst, 85- *Mem:* Chem Inst Can; Am Chem Soc; Sigma Xi. *Res:* Fundamental studies on nucleation and coprecipitation from solution and on mesophase transformations in lyotropic liquid crystalline systems; liquid redox desulfurization systems; electron transfer processes. *Mailing Add:* Dept Chem McMaster Univ Hamilton ON L8S 4M1 Can

HILEMAN, ROBERT E, b Bloomsburg, Pa, Aug 16, 29; m 51; c 5. ORGANIC CHEMISTRY. *Educ:* Bloomsburg Univ, BS, 51; Rutgers Univ, BS, 56; Univ Minn, PhD(org chem), 60. *Prof Exp:* From sr chemist to res chemist, 60-69, group leader, 69-80, technologist, 80-85, sr technologist, Beacon Res Lab, Texaco Inc, 85-87; CONSULT, 87- *Mem:* Am Soc Testing & Mat; Sigma Xi. *Res:* Mechanisms of organic chemistry; radioactive tracer studies; synthesis of fuel additives; liquid rocket propellant fuels; dispersions of solids in gelled liquids; automobile exhaust emissions; gasoline blending; fuel additives. *Mailing Add:* 25 Bates Lane Wallkill NY 12589

HILER, EDWARD ALLAN, b Hamilton, Ohio, May 14, 39; m 60; c 3. AGRICULTURAL ENGINEERING. *Educ:* Ohio State Univ, BS & MS, 63, PhD(agr eng), 66. *Prof Exp:* Res asst agr eng, Ohio State Univ, 62-63, instr, 63-64 & 65-66; from asst prof to prof agr eng, 66-68, head dept, 74-88, DEP CHANCELLOR ACAD PROG PLANNING & RES, TEX A&M UNIV, 89- *Concurrent Pos:* Consult. *Mem:* Fel AAAS; Am Soc Agr Engrs; Am Soc Eng Educ; Nat Soc Prof Engrs; Am Geophys Union; Sigma Xi. *Res:* Irrigation and drainage design criteria; water use efficiency; energy conservation; biomass utilization; liquid fuels from agricultural crops. *Mailing Add:* Dept Agr Eng Tex A&M Univ College Station TX 77843

HILES, MAURICE, b London, Eng, Oct 14, 36; m 60; c 1. THE USE OF POLYMERS IN THE CONTAINMENT OF MECHANICALLY INDUCED VIBRATION. *Educ:* Oxford Univ, BA, 57; London Univ, BSc, 59, MSc, 65, PhD, 69. *Prof Exp:* Res chemist org, Procter & Gamble, 65-68; training mgr, Thorn Elec Industs, 68-72; pres polymer res, Advan Polymer Technol, 72-81; prof biomat sci, Univ Akron, 81-83; CONSULT, 83- *Concurrent Pos:* Consult, BTR Indust, 74-88, Brit Technol Group & Ministry Defence, 76-81. *Mem:* Chartered Chemist, fel of the Royal Soc of Chem; fel Am Inst Chem; Am Col Sports Med; Am Asn Orthop Med; Aerospace Med Asn; Plastics & Rubber Inst. *Res:* Development and formulation of Lurex and Abraflex; development and formulation of visco-elastomers; sorbothane, sarathon, SIN4 and SIN6; the development of simultaneous interpenetrating; network (SIN) technology. *Mailing Add:* 234 Silver Valley Blvd Munroe Falls OH 44262

HILES, RICHARD ALLEN, b Asheville, NC, Aug 31, 43; m 66. BIOCHEMISTRY, TOXICOLOGY. *Educ:* Clemson Univ, BS, 65; Mich State Univ, PhD(biochem), 70. *Prof Exp:* Fel, Gortner Lab, Univ Minn, St Paul, 70-71; toxicologist, Procter & Gamble Co, 71-78; vpres toxicol, Springborn Inst Biores, 78-86; prog mgr, Hazleton Labs, Madison, Wis, 86-91; MGR PHARMACOL & TOXICOL, FUJI SAWA, 91- *Concurrent Pos:* Adj prof, Ohio Northern Univ, 80- *Mem:* Am Chem Soc; AAAS; Soc Toxicol; Soc Cosmetic Chemists. *Res:* Metabolism of natural and man-made compounds in animals; biochemical toxicology of metabolism and pharmacokinetics. *Mailing Add:* Fuji Sawa Parkway North Ctr Three Parkway N Deerfield IL 60015-2548

HILF, RUSSELL, b Brooklyn, NY, Aug 13, 31; m 55; c 3. BIOCHEMISTRY. *Educ:* City Col New York, BS, 52; Rutgers Univ, MS, 53, PhD(biochem), 55. *Prof Exp:* Head biochem lab, QM Food & Container Inst, 58-59; head cancer endocrine sect, Squibb Inst Med Res, Olin Mathieson Chem Corp, 59-69; PROF BIOCHEM, UNIV ROCHESTER, 69- *Concurrent Pos:* Mem, Breast Cancer Task Force, Nat Cancer Inst, 76-80, Merit Review Bd Oncol, Vet Admin, 80-83 & Sci Adv Bd, Univ Wis Cancer Ctr, 81-84; Reproduction Endocrinol Study Sect, NIH, 86-89; Biochem Chem Carcinogen Adv Comt & Biochem Endocrinol Adv Comt, Am Cancer Soc, 87- *Mem:* Fel AAAS; Am Soc Biol Chemists; Am Asn Cancer Res; Endocrine Soc; fel NY Acad Sci; Am Soc Photobiol. *Res:* Biochemistry and metabolism of tumors and normal tissue; response of target tissue to hormones; photodynamic therapy of tumors. *Mailing Add:* Dept Biochem Univ Rochester Med Ctr Box 607 Rochester NY 14642

HILFER, SAUL ROBERT, b Quakertown, Pa, June 12, 31; m 59; c 3. ORGANOGENESIS, CELL CULTURE. *Educ:* Queens Col, BS, 54; Amherst Col, MA, 55; Yale Univ, PhD(zool), 60. *Prof Exp:* NIH trainee anat, Harvard Med Sch, 59-60, fel, 60-61; from asst prof to assoc prof, 61-74, PROF BIOL, TEMPLE UNIV, 74- *Mem:* Int Soc Develop Biol; Am Soc Cell Biol; Soc Develop Biol. *Res:* Control of cell shape and cell-matrix interactions during early embryonic development; morphogenetic movements during organogenesis; role of extracellular matrix in differentiation of organs; control of shape changes during organ formation; role of cytoskeleton. *Mailing Add:* Dept Biol Temple Univ Philadelphia PA 19122

HILFERTY, FRANK JOSEPH, b Medway, Mass, July 1, 20. BRYOLOGY. *Educ:* Bridgewater State Col, BS, 42; Cornell Univ, PhD(biol), 52. *Prof Exp:* Instr pub sch, Mass, 42-46; asst prof biol, Univ Maine, 46-52; assoc prof biol, Salem State Col, 52-54; chmn, Dept Biol Sci, Bridgewater State Col, 54-65, commonwealth prof bot, 54-81, dir, Div Sci & Math, 65-78, dean, Grad Sch, 65-81; RETIRED. *Concurrent Pos:* NSF grant, Inst Bot, Cornell Univ, 56; NSF fac & res fels, Gray Herbarium, Harvard Univ, 59-60. *Mem:* Sigma Xi. *Res:* Taxonomy of New England mosses and vascular plants; teaching biology at the college level. *Mailing Add:* 20 Aspen Dr Bridgewater MA 02324-1244

HILFIKER, FRANKLIN ROBERTS, b Warsaw, NY, June 7, 43; m 67; c 3. ORGANIC CHEMISTRY. *Educ:* Univ Rochester, BS, 65; State Univ NY Buffalo, PhD(org chem), 69. *Prof Exp:* Turner fel chem, Rice Univ, 69-70; res chemist, 70-73, SUPVR, JACKSON LAB, E I DU PONT DE NEMOURS & CO, INC, 73-, TECH SERV SUPVR, CHESTNUT RUN, 78-, TECH SERV CONSULT, 80- *Mem:* Am Chem Soc. *Res:* Dye chemistry; natural products chemistry. *Mailing Add:* 103 Venus Dr Newark DE 19711

HILGAR, ARTHUR GILBERT, b Butler Co, Pa, Dec 21, 26; m 66; c 2. MEDICAL RESEARCH, MEDICAL ADMINISTRATION. *Educ:* Eastern Nazarene Col, AB, 50. *Prof Exp:* Biologist, Bur Entom & Plant Quarantine, USDA, 52-53; biologist, Path Lab, Nat Cancer Inst, 53-56; biologist endocrinol sect, Cancer Chemother Nat Serv Ctr, 56-61; head biol activities hormones sect, Endocrine Eval Br, Nat Cancer Inst, 61-67; dir data mgt, Mason Res Inst, Mass, 67-71; from asst dir to assoc dir, Nat Bladder Cancer Inst, St Vincent Hosp, 71-84; sci adminr Organ Systs Coord Ctr, Roswell Park Mem Inst, 84-89; CONSULT, 89- *Mem:* AAAS; Am Asn Cancer Res; NY Acad Sci. *Res:* Cancer research administration; endocrine bioassays; experimental designs and data evaluation. *Mailing Add:* 37 Cherlyn Dr Northboro MA 01532

HILGARD, HENRY ROHRS, CELLULAR IMMUNOLOGY. *Educ:* Stanford Univ, MD, 62; Univ Minn, PhD(microbiol), 70. *Prof Exp:* PROF BIOL, UNIV CALIF, SANTA CRUZ, 67- *Res:* Graft-versus-host reactions. *Mailing Add:* Dept Biol Univ Calif Santa Cruz CA 95064

HILGER, ANTHONY EDWARD, b Louisville, Ky, Aug 1, 44; m 69; c 1. MEDICAL MYCOLOGY, MYCOLOGY. *Educ:* Bellarmine Col, BA, 66; Univ SC, MS, 69, PhD(mycol), 74. *Prof Exp:* Instr biol, Univ SC, 68-69; biologist asst, Letterman Army Inst Res, 70-71; asst prof biol, Presby Col, 73-75; asst prof microbiol, Univ Health Sci, Chicago Med Sch, 75-80; asst prof, 80-84, ASSOC PROF MED TECHNOL, UNIV NC, CHAPEL HILL, 84- *Mem:* Int Soc Human & Animal Mycol; Am Soc Microbiol; Med Mycol Soc Am; Am Soc Clin Path. *Res:* Fungal susceptibility testing. *Mailing Add:* Dept Med Allied Health Prof Div Med Tech Univ NC Med Sch CB 7145 Chapel Hill NC 27514

HILGER, JAMES EUGENE, b Wilmington, Del. IMAGE PROCESSING, PATTERN RECOGNITION. *Educ:* Salisbury State Univ, BS, 85; Univ Tenn, MS, 88. *Prof Exp:* Teaching asst, Elec Eng Lab, Univ Tenn, 86-87; res asst, Oak Ridge Nat Lab, 87-89; ELECTRONICS ENGR, CTR NIGHT VISION, 89- *Res:* Developing high speed processors in miniaturized packages for image processing applications. *Mailing Add:* 13737 Mahoney Dr Woodbridge VA 22193

HILIBRAND, J(ACK), b New York, NY, Sept 15, 30; m 56; c 3. ELECTRICAL ENGINEERING. *Educ:* City Col New York, BEE, 51; Mass Inst Technol, ScD(elec eng), 56. *Prof Exp:* Mem tech staff, labs, Radio Corp Am, 56-60, mgr advan devices, solid state div, 60-62, mgr indust transistor design, 62-65, mgr indust semiconductor eng, 65-66; opers mgr transistors & semiconductors, Transitron Electronics Corp, 66; mgr integrated circuit technol, Solid State Div, 67-68, mgr digital integrated circuit design, 68-69, mgr advan projs, 69-70, mgr adv mat & processes, 70-71, staff tech adv, Govt Systs Div, 71-81, mgr, integrated circuit planning, RCA Corp, 81-85; prin staff scientist, GE Aerospace, 85-90; PRIN ENGR, MICROELECTRONICS ASSOCS, 91- *Concurrent Pos:* Lectr, grad sch technol, City Col New York, 57-59. *Mem:* Fel Inst Elec & Electronics Engrs; Am Inst Physics; Sigma Xi. *Res:* Integrated circuit design, development and applications; techniques for very large scale integration; complementary metal-oxide semiconductor technology; bulk and SOS technology; very large scale integration packaging. *Mailing Add:* Microelectronics Assocs 1037 Owl Lane Suite 200B Cherry Hill NJ 08003-2934

HILKER, DORIS M, b Aurora, Ind, Dec 23, 23. NUTRITION, BIOCHEMISTRY. *Educ:* Univ Chicago, BS, 49; Loyola Univ, MS, 55; Tulane Univ, PhD(biochem), 58. *Prof Exp:* Res biochemist, Vet Admin Ctr, WVa, 59-60; asst prof nutrit & chmn dept, Univ Hawaii, 60-68, assoc prof food & nutrit sci, 68-75, prof, 75-; RETIRED. *Mem:* Am Chem Soc; Inst Food Technol; NY Acad Sci. *Res:* Toxic and deleterious substances in food; bioavailability of nutrients. *Mailing Add:* Rte 5 Box 7090 Lexington TN 38351

HILKER, HARRY VAN DER VEER, JR, b Hamilton, Ohio, June 19, 25; m 47; c 3. PHYSICS. *Educ:* Kalamazoo Col, AB, 47. *Prof Exp:* Advert asst, Atlas Press Co, 47-48; assoc ed, Modern Photog, 48-51; physicist, US Naval Ord Test Sta, 51-52, missile proj engr, 52-54, sr physicist, 54-55, res scientist, 55-59, consult, 59-61; vpres, Decision Control, Inc, 61-67 & Varian Data Mach, Inc, 67-70; pres, Accra-Point Arrays, 70-75; CONSULT, 75- *Mem:* Sigma Xi; Inst Elec & Electronics Engrs. *Res:* Applied physics; sequential machines; logic design and switching theory; digital computer systems; electronics; missile test instrumentation; analog-digital techniques; robotic control systems; design of high-speed data processors. *Mailing Add:* 1118 Pescador Dr Newport Beach CA 92660

HILL, ALFRED, JR, b Henderson, Tex, Feb 17, 19; m 47; c 3. MEDICAL ENTOMOLOGY. *Educ:* Prairie View Agr & Mech Col, BS, 47; Colo State Univ, MS, 51; Kans State Univ, PhD(entom), 62. *Prof Exp:* From instr to asst prof, 52-62, PROF BIOL, A&T STATE UNIV, NC, 62- *Mem:* Entom Soc Am. *Res:* Systemic insecticides for control of insect pests on plants and on animals. *Mailing Add:* Dept Biol A&T State Univ 1601 E Market St Greensboro NC 27411

HILL, ANN GERTRUDE, b Cleveland, Ohio, Mar 15, 22. INORGANIC CHEMISTRY, COMPUTER SCIENCES. *Educ:* Ursuline Col, BS, 44; Univ Notre Dame, MS, 52, PhD, 57. *Prof Exp:* Teacher, Villa Angela Acad, 44-46; prof chem, Ursuline Col, 46-52, head dept, 52-88, chmn, Div Natural Sci, 77-85, dir Comput Ctr, 81-88, PROF COMPUTER SCI, URSULINE COL, 88- *Concurrent Pos:* Res assoc, Atomic Energy Comn, Univ Notre Dame, 55-57. *Mem:* Am Chem Soc. *Res:* Absorption spectra of coordination compounds; ligand exchange on ion exchange resins. *Mailing Add:* Dept Computer Sci Ursuline Col Cleveland OH 44124

HILL, ANNLIA PAGANINI, b San Mateo, Calif, Aug 22, 46; m 72. PUBLIC HEALTH & EPIDEMIOLOGY, BIOMETRICS-BIOSTATISTICS. *Educ:* Univ Calif Los Angeles, BA, 68, MS, 70, PhD(biostatist), 74. *Prof Exp:* Fel ment retardation, Univ Calif, Los Angeles, 74-76; instr community med & pub health, 76-77, asst prof, family & prev med, 77-82, assoc prof, 82-89, PROF PREV MED, UNIV SOUTHERN CALIF, 90- *Concurrent Pos:* Statist consult, HEAR Found, 72-73; data coordr, Univ Southern Calif Comprehensive Cancer Ctr, 76-77; prin investr, Nat Cancer Inst, 85- *Mem:* Am Statist Asn; Soc Epidemiol Res. *Res:* Prospective cohort study designed to measure the risks and benefits of menopausal estrogen replacement therapy in terms of incident disease and mortality and to evaluate the effects of other health-related practices in older adults; premenopausal bilateral breast cancer family registry to study the genetic component in breast cancer etiology and to provide a comprehensive breast cancer screening program. *Mailing Add:* Dept Prev Med Univ Southern Calif Sch Med 1441 Eastlake Ave No 802 Los Angeles CA 90033

HILL, ARCHIBALD G, b Louisville, Ky, Jan 5, 50; m 87; c 2. PROCESS CONTROL. *Educ:* Univ Louisville, MEng, 73; La Tech Univ, PhD(chem eng), 80. *Prof Exp:* Res assoc chem eng, La Tech Univ, 77-80; asst prof chem eng, Okla State Univ, 80-86; ASSOC PROF CHEM ENG, UNIV SOUTHWESTERN LA, 86- *Mem:* Am Inst Chem Engrs; Instrument Soc Am; Int Ozone Asn; Sigma Xi; Am Soc Eng Educ; Soc Indust & Appl Math. *Res:* Process instrumentation and control; process identification; adaptive control; control of multivariable processes. *Mailing Add:* Dept Chem Eng Univ SWestern La Lafayette LA 70504-4130

HILL, ARCHIE CLYDE, b Salt Lake City, Utah, July 18, 22; m 46; c 6. SOIL CHEMISTRY. *Educ:* Brigham Young Univ, BS, 50; Rutgers Univ, PhD(soil chem), 52. *Prof Exp:* Plant physiologist, Monsanto Chem Co, 52-54; supvr agr res & technol, Columbia-Geneva Steel Div, US Steel Corp, 54-64; assoc res prof bot, Univ Utah, 64-75, prof biol, 75-90, dir, Environ Studies Lab, Res Inst, 85-90; RETIRED. *Mem:* Am Soc Plant Physiol; Am Chem Soc; Am Soc Agron; Soil Sci Soc Am; Air Pollution Control Asn. *Res:* Air pollution effects on plants, acid rain and visibility; soil and plant relationships; trace elements; herbicides. *Mailing Add:* 310 S 1500 E Pleasant Grove UT 84062

HILL, ARMIN JOHN, b Riverdale, Idaho, June 7, 12; m 33, 75; c 6. PHYSICS. *Educ:* Mont State Col, BS, 32, MS, 38; Calif Inst Technol, MS, 49, PhD, 50. *Prof Exp:* Asst prof math & eng, NDak Sch Forestry, 32-37; asst elec eng, Mont State Col, 37-38, electrification specialist, exten serv & rural electrification engr, Agr Exp Sta, 38-41, asst prof physics, 41-50; physicist, Motion Picture Res Coun, Calif, 50-57; prof physics & dean Col Phys & Eng Sci, 57-72, dean Col Eng Sci & Technol, 72-77, EMER DEAN, BRIGHAM YOUNG UNIV, 77- *Mem:* AAAS; Fel Soc Motion Picture & TV Engrs; Optical Soc Am; Am Soc Eng Educ. *Res:* Special light sources; stereoscopic photography; light diffusion through translucent screens; characteristics of reflecting screens. *Mailing Add:* 4750 Eastcliff Ave Provo UT 84604

HILL, ARTHUR JOSEPH, JR, b New Haven, Conn, Oct 24, 18; m 44; c 4. ORGANIC CHEMISTRY. *Educ:* Yale Univ, BS, 41, PhD(org chem), 44. *Prof Exp:* Res chemist, 44-53, sr res supvr, 53-60, sr res chemist, 60-74, STAFF CHEMIST, E I DU PONT DE NEMOURS & CO, INC. 74- *Mem:* Fel AAAS; Am Chem Soc; Am Nuclear Soc; fel Am Inst Chem; Sigma Xi. *Res:* Organic syntheses; chemotherapeutics; design, development and operation of facilities for research work with radioactive materials; decontamination of equipment and processing of radioactive waste products; fire prevention and protection in atomic energy facilities. *Mailing Add:* PO Box 42 Aiken SC 29802-0042

HILL, ARTHUR S, b McKeesport, Pa, Apr 7, 41; m 64; c 3. CHEMICAL ENGINEERING, POLYMER ENGINEERING. *Educ:* Mass Inst Technol, SB, 63; Princeton Univ, MSE, 64, PhD(chem eng), 68. *Prof Exp:* Res engr polymers, Harry Diamond Lab, Washington, DC, 67-69; res engr plastics, Exp Sta, E I du Pont de Nemours & Co Inc, Wilmington, Del, 69-75; sr proj leader med prod, 75-80, SR RES ASSOC & ASST MGR, JOHNSON & JOHNSON MED RES CTR, 80- *Concurrent Pos:* Adj asst prof, Philadelphia Col Textiles & Sci, 71-73. *Mem:* Am Chem Soc; Soc Plastics Engrs; Adhesion Soc. *Res:* Pressure sensitive adhesives, rheology and processing. *Mailing Add:* 6619 Nantucket Lane Arlington TX 76017

HILL, ARTHUR THOMAS, b Carman, Man, Jan 15, 20; m 48; c 5. POULTRY HUSBANDRY, POULTRY PHYSIOLOGY. *Educ:* Univ Man, BSA, 43; Univ BC, MSA, 47; Tex A&M Univ, PhD(animal genetics), 64. *Prof Exp:* Instr poultry, Univ Sask, 47-48; geneticist, Can Dept Agr, Ottawa, 49-53 & Dereen Poultry Farm Ltd, Sardis, 53-54; poultry farmer, 54-56; mem, Poultry Mgt Animal Sect, Res Sta, Can Dept Agr, 56-79; CAN EXCHANGE SERV OVERSEAS VOLUNTEER, 80- *Concurrent Pos:* Transfer, Poultry Res Centre, Edinburgh, Scotland, 72-73; broiler prod proj, Peru, 81, Agr Educ Panama, 84; secy-treas, BC Div, Can Feed Indust, 82- *Mem:* Poultry Sci Asn; World Poultry Sci Asn; Agr Inst Can. *Res:* Population stress factors affecting the performance of laying and broiler stock and the quality of the resultant eggs and meat. *Mailing Add:* 5410 Huston Rd RR Two Sardis BC V2R 1B1 Can

HILL, BENNY JOE, b Cordell, Okla, Feb 20, 35; m 53; c 4. PHYSICS. *Educ:* Okla State Univ, BS, 57; Tex A&M Univ, PhD(physics), 69. *Prof Exp:* Res asst physics, Los Alamos Sci Lab, 57-59; instr, Southwestern Okla State Univ, 59-60; mem staff, Los Alamos Sci Lab, 60-64; asst prof & head dept, 64-66, PROF & HEAD DEPT PHYSICS, SOUTHWESTERN OKLA STATE UNIV, 68- *Concurrent Pos:* Res assoc, Theoret Physics Group, Tex A&M Univ, 66-68; vis staff mem, Los Alamos Sci Lab, 69; prin staff mem, The BDM Corp, 84- *Mem:* Am Phys Soc; Am Asn Physics Teachers. *Res:* Theoretical physics; nuclear physics; computers in physics education; plasma physics; electromagnetics. *Mailing Add:* Dept Physics Southwestern Okla State Univ 100 Campus Dr Weatherford OK 73096

HILL, BERNARD DALE, b Brandon, Man, Aug 10, 48; m 73. PESTICIDE SCIENCE, MODELING. *Educ:* Brandon Univ, BSc, 69; Univ Man, MSc, 74, PhD(pesticide biochem), 78. *Prof Exp:* Res assoc pesticide biochem, Univ Man, 77; RES SCIENTIST PESTICIDE RESIDUE CHEM, AGR CAN, 78- *Mem:* Am Chem Soc. *Res:* Determination of the fate, persistence and environmental acceptability of pesticides; residue data for new insecticides on soils and crops. *Mailing Add:* Agr Can Res Sta PO Box 3000 Main Lethbridge AB T1J 4B1 Can

HILL, BRIAN, b Carmarthen, Wales, May 11, 38; m 62; c 2. CERAMICS. *Educ:* Univ Wales, BSc, 58; Univ London, PhD(ceramics) & DIC, 63. *Prof Exp:* Res engr, E I du Pont de Nemours & Co, Inc, 63-67; asst prof, Drexel Inst Technol, 67-69; res metallurgist, Paul D Merica Res Labs, Int Nickel Co, Inc, 69-78; sales mgr, Zircar Prods, Inc, 78-80; RES MGR, BRUSH WELLMAN, INC, 80- *Mem:* Am Ceramic Soc; Nat Inst Ceramic Engrs; Brit Inst Metals; Int Soc Hybrid Microelectronics. *Res:* Properties of alumina; processing of ceramics; electronic ceramics. *Mailing Add:* Brush Wellman Inc 17876 St Clair Ave Cleveland OH 44110

HILL, BRIAN KELLOGG, b Boise, Idaho, July 11, 43; m 65. REFLECTIVE MATERIALS, OPTICAL SYSTEMS. *Educ:* Univ Idaho, BS, 65; Mont State Univ, PhD(chem), 69. *Prof Exp:* Sr res chemist, 3M Co, 69-74, supvr, Nuclear Med Dept, 74-76, supvr, New Prod Traffic Control Mat, 76-80, mgr, New Bus Develop, 80-89, lab mgr, Energy Control Prods, 83-89, TECH DIR, COM OFF, SUPPLY DIV, 3M CO, 89- *Res:* Photo and thermal chemistry of transition metals; catalytic activity of transition metal complexes; development of nuclear diagnostic pharmaceuticals; development of products for new business opportunities in reflective materials; large scale imaging; optical material for controlling energy. *Mailing Add:* Bldg 207-1W-08 3M Co St Paul MN 55144

HILL, BRUCE COLMAN, b Houston, Tex, Jan 24, 48. BIOMEDICAL ENGINEERING, BIOPHYSICS. *Educ:* Rice Univ, BA, 69; Stanford Univ, MS, 72, PhD(appl physics), 77. *Prof Exp:* SR INVESTR BIOENG, PALO ALTO MED FOUND RES INST, 78- *Concurrent Pos:* Consult, Vets Admin Hosp, Palo Alto, 78-82. *Mem:* Biophys Soc. *Res:* Interferometric measurements of axon structural changes; biomedical instrumentation for blood flow measurements; electro-optical system design for eardrum displacement measurements; optical measurements of action potential propagation in cardiac tissue. *Mailing Add:* 610 Ashton Ave Palo Alto CA 94306

HILL, BRUCE M, b Chicago, Ill, Mar 13, 35; m 58; c 3. STATISTICS. *Educ:* Univ Chicago, BS, 56; Stanford Univ, MS, 58, PhD(statist), 61. *Prof Exp:* Lectr bio-statist, 60-61, from asst prof to prof math, 61-70, PROF STATIST, UNIV MICH, ANN ARBOR, 70- *Concurrent Pos:* Vis assoc prof, Harvard Univ, 64-65; vis prof, Univ Lancaster, UK, 68-69, Univ Utah, 79, Univ Milan, Italy, 89 & Univ Rome, Italy, 89; consult var indust firms, 61-81; prin investr, NSF grants, 81-85; pres, Ann Arbor Chap, Am Statist Asn, 86-; USAF, 87-88. *Honors & Awards:* Fel, Am Statist Asn; fel, Inst Math Statist. *Mem:* Inst Math Statist; Am Statist Asn. *Res:* Statistical theory and applications to science and industry; Bayesian inference and decision-making; subjective probability models; Zipf's law. *Mailing Add:* Dept Statist Univ Mich Ann Arbor MI 48109

HILL, CARL MCCLELLAN, b Norfolk, Va, July 27, 07; m 27, 70; c 2. ORGANIC CHEMISTRY. *Educ:* Hampton Inst, BS, 31; Cornell Univ, MS, 35, PhD(org chem), 41. *Hon Degrees:* LLD, Univ Ky; DSc, East Ky Univ, 75. *Prof Exp:* Instr, High Sch, Hampton Inst, 31-39, asst prof chem, 39-40, prin, Lab Sch, 40-41; assoc prof chem, Agr & Tech Univ NC, 41-44; head dept, Tenn State Univ, 44-52, prof, 44-62, chmn dept, 52-58, dean fac & sch arts & sci, 58-62; prof chem & pres, Ky State Univ, 62-75; pres, Hampton Inst, 76-78. *Concurrent Pos:* Gen Educ Bd fel; Rosenwald fel. *Honors & Awards:* Col Chem Teachers Award, Mfg Chemists Asn, 62. *Mem:* Fel AAAS; fel Am Chem Soc; fel Nat Inst Sci (pres, 46); fel Am Inst Chemists; Sigma Xi. *Res:* Organic chemical reserach with ketenes, aliphatic, alicyclic and arylunsaturated ethers and Grignard reagents, quality levels of fruits and vegetables; low temperature treatment of soft coals; aryloxy acids and plant hormones; lithium aluminum hydride reduction studies of dimetric ketenes. *Mailing Add:* 431 Elizabeth Lake Dr Hampton VA 23669

HILL, CHARLES EARL, b Baltimore, Md, July 12, 35; m 57; c 4. FAMILY MEDICINE. *Educ:* Loyola Col, Md, BSc, 56; Univ Md, Baltimore, MD, 60. *Prof Exp:* Intern, St Agnes Hosp, Baltimore, 60-61; pvt pract, 61-62 & 64-72; consult, Md House Correction, Jessup, 68-72; asst dir, 72-75, EXEC DIR FAMILY PRACT PROG, SCH MED, UNIV MD, BALTIMORE, 75-, MEM FAC FAMILY MED, 72-, DIR, FAMILY PRACT RESIDENT PROG, 78- *Concurrent Pos:* Consult, Resident Asst Prog, 80-86; mem, Accreditation Rev Comt, ACCME, 87- *Mem:* Soc Teachers Family Med; Asn Am Med Cols. *Res:* Color vision testing; family practice library. *Mailing Add:* 1677 Thorpe Rd Pasadena MD 21122-6104

HILL, CHARLES GRAHAM, JR, b Elmira, NY, July 28, 37; m 64; c 3. CHEMICAL ENGINEERING. *Educ:* Mass Inst Technol, SB, 59, SM, 60, ScD(chem eng), 64. *Prof Exp:* Asst prof chem eng, Mass Inst Technol, 64-65; from asst prof to assoc prof, 67-76, PROF CHEM ENG, UNIV WIS-MADISON, 76-, CHMN DEPT CHEM ENG & PROF FOOD SCI, 89- *Concurrent Pos:* Consult, Arthur D Little, Inc, 63-65, Joseph Schlitz Brewing Co, 73-76 & Tomah Prod, 76 & Nat Bur Standards, 79-87; Ford Found fel eng, 64-65. *Mem:* AAAS; Am Inst Chem Engrs; Am Chem Soc. *Res:* Chemical kinetics; catalysis; reverse osmosis and membrane separation processes; thermodynamics; laser Raman spectroscopy. *Mailing Add:* Dept Chem Eng Univ Wis Madison WI 53706

HILL, CHARLES HORACE, JR, b Indianapolis, Ind, Aug 2, 21; m 48; c 2. NUTRITION, BIOCHEMISTRY. *Educ:* Colo Agr & Mech Col, BS, 48; Cornell Univ, MS, 49, PhD, 51. *Prof Exp:* Asst chemist biochem & nutrit, Univ Nebr, 51-52; assoc prof poultry nutrit, 52-70, PROF POULTRY SCI, NC STATE UNIV, 70- *Mem:* Am Chem Soc; Poultry Sci Asn. *Res:* Unidentified factors and role of antibiotics in poultry nutrition; nutrition in resistance to disease; inhibition of growth due to unheated soybean meal; bacterial nutrition. *Mailing Add:* Dept Poultry Sci NC State Univ Raleigh NC 27695

HILL, CHARLES WHITACRE, b Danville, Va, Mar 6, 40; m 62; c 2. BIOCHEMISTRY, GENETICS. *Educ:* Univ Va, BSChem, 62; Univ Wis, PhD(biochem), 67. *Prof Exp:* From asst prof to assoc prof, 68-81, PROF BIOL CHEM, HERSHEY MED CTR, PA STATE UNIV, 81- *Concurrent Pos:* NSF fel biochem, Stanford Univ, 66-68. *Mem:* Am Soc Biol Chemists; Am Soc Microbiol; Genetics Soc Am. *Res:* Genetics and biochemistry of transfer RNA; normal and illegitimate genetic recombination; chromosomal rearrangement. *Mailing Add:* Dept Biol Chem Pa State Univ Hershey Med Ctr PO Box 850 Hershey PA 17033

HILL, CHRISTOPHER T, b Neenah, Wis, June 19, 51; m 78; c 2. ELEMENTARY PARTICLE PHYSICS. *Educ:* Mass Inst Technol, BS & MS, 72; Calif Inst Technol, PhD(physics), 77. *Prof Exp:* Fel physics, Univ Chicago, 77-79; fel, 79-81, assoc scientist physics, 81-85, Scientist I, 85-89, SCIENTIST II, FERMI NAT ACCELERATOR LAB, 89- *Concurrent Pos:* CERN assoc, 87-88; vis scholar, Univ Chicago, 88- *Honors & Awards:* Arthur H Compton lectr, Univ Chicago, 79. *Mem:* Fel Am Phys Soc. *Res:* Problems of high energy theoretical physics including construction of realistic grand unified field theories and studies of geometry and topology in relation to quantum field theory. *Mailing Add:* Fermi Nat Accelerator Lab PO Box 500 Batavia IL 60510

HILL, CHRISTOPHER THOMAS, b Clarksburg, WVa, Aug 29, 42; m 65. CHEMICAL ENGINEERING, POLICY ANALYSIS. *Educ:* Ill Inst Technol, BS, 64; Univ Wis-Madison, MS, 66, PhD(chem eng), 69. *Prof Exp:* Res engr, Uniroyal, Inc, NJ, 68-70; asst prof chem eng, Wash Univ, 70-74, assoc prof chem eng, 74-76, assoc prof technol & human affairs, 76-78; sr res assoc, Ctr Policy Alternatives, Mass Inst Technol, 78-83; sr specialist sci & tech policy, Libr Cong, 83-90; EXEC DIR MFG FORUM, NAT ACAD ENG & SCI, 90- *Mem:* Fel AAAS; Am Inst Chem Engrs; Am Chem Soc; Am Economics Asn. *Res:* Policy analysis; technological innovation and assessment; manufacturing energy, resources and environment; government regulation; chemical industry; multiphase rheology. *Mailing Add:* Mfg Forum Nat Acad Eng & Sci 2101 Constitution Ave NW Washington DC 20418

HILL, CLEMENT JOSEPH, pedodontics, for more information see previous edition

HILL, CLIFF OTIS, b Glendale, Calif, Sept 10, 41; m 64; c 2. PHYSICAL CHEMISTRY, PHOTOCHEMISTRY. *Educ:* Wichita State Univ, BS, 64, MS, 65; Ari State Univ, PhD(phys chem), 69. *Prof Exp:* SR RES CHEMIST, EASTMAN KODAK CO, 69- *Mem:* Am Chem Soc. *Res:* GaAs integrated circuit design; Photochemistry of transition metal complexes; theoretical studies of luminescence and luminescence quenching. *Mailing Add:* 1254 Old Farm Circle Webster NY 14580

HILL, DALE EUGENE, b Struthers, Ohio, Mar 1, 31; m 53; c 3. PHYSICS, DEFECTS SILICON. *Educ:* Case Inst Technol, BS, 53; Purdue Univ, MS, 55, PhD(physics), 59. *Prof Exp:* Res physicist, 58-70, SCI FEL, MONSANTO CO, 70- *Honors & Awards:* IR 100 Award, 84. *Mem:* AAAS; Am Phys Soc; Sigma Xi. *Res:* Semiconductor physics, especially process induced defects in silicon wafers, also electrical, optical and luminescence properties of III-V compound semiconductors; light emitting devices. *Mailing Add:* 940 Wood Ave St Louis MO 63122

HILL, DAVID ALLAN, b Cleveland, Ohio, Apr 21, 42; m 71. ELECTROMAGNETICS, ANTENNAS. *Educ:* Ohio Univ, BS, 64, MS, 66; Ohio State Univ, PhD(elec eng), 70. *Prof Exp:* Vis fel, Coop Inst Res Environ Sci, 70-71; electronics engr, Inst Telecommun Sci, 71-82; ELECTRONICS ENGR, NAT INST STANDARDS & TECHNOL, 82- *Concurrent Pos:* Adj prof, Univ Colo, 79-; ed, Inst Elec & Electronics Engrs Trans Geosci & Remote Sensing, 80-84 & Inst Elec & Electronics Engrs Trans Antennas & Propagation, 86-89. *Mem:* Fel Inst Elec & Electronics Engrs; Int Union Radio Sci; fel Electromagnetics Acad. *Res:* Electromagnetic theory with applications to antennas, propagation, electromagnetic compatibility, and remote sensing; author of more than 100 publications. *Mailing Add:* Nat Inst Standards & Technol 325 Broadway Boulder CO 80303

HILL, DAVID BYRNE, b Brooklyn, NY, Nov 3, 38; m 60; c 4. MATHEMATICS, STATISTICS. *Educ:* Stevens Inst Technol, ME, 60, MS, 61, PhD(math), 63. *Prof Exp:* Instr math, Stevens Inst Technol, 63-64; USPHS fel biomath, Stanford Univ, 64-65; from asst prof to assoc prof math & community med, 69-78, dir, Acad Comput Ctr, 66-74, PROF & CHMN DEPT COMPUTER SCI, UNIV VT, 74-, PROF SURG, 78- *Concurrent Pos:* Consult, Vt Dept Taxes, 65-, Health Serv & Ment Health Admin, 69-, NSF, 69- & NIH, 71-; vpres, Geomet Inc, 72-74. *Mem:* Am Math Soc; Am Inst Math Statist; Asn Comput Mach. *Res:* Computer architecture. *Mailing Add:* RD 1 Box 362 Grand Isle VT 05458

HILL, DAVID EASTON, b Orange, NJ, Apr 26, 29; m 57; c 3. SOIL SCIENCE. *Educ:* Rutgers Univ, BS, 52, PhD(soil sci), 58. *Prof Exp:* Asst soil scientist, 57-60, assoc, 61-80, SOIL SCIENTIST, CONN AGR EXP STA, 80- *Mem:* Am Soc Agron. *Res:* Soil morphology and genesis; land classification and survey; soil survey interpretations; horticulture. *Mailing Add:* Conn Agr Exp Sta PO Box 1106 New Haven CT 06504

HILL, DAVID G, b Tarentum, Pa, Dec 1, 37; m 61; c 2. MEDICAL PHYSICS, HIGH ENERGY PHYSICS. *Educ:* Carnegie Inst Technol, BS, 59, MS, 60, PhD(physics), 64. *Prof Exp:* Physicist, Brookhaven Nat Lab, 64-72; asst prof, Univ Md, College Park, 72-76; develop scientist, Pfizer Med Systs, 76-82; MGR, CTR RES & DEVELOP, SIEMENS CORP, 82- *Mem:* Am Phys Soc; Am Asn Phys Med. *Res:* High energy hadron physics; development of x-ray computerized axial tomographic scanners. *Mailing Add:* Siemens Med Syst Inc 186 Wood Ave S Iselin NJ 08830

HILL, DAVID LAWRENCE, b Booneville, Miss, Nov 11, 19; m 50; c 7. NUCLEAR PHYSICS, INSTRUMENTATION. *Educ:* Calif Inst Technol, BS, 42; Princeton Univ, PhD(nuclear physics), 51. *Prof Exp:* Asst, Metall Lab, Chicago, 42-43, jr physicist, 43-44; assoc physicist & group leader, Argonne Nat Lab, 44-46; from asst prof to assoc prof physics, Vanderbilt Univ, 49-54; group leader, Los Alamos Sci Lab, 54-58, consult theoret physics, 58-59; pres Southport Comput, Inc, 73-80; PRES, HARBOR RES CORP, 78-, VALUTRON NV, 80-, PATENT ENFORCEMENT FUND, INC, 90- *Concurrent Pos:* Consult, Los Alamos Sci Lab, 52-54; mem fel panel, Nat Res Coun, 54. *Mem:* AAAS; Fedn Am Scientists (chmn, 53-54); fel Am Phys Soc; Inst Elec & Electronics Engrs. *Res:* Nanosecond pulse generators; 100 and 200 megacycle logic systems and counters; instrumentation; fission theory; nuclear structure; relation of nuclear shapes to transition rates and energy levels. *Mailing Add:* 1074 Harbor Rd Southport CT 06490

HILL, DAVID PAUL, b Livingston, Mont, June 18, 35; m 61; c 1. GEOPHYSICS. *Educ:* San Jose State Col, BS, 58; Colo Sch Mines, MS, 61; Calif Inst Technol, PhD(geophys), 71. *Prof Exp:* Geophysicist, Crustal Studies Br, 59-64 & Hawaiian Volcano Observ, 64-66, geophysicist, Off Earthquake Studies, 71-78, chief, Seismol Br, 78-82, CHIEF SCIENTIST, LONG VALLEY CALDERA STUDIES, US GEOL SURV, 83- *Concurrent Pos:* Assoc ed, J Geophys Res, 85-88. *Mem:* AAAS; Am Geophys Union; Seismol Soc Am. *Res:* Elastic wave propagation with applications to earth structure and earthquake studies; application of geophysical methods to the study of earthquake and volcanic processes. *Mailing Add:* 3794 Redwood Circle Palo Alto CA 94306

HILL, DAVID THOMAS, b Griffin, Ga, Oct 11, 47; m 67; c 3. AGRICULTURAL WASTE UTILIZATION, ENERGY PRODUCTION. *Educ:* Univ Ga, BSAE, 70, MS, 71; Clemson Univ, PhD(agr eng), 75. *Prof Exp:* Instr agr eng, Clemson Univ, 73-75; asst prof agr eng, Univ Fla, 75-79; assoc prof, 79-86, ALUMNI PROF AGR ENG, AUBURN UNIV, 86- *Honors & Awards:* Nat Young Researcher, Am Soc Agr Engrs, 85. *Mem:* Am Soc Agr Engrs; Nat Soc Prof Engrs; Am Soc Eng Educ. *Res:* Energy recovery from agricultural wastes through anaerobic fermentation; novel utilization systems development and design refinement of animal waste systems. *Mailing Add:* Agr Eng Dept Auburn Univ Auburn AL 36849-5417

HILL, DAVID W(ILLIAM), b Orange, NJ, Sept 24, 36; m 60; c 3. GROUNDWATER GEOCHEMISTRY, HAZARDOUS WASTE SITE ANALYSIS. *Educ:* Cornell Univ, AB, 58; Stanford Univ, MS, 63, PhD(sanit eng), 66. *Prof Exp:* Res sanit engr, Fed Water Pollution Control Admin, 66-71, supv sanit engr, 71-83, environ engr, Environ Protection Agency, Southeast Water Lab, 83-86, ENVIRON ENGR, ENVIRON PROTECTION AGENCY, GROUNDWATER BR, 86- *Honors & Awards:* Spec Achievement Award, Environ Protection Agency, 88 & 89; Notable Achievement Award, Environ Protection Agency, 90. *Mem:* Am Geophys Union. *Res:* Anaerobic degradation of pesticides; waste treatment and characterization; environmental monitoring; remote sensing; environmental assessments; groundwater contamination. *Mailing Add:* Water Mgt Div Region 4 Environ Protection Agency 345 Cortland St NE Atlanta GA 30365

HILL, DEREK LEONARD, b Croydon, Eng, Dec 6, 30; m 58; c 4. PHYSICAL CHEMISTRY, ELECTROCHEMISTRY. *Educ:* Univ London, BSc, 53, PhD(electrochem) & DIC, 57. *Prof Exp:* Res assoc, Rensselaer Polytech Inst, 57-59, asst prof chem, 59-61; sr lectr phys chem, Univ Hong Kong, 61-68; assoc prof, 68-71, chmn dept, 71-79, dean fac nat & math sci, 79-82, PROF CHEM, STATE UNIV NY COL BROCKPORT, 71- *Mem:* Am Chem Soc; Sigma Xi; The Royal Soc Chem. *Res:* Electrochemistry in molten salts; thermochemistry in nonaqueous solutions. *Mailing Add:* 123 Sherwood Dr Brockport NY 14420-1453

HILL, DONALD GARDNER, b Lansing, Mich, May 17, 41; m 65; c 1. GEOLOGY, GEOPHYSICS. *Educ:* Mich State Univ, BS, 63, PhD(geol, geophys), 69. *Prof Exp:* Res geophysicist, Chevron Oil Field Res Co, 69-78, sr geophysicist, Chevron Resources Co, 78-81, SR DEVELOP GEOLOGIST, CHEVRON OVERSEAS PETROL, INC, STANDARD OIL CO CALIF, 81- *Mem:* Am Geophys Union; Soc Explor Geophys; Soc Petrol Eng; Europ Asn Explor Geophys; Soc Prof Well Log Analysists. *Res:* Petrophysics; well log analysis; seismic interpretation research; mining geophysics and geothermal research. *Mailing Add:* Chevron Overseas Petrol Inc Box 5046 San Ramon CA 94583-0946

HILL, DONALD LOUIS, b Klaber, Wash, Feb 1, 21; m 43; c 2. ANIMAL PHYSIOLOGY. *Educ:* State Col Wash, BS, 43; Univ Minn, PhD, 52. *Prof Exp:* Asst dairy husb, Univ Minn, 46-49; asst prof, 49-59, ASSOC PROF ANIMAL SCI, PURDUE UNIV, WEST LAFAYETTE, 59- *Mem:* Am Dairy Sci Asn; Sigma Xi. *Res:* Hormonal control of lactation and reproduction in bovine; forage utilization by dairy cattle. *Mailing Add:* Dept Animal Sci Poultry Bldg Purdue Univ Lafayette IN 47907

HILL, DONALD LYNCH, b Decherd, Tenn, June 24, 37; m 60; c 3. BIOCHEMICAL PHARMACOLOGY, CANCER. *Educ:* Mid Tenn State Col, BS, 60; Vanderbilt Univ, PhD(biochem), 64. *Prof Exp:* Res asst, Univ Calif, 64-65; sr biochemist, 65-71, head membrane biochem sect, 71-76, assoc dir, 84-85, head biochem pharmacol div, 76-85, dir, 86-90, VPRES, BIOCHEM RES, SOUTHERN RES INST, 90- *Concurrent Pos:* Assoc adj prof pharmacol, Univ Ala, Birmingham, 77- *Mem:* AAAS; Am Asn Biol Chem; Am Soc Pharmacol & Exp Therapeut; Am Asn Cancer Res; Soc Toxicol. *Res:* Dispostion metabolism and sites of action of antitumor agents and carcinogens; prevention of cancer. *Mailing Add:* Southern Res Inst Box 55305 Birmingham AL 35255-5305

HILL, DONALD P, b Orillia, Ont, Nov 9, 29; m 54; c 8. PATHOLOGY. *Educ:* Univ Ottawa, MD, 54. *Prof Exp:* From asst prof to assoc prof, 60-70, actg chmn dept, 78-79, PROF PATH, UNIV OTTAWA, 70- *Concurrent Pos:* Res fel, Ont Cancer Treatment & Res Found, 59-64; fel cytol, Mem Hosp, NY, 60; pathologist, Ottawa Gen Hosp, 58-, assoc dir labs, 69-75, dir, 75-79; asst registr tumor path, Can Tumor Registry, 60-66, assoc registr, 67-73, actg registr, 73-76; asst, Can Asn Path, 65-73; mem assoc comt dent res, Nat Res Coun Can, 65-67. *Mem:* Can Asn Pathologists (secy-treas, 73-76, vpres, 76-77, pres-elect, 77-78, pres, 78-79); Can Soc Forensic Sci (vpres, 70-, pres, 71). *Res:* Exfoliative cytology; tumor, forensic and anatomic pathology. *Mailing Add:* Ottawa Gen Hosp 501 Smyth Rd Ottawa ON K1H 8L6 Can

HILL, DOUGLAS WAYNE, b Bountiful, Utah, Aug 6, 27; m 50; c 2. MEDICAL MICROBIOLOGY. *Educ:* Univ Utah, BS, 50, MS, 52, PhD(bact), 59. *Prof Exp:* Asst mycol, 50-52, res assoc virol, 53-55, instr, 55-70, ASSOC PROF MICROBIOL, UNIV UTAH, 70- *Concurrent Pos:* Consult, US Army Chem Corps, 59. *Mem:* Am Soc Microbiol; Reticuloendothelial Soc. *Res:* Experimental immunology of viral and mycotic diseases. *Mailing Add:* Dept Cell-Viral-Molecular Biol Univ Utah Salt Lake City UT 84112

HILL, DOYLE EUGENE, b Anthony, Kans, Oct 15, 46; m 66; c 3. BIOCHEMISTRY, CLINICAL CHEMISTRY. *Educ:* Northwestern Okla State Univ, BS, 68; Okla State Univ, PhD(biochem), 72. *Prof Exp:* NIH fel biochem, Cornell Univ, 73-74; res chemist, Res Labs, 74-79, supvr med prod eval, 79-89, MGR, STRATEGIC RESOURCES, CLIN PROD DIV, EASTMAN KODAK CO, 89- *Mem:* Sigma Xi; Am Chem Soc; AAAS; Am Asn Clin Chem. *Res:* Enzyme kinetics in solution and diffusion limited situations; physical chemistry of enzymes to delineate structure-function problems; use of enzymes for analytical determinations; in vitro diagnostic product development. *Mailing Add:* Kodak Co 343 State St Rochester NY 14650-0546

HILL, EDDIE P, b Allen, Nebr, June 9, 30; m 59; c 2. BOTANY, MYCOLOGY. *Educ:* Nebr State Teachers Col, Wayne, BA, 52; Colo State Col, MA, 57; Univ Nebr, PhD(bot), 62. *Prof Exp:* Res assoc bot, Univ Mich, 62-64; from asst prof to assoc prof, 64-76, PROF BIOL, MACALESTER COL, 75- *Mem:* AAAS; Am Soc Microbiol. *Res:* Fungal physiology; spore dormancy; growth and development. *Mailing Add:* Dept Biol Macalester Col 1600 Grand Ave St Paul MN 55105

HILL, EDWARD C, b Pocatello, Idaho, Dec 18, 20; m; c 2. OBSTETRICS & GYNECOLOGY. *Educ:* George Washington Univ, MD, 46. *Prof Exp:* Intern, King County Hosp, Seattle, 46-47; resident obstet & gynec, Univ Calif Hosps, 50-54; from instr to assoc prof, 54-73, PROF OBSTET & GYNEC, SCH MED, UNIV CALIF, SAN FRANCISCO, 73- *Mailing Add:* 350 Parnassus Ave San Francisco CA 94117

HILL, EDWARD ORSON, medical microbiology, for more information see previous edition

HILL, EDWARD T, b Minneapolis, Minn, Mar 1, 38; m 64; c 2. MATHEMATICS. *Educ:* Luther Col, Iowa, BA, 63; Vanderbilt Univ, MA, 65, PhD(math), 68. *Prof Exp:* Asst prof math, Calif Lutheran Col, 67-69; asst prof math, 69-75, actg dean, 75-76, PROF MATH, CORNELL COL, 78- *Mem:* Sigma Xi; AAAS; Math Asn Am. *Res:* Modular group rings of p-groups. *Mailing Add:* Cornell Col Mt Vernon IA 52314

HILL, ELDON G, b Hudson, Colo, Apr 30, 18; m 44; c 2. NUTRITIONAL BIOCHEMISTRY. *Educ:* Colo State Univ, BS, 40; Univ Minn, MS, 50, PhD, 53. *Prof Exp:* Instr chem, Colo State Univ, 45-48; res asst poultry husb, Univ Minn, 48-53, res fel, Hormel Inst, 53-54, asst prof, 55-61, assoc prof, Hormel Inst, Univ Minn, 62. *Mem:* Am Inst Nutrit; Soc Exp Biol & Med; Poultry Sci Asn; Am Soc Animal Sci; Am Oil Chem Soc. *Res:* Poultry and swine nutrition; polyunsaturated fatty acid metabolism. *Mailing Add:* Dept Nutrit Univ Guelph Guelph ON N1G 2W1 Can

HILL, ELGIN ALEXANDER, b Pittsburgh, Pa, May 16, 35; m 58; c 3. PHYSICAL ORGANIC CHEMISTRY, ORGANOMETALLIC COMPOUNDS. *Educ:* Allegheny Col, BS, 57; Calif Inst Technol, PhD(chem), 61. *Prof Exp:* NSF fel, Pa State Univ, 60-61; asst prof org chem, Univ Minn, Minneapolis, 61-66; chmn dept, 70-72, from asst prof to assoc prof, 66-80, PROF ORG CHEM, UNIV WIS-MILWAUKEE, 80- *Mem:* Am Chem Soc. *Res:* Physical organic chemistry; organometallic structure and rearrangements; structure and reactivity correlations; isotope effects; free radical rearrangements. *Mailing Add:* Dept Chem Univ Wis-Milwaukee Milwaukee WI 53201

HILL, ELWOOD FAYETTE, b Tonopah, Nev, Feb 18, 39; m 63; c 3. TOXICOLOGY. *Educ:* San Jose State Univ, BA, 61; Univ Md, MS, 72, PhD(avian physiol), 81. *Prof Exp:* Wildlife mgr II, Nev Fish & Game Dept, 61-66; wildlife biologist, Tech Develop Labs, US Pub Health Serv, Fla, 66-69; LEADER, WILDLIFE TOXICOL GROUP & RES TOXICOLOGIST, PATUXENT WILDLIFE RES CTR, US FISH & WILDLIFE SERV, MD, 69- *Mem:* Soc Environ Toxicol & Chem; Sigma Xi; Wildlife Soc; Am Col Toxicol; AAAS; Soc Toxicol. *Res:* Toxicologic research on effects of pollutants and pesticides on wild birds and mammals; emphasis on biochemical indicators of toxicity and development of toxicity testing protocols. *Mailing Add:* Patuxent Wildlife Res Ctr Laurel MD 20708

HILL, ERIC STANLEY, b Liverpool, Eng, Feb 12, 25; m 55; c 3. PHYSICAL CHEMISTRY. *Educ:* Univ Liverpool, BS, 50, PhD(phys chem), 55. *Prof Exp:* Res chemist, Imp Chem Industs, Eng, 53-56, sect head man made fibers, 56-58, qual control mgr, 58-60; asst tech dir, Fiber Industs, Inc, 60-67, mfg tech dir, 67-70, tech dir, 70-, vpres, 71-; RETIRED. *Concurrent Pos:* Consult, Montefibres, Milan, Italy, 84-86. *Res:* Nylon and polyester fibers including new products and development of new design data for new plant facilities. *Mailing Add:* 3953 Abingdon Rd Charlotte NC 28211

HILL, ERIC VON KRUMREIG, mechanical engineering, for more information see previous edition

HILL, ERNEST E(LWOOD), b Oakland, Calif, May 15, 22; m 42; c 3. NUCLEAR ENGINEERING. *Educ:* Univ Calif, BS, 43, MS, 59. *Prof Exp:* Design engr, Daco Develop Co, 46-47; indust engr & prod supt, Fed Pac Elec Co, 47-55; supvr opers res reactor, Lawrence Radiation Lab, Univ Calif, 55-64; chief, reactor safety br, San Francisco Opers Off, AEC, 64-67; head safety systs anal group, Lawrence Livermore Lab, Univ Calif, 67-74, assoc div leader, Nuclear Test Eng div, 74-77, div leader, Magnetic Fusion Eng Div, 77-82; PRES, HILL ASSOCS, 82- *Concurrent Pos:* Admin judge, Atomic Safety Licensing Bd, Panel Nuclear Regulatory Comn, 73- *Mem:* Am Nuclear Soc. *Res:* Nuclear engineering and reactor operation; nuclear safety; human factors engineering; probabilistic risk assessment. *Mailing Add:* 210 Montego Dr Danville CA 94526

HILL, FLOYD ISOM, b Memphis, Tenn, Aug 10, 21; m 43; c 3. MECHANICAL ENGINEERING. *Educ:* Univ Tenn, BS, 43; Johns Hopkins Univ, MS, 52. *Prof Exp:* Analyst, Pratt & Whitney Aircraft Div, United Aircraft Corp, 43-45; ord engr recoil systs anal, US Army Ballistic Res Labs, 46-49, sect head weapons systs anal, 49-51, br chief, 51-55; sr engr, Opers Res, Inc, 56-57; assoc dir oper testing, Tech Opers, Inc, 57-58; prin engr combat surveillance, Cornell Aeronaut Lab, Inc, 58-60, dep dir, 60-62, head Wash projs dept, 62-71; CONSULT, 71- *Mem:* AAAS; Opers Res Soc Am; NY Acad Sci; Sigma Xi. *Res:* Analysis of unsteady compressible flow of liquid; mathematical analysis of ground warfare; design and analysis of operational field tests and war games. *Mailing Add:* 6338 Cross Woods Dr Falls Church VA 22044

HILL, FRANK B(RUCE), b Washington, DC, Aug 9, 24; m 49; c 3. PHOTOREACTOR ENGINEERING. *Educ:* Cath Univ, BChE, 49; Princeton Univ, PhD(chem eng), 59. *Prof Exp:* Instr chem eng, Princeton Univ, 52-53, asst, 53-54; assoc chem engr, Nuclear Eng Dept, 54-57, supvr, Processing Sect, 57-58, proj chem engr, 58-61, chem engr, 61-62, supvr radiation chem eng, Dept Appl Sci, 62-72, prin investr environ sci, Dept Appl Sci, 72-75, prin investr chem eng sci, Dept Energy & Environ, 75-80, GROUP LEADER SEPARATION SCI & TECHNOL, BROOKHAVEN NAT LAB, 80- *Concurrent Pos:* Adj res prof & consult, NJ Inst Technol, 73-; adj prof chem eng, NC State Univ, 73-75; adj prof nuclear eng, Polytechnic Inst NY, 79. *Mem:* Am Chem Soc; Am Inst Chem Engrs. *Res:* Separation of industrial gases, isotopes, biological substances, hazardous wastes; engineering of photochemical and radiation chemical reactors; modelling of acid rain production; emission of biogenic atmospheric sulfur compounds. *Mailing Add:* 13097 SE Point O Woods Ct Hobe Sound FL 33455

HILL, FRANKLIN D, b Boley, Okla, May 15, 33; m 55; c 3. BIOCHEMISTRY. *Educ:* Langston Univ, BS, 56; Iowa State Univ, MS, 58, PhD(biochem), 60. *Prof Exp:* Prof chem, Langston Univ, 60-62; assoc biochem, Iowa State Univ, 62-63; assoc prof, 63-65, PROF CHEM, GRAMBLING STATE UNIV, 65- *Mem:* Am Chem Soc; Am Oil Chem Soc. *Res:* Autoxidation and metabolism of lipids. *Mailing Add:* Dept Chem Grambling State Univ Grambling LA 71245

HILL, FREDERICK BURNS, JR, b Portsmouth, Va, Aug 2, 13; m 45; c 3. FLUORINE CHEMISTRY, POLYMER CHEMISTRY. *Educ:* Col of William & Mary, BS, 34; Univ Va, PhD(org chem), 40. *Prof Exp:* Chemist, Corn Prod Refining Co, Ill, 34-36; res chemist, Org Chem Dept, Jackson Lab, E I du Pont de Nemours & Co, Inc, 40-55, tech sales serv, Freon Prod Div, 55-61, mkt assoc, 61-69, mgr blowing agent develop, 69-71, sales mgr blowing agents, 71-76; RETIRED. *Mem:* Am Chem Soc; Sigma Xi. *Res:* Elastomers; resins; fluorocarbons. *Mailing Add:* 1904 Greenbriar Dr Westwood Manor Wilmington DE 19810

HILL, FREDERICK CONRAD, b Waukegan, Ill, Apr 6, 43; m 75; c 3. VERTEBRATE ZOOLOGY, ARCHAEOZOOLOGY. *Educ:* Ill State Univ, BS, 65, MS, 70; Univ Louisville, PhD(biol), 75. *Prof Exp:* Instr biol, Jefferson Col, 67-70; PROF ZOOL, BLOOMSBURG UNIV, PA, 75- *Mem:* Sigma Xi; Am Malacol Union; Soc Am Archaeol. *Res:* Archaeozoology; fish and mollusc remains. *Mailing Add:* Dept Biol Bloomsburg State Col Bloomsburg PA 17815

HILL, FREDRIC WILLIAM, b Erie, Pa, Sept 2, 18; m 44; c 3. ANIMAL SCIENCE & NUTRITION, AGRICULTURAL & FOOD CHEMISTRY. *Educ:* Pa State Univ, BS, 39, MS, 40; Cornell Univ, PhD(animal nutrit), 44. *Prof Exp:* Asst poultry husb, Pa State Univ, 39-40 & Cornell Univ, 40-43; head nutrit div, Res Labs, Western Condensing Co, 44-48; assoc prof animal nutrit & poultry husb, Cornell Univ, 48-53, prof, 53-59; prof & chmn dept poultry husb, 59-65, assoc dean, Cl Agr, 65-66, chn dept nutrit, 65-73, assoc dean res & int prog, 76-80, PROF NUTRIT, UNIV CALIF, DAVIS, 65- *Concurrent Pos:* Mem, Animal Nutrit Subcomt on Hormonal Relationships & Applns, Nat Res Coun, 53, Animal Nutrit Subcomt on Poultry Nutrit, 53-74 & Food & Nutrit Bd, 75-78; mem task force livestock res, USDA; comnr, Calif Poultry Improvement Comn, 59-65; lectr, Univ Nottingham, Eng, 61; deleg, World Conf Animal Prod, Rome, 63; Guggenheim fel, Nat Inst Med Res, London, 66; vis scientist, Hebrew Univ, Jerusalem, , 67, Univ Nagoya, Japan, 74-75, USDA, 75 & US Food & Drug Admin, 75 & 88; ed, J Nutrit, 69-79; mem, USAID-Nat Acad Sci Sem on Protein Foods, Bangkok, Thailand, 70 & US Info Agency Sem on Food, Pop & Energy, Philippines, S Vietnam, Malaysia, Indonesia, Japan, India & Bangladesh, 74-75; consult, Nat Inst Agron Res, France, 82; alumni fel, Pa State Univ, 83; plenary lectr, Int Cong Asian-Australasian Animal Prod Socs, Seoul, Korea, 85. *Honors & Awards:* Res Prize, Poultry Sci Asn, 57; Nutrit Res Award, Am Feed Mfrs Asn, 58; Newman Mem Int Res Award, 59; Borden Award, 61. *Mem:* Fel AAAS; Am Chem Soc; fel Poultry Sci Asn; Am Soc Animal Sci; fel Am Inst Nutrit; fel Japan Soc Promotion Sci; Nutrit Soc Gt Brit. *Res:* Experimental nutrition, especially energetics, quantitative nutrient requirements; nutritional properties of fats; carbohydrates. *Mailing Add:* Dept Nutrit Univ Calif Davis CA 95616

HILL, FREDRICK J, b Las Vegas, Nev, Sept 2, 36; m 57; c 3. ELECTRICAL ENGINEERING. *Educ:* Univ Utah, BS, 58, MS, 60, PhD(elec eng), 63. *Prof Exp:* Teaching asst elec eng, Univ Utah, 58-61; assoc prof, 63-70, PROF ELEC ENG, UNIV ARIZ, 70- *Concurrent Pos:* Consult, Motorola Integrated Circuits Ctr, Rockwell Microelectronics & Gen Instrument Res & Develop; mem, Int Working Group on a consensus hardware description lang, 75- *Mem:* Inst Elec & Electronics Engrs. *Res:* Test sequence generation for sequential circuits; hardware description languages. *Mailing Add:* Dept Elec Eng Univ Ariz Tucson AZ 85721

HILL, GALE BARTHOLOMEW, b Atlanta, Ga, Sept 20, 36. MICROBIOLOGY, INFECTIOUS DISEASES. *Educ:* Fla State Univ, BS, 60; Duke Univ, PhD(microbiol), 66. *Prof Exp:* USPHS fel, Dept Microbiol, Med Sch, 66-67, res assoc microbiol & immunol, Med Ctr, 66-71, assoc radiol, Div Radiol, 67-72, instr microbiol, 71-73, asst prof radiol, 72-73, asst prof obstet & gynec, 73-78, ASST PROF MICROBIOL, MED CTR, DUKE UNIV, 73-, ASSOC PROF OBSTET & GYNEC, 78-, DIR DIV ANAEROBIC MICROBIOL, 75- *Mem:* Am Soc Microbiol; Infectious Dis Soc Obstet & Gynec; AAAS; Sigma Xi. *Res:* Infectious diseases primarily anaerobic bacterial infections and pathogenesis; anaerobic infections in obstetrics and gynecology; therapeutic agents for infection and evaluation in animal models. *Mailing Add:* Dept Reproductive Biol Box 3172 Duke Univ Med Ctr Durham NC 27710

HILL, GARY MARTIN, b Hendersonville, NC, Nov 11, 46; m 68; c 3. RUMINANT NUTRITION, ANIMAL SCIENCE. *Educ:* Clemson Univ, BS, 69; Univ Kent, MS, 74, PhD(animal sci), 77. *Prof Exp:* Res asst animal sci, Univ Kent, 72-77; asst prof animal sci, 77-80, assoc prof, La State Univ, 80-83; asst prof, 83-87, ASSOC PROF, ANIMAL SCI DEPT, COASTAL PLAIN STA, TIFTON, GA, 87- *Honors & Awards:* Merit Cert, Am Forage & Grassland Coun, 90. *Mem:* Am Soc Animal Sci; Am Dairy Sci Asn; Am Forage & Grassland Coun; Sigma Xi; Am Registry Prof Animal Scientists. *Res:* Ruminant nutrition as it affects preweaning and postweaning beef animals; forage utilization by ruminants and forage management systems. *Mailing Add:* Animal Sci Dept Coastal Plain Exp Sta PO Box 748 Tifton GA 31793-0748

HILL, GEORGE CARVER, b Moorestown, NJ, Feb 19, 39; m 65; c 2. PARASITOLOGY, BIOCHEMISTRY. *Educ:* Rutgers Univ, BA, 61; Howard Univ, MS, 63; NY Univ, PhD(biol), 67. *Prof Exp:* NIH fel, Med Ctr, Univ Ky, 67-69; res investr biochem parasitol, Squibb Inst Med Res, 69-71; NIH spec res fel, Molteno Inst, Univ Cambridge, 71-73; mem staff, Squibb Inst Med Res, 73-77; assoc prof path, Colo State Univ, Ft Collins, 77-; ASSOC PROF, DIV BIOMED SCI, MEBERRY MED COL. *Concurrent Pos:* Dept Army res grant, 67-68. *Mem:* Soc Protozoologists; Am Soc Microbiol. *Res:* Purification and properties of Trypanosomatid cytochromes C; characterization of electron transport systems in Trypanosomatids; ultrastructure and function of mitochondria in Trypanosomatids. *Mailing Add:* Div Biomed Sci Meberry Med Col 1005 D B Todd Jr Blvd Nashville TN 37208

HILL, GEORGE JAMES, II, b Cedar Rapids, Iowa, Oct 7, 32; m 60; c 4. SURGERY, CANCER. *Educ:* Yale Univ, AB, 53; Harvard Med Sch, MD, 57. *Prof Exp:* Clin assoc med, NIH, Bethesda, Md, 61-63; instr surg, Univ Colo Sch Med, 66-67, from asst prof to assoc prof, 67-73; prof, Wash Univ Sch Med, 73-76; prof, Surg & Chmn Dept, Marshall Univ Sch Med, 76-81;

PROF SURG & DIR ONCOL, NJ MED SCH, UNIV MED & DENT, NJ, 81-; CLIN PROF SURG, UNIFORMED SERVS UNIV HEALTH SCIS, BETHESDA, MD, 89- *Concurrent Pos:* Jr fac fel, Am Cancer Soc, 70-73; vis prof, Univ Saigon, Repub Vietnam, 72-73; chief surg, St Louis City Hosp, 73-76; assoc dean clin affairs, Marshall Univ Sch Med, 76-78; chmn, Nat Cancer Inst Clin Cancer Educ Comt, 78-80; pres, Am Cancer Soc, WVa Div, 80-81; pres, Am Cancer Asn, Cancer Educ, 85-86; pres, Am Cancer Soc, NJ Div, 87-88; vis fel, Molecular Biol, Princeton Univ, 88; dir at large, Am Cancer Soc Nat Bd, 89-91; mem bd trustees, Sterling Col, Craftsburg Common, Vt, 90-93. *Honors & Awards:* Lederle Med Fac Award, 67-70. *Mem:* Am Asn Cancer Res; fel Am Col Surgeons; Soc Surg Oncol; Soc Univ Surgeons; Sigma Xi. *Res:* Cancer research, surgery; chemotherapy; immunotherapy; radiation therapy; cancer education-new programs, trials and evaluation; medical education programs in developing countries and rural America. *Mailing Add:* Three Silver Spring Rd West Orange NJ 07052

HILL, GEORGE RICHARD, b Ogden, Utah, Nov 24, 21; m 41; c 7. CHEMISTRY, FUEL SCIENCE. *Educ:* Brigham Young Univ, AB, 42; Cornell Univ, PhD(inorg chem), 46. *Hon Degrees:* Brighan Young Univ, DSc, 80. *Prof Exp:* Asst chem, Cornell Univ, 42-43, instr, 43-46; from instr to prof, Univ Utah, 46-72; dir, US Off Coal Res, 72-73; dir, Fossil Fuel Dept, Elec Power Res Inst, 73-77; envirotech prof chem eng, 77-87, EMER PROF, UNIV UTAH, 87- *Concurrent Pos:* Head dept fuel technol, Univ Utah, 53-60, chmn dept fuels eng, 60-66, dean, Col Mines & Mineral Indust, 65-72; consult, NSF, 60-; mem comt mineral sci & technol, Nat Res Coun; mem coal sci team, Europ Econ Community Group, 67 & 71; mem comt mineral resources, Nat Asn State Univs & Land-Grant Cols, 70-; chmn, Gordon Res Conf Coal Sci, 71; mem, Energy Res & Develop Admin-FE Gen Tech Adv Comt, 73-; mem bd energy studies, Nat Acad Sci-Nat Acad Eng, 74-76; mem bd mineral & energy resources, Nat Res Coun, 76- *Honors & Awards:* Henry H Storch Award, 71. *Mem:* Nat Acad Eng; fel Am Inst Chem; Am Inst Mining, Metall & Petrol Eng; Sigma Xi; Am Chem Soc. *Res:* Kinetics of coal conversion reactions and in situ production of shale oil. *Mailing Add:* 1430 Yale Ave Salt Lake City UT 84105

HILL, GIDEON D, b Winston-Salem, NC, Nov 2, 22; m 47; c 2. AGRONOMY, WEED SCIENCE. *Educ:* Berea Col, BS, 48; NC State Col, MS, 51; Ohio State Univ, PhD(agron), 53. *Prof Exp:* Instr agron, NC State Col, 49-50; res supvr agr chem, 53-65, prod develop mgr, 65-68, mgr prod develop, 68-78, assoc div biol res, Exp Sta, E I Du Pont De Nemours & Co, 78-87; CONSULT, 87- *Mem:* Hon mem Weed Sci Soc Am. *Res:* Weed control in tobacco plant beds with methyl bromide, allyl alcohol and uramon plus cyanamide; control of Canada thistle with various herbicides and adjuvants; role of soil factors in pre-emergence weed control in economic crops; herbicides; insecticides; fungicide; agricultural chemicals. *Mailing Add:* 4915 Threadneedle Rd Wilmington DE 19807

HILL, GRETCHEN MYERS, b July 28, 42; c 1. COPPER, ZINC. *Educ:* Mich State Univ, PhD(nutrit), 81. *Prof Exp:* Asst prof clin nutrit & nutrit sci, Community Health & Nutrit Prog, Univ Mich, 84-86; asst prof, 86-88, ASSOC PROF, DEPT HUMAN NUTRIT, UNIV MO, 88- *Concurrent Pos:* Supvr, res in nutrit, Community Health & Nutrit Prog, Mich State, 84- *Mem:* Am Inst Nutrit; Am Soc Clin Nutrit. *Res:* Copper and factors affecting its metabolism; impact of breakfast in dietary adequacy. *Mailing Add:* Dept Human Nutrit Univ Missouri 217 Gwynn Hall Columbia MO 65211

HILL, HAMILTON STANTON, b Brookline, Mass, Jan 19, 11; m 36; c 2. MINERALOGY. *Educ:* Pomona Col, BA, 33; Claremont Grad Sch, MA, 40. *Prof Exp:* Prof, 35-72, EMER PROF GEOL & MINERAL, PASADENA CITY COL, 72- *Concurrent Pos:* Cur minerals, Calif Inst Technol, 79-88. *Honors & Awards:* Neil Miner Award, Nat Asn Geol Teachers, 71. *Mem:* Fel Geol Soc Am; Mineral Soc Am; Mineral Soc Gt Brit & Ireland; fel Geol Soc, London; Hist Sci Soc. *Res:* History of geological sciences; development of early geologic ideas. *Mailing Add:* 2661 Bowring Dr Altadena CA 91001

HILL, HARRY RAYMOND, b Slidell, La, Dec 18, 41; m 66; c 2. PEDIATRICS, PATHOLOGY. *Educ:* Baylor Col Med, MD, 66. *Prof Exp:* Intern, Grady Mem Hosp, Emory Univ, 66-67; epidemic intel serv officer, Ctr Dis Control, USPHS, 67-69; resident pediat, Univ Wash, 69-71; fel infectious dis, Univ Minn, 71-73, from instr to asst prof pediat, 73-74; asst prof, 74-76, ASSOC PROF PEDIAT & PATH, UNIV UTAH, 76- *Concurrent Pos:* Investr, Howard Hughes Med Inst, 75, prof path & pediat, 81. *Honors & Awards:* Outstanding Young Investr, Western Soc Pediat Res, 80. *Mem:* Soc Pediat Res; Am Fedn Clin Res; Reticuloendothelial Soc; Am Soc Microbiol; Western Soc Clin Res. *Res:* The role of cyclic nucleotides and pharmacologic agents in regulating the host defense mechanism; host defense mechanisms in group B streptococcal infection. *Mailing Add:* Dept Path & Pediat Univ Utah Col Med Salt Lake City UT 84132

HILL, HELENE ZIMMERMANN, b Philadelphia, Pa, Apr 10, 29; m 60; c 4. CANCER BIOLOGY, GENETICS. *Educ:* Smith Col, AB, 50; Brandeis Univ, PhD(biol), 64. *Prof Exp:* Instr biol, Brandeis Univ, 63-64; from instr biophys to asst prof biophys & genetics, Univ Colo Med Ctr, Denver, 67-72; assoc prof radiol, Wash Univ Sch Med, 73-76; from assoc prof to prof biochem, Sch Med, Marshall Univ, 76-81; PROF & HEAD CANCER SECT, DEPT RADIOL, NJ MED SCH, 81- *Concurrent Pos:* Fel, Harvard Med Sch, 64-66; NIH fels, 64-68; prin investr, NIH grant, 71-75; adj assoc prof biol, Wash Univ. *Mem:* Am Soc Cell Biol; Am Soc Photobiol; Radiation Res Soc; Am Asn Cancer Res; Biophys Soc; Sigma Xi. *Res:* Ultraviolet and photoreactivation effects in Euglena; protein synthesis patterns in normal and regenerating rat liver; cancer biology and malignant transformation; gene expression and enzyme kinetics in mammalian cells; DNA damage and repair and its role in carcinogenesis annd chemotherapy. *Mailing Add:* Three Silver Spring Rd West Orange NJ 07052

HILL, HENRY ALLEN, b Port Arthur, Tex, Nov 25, 33; m 54; c 3. ASTROPHYSICS, OPTICS & NUCLEAR STRUCTURE. *Educ:* Houston Univ, BS, 53; Univ Minn, MS, 56, PhD(nuclear physics), 57. *Hon Degrees:* MA, Wesleyan Univ, 66. *Prof Exp:* Asst nuclear physics, Princeton Univ, 57-58, from instr to asst prof physics, 58-64; from assoc prof to prof, Wesleyan Univ, 64-74, chmn dept, 69-71; PROF PHYSICS, UNIV ARIZ, 66- *Concurrent Pos:* Sloan fel, 66-68. *Mem:* Fel Am Phys Soc; Am Astron Soc; Royal Astron Soc; Optical Soc Am; Int Astron Union. *Res:* Major research in experimental relativity and the solar interior via the study of global oscillations of the sun. *Mailing Add:* Dept Physics Univ Ariz Tucson AZ 85721

HILL, HERBERT HENDERSON, JR, b Helena, Ark, Nov 25, 45; m 68; c 2. ANALYTICAL CHEMISTRY, CHROMATOGRAPHY. *Educ:* Rhodes Col, Memphis, BS, 70; Univ Mo, Columbia, MS, 73; Dalhousie Univ, NS, PhD(chem), 75. *Prof Exp:* From asst prof to assoc prof, 76-85, PROF CHEM, WASH STATE UNIV, 85- *Concurrent Pos:* Killam fel, 73-74, Japan Soc for Prom Sci fel, 83-84; dir, Off Grant & Res Develop, Wash State Univ, 85-87. *Honors & Awards:* Keene P Dimick Award Chromatography, 89. *Mem:* Am Chem Soc. *Res:* Development of analytical methods and instrumentation for trace organic analysis of environmental, biological and industrial samples. *Mailing Add:* Dept Chem Wash State Univ Pullman WA 99164

HILL, HERBERT HEWETT, b Tampa, Fla, Aug 31, 33; m 55. MECHANICAL ENGINEERING. *Educ:* Ga Inst Technol, BSME, 55, MSME, 61; NC State Univ, PhD, 69. *Prof Exp:* Engr, Lockheed Aircraft Corp, 59-61; res engr, Astra, Inc, NC, 62; systs analyst, Res Triangle Inst, 63-67; PRES, RES ENGRS, INC, 68- *Mem:* Nat Soc Prof Engrs. *Res:* Vehicle and traffic research; kinematics of accident reconstruction; mechanical machine design; vibrational and stress analysis. *Mailing Add:* Res Engrs Inc PO Box 12072 Research Triangle Park NC 27709

HILL, HUBERT MACK, b Sheridan, Mich, Aug 3, 18; m 44; c 4. SYSTEMS ANALYSIS, APPLIED STATISTICS. *Educ:* Alma Col, BS, 40; Univ Conn, MS, 42; Purdue Univ, PhD(org chem), 47. *Prof Exp:* Asst chem, Univ Conn, 40-42; Bristol res fel, Purdue Univ, 42-46; Bristol res assoc, Ohio State Univ, 46-47; sr chemist, Tenn Eastman Co, Eastman Kodak Co, 47-58, sr mathematician, 58-65, consult, 65-81; CONSULT, INNOVATIVE PROBLEM SOLVING, 81- *Concurrent Pos:* Vis prof, Alma Col, Mich, 82. *Mem:* Am Chem Soc; fel Am Soc Qual Control. *Res:* Design of experiments; quality control; information systems; systems planning; problem solving strategies; statistics. *Mailing Add:* 1225 Buchelew Dr Kingsport TN 37663

HILL, JACK DOUGLAS, b Fort Frances, Ont, Nov 28, 37; m 59; c 2. ELECTRICAL ENGINEERING, MATHEMATICS. *Educ:* Univ Man, BSc, 59, MS, 60; Purdue Univ, PhD(elec eng), 65. *Prof Exp:* Sci officer, Defence Res Bd Can, Ottawa, 60-61; instr & res engr, Purdue Univ, 61-64; mem tech staff, Bellcomm Inc, 64-65; MGR, ELECTRONICS DEPT, BATTELLE MEM INST, 65- *Mem:* AAAS; Inst Elec & Electronics Engrs. *Res:* Development and application of system methodologies for planning, managing and assessing complex interdisciplinary research programs in both social and engineering sciences; basic research in optimization, modeling, technology assessment and management. *Mailing Add:* Battelle Mem Inst 505 King Ave Columbus OH 43201

HILL, JACK FILSON, b Hamilton, Mo, Feb 23, 26; m 49; c 6. POULTRY BREEDING, GENETICS. *Educ:* Univ Mo, BS, 49, MS, 50; Iowa State Univ, PhD(poultry breeding), 57. *Prof Exp:* Geneticist, Hy-Line Poultry Farms, 50-59; vpres & dir res, Babcock Industs, 59-76 & Colonial Poultry Farms, 76-80; consult geneticist, Tatum Farms, 81-82; consult geneticist, H & N, Inc, 82-85; CONSULT, TATUM FARMS, 85- *Concurrent Pos:* Geneticist, Tokai Breeding Farms Ltd, 77-84. *Honors & Awards:* Gold Cup, Flour Adv Bd, UK, 66. *Mem:* Fel AAAS; Poultry Sci Asn; World Poultry Sci Asn; Sigma Xi. *Res:* Development and improvement of commercial egg laying crosses of chickens by hybridization. *Mailing Add:* 1203 Crest Dr Pleasant Hill MO 64080

HILL, JAMES CARROLL, b Manila, Ark, June 30, 41; div. MICROBIOLOGY, BACTERIOLOGY. *Educ:* Ark State Col, BS, 62; Univ Ark, MS, 65, PhD(microbiol), 67. *Prof Exp:* Bacteriologist, Naval Biol Lab, Naval Supply Ctr, US Navy, 67-69, microbiologist, Naval Med Res Inst, Nat Naval Med Ctr, 69-74; bact vaccines prog officer, Develop & Applications Br, Nat Inst Allergy & Infectious Dis, 74-83, assoc dir, Intramoral Res Prog, 83-84, asst to dir, 84-87, DEP DIR, NAT INST ALLERGY & INFECTIOUS DIS, 87- *Mem:* Am Soc Microbiol; Infectious Dis Soc Am; Sigma Xi; Am Venereal Dis Asn; Sigma Xi. *Res:* Pigment production by Aspergillis species and Pseudomonas species; serology and antigenicity of Neisseria meningitidis, physiology and basic biology of the Neisseria; science administration. *Mailing Add:* Nat Inst Allergy & Infect Dis-NIH Bldg 31-Rm 7A03 Bethesda MD 20892-0031

HILL, JAMES EDWARD, b Bluefield, WVa, Jan 19, 42; m 63; c 3. BUILDING RESEARCH. *Educ:* Va Polytech Inst, BS, 63; Ga Inst Technol, MS, 66, PhD(mech eng), 68. *Prof Exp:* Res asst, Ga Inst Technol, 63-67; engr, Mission Planning & Anal Div, Manned Spacecraft Ctr, NASA, 67-68; aerospace res engr, Air Force Dynamics Lab, 68-69; asst prof mech eng, Univ Md, College Park, 69-72; mech engr thermal eng, Nat Bur Standards, 72-77, leader thermal solar group, Ctr Bldg Technol, 78-80, chief bldg equip div, 80-86, CHIEF BLDG ENVIRON DIV, CTR BLDG TECHNOL, NAT BUR STANDARDS, 86- *Honors & Awards:* Dept Com Spec Achievement Award, 74; Crosby Field Award, Am Soc Heating, Refrig & Air-Conditioning Engrs, 75; Dept Com Silver Medal, 76. *Mem:* Am Soc Mech Engrs; Am Soc Heating, Refrig, & Air-Conditioning Engrs. *Res:* Air-conditioning; refrigeration; solar heating and cooling of buildings. *Mailing Add:* No 2 Diller Ct Boyds MD 20841

HILL, JAMES LAFE, b Clinton, Okla, Feb 23, 37; m 61, 81. ENGINEERING MECHANICS, MECHANICAL ENGINEERING. *Educ:* Univ Okla, BS, 59; Univ Ill, MS, 62, PhD(theoret & appl mech), 63. *Prof Exp:* Assoc prof, 63-76, PROF ENG MECH, UNIV ALA, 76- *Honors & Awards:* AT&T Found Award, 85. *Mem:* Am Soc Mech Engrs; Am Soc Eng Educ; Am Acad Mech; Sigma Xi. *Res:* FEM; expert systems; CAD; dynamics; vibrations; solid mechanics. *Mailing Add:* Box 870278 Tuscaloosa AL 35487-0278

HILL, JAMES LESLIE, b Calgary, Alta, June 11, 40; m 66; c 2. BEHAVIORAL BIOLOGY. *Educ:* Univ Alta, Calgary, BSc, 63, MSc, 66; Mich State Univ, PhD(zool), 70. *Prof Exp:* Asst prof biol, Richmond Col, City Univ New York, 70-75; sr res assoc zool, Mich State Univ, 75-77, Dept Biomech, 76-77; vis assoc, unit res behav systs, NIMH, 77-84, expert consult, comp studies of brain & behav, 84-85, BIOSTATISTICIAN, CLIN SCI LAB, NIMH, 85- *Concurrent Pos:* Statist consult, Lamar Res Group, 74-75. *Mem:* AAAS; Am Soc Mammalogists; Animal Behav Soc. *Res:* Social behavior of mammals as a regulator of spacing patterns, resource utilization, mate selection and gene flow within and between populations. *Mailing Add:* Lab Clin Sci NIMH Bldg 10 Rm 3D41 Bethesda MD 20892

HILL, JAMES MILTON, b Pascagoula, Miss, Feb 20, 44; m 68; c 2. BIOLOGICAL CHEMISTRY. *Educ:* Spring Hill Col, BS, 64; Univ Miss, MS, 67; Baylor Col Med, PhD(pharmacol), 71. *Prof Exp:* Res asst microbiol, Univ Miss Sch Med, 67; fel biochem, Harvard Med Sch, 71-73; from asst prof to assoc prof biol, microbiol & pharmacol, Med Col Ga, 73-85; PROF OPHTHAL, PHARMACOL & MICROBIOL, LA STATE UNIV EYE CTR, 85- *Concurrent Pos:* NIH fel, 71-73; NSF grant, 73-; Am Lung Asn fel, 77-; consult, Warner-Lambert/Parke-Davis & Co, 76- & Burton-Parsons & Co, Inc, 78-; Nat Inst Dent Res fel, 78-; grant, Nat Eye Inst, 79- *Mem:* Am Soc Cell Biol; Am Asn Cancer Res; Am Soc Microbiol; Soc Exp Biol & Med. *Res:* Biochemical regulation of stimulated growth of liver, kidney, and lung; control mechanisms for protein degradation; polyamine metabolism; ribosomal RNA metabolism; Herpes virus forms of infections; antiviral chemotherapy; iontophoresis for drug delivery. *Mailing Add:* Dept Ophthal La State Univ Eye Ctr 2020 Gravier St Suite B New Orleans LA 70112-2234

HILL, JAMES STEWART, b Washington, DC, Dec 2, 12; m 36; c 4. ELECTROMAGNETIC COMPATIBILITY, ELECTROMAGNETIC ENVIRONMENT. *Educ:* Case Sch Appl Sci, BSc, 34. *Prof Exp:* Chief engr, United Broadcasting Co, 42-53; vpres, Smith Electronics Inc, 53-59; proj engr, Jansky & Bailey Div, Atlantic Res Corp, 59-64; sr res engr, Genisco Technol Inc, 65-69 & RCA Corp, 69-78; PRIN ENGR, EMXX CORP, 78- *Concurrent Pos:* mem comt G-46, Electronics Indust Asn, 67-; dir, Inst Elec & Electronics Engrs Electromagnetic Compatibility Soc, 64- *Honors & Awards:* Laurence Cumming Award, Inst Elec & Electronics Engrs, 83, Centennial Medal, 84. *Mem:* Fel Inst Elec & Electronics Engrs; Inst Elec & Electronics Engrs Electromagnetic Compatibility Soc. *Res:* Electromagnetic environment measurement and evaluation; electromagnetic compatibility design principles; author and editor of books written on sub-speciality. *Mailing Add:* 263 N Main St Hudson OH 44236

HILL, JAMES WAGY, b Southgate, Calif, Apr 7, 42; m 70. ORGANIC CHEMISTRY, ANALYTICAL CHEMISTRY. *Educ:* Kans State Teachers Col, BS, 63; Southern Ill Univ, MS, 66, PhD(chem), 68. *Prof Exp:* Asst prof chem, Emporia Kans State Col, 68-74; asst prof, 74-, PROF CHEM, PANHANDLE STATE UNIV, 88- *Mem:* Am Chem Soc. *Res:* mercury in minerals by atomic absorption spectroscopy; superpure gold and superpure silver. *Mailing Add:* Dept Chem Panhandle State Univ Goodwell OK 73939

HILL, JANE VIRGINIA FOSTER, b Portland, Ore, Apr 15, 46; m 71. PLANT MORPHOGENESIS & FUNCTION. *Educ:* Carleton Col, BA, 68; George Washington Univ, MS, 76, PhD(bot), 80. *Prof Exp:* Botanist, Water Resources Div, US Geol Surv, Reston, Va, 78-80; vol plant physiologist, Forest Physiol Lab, 80-82, VOL PLANT PHYSIOLOGIST, CLIMATE STRESS LAB, USDA BARC-W, BELTSVILLE, MD, 83- *Mem:* Sigma Xi; Bot Soc Am. *Res:* Study of functional and developmental plant anatomy (structure and morphogensis) with special reference to trees; identification of environmental and endogenous factors which influence and direct the course of wood-cell differentiation. *Mailing Add:* 8211 Hawthorne Rd Bethesda MD 20817

HILL, JESSE KING, b Billings, Mont, Dec 31, 47. ASTRONOMY. *Educ:* Stanford Univ, BS, 70; Univ Calif, Berkeley, MA, 71, PhD(astron), 75. *Prof Exp:* Res assoc astron, Nat Radio Astron Observ, 75-78; analyst astron, Systs & Appl Sci Corp, 78-86, ST SYSTS CORP, 86- *Mem:* Am Astron Soc. *Res:* Interstellar gas dynamics; observation of interstellar absorption lines; ultraviolet astronomy; galaxies; globular clusters. *Mailing Add:* 6206 87th Ave Hyattsville MD 20784

HILL, JIM T, b Cushing, Okla, Apr 27, 39; m 63; c 3. METABOLISM, BIOCHEMISTRY. *Educ:* Abilene Christian Univ, BS, 61; Univ Tenn, MS, 64, PhD(biochem), 68. *Prof Exp:* Sr res scientist, E R Squibb & Sons, 68-69; sr res biochemist, Lakeside Labs, 69-75; sr res specialist, Monsanto Co, 75-78; dir chem, Hazleton Labs Am Inc, 78-80; mgr toxicol, Phelps Dodge Corp, 80-81; dir sci affairs, 81-86, DIR PESTICIDE INGREDIENT REV PROG, CHEM SPEC MFR ASN, 86- *Mem:* Am Chem Soc; Environ Mutagen Soc; Sigma Xi; Am Soc Pharmacol & Exp Therapeut; Am Col Toxicol; Soc Toxicol. *Res:* Biochemical pharmacology; chemical carcinogenesis; microbial fermentation of hydrocarbons; development of analytical techniques for identification of biological products; environmental and mammalian safety, toxicology and metabolism. *Mailing Add:* 2477 Freetown Dr Reston VA 22091

HILL, JOHN CAMPBELL, b Detroit, Mich, May 30, 38; m 66; c 3. SOLID STATE PHYSICS. *Educ:* Wayne State Univ, BS, 60; Yale Univ, MS, 61, PhD(far infrared spectros), 66. *Prof Exp:* Res asst, Yale Univ, 62-66; from assoc sr res physicist to sr res physicist, 66-73, supvry res physicist, 73-77, dept res scientist, 77-79, asst dept head, 79-87, PRIN RES SCIENTIST, GEN MOTORS RES LABS, 87- *Mem:* Am Phys Soc; Inst Elec & Electronics Engrs; Soc Automotive Engrs; Sigma Xi. *Res:* Infrared spectroscopy of crystalline solids; infrared diode laser spectroscopy of gases; tungsten-halogen light bulb technology; light emitting diodes; metal physics, metal deformation; combustion physics. *Mailing Add:* Gen Motors Res Labs Physics Dept 12 Mile & Mound Rds Warren MI 48090-9055

HILL, JOHN CHRISTIAN, b Blacksburg, Va, Apr 13, 36; m 67; c 1. NUCLEAR PHYSICS, HEAVY ION PHYSICS. *Educ:* Davidson Col, BS, 57; Purdue Univ, MS, 62, PhD(physics), 66. *Prof Exp:* Fel physics, Univ Mich, 66-68; asst prof, Tex A&M Univ, 68-75; from asst prof to assoc prof, 75-81, PROF PHYSICS, IOWA STATE UNIV, 81- *Concurrent Pos:* Res assoc, Univ Mich, 66-68; res grant, Robert A Welch Found, 69-75 & NSF, 74-78; prog dir, Ames Lab, 77-; vis scientist, KFA Jülich, WGer, 80-81. *Mem:* Am Phys Soc; Am Chem Soc; AAAS; Sigma Xi. *Res:* Structured neutron-rich fission products using the mass separator Tristan search for new isotopes; electromagnetic dissociation processes with relativistic heavy ions at the Bevalac, AGS & CERN-SPS accelerators. *Mailing Add:* Dept Physics Iowa State Univ Ames IA 50011

HILL, JOHN DONALD, b Santa Monica, Calif, Dec 5, 30. VETERINARY MEDICINE, CARDIOLOGY. *Educ:* Wash State Univ, BS, 58, DVM, 60; Univ Pa, MMS, 67. *Prof Exp:* Pvt pract, Calif, 60-62; instr med & surg, Sch Vet Med, Univ Calif, Davis, 62-63; instr cardiol, Univ Pa, 63-64, Nat Heart Inst fel, 64-67, asst prof med & USPHS fel physiol, 67-68; assoc scientist, Ore Regional Primate Res Ctr, 68-76, ASST PROF MED, MED SCH, ORE HEALTH SCI UNIV, 68-; scientist, Ore Regional Primate Res Ctr, 76-, chmn, Dept Surg, 68-; RETIRED. *Mem:* AAAS; Am Heart Asn; Am Vet Med Asn. *Res:* Experimental surgery; comparative cardiology; electrocardiography; experimental surgery in nonhuman primates; electrophysiology and cardiovascular physiology of spontaneous heart diseases in dogs; cardiovascular physiology in monkeys including pregnant animals; fetal surgery in monkeys. *Mailing Add:* 160 NW 11th Ave Canby OR 97013

HILL, JOHN HAMON MASSEY, b Belfast, Northern Ireland, June 12, 32; m 60. ORGANIC CHEMISTRY. *Educ:* Queen's Univ, Belfast, BSc, 53, PhD(org chem), 56. *Prof Exp:* Fel org chem, Univ Munich, 56-57; res assoc, Johns Hopkins Univ, 57-59; res chemist, Du Pont of Can, 59-60; lectr chem, Royal Mil Col, Ont, 60-61; assoc prof, 61-66, PROF CHEM, HOBART & WILLIAM SMITH COLS, 66-, CHMN DEPT, 69- *Concurrent Pos:* F G Cottrell res grant, 63-65. *Mem:* Am Chem Soc. *Res:* Nucleophilic heteroaromatic substitution; novel heteroaromatic compounds; chemiluminescence; carbanion rearrangements. *Mailing Add:* Hobart & William Smith Cols Geneva NY 14456

HILL, JOHN JOSEPH, b Kansas City, Mo, Jan 11, 21; wid; c 2. PHYSICS. *Educ:* Rockhurst Col, AB, 42; Johns Hopkins Univ, MA, 48; St Louis Univ, PhD(physics), 64. *Prof Exp:* Meteorologist, US Weather Bur, Washington, DC, 42-44; instr math, Loyola Col, Md, 47-48; physicist, Lewis Lab, NASA, Ohio, 48-53; instr math & physics, 53-56, from asst prof to prof physics, 57-87, EMER PROF, ROCKHURST COL, 87- *Concurrent Pos:* Res partic, Oak Ridge Nat Lab, 68 & Argonne Nat Lab, 65-66, 72-73, 79-80, 83-84; NSF fac fel, 60. *Mem:* Am Phys Soc. *Res:* Dynamic polarization of nuclei; electron and nuclear magnetic resonance; thermoluminescence; biophysical research in renal flow. *Mailing Add:* 229 Ward Pkwy Apt 801A Kansas City MO 64112

HILL, JOHN LEDYARD, b El Paso, Tex, Aug 11, 19; m 44; c 3. WOOD SCIENCE & TECHNOLOGY. *Educ:* Colo State Univ, BS, 42; Yale Univ, MS, 47, DFor, 54. *Prof Exp:* Instr wood technol, Mich State Univ, 50-55; prof, Auburn Univ, 56-58; wood technologist, Nat Lumber Mfrs Asn, Washington, DC, 58-64; assoc prof forest resources, 64-70, PROF WOOD SCI & TECHNOL, UNIV NH, 70- *Honors & Awards:* Wood Award, Forest Prod Res Soc, 49. *Mem:* Soc Wood Sci & Technol; Forest Prod Res Soc; Am Soc Testing & Mat; Sigma Xi. *Res:* Relation of tree growth to wood quality; drying stress analysis and automatic dry kiln control. *Mailing Add:* Pettee Hall Univ NH Durham NH 03824

HILL, JOHN MAYES, JR, b San Antonio, Tex, Mar 4, 38; c 1. BIOCHEMISTRY. *Educ:* Rice Univ, BA, 61, PhD(biol), 65. *Prof Exp:* USPHS res fel, Univ Calif, Los Angeles, 65-66; asst prof chem & genetics, Wash State Univ, 66-70, asst prof genetics, 70-71; asst prof, 71-74, ASSOC PROF FOOD SCI & NUTRIT, BRIGHAM YOUNG UNIV, 71- *Mem:* AAAS; Genetics Soc Am. *Res:* Genetic control and metabolic regulation of cells, especially regulatory enzymes in the metabolic pathways which supply energy to the cell. *Mailing Add:* Dept Food Sci SFLC 2218C Brigham Young Univ Provo UT 84602

HILL, JOHN ROGER, b Reinersville, Ohio, Mar 5, 12; m 37; c 3. COLON & RECTAL SURGERY. *Educ:* Ohio State Univ, AB, 33, MD, 36; Univ Minn, MS, 41. *Prof Exp:* Assoc prof, 65-70, PROF PROCTOL, MAYO MED SCH, UNIV MINN, 71-, MEM STAFF SECT PROCTOL, MAYO CLIN, 46-, SR CONSULT SECT PROCTOL, 67- *Concurrent Pos:* Ed-in-chief, Dis Colon & Rectum, 67-87. *Mem:* Fel Am Proctol Soc; fel Am Col Surgeons; Am Gastroenterol Asn. *Res:* Colloid adenocarcinoma; polypoid lesions of the terminal portion of the colon; annular rectal structure resulting from internal fistula in the ano; abscesses, fistulas and pain in the anorectal region; anorectal rings in infancy; multiple polyposis; infections and sinuses other than fistulas in the perianal region; ulceration of the rectum and colon; physiology of the large intestine. *Mailing Add:* 1010 SW Tenth St Rochester MN 55902

HILL, JOHN WILLIAM, b Decherd, Tenn, Aug 20, 33; m 56; c 2. ORGANIC CHEMISTRY. *Educ:* Mid Tenn State Col, BS, 57; Univ Ark, PhD(org chem), 60. *Prof Exp:* Asst prof chem, Northeast La State Col, 60-63; from asst prof to assoc prof, 63-69, chmn dept, 70-72 & 74-84, PROF CHEM, UNIV WIS-RIVER FALLS, 69- *Concurrent Pos:* Vis prof, Univ Ariz, 71 &

Cornell Univ, 81, Murray State Univ, 90. *Mem:* Am Chem Soc; AAAS; Soc Chem Indust; Nat Sci Teachers Asn; fel Am Inst Chemists. *Res:* Organic sytheses using phase transfer catalysts; author: Introductory Chemistry books. *Mailing Add:* Dept Chem Univ Wis-River Falls River Falls WI 54022

HILL, JOSEPH MACGLASHAN, b Buffalo, NY, Mar 26, 05; m 33; c 5. PATHOLOGY. *Educ:* Univ Buffalo, BS & Md, 28; Am Bd Path, dipl, 37. *Hon Degrees:* Dr, Univ Guadalajara, 44; DSc, Baylor Univ, 45. *Prof Exp:* Assoc pathologist, Buffalo City Hosp, 31-32; asst prof path, Col Med, Univ Okla, 32-33; from asst prof to assoc prof path, Baylor Col Dent, 34-43, prof, 45-69, dean, Grad Res Inst, 48-68, dir labs, Baylor Univ Hosp, 34-59, chief hemat sect, 54-71; exec dir, 69-76, pres, 76-77, DIR RES & CHMN, WADLEY INSTS MOLECULAR MED, 77- *Concurrent Pos:* Instr, Sch Med, Univ Buffalo, 29-32, asst, 31-32; dir, Buchanan Blood, Plasma & Serum Ctr, Dallas, 43-44; prof clin path, Univ Tex Southwestern Med Sch Dallas, 43-51, clin prof path, 57-; hon prof, Univ Guadalajara, 44-; consult, Brooks Gen Hosp, San Antonio, 47-; dir, Wadley Insts Molecular Med, 51-69; res consult, Baylor Univ Med Ctr, 59-; exec comt bd trustees, Wadley Insts Molecular Med, 77- *Honors & Awards:* Marchman Award, AMA, 47; Res Medal, Southern Med Asn, 57. *Mem:* AAAS; fel Am Col Physicians; fel Col Am Path; fel Int Soc Hemat (pres, 48-49); Am Soc Clin Oncol. *Res:* Immuno-hematology; desiccation of plasma from frozen state; blood coagulation and leukemia. *Mailing Add:* 4339 Shady Hill Dr Dallas TX 75229

HILL, JOSEPH PAUL, b Colver, Pa, Feb 5, 47; m 69; c 1. GENETICS, EPIDEMIOLOGY. *Educ:* Pa State Univ, BS, 69, MA, 75, PhD(plant path), 79. *Prof Exp:* Res aide, Dept Plant Path, Pa State Univ, 73-78; asst prof, 78-84, ASSOC PROF PLANT PATH & WEED SCI, COLO STATE UNIV, 84- *Mem:* Am Soc Agron; Am Phytopath Soc. *Res:* Diseases of cereals; genetics of host parasite interactions; fungal genetics; epidemiology; population genetics. *Mailing Add:* Dept Plant Path & Weed Sci Colo State Univ Ft Collins CO 80523

HILL, KENNETH LEE, b Moundsville, WVa, Dec 19, 31; m 51; c 3. PLANT PHYSIOLOGY, ENTOMOLOGY. *Educ:* Marshall Univ, BS & MS, 55; NC State Univ, PhD(field crops), 60. *Prof Exp:* Proj leader herbicide invests, FMC Corp, 60-67; prin biologist, Air Prod & Chem Inc, 67-68, mgr biol res, 68-70; mgr pesticide develop, 70-71, DIR PLANT INDUST RES & DEVELOP, AM CYANAMID CO, 71- *Mem:* Weed Sci Soc Am; Sigma Xi; Entom Soc Am. *Res:* Research and development of agricultural chemicals. *Mailing Add:* 138 Sandy Knoll Dr Doylestown PA 18901

HILL, KENNETH RICHARD, b Hillsboro, Ill, May 12, 30; m 55; c 2. PESTICIDE RESIDUE ANALYSIS. *Educ:* Purdue Univ, BS, 52, MS, 58, PhD(org chem), 61. *Prof Exp:* Chemist, Pitman-Moore Co Div, Dow Chem Co, 54-56; staff scientist org electrochem, P R Mallory & Co, Inc, 61-67; supvry chemist, Agr Res Serv, USDA, 67-73, res leader, Anal Chem Lab, Agr Environ Qual Inst, 73-89; SR SCI FEL, EPL/BIO-ANALYTIC SERV, INC, 89- *Concurrent Pos:* Mem, Terminal Pesticide Residues Comn, Int Union Pure & Appl Chem, 68-77, secy, 68-73, chmn, 73-75; mem, Food & Agr Orgn-WHO Joint Meeting Pesticide Residue, 70-73 & 78-83, vchmn, 71-73, chmn, 72 & 82; co-organizer & chmn workshop, Fourth Int Cong Pesticide Chem, Int Union Pure & Appl Chem, 78 & chmn, Comt Instrumental Methods & Data Handling, Asn Off Anal Chemists, 83-87; vchmn, Pilot Secretariat No 17 Int Orgn Legal Metrol, 88-89; invited lectr, residue anal & int regulations, Training Course Control Environ Contaminants in Food, Food & Agr Orgn, UN Environ Prog, Mysore, India, 78, 79, Int Course Prin & Methods Modern Toxicol, Belgirate, Italy, 79 & World Bank, Regional Training Prog on Qual Control of Pesticide Formulations, Gurgaon, India, 88; consult, AID, Islamabad, Pakistan, 80. *Honors & Awards:* Bronze Medallion, Organizing Comt, Third Int Cong Pesticide Chem, Helsinki, Finland, 74, Engraved Silver Teaspoon, Sci Prog Comt, Fourth Int Cong Pesticide Chem, Zurich, 78. *Mem:* NY Acad Sci; AAAS; Am Chem Soc; Asn Off Anal Chem; Sigma Xi; fel Am Inst Chemists. *Res:* Develop methods of analysis for pesticide chemicals and their residues; new instrumentation for residue analysis. *Mailing Add:* 4166 Christmas Tree Rd Decatur IL 62521

HILL, KENNETH WAYNE, b Winston-Salem, NC, Apr 23, 45; m 66; c 2. ATOMIC SPECTROSCOPY, ATOMIC PHYSICS. *Educ:* Drexel Inst Technol, BS, 68; Univ NC, MS, 69, PhD(physics), 74. *Prof Exp:* Instr physics, Surry Community Col, 69-71; Nat Res Coun resident res assoc atomic collision & x-ray spectros, Naval Res Lab, 74-76; physicist, Fusion Energy Div, Oak Ridge Nat Lab, 76-78; PHYSICIST, PRINCETON PLASMA PHYSICS LAB, 78- *Mem:* Am Phys Soc. *Res:* Study of ion-atom collision physics and atomic structure by x-ray and vacuum-ultraviolet spectroscopy of ion beam-target interactions and of plasmas; physics studies and design of x-ray diagnostics for Tokamak Fusion Test Reactor; plasma diagnostics. *Mailing Add:* 2651 Princeton Pike Lawrenceville NJ 08648

HILL, L(OUIS) LEIGHTON, b Baton Rouge, La, Dec 19, 28; m 58; c 3. PEDIATRICS, NEPHROLOGY. *Educ:* La State Univ, MD, 52. *Prof Exp:* Intern med, Brooke Army Hosp, San Antonio, 52-53; resident pediat, Affil Hosps, 55-57, from instr to assoc prof, 59-71, PROF PEDIAT, BAYLOR COL MED, 71-; HEAD RENAL & METAB SERV & CLINS, TEX CHILDREN'S HOSP, 60- *Concurrent Pos:* Res fel metab & renalogy, Baylor Col Med, 57-58; res fel metab & renalogy, Case Western Reserve Univ, 58-59; pres, Southern Soc Pediat Res, 67. *Mem:* Soc Pediat Res; Am Soc Nephrology; Am Diabetes Asn; Am Fedn Clin Res; Am Pediat Soc. Res; Am Acad Pediat; hon mem Pediat Asn Mex. *Res:* Water and electrolyte metabolism; renal diseases. *Mailing Add:* Dept Pediat Baylor Col Med Houston TX 77030

HILL, LEMMUEL LEROY, b Ithaca, NY, Mar 17, 33; m 60; c 4. NUCLEAR PHYSICS. *Educ:* Rensselaer Polytech Inst, BS, 59; Cath Univ Am, PhD(physics), 67. *Prof Exp:* Physicist, US Naval Ord Lab, Naval Surface Warfare Ctr, 59-62, res physicist, 62-64, supvry res physicist, 64-68, supvry physicist, 68-74, sci adv to Comdr, US Naval Surface Fleet, Atlantic, 74-75, head radiation div, 75-76, head res dept, 76-78, head weapon systs dept, 78-80, tech dir, Off Naval Technol, 80-84, dir, 84-89, DIR, SCI & TECHNOL DIV, INST DEFENSE ANALYSIS, NAVAL SURFACE WARFARE CTR, 89- *Mem:* Am Phys Soc. *Res:* Plasma, reentry, theoretical nuclear physics and nuclear weapon effects; electromagnetics; electrooptics; research administration. *Mailing Add:* 12307 Keel Turn Bowie MD 20715

HILL, LOREN GILBERT, b Tulia, Tex, Mar 23, 40; m 61; c 2. ICHTHYOLOGY, ECOLOGY. *Educ:* WTex State Univ, BS, 61; Univ Ark, MS, 63; Univ Louisville, PhD(ichthyol), 66. *Prof Exp:* Instr zool, Catherine Spalding Col, 65-66; from asst prof to assoc prof zool, 66-75, asst dir, Biol Sta, 67-69, dir, Biol Sta, 69-74, PROF ZOOL & DIR BIOL SURV, UNIV OKLA, 75-, CHMN, DEPT ZOOL, 80- *Concurrent Pos:* Pres, Southwestern Asn Naturalists, 77-79. *Mem:* Am Fisheries Soc; Am Soc Ichthyologists & Herpetologists. *Res:* Organismic responses of fishes to water quality parameters. *Mailing Add:* Dept Zool Univ Okla 730 Van Vleet Oval Norman OK 73019

HILL, LOREN WALLACE, b New Leipzig, NDak, May 1, 39; m 62; c 2. POLYMER CHEMISTRY. *Educ:* NDak State Univ, BS, 61; Pa State Univ, PhD(phys chem), 65. *Prof Exp:* From asst prof to prof phys chem, NDak State Univ, 65-71, prof polymers & coatings, 71-80; sci fel, Monsanto Polymer Prod Co, 80-86, SR FEL, MONSANTO CAEM CO, 86- *Concurrent Pos:* Chmn, Gordon Res Conf on Coatings & Films, 85. *Honors & Awards:* Mattilleo Lectr, Fedn Soc Coatings Technol, 91. *Mem:* Am Chem Soc; Sigma Xi; Fedn Soc Coatings Technol. *Res:* Polymer chemistry; weathering of industrial coatings; water soluble coatings; adhesion; kinetics; catalysis; surface chemistry. *Mailing Add:* Monsanto Chem Co 730 Worcester St Springfield MA 01151

HILL, LOUIS A, JR, b Okemah, Okla, May 18, 27; m 50; c 3. CIVIL & STRUCTURAL ENGINEERING. *Educ:* Okla State Univ, BA, 49, BSCE, 54, MSCE, 55; Case Inst Technol, PhD(civil struct eng), 65. *Prof Exp:* Engr struct & hwy, Lee Hendrix, Consult, 55-57; struct eng, Hudgins, Thompson, Ball & Assoc, 57-58; dean eng, Univ Akron, Ohio, 81, assoc vpres, res & grad studies, 88; from asst prof to assoc prof struct eng, 58-70, chmn, Dept Civil Eng, 74-81, PROF STRUCT ENG, ARIZ STATE UNIV, 70-; CONSULT, 88- *Honors & Awards:* Western Elec Fund Award for Excellence in Instr Eng Studies. *Mem:* Nat Soc Prof Engrs; Am Soc Civil Engrs; Am Soc Eng Educ; Sigma Xi. *Res:* Application of electronic computers to structural engineering problems, especially automated optimum cost design; integrated approach to study of structural engineering. *Mailing Add:* 3208 N 81st Pl Scottsdale AZ 85251

HILL, LYNN MICHAEL, b Washington, DC, Aug 12, 41; m 69; c 2. PLANT SCIENCES. *Educ:* Ala Col, BS, 63; Tenn Technol Univ, MS, 65; Univ NH, PhD(plant sci), 72. *Prof Exp:* Instr biol, Univ Tenn, Martin, 66-69; from asst prof to assoc prof, 72-78, PROF BIOL, 78-, CHMN, BIOL DEPT, BRIDGEWATER COL, 81- *Mem:* AAAS; Bot Soc Am; Int Asn Plant Taxon & Nomenclature. *Res:* Chromosome numbers of angiosperms. *Mailing Add:* Dept Biol Bridgewater Col Bridgewater VA 22812

HILL, MARION ELZIE, b Pawnee, Okla, Jan 18, 20; m 45; c 3. ORGANIC CHEMISTRY. *Educ:* Univ Ore, BA, 48, MA, 50. *Prof Exp:* Res assoc phys chem, Nat Bur Standards, 49-51; sr res assoc org chem, Naval Ord Lab, 51-60; sr res chemist, 60-61, mgr synthesis res, 62-65, chmn synthesis res dept, 65-67, dir org & polymer chem div, 67-68, dir chem lab, 68-84, SR SCI ADV, STANFORD RES INST, 84- *Mem:* Am Chem Soc; AAAS; Sigma Xi. *Res:* Organic reaction mechanisms; synthesis; catalysis of alcohol reactions; reactions in sulfuric acid; explosives; propellants; fluorine chemistry. *Mailing Add:* 4270 Pomona Ave Palo Alto CA 94306-4337

HILL, MARQUITA K, b Hickory, NC, Dec 18, 38; m 76; c 1. RESEARCH ADMINISTRATION, WOOD & PULPING CHEMISTRY. *Educ:* Mich State Univ, BA, 60; Univ Calif, Davis, PhD(biochem), 66. *Prof Exp:* Assoc fel, Univ Calif, Los Angeles, 66-68; from asst prof to assoc prof biochem, Va Polytech Inst, 68-77; instr, 78-79, prog develop specialist, 79-81, ASSOC RES PROF, DEPT CHEM ENG, UNIV MAINE, 81- *Concurrent Pos:* Prin investr, Dept Energy contract, Univ Maine, 84-87. *Mem:* Fedn Am Soc Exp Biol; Am Chem Soc; Tech Asn Pulp & Paper Indust. *Res:* Ultrafiltration of industrial streams, especially of kraft black liquor; wood and pulping chemistry; utilization of biomass. *Mailing Add:* Dept Chem Eng 207 Jenness Hall Univ Maine Orono ME 04469

HILL, MARTHA ADELE, b London, Ont, Apr 13, 22; nat Can. ANALYTICAL CHEMISTRY, BIOCHEMICAL ENGINEERING. *Educ:* Wheaton Col, BS, 43; Rutgers Univ, MSc, 45; Univ Toronto, PhD(chem), 49. *Prof Exp:* Asst chem, Wheaton Col, 42-43 & NJ Col Women, Rutgers Univ, 43-45; demonstr, Univ Toronto, 45-48; res chemist, Eastman Kodak Co, 49-57, res assoc, 57-64; res assoc, Univ Western Ont, 73-80; consult, Univ Tex, El Paso, 81, Univ Western Ont, 82; RETIRED. *Concurrent Pos:* Exec asst, VI Int Fermentation Symp, 78-80. *Mem:* Am Chem Soc. *Res:* Study of biodegradation of sulfite mill waste for pollution control of pulp and paper industry; chemical literature searches; photographic chemistry. *Mailing Add:* 103 Wychwood Court London ON N6G 1S6 Can

HILL, MARVIN FRANCIS, b Missoula, Mont, Feb 4, 25; m 55; c 4. ANATOMY, PHYSIOLOGY. *Educ:* Ore State Col, BS, 49, MS, 51, PhD, 55. *Prof Exp:* Lectr anat, Ore State Col, 54-55; lectr, Univ Heidelberg, 55; instr, Med Sch, Tulane Univ, 56-58; from asst prof to assoc prof, Sch Med, 58-70, PROF ANAT, SCH DENT, CREIGHTON UNIV, 70- *Mem:* AAAS; Am Asn Anatomists; Soc Exp Biol & Med; Soc Cryobiol. *Res:* Gross neuro-anatomy; developmental histochemistry; regeneration in mammals; histochemical calcium stains. *Mailing Add:* Dept Oral Biol Creighton Univ 2500 Cal St Omaha NE 68178

HILL, MARY RAE, b Great Falls, Mont, Sept 2, 23. GEOMORPHOLOGY. *Educ:* Univ Colo, BA, 44; San Francisco State Univ, MA, 70. *Prof Exp:* From geol aide to geologist & geol data officer, Calif Div Mines & Geol, 49-75; info officer, US Geol Surv, Menlo Park, 75-79; RETIRED. *Concurrent Pos:* Ed, Calif Geol, 53-74; adj prof, San Francisco State Univ, 79- *Mem:* Asn Earth Sci Ed (vpres, 73, pres, 74). *Res:* Environmental geology; shore processes; cinematography as an earth science tool; science for the layman. *Mailing Add:* Rte 7 Box 124-MU Santa Fe NM 87505

HILL, MASON LOWELL, b Pomona, Calif, Jan 17, 04; m 42; c 5. GEOLOGY. *Educ:* Pomona Col, AB, 26; Claremont Col, AM, 29; Univ Wis, PhD, 32. *Hon Degrees:* ScD, Pomona Col, 71. *Prof Exp:* Geologist, Shell Oil Co, 27-28, 29-30; instr geol, Coalinga Jr Col, Calif, 33-35; geologist, Shell Oil Co, 35-37 & Richfield Oil Co, 37-65, explor mgr, Int Div, Atlantic Richfield Co, 65-70; res assoc, Univ Calif, Los Angeles, 70-73; instr geol, Univ Calif, Irvine, 73-; RETIRED. *Concurrent Pos:* Consult geologist, 70-79. *Mem:* Fel Geol Soc Am; Am Asn Petrol Geologists (pres, 62-63); Asn Eng Geologists. *Res:* Structural geology and petroleum; seismic hazards, fault tectonics and faulting in California. *Mailing Add:* 14067 E Summit Dr Whittier CA 90602

HILL, MAX W, b Payson, Utah, May 20, 30; m 54; c 7. NUCLEAR PHYSICS. *Educ:* Brigham Young Univ, BA, 54; Univ Calif, Berkeley, PhD(nuclear chem), 59. *Prof Exp:* From asst prof to assoc prof, 58-68, PROF PHYSICS, BRIGHAM YOUNG UNIV, 68-, ASST DEPT CHMN, 79- *Concurrent Pos:* Res asst chem, Univ Calif, Berkeley, 55-58; res scientist, Gen Elec Co & Pac Northwest Labs, Battelle Mem Inst, 64-65. *Res:* Nuclear spectroscopy; trace element analysis of environmental samples. *Mailing Add:* Dept Physics Brigham Young Univ Provo UT 84602

HILL, MERTON EARLE, III, b Long Beach, Calif, Oct 17, 47; m 72. EARTH SCIENCES, PALEONTOLOGY. *Educ:* Univ Redlands, BS, 69; Univ Calif, Los Angeles, PhD(geol), 75. *Prof Exp:* asst prof geol, Calif State Univ, 76-79; SR RES MICROPALEONTOLOGIST, UNION SCI & TECHNOL DIV, UNION OIL CO OF CALIF, 79- *Concurrent Pos:* Lectr geol, Santa Monica Col, 75-76; lectr paleont, Univ Southern Calif, 76-; explor res geologist, 79- *Mem:* Int Nannofossil Asn. *Res:* Calcareous nannofossil (coccolith) biostratigraphy of the North Sea, North Africa, Ecuador, Colombia, Western Interior of the United States, and Gulf of Mexico. *Mailing Add:* 842 Manzita Dr Laguna Beach CA 92651-1960

HILL, ORVILLE FARROW, b Decatur, Ill, Jan 6, 19; m 44; c 3. NUCLEAR SPENT FUEL REPROCESSING. *Educ:* Millikin Univ, BS, 40; Univ Ill, MS, 41, PhD(inorg chem), 48. *Hon Degrees:* DSc, Millikin Univ, 63. *Prof Exp:* Res chemist, Manhattan Dist, 42-46; res chemist, Hanford Atomic Prod Opers, Gen Elec Co, Richland, Wash, 48-49, mgr chem res, 50-53, mgr plant processes, 53-56, mgr chem develop, 56-64, prin chem engr, 65-66; staff scientist, Pac Northwest Labs, Battelle Mem Inst, 64-65; prin chem engr, Isochem, Inc, 66-67 & Atlantic Richfield Hanford Opers, Atlantic Richfield Co, Wa, 67-77; staff engr, Pac Northwest Lab, Battle Mem Inst, 77-84; CONSULT NUCLEAR FUEL CYCLE, 84- *Concurrent Pos:* Engr-on-loan, Atomic Energy Comn Combined Oper Planning, Tenn, 67-69, Div Regulation, 72-74; vis prof environ chem, Prairie View A&M, Tex, 69-70. *Honors & Awards:* Richland Sect Award, Am Chem Soc, 82. *Mem:* Am Chem Soc; Am Nuclear Soc; fel Am Inst Chem Engrs; fel AAAS; Sigma Xi. *Res:* Chemistry of the nuclear fuel cycle; science and engineering of spent fuel reprocessing and nuclear waste management; environmental surveys and impact statements. *Mailing Add:* 1510 SE 127th Ave Vancouver WA 98684-6473

HILL, PATRICK ARTHUR, b Calcutta, India, Jan 8, 22. ECONOMIC GEOLOGY, ENVIRONMENTAL GEOLOGY. *Educ:* Univ London, BSc, 48; Columbia Univ, PhD(geol), 57. *Prof Exp:* Instr geol, Brooklyn Col, 50-51; instr phys & hist geol, Wellesley Col, 51-52; lectr, Carlton Univ, 53-54, Geol Dept, 53-59, from asst prof to prof, 54-87, ADJ PROF GEOL, CARLTON UNIV, 87- *Concurrent Pos:* Consult to govt & indust agencies, 43-; vis res prof, Univ Tasmania, 60-61; visitor, Univ Reading, 67 & Bristol Univ, 71; res assoc, Univ Ireland, Cork, 86- *Mem:* Fel Royal Geog Soc; Quaternary Res Asn; Geol Soc London. *Res:* Archaeological and applied geology; coal geology; medical geology. *Mailing Add:* Dept Geol Carleton Univ Ottawa ON K1S 5B6 Can

HILL, PAUL DANIEL, b Dadeville, Ala, May 8, 33; m 58; c 2. MATHEMATICS. *Educ:* Auburn Univ, BS, 56, MS, 58, PhD(math), 60. *Prof Exp:* NSF fel math, Inst Advan Study, 60-61; asst prof, Auburn Univ, 61-63; assoc prof, Emory Univ, 63-65; prof, Univ Houston, 65-68; prof math, Fla State Univ, 68-75; PROF MATH, AUBURN UNIV, 75- *Mem:* Am Math Soc; Math Asn Am. *Res:* Semigroups and groups. *Mailing Add:* Dept Math Auburn Univ Auburn AL 36849

HILL, PERCY HOLMES, b Norfolk, Va, Feb 19, 23; m 46; c 2. ERGONOMICS, CONSUMER PRODUCT RESEARCH & DEVELOPMENT. *Educ:* Rensselaer Polytech Inst, BME, 44; Harvard Univ, SM, 51. *Prof Exp:* Instr eng & math, Va Polytech Inst, 46-48; prof, 48-83, EMER PROF ENG DESIGN, TUFTS UNIV, 83-; PRES, APPL ERGONOMICS CO, 76- *Concurrent Pos:* Dir & prin investr, Biodent Eng Res, Tufts Univ, 64-70, chmn comt decision & policy making, 77-81; pres, Stratford Labs, Inc, 82-; mem bd dirs, Am Soc Eng Educ, 85-87; vpres res & develop, Powell Technol Prod, 89- *Honors & Awards:* Oppenheimer Award Design Graphics, Am Soc Eng Educ, 68, Distinguished Serv Award, 72; Fred Merrifield Award, Am Soc Eng Educ, 82. *Mem:* Am Soc Eng Educ; Human Factors Soc. *Res:* Author of 9 textbooks in area of descriptive geometry, engineering graphics, kinematics and creative problem solving; holder of 6 patents in the area of dental health. *Mailing Add:* East Shore Dr Silver Lake NH 03875

HILL, PHILIP G(RAHAM), b Vancouver, BC, July 18, 32; m 59. MECHANICAL ENGINEERING. *Educ:* Queen's Univ, Ont, BSc, 53; Univ Birmingham, MSc, 55; Mass Inst Technol, ScD, 58. *Prof Exp:* From asst prof to prof mech eng, Queen's Univ, 68-73; Commonwealth vis prof, Univ Sheffield, Cambridge, 73-74; prof, Queen's Univ, 74-75; head dept, 78-83, PROF MECH ENG, UNIV BC, 76- *Concurrent Pos:* Guggenheim fel, 62-63; consult, various indust co. *Res:* Fluid mechanics; propulsion; power. *Mailing Add:* Dept Mech Eng Univ BC 2075 Wesbrook Pl Vancouver BC V6T 1W5 Can

HILL, RAY ALLEN, b Houston, Tex, Sept 16, 42. CELL BIOLOGY, ANATOMY. *Educ:* Howard Univ, BS, 64, MS, 65; Univ Calif, PhD(bot), 77. *Prof Exp:* Instr bot, Southern Univ, 65-66 & Howard Univ, 66-73; asst prof, Fisk Univ, 77-80; assoc prof bot, Col Baltimore, 80-82; assoc prof biol, Morgan State Univ, 82-85; SCI TEACHER, LOWELL COL PREP SCH, SAN FRANCISCO, 86- *Concurrent Pos:* NSF fac fel, Fac Sci Develop Prog, NSF, 74-75; vis scientist, Environ Protection Agency, 78, Nat Aeronautics & Space Admin, 79 & Lawrence Berkeley Lab, 80; panelist, Comprehensive Assistance to Undergrad Sci, NSF, 79 & Instrnl Sci Equip Prog, 81; vis scholar, Stanford Univ & Asn Marine Eng Sch Res Facil, NASA, 83-84, vis res assoc prof, Univ Calif, San Francisco, 85; fel, NSF, jr fac fel Howard Univ & diss fel Ford Found; vis prof, bot & plant path, Purdue Univ. *Mem:* Inst Med Nat Acad Sci; Bot Soc Am; NY Acad Sci; AAAS. *Res:* Ultrastructure and histochemistry of ovules in cotton; developmental approach through maturation and senescence. *Mailing Add:* 8751 Fehler Lane Cotati CA 94931

HILL, REBA MICHELS, b Houston, Tex, Oct 8, 30; m 58; c 3. PEDIATRICS, NEONATOLOGY. *Educ:* Baylor Univ, BS, 53, MD, 55. *Prof Exp:* Fel pediat, Case Western Reserve, 58-59; intern & res, Baylor Affil Prog, 55-58, fel, Baylor Col Med, 59-60, from instr to assoc prof, 60-79, PROF PEDIAT, BAYLOR COL MED, 79-, PROF OBSTET, 85- *Concurrent Pos:* Dir, neurol clin, Jefferson Davis Hosp, 60-62; chief, newborn res, St Lukes Episcopal Hosp, 62-; consult, NIH Automation Med Lab, 72-75; mem, Drug Comt, Am Acad Pediat, 77-83, Ad Hoc Comt Adolescence & Drug Abuse, 84; vis prof, Michael Reese Hosp & Chicago Perinatology Soc, 74, Ariz Health Sci Ctr, 79, Univ Southern Calif Med Sch, 80, Med Sch Costa Rica, 83. *Mem:* Am Soc Pharmacol & Exp Therapeut; Soc Pediat Res; Am Pediat Soc; Am Acad Pediat. *Res:* Effect of maternal drugs on the fetus and newborn; transfer of drugs into breast milk; effect of intrauterine malnutrition on the infant. *Mailing Add:* Dept Pediat Baylor Col Med 6720 Bertner Dr Houston TX 77030

HILL, RICHARD A, b Springfield, Ohio, Aug 14, 32; m 56; c 3. CHEMICAL ENGINEERING. *Educ:* Univ Fla, BSPh, 56, MS, 58, BChE & PhD(pharm), 60. *Prof Exp:* Sr res pharmacist, Squibb Inst Med Res, 60-64, res supvr, 64-67, mgr res eng, 67-69, dir res eng, 69-71, dir eng & tech servs, 71-75, dir tech admin, 75-76; dir develop, Norwich Eaton Pharmaceut, 76-78, vpres develop, 78-80, vpres res & develop, 80-84, vpres prod develop-int, 84-90; RETIRED. *Concurrent Pos:* Fel, Am Found Pharmaceut Educ & NSF. *Mem:* Am Pharmaceut Asn; Sigma Xi; Am Soc Hosp Pharmacists. *Res:* Directing research and development; new product development, licensing and acquisition, computer systems, statistical services, corporate management committee, international liaison. *Mailing Add:* 1610 NW 19th Circle Gainesville FL 32605

HILL, RICHARD C(ONRAD), b Schenectady, NY, Oct 15, 18; m 44; c 5. MECHANICAL ENGINEERING. *Educ:* Syracuse Univ, BS, 41. *Prof Exp:* Engr gas turbine, Gen Elec Co, 41-46; actg dean col technol, 67-69, PROF MECH ENG, UNIV MAINE, ORONO, 46-, DIR DEPT INDUST COOP, 69- *Mem:* Am Soc Eng Educ; Am Soc Mech Engrs. *Res:* Thermodynamics; steam and gas turbines; compressible fluid flow. *Mailing Add:* 501 College Rd Orono ME 04469

HILL, RICHARD KEITH, b Erie, Pa, June 1, 28; m 54; c 4. ORGANIC CHEMISTRY. *Educ:* Pa State Univ, BS, 49; Harvard Univ, MA, 50, PhD(chem), 54. *Prof Exp:* From instr to assoc prof chem, Princeton Univ, 53-68; actg head dept, 69-71, 77-78, PROF CHEM, UNIV GA, 68- *Concurrent Pos:* Consult, Schering Corp, 56-57; Sloan Found fel, 61-65; res assoc, Textile Res Inst, 64-68; NSF sr fel, 65-66; mem, Med Chem Study Sect, NIH, 68-72; mem chem panel, Sr Fulbright-Hays Prog, 70-73; consult, Burroughs Wellcome Co, 74-77; NATO sr fel sci, NSF, 76; mem adv bd, Petrol Res Fund, 80-85. *Mem:* Fel AAAS; Am Chem Soc; Royal Soc Chem; Am Soc Pharmacog. *Res:* Organic natural products; stereochemistry; molecular rearrangements; synthesis. *Mailing Add:* Dept Chem Univ Ga Athens GA 30602

HILL, RICHARD M, b San Francisco, Calif, Dec 16, 34; m 57; c 1. PHYSIOLOGICAL OPTICS. *Educ:* Univ Calif, Berkeley, BS, 57, MOpt, 58, PhD(physiol optics), 61. *Prof Exp:* Clin instr optom, Univ Calif, Berkeley, 58-61, asst prof physiol optics, 61-64; assoc prof, 64-68, PROF PHYSIOL OPTICS & BIOPHYS, COL OPTOM, OHIO STATE UNIV, 68-, ASSOC DEAN, COL OPTOM, 78- *Mem:* Fel AAAS; Am Physiol Soc; Sigma Xi. *Res:* Neurophysiology of visual system; single cell responses of visual pathways and gas exchange properties of cornea. *Mailing Add:* 1601 Lafayette Dr Upper Arlington OH 43220

HILL, RICHARD NORMAN, b Stockton, Calif, May 30, 37; m 61; c 2. ENVIRONMENTAL HEALTH, GENETICS. *Educ:* Johns Hopkins Univ, BA, 59; Univ Minn, MD, 67, PhD(genetics), 68. *Prof Exp:* Asst prof pharmacol & genetics, Hershey Med Ctr, Pa State Univ, 69-76; mem staff, Off Asst Admin Pesticides & Toxic Substances, 76-80, SR SCI ADV, PESTICIDES & TOXIC SUBSTANCES, US ENVIRON PROTECTION AGENCY, WASH, DC, 80- *Concurrent Pos:* NIH fel biol, Univ Colo, 67-69. *Res:* Behavior of cells at the molecular level using tissue culture; biochemical toxicology; cytogenetics. *Mailing Add:* US Environ Protection Agency TS-788 Washington DC 20460

HILL, RICHARD PETER, b Belfast, Northern Ireland, Dec 27, 42; m 66; c 3. RADIATION BIOLOGY, TUMOR BIOLOGY. *Educ:* Oxford Univ, Eng, BA, 64; London Univ, PhD(radiation biol), 67. *Prof Exp:* Staff biophysics dept, Inst Cancer Res, London, 71-73; MEM STAFF, RES DIV, ONT CANCER INST, TORONTO, 73-; PROF, DEPT MED BIOPHYS, UNIV TORONTO, 79- *Concurrent Pos:* Fel radiol res, James Picker Found, 67-71. *Mem:* Brit Inst Radiol; Radiation Res Soc; Am Asn Cancer Res; Metastasis Res Soc. *Res:* Radiation biology and tumor biology in their application to the treatment of cancer; mechanisms of metastasis and tumor progression. *Mailing Add:* Res Div Ont Cancer Inst Toronto ON M4X 1K9 Can

HILL, RICHARD RAY, JR, b Louisburg, NC, Oct 18, 36; m 60; c 2. PLANT BREEDING, GENETICS. *Educ:* NC State Univ, BS, 59, MS, 61; Cornell Univ, PhD(plant breeding), 64. *Prof Exp:* RES AGRONOMIST ALFALFA BREEDING, AGR RES SERV, USDA, 64- *Concurrent Pos:* Assoc ed, Crop Sci, 75-76, ed, 79-81. *Mem:* Fel Am Soc Agron; Crop Sci Soc Am; Biomet Soc; Sigma Xi. *Res:* Breeding and genetics of alfalfa. *Mailing Add:* 227 Hickory Rd State College PA 16801

HILL, RICHARD WILLIAM, b Paterson, NJ, Mar 5, 25; m 48; c 3. NUCLEAR PHYSICS, RESEARCH ADMINISTRATION. *Educ:* NY Univ, BS, 48; Univ Wis, MS, 49, PhD(physics), 53. *Prof Exp:* Res physicist, Res Labs, Westinghouse Elec Corp, 53-58; physicist, Lawrence Livermore Lab, Univ Calif, 58-88; CONSULT, IN-SITU COAL GASIFICATION, 88- *Mem:* Am Phys Soc. *Res:* In-situ coal gasification. *Mailing Add:* 1945 Heidelberg Livermore CA 94550

HILL, RICHARD WILLIAM, b Philadelphia, Pa, July 11, 42; m 68; c 2. COMPARATIVE PHYSIOLOGY, ENVIRONMENTAL BIOLOGY. *Educ:* Univ Del, BA, 64; Univ Mich, Ann Arbor, AM, 66, PhD(zool), 70. *Prof Exp:* Lectr zool, Univ Mich, Ann Arbor, 69-70; asst prof biol sci, Univ Del, 70-72; from asst prof to assoc prof, 72-82, CUR LIVING VERT, THE MUSEUM, 72-, PROF ZOOL, MICH STATE UNIV, 82- *Mem:* AAAS; Am Soc Mammal; Sigma Xi; Am Soc Zoologists. *Res:* Physiological ecology; mammalian and avian thermoregulation; ontogeny of thermoregulation; energetics. *Mailing Add:* Dept Zool & Museum Mich State Univ East Lansing MI 48824

HILL, ROBERT, b New York, NY, Apr 6, 22; m 51; c 3. PHYSIOLOGY, BIOCHEMISTRY. *Educ:* Univ Calif, AB, 49, PhD(physiol), 53. *Prof Exp:* Assoc res physiologist, Univ Calif, 53-60; pres & sci dir, Diablo Labs, Inc, 58-64; pres & sci dir, Hill Res Inst, 64-65; mgr toxicol & path, Syntex Corp, 65-69, dir dept toxicol, 69-77, asst dir, Inst Clin Med, 73-77, dir toxicol & path, 77-81, vpres & dir, Inst Toxicol Sci, 81-85; PRES, HILL RES ASSOCS, INC, 85- *Concurrent Pos:* Advan res fel, Am Heart Asn, 58-60; abstractor, Chem Abstr, 59-60; vis prof environ toxicol, Univ Calif, Santa Cruz, 77-78; expert toxicol & pharmacol, French Minister Health, 78; adj prof toxicol, San Jose State Univ, 81. *Mem:* Sigma Xi; Am Physiol Soc; Soc Toxicol; Environ Mutagenesis Soc; Europ Soc Toxicol. *Res:* Endocrines; nutrition; intermediary metabolism; toxicology. *Mailing Add:* 150 Robin Way Los Gatos CA 95032

HILL, ROBERT BENJAMIN, b New York, NY, July 23, 30. COMPARATIVE PHYSIOLOGY. *Educ:* Tufts Univ, SB, 52; Harvard Univ, PhD(biol), 57. *Prof Exp:* Instr zool, Univ Maine, 56-59; USPHS fel, Dept Zool, Glasgow Univ, 59-61; from instr to asst prof physiol, Med Sch, Dartmouth Col, 61-68; assoc prof, 68-75, PROF ZOOL, UNIV RI, 75- *Concurrent Pos:* Corp mem, Bermuda Biol Sta & Marine Biol Lab. *Mem:* Fel AAAS; Am Soc Zoologists; Soc Exp Biol; Soc Gen Physiol; Am Physiol Soc; Soc Neurosci; Biophys Soc. *Res:* Neurohumoral transmission in invertebrates; nervous control of molluscan hearts; active state ionic pumps and excitation-contraction coupling in molluscan muscle. *Mailing Add:* Dept Zool Univ RI Kingston RI 02881

HILL, ROBERT D, b Winnipeg, Man, Feb 23, 37; m 59; c 2. AGRICULTURAL BIOCHEMISTRY. *Educ:* Univ Man, BSc, 58, MSc, 63, PhD(biochem), 65. *Prof Exp:* Res asst chem, plant sci dept, Univ Man, 59-63; fel biochem, Univ Calif, Los Angeles, 65-67; from asst prof to assoc prof, 67-77, PROF PLANT SCI, UNIV MAN, 77- *Mem:* Can Soc Plant Physiologists; Can Biochem Soc; Am Soc Plant Physiol; Am Asn Cereal Chemists. *Res:* Carbohydrates in cereal grains; germination physiology. *Mailing Add:* Dept Plant Sci Univ Man Winnipeg MB R3T 2N2 Can

HILL, ROBERT DICKSON, b Melbourne, Australia, July 3, 13; m 42; c 1. PHYSICS. *Educ:* Cambridge Univ, 37-38; Univ Ill, 38-39; Univ Melbourne, DSc, 46. *Prof Exp:* Sr lectr physics, Univ Melbourne, 41-47; from asst prof to prof physics, Univ Ill, Urbana, 47-65; CONSULT PHYSICIST, 66-; RES PHYSICIST, UNIV CALIF, SANTA BARBARA, 78- *Concurrent Pos:* Fulbright sr fel, 60; Guggenheim fel, 61. *Mem:* Fel Am Phys Soc; fel Royal Meteorol Soc. *Mailing Add:* 612 Alston Rd Santa Barbara CA 93108

HILL, ROBERT F, b Montrose, Colo, June 28, 29; m 50; c 4. NUCLEAR & CHEMICAL ENGINEERING. *Educ:* Purdue Univ, BSChE, 53, MSChE, 54; Oak Ridge Sch Reactor Technol, BSChE, 55. *Prof Exp:* Nuclear engr, Naval Reactors Br, Atomic Energy Comn, 54-57; sr res engr, 57-65, supvr appl tech group, Physics Dept, 65-72, SUPVR RADIATION SAFETY RES, RES LABS, GEN MOTORS CORP, 66-, SUPVR RADIOISOTOPIC METHODS GROUP, ANALYTICAL CHEM, 72- *Concurrent Pos:* Mem fac, Univ Detroit, 58-60. *Mem:* Sigma Xi. *Res:* Chemical thermodynamics; radioisotope applications, radioisotopic x-ray fluorescence and heat transfer. *Mailing Add:* 1207-C Kirts Rd Troy MI 48084

HILL, ROBERT GEORGE, JR, b Silver Spring, Md, Jan 11, 22; m 48; c 2. HORTICULTURE. *Educ:* Univ Md, BS, 43, MS, 48, PhD(hort), 50. *Prof Exp:* From asst prof to prof, 50-60, assoc chmn dept, 70-83, EMER PROF HORT, OHIO STATE UNIV & OHIO AGR RES DEVELOP CTR, 83- *Mem:* Fel Am Soc Hort Sci; Am Soc Plant Physiol; Weed Sci Soc Am; Am Pomol Soc. *Res:* Stone and small fruit cultural practices; cultivar evaluation; chemical weed control. *Mailing Add:* 473 E Beverly Rd Wooster OH 44691

HILL, ROBERT JAMES, b Ann Arbor, Mich, Dec 1, 29; m 56; c 2. BIOCHEMISTRY. *Educ:* Stanford Univ, BS, 52; Univ Tenn, MS, 54, PhD, 56. *Prof Exp:* Res assoc, Rockefeller Inst, 57-61; asst prof, 61-65, assoc prof, 65-71, PROF BIOCHEM, CTR HEALTH SCI, UNIV TENN, MEMPHIS, 71- *Concurrent Pos:* USPHS res fel, Univ Tenn, 56-57; USPHS career develop award, 61-70. *Mem:* AAAS; Am Chem Soc; Am Soc Biol Chemists. *Res:* Isolation and determination of the structure of proteins. *Mailing Add:* 1207 Central Ave Memphis TN 38104

HILL, ROBERT JOE, b Tipton, Ind, Feb 25, 30; m 53; c 3. MATHEMATICS. *Educ:* Auburn Univ, BS, 57, MS, 58; Univ Ala, PhD, 70. *Prof Exp:* Mem staff, NASA, 60-70; ASSOC PROF MATH, PURDUE UNIV, CALUMET CAMPUS, 70- *Res:* Linear algebra. *Mailing Add:* Dept Math Sci Purdue Univ Calumet 2233 171st St Hammond IN 46323

HILL, ROBERT LEE, b Kansas City, Mo, June 8, 28; m 48; c 4. PROTEIN CHEMISTRY, ENZYME CHEMISTRY. *Educ:* Univ Kans, AB, 49, MA, 51, PhD(biochem), 54. *Prof Exp:* Asst instr biochem, Univ Kans, 49-51, instr, 51-53; instr, Col Med, Univ Utah, 56-57, asst res prof, 57-60, assoc prof, 60-61; from assoc prof to prof, 61-74, J B DUKE PROF BIOCHEM, MED CTR, DUKE UNIV, 74-, CHMN DEPT, 69- *Concurrent Pos:* Consult biochem training comt, Nat Inst Gen Med Sci, 66-69; mem biochem training comt, NIH, 66-69, chmn, 67-69, mem physiol chem study sect, 69-73, chmn, 70-73; mem, Nat Bd Med Examr, 68-, mem biochem test comt, 66-70, chmn, 68-70; mem bd, Fedn Am Socs Exp Biol, 71-74. *Mem:* Nat Acad Sci; Inst Med Nat Acad Sci; AAAS; Am Soc Biol Chemists (secy, 72-75, pres, 76-77); fel Am Acad Arts & Sci. *Res:* Structural aspects of protein chemistry; mechanism of action of enzyme. *Mailing Add:* Dept Biochem Med Ctr Duke Univ Durham NC 27710

HILL, ROBERT MATHEW, b Caledonia, Minn, Mar 21, 22; m 46; c 2. BIOCHEMISTRY. *Educ:* St Mary's Col, Minn, BS, 44; Univ Minn, MS, 54, PhD(agr biochem), 55. *Prof Exp:* Instr chem & math, St Mary's Col, Minn, 46-47; from asst to instr agr biochem, Univ Minn, 47-55; from asst biochemist to assoc biochemist, Univ Nebr, 55-60, asst prof chem, 56-60, assoc prof biochem, 60-87; RETIRED. *Mem:* Am Chem Soc. *Res:* Chemical and physical properties of proteins; analysis of biochemical materials. *Mailing Add:* 308 S 54th Lincoln NE 68510

HILL, ROBERT MATTESON, b New York, NY, Sept 20, 26; m 47; c 4. PHYSICS. *Educ:* Cornell Univ, AB, 49; Duke Univ, PhD(physics), 53. *Prof Exp:* Physicist, Sylvania Elec Prod Inc, Gen Tel & Electronics Corp, 53-63 & Lockheed Res Lab, 63-70; mem staff, 70-83, SR SCI ADV, STANFORD RES INST INT, 83- *Concurrent Pos:* Prog officer, Nat Sci Found, 80-81. *Mem:* AAAS; fel Am Phys Soc; NY Acad Sci; Sigma Xi. *Res:* Gaseous electronics; plasma physics; ferroelectricity; echos in ferri and anti-ferri magnetic resonance and in electron cyclotron resonance; Rydberg atoms, kinetics of laser media; chemical physics. *Mailing Add:* Stanford Res Inst 333 Ravenwood Ave PN071 Menlo Park CA 94025

HILL, ROBERT NYDEN, b Evanston, Ill, Mar 5, 35; m 62; c 1. THEORETICAL PHYSICS. *Educ:* Carleton Col, BA, 56; Yale Univ, MS, 57, PhD(physics), 62. *Prof Exp:* NSF fel physics, Princeton Univ, 62-63 & Yale Univ, 64; from asst prof to assoc prof, 64-73, PROF PHYSICS, UNIV DEL, 73- *Mem:* Am Phys Soc. *Res:* Mathematical physics of atoms and molecules; eigenvalue estimation; relativistic classical particle dynamics; miscellaneous theoretical physics. *Mailing Add:* 40 Lynn Dr Newark DE 19711

HILL, ROBERT WILLIAM, b Chicago, Ill, Dec 13, 27; m 51; c 4. POLYMER CHEMISTRY. *Educ:* Augustana Col, BA, 50; Univ Ill, PhD(chem), 54. *Prof Exp:* Res chemist, Spencer Chem Div, Gulf Oil Corp, 54-59, group leader, Gulf Res & Develop Co, 59-68, sect supvr, 68-75, mgr Polymer Res, Gulf Oil Chem Co, 76-83; RETIRED. *Mem:* Am Chem Soc; Sigma Xi; The Chem Soc; AAAS. *Res:* Organometallic chemistry; coordination polymerization; polyolefins; surface coatings; aminoplast and phenolic resins; polyamides; aromatic polymers; composites. *Mailing Add:* 423 Green Park Dr Houston TX 77079

HILL, ROGER W(ARREN), b Fruitdale, SDak, June 11, 19; m 45. CHEMICAL ENGINEERING. *Educ:* SDak Sch Mines & Technol, BS, 42; Va Polytech Inst, MS, 48. *Prof Exp:* Shift supvr, Hercules Powder Co, 42-43; foreman prod opers, Tenn Eastman Corp, 43-44; asst chem eng, Va Polytech Inst, 46-47; engr, Cent Res Labs, Gen Aniline & Film Corp, 47-52; res engr, Calif Res Corp, 52-59; sr chemist, Iranian Oil Ref Corp, 59-61; group supvr polymer process res, 62-69, admin asst to vpres, 69-72, prod specialist, 72-73, SR RES ENGR, CHEVRON RES CORP, 73- *Mem:* AAAS; Am Chem Soc; Am Inst Chem Engrs. *Res:* Process research and development; polyolefin polymerization and petrochemical manufacturing processes. *Mailing Add:* 1231 Grove St No 10 San Francisco CA 94117

HILL, ROLLA B, JR, b Baltimore, Md, June 11, 29; m 51; c 3. PATHOLOGY, DEVELOPMENTAL BIOLOGY. *Educ:* Univ Rochester, BA, 50, MD, 55; Am Bd Path, cert anat & clin path, 59. *Prof Exp:* Resident path, Yale Univ, 58-59; pathologist, Bridgeport Hosp, Conn, 59-61; asst prof path, Univ Colo, 61-65, assoc prof, 65-68, actg chmn dept, 67-68; prof & vchmn dept, Univ Calif, Davis, 68-69; PROF PATH & CHMN DEPT, STATE UNIV NY UPSTATE MED CTR, 69- *Concurrent Pos:* Fel path, Yale Univ, 58-59; mem coun, Am Asn Path & Bact, 71-; mem coun grad & undergrad educ, Am Soc Clin Path, 73-; chmn coun, Asn Path, 70-, pres, 74-75. *Mem:* AAAS; Am Soc Exp Path; Am Asn Path & Bact; Am Soc Cell Biol; Am Soc Clin Path. *Res:* Biochemical basis of human disease; liver physiology and pathology. *Mailing Add:* 2151 Hwy 128 Philo CA 95466

HILL, RONALD AMES, b Caro, Mich, June 19, 34; div; c 2. PHOTOMETRICS AND OPTICAL DEVELOPMENT. *Educ:* Mich State Univ, BS, 57, MS, 58, PhD(molecular struct), 63. *Prof Exp:* Staff mem plasma spectros, 63-69, supvr exp aerophys, 69-77, staff mem laser applns & spectros, 77-79, staff mem neutron devices, 79-87, SUPVR PHOTOM & OPTICAL

DEVELOP, SANDIA LABS, 87- *Mem:* Am Phys Soc; Optical Soc Am. *Res:* High resolution infrared spectroscopy and molecular structure; time resolved plasma spectroscopy; plasma diagnostics; raman spectroscopy; optical devices. *Mailing Add:* Sandia Labs Org 7556 Albuquerque NM 87115

HILL, RONALD STEWART, b Ironwood, Mich, Aug 20, 51; m 82; c 1. ENDOCRINOLOGY, CELL BIOLOGY. *Educ:* Univ Ill, Urbana, BS, 74, PhD(physiol), 82; Northern Ill Univ, DeKalb, MS, 77. *Prof Exp:* Instr gross anat, NJ Med Sch, Univ Med & Dent NJ, Newark, 80-81, res assoc, 81-82; teaching fel, 82-85, INSTR, BAYLOR COL MED, HOUSTON, TEX, 85-, ACTG DIR, RIA CORE LAB, DIABETES & ENDOCRIN RES CTR, 85- *Mem:* AAAS; Am Diabetes Asn; Am Asn Anatomists. *Res:* Mechanisms of insulin secretion. *Mailing Add:* 903 W Red Oak Blvd Red Oak TX 75154

HILL, RUSSELL JOHN, b Ilford, Eng, May 7, 34; US citizen; m 58; c 3. PHYSICAL CHEMISTRY, HIGH TEMPERATURE CHEMISTRY. *Educ:* Univ Wales, BSc, 56; Univ London, PhD(phys chem) & DIC, 59. *Prof Exp:* Sci officer, UK Atomic Energy Res Estab, 59-62; adv develop engr, Radio Corp Am, 62-65; actg sect chief metals res, Space Systs Div, Avco Corp, Mass, 65-71; asst dir res, 71-73, dir res, 73-76, VPRES TEMESCAL, HIGH VACUUM INT, BERKELEY, 76- *Concurrent Pos:* Lectr, West Ham Col Technol, Eng, 59-61 & Ctr Continuing Educ, Northeastern Univ, 67-69. *Mem:* Am Ceramic Soc; Am Soc Metals; Am Vacuum Soc. *Res:* Metals and ceramic materials science; surface chemistry; ultra-high vacuum technology. *Mailing Add:* Temescal Edwards High Vacuum Int 2850 7th St Berkeley CA 94710

HILL, SAMUEL RICHARDSON, JR, b Greensboro, NC, May 19, 23; m 50; c 4. INTERNAL MEDICINE. *Educ:* Duke Univ, BA, 43; Wake Forest Univ, MD, 46; Am Bd Internal Med, dipl, 54. *Hon Degrees:* DSc, Univ Ala, 75, Wake Forest Univ, 79. *Prof Exp:* Intern med, Peter Bent Brigham Hosp, Boston, 47-48, asst resident, 48-49, asst, 49-50; instr, Bowman Gray Sch Med, 50-51; asst, Peter Bent Brigham Hosp, Boston, 53-54; from asst prof to assoc prof med, 54-62, dir metab & endocrine sect, 54-62, dean med col, vpres health affairs & dir med ctr, 62-77, pres, 77-87, PROF MED, SCH MED, UNIV ALA, BIRMINGHAM, 62-, DISTINGUISHED PROF, 87- *Concurrent Pos:* Teaching fel, Harvard Med Sch, 48-49, Dazian Med Found fel, 49-50; asst, Harvard Med Sch, 53-54; chief resident, NC Baptist Hosp, 50-51; chief metab sect, Birmingham Vet Admin Hosp, 54-57; dir metab & endocrine sect, Med Col Ala, 57-62; dir syst med educ prog, Univ Ala, 72-79. *Mem:* Inst Med-Nat Acad Sci; Endocrine Soc; Am Diabetes Asn; Am Thyroid Asn; Am Col Physicians; Am Fedn Clin Res (pres, 61-62). *Res:* Metabolic and endocrine diseases. *Mailing Add:* Univ Ala at Birmingham UAB Sta Birmingham AL 35294

HILL, SHARON WILKINS, immunology, for more information see previous edition

HILL, SHIRLEY ANN, b Kansas City, Mo, Aug 26, 27. MATHEMATICS EDUCATION. *Educ:* Univ Mo, BA, 48; Univ Kansas City, MA, 56; Stanford Univ, PhD(math educ), 61. *Prof Exp:* Res assoc math, Stanford Univ, 60-63; assoc prof, Univ Mo, Kansas City, 63-65, prof math & educ, 65-80; CONSULT. *Concurrent Pos:* Chairperson, Nat Adv Comt Math Educ, 74-76 & US Comn Math Instr, 75-; mem, US Nat Comt Math, 75- sr res assoc, Stanford Univ, 76; pres, Nat Coun Teachers Math, 78-80. *Mem:* Math Asn Am; Nat Coun Teachers Math; Am Math Soc; Asn Symbolic Logic; Philos Educ Soc. *Res:* Mathematics learning; the development of logical and reasoning abilities. *Mailing Add:* Dept Educ Univ Mo 5100 Rockhill Rd Kansas City MO 64110

HILL, SHIRLEY YARDE, b Galesburg, Ill, Nov 2, 41; c 1. PSYCHOPHYSIOLOGY. *Educ:* Grinnell Col, BA, 63; Washington Univ, PhD(psychol), 71. *Prof Exp:* Res asst, psychiat, Washington Univ Sch Med, 63-66, res fel, 66-71, res instr, 71-74, asst prof, neuropsychol & psychiat, 74-77; assoc prof psychiat, 77-88, ASSOC PROF PSYCHOL, UNIV PITTSBURGH SCH MED, 84-, PROF PSYCHIAT, 88- *Concurrent Pos:* Dir, Substance Abuse Treatment Serv, 80-86 & Alcoholism & Genetics Res Prog, Univ Pittsburgh, 86- *Mem:* Am Col Neuropsychopharmacol; Am Psychol Asn; Am Psychopathol Asn; Int Soc Biomed Res Alcoholism; Sigma Xi. *Res:* Biological markers for risk development of alcoholism. *Mailing Add:* W Psychiat Inst & Clin 3811 O'Hara St Pittsburgh PA 15213

HILL, STEPHEN JAMES, b Des Moines, Iowa, July 20, 43; m 65; c 2. GEOPHYSICS, ASTROPHYSICS. *Educ:* Iowa State Univ, BS, 65; Univ Colo, PhD(physics, astrophys), 71. *Prof Exp:* From asst prof to assoc prof astrophys, Mich State Univ, 71-78; assoc geophysicist, Continental Oil Corp, 78-80; STAFF GEOPHYSICIST, CONOCO INC, 80- *Mem:* Soc Explor Geophysics. *Res:* Seismic migration theory; hydrodynamics in variable stars; numerical seismic reduction techniques; computer automation in optical telescopes; seismic data reduction to modeling. *Mailing Add:* Conoco Inc Ponca City OK 74601

HILL, STUART BAXTER, b Aylesbury, Eng, Apr 11, 43; m 63, 66; c 2. ECOLOGY, ENTOMOLOGY. *Educ:* Univ Wales, Swansea, BSc, 65; Univ WI, Trinidad, PhD(ecol), 69. *Prof Exp:* Demonstr zool, Univ WI, Trinidad, 65-67 & Univ Reading, 67-69; Nat Res Coun Can res assoc entom, 69-72, lectr, 72-73, asst prof, 73-79, ASSOC PROF ENTOM, MACDONALD COL, MCGILL UNIV, 79- *Honors & Awards:* Queen's Silver Jubilee Medal, 77. *Mem:* Brit Ecol Soc; Entom Soc Can. *Res:* Orchard pest management; tropical cave ecology; ecological agriculture; soil mite ecology, taxonomy, extraction and effects of agricultural and forestry practices on them; food policy and agroecosystem design. *Mailing Add:* Dept Entom Macdonald Col McGill Univ 21,111 Lakeshore Ste Anne de Bellevue PQ H9X 1C0 Can

HILL, SUSAN DOUGLAS, b Toronto, Ont, Can, Sept 4, 40; m 68; c 2. EFFECTS OF ENVIRONMENT ON DEVELOPMENT. *Educ:* Queen's Univ, Kingston, Ont, Can, BSc, 63; Univ Mich, Ann Arbor, MS, 65, PhD(zool), 69. *Prof Exp:* Postdoctoral, Dept Anat, Univ Mich, 69-70; asst prof, Dept Health Sci, Univ Del, 70-72; res assoc, Dept Zool, Mich State Univ, 73-81; instr, Lansing Community Col, 79-80; ASST PROF, DEPT ZOOL, MICH STATE UNIV, 81- *Concurrent Pos:* Mem, Bermuda Biol Sta, 70-, Marine Biol Lab, 87- *Mem:* Soc Develop Biol; Am Soc Zoologists; Sigma Xi. *Res:* Regeneration in polychaete worms; effects of tail loss and regeneration on fitness; effects of extra low frequency ultramagnetic fields on embryological development. *Mailing Add:* Dept Zool Mich State Univ East Lansing MI 48824

HILL, TERRELL LESLIE, b Oakland, Calif, Dec 19, 17; m 42; c 3. BIOPHYSICS, PHYSICAL BIOCHEMISTRY. *Educ:* Univ Calif, Berkeley, AB, 38, PhD(chem), 42. *Prof Exp:* Instr chem, Western Reserve Univ, 42-44; res chemist, Radiation Lab, Univ Calif, 44-45; res assoc, Univ Rochester, 45-46, asst prof chem, 46-49; chemist, Naval Med Res Inst, 49-57; prof chem, Univ Ore, 57-67 & Univ Calif, Santa Cruz, 67-71, vchancellor sci, 68-69; res chemist, 71-88, EMER SCIENTIST, LAB MOLECULAR BIOL, NAT INST ARTHRITIS, METAB & DIGESTIVE DIS, 88- *Concurrent Pos:* Guggenheim fel, Yale Univ, 52-53; Sloan fel, 58-62; adj prof, 77-88, emer prof, Univ Calif, Santa Cruz, 88- *Honors & Awards:* Flemming Award, US Govt, 54; Kendall Award, Am Chem Soc, 69. *Mem:* Nat Acad Sci; Am Chem Soc; Biophys Soc. *Res:* Thermodynamics; statistical mechanics; biophysics; physical biochemistry. *Mailing Add:* 433 Logan St Santa Cruz CA 95062

HILL, THOMAS WESTFALL, b Olean, NY, July 28, 45; m 76; c 3. PLANETARY MAGNETOSPHERES. *Educ:* Rice Univ, BA, 67, MS, 71, PhD(space sci), 73. *Prof Exp:* Res assoc, 72-74, sr res assoc, 74-76, asst prof, 76-80, assoc res scientist, 80-85, SR RES SCIENTIST, SPACE PHYSICS, RICE UNIV, 86- *Concurrent Pos:* Mem, Upper Atmosphere Panel, Nat Res Coun, 74-77; resident res assoc, Nat Oceanic & Atmospheric Admin, 75-76; assoc ed, J Geophys Res, 78-80 & Geophys Res Lett, 83-88; consult, Southwest Res Inst, 79; mem, Associateship Panel, Nat Res Coun, 84-89, steering comt, Geospace Environ Modeling Prog, NSF, 89-, Mgt Opers Working Group, NASA, 89- *Honors & Awards:* James B Macelwane Award, Am Geophys Union, 80. *Mem:* Fel Am Geophys Union; Int Asn Geomagnetism & Aeronomy; AAAS. *Res:* Theoretical research in space plasma physics, particularly solar-wind interactions with the terrestrial magnetosphere and the dynamics of rapidly rotating planetary and astrophysical magnetospheres. *Mailing Add:* Space Physics & Astron Dept Rice Univ Houston TX 77251-1892

HILL, TREVOR BRUCE, b St Catharines, Ont, May 23, 28; m 48. ORGANIC CHEMISTRY. *Educ:* Univ Alta, BSc, 52; Cornell Univ, PhD(chem), 58. *Prof Exp:* Res chemist, E I du Pont de Nemours & Co, Inc, 57-63; assoc prof org chem, 63-71, PROF CHEM, COL WILLIAM & MARY, 71- *Mem:* Am Chem Soc. *Res:* Synthesis of polynuclear aromatic hydrocarbons; organic coatings. *Mailing Add:* 228 Longhill Rd Williamsburg VA 23185

HILL, VICTOR ERNST, IV, b Pittsburgh, Pa, Nov 3, 39; div; c 2. GROUP THEORY, HISTORY MATHEMATICS. *Educ:* Carnegie-Mellon Univ, BS, 61; Univ Wis, MA, 62; Univ Ore, PhD(math), 66. *Prof Exp:* From asst prof to assoc prof, 66-78, prof, 78-89, THOMAS T READ PROF MATH, WILLIAMS COL, 89- *Concurrent Pos:* Vis prof, Ga Tech, 88 & 91-92. *Mem:* Math Asn Am; Sigma Xi. *Res:* Arithmetical structure and extension theory of finite groups; history of mathematics. *Mailing Add:* Dept Math Williams Col Williamstown MA 01267

HILL, W(ILLIAM) RYLAND, electrical engineering; deceased, see previous edition for last biography

HILL, WALTER ANDREW, b New Brunswick, NJ, Aug 9, 46; m 84; c 3. AGRONOMY, HORTICULTURE. *Educ:* Lake Forest Col, BA, 68; Univ Chicago, MAT, 70; Univ Ariz, MS, 73; Univ Ill, Urbana, PhD(soil chem & fertil), 78. *Prof Exp:* Res asst chem, Lake Forest Col, 67-68; teacher chem, Chicago, 69-71; res asst soil chem/fertil, Univ Ariz, 71-73; irrig/fertil specialist, Univ Ariz Exp Sta, 73-74; teaching asst soils, Univ Ill, 76-77; res asst soil chem /fertil, Univ Ill, 74-77; from asst prof to assoc prof, 78-84, PROF SOIL SCI, TUSKEGEE UNIV, 84-, DIR AGR EXP STA, 86- *Concurrent Pos:* Dean, Res Dir & Exten Adminr, Tuskegee Univ, 87-; prin investr, sweet potato plant-environ relationships, NASA, 85-, USAID, 86- & US Dept Agr/CSRS, 79-, environ technol & waste mgt, DOE, 90-; bd trustees, Lake Forest Col; bd dir, Agr Satellite Corp, 90-93; mem, Nat Res Coun Comt Syst Ag & Environ in the Tropics, 90-; sci liaison officer, Asian Veg Res & Develop Ctr, 89-93; consult, USAID/Univ Md-ES & Ark Comn Higher Educ; chmn, Prof Agr Workers Conf, 88- & Int Symp Sweet Potato Tech 21st Centennial, 91; counr, Int Soc Trop Root Corps, 85-88; vis scientist, Purdue Univ, 82, Int Inst Trop Agr, 85 & NASA, 87; Kellogg fel, NCFAP/RFF, 88. *Honors & Awards:* Distinguished Serv Award, Carver Plant & Soil Sci Club, 82. *Mem:* Soil Sci Soc Am; Am Soc Agron; Int Soc Soil Sci; Int Soc Trop Root Crops; Am Soc Hort Sci; Sigma Xi. *Res:* Administers interdisciplinary research/extension teams and teaching faculty in agriculture, food, environment, rural development and forestry; uses the sweet potato as a model crop to assess plant- microbiol-soil-nutrient-water interactions and crop production for controlled ecological life support systems for space missions. *Mailing Add:* Dept Agr Sci Milbank Hall Tuskegee Inst Tuskegee Institute AL 36088

HILL, WALTER EDWARD, JR, b Moberly, Mo, June 4, 31; m 51; c 5. GEOCHEMISTRY. *Educ:* Univ Kans, AB, 55, MA, 64. *Prof Exp:* Chemist & geochemist, Kans Geol Surv, 54-67; asst dir process res, C F & I Steel Corp, 67-70; chief chemist, Amax Explor Inc, 70-74; mgr, Standards Hazen Res Inc, 74-79; lab dir, Earth Sci Inc, 79-80; tech serv supt, Tex Gulf Metals, 80-82; gen mgr, Calmet of Colo, 83-84; tech serv mgr, Marathon Gold, 84-85; consult precious metals & geochem, 85-87; chief chemist, Nev Gold Mining Inc, 87-89; OPERS MGR, APACHE ENERGY & MINERALS, 89- *Concurrent Pos:* C F & I Co rep, nondestruct testing comt, Am Iron & Steel Inst, 67-69. *Mem:* Fel Am Inst Chemists; Asn Explor Geochemists. *Res:* Methods of analytical geochemistry; exploration geochemistry; development of rock and mineral analytical standards, metallurgical processing; precious metals. *Mailing Add:* 1486 S Wright St Lakewood CO 80228

HILL, WALTER ENSIGN, b Bottineau, NDak, July 25, 37; m 61; c 7. BIOPHYSICS, BIOCHEMISTRY. *Educ:* Brigham Young Univ, BS, 61; Univ Wis, MS, 64, PhD(biophys), 67. *Prof Exp:* Res assoc biochem & biophys, Ore State Univ, 67-69; from asst prof to assoc prof, 69-77, assoc dean grad sch, 78-79, PROF CHEM, UNIV MONT, 77- *Concurrent Pos:* NIH career develop award, 72-77; prog dir, NSF, 89-90. *Mem:* Biophys Soc; Am Soc Biol Chem; fel AAAS; Am Soc Microbiol. *Res:* Probing structure and function of ribosomes using short complementary DNA oligonucleotides; physical study of ribosomes and other macromolecules using analytical utlracentrifugation, density measurements, diffusion measurements and viscometry. *Mailing Add:* Dept Biol Sci Univ Mont Missoula MT 59812

HILL, WALTER ERNEST, b San Francisco, Calif, July 24, 45. GENE PROBES, BACTERIAL VIRULENCE TESTING. *Educ:* Univ Calif, Berkeley, BA, 67; Univ Wash, Seattle, PhD(genetics), 72. *Prof Exp:* Fel molecular biol, Univ Wis-Madison, 72-74, asst prof biol genetics, Whitewater, 74; fel virol, Univ Chicago, 74-75; asst prof microbiol, Univ Southwestern La, 75-78; staff fel microbiol, Bur Foods, 78-82, RES MICROBIOLOGIST, DIV MICROBIOL, CTR FOOD SAFETY & APPL NUTRIT, FOOD & DRUG ADMIN, 82- *Mem:* Am Soc Microbiol; Genetics Soc Am; Sigma Xi; AAAS; Asn Off Anal Chemists; Inst Food Technologists. *Res:* Development of DNA hybridization probes to detect pathogenic bacteria in foods; application of the polymease chair reaction to safety testing of foods. *Mailing Add:* HFF-235 Food & Drug Admin 200 C St SW Washington DC 20204

HILL, WARNER MICHAEL, food microbiology, food science, for more information see previous edition

HILL, WILLIAM FRANCIS, JR, virology, environmental biology, for more information see previous edition

HILL, WILLIAM JOSEPH, b St Catharines, Ont, Jan 20, 40; m 62; c 2. STATISTICS. *Educ:* Princeton Univ, BSE, 63; Univ Wis, MS, 64, PhD(statist), 66. *Prof Exp:* Sr engr, Allied Corp, 66-72, mgr math sci, 73-84, dir, Ctr Appl Math, 84-89; dir, Math & Simulation Sci Dept, Allied-Signal, 89-91; EMERSON ELECTION & WIS DISTINGUISHED PROF, INDUST ENG, UNIV WIS, 90-, DIR, CTR QUAL & PRODUCTIVITY IMPROV, 91- *Concurrent Pos:* Adj assoc prof, State Univ NY, Buffalo, 74-77; mem epidemiol subcomt, Chem Indust Inst Toxicol, 78-80; chair, Gordon Conf on Statists, 80; vchair, continuing educ adv comt, Am Statist Asn, 81-84, mem bd dirs, 83, UN Environ Programme Comt on the Sci Assessment Stratospheric Ozone, 89, steering comt, Network for the Detection of Stratospheric Change, 90- *Honors & Awards:* William G Hunter Award, Am Soc Qual Control. *Mem:* Fel Am Statist Asn; Am Soc Qual Control; fel AAAS. *Res:* Design and analysis of experiments for model discrimination, parameter estimation and chemical process improvement; time series analysis of environmental data; probability analysis in epidemiological and occupational health studies; statistical process control and quality improvement. *Mailing Add:* Ctr Qual & Productivity Improv Univ Wis 610 Walnut St Madison WI 53705

HILL, WILLIAM T, b Covington, Ky, Apr 8, 25; m 50; c 3. GEOLOGY. *Educ:* Univ Tenn, BA, 50, MS, 51, PhD, 71. *Prof Exp:* Geologist, Am Zinc Co, 51-53; teaching asst geol, Univ Tenn, 53-55; geologist, Tenn Div Geol, 55-56; explor supt, B H Putnam & Assocs, 56-59; geologist I, NJ Zinc Co, Jefferson City, Tenn, 59-60, res geologist, Treadway, 60-65, distr geologist, 65-66, asst to mgr, 66-69, asst to div mgr mines, 69-70, proj mgr, 70-74, mgr, Elmwood Mine, vpres, 76-79, pres, 79-84; assoc, Condor Mine Mgt, 84-85; state geologist, Tenn, 85-90; CONSULT, 90- *Concurrent Pos:* Dir, Enuex Mineracao, Brazil, 74-76. *Mem:* Soc Econ Geol; Am Inst Mining, Metall & Petrol Eng. *Res:* Minerals exploration; mine and production planning; geological research. *Mailing Add:* 100 Jackson Lane Hendersonville TN 37075

HILLAIRE-MARCEL, CLAUDE, b Salies de Bearn, France, April 1, 44; m 75; c 1. ISOTOPIC GEOCHEMISTRY, QUATERNARY GEOLOGY. *Educ:* Univ Paris at the Sorbonne, France, LLSci, 67, DES, 67; Univ Pierre & Marie Curie, Paris, DSci, 79. *Prof Exp:* PROF GEOL, UNIV QUE, MONTREAL, 69- *Concurrent Pos:* Vis prof, Ctr Nat Res Sci, France, 80; assoc prof, Univ Orsay, Paris, 81. *Mem:* Soc Geol France; Geol Soc Am; Can Nat Comt Inqua; NY Acad Sci; Geol Asn Can. *Res:* Isotope geochemistry; paleoclimatology; quaternary geology (Eastern Canada-Tropical Africa); isotopic biochemistry. *Mailing Add:* GEOTOP BP8888-SuccA Univ Que Montreal PQ H3C 3P8 Can

HILLAKER, HARRY J, b Flint, Mich, May 9, 19. ENGINEERING. *Educ:* Univ Mich, AE. *Prof Exp:* Vpres & dep prog dir, F-16 Gen Dynamics, Ft Worth; AEROSPACE CONSULT. *Concurrent Pos:* Mem, sci adv bd & adv group, Aeronaut Systs Div, USAF; chmn, Aerospace Vehicles Panel. *Honors & Awards:* Aircraft Design Award, Am Inst Aeronaut & Astronaut. *Mem:* Nat Acad Eng; Am Inst Aeronaut & Astronaut. *Res:* Aircraft design and technologies. *Mailing Add:* 4001 Shannon Dr Ft Worth TX 76116

HILLAM, BRUCE PARKS, b Salt Lake City, Utah, May 23, 45. MATHEMATICAL ANALYSIS. *Educ:* Univ Calif, Riverside, BA, 68, MA, 69, PhD(math), 73. *Prof Exp:* Asst prof math, Univ Calif, Riverside, 73; asst prof, 73-77, assoc prof math & comput sci, 77-81, assoc prof comput sci, 81-83, PROF & CHAIR, DEPT COMPUT SCI, CALIF STATE POLYTECH UNIV, POMONA, 83- *Mem:* Soc Indust & Appl Math; Math Asn Am; Asn Comput Mach; AAAS. *Res:* Existence of fixed point solutions of functional equations with applications to modeling and simulation on computers. *Mailing Add:* Dept Comput Sci Calif State Polytech Univ Pomona CA 91768

HILLAM, KENNETH L, b Salt Lake City, Utah, July 15, 27; m 50; c 3. MATHEMATICS. *Educ:* Univ Utah, BS, 49, MS, 56; Univ Colo, PhD(math), 62. *Prof Exp:* Instr appl math, Univ Colo, 56-57; from instr to assoc prof, 57-69, PROF MATH, BRIGHAM YOUNG UNIV, 69-, CHMN DEPT, 63- *Mem:* Am Math Soc; Math Asn Am. *Res:* Complex analysis; convergence criteria for continued fractions; real variables. *Mailing Add:* 11017 S Surrey Dr Covered Bridge Canyon Spanish Fork UT 84660

HILLAR, MARIAN, b Bydgoszcz, Poland, Mar 22, 38; US citizen; m 70; c 2. BIOCHEMISTRY, MOLECULAR BIOLOGY. *Educ:* Univ Med Sch, Gdansk, Poland, MD, 62, PhD(biochem), 66. *Prof Exp:* Instr biochem, Univ Med Sch, Gdansk, Poland, 58-62, asst prof, 62-69; res assoc, Baylor Col Med, 69-70; from asst prof to assoc prof, 71-81, PROF BIOCHEM, TEX SOUTHERN UNIV, 81- *Concurrent Pos:* NIH fel, Nat Cancer Inst, 72-84; Fulbright award, 80; prof, Ponce Sch Med, PR, 85-86. *Mem:* Biophys Soc; Biochem Soc London. *Res:* Control of bioenergetic reactions in mitochondria; mechanisms of control of gene expression in normal and cancer cells by endogenous peptides associated with chromatin and poly(A)-mRNA; glutamate dehydrogenase biosynthesis. *Mailing Add:* 9330 Bankside Houston TX 77031

HILLARD, CECILIA JANE, m 80; c 1. DRUG ABUSE, NEUROCHEMISTRY. *Educ:* Med Col Wis, PhD(pharmacol), 83. *Prof Exp:* ASST PROF PHARMACOL, MED COL WIS, 85- *Mem:* Soc Neurosci; Am Soc Pharmacol & Exp Therapeut. *Mailing Add:* Med Col Wis 8701 Watertown Plank Rd Milwaukee WI 53226

HILLCOAT, BRIAN LESLIE, b Rockhampton, Australia, July 26, 32; m 57; c 4. CANCER CHEMOTHERAPY. *Educ:* Univ Queensland, MB, BS, 55, BSc, 62, MD, 63; Australian Nat Univ, PhD(biochem), 65. *Prof Exp:* USPHS fel biochem pharmacol, Sch Med, Yale Univ, 65-67, asst prof, 67-68; from assoc prof to prof biochem, McMaster Univ, 68-78; exp, Nat Cancer Inst, NIH, 78; head, Solid Tumor Chemother Unit, Cancer Inst, Melbourne, 78-84, prof cancer med, 84-89; MED SERV ADV, DEPT COMMUN SERV & HEALTH, DRUG EVAL BR, 89- *Concurrent Pos:* Vis prof med, Yale Univ, 75; Nuffield travelling fel, 75; outpatient physician, Cancer Clin, Henderson Hosp, 75-78. *Mem:* AAAS; Clin Oncol Soc Australia; NY Acad Sci; Am Soc Clin Oncol; Am Asn Cancer Res. *Res:* Mechanism of action of drugs that interact with DNA; effects of anticancer drugs on gene expression. *Mailing Add:* Drug Eval Br Therapeut Goods Admin PO Box 100 Woden ACT 2606 Australia

HILLDRUP, DAVID J, b Manchester, Eng, Apr 5, 36. MATHEMATICS, OPERATIONS RESEARCH. *Educ:* Univ London, BSc, 59, MSc, 60; Univ Hawaii, MBA, 73; Laurentian Univ, BA, 75. *Prof Exp:* Head, dept appl math, N Manchester Grammar Sch, 59-61; asst prof math, Bishop's Univ, 61-65; dir, Sch Com & Admin & dean, Fac Prof Schs, 78-81, ASSOC PROF COM, LAURENTIAN UNIV, 65- *Mem:* Opers Res Soc Am; Can Opers Res Soc; Inst Mgt Sci. *Res:* Inventory, forecasting and queueing models. *Mailing Add:* Sch Commerce Laurentian Univ Ramsey Lake Rd Sudbury ON P3E 2C6 Can

HILLE, BERTIL, b New Haven, Conn, Oct 10, 40; m 64; c 2. PHYSIOLOGY, BIOPHYSICS. *Educ:* Yale Univ, BS, 62; Rockefeller Univ, PhD(life sci), 67. *Prof Exp:* Helen Hay Whitney fel physiol, Cambridge Univ & Agr Res Coun Inst Animal Physiol, Babraham, 67-68; from asst prof to assoc prof, 68-74, PROF PHYSIOL & BIOPHYS, UNIV WASH, 74- *Concurrent Pos:* Biophys Soc Counr, 77-79; vis prof, Univ Saarland, Fed Repub Germany, Nat Inst Health, Physiol Study Sect, 80-84; Neurosci res prog assoc, 85- *Honors & Awards:* Sr Scientist Award, Alexander von Humboldt Found, 75-76; Bristol-Myers Squibb Award, Neurosci Res. *Mem:* Nat Acad Sci; Biophys Soc; Soc Gen Physiologists; Soc Neurosci. *Res:* Permeability changes in axon and muscle membranes; conduction of nervous impulses; ionic channels of excitable membranes; neurotransmitters and second messengers. *Mailing Add:* Dept Physiol & Biophys SJ-40 Univ Wash Seattle WA 98195

HILLE, KENNETH R, b New York, NY, Sept 16, 27; m 56; c 2. LIMNOLOGY, AQUATIC TOXICOLOGY. *Educ:* Wagner Col, BS, 52; Bowling Green State Univ, MA, 56; Ohio State Univ, PhD(aquatic ecol), 69. *Prof Exp:* Instr, Fremont City Schs, 54-65; assoc prof biol, 69-81, ASSOC PROF BIOL & CHMN NATURAL & SOCIAL SCI, FIRELANDS COL, BOWLING GREEN STATE UNIV, 81- *Mem:* Am Soc Limnol & Oceanog; Am Fisheries Soc; Sigma Xi. *Res:* Toxicity of chlorinated hydrocarbons, as dieldrin, to fish; toxicity of surfactants, as detergents to fish; stream macro-invertebrate community structure; water quality criteria of streams. *Mailing Add:* Firelands Col 901 Rye Beach Rd Huron OH 44839

HILLE, MERRILL BURR, b Feb 15, 39; m; c 2. EMBROYOLOGY. *Educ:* Rockefeller Univ, PhD(life sci), 65. *Prof Exp:* ASSOC PROF ZOOL, UNIV WASH, SEATTLE, 76- *Mem:* Soc Develop Biol; Am Soc Cell Biol. *Res:* Translational control in eggs; embryos of Echinoderms; cell adhesion during gastrulation. *Mailing Add:* Dept Zool NJ-15 Univ Wash Seattle WA 98195

HILLEGAS, WILLIAM JOSEPH, b Quakertown, Pa, Aug 31, 37; m 82; c 5. MATERIALS SCIENCE, BIOENGINEERING & VACUUM TECHNOLOGY. *Educ:* Drexel Inst Technol, BSChE, 60; Northwestern Univ, MS, 67, PhD(mat sci), 68. *Prof Exp:* Jr engr, Lansdale Div, Philco Corp, 60-61, engr, 61-63; assoc scientist, Res Div, Xerox Corp, 68-69, scientist, 69-78; res scientist, KMS Fusion Inc, 78-85; VPRES, CHIEF TECH OFFICER, DIR & CO-FOUNDER, SOLO HILL ENG INC, 84- *Concurrent Pos:* Prin investr, Small Bus Innovative Res Phase I & II; Nat Cancer Inst res award, 84-86, 87-90; res fund grants, Mich State, 87-90. *Mem:* Tissue Cult Asn; AAAS; Mat Res Soc. *Res:* Bioengineering process development for large scale cell cultures; fabrication of inertial fusion targets; materials process technologies associated with semiconducting and amorphous photoreceptive materials; physics of defects in solids; thin film vacuum technologies. *Mailing Add:* Solo Hill Eng Inc 1919 Green Rd Ann Arbor MI 48105

HILLEL, DANIEL, b Los Angeles, Calif, Sept 13, 30; m 84; c 5. SOIL & ENVIRONMENTAL PHYSICS, HYDROLOGY. *Educ:* Univ Ga, BS, 50; Rutgers Univ, MS, 51; Hebrew Univ, PhD(soil physics), 58. *Prof Exp:* Soil survr, Soil Conserv Serv, Israel, 51-52; founding mem, Kibbutz Sde-Boker, Negev, Israel, 52-55; adv land develop, Govt Burma, 56-58; res fel soil physics, Univ Calif, 59-61; head, soil technol div, Agr Res Ctr, Rehovot, Israel, 61-65; prof & head soil & water sci, Hebrew Univ, Jerusalem, 66-74; vis prof soil physics, Tex A&M Univ, 74-75; vis prof environ sci, Univ Va,

75-77; PROF SOIL PHYSICS & HYDROL, UNIV MASS, 77- *Concurrent Pos:* Consult water conserv, Govt Calif, 79-80; irrig consult, World Bank, 82-87, environ sci adv, 87-, US AID, Govt Egypt, 84; ed, Advan in Irrig, 82-; vis prof, Israel Inst Technol, 83-84 & Haifa Univ, 85-86; res consult, Commonwealth Sci & Indust Res Orgn, Australia, 84; consult ed, Soil Sci & Hydrol Processes, 84-; vchmn, soil physics, Int Soil Sci Soc, 64-68 & 82-86, chmn, Soil Sci Soc of Am, 88-89; prin investr, NSF Res Projs, 79-83 & 86-89; nat lectr, Sigma Xi Sci Honor Soc, 87-89. *Mem:* fel Soil Sci Soc Am; fel Am Soc Agron; Am Geophys Union; AAAS; Am Soc Agr Engrs; Int Soil Sci Soc (vpres, 64-68 & 82-86). *Res:* Soil-water dynamics; soil-plant-water relations; the water regime of natural habitats and agricultural fields; irrigation efficiency; drainage; water conservation; land reclamation; environmental physics; pollution control; mathematical modeling and field measurements of hydrological processes; author of over 150 scientific papers and of 14 books in soil and environmental physics. *Mailing Add:* 58 High Point Rd Amherst MA 01002

HILLEMAN, MAURICE RALPH, b Miles City, Mont, Aug 30, 19; m 43, 63; c 2. VIROLOGY. *Educ:* Mont State Col, BS, 41; Univ Chicago, PhD(microbiol & virol), 44. *Hon Degrees:* ScD, Mont State Col, 66; DSc, Univ Md, 68. *Prof Exp:* Asst bacteriologist, Univ Chicago, 42-44; res assoc virus labs, E R Squibb & Sons, NJ, 44-47, chief virus dept, 47-48; chief respiratory & virus res & diag sect, Army Med Serv Grad Sch, Walter Reed Army Med Ctr, 48-56, asst chief lab affairs, 53-56, chief respiratory dis, Walter Reed Army Inst Res, 56-57; from dir to exec dir virus & cell biol res, 57-71, vpres, Merck Inst Therapeut Res, 72-78, dir virus & cell biol res & vpres, Merck Sharp & Dohme Res Labs, 70-78, SR VPRES, MERCK INST THERAPEUT RES, MERCK SHARP & DOHME RES LABS, MERCK & CO, INC, 78. *Concurrent Pos:* Vis instr, Rutgers Univ, 47; vis investr, Rockefeller Inst Med Res, 51; mem expert adv panel virus dis, WHO, 52-, mem expert comt influenza, 52, mem respiratory dis, 58, mem sci group measles vaccine studies, 63, viruses & cancer, 64 & virus vaccines, 65; vis prof, Univ Md, 53-57; mem study sect microbiol & immunol, Grants-in-aid Prog, USPHS, 53-61, spec consult, Panel Respiratory & Related Viruses, 60-64; assoc mem comn influenza, US Armed Forces Epidemiol Bd, 55-58 & comn respiratory dis, 56-58; consult, Surgeon Gen, US Army, 58-63; mem primate study group, Nat Cancer Inst, 64-70, mem working group immunol & epidemiol, Spec Virus Cancer Prog, 69-70; adj prof virol in pediat, Sch Med, Univ Pa & consult, Children's Hosp, Philadelphia, 68-; John Herr Musser lectr, Tulane Univ, 69; mem permanent sect microbiol standardization, Int Asn Microbiol Socs; mem coun anal & projection, Am Cancer Soc, 71-76; mem investr on animal models for res on immunity to cancer, WHO, UN, 73, sci group on immunol adjuvants, 75 & study group on cerebrospinal meningitis control, 75; trustee, Merck Inst Therapeut Res, 76-; mem, Coun Div Biol Sci & Pritzger Sch Med, Univ Chicago, 77-; mem, Overseas Med Res Lab Comt, US Dept Defense, 80. *Honors & Awards:* Howard Taylor Ricketts Prize, 45; Flemming Award, 56; Golden Plate Award, Am Acad Achievement, 75; 19th Graugnard Lectr, Tulane Univ, 78. *Mem:* Nat Acad Sci; Am Soc Microbiol; Soc Exp Biol & Med; fel Am Acad Microbiol; Am Asn Immunol; Am Asn Cancer Res; Tissue Culture Asn; Infectious Dis Soc. *Res:* Medical virology, especially adenoviruses, influenza, psittacosis-lymphogranuloma venereum group, poliomyelitis, measles and carcinogenic viruses; virus in cancer and immunology in cancer; epidemiology and preventive medicine; interferon and host resistance, chemotherapy in viral infections; viral vaccines and immunologic adjuvants. *Mailing Add:* Merck Inst Therapeut Res Merck Sharp & Dohme Res Labs West Point PA 19486

HILLER, DALE MURRAY, b Chicago, Ill, Oct 13, 24; m 52; c 4. PHYSICAL CHEMISTRY. *Educ:* Univ Nebr, BSEE, 47; Miami Univ, AB, 48; Iowa State Col, PhD(chem), 52. *Prof Exp:* Res chemist, Pigments Dept, E I Du Pont de Nemours & Co, Inc, 52-65, sr res chemist, Photo Dept, 65-69, 72-77, qual control supvr, 69-72, res assoc, Electronics Dept, 77-90; INDEPENDENT CONSULT, 90- *Mem:* Am Chem Soc; Inst Elec & Electronics Eng; Sigma Xi. *Res:* Photofission of thorium; titanium dioxide pigments; extractive metallurgy; photographic film; magnetic tape. *Mailing Add:* Westgate Farms 151 Oldbury Dr Wilmington DE 19808-1433

HILLER, FREDERICK CHARLES, b Kansas City, Kans, Jan 30, 42; m 65; c 4. MEDICINE. *Educ:* Univ Kans, Lawrence, AB, 64, Kansas City, MD, 68. *Prof Exp:* Intern, Med Ctr, Univ Kans, Kansas City, 68-69, resident internal med, 71-73, fel, 73-75; captain, Med Corps, US Army Reserves, 69-71; from asst prof to assoc prof, dept med, 75-83, assoc prof, dept pharmacol, 82-83, PROF MED & INTERDISCIPLINARY TOXICOL, UNIV ARK MED SCI, 83-, MED DIR, RESPIRATORY THER DEPT, 80-, SLEEP DIS CTR & MED & INTENSIVE CARE UNIT, 84-, DIR, DIV PULMONARY & CRIT CARE MED, 85- *Concurrent Pos:* Dir, pulmonary function lab, Univ Ark Med Sci, 75-; chest consult, Ark State Health Dept, 75-; consult, Little Rock Vet Admin Med Ctr, 75-; med adv, Ark State Respiratory Ther Asn, 81-; chief, Pulmonary & Critical Care Med Div, Univ Ark Med Sci & Little Rock Vet Admin Med Ctr, 85- *Mem:* Am Thoracic Soc; fel Am Col Chest Physicians; Am Fedn Clin Res; Sigma Xi. *Res:* Aerosol physical properties, deposition in respiratory tract; sleep apnea and effects of alcohol on sleep and sleep apnea. *Mailing Add:* Pulmonary Div Slot 555 Univ Ark Med Sci 4301 W Markham Little Rock AR 72206

HILLER, FREDERICK W, b Melstone, Mont, Feb 28, 27; m 56, 77; c 3. INORGANIC CHEMISTRY, PHYSICAL CHEMISTRY. *Educ:* Ore State Univ, BS, 50, PhD(inorg chem), 67. *Prof Exp:* Instr chem, SDak State Col, 56-57, Idaho State Col, 57-64 & Ore State Univ, 64-66; from asst prof to assoc prof, 66-74, chmn dept, 80-83, PROF CHEM, CALIF STATE UNIV, CHICO, 74- *Mem:* AAAS; Am Chem Soc; Sigma Xi. *Res:* Inorganic kinetics and reaction mechanisms. *Mailing Add:* Four El Verta Circle Chico CA 95926

HILLER, JACOB MOSES, b New York, NY, Dec 12, 39; m 65; c 3. NEUROPHARMACOLOGY. *Educ:* City Col New York, BS, 61; NY Univ, MS, 67, PhD(biol), 70. *Prof Exp:* From asst res scientist to res scientist biol, Grad Sch Art & Sci, NY Univ, 69-70; from asst res scientist to assoc res scientist med, 70-74, res asst prof med, 74-80, RES ASSOC PROF PSYCHIATRY, MED CTR, NY UNIV, 80- *Concurrent Pos:* Prin investr grant, Nat Inst Aging, 89-92. *Mem:* AAAS; Am Soc Microbiol; Sigma Xi; NY Acad Sci. *Res:* Molecular basis of the pharmacological effects of narcotics, including analgesia and development of tolerance and physical dependence; localization, solubilization and purification of stereospecific opiate receptors in animal and human brain. *Mailing Add:* Dept Psychiatry NY Univ Med Ctr New York NY 10016

HILLER, LARRY KEITH, b Morning Sun, Iowa, Apr 28, 41; m 63; c 2. HORTICULTURE, VEGETABLE CROPS. *Educ:* Iowa State Univ, BS, 63, MS, 64; Cornell Univ, PhD(veg crops), 74. *Prof Exp:* Tech adv, Iowa State, USAID Prog, Uruguay, 64-67; res asst, Cornell Univ, 68-73; asst prof & asst hort, 73-79, ASSOC PROF & ASSOC HORT, WASH STATE UNIV, 79- *Concurrent Pos:* Kellogg Found Nat Fel Prog, 80-83; consult vegatable, Ecuador, Peru & Columbia, 79-82. *Mem:* Am Soc Hort Sci; Am Soc Plant Physiologists; Coun Agr Sci & Technol; Potato Asn Am; Weed Sci Soc Am. *Res:* Production practices, physiological tuber disorders and quality factors affecting fresh market and processing grades of Washington potatoes; production and quality factors in irrigated vegetable crops. *Mailing Add:* Hort & Landscape Archit Dept Wash State Univ Pullman WA 99164-6414

HILLER, LEJAREN ARTHUR, b New York, NY, Feb 23, 24; m 45; c 2. ACOUSTICS. *Educ:* Princeton Univ, AB, 44, MA, 46, PhD(chem), 47; Univ Ill, MMus, 58. *Prof Exp:* Asst chem, Princeton Univ, 44-45; res chemist, E I du Pont de Nemours & Co, Inc, 47-52; res assoc, Univ Ill, 52-55, asst prof chem, 55-58, from res asst prof to prof music, 58-68; Slee prof compos, 68-81, BIRGE-CARY PROF COMPOS, STATE UNIV NY BUFFALO, 81- *Concurrent Pos:* Sr Fulbright lectr, Poland, 73-74 & Brazil, 80. *Mem:* Am Soc Composers, Auth & Publ. *Res:* Application of high-speed computers to scientific problems; acoustics and communication theory; electronic and experimental music. *Mailing Add:* Dept Music Baird Hall State Univ NY Buffalo NY 14260

HILLER, ROBERT ELLIS, b Traveler's Rest, Sc, Dec 14, 27; m 52; c 2. OPERATIONS RESEARCH. *Educ:* Clemson Col, BS, 50; NC State Col, MS, 53; Univ NC, PhD(physics), 56. *Prof Exp:* Asst, NC State Col, 51-53; instr physics, Univ NC, 53-56, asst prof & res assoc, 56-57; opers analyst, USAF, Europe, 57-61, chief opers analyst, Pac Air Forces, 61-87; RETIRED. *Mem:* Opers Res Soc Am. *Res:* Mathematics. *Mailing Add:* 44-404 Kaneohe Bay Dr Kaneohe Oahu HI 96744

HILLERS, JOE KARL, b Centerville, Iowa, July 20, 38; m 63; c 2. GENETICS, ANIMAL SCIENCE. *Educ:* Northwest Mo State Col, BA, 60; Iowa State Univ, MS, 62, PhD(animal breeding), 65. *Prof Exp:* From asst prof to assoc prof, 71-82, PROF ANIMAL SCI, WASH STATE UNIV, 82- *Mem:* Am Dairy Sci Asn; Am Soc Animal Sci; Biomet Soc. *Res:* Selection programs and computer systems analysis of livestock production traits, particularly dairy. *Mailing Add:* Dept Animal Sci Clark Hall 116 Wash State Univ Pullman WA 99163

HILLERY, HERBERT VINCENT, b Lima, Ohio, Dec 8, 24; m 56, 90; c 2. ART CONSERVATION, BAMBOO RESEARCH. *Educ:* Oberlin Col, AB, 47. *Prof Exp:* Mem staff acoust, 50-67, spec res assoc acoust, Appl Res Lab, Univ Tex, Austin, 67-78; MEM STAFF, HILLERY ENTERPRISES, 78- *Concurrent Pos:* NSF res grant, 72- *Mem:* Am Bamboo Soc; assoc mem Am Inst Conserv; Sigma Xi. *Res:* Art conservation & restoration; research on bamboo agriculture for Texas; noise vibration in solids; organic adhesives. *Mailing Add:* 1909 Richcreek Rd Austin TX 78757

HILLERY, PAUL STUART, b Morristown, NJ, June 19, 41; m 69; c 3. FEDERAL DRUG REGULATIONS. *Educ:* Johns Hopkins Univ, BA, 63; Univ Va, PhD(phys org chem), 68. *Prof Exp:* Lab instr, Univ Va, 63-65; res assoc, Inst Environ Med, NY Univ, 69-70; res chemist, Biochem Mech Sect, Nat Inst Arthritis, Metab & Digestive Dis, NIH, 71-77; chemist, Div Oncol & Radiopharmaceut Drug Prod, 78-89, REVIEWING CHEMIST, PILOT DRUG EVAL STAFF, FOOD & DRUG ADMIN, CTR DRUG EVAL & RES, 89- *Concurrent Pos:* Vis scientist, Lab Analytical Chem, Nat Cancer Inst, 79-80 & Lab Med Chem, Nat Inst Diabetes, Digestive & Kidney Dis, NIH, 81- *Mem:* Am Chem Soc, Org Div, Med Chem Div. *Res:* Intramolecular catalysis; solution kinetics and reaction mechanisms; heterocyclic synthesis, including phencyclidine analogs. *Mailing Add:* Food & Drug Admin Rm 9B45 5600 Fishers Lane Rockville MD 20857

HILLIARD, JOHN E, materials science, stereology; deceased, see previous edition for last biography

HILLIARD, JOHN ROY, JR, b Irving, Tex, Feb 7, 24; m 50; c 2. ENTOMOLOGY. *Educ:* Trinity Col, Tex, BA, 47; Univ Colo, MA, 51; Univ Tex, PhD(zool), 59. *Prof Exp:* Instr, Trinity Col, Tex, 48-51; prof, McMurry Col, 51-54, 57-68; assoc prof, 68-71, PROF BIOL, SAM HOUSTON STATE UNIV, 72- *Mem:* Pan-Am Acridological Soc; Entom Soc Am; Southwestern Entom Soc. *Res:* Ecology and taxonomy of Orthopterous insects; systematics of the eggs and egg pods of grasshoppers. *Mailing Add:* Dept Life Sci Sam Houston State Univ Huntsville TX 77341

HILLIARD, RONNIE LEWIS, b Havre, Mont, July 17, 37; div; c 3. PHYSICS. *Educ:* Whitman Col, BA, 59; Univ Sask, PhD(physics), 64. *Prof Exp:* Res asst, Steward Observ, Univ Ariz, 64-66, asst prof astron & asst astronr, 66-70, res assoc instrumentation, 70-80; PRES, OPTOMECHANICS RES, INC, 81- *Mem:* Int Astron Union. *Res:* Design and development of systems and techniques for astronomical optical instrumentation. *Mailing Add:* PO Box 87 Vail AZ 85641

HILLIARD, ROY C, b Middlesex, NC, May 19, 31; m 57; c 2. CHEMISTRY. *Educ:* Duke Univ, AB, 53. *Prof Exp:* Chemist, 53-65, supvr flavor res, 65-72, asst mgr prod develop div, 72-75, mgr prod develop div, 75-87, DIR RES, LIGGETT & MYERS TOBACCO CO, 88- *Mem:* Inst Food Technologists. *Res:* Flavor chemistry of tobacco and related products; panel evaluation of tobacco products; organoleptic chemistry. *Mailing Add:* Liggett & Myers Tobacco Co 710 W Main St PO Box 1572 Durham NC 27702-1572

HILLIARD, STEPHEN DALE, b Greenfield, Ohio, May 30, 39; m 62; c 3. EMBRYOLOGY, PHYSIOLOGY. *Educ:* Univ Cincinnati, BS, 64, MS, 66, PhD(physiol), 68. *Prof Exp:* Technician, Gastric Lab, Cincinnati Gen Hosp, Ohio, 58-65; lectr biol, Univ Cincinnati, 67-68; from asst prof to assoc prof, Col of the Virgin Islands, 68-72; PROF BIOL, BALDWIN-WALLACE COL, 72- *Mem:* Soc Develop Biol; Nat Soc Teachers Asn; Am Inst Biol Sci; Sigma Xi. *Res:* Development and electron microscopy. *Mailing Add:* 32611 Carriage Lane Avon Lake OH 44012

HILLIDGE, CHRISTOPHER JAMES, b Halton, Eng, Nov 11, 44. EQUINE MEDICINE, EQUINE SURGERY. *Educ:* Univ London, BS, 66, BVet Med, 69, PhD(anesthesiol), 73; FRCVS, 81. *Prof Exp:* Univ lectr cardiopulmonary med, Royal Vet Col, Univ London, 75-78; asst prof equine med, Sch Vet Med, Univ Ill, 79-81; ASSOC PROF, COL VET MED, UNIV FLA, 81- *Mem:* Physiol Soc London; Am Asn Equine Practr; Brit Equine Vet Asn; Vet Cardiovasc Soc; Comp Respiratory Soc. *Res:* Non-invasive methods for the assessment of equine cardiopulmonary function; cardiopulmonary and metabolic function in neonatal foals. *Mailing Add:* 223 Lac Col Vet Med Urbana IL 61801

HILLIER, FREDERICK STANTON, b Aberdeen, Wash, Mar 4, 36; m 58; c 3. OPERATIONS RESEARCH. *Educ:* Stanford Univ, BS, 58, MS, 59, PhD(indust eng), 61. *Prof Exp:* From asst prof to assoc prof indust eng, 61-68, PROF OPERS RES, STANFORD UNIV, 68- *Concurrent Pos:* Vis asst prof, Cornell Univ, 62-63; Inst Mgt Sci-Off Naval Res grant, Stanford Univ, 65-66; vis prof, Carnegie-Mellon Univ, 69-70 & Tech Univ Denmark, 76-77; Erskine fel, Univ Canterbury, NZ, 89. *Mem:* Opers Res Soc Am (treas, 74-76); Inst Mgt Sci (vpres, 81-84); Am Inst Indust Eng; Math Prog Soc. *Res:* Queueing theory; mathematical programming; capital budgeting; textbooks in operations research. *Mailing Add:* Dept Opers Res Stanford Univ Stanford CA 94305-4022

HILLIER, JAMES, b Brantford, Ont, Aug 22, 15; nat US; m 36; c 2. PHYSICS, RESEARCH ADMINISTRATION. *Educ:* Univ Toronto, BA, 37, MA, 38, PhD(physics), 41. *Hon Degrees:* Univ Toronto, DSc, 78, NJ Inst Technol, 81. *Prof Exp:* Demonstr physics, Univ Toronto, 37-39, asst, Banting Inst, 40; res engr, labs, Radio Corp Am, 40-53; dir res dept, Melpar, Inc, Westinghouse Air Brake Co, 53-54; admin engr, res & eng, RCA Corp, 54-55, chief engr, commercial electronic prod, 55-57, gen mgr, labs, 57-58, vpres, 58-68, vpres, res eng, 68-69, exec vpres res eng, 69-76, exec vpres & sr scientist, 76-78; CONSULT, 78- *Concurrent Pos:* Assoc, Sloan Kettering Inst, 49-51; vis lectr, dept biol, Princeton Univ, 51-53, chmn adv coun, dept elec eng, 65-69; pres, Indust Reactor Labs, 64-65. Mem, indust adv comt, NASA, 62-64, gov bd, Am Inst Physics, 62-65 & Gov NJ Higher Educ Study Comt, 63-64; pres, Indust Res Inst, 63-64; mem commerce tech adv bd, US Dept Commerce, 64-70; mem adv coun, Col Eng, Cornell Univ, 65-; mem joint consult comt, Acad Sci Res & Technol, Egypt & Nat Acad Sci, 77-85. *Honors & Awards:* Albert Lasker Award, Am Pub Health Asn, 60; David Sarnoff Award, Inst Elec & Electronics Engrs, 67, Founders Medal, 81; IRI Medal, Indust Res Inst, 75; Nat Inventors Hall Fame, 80; Common Wealth Award Laureate, 80. *Mem:* Nat Acad Eng; fel AAAS; fel Am Phys Soc; Electron Micros Soc Am (vpres, 44, pres, 45); fel Inst Elec & Electronics Engrs; Indust Res Inst (pres, 63). *Res:* Electron microscopy; instrumentation and applications; electron diffraction, instrumentation and applications; electronic methods of microanalysis; biophysics; research management. *Mailing Add:* 22 Arreton Rd Princeton NJ 08540-1402

HILLIER, RICHARD DAVID, b Boulder, Colo, Sept 3, 38; m 67; c 1. ECOLOGY, BOTANY. *Educ:* Haverford Col, AB, 60; Duke Univ, PhD(bot), 70. *Prof Exp:* Asst prof biol, Erskine Col, 69-71; asst prof, 71-76, ASSOC PROF BIOL, UNIV WIS-STEVENS POINT, 76- *Mem:* Ecol Soc Am; Am Inst Biol Sci; Sigma Xi. *Res:* Plant microenvironments and physiological ecology; vegetation of central Wisconsin; alpine plant distribution. *Mailing Add:* Dept Biol Univ Wisc Stevens Point Stevens Point WI 54481

HILLIG, KURT WALTER, II, b Schenectady, NY, Sept 12, 54; m 81. MOLECULAR SPECTROSCOPY. *Educ:* Union Col, NY, BS, 75; Univ Mich, Ann Arbor, MS, 79, PhD(chem), 81. *Prof Exp:* Postdoctoral scholar, 81-83, ASST RES SCIENTIST, UNIV MICH, 83- *Concurrent Pos:* Fac fel, NASA/Am Soc Eng Educ/Jet Propulsion Lab, 83 & 84. *Mem:* Am Chem Soc. *Res:* Microwave and millimeter wave spectroscopy to study molecular structures and hyperfine interactions in molecular spectra; instrumentation development for molecular beam; pulsed fourier transform microwave spectroscopy. *Mailing Add:* Dept Chem Univ Mich Ann Arbor MI 48109

HILLIG, WILLIAM BRUNO, b Detroit, Mich, Oct 3, 24; m 49; c 3. THEORY OF FRACTURE, COMPOSITES. *Educ:* Univ Mich, BS, 44, MS, 48, PhD(phys chem), 54. *Prof Exp:* Res asst inorg mat, Mass Inst Technol, 44-47; instr chem, Univ Mich, 49-52; scientist phys chem, Gen Elec Res & Develop Ctr, 53-63, liaison scientist mat, 63-65, scientist phys chem, 65-69, mgr composites, 69-80, scientist phys chem, 80-90; CONSULT & VIS PROF MAT ENG, RENSSELAER POLYTECHNIC INST, 90- *Concurrent Pos:* Mem bd assessment, Nat Bur Standards Progs, Mat Sci, 81-87; vol, Tech Assistance, 59-; mem, Assessment Team, Japanese Composite Technol, 90. *Honors & Awards:* S B Meyer Award, Am Ceramic Soc, 63. *Mem:* Fel Am Ceramic Soc; fel AAAS; Am Chem Soc; NY Acad Sci. *Res:* Theory of strength and fracture of materials; ceramic composites; impact phenomena; solidification; kinetics of phase transformations; vitreous state; high temperature materials and coatings; solar reflector devices; magnetochemistry. *Mailing Add:* Rensselaer Polytechnic Inst Troy NY 12180-3590

HILLIKER, DAVID LEE, b Los Angeles, Calif, Sept 1, 35. SYSTEMS PROGRAMMING. *Educ:* Univ Calif, Los Angeles, AB, 59, MA & PhD(math), 65. *Prof Exp:* Asst prof math, Univ Calif, Irvine, 66-68; asst prof, Pa State Univ, 68-70; assoc prof math, Cleveland State Univ, 70-75; mem staff comput sci, Calif State Univ, Fullerton, 75-79; mem staff math, Univ Calif, Los Angeles, 79-81; mem staff, Dominguez Hills, 81-84, MEM STAFF COMPUT SCI, CALIF STATE UNIV, LONG BEACH, 84- *Concurrent Pos:* NSF grant, 68-69. *Mem:* Am Math Soc; Math Asn Am; Asn Comput Mach. *Res:* Diophantine equations; transcendental numbers; analytic number theory. *Mailing Add:* 2415 Fordham Dr Costa Mesa CA 92626

HILLIS, ARGYE BRIGGS, b Borger, Tex, July 27, 33; m 52; c 3. MEDICAL STATISTICS. *Educ:* Towson State Col, BS, 68; Johns Hopkins Univ, PhD(biostatist), 74. *Prof Exp:* Asst prof social & prev med, Univ Md, 74-78; asst prof ophthal & statist, Johns Hopkins Univ, 78-83; asst prof, 82-84, ASSOC PROF MED MICROBIOL & IMMUNOL STATIST, TEX A&M UNIV, 84- *Concurrent Pos:* Prin investr, NIH, 78-82 & 88-; ad hoc consult, Nat Eye Inst, NIH, 78; mem panel, NIH Consensus Conf, 88; chmn Biostatist Comt Intraocular Lenses, Food & Drug Admin, 80-82; mem Coun Epidemiol, Am Diabetes Asn; mem, Vision Res Rev Comt, Nat Eye Inst, NIHV, 89-90. *Mem:* Am Statist Asn; Biometrics Soc; Asn Res Vision & Ophthal; Am Pub Health Asn; Soc Clin Trials. *Res:* Design, organize and coordinate collaborative clinical trials, primarily in ophthalmology; clinically oriented research in collaboration with physicians; develop statistical methodology for these studies; mathematical modeling of disease processes and epidemiological phenomena. *Mailing Add:* Scott & White Hosp Res & Educ 2401 S 31st St Temple TX 76508

HILLIS, DAVID MARK, b Copenhagen, Denmark, Dec 21, 58; US citizen; m 80; c 2. EVOLUTION, SYSTEMATICS. *Educ:* Baylor Univ, BS, 80; Univ Kans, MA, 83, MPh, 84, PhD(biol sci), 85. *Prof Exp:* Asst prof biol, Univ Miami, 85-87; asst prof, 85-87, ASSOC PROF ZOOL, UNIV TEX, AUSTIN, 89- *Concurrent Pos:* NSF presidential young investr award, 87; mem, adv panel, Biol Res Ctr Prog, NSF, 87-88, adv panel, Res Training Groups Prog, 90-91 & Species Survival Comn, World Conserv Union, 90-; ed, Syst Zool, 89- *Mem:* Soc Syst Zoologists; Am Soc Ichthyologists & Herpetologists; Soc Study Evolution; AAAS; Herpetologists League; Am Malacological Union. *Res:* Molecular aspects of evolutionary biology; phylogenetic analysis; molecular mechanisms of evolution; systematic theory. *Mailing Add:* Dept Zool Texas Univ Austin TX 78712-1064

HILLIS, LLEWELLYA, b Windsor, Ont, Jan 17, 30; m 61; c 2. CORAL REEFS, MOLECULAR EVOLUTION. *Educ:* Queen's Univ, Ont, BA Hons, 52; Univ Mich, MS, 53, PhD(bot), 57. *Hon Degrees:* DSc, Queen's Univ, 85. *Prof Exp:* Nat Res Coun Can fel, Univ NB, 57-59; asst prof bot & biol, Univ Victoria, 59-61; vis instr bot, Duke Univ, 61-62 & Queen's Univ, Belfast, 62-63; phycologist, Int Indian Ocean Exped Cruise, 63; guest investr biol, Yale Univ, 64; asst prof bot, 64-69, adj prof zool, 69-72, asst prof, 72-79, ASSOC PROF ZOOL, OHIO STATE UNIV, 79- *Concurrent Pos:* Consult, NY Bot Gardens, 62-65; Founder's fel, Am Asn Univ Women, 71-72 & sci fel, Bunting Inst, Harvard Univ, 85-87; mem ocean sci panel, Comt Instnl Coop, 74-; nat scholar, People's Repub China, 85; corp mem, Marine Biol Lab, Woods Hole. *Mem:* AAAS; Phycol Soc Am; Marine Biol Asn UK; Brit Phycol Soc; Int Phycol Soc; Int Soc Reef Studies; Biomineral Asn. *Res:* Marine biology with emphasis on primary and carbonate productivity, coral reefs, ecology and taxonomy; marine algal genus Halimeda; marine phycology, particularly ecology, culturing, taxonomy. *Mailing Add:* Dept Zool Ohio State Univ 1735 Neil Ave Columbus OH 43210

HILLIS, MARY OLIVE, analytical chemistry; deceased, see previous edition for last biography

HILLIS, WILLIAM DANIEL, SR, b Paris, Ark, June 12, 33; m 52; c 3. VIROLOGY, NEPHROLOGY. *Educ:* Baylor Univ, BS, 53; Johns Hopkins Univ, MD, 57. *Prof Exp:* Intern internal med, Johns Hopkins Hosp, Baltimore, Md, 57-58; virologist, USAF Sch Aerospace Med, 60-61, chief epidemiol, Air Force Epidemiol Lab, Lackland AFB, Tex, 62-65; asst prof to assoc prof med, Sch Med, 72-82, ASSOC PROF PATHOBIOL, SCH HYG, JOHNS HOPKINS UNIV, 68-; PROF & CHMN BIOL, BAYLOR UNIV, 82- *Concurrent Pos:* Fel pathobiol, Sch Hyg, Johns Hopkins Univ, 58-60; vis scientist, State Serum Inst, Copenhagen, Denmark, 58-60, Yerkes Labs Primate Biol, Orange Park, Fla, 61-62 & Delta Regional Primate Res Ctr, Tulane Univ, 64-65; chief virol, Johns Hopkins Univ Ctr for Med Res & Training, Calcutta, India, 68-70, resident coordr, 70; resident renology, Johns Hopkins Hosp, Baltimore, 71-72, dir Outpatient Dept, Clin Res Ctr, 73-82; outstanding prof, Baylor Univ, 85, exec vpres, 85- *Honors & Awards:* Louis Livingston Seaman Prize, Asn Mil Surgeons US. *Mem:* AAAS; Am Soc Microbiol; Aerospace Med Asn; Soc Exp Biol & Med; NY Acad Sci; Sigma Xi. *Res:* Epidemiology of virus infections of man and animals; in vitro growth and characterization of viruses; serological and immunological aspects of virus infections; pathology of renal diseases of man; clinical aspects of hemodialysis and transplantation. *Mailing Add:* PO Box 97016 Baylor Univ Waco TX 76798-7016

HILLMAN, ABRAHAM P, b Brooklyn, NY, Dec 18, 18; m 55. MATHEMATICS. *Educ:* Brooklyn Col, BA, 39, MA, 40; Princeton Univ, PhD(math), 50. *Prof Exp:* Asst mathematician, Nat Bur Standards, 41-43, mathematician, 45-48; asst, Princeton Univ, 43-44, instr, 44-45; instr, Columbia Univ, 50-56, asst prof, Wash State Univ, 56-57; from asst prof to assoc prof, Univ Santa Clara, 57-65; assoc prof, 65-67, PROF MATH, UNIV NMEX, 67- *Mem:* Math Asn Am; Am Math Soc; Fibonacci Asn. *Res:* Differential algebra; combinatorial analysis. *Mailing Add:* 709 Solano Dr SE Albuquerque NM 87108

HILLMAN, DEAN ELOF, b Shell Lake, Wis, Nov 26, 36; m 59; c 3. NEUROBIOLOGY. *Educ:* Gustavus Adolphus Col, BS, 59; Univ NDak, MS, 62, PhD(anat), 64. *Prof Exp:* NIH fel, Marquette Univ, 64-65; asst prof anat, Sch Med, Wayne State Univ, 65; asst mem neurobiol, Inst Biomed Res, AMA, 67-70; from assoc prof to prof physiol & biophys, Univ Iowa, 70-76; PROF PHYSIOL, SCH MED, NY UNIV, 76- *Concurrent Pos:* Mem study sect neurol B, NIH, 71-75. *Mem:* Soc Neurosci; AAAS; Am Asn Anatomists; NY Acad Sci. *Res:* Neuronal form; components and connections of the cerebellum and vestibular system studied by use of Golgi impregnations, electron microscopy, degeneration techniques and three dimensional reconstruction. *Mailing Add:* Dept Physiol NY Univ Sch Med 550 First Ave New York NY 10016

HILLMAN, DONALD, b Deckerville, Mich, Jan 10, 29; m 51; c 7. DAIRY NUTRITION, ANIMALS. *Educ:* Mich State Univ, BS, 51, MS, 56, PhD(dairy nutrit), 59. *Prof Exp:* Instr voc agr, Pub Schs, Mich, 51-53; 4-H Club Agent Genesee County, Mich State Univ, 53-55, exten dairy specialist, 55-82; PVT CONSULT RES & SALES, 82- *Concurrent Pos:* Int consult fluoride & iodide toxicities, animal nutrition, agr educ, animal waste mgt & odor control. *Mem:* Am Dairy Sci Asn; Am Soc Animal Sci. *Res:* Feed preservation, processing and feeding value; nutrient requirements for growth and milk production, computer ration formulation; cattle feeding and management; fluoride and iodide toxicities; animal waste management and odor control. *Mailing Add:* 750 Berkshire Lane E Lansing MI 48823-2745

HILLMAN, DONALD ARTHUR, b Montreal, Que, June 25, 25; m 56; c 5. PEDIATRICS, ENDOCRINOLOGY. *Educ:* McGill Univ, BSc, 49, MD, 51, PhD(endocrinol), 61. *Prof Exp:* Assoc prof pediat, McGill Univ, 58-76, assoc dean, 71-74; CHMN DEPT & PROF PEDIAT, MEM UNIV NFLD, 76- *Concurrent Pos:* Chief physician, Janeway Child Health Ctr, 76- *Mem:* Soc Pediat Res. *Res:* Pediatric endocrinology. *Mailing Add:* Dept Pediat Mem Univ Prince Philip Dr St Johns NF A1B 3V6 Can

HILLMAN, ELIZABETH S, b Ont; m 55; c 5. PEDIATRICS, TOXICOLOGY. *Educ:* Univ Western Ont, MD, 51. *Prof Exp:* Dir ambulatory pediat, Montreal Children's Hosp, 57-69, dir poison ctr, 58-69; asst prof pediat, McGill Univ, 60-69; sr lectr, Univ Nairobi, 69-71; assoc prof pediat, McGill Univ, 71-74; dir, Ambulatory Pediat & Poison Ctr, Montreal Children's Hosp, 71-74; sr lectr pediat, Univ Nairobi, Kenya, 74-76; PROF PEDIAT, MEM UNIV, ST JOHNS, NFLD, 76- *Concurrent Pos:* Pres, Med Coun Can, 81-82. *Mem:* Can Paediat Soc; Am Acad Pediat. *Res:* Community pediatrics; prevention of poisoning. *Mailing Add:* Dept Pediat Mem Univ Prince Philip Dr St Johns NF A1B 3V6 Can

HILLMAN, GILBERT R, b New Haven, Conn, May 1, 43; m 65; c 1. NEUROPHARMACOLOGY, BIOCHEMISTRY. *Educ:* Harvard Univ, BA, 65; Yale Univ, PhD(pharmacol), 69. *Prof Exp:* Conn Heart Asn fel, Yale Univ, 69-70; instr, 70-71, asst prof med sci, Brown Univ, 71-76; assoc prof, 76-85, PROF PHARMACOL, UNIV TEX MED BR, GALVESTON, 85-, ASSOC DIR ACAD COMPUT, 89- *Concurrent Pos:* Res grants, NIH & Dept Defense. *Mem:* Am Soc Trop Med & Hyg; AAAS; Am Soc Pharmacol & Exp Therapeut; NY Acad Sci; Inst Elec & Electronics Engrs. *Res:* Pharmacology of synapses; computer applications in histochemistry; computer image analysis. *Mailing Add:* Dept Pharmacol Univ Tex Med Br Galveston TX 77550

HILLMAN, MANNY, b St Louis, Mo, Nov 8, 28; m 53; c 2. ORGANOMETALLIC & NUCLEAR CHEMISTRY. *Educ:* Brooklyn Col, AB, 49; Wash Univ, PhD(chem), 53. *Prof Exp:* Instr chem, Brooklyn Col, 49-50; asst, Wash Univ, 50-52, instr, 53; res fel, Wayne State Univ, 55-56; sr res chemist, Olin Mathieson Chem Co, 56-59; chemist, 59-62, group leader, Brookhaven Nat Lab, 62-86, CONSULT, 86- *Concurrent Pos:* Asst, Brookhaven Nat Lab, 51; Nat Acad Sci exchange fel, US & Hungary, 74; adj prof, Southampton Col, 74-; vis prof, Weizmann Inst Sci, 75-76; adj prof, State Univ NY, Binghamton, 81- *Mem:* AAAS. *Res:* Metallocene chemistry; organic chemical reactions; chemistry of boranes; numerical analysis; nuclear structure and reactions; hot atom chemistry. *Mailing Add:* 22 Bayview Ave Blue Point NY 11715

HILLMAN, RALPH, b Bridgeport, Conn, Apr 9, 29; m 60; c 1. GENETICS, MOLECULAR BIOLOGY & EMBRYOLOGY. *Educ:* Stanford Univ, BA, 51; Yale Univ, PhD, 57. *Prof Exp:* Asst zool, Yale Univ, 51-56; instr, Univ Conn, 56-57; instr biol, Univ Pa, 57-60; from asst prof to assoc prof, 60-71, chmn dept, 66-67 & 85-88, PROF BIOL, TEMPLE UNIV, 71- *Concurrent Pos:* Vis scientist, Univ Sussex, 70. *Mem:* Genetics Soc Am; Am Soc Zool; Sigma Xi; fel AAAS. *Res:* Developmental and physiological genetics. *Mailing Add:* Dept Biol Temple Univ Philadelphia PA 19122

HILLMAN, RICHARD EPHRAIM, b Pawtucket, RI, Oct 6, 40; m 70; c 6. METABOLISM, MEDICAL GENETICS. *Educ:* Brown Univ, AB, 62; Yale Univ, MD, 65; Am Bd Pediat, dipl, 71. *Prof Exp:* Fel genetics & metab, Yale Univ, 67-69; staff pediat, metab & genetics, Nat Naval Med Ctr, 69-71; from asst prof to assoc prof pediat, Sch Med Wash Univ,71-78 assoc prof genetics, 78-81, dir, Endocrinol & Metab, 79-85, prof pediat & genetics 78-87, prof genetics, 81-87, dir genetics, 85-87; PROF CHILD HEALTH & BIOCHEM, UNIV MO, COLUMBIA, 87- *Concurrent Pos:* Dir, Metab Screening Labs, St Louis Children's Hosp, 71-87; dir metab genetics, Univ Mo, Columbia, 87- *Mem:* Soc Pediat Res; Am Acad Pediat; Am Soc Human Genetics; Am Soc Clin Invest; Soc Inherited Metab Disorders (pres-elect, 81, pres, 82). *Res:* Study of metabolism of amino acids by cultured cells. *Mailing Add:* Dept Child Health Univ Mo One Hosp Dr Columbia MO 65212

HILLMAN, ROBERT B, b Hancock, NY, Jan 23, 30; m 64; c 3. VETERINARY MEDICINE. *Educ:* Syracuse Univ, BA, 51; Cornell Univ, DVM, 55, MS, 61. *Prof Exp:* Pvt & mixed pract, NY, 57-58; med intern large animal med, ambulatory clin, 58-60, Nat Inst Allergy & Infectious Dis res fel, 60-61, from asst prof to assoc prof large animal med, 61-71, SR CLINICIAN, NY STATE VET COL, CORNELL UNIV, 71- *Mem:* Am Vet Med Asn; NY Acad Sci; Am Asn Equine Practrs; Am Asn Bovine Practrs. *Res:* Bovine mycotic abortion; plant poisonings in large animals; artificial insemination in mares; cyclic and seasonal changes in hormone levels in mare; bovine fetal serology; induction of parturition in mares. *Mailing Add:* NY State Vet Col Cornell Univ Ithaca NY 14850

HILLMAN, ROBERT EDWARD, b Brooklyn, NY, Nov 5, 33; div; c 2. INVERTEBRATE ZOOLOGY, MARINE BIOLOGY. *Educ:* Hofstra Univ, BA, 55, MA, 58; Univ Del, PhD(zool), 62. *Prof Exp:* Assoc invert zool, Marine Biol Lab, Univ Del, 61-62; res assoc shellfish genetics, Natural Resources Inst, Univ Md, 62-64, res asst prof, 64-65; marine biologist, 65-68, div chief, 68-73, sr marine biologist, William F Clapp Labs, 73-78, res leader, 78-84, SR RES SCIENTIST, BATTELLE MEM INST, 84- *Concurrent Pos:* NSF grant, Univ Md, 64-66; ed, Proc Nat Shellfisheries Asn, 76-81, J Shellfish Res, 81-83. *Mem:* AAAS; Am Soc Zoologists; Soc Invert Path; Nat Shellfisheries Asn (vpres, 83-84, pres-elect, 84-85, pres, 85-86); NY Acad Sci. *Res:* Genetics, histology, histochemistry and histopathology of marine invertebrates; oceanography and estuarine ecology. *Mailing Add:* Battelle Mem Inst Duxbury MA 02332

HILLMAN, STANLEY SEVERIN, b Chicago, Ill, Feb 4, 48; m 76; c 2. COMPARATIVE PHYSIOLOGY, PHYSIOLOGICAL ECOLOGY. *Educ:* Calif State Univ, Fullerton, BA, 70, MA, 72; Univ Calif, Los Angeles, PhD(biol), 77. *Prof Exp:* Lectr biol, Univ Calif, Riverside, 76-77; from asst prof to assoc prof, 77-85, PROF BIOL, PORTLAND STATE UNIV, 85- *Mem:* Am Soc Ichthyologists & Herpetologists; Am Physiol Soc; Am Soc Zoologists. *Res:* Cardiovascular consequences of hyperosmolality and hypovolemia; physiology of activity metabolism in lower vertebrates. *Mailing Add:* Dept Biol Portland State Univ Portland OR 97207

HILLMAN, WILLIAM SERMOLINO, plant physiology; deceased, see previous edition for last biography

HILLNER, EDWARD, b Bronx, NY, Apr 1, 29; m 51; c 2. METALLURGICAL CHEMISTRY, METALLURGICAL ENGINEERING. *Educ:* City Col New York, BS, 51; NY Univ, MS, 53, PhD(chem), 55. *Prof Exp:* Metall engr, 57-59, sr metall engr, 59-71, fel engr, 71-82, ADV ENGR, BETTIS ATOMIC POWER DIV, WESTINGHOUSE ELEC CORP, 82- *Concurrent Pos:* Consult, Elec Power Res Inst, 78-80. *Mem:* Nat Asn Corrosion Engrs. *Res:* High temperature oxidation and corrosion of refractory metals and plant materials. *Mailing Add:* PO Box 79 West Mifflin PA 15122

HILLOOWALA, RUMY ARDESHIR, b Surat, India, Dec 20, 35; m 65; c 2. ANATOMY, PHYSICAL ANTHROPOLOGY. *Educ:* Univ Bombay, BDent Surg, 60; Univ Pa; Howard Univ, MS, 65; Univ Ala, Birmingham, PhD(anat), 70. *Prof Exp:* Intern oral surg, Goculdas Tejpal Hosp, Bombay, 60; resident anesthesiol, Hosp, George Washington Univ, 61-62; scientist, Bionetics Res Labs, 64-66; instr anat, Med Ctr, Univ Ala, Birmingham, 66-69; from instr to asst prof, 70-75, ASSOC PROF ANAT & ANTHROP, WVA UNIV, 75- *Mem:* Am Asn Anat; Am Asn Phys Anthrop; Int Primatological Soc; Am Asn Hist Med; Smithsonian Inst; Am Asn Dent Schs. *Res:* Evolution and phylogeny in relation to human anatomy; primate anatomy; cranio-facial development (post-natal) in fossil man, apes and modern human; history of anatomy especially in relation to Renaissance art; XVIII century anatomical wax models. *Mailing Add:* Dept Anat WVa Univ Med Ctr Morgantown WV 26506

HILLS, CLAUDE HIBBARD, b Kirksville, Mo, June 21, 12; m 40; c 4. FOOD SCIENCE. *Educ:* Northeast Mo State Teachers Col, BS, 33; Univ Mo, MA, 35; Univ Minn, PhD(agr biochem), 37. *Prof Exp:* Asst horticulturist, Univ Mo, 34-35; asst biochemist, Univ Minn, 36-37; res chemist, Borden Co, NY, 37-38; Hormel res fel, Univ Minn, 38-40; assoc chemist, protein div, Bur Agr Chem & Eng, 41-42 & Bur Agr & Indust Chem, Sci & Educ Admin, Agr Res, USDA, 42-45, chemist, 45-53, chemist, Eastern Regional Res Lab, 53-77, head fruit prod sect, 48-73, head maple syrup sect, 73-77; RETIRED. *Honors & Awards:* Superior Serv Award, US Dept Agr, 67. *Mem:* Am Chem Soc; Am Soc Hort Sci; assoc mem NAm Maple Syrup Coun; Inst Food Technologists. *Res:* Tobacco mosaic virus; amylases of barley; milk solids in baked products; meat processing; pectin; fruit products; maple syrup. *Mailing Add:* 510 E Valley Green Rd Flourtown PA 19031

HILLS, F JACKSON, b Oakland, Calif, Feb 27, 19; m 43; c 4. SUGARBEET AGRONOMY & PATHOLOGY, EXPERIMENTAL DESIGN. *Educ:* Univ Calif, Berkeley, BS, 41, Univ Calif, Davis, MS, 51, PhD(plant path), 61. *Prof Exp:* Asst field supt, sugar beet prod, Spreckels Sugar Co, 41-42, field supt, 46-48, agronomist, 48-51; EXTEN AGRONOMIST, UNIV CALIF, DAVIS, 51- *Honors & Awards:* Distinguished Serv Award, Am Soc Sugar Beet Technologists. *Mem:* Am Soc Agron; Am Phytopath Soc; Am Soc Sugar Beet Technol. *Res:* Sugar beet agronomy and pathology; design and analysis of experiments. *Mailing Add:* Agron & Range Sci Exten Univ Calif Davis CA 95616

HILLS, FRANCIS ALLAN, b Charleston, SC, Aug 17, 34; m 62; c 3. URANIUM GEOCHEMISTRY, PETROLOGY GRANITE. *Educ:* Univ NC, BS, 56; Yale Univ, PhD(geol), 65. *Prof Exp:* Res fel geol, Univ Minn, 62-65; res staff geologist, Yale Univ, 65-68; asst prof geol, State Univ NY Buffalo, 68-75; GEOLOGIST, US GEOL SURV, DENVER, COLO, 75- *Mem:* AAAS; Geol Soc Am. *Res:* Uranium in Precambrian rocks; geochronology and tectonics of Proterozolc-Archean boundary; origin of Anorthosite-syenite; uranium and gold in Precambrian conglomerate; geochemistry of granite; geochronology. *Mailing Add:* US Geol Surv Stop 939 Box 25046 Denver Fed Ctr Denver CO 80225

HILLS, GRAHAM WILLIAM, b Slough, Eng, May 19, 49; m 72; c 1. CHEMICAL PHYSICS, PHYSICAL CHEMISTRY. *Educ:* Cambridge Univ, BA, 71, PhD(phys chem), 75. *Prof Exp:* Res assoc phys chem, Rice Univ, 74-77; res assoc astrophysics, HIA, NRC, Ottawa, Ont, 77-78; asst prof chem physics, Univ NC, 78-; AT AT&T BELL LAB, ALLENTOWN, PA. *Concurrent Pos:* Lindemann trust fel, Eng Speaking Union, Rice Univ, 74-76. *Mem:* Am Chem Soc. *Res:* Laser spectroscopy of small gas molecules using saturation, double resonance and fluorescence techniques; free radicals and molecular ions. *Mailing Add:* 100 Vasona Oaks Dr Los Gatos CA 95030-2320

HILLS, HOWARD KENT, b Mt Pleasant, Iowa, Sept 21, 38; m 60; c 3. SPACE SCIENCE. *Educ:* Univ Iowa, BA, 61, MS, 64, PhD(space physics), 67. *Prof Exp:* Asst prof physics, Iowa Wesleyan Col, 67-68; res assoc, Rice Univ, 68-71, sr res assoc space sci, 71-78; SR SCIENTIST, SIGMA DATA SERV CORP, 78- *Mem:* Am Geophys Union. *Res:* Particles and fields in the earth's magnetosphere, in interplanetary space and in the lunar surface environment. *Mailing Add:* Nat Space Sci Data Ctr Goddard Space Flight Ctr Code 633 Greenbelt MD 20771

HILLS, JACK GILBERT, b Keflavik, Iceland, May 15, 43; US citizen; m; c 1. THEORETICAL ASTROPHYSICS. *Educ:* Univ Kans, AB, 66, MA, 67; Univ Mich, Ann Arbor, MS, 67, PhD(astron), 69. *Prof Exp:* From instr to asst prof astronm Univ Mich, Ann Arbor, 69-76; asst prof, Univ Ill, Urbana, 76-77; from assoc prof astron-astrophys to prof physics & astron, Mich State Univ, East Lansing, 77-83; Dep Group Leader, 83-90, MEM STAFF, 81-, ACTG GROUP LEADER, THEORET ASTROPHYS GROUP, LOS ALAMOS NAT LAB, 91- *Mem:* Am Astron Soc; Royal Astron Soc; Int Astron Union; fel Am Phys Soc. *Res:* Stellar dynamics; quasi-stellar objects. *Mailing Add:* Los Alamos Nat Lab PO Box 1663 Mail Stop B288 Los Alamos NM 87545

HILLS, JOHN MOORE, b Oak Park, Ill, Mar 15, 10; c 3. GEOLOGY. *Educ:* Lafayette Col, BS, 31; Univ Chicago, PhD(geol), 34. *Prof Exp:* Asst geologist, Amerada Petrol Corp, 34-41; prof, Univ Tex, El Paso, 67-80; CONSULT GEOLOGIST, 41-; EMER PROF GEOL, UNIV TEX, EL PASO, 80- *Concurrent Pos:* Vis lectr, Univ Tex, 59-60; partner, Penn, Hills & Turner, Consult Geologists, 67- *Mem:* Fel AAAS; fel Geol Soc Am; Am Inst Mining, Metall & Petrol Engrs; hon mem Am Asn Petrol Geologists; Sigma Xi. *Res:* Permian stratigraphy; oil field structure; microscopic examination of well cuttings; petroleum geology; oil and gas property exploration and appraisal; structural geology of West Texas and New Mexico. *Mailing Add:* Dept Geol Sci Univ Tex El Paso TX 79968

HILLS, LEONARD VINCENT, b Judah, Alta, Jan 3, 33; m 59; c 2. GEOLOGY. *Educ:* Univ BC, BSc, 60, MSc, 62; Univ Alta, PhD, 65. *Prof Exp:* Explor palynologist, Shell Can, 65-66; asst prof to assoc prof, 66-74, PROF GEOL, UNIV CALGARY, 74- *Concurrent Pos:* Ed, Arctic, Jour of Arctic Inst NAm; co-ed, Arctic Profiles, 88-; adj res scientist, Tyrrell Mus Palaeont, 88-; adj prof, fac environ design, Univ Calgary. *Mem:* Can Soc Petrol Geologists (pres, 79); Am Asn Stratig Palynologists; fel Arctic Inst NAm. *Res:* Devonian-Tertiary-Pleistocene palynology; stratigraphy. *Mailing Add:* Dept Geol & Geophys Univ Calgary Calgary AB T2N 1N4 Can

HILLS, LORAN C, b Oacoma, NDak, May 15, 25; m 47; c 3. BIOCHEMISTRY. *Educ:* Black Hills State Col, BS, 50; Stanford Univ, MA, 58; SDak State Univ, 63-64. *Prof Exp:* Teacher, high schs, Wyo, 50-51, SDak, 51-61; from instr to assoc prof, 61-70, dir, Med Lab Technician Prog, 71-81, PROF CHEM, DAKOTA WESLEYAN UNIV, 70-, HEAD DEPT, 65-, HEAD, NAT SCI DIV, 70-, DIR, COMPUT SCI PROG, 81- *Mem:* Sigma Xi. *Res:* Selenium metabolism. *Mailing Add:* 1313 E Fifth Ave Mitchell SD 57301

HILLS, ROBERT, JR, optics; deceased, see previous edition for last biography

HILLS, STANLEY, b New York, NY, June 4, 30; m 53; c 2. ELECTROCHEMISTRY. *Educ:* City Col New York, BS, 52; Duke Univ, MA, 54, PhD(phys chem), 56. *Prof Exp:* Res electrochemist, Reynolds Metals Co, 55-56 & Yardney Elec Co, 56-61; instr chem, City Col New York, 58-60; head, silver cadmium lab, Elec Storage Battery Co, 61-66, head basic develop chem, 66-69; sr electrochemist, Switchgear Bus Div, 69-83, PROCESS DEVELOP ENGR, SPACE SYSTS DIV, GEN ELEC CO, 83- *Concurrent Pos:* Co-adj assoc prof chem, Rutgers Univ, 71-74. *Mem:* Am Chem Soc; Electrochem Soc; Sigma Xi. *Res:* Plating processes; corrosion studies; pollution control; electrodeposition of alloys and organic polymers. *Mailing Add:* 1224 Heartwood Dr Cherry Hill NJ 08003

HILLSMAN, MATTHEW JEROME, b Crucible, Pa, Oct 1, 35; m 61; c 5. PHYSICS, ADMINISTRATIVE MANAGEMENT. *Educ:* Pa State Univ, BS, 61. *Prof Exp:* Physicist optical systs, Naval Underwater Sound Lab, Conn, 61-67; physicist & bio-instrumentation technologist biomed measurements & instrumentation systs technol, Electronics Res Ctr, NASA, 67-70; PHYSICIST & ADMIN MGR OPTICAL SYSTS & ADMIN MGT, NAVAL UNDERWATER SYSTS CTR, CONN, 71- *Concurrent Pos:* Physicist & consult biomed systs, 69-71. *Mem:* AAAS; Optical Soc Am; Am Inst Physics. *Res:* Submarine optical and electromagnetic communication systems, including near infrared communications, low light level telescopes and television, periscopes and radio communications. *Mailing Add:* PO Box 2886 Portales NM 88130

HILLSON, CHARLES JAMES, b Monroeville, Ohio, Dec 18, 26; m 58; c 1. PLANT ANATOMY, PLANT MORPHOLOGY. *Educ:* Bowling Green State Univ, BS(educ) & BS(biol), 50; Univ Miami, MS, 52; Pa State Univ, PhD(bot), 57. *Prof Exp:* From instr to prof bot, Buckhout Lab, 53-89, EMER PROF BOT, SOUTH FREAR LAB, PA STATE UNIV, 89- *Mem:* AAAS; Bot Soc Am; Phycol Soc Am; Int Oceanog Found. *Res:* Developmental anatomy; morphology, particularly the phylogenetic specialization of vascular tissue; morphology of lower plants. *Mailing Add:* Dept Biol 302 S Frear Lab Pa State Univ University Park PA 16802

HILLSTROM, WARREN W, b Chicago, Ill, July 29, 35; m 63. SURFACE CHEMISTRY, ORGANIC CHEMISTRY. *Educ:* Earlham Col, BA, 58; Univ Cincinnati, PhD(org chem), 63. *Prof Exp:* Asst proj chemist, Am Oil Co, 63-64, develop chemist, Amoco Chem Corp, 64-68; PROJ LEADER, BALLISTIC RES LAB, ABERDEEN PROVING GROUND, 68- *Mem:* Am Chem Soc; Sigma Xi. *Res:* High temperature metal oxidation; solid-solid reactions; lubricant additives; liquid surface effects; liquid pool fires; explosive formulation. *Mailing Add:* 808 Stiles Ct Joppa MD 21085

HILLYARD, IRA WILLIAM, b Richmond, Utah, Mar 23, 24; m 70; c 3. TOXICOLOGY, TERATOLOGY. *Educ:* Idaho State Col, Pocatello, BS, 49; Univ Nebr, Lincoln, MS, 51; St Louis Univ, PhD(pharmacol), 57. *Prof Exp:* Instr, Col Pharm, Univ Nebr, 49-51; dir, Pharm Serv, US Naval Hosp, Oakland, Calif, 51-53; instr, pharmacol dept, Med Sch, St Louis Univ, 53-57; sr pharmacologist, dept pharmacol, Mead Johnson & Co, 57-59; sr res assoc, dept pharmacol, Warner-Lambert Res Inst, 59-69; assoc prof pharmacol, Col Pharm, Idaho State Univ, 69-73; dir, Pharmacol & Toxicol Res, ICN Pharmaceut, Inc, Irvine, Calif, 73-77; assoc prof & dir, poison control & drug info, Idaho Drug Info & Regional Poison Control Ctr, 77-79, dean Col Pharm, 79-86, PROF PHARMACOL & TOXICOL, COL PHARM, IDAHO STATE UNIV, 79- *Concurrent Pos:* Continuing ed lectr, Am Optom Asn, 70-77; sci rev, Drug Interaction Panel, Am Pharmaceut Asn, 73-76; consult, ICN Pharmaceut, Inc, Irvine, Calif, 78-80, Pennwalt Corp, Rochester, NY, 79-82. *Mem:* Am Soc Pharmacol & Exp Therapeut; NY Acad Sci; Am Pharmaceut; Am Asn Col Pharm; Am Acad Pharmaceut Sci. *Res:* Assessment of adverse effects caused by maternal drug-taking on the developing embryo or surving newborn; terato logic evaluations of commonly used-and abused-over-the counter and social drugs. *Mailing Add:* Col Pharm Univ Idaho Pocatello ID 83209

HILLYARD, LAWRENCE R(OBERTSON), b Dows, Iowa, Feb 11, 09; m 34; c 2. INDUSTRIAL ENGINEERING. *Educ:* Iowa State Univ, BS, 32, MS, 36. *Prof Exp:* Teacher pub sch, Iowa, 32-36; from instr to assoc prof gen eng, 36-55, eng personnel officer, 46-59, eng & sci personnel officer, 59-63, placement dir eng, sci & humanities, 63-74, prof indust eng, 56-79, EMER PROF INDUST ENG, IOWA STATE UNIV, 56- *Mem:* Am Soc Eng Educ. *Res:* Technical placement. *Mailing Add:* 1006 Roosevelt Ames IA 50010

HILLYARD, STANLEY DONALD, EPITHELIAL TRANSPORT. *Educ:* Univ Calif, Riverside, PhD(biol), 74. *Prof Exp:* ASSOC PROF BIOL, UNIV NEV, 76- *Res:* Changes in the ionic channels in cell membranes of amphibian skin during metamorphosis. *Mailing Add:* Dept Biol Sci Univ Nev Las Vegas Las Vegas NV 89154

HILLYARD, STEVEN ALLEN, b Long Beach, Calif, Oct 8, 42; m 66; c 2. NEUROPSYCHOLOGY. *Educ:* Calif Inst Technol, BS, 64; Yale Univ, PhD(psychol), 68. *Prof Exp:* From asst prof to assoc prof, 71-80, PROF NEUROSCI, UNIV CALIF, SAN DIEGO, 80- *Concurrent Pos:* USPHS grant neurosci, Univ Calif, San Diego, 68-70; consult, NIH, NSF & study sect, NIMH. *Mem:* Fel AAAS; Soc Neurosci. *Res:* Brain electrophysiology and behavior; perception and information processing. *Mailing Add:* Dept Neurosci M-008 Univ Calif San Diego La Jolla CA 92093

HILLYER, GEORGE VANZANDT, b Santurce, PR, Dec 8, 43; US citizen; m 68; c 2. PARASITOLOGY, IMMUNOLOGY. *Educ:* Univ PR, BS, 67; Univ Chicago, PhD(parasite immunol), 72. *Prof Exp:* From asst prof to prof parasitol & immunol, Univ PR, 72-80, assoc prof path, 76-80, prof parasitol immunol, 80-87, prof & chmn dept biol, Col Natural Sci, 81-87, PROF PATH, MED SCI CAMPUS, UNIV PR, RIO PIEDRAS, 80- *Concurrent Pos:* Sr scientist (ad honorem), PR Nuclear Ctr, 72-; consult immunologist, parasitologist, Vet Admin Ctr, San Juan, PR, 73-, Div Res Resources, NIH, 75- & Trop Med & Parasitol Study Sect, NIH, 80-85; dir training prog, Fac Nat Sci, Univ PR, Rio Piedras, 74-; adj assoc prof trop med & parasitol, Sch Pub Health & Trop Med, Tulane Univ, 79-81, prof, 81-; James W McLaughlin vis prof, Univ Tex Med Br, Galveston, 85; Rockefeller Found fel, dept path, Univ Cambridge, Eng, 85; dir parasitic dis res prog, PR, 85-, Lab Parasite Immunol & Path, 87-; mem microbiol & infectious dis rev comt, NIH-NIAID, 89-; pres, Caribbean div, AAAS, 90-91. *Honors & Awards:* Henry Baldwin Ward Medal, Am Soc Parasitologists, 82; Bailey K Ashford Award, Am Soc Trop Med & Hyg, 86; Carrion Award, Soc Microbiol PR. *Mem:* Am Asn Immunologists; Am Soc Parasitologists; Am Soc Trop Med & Hyg; fel Royal Soc Trop Med & Hyg; Am Soc Microbiol; Soc Exp Biol & Med; Soc Microbiol PR (pres, 79-80); Soc Allergy & Immunol PR (pres, 87-91); Sigma Xi; AAAS. *Res:* Immunology of parasitic infections. *Mailing Add:* Dept Path Univ PR Sch Med San Juan PR 00936-5067

HILLYER, IRVIN GEORGE, b Thief River Falls, Minn, Dec 1, 27. HORTICULTURE. *Educ:* NDak Agr Col, BS, 51; Univ Idaho, MS, 53; Mich State Univ, PhD(hort), 56. *Prof Exp:* Asst hort, Mich State Univ, 53-56; from asst prof to assoc prof, 56-72, PROF AGR, SOUTHERN ILL UNIV, CARBONDALE, 72- *Mem:* Am Soc Hort Sci; Am Soc Plant Physiol. *Res:* Physiology of vegetable crops; growth regulators and their effects on flowering and fruiting of plants. *Mailing Add:* Southern Ill Univ Soil Sci & Agr Bldg 0176B Carbondale IL 62901

HILMAS, DUANE EUGENE, b Virginia, Minn, Jan 6, 38; m 61; c 5. RADIATION BIOLOGY, VETERINARY MEDICINE. *Educ:* Univ Minn, BS, 59, DVM, 61; Univ NC, MSPH, 64; Colo State Univ, PhD(radiation biol), 72; Am Bd Vet Pub Health, dipl, 75. *Prof Exp:* Vet pract, Cloquet, Minn, 61-62; asst vet, Martin Army Hosp, Ft Benning, Ga, 62-63; asst food irradiation, Natick Labs, US Army, 64-66, asst chief Med Res Div, Med Res & Develop Command, 66-69, Chief Animal Assessment Div, Med Res Inst Infectious Dis, 72-77, chief Off Irradiated Food, 77-79, res mgr med defense against chem agents, Med Res & Develop Command, 79-82, chief Drug Assessment Div, Med Res Inst Chem Defense, 82-83, dir & comdr, 83-85; SR PROG MGR, BIOL & CHEM TECH CTR, BATTELLE, 85- *Concurrent Pos:* Consult radiobiol, Surg Gen, Dept Army, 75-; clin asst prof, Hershey Med Ctr, Pa State Univ, 76-79; Army Med Dept Liaison Radiation Study Sect, NIH, 78-85. *Honors & Awards:* Army Commendation Medal, US Army Labs, 66; Army Commendation Medal First Oak Leaf Cluster, US Army Med Res & Develop Command, 69; Meritorious Serv Medal, US Army Res Inst Infectious Dis, 77; Legion of Merit, US Army Med Res & Develop Command, 85. *Mem:* Am Vet Med Asn; Sigma Xi; Health Physics Soc; Radiation Res Soc; Soc Exp Biol & Med; Risk Assessment Soc. *Res:* Radiation oncology and experimental therapy; in vivo and in vitro immunobiology; antiviral chemotherapy and combined therapy modalities; radiation and infection; comparative toxicologic assessments in experimental animals; chemical agent pretreatment, therapy and protection; toxicology. *Mailing Add:* 1608 Buckshot Ct Worthington OH 43085

HILMOE, RUSSELL J(ULIAN), b Colman, SDak, Jan 26, 21; m 47; c 4. BIOCHEMISTRY. *Educ:* SDak State Univ, BS, 43; Georgetown Univ, MS, 50, PhD(biochem), 56. *Prof Exp:* Analyst, SDak State Univ, 43-44; bacteriologist, US Army Chem Corps, Ft Detrick, Md, 46-48; biochemist, Nat Inst Arthritis & Metab Diseases, 48-56, res biochemist, 56-64; sci adminr, Nat Inst Gen Med Sci, 64-70, chief biochem sci training grants, 70-74, assoc prog dir, Cellular & Molecular Basis Disease Prog, 74-77; exec officer, Am Soc Biol Chemists, Inc, 78-80, mgr, J Biol Chem, 78-80; staff officer, Comn Human Resources, Nat Res Coun-Nat Acad Sci, 80-82; CONSULT SCI ADMIN, 80- *Mem:* Am Soc Biol Chemists. *Res:* Nucleic acid chemistry, metabolism and enzymology; general enzymology; science administration. *Mailing Add:* 6309 Kirby Rd Bethesda MD 20817

HILPMAN, PAUL LORENZ, b New York, NY, Feb 3, 32; div; c 3. GENERAL ENVIRONMENTAL SCIENCES. *Educ:* Brown Univ, AB, 54; Univ Kans, AM, 56, PhD, 69. *Prof Exp:* Geologist, oil & gas div, State Geol Surv Kans, 56-62, div head, 62-65, sect chief environ geol, 65-74, assoc prof geol, Univ Kans, 70-74; PROF GEOSCI, UNIV MO, KANSAD CITY, 74- *Concurrent Pos:* Dir, Ctr Undergrad Studies, Univ Mo, Kansas City, 87- *Mem:* Fel Geol Soc Am; Am Asn Petrol Geologists; Am Inst Prof Geologists; Asn Eng Geologists; fel AAAS. *Res:* Engineering and environmental geology; urban planning; geology of radon. *Mailing Add:* Geosci Dept Univ Mo 5100 Rock Hill Rd Kansas City MO 64110

HILSENHOFF, WILLIAM LEROY, b Madison, Wis, July 13, 29; m 52; c 3. ENTOMOLOGY. *Educ:* Univ Wis, BS, 51, MS, 52, PhD(entom), 57. *Prof Exp:* Proj assoc entom, 57-60, from asst prof to assoc prof, 60-72, PROF ENTOM, UNIV WIS-MADISON, 72- *Mem:* Entom Soc Am; NAm Benthological Soc; Coleopterists Soc; Am Mosquito Control Asn. *Res:* Biology, ecology and taxonomy of aquatic insects and their use as indicators of water quality. *Mailing Add:* Entom-445 Russell Lab Univ Wis Madison WI 53706

HILSINGER, HAROLD W, b Midland, Mich, May 31, 32; m 75. THEORETICAL PHYSICS. *Educ:* Univ Mich, BS, 54; Univ Conn, MS, 57, PhD(physics), 64. *Prof Exp:* Asst instr physics, Univ Conn, 57-59; asst prof, 62-67, ASSOC PROF PHYSICS, WORCESTER POLYTECH INST, 67- *Res:* Theory of relativity and philosophy of physics. *Mailing Add:* Dept Physics Worcester Polytech Inst Worcester MA 01609

HILST, ARVIN RUDOLPH, b Manito, Ill, June 14, 24; m 44; c 2. AGRONOMY. *Educ:* Univ Ill, BS, 46, MS, 49; Purdue Univ, PhD, 55. *Prof Exp:* Asst, Univ Ill, 48; instr, 49-54, from asst prof to assoc prof, 55-61, asst head dept agron, 66-69, PROF AGRON, PURDUE UNIV, 61-, DIR RESIDENT INSTR & ASSOC DEAN SCH AGR, 69- *Mem:* Am Soc Agron. *Res:* Herbicides; morphology. *Mailing Add:* Sch Agr Rm 125 Purdue Univ West Lafayette IN 47907

HILST, GLENN RUDOLPH, b Meade, Kans, May 1, 23; m 86; c 3. METEOROLOGY. *Educ:* Mass Inst Technol, SB, 48, SM, 49; Univ Chicago, PhD(meteorol), 57. *Prof Exp:* Asst radar meteorol, Mass Inst Technol, 48-49; res meteorologist, Gen Elec Co, 49-52, mgr atmospheric physics, Wash, 54-60; assoc meteorologist, Argonne Nat Lab, 52-54; asst dir weather systs div, Travelers Ins Co, 60, vpres, Travelers Res Ctr, Inc, 61-68, exec vpres, Travelers Res Corp, 68-70; vpres environ res, Aeronaut Res Assocs of Princeton, Inc, 70-75; chief scientist, The Res Corp of New Eng, 75-77; sr mem tech staff, Elec Power Res Inst, 77-80, prog mgr environ physics & chem, 80-87; CONSULT, 87- *Concurrent Pos:* Spec consult to surgeon gen, USPHS, 61; consult, NIH, 63-; mem, Conn Air Pollution Comn, 67-69; mem, US Nat Comt, Int Biol Prog, 67-70; chmn ad hoc comt, Global Network Environ Monitoring; mem environ sci adv comt, NSF, 68-69, chmn, 70; study dir, Nat Acad Sci, 70, mem, Monitoring Panel, Comt Int Environ Progs, 74-77; vis scientist, Am Meteorol Soc, 70; mem, Nat Acid Precipitation Assessment Prog, ORB, 89-91. *Honors & Awards:* Charles Franklin Brooks Award, Am Meteorol Soc, 73. *Mem:* Fel AAAS; fel Am Meteorol Soc; fel Explorers Club. *Res:* Atmospheric turbulence and diffusion; earth-air exchange phenomena; micrometeorology; physical radar meteorology; environmental effects of atmosphere and hydrologic phenomena. *Mailing Add:* W 1404 Country Club Ct Spokane WA 99218-2964

HILT, RICHARD LEIGHTON, b Detroit, Mich, Apr 3, 36; m 65. PHYSICS. *Educ:* Oberlin Col, AB, 58; Univ NC, PhD(physics), 64. *Prof Exp:* Asst prof physics, NC State Univ, 63-64; from asst prof to assoc prof, 64-80, PROF PHYSICS, COLO COL, 80- *Mem:* Am Asn Physics Teachers; Am Geophys Union; Sigma Xi. *Res:* Geophysics. *Mailing Add:* Dept Physics Colo Col Colorado Springs CO 80903

HILTBOLD, ARTHUR EDWARD, JR, b Brooklyn, NY, Apr 24, 25; m 49; c 3. SOIL MICROBIOLOGY. *Educ:* Cornell Univ, BS, 48, PhD, 55; Iowa State Col, MS, 50. *Prof Exp:* Asst agron, Iowa State Col, 48-50; instr, Cornell Univ, 50-54; assoc prof agron, 55-68, PROF AGRON & SOILS, AUBURN UNIV, 68- *Concurrent Pos:* Assoc ed, Agron J; reviewer, Weed Sci; assoc Soil Microbiologist, 55- *Mem:* Am Soc Agron; Soil Sci Soc Am; Weed Sci Soc Am; Am Peanut Res & Educ Soc. *Res:* Microbiological transformations of nitrogen in soils; movement and persistence of herbicides in soil; nitrogen fixation in soybeans and peanuts. *Mailing Add:* Dept Agron Auburn Univ Auburn AL 36849

HILTERMAN, FRED JOHN, b Pittsburgh, Pa, Aug 30, 41; m 61; c 2. EXPLORATION GEOPHYSICS. *Educ:* Colo Sch Mines, Prof Engr, 63, PhD(geophys), 70. *Prof Exp:* Engr, Geophys Serv Ctr, Mobil Oil Corp, 63-66, assoc geophysicist, Explor Serv Ctr, 70-72, activ leader, Field Res Lab, 72-73; ASSOC PROF GEOPHYS, UNIV HOUSTON, 73- *Concurrent Pos:* Consult res, GeoQuest Int, Ltd, 74-, educ, Prof Geophys Inc; assoc ed, Soc Explor Geophysicists, 75- *Mem:* Soc Explor Geophysicists; Europ Asn Explor Geophysicists; Sigma Xi. *Res:* Application of wave propagation models to seismic exploration data processing techniques and interpretational techniques. *Mailing Add:* Geophys Develop Corp 8401 Westheimer Suite 150 Houston TX 77063

HILTIBRAN, ROBERT COMEGYS, b Urbana, Ohio, Sept 24, 20; m 45; c 3. BIOCHEMISTRY. *Educ:* Denison Univ, BS, 48; Univ Kans, MS, 51, PhD(biochem), 54. *Prof Exp:* Instr biochem, Med Sch, Northwestern Univ, 53-55; Herman Frasch Found fel protein chem, Dept Agr Biochem, Ohio State Univ, 55-57; assoc biochemist, ILL Natural Hist Surv, 57-69, biochemist, aquatic biol sect, 69-82; ASSOC PROF AGRON, UNIV ILL, URBANA, 69- *Concurrent Pos:* Res biochemist, Vet Admin Hosp, Hines, Ill, 53-55. *Mem:* Am Chem Soc; Weed Sci Soc Am; Sigma Xi. *Res:* Fish; enzymes of fish liver; muscle tissues; aquatic weed control; biochemistry of brain tissue; isolation and characterization of natural products; water pollution and its effect. *Mailing Add:* 608 E Washington St Urbana IL 61801

HILTNER, WILLIAM ALBERT, b Continental, Ohio, Aug 27, 14; m 39; c 4. ASTROPHYSICS. *Educ:* Univ Toledo, BS, 37; Univ Mich, MS, 38, PhD(astrophys), 42. *Hon Degrees:* DSc, Univ Toledo, 67. *Prof Exp:* Nat Res Coun fel, Univ Chicago, 42-43, from instr to prof astrophys, Yerkes Observ, 43-70; prof, 70-85, chmn dept, 70-83, EMER PROF ASTRON, UNIV MICH, ANN ARBOR, 85-; STAFF ASTROM & PROJ MGR, OBSERV, CARNEGIE INST WASH, PASADENA, 86- *Concurrent Pos:* Mem bd, Asn Univs Res in Astron, 59-71, 74-83, pres bd, 68-71; dir, Yerkes Observ, Univ Chicago, 63-66. *Mem:* Am Astron Soc; Int Astron Union. *Res:* Stellar spectroscopy; photoelectric photometry; electronic image intensification; interstellar polarization of starlight; x-ray astronomy; instrumentation. *Mailing Add:* 801 Berkshire Rd Ann Arbor MI 48104

HILTON, ASHLEY STEWART, b St Marys, Pa, Apr 28, 40; m 62; c 2. ANALYTICAL CHEMISTRY. *Educ:* Pa State Univ, BS, 62; Ohio State Univ, PhD(anal chem), 68. *Prof Exp:* Res scientist, Aerospace Res Labs, USAF, 67-70; res scientist, Firestone Tire & Rubber Co, 70-80, sr res scientist, 80-87, ASSOC SCIENTIST, BRIDGESTONE/FIRESTONE, INC, AKRON, 87- *Mem:* Am Chem Soc. *Res:* Application of mass spectrometry, high performance liquid chromatography and other instrumental analytical techniques to research in polymer chemistry. *Mailing Add:* 7431 Shadyview NW Massillon OH 44646

HILTON, DONALD FREDERICK JAMES, b Calgary, Alta, May 9, 44; m 66; c 2. INSECT ECOLOGY, BEHAVIORAL ECOLOGY. *Educ:* Univ Alta, BSc, 65, PhD(parasitol), 71; Univ Kans, MA, 67; Univ London, DIC, 72. *Prof Exp:* Bacteriologist, Royal Alexandria Hosp, 67-68; NATO fel med entom, Imperial Col, Univ London, 71-73; from asst prof to assoc prof, 73-85, PROF ZOOL, BISHOP'S UNIV, 85- *Mem:* Entom Soc Can; Soc Int Odonatologica; Can Nature Fedn; Entom Soc Am; Royal Entom Soc London. *Res:* Behavioral ecology of insects, especially reproductive behavior of dragonflies. *Mailing Add:* Dept Biol Sci Bishop's Univ Lennoxville PQ J1M 1Z7 Can

HILTON, FREDERICK KELKER, b Harrisburg, Pa, Feb 8, 26; m 50; c 4. ZOOLOGY. *Educ:* Cornell Univ, BS, 50; Johns Hopkins Univ, DSc(ecol), 58. *Prof Exp:* Asst pathobiol, Johns Hopkins Univ, 55-58; from instr to assoc prof, 58-74, PROF ANAT, SCH MED, UNIV LOUISVILLE, 74- *Mem:* Am Asn Anatomists; Soc Study Reproduction; Animal Behav Soc. *Res:* Biochemistry and endocrinology of animal behavior; physiology and endocrinology of reproduction. *Mailing Add:* Dept Anat Univ Louisville Health Sci Ctr Louisville KY 40292

HILTON, H WAYNE, b Salt Lake City, Utah, Aug 24, 23; m 48. ORGANIC CHEMISTRY. *Educ:* Univ Utah, BA, 49; Ohio State Univ, PhD(org chem), 52. *Prof Exp:* Res chemist high polymers, Visking Corp, 52-55; Prin chemist, Exp Sta, Hawaiian Sugar Planter's Asn, 55-76, head, Dept Crop Sci, 76-88; RETIRED. *Concurrent Pos:* Consult, 88- *Mem:* Am Chem Soc; AAAS; Am Soc Agron; Weed Sci Soc. *Res:* Carbohydrates; vinyl high polymers; agricultural chemicals. *Mailing Add:* PO Box 1057 Aiea HI 96701

HILTON, HENRY H, physics, for more information see previous edition

HILTON, HORACE GILL, b Delta, Utah, June 28, 32; m 56; c 9. STATISTICS. *Educ:* Brigham Young Univ, BS, 57; NC State Univ, MS, 60, PhD, 63. *Prof Exp:* From asst prof to assoc prof, 62-75, chmn dept, 69-80, PROF STATIST, BRIGHAM YOUNG UNIV, 75- *Concurrent Pos:* Nat Res Coun fel, 68-69. *Mem:* Am Statist Asn; Biomet Soc. *Res:* Design and analysis of scientific experiments. *Mailing Add:* Dept Statist Brigham Young Univ Provo UT 84602

HILTON, JAMES GORTON, b Baltimore, Md, Sept 21, 23; m 46; c 2. PHARMACOLOGY. *Educ:* Va Polytech Inst, BS, 47; Univ Tenn, MS, 52, PhD(pharmacol), 54. *Prof Exp:* Res assoc pharmacol, Univ Va, 48-50; from actg asst prof to assoc prof, Univ Miss, 53-58; assoc prof, Sch Med, Marquette Univ, 59-61; assoc prof, 61-63, actg chmn, 79-82, PROF PHARMACOL, UNIV TEX MED BR GALVESTON, 63-; CHIEF PHARMACOL, SHRINERS BURNS INST, 76- *Concurrent Pos:* Fulbright res scholar, Biol Ctr, Gulbenkian Inst Sci, Portugal, 68-69; Fulbright lectr, Univ San Agustin, Peru, 59; mem pharmacol & endocrinol fel rev comt, NIH, 64-67; mem comt persistent pesticides, Nat Acad Sci, 67-69; consult, Environ Protection Agency, 70-72. *Mem:* AAAS; Am Physiol Soc; Am Soc Pharmacol & Exp Therapeut; Am Heart Asn; Am Burn Asn. *Res:* Sympathetic ganglion transmission; factors regulating vascular permeability; cardiovascular changes after thermal injury. *Mailing Add:* 2626 Gerol Ct Galveston TX 77551

HILTON, JAMES LEE, b Bristol, Va, Apr 14, 30; m 58; c 3. PESTICIDE CHEMISTRY. *Educ:* Duke Univ, AB, 52; Iowa State Univ, MS, 54, PhD(plant physiol), 55. *Prof Exp:* Plant physiologist, NC State Col exp sta, USDA, 56, plant indust sta, Plant Sci Res Div, 56-72, chief, Pesticide Action Lab, 72-76, dir, Agr Environ Qual Inst, 76-88, ASSOC DIR S ATLANTIC AREA, AGR RES SERV, USDA, 88- *Concurrent Pos:* Ed-in-chief, Weed Sci Soc Am, 78- *Honors & Awards:* Res Award, Wash Acad Sci, 66, Weed Sci Soc Am, 76; Agr Award, Ciba-Geigy, 76. *Mem:* AAAS; Am Soc Plant Physiol; fel Weed Sci Soc Am. *Res:* Mechanism of herbicide action. *Mailing Add:* Richard D Russel Agr Res Ctr Agr Res Serv USDA Box 5677 Athens GA 30613

HILTON, JAMES LEE, b Bristol, Tenn, Apr 14, 30; m 58; c 3. HERBICIDES. *Educ:* Duke Univ, AB, 52; Iowa State Univ, MS, 55, PhD(plant physiol), 56. *Prof Exp:* Res scientist, Agr Res Serv, USDA, Beltsville, Md, 56-72, res leader, 72-76, inst dir, Beltsville, 76-88, ASSOC AREA DIR, AGR RES SERV, USDA, ATHENS, GA, 88- *Concurrent Pos:* Assoc ed, Weed Sci Soc Am, 73-77, ed-in-chief, 78-87. *Honors & Awards:* Spec Award, Weed Sci Soc Am, 87. *Mem:* Fel Weed Sci Soc Am. *Res:* Mechanisms of action of herbicides. *Mailing Add:* Russel Res Ctr USDA PO Box 5677 Athens GA 30613

HILTON, JOHN L, b Woodruff, Ariz, May 25, 27; m 50; c 8. PLASMA PHYSICS. *Educ:* Brigham Young Univ, BA, 52. *Prof Exp:* Proj engr, Sandia Corp, NMex, 52-57; dept head, Aerojet-Gen Nucleonics Div, Gen Tire & Rubber Co, 57-70; proj engr, Cyclotron Corp, 70-84; consult, Univ Pa & Fox Chase Cancer Ctr, 84-86; staff physicist, Physics Int Co, 86-88; ADJ PROF, DEPT STATIST, BRIGHAM YOUNG UNIV, PROVO, UTAH, 86- *Mem:* AAAS; Am Phys Soc. *Res:* Ion sources; accelerators; statistical studies of word patterns. *Mailing Add:* 604 Sagewood Ave Provo UT 84604

HILTON, MARY ANDERSON, b State College, Pa, July 6, 26; m 50; c 4. BIOCHEMISTRY. *Educ:* Pa State Col, BS, 47; Cornell Univ, PhD(biochem), 51. *Prof Exp:* Asst biochem, Cornell Univ, 47-50; asst med, Johns Hopkins Univ, 51-55, instr chem, Hosp Sch Nursing, 57-58; from instr to asst prof, 63-73, ASSOC PROF BIOCHEM, SCH MED, UNIV LOUISVILLE, 74- *Res:* Cell biology; amino acid metabolism. *Mailing Add:* Health Sci Ctr Univ Louisville Sch Med Louisville KY 40292

HILTON, PETER JOHN, b London, Eng, Apr 7, 23; m 49; c 2. MATHEMATICS. *Educ:* Oxford Univ, MA & DPhil, 50; Cambridge Univ, PhD(topology), 52. *Hon Degrees:* DHum, NMich Univ, 77; DSc, Mem Univ, 83; DSc, Autonomous Univ Barcelona, 89. *Prof Exp:* Prof math, Cornell Univ, 62-71 & Univ Wash, 71-73; Beaumont Univ Prof Math, Case Western Reserve Univ, 73-82; DISTINGUISHED PROF, DEPT MATH SCI, STATE UNIV NY, BINGHAMTON. *Concurrent Pos:* Co-chmn, Cambridge Conf Sch Math, 65; fel, Battelle Res Ctr, Seattle, 71-; chmn, US Comn Math Instr, 71-74; chmn comt appl math training, Nat Res Coun, 77-; corresp mem, Brazil Acad Sci, 79. *Honors & Awards:* Silver Medal, Univ Helsinki, 75; Centenary Medal, John Carroll Univ, 85. *Mem:* Am Math Soc; London Math Soc; Royal Statist Soc; hon mem Math Soc Belg; Math Asn Am (first vpres, 78-80). *Res:* Algebraic topology; homological algebra; category theory. *Mailing Add:* Dept Math Sci State Univ NY, Binghamton Binghamton NY 13902-6000

HILTON, RAY, b Houston, Tex, Sept 3, 30; m 51; c 4. PHYSICAL CHEMISTRY. *Educ:* Tex A&M Univ, BS, 53, MS, 57, PhD(chem), 59. *Prof Exp:* Anal chemist, Sinclair Oil & Refining Co, 55-56; res engr, Prod Res Div, Humble Oil & Refining Co, 59-60; SR SCIENTIST, PHYS SCI RES LAB, TEX INSTRUMENTS INC, 60-, MGR INFRARED GLASS LAB, ELECTROOPTICS DIV, 74- *Mem:* Soc Appl Spectros; Optical Soc Am. *Res:* Optical properties of solids, especially semiconductors; solid state chemistry of chalcogen based infrared transmitting glasses; infrared optical materials and electrooptic devices. *Mailing Add:* Amorphous Mat 3130 Benton St Garland TX 75042

HILTON, WALLACE ATWOOD, b Hardin, Mo, Aug 17, 11; m 40; c 2. PHYSICS. *Educ:* William Jewell Col, AB, 33; Univ Mo, AM, 39, EdD, 41; Univ Ark, MS, 48. *Prof Exp:* Teacher, pub sch, Mo, 38-42; asst physics, Univ Ark, 46; prof physics, 46-80, head dept, 46-80, EMER PROF, WILLIAM JEWELL COL, 80- *Concurrent Pos:* Supv res, optical properties of thin films. *Honors & Awards:* Distinguished Serv Citation, Am Asn Physics Teachers, 76, Oersted Medal, 78. *Mem:* AAAS; fel Optical Soc Am; Sigma Xi; Am Asn Physics Teachers. *Res:* Acoustical impedance of acoustical tile; some functions of education at the junior college level; sun spot activity at radio frequencies. *Mailing Add:* 322 Arthur St Liberty MO 64068

HILTUNEN, JARL KALERVO, b Sault Ste Marie, Mich, Apr 14, 33. FRESH WATER ECOLOGY. *Educ:* Wayne State Univ, BA, 59, teaching cert, 62. *Prof Exp:* Fishery biologist, Great Lakes Fishery Lab, US Fish & Wildlife Serv, Ann Arbor, Mich, 62-85; RETIRED. *Mem:* Int Asn Great Lakes Res; NAm Benthological Soc (pres, 77-78). *Res:* Vascular flora and benthic invertebrate fauna of the St Lawrence Great Lakes region. *Mailing Add:* HC57-Box 335 Sault Ste Marie MI 49783

HILTY, JAMES WILLARD, b Wadsworth, Ohio, Oct 11, 36; m 59; c 2. PLANT PATHOLOGY. *Educ:* Ohio State Univ, BSc, 58, MSc, 60, PhD(plant path), 64. *Prof Exp:* From asst prof to assoc prof, 65-76, PROF PLANT PATH, UNIV TENN, KNOXVILLE, 76- *Mem:* Am Phytopath Soc; Sigma Xi; Am Inst Biol Sci. *Res:* Diseases of field crops. *Mailing Add:* Agric Exp Sta Univ Tenn Knoxville TN 37916

HILTZ, ARNOLD AUBREY, b PEI, July 31, 24; nat US; m 46; c 2. PHYSICAL CHEMISTRY. *Educ:* Acadia Univ, BS, 47; McGill Univ, PhD(chem), 52. *Prof Exp:* Defense res sci officer, Defense Res Bd Can, 51-53; res chemist, Am Viscose Corp, 53-59, group leader, 59-60, group leader, Avisun Corp, 60-65; res chemist, Cent Res Lab, Borden Chem Co, 65-66; sr scientist, Reentry Systs Dept, 66-79, mgr mat appln, Space Systs Div, 79-87, ASTRO SPACE DIV, GEN ELEC CO, KING OF PRUSSIA, 87- *Res:* Determination of polymer properties; effects of molecular structure on properties; thermal protection for re-entry vehicles; materials and materials applications for satellites. *Mailing Add:* 524 Cedar Lane Swarthmore PA 19081

HILU, KHIDIR WANNI, b Baghdad, Iraq, Jan 1, 46; m 77; c 2. GRASS SYSTEMATICS, CROP EVOLUTION. *Educ:* Univ Baghdad, BS, 66, MS, 71; Univ Ill, PhD(bot), 76. *Prof Exp:* Asst prof bot, Okla City Univ, 77-78; res assoc biosysts, Univ Calif, Riverside, 78-79, Univ Ill, 79-81; ASSOC PROF BOT, VA POLYTECH INST & STATE UNIV, 81- *Concurrent Pos:* NSF travel grant, 81; assoc curator, Crop Evolution Lab herbarium; consult, UN Int Bd Plant Genetic Resources. *Mem:* Bot Soc Am; Int Asn Plant Taxon; AAAS; Am Soc Naturalists; Soc Econ Bot; Am Soc Plant Taxonomists; Int Soc Plant Molecular Biol. *Res:* Biosystematic and evolution of higher plants in general and grasses in particular with emphasis on using molecular approaches; origin, evolution and systematics of domesticated plants; genetics of morphological characters and macroevolution in flowering plants. *Mailing Add:* Dept Biol Va Polytech Inst & State Univ Blacksburg VA 24061

HIME, WILLIAM GENE, b Dayton, Ohio, Dec 31, 25; m 50; c 3. ANALYTICAL CHEMISTRY. *Educ:* Heidelberg Col, BS, 48. *Prof Exp:* Assoc res chemist, Portland Cement Asn, 51-54; asst prof chem, La Polytech Inst, 54-55; res chemist, Portland Cement Asn, 55-58, supvr anal chem, 58-66, mgr, chem & petrog res sect, 66-71; VPRES, ERLIN, HIME ASSOCS, 71- *Mem:* Am Soc Testing & Mat; Am Chem Soc; Soc Appl Spectros; Am Concrete Inst. *Res:* Silicate analysis; infrared, flame and x-ray emission spectroscopy; x-ray diffraction analyses. *Mailing Add:* 2701 Fontana Dr Glenview IL 60025-4707

HIMEL, CHESTER MORA, b Des Plaines, Ill, Mar 10, 16; m 43; c 2. BIOPHYSICS, ENTOMOLOGY. *Educ:* Univ Chicago, BS, 38; Univ Ill, PhD(org chem), 42. *Prof Exp:* Chemist, ammonia dept, E I du Pont de Nemours & Co, 42; asst group leader, Barrett Div, Allied Chem Corp, 42-44; group leader, Phillips Petrol Co, 44-49; dir org div, Stanford Res Inst, 49-65; RES PROF ENTOM, UNIV GA, 65- *Concurrent Pos:* Fel, Intra-Sci Res Found. *Mem:* Am Chem Soc; Entom Soc Am. *Res:* Fluorescence spectroscopy; molecular fluorescence; cholinergic transmission and receptor sites; fluorescent substrate analogs; narcotic receptors; insect biochemistry; insecticides; agricultural physics and biology; controlled release pesticide systems; pesticide spray physics. *Mailing Add:* 2005 E View Dr Sun City Center FL 33573

HIMELICK, EUGENE BRYSON, b Summitville, Ind, Feb 11, 26; m 51; c 3. PLANT PATHOLOGY. *Educ:* Ball State Univ, BS, 50; Purdue Univ, MS, 52; Univ Ill, PhD(plant path), 59. *Prof Exp:* From asst plant pathologist to assoc plant pathologist, Ill Natural Hist Surv, 52-65, plant pathologist, 65-87; prof, 72-87, EMER PROF PLANT PATH, UNIV ILL, URBANA, 87- *Concurrent Pos:* Assoc ed, Plant Dis Reporter, USDA, 77-79 & Plant Dis, Am Phytopath Soc, 79-81 & ed, Int Wood Collectors Soc, 83-86; exec dir, Int Soc Arboricult, 69-79; pres, Int Wood Collectors Soc, 90- *Mem:* Am Phytopath Soc; Am Soc Consult Arborists; hon mem Int Soc Arboricult; Sigma Xi. *Res:* Forest and shade tree diseases; urban tree maintenance and selection. *Mailing Add:* Ill Natural Hist Surv 607 Peabody Champaign IL 61820

HIMES, CRAIG L, b Homestead Park, Pa, June 11, 27; m 61; c 1. BIOLOGY. *Educ:* Clarion State Col, BSEd, 49; Univ Pittsburgh, MS, 57, PhD(biol & higher educ), 71. *Prof Exp:* Teacher sci & biol, S Butler County Joint Schs, 49-57; teacher biol, Butler Area Sr High Sch, 57-58 & Am Dependent Sch, Orleans, France, 58-60; res asst, Univ Pittsburgh, 60-61; chmn, Bloomsburg State Col, 73-79, emer prof, Bloomsburg Univ, 61-83; RETIRED. *Concurrent Pos:* Reviewing textbook manuscripts in biol & human sexuality. *Res:* Freshwater biology; limnology and pollution. *Mailing Add:* Bloomsburg State Col Bloomsburg PA 17815

HIMES, FRANK LAWRENCE, b Crawfordsville, Ind, July 30, 27; m 51; c 3. SOIL CHEMISTRY. *Educ:* Wabash Col, AB, 49; Purdue Univ, MS, 51, PhD(soil fertil), 56. *Prof Exp:* Asst prof agron, Mid Tenn State Col, 56-57; from asst prof to assoc prof, 57-67, asst chmn, 85-88, PROF AGRON, OHIO STATE UNIV, 67- *Concurrent Pos:* Res assignment, Rothamsted Exp Sta, Harpenden, Eng, 66 & Dept Chem, Univ Birmingham, Eng, 81. *Mem:* Fel Soil Sci Soc Am; fel Am Soc Agron; Am Chem Soc; Sigma Xi; Int Soil Sci Soc; Soil Sci Soc Am; Am Soc Agron. *Res:* Fertility; movement and availability of essential elements for growth which form chelates with organic compounds found in soil. *Mailing Add:* Dept Agron Ohio State Univ 2021 Coffey Rd Columbus OH 43210

HIMES, JAMES ALBERT, b Lucas, Ohio, Aug 12, 19; m 73; c 2. VETERINARY PHARMACOLOGY, VETERINARY PHYSIOLOGY. *Educ:* Muskingum Col, BS, 41; Univ Pa, VMD, 50; Cornell Univ, PhD(physiol), 65. *Prof Exp:* Asst zool, Univ Nebr, 41-42, 46; NIH trainee physiol & pharmacol, NY State Vet Col, Cornell Univ, 62-65; asst pharmacologist, 65-69, assoc pharmacologist, 69-76, dir vet med educ, 75-77, pharmacologist, 76-90, assoc dean student activities, Col Vet Med, 77-90, EMER PROF, UNIV FLA, 90- *Mem:* AAAS; Am Vet Med Asn; Am Soc Vet Physiol & Pharmacol. *Res:* Plasma cholinesterase; bile flow and related liver physiology; hepatic metabolism of drugs and enzyme induction. *Mailing Add:* Off Student Serv Col Vet Med Univ Fla Gainesville FL 32610

HIMES, JOHN HARTER, b Salt Lake City, Utah, July 25, 47; m 72; c 3. MATERNAL & CHILD HEALTH, CHILD GROWTH & NUTRITION. *Educ:* Ariz State Univ, BS, 71; Univ Tex, Austin, PhD(human biol), 75; Harvard Univ, MPH, 82. *Prof Exp:* Res scientist, Fels Res Inst, 76-79; asst prof pediat, Sch Med, Wright State Univ, 77-79; proj dir, Abt Assocs, Inc, 79-82; ASSOC PROF HEALTH & NUTRIT SCI, BROOKLYN COL, CITY UNIV NEW YORK, 82- *Concurrent Pos:* Vis scientist, Inst Nutrit Cent Am & Panama, 73-74; bk rev ed, Human Biol, 78-81; NIH trainee, Harvard Sch Pub Health, 80-82. *Honors & Awards:* Nathalie Masse Mem Prize, Int Childrens's Ctr, 80. *Mem:* Am Soc Clin Nutrit; Human Biol Coun; Am Pub Health Asn; Soc Study Human Biol; Am Asn Phys Anthropologists; Int Asn Human Auxology. *Res:* Physical growth and maturation of children; anthropometry, growth and nutrition; anthropometric assessment of nutritional status; body composition; subcutaneous fat; obesity; secular change; analysis of serial data; skeletal growth and maturation; clinical assessment of child growth. *Mailing Add:* 11 Geyser Dr Staten Island NY 10312

HIMES, MARION, b New York, NY, Mar 22, 23; c 2. BIOLOGY, CYTOLOGY. *Educ:* Columbia Univ, AB, 44, PhD, 51. *Prof Exp:* Instr biol, Queens Col, 61-63; asst prof, Rutgers Univ, 63-64; asst prof, 65-70, assoc prof, 70-81, PROF BIOL, BROOKLYN COL, 81- *Mem:* Am Soc Cell Biol; Histochem Soc. *Res:* Cytochemistry. *Mailing Add:* 202 Marlborough Rd Brooklyn NY 11226

HIMES, RICHARD H, b Philadelphia, Pa, July 2, 35; m 59; c 3. BIOCHEMISTRY. *Educ:* Univ Pa, AB, 56; Univ Calif, Berkeley, PhD(biochem), 61. *Prof Exp:* NIH fel, 61-63; from asst prof to assoc prof, 63-71, PROF BIOCHEM, UNIV KANS, 71-, CHMN DEPT, 78- *Mem:* AAAS; Am Chem Soc; Am Soc Biol Chemists. *Res:* Structure and function of proteins. *Mailing Add:* Dept Biochem Univ Kans Lawrence KS 66045

HIMMEL, LEON, b Mount Vernon, NY, Jan 24, 21; m 45; c 1. ELECTRICAL ENGINEERING, MANAGEMENT. *Educ:* City Col New York, BEE, 42. *Prof Exp:* From engr to pres, avionics, prod assurance & value eng, Fed Labs, Int Tel & Tel Corp, 42-67, pres, Avionics Div, 67-70, ASST TO PRES, INT TEL & TEL CORP, 70-, DIR CAPITAL ASSETS, 88-, DIR MIS, 90- *Concurrent Pos:* Mgt consult, eng, info systs, commun & financial. *Mem:* Fel Inst Elec & Electronics Engrs. *Res:* Instrument landing and radio navigation, antisubmarine warfare, range instrumentation and simulation; electronic defense, countermeasures and reconnaissance; communications. *Mailing Add:* Himmel Assocs 39 Club Rd Upper Montclair NJ 07043

HIMMELBERG, CHARLES JOHN, b North Kansas City, Mo, Nov 12, 31; m; c 5. TOPOLOGY, MATHEMATICAL ANALYSIS. *Educ:* Rockhurst Col, BS, 52; Univ Notre Dame, MS, 54, PhD(math), 57. *Prof Exp:* From asst prof to assoc prof, 59-65, PROF MATH, UNIV KANS, 68-, CHMN DEPT, 78- *Concurrent Pos:* Assoc analyst, Midwest Res Inst, 57-59; vis prof, Inst of Math, Univ Florence, 75. *Mem:* Am Math Soc; Math Asn Am; Sigma Xi. *Res:* Generalized differential equations; set-valued functions; stochastic decision theory and fixed point theory. *Mailing Add:* Dept Math Univ Kans Lawrence KS 66045

HIMMELBERG, GLEN RAY, b Glasgow, Mo, Oct 31, 37; m 57; c 3. PETROLOGY. *Educ:* Tex Tech Col, BS, 59, MS, 60; Univ Minn, PhD(geol), 65. *Prof Exp:* Geologist, US Geol Surv, 65-69; assoc prof, 69-73, PROF GEOL, UNIV MO, COLUMBIA, 73- *Mem:* Mineral Soc Am; Am Geophys Union; Geol Soc Am. *Res:* Studies of phase equilibrium of coexisting minerals in igneous and metamorphic rocks to determine the physical and chemical environment of crystallization. *Mailing Add:* Dept Geol Univ Mo Columbia MO 65211

HIMMELBLAU, DAVID M(AUTNER), b Chicago, Ill, Aug 29, 23; m 48; c 2. CHEMICAL ENGINEERING. *Educ:* Mass Inst Technol, BS, 47; Northwestern Univ, MBA, 50; Univ Wash, Seattle, MS, 56, PhD, 57. *Prof Exp:* Instr chem eng, Univ Wash, Seattle, 55-57; PROF, UNIV TEX, AUSTIN, 57- *Concurrent Pos:* Pres, Cache Corp, 78-80; pres, Ramad Corp, 79- *Mem:* Am Chem Soc; Am Inst Chem Engrs; Sigma Xi. *Res:* Fault detection; optimization; process analysis and simulation. *Mailing Add:* Dept Chem Eng Univ Tex Austin TX 78712

HIMMELSBACH, CLIFTON KECK, b Philadelphia, Pa, Mar 17, 07; m 28, 61; c 2. PHARMACOLOGY. *Educ:* Univ Va, MD, 31. *Prof Exp:* Intern, USPHS Hosp, New Orleans, La, 31-32; clin researcher drug addiction, Kans, 33-34, Mass, 34-35, Ky, 35-44; clin researcher, NIH, 44, off voc rehab, 45-46, fed employee, Health Div, 47, med officer chg outpatient clin, 48-53, asst chief, Div hosps, 53-54, chief, 54-57, dir spec prog, Div Res Grants, 57-59, assoc dir, Clin Ctr, 59-65; assoc dean, Schs Med & Dent, 65-77, adminr sponsored progs, 71-72, prof, 66-77, EMER PROF COMMUNITY MED & PHARMACOL, GEORGETOWN UNIV, 77- *Concurrent Pos:* Fel pharmacol, Western Reserve Univ, 33; mem bd dirs, Washington Home, Iona House Sr Serv Ctr, Am Cancer Soc, Retired Officers Asn; mem bd dirs, Acad Med of Washington DC, pres, 90-91. *Honors & Awards:* Meritorious Serv Award, USPHS, 65; Nathan B Eddy Mem Award, 87. *Mem:* Fel AMA; fel Am Col Physicians; Am Hosp Asn; Sigma Xi. *Res:* Addiction; opiates in man; control of hospital-acquired infections; human experimentation; hospice care. *Mailing Add:* 3731 Harrison St NW Washington DC 20015

HIMMELSTEIN, SYDNEY, b New York, NY, Oct 1, 27; m 66. ELECTRICAL ENGINEERING. *Educ:* NY Univ, BEE, 47. *Prof Exp:* Proj engr, US Naval Ord Lab, 47-50; chief, Vehicular Equip Sect, Eng Res & Develop Labs, 50-56; exec engr, Res Labs, Cook Elec Co, 56-57, tech dir, Data Stor Div, 57-60; dir eng, 60-80, PRES, S HIMMELSTEIN & CO, 80- *Concurrent Pos:* Lectr, Cath Univ Am, 52-53; pres, Electronic Applications, Inc, 53-56. *Mem:* Inst Elec & Electronics Engrs; Audio Eng Soc; Nat Soc Prof Engrs. *Res:* Magnetic recording systems and computer peripheral equipment; design and manufacture of control and monitoring systems for rotating machinery. *Mailing Add:* 2490 Pembroke Ave Hoffman Estates IL 60195

HIMMS-HAGEN, JEAN, b Oxford, Eng, Dec 18, 33; m 56; c 2. BIOCHEMISTRY, METABOLISM. *Educ:* Univ London, BSc, 55; Oxford Univ, DPhil(pharmacol), 58. *Prof Exp:* Fel biochem, Harvard Univ Med Sch, 58-59; asst prof physiol, Univ Man, 59-64; asst prof biochem, Queen's Univ, Ont, 64-67; assoc prof, 67-71, actg chmn dept, 75-77, 88, chmn dept, 77-82, PROF BIOCHEM, UNIV OTTAWA, 71- *Concurrent Pos:* Assoc, Med Res Coun, Can, 66-77, mem, 70-75. *Honors & Awards:* Bond Award, Am Oil Chemists Soc, 72; Ayerst Award, Can Biochem Soc, 73. *Mem:* Fel, Royal Soc Can; Am Soc Pharmacol & Exp Therapeut; Brit Biochem Soc; Am Inst Nutrit; Brit Nutrit Soc; Can Biochem Soc. *Res:* Brown adipose tissue metabolism; disordered brown adipose tissue metabolism in obesity. *Mailing Add:* Dept Biochem Univ Ottawa Ottawa ON K1H 8M5 Can

HIMOE, ALBERT, b Albany, Calif, Oct 23, 38. BIOCHEMISTRY. *Educ:* Reed Col, BA, 59; Univ Chicago, PhD(chem), 64. *Prof Exp:* Asst prof biochem, Baylor Col Med, 67-75; res assoc agron, Univ Ill, 76-77. *Concurrent Pos:* Fel biochem, Cornell Univ, 64-67. *Mem:* AAAS; Am Chem Soc. *Res:* Enzyme kinetics; energy metabolism. *Mailing Add:* 811 Oakland Ave Urbana IL 61801

HIMPSEL, FRANZ J, b Rosenheim, Ger, Oct 30, 49; m 80; c 2. SOLID STATE PHYSICS, TECHNICAL MANAGEMENT. *Educ:* Munich Univ, PhD(physics), 76. *Prof Exp:* MEM RES STAFF, IBM RES DIV, 80- *Concurrent Pos:* Vis prof, Univ Munich, 84-85; sr mgr surface sci, IBM Res, 85-89. *Honors & Awards:* Peter Mark Award, Am Vacuum Soc, 85. *Mem:* Fel Am Phys Soc; Am Vacuum Soc. *Res:* Surface science, using photoelectron spectroscopy with synchrotron radiation and inverse photoemission. *Mailing Add:* IBM Res PO Box 218 Yorktown Heights NY 10598

HIMWICH, WILLIAMINA (ELIZABETH) ARMSTRONG, b Spokane, Wash, Mar 4, 12; m 43; c 3. PHYSIOLOGY. *Educ:* Univ Idaho, BS, 33, MS, 34; Iowa State Col, PhD(nutrit), 39. *Prof Exp:* Instr home econ, Univ Ill, 39-41; prof home econ res, Okla Agr & Mech Col, 41-43; assoc physiol & pharm, Albany Med Col, Union Univ, NY, 43-46; toxicologist, Army Chem Ctr, 46-49; res med bibliogr, Sch Med, Johns Hopkins Univ, 49-51; med res assoc, Thudichum Psychiat Res Lab, Galesburg State Res Hosp, 52-69, admin res scientist, 69-72; res prof, Col Med, Univ Nebr Med Ctr, Omaha, 72-77, expert, Nat Libr Med, 79-81, emer prof psychiat & prof biochem, 77; RETIRED. *Mem:* Am Physiol Soc; Am Soc Pharmacol & Exp Therapeut. *Res:* Glutamic acid; metabolism; blood-brain barrier; neuropharmacology; cerebral circulation; developing brain. *Mailing Add:* 11623 Little Patuxent Pkwy Columbia MD 21044

HINATA, SATOSHI, b Tokyo, Japan, Aug 6, 44; m 69; c 3. PLASMA ASTROPHYSICS, SOLAR PHYSICS. *Educ:* Univ Tokyo, BE, 67; Univ Ill, MS, 69, PhD(physics), 73. *Prof Exp:* Res assoc, Univ Ill, 73-74; res fel, Harvard Univ, 74-76; res instr, Yale Univ, 76-78; res scientist, Solar Nat Observ, 78-80; CONSULT. *Mem:* Am Astron Soc. *Res:* Plasma astrophysics of stars and sun; nonlinear dynamos of solar type stars; stellar atmosphere & magnetic field. *Mailing Add:* Dept Physics Auburn Univ Auburn AL 36849

HINCHEN, JOHN J(OSEPH), b North Bergen, NJ, Sept 18, 26; m 53; c 2. PHYSICAL & LASER CHEMISTRY. *Educ:* Seton Hall Univ, BS, 50; Rensselaer Polytech Inst, PhD(phys chem), 62. *Prof Exp:* Chemist, Wright Aeronaut Corp, 50-51 & Gen Elec Co, 54-59; fel & res assoc phys chem Rensselaer Polytech Inst, 62-63; res scientist, United Technol Corp, 63-64, sr res scientist; MEM STAFF, UNITED TECH RES CTR, 64- *Concurrent Pos:* Consult, Kearfott Div, Gen Precision Inc, 60-61; adj prof, Univ Hartford, 75-85. *Mem:* Am Chem Soc; Am Phys Soc; Sigma Xi. *Res:* Gas-solid absorption; molecular beam studies of heterogeneous catalysis; kinetics; laser chemistry, laser spectroscopy, development of chemical lasers; laser studies of excited state lifetimes. *Mailing Add:* United Tech Res Ctr 400 Main St East Hartford CT 06108

HINCHMAN, RAY RICHARD, b Chicago, Ill, Aug 20, 37; m 63; c 2. PLANT ECOLOGY, ENVIRONMENTAL BIOLOGY. *Educ:* Univ Ill, Urbana, BS, 61; Univ Wash, MS, 64; Univ Chicago, PhD(bot), 71. *Prof Exp:* Sci asst plant physiol & morphol, Argonne Nat Lab, 64-70, appointee, 71-74, staff scientist plant ecol & environ biol, 74 & dep dir, land reclamation prog, 77-82, PROJ LEADER & PRIN INVESTR, ENERGY SYSTS DIV, ARGONNE NAT LAB, 82- *Concurrent Pos:* Adv, Argonne Nat Lab, Northern Ill Univ, 75-83; prin investr, reclamation & revegetation of disturbed land, 79-; consult, Utah State Univ, 80-83. *Mem:* Bot Soc Am; Am Soc Plant Physiologists; AAAS; Sigma Xi. *Res:* Rehabilitation and revegetation of disturbed land in the US, Germany, Nepal and China; environmental impact assessment; effects of pollutants on plants; gravity sensing mechanisms in plants; ultrastructure of plant plastids; plant morphology; plant and algal physiology; environmental fate of hazardous wastes in soils and vegetation. *Mailing Add:* Energy Systs Div Argonne Nat Lab Bldg 362 Argonne IL 60439-4815

HINCK, LAWRENCE WILSON, b Cape Girardeau, Mo, Dec 12, 40; m 64; c 2. MICROBIOLOGY, PARASITOLOGY. *Educ:* Southeast Mo State Univ, BS, 63; Med Ctr, Univ Mo, MS, 65; Med Ctr, Univ Okla, PhD(parasitol), 71. *Prof Exp:* PROF BIOL, ARK STATE UNIV, 69- *Concurrent Pos:* Consult microbiol, Helena Chem Co, 74-, Foremost Dairy, 76, Joseph A Coors Brewing Co, 84. *Mem:* Am Soc Microbiol; Am Soc Parasitologists; Sigma Xi. *Res:* Physiology of infection mechanisms of nematodes; bacterial and parasitic infections of catfish; trypanosomes of wild mammals; influence of fever on acquired immunity; lyme disease in Ark; motility of gliding bacteria. *Mailing Add:* Dept Biol Sci Ark State Univ Box 1561 State University AR 72467

HINCK, VINCENT C, b New York, NY, Feb 7, 26; m 59; c 1. RADIOLOGY. *Educ:* Cornell Univ, AB, 48; New York Med Col, MD, 53; Am Bd Radiol, dipl. *Prof Exp:* Intern, Grace-New Haven Community Hosp, Conn, 53-54; resident radiol, Hosp St Raphael, New Haven, 54-56; chief resident, Med Sch Hosps, Univ Ore, 56-57, from instr to prof, 57-69; prof radiol & head div diag radiol, Med Ctr, Univ Wis-Madison, 69-73; PROF RADIOL, BAYLOR COL MED, 73- *Mem:* Fel Am Col Radiol. *Res:* Neuroradiology; radiographic measurement criteria for diagnosis of intraspinal tumor and spinal canal stenosis; femorocerebral angiographic technic. *Mailing Add:* Dept Radiol Baylor Col Med 1200 Moursund Ave Houston TX 77030

HINCKLEY, ALDEN DEXTER, b New York, NY, Nov 2, 31; m 59; c 5. APPLIED ECOLOGY. *Educ:* Harvard Univ, AB, 53; Univ Hawaii, MS, 56, PhD, 60. *Prof Exp:* Entomologist, Fiji, 60-63, ecologist, Apia, WSamoa, 64-67; vis ecologist, Biol Dept, Brookhaven Nat Lab, 67-70; assoc prof environ sci, Univ Va, 70-74, asst dean, Col Arts & Sci, 73-74; assoc dir, Inst Ecol, 74-76, actg dir, 77; tech staff, MITRE Corp, 78; sr ecologist, Gen Res Corp, 79-81; environ protection specialist, Off Res & Develop, 82, Off Solid Waste, 83-86, ENVIRON PROTECTION SPECIALIST, OFF POLICY ANALYSIS, ENVIRON PROTECTION AGENCY, 87- *Mem:* Am Inst Biol Sci; Ecol Soc Am; AAAS. *Res:* Application of ecological principles to the management of renewable resources; ecological risk assessment; ecological effects of rapid climate change. *Mailing Add:* 5604 Bloomfield Dr No 102 Alexandria VA 22312

HINCKLEY, CONRAD CUTLER, b Ft Worth, Tex, May 8, 34; m 56; c 2. PHYSICAL CHEMISTRY. *Educ:* NTex State Univ, BS, 59, MS, 60; Univ Tex, PhD(chem), 64. *Prof Exp:* Fel chem, Univ Tex, 64-66; from asst prof to assoc prof, 66-76, PROF CHEM, SOUTHERN ILL UNIV,

CARBONDALE, 76- *Concurrent Pos:* Fel, A P Sloan Found, 73. *Honors & Awards:* Kaplan Res Award. *Mem:* AAAS; Am Chem Soc; Chem Soc London; Sigma Xi. *Res:* Applications of lanthanide shift reagents in nuclear magnetic resonance spectroscopy; inorganic chemistry; osmium chemistry; coal chemistry. *Mailing Add:* Dept Chem Southern Ill Univ Carbondale IL 62901

HIND, ALFRED THOMAS, JR, b Waleska, Ga, Aug 26, 14; m 40; c 2. MATHEMATICS. *Educ:* Emory Univ, BA, 34, MA, 36; Univ Ga, PhD(math), 52. *Prof Exp:* Teacher pub schs, Ga, 35-40; loan & collection off, Farm Security Admin, 40-42; sales mgr, Pope Distribution Co, Ga, 46-47; instr math, Clemson Univ, 47-49, assoc prof, 52-54, prof, 54-78; RETIRED. *Concurrent Pos:* Consult, Opers Res Br, Applns Res Div, Naval Res Lab, 52- *Mem:* Am Math Soc; Math Asn Am; Inst Elec & Electronics Eng. *Res:* Calculus of finite differences; numerical analysis, especially high speed computer applications. *Mailing Add:* 203 Wyatt St Clemson SC 29631

HIND, GEOFFREY, b Bolton, Eng, Apr 7, 37; m 85. BIOCHEMISTRY, PLANT PHYSIOLOGY. *Educ:* Cambridge Univ, BA, 58; Univ London, PhD(biochem), 61. *Prof Exp:* Res assoc biol, McCollum-Pratt Inst, Johns Hopkins Univ, 61-64; res assoc, Brookhaven Nat Lab, 64-65, assoc plant physiologist, 65-67, plant physiologist, 67-78, actg chmn, 77-78, dep chmn, 78-87, chmn, Biol Dept, 87-90, SR BIOCHEMIST, BROOKHAVEN NAT LAB, 87- *Mem:* Am Soc Plant Physiol; Am Soc Biol Chem; Am Soc Photobiol; Brit Soc Photobiol; Biophys Soc. *Res:* Photosynthetic electron transport; ion transport and phosphorylation in higher plants. *Mailing Add:* Biol Dept Brookhaven Nat Lab Upton NY 11973-5000

HIND, JOSEPH EDWARD, b Chicago, Ill, Apr 2, 23; m 47; c 3. AUDITORY NEUROPHYSIOLOGY, PHYSIOLOGICAL ACOUSTICS. *Educ:* Ill Inst Technol, BS, 44; Univ Chicago, PhD(psychol), 52. *Prof Exp:* Asst mfg engr, Western Elec Co, Ill, 44-45; radio engr, Naval Res Lab, 45-46; asst otolaryngol & physiol acoust, Univ Chicago, 47-53; res assoc auditory electrophysiol, Cent Inst for Deaf, Mo, 53-54; proj assoc, Dept Neurophysiol & Psychiat Inst, Univ Wis-Madison, 54-56, from asst prof to assoc prof physiol & neurophysiol, 56-64, chmn dept, 73-88, PROF NEUROPHYSIOL, UNIV WIS-MADISON, 64- *Concurrent Pos:* Mem, NIH Commun Sci Study Sect, 62-66 & Comput & Biomath Sci Study Sect, 69-73; trustee, Beltone Inst for Hearing Res, 63-86. *Mem:* AAAS; Am Physiol Soc; fel Acoust Soc Am; Soc Neurosci; Int Brain Res Orgn; Asn Res Otolaryngol. *Res:* Auditory neurophysiology, especially the acoustical and neural mechanisms of sound localization; electrophysiological and electroacoustical instrumentation, including computer techniques for processing physiological data. *Mailing Add:* Dept Neurophysiol Univ Wis Med Sch 1300 University Ave Madison WI 53706

HINDAL, DALE FRANK, b Ladysmith, Wis, May 18, 38; m 65; c 1. PLANT PATHOLOGY, MYCOLOGY. *Educ:* Univ Wis, Eau Claire, BS, 67; Iowa State Univ, PhD(plant path), 73. *Prof Exp:* Fel plant path, 73-75, asst prof mycol, 76-81, ASSOC PROF MYCOL & ASSOC MYCOLOGIST, WVA UNIV, 81- *Mem:* Am Phytopath Soc; Am Mycol Soc; Sigma Xi. *Res:* Host parasite interaction of woody plant wilt disease; fungus physiology; mycoparasite relationships. *Mailing Add:* 401 Brooks Hall WVa Univ Morgantown WV 26506

HINDER, RONALD ALBERT, b Johannesburg, SAfrica, Jan 14, 42; Swiss citizen; m 68; c 3. SURGERY. *Educ:* Witwatersrand Univ, MB Bch, 65, PhD(surg), 76. *Prof Exp:* Lectr surg, Witwatersrand Univ, 70-82, assoc prof, 82-87; assoc prof, 87-91, PROF SURG, CREIGHTON UNIV, 91- *Mem:* Am Col Surgeons; Cent Surg Asn; Am Gastroenterol Asn. *Res:* Foregut physiology and pathophysiology; gastroesophageal and duodenogastric reflex. *Mailing Add:* Creighton Univ Medical Ctr 601 N 30th St Omaha NE 68131

HINDERSINN, RAYMOND RICHARD, b Central Falls, RI, July 24, 18; m 44; c 1. ORGANIC POLYMER CHEMISTRY. *Educ:* Brown Univ, ScB, 49; Univ Wis, PhD(org chem), 54. *Prof Exp:* Res chemist, Merck & Co, Inc, 49-51; res chemist, Hooker Chem Corp, 54-58, res supvr polymer chem, 58-68, supvr polymer testing & eval, 68-70, res coordr, 70-73, mgr polymer res, Hooker Chems & Plastics Corp, 73-75, mgr polymer appln & test, 75-78, sr scientist polymer res, 78-81; CONSULT, 81- *Concurrent Pos:* Mem ad hoc comt on fire safety aspects of polymeric materials, Nat Mat Adv Bd-Nat Acad Sci, 72-78; consult polymer indust, 81- *Honors & Awards:* Schoellkopf Medal, Am Chem Soc, 82. *Mem:* Am Chem Soc; Soc Plastic Eng; Sigma Xi. *Res:* Polymer synthesis; polymer testing and evaluation; organophosphorous chemistry; polymer fire retardance and fire retardant mechanisms; unsaturated polyesters; polyurethane foams; engineering thermoplastics. *Mailing Add:* 4288 Lower River Rd Youngstown NY 14174

HINDLE, BROOKE, b Drexel Hill, Pa, Sept 28, 18; m 43; c 2. EARLY AMERICAN SCIENCE & TECHNOLOGY. *Educ:* Brown Univ, BA, 40; Univ Pa, MA, 42, PhD(hist sci), 49. *Prof Exp:* Asst hist, Univ Pa, 41-42 & 45-48; res assoc hist, Inst Early Am Hist & Cult, 48-50; from assoc prof to prof hist, NY Univ, 50-74, chmn dept hist, Univ Col, 65-67, dean, 67-69, head dept, 70-74; dir Nat Mus Hist & Technol, 74-78, sr historian, 78-85, EMER HISTORIAN, US NAT MUS AM HIST, SMITHSONIAN INST, 85- *Concurrent Pos:* Lectr hist, Col William & Mary, 48-50, Northwestern Univ, 50; sr res scholar, Eleutherian Mills-Hagley Found, 69-70; Killian vis prof hist sci & technol, Mass Inst Technol, 71-72; lectr, NY State Hist Asn, Cooperstown, 80; vis prof hist technol, Univ Central Fla, 81; distinguished vis prof hist technol, Univ Del, 82. *Honors & Awards:* Anson Phelps lectr, New York Univ, 79; Fel's Award Medal, Early Am Industs Asn, 83; Leonardo Da Vinci Medal, Soc Hist Technol, 84. *Mem:* Fel AAAS; Soc Hist Technol (vpres, 77-80, pres, 81-82); Soc Indust Archaeol; Hist Sci Soc (secy, 58-60); Am Philos Soc; Int Acad Hist Sci. *Res:* American technology and science from the Revolutionary to the Ante-bellum era; internal and external history; creativity in technology. *Mailing Add:* 5114 Dalecarlia Dr Bethesda MD 20816

HINDMAN, EDWARD EVANS, b Los Angeles, Calif, Sept 26, 42; div; c 3. ATMOSPHERIC SCIENCES. *Educ:* Univ Utah, BSc, 65; Colo State Univ, MSc, 67; Univ Wash, PhD(atmos sci), 75. *Prof Exp:* Res meteorologist, Navy Weather Res Facil, 67-71, actg head phys meteor and weather modification br, 70-71; head plans, anal & reports sect, Naval Weapons Ctr, 71-74, head atmospheric interactions sect, 74-79; res assoc atmos sci, Colo State Univ, 79-88; ASSOC PROF, EARTH & PLANETARY SCI, CITY COL, CITY UNIV NY, 88- *Concurrent Pos:* Naval Weapons Ctr fel, Univ Wash, 72-74; vis prof, US Naval Acad, 84-87, Drexel Univ, 88. *Honors & Awards:* Father J B Macelwane Award, Am Meteorol Soc, 66. *Mem:* Am Meteorol Soc; Royal Meteorol Soc; Am Geophys Union; Int Tech & Sci Orgn Soaring Flight. *Res:* Micro-structure of winter snowstorms, hurricanes and fogs; impact of industrial effluents on rainfall; physical, chemical, optical properties of clouds. *Mailing Add:* Dept Earth & Planetary Sci City Col New York City NY 10031

HINDMAN, JOSEPH LEE, b Baker, Ore, Dec 24, 33; m 55; c 4. PLANT MORPHOLOGY, DEVELOPMENTAL ANATOMY. *Educ:* Harvard Univ, AB, 56; Univ Ore, MA, 58, PhD(plant morphol), 62. *Prof Exp:* Asst prof biol, La State Univ, New Orleans, 61-63 & State Univ NY Buffalo, 63-67; assoc prof bot, Univ Ga, 67-70; prof & chmn prog gen biol, 70-81, DIR ADV, WASH STATE UNIV, 81- *Mem:* Bot Soc Am; Am Inst Biol Sci; Sigma Xi. *Res:* Floral morphogenesis; culture of floral buds in vitro. *Mailing Add:* SALC Wash State Univ Pullman WA 99164-2105

HINDS, DAVID STEWART, b LaJolla, Calif, Dec 3, 39; c 3. PHYSIOLOGICAL ECOLOGY. *Educ:* Pomona Col, BA, 62; Univ Ariz, MS, 64, PhD(zool), 69. *Prof Exp:* Res assoc physiol ecol, Duke Univ, 64 & Univ Ariz, 69-70; from asst prof to assoc prof, 70-76, chmn dept biol, 86-88, PROF BIOL, CALIF STATE UNIV, BAKERSFIELD, 76- *Concurrent Pos:* Fulbright sr scholar, Flinders Univ, SAustralia, 85-86, vis prof, 85-86 & 90; vis researcher, Dept Ecol & Evolutionary Biol, Univ Calif, Irvine, 80-82. *Mem:* Am Inst Biol Sci; AAAS; Ecol Soc Am; Am Asn Biol Sci; Nat Asn Biol Teachers. *Res:* Adjustment of physiology of vertebrates in response to the changing environment; seasonal changes in thermoregulation of vertebrates and their energy exchange in the natural microclimate. *Mailing Add:* Dept Biol Calif State Univ Bakersfield CA 93311-1099

HINDS, EDWARD C, b Park Rapids, Minn, May 10, 17; m 43; c 4. ORAL SURGERY. *Educ:* Baylor Univ, BA & DDS, 40, MD, 45; Am Bd Oral Surg, dipl, 52; Am Bd Surg, dipl, 56. *Prof Exp:* Instr oral surg, Col Dent, Baylor Univ, 40-42, clin asst surg, 49-51; assoc prof, Dent Br, Univ Tex, Houston, 48, resident surg, M D Anderson Hosp & Tumor Inst, 51-52, prof surg, dent br, 52-87; RETIRED. *Concurrent Pos:* J Jones cancer res fel, M D Anderson Hosp & Tumor Inst, 51-52; consult oral surg, Vet Admin Hosp, Houston, 52-, USAF Hosp, Lackland AFB, 56-, Wilford Hall Air Force Hosp, San Antonio, 57, William Beaumont Army Hosp, El Paso, 58, 58, Brooke Army Hosp, San Antonio, 65 & Cent Off, Vet Admin; nat consult to Surgeon Gen, USAF, DC, 64-69; mem adv comt, Am Bd Oral Surg, 56-60. *Mem:* Am Dent Asn; AMA; Am Soc Oral Surg; fel Am Col Surgeons; fel Am Col Dent. *Mailing Add:* Valley Ledge Dr Houston TX 77078

HINDS, FRANK CROSSMAN, b Sioux Falls, SDak, Nov 30, 30; m 53; c 3. ANIMAL NUTRITION. *Educ:* Ill State Norm Univ, BS, 52; Univ Ill, MS, 57, PhD(animal sci), 59. *Prof Exp:* Asst animal sci, Univ Ill, Urbana, 55-59, asst prof, 59-70, assoc prof, 70-74, prof, 74-80, res assoc animal nutrit, Dixon Springs Exp Sta, 59-80; PROF & HEAD, DEPT ANIMAL SCI, UNIV WYO, 80- *Mem:* AAAS; Am Soc Animal Sci; Soc Range Mgt. *Res:* Nutrition and reproductive physiology of ruminants, especially the influence of anabolic agents on ruminants and factors related to utilization of forages by ruminants. *Mailing Add:* Dept Animal Sci Univ Wyo Laramie WY 82071

HINDS, HORACE, JR, chemical engineering, for more information see previous edition

HINDS, JAMES WADSWORTH, b Waterville, Maine, May 20, 41; m 67; c 1. ANATOMY. *Educ:* Williams Col, AB, 63; Harvard Univ, PhD(anat), 68. *Prof Exp:* From instr to asst prof, 67-75, ASSOC PROF ANAT, SCH MED, BOSTON UNIV, 75-; ACCT, LOISELLE & BEATHAM, CPA'S. *Concurrent Pos:* USPHS res grants, Sch Med, Boston Univ, 69-74, 74-80 & 80-85. *Mem:* AAAS; Am Asn Anat; Soc Neurosci. *Res:* Neuroembryology; neurocytology; neuroanatomy; aging. *Mailing Add:* Loiselle & Beatham CPA's PO Box 62 Bangor ME 00411-0062

HINDS, MARVIN HAROLD, b Brookville, Ind, Oct 25, 28; m 52; c 2. PHYSIOLOGY, BIOMEDICAL ENGINEERING. *Educ:* Marion Col, BS, 51; Valparaiso Tech Inst, BS, 57; Tex A&M Univ, PhD(vet physiol), 71. *Prof Exp:* TV engr, M D Anderson Hosp, Univ Tex, 57-59; electronics supvr, Dept Physiol, Baylor Col Med, 59-67; asst prof physiol, Dept Vet Physiol & Pharmacol, Tex A&M Univ, 69-73; asst prof, 73-80, PROF, BIOL DEPT & COORDR, IND WESLEYAN UNIV, 80- *Mem:* Med Electronics & Data Soc; Am Sci Affil. *Res:* New instrumentation for cardiac defibrillation; use of skeletal muscle for cardiac assist. *Mailing Add:* Dept Biol Ind Wesleyan Univ Marion IN 46952

HINDS, NANCY WEBB, b Milan, Tenn, Jan 14, 47; m 75; c 1. PHYSICAL CHEMISTRY. *Educ:* Univ Tenn, AB, 69; Memphis State Univ, PhD(phys chem), 73. *Prof Exp:* From res asst to res assoc phys biochem, Ctr Health Sci, Univ Tenn, 73-75; asst prof chem, Christian Brothers Col, 75-80; ASST PROF CHEM, UNIV TENN, MARTIN, 80- *Mem:* Am Chem Soc; Sigma Xi. *Res:* Physical parameters of drug interactions. *Mailing Add:* 403 McGill St Martin TN 38237

HINDS, THOMAS EDWARD, b Albuquerque, NMex, Nov 30, 22; m 56; c 3. PLANT PATHOLOGY. *Educ:* Colo State Univ, BS, 58, MS, 68. *Prof Exp:* Plant pathologist, Rocky Mountain Forest & Range Exp Sta, 59-70, res plant pathologist tree dis, 70-84; RETIRED. *Mem:* Am Phytopath Soc; Mycol Soc Am; AAAS. *Res:* Diseases of forest trees in general; aspen diseases in particular; decays and stains in forest trees. *Mailing Add:* 1212 Emigh St Ft Collins CO 80524

HINE, FREDERICK ROY, b Cincinnati, Ohio, Nov 16, 25; m 48; c 1. PSYCHIATRY. *Educ:* Yale Univ, BS, 46, MD, 49. *Prof Exp:* Intern, Charity Hosp of La, New Orleans, 49-50; resident fel psychiat, Sch Med, Tulane Univ, 50-53; psychiatrist, dir training & res & clin dir, Southeast La Hosp, 53-59; from asst prof to assoc prof, 59-71, PROF PSYCHIAT, MED CTR, DUKE UNIV, 71- *Concurrent Pos:* Instr psychiat, Sch Med, Tulane Univ, 52-59; attend psychiat, Durham Vet Admin Hosp, 59-74, consult psychiat, 74- *Mem:* Am Psychiat Asn; Asn Dir Med Student Educ Psychiat. *Res:* Medical education. *Mailing Add:* Med Ctr Duke Univ Box 2915 Durham NC 27710

HINE, GERALD JOHN, health sciences, nuclear medicine; deceased, see previous edition for last biography

HINE, JACK, b Coronado, Calif, July 2, 23; m 46, 78; c 1. CHEMISTRY. *Educ:* Univ Ark, BS, 43; Univ Ill, PhD(chem), 47. *Hon Degrees:* LLD, Lewis Col, 65. *Prof Exp:* Asst res chemist, Cities Serv Oil Co, 43-45; res assoc org chem, Mass Inst Technol, 47-48; du Pont fel, Harvard Univ, 48-49; from asst prof to prof chem, Ga Inst Technol, 49-58, Regent's prof, 58-65; PROF CHEM, OHIO STATE UNIV, 65- *Concurrent Pos:* Sloan fel, 56-60; lectr, Western Reserve Univ, 59; Phillips visitor, Haverford Col, 60; John W Watson lectr, Va Polytech Inst & State Univ, 62; Kelly lectr, Purdue Univ, 64. *Honors & Awards:* Ga Inst Technol Res Award, 58; Fla Award, Am Chem Soc, 62, Herty Medal, Ga Sect, 63. *Mem:* Fel AAAS; Am Chem Soc; Am Phys Soc; The Chem Soc; Chem Soc France. *Res:* Effect of structure on rate and equilibrium constants; polyfunctional catalysis; imines and iminium ions; enzyme model studies. *Mailing Add:* 2930 N Star Rd Columbus OH 43221

HINE, JOHN MAYNARD, b Tacoma, Wash, Aug 1, 20; m 44; c 2. ORGANIC POLYMER CHEMISTRY. *Educ:* Col Puget Sound, BS, 44. *Prof Exp:* Asst chem, Univ Ore, 43-44; resin chemist, Borden Inc, 46-53, lab mgr, 53-54, West Coast develop mgr, 54-63, develop mgr, Thermosetting Prod Dept, Western Opers, 63-70, Borden Chem Co Div, 70-80, dir res & develop, Resins & Chem Div, 80-87; RETIRED. *Mem:* Am Inst Chemists. *Res:* Synthetic resins; plastics. *Mailing Add:* 10515 NE 60th Kirkland WA 98033

HINE, MAYNARD KIPLINGER, b Waterloo, Ind, Aug 25, 07; m 32; c 3. DENTISTRY, DENTAL HISTORY. *Educ:* Univ Ill, DDS, 30, MS, 32. *Hon Degrees:* DSc, Case Western Reserve Univ, 67, Univ Ill, 69, Boston Univ, 69, Ohio State Univ, 70, Marian Col, 73, Temple Univ, 73, Ind Univ, 79. *Prof Exp:* From instr to assoc prof oral path & periodontia, Col Dent, Univ Ill, 36-44; prof oral histopath & periodontia & head dept, Sch Dent, Ind Univ, Indianapolis, 44-78; dean, Sch Dent, Ind Univ, Indianapolis, 45-69; chancellor, 69-73, exec assoc, Ind Univ Found, 73-78, EMER CHANCELLOR, IND UNIV-PURDUE UNIV, INDIANAPOLIS, 78- *Concurrent Pos:* Rockefeller fel, Univ Rochester, 34-35, Carnegie fel, 35-36; pres, Am Asn Dent Sch, 53; mem, US Adv Panel Med Sci, 54; mem bd regents, Nat Libr Med, 59-63; mem, Nat Adv Dent Res Coun, 48-50 & 64-68 & Dept Defense Dent Adv Comt, 65-68; ed, Jour, Am Acad Periodont, 50-70; pres, Bd Cent Ind Coun on Aging, 80-83; pres, Am Asn; dent ed, Indianapolis Dist Dent Soc, Ind State Dent Asn, 58. *Honors & Awards:* Presidental Award, Fedn Dentaire Int, 87; Hayden Harris Award, Am Acad Hist Dent, 83. *Mem:* AAAS; Am Dent Asn (pres elect, 64, pres, 65); Am Acad Periodont (pres, 64); Am Acad History Dent (pres, 81); Fedn Dent Int (pres, 75); Am Asn Dent Schs (pres, 53); Am Asn Endodontists (pres, 47); hon fel Royal Col Surgeons, Ireland; hon fel Royal Col Surgeons, Can. *Res:* Oral pathology; periodontia; dental hygiene; dental history. *Mailing Add:* 1121 Michigan St Indianapolis IN 46202

HINE, RICHARD BATES, b Los Angeles, Calif, June 2, 29; m 51; c 4. PLANT PATHOLOGY. *Educ:* Univ Calif, Los Angeles, BA, 52; Univ Calif, Davis, PhD(plant path), 58. *Prof Exp:* Asst, Univ Calif, 56-58; tech rep, E I du Pont de Nemours & Co, 59-61; assoc prof, Univ Hawaii, 61-67, chmn dept, 63-67; PROF PLANT PATH & PLANT PATHOLOGIST, UNIV ARIZ, 67- *Concurrent Pos:* Staff mem, Calif Packing Corp, Philippines, 65-67. *Mem:* Am Phytopath Soc; Mycol Soc Am. *Res:* Soil fungi; fungus diseases of plants; agricultural chemicals. *Mailing Add:* Dept Plant Path Univ Ariz Tucson AZ 85721

HINE, RUTH LOUISE, b Columbus, Ohio, Aug 19, 23. WILDLIFE CONSERVATION. *Educ:* Conn Col, BA, 44; Univ Wis, MA, 47, PhD, 52. *Prof Exp:* Asst biol, Wesleyan Univ, 44-46; asst zool, Univ Wis, 46-48, asst wildlife mgt, 48-49; conserv aid, Dept Natural Resources, 49-52, biologist, 52-57, ed res publ, Natural Resources, Ecol & Environ, 57-74, natural resource specialist, 74-86; NATURALIST, BETHEL HORIZONS NATURE CTR, 86- *Concurrent Pos:* Mem conf biol eds, Nat Conservancy; Consult naturalist Lutheran Outdoor Ministries of Wis & Upper Mich, 84-85. *Honors & Awards:* Am Motors Conserv Award, 63; Commendation Award, Am Soil Conserv Soc, 64; Steven Mather Award, Nat Park & Conserv Assoc, 84. *Mem:* Fel Am Soil Conserv Soc; Wildlife Soc; Nat Audubon Soc; Nature Conservancy. *Res:* Conservation of natural resources, especially game, fish, natural areas and endangered species; surveys and investigations on nongame, threatened and endangered species. *Mailing Add:* 3609 Nakoma Rd Madison WI 53711

HINEGARDNER, RALPH, b New York, NY, Feb 18, 31; m 56; c 1. DEVELOPMENTAL BIOLOGY. *Educ:* Denison Univ, BA, 53; Univ Southern Calif, MS, 57; Univ Hawaii, PhD(zool), 61. *Prof Exp:* USPHS res fel, Univ Hawaii, 60-63; from asst prof to assoc prof zool, Columbia Univ, 63-68; assoc prof natural sci, 68-72, PROF BIOL, CROWN COL, UNIV CALIF, SANTA CRUZ, 72- *Mem:* Fel AAAS; Soc Gen Physiol; Soc Develop Biol; Sigma Xi. *Res:* Theoretical biology; invertebrate development; nucleic acid synthesis and evolution. *Mailing Add:* Div Natural Sci Univ Calif Santa Cruz CA 95060

HINERMAN, CHARLES OVALEE, science education; deceased, see previous edition for last biography

HINERMAN, DORIN LEE, b Huntington, WVa, Apr 19, 14; m 44; c 2. PATHOLOGY. *Educ:* Marshall Col, AB, 37; Univ Mich, MD, 42; Am Bd Path, dipl, 48. *Prof Exp:* Intern, Milwaukee Hosp, 42-43; resident, Hosp, 43-45, from instr to prof, Med Sch, 45-84, med sch counsr, 63-75, EMER PROF PATH, UNIV MICH, ANN ARBOR, 84- *Concurrent Pos:* Consult, Vet Admin Hosp, Ann Arbor, 54- & Mich Tumor Registry, 54-61; chmn bd trustees, Mich Cancer Found, 81-84; chmn, Mich Cancer Registry Comt; chmn spec admis comt, Univ Mich, Ann Arbor, mem exec comt, Fac Gov Body & mem intercol athletics control bd. *Mem:* Fel Am Soc Clin Path; AMA; Am Asn Path & Bact; Int Acad Path. *Res:* Endocrine pathology; oncology; medical education. *Mailing Add:* Dept Path Box 45 Univ of Mich Ann Arbor MI 48109

HINES, ANSON HEMINGWAY, b Honolulu, Hawaii, Jan 4, 47. ZOOLOGY, MARINE BIOLOGY. *Educ:* Pomona Col, BA, 69; Univ Calif, Berkeley, PhD(zool), 76. *Prof Exp:* Res fel zool, Univ Calif, Santa Cruz, 75-78; sr marine biologist, Tera Corp, 78-79; RES ECOLOGIST, SMITHSONIAN INST, 79- *Concurrent Pos:* Seven hundred & twenty-five grants & awards. *Mem:* Am Soc Zoologists; Sigma Xi; Ecol Soc Am. *Res:* Marine invertebrate zoology; population and community ecology. *Mailing Add:* 410 Fernwood Dr Severna Park MD 21146

HINES, ANTHONY LORING, b Sept 19, 1941; m 63; c 7. MASS TRANSFER, SYNTHETIC FUELS. *Educ:* Univ Okla, BS, 67; Okla State Univ, MS, 69; Univ Tex, Austin, PhD(mech eng), 73. *Prof Exp:* Chem process & plant engr, Gulf Oil Co, 67-69; sr assoc chem engr, IBM, 69-73; instr mech eng, Univ Tex, Austin, 72-73; asst prof chem eng, Ga Tech, 73-75; from asst to assoc prof, Colo Sch Mines, 75-80; dept head & prof, Chem Eng, Univ Wyo, 80-83; assoc dean & prof Chem Eng, Okla State Univ, 83-87; DEAN & PROF, COL ENG, UNIV MO, COLUMBIA, 87- *Concurrent Pos:* Consult, Laramie Energy Res Ctr, Dept Energy, 76-83, Solar Energy Res Inst, 79-81 & Western Oil Sands, Ltd, 77-81; proj dir synthetic fuel studies, US Dept Energy, 78; proj dir adsorption studies, Solar Energy Res Inst, 80-81; vis prof, Assoc Western Univs, US Dept Energy, 82; prin investr weapon studies, Dept Air Force, 83- *Mem:* Am Inst Chem Engrs; Sigma Xi; AAAS; Am Soc Eng Educ; Nat Soc Prof Engrs; Int Adsorption Soc. *Res:* Oil shale and tar sands; adsorption of gases and liquids on solids; mass transfer and diffusion; indoor air quality. *Mailing Add:* Col Eng Univ Mo Columbia MO 65211

HINES, DOUGLAS P, nuclear & mechanical engineering, for more information see previous edition

HINES, HAROLD C, b Ashville, Ohio, Dec 12, 37; m 63; c 3. DAIRY SCIENCE, IMMUNOGENETICS. *Educ:* Ohio State Univ, BS, 59, MS, 61, PhD(dairy sci), 65. *Prof Exp:* Asst prof, 65-70, ASSOC PROF DAIRY SCI, OHIO STATE UNIV, 70- *Mem:* AAAS; Am Soc Animal Sci; Int Soc Animal Blood Group Res; Am Dairy Sci Asn; Sigma Xi. *Res:* Cattle blood antigens; serum and milk genetic polymorphisms. *Mailing Add:* Dairy Sci Dept Ohio State Univ 625 Stadium Dr Columbus OH 43210

HINES, JAMES R, b Kewanna, Ind, Sept 16, 23; m 57; c 5. SURGERY. *Educ:* Univ Ill, BS, 44, MD, 46; Northwestern Univ, MS, 51. *Prof Exp:* PROF SURG, MED SCH, NORTHWESTERN UNIV, CHICAGO, 51-; CHIEF GEN SURG & VCHMN DEPT SURG, MED SCH, NORTHWESTERN UNIV, 73- *Mem:* Soc Surg Alimentary Tract; Int Soc Gastrointestinal Surg. *Res:* Mammary carcinoma; peptic ulcer disease. *Mailing Add:* 251 E Chicago Ave Chicago IL 60611

HINES, LEONARD RUSSELL, b Mechanicsville, Iowa, Feb 8, 13; m 41; c 3. BIOCHEMISTRY, PHARMACOLOGY. *Educ:* Cornell Col, BS, 34; Univ Iowa, PhD(biochem), 43. *Prof Exp:* Res chemist, Wilson & Co, Ill, 34-37; asst biochem, Univ Chicago, 37-40; asst, Univ Iowa, 40-42; head dept biochem, E R Squibb & Sons, 43-52; chief chem pharmacol, Lederle Labs Div, Am Cyanamid Co, 52-58; clin investr, Hoffmann-La Roche, Inc, 58-59, assoc dir biol res, 59-62, dir biol res, 62-78, asst vpres, 73-78; RETIRED. *Mem:* Fel AAAS; Am Fedn Clin Res; Am Col Clin Pharmacol; Am Col Neuropsychopharmacol; NY Acad Sci. *Res:* Drug metabolism; toxicology; chemotherapy; clinical investigation. *Mailing Add:* 17233 Cuvee Ct Poway CA 92064

HINES, MARGARET H, b Cincinnati, Ohio, Sept 29, 23; m 49; c 2. ANATOMY. *Educ:* Univ Cincinnati, BSc, 45; Ohio State Univ, MSc, 42, PhD(anat and student personnel admin), 74. *Prof Exp:* Phys therapist, US Army, 46-48 & Elizabeth Kenny Inst, Minneapolis, 49-50; asst instr anat, 62-63, instr, 63-74, asst prof, 75-81, ASSOC PROF, OHIO STATE UNIV, 81- *Concurrent Pos:* Phys therapist, Children's Hosp, Columbus, Ohio, 48-, asst dir outpatient dept, 51-; fel, Elizabeth Kenny Inst, 49-51; guest lectr, Bowling Green State Univ, 73, Ohio Dominican Col, 73 & Dept Surg, Beaumont Gen Hosp, Royal Oak, Mich, 75. *Mem:* Sigma Xi; Asn Am Women Sci; Asn Am Med Cols; Am Asn Clin Anatomists. *Res:* Control, computer simulation and clinical investigation of human and biped robot locomotion; medical college admissions process and student advising systems; clinical manifestations of vascular pain. *Mailing Add:* 5970 Winstead Dr Worthington OH 43085

HINES, MARK EDWARD, b Pasadena, Tex, Apr 30, 50; m 69; c 2. MARINE MICROBIOLOGY, BIOGEOCHEMISTRY. *Educ:* Ohio State Univ, BS, 73; Univ Conn, MS, 78; Univ NH, PhD(microbiol), 81. *Prof Exp:* Res scientist marine biol, 81-84, res asst prof, Earth Sci, 84-89, RES ASSOC PROF, EARTH SCI, UNIV NH, 89- *Concurrent Pos:* Prin investr sea grant, Nat Oceanic & Atmospheric Admin, Univ NH, 81-84, NSF grant, 81-84 & 85-87 & 86-88, Nat Aeronaut & Space Admin grant, 83-88 & 88- & Am Chem Soc, 88-90; instr, Sch Life Long Learning, 82-85. *Mem:* Am Soc Microbiol; Am Soc Limnol & Oceanog; Sigma Xi; Estuarine Res Fedn; AAAS. *Res:* Microbial biogeochemistry of marine sediments with emphasis on microbial activity rates; interactions with trace metals; anaerobic nutrient regeneration; hypersaline ecology; wetlands ecology; production and emission of biogenic trace gases. *Mailing Add:* Inst Study Earth, Oceans & Space Univ NH Durham NH 03824

HINES, PAMELA JEAN, b Detroit, Mich, July 19, 52. DEVELOPMENTAL BIOLOGY. *Educ:* Oberlin Col, AB, 74; Univ Wis, MS, 77; Johns Hopkins Univ, PhD(biol), 83. *Prof Exp:* Instr develop biol, Purdue Univ, 77-79; res assoc molecular biol, Univ Wash, 84-89; ASSOC ED DEVELOP & PLANT BIOL, SCIENCE MAG, AAAS, 89- *Mem:* Soc Develop Biol; Am Soc Cell Biol; Am Soc Plant Physiol; Asn Women Sci. *Res:* Developmental biology; human genetics; chromosomes; plant development and molecular biology; hematopoiesis. *Mailing Add:* 1333 H St NW Washington DC 20005

HINES, PAUL STEWARD, b New Brighton, Pa, Aug 31, 40; m 72; c 4. ORGANIC CHEMISTRY. *Educ:* Geneva Col, BSCh, 62; Univ Pa, PhD(org chem), 69. *Prof Exp:* Asst prof, 68-74, assoc prof, 74-78, Chmn Dept, 71-80, PROF CHEM, WESTERN CONN STATE UNIV, 78- *Concurrent Pos:* Adj prof, New Eng Inst, Inc, 70- *Mem:* Am Chem Soc. *Res:* Synthesis, physical properties, and medicinal structure-function relationships of heterocyclic compounds, especially nitrogen bridgehead compounds and porphyrins. *Mailing Add:* Dept Chem Western Conn State Univ 181 White St Danbury CT 06810

HINES, RODERICK LUDLOW, b Cleveland, Ohio, Nov 20, 25; m 49; c 5. SOLID STATE PHYSICS. *Educ:* Oberlin Col, BA, 47; Univ Mich, MS, 49, PhD(physics), 54. *Prof Exp:* Trainee, Gen Elec Co, NY, 47-48; physicist, sci lab, Ford Motor Co, Mich, 53-57; from asst prof to assoc prof, 57-66, PROF PHYSICS, NORTHWESTERN UNIV, EVANSTON, 66- *Mem:* Am Phys Soc. *Res:* Irradiation effects; ion ranges in solids; imperfections in solids; electrostatic atomization; high resolution electron microscopy. *Mailing Add:* Dept Physics Northwestern Univ Evanston IL 60201

HINES, WILLIAM CURTIS, b Greenwood, Miss, Mar 23, 40; m 61; c 2. ELECTRICAL ENGINEERING. *Educ:* Miss State Univ, BS, 62, MS, 64; Auburn Univ, PhD(elec eng), 67. *Prof Exp:* Mem tech staff, 67-69, supvr systs studies div II, 69-82, DEPT MGR, SANDIA NAT LABS, 82- *Mem:* Inst Elec & Electronics Engrs. *Res:* Optical control theory. *Mailing Add:* Dept 9410 Sandia Nat Labs PO Box 5800 Albuquerque NM 87185

HINES, WILLIAM GRANT, b Hamilton, Ont, June 22, 16; m 39; c 3. PHYSICAL CHEMISTRY. *Educ:* Univ Toronto, BA, 37, MA, 38, PhD, 42. *Prof Exp:* Asst corrosion, Nat Res Coun Can, 38-39, asst chem warfare, 39-40; demonstr, Univ Toronto, 40-42; inspector, Res Enterprises Ltd, 42-43, head, approvals lab, 43-46; sr res assoc indust chem res, BC Res Coun, 46-49; dir, Sch Indust Chem, Ryerson Inst Technol, 49-56; develop chemist surface chem, Steel Co Can, 56-58, div metallurgist chem activities, 58-64, gen supv metallurgist, Labs, 64-81; RETIRED. *Mem:* Fel Chem Inst Can. *Res:* Surface chemistry; electrochemistry; analytical chemistry; education. *Mailing Add:* 225 Bamburgh Circle Suite 201 Scarborough ON M1W 3X9 Can

HINES, WILLIAM W, b Tampa, Fla, Dec 12, 32; m 59; c 2. INDUSTRIAL & SYSTEMS ENGINEERING. *Educ:* Memphis State Univ, BS, 54; Ga Inst Technol, MSIE, 58, PhD(opers res), 64. *Prof Exp:* From instr to asst prof indust eng, 59-64, res asst prof comput appln res, 59-64, assoc prof indust eng, 64-67, PROF INDUST ENG, GA INST TECHNOL, 67-, ASSOC DIR, SCH INDUST & SYSTS ENG, 68- *Concurrent Pos:* Asst ed & tech notes ed, J Indust Eng, 59-67; lead engr, assoc prin engr & consult, Radiation Inc, 65-70; mem, Food & Drug Admin adv comt on med device good mfg pract, 77-80; pres, Enstat, Inc. *Mem:* Opers Res Soc Am; Am Statist Asn; Inst Indust Engrs; Am Soc Eng Educ. *Res:* Sampling methods; engineering statistics; simulation; reliability engineering; power systems load research and management. *Mailing Add:* 1989 Tall Tree Dr NE Atlanta GA 30324

HINESLEY, CARL PHILLIP, b Muncie, Ind, Aug 13, 44; m 66; c 3. MECHANICAL METALLURGY, ENERGY CONSERVATION TECHNOLOGY. *Educ:* Univ Ky, BS, 67, MS, 69, PhD(mat sci), 72. *Prof Exp:* Proj mgr mech metall, Brush-Wellman, Inc, 72-75; proj mgr, Gould Labs, Gould Inc, 75-76, mgr prod develop energy conserv, Elec Heat Div, 76-77, prog mgr elec mat, 77-78; dir prod develop, Energy Conserv, Chromalox Div, Emerson Elec, 78-86; VPRES ENG, MARLEY ELEC HEATING, 86- *Mem:* Inst Elec & Electronics Engrs; Am Soc Metals; Am Inst Mining, Metall & Petrol Engrs. *Res:* Reciprocal relations effecting the structure, properties and elevated temperature formability of materials. *Mailing Add:* PO Box 1298 Pinehurst NC 28374

HINGERTY, BRIAN EDWARD, b Brooklyn, NY, July 30, 48. BIOPHYSICS, MOLECULAR BIOLOGY. *Educ:* Brooklyn Col, BS, 69; Princeton Univ, MA, 71, PhD(physics), 74. *Prof Exp:* NATO fel chem, Max Planck Inst Exp Med, Gottingen, WGer, 74-75; NIH fel, MRC Lab Molecular Biol, Cambridge, Eng, 75-78; Eugene P Wigner fel biol, 78-80, sci staff biol, 80-81, SCI STAFF HEALTH & SAFETY, OAK RIDGE NAT LAB, 81- *Concurrent Pos:* Consult Biol Dept, New York Univ, 77-; instr, Physics Dept & asst prof, Math Dept, Univ Tenn, Knoxville, 81- *Mem:* Fel Am Phys Soc; Am Crystallog Asn; Biophys Soc; Sigma Xi. *Res:* X-ray structure of biologically important compounds including both small and macromolecular problems; semi-empirical potential energy methods; nucleic acid structure and function; protein crystallography; supercomputers. *Mailing Add:* Oak Ridge Nat Lab-HASRD PO Box 2009 MS 8077 Oak Ridge TN 37831-8077

HINGORANI, NARAIN G, b Karachi, Pakistan, June 15, 31; US citizen; m; c 2. HIGH VOLTAGE DIRECT CURRENT TRANSMISSION. *Educ:* Baroda Univ, BE, 54; Manchester Univ, MSc, 57, PhD(elec eng), 60. *Prof Exp:* Res fel, Inst Sci & Technol, Univ Manchester, 57-61; lectr, Loughborough Univ, Eng, 61-63; sr lectr, Univ Salford, Eng, 63-68; consult, Bonneville Power Admin, 68-74; prog mgr, AC/DC substations, 74-83, dir Transmission Dept, 83-86, VPRES ELEC SYSTS DIV, ELEC POWER RES INST, 86- *Concurrent Pos:* Mem, Joint US-USSR Subcomt Res Design & Oper Ultrahigh Voltage Transmission, 70-76; mem, Coord Comt High Power Electronics Res & Develop, Elec Power Res Inst, 85- & co-chmn, Superconducting Coord Comt, 87-; mem bd dir, Power Eng Soc, Inst Elec & Electronics Engrs; spec reporter, CIGRE SC14, 84-86, chmn, 89- *Honors & Awards:* Uno Lamm Award, Power Eng Soc, Inst Elec & Electronics Engrs, 85; ASM Inst Eng Mat Achievement Award, 87. *Mem:* Nat Acad Eng; fel Inst Elec & Electronics Engrs. *Res:* Electric utility power transmission; application of amorphous steel in transformer cores; zinc oxide technology for overvoltage control; advancements in power semiconductor technology; performance & extended application of direct current transmission; flexibile AC transmission systs (FACTS); custom power. *Mailing Add:* 3412 Hillview Ave Palo Alto CA 94304

HINISH, WILMER WAYNE, b Curryville, Pa, Apr 1, 25; m 46; c 2. AGRONOMY. *Educ:* Pa State Univ, BS, 51, MS, 53, PhD(agron), 55. *Prof Exp:* Asst agron, 51-54, from instr to assoc prof, 54-68, prof coordr agron ext prog, 76-80, asst dean coop exten, 80-82, assoc dean, 82-88, prof agron, Pa State Univ, 68-, actg dean agr, 85-86; RETIRED. *Concurrent Pos:* Chief agriculturist, Tipton & Kalmbach, Inc, consult water & power develop authority, WPakistan, 63-65. *Mem:* Am Soc Agron; Coun Soil Testing & Plant Anal. *Res:* Relationships between chemical soil test and the potash content of timothy and red clover; amino acid content of grasses as affected by nitrogen fertilizers and time of cutting; agronomy extension; soil fertility; plant nutrition. *Mailing Add:* 1851 Park Forest Ave State Col University Park PA 16803

HINK, WALTER FREDRIC, b Alhambra, Calif, June 29, 38; m 61; c 2. CELL BIOLOGY, NATURAL PRODUCT INSECTICIDES. *Educ:* Ashland Col, BS, 61; Ohio State Univ, MS, 63, PhD(entom), 65. *Prof Exp:* Res entomologist, Int Minerals & Chem Corp, 66-68; from asst prof to assoc prof, 68-76, PROF ENTOM & MICROBIOL, OHIO STATE UNIV, 77 - *Concurrent Pos:* Res grants, NSF & Environ Protection Agency, 73, WHO, 78. *Mem:* AAAS; Tissue Cult Asn (treas, 76); Entom Soc Am; Soc Invert Path; Am Soc Microbiol. *Res:* Invertebrate cell culture; characterization of cultured cells; cell nutrition; new approaches to insect control, in vitro culture of differentiated cells and cell hybridization; insecticides from natural products; biological activity of venoms; flea control with natural products and systemics. *Mailing Add:* Dept Entom Ohio State Univ 103 Bot/Zool Bldg Columbus OH 43210

HINKAMP, JAMES BENJAMIN, b Holland, Mich, Mar 18, 19; m 43; c 3. CHEMISTRY. *Educ:* Hope Col, AB, 40; Ohio State Univ, PhD(org chem), 43. *Prof Exp:* Asst, Ohio State Univ, 40-42; res chemist, 43-48, sect head, 48-52, res supvr, 52-63, dir chem tech serv, 63-65, asst dir res & develop, 65-83, CONSULT, ETHYL CORP, 83- *Honors & Awards:* Midgley Award, 85. *Mem:* AAAS; Am Chem Soc. *Res:* Development, testing and analysis of fuel and lubricant additives for diesel and gasoline engines. *Mailing Add:* 1444 S Bates St Birmingham MI 48009-1903

HINKAMP, PAUL EUGENE, (II), b Mich, Jan 29, 25; m 50; c 3. ORGANIC CHEMISTRY. *Educ:* Hope Col, AB, 49; Ohio State Univ, PhD, 53. *Prof Exp:* Res chemist & proj leader petrochem res, Dow Chem Co, 54-65, res chemist, Plastics Prod Res Dept, 65-67, prod specialist saran coating resins, 67-70, sr res chemist designed polymer res, 70- 73, res specialist designed latex & resins, 74-80, res leader, 80-86; RETIRED. *Mem:* Am Chem Soc; Sci Res Soc Am. *Res:* Polymer chemistry; surfactants; latex chemistry; latex polymerizations. *Mailing Add:* 1304 Ashly Midland MI 48640

HINKE, JOSEPH ANTHONY MICHAEL, b Vancouver, BC, Nov 6, 31; m 57; c 3. ANATOMY, BIOPHYSICS. *Educ:* Univ BC, MD, 57. *Prof Exp:* Asst prof, Univ BC, 60-65, assoc prof, 65-68, prof anat, 68-77; PROF ANAT & CHMN DEPT, SCH MED, UNIV OTTAWA, 77- *Concurrent Pos:* Am Life Ins res fel, Univ Col, Univ London, 58-60; Markle scholar, Univ BC, 60-65; mem adv, BC Heart Found, 65-69 & Can Heart Found, 66-69; mem scholar selection comt, Med Res Coun Can, 69-74, mem grants comt physiol & pharmacol, 73-; assoc ed, Can J Physiol & Pharmacol, 69-74. *Mem:* Can Asn Anat; Can Physiol Soc; Biophys Soc; Am Physiol Soc; Am Asn Anat. *Res:* Physico-chemical state of water and ions in muscle; electrolyte permeability, transport and contraction; ion-sensitive microelectrodes. *Mailing Add:* Dept Anat Univ Ottawa Ottawa ON K1N 6N5 Can

HINKEBEIN, JOHN ARNOLD, b Leopold, Mo, Mar 5, 31; m 55; c 6. PHYSICAL CHEMISTRY. *Educ:* Southeast Mo State Col, BS, 54; Iowa State Univ, PhD, 58. *Prof Exp:* Res chemist, 58-67, res specialist, 67-85, SR RES SPECIALIST, MONSANTO CO, 85- *Mem:* Am Chem Soc. *Res:* Phosphorus compounds. *Mailing Add:* 2450 Barrett Station Rd Manchester MO 63011

HINKEL, ROBERT DALE, b York, Pa, Mar 11, 14; m 38; c 2. CHEMISTRY. *Educ:* Pa State Col, BS, 36, PhD(tech of fuels), 46; Lawrence Col, MS, 38. *Prof Exp:* Asst petrol ref, Pa State Col, 38-43 & fuel tech, 43-47; chief anal chem sect, off synthetic liquid fuels, Res & Develop Br, US Bur Mines, 47-53; pres, R D Hinkel & Co, 53-; mgr anal res & develop, Koppers Co, Inc, 54-62; sr proj res engr, Arnold Eng Develop Ctr, 63; assoc prof, 63-64, prof 64-79, EMER PROF CHEM, UNIV PITTSBURGH, 79- *Res:* Chemical and instrumental methods of analysis of hydrocarbons, synthetic fuels, gases, polymers, coal tar oils, catalysts, latices, dyestuffs and fine organic chemicals. *Mailing Add:* 284 Streets Run Rd Pittsburgh PA 15236-2005

HINKELMANN, KLAUS HEINRICH, b Bad Segeberg, Ger, June 6, 32; m 66; c 1. STATISTICS. *Educ:* Univ Hamburg, dipl, 58; Iowa State Univ, PhD(statist), 63. *Prof Exp:* Sci asst statist, Univ Freiburg, 64-66; assoc prof, 66-72, PROF STATIST, VA POLYTECH INST & STATE UNIV, 72-, HEAD DEPT, 82- *Concurrent Pos:* Ed, Biomet, 90-93. *Mem:* Biomet Soc; Inst Math Statist; fel Am Statist Asn; Sigma Xi; fel AAAS. *Res:* Experimental design; genetic statistics; linear models. *Mailing Add:* Dept Statist Va Polytech Inst & State Univ Blacksburg VA 24061-0439

HINKES, THOMAS MICHAEL, b Watertown, Wis, Aug 14, 29; m 53, 84; c 3. TECHNOLOGY TRANSFER, PATENT LICENSING. *Educ:* Univ Notre Dame, BS, 51. *Prof Exp:* Applns res chemist, Victor Chem Works, Ill, 51-58; sr res chemist, Trionics Corp, 58-62; vpres & gen mgr chem div, Madison Res & Develop Labs, Inc, 62-67, pres & sales mgr, Madison Res, Inc, 67-68, dir coating develop lab, 68-77, ASSOC LICENSING DIR & SR LICENSING ASSOC, WIS ALUMNI RES FOUND, 77- *Concurrent Pos:* Instr, Edgewood Col, 59-60; vpres cent region, Asn Univ Technol Mgrs, 91-; consult, expert witness, lectr-intelectual prop. *Mem:* Licensing Execs Soc; Asn Univ Technol Mgrs. *Res:* Technical management; technology valuation. *Mailing Add:* 4934 Raymond Rd Madison WI 53711

HINKLE, BARTON L(ESLIE), b Miami Beach, Fla, Nov 2, 25; m 49, 59; c 5. CHEMICAL ENGINEERING. *Educ:* Purdue Univ, BSChE, 49; Inst Textile Technol, MS, 51; Ga Inst Technol, PhD(chem eng), 53. *Prof Exp:* Asst, eng exp sta, Ga Inst Technol, 51-53; res engr, E I Du Pont De Nemours & Co, Inc, 53-55, supvr, 55-58, tech supt, 58-63, asst plant mgr, 63-64, plant supt, 64-69, res mgr, 69-75, admin asst, 75-77, personnel mgr, 77-84; vpres Human Resources, Electromagnetic Sci, Inc, 84-87, CONSULT, 87- *Mem:* Am Inst Chem Engrs. *Res:* Cellulose chemistry; rheology of cellulose dispersions; heat-transfer characteristics of cellulose dispersions; fine particle technology; aggregation of aerosols, pneumatic conveying. *Mailing Add:* 5773 Huntonwood Dr Broad Run VA 22014

HINKLE, CHARLES N(ELSON), b Lafayette, Ind, Sept 12, 30; m 51; c 4. AGRICULTURAL ENGINEERING, MICROPROCESSOR CONTROLS. *Educ:* Purdue Univ, BSAE, 51; Mich State Univ, MS, 53; Univ Mo, PhD(agr eng), 57. *Prof Exp:* Asst farm struct, Mich State Univ, 51-52, instr, 52-53; agr market specialist, Armco Steel Corp, 53; asst farm struct, Univ Mo, 55-56, instr, 56-57; assoc prof, SDak State Univ, 57-65; assoc prof, 65-69, PROF AGR ENG, PURDUE UNIV, 69- *Concurrent Pos:* Adv, Freshman Eng Dept, Purdue Univ, 68-; vis prof, NC State Univ, 79. *Mem:* Fel Am Soc Agr Engrs; Am Soc Eng Educ; Am Soc Heating, Refrig & Air-Conditioning Engrs. *Res:* Environmental control for animals, plants and their products; instruments for measuring man's environment; microprocessor control systems for agricultural equipment. *Mailing Add:* 401 N Sharon Chapel Rd West Lafayette IN 47906

HINKLE, DALE ALBERT, b Perkins, Okla, Feb 2, 11; m 36; c 1. AGRONOMY. *Educ:* Okla State Univ, BS, 34, MS, 36; Univ Mass, PhD(agron), 49. *Prof Exp:* Asst prof agron, NMex State Col, 36-46; actg head dept, 49-50, from assoc prof, to prof agron, 48-77, head dept, 50-77, EMER PROF AGRON, UNIV ARK, FAYETTEVILLE, 77. *Concurrent Pos:* Sabbatical, CIMMYT, El Batan, Mex. *Mem:* Am Soc Agron; Soil Sci Soc Am. *Res:* Soil fertility; soil management; plant nutrition. *Mailing Add:* 1648 E Shadowridge Fayetteville AR 72701

HINKLE, DAVID CURRIER, b Brattleboro, Vt, June 7, 44; m 68; c 3. ENZYMOLOGY. *Educ:* Harvard Univ, BA, 66; Univ Calif, Berkeley, PhD(biochem), 71. *Prof Exp:* Fel biochem, Harvard Med Sch, 71-74; asst prof, 74-80, ASSOC PROF BIOL UNIV ROCHESTER, 80- *Concurrent Pos:* Fel, Jane Coffin Childs Mem Fund Med Res, Harvard Med Sch, 71-73. *Mem:* Am Soc Biol Chemists. *Res:* Mechanisms and biological functions of nucleic acid enzymes with a current emphasis on proteins involved in DNA replication, recombination and repair. *Mailing Add:* Dept Biol Univ Rochester Rochester NY 14627

HINKLE, GEORGE HENRY, b Renssalear, Ind, May 7, 52; m 72; c 2. NUCLEAR MEDICINE IMAGING, RADIOIMMUNODIAGNOSIS. *Educ:* Purdue Univ, BS, 75, MS, 77. *Prof Exp:* Grad instr nuclear pharm, Sch Pharm, Purdue Univ, 75-77, instr, 77-78; asst prof nuclear pharm, Col Pharm, Univ Okla, 78-81; CLIN ASST PROF RADIOL & PHARM, COL MED, OHIO STATE UNIV, 81- *Concurrent Pos:* Dir, Nuclear Pharm Serv, Ohio State Univ Med Ctr, 81-; chmn, Sect Nuclear Pharm, Acad Pharm Pract, 85-86 & Specialty Coun Nuclear Pharm, Bd Pharmaceut Specialties, 85-91. *Mem:* Soc Nuclear Med; Am Pharmaceut Asn; Sigma Xi. *Res:* Radionuclide methodology and nuclear pharmacy practice; design, development preparation, quality assessment, clinical use and disposal of radiopharmaceuticals; development of new diagnostic and therapeutic radiopharmaceuticals; development of radiolabeled antibodies for diagnosis and therapy of cancer. *Mailing Add:* 1795 Upper Chelsea Rd Columbus OH 43212

HINKLE, LAWRENCE EARL, JR, b Raleigh, NC, Feb 7, 18; m 42; c 6. CARDIOVASCULAR EPIDEMIOLOGY, HUMAN ECOLOGY. *Educ:* Univ NC, AB, 38; Harvard Univ, MD, 42; Am Bd Internal Med, dipl, 53. *Prof Exp:* Intern, Peter Bent Brigham Hosp, Boston, 42-43; asst resident internal med, New York Hosp, 46-48; instr, 50-51, asst prof clin med, 51-56, assoc prof, 56-61, clin assoc prof med, 61-64, assoc prof, 64-71, PROF MED, MED COL, CORNELL UNIV, 71-, DIR HUMAN ECOL, 64- *Concurrent Pos:* Commonwealth fel med, New York Hosp, 48-50; asst physician, Out-Patient Dept, New York Hosp, 50-51, physician, 51-58, from asst attend physician to assoc attend physician, 58-71; consult, Ed Dept, Time, Inc, 54-; consult, Med Dept, Am Tel & Tel Co, 56-; mem environ sci rev comt, NIH, 64-66; mem environ sci training comt, Nat Inst Environ Health Sci, 66-70, chmn, 68-71; consult, Prog Resources & Environ, Ford Found, 68-70; chmn, Interuniv Bd Collabr, Bur Community Environ Mgt, Dept Health, Educ & Welfare, 69-70; mem bldg res adv bd, Comn on Sociotech Systs, Nat Res Coun, 75-81; mem policy bd, Multictr Invest Limitation Infarction Size, Nat Heart, Lung & Blood Inst, 78-; mem coun epidemiol, Am Heart Asn. *Mem:* Am Fedn Clin Res (secy-treas, 52-54, vpres, 55, pres, 56); Am Col Physicians; Am Psychosom Soc (pres, 56); fel Am Heart Asn; fel Am Pub Health Asn. *Res:* Relation of social and man-made environment to health; natural history of coronary heart disease and sudden death; epidemiology of cardiac dysrhythmias; human ecology. *Mailing Add:* 198 Bridle Path Lane New Cedars CT 06848

HINKLE, PATRICIA M, b Glen Ridge, NJ, Oct 12, 43; m 68; c 3. ENDOCRINOLOGY. *Educ:* Mt Holyoke Col, BA, 65; Univ Calif, Berkeley, PhD(biochem), 70. *Prof Exp:* Fel, Harvard Univ, 71-74; from asst prof to assoc prof, 75-87, PROF PHARMACOL, UNIV ROCHESTER MED SCH, 87- *Concurrent Pos:* Mem endocrinol study sect, 87-91; ed, Endocrinol, 90. *Mem:* Endocrine Soc; Am Soc Biochem & Molecular Biol; Am Soc Pharmacol & Exp Therapeut; Am Thyroid Asn. *Res:* Molecular mechanisms of hormone action with emphasis on the role of receptors and signal transduction pathways in the anterior pituitary. *Mailing Add:* Dept Pharmacol Univ Rochester Med Sch Rochester NY 14642

HINKLE, PETER CURRIER, b Keene, NH, Nov 13, 40; m 66; c 3. BIOCHEMISTRY. *Educ:* Harvard Univ, BS, 62; NY Univ, PhD(biochem), 67. *Prof Exp:* NIH fel, Glynn Res Labs, Bodmin, Eng, 67-68; asst prof, 70-80, PROF BIOCHEM & CHMN DEPT, CORNELL UNIV, 82- *Res:* Oxidative phosphorylation; ion transport and membrane structure. *Mailing Add:* Biotech Bldg Cornell Univ Ithaca NY 14853

HINKLEY, DAVID VICTOR, b Kent, Eng, Sept 10, 44; m 70; c 2. STATISTICS. *Educ:* Birmingham Univ, BSc, 65; London Univ, PhD(statist), 69; Imp Col, DIC, 69. *Prof Exp:* Lectr statist, Imp Col Sci & Technol, 67-69; vis asst prof, Stanford Univ, 69-71; lectr, Imp Col, 71-73; assoc prof, 73-76, chmn dept appl statist, 78-80, prof statist, Univ Minn, 76-; AT DEPT MATH, UNIV TEX. *Concurrent Pos:* Assoc ed, Annals Statist, 76-79, ed, 80-82; assoc ed, Biometrika, 76-80; ed, Annals Statist, 80-82. *Mem:* Royal Statist Soc; Int Statist Inst; fel Inst Math Statist; fel Am Statist Asn; fel AAAS. *Res:* Statistical methods and inference. *Mailing Add:* 1808 Intervail Dr Austin TX 78746

HINKLEY, EVERETT DAVID, b Augusta, Maine, Nov 19, 36; m 60; c 4. PHYSICS. *Educ:* Wash Univ, BS, 58; Northwestern Univ, MS, 61, PhD(physics), 63. *Prof Exp:* Res physicist, Automatic Elec Labs, Inc, 58-59; asst physics, Northwestern Univ, 59-60, res asst & assoc, 60-63; physicist, Lincoln Lab, Mass Inst Technol, 63-76; vpres, Laser Anal Inc, 76-77; group leader, Calif Inst Technol, 77-80, prog mgr, 80-82, mgr, Atmospheric Sci Sect, Jet Propulsion Lab, 82-84, mgr, Sensor Tech Prog, 84-86; chief electronics scientist, Lockheed Corp, 86-87; chief scientist, Hughes Aircraft Corp, 87-89; CHIEF SCIENTIST, GLOBAL CHANGE PROG, TRW, 89- *Concurrent Pos:* Sr res fel, Univ Calif, Los Angeles, 91-; mem, Tech Adv Comt, NASA. *Mem:* Sr fel Inst Elec & Electronics Engrs; fel Optical Soc Am. *Res:* Semiconductor heterojunctions; band structure; quantum effects; computer memory systems and components; communications techniques; tunable lasers; laser spectroscopy; air pollution monitoring; ion implantation; planetary atmospheres; infrared techniques; applications of lasers to pollution detection; remote sensing; space technology; author of one book. *Mailing Add:* One Space Park Redondo Beach CA 90278

HINKLEY, ROBERT EDWIN, JR, b Kansas City, Mo, Mar 10, 43; m 66; c 1. CELL BIOLOGY, ANESTHESIOLOGY. *Educ:* Tulane Univ, BS, 65, MS, 66; Univ Kans, PhD(comp biochem & physiol), 71. *Prof Exp:* Res assoc anesthesia, Med Sch, Northwestern Univ, Chicago, 71-72, asst prof anesthesia & anat, 72-77; from asst prof to assoc prof, 77-87, PROF ANAT, UNIV MIAMI, 87-, ASSOC DEAN, 89- *Concurrent Pos:* Grant, Med Sch, Northwestern Univ, Chicago, 72-77; prin investr, Dept Health & Human Serv grant, Univ Miami, 77-; vis worker Anesthesia, Northwick Park Hosp, Harrow, 79-90; dir & chmn Honors Prog Med Educ, Univ Miami Med Sch, 80-, assoc dean Honors Med Prog, 90- *Mem:* Am Soc Cell Biol. *Res:* Effects of general anesthetics on reproduction; toxicity, teratology; medical education. *Mailing Add:* Dept Anat Univ Miami Med Sch Miami FL 33101

HINKS, DAVID GEORGE, b Neenah, Wis, July 11, 39; m 76; c 2. SUPERCONDUCTIVITY, MAGNETISM. *Educ:* Ill Inst Technol, BS, 62; Ore State Univ, PhD(chem), 68. *Prof Exp:* asst chemist, 68-75, CHEMIST, ARGONNE NAT LAB, 75- *Mem:* Am Chem Soc; Am Phys Soc. *Res:* Synthesis and characterization of superconductivity materials; crystal growth. *Mailing Add:* Argonne Nat Lab Bldg 223 9700 S Cass Ave Argonne IL 60439

HINKSON, JIMMY WILFORD, b Los Angeles, Calif, June 18, 31; m 54; c 6. BIOCHEMISTRY. *Educ:* Brigham Young Univ, BSc, 56, MSc, 58; Ind Univ, PhD(biol chem), 62. *Prof Exp:* NIH fel biochem, Univ Minn, 62-63; res fel, Charles F Kettering Res Labs, 63-65; asst prof, Univ Ariz, 65-68; mem staff, Xerox Med Diag Opers, 68-70; PROF CHEM, CALIF STATE UNIV, STANISLAUS, 70- *Concurrent Pos:* NSF res grant, 66-68. *Mem:* AAAS; Am Chem Soc. *Res:* Enzymology; kinetics of enzyme reactions. *Mailing Add:* 801 W Monte Vista Ave Turlock CA 95380

HINKSON, THOMAS CLIFFORD, b Winston-Salem, NC, July 8, 39; m 60; c 3. PHYSICAL CHEMISTRY. *Educ:* Midwestern Univ, BS, 60; Univ Fla, MS, 62; Univ Ark, Fayetteville, PhD(phys chem), 70. *Prof Exp:* Sr nuclear engr, Ling-Temco-Vought, Inc, 62-64; instr chem, Tex A&I Univ, 64-65; from asst prof to assoc prof, 68-75, prof & head dept, 75-83, PROF CHEM & ASST VPRES ACAD AFFAIRS, TARLETON STATE UNIV, 83- *Mem:* Am Chem Soc. *Res:* Nuclear emulsions; hot atom recoil chemistry, photochemistry. *Mailing Add:* IBM Corp 1503 LBJ Freeway Dallas TX 75234-6032

HINMAN, ALANSON, b San Francisco, Calif, Apr 18, 21; m 50; c 4. PEDIATRICS. *Educ:* Stanford Univ, BA, 42; Johns Hopkins Univ, MD, 46; Am Bd Pediat, dipl, 55. *Prof Exp:* Intern, San Francisco Hosp, 46; jr asst resident, Outpatient Dept, Johns Hopkins Hosp, 49-50, house officer pediat, 50-51; instr pediat, Bowman Gray Sch Med, 52-56, assoc neuropsychiat, 52-60, asst prof pediat, 56-60, from asst prof to assoc prof pediat neurol, 68-86, dir Develop Eval Clin, 61-86, emer assoc prof pediat, 86-; RETIRED. *Concurrent Pos:* USPHS pediat fel child psychiat, Stanford Univ Hosps, San Francisco, 51-52; Markle scholar med sci, 55; Nat Inst Neurol Dis & Blindness vis fel neurol, Columbia-Presby Med Ctr, 57-60; med dir, Amos Cottage Rehab Hosp, 60-86; dep spec asst for ment retardation to President, 64; mem adv bd ment retarded & law, NIMH Proj grant, Inst Law & Criminol,

George Washington Univ, 64-68; consult, Ment Retardation Div, Bur Social & Rehab Serv, Dept Health, Educ & Welfare, 64-69; mem adv coun, NC Comt Children & Youth, 68-70; mem adv bd, NC Neurol & Sensory Dis Proj, Sch Pub Health, Univ NC; consult subcomt on prev & eval, NC Coun Develop Disabilities, mem, 74-76. *Honors & Awards:* Hadassah Humanitarian Award, 63. *Mem:* AMA; Am Acad Pediat; Am Asn Ment Deficiency. *Res:* Pediatric neurology. *Mailing Add:* 1244 Arbor Rd No 426 Winston-Salem NC 27104

HINMAN, CHANNING L, b Portland, Ore, Sept 3, 43; m 76; c 2. NEUROSCIENCE, NEUROIMMUNOLOGY. *Educ:* Brigham Young Univ, BS, 72; Univ Calif, Los Angeles, PhD(neurosci), 77. *Prof Exp:* Fel auditory physiol, Brain Res Inst, Univ Calif, Los Angeles, 77-79; FEL IMMUNOL, DEPT IMMUNOL & MICROBIOL, WAYNE STATE UNIV, 79- *Res:* Amelioration of antibody-mediated Myasthenia Gravis; autoimmune and viral mechanisms involved in multiple sclerosis; single neuron and evoked potential recording of brain activity; functional power series analysis of evoked potentials. *Mailing Add:* 2334 Chriswood Toledo OH 43615-1258

HINMAN, CHARLES WILEY, b Boone, Iowa, Feb 9, 27; m 47; c 1. ORGANIC CHEMISTRY. *Educ:* Grinnell Col, BA, 51; Univ Ill, PhD(org chem), 54. *Prof Exp:* Asst, Rubber Res Lab, Univ Ill, 51-54; res chemist, Dow Chem Co, 54-55, group leader, Agr Chem Res Dept, 57-61, dir res & develop, Pitman-Moore Div, 61-64, sci adv, Dow-Ledogo S P A, Milan, 64-68, asst dir res & develop, 68-72, DIR PHARMACEUT RES & DEVELOP, DOW CHEM CO, USA, 72- *Concurrent Pos:* Dir life scis res & develop, Diamond Shamrock Corp, 76-78, dir exploratory technol, 78-; pres, Hinman Assoc, 83-; vpres, Vega Biotechnol, Inc, 85- *Mem:* AAAS; Am Chem Soc. *Res:* Scientific adviser in matters of chemistry, drug research and technical organization. *Mailing Add:* 4211 N Camino Ferreo Tucson AZ 85715-6360

HINMAN, CHESTER ARTHUR, b Bremerton, Wash, Apr 22, 39; m 62; c 1. SOLID STATE CHEMISTRY, NUCLEAR REACTOR SAFETY. *Educ:* Univ Wash, BS, 61, MS, 63, PhD(ceramic eng), 73. *Prof Exp:* Engr mat res & develop, Gen Elec Corp, 62-65; scientist, Battelle Northwest Lab, 65-70; SR SCIENTIST MAT RES & DEVELOP, HANFORD ENG DEVELOP LAB, WESTINGHOUSE- HANFORD, 70- *Mem:* Am Ceramic Soc; Nat Inst Ceramic Eng; AAAS. *Res:* Equilibrium and kinetic processes in non-metallic, inorganic solids, liquids and vapors; vapor pressure; diffusion kinetics; fission gas release from ceramic nuclear fuels, transient ceramic nuclear fuels behavior. *Mailing Add:* 2214 Camas Ave Richland WA 99352

HINMAN, EDWARD JOHN, b New Orleans, La, Nov 10, 31; m 54; c 3. HEALTH CARE ADMINISTRATION, INTERNAL MEDICINE. *Educ:* Univ Okla, BA, 51; Tulane Univ, MD, 55; Johns Hopkins Univ, MPH, 71; Am Bd Prev Med, cert, 71. *Prof Exp:* Dir, Prof & Tech Develop Div, Regional Med Progs Serv, 71-73; dir res admin, Spec Res & Develop Projs, Nat Ctr Health Serv Res, USPHS, 73-74, dep dir admin, Off Prog Oper, 74, dir, Div Hosps & Clins, 74-78; EXEC DIR ADMIN, GROUP HEALTH ASN, INC, 78- *Concurrent Pos:* Intern, USPHS Hosp, New Orleans, La, 55-56, chief prof serv, 66-67; residency internal med, USPHS Hosp, Baltimore, Md, 58-61, dir, 68-71; asst med, Johns Hopkins Univ, 63-66, lectr pub health admin, 68-79; instr med, Tulane Univ, 66-67; renal consult, Greater Baltimore Med Ctr, 66; pres, Kidney Found Md, 68-70, mem med adv bd, 68-71, chmn dialysis & transplantation comt, 69-71; nat secy, USPHS Clin Soc, 65-66, nat pres, 66-67, pres Baltimore Chap, 68-70; fel, Am Col Prevent Med, 71, Am Pub Health Asn, Am Col Physicians & charter fel Soc Adv Med Systs, 71; mem coun, Alliance for Eng Med & Biol, 75, chmn proj adv comt, 75-77, vpres, 76-77, pres, 77-79; vpres pub health, Am Col Prev Med, 76-77; mem bd, Symp Computs Applns Med Care, 80-; assoc clin prof, Health Care Sci, George Wash Univ, 79-; exec comt, Group Health Asn Am, 79- *Mem:* Soc Adv Med Syst (treas, 74, pres-elect, 75-76, pres, 76-77); Am Fedn Clin Res; Am Soc Nephrology. *Mailing Add:* 2225 Cape Leonard Dr St Leonard MD 20685

HINMAN, EUGENE EDWARD, b Dubuque, Iowa, Jan 29, 30; m 52; c 2. STRATIGRAPHY, PALEONTOLOGY. *Educ:* Cornell Col, BA, 52; Wash State Univ, MS, 54; Univ Iowa, PhD(geol), 63. *Prof Exp:* Instr, 56-58, from asst prof to assoc prof, 60-74, PROF GEOL, CORNELL COL, 74- *Mem:* AAAS; Am Asn Petrol Geol; Soc Econ Paleont & Mineral; Sigma Xi. *Res:* Jurassic Carmel-Twin Creek facies of northern Utah; Silurian bioherms of Iowa. *Mailing Add:* 208 Seventh Ave N Mt Vernon IA 52314

HINMAN, GEORGE WHEELER, b Evanston, Ill, Nov 7, 27; m 52; c 3. ENERGY, TECHNOLOGY ASSESSMENT. *Educ:* Carnegie Inst Technol, BS, 47, MS, 50, DSc(physics), 52. *Prof Exp:* Instr physics, Carnegie Inst Technol, 51-52, res physicist, 52-54, from asst prof to assoc prof physics, 54-63; chmn, Exp Physics Dept, Gen Atomic Div, Gen Dynamics Corp, 63-66, chmn, Dept Physics, 66-69; dir, Nuclear Radiation Ctr, 69-79, dir, Environ Res Ctr, 69-79, PROF PHYSICS & DIR OFF APPL ENERGY STUDIES, WASH STATE UNIV, 69- *Concurrent Pos:* Consult, US Gen Acct Off, 77-; dir, NMex Energy Res & Develop Inst, 82-83. *Mem:* Fel Am Phys Soc; Am Nuclear Soc; Am Soc Eng Educ. *Res:* Nuclear and reactor physics; systems studies on energy resources and environmental quality; technology assessments. *Mailing Add:* 305 Troy Hall Wash State Univ Pullman WA 99164-4430

HINMAN, JACK WILEY, b Pilot Mound, Iowa, June 27, 19; m 42; c 6. BIOCHEMISTRY, RESEARCH ADMINISTRATION. *Educ:* Grinnell Col, AB, 41; Univ Vt, MS, 43; Univ Ill, PhD(biochem), 47. *Prof Exp:* Res chemist, Container Corp Am, Ill, 42; instr chem, Univ Vt, 42-43; spec asst, Univ Ill, 43-44; res chemist, Upjohn Co, 44-57, head natural prod res sect, 57-73, chmn, Calcium Metab Res Proj Team & clin res monitor exp med res, 73-77; clin res scientist, Hoffmann-LaRoche, Inc, Nutley, NJ, 78-82; CONSULT, 82- *Honors & Awards:* W E Upjohn Prize, 63. *Mem:* Fel Am Inst Chem; Am Chem Soc; Am Soc Bone & Mineral Res; Marine Technol Soc. *Res:* Chemistry of penicillin; amino acid derivatives; extraction, purification and chemistry of antibiotics, protein hormones, hypothalamic factors, prostaglandins, antihypertensive kidney lipids and vitamin D metabolites. *Mailing Add:* 717 Chalfonte Ave Kalamazoo MI 49008

HINMAN, NORMAN DEAN, b Greeley, Colo, June 4, 44; m 66; c 2. BIOPROCESS DEVELOPMENT. *Educ:* Johns Hopkins Univ, BA, 66; Univ Conn, PhD(biochem), 72; Colo Sch Mines, MS, 80. *Prof Exp:* Res assoc biophys, Med Sch, Univ Colo, 72-75; asst prof biochem, Eastern Va Med Sch, 75-77; instr, Colo Sch Mines, 79-80; sales forecast coordr, Gates Rubber Co, 80; process engr, Stearns Catalytic Corp, 80-86, sr biochem engr, 86-87; SR BIOCHEM ENGR, SOLAR ENERGY RES INST, 87- *Mem:* AAAS. *Res:* Bioprocess development and demonstration; conceptual design and economic analysis of bioprocesses; bench scale and pilotscale bioprocess research. *Mailing Add:* Solar Energy Res Inst 1617 Cole Blvd Golden CO 80401

HINMAN, PETER GREAYER, b Chicago, IL, Dec 8, 37; US citizen. RECURSION THEORY. *Educ:* Univ Calif Berkeley, PhD(math), 66. *Prof Exp:* From asst prof to assoc prof, 66-84, PROF MATH, UNIV MICH, 84- *Concurrent Pos:* Vis prof, Univ Oslo, 71-72, Univ Heidelberg, 76-77; res fel, Math Inst, ETH, Zurich, 83-84. *Res:* Recursion theory (classical and generalized); complexity theory. *Mailing Add:* Dept Math Univ Mich Ann Arbor MI 48109-1003

HINMAN, RICHARD LESLIE, b Utica, NY, Mar 8, 27; m 67; c 4. BIO-ORGANIC CHEMISTRY, BIOTECHNOLOGY. *Educ:* Columbia Col, AB, 49; Univ Ill, Urbana, PhD, 52. *Prof Exp:* Merck fel, Cambribge Univ, Eng, 52-53; instr orgn chem, State Univ Iowa, 53-57; res chemist & group leader, Union Carbide Res Inst, 58-65, dir med chem, 65-67, mgr pharm technol, 67-69, prog mgr, chem sci, 69-72; vpres, 72-81, SR VPRES, CHEM PROD RES & DEVELOP, PFIZER INC, GROTON, CONN, 81- *Concurrent Pos:* Mem Energy Res Adv Bd, Wash, DC, 79-82; vis comt dept chem, Mass Inst Technol, 81-, Indust Adv Bd, Dept Biochem Eng, 85-; chmn, Steering Comt, Old Lyme Parents Network, Conn, 86-87; mem, Educ Comt, Conn Acad Sci & Eng, 91- *Mem:* Am Chem Soc; Chem Soc London; Sigma Xi. *Res:* Fermentation research biotechnology, bioorganic chemistry; author of over 50 scientific papers. *Mailing Add:* Pfizer Inc Eastern Point Rd Groton CT 06340

HINNERS, NOEL W, b Brooklyn, NY, Dec 25, 35; m 62; c 2. GEOCHEMISTRY, GEOLOGY. *Educ:* Rutgers Univ, BSc, 58; Calif Inst Technol, MSc, 60; Princeton Univ, MA, 61, PhD(geochem, geol), 63. *Prof Exp:* Mem tech staff lunar sci, Bellcomm, Inc, Washington, DC, 63-65, supvr, 65-70, dept head, 70-72; dir, Lunar Progs, NASA, 72-74, assoc adminr, Space Sci, 74-79, dir, Nat Air & Space Mus, 79-82, dir, Goddard Space Flight Ctr, 82-87, ASSO DEP ADMINR, NASA, 87- *Mem:* Am Geophys Union; AAAS; Am Inst Aeronauts & Astronauts. *Res:* Geochemistry of ore-forming fluids; lunar geology and mission planning; physics and chemistry of meteorites. *Mailing Add:* Seven Greyswood Ct Rockville MD 20854

HINNOV, EINAR, b Estonia, Mar 17, 30; nat US; m 56; c 1. PHYSICS. *Educ:* St Olaf Col, BA, 52; Duke Univ, PhD(physics), 56. *Prof Exp:* Res assoc physics, Univ Md, 56-58, asst res prof, 58-59; res physicist, Proj Matterhorn, 59-63 & Plasma Physics Lab, 63-74, sr res physicist, 74-80, PRIN RES PHYSICIST, PLASMA PHYSICS LAB, PRINCETON UNIV, 80- *Concurrent Pos:* Lectr, Dept Astrophys Sci, Princeton Univ, 65-74. *Mem:* Fel Am Phys Soc; Sigma Xi. *Res:* Atomic physics; plasma physics. *Mailing Add:* Plasma Physics Lab Princeton Univ Princeton NJ 08540

HINRICHS, CLARENCE H, b Basin, Wyo, Jan 21, 35; m 55; c 3. PHYSICS. *Educ:* Linfield Col, BA, 57; Iowa State Univ, MS, 60; Washington State Univ, PhD(physics), 66. *Prof Exp:* Res asst, Ames Lab, AEC, 57-60; res physicist, Linfield Res Inst, 60-62; instr, Wash State Univ, 62-64; assoc prof physics & registr, 66-77, actg dean fac, 74-77, PROF PHYSICS, LINFIELD COL, 77- *Mem:* Am Asn Physics Teachers. *Res:* Vacuum, low temperature and high pressure physics; superconductivity; field emission and surface physics; geophysics. *Mailing Add:* 726 Villard St McMinnville OR 97128

HINRICHS, KARL, b Port Washington, NY, Aug 11, 25; m 46; c 3. ELECTRONICS, COMPUTER SCIENCE. *Educ:* Swarthmore Col, BS, 45; Harvard Univ, MS, 47; Univ Calif, Berkeley, PhD(elec eng), 55. *Prof Exp:* Lectr, Univ Calif, Berkeley, 47-51, asst prof, 53-55; tech staff mem, Thompson-Ramo-Wooldridge, 55-56; sr engr, Systs Div, Beckman Instruments, Inc, 56-57, chief develop engr, 57-58, asst dir res & develop, 58-60, chief scientist, 60-64, dir res, 64-65, mgr comput opers, 65-66; dir res, Zeltex, Inc, 66-67; vpres & dir eng, Comcor/Astrodata, 67-68, Astrodata, 68-69; dir eng, Data Prod Div, Lockheed Electronics Co, 69-77; VPRES ENG, NEWPORT LABS, 77- *Mem:* AAAS; Inst Elec & Electronics Engrs; Instrument Soc Am; Sigma Xi. *Res:* Invention and application of large scale integrated circuits to computer memories and central processing unit, including minicomputer systems. *Mailing Add:* 7230 Rio Flora Pl Downey CA 90241-2030

HINRICHS, KATRIN, b Oakland, Calif, Nov 13, 54; m 88; c 1. OOCYTE MATURATION, EMBRYO TRANSFER. *Educ:* Univ Calif, Davis, BS, 76, DVM, 78; Univ Pa, PhD(comp med sci), 88. *Prof Exp:* Vet, pvt pract, 78-82; resident reproduction, New Bolton Ctr, Univ Pa Sch Vet Med, 82-84, lectr, 84-88; ASST PROF REPRODUCTION, TUFTS SCH VET MED, 88- *Honors & Awards:* Hamilton-Thorn Award, 90. *Mem:* Am Vet Med Asn; Soc Study Reproduction; Soc Theriogenol; Am Col Theriogenologists; Int Embryo Transfer Soc; Am Asn Equine Practitioners. *Res:* Early pregnancy in the mare including hormonal requirements for maintenance of pregnancy in ovaviectomized mares; oocyte retrieval, maturation, fertilization; early embryonic development in the horse. *Mailing Add:* Tufts Univ Sch Vet Med 200 Westboro Rd North Grafton MA 01536

HINRICHS, LOWELL A, b Hillsboro, Ore, June 16, 35; m 58; c 1. MATHEMATICS. *Educ:* Univ Ore, BA, 58, MA, 60, PhD(math), 62. *Prof Exp:* Asst prof math, Duke Univ, 62-67; ASSOC PROF MATH, UNIV VICTORIA BC, 67- *Res:* Symmetry geometry, polyehedra and tensigrids. *Mailing Add:* Dept Math Univ Victoria Box 1700 Victoria BC V8W 2Y2 Can

HINRICHS, PAUL RUTLAND, b Tulsa, Okla, July 17, 28; m 49; c 1. AERONAUTICAL & ASTRONAUTICAL ENGINEERING. *Educ:* Univ Okla, BS, 60, MS,61, PhD(elec eng),63. *Prof Exp:* Aero engr, Gen Dynamics, Ft Worth, 63-65; prof dir, Gorland Div, E-Systs, Inc, 65-78; chief engr, Missile Systs Div, Rockwell Int, 78-88; vpres eng, BEI Electronics, 88-91; RETIRED. *Mem:* Inst Elec & Electronics Engrs; Am Inst Aeronaut & Astronaut. *Res:* Integrated position updating navigation systems using terrain correlation, for use in navigating long range cruise missiles. *Mailing Add:* 2317 Meandering Way Arlington TX 76011-2619

HINRICHSEN, JOHN JAMES LUETT, b Davenport, Iowa, July 21, 03; m 29, 38; c 4. MATHEMATICS. *Educ:* Iowa State Col, BS, 25; Harvard Univ, AM, 27, PhD(math), 29. *Prof Exp:* From instr to prof, 29-74, asst to dean grad col, 35-38, assoc dean col sci & humanities, 60-61 & 62-69, actg dean, 61-62, actg head dept sociol & anthrop, 66-68, head dept math, 54-60, budget & personnel supvr col sci & humanities, 69-74, EMER PROF MATH, IOWA STATE UNIV, 74- *Mem:* AAAS; Am Math Soc; Soc Indust & Appl Math; Am Soc Eng Educ; Math Asn Am. *Res:* Problem of n-bodies; potential theory; analysis. *Mailing Add:* 321 Pearson Ave Ames IA 50010

HINSCH, GERTRUDE WILMA, b Chicago, Ill, Oct 20, 32; wid. ZOOLOGY, EMBRYOLOGY. *Educ:* Northern Ill Univ, BS, 53; Iowa State Univ, MS, 55, PhD, 57. *Prof Exp:* Instr zool, Mt Holyoke Col, 57-60; from asst prof to assoc prof, Mt Union Col, 60-66; res scientist, Inst Molecular Evolution, Univ Miami, 66-69, assoc prof, Inst Molecular Evolution & Dept Biol, 69-74; assoc prof, 74-80, PROF, UNIV SFLA, 80- *Concurrent Pos:* Corp mem, Marine Biol Lab. *Mem:* AAAS; Am Soc Zool; Soc Develop Biol; Am Soc Cell Biol; Int Soc Develop Biol. *Res:* Ultrastructure of fertilization and gamete structure in coelenterates and crustaceans; hormonal control of reproduction in crustaceans. *Mailing Add:* jept Biol Univ SFla Tampa FL 33620

HINSCH, JAMES E, b Pontiac, Mich, Oct 18, 37; m 58; c 4. CHEMICAL ENGINEERING. *Educ:* Univ Detroit, BChE, 60; Wayne State Univ, MSChE, 63. *Prof Exp:* Res scientist, sci res staff, Ford Motor Co, 60-66, sr res scientist, 66-67, sect supvr, Indust & Chem Prod Div, 67-71, sales & mkt mgr, electrocure opers, 71-72, asst mgr res & develop, 72-75, mgr, Res & Develop, 75-80, mfg mgr, Mt Clemens Paint Oper, 80-83 & Milan Plastics Plant, 83-85, mfg mgr, Mt Clemens Paint & Vinyl Opers, 85-86; Qual Mgr, 86-87, lab mgr, 87-90, SUPPLY CHAIN MGR, DU PONT CO, 90- *Mem:* Am Chem Soc; Sigma Xi. *Res:* Automotive coatings development including body enamels; electrodeposition and low emission systems. *Mailing Add:* Du Pont Automotive Prod 950 Stephenson Hwy Troy MI 48007-7013

HINSDILL, RONALD D, b Chicago, Ill, Dec 6, 33; m 60; c 3. IMMUNOTOXICOLOGY. *Educ:* Univ Wis-Madison, PhD(bact), 63. *Prof Exp:* PROF BACT, UNIV WIS-MADISON, 66-, CHMN DEPT, 82- *Concurrent Pos:* Dir, Environ Toxicol Ctr, 74-83. *Mem:* Am Soc Microbiol; Am Asn Immunol; Sigma Xi. *Mailing Add:* 411 Bact Bldg Univ Wis 1550 Linden Dr Madison WI 53706

HINSHAW, ADA SUE, b Arkansas City, Kans, May 20, 39. NURSING. *Educ:* Univ Kans, BS, 61; Yale Univ, MSN, 63; Univ Ariz, MA, 73, PhD(sociol), 75. *Hon Degrees:* DSc, Univ Md, 88, Med Col Ohio, 88, Marquette Univ, 90. *Prof Exp:* Instr nursing, Univ Kans, 63-65; instr, Sch Nursing, Univ Calif, San Francisco, 66-67, asst prof, 67-71; assoc dir nursing res, Dept Nursing, Univ Med Ctr, Tucson, Ariz, 75-87; DIR, NAT CTR NURSING RES, DEPT HEALTH & HUMAN SERV, NIH, BETHESDA, MD, 87- *Concurrent Pos:* Nurse-scientist fel, Univ Ariz, 71-75; green chair vis prof, Sch Nursing, Tex Christian Univ, 80; comnr, Comn Nursing Res, Am Nurses Asn, 80; vis prof, Sch Nursing, Univ Tex, 86, Univ Mich, Ann Arbor, 87; mem, Comt Advise Pub Health Serv Med Pract Guidelines, Inst Med, 89-90; co-chair, Int Task Force Nursing Res, Nat Ctr Nursing Res & Int Coun Nurses, Geneva, Switz, 90. *Honors & Awards:* Nurse Scientist Year Award, Am Nurses' Asn, 85; Helen Denne Schulte Vis Lectr, Univ Wis-Madison, 86. *Mem:* Inst Med; Sigma Xi; fel Am Acad Nursing; Am Nurses Asn. *Res:* Clinical nursing and nursing administration; theory construction and testing; instrument development and testing; field designs and issues. *Mailing Add:* Nat Ctr Nursing Res NIH Bldg 31-Rm 5B-03 Bethesda MD 20892

HINSHAW, CHARLES THERON, JR, b Wichita, Kans, Jan 31, 32; m 88; c 3. CLINICAL PATHOLOGY, NUCLEAR MEDICINE. *Educ:* Univ Kans, BA, 54, MD, 58. *Prof Exp:* Intern, Med Ctr, Univ Kans, 58-59, resident fel path, 61-65; assoc pathologist, Hutchinson Hosp, 65-79, dir, Biol Ctr Lab, 79-83; clin asst prof path, 81-84, PRECEPTOR, UNIV KANS SCH MED, 85- *Concurrent Pos:* Assoc dir, Lettner-Hinshaw Labs, 72-79; med dir, Sci Labs, 80-83, Ice Dream Int, 85-90. *Mem:* Am Soc Clin Pathologists; Col Am Pathologists; AMA; AAAS; Am Acad Environ Med. *Res:* Food allergy. *Mailing Add:* 1133 E Second Wichita KS 67214

HINSHAW, DAVID B, b Whittier, Calif, Nov 24, 23; m 43; c 3. SURGERY. *Educ:* Loma Linda Univ, MD, 47. *Prof Exp:* From instr to asst prof surg, 54-61, dean sch med, 62-74, prof surg, Loma Linda Univ, 61-, chmn dept, 74-; DEAN, SCH MED, ORAL ROBERTS UNIV. *Mem:* Am Surg Asn; Am Col Surgeons; Soc Univ Surg; Int Soc Surg. *Res:* Dumping syndrome; homograft rejection factor; metabolic and hemodynamic studies in surgical diseases. *Mailing Add:* Vet Admin Hosp Loma Linda CA 92357

HINSHAW, J RAYMOND, b Butler, Okla, Oct 8, 23. SURGERY. *Educ:* Univ Okla, BA, 43, MD, 46; Oxford Univ, DPhil(exp anat), 51. *Prof Exp:* From instr to prof, 54-89, EMER PROF SURG, SCH MED & DENT, UNIV ROCHESTER, 89-; EMER CHMN, DEPT SURG, ROCHESTER GEN HOSP, 89- *Concurrent Pos:* Chief flashburn sect, Atomic Energy Proj, Univ Rochester, 55-60; attend surgeon, Willard State Hosp, 63-; consult, Batavia Vet Admin Hosp, 63-; assoc chief surg, Rochester Gen Hosp, 64-67; mem, Plastic Surg Res Coun, 65-; mem bd govs, Am Col Surgeons, 75; NY State Bd Med, 77-80; dir surg educ, Rochester Gen Hosp, 64-, chief surg, 67- *Mem:* Am Soc Exp Path; Soc Univ Surg; fel Am Col Surg; Int Soc Surg. *Res:* Experimental anatomy; thermal burns; wound healing; splenic tissue autotransplantation; laser surgery. *Mailing Add:* 748 Quaker Rd Scottsville NY 14546

HINSHAW, JERALD CLYDE, b Weiser, Idaho, May 6, 44. ORGANIC CHEMISTRY. *Educ:* Ore State Univ, BS, 66; Univ Utah, PhD(org chem), 70. *Prof Exp:* Sr chemist, 70-76, res assoc, res labs, Eastman Kodak Co, 76-78; scientist, Res & Develop Lab, 78-84, supvr, 84-90, MGR APPL RES DEPT, RES & DEVELOP LAB, THIOKOL CORP, 90- *Mem:* Am Chem Soc; AAAS. *Res:* Synthesis and properties of unusual and or theoretically interesting molecules; high energy compounds; applications of chemistry to problems in geology and biology; research management. *Mailing Add:* Res & Develop Lab Thiokol Corp PO Box 707 MS 244 Brigham City UT 84302-0707

HINSHAW, LERNER BRADY, b San Diego, Calif, June 9, 21; m 46; c 4. PHYSIOLOGY & PATHOPHYSIOLOGY. *Educ:* Univ Southern Calif, AB, 49, MS, 50, MA, 52, PhD(physiol), 55. *Prof Exp:* From instr to asst prof physiol, Med Sch, Univ Minn, 58-61; supvry res physiologist, Civil Aeromed Res Inst, 61-65; from assoc prof to prof physiol, 61-71, LYNN CROSS RES PROF, UNIV OKLA, 85- *Concurrent Pos:* Fel, Life Ins Med Res Fund, Med Sch, Univ Minn, 55-58; Lederle med fac award, 59-62; adj prof surg, med sch, Univ Okla, 66-; chmn, res & develop comt, Vet Admin Med Ctr, Oklahoma City, 71-73 & 80-82, prin proponent & co-chmn, Vet Admin coop study No 209 (nat clin trial study); med investr, Vet Admin Hosp, Oklahoma City, 73-82; res career scientist, Vet Admin Med Ctr, Oklahoma City, 83-87; mem, Okla Med Res Found, 87- *Mem:* West Soc Clin Res; Soc Critical Care Med; Soc Exp Biol & Med; Am Physiol Soc; Shock Soc (pres, 85); fel Am Col Critical Care Med; Am Soc Exp Therapeuts. *Res:* Cardiovascular physiology; shock. *Mailing Add:* Okla Med Res Found 825 NE 13th St Oklahoma City OK 73104

HINSHAW, VIRGINIA SNYDER, b Oak Ridge, Tenn, Mar 25, 44; m 63; c 2. VIROLOGY, IMMUNOLOGY. *Educ:* Auburn Univ, BS, 66, MS, 67, PhD(microbiol), 72. *Prof Exp:* Clin & res microbiologist bacteriol, Med Col Va, 67-68; res virologist oncogenic viruses, Univ Calif, Berkeley, 74-75; res assoc influenza, St Jude Children's Res Hosp, 75-77, asst mem, Influenza, 77-80, assoc mem, 81-85; assoc prof, 85-88, PROF VIROL, SCH VET MED, UNIV WIS-MADISON, 88- *Mem:* Am Soc Microbiol; AAAS. *Res:* Ecology; structure and immunology of influenza viruses in humans, lower animals and birds. *Mailing Add:* Dept Pathobiol Sci Sch Vet Med Univ Wis 2015 Linden Dr W Rm 3174 Madison WI 53706

HINSMAN, EDWARD JAMES, b Wyandotte, Mich, Aug 5, 34; m 59; c 3. VETERINARY ANATOMY, COMPARATIVE NEUROLOGY. *Educ:* Mich State Univ, BS, 56, DVM, 58, MS, 60; Purdue Univ, PhD(vet anat), 63. *Prof Exp:* From instr to assoc prof, 59-70, PROF VET ANAT, PURDUE UNIV, 70- *Mem:* Am Asn Anatomists; World Asn Vet Anatomists; Am Asn Vet Anat (pres, 82-83). *Res:* Use of electron microscope to study the nervous system. *Mailing Add:* 2450 Lindberg Rd West Lafayette IN 47906

HINSON, DAVID CHANDLER, b Millen, Ga, Apr 17, 39; m 73; c 2. PHYSICS. *Educ:* Ga Inst Technol, BS, 61, MS, 64, PhD(physics), 66. *Prof Exp:* Sr res physicist, ITT Electron Tube Div, 67-70; mgr new prod develop, Owens-Ill Electro-Optical Displays, 70-76; mgr phys electronics, Gould Labs, 76-80; asst dept mgr, Litton Electron Tube Div, 80-; TECH DIR, MAT RES CORP. *Mem:* Inst Elec & Electronics Engrs; Am Phys Soc; Am Vacuum Soc; Sigma Xi. *Res:* Thin film electronic materials and devices; gas discharge display devices; surface and gas discharge physics; ultraviolet optical properties of solids. *Mailing Add:* 35 Jefferson St Pearl River NY 10965

HINSON, JACK ALLSBROOK, b Aug 18, 44; m 71; c 2. TOXICOLOGY, DRUG METABOLISM. *Educ:* Col of Charleston, BS(biology), 66, Univ S Carolina, MS(biol), 68, Vanderbilt Univ, PhD(biochem), 72. *Prof Exp:* Sr staff fel, Nat Heart, Lung & Blood Inst, 75-80; chief, Biochem Mechanisms Br, Nat Ctr Toxicol Res, 80-90; PROF & DIR, DIV TOXICOL, UNIV ARK MED SCI, 90- *Concurrent Pos:* Postdoctoral fel, NIH, 72-75; adj prof toxicol, Sch Med, Univ Ark, 80-90; vis fel, Can Res Campaign, Middlesex Hosp Med Sch, London, 82; adj assoc prof, Col Pharm, Univ Tenn, 82-90; vis prof, Div Toxicol, Univ Leiden, Neth, 86. *Mem:* Soc Toxicol (pres, 90-); Am Soc Pharmacol & Exp Therapeut. *Res:* Toxicology; drug metabolism carcinogenesis. *Mailing Add:* Div Toxicol Univ Ark Med Sci 4301 W Markham Little Rock AR 72205

HINSON, KUELL, b Moss, Tenn, Jan 8, 24; m 55; c 4. GENETICS, PLANT BREEDING. *Educ:* Tenn Polytech Inst, BS, 49; Univ Wis, MS, 51, PhD(genetics, agron), 54. *Prof Exp:* Geneticist, 54-73, RES AGRONOMIST, AGR RES SERV, USDA, 73- *Mem:* Am Soc Agron; Crop Sci Soc Am. *Res:* Soybean breeding and genetics; pyramiding pest-resistance factors into high yielding genotypes; adapting genotypes to broader bands of latitude; soybean production research. *Mailing Add:* Dept Agron Univ Fla Gainesville FL 32611

HINTERBERGER, HENRY, b Melk, Austria, June 27, 21; US citizen; m 50; c 3. SOLAR ENERGY, ENERGY ENGINEERING. *Educ:* City Col New York, BS, 48. *Prof Exp:* Chief engr synchrocyclotron, Univ Chicago, 48-68; assoc dir res, Fermi Nat Accelerator Lab, 68-85; SR SCIENTIST, RICE UNIV, 85- *Concurrent Pos:* Consult, Dept Energy, Solar Div, 77- *Mem:* Int Solar Energy Soc. *Res:* Superconducting synchrotron design; non-tracking solar concentrator design. *Mailing Add:* 88 Rolling Hills Dr Conroe TX 77304

HINTERBUCHNER, CATHERINE NICOLAIDES, b Corfu, Greece, Nov 22, 26; US citizen; m 55. PHYSICAL MEDICINE & REHABILITATION. *Educ:* Nat Univ Athens, MD, 51; Am Phys Med & Rehab, dipl, 62. *Prof Exp:* Asst instr, NY Med Col, 59-60; instr, State Univ NY Downstate Med Ctr, 60-64; from asst prof to assoc prof, 64-71, PROF REHAB MED & CHMN DEPT, NY MED COL, 71-, DIR METROP HOSP, NY, 64- *Concurrent Pos:* Fel rehab med, Jewish Chronic Dis Hosp, Brooklyn, NY, 56-57 & Metrop Hosp, 59-60; vis physician, Kings County Hosp, Brooklyn, 60-64; asst attend, Jewish Chronic Dis Hosp, 60-66; attend Vet Admin Hosp, 61-64; assoc attend, Flower & Fifth Ave, 64-71, attend, 71-78, Metrop, 64- & Bird S Coler

Hosps, 64-88; attend, Westchester County Med Ctr, Valhalla, NY, 71-, Lincoln Hosp, 79-; mem, Am Bd Phys Med & Rehab & Accreditation Coun Grad Med Ed. *Mem:* AMA; fel Am Col Physicians; NY Acad Med; Am Acad Phys Med & Rehab; Am Cong Rehab Med; NY Acad Sci. *Mailing Add:* Polly Park Rd Rye NY 10580

HINTERBUCHNER, LADISLAV PAUL, b Slovakia, Dec 10, 22; m 55. NEUROLOGY. *Educ:* Univ Bratislava, MD, 47. *Prof Exp:* From asst instr to instr neurol, State Univ NY Downstate Med Ctr, 55-57; USPHS trainee neurochem, Col Physicians & Surgeons, Columbia Univ, 57-59; asst prof, 59-61, from clin asst prof to clin assoc prof, 61-76, chmn neurol dept, Brooklyn Cumberland Med Ctr, 71-85, PROF NEUROL, STATE UNIV NY DOWNSTATE MED CTR, 76- *Concurrent Pos:* Dir neurol & psychiat, Jewish Chronic Dis Hosp, 61-66. *Mem:* Asn Res Nerv & Ment Dis; fel Am Acad Neurol; NY Acad Sci; fel Am Col Med; AMA. *Res:* Muscle physiology and its application to clinical neurology. *Mailing Add:* Polly Park Rd Rye NY 10580

HINTEREGGER, HANS ERICH, b Waidhofen, Austria, Sept 3, 19; nat US; m 42; c 6. PHYSICS. *Educ:* Vienna Inst Technol, Dipl Ing, 44, Dr Tech(physics), 47. *Prof Exp:* Asst physics, Univ Gottingen, 44-46; asst prof, Vienna Inst Technol, 46-51; physicist, 51-53, SR SCIENTIST, AIR FORCE GEOPHYS LAB, 61- *Concurrent Pos:* Consult, Los Alamos Sci Labs, 60-64. *Mem:* Int Astron Union; Int Union Geod & Geophys; fel Optical Soc Am; Am Geophys Union. *Res:* Physical electronics; extreme ultraviolet radiation; ionospheric physics; solar physics; space research; aeronomy. *Mailing Add:* 140 Newtonville Ave Newton MA 02158

HINTHORNE, JAMES ROSCOE, b Los Angeles, Calif, Dec 23, 42; m 64; c 2. MINERALOGY, COMPUTER SOFTWARE. *Educ:* Univ Calif, BA, 65, PhD(geol), 74; Univ Mass, MS, 67. *Prof Exp:* Assoc scientist microprobe res, 69-73, scientist, 73-75, res scientist, 75-76, mgr methods & appl div, Appl Res Labs, 76-80; PROF GEOL, CENT WASH UNIV, 80- *Concurrent Pos:* Consult mineral & comput appln. *Mem:* Geol Soc Am; Mineral Soc Am; Microbeam Anal Soc. *Res:* Ion microprobe; electron microprobe; trace element contents of minerals; U-Pb ages in lunar minerals and very old terrestrial rocks; control of analytical instruments by minicomputer; applications software for geographic information systems. *Mailing Add:* Dept Geol Cent Wash Univ Ellensburg WA 98926

HINTON, BARRY THOMAS, b 1950. MALE REPRODUCTIVE CELL & MOLECULAR BIOLOGY, SPERM PHYSIOLOGY. *Educ:* Inst Animal Physiol, Cambridge, Eng, PhD(male reproductive physiol), 79. *Prof Exp:* ASSOC PROF GROSS ANAT & REPRODUCTIVE BIOL, SCH MED, UNIV VA, 83- *Mem:* Soc Study Reproduction; Am Soc Cell Biol; Am Asn Anatomists. *Res:* Cell and molecular biology of testicular and epididymal function; role of epididymal epithelium in regulating the luminal microenvironment necessary for sperm development, maturation and survival. *Mailing Add:* Dept Anat & Cell Biol Sch Med Univ Va Box 439 Charlottesville VA 22908

HINTON, CLAUDE WILLEY, b Gatesville, NC, Aug 1, 28; m 52; c 2. GENETICS. *Educ:* Univ NC, AB, 48, MA, 50; Calif Inst Technol, PhD(biol), 54. *Prof Exp:* Res assoc, Biol Div, Oak Ridge Nat Lab, 54-55; from asst prof to prof zool, Univ Ga, 55-68; MATEER PROF BIOL, COL WOOSTER, 68- *Concurrent Pos:* Res grants, NIH, 55-; consult, Oak Ridge Nat Lab, 59-68; guest investr, Div Plant Indust, Commonwealth Sci & Indust Res Orgn, Canberra, Australia, 72-73; vis res biologist, Univ Calif, San Diego, 77-78, Univ NC, Chapel Hill, 82-83. *Mem:* AAAS; Genetics Soc Am; Am Soc Naturalists; Am Genetic Asn. *Res:* Chromosome behavior in Drosophila; hybrid dysgenesis. *Mailing Add:* Dept Biol Col Wooster Wooster OH 44691

HINTON, DAVID EARL, b Hattiesburg, Miss, Jan 4, 42; m 63; c 1. COMPARATIVE PATHOLOGY. *Educ:* Miss Col, BS, 65; Univ Miss, MS, 67, PhD(anat), 69. *Prof Exp:* From instr to asst prof gross anat, Sch Med, Univ Louisville, 70-73, res assoc water resources biol, 70-73; res assoc path, 73-77, ASST PROF PATH, SCH MED, UNIV MD, BALTIMORE, 73- *Res:* Electron microscopy; toxicology; comparative and environmental pathobiology; fishes; aquatic pollutants. *Mailing Add:* Dept Med Sch Vet Med Univ Calif Davis CA 95616

HINTON, DEBORAH M, b Greenville, SC, July 7, 53. CHEMISTRY. *Educ:* Univ NC, Chapel Hill, BS, 74; Univ Ill, MS, 76, PhD(biochem), 80. *Prof Exp:* Res Asst, Dept Biochem, Univ Ill, 75-80; postdoctoral fel, Cancer Biol Prog, Nat Cancer Inst, NIH, 80-82, staff fel, Sect Pharmacol, Lab Biochem Pharmacol, Nat Inst Arthritis, Diabetes & Digestive & Kidney Dis, 82-83, sr staff fel, 83-86, RES CHEMIST, SECT NUCLEIC ACID BIOCHEM, LAB BIOCHEM PHARMACOL, NAT INST DIABETES & DIGESTIVE & KIDNEY DIS, NIH, 86- *Concurrent Pos:* Postdoctoral fel, Am Cancer Soc, 80-82. *Mem:* Sigma Xi; AAAS; Am Soc Biol Chemists. *Res:* Biochemistry; numerous technical publications. *Mailing Add:* Nat Inst Diabetes Digestive & Kidney Dis NIH Bldg 8 Room 225 9000 Rockville Pike Bethesda MD 20892

HINTON, DON BARKER, b Savannah, Tenn, Aug 11, 37; m 58; c 1. MATHEMATICS. *Educ:* Univ Tenn, BSEE, 60, PhD(math), 63. *Prof Exp:* Asst prof math, Univ Tenn, 63-64; from asst prof to assoc prof, Univ Ga, 64-70; assoc prof, 70-74, PROF MATH, UNIV TENN, KNOXVILLE, 74- *Mem:* Am Math Soc; Soc Indust & Appl Math. *Res:* Differential equations. *Mailing Add:* Dept Math Univ Tenn Knoxville TN 37996

HINTON, FREDERICK LEE, b Yale, Mich, Aug 4, 39. PLASMA PHYSICS. *Educ:* Univ Mich, BSE, 62, MS, 63; Calif Inst Technol, PhD(physics), 67. *Prof Exp:* Res assoc physics, Plasma Physics Lab, Princeton Univ, 67-69; from asst prof to assoc prof physics, Univ Tex, Austin, 73-81; SR TECH ADV, GEN ATOMIC, 81- *Concurrent Pos:* Alfred P Sloan Found res fel, 74-76; assoc ed, Physics of Fluids, 75-77. *Mem:* Am Phys Soc. *Res:* Stability and transport properties of magnetically confined plasmas. *Mailing Add:* Gen Atomics PO Box 85608 San Diego CA 92138

HINTON, GEORGE GREENOUGH, b Shipley, Eng, July 31, 25; Can citizen; m 56; c 2. PEDIATRIC NEUROLOGY. *Educ:* McGill Univ, BSc, 51, MD, CM, 55; Univ London, dipl child health, 60; FRCP(C). *Prof Exp:* PROF PEDIAT, FAC MED, UNIV WESTERN ONT, 62-, PEDIAT NEUROLOGIST, WAR MEM CHILDREN'S HOSP, 62- *Mem:* Can Med Asn; Can Pediat Soc; Can Neurol Soc; Can Psychiat Asn; Am Acad Cerebral Palsy. *Res:* Mental retardation; epilepsy; learning disability. *Mailing Add:* Dept Pediat Univ Western Ont London ON N6C 2V5 Can

HINTON, JAMES FAULK, b Bessemer, Ala, May 5, 38; m 61; c 1. PHYSICAL CHEMISTRY. *Educ:* Univ Ala, BS, 60; Univ Ga, MS, 62, PhD(chem), 64. *Prof Exp:* From asst prof to assoc prof, 67-75, PROF PHYS CHEM, UNIV ARK, FAYETTEVILLE, 75- *Concurrent Pos:* Res fels, Univ Ga, 64-65 & Univ Ark, 65-67. *Mem:* Am Chem Soc; NY Acad Sci. *Res:* Kinetics of reactions in solution kinetic salt and solution effects; photochemically induced reactions in solution; magnetic resonance applications to solution phenomena. *Mailing Add:* Dept Chem Univ Ark Fayetteville AR 72701

HINTON, JONATHAN WAYNE, b Canton, Ohio, Apr 12, 44; m 66. CERAMIC ENGINEERING. *Educ:* Ohio State Univ, BS, 66, MS, 68, PhD(ceramic eng), 70. *Prof Exp:* Staff engr ceramic eng, 70-75, SECT HEAD, CERAMIC DIV, CHAMPION SPARK PLUG CO, 75- *Concurrent Pos:* Trustee, Edward Orton Jr Ceramic Found. *Mem:* Fel Am Ceramic Soc; Nat Inst Ceramic Engrs; Fed Mats Soc. *Res:* Chemical vapor deposition; sonic compaction; bioceramics; refractory, dielectric, and high temperature ceramics; materials for energy conservation; environmental effects of ceramic components and methods; ceramics for spark plug components. *Mailing Add:* Champion Spark Plug Co 20000 Conner Ave Detroit MI 48234

HINTON, RAYMOND PRICE, b Hereford, Gt Brit, Jan 18, 25; US citizen; m 49; c 4. TELECOMMUNICATION SYSTEMS, TELEPHONE & DATA SWITCHING. *Educ:* Univ Bristol, Gt Brit, BSc, 45. *Prof Exp:* Engr, Standard Tel & Cables, 46-53; lab dir, ITT, 53-65; vpres, Computer Sci Corp, 65-67; tech dir, Ford Aerospace, 67-77; chief engr, Sperry Univac, 78-81; pres, Telecommun Res Inc, 82-85; PRIN ENGR, GEN ELEC, 85- *Mem:* Fel Inst Elec & Electronics Engrs. *Res:* Electronic telecommunications; digital devices for data switching systems and telephone switching equipment; awarded 22 US and British patents. *Mailing Add:* 901 Carroll Rd Wynnewood PA 19096

HINTZ, HAROLD FRANKLIN, b Frank, Ohio, Oct 28, 37; m 59. ANIMAL NUTRITION. *Educ:* Ohio State Univ, BS, 59; Cornell Univ, MS, 61, PhD(animal nutrit), 64. *Prof Exp:* Asst prof animal husb, Univ Calif, Davis, 64-67; asst prof animal sci, 67-70, assoc prof, 70-78, PROF ANIMAL NUTRIT, CORNELL UNIV, 78- *Mem:* Am Soc Animal Sci; Am Inst Nutrit. *Res:* Horse nutrition; Ca and P metabolism. *Mailing Add:* Animal Sci Cornell Univ 149 Morrison Hall Ithaca NY 14853

HINTZ, HOWARD WILLIAM, b Dubuque, Iowa, Nov 27, 21; m 52; c 3. ZOOLOGY, ENTOMOLOGY. *Educ:* Iowa State Col, BS, 47; Ohio State Univ, MS, 49, PhD, 52. *Prof Exp:* Asst, Ohio State Univ, 47-51; from asst prof to assoc prof, 51-64, PROF BIOL, HEIDELBERG COL, 64- *Mem:* Entom Soc Am; Sigma Xi. *Res:* Insect ecology and taxonomy. *Mailing Add:* 229 Coe St Tiffin OH 44883

HINTZ, MARIE I, antiviral research, for more information see previous edition

HINTZ, NORTON MARK, b Milwaukee, Wis, Nov 22, 22; m 43, 90; c 3. NUCLEAR REACTIONS. *Educ:* Univ Calif, Los Angeles, BA, 43; Harvard Univ, PhD, 51. *Prof Exp:* AEC fel, Cavendish Lab, Eng, 51-52; from asst prof to assoc prof, 52-61, PROF PHYSICS, UNIV MINN, MINNEAPOLIS, 61- *Concurrent Pos:* NSF sr fel, Inst Theoret Physics, Copenhagen, 59-60; Guggenheim award, Israel & Copenhagen, 64; NSF physics adv panel, Los Alamos Meson Physics Faculty, 69-; NATO sr fel, Europ Orgn Nuclear Res, Geneva, 72; US-Japan Co-op Sci Fel, Kyoto Univ, 74-75; vis scientist, IPN, Onsay, 89. *Mem:* Am Phys Soc. *Res:* Nuclear reactions and spectroscopy. *Mailing Add:* Sch Physics 165 Physics Univ Minn Minneapolis MN 55455

HINTZ, RICHARD LEE, b Norwalk, Ohio, Mar 30, 49. BIOSTATISTICS. *Educ:* Ohio State Univ, BS, 71; Cornell Univ, PhD(animal breeding), 77. *Prof Exp:* ASST PROF SYSTS ANALYSIS & ANIMAL BREEDING, OKLA STATE UNIV, 77-; STATISTICIAN, MONSANTO INC, ST LOUIS, MGR, STATIST. *Mem:* Am Soc Animal Sci; Am Soc Dairy Sci; Biomet Soc; Am Statist Asn. *Res:* Manage a statistics group with an animal science division of Monsanto. *Mailing Add:* 700 Chesterfield Village Pkwy St Louis MO 63198

HINTZE, LEHI FERDINAND, b Denver, Colo, Apr 14, 21; m 42; c 4. GEOLOGY. *Educ:* Univ Utah, AB, 41; Columbia Univ, AM, 48, PhD(geol), 51. *Prof Exp:* From instr to asst prof geol, Ore State Col, 49-54; assoc prof geol, Brigham Young Univ, 55-59, chmn dept, 60-69, prof, 60-86; REGIONAL GEOLOGIST, UTAH GEOL SURV, 81- *Concurrent Pos:* Consult, US Geol Surv, 60-86. *Mem:* Geol Soc Am; Am Asn Petrol Geol. *Res:* Environmental geology; Ordovician stratigraphy of Utah and Nevada; Ordovician trilobites; geologic maps of Oregon and Utah; geologic mapping; structure and stratigraphy of western Utah. *Mailing Add:* 1835 N 1450 E Provo UT 84604

HINTZE, THOMAS HENRY, PHYSIOLOGY OF ATREAL ANTRURETIC FACTORS. *Educ:* NY Med Col, PhD(physiol), 79. *Prof Exp:* ASST PROF PHYSIOL, NY MED COL, 82- *Mailing Add:* Dept Physiol NY Med Col Elmwood Hall Valhalla NY 10595

HINTZEN, PAUL MICHAEL, b Glenwood, Minn, Feb 17, 50. ASTRONOMY. *Educ:* Univ Minn, BSc, 71; Univ Ariz, PhD(astron), 75. *Prof Exp:* Res asst astron, Univ Ariz, 74-75; fel astron, Kitt Peak Nat Observ, 75-; AT GODDARD SPACE FLIGHT CTR, NASA. *Res:* Observations and atmospheric analysis of white dwarfs. *Mailing Add:* Goddard Space Flight Ctr NASA Greenbelt Rd Greenbelt MD 20771

HINZ, CARL FREDERICK, JR, b Cleveland, Ohio, Apr 9, 27; m 53; c 4. INTERNAL MEDICINE, HEMATOLOGY. *Educ:* Western Reserve Univ, BS, 48, MD, 51. *Prof Exp:* From instr to assoc prof med, Western Reserve Univ, 57-67; PROF MED & ASSOC DEAN, SCH MED, UNIV CONN, 67- *Concurrent Pos:* Res fel med, Western Reserve Univ, 53-57; Markle scholar, 59-64. *Mem:* Am Fedn Clin Res; Am Soc Clin Invest; Am Asn Immunol; Am Soc Hemat. *Res:* Immunochemistry; mechanisms of hemolysis of red cells; complement and properdin; lymphocyte mediated cytotoxicity. *Mailing Add:* 11 Highwood Dr Avon CT 06001

HINZ, PAUL NORMAN, b Tarentum, Pa, Dec 23, 35; m 61; c 3. STATISTICS. *Educ:* Pa State Univ, BS, 57; NC State Univ, MS, 60; Univ Wis, MS, 63, PhD(statist), 67. *Prof Exp:* Wood technologist, US Forest Prod Lab, 59-62, math statistician, 63-64 & 67-68; asst prof, 68-72, assoc prof statist, 72-77, PROF STATIST & FORESTRY, IOWA STATE UNIV, 78- *Mem:* Biomet Soc; Am Statist Asn. *Res:* Statistical methods. *Mailing Add:* Dept Statist Iowa State Univ 122 Snedecor Ames IA 50011

HINZE, HARRY CLIFFORD, b Middleton, Wis, July 18, 30; m 57; c 3. VIROLOGY, VIRAL IMMUNOLOGY. *Educ:* Univ Wis, BS, 53, MS, 55, PhD(med microbiol), 58. *Prof Exp:* Proj assoc virol, 58-59, asst prof, 61-72, ASSOC PROF MED MICROBIOL, SCH MED, UNIV WIS-MADISON, 72- *Concurrent Pos:* USPHS trainee, Children's Hosp, Philadelphia, 59-61; prin investr, Meharry Med Sch, 65-, vis prof virol, 79-81. *Mem:* Am Soc Microbiol; Am Asn Immunol; Am Soc Virol. *Res:* Oncogenic herpes viruses; tumor virus-host cell interactions; tumor virus immunology. *Mailing Add:* Dept Med Microbiol Univ Wis Med Sch Madison WI 53706

HINZE, WILLIAM JAMES, b Milwaukee, Wis, July 26, 30; m 56; c 2. GEOPHYSICS. *Educ:* Univ Wis, BS, 51, PhD(geol), 57. *Prof Exp:* Staff geophysicist, Jones & Laughlin Steel Corp, 53 & 56-58; from asst prof to prof geol, Mich State Univ, 58-72; PROF GEOPHYS, PURDUE UNIV, 72- *Mem:* Soc Explor Geophys; Am Geophys Union. *Res:* Regional geophysical studies; mining and engineering geophysics; geoscience data management. *Mailing Add:* Dept Earth & Atmospheric Sci Purdue Univ W Lafayette IN 47907

HINZE, WILLIE LEE, b Burton, Tex, Jan 17, 49; m 80. ANALYTICAL & MEMBRANE MIMETIC CHEMISTRY, SEPARATION SCIENCE. *Educ:* Sam Houston State Univ, BS, 70, MA, 72; Tex A&M Univ, PhD(chem), 74. *Prof Exp:* Instr chem, Blinn Col, 74-75; from asst prof to prof, 75-89, UNIV PROF CHEM, WAKE FOREST UNIV, 89- *Concurrent Pos:* Lectr & NIH fel chem, Tex A&M Univ, 74-75. *Mem:* Am Chem Soc; Am Inst Chemists; Soc Appl Spectroscopy; Sigma Xi; Royal Soc Chem; Assoc Off Anal Chemists. *Res:* Use of membrane mimetic agents in analytical chemistry; luminescence methods; separations; enzymology. *Mailing Add:* Dept Chem Wake Forest Univ Box 7486 Winston-Salem NC 27109-7486

HIPEL, KEITH WILLIAM, b Kitchener, Ont, March 15, 46; m 71; c 4. STOCHASTIC MODELLING, CONFLICT ANALYSIS. *Educ:* Univ Waterloo, BASc, 70, MASc, 72, PhD(civil eng), 75. *Prof Exp:* Teaching asst eng, Univ Waterloo, 71-75; vis prof civil eng, Fed Univ, Brazil, 75-76; PROF SYSTS DESIGN, UNIV WATERLOO, 76- *Concurrent Pos:* Vis prof, Univ Sao Paulo, Brazil, 80 & State Sci & Technol, China, 82; consult, Environ Can & Elec Utilities, 77-; Japan Soc Prom Sci res fel, dept appl math & physics, Kyoto Univ, 84. *Mem:* Am Geophys Union; Am Water Resources Asn; Asn Prof Engrs Ont. *Mailing Add:* Dept Syst Sci Univ Waterloo Waterloo ON N2L 3G1 Can

HIPKIN, HOWARD G(EORGE), chemical engineering, thermodynamics, for more information see previous edition

HIPP, BILLY WAYNE, b Blackwell, Tex, June 18, 33; m 54; c 2. SOIL FERTILITY. *Educ:* Tex Tech Univ, BS, 55; Tex A&M Univ, MS, 63, PhD(soil chem), 66. *Prof Exp:* Field rep fertilizer, Wood Chem Co, 55-56 & 58-60; instr soils, 60-65, PROF SOILS, TEX A&M UNIV, AGR RES & EXTEN CTR, DALLAS, 65- *Mem:* Am Soc Agron; Am Hort Soc. *Res:* Soil fertility; nutrient movement in soils. *Mailing Add:* Tex A&M Univ Res & Exten Ctr 17360 Coit Rd Dallas TX 75252

HIPP, SALLY SLOAN, b Philadelphia, Pa, Apr 5, 37; c 1. SEXUALLY TRANSMISSIBLE DISEASE. *Educ:* Pa State Univ, BS, 58; Albany Med Col, PhD(biochem), 69. *Prof Exp:* res scientist, NY State Dept Health, 62-70; sr res scientist, London Sch Hyg & Trop Med, 70-71; res scientist, Div Labs & Res, 71-87, STD CONTROL PROG, NY STATE DEPT HEALTH, 87- *Concurrent Pos:* Adj asst prof, Dept Obstet & Gynec, Albany Med Col, 81-; vis lectr, Dept Biol, Skidmore Col, 72-73; WHO fel, UK, 77. *Mem:* Int Soc Human & Animal Mycol; Am Soc Microbiol; Sigma Xi; Am Venereal Dis Asn; NY Acad Sci. *Res:* Chlamydea; mycoplasma; sexually transmissible diseases of women; identification of causes of chronic vaginitis; role of microorganisms in reproductive failure; transport systems for genital microorganisms. *Mailing Add:* 15 Pineridge Pl Delmar NY 12054

HIPPS, KERRY W, b El Paso, Tex, Mar 16, 48; c 2. CHEMICAL PHYSICS, PHYSICAL CHEMISTRY. *Educ:* Univ Tex, BS, 70; Wash State Univ, PhD(chem physics), 76. *Prof Exp:* Lectr phys chem, Univ Mich, 76-78; from asst prof to assoc prof, 78-84, PROF CHEM & CHEM PHYSICS, WASH STATE UNIV, 84- *Concurrent Pos:* NSF fel, Univ Mich, 76-78; Alfred P Sloan Fel, 80-82. *Mem:* Am Chem Soc; Am Phys Soc; Sigma Xi. *Res:* Spectroscopy of solids and surfaces; chemical physics of inorganic materials. *Mailing Add:* Chem Dept Wash State Univ Pullman WA 99164-4630

HIRAMOTO, RAYMOND NATSUO, b Honolulu, Hawaii, June 12, 30; m 52; c 5. BACTERIOLOGY, IMMUNOLOGY. *Educ:* Univ Mich, Ann Arbor, BA, 52, MS, 53, PhD(bact), 56. *Prof Exp:* Teaching asst, Univ Mich, Ann Arbor, 55-56; cancer res scientist biochem, Roswell Park Mem Inst, NY, 56-59, sr cancer res scientist, 59-62; from asst prof to assoc prof microbiol, Univ Tenn, Memphis, 61-66; PROF MICROBIOL, MED CTR, UNIV ALA, BIRMINGHAM, 66- *Concurrent Pos:* Chief immunol sect, St Jude Children's Res Hosp, Memphis, 62-66, dir blood bank, 63-66. *Mem:* Fel Am Acad Microbiol; Am Asn Immunol; Am Asn Cancer Res; Soc Exp Biol & Med; NY Acad Sci. *Res:* Site of fixation in various organs of antikidney antibodies; penetration of antibodies and globulins into viable cells; immunohistochemical studies on human tumors and related tissues; antibody formation; Pavlovian conditioning of immune resistance. *Mailing Add:* Dept Microbiol Sch Med Univ Ala Univ Sta Birmingham AL 35233

HIRANO, ASAO, b Tomioka, Gunma, Japan, Nov 26, 26; m 59; c 4. NEUROPATHOLOGY. *Educ:* Univ Niigata, BS, 47; Univ Kyoto, MD, 52. *Prof Exp:* Intern, Osaka US Army Hosp, Japan, 52-53 & Harlem Hosp, New York, 53-54; resident neurol, Bellevue Hosp, New York, 54-55 & Montefiore Hosp, 55-56, chief resident, 57-58; vis scientist, Nat Inst Neurol Dis & Blindness, 59-65; HEAD DIV NEUROPATH, MONTEFIORE HOSP, 65- *Concurrent Pos:* Fel neuropath, Montefiore Hosp, 56-69; asst, Columbia Univ, 59; assoc prof, Albert Einstein Col Med, 65-71, prof, 71-; vis prof, Kansai Med Univ, Osaka, Japan, 85; vis prof, Univ Western Australia, 86. *Honors & Awards:* Billings Silver Medal, AMA, 59; Annual Prize, Am Asn Neuropath, 61; Henry Moses Res Award, 68; Royal Col lectr, Can Asn Neuropathologists, Royal Col Physicians & Surgeons Can, 80; Jack Prichard Mem Lectr, Queen's Univ, Belfast, Ireland, 81; Rajan Bharati Mem Lectr, Madras Med Col, India, 84. *Mem:* Am Asn Neuropath (pres, 77-78); Am Acad Neurol; Asn Res Nervous & Ment Dis; NY Soc Electron Microscopists; Am Neurol Asn; Japanese Neuropath Asn. *Res:* Fine structure of normal and pathological nervous tissue; pathology of motor neuron diseases. *Mailing Add:* Montefiore Med Ctr 111 E 210th St Bronx NY 10467

HIRANO, TOSHIO, b Osaka, Apr 17, 47; m 74; c 2. SIGNAL TRANSDUCTION, GENE EXPRESSION. *Educ:* Osaka Univ, med degree, 72, PhD, 79. *Prof Exp:* Assoc prof, Kumamato Univ Med Sch, 80-84; assoc prof, 84-89, PROF, INST MOLECULAR CELL BIOL, OSAKA UNIV MED SCH, 89- *Honors & Awards:* Erwin von Belz prize, 86; Rheumatism Prize, Japanese Found Rheumatism, 90. *Mem:* Am Asn Immunologists. *Res:* Molecular mechanism(s) of growth and differentiation of hematopoietic lineage cells; gene expression and signal transduction mechanism(s) of interleukin 6; molecular mechanism(s) of autoimmune diseases and oncogenesis of lymphoid cells. *Mailing Add:* Div Molecular Oncol Biomed Res Ctr Osaka Univ Med Sch 2-2 Yamada-oka Suita Osaka 565 Japan

HIRAOKA, HIROYUKI, b Kyoto, Japan, Mar 20, 32; m 63; c 2. PHYSICAL CHEMISTRY. *Educ:* Kyoto Univ, BS, 54, MS, 56, PhD(phys chem), 59. *Prof Exp:* Res staff chem, Teijin Ltd, Japan, 59-60; jr res chemist, Univ Calif, Los Angeles, 60-62; jr res chemist, Univ Calif, Berkeley, 62-63; asst res chemist, Univ Calif, Los Angeles, 63-64; res staff chem, Teijin Ltd, 64-66, sr res staff, 66; res staff, Thomas J Watson Res Ctr, 66-71, RES STAFF CHEM, SAN JOSE RES LAB, IBM CORP, 71- *Mem:* Am Chem Soc; Am Inst Chem; Chem Soc London; Electrochem Soc; Chem Soc Japan. *Res:* Phase equilibrium and diffusion process between/in gases and liquids; chemical kinetics in shock waves and in flow systems; radiation chemistry; photochemistry. *Mailing Add:* IBM Res Lab K91 San Jose CA 95120-6099

HIRASAKI, GEORGE J, b Beaumont, Tex, Sept 26, 39. OIL RESERVOIR RESEARCH. *Educ:* Lamar State Col Technol, BS, 63; Rice Univ, PhD(chem eng), 67. *Prof Exp:* Res assoc, Shell Develop Co, Bellaire Res Ctr, 67-72 & 74-84, staff reservoir engr, Shell Oil Co, W Coast Div, 72-74, RES ADV, SHELL DEVELOP CO, BELLAIRE RES CTR, 84- *Concurrent Pos:* Lectr, Chem Eng Dept, Rice Univ, 77-; mem, Comt Enhanced Oil Recovery, Nat Petrol Coun. *Honors & Awards:* Lester C Uren Award, Soc Petrol Engrs, 89. *Mem:* Nat Acad Eng; Soc Petrol Engrs; Am Inst Chem Engrs; Am Chem Soc; Soc Core Analyst. *Res:* Interfacial phenomena; numerical simulation; multiphase flow in porous media; wettability; foam in porous media; surfactant flooding; author of various publications; granted 3 patents. *Mailing Add:* Shell Develop Co PO Box 481 Houston TX 77001

HIRASUNA, ALAN RYO, b Fresno, Calif, Sept 27, 39; m 86; c 1. MECHANICAL ENGINEERING. *Educ:* Univ Calif, Berkeley, BS, 62; Univ SCalif, MS, 69. *Prof Exp:* Refinery engr, Chevron, 62-64; heat transfer analyst, NAm Rockwell, 64; prog engr, Ford Aerospace, 64-71; VPRES, L'GARDE, INC, 71- *Concurrent Pos:* Dir, bd L'Garde Inc, 71-; dir, bd Am Assoc SML Res Cos, 84-; comt mem, Cont Elastomers, Geothermal Resources Coun, 84. *Mem:* Am Chem Soc; Am Assoc Mech Engrs; Geothermal Res Coun; Soc Petrol Engrs; Am Defense Preparedness Assoc. *Res:* Develop new elastomer compounds for extremely hostile environments; developed Y267 EPDM which has become a standard in the geothermal industry for wells up to 320 ,; developed compounds for army tank track and strategic missile applications which must survive a nuclear encounter. *Mailing Add:* L'Garde Inc 15181 Woodland Ave Tustin CA 92680

HIRATA, ARTHUR ATSUNOBU, immunology; deceased, see previous edition for last biography

HIRATA, FUSAO, b Ube, Yamagulhi, Japan, Aug 28, 41; m 68; c 3. MOLECULAR & BIOCHEMICAL PHARMACOLOGY, DEVELOPMENT PHARMACEUTICS BY BIOTECHNOLOGY. *Educ:* Tokyo Med & Dent Univ, MD, 67; Kyoto Univ, PhD(biochem), 72. *Prof Exp:* Asst prof biochem, Fac Med, Kyoto Univ, 72-80; vis scientist pharmacol, NIMH, 77-84, unit chief cell biol, 84-86; assoc prof toxicol, Johns Hopkins Univ, 86-90; PROF PHARMACOL, FAC PHARM, WAYNE STATE UNIV, 90-, PROF TOXICOL, INS CHEM TOXICOL, 90- *Concurrent Pos:* State-of-the-Art lect, Am Asn Endocrinol, 82; adj fac, Sch Hyg & Pub Health, Johns Hopkins Univ, 90-; prof pharmacol, Sch Med, Wayne State Univ, 91- *Honors & Awards:* Distinguished Young Scientist Award, Japanese Biochem Soc, 77. *Mem:* AAAS; Am Soc Biochem & Molecular Biol; NY Acad Sci. *Res:* Pathophysiology of hyperreactive airways with respect to dysfunction of neuroreceptors induced by inflammatory cytokine networks and mechanisms of therapeutic action of glucocorticoids. *Mailing Add:* Dept Pharmaceut Sci Wayne State Univ Detroit MI 48202

HIRATSUKA, YASUYUKI, b Tottori, Japan, Dec 27, 33; Can citizen; m 59; c 3. MYCOLOGY, PLANT PATHOLOGY. *Educ:* Int Christian Univ, Tokyo, BS, 57; Hokkaido Univ, MS, 59; Purdue Univ, PhD(mycol, plant path), 62. *Prof Exp:* RES SCIENTIST MYCOL, CAN FORESTRY SERV, CAN GOVT, 64- *Concurrent Pos:* Nat Res Coun Can fel, 62-65; adj prof bot, Univ Man, 83-; adj prof forest & plant sci, Univ Alta, 81- *Honors & Awards:* Dr & Mrs D L Bailey Award, Can Phytopath Soc, 88; Distinguished Award, Soc Tech Commun, 88. *Mem:* Am Mycol Soc; Phytopath Soc Japan; Mycol Soc Japan; Int Soc Plant Taxon; Can Bot Asn; Am Phytopath Soc. *Res:* Taxonomy and biology of uredinales; taxonomy and morphology of forest disease pathogens; biochemical study of metabolites of tree disease pathogens; pine stem rusts of the world; armillaria root rot of conifers; aspen decay and stain. *Mailing Add:* Northern Forestry Ctr 5320 122nd St Edmonton AB T6H 3S5 Can

HIRAYAMA, CHIKARA, b Honolulu, Hawaii, Feb 12, 24; m 56; c 5. PHYSICAL CHEMISTRY, INORGANIC CHEMISTRY. *Educ:* Univ Hawaii, BS, 46, MS, 48; Univ Minn, PhD(chem), 57. *Prof Exp:* Chemist, Coca Cola Co, 48-50, Wailuku Sugar Co, 51-52 & US Fish & Wildlife Serv, 52-53; res chemist, Res Labs, Westinghouse Elec Corp, 57-73, consult scientist, 85-86; CONSULT, 86- *Mem:* AAAS; Am Chem Soc. *Res:* Dielectric properties of inorganic electrical insulators; physics and chemistry of glasses; chemistry of coordination compounds, kinetics and optical properties; lasers; electroluminescence; high temperature chemistry and thermodynamics; mass spectrometry; solid electrolyte sensors; thin dielectric films. *Mailing Add:* PO Box 318 Kaunakakai HI 96748

HIRD, DAVID WILLIAM, b Newport News, Va, June 30, 42; m 73; c 2. VETERINARY MEDICINE, PUBLIC HEALTH & EPIDEMIOLOGY. *Educ:* Stanford Univ, AB, 64; Univ Calif-Davis, DVM, 68, MPVM, 73; Univ Minn, PhD(epidemiol), 80. *Prof Exp:* Res veterinarian, Univ Chile, Santiago, 68-70; clin veterinarian, Australia, 71-72; livestock prod specialist, Food & Agr Orgn, Venezuela, 70-71; county veterinarian, Imperial County, Calif, 73-77; asst prof, 80-84, ASSOC PROF, UNIV CALIF, DAVIS, 84- *Concurrent Pos:* Fulbright res grant, 87. *Res:* Quantitative epidemiologic approaches to livestock, health and productivity; food safety. *Mailing Add:* Dept Epidemiol & Prev Med Univ Calif Davis CA 95616

HIREGOWDARA, DANANAGOUD, MEDICINE. *Prof Exp:* POSTDOCTORAL FEL, SCH MED, BOSTON UNIV, 89- *Mailing Add:* Dept Med Renal Unit E428 Sch Med Boston Univ 75 E Newton St Boston MA 02118

HIREMATH, SHIVANAND T, b Bijapur, India, Jan 26, 52; m 80; c 1. MOLECULAR ENTOMOLOGY, BIOLOGICAL CONTROL OF FOREST PESTS. *Educ:* Karnatak Univ, Dharwar, India, BS, 71, MS, 73; Poona Univ, India, PhD(biochem), 78. *Prof Exp:* Res assoc, State Univ NY, Buffalo, 78-82, Univ Ky, Lexington, 83-87; res scientist, Univ Calif, Riverside, 87-88; RES BIOLOGIST, USDA FOREST SERV, 88- *Mem:* Am Soc Biochem & Molecular Biol; Am Soc Microbiologists; AAAS. *Res:* Safe and effective microbial agents for use in integrated forest pest management systems. *Mailing Add:* 60 Trail Edge Circle Powell OH 43065

HIRES, RICHARD IVES, b Camden, NJ, Aug 19, 39; m 63; c 4. PHYSICAL OCEANOGRAPHY. *Educ:* Rensselaer Polytech Inst, BS, 61; Johns Hopkins Univ, MA, 64, PhD(oceanog), 68. *Prof Exp:* Asst prof, 67-73, ASSOC PROF OCEAN ENG, STEVENS INST TECHNOL, 73- *Concurrent Pos:* Instr, NJ Marine Sci Consortium, 69-; consult, Woodward Envicon Inc, 72-74 & Enviro Sci Inc, 74- *Mem:* Soc Naval Architects & Marine Engrs; Am Geophys Union. *Res:* Wind generation of waves; interactions across the air-sea interface; diffusion of contaminants in coastal waters; estuarine circulation; wind engineering. *Mailing Add:* 14 Farragut Pl Morristown NJ 07960

HIRKO, RONALD JOHN, b Cleveland, Ohio, Mar 3, 43; m 65; c 3. PHYSICAL CHEMISTRY. *Educ:* Kent State Univ, BS, 65; Utah State Univ, PhD(chem), 69. *Prof Exp:* Sect head chem, 69-79, MGR, QUALITY ASSURANCE, OCCIDENTAL CHEM CO, WHITE SPRINGS, 79- *Mem:* Am Chem Soc. *Res:* Thermodynamics and statistical mechanics of electrolytic solutions; electrochemistry; pollution abatement and control; fluorine chemistry. *Mailing Add:* Rt 6 Box 446D-D Lake City FL 32055-3299

HIRLEMAN, EDWIN DANIEL, JR, b Wichita, Kans, Dec 1, 51; m 75; c 3. THERMAL SCIENCES. *Educ:* Purdue Univ, BSME, 72, ME, 74, PhD(mech eng), 77. *Prof Exp:* Res assoc laser diag, Tech Univ Denmark, 74-75; staff engr, Hughes Aircraft Co, 74-77; from asst prof to assoc prof mech eng, 81-87, PROF MECH ENG, ARIZ STATE UNIV, 88-, PROF & VCHAIR AEROSPACE, 89- *Concurrent Pos:* Prin investr, various grants founded by NSF, Off Naval Res, Air Force Off Sci Res, Semiconductor Res Corp, IBM, Allied-Signal Aerospace, McDonnell-Douglas Helicopter. *Mem:* Combustion Inst; Am Soc Mech Engrs; Am Inst Aeronaut & Astronaut; Optical Soc Am. *Res:* Laser optical diagnostics for thermo-fluid science; optical sensors for manufacturing; optical characterization of surface microfeatures design theory; design methodology. *Mailing Add:* Mech Aero Eng Dept Ariz State Univ Tempe AZ 85287-6106

HIRNING, LANE D, b Twin Falls, Idaho, Oct 1, 58. MOLECULAR BIOLOGY. *Educ:* Idaho State Univ, BS, 81; Univ Ariz, PhD(pharmacol), 85. *Prof Exp:* Postdoctoral fel, Dept Pharmacol & Phys Sci, Univ Chicago, 85-89; SR SCIENTIST, NATURAL PROD SCI INC, 89- *Mem:* Am Soc Pharmacol & Exp Therapeut; AAAS; Soc Neurosci. *Mailing Add:* Dept Neurophysiol Natural Prod Sci Inc 420 Chipeta Way Suite 240 Salt Lake City UT 84108

HIROKAWA, KATSUIKU, b Tokyo, Japan, Nov 5, 39; m 68; c 3. STRUCTURE & IMMUNOLOGICAL FUNCTION OF THE THYMUS, BIOLOGY OF AGING. *Educ:* Tokyo Med & Dent Univ, MD, 64, DMed Sci, 69. *Prof Exp:* Sr instr path, Tokyo Med & Dent Univ, 69-76, assoc prof, 76-81; chief, 81-85, DIR PATH, TOKYO METROP INST GERONT, 85- *Mem:* Am Asn Immunologists; NY Acad Sci. *Res:* Ontogenic development and aging of the thymus and T cell dependent immune functions; mechanism of the age-related alteration of the immune system; neuroendocrine control of the thymus and the immune system. *Mailing Add:* Dept Path Tokyo Metro Inst Geront 35-2 Sakaecho, Itabashi-Ku Tokyo 173 Japan

HIROSE, AKIRA, b Nagano Prefecture, Japan, Aug 16, 41; m 69; c 2. PHYSICAL MATHEMATICS. *Educ:* Yokohama Nat Univ, BS, 65, MS, 67; Univ Tenn, PhD(elec eng), 69. *Prof Exp:* Consult, Oak Ridge Nat Lab, 68-69, mem res staff, Thermonuclear Div, 69-71; res assoc, Univ Sask, 71-74, res scientist, 75-77, assoc prof, 77-79, PROF, DEPT PHYSICS, UNIV SASK, 79- *Concurrent Pos:* Res fel, Japan Soc Promotion Sci, 84. *Mem:* Fel Am Phys Soc; fel Inst Elec & Electronics Engrs; Can Asn Physicists. *Res:* Plasma turbulence; linear and nonlinear waves and instabilities; turbulent heating; toroidal confinement. *Mailing Add:* Dept Physics Univ Sask Saskatoon SK S7N 0W0 Can

HIROTA, JED, b Honolulu, Hawaii, Mar 2, 43; m 70; c 2. BIOLOGICAL OCEANOGRAPHY. *Educ:* Univ Wash, BSc, 65, MSc, 67; Scripps Inst Oceanog, PhD(biol oceanog), 73. *Prof Exp:* Res asst biol oceanog, Scripps Inst & Dept Oceanog, Univ Wash, 65-72; fel, 73-74, asst prof, 74-79, ASSOC PROF OCEANOG, HAWAII INST MARINE BIOL, 79- *Honors & Awards:* Calif Acad Sci Award, Am Soc Limnol & Oceanog, 71. *Mem:* Am Soc Limnol & Oceanog; Ecol Soc Am. *Res:* Ecology of zooplankton distribution, faunal assemblages, metabolism, secondary production, natural history, feeding behavior; biological and chemical oceanographic methodology. *Mailing Add:* 186 Kuukama St Kailua HI 96734

HIRS, CHRISTOPHE HENRI WERNER, b Bern, Switz, Apr 25, 23; nat US; m 49. ORGANIC CHEMISTRY. *Educ:* Manchester Univ, BSc, 44; Columbia Univ, PhD(biochem), 49. *Prof Exp:* Assoc in biochem, Rockefeller Inst Med Res, 49-58; from assoc biochemist to biochemist, Brookhaven Nat Lab, 58-63, sr biochemist, 63-65, chmn dept biol, 65-69; dir, Div Biol Sci, Ind Univ, Bloomington, 69-77, prof biol, 69-78; PROF & CHMN, DEPT BIOCHEM, BIOPHYS & GENETICS, SCH MED, UNIV COLO, DENVER, 78- *Concurrent Pos:* Mem Am Inst Biol Sci adv comt biochem, Off Naval Res, 60-63 & physiol chem study sect, Div Res Grants, NIH, 63-67; exec ed, Arch Biochem & Biophys, 63-72; mem adv panel molecular biol, NSF, 71-74; assoc ed, J Biol Chem, 72- *Mem:* AAAS; Am Chem Soc; Am Soc Biochem & Molecular Biol. *Res:* Protein chemistry. *Mailing Add:* Dept Biochem Biophys Genetics Sch Med Univ Colo Denver CO 80262

HIRSCH, ALBERT EDGAR, b Buffalo, NY, Nov 27, 24; m 48; c 1. CHEMICAL ENGINEERING. *Educ:* Univ Mich, BS, 48, Univ Tenn, MS, 62, PhD(chem eng), 64. *Prof Exp:* Engr chem eng, E I du Pont de Nemours & Co Inc, 48-57, proc control supt, 51-61, res engr chem eng, 64-69, res engr mat sci, 69-70; RETIRED. *Concurrent Pos:* Consult, 90- *Mem:* Am Chem Soc. *Res:* Engineering and materials science; chemical engineering; materials science engineering. *Mailing Add:* 2410 Raven Rd Wilmington DE 19810

HIRSCH, ALLEN FREDERICK, b New York, NY, May 24, 35; m 57; c 3. MEDICINAL CHEMISTRY. *Educ:* Fordham Univ, BS, 57; Univ NC, MS, 62, PhD(med chem), 64. *Prof Exp:* Res assoc org chem, Sloan-Kettering Inst Cancer Res, 63-65; fel, Albert Einstein Col Med, 65-67; group leader, Ortho Pharmaceut Corp, 67-77, dir res qual assurance, 77-89; SR DIR, REGULATORY AFFAIRS ADMIN WORLDWIDE, R W JOHNSON PHARMACEUT RES INST, 89- *Concurrent Pos:* Course Dir, Ctr Prof Advan. *Mem:* Am Chem Soc; Sigma Xi; Drug Info Asn. *Res:* Organic synthesis in plasmalogens, thiosulfinates, nitrogen mustards, pseudouridine, steroids and nucleic acids; male and female antifertility agents. *Mailing Add:* 4 N Cadillac Dr Somerville NJ 08876

HIRSCH, ANN MARY, b Milwaukee, Wis, June 2, 47; m 70. PLANT MOLECULAR BIOL, PLANT DEVELOPMENT. *Educ:* Marquette Univ, BS, 69; Univ Calif, PhD(bot), 74. *Prof Exp:* Asst prof bot, Univ Minn, 74-76; res assoc, Harvard Univ, 76-78; from asst prof to assoc prof biol, Wellesley Col, 78-88, dir greenhouses, 84-87; ASSOC PROF BIOL, UNIV CALIF, LOS ANGELES, 88- *Concurrent Pos:* Biol tutor, Harvard Univ, 77-80, vis scientist, 81-82; vis scientist, Stanford Univ, 83; panel mem, comt res grants, Nat Acad Sci-Nat Res Coun, 84-87 & nitrogen fixation prog, USDA, 86, NSF, 86. *Mem:* Am Asn Plant Physiologists; Am Soc Microbiol; Am Soc Cell Biol; Sigma Xi; Bot Soc Am. *Res:* Symbiotic nitrogen fixation; plant and rhizobial genes expressed during the early stages of nodule development. *Mailing Add:* Dept Biol Univ Calif Los Angeles CA 90024

HIRSCH, ARTHUR, b Vienna, Austria, Dec 10, 21; US citizen; m 49; c 1. ORGANIC CHEMISTRY. *Educ:* City Col New York, BS, 46; Brooklyn Col, MS, 56; NY Univ, PhD, 60. *Prof Exp:* Chemist, Pease Labs, 46-52; chief chemist, Atlantic Gummed Paper Co, 52-54 & Swingline Indust Corp, 54-55; tech dir, Gen Gummed Prod, 55-60; dir res, Can Tech Tape, Ltd, 62-68; dir res, Standard Packaging Corp, Clifton, 68-78; tech dir, Arvey Corp, Chicago & Cedar Grove, 78-89; CONSULT, 89- *Concurrent Pos:* Lectr, Wagner Lutheran Col, 56-58; Nat Res Coun Can & Defense Res Bd Can, grants, 63, 64 & 65; chmn educ comt, Nat Flexible Packaging Asn; mem comt adhesives, Am Soc Testing & Mat; adj prof, BCCC, 89- *Honors & Awards:* NY Univ Founders Day Award, 60. *Mem:* Am Chem Soc; Tech Asn Pulp & Paper Indust; Chem Inst Can; fel Am Inst Chem; sr mem Inst Elec & Electronics Engrs. *Res:* Polymers; adhesives; functional paper coatings; heterocyclic organic chemistry; packaging; extrusion and lamination; paper saturation. *Mailing Add:* 2000 Linwood Ave Ft Lee NJ 07024

HIRSCH, CARL ALVIN, b Los Angeles, Calif, Sept 21, 29. CHEMISTRY. *Educ:* Calif Inst Technol, BS, 51; Wash Univ, MD, 59; Am Bd Path, cert, 75. *Prof Exp:* Chemist, Army Chem Corps Biol Labs, Md, 53-55; intern, Eastern Marine Gen Hosp, 59-60; res fel, Dept Bact, Harvard Med Sch, 60-63, instr bact, 63-64, instr med, 64-68, assoc, 68-69, asst prof, 69-72; asst clin prof lab med & clin path, Sch Med, Univ Calif, San Francisco, 72-74; DIR CLIN CHEM, NEW ENG MED CTR HOSP, BOSTON, 75-; ASSOC CLIN

PROF PATH, SCH MED, TUFTS UNIV, 75- *Concurrent Pos:* Res assoc med, Beth Israel Hosp, Boston, 64-72; NSF grant, 71-73; asst chief, Clin Path Serv, San Francisco Vet Admin Hosp, 72-74, actg chief, 74; pathologist, New Eng Deaconess Hosp, Boston, 74-75, New Eng Baptist Hosp, Boston, 74-75. *Mem:* AAAS; Am Soc Biochem & Molecular Biol; Col Am Pathologists; Am Asn Clin Chem; Acad Clin Lab Physicians & Scientists. *Res:* Clinical pathology. *Mailing Add:* Clin Chem Lab Tufts-New Eng Med Ctr Box 830 Boston MA 02111

HIRSCH, DONALD EARL, b Erie, Pa, Sept 17, 24; m 46; c 4. CHEMICAL ENGINEERING. *Educ:* Univ Pittsburgh, BS, 47, MS, 48, PhD(chem eng), 54. *Prof Exp:* Chemist, Los Alamos Sci Lab, 45-46; instr chem eng, Univ Pittsburgh, 48-50; chemist, E I du Pont de Nemours & Co, 50-57, proj mgr, 57-58, sr assoc engr, 59-62; adv engr, Am Potash & Chem Co, 63-65, mgr plant tech serv, 65-68; consult engr, Kerr McGee Co, 68-69, tech asst to pres, 69-72, chief process design engr, 72-77, dir mfg servs, 77-80, mgr process eng, 80-81; DIR TECH SERV, IMC FERTILIZER, INC, 81- *Mem:* Nat Soc Prof Engrs; Am Chem Soc; Sigma Xi; Am Inst Chem Engrs. *Res:* High temperature kinetics and thermodynamics; phase chemistry; process design; plant optimization; winning of metals by chlorine technology. *Mailing Add:* 4852 Leisurewood Lane Lakeland FL 33811-1592

HIRSCH, HELMUT V B, b Chicago, Ill, Sept 22, 43; div; c 1. NEUROBIOLOGY. *Educ:* Univ Chicago, BA, 65; Stanford Univ, PhD(psychol), 70. *Prof Exp:* Fel, Johns Hopkins Univ, 70-72; asst prof, 72-77, ASSOC PROF BIOL, STATE UNIV NY ALBANY, 77- *Concurrent Pos:* Alfred P Sloan Found fel, 75; adj assoc prof psychol, State Univ NY, Albany, 77- *Mem:* AAAS; Soc Neurosci; Sigma Xi; Asn Res Vision & Ophthalmol; Sigma Xi. *Res:* Role of early visual experience in the development of the mammalian visual system; neuronal mechanisms underlying the perception of form and pattern. *Mailing Add:* Dept Biol State Univ NY 1400 Washington Ave Albany NY 12222

HIRSCH, HENRY RICHARD, b New York, NY, Mar 27, 33; m 54; c 2. GERONTOLOGY. *Educ:* Mass Inst Technol, SB, 54, PhD(physics), 60. *Prof Exp:* Elec engr, Bell Tel Labs, 54-57; physicist, NIH, 61-63; asst prof, 63-67, assoc prof, 67-76, PROF PHYSIOL & BIOPHYS, UNIV KY, 76- *Concurrent Pos:* Reviewer, Zentralblatt fur Mathematik, 67-; coordr nat corresp, Fed Am Soc Exp Biol, 71-79; guest assoc biophysicist, Brookhaven Nat Lab, 71; mem pub affairs comt, Fedn Am Soc Exp Biol, 79-85, pub affairs adv comt, Am Physiol Soc, 83-, pub affairs exec comt, 91- *Mem:* Am Physiol Soc; Am Phys Soc; Biophys Soc; fel Geront Soc; sr mem Inst Elec & Electronics Engrs; AAAS. *Res:* Theoretical biology; physiology. *Mailing Add:* Dept Physiol & Biophys Univ Ky Col Med Lexington KY 40536-0084

HIRSCH, HORST EBERHARD, b Woelsendorf, W Germany, July 26, 33; Can citizen; m 61; c 3. EXTRACTIVE METALLURGY, SEMICONDUCTORS. *Educ:* Univ Karlsruhe, W Germany, dipl chem, 58, Dr rer nat(chem tech), 60. *Prof Exp:* Res engr extractive metall, 62-64, group leader mineral process, 65-68, develop supvr process develop, 69-70, supt elec mat, 71-76, asst mgr, 76-79, mgr tech res, 79-81, gen mgr, elec mat div, 82-84, PRES ELEC MAT DIV, COMINCO LTD, 84- *Concurrent Pos:* Nat Res Coun Can fel, 61-62; pres & chief exec officer, Comt Emergency Med Identification, 84- *Mem:* Chem Inst Can; Am Inst Mining & Metall Engrs; Am Soc Metals; German Soc Mining Metals Eng. *Res:* Extractive metallurgy process development; new compound semiconductors; ultrapurification of metals. *Mailing Add:* 1005 E 54th Ave Spokane WA 99223

HIRSCH, JACOB IRWIN, b New York, NY, Aug 27, 26; m 50; c 3. CARDIOVASCULAR DISEASE, BIOMEDICAL ENGINEERING. *Educ:* Brooklyn Col, AB, 48; NY Univ, MD, 52. *Prof Exp:* From instr to asst prof, 59-71, ASSOC PROF MED, SCH MED, NY UNIV, 71- *Concurrent Pos:* Consult, Biomed Eng Training Comt, Nat Inst Gen Med Sci, 62- *Mem:* AAAS; AMA; Am Heart Asn; NY Acad Sci; Inst Elec & Electronics Engrs. *Res:* Volume conductor characteristics of human torso; application of computers to monitoring of patient electrocardiograms on line; biomedical engineering education; experimental myocardial infarction. *Mailing Add:* 530 First Ave New York NY 10016

HIRSCH, JAY G, b Cleveland, Ohio, Aug 6, 30; m 62; c 4. CHILD PSYCHIATRY. *Educ:* Univ Cincinnati, BS, 50, MD, 54; Am Bd Psychiat & Neurol, dipl, psychiat, 63 & child psychiat, 64. *Prof Exp:* Intern, Philadelphia Gen Hosp, Pa, 54-55; res psychiat, Univ Ill, 57-60; fel child psychiat, Inst Juv Res, 60-62, res child psychiatrist, 62-66, chief div prev psychiat, 66-71; assoc prof, 71-74, PROF PSYCHIAT, ABRAHAM LINCOLN SCH MED, UNIV ILL COL MED, 74- *Concurrent Pos:* Dir prof educ, Inst Juv Res, Chicago, 73-77. *Mem:* Fel Am Acad Child Psychiat; fel Am Psychiat Asn; fel Am Orthopsychiat Asn; Soc Res Child Develop. *Mailing Add:* 1971 Second St Highland Park IL 60035

HIRSCH, JERRY, b New York, NY, Sept 20, 22; m 50; c 1. PSYCHOLOGY GENETICS, RACISM. *Educ:* Univ Calif, BA, 52, PhD(psychol), 55. *Hon Degrees:* Doctorat Honoris Causa, Univ Rene Descartes, Paris, France, 87. *Prof Exp:* Asst prof psychol, Columbia Univ, 56-60; assoc prof psychol, Univ Ill, 60-63, prof zool, 67-76, PROF PSYCHOL, UNIV ILL, URBANA-CHAMPAIGN, 63-, PROF ECOLOGY, ETHOLOGY & EVOLUTION, 76- *Concurrent Pos:* NSF fel, 55-57; NIMH fel, Ctr Advan Study Behav Sci, 60-61; dir, NIMH pre- & postdoctoral Biopsych Res training prog, 66-78; vis res scholar zool dept, Univ Edinburgh, 68; ed, Animal Behav, 68-72, J Comp Psychol, 83-88; mem, US Nat Comt, Int Union Biol Sci, 75-81, Int Ethological Conf Comt, 75-83; co-dir, NIMH pre- & postdoctoral training prog for res on Instnl Racism, Univ Ill, 77-86. *Honors & Awards:* Auxillary Res Award, Soc Sci Res Coun, 62; Robert Choate Tryon Mem Lectr, dep psych, Univ Calif, Berkeley, 87. *Mem:* AAAS; Animal Behav Soc (pres, 75); Am Psychol Asn; Sigma Xi. *Res:* Heredity-environment analysis of behavior, measuring individual differences in tropisms, excitatory states, conditioning, etc; behavior-genetic component analyses of species; methodological, theoretical and scholarly analyses of institutional and scientific racism. *Mailing Add:* Psychol Dept Univ Ill 603 E Daniel Champaign IL 61820

HIRSCH, JERRY ALLAN, b Louisville, Ky, June 6, 41; m 69; c 3. ORGANIC CHEMISTRY. *Educ:* Vanderbilt Univ, AB, 62; Stanford Univ, PhD(org chem), 66. *Prof Exp:* Res assoc chem, Wayne State Univ, 66-67; from asst prof to assoc prof, 67-75, PROF CHEM, SETON HALL UNIV, 75- *Concurrent Pos:* Vis prof, State Univ Leiden, Neth, 74-75; vis prof, Weizmann Inst, Israel, 81-82. *Mem:* Am Chem Soc. *Res:* Stereochemistry and conformational analysis; ionene polymers; nonconjugated chromophores; heterocyclic systems related to trans-decalin; medium-ring carbocyclic systems. *Mailing Add:* Dept Chem Seton Hall Univ South Orange NJ 07079

HIRSCH, JOHN MICHELE, b Cleveland, Ohio, Feb 1, 47; m 69. EXPERIMENTAL PHYSICS, PETROLEUM PHYSICS. *Educ:* Mass Inst Technol, SB, 69; Harvard Univ, MA, 72, PhD(physics), 74. *Prof Exp:* Physicist, 75-77, RES PHYSICIST, SHELL DEVELOP CO, 77- *Mem:* Am Inst Physics; Am Asn Physics Teachers; Soc Petrol Engrs. *Res:* Application of physical techniques to the location of petroleum in the subsurface. *Mailing Add:* 1502 Cedarbrook Dr Houston TX 77055

HIRSCH, JORGE E, b Buenos Aires, Arg, Aug 5, 51; c 3. PHYSICS. *Educ:* Univ Chicago, MSc, 77, PhD(physics), 80. *Prof Exp:* Postdoctoral res assoc, Univ Calif, Santa Barbara, Inst Theoret Physics, 80-82; from asst prof to assoc prof, 83-87, PROF PHYSICS, UNIV CALIF, SAN DIEGO, 87- *Concurrent Pos:* Sloan fel, Sloan Found, 84. *Mem:* Fel Am Phys Soc. *Res:* Superconductivity; magnetism; theory; mechanisms. *Mailing Add:* Dept Physics 0319 Univ Calif San Diego La Jolla CA 92093

HIRSCH, JUDITH ANN, CARDIOPULMONARY PHYSIOLOGY, PEDIATRIC PULMONARY MEDICINE. *Educ:* State Univ NY, Buffalo, PhD(physiol), 79. *Prof Exp:* ASST RES PROF PEDIAT PULMONARY MED, STATE UNIV NY, BUFFALO, 79- *Mailing Add:* Dept Physiol & Pediat SUNY Buffalo 51 Lennox Ave Amherst NY 14226

HIRSCH, JULES, b New York, NY, Apr 6, 27. MEDICINE. *Educ:* Rutgers Univ, 43-45; Univ Tex, MD, 48. *Prof Exp:* Intern path & med, Duke Hosp, NC, 48-50; from asst resident to resident, State Univ NY Col Med, Syracuse, 50-52; asst prof biochem & assoc physician, 54-60, assoc prof & physician, 60-67, PROF & SR PHYSICIAN, ROCKEFELLER UNIV, 67- *Mem:* AAAS; Am Soc Clin Invest; Harvey Soc; Asn Am Physicians; Am Fedn Clin Res. *Res:* Obesity; human behavior; internal medicine; biochemistry and physiology of lipids; lipid metabolism and nutrition. *Mailing Add:* Lab Human Behav & Med Rockefeller Univ New York NY 10021

HIRSCH, LAWRENCE LEONARD, b Chicago, Ill, Aug 20, 22. FAMILY MEDICINE, MEDICAL SOCIOECONOMICS. *Educ:* Univ Ill, BS, 43, MD, 50. *Prof Exp:* Pvt pract, 51-70; dir ambulatory care, Ill Masonic Med Ctr, 70-75, dir family pract residency, 71-75; PROF FAMILY MED & CHMN DEPT, CHICAGO MED SCH, 75- *Concurrent Pos:* Vis fac, Great Lakes Naval Hosp, 75-; trustee, Cook County Grad Med Sch, 84- *Mem:* Fel AAAS; fel Am Acad Family Physicians; fel Am Geriat Soc; AMA; Am Geront Soc; Soc Teachers Family Med. *Res:* Family physicians. *Mailing Add:* Dept Family Med Univ Health Sci Chicago Med Sch 3333 Green Bay Rd North Chicago IL 60064

HIRSCH, MARTIN STANLEY, b Cortland, NY, Apr 16, 39; m 64; c 2. IMMUNOLOGY, VIROLOGY. *Educ:* Hamilton Col, AB, 60; Johns Hopkins Univ, MD, 64; Am Bd Internal Med, dipl, 72. *Prof Exp:* Intern med, Univ Chicago Clins & Hosps, 64-65, resident, Univ Chicago Clins, 65-66; med officer virol, Nat Commun Dis Ctr, 66-68; asst prof, 71-76, assoc prof, 76-88, PROF, DEPT MED, HARVARD MED SCH, 88- *Concurrent Pos:* Fel immunol, Nat Inst Med Res, London, 68-69; fel infectious dis, Mass Gen Hosp, 69-71. *Mem:* Infectious Dis Soc Am; Am Asn Immunol; Am Soc Clin Investr; NIH AIDS Adv Comt; Am Found AIDS Res Sci Adv Comt. *Res:* Host response, both cellular and humoral, to viral infections; tumor viruses; autoimmunity; AIDS. *Mailing Add:* Mass Gen Hosp Dept Med Harvard Med Sch Boston MA 02114

HIRSCH, MORRIS WILLIAM, b Chicago, Ill, June 28, 33; m 57; c 2. TOPOLOGY OF MANIFOLDS, DYNAMICAL SYSTEMS. *Educ:* Univ Chicago, PhD(math), 58. *Prof Exp:* NSF fel math, Inst Adv Study, 58-60 & 83-84; asst prof, 60-65, vchmn dept, 66-67, chmn dept, 81-83, PROF MATH, UNIV CALIF, BERKELEY, 65- *Concurrent Pos:* Vis prof, Cambridge Univ, 63-64, Geneva Univ, 68-69 & Harvard Univ, 75-76; Sloan fel, 64-66; fel, Miller Inst Basic Res, in Sci, 71-72; Zyskind prof, Brandeis Univ, 76-77. *Mem:* Am Math Soc; Soc Indust & Appl Math. *Res:* Neural computation; dynamical systems; mathematical models in applied sciences; history and philosophy of mathematics. *Mailing Add:* Dept Math Univ Calif Berkeley CA 94720

HIRSCH, PETER M, b Cortland, NY, Apr 12, 39; m 61; c 3. MATHEMATICS, COMPUTER SCIENCE MANAGEMENT. *Educ:* Univ Wis, BS, 61, MS, 63, PhD(math), 66. *Prof Exp:* Teaching asst math, Univ Wis, 61-65; instr, Pa State Univ, 65-66; MEM STAFF, IBM CORP, 66- *Res:* Numerical analysis; holography; electric power simulation; object oriental programming. *Mailing Add:* IBM Sci Ctr 1530 Page Mill Rd Palo Alto CA 94303

HIRSCH, PHILIP FRANCIS, b Stockton, Calif, June 24, 25; m 56; c 4. PHARMACOLOGY. *Educ:* Univ Calif, Berkeley, BS, 50, PhD(physiol), 54. *Prof Exp:* Asst physiol, Univ Calif, Berkeley, 53-54, lectr & jr res physiol, 54-55; instr pharmacol, Sch Dent Med, Harvard Univ, 55-57, assoc, 57-64, asst prof, 64; physiologist, Lawrence Radiation Lab, Univ Calif, Livermore, 64-66; assoc prof, 66-70, dir dent res ctr, 75-83, PROF PHARMACOL, SCH MED, UNIV NC, CHAPEL HILL, 70-, PROF, DEPT DENT ECOL, SCH DENT, 88- *Concurrent Pos:* Res fel biochem, Brandeis Univ, 58-59. *Mem:* AAAS; Endocrine Soc; Am Soc Pharmacol & Exp Therapeut; Am Soc Bone & Mineral Res. *Res:* Carbohydrate and lipid metabolism of mammary gland and liver; physiological control of serum calcium; pharmacological action of parathyroid hormone, calcitonin and glucocorticoids. *Mailing Add:* Dent Res Ctr Univ NC Chapel Hill NC 27599-7455

HIRSCH, ROBERT GEORGE, b Baltimore, Md, Nov 25, 46; m 70; c 2. ATOMIC & MOLECULAR PHYSICS. *Educ:* Univ Scranton, Pa, BS, 69; Harvard Univ, MA, 70; Univ Va, PhD(physics), 74. *Prof Exp:* Res physicist appl physics, E I Du Pont de Nemours Co, Inc, 74-76, res supvr, 76-78, div prog coordr, Eng Res & Develop, 78-79, sr res supvr, Mat Res, 79-81, prin consult, Corp Res & Develop Planning, 81-84, mkt mgr adv electronics systs, 84-85, dir & site mgr, Electronics Tech Lab, 85-88, dir IBM account, 89-90, TECH DIR POLYESTER FILM ENTERPRISE, E I DU PONT DE NEMOURS CO, INC, 91- *Concurrent Pos:* Woodrow Wilson Found fel, 69-74; NSF trainee, Harvard Univ, 69-70; consult, Teledyne Avionics, 73-74; mem, Indust Adv Bd, NC A&T Univ, 85-88; mem, Duke Univ Bd Visitors Eng, 87-; mem, NC State Univ Physics & Math Sci Found, 87- *Mem:* Am Phys Soc; Am Inst Physics; Sigma Xi; Am Inst Chem Engrs. *Res:* Atomic and molecular physics; ion-molecule reactions; atmospheric physics; mass spectroscopy; plasma chromatography; trace substance analysis; research and development strategy; microelectronic packaging; tech ceramics; high-density ceramic packaging; polyester film technology; research and development management. *Mailing Add:* Two Oakhurst Lane Raleigh NC 15215-1572

HIRSCH, ROBERT L, b Boston, Mass. NEUROIMMUNOLOGY, VIROLOGY. *Educ:* Brandeis Univ, AB, 73; Georgetown Univ, PhD(immunobiol), 77. *Prof Exp:* Fel neurovirol, Sch Med, Johns Hopkins Univ, 77-79, asst prof, 79-82; asst prof neurol & microbiol, Sch Med, Univ Md, 82-85; mgr med info serv, Ortho Pharmaceut Corp, 85-86; group mgr clin develop, 86-87, asst dir clin invest, 88-89, DIR CLIN INVEST, IMMUNOBIOL RES INST, 90- *Concurrent Pos:* Instr neurol, Sch Med, Johns Hopkins Univ, 79; res assoc, Howard Hughes Med Inst, 79-82. *Mem:* Am Asn Immunologists; Am Soc Microbiol; AAAS; Sigma Xi. *Res:* Autoimmunity; AIDS; peptides-immunomodulators in HIV infection; role of complement in recovery from acute viral infections; immunobiology of multiple sclerosis; therapeutic trials in multiple sclerosis; monoclonal antibodies in organ transplantation in man. *Mailing Add:* Immunobiol Res Inst Rt 22 E PO Box 999 Annandale NJ 08801-0999

HIRSCH, ROBERT LOUIS, b Evanston, Ill, Mar 6, 35; m 59; c 3. NUCLEAR ENGINEERING & PHYSICS. *Educ:* Univ Ill, BS, 58, PhD(nuclear eng), 64; Univ Mich, MS, 59. *Prof Exp:* Sr engr, Atomics Int Div, NAm Aviation, Inc, 59-60; res asst elec eng, Univ Ill, 63-64; sr engr, ITT Farns Res Corp, 64-67, dir plasma physics, ITT Indust Labs Div, Int Tel & Tel Corp, 67-68; 00050593x, Controlled Themonuclear Prog, Div Res, US Atomic Energy Comn, 68-72, dir, Div Magnetic Fusion Energy, US Energy Res & Develop Admin, 72-76, asst admin solar, geothermal & advan energy systs, 76-77; gen mgr petrol explor res, Exxon Res & Eng Co, 77-80, mgr, Baytown Synthetic Fuels Res, 80-82; VPRES & MGR EXPLOR & PROD RES, ARCO EXPLOR & TECHNOL CO, 83- *Concurrent Pos:* Lectr, Purdue Univ, 65-66; mem, Atomic Energy Comn Laser Fusion Coord Comt; US rep, Int Atomic Energy Agency Int Fusion Res Coun; head, US mem, US-USSR Joint Fusion Power Coord Comt; chmn, Govts Interagency Geothermal Adv Coun; consult, Dept Energy, Gen Acctg Off, Lawrence Livermore Lab, Oak Ridge Nat Lab, Princeton Univ, Gas Res Inst & Nat Res Coun; distinguished fel, Ctr Theoret Studies, Univ Miami, 83; chmn bd, Fusion Power Asn; mem, Energy Res Adv Bd, Dept Energy & Int Sci Forum Energy. *Mem:* Fel AAAS; Am Nuclear Soc; Am Phys Soc; NY Acad Sci; Am Petrol Inst. *Res:* Experimental plasma physics; ion gun and vacuum pump development; experimental research in thermionic energy conversion; space nuclear reactor engineering; reactor analysis for aircraft propulsion; research administration. *Mailing Add:* ARCO Oil & Gas Co 2300 W Plano Pkwy Plano TX 75075

HIRSCH, ROLAND FELIX, b Rhinebeck, NY, Nov 30, 39; m 71; c 3. ANALYTICAL CHEMISTRY. *Educ:* Oberlin Col, BA, 61; Univ Mich, MS, 63, PhD(anal chem), 65. *Prof Exp:* From instr to assoc prof, 65-83, from asst chmn to assoc chmn dept, 71-75, assoc dean, Arts & Sci, 81-86, prof chem, Seton Hall Univ, 83-88; prog mgr, US Dept Energy, 84-88, PROG MGR, NIH, 88- *Concurrent Pos:* Vis scientist, Inorg Chem Labs, Oxford Univ, 75-76; adv bd, Advan Chem, 85-88; chair, Comt Int Activ, Am Chem Soc, 90- *Mem:* Am Chem Soc; ASTM. *Res:* Gas and liquid chromatography; statistical analysis; ion-sensitive electrodes. *Mailing Add:* Biomed Res Tech Prog NIH Westwood Bldg Rm 8A11 Bethesda MD 20892

HIRSCH, SAMUEL ROGER, b Chicago, Ill, Dec 19, 30; m 54; c 4. ALLERGY, IMMUNOLOGY. *Educ:* Univ Wis, BA, 53, MD, 56. *Prof Exp:* res assoc, Vet Admin Ctr, Milwaukee, 64-73, chief, Allergy & Pulmonary Res Lab, 75-82; assoc clin prof, 73-87, CLIN PROF MED, MED COL WIS, 87-, ASSOC CLIN PROF MED, UNIV WIS, 76- *Concurrent Pos:* Pvt pract, Allergy Assocs, Ltd, 63-80; attending physician, Milwaukee County Med Complex, 63-78; consult staff mem, Milwaukee Children's Hosp, 63-, Samaritan Hosp, Milwaukee, 65- & St Michael's Hosp, 65-80; assoc attending physician, Mt Sinai Hosp, 67-; consult, Panel on Diphenhydramine, FDA, 76-77. *Mem:* Fel Am Acad Allergy, Am Col Chest Physicians; fel Am Col Physicians; fel Am Col Allergy. *Res:* Pulmonary disease. *Mailing Add:* 5020 W Oklahoma Ave Milwaukee WI 53219

HIRSCH, SOLOMON, b Sydney, NS, Jan 24, 26; m 54; c 3. MEDICINE, PSYCHIATRY. *Educ:* Dalhousie Univ, BSc, 45, MD, 49; Royal Col Physicians & Surgeons Can, cert, 53. *Prof Exp:* Resident psychiat, Dalhousie Univ, 49-50, asst prof, 53-54; asst resident, Johns Hopkins Univ, 50-52; Head dept psychiat, Victoria Gen Hosp, 75-85; assoc prof, 58-64, PROF PSYCHIAT, DALHOUSIE UNIV, 64-, DEP HEAD DEPT, 81- *Concurrent Pos:* Asst psychiatrist, Victoria Gen Hosp, Can, 43-54, assoc psychiatrist, 59-; psychiatrist, NS Hosp, 53-54; consult psychiatrist, Grace Maternity Hosp, 54- *Mem:* Am Psychiat Asn; fel Am Col Psychiat; Can Psychiat Asn. *Res:* suicidology. *Mailing Add:* Dept Psychiat Victoria Gen Hosp 1278 Tower Rd Halifax NS B3H 2Y9 Can

HIRSCH, STEPHEN SIMEON, b New York, NY, Mar 25, 37; m 59; c 2. ORGANIC CHEMISTRY, POLYMER CHEMISTRY. *Educ:* Polytech Inst Brooklyn, BS, 58; Univ Md, PhD(chem), 62. *Prof Exp:* Sr res chemist, Goodyear Tire & Rubber Co, Ohio, 62-63; res chemist, Monsanto Co, 64-66, group leader, 67-68; res group leader, Allied Chem Corp, 68-69; assoc dir, Res Dept, Ciba-Geigy Corp, 69-71, prod mgr, Resins Dept, 71-72, mgr bus develop plastics, 72-75; dir res & develop, Hooker Chem & Plastics Corp, 75-78; vpres technol & bus develop, Plastics & Chem Spec Group, Hooker Chem Co, 78-81; dir chem res, Armand Hammer Tech Ctr, 81-84, dir, Oper Anal, Occidental Petrol Corp, 84-85; DIR NEW BUS DEVELOP & PLANNING, GAF CHEM, 85- *Mem:* Am Chem Soc; Am Mgt Asn; Com Develop Asn. *Res:* Plastics; fiber; composites; advanced materials; high temperature chemistry; flammability; organics; metal plating chemicals; new business development. *Mailing Add:* Atochem N America Three Parkway Philadelphia PA 19102

HIRSCH, TEDDY JAMES, b Beaumont, Tex, Aug 11, 29; m 55; c 4. CIVIL ENGINEERING. *Educ:* Tex A&M Univ, BS, 52, MEng, 53, PhD(civil eng), 61. *Prof Exp:* Struct engr, Stone & Pitts Architects & Engrs, 55-56; from instr & res asst civil eng to PROF CIVIL ENG, TEX A&M UNIV, 67-, RES ENGR, 67- *Concurrent Pos:* Assoc head, Struct Eng Div, Tex Transp Inst, 64-65, head, 65-83; mem grad fac, 63-, head structural engr div, 66-, assoc head dept, 80-83, interim head, Dept Civil Eng, Texas A&M Univ, 83-85. *Honors & Awards:* Paul Gray Hoffman award, Automobile Safety Found, 68. *Res:* Structural engineering and materials; motor vehicle impact; dynamic behavior of piling; dynamic loads; heavy truck and railroad barriers; foundations of structures; highway safety appurtenances; human tolerance to impact free fall life boats. *Mailing Add:* Dept Civil Eng & Tex Transp Inst Tex A&M Univ College Station TX 77843

HIRSCH, WARREN MAURICE, b New York, NY, Aug 3, 23; m 60; c 2. MATHEMATICS. *Educ:* NY Univ, AB, 47, MS, 48, PhD(math), 52. *Prof Exp:* Asst, 47-52, instr, 48-51, lectr, grad dept, 52, vis lectr, 52-53, from asst prof to assoc prof, 53-60, PROF MATH, COURANT INST MATH SCI, NY UNIV, 57- *Concurrent Pos:* Lectr, Columbia Univ, 52-53; consult Rand Corp, 53- *Mem:* AAAS; Am Math Soc; Math Asn Am; Am Statist Asn; Sigma Xi. *Res:* Theory of probability; mathematical statistics. *Mailing Add:* 51 Kettle Creek Rd Weston CT 06883

HIRSCHBERG, ALBERT I, b Brooklyn, NY, May 9, 34; m 60; c 2. ORGANIC CHEMISTRY. *Educ:* Brooklyn Col, BS, 54, Polytech Inst New York, MS, 56, PhD(chem), 60. *Prof Exp:* NIH fel, 60-61; from asst prof to assoc prof, 62-71, dept chmn, 79-85, PROF CHEM, LONG ISLAND UNIV, 71- *Concurrent Pos:* Fac res fel, Seiler Res Lab, US Air Force Acad, 87. *Mem:* Am Chem Soc; Royal Soc Chem. *Res:* Heterocyclic synthesis; chemotherapy; reaction mechanisms; chemical education. *Mailing Add:* Dept Chem Long Island Univ University Plaza Brooklyn NY 11201

HIRSCHBERG, CARLOS BENJAMIN, b Santiago, Chile, Feb 1, 43; m 73. CELL BIOLOGY, GENETICS. *Educ:* Rutgers Univ, MS, 66; Univ Ill, Urbana, PhD(chem), 70. *Prof Exp:* Res fel biol chem, Harvard Univ, 70-72; res assoc biol, Mass Inst Technol, 72-74; from asst prof to assoc prof, Sch Med, St Louis Univ, 74-82, prof biochem, 82-87; PROF BIOCHEM & MOLECULAR BIOL, UNIV MASS MED CTR, 87- *Concurrent Pos:* Jane Coffin Childs fel, Harvard Med Sch, 70-72 & Mass Inst Technol, 72-73; mem, Cell Biol Study Sect, NIH, 78-82; mem adv comt biochem & carcinogenesis, Am Cancer Soc, 84-87. *Mem:* AAAS; Am Soc Cell Biol; Am Soc Biochem & Molecular Biol. *Res:* Metabolism and function of glycoproteins and lipids in membranes and cell surfaces; biosynthesis of glycosaminolycans. *Mailing Add:* Dept Biochem & Molecular Biol Univ Mass Med Ctr 55 Lake Ave N Worcester MA 01655

HIRSCHBERG, ERICH, b Munich, Germany, Oct 24, 21; nat US; m 45; c 3. BIOCHEMISTRY. *Educ:* Univ Wis, PhD(physiol chem), 50. *Prof Exp:* Chemist, Hoffmann-La Roche, NJ, 43-47; asst, Univ Wis, 47-49; res assoc, Columbia Univ, 51-55, asst prof biochem, 55-68; assoc prof, 68-70, PROF BIOCHEM, NJ MED SCH, UNIV MED & DENT, 70-, ASSOC DEAN RES, 68- *Concurrent Pos:* USPHS fel, Columbia Univ, 50-51; assoc ed, Cancer Res, Am Asn Cancer Res. *Mem:* AAAS; Am Chem Soc; Am Soc Biol Chem; Am Asn Cancer Res; Am Asn Cancer Educ. *Res:* Experimental cancer research; biochemistry of differentiation and morphogenesis; clinical biochemistry. *Mailing Add:* NJ Med Sch 185 S Orange Ave Newark NJ 07103-2757

HIRSCHBERG, JOSEPH GUSTAV, b Chicago, Ill, Apr 13, 21; m 47; c 4. PHYSICS. *Educ:* Dartmouth Col, AB, 43; Univ Wis, PhD(physics), 52. *Prof Exp:* Fulbright scholar physics, Ecole Normale Superieure, Paris, 52-53; res assoc, Univ Wis, 53-58; head, Optical Sect, Plasma Physics Lab, Princeton Univ, 58-65, res physicist, 62-65; chmn dept, 65-72, dir, Optical Physics Lab, 67-77, PROF PHYSICS, UNIV MIAMI, 65- *Concurrent Pos:* Pres, Fed Eng Co, 54-; exchange prof, Univ Paris, 64; res assoc, French Asn Atomic Energy Comn-Europ Atomic Energy Comn, Paris, 64; consult, Plasma Physics Lab, Princeton Univ, 65-68 & Langley Res Ctr, NASA, 66-67; coun, Oak Ridge Assoc Univs, 70-72; fel, Papanicolaou Cancer Res Inst, Miami; vis astron, Sacramento Peak Observ, 77; vis scientist, Princeton Univ, 81-; exchange researcher, Museum Natural History, Paris, France, 83, 87 & 88. *Mem:* fel Am Phys Soc; fel Am Optical Soc; Sigma Xi; Europ Acad Sci, Arts & Humanities. *Res:* Meteorology; atomic spectra; physical optics; Fabry-Perot interferometer; plasma spectroscopy; holographic interferometry; solar physics; infrared imaging; fluoromicroscopy of living cells; optical oceanography. *Mailing Add:* Dept Physics Univ Miami Coral Gables FL 33124

HIRSCHBERG, NELL, microbiology, for more information see previous edition

HIRSCHBERG, RONA L, b Gary, Ind, Mar 26, 43. GENE REGULATION, NITROGEN FIXATION. *Educ:* Purdue Univ, BS, 65; Univ Wis-Madison, MS, 68, PhD(bacteriol), 70. *Prof Exp:* Fel molecular biol & biochem, Albert Einstein Col Med, 69-73; instr biol, Hunter Col, City Univ NY, 73-75; asst prof biol, 75-81, ASSOC PROF BASIC LIFE SCI, UNIV MO-KANSAS CITY, 81-, ASSOC DEAN BASIC LIFE SCI, 89- *Concurrent Pos:* Vis scientist, Univ Wis-Madison, 81. *Mem:* Am Soc Microbiol; Soc Gen Microbiol; AAAS; Am Soc Biochem & Molecular Biol. *Res:* Regulation of gene expression in procaryotes; molecular genetics of microbial pathogens. *Mailing Add:* Sch Basic Life Sci Univ Mo Kansas City MO 64110

HIRSCHFELD, RONALD COLMAN, b Amsterdam, NY, Nov 23, 30; m 64; c 2. CIVIL ENGINEERING. *Educ:* Union Col, NY, BSCE, 50; Harvard Univ, SM, 51, PhD, 58. *Prof Exp:* From instr to asst prof soil mech, Harvard Univ, 58-64; assoc prof, Mass Inst Technol, 64-72; PRIN, GEOTECH ENGRS, INC, 70- *Concurrent Pos:* Mem, US Comt Large Dams; nat dir, Asn Eng Geol, 72-73, Am Soc Civil Engrs, 72-73 & Am Consult Engrs Coun, 83-84. *Mem:* Am Soc Civil Engrs; Geol Soc Am; Int Soc Soil Mech & Found Eng; Asn Eng Geol. *Res:* Engineering geology; soil and rock mechanics. *Mailing Add:* GEI Consult Inc 1021 Main St Winchester MA 01890

HIRSCHFELD, SUE ELLEN, b Ossining, NY, Jan 12, 41; div. GEOLOGY, PALEONTOLOGY. *Educ:* Univ Fla, BS, 63, MS, 65; Univ Calif, Berkeley, PhD(paleont), 71. *Prof Exp:* Instr earth sci, St Petersburg Jr Col, 65-66; instr, 71-75, assoc prof earth sci, 75-80, PROF GEOL SCI, CALIF STATE UNIV, HAYWARD, 80-, DEPT CHAIR, 88- *Concurrent Pos:* Consult. *Mem:* AAAS; hon mem, Golden Key Soc; Asn Women Geoscientists (nat secy, 79-80); Geol Soc Am; Soc Econ Paleontologists & Mineralogists. *Res:* Earthquake hazards of the San Francisco Bay area. *Mailing Add:* Dept Geol Sci Calif State Univ Hayward CA 94542

HIRSCHFELD, TOMAS BENO, molecular spectroscopy, optics; deceased, see previous edition for last biography

HIRSCHFELDER, JOHN JOSEPH, b Ft Wayne, Ind, Sept 12, 43; m 66; c 3. MATHEMATICS. *Educ:* Univ Notre Dame, BS, 65, MS, 66, PhD(math), 68. *Prof Exp:* Asst prof math, Univ Wash, 68-75; SR PRIN ENGR, HONEYWELL INC, 75- *Mem:* Asn Comput Mech; Am Math Soc. *Res:* Several complex variables; complex manifolds. *Mailing Add:* 6222 54th Ave NE Seattle WA 98115

HIRSCHFELDER, JOSEPH OAKLAND, theoretical chemistry, theoretical physics; deceased, see previous edition for last biography

HIRSCHHORN, JOEL S(TEPHEN), b New York, NY, Sept 8, 39; m 61; c 2. SCIENCE POLICY, TECHNICAL MANAGEMENT. *Educ:* Polytech Inst Brooklyn, BMetE, 61, MS, 62; Rensselaer Polytech Inst, PhD(mat eng), 65. *Prof Exp:* Res metallurgist, adv mat res & develop lab, Pratt & Whitney Aircraft Div, United Aircraft Corp, 62-63; from asst prof to prof metall eng, Univ Wis-Madison, 65-78; sr assoc, Off Technol Assessment, US Cong, 78-90; PRES, HIRSCHHORN & ASSOCS, INC, 90- *Concurrent Pos:* Consult, Friction Prod Co, 67-68, dir res, 68-72; consult, Stellite Div, Cabot Corp, 70-71, Advan Prod Corp, 71-73 & Que Metal Powders, 75-76. *Mem:* AAAS. *Res:* Indust waste reduction; environmental technology; hazardous waste; strategic planning; environmental regulations; technology assessment; science and technology policy. *Mailing Add:* Hirschhorn & Assocs Inc 2400 Virginia Ave Suite C103 Washington DC 20037

HIRSCHHORN, KURT, b Vienna, Austria, May 18, 26; nat US; m 52; c 3. HUMAN GENETICS, PEDIATRICS. *Educ:* NY Univ, BA, 50, MD, 54, MS, 58. *Prof Exp:* Intern internal med, Bellevue Hosp, NY, 54-55, asst resident, 55-56; from instr to assoc prof med, Sch Med, NY Univ, 55-66; prof pediat & chief div med genetics, 66-76, Arthur J & Nellie Z Cohen prof pediat, 68-76, HERBERT H LEHMAN PROF & CHMN DEPT PEDIAT, MT SINAI SCH MED, 76- *Concurrent Pos:* USPHS clin trainee metab dis, 56-57; Berquist & Pop Coun fels, Inst Med Genetics, Uppsala, 57-58; Am Heart Asn res fel, NY Univ, 58-60, estab investr, 60-65; from asst vis physician to assoc vis physician, Bellevue Hosp, 56-66; from asst attend physician to assoc attend physician, NY Univ Hosp, 56-66; career scientist, NY City Health Res Coun, 65-75; attend pediat, Mt Sinai Hosp, 66-; vis prof, Galton Lab, Dept Human Genetics, Univ Col, Univ London, 71-72; mem coun arteriosclerosis, Am Heart Asn; vis prof, Harvard Med Sch, 86- *Mem:* Fel AAAS; Am Soc Human Genetics (dir, 64-65 & 68-71, pres, 69); Harvey Soc (vpres, 79-80, pres, 80-81); Am Asn Immunol; Asn Am Physicians; Am Pediat Soc (coun, 81-84). *Res:* Immunogenetics; human biochemical, molecular and cytogenetics. *Mailing Add:* Mt Sinai Sch Med One Gustave L Levy Pl New York NY 10029-6574

HIRSCHHORN, ROCHELLE, b New York, NY, Mar 19, 32; m 52; c 3. IMMUNOBIOLOGY, HUMAN GENETICS. *Educ:* Barnard Col, Columbia Univ, BA, 53; NY Univ, MD, 57. *Prof Exp:* Intern, Bellevue IV Med Div, NY Univ, 58-59, assoc res scientist, 65-66, from instr to assoc prof, 66-79, PROF MED, DEPT MED, SCH MED, NY UNIV, 79- HEAD, DIV MED GENETICS, 84- *Concurrent Pos:* Fel, Arthritis Found, 66-69; sr investr, New York Heart Asn, 69-72; res career develop award, NIH, 72-77; hon fel, Dept Human Genetics & Biomet, Univ Col, Univ London, 71-72; mem, NIH Study Sect, Allergy & Immunol Clin Res Adv Comt, Nat Found March Dimes; vis prof, Harvard Med Sch, 85-86; fel rev comt, Arthritis Found, 86-88; mem Inst Allergy & Infectious Dis Bd Sci Counselors, NIH, 87- *Honors & Awards:* Jeffrey Modell Found Lifetime Achievement Award, 90. *Mem:* Am Soc Clin Invest; Am Asn Immunologists; Am Soc Human Genetics; Am Soc Cell Biol; Am Asn Physicians; Am Fedn Clin Res; fel AAAS; fel Arthritis & Rheumatism Asn; Harvey Soc. *Res:* Control of differentiation and activation of immunocompetent cells; inherited disorders of man and genetic polymorphism. *Mailing Add:* Dept Med Sch Med NY Univ 550 First Ave New York NY 10016

HIRSCHMAN, ALBERT, b New York, NY, Oct 20, 21; c 3. BIOCHEMISTRY, HISTOLOGY. *Educ:* City Col New York, BS, 42; Polytech Inst Brooklyn, MS, 46, PhD(biochem), 52. *Prof Exp:* Chem technician, Jewish Hosp, Brooklyn, 42-43; asst chemist, 43-47; instr histol, 48-54, asst prof anat, 54-68, ASSOC PROF ANAT, STATE UNIV NY DOWNSTATE MED CTR, 68- *Concurrent Pos:* Adj prof, Ctr Biomed Educ, Touro Col, Huntington, NY, 82- *Mem:* Am Chem Soc; Am Asn Anat; Am Crystallog Asn; NY Acad Sci. *Res:* Physical, chemical and histochemical studies of bone and the mechanism of calcification; x-ray diffraction of mineralized tissue; biochemical studies of enzymes and proteoglycans in bone and in calcifying and non-calcifying cartilages. *Mailing Add:* Downstate Med Ctr Box 5 Brooklyn NY 11203

HIRSCHMAN, ISIDORE ISAAC, JR, b Washington, DC, Nov 22, 22; m 43; c 3. MATHEMATICS. *Educ:* Harvard Univ, BA, 42; Brown Univ, ScM, 43; Harvard Univ, PhD(math), 47. *Prof Exp:* Naval res fel, Harvard Univ, 47-49; from asst prof to assoc prof, 49-55, chmn dept, 61-66, PROF MATH, WASHINGTON UNIV, 55- *Concurrent Pos:* Guggenheim fel, 52-53; Fulbright res grant, France, 55-56; Air Force res grant, Switz, 60-61; vis prof, Stanford Univ, 64-65, Duke Univ, 68-69 & Univ Montpellier, France, 74-75. *Mem:* Am Math Soc. *Res:* Integral equations; convolution transforms; quasi-analytic functions; Fourier analysis; harmonic analysis, Wiener-Hopf Toeplitz operators. *Mailing Add:* Dept of Math Washington Univ Lindell-Skinker Blvd St Louis MO 63130

HIRSCHMAN, LYNETTE, b Huntington, WVa, Nov 22, 45. NATURAL LANGUAGE PROCESSING, LOGIC PROGRAMMING. *Educ:* Oberlin Col, BA, 66; Univ Calif, Santa Barbara, MA, 68; Univ Pa, PhD(linguistics), 72. *Prof Exp:* Res scientist, NY Univ, 75-82; mgr, 82-86, TECH DIR, LOGIC-BAND SYSTS, PAOLI RES CTR, UNISYS DEFENSE SYSTS, 86- *Concurrent Pos:* Adj asst prof, Dept Comput Sci, NY Univ, 80-81; adj prof, Dept Comput Sci, Univ Pa, 87- *Mem:* Am Asn Artificial Intelligence; Asn Comput Mach; Asn Logic Prog; Asn Comput Ling. *Res:* Natural language processing and applications of logic programming to natural language; parallel processing and biomedical information processing. *Mailing Add:* 264 W Harvey St Philadelphia PA 19144

HIRSCHMAN, SHALOM ZARACH, b Troy, NY, Aug 5, 36; m 65; c 2. INFECTIOUS DISEASE, MOLECULAR BIOLOGY. *Educ:* Yeshiva Univ, BA, 57; Albert Einstein Col Med, MD, 61; FRSTM. *Prof Exp:* Intern, Mass Gen Hosp, Harvard Med Sch, Boston, 61-62, asst resident, 62-63; NIH res assoc, Nat Inst Arthritis & Metab Dis, Md, 63-66; sr investr molecular virol, Nat Cancer Inst, Md, 67-69; assoc prof, 69-71, assoc attend physician, Hosp, 69-71, PROF MED, MT SINAI SCH MED, 71-, DIR DIV INFECTIOUS DIS, MED CTR, 69-, ATTEND PHYSICIAN, HOSP, 71- *Concurrent Pos:* NIH spec fel, 64 & fel, Columbia-Presby Med Ctr, New York, 66-67; vchmn microbiol sect, NY Acad Sci, 73-, chmn, 75; mem, Merit Rev Bd, Vet Admin, 76-79, & Microbiol Rev Bd, Am Cancer Soc, 80-85. *Mem:* AAAS; fel Am Col Physicians; NY Acad Sci; Am Soc Clin Invest; Soc Exp Biol Med; Asn Am Physicians; Biophys Soc; Infectious Dis Soc Am; Am Soc Microbiol; Royal Soc Trop Hyg Med; Am Col Pharamacol; Liver Soc; Harvey Soc. *Res:* Molecular biology of bacteria and viruses; biology of hepatitis B virus; internal medicine. *Mailing Add:* Mt Sinai Sch Med Fifth Ave 100th St New York NY 10029

HIRSCHMANN, ERWIN, b Vienna, Austria, July 17, 24; m 63; c 2. MICROWAVE PHYSICS. *Educ:* Vienna Tech Univ, dipl, 49. *Prof Exp:* Prod technician, Siemens & Halske Co, Inc, Austria, 42-45, prod engr, 50-52; mem staff, Sound-Frequency Lab, Hasler Co, Inc, Switz, 54-55; mem staff develop transistor field, Siemens & Halske, Ger, 56-58; researcher semiconductors, Diamond Ord Fuze Labs, 58-62; researcher laser commun & millimeter propagation, 62-80, COMMAND & CONTROL-COMMUN & DATA HANDLING SUBSYSTS TV INFRARED OBSERV SATELLITE WEATHER SATELLITES, GODDARD SPACE FLIGHT CTR, NASA, 81- *Mem:* Inst Elec & Electronics Engrs; Am Phys Soc. *Res:* Microwave and laser propagation and communication; semiconductor, solid state physics; theoretical physics. *Mailing Add:* 35 Lakeside Dr Greenbelt MD 20770-1973

HIRSCHMANN, HANS, b Bavaria, Ger, July 1, 09; US citizen; m 38; c 1. BIOLOGICAL CHEMISTRY. *Educ:* Univ Basel, MD, 34; Columbia Univ, PhD(biochem), 38. *Prof Exp:* Res fel, Univ Pa, 38-42; from asst prof to prof biochem, 42-78, prof chem, 68-78, EMER PROF, CASE WESTERN RESERVE UNIV, 78- *Honors & Awards:* Morley Award, Am Chem Soc, 84. *Mem:* AAAS; Am Chem Soc; Endocrine Soc; Am Soc Biochem & Molecular Biol. *Res:* Chemistry of steroids, adrenal steroids; stereochemistry. *Mailing Add:* 14018 Becket Rd Cleveland OH 44120-2823

HIRSCHMANN, RALPH FRANZ, b Bavaria, Ger, May 6, 22; nat US; m 51; c 2. ORGANIC CHEMISTRY. *Educ:* Oberlin Col, AB, 43; Univ Wis, MA, 48, PhD(org chem), 50. *Hon Degrees:* DSc, Oberlin Col, 69. *Prof Exp:* Chemist process res, Res Labs, Merck Sharp & Dohme, Inc, Rahway, 50-54, sr chemist, 54-56, sect head, 57-58, res assoc fundamental res, 58-63, res assoc explor, 63-64, asst dir explor res, 64-68; dir peptide res, Merck Sharp & Dohme Res Labs, Rahway, NJ, 68-69, dir protein chem, 69-71, sr dir new lead discovery, 71-72, sr dir med chem, 72-74, exec dir, med chem, West Point, Pa, 74-76, vpres, 76-78, sr vpres basic res, 78-84, sr vpres chem, 84-87; RES PROF CHEM, UNIV PA, 87-; UNIV PROF BIOMED RES, MED UNIV SC, CHARLESTON, 87- *Concurrent Pos:* Vis prof, Univ Wis, 73; mem, NIH Med Chem A Study Sect, 79-82; distinguished lectr, Wesleyan Univ, 79; cochmn, Gordon Res Conf, 78 & bd trustees, 81-; mem, Comt Surv Opportunities Chem Sci, Nat Res Coun, 82-; adv comt chem, NSF, 85-; med prof, Univ SC, 87-; consult, expert witness-patents. *Honors & Awards:* Intrasci Res Found Award, 70; Alan E Pierce Award, 83; Romanes Lectr, Univ Edinburgh, 85; Agnes Borrowman Spec Lectr in Pharm, London Univ, 85; Hurd Lectr, Northwestern Univ, 85; Nichol medal, Darmouth Lectr, 88. *Mem:* Fel AAAS; NY Acad Sci; Am Soc Biol Chemists; Am Acad Arts & Sci; Am Chem Soc. *Res:* Organic synthesis; steroids; vitamin K; first total synthesis of ribonuclease; chemistry and biology relating to human and animal health; peptides; peptidomimetics. *Mailing Add:* 740 Palmer Pl Blue Bell PA 19422-1725

HIRSCHMANN, ROBERT P, b New York, NY, Dec 26, 34; m 57; c 4. APPLIED CHEMISTRY. *Educ:* City Col New York, BS, 56; Iowa State Univ, PhD(phys chem), 63; Wichita State Univ, MBA, 82. *Prof Exp:* Res chemist, Gen Chem Div, Allied Chem Corp, NJ, 56-58, spectroscopist, Ill, 58-59, sr res chemist, Morristown, 63-66, sr res scientist, Indust Chem Div, 66-70; group leader, 70-82, asst mgr chem res, 82-88, DIR RES & DEVELOP, VULCAN MAT CORP, 88- *Mem:* Am Chem Soc. *Res:* Vibrational spectra of -X-Y-Z compounds such as alkyl isocyanates and isothiocyanates; the study of oxidative chlorination of simple aliphatic molecules. *Mailing Add:* 9919 W 12th Wichita KS 67212

HIRSCHOWITZ, BASIL ISAAC, b Bethal, SAfrica, May 29, 25; m 58. MEDICINE. *Educ:* Univ Witwatersrand, BSc, 43, MB, ChB, 47, MD, 54; FRCP(Ed); FRCP. *Prof Exp:* House officer med, Johannesburg Gen Hosp, SAfrica, 48-50; house officer, Postgrad Med Sch, London, 50; registr, Cent Middlesex Hosp, 51-53; from instr to asst prof internal med, Med Sch, Univ Mich, 53-56; asst prof, Sch Med, Temple Univ, 57-59; assoc prof med, 59-64, physiol, 68-71, PROF MED, MED COL ALA, 64-, PHYSIOL, 71-, DIR DIV GASTROENTEROL, 59- *Honors & Awards:* Schindler Medal, Am Soc Gastrointestinal Endoscopy, 73; Kettering Prize, Gen Motors Cancer Res Found, 87. *Mem:* AAAS; Am Physiol Soc; Sigma Xi; NY Acad Sci; Med Res Soc Gt Britain; Am Gastroenterol Asn. *Res:* Diseases of the gastrointestinal tract; physiology of the stomach. *Mailing Add:* 3200 E Briarcliff Birmingham AL 35223

HIRSH, ALLEN GENE, b Pinewald, NJ, Nov 26, 47; m 87. STRESS PHYSIOLOGY OF FREEZING IN WOODY PLANTS. *Educ:* Calif Inst Technol, BS, 70; Univ Md, PhD, 85. *Prof Exp:* Nurseryman, Calgo Gardens Nursery, Toms River, NJ, 73-77, Potomac Garden Ctr, Potomac, Md, 77-79; res technologist, Bethesda, Md, 79-86, RES SCIENTIST, BIOMED RES & DEVELOP LAB, AM RED CROSS, ROCKVILLE, MD, 86- *Mem:* Am Soc Plant Physiologists; Biophys Soc; Int Soc Cryobiol; NAm Thermal Anal Soc. *Res:* Biophysical analysis of the affects of freezing on living tissue; use of electron microscopy, scanning calorimetry, and dynamic mechanical analysis to characterize phase changes in cells at low temperature; concerned with the natural resistance to extreme freezing manifested by certain artic tree species; author or co-author of 11 journal articles and book chapters. *Mailing Add:* Transplantation Lab 15601 Crabbs Br Way Rockville MD 20855

HIRSH, DWIGHT CHARLES, III, b Los Angeles, Calif, Oct 5, 38; m 67; c 2. VETERINARY MICROBIOLOGY. *Educ:* Loyola Univ, Calif, BS, 60; Univ Calif, Davis, DVM, 66; Stanford Univ, PhD(med microbiol), 71. *Prof Exp:* Fel med microbiol, Stanford Univ, 66-68; asst prof vet microbiol, Univ Mo-Columbia, 71-73; from asst prof to assoc prof, 73-83, PROF MICROBIOL, UNIV CALIF, DAVIS, 83- *Mem:* Am Soc Microbiol; Sigma Xi; Am Asn Immunologists. *Res:* Host-parasite relationships, especially immunological and ecological aspects; plasmid mediated resistance and virulence determinants in Pasteurella. *Mailing Add:* Dept Vet Microbiol Univ Calif Sch Vet Med Davis CA 95616

HIRSH, EVA MARIA HAUPTMANN, b Freiburg, Ger, July 14, 28; US citizen; m 62; c 2. PHARMACOLOGY, PHYSIOLOGY. *Educ:* Bryn Mawr Col, AB, 50; NY Univ, PhD(physiol-pharmacol), 73. *Prof Exp:* Instr physiol & pharmacol, NY Univ Col Dent, 60-68; lectr Philadelphia Col Osteop Med, 69-72; fel pharmacol & psychiat, NY Med Col, 73-74; res assoc psychopharmacol, Med Sch Thomas Jefferson Univ, 74-75; ASST PROF PHARMACOL, NY MED COL, 75- *Mem:* NY Acad Sci; AAAS; Am Asn Electroencephalographers; Sigma Xi. *Res:* Mechanisms involved in production of experimental myocardial ischemia and necrosis by means of cardiotoxic adrenergic amines; measurement of blood flow by electrical impedance plethysmography; neuropsychopharmacology of hallucinogens. *Mailing Add:* James St Norwalk CT 06850

HIRSH, IRA JEAN, b New York, NY, Feb 22, 22; m 43; c 4. PSYCHOPHYSICS. *Educ:* NY State Col Teachers, AB, 42; Northwestern Univ, MA, 43; Harvard Univ, PhD(exp psychol), 48. *Prof Exp:* Asst, Psycho-acoustic Lab, Harvard Univ, 46-48, res fel, 48-51; res assoc, Cent Inst Deaf, 51-60, from asst dir res to dir res 60-83; from assoc prof to prof, 52-85, dean, Fac Arts & Sci, 69-73, MALINCKROOT DISTINGUISHED PROF PSYCHOL, WASH UNIV, 85-; SR SCIENTIST, CENT INST DEAF, 83- *Concurrent Pos:* Vis prof, Sorbonne, 62-63 & Tsukuba Univ, Japan, 81-82; consult, USAF, 55-61, Vet Admin, 56-65, NIH, 63-66 & Dept Trans, 67-73; mem comt hearing & bio-acoustics, Armed Forces-Nat Acad Sci, 56-59 & 63-66, chmn, 64-65; chmn eval panel acoustics, Nat Bur Standards, 71-74. *Honors & Awards:* Biennial Award, Acoust Soc Am, 56; Honors Award, Am Speech & Hearing Asn, 68. *Mem:* Nat Acad Sci; fel Acoust Soc Am (pres, 67-68); fel Am Psychol Asn; fel Am Speech & Hearing Asn; Soc Exp Psychol. *Res:* Hearing, auditory perception, communication, speech and language, communication disorders. *Mailing Add:* Cent Inst Deaf 818 S Euclid St Louis MO 63110

HIRSH, KENNETH ROY, pharmacology, technical management, for more information see previous edition

HIRSH, MERLE NORMAN, b New York, NY, Apr 27, 31; m 54; c 4. MATERIALS SCIENCE, ATOMIC PHYSICS. *Educ:* Univ Pittsburgh, BS, 52; Johns Hopkins Univ, PhD(physics), 58. *Prof Exp:* Proj dir maser res, ITT Fed Labs, 58-61; tech dir space physics, G C Dewey Corp, 61-71 & Dewey Electronics Corp, 71-72; chmn, Sci-Math Div, Univ Minn, Morris, 72-78, prof physics, 72-80; sr scientist, Lab Laser Energetics, 80-81, PRES, PLASMA RESOURCES, UNIV ROCHESTER, 81- *Concurrent Pos:* Mem staff, Physics Discharge Lab, Superior Sch Elec, Gif-sur-Yvette, France, 76-77. *Mem:* Am Asn Physics Teachers; Inst Elec & Electronics Engrs; Indust Appln Soc; Am Phys Soc; Am Vacuum Soc. *Res:* Plasma modification of surfaces; plasma chemistry; gaseous electronics, particularly electron and ion production processes in the upper atmosphere; laboratory investigation of ionospheric reactions. *Mailing Add:* 1600 East Ave No 810 Rochester NY 14610

HIRSHAUT, YASHAR, b Berlin, Ger, Feb 27, 38; US citizen; m 64; c 8. INTERNAL MEDICINE, ONCOLOGY. *Educ:* Yeshiva Univ, BA, 59, Albert Einstein Col Med, MD, 63; Am Bd Internal Med, dipl, 72, dipl med oncol, 75. *Prof Exp:* Intern, Montefiore Hosp, 63-64, asst resident, 64-65; clin assoc, Med Br, Nat Cancer Inst, 65-68; res assoc, Inst, 68-72, assoc, 72-75, asst prof, 72-79, ASSOC PROF MED, MED COL, CORNELL UNIV, 79-, HEAD LAB IMMUNODIAG, SLOAN-KETTERING INST CANCER RES, 75-, ASST MEM, INST, 81- *Concurrent Pos:* Fel med oncol, Dept Med, Mem Hosp, New York, 69-70; fel, Dept Med, Cornell Univ Med Sch, 69-70, instr med, 70-72; clin asst physician, Clin Immunol Serv, Mem Hosp, New York, 70-73, asst attend physician, 73-79, assoc attend physician, 79- *Mem:* Am Asn Immunol; Am Asn Cancer Res; Am Soc Clin Oncol; NY Acad Sci; AMA. *Res:* Tumor immunology; clinical immunology; cancer chemotherapy and immunotherapy. *Mailing Add:* 425 E 67th St New York NY 10021

HIRSHBURG, ROBERT IVAN, b Dallas, Tex, June 19, 49; m 74; c 1. MECHANICAL ENGINEERING. *Educ:* Auburn Univ, BME, 72, MS, 74; Ariz State Univ, PhD(mech eng), 80. *Prof Exp:* Proj engr, Sperry Flight Systs, 74-75; res asst, Ariz State Univ, 75-79; RES ENGR, E I DU PONT DE NEMOURS & CO, INC, 80- *Mem:* Am Soc Mech Engrs. *Res:* Theoretical and experimental investigations of free-surface liquid flows, particularly analysis and development of precision coating equipment. *Mailing Add:* E I du Pont de Nemours & Co Inc PO Box 267 Brevard NC 28712

HIRSHFELD, FRED LURIE, b Brooklyn, NY, Apr 3, 27; m 56; c 5. CHEMICAL CRYSTALLOGRAPHY. *Educ:* Columbia Univ, BA, 47, MA, 48; Hebrew Univ, Israel, PhD(physics), 56. *Prof Exp:* Jr scientist, Weizmann Inst, 50-56; fel, Univ Minn, 56-57; res worker, Columbia Univ, 57-58; res assoc, 58-64, sr scientist, 64-72, assoc prof, 72-79, PROF, WEIZMANN INST SCI, 79- *Concurrent Pos:* Chmn, Israel Crystallog Soc, 71-73. *Mem:* Am Phys Soc; Israel Crystallog Soc. *Res:* Crystallography of organic compounds; electron-density distributions. *Mailing Add:* Weizmann Inst Sci PO Box 26 Rehovot Israel

HIRSHFIELD, JAY LEONARD, b Washington, DC, Oct 24, 31; m 57; c 4. PLASMA PHYSICS. *Educ:* Univ Md, BS, 52; Ohio State Univ, MS, 56; US Air Force Inst Technol, MS, 58; Mass Inst Technol, PhD(physics), 60. *Prof Exp:* Proj engr, Air Develop Ctr, Wright-Patterson Air Force Base, Ohio, 54-57; res assoc physics, Mass Inst Technol, 60; NATO fel, Inst Plasma Physics, Jutphaas, Neth, 61-62; from instr to asst prof physics, 62-63, from asst prof to assoc prof appl sci, 63-68, PROF APPL PHYSICS, YALE UNIV, 68- *Concurrent Pos:* Guggenheim fel, Lab Gas Ionization, Frascati, Italy, 68; vis prof, Racah Inst Physics, Hebrew Univ, Jerusalem, Israel, 72 & 77-78. *Mem:* Fel Am Phys Soc. *Res:* Microwave optics; microwave gas discharge and plasma physics; interactions of radiation with relativistic electron beams, including gyrotrons, free electron lasers, slow cyclotron wave amplifiers and related mechanisms; laser-produced plasmas; plasma diagnostics. *Mailing Add:* 55 Killdeer Rd Hamden CT 06517

HIRSHMAN, CAROL A, b Montreal, Can, Aug 12, 44; m 70; c 1. ANESTHESIOLOGY, PHARMACOLOGY. *Educ:* McGill Univ, BSc, 65, MD, 69. *Prof Exp:* Asst prof anesthesiol, Univ Colo Med Ctr, 74-75; asst prof anesthesiol, Ore Health Sci Univ, 76-80, from assoc prof to prof anesthesiol & pharmacol, 80-86; PROF ANESTHESIOL MED & ENVIRON HEALTH SCI, JOHNS HOPKINS UNIV, 86- *Concurrent Pos:* Prin investr, NIH, 90-95. *Mem:* Am Soc Anesthesiologists; Am Physiol Soc; Am Thoracic Soc; Asn Univ Anesthetists. *Mailing Add:* Physiol Div Rm 7006 615 N Wolfe St Baltimore MD 21205

HIRSHMAN, JUSTIN LEONARD, b New York, NY, Dec 13, 30; m 59; c 2. ORGANIC & ORGANOMETALLIC CHEMISTRY. *Educ:* City Col, City Univ New York, BS, 53; Brooklyn Col, City Univ New York, MA, 59; Fairleigh Dickinson Univ, MBA, 77. *Prof Exp:* Develop chemist, US Vitamin & Pharmaceut Corp, 55-59 & Specific Pharmaceut, Chemetron Co, 59-63; res chemist, M&T Chem, Inc, 63-68, res supvr, 68-72, res mgr, 72-74, res dir, Chem Div, 74-80, res dir chem, 80-86, dir, comm eval & develop, 86-89; DIR, NEW BUS DEVELOP, ATOCHEM NAM, INC, 90- *Mem:* Am Chem Soc; CDA; Licensing Exec Soc. *Res:* Product and process research and development; organic and organometallic chemicals; condensation catalysis; polymer stabilization and biocidal activity of organometallic chemicals. *Mailing Add:* Atochem NAm Inc PO Box 195 Somerville NJ 08876-1295

HIRSHON, JORDON BARRY, b Brooklyn, NY, Feb 26, 39; m 60; c 2. BIOLOGICAL RHYTHMS, DNA SEQUENCES. *Educ:* City Col New York, BS, 60; Rutgers Univ, PhD(bot), 64. *Prof Exp:* From asst prof to assoc prof, 64-76, PROF BIOL, LONG ISLAND UNIV, 76- *Concurrent Pos:* Vis prof, Chungang Univ, Korea, 65-66; partic, NSF Summer Inst Math. *Mem:* Sigma Xi. *Res:* Circadian rhythms in Neurospora; computer analysis of DNA sequences. *Mailing Add:* Dept Biol Long Island Univ Brooklyn Ctr Brooklyn NY 11201

HIRSHON, RONALD, b May 17, 39; US citizen; m 62; c 2. MATHEMATICS. *Educ:* Brooklyn Col, BS, 61, MA, 63; Adelphi Univ, PhD(math), 67. *Prof Exp:* Instr math, State Univ NY Stony Brook, 65-66; asst prof, Staten Island Community Col, 66-67; from asst prof to assoc prof, 67-79, PROF MATH, POLYTECH INST NEW YORK, 80- *Mem:* Am Math Soc; Math Asn Am. *Res:* Group theory, particularly hopfian groups. *Mailing Add:* 90 Shiloh St Staten Island NY 10314

HIRST, ALBERT EDMUND, b Riverside, Calif, June 13, 15; m 52; c 3. MEDICINE. *Educ:* Pacific Union Col, BA, 38; Loma Linda Univ, MD, 42; Am Bd Path, dipl, 47. *Prof Exp:* From instr to assoc prof, 43-63 chmn dept, 64-73, PROF PATH, SCH MED, LOMA LINDA UNIV, 63- *Concurrent Pos:* Jr attend physician, Los Angeles County Hosp, 46-64; consult pathologist, San Antonio Community Hosp, Upland, 50-52; dir labs, John Wesley County Hosp, 59-60; res assoc, Harvard Univ, 61-62; vis prof, Univ Ala, Birmingham, 71-72; chief Lab Serv, Loma Linda Vet Admin Hosp, 77-87; assoc pathologist, San Bernardino Community Hosp, 83-84 & Loma Linda Community Hosp, 85. *Mem:* Am Soc Clin Path. *Res:* Arterial disease, especially arteriosclerosis and dissecting aneurysm. *Mailing Add:* Dept Path Loma Linda Univ Sch Med Loma Linda CA 92354

HIRST, GEORGE KEBLE, b Eau Claire, Wis, Mar 2, 09; m 37; c 5. VIROLOGY. *Educ:* Yale Univ, BS, 30, MD, 33. *Prof Exp:* Intern psychiat, New Haven Hosp, Conn, 33-34; intern med, Univ Hosps, Cleveland, 34-35, asst resident, 35-36; asst resident, Rockefeller Inst Hosp, 36-40, mem int health div, Rockefeller Found, 40-46; res prof microbiol, Post-Grad Med Sch, NY Univ, 68-80; chief virol, Pub Health Res Inst, 46-80, dir, 56-80, emer dir virol, 80-83; RETIRED. *Concurrent Pos:* Ed-in-chief, Virol, 55-76; adj prof, Post-Grad Med Sch, NY Univ, 57-68. *Mem:* Nat Acad Sci; Am Acad Arts & Sci; Harvey Soc. *Res:* Virus diseases; streptococci; rheumatic fever; influenza. *Mailing Add:* 18600 Castle Lake Dr Morgan Hill CA 95037

HIRST, MAURICE, b Leeds, Eng, Aug 23, 40; m 65; c 3. CHEMICAL PHARMACOLOGY. *Educ:* Univ Liverpool, BSc, 62, PhD(org chem), 65; Univ Western Ont, PhD(pharmacol), 69. *Prof Exp:* NIH fel org chem, 65-66, from asst prof to assoc prof, 67-79, PROF PHARMACOL, UNIV WESTERN ONT, 79- *Concurrent Pos:* Sci review mem, Comt Biomed Res, Health & Welfare, Med Res Coun Can, 75-78; mem fel comt, Ont Ment Health Found, 81-86; mem, biol, chem & toxicol comt, Health & Welfare, Can, 84- *Mem:* NY Acad Sci; Soc Toxicol Can; Can Col Neuropsychopharmacol; Pharmacol Soc Can. *Res:* Dependence producing drugs and their mechanisms of action. *Mailing Add:* Dept Pharmacol & Toxicol Health Sci Ctr Univ Western Ont London ON N6A 5C1 Can

HIRST, ROBERT CHARLES, b Chillicothe, Mo, Nov 2, 32; m 61; c 2. PHYSICAL CHEMISTRY. *Educ:* Univ Calif, Berkeley, BS, 54; Univ Utah, PhD(phys chem), 63. *Prof Exp:* Fel, Div Pure Chem, Nat Res Coun Can, 63-64; res chemist, Cent Res Div Lab, Socony Mobil Oil Co, 64-68, sr res chemist, Mobil Res & Develop Corp, 68-69; sr res chemist res dept, 69-75, group leader, NMR Spectros, 75-77, SECT HEAD SPECTROS, GOODYEAR TIRE & RUBBER CO, 77- *Mem:* Am Phys Soc; Am Chem Soc. *Res:* Polymer characterization and analysis by nuclear magnetic resonance spectroscopy, including high resolution, pulse and carbon-13 nuclear magnetic resonance of solids; characterization of polymers and surfaces by Fourier transform infrared spectroscopy. *Mailing Add:* 1082 Greenvale Ave Akron OH 44313-6702

HIRSTY, SYLVAIN MAX, b Antwerp, Belg, Apr 11, 25; nat US; m 46; c 3. RUBBER CHEMISTRY, POLYMER CHEMISTRY. *Educ:* Rensselaer Polytech Inst, BS, 46, PhD(org chem), 53. *Prof Exp:* Instr, NY State Col Teachers, Albany, 52; res chemist rubber chem, E I du Pont de Nemours & Co, 53-71; TECH DIR, GATES ENG, 71- *Mem:* Am Chem Soc. *Res:* Elastomers technology; rubber chemicals; fluorine and neoprene chemistry; elastomeric coatings. *Mailing Add:* 1124 Grinnel Rd Wilmington DE 19803

HIRT, ANDREW MICHAEL, b Youngstown, Ohio, Mar 30, 55; m 83; c 1. SURFACE & INTERFACE CHEMISTRY, MATERIALS CHARACTERIZATION. *Educ:* Case Inst Technol, BS, 77. *Prof Exp:* Sr scientist, Structure Probe, Inc, 77-80, lab mgr, 82-88; lab scientist, Phys Electronics Div, Perkin-Elmer, 80-82; PRES & SR SCIENTIST, MAT RES LABS, INC, 88-; VPRES RES, LASIR SYSTS, INC, 90- *Concurrent Pos:* Publicity chmn, Int Soc Hybrid Microelectronics, Garden State, 85-, prog comt, Annual Joint Symp, 89-; course instr, Instrumental Surface Anal, Ctr Prof Advan, 89; consult electron spectros, Dept Physics, Rutgers Univ, 90- *Mem:* Am Vacuum Soc; Am Soc Metals Int; Int Soc Hybrid Microelectronics (pres, 88-89); Soc Appl Spectros; Catalysis Soc Am; Eastern Electron Spectros Soc; Microbeam Anal Soc. *Res:* Surface and interface chemistry; interaction of solid-solid and solid-gas interfaces; thermal transport mechanisms; materials characterization; experimental design; evaporation and sputtering techniques for metallization; metallographic preparation petrography and fractography; computer programming. *Mailing Add:* 720 King Georges Post Rd Suite 1E Fords NJ 08863-1974

HIRT, CYRIL WILLIAM, JR, b Flushing, NY, Dec 20, 36; m; c 2. THEORETICAL PHYSICS. *Educ:* Univ Mich, BS, 58, MS, 60, PhD(plasma physics), 63. *Prof Exp:* Staff mem, Los Alamos Nat Lab, 63-80; PRES, FLOW SCI, INC, 80- *Concurrent Pos:* Sr scientist, Sci Appl Inc, 72-73; consult. *Res:* Fluid mechanics; numerical fluid dynamics. *Mailing Add:* Flow Sci Inc 1325 Trinity Dr Los Alamos NM 87544

HIRT, THOMAS J(AMES), b Cincinnati, Ohio, Mar 24, 31; m 52; c 3. PHYSICAL CHEMISTRY, FUEL TECHNOLOGY. *Educ:* Xavier Univ, Ohio, BS, 56; Pa State Univ, PhD(fuel tech), 62. *Prof Exp:* Chemist, Carbon & Ribbon Div, Interchem Corp, 56-57; res asst fuel tech, Pa State Univ, 57-62; sr res chemist, Allison Div, Gen Motors Corp, 62-64; res scientist, HNG/Internorth Inc, 64-67, mgr, Corp Res & Develop, Northern Natural Gas Co, 67-75, vpres res, HNG/Internorth, 75-86; RETIRED. *Mem:* Am Chem Soc; Electrochem Soc; Indust Res Inst. *Res:* Chemical kinetics in the middle and high temperature ranges; combustion phenomena reactions of element carbon; structure of carbon; electrochemistry; energy conversion. *Mailing Add:* Sun Air Country Club One Cypress Run Haines City FL 33844

HIRTH, HAROLD FREDERICK, b Rockville, Conn, Dec 6, 32; m 62; c 2. ANIMAL ECOLOGY. *Educ:* Univ Conn, BA, 54, MS, 58; Univ Fla, PhD(biol), 62. *Prof Exp:* Asst prof biol, Univ Fla, 62-63; asst prof zool, 63-68, assoc prof biol, 68-71, PROF BIOL, UNIV UTAH, 72- *Concurrent Pos:* NIH, Food & Agr Org & AEC res grants; Fulbright scholar. *Mem:* Ecol Soc Am; Animal Behav Soc; Sigma Xi. *Res:* Ecology and behavior of animals. *Mailing Add:* Dept Biol Univ Utah Salt Lake City UT 84112

HIRTH, JOHN P(RICE), b Cincinnati, Ohio, Dec 16, 30; m 53; c 4. PHYSICAL METALLURGY. *Educ:* Ohio State Univ, BMetE & MS, 53; Carnegie Inst Technol, PhD(metall eng), 58. *Prof Exp:* Fulbright res fel, Univ Bristol, 57-58; asst prof metall eng, Carnegie Inst Technol, 58-61; Mershon assoc prof, Ohio State Univ, 61-64, prof phys metall, 64-88; PROF MAT ENG, WASH STATE UNIV, 88- *Concurrent Pos:* Consult, Crucible Steel Co. *Honors & Awards:* Hardy Medal, Am Inst Mining, Metall & Petrol Engrs, 61, Mehl Medal, 80, Mathewson Medal, 82; Stoughton Award, Am Soc Metals, 64; Curtis McGraw Award, Am Soc Eng Educ, 67. *Mem:* Nat Acad Eng; Am Inst Mining, Metall & Petrol Engrs; Am Soc Eng Educ; AAUP; fel Am Soc Metals; fel Metall Soc; fel Am Inst Mining Engrs. *Res:* Crystal growth; dislocation theory. *Mailing Add:* Dept Mech & Mat Eng Wash State Univ Pullman WA 99164

HIRTH, ROBERT STEPHEN, b Rockville, Conn, Jan 29, 35; m 59; c 1. VETERINARY PATHOLOGY. *Educ:* Univ Conn, BA, 57, PhD(vet path), 68; Univ Ill, DVM, 61. *Prof Exp:* Sr res asst path, Univ Conn, 63-68; asst dir, 68-80, DIR PATH, BRISTOL LABS, 80- *Mem:* Sigma Xi; Am Col Vet Pathologists; Am Vet Med Asn; Soc Toxicol Pathologists; Int Acad Path. *Res:* Investigation of toxic effects of new compounds on experimental animals and also spontaneous diseases of laboratory animals. *Mailing Add:* Bristol Meyers Squibb Co Bldg 32 315 N Thompson Rd PO Box 4755 Syracuse NY 13221

HIRTHE, WALTER M(ATTHEW), metallurgy, materials science; deceased, see previous edition for last biography

HIRTZEL, CYNTHIA S, b Ind. APPLIED MATHEMATICS, SYSTEMS SIMULATION. *Educ:* Wash Univ, AB, 73; Northwestern Univ, MS, 77, PhD(civil & environ eng), 80. *Prof Exp:* Vis scholar environ eng, Northwestern Univ, 80-81; asst prof chem & environ eng, Rensselaer Polytech Inst, 81-86; assoc prof, 86-90, PROF CHEM ENG & MAT SCI, SYRACUSE UNIV, 90-, CHAIRPERSON DEPT, 90- *Concurrent Pos:* Consult, various environ, indust & other consult firms, 82- *Mem:* AAAS; Am Inst Chem Engrs; Am Soc Eng Educ; Am Phys Soc; Am Chem Soc; Soc Insust & Appl Math; Asn Women Sci. *Res:* Adsorption/desorption phenomena; colloidal and interfacial phenomena; stochastic processes and applications to engineering systems; environmental analysis and modeling (including air pollution); computer experiments for colloidal systems; atmospheric sciences and modeling. *Mailing Add:* Dept Chem Eng & Mat Sci Syracuse Univ Syracuse NY 13244-1190

HIRWE, ASHALATA SHYAMSUNDER, b Goa, India, Aug 20, 38; m 62; c 2. SYNTHETIC ORGANIC CHEMISTRY. *Educ:* Wilson Col, India, BSc, 59; Univ Col, Ireland, PhD(org chem), 65. *Prof Exp:* Sr demonstr chem, Univ Col, Dublin, Ireland, 62-65; sr res chemist org chem, Fraco-India Pharmaceut, 65-68; res assoc radiochem, Dept Chem, Univ Ill, 68-69, res assoc synthetic org chem, Dept Entom, 70-74; res chemist, 74-80, SR RES CHEMIST, RES CTR, HERCULES, INC, 80- *Concurrent Pos:* Mgr, Tech Info Div, Hercules, Inc. *Mem:* Am Chem Soc. *Res:* Synthesis of bioactive compounds; application of radioisotopes to industrial problems and surface chemistry. *Mailing Add:* 210 Marion Ct Hockessin DE 19707-9781

HIRZY, JOHN WILLIAM, b St Louis, Mo, Sept 19, 36; m 87; c 3. PLASTICS CHEMISTRY, RISK ASSESSMENT. *Educ:* Univ Mo, BS, 58, PhD(org chem), 62. *Prof Exp:* Res specialist plasticizers, Monsanto Indust Chem Co, 62-81; SR SCIENTIST, ASSESSMENT DIV, US ENVIRON PROTECTION AGENCY, 81- *Concurrent Pos:* Instr, St Louis-St Louis County Jr Col, 63-68, 70-71 & 75-76; asst prof, Univ Mo, St Louis, 68-70; pres, local 2050, Nat Fedn Fed Employees; consult, Community Ecol, Inc. *Honors & Awards:* Bronze & Silver Medals, Environ Protection Agency. *Mem:* Am Chem Soc; AAAS; Soc Risk Analysis; NY Acad Sci. *Res:* Reduction of nitroaromatic compounds; effect of substituents on ultraviolet spectra of aromatic compounds; chemical modification of polymers; plasticizer theory and technology; environmental effects of plasticizers; flame retardants synthesis; risk assessment; industrial chemicals. *Mailing Add:* US Environ Protection Agency 401 M St SW TS 778 Washington DC 20460

HISADA, MITUHIKO, b Nagoya, Japan, Oct 30, 29; m 56; c 1. NEUROSCIENCES. *Educ:* Univ Tokyo, BSc, 52; Hokkaido Univ, DSc, 60. *Prof Exp:* Res assoc zool, Univ Tokyo, 52-54; res assoc zool, Hokkaido Univ, 54-56, lectr physiol, 56-60, assoc prof, 60-76, PROF PHYSIOL, HOKKAIDO UNIV, 76- *Concurrent Pos:* Coun, Physiol Soc Japan, 61-, Biol Soc Japan, 77- & Jap Soc Comp Physiol & Biochem, 91-; vis lectr, Univ Pa, 62-64, vis assoc prof, 71-73; dir, Ctr Exp Plants & Animals, Hokkaido Univ, 81-91; lectr, Sapporo Med Col, 78-, Asahikawa Med Col, 85- *Honors & Awards:* Soc prize, Zool Soc Japan, 88. *Mem:* Biol Soc Japan; Physiol Soc Japan; Japanese Soc Comp Physiol & Biochem; Int Soc Neuroethology (secy, 89-); Am Physiol Soc; Soc Neurosci. *Res:* Neuronal substrates of invertebrate behavior; anthropod interneurons; nonspiking integration of sensory input and motor output; neural network subserving gravitational responses of crustacea; neural circuit simulation study. *Mailing Add:* Zool Inst Fac Sci Hokkaido Univ Sapporo Hokkaido 060 Japan

HISATSUNE, ISAMU CLARENCE, b Stockton, Calif, Jan 3, 24; m 51. PHYSICAL CHEMISTRY. *Educ:* Univ Calif, BS, 51; Univ Wash, PhD(chem), 54. *Prof Exp:* Asst aeronaut eng, Aeronaut Res Inst, Tokyo, 44-45; civil censor, US Army, 45-47; asst chem, Univ Calif, 51 & Univ Wash, 51-52; res fel, 52-54; res fel, Univ Minn, 54-56; asst prof, Kans State Univ, 56-60; from asst prof to prof, 60-84, EMER PROF CHEM, PA STATE UNIV, 84- *Concurrent Pos:* Vis prof, Inst Org Chem, Univ Florence, 66 & Dept Chem, Fac Sci, Univ Tokyo, 68 & 75 & Inst Buddhist Studies, Berkeley, Calif, 88- *Mem:* Am Chem Soc; Am Phys Soc. *Res:* Molecular structure; kinetics. *Mailing Add:* 888 O'Farrell St Apt W-1206 San Francisco CA 94109

HISAW, FREDERICK LEE, JR, b Madison, Wis, Mar 26, 27; m 48; c 6. ZOOLOGY, ENDOCRINOLOGY. *Educ:* Univ Mo, BS, 50, MS, 52; Harvard Univ, PhD(zool), 55. *Prof Exp:* Res fel endocrinol, Biol Labs, Harvard Univ, 55-58; asst prof zool, 58-61, ASSOC PROF ZOOL, ORE STATE UNIV, 61- *Mem:* Am Soc Zool; Endocrine Soc; Soc Gen Physiol. *Res:* Endocrinology and the comparative physiology of reproduction. *Mailing Add:* 5925 SW Plymouth Dr Corvallis OR 97330

HISCOCK, WILLIAM ALLEN, b Santa Monica, Calif, Oct 31, 51; m 75; c 2. THEORETICAL GENERAL RELATIVITY, QUANTUM FIELD THEORY. *Educ:* Calif Inst Technol, Bs, 73; Univ Md, MS, 75, PhD(physics), 79. *Prof Exp:* Res assoc, dept physics, Yale Univ, 79-81; res assoc, dept physics, Ctr Relativity, Univ Tex, 81-82, lectr, 82-83; res assoc, dept physics,

Univ Calif, Santa Barbara, 83-84; asst prof, 84-88, ASSOC PROF PHYSICS, DEPT PHYSICS, MONT STATE UNIV, 88- *Concurrent Pos:* Mem, Quantum Gravity Group, Nat Inst Theoret Physics, Univ Calif, Santa Barbara, 80. *Honors & Awards:* Wiley Res Award, 90. *Mem:* Am Phys Soc; Sigma Xi. *Res:* Quantum field theory in curved spacetimes, especially black hole evaporation; relativistic dissipative fluid mechanics; early universe cosmology; physics of neutron stars and black holes; quantum gravity. *Mailing Add:* Dept Physics Mont State Univ Bozeman MT 59717

HISCOE, HELEN BRUSH, b New London, Conn, Mar 10, 19; m 46; c 4. REPRODUCTIVE BIOLOGY. *Educ:* Vassar Col, AB, 39; Brown Univ, MS, 40; Univ Calif, Los Angeles, PhD(zool), 43. *Prof Exp:* Asst zool, Vassar Col, 40-41, instr, 43-46; res assoc for Dr Paul Weiss, Univ Chicago, 46-47; lectr, 59-68, from asst prof to prof, 68-86, EMER PROF, DEPT NATURAL SCI, MICH STATE UNIV, 86- *Mem:* AAAS; Asn Gen & Liberal Studies; Soc Col Sci Teachers. *Res:* General education in natural sciences. *Mailing Add:* 1817 Walnut Heights Dr East Lansing MI 48823

HISCOTT, RICHARD NICHOLAS, b St Catharines, Ont, Feb 12, 51; m 80; c 1. DEEP-SEA CLASTIC SEDIMENTOLOGY, BASIN ANALYSIS. *Educ:* Brock Univ, BSc, 74; McMaster Univ, PhD(geol), 77. *Prof Exp:* ASSOC PROF EARTH SCI, MEM UNIV NFLD, 77- *Concurrent Pos:* Res fel, Husky Oil Opers, 85-87. *Mem:* Soc Econ Paleontologists & Mineralogists; Am Asn Petrol Geologists; Can Soc Petrol Geologists; Int Asn Sedimentologists; Geol Asn Can. *Res:* Sedimentology of ancient deep-sea fans, slopes and shallow-marine clastics; comparison of upper Jurassic and lower Cretaceous sequences from the Grand Banks and the Iberian Peninsula. *Mailing Add:* Dept Earth Sci Mem Univ Nfld St John's NF A1C 5S7 Can

HISER, HOMER WENDELL, b Ava, Ill, Nov 21, 24; m 53. METEOROLOGY, ENVIRONMENTAL ENGINEERING. *Educ:* Univ Ill, BS, 51, MS, 54; Wash Univ, St Louis, DSc(environ eng), 72. *Prof Exp:* Res assoc & res radar meteorologist, Ill State Water Surv, Univ Ill, 50-55; prof & dir, Radar Meteorol Lab, Rosenstiel Sch Marine & Atmospheric Sci, 55-73, PROF & DIR REMOTE SENSING LAB, SCH ENG & ARCHIT, UNIV MIAMI, 74- *Concurrent Pos:* Consult, Battelle Columbus Labs, 74 & adv panel, NASA, 75. *Honors & Awards:* Cert Appreciation, Nat Weather Serv, US Dept Commerce, 71. *Mem:* Am Meteorol Soc; Am Geophys Union; sr mem Inst Elec & Electronics Engrs; Air Pollution Control Asn; Sigma Xi. *Res:* Applications of satellites, radar, and other remote sensing systems to meteorology, atmospheric physics, solar energy, microwave propagation, and air pollution problems. *Mailing Add:* 4705 Univ Dr Coral Gables FL 33146

HISEY, ROBERT WARREN, b Chicago, Ill, May 7, 31; c 5. ORGANIC CHEMISTRY, CHEMICAL ENGINEERING. *Educ:* Middlebury Col, AB, 52; Inst Paper Chem, MS, 54, PhD(chem eng), 55. *Prof Exp:* Tech serv dir, S D Warren Co, Cumberland Mills, 55-60, chief engr, 60-64; vpres, tech dir & div mgr, Brown Co, NH, 64-67; vpres-gen mgr, Bleached Prod, Continental Can Co, 67-77; exec vpres, Bleached Syst Oper, Continental Forest Industs, 77-86; CONSULT, 86- *Honors & Awards:* Westbrook Steel Gold Medal, 55. *Mem:* Tech Asn Pulp & Paper Indust. *Res:* Filtration; pulp and paper processing. *Mailing Add:* 7227 Sparta Rd Sebring FL 33872

HISKES, JOHN ROBERT, b Chicago, Ill, May 30, 28; m 51; c 2. THEORETICAL PHYSICS. *Educ:* Univ Calif, AB, 51, MA, 52, PhD, 60. *Prof Exp:* Physicist, Calif Res & Develop Co, 52-54, Lawrence Livermore Lab, Univ Calif, 54-55 & Univ Calif, Berkeley, 55-60, PHYSICIST, LAWRENCE LIVERMORE LAB, UNIV CALIF, 60- *Concurrent Pos:* Consult, Gen Elec Co, 62-63; vis physicist, Culham Lab, Berkshire, Eng, 63-64, Ecole Polytecnic, France, 80, Japan At Eu R Inst, NAKA, Ibara Ki-Ken, 87-89, Nat Inst Fusion Sci, Nagoya, Japan, 91. *Mem:* Fel Am Phys Soc; Sigma Xi. *Res:* Atomic and molecular physics; controlled fusion; particle accelerators; liquid drop model of nuclear fission; negative ions. *Mailing Add:* Lawrence Radiation Lab Livermore CA 94550

HISKES, RONALD, b Evergreen Park, Ill, Jan 30, 41. SOLID STATE PHYSICS. *Educ:* Calvin Col, BS, 63; Univ Mich, BSE(chem eng) & BSE(metal eng), 63; Stanford Univ, MS, 64, PhD(mat sci), 68. *Prof Exp:* Instr mat sci, Stanford Univ, 68-70; mem tech staff mat res, 70-77, PROF MGR, HEWLETT-PACKARD LAB, 77- *Concurrent Pos:* Res assoc & staff scientist crystal growth, Ctr Mat Res, Stanford Univ, 68-70; consult, Battelle Mem Inst, 68-70. *Mem:* Am Asn Crystal Growth. *Res:* Crystal growth and characterization of magnetic oxides; materials research; molecular beam epitaxy; fiber optics. *Mailing Add:* 3484 Waverley Palo Alto CA 94306

HISKEY, CLARENCE FRANCIS, b Milwaukee, Wis, June 5, 12;; c 2. ANALYTICAL CHEMISTRY. *Educ:* Univ Wis, BA, 35, MA, 36, PhD(inorg & phys chem), 39. *Prof Exp:* Instr chem, Univ Tenn, 39-41; instr, Columbia Univ, 41-42, res sect mgr, Manhattan Dist Proj, s a m labs, 42-43; sect chief metall lab, Univ Chicago, 43-44; assoc prof chem, Polytech Inst Brooklyn, 46-53; vpres, Transition Metals & Chems, 55-58; DIR ANALYTICAL RES, ENDO LABS, INC, 58- *Mem:* AAAS; Am Chem Soc; Optical Soc Am; NY Acad Sci. *Res:* Analytical and metallurgical chemistry of niobium, tantalum, rhenium, molybdenum and manganese; spectrophotometry; isotope fractionation; instrumental techniques of analysis; pharmaceutical analysis; development of analytical methods of a pharmaceutical and inorganic chemical character. *Mailing Add:* 49 Long Hill Farm Guilford CT 06437

HISKEY, J BRENT, b Salina, Utah, Aug 18, 44; m 67; c 2. CHEMICAL METALLURGY, HYDROMETALLURGY. *Educ:* Univ Utah, BS, 67, MS, 71, PhD(metall), 73. *Prof Exp:* Extractive metallurgist res, Alcoa Labs, 73-74; asst prof extractive metall, NMex Inst Mining & Technol, 74-77; res scientist, US Steel Res Labs, 77-80; mgr, Metall Res, Kennecott Copper Corp, 80-84; PROF EXTRACTIVE METALL, DEPT MAT SCI & ENG, UNIV ARIZ, 84-, DIR, ARIZ MINING & MINERAL RESOURCES RES INST, 85-, DIR COPPER RES CTR, 89- *Concurrent Pos:* Lectr, Carnegie Mellon Univ, 77-79; consult, E I du Pont de Nemours & Co, Inc, 84-, Phelps Dodge, 85-, Newmont Gold Co, 85-, Kennecott Corp, 85-; chmn, Nat Asn Mineral Inst Dirs, 90-91; chmn, Mineral & Metall Processing Div, Soc Mining, Metall & Explor Inc, 91-92. *Honors & Awards:* Taggart Award, Soc Mining Engrs, Am Inst Mining, Metall & Petrol Engrs, 74. *Mem:* Soc Mining Metall & Explor Inc; Nat Asn Mineral Inst Dirs; Sigma Xi. *Res:* Hydrometallurgy and mineral processing, specifically the physical chemistry of leaching, ion exchange, solvent extraction and metal recovery; kinetics of hydrometallurgical reactions with special emphasis on electrochemical processes; extraction of copper, gold and silver, and critical and strategic metals. *Mailing Add:* Ariz Mining & Mineral Resources Res Inst Univ Ariz Geol 103 Tucson AZ 85721

HISKEY, RICHARD GRANT, b Emporia, Kans, May 21, 29; m 53; c 5. SYNTHETIC ORGANIC CHEMISTRY, PROTEIN CHEMISTRY. *Educ:* Emporia State Univ, AB, 51; Kans State Col, MS, 53; Wayne State Col, PhD(chem), 55. *Prof Exp:* Res assoc org chem, Polytech Inst Brooklyn, 55-58; from asst prof to prof, 58-75, chmn dept, 70-75, ALUMNI PROF CHEM, UNIV NC, CHAPEL HILL, 75- *Concurrent Pos:* Guggenheim fel, 70-71; consult, NIH, 74- *Honors & Awards:* Standard Oil Award, 69. *Mem:* fel AAAS; Am Chem Soc; fel Japan Soc Prom Sci; Sigma Xi. *Res:* Protein-metalion-lipid interactions; peptide synthesis; enzyme modification; model enzyme systems; role of gamma-carboxy glutamic acid in blood coagulation. *Mailing Add:* Dept Chem Univ NC Chapel Hill NC 27599-3290

HISLOP, HELEN JEAN, b Linden, NJ, Mar 13, 29. MEDICAL PHYSIOLOGY, EXERCISE PHYSIOLOGY. *Educ:* Cent Col Iowa, BA, 50, cert phys ther, 51; Univ Iowa, MS, 53, PhD, 60. *Hon Degrees:* ScD, Cent Col, 79. *Prof Exp:* Instr phys ther, Sch Med, Univ Minn, 53-55; asst to dir prof educ, Nat Found, 55-58; res assoc physiol & pharmacol, Sch Med, Univ Pittsburgh, 60-62; ed j, Am Phys Ther Asn, 62-68; chmn dept, 76-87, PROF PHYS THER, UNIV SOUTHERN CALIF, 68- *Concurrent Pos:* Sr res consult, Inst Rehab Med, NY Med Ctr, 66-68; consult phys ther, Surg Gen, US Army, 67-69; dir phys ther, Rehab Res & Training Ctr, Univ Southern Calif, 68-74; Worthingham fel, Am Phys Ther Asn, 82; consult, S Ill Univ Sch Med, 86-, SUNY, Buffalo, 86-87. *Honors & Awards:* Golden Pen Award, Am Phys Ther Asn, 69; John Stanley Coulter Lectr, Am Congress Rehab Med, 77; McMillan Lect, Am Phys Ther Asn, 75. *Mem:* Fel AAAS; fel Am Col Sports Med; Am Phys Ther Asn; Am Cong Rehab Med; Sigma Xi. *Res:* Physiology of exercise; nerve-muscle physiology; adaptations to exercise; pathokinesiology; physical therapy. *Mailing Add:* 12313 Brock Ave Downey CA 90242

HISSERICH, JOHN CHARLES, b Los Angeles, Calif, Apr 29, 39; m 73; c 4. HEALTH ADMINISTRATION. *Educ:* Calif State Univ, Los Angeles, BA, 65; Univ Calif, Los Angeles, MPH, 66, DrPH, 70. *Prof Exp:* Adminr, Charles R Drew Med Sch, Univ Calif, Los Angeles, 70-71; admin dir, COMPREHENSIVE CANCER CTR, 71-86, ASSOC VPRES, HEALTH AFFAIRS, UNIV SOUTHERN CALIF, 86- *Concurrent Pos:* Assoc clin prof preventive med & pharm, Sch Med, Univ Southern Calif, 72-; WHO study fel, 75; Nat Cancer Inst Interventions Comt, NIH, 77-79. *Mem:* Am Pub Health Asn; Am Asn Cancer Educ; Inst Soc, Ethics & Life Sci; Soc Health & Human Values; AAAS; Am Soc Prev Oncol. *Res:* Cancer control; technology transfer; biomedical ethics; confidentiality of medical data. *Mailing Add:* Off VPres Health Affairs Uni Southern Calif 1985 Zoral Ave Los Angeles CA 90033

HISTAND, MICHAEL B(ENJAMIN), b Ft Lewis, Wash, Oct 31, 42; m 67; c 2. BIOENGINEERING, BIOMECHANICS. *Educ:* Lehigh Univ, BS, 64; Stanford Univ, MS, 65, PhD(aeronaut, astron), 69. *Prof Exp:* asst prof, 69-73, assoc prof, 73-80, PROF MECH ENG, PHYSIOL & BIOPHYS, COLO STATE UNIV, 80- *Mem:* Am Soc Mech Engrs. *Res:* Medical ultrasound; hemodynamics; medical instrumentation. *Mailing Add:* Dept Mech Eng Colo State Univ Ft Collins CO 80523

HISTED, JOHN ALLAN, b Hamilton, Ont, Feb 14, 29; m 53; c 3. PULP CHEMISTRY, PULP BLEACHING. *Educ:* McMaster Univ, BSc, 53. *Prof Exp:* Res chemist, CIP Res Ltd, 53-66, group leader specialty pulps & bleaching, 66-73, res assoc, 73-75, SR RES ASSOC, C P FOREST PROD RES LTD, 75- *Concurrent Pos:* Tappi fel, Tech Asn Pulp & Paper Indust, 85. *Honors & Awards:* Douglas Jones Award, Tech Sect, Can Pulp & Paper Asn, 73, I H Weldon Medal, 83. *Mem:* Can Pulp & Paper Asn; sr mem Tech Asn Pulp & Paper Indust; Chem Inst Can. *Res:* Optimization and simplification of bleach sequences; development of bleachery control systems; reduction of pollution from bleacheries; development of new bleaching process. *Mailing Add:* 179 Main St W Hawkesbury ON K6A 2H4 Can

HITCHCOCK, ADAM PERCIVAL, b Hamilton, Ont, June 27, 51; m 72; c 2. ELECTRON AND X-RAY ABSORPTION. *Educ:* McMaster Univ, BSc, 74; Univ British Columbia, PhD(chem), 78. *Prof Exp:* postdoctoral fel, Univ BC, 78-79; from asst prof to assoc prof, 79-89, PROF CHEM, MCMASTER UNIV, 89- *Concurrent Pos:* Post doc fellow, Univ British Columbia, 78-79; prin investr, Ont Ctr Mat Res, 88- *Honors & Awards:* Rutherford Medal, Royal Soc Can, 78; Noranda Award, Can Soc Chem, 89. *Mem:* Am Chem Soc; Am Vacuum Soc; Can Soc Chem; Can Inst Physics. *Res:* Inner-shell excitations and ionization phenomina in molecules; surfaces and solids studied by electron impact and X-ray absorption. *Mailing Add:* Inst Mat Res McMaster Univ Hamilton ON L8S 4M1 Can

HITCHCOCK, CLAUDE RAYMOND, b Minneapolis, Minn, Oct 14, 17; m 49; c 2. SURGERY. *Educ:* Univ Minn, BA, 40, BS, 43, MB, 44, MD, 45, PhD(surg), 54; Am Bd Surg, dipl, 54. *Prof Exp:* Intern surg, 52-54, from asst prof to assoc prof, 54-61, dir cancer detection ctr, 52-55, PROF SURG, MED SCH, UNIV MINN, MINNEAPOLIS, 61- *Concurrent Pos:* Trainee, Am Cancer Soc, 51-52; sr investr, Nat Cancer Inst, 53-55; chief surg dept, Hennepin Co Med Ctr, Minneapolis, 55- *Mem:* AAAS; AMA; fel Am Col Surg; Am Chem Soc; Sigma Xi. *Res:* Cancer biology transfusion problems and intra-arterial transfusions; carcinogenesis studies in animals; vascular surgery and physiology; epidemiology of wound infections; organ grafting; hyperbaric medicine and surgery. *Mailing Add:* Surg 5th Portland Hennepin County Gen Hosp Minneapolis MN 55415

HITCHCOCK, DANIEL AUGUSTUS, b Cleveland, Ohio, Oct 14, 47; m 73. PLASMA PHYSICS, APPLIED MATHEMATICS. *Educ:* Univ Calif, Berkeley, AB, 69; Princeton Univ, MA, 71, PhD(appl math), 75. *Prof Exp:* RES ASSOC PLASMA PHYSICS, FUSION RES CTR, UNIV TEX, AUSTIN, 75- *Mem:* Am Phys Soc. *Res:* Stability theory for two component tokomaks, nonlinear plasma theory, especially resonance broadening techniques; singular perturbation theory for integro-differential equations. *Mailing Add:* 7802 Marion Ln Bethesda MD 20814

HITCHCOCK, DOROTHY JEAN, b Charlotte, Mich. BIOLOGY. *Educ:* Mich State Univ, BS, 43, MS, 47, PhD, 54. *Prof Exp:* Med technologist, Parke, Davis & Co, 43-45; med technologist, Mich State Univ, 45-47, instr parasitol, 47-55; asst prof, Clemson Col, 55-56; bacteriologist, Mt Carmel Mercy Hosp, 56-63; PROF BIOL, MERCY COL MICH, 63- *Honors & Awards:* Sayer Bact Award. *Mem:* Am Soc Microbiol. *Res:* Parasitology, bacteriology. *Mailing Add:* Dept Math & Sci Mercy Col 8200 W Outer Dr Detroit MI 48219

HITCHCOCK, ELDON TITUS, b Leonard, Mich, Feb 2, 24; m 54; c 3. ANALYTICAL CHEMISTRY. *Educ:* Western Mich Univ, BS, 46; Univ Mich, PhD, 61. *Prof Exp:* Du Pont instr chem, Univ Mich, 56-57; chmn dept, Bob Jones Univ, 48-54; from asst prof to assoc prof, 57-69, chmn dept, 79-85, PROF CHEM, COLO COL, 69- *Concurrent Pos:* Vis staff mem, Los Alamos Sci Lab, 65- *Mem:* Am Chem Soc; Am Sci Affil; Sigma Xi; fel AAAS. *Res:* Nonaqueous and high frequency titrimetry; acid-base theory; electroanalytical methods; teaching techniques and curricula in undergraduate chemistry training. *Mailing Add:* Dept Chem Colo Col Colorado Springs CO 80903

HITCHCOCK, HAROLD BRADFORD, b Hartford, Conn, June 23, 03; m 42; c 3. SCIENCE EDUCATION. *Educ:* Williams Col, AB, 26; Harvard Univ, MA, 32, PhD(biol), 38. *Prof Exp:* Asst biol, Williams Col, 32-34 & Harvard Univ, 34-38; from instr to asst prof zool, Univ Western Ont, 39-43; from asst prof to prof biol, Middlebury Col, 43-68, chmn dept, 45-64, Mead prof, 64-68; Dana prof & chmn dept, 69-72, EMER PROF BIOL, BATES COL, 72-; EMER PROF BIOL, MIDDLEBURY COL, 68- *Concurrent Pos:* Soil Scientist soil conserv serv, USDA, 45-46; grants, Am Acad Arts & Sci, 47 & Am Philos Soc, 49, 51 & 52; Fulbright res scholar, Max Planck Inst, Wilhelmshaven, Ger, 54; grant, NSF, 64; vis prof, Boston Univ, 68, Norwich Univ, 68 & Univ Hawaii, 69. *Mem:* Am Soc Mammal. *Res:* Life histories, hibernation and homing of bats; homing of pigeons; puma, Felis concolor, in the Northeast. *Mailing Add:* One Locust Lane Middlebury VT 05753

HITCHCOCK, JOHN PAUL, b Muscatine, Iowa, Apr 13, 45; m 75. ANIMAL NUTRITION. *Educ:* Iowa State Univ, BS, 68; Pa State Univ, MS, 70; Mich State Univ, PhD(animal husb), 75. *Prof Exp:* ASST PROF ANIMAL SCI, UNIV TENN, KNOXVILLE, 75- *Mem:* Am Soc Animal Sci. *Res:* Vitamin and trace mineral research; amino acid and fatty acid requirements of swine. *Mailing Add:* 205 Derby Hall 154 N Oval Mall Columbus OH 43210

HITCHCOCK, MARGARET, b Harrow, Eng, July 17, 40. PHARMACOLOGY, METABOLISM. *Educ:* Univ London, BSc, 62, PhD(biochem), 65. *Prof Exp:* Res specialist air pollution, Calif Dept Pub Health, 67-69; from asst prof to assoc prof pharmacol med & environ health, 69-79, sr res assoc & lectr pharmacol, 79-83, SR ADMIN ASST TO CHMN, DEPT PHARMACOL, SCH MED, YALE UNIV, 84- *Concurrent Pos:* Res fel physiol, Harvard Sch Pub Health, 65-67; asst fel, John B Pierce Found Lab, 69-, assoc fel, 76- *Mem:* Soc Toxicol; Am Soc Pharmacol & Exp Therapeut; NY Acad Sci; Soc Occup & Environ Health. *Res:* Pulmonary pharmacology; immunotoxicology; biochemical mechanisms underlying bronchial reactivity. *Mailing Add:* 333 Cedar St New Haven CT 06510

HITCHCOCK-DEGREGORI, SARAH ELLEN, b Washington, DC, Nov 19, 43. BIOLOGY, BIOCHEMISTRY. *Educ:* Smith Col, AB, 65; Wesleyan Univ, MA, 67; Case Western Reserve Univ, PhD(biol), 70. *Prof Exp:* Fel Brandeis Univ, 70-73 & MRC Lab Molecular Biol, Cambridge, Eng, 73-76; asst prof biol sci, Carnegie-Mellon Univ, 76-81, assoc prof, 80-85; ASSOC PROF ANAT, ROBERT WOOD JOHNSON MED SCH, UNIV MED & DENT NJ, 85- *Concurrent Pos:* Fel, Muscular Dystrophy Asn, 71-73, Brit-Am Heart Asn, 73-75 & NIH, 75-76; Res Career Develop Award, NIH, 82- *Mem:* Am Soc Cell Biol; Biophys Soc. *Res:* Molecular biology of contractile proteins in muscle and non-muscle cells. *Mailing Add:* Dept Anatomy UMDNJ Robert Wood Johnson Med Sch 675 Hoes Lane Piscataway NJ 08854

HITCHINGS, GEORGE HERBERT, b Hoquiam, Wash, Apr 18, 05; m 33, 89; c 2. BIOCHEMISTRY. *Educ:* Univ Wash, BS, 27, MS, 28; Harvard Univ, PhD(biol chem), 33. *Hon Degrees:* DSc, Univ Mich, Ann Arbor, 71, Univ Strathclyde, 77, NY Med Col & Emory Univ, 81, Duke Univ & Univ NC, 82, Mt Sinai Sch Med, 83. *Prof Exp:* Instr & tutor biochem sci, Harvard Univ, 32-36, assoc, 36-39; sr instr biochem, Western Reserve Univ, 39-42; biochemist, Burroughs Wellcome Co, 42-46, chief biochemist, 46-55, assoc res dir, 55-63, res dir, Chemother Div, 63-67, vpres chg res, 67-75, dir, 68-77, EMER SCIENTIST & CONSULT, BURROUGHS WELLCOME CO, 75-, DIR, BURROUGHS WELLCOME FUND, 68- *Concurrent Pos:* Res fel, Harvard Univ, 34-36; mem comt growth, Nat Res Coun, 52-53; chem panel, Cancer Chemother Nat Serv Ctr, 55-57; indust subcomt, Cancer Chemother Nat Comt, 55-57; consult, Cancer Chemother Study Sect, USPHS, 56-60; mem ther comt, Am Cancer Soc, 63-66; Clowes lectr, 68; vis prof, Brown Univ, 68-80; Charles E Dohme lectr, 69; mem bd trustees & mem med & sci adv comt, Leukemia Soc, 69-73; adj prof exp med & pharmacol, Duke Univ, 70-85, pharmacol, Univ NC, Chapel Hill, 72-85; Walter H Hartung mem lectr, Univ NC, Chapel Hill, 72, adj prof pharmacol, 72-; Michael Cross mem lectr, Cambridge Univ, 74; mem vis comt, Drug Res Bd, Nat Res Coun-Nat Acad Sci, 74-75, mem bd sci technol, Nat Res Coun, 75-79; mem drug develop comt, Nat Cancer Inst, 75-79; mem external adv comt, Duke Cancer Ctr, 77-83. *Honors & Awards:* Nobel Prize Med & Physiol, 88; Clowes Mem Lect & Award, Am Asn Cancer Res, 68, Bruce F Cain Award, 84; Charles E Dohme Lect & Award, Johns Hopkins, 69; Robert de Villier Sci Award, Leukemia Soc Am, 69; Walter H Hartung Mem Lectr, Univ NC, 72; Med Chem Award, Am Chem Soc, 72, Alfred Burger Award Med Chem, 84; Michael Cross Mem Lect, Cambridge Univ, 74; Pfizer Lect, State Univ NY, Stoneybrook, 81; Oscar B Hunter Award, Am Soc Clin Pharmacol, 84; Matthia Lect Chemother, 84; Award Sci Achievement, Nat Cancer Soc, 78; Papanicolaou Award, 78; C Chester Stock Medal, 81; Pfizer lectr, State Univ NY Stony Brook, 81; Burroughs Wellcome Distinguished Lectr, Philadelphia Col Pharm & Sci, 81 515 Oscar B Hunter Award, Bruce F Cain Award, 84. *Mem:* Nat Acad Sci; fel AAAS; foreign mem Royal Soc, London; hon fel Royal Soc Med, London; hon mem Am Asn Cancer Res; hon fel Royal Soc Chem; Sigma Xi; Am Chem Soc; Am Soc Biol Chem; Soc Exp Biol Chem. *Res:* Chemistry of condensed pyrimidine systems; purine and pyrimidine antagonists; nutrition and disease; antimalarials; antibacterials; cancer chemotherapy; immunosuppressive drugs; antimetabolites; organic chemistry of heterocycles; nucleic acids; antitumor, antimalarial and antibacterial drugs; medical education and care. *Mailing Add:* Burroughs Wellcome Co 3030 Cornwallis Rd Research Triangle Park NC 27709

HITCHNER, STEPHEN BALLINGER, b Daretown, NJ, Feb 4, 16; m 43; c 5. POULTRY PATHOLOGY. *Educ:* Rutgers Univ, BS, 39; Univ Pa, VMD, 43. *Prof Exp:* Instr avian path, Col Vet Med, Univ Ill, 46; assoc animal & poultry path, Agr Exp Sta, Va Polytech Inst, 47-49; res prof poultry path, Agr Exp Sta, Mass State Col, 49-53; res vet, Am Sci Labs, 53-58, dir res, 59-60; dir res, L&M Labs, 60-64, Amdal Co, 64-65 & infectious dis res div, Abbott Labs, 65-66; chmn dept, 66-76, prof, 66-81, EMER PROF AVIAN DIS, NY STATE COL VET MED, CORNELL UNIV, 81- *Concurrent Pos:* Lectr, VOCA, Bolivia, 87, 89; vaccine consult, Int Execs Serv Corps, Peru, 87-88. *Honors & Awards:* Special Serv Award, Am Asn Avian Path, 81. *Mem:* Am Vet Med Asn; Poultry Sci Asn; Am Asn Avian Path (pres, 60-61); NY Acad Sci. *Res:* Virology. *Mailing Add:* Rt 1 K-54 Worton MD 21678

HITCHON, BRIAN, b St John, NB, Can, Dec 13, 30; div. GEOCHEMISTRY. *Educ:* Univ Manchester, BSc, 52, PhD(geol), 55. *Prof Exp:* Geologist, Northern Rhodesia Geol Surv, 55-57; EMER RES OFFICER, RES COUN ALTA, 57- *Mem:* Am Asn Petrol Geol; Am Geochem Soc; Asn Explor Geochem; Int Asn Geochem Cosmochem. *Res:* Geochemistry of formation fluids, especially their origin, migration and accumulation; organic geochemistry of crude oil; geochemistry of recent and ancient sediments; hydrodynamics; hydrogeochemistry of surface waters. *Mailing Add:* Alta Res Council PO Box 8330, Station F Edmonton AB T6H 5X2 Can

HITE, GILBERT J, b Boston, Mass, Jan 24, 31; m 60; c 2. MEDICINAL CHEMISTRY. *Educ:* New Eng Col Pharm, BS, 54; Univ Wis, MS, 57, PhD(pharmaceut chem), 59. *Prof Exp:* Assoc prof pharmaceut chem & chmn dept, Col Pharm, Howard Univ, 59-61; from asst prof to prof pharmaceut chem, Col Pharm, Columbia Univ, 61-73, chmn dept chem, 67-72; PROF MED CHEM, SCH PHARM, UNIV CONN, 73-, HEAD SECT MED CHEM & PHARMACOGNOSY, 76- *Concurrent Pos:* Vis prof, Dept Crystallog, Univ Pittsburgh, 71-73. *Mem:* Am Pharmaceut Asn; Am Chem Soc; Am Asn Col Pharm; Am Asn Pharmaceut Scientists; Am Crystallog Asn. *Res:* Mechanisms of organic reactions; structure-activity relationships; modes of drug action; stereochemical aspects of drug action. *Mailing Add:* Sch Pharmrm Univ Conn U-92 372 Fairfield Rd Storrs CT 06268

HITE, J ROGER, b Houston, Tex, Aug 18, 39; m 60; c 3. CHEMICAL ENGINEERING. *Educ:* Tulane Univ, BS, 61; Princeton Univ, MA, 63, PhD(chem eng), 65. *Prof Exp:* NATO fel, Phys Chem Inst, Munich, 65-66; chem engr, Shell Develop Co, 66-69, res engr, 69-74, res engr, 75-79, mgr engr, Shell Oil Co, 80-90, DIR PROD RES, SHELL DEVELOP DIV, 90- *Mem:* Am Chem Soc; Am Inst Chem Engrs; Soc Petrol Engrs. *Res:* Reservoir engineering. *Mailing Add:* 3734 Ella Lee Lane Houston TX 77027

HITE, MARK, b Cambridge, Mass, Feb 7, 35; m 60; c 2. TOXICOLOGY, SAFETY ASSESSMENT. *Educ:* Rensselaer Polytech Inst, BCE, 56; Univ Rochester, 56-57; Univ Cincinnati, SCD(indust toxicol), 60; Am Bd Toxicol, dipl. *Prof Exp:* Sr asst scientist, USPHS, 60-63; toxicologist, Dow Chem Co, 63-64; sr res toxicologist, Merck Inst Therapeut Res, 64-70, dir toxicol, 67-70, dir toxicol & path, 70-85; DIR, DRUG SAFETY EVAL, WYETH-AYERST RES, 85- *Mem:* AAAS; Environ Mutagen Soc; Sigma Xi; Soc Toxicol; Am Col Toxicol; fel Acad Toxicol Sci. *Res:* Pharmaceutical and chemical safety evaluations; mutagenic and carcinogenic effects of chemicals and drugs; genetic toxicology; safety assessment; risk analysis. *Mailing Add:* Wyeth-Ayerst Res PO Box 8299 Philadelphia PA 19101

HITE, S(AMUEL) C(HARLES), b Ft Wayne, Ind, Aug 20, 22; m 53; c 1. CHEMICAL ENGINEERING. *Educ:* Purdue Univ, BSChE, 43, PhD(chem eng), 51. *Prof Exp:* From instr to assoc prof chem eng, Purdue Univ, 43-57; prof & chmn dept, Univ Ky, 57-70; PROF CHEM ENG & CHMN DEPT, ROSE-HULMAN INST TECHNOL, 70- *Concurrent Pos:* Res assoc, Purdue Univ, 46-51; consult, Com Solvents Corp & Eli Lilly & Co. *Mem:* Am Chem Soc; Am Inst Chem Engrs. *Res:* Gas engineering; unit process and operations. *Mailing Add:* Dept Chem Eng Rose-Hulman Inst Technol 5500 Wabash Ave Terre Haute IN 47803

HITES, RONALD ATLEE, b Jackson, Mich, Sept 19, 42; m 64; c 3. ANALYTICAL CHEMISTRY, ENVIRONMENTAL CHEMISTRY. *Educ:* Oakland Univ, BA, 64; Mass Inst Technol, PhD(anal chem), 68. *Prof Exp:* Fel chem, Arg Res Serv, 68-69; res staff chem, Mass Inst Technol, 69-72, from asst prof to assoc prof chem eng, 72-79; prof, 79-89, DISTINGUISHED PROF PUB & ENVIRON AFFAIRS & CHEM, INDIANA UNIV, 89- *Concurrent Pos:* Mem, Diesel Impacts Study Comt, Nat Res Coun, 79; mem, Comt Environ Improv, Am Chem Soc, 80-83; pres, Am Soc Mass Spec, 88-90. *Honors & Awards:* Award for Creative Advan in Environ Sci & Technol, Am Chem Soc, 91. *Mem:* Am Chem Soc; Am Soc Mass Spectrometry; AAAS; Int Asn Great Lakes Res; Sigma Xi. *Res:* Organic analytical chemistry; organic environmental chemistry; mass spectrometry. *Mailing Add:* Sch Public & Environ Affairs Indiana Univ Bloomington IN 47405

HITLIN, DAVID G, b Brooklyn, NY, Apr 15, 42; m 66, 82. ELEMENTARY PARTICLE PHYSICS. *Educ:* Columbia Univ, BA, 63, MA, 65, PhD(physics), 68. *Prof Exp:* Instr physics, Columbia Univ, 67-69; res assoc, Stanford Linear Accelerator Ctr, 69-72; asst prof, Stanford Univ, 72-75; asst prof, Stanford Linear Accelerator Ctr, 75-79; assoc prof, 79-85, PROF PHYSICS, CALIF INST TECHNOL, 85- *Mem:* Fel Am Phys Soc. *Res:* Experimental high energy physics; with emphasis on e plus e minus annihilation; weak decays of elementary particles. *Mailing Add:* 356-48 Lauritsen Lab Calif Inst Technol Pasadena CA 91125

HITNER, HENRY WILLIAM, b Bethlehem, Pa, Oct 9, 39; m 66; c 2. PHARMACOLOGY, PHYSIOLOGY. *Educ:* Moravian Col, BS, 65; Hahnemann Med Col, MS, 72, PhD(pharmacol), 74. *Prof Exp:* Biologist toxicol, Pharmachem Corp, 66; res asst toxicol, Wyeth Labs, Inc, 66-69; pharmacologist, Nat Drug Co, 69-70; instr biol, Montgomery Co Community Col, 74-77; from asst prof to assoc prof, 77-88, PROF PHARMACOL, PHILADELPHIA COL OSTEOPATH, 88-, VCHMN, DEPT PHYSIOL & PHARMACOL, 82- *Mem:* Sigma Xi; Mid Atlantic Reproduction & Teratology Asn. *Res:* Autonomic nervous system; shock; drug abuse; hallucinogens; teratology. *Mailing Add:* Philadelphia Col Osteopath 506 Madison Ave Ft Washington PA 19034

HITSCHFELD, WALTER, meteorology, atmospheric physics; deceased, see previous edition for last biography

HITT, JOHN BURTON, b Springfield, Ohio, June 15, 31; m 57, 85; c 7. CELL BIOLOGY. *Educ:* Wittenberg Univ, BS, 57; Ohio State Univ, MS, 60, PhD(zool), 69. *Prof Exp:* Teacher gen sci, High Sch, Ohio, 60-61; from instr to asst prof, 61-73, ASSOC PROF BIOL, WITTENBERG UNIV, 73- *Mem:* AAAS; Sigma Xi. *Res:* Studies of calcium transport in mitochondria of the earthworm calciferous gland; glucose as a depressor of menstrual cramps. *Mailing Add:* Dept Biol Wittenberg Univ PO Box 720 N Wittenberg Ave Springfield OH 45501

HITTELMAN, ALLEN M, b New York, NY, Dec 21, 45; div; c 2. DATA MANAGEMENT, MAPPING. *Educ:* Queen's Col, BS, 67; Univ Chicago, MS, 69. *Prof Exp:* Geophysicist oceanog, US Naval Oceanog Off, 67-75; sect chief geophys, Nat Geophys & Solar Terrestrial Data Ctr, 75-82; mgr, Data Mgt Sect, Arco Explor & Technol Co, 82-86; SECT CHIEF, NAT GEOPHYS DATA CTR, 87- *Concurrent Pos:* Res asst, Advan Res Proj Admin, 67-69; chmn, Marine Geophys Data Exchange Format Task Group, 77. *Honors & Awards:* Super Accomplishment Award, US Naval Oceanog Off, 73. *Mem:* Sigma Xi; Am Geophys Union; Soc Explor Geophysicists; Am Asn Petrol Geologists; Am Petrol Inst. *Res:* Geophysical database systems; potential field continuation. *Mailing Add:* NOAA Mail Code E/GC 325 Broadway Boulder CO 80303

HITTELMAN, WALTER NATHAN, b Fontana, Calif, May 22, 44; div; c 3. CELL BIOLOGY, CYTOGENETICS. *Educ:* Univ Calif, Berkeley, BS, 66, MS, 68, PhD(biophys), 72. *Prof Exp:* Fel, 72-74, res assoc, 74-75, from asst prof to assoc prof, 75-85, PROF CELL BIOL, M D ANDERSON HOSP & TUMOR INST, UNIV TEX, HOUSTON, 85- *Mem:* Am Soc Cell Biol; Fedn Am Scientists; Am Asn Cancer Res. *Res:* Growth regulation and cytogenetics of mammalian cells with special regard to the process of malignancy in humans. *Mailing Add:* M D Anderson Cancer Ctr Univ Tex 1515 Holcombe Blvd Houston TX 77030

HITTINGER, WILLIAM C(HARLES), b Bethlehem, Pa, Nov 10, 22; m 44; c 4. PHYSICAL METALLURGY. *Educ:* Lehigh Univ, BS, 44. *Hon Degrees:* Dr, Lehigh Univ, 74. *Prof Exp:* Mat Engr, Western Elec Co, 46-52; prod mgr, Nat Union Radio Corp, 52-54; mem tech staff semiconductor res & develop, Bell Tel Labs, 54-57, dept head, 57-60, dir, 60-62, exec dir, NJ, 62-66; pres, Bellcomm, Inc, DC, 66-68; pres, Gen Instrument Corp, 68-70; gen mgr & vpres, RCA Solid State Div, RCA Corp, 70-73, exec vpres consumer elec & solid state electronics, 73-75, exec vpres, 74-86; RETIRED. *Concurrent Pos:* Corp dir. *Honors & Awards:* Frederik Philips Medal, Inst Elect & Eng. *Mem:* Nat Acad Eng; fel Inst Elec & Electronics Engrs; Am Inst Mining Metall & Petrol Engrs; fel Royal Soc Arts. *Res:* Germanium crystal growth; diffused silicon transistors. *Mailing Add:* 149 Bellview Ave Summit NJ 07901

HITTLE, CARL NELSON, b Hartman, Colo, Mar 18, 22; m 45; c 4. CROP BREEDING. *Educ:* Colo State Univ, BS, 47; Cornell Univ, PhD(plant breeding), 51. *Prof Exp:* Res agronomist & regional geneticist, Latin Am, USDA, 51-53; from asst prof to assoc prof plant genetics, 53-64, prof plant breeding, 64-77, prof agron & plant breeding, Univ Ill, Urbana, 77-82; prof int agr, Int Agr Prog & Dept Agron, 81-82; consult in Nepal for Winrock Int, 82-86, FREELANCE CONSULT AGR DEVELOP OVERSEAS EXP AS RES STAFF, COSTA RICA, INDIA, SCI LANKA, NEPAL, 87- *Concurrent Pos:* NSF fel, Plant Breeding Inst, Cambridge, 62-63; adv agron & plant breeding, Univ Ill-US AID Contract Team, Jawaharlal Nehru Univ, 69-72 & adv & coord, All India Soybean Proj, 69-72; coordr & agronomist of BOA-73-30-Task Order to assist Govt Guyana, 73-75; prof plant breeding & consult, Int Soybean Prog, 73-82; coordr, Sri Lanka Soybean Prog, 76-81. *Honors & Awards:* Hon prof agron, Col Agr Sci, Univ PR, Mayaguez Campus, 74. *Mem:* Am Soc Agron; Crop Sci Soc Am. *Res:* Improvement of Hevea brasiliensis by methods of plant breeding; forage crop breeding; breeding tropical soybeans. *Mailing Add:* 22 Sangaraja Mawath Kandy Sri Lanka

HITTLE, DOUGLAS CARL, b Ft Collins, Colo, Apr 16, 47; m 69; c 2. ENERGY CONSERVATION, HEAT TRANSFER & THERMODYNAMICS. *Educ:* Univ Ill, BS, 69, MS, 75, PhD(mech eng), 81. *Prof Exp:* Mech engr, Chanute AFB, 69-73; prin investr, US Army Construct Eng Res Lab, 73-81, res team leader, 81-86; assoc prof, Purdue Univ, 86-89; DIR, SOLAR ENERGY APPLICATIONS LAB, COLO STATE UNIV, 89- *Concurrent Pos:* Consult, Architect US Capitol, 77-79 & Elec Power Res Inst, 87-; vis researcher, Bldg Res Estab, UK, 84. *Mem:* Fel Am Soc Heating Refrig & Air-Conditioning Engrs; Am Soc Mech Engrs; Int Solar Energy Soc; Sigma Xi; AAAS. *Res:* Building energy conservation; simulation of building energy systems; control of heating and air-conditioning systems; renewable solar energy system development; expert systems for hvac design; air flow and ventilation; heat transfer and thermodynamics. *Mailing Add:* Solar Energy Applications Lab Colo State Univ Ft Collins CO 80523

HITZ, CHESTER W, b Fortescue, Mo, Dec 13, 12; m 41. POMOLOGY. *Educ:* Univ Mo, BSAgr, 36; Univ Md, MS, 38, PhD(pomol), 41. *Prof Exp:* Asst pomol, Univ Md, 36-40, asst, 40-42; head dept hort & exten specialist, Agr Exten Serv & Exp Sta, Univ Maine, 46-48; assoc horticulturist, Univ Del, 48-57; prof, 57-76, EMER PROF POMOL, PA STATE UNIV, 76- *Concurrent Pos:* Coordr, Pa State Univ Five-Year Prog for Aiding Hort Res & Practs in Argentina, 63-68; horticulturist, Int Develop Serv Uruguay & chief-of-party & pomologist, Tri-Univ Consortium Int Agr Uruguay, 70-76. *Mem:* Am Soc Hort Sci; Am Pomol Soc; Int Soc Hort Sci. *Res:* Fruit physiology; small fruit propagation; tree fruit nutrition and soil management; fruit storage. *Mailing Add:* 10402 Tumblewood Dr Sun City AZ 85351

HITZEMAN, JEAN WALTER, b Chicago, Ill, Oct 29, 26. CELL BIOLOGY. *Educ:* Barry Col, BS, 50; DePaul Univ, MS, 56; Univ Mich, PhD(zool), 62. *Prof Exp:* Teacher parochial elem schs, Fla, 45-53; teacher parochial sec schs, Mich, 53-59; from instr to prof biol, Siena Heights Col, 62-70; prof & chmn dept biol, J C Smith Univ, 70-73; dir of studies, Adrian Dominican Generalate, 73-75; prof biol & chmn div natural sci & math, Johnson State Col, Vt, 75-77; PROF ANAT & PHYSIOL, DEPT BIOL SCI, STATE UNIV NY, BROCKPORT, 77-; AT ZOOL DEPT, WASHINGTON STATE UNIV, PULLMAN. *Concurrent Pos:* NIH res grant gen med, 63-70; chmn dept biol, Siena Heights Col, 68-70. *Mem:* AAAS; Sigma Xi. *Res:* Pathways of metabolism in Leydig cells of the mouse; hormonal control of sperm differentiation. *Mailing Add:* Dept Biol Sci State Univ NY Brockport NY 14420

HITZMAN, DONALD OLIVER, b Milwaukee, Wis, Dec 2, 26; m 49; c 2. BACTERIOLOGY, GEOCHEMISTRY. *Educ:* Carleton Col, AB, 48; Univ Ill, MS, 50, PhD(bact), 56. *Prof Exp:* Asst, Univ Ill, 48-50 & 52-54; res bacteriologist, Tex Co, 50; asst, Univ Southern Calif, 50-51; res bacteriologist, Phillips Petrol Co, 54-85; RES DIR, GEO-MICROBIAL TECHNOL INC, 85-; VPRES RES, INJECTECH INC, 85- *Concurrent Pos:* Consult, petrol microbiol & fermentation, 85- *Mem:* Am Chem Soc; Am Soc Microbiol; Soc Indust Microbiol; AAAS; Geochem Soc. *Res:* Bacteriological aspects of petroleum and petroleum products; microbial, prospection; microbial enhanced oil recovery; geochemical prospection for oil and gas. *Mailing Add:* 1717 Church Ct Bartlesville OK 74006

HIU, DAWES NYUKLEU, b Honolulu, Hawaii, Feb 14, 27; m 55; c 5. ORGANIC CHEMISTRY. *Educ:* Univ Southern Calif, BA, 50; Univ Hawaii, PhD(chem), 59. *Prof Exp:* Chemist, Res Div, Allis-Chalmers Mfg Co, Wis, 52-54; asst chem, Univ Hawaii, 54-56, from instr to asst prof, 57-60; from asst prof to prof chem, Chaminade Col, Honolulu, 60-67, acad dean, 68-76, vpres acad affairs, 78-89; RETIRED. *Mem:* Am Chem Soc; Sigma Xi. *Res:* Analysis and structural determination of volatiles and pigments from natural products, especially by chromatographic means; infra-red spectroscopy. *Mailing Add:* 3276 Ala Laulani Honolulu HI 96818

HIVELY, ROBERT ARLAND, b Salem, Ohio, Apr 27, 21; m 48; c 2. ANALYTICAL CHEMISTRY. *Educ:* Mt Union Col, BS, 43; Univ Akron, MS, 53. *Prof Exp:* Control chemist, Goodyear Tire & Rubber Co, 45-53, from res chemist to sr res chemist, 53-65, sect head chromatography, 65-74, sect head quant anal, 74-77, sect head separations & identifications, 77-80, res scientist, 80-81, res & develop Assoc, 81-87; RETIRED. *Mem:* Am Chem Soc. *Res:* Separations and identifications; gas and liquid partition chromotography; mass spectroscopy; ion exchange; organic qualitative analysis; ultraviolet spectrophotometry; polarography; atomic absorption spectroscopy. *Mailing Add:* 2431 27th St Cuyahoga Falls OH 44223

HIX, ELLIOTT LEE, b Rock Hill, SC, July 28, 25; m 50; c 4. PHARMACOLOGY. *Educ:* Univ Ga, BS, 49; Kans State Univ, MS, 50, PhD(mammal physiol, biochem), 53. *Prof Exp:* Asst physiol, Kans State Univ, 49-50, asst animal sci, 50-52, instr nutrit physiol, 52-53; from instr to assoc prof, 53-67, chmn dept, 67-75, PROF PHARMACOL, KIRKSVILLE COL OSTEOP MED, 67- *Mem:* Am Physiol Soc; Sigma Xi. *Res:* Renal physiology; autonomic innervation of kidney and its influence on kidney function; somato-renal and viscero-renal reflexology; trophic functions of kidney innervation; biochemical synthesis in transplanted kidney; experimental surgery. *Mailing Add:* 811 S Cottage Grove Kirksville MO 63501

HIX, HOMER BENNETT, b Denver, Colo, Mar 13, 20; m 45; c 3. ORGANIC CHEMISTRY. *Educ:* Nebr Wesleyan Univ, AB, 41; La State Univ, MS, 43, PhD(chem), 48. *Prof Exp:* Instr chem, La State Univ, 43-44; sr res chemist, Field Res Lab, Socony Mobil Oil Co, Inc, 48-52, supvr employee rels, 52-61, employee rels mgr, Socony Mobil Can, Ltd, Alta, 61-62, Los Angeles Explor & Prod Div, Mobil Oil Co, 62-64, sr compensation adv, Mobil Oil Corp, 64-67, mgr, corp compensation, 67-73, ADV, HUMAN RESOURCES PLANNING & ANALYSIS, MOBIL CORP, 73- *Mem:* AAAS; Am Chem Soc; Sigma Xi; Am Inst Chemists. *Res:* Fatty acids; petroleum chemistry; organic soil analysis; employee relations; wage and salary administration; human resources planning. *Mailing Add:* 61 Cedar Hills Circle Chapel Hill NC 27514

HIXON, MARK A, b San Diego, Calif, June 10, 51; m 88; c 1. MARINE ICHTHYOLOGY. *Educ:* Univ Calif, Santa Barbara, BA, 73, MA, 74, PhD(pop & aquatic biol), 79. *Prof Exp:* Seasonal aide, Calif Dept Fish & Game, 73-74; teaching asst, Univ Calif, Santa Barbara, 74-76, res asst, 77-79; NSF fel, Univ Calif, Irvine, 79 & Univ Hawaii, 79-80; vis prof, Univ Hawaii, 80-81; lectr, Univ Calif, Irvine, 81-84; asst prof, 84-89, ASSOC PROF, ORE STATE UNIV, 89- *Concurrent Pos:* vis prof, Univ VI, 83- *Mem:* AAAS; Am Soc Zoologists; Animal Behav Soc; Ecol Soc Am; Soc Study Evolution. *Res:* Community ecology of marine reef fishes and benthos; behavioral ecology of reef fishes and hummingbirds; theories of territoriality, foraging and mechanisms maintaining local species diversity. *Mailing Add:* Dept Zool Ore State Univ Corvallis OR 97331

HIXON, SUMNER B, b La Junta, Colo, Sept 7, 30; m 61; c 3. IMAGE INTERPRETATION, PETROGRAPHY. *Educ:* Univ Colo, BA, 52; Univ Tex, MA, 59; Univ Mich, PhD(chem metall), 64. *Prof Exp:* Asst, Univs Tex, Mich & Eastern Mich; asst prof geol & geol eng, Univ Miss, 64-67; engr sr, Lockheed Electronics Co, Houston, 67-81; sr geologist, geophysicist & image interpreter, Aero Serv, Houston, 81-83; consult, Monarch Int, 83-85, chief geologist, 85-; PROGRAMMER, UNISYS, 88- *Concurrent Pos:* Jr geologist, Pa State Geol Surv, 59 & Great Lakes Res, Mich, 61; geologist, Que Dept Mines, 60 & 64 & NASA, Ala, 65; consult, Consumers Power Co, Mich, 62 & Dow Chem Co, 63; fac fel, NASA, Johnson Space Ctr, Houston, 67. *Mem:* Am Asn Petrol Geol; Am Soc Photogram. *Res:* Sedimentary petrography; computer programming applications to photometry, to earth, moon, and martian photography and to data storage and retrieval; lunar geology; remote sensing applications from space and aircraft. *Mailing Add:* 504 Misty Lane Friendswood TX 77546-4532

HIXSON, ELMER LAVERNE, b Arlington, Calif, Sept 29, 24; m 45; c 5. ELECTROACOUSTICS. *Educ:* Univ Tex, BS, 47, MS, 48, PhD(elec eng), 60. *Prof Exp:* Res engr, elec eng res lab, Univ Tex, 47-48; electronic scientist, electronics lab, US Navy, 48-54; from asst prof to assoc prof elec eng, 54-70, PROF ELEC ENG, UNIV TEX, AUSTIN, 71- *Mem:* Fel Acoust Soc Am; sr mem Inst Elec & Electronics Engrs; Inst Noise Control Eng; Am Soc Eng Educ (vpres, 78-79). *Res:* Acoustics; transducers; mechanical vibrations. *Mailing Add:* Dept Elec & Comput Eng Univ Tex Austin TX 78712

HIXSON, FLOYD MARCUS, b Holdenville, Okla, May 15, 18; m 41; c 5. ANIMAL BREEDING, ENVIRONMENTAL PHYSIOLOGY. *Educ:* Okla State Univ, BS, 41; Kans State Univ, MS, 48, PhD(genetics), 60. *Prof Exp:* Assoc prof poultry husb, Okla State Univ, 49-50; prof animal sci, Calif State Univ, Fresno, 51-80; RETIRED. *Mem:* Poultry Sci Asn. *Mailing Add:* 1712 Harvard Clovis CA 93612

HIXSON, JAMES ELMER, b San Diego, Calif, Nov 24, 52; m 87. GENETICS. *Educ:* Univ Tex Austin, BA, 78; Univ Mich Ann Arbor, PhD(human genetics), 83. *Prof Exp:* Teaching asst human genetics, Univ Mich, 79-80; postdoctoral fel, Sch Med, Stanford Univ, 83-85; asst scientist, 85-89, ASSOC SCIENTIST, SOUTHWEST FOUND, BIOMED RES, 90- *Concurrent Pos:* Adj asst prof, Univ Tex Health Sci Ctr, San Antonio, 85-90, adj assoc prof, 90-; prin investr, NIH grant, 88-93. *Mem:* Am Heart Asn; Int Soc Heart Res; Am Soc Human Genetics. *Res:* Identification of genetic factors that cause predisposition to heart disease; genetic analysis of atherosclerosis in human populations; lipid metabolism in non-human primates. *Mailing Add:* Dept Genetics Southwest Found Biomed Res PO Box 28147 San Antonio TX 78228-0147

HIXSON, STEPHEN SHERWIN, b Philadelphia, Pa, Sept 4, 43; m 67, 86; c 3. ORGANIC CHEMISTRY. *Educ:* Univ Pa, BA, 65; Univ Wis, PhD(chem), 70. *Prof Exp:* Fel chem, Harvard Univ, 69-70; from asst prof to assoc prof, 70-80, PROF CHEM, UNIV MASS, 80- *Concurrent Pos:* NSF fel, 69-70. *Mem:* Am Chem Soc. *Res:* Organic photochemistry; photochemical probes of the structure of biological systems. *Mailing Add:* Dept Chem Univ Mass Amherst MA 01003

HIXSON, SUSAN HARVILL, b Orange, NJ, Sept 26, 44; c 2. BIO-ORGANIC CHEMISTRY. *Educ:* Univ Mich, Ann Arbor, BSCh, 65; Univ Wis-Madison, PhD(biochem), 70. *Prof Exp:* Instr chem, Boston Univ, 69-70; res assoc, Univ Mass, Amherst, 70-73; from asst prof to assoc prof, 73-86, PROF CHEM, MT HOLYOKE COL, 86- *Concurrent Pos:* NIH fel, 71-72; vis prof, dept biochem, Univ NC, Chapel Hill, 80; vis scientist, dept biochem & molecular biol, Univ Tex Health Sci Ctr, Houston; coun, Council Undergrad Res, AAAS, 82-88. *Mem:* Am Chem Soc; Sigma Xi; Asn Women Sci; Am Soc Biol Chemists; AAAS. *Res:* Photoaffinity labeling of enzymes; enzyme chemistry; glycoprotein biochemistry. *Mailing Add:* Dept Chem Mt Holyoke Col South Hadley MA 01075-1483

HIYAMA, TETSUO, b Tokyo, Japan, Jan 28, 39; m 80; c 2. BIOCHEMISTRY, PLANT PHYSIOLOGY. *Educ:* Univ Tokyo, BS, 62, MA, 64, PhD(biochem), 67. *Prof Exp:* Asst instr microbiol, Univ Tokyo, 67; NIH fel, Johnson Found, Univ Pa, 67-69; NSF fel, C F Kettering Lab, Ohio, 69-71; Carnegie Inst res fel, Carnegie Inst, Dept Plant Biol, 71-74; asst res biochemist, Univ Calif, Berkeley, 74-79; assoc prof, Nat Inst Basic Biol, Japan, 79-82; assoc prof, 79-87, PROF, SAITAMA UNIV, JAPAN, 87-, DEPT CHMN, 89- *Mem:* AAAS; Am Soc Plant Physiologists; Am Soc Photobiol; Int Soc Plant Molecular Biol. *Res:* Photosynthesis and bioenergetics; photosynthesis. *Mailing Add:* Dept Biochem Saitama Univ Urawa Saitama 338 Japan

HIZA, MICHAEL JOHN, JR, b Pueblo, Colo, Sept 8, 31; m 53; c 2. PHASE EQUILIBRIA, GAS SEPARATION & PURIFICATION. *Educ:* Univ Colo, BS, 53, MS, 60. *Prof Exp:* Res chem engr properties mat, Cryogenics Div, 56, proj engr hydrogen & helium liquefaction, 56-59, res chem engr gas purification, 59-60, proj leader properties cryogenic fluid mixtures, 60-78, res chem engr properties fluid mixtures, 78-80, leader, unit opers & process group, Chem Eng Sci Div, 80-83, actg chief, Chem Process Metrol Div, 83, actg chief, Chem Eng Sci Div, Nat Bur Standards, 84-86; PRES, DELTA-MH INC, BOULDER, COLO, 87- *Concurrent Pos:* Res assoc chem eng, Univ Colo, 62-68; consult hydrogen safety, Lockheed Calif Co & NAm Aviation, Inc, 63-64; mem, task group, Comt Data Sci & Technol Chem Indust, 78-86; adj prof chem eng, Univ Wyo, 77-85. *Honors & Awards:* Russell B Scott Mem Award, Cryogenic Eng Conf Bd, 67. *Mem:* Am Inst Chem Engrs; Am Chem Soc; AAAS; Sigma Xi. *Res:* Low temperature phase equilibria properties of fluid mixtures; physical adsorption of pure fluids and of fluid mixtures at high pressure and low temperatures; measurement and prediction of liquified natural gas densities for equity in trade. *Mailing Add:* PO Box 490 Story WY 82842

HJELLE, JOSEPH THOMAS, b LaCrosse, Wis, Aug 26, 49; m 71; c 2. CELLULAR PHARMACOLOGY, DRUG TARGETING. *Educ:* Univ Wis-LaCrosse, BS, 71; Univ Ariz, PhD(pharmacol), 76. *Prof Exp:* Fel pharmacol, Roche Inst Molecular Biol, 76-78; fel cell biol, Pharmaceutical Mfrs Asn, Found Inst Cellular & Molecular Path, 78-80; asst prof, 80-86, ASSOC PROF PHARMACOL, COL MED, UNIV ILL, PEORIA, 87- *Concurrent Pos:* Prin investr, Univ Ill, 80-; consult, 85- *Mem:* Soc Exp Biol & Med; Sigma Xi; NY Acad Sci; Am Soc Pharmacol & Exp Therapeut; AAAS; Int Soc Peritoneal Dialysis. *Res:* Studies of drug localization and action on subcellular organelles; mesothelial cell biology; peritoneal dialysis; polycystic kidney disease; renal toxicology; effects of endotoxin in disease and toxicology. *Mailing Add:* Univ Ill Col Med One Illini Dr PO Box 1649 Peoria IL 61656

HJELME, DAG ROAR, b Valldal, Norway, Mar 25, 59. ELECTRICAL ENGINEERING. *Educ:* Norweg Inst Technol, dipl eng, 82; Univ Colo, PhD(elec eng), 88. *Prof Exp:* Res asst, Norweg Inst Technol, 83-84; res asst, 84-88, RES ASSOC, UNIV COLO, 88- *Mem:* Optical Soc Am; Inst Elec & Electronics Engrs; Int Soc Optical Eng. *Res:* Semiconductor lasers; spectral and dynamic properties, stabilization, optical feedback and noise; injection locking; ultrafast lasers; noise analysis and application to optical sampling; microwave optics; application to integrated optics and microwave device characterization. *Mailing Add:* Univ Colo Campus Box 425 Boulder CO 80309

HJELMELAND, LEONARD M, GROWTH FACTORS, FIBROSIS. *Educ:* Stanford Univ, PhD(phys biochem), 79. *Prof Exp:* ASSOC PROF OPHTHAL, UNIV CALIF, DAVIS, 86- *Mailing Add:* Dept Ophthal & Biol Chem Univ Calif Davis CA 95616

HJELMFELT, ALLEN T, JR, b Holdrege, Nebr, Oct 21, 37; m 80; c 4. HYDROLOGY, CIVIL ENGINEERING. *Educ:* Kans State Univ, BS, 59; Univ Kans, MS, 61; Northwestern Univ, PhD(fluid mech), 65. *Prof Exp:* Jr civil engr, State Calif Dept Water Resources, 59; asst instr eng mech, Univ Kans, 59-62; assoc develop engr, Nuclear Div, Union Carbide Corp, 62-63; res engr, Northwestern Univ, 63-64; from asst prof to prof civil eng, Univ Mo-Columbia, 65-77; RES HYDRAUL ENGR & RES LEADER, AGR RES SERV, USDA, 78-; PROF AGR ENG, UNIV MO-COLUMBIA, 78- *Concurrent Pos:* Consult, Mineracao Vera Cruz, 76; vis prof, Univ Fed do Para, Belem, Brazil, 76. *Honors & Awards:* J C Stevens Award, Am Soc Civil Engrs, 83; Fel, Am Soc Civil Engrs, 85. *Mem:* Fel Am Soc Civil Engrs; Am Geophys Union; Int Asn Hydraul Res; AAAS; Am Water Resources Asn; Asn Groundwater Scientists & Engrs. *Res:* Hydrologic processes on small watersheds. *Mailing Add:* 269 Agr Engr Bldg Univ Mo Columbia MO 65211

HJERTAGER, BJORN HELGE, applied fluid dynamics, combustion, for more information see previous edition

HLASTA, DENNIS JOHN, b Youngstown, Ohio, July 13, 53; m 75; c 4. MEDICINAL CHEMISTRY. *Educ:* Ohio State Univ, BS, 75; Yale Univ, MS, 77, MPh, 78, PhD(org chem), 79. *Prof Exp:* Sr res chemist, Sterling Winthrop Res Inst, 79-83, group leader, 83-90, RES LEADER, STERLING RES GROUP, 90- *Mem:* Am Chem Soc. *Res:* Design and synthesis of organic compounds as potential therapeutic agents. *Mailing Add:* Sterling Res Group 81 Columbia Turnpike Rensselaer NY 12144

HLAVACEK, ROBERT ALLEN, b Bridgeport, Conn, June 14, 49; m 72; c 2. MATERIAL SCIENCE. *Educ:* Univ Conn, BS, 71, MS, 74. *Prof Exp:* Development engr, Uniroyal Chem, 71-79; sr proj engr, Raymark Corp, 79-82; SR RES SCIENTIST, DAVIS & GECK, DIV AM CYANAMID, 82- *Mem:* AAAS. *Res:* Development of biodegradeable implantable and surgical devices. *Mailing Add:* 42 Visconti Dr Naugatuck CT 06770

HLAVACEK, ROBERT JOHN, b Chicago, Ill, Apr 18, 23; m 51; c 3. FOOD CHEMISTRY, QUALITY SYSTEMS-FOODS. *Educ:* Lawrence Col, BA, 44; Northwestern Univ, PhD(org chem), 49. *Prof Exp:* Chemist, Swift & Co, Ill, 48-50, head indust oils res, 50-53, asst to dir labs, 53, spec assignment, 53-54, head finished prod qual div, 54-56, asst head fresh meats res, 56-59, head table ready meats res, 59-63; asst to tech dir, Hunt-Wesson Foods, 63-64, assoc dir res & develop, 64-73; vpres res & qual control, Best Foods Div, CPC Int, Inc, Union, NJ, 73-76; asst vpres, 76-77, vpres qual assurance, Thomas J Lipton, 77-88; RETIRED. *Mem:* Am Chem Soc; Am Oil Chemists Soc; fel Inst Food Technologists. *Res:* Fats and oils chemistry and processing, sensory evaluation methodology; packaging; analytical methods; quality control methodology. *Mailing Add:* 16 Oechsner Ct Berkeley Heights NJ 07922

HLAVACEK, VLADIMIR, b Prague, Czech, Feb 28, 39; US citizen; m 70. CHEMICAL ENGINEERING. *Educ:* Inst Chem Technol, Prague, MS, 61; Charles Univ, Prague, BS, 67. *Prof Exp:* Prof & sr res scientist, chem eng, Inst Chem Technol, Prague, 67-80; prof chem eng, Catholic Univ, Leuven, Belgium, 80-81; PROF CHEM ENG, STATE UNIV NY, BUFFALO, 81- *Concurrent Pos:* Ed, Chem Eng Sci, 75-81; vis prof, Catholic Univ, Leuven, Belgium, 74 & 80-81, Leningrad, USSR, 73 & Zurich, Switz, 77; consult, var co. *Honors & Awards:* Wilhelm Award. *Mem:* Am Inst Chem Engrs; Am Ceramic Soc; Soc Indust & Appl Math. *Res:* Chemical reactor theory; flames; separation process; nonlinear boundary value problems; bifurcation theory; synthesis of ceramic materials. *Mailing Add:* Chem Eng Dept Furnas Hall State Univ Buffalo Amherst NY 14260

HLAVKA, JOSEPH JOHN, b Franklin, NJ, Feb 13, 27; m 54; c 2. ORGANIC CHEMISTRY. *Educ:* Rutgers Univ, 50; Mass Inst Technol, PhD(org chem), 57. *Prof Exp:* RES CHEMIST, LEDERLE LABS, AM CYANAMID CO, 57- *Concurrent Pos:* Am Cyanamid sr ed award, Imp Col, Univ London, 62. *Mem:* Am Chem Soc. *Res:* Photochemistry; peptide synthesis; chemistry of antibiotics. *Mailing Add:* Org Chem Res Sect Lederle Labs Pearl River NY 10965

HNATIUK, BOHDAN T, b Zaliszczyki, Ukraine, July 25, 15; nat US; m 44; c 3. AERONAUTICAL ENGINEERING. *Educ:* State Sem, Ukraine, BSEd, 35; Danzig Tech Univ, Dipl, 43, DrIng, 45. *Prof Exp:* Sci asst aeronaut eng, Danzig Tech Univ, 42-44; res scientist & asst to Prof Schrenk & Prof Katzmayer, Vienna Tech Univ, 44-45; anal engr, Dornier Works, Ger, 45-47; teacher & dir UNRRA, Mech Training Sch, Ger, 47-48; with Displaced Persons Serv, Tettnang Dist, French Zone, Ger, 48-49; engr, Link Belt Co, Pa, 50-51; from asst prof to assoc prof aeronaut eng, Univ Notre Dame, 51-57; prof aerospace eng, Univ WVa, 57-60; PRES AEROSPACE ENG, DREXEL UNIV, 60- *Concurrent Pos:* Consult, Guided Missile Sect, Bendix Aviation Corp, Ind, 55-57, Pneumo Dynamics Corp, Systs Eng Div, Md, 61-63 & George C Marshall Space Flight Ctr, Ala, 67-69; lectr & consult, Allegany Ballistics Lab, Dept of Navy, 59-; fel Am Soc Eng Educ-NASA; Alexander von Humboldt scholar. *Mem:* AAAS; Am Ord Asn; AAUP; Shevchenko Sci Soc; Am Inst Aeronaut & Astronaut. *Res:* Thermodynamics, heat transfer; fluid mechanics; aerodynamics; gas dynamics; aircraft and missile propulsion; gas turbines; space flight dynamics. *Mailing Add:* Dept Mech Eng Drexel Univ Philadelphia PA 19104

HNATOW, MIQUEL ALEXANDER, b Arg, Feb 26, 42; US citizen; m 66; c 3. CHEMICAL ENGINEERING. *Educ:* Univ Mass, BS, 65; NY Univ, MS, 67, PhD(chem eng), 70. *Prof Exp:* Engr, Esso Res & Eng Co, 69-70; asst prof chem eng, NY Univ, 70-73; asst prof, Polytech Inst NY, 73-75; VPRES & TREAS, CATALYSIS RES CORP, 73- *Mem:* Am Inst Chem Engrs; Am Chem Soc; NY Acad Sci; Catalysis Soc; Sigma Xi. *Res:* Kinetics and catalysis; use of tracers in elucidating structure and mechanisms in heterogeneous catalysis; use of complex oxides as catalysts; catalyst synthesis; hydrates, kinetics of formation and decomposition. *Mailing Add:* 45 Cedar Rd Caldwell NJ 07006

HNATOWICH, DONALD JOHN, b Winnipeg, Man, Apr 20, 40; m 68; c 2. NUCLEAR MEDICINE. *Educ:* Univ Man, BSc, 63, MSc, 64; Mass Inst Technol, PhD(chem), 68. *Prof Exp:* Res assoc, Europ Inst Nuclear Res, 68-69; res collabr, Brookhaven Nat Lab, 69-70; Mem res staff nuclear eng, Mass Inst Technol, 71-78; PROF NUCLEAR MED, MED SCH, UNIV MASS, WORCESTER, 79- *Concurrent Pos:* Nat Res Coun Can fels, Europ Inst Nuclear Res, 68-69 & Brookhaven Nat Lab, 69-70; asst chemist, Dept Radiol, Mass Gen Hosp, 71-78, Hammersmith Hosp, London, 87. *Mem:* Soc Nuclear Med; Int Asn Radiopharmacol. *Res:* Diagnostic nuclear medicine; development of new radioactive imaging agents. *Mailing Add:* Dept Nuclear Med Univ Mass Med Ctr Worcester MA 01655

HO, ANDREW KONG-SUN, b Apr 22, 39. PHARMACOLOGY, MOLECULAR BIOLOGY OF MUSCARINIC ACETYLCHOLINE RECEPTOR & ALCOHOL STUDIES. *Educ:* Monash Univ, Victoria, Australia, PhD(pharmacol & cell biol), 67. *Prof Exp:* PROF PHARMACOL, COL MED, UNIV ILL, PEORIA & URBANA, 75- *Concurrent Pos:* Prin investr grants, NIH, Am Heart Asn & Am Fedn on Aging; vis prof, Fac Med, Univ Calgary, 84-85. *Mem:* Am Soc Pharmacol & Exp Therapeut; Int Soc Biomed Res Alcoholism. *Res:* Calcium binding proteins; muscarinic acetylcholine receptors; molecular pharmacology of the muscarinic acetylcholine receptor regulation and Ca2plus-binding proteins; role of receptor mechanisms in aging and alcoholism. *Mailing Add:* Dept Pharmacol Col Med Uni Ill One Illini Dr Peoria IL 61605

HO, BEGONIA Y, b Hong Kong, July 8, 62. RECEPTOR PHARMACOLOGY, DRUG DESIGN. *Educ:* Chinese Univ Hong Kong, BS, 84; Univ Minn, Minneapolis, PhD(pharmacol), 89. *Prof Exp:* Res asst, Behav Pharmacol Lab, Univ Minn, Minneapolis, 84-85, res asst pharmacol, 85-89; RES FEL BIOL, CALIF INST TECHNOL, 89- *Mem:* Fel Am Soc Pharmacol & Exp Therapeut; Soc Neurosci. *Res:* Expression of neurotransmitter, hormone, drug receptors and ion channels in mammalian cells and characterization of their structure-function relationship in ligand binding and signal transduction by molecular biological, biochemical and electrophysiological approaches. *Mailing Add:* Div Biol 156-29 Calif Inst Technol Pasadena CA 91125

HO, BENG THONG, b Malacca, Malaysia, Sept 3, 32; US citizen; m 62. BIOCHEMICAL PHARMACOLOGY. *Educ:* Nat Taiwan Univ, BS, 55; Univ Ore, MS, 59; Univ Wash, PhD(pharmaceut chem), 62. *Prof Exp:* Res assoc med chem, Sch Pharm, State Univ NY Buffalo, 62-64, asst res prof, 64-65; asst prof biochem & psychiat, Col Med, Baylor Univ, 65-68; from asst prof to assoc prof, 68-71, PROF MENT SCI, GRAD SCH BIOMED SCI, UNIV TEX, HOUSTON, 71-, HEAD NEUROCHEM & NEUROPHARMACOL RES SECT, TEX RES INST MENT SCI, 69-; AT UNIV MD CANCER CTR. *Concurrent Pos:* Fel, Sch Pharm, State Univ NY Buffalo, 62-64; chief chem res sect, Tex Res Inst Ment Sci, 65-68, chief neurochem & chem res sect, 68-69. *Mem:* Fel NY Acad Sci; fel Acad Pharmaceut Sci; Am Soc Neurochem; Am Chem Soc; Sigma Xi. *Res:* Neurochemical adaptation to chronic treatment of psychotherapeutic agents; evaluation of agents capable of selectively altering the actions of a specific biogenic amine; behavioral correlates of drug actions at the subcellular level; effects of nutritional state on bioavailability of psychotherapeutic agents, and on alcohol dependency and withdrawal. *Mailing Add:* Dept Neuro Oncol M D Anderson Hosp 6723 Bertner Ave Houston TX 77030

HO, BONG, b Hong Kong, Dec 23, 31; m 59; c 2. ELECTRICAL ENGINEERING. *Educ:* Chung Kung Univ, BS, 56; SDak State Univ, MS, 61; Univ Mich, PhD(elec eng), 67. *Prof Exp:* Asst prof, 67-76, ASSOC PROF ELEC ENG, MICH STATE UNIV, 76- *Mem:* Inst Elec & Electronics Engrs. *Res:* Microwave electronics. *Mailing Add:* Dept Elec Eng Mich State Univ East Lansing MI 48823

HO, CHIEN, b Shanghai, China, Oct 23, 34; US citizen; m 63; c 2. BIOCHEMISTRY, BIOPHYSICS. *Educ:* Williams Col, BA, 57; Yale Univ, PhD(phys chem), 61. *Prof Exp:* Res chemist, Linde Co, Union Carbide Corp, 60-61; res assoc chem, Mass Inst Technol, 61-62, res assoc biol, 62-64; asst prof biophys, Univ Pittsburg, 64-67, assoc prof molecular biol, 67-70, chmn dept biophys & microbiol, 67-69, prof molecular biol & biochem, 71-79; head, 79-86, PROF, DEPT BIOL SCI, CARNEGIE MELLON UNIV, 79- *Concurrent Pos:* Guggenheim fels, 70-71; mem, Biophysics & Biophys Chem B Study Sect, NIH, 74-78; mem, Biotechnol Resources Rev Comt, NIH, 79-83; mem, Molecular Biol Adv Panel, NSF, 80-83; mem Cellular & Molecular Basis Dis Rev Comt, NIH, 87- *Honors & Awards:* Merit Award, Nat Heart, Lung & Blood Inst, 86. *Mem:* AAAS; Am Chem Soc; Biophys Soc; Am Soc Biochem & Molecular Biol; Soc Magnetic Resonance Med; Protein Soc. *Res:* Structure-function relationships in biological systems; application of nuclear magnetic resonance to biological and biomedical problems; conformations of macromolecules in solution; normal and abnormal hemoglobins; biological membranes. *Mailing Add:* Dept Biol Sci Carnegie Mellon Univ Pittsburgh PA 15213

HO, CHI-TANG, b Fuzhou, Fujian, China, Dec 26, 44; m 74; c 2. FOOD SCIENCE & TECHNOLOGY. *Educ:* Nat Taiwan Univ, BS, 68; Wash Univ, St Louis, MA, 71 & PhD(chem), 74. *Prof Exp:* Fel Sch Chem, Rutgers Univ, 75-76; asst res fel, dept food sci, Rutgers Univ, 76-78, asst prof; 78-83, assoc res prof, 83-87, RES PROF, 87- *Concurrent Pos:* Ed bd mem Food Reviews Int, 87, ed bd adv, Reviews in Food Sci & Nutrit, 87; chmn Flavor subdiv, Agr Food Div, Am Chem Soc, 90. *Mem:* Am Chem Soc; Am Oil & Chemists Soc; Inst Food Technologists. *Res:* Mechanisms of the formulation of flavor compounds in foods; antimutagenic and anticarcinogenic compounds from natural products; antoxidative mechanism of lipids. *Mailing Add:* Eight Newman St East Brunswick NJ 08816

HO, CHONG CHEONG, b Singapore, July 5, 38; m 68; c 2. POLYMER SCIENCE. *Educ:* Univ Idaho, BS, 59, MS, 61; Rensselaer Polytech Inst, PhD(chem eng), 64. *Prof Exp:* NASA res asst, Rensselaer Polytech Inst, 64; res engr, Uniroyal Res Ctr, 64-66, sr res engr, Uniroyal Chem, Naugatuck, 66-71, sr group leader, Uniroyal Chem, Uniroyal Inc, Naugatuck, 71-85, SR RES SCIENTIST, UNIROYAL CHEM, GEISMAR, LA, 86- *Mem:* Am Chem Soc. *Res:* Relationship between molecular structure and properties of polymers; melt rheology and polymer processing; physical properties and morphology of polymer blends; rubber-reinforced plastics; physical testing. *Mailing Add:* Uniroyal Chem Co PO Box 397 Geismar LA 70734

HO, CHO-YEN, b Kweiping, China, Aug 11, 28; nat US; m 63; c 2. THERMAL PHYSICS, RESEARCH ADMINISTRATION. *Educ:* Nat Taiwan Univ, BS, 55; Univ Ky, MSME, 60; Purdue Univ, PhD(thermodyn, heat transfer), 64. *Prof Exp:* Asst, Univ Ky, 58-60; res asst, Thermophys Properties Res Ctr, 60-64, from asst sr researcher to assoc sr researcher, 64-69, HEAD REF DATA TABLES DIV, PURDUE UNIV, WEST LAFAYETTE, 67-, ASST DIR, CTR INFO & NUMERICAL DATA ANALYSIS & SYNTHESIS, 73-, SR RESEARCHER, 74-, INTERIM DIR, 81- *Concurrent Pos:* Indexer, Appl Mech Rev, 67-72; reviewer & referee, Int J Heat & Mass Transfer & Am Soc Mech Eng J Heat Transfer, 68-; co-chmn, Eighth Int Thermal Conductivity Conf, 68; mem, Standing Comt on Thermophys Properties, Am Soc Mech Engrs, 68-; short course instr, NSF, 73; treas & mem, Gov Bd, Int Thermal Conductivity Conf, 73- *Mem:* Am Soc Mech Engrs; Am Phys Soc; Sigma Xi; Am Soc Metals; AAAS. *Res:* Thermophysics, thermophysical and electrical properties of materials, thermoelectric power generation and conversion, thermodynamics and heat transfer; engineering physics; mechanical engineering. *Mailing Add:* 606 Riley Lane West Lafayette IN 47906

HO, CHUNG YOU (PETER), b China, Sept 12, 33; US citizen; m 60; c 3. APPLIED MATHEMATICS, COMPUTER SCIENCE. *Educ:* Univ Taiwan, BS, 56; Univ Tex, MS, 59; Rensselaer Polytech Inst, PhD, 62. *Prof Exp:* Sr engr, Bendix Corp, NY, 62-63; syst analyst, Int Bus Mach, Minn & Vt, 63-67; assoc prof comput sci & mech eng, 67-81, PROF COMPUT SCI, UNIV MO, ROLLA, 82- *Concurrent Pos:* Asst prof, St Mary's Col, Minn, 63-65. *Mem:* Inst Elec & Electronics Engrs; Am Soc Mech Engrs. *Res:* Operations research; spatial kinematics; robotics; computer integrated manufacturing. *Mailing Add:* Dept Comput Sci Univ Mo Rolla MO 65401

HO, CHUNG-WU, b Hankow, China; m 64; c 2. GEOMETRIC TOPOLOGY, DYNAMICAL SYSTEMS. *Educ:* Univ Wash, BS(physics) & BS(math), 64, MA, 65; Mass Inst Technol, PhD(math), 70. *Prof Exp:* Asst prof, 70-74, assoc prof, 74-78, PROF MATH, 78-, CHMN DEPT, SOUTHERN ILL UNIV, 88- *Concurrent Pos:* Hon prof, Hefei Educ Inst, Hefei, Anhui, China; Res Scholar Award, So Ill Univ, 81-82; vis prof, Wash Univ, 84-85. *Mem:* Am Math Soc; Math Asn Am. *Res:* Differential and geometric topology; differential geometry; topological dynamics. *Mailing Add:* Dept Math Southern Ill Univ Edwardsville IL 62026

HO, CLARA LIN, b Shantung, China, Jan 16, 32; US citizen; m 60; c 2. GEOCHEMISTRY. *Educ:* Nat Taiwan Univ, BS, 54; Ore State Col, MS, 56; NC State Col, PhD(soil chem), 59. *Prof Exp:* assoc fel soil biochem, Iowa State Univ, 59-62 & civil eng, 62-65; asst prof geochem, Coastal Studies Inst, La State Univ, 66-69, assoc prof, Dept Marine Sci, 69-78; res scientist & dir, Mineral Studies Lab, Anal Geochem, Bur Econ Geol, Univ Tex, Austin, 78-84. *Mem:* Fel Am Inst Chemists; Int Geochem Soc; Am Geochem Soc; Am Lab Mgrs Asn; Am Soc Testing & Mat. *Res:* Clay mineralogy and clay chemistry; soil engineering and soil stabilization. *Mailing Add:* 4114 First View Austin TX 78731

HO, DAH-HSI, b Shanghai, China, June 22, 31; m 63. BIOCHEMICAL PHARMACOLOGY. *Educ:* Nat Taiwan Univ, BS, 55; Univ Ore, MS, 59, PhD(physiol), 62. *Hon Degrees:* MD. *Prof Exp:* Cancer res scientist biochem, Roswell Park Mem Inst, 64-65; ASST BIOCHEMIST, UNIV TEX SYST CANCER CTR, M D ANDERSON HOSP & TUMOR INST, 65-, ASST PROF BIOCHEM, 72-; BIOCHEMIST & PROF BIOCHEM, GRAD SCH SCI, UNIV TEX HEALTH SCI CTR, HOUSTON, 81- *Concurrent Pos:* Assoc & assoc prof biochem, Grad Sch Sci, Univ Tex Health Sci Ctr, Houston, 68-81. *Mem:* Am Asn Cancer Res; Am Soc Pharmacol & Exp Therapeut; Am Soc Clin Oncol. *Res:* Drug metabolism mechanism of action of anticancer drugs; clinical and biochemical pharmacology of anticancer drugs; radioimmunoassay for phase I anticancer drugs. *Mailing Add:* M D Anderson Hosp & Tumor Inst 1515 Holcombe Box 52 Houston TX 77030

HO, DAR-VEIG, b Peiping, China, Jan 3, 31; US citizen; m 55; c 3. APPLIED MATHEMATICS. *Educ:* Nat Taiwan Univ, BS, 53; Univ Tenn, MS, 55; Brown Univ, PhD(appl math), 62. *Prof Exp:* Asst prof math, 62-67, ASSOC PROF MATH, GA INST TECHNOL, 67- *Mem:* Am Math Soc; Soc Indust & Appl Math. *Res:* Fluid dynamics; partial differential equations. *Mailing Add:* Dept Math Ga Inst Technol 225 N Ave Atlanta GA 30332

HO, DAVID TUAN-HUA, b Nanking, China, Oct 1, 48; c 1. PLANT BIOCHEMISTRY, DEVELOPMENTAL BIOLOGY. *Educ:* Nat Taiwan Univ, BS, 70; Mich State Univ, PhD(biochem), 76. *Prof Exp:* Res asst plant biochem, Mich State Univ, 71-75; fel plant biochem, Wash Univ, 75-76; fel molecular biol, Mass Inst Technol, 76-78; asst prof bot, Univ Ill, 78-84; ASSOC PROF, DEPT BIOL, WASH UNIV, ST LOUIS, MO, 84- *Concurrent Pos:* Fel, Jane Coffin Child Mem Fund for Med Res, 76-78. *Mem:* Am Soc Plant Physiologists; AAAS; Am Soc Biochem & Molecular Biol; Int Asn Plant Molecular Biol. *Res:* Plant biochemistry; action of plant hormones; molecular biology; plant physiology. *Mailing Add:* Dept Biol Wash Univ Lindell & Skiner St Louis MO 63130

HO, FANGHUAI H(UBERT), b Nanking, China, Dec 25, 34; m 61; c 2. ENGINEERING MECHANICS, MECHANICAL SYSTEMS. *Educ:* Nat Cheng Kung Univ, Taiwan, BS, 56; Univ Fla, PhD(eng mech), 64. *Prof Exp:* Asst prof eng sci & mech, Univ Fla, 64-65; res engr, 65-69, res assoc, 69-81, RES FEL, B F GOODRICH RES CTR, 81- *Concurrent Pos:* Lectr, Cleveland State Univ, 80- *Mem:* Am Inst Aeronaut & Astronaut. *Res:* Mechanics and materials; research of high-temperature composite materials and structures. *Mailing Add:* B F Goodrich Res Ctr 9921 Brecksville Rd Brecksville OH 44141

HO, FLOYD FONG-LOK, b China; nat US. ANALYTICAL CHEMISTRY, CATALYSIS. *Educ:* Hong Kong Baptist Col, BS, 63; Stetson Univ, Fla, MS, 64; Univ NC, PhD(chem), 69. *Prof Exp:* res chemist, 69-76, SR RES CHEMIST, HERCULES, INC, 76- *Mem:* Am Chem Soc. *Res:* Research and development of analytical methods and technology, mainly in the fields of nuclear magnetic resonance, electron spin resonance and ESCA, and applied mostly to polymers and polymerization catalysts; catalysis research involving Zeigler-Natta type catalysis. *Mailing Add:* 715 Pleasant Run W Chester PA 19380

HO, GRACE PING-POO, b Surabaia, Indonesia. MATHEMATICS. *Educ:* Ore State Univ, BA, 57; Iowa State Univ, MA, 60; Pa State Univ, PhD(math), 68. *Prof Exp:* Instr physics, Winona State Col, 60-61; asst prof, 67-72, ASSOC PROF MATH, UNIV BRIDGEPORT, 72- *Mem:* Am Math Soc; Math Asn Am; Soc Indust & Appl Math. *Res:* Partial differential equation. *Mailing Add:* 109 Village Dr Shelton CT 06484-1732

HO, HENRY SOKORE, b Fukien, China, Sept 8, 42; US citizen; m 72; c 2. BIOPHYSICS, ELECTRONICS ENGINEERING. *Educ:* Univ Wash, BS, 65, MS, 67, PhD(elec eng), 70. *Prof Exp:* RES MICROWAVE ENGR BIOL EFFECTS, BUR RADIOL HEALTH, FOOD & DRUG ADMIN, 71- *Concurrent Pos:* Res medical device electronics engr. *Mem:* Sigma Xi; Inst Elec & Electronics Engrs. *Res:* Biological effects and dosimetry of microwaves; medical devices. *Mailing Add:* Fed Drug Admin/CDRH 5600 Fishers Lane HFZ 133 Rockville MD 20857

HO, HON HING, b Hong Kong, May 3, 39; m 68; c 2. MYCOLOGY, PHYTOPATHOLOGY. *Educ:* Univ Hong Kong, BSc, 62, Hons, 63; Univ Western Ont, PhD(bot), 66. *Prof Exp:* Can fel, 66-67; asst prof bot & bact, Ohio Wesleyan Univ, 67-68; from asst prof to assoc prof, 68-79, PROF BIOL, STATE UNIV NY COL NEW PALTZ, 79- *Concurrent Pos:* Vis prof, Inst Microbiol, Acad Sinica, Beijing & Dept Plant Protection, Nanjing Agr Univ, China, 82-83, Inst Bot, Acad Sinica, Taipei, 89-90. *Mem:* Mycol Soc Am; Am Phytopath Soc. *Res:* Biology of plant-pathogenic phycomycetes; morphology and taxonomy of phytophthora. *Mailing Add:* Dept Biol State Univ NY Col New Paltz NY 12561

HO, HUNG-TA, b Anking, Anhui, China, Oct 14, 21; m 61; c 2. APPLIED MATHEMATICS, FLUID DYNAMICS. *Educ:* Ord Eng Col, China, BSc, 45; Va Polytech Inst, MSc, 56; Brown Univ, PhD(appl math), 61. *Prof Exp:* Res engr, Chinese Ord Bur, 45-49, res assoc mech eng, Chinese Ord Res Inst, 49-54; res asst, Va Polytech Inst, 54-55; appl math & hypersonic aerodyn, Brown Univ, 55-60; res scientist, Hydronautics Inc, 60-63, asst prof aerospace, Univ Okla, 63-66; from asst prof to assoc prof, 66-72, PROF MATH, SAN DIEGO STATE UNIV, 72- *Concurrent Pos:* Instr, Taiwan Prov Taipei First Tech Col, 49-54; consult, ARO, Inc, Arnold Air Force Sta, Tenn, 66-68. *Mem:* Am Math Soc. *Res:* Aerodynamics, hydrodynamics. *Mailing Add:* Dept Math 5300 Campanile Dr San Diego State Univ San Diego CA 92182-0763

HO, ING KANG, b Taiwan, China, May 7, 39; m 65; c 3. NEUROCHEMISTRY, PHARMACOLOGY. *Educ:* Nat Taiwan Univ, BS, 62; Univ Calif, San Francisco, PhD(biochem), 68. *Prof Exp:* Res assoc anesthesiol, Baylor Col Med, 68-69, instr, 69-70; res pharmacologist, Univ Calif, San Francisco, 70-71, asst res pharmacologist, 71-73, asst prof, 73-75; assoc prof, 75-78, PROF PHARMACOL, MED CTR, UNIV MISS, 78-, CHMN, 82- *Mem:* Soc Neurosci; Am Soc Pharmacol & Exp Therapeuts; Soc Toxicol. *Res:* Drug addiction; mechanism of development of tolerance to and dependence on psychoactive agents; neurotoxicity. *Mailing Add:* Dept Pharmacol/Toxology Univ Miss Med Ctr 2500 N State St Jackson MS 39216-4505

HO, IWAN, b Souzhou, Kiangsu, China, Apr 15, 25; US citizen; m 75; c 1. MYCORPHIZAE PHYSIOLOGY, ENZYMES. *Educ:* Nat Shanghai Univ, BS, 46; La State Univ, MS, 56; Ore State Univ, PhD(forest sci), 84. *Prof Exp:* Specialist pest mgt, Taiwan Agr Sta, Taiwan, 47-51; supvr res, Taichung Sugarcane Exp Sta, 51-56; microbiologist diagnostics, Seattle-King County Health Dept, 62-67; RES PLANT PATHOLOGIST, FORESTRY SCI LAB, US FORESTRY SERV, 70- *Concurrent Pos:* Grant, Sigma Xi, 72; vis prof, Academia Sinica, 85; courtesy adj prof, Col Forestry, Ore State Univ. *Mem:* Sigma Xi; Mycol Soc Am; Am Soc Plant Physiologists; Int Soc Molecular Biol. *Res:* Physiological and biochemical process of mycorrhizae and their ecological significance in relationship with their hosts and site condition; genetic variations in enzyme activity and phytohormone production of various species. *Mailing Add:* Forestry Sci Lab US Forestry Serv 3200 SW Jefferson Corvallis OR 97331

HO, JAMES CHIEN MING, b China, July 31, 37; US Citizen; m 64; c 2. PHYSICAL CHEMISTRY, SOLID STATE PHYSICS. *Educ:* Nat Taiwan Univ, BS, 59; Univ Calif, Berkeley, MS, 63, PhD(chem), 66. *Prof Exp:* Fel chem, Lawrence Radiation Lab, Univ Calif, Berkeley, 65-67; sr scientist, Battelle Mem Inst, 67-71; from assoc prof to prof physics, 71-85, PROF PHYSICS & CHEM, WICHITA STATE UNIV, 85-, SR STAFF SCIENTIST, NAT INST AVIATION RES, 87- *Concurrent Pos:* Vis prof chem, Univ Calif, Berkeley, 77-78. *Mem:* AAAS; Am Phys Soc; Am Soc Testing & Mat; Am Chem Soc; Sigma Xi; Mat Res Soc. *Res:* Low-temperature calorimetry; superconductivity; metals and alloys. *Mailing Add:* Dept Physics Wichita State Univ Wichita KS 67208

HO, JOHN TING-SUM, b Hong Kong, July 5, 42; US citizen; m 70. EXPERIMENTAL SOLID STATE PHYSICS. *Educ:* Univ Hong Kong, BSc Gen, 64, BSc Spec, 65; Mass Inst Technol, PhD(physics), 69. *Prof Exp:* Asst prof physics, Univ Pa, 69-74; assoc prof, Univ Houston, 74-75; assoc prof, 75-83, interim dean, 84-85, PROF PHYSICS & ASSOC DEAN NATURAL SCI & MATH, STATE UNIV NY, BUFFALO, 83- *Concurrent Pos:* Vis prof physics, Haverford Col, 84; Guggenheim fel, 90-91; vis scientist, AT&T Bell Labs, 91. *Mem:* Fel Am Phys Soc; Biophys Soc. *Res:* Critical phenomena; biophysics; light scattering; liquid crystals. *Mailing Add:* Dept Physics & Astron State Univ NY Buffalo NY 14260

HO, JU-SHEY, b Taipei, Formosa, Dec 20, 35; m 65; c 2. ZOOLOGY, PARASITOLOGY. *Educ:* Nat Taiwan Univ, BS, 58; Boston Univ, MA, 65, PhD(zool), 69. *Prof Exp:* Lab instr zool, Nat Taiwan Univ, 60-62; NSF fel, Boston Univ, 69-70; from asst prof to assoc prof, 70-79, PROF, CALIF STATE UNIV, LONG BEACH, 79- *Mem:* AAAS; Soc Syst Zool; Crustacean Soc. *Res:* Crustacean parasites of fish; copepod parasites of marine invertebrates. *Mailing Add:* Dept Biol Calif State Univ Long Beach CA 90840

HO, KANG-JEY, b Tainan, Taiwan, China, Aug 2, 37; m 69; c 2. PATHOLOGY. *Educ:* Nat Taiwan Univ, MD, 63; Northwestern Univ, PhD(path), 68. *Prof Exp:* From instr to asst prof path, Med Sch, Northwestern Univ, 68-70; from asst prof to assoc prof, 70-75, PROF PATH, MED CTR, UNIV ALA, BIRMINGHAM, 76-; CHIEF PATH RES, VET ADMIN HOSP, BIRMINGHAM, 76- *Concurrent Pos:* Res fel, Evanston Hosp, Ill, 68-69; Schweppe Found fel, 73; dir path res, Evanston Hosp, 69-70; actg chief path serv, Vet Admin Hosp, Birmingham, 75-76; hon prof, Guiyang Med Col, Guiyang, China, 87. *Mem:* Am Heart Asn; Am Soc Clin Nutrit; Am Inst Nutrit; Soc Exp Biol & Med; Int Soc Chronobiol. *Res:* Lipid metabolism and its relationship to atherosclerogenesis and cholelithiasis. *Mailing Add:* Dept Path Univ Ala Med Ctr Birmingham AL 35294

HO, KEH MING, b Ping-Tong, Taiwan, Oct 29, 36; m 72. CROP BREEDING, CYTOGENETICS. *Educ:* Chung-Hsing Univ, Taiwan, BS, 60; Univ Guelph, Ont, MS, 70, PhD(cytogenetics), 74. *Prof Exp:* Res & teaching asst genetics, Agron Dept, Chung-Hsing Univ, Taichung, Taiwan, 62-68, assoc prof, 79-80; plant breeder barley breeding, Ciba-Geigy Seeds Ltd, Ailsa Craig, Ont, 73-81; BARLEY BREEDER, OTTAWA RES STA, OTTAWA, 81- *Concurrent Pos:* Mem, Subcomt Barley, Expert Comt Grain Breeding, 74. *Mem:* Genetic Soc Can; Am Soc Agron; Sigma Xi. *Res:* Development of the haploid method in barley breeding; the techniques for haploid production and doubling haploids; development of feed barley cultivars; genetics and cytogenetics studies on barley. *Mailing Add:* Bldg 75 Ottawa Res Sta Ottawa ON K1A 0C6 Can

HO, LOUIS T, b Shanghai, China, July 22, 30; m 61; c 2. ACOUSTICS. *Educ:* Cath Univ Am, BS, 53, MS, 61, PhD(acoustics), 72. *Prof Exp:* Res assoc molecular physics lab, Univ Md, 55-61; res physicist, Gen Kinetics, Inc, 61-63; res physicist, Marine Eng Lab, 63-67, sr proj engr, Naval Ship Res & Develop Ctr, 67-87,; res scientist, 87-89, BR HEAD, MACH ACOUSTICS BR, DAVID TAYLOR RES CTR, ANNAPOLIS, 89- *Mem:* Am Phys Soc; Acoust Soc Am. *Res:* Sound radiation and propagation; noise reduction techniques; acoustic measurements; fluid-structure interaction. *Mailing Add:* 11607 Candor Dr Mitchellville MD 20721

HO, LYDIA SU-YONG, b China, Dec 16, 39; US citizen; m 64; c 2. CHEMISTRY, STATISTICAL QUALITY CONTROL. *Educ:* Nat Taiwan Univ, BS, 61; Univ Calif, Berkeley, MS, 64, PhD(nutrit), 67. *Prof Exp:* Asst ed chem & biochem, Chem Abstracts Serv, Am Chem Soc, 67-70; clin chemist, St Francis Hosp, 74-78; mgr & sr chemist, Kans Energy & Environ Lab, Inc, 78-79; res assoc, 79-85, res asst prof, Internal Med, Univ Kansas Med Ctr, Wichita Br, 85-86; STATISTICAN QUAL CONTROL, BOEING CO, 86- *Mem:* Am Chem Soc; Am Soc Qual Control. *Mailing Add:* SE 65th St Bellevue WA 98006

HO, MAT H, b May 29, 51; m 81; c 2. ANALYTICAL BIOCHEMISTRY, CLINICAL CHEMISTRY. *Educ:* Univ New Orleans, PhD(chem), 81. *Prof Exp:* Asst prof chem, 81-77, ASSOC PROF CHEM, UNIV ALA, BIRMINGHAM, 87- *Concurrent Pos:* From asst prof forensic sci to assoc prof, Univ Ala Birmingham, 83-87; from asst prof biomed eng, to assoc prof, Univ Ala, Birmingham, 87-88; consult, Southern Res Inst, 82-85; chmn, Symposium Biol Montitoring of Exposure to Chem, Am Chem Soc, 84-85, Symposium Methods Forensic Chem, 85. *Mem:* Am Chem Soc; Sigma Xi. *Res:* Fundamental studies of bio-electro chemistry and development of biosensor based on electrochemical, piezoelectric and fiber optical detections; study of immobilized enzymes, antibodies, and DNA and use them as analytical reagents in clinical, environmental and forensic chemistry; investigation of flow injection analysis, biosensors, and chemical sensors. *Mailing Add:* 1744 Napier Dr Birmingham AL 35226

HO, MAY-KIN, b Hong Kong, Oct 16, 52. IMMUNOLOGY. *Educ:* State Univ NY, Downstate Med Ctr, BS, 75, PhD(immunol), 80. *Prof Exp:* SUPVR IMMUNOL RES, E I DU PONT DE NEMOURS CO, 82- *Mem:* Am Asn Immunologists; Reticuloendothelial Soc; AAAS; Am Asn Women Sci; NY Acad Sci. *Mailing Add:* Immunol Res Bldg 500-2 E I du Pont de Nemours Co 331 Treble Cove Rd Billerica MA 01862

HO, MONTO, b Hunan, China, Mar 28, 27; m 52; c 2. VIROLOGY, MEDICINE. *Educ:* Harvard Univ, AB, 49; Harvard Med Sch, MD, 54; Am Bd Internal Med, dipl; Am Bd Microbiol, dipl. *Prof Exp:* Intern & resident, Boston City Hosp, 54-56; asst prof med & microbiol, 59-62, PROF MICROBIOL, 64-, CHMN DEPT INFECTIOUS DIS & MICROBIOL, GRAD SCH PUB HEALTH, UNIV PITTSBURGH, 69- *Concurrent Pos:* Res fel med & bact, Harvard Med Sch, 56-59; prof med & chief infectious dis, Med Sch, Univ Pittsburgh, 70- *Mem:* Fel Am Acad Microbiol; Am Asn Immunol; Am Soc Clin Invest; fel Am Col Physicians; Asn Am Physicians; Academia Sinica. *Res:* Interferon, cytomegalovirus, pathogenesis of infectious diseases. *Mailing Add:* Dept Infectious Dis & Microbiol Univ Pittsburgh Grad Sch Pub Health Crabtree A-427 Pittsburgh PA 15261

HO, NANCY WANG-YANG, b Tai-City, Kiangsu, China; US citizen; m 63; c 2. MOLECULAR BIOLOGY, BIOCHEMISTRY. *Educ:* Nat Taiwan Univ, BS, 57; Temple Univ, MA, 60; Purdue Univ, PhD(molecular biol), 68. *Prof Exp:* Res asst biochem, LaRabida-Univ Chicago Inst, 62-63; res assoc, Dept Biol Sci, 68-69, NIH res fel, 69-71, res assoc molecular biol, 71-77, ASST RES SCIENTIST & ASST PROF, DEPT BIOL SCI, PURDUE UNIV, 77- *Concurrent Pos:* Co-prin investr, NIH res grants, 71- *Mem:* Am Soc Biol Chemists; Am Inst Biol Sci; Sigma Xi. *Res:* Nucleic acids; organic chemistry. *Mailing Add:* LORRE Purdue Univ West Lafayette IN 47907

HO, PATIENCE CHING-RU, b Peiping, China; US citizen. PHYSICAL CHEMISTRY, ANALYTICAL CHEMISTRY. *Educ:* Nat Taiwan Norm Univ, BS, 63; Univ Calif, MS, 66; Univ Bonn, WGermany, PhD(org chem), 71. *Prof Exp:* Chemist, Hogan Faximile Corp, 66-67; res asst, Upjohn Co, 68; assoc org synthesis, Stanford Univ, 72; assoc med res, State Univ NY Buffalo, 72-74; MEM RES STAFF NUCLEAR WASTE CHEM, OAK RIDGE NAT LAB, 76- *Mem:* Am Chem Soc. *Res:* Solution chemistry. *Mailing Add:* Oak Ridge Nat Lab PO Box 2008 Oak Ridge TN 37831-6201

HO, PAUL SIU-CHUNG, b Canton, China, Oct 6, 36; m 67; c 2. MATERIALS SCIENCE ENGINEERING. *Educ:* Nat Chengkung Univ, Taiwan, BS, 57; Nat Tsinghue Univ, Taiwan, MS, 59; Rensselaer Polytech, NY, PhD(physics), 65. *Prof Exp:* From asst prof to assoc prof mat sci, Cornell Univ, 67-72; res staff mem, 72-75, SR MGR, IBM RES CTR, 85- *Mem:* Fel Am Phys Soc; Am Vacuum Soc; Mat Res Soc; Am Soc Metals. *Res:* Thin films and interfaces in multilayeral structures for ULSI applications; published articles in numerous journals. *Mailing Add:* IBM T J Watson Res Ctr Yorktown Heights NY 10598

HO, PETER PECK KOH, b Fukien, China, Sept 12, 37; US citizen; m 62; c 2. BIOCHEMISTRY, IMMUNOLOGY. *Educ:* Chung Chi Col, Hong Kong, BS, 58; Univ Wash, PhD(biochem), 63. *Hon Degrees:* JD, Ind Univ, 84. *Prof Exp:* Res assoc biochem, Univ Calif, Berkeley, 63-64; sr biochemist virol, 64-69, from res scientist to sr res scientist, 69-82, RES ADV, RES LABS, ELI LILLY & CO, 82- *Mem:* Am Chem Soc; Am Soc Biol Chemists; NY Acad Sci; Sigma Xi. *Res:* Virus-induced enzyme systems; biochemical studies on virus replication; mechanism of antibiotics; metabolic inhibitors; prostaglandin and leukotriene biosynthesis and metabolism; asthma; cellular immunology, arthritis and inflammation. *Mailing Add:* 1228 Ironwood Drive W Carmel IN 46032

HO, PING-PEI, b Kao-Hsiung, Taiwan, May 19, 49; m 73, 88; c 3. PHYSICS, LASERS. *Educ:* Nat Tsing-Hua Univ, Taiwan, BS, 70; City Col New York, MA, 74, PhD(physics), 78; Kent State Univ, Ohio, MBA, 83. *Prof Exp:* Sr res assoc physics, City Col New York, 78-79; sr engr, NCR Corp, 79-83; PROF, DEPT ELEC ENG, CITY COL NY, 83- *Mem:* Optical Soc Am. *Res:* Ultrafast phenomena in solid state physics; chemical reaction of biological systems; nonlinear optical research in liquids and solids; state of arts laser research and development. *Mailing Add:* Dept Elec Eng City Col NY New York NY 10031

HO, REN-JYE, ADENLYLATE CYCLASE, ADIPOSE TISSUES. *Educ:* Vanderbilt Univ, PhD(physiol), 63. *Prof Exp:* PROF BIOCHEM, SCH MED, UNIV MIAMI, 73- *Mailing Add:* Dept Biochem Sch Med Univ Miami Sch Med PO Box 520875 Miami FL 33152

HO, S(HUI), b Macao, Sept 16, 31; US citizen; m 59; c 1. OPERATIONS RESEARCH. *Educ:* Univ Adelaide, BE, 55; Mass Inst Technol, SM, 57, ScD(struct eng), 63. *Prof Exp:* Res asst struct dynamics & comput appln, Mass Inst Technol, 55-57 & 60-62; designer, Jackson & Moreland Inc, Engrs, Mass, 57-58, jr engr, 58-60; mem tech staff, Mitre Corp, 62-66; mem prof staff, Ctr Naval Anal, 66-83, opers res analyst, Naval Res Lab, 83-84 & US Army Concept Anal Agency 84-88; SCI ADV, US NCCS SUPPORT STAFF, 88- *Concurrent Pos:* Adj asst prof, Howard Univ, 67-68, adj prof, 68-69; fel opers res, Stanford Univ, 71-72. *Mem:* Opers Res Soc Am. *Res:* Command and control; nuclear weapons effects; operations research; computer applications. *Mailing Add:* 9114 Ashmeade Dr Fairfax VA 22032

HO, SHIH MING, solid state chemistry, for more information see previous edition

HO, THOMAS INN MIN, b Honolulu, Hawaii, Oct 17, 48; div; c 1. TELECOMMUNICATIONS, INFORMATION NETWORKING. *Educ:* Purdue Univ, BS, 70, MS, 71, PhD(computer sci), 74. *Prof Exp:* From asst prof to assoc prof computer sci, Purdue Univ, 75-78, prof info systs, 84-90; DIR, INFO NETWORKING INST, CARNEGIE MELLON UNIV, 90- *Concurrent Pos:* Bd regents, Data Processing Mgt Asn Educ Found, 85-89; exec dir, Intelenet Comn, 86-88. *Mem:* Asn Comput Mach; Inst Elec & Electronics Engrs. *Res:* Relationship between telecommunications and economic development with particular emphasis on educational applications and the infrastructure necessary to support educational applications such as distance education. *Mailing Add:* Info Networking Inst Carnegie Mellon Univ 5000 Forbes Ave Pittsburgh PA 15213-3890

HO, THOMAS TONG-YUN, b Taiwan, China, July 2, 31; m 63; c 2. ORGANIC GEOCHEMISTRY. *Educ:* Nat Taiwan Univ, BS, 54; Univ Kans, MA, 61, PhD(paleont), 64. *Prof Exp:* Teaching asst zool, Nat Taiwan Univ, 56-58; res assoc org geochem, Univ Ariz, 64-67; sr geologist, Exxon Prod Res Co, 67-75; group leader, Continental Oil Co, 75-80, sr res assoc, 80-81, DUPONT FEL, CONOCO, INC, 81- *Concurrent Pos:* Vis scientist, Univ Calif, Los Angeles, 67; UN consult, 80. *Mem:* AAAS; fel Geol Soc Am; Am Geophys Union; Geochem Soc; Am Asn Petrol Geologists. *Res:* Fishery biology; paleontology; organic geochemistry; Quaternary, marine and petroleum geology; computer application in geology. *Mailing Add:* 2409 Hummingbird St Ponca City OK 74604

HO, YEW KAM, b Macau, June 5, 46; Can citizen; m 72; c 1. ATOMIC SCATTERING, POSITRON PHYSICS. *Educ:* Sir George Williams Univ, Can, BSc, 69; Univ Western Ont, MSc, 71, PhD(appl math), 75. *Prof Exp:* Res assoc, NASA-Goddard Space Flight Ctr, 75-77; asst prof, physics, Wayne State Univ, 77-80; res asst prof, 84-87, RES ASSOC PROF, LA STATE UNIV, 87- *Mem:* Am Phys Soc. *Res:* Theoretical studies of atomic resonance phenomena; interactions between positrons and positroniums with atoms; electron-ion interaction phenomena related to astrophysics and laboratory plasmas. *Mailing Add:* Dept Physics & Astron LA State Univ Baton Rouge LA 70803-4001

HO, YU-CHI, b China, Mar 1, 34; US citizen; m 59; c 3. APPLIED MATHEMATICS, SYSTEMS ENGINEERING. *Educ:* Mass Inst Technol, SB, 53, SM, 55; Harvard Univ, PhD(appl math), 61. *Prof Exp:* Sr engr, Res Lab Div, Bendix Corp, 55-58; asst prof appl physics, 61-65, assoc prof eng & appl math, 65-69, GORDON MCKAY PROF, ENG & APPL MATH, HARVARD UNIV, 69- *Concurrent Pos:* Mem, Army Sci Adv Panel, 68-; Guggenheim fel, 70. *Honors & Awards:* Field Award, Control Eng, Inst Elec & Electronics Engrs. *Mem:* fel Inst Elec & Electronics Engrs; Nat Acad Eng. *Res:* Control and information science; manufacturing automation. *Mailing Add:* Pierce Hall Harvard Univ Cambridge MA 02138

HO, YUK LIN, microbiology, for more information see previous edition

HOA, SUONG VAN, b Longan, Viet Nam, Mar 20, 48; m 83; c 3. STRESS ANALYSIS, ACOUSTIC EMISSION. *Educ:* Calif State Univ, San Luis Abispo, BEng, 71; Univ Toronto, MS, 73 & PhD(mech eng), 76. *Prof Exp:* Design Engineer, Canadian Fram Ltd, 76-77, asst prof mech eng, 77-81, PROF MECH ENG, CONCORDIA UNIV, 81- *Concurrent Pos:* Consult, CPF Dualan Ltd, 79-, SPAR Aerospace, 81-87, comt mem, Am Soc Mech Engr Boiler & Pressure Vessel Code, 83- *Honors & Awards:* Ralph R Teetor Award, Soc Automotive Engr, 80. *Mem:* Am Soc Mech Eng; Am Soc Mat; Am Soc for Testing & Mat; Soc Exp Mech. *Res:* Composite structures and materials; graphite, glass, keular fibers; epoxy, polyester and thermoplastic matrix materials. *Mailing Add:* 1455 de Maisonneuve W #929 Montreal PQ H3G 1M8 Can

HOADLEY, ALFRED WARNER, b Roxbury, Mass, Dec 26, 34; m 59, 89; c 1. ENVIRONMENTAL ENGINEERING. *Educ:* Harvard Univ, AB, 57, SB, 58; Univ Wis, MS, 63, PhD(bact, civil eng), 67; Johns Hopkins Univ, MPH, 78. *Prof Exp:* Sanit engr, USPHS, 57-66; aquatic biologist, Nat Water Qual Lab, US Dept Interior, 66-68; assoc prof environ eng & biol, Sch Civil Eng & Sch Biol, Ga Inst Technol, 68-77, prof, 77-78; sanitary engr, WHO, Dacca, Bangladesh, 79-81; pub health engr, Acad Educ Develop, Rural Water Supply Bd, Mbabane, Swaziland, 81-86, pub health engr, 86-89; consult, Burlington, Vt, 90-91; CHIEF ENVIRON HEALTH, UN RELIEF & WORKS AGENCY FOR PALESTINE REFUGEES IN THE NEAR EAST, AMMAN, JORDAN, 91- *Concurrent Pos:* Assoc prof environ sci & eng, Univ Fla, 72-73. *Mem:* Sigma Xi; Am Pub Health Asn. *Res:* Aquatic microbiology; microbial ecology; effects of pollution on the biology of natural surface waters; ecology of waste treatment; behavior of pathogenic bacteria and viruses in natural surface waters. *Mailing Add:* WHO Rep PO Box 811547 Amman Jordan

HOADLEY, GEORGE B(URNHAM), b Swarthmore, Pa, June 24, 09; m 34; c 3. ELECTRICAL ENGINEERING. *Educ:* Swarthmore Col, BSc, 30; Mass Inst Technol, MSc, 32, DSc, 37. *Prof Exp:* Asst elec eng, Mass Inst Technol, 30-31, instr, 31-40; from asst prof to assoc prof, Polytech Inst Brooklyn, 40-48; prof, 48-74, head dept, 54-74, EMER PROF ELEC ENG, NC STATE UNIV, 74- *Concurrent Pos:* Ed, Transactions Instrumentation & Measurement, Inst Elec & Electronics Engrs, 55-78. *Honors & Awards:* Inst Elec & Electronics Engrs Centennial Award. *Mem:* Am Soc Eng Educ; fel Inst Elec & Electronics Engrs. *Res:* Electric networks; electrical measurements and control devices. *Mailing Add:* 3213 Leonard Raleigh NC 27607

HOADLEY, ROBERT BRUCE, b Waterbury, Conn, July 24, 33; m 57; c 2. WOOD SCIENCE & TECHNOLOGY. *Educ:* Univ Conn, BS, 55; Yale Univ, MF, 57, DFor, 62. *Prof Exp:* Lab supvr & res technician wood tech, Yale Univ, 57-62; asst prof, 62-71, ASSOC PROF WOOD SCI & TECH, UNIV MASS, AMHERST, 71- *Mem:* Forest Prod Res Soc; Soc Wood Sci & Technol; Sigma Xi. *Res:* Wood machining; veneer cutting; fundamental properties of wood. *Mailing Add:* Holdsworth Nat Res Ctr Univ Mass Amherst MA 01002

HOAG, ARTHUR ALLEN, b Ann Arbor, Mich, Jan 28, 21; m 49; c 2. ASTRONOMY. *Educ:* Brown Univ, AB, 42; Harvard Univ, PhD(astron) 51. *Prof Exp:* Physicist, US Naval Ord Lab, 42-44; astronomer, US Naval Observ, 50-65; astronomer, Kitt Peak Nat Observ, 65-77; dir, Lowell Observ, 77-86; RETIRED. *Mem:* AAAS; Am Astron Soc; Int Astron Union; Fel Royal Astron Soc. *Res:* Photoelectric and photographic photometry; astronomical sites and instruments; quasi-stellar sources. *Mailing Add:* 4410 E 14th St Tucson AZ 85711

HOAG, DAVID GARRATT, b Boston, Mass, Oct 11, 25; m 52; c 5. SYSTEM ENGINEERING. *Educ:* Mass Inst Technol, BS, 46, MS, 50. *Prof Exp:* Tech dir polaris missile guid, Mass Inst Technol, 58-61, prog mgr, Apollo guid & navigation, 61-73; dept head, adv systs, C S Draper Lab, Inc, 73-85, sr tech adv, 85-89; CONSULT, 89- *Honors & Awards:* Thurlow Award, Inst Navigation, 70; Lewis W Hill Space Transp Award, Am Inst Aeronaut & Astronaut, 72. *Mem:* Nat Acad Eng; Int Acad Astronaut; Inst Navigation (pres, 78-79); fel Am Inst Aeronaut & Astronaut. *Res:* System engineering for aerospace vehicles specializing on guidance, navigation, control, and precision pointing with emphasis on inertial and electro-optical sensing; real time computing; trajectory analysis and accuracy. *Mailing Add:* 116 Winthrop St Medway MA 02053

HOAG, ROLAND BOYDEN, JR, b Boston, Mass, Sept 3, 45; m 68; c 2. EXPLORATION GEOCHEMISTRY. *Educ:* Univ NH, BS, 67; McGill Univ, PhD(geol), 75. *Prof Exp:* Geologist stream sediment geochem, NAm Explor, 67; consult geologist, Scandia Mining & Explor Co & Labrador Mining & Explor Co, 71; res assoc, McGill Univ, 76; consult geologist, 76-77; VPRES, BCI GEONETICS, INC, 77- *Mem:* Sigma Xi; Geol Soc Am; Soc Explor Geochemists; Soc Econ Geologists; Newcomen Soc. *Res:* Innovation techniques applying remote sensing, geophysics, geochemistry, and hydrogeology for the exploration and evaluation of municipal groundwater supplies in fractured crystalline rock. *Mailing Add:* BCI Geonetics Inc Water Resources Div PO Box 529 Laconia NH 03247

HOAGBERG, RUDOLPH KARL, b Ironwood, Mich, Mar 11, 30; m 53; c 5. GEOLOGY. *Educ:* Mich State Univ, BS, 52, MS, 57. *Prof Exp:* Geophysicist, Humble Oil Div, Standard Oil Co, NJ, Ill & Mont, 52-56; mining geologist, Properties & Indust Develop Dept, Northern Pac Rwy Co, Minn, 57-62; instr phys geol, Sch Earth Sci, Univ Minn, 62-63; asst to dir indust minerals & environ geol, Minn Geol Surv, 63-69; managing assoc, Lindgren Explor Co, 69-70; geologist, Minn Geol Surv, 70-76; sr geologist, E K Lehmann & Assocs, Inc, 76-82; PRES, R K HOAGBERG ASSOCS, INC, 82- *Mem:* AAAS; Geol Soc Am; Asn Prof Scientists; Am Inst Prof Geologists. *Res:* Glacial geology studies of industrial minerals; environmental geology and hydrogeology of glaciated and Karst terranes. *Mailing Add:* R K Hoagberg Assoc Inc Suite 7N90 430 First Ave N Minneapolis MN 55401

HOAGE, TERRELL RUDOLPH, b Fairfield, Iowa, Aug, 12, 34; m 57; c 2. CELL BIOLOGY, REPRODUCTIVE ULTRASTRUCTURE. *Educ:* Parsons Col, BS, 57; Iowa State Univ, MS, 63, PhD(entom), 64. *Prof Exp:* Res entomologist, Bee Mgt Invests, USDA, 64-65; asst prof zool, Parsons Col, 65-68; assoc prof, 68-75, Path Serv Prog, Nat Ctr Toxicol Res, 81-91, PROF BIOL, SAM HOUSTON STATE UNIV, 75- *Concurrent Pos:* Dir, BAPC, Inc & Div Microeng Cellular Graphics; postdoctoral fel, Univ Iowa, 67. *Res:* Study of insect gonad tissue culture and ultrastructure; spermatogenesis and spermstorage mechanisms; metabolic DNA in post-mitotic cells; molecular genetics; genetics of cloned plants; rege nerative tissue toxicology; lyme disease. *Mailing Add:* Dept Biol Sam Houston State Univ Huntsville TX 77341

HOAGLAND, A(LBERT) S(MILEY), b Berkeley, Calif, Sept 13, 26; m 50; c 3. ELECTRICAL ENGINEERING. *Educ:* Univ Calif, BS, 47, MS, 48, PhD(elec eng), 54. *Prof Exp:* Res engr elec eng, Univ Calif, 47-52, from instr to asst prof, 52-56; mgr, Magnetic Record Technol Res Lab, IBM Corp, Calif, 56-59, mgr eng sci, 59-61, systs mgr, Adv Systs Div, 61-62, sr tech consult, World Trade Corp, 62-64, mgr eng sci, Res Ctr Yorktown Heights, 64-68, dir tech planning, 68-71, corp prog coordr, Boulder, 71-76, mgr explor magnetic recording, 76-80; tech adv pres, GPD, 80-84, DIR, INST INFO STORAGE TECHNOL, SANTA CLARA UNIV, 84- *Concurrent Pos:* Consult, IBM Corp, 54-56 & State Calif, 56; dir bd, Inst Elec & Electronics Engrs, 73-75; adj prof, Harvey Mudd Col, 78-; trustee, Charles Babbage Found, 79- *Honors & Awards:* Centennial Medal, Inst Elec & Electronics Engrs, 84. *Mem:* Fel Inst Elec & Electronics Engrs(pres comput soc, 71-73); Am Fedn Info Processing Soc (vpres, 75-78, pres, 78-80); Sigma Xi. *Res:* Digital computers, especially technologies exploiting magnetic recording for storage; magnetic recording theory; memory and storage for data processing systems; technology and design. *Mailing Add:* Inst Info Storage Technol EECS Santa Clara Univ Santa Clara CA 95053

HOAGLAND, ALAN D, b Winnetka, Ill, Mar 25, 11; m 39; c 3. GEOLOGY. *Educ:* Northwestern Univ, BSc, 33, MA, 35. *Prof Exp:* Chief geologist, NJ Zinc Co, 41-70, mgr eastern explor, 70-76; CONSULT GEOLOGIST, 76- *Mem:* Geol Soc Am; Soc Econ Geol. *Res:* Metal and industrial mineral exploration. *Mailing Add:* 3749 Amesbury Lane Sarasota FL 34232

HOAGLAND, GORDON WOOD, b Nampa, Idaho, Oct 22, 36; m 62; c 3. NUMERICAL ANALYSIS, PARTIAL DIFFERENTIAL EQUATIONS. *Educ:* Brigham Young Univ, BS, 66, MS, 68. *Prof Exp:* Res consult, Ore State Univ, 68-69; chmn dept, 79-84, PROF MATH & COMPUT SCI, RICKS COL, 69- *Mem:* Soc Indust & Appl Math; Am Math Soc. *Res:* Gamma ray spectroscopy; numerical forestry analysis; applied mathematics. *Mailing Add:* Dept Math Ricks Col Rexburg ID 83460-0515

HOAGLAND, JACK CHARLES, communications, systems; deceased, see previous edition for last biography

HOAGLAND, K ELAINE, b Los Angeles, Calif, Dec 10, 47; m 75. ECOLOGY, EVOLUTION. *Educ:* Pomona Col, AB, 70; Harvard Univ, AM, 73, PhD(biol), 75. *Prof Exp:* Asst prof biol, 75-82, sr res scientist, ctr marine & environ studies, Lehigh Univ, 83-87; EXEC DIR, ASN SYSTS COLLECTIONS, 87- *Concurrent Pos:* Res assoc, Acad Natural Sci Philadelphia, 75- *Mem:* Ecol Soc Am; AAAS; Am Malacol Union; Soc Conserv Biol. *Res:* Population dynamics and population genetics of marine invertebrates; ecology of colonizing species; molluscan systematics; science policy. *Mailing Add:* Asn Systs Collections 730 11th St NW Washington DC 20001

HOAGLAND, LAWRENCE CLAY, b Marion, Ohio, Nov 15, 31; m 51; c 4. MECHANICAL ENGINEERING. *Educ:* Gen Motors Inst, BMech Eng, 54; Mass Inst Technol, MS, 55, ME, 56, ScD(mech eng), 60. *Prof Exp:* Jr engr, Small Gas Turbine Dept, Gen Motors Corp, 53-54; from instr to asst prof heat transfer & thermodyn, 56-59; vpres eng res, Dynatech Corp, 60-68, vpres & dir res, Steam Engine Systs Corp, Newton, 68-75; dir develop, CTI-Cryogenics Div, Helix Tech Corp, 75-77; dir res & develop, Amtech Inc, Newton, Mass, 77-80; VPRES & DIR RES, AIRXCHANGE INC, HINGHAM, MASS, 80- *Concurrent Pos:* Consult heat transfer, 85- *Mem:* Am Soc Mech Engrs; Soc Automotive Engrs; Am Soc Heating, Refrig & Air-Conditioning Engrs. *Res:* Development of heat engines, heat exchanger applications and heat transfer and flow processes of interest in heat exchange equipment; heat transfer and flow in non circular ducts. *Mailing Add:* Star Rte 62 Box 412A Center Harbor NH 03226

HOAGLAND, MAHLON BUSH, b Boston, Mass, Oct 5, 21; m 43, 61; c 3. MOLECULAR BIOLOGY, RESEARCH ADMINISTRATION. *Educ:* Harvard Med Sch, MD, 48. *Hon Degrees:* ScD, Worcester Polytech Inst, 73; ScD, Univ Mass, 84. *Prof Exp:* From asst to asst prof med, Huntington Lab, Mass Gen Hosp, Harvard Med Sch, 48-60, assoc prof bact & immunol, 60-67, tutor, 53-60, exec secy comt res, Hosp 54-57; prof biochem & chmn dept, Dartmouth Med Sch, 67-70; founder & spokesman, Deleg Basic Biomed Res, 78-87; pres & sci dir, 70-85, EMER PRES, WORCESTER FOUND EXP BIOL, 85- *Concurrent Pos:* Res fel med, Huntington Lab, Mass Gen Hosp, 48-51; fel, Am Cancer Soc, Carlsberg Lab, Copenhagen, 51-52; fel, Biochem Res Lab, Mass Gen Hosp, 52-53; scholar cancer res, Am Cancer Soc, 53-58; mem med res coun unit molecular biol, Cavendish Lab, Cambridge Univ, 57-58; mem biochem study sect, NIH, 61-64; mem comt etiol cancer, Am Cancer Soc, 65-68, chmn med res comt, Mass Div, 74-76; res prof, Med Sch Univ Mass, 70-85; mem bd sci counrs, Nat Heart & Lung Inst, 72-74; mem, Corp Woods Hole Oceanog Inst, 74-85, trustee, 82-87; vis scholar, Biochem Prog & Molecular Genetics Ctr, Dartmouth Med Sch, 86-; trustee, Montshire Mus Sci, 90- *Honors & Awards:* Franklin Medal, 76. *Mem:* Nat Acad Sci; Am Acad Arts & Sci. *Res:* Carcinogenic and biological effects of beryllium; biosynthesis of co-enzyme A; biosynthesis of protein; biochemistry; liver regeneration. *Mailing Add:* Box 153 Thetford VT 05074

HOAGLAND, ROBERT EDWARD, b Cleveland, Ohio, Feb 3, 42. BIOCHEMISTRY, PLANT PHYSIOLOGY. *Educ:* Youngstown State Univ, BS, 64; Univ Cincinnati, MS, 68; NDak State Univ, PhD(biochem), 70. *Prof Exp:* Res assoc chem, NDak State Univ, 67-69, res chemist herbicide metab, 70-72; res chemist, Metab & Radiation Res Lab, Fargo, NDak, 72, RES BIOCHEMIST HERBICIDE ACTION, SOUTHERN WEED SCI LAB, SCI & EDUC ADMIN-AGR RES, USDA, STONEVILLE, MISS, 72- *Mem:* Sigma Xi; Am Chem Soc; Am Soc Plant Physiologists; Weed Sci Soc Am; Bot Soc Am. *Res:* Plant biochemistry; physiology and enzymology; aspects of phenolic metabolism; mode and mechanisms of herbicide action. *Mailing Add:* Southern Weed Sci Lab Stoneville MS 38776

HOAGLAND, VINCENT DEFOREST, JR, b Waltham, Mass, June 10, 40; m 70. BIOCHEMISTRY. *Educ:* Wesleyan Univ, BA, 62; Fla State Univ, PhD(chem), 67. *Prof Exp:* NIH res fel biochem, Univ Wash, 67-69; asst prof, 69-72, assoc prof, 73-78, chmn dept, 74-77, PROF CHEM, SONOMA STATE UNIV, 78- *Mem:* Am Chem Soc; AAAS; Sigma Xi. *Res:* Enzymology; quaternary structure of enzymes; enzyme kinetics. *Mailing Add:* Dept Chem Sonoma State Univ Rohnert Park CA 94928

HOAGLIN, DAVID CASTER, b Charleston, WVa, Mar 4, 44; m 68; c 1. DATA ANALYSIS, ROBUSTNESS. *Educ:* Duke Univ, BS, 66; Princeton Univ, PhD(statist), 71. *Prof Exp:* From instr to asst prof statist, 70-72, lectr statist, 72-77, RES ASSOC STATIST, HARVARD UNIV, 77- *Concurrent Pos:* Sr analyst, Abt Assocs Inc, Cambridge, Mass, 77-78, sr scientist, 78-; biostatistician, Harvard Anaesthesia Ctr, Mass Gen Hosp, 83-88. *Mem:* Fel AAAS; Soc Indust & Appl Math; fel Am Statist Asn; Inst Math Statist; Int Statist Inst; Asn Comput Mach. *Res:* Exploratory data analysis; applied robust methods; statistical computation; applications of statistics to policy problems. *Mailing Add:* Dept Statist Harvard Univ One Oxford St Cambridge MA 02138

HOAGSTROM, CARL WILLIAM, b Holdrege, Nebr, Aug 24, 40; m 65; c 3. BIOLOGY. *Educ:* Kearney State Col, Nebr, BS, 66; Purdue Univ, MS, 68; Univ Ariz, PhD(zool), 78. *Prof Exp:* Instr sci, Arcadia Nebr Sch, 68-69; instr biol & chem, Grand Island Nebr Sch, 69-71; PROF, OHIO NORTHERN UNIV, ADA, 75- *Mem:* AAAS; Am Inst Biol Sci; Ecol Soc Am; Am Soc Mammalogists. *Res:* Monitoring population fluctuations in small mammals. *Mailing Add:* Dept Biol Sci Ohio Northern Univ Ada OH 45810

HOAK, JOHN C, BLOOD DISEASE. *Prof Exp:* DIR DIV BLOOD DIS & RESOURCES, NAT HEART LUNG & BLOOD INST, NIH, 89- *Mailing Add:* Div Blood Dis & Resources Nat Heart Lung & Blood Inst NIH Fed Bldg Rm 516 Bethesda MD 20892

HOAR, RICHARD MORGAN, b Boston, Mass, Nov 22, 27; m 49; c 1. DEVELOPMENTAL ANATOMY, REPRODUCTIVE BIOLOGY. *Educ:* Dartmouth Col, AB, 50; Univ Kans, PhD(anat), 56. *Prof Exp:* Instr anat, Col Med, Univ Cincinnati, 56-58, from asst prof to assoc prof, 58-69; head teratology, Res Div, Hoffmann-La Roche Inc, 69-77, from asst dir to assoc dir, dept toxicol & path, 77-85; Findley Res, Inc, Fall River, Mass, 85-86; CONSULT DEVELOP TOXICOL, ARGUS INT INC, HORSHAM, PA, 86- *Concurrent Pos:* Consult, Nat Inst Environ Health Sci, NIH, 70-72 & 76-78 & Environ prof agency, 80-82; adj assoc prof anat, Col Physicians & Surgeons, Columbia Univ, 73-85; lectr anat, Mt Sinai Sch Med, 76-85. *Mem:* AAAS; Am Asn Anat; Teratology Soc (treas, 74-78, pres, 80-81); Soc Study Reproduction; Am Asn Lab Animal Sci; Sigma Xi; Am Col Toxicol (pres, 88-89). *Res:* Endocrinology of reproduction, teratology, and biochemistry of development in guinea pigs; reproduction and teratology in ferrets. *Mailing Add:* Box 9355 North Dartmouth MA 02747

HOAR, WILLIAM STEWART, b Moncton, NB, Aug 31, 13; m 41; c 4. COMPARATIVE PHYSIOLOGY. *Educ:* Univ NB, BA, 34; Univ Western Ont, MA, 36; Boston Univ, PhD(med sci), 39, FRSC, 55. *Hon Degrees:* DSc, Univ NB, 65, Memorial Univ, 67; St Francis Xavier Univ, 76 & Univ Western Ont, 78; LLD, Simon Fraser, 80, Toronto, 81. *Prof Exp:* Demonstr zool, Univ Western Ont, 34-36; asst histol & embryol, Med Sch, Boston Univ, 36-39; asst prof biol, Univ NB, 39-42, prof zool, 42-45; prof zool & fisheries, Univ BC, 45-64, head dept, 64-71, prof, 64-79, EMER PROF ZOOL, UNIV BC, 79- *Concurrent Pos:* Guggenheim fel, Oxford Univ, 58-59. *Honors & Awards:* Flavelle Medal, Royal Soc Can, 65; Officer, Order of Can, 75; Fry Medal, Can Soc Zool, 74. *Mem:* Can Soc Zoologists; Am Soc Zoologists; Sigma Xi. *Res:* Physiology; aspects of histology, embryology behavior and physiology of salmon; fish endocrinology especially pituitary and reproductive hormones. *Mailing Add:* Dept Zool Univ BC Vancouver BC V6T 2A9 Can

HOARD, DONALD ELLSWORTH, b Seattle, Wash, Apr 14, 28; m 54; c 4. BIOCHEMISTRY. *Educ:* Univ Wash, BS, 52; Univ Calif, PhD(biochem), 57. *Prof Exp:* Res chemist, Pabst Labs, Pabst Brewing Co, 60-63; mem staff, 63-82, biomed res group, 63-82, data anal, Los Alamos Sci Lab, Univ Calif, 82-86; AMPARO CORP, 86- *Concurrent Pos:* Res fel biochem, Calif Inst Technol, 57-60. *Mem:* AAAS; Am Chem Soc; Sigma Xi. *Res:* Organic and radiation chemistry of nucleotides; oligonucleotides and nucleic acids. *Mailing Add:* 11 Los Arboles Dr Los Alamos NM 87544

HOARD, JAMES LYNN, b Elk City, Okla, Dec 28, 05; m 35; c 3. STRUCTURAL CHEMISTRY, CRYSTALLOGRAPHY. *Educ:* Univ Wash, Seattle, BS, 27, MS, 29; Calif Inst Technol, PhD(chem), 32. *Prof Exp:* Anal chemist, Falkenburg & Co, Wash, 27-28; asst chem, Calif Inst Technol, 29-32; instr, Stanford Univ, 32-35 & Ohio State Univ, 35-36; from instr to prof, 36-42, EMER PROF CHEM, CORNELL UNIV, 71- *Concurrent Pos:* Guggenheim fel, 46, 60 & 66; civilian with Manhattan proj subcontract, Cornell Univ, 43-44 & Off Sci Res & Develop, 42-45; prin investr grants, NSF, Cornell Univ, 56-73 & NIH, 62-81. *Honors & Awards:* Award Distinguished Serv in Advan of Inorg Chem, Am Chem Soc. *Mem:* Nat Acad Sci; fel AAAS; Am Chem Soc; Am Phys Soc; Am Crystallog Asn. *Res:* Structure and properties of complex inorganic substances and of tetrapyrrole systems, especially the metalloporphyrins. *Mailing Add:* 42 Cornell St Ithaca NY 14850

HOARE, JAMES PATRICK, b Denver, Colo, Jan 9, 21; m 53; c 3. ELECTROCHEMISTRY. *Educ:* Regis Col, BS, 43; Cath Univ, MS, 48, PhD(phys chem), 49. *Prof Exp:* Asst chem, Cath Univ, 46-49; from instr to asst prof physics & chem, Trinity Col, DC, 49-54; phys chemist, Naval Res Lab, Washington, DC, 54-57; prin res engr, Sci Lab, Ford Motor Co, 57-60; sr res chemist, Gen Motors Corp, 60-70, mem sr tech staff electrochem dept, Res Labs Div, 70-77, dept res scientist, 77-79, res fel, 79-87; CONSULT, 87- *Concurrent Pos:* Bonfils scholar, 39-43; Mullen fel, 43-44; electronic technician, US Navy, 44-46; grant, Res Corp, 52-54. *Honors & Awards:* John Campbell Award, Gen Motors Corp, 80, 86; Silver Medal Award, Am Electroplaters Soc, 81 & 83, Gold Medal Award, 85, Sci Achievement Award, 89; Res Award, Electrochem Soc, 87. *Mem:* Am Chem Soc; Electrochem Soc; Int Soc Electrochem; NY Acad Sci; Am Electroplaters Soc; Catalyst Soc. *Res:* Chemical kinetics; electrochemistry; mechanisms of electrode reactions; catalysts; surface chemistry; fuel cells; electrochemical machining; mechanisms of chromium electrodeposition. *Mailing Add:* 24860 Davenport Novi MI 48374

HOARE, RICHARD DAVID, b Rosiclare, Ill, May 14, 27; m 47, 71; c 4. INVERTEBRATE PALEONTOLOGY. *Educ:* Augustana Col, AB, 51; Univ Mo, MA, 53, PhD, 57. *Prof Exp:* Instr geol, Univ Mo, 53-57; chmn dept, 67-70, 71-75 & 80-84, assoc vpres acad affairs, 84-88, PROF GEOL, BOWLING GREEN STATE UNIV, 57- *Concurrent Pos:* Geologist, Cartier Mining Co, Can, 56; asst prof, Univ Mo, 57-58; managing ed, J Paleontology, 85- *Mem:* Paleont Soc; Brit Palaeont Asn. *Res:* Upper Paleozoic invertebrate paleontology; Pennsylvanian gastropods and polyplacophora. *Mailing Add:* Dept Geol Bowling Green State Univ Bowling Green OH 43403

HOBACK, JOHN HOLLAND, b Huntington, WVa, Sept 6, 20; m 45; c 2. PHYSICAL CHEMISTRY. *Educ:* Marshall Co, AB, 41, MS, 42; WVa Univ, PhD(phys chem), 47. *Prof Exp:* From instr to assoc prof, 45-64, PROF CHEM, MARSHALL UNIV, 64- *Concurrent Pos:* Consult, Vet Admin, 57-64; vpres res & develop, Continental Tobacco Co, 59-62, consult, 62-63; pre med adv, Marshall Univ, 70- *Mem:* Fel Am Inst Chem; Am Chem Soc; Sigma Xi. *Res:* Relationships between physical properties and molecular structure; cellulose and salt solutions; paper electrophoresis and pulmonary function. *Mailing Add:* 2658 Third Ave Huntington WV 25702

HOBART, DAVID EDWARD, b Middletown, Ohio, Jan 5, 49; m 86; c 2. SPECTROELECTROCHEMISTRY, ACTINIDE SOLUTION AND SOLID STATE CHEMISTRY. *Educ:* Rollins Col, BA, 71; Univ Tenn, Knoxville, PhD(chem), 81. *Prof Exp:* Teaching asst chem, Univ Tenn, Knoxville, 75-77, res asst, 77-81, res assoc, 81-83; STAFF MEM, ISOTOPE NUCLEAR CHEM DIV, LOS ALAMOS NAT LAB, 83- *Concurrent Pos:* Reviewer, Nat Sci Found, Inorg Chem, Chem Reviews, Lanth Act Res, Radiochim Acta. *Mem:* Am Chem Soc; Sigma Xi; Soc Electroanal Chem. *Res:* Less stable oxidation states of lanthanide and actinide elements via spectroelectrochemistry in complexing aqueous media; preparation and analysis of solid state actinides and actinides in near-neutral aqueous media as pertains to nuclear waste isolation and the environment. *Mailing Add:* INC-11 Mail Stop G739 Los Alamos Nat Lab Los Alamos NM 87545

HOBART, EVERETT W, b Cincinnati, Ohio, Dec 15, 31; m 56; c 5. ANALYTICAL CHEMISTRY. *Educ:* Mass Inst Technol, BS, 53. *Prof Exp:* Jr chemist, Am Cyanamid Co, Conn, 53-55, chemist, 55-56; sr chemist, Pratt & Whitney Aircraft Div, United Aircraft Corp, Conn, 56-60, supvr anal chem, 60-65; asst res dir, Ledoux & Co, 65-71, lab dir, 71-76; tech dir, Spectrochem Labs, Inc, 76-80; LAB DIR, LEDOUX & CO, 80- *Mem:* Am Chem Soc; Am Soc Test & Mat. *Res:* Chemical and instrumental analysis. *Mailing Add:* Ledoux & Co 359 Alfred Ave Teaneck NJ 07666-5754

HOBART, PETER MERRILL, b Nov 13, 46; US citizen. MOLECULAR BIOLOGY. *Educ:* Colby Col, AB, 68; Wesleyan Univ, PhD(molecular biol), 78. *Prof Exp:* Fel molecular genetics, Univ Calif, San Francisco, 77-81; ADJ PROF, CONN COL, 83- *Mem:* Am Soc Biochem & Molecular Biol; NY Acad Sci. *Res:* Molecular mechanisms of cellular regulation; contribution of primary structure of macromolecules and deriving information from primary structure and levels of expression of the macromolecules which may reveal regulating mechanisms. *Mailing Add:* Cent Res Div Pfizer Inc Groton CT 06340

HOBART, ROBERT H, b New York, NY, Feb 2, 32. PHYSICS. *Educ:* Mass Inst Technol, BS, 54; Stanford Univ, MS, 55; Univ Ill, PhD, 61. *Prof Exp:* Asst prof physics, Dalhousie Univ, 61-63; sr physicist, Battelle Mem Inst, 63-70; lectr, Mid East Tech Univ, Ankara, 70-71; hons tutor, St Andrews Univ, 71-72; prof physics & head dept, Ft Hare Univ, 73-75 & Kenyatta Univ Col, 76-78; adj prof physics, Western Ky Univ, 78-79; prof physics & math, Mid Ga Col, 80-82; PHYSICIST, FOREIGN SCI & TECHNOL CTR, 83- *Mem:* Am Phys Soc; Am Asn Univ Prof. *Res:* Nonlinear modeling; electrodynamics; nonlinear field theory. *Mailing Add:* 701 E High St Apt 306 Charlottesville VA 22901

HOBBIE, JOHN EYRES, b Buffalo, NY, June 5, 35; m 59; c 3. HYDROBIOLOGY. *Educ:* Dartmouth Col, BA, 57; Univ Calif, Berkeley, MA, 59; Ind Univ, PhD(zool), 62. *Prof Exp:* Res assoc zool, Univ Calif, Davis, 62-63; NIH fel, Inst Limnol, Uppsala, 63-65; from asst prof zool to prof, NC State Univ, 65-75; SR SCIENTIST, MARINE BIOL LAB, 76-, CO-DIR, ECOSYSTS CTR, 85- *Concurrent Pos:* NSF sr fel, Norweg Inst Water Res, Oslo, 71-72; dir, Tundra Biome Aquatic Prog, US Int Biol Prog, 71-74; Tage Erlander professorship, Sweden, 88-89. *Honors & Awards:* Hutchinson Medal, Am Soc Limnol & Oceanog, 83. *Mem:* Ecol Soc Am; Am Soc Limnol & Oceanog (pres, 84-86); Am Soc Microbiol. *Res:* Arctic and antarctic limnology; heterotrophic bacteria in aquatic ecosystems; estuarine ecology; global carbon cycle. *Mailing Add:* Marine Biol Lab Woods Hole MA 02543

HOBBIE, RUSSELL KLYVER, b Albany, NY, Nov 3, 34; m 57; c 4. PHYSICS. *Educ:* Mass Inst Technol, BS, 56; Harvard Univ, AM, 57, PhD(physics), 60. *Prof Exp:* Res assoc, 60-62, from asst prof to assoc prof, 62-72, dir, Space Sci Ctr, 79-84, PROF PHYSICS, UNIV MINN, MINNEAPOLIS, 72-, ASSOC DEAN, 84- *Mem:* Am Phys Soc; Am Asn Physics Teachers; Am Asn Physicists Med; AAAS; Inst Elec & Electronics Engrs. *Res:* Radiological physics; biophysics; education administration. *Mailing Add:* Sch Physics Univ Minn Minneapolis MN 55455

HOBBINS, JOHN CLARK, b New York, NY, July 13, 36. MATERNAL-FETAL MEDICINE, DIAGNOSTIC ULTRASONOGRAPHY. *Educ:* Hamilton Col, MS, 58; NY Med Col, MD, 63; Yale Univ, MS, 80. *Prof Exp:* Instr, 70-71, asst prof, 74-76, assoc prof, 76-80, PROF OBSTET & GYNEC, YALE UNIV SCH MED, 80-, DIR OBSTET, NEW HAVEN MED CTR, 76- *Concurrent Pos:* Ed, Clin Diag Ultrasound & Ultrasound Med & Biol; examr, Maternal-Fetal Med Bds. *Mem:* Fel Am Col Obstet & Gynec; Soc Gynec Invest; Am Inst Ultrasound Med; Soc Perinatal Obstetricians (vpres). *Res:* Prenatal diagnosis of congenital anomalies by ultrasound and fetal treatment of some of these anomalies such as hydrocephalus and posterior urethral obstruction. *Mailing Add:* Diag Radiology Obstet & Gynec Yale Univ Sch Med New Haven CT 06510

HOBBS, ANN SNOW, b Washington, DC, Nov 20, 45. ION TRANSPORT, ENZYME KINETICS. *Educ:* Univ Md, BS, 68, PhD(biophysics), 73. *Prof Exp:* Res assoc, Dept Biol, Syracuse Univ, 73-75; fel, Lab Biophysics & Neurochem, NIH, 75-77, staff fel, 77-79, sr staff fel, Lab Neurochem, Neurol Commun Disorders & Strokes, 79-84; RES ASST PROF, DEPT PHYSIOL, SCH MED, UNIV MD, 84- *Concurrent Pos:* Fel Grass Found, Marine Biol Lab, 74; adj asst prof, Dept Physiol, Uniformed Serv Univ Health Sci, 78- *Mem:* Biophys Soc; AAAS. *Res:* Enzyme mechanisms and regulatory processes for active ion transport across cell membranes, particularly the transport of sodium and potassium. *Mailing Add:* Bldg 36 Rm 4D20 NIH NINDS Bethesda MD 20205

HOBBS, ANSON PARKER, b Greensburg, Ind, Mar 4, 15; m; c 3. PHYSICAL CHEMISTRY. *Educ:* Univ Ind, AB, 36. *Prof Exp:* Chemist, Dow Chem Co, 37-44, group leader anal chem, 44, asst dir spec serv lab, 44-69; res chemist, Helium Res Ctr, Tex, 69-70, res chemist, Pittsburgh Energy Res Ctr. Energy Res & Develop Admin, 70-77; RETIRED. *Concurrent Pos:* Rep, Natural Gas Processors Asn, 55-; mem, Gas Asn. *Mem:* AAAS; fel Am Inst Chem; Am Chem Soc; Alzheimers Found (vpres). *Res:* Gas analysis; catalyst research; physical properties and other reactions of gaseous materials. *Mailing Add:* 8555 S Lewis 24C Tulsa OK 74137

HOBBS, BENJAMIN F, b Summit, NJ, July 31, 54; m 80; c 2. ENVIRONMENTAL SYSTEMS ENGINEERING, ELECTRIC UTILITY SYSTEMS PLANNING. *Educ:* SDak State Univ, BS, 76; State Univ NY, Syracuse, MS, 78; Cornell Univ, PhD(environ systs eng), 83. *Prof Exp:* Econ assoc, Brookhaven Nat Lab, 77-79; Wigner fel, Oak Ridge Nat Lab, 82-84; asst prof, 84-88, ASSOC PROF SYSTS ENG, CASE WESTERN RESERVE UNIV, 88- *Concurrent Pos:* NSF presidential young investr, 86. *Mem:* Am Soc Civil Engrs; Opers Res Soc Am; Am Geophys Union; Air & Waste Mgt Asn; Am Water Resources Asn. *Res:* Energy and environmental engineering problems using systems analysis and economics; energy conservation planning; acid rain compliance planning; multiobjective decision analysis method development. *Mailing Add:* Dept Systs Eng Case Western Reserve Univ Cleveland OH 44106

HOBBS, BILLY FRANKEL, b Hiltons, Va, Mar 5, 31; m 53; c 2. MATHEMATICS. *Educ:* Ball State Univ, BS, 54, MA, 55; Purdue Univ, PhD, 67. *Prof Exp:* Teacher pub sch, Ind, 54-57; asst prof math, Olivet Nazarene Col, 57-66, chmn dept, 60-66; asst prof, Purdue Univ, 66-67; prof, Olivet Nazarene Col, 67-70, prof & chmn dept, Pasadena Col, 70; chmn dept, 70-80, PROF, POINT LOMA COL, 70-, CHMN, 87- *Concurrent Pos:* NSF sci fac fel, 62-63, grad fel, 64-65, grad trainee, 65-66; mem, deleg math educ to China, 83, Nat Coun Teacher Math. *Mem:* Am Math Soc; Math Asn Am. *Res:* Topological and commutative algebra; general topology; use of logo in problem solving. *Mailing Add:* Dept Math Point Loma Col 3900 Lomaland Dr San Diego CA 92106

HOBBS, CHARLES CLIFTON, JR, b Wewoka, Okla, Mar 23, 27; m 50; c 3. ORGANIC CHEMISTRY. *Educ:* Univ Okla, BS, 49, MS, 51, PhD(chem), 53. *Prof Exp:* From res chemist to sr res chemist, Celanese Corp Am, 52-58, from res assoc, Explor Res Dept to group leader, basic process res sect, Chem Div, 58-67, sr res assoc, Chem Res Dept, Celanese Chem Co, 67-85; CONSULT LIQUID PHASE OXIDATION, 85- *Honors & Awards:* Chem Pioneer Award, Am Inst Chemists, 74. *Mem:* Am Chem Soc; Am Inst Chemists. *Res:* Chromyl chloride oxidation of saturated hydrocarbons; derivatives of petroleum chemicals; organic electrochemistry; peroxygen chemicals; free radical reactions; liquid phase oxidation; mathematical modeling; vapor phase oxidation; organic syntheses. *Mailing Add:* 1437 Casa Verde Dr Corpus Christi TX 78411-3331

HOBBS, CHARLES FLOYD, b Lebo, Kans, Jan 17, 35; m 55; c 2. ORGANIC CHEMISTRY. *Educ:* Emporia Kans State Univ, AB, 56; Univ Kans, PhD(org chem), 60. *Prof Exp:* Res asst org chem, Univ Kans, 56-60; sr res chemist, 60-68, res specialist, 68-73, sr res specialist org chem, Corp Res Labs, 73-79, sr res specialist, Monsanto Chem Intermediates Co, 79-83, sr res specialist, 83-88, GROUP LEADER, MONSANTO AGR CHEM CO, 88- *Mem:* Am Chem Soc; fel Am Inst Chemists; Sigma Xi. *Res:* Homogeneous catalysis; organic nitrogen chemistry; organometallic chemistry; olefin metathesis; heterocyclic chemistry. *Mailing Add:* Monsanto Co 800 N Lindbergh Blvd St Louis MO 63167

HOBBS, CHARLES HENRY, b Longmont, Colo, Oct 4, 42; div; c 2. TOXICOLOGY & VETERINARY TOXICOLOGY. *Educ:* Colo State Univ, DVM, 66; Am Bd Vet Toxicol, dipl, 72; Am Bd Toxicol, cert, 81. *Prof Exp:* Toxicologist, Inhalation Toxicol Res Inst, Lovelace Found Med Educ & Res, 69-75, asst dir, 74-75; toxicologist, Div Biomed & Environ Res, Us Energy Res & Develop Admin, 75-76; ASST DIR, INHALATION TOXICOL RES INST, LOVELACE BIOMED & ENVIRON RES INST, 77- *Mem:* Am Acad Vet & Comp Toxicol (secy-treas, 76-80); Am Vet Med Asn; Radiation Res Soc; Soc Toxicol. *Res:* Inhalation toxicology; toxicology of inhaled chemicals; late effects of radiation; long-term toxcity studies. *Mailing Add:* Lovelace Inhalation Toxicol Res Inst PO Box 5890 Albuquerque NM 87185

HOBBS, CHARLES RODERICK BRUCE, JR, b Englewood, NJ, Dec 24, 29; m 62. GEOLOGY. *Educ:* Va Polytech Inst, BS, 52, MS, 53, PhD(geol), 57. *Prof Exp:* Deputy state geologist & mineral res, Div Mineral Resources, VA Dept Conserv & Econ Develop, 57-85; VA DEPT MINES, MINERALS & ENERGY, 85- *Mem:* Geol Soc Am; Am Asn Petrol Geol; Sigma Xi. *Res:* Carbonate petrology. *Mailing Add:* 717 Cargil Lane Charlottesville VA 22901

HOBBS, CLIFFORD DEAN, b Atlanta, Tex, June 9, 33; m 56; c 3. PLANT PATHOLOGY, GENERAL AGRICULTURE. *Educ:* Tex A&M Univ, BS, 55, MS, 60, PHD(plant path), 64. *Prof Exp:* Res agronomist, Crops Res Div, Agr Res Serv, USDA, 58-61, res plant pathologist, 61-65; exten plant pathologist, Tex Agr Exten Serv, 65-67; plant sci rep, Eli Lilly & Co, 67-74; consult, Agr Mgt & Consult Serv, 75-77; pres, Valley Garden Ctr, Inc, 76-85; AGR MISSIONARY, FOREIGN MISSION BD, SOUTHERN BAPTIST CONV, 85- *Res:* Manage garden center; educational seminars; consultant on ornamental crops; evaluation of mung bean varieties for use as high protein food in Liberia and other West African countries; work has included varietal selection, cultural practices, processing and utilization, seed distributions and agricultural extension type activities. *Mailing Add:* 10806 Burnwood Dr Austin TX 78758

HOBBS, CLINTON HOWARD, b Indianapolis, Ind, Jan 16, 15; m 44; c 4. BOTANY. *Educ:* Purdue Univ, BSA, 36, MS, 39, PhD(plant physiol), 42. *Prof Exp:* Asst biol, Purdue Univ, 36-42; teacher pub sch, Ind, 43; asst plant physiologist, Guayule Rubber Res Proj, USDA, Calif, 43-45; from asst prof to assoc prof, 45-58, prof biol, 58-80, EMER PROF BIOL SCI, KENT STATE UNIV, 80- *Mem:* AAAS; Am Inst Biol Sci; Am Nature Study Soc. *Res:* Plant respiration; mineral deficiency in pine; polygonaceae of Ohio. *Mailing Add:* Dept Biol Kent State Univ Kent OH 44242

HOBBS, DONALD CLIFFORD, b Edmonton, Alta, Apr 12, 30; m 52; c 3. BIOCHEMISTRY. *Educ:* Univ Alta, BSc, 52, MSc, 55; Univ Tenn, PhD(biochem), 57. *Prof Exp:* Lectr, Univ Alta, 52-54; res biochemist, 57-73, mgr, 73-78, asst dir, 78-81, DIR DRUG METAB DEPT, PFIZER, INC, 81- *Mem:* AAAS; Am Soc Pharmacol & Therapeut. *Res:* Intermediary metabolism; peptide chemistry; isolation and chemistry of antibiotics and other natural products; drug disposition and biotransformation. *Mailing Add:* Seven Ledge Rock Rd Niantic CT 06357

HOBBS, GEORGE EDGAR, psychiatry; deceased, see previous edition for last biography

HOBBS, HERMAN HEDBERG, b Dallas, Tex, Jun 25, 27; m 48. SOLID STATE PHYSICS. *Educ:* George Washington Univ, BS, 53, MS, 55; Univ Va, PhD(physics), 58. *Prof Exp:* Physicist, Nat Bur Standards, 53; instr physics, George Washington Univ, 54-55; instr, Univ Va, 56-58, res fel, 58-59; chmn dept, 61-68, 74-77, PROF PHYSICS, GEORGE WASHINGTON UNIV, 59- *Mem:* Am Phys Soc; Am Asn Physics Teachers; Sigma Xi; AAAS; Am Asn Univ Prof; Exp Aircraft Asn. *Res:* Growth and perfection of metal crystals; electrical conductivity and piezo-resistance effects in metal crystals; microwave electronics; studies of growth of metal whiskers under conditions which resemble those on an orbiting space station sponsored by NASA. *Mailing Add:* Dept Physics George Washington Univ Washington DC 20052

HOBBS, HORTON HOLCOMBE, JR, b Alachua, Fla, Mar 29, 14; m 39; c 2. SYSTEMATIC ZOOLOGY. *Educ:* Univ Fla, BS, 35, MS, 36, PhD(biol), 40. *Prof Exp:* Instr biol, Univ Fla, 37-41, from asst prof to assoc prof, 41-46; from assoc prof to prof, Univ Va, 46-62, dir Mt Lake Biol Sta, 56-60; head cur dept zool, 62-64, sr zoologist, 64-84, EMER ZOOLOGIST, DEPT INVERT ZOOL, US NAT MUS, 84- *Concurrent Pos:* Vis lectr & investr, Nat Univ Mex, Int Educ Exchange Serv, US Dept State, 57. *Mem:* Int Asn Astacology; Crustacean Soc; Am Micros Soc (pres, 63-64); Am Soc Zool. *Res:* Taxonomy, ecology and geographic distribution of freshwater decapod crustaceans. *Mailing Add:* Dept Invert Zool Smithsonian Inst Washington DC 20560

HOBBS, HORTON HOLCOMBE, III, b Gainesville, Fla, Dec 17, 44; m 67; c 2. LIMNOLOGY, BIOSPELEOLOGY. *Educ:* Univ Richmond, BA, 67; Miss State Univ, MS, 69; Ind Univ, PhD(zool), 73. *Prof Exp:* Instr biol, Christopher Newport Col, 73-75; asst prof biol, George Mason Univ, 75-76; PROF BIOL, WITTENBERG UNIV, 76- *Concurrent Pos:* Dir Nat Speleol Soc, bd gov, Ohio Valley Region; bd mem, Cave Conserv, Va; bd trustees, Island Cave Res Ctr. *Mem:* Am Soc Limnol & Oceanog; Am Inst Biol Sci; fel Nat Speleol Soc; Sigma Xi; Cave Res Found; Crustacean Soc. *Res:* Cave ecology; systematics, evolution and ecology of aquatic decapods and entocytherid ostracods. *Mailing Add:* Dept Biol Wittenberg Univ PO Box 720 Springfield OH 45501-0720

HOBBS, JAMES ARTHUR, b East Kildonan, Can, Dec 5, 14; nat US; m 40; c 3. AGRONOMY, SOILS. *Educ:* Univ Man, BSA, 35, MSc, 40; Purdue Univ, PhD(soils), 48. *Prof Exp:* Asst, Dom Exp Farms, Ottawa, 35-41; agr rep exten serv, Dept Agr, Man, 41-46; asst prof soils, Univ Man, 48-49; asst agronomist & asst prof, Kans State Univ, 50-52, assoc agronomist & assoc prof, 52-58, agronomist & prof, 58-85, EMER PROF SOILS, EXP STA, KANS STATE UNIV, 85- *Concurrent Pos:* Head dept soil sci, Ahmadu Bello Univ, Nigeria, 64-66 & 70-74, dean agr, 65-66 & 71-72; consult, Dept Agr, Sri Lanka, 79, Min Agr, Tanzania, 79, Fac Agr, Univ Zimbabwe, 82 & INRA, Min Agr, Morocco, 87; res exten liaison officer, Min Agr, Botswana, 82-85. *Mem:* Am Soc Agron; Soil Sci Soc Am; Int Soc Soil Sci; Soil Water Conserv Soc. *Res:* Soil management, soil conservation. *Mailing Add:* 571 Jessie Ave Winnipeg MB R3L 0R1 Can

HOBBS, JOHN ROBERT, b Plainfield, Ind, Aug 22, 41; m 69. PHYSICAL CHEMISTRY, ANALYTICAL CHEMISTRY. *Educ:* Rose Polytech Inst, BS, 63; Univ NH, PhD(phys chem), 68. *Prof Exp:* Nat Acad Sci-Nat Res Coun fels, Army Mat & Mech Res Ctr, 68-69; NSF fel chem, Cornell Univ, 69-70; RES CHEMIST, SAFETY & ENVIRON TECHNOL DIV, VOL PE NAT TRANSP SYST CTR, RES & SPEC PROG ADMIN, US DEPT TRANSP, 70- *Mem:* Am Chem Soc; NY Acad Sci; Am Soc Mass Spectrometry; Am Soc Testing & Mat. *Res:* Mass spectroscopy applied to trace analysis; trace analysis of impurities in materials; high energy chemical kinetics; trace gas analysis; explosive vapor detection methods; security in transportation systems. *Mailing Add:* Volpe Nat Transp Syst Ctr Code 75 US Dept Transp Kendall Sq Cambridge MA 02142

HOBBS, LEWIS MANKIN, b Upper Darby, Pa, May 16, 37; m 62; c 3. ASTRONOMY, ASTROPHYSICS. *Educ:* Cornell Univ, BEngPhys, 60; Univ Wis, MS, 62, PhD(physics), 66. *Prof Exp:* Jr astronomer, Univ Calif, Santa Cruz, 65-66; from asst prof to assoc prof, 66-76, dir, Yerkes Observ, 74-82, PROF ASTRON & ASTROPHYS, UNIV CHICAGO, 76- *Concurrent Pos:* Alfred P Sloan scholar, 56-60; mem, bd dirs, Asn Univs for Res Astron, Inc, Washington, DC, 74-85; mem, Astron Comt of Bd Trustees, Univs Res Asn, Inc, Washington, DC, 79-83, chmn, 79-81; coun mem, Space Telescope Inst, Asn Univs for Res Astron, Inc, 82-87; mem, bd govs, Astrophys Res Consortium, Inc, ARC, Seattle, Wash, 84-91; mem, Hubble Space Telescope Users' Comt, NASA, 90- *Mem:* Am Astron Soc; Am Phys Soc; Int Astron Union. *Res:* Interstellar medium; galactic structure; observational cosmology. *Mailing Add:* Yerkes Observ Univ Chicago Williams Bay WI 53191-0258

HOBBS, M FLOYD, b San Jose, Calif, Dec 19, 24; m 47; c 5. PHYSICAL CHEMISTRY, ORGANIC CHEMISTRY. *Educ:* Univ Calif, Berkeley, BA, 48. *Prof Exp:* Anal chemist, Marine Magnesium Div, Merck & Co, 48-51; chemist, Bioferm Div, Int Minerals & Chem Corp, 51-54, proj chemist, 54-61, head org chem, 61-65; sr res chemist, FMC Corp, 65-82; PRES, M FLOYD HOBBS & ASSOC INC, 83- *Mem:* Am Chem Soc; fel Am Inst Chem; Am Soc Enol & Viticult. *Res:* Isolation of organic chemicals from fermentation media; ion exchange; liquid ion exchange; kinetics; phase relationships; solvent extraction; waste and waste water chemistry. *Mailing Add:* 15604 Kavin Lane Monte Sereno CA 95030

HOBBS, MARCUS EDWIN, b Chadbourn, NC, Aug 11, 09; m 37; c 2. PHYSICAL CHEMISTRY. *Educ:* Duke Univ, AB, 32, MA, 34, PhD(chem), 36. *Prof Exp:* Asst indust chem lab, 31-35, from instr to asst prof, 35-45, assoc prof, 45-49, chmn dept, 51-54, dean, Grad Sch Arts & Sci, 54-59, dean univ, 58-63, provost, 69-70, prof chem, 49-79, UNIV DISTINGUISHED SERV EMER PROF CHEM, DUKE UNIV, 79- *Concurrent Pos:* Civilian with Off Sci Res & Develop; actg chief scientist, Off Ord Res, 51-52; mem predoctoral fels bd, Nat Res Coun, NSF, 51-53, mem adv comt spec projs sci ed, 58-65; mem adv comt utilization res & develop, USDA, 64-70; chmn exec comt, Res Triangle Inst, 64-69 & 77- *Honors & Awards:* Civilian Serv Medal, Dept Army, 59. *Mem:* AAAS; Am Chem Soc; Am Inst Physics. *Res:* Dielectric constants and dipole moments; infrared and ultraviolet absorption spectra; colloids; chemistry of tobacco and smoke; nuclear magnetic resonance; relaxation times. *Mailing Add:* Dept Chem Duke Univ Durham NC 27706

HOBBS, MILFORD LEROY, b Waynesville, Mo, Dec 28, 04; m 31; c 2. CLINICAL PATHOLOGY. *Educ:* Wash Univ, MD, 31. *Prof Exp:* Instr internal med, Col Med, Univ Vt, 38-39, from instr to asst prof path, 39-45; pathologist & dir labs, Fairmont Gen Hosp, WVa, 45-47; prof path, Sch Med, Univ WVa, 47-61; pathologist & dir labs, St Mary's Hosp, Huntington, WVa, 61-63; prof, Med Col Ga, 63-74, emer prof path, 74; RETIRED. *Concurrent Pos:* Chief lab serv, Vet Admin Hosp, 63-74. *Mem:* AMA; fel Am Col Physicians; fel Col Am Path. *Res:* Diagnostic pathology. *Mailing Add:* 4275 Owens Rd Evans GA 30809

HOBBS, PETER VICTOR, b London, Eng, May 31, 36; US citizen; m 63; c 3. ATMOSPHERIC PHYSICS. *Educ:* Univ London, BSc, 60, DIC & PhD(cloud physics), 63. *Prof Exp:* Asst prof, 63-65, assoc prof, 65-70, PROF ATMOSPHERIC SCI, UNIV WASH, 70-, DIR CLOUD PHYSICS GROUP, 63-, ADJ PROF GEOPHYS, 74- *Mem:* AAAS; Am Geophys Union; Am Meteorol Soc; Royal Meteorol Soc. *Res:* Physics of clouds, rain, snow, thunderstorms and weather modification; air pollution; atmospheric chemistry; mesometeorology. *Mailing Add:* Dept Atmospheric Sci AK-40 Univ Wash Seattle WA 98195

HOBBS, ROBERT WESLEY, b Chester, WVa, Jan 28, 38; m 62; c 2. RADIO ASTRONOMY. *Educ:* Case Inst Technol, BS, 60; Univ Mich, MS, 62, PhD(astron), 64. *Prof Exp:* Sect head, Radio Astron Br, US Naval Res Lab, 64-69; br head, Lab Astron & Solar Physics, Goddard Space Flight Ctr, NASA, 69-82; DIR ENG, CTA INC, 82- *Concurrent Pos:* Guest lectr, Univ Md, 64. *Mem:* Am Astron Soc; Int Sci Radio Union; Int Astron Union. *Res:* Radio, ultraviolet and optical astronomy; polarization of radio sources; flux density variations in quasars; solar active regions; radio emission from comets; author of 120 technical publications. *Mailing Add:* 737 Walnut Ave North Beach MD 20714

HOBBS, SAMUEL WARREN, b Des Moines, Iowa, Nov 20, 11; m 47; c 2. GEOLOGY. *Educ:* Univ Wash, Seattle, BS, 35; Yale Univ, PhD(geol), 48. *Prof Exp:* Asst geol, Yale Univ, 37-38, instr, 38-40; from geologist to chief, Mineral Deposits Br, 40-60, asst chief geologist econ geol, 60-63, RES GEOLOGIST, US GEOL SURV, 63- *Mem:* Fel Geol Soc Am; Soc Econ Geol; Geochem Soc. *Res:* Areal geology; geology of tungsten and base-metal deposits; igneous and metamorphic petrology and petrography. *Mailing Add:* 29663 Paint Brush Dr Evergreen CO 80439

HOBBS, STANLEY YOUNG, b Gloversville, NY, June 1, 44; m 68; c 1. POLYMER PHYSICS. *Educ:* Dartmouth Col, BA, 66; Rensselaer Polytech Inst, PhD(chem), 69. *Prof Exp:* MEM STAFF POLYMER PHYSICS, GEN ELEC CORP RES & DEVELOP, 69-, TECH COORDR LAB ADMIN, 75-, MGR STRUCT PLASTICS PROJ, 76-, MGR POLYMER PHYSICS UNIT, 78-, STAFF SCIENTIST, 88- *Mem:* Am Phys Soc; Electron Micros Soc Am; Sigma Xi. *Res:* Polymer crystallization and morphology; nucleation phenomena; polymer blends. *Mailing Add:* 33 Gould Dr Scotia NY 12302

HOBBY, CHARLES R, b Tyler, Tex, Sept 5, 30; m 53; c 2. MATHEMATICS. *Educ:* Univ Calif, Berkeley, BA, 53; Univ Houston, MS, 57; Calif Inst Technol, PhD(math), 60. *Prof Exp:* Mathematician, Inst Defense Anal, 60-61; from asst prof to prof math, Univ Wash, 61-73; chmn dept, 73-84, PROF MATH, UNIV ALA, 84- *Mem:* Am Math Soc. *Res:* Group theory; probability. *Mailing Add:* Box 870350 Tuscaloosa AL 35487-0350

HOBBY, GLADYS LOUNSBURY, b New York, NY, Nov 19, 10. MICROBIOLOGY, MEDICAL SCIENCES. *Educ:* Vassar Col, AB, 31; Columbia Univ, MA, 32, PhD(bact), 35; Am Bd Microbiol, Dipl, 62. *Prof Exp:* Res assoc, Presby Hosp & Col Physicians & Surgeons, Columbia Univ, 34-43; microbiologist, Chas Pfizer & Co, Inc, 44-59; sci dir, Infectious Dis Res Inst, 59-82; SCI WRITER, 77- *Concurrent Pos:* Consult, East Orange Vet Admin Hosp, 54-59, chief vet admin spec res lab, 59-77; clin instr, Med Col, Cornell Univ, 59-73, clin asst prof, 74-77, emer clin asst prof, 77-82; ed, Antimicrobial Agents & Chemother, 65-70; ed-in-chief, Antimicrobial Agents & Chemother J, 72-80. *Honors & Awards:* Trudeau Medal, Am Lung Asn, 83. *Mem:* emer fel Am Acad Microbiol; hon mem Am Soc Microbiol; emer mem Soc Exp Biol & Med; hon fel Am Thoracic Soc; fel AAAS; Am Pub Health Asn; Infectious Dis Soc Am. *Res:* Bacteriophage; bacterial variation and enzymes; streptococci; pneumococci; tubercle bacilli; rat-bite fever; experimental tuberculosis; rheumatic diseases; sulfonamides; immunizing agents; germ-free life; antimicrobial drug chemotherapy. *Mailing Add:* 25 Crosslands Kennett Square PA 19348-2099

HOBEIKA, ANTOINE GEORGE, b Homsieh, Lebanon, Sept 28, 43; US citizen. TRANSPORTATION ENGINEERING. *Educ:* Am Univ Beirut, Lebanon, BS, 67, MS, 70; Purdue Univ, PhD(civil eng), 73. *Prof Exp:* Training engr construct & design, Victoria Line, London, 66; site engr construct, Esso Gas Liquification Plant, Libya, 67; teaching asst, Am Univ Beirut, 69-70; supv engr civil works, Off Reconstruct, Beirut, 68-70; res asst, Purdue Univ, 70-73; from asst prof to assoc prof, 73-83, PROF TEACHING & RES, VA POLYTECH INST & STATE UNIV, 83- *Concurrent Pos:* Consult, Ministry Planning, Iraq, 74-75 & RR Orgn, Saudi-Arabia, 78; prin investr, Va Div Motor Vehicles, 77-82, NASA, Langley, 78-81, Dept Energy, 79-81 & Urban Mass Transp Admin, 82-85, NSF, Va Elec & Power Co, Va Dept Emergency Serv, Va Dept Highways & Transp. *Mem:* Am Soc Civil Engrs; Inst Transp Engrs; Am Rd & Transp Builders Asn; Sigma Xi. *Res:* Transportation evacuation planning particularly in emergency management; airfield runway geometrics under future air traffic control systems; transportation and development in less developed countries; increasing productivity of mass transit systems; shipment of hazardous materials. *Mailing Add:* Dept Civil Eng Va Polytech Inst Blacksburg VA 24061

HOBERMAN, ALFRED ELLIOTT, b Lock Haven, Pa, Aug 24, 39; m 67; c 2. EDUCATION. *Educ:* Lock Haven Univ, BS, 62; Univ Miss, MSc, 67; Louisiana State Univ, ABD. *Prof Exp:* Teacher chem, Spec Sch Dist, Dover, Del, 62-65 & Chichester Sch Dist, 65-66; PROF & CHMN DEPT CHEM, PHYSICS & GEOSCI, LOCK HAVEN UNIV, 67- *Concurrent Pos:* Instr chem, Louisiana State Univ, 72-74; prin investr, stream acidification, Lock Haven Univ, 85- *Mem:* Am Chem Soc. *Res:* Stream acidification in central Pennsylvania; the effect of acid precipitation on limestone and free stone streams. *Mailing Add:* Dept Chem Physics & Geosci Lock Haven Univ Lock Haven PA 17745

HOBERMAN, HENRY DON, b Bridgeport, Conn, Apr 23, 14; m 43, 74; c 4. BIOCHEMISTRY. *Educ:* Columbia Univ, AB, 36, PhD(biochem), 43; Harvard Univ, MD, 47. *Prof Exp:* Asst trop med & tutor biochem sci, Harvard Univ, 43-44; asst prof internal med, Yale Univ, 47-48, asst prof physiol chem, 48-53; assoc prof, 53-58, PROF BIOCHEM, ALBERT EINSTEIN COL MED, 58-; ASSOC ATTEND PHYSICIAN, BRONX MUNIC HOSP CTR, 81- *Concurrent Pos:* Markle scholar, Yale Univ, 48-53. *Mem:* Am Soc Biol Chemists; Harvey Soc; Am Chem Soc. *Res:* Regulation of intermediary metabolism of amino acids, coupling of intracellular oxidation-reduction reactions; alcohol metabolism; post-translational modifications of human hemoglobin. *Mailing Add:* Albert Einstein Col Med Yeshiva Univ New York NY 10461

HOBEY, WILLIAM DAVID, b Lynn, Mass, Aug 8, 35; m 61; c 3. BIOMEMBRANE STRUCTURE, DYNAMICS. *Educ:* Tufts Univ, BS, 57; Calif Inst Technol, PhD(theoret chem), 62. *Prof Exp:* NSF res fel, Trinity Col, Cambridge Univ, 62-63; asst prof, 63-67, ASSOC PROF CHEM, WORCESTER POLYTECH INST, 67- *Concurrent Pos:* Gen Elec found grant, 72-76; NSF microcomput grant, 81-83. *Mem:* Fel Am Inst Chem; Sigma Xi; Am Chem Soc. *Res:* Kinetics and mechanisms of passive cation transport in human erythrocytes; modes of action of beta adrenergic blockers and related compounds on human erythrocytes; mechanism of activation of erythrocytic acetylcholinesterase. *Mailing Add:* Dept Chem Worcester Polytech Inst 100 Inst Rd Worcester MA 01609-2247

HOBGOOD, RICHARD TROY, JR, b Calhoun, Ga, Jan 15, 39. PHYSICAL CHEMISTRY. *Educ:* Emory Univ, AB, 60, MS, 61, PhD(phys chem), 63. *Prof Exp:* Res chemist, ICI Americas Inc, 63-77, sr res chemist, 77-81, supvr, Corp Res Dept, 81-89; CONSULT, 89- *Mem:* Am Chem Soc; Soc Appl Spectros; Sigma Xi. *Res:* Theoretical and applied spectroscopy, particularly nuclear magnetic resonance; studies of conjugated systems; unsaturated hetero atom containing hydrocarbons; carbohydrates and their dehydrated derivatives; heterocyclic ring systems. *Mailing Add:* 2102 Baynard Blvd Wilmington DE 19802-3937

HOBKIRK, RONALD, b Peebles, Scotland, June 13, 30; m 54; c 3. BIOCHEMISTRY, ENDOCRINOLOGY. *Educ:* Univ Edinburgh, BSc, 52, PhD(biochem), 55, DSc, 73. *Prof Exp:* Asst lectr biochem, Univ Edinburgh, 52-55; asst lectr biochem & surg, Univ Glasgow, 55-57; res asst, McGill Univ, 57-58, from asst prof to prof exp med, 60-71; CAREER INVESTR, MED RES COUN CAN, 66-; PROF BIOCHEM, UNIV WESTERN ONT, 72- *Concurrent Pos:* Brit Empire Cancer Campaign jr fel, 55-57; Med Res Coun Can grants, 59-; Banting Res Found grants, 60-; res asst biochem, Div Metab Dis, Montreal Gen Hosp, 58-60, lab dir endocrinol, 60-66; prof path chem, Univ Western Ont, 71-72. *Mem:* Am Soc Biol Chemists; Endocrine Soc; Can Biochem Soc. *Res:* Biochemistry and regulation of steroid sulfotransferases, sulfohydrolases, glucuronyltransferases and B-glucuronidases; their roles in controlling steroid hormone concentrations in uterus and intrauterine tissues in pregnancy; steroid sulfate hydroxylation. *Mailing Add:* Dept Biochem Univ Western Ont London ON N6A 5C1 Can

HOBLIT, LOUIS DOUGLAS, b Cleburne, Tex, Dec 15, 19; m 41; c 3. ORGANIC CHEMISTRY. *Educ:* Tex Christian Univ, BA, 41; Univ Chicago, dipl, 43. *Prof Exp:* Group leader, Dow Chem Co, 41-58, dir res lab div, 58-69, tech dir, Dow Badische Co, 69-71, vpres & tech dir, 71-77, RES SCIENTIST, DOW CHEM CO, 77- *Mem:* Am Chem Soc. *Res:* Organic chemical productions; pilot plant studies; fiber processes and uses. *Mailing Add:* 217 Rabbit Trail Lake Jackson TX 77566

HOBLITT, RICHARD PATRICK, b Stockton, Calif, Oct 14, 44; m 67. GEOLOGY, GEOPHYSICS. *Educ:* Univ Wash, BS, 67; Univ Colo, MS, 70, PhD(geol), 78. *Prof Exp:* Geophysicist paleomagnetism, 75-77, GEOLOGIST, US GEOL SURV, 77- *Mem:* Geol Soc Am; Am Geophys Union; Sigma Xi. *Res:* Volcanology; volcanic hazards; paleomagnetism; petrology; organic chemistry. *Mailing Add:* US Geol Surv 5400 Macarthur Blvd Vancouver WA 98661

HOBROCK, DON LEROY, b Arenzville, Ill, Jan 28, 36; m 55; c 3. PHYSICAL CHEMISTRY, INORGANIC CHEMISTRY. *Educ:* Western Ill Univ, BS, 58, MS, 60; Kans State Univ, PhD(phys & inorg chem), 65. *Prof Exp:* Instr high sch, Ill, 58-62; from sr res chemist to group leader mass spectrometry, Monsanto Res Corp, 66-72, prog mgr, 72-75, sr res specialist, Mound Lab, 75-78, fel, 78-86, sr fel, 86-88, sr fel, EG&G, Mound Lab, 88-91; RETIRED. *Concurrent Pos:* Res assoc, Univ Kans, 65-66. *Mem:* AAAS; Am Chem Soc. *Res:* Mass spectrometry, molecular structure and appearance potentials; dynamics of gaseous isotopes, their thermodynamics and analyses; development of measurement instrumentation, research and development for surveillance and production of nuclear weapons components; high precision measurements; experimental design and systems evaluation, especially with respect to tritium; program management; tritium containment and engineering. *Mailing Add:* 8736 Maltbie Rd Centerville OH 45459

HOBSON, ARTHUR STANLEY, b Philadelphia, Pa, Nov 27, 34; m 59; c 1. THEORETICAL PHYSICS. *Educ:* NTex State Univ, BMusic, 55; Kans State Univ, BSc, 60, PhD(physics), 64. *Prof Exp:* Assoc prof, 64-74, PROF PHYSICS, UNIV ARK, FAYETTEVILLE, 74- *Mem:* AAAS; Am Phys Soc; Am Assoc Physics Teachers; Fedn Am Sci. *Res:* Foundations of statistical mechanics. *Mailing Add:* Dept Physics Univ Ark Fayetteville AR 72701

HOBSON, EDMUND SCHOFIELD, b White Plains, NY, Apr 7, 31; m 59; c 3. ICHTHYOLOGY. *Educ:* Univ Hawaii, BA, 58, MA, 61; Univ Calif, Los Angeles, PhD(zool), 67. *Prof Exp:* RES FISHERY BIOLOGIST, US NAT MARINE FISHERIES SERV, 62- *Concurrent Pos:* Res assoc, Scripps Inst Oceanog, Univ Calif, San Diego, 68-76. *Honors & Awards:* Publ Award, Am Inst Fishery Res Biologists, 64; Silver Medal Award, US Dept Com, 76; Fisheries Publ Award, Wildlife Soc, 76; Publ Award, US Nat Marine Fisheries Serv, 76, 78, 81, 89. *Mem:* Am Soc Ichthyologists & Herpetologists; Am Fisheries Soc; Am Inst Fishery Res Biologists; West Soc Naturalists. *Res:* Behavior and ecology of marine fishes. *Mailing Add:* Tiburon Lab 3150 Paradise Dr Tiburon CA 94920

HOBSON, JOHN ALLAN, b Hartford, Conn, June 3, 33; m 56; c 3. PSYCHIATRY, NEUROPHYSIOLOGY. *Educ:* Wesleyan Univ, AB, 55; Harvard Med Sch, MD, 59. *Prof Exp:* Intern med, Bellevue Hosp, New York, 59-60; resident psychiat, Mass Ment Health Ctr, Boston, 60-61; clin assoc exp psychiat, NIMH, 61-63; resident psychiat, Mass Ment Health Ctr, Boston, 64-66; res assoc neurophysiol, Lyon, 63-64; res assoc, 64-66, from asst prof to assoc prof, 69-78, PROF PSYCHIAT, HARVARD MED SCH, 78-; PRIN PSYCHIATRIST & DIR LAB NEUROPHYSIOL, MASS MENT HEALTH CTR, 67- *Concurrent Pos:* NIMH spec fel, 63-64, grant, 67-; psychiat consult, Mass Rehab Comn, 66-; chmn Intramural Rev Panel, NIMH, 82-84; vis comt, Max Planck Inst, Munich, 85- *Honors & Awards:* Benjamin Rush Gold Medal, Am Psychiat Asn, 78. *Mem:* Asn Psychophysiol Study Sleep; Soc Neurosci; Int Brain Res Orgn. Res; Physiological correlates of behavior, especially sleep, in man and animals. *Res:* Cellular and molecular basis of sleep cycle control with special reference to REM sleep and dreaming. *Mailing Add:* 74 Fenwood Rd Boston MA 02115

HOBSON, JOHN PETER, b India, Aug 25, 24; nat Can; m 48; c 4. ELECTRON PHYSICS. *Educ:* Univ BC, BASc, 49, MASc, 50; Univ Calif, PhD(physics), 54. *Prof Exp:* From asst res officer to assoc res officer, Nat Res Coun Can, 54-62, sr res officer, 62-73, prin res officer, 73-81, asst dir, Elec Eng Div, 81-86; PRES, NAT VACUUM TECHNOLOGIES, INC, 86- *Honors & Awards:* Albert Nerken Award, Am Vacuum Soc, 90. *Mem:* Am Phys Soc; hon mem Am Vacuum Soc; Can Asn Physicists. *Res:* Molecular beams; electron physics; physical adsorption; vacuum physics. *Mailing Add:* Nat Vacuum Technologies Inc PO Box 4160 Postal Sta E Ottawa ON K1S 5B2 Can

HOBSON, JOHN PETER, JR, geology, for more information see previous edition

HOBSON, LELAND S(TANFORD), research administration, for more information see previous edition

HOBSON, MELVIN CLAY, JR, b Ft Myers, Fla, Aug 7, 26; m 56; c 3. PHYSICAL CHEMISTRY. *Educ:* Va Polytech Inst, BS, 48; Univ Pa, MS, 51, PhD, 57. *Prof Exp:* With Nat Paint, Varnish & Lacquer Asn, 48-49; asst, Univ Pa, 51-53, res assoc, 54-56; res chemist, Gen Elec Co, 56-58; chemist, M W Kellogg Co, 58-60; res assoc, Univ Pa, 60-62; sr res chemist, Va Inst Sci Res, 62-66, head, Surface Chem Div, 66-72; sr chemist, Consol Lab Servs, 72-74; sr res scientist, United Technol Res Ctr, 74-80; RES SECT HEAD, ENGELHARD INDUST, 80- *Concurrent Pos:* Adj prof, Va Commonwealth Univ, 64-74; consult, Phillip Morris, Inc, 68-71. *Mem:* Am Chem Soc; The Chem Soc; The Catalysis Soc. *Res:* Heterogenous catalysis; free radical chemistry. *Mailing Add:* Dept Phys Chem Engelhard Corp Menlo Park Edison NJ 08818

HOBSON, ROBERT MARSHALL, physics; deceased, see previous edition for last biography

HOBSTETTER, JOHN NORMAN, b Dayton, Ohio, Feb 19, 17. MATERIALS SCIENCE. *Educ:* Mass Inst Technol, SB, 39; Harvard Univ, SD(phys metall), 46. *Prof Exp:* Instr metall, Harvard Univ, 41-46, instr eng sci, 46-47, asst prof, 47-52; mem tech staff, Bell Tel Labs, 52-58; from assoc prof to prof metall, 58-69, dir lab res struct of matter, 60-76, vprovost res, 67-71, dean grad sch arts & sci, 68-69, prof metall & mat sci, 69-80, PROF MAT SCI & ENG, UNIV PA, 80-, ASSOC PROVOST ACAD PLANNING, 71- *Mem:* AAAS; Am Phys Soc; Am Inst Mining, Metall & Petrol Engrs; Sigma Xi. *Res:* Erosion of materials under extreme conditions; phase transformations in solids; physics of solids; defect structures in metals and semiconductors; biomaterials. *Mailing Add:* Cape May Ave Sewell NJ 08080

HOBURG, JAMES FREDERICK, b Pittsburgh, Pa, Dec 30, 46; m 78; c 2. APPLIED ELECTROMAGNETICS, CONTINUUM ELECTROMECHANICS. *Educ:* Drexel Univ, BS, 69; Mass Inst Technol, SM, 71, PhD(elec eng), 75. *Prof Exp:* Instr, Mass Inst Technol, 73-75; from asst prof to assoc prof, 75-84, PROF ELEC & COMP ENG, CARNEGIE-MELLON UNIV, 84-, ASSOC DEPT HEAD, 85- *Concurrent Pos:* Consult, Westinghouse Elec Corp, 79-82, Midland Ross Corp, 85, Mentor Robotics, 85, US Steel Corp, 86, McGraw-Hill Book Co, 86-87, Bethlehem Steel, 87. *Mem:* Inst Elec & Electronics Engrs; Sigma Xi; Electrostatics Soc; Am Soc Eng Educ. *Res:* Electromagnetic field theory and engineering; continuum electromechanics; electrohydrodynamics; electrostatic precipitation; magnetohydro dynamics; magnetic confinement and shaping. *Mailing Add:* 1000 Oak Creek Lane Baden PA 15005

HOCH, FREDERIC LOUIS, b Vienna, Austria, Apr 14, 20; nat US; m 61. ENDOCRINOLOGY, BIOCHEMISTRY. *Educ:* City Col New York, BS, 39; NY Univ, MD, 43; Mass Inst Technol, MS, 52. *Prof Exp:* Intern, Michael Reese Hosp, 43-44; resident & teaching fel, Dept Path, Tufts Med Sch & Mt Auburn Hosp, 47-48; res assoc, dept biol, Mass Inst Technol, 48-51; res fel biochem, Mass Gen Hosp, 51-53; from res assoc to asst prof med, Harvard Med Sch, 53-67; from assoc prof to prof internal med, 68-86, from assoc prof to prof biochem, 70-86, EMER PROF INTERNAL MED & BIOCHEM, UNIV MICH MED SCH, 87- *Concurrent Pos:* 2nd, 1st Lt, Med Corps, US Army, 41-46, capt, 44-46; fel clin physiol, McLean Hosp, 48; from jr assoc to sr assoc med, Peter Bent Brigham Hosp, 53-67; tutor med sci, Harvard Med Sch, 57-66; vis scientist, Biophys Res Div, Inst Sci & Technol, Univ Mich, 67-68; vis prof, Arrhenius Lab, Univ Stockholm, Sweden, 77-78, Lab Nuclear Med & Radiation Biol, Univ Calif, Los Angeles, 78. *Mem:* AAAS; Am Chem Soc; Am Soc Biol Chem; Brit Biochem Soc. *Res:* Mechanism of action of hormones; cellular intermediary metabolism and energy transfer. *Mailing Add:* 3455 Woodland Rd Ann Arbor MI 48104-4257

HOCH, GEORGE EDWARD, b Brookings, SDak, Mar 11, 31; m 53; c 4. BIOCHEMISTRY. *Educ:* SDak State Col, BS, 53; Univ Wis, MS, 56, PhD(biochem), 58. *Prof Exp:* Staff scientist, Res Inst Adv Studies, Martin Co, 58-64; mem staff, Univ Pa, 64-65; assoc prof, 65-74, PROF BIOL, UNIV ROCHESTER, 74-, CHMN DEPT, 80- *Res:* Photosynthesis; biological nitrogen fixation. *Mailing Add:* Dept Biol Univ Rochester Wilson Blvd Rochester NY 14627

HOCH, HARVEY C, FUNGAL ULTRA-STRUCTURE, CYTOLOGY. *Educ:* Univ Wis, PhD(plant path). *Prof Exp:* ASSOC PROF, NY STATE AGR EXP STA, CORNELL UNIV, 86- *Mailing Add:* Dept Plant Path Cornell Univ NY State Agr Exp Sta Geneva NY 14853

HOCH, JAMES ALFRED, b Brookings, SDak, Jan 22, 39; m 69; c 2. MICROBIOLOGY, BIOCHEMISTRY. *Educ:* SDak State Univ, BSc, 61; Univ Ill, PhD(microbiol), 65. *Prof Exp:* USPHS fel, Ctr Molecular Genetics, Gif-sur-Yvette, France, 65-67; fel, 67-68, from asst to assoc, 68-72, assoc mem microbiol, 72-77, assoc mem cellular biol, 78-80, MEM CELLULAR BIOL, SCRIPPS CLIN & RES FOUND, 80- *Concurrent Pos:* Am Cancer Soc fac res assoc awardee, 69-74; mem ed bd, Bact Reviews, 73-79 & J of Bact, 75-78; mem, microbiol chem, NIH, 77-81, microbiol physiol & genetics, 85-88. *Mem:* AAAS; Am Soc Microbiol; Genetics Soc Am; Am Soc Biol Chemists. *Res:* Genetics and regulation of development in microorganisms. *Mailing Add:* Dept Molecular & Exp Med Res Inst Scripps Clin 10666 N Torrey Pines Rd La Jolla CA 92037

HOCH, MICHAEL, b Budapest, Hungary, Feb 9, 23; nat US; m 56. MATERIALS SCIENCE. *Educ:* Swiss Fed Inst Technol, DSc, 47. *Prof Exp:* Res assoc, Univ Zurich, 48-50; res assoc, Res Found, Ohio State Univ, 51-53, assoc supvr, 53-56; from asst prof to assoc prof metall eng, 56-63, head dept, 68-75, PROF MAT SCI & METALL ENG, UNIV CINCINNATI, 63- *Concurrent Pos:* Consult, Solid State Sci Div, Argonne Nat Lab, 58-59, NMPO, Gen Elec Co, 62- & Mound Lab, Monsanto Chem Corp, 62- *Mem:* Am Soc Metals; Am Chem Soc; Am Phys Soc; Am Ceramic Soc; Am Inst Mining, Metall & Petrol Engrs. *Res:* High temperature chemistry and metallurgy; thermophysical properties; alloys of transition metals; deviation from stoichiometry. *Mailing Add:* Dept Mat Sci & Eng Univ Cincinnati Cincinnati OH 45221-0012

HOCH, ORION L, b Canonsburg, Pa, Dec 21, 28; m 52; c 3. PHYSICS, ELECTRICAL ENGINEERING. *Educ:* Carnegie Inst Technol, BS, 52; Univ Calif, Los Angeles, MS, 54; Stanford Univ, PhD(elec eng), 57. *Prof Exp:* Electronics engr, Res & Develop Labs, Hughes Aircraft Corp, 52-54; res engr microwave tubes, Electronics Labs, Stanford Univ, 54-57; sr engr & dept mgr, Electron Tube Div, Litton Industs, Inc, 57-63, pres & gen mgr, 63-68, vpres, Components Group, 68-70, corp sr vpres, 70-77; pres & chief exec officer, Advan Memory Systs, 77-; PRES, LITTON INDUSTS INC. *Mem:* Inst Elec & Electronics Engrs. *Mailing Add:* 707 N Palm Dr Beverly Hills CA 90210

HOCH, PAUL EDWIN, b Philadelphia, Pa, Nov 6, 20; m 54; c 3. ORGANIC CHEMISTRY. *Educ:* Ohio Wesleyan Univ, BA, 42; Syracuse Univ, MS, 44; Univ Ill, PhD(org chem), 48. *Prof Exp:* Res chemist biochem res, Manhattan Proj, Univ Rochester, 44-45; org synthetic res, Gen Aniline & Film Corp, 48-53; electrochem res, Carwin Co, 53-54 & Hooker Chem Co, 54-66; mgr chem res, Richmond Res Ctr, Stauffer Chem Co, 66-74, dir, 74-77, 82-85, chief scientist, 77-86; RETIRED. *Mem:* Am Chem Soc; Entom Soc Am. *Res:* Biochemical and synthetic organic electrochemistry; agricultural chemistry; pest management. *Mailing Add:* 368 Tharp Dr Moraga CA 94556

HOCH, RICHMOND JOEL, b Oak Park, Ill, Oct 5, 35; m 53; c 2. AERONOMY, SPACE SCIENCE. *Educ:* Harvey Mudd Col, BS, 63; Univ Wash, MS, 65, PhD(physics), 70. *Prof Exp:* Instrument designer, Davidson Optronics Inc, 58-63; res scientist nuclear reactor physics, Gen Elec Co, 63-65; res scientist, Battelle-Northwest Lab, Wash, 65-66, sr res scientist & mgr aeronomy & space sci, 66-74; chmn, bd dirs & chief exec officer, Sigma Res Inc, 74-88, PRES, SIGMA FINANCIAL GROUP, INC, 88-; PRES, BENTON-FRANKLIN VENTURES, INC, 89- *Concurrent Pos:* Prof math & physics, Columbia Basin Col, 88-89; chmn, Pac Rim Technol Conf, 89; lectr physics, Wash State Univ, Tri Cities, 90- *Mem:* AAAS; Am Geophys Union; Sigma Xi. *Res:* Photometric and spectroscopic research on auroras; airglow; zodiacal light and astronomical objects; theoretical studies in plasma physics; hydromagnetics and atomic physics; design of automated measuring instruments. *Mailing Add:* 2921 S Auburn Pl Kennewick WA 99337

HOCH, SALLIE O'NEIL, b Mineola, NY, Feb 15, 41; m 69; c 2. CELLULAR BIOLOGY, BIOCHEMISTRY. *Educ:* St Bonaventure Univ, BS, 64; Univ Ill, MS, 66, PhD(microbiol), 69. *Prof Exp:* Am Cancer Soc fel cellular biol, Scripps Clin & Res Found, 69-71, asst, 71-72, assoc, 72-83; PRIN SCIENTIST, AGOURON INST, 83-; SR SCIENTIST, AGOURON PHARMACEUT INC, 85- *Concurrent Pos:* Fel, Am Cancer Soc, 69-71; Res Career Develop Award, Nat Cancer Inst, NIH, 75-80; prin investr res grants, NIH, Nat Sci Found, Lupus Found Am, 71-; mem spec rev comts, NIH, 81, 90; mem, Fel Comt, Arthritis Found, 88-91. *Mem:* Am Soc Microbiol; Am Soc Biochem & Molecular Biol; Am Soc Cell Biol; Sigma Xi; Am Col Rheumatology. *Res:* Structure and function of autoimmunity-associated antigens; immunodiagnostic markers; ribonucleoprotein complexes. *Mailing Add:* Agouron Inst 505 Coast Blvd S La Jolla CA 92037

HOCHACHKA, PETER WILLIAM, b Therien, Alta, Mar 9, 37; c 3. COMPARATIVE BIOCHEMISTRY. *Educ:* Univ Alta, BSc, 59; Dalhousie Univ, MSc, 61; Duke Univ, PhD(zool), 65. *Prof Exp:* Asst prof biol, Univ Toronto, 64-65; fel biochem, Duke Univ, 65-66; from asst prof to assoc prof, 66-75, PROF BIOL & ZOOL, UNIV BC, 75- *Concurrent Pos:* Vis investr, Inst Arctic Biol, Univ Alaska, 71, Univ Hawaii, 73, 75 & 76, Plymouth Marine Lab, 78, Nat Marine Fisheries, Honolulu, 81, 82 & 84 & Concord Field Sta, Harvard Univ, 84; vis prof, Friday Harbor Marine Lab, Univ Wash, 75, Med Sch, Harvard Univ, 76-77 & Monash Univ, 83; NSF grant, 78-79; Nat Sci & Eng Coun Can grants, 78-81, 82-85 & 85- *Honors & Awards:* Killiam Res Prizes, 87 & 88. *Mem:* Fel AAAS; NY Acad Sci; Am Soc Zoologists; Am Soc Biol Chemists; Soc Exp Biol; fel Royal Soc Can; Can Soc Zoologists; Sigma Xi. *Res:* Unique problems of metabolic control faced by animals unable to maintain classical homeostasis; examination of three levels of organization whole animal, multienzyme preparations and enzyme systems. *Mailing Add:* Dept Zool Univ BC Vancouver BC V6T 1W5 Can

HOCHANADEL, CLARENCE JOSEPH, b Gibsonburg, Ohio, Nov 25, 16; m 44; c 8. PHYSICAL CHEMISTRY, ATMOSPHERIC CHEMISTRY. *Educ:* Bowling Green State Univ, AB, 40; Ind Univ, MA, 41, PhD(phys chem), 43. *Prof Exp:* Res chemist, Metall Lab, 43-46; sr chemist, Oak Ridge Nat Lab, 46-80; RETIRED. *Mem:* Am Chem Soc; Radiation Res Soc. *Res:* Radiation chemistry; photochemistry; chemical kinetics. *Mailing Add:* 101 Endicott Lane Oak Ridge TN 37830

HOCHBERG, IRVING, b Brooklyn, NY, Apr 17, 34; c 2. AUDIOLOGY. *Educ:* Brooklyn Col, BA, 55; Teachers Col, Columbia, MA, 57; Pa State Univ, PhD(audiol), 62. *Prof Exp:* Prof audiol, NY Univ, 62-70; PROF AUDIOL, CITY UNIV NY BROOKLYN COL, 70-, EXEC OFFICER SPEECH & HEARING SCI, GRAD SCH, 74- *Concurrent Pos:* Danforth Assoc, 68-71; consult audiol, Goldwater Mem Hosp & Mt Sinai Hosp, 70-; fel, Am Speech & Hearing Asn, 71; training proj dir, Off Special Educ, US Dept Educ, 74-; consult audiol, Vet Admin Hosp, East Orange, NJ & Bronx, 77-; dir, Ctr Res Speech & Hearing Sci, City Univ NY Grad Ctr, 79- *Mem:* Fel Am Speech & Hearing Asn; Acoust Soc Am; Inst Soc Audiol; Acad Rehab Audiol. *Res:* Clinical audiology; psychoacoustic behavior of children and the aged; adaptive testing in audition. *Mailing Add:* City Univ NY Grad Sch 33 W 42nd St New York NY 10036

HOCHBERG, JEROME, b New York, NY, Dec 21, 25; m 48; c 2. CHEMICAL ENGINEERING. *Educ:* City Col New York, BChE, 48; NY Univ, MChE, 51. *Prof Exp:* Res chem engr, Laurel Hill Res Lab, Gen Chem Div, Allied Chem & Dye Corp, 48-50; res engr coated fabrics, 50-63, res assoc, 63-65, res supvr, 65-68, RES FEL, EXP STA, E I DU PONT DE NEMOURS & CO, 68- *Mem:* Am Chem Soc. *Res:* Coated fabrics, finishes, poromeric materials; polymer manufacture; air pollution control. *Mailing Add:* 2413 Chatham Dr Wilmington DE 19803

HOCHBERG, KENNETH J, b New York, NY, July 26, 50; m 71; c 4. STOCHASTIC PROCESSES, MATHEMATICAL BIOLOGY. *Educ:* Yeshiva Univ, BA, 71; New York Univ, MS, 73, PhD(math), 76. *Prof Exp:* Asst prof math, Carleton Univ, 76-78; asst prof, 78-82, ASSOC PROF MATH & STATIST, CASE WESTERN RESERVE UNIV, 82- *Concurrent Pos:* Vis asst prof, Northwestern Univ, 78; prin investr res grants, NIH, 80-83, Nat Sci Found, 84-87, Nat Sec Agency, 88-89; vis prof, Hebrew Univ, 83-84; Fulbright scholar, 86-87. *Mem:* Am Math Soc; Inst Math Statist. *Res:* Mathematical areas of probability theory and stochastic processes; measure-valued diffusion processes with applications to epidemiology and population genetics. *Mailing Add:* Dept Math & Comput Sci Bar Ilan Univ Ramatt Gan 52900 Israel

HOCHBERG, MELVIN, b New York, NY, Feb 12, 20; m 45; c 2. CHEMISTRY. *Educ:* City Col New York, BS, 40; Columbia Univ, AM, 43; NY Univ, PhD(phys chem), 49. *Prof Exp:* Prin res chemist, Food Res Labs, Inc, 41-46; dir, Vitamin Prod Lab, Nopco Chem Co, 47-53, dir res & prod, 53-57, vpres, 58-59, dir, 59, mem exec comt, 62; vpres & gen mgr biochem div, Diamond Shamrock Chem Co, 68-70, vpres develop life sci, 69-70; vpres, 70-72, EXEC VPRES, INT RECTIFIER CORP, 72-, PRES, RACHELLE LABS INC DIV, 70- *Mem:* AAAS; Am Chem Soc; fel NY Acad Sci; Am Inst Chem; Sigma Xi. *Res:* Nutritional and pharmaceutical chemicals for humans and animals. *Mailing Add:* 30035 Avenida Elegante Rancho Palos Verdes CA 90274

HOCHBERG, MURRAY, b Brooklyn, NY, May 30, 43; m 67; c 2. MATHEMATICS. *Educ:* Yeshiva Univ, BA, 64; NY Univ, MS, 66, PhD(math), 69. *Prof Exp:* Asst prof, 68-70, ASSOC PROF MATH, BROOKLYN COL, 70- *Mem:* Am Math Soc; Math Asn Am. *Res:* Probability; combinatorial analysis; reliability theory and biostatistics. *Mailing Add:* Dept Math Brooklyn Col Brooklyn NY 11210

HOCHEL, ROBERT CHARLES, b Chicago, Ill, Oct 10, 44; m 69; c 3. INSTRUMENTATION, SYSTEMS DEVELOPMENT. *Educ:* Univ Ill, BS, 67; Purdue Univ, PhD(nuclear chem), 73. *Prof Exp:* Chemist, Babcock & Wilcox Nuclear Develop Ctr, 67-68; res chemist, 73-76, staff chemist, 76-77, res supvr, 78-80, RES STAFF CHEMIST, SAVANNAH RIVER LAB, WESTINGHOUSE INC, 80- *Mem:* Am Nuclear Soc. *Res:* Development of nuclear and radio chemical techniques; neutron activation analysis; automated or specialized counting systems. *Mailing Add:* 1058 Kismet Dr Aiken SC 29803

HOCHELLA, NORMAN JOSEPH, b Hazleton, Pa, June 28, 30; m 66; c 2. ANALYTICAL CHEMISTRY, BIOCHEMISTRY. *Educ:* Lafayette Col, AB, 52. *Prof Exp:* Res chemist, Univ Pa, 52-56 & Thomas Jefferson Univ, 56-58; res assoc, Inst Cancer Res, 58-60, Fels Res Inst, Temple Univ, 60-66 & Sch Med, Univ NC, 66-70; sr res chemist, Damon Corp, 70-72; res assoc, Mass Inst Technol, 72-75; CHIEF CHEMIST, PRECISION SYSTS, INC, 75- *Concurrent Pos:* Consult, Women's Med Col Pa, 58-60. *Mem:* Am Chem Soc; fel Am Inst Chemists; Sigma Xi; AAAS. *Res:* Boron chemistry; endocrine chemistry; automated chemical analysis; glucose metabolism; chemistry of bladder cancer; inborn errors of metabolism; mitogenic agents; calcium analysis; osmometry; molecular weight characterization. *Mailing Add:* 4 Milford St West Medway MA 02053

HOCHHAUS, LARRY, b Britt, Iowa, Mar 9, 42; div; c 2. COGNITIVE PSYCHOLOGY, APPLIED STATISTICS & EXPERIMENTAL DESIGN. *Educ:* Iowa State, BS, 67, MS, 69, PhD(psychol), 70. *Prof Exp:* Asst prof psychol, Cent Col, 70-71; asst prof, 71-73, ASSOC PROF PSYCHOL, OKLA STATE UNIV, 73- *Concurrent Pos:* Vis assoc prof res, Okla Univ Health Sci Ctr, 80-81; assoc ed, Behav Res Methods, Instr & Comput, 82- & J Exp Psych: Human Perception & Performance, 83- *Mem:* Am Psychol Asn; Psychonomic Soc. *Res:* Cognitive psychology; perception, memory and language processes; effects of alcohol and drugs on human information processing. *Mailing Add:* Psychol Dept Okla State Univ Stillwater OK 74078

HOCHHEIMER, BERNARD FORD, b Rochester, NY, May 24, 29; m 55; c 6. OPTICS, SPECTROSCOPY. *Educ:* St Bonaventure Univ, BS, 51; Univ Rochester, MS, 53. *Prof Exp:* Tech assoc optics, Univ Rochester, 53-54; assoc physicist, Appl Physics Lab, Johns Hopkins Univ, 54-56; scientist, Hayes Aircraft Corp, 56-60; PHYSICIST, APPL PHYSICS LAB, JOHNS HOPKINS UNIV, 60-, INSTR OPHTHAL HOSP, 70- *Mem:* Optical Soc Am. *Res:* Gas dynamics and gas lasers; infrared spectroscopy and physics; medical instrumentation and research; applications of lasers to medicine. *Mailing Add:* Appl Physics Lab Johns Hopkins Rd Laurel MD 20723

HOCHMAN, BENJAMIN, b New York, Apr 8, 25; m 69. GENETICS, ZOOLOGY. *Educ:* Univ Calif, AB, 49, MA, 52, PhD, 56. *Prof Exp:* Asst zool, Univ Calif, 52-54, assoc, 54-55; res assoc genetics, City of Hope Med Ctr, 56-57; asst prof, Univ Utah, 57-63; assoc prof, 63-72, PROF GENETICS, UNIV TENN, KNOXVILLE, 72- *Concurrent Pos:* Res partic biol div, Oak Ridge Nat Lab, 60-61. *Mem:* Genetics Soc Am; Soc Study Evolution; NY Acad Sci; Am Asn Univ Prof. *Res:* Genetics of Drosophila; cytogenetics & developmental effects of chromosome 4. *Mailing Add:* 8116 W Cliff Dr Knoxville TN 37909

HOCHMAN, JACK M(ARTIN), b Brooklyn, NY, Jan 26, 39; m 69; c 2. CHEMICAL ENGINEERING. *Educ:* Polytech Inst Brooklyn, BChE, 59; Yale Univ, MEng, 60; Columbia Univ, DEngSc, 64. *Prof Exp:* Sr res engr, Atomics Int Div, NAm Aviation, Inc, 64-65; sr res engr, Esso Res & Eng Co, 65-88, DEPT MGR, EQUIP OIL PROCESSING, EXXON RES & ENG CO, 86- *Concurrent Pos:* Adj asst prof, NY Univ, 68-69. *Mem:* Am Inst Chem Engrs; Am Chem Soc. *Res:* Petroleum refinery planning; chemical reactor engineering; conversion of coal to synthetic fuels via liquefaction/gasification; heavy oil processing. *Mailing Add:* Exxon Res & Eng Co PO Box 101 Florham Park NJ 07932

HOCHMAN, PAULA S, b Brooklyn, NY, Feb 14, 53. IMMUNOGENETICS. *Educ:* State Univ NY, Buffalo, PhD(microbiol & immunol), 79. *Prof Exp:* Fel, 81-83, instr, 83-85, RES ASST PROF PATH, TUFTS UNIV, 85- *Mem:* Am Asn Immunologists. *Mailing Add:* Tufts Univ Sch Vet Med 136 Harrison Ave Boston MA 02111

HOCHMAN, ROBERT F(RANCIS), b Chicago, Ill, May 1, 28; m 60; c 2. METALLURGY. *Educ:* Univ Notre Dame, BS, 50, MS, 54, PhD(metall), 59. *Prof Exp:* Foundry metallurgist, Dodge Mfg Corp, Ind, 50-51; metallurgist, Bendix Aviation Corp, 54-55; spec instr, Eve Div, Mich State Univ, 58-59; from asst prof to prof & assoc dir metall, Sch Chem Eng, 59-82, PROF METALL, SCH MAT ENG, GA INST TECHNOL, 80- *Concurrent Pos:* Consult, Zimmer Mfg Co, 57-, Lockheed Ga Ford Motor Co, 64-, Howmedica Inc, 85- *Honors & Awards:* Res Award, Sigma Xi, 77. *Mem:* Fel Am Soc Metals; Nat Asn Corrosion Engrs; Am Soc Nondestructive Testing; Am Soc Testing & Mat; fel Am Soc Met, 77. *Res:* Physical metallurgy surface modification, ion implantation and plating; basic and applied corrosion; biomedical materials. *Mailing Add:* Sch Mat Eng Ga Inst Technol 225 N Ave NW Atlanta GA 30332-0245

HOCHMUTH, ROBERT MILO, b Berkeley, Calif, May 29, 39; m 64; c 1. CELL BIOMECHANICS, BIORHEOLOGY. *Educ:* Univ Colo, BS, 61; Ohio State Univ, MS, 62; Brown Univ, PhD(eng), 67. *Prof Exp:* From asst prof to prof chem eng, Wash Univ, 67-78; PROF BIOMED ENG, DUKE UNIV, 78-, PROF MECH ENG & CHAIR DEPT, 86- *Concurrent Pos:* NIH res career develop award, 73-78; vis prof, Dept Biophys, Univ Rochester, 85-86; assoc ed, J Biomech Eng, 87-; mem bd dirs, Biomed Eng Soc, 87-91; chmn, Bioprocess Eng Prog, Am Soc Mech Engrs, 89-91; consult, NIH & NSF. *Mem:* Am Soc Mech Engrs; Biophys Soc; Soc Rheology; Biomed Eng Soc; AAAS. *Res:* Elastic and viscous properties of cells and cell membranes; biorheology. *Mailing Add:* Dept Mech Eng & Mat Sci Duke Univ Durham NC 27706

HOCHSCHILD, GERHARD P, MATHEMATICS. *Prof Exp:* Emer prof math, Univ Calif, Berkely, 86; RETIRED. *Mem:* Nat Acad Sci. *Mailing Add:* Dept Math Univ Calif Berkeley CA 94720

HOCHSTADT, HARRY, b Vienna, Austria, Sept 7, 25; nat US; m 53; c 2. MATHEMATICS. *Educ:* Cooper Union, BChE, 49; NY Univ, MS, 50, PhD(math), 56. *Prof Exp:* Res engr, W L Maxson Corp, 51-57; assoc prof, Polytech Inst NY, 57-59, head dept, 63-90, dean arts & sci, 74-75, PROF MATH, POLYTECH INST NY, 59- *Concurrent Pos:* Adv ed, Intersci Series Pure & Appl Math, 72- *Mem:* Am Math Soc; Math Asn Am; Soc Indust & Appl Math; Sigma Xi. *Res:* Special functions; partial differential equations; problems of wave propagation. *Mailing Add:* 126 Joralemon St Brooklyn NY 11201

HOCHSTADT, JOY, b New York, NY, May 6, 39; m 60. BIOCHEMISTRY, MICROBIOLOGY. *Educ:* Columbia Univ, AB, 60; Stanford Univ, AM 63; Georgetown Univ, PhD(microbiol), 69. *Prof Exp:* USPHS res fel, Lab Biochem, Nat Heart & Lung Inst, 68-70, guest worker, 70-72; sr scientist molecular biol, Worcester Found Exp Biol, 72-76; vis prof, Weizmann Inst, Israel, 76; vis prof biochem & biophysics, Univ RI, 76-77; res prof microbiol, NY Med Col, 77-81; DIR CELL GENETICS RES LAB & DIV CLIN BIOCHEM, CATHOLIC MED CTR, WOODHAVEN, NY, 81- *Concurrent Pos:* Estab investigatorship, Am Heart Asn, 70-75, mem coun, Basic Sci, 71-; adj prof biochem, Cent New England Col, 74-75; mem NSF fel educ panel, Nat Res Coun, 75-; prin investr res grants, Nat Inst Gen Med Sci, 73-76, & Nat Cancer Inst, 73- & Nat Sci Found, 78-81; mem NATO fel panel, 78 & cell biol study sect, 79-; vis scholar, Columbia Univ, 84- *Mem:* Am Soc Microbiol; fel Am Inst Chemists; Am Soc Biol Chemists; fel Am Acad Microbiol; Am Soc Cell Biol. *Res:* Cell division and cell differentiation; molecular biology of regulation of differentiation in lens cells and hepatoma cells; cell genetics including gene isolation for cell division specific genes. *Mailing Add:* Princeton Polymer Labs 501 Plainsboro Rd Plainsboro NJ 08536

HOCHSTEIN, HERBERT DONALD, b Johnstown, Pa, Sept 12, 29; m 54; c 3. MICROBIOLOGY. *Educ:* Univ Md, BS, 56; George Washington Univ, MS, 59; Univ NC, MPH, 68, DrPH, 70. *Prof Exp:* Med technician med microbiol, Clin Path Dept, NIH, 56-67, microbiologist, Div Biol Stands, 70-72; MICROBIOLOGIST, BUR BIOLOGICS, FOOD & DRUG ADMIN, 72- *Mem:* Am Pub Health Asn; Am Soc Microbiol. *Res:* Detecting endotoxin in biological products using both the rabbit pyrogen test and the Limulus amebocyte lysate test. *Mailing Add:* 11313 Orleans Way Kensington MD 20795

HOCHSTEIN, LAWRENCE I, b Chicago, Ill, June 16, 28; m 55, 73; c 2. BACTERIOLOGY, MICROBIAL PHYSIOLOGY. *Educ:* Univ Southern Calif, BA, 53, MS, 55, PhD(bact, biochem), 58. *Prof Exp:* Instr biochem, Univ Southern Calif, 55-56, res assoc bact, 57-58, lectr, 58, res assoc dept med microbiol, 58-59; res assoc, Elgin State Hosp, 59-60; instr microbiol, Col Med, Univ Ill, 60-61; res assoc dept genetics, Sch Med, Stanford Univ, 61-63; SR SCIENTIST, AMES RES CTR, 63- *Mem:* Am Soc Microbiol; Brit Soc Gen Microbiol. *Res:* Physiology of halophilic bacteria and sulfolobus; denitrification. *Mailing Add:* Ames Res Ctr 239-10 Moffett Field CA 94035

HOCHSTEIN, PAUL EUGENE, b New York, NY, Feb 7, 26; m 56; c 2. PHARMACOLOGY. *Educ:* Rutgers Univ, BS, 50; Univ Md, MS, 52, PhD(phytopath), 54. *Hon Degrees:* PhD, Univ Stockholm, 86. *Prof Exp:* Assoc biochem, Col Physicians & Surgeons, Columbia Univ, 57-62; from asst prof to assoc prof pharmacol, Duke Univ, 63-69; prof pharmacol, 69-80, PROF TOXICOL & BIOCHEM, DIR INST TOXICOL & ASSOC DEAN, UNIV SOUTHERN CALIF, 80- *Concurrent Pos:* NIH res fel, 54-57, career award, 65-69; NSF sr fel, Wenner-Gren Inst, 62-63; vis prof, Univ Stockholm, 78-79, Univ Genoa, 86. *Mem:* AAAS; Am Asn Cancer Res; Am Soc Biol Chemists; Soc Gen Physiol; Am Soc Pharmacol & Exp Therapeut; Soc Toxicol. *Res:* Toxicological mechanisms; hydrogen peroxide and organic peroxide formation and detoxification; oxygen toxicity; hemolytic mechanisms. *Mailing Add:* Inst Toxicol Univ Southern Calif Los Angeles CA 90033

HOCHSTER, MELVIN, b Brooklyn, NY, Aug 2, 43. COMMUTATIVE ALGEBRA, ALGEBRAIC GEOMETRY. *Educ:* Harvard Univ, BA, 64; Princeton Univ, MA, 66, PhD(math), 67. *Prof Exp:* From asst prof to assoc prof math, Univ Minn, 67-73; prof, Purdue Univ, 73-77; prof, 77-84, R L WILDER PROF MATH, UNIV MICH, 84- *Concurrent Pos:* Guest prof, Math Inst Aarhn, Denmark, 73-74; fel, John Simon Guggenheim Mem Found, 82; bd gov, Inst Math & Applns, 85-; bd trustees, Math Sci Res Inst, Berkeley, 85- *Honors & Awards:* Frank Nelson Cole Prize, Am Math Soc, 80. *Mem:* Am Math Soc; Math Asn Am. *Res:* Commutative Noetherian rings including Cohen-Macaulay rings and modules; invariant theory; algebraic geometry. *Mailing Add:* Math Dept Univ Mich Ann Arbor MI 48109-1003

HOCHSTETLER, ALAN RAY, b Nappanee, Ind, May 7, 39; m 81. SYNTHETIC ORGANIC CHEMISTRY. *Educ:* Goshen Col, BA, 64; Northwestern Univ, PhD(org chem), 68. *Prof Exp:* Instr org chem, Northwestern Univ, 68-69; sr res chemist, 69-81, DIR CHEM DEVELOP, GIVAUDAN CORP, 81- *Mem:* Am Chem Soc. *Res:* Natural products chemistry. *Mailing Add:* Givaudan Corp 125 Delawanna Ave Clifton NJ 07014

HOCHSTIM, ADOLF R, b Krakow, Poland, Nov 13, 28; US citizen; div; c 2. PHYSICS. *Educ:* Univ Miami, Fla, BS, 51; Fla State Univ, MS, 53; Univ Fla, PhD(physics), 67. *Prof Exp:* Sr staff scientist, Gen Dynamics/Convair, Calif, 56-62; mem tech staff, Inst Defense Anal, 62-68; dir Res Inst Eng Sci, Wayne State Univ, 68-74, prof eng sci, 68-75; PRES, PHYS DYNAMICS, INC, 69- *Concurrent Pos:* Pres, La Jolla Inst, 76-; adj prof, Univ Calif, San Diego, 79- *Mem:* AAAS; Am Phys Soc; NY Acad Sci; Sigma Xi. *Res:* Thermodynamics of high temperature gases and properties behind shocks; physics of reentry of space vehicles; transport processes in ionized gases; microwave scattering from ionized gases and random media; hypersonic chemisynthesis and origin of life; origin of asymmetry of biological molecules. *Mailing Add:* 1672 Via Corona La Jolla CA 92037-1883

HOCHSTRASSER, DONALD LEE, b Taylorsville, Ky, June 10, 27; m 60; c 3. MEDICAL ANTHROPOLOGY, COMMUNITY HEALTH. *Educ:* Univ Ky, BA, 52, MA, 55; Univ Ore, PhD(anthrop), 63; Univ Calif, Berkeley, MPH, 69. *Prof Exp:* Instr anthrop, Univ Ky, 56-57; instr, Univ Ore, 58-59; from instr to prof anthrop & community med, Col Arts & Sci & Col Med, 61-80, assoc dir, Ctr Develop Change, 70-73, PROF ANTHROP, COMMUNITY HEALTH, & PUB ADMIN, COL ARTS & SCI, COL ALLIED HEALTH, & SCH PUB ADMIN, UNIV KY, 80- *Concurrent Pos:* Nat Tuberc Asn res grant, Cent Ky, Univ Ky, 62-65, USPHS proj grant, Appalachia, 64-65, Nat Inst Child Health & Human Develop res grant, Appalachia, 71-74, US AID proj grant, 72, US Agency Action prog grant, 72-73 & Robert Wood Johnson Found res & prog eval grant, 74-77; mem, Nat Guid Comt, TB Control Prog, Nat Asn-Am Thoracic Soc, 64-67; NIMH spec res fel, Univ Calif, Berkeley, 68-69, vis scholar, Sch Pub Health, 79-80; mem, Adv Bd & consult panel, Small Towns Inst, Illensburg, Wash, 69-; chmn, State Family Planning Prog Rev Comt, Ky State Comprehensive Health Planning Coun, 72-74; consult regionalization effectiveness study, Southeastern Ky Regional Health Demonstration, Inc, Lexington, 73-74; res consult, Infant-Child Care Eval Prog, Hazard Hosp, Appalachian Hosps, Lexington, 74-78; primary care eval proj, Ohio Valley Regional Med Prog Spec Res Contract, 74-75; sickle cell prog eval study res grant, Ky State Bur Health Serv, 75-78; Appalachian fertil study contract res proj, Nat Inst Child Health & Human Develop, Pop Res Ctr, 80-83; Appalachian Fertil Study, Nat Inst Child Health & Develop res proj, Univ Ky Res Found, 83-85; res consult, State Family Planning Prog, Eval Study, Access to Family Planning Serv in Appalachia, Ky, State Dept Health Serv, 88-89. *Mem:* Am Pub Health Asn; fel Soc Appl Anthrop; fel Am Anthrop Asn; AAAS; Am Acad Polit & Soc Sci; Sigma Xi. *Res:* Medical anthropology, cultural and social change, community study research; health services research and program evaluation, population and fertility study, social and behavioral aspects of health behavior; health services utilization and patient compliance; rural health care. *Mailing Add:* Dept Health Serv Univ Ky 131 Annex Two Chandler Med Ctr Lexington KY 40536-0080

HOCHSTRASSER, ROBIN, b Edinburgh, Scotland, Jan 4, 31; US citizen; m 60; c 2. CHEMISTRY, CHEMICAL PHYSICS. *Educ:* Heriot-Watt Univ, BSc, 52; Univ Edinburgh, PhD(phys chem), 55. *Hon Degrees:* DSc, Heriot-Watt Univ. *Prof Exp:* Instr, Univ BC, 57-60, asst prof, 60-62; from assoc prof to prof, 63-69, Blanchard prof, 69-83, DONNER PROF CHEM, UNIV PA, 83- *Concurrent Pos:* Alfred P Sloan fel, 63-67; Guggenheim fel, 71-72; vis prof, Cambridge Univ, Eng, 72, Australian Nat Univ, 73, Univ Grenoble, 74, Calif Inst Technol, 75; Alexander von Humboldt sr fel, 78. *Honors & Awards:* Sherwin-Williams Lectr, Univ Ill, 77; Phillips Lectr, Haverford Col, 78; Bourke Lectr, Faraday Soc, Royal Inst, 81; Noyes Distinguished Lectr, Oxford Univ, 82; Cyril Hinshelwood Lectr, Oxford Univ, 82; Reilly Lectr, Notre Dame Univ, 86. *Mem:* Nat Acad Sci; AAAS; Am Chem Soc; Am Acad Arts & Scis; Royal Inst Chem; Am Phys Soc; Am Inst Physics; Am Asn Univ Profs. *Res:* Molecular spectroscopy with fentosecond and other laser methods; photo properties of large molecules and solids with emphasis on primary processes. *Mailing Add:* 242 S Phillip Pl Philadelphia PA 19106

HOCHULI, URS ERWIN, b Biel, Switz, Feb 9, 27; m 54; c 3. ELECTRICAL ENGINEERING, PHYSICS. *Educ:* Univ Md, MS, 55; Cath Univ, PhD, 61. *Prof Exp:* Elec engr, Brown Boveri Co, Switz, 50-52; asst inst fluid dynamics, Univ Md, 52-55; from asst prof to assoc prof elec eng, 55-71, PROF ELEC ENG, UNIV MD, COLLEGE PARK, 71- *Honors & Awards:* Cert Recognition, NASA. *Mem:* Am Phys Soc; Inst Elec & Electronic Engrs. *Res:* Quantum electronics; laser technology. *Mailing Add:* Dept Elec Eng Univ Md College Park MD 20742

HOCHWALD, GERALD MARTIN, b New York, NY, June 13, 32; m 59; c 1. NEUROLOGY, IMMUNOLOGY. *Educ:* Alfred Univ, BA, 53; State Univ Leiden, MD, 59. *Prof Exp:* Intern, Bronx Hosp, NY, 59-60; resident neurol, 60-64, from instr to prof neurol, 64-77, PROF PHYSIOL & BIOPHYS, MED CTR, NY UNIV, 77- *Concurrent Pos:* Fel path, Med Ctr, NY Univ, 61-62; Nat Inst Neurol Dis & Stroke fel, 61-62, spec fel & res grant, 64-67, career develop award, 67-; assoc attend physician, Bellevue & Univ Hosps, 65. *Mem:* Am Acad Neurol; Am Asn Immunol; Am Physiol Soc; Brit Soc Res Hydrocephalus & Spina Bifida; Am Neurol Asn. *Res:* Origin of serum and spinal fluid proteins; blood-brain barrier for proteins; experimental hydrocephalus. *Mailing Add:* Dept Neurol NY Univ Med Ctr New York NY 10016

HOCHWALT, CARROLL ALONZO, organic chemistry; deceased, see previous edition for last biography

HOCK, ARTHUR GEORGE, b Hamilton, Ohio, Nov 20, 41; m 70. AGRONOMY, SOIL MORPHOLOGY. *Educ:* Ohio State Univ, BS, 67, MS, 71. *Prof Exp:* Soil scientist, Div Lands & Soils, Ohio Dept Natural Resources, 66-68; soils consult, Hock & Assocs, 69-71; agronomist, 71-77, dir res & agron serv, 77-79, MGR SEED AGRON SERV & RES, COUNTRYMARK, INC, 79- *Mem:* Am Soc Agron; Soil Sci Soc Am; Crop Sci Soc Am; Coun Agr Sci & Technol. *Res:* Development of soybean, corn and forage varieties; soil fertility; use of sewage sludge as a fertilizer. *Mailing Add:* Countrymark Inc 4565 Columbus Pike Delaware OH 43015

HOCK, DONALD CHARLES, b North Tonawanda, NY, Apr 18, 29; m 57; c 2. PHYSICS. *Educ:* Univ Buffalo, BA, 51; Agr & Mech Col, Tex, MS, 52; Pa State Univ, PhD(physics), 56. *Prof Exp:* Sr staff mem, Appl Physics Lab, Johns Hopkins Univ, 55-58; res physicist, Res Div, Radiation, Inc, 58-61; design engr, Orlando Div, Martin Co, 61-64; sr mem tech staff, Data Dynamics, Inc, 64-65; SR MEM PROF STAFF, ORLANDO DIV, MARTIN MARIETTA CORP, 65- *Mailing Add:* 2480 Temple Dr Winter Park FL 32789

HOCKEL, GREGORY MARTIN, b Philadelphia, Pa, Nov 10, 50; m 76; c 2. REGULATORY AFFAIRS, MEDICINAL DRUG DEVELOPMENT. *Educ:* Calif State Univ, Long Beach, BA, 72; Ind Univ, PhD(physiol), 77. *Prof Exp:* From instr to asst prof physiol, Sch Med, Univ Miss, 77-80; res scientist, 80-83, sr res scientist, 83-85, SR REGULATORY AFFAIRS SCIENTIST, PFIZER CENT RES, 85- *Mem:* Am Physiol Soc; Regulatory Affairs Prof Soc; NY Acad Sci. *Res:* Role of the kidney and renal prostaglandins in the control of arterial blood pressure; renin-angiotensin-aldosterone system; design of renin inhibitors for treatment of hypertension. *Mailing Add:* G H Besselaar Assocs 103 College Rd E Princeton NJ 08540

HOCKEN, ROBERT JOHN, b Dinuba, Calif, Feb 20, 44; m 80; c 3. PHYSICS, METROLOGY. *Educ:* Ore State Univ, BA, 68; State Univ NY, Stony Brook, MA, 69, PhD(physics), 73. *Prof Exp:* Fel, Equation State Sect, 73-75, physicist dimensional technol, 75-76, group leader dimensional metrol, 76-80, chief, automated prod technol, 80-85, CHIEF, PRECISION ENG DIV, NAT BUR STANDARDS, 85- *Honors & Awards:* Silver Medal, Dept Com, 78; F W Taylor Medal, Int Inst Prod Eng Res, 79; IR-100 Award, 80; Nat Bur Standards Appld Res Award, 80; Fredrick W Taylor Res Award, Soc Mfg Engrs, 85. *Mem:* Am Phys Soc; Am Soc Testing & Mat; Int Inst Prod Eng Res; Soc Mfg Engrs. *Res:* Laser interferometry; three-dimensional metrology; machine tool metrology; dimensional measurement of large structures; critical phenomena; machine tool automation; robotics; computer aided design/computer aided manufacturing. *Mailing Add:* Dept Mech Eng Univ NC Charlotte NC 28223

HOCKENBERRY, TERRY OLIVER, b Butler, Pa, Apr 13, 38. ELECTRICAL ENGINEERING. *Educ:* Univ Cincinnati, Elec Eng, 61; Carnegie Inst Technol, MSc, 62, PhD(elec eng), 64. *Prof Exp:* Asst elec eng, Particle Accelerator Div, Argonne Nat Lab, 61; asst prof elec eng, Carnegie-Mellon Univ, 64-70; consult engr, 70-90; PRES, TOH LABS, 90- *Concurrent Pos:* Sci consult, Siltronics, Inc, 64-65, dir res, 66-70; expert witness, var legal firms, 66-78; dir res, Krux, Inc, 70-73; lectr, Carnegie-Mellon Univ, 70-79. *Mem:* Inst Elec & Electronic Engrs; NY Acad Sci; AAAS; Soc Mfg Engrs; Nat Fire Protection Asn; Sigma Xi. *Res:* Electrical discharge machining; basic research in spark erosion of metallic conductors; effects of electrical arcs on materials; electrical distress conditions; fire causation research. *Mailing Add:* 112 Wynnwood Rd Pittsburgh PA 15215

HOCKENBURY, ROBERT WESLEY, b Springfield, Mass, July 31, 28; m 55; c 3. NUCLEAR ENGINEERING. *Educ:* Union Col, BS, 52; Rensselaer Polytech Inst, MS, 59, PhD(nuclear sci), 67. *Prof Exp:* Nuclear analyst, Knolls Atomic Power Lab, Gen Elec Co, 56-58; res asst, Nuclear Eng & Sci Div, 58-67, res assoc, 67-74, ASSOC PROF NUCLEAR ENG & SCI DIV, RENSSELAER POLYTECH INST, 74- *Concurrent Pos:* Consult reactor safety, NY risk assessment, availability; mem, Exec Coun, Soc Risk Anal. *Mem:* Am Nuclear Soc; Am Phys Soc; Soc Risk Anal. *Res:* Reliability; risk analysis; failure data analysis; probabilistic methods; radiation induced curing of polymers. *Mailing Add:* Eight Oak Dr Albany NY 12203

HOCKER, GEORGE BENJAMIN, b Harrisburg, Pa, Sept 11, 42; m 72; c 2. SENSORS. *Educ:* Cornell Univ, BEP, 65, PhD(appl physics), 70. *Prof Exp:* Asst prof elec eng, Univ Minn, Minneapolis, 70-76; sect head, 76-85, dept mgr, 85-87, PRIN RES FEL, HONEYWELL SENSOR & SYST DEVELOP CTR, 87- *Concurrent Pos:* Prog chmn, 88, gen chmn, Inst Elect & Electronic Engrs Solid State Sensor & Actuator Works Lab, 90. *Mem:* Inst Elec & Electronics Engrs. *Res:* Sensor technology and applications, especially silicon sensors, micromachining, and micro-structure devices. *Mailing Add:* Honeywell Sensor & Syst Develop Ctr 10701 Lyndale Ave S Bloomington MN 55420

HOCKETT, ROBERT C(ASAD), biochemistry, for more information see previous edition

HOCKING, DRAKE, b Kurseong, India, Oct 14, 40; Can citizen; m 61; c 3. AGROFORESTRY, SOCIAL FORESTRY. *Educ:* Univ Alta, BSc, 60; Univ St Andrews, Scotland, MSc, 61; Durham Univ Eng, PhD(biochem genetics), 63; Harvard Univ, MPA, 71. *Prof Exp:* Res officer phytopath, EAfrican Conserv Serv, 63-67; res scientist, Can Forestry Serv, 67-75; chief Wildlife Mgt, Can Wildlife Serv, 75-78; first secy, Natural Resources, Brit High Comn, Dhaka Bangladesh, 78-80; tech adv agroforestry, Indian Coun Agr Res, 83-85; assoc agroforestry, Oxford Forestry Inst, 85-86; monitoring specialist, INFRAS Consult Int Develop, 87-90; PRIN CONSULT AGROFORESTRY, STEEP ASSOC, 90- *Concurrent Pos:* Consult, ICRAF, Nairobi, 89, CIDA, O'Hara, 80-83, Intercoop, Berne, Switz, 87-, FAO, Rome, 89. *Res:* Agroforestry, particularly tropical and covering arid to humid environments; related social forestry and range management; social organization; management of common property resources and rural development. *Mailing Add:* Thrushes Stroud Lane Steep Petersfield Hants GU32 1AL England

HOCKING, GEORGE MACDONALD, b Newquay, Eng, Mar 21, 08; US citizen; m 33; c 1. PHARMACOGNOSY. *Educ:* Univ Wash, BSP, 31; Univ Fla, MSP, 32, PhD(pharmacog), 42. *Prof Exp:* Instr, Col Pharm, George Washington Univ, 33-35; chief pharmacognosist, S B Penick & Co, New York, 42-46; prof pharmacog, Col Pharm, Univ Buffalo, 46-48; prof pharmacog & pharmacol, Col Pharm, Univ NMex, 48-51; PROF PHARMACOG, SCH PHARM, AUBURN UNIV, 51- *Concurrent Pos:* Instr, Col Pharm, Ohio Northern Univ, 37-40; abstr, Biol Abstracts, Excerpta Botanica Sect A & Phytologia, 46-; mem, Comt Nat Formulary & chmn, Subcomt Pharmacog, 47-60; vis prof, Univ Miss, 47-51; consult med plants, Food & Agr Orgn, Govt Pakistan, 51; mem, Comn Study Med & Toxic Plants Brazil, 52 & Int Comn Plant Raw Mat, 65; ed, Int J Crude Drug Res, Lisse, Netherlands, 75- *Honors & Awards:* ACS Award, Am Chem Soc, 26. *Mem:* Am Pharmaceut Asn; Am Soc Pharmacog; Nat Asn Stand Med Vocabulary. *Res:* Crude plant drug materials and drug plants, including plants belonging to genera Liatris, Mentha and Iris; terminology drugs. *Mailing Add:* Sch Pharm Auburn Univ Auburn AL 36849-5503

HOCKING, JOHN GILBERT, b Caspian, Mich, Sept 26, 20; m 44; c 4. MATHEMATICS. *Educ:* Univ Mich, PhD(math), 53. *Prof Exp:* Asst, Off Naval Res, Univ Mich, 48-51; from instr to assoc prof, 51-63, PROF MATH, MICH STATE UNIV, 63- *Concurrent Pos:* Fulbright guest prof, Univ T06bingen, 62-63. *Mem:* Am Math Soc; Math Asn Am. *Res:* Topology of manifolds; game theory; logic. *Mailing Add:* Mich State Univ East Lansing MI 48824r

HOCKING, MARTIN BLAKE, b London, Nov 25, 38; Can citizen; m 62; c 3. ORGANIC REDOX RATES & MECHANISMS. *Educ:* Univ Alta, BSc, 59; Univ Southampton, PhD(org chem), 63. *Prof Exp:* Res chemist, Dow Chem Can, Ltd, 63-71; asst prof, 71-75, ASSOC PROF CHEM, UNIV VICTORIA, 75- *Concurrent Pos:* Vis prof, McGill Univ, Montreal, 78-79; vis scientist, Pulp & Paper Res Inst Can, 78-79; Harold Hibbert Mem fel, 78-79; consult, MacMillan Bloedel Ltd, 85-87; res fel, Univ Col London, 87-88. *Mem:* Fel Royal Inst Chem; fel Chem Inst Can; Royal Soc Chem; Am Chem Soc; Can Soc Chem & Biochem Technol. *Res:* Application of principles of aqueous aromatic oxidation chemistry to development of peroxide bleaching of mechanical pulps; synthesis and stereochemistry of bridged ring phosphorus heterocycles; environmental chemistry. *Mailing Add:* Dept Chem Univ Victoria PO Box 3055 Victoria BC V8W 3P6 Can

HOCKING, RONALD RAYMOND, b Ishpeming, Mich, June 4, 32; m 54; c 3. MATHEMATICS, STATISTICS. *Educ:* Mich Tech Univ, BS, 54; Univ Mich, MS, 57; Iowa State Univ, PhD(math, statist), 62. *Prof Exp:* Res assoc, Boeing Airplane Co, 57-58; asst prof math, Mich Tech Univ, 58-63; from assoc prof to prof statist, Tex A&M Univ, 63-71; prof mgt sci, Univ Houston, 71-76; prof statist & head dept, Miss State Univ, 76-80; PROF STATIST, TEX A&M UNIV, 80- *Honors & Awards:* Snedecor Award, Iowa State Univ, 62; Wilcoxon Award, 71 & 76; Shewell Prize, 78 & 80. *Mem:* Fel Am Statist Asn. *Res:* Linear models; regression; statistical estimation theory, especially minimum variance estimation; multivariate analysis. *Mailing Add:* Dept Statist Tex A&M Univ College Station TX 77843

HOCKMAN, CHARLES HENRY, b Montreal, Que, Mar 15, 23; m 52; c 3. NEUROPHYSIOLOGY, NEUROPHARMACOLOGY. *Educ:* Queen's Univ, Ont, BA, 58; Brown Univ, ScM, 60, PhD(exp psychol, electrophysiol), 63. *Prof Exp:* Asst prof neurophysiol, Med Col Va, 62-66; assoc prof pharmacol, Fac Med, Univ Toronto, 66-72; assoc prof neurosci, Sch Basic Med Sci, Univ Ill Col Med, 72-82; PROF NEUROSCI, SCH MED, MERCER UNIV, MACON, GA, 82- *Concurrent Pos:* Consult, Hoffman-LaRoche, US, 62-66, Can, 66-71; sci officer, Dept Health & Welfare, Can, 72-73. *Mem:* Am Physiol Soc; Can Physiol Soc; Soc Neurosci; Sigma Xi; World Fedn Neurol; Int Brain Res Orgn. *Res:* Neurophysiol and neuropharmacol relationships between basal forebrain and bulbar motor functions; CNS regulation of autonomic function; effects of psychotropic drugs on cortical and subcortical regions of the brain. *Mailing Add:* Sch Med Mercer Univ Macon GA 31207

HOCKMAN, DEBORAH C, b Sept 1, 55; c 2. RESEARCH & DEVELOPMENT. *Educ:* Northeastern Ill Univ, BS, 77; Loyola Univ, Chicago, PhD(anal Chem), 82. *Prof Exp:* Spectroscopist & lab coordr, Anal Serv Lab, Northwestern Univ, 82; res scientist, Prod Develop Anal Dept, G D Searle & Co, 82-84; group leader formulations res, Med Res Div, Am Cyanamid Co, 84-87; ASST DIR, WMI ENVIRON MONITORING LABS, INC, 87- *Concurrent Pos:* Schmitt fel. *Mem:* Am Chem Soc; Soc Appl Spectros (pres, 90-91); Sigma Xi; Asn Women Sci. *Res:* Lab robotics; chromatography; spectroscopy. *Mailing Add:* WMI Environ Monitoring Lab Inc 2100 Cleanwater Dr Geneva IL 60134-4101

HOCKNEY, RICHARD L, POWER ELECTRONICS, CONTROL SYSTEMS. *Educ:* Northeastern Univ, BS, 72, MS, 80. *Prof Exp:* Design engr, C S Draper Lab, 72-75, sr engr, 76-79, prin engr, 80-85; div leader, 86-90, VPRES ENG, SATCON TECHNOL CORP, 91- *Mem:* Sr mem Inst Elec & Electronic Engrs. *Res:* Electro-mechanical control systems including magnetic bearings, magnetic suspensions, vibration isolators; power electronics for actuator and motor drives; low-level signal interface for various types of sensors. *Mailing Add:* Sat Con Technol Corp 12 Emily St Cambridge MA 02139

HOCKSTRA, DALE JON, b San Diego, Calif, Mar 29, 48; m 72. MANAGEMENT SCIENCE, OPERATIONS MANAGEMENT. *Educ:* Occidental Col, BA, 69; Stanford Univ, MS, 70, PhD(opers res), 74. *Prof Exp:* Asst prof mgt sci, Wash State Univ, 72-76; assoc prof mgt, Univ Evansville, 76-77, asst dean, 77-80, dean, 81-90, PROF MGT, UNIV EVANSVILLE, 90- *Concurrent Pos:* Assoc prof, Univ Denver, 80-81. *Mem:* Inst Mgt Sci; Decision Sci Inst. *Res:* Mathematical models in production/operations management. *Mailing Add:* Sch Bus Admin Univ Evansville 1800 Lincoln Ave Evansville IN 47722

HOCOTT, C(LAUDE) R(ICHARD), b Excelsior, Ark, Nov 16, 09; m 37; c 2. PETROLEUM ENGINEERING. *Educ:* Univ Tex, BS, 33, MS, 34, PhD(chem eng), 37. *Prof Exp:* Tutor chem, Univ Tex, 33-34, instr chem eng, 34-35, chem, 35-37; res engr, Humble Oil & Refining Co, Standard Oil Co, NJ, 37-42, asst head prod res, 42-54, div head, 54-58, res mgr, 58-64, vpres, Esso Prod Res Co, 64-71, exec vpres, 71-77; dir, Petrol Res Comt, 77-79, prof, 77-80, EMER PROF ENG, UNIV TEX, AUSTIN, 80- *Mem:* Nat Acad Eng; Am Asn Petrol Geol; Am Inst Mining, Metall & Petrol Engrs; Am Inst Chem Engrs; Soc Petrol Eng. *Res:* Occurrence and recovery of petroleum; cooling tower for suspensions; geochemical prospecting; well logging; decomposition of methane in electrical discharge. *Mailing Add:* Col Eng Univ Tex Austin TX 78712

HOCUTT, CHARLES H, b Florence, Ala, Feb 25, 44; m; c 2. AQUATIC ECOLOGY, ICHTHYOLOGY. *Educ:* Va Polytech Inst & State Univ, BS, 68, PhD, 74; Southern Conn State Col, MS, 70. *Prof Exp:* Lead scientist ecol, Environ Res & Technol, 74-75; res assoc, Appalachian Environ Lab, Univ Md, 75-77, asst prof, 77-81, assoc prof, 81-88, prof, 88; MEM STAFF, HORN POINT ENVIRON LAB, UNIV MD, 81- *Concurrent Pos:* Prof fisheries scientist, 74-; mem comt exotic & ornamental fishes, Am Fisheries Soc, 77-78 & mem environ concerns comt, 78-79; mem comt rare & endangered fishes, Pa, Va & WVa, 78-88; mem, Prof Cert Comt, 87-91. *Honors & Awards:* Fulbright recipient, 86-87. *Mem:* Am Fisheries Soc; Ecol Soc Am; Soc Int Limnol; Soc Ichthyologists & Herpetologists; Sigma Xi. *Res:* Effects of natural and man-related stress on aquatic ecosystems; succession in aquatic communities; ecology, systematics and zoogeography of fishes; fisheries development in subsaharan Africa. *Mailing Add:* Horn Pt Environ Lab Box No 775 Univ Md Cambridge MD 21613

HODDER, ROBERT WILLIAM, b Ottawa, Ont, Apr 10, 32; m 65; c 2. GEOLOGY. *Educ:* Queen's Univ, Ont, BA, 55; Univ Calif, Berkeley, PhD(geol), 59. *Prof Exp:* Geologist, Am Metal Climax Inc, 59-64; geologist & vpres, Callahan Mining Corp, 64-70; from asst prof to assoc prof, 70-75, PROF GEOL, UNIV WESTERN ONT, 75-, CHMN, DEPT GEOL, 83- *Concurrent Pos:* Consult, Callahan Mining Corp, 70- *Mem:* Am Inst Prof Geologists; Am Inst Mining, Metall & Petrol Engrs; Soc Econ Geologists; Geol Asn Can. *Res:* Metallogenetic relationships of mineral deposits and their economic significance; physical volcanolgy and its role in metal concentration. *Mailing Add:* Dept Geol Univ Western Ont London ON N6A 5B9 Can

HODDER, VINCENT MACKAY, b Harbour Breton, Newf, Jan 24, 27; m 61; c 2. FISHERIES SCIENCE. *Educ:* McGill Univ, BSc, 56; Memorial Univ, MSc, 62. *Prof Exp:* Asst scientist biol sta, Fisheries Res Bd Can, 56-60, haddock invests, 60-64, actg asst dir, 64-65, scientist in charge pelagic invests, 64-71, consult math statists, 56-71; mem assessment subcomt & co-ed assessment report, Northwest Atlantic Fisheries Orgn, 59-64, asst exec secy, Int Comn for the Northwest Atlantic Fisheries, 71-79, asst exec secy, 79-; RETIRED. *Concurrent Pos:* Ed, Jour Northwest Atlantic Fisheries Sci & NAFO Sci Coun Studies, 80- *Res:* Biology of haddock and herring; application of fish population dynamics models; selectivity of trawls with regard to bottom fishes; assessment of fish stocks, conservation. *Mailing Add:* Ten Cathycross Dr Dartmouth NS B2W 2R5 Can

HODEL, MARGARET JONES, b Lewistown, Pa, July 6, 41; m 70; c 2. COMBINATORIAL ANALYSIS, NUMBER THEORY. *Educ:* Vassar Col, AB, 63; Duke Univ, PhD(math), 72. *Prof Exp:* Math teacher, Am Community Sch, Athens, Greece, 63-65 & St Stephens Sch, Rome, Italy, 65-68; INSTR MATH, DUKE UNIV, 74- *Mem:* Am Math Soc. *Res:* Enumeration of sequences of nonnegative integers, of weighted sequences, of weighted rectangular arrays; Dedekind sums. *Mailing Add:* Dept Math Duke Univ Durham NC 27706

HODEL, RICHARD EARL, b Winston-Salem, NC, Sept 24, 37; m 70; c 2. MATHEMATICS. *Educ:* Davidson Col, BS, 59; Duke Univ, PhD(math), 62. *Prof Exp:* Asst prof, 65-70, ASSOC PROF MATH, DUKE UNIV, 70- *Mem:* Am Math Soc; Math Asn Am; Symbolic Logic Asn. *Res:* Set-theoretic topology. *Mailing Add:* Dept Math Duke Univ Durham NC 27706

HODES, LOUIS, b New York, NY, June 19, 34; m 67. APPLICATIONS IN CHEMISTRY, STRUCTURE ACTIVITY RELATIONSHIPS. *Educ:* Polytech Inst Brooklyn, BEE, 56, MS, 58; Mass Inst Technol, PhD(math), 62. *Prof Exp:* Staff mem math, IBM Res Ctr, 61-65; vis mem, Courant Inst Math Sci, NY Univ, 65-66; res mathematician, Div Comp Res & Technol, NIH, 66-74, RES MATHEMATICIAN, DIV CANCER TREATMENT, NAT CANCER INST, 74- *Concurrent Pos:* Vis mathematician, Appl Logic Group, Hebrew Univ, Jerusalem, 64. *Mem:* Am Chem Soc; Soc Indust & Appl Math; Asn Comput Mach; Inst Elec & Electronics Engrs Comput Soc; Am Asn Cancer Res. *Res:* Theory of pattern recognition on a discrete space; biomedical applications of computers, radiation treatment planning; information theory applied to chemical structure searching; quantitative structure-activity relationships in biochemistry. *Mailing Add:* 10201 Grosvenor Place Rockville MD 20852

HODES, MARION EDWARD, b New York, NY, Aug 6, 25; m 49; c 4. MEDICAL GENETICS, MOLECULAR GENETICS. *Educ:* Univ Buffalo, MD, 47; Columbia Univ, PhD(biochem), 55; Am Bd Clin Chem, dipl, 52, Am Bd Med Genetics, dipl, 82. *Prof Exp:* Asst med, Columbia Univ, 55-56, assoc, 56; from asst prof to assoc prof, 56-66, prof med & biochem, 66-72, PROF MED GENETICS & MED, SCH MED, IND UNIV, INDIANAPOLIS, 72- *Concurrent Pos:* Consult, Regional Genetics Counseling Clin, South Bend, Ind, 80-, Evansville Ind, 83-86 & Med Genetics Clin, Richmond, Ind, 81-82; chmn, Int Sci Coun, Israel Cancer Res Fund, 88- *Mem:* Am Soc Biochem & Molecular Biol; Am Chem Soc; Am Asn Clin Chem; Am Asn Cancer Res; Am Soc Human Genetics. *Res:* Genetic diseases (molecular and clinical); recombinant DNA; preservation of DNA. *Mailing Add:* Dept Med Genetics Ind Univ Med Ctr Indianapolis IN 46202-5251

HODES, PHILIP J, b New York, NY, Apr 15, 06; m 35; c 2. RADIOLOGY. *Educ:* Univ Pa, BS, 28, MD, 31, DSc(radiol), 40; Am Bd Radiol, dipl. *Prof Exp:* Assoc, Hosp Univ Pa, 35-38; prof radiol, Med Sch, Univ Pa, 52-58; prof & chmn dept, Jefferson Med Col, 58-73; prof radiol, 73-87, DISTINGUISHED PROF RADIOL, SCH MED, UNIV MIAMI, 87- *Concurrent Pos:* Fel radiol, Hosp Univ Pa, 40; consult, Jeanne's Hosp, Philadelphia, 52-58; chmn US deleg, Inter-Am Cong Radiol, Peru, 58; consult, Vet Admin AMA II & US Pharmacopoeia; consult & lectr, US Naval Hosp, Philadelphia, Walter Reed Hosp, Washington, DC & Armed Forces Inst Path; mem, Adv Bd Hist Radiol, World War II; mem hist unit, US Army Med Serv. *Honors & Awards:* Gold Medal, Inter-Am Col Radiol; Humanitarian Award, Am Cancer Soc; Gold Medal, Am Col Radiol; Gold Medal, Radiol Soc NAm. *Mem:* Am Roentgen Ray Soc (vpres, 55); Radiol Soc NAm (vpres, 57); fel Am Col Radiol (vpres, 58). *Res:* Neuroradiology; osteoarticular; gastrointestinal tract. *Mailing Add:* Three Island Ave Miami Beach FL 33139

HODES, WILLIAM, b Newark, NJ, May 25, 25; m 55; c 4. ORGANIC CHEMISTRY, POLYMER CHEMISTRY. *Educ:* NY Univ, AB, 48; Ind Univ, MA, 50, PhD(chem), 52. *Prof Exp:* Asst proj chemist, Standard Oil Co, Ind, 52-56; res chemist, Am Cyanamid Co, 56-60; mgr chem dept, Res Div, Am-Standard Corp, 60-69; mgr chem develop lab, 69-80, dir chem develop & gen mgr, chem oper, 80-87, GEN MGR COM CHEM, POLAROID CORP, 87- *Mem:* AAAS; Am Chem Soc. *Res:* Synthetic chemicals-process research and development; polymerization processes; polyelectrolyte properties and applications; membrane separations; steric control of organic reactions; catalysis; manufacture of fine organics and chemical specialties; marketing. *Mailing Add:* 21 Oxford St Winchester MA 01890

HODGDON, F(RANK) E(LLIS), b Malden, Mass, Aug 21, 15; m 47; c 4. MECHANICAL ENGINEERING. *Educ:* Ga Sch Technol, BS, 37. *Prof Exp:* Engr, Gen Elec Co, Ind, 37-38 & Hydril Co, Pa, 38-39; res engr, labs, Am Gas Asn, 39-42, asst chief res engr, 46-49, chief methods engr, 49-52, asst to dir, 52-55, asst dir, 55-56, dir, 56-71, asst managing dir, 62-71, vpres & dir labs, 71-77, sr vpres consumer affairs & safety, 77-80; RETIRED. *Mem:* Int Gas Union; Am Soc Testing & Mat; Newcomen Soc; Nat Soc Prof Engrs. *Res:* Gas appliances and accessories. *Mailing Add:* 1152 Verdon Dr Dunwoody GA 30338

HODGDON, HARRY EDWARD, b Brattleboro, Vt, Sept 4, 46; m 67; c 3. SCIENTIFIC SOCIETY ADMINISTRATION. *Educ:* Univ Maine, Orono, BS, 68; Univ Mass, Amherst, MS, 71, PhD(wildlife biol), 78. *Prof Exp:* Mgr, conserv activ dept, Nat Rifle Asn, 75-76, dir, hunting & conserv div, 76-77; field dir, 77-82, EXEC DIR, WILDLIFE SOC, 82- *Concurrent Pos:* Mem, Nat Nongame Bird Steering Comt, Forest Serv, USDA, 76-80 & Rangeland Policy Comt, Off Secy, 78-80; mem, Exec Comt, Renewable Natural Resources Found, 81-; ed, The Wildlifer, Wildlife Soc, 82-; mem, conserv adv coun, Sen Roger W Jepson, Chmn, US Sen Subcomt Soil, Conserv, Forestry & Environ, 83-84; mem, fish & wildlife adv comt, Conserv Found, 83-84; mem, Info & Commun Study Team, President's Comn Americans Outdoors, 85-86; mem, Nat Threatened & Endangered Species Task Force, Forest Serv, USDA, 88-89, US Implementation Bd, NAm Waterfowl Mgt Plan, 88- & Nat Adv Comt for Eval of the Conserv Title of the Food Security Act of 1985, Soil & Water Conserv Soc, 88- *Mem:* Sigma Xi; AAAS; Am Inst Biol Sci; Wildlife Soc; Am Soc Mammalogists; Soc Range Mgt. *Res:* Integration of wildlife research findings into the political decisionmaking process; improvement of the status of wildlife professionals; enhancement of wildlife education programs; furbearer management; behavior, ecology and population dynamics of beaver. *Mailing Add:* Wildlife Soc 5410 Grosvenor Lane Bethesda MD 20814

HODGDON, RUSSELL BATES, JR, b Medford, Mass, Nov 9, 24; m 58; c 3. PHYSICAL ORGANIC CHEMISTRY. *Educ:* Mass Inst Technol, BS, 51; Columbia Univ, MS, 54, PhD(phys org chem), 57. *Prof Exp:* Res chemist, Monsanto Chem Co, 59-60; sr scientist, Gen Elec Co, Mass, 60-69; MGR MEMBRANE RES & DEVELOP, IONICS INC, 69- *Mem:* Am Chem Soc. *Res:* Polyelectrolyte chemistry; ion exchange membranes synthesis, rheological behavior and solution thermodynamics. *Mailing Add:* 15 Partridge Lane Sudbury MA 01776-2526

HODGE, BARTOW, b Winnfield, La, Jan 11, 20; m 46; c 3. INFORMATION SCIENCE. *Educ:* La State Univ, BS, 47, MS, 48, PhD, 54. *Prof Exp:* Asst physics, La State Univ, 47-51; physicist, Esso Labs, Esso Standard Oil Co, 51-55; appl scientist, IBM Corp, Tex, 55-56, appl sci mgr, La, 57-58, mgr petrol industs, NY, 58-60, mgr control systs, Ill, 60, mgr advan systs, 61-64, prog adminr process industs, White Plains, 64-68, sr prog adminr sensor based syst, 68-73; dis lectr info syst, Col Bus Admin, Univ SC, 73-82; PROF INFO SYSTS, SCH BUS, VA COMMONWEALTH UNIV, 82- *Concurrent Pos:* Consult, Cert Syst Prof, 86. *Mem:* Am Phys Soc; Data Processing Mgt Asn; Asn Syst Mgr; Acad Mgt; Asn Comput Mach. *Res:* Information theory and sciences; application to management information and organization; health care administration. *Mailing Add:* 11220 TurnleyLane Midlothian VA 23113-1373

HODGE, DANIEL B, b Warsaw, Ind, Nov 29, 38; m 64; c 1. ELECTRICAL ENGINEERING. *Educ:* Purdue Univ, BSEE, 60, MSEE, 61, PhD(elec eng), 65. *Prof Exp:* Vis asst prof elec eng, Univ Ill, 64-65; Royal Norweg Coun Sci & Indust res fel, 65-66; asst prof, 66-80, prod elec eng, Ohio State Univ, 80-; CHMN, DEPT ELEC ENG, VA POLYTECH INST & STATE UNIV. *Mem:* Inst Elec & Electronics Engrs. *Res:* Electromagnetic theory; wave propagation and scattering. *Mailing Add:* Elec Eng Dept Ohio State Univ 2015 Neil Ave Columbus OH 43210

HODGE, DAVID CHARLES, b Ft Worth, Tex, July 5, 31; m 57. HUMAN FACTORS ENGINEERING, ROBOTICS. *Educ:* Univ Rochester, PhD(exp psychol), 63. *Prof Exp:* Res psychologist, 62-85, ENG PSYCHOLOGIST, HUMAN ENG LAB, US ARMY, ABERDEEN PROVING GROUND, MD, 85- *Mem:* Sigma Xi; Human Factors Soc. *Res:* Robotics for field applications; technology assessment of foreign robotics capabilities; development and management of robotics cooperation programs with allied governments. *Mailing Add:* 3103 Woolsey Dr Churchville MD 21028

HODGE, DENNIS, b Chicago, Ill, Jun 17, 39; m 62; c 2. GEOPHYSICS, PETROLOGY. *Educ:* Beloit Col, BS, 61; Univ Wyo, MS, 63, PhD(geol, geophys), 66. *Prof Exp:* From asst prof to assoc prof, 66-81, chmn dept, 73-75, PROF GEOL SCI, STATE UNIV NY BUFFALO, 81- *Mem:* Am Geophys Union; Soc Expl Geophysicists. *Res:* Field geophysical studies of New England; heat flow studies of New York; earth molds of flexine. *Mailing Add:* Dept Geol Sci Rm 42B State Univ NY 4240 Ridge Lea Rd Amherst NY 14226

HODGE, DONALD RAY, b Springfield, Mo, Aug 22, 39; m 79; c 2. INFORMATION ENGINEERING, INTEROPERABILITY. *Educ:* Drury Col, BS, 61; Univ Wis, Madison, MS, 63, PhD(physics), 68. *Prof Exp:* Analyst opers res, Ctr Naval Anal, 68-71; opers res analyst, US Army Commun, Electronics Comput Appln Agency, 71-73; sr scientist, BDM Corp, 73-77; sr staff, TRW Defense Systs Group, TRW Corp, 77-78; sr anal, Jaycor, 88-89; EXEC STAFF, COMPUTER SCI CORP, 89- *Concurrent Pos:* Staff mem physics, Nat Acad Sci, 71- *Mem:* Am Phys Soc; Opers Res Soc Am; Inst Elec & Electronic Engr Computer Soc. *Res:* Information systems architecture; systems engineering; command and control systems; standards and piotocols; information interfaces; communications compatibility and interoperability; system development methodologies; measures of effectiveness; simulation and modeling. *Mailing Add:* 2907 Farm Rd Alexandria VA 22302-2411

HODGE, FREDERICK ALLEN, b Durango, Colo, Jan 10, 39; m 63, 85; c 4. RADIOBIOLOGY, ENVIRONMENTAL SCIENCE. *Educ:* Colo State Univ, BS, 64, MS, 65; Ore State Univ, PhD(radiation biol), 71. *Prof Exp:* Microbiologist, Naval Radiol Defense Lab, 65-68 & Naval Biomed Res Lab, Calif, 71-73; head bact div & head med physics dept, Naval Med Res Unit 2, Taipei, Taiwan, 73-76; radiobiologist, Spec Effects Biomed Res Detachment, Naval Med Res Inst, 76-77; RADIOBIOLOGIST & ACTG CHIEF, BIOEFFECTS ANAL BR, OFF RADIATION PROGS, US ENVIRON PROTECTION AGENCY, 77- *Concurrent Pos:* Med Serv Corps, US Navy, 65-76; capt, USPHS, 77- *Mem:* Radiation Res Soc; AAAS; Sigma Xi; NY Acad Sci; Asn Mil Surgeons US. *Res:* Effects of ionizing and non-ionizing radiation on cells; role of lysosomes in cellular responses to various physical, chemical and biological agents; mechanisms of division delay in synchronized cells; radiation protection of man and his environment; radiation risk assessment. *Mailing Add:* 1206 Governor Bridge Rd Davidsonville MD 21035

HODGE, HAROLD CARPENTER, b Chicago, Ill, Dec 19, 04; m 28; c 2. PHARMACOLOGY, TOXICOLOGY. *Educ:* Ill Wesleyan Univ, BS, 25, Univ Iowa, MS, 27, PhD(phys chem), 30; Am Bd Indust Hyg, cert toxicol. *Hon Degrees:* DSc, Ill Wesleyan Univ, 49 & Case Western Reserve Univ, 67. *Prof Exp:* From asst prof to assoc prof biochem & pharmacol, 37-46, prof pharmacol & toxicol, 46-58, prof pharmacol & radiation biol & chmn dept pharmacol, 58-70, consult dent res, 40-67, from asst prof to prof dent res, 36-70, EMER PROF PHARMACOL & RADIATION BIOL & DENT, UNIV ROCHESTER, 70-; PROF ENVIRON TOXICOL, MED PHARMACOL & THERAPEUT, UNIV CALIF, IRVINE, 72- *Concurrent Pos:* Rockefeller

Found fel, Univ Rochester, 31-36; chief pharmacologist, Atomic Energy Proj, 43-58; mem, Nat Adv Dent Res Coun, 48-52 & 57-58; observer, Bikini tests, 46; chmn comt toxicol, Nat Res Coun, 51-58; mem expert adv panel, Dent Health, WHO, 56-75, consult; prof pharmacol, Univ Calif, San Francisco, 70-72. *Honors & Awards:* Award in Oral Therapeut, Int Asn Dent Res, 65; Merit Award, Soc Toxicol, 69, Toxicol Educ Award, (Forum Adv Toxicol), 75; Dean Mem Award, Int Asn Dent Res, 76; Kammer Merit Authorship Award, Am Occup Med Asn, 78. *Mem:* Am Soc Pharmacol & Exp Ther (pres, 66-67); Soc Toxicol (pres, 61-62); fel Am Col Dent; Int Asn Dent Res (pres, 47); fel Royal Soc Med; Am Chem Soc. *Res:* Fluorine and caries; toxicology of uranium and fluorine compounds; toxicology of commercial products. *Mailing Add:* Nine Commonwealth Ave Boston MA 02116-2106

HODGE, IAN MOIR, b Auckland, NZ, Jan 28, 46; m. PHYSICAL CHEMISTRY, POLYMER PHYSICS. *Educ:* Univ Auckland, BSc, 67, MSc, 68; Purdue Univ, PhD(phys chem), 74. *Prof Exp:* Res chemist, B F Goodrich Co, 78-85; AT EASTMAN KODAK RES LABS, ROCHESTER, NY, 85- *Concurrent Pos:* Fel, Univ Aberdeen, UK, 74-75, McGill Univ, Montreal, 75-76 & Purdue Univ, 77-78. *Mem:* Am Chem Soc; Am Physical Soc. *Res:* Conductivity and dielectric relaxation; glass transition; physical aging; diffusion polymer physics. *Mailing Add:* Eastman Kodak Res Labs 1669 Lake Ave Rochester NY 14650-2116

HODGE, JAMES DWIGHT, b Mechanicsburg, Ohio, July 14, 33; m 55; c 3. POLYMER CHEMISTRY. *Educ:* Bob Jones Univ, BS, 56; Univ Mich, MS, 59; Pa State Univ, PhD(org chem), 63. *Prof Exp:* Chemist, Plastics & Resins Div, Shell Chem Co, Tex, 59-61; res chemist, 63-66, sr res chemist, 66-87, RES ASSOC, DACRON RES LAB, TEXTILE FIBERS DEPT, E I DU PONT DE NEMOURS & CO, INC, 87- *Mem:* Am Chem Soc. *Res:* Organic reaction mechanisms; carbonium ions; epoxy compounds; polyesters. *Mailing Add:* 2101 Oakengate Lane Midlothian VA 23113-4047

HODGE, JAMES EDGAR, b Corsicana, Tex, Apr 19, 29; m 48; c 2. MATHEMATICS. *Educ:* NTex State Univ, BS, 50, MS, 57; Univ Ga, PhD(math), 65. *Prof Exp:* Teacher pub sch, Tex, 50-52, prin high sch, 54-56; instr math, NTex State Univ, 57-60, asst prof, 65-71; assoc prof, 71-77, PROF MATH, ANGELO STATE UNIV, 77- *Mem:* Am Math Soc; Math Asn Am. *Res:* Mathematical topology. *Mailing Add:* Dept Math Angelo State Univ PO Box 10900 San Angelo TX 76909

HODGE, JOHN DENNIS, engineering, for more information see previous edition

HODGE, JOHN EDWARD, b Oct 12, 14; US citizen; m 48; c 4. CARBOHYDRATE CHEMISTRY. *Educ:* Univ Kans, AB, 36, MA, 40. *Prof Exp:* Prin chemist gasoline oils, State Kans Dept Inspections & Regist, Topeka, 37-39; teacher chem, State Indust Dept, Western Univ, Kansas City, Kans, 39-41; org chemist carbohydrates, Northern Regional Res Lab, Sci & Educ Admin, Agr Res, USDA, Peoria, Ill, 41-73, supvry res chemist, 73-81; ADJ PROF CHEM, BRADLEY UNIV, PEORIA, ILL, 84- *Concurrent Pos:* Mem comt food stability, Adv Bd Mil Personnel Supplies, Nat Res Coun, Nat Acad Sci, 77-, supvr grant res prog; consult chemist carbohydrates, 81- *Honors & Awards:* Super Serv Award, USDA, 53. *Mem:* Emer mem Am Chem Soc; emer mem Am Asn Cereal Chemists. *Res:* Organic, food and cereal chemistry; sweeteners; gums. *Mailing Add:* 1107 W Groveland Ave Peoria IL 61604-3131

HODGE, PAUL WILLIAM, b Seattle, Wash, Nov 8, 34; m 62; c 3. ASTRONOMY. *Educ:* Yale Univ, BS, 56; Harvard Univ, PhD(astron), 60. *Prof Exp:* Lectr astron, Harvard Univ, 60-61; asst prof, Univ Calif, Berkeley, 61-65; assoc prof, Univ Wash, 65-69, assoc dean, Grad Sch, 70-72, assoc dean arts & sci, 78-79, chmn astron dept, 87-90, PROF ASTRON, UNIV WASH, 69- *Concurrent Pos:* Physicist, Astrophys Observ, Smithsonian Inst, 56-65; NSF fel, 60-61; fel, Mt Wilson & Palomar Observs & Calif Inst Technol, 60-61; ed, Astron J, 84- *Honors & Awards:* Bok Astron Prize; Beckwith Astron Prize. *Mem:* Am Astron Soc (vpres, 90-93); Int Astron Union; Meteoritical Soc; AAAS. *Res:* Structure of galaxies; evolution of galaxies. *Mailing Add:* Dept Astron FM-20 Univ Wash Seattle WA 98195

HODGE, PHILIP G(IBSON), JR, b New Haven, Conn, Nov 9, 20; m 43; c 3. PLASTICITY, COMPUTATIONS. *Educ:* Antioch Col, AB, 43; Brown Univ, PhD(appl math), 49. *Prof Exp:* Res assoc appl math, Brown Univ, 47-49; asst prof math, Univ Calif, Los Angeles, 49-53; from assoc prof to prof appl mech, Polytech Inst Brooklyn, 53-57; prof mech, Ill Inst Technol, 57-71; prof mech, Univ Minn, Minneapolis, 71-91; CONSULT, 91- *Concurrent Pos:* Consult, NAm Aviation, Inc, 51-52, AiResearch Mfg Co, 52-53 & IIT Res Inst, 57-71; Nat Sci Found sr fel, 63-64; tech ed, J Appl Mech, 71-76; secy, US Nat Comn Theo & Appl Mech, 82-; asst treas, Int Union Theo & Appl Mech, 84- *Honors & Awards:* Worcester Reed Warner Medal, Am Soc Mech Engrs, 75, Medal, 87; Theodore von Karman Medal, Am Soc Civil Engrs, 85; Distinguished Serv Award, Am Acad Mech, 85. *Mem:* Nat Acad Eng; Am Soc Mech Engrs. *Res:* Theory of plasticity; plastic analysis of structures; numerical analysis. *Mailing Add:* Dept Aerospace Eng & Mech Univ Minn Minneapolis MN 55455

HODGE, RAYMOND, air transport, airport design; deceased, see previous edition for last biography

HODGE, STEVEN MCNIVEN, b London, Eng, Oct 15, 42; US citizen; div; c 2. GLACIOLOGY. *Educ:* Univ BC, BSc, 64; Univ Wash, PhD(geophys), 72. *Prof Exp:* GEOPHYSICIST, WATER RESOURCES DIV, US GEOL SURV, 72- *Concurrent Pos:* Mem, Comt Snow & Ice, Am Geophys Union, 74-77; mem, Glaciol Comt, Polar Res Bd, US Nat Acad Sci, 76-79. *Mem:* Int Glaciol Soc; Am Geophys Union. *Res:* Numerical modeling of the flow of ice sheets and glaciers; radar sounding of ice sheets and glaciers; the interaction between glaciers and climate. *Mailing Add:* US Geol Surv Univ Puget Sound 113 Thompson Hall Tacoma WA 98416

HODGEN, GARY DEAN, b Frankfort, Ind, May 10, 43; m 64; c 2. ENDOCRINOLOGY. *Educ:* Purdue Univ, BS, 65, MS, 66; Ohio State Univ, PhD(endocrinol), 69. *Prof Exp:* Staff fel endocrinol, NIH, 69-71, sr staff fel, 71-73, chief sect endocrinol, Reproduction Res Br, 73-76, CHIEF, PREGNANCY RES BR, NAT INST CHILD HEALTH & HUMAN DEVELOP, 77- *Concurrent Pos:* Res award, Soc Study Reproduction, 81. *Honors & Awards:* Ayerst Lectr, Am Fertility Soc, 82; Gregory Pincus Mem Lectr, Laurentian Hormone Conf, 82. *Mem:* Endocrine Soc; Soc Study Reproduction; Soc Gynec Invest; Am Fertil Soc. *Res:* Primate reproductive physiology, including intrauterine fetal development and ovarian and testicular function during puberty and adulthood. *Mailing Add:* Jones Inst Reproductive Med EVa Med Sch Lewis Hall Rm 2049 Norfolk VA 23501

HODGES, BILLY GENE, b Wayne, Okla, Nov 2, 29; m 60; c 3. MATHEMATICS. *Educ:* Univ Okla, BA, 52, MA, 58, PhD(differential geom), 64. *Prof Exp:* Spec instr math, Univ Okla, 58-62; asst prof, Winthrop Col, 62-64; assoc prof, Ill State Univ, 64-65; assoc prof, 65-74, chmn dept, 74-77, PROF MATH, WINTHROP COL, 74- *Mem:* Am Math Soc; Math Asn Am. *Res:* Riemannian geometry. *Mailing Add:* Dept Math Winthrop Col 701 Oakland Ave Rock Hill SC 29733

HODGES, CARL NORRIS, b New Braunfels, Tex, Mar 19, 37; m 60, 87; c 4. CONTROLLED ENVIRONMENTS, FOOD PRODUCTION. *Educ:* Univ Ariz, BS, 59. *Prof Exp:* Res assoc, Univ Ariz, 61-63, supvr, Solar Energy Res Lab, 63-67, DIR, ENVIRON RES LAB, UNIV ARIZ, 67-; PRES, PLANETARY DESIGN CORP, 85- *Concurrent Pos:* Mem, Comt Arid Lands, AAAS, 67-75 & Desertification Adv Comt, 76-78; mem bd trustees, Oceanic Found, Waimanalo, Hawaii, 72-82, Western Behav Sci Inst, La Jolla, Calif, 85-; pres, Desert Develop Found, Puerto Penasco, Mex, 75-; mem, Comt Aquacult, Nat Acad Sci, Washington, DC, 77, Panel Aquacult Eng & Simulation, 84-85, Adv Comt Technol Innovation & Comt Climate & Weather Fluctuations & Agr Prod; mem, Technol Assessment Adv Coun, Off Technol Assessment, US Cong, Washington, DC, 79-87, Adv Panel Technol Transfer to Middle East, 83 & Adv Panel Energy, Technol & Environ Develop Countries, 90; mem, Tarrytown One Hundred, NY, 83-85 & PRONATURA (Mex Asn Conserv Nature) Hermosillo, Mex, 84-; mem adv bd, Ariz Cancer Ctr, Tucson, 87-; mem bd dirs, Nat Asn Homebuilders Res Ctr, Washington, DC, 88-89, Univ Phoenix, 90-; mem adv coun, Kino Learning Ctr, Tucson, Ariz, 89-; chmn, Ariz Solar Energy Comm, Phoenix, Ariz; first Pierson vis fel, Pierson Col, Yale Univ, 90. *Mem:* Fel World Acad Art & Sci; fel AAAS; Int Solar Energy Soc/Am Solar Energy Soc; World Maricult Soc. *Res:* Controlled environments and growing systems, future food production including greenhouses, aquatic animals and halophytes; development of totally closed ecological "biospheres" for the study of environmental problems on earth and ultimately use in space colonies; integrating biology into urban environments; seawater technology for seawater irrigation of coastal desert areas and its effect on global warming. *Mailing Add:* Environ Res Lab 2601 E Airport Dr Tucson AZ 85706

HODGES, CARROLL ANN, b Pomona, Calif, June 30, 36. PLANETARY GEOLOGY, GEOMORPHOLOGY. *Educ:* Univ Tex, Austin, BA, 58; Univ Wis, MS, 60; Stanford Univ, PhD(geol), 66. *Prof Exp:* Geologist, Shell Oil Co, 60-62; asst prof geol, Colo State Univ, 65-66; geologist, Utah Construct & Mining Co, 66-69; asst prof geol, San Jose State Univ, 69-70; asst chief geologist, Western Region, 84-88, GEOLOGIST, US GEOL SURV, 70- *Concurrent Pos:* Lectr, San Francisco State Univ, 78; Am Geophys Union Cong Sci fel, Washington, DC, 80-81. *Mem:* Geol Soc Am; fel AAAS; fel Am Geophys Union. *Res:* Analysis and interpretation of lunar and Martian landforms of volcanic and impact origin; mineral resource assessment. *Mailing Add:* US Geol Surv MS 984 345 Middlefield Rd Menlo Park CA 94025

HODGES, CHARLES THOMAS, b Birmingham, Ala, Apr 28, 51; m 75; c 2. CHEMISTRY. *Educ:* Univ South, BA, 73; Univ NC, Chapel Hill, PhD(biochem), 78. *Prof Exp:* develop biochemist, 78-84, RES SUPVR, E I DU PONT DE NEMOURS & CO, INC, 84- *Mem:* Am Asn Clin Chemists. *Res:* Chemical modification of proteins; development of clinical chemistry assays; immunoassay techniques. *Mailing Add:* 28 Pierson Dr Hockessin DE 19707-9801

HODGES, CLINTON FREDERICK, b Danville, Ill, Apr 20, 39; m 60; c 3. PLANT PHYSIOLOGY. *Educ:* Univ Ill, Urbana, BS, 62, MS, 64, PhD(plant path), 67. *Prof Exp:* Assoc prof agron & hort, 67-76, PROF HORT, AGRON & PLANT PATH, IOWA STATE UNIV, 76- *Mem:* Am Phytopath Soc; Am Soc Plant Physiologists. *Res:* Physiology of lost-pathogen interactions. *Mailing Add:* Dept Plant Path Iowa State Univ 351 Bessey Ames IA 50011

HODGES, DAVID A(LBERT), b Hackensack, NJ, Aug 25, 37; m 65; c 2. ELECTRICAL ENGINEERING. *Educ:* Cornell Univ, BEE, 60; Univ Calif, Berkeley, MS, 61, PhD(elec eng), 66. *Prof Exp:* Mem tech staff, Functional Integration Group, Bell Tel Labs, Inc, 66-68, tech supvr, 68-69, head, Systs Elements Res Dept, 69-70; assoc prof, 70-74, PROF ELEC ENG & COMPUT SCI, UNIV CALIF, BERKELEY, 74-, DEAN, COL ENG, 90- *Concurrent Pos:* Indust consult, 70- *Mem:* Nat Acad Eng; Inst Elec & Electronic Engrs. *Res:* Semiconductor; integrated circuits; computer integrated manufacturing. *Mailing Add:* Col Eng Univ Calif 320 McLaughlin Hall Berkeley CA 74720

HODGES, DEAN T, JR, b Bremerton, Wash, Mar 28, 44; m 76. QUANTUM ELECTRONICS. *Educ:* Calif State Univ, Humboldt, BS, 66; Cornell Univ, PhD(appl physics), 71. *Prof Exp:* Mgr, Lasers & Optics Dept, Electronics Res Lab, Aerospace Corp, 71-80, eng mgr, Laser Prod, Spectra Physics, 80-84; SR VPRES, NEWPORT CORP, 84- *Concurrent Pos:* Adj prof, Univ Calif, Los Angeles, 76-77. *Mem:* Optical Soc Am; fel Inst Elec & Electronics Engrs. *Res:* Quantum electronic processes of rare-gas and metal-vapor ion lasers, discharge molecular lasers, and optically pumped infrared and far infrared molecular lasers; development of near millimeter wave detectors and radiometers. *Mailing Add:* Newport Corp 18235 Mt Baldy Circle PO Box 8020 Fountain Valley CA 92708

HODGES, FLOYD NORMAN, b San Antonio, Tex, Sept 16, 39; m 66; c 2. ENVIRONMENTAL GEOCHEMISTRY, GEOHYDROLOGY. *Educ:* Univ Tex, Austin, BS, 69, PhD(geol), 75. *Prof Exp:* Fel Geophys Lab, Carnegie Inst, Washington, 72-74; fel, State Univ NY, Stony Brook, 74-76; asst prof geol, Furman Univ, 76-79; staff scientist geochem, Rockwell Int, 79-81; sr scientist geochem, Battelle Northwest Lab, 81-88; SR SCIENTIST, WESTINGHOUSE, 89- *Concurrent Pos:* Lectr chem, Joint Ctr Grad Studies, Washington State Univ, Richland, Wash, 80- *Mem:* Geol Soc Am; Mineral Soc Am; Am Geophys Union; Geochem Soc; AAAS. *Res:* Application of geochemistry to problems of hazardous waste disposal and remediation; igneous and metamorphic petrogenesis and the role of these processes in the evolution of planetary crusts. *Mailing Add:* 8130 W Falls Pl Kennewick WA 99336

HODGES, GLENN R(OSS), b Cleveland, Ohio, Aug 28, 41; m 64; c 4. INFECTIOUS DISEASES, INTERNAL MEDICINE. *Educ:* Muskingum Col, BA, 63; Univ Chicago, MD, 67; Ohio State Univ, MS, 72. *Prof Exp:* Clin instr med, Col Med, Ohio State Univ, 70-72; from asst prof to assoc prof, 74-84, PROF MED, COL MED, UNIV KANS, 84- *Concurrent Pos:* Head, sect infectious dis, Naval Hosp, Great Lakes, Ill, 72-74; chief sect, infectious dis, Vet Admin Med Ctr Kansas City, Mo, 74-90, chief of staff, 87- *Mem:* Am Col Physicians; Am Soc Microbiol; Infectious Dis Soc Am. *Res:* Antimicrobial therapy; nosocomial infections. *Mailing Add:* Chief of Staff VA Med Ctr 4801 Linwood Blvd Kansas City MO 64128

HODGES, HARDY M, b Tampa, Fla, Nov 26, 63. COSMOLOGY. *Educ:* Fla State Univ, BS, 83; Univ Chicago, PhD(physics), 88. *Prof Exp:* Res assoc, Santa Cruz Inst Particle Physics, Univ Calif, 88-90; SMITHSONIAN POSTDOCTORAL FEL, HARVARD-SMITHSONIAN CTR ASTROPHYSICS, 90- *Res:* Particle physics and astrophysics interface; cosmology; inflation; topological defects; phase transitions; origin of large-scale structure; stochastic processes in cosmology. *Mailing Add:* Harvard Ctr Astrophys MS 51 60 Garden St Cambridge MA 02138

HODGES, HARRY FRANKLIN, b Bedford, Ind, Nov 17, 33; m 55; c 3. PLANT PHYSIOLOGY. *Educ:* Purdue Univ, BS, 56, MS, 58, PhD(plant physiol), 64. *Prof Exp:* Instr res agron, Purdue Univ, 56-65, from asst prof to assoc prof plant physiol, 65-75; PROF AGRON, MISS STATE UNIV, 75- *Concurrent Pos:* Res asst, Pa State Univ, 64-65. *Mem:* Am Soc Agron; Crop Sci Soc Am; Am Soc Plant Plysiologists; Sigma Xi. *Res:* Regulation of physiological development. *Mailing Add:* Agron Miss State Univ Box 5248 Mississippi State MS 39762

HODGES, HELEN LESLIE, b Alexandria, La, Feb 7, 45. BIOINORGANIC CHEMISTRY. *Educ:* Univ Mich, BS, 66, MS, 68, PhD(chem), 71. *Prof Exp:* Fel chem, Calif Inst Technol, 72-74; ASST PROF CHEM, UNIV CALIF, SANTA CRUZ, 74- *Mem:* Am Chem Soc. *Res:* Mechanistic studies of electron transfer of metal ions on micelle surfaces. *Mailing Add:* 100-1B Pickering Pl Cary NC 27513-5209

HODGES, JOHN DEAVOURS, b Laurel, Miss, Apr 12, 37; m 58; c 2. PLANT PHYSIOLOGY, SILVICULTURE. *Educ:* Miss State Univ, BS, 59; Univ Wash, MS, 62, PhD(tree physiol), 65. *Prof Exp:* Res forester, Southern Forest Exp Sta, USDA, 59-61, res plant physiologist, 65-75; SILVICULTURIST, MISS STATE UNIV, 75- *Mem:* Soc Am Foresters; Am Soc Plant Physiol. *Res:* Silviculture of southern hardwoods; plant water relations; plant chemistry; environmental physiology; physiology of disease and insect resistance in plants. *Mailing Add:* Dept Forestry Miss State Univ PO Drawer F R Mississippi State MS 39762

HODGES, JOHN HENDRICKS, b Harpers Ferry, WVa, Aug 1, 14; m 40; c 1. MEDICINE. *Educ:* Catholic Univ, BS, 35; Jefferson Med Col, MD, 39; Am Bd Internal Med, dipl, 48, recertification, 77. *Prof Exp:* Asst demonstr, Jefferson Med Col, 44-53, asst prof, 53-63, prof clin med, 63, prof, 64-79, bd trustees, Children's Rehab Hosp, 78-89, LUDWIG A KIND EMER PROF MED, JEFFERSON MED COL, 79-, BD TRUSTEES, MAGEE REHAB HOSP, 87- *Concurrent Pos:* Fel med, Jefferson Med Col, 42-44; Mary Markle fel in trop med, Walter Reed Hosp & Cent Am, 43; consult hemat, Lankenau Hosp, 64-88; mem bd dir, Mercy Cath Med Ctr, 71-81; assoc, Cardeza Found, 69-79; mem bd trustees, Thomas Jefferson Univ, 78- *Mem:* Fel Am Col Physicians; Am Soc Hemat; Am Soc Trop Med & Hyg; Am Heart Asn; Int Soc Hemat; AMA; Sigma Xi. *Res:* Original clinical and laboratory studies on the origin of viral dysentery; studies on sickle cell disease, bone marrow and coagulation. *Mailing Add:* 436 Sabine Ave Wynnewood PA 19096

HODGES, JOHN HERBERT, b Pittsburgh, Pa, Sept 30, 28; m 55; c 3. MATHEMATICS. *Educ:* Westminster Col, Pa, BS, 51; Duke Univ, MA, 53, PhD, 55. *Prof Exp:* Instr math, Duke Univ, 54-55; asst prof, Univ Buffalo, 55-58; res mathematician, Cornell Aeronaut Lab, Inc, 58-60; from asst prof to assoc prof, 60-66, PROF MATH, UNIV COLO, BOULDER, 66- *Concurrent Pos:* NSF res grants, 56-58 & 61-71 & sci fel, 64-65. *Mem:* Am Math Soc; Math Asn Am. *Res:* Number-theoretic properties of matrices over a finite field; matrix equations; distribution of matrices with given properties; weighted partitions; exponential sums. *Mailing Add:* Campus Box 426 Univ Colo Boulder CO 80309

HODGES, JOSEPH LAWSON, JR, b Shreveport, La, Apr 10, 22; m 47; c 5. THEORETICAL STATISTICS. *Educ:* Univ Calif, BA, 42, PhD(math), 48. *Prof Exp:* From instr to assoc prof math, 48-57, PROF STATIST, UNIV CALIF, BERKELEY, 57- *Concurrent Pos:* Asst prof, Univ Chicago, 51-52; Guggenheim fel, 56-57. *Mailing Add:* Dept Statist Univ Calif 2120 Oxford St Berkeley CA 94720

HODGES, LANCE THOMAS, b Windsor, Ont, May 20, 34. SEDIMENTARY PETROLOGY, PALEOECOLOGY. *Educ:* Andrews Univ, BA, 57; Univ Waterloo, MSc, 63; Loma Linda Univ, PhD(geol), 77. *Prof Exp:* Instr phys sci, Andrews Univ, 65-66; engr elec eng, Calif Inst Technol, 67-68; instr eng, Grantham Eng Sch, 69-71; asst prof geol, NMex Highlands Univ, 77; staff geologist, CER Corp, 78-79; res geologist, C K Geoenergy Corp, 79-86; SR RES ASSOC, PHYSIOL & PHARMACOL, LOMA LINDA UNIV, 87. *Mem:* Am Asn Petrol Geologists; Geol Soc Am; Int Asn Sedimentologists; Int Asn Study Fossil Cnidaria; Soc Econ Paleontologists & Mineralogists. *Res:* Fossil reef paleoecology and sedimentology; carbonate petrology and petrography; bioherm structure; structure, stratigraphy and natural gas of the Green River Basin. *Mailing Add:* 11644 Ciello Lane Loma Linda CA 92354

HODGES, LAURENT, b Houston, Tex, Jan 16, 40; m 66; c 2. SOLAR ENERGY, GENERAL PHYSICS. *Educ:* Harvard Univ, AB, 60, AM, 61, PhD(physics), 66. *Prof Exp:* Assoc, 66-68, from asst prof to assoc prof, 68-78, PROF PHYSICS, IOWA STATE UNIV, 78- *Concurrent Pos:* From asst physicist to sr physicist, Ames Lab, US Dept Energy, 66-81; energy specialist, Iowa Energy Exten Serv, 81- *Mem:* AAAS; Am Phys Soc; Am Solar Soc; Am Asn Phys Teachers; Am Math Soc. *Res:* Energy conservation and passive solar energy; environmental pollution. *Mailing Add:* Dept Physics Iowa State Univ Ames IA 50011

HODGES, LAWRENCE H, AGRICULTURAL ENGINEERING. *Prof Exp:* CONSULT AGR ENG. *Mem:* Nat Acad Eng. *Mailing Add:* Lake Forum Bldg 840 Lake Ave PO Box 307 Racine WI 53401

HODGES, LINDA CAROL, b Covington, Ky, Jan 24, 51. BIOCHEMISTRY. *Educ:* Centre Col, Ky, BS, 72; Univ Ky, PhD(biochem), 79. *Prof Exp:* Instr chem, Georgetown Col, 77, Berea Col, 78-79; vis prof, Centre Col, 79-80; ASST PROF CHEM, KENNESAW COL, 80- *Concurrent Pos:* Res scientist, Ga Inst Technol, 81- *Mem:* Sigma Xi; Am Chem Soc. *Res:* Glycoprotein structure and function; carbohydrate structures and sites of glycosidic linkage in protease-inhibitor; structural studies on a polysaccharides from Cannabis that lower inner ocular pressure. *Mailing Add:* Dept Chem Kennesaw Col Marietta GA 30061

HODGES, NEAL HOWARD, II, b Tacoma, Wash, Oct 22, 48; m 78; c 2. INTELLIGENT SYSTEMS. *Educ:* Black Hills State Univ, BS, 74. *Prof Exp:* DATA PROCESSING SPECIALIST, SDAK SCH MINES & TECHNOL, 77- *Concurrent Pos:* Adj assoc prof, SDak Sch Mines & Technol, 80-91. *Mem:* Am Asn Artificial Intel. *Res:* Mobile semi-autonomous robots for exploration and military applications. *Mailing Add:* 2327 S Valley Dr Rapid City SD 57701-5997

HODGES, RALPH RICHARD, JR, b St Louis, Mo, Aug 18, 33; m 57; c 4. PLANETARY ATMOSPHERES. *Educ:* Univ Ill, BS, 55, MS, 57, PhD(elec eng), 64. *Prof Exp:* Res asst antenna lab, Univ Ill, 55-56, res assoc, 60-64; proj engr, Collins Radio Co, 58-60, head appl sci dept, 64-65; res scientist, 65-67, asst prof atmospheric & space sci, 67-72, RES SCIENTIST, UNIV TEX, DALLAS, 72- *Mem:* Am Geophys Union; Inst Elec & Electronics Engrs; Sigma Xi. *Res:* Planetary atmospheres and ionospheres; electromagnetic theory. *Mailing Add:* 7620 Tophill Lane Dallas TX 75240

HODGES, ROBERT EDGAR, b Marshalltown, Iowa, July 30, 22; m 46; c 4. INTERNAL MEDICINE. *Educ:* Univ Iowa, BA, 44, MD, 47, MS, 49; Am Bd Internal Med, dipl, 54; Am Bd Nutrit, cert. *Prof Exp:* Intern, Mem Hosp, Pa, 47-48; resident internal med, Univ Iowa, 49-52, instr, 52-53, assoc, 53-54, from asst prof to prof 54-71, dir metab ward, Hosp, 52-54 & 56-71; prof, Sch Med, Univ Calif, Davis, 71-; at Dept Biochem & Internal Med, Med Ctr, Univ Nebr; RETIRED. *Concurrent Pos:* Fel coun arteriosclerosis & coun epidemiol, Am Heart Asn. *Mem:* AMA; Am Heart Asn; fel Am Col Physicians; Am Fedn Clin Res; Am Soc Clin Nutrit (pres, 66). *Res:* Nutrition; physiology; diet in atherosclerosis; experimental human deficiencies of pantothenic acid, ascorbic acid and vitamin A. *Mailing Add:* 1323 E Fairway Orange CA 92666

HODGES, ROBERT MANLEY, clinical pharmacology, gynecology, for more information see previous edition

HODGES, ROBERT STANLEY, b Saskatoon, Sask, Dec 30, 43; m 62; c 2. PROTEIN CHEMISTRY. *Educ:* Univ Sask, BSc, 65; Univ Alta, PhD(biochem), 71. *Prof Exp:* Defense sci serv officer, Dept Nat Defense, Defense Res Bd, Defense Res Estab, Suffield, 65-67; Med Res Coun Can fel, Rockefeller Univ, 71-73, res assoc, 73-74; from asst prof to assoc prof, 74-84, PROF BIOCHEM, UNIV ALTA, 84- *Mem:* Can Biochem Soc; Biophys Soc; Am Soc Biol Chemists; NY Acad Sci; Am Chem Soc. *Res:* Structure and function of muscle protein systems involved in contraction and relaxation (tropomyosin, troponin I, troponin T, troponin C and calmodulin); chemical modification and sequence determination; chemical synthesis of peptides and proteins; synthesis and use of heterobifunctional crosslinking reagents; immunogenicity-antigenicity of proteins; synthetic vaccines. *Mailing Add:* Dept Biochem Univ Alta Edmonton AB T6G 2H7 Can

HODGES, RONALD WILLIAM, b Lansing, Mich, Aug 7, 34; m 67; c 2. SYSTEMATIC ENTOMOLOGY. *Educ:* Mich State Univ, BSc, 56, MSc, 57; Cornell Univ, PhD(entom), 61. *Prof Exp:* Res asst, Mich State Univ, 56-57; NSF jr fel, 61-62; res entomologist, 62-76, chief, syst entom lab, 76-80, RES ENTOMOLOGIST, SCI & EDUC ADMIN, AGR RES, USDA, 80- *Concurrent Pos:* Ed-in-chief, Moths of Am N Am. *Mem:* Soc Syst Zool; Lepidop Soc (pres, 75-76); Entom Soc Can; Entom Soc Am. *Res:* Taxonomy of Lepidoptera, particularly Gelechioidea. *Mailing Add:* Syst Entom Lab USDA US Nat Mus NHB 127 Washington DC 20560

HODGES, SIDNEY EDWARD, b Booneville, Miss, Feb 11, 24; m 47; c 5. PHYSICS. *Educ:* Southern Methodist Univ, BS, 49; Agr & Mech Col, Univ Tex, PhD(physics), 58. *Prof Exp:* Instr physics, ETex State Col, 50-54, asst prof, 56-58; prof & head dept, Stephen F Austin State Col, 58-65 & Howard Payne Col, 65-67; prof physics, Southeast Mo State Univ, 67-88; RETIRED. *Mem:* Am Asn Physics Teachers; Sigma Xi. *Res:* Molecular spectroscopy; acoustics; laboratory standard for absolute intensities; optics. *Mailing Add:* 1821 Emery St Denton TX 76201

HODGES, THOMAS KENT, b Bedford, Ind, Oct 18, 36; c 3. PLANT PHYSIOLOGY. *Educ:* Purdue Univ, BS, 58; Univ Calif, Davis, MS, 60, PhD(plant physiol), 62. *Prof Exp:* Res assoc plant physiol, Univ Ill, 62-63, from asst prof to assoc prof, 63-68; assoc prof, 68-74, head dept bot & plant path, 77-82, PROF PLANT PHYSIOL, PURDUE UNIV, 71- *Honors & Awards:* Charles Albert Shull Award, Am Soc Plant Physiol, 75. *Mem:* Am Soc Plant Physiol. *Res:* Plant biochemistry; membrane transport; plant regeneration. *Mailing Add:* Dept Bot & Plant Path Purdue Univ West Lafayette IN 47906

HODGETTS, ROSS BIRNIE, b Ont, May 27, 41; m 64. DEVELOPMENTAL GENETICS. *Educ:* Queen's Univ, Ont, BSc, 63; Yale Univ, MS, 65, PhD(biophysics), 67. *Prof Exp:* Jane Coffin Childs fel, 67-69; asst prof, 70-75, ASSOC PROF GENETICS, UNIV ALTA, 75- *Mem:* Genetics Soc Can. *Res:* Gene-enzyme studies on dope decarboxylase in Drosophila. *Mailing Add:* Dept Genetics Univ Alta Edmonton AB T6G 2E2 Can

HODGIN, EZRA CLAY, b Greensboro, NC, Apr 1, 47; m 74; c 1. IMMUNOLOGY, VETERINARY PATHOLOGY. *Educ:* NC State Univ, BS, 68; Okla State Univ, DVM, 72; Wash State Univ, PhD(path), 76. *Prof Exp:* asst prof immunopath, NC State Univ, 76-79; MEM STAFF, DEPT PATH, MICH STATE UNIV, 79- *Mem:* Asn Am Vet Med Cols. *Res:* Modulation of immune responses; lymphocyte function; mechanisms of host defense and disease; eosinophil function. *Mailing Add:* La State Univ Sch Vet Med Vet Diag Lab PO Box 25070 Baton Rouge LA 70894

HODGINS, DANIEL STEPHEN, b Scranton, Pa, June 21, 39; m 65; c 3. BIOCHEMISTRY. *Educ:* Elizabethtown Col, BS, 61; Univ Del, PhD(chem), 66; Okla City Univ, JD, 81. *Prof Exp:* NIH fel, Brandeis Univ, 65-67; from asst prof to assoc prof biochem & molecular biol, Med Sch, Univ Okla, 67-81; assoc, Dunlap & Codding Patent Attorneys, 81-84; assoc, 84-90, PARTNER, ARNOLD, WHITE & DURKEE, 91- *Mem:* Am Soc Biol Chemists; Am Chem Soc; Soc Exper Biol Med; AAAS; Am Bar Asn. *Res:* Enzyme mechanism; cancer therapy; cyclic adenosine monophosphate and hypertension; cardiac hypertrophy. *Mailing Add:* Arnold White & Durkee 2300 One American Ctr 600 Congress Ave Austin TX 78701

HODGINS, GEORGE RAYMOND, b East Rockaway, NY, May 24, 26; m 52; c 3. PHOTOGRAPHIC CHEMISTRY. *Educ:* Hofstra Col, BA, 50; Adelphi Col, MS, 56. *Prof Exp:* Dir res, Litho Chem & Supply Co, 50-59; group leader basic res, Technifax Corp, 59-64, res supvr, 64-70; mgr contract develop, 70-73, RES MGR, JAMES RIVER GRAPHICS INC, 73- *Mem:* Am Chem Soc; Soc Photog Scientists & Engr; Tech Asn Pulp & Paper Indust; Nat Microfilm Asn; Tech Asn Graphic Arts. *Res:* Light sensitive materials used in the reproduction of printed materials; diazotypes; lithography; photography; photopolymerization; organic synthesis; electrophotographic imaging materials and processes; non-silver photoimaging materials and processes; photopolymers. *Mailing Add:* 666 Amherst Rd South Hadley Ctr South Hadley MA 01075

HODGKIN, BRIAN CHARLES, b Lewiston, Maine, May 1, 41; m 69; c 2. BIOENGINEERING & BIOMEDICAL ENGINEERING. *Educ:* Univ Maine, BSc, 63 & 64; Johns Hopkins Univ, PhD(biomed eng), 69. *Prof Exp:* Instr surg, Sch Med, Johns Hopkins Univ, 69-72; res assoc electrocardiology, Maine Med Ctr, 72-75; res assoc med res, Biomed Res Inst, 76-82, coordr elec eng, 82-85, DEAN, SCH APPL SCI, UNIV SOUTHERN MAINE, 85- *Concurrent Pos:* Res assoc, Maine Med Ctr, 76- *Mem:* Biomed Eng Soc; Inst Elec & Electronic Engrs. *Mailing Add:* Dept Elec Eng Univ Southern Maine 96 Falmouth St Portland ME 04103

HODGKINS, EARL JOSEPH, b Evansville, Ind, Jan 16, 16. FOREST ECOLOGY. *Educ:* Mich State Univ, BS, 38, PhD, 56; Univ Calif, MS, 40. *Prof Exp:* Asst, Univ Calif, 38-40 & US Forest Serv, 40-41; farm forester, US Soil Conserv Serv, 41-45; forester, Scott Lumber Co, 45-47; asst prof, Mich State Col, 47-51; assoc prof, 51-58, prof, 58-78, EMER PROF FORESTRY, AUBURN UNIV, 78- *Res:* Forest habitat and vegetation studies; applied forest ecology. *Mailing Add:* 206 Pier Ave Fairhope AL 36532

HODGMAN, JOAN ELIZABETH, b Portland, Ore, Sept 7, 23; m 49; c 2. MEDICINE. *Educ:* Stanford Univ, BA, 43; Univ Calif, MD, 46; Am Pediat Bd, dipl, 56; Am Bd Neonatal-Prenatal Med, cert, 75. *Prof Exp:* Intern pediat, Univ Calif Hosp, 46-47; resident, Los Angeles County Hosp, 48-50; from instr to assoc prof, 51-69, PROF PEDIAT, SCH MED, UNIV SOUTHERN CALIF, 69-; MEM ATTEND STAFF, LOS ANGELES COUNTY-UNIV SOUTHERN CALIF MED CTR, 50-, PHYSICIAN SPECIALIST, 52- *Concurrent Pos:* Resident, Harbor Gen Hosp, Torrance, 48; mem attend staff, Rancho Los Amigos, 53-; consult nursery, Booth Mem Hosp, 50-63. *Mem:* AMA; Am Pediat Soc; Am Acad Pediat. *Res:* Problems of the newborn, especially the premature infant. *Mailing Add:* Los Angeles County-Univ Southern 1240 Mission Rd Los Angeles CA 90033

HODGSON, DEREK JOHN, b Watford, Eng, July 1, 42; m 78; c 1. STRUCTURAL CHEMISTRY, INORGANIC CHEMISTRY. *Educ:* Harvard Univ, AB, 65; Northwestern Univ, Ill, MS, 68, PhD(chem), 69. *Prof Exp:* From asst prof to prof chem, Univ NC, Chapel, 69-87; prof & head chem, 87-90, VPRES RES, UNIV WYO, 90-, PRES, UNIV WYO TECH TRANSFER FOUND, INC, 91- *Concurrent Pos:* Acad vis, Oxford Univ, 80; vis prof, Univ Copenhagen, 80 & Univ Aukland, 87; dir, Western Res Inst, 91- *Honors & Awards:* Tanner Award, 85. *Mem:* Am Chem Soc; Am Crystallog Asn; Coun Chem Res. *Res:* Structures of magnetically condensed materials; redox-active dimers and oligomers; aza analogs of nucleic acid constituents; calcium binding in biological systems. *Mailing Add:* Off Res Univ Wyo Laramie WY 82071-3355

HODGSON, EDWARD SHILLING, b Wilmington, Del, Oct 28, 26; m 49; c 2. COMPARATIVE PHYSIOLOGY. *Educ:* Allegheny Col, BS, 47; Johns Hopkins Univ, PhD(biol), 51. *Prof Exp:* Jr instr, Johns Hopkins Univ, 47-51; from instr to prof zool, Barnard Col, Columbia Univ, 51-69; chmn dept, 69-76, prof, 69-89, EMER PROF BIOL, TUFTS UNIV, 90- *Concurrent Pos:* USPHS fel, 54-55; Fulbright res fel, Australia, 59-60; hon fel, Australian Nat Univ. *Mem:* AAAS; Am Soc Zool; Am Physiol Soc; NY Acad Sci. *Res:* Sensory physiology; invertebrate physiology; neurophysiology; neurosecretion; animal behavior. *Mailing Add:* Dept Biol Tufts Univ Medford MA 02155

HODGSON, ERNEST, b Hetton-le-Hole, Eng, July 26, 32; US citizen; m 57; c 4. BIOCHEMICAL & COMPARATIVE TOXICOLOGY. *Educ:* Univ Durham, BSc, 54; Ore State Univ, PhD, 59. *Prof Exp:* Demonstr zool, Kings Col, Univ Durham, 54-55; fel entom, Univ Wis, 59-61; from asst prof to prof, NC State Univ, 61-77, William Neal Reynolds prof toxicol, 77, chmn toxicol prog, 83-89, HEAD, DEPT TOXICOL, NC STATE UNIV, 89- *Concurrent Pos:* Ed, J Biochem Toxicol Reviews Biochem Toxicol, Environ Toxicol. *Honors & Awards:* Educ Award, Soc Toxicol, 84; Burdick & Jackson Award, Am Chem Soc, 89. *Mem:* AAAS; Am Soc Pharmacol & Exp Therapeut; Soc Toxicol; Am Chem Soc; Int Soc Study Xenobiotics. *Res:* Enzymatic aspects of toxicology; comparative toxicology. *Mailing Add:* Dept Toxicol Box 7633 NC State Univ Raleigh NC 27695-7633

HODGSON, GORDON WESLEY, b Alta, May 25, 24; m 53; c 5. PHYSICAL CHEMISTRY. *Educ:* Univ Alta, BSc, 46, MSc, 47; McGill Univ, PhD(phys chem), 49. *Prof Exp:* Chief chemist, Oil Sands Proj, 49; res chemist, Res Coun Alta, 49-58, head petrol res, 58-67; Nat Acad Sci res assoc, NASA Ames Res Ctr, 67-68; NASA res assoc, Sch Med, Stanford Univ, 68-69; fac prof, 69-72, Kananaskis Ctr, Univ Calgary, 69-72, prof environ sci, 72-73, prof & dir, prof fac eng, 82-85; RETIRED. *Concurrent Pos:* Vis prof, Tohoku, Japan, 62; ed, Univ Calgary, Arctic Inst, 85-89. *Mem:* Am Chem Soc; Geochem Soc; Am Geophys Union; Am Astron Soc; Chem Inst Can. *Res:* Organic geochemistry; solids pipelining; exobiology; cosmochemistry; chemical oceanography; environmental science; environmental management; technology transfer. *Mailing Add:* Dept Common Eng Univ Calgary 2500 University Dr NW Calgary AB T2N 1N4 Can

HODGSON, J(EFFREY) WILLIAM, b Stockport, Eng, Nov 15, 40; US citizen; m 63; c 2. MECHANICAL ENGINEERING. *Educ:* Ga Inst Technol, BME, 63, MSME, 65, PhD(mech eng), 67. *Prof Exp:* Asst mech eng, Ga Inst Technol, 63-67; from asst prof to assoc prof, 67-76, PROF MECH ENG, UNIV TENN, KNOXVILLE, 76- *Mem:* Am Soc Mech Engrs; Soc Automotive Engrs. *Res:* Heat transfer; fluid mechanics; thermodynamics; automotive safety; dynamic sealing. *Mailing Add:* Dept Mech Eng Univ Tenn Knoxville TN 37996

HODGSON, JAMES B, JR, b Cedar Rapids, Iowa, July 29, 20; m 46; c 5. INFORMATION SCIENCE. *Educ:* Coe Col, BA, 42. *Prof Exp:* Historian mil hist, Off Chief Mil Hist, 53-55; sr opers analyst mil opers res, Tech Opers, Combat Opers Res Group, 55-58; proj eng data systs, Syst Anal Group, Melpar, Inc, WABCO, 58-61; sect head data systs, Washington Opers, Syst Develop Corp, 61-65; pvt consult syst eng & opers res, 65-68; pres syst eng & opers res, Ability Develop Serv, Inc, 68-76; consult, Auerbach Assocs, Inc, 76-77; mgt systs consult, Comput Sci Corp, 77-82; Sr comput scientist, Sterling Systs Inc, Plan Res Corp, 82-84; RETIRED. *Concurrent Pos:* Consult, Page Communs Engrs, 66-72, Inst Comput Technol, 66-68, TRW Systs Group, 67, Info Syst Off, Libr Cong, 67-68, Admin Urban Mass Transp, 68-69 & Indust Col Armed Forces, 68-69; dir transp, Urban Design & Develop Corp, 69-71. *Res:* Information management systems and technologies; language and linguistic research; teaching of languages. *Mailing Add:* 506 A St SE Washington DC 20003

HODGSON, JAMES RUSSELL, b Sept 1, 42; US citizen; m 67; c 2. ZOOLOGY. *Educ:* Univ Wis, BS, 65; Western Mich Univ, MA, 69; Mont State Univ, PhD(zool), 70. *Prof Exp:* Asst prof, 70-76, ASSOC PROF BIOL, ST NORBERT COL, 76- *Concurrent Pos:* Environ consult, 77-; vis prof zool & hon fel, Univ Wis-Madison, 80-81. *Mem:* AAAS; Am Soc Mammalogists. *Res:* Habitat preference and niche utilization of small mammals; prey-predator interactions in aquatic ecosystems. *Mailing Add:* Div Natural Sci St Norbert Col De Pere WI 54115

HODGSON, JOHN HUMPHREY, seismology, for more information see previous edition

HODGSON, JONATHAN PETER EDWARD, b Sheffield, Eng, Sept 3, 42; m 67. GEOMETRIC TOPOLOGY, MATHEMATICAL BIOLOGY. *Educ:* Cambridge Univ, BA, 63, MA & PhD(math), 67. *Prof Exp:* Instr math, Univ Pa, 66-68; Sci Res Coun fel, Univ Warwick, 68-69; asst prof, Univ BC, 69-70 & Univ Pa, 71-74; assoc prof, 74-80, PROF MATH, ADELPHI UNIV, 80-, CHMN, DEPT MATH & CORP SCI, 81- *Mem:* Am Math Soc; London Math Soc; NY Acad Sci. *Res:* Automorphisms of manifolds, structure of manifolds and poincare spaces; diffusion in ecological systems. *Mailing Add:* 533 Riverview Rd Swarthmore PA 19081

HODGSON, KEITH OWEN, b Sept 28, 47; US citizen; m 69. BIOINORGANIC CHEMISTRY, STRUCTURAL CHEMISTRY. *Educ:* Univ Va, BS, 69; Univ Calif, Berkeley, PhD(chem), 72. *Prof Exp:* Fel chem, Swiss Fed Inst Technol, 72-73; asst prof, 73-78, ASSOC PROF CHEM, STANFORD UNIV, 79- *Mem:* Am Chem Soc; Am Crystallog Soc; AAAS. *Res:* Synthetic and structural organometallic chemistry; application of x-ray absorption spectroscopy for investigation of molecular structure; anomalous scattering in protein crystallography and use of synchrotron radiation in such studies. *Mailing Add:* Dept Chem 101 Org Bldg Stanford Univ Stanford CA 94305

HODGSON, LYNN MORRISON, b Atlanta, Ga, July 30, 48. MARINE PHYCOLOGY. *Educ:* Col William & Mary, BS, 70; Univ Wash, MS, 72; Stanford Univ, PhD(biol), 79. *Prof Exp:* Asst res scientist aquatic ecol, Ctr Aquatic Weeds, Univ Fla, 79-81; asst prof biol & bot, Univ Ark, 81-82; asst res scientist aquacult, Div Appl Biol, Harbor Br Found, 83-85; asst prof, 85-89, CHAIR BIOL & BOT, DEPT MATH & NATURAL SCI, NORTHERN STATE UNIV, 88-, ASSOC PROF, 89- *Concurrent Pos:* Nominating comt chair, Phycological Soc Am, 87-89. *Mem:* Int Phycological Soc; Am Soc Limnol & Oceanog; Sigma Xi; Asn Biol Lab Educ; Western Soc Naturalists. *Res:* Nearshore marine algae of the Ahihi-Kinau Natural Area Reserve on Maui in Hawaii; biodiversity of seaweeds. *Mailing Add:* Dept Math & Natural Sci Northern State Univ Aberdeen SD 57401

HODGSON, PAUL EDMUND, b Wilwaukee, Wis, Dec 14, 21; m 45; c 2. SURGERY. *Educ:* Univ Mich, MD, 45; Am Bd Surg, dipl, 53. *Prof Exp:* Resident, Univ Mich Hosp, Ann Arbor, 48-52, from instr to assoc prof surg, Med Sch, 52-62; asst dean curric, Col Med, Univ Nebr Med Ctr, Omaha, 69-72, chmn dept, 72-84, prof, 62-88, EMER PROF SURG, COL MED, UNIV NEBR MED CTR, 88- *Concurrent Pos:* Chief surg serv, Ann Arbor Vet Admin Hosp, 56-60. *Mem:* Am Col Surg; Soc Univ Surg; Am Surg Asn; Soc Surg Alimentary Tract. *Res:* Blood coagulation; metabolism; vascular disease. *Mailing Add:* 600 S 42nd St Omaha NE 68198-3280

HODGSON, RICHARD HOLMES, b Orange, NJ, Aug 30, 29; m 51; c 1. BOTANY. *Educ:* Duke Univ, BS, 51, MA, 54; Ohio State Univ, PhD(bot), 59. *Prof Exp:* Plant physiologist, Denver Fed Ctr, 59-64, res plant physiologist, Metab & Radiation Res Lab, USDA, Fargo, NDak, 64-78, RES LEADER, WEED PHYSIOL & GROWTH REGULATOR RES LAB, SCI & EDUC ADMIN, USDA, FREDERICK, MD, 78- *Mem:* AAAS; Weed Sci Soc Am; Am Soc Plant Physiol; Scand Soc Plant Physiol; Am Soc Hort Sci. *Res:* Pesticide and growth regulator metabolism; organic translocation in plants; radiobiology; general plant physiology; weed science; air pollution; environmental physiology. *Mailing Add:* Bldg 1301 Ft Detrick Frederick MD 21701

HODGSON, RICHARD JOHN WESLEY, b Edmonton, Alta, Dec 10, 41; m 66; c 2. THEORETICAL PHYSICS, INVERSE SCATTERING. *Educ:* Univ Alta, BSc, 63; Univ Sydney, PhD(physics), 68. *Prof Exp:* Fel theoret physics, Univ Alta, 67-69; asst prof, 69-73, ASSOC PROF THEORET PHYSICS, UNIV OTTAWA, 73- *Mem:* Can Asn Physicists; Am Asn Physics Teachers; Am Phys Soc. *Res:* many-body theory; scattering theory; heavy-ion scattering; inverse scattering. *Mailing Add:* Dept Physics Univ Ottawa Ottawa ON K1N 6N5 Can

HODGSON, ROBERT ARNOLD, b Orange, NJ, Sept 22, 24; m 48; c 1. ENVIRONMENTAL POLLUTION BY HYDROCARBONS. *Educ:* Antioch Col, BA, 50; Brigham Young Univ, MS, 51; Yale Univ, MS, 57, PhD(struct geol), 58. *Prof Exp:* Sr scientist, Gulf Res & Develop Corp, 58-83; CONSULT OIL & GAS EXPLOR, RAH GEOL CONSULT SERV, 83- *Concurrent Pos:* Geologist, Ohio Oil Co, 51-54. *Mem:* Fel Geol Soc Am; fel AAAS; fel Am Cong Surveying & Mapping; NY Acad Sci; Sigma Xi; Explorers Club. *Res:* Natural rock fracturing and lineaments and their implications with respect to global structure and modes of origin. *Mailing Add:* PO Box 531 Jamestown PA 16134

HODGSON, RODNEY, b Salmon Arm, BC, Jan 6, 39; m 62; c 3. PHYSICS. *Educ:* Univ BC, BSc, 60, PhD(physics), 64. *Prof Exp:* Scientist, Max Planck Inst Plasma Physics, 64-67; asst prof mech eng, Colo State Univ, 67-68; MEM RES STAFF, RES DIV, IBM CORP, 68- *Concurrent Pos:* Nat Res Coun Can fel, 64; vis fel Imperial Col, 74. *Mem:* Fel Am Phys Soc; fel Am Optical Soc. *Res:* Plasma instabilities; non-equilibrium ionization in plasmas; ion engines; atomic and electronic collision phenomena; vacuum ultraviolet lasers; four wave parametric generation; laser and rapid annealing of semiconductors; laser welding; high power transmission of laser light through fibers; holography. *Mailing Add:* T J Watson Res Ctr IBM PO Box 218 Yorktown Heights NY 10598-0218

HODGSON, THOM JOEL, b Hillsdale, Mich, Mar 6, 38; m 65; c 2. OPERATIONS RESEARCH, INDUSTRIAL ENGINEERING. *Educ:* Univ Mich, Ann Arbor, BSE, 61, MBA, 65, PhD(indust eng), 70. *Prof Exp:* Opers res analyst, Ford Motor Co, 66-67; from asst prof to prof indust & systs eng, Univ Fla, 70-78; prof & head dept indust eng, 83-90, PROF INDUST ENG DEPT, NC STATE UNIV, 90- *Concurrent Pos:* Consult, Ford Motor Co, 68-71. *Mem:* Opers Res Soc Am; Inst Mgt Sci; Am Inst Indust Engrs. *Res:* Analytic models of production-inventory and manufacturing systems. *Mailing Add:* Dept Indust Eng NC State Univ Raleigh NC 27695-7906

HODGSON, VOIGT R, b Turtle Creek, Pa, June 7, 23; m 56; c 5. BIOMECHANICS. *Educ:* Wayne State Univ, BS, 55, MS, 60, PhD(eng mech), 68. *Prof Exp:* Engr, Missile Div, Chrysler Corp, Mich, 56-59; asst eng mech, Col Eng, 59-60, res assoc, 60-62, asst prof, 62-63, res assoc, Sch Med, 63-65, res asst biomech, 65-68, from instr to asst prof, 68-69, asst prof, 69-71, assoc prof, 71-77, PROF NEUROSURG, SCH MED, WAYNE STATE UNIV, 77-, DIR, GURDJIAN-LISSNER BIOMECH LAB, 72- *Concurrent Pos:* Co-prin investr head injury res projs, Dept Transp, Gen Motors Corp & Nat Oper Comt Stand in Athletic Equip; mem fac, Dept Phys Ther, Wayne State Univ, 67- *Mem:* Am Soc Testing & Mat; Soc Automotive Eng; Soc Exp Stress Anal. *Res:* Head injury. *Mailing Add:* Dept Neurosurg/116 Lande Wayne State Univ 5950 Cass Ave Detroit MI 48202

HODGSON, WILLIAM GORDON, b Rochdale, Eng, Nov 7, 29. ANALYTICAL CHEMISTRY. *Educ:* Univ Sheffield, Eng, BSc, 51, PhD(chem), 56. *Prof Exp:* Sci officer, Royal Naval Sci Serv, Eng, 54-58; res chemist, Lederle Labs Div, 58-64, group leader infrared & electron spin resonance, 64-66, group leader molecular spectros, 66-70, mgr spectros & phys chem sect, Sci Serv Dept, Cent Res Div, 70-75, dir pharmacodynamics & anal res, 75-78; dir, sci serv, 78-87, dir res serv, Med Res Div, 87-88, EXEC DIR CLIN ADMIN, AM CYANAMID CO, 88- *Mem:* Am Chem Soc; NY Acad Sci. *Res:* Infrared and Raman spectroscopy; electron spin resonance. *Mailing Add:* Med Res Div Am Cyanamid Co N Middletown Rd Pearl River NY 10965

HODNETT, ERNEST MATELLE, b Hartford, Ala, Feb 6, 14; m 39; c 1. MEDICINAL CHEMISTRY. *Educ:* Univ Fla, BS, 36, MS, 39; Purdue Univ, PhD(org chem), 45. *Prof Exp:* Asst chem, Univ Fla, 36-39; state food chemist, Fla, 40-43; from asst prof to assoc prof, 45-57, prof, 57-79, EMER PROF CHEM, OKLA STATE UNIV, 79- *Concurrent Pos:* Assoc chemist, Argonne Nat Lab, 52-53; res partic, Oak Ridge Nat Lab, 54; sr scientist, NIH, 62-63; scientist dir, USPHS Res Corps. *Mem:* Am Chem Soc; Sigma Xi. *Res:* Chemotherapy of cancer; synthesis and study of drugs used for treatment of cancer. *Mailing Add:* Dept Chem Okla State Univ Stillwater OK 74078

HODSON, CHARLES ANDREW, b Mason City, Iowa, Aug 9, 47; m 84; c 1. NEUROENDOCRINOLOGY, REPRODUCTIVE PHYSIOLOGY. *Educ:* Iowa State Univ, BS, 69, MS, 71, PhD(zool), 75. *Prof Exp:* Res assoc physiol, Mich State Univ, 75-78; asst prof, 78-84, ASSOC PROF OBSTET & GYNEC, E CAROLINA UNIV, 84- *Mem:* Endocrine Soc; Am Physiol Soc; Soc Study Reproduction; Sigma Xi; AAAS; Am Soc Zoologists. *Res:* Neuroendocrinology of the female reproductive process; neural regulation of gonadotropin and prolactin and mammary tumor growth. *Mailing Add:* Dept Obstet & Gynec E Carolina Univ Sch Med Greenville NC 27834

HODSON, HAROLD H, JR, b Grand Forks, NDak, Mar 20, 39; m 59; c 3. ANIMAL NUTRITION. *Educ:* Iowa State Univ, BS, 61, PhD(animal nutrit), 65. *Prof Exp:* Instr animal sci, Iowa State Univ, 63-65; instr night sch, Alamogordo Commun Col, 65-68; asst prof animal sci, Iowa State Univ, 68-71; assoc prof, 71-75, PROF ANIMAL INDUSTS, SOUTHERN ILL UNIV, CARBONDALE, 75-, CHMN DEPT, 73- *Mem:* Am Soc Animal Sci; Sigma Xi. *Res:* Nutritional requirements and general production problems of swine. *Mailing Add:* Rte 4 Ames IA 50010

HODSON, JEAN TURNBAUGH, b Anthony, Kans, July 10, 20. RESTORATIVE DENTISTRY. *Educ:* Univ Wash, BA, 52, MS, 58. *Prof Exp:* From instr to assoc prof oral anat & dent ceramics, 52-69, prof oper dent, 69-70, PROF RESTORATIVE DENT, UNIV WASH, 70- *Mem:* Am Asn Dent Schs; assoc Am Dent Asn; Int Asn Dent Res; hon mem Am Acad Gold Foil Opers; Sigma Xi. *Res:* Human tooth morphology teaching methods; microstructure, strength and physical properties of dental porcelains; dental amalgam and pure gold dental restorative materials; dental anatomy and occlusion. *Mailing Add:* Restorative Dent SM-56 Univ Wash Seattle WA 98195

HODSON, PHILLIP HARVEY, b Springville, Utah, July 25, 31; m 55; c 4. INDUSTRIAL MICROBIOLOGY. *Educ:* Brigham Young Univ, BS, 53, MS, 57; Univ Tex, PhD(bact), 62. *Prof Exp:* Res biochemist, Cent Res Dept, Monsanto Co, 62-69, group leader, Inorg Res & Develop Div, 69-72; TEAM LEADER & RES SCIENTIST, ELI LILLY & CO, 72- *Mem:* Am Soc Microbiol; Sigma Xi; Am Chem Soc. *Res:* Bacterial endospores; production of dipicolinic acid by fungi; Bacillus exo-enzymes amylase, protease; citric acid; cellulase, microbial detection of chemical carcinogen; large scale production of antibiotics monensin, tylosin, penicillin, vancomycin and hygromycin; use of genetically engineered microorganisms to commercially produce human insulin (Humulin R). *Mailing Add:* 8928 Dandy Creek Dr Indianapolis IN 46234

HODSON, ROBERT CLEAVES, b St Paul, Minn, Apr 21, 37; m 64; c 2. GENE STRUCTURE & REGULATION. *Educ:* Univ Minn, BA, 59; Cornell Univ, MS, 63, PhD(plant physiol), 66. *Prof Exp:* NIH fel, 65-67; NIH traineeship biol, Brandeis Univ, 67-69; asst prof biol sci, 69-75, ASSOC PROF, SCH LIFE & HEALTH SCI, UNIV DEL, 75- *Concurrent Pos:* Prin investr, Crop Res Grants Off, USDA, 81-83. *Mem:* AAAS; Am Soc Plant Physiol; Int Soc Plant Molecular Biol; Am Soc Microbiol; Sigma Xi. *Res:* Nitrogen metabolism in algae; regulation of genes for nitrogen assimilation in chlamydomonas; regulation of gametogenesis in chlamydomonas; plant cell transformation; plant gene cloning. *Mailing Add:* Sch Life & Health Sci Univ Del Newark DE 19716

HODSON, WILLIAM MYRON, b Redlands, Calif, Feb 22, 43; m 66; c 3. GEOGRAPHIC ANALYSIS, COMPUTER MODELING. *Educ:* Univ Redlands, BS, 65; Univ Calif, Los Angeles, MA, 66; Univ SC, PhD(biol), 71. *Prof Exp:* From instr to asst prof, Talladega Col, 67-74; Fulbright lectr biol, Univ Jordan, 74-76; environ consult, Environ Systs Res Inst, 77-82; assoc prof remote sensing, Asian Inst Technol, 82-84; HEAD, TRW GEOG MGT INFO SYSTS, 84- *Concurrent Pos:* Consult curric develop, Univ Jordan, 74-76, Asian Inst Technol, 82-84, remote sensing data base, UN, Bangkok, 83-84. *Mem:* AAAS; Am Ecol Soc; Am Soc Photogram. *Res:* Remote sensing and computer modeling applied to geographic and environmental analysis. *Mailing Add:* 835 College Ave Redlands CA 92374

HOEFELMEYER, ALBERT BERNARD, b San Antonio, Tex, Mar 27, 28; div; c 2. COSMETIC CHEMISTRY. *Educ:* St Mary's Univ, BS, 49; Univ Tex, MA, 50; Tex A&M Univ, PhD, 54. *Prof Exp:* Res chemist, Petrochem Res Lab, Celanese Corp Am, 53-55; asst prof chem, Tex Col Arts & Indust, 55-57; sr nuclear engr, Radiation Effects Group, Nuclear Aerospace Res Facility, Gen Dynamics Corp, 57-63, sr res scientist, Solid State Physics Group, Appl Sci Lab, Tex, 63-71, sr res scientist, Mat Group, Appl Res Lab, 71; chem consult, 71-74; sr testing engr, Bell Helicopter Co, 74-81; PRES & OWNER, EAGLE BEAUTY LABS INC, 75- *Concurrent Pos:* Adj prof, Tex Christian Univ, 59- *Mem:* Am Chem Soc; Radiation Res Soc; Soc Cosmetic Chemists; Sigma Xi; NY Acad Sci. *Mailing Add:* 7355 Greenacres Dr Ft Worth TX 76112-3533

HOEFER, JACOB A, b Lafayette, Ind, Nov 21, 15; m 42; c 5. ANIMAL HUSBANDRY. *Educ:* Purdue Univ, BSA, 37, MSA, 40, PhD(animal nutrit), 46. *Prof Exp:* From asst to instr animal husb, Purdue Univ, 37-43; asst prof, Okla Agr & Mech Col, 44-46; from asst prof to assoc prof, Purdue Univ, 46-50; prof, 50-66, actg chmn dept animal husb, 66-68 & dept food sci & human nutrit, 70-71, actg dean, Col Agr & Natural Resources, 75-77, ASSOC DIR, AGR EXP STA, MICH STATE UNIV, 67-, ASST DEAN, COL AGR & NATURAL RESOURCES & COL NATURAL SCI, 78- *Mem:* Am Inst Nutrit. *Res:* Animal nutrition; swine, cattle and sheep. *Mailing Add:* 1337 Albert East Lansing MI 48823

HOEFER, RAYMOND H, b Sutherland, Nebr, Mar 30, 52; m 79; c 2. TECHNICAL MANAGEMENT, PLANT PHYSIOLOGY. *Educ:* Hastings Col, BS, 74; Univ Nebr, MS, 77, PhD(agron), 81. *Prof Exp:* Plant sci rep, Eli Lilly & Co, 81-85, regional res rep, 85, regulatory mgr, 85-86, proj mgr, 86-87, head, agr anal chem, 87- 89; head, plant sci chem res, 89-90, MGR TECH SERV & DEVELOP, DOWELANCO, 90- *Mem:* Weed Sci Soc Am; Am Chem Soc. *Res:* Crop protection chemicals on major and minor US crops; pesticide registration. *Mailing Add:* DowElanco 9002 Purdue Rd Indianapolis IN 46268

HOEFER, WOLFGANG JOHANNES REINHARD, b Urmitz Rhein, Ger, Feb 6, 41; m 70; c 2. ELECTRICAL ENGINEERING. *Educ:* Aachen Tech Univ, Dipl Ing, 65; Univ Grenoble, Dr Ing, 68. *Prof Exp:* Asst elec eng, Univ Grenoble, 68-69; asst prof, 69-75, assoc prof, 75-80, chmn dept, 78-81, PROF ELEC ENG, UNIV OTTAWA, 80 - *Concurrent Pos:* Nat Res Coun Can fel, 70-81; consult, Dept Commun, Govt Can, 74-85, AEG-Telfunken, 76 & Can Int Develop Agency, 78; vis prof, Univ Grenoble, 77, Univ Rome II, 90, Univ Nice, 91; managing ed, Int J Numerical Modelling, 88- *Mem:* Fel Inst Elec & Electronic Engrs. *Res:* Microwave and millimeter-wave integrated circuits; planar microwave structures; fin lines; microwave and millimeter-wave communication systems; numerical techniques. *Mailing Add:* Dept Elec Eng Univ Ottawa Ottawa ON K1N 6N5 Can

HOEFERT, LYNN LUCRETIA, b Billings, Mont, July 7, 35. BOTANY. *Educ:* Mont State Univ, BS, 58, MS, 61; Univ Calif, Davis, PhD(bot), 65. *Prof Exp:* Lab technician, Univ Calif, Davis, 64-65; res botanist, Univ Calif, Santa Barbara, 65-66; BOTANIST, AGR RES SERV, USDA, 66- *Mem:* Bot Soc Am; Electron Microscopy Soc Am; Am Soc Sugar Beet Technol; Am Inst Biol Sci; Am Phytopathol Soc; Sigma Xi. *Res:* Pathological plant anatomy and cytology. *Mailing Add:* US Dept Agr 1636 E Alisal St Salinas CA 93905

HOEFFDING, WASSILY, mathematical statistics; deceased, see previous edition for last biography

HOEFLE, MILTON LOUIS, b Manhattan, Kans, Feb 19, 22; m 45; c 5. ORGANIC CHEMISTRY. *Educ:* Univ Ill, BS, 43; Univ Minn, MS, 44, PhD(org chem), 49. *Prof Exp:* Res chemist, Gen Aniline & Film Corp, 49-53; sr res chemist, Parke, Davis & Co, 53-70, sect dir chem, 70-77, sect dir chem, Pharmaceut Res Div, Warner-Lambert/Parke-Davis, 77-; RETIRED. *Mem:* Am Chem Soc; AAAS. *Res:* Pharmaceutical chemistry; development of cardiovascular agents. *Mailing Add:* 1020 Belmont Ann Arbor MI 48104

HOEFT, LOTHAR OTTO, b Antigo, Wis, Sept 17, 31; m 58; c 3. APPLIED PHYSICS, ELECTROMAGNETISM. *Educ:* Univ Wis, BS, 53, MS, 54; Pa State Univ, PhD(physics, biophys), 61. *Prof Exp:* Asst chief phys acoust, Aero Med Lab, Wright Air Develop Ctr, Wright-Patterson AFB, Ohio, 54-58 & neurophysiol br, Aerospace Med Res Lab, 61-65; group leader simulation group, Air Force Weapons Lab, Kirtland AFB NMex, 66-69, dep chief technol div, 69-71, sci & tech coordr, 76-79; physicist, European Off Aerospace Res & Develop, London Eng, 71-75, spec asst for spec studies, Anal Div, 75-76; SR PRIN EM EFFECTS, BDM INT, INC, ALBUQUERQUE, NMEX, 79- *Mem:* AAAS; Int Solar Energy Soc; Inst Elec & Electronic Engrs; Soc Aerospace Engrs. *Res:* Electromagnetic compolibility; pulse power technology; alternative energy technology. *Mailing Add:* BDM Int Inc 1801 Randolph Rd SE Albuquerque NM 87106

HOEFT, ROBERT GENE, b David City, Nebr, May 21, 44; m 63; c 2. SOIL SCIENCE, AGRONOMY. *Educ:* Univ Nebr, Lincoln, BSc, 65, MSc, 67; Univ Wis-Madison, PhD(soils), 72. *Prof Exp:* Fel, Univ Wis-Madison, 71-72; asst prof soil & plant sci, SDak State Univ, 72-73; from asst prof to assoc prof, 73-81, PROF AGRON, SOIL FERTILITY EXTEN, UNIV ILL, 81- *Honors & Awards:* CIBA-Geigy Award, Am Soc Agron, 78. *Mem:* Am Soc Agron; Soil Sci Soc Am; Coun Agr Sci & Technol; Sigma Xi; fel Soil Sci Soc Am; fel Am Soc Agron. *Res:* Soil fertility, macro and micro-nutrients; municipal and industrial waste disposal; air and water pollution; nitrification inhibitors. *Mailing Add:* Agron/N305 Turner-MC-046 Univ Ill Urbana IL 61801

HOEG, DONALD FRANCIS, b Brooklyn, NY, Aug 2, 31; m 52; c 5. PHYSICAL ORGANIC CHEMISTRY. *Educ:* St John's Univ, NY, BS, 53; Ill Inst Technol, PhD(chem), 57. *Prof Exp:* Sr res chemist, Polymer Res Dept, Res Div, W R Grace & Co, 56-61; from group leader addition polymers to mgr polymer chem, Borg-Warner Corp, 61-66, assoc dir, 66-75, dir, res ctr, 75-88; CONSULT, DFH ASSOCS, 88- *Concurrent Pos:* Adj prof, Ill Inst Technol Bus Sch, 87. *Mem:* AAAS; Am Chem Soc; Am Mgt Asn; Sigma Xi. *Res:* Organic and polymer chemistry; reaction kinetics and mechanisms; isotope effects; organometallic chemistry; stereospecific polymerizations; chemical reactions of macromolecules. *Mailing Add:* 313 S Elmhurst Ave Mt Prospect IL 60056

HOEG, JEFFREY MICHAEL, b Gary, Ind; m 73; c 2. MOLECULAR BIOLOGY. *Educ:* Ind Univ, AB, 74, MD, 77. *Prof Exp:* Intern, Barnes Hosp, St Louis, 77-78, resident, 78-79, chief resident, 79-80; res assoc, Wash Univ Vet Admin Hosp, 80-83; med staff fel, 83-87, SR INVESTR, MOLECULAR DIS BR, NIH, 87- *Concurrent Pos:* Head, Sect Cellular Biol, NIH. *Mem:* Am Fedn Clin Res; Am Col Physicians; AAAS; Am Col Nutrit; Am Col Cardiol. *Res:* Apolipoproteins recognized by specific membrane-associated receptors within the human liver; regulation of hepatic apolipoprotein output; posttranslational modifications of apolipoproteins A-I and B including glycosylation, phosphorylation, and fatty acid acylation. *Mailing Add:* 10-7N117 9000 Rockville Pike Bethesda MD 20892

HOEGERMAN, STANTON FRED, b Brooklyn, NY, May 13, 44; div; c 2. CYTOGENETICS, RADIOBIOLOGY. *Educ:* Cornell Univ, BS, 65; NC State Univ, MS, 68, PhD(genetics), 72; Am Bd Med Genetics, dipl Clin Cytogenetics. *Prof Exp:* Instr biol, Lincoln Univ, Pa, 70-72; asst biologist, Radiol & Environ Res Div, Argonne Nat Lab, 72-76; asst prof, 76-80, ASSOC PROF BIOL, COL WILLIAM & MARY, 80- *Concurrent Pos:* Adj assoc prof, Dept Pediat, Eastern Va Med Sch. *Mem:* AAAS; Genetics Soc Am; Bot Soc Am; Am Soc Human Genetics. *Res:* Cytogenetics of fragile sites in humans; general human cytogenetics. *Mailing Add:* Dept Biol Col William & Mary Williamsburg VA 23185

HOEGGER, ERHARD FRITZ, b Baden, Switz, July 3, 24; nat US; m 56; c 1. CHEMICAL CONSULTING, GENERAL CHEMISTRY. *Educ:* Basel Univ, PhD(chem), 52. *Prof Exp:* Asst lectr, Univ Basel, 50-54; res fel, Univ Colo, 54-55; res assoc, L Givaudan & Co SA, Switz, 56; sr res & staff chemist, Film & Plastic Prods & Resins Depts, E I du Pont de Nemours & Co, Inc, 56-88; INDEPENDENT CONSULT, 85- *Concurrent Pos:* Mem gov bd, Fedn Anal Chemists & Spectros Socs, 86- *Mem:* Am Chem Soc; Sigma Xi; Fedn Anal Chemists & Spectros Socs. *Res:* Polymer synthesis, characterization and new applications; films and foams; liquid and gas chromatography; gel permeation chromatography. *Mailing Add:* 1513 Harvey Rd Wilmington DE 19810

HOEHN, A(LFRED) J(OSEPH), b Ft Dodge, Iowa, Mar 31, 19; m 71. ELECTRICAL ENGINEERING, MATHEMATICS. *Educ:* Iowa State Univ, BS, 41; Univ Minn, MS, 48; Univ Ariz, PhD(elec eng), 60. *Prof Exp:* Jr engr, Commonwealth Edison Co, Ill, 41-42; instr rotating elec mach, Naval Training Sch, Iowa State Col, 42-43; elec engr, Stromberg Carlson Co, 43-46; instr elec eng & radio, Univ Minn, 46-47; elec engr & supvr elec eng res, Armour Res Found, 47-52, asst mgr, 52-55, mgr southwestern labs, 55-57; instr radio & power, Univ Ariz, 57-60; mgr theoret anal, Bell Aerosysts Co, Ariz, 60-61, res dir, 61-62; consult, pvt pract, 62-64; vpres & res dir, Electro Tech Anal Corp, 64-69, pres & res dir, 69-71; consult elec eng, 71-85; PROPRIETOR, ETA ENGRS, 85- *Concurrent Pos:* Consult, Fed Aviation Admin, 64-65, 74-75. *Mem:* Inst Elec & Electronics Engrs; Sigma Xi. *Res:* Nonlinear phenomena in electrical engineering; electromagnetic engineering; modulation theory; frequency region from power applications through radar. *Mailing Add:* 3325 N Forgeus Ave Tucson AZ 85716-1165

HOEHN, G(USTAVE) L(EO), b Dallas, Tex, June 27, 25; m 50; c 2. ELECTRICAL ENGINEERING. *Educ:* Southern Methodist Univ, BS, 47; Stanford Univ, MS, 49, PhD(elec eng), 58. *Prof Exp:* Prod engr, geophys instruments, Geotronic Labs, 47-50; vpres, Radian Instrument Co, 50-52; res technologist, Field Res Labs, Magnolia Petrol Co, Socony Mobil Oil Co, Inc, 52-54, sr res technologist, 54-59, res assoc, Mobil Res & Develop Corp, 59-85; PVT CONSULT, 85- *Mem:* Sigma Xi; Inst Elec & Electronics Engrs; AAAS. *Res:* Geophysical electromagnetics, geophysical instrumentation; data processing. *Mailing Add:* 3512 S Franklin St Dallas TX 75233

HOEHN, HARVEY HERBERT, b Fall Creek, Wis, June 25, 14; m 35; c 4. ORGANIC POLYMER CHEMISTRY. *Educ:* Univ Wis-Eau Claire, BEd, 35; Univ Minn, PhD(org chem), 40. *Prof Exp:* Asst, Univ Minn, 37-40; res scientist, Cent Res & Develop Dept, E I Du Pont de Nemours & Co, Inc, 40-84; PRES, MERT, INC, HOCKESSIN, DEL, 84- *Mem:* Am Chem Soc. *Res:* Polymer synthesis; polymer structure; permeability properties of polymers; polymer physics; membrane structure; membrane properties; engineering resins. *Mailing Add:* PO Box 1323 Hockessin DE 19707-5323

HOEHN, MARTHA VAUGHAN, b Lubbock, Tex, Jan 8, 51; m 73. NUCLEAR PHYSICS, NUCLEAR CHEMISTRY. *Educ:* Hardin-Simmons Univ, BS, 72; Auburn Univ, MS, 74; Fla State Univ, PhD(nuclear chem), 77. *Prof Exp:* STAFF MEM MESON PHYSICS, LOS ALAMOS NAT LAB, UNIV CALIF, 78- *Mem:* Am Phys Soc. *Res:* Nuclear structure; nuclear moments; mesic x-ray studies. *Mailing Add:* 963 Otowi Pl Los Alamos NM 87544-3206

HOEHN, MARVIN MARTIN, b Fall Creek, Wis, May 4, 20; m 44; c 3. MICROBIOLOGY. *Educ:* Wis State Col, BS, 42; Univ Wis, MS, 49. *Prof Exp:* Sr microbiologist, Eli Lily & Co, 49-68, res scientist, 68-74, res assoc, 74-84; RETIRED. *Mem:* Am Soc Microbiol; Soc Indust Microbiol; US Fedn Cult Collections. *Res:* Isolation and study of actinophages; fermentation and development of new antibiotics. *Mailing Add:* 4975 E 79th St Indianapolis IN 46250

HOEHN-SARIC, RUDOLF, b Graz, Austria, Feb 5, 29; US citizen; m 60; c 3. PSYCHIATRY. *Educ:* Graz Univ, MD, 54; McGill Univ, dipl psychiat, 59. *Prof Exp:* Fel med, Johns Hopkins Univ, 58-59; from instr to asst prof, 60-72, ASSOC PROF PSYCHIAT, SCH MED, JOHNS HOPKINS UNIV, 72- *Concurrent Pos:* Res assoc, Springfield State Hosp, Md, 60-61. *Mem:* Am Psychiat Asn; AAAS; Am Psychopath Asn. *Res:* Psychopharmacology and psychophysiology of anxiety disorders. *Mailing Add:* 115 Meyer Bldg Johns Hopkins Hosp Baltimore MD 21205

HOEK, JOANNES (JAN) BERNARDUS, b Haarlem, Netherlands, Jan 25, 42; m 66; c 3. CALCIUM HOMEOSTASIS, MEMBRANE BIOCHEMISTRY. *Educ:* Univ Amsterdam, Netherlands, PhD(biochem), 72. *Prof Exp:* Assoc prof path & cell biol, 86-88, PROF PATHOLOGY & CELL BIOL, THOMAS JEFFERSON UNIV, 88- *Concurrent Pos:* Lectr, biochem, Univ Nairobi, Kenya, 74-77; asst prof, Univ Pa, 77-81; assoc prof, pathology, Hahnemann Univ, Pa, 81-86. *Mem:* AAAS; Am Soc Biochem & Molecular Biol; NY Acad Sci; Res Soc Alcoholism; Int Soc Biomed Res Alcoholism. *Res:* Intracellular signalling; control of intracellular calcium homeostasis; mechanisms of adaptation to ethanol; hormonal control of metabolism in liver cells; alcoholism. *Mailing Add:* Dept Path & Cell Biol Thomas Jefferson Univ Philadelphia PA 19107

HOEKENGA, MARK T, b Ripon, Calif, Apr 6, 20; m 44; c 5. INTERNAL MEDICINE, RESEARCH ADMINISTRATION. *Educ:* Stanford Univ, AB, 41, MD, 43. *Prof Exp:* Intern, Gorgas Hosp, CZ, Stanford Univ, 44, sr res med, 48-49; USPHS sr res fel, Johns Hopkins Hosp, 47-48; physician & chief res sect, United Fruit Co, Honduras, 49-56, med supt, Chiriqui Land Co, Panama, 56-61; assoc dir clin res, Merrell-Nat Labs, Div Richardson-Merrell Inc, 61-63, dir med & sci coord, 63-70, vpres res, 70-80, VPRES, MERRELL DOW PHARMACEUT INC, 80- *Concurrent Pos:* Physician, Kuwait Oil Co Ltd, 49; vpres, Col Med Chiriqui, 57-60; vis prof, Sch Med, Univ Miami, 60-72; asst prof, Sch Med, Univ Cincinnati, 61-; clinician, Out-Patient Dept, Cincinnati Gen Hosp; attend physician, Daniel Drake Mem Hosp, Cincinnati, 61-, pres med staff, 73-, med dir, 76-78; Rockefeller Found consult, Valle, Colombia, 63; vchmn sect clin & path malaria, Int Cong Trop Med & Malaria,

Rio de Janeiro, Brazil, 63; clin prof, Sch Med, Univ Ky, 64- *Honors & Awards:* Commendation, Minister Health & Soc Assistance, Honduras, 61. *Mem:* Fel Am Col Physicians; Am Soc Trop Med & Hyg (secy-treas, 75-80, pres elect, 80-81 & pres, 81-82); Am Soc Clin Pharmacol & Therapeut; Am Acad Clin Toxicol; NY Acad Sci. *Res:* Chemotherapy of tropical diseases; tropical public health; pharmaceutical scientific administration. *Mailing Add:* 925 Congress Ave No GL Cincinnati OH 45246

HOEKMAN, THEODORE BERNARD, b Lethbridge, Alta, Nov 1, 39; m 63; c 5. PHARMACOLOGY, BIOPHYSICS. *Educ:* Hope Col, AB, 62; Univ Ill, Urbana, PhD(biophys), 68. *Prof Exp:* Res assoc pharmacol, Vanderbilt Univ Sch Med, 68-70, instr, 70-71; from instr to asst prof, Georgetown Unic, Sch Med, 71-77; ASSOC PROF BIOPHYS, MEM UNIV NFLD, 77- *Concurrent Pos:* Prin investr, NIH res grant, 74-77, Med Res Coun res grant, 78-82 & Muscular Dystrophy Asn Can res grant, 80-82. *Mem:* Am Soc Pharmacol & Exp Therapeut; Soc Neurosci. *Res:* Physiology and pharmacology of neuromuscular junction; muscle; neurotrophic regulation of muscle in relation to muscular dystrophy; expert testimony in drug and alcohol cases; developed computer assisted medical education software for microcomputers. *Mailing Add:* Fac Med Mem Univ Nfld St John's NF A1B 3V6 Can

HOEKSEMA, HERMAN, JR, b Grand Rapids, Mich, Mar 21, 20; m 43; c 4. ORGANIC CHEMISTRY. *Educ:* Calvin Col, AB, 42; Univ Nebr, MS, 44, PhD(org chem), 48. *Prof Exp:* Res chemist, Standard Oil Co Ind, 48-51; RES ASSOC, UPJOHN CO, 51- *Mem:* Am Chem Soc. *Res:* Chemistry of antibiotics. *Mailing Add:* 6898 Blue Star Hwy Coloma MI 49038-9316

HOEKSEMA, WALTER DAVID, b Holland, Mich, Apr 22, 42; m 66; c 1. MICROBIOLOGY, BIOCHEMISTRY. *Educ:* Calvin Col, BS, 64; Western Mich Univ, MA, 66; Mich State Univ, PhD(microbiol), 71. *Prof Exp:* Teaching asst biol, Western Mich Univ, 64-66; asst prof, 71-80, ASSOC PROF BIOL, FERRIS STATE COL, 80- *Mem:* Am Soc Microbiol. *Res:* Physiology, genetics and biochemistry of the synthesis of amino acids and proteins in the bacteria; control mechanisms in catabolic and anabolic pathways. *Mailing Add:* Dept Biol Ferris State Col Sci Bldg Rm 221 Big Rapids MI 49307

HOEKSTRA, JOHN JUNIOR, b Grand Rapids, Mich, Jan 18, 29; m 57; c 3. PHYSICAL CHEMISTRY. *Educ:* Calvin Col, BS, 51; Wayne State Univ, PhD(phys chem), 58. *Prof Exp:* Asst, Wayne State Univ, 51-52, res assoc, 52-54; chemist, Dow Chem Co, 57-63, res chemist, 64-69; assoc prof chem, Cent Col, Iowa, 69-72; assoc prof, 72-77, PROF CHEM, UNIV DUBUQUE, 77-, ACTG ACAD DEAN, 80- *Mem:* Am Chem Soc; Electrochem Soc. *Res:* Electrochemistry; batteries; electro-deposition; single crystal studies; short contact time hydrocarbon cracking studies. *Mailing Add:* Univ Dubuque Dubuque IA 52001

HOEKSTRA, KARL EGMOND, b Battle Creek, Mich, Oct 9, 35; m 57; c 3. MINERALOGY, GLASS TECHNOLOGY. *Educ:* Miami Univ, BA, 58, MS, 65; Ohio State Univ, PhD(mineral), 69. *Prof Exp:* Res engr, Ferro Corp; mgr mat eval lab, Precision Furnace Div, Harrop Ceramic Serv Co, 64-67, dir lab div, 67-70; consult, Tradet, Inc, 70-71; sr res scientist, Corning Glass Works, 72-; PRES & GEN MGR, RANSOM & RANDOLPH, 91- *Concurrent Pos:* Res fel, Corning Glass Works, 71-72. *Mem:* Mineral Soc Am; Am Ceramic Soc. *Res:* Diffusion and crystallization in the glassy phase, glass-glass and glass-plastic composites. *Mailing Add:* Ransom & Randolph 120 W Wayne St Maumee OH 43537

HOEKSTRA, PIETER, geophysics, geotechnical engineering, for more information see previous edition

HOEKSTRA, WILLIAM GEORGE, b Golden, Colo, Aug 14, 28; m 55; c 3. NUTRITIONAL BIOCHEMISTRY. *Educ:* Colo State Univ, BS, 50; Univ Wis, MS, 52, PhD(biochem), 54. *Prof Exp:* From instr to assoc prof, 54-64, PROF BIOCHEM, UNIV WIS-MADISON, 64-, PROF NUTRIT SCI, 69- *Concurrent Pos:* Merck sr fel, Nat Inst Res Dairying, Reading, Eng, 57-58; mem nutrit study sect, NIH, 70-74; NIH fel, Rowett Res Inst, Aberdeen, Scotland, 75-76. *Honors & Awards:* Gustav Bohstedt Award, Am Soc Animal Sci, 67; Borden Award, Am Inst Nutrit, 75. *Mem:* Am Inst Nutrit (pres, 80); Am Soc Animal Sci; Am Chem Soc; Soc Exp Biol & Med; Am Soc Biol Chemists; Sigma Xi. *Res:* General animal metabolism and nutrition; mineral nutrition of animals; biochemical function and metabolism of trace elements. *Mailing Add:* Dept Biochem Univ Wis Madison WI 53706

HOEL, DAVID GERHARD, b Los Angeles, Calif, Nov 18, 39; m 61; c 3. PUBLIC HEALTH & EPIDEMIOLOGY. *Educ:* Univ Calif, Berkeley, AB, 61; Univ NC, Chapel Hill, PhD(statist), 66. *Prof Exp:* Fel, Stanford Univ, 66-67; sr mathematician, Westinghouse Res Labs, 67-68; res statistician, Oak Ridge Nat Labs, 68-70; chief, Environ Biomet Br, 73-81, math statistician, 70-73, DIR, BIOMET & RISK ASSESSMENT PROG, NAT INST ENVIRON HEALTH SCI, NIH, 81-, ACTG DIR, 90- *Concurrent Pos:* Assoc dir, Radiation Effects Res Found, Hiroshima, 84-86. *Mem:* Inst Med-Nat Acad Sci; Am Statist Asn; Royal Statist Soc; Biomet Soc; Int Statist Inst; Soc Risk Anal. *Res:* Statistical inference particularly sequential procedures; statistical and mathematical applications in biology and medicine; epidemiology; radiation health effects. *Mailing Add:* Nat Inst Environ Health Sci PO Box 12233 Research Triangle Park NC 27709

HOEL, LESTER A, b Feb 26, 35. CIVIL ENGINEERING. *Educ:* City Col NY, BCE, 57; Polytech Inst NY, MCE, 60; Univ Calif, Berkeley, DEng, 63. *Prof Exp:* Lectr civil eng, City Col, NY, 58-60; asst prof eng, San Diego State Col, Calif, 62-64; Fulbright res scholar & vis prof transp, Norweg Tech Univ, Trondheim & Inst Transp Econ, Oslo, Norway, 64-65; prin engr, Wilbur Smith & Assocs, 65-66; from assoc prof to prof civil eng, Carnegie-Mellon Univ, Pittsburgh, Pa, 66-74, assoc dir, Transp Res Inst, 66-74; chmn & Hamilton prof civil eng, 74-89, FAC RES SCIENTIST, VA TRANSP RES COUN, UNIV VA, CHARLOTTESVILLE, 75-, HAMILTON PROF CIVIL ENG, 91- *Concurrent Pos:* Consult var firms; mem, Transp Res Coun, Am Soc Civil Engrs, chmn, Transp Educ Comt, Hwy Safety Comt & Nat Conf Educ; chmn, Comt Transp Prof Needs, Nat Res Coun & Truck Weight Study Comt; dir, Transp Exec Inst, Am Asn State Hwy & Transp Officials; vis prof civil eng, Univ Calif, Irvine & Inst Road & Railroad Eng, Norway, 90-91. *Honors & Awards:* Walter L Huber Civil Eng Res Prize, Am Soc Civil Engrs, Frank M Masters Transp Eng Award; Pyke Johnson Award, Transp Res Bd; Stanley W Gustafson Leadership Award, Hwy Users Fedn. *Mem:* Nat Acad Eng; fel Am Soc Civil Engrs; Inst Transp Engrs; Transp Res Bd; Am Soc Eng Educ; Transp Res Forum; Am Road & Transp Builders Asn; fel Urban Land Inst; Sigma Xi. *Res:* Author of various publications. *Mailing Add:* Thornton Hall Univ Va Charlottesville VA 22901

HOEL, PAUL GERHARD, b Iola, Wis, Mar 23, 05; m 32; c 3. MATHEMATICS. *Educ:* Luther Col, BA, 26; Univ Minn, MA, 29, PhD(math), 33. *Prof Exp:* Instr math, Rose Polytech Inst, 31-35, asst prof, 35-36; Am-Scandinavian Found fel, 36-37; instr math, Ore State Col, 37-39; from instr to assoc prof, 39-48, PROF MATH, UNIV CALIF, LOS ANGELES, 48- *Concurrent Pos:* Fulbright fel, Norway, 53-54. *Mem:* Am Math Soc; Math Asn Am; Am Statist Asn; fel Inst Math Statist. *Res:* Mathematical statistics; theory of computational errors; theoretical statistics; approximation theory. *Mailing Add:* Dept Math 6364 Math Sci Bldg Univ Calif Los Angeles CA 90024

HOELZEL, CHARLES BERNARD, b St Paul, Minn, Nov 21, 32; m 56; c 2. TECHNICAL MANAGEMENT. *Educ:* Hamline Univ, BS, 54; Vanderbilt Univ, PhD(chem), 60. *Prof Exp:* Chemist, Tenn Prod & Chem Corp, 56; sr res chemist, Hercules Powder Co, 59-65; res & develop engr, Fiber Industs, Inc, 65-67, res chemist, Halocarbon Prod Corp, 67-70; res scientist, Philip Morris Inc, 70-78, assoc sr scientist, 78-84; lectr chem, Univ NC, Wilmington, 84-89; LECTR GEN SCI, ECSU, ELIZABETH CITY, 89- *Mem:* Am Chem Soc; Sigma Xi. *Res:* Organosulfur compounds; disulfides and sulfinate esters; nitrogen heterocyclics; propellant binder systems; tobacco properties; liquid adhesive rheology. *Mailing Add:* 610 Alton St Elizabeth City NC 27909-9010

HOELZEMAN, RONALD G, b Pittsburgh, Pa, Oct 6, 40. ELECTRICAL & COMPUTER ENGINEERING. *Educ:* Univ Pittsburgh, BSEE, 64, MSEE, 67, PhD(elec eng), 70. *Prof Exp:* Engr, Westinghouse Elec Co, 64-67; instr elec eng, Univ Pittsburgh, 67-69; instr, Am Univ Beirut, 69-70; asst prof, 70-77, dir continuing eng educ, 77-80, ASSOC PROF ELEC ENG & ASSOC CHMN DEPT, UNIV PITTSBURGH, 80- *Mem:* Fel Inst Elec & Electronics Engrs; Am Soc Eng Educ. *Res:* Optimization; computer design; mathematical programming; computer graphics. *Mailing Add:* Dept Elec Eng 348 BEH Pittsburgh PA 15261

HOENER, BETTY-ANN, b Dayton, Ohio, Nov 25, 46. BIOPHARMACEUTICS. *Educ:* Univ Cincinnati, BS, 70; Ohio State Univ, PhD(pharmaceut chem), 74. *Prof Exp:* Asst prof, 74-81, VCHMN, DEPT PHARM, UNIV CALIF, SAN FRANCISCO, 78-, ASSOC PROF PHARM & PHARMACEUT CHEM, SCH PHARM, 81- *Concurrent Pos:* Secy, Pharmaceut Sci Sect, AAAS, 85-; chmn, Basic Sci Sect, Acad Pharmaceut Res & Sci, 91- *Mem:* AAAS; Am Asn Cols Pharm; Am Pharmaceut Asn; Am Chem Soc; Am Asn Pharmaceut Scientists. *Res:* Reductive metabolism and covalent binding to DNA of the mutagenic carcinogenic 5-nitrofurans; disposition of 5-nitrofurans in isolated perfused rat livers and kidneys. *Mailing Add:* Dept Pharm Univ Calif San Francisco CA 94143

HOENIG, ALAN, b Brooklyn, NY, June 18, 49; m 80; c 2. FRACTURE MECHANICS, COMPUTER TYPESETTING. *Educ:* Yale Univ, BA, 71; Harvard Univ, MA, 72, PhD(appl math), 77. *Prof Exp:* From asst prof to assoc prof, 79-90, PROF MATH, JOHN JAY COL, CITY UNIV NY, 91- *Mem:* NY Acad Sci; Math Asn Am; Am Math Soc; Am Acad Mech. *Res:* Mathematical analysis of fracture and crack analysis; effects of cracking on large-scale physical properties of a specimen. *Mailing Add:* 17 Bay Ave Huntington NY 11743

HOENIG, CLARENCE L, b Chester, Ill, Feb 8, 31; m 57; c 3. CERAMIC ENGINEERING. *Educ:* Univ Ill, BS, 55, MS, 56; Univ Calif, Berkeley, PhD(eng sci), 64. *Prof Exp:* Asst ceramic coatings, Univ Ill, 55-56; asst ceramic engr, Argonne Nat Lab, 56-58; asst mat sci, 58-59, RES CHEMIST NUCLEAR FUELS, LAWRENCE LIVERMORE LAB, UNIV CALIF, 59- *Mem:* Am Ceramic Soc. *Res:* Thermodynamics; ceramics; metallurgy; nuclear fuels. *Mailing Add:* 588 Tyler Ave Livermore CA 94550

HOENIG, STUART ALFRED, b Bridgeport, Conn, May 2, 28; m 52; c 2. ELECTRICAL ENGINEERING, AIR POLLUTION. *Educ:* Univ Mich, BS, 51; Univ Md, MS, 54; Univ Calif, Berkeley, PhD(eng sci), 61. *Prof Exp:* Physicist, US Naval Ord Lab, Md, 51-52; asst, Inst Fluid Dynamics & Appl Math, Univ Md, 52-54; asst engr, Sperry Gyroscope Co, NY, 54-55; assoc engr, Armour Res Inst, 55-57; asst, Low Pressure Lab, Univ Calif, Berkeley, 57-79; assoc prof eng, 59-71, PROF ELEC ENG, UNIV ARIZ, 71- *Concurrent Pos:* Mem, Nat Acad Sci Comts, 77- *Mem:* AAAS; Inst Elec & Electronics Engrs; Electrostatic Soc Am. *Res:* New technology for control of fugitive dust pollution; sulfur dioxide and diesel engine exhaust particulates; new medical devices for operating room and burn unit use; apparatus for detection of under ground rock failure. *Mailing Add:* Dept Elec & Computer Eng Univ Ariz Tucson AZ 85721

HOEPFINGER, LYNN MORRIS, b Neligh, Nebr, Aug 7, 41; m 65; c 2. QUALITY ASSURANCE, FOOD CHEMISTRY. *Educ:* Hastings Col, AB, 63; Purdue Univ, PhD(biochem), 68. *Prof Exp:* Asst prof chem, Hope Col, 67-71, assoc prof, 71-74; mgr, Mead Johnson & Co, 74-84; MGR QUAL ASSURANCE, HENKEL CORP, 84- *Mem:* Am Soc Qual Control; Am Chem Soc. *Res:* Enzyme isolation and inhibition studies; biochemistry of ascorbic acid. *Mailing Add:* 5 Bristol Green Bourbonnais IL 60914

HOEPPNER, CONRAD HENRY, b Spooner, Wis, March 12, 28; m 52; c 2. ELECTRO-OPTICS, RADAR ENGINEERING. *Educ:* Univ Wis, BS, 49, MS, 50; Mass Inst Technol, EE, 57. *Prof Exp:* Dept mgr, Raytheon Corp, 60-65; chief scientist, Harris Corp, 65-70; pres, Indust Electronetics Corp, 70-80; CONSULT, HOEPPNER ASSOC, 80- *Concurrent Pos:* Mem, US Govt Res & Develop Bd Comt Guided Missiles, Dept Defense, 50-53; deleg, Int Space Conf, London, Paris, Vienna & Tokyo; deleg, UN Int Space Conf, Vienna. *Mem:* Fel Inst Radio Engrs; fel Inst Elec & Electronics Engrs; fel Am Inst Aeronaut & Astronaut; Instrument Soc Am; Soc Exp Stress Anal; Soc Mfg Engrs; Soc Automotive Engrs; Sigma Xi. *Res:* High field electron emission; radar target characterization; high performance radio telemetry; author of numerous publications. *Mailing Add:* 320 12th Terr Indialantic FL 32903

HOEPPNER, DAVID WILLIAM, b Waukesha, Wis, Dec 17, 35; m 59; c 3. FATIGUE DESIGN, FRACTURE MECHANICS. *Educ:* Marquette Univ, BME, 58; Univ Wis, Madison, MS, 60, PhD(metall & mech eng), 63. *Prof Exp:* Asst prof mat eng, Calif State Univ, 63; asst prof mat & mech eng, Univ Wis, Milwaukee, 63-64; res metallurgist struct mat eng, Battelle Mem Inst, 64-69; group engr fatigue & fracture mech, Lockheed Calif Co, 69-74; prof mech & aerospace eng, Univ Mo, Columbia, 74-78; Cockburn prof eng design, Univ Toronto, 78-85; PROF & CHMN, MECH ENG, UNIV UTAH, 85- *Concurrent Pos:* Prin investr, US Navy, US Air Force, 70-78 & Elec Power Res Inst, 75-78; consult, Metal Properties Coun, 78-82, Detroit Diesel Allison, 79-81, Woodward Assoc, US Bur Mines, 78- & Rolls Royce Aero Engine Div, 80-, Lockheed Corp, Boeing Corp, McDonnell-Douglas Corp, Garrett Turbine, Pratt & Whitney & NIH, 85- *Mem:* Am Soc Mech Engrs; Am Soc Testing & Mat; Soc Automotive Engrs; Am Soc Metals; Am Soc Eng Educ; Am Inst Aeronaut & Astronaut; Nat Soc Prof Engrs. *Res:* Fatigue design; fracture mechanics; fretting fatigue; corrosion fatigue; structural integrity based design; mechanical design; design of mechanical joints; physics of fatigue; elevated temperature fatigue. *Mailing Add:* Mech Eng Col Eng Univ Utah Salt Lake City UT 84112

HOEPRICH, PAUL DANIEL, b Alliance, Ohio, Jan 3, 24; m 48; c 4. INTERNAL MEDICINE, INFECTIOUS DISEASES. *Educ:* Harvard Med Sch, MD, 47; Am Bd Internal Med, dipl; Am Bd Med Microbiol, dipl. *Prof Exp:* Med house officer, Peter Bent Brigham Hosp, 47-48; res fel bact & immunol, Harvard Med Sch, 49-51; asst resident internal med, New Haven Hosp, 53-54; resident infectious dis, Western Reserve Univ Hosp, 54-55; instr med, Sch Med & instr epidemiol, Sch Hyg & Pub Health, Johns Hopkins Univ, 56-57; asst prof med & asst res prof path, Col Med, Univ Utah, 57-62, assoc prof med & assoc res prof path, 62-67; prof path, 68-86, prof med, 67-91, EMER PROF, SCH MED, UNIV CALIF, DAVIS, 91- *Concurrent Pos:* Asst physician, Outpatient Dept, Johns Hopkins Hosp, 55-57; prof dir clin microbiol lab & assoc physician, Salt Lake County Gen Hosp, 57-65, Univ Hosp, 65-67; consult, Vet Admin Hosp, 56-67; consult med specialist, David Grant US Air Force Med Ctr, Travis Air Force Base, 68-91; consult, Letterman Gen Hosp, San Francisco, 70-76; Alice Sneed West lectr & vis prof med, St Joseph Hosp & Univ Tex Med Sch, Houston, 73; Paul A Rhoads lectr & vis prof, Med Sch, Northwestern Univ, Chicago, 75; lectr, Found Microbiol, 74-75; Fogerty Sr Int fel, 76; H Wm Harris vis prof med, Am Bureau Med Advan China, 89. *Honors & Awards:* Mod Med Monographs Award, 58. *Mem:* Am Acad Microbiol; Fel Am Col Physicians; Am Fedn Clin Res; Am Soc Clin Invest; Am Soc Exp Path; Am Soc Microbiol; Am Thoracic Soc; Fel Infectious Dis Soc Am; AAAS; Am Asn Pathol Bacteriol; Am Asn Univ Prof; Asn Am Physicians; Sigma Xi. *Res:* Clinical investigation in infectious diseases; laboratory investigation of methods of microbiology; mechanisms of action of antimicrobial agents; antifungal chemotherapy. *Mailing Add:* Sect Med Mycol/Div Infectious & Immunol Dis Univ Calif-Davis Sch Med 4301 X St Sacramento CA 95817

HOERCHER, HENRY E(RHARDT), b New York, NY, Oct 19, 30; m 57; c 2. THERMODYNAMICS. *Educ:* Polytech Inst Brooklyn, BME, 52. *Prof Exp:* Asst, Mech Eng Lab, Polytech Inst Brooklyn, 51-52; res scientist, Rocketdyne Div, NAm Aviation, Inc, 53-55; sr scientist, Systs Div, Avco Corp, 55-64, CHIEF ENERGY TRANSFER SECT, SYSTS DIV, AVCO CORP, 64- *Mem:* Am Inst Aeronaut & Astronaut; Am Soc Mech Engrs; Sigma Xi. *Res:* Heat transfer and thermodynamics of reentry vehicles and nozzles; laboratory simulation devices for reentry vehicles and rocket nozzles; space propulsion. *Mailing Add:* One Candlewood Rd Lynnfield MA 01940

HOERGER, FRED DONALD, b Wadsworth, Ohio, Jan 3, 29; m 51; c 5. ENVIRONMENTAL SCIENCE POLICY, REGULATORY AFFAIRS. *Educ:* Heidelberg Col, BS, 51; Purdue Univ, MS, 53, PhD(chem), 54. *Prof Exp:* Chemist, Strosackers Lab, Dow Chem Co, 54-57, asst lab dir, Ethylene Res Lab, 57-62, lab dir, 62-68, prod registr mgr, 68-74, opers mgr, Health-Environ Res, 74-77, dir regulatory & legis issues, Health-Environ Sci, 77-84, REGULATORY & POLICY CONSULT, DOW CHEM CO, 84- *Mem:* AAAS; Am Chem Soc; Sigma Xi; NY Acad Sci; Soc Risk Anal. *Res:* Agricultural chemicals; toxicology of biologically active compounds; safety and environmental fate of chemicals and pesticides; regulatory policy in occupational health, toxic substances and environmental affairs; science policy; risk assessment, risk communication. *Mailing Add:* 4415 Andre Midland MI 48642

HOERING, THOMAS CARL, b Alton, Ill, May 4, 25; m 50; c 2. GEOCHEMISTRY, PHYSICAL CHEMISTRY. *Educ:* Wash Univ, BA, 48, MA, 50, PhD(chem), 52. *Prof Exp:* From asst prof to assoc prof chem, Univ Ark, 52-58; STAFF MEM, CARNEGIE INST WASH, 59- *Honors & Awards:* Treibs Medal, Geochemical Soc. *Mem:* Am Chem Soc; AAAS; Geochemical Soc; Am Soc Limnol & Oceanog. *Res:* Geochemistry of stable isotopes; geochemistry of organic substances. *Mailing Add:* Carnegie Inst Wash 5251 Broad Branch Rd NW Washington DC 20015-1305

HOERL, BRYAN G, b Kiel, Wis, Feb 7, 21; m 49; c 4. MICROBIOLOGY, PATHOLOGY. *Educ:* St Francis Sem, BA, 43; Marquette Univ, MS, 53, PhD(exp path), 61. *Prof Exp:* Asst microbiol, Sch Med, 53-57, from instr to asst prof, 57-63, ASSOC PROF MICROBIOL & PATH, SCH DENT, MARQUETTE UNIV, 63-, CHMN DIV MICROBIOL, 78- *Concurrent Pos:* USPHS career res develop award, 61-66. *Mem:* Am Soc Microbiol. *Res:* Tumor immunology; basic medical microbiology. *Mailing Add:* 6004 Saddle Ridge Portage WI 53901-9102

HOERLEIN, ALVIN BERNARD, b Ft Collins, Colo, July 4, 17; m 42; c 2. ANIMAL PATHOLOGY. *Educ:* Colo Agr & Mech Col, DVM, 40; Cornell Univ, PhD(vet bact), 45. *Prof Exp:* Instr vet bact, Colo Agr & Mech Col, 40-41 & Cornell Univ, 41-45; assoc prof vet path, SDak State Col, 45-46; vet researcher, Iowa State Col, 47-53; prof animal path, Univ Nebr, 53-57; prof microbiol & hyg, Univ Ill, 57-59; prof vet bact, 59-77, EMER PROF VET MICROBIOL, COLO STATE UNIV, 77- *Honors & Awards:* Borden Award, Am Vet Med Asn, 69. *Mem:* Am Soc Microbiol; Am Asn Immunol; Am Vet Med Asn; US Animal Health Asn; Conf Res Workers Animal Dis. *Res:* Pathology and pathogenesis of diseases of domestic animals; the etiology, pathogenesis, and immunology of infectious diseases of cattle; the practical control and diagnosis of these diseases is implied. *Mailing Add:* Colo State Univ Col Vet Med PO Box 3636 Estes Park CO 80517

HOERMAN, KIRK CONKLIN, b Lyndon, Kans, Nov 26, 24; m 62; c 4. BIOCHEMISTRY. *Educ:* Univ Mo, DDS, 47, MS, 54. *Prof Exp:* Res assoc biochem & head div, Naval Dent Res Facil, Ill, 54-57, head dent res dept, Naval Med Res Unit 3, Cairo, Egypt, 57-60, dir dept biochem, 59-60, head dent dept, Chem Div, Naval Med Res Inst, Md, 60-69; chief biophys div & commanding officer, Math Statistician, Dent Res Inst, Naval Training Ctr, 69-72; dir, Res Inst, Am Dent Asn Health Found, 72-75; assoc prof, 75-80, PROF PREV DENT, SCH DENT, LOYOLA UNIV CHICAGO, 80- *Concurrent Pos:* Consult, Interdept Comt Nutrit Nat Defense, 60-63. *Mem:* Am Dent Asn; Biophys Soc; Int Asn Dent Res; fel Am Col Dent; fel Am Inst Nutrit. *Res:* Protein chemistry; connective tissue and aging; biophysical characterization of parotid gland protein and abnormal hemoglobins. *Mailing Add:* 566 Oakwood Lake Forest IL 60045

HOERNER, GEORGE M, JR, b Hershey, Pa, Feb 12, 29; m 57; c 2. CHEMICAL ENGINEERING. *Educ:* Lafayette Col, BS, 51; Univ Rochester, MEd, 58; Lehigh Univ, PhD(chem eng), 63. *Prof Exp:* Shift supvr, E I du Pont de Nemours & Co, Inc, 51-54 & 56-57; from asst prof to assoc prof, 58-75, exec officer, 77-81, PROF CHEM ENG, LAFAYETTE COL, 76-, DEPT HEAD, 81- *Concurrent Pos:* Res assoc, Lewis Res Ctr, NASA, summers 64 & 65; resident process engr, Sun Oil Co, 66-67; eng consult, J T Baker Chem Co, 68-83. *Honors & Awards:* Jones Lectr Award, Lafayette Col, 64. *Mem:* Fel Am Inst Chem Engrs; Am Soc Eng Educ; Nat Soc Prof Engrs; Instrument Soc Am. *Res:* Dynamics of packed-tower gas absorption; chemical process instrumentation; chemical process design and economics. *Mailing Add:* Dept Chem Eng Lafayette Col Easton PA 18042

HOERR, FREDERIC JOHN, b Peoria, Ill, May 2, 52; m 75; c 2. AVIAN PATHOLOGY. *Educ:* Purdue Univ, DVM, 76, MS, 77, PhD(vet path), 81. *Prof Exp:* VET PATHOLOGIST, ALA DEPT AGR & INDUSTS & C S ROBERTS VET DIAG LAB, 87- *Concurrent Pos:* Assoc prof, Dept Pathobiology, Col Vet Med, Auburn Univ, 80- *Mem:* Am Vet Med Asn; Am Asn Avian Pathologists; World Vet Poultry Asn; Am Col Vet Path. *Res:* Respiratory diseases of poultry; cryptosporidiosis; trichothecene mycotoxicoses; aflatoxicosis; applied research in veterinary pathology and toxicology. *Mailing Add:* C S Roberts Vet Diag Lab PO Box 2209 Auburn AL 36831-2209

HOERTZ, CHARLES DAVID, JR, b Louisville, Ky, Aug 5, 26; m 45; c 4. CHEMICAL ENGINEERING. *Educ:* Univ Louisville, BChE, 48. *Prof Exp:* Refining engr, Aetna Oil Co, Ky, 48-54; sales engr, Ashland Oil Co, 54-55, chem engr, 55-60, group leader res & develop, 60-66, admin asst, 66-67, mgr res & develop, 67-80, PRES, ASHLAND SYNTHETIC FUELS, 80- *Mem:* Am Chem Soc; Am Inst Chem Engrs. *Res:* Process research and development in all phases of petroleum refining and petrochemical manufacturing. *Mailing Add:* 420 Bellefonte Princes Ashland KY 41101

HOESCHELE, GUENTHER KURT, b Stuttgart, Ger, Mar 25, 25; nat US; m 55; c 2. ORGANIC POLYMER CHEMISTRY. *Educ:* Stuttgart Tech Inst, dipl, 49, Dr rer nat, 52. *Prof Exp:* Asst, Stuttgart Tech Inst, 51-53; res assoc, Polytech Inst Brooklyn, 53-54; res chemist, E I du Pont de Nemours & Co, Inc, 54-79, res fel, 79-83, sr res fel, 84-91; RETIRED. *Mem:* Am Chem Soc; Ger Chem Soc. *Res:* Organic preparative chemistry; polymer chemistry; thermoplastic elastomers. *Mailing Add:* 2007 Dogwood Lane Wilmington DE 19810-3606

HOESE, HINTON DICKSON, b San Antonio, Tex, June 26, 35; m 58; c 3. MARINE BIOLOGY. *Educ:* Agr & Mech Col, Tex, BS, 56, MS, 59; Univ Tex, PhD(zool), 65. *Prof Exp:* Marine biologist, Marine Div, Tex Game & Fish Comn, 57-59; asst marine scientist, Va Fisheries Lab, Eastern Shore Sta, 59-62; res assoc, Marine Inst, Univ Ga, 65-66; asst prof marine biol, Univ Ala, 66-67; from asst prof to assoc prof, 67-76, PROF BIOL, UNIV SOUTHWESTERN LA, 76- *Concurrent Pos:* Interim Dir, La Univ Marine Consortium, 79-80. *Mem:* Am Soc Ichthyol & Herpet; Nat Shellfisheries Asn; Am Fisheries Soc; Sigma Xi. *Res:* Distribution, ecology and taxonomy of marine fishes; ecology of marine invertebrates. *Mailing Add:* 213 Des Jardin Lafayette LA 70507

HOEVE, CORNELIS ABRAHAM JACOB, b Landsmeer, Neth, Aug 31, 24; m 53; c 1. PHYSICAL CHEMISTRY. *Educ:* Univ Amsterdam, cand, 47, MS, 50; Univ Pretoria, DSc, 55. *Prof Exp:* Res chemist, Toegepast Natuurwetenschappelijk Onderzoek, Neth, 50-52 & coun sci & indust res, SAfrica, 52-55; fel, Nat Res Coun Can, 55-56; res assoc, Cornell Univ, 56-57; sr fel, Mellon Inst, 57-64; mem staff, Nat Bur Stand, 64-69, sect chief, 69; PROF CHEM, TEX A&M UNIV, 69- *Mem:* Am Chem Soc; Sigma Xi. *Res:* Polymer chemistry. *Mailing Add:* 707 Llano Pl College Station TX 77840-4251

HOF, LISELOTTE BERTHA, b Cologne, Ger, Jan 1, 37; m 72. BIOCHEMISTRY. *Educ:* Univ Cologne, BS, 59, MS, 63, PhD(physiol chem), 65. *Prof Exp:* Res asst glycolipid metab, Ment Health Res Inst, Univ Mich, 67-68; NIH fel, Dept Pediat, Univ Chicago, 68-70; res asst, Univ Bochum, 70-72; instr, 72-73, asst prof, 73-78, ASSOC PROF BIOCHEM, ALBANY MED COL, 78- *Mem:* Soc Complex Carbohydrates; Asn Women in Sci; Am Chem Soc; Am Soc Biol Chemists; Am Soc Neurochemistry. *Res:* Glycolipids in brain development; membrane glycoconjugates. *Mailing Add:* Dept Biochem Albany Med Col 47 New Scotland Ave Albany NY 12208

HOFELDT, FRED DAN, b Rock Springs, Wyo, Sept 11, 36; m 77; c 4. INTERNAL MEDICINE, ENDOCRINOLOGY-METABOLISM. *Educ:* Col Idaho, BS, 59; Univ Wash, MD, 63. *Prof Exp:* Chief endocrinol, Fitzsimons Army Med Ctr, 71-84, dir fel prog endocrinol, 78-84; CHIEF ENDOCRINOL, DENVER GEN HOSP, DENVER, COLO, 84-, ACTG CHIEF MED, 89- *Concurrent Pos:* Prof med, Univ Colo Health Sci Ctr, Denver, 84-; assoc dir med, Denver Gen Hosp, 89- *Mem:* Fel Am Col Physicians; Am Diabetes Asn; Endocrine Soc; Am Inst Nutrit; Am Soc Clin Nutrit; Am Soc Bone & Mineral Resources. *Res:* Diabetes mellitus; hypoglycemia; thyroid cancer; metabolic bone disease and dyslipidemia. *Mailing Add:* 6680 E Mansfield Ave Denver CO 80237

HOFER, KENNETH EMIL, b Chicago, Ill, Sept 26, 34; m 75; c 6. COMPOSITES FATIGUE & FRACTURE, UNDERGROUND STRUCTURES. *Educ:* Ill Inst Technol, BS, 56, MS, 60. *Prof Exp:* Mgr mat sci, IIT Res Inst, 65-77, sr eng, 77-82; VPRES MAT, L J BROUTMAN & ASSOC, LTD, 82-; VPRES MFG, TEST SYSTS TECHNOL, 83- *Concurrent Pos:* Mem, Am Soc Testing Mat Subcomts; expert, fatigue & fracture composites, NASA. *Mem:* Soc Exp Mech Inc; Soc Advan Mat & Process Eng; Soc Mfg Eng; Am Soc Metals Int; Am Water Works Asn; Soc Plastics Indust. *Res:* Fatigue and fracture of composite materials and reinforced plastics; materials and structural, stress analysis and failure analysis; developed several testing methods and fixtures which have been adopted by the National Association for Testing. *Mailing Add:* 10601 Orchard Lane Chicago Ridge IL 60415

HOFER, KURT GABRIEL, b Feldkirchen, Austria, Mar 2, 39; m 65; c 2. CELL BIOLOGY, RADIATION BIOLOGY. *Educ:* Teacher Training Col, Austria, MEd, 59; Univ Vienna, PhD(biol), 65. *Prof Exp:* Asst radiobiol, Univ Vienna, 65-66; res assoc cancer biol, Sch Med, Tufts Univ, 66-70; asst prof radiobiol & Bremer Found Fund res grant, Ohio State Univ, 70-71; asst prof, 71-75, assoc prof & USPHS res grant, 75-79, PROF RADIOBIOL, FLA STATE UNIV, 79- *Concurrent Pos:* Mem res comt, Fla Div, Am Cancer Soc, 72- *Mem:* AAAS; Radiation Res Soc; Am Asn Cancer Res; Sigma Xi. *Res:* Effects of radiation and antimetabolites on biological systems; damage from intracellular radioisotope decay; radiation repair; diurnal and annual rhythms of radiosensitivity; growth kinetics of normal and neoplastic cell populations. *Mailing Add:* Inst Molecular Biophys Fla State Univ Tallahassee FL 32306

HOFER, LAWRENCE JOHN EDWARD, b Salt Lake City, Utah, June 26, 15; m 41; c 1. INORGANIC CHEMISTRY. *Educ:* Univ Utah, BA, 37, MA, 39; Univ Rochester, PhD(phys chem), 41. *Prof Exp:* Sherman Clark fel, Univ Rochester, 38-40; res assoc propellants, Nat Defense Res Comt, Carnegie Inst Technol, 41-43; actg chief, Phys Chem Res & Develop Br, Off Synthetic Liquid Fuels, US Bur Mines, 43-50, chief, Phys Chem Sect, Synthetic Fuels Res Br, 50-55 & Phys & Catalytic Chem Sect, Coal-to-Oil Res Br, 55-61; supvry phys chemist adsorption, Pittsburgh Coal Res Ctr, 61-65; sr fel, Mellon Inst, 65-71, head, Adsorption Fel, 67-71; mem staff anal toxicol in mining res, US Bur Mines, 71-76; chief, Mat Toxicol Br, Pittsburgh Tech Support Ctr, 76-78, Toxic Mat Div, 78-86, SR SCIENTIST, PITTSBURGH HEALTH TECHNOL CTR, MINE SAFETY & HEALTH ADMIN, US DEPT LABOR, 86- *Concurrent Pos:* Consult, 71- *Mem:* Fel Am Inst Chem; Catalysis Soc; Am Chem Soc; Air pollution Control Asn; fel AAAS. *Res:* Catalysis; Fischer-Tropsch synthesis; auto exhaust oxidation; carbides of iron, cobalt and nickel; photochemistry; fluorescence of cyanine dyes; activated carbon structure; adsorption; toxicology; mine atmospheres; structure chemistry of coal. *Mailing Add:* Mine Safety & Health Admin 4800 Forbes Ave Pittsburgh PA 15213

HOFERT, JOHN FREDERICK, b Oak Park, Ill, Oct 13, 35; m 62; c 2. ENDOCRINOLOGY. *Educ:* NCent Col, Ill, BA, 57; Mich State Univ, MS, 59; Univ Wis, PhD(oncol), 63. *Prof Exp:* Res asst biochem, Mich State Univ, 57-59; res asst oncol, Univ Wis, 59-63; res fel biochem, Albert Einstein Col Med, 63-64, instr, 64-66; asst prof, 66-71, ASSOC PROF BIOCHEM, UNIV NEBR MED CTR, OMAHA, 71- *Concurrent Pos:* Vis scholar, Dept Internal Med, Sch Med, Emory Univ, Atlanta, Ga, 86. *Mem:* AAAS; Sigma Xi; Am Soc Biochem & Molecular Biol; Endocrine Soc. *Res:* metabolic control mechanisms. *Mailing Add:* Dept Biochem Univ Nebr Med Ctr 600 S 42nd St Omaha NE 68198-4525

HOFF, BERT JOHN, b Windthorst, Tex, Nov 26, 34; m 71, 86; c 2. PLANT BREEDING, RICE GENETICS. *Educ:* Tex A&M Univ, BS, 57, MS, 63; Univ Ariz, PhD(agron), 67. *Prof Exp:* ASSOC PROF AGRON, LA STATE UNIV, BATON ROUGE, 67- *Mem:* Am Soc Agron; Crop Sci Soc Am. *Res:* Cytology and reproductive behavior of Cynodon dactylon; reproductive behavior of Zizania aquatica in Louisiana; male sterility in Oryza sativa; inheritance of disease resistance in rice; inheritance of aroma in rice. *Mailing Add:* Dept Agron La State Univ Baton Rouge LA 70803

HOFF, CHARLES JAY, b Newark, NY, Oct 28, 37; m 77. MEDICAL GENETICS. *Educ:* Univ Oxford, dipl human biol, 68; Pa State Univ, BS, 61, PhD(phys anthrop), 72. *Prof Exp:* Comput programmer, IBM, 62-64; asst prof & phys anthrop, Dept Anthrop, Univ Ore, 70-76; ASSOC PROF MED GENETICS, COL MED, UNIV SOUTH ALA, 76- *Concurrent Pos:* Partic growth & develop, Advan Study Inst, NATO, 79; prin investr res grants, NIH, 79-81, 86-88. *Mem:* Am Soc Human Genetics; Soc Study Human Biol; Human Biol Coun; Soc Epidemiol Res; Am Asn Phys Anthropologists; fel Am Col Epidemiol. *Res:* Human population genetics; demography; physical growth and development; risk factors associated with pregnancy outcome. *Mailing Add:* Dept Pediat Univ South Ala Med Col Mobile AL 36688

HOFF, DALE RICHARD, b Galesburg, Ill, Nov 16, 27; m 59; c 5. MEDICINAL CHEMISTRY. *Educ:* Univ Ill, BS, 52; Mass Inst Technol, PhD(org chem), 56. *Prof Exp:* From sr chemist to asst dir, Merck Sharp & Dohme Res Labs, Rahway, 56-68, assoc dir, 68-69, dir, 70-78, sr investr, 78-88; RETIRED. *Mem:* Am Chem Soc; The Chem Soc; AAAS. *Res:* Antiparasitic chemotherapy; computer applications to chemistry. *Mailing Add:* 2 King's Ridge Rd RD 1 Basking Ridge NJ 07920

HOFF, DARREL BARTON, b Viroqua, Wis, Dec 17, 32; m 54; c 2. ASTRONOMY, EARTH SCIENCE. *Educ:* Luther Col, Iowa, BA, 55; Univ Northern Iowa, MA, 63; Univ Iowa, PhD(sci educ & astron), 70. *Prof Exp:* Teacher chem & physics, Westby High Sch, 58-61; teacher phys sci & astron, 64-67, PROF ASTRON & SCI EDUC, UNIV NORTHERN IOWA, 70- *Mem:* Am Astron Soc; Am Asn Physics Teachers; Sigma Xi; Nat Sci Teachers Asn; Int Astron Union. *Res:* Astronomy education; UBV photometry-variable stars. *Mailing Add:* Harvard Col Observ 60 Garden St MS 71 Cambridge MA 02138

HOFF, GLORIA THELMA (ALBUERNE), b Cienfuegos, Cuba, May 10, 30; m 59; c 3. HIGH ENERGY PHYSICS. *Educ:* Univ Havana, DCFM(physics, math), 54; Univ Chicago, MS, 57, PhD(physics), 65. *Prof Exp:* Asst prof physics, Sec Educ Inst, Cienfuegos, Cuba, 51-54, Sec Educ Inst, Vibora, Cuba, 55-56 & Univ Villanueva, Cuba, 57-58; asst, Univ Chicago, 58-63; from instr to asst prof, 64-68, ASSOC PROF PHYSICS, UNIV ILL, CHICAGO CIRCLE, 68- *Concurrent Pos:* Referee, Phys Rev & Phys Rev Letters, 65- & Am J Physics, 75-; Res Bd grant, Univ Ill, Chicago Circle, 75-76. *Mem:* NY Acad Sci; AAAS; Am Phys Soc. *Res:* Phenomenology of pion photoproduction with polarized photons; analysis of lambda-kaon production data; interpretation of pion-nucleon differential cross section and polarization data; strongly correlated resonances model. *Mailing Add:* Dept Physics M/C 273 Univ Ill Box 4348 Chicago IL 60680

HOFF, HENRY FREDERICK, b Vienna, Austria, Mar 22, 38; US Citizen; m 67; c 1. PATHOLOGY, BIOPHYSICS. *Educ:* Dartmouth Col, BA, 59; Univ Calif, Berkeley, PhD(biophys), 65. *Prof Exp:* Teaching asst physiol, Univ Calif, Berkeley, 61-63; Alameda County Heart Asn fel histol, Univ Vienna, 65-66; res assoc neuropath, Max-Planck Inst Psychiat, 66-70; asst prof neurol, Baylor Col Med, 70-74, assoc prof, 74-78, assoc prof exp med, 78-80; HEAD ATHEROSCLEROSIS SECT, DEPT CARDIOVASC RES, CLEVELAND CLIN FOUND, 80-, DIR, PROG PROJ ATHEROSCLEROSIS, 83- *Concurrent Pos:* Fel arteriosclerosis coun, Am Heart Asn; Am Heart Asn advan res fel atherosclerosis, Max-Planck Inst Psychiat, Munich, 68-70; proj in cardiovasc & stroke ctr grants, Baylor Col Med, 70-80; estab investr award, Am Heart Asn, 76-81 & mem, Prog Rev Comt, 80-81; ad hoc mem, Metabolism Study Sect, NIH, 84; mem, Parent Review Comt, Arteriosclerosis Spec Ctr Res Prog, Nat Heart, Lung & Blood Inst, 85-86; ed, Artery, 77-84. *Mem:* Am Asn Exp Path; Am Heart Asn; Am Soc Cell Biol; NY Acad Sci. *Res:* Vascular pathology, injury and repair; cerebral atherosclerosis; lipoproteins and atherosclerosis in experimental animals and humans using quantitative immunochemical, enzyme and immunohistochemical and electron microscope techniques; interaction of lipoproteins extracted from human arteries with macrophages and arterial cells in culture-mechanism of foam cell formation; cell biology. *Mailing Add:* Cleveland Clin Found Dept Atherosclerosis 9500 Euclid Ave Cleveland OH 44106

HOFF, JAMES GAVEN, b Philadelphia, Pa, Mar, 35; m 56; c 2. ECOLOGY, ICHTHYOLOGY. *Educ:* East Stroudsburg State Col, BS, 60; Rutgers Univ, MS, 62, PhD(fish ecol), 65. *Prof Exp:* Res asst ichthyol, Rutgers Univ, 60-62, res assoc, 62-65; from asst prof to assoc prof, 65-71, PROF ICHTHYOL & ECOL, SOUTHEASTERN MASS UNIV, 71- *Mem:* Am Fisheries Soc; Wildlife Soc; Am Soc Ichthyol & Herpet. *Res:* Distribution of estuarine fishes, especially temperature-salinity-oxygen relationships as limiting factors; sturgeon biology; endangered species. *Mailing Add:* Dept Biol Southeastern Mass Univ Old Westbury Rd North Dartmouth MA 02747

HOFF, JOHAN EDUARD, food science, analytical chemistry, for more information see previous edition

HOFF, JOHN C, b Billings, Mont, Mar 9, 28; m 57; c 3. MICROBIOLOGY. *Educ:* Mont State Col, BS, 50; Utah State Univ, MS, 56; Wash State Univ, PhD(bact), 62. *Prof Exp:* Bacteriologist, Mont State Bd Health, 50, 52-54, virologist, 56-58; res microbiologist, Northwestern Water Supply Res Lab, Environ Protection Agency, 62-69, dir, 70-73, res microbiologist, Risk Reduction Eng Lab, 73-90; CONSULT, ENVIRON MICROBIOL, 90- *Mem:* Am Soc Microbiol; Sigma Xi; Am Water Works Asn. *Res:* Sanitary and environmental bacteriology and virology; bacteriophage typing; bacteriological and virological methods for examination of water; microbiological aspects of water disinfection. *Mailing Add:* 6962 Gammwell Dr Cincinnati OH 45230

HOFF, KENNETH MICHAEL, b Dickinson, NDak, July 6, 34. NEUROCHEMISTRY, COMPARATIVE ANATOMY. *Educ:* Carroll Col, Mont, AB, 61; Wash State Univ, PhD(zool), 66. *Prof Exp:* From asst prof to assoc prof, 66-77, PROF BIOL, CLEVELAND STATE UNIV, 77- *Mem:* AAAS; Am Soc Zoologists. *Res:* Maturation and control of indoleamine metabolism during mouse brain development; interaction of various drugs with indoleamines during mouse brain development. *Mailing Add:* Dept Biol Cleveland State Univ Euclid Ave at E 24th Cleveland OH 44115

HOFF, MARCIAN EDWARD, JR, b Rochester, NY, Oct 28, 37; m 62, 77; c 1. ELECTRICAL ENGINEERING. *Educ:* Rensselaer Polytech Inst, BEE, 58; Stanford Univ, MS, 59, PhD(elec eng), 62. *Prof Exp:* Res assoc elec eng, Stanford Univ, 62-68; mgr applns res, Intel Corp, 68-83; vpres technol, Atari, Inc, 83-84; VPRES & CHIEF TECH OFFICER, TEKLICON INC, 90- *Concurrent Pos:* Consult, 84- *Honors & Awards:* Stuart Ballantine Medal, Franklin Inst, 79; Cledo Brunetti Award & Centennial Medal, Inst Elec & Electronics Engrs, 84. *Mem:* Fel Inst Elec & Electronics Engrs. *Res:* Electronic and adaptive computers; telecommunications. *Mailing Add:* Teklicon Inc 444 Castro St Mountain View CA 94041-9909

HOFF, N(ICHOLAS) J(OHN), b Magyarovar, Hungary, Jan 3, 06; nat US; m 40, 72; c 1. AERONAUTICAL ENGINEERING, MECHANICS. *Educ:* Fed Polytech Inst, Zurich, Switz, Dipl Ing, 28; Stanford Univ, PhD(eng mech), 42. *Hon Degrees:* DSc Eng, Technion, Haifa, Israel, 81. *Prof Exp:* Instr gliding & soaring, Hungarian Soaring Soc, 28; struct designer, Aeroplane Div, Manfred Weiss Aeroplane & Motor Works, Ltd, Budapest, Hungary, 29-39; res asst mech eng, Stanford Univ, 39-40; from instr to prof aeronaut eng & appl mech, Polytech Inst Brooklyn, 40-57, chmn dept, 41-57; prof & head dept, 57-71, EMER PROF AERONAUT & ASTRONAUT, STANFORD UNIV, 71- *Concurrent Pos:* Vis prof, Univ Tokyo, 63, Univ Paris, 64, Univ New S Wales, Sydney, 66, Monash Univ, 71, Ga Inst Technol, 73, Cranfield Inst Technol, 73-74, Fed Polytech Inst, Zurich, 74 & Rensselaer Polytech Inst, 76-81; mem subcomt aircraft structures, Nat Adv Comt Aeronaut, 48-57, mem struct design comt, 58; mem, Gen Assembly Int Union for Theoret & Appl Mech, 49-; mem, US Nat Comt Theoret & Appl Mech, 49-, chmn, 57-60; mem ship struct design comt, Nat Res Coun, 53-63, chmn, 60-63; mem res adv comt aircraft struct, NASA, 59-67 & res & technol adv coun, 67-70; invited lectr, 4th Int Conf Composite Mats, Tokyo, Japan, 82 & Hungarian Acad Sci, 83. *Honors & Awards:* William M Murray lectr, Soc Exp Stress Anal, 58; Theodore von Karman Mem lectr, Israel Soc Aeronaut Sci, Tel Aviv, 66; von Karman lectr, Am Inst Aeronaut & Astronaut, 66, G Edward Pendray Award & Struct Dynamics & Mat Award, 71; Worcester Reed Warner Medal, Am Soc Mech Engrs, 67, ASME Medal, 74, Centennial Medal, 80; Theodore von Karman Medal, Am Soc Civil Engrs, 72; Lester D Gardner lectr, Mass Inst Technol, 73; Monie A Ferst Award, Sigma Xi, 82; I B Laskowitz Award, NY Acad Sci, 83; Daniel Guggenheim Medal, 83. *Mem:* Nat Acad Eng; Am Acad Arts & Sci; hon fel Am Inst Aeronaut & Astronaut; hon mem Aeronaut Soc India; hon mem & fel Am Soc Mech Engrs; hon mem, Hungarian Acad Sci, Budapest, Hungary; hon mem, Japanese Rocket Soc, Tokyo, Japan; hon mem Nat Acad Air & Space, Toulouse, France; assoc mem Acad Sci, Paris, France. *Res:* Applied mechanics; structural analysis and airplane analysis; shell theory; static and dynamic stability; high temperature effects on structures; applied mechanics. *Mailing Add:* 782 Esplanada Way Stanford CA 94305

HOFF, RAYMOND E, b Norfolk Va, Mar 24, 34; m 62; c 2. POLYMER CHEMISTRY. *Educ:* Beloit Col, BS, 56; Univ Utah, PhD(org chem), 64. *Prof Exp:* Technician, Westinghouse Res Lab, 56-60; res chemist, B F Goodrich Res Ctr, 64-66; res chemist, Chemplex Co, 66-85; sr res scientist, Norchem, Inc, 85-86; assoc scientist, USI Chem Co, 86-88; RES SCIENTIST, QUANTUM CHEM CORP, 88- *Mem:* Am Chem Soc. *Res:* Catalysis of polymerization. *Mailing Add:* 244 S Oak St Palatine IL 60067

HOFF, RICHARD WILLIAM, b Duluth, Minn, Jan 25, 30; m 53; c 2. NUCLEAR CHEMISTRY. *Educ:* Univ Minn, AB, 50; Univ Calif, PhD(chem), 54. *Prof Exp:* Asst, 50-53, group leader, 59-69, assoc leader, Nuclear Chem Div, 69-79, RES CHEMIST, LAWRENCE LIVERMORE LAB, UNIV CALIF, 53- *Mem:* Am Chem Soc; Am Phys Soc. *Res:* Radiochemical and nuclear properties of actinide elements; neutron capture gamma-ray spectroscopy. *Mailing Add:* Nuclear Chem Div Lawrence Livermore Nat Lab Livermore CA 94550

HOFF, VICTOR JOHN, b Cincinnati, Ohio, Oct 18, 29; m 62. CYTOGENETICS. *Educ:* Univ Ky, BS, 53, MS, 58; Univ Ind, PhD(bot), 61. *Prof Exp:* Asst prof bot, Univ Ark, 61-66; assoc prof, 66-71, PROF BIOL, STEPHEN F AUSTIN STATE UNIV, 71- *Mem:* Bot Soc Am; Am Genetic Asn; Am Inst Biol Sci. *Res:* Cytogenetics of Onagraceae. *Mailing Add:* Dept Biol Stephen F Austin State Univ Nacogdoches TX 75962

HOFF, WILFORD J, JR, b Walterboro, SC, Oct 3, 28; m 50; c 3. ORGANIC CHEMISTRY. *Educ:* The Citadel, BS, 50; Princeton Univ, MA, 55, PhD(chem), 62. *Prof Exp:* From instr to assoc prof, 58-67, dept head, 67-77, PROF CHEM & HEAD DEPT, US MIL ACAD, 77- *Mem:* Am Chem Soc; AAAS; Sigma Xi. *Res:* Chemical education. *Mailing Add:* Dept Chem US Mil Acad West Point NY 10996

HOFFBECK, LOREN JOHN, b Ortonville, Minn, Aug 26, 32; m 60; c 5. AGRONOMY. *Educ:* SDak State Univ, BS, 54; Univ Wis, MS, 59, PhD(agron), 62. *Prof Exp:* Res agronomist, USDA, 61-66; res coordr, 77-86, regist res dir, 86-90, CORN BREEDER, RES DEPT, PIONEER HI-BRED INT, INC, 66-, RES FEL, 90- *Mem:* Am Soc Agron; Crop Sci Soc. *Res:* Crop breeding; development of improved corn hybrids. *Mailing Add:* Pioneer Hi-Bred Int Inc 1000 W Jefferson Tipton IN 46072

HOFFEE, PATRICIA ANNE, b Columbiana, Ohio, Oct 1, 37. BIOCHEMISTRY, MICROBIOLOGY. *Educ:* Univ Pittsburgh, BS, 59, MS, 61, PhD(microbiol), 63. *Prof Exp:* NIH fel, Albert Einstein Col Med, 63-65, from instr to asst prof molecular biol, 65-67; from asst prof to assoc prof, 67-78, PROF MICROBIOL, SCH MED UNIV PITTSBURGH, 78- *Concurrent Pos:* Career develop award, Albert Einstein Col Med, 65-67. *Mem:* AAAS; Am Soc Biol Chem & Molec Biol; Am Soc Microbiol. *Res:* Genetic regulation; enzyme structure and function; immuno deficiency diseases-adenosine deaminase, purine nucleaside phosphorylase; purine and pyrimidine nucleotide metabolism. *Mailing Add:* Dept Molec Genet Biochem Univ Pittsburgh Sch Med Pittsburgh PA 15261-2072

HOFFELD, J TERRELL, b Oct 18, 46; m 73; c 2. IMMUNOLOGY, DENTAL RESEARCH. *Educ:* Univ Cincinnati, BS, 68; Ohio State Univ, DDS, 72; Univ Rochester, PhD(microbiol & immunol), 77. *Prof Exp:* Investr, Lab Microbiol & Immunol, Nat Inst Dent Res, NIH, 77-81, sr investr, Clin Invests & Patient Care Br, 81-85, EXEC SECY, ORAL BIOL & MED STUDY SECT, DIV RES GRANT, NIH, 85-; DENT OFFICER, COMN CORPS, US PUB HEALTH SERV, DEPT HEALTH & HUMAN SERV, 78- *Concurrent Pos:* Adj res fac, dept microbiol, Baltimore Col Dent Surg, Univ Md, 82-; vis lectr, dept periodont, Naval Dent Sch, 83-; co-founder, NIH Oxygen Radicals Forum, 81, chmn, 81-87; mem, Int Asn Den Res, Wash Acad Sci liasion, 85-, fel, 91- *Mem:* Int Asn Dent Res (secy-treas, 86-87, vpres, 87-88, pres, 88-, secy-treas, Microbiol & Immunol Res Group, 86-); Am Asn Immunologists; Am Soc Microbiol; Am Soc Bone & Mineral Res. *Res:* Dental research and oral biology. *Mailing Add:* Oral Biol & Med Study Sect Div Res Grant NIH Bethesda MD 20892

HOFFELDER, ANN MCINTOSH, b Creedmoor, NC, Feb 3, 35; m 59; c 2. CHEMICAL EDUCATION. *Educ:* Univ NC, Greensboro, BA, 57; Univ NC, Chapel Hill, MAT, 63; Univ Md, College Park, PhD(chem educ), 73. *Prof Exp:* Teacher & counr gen sci, English & soc studies, Pac Beach Jr High, San Diego, Calif, 57-61; teacher chem, biol & gen sci, Chapel Hill High Sch, NC, 61-62; from instr to assoc prof, 63-78, PROF GEN & ORG CHEM & DEPT HEAD, CUMBERLAND COL, KY, 78- *Concurrent Pos:* Mem bd dir, Ky Asn Progress Sci, 72-; mem coun, Ky State Sci Adv Coun, 73- *Mem:* Am Chem Soc; Nat Sci Teachers Asn. *Res:* Curriculum development in chemical education. *Mailing Add:* Dept Math & Sci Cumberland Col Walnut St Williamsburg KY 40769

HOFFENBERG, PAUL HENRY, color science, for more information see previous edition

HOFFER, ABRAHAM, b Budapest, Hungary; US citizen. EARTH SCIENCES, GEOPHYSICS. *Educ:* Univ Toronto, BASc, 56, MASc, 57; Univ Chicago, PhD(geophys sci), 59. *Prof Exp:* Res assoc solid state physics, Pa State Univ, 60-62; asst prof, Loyola Univ, 62-64 & Sir George Williams Univ, 65-66; assoc prof, Marquette Univ, 66-68; assoc ed & staff writer, Encycl Britannica, Inc, 68-71; teacher phys sci, Chicago City Col, 72-88. *Mem:* AAAS; Am Geophys Union; Am Phys Soc; Nat Soc Prof Engr. *Res:* Causes of intra-terrestrial unrest; origin of geomagnetic field; solar-planetary relationships. *Mailing Add:* 992 S Park Terrace Chicago IL 60605-2018

HOFFER, ALAN R, b Chicago, Ill. MATHEMATICS. *Educ:* Univ Calif, BA, 58; Univ Notre Dame, MS, 63; Univ Mich, PhD(math), 69. *Prof Exp:* Teacher high schs, Calif, 58-67; asst prof math, Univ Mont, 69-71; assoc prof, 71-76, PROF MATH, UNIV ORE, 76- *Concurrent Pos:* Chmn, Comn Educ Teachers Math, Nat Coun Teachers Math, 74-76; dir, Math Resource Proj, NSF, 74-78. *Mem:* Math Asn Am. *Res:* Mathematics education; group theory and geometry; combinatorics; mathematics combinatorics. *Mailing Add:* Dept Curriculum & Instruction Boston Univ 605 Commonwealth Ave Boston MA 02215-3701

HOFFER, J(OAQUIN) A(NDRES), b Montevideo, Uruguay, Aug 4, 46; c 1. NEUROPHYSIOLOGY, SENSORIMOTOR SYSTEMS. *Educ:* Harvey Mudd Col, BS, 70; Johns Hopkins Univ, PhD(biophysics), 75. *Prof Exp:* Fel nerve & muscle physiol, Dept Physiol, Univ Alberta, 75-78; staff fel integrative neurophysiol, Lab Neural Control, Nat Inst Neurol & Commun Dis & Stroke, Bethesda, Md, 78-80, sr staff fel, 80-82; from asst prof to assoc prof, 82-91, PROF, DEPT CLIN NEUROSCI, UNIV CALGARY, 91- *Concurrent Pos:* Vis prof, Nat Res Inst Physiol Cent Nerv Syst, Milano Italy, 82; med scholar, Alta Heritage Found Med Res, 82-87 & 87-92. *Mem:* Soc Neuroscience; Behav & Brain Sci Asn; Inst Elec & Electronic Engrs; Emba Mink Breeders Asn. *Res:* Central control and reflex regulation of movements; activity of single sensory and motor nerve cells in freely moving animals; design of implantable electrodes for the control of prosthetic limbs. *Mailing Add:* Dept Clin Neurosci Univ Calgary 2500 University Dr NW Calgary AB T2N 1N4 Can

HOFFER, JERRY M, b Marshalltown, Iowa, Aug 10, 34; m 57; c 2. GEOLOGY. *Educ:* Univ Iowa, BA, 56, MS, 58; Wash State Univ, PhD(geol), 65. *Prof Exp:* Instr high sch, 58-59; teaching asst geol, Wash State Univ, 60-64; asst prof, Western Ill Univ, 65 & Tex Western Col, 65-68; assoc prof, 68-72, PROF GEOL, UNIV TEX, EL PASO, 72- *Mem:* AAAS; Geol Soc Am; Can Mineral Soc; Mineral Soc Am; Sigma Xi. *Res:* Vulcanism; x-ray mineralogy. *Mailing Add:* 8022 Antero Pl El Paso TX 79968

HOFFER, PAUL B, b New York, NY, Apr 9, 39; m 64; c 5. RADIOLOGY, NUCLEAR MEDICINE. *Educ:* Univ Chicago, MD, 63; Am Bd Radiol, dipl, 70, cert nuclear imaging, 74; Am Bd Nuclear Med, dipl, 72; Yale, MA, 77. *Prof Exp:* Intern med, Mary Hitchcock Mem Hosp, 63-64; resident radiol, Univ Chicago, 66-69, from instr to assoc prof, 69-74, dir sect nuclear med, Pritzker Sch Med, 70-74; prof radiol & dir, Sect Nuclear Med, Med Ctr, Univ Calif, San Francisco, 74-77; PROF RADIOL & DIR SECT NUCLEAR MED, YALE UNIV, 77- *Concurrent Pos:* James Picker Found res scholar, Pritzker Sch Med, Univ Chicago, 69-72; proj leader, Argonne Cancer Res Hosp, 69-74. *Mem:* AAAS; Soc Nuclear Med (vpres, 80-81); Am Col Radiol; Asn Univ Radiol; Am Col Nuclear Physicians. *Res:* Nuclear medicine. *Mailing Add:* Sect Nuclear Med Yale Univ Sch Med 333 Cedar St New Haven CT 06510

HOFFER, ROGER M(ILTON), b Rogers City, Mich, Dec 5, 37; m 58; c 3. REMOTE SENSING, PHOTO INTERPRETATION. *Educ:* Mich State Univ, BS, 59; Colo State Univ, MS, 60, PhD(watershed mgt), 62. *Prof Exp:* Watershed specialist, US Forest Serv, 62; res assoc remote sensing, Purdue Univ, assoc prof, 68-75, prof forestry, 75-88, leader, Ecosysts Res Prog, Lab Appln Remote Sensing, 66-88; PROF FORESTRY & DIR REMOTE SENSING, GIS PROG, COLO STATE UNIV, 88- *Concurrent Pos:* NASA grant, 66-72; consult, AID, 71 & 76-77, US Geol Surv & US Army Corps Engrs, 73, UN, 75 & NASA, 78 & 80; prin investr, Landsat, 72-76 & Skylab, 73-75; assoc ed, Photogram Eng & Remote Sensing, 74-75. *Honors & Awards:* Pecora Award, NASA & US Geol Surv, 76. *Mem:* Soc Am Foresters; Am Soc Photogram (vpres, 87-89, pres, 89-90); Sigma Xi. *Res:* Interpretation of remote sensor imagery and color infrared photography; spectral characteristics of vegetation, soil, and water; development and evaluation of computer-aided analysis techniques for mapping earth surface features using satellite multispectral scanner data; author of over 100 scientific publications. *Mailing Add:* Dept Forestry & Wood Sci Colo State Univ Ft Collins CO 80523

HOFFER, THOMAS EDWARD, b Salt Lake City, Utah, Aug 1, 27; m 76; c 2. PHYSICAL METEOROLOGY. *Educ:* Univ Utah, BS, 50, MS, 52; Univ Chicago, PhD(meteorol), 61. *Prof Exp:* Lectr meteorol, Univ NMex, 51-52 & Univ Utah, 52-56; res assoc, Univ Chicago, 56-61; sr res scientist, Lockheed-Calif Co, 61-64; assoc res prof, 64-71, RES PROF ATMOSPHERIC PHYSICS & PROF PHYSICS, UNIV NEV, RENO, 71-

Mem: Sigma Xi; Am Meteorol Soc; AAAS; NY Acad Sci; Air Pollution Control Asn. *Res:* Study of aerosols in pollution and nucleation; air pollution monitoring and research; aircraft sampling and analysis of particulates gases and wind field; visibility research and laser tranmissometer development. *Mailing Add:* Desert Res Inst Univ Nev Reno NV 89507

HOFFERT, JACK RUSSELL, comparative physiology, ophthalmology; deceased, see previous edition for last biography

HOFFERTH, BURT FREDERICK, b Kouts, Ind, Nov 6, 22; m 43; c 5. RESEARCH ADMINISTRATION. *Educ:* Purdue Univ, BSCHE, 43, Iowa State Univ, PhD(chem), 50. *Prof Exp:* Chem engr fatty acids & derivatives, Armour & Co, 43-45; instr org chem, Iowa State Univ, 46-50; mgr, Armstrong World Indust, Inc, 61-68, gen mgr, Armstrong Cork Co, 68-80, sr res assoc, 81-86; RETIRED. *Mem:* Am Chem Soc. *Res:* Fatty chemicals; organometallic chemistry; protective coatings; polymer processing. *Mailing Add:* 125 White Oak Dr Lancaster PA 17601

HOFFLEIT, ELLEN DORRIT, b Florence, Ala, Mar 12, 07. ASTRONOMY. *Educ:* Radcliffe Col, AB, 28, MA, 32, PhD(astron), 38. *Hon Degrees:* DSc, Smith Col, 84. *Prof Exp:* Asst, Observ, Harvard Univ, 29-38, res assoc, 38-43; mathematician, Ballistic Res Lab, Aberdeen Proving Ground, MD, 43-48; astronr, Observ, Harvard Univ, 48-56; dir, Maria Mitchell Observ, Mass, 57-78; res assoc, 56-69, SR RES ASTRONR, OBSERV, YALE UNIV, 69- *Concurrent Pos:* Lectr, Wellesley Col, 55-56; tech expert, Ballistic Res Lab, Aberdeen Proving Ground, 48-61; astronr adjoint, Astron Observ, Pasteur Inst, Strasbourg, France, 76. *Honors & Awards:* George van Biesbroeck Award, 88. *Mem:* AAAS; Meteoritical Soc; Am Asn Variable Observers (vpres, 58-60, pres, 61-63); Int Astron Union; Am Astron Soc; Am Defense Preparedness Asn. *Res:* Variable stars; stellar spectra; proper motions; meteors; galactic structure; bright stars; updating star catalogues; 19th-20th century topics in the history of astronomy. *Mailing Add:* Yale Univ Observ Box 6666 New Haven CT 06511

HOFFMAN, ALAN BRUCE, b Canton, Ill, Sept 22, 41; m 64; c 4. PHYSICAL INORGANIC CHEMISTRY. *Educ:* Univ Ill, BS, 63; Stanford Univ, PhD(inorg chem), 67. *Prof Exp:* NIH res fel crystallog, Cornell Univ, 67-69; sr res chemist phys chem, Monsanto Co, St Louis, 69-71; ASSOC PROF CHEM, LEWIS & CLARK COMMUNITY COL, 71- *Concurrent Pos:* Lectr, Southern Ill Univ, Edwardsville, Ill, 70- *Mem:* Sigma Xi; Am Chem Soc. *Res:* Kinetics and mechanism of reactions of complex ions; structures and reaction mechanisms of complex ions involved in oxygen-carrying and similar phenomena. *Mailing Add:* Dept Health/Life Sci Lewis & Clark Community Col 5800 Godfrey Rd Godfrey IL 62035

HOFFMAN, ALAN JEROME, b New York, NY, May 30, 24; m 47, 90; c 2. MATHEMATICS. *Educ:* Columbia Univ, AB, 47, PhD, 50. *Hon Degrees:* DSc, Technion, 86. *Prof Exp:* Lectr & instr math, Columbia Univ, 46-50; mem, Inst Adv Study, 50-51; mathematician, Nat Bur Standards, 51-56; sci liaison officer, Off Naval Res, London, 56-57; consult, Gen Elec Co, 57-61; RES STAFF MEM, IBM RES CTR, 61- *Concurrent Pos:* Adj prof, City Univ New York, 65-76, Yale Univ, 76-80, Stanford, Univ, 80- *Mem:* Nat Acad Sci; Math Asn Am; fel Am Acad Arts & Sci; fel NY Acad Sci; Am Math Soc; Soc Indust Appl Math. *Res:* Combinatorial mathematics; graph theory; experimental designs; linear inequalities and programming; estimation of eigenvalues. *Mailing Add:* IBM Res Ctr PO Box 218 Yorktown Heights NY 10598

HOFFMAN, ALBERT CHARLES, b Northumberland, Pa, July 8, 38; m 66; c 2. GENETICS, BIOLOGY. *Educ:* Bloomsburg State Col, BSEd, 64; NC State Univ, MS, 68, PhD(genetics), 70. *Prof Exp:* From asst prof to assoc prof, Millersville Univ, 70-77, prof biol genetics, 77-81, actg dean, 81-82, DEAN, SCH SCI & MATH, MILLERSVILLE UNIV, 82- *Mem:* Coun Cols Arts & Sci. *Res:* Use of insects as assay organisms to determine mutagenic effectiveness of chemical agents; repair of induced genetic damage in insect oocytes. *Mailing Add:* Dean Sci/Math Millersville State Col Millersville PA 17551

HOFFMAN, ALEXANDER A J, b Suffern, NY, Dec 3, 31; m 56; c 3. THEORETICAL PHYSICS, SOLAR ENERGY. *Educ:* Univ Tex, Austin, BA, 55, MA, 57, PhD(physics), 62. *Prof Exp:* Teaching asst math, Univ Tex, Austin, 56-58, spec instr, 58-60, res physicist, Defense Res Labs, 60-62; from asst prof to prof math & physics, 62-81, Tex Christian Univ, dir comput sci prog, 73-81; INDEPENDENT PRACT, 81- *Mem:* Geoscience Electronics Soc (pres, 78); AAAS; Asn Comput Mach; Inst Elec & Electronics Engrs; Sigma Xi. *Res:* Insulation and climatology; solar energy; computer based mathematical models. *Mailing Add:* 4201 Hildring Dr E Ft Worth TX 76109

HOFFMAN, ALLAN JORDAN, b Hicksville, NY, June 27, 25; m 52; c 2. PHARMACEUTICAL CHEMISTRY, BIOCHEMISTRY. *Educ:* Columbia Univ, BS, 49; Rutgers Univ, PhD(pharmaceut chem), 69. *Prof Exp:* Supvr anal chem, Carter-Wallace Inc, 54-59; anal res supvr, White Labs, Inc, Div Schering Corp, 59-66; sr scientist, Am Cyanamid Co, 66-70; head bioanal res, Hoechst-Roussel Pharmaceut Inc, Div Am Hoechst Corp, 70-81; RADIATION PHYSICIST I, DEPT ENVIRON PROTECTION, NJ STATE, 84- *Mem:* AAAS; Am Chem Soc; Acad Pharmaceut Sci. *Res:* Analysis of drugs in biological matrices; pharmaceutical analysis; drug-sensitized photodynamic oxidation; analytical instrumentation; analysis and stability studies of drug formulations; thorium, uranium, radon studies on drinking water. *Mailing Add:* 223 S Fourth Ave Highland Park NJ 08904

HOFFMAN, ALLAN RICHARD, b New York, NY, May 31, 37; m 59; c 2. SOLID STATE PHYSICS. *Educ:* Cornell Univ, BEP, 59; Univ Ill, Urbana, MS, 61; Brown Univ, PhD(physics), 67. *Prof Exp:* Res physicist, Tex Instruments, Inc, 61-63; res assoc physics, Brown Univ, 66-68; asst prof, Univ Mass, Amherst, 68-75; staff scientist, US Senate Comt on Com, Sci & Transp, 75-78; dir, Advanc Energy Systs Policy Div, US Dept Energy, 78-79; asst dir indust prog, Energy Prod Ctr/Mellon Inst, 79-81; EXEC DIR, COMT SCI, ENG & PUB POLICY, NAT ACAD SCI, 82- *Concurrent Pos:* Cong fel, Am Phys Soc, 74-75. *Mem:* AAAS; Fedn Am Sci; Am Phys Soc. *Res:* Low temperature thermometry; radiation detection; transport properties of metals; ultrasonic studies in metals; solar energy; energy conservation. *Mailing Add:* 1621 Apricot Ct Reston VA 22090

HOFFMAN, ALLAN SACHS, b Chicago, Ill, Oct 27, 32; m 62; c 2. CHEMICAL ENGINEERING, BIOENGINEERING. *Educ:* Mass Inst Technol, SB, 53, SM, 55, ScD, 57. *Prof Exp:* Asst dir, Oak Ridge Eng Practice Sch, Mass Inst Technol, 54-55, instr chem eng, 55-56, asst prof, 58-60; res staff mem, Calif Res Corp, 60-63; assoc dir res, Amicon Corp, 63-65; assoc prof chem eng, Mass Inst Technol, 65-70; asst dir, Ctr Bioeng, 72-83, PROF CHEM ENG & BIOENG, UNIV WASH, 70- *Concurrent Pos:* Kimberly-Clark fel, 56-57; Fulbright fel, 57-58 & Battelle fel, 70-72. *Honors & Awards:* Biomat Sci Prize, Japanese Soc Biomat, 90- *Mem:* Am Chem Soc; Am Inst Chem Engrs; Am Soc Artificial Internal Organs; Soc Biomat (pres, 83-84); Int Soc Artificial Organs; Controlled Release Soc. *Res:* Materials science of surfaces, polymers and biomaterials; applied radiation and plasma chemistry. *Mailing Add:* Ctr Bioeng FL-20 Univ Wash Seattle WA 98195

HOFFMAN, ALLEN HERBERT, b Salem, Mass, Mar 23, 42; m 68; c 2. BIOMECHANICS. *Educ:* Worcester Polytech Inst, BS, 63, MS, 67; Univ Colo, PhD(mech eng), 70. *Prof Exp:* From asst prof to assoc prof, 70-83, PROF MECH ENG, WORCESTER POLYTECH INST, 83- *Concurrent Pos:* Vis prof, Univ Mass Med Sch, 78- *Honors & Awards:* Elizabeth W Lanir Award, Am Acad Orthop Surgeons, 88. *Mem:* Am Soc Mech Engrs; AAAS; Am Soc Eng Educ; Sigma Xi. *Res:* Constitutive equations to describe soft biological tissue; use of finite element methods to determine stresses in soft tissue. *Mailing Add:* Mech Eng Dept Worcester Polytech Inst 100 Inst Rd Worcester MA 01609

HOFFMAN, ARLENE FAUN, b New York, NY, Nov 23, 41. VASCULAR EVALUATION. *Educ:* Queens Col, BS, 62; State Univ NY Downstate Med Ctr, PhD(physiol), 67; Calif Col Podiatric Med, DPM, 76. *Prof Exp:* Bact technician, Maimonides Hosp, 61; high sch teacher biol, New York City Bd Educ, 62; teaching asst physiol, State Univ NY Downstate Med Ctr, 63-66; USPHS fel immunophysiol, Depts Med & Physiol, Sch Med, Stanford Univ, 66-67; assoc prof, Basic Sci Dept, Calif Col Podiatric Med, 67-69, dir basic sci, 68-75, assoc dean, curric affairs, 72-75, from asst prof to assoc prof, Dept Podiatric Med, 78-85, PROF BASIC SCI, CALIF COL PODIATRIC MED, 69-, PROF PODIATRIC MED, 85- *Concurrent Pos:* Instr biol & physiol, City Univ New York, 64-66; mem, Nat Bd Podiatry Examrs, 69-; spec ed basic sci in med, J Am Podiatry Asn, 72-; eval consult, Mt Zion Hosp, 74-75; mem rev comt spec proj grants, Bur Health Manpower, 75-77; resident surg, Jewish Mem Hosp, NY, 76-77; mem rev comt, Nat Heart & Lung Inst instnl res fel grant appln, 76-77; mem, Dept Health, Educ & Welfare, Women's Action Prog Task Force Comt; mem rev comt biomed res develop grants, NIH, 77-78; assoc, Med Ctr Podiatry Group, Augusta, Ga, 77-78; pvt pract, 78; gen ed, J Am Podiatric Med Asn, 80-; chief, Vascular Clin, 84- *Mem:* Am Asn Col Podiatric Med; Am Educ Res Asn; Am Pub Health Asn; Am Podiatric Med Asn; Asn Women in Sci; Am Asn Women Podiatrists (pres, 82-84); Nat Acad Pract; fel Am Soc Podiatric Med; fel Am Soc Podiatric Dermat; fel Am Soc Ambulatory Surg. *Res:* Podiatric medicine; medical curriculum development, implementation and evaluation with special emphasis upon developing curricula compatible with health delivery systems; evaluation of blood flow in the lower extremities; utilization of physiological principles in the treatment of podiatric problems. *Mailing Add:* Calif Col Podiatric Med 1210 Scott St San Francisco CA 94115

HOFFMAN, BRIAN MARK, b Chicago, Ill, Aug 7, 41; m 80. BIOINORGANIC CHEMISTRY. *Educ:* Univ Chicago, BS, 62; Calif Inst Technol, PhD(chem), 66. *Prof Exp:* Air Force Off Spec Res, Nat Acad Sci-Nat Res Coun res fel biol, Mass Inst Technol, 66-67; from asst prof to assoc prof chem, 67-74, prof chem & biochem, 74-80, PROF CHEM, BIOCHEM & MOLECULAR & CELL BIOL, NORTHWESTERN UNIV, 80- *Concurrent Pos:* Alfred P Sloan Found fel, 71-73; career develop award, NIH, 71-73; F Leon Watkins vis prof, Wichita State Univ, 81; US/USSR acad seminar on catalysis, 85; metals in biol, Gordon Res Conf, 87; mem adv bd, Nat Biomed Electron Spin Resonance Ctr, 80- *Honors & Awards:* FMC lectr, Princeton Univ, 88. *Mem:* Fel AAAS; Am Chem Soc; Am Soc Biol Chemists; Biophys Soc. *Res:* Electron paramagnetic resonance and electron-nuclear double resonance of metalloenzymes; long-range electron transfer within protein complexes; magnetism and metallic conductivity in molecular crystals. *Mailing Add:* 733 Milburn Evanston IL 60201-2408

HOFFMAN, CHARLES JOHN, b Cleveland, Ohio, July 9, 18; m 55; c 1. PHYSICAL INORGANIC CHEMISTRY. *Educ:* Case Inst Technol, BSc, 42, MS, 48; Univ Ill, PhD(chem), 51. *Prof Exp:* Res org chem, Diamond Alkali Co, 42-43; asst chem, Case Inst Technol, 47; phys chemist, Univ Ill, 48-50; cryogenist & mem staff, Los Alamos Nat Lab, 51-53; phys chemist, Merrill Co Div, Arthur D Little, Inc, 53-55; chemist, Lawrence Radiation Lab, Calif, 55-58; staff scientist, Lockheed Missiles & Space Co, 58-78; CONSULT, 78- *Res:* Inorganic-physical rocket propulsion chemistry. *Mailing Add:* 264 Andanada St Sunnyvale CA 94086

HOFFMAN, CLARK SAMUEL, JR, b Hershey, Pa, Nov 26, 38; div; c 2. ENVIRONMENTAL SCIENCES, ANALYTICAL CHEMISTRY. *Educ:* Lebanon Valley Col, BS, 60; Mont State Univ, MS, 64, PhD(chem), 67. *Prof Exp:* Res chemist, 66-73, res supvr, 73-75, environ mgr, 75-77, mgf mgr, 78-81, EMPLOYEE REINS MGR, E I DU PONT DE NEMOURS & CO, INC, 81- *Mem:* Am Chem Soc. *Res:* Vehicle emission research; air pollution; polycyclic aromatic hydrocarbons; halocarbon and fluorocarbon analyses; water conservation; pilot plant in-line instrumentation; gas, liquid, thin-layer and paper chromatography; moisture and aerosol propellant analyses. *Mailing Add:* C & P Dept Chambers Works Dupont Co Deepwater NJ 08023

HOFFMAN, CLYDE H(ARRIS), electrical engineering, for more information see previous edition

HOFFMAN, CYRUS MILLER, b New York, NY, Feb 20, 42; m 63; c 4. PARTICLE PHYSICS. *Educ:* Brown Univ, ScB, 62; Harvard Univ, MA, 64, PhD(physics), 69. *Prof Exp:* Res asst physics, Princeton Univ, 69, from instr to asst prof, 69-74; staff mem, 74-84, ASSOC GROUP LEADER, LOS ALAMOS NAT LAB, 84- *Mem:* Fel Am Phys Soc. *Res:* Quantum electrodynamics, including muon inelastic scattering and properties of vector mesons; weak interactions including CP violation, hyperon beta decay and rare pion and muon decays. *Mailing Add:* 660 Totavi Los Alamos NM 87544

HOFFMAN, DALE A, b Denver, Colo, Feb 25, 30; m 58; c 1. LIMNOLOGY. *Educ:* Colo State Univ, BS, 52, MS, 56, PhD(limnol), 62. *Prof Exp:* Aquatic biologist, US Fed Water Pollution Control Admin, 61-65; res scientist, US Bur Reclamation, 65-73; WATER QUAL SPECIALIST, BUR LAND MGT, 73- *Concurrent Pos:* Affiliate fac mem, Grad Sch, Colo State Univ, 70- *Mem:* Ecol Soc Am; Am Soc Limnol & Oceanog; Am Fisheries Soc; Sigma Xi. *Res:* Water quality and changes; regimen in reservoir; bottom fauna reactions to water quality changes. *Mailing Add:* 21861 Pleasant Pk Rd Conifer CO 80433

HOFFMAN, DANIEL LEWIS, b La Salle, Ill, June 23, 38; m 80; c 3. MARINE INVERTEBRATE ZOOLOGY. *Educ:* Univ Calif, Berkeley, BA, 61, MA, 63; Univ Wash, PhD(zool), 68. *Prof Exp:* NIH fel zool, Northwestern Univ, 68-69, instr biol sci, 69-70; asst prof, 70-77, ASSOC PROF BIOL, BUCKNELL UNIV, 77- *Concurrent Pos:* Vis scholar, Scripps Inst Oceanog, 83-84. *Mem:* Am Soc Zoologists; Crustacean Soc. *Res:* Life history strategies of benthic marine invertebrates; predatory-prey interactions among benthic invertebrates; settlement, recruitment and growth of pedunculate barnacles. *Mailing Add:* Dept Biol Bucknell Univ Lewisburg PA 17837

HOFFMAN, DARLEANE CHRISTIAN, b Terril, Iowa, Nov 8, 26; m 51; c 2. NUCLEAR CHEMISTRY. *Educ:* Iowa State Col, BS, 48, PhD(phys chem), 51. *Prof Exp:* Asst, Ames Lab, Atomic Energy Comn, 47-51; chemist, Oak Ridge Nat Lab, 51-52; chemist, Los Alamos Nat Lab, 52-71, assoc group leader, 71-78, div leader, chem-nuclear chem div, 79-81, leader, Isotope Nuclear Chem Div, 81-84; PROF CHEM, NUCLEAR SCI DIV, LAWRENCE BERKELEY LAB, UNIV CALIF, 84- *Concurrent Pos:* Sr NSF fel, Oslo, Norway, 64-65; pres, NMex Inst Chemists, 76-78; Guggenheim fel mechanisms nuclear fission, Lawrence Berkeley Lab, Univ Calif, 78-79. *Honors & Awards:* John Dustin Clark Award, Am Chem Soc, 76, Award for Nuclear Chem, 83, Garvan Medal, 90. *Mem:* fel Am Inst Chem; Am Chem Soc; fel Am Phys Soc; AAAS; dirs fel Los Alamos Nat Lab, 90. *Res:* Low energy and spontaneous fission; radionuclide migration in the environment; nuclear waste management; production mechanisms for heavy element isotopes; chemical and nuclear properties of heaviest elements. *Mailing Add:* Nuclear Sci Div 70A-3307 Lawrence Berkeley Lab Berkeley CA 94720

HOFFMAN, DAVID ALLEN, b Far Rockaway, NY, July 21, 44; m; c 2. GEOMETRY, COMPUTATION. *Educ:* Univ Rochester, AB, 66; Stanford Univ, MS, 69, PhD(math), 72. *Prof Exp:* Lectr math, Univ Durham, 72; asst prof, Univ Mich, 72-73; from asst prof to assoc prof, 74-84, PROF MATH, UNIV MASS, AMHERST, 84- *Concurrent Pos:* Co-prin investr, NSF Res Grants Geometry & Anal, 73-78, 80-85, prin investr, 85-; vis prof, Stanford Univ, 78, Inst Pure & Appl Math, Rio de Janeiro, 81, Univ Paris VII, 89 & Ecole Technique, 90; Danforth fel; prin investr, US Dept Energy, 86- *Honors & Awards:* Chauvenet Prize, Math Asn Am, 90. *Mem:* Am Math Soc; Math Asn Am; Inst Elec & Electronics Engrs; Asn Comput Mach. *Res:* Geometry of immersed submanifolds, including problems concerning minimal surfaces, variational inequalities, isoperimetric inequalities in Riemannian geometry; transformation groups and Riemmanian geometry; computational methods in mathematical research; computer graphics; generalized Gauss map; harmonic mappings. *Mailing Add:* Dept Math Univ Mass Amherst MA 01003

HOFFMAN, DAVID J, b New London, Conn, Sept 22, 44; m 66; c 1. DEVELOPMENTAL PHYSIOLOGY, BIOCHEMISTRY. *Educ:* McGill Univ, BS, 66; Univ Md, PhD(develop biol), 71. *Prof Exp:* Instr biol, Univ Nebr-Lincoln, 66-67; instr embryol & genetics, Univ Md, College Park, 67-70; NIH fel biochem, Oak Ridge Nat Lab, Tenn, 71-73; mem fac biol, Boston Col, 73-74; sr staff physiologist, Toxicol Div, Health Effects Lab, Environ Protection Agency, Cincinnati, Ohio, 74-76; SR STAFF PHYSIOLOGIST, ENVIRON PHYSIOL & TOXICOL, PATUXENT RES CTR, US DEPT INTERIOR, 76- *Honors & Awards:* Achievement Award, US Dept Interior, 90. *Mem:* AAAS; Soc Environ Toxicol & Chem; Teratology Soc; Toxicol Soc; Soc Exp Biol & Med. *Res:* Developmental physiology, teratology and environmental health, nucleic acid regulation, oxidant stress, avian and wildlife toxicology. *Mailing Add:* Patuxent Res Ctr US Dept of Interior Laurel MD 20708

HOFFMAN, DENNIS MARK, b Huntingdon, Pa, July 22, 47; m 81; c 2. POLYMER CRYSTALLIZATION, POLYMER CHARACTERIZATION. *Educ:* Juniata Col, Huntingdon, Pa, BS, 69; Univ Mass, Amherst, PhD(polymer sci), 79. *Prof Exp:* RES SCIENTIST, LAWRENCE LIVERMORE NAT LAB, 79- *Mem:* Am Chem Soc; Am Phys Soc; AAAS; Soc Plastics Engrs. *Res:* Polymer blends, morphology, crystallization, solution properties, cure kinetics; thermal analysis, mechanical and rheological properties, electron and optical microscopy and molecular weight measurements; polymer explosives, polyolefins, epoxy prepregs and foams. *Mailing Add:* Lawrence Livermore Nat Lab L-282 Livermore CA 94550

HOFFMAN, DONALD BERTRAND, b New York, NY, May 3, 39. OTHER MEDICAL & HEALTH SCIENCES. *Educ:* NY Univ, BA, 59; Columbia Univ, MA, 60, PhD(chem), 67; Am Bd Forensic Toxicol, dipl, 88. *Prof Exp:* Res assoc biochem, Jewish Chronic Dis Hosp, New York, 67-68; NIMH training fel psychopharmacology, Dept Psychiat, NY Univ Med Ctr, 68-69; sr chemist toxicol, 69-87, RES SCIENTIST, DEPT TOXICOL, OFF CHIEF MED EXAMR, NEW YORK, 87-; ASST PROF FORENSIC MED, MED CTR, NY UNIV, 77- *Concurrent Pos:* Sr res scientist forensic med, Med Ctr, NY Univ, 73-76; expert testimony in court; guest lectr med, legal, law enforcement & forensic groups; adj asst prof, Dept Forensic sci, John Jay Col Criminal Justice, 76-84; consult forensic toxicol. *Mem:* Am Acad Forensic Sci; NY Acad Sci; Int Asn Forensic Toxicol; Soc Med Jurisp; Soc Forensic Toxicologists; Sigma Xi. *Res:* Methodology of forensic chemistry; new techniques for isolation of drugs from tissues and biological fluids; mass urine screening procedures for detection of drugs of abuse; forensic toxicology. *Mailing Add:* Dept Toxicol Off Chief Med Examr 520 First Ave New York NY 10016

HOFFMAN, DONALD OLIVER, b Mt Pleasant, Iowa, June 14, 16; m 44; c 3. INDUSTRIAL CHEMISTRY. *Educ:* Northwestern Univ, BS, 37; Univ Minn, MS, 41; Ohio State Univ, PhD(org chem), 48. *Prof Exp:* Chemist, Columbia Chem Div, Pittsburgh Plate Glass Co, Ohio, 37-39; asst chem, Univ Minn, 41-43; res assoc, Nat Defense Res Comt Proj, Columbia Univ, 43-44; res assoc, Ohio State Univ, 44-46, asst instr org chem, 47-48, asst prof, 48-49; from biochemist to head dept biochem, US Naval Med Res Unit No 3, Egypt, 49-53; sr res chemist, Am Optical Corp, 54-65, mgr mat res, 66-68, chief surface res, 69-76; RETIRED. *Res:* Chromatography of sugars and related compounds; molluscicides for control of snail intermediate hosts of bilharziasis; infrared absorbing plastics; mechanism of glass polishing; surface chemistry. *Mailing Add:* Seven Ridgeview Rd Sturbridge MA 01566

HOFFMAN, DONALD RICHARD, b Boston, Mass, Aug 25, 43; m 71; c 3. IMMUNOLOGY. *Educ:* Harvard Col, AB, 65; Calif Inst Technol, PhD(chem), 70. *Prof Exp:* Cancer res scientist immunol, Roswell Park Mem Inst, NY State Health Dept, 70-71; asst prof pediat, Sch Med, Univ Southern Calif, 71-75; assoc prof path, Sch Med, Creighton Univ, 75-77; assoc prof path & lab med, 77-82, PROF PATHO, SCH MED, ECAROLINA UNIV, 82- *Concurrent Pos:* Consult, Allermed Labs Inc, 73-80, Pharmacia Labs Inc, 74-86, Allergenic Prod Adv, US Food & Drug Admin, 90-93. *Mem:* Am Asn Immunologists; Am Acad Allergy & Immunol; Sigma Xi; NY Acad Sci; Clin Immunol Soc. *Res:* Human immunoglobulin E antibodies and allergy; allergens; antibody sites and structure; immunopathology; autoimmune diseases; immune response in children; insect venoms. *Mailing Add:* Dept Path ECarolina Univ Sch Med Greenville NC 27858-4354

HOFFMAN, DOROTHEA HEYL, b Easton, Pa, May 30, 19; m 46; c 2. ORGANIC CHEMISTRY. *Educ:* Bryn Mawr Col, BA, 39, MA, 40, PhD(org chem), 42. *Prof Exp:* Demonstr chem, Bryn Mawr Col, 39-40; res chemist, Merck & Co, 42-52; teacher, Westfield High Sch, 65-84; RETIRED. *Concurrent Pos:* Chem consult, 63-65; sci writing, 52- *Honors & Awards:* James Bryant Conant Award, High Sch Chem Teaching, Am Chem Soc, 76. *Mem:* AAAS; Am Chem Soc. *Res:* Substituted malonic esters and barbituric acids; biotin; vitamin B6 group; codecarboxylase; vitamin B12. *Mailing Add:* Four Cowperthwaite Sq Westfield NJ 07090-4048

HOFFMAN, DOUGLAS WEIR, b Toronto, Ont, Feb 16, 20; m 44; c 4. AGRICULTURE, EARTH SCIENCE. *Educ:* Univ Toronto, BSA, 46, MSA, 49; Univ Waterloo, PhD(planning), 73. *Prof Exp:* Soils specialist taxon, Ont Dept Agr, 46-48; pedologist soil taxon, Can Agr Soil Res Inst, 48-62; from assoc prof to prof soil surv & land use, Univ Guelph, 62-78, dir resource develop, 73-78; dir planning & resource mgt, Sch Urban Planning, Univ Waterloo, 78-84, prof planning & resource mgt, 85-88; RETIRED. *Concurrent Pos:* Mem, Can Soil Survey Comt, 55-73, Grand River Conserv Authority, 70-74, Peatlands Study Comt Nat Forest Lands, 71-75 & Land Use Comt, Soil Conserv Soc Am, 74-76; dir, Ecoplans Ltd, 76-; pres, Asn Urban & Regional Planning Progs Can Univs, 82-85. *Mem:* Agr Inst Can; Geol Asn Can; fel Can Soil Sci Soc; Int Soil Sci Soc; Can Asn Geol; Can Asn Geog; Sigma Xi; Can Inst Planners. *Res:* Soil classification, genesis and morphology and their relationship to land evaluation and rural planning. *Mailing Add:* 45 Winston Cres Guelph ON N2L 3G1 Can

HOFFMAN, DOYT K, JR, b Gastonia, NC, Nov 29, 43; m 66; c 2. ELECTRO-CHEMISTRY. *Educ:* Wake Forest Univ, BS, 67. *Prof Exp:* Res chemist, Lithium Corp Am, 67-79, chemist-group leader, 67-80, proj dir, 80-87, SR DEVELOP ENG, LITHIUM CORP AM, 88- *Mem:* Am Chem Soc; Sigma Xi. *Res:* Improvement of extractive metallurgical plant processes connected with efficient and economical recovery of lithium values from spodumene ore; production of battery quality lithium metal, alloys and salts; development of new and improved battery separators. *Mailing Add:* 104 Windsong Ct Gastonia NC 28056

HOFFMAN, EDWARD JACK, b Marion, Kans, June 10, 25; m 53; c 3. COAL & ENERGY CONVERSION, PETROLEUM PRODUCTION & RESERVOIR ENGINEERING. *Educ:* Okla State Univ, BS, 44; Univ Mich, MS, 50. *Prof Exp:* Refinery control, petrol refining, Continental Oil Co, 46-48; petrol prod res, Carter Oil Co, 48-49; process engr oil & gas equip, Black, Sivalls & Bryson, 51-52; instr chem eng, Okla State Univ, 52-54; asst prof, Univ Colo, 54-57; assoc prof, Univ Tulsa, 57-62; res engr, Heat Transfer Res Inc, 63-65; assoc prof & res engr coal conversion, Univ Wyo, 65-72; consult, King-Wilkinson, Inc, 77-84 & Gas Res Inst, 78-85; RETIRED. *Concurrent Pos:* Prin investr, FMC Corp, 56-57; vis prof, Cities Serv Co, 57, Phillips Petrol Co, 60-61 & Humble Oil Co, 62; consult fuels & energy, 72- *Res:* Coal and energy conversion; thermodynamic behavior of non-ideal systems; net energy analysis; distillation; energy concepts; heat transfer; coal gasifiers; petroleum production; synthetic fuels; process development; hazardous wastes. *Mailing Add:* PO Box 1352 Laramie WY 82070

HOFFMAN, ERIC ALFRED, b Rochester, Minn, May 9, 51; m 87. CARDIAC MECHANICS, PULMONARY VENTILATION-PERFUSION & IMAGING. *Educ:* Antioch Col, Ohio, BA, 74; Univ Minn, PhD(physiol), 81. *Prof Exp:* Res asst cardiopulmonary physiol, Cardiovasc Pulmonary Res Lab, Univ Colo, Denver, 74-75 & physiol, dept physiol & biophys, Colo State Univ, Ft Collins, 75; fel, Mayo Clin, 75-81, res fel, 81-82, sr res fel, 83-84, instr, 82-84, assoc consult & asst prof physiol, Biodynamics Res Unit, 84-87; asst prof, 87-89, ASSOC PROF, RADIOL SCI & PHYSIOL, UNIV PA, 89-, CHIEF, SECT CARDIOTHORACIC IMAGING RES, DEPT RADIOL, 88- *Concurrent Pos:* John G Searle-Mayo scholar, 83-86; new investr, Heart, Lung & Blood Inst, NIH, 83-86; mem exec comt, Coun Cadiopulmonary Dis, Am Heart Asn, 87-, Coun Cardiac Radiol, 90-, estab investr, 87-91; study sect mem, NIH, 90- *Mem:* Am Physiol Soc; Sigma Xi; Am Heart Asn; Am Thoracic Soc; AAAS; Union Concerned Scientists; NAm Soc Cardiac Imaging. *Res:* New imaging techniques associated with synchronous volumetric x-ray scanning computed tomography and magnetic resonance imaging; the intrathoracic determinants of cardiac and pulmonary geometry to function relationships. *Mailing Add:* Dept Radiol Hosp Univ Pa 3400 Spruce St Philadelphia PA 19104

HOFFMAN, EUGENE JAMES, b Peoria, Ill, Dec 1, 35; m 62; c 2. CELL BIOLOGY. *Educ:* Univ Chicago, AB, 55, SB, 57; Iowa State Univ, PhD(biophys), 63. *Prof Exp:* Res assoc molecular biol, Inst Therapeut Biochem, Univ Frankfurt, 64-65 & cell biol, Yale Univ, 66-67; vis asst prof biol, Inst Biol, Hacettepe Univ Turkey, 67-71; vis assoc prof, Kans State Univ, 72; vis specialist biophys, Istanbul Univ, 72-76; res assoc enzym, Univ Kans, 75-77; assoc prof biophysics, Western Ky Univ, 77-; RES ASSOC, DEPT BIOL SCI, UNIV MD, BALTIMORE COUNTY, CATONSVILLE. *Concurrent Pos:* Res fel, Nat Cancer Inst, NIH, 65-67; Fulbright lectr, US Dept State, 67-69. *Mem:* Am Soc Cell Biol; Soc Protozoologists; AAAS. *Res:* Morphogenesis and function of motility organelles in protozoan and other specialized cells. *Mailing Add:* Physics Dept Western Ky Univ Bowling Green KY 42101

HOFFMAN, EVERETT JOHN, b Spring Valley, Minn, Dec 31, 08; m 42; c 1. PHYSICAL CHEMISTRY. *Educ:* Univ Minn, BChE, 31, PhD(phys chem), 35. *Prof Exp:* Asst chem, Univ Minn, 34-35; res chemist, Armour & Co, Ill, 35-46, chief librn, Chem Res & Develop Dept, 46-50; ed, Nuclear Sci Abstracts, Div Tech Info Exten, US Atomic Energy Comn & Environ Res & Develop Admin, 50-56, sci ed, 56-58, asst to chief, Cataloging Br, 58-61, sr sci asst staff officer to chief, Catalog Opers Sect, 61-66, chief, Chem Sect, Sci & Technol Br, 66-74, asst to chief, Sci & Technol Br, 74-77; RETIRED. *Mem:* Am Chem Soc. *Mailing Add:* 103 Nixon Rd Oak Ridge TN 37830-5224

HOFFMAN, F(RANK) E(DWARD), chemical engineering; deceased, see previous edition for last biography

HOFFMAN, FREDERICK, b Cleveland, Ohio, Nov 10, 37; m 68; c 4. EXPERT SYSTEMS, COMBINATORICS & FINITE GROUPS. *Educ:* Georgetown Univ, BS, 58; Univ Va, PhD(math), 63. *Prof Exp:* From instr asst prof math, Univ Ill, Urbana, 62-66; mathematician, Commun Res Div, Inst Defense Anal, 66-67; asst prof math, Drexel Inst Technol, 67-68; asst prof, 68-70, assoc prof, 70-77, PROF MATH, FLA ATLANTIC UNIV, 77- *Concurrent Pos:* Vis assoc prof combinatorics & optimization, Univ Waterloo, 75; fac visitor, IBM, 82-83; mem bd dirs, Teletimer Int, 87- *Mem:* Am Asn Artificial Intel; Am Math Soc; Math Asn Am; Soc Indust & Appl Math; Inst Elec & Electronics Engrs; Asn Comput Mach; Inst Elec & Electronics Engrs Computer Soc; Sigma Xi; NY Acad Sci; fel Inst Combinatorics & Appln. *Res:* Development of expert systems for configuration and experimental design; interplay of combinatiorics and algebra, especially finite groups and vector spaces. *Mailing Add:* Dept Math Fla Atlantic Univ Boca Raton FL 33431-0991

HOFFMAN, G(RAHAM) W(ALTER), b Detroit, Mich, Jan 10, 28; wid; c 4. ELECTRICAL ENGINEERING. *Educ:* Lafayette Col, BS, 50; Harvard Univ, SM, 51, PhD(elec eng), 55. *Prof Exp:* Scientist, Bettis Atomic Power Div, Westinghouse Elec Corp, 55-59; assoc prof, 59-65, PROF ELEC ENG, UNIV TENN, KNOXVILLE, 65- *Concurrent Pos:* Consult, Oak Ridge Nat Lab, 60-; vis prof, Univ Nottingham, 70-71. *Mem:* Am Phys Soc; Am Nuclear Soc; Am Soc Eng Educ; Inst Elec & Electronics Engrs. *Res:* Physical electronics; conduction in gases; nuclear reactor physics; electron optics; plasma. *Mailing Add:* Dept Elec Eng Univ Tenn Knoxville TN 37996-2100

HOFFMAN, GEORGE R, b Hastings, Nebr, Feb 5, 35; m 60; c 2. PLANT ECOLOGY. *Educ:* Hastings Col, BA, 57; Wash State Univ, PhD(bot), 62. *Prof Exp:* Instr bot, Univ Denver, 62-64; from asst prof to assoc prof, 64-70, PROF BIOL, UNIV SDAK, 70-, CHMN DEPT BIOL, 78- *Concurrent Pos:* Res grant, 65-67; vis scientist, Oak Ridge Nat Lab, 67-68; res assoc, Ctr Study Natural Systs, Mo Bot Gardens, 68-69; res grants, Nat Air Pollution Control Admin, 68-71, US Dept Interior 71-73, US Forest Serv, 72-74, 76-81 & 81-, US Army Corps of Engrs, 73-82, NSF, 74-76 & 76-79 & Nat Park Serv, 78-80, 84-85 & 88- *Mem:* Soc Range Mgt; Torrey Bot Club; Ecol Soc Am; Am Bryol & Lichenological Soc; Brit Ecol Soc. *Res:* Autecologic investigations of bryophtes; ecology of epiphytic bryophytes and lichens and influence of air pollution on them; studies on revegetation of reservoir shorelines; habitat type studies in Rocky Mountains and grasslands of the Great Plains. *Mailing Add:* Dept Biol Univ SDak 414 E Clark St Vermillion SD 57069

HOFFMAN, GERALD M, b Chicago, Ill, Oct 31, 26; m 60; c 2. OPERATIONS RESEARCH. *Educ:* Purdue Univ, BS, 46, MS, 49; Northwestern Univ, PhD, 78. *Prof Exp:* Res physicist, Motorola, Inc, 49-51; pres, Panoramic Builders, Inc, 51-65; mgr opers res, Standard Oil Co Ind, 68-79; exec vpres & gen mgr, Amoco Computer Serv, 79-84; PRES, GERALD HOFFMAN CO, 85- *Mem:* Inst Mgt Sci (pres); Soc Mgt Info Systs (pres, secy); Opers Res Soc Am; fel AAAS; Sigma Xi. *Res:* Practical applications of operations research methodology; management of information systems. *Mailing Add:* 212 E Ontario St Chicago IL 60611

HOFFMAN, GLENN LYLE, b Aurora, Iowa, Dec 28, 18; m 48; c 4. PARASITOLOGY. *Educ:* Univ Iowa, BA, 42, PhD(zool, parasitol), 50. *Prof Exp:* Asst prof parasitol & bact, Med Sch, Univ NDak, 50-58; parasitologist, Eastern Fish Dis Lab, US Fish & Wildlife Serv, 58-74, res fish parasitologist, Fish Farming Exp Sta, 74-85; RETIRED. *Concurrent Pos:* Assoc ed, J Wildlife Dis, 70-71; asst ed, Int J Fish Dis, 78-; adj prof, Memphis State Univ, 75-85; bd rev, J Protozool, 80-82. *Honors & Awards:* Distinguished Serv Award, Wildlife Dis Asn, 74; Distinguished Serv Award, Fish Health Sect, Am Fish Soc, 82, Fish Culture Sect, 85. *Mem:* Am Soc Parasitol; Am Fisheries Soc; Am Micros Soc; Soc Protozool; Wildlife Dis Asn; Sigma Xi; hon mem, Czechoslovakian Fishermens Guild, Ceske Budejovice; Europ Asn Fish Path; Am Fish Soc (pres, 84). *Res:* Parasites and diseases of freshwater fish, especially strigeoid trematodes; Myxosporida; ciliates. *Mailing Add:* Rte 3 Box 36 Kearneysville WV 25430

HOFFMAN, H(ERBERT) W(ILLIAM), b Baltimore, Md, Jan 20, 23; m 47; c 4. CHEMICAL ENGINEERING. *Educ:* Johns Hopkins Univ, BE, 43, DrEng, 51. *Prof Exp:* Res engr distillation, Res & Develop Div, Socony-Vacuum Oil Co, 43-44; res assoc gaseous separations, SAM Labs, Columbia Univ, 44; engr thermal separations, Fercleve Corp, 44-45; res assoc explosives, Los Alamos Sci Labs, Univ Calif, 45-46; res assoc heat transfer, Johns Hopkins Univ, 47-50; HEAD DEPT HEAT TRANSFER & FLUID DYNAMICS, OAK RIDGE NAT LAB, 50- *Concurrent Pos:* Lectr, Univ Tenn, 64- *Mem:* AAAS; Am Nuclear Soc; Am Inst Chem Engrs; Sigma Xi. *Res:* Heat transfer and fluid dynamics research with molten salts, liquid metals and gases as nuclear reactor coolants; desalination; water reactor thermal hydraulics. *Mailing Add:* 116 Canterbury Rd Oak Ridge TN 37830

HOFFMAN, HANNA J, solid state electronics, non-linear optics, for more information see previous edition

HOFFMAN, HAROLD A, b Grand Rapids, Mich, Mar 21, 30; m 55; c 2. GENETICS. *Educ:* Univ Calif, BS, 57, MA, 60, PhD(genetics), 62. *Prof Exp:* Geneticist, Div Res Serv, NIH, 61-63; GENETICIST, NAT CANCER INST, 63- *Concurrent Pos:* Lectr, George Washington Univ, 62, Am Univ, 66 & Univ Calif, Berkeley, 69; assoc ed, J Nat Cancer Inst, 74- *Mem:* AAAS; Am Genetic Asn; Am Soc Human Genetics. *Res:* Biochemical genetics; genetic control of protein poly-morphism and regulatory mechanisms. *Mailing Add:* 11519 College View Dr Wheaton MD 20922

HOFFMAN, HEINER, b Cleveland, Ohio, Mar 20, 19; m 55. ORAL MICROBIOLOGY. *Educ:* Case Western Reserve Univ, BS, 41; Univ Louisville, DMD, 45; Ohio State Univ, MSc, 49, PhD(microbiol), 50. *Prof Exp:* Grad asst microbiol, Ohio State Univ, 48-49, res assoc, Res Found, 50-51; instr, Col Med, Univ Nebr, 51-53; from instr to prof, 54-84, EMER PROF MICROBIOL, COL DENT, NY UNIV, 84- *Concurrent Pos:* NIH fel, Ohio State Univ, 49-50; NIH res grants bact, NY Univ, 57-68; adj prof, York Col, NY, 74-75; vis lectr, St Lukes Hosp Roosevelt Med Ctr, 82-85, NY Col Osteop, 85-86. *Mem:* Am Soc Microbiol. *Res:* Cytology and cytochemistry of bacteria; morphogenesis of bacterial aggregations; microbial ecology of human body, especially oral cavity. *Mailing Add:* 10-44 Totten St Whitestone NY 11357

HOFFMAN, HENRY ALLEN, JR, organic chemistry; deceased, see previous edition for last biography

HOFFMAN, HENRY HARLAND, b Hughesville, Mo, Oct 30, 22; m 55; c 3. ANATOMY. *Educ:* Mo Valley Col, BS, 50; St Louis Univ, MS, 54, PhD(anat), 56. *Prof Exp:* Instr anat, St Louis Univ, 56-57; from instr to prof anat, Univ Ala, Birmingham, 57-88, dir med admin, Med Ctr, 70-88; RETIRED. *Mem:* AAAS; Sigma Xi; Acad Educ Develop; Am Asn Anat. *Res:* Neuroanatomy; comparative neuroanatomy; autonomic nervous system. *Mailing Add:* 1244 Lincoya Dr Birmingham AL 35216

HOFFMAN, HENRY TICE, JR, b Lake City, Colo, Sept 14, 25; m 56; c 3. ANALYTICAL CHEMISTRY. *Educ:* Western State Col Colo, AB, 46; Univ Iowa, PhD(phys chem), 50. *Prof Exp:* Asst, Univ Iowa, 46-49; from chemist to sr res chemist, Res Div, Am Can Co, 50-64, res assoc, 64-83; anal chemist, Environ & Health Sci Lab, Mobil Oil Corp, 83-86; RES SCIENTIST, DEPT ENVIRON PROTECTION, STATE NJ, 86- *Mem:* Am Chem Soc; Am Soc Mass Spectrometry; Sigma Xi. *Res:* Molecular spectroscopy; gas chromatography; mass spectrometry. *Mailing Add:* 22 Vanderveer Dr Lawrenceville NJ 08648-3151

HOFFMAN, HERBERT I(RVING), b Uniontown, Pa, Oct 22, 25; m 48; c 3. MECHANICAL ENGINEERING. *Educ:* A&M Univ, Tex, BS, 47. *Prof Exp:* Mech engr, Beavers & Lodal, Consult Engrs, 47-48; proj engr, Friedrich Refrigerators, Inc, 49-51; from res engr to sr res engr, Southwes Res Inst, 51-66, dir, prog develop, 66-68; CONSULT, 88- *Concurrent Pos:* Instr physics & eng, eve div, San Antonio Col, 53-82; bus develop eng & tech firms, 75- *Mem:* Sigma Xi. *Res:* Coordination for planning of multidisciplinary research; managing capabilities information network for proposals; energy conservation and thermal systems engineering. *Mailing Add:* 514 W Crestline San Antonio TX 78228

HOFFMAN, HOWARD EDGAR, b New York, NY, Nov 10, 26; m 51; c 3. DRUG DEVELOPMENT, RESEARCH ADMINISTRATION. *Educ:* Univ Calif, Los Angeles, BA, 49, MA, 51; Stanford Univ, PhD(chem), 56. *Prof Exp:* Asst chem, Univ Calif, Los Angeles, 49-53; asst biochem, Stanford Univ, 53-55; res assoc, Stine-Haskell Pharmaceut Lab, Newark, Del, 55-86, admin supvr, 86-87, asst dir infectious dis clin invests, E I Du Pont de Nemours & Co, Barley Mill Plaza, Wilmington, Del, 87-88; RETIRED. *Concurrent Pos:* Asst, Palo Alto Med Res Fedn, 54-55; instr, Univ Del, 89-; sci consult, 89- *Res:* Studies on metabolism and pharmacokinetics of drug candidates, both preclinical and clinical; clinical development of new drugs. *Mailing Add:* 2220 Pennington Dr Wilmington DE 19810

HOFFMAN, HOWARD TORRENS, b East St Louis, Ill, Dec 30, 23; m 47; c 3. ELECTRONICS ENGINEERING, SCIENCE ADMINISTRATION. *Educ:* Thomas Univ, MSEE, 72, PhD(mgt sci), 77. *Prof Exp:* Head, Eng Sect, Joy Mfg Co, St Louis, 50-55; missile systs engr, McDonnell Aircraft Corp, St Louis, 55-57; exec engr, IT&T Labs, Ft Wayne, Ind, 57-59; mgr missile systs, Litton Industs, College Park, Md, 59-60; prog mgr, chief mgr & div mgr, Teledyne Ryan Co, San Diego, 60-69; PRES & CHIEF EXEC OFFICER, HOFFMAN ASSOCS, SAN DIEGO, 69-; CHIEF EXEC OFFICER & CHIEF FINANCIAL OFFICER, HOFFMAN GROUP, SAN DIEGO, 87- *Concurrent Pos:* Chmn bd, exec dir, asset mgr, mgt conserv, H&R Assocs, San Diego, 70- *Mem:* Am Inst Aeronaut & Astronaut; Inst Elec & Electronics Engrs; AAAS; Nat Soc Prof Engrs; Am Mgt Asn; Nat Mgt Asn. *Res:* Military, space and computer areas; patentee in field. *Mailing Add:* 5545 Stresemann St San Diego CA 92122

HOFFMAN, JACOB MATTHEW, JR, b Pittsburgh, Pa, Dec 10, 44. MEDICINAL CHEMISTRY, SYNTHETIC ORGANIC CHEMISTRY. *Educ:* Carnegie Inst Technol, BSc, 66; Univ Rochester, PhD(org chem), 71. *Prof Exp:* Res assoc, Columbia Univ, 70-72; res chemist, Univ Calif, Los Angeles, 72-73; sr res chemist, 73-79, res fel, 79-90, SR RES FEL, MERCK SHARP & DOHME RES LAB, 91- *Mem:* Am Chem Soc; AAAS. *Res:* Synthetic methods; drug design and development. *Mailing Add:* Merck Sharp & Dohme Res Labs W26-410 Sumneytown Pike West Point PA 19486

HOFFMAN, JACQUELINE LOUISE, b Los Angeles, Calif, 1952; m 82; c 2. GENETICS MICROBIOLOGY, BACTERIOLOGY. *Educ:* Univ Calif, Davis, BS, 73; Harvard Univ, AM, 74, PhD(biol), 79. *Prof Exp:* LECTR & LAB COURSE COORDR, GENETICS & MICROBIOL, DEPT BIOL, WASH UNIV, 84-, SCI OUTREACH COORDR, 90- *Mem:* Am Soc Cell Biol; Soc Develop Biol; Sigma Xi. *Mailing Add:* Dept Biol Wash Univ Box 1137 St Louis MO 63130

HOFFMAN, JAMES IRVIE, b Trenton, NJ, Dec 23, 41; m 68; c 1. GEOLOGY. *Educ:* Allegheny Col, BS, 63; Mich State Univ, MS, 65, PhD(geol), 69. *Prof Exp:* Explor field geologist, Anaconda Am Brass, Ltd, Can, 67; instr geol, Mich State Univ, 68-69; from asst prof to assoc prof, 69-79, chmn dept, 75-78, assoc dean, 79-84, PROF GEOL, UNIV WIS-OSHKOSH, 79-, DEAN LETTERS & SCI, 84- *Concurrent Pos:* Consult geologist, Earth Studies Inc, 72- *Mem:* Geochemistry Soc; Int Asn Geochemistry & Cosmohemistry; AAAS; Nat Asn Geol Teachers; Sigma Xi; Am Water Resources Asn. *Res:* Geochemical mass balance of transition metals with sediments and associated waters, with special attention to base metals in organic rich sediments. *Mailing Add:* Dean Col Letters & Sci Univ Wis 800 Algoma Blvd Oshkosh WI 54901

HOFFMAN, JAMES TRACY, b Lancaster, Ohio, July 18, 35; m 57; c 2. ELECTRONICS, SYSTEMS ENGINEERING. *Educ:* Univ Kans, BS, 61, PhD(physics), 70. *Prof Exp:* Physicist nuclear optics, Autonetics Div, Rockwell Int, 67-70; sr scientist optics, Systs Res Lab, 70-73; dir electronics, HRB Div, Singer Co, 73-75; exec vpres, Interstate Electronics Div, Figgie Int, 75-84, pres, Musys, 84-86; VPRES PROG, TELEDYNE SYSTS, 86- *Mem:* Am Phys Soc. *Res:* Signal processing technology and techniques; theory of distributed and array processing. *Mailing Add:* Teledyne Systs Co 19601 Nordhoff Northridge CA 91324

HOFFMAN, JEFFREY ALAN, b New York, NY, Nov 2, 44. X-RAY ASTRONOMY. *Educ:* Amherst Col, BA, 66; Harvard Univ, PhD(astron), 71. *Prof Exp:* Nat Acad Sci fel astron, Smithsonian Astrophys Observ, 71-72; Harvard-Sheldon Int Fel, Univ Leicester, 72-73, NATO fel, 73-74, EXOSAT proj scientist, Physics Dept, X-ray Astronomy Group, 74-75; head-A4 proj scientist, Mass Inst Technol, 75-78; NASA SPACE SHUTTLE ASTRONAUT/MISSION SPECIALIST, JOHNSON SPACE CTR, 78- *Mem:* Am Astron Soc; Int Astron Union; Sigma Xi. *Res:* X-ray; gamma ray; space-based observations; high energy astrophysics; x-ray astronomy; high-energy astrophysics. *Mailing Add:* 6527 Rutgers Houston TX 77005

HOFFMAN, JERRY C, b Madisonville, Ky, Apr 15, 43; m 66; c 4. HAZARDOUS MATERIAL MANAGEMENT, HAZARDOUS WASTE MANAGEMENT. *Educ:* Western Ky Univ, BS, 72. *Prof Exp:* Proj engr, 72-78, dept head, 78-88, PROJ MGR, BURNS & MCDONNELL ENG CO, 88- *Concurrent Pos:* Chmn, Air Toxic Subcomt, Mid Am Regional Coun, 89- *Mem:* Am Soc Civil Engrs; Am Soc Testing & Mat; Nat Soc Prof Engrs. *Res:* Hazardous material management; investigation of hazardous waste disposal sites. *Mailing Add:* Box 419173 Kansas City MO 64141-6173

HOFFMAN, JOE DOUGLAS, b Memphis, Tenn, Aug 9, 34; m 56; c 3. FLUID DYNAMICS, PROPULSION. *Educ:* Tex A&M univ, BS, 58, MS, 59; Purdue Univ, PhD(mech eng), 63. *Prof Exp:* From asst prof to assoc prof, 63-73, PROF MECH ENG, PURDUE UNIV, 73- *Concurrent Pos:* Design engr, Aerojet-Gen Corp, 61-63; vis prof, Univ Colo, 74. *Mem:* Am Inst Aeronaut & Astronaut; Am Soc Eng Educ; Combustion Inst; Am Soc Mech Engrs. *Res:* Gas dynamics; computational fluid dynamics. *Mailing Add:* Dept Mech Eng Purdue Univ West Lafayette IN 47907

HOFFMAN, JOHN D, b Washington, DC, Nov 26, 22. CRYSTALLOGRAPHY. *Educ:* Franklin & Marshall Col, BS, 42; Princeton Univ, MS, 48, PhD, 49. *Prof Exp:* Dir, Inst Mat Res, Nat Bur Standards, 67-78, dir, 78-82; prof & chmn nuclear eng, Univ Md, 82-85; dir & chief exec, Mich Molecular Inst, 85-90; RES PROF, DEPT MATH SCI & ENG, JOHNS HOPKINS UNIV, 90- *Concurrent Pos:* Distinguished res fel, Mich Molecular Inst, 90. *Honors & Awards:* High Polymer Physics Prize, Am Phys Soc, 71; Presidents Meritorious Excellence Award, 80. *Mem:* Nat Acad Eng; fel Am Phys Soc; Sigma Xi; NY Acad Sci; hon mem Am Dent Asn. *Mailing Add:* 6121 Maiden Lane Bethesda MD 20817

HOFFMAN, JOHN HAROLD, b Winona, Minn, Sept 7, 29; m 59; c 4. ATMOSPHERIC PHYSICS, MASS SPECTROMETRY. *Educ:* St Mary's Col, Minn, BS, 51; Univ Minn, MS, 54, PhD(physics), 58. *Prof Exp:* Asst physics, Univ Minn, res assoc, 58-59; physicist, US Naval Res Lab, Wash, DC, 59-66; assoc prof physics, Southwest Ctr Advan Studies, 66-69, assoc prof, 69-78, PROF PHYSICS & HEAD DEPT, UNIV TEX, DALLAS, 78- *Honors & Awards:* Group Achievement Award, NASA. *Mem:* Am Geophys Union; Sigma Xi. *Res:* Mass spectrometric studies of cosmogenic helium in iron meteorites; gas and ionic composition of the upper atmosphere; lunar and planetary atmospheric composition; stratospheric ion composition; cometary coma studies. *Mailing Add:* Dept Physics Univ Tex Dallas PO Box 688 Richardson TX 75083-0688

HOFFMAN, JOHN RALEIGH, b Evansville, Ind, July 7, 26; m 50; c 2. PHYSICS ENGINEERING. *Educ:* Univ Richmond, BS, 49; Univ Fla, MS, 51, PhD(physics), 54. *Prof Exp:* Asst, Univ Fla, 49-53, asst physics, 53-54; mem staff, Sandia Corp, 54-57; proj scientist, Kaman Nuclear, 57-68, from vpres to sr vpres, 74-90, EXEC VPRES, KAMAN SCI CORP, 90- *Mem:* Am Phys Soc; Inst Elec & Electronics Engrs; Sigma Xi; AIAA. *Res:* Nuclear engineering; nuclear physics; atomic physics. *Mailing Add:* 5020 Lyda Lane Colorado Springs CO 80904

HOFFMAN, JOSEPH ELLSWORTH, JR, b Somerville, NJ, July 9, 35. ELECTRICAL ENGINEERING, PHYSICS. *Educ:* Rensselaer Polytech Inst, BEE, 57; Univ Pa, MSEE, 59; Northeastern Univ, PhD(spectros instrumentation), 68; Univ Conn, MBA, 75. *Prof Exp:* Elec engr, Radio Corp Am, 57-59; electronics officer & engr, US Air Force, 59-63; asst res engr, Utah State Univ, 63-66; electronics engr, USAF, 66-70, Canberra Industs, 70-71 & Hoffman Elec, 71; asst chief engr, US Vet Admin Med Ctr, Newington, Conn, 71-75, chief eng, Providence, RI, 75-84, electronics eng prog mgr, Vet Admin Cent Off, Wash, DC, 84-86, proj mgr, 86-88, ASST CHIEF ENG, VA MEDICAL CTR, BOSTON, MASS, 88- *Concurrent Pos:* Consult instrumentation, 71-; adj asst prof, Univ Hartford, 72-74; vis lectr, Providence Col & Bryant Col, 75-84, Capitol Col & Prince George Community Col, 85-88. *Mem:* Inst Elec & Electronics Engrs; Nat Soc Prof Engrs; Am Mgt Asn. *Res:* Improving instrumentation and techniques of Fourier spectroscopy; lasers and their applications. *Mailing Add:* 304 N London Ave West Warwick RI 02893

HOFFMAN, JOSEPH FREDERICK, b Oklahoma City, Okla, Mar 7, 25; m 74. PHYSIOLOGY. *Educ:* Univ Okla, BS, 47, MS, 48; Princeton Univ, MA, 51, PhD, 52. *Hon Degrees:* MA, Yale Univ, 65. *Prof Exp:* Lab asst, Univ Okla, 47; asst instr biol, Princeton Univ, 49-51, asst, 52-53, res assoc, 53-54, lectr, 54-56; USPHS spec res fel, 56-57; from physiologist to head sect membrane physiol, Nat Heart Inst, 57-65; prof, 65-74, chmn dept, 73-79, EUGENE HIGGINS PROF CELLULAR & MOLECULAR PHYSIOL, SCH MED, YALE UNIV, 74- *Concurrent Pos:* Lectr, George Washington Univ, 58-65; mem fac, NIH Grad Sch, 59-65; mem, Woods Hole Marine Biol Lab. *Mem:* Nat Acad Sci; Soc Gen Physiol (pres, 75-76); Biophys Soc (pres, 85-86); fel AAAS; Am Physiol Soc; Sigma Xi; Fel Am Acad Arts & Sci; hon mem Argentine Soc Physiol Sci. *Res:* Membrane transport processes in red blood cells. *Mailing Add:* Dept Cellular & Molecular Physiol Yale Univ Sch Med New Haven CT 06510

HOFFMAN, JULIEN IVOR ELLIS, b Salisbury, SRhodesia, July 26, 25; US citizen; m; c 2. PEDIATRIC CARDIOLOGY. *Educ:* Univ Witwatersrand, BSc, 44, Hons, 45, MB & BCh, 49, MD, 70; Am Bd Pediat, dipl pediat cardiol, 68 & dipl pediat intensive care, 78; FRCP, 73. *Prof Exp:* Vol res asst med, Postgrad Med Sch, London, Eng, 56-57; fel pediat, Children's Hosp, Boston, 57-59; sr fel cardiol, Cardiovasc Res Inst, San Francisco, 59-60; asst prof med & pediat, Albert Einstein Col Med, 61-66; assoc prof, pediat, 66-70, prof physiol, 81-88, PROF PEDIAT, UNIV CALIF, SAN FRANCISCO, 70-, MEM CARDIOVASC RES INST, 66- *Concurrent Pos:* Estab investr, Am Heart Asn, 63-68; asst physician & pediatrician, Bronx Munic Hosp Ctr, 65-66; sub-bd Pediat Intensive Care, 85-87; distinguished physiol lectr, Am Col Chest Physicians, 85. *Honors & Awards:* George Brown Mem lectr, Am Heart Asn, 77; Lilly lectr, Royal Col Physicians, London, 81; George Alexander Gibson lectr, Royal Col Physicians, Edinburgh, 78; Isaac Starr lectr, Cardiac Systs Dynamics Soc, 82; Janet Baldwin Mem lectr, 84; John Keith lectr, Can Cardiovasc Soc, 85; First Nadas lectr, Am Heart Asn, 87; Bayer Award, 89; First Donald C Fyler lectr, Boston Children's Hosp, 90. *Mem:* Am Heart Asn; Soc Pediat Res; Am Pediat Soc; Am Physiol Soc. *Res:* Pathophysiology and natural history of congenital heart disease; pathophysiology of regional coronary blood flow; indicator dilution studies; natural history of ventricular septal defects in infancy. *Mailing Add:* Box 0544 Univ Calif San Francisco CA 94143-0544

HOFFMAN, JULIUS, b New York, NY, Feb 4, 21; m 43; c 3. PSYCHIATRY, NEUROLOGY. *Educ:* NY Univ, BA, 41, MD, 44; Ohio State Univ, MA, 58, MMSc, 62; Southern Ill Univ, MS, 72; Am Bd Psychiat & Neurol, dipl neurol, 50, dipl psychiat, 53, dipl child neurol, 68; Am Bd Pediat, dipl, 64. *Prof Exp:* Intern, Kings County Hosp, NY, 44-45, resident neurol, 45-46 & 48-49; resident psychiat, Winter Vet Admin Hosp, Kans, 49-51; sr psychiat physician, Fairfield State Hosp, Conn, 51; assoc prof psychiat, Grad Sch, Ohio State Univ, 51-53, social admin, 53-63; assoc prof neurol & pediat, Georgetown Univ, 63-65; PROF PEDIAT NEUROL, SCH MED, HOWARD UNIV, 65- *Concurrent Pos:* Nat Inst Neurol Dis & Blindness award pediat neurol, 61-62; staff psychiatrist, Columbus Receiving Hosp, 51-63; consult, Children's Hosp, Columbus, Ohio, St Anthony, Grant & Mt Carmel Hosps, Columbus State Sch & Vet Admin, 53-56, Old Age Survivors Ins, 58-63, Columbus Receiving Hosp Children, 59-63, USAF, 59-63, Freedman's Hosp, DC & Howard Univ Hosp, 66-, Suburban Hosp, Md, 67-68; resident, Children's Hosp, Columbus, Ohio, 61-62; mem staff, Georgetown Univ, DC Gen, Children's Convalescent & Children's Hosps, Washington, DC, 63-65, Sibley Mem Hosp, 64-70, Wash Hosp Ctr, 71-; consult to Presidential Spec Asst on Ment Retardation, 63; consult, Div Res Facilities & Resources, NIH, 64-, Neurol & Sensory Dis Prog, USPHS, 65- & Civil Serv Comn, 71-; prof-in-residence, Lynchburg Training Sch & Hosp, Va, 66 & 67; spec lectr, George Washington Univ, 69-, Found Advan Educ Sci, NIH, 71-74. *Honors & Awards:* Hughlings-Jackson Found Nuerol Res Award, 62. *Mem:* AMA; fel Am Psychiat Asn; fel Am Asn Ment Deficiency; fel Am Acad Neurol; fel Am Acad Pediat. *Res:* Pediatric neurology; basic and clinical neurological sciences; neurophysiology; psychotherapy. *Mailing Add:* 5410 Connecticut Ave NW Washington DC 20015

HOFFMAN, JULIUS R, b Toledo, Ohio, Aug 1, 19; m 43; c 4. ENTOMOLOGY. *Educ:* Ohio State Univ, BA, 40; Cornell Univ, MS, 47, PhD(entom), 49. *Prof Exp:* From asst res prof to assoc res prof entom, 49-63, assoc prof natural sci & entom, 63-69, asst to dean undergrad coun, 65-69, PROF ENTOM & ASST DEAN COL NATURAL SCI, MICH STATE UNIV, 69- *Mem:* AAAS; Entom Soc Am; Sigma Xi. *Res:* Virus and disease transmission by insects; greenhouse and ornamental pests control; leaf-hoppers and mites. *Mailing Add:* 246 Clarendon Rd East Lansing MI 48823

HOFFMAN, KARLA LEIGH, b Paterson, NJ, Feb 14, 48; m 71; c 1. OPERATIONS RESEARCH, COMBINATORIAL OPTIMIZATION. *Educ:* Rutgers Univ, BA, 69; George Wash Univ, MBA, 71, DSc, 75. *Prof Exp:* Oper res analyst, Internal Revenue Serv, 71-73; res fel, NSF & Nat Acad Sci, 75-76; mathematician, Nat Bur Standards, 76-85; assoc prof, 85-89, PROF OPERS RES, GEORGE MASON UNIV, 90- *Concurrent Pos:*

Consult var govt agencies, 76-86, airline indust & telecommun corp, 86-; distinguished fac award, George Mason Univ, 89. *Honors & Awards:* Silver Medal, US Dept Com, 84; Appl Res Award, Nat Bur Standards, 84. *Mem:* Opers Res Soc Am; Math Prog Soc; Soc Indust & Appl Math. *Res:* Testing of large-scale mathematical models; combinatorial optimization; nonconvex optimization. *Mailing Add:* 6921 Clifton Rd Clifton VA 22024

HOFFMAN, KENNETH CHARLES, b Long Island City, NY, Nov 29, 33; m 55; c 3. MECHANICAL ENGINEERING, SYSTEM SCIENCE. *Educ:* NY Univ, BME, 54; Adelphi Univ, MS, 68; Polytech Inst Brooklyn, PhD(systs eng), 72. *Prof Exp:* Sci staff engr, Brookhaven Nat Lab, 56-79, head, Nat Ctr Anal Energy Systs, 76-79, chmn, Dept Energy & Environ, 77-79; sr vpres, Mathtech, Inc, 79-83, pres, 84-88; vpres, Martin Marietta Data Systs, 87-; PRIN ENGR, MITRE CORP. *Concurrent Pos:* Consult, Systs Europ, Inc, 76-; mem, Commerce Tech Adv Bd, Panel Proj Independence, 76-77, Comt Renewable Resources for Indust Mats, Nat Acad Sci, 76-77, Comt Adv Energy Storage Systs, 77- & Res Coord Panel, Gas Res Inst, 77-79. *Mem:* Am Soc Mech Engrs; Opers Res Soc Am; AAAS. *Res:* Development of information technology; development of engineering, materials, and advanced energy conversion systems; development of analytical models of the energy and materials system with applications to industry and national policy. *Mailing Add:* 2007 Swans Neck Way Reston VA 22091

HOFFMAN, KENNETH MYRON, b Long Beach, Calif, Nov 30, 30; div; c 3. MATHEMATICS. *Educ:* Occidental Col, AB, 52; Univ Calif, Los Angeles, MA, 54, PhD(math), 56. *Prof Exp:* Instr math, 56-57, C L E Moore instr, 57-59, from asst prof to assoc prof, 59-64, chmn comn educ, 69-71, head dept, 71-80, PROF MATH, MASS INST TECHNOL, 64- *Concurrent Pos:* Fel, Sloan Found, 64-66; co-chmn, Mass Inst Technol/Wellesley Exchange Prog, 74-79; exec dir, Nat Res Coun Comn Resources Math Sci, 81-83; dir fed rel, Joint Policy Bd Math, 84-; chmn adv comt, NSF Sci & Eng Educ Directorate, 84-85; exec dir, Math Sci Educ Bd, 89-91; assoc exec officer, Nat Res Coun, 91- *Mem:* Am Math Soc; Math Asn Am; Soc Indust & Appl Math; AAAS; Oper Res Soc Am; Nat Coun Teachers Math. *Res:* Functional analysis. *Mailing Add:* Nat Res Coun 2101 Constitution Ave NW Washington DC 20418

HOFFMAN, KENNETH WAYNE, b Irvington, NJ, June 9, 55; m 76; c 1. APPLICATION OF ANALYTICAL INSTRUMENTATION TECHNIQUES, POLYMER ANALYSIS & PROBLEM RESOLUTION. *Educ:* Tenn Technol Univ, BS, 80. *Prof Exp:* Forensic chemist, Ga Bur Invest, 80-82, forensic chemist prin, 82-84; plating chemist, Siegel Robert Inc, 84-85, res chemist, 85-86, chief res & develop, 86-90; RES CHEMIST & LAB MGR, RH O05E POULENC, NASHVILLE, 90- *Mem:* Am Chem Soc; Soc Appl Spectros. *Res:* Provide analytical instrumentation expertise for support of any research projects or troubleshooting activities of current and future production processes. *Mailing Add:* 4600 Centennial Blvd Box 1130 Nashville TN 37202

HOFFMAN, LANCE J, b Pittsburgh, Pa, Dec 8, 42; div; c 1. AUTOMATED RISK ANALYSIS, COMPUTER SECURITY. *Educ:* Carnegie Inst Technol, BS, 64; Stanford Univ, MS, 67, PhD(comput sci), 70. *Prof Exp:* Asst prof comput sci, Univ Calif, Berkeley, 70-77; PROF COMPUT SCI, GEORGE WASHINGTON UNIV, 77- *Concurrent Pos:* Pres, Info Policy, Inc, 81-84, Hoffman Bus Soc, Inc, 85-; distinguished lectr, Inst Elec & Electronics Engrs; vis lectr, Asn Comput Mach. *Mem:* Asn Comput Mach; AAAS; Inst Elec & Electronics Engrs. *Res:* Computer security, risk analysis; social impact analysis for computer systems; automated risk analysis; computer viruses. *Mailing Add:* Dept Elec Eng & Comput Sci George Washington Univ Washington DC 20052

HOFFMAN, LARRY RONALD, b Sigourney, Iowa, May 12, 36; m 59; c 3. PHYCOLOGY. *Educ:* Iowa State Univ, BS, 58; Univ Tex, PhD(bot), 61. *Prof Exp:* NSF res fel, Prof Manton's Lab, Univ Leeds, 61-62; from asst prof to assoc prof, 62-75, PROF BOT, UNIV ILL, URBANA, 75- *Concurrent Pos:* Vis prof, Witts Univ, Johannesburg, S Africa, 83. *Honors & Awards:* The Darbaker Prize, Bot Soc Am, 75. *Mem:* Bot Soc Am; Phycol Soc Am (vpres, 75 & 80, pres, 81); Int Phycol Soc; Brit Phycol Soc; Am Microscop Soc (secy, 88-). *Res:* Morphology; fine structure; eukaryotic algal viruses; sexuality of algae, especially Sphaeropleales, Oedogoniales and other oogamous members of the Chlorophyta. *Mailing Add:* Dept Plant Biol Univ Ill 505 S Goodwin Urbana IL 61801-3793

HOFFMAN, LEWIS CHARLES, inorganic chemistry, for more information see previous edition

HOFFMAN, LINDA M, b New York, NY, Dec 18, 39; m 58; c 1. NEUROCHEMISTRY, CANCER RESEARCH. *Educ:* Queens Col, BS, 59; NY Univ, MS, 67, PhD(org chem), 70. *Prof Exp:* Fel biochem, Sloan Kettering Inst Cancer Res, 72-73; res assoc neurochemistry, Kingsbrook Jewish Med Ctr, 73-77; from asst prof to assoc prof, 77-81, PROF CHEM, BARUCH COL, 82- *Concurrent Pos:* Res assoc, Kingsbrook Jewish Med Ctr, 78- *Honors & Awards:* Moore Award, Am Soc Neuropathologists, 80, 84. *Mem:* Am Chem Soc; AAAS; Sigma Xi; NY Acad Sci. *Res:* Relationship of glycosphingolipids to surface properties and disease in cultured cells derived from Tay-Sachs diseased brains; glycosphingolipids in neurological tumors and cell cultures derived from these tumors; porphyrins as tumor-localizing agents. *Mailing Add:* Baruch Col Box 291 17 Lexington Ave New York NY 10010

HOFFMAN, LOREN HAROLD, b Los Angeles, Calif, Dec 18, 41; m 63; c 2. HISTOLOGY, REPRODUCTIVE BIOLOGY. *Educ:* Jamestown Col, BA, 63; Cornell Univ, PhD(zool), 68. *Prof Exp:* From instr to asst prof anat, 69-84, PROF CELL BIOL, SCH MED, VANDERBILT UNIV, 84- *Concurrent Pos:* USPHS res fel, Sch Med, Vanderbilt Univ, 68-70. *Mem:* Am Asn Anat; Soc Study Reproduction; Sigma Xi; Am Soc Cell Biol. *Res:* Female reproductive biology; uterus, ovo-implatation and placentation. *Mailing Add:* Dept Cell Biol Vanderbilt Sch Med Nashville TN 37232

HOFFMAN, MARK PETER, b West Reading, Pa, Feb 4, 41; m 69; c 2. BEEF & SWINE PRODUCTION. *Educ:* Delaware Valley Col, BS, 63; Iowa State Univ, MS, 67, PhD(animal sci), 69. *Prof Exp:* From asst prof to assoc prof, 69-80, PROF ANIMAL SCI, IOWA STATE UNIV, 80- *Honors & Awards:* Animal Mgt Award, Am Soc Animal Sci. *Mem:* AAAS; Am Dairy Sci Asn; Am Soc Animal Sci; Am Inst Biol Sci; Coun Agr Sci & Technol; Sigma Xi; Am Regist Prof Animal Sci. *Res:* Bovine production; effects of genotype and environment on productivity. *Mailing Add:* 119 Kildee Hall Iowa State Univ Ames IA 50011

HOFFMAN, MARVIN MORRISON, b Kan, Jan 16, 25; m 51; c 2. PHYSICS. *Educ:* Greenville Col, AB, 47; Iowa State Col, PhD(physics), 52; Univ NMex, MBA, 79. *Prof Exp:* Asst, Ames Lab, Atomic Energy Comn, 48-52; physicist, Los Alamos Nat Lab, 52-80; CONSULT, AMPARO CORP, 80- *Mem:* Am Phys Soc; Am Asn Physics Teachers; Inst Elec & Electronics Engrs. *Res:* Photonuclear reactions; fission; nuclear structure. *Mailing Add:* 2277 Manzanita Dr Oakland CA 94611

HOFFMAN, MICHAEL G, b Sunnyside, Wash, Oct 2, 52; m 83. ELECTRICAL ENGINEERING, HARDWARE SYSTEMS. *Educ:* Univ Portland, BSEE, 78. *Prof Exp:* Distrib engr, Pac Power & Light, 78-79; proj engr, Saudi Consol Elec Co, 79-81, 83-84; scheduler, Alusuisse Serv, 81-82; elec engr, Pac Eng, 84-86; syst engr, Computer Tools, Inc, 87-89; scheduler, Wright Schuchart Harbor, 89-90; SCHEDULER, HOFFMAN CONSTRUCT, 91- *Concurrent Pos:* Founder Saudi Arabian Inst Elec & Electronics Engrs Sect, 80-81, sect chmn, 81, chmn, Prof Activ Comt, Portland, Ore Sect, 83-85; ed, Computer Tools Mag, 87-89; publ, Productivity Solutions Mag, 91- *Honors & Awards:* Centennial Medal, Inst Elec & Electronics Engrs, 83. *Mem:* Inst Elec & Electronics Engrs. *Res:* Integration of off the shelf software and hardware for PC based project management systems running on local area networks. *Mailing Add:* 7466 N Fiske St Portland OR 97203

HOFFMAN, MICHAEL K, b Philadelphia, Pa, Sept 7, 41; div; c 2. PHYSICAL ORGANIC CHEMISTRY, MASS SPECTROMETRY. *Educ:* Univ Pa, BS, 63; Bryn Mawr Col, PhD(phys org chem), 68. *Prof Exp:* Res assoc, Univ Pa, 67-68 & Univ NC, 68-71; asst prof chem, Kans State Univ, 71-75; asst prof dept biochem, Sch Med, Wash Univ, 75-79; chemist, 79-85, staff officer, 85-89, BRANCH CHIEF, FOOD SAFETY & INSPECTION SERV, USDA, 89- *Concurrent Pos:* Sr fel sci, NATO, 73; Congressional fel, 88. *Mem:* Am Soc Mass Spectrom. *Res:* Application of mass spectrometry to environmental problems; food safety, risk assessment. *Mailing Add:* USDA Food Safety & Inspection Serv Chem 300 12th St SW Rm 521 Washington DC 20250

HOFFMAN, MORTON Z, b New York, NY, Apr 22, 35; m 65; c 2. PHYSICAL INORGANIC CHEMISTRY. *Educ:* Hunter Col, AB, 55; Univ Mich, MS, 57, PhD(photochemistry), 60. *Prof Exp:* Res assoc phys chem, Univ Sheffield, 60-61; from asst prof to assoc prof, 61-71, PROF PHYS CHEM, BOSTON UNIV, 71- *Concurrent Pos:* Rackham fel, 60-61; Nat Acad Sci-Nat Res Coun sr postdoctoral res assoc, 69-70; vis scientist, US Army Natick Labs, 69-70. *Mem:* AAAS; Am Chem Soc; Am Asn Univ Professors; Sigma Xi; Inter-Am Photochem Soc; Radiation Res Soc. *Res:* Photo and radiation chemistry of transition metal coordination complexes; photochemical conversion and storage of solar energy. *Mailing Add:* Dept Chem Boston Univ 590 Commonwealth Ave Boston MA 02215

HOFFMAN, MYRON A(RNOLD), b Chicago, Ill, Nov 15, 30; m 63; c 2. ENGINEERING. *Educ:* Mass Inst Technol, SB & SM, 52, ScD(aeronaut eng), 55. *Prof Exp:* Assoc prof aeronaut & astronaut, Mass Inst Technol, 55-56, 59-68; PROF MECH ENG, UNIV CALIF, DAVIS, 68- *Mem:* Am Nuclear Soc. *Res:* Electric power generation; heat transfer; fusion energy conversion. *Mailing Add:* Dept Mech Eng Univ Calif Davis CA 95616

HOFFMAN, NELSON MILES, III, b Manhattan, Kans, Oct 4, 48; c 2. RADIATIVE TRANSFER, HYDRODYNAMICS. *Educ:* Rice Univ, BA, 70; Univ Wis, PhD(astron), 74. *Prof Exp:* Staff mem, 75-80, asst group leader, 80-81, assoc group leader, Diag Physics Group, 81-85, STAFF MEM, INERTIAL FUSION & PLASMA THEORY GROUP, APPL THEORET PHYSICS DIV, LOS ALAMOS NAT LAB, 85- *Mem:* Am Phys Soc. *Res:* Inertial fusion target physics. *Mailing Add:* M/S E531 Los Alamos Nat Lab PO Box 1663 Los Alamos NM 87545

HOFFMAN, NORMAN EDWIN, b Chicago, Ill, Oct 26, 28; m 53. CLINICAL CHEMISTRY. *Educ:* Loyola Univ, BS, 50; Northwestern Univ, PhD(chem), 54. *Prof Exp:* Res chemist, Standard Oil Co, 53-56; from instr to assoc prof, 56-67, chmn dept, 72-81, PROF CHEM, MARQUETTE UNIV, 67- *Concurrent Pos:* Vis prof, Univ Wis-Madison, 85. *Mem:* AAAS; Am Chem Soc; Am Asn Clin Chem; Sigma Xi; NY Acad Sci. *Res:* Development of gas and liquid chromatographic methods for use in clinical chemistry; chromatographic theory; chromatographic techniques for determination of constituents such as therapeutic drugs and metabolites in biological fluids, usually urine and blood or plasma; evaluation of techniques as biomedical predictors. *Mailing Add:* Chem Dept Marquette Univ Milwaukee WI 53233

HOFFMAN, PAUL NED, b Cornwall, NY, Sept 16, 47; m 83. NEURO-OPHTHALMOLOGY. *Educ:* Cornell Univ, BS, 69; Case Western Reserve Univ, PhD(neurobiol), 74, MD, 76. *Prof Exp:* Fel neuropath, 76-77, resident ophthal, 77-80, fel ophthal, 80-81, asst prof neurol, 81-87, ASSOC PROF OPHTHAL, JOHNS HOPKINS UNIV, 87- *Mem:* Soc Neurosci; Am Soc Cell Biol; Am Soc Neurochem; Am Acad Ophthal. *Res:* The role of the neuronal cytoskeleton in axonal growth and the response of neurons to axonal injury; the control of neuronal gene expression. *Mailing Add:* Wilmer Ophthal Inst Johns Hopkins Univ Sch Med Baltimore MD 21205

HOFFMAN, PAUL ROGER, b New York, NY, May 10, 34; c 2. CHEMICAL ENGINEERING. *Educ:* Yale Univ, BE, 55, MS, 58, PhD(chem eng), 59. *Prof Exp:* Mkt res analyst, Union Carbide Chem Co, 59; sect chief mat sci, Res & Adv Develop Div, 62-63, mgr res projs staff, 63-64, mgr applns dept, 64-65, dir mat technol, 65-69, vpres res & eng, Avco Space Systs Div, 69-70, vpres mat & chem processes, Avco Systs Div, 70-77, VPRES & GEN MGR, AVCO SPECIALTY MAT DIV, AVCO CORP, 77- *Mem:* Am Inst Aeronaut & Astronaut; Am Inst Chem. *Res:* Materials research, development and production; chemical market research. *Mailing Add:* Textron Specialty Mat Div Two Industrial Ave Lowell MA 01851

HOFFMAN, PHILIP, biochemistry, for more information see previous edition

HOFFMAN, RICHARD BRUCE, b Manheim, Pa, Nov 29, 36; m 57; c 3. PHYSICS. *Educ:* Lehigh Univ, BS, 62, MS, 63, PhD(physics), 67. *Prof Exp:* Instr physics, Temple Univ, 66-67, res assoc, 67-68; asst prof physics, 68-72, VPRES & LECTR PHYSICS, FRANKLIN & MARSHALL COL, 72- *Mem:* Am Phys Soc; Am Asn Physics Teachers; NY Acad Sci; Am Asn Univ Professors. *Res:* Double crystal spectrometry; celestial mechanics; general relativity. *Mailing Add:* 173 N Linden St Manheim PA 17545

HOFFMAN, RICHARD LAIRD, b Cincinnati, Ohio, Apr 30, 39; m 63; c 2. RHEOLOGY, COLLOID CHEMISTRY. *Educ:* Ohio State Univ, BChE, 62, MS, 62; Princeton Univ, MA & PhD(chem eng), 68. *Prof Exp:* Res specialist, 68-78, sci fel, 78-86, SR SCI FEL, MONSANTO CO, 86- *Mem:* Am Inst Chem Engrs; Soc Rheology. *Res:* Fluid mechanics; flow behavior of concentrated suspensions of solids in liquid, especially dilatant flow; characterization of flow near interfaces; polymer rheology; particle size measurement. *Mailing Add:* Monsanto Co 730 Worcester St Indian Orchard MA 01151

HOFFMAN, RICHARD LAWRENCE, b Clifton Forge, Va, Sept 25, 27; m 54; c 3. SYSTEMATIC ZOOLOGY. *Educ:* Cornell Univ, MS, 58; Va Polytech Inst, PhD(zool), 60. *Prof Exp:* Chemist-supvr, Hercules Powder Co, Radford Arsenal, 50-57; NSF res asst arthropod taxon, US Nat Mus, DC, 59-60; assoc prof, 60-62, prin investr, NSF grants, 62-78, PROF BIOL, RADFORD UNIV, 62- *Res:* Evolution and zoogeography of arthropods, chiefly Diplopoda and Arachnida; origin and distribution of the fauna of eastern North America. *Mailing Add:* PO Box 501 Blackburg VA 24060

HOFFMAN, RICHARD OTTO, b Chicago, Ill, Aug 4, 39; m 61; c 2. INDUSTRIAL ENGINEERING, MANUFACTURING ENGINEERING. *Educ:* Iowa State Univ, BS, 63, MS, 66; Va Polytech Inst & State Univ, PhD(indust eng), 71. *Prof Exp:* Res asst indust eng, Iowa State Univ, 62-63, grad asst, 63-64, instr, 64-66; indust engr, Hosp Systs Res Group, Univ Mich-Ann Arbor, 66-67; indust engr, Community Systs Found, 67-68; instr indust eng, Va Polytech Inst & State Univ, 68-70; asst prof indust eng, 70-76, assoc prof, 76-81, PROF INDUST & MGT ENG, UNIV NEBR, LINCOLN, 81- *Mem:* Am Inst Indust Engrs; Am Soc Eng Educ; Soc Mfg Engrs. *Res:* Facility design; computer-aided manufacturing; micro-mini computer software design; quality control systems. *Mailing Add:* Dept Indust Eng Univ Nebr Lincoln NE 68588

HOFFMAN, RICHARD WAGNER, b Cleveland, Ohio, Apr 9, 27; m 50; c 3. SOLID STATE PHYSICS. *Educ:* Case Inst, BS, 47, MS, 49, PhD(physics), 52. *Prof Exp:* Res assoc, Case Western Reserve Univ, 49-52, from instr to prof, 52-81, AMBROSE SUASEY PROF PHYSICS, CASE WESTERN RESERVE UNIV, 81- *Concurrent Pos:* Mem tech staff, Bell Labs, 55-56, dep to George Kelley Reader, Oxford, 62-63. *Mem:* Am Phys Soc; Am Vacuum Soc (pres, 77). *Res:* Structure-property relations in thin film and solid surfaces, especially magnetic and mechanical properties; surface analysis, including specialized techniques of optical, electron and x-ray spectroscopy. *Mailing Add:* Dept Physics Case Western Reserve Univ Cleveland OH 44106

HOFFMAN, ROBERT, b Brooklyn, NY, Nov 27, 28; m 52; c 3. DENTISTRY, ORTHODONTICS. *Educ:* Colgate Univ, BA, 50; NY Univ, DDS, 55. *Prof Exp:* Staff scientist, Waldemar Med Res Found, 63-67, dir dent mat res, 67-76, dir micros, 74-76; DIR RES, MODULATION OPTICS INC, 76- *Concurrent Pos:* Adj prof, Adelphi Univ, Garden City, NY, 76-; consult. *Honors & Awards:* Albert Joachim Int Award, 67. *Mem:* Soc Photo-Optical Instrumentation Engr; Am Dent Asn; Am Asn Orthod; Int Asn Dent Res; Int Dent Fedn. *Res:* Ultrasonic bonding of metal to hard tissue; ultrasonic casting; invention and development of the modulation contrast microscope for imaging phase gradient objects; differential interference microscopy; advanced optical techniques for hard tissues study; electro-optic methods for detecting and measuring motion; microstructures of dental enamel; invention of "HMOS" precision measuring device. *Mailing Add:* Modulation Optics Inc 100 Forest Dr Greenvale NY 11548

HOFFMAN, ROBERT A, b Winona, Minn, Jan 30, 34; m 59; c 2. SPACE PHYSICS. *Educ:* St Mary's Col, Minn, BS, 56; Univ Minn, MS, 61, PhD(physics), 62. *Prof Exp:* MEM STAFF AEROSPACE TECHNOL, LAB PLANETARY ATMOSPHERES, GODDARD SPACE FLIGHT CTR, NASA, 61-, MEM STAFF, LAB EXTRATERRESTRIAL PHYSICS, 80- *Concurrent Pos:* Nat Acad Sci-NASA res fel, 61-63; lectr, Cath Univ Am, 65-68. *Honors & Awards:* Medal Except Sci Achievement, NASA. *Mem:* Am Geophys Union. *Res:* Magnetospheric physics; auroral physics and high latitude ionospheric phenomena; magnetosphere-ionosphere coupling. *Mailing Add:* Code 696 Goddard Space Flight Ctr Greenbelt MD 20771

HOFFMAN, ROBERT A, b Portland, Ore, June 30, 19; m 43; c 2. AGRICULTURAL ADMINISTRATION. *Educ:* Ore State Univ, BS, 47; Okla State Univ, MS, 65, PhD, 68; Am Registry Prof Entomologists, dipl. *Prof Exp:* Fishery biologist, State Game Comn, Ore, 47; res asst, Entom Res Div, 47-49, jr entomologist, 50-56, entomologist & substa leader, Miss, 56-60, res entomologist, Tex, 60-68, agr adminr, Plant Indust Sta, Agr Res Serv, Md, 68-72, invest leader & res entomologist, Vet Toxicol & Entom Res Lab, 72-73, AGR ADMINR, OKLA-TEX AREA, AGR RES SERV, USDA, 73- *Concurrent Pos:* Mem NSF seminar, NC State Univ, 65. *Mem:* Entom Soc Am; Am Mosquito Control Asn; Sigma Xi. *Res:* Medical and veterinary entomology, especially biology and control of the blood feeding insects. *Mailing Add:* Rte 3 Box L-312 Franklin TX 77856

HOFFMAN, ROBERT FRANK, b Riverside, NJ, Apr 2, 35; m 60; c 2. CHEMICAL ENGINEERING, PHYSICAL CHEMISTRY. *Educ:* Drexel Univ, BS, 57; Princeton Univ, MSE, 58; Univ Pa, PhD(chem eng), 67. *Prof Exp:* Process engr chem eng, Thiokol Corp, Trenton NJ, 58-62, pilot plant supvr, 62-67, mgr process develop, 67-74, res & develop dir, 74-77, dir tech opers chem eng & admin, 77-79; TECH DIR, RUBICON CHEM INC, WILMINGTON, 79- *Concurrent Pos:* Adj asst prof, Drexel Univ, 68-74. *Mem:* Am Inst Chem Engrs; Sigma Xi. *Res:* Development of commercial processes for chemicals and polymers; physical chemistry of polymers; polymer processing. *Mailing Add:* 25 Horse Shoe Curve Medford NJ 08055

HOFFMAN, ROBERT M, b Greenwich, Conn, June 19, 44. CANCER BIOLOGY. *Educ:* Univ Buffalo, BA, 65; Harvard Univ, PhD(biol), 71. *Prof Exp:* PROF CANCER BIOL, UNIV CALIF, SAN DIEGO, 79- *Concurrent Pos:* Pres, Anticancer Inc, 84- *Honors & Awards:* Belozersky Medal, Moscow State Univ, 90. *Mem:* Am Asn Cancer Res; Am Soc Clin Oncol; Tissue Cult Asn; Am Soc Cell Biol; Japanese Cancer Asn. *Res:* Cancer biology; biochemistry of cancer; development of preclinical models of human cancer, in vitro and in vivo. *Mailing Add:* Lab Cancer Biol Univ Calif San Diego 0609F La Jolla CA 92037-0609

HOFFMAN, ROBERT VERNON, b Canton, Ohio, Nov 11, 44; m 69. ORGANIC CHEMISTRY. *Educ:* Case Western Reserve Univ, BS, 66, PhD(chem), 70. *Prof Exp:* Res assoc org chem, Ohio State Univ, 69-74; mem fac, 74-80, ASSOC PROF CHEM, NMEX STATE UNIV, 80- *Mem:* Am Chem Soc; Sigma Xi. *Res:* Electrophilic aromatic substitution; peroxide decompositions and oxidation; intermolecular and intramolecular carbenic reactions. *Mailing Add:* 605 College Pl Las Cruces NM 88005

HOFFMAN, ROGER ALAN, b Willimantic, Conn, Feb 23, 24; m 46; c 3. NEUROENDOCRINOLOGY, REPRODUCTIVE BIOLOGY. *Educ:* Univ Conn, BS, 50; Purdue Univ, MS, 52, PhD(physiol, endocrinol), 56. *Prof Exp:* Asst wildlife mgt, Purdue Univ, 50-52, asst zool, 53-54, wildlife biologist, Maine Dept Inland Fish & Game, 52-53; biologist, physiol & endocrinol, Army Chem Ctr, 56-65; assoc prof biol, 65-70, PROF BIOL, COLGATE UNIV, 70- *Concurrent Pos:* Res grant, NSF, Purdue Univ, 59; vis lectr, Loyola Col, Md, 59-; NIMH fel, Univ Rochester, 69-70; Pop Coun fel, Univ Del Valle, Colombia, 71-72; dir reg biol, NSF, 75-76; vis scientist, Bur Drug Biol, Food & Drug Admin, Washington, DC, 78-79. *Mem:* AAAS; Am Soc Zoologists; Am Physiol Soc; Am Soc Mammal; Soc Study Reproduction. *Res:* Male reproduction in seasonal breeders; thyroid-gonad relationships; psychopharmacology and temperature regulation; hormonal aspects of temperature regulation; hibernation; comparative endocrinology; pineal gland-endocrine system relationships. *Mailing Add:* Dept Biol Colgate Univ Hamilton NY 13346

HOFFMAN, ROGER ALLEN, b LaCrosse, Wis, Sept 6, 15; m 39; c 2. ORGANIC CHEMISTRY. *Educ:* Univ Wis, BS, 39; Univ Mich, MS, 40, PhD(org chem), 42. *Prof Exp:* Res chemist, Merck, Sharp & Dohme, 42-50, mgr develop res group, Cherokee Plant, 50-55, mgr, Rahway Process Develop Group, 55-58, prod mgr, India, 58-63, managing dir, Pakistan Ltd, 63-67, admin mgr, Int Div, Southeast Asia, 67-68 & Europe, 69-74, sr dir subsidiary serv Europe, 74-79, exec dir tech opers, 80-85; RETIRED. *Mem:* Am Chem Soc. *Res:* Developmental research on pharmaceuticals; dissociation on pentaarylethanes. *Mailing Add:* 20 Dogwood Dr Madison NJ 07940

HOFFMAN, RUTH I, b Denver, Colo, Mar 23, 25. MATHEMATICS, EDUCATION IN COMPUTERS & MATHEMATICS. *Educ:* Univ Colo, BA, 46, MA, 47; Univ Denver, EdD(math educ), 53. *Prof Exp:* Teacher math, Denver Pub Schs, 47-53, dean admin, 53-58, prin, 58-63; lectr, 55-63, assoc prof, 63-70, PROF MATH, UNIV DENVER, 70- *Concurrent Pos:* Lectr, Regis Col, Colo, 55-58 & Univ Colo, 55-63; consult, Addison Wesley Pub Co, 63- & Sci Res Assocs, 72-; author and math systems dir, Educ Develop Lab, McGraw Hill, 73- *Mem:* Math Asn Am; Am Math Soc; AAAS; NY Acad Sci. *Res:* Mathematics education; geometry; topology. *Mailing Add:* Dept Math & Comput Sci Univ Denver Denver CO 80208

HOFFMAN, T(ERRENCE) W(ILLIAM), b Kitchener, Ont, Jan 3, 31; m 54; c 4. CHEMICAL ENGINEERING. *Educ:* Royal Mil Col, Can, dipl, 52; Queen's Univ, Ont, BSc, 53, MSc, 55; McGill Univ, PhD(chem eng), 59. *Prof Exp:* Lectr chem eng, Royal Mil Col, Can, 53-55; lectr, McGill Univ, 55-58; from asst prof to prof chem, McMaster Univ, 58-90; CONSULT, DYNAMIC MATRIX COMPTROL CORP, 90- *Concurrent Pos:* Chmn dept, McMaster Univ, 65-71. *Honors & Awards:* Erco Award, Can Soc Chem Eng, 70. *Mem:* Am Inst Chem Engrs; Chem Inst Can. *Res:* Polymer reaction engineering, advanced process control; chemical reaction engineering; simulation of chemical processes. *Mailing Add:* Dynamic Matrix Comptrol Corp 1220 Blalock Houston TX 77055

HOFFMAN, THEODORE P, b Jersey City, NJ, Mar 9, 29; m 81. ORGANIC CHEMISTRY. *Educ:* Rutgers Univ, BSc, 50, MSc, 52, PhD(org chem), 55. *Prof Exp:* Develop chemist, Am Cyanamid Co, 55-57; res chemist, Esso Res & Eng Co, 57-61; res scientist, Am Standard Co, 61-65; ed chem, Acad Press, Inc, 65-66; ed, 66-84, SR ED, WILEY-INTERSCI, JOHN WILEY & SONS, 85- *Mem:* AAAS; Am Chem Soc; NY Acad Sci. *Res:* Organic, inorganic, physical chemistry; biochemistry; macromolecular chemistry. *Mailing Add:* 1311 Lexington Ave New York NY 10128

HOFFMAN, THOMAS, b Jan 31, 47. CELL BIOLOGY. *Educ:* City Col New York, BS, 67; Univ Pittsburgh, MD, 71. *Prof Exp:* Pediat internship & residency, Johns Hopkins Hosp, 71-74; pediat immunol fel, Mem Sloan Kettering Cancer Ctr & immunol res fel, Rockefeller Univ, 74-79; sr investr,

Lab Immunodiag & Biol Response Modifiers Prog, Frederick Cancer Res Ctr, Nat Cancer Inst, 79-83; sr investr, Div Biochem & Biophys, Ctr Drugs & Biologics, 83-84, CHIEF, LAB CELL BIOL, DIV HEMAT, CTR BIOLOGICS EVAL & RES, US FOOD & DRUG ADMIN, NIH, 84- *Concurrent Pos:* Med officer, USPHS Comn Corp, comd, 79-88, capt, 88-, comn corps liason rep CBER, 87-89, res officer group, 90- *Mem:* Am Asn Immunologists; Am Acad Pediat. *Res:* Cell biology; immunology; pediatrics. *Mailing Add:* Ctr Biol Eval & Res NIH Lab Cell Biol Bldg 29 Rm 225 8800 Rockville Pike Bethesda MD 20892

HOFFMAN, THOMAS R(IPTON), b White Plains, NY, Oct 18, 23; m 45; c 4. VLSI DESIGN. *Educ:* Union Col, NY, BS, 44, MS, 50. *Prof Exp:* From instr to asst prof elec eng, Union Univ, NY, 46-50; mem tech staff, Bell Tel Labs, 50-54; from assoc prof to prof elec eng, Union Col, NY, 54-79; consult engr, Gen Elec Co, Fla, 79-84, mgr microelectronic, 84-89; RETIRED. *Concurrent Pos:* Consult, Gen Elec Co, 54-79 & Bell Tel Labs, 70-79; Fulbright lectr, Alexandria, 60-61, Taiwan, 67-68; vis prof, Univ Canterbury, Christchurch, NZ, 74. *Honors & Awards:* Centennial Medal, Inst Elec & Electronics Engrs. *Mem:* Sr mem Inst Elec & Electronics Engrs; Sigma Xi. *Res:* Digital computers; logic design and programming; digital circuits; semiconductor devices and circuits; microprocessors--applications; electrical engineering. *Mailing Add:* 1115 Jacaranda Ave Daytona Beach FL 32118-3615

HOFFMAN, WARREN E, b Buffalo, NY, Apr 5, 23; m 48; c 3. ORGANIC CHEMISTRY. *Educ:* Union Col, NY, BS, 51; Univ Buffalo, PhD, 55. *Prof Exp:* Payroll accountant, Hanna Furnace Corp, 46-48; res & develop chemist, Nat Analine Div, Allied Chem & Dye Corp, 54-60; from asst prof to prof chem, Ind Inst Technol, 60-83, chmn dept, 65-76, acad dean, 75-76; RETIRED. *Concurrent Pos:* Qual control engr, Western Elec Corp, 51-52; exp control chemist, E I du Pont de Nemours & Co, 53; consult, Protective Coatings Inc, Tokheim Corp & ITT. *Mem:* AAAS; Am Chem Soc; Am Inst Chem; Am Inst Chem Eng; Soc Appl Spectros; Sigma Xi. *Res:* Nitrogen organic compounds including isocyanates and polymers; oxamidines and cyanogen; physical organic and polymer chemistry; education and medicinal chemistry. *Mailing Add:* 1619 Lee Way Orlando FL 32810

HOFFMAN, WAYNE LARRY, b Oshkosh, Wis, Aug 12, 37; m 63; c 2. ENVIRONMENTAL SCIENCE, EARTH SCIENCE. *Educ:* Univ Wis-Oshkosh, BS, 62; Ohio State Univ, MA, 65; Univ Fla, PhD(geog), 70. *Prof Exp:* Instr geog, Univ Wis-Oshkosh, 66-67 & 69-70; PROF GEOG & HEAD DEPT GEOG & GEOL, WESTERN KY UNIV, 70- *Mem:* Asn Am Geogrs; Am Planning Asn; Am Geog Soc. *Res:* Land use and socio-economic areas of geography. *Mailing Add:* Dept Geog & Geol Western Ky Univ Bowling Green KY 42101

HOFFMAN, WILLIAM ANDREW, JR, b Hughesville, Mo, July 19, 28; m 51; c 5. ANALYTICAL CHEMISTRY. *Educ:* Mo Valley Col, BS, 50: Purdue Univ, MS, 52, PhD(chem), 55. *Prof Exp:* Instr Wabash Col, 54-55; asst prof chem, Wesleyan Univ, 55-60; from asst prof to assoc prof, 60-67, chmn dept, 64-67 & 68-69, dean admis, 73-77, PROF CHEM, DENISON UNIV, 67- *Mem:* AAAS; Am Chem Soc. *Res:* Solution and atmospheric chemistry. *Mailing Add:* Dept Chem Denison Univ Granville OH 43023

HOFFMAN, WILLIAM CHARLES, b Portland, Ore, Aug 11, 19; m 50, 75; c 3. PHYSICAL MATHEMATICS, MATHEMATICAL STATISTICS. *Educ:* Univ Calif, Berkeley, BA, 43; Univ Calif, Los Angeles, MA, 47, PhD(math), 53. *Prof Exp:* Head anal staff, USN Electronics Lab, 47-49; asst probability proj, Off Naval Res, 49-50; res physicist, Hughes Aircraft Co, 51-52; head anal staff, USN Electronics Lab, 52-55; consult, US Naval Air Missile Test Ctr, 55; mathematician & consult, Rand Corp, 55-59; res mathematician, Hughes Aircraft Co, 59; sr lectr physics, Univ Queensland, 60-61; res mathematician, Boeing Sci Res Labs, 61-65; prof math, Ore State Univ, 66-69; prof 69-85, EMER PROF MATH, OAKLAND UNIV, 85- *Concurrent Pos:* Vis prof math, NMex State Univ, 85-86; fel, math psychol, Melbourne Univ, 75-76. *Honors & Awards:* Wilson Award, 85. *Mem:* Am Math Soc; Psychometric Soc; Am Educ Res Asn; Soc Indust & Appl Math; Am Statist Asn. *Res:* Lie group theory of neuropsychology; geometric psychology; electromagnetic wave propagation; statistical methods. *Mailing Add:* 1430 Camino Amapola Sierra Vista AZ 85635

HOFFMAN, WILLIAM E, b Sheridan, Iowa, Jan 9, 46. ANESTHESIOLOGY, BRAIN FUNCTION. *Educ:* Univ Iowa, PhD(physiol), 75. *Prof Exp:* Asst prof, 79-85, ASSOC PROF ANESTHESIOL, MICHAEL REESE HOSP, 85- *Concurrent Pos:* Dir anesthesia res, Michael Reese Hosp, 79- *Mem:* Am Soc Anesthesia; Am Physiol Soc; Int Soc Anesthesia Res. *Mailing Add:* Dept Anesthesiol Michael Reese Hosp Chicago IL 60616

HOFFMAN, WILLIAM F, b Alvo, Nebr, Jan 16, 29; m 68; c 1. ANIMAL SCIENCE, DAIRY SCIENCE. *Educ:* Iowa State Univ, BS, 60; Univ Mo, MS, 62, PhD(dairy sci), 65. *Prof Exp:* Inspector processed foods, USDA, 60; res asst, Univ Mo, 61-64; from asst prof to assoc prof, 64- 68, PROF AGR, COL AGR, UNIV WIS-PLATTEVILLE, 68- *Mem:* Am Soc Animal Sci; Am Dairy Sci Asn; Am Inst Biol Sci. *Res:* Lyophilization of bovine spermatozoa; GnRH and its relationship to dairy cows with retained placentas; currently funded for a joint research project on The Effects of Relaxin on Retained Placentas in Dairy Cows with Dr Roy Ax and Wm Madison. *Mailing Add:* Dept Agr Univ Wis One University Plaza Platteville WI 53818

HOFFMAN, WILLIAM HUBERT, b Toledo, Ohio; m 65; c 3. PEDIATRIC ENDOCRINOLOGY. *Educ:* Marquette Univ, BA, 61, MD, 65. *Prof Exp:* Chief, Childs Hosp Mich, 73-83; Med res officer, NIH, 83-84; CHIEF PEDIAT ENDOCRINOL, MED COL GA, 84- *Mem:* Am Diabetes Asn; Am Thyroid Asn; Am Col Nutrit; Am Soc Human Genetics; Endocrinol Soc; Pediat Endocrinol Soc. *Mailing Add:* 718 Somerset Way Augusta GA 30909

HOFFMAN-GOETZ, LAURIE, NUTRITIONAL IMMUNOLOGY, THERMOREGULATION. *Educ:* Univ Mich, PhD(biol anthrop & physiol), 79. *Prof Exp:* Asst prof, 80-86, ASSOC PROF HEALTH STUDIES, UNIV WATERLOO, 86- *Mailing Add:* Dept Health Studies Univ Ave Univ Waterloo Waterloo ON N2L 3G1 Can

HOFFMANN, CONRAD EDMUND, b Lawrence, Kans, Apr 15, 20; m 42; c 3. MICROBIOLOGY & DRUG METABOLISM. *Educ:* Cornell Univ, BS, 42, MS, 43; Western Reserve Univ, PhD(microbiol), 52. *Prof Exp:* Res assoc, Owens-Ill Glass Co, 45-46; res scientist nutrit & physiol, Lederle Labs, 46-49; asst, Western Reserve Univ, 49-52; res microbiologist animal med, Stine Lab, 52-58, res supvr, Therapeut Chem Sect, 58-61, mgr microbiol sect, 61-77, MGR MICROBIOL & DRUG METAB, PHARMACEUT DIV, STINE LAB, E I DU PONT DE NEMOURS & CO, INC, 77-, COORDR, PHARMACEUT DIV TASK FORCE, 80-, MGR, LIFE SCI CONSTRUCT, 80- *Mem:* Am Soc Microbiol; NY Acad Sci. *Res:* Microbiology; metabolism. *Mailing Add:* 701 Cruthers Rising Sun MD 21911

HOFFMANN, DIETRICH, b Danzig, Ger, Dec 10, 24; m 60; c 2. BIOCHEMISTRY, ORGANIC CHEMISTRY. *Educ:* Univ Kiel, BS, 52, MS, 55; Max Planck Inst, PhD(biochem), 57. *Prof Exp:* Res assoc, 57-60, assoc, 60-66, assoc mem, 66-70, ASSOC SCIENTIST, SLOAN-KETTERING INST CANCER RES, 70-; CHIEF, DIV ENVIRON CARCINOGENESIS, AM HEALTH FOUND, VALHALLA, NY, 70- *Concurrent Pos:* Asst prof, Sloan-Kettering Div, Med Col, Cornell Univ, 61-69; assoc dir, chem & biochem, Am Health Found, Valhalla, NY, 80- *Mem:* Am Chem Soc; NY Acad Sci; Am Asn Cancer Res; fel Am Inst Chem; Soc Toxicol; Am Soc Prev Oncol; Phytochemical Soc. *Res:* Experimental and environmental carcinogenesis; environmental analysis; tobacco sciences. *Mailing Add:* 29 Holly Pl Larchmont NY 10538

HOFFMANN, EDWARD MARKER, b Mar 1, 34; US citizen; m 57; c 2. IMMUNOLOGY. *Educ:* Fla State Univ, BS, 56; Univ Miami, MS, 62, PhD(microbiol), 67. *Prof Exp:* Lab asst bact, Fla State Bd Health, 56; res asst immunol, Univ Fla, 59; res scientist immunochem, Sch Med, Univ Miami, 66-68; from asst prof to assoc prof, 68-77, prof microbiol, 77-80, PROF MICROBIOL, CELL SCI, IMMUNOL & MED MICROBIOL, INST FOOD & AGR SCI, UNIV FLA, 80-, CHMN, DEPT MICROBIOL & CELL SCI, 89- *Concurrent Pos:* USPHS trainee, 59-60. *Mem:* Am Soc Microbiol; Am Asn Immunologists. *Res:* Isolation of an anticomplementary substance from human erythrocytes; complement deficiencies in lymphoproliferative diseases; phylogenetic aspects of complement system; role of complement in virus elimination. *Mailing Add:* Dept Microbiol & Cell Sci Univ Fla Gainesville FL 32611

HOFFMANN, GEOFFREY WILLIAM, b Hamilton, Victoria, Australia, Oct 20, 44; m 71; c 2. IMMUNE SYSTEM NETWORK THEORY, NEURAL NETWORK THEORY. *Educ:* Univ Melbourne, BSc, 67, MSc, 68; Max Planck Inst, PhD(biophys chem), 72. *Prof Exp:* Postdoctoral phys chem, IBM Res Labs, San Jose, Calif, 72-73; scientist theoret biol, Max Planck Inst Biophys Chem, Göttingen, Ger, 73-74; mem theoret immunol, Basel Inst Immunol, Switz, 74-79; ASSOC PROF PHYSICS & MICROBIOL, UNIV BC, VANCOUVER, CAN, 79- *Concurrent Pos:* Collabr theoret biol group, Los Alamos Nat Lab, 85-86. *Mem:* Can Soc Immunologists; Can Asn Physicists. *Res:* Theoretical biology; stochastic theory of origin of life; symmetric network theory of regulation of immune system; neural network theory. *Mailing Add:* Dept Physics Univ BC Vancouver BC V6T 2A6 Can

HOFFMANN, GEORGE ROBERT, b Jersey City, NJ, June 6, 46; m 69; c 1. MICROBIAL GENETICS, GENETIC TOXICOLOGY. *Educ:* Rutgers Univ, AB, 67; Univ Tenn, Knoxville, MS, 69, PhD(bot), 72. *Prof Exp:* Instr biol, Roane State Community Col, Harriman, Tenn, 72-73; fel genetics, Nat Inst Environ Health Sci, Res Triangle Park, NC, 73-74; asst prof biol, Meredith Col, Raleigh, NC, 74-77; geneticist, Nat Inst Environ Health Sci, 77-78; sr staff off toxicol, Nat Acad Sci, Washington, DC, 79-81; from asst prof to assoc prof, 81-89, CHMN BIOL, HOLY CROSS COL, WORCESTER, MASS, 85-, PROF BIOL, 89- *Concurrent Pos:* Mem, Nat Res Coun Panel Irritant Chem, 82-84; consult, L'Oreal Res Lab, Aulnay-sous-Bois, Fr, 81-, US Environ Protection Agency, Washington, DC, 82-; assoc ed genetic toxicol, J Am Col Toxicol, 82-, ed in chief, Environ Molecular Mutagenesis, 83-88; vis scientist, Nat Ctr Social Republicans, Strasbourg, France, 88-89. *Mem:* Genetics Soc Am; Environ Mutagen Soc (pres, 90-91); Am Soc Microbiol; Bot Soc Am; Am Col Toxicol; Coun Biol Ed. *Res:* Bacterial and fungal genetics; mutagenesis; genetic toxicology testing. *Mailing Add:* Dept Biol Holy Cross Col Worcester MA 01610

HOFFMANN, HARRISON ADOLPH, b Breese, Ill, July 16, 10; m 48; c 1. MICROBIAL ECOLOGY. *Educ:* McKendree Col, BS, 34; Univ Ill, MS, 39; Univ Wash, St Louis, PhD(microbiol), 51. *Prof Exp:* Dir microbiol res group, Anheuser-Busch, Inc, 51-62; assoc prof biol, 62-73, chmn dept, 70-73, prof microbiol, 73-75, PROF EMER MICROBIOL, FLA ATLANTIC UNIV, 75- *Mem:* Am Soc Microbiol; Inst Biol Sci; Sigma Xi. *Res:* Microbial ecology; marine pollution. *Mailing Add:* 5659 Nassau Dr Boca Raton FL 33487-4158

HOFFMANN, JAMES ALLEN, plant pathology, for more information see previous edition

HOFFMANN, JOAN CAROL, b Cedarburg, Wis, Feb 20, 34. ENDOCRINOLOGY, REPRODUCTIVE BIOLOGY. *Educ:* Univ Wis-Madison, BS, 59; Univ Ill, Chicago, PhD(physiol), 65. *Prof Exp:* Asst prof nursing & physiol, Sch Med & Dent, Univ Rochester, 65-70; from assoc prof to prof anat & reproductive biol, Sch Med, Univ Hawaii, 74-83; ASSOC DEAN, OFF STUDENT AFFAIRS, UNIV MASS MED SCH, 83- *Concurrent Pos:* Nat Inst Child Health & Human Develop res grant, 66-75. *Mem:* Sigma Xi; Am Physiol Soc; Endocrine Soc; Soc Study Reproduction; Am Asn Anat. *Res:* Factors controlling the timing of ovulation; effects of the environment, particularly light, on reproduction in mammals; control mechanisms of neuro-endocrine function. *Mailing Add:* Off Student Affairs Univ Mass Med Sch 55 Lake Ave N Worcester MA 01605

HOFFMANN, JON ARNOLD, b Wausau, Wis, Jan, 13, 42; m 73. FLUID DYNAMICS. *Educ:* Univ Wis, BS, 64, MS, 66. *Prof Exp:* Res engr, Trane Co, 66-68; PROF AERONAUT ENG, CALIF POLYTECH STATE UNIV, 68- *Concurrent Pos:* Res engr, Stanford Univ, 70; res fel, NASA Ames Res Ctr, 74-75, consult, 75-77, prin investr, 81-89. *Mem:* Am Soc Mech Engrs. *Res:* Author of numerous publications. *Mailing Add:* Dept Aeronaut Eng Calif Polytech State Univ San Luis Obispo CA 93407

HOFFMANN, JOSEPH JOHN, b Waukesha, Wis, April 26, 50; m 85; c 1. RENEWABLE RESOURCES, PHYTOCHEMISTRY. *Educ:* St Norbert Col, BS, 71; Univ Ariz, PhD(med chem), 75. *Prof Exp:* Teaching asst chem, St Norbert Col, 69-71; teaching asst, Univ Ariz, 72-75, res assoc, 75-79, asst prof med chem, 80-87, DIR, BIORESOURCES RES FACIL, 81-, ASSOC PROF, UNIV ARIZ, 87- *Concurrent Pos:* Fel, Am Found Pharm Educ, 72; vis lectr, Univs Venice & Florence, Italy, 78. *Mem:* Am Chem Soc; Soc Econ Bot; Phytochem Soc NAm; Am, Soc Pytochem. *Res:* Renewable resources, especially chemicals and fuels from arid lands plants; isolation and identification of medicinal agents from plants. *Mailing Add:* 250 E Valencia Rd Tucson AZ 85706

HOFFMANN, LOUIS GERHARD, b Bloemendaal, Neth, July 12, 32; US citizen; m 55; c 2. IMMUNOLOGY. *Educ:* Wesleyan Univ, BA, 53; Johns Hopkins Univ, ScM, 58, ScD(microbiol), 60. *Prof Exp:* NSF fel, Univ Calif, Berkeley, 60-62; from instr to asst prof microbiol, Sch Med, Johns Hopkins Univ, 62-64; from asst prof to assoc prof, 64-73, PROF MICROBIOL, COL MED, UNIV IOWA, 73- *Concurrent Pos:* USPHS fel, 62-63, res grant, 64-67; NSF res grants, 67-73; Iowa Heart Asn res grant, 69-72; Damon Runyon-Walter Winchell Cancer Fund res grant, 71-73; Roche Res Found fel, 75; res grants, Iowa Heart Asn, 77-78 & NIH, 80-83. *Mem:* AAAS; Soc Exp Biol & Med; NY Acad Sci; Am Asn Immunol. *Res:* Chemical basis of antibody activity as manifested in cytotoxic effects in conjunction with complement. *Mailing Add:* Dept Microbiol Univ Iowa Col Med Iowa City IA 52241

HOFFMANN, MICHAEL K, b Breslau, Ger, Mar 6, 39; wid. MICROBIOLOGY. *Educ:* Univ Heidelberg, Physikum, 63; Univ Tübingen, MD, 66. *Prof Exp:* Postdoctoral sci teaching asst, Max-Planck Inst für Virusforschung, Ger, 66-70; postdoctoral fel, Univ Calif, San Diego, 70-71; res asst, Sloan-Kettering Inst Cancer Res, NY, 71-72, assoc, 73-78, assoc mem, 78-89; assoc prof, Cornell Univ Med Col, 73-89; PROF, NY MED COL, VALHALLA, 89- *Mailing Add:* Dept Microbiol & Immunol NY Med Col Valhalla NY 10595

HOFFMANN, MICHAEL ROBERT, b Fond du Lac, Wis, Nov 13, 46; m; c 2. APPLIED CHEMICAL & MICROBIAL CATALYSIS, ENVIRONMENTAL, CLOUD & FOG CHEMISTRY. *Educ:* Northwestern Univ, BA, 68; Brown Univ, PhD(chem), 73. *Prof Exp:* Asst prof civil eng, Univ Minn, 75-79, assoc prof, 79-80; from assoc prof to prof environ eng, 80-90, PROF ENVIRON CHEM, CALIF INST TECHNOL, 90- *Concurrent Pos:* USPHS res fel, Calif Inst Technol, 73-75; ed, Environ Div, Am Chem Soc, 78-84 & assoc ed, J Geophys Res, 84-86; chmn, Gordon Res Conf, 88. *Mem:* AAAS; Am Chem Soc; Am Geophys Union; Asn Environ Eng Prof; Soc Limnol & Oceanog. *Res:* Air and water pollution; atmospheric and aquatic chemistry; photocatalysis, microbial and chemical catalysis; chemistry and physics of clouds and fogs; reaction kinetics. *Mailing Add:* Environ Eng Sci Calif Inst Technol Pasadena CA 91125

HOFFMANN, PHILIP CRAIG, b Evanston, Ill, June 18, 36. PHARMACOLOGY. *Educ:* Univ Chicago, BS, 57, PhD(pharmacol), 62. *Prof Exp:* NSF fel, Pharmacol Inst, Univ Marburg, 62-63; guest res assoc, Royal Vet Col Sweden, 63; from instr to assoc prof, 63-80, PROF PHARMACOL, SCH MED, UNIV CHICAGO, 80- *Concurrent Pos:* Prin investr, USPHS res grant, 64- *Honors & Awards:* Quantrell Prize, 71. *Mem:* AAAS; Am Soc Pharmacol & Exp Therapeut; Sigma Xi. *Res:* Neurochemical control of metabolism; neurochemical aspects of development. *Mailing Add:* Pharmacol & Physiol Sci Univ Chicago 947 E 58th St Chicago IL 60637

HOFFMANN, RICHARD JOHN, b Ames, Iowa, Nov 8, 46; m 69; c 2. POPULATION BIOLOGY, MARINE BIOLOGY. *Educ:* Col William & Mary, BS, 69; Stanford Univ, MA, 71, PhD(biol sci), 74. *Prof Exp:* Scholar biol oceanog, Woods Hole Oceanog Inst, 74-75; asst prof biol sci, Univ Pittsburgh, 75-79; assoc prof zool & genetics, 80-87, PROF ZOOL & GENETICS, IOWA STATE UNIV, 87- *Concurrent Pos:* Woodrow Wilson fel, 69; instr invert zool, Marine Biol Lab, 77; prin investr, NSF, 78-79, 82-86, 89-90, 90-93 & NIH, 79-82. *Mem:* Fel AAAS; Soc Study Evolution. *Res:* Function of genetic variation at the enzyme level in natural populations; ecological genetics; microevolution; biology of coelenterates; population structure of asexual organisms. *Mailing Add:* Dept Zool Iowa State Univ Ames IA 50011-3223

HOFFMANN, ROALD, b Zloczow, Poland, July 18, 37; US citizen; m 60; c 2. THEORETICAL CHEMISTRY. *Educ:* Columbia Univ, BA, 58; Harvard Univ, MA, 60, PhD(chem Physics), 62. *Hon Degrees:* Fifteen from US & foreign univs, 78-90. *Prof Exp:* Jr fel, Soc Fels, Harvard Univ, 62-65; from assoc prof to prof chem, 68-74, JOHN A NEWMAN PROF PHYS SCI, CORNELL UNIV, 74- *Honors & Awards:* Nobel Prize in Chem, 81; Int Acad Quantum Molecular Sci Award, 71; Pure Chem Award, Am Chem Soc, 69, Harrison Howe Award, 70, Arthur C Cope Award, 73, Pauling Award, 74 & Inorg Chem Award, 82, Priestly Medal, 90; Nichols Medal, 81; Inorg Chem Award, 82; Nat Medal Sci, 83. *Mem:* Nat Acad Sci; AAAS; Int Acad Quantum Molecular Sci; Am Phys Soc; foreign mem Royal Swed Acad Sci; foreign mem Royal Soc; foreign fel Indian Nat Sci Acad; Am Philos Soc. *Res:* Molecular orbital calculations of electronic structure of ground and excited states of molecules; theoretical studies of transition states and intermediates in organic and inorganic reactions. *Mailing Add:* Dept Chem Cornell Univ Ithaca NY 14853

HOFFMANN, ROBERT SHAW, b Evanston, Ill, Mar 2, 29; m 51; c 4. SYSTEMATICS, ECOLOGY. *Educ:* Utah State Univ, BS, 50; Univ Calif, MA, 54, PhD(zool), 55. *Hon Degrees:* DSc, Utah State Univ, 88. *Prof Exp:* From instr to prof zool, Univ Mont, 55-68; chmn dept systs & ecol, Univ Kans Mus Natural Hist, 69-72, chmn div biol sci, 75-76, assoc dean & actg dean, Col Liberal Arts & Sci, 78-82, Summerfield distinguished prof systs & ecol, 68-86; dir, Nat Mus Natural Hist, 86-88; ASST SECY RES, SMITHSONIAN INST, WASHINGTON, DC, 88- *Concurrent Pos:* Nat Acad Sci exchange fel, USSR, 63-64; mem & chmn, US Nat Comt Int Union Quaternary Res, 78-82; mem, Nat Acad Sci & US-USSR joint Comt Sci Policy. *Mem:* Brit Mammal Soc; Ecol Soc Am; Am Soc Mammal (pres, 78-80); Soc Study Evolution; Soc Syst Zoologists. *Res:* Systematics, evolution and biogeography of birds and mammals; population and community ecology. *Mailing Add:* Asst Secy Res Smithsonian Inst Washington DC 20560

HOFFMANN, RONALD LEE, analytical chemistry, physical chemistry, for more information see previous edition

HOFFMANN, THOMAS RUSSELL, b Milwaukee, Wis, Sept 10, 33; m 57; c 1. TECHNICAL MANAGEMENT, INDUSTRIAL & MANUFACTURING ENGINEERING. *Educ:* Univ Wis, BS, 55, MS, 56, PhD(indust eng), 59. *Prof Exp:* Engr, Allis-Chalmers Mfg Co, 56-59; asst prof prod mgt, Univ Wis, 59-63; assoc prof, 63-65, PROF OPERS MGT, UNIV MINN, 65- *Concurrent Pos:* Chmn, Mgt Sci Dept, Univ Minn, 69-78, dir, W Bank Computer Serv, 71-87. *Mem:* Am Prod & Inventory Control Soc; Inst Mgt Sci; Asn Comput Mach. *Res:* Computing technology in the field of operations management; assembly line balancing, shop floor scheduling and linear programming; author or co-author of five books and 50 articles. *Mailing Add:* 4501 Sedum Lane Edina MN 55435

HOFFMANN, WILLIAM FREDERICK, b Manchester, NH, Feb 26, 33; m 65; c 2. INFRARED ASTRONOMY, ASTRONOMICAL INSTRUMENTATION. *Educ:* Bowdoin Col, AB, 54; Princeton Univ, PhD(physics), 62. *Prof Exp:* Instr physics, Princeton Univ, 58-61; resident res assoc, NASA Goddard Inst Space Studies, 62; instr, Yale Univ, 62-65; physicist, NASA Goddard Inst Space Studies, 65-73; PROF ASTRON, STEWARD OBSERV, UNIV ARIZ, 73- *Concurrent Pos:* Adj assoc prof astron, Columbia Univ, 69-73; NSF fel, 54; Danfull fel, 54-58. *Honors & Awards:* Exceptional Sci Achievement Medal, NASA, 72. *Mem:* AAAS; Am Phys Soc; Am Astron Soc; AAAS; Sigma Xi. *Res:* Infrared astronomy; astronomical telescope construction. *Mailing Add:* Steward Observ Univ Ariz Tucson AZ 85721

HOFFMANN-PINTHER, PETER HUGO, b Chicago, Ill, Feb 14, 35; m 62. NUCLEAR STRUCTURE CALCULATION. *Educ:* St Mary's Univ, BS, 58, Ind Univ, MS, 64; Ohio Univ, PhD(physics), 73. *Prof Exp:* Res Scientist US Tank Automotive Command, 60-61; instr physics, Ohio State Univ, 71-73, vis fel, 74-77; asst prof, Marshall Univ, 77-78 & Ohio State Univ, 79-81. *Mem:* Am Phys Soc; Sigma Xi. *Res:* Theory fructuation nuclear cross-section and nuclear structure calculation; science in pre-Columbian America. *Mailing Add:* Dept Nat Sci Univ Houston One Main St Houston TX 77002

HOFFMASTER, DONALD EDEBURN, b New Castle, Pa, July 14, 11; m 37; c 2. BOTANY, CONSERVATION. *Educ:* Pa State Teachers Col, BS, 31; Cornell Univ, MS, 36; Univ WVa, PhD(plant path), 40. *Prof Exp:* Instr pub sch, Pa, 32-36; asst, Univ WVa, 36-40; instr bot, Okla Agr & Mech Col, 40-46; asst prof bot, bact & zool, Westminster Col, 46-47; prof, 47-71, chmn dept, 65-69, EMER PROF BOT, IND UNIV PA, 71- *Concurrent Pos:* Asst plant pathologist, Exp Sta, Okla Agr & Mech Col, 40-45. *Res:* Conservation. *Mailing Add:* 250 School St Indiana PA 15701

HOFFMEISTER, DONALD FREDERICK, b San Bernardino, Calif, Mar 21, 16; m 38; c 2. ZOOLOGY. *Educ:* Univ Calif, AB, 38, MA, 40, PhD(zool), 44. *Prof Exp:* Tech asst, Mus Vert Zool, Univ Calif, 40-41, asst, 41-42, asst zool, Univ, 42-44, assoc, 44; asst prof zool & asst cur mus natural hist, Univ Kans, 44-46; from asst prof to assoc prof, 46-59, prof zool, 59-84, EMER PROF ZOOL, UNIV ILL, URBANA, 84- *Concurrent Pos:* Curatorial asst, Mus Vert Zool, Univ Calif, 43-44; mem natural hist field expeds, Kans, 45, Ill, Ariz & Calif; res assoc, Mus N Ariz, 69-; cur mus, Univ Mus Nat Hist, Univ Ill, Urbana, 46-64, dir, 64-84. *Honors & Awards:* Hartley H T Jackson Award, 86. *Mem:* Am Soc Mammal (secy, 47-52, pres, 64-66); Soc Syst Zool; Asn Sci Mus Dirs; Am Asn Mus. *Res:* Mammalogy, especially phylogeny; taxonomy and speciation in rodents, particularly cricetine rodents, both fossil and recent; southwestern American mammals. *Mailing Add:* 438 Natural Hist Bldg Univ Ill Urbana IL 61801

HOFFSOMMER, JOHN C, b Montgomery, Ala, Jan 23, 32; m 56; c 2. ORGANIC CHEMISTRY. *Educ:* Univ Pa, AB, 54; Univ Md, MS, 60; George Washington Univ, PhD(chem), 64. *Prof Exp:* RES ASSOC CHEMIST, US NAVAL ORD LAB, 59- *Mem:* Am Chem Soc. *Res:* Explosives chemistry research; synthesis; kinetics. *Mailing Add:* 12805 Lacy Dr Silver Spring MD 20904-2917

HOFFSTEIN, VICTOR, b Soviet Union, Dec 7, 42; nat US. RESPIRATORY MEDICINE, RESPIRATORY PHYSIOLOGY. *Educ:* Polytech Inst Brooklyn, 65, MS, 67, PhD(physics), 73; Univ Miami, MD, 73, Royal Col Physicians & Surgeons Can, 82. *Prof Exp:* Researcher physics, RCA Labs, Princeton, 65-67, Nat Sci Res Ctr, France, 70-73; STAFF RESPIROLOGIST MED, ST MICHAELS HOSP, UNIV TORONTO, 80-, ASSOC PROF MED, 87- *Mem:* Am Thoracic Soc; Can Thoracic Soc. *Res:* Upper airways, asthma, obstructive sleep breathing. *Mailing Add:* Dept Med St Michaels Hosp 30 Bond St Toronto ON M5B 1W8 Can

HOFMAN, EMIL THOMAS, b Paterson, NJ, June 24, 21; m 57; c 3. INORGANIC CHEMISTRY. *Educ:* Univ Miami, AB, 49; Univ Notre Dame, MS, 53, PhD(chem), 62. *Prof Exp:* From instr to assoc prof, 53-68, asst dean, Col Sci, 65-71, PROF CHEM, UNIV NOTRE DAME, 68-, DEAN FRESHMAN YEAR STUDIES, 71- *Mem:* AAAS; Am Chem Soc; Nat Sci Teachers Asn; Sigma Xi. *Res:* Metal chelates; teacher training. *Mailing Add:* Freshman Year Studies Univ Notre Dame Notre Dame IN 46556

HOFMAN, WENDELL FEY, b Grand Rapids, Mich, Oct 28, 42; m 63; c 3. PHYSIOLOGY. *Educ:* Western Mich Univ, BS, 64, MA, 66; Mich State Univ, PhD(physiol), 70. *Prof Exp:* Teaching asst biol, Western Mich Univ, 64-66; teaching asst physiol, Mich State Univ, 66-69, spec res assoc, Endocrine Res Unit, 69-70; asst prof, 70-77, ASSOC PROF PHYSIOL, MED COL GA, 77- *Concurrent Pos:* Prin investr, NIH, 87-92, mem study sects, 89-90; mem, Ga Lung Asn Res Grants Panel, 89-, Ga Heart Asn Res Comt, 89- *Mem:* Sigma Xi; Am Lung Asn; Am Physiol Soc; Am Heart Asn. *Res:* Pulmonary edema and vasoreactivity dogs to serotorin and prostaglandins. *Mailing Add:* Dept Physiol Med Col Ga Augusta GA 30912

HOFMANN, ALAN FREDERICK, b Baltimore, Md, May 17, 31; m 59, 78; c 2. GASTROENTEROLOGY. *Educ:* Johns Hopkins Univ, AB, 51, MD, 55; Univ Lund, PhD, 65. *Hon Degrees:* MD, Univ Bologna, Italy, 88. *Prof Exp:* Intern & res, Dept Med, Columbia-Presby Med Ctr, 55-57; clin assoc internal med, Nat Heart Inst, 57-59; res assoc, Rockefeller Univ, 62-64, asst prof med & biochem & assoc physician, 64-67; assoc prof physiol & med, Mayo Med Sch, Univ Minn, 67-70, prof med & physiol, 70-77, assoc dir, Gastroenterol Unit, Mayo Clin & Mayo Found, 66-77; PROF MED, UNIV CALIF, SAN DIEGO, 77-, DIR RES & TRAINING PROG, DIV GASTROENTEROL, 80- *Concurrent Pos:* Nat Found fel, Univ Lund, 59-61, NIH fel, 61-62; vis prof, var univs & clins US, Scotland & Australia, 69-85; mem, Gen Med A Study Sect, NIH, 71-75; distinguished prof lectr, Univ Cincinnati, 76; attend physician, San Diego Med Ctr, Univ Calif, 77-; Fogarty Int sr fel, NIH, 86; vis prof, Royal Soc Med, UK, 87. *Honors & Awards:* Snell Mem Lectr, Palo Alto, Calif, 73; Beaumont Mem Lectr, USAF, 73; Sir Arthur Hurst Mem Lectr, Brit Soc Gastroenterol, 74; Schorstein Mem Lectr, London Hosp, 74; Solomon Meyers Lectr, Sch Med, Wayne State Univ, 75; Eppinger Prize, Falk Found, WGer, 76; Humboldt Found Sr Scientist Award, Ger, 76; H M Pollard Lectr, Univ Mich, 77; Christian Johann Berger Lectr, Albany Hosp, Denmark, 77; William Beaumont Prize, Am Gastroenterol Asn, 78; R D McKenna Lectr, Can Soc Gastroenterol, 79; Samuel D Kushlan Lectr, Yale Univ, 83; Leon Schiff Lectr, Univ Cincinnati, 84; Lepow Lectr, Univ Conn, 88; Walter B Cannon Lectr, Soc Gastroenterol Radiol; Bengt Ihre Lectr, Swed Med Asn, 88. *Mem:* Am Asn Study Liver Dis; Am Gastroenterol Asn; Am Physiol Soc; Am Soc Clin Invest; Asn Am Physicians; Biochem Soc; Int Asn Study Liver; hon mem Brit Soc Gastroenterol; fel AAAS; Am Fedn Clin Res; corres mem Royal Flemish Acad Med. *Res:* Bile acid and lipid metabolism; cholelithiasis; biliary pathobiology, pathochemistry and physiology; diagnosis and treatment of digestive and hepatobiliary disease; enterohepatic circulation of endo and xenobiotics. *Mailing Add:* Dept Med 0813 Univ Calif San Diego La Jolla CA 92093-0813

HOFMANN, ALBRECHT WERNER, b Zeitz, Germany, Mar 11, 39; m 66; c 2. GEOCHEMISTRY, GEOCHRONOLOGY. *Educ:* Univ Freiburg, Vordiplom, 62; Brown Univ, MSc, 65, PhD(geochemistry), 69. *Prof Exp:* Asst geochronology, Univ Heidelberg, 68-70; Carnegie fel geochem, Carnegie Inst, 70-72; mem staff geochem, Wash Dept Terrestrail Magnetism, 72-80; DIR, MAX PLANCK INST CHEM, MAINZ, WGER, 80- *Mem:* Am Geophys Union; Geochem Soc; Sigma Xi. *Res:* Isotopic and trace element geochemistry; diffusion in solid and liquid silicates. *Mailing Add:* Max Planck Inst Fuer Chemie 6500 Mainz Saarstr 23 Postfach 3060 Germany

HOFMANN, BO, b Copenhagen, Denmark, Sept 14, 53. AIDS PATHOGENESIS, LYMPHOCYTE ACTIVATION. *Educ:* Copenhagen Univ, MD, 81, DSc(immunol), 91. *Prof Exp:* Resident med & surg, Copenhagen Univ Hosp, 81-83, resident clin immunol, 83-86, res fel immunol, 86-88; res fel, 88-90, ASST RESEARCHER IMMUNOL, DEPT MICROBIOL-IMMUNOL, SCH MED, UNIV CALIF, LOS ANGELES, 90- *Concurrent Pos:* Teaching asst anat, Copenhagen Univ, 79-81 & 84-85. *Mem:* Am Asn Immunologists; Int AIDS Soc. *Res:* Early events in lymphocyte activation; AIDS pathology; HIV's effect on the immune system; HIV epidemiology. *Mailing Add:* 740 W Kings Rd No 119 Los Angeles CA 90069

HOFMANN, CHARLES BARTHOLOMEW, b New York, NY, Sept 5, 39; m 65; c 2. ELECTRICAL ENGINEERING. *Educ:* Lehigh Univ, BSEE, 61; Columbia Univ, MBA, 63. *Prof Exp:* Engr electronic warfare, Airborne Instruments Lab, 63-66; sr engr, Maxson Electronics, 66-69; prin engr, 69-71, dir eng proj, 71-75, vpres eng, 75-, PRES, AMECOM DIV, LITTON SYSTS, INC. *Mem:* Inst Elec & Electronics Engrs; Armed Forces Commun & Electronics Asn; Soc Photooptical Instrumentation Engrs; Asn Old Crows. *Res:* Electronic warfare; electronic support measures equipment; air traffic control voice switching system, high frequency communications, and Loran navigation. *Mailing Add:* 13517 Middlevale Lane Silver Spring MD 20906

HOFMANN, CORRIS MABELLE, b Plainville, Mass, Sept 18, 15. PHARMACEUTICAL CHEMISTRY. *Educ:* Univ Ill, BS, 37; Bryn Mawr Col, PhD(org chem), 41. *Prof Exp:* Demonstr chem, Bryn Mawr Col, 37-39; res chemist, Calco Chem Div, 41-55, RES CHEMIST, LEDERLE LABS, AM CYANAMID CO, 55- *Mem:* AAAS; Am Chem Soc; Am Inst Chem. *Res:* Synthesis of unsaturated esters; molecular rearrangements; azo dyestuffs; pharmaceuticals. *Mailing Add:* 258 Ackerman Ave Ho Ho Kus NJ 07423

HOFMANN, DAVID JOHN, b Albany, Minn, Jan 3, 37; m 59; c 3. ATMOSPHERIC PHYSICS. *Educ:* Univ Minn, BS, 61, MS, 63, PhD(physics), 66. *Prof Exp:* Res assoc cosmic ray physics, Univ Minn, 65-66; from asst prof to prof physics, Univ Wyo, 66-91; CHIEF SCIENTIST, NAT OCEANIC & ATMOSPHERIC ADMIN/CMOL, 90- *Concurrent Pos:* Res grants, NSF, 67-88, Off Naval Res, 70-85 & NASA, 76-88 & Environ Protection Agency, 71-72. *Honors & Awards:* Alexander von Humboldt Sr Scientist Award, 82. *Mem:* Am Geophys Union; Sigma Xi. *Res:* Stratospheric constituents; stratospheric aerosol measurements. *Mailing Add:* Climate Monitoring & Diag Lab NOAA/ERL 325 Broadway Boulder CO 80303

HOFMANN, FREDERICK GUSTAVE, b Detroit, Mich, May 25, 23; m 56; c 2. ENDOCRINOLOGY. *Educ:* Univ Mich, AB, 43; Harvard Univ, PhD(physiol), 52. *Prof Exp:* Nat Res Coun fel med sci & res fel endocrinol, Harvard Univ, 52-53; from assoc to assoc prof, 53-68, PROF PHARMACOL, COL PHYSICIANS & SURGEONS, COLUMBIA UNIV, 68-, ASSOC DEAN, 70- *Concurrent Pos:* Markle scholar, 55-60; ed in chief, Endocrinol, Endocrine Soc, 63-67. *Mem:* Endocrine Soc; Am Soc Pharmacol & Exp Therapeut; Harvey Soc. *Res:* Adrenocortical insufficiency; pituitary-adrenal interrelationships; biogenesis of steroid hormones; drug abuse. *Mailing Add:* 923 Van Houten B-1 Clifton NJ 07013

HOFMANN, GUNTER AUGUST GEORGE, b Munich, Ger, Sept 9, 35. PLASMA PHYSICS. *Educ:* Munich Tech Univ, Dipl Ing, 59, Dr rer nat(plasma physics), 62. *Prof Exp:* Res scientist, Max Planck Inst Plasma Physics, 62-67; sr staff physicist, 67-80, MGR PROD DEVELOP, HUGHES RES LABS, 80- *Mem:* Am Phys Soc; Europ Phys Soc; Ger Phys Soc. *Res:* Thermonuclear fusion; fast discharges; plasma diagnostics; crossed field devices; direct current interrupters; ionizing radiation; radiation transport. *Mailing Add:* 3750 Riviera Dr San Diego CA 92109

HOFMANN, KARL HEINRICH, b Heilbribbm Ger, Oct 3, 32; m 63; c 2. MATHEMATICS. *Educ:* Univ Tibingen, Dr rer nat, 58. *Prof Exp:* Asst math, Univ Tibingen, 58-59, res assoc math statist, 59-60, docent math, 61-63; res assoc math, Tulane Univ, 60-61, assoc prof, 63-65, prof, 65-, W R Irby prof, 79-; PROF MATH, TECHNISCHE HOCHSCHULE DARMSTADT, GER, 82- *Concurrent Pos:* Fel, Alfred P Sloan Found, 67-68; vis prof, Univ Paris, 73-74 & Tech Univ, Darmstadt, 80-81; adj prof, Tulane Univ. *Honors & Awards:* E Harris Harbison Award, Danforth Found, 70. *Mem:* Am Math Soc; Australian Math Soc; Ger Math Soc; Math Soc France; Sigma Xi. *Res:* Topological algebraic structures; functional analysis. *Mailing Add:* Fachbereich Math Tech Hochsch Schlossgartenstr 7 D-6100 Darmstadt Germany

HOFMANN, KLAUS HEINRICH, b Karlsruhe, Ger, Feb 21, 11; US citizen; m 36, 64; c 1. BIOCHEMISTRY. *Educ:* Swiss Fed Inst Technol, PhD(org chem), 36. *Prof Exp:* Fel, Rockefeller Found, 38-40; asst biochem, Med Col, Cornell Univ, 40-42; sci guest, Ciba Pharmaceut Prod, Inc, NJ, 42-44; from asst res prof to assoc res prof org biochem, 44-47, res prof chem, 47-52, chmn dept biochem, Sch Med, 53-64, PROF BIOCHEM, SCH MED, UNIV PITTSBURGH, 52-, SALK COMMONWEALTH PROF EXP MED & DIR PROTEIN RES LAB, 64-, UNIV PROF EXP MED & BIOCHEM. *Honors & Awards:* Sr Scientist Award, Alexander Von Humboldt Found, Bonn, WGer, 76. *Mem:* Nat Acad Sci; AAAS; Am Chem Soc; Am Soc Biochemists; Endocrine Soc; Swiss Chem Soc. *Res:* Degradation and synthesis of natural products; chemistry of biotin and its antagonists; synthesis of peptides; mechanism of action of adrenocorticotropic hormone; structure and metabolism of bacterial fatty acids; synthesis of pituitary hormones. *Mailing Add:* Protein Res Lab Univ Pittsburgh Sch Med Pittsburgh PA 15261

HOFMANN, LENAT, b Cedarburg, Wis, May 25, 34; m 59; c 3. GRAZING SEASONALITY. *Educ:* Univ Wis-River Falls, BS, 62; NDak State Univ, MS, 68, PhD(agron), 69. *Prof Exp:* Soil scientist, US Army CEngr, Ill, 62-65; asst prof forage crops, Univ Md, College Park, 69-74; RES AGRONOMIST RANGE MGT, AGR RES SERV, USDA, 74- *Mem:* Am Soc Agron; Soil Water Conserv Soc; Soc Range Mgt. *Res:* Develop use of cool-season grass species for season-long grazing in the Northern Great Plains; range and pasture management systems. *Mailing Add:* Northern Great Plains Res Lab PO Box 459 Mandan ND 58554

HOFMANN, LORENZ M, b Chicago, Ill, Jan 10, 37; m 68. PHARMACOLOGY, PHYSIOLOGY. *Educ:* Univ Ill, BS, 59, MS, 61, PhD(pharmacol), 64. *Prof Exp:* Sr investr renal pharmacol, Div Biol Res, G D Searle & Co, 64-77; assoc dir clin pharmacol res, Ross Labs, 77-; ASSOC DIR CLIN DEVELOP, DEPT MED, ADRIA LABS. *Concurrent Pos:* Assoc dir clin sci, Parke-Davis. *Mem:* AAAS; Am Pharmaceut Asn; Am Soc Pharmacol & Exp Therapeut; Am Soc Clin Pharmacol & Therapeut. *Res:* Renal and cardiovascular pharmacology; application of statistics in design of experiments; clinical research in antibiotics, analgesics, dermatologicals, antihypertensives, cardiotonics and devices. *Mailing Add:* Clin Res Int Inc 2224 Chapel Hill Nelson Hwy Durham NC 27713

HOFMANN, PETER L(UDWIG), b Vienna, Austria, Jan 25, 25; nat US; m 50; c 2. TECHNICAL MANAGEMENT. *Educ:* Cooper Union, BEE, 50; Union Col, NY, MS, 54; Rensselaer Polytech Inst, DEngSci, 60. *Prof Exp:* Test engr, Gen Elec Co, 50-51; nuclear engr & mgr nuclear design & anal, Knolls Atomic Power Lab, 51-61; tech consult, Hanford Lab, 61-63, mgr eng physics, 63-66, mgr fast flux test facil reactor physics, 66-68, mgr reactor & plant technol, 68-70, mgr systs anal, 70-74; assoc dir planning & anal, Battelle Corp Staff, 74-79, various mgt positions, Battelle Proj Mgt Div, 79-90; RETIRED. *Concurrent Pos:* Adj assoc prof, Rensselaer Polytech Inst, 60-61; assoc prof & prog chmn, Joint Ctr Grad Study, Univ Wash, 62-74. *Mem:* Am Phys Soc; Am Nuclear Soc; sr mem Inst Nuclear Mat Mgt; Sigma Xi. *Res:* Analysis and engineering of power reactors; nuclear strategies, fuel cycles and waste management including nuclear waste transport; environmental aspects of reactor siting; energy policy. *Mailing Add:* 5080 Dublin Rd Dublin OH 43017

HOFMANN, THEO, b Zurich, Switz, Feb 2, 24; Can citizen; m 53; c 3. BIOCHEMISTRY, MOLECULAR BIOLOGY. *Educ:* Swiss Fedn Inst Technol, dipl Ing Chem, 47, DrScTech(pharmaceut chem), 50. *Prof Exp:* Agr Res Coun grant biochem, Aberdeen Univ, 50-52; from sci officer to sr sci officer, Hannah Dairy Res Inst, Ayr, Scotland, 52-56; lectr biochem, Univ Sheffield, 56-64; assoc prof biochem, 64-66, prof, 66-89, EMER PROF BIOCHEM & CHEM, UNIV TORONTO, 89- *Concurrent Pos:* Vis assoc prof, Univ Wash, 62-63; mem group comt, Med Res Coun Can, 70-74; vis scientist, Div Animal Genetics, Commonwealth Sci & Indust Res Orgn, Sydney, Australia, 72; vis prof, Dept Chem, Univ Calif, Santa Cruz, 81; vis scientist, Univ Lund, Sweden, 84, 87. *Mem:* Am Soc Biol Chem; Can Biochem Soc; Brit Biochem Soc. *Res:* Structure and function of proteolytic enzymes and calcium-binding proteins. *Mailing Add:* Dept Biochem Univ Toronto Fac Med Toronto ON M5S 1A8 Can

HOFMASTER, RICHARD NAMON, b Fostoria, Ohio, Apr 12, 15; m 46. ENTOMOLOGY. *Educ:* Ohio State Univ, BSc, 37, MSc, 40, PhD(entom), 48. *Prof Exp:* Field aide, Sugar Beet Leafhopper Lab, Bur Entom & Plant Quarantine, USDA, Idaho, 38-41; asst entom, Ohio State Univ, 46-47; entomologist, Va Truck Exp Sta, 47-81; RETIRED. *Honors & Awards:* L O Howard Award, ENT, 74. *Mem:* AAAS; Entom Soc Am; Sigma Xi. *Res:* Nuclear polyhedrosis virus studies on cabbage looper, its effectiveness and environmental relationships; ecology of the sugar beet leafhopper; precise insecticidal tests; biology and control of the potato tubermoth and the corn earworm; mites with special emphasis on joint insecticide-herbicide treatments; potato, tomato and fall cucurbit insects; general vegetable insects and control. *Mailing Add:* Box 327 Belle Haven VA 23306

HOFREITER, BERNARD T, b Peoria, Ill, July 22, 24; m 47; c 4. ORGANIC CHEMISTRY, PAPER TECHNOLOGY. *Educ:* Bradley Univ, BS, 48, MS, 50; NC State Univ, PhD, 73. *Prof Exp:* Chemist, Corn Prod Ref Co, 50; chemist, Northern Regional Res Ctr, USDA, 51-60, res leader, 60-80; mem staff, Peoria Sch Med, Univ Ill, 80-83; CONSULT, INT EXEC SERV CORPS, KOREA, SRI LANKA, US, 83- *Concurrent Pos:* Fac chem, Bradley Univ, 85- *Mem:* Am Chem Soc; Tech Asn Pulp & Paper Indust. *Res:* Industrial utilization of cereal derived products; physical and chemical modification of starch and related carbohydrate material to produce papermaking adjuncts. *Mailing Add:* 2516 N Woodbine Terr Peoria IL 61604

HOFRICHTER, CHARLES HENRY, b Cleveland, Ohio, Sept 3, 14; m 38; c 4. CHEMISTRY. *Educ:* Hiram Col, BA, 37; Univ Buffalo, PhD(org chem), 40. *Prof Exp:* Off Sci Res & Develop, Northwestern Univ, 40-41; res chemist, Durez Plastics, 41-43; res chemist, E I du Pont de Nemours & Co, 43-51; chief film res sect, Olin Industs, Inc, 51-54, mgr pesticides res & develop dept, 54-58, tech adv to vpres res, 58-62, tech dir appln res chem div, 62-69, tech dir urethanes, 69-75, dir urethane chem safety, Olin Corp, 75-78; RETIRED. *Concurrent Pos:* Mem safety group, Int Isocyanates Inst, 72-; chmn pub affairs tech info comt, SPI Coordinating Comt on Consumer Safety, 73-78; consult urethane & fire technol, 78- *Mem:* Fel AAAS, Am Chem Soc; fel Am Inst Chem; NY Acad Sci; Soc Fire Protection Engrs; Sigma Xi. *Res:* Silicon orthoesters; cellophane coatings; mylar and polyethylene film; agricultural chemicals; urethane; automotive fluids; biocides; plastics combustibility; fire sciences. *Mailing Add:* Six Stone Pasture Lane Killingworth CT 06417

HOFRICHTER, HARRY JAMES, b Buffalo, NY, Aug 21, 43; m 65; c 2. PHYSICAL CHEMISTRY, BIOPHYSICS. *Educ:* Dartmouth Col, BA, 65; Univ Ore, PhD(phys chem), 71. *Prof Exp:* Fel phys chem, Univ Ore, 71-72; staff fel phys chem, 72-75, sr staff fel, 75-77, RES CHEMIST PHYS CHEM, LAB CHEM PHYSICS, NIH, 77- *Res:* Polymer physical chemistry; linear and circular dichroism of biological molecules; optical spectroscopy; microspectrophotometry; protein assembly reactions; sickle cell hemoglobin; kinetics of nucleated reactions. *Mailing Add:* Chem Physics Lab Bldg 2 Rm B1-22 NIH Bethesda MD 20892

HOFSLUND, PERSHING BENARD, b Jeffers, Minn, Apr 13, 18; m 40; c 2. ZOOLOGY. *Educ:* Mankato State Teachers Col, Minn, BS, 40; Univ Mich, MS, 47, PhD(zool), 54. *Prof Exp:* Pub sch teacher, Minn, 40-45; from instr to prof, 49-82, EMER PROF ZOOL, UNIV MINN, DULUTH, 82- *Honors & Awards:* Thomas Sadler Roberts Award. *Mem:* Wilson Ornith Soc (secy, 63-71, pres, 71-73); Am Ornithologists Union; Am Birding Asn. *Res:* Ornithology, behavior and migrations of hawks and wood warblers. *Mailing Add:* 4726 Jay St Duluth MN 55804

HOFSTADTER, ROBERT, physics; deceased, see previous edition for last biography

HOFSTETTER, EDWARD, b New York, NY, Nov 17, 32; m 55; c 2. ELECTRICAL ENGINEERING, MATHEMATICS. *Educ:* Mass Inst Technol, SB & SM, 55, ScD(elec eng), 59. *Prof Exp:* Asst prof elec eng, 59-63, STAFF MEM, LINCOLN LAB, MASS INST TECHNOL, 63- *Concurrent Pos:* Consult, Tex Instruments Inc, Tex, 60-61 & Raytheon Co, Mass, 61-63. *Mem:* Inst Elec & Electronics Engrs; Sigma Xi. *Res:* Digital signal processing; probability theory and its application to signal detection and parameter estimation; speech recognition; speech bandwidth compression. *Mailing Add:* 98 Wolf Rock Rd Carlisle MA 01741

HOFSTETTER, HENRY W, b Windsor, Ohio, Sept 10, 14; m 41; c 2. VISUAL PHYSIOLOGY, OPTOMETRY. *Educ:* Ohio State Univ, BSc, 39, MSc, 40, PhD(physiol optics), 42. *Hon Degrees:* DOS, Los Angeles Col Optom, 54 & Mass Col Optom, 68; ScD, Pa Col Optom, 69, Univ Waterloo, 77, State Univ NY, 91. *Prof Exp:* From instr to assoc prof optom, Ohio State Univ, 42-49; dean, Los Angeles Col Optom, 49-52; dir, Div Optom, 52-70, prof, 52-74, Rudy prof, 74-79, EMER PROF OPTOM, IND UNIV, 80- *Concurrent Pos:* Mem, Adv Res Coun, Am Optom Found, 49-70, Nat Adv Coun Educ for Health Professions & Armed Forces-Nat Res Coun Comt Vision; chmn, Tech Adv Comt Light & Vision, Illum Eng Res Inst, 77-81; vis prof optom, Univ Waterloo, Can, 80; actg dean optom, Int Am Univ, PR, 81; mem adv comt, Lighting Res Inst, 88- *Honors & Awards:* Int Optical League Medal, 68; Apollo Award, Am Optom Asn, 73; Orion Award, Armed Forces Optom Soc, 74; Prentice Award, Am Acad Optom, 76. *Mem:* AAAS; Optical Soc Am; Am Optom Asn (pres, 68-69); Am Acad Optom; Asn Schs & Cols Optom (pres, 53). *Res:* Ocular accommodation and convergence; ocular findings in twins; validity of clinical optometric tests; physiology of concomitant squint; environmental optics; optometric and visual science history. *Mailing Add:* 2615 Windermere Woods Dr Bloomington IN 47401

HOFSTETTER, KENNETH JOHN, b Moline, Ill, Aug 18, 40; m 65; c 2. RADIOANALYTICAL CHEMISTRY, RADIOACTIVE WASTE PROCESSING. *Educ:* Augustana Col, AB, 62; Purdue Univ, PhD(chem), 67. *Prof Exp:* Post-doctoral fel nuclear chem, Tex A&M Univ, 67-69; asst prof chem, Univ Ky, 69-74; radiochem supvr, Allied Gen Nuclear Servs, 74-80; radiochem eng supvr, Gen Pub Utilities-TMI-2, 80-87; res staff chemist, E I Du Pont, Savannah River Lab, 87-89; SR RES SCIENTIST, WESTINGHOUSE SAVANNAH RIVER CO, 89- *Concurrent Pos:* Consult, Int Atomic Energy Agency, 85-87; lectr, Am Chem Soc Speakers Tour, 70-74. *Mem:* Am Chem Soc; Am Nuclear Soc; AAAS. *Res:* Nuclear decay scheme spectroscopy and the study of nuclear reaction mechanisms; development of charged particle induced x-ray fluorescence and other radioanalytical methods; developed real-time monitors for special nuclear materials; developed liquid radiowaste processing techniques; monitoring of radioactive material at environmental levels. *Mailing Add:* Westhouse Savannah River Co Savannah River Lab Bldg 735-A Aiken SC 29808

HOFSTETTER, RONALD HAROLD, b Kitchener, Ont, Jan 25, 39; m 62; c 3. ECOLOGY. *Educ:* McMaster Univ, BSc, 62, MSc, 64; Univ Minn, Minneapolis, PhD(bot), 69. *Prof Exp:* Asst prof, 68-74, ASSOC PROF BIOL, UNIV MIAMI, 74- *Mem:* Am Inst Biol Sci; Ecol Soc Am; Soc Wetland Scientists. *Res:* Ecology of wetlands and forests; plant ecology of subtropical ecosystems particularly the role of fire. *Mailing Add:* Dept Biol Univ Miami Coral Gables FL 33124

HOFSTRA, GERALD, b Heeg, Neth, Dec 11, 38; Can citizen; m 62; c 3. PLANT PHYSIOLOGY, PLANT ECOLOGY. *Educ:* Univ Toronto, BSA, 63, MSA, 64; Simon Fraser Univ, PhD(plant physiol), 67. *Prof Exp:* Nat Res Coun Can overseas fel, 67-68; asst prof, 68-73, assoc prof, 73-81, PROF ENVIRON BIOL, UNIV GUELPH, 81- *Mem:* Can Sci & Christian Affil; Can Soc Plant Physiol. *Res:* Environmental physiology; assessment of economic losses from air pollution; low temperature stress and drought stress in crops and conifer seedlings; stress alleviation. *Mailing Add:* Dept Environ Biol Univ Guelph Guelph ON N1G 2W1 Can

HOFT, RICHARD GIBSON, b Wall Lake, Iowa, Dec 4, 26; m 46; c 4. ELECTRICAL ENGINEERING. *Educ:* Iowa State Univ, BS, 48, PhD(elec eng), 65; Rensselaer Polytech Inst, MEE, 54. *Prof Exp:* Develop engr, Gen Eng Lab, Gen Elec Co, 49-56, mgr elec control, 56-60, mgr converter circuits, Adv Tech Labs, 60-63; assoc prof, 65-68, PROF ELEC ENG, UNIV MO-COLUMBIA, 68- *Mem:* Fel Inst Elec & Electronics Engrs; Sigma Xi; Nat Soc Prof Engrs; Am Soc Elec Engr. *Res:* Applied research and development on power electronics, automatic control, and electrical machines. *Mailing Add:* Dept Elec Eng Univ Mo Columbia MO 65211

HOGABOAM, GEORGE JOSEPH, b Lewiston, Idaho, Nov 13, 19; m 43; c 3. PLANT BREEDING. *Educ:* Univ Idaho, BS, 47; Mich State Univ, MS, 48, PhD(plant breeding), 56. *Prof Exp:* Marketing specialist, Fed Pea & Bean Inspector, Idaho, 46-47; asst farm crops, Mich State Col, 47-48; res agronomist & res leader, USDA & Mich State Univ, 48-84; RETIRED. *Mem:* Am Soc Sugar Beet Technol; Am Soc Agron; Crop Sci Soc Am. *Res:* Breeding improved varieties of sugar beets with special emphasis on resistance to black root, leaf spot diseases and rhizoctonia. *Mailing Add:* 1778 Dogwood Dr Holt MI 48842-1529

HOGAN, ALOYSIUS JOSEPH, JR, b Albany, NY, Sept 18, 28; m 52; c 5. NUCLEAR ENGINEERING, PHYSICAL CHEMISTRY. *Educ:* Col Holy Cross, BS, 50; Rensselaer Polytech Inst, MS, 52. *Prof Exp:* Jr engr, Philadelphia Elec Co, 54-55; proj engr, Atomic Power Develop Assoc, Inc, 55, asst to tech dir, 55-57, asst Fermi coordr, 57-61, field engr, 61-62, proj engr, 62-63; sr engr, Philadelphia Elec Co, 63-75, group leader, 75-76, staff engr, 76-90; RETIRED. *Concurrent Pos:* Mem Adv Comt Radioactive Mat, Hazardous Transp Bd, Commonwealth Pa, 66-73; mem Task Force Reactor Safety Res, Elec Res Coun, 69-73, Atomic Indust Forum, 66-71 & Elec Power Res Inst, 74-80; mem Environ Res Guid Comt, Md Acad Sci, Power Plant Siting Prog, Md Dept Natural Resources, 73-82, vchmn, 77-82; mem, Am Nuclear Soc Standards Comt Environ Stand, 72-84. *Mem:* Am Chem Soc; Am Nuclear Soc; fel Am Inst Chem. *Res:* Research in sodium chemistry and nuclear chemistry using tracer techniques research, design and development of sodium-cooled fast breeder and high temperature gas cooled power reactors; design of boiling and pressurized water reactors; environmental studies in meteorology and radiation monitoring. *Mailing Add:* 1221 Concord Ave Drexel Hill PA 19026

HOGAN, CHRISTOPHER JAMES, b Tehran, Iran, Jan 25, 57; m 82; c 2. PLANT CELL BIOLOGY. *Educ:* Colo State Univ, BS, 80, PhD(bot & cell biol), 86. *Prof Exp:* NSF POSTDOCTORAL FEL CELL BIOL MITOSIS, DEPT MOLECULAR & CELL BIOL, UNIV CALIF, BERKELEY, 87- *Mem:* Am Soc Cell Biol. *Res:* Higher plant cytoskeleton as it relates to division polarity and plant development; mitosis and meiosis, specifically the mechanism of the polypeptides involved in, and the controlling mechanisms that regulate anaphase B (spindle elongation). *Mailing Add:* Dept Molecular & Cell Biol 345 LSA Univ Calif Berkeley CA 94720

HOGAN, CLARENCE LESTER, b Great Falls, Mont, Feb 8, 20; m 46; c 1. APPLIED PHYSICS. *Educ:* Mont State Univ, BS, 42; Lehigh Univ, MS, 47, PhD(physics), 50; Harvard Univ, MA, 54. *Hon Degrees:* DEng, Mont State Univ, 68 & Lehigh Univ, 71; DSc, Worcester Polytech Inst, 69. *Prof Exp:* Res engr, Anaconda Copper Mining Co, 42-43; instr physics, Lehigh Univ, 47-50; mem tech staff, Bell Tel Labs, 50-53, sub dept head, 53; assoc prof appl physics, Harvard Univ, 54-56, Gordon McKay prof, 56-58; vpres & gen mgr, Motorola Inc, 58-68; pres & chief exec off, 68-74, Fairchild Camera & Instrument Corp, vchmn bd, 75-79 consult to pres & dir, 80-85; RETIRED. *Concurrent Pos:* Mem comt undersea warfare, Torpedo Study Group, Nat Acad Sci, 63-64; res coun comt adv, Nat Bur Standards for Radio Standards Lab, 65-67; vis comt, Lehigh Univ, 65- *Mem:* Nat Acad Eng; fel Inst Elec & Electronics Engrs; Am Phys Soc. *Res:* Thermal conductivity of metals; application of ferrites and semiconductors to microwave transmission systems; inventor of the microwave gyrator, circulator, and isolator. *Mailing Add:* 36 Barry Lane Atherton CA 94027

HOGAN, EDWARD L, b Arlington, Mass, July 26, 32; m 61; c 5. NEUROLOGY, NEUROCHEMISTRY. *Educ:* Tufts Univ, BS, 53, MD, 57. *Prof Exp:* Res fel neurol, Harvard Med Sch, 63-64; res fel biochem, Sch Med, Tufts Univ, 64-66; from asst prof to assoc prof neurol & from asst prof to assoc prof biochem, Sch Med, Univ NC, Chapel Hill, 66-73; PROF NEUROL & BIOCHEM & CHMN DEPT NEUROL, MED UNIV SC, 73- *Concurrent Pos:* Nat Inst Neurol Commun Dis & Stroke res grants, 66-71 & 74- *Mem:* Asn Univ Prof Neurol; Am Soc Biol Chemists; Soc Neurosci; Am Acad Neurol; Asn Res Nerv & Ment Dis. *Res:* Pathological biochemistry of developmental disorders of the brain, particularly the process of myelination and genetic disorders of myelination. *Mailing Add:* Dept Neurol Med Sch SC Charleston SC 29425

HOGAN, GARY D, b Hayes, Eng, Nov 11, 49; Can citizen; m 71; c 2. PLANT MINERAL NUTRITION. *Educ:* Univ NB, BSc, 71; Laurentian Univ, MSc, 75; Univ Guelph, PhD(plant physiol), 78. *Prof Exp:* RES SCIENTIST, CAN FORESTRY SERV, 77- *Mem:* Can Soc Plant Physiol. *Res:* The effects of environmental chemical factors on physiological function of native trees through measurement of growth, carbon utilization and nutrient relations. *Mailing Add:* Great Lakes Forestry Center Forestry Can Ont Region Sault St Marie ON P6A 5M7 Can

HOGAN, GUY T, b La Boca, CZ, May 21, 33; US citizen; m 61; c 3. MATHEMATICS. *Educ:* Talladega Col, BA, 57; Univ Chicago, MS, 59; Ohio State Univ, PhD(math), 69; Suffolk Univ Law Sch, JD, 85. *Prof Exp:* Instr natural sci & math, Talladega Col, 59-61; asst prof math, Cent State Univ, Ohio, 61-66; assoc prof, State Univ NY Col Oneonta, 69-73; ASSOC PROF MATH, UNIV MASS, BOSTON, 73- *Mem:* AAAS; Am Math Soc. *Res:* Structure of finite p-groups; existence and construction of social welfare functions. *Mailing Add:* Dept Math Univ Mass Boston MA 02125

HOGAN, JAMES C, b Milledgeville, Ga, Jan 3, 39; m; c 3. CELL BIOLOGY. *Educ:* Brown Univ, PhD(cell biol), 72. *Prof Exp:* Dir, Minority Student Affairs, 83-87, ASSOC PROF ALLIED HEALTH, UNIV CONN HEALTH SCI CTR, 80-; CHIEF, CLIN CHEM & HEMAT, CONN DEPT HEALTH, 87- *Concurrent Pos:* Josiah Macy fel, 78-80; Ford Found fel, 80-81; vis fac fel, Yale Univ, 84-85. *Mem:* Am Soc Cell Biol; AAAS; NY Acad Sci; Am Inst Biol Sci; Sigma Xi. *Mailing Add:* 51 Pool Rd PO Box 146 North Haven CT 06473

HOGAN, JAMES J(OSEPH), b Cleveland, Ohio, Oct 31, 37; m 65; c 4. ELECTRICAL ENGINEERING. *Educ:* Univ Dayton, BEE, 59; Univ Ill, MS, 61, PhD(elec eng), 64. *Prof Exp:* Instr elec eng, Univ Ill, 62-64; eng specialist, 64-69, head electro-optical sensors, 69-76, eng prin & head advan guid technol, 76-80, mgr, Weapon Syst Eng, 80-82, prog dir, Missile FSD prog, 82-85, DIR SYSTS ENGR, LORAL DEFENSE SYSTS, AKRON, 85- *Mem:* Fel Inst Elec & Electronics Engrs. *Res:* Electrically sprayed liquid particles; high resolution radar sensors; area correlation, map-matching and image processing for guidance and tracking systems; missile navigation and control; digital image processor simulation and development; underwater acoustics, transducers and signal processing; antisubmarine warfare system development, smart weapons. *Mailing Add:* Loral Systs Group Defense Systs Div Group 1210 Massillion Rd Akron OH 44315

HOGAN, JAMES JOSEPH, b New Haven, Conn, May 30, 41; m 63; c 3. NUCLEAR CHEMISTRY. *Educ:* Rensselaer Polytech, BS, 62; Univ Chicago, MSc, 66, PhD(chem), 68. *Prof Exp:* Res assoc chem, Univ Chicago, 68; from asst prof to assoc prof, 68-80, PROF CHEM, MCGILL UNIV, 81- *Concurrent Pos:* Vis scientist, Lawrence Berkeley Lab, Berkeley, Calif, 78-79 & UK Atomic Energy, Harwell, 83-90. *Mem:* AAAS; Am Chem Soc; Am Phys Soc; Sigma Xi; Chem Inst Can. *Res:* High energy nuclear fission; nuclear reactions; heavy ion induced reactions; fissionability of heavy nuclides. *Mailing Add:* Dept Chem McGill Univ 801 Sherbrooke St W Montreal PQ H3A 2K6 Can

HOGAN, JOHN PAUL, b Lowes, Ky, Aug 7, 19; m 43; c 3. POLYMER CHEMISTRY, CATALYSIS. *Educ:* Murray State Col, BS, 42. *Hon Degrees:* DSc, Murray State Col, 71. *Prof Exp:* Teacher pub sch, 42-43; instr physics, Okla State Univ, 43-44; res chemist, Phillips Petrol Co, 44-54, group leader to sr group leader, 54-61, sect mgr catalysis, 61-77, sr res assoc, 77-85; RETIRED. *Concurrent Pos:* Consult, 85- *Honors & Awards:* Nat Award, Am Chem Soc, 69; Pioneer Chemist Award, Am Inst Chemists, 72, Soc Plastics Eng, 81; Perkin Medal, Soc Chem Indust, 87. *Mem:* Am Chem Soc; Am Inst Chemists. *Res:* Isomerization; alkylation; Fischer-Tropsch synthesis; polymer chemistry; catalysis. *Mailing Add:* 3908 Lester Ave Bartlesville OK 74006-7111

HOGAN, JOHN THOMAS, b East St Louis, Ill, Aug 18, 41; m 67; c 2. PLASMA PHYSICS, NUCLEAR FUSION. *Educ:* St Louis Univ, BS, 62; Northwestern Univ, PhD(mech eng), 67. *Prof Exp:* Vis prof mech, State Univ NY, Stony Brook, 67-68; res assoc magneto-fluid dynamics, Courant Inst NY Univ, 68-70; RES SCIENTIST & GROUP LEADER NUCLEAR FUSION, OAK RIDGE NAT LAB, 70- *Honors & Awards:* Fel, Am Phys Soc; Distinguished Assoc, Dept Energy. *Mem:* AAAS; Am Phys Soc. *Res:* Controlled nuclear fusion by magnetic confinement. *Mailing Add:* 103 Pomona Rd Oak Ridge TN 37830

HOGAN, JOHN WESLEY, b Monterville, WVa, Jan 20, 36; m 78; c 4. MATHEMATICS. *Educ:* Berea Col, BA, 57; Univ Wis, Madison, MS, 59; Va Polytech Inst, PhD(math), 69. *Prof Exp:* Instr math, Berea Col, 59-60; instr math, WVa Univ, 62-67; instr math, Va Polytech Inst, 67-69; assoc prof, 69-72, chmn dept, 72-77, PROF MATH, MARSHALL UNIV, 72- *Mem:* Math Asn Am; Am Math Soc. *Res:* Algebraic and topological theory of semigroups. *Mailing Add:* Dept Math Marshall Univ Huntington WV 25701

HOGAN, JOSEPH C(HARLES), b St Louis, Mo, May 26, 22; m 44; c 8. ELECTRICAL ENGINEERING. *Educ:* Wash Univ, BS, 43; Univ Mo, MS, 49; Univ Wis, PhD(elec eng), 53. *Prof Exp:* Elec engr, Knapp-Monarch Co, 45-46 & McQuay-Norris Co, 46-47; instr elec eng, Univ NDak, 47; from instr to prof, Univ Mo, 47-67, dean eng, 61-67; dean eng, 67-81, prof elec eng, 67-87, EMER DEAN ENG, UNIV NOTRE DAME, 81-, EMER PROF ELEC ENG, 87- *Concurrent Pos:* Instr, Univ Wis, 52-53; engr, Commonwealth Assoc, Inc, 53-54; consult, Syst Analyzer Corp, 54, Henry Elec Co, 57, Cent Elec Power Corp, 59, Aerojet-Gen Corp, 61-65, North Cent Asn, 67- & St Louis Univ, 68-69; bd dirs, Weldun Int, 72- & TII Corp, 73- *Mem:* Am Soc Eng Educ (pres-elect, 81-82, pres, 82-83); Inst Elec & Electronics Engrs; Nat Soc Prof Engrs; Am Asn Eng Socs Inc; Sigma Xi. *Res:* Analysis of unsymmetrical induction machines by means of derived equivalent circuits; shaded pole induction machines; analysis of power systems, electrical machinery; control systems. *Mailing Add:* 12 Sandhill Crane Hilton Head Island SC 29928

HOGAN, LE MOYNE, b Choudrant, La, Apr 24, 26; m 47; c 4. HORTICULTURE. *Educ:* La State Univ, BS, 53, MS, 57; Univ Md, PhD(hort), 62. *Prof Exp:* Nursery supt, La State Forestry Comn, 53-54; res assoc hort, La State Univ, 54-56; gen mgr, Fred's Greenhouses & Nursery, 56-58; asst hort, Univ Md, 58-59, instr, 59-62; asst horticulturist, Univ Ariz, 62-63, assoc prof & assoc horticulturist, 63-66, prof hort & bot, US Agency Int Develop, Brazil, 66-70, prof hort & chief of party, Univ Ariz, 70-72, horticulturist, 72-81, prof hort, 72-86, dept head, 81-84; dep proj dir, Western Sudan Agr Res Proj, 84-85; dep head, Sultan Qaboos Univ, Oman, 86-87, DEAN, COL AGR, SULTAN QABOOS, UNIV OMAN, 87- *Mem:* Am Soc Hort Sci. *Res:* Nutrition of landscape and floricultural plants; domestication of jojoba (Simmondsia chinensis). *Mailing Add:* Col Agr Sultan Qaboos Univ AC-KHOD PO Box 32484 Sultanate Oman

HOGAN, MICHAEL EDWARD, b Phoenix, Ariz, Dec 7, 51; m 78. BIOPHYSICS. *Educ:* Dartmouth Col, BA, 73; Yale Univ, PhD(biochem), 78. *Prof Exp:* ASST PROF PHYS BIOCHEM, DEPT BIOCHEM SCI, PRINCETON UNIV, 80-; ASSOC PROF BIOTECHNOL, BAYLOR COL MED, 88- *Res:* Structural studies of chromatin and its complexes with toxins; studies of the mechanism by which hydrocarbons induce tumors; physical studies of molecular flexibility; structural origins of eucharyotic gene control. *Mailing Add:* Baylor Col Med 4000 Research Forest Dr The Woodlands TX 77381

HOGAN, PHILIP, b Chicago, Ill, Oct 22, 25; m 69; c 2. ORGANIC CHEMISTRY. *Educ:* St Mary's Col, Minn, BS, 48; Univ Mo, MS, 53; Loyola Univ, PhD(org chem), 59. *Prof Exp:* Asst prof chem, St Mary's Col, 59-61, chmn, 61-63; assoc prof, 63-64, chmn dept, 64-76, PROF CHEM, LEWIS UNIV, 65- *Concurrent Pos:* Res grant, NSF, Kans State Univ, 64; fac fel, Pa State Univ, 67-68, vis prof, 68. *Mem:* Am Chem Soc; NY Acad Sci; The Chem Soc. *Res:* Organic reaction mechanics; physical organic chemistry. *Mailing Add:* Dept Chem Lewis Univ Romeoville IL 60441

HOGAN, WILLIAM, b New York, NY, June 27, 23. PHYSICS. *Educ:* Cath Univ, BS, 46, PhD(physics), 62; Fordham Univ, MS, 52. *Prof Exp:* Teacher parochial sch, NY, 46-48, high schs, NH, 48-52, NY, 52-53 & Mich, 53-56; instr physics, De La Salle Col, 56-62; asst prof, St Mary's Col, Minn, 62-64; from asst prof to assoc prof physics, Manhattan Col, 64-, dir comput ctr, 69-; RETIRED. *Mem:* Am Phys Soc; Am Asn Physics Teachers; Asn Comput Mach. *Res:* Nuclear particle physics, using liquid scintillator detectors. *Mailing Add:* 4415 Post Rd Riverdale NY 10471

HOGAN, WILLIAM ALFRED, theoretical physics, for more information see previous edition

HOGAN, WILLIAM JOHN, b Omaha, Nebr, Oct 8, 40; m 62; c 3. APPLIED PHYSICS. *Educ:* Calif Inst Technol, BS, 62; Princeton Univ, MA, 64, PhD(physics), 66. *Prof Exp:* Res & teaching assoc physics, Princeton Univ, 62-66; sr scientist, Lawrence Livermore Nat Lab, 66-73, assoc div leader, Spec Proj Div, 73-77, prog leader, Liquefied Gaseous Fuels Safety Prog, 78-83, leader, 83-86, DEP PROG LEADER, INERTIAL CONFINEMENT FUSION, LAWRENCE LIVERMORE NAT LAB, 86- *Concurrent Pos:* Mem adv bd, US Air Force Sci, 72-78 & Gas Res Inst, 80-83. *Mem:* AAAS; Am Nuclear Soc; Am Phys Soc. *Res:* Theoretical work on the physics of nuclear explosions; interaction of radiation and matter at large temperatures and densities; non-equilibrium processes; system studies and operations research on nuclear weapons systems, heavy gas dispersion, combustion and detonation phenomena; inertial confinement fusion physics; fusion reactor design; fusion program management. *Mailing Add:* Lawrence Livermore Nat Lab P O Box 5508 Livermore CA 94550

HOGARTH, JACKE EDWIN, b Huntsville, Ont, Sept 22, 23; m 44; c 3. APPLIED MATHEMATICS, COSMOLOGY. *Educ:* Univ Toronto, BASc, 49, MA, 50; Univ London, PhD(theoret physics), 53. *Prof Exp:* Defence Res Sci Officer, Defence Res Bd, Dept Nat Defence, Can, 50-59; assoc dean, Sch Grad Studies & Res, Queens Univ, Ont, 70-76, Abitibi prof eng math, 59-86; RETIRED. *Concurrent Pos:* Math consult, Ont Dept Educ, 66-68. *Res:* Relativity and cosmology; algebraic coding theory. *Mailing Add:* 522 King W Kingston ON K7L 2X8 Can

HOGBEN, CHARLES ADRIAN MICHAEL, b Eng, Nov 12, 21; nat US; m 48; c 2. PHYSIOLOGY. *Educ:* Univ Wis, BS, 41, MD, 43; Univ Minn, PhD(med), 50. *Prof Exp:* Med officer, Sect Kidney & Electrolyte Metab, Nat Heart Inst, Md, 51-57; prof physiol & exec off dept, George Washington Univ, 57-61; head dept, 61-73, PROF PHYSIOL, COL MED, UNIV IOWA, 61- *Concurrent Pos:* Fel med sci, Nat Res Coun, Copenhagen, 50-51; mem res career award comt, NIH, 58-62, med study res training prog, 61-63, physiol study sect, 63-67; mem steering comt, Gastroenterol Res Group, 60-64; trustee, Mt Desert Island Biol Lab, 60-66, mem exec comt, 64-66; mem comt to eval appln for NSF postdoctoral fels, Nat Acad Sci-Nat Res Coun, 61-63 & sr postdoctoral fels, 65, mem comt on NIH training, 72-75. *Mem:* Am Physiol Soc; Am Soc Pharmacol & Exp Therapeut; Biophys Soc; Am Gastroenterol Asn. *Res:* Mechanisms of gastric hydrochloric acid secretion; gastrointestinal physiology; membrane transport; kidney and electrolyte physiology. *Mailing Add:* Box 243 RFD 1 Ellsworth ME 04605

HOGBEN, DAVID, b Cape Town, SAfrica, Feb 28, 29; US citizen; div. STATISTICS, SOFTWARE SYSTEMS. *Educ:* Univ Minn, BA, 51; Rutgers Univ, MS, 58, PhD(statist), 63. *Prof Exp:* Actuarial asst, NAm Life & Casualty Co, 52-53; qual control analyst, Ford Motor Co, 53-55; qual control develop engr, Western Elec Co, Inc, 55-59; asst statist, Rutgers Univ, 59-61, from instr to asst prof, 61-65; mathematician, Nat Bur Standards, 65-81; CONSULT, 81- *Concurrent Pos:* Consult, US Naval Res Supply & Develop Facil, NJ, 65; lectr, Univ Md, 66-67. *Honors & Awards:* Silver Med, Dept Com. *Mem:* Fel Am Statist Asn. *Res:* Statistical research in the computing physical sciences. *Mailing Add:* 10407 46th Ave No 304A Beltsville MD 20705-2416

HOGBEN, LESLIE, b Washington, DC, Feb 10, 52; m 78. ABSTRACT ALGEBRA. *Educ:* Swarthmore Col, BA, 74; Yale Univ, PhD(math), 78. *Prof Exp:* Instr, 78-80, ASSOC PROF MATH, IOWA STATE UNIV, 80- *Mem:* Am Math Soc. *Res:* Non-associative algebras, especially Jordan algebras; radical classes. *Mailing Add:* Dept Math 445 Carver Iowa State Univ Ames IA 50011

HOGE, HAROLD JAMES, b Springville, Iowa, Mar 15, 07; m 34; c 1. THERMAL SCIENCES, FLUID MECHANICS. *Educ:* Univ Iowa, AB, 32; Yale Univ, PhD(physics), 35. *Prof Exp:* Asst physics, Yale Univ, 32-35; res physicist, Nat Bur Standards, 35-50 & Leeds & Northrup Co, 50-55; head, Thermodyn Lab, US Army Natick Labs, 55-74; VIS RES ASSOC, TUFTS UNIV, 74- *Mem:* Am Phys Soc. *Res:* Thermodynamics; thermal properties; temperature measurement. *Mailing Add:* Dept Mech Eng Tufts Univ Medford MA 02155

HOGE, HARRY PORTER, b Coshocton, Ohio, Jan 12, 36; m 59; c 2. SEDIMENTARY PETROLOGY. *Educ:* Ohio Univ, BSc, 61, MSc, 63; Univ NMex, PhD(geol), 71. *Prof Exp:* Explor geologist, Texaco, Inc, 63-66; instr geol, Morehead State Univ, 66-69; asst field geol, Univ NMex, 69-71; from asst prof to prof geol, Eastern Ky Univ, 71-79, chmn geol, 78-79; prof & chmn geol, Stephen F Austin State Univ, 79-85; PROF GEOL, LEE COL, HUNTSVILLE, TEX, 86- *Concurrent Pos:* Field geologist, Ky Geol Surv-US Geol Surv, 67-69 & 71-78; secy-treas, Keystone Geol Consults Inc, 75-78; vis prof geol, San Diego State Univ, 86-88. *Mem:* Soc Econ Paleontologists & Mineralogists; Am Asn Petrol Geologists; Sigma Xi. *Res:* Sedimentology and stratigraphy of Cenozoic deposits in alluvial and paralic environments; petrology and stratigraphy of Upper Paleozoic deposits in Eastern and Southeastern United States. *Mailing Add:* 14000 Ella Blvd Apt 1703 Houston TX 77014

HOGENBOOM, DAVID L, b Pikeville, Ky, Nov 29, 36; m 60; c 2. HIGH PRESSURE PHYSICS, PLANETARY SCIENCE. *Educ:* Col Wooster, BA, 57; Pa State Univ, MS, 61, PhD(physics), 63. *Prof Exp:* Nat Acad Sci-Nat Res Coun res assoc & physicist, Nat Bur Standards, 63-65; from asst prof to assoc prof, 65-81, PROF PHYSICS, LAFAYETTE COL, 81- *Mem:* Am Phys Soc; Am Asn Physics Teachers; Sigma Xi; AAAS. *Res:* Phase transitions at elevated pressures; density and phase diagrams of solutions to 4000 bars and 130K that are important constituents of the icy moons of the outer planets. *Mailing Add:* 165 Parker Ave Easton PA 18042

HOGEN-ESCH, THIEO ELTJO, b Terneuzen, Neth, Feb 22, 36; m 66; c 3. ORGANIC POLYMER CHEMISTRY, PHYSICAL ORGANIC CHEMISTRY. *Educ:* State Univ Leiden, BSc, 58, MCSc, 61, PhD(phys org chem), 67. *Prof Exp:* Adv technologist, Shell Chem Co, NV, 67-68; res assoc polymer chem, State Univ NY Col Forestry, Syracuse, 68-69, asst prof, 69-70; asst prof polymer chem, Univ Fla, 70-75, assoc prof, 75-79, prof, 79-88; PROF, DEPT CHEM & LOKER HYDROCARBON INST, UNIV SOUTHERN CALIF, LOS ANGELES, 88- *Concurrent Pos:* Vis prof, Univ Louvain, Belgium, 77, Univ Pierre Et Marie Curie, Paris, 77 & Johannes Gutenberg Univ, Mainz, WGer, 82. *Mem:* Sigma Xi; Am Chem Soc. *Res:* Mechanisms of stereo regular vinyl polymerization; interactions of ions and macromolecules; synthesis of water-soluble polymers; synthesis of macrocyclic and star polymers. *Mailing Add:* Dept Chem & Loker Hydrocarbon Inst Univ Southern Calif Los Angeles CA 90007

HOGENKAMP, HENRICUS PETRUS C, b Doesburg, Netherlands, Dec 20, 25; US citizen; m 53; c 3. BIOCHEMISTRY. *Educ:* Univ BC, BSA, 57, MSc, 58; Univ Calif, Berkeley, PhD(biochem), 61. *Prof Exp:* Jr res biochemist, Univ Calif, 61-62; assoc scientist, Fisheries Res Bd Can, 62-63; from asst prof to prof chem, Univ Iowa, 63-76; PROF & HEAD DEPT BIOCHEM, UNIV MINN, MINNEAPOLIS, 76- *Concurrent Pos:* Vis prof, John Curtin Sch Med Res, Australian Nat Univ, 66-67, Pilipps Univ, Marburg, Ger, 86-87; guest scientist, Los Alamos Sci Labs, 74-75; fel John Simon Guggenheim Mem Found, 74-75; mem ed bd, J Biol Chem, 75-80; Alexander von Humboldt-Stiftung Award, 86-87. *Mem:* Am Soc Biol Chem; Am Chem Soc. *Res:* Chemistry and biochemistry of cobamide coenzymes. *Mailing Add:* Dept Biochem Univ Minn Minneapolis MN 55455

HOGG, ALAN MITCHELL, b Edinburgh, Scotland, Aug 8, 36; m 63; c 3. MASS SPECTROMETRY. *Educ:* Univ Edinburgh, BSc, 58; Univ Alta, PhD(chem), 64. *Prof Exp:* Nat Res Coun Can fel chem, 64-65, res assoc mass spectrometry, 65-69, prof officer, 69-76, fac serv officer mass spectrometry, 76-80, FAC SERV OFFICER CHEM, UNIV ALTA, 80- *Mem:* Am Soc Mass Spectrometry; Chem Inst Can. *Res:* Analytical mass spectrometry. *Mailing Add:* Dept Chem Univ Alta Edmonton AB T6G 2G2 Can

HOGG, DAVID CLARENCE, b Sask, Can, Sept 5, 21; m 47; c 2. RADIO PHYSICS, ATMOSPHERIC SCIENCES. *Educ:* Univ Western Ont, BSc, 49; McGill Univ, MSc, 50, PhD, 53. *Prof Exp:* Mem tech staff, radio res, Bell Labs, 53-66, head antenna & propagation dept, 66-77; chief radio meteorol, Wave Propagation Lab, Environ Res Lab, Nat Oceanic & Atmospheric Admin, 77-86; sci assoc, Colo Inst Res Environ Sci, 86-89, LECTR, UNIV COLO, 89- *Concurrent Pos:* Distinguished lectr, Antenna & Prop Soc, IEEE. *Honors & Awards:* Silver Medal Award, US Dept Com. *Mem:* Nat Acad Eng; AAAS; fel Inst Elec & Electronics Engrs; Antenna & Prop Soc; Geosci & Remote Sensing Soc. *Res:* Radio propagation; microwave diffraction; electronics, research and design of remote sensors. *Mailing Add:* 4978 Carter Ct Boulder CO 80301-3804

HOGG, DAVID EDWARD, b Newmarket, Ont, Jan 18, 36; m 59; c 2. ASTRONOMY. *Educ:* Queen's Univ, Ont, BA, 57, MSc, 59; Univ Toronto, PhD(radio astron), 62. *Prof Exp:* Asst astronr, 61-64, assoc astronr, 64-69, asst dir, 70-74, SCIENTIST NAT RADIO ASTRON OBSERV, 69-, ASSOC DIR, 74- *Mem:* Am Astron Soc; Royal Astron Soc Can; Can Asn Physicists; Int Astron Union. *Res:* Radio emission from supernova remnants; the origin of cosmic rays; structure of radio sources by means of radio interferometry. *Mailing Add:* Nat Radio Astron Observ Edgemont Rd Charlottesville VA 22903

HOGG, GARY LYNN, b Olney, Ill, July 3, 46; m 66; c 2. INDUSTRIAL ENGINEERING, OPERATIONS RESEARCH. *Educ:* Tex A&M Univ, BS, 68; Univ Tex, Austin, MS, 70, PhD(mech eng), 71. *Prof Exp:* Teaching asst, Univ Tex, 69-71; asst prof, Univ Ill, 71-76; assoc prof, 76-81, PROF INDUST ENG, TEX A&M UNIV, 82-, ASSOC DEPT HEAD, 90- *Concurrent Pos:* Dir simulation res, Nat Clearinghouse Criminal Justice Planning, 72-76; eng consult, pres Lodestane II Inc; dir, Opers Res Div, Inst Indust Engrs. *Mem:* Inst Indust Engrs. *Res:* Mathematical programming; simulation languages and techniques; power generation systems; production control systems and manufacturing. *Mailing Add:* Dept Indust Eng Tex A&M Univ College Station TX 77843

HOGG, HELEN (BATTLES) SAWYER, b Lowell, Mass, Aug 1, 05; m 30; c 3. ASTRONOMY. *Educ:* Mt Holyoke Col, AB, 26; Radcliffe Col, AM, 28, PhD(astron), 31. *Hon Degrees:* DSc, Mt Holyoke Col, 58, Univ Waterloo, 62, McMaster Univ, 76 & Univ Toronto, 77; LittD, St Mary's Univ, 81. *Prof Exp:* Instr, Smith Col, 27, Mt Holyoke Col, 30-31 & Dominion Astrophys Observ, BC, 31-34; asst, David Dunlap Observ, Univ Toronto, 35-37, res assoc, 38-41, res assoc, Univ, 41-51, from asst prof to prof astron, 51-76, EMER PROF ASTRON, UNIV TORONTO, 76- *Concurrent Pos:* Asst prof & actg chmn dept, Mt Holyoke Col, 40-41; astron columnist, Toronto Daily Star, 51-81; prog dir astron, NSF, Washington, DC, 55-56; dir, Bell Tel Co, Can, 68-78. *Honors & Awards:* Cannon Prize, Am Astron Soc, 50; Rittenhouse Medal, Rittenhouse Astron Soc, 67; Centennial Medal Can, 67. *Mem:* Can Astron Soc (first pres, 71-72); Am Asn Variable Stars Observ (pres, 39-41); Royal Astron Soc Can (pres, 57-59); fel Royal Soc Can (pres sect III, 60-61); Royal Can Inst (pres, 64-65); Sigma Xi. *Res:* Globular star clusters; variable stars. *Mailing Add:* 98 Richmond St Richmond Hill ON L4C 3Y4 Can

HOGG, JAMES CAMERON, PATHOLOGY, PHYSIOLOGY. *Educ:* McGill Univ, Montreal, PhD(exp med), 69. *Prof Exp:* PROF PATH & PHYSIOL, UNIV BC, 77- *Mailing Add:* Dir Pulmonary Res St Paul's Hosp 1081 Burrard St Vancouver BC V6Z 1Y6 Can

HOGG, JAMES FELTER, b Denison, Tex, Dec 11, 18; m 45; c 2. ANALYTICAL CHEMISTRY. *Educ:* Rice Inst, BA, 41; Univ Tex, MA, 42, PhD(biol org chem), 48. *Prof Exp:* Instr chem, Univ Tex, 41-44 & 46-47; res chemist & biochemist, Chas Pfizer & Co, Inc, NY, 44-46; from instr to asst prof biol chem, Univ Mich, 47-64; assoc prof, Queen's Col, 64-67, chmn dept, 68-72, prof chem, 67-89, prof biochem, 80-89, EMER PROF BIOCHEM, QUEENS COL, 89- *Concurrent Pos:* Exec officer biochem, City Univ New York, 66-70; chmn educ comt, Div Biol Chem, Am Chem Soc, 73-76. *Mem:* Am Chem Soc; Am Soc Biochem & Molecular Biol; Biochem Soc Eng. *Res:* Adsorption methods; biochemistry of microorganisms, especially Protozoa; hexose transport; enzyme cytology; metabolism in peroxisomes and autophagic vacuoles; gluconeogenesis. *Mailing Add:* Dept Chem & Biochem Queens Col Flushing NY 11367

HOGG, JOHN LESLIE, b Mangum, Okla, Aug 18, 48; m 69. BIO-ORGANIC CHEMISTRY. *Educ:* Southwestern State Col, BS, 70; Univ Kans, PhD(chem), 74. *Prof Exp:* Res assoc chem, Dept Biochem, Brandeis Univ, 74-75; asst prof, 75-81, ASSOC PROF CHEM, TEX A&M UNIV, 81- *Mem:* Am Chem Soc. *Res:* Bioorganic and physical organic chemistry; mechanism of enzyme action; isotope effects; solution catalysis. *Mailing Add:* Dept Chem Tex A&M Univ College Station TX 77843-5000

HOGG, RICHARD, b Redcar, Eng, Jan 6, 38; m. MINERAL PROCESSING, FINE PARTICLE TECHNOLOGY. *Educ:* Univ Leeds, BSc, 63; Univ Calif, Berkeley, MS, 65, PhD(mineral technol), 70. *Prof Exp:* Asst specialist mineral technol, Univ Calif, Berkeley, 66-69; from asst prof to assoc prof, 69-79, PROF MINERAL PROCESSING, PA STATE UNIV, 79- *Concurrent Pos:* Mem comt, Comminution and Energy Consumption, Nat Acad Sci, 78-80. *Mem:* Am Inst Mining, Metall & Petrol Engrs; Fine Particle Soc; Sigma Xi. *Res:* Fundamental basis of mineral processing operations; colloid and surface chemistry; mixing, segregation and flow of particulate solid materials. *Mailing Add:* 1232 S Garner St State College PA 16801

HOGG, ROBERT VINCENT, JR, b Hannibal, Mo, Nov 8, 24; m 56; c 4. STATISTICS. *Educ:* Univ Ill, AB, 47; Univ Iowa, MS, 48, PhD(math), 50. *Prof Exp:* From instr to assoc prof, 47-62, chmn dept, 65-83, PROF MATH, UNIV IOWA, 62-, PROF STATIST, 83- *Concurrent Pos:* NIH grants, 67-69 & 75-78, NSF grants, 69-74. *Mem:* Math Asn Am; fel Inst Math Statist; fel Am Statist Asn (pres, 88); Sigma Xi; Int Statist Inst. *Res:* Mathematical statistics. *Mailing Add:* Dept Statist Univ Iowa Iowa City IA 52242

HOGG, ROBERT W, b New York, NY, June 30, 38; m 60; c 2. MICROBIOLOGY, GENETICS. *Educ:* Univ BC, BSA, 60, MSA, 62; Univ Ill, Urbana, PhD(dairy biochem), 67. *Prof Exp:* Scholar, Univ Calif, Santa Barbara, 67-68; asst prof, 69-73, assoc prof microbiol, Sch Med, 73-, PROF, DEPT MOLECULAR BIOL & MICROBIOL, CASE WESTERN RESERVE UNIV. *Concurrent Pos:* USPHS res grant, 69, 72, 78, 84, res career develop award, 70, NSF, 90. *Mem:* Am Soc Biol Chem; Am Soc Microbiol; Protein Soc. *Res:* Genetic and molecular studies on the process of active transport of small molecules into microbial cells. *Mailing Add:* Dept Molecular Biol & Microbiol Case Western Reserve Univ Cleveland OH 44106

HOGGAN, DANIEL HUNTER, b Lorenzo, Idaho, Sept 25, 29; m 60; c 4. CIVIL ENGINEERING, HYDROLOGY & WATER RESOURCES. *Educ:* Utah State Univ, BS, 52, PhD(civil eng), 69; Stanford Univ, MS, 53. *Prof Exp:* Asst civil engr, Fluor Corp Ltd, 53-54; detailer & designer, US Steel Corp, 56-60; asst proj engr, US Atomic Energy Comn, 60-61; struct engr, Ralph M Parsons Co, 61-63, proj engr, 63-65; staff specialist water resources planning, US Water Resources Coun, 67-68; assoc prof civil eng, 68-75, asst dir, Utah Water Res Lab, 71-76, PROF CIVIL & ENVIRON ENG, UTAH STATE UNIV, 75- *Concurrent Pos:* Consult, Pac Northwest River Basins Comn, 69, US Water Resources Coun, 70-71, Sergant, Hauskins & Beckwith, 88, Indo-Am Eng Co, 89, EG&G Idaho, Inc, 90-91; Off Water Resources Res grant, 71-73; advan technol specialist, US Energy Res & Develop Admin, 76-77; hydraul engr, Hydrol Eng Ctr, US Army CEngr, 84-86, Sacramento Dist, 89-90. *Mem:* Am Soc Civil Engrs; Am Water Resources Asn; Am Soc Engr Educ; Am Inst Hydraul. *Res:* State organizational patterns for water planning; state and local government capability to finance water development; effectiveness of river basin water planning groups; legal and institutional aspects of low head hydroelectric energy development; urban drainage and flood control; computer simulation models of hydrologic systems; real time data systems. *Mailing Add:* 955 Foothill Dr Providence UT 84332

HOGGAN, MALCOLM DAVID, b Salt Lake City, Utah, Feb 28, 30; m 50; c 2. MICROBIOLOGY. *Educ:* Univ Utah, BS, 52, MS, 53; Johns Hopkins Univ, ScD(microbiol), 59. *Prof Exp:* Asst chief bact, Br, Sixth Army Area Med Lab, Ft Baker, Calif, 53-56; virologist, Histobact Sect, Armed Forces Inst Path, 59-63; VIROLOGIST, LAB VIRAL DIS, NAT INST ALLERGY & INFECTIOUS DIS, 63- *Mem:* Am Soc Microbiol; Am Asn Immunol; Electron Micros Soc Am. *Res:* Tumor virology and immunology; electron microscopy; parvoviruses. *Mailing Add:* Nat Inst Allergy & Infectious Dis Bldg 4 Rm 10 NIH Bethesda MD 20892

HOGGAN, ROGER D, b Ogden, Utah, Jan 3, 40; m 65; c 2. GEOLOGY. *Educ:* Weber State Col, BS, 66; Brigham Young Univ, MS, 69, PhD(paleont), 71. *Prof Exp:* INSTR GEOL, RICKS COL, 71- *Concurrent Pos:* Mem staff, US Geol Surv, 74-81, geol dept chmn, 81-86, chmn, Nat Sci Div, 86- *Mem:* Nat Asn Geol Teachers; Paleont Soc; assoc Sigma Xi. *Res:* Geologic mapping in eastern Idaho; paleoecologic and stratigraphic investigations of paleozoic rock units in eastern Idaho. *Mailing Add:* Ricks Col Rexburg ID 83440

HOGGARD, PATRICK EARLE, b Los Angeles, Calif, July 5, 44; m 67; c 2. PHYSICAL INORGANIC CHEMISTRY. *Educ:* Univ Calif, Berkeley, BS, 65; Wash State Univ, PhD(chem), 70. *Prof Exp:* Fel chem, Inst Phys Chem, Frankfurt, Ger, 70-73 & Univ BC, 73-75; asst prof chem, Polytech Inst NY, 75-81; asst prof, 81-86, ASSOC PROF CHEM, NDAK STATE UNIV, 86- *Mem:* Am Chem Soc. *Res:* Sharp line luminescence and excitation spectroscopy of transition metal complexes. *Mailing Add:* Chem Dept NDak State Univ Fargo ND 58105

HOGLE, DONALD HUGH, b Ticonderoga, NY, Dec 11, 24; m 45; c 3. PHYSICAL CHEMISTRY, DIRECT PROCESS NON WOVENS. *Educ:* St Lawrence Univ, BS, 49, MS, 50; Iowa State Univ, PhD(phys org chem), 54. *Prof Exp:* Control chemist, Int Paper Co, 46; lab asst, Eastman Kodak Co, 46-47; res engr, Res Lab, Westinghouse Elec Corp, 54-61; supvr elec prod div, Minn Mining & Mfg Co, 61-68, supvr, 70-74, mgr, Bldg Serv & Cleaning Prod Div, (3M Co,) 75-81, staff scientist, Non Wovens Technol Ctr, 82-86, staff scientist, Life Sci Sector Lab 3M Co, 86-89; CONSULT, 89- *Mem:* Am Chem Soc. *Res:* Relationship of chemical structure to dielectric properties of liquids, polymers and inorganic insulators; dielectrics; semiconductors; synthetic fibers; non wovens. *Mailing Add:* 21509 Lofton Ave N Scandia MN 55073

HOGNESS, DAVID SWENSON, b Oakland, Calif, Nov 17, 25; m 48; c 2. MOLECULAR GENETICS. *Educ:* Calif Inst Technol, BS, 49, PhD(biol & chem), 52. *Hon Degrees:* Univ Crete, Greece & Univ Basel, Switz, 86. *Prof Exp:* Nat Res Coun Lilly fel, Inst Pasteur, Paris, France, 52-54; Nat Found fel, NY Univ, 54-55; from instr to asst prof microbiol, Sch Med, Wash Univ, St Louis, Mo, 55-59; from asst prof to assoc prof, 59-66, chmn, 86-89, PROF, DEPT BIOCHEM, STANFORD UNIV, CALIF, 66-, PROF DEVELOP BIOL & BIOCHEM, 89- *Concurrent Pos:* Guggenheim Found fel, 68-69; co-chmn, Gordon Conf, Nucleic Acids, 70, Develop Biol, 81; mem, Nat Cancer Adv Bd, 77-78; Harvey Soc Lectr, 79; mem, Alan T Waterman Award Comt, NSF, 82; Rudy J & Daphne Donohue Munzer prof, 91. *Honors & Awards:* Newcomb Cleveland Prize, AAAS, 66 & 88; Genetics Soc Am Medal, 84; Sonneborn Lectr, Ind Univ, 85. *Mem:* Nat Acad Sci; Am Acad Arts & Sci. *Res:* Genetic organization and regulation of eukaryotic chromosomes. *Mailing Add:* Dept Develop Biochem Sch Med Stanford Univ Stanford CA 94305-5427

HOGNESS, JOHN RUSTEN, b Oakland, Calif, June 27, 22. ACADEMIC ADMINISTRATION. *Educ:* Univ Chicago, BS, 43, MD, 46; Am Bd Internal Med, dipl, 54. *Hon Degrees:* DSc, Med Col Ohio, 72 & Haverford Col, 73; LLD, George Washington Univ, 73; DLitt, Thomas Jefferson Univ, 80. *Prof Exp:* From asst prof to prof med, Univ Wash, 58-69, exec vpres, 69-70, dir, Health Sci Ctr, 70-71; pres, Inst Med, Nat Acad Sci, 71-74; pres, Univ Wash, 74-79; pres acad admin, Asn Acad Health Sci, 79-80; RETIRED. *Concurrent Pos:* From asst dean to dean, Sch Med, Univ Wash, 59-69; mem nat adv coun regional med progs, USPHS, 66-70; trustee, China Med Bd, NY, 67-; mem adv comt exempt orgn, Internal Revenue Serv, 69-71; mem task force medicaid & related progs, HEW, 69-70, adv comt to dir, NIH & adv comt to adminr, Health Sci & Ment Health Admin, 70-71; vchmn adv comt environ sci, NSF, 70-71; mem, Nat Cancer Adv Bd, Nat Cancer Inst, 72-76; chmn med injury compensation study comt, Inst Med, Nat Acad Sci, 76-78; mem selection comt, Rockefeller Pub Serv Award, 76-81 & Tyle Ecol Award, 76-79; mem, Nat Sci Bd, 76-82; mem, Coun Biol Sci & Pritzker Sch Med, Univ Chicago, 77-89; mem, Nat Coun Health Care Technol, 79- *Honors & Awards:* Cartwright Medal, Columbia Univ, 78. *Mem:* Inst Med-Nat Acad Sci; fel Am Acad Arts & Sci; Am Soc Internal Med; Archeol Inst Am; Asn Am Physicians; Am Col Physicians. *Res:* Endocrinology. *Mailing Add:* 4919 102nd Lane NE Kirkland WA 98033

HOGNESTAD, EIVIND, b Time, Norway, July 17, 21; nat US; m 49; c 3. ENGINEERING MECHANICS. *Educ:* Norweg Inst Technol, MSc, 47, DSc, 52; Univ Ill, MS, 49. *Prof Exp:* Asst concrete technol, Norweg Inst Technol, 46-47; reinforced concrete technol, Univ Ill, 47-49, res assoc, 49-50, res asst prof, 50-52, res assoc prof, 52-53; mgr struct develop sect, 53-66, dir develop, 66-69, dir eng res, 69-77, dir technol & sci develop, Portland Cement Asn, 77-87; PRIN CONSULT, CONSTRUCT TECH LAB, 87- *Mem:* Nat Acad Eng; Am Soc Civil Engrs; Am Soc Testing & Mat; Am Concrete Inst; Int Asn Bridge & Struct Engrs. *Res:* Concrete technology; reinforced concrete members and structures; structural design. *Mailing Add:* Construct Technol Lab 5420 Old Orchard Rd Skokie IL 60077

HOGUE, CAROL JANE ROWLAND, b Springfield, Mo, Dec 11, 45; m 66; c 1. EPIDEMIOLOGY, BIOSTATISTICS. *Educ:* William Jewell Col, AB, 66; Univ NC, MPH, 71, PhD(epidemiol), 73. *Prof Exp:* Res assoc epidemiol, Univ NC, 73-74, asst prof biostatist, 74-77; from asst prof to assoc prof biomet, Univ Ark Med Sci, 77-82; chief, Pregnancy Epidemiol Br, 82-88, DIR, DIV REPRODUCTIVE HEALTH CTRS DIS CONTROL, ATLANTA, 88- *Concurrent Pos:* Prin investr grant, Ford & Rockefeller Found, 72-73; pos coordr, Sch Pub Health & Area Health Educ Ctrs, 75-76; res fel, Carolina Pop Ctr, 76-77; res assoc, Health Serv Res Ctr, Univ NC, 76-77; consult, Int Fertil Res Prog, 73-78, Div Biomet, Nat Ctr Toxicol Res, 78-80 & Bur Med Devices, FDA, 78-81; scholar in residence, Ctrs Dis Control, 81; prin investr grant, Nat Inst Child Health & Human Develop, 80-82. *Honors & Awards:* First Ann Award Excellence, Statist & Epidemiol Sect, NC Pub Health Asn, 73. *Mem:* Int Epidemiol Asn; Soc Epidemiol Res (pres, 87-88); Pop Asn Am; Am Pub Health Asn; Sigma Xi. *Res:* Long-term complications of induced abortion; methodology of epidemiologic studies; epidemiology of reproduction. *Mailing Add:* Div Reproductive Health Ctrs Dis Control Atlanta GA 30333

HOGUE, CHARLES LEONARD, b Caruthersville, Mo, Feb 4, 35; m 56; c 3. ENTOMOLOGY, TAXONOMY. *Educ:* Univ Calif, Berkeley, BS, 57, Los Angeles, PhD(zool), 62. *Prof Exp:* Sci illustrator entom, Univ Calif, Los Angeles, 57-61, jr res entomologist, 62; CUR ENTOM, LOS ANGELES COUNTY MUS, 62- *Concurrent Pos:* Adj prof, Univ Southern Calif, 62-; mem, Lepidop Found, 63-; fel, Int Cong Entom, London, 64 & Hamburg, 84. *Mem:* Soc Syst Zool; Entom Soc Am; Asn Trop Biol. *Res:* Blephariceridae of the world; primitive aquatic Diptera of New World; insect ecology and evolution; taxonomy and biology of primitive aquatic Diptera; entomological illustration; ethnoentomology. *Mailing Add:* Natural Hist Mus Los Angeles County 900 Exposition Blvd Los Angeles CA 90007-4057

HOGUE, DOUGLAS EMERSON, b Holdrege, Nebr, Aug 8, 31; m 55; c 2. ANIMAL NUTRITION. *Educ:* Univ Calif, BS, 53; Cornell Univ, MS, 55, PhD(animal nutrit), 57. *Prof Exp:* From asst prof to assoc prof, 57-73, PROF ANIMAL HUSB & NUTRIT, CORNELL UNIV, 73- *Mem:* Am Soc Animal Sci; Brit Soc Animal Prod. *Res:* Animal breeding. *Mailing Add:* Morrison Hall Cornell Univ Ithaca NY 14853

HOGUE, RAYMOND ELLSWORTH, physical medicine, for more information see previous edition

HOGUET, ROBERT GERARD, b Darby, Pa, Apr 2, 48; m 70; c 2. POLYMER CHEMISTRY. *Educ:* Dayton Univ, BS, 70; Syracuse Univ, PhD(chem), 76. *Prof Exp:* Res chemist, GAF Corp, 75-78; RES CHEMIST, VERONA DIV, MOBAY CORP, 78- *Mem:* Am Chem Soc; Sigma Xi. *Res:* Dispersion of powders in liquids. *Mailing Add:* 200 Duncannon Rd Bel Air MD 21014

HOH, GEORGE LOK KWONG, b Canton, China, Feb 22, 35; US citizen; m 62; c 2. ORGANIC CHEMISTRY, POLYMER CHEMISTRY. *Educ:* Pomona Col, BA, 56; Univ Kans, PhD(org chem), 61. *Prof Exp:* Res asst ferrocene chem, Univ Kans, 58-60; res assoc, Brandeis Univ, 60-61; res chemist, Electrochem Dept, 62-71, sr res chemist, 72-81, RES ASSOC, PLASTIC PROD & RESINS DEPT, E I DU PONT DE NEMOURS & CO INC, 81- *Mem:* AAAS; Am Chem Soc; Sigma Xi. *Res:* Ferrocene chemistry; amine oxidation and olefin epoxidation with hydrogen peroxide; ethylene-vinyl acetate, polyamide, and polyester thermoplastics; hot melt adhesives; polymer research and development. *Mailing Add:* 1404 Fresno Rd Wilmington DE 19803-5122

HOHAM, RONALD WILLIAM, b Omaha, Nebr, July 10, 42; div; c 2. PHYCOLOGY, SNOW. *Educ:* Munic Univ Omaha, BA, 64; Mich State Univ, MS, 66; Univ Wash, PhD(bot), 71. *Prof Exp:* From asst prof to assoc prof, 71-83, PROF BIOL, COLGATE UNIV, 83- *Concurrent Pos:* Res assoc, Northern Ariz Univ, 78; vis prof bot, Biol Sta, Univ Mont, 80, 82, 84 & 86; chair, Colgate Univ Res Coun, 80-84; assoc ed, Phycologia, 78-83. *Mem:* Phycol Soc Am; Int Phycol Soc; Bot Soc Am; Sigma Xi; Inst Arctic & Alpine Res. *Res:* Snow algae, emphasizing physiological ecology, distribution, nutrional requirements for growth; acid precipitation, interactions with snow chemistry and life histories; culture collections; reviewer and contributor of photographic material of textbooks in biology, botany and phycology. *Mailing Add:* Dept Biol Colgate Univ Hamilton NY 13346

HOHENBERG, CHARLES MORRIS, b Selma, Ala, May 19, 40; div; c 2. PHYSICS. *Educ:* Princeton Univ, BA, 62; Univ Calif, Berkeley, PhD(physics), 68. *Prof Exp:* Res physicist, Univ Calif, Berkeley, 68-69; from asst prof to assoc prof, 70-78, PROF PHYSICS, WASH UNIV, 78- *Mem:* Fel Meteoritical Soc. *Res:* Space physics; rare gas mass spectroscopy. *Mailing Add:* Dept Physics Wash Univ St Louis MO 63130

HOHENBERG, PIERRE CLAUDE, b Neuilly-sur-Seine, France, Oct 3, 34; US citizen. THEORETICAL & SOLID STATE PHYSICS. *Educ:* Harvard Univ, AB, 56, PhD(physics), 62. *Prof Exp:* Nat Acad Sci exchange fel physics, Inst Phys Probs, Moscow, USSR, 62-63; NATO fel, Ecole Normale Superieure, Paris, 63-64; MEM TECH STAFF, BELL LABS, 64- *Concurrent Pos:* Vis prof, Tech Univ Munich, 71-72, prof, 74-77; Lorentz prof, Univ

Leiden, 91. *Honors & Awards:* Fritz London Prize Low Temperature Physics, 90. *Mem:* Nat Acad Sci; fel Am Acad Arts & Sci; fel NY Acad Sci; fel AAAS; fel Am Phys Soc. *Res:* Theory of superfluids and superconductors; electronic theory of solids; magnetism, critical phenomena and phase transitions; nonlinear and nonequilibrium phenomena; statistical mechanics; instabilities in fluids. *Mailing Add:* AT&T Bell Labs Murray Hill NJ 07974-2070

HOHENBOKEN, WILLIAM DANIEL, b Davenport, Iowa, Nov 13, 41; div; c 1. ANIMAL BREEDING, ANIMAL SCIENCE. *Educ:* Okla State Univ, BS, 63; Colo State Univ, MS, 69, PhD(animal breeding & genetics), 70. *Prof Exp:* USDA Agr Res Serv fel, Univ Wis-Madison, 70-71; from asst prof to prof animal breeding & genetics, Ore State Univ, 71-87; PROF ANIMAL BREEDING & GENETICS, VA POLYTECH INST & STATE UNIV, 87- *Concurrent Pos:* NZ Nat Res Adv Coun fel, Ruakura Animal Res Ctr, Hamilton, NZ, 85-86. *Mem:* Am Soc Animal Sci. *Res:* Quantitative animal genetics; livestock production. *Mailing Add:* Dept Animal Sci Va Polytech Inst State Univ Blacksburg VA 24061

HOHENEMSER, CHRISTOPH, b Berlin, Ger, May 29, 37; US citizen; m 60; c 1. PHYSICS. *Educ:* Swarthmore Col, BA, 58; Wash Univ, St Louis, PhD(physics), 63. *Prof Exp:* Res assoc physics, Wash Univ, St Louis, 63-64; from instr to asst prof physics, Brandeis Univ, 64-71; assoc prof, 71-77, PROF PHYSICS, CLARK UNIV, 77- *Concurrent Pos:* Vis scientist, Natuurkundig Laboratorium, Univ Groningen, Neth, 73-74 & 78-79. *Mem:* AAAS; Am Phys Soc. *Res:* Use of nuclear physics techniques in solid state physics, such as Mossbauer effect; positron annihilation in solids; perturbed angular correlations; x-ray spectroscopy; technological hazard management; regulation of hazardous technology. *Mailing Add:* Dept Physics Clark Univ 950 Main St Worcester MA 01610

HOHENEMSER, KURT HEINRICH, b Berlin, Ger, Jan 3, 06; nat US; m 33; c 2. MECHANICS. *Educ:* Darmstadt Tech Univ, dipl, 27, DrIng, 29. *Prof Exp:* Privatdocent appl mech, Univ Gottingen, 32-34; consult helicopter develop, Anton Flettner Aircraft Corp, 35-45; chief aeromech eng, McDonnell Aircraft Corp, 47-65; vis prof, 57-65, prof, 65-76, EMER PROF AEROSPACE ENG, WASH UNIV, 76- *Honors & Awards:* Bell Award, 57; Alexander Klemin Award, 64. *Mem:* Fel Am Helicopter Soc; assoc fel Am Inst Aeronaut & Astronaut. *Res:* Helicopters; structural dynamics; aeromechanics; applied mechanics. *Mailing Add:* 2421 Remington Lane St Louis MO 63144

HOHIMER, A ROGER, PERINATAL PHYSIOLOGY. *Educ:* Univ Wash, Seattle, PhD(physiol & biophys), 77. *Prof Exp:* ASST PROF PHYSIOL, DEPT OBSTET & GYNEC, ORE HEALTH SCI UNIV, 81- *Res:* Fetal-cerebral blood flow and metabolism. *Mailing Add:* Dept Obstet & Gynec Ore Health Sci Univ 3181 SW Sam Jackson Park Rd Portland OR 97201

HOHL, FRANK, physics, astronomy; deceased, see previous edition for last biography

HOHL, JAKOB HANS, b Heiden, Switz, Feb 18, 30; m 55; c 3. SOLID STATE ELECTRONICS, RELIABILITY ENGINEERING. *Educ:* Tech Winterthur, Switz, dipl, 53; Univ Vt, MS, 70; Univ Ariz, MS, 72, PhD(elec eng), 76. *Prof Exp:* Engr tel switch circuits, Standard Tel & Radio, Int Tel & Tel Corp, 53-56; res staff mem electronics, IBM Corp, Switz, 56-66; mem staff & adv engr solid state circuits, IBM Corp, Vt, 66-78; adv engr large scale integrated circuits, IBM Corp, Tucson, Ariz, 78-85; res engr, 85-88, VIS PROF, UNIV ARIZ, 88- *Concurrent Pos:* Adj assoc prof, Univ Vt, 77-78 & adj prof, Univ Ariz, 81-88. *Mem:* Sr mem Inst Elec & Electronics Engrs; Int Soc Hybrid Microelectronics. *Res:* Evaluation and modeling of integrated solid state electronic devices; development, simulation and analysis of very large-scale integrated memory logics circuits; techniques for reliability and functional evaluation of such circuits; modeling of radiation effect in very large-scale integrated circuits. *Mailing Add:* Dept Elec & Comp Eng Univ Ariz Tucson AZ 85721

HOHMAN, WILLIAM H, b Canton, Ohio, May 1, 39; m 70; c 2. INORGANIC CHEMISTRY. *Educ:* Mt Union Col, BS, 61; Univ Ohio, PhD(chem), 66. *Prof Exp:* From asst prof to assoc 65-80, PROF INORG CHEM, MARIETTA COL, 80- *Mem:* Am Chem Soc. *Res:* Preparation and identification of coordination compounds; transition metals. *Mailing Add:* Dept Chem Marietta Col Marietta OH 45750-3031

HOHMANN, PHILIP GEORGE, b Harvey, Ill, Jan 25, 43; m 71; c 1. BIOCHEMISTRY. *Educ:* Univ Ill, Urbana, BS, 64; Univ Calif, Berkeley, PhD(biochem), 69. *Prof Exp:* Damon Runyon Mem fel, Univ Calif, Berkeley, 69-70; Am Cancer Soc fel, Nat Inst Med Res, Eng, 70-71; res assoc biochem, Col Med, Univ Iowa, 71-72; NIH fel, Dept Pharmacol, Med Sch, Univ Colo, Denver, 72-74; staff mem, Cellular & Molecular Radiobiol Group, Los Alamos Sci Lab, Los Alamos, NMex, 74-75; sr scientist, 75-78, asst res prof, Dept Exp Biol, 78-81, ASSOC CANCER RES SCIENTIST, ROSWELL PARK MEM INST, 78- *Mem:* AAAS; Am Soc Cell Biol; Am Soc Biol Chemists. *Res:* H1 histone synthesis and phosphorylation during cell replication; protein kinases. *Mailing Add:* 43 Hawthorne Ct Orchard Park NY 14127

HOHN, ARNO R, b Paterson, NJ, Aug 4, 31; m 61; c 2. PEDIATRICS, CARDIOLOGY. *Educ:* Rutgers Univ, New Brunswick, BS, 52; NY Med Col, MD, 56. *Prof Exp:* Fels, Buffalo Children's Hosp, NY, 58-59 & 62-63; from instr to asst prof, State Univ NY, Buffalo, 63-67; from assoc prof to prof pediat, Med Univ SC, 67-83, dir div pediat cardiol, 67-83; PROF PEDIAT & HEAD DIV PEDIAT CARDIOL, UNIV SOUTHERN CALIF, 84- *Concurrent Pos:* Dir cardiac clins, Buffalo Children's Hosp, NY, 63-67, actg dir sect cardiol, 66-67, dir resident training, 67; consult & chmn comt hosp care, Am Acad Pediat, 69-81; actg chmn, Dept Pediat, Med Univ SC, 81-83; head, Div Cardiol, Children's Hosp Los Angeles, 84- *Mem:* Fel Am Acad Pediat; fel Am Col Cardiol; fel Am Col Chest Physicians. *Res:* Improvement of heart catheterization techniques in children; hypertension and exercise testing; cardiac function in noncardiac diseases in infants and children. *Mailing Add:* Dept Pediat Univ Southern Calif 2025 Zonal Ave Los Angeles CA 90033

HOHN, EMIL OTTO, b Basel, Switz, Mar 14, 19; m 44; c 2. ENDOCRINOLOGY. *Educ:* Univ London, MB & BS, 43, MSc, 46, PhD(physiol), 51. *Prof Exp:* From demonstr to asst lectr physiol, Guy's Hosp, Med Sch, Univ London, 44-47; from asst prof to prof physiol, Univ Alta, 47-83; RETIRED. *Mem:* Arctic Inst NAm; Can Physiol Soc; Brit Physiol Soc; Brit Soc Endocrinol; Am Ornithologists Union. *Res:* Ornithology; endocrinology of mammals and birds. *Mailing Add:* 11511 78th Ave Edmonton AB T6G 0N4 Can

HOHN, MATTHEW HENRY, b New Stanton, Pa, Nov 9, 20; m 47; c 4. ECONOMIC BOTANY, LIMNOLOGY. *Educ:* Pa State Teachers Col, Ind, BS, 43; Cornell Univ, MS, 49, PhD(econ bot), 51. *Prof Exp:* Asst bot, Cornell Univ, 48-50, instr, 50-59; assoc prof biol, Bloomsburg State Col, 59-61; assoc prof, Beaver Island, Cent Mich Univ, 61-67, prof biol & dir biol sta, 67-83; RETIRED. *Mem:* Am Soc Limnol & Oceanog; Phycol Soc Am; Am Micros Soc; Int Phycol Soc. *Res:* Diatom taxonomy and population structure with reference to stream pollution. *Mailing Add:* East Shore Dr St James MI 49782

HOHNADEL, DAVID CHARLES, b Englewood, NJ, Dec 12, 41; m 62; c 2. CLINICAL CHEMISTRY, BIOCHEMISTRY. *Educ:* Antioch Col, BS, 63; Case Western Reserve Univ, PhD(biochem), 70. *Prof Exp:* IBM fel, Dept Lab Med, Univ Conn, 69-71; clin chemist, Bristol Hosp, 71-73; dir clin chem, Dept Lab Med, Univ Conn, 73-78; DIR CLIN CHEM, DEPT LAB MED, CHRIST HOSP, 78- *Concurrent Pos:* Consult, Int Bus Mach Co, NY, 69-73, US Vet Admin Hosp, Newington, Conn, 70-73 & New Brit Gen Hosp, Conn, 71-73. *Honors & Awards:* B J Katchman Award, 87. *Mem:* Sigma Xi; Am Asn Clin Chemists; AAAS; Asn Clin Scientists; Am Chem Soc; Nat Acad Clin Biochem. *Mailing Add:* Christ Hosp 2139 Auburn Ave Cincinnati OH 45219

HOHNKE, DIETER KARL, b Ger, Dec 22, 36. MATERIALS SCIENCE. *Educ:* Univ Hamburg, BS, 60; Univ Pa, MS, 65, PhD(mat sci), 68. *Prof Exp:* Sr res scientist, 68-73, PRIN RES SCI ASSOC, DEPT PHYSICS, FORD MOTOR CO, 73- *Mem:* Am Phys Soc; Electrochem Soc. *Res:* Electrical transport processes and defects in ionic solids and semiconductors; interaction of gases with solid surface (solid state chemical sensors). *Mailing Add:* 1201 Harbrooke Ann Arbor MI 48103

HOHNSTEDT, LEO FRANK, b Alton, Ill, June 12, 24; m 60. BORON COMPOUNDS. *Educ:* St Louis Univ, BS, 49; Univ Chicago, PhD(chem), 55. *Prof Exp:* From instr to asst prof chem, St Louis Univ, 54-60; assoc prof chem, Polytech Inst Brooklyn, 60-61; mem weapon systs anal div, Inst Defense Anal, 61-63; from assoc prof to prof chem, St Louis Univ, 63-83, chmn dept, 66-77; RETIRED. *Mem:* Am Inst Chemists. *Res:* Boron-nitrogen and boron-sulfur compounds. *Mailing Add:* 4811 Kaskaskia Trail Godfrey IL 62035

HOINESS, DAVID ELDON, b Decorah, Iowa, Aug 14, 35; wid; c 1. ORGANIC CHEMISTRY, TECHNICAL MANAGEMENT. *Educ:* St Olaf Col, BA, 57; Purdue Univ, MS, 61; Baylor Univ, PhD(org chem), 64. *Prof Exp:* Fel & res assoc, Southern Regional Res Lab, USDA, New Orleans, 64-66; RES ASSOC, TEXTILE FIBERS DEPT, E I DU PONT DE NEMOURS & CO, INC, 66- *Mem:* AAAS; Am Chem Soc; Soc Advan Mat & Process Eng; Soc Automotive Engr. *Res:* Electrical insulation materials; organic polymer chemistry, paper and fiber chemistry; friction materials; materials engineering; non-structural composites. *Mailing Add:* Textile Fibers Dept Chestnut Run Plaza E I du Pont de Nemours & Co Inc Wilmington DE 19880-0701

HOISIE, ADOLFY, b Botosani, Romania, May 15, 57; US citizen; m 83; c 2. COMPUTATIONAL LINEAR ALGEBRA, PARALLEL COMPUTING & ALGORITHMS. *Educ:* Bucharest Univ, Romania, BS, 80, MS, 81. *Prof Exp:* Res assoc, Nuclear Power Reactors Inst, Romania, 81-88; sci software analyst, 88-91, MGR, CORNELL THEORY CTR, CORNELL UNIV, 91- *Concurrent Pos:* Consult, New York City Bd Educ, 88. *Mem:* Soc Indust & Appl Math. *Res:* Blocking algorithms for dense linear algebra; parallel algorithms for shared and distributed computing; performance programming in distributed settings; modal techniques for three dimensional nuclear reactor dynamics. *Mailing Add:* Cornell Univ 623 E&TC Bldg Ithaca NY 14853

HOISINGTON, DAVID B(OYSEN), b Philadelphia, Pa, Feb 11, 20; m 40; c 3. ELECTRONICS SYSTEMS DESIGN & SYSTEMS SCIENCE. *Educ:* Mass Inst Technol, SB, 40; Univ Pa, MS, 41. *Prof Exp:* Develop engr, Hazeltine Electronics Corp, NY, 41-46; proj engr in charge develop radar equip, Sperry Gyroscope Co, 46-47; from asst prof to prof elec eng, Naval Postgrad Sch, 57-80, chmn, Electronic Warfare Acad Group, 76-80; PARTNER, E W ENTERPRISES, FRIDAY HARBOR, WASH, 80- *Concurrent Pos:* Instr night sch, City Col New York, 45-46; consult engr, Carmel Valley, Calif, 51-80. *Honors & Awards:* Straton Prize, Mass Inst Technol, 40. *Mem:* Inst Elec & Electronics Engrs; Sigma Xi; Asn Old Crows; Armed Forces Commun & Electronics Asn. *Res:* Design of radar equipment; antijamming radar receiver; detector for antijamming receiver; electronic warfare; electronic countermeasures; direction finding systems. *Mailing Add:* E W Enterprises 5661 Davison Head Dr Friday Harbor WA 98250-9728

HOITINK, HARRY A J, b Neth, July 28, 38; m 64; c 2. PLANT PATHOLOGY, BACTERIOLOGY. *Educ:* Univ Wis, PhD(plant path), 67. *Prof Exp:* Assoc prof, 67-74, PROF PLANT PATH, OHIO AGR RES & DEVELOP CTR, 74- *Mem:* Am Phytopath Soc; AAAS. *Res:* Soil microbiology; ornamental pathology. *Mailing Add:* 946 Linden Dr Wooster OH 44691

HOJNACKI, JEROME LOUIS, b Stamford, Conn, Mar 9, 47; m 78. BIOCHEMISTRY, ATHEROSCLEROSIS. *Educ:* Southern Conn State Col, BS, 69; Univ Bridgeport, MS, 71; Univ NH, PhD(lipid biochem), 75; Clark Univ, MHA, 82. *Prof Exp:* Instr biol & physiol, Stamford Pub Sch, Stamford, Conn, 69-71; PROF BIOL SCI, UNIV LOWELL, 77- *Concurrent Pos:* Res asst, Univ NH, 71-75; Am Heart Asn res fel, Harvard Univ Sch Pub Health, 75-76 & NIH res fel, 76-77; res consult, Beth Israel Hosp, Harvard Med Sch, 77-; res grant-in-aid, Am Heart Asn, 78-86, Coun Tobacco Res, USA Inc, US Brewer's Asn & Distilled Spirits Coun US Inc & NIH, Alcoholic Beverage Med Res Found, Johns Hopkins Univ; fel, Coun Arteriosclerosis. *Honors & Awards:* Stamford Med Asn Award, 65. *Mem:* Soc Exp Biol & Med; Am Heart Asn; AAAS; Sigma Xi; NY Acad Sci; Am Inst Nutrit; Am Fedn Clin Res. *Res:* Lipid chemistry; lipoprotein metabolism; pathology. *Mailing Add:* One Sunset Rd Groton MA 01450

HOJVAT, CARLOS F, b Buenos Aires, Arg, May 5, 39; Can citizen; m 67; c 1. ACCELERATOR PHYSICS. *Educ:* Inst Balseiro, Arg, Lic physics, 62; Univ BC, PhD(physics), 67. *Prof Exp:* Fel, Univ BC, 68-69; res assoc, Queen Mary Col, 70-74 & McGill Univ, 74-77; head Antiproton Prod Group, 80-85, SCIENTIST II, FERMI NAT ACCLERATOR LAB 77-, HEAD, PHYSICS DEPT, 86- *Mem:* Can Asn Physics; Am Phys Soc; AAAS. *Res:* Elementary particle physics; accelerator physics. *Mailing Add:* Fermi Nat Accelerator Lab MS 122 PO Box 500 Batavia IL 60510

HOKAMA, TAKEO, b Hilo, Hawaii, June 20, 31. ORGANIC CHEMISTRY. *Educ:* Univ Hawaii, BA, 53; Purdue Univ, MS, 56, PhD(chem), 58. *Prof Exp:* Res chemist, Koppers Chem Co, 58-65; res chemist, 65-71, group leader polymer synthesis, 71-78, GROUP LEADER AGR CHEM SYNTHESIS, VELSICOL CHEM CORP, 78- *Mem:* Am Chem Soc. *Res:* Organic synthesis in nitroparaffins, phenols and hydrocarbons. *Mailing Add:* 715 Quetta Ave No E Sunnyvale CA 94087-1249

HOKAMA, YOSHITSUGI, b Kohala, Hawaii, Oct 25, 26; m 51; c 2. IMMUNOLOGY, PATHOLOGY. *Educ:* Univ Calif, Los Angeles, BA, 51, MA, 53, PhD(microbiol), 57. *Prof Exp:* Asst, Univ Calif, Los Angeles, 52-53, from microbiologist to asst res microbiologist, 53-66; assoc prof, 66-68, PROF PATH, MED SCH, UNIV HAWAII, 68- *Concurrent Pos:* Consult, Courtland Labs, 62-66 & Accupath Lab, 74-; assoc prof microbiol, Calif State Univ, Los Angeles, 63-66. *Mem:* Sigma Xi; Am Asn Immunologists; Am Soc Microbiol; NY Acad Sci; Am Asn Pathologists; Int Soc Toxinology. *Res:* Medical bacteriology; mycology; physiology and biochemistry of acute phase proteins; C-reactive protein; influenza sialidases; catalase biochemistry and physiology; prostaglandins and drugs in immune response; peroxisome-functions; ciguatoxin (fish toxin) assay. *Mailing Add:* Dept Path Med Sch Univ Hawaii Honolulu HI 96822

HOKANSON, GERARD CLIFFORD, b Elgin, Ill, Dec 19, 49; m 76. MEDICINAL CHEMISTRY. *Educ:* DePauw Univ, BA, 71; Purdue Univ, PhD(med chem), 75. *Prof Exp:* asst prof pharmacognosy, 75-79, assoc prof pharmaceutical chem, Col Pharm & Allied Health Professions, Wayne State Univ, 79-81; sr scientist, Parke-Davis Pharmaceut, 81-83, res assoc, 83-85, sr res assoc, 85-90, SECT DIR, PARKE-DAVIS PHARMACEUT, 90- *Concurrent Pos:* Res Div, Warner-Lambert Co & Parke-Davis Pharmaceut. *Mem:* Am Chem Soc; Am Asn Pharmaceut Scientists. *Res:* Analytical methods development; stability of drug substances in pharmaceutical dosage forms. *Mailing Add:* One Hunter Dr Long Valley NJ 07853

HOKANSON, KENNETH ERIC FABIAN, b Minneapolis, Minn, Nov 10, 41; m 70; c 2. FISH BIOLOGY. *Educ:* Univ Minn, St Paul, BS, 63, PhD(fishery biol), 68. *Prof Exp:* Chief, Monticello Field Sta, 73-78, RES AQUATIC BIOLOGIST, NAT WATER QUAL LAB, 78-, ENVIRON RES LAB-DULUTH, ENVIRON PROTECTION AGENCY, 68-, RES AQUATIC BIOLOGIST,NAT WATER QUAL LAB, 78- *Concurrent Pos:* Adj assoc prof, Univ Minn, Duluth, 71- *Mem:* Am Fisheries Soc; Am Inst Fishery Res; Sigma Xi. *Res:* Thermal requirements of fishes; population dynamics; water quality effects; toxicology; bioassay methods; aquatic entomology; biometrics; experimental design. *Mailing Add:* Environ Res Lab-Duluth US Environ Protection Agency 6201 Congdon Blvd Duluth MN 55804

HOKE, CAROLYN KAY, medical technology, for more information see previous edition

HOKE, DONALD I, b Elgin, Ill, Dec 19, 30; m 60; c 2. POLYMER CHEMISTRY, ORGANIC CHEMISTRY. *Educ:* Ill Inst Technol, BS, 53; Purdue Univ, MS, 56, PhD(org chem), 58. *Prof Exp:* Res chemist, Lubrizol Corp, 57-60, group leader develop lubricant additives, 60-62, staff asst to vpres res & develop, 63-64, supvr chem prod res dept, 64-71, supvr lubricant additive res, 71-73, dept head additive res, 73-81, DIR, CORP PROD SAFETY & COMPLIANCE, LUBRIZOL CORP, 81- *Honors & Awards:* Outstanding Contrib Chem, Am Chem Soc, 88. *Mem:* Am Chem Soc. *Res:* Lubricant additives; chemical process development; synthesis of monomers and polymers. *Mailing Add:* 15440 Dale Rd Chagrin Falls OH 44022

HOKE, GLENN DALE, b Nashville, Tenn, Apr 18, 52; m; c 2. MITOCHONDRIAL TOXICITY, ANTISENSE OLIGONUCLEOTIDES. *Educ:* Va Polytech Inst & State Univ, BS, 74; Va Commonwealth Univ, MS, 79, PhD(biochem), 86. *Prof Exp:* Postdoctoral sci, Smith Kline & French, 86-89; SR SCIENTIST, ISIS PHARMACEUT, 89- *Mem:* Am Soc Biochem & Molecular Biol; AAAS. *Res:* Biochemical evaluation of novel oligonucleotide modifications for nuclease resistance; hybridization quality; RNase H substrate activity, uptake, and toxicity as relates to antisense therapy. *Mailing Add:* Dept Molecular & Cell Biol ISIS Pharmaceut 2280 Faraday Ave Carlsbad CA 92008

HOKE, JOHN HENRY, b Greencastle, Pa, July 24, 22; m 44; c 5. METALLURGY. *Educ:* Pa State Univ, BS, 46, MS, 48; Johns Hopkins Univ, DEng(mech eng), 55. *Prof Exp:* Res asst, Armco Steel Corp, 47-50; instr, Johns Hopkins Univ, 50-54; res metallurgist, Babcock & Wilcox Co, 54-57; res supvr stainless steel, Crucible Steel Co Am, 57-60; from asst prof to prof, 60-85, EMER PROF METALL, PA STATE UNIV, 85- *Mem:* Am Soc Metals; Am Inst Mining, Metall & Petrol Engrs; Nat Asn Corrosion Eng; Am Soc Testing & Mat; Am Soc Eng Educ. *Res:* Stainless steels and high temperature materials; mechanical properties, especially brittle fracture and creep. *Mailing Add:* 285 Spring St State College PA 16801

HOKE, JOHN HUMPHREYS, b Toledo, Ohio, Aug 5, 26; m 51, 81; c 3. GEOLOGY, GEOPHYSICS. *Educ:* Duke Univ, BS, 46; Univ Mich, MS, 51. *Prof Exp:* Res org chemist, Plaskon Div, Libbey-Owens-Ford Glass Co, 47-49; geologist & geophysicist, Standard Oil Co Calif, 51-54; seismograph geologist, Saudi Arabia, 54-56, sr geophysicist, NY & Tex, 56-59, sr geologist, Hq Staff, Saudi Arabia, 59-64, sr staff geophysicist, Explor Dept, 64-66, chief geophysicist, Explor Dept, 66-80, MGR GEOPHYS DEPT, ARABIAN AM OIL CO, SAUDI ARABIA, 80- *Mem:* Geol Soc Am; Soc Explor Geophys; Geochem Soc; Seismol Soc Am; Am Asn Petrol Geol. *Res:* Geophysical interpretation; organic geochemistry. *Mailing Add:* PO Porto Cervo Costa Smeralda Sardinia 07020 Italy

HO-KIM, QUANG, b Saigon, Viet Nam; m; c 2. THEORETICAL PHYSICS. *Educ:* Syracuse Univ, BS, 61; Mass Inst Technol, MS, 63, PhD(physics), 65. *Prof Exp:* Res asst physics, Mass Inst Technol, 63-65; Nat Res Coun Can fel & res assoc, McGill Univ, 65-66; from asst prof to assoc prof, 66-75, PROF PHYSICS, LAVAL UNIV, 75- *Mem:* Am Phys Soc; Can Asn Physicists. *Res:* Particles and field theory. *Mailing Add:* Dept Physics Laval Univ Quebec PQ G1K 7P4 Can

HOKIN, LOWELL EDWARD, b Chicago, Ill, Sept 20, 24; m 78; c 4. PHARMACOLOGY. *Educ:* Univ Louisville, MD, 48; Univ Sheffield, PhD, 52. *Prof Exp:* From lectr to asst prof pharmacol, McGill Univ, 53-57; from asst prof to assoc prof, 57-61, PROF PHYSIOL CHEM, UNIV WIS-MADISON, 61-, CHMN DEPT PHARMACOL, 68- *Mem:* Am Soc Biol Chem; Brit Biochem Soc; Am Chem Soc; Am Soc Pharmacol & Exp Therapeut; NY Acad Sci; AAAS. *Res:* Phosphoinositide metabolism and function; neurochemistry; ion transport. *Mailing Add:* 383 Med Sci Bldg Univ Wis Madison WI 53706

HOKIN-NEAVERSON, MABEL, PSYCHOPHARMACOLOGY. *Educ:* Univ Sheffield, Eng, PhD(biochem), 52. *Prof Exp:* PROF PSYCHOPHARMACOL, UNIV WIS, 57- *Mailing Add:* Dept Psychiat Univ Wis 589 Med Sci Bldg Madison WI 53706

HOLADAY, DUNCAN ASA, b Denver, Colo, July 22, 16; m 41; c 5. ANESTHESIOLOGY. *Educ:* Univ Chicago, BS, 40, MD, 43; Am Bd Anesthesiol, dipl. *Prof Exp:* Intern, St Luke's Hosp, New York, 43-44, resident anesthesiol, 44-45; Sharp & Dohme fel pharmacol & exp therapeut, Sch Med, Johns Hopkins Univ, 47-50; from asst prof to assoc prof, Col Physicians & Surgeons, Columbia Univ, 50-59; prof surg & anesthesiol, Sch Med, Univ Chicago, 59-68; prof anesthesiol, Sch Med, Univ Miami, 68-80; PROF ANESTHESIOL, SCH MED, VANDERBILT UNIV, 80- *Concurrent Pos:* Asst attend anesthetist, Presby Hosp, NY, 50-59. *Mem:* Am Physiol Soc; Am Soc Pharmacol & Exp Therapeut; Am Soc Anesthesiol; Asn Univ Anesthetists; fel NY Acad Sci; Sigma Xi. *Res:* Physiology of respiration and acid-base balance; pharmacology of curariform drugs and other anesthetic agents; respiratory therapy; management of acute respiratory failure; biotransformation of volatile halogenated hydrocarbons; medical instrumentation. *Mailing Add:* Dept Anesthesiol Vanderbilt Univ Sch Med Nashville TN 37232

HOLADAY, JOHN W, b New York, NY, June 9, 45; div. NEUROPHARMACOLOGY, PSYCHOSOMATIC MEDICINE. *Educ:* Univ Ala, Tuscaloosa, BS, 66, MS, 69; Univ Calif, San Francisco, PhD(pharmacol), 77. *Prof Exp:* Chief, Neuropharmacol Br, Walter Reed Army Inst Res, 81-89; SR VPRES RES & DEVELOP, MEDICIS PHARMACEUT CORP, 89- *Concurrent Pos:* Prof pharmacol & psychiat, Uniformed Serv Univ Health Sci. *Honors & Awards:* Dean Calvert Award, Am Soc Pharmacol & Exp Therapeut, 77; US Army Res & Develop Award & Bronze Medallion, 80 & 84; US Army Serv Conf Award & Bronze Medallion Outstanding Sci Achievement, 80. *Mem:* Am Soc Pharmacol & Exp Therapeut; Am Col Neuropsychopharmacol; Shock Soc (treas, 83-); Soc Crit Care Med; Soc Neurosci; Sigma Xi. *Res:* Pathophysiological mechanisms and therapeutics in circulatory shock and central nervous system injury; biochemical basis of antidepressants; mechanisms of seizures and anticonvulsant pharmacology; immune correlates of endocrine function; stress and immunity relationships to neuroendocrine status; pain and antonomic regulation; peptide physiology; opiate receptor coupling through allostene mechanisms. *Mailing Add:* Medicis Pharmaceut Corp 100 E 42nd St 15th Floor New York NY 10017

HOLADAY, WILLIAM J, b Chicago, Ill, June 23, 28; m 52; c 4. MEDICINE, SURGICAL PATHOLOGY. *Educ:* Hobart Col, BS, 50; NY Med Col, MD, 55. *Prof Exp:* Asst prof path, Med Br, Univ Tex, 64-68; from assoc prof to prof path, Col Med, Ohio State Univ, 71-78; instr div radiations oncol, Univ Tex, San Antonio, 78-; AT DEPT PATHOL, ST JOHNS HOSP, LONGVIEW, WA. *Concurrent Pos:* Consult, USPHS Hosp, Galveston, Tex, 65-68 & Childrens Hosp, Columbus, Ohio, 68- *Mem:* Fel Col Am Path; fel Am Soc Clin Path. *Mailing Add:* PO Box 998 Longview CA 98632

HOLBERT, DONALD, b NSW, Australia, Aug 27, 41; m 73. APPLIED STATISTICS. *Educ:* Univ Ore, BS, 67; Wash State Univ, MA, 69; Okla State Univ, PhD(statist), 73. *Prof Exp:* Teacher math, Australian Sec Schs, 61-65; asst prof statist, Med Univ SC, 73-74; ASST PROF STATIST, OKLA STATE UNIV, 74- *Mem:* Sigma Xi; Am Statist Asn; Biomet Soc. *Res:* Methods in applied statistics. *Mailing Add:* Sch Allied Health ECarolina Univ Greenville NC 27858

HOLBERT, GENE W(ARWICK), b Honesdale, Pa, Aug 8, 48; m 72; c 2. CHEMISTRY. *Educ:* Pa State Univ, BS, 70; Cornell Univ, MS, 77, PhD(chem), 79. *Prof Exp:* Res chemist, Merrell-Nat Labs, 79-81, SR RES CHEMIST, MARION MERRELL DOW RES INST, 81- *Mem:* Am Chem Soc. *Res:* Design and synthesis of potentially useful therapeutic agents, especially suicide enzyme inhibitors; natural products synthesis and synthetic methods. *Mailing Add:* 266 Stockton Dr Loveland OH 45140

HOLBROOK, DAVID JAMES, JR, b Norfolk, Va, Apr 16, 33; m 65; c 1. BIOCHEMISTRY. *Educ:* Col William & Mary, BS, 55; Univ NC, PhD(biochem), 59. *Prof Exp:* Res assoc, 59-60, from instr to asst prof, 60-71, ASSOC PROF BIOCHEM, SCH MED, UNIV NC, CHAPEL HILL, 71-, DIR GRAD STUDIES TOXICOL, 79- *Mem:* Am Soc Biol Chem; Am Soc Pharmacol & Exp Therapeut; Soc Toxicol. *Res:* Metabolism and chemistry of nucleic acids; biochemical pharmacology and toxicology. *Mailing Add:* Dept Biochem Sch Med Univ NC Chapel Hill NC 27599

HOLBROOK, FREDERICK R, b Warwick, Mass, Nov 15, 35; div; c 2. INSECT ECOLOGY. *Educ:* Univ NH, BS, 61; Univ Mass, Amherst, MS, 64, PhD(entom, Simuliidae), 67. *Prof Exp:* Plant pest control inspector, Plant Pest Control Div, 61-64, entomologist, Entom Res Div, Hawaii, 67-68, res entomologist, Fla, 68-71, res entomologist, Maine, 71-73, res leader, Maine, 73-78, res entomologist, Colo, 78-85, & ARTHROPOD-BORNE ANIMAL DIS RES LAB, LARAMIE, WYO, 85- *Concurrent Pos:* Consult, Off Technol Assessment, 77-80. *Honors & Awards:* Cert of Merit, USDA, 89. *Mem:* Entom Soc Am; Am Mosquito Control Asn; Soc Invertebrate Path; AAAS; Sigma Xi; Soc Vector Ecol; Am Benthological Soc. *Res:* Ecology of gypsy moth and Simuliidae; ecology and control of alfalfa weevil; Mediterranean, Oriental, Caribbean and melon flies; aphids on potatoes; development of mycological insect control; biology and control of insect vectors of animal virus disease. *Mailing Add:* USDA-Agr Res Serv PO Box 3965-Univ Sta Laramie WY 82071-3965

HOLBROOK, GABRIEL PETER, b Ilfracombe, Eng, Nov 20, 57; m 88; c 1. PHOTOSYNTHESIS, CARBON METABOLISM. *Educ:* Univ York, Eng, BS, 79, PhD(plant biochem), 86. *Prof Exp:* Res assoc, Biochem Dept, Univ Neb, 83-85; postdoctoral assoc, Bot Dept, Univ Fla, 85-87; ASST PROF, DEPT BIOL SCI, NORTHERN ILL UNIV, 87- *Mem:* Am Soc Plant Physiol. *Res:* Photosynthesis; carbon metabolism; biochemistry and enzymology of photosynthetic reactions; enzyme regulation; mechanisms for reducing photorespiration in higher plants and algae. *Mailing Add:* Dept Biol Sci Northern Ill Univ De Kalb IL 60115

HOLBROOK, GEORGE W, electrical engineering, for more information see previous edition

HOLBROOK, KAREN ANN, b Des Moines, Iowa, Nov 6, 42; m 73; c 1. BIOLOGICAL STRUCTURE. *Educ:* Univ Wis, BS, 63, MS, 66; Univ Wash, PhD(biol struct), 72. *Prof Exp:* Instr biol, Ripon Col, 66-69; instr, 72-75, from asst prof biol struct to assoc prof biol struct, 75-79 VCHMN DEPT, 81-PROF, 84-, ASSOC DEAN SCI AFFAIRS, SCH MED, UNIV WASH, 85- *Concurrent Pos:* NIH trainee, Univ Wash, 69-72, NIH trainee & sr fel dermat, 76-78, mem, Study Sect, Gen Med, NIH, 80-; adj assoc prof med dermat, Sch Med, Univ Wash, 79-; Nat Inst Arthritis & Metab Dis, Nat Inst Arthritis, Diabetes & Digestive Kidney Dis Spec Study Sect, 85-88; adj prof, Med Dermat, 84- *Mem:* Am Asn Anatomists; Soc Invest Dermat; AAAS; Sigma Xi; Am Soc Cell Biol; Soc Pediat Dermat. *Res:* Fine structural & biochemical analysis of human skin including development of the human epidermis and dermis in vivo prenatal diagnosis of inherited skin diseases, and structural abnormalities of the dermis in individuals with inherited disorders of connective tissue metabolism and of epidermis in inherited disorders of keratinization. *Mailing Add:* Dept Biol Struct SM20 Univ Wash Seattle WA 98195

HOLBROOK, NIKKI J, MOLECULAR GENETICS. *Prof Exp:* CHIEF UNIT GENE EXPRESSION & AGING, LAB MOLECULAR GENETICS, NAT INST AGING, 83- *Mailing Add:* Molecular Genetics Lab Nat Inst Aging Geront Res Ctr 4940 Eastern Ave Baltimore MD 21224

HOLBROW, CHARLES H, b Melrose, Mass, Sept 23, 35; m 56; c 5. EXPERIMENTAL NUCLEAR PHYSICS. *Educ:* Univ Wis, BA, 55; Columbia Univ, MA, 57; Univ Wis, MS, 60, PhD(physics), 63. *Prof Exp:* Asst prof physics, Haverford Col, 62-65; res investr, Univ Pa, 65-67; assoc prof, Colgate Univ, 67-75, assoc dir, Comput Ctr, 68-69, sr assoc, 70-72, chmn, Dept Physics & Astron, 70-71, 77-80 & 82-84, prof physics, 75-86, dir, Div Natural Sci & Math, 85-88, CHARLES A DANA PROF PHYSICS, COLGATE UNIV, 86- *Concurrent Pos:* Am Coun Educ acad admin intern, Stanford Univ, 72-73; vis assoc, Calif Inst Technol, 75-76; vis physicist, Brookhaven Nat Lab, 80-81; vis prof, Spectros Lab, Mass Inst Technol, 82 & 88-89; guest scientist, dept physics, State Univ NY, Stony Brook, 83- *Mem:* Am Phys Soc; Am Asn Physics Teachers; Sigma Xi; Am Asn Univ Professors. *Res:* Charged-particle nuclear spectroscopy with magnetic spectrograph; laser-based atomic spectroscopy relevant to nuclear properties; history of nuclear physics; information science. *Mailing Add:* Dept Physics & Astron Colgate Univ Hamilton NY 13346-1398

HOLCENBERG, JOHN STANLEY, b San Francisco, Calif, Oct 9, 35; m 58; c 2. CLINICAL PHARMACOLOGY, ONCOLOGY. *Educ:* Harvard Univ, AB, 56; Univ Wash, MD, 61. *Prof Exp:* Intern & resident ward med, Barnes Hosp, St Louis, 61-63; surgeon, Clin Endocrine Br, Nat Inst Arthritis & Metab Dis, 63-65; Am Cancer Soc res fel, Lab Biochem, Nat Heart Inst, 65-67; from asst prof to assoc prof med & pharmacol, Sch Med, Univ Wash, 67-76; prof pharmacol, med & pediat, Med Col Wis & dir pharmacol, Midwest Childhood Cancer Ctr, Milwaukee, 76-82; prof pediat & biochem, Dept Pediat, Univ Southern Calif Children's Hosp, 82-90; DIR, CLIN AFFAIRS, IMMUNEX CORP, SEATTLE, WASH, 91- *Concurrent Pos:* Pharmaceut Mfrs Asn Found career develop award, 67-70; Leukemia Soc scholar, 70; NIH career develop award, 71-76; Burroughs Wellcome Clin Pharmacol Scholar, 78-83. *Mem:* AAAS; Am Soc Biol Chemists; Am Soc Pharmacol & Exp Therapeut; Am Soc Clin Pharmacol & Therapeut; Am Asn Cancer Res; Am Soc Clin Invest. *Mailing Add:* Immunex Corp 51 University St Seattle WA 98101

HOLCK, GARY LEROY, b Paullina, Iowa, Dec 30, 38; m 60; c 3. ANIMAL NUTRITION. *Educ:* Iowa State Univ, BS, 60; Univ Vt, MS, 62; Univ Mo, PhD(animal nutrit), 66. *Prof Exp:* Swine res specialist, Cent Soya Co, Inc, 65-72; ANIMAL NUTRITIONIST, HOLCO AGRI-PROD, INC, 72- *Mem:* Am Soc Animal Sci. *Res:* Calcium and phosphorus nutrition of growing dairy calves; the relationship between lysine and tryptophan and their effect upon growth; nitrogen retention and plasma levels of these amino acids in swine. *Mailing Add:* 1406 Saint Lukes Dr Spencer IA 51301

HOLCOMB, CHARLES EDWARD, b Colorado Springs, Colo, July 17, 24; m 46; c 2. DRYING, HEAT TRANSFER. *Educ:* Univ Colo, BS, 49; Univ Wis, MS, 51, PhD(chem eng), 53. *Prof Exp:* Instr chem eng, Univ Wis, 49-53; RETIRED. *Concurrent Pos:* Res engr, E I du Pont de Nemours & Co, 53-58, staff engr, 58-60, res assoc, 60-74, res fel, 74-85. *Mem:* Am Inst Chem Engrs. *Res:* Process development for plastic films; photographic films from emulsion making to coating; drying of photographic films. *Mailing Add:* 3210 Laurel Park Hwy Hendersonville NC 28739

HOLCOMB, DAVID NELSON, b Sioux City, Iowa, Sept 12, 36; m 65; c 4. PHYSICAL CHEMISTRY. *Educ:* Univ Nebr, BS, 58; Univ Ill, PhD(chem), 62; Roosevelt Univ, MBA, 77. *Prof Exp:* NSF fel chem, Univ Calif, Berkeley, 62-64; res chemist, Agr Res Serv, USDA, 64-66; group leader, 66-80, mgr res & develop, 80-87, TECHNOL MGR, BASIC FOOD SCI LAB, KRAFT, INC, 87- *Concurrent Pos:* Ed, Food Microstruct. *Mem:* AAAS; Am Chem Soc; Electron Micros Soc Am; Am Oil Chemists Soc; Sigma Xi. *Res:* physical chemistry; thermodynamics, rheology and microscopy of foods and food components; research management; analytical data quality control. *Mailing Add:* Kraft Res & Develop 801 Waukegan Rd Glenview IL 60025

HOLCOMB, DONALD FRANK, b Chesterton, Ind, Nov 8, 25; m 50; c 3. PHYSICS EDUCATION. *Educ:* DePauw Univ, AB, 49; Univ Ill, MS, 50, PhD(physics), 54. *Prof Exp:* From instr to assoc prof physics, 54-62, chmn dept, 69-74 & 82-86, dir lab atomic & solid state physics, 64-68, PROF PHYSICS, CORNELL UNIV, 62- *Concurrent Pos:* Guggenheim fel, 68-69; Sci Res Coun sr vis fel, 78. *Mem:* Fel Am Phys Soc; Am Asn Physics Teachers (pres,87); fel AAAS. *Res:* Spin resonance phenomena; physics solids; studies of electron localization phenomena in disordered systems such as heavily-doped, compensated semiconductors; strong engagement in modernizing undergraduate physics curricula. *Mailing Add:* Dept Physics Clark Hall Cornell Univ Ithaca NY 14853-2501

HOLCOMB, GEORGE RUHLE, b Kankakee, Ill, Oct 25, 27; m 52; c 3. PHYSICAL ANTHROPOLOGY, RESEARCH ADMINISTRATION. *Educ:* Univ Wis, BA, 50, MA, 52, PhD, 56. *Prof Exp:* Asst anthrop, Univ Wis, 51-54; from instr to asst prof anat, Sch Med, Creighton Univ, 54-57; asst prof, Univ NC, Chapel Hill, 57-68, assoc dean grad sch, 62-65, dean res admin, 65-82, chmn dept, 85-90, PROF ANTHOP, UNIV NC, CHAPEL HILL, 68- *Mem:* AAAS; Am Anthrop Asn; Am Asn Phys Anthrop; Am Asn Anatomists; Nat Coun Univ Res Adminrs (pres, 67-68); Sigma Xi. *Res:* Human morphology, both descriptive and functional; skeletal morphogenesis. *Mailing Add:* Dept Anthrop Univ NC Chapel Hill NC 27599-3115

HOLCOMB, GORDON ERNEST, b Monroe, Wis, July 6, 32; m 64; c 2. DISEASES OF ORNAMENTALS & TURFGRASSES, PLANT TISSUE CULTURE. *Educ:* Univ Wis, Platteville, BS, 59; Univ Wis, Madison, PhD(plant path), 66. *Prof Exp:* From asst prof to assoc prof, 65-70, PROF PLANT PATH, LA STATE UNIV, BATON ROUGE, 78- *Honors & Awards:* Dave Feathers Award, Am Camellia Soc, 82-83, 85. *Mem:* AAAS; Am Phytopath Soc; Am Inst Biol Sci; Sigma Xi. *Res:* Diseases of ornamental plants and turfgrasses; disease resistance in plants regenerated from cell culture; mycology. *Mailing Add:* Dept Plant Path & Crop Phys La State Univ Baton Rouge LA 70803-1720

HOLCOMB, HERMAN PERRY, b Jonesville, NC, Dec 30, 34; m 90; c 2. ANALYTICAL CHEMISTRY, RADIOCHEMISTRY. *Educ:* Duke Univ, BS, 56; Univ Va, MS, 58, PhD(anal Chem), 60. *Prof Exp:* Res chemist, Separations Chem Div, E I du Pont de Nemours & Co, Inc, 60-75, staff chemist, Environ Effects Div, Savannah River Lab, 75-80, process staff chemist, Separations Technol Dept, Savannah River Plant, 80-89, FEL SCIENTIST, SEPARATIONS TECHNOL DEPT, SAVANNAH RIVER SITE, WESTINGHOUSE SAVANNAH RIVER CO, 89- *Concurrent Pos:* Asst res prof, Med Col Ga, 65-68. *Mem:* Am Chem Soc; Sigma Xi. *Res:* Spectrophotometry; inorganic separations; analytical chemistry of transuranium elements; analytical applications of noble gas compounds; separation and purification of actinide elements; radioactive waste studies. *Mailing Add:* 1891 Green Forest Dr N Augusta SC 29841-2157

HOLCOMB, IRA JAMES, b Flint, Mich, June 10, 34; m 60; c 3. ANALYTICAL CHEMISTRY. *Educ:* Wayne State Univ, BS, 59, MS, 61, PhD(anal chem), 64, MBA, 80. *Prof Exp:* Instr chem, Detroit Inst Technol, 60-62; res fel, Wayne State Univ, 62-64; control chemist, 64-78, SR RES ASSOC, PARKE DAVIS & CO, DIV WARNER LAMBERT CO, 78- *Mem:* Am Chem Soc; Coblentz Soc; fel Asn Anal Chemists; Fed Anal Chem & Spectros Soc. *Res:* Thin layer and paper chromatography; ultraviolet and infrared spectrophotometry; non-aqueous titrations; functional group analysis; liquid chromatography; chemical microscopy; quality control. *Mailing Add:* 3603 28 M Rd Romeo MI 48065

HOLCOMB, ROBERT M(ARION), b Sapulpa, Okla, Mar 7, 16; m 38; c 3. CIVIL ENGINEERING. *Educ:* Univ Ariz, BS, 36; Iowa State Col, MS, 41, PhD(struct eng), 56. *Prof Exp:* Jr engr, Ariz Hwy Dept, Phoenix, 36-37 & US Bur Reclamation, 37-38; instr civil eng, Univ Ariz, 38-40; instr gen eng, Iowa State Col, 40-41; civil engr & designer, Donald R Warren, Los Angeles, 41-43

& Barrett & Hilp, San Francisco, 43; designer, Radiation Lab, Calif, 43-47; prof struct eng, Tex A&M Univ, 47-79; RETIRED. *Concurrent Pos:* Consult engr, 47-; dean, Seato Grad Sch Eng, Bangkok, 61-63. *Mem:* Am Soc Civil Engrs; Am Soc Eng Educ; Am Concrete Inst. *Res:* Structural engineering; comparative economy of various structural types. *Mailing Add:* 1100 Ashburn College Station TX 77840

HOLCOMB, ROBIN TERRY, b Sterling, Ill, Mar 11, 43; m 65; c 2. VOLCANOLOGY, MARINE GEOLOGY. *Educ:* Cornell Col, BA, 65; Univ Ariz, MS, 75; Stanford Univ, PhD(geol), 81. *Prof Exp:* RES GEOLOGIST, US GEOL SURV, 71- *Concurrent Pos:* Affil prof, Univ Wash, 88- *Mem:* Am Geophys Union; Geol Soc Am; AAAS; Marine Technol Soc; Am Soc Photogram & Remote Sensing. *Res:* Volcanic geomorphology, including studies of active volcanic eruption processes; morphologic interpretation of subaerial and submarine lava flows; long-range forecasting of volcanic eruptions. *Mailing Add:* Sch Oceanog WB-10 US Geol Surv Univ Wash Seattle WA 98195

HOLCOMBE, CRESSIE EARL, JR, b Anderson, SC, Dec 18, 45; m 66; c 2. REFRACTORY CERAMICS & METALS, HIGH-TEMPERATURE INTERACTIONS & COATINGS. *Educ:* Clemson Univ, BS, 66, MS, 67. *Prof Exp:* Assoc develop engr, Union Carbide Nuclear Div, Oak Ridge Y-12 Plant, Tenn, 67-72, develop engr, 72-76, develop staff, 77-80, advan develop staff, 80-84; advan develop staff, 84-88, sr develop staff, 88-90, ADVAN SR DEVELOP STAFF, MARTIN MARIETTA ENERGY SYSTS INC, OAK RIDGE Y-12 PLANT, TENN, 90-; PRES & DEVELOP ENGR, ZYP COATINGS, INC, 85- *Concurrent Pos:* Pres & develop engr, Orpac, Inc, 89- *Mem:* Fel Am Ceramic Soc; Nat Inst Ceramic Engrs; Sigma Xi. *Res:* Paintable refractory coatings; yttria cements, paints and concretes; low expansion ceramics; diamond syntheses; specialty refractory insulations; microwave sintering/processing; compatibility-interaction studies; specialty refractory metal and ceramic crucibles and foundry materials; 75 publications and 34 issued US patents. *Mailing Add:* ZYP Coatings Inc 120 Valley Court Rd Oak Ridge TN 37830

HOLCOMBE, JAMES ANDREW, b Denver, Colo, Nov 12, 48; m 69; c 2. ANALYTICAL CHEMISTRY. *Educ:* Colo Col, BA, 70; Univ Mich, MS, 72, PhD(chem), 74. *Prof Exp:* PROF CHEM, UNIV TEX, AUSTIN, 74- *Mem:* Am Chem Soc; Soc Appl Spectros. *Res:* Graphite furnace atomic absorption spectrometry; trace and ultratrace metal analysis; surface studies using sims; computer simulation and interfacing analytical instrumentation to computers; software development. *Mailing Add:* Dept Chem Univ Tex Austin TX 78712-1104

HOLCOMBE, TROY LEON, b Roxton, Tex, Mar 8, 40; m 71; c 3. OCEAN MAPPING, SCIENTIFIC DATA MANAGEMENT & ADMINISTRATION. *Educ:* Hardin-Simmons Univ, BA, 61; Univ Mo, AM, 64; Columbia Univ, PhD(geol), 72. *Prof Exp:* Res oceanogr, US Naval Oceanog Off, Chesapeake Beach, Md, 68-75; head, Geol Br, Naval Ocean Res & Develop Activ, NSTL, Miss, 75-84; SR SCIENTIST, MARINE GEOL & GEOPHYS DIV, NAT GEOPHYS DATA CTR, NAT OCEANIC & ATMOSPHERIC ADMIN, BOULDER, COLO, 84- *Concurrent Pos:* Vis assoc prof oceanog, Dept Oceanog, Tex A&M Univ, College Station, 80-81; assoc prof, Ctr Wetland Resources, La State Univ, Baton Rouge, 82-83; prog mgr & prin investr, NGDC Prog Paleoclimatol, Nat Oceanic & Atmospheric Admin, 87-91; US sci coordr, Int Bathymetric Chart Caribbean Sea & Gulf Mex proj, Intergovt Oceanog Comn, UNESCO, 87-; mem, Panel Int Progs & Intergovt Coop Ocean Affairs, US State Dept, 87-; reviewer, Caribbean & Gulf Mex, Gen Bathymetric Chart Oceans, 89- *Mem:* Geol Soc Am; Am Asn Petrol Geologists. *Res:* Origin and evolution of seafloor fabric and the accumulation of ocean sediments in space and time-supported by ocean mapping and studies of acoustic stratigraphy; areas of concentration, Caribbean Sea, Gulf of Mexico and Laurentian Great Lakes. *Mailing Add:* Nat Geophys Data Ctr Nat Oceanic & Atmospheric Admin 325 Broadway Boulder CO 80303

HOLCSLAW, TERRY LEE, b Hammond, Ind, July 1, 46; m 67; c 1. PHARMACOLOGY. *Educ:* Univ Ariz, BS, 68; Purdue Univ, MS, 72, PhD(pharmacol), 74. *Prof Exp:* ASST PROF PHARMACOL, UNIV NEBR, 74- *Mem:* AAAS; Am Asn Cols Pharm. *Res:* Overall regulation of drug metabolizing enzyme activity as well as environmental and pharmacological agent which modifies drug metabolism; biochemical mechanisms in the development and maintenance of essential hypertension. *Mailing Add:* Clin Res & Develop Dept Smith Kline & French 709 Swedeland Rd L-223 King of Prussia PA 19479

HOLDAWAY, MICHAEL JON, b Canberra, Australia, Jan 1, 36; US citizen; m 58; c 3. PETROLOGY, MINERALOGY. *Educ:* Yale Univ, BS, 58; Univ Calif, Berkeley, PhD(geol), 62. *Prof Exp:* From asst prof to assoc prof, 62-74, chmn, 77-82, PROF GEOL, SOUTHERN METHODIST UNIV, 74-, CHMN, 86- *Concurrent Pos:* Res grants, NSF, 63-65, 72-75, 76-79, 81-83, 83-85 & NASA, 68-69; ed, Am Mineralogist, 80-84. *Mem:* Mineral Soc Am; Sigma Xi. *Res:* Stability relations of epidote, aluminum silicates, cordierite, almandine, staurolite; pelitic metamorphic rocks in Maine and New Mexico; crystal chemistry of staurolite. *Mailing Add:* Dept Geol Sci Southern Methodist Univ Dallas TX 75275

HOLDBROOK, RONALD GAIL, mechanical engineering, for more information see previous edition

HOLDEMAN, JONAS TILLMAN, JR, b Baton Rouge, La, Feb 9, 37; m 64; c 3. PHYSICS, COMPUTER SCIENCE. *Educ:* La State Univ, BS, 58, MS, 60; Case Western Reserve Univ, PhD(physics), 66. *Prof Exp:* Mem staff physics, Los Alamos Sci Lab, 66-68; res assoc, Mich State Univ, 69-71, asst prof, 71-72; asst prof, Univ Southwestern La, 72-74; SECT HEAD-TECH COMPUT, OAK RIDGE NAT LAB, 74- *Mem:* Am Phys Soc; Sigma Xi; AAAS. *Res:* Numerical analysis; mathematical physics; environmental modeling; optimization. *Mailing Add:* 1056 Lovell Rd Knoxville TN 37932

HOLDEMAN, LOUIS BRIAN, b Baton Rouge, La, Aug 9, 38; m 59; c 2. CRYOGENICS, MICROELECTRONICS. *Educ:* La State Univ, BS, 62; Stanford Univ, MS, 65, PhD(physics), 73. *Prof Exp:* Planetarium lectr & instr astron, La State Univ, 62; res assoc space sci, George C Marshall Space Flight Ctr, NASA, 73-75; physicist, Elec Div, 75-78, Nat Bur Standards, 75-83; AT COMSAT LABS, CLARKSBURG, MD. *Concurrent Pos:* Res assoc, Nat Acad Sci/Nat Res Coun, 73-75. *Honors & Awards:* Bronze Medal, Dept Comn, 81. *Mem:* Am Phys Soc; Sigma Xi; Inst Elec & Electronics Engrs; Am Vacuum Soc. *Res:* Microelectronics fabrication technology for GaAs field; effect transistors; monolithic microwave intigrated circuits. *Mailing Add:* Comsat Labs 22300 Comsat Dr Clarksburg MD 20871

HOLDEN, ALISTAIR DAVID CRAIG, b Lochgilphead, Argyll, Scotland, Nov 8, 28; US citizen; m 59; c 2. ELECTRICAL ENGINEERING, COMPUTER SCIENCE. *Educ:* Glasgow Univ, BSc, 55; Yale Univ, MEng, 58; Univ Wash, PhD(elec eng), 64. *Prof Exp:* Grad apprentice, Brit Broadcasting Corp, London, 55-57; res engr, Boeing Co, 58-60; from asst prof to prof elec eng & computer sci, 64-89, PROF COMPUTER SCI & ENG & ELEC ENG, UNIV WASH, 89- *Concurrent Pos:* Gen chmn, 1st Int Joint Conf Artificial Intel, 69; sect ed, artificial intel, Int J Cybernetics & Gen Systs, 72-; consult, expert systs & artificial intel. *Mem:* Asn Comput Mach; Inst Elec & Electronics Engrs. *Res:* Artificial intelligence, the attempt to use digital computers to augment human itelligence, to simulate human intelligent activities and to shed light on the learning, problem-solving abilities and development of concepts in children; man and machine communication involving natural language, speech and vision. *Mailing Add:* Dept Elec Eng FT-10 Univ Wash Seattle WA 98195

HOLDEN, FRANK C(HARLES), b Bangor, Maine, Nov 28, 21; m 59; c 1. PHYSICAL METALLURGY. *Educ:* Univ Maine, BS, 43; Harvard Univ, MS, 47, MES, 49. *Prof Exp:* Jr engr, Explosives Div, Los Alamos Labs, 44-46; instr mech eng, Univ Maine, 49-51; prin res metallurgist, 51-56, from asst div chief to div chief, 56-67, assoc mgr, 67-70, mgr process & phys metall, 70-73, mgr, Metall Dept, 73-78, mgr metall sci, 78-81, mgr phys metall, 81-83, mgr ceramics & glass, 83-85, mgr corrosion & electrochem, Battelle Mem Inst, 85-86; RETIRED. *Mem:* Am Soc Metals; Am Inst Mining, Metall & Petrol Engrs. *Res:* Titanium alloys; mechanical metallurgy. *Mailing Add:* 1763 Merriweather Dr Columbus OH 43221

HOLDEN, FREDERICK THOMPSON, b New York, NY, June 23, 15. PETROLEUM GEOLOGY. *Educ:* Denison Univ, AB, 37; Univ Chicago, PhD(geol), 41. *Hon Degrees:* DSc, Denison Univ, 80. *Prof Exp:* Field geologist, US Geol Surv, Kans, 38-39; subsurface geologist, Carter Oil Co, La, 41-46, dist geologist, Miss, 46-52, staff geologist, Okla, 52-53, asst to div geologist, southern US, 53-56, sr geologist, 56-60; sr geologist, Humble Oil & Ref Co, 60-67, sr prof geologist, 67-71, explor geologist, 71-73; sr explor geologist, Exxon Corp, CO, USA, 73-75, geol scientist, 75-78, sr geol scientist, 78-85; RETIRED. *Concurrent Pos:* Mem, geol assocs adv bd, Univ Kans, 90- *Mem:* Fel Geol Soc Am; Am Asn Petrol Geol; Am Inst Prof Geologists. *Res:* Mississippian stratigraphy of Ohio; stratigraphy and structure of eastern Gulf Coast and western Mid-Continent; stratigraphy and geologic history of Mid-Continent. *Mailing Add:* 118 Bridgewater Circle Midland TX 79707-6112

HOLDEN, HAROLD WILLIAM, b Toronto, Ont, Sept 29, 29; m 54; c 6. PHYSICAL CHEMISTRY. *Educ:* Univ Toronto, BA, 51, PhD(phys chem), 55. *Prof Exp:* UK Nat Coal Bd res fel, Univ Birmingham, 55, univ res fel, 56-58; Nat Res Coun Can res fel photochem, Nat Res Coun Div Pure Chem, Ottawa, 58-60; res chemist, Cent Res Lab, Can Industs Ltd, 60-71, group leader pioneering, 71-73, group mgr pioneering res, Explosives Res Lab, 73-75, group mgr prod res & develop, 75-79, group mgr, explosives physics, Explosives Res Lab, Can Industs Ltd, 79-84, tech mgr, New Ventures, Explosives Div, C-I-L Inc, 84-90; EXPLOSIVES TECH CTR, ICI EXPLOSIVE CAN, 90- *Mem:* Chem Inst Can. *Res:* High explosives and blasting agents; polymer physics and physical chemistry; physical properties of polymers and composites; emulsions and gels; chemical kinetics; explosives formulation, sensitization and hazards; energetic propellant constituents. *Mailing Add:* c/o Explosives Tech Ctr ICI Explosives Can McMasterville PQ J3G 1T9 Can

HOLDEN, HOWARD T, IMMUNOLOGY. *Educ:* Univ Miami, Fla, PhD(microbiol), 73. *Prof Exp:* Sr investr, Nat Cancer Inst, 75-88; STAFF, WARNER LAMBERT, ANN ARBOR, 88- *Res:* Drug regulatory affairs. *Mailing Add:* Parke-Davis Pharmaceut Res Div Warner-Lambert Co 2800 Plymonth Rd Ann Arbor MI 48105-2430

HOLDEN, JAMES EDWARD, b Yonkers, NY, Oct 19, 44; m 68; c 2. MEDICAL PHYSICS. *Educ:* Providence Col, RI, BS, 66; Univ Pa, PhD(physics), 71. *Prof Exp:* Res assoc nuclear physics, Nuclear Physics Lab, Univ Pittsburgh, 71-74; asst prof radiol, 74-79, assoc prof med phys & radiol, 79-80, PROF MED PHYS & RADIOL, UNIV WIS-MADISON, 80- *Concurrent Pos:* James Picker Found scholar, 76-80. *Mem:* Soc Nuclear Med; Soc Cerebral Blood Flow & Metab. *Res:* Linear systems analysis of medical imaging processes and of tracer kinetics for investigation of cerebral and cardiac physiology; radioactive tracer dynamics and compartmental analysis; computer assisted transverse-axial reconstruction algorithms. *Mailing Add:* Dept Radiol Med Physics Univ Wis 1530 Med Sci Ctr 1300 University Ave Madison WI 53706-1532

HOLDEN, JAMES RICHARD, b Omaha, Nebr, Aug 1, 28; m 58; c 2. PHYSICAL CHEMISTRY. *Educ:* Univ Nebr, BSc, 50, MSc, 51; Univ Iowa, PhD(phys chem), 55. *Prof Exp:* Phys chemist, US Naval Surface Warfare Ctr, 55-88; RETIRED. *Mem:* Am Chem Soc; Am Crystallog Asn. *Res:* Study of the physical properties of explosives and their relation to chemical constitution; determination of the crystal structure of organic compounds by x-ray crystallography. *Mailing Add:* HCR 31 Box 347 Bath ME 04530-9609

HOLDEN, JOHN B, JR, b Washington, DC, May 26, 35; m 64; c 3. ANALYTICAL CHEMISTRY. *Educ:* Cath Univ Am, BChE, 57; Princeton Univ, MA, 59, PhD(chem), 61. *Prof Exp:* Sr res chemist, Printing Prod Div, Minn Mining & Mfg Co, 61-68; from asst prof to assoc prof chem, 68-74, prof chem & dir Environ Studies Inst, 74-86, PROF CHEM, MANKATO STATE UNIV, 86- *Mem:* Am Chem Soc. *Res:* Iron III chemistry; experimentation for instructional chemistry; water chemistry. *Mailing Add:* Chem Dept Mankato State Univ PO Box 40 Mankato MN 56001

HOLDEN, JOSEPH THADDEUS, b Brooklyn, NY, Jan 18, 25; m 48; c 4. MICROBIOLOGY. *Educ:* Polytech Inst Brooklyn, BS, 44; Univ Wis, MS, 48, PhD(biochem), 51. *Prof Exp:* Asst biochem, Univ Wis, 47-51; USPHS res fel biochem & genetics, Calif Inst Technol, 51-53; biochemist exp sta, Chem Dept, E I du Pont de Nemours & Co, 53-54; res scientist, assoc div chmn & head, Cell Physiol Sect, Neurosci Div, City of Hope Res Inst, 54-80, asst dir res, 80-87; ACTG DIR, BECKMAN RES INST, 87- *Concurrent Pos:* Nat Inst Neurol Dis & Stroke spec trainee, Col France, Paris, 74. *Mem:* Am Soc Microbiol; Am Soc Biochemists; Am Soc Cell Biol; Am Soc Neurochem; Soc Neurosci. *Res:* Microbial amino acid and vitamin nutrition and metabolism; genetic factors in nutrition and metabolism; active transport of cell nutrients; biosynthesis of bacterial cell wall; structure and function of cell membranes. *Mailing Add:* City of Hope Beckmann Res Inst 1450 E Duarte Rd Duarte CA 91010

HOLDEN, KENNETH GEORGE, b Summit, NJ, Dec 19, 34; m 60; c 1. MEDICINAL CHEMISTRY. *Educ:* Brown Univ, BS, 57; Univ Calif, Berkeley, PhD(chem), 61. *Prof Exp:* Sr med chemist, 61-67, asst dir chem, 67-72, sr investr, res & develop div, 72-79, assoc sci dir, 79-80, DEP DIR MED CHEM, SMITH KLINE & FRENCH LABS, 80- *Mem:* AAAS; Am Chem Soc; NY Acad Sci; The Chem Soc. *Res:* Natural products chemistry; synthesis and structure determination of alkaloids, microorganism metabolites, steroids, prostaglandins, glycoproteins, B-lactam antibiotics and enzyme inhibitors. *Mailing Add:* RD 5 Box 336 Horseshoe Trail Malvern PA 19355-9501

HOLDEN, NORMAN EDWARD, b New York, NY, Feb 1, 36; m 62; c 9. NUCLEAR PHYSICS. *Educ:* Fordham Univ, BS, 57; Cath Univ Am, MS, 59, PhD(physics), 64. *Prof Exp:* Officer nuclear physics, Air Force Tech Appln Ctr, 62-65; physicist, G E Knolls Atomic Power Lab, 65-74; PHYSICIST NUCLEAR PHYSICS, BROOKHAVEN NAT LAB, 74- *Concurrent Pos:* Mem Int Comn Atomic Weights, 71-, secy, 75-79, pres, 79-83; mem Subcomt Assessment Isotopic Compos, Int Union Pure & Appl Chem, 73-; US adv nuclear struct & decay data, Int Atomic Energy Agency, 76-78; mem Int Comt Radionuclide Metrol, 77- & Int Comt Radiochem & Nuclear Tech, 83- *Mem:* Am Phys Soc; Am Nuclear Soc; NY Acad Sci; AAAS. *Res:* Atomic weights; isotopic abundances of elements; neutron cross sections; radioactivity and nuclear data. *Mailing Add:* High Flux Beam Reactor Brookhaven Nat Lab Upton NY 11973

HOLDEN, PALMER JOSEPH, b Apr 10, 43; US citizen; m 74; c 1. ANIMAL NUTRITION, ANIMAL HUSBANDRY. *Educ:* NDak State Univ, BS, 63; Iowa State Univ, MS, 67, PhD(animal nutrit), 70. *Prof Exp:* Exten adv agr, US Army, Vietnam, 70-71; EXTEN SPECIALIST SWINE NUTRIT, IOWA STATE UNIV, 72- *Concurrent Pos:* Swine consult, Monterey Farms, Philippines, 73-86, Univ Costa Rica, 78, 81 & 89, Poland, Bulgaria & Ger Dem Repub, 86, Honduras, 87, Sardenia, Malaysia, 89, Sweden, Denmark & Holland, 89. *Mem:* Am Soc Animal Sci. *Res:* Swine nutrition research of practical problems related directly to production. *Mailing Add:* Dept Animal Sci Iowa State Univ 109 Kildee Hall Ames IA 50011

HOLDEN, THOMAS MORE, b Ilkley, Yorkshire, Eng; Can citizen. METALLIC MAGNETISM, APPLIED PHYSICS. *Educ:* Univ Leeds, Eng, BSc, 61, PhD(physics), 65. *Prof Exp:* Indust fel, Atomic Energy Res Estab, Harwell, UK, 64-66; Nat Res Coun fel, 66-67, asst res officer, 67-75, assoc res officer, 75-84, SR RES OFFICER, CHALK RIVER NUCLEAR LABS, ATOMIC ENERGY CAN LTD, ONT, 84- *Mem:* Can Asn Physicists. *Res:* Application of thermal neutrons to the study of the structure and dynamics of magnetic materials, particularly rare-earth and actinide compounds; engineering applications of thermal neutron diffraction methods. *Mailing Add:* Neutron & Solid State Physics Br Atomic Energy Can Ltd Chalk River ON K0J 1J0 Can

HOLDEN, WILLIAM DOUGLAS, b Pittsfield, Mass, Aug 25, 12; m 36; c 3. SURGERY. *Educ:* Cornell Univ, AB, 34, MD, 37. *Prof Exp:* Asst surgeon, Univ Hosps, Cleveland, 41-42 & 46-50; from instr to assoc prof, 46-50, Payne Prof surg & chmn, 50-77, prof, 77-82, EMER PROF SURG, SCH MED, CASE WESTERN RESERVE UNIV, 82- *Concurrent Pos:* Dir surg, 50-77, assoc surg, 77-85, hon staff, Univ Hosps, Cleveland, 85-; hon mem, Nat Bd Med Examiners. *Honors & Awards:* Distinguished Serv Award, AMA & Am Col Surgeons; Special Award, Am Bd Med Specialties. *Mem:* Soc Vascular Surg; Soc Univ Surg; Am Surg Asn; AMA; fel Am Col Surg. *Res:* Author of 180 publications. *Mailing Add:* 9775 Country Scene Lane Concord Township OH 44060

HOLDEN, WILLIAM R(OBERT), b Ft Worth, Tex, Dec 20, 28; m 57; c 1. ENGINEERING PHYSICS. *Educ:* Univ Okla, BS, 53, MS, 55. *Prof Exp:* From asst prof to assoc prof, 55-77, PROF PETROL ENG, LA STATE UNIV, BATON ROUGE, 77- *Mem:* Am Inst Mining, Metall & Petrol Engrs; Soc Petrol Engrs. *Res:* Turbulent flow properties of non-Newtonian fluids. *Mailing Add:* Dept Petrol Eng La State Univ 3526 Cera Bldg Baton Rouge LA 70803

HOLDER, CHARLES BURT, JR, b Berea, Ky, Aug 29, 14; m 53; c 5. CHEMISTRY. *Educ:* Transylvania Col, AB, 36; Ga Inst Technol, MS, 38; Univ Tex, PhD(org chem), 41. *Prof Exp:* Chemist, Texaco, Inc, 41-65, res chemist petrol prod res, 65-80; RETIRED. *Mem:* Am Chem Soc; Sigma Xi. *Res:* Polymers; synthesis of hydantoin derivatives; petroleum derivatives; organic chemistry. *Mailing Add:* 33 Edge Hill Dr Wappingers Falls NY 12590

HOLDER, DAVID GORDON, b Louisville, Miss, Oct 18, 43; m 67; c 1. PLANT BREEDING. *Educ:* Miss State Univ, BS, 65, MS, 67; Purdue Univ, PhD(plant genetics), 71. *Prof Exp:* Rockefeller Found res grant entom, Miss State Univ, 71-73; GENETICIST, US SUGAR CORP, 73- *Mem:* Am Soc Agron; Crop Sci Soc Am; Int Soc Sugarcane Technologists; Am Soc Sugarcane Technologists. *Res:* Sugarcane genetics and breeding, sugarcane pathology. *Mailing Add:* PO Drawer 1207 Clewiston FL 33440

HOLDER, GERALD D, b Los Angeles, Calif, July 29, 50; m 68; c 2. NATURAL GAS RECOVERY, THERMODYNAMICS. *Educ:* Kalamazoo Col, BA, 72; Univ Mich, BSE, 73, MSE, 74, PhD(chem eng), 76. *Prof Exp:* Asst prof, Columbia Univ, 76-79; from asst prof to assoc prof chem eng, 79-86, PROF CHEM ENG, UNIV PITTSBURGH, 86-, CHMN, 87- *Concurrent Pos:* Consult, Exxon Res & Eng, 77, Gulf Sci & Technol, 80-81, Norks-Nyddo, 87, Petrobas, 87. *Mem:* Am Chem Soc; Am Inst Chem Engrs; Soc Petrol Engrs. *Res:* Thermodynamics of natural gas hydrates and their potential as an alternative energy source; molecular dynamics; extraction and reaction of substances using supercritical fluids. *Mailing Add:* Dept Chem Eng Univ Pittsburgh Pittsburgh PA 15261

HOLDER, HAROLD DOUGLAS, b Raleigh, NC, Aug 9, 39; m 80; c 3. COMPUTER STIMULATION COMMUNITY SYSTEMS. *Educ:* Samford Univ, AB, 61; Syracuse Univ, MA, 62, PhD(commun sci), 65. *Prof Exp:* Asst prof jour & sociol, Baylor Univ, 65-69; assoc prof sociol, NC State Univ, 69-74; sr scientist, Human Ecol Inst, Raleigh, NC, 74-80, dir, Chapel Hill, NC, 80-86; DIR & SR SCIENTIST, PREV RES CTR, BERKELEY, CALIF, 86- *Concurrent Pos:* Vis prof, Univ NC, Chapel Hill, 80-86 & Sch Pub Health, Univ Calif, Berkeley, 86-91. *Mem:* Res Soc Alcoholism; Am Pub Health Asn. *Res:* Conduct cost/benefit studies of the alcoholism treatment; policy study of prevention strategies for the reduction of alcohol problems. *Mailing Add:* Prevention Res Ctr Pac Inst Res & Eval Nat Inst Alcohol Abuse & Alcoholism 2532 Durant Ave Berkeley CA 94704

HOLDER, IAN ALAN, b Brooklyn, NY, July 23, 34. MICROBIOLOGY. *Educ:* Brooklyn Col, BS, 58, MA, 63; Univ Kans, PhD(microbiol), 67. *Prof Exp:* Res asst microbiol, Brooklyn Col, 58-59; res & develop biologist, Res & Develop Lab, Schieffelin & Co, 59-61; res assoc, Bellevue Hosp, 61-63; from asst prof to prof exp surg, Col Med, Univ Cincinnati, 67-79, from asst prof to prof microbiol, 70-80, actg dir, Div Microbiol, Med Ctr, 73-75; DIR, DEPT MICROBIOL, SHRINERS BURNS INST, CINCINNATI, 67- *Concurrent Pos:* Vis prof, Kitasato Inst, Tokyo, Japan, 84, dept surg & pediat, Univ Tex Health & Sci Ctr, Houston, 85 & dept biol, Youngstown State Univ, Youngstown, Ohio, 88. *Honors & Awards:* Am Burn Asn Award, 85; Tanner Vandeput Prize, 86. *Mem:* Am Soc Microbiol; Am Burn Asn; Sigma Xi; fel Am Acad Microbiol; Int Soc Burn Injuries; Japan Pseudomonas Aeruginosa Soc. *Res:* Host parasite interactions, virulence factors, mechanisms of pathogenesis and clinical aspects of Pseudomonas aeruginosa and Candida albicans infections. *Mailing Add:* Shriners Burns Inst 202 Goodman St Cincinnati OH 45219

HOLDER, LEONARD IRVIN, b Ft Worth, Tex, Nov 30, 23; m 49; c 2. MATHEMATICS. *Educ:* Agr & Mech Col Tex, BS, 47, MS, 51; Purdue Univ, PhD(math), 55. *Prof Exp:* Asst prof math, Arlington State Col, 47-49; instr, Agr & Mech Col Tex, 51-52; engr-designer, Boeing Co, Kans, 52; asst math, Purdue Univ, 52-55; assoc prof, San Jose State Col, 55-64; chmn dept, 64-75, PROF MATH, GETTYSBURG COL, 64-, DEAN, 75-, CHMN DEPT, 80- *Mem:* Am Math Soc; Math Asn Am. *Res:* Infinite series, particularly summability of series as applied to Fourier series. *Mailing Add:* Dept Math Gettysburg Col Gettysburg PA 17325

HOLDER, THOMAS M, b Corinth, Miss, Sept 1, 26; m; c 3. PEDIATRIC SURGERY. *Educ:* Bowman Gray Sch Med, MD, 52. *Prof Exp:* Intern, Jefferson Med Col Hosp, Philadelphia, 52-53; resident, Children's Hosp, Boston, 53-54, 58 & 59; chmn dept surg, Children's Mercy Hosp, Kansas City, Mo, 60-63; from assoc prof to prof surg & head sect pediat surg, Univ Kans, 63-72; CLIN PROF SURG, UNIV MO-KANSAS CITY, 72-; HEAD CARDIOTHORACIC SURG, CHILDREN'S MERCY HOSP, 72- *Concurrent Pos:* Vis scientist, McIndoe Mem Labs, Eng, 69-70. *Mem:* Am Col Surg; Am Asn Thoracic Surg; Am Pediat Surg Asn (pres, 75-76); Am Acad Pediat; Am Surg Asn. *Res:* Production of congenital anomalies by in utero operative procedures; effect of vasodilators on blood flow and systemic pulmonary shunts and physiologic cardiovascular studies. *Mailing Add:* 5520 College Blvd Suite 206 Overland Park KS 66211

HOLDERBAUM, DANIEL, VASCULAR & CONNECTIVE TISSUE. *Educ:* Cleveland State Univ, PhD(regulatory biol), 80. *Prof Exp:* ASST STAFF, CLEVELAND CLIN FOUND, 82- *Mailing Add:* Biol Dept Case Western Reserve Univ Cleveland OH 44106

HOLDHUSEN, JAMES S(TAFFORD), b Houghton, SDak, Apr 2, 25; m 55; c 3. AERONAUTICAL ENGINEERING. *Educ:* Univ Minn, BCE, 45, MS, 47, PhD(fluid mech), 52. *Prof Exp:* Instr civil eng, Univ Minn, 49-52; chief res engr, 52-57, vpres res & planning, 57-71, EXEC VPRES, FLUIDYNE ENG CORP, 71- *Honors & Awards:* J C Stevens Award, Am Soc Civil Engrs, 51. *Mem:* Am Inst Aeronaut & Astronaut; Am Soc Mech Engrs. *Res:* Propulsion aerodynamics; wind tunnel development; fluidized bed combustors; aerodynamics. *Mailing Add:* 4520 Sunset Ridge Golden Valley MN 55427

HOLDREDGE, RUSSELL M, b Paonia, Colo, July 22, 33; m 55; c 4. MECHANICAL ENGINEERING. *Educ:* Univ Colo, BS(mech eng) & BS(bus admin), 56, MS, 59; Purdue Univ, PhD(mech eng), 65. *Prof Exp:* Systs engr, Sandia Corp, 56-57; instr mech eng, Univ Colo, 57-59; from asst prof to assoc prof, 59-70, head dept, 70-76, PROF MECH ENG, 70-, ASSOC DEAN, COL ENG, UTAH STATE UNIV, 76- *Concurrent Pos:* NSF fac fel, Purdue Univ, 65-66. *Mem:* Am Soc Mech Engrs; Am Soc Eng Educ. *Res:* Cryogenic heat transfer, primarily transient cooling and pool boiling; reactor heat transfer, primarily transient cooling and conduction in nonhomogeneous materials. *Mailing Add:* Col Mech Eng Utah State Univ Logan UT 84322

HOLDREN, JOHN PAUL, b Sewickley, Pa, Mar 1, 44; m 66; c 2. ENERGY CONVERSION, ENVIRONMENTAL PHYSICS. *Educ:* Mass Inst Technol, BS, 65, MS, 66; Stanford Univ, PhD(aeronaut, astronaut & elec eng), 70. *Hon Degrees:* ScD, Univ Puget Sound, 75. *Prof Exp:* Physicist, Magnetic Fusion energy, Lawrence Livermore Nat Lab, Univ Calif, 70-73; from asst prof to assoc prof, 70-78, PROF ENERGY & RESOURCES, UNIV CALIF, BERKELEY, 78-; SR INVESTR, ROCKY MOUNTAIN BIOL LAB, 76- *Concurrent Pos:* Mem, Int Environ Prog Comt, Nat Acad Sci-Nat Acad Eng, 70-75, mem comt nuclear & alt energy systs & comt to rev lit nuclear risks, 75-; lectr, Univ Calif, Berkeley, 71-72; sr res fel, Pop Prog, Calif Inst Technol, 72-73; consult, Magnetic Fusion Energy Div, Lawrence Livermore Lab, 73-, Energy & Environ Div, Lawrence Berkeley Lab, 74- & Int Inst Appl Systs Anal, 75-; mem, Energy Res Adv Bd, US Dept Energy, 78-79; vis fel, Resource Systs Inst & Environ & Policy Inst, East West Ctr, Honolulu, 79 & 80; vchair, Comt Int Security Studies, Am Acad Arts & Sci, 83-; MacArthur Found Prize fel, 81-86; chmn, Fedn Am Scientists, 84-86; chmn, Sr Comt Environ Safety & Econ Aspects Magnetic Fusion Energy, US Dept Energy, 85-; vis prof, Univ Rome, 87 & vis scholar, Ctr Int Studies, Mass Inst Technol, 88; Kistiakowsky vis scholar, AAAS, 86-87; guest res, Max-Plank Soc, WGer, 87-88. *Honors & Awards:* Gustavsen lectr, Univ Chicago, 78. *Mem:* Nat Acad Sci; Fel Am Phys Soc; Fedn Am Scientists; Am Nuclear Soc; Sigma Xi; fel Am Acad Arts & Sci; fel AAAS. *Res:* Environmental aspects of coal, fusion and fission; national and international problems in energy and environment nuclear weapons and arms control; plasma physics, energy technology and policy, environmental problems, population and arms control. *Mailing Add:* Energy & Resources Group Univ Calif 100 T-4 Berkeley CA 94720

HOLDSWORTH, ROBERT POWELL, b Stoughton, Mass, Jan 12, 15; m 40; c 3. ENTOMOLOGY. *Educ:* Univ Mass, BS, 37; Harvard Univ, MA, 38, PhD(zool), 41. *Prof Exp:* Agt entom, USDA, 46-47; salesman, E I du Pont de Nemours & Co, 47-56, prod mgr, 56-58; exten entomologist, Ohio State Univ, 58-65, resident teacher entom & researcher, 65-81; RETIRED. *Mem:* Entom Soc Am. *Res:* Integrated pest management of apple arthropods. *Mailing Add:* 2739 Westmont Blvd Columbus OH 43221

HOLE, FRANCIS DOAN, b Muncie, Ind, Aug 25, 13; m 41; c 2. SOIL ZOOLOGY, SOIL LANDSCAPE DYNAMICS. *Educ:* Earlham Col, AB, 33; Haverford Col, MA, 34; Univ Wis-Madison, PhD(geol soil sci), 43. *Prof Exp:* Instr Ger, Friends Cent Country Day Sch, 34-35; instr French, Ger, music, & physiography, Westtown Friends Boarding Sch, 35-38; instr geol & soil sci, Earlham Col, 40-41 & 43-44; asst prof soil sci, 46-51, assoc prof, 52-57, prof, 58-83, EMER PROF SOIL SCI & GEOG, UNIV WIS-MADISON, 83- *Concurrent Pos:* Ed, Soil Surv Horizons, 60-63; vis prof, Dept Soils, Water & Eng, Univ Ariz, 81. *Mem:* Soil Sci Soc Am; Geol Soc Am; Asn Am Geographers; Sigma Xi; Soc Social Responsibility in Sci. *Res:* Influence of animals on soils; factors affecting the incorporation of organic matter into soil; numerical classification of soils; soil landscape analysis. *Mailing Add:* 2201 Center Ave Madison WI 53704

HOLECHEK, JERRY LEE, b Lebanon, Ore, Mar 6, 48. RANGE NUTRITION. *Educ:* Mont State Univ, MS, 76; Ore State Univ, BS, 71, PhD(animal sci), 80. *Prof Exp:* PROF RANGE SCI, DEPT ANIMAL & RANGE SCI, NMEX STATE UNIV, 79- *Mem:* Soc Range Mgt; Am Soc Animal Sci; Soil Conserv Soc Am; Wildlife Soc. *Res:* Nutrition of free ranging animals; reclamation of deteriorated ranges; range/wildlife interactions; techniques for studying food habits and nutrition of wild and domestic ruminant animals. *Mailing Add:* Dept Animal & Range Sci NMex State Univ Las Cruces NM 88003

HOLEMAN, DENNIS LEIGH, b Portland, Ore, May 10, 46; m 81. MAN-MACHINE INTELLIGENCE. *Educ:* Harvey Mudd Col, BSE, 68; Univ Calif, Berkeley, MEME, 70. *Prof Exp:* Sr engr, Oceanic Div, Westinghouse Elec Corp, 68-75; co-founder & coordr eng, IMSAI Mfg Corp, 75-78; PRIN ENGR, SRI INT, 78- *Concurrent Pos:* Prin, Heuristica: A Systs Consultancy, 72-75. *Mem:* Am Asn Artificial Intel; Inst Elec & Electronics Engrs; Am Inst Aeronaut & Astronaut; AAAS. *Res:* Development of systems integrating human and machine intelligence capabilities in a synergistic fashion for defense, aerospace and other complex system control functions. *Mailing Add:* SRI Int MS BN390 333 Ravenswood Ave Menlo Park CA 94061

HOLESKI, PAUL MICHAEL, b Akron, Ohio, July 7, 43. INSECT ECOLOGY, TAXONOMY. *Educ:* Wilmington Col, AB, 66; Univ Akron, MS, 69; Bowling Green State Univ, PhD(biol), 76. *Prof Exp:* Teacher biol, Wayne Trace Local Sch, 68-72; fel, Bowling Green State Univ, 75-76, asst prof biol, Firelands Br, 76-77; asst prof biol, Agr Tech Inst, Ohio State Univ, 79-80; ASSOC PROF BIOL, RIO GRANDE COL, 80- *Concurrent Pos:* Vpres elect, Zool Sect, Ohio Acad Sci. *Mem:* Sigma Xi; Entom Soc Am; Coleopterists Soc. *Res:* Adaptions and changes in the composition of shore-inhabiting insect communities in response to environmental change; taxonomy of Coleoptera which inhabit the shore. *Mailing Add:* 628 Linwood PO Box 268 Rio Grande OH 45674

HOLFELD, WINFRIED THOMAS, b Essen, Ger, June 1, 26; nat US; m 53; c 3. ORGANIC CHEMISTRY. *Educ:* Fordham Univ, BS, 47, MS, 49; Rutgers Univ, PhD(chem), 57. *Prof Exp:* Asst supvr biochem lab, St Vincent's Hosp, NY, 49-51; biochemist, Carter Prod Inc, NJ, 51-54; asst instr, Rutgers Univ, 54-55; sr res chemist, Tech Div, 56-69, tech specialist, Mkt Div, 69-78, DEVELOP ASSOC, TECH DIV, E I DU PONT DE NEMOURS & CO, INC, 78- *Mem:* Am Asn Textile Chemists & Colorists. *Res:* Natural products; textiles. *Mailing Add:* 700 Thornby Rd Sharpley DE 19899

HOLFORD, RICHARD L, b London, Eng, Apr 1, 37; wid; c 2. APPLIED MATHEMATICS. *Educ:* Cambridge Univ, BA, 58, PhD(math), 64. *Prof Exp:* Res fel math, Manchester Univ, 62-64; lectr, Stanford Univ, 64-65; res scientist, NY Univ, 65-66; sr scientist, TRG, Inc Div, Control Data Corp, 66-68; MEM TECH STAFF, BELL LABS, 68- *Mem:* Assoc Acoust Soc Am. *Res:* Water waves; rough surface scattering; asymptotic methods in acoustics. *Mailing Add:* AT&T Bell Labs 15F-310 Whippany NJ 07981

HOLFORD, RICHARD MOORE, b Cheltenham, UK, June 25, 38; m 61; c 3. HEALTH PHYSICS, NUCLEAR INSTRUMENTATION. *Educ:* Cambridge Univ, BA, 61, PhD, 66. *Prof Exp:* Res asst radiotherapeut, Cambridge Univ, 61-62; from asst res officer to res officer, Atomic Energy Can Ltd, 66-89; PROG SYSTS ANALYST, HAMILTON DIGITAL DESIGNS, 89- *Mem:* Can Radiation Protection Asn. *Res:* Design of health physics instruments; biomathematics. *Mailing Add:* Hamilton Digital Designs 3180 Main Way Dr Burlington ON L7M 1A5 Can

HOLFORD, THEODORE RICHARD, b Columbus, Ohio, May 19, 47; m 69; c 2. BIOSTATISTICS. *Educ:* Andrews Univ, BA, 69; Yale Univ, MPhil, 72, PhD(biomet), 73. *Prof Exp:* Mem res biomet, 72-73, from asst prof to assoc prof pub health, 74-89, PROF PUB HEALTH, MED SCH, YALE UNIV, 89- *Concurrent Pos:* Eleanor Roosevelt int fel, Am Union Int Contre le Cancer, Univ Oxford; Epidemiol & Dis Control Study Sect, NIH, 86-89; assoc ed, Biomet, 84-88, Am J Epidemiol, 89-92. *Honors & Awards:* Wakeman Award, 89. *Mem:* Am Statist Soc; Soc Epidemiol Res; Biomet Soc. *Res:* Formulation of mathematical models to describe and explain the underlying mechanics of fundamental processes and the development and application of statistical methods in the field of health. *Mailing Add:* Dept Epidemiol & Pub Health Yale Univ Sch Med PO Box 3333 New Haven CT 06510

HOLGUIN, ALFONSO HUDSON, b El Paso, Tex, April 3, 31; m 54; c 6. EPIDEMIOLOGY. *Educ:* Univ Tex, El Paso, BA, 51, Galveston, MD, 57; Harvard Univ, MPH, 64. *Prof Exp:* Med officer, USPHS, Seattle, 57-58; med officer virus res, Lab Br, Ctr Dis Control, USPHS, 58-62, asst to chief, 62-64, chief, Tuberc Prog, 64-68, dir, Bur Training, 70-74; vis prof, 74-78, PROF EPIDEMIOL, SCH PUB HEALTH, UNIV TEX, HOUSTON, 78- *Concurrent Pos:* Prin investr, Epidemiol Res Univ, 79-; consult, Southwest Res Inst, 79- & Nat Inst Heart, Lung & Blood, 80- *Res:* Epidemiology of infectious diseases, especially tuberculosis and nosocomial infections; epidemiology of environments and occupationally related health problems. *Mailing Add:* Sch Pub Health HSC-SA Univ Tex 7703 Floyd Curl Dr San Antonio TX 78284

HOLICK, MICHAEL FRANCIS, b Mar 15, 46; c 1. ENDOCRINOLOGY, NUTRITIONAL BIOCHEMISTRY. *Educ:* Seton Hall Univ, BS, 68; Univ Wis, MS, 70, PhD, 71, MD, 76. *Prof Exp:* Clin fel, 76-78, asst prof, 78-81, ASSOC PROF MED, HARVARD MED SCH, 81-; ASSOC PROF NUTRIT BIOCHEM, MASS INST TECHNOL, 81- *Concurrent Pos:* Clin asst med, Mass Gen Hosp, 78-81, assoc prof, 81-; lectr, Mass Inst Technol, 79-81. *Honors & Awards:* Wilson S Stone Mem Award, M D Anderson Hosp, 72; Fuller Albright Young Investr Award, 80. *Mem:* Am Soc Bone & Mineral Res; Am Soc Clin Invest; Endocrine Soc; Am Soc Photobiol; Am Col Physicians. *Res:* Photobiology of vitamin D; chemical synthesis of vitamin D analogs; role of caffeine on fetal bone growth and development; metabolism of vitamin D in health and disease. *Mailing Add:* Boston Univ Sch Med Boston City Hosp 80 E Concord St M-1013 Boston MA 02119

HOLICK, SALLY ANN, b Harrisburg, Pa, Dec 11, 48; m 72; c 1. VITAMIN D CHEMISTRY. *Educ:* Bloomsburg Col, Pa, BA, 70; Univ Wis, PhD(biochem), 74. *Prof Exp:* Fel biochem, Univ Wis, 74-75; instr chem, Boston Univ, 75-77; res fel med, Harvard Med Sch & Mass Gen Hosp, 77-81; instr med, Harvard Med Sch, 77-82; res scientist nutrit, Mass Inst Technol, 82-85 & Tufts Univ, 85-88; CONSULT, VITAMIN D, 88- *Mem:* Am Chem Soc; Am Soc Biol Chemists; Sigma Xi. *Res:* Metabolism of vitamin D; structure function relationships; photobiology. *Mailing Add:* 31 Bishop Lane Sudbury MA 01776

HOLIFIELD, CHARLES LESLIE, b Minneapolis, Minn, May 16, 40; m 65; c 2. ANALYTICAL CHEMISTRY, INORGANIC CHEMISTRY. *Educ:* Univ Tex, BS, 62; Univ Minn, MS, 65. *Prof Exp:* RES CHEMIST ANALYTICAL, CHEM DIV, PPG INDUSTS, INC, 65-, RES ASSOC, ANALYTICAL INSTRUMENTAL ANALYSIS. *Mem:* Am Chem Soc; Electron Micros Soc Am. *Res:* X-ray fluorescence and diffraction; atomic absorption spectroscopy; emission spectroscopy; thermal analysis; scanning electron microscopy; transmission electron microscopy. *Mailing Add:* PPG Indust Chem Div 440 College Park Dr Monroeville PA 15146

HOLKEBOER, PAUL EDWARD, b Holland, Mich, Jan 16, 28; m 51; c 4. ANALYTICAL CHEMISTRY. *Educ:* Hope Col, AB, 51; Purdue Univ, MS, 53, PhD(chem), 56. *Prof Exp:* From asst prof to prof chem, 55-88, Western Mich Univ, Coordr Grad Sci Educ Prog, 66-88, Coordr Acad Adv, 80-88; RETIRED. *Mem:* Am Chem Soc. *Res:* Inorganic chemistry; coordination compounds and their application to analytical chemistry; science education. *Mailing Add:* 2381 W Lakewood Blvd Holland MI 49424

HOLL, FREDERICK BRIAN, b Winnipeg, Man, 1944. PLANT GENETICS, SUSTAINABLE AGRICULTURE. *Educ:* Univ Man, BSA, 66, MSc, 68; Univ Cambridge, Eng, PhD(genetics), 71. *Prof Exp:* Res officer plant genetics, Prairie Regional Lab, Nat Res Coun, 72-78; ASSOC PROF, DEPT PLANT SCI, UNIV BC, 78- *Concurrent Pos:* Sr consult, Lamorna Enterprises Ltd. *Mem:* Genetics Soc Can; Genetical Soc Brit. *Res:* Inheritance of dinitrogen fixation in legumes; use of rhizosphere microorganisms in sustainable agriculture and forestry. *Mailing Add:* Dept Plant Sci Univ BC Vancouver BC V6T 2A2 Can

HOLL, J(OHN) WILLIAM, b Danville, Ill, Feb 20, 28; m 50; c 7. MECHANICAL & AERONAUTICAL ENGINEERING. *Educ:* Univ Ill, BS, 49, MS, 51; Pa State Univ, PhD(mech eng), 58. *Prof Exp:* Asst mech eng, Univ Ill, 49-51; res assoc ord res lab, Pa State Univ, 51-55; res engr subsonic wind tunnel, Chem Warfare Lab, Chem Ctr, US Army, Md, 55-56; res assoc Appl Res Lab, Pa State Univ, 56-58, asst prof, 58-59; assoc prof mech eng, Univ Nebr, 59-63; assoc prof, 63-67, PROF AEROSPACE ENG, PA STATE UNIV, 67- *Honors & Awards:* R T Knapp Award, Am Soc Mech Engrs, 70, Melville Medal, 70. *Mem:* Assoc fel Aeronaut & Astronaut; fel Am Soc Mech Engrs; Sigma Xi. *Res:* Fluid mechanics; two phase flow; cavitation. *Mailing Add:* Dept Aerospace Studies University Park PA 16802

HOLL, MANFRED MATTHIAS, b Cologne, Ger, Nov 12, 28; nat US; m 52; c 2. MARINE WEATHER ANALYSIS & PREDICTION. *Educ:* McGill Univ, BSc, 50; Univ Toronto, MA, 51; Univ Calif, Los Angeles, PhD(meteorol), 57. *Prof Exp:* Meteorologist, Meteorol Serv Can, 50-52; instr meteorol & jr res meteorologist, Univ Calif, Los Angeles, 52-57; proj scientist & sect chief geophys res directorate, Air Force Cambridge Res Ctr, 57-58; head weather dynamics prog, Stanford Res Inst, 58-61; pres & dir res, Meteorol Int, Inc, 61-84; sr scientist, Nat Ocean Serv, Nat Oceanic & Atmospheric Admin, US Dept Com, 84-90; CONSULT, 90- *Mem:* Am Meteorol Soc. *Res:* Numerical models of atmosphere and ocean structures for analysis and forecasting of significant variabilities; environmental-data evaluation; processing and objective assembly; inherent-scale techniques; mathematical-physical, statistical and probabalistic approaches to weather studies; related exploitation of computers. *Mailing Add:* 25526 Carmel Knolls Dr Carmel CA 93923

HOLL, RICHARD JACOB, nuclear science, for more information see previous edition

HOLLAAR, LEE ALLEN, b Litchfield, Minn, Mar 9, 47; m 68; c 1. INFORMATION RETRIEVAL, SYSTEMS ENGINEERING. *Educ:* Ill Inst Technol, BS, 69; Univ Ill, Urbana-Champaign, MS, 74, PhD(comput sci), 75. *Prof Exp:* Sr syst programmer, First Nat Bank Chicago, 67-69; eng mgr, Datalogics, Inc, 69-74; asst prof comput sci, Univ Ill, 75-80; assoc prof comput sci, 80-86, prof comput sci & elec eng, Univ Utah, 86-, DIR, CAMPUS NETWORKING, 86-; PRES, CONTEXTURE, INC, 83- *Concurrent Pos:* Distinguished visitor, Inst Elec & Electronics Engrs Comput Soc, 85-86. *Mem:* Sr mem Inst Elec & Electronics Engrs; Asn Comput Mach. *Res:* Hardware and software system for the retrieval and handling of text; development of special purpose backend processors; high-speed text search engine implemented in custom very-large-scale integration; design and implementation of a distributed software architecture for text handling; very-large-scale integration design tools and databases. *Mailing Add:* Dept Comput Sci 3160 Merrill Eng Bldg Univ Utah Salt Lake City UT 84112

HOLLAHAN, JOHN RONALD, physical chemistry; deceased, see previous edition for last biography

HOLLAND, ANDREW BRIAN, b Nashville, Tenn, Feb 24, 40; m 63; c 2. SOLID STATE PHYSICS, PHOTOGRAPHIC CHEMISTRY. *Educ:* Vanderbilt Univ, AB, 61; Univ Wis, MS, 64, PhD(physics), 69. *Prof Exp:* Res physicist, Photo Prod Dept, Exp Sta, E I du Pont de Nemours & Co, 69-75; SCIENTIST, POLAROID CORP, 75- *Mem:* Am Phys Soc; Soc Photog Scientists & Engrs. *Res:* Fermi surfaces of solids; electroluminesence; solid state electronics; electrooptics; photographic science; particle sizing; sensitometry. *Mailing Add:* Seven Autumn Lane Wayland MA 01778

HOLLAND, CHARLES D(ONALD), b Iredell Co, NC, Oct 9, 21; m 45; c 3. CHEMICAL ENGINEERING. *Educ:* NC State Col, BS, 43; Agr & Mech Col, Tex, MS, 49, PhD(chem eng), 53. *Prof Exp:* Engr, Burlington Mills Corp, 47-48; from asst prof to prof, 52-87, head dept, 84-87, EMER PROF CHEM ENG, TEX A&M UNIV, 87-; PRES, TEX INST ADVAN CHEM TECHNOL, 86- *Mem:* Fel Am Inst Chem Engrs; Am Chem Soc; Am Soc Eng Educ; AAAS; Sigma Xi; fel Am Inst Chemists. *Res:* Design of distillation columns; catalysis and kinetics; reactor design. *Mailing Add:* Dept Chem Eng Tex A&M Univ College Station TX 77843

HOLLAND, CHARLES JORDAN, b Dublin, Ga, Mar 28, 48; m 86; c 2. APPLIED MATHEMATICS, COMPUTER SCIENCE. *Educ:* Ga Inst Technol, BS, 68, MS, 69; Brown Univ, PhD(appl math), 72. *Prof Exp:* From asst prof to assoc prof math, Purdue Univ, 72-81; sci officer math, 81-84, liaison scientist, London, 84-85, head, computer sci div, Off Naval Res, 85-88; DIR, MATH & INFO SCI, AIR FORCE OFF SCI RES, 88- *Concurrent Pos:* Vis mem, Courant Inst Math Sci, 76-78; mem staff, Off Technol Assessment, US Cong, 79-81. *Mem:* Soc Indust & Appl Math; sr mem Inst Elec & Electronics Engrs. *Res:* Deterministic and stochastic optimal control, filtering; probabilistic methods in applied mathematics; nonlinear diffusion equations; stochastic control theory. *Mailing Add:* Air Force Off Sci Res/NM Bolling AFB Bldg 410 Washington DC 20332-6448

HOLLAND, CHRISTIE ANNA, b 1950; m 89; c 1. VIROLOGY, RECOMBINATE DNA. *Educ:* Univ Tenn, PhD(biomed sci), 77. *Prof Exp:* ASST PROF, UNIV MASS, 85- *Mem:* AAAS; Am Soc Cell Biol. *Res:* Oncogenesis. *Mailing Add:* Dept Radiation Oncol Univ Mass 55 Lake Ave Worcester MA 01605

HOLLAND, DEWEY G, b New York, NY, Apr 15, 37; m 59; c 4. POLYMER SYNTHESIS, POLYMER APPLICATIONS. *Educ:* Fordham Univ, BS, 58; Lehigh Univ, PhD(chem), 62. *Prof Exp:* Res chemist, mil res & develop, polymer br, Air Force Mat Lab, 62-65; sect head, corp res & develop, Air Prod & Chem, Inc, 65-71, dir, indust res & develop, 71-78; vpres technol, Eschem, Inc, 78-85 & Reichhold Chem, Inc, 85; PRES, RD&E CONSULT. *Mem:* Am Chem Soc; Am Asn Consult Chemists & Chem Engrs. *Res:* Monomer and polymer synthesis; polymerization; electrochemistry adhesives; coatings and adhesives; product formulation and applications; surface chemistry; process development; reaction catalysts; foamed plastics. *Mailing Add:* 6870 S Tenth St Oak Creek WI 53154

HOLLAND, EUGENE PAUL, b Niagara Falls, NY, Apr 3, 35; c 3. STRUCTURAL ENGINEERING. *Educ:* Valparaiso Univ, BS, 57. *Prof Exp:* Struct engr, Ketchum & Konkel, Denver, 57-60; prin engr, Adv Eng Group, Portland Cement Asn, Chicago, 60-65; consult engr, 65-66; pres, Wiesinger-Holland LTD, 66-78; pres & chief exec officer, Coder Taylor Assoc, Inc, 78-81; CO-FOUNDER & PRES, EUGENE HOLLAND & ASSOCS LTD, 82- *Concurrent Pos:* Lectr, thin shell sem, Clemson Univ, 63, Exten Div, Univ Ill, 66-, Struct Engrs State Bd Exam Refresher Course, 66-, Architects State Bd Exam Refresher Course, 68- & Hawaii Struct Engrs, Contractors & Planners, Honolulu, 71; adj prof, Dept Archit, Univ Ill, 65- *Honors & Awards:* Del R Bloom Award, Am Concrete Inst, 78. *Mem:* Am Soc Civil Engrs; fel Am Concrete Inst; Nat Soc Prof Engrs; Prestressed Concrete Inst (pres, 76). *Res:* Seismology structural design; ultimate strength design. *Mailing Add:* 923 N Plum Grove Rd Suite C Schaumburg IL 60173

HOLLAND, F D, JR, b Leavenworth, Kans, Mar 6, 24; m 45; c 2. PALEONTOLOGY. *Educ:* Univ Kans, BS, 48; Univ Mo, AM, 50; Univ Cincinnati, PhD(geol), 58. *Prof Exp:* Asst to cur geol mus, Univ Kans, 47-48; asst, Univ Cincinnati, 50-51, cur mus, 51-54; from asst prof to prof, 54-89, EMER PROF GEOL, UNIV NDAK, 89- *Concurrent Pos:* Dir coun educ geol sci & AGI dir educ & manpower, Am Geol Inst, 70-72. *Mem:* Paleont Res Inst; AAAS; Paleont Soc; Soc Econ Paleontologists & Mineralogists; Geol Soc Am. *Res:* Paleozoic invertebrates; Devonian and Mississippian stratigraphy and paleontology; stratigraphy and paleontology of North Dakota; earth science education. *Mailing Add:* Dept Geol & Geol Eng Univ Sta Box 8068 Grand Forks ND 58202

HOLLAND, GEORGE PEARSON, entomology, for more information see previous edition

HOLLAND, GERALD FAGAN, b Boston, Mass, Aug 16, 31; m 62; c 4. MEDICINAL CHEMISTRY. *Educ:* Boston Col, BS, 52; Mass Inst Technol, PhD(chem), 56. *Prof Exp:* Sr asst scientist, NIH, 56-58; chemist, Med Res Labs, Pfizer Inc, 58-74; ysr res investr, 74-87, SR PROJ ANALYST, PFIZER CENT RES, 87- *Mem:* Am Chem Soc. *Res:* Atherosclerosis and lipid metabolism; metabolic diseases; diabetes; organic chemistry; amino acids and proteins; heterocyclic compounds. *Mailing Add:* Pfizer Ctr Res Groton CT 06340

HOLLAND, GRAHAM REX, b Rainford, Eng, Dec 3, 46; m 69; c 3. ANATOMY, DENTISTRY. *Educ:* Univ Bristol, BSc, 68, BDS, 71, PhD(anat), 73. *Prof Exp:* Res asst, Med Res Coun Gt Brit, 72-75; asst prof endodontics, Univ Iowa, 75-77; asst prof anat, Univ Man, 77-80, assoc prof, 80-81; assoc prof, dept endodontics, Col Dent, Univ Iowa, 81-83; assoc prof, 83-85, PROF, & CHMN DIV ENDODONTICS, UNIV ALTA, 85- *Concurrent Pos:* McCalla prof, 89-90. *Honors & Awards:* Murray L Barr Jr Scientist Award, 80. *Mem:* Brit Asn Clin Anatomists; Int Asn Dent Res; Anat Soc Gt Brit & Ireland. *Res:* Structural basis of pain; neural damage and recovery. *Mailing Add:* Div Endodontics Fac Dent Univ Alta Edmonton AB T6C 2N8 Can

HOLLAND, HANS J, b Mannheim, Ger, July 1, 29; US citizen; m 54; c 5. INORGANIC CHEMISTRY. *Educ:* Houghton Col, BS, 50; Columbia Univ, MA, 52; Univ Utah, PhD(chem), 63. *Prof Exp:* Phys scientist, Dugway Proving Ground, Utah, 51-52; res chemist, Continental Oil Co, 54-59; RES ASSOC, CORNING GLASS WORKS, 63- *Mem:* Am Chem Soc; Am Crystallog Asn; Mineral Soc Am. *Res:* Carbon dating; coprecipitation of cations with calcium carbonate; structure of alkali earth hexammoniates; glass devitrification products and their structure and mechanism of formation; x-ray diffraction instrumentation and automation; x-ray crystallography. *Mailing Add:* Res & Develop FR-18 Corning Glass Works Corning NY 14831

HOLLAND, HEINRICH DIETER, b Mannheim, Ger, May 27, 27; US citizen; m 53; c 4. GEOCHEMISTRY. *Educ:* Princeton Univ, BA, 46; Columbia Univ, MS, 48, PhD(geol), 52, Harvard Univ, MA, 72. *Prof Exp:* Asst, Columbia Univ, 47-50; from instr to prof geochem, Princeton Univ, 50-72, dir summer studies, 62-66; prof geochem, 72-80, PROF GEOL, HOFFMAN LAB, HARVARD UNIV, 80- *Concurrent Pos:* Instr, Shelton Col, 48-50; NSF fel, Oxford Univ, 56-57; Fulbright lectr, Univ Durham, 63-64; vis prof, Univ Hawaii, 68-69 & 81, Pa State Univ, 85-86; Guggenheim fel, 75-76; dir, Ctr Earth & Planetary Physics, Harvard Univ, 78-; sr scientist award, Von Humboldt Found, 80-81. *Mem:* Nat Acad Sci; Geol Soc Am; Geochem Soc (vpres, 69-70, pres, 70-71); Am Geophys Union. *Res:* Chemistry of ore forming fluids; chemical evolution of the atmosphere and ocean; planetary geology. *Mailing Add:* 306 Hoffman Labs Harvard Univ 20 Oxford St Cambridge MA 02138

HOLLAND, HERBERT LESLIE, b Bolton, Eng, Aug 2, 47; m 73; c 2. ORGANIC CHEMISTRY. *Educ:* Univ Cambridge, BA, 68, MA, 71; Univ Warwick, MSc, 69; Queen's Univ Belfast, PhD(chem), 72. *Prof Exp:* Univ demonstr chem, Queen's Univ Belfast, 69-72; fel, McMaster Univ, 72-74, teaching fel, 74-76; from asst prof to assoc prof, 76-85, PROF CHEM, BROCK UNIV, 85- *Concurrent Pos:* Res grants, Nat Res Coun Can, 76-, Imp Oil, 78-79, Ont Ministry Environ, 79 & 80 & Petrol Res Fund, Am Chem Soc, 82. *Mem:* Royal Soc Chem; Can Inst Chem. *Res:* Bioorganic chemistry; mechanistic chemistry of biological oxidation processes; natural products chemistry. *Mailing Add:* Dept Chem Brock Univ St Catharines ON L2S 3A1 Can

HOLLAND, ISRAEL IRVING, b Houston, Tex, Nov 25, 15; m 42; c 3. FOREST ECONOMICS. *Educ:* Univ Calif, BS, 40, MS, 41, PhD, 55. *Prof Exp:* Field asst range mgt res, Rocky Mountain Forest & Range Exp Sta, US Forest Serv, 41, logging time studies, Calif, 41, chief war mapping proj party, 42, forester, 46-49, forest economist, Div Forest Econ Res, Washington, DC, 52-57; asst prof forestry, Iowa State Univ, 57-59; assoc prof forest econ, Univ Ill, Urbana, 59-64, assoc dept head, 71-74, prof, 64-80, dept head, 74-81; RETIRED. *Mem:* Soc Am Foresters. *Mailing Add:* 1004 Harding Dr Urbana IL 61801

HOLLAND, JACK CALVIN, b Alameda, Calif, Mar 11, 25; m 49; c 3. BIOCHEMISTRY, CLINICAL CHEMISTRY. *Educ:* Mich Col Mining & Technol, BS, 48, MS, 49; Mich Technol Univ, PhD(chem), 68. *Prof Exp:* Res engr chem eng, Copper Range Mining Co, Houghton, Mich, 49-50; clin biochemist, St Luke's Hosp, Duluth, Minn, 50-54; instr chem, Univ Minn, 50-54; dir med lab, Duluth Clin, Minn, 54-63; prof med tech, Mich Technol Univ, 65-; RETIRED. *Concurrent Pos:* Res assoc, Am Cancer Soc, Mich Technol Univ, 63-65; consult, Allied Health Planning Asn, 73-76 & Mich Dept Pub Health-Continuing Educ, 76- *Mem:* Am Chem Soc; Sigma Xi; Am

Asn Clin Chem. *Res:* Hepatic cancer; fluorescent analysis of cervical smears; adrenalin levels in football players; femur fat in deer bone marrow; hyperthermia; medicinal value of the Finnish sauna; hibernation phenomenon; comparative biochemistry; effects of nutritional variation in deer herds of Michigan; high density lipoprotein cholesterol variations in Michigan populations. *Mailing Add:* 1624 Jasberg St Hancock MI 49930

HOLLAND, JAMES FREDERICK, b Morristown, NJ, May 16, 25; m 56; c 6. MEDICINE. *Educ:* Princeton Univ, AB, 44; Columbia Univ, MD, 47. *Prof Exp:* House staff, Presby Hosp, NY, 47-49; resident med, Francis Delafield Hosp, 51-52; asst, Columbia Univ, 52-53; sr asst surgeon, Nat Cancer Inst, 53-54; from assoc chief to chief med, Roswell Park Mem Inst, 54-73, from assoc res prof to res prof med, State Univ NY Buffalo, 62-73, dir, Cancer Clin Res Ctr, 63-73; PROF NEOPLASTIC DIS & CHMN DEPT & PROF MED, MT SINAI SCH MED, 73-, DIR CANCER CTR, UNIV & HOSP, 73- *Concurrent Pos:* Consult, Cancer World Health Orgn; mem, Bd Sci Counselors, Div Cancer Prev & Control, Nat Cancer Inst; cancer res award, Milken Family Med Found, 86- *Honors & Awards:* Albert Lasker Award Cancer Chemother, 72; Nat Annual Award, Am Cancer Soc, 81; David A Karnofsky Mem Lectr, 82; Katherine Berkan Judd Award, Mem Sloan-Kettering Cancer Ctr, 84; Return of the Child Award, Leukemia Soc Am, 86. *Mem:* Am Soc Clin Invest; Am Asn Cancer Res (pres, 70); Am Fedn Clin Res; Am Soc Hemat; Am Soc Clin Oncol (pres, 76); Sigma Xi; Asn Am Physicians. *Res:* Internal medicine; neoplastic diseases. *Mailing Add:* Dept Neoplastic Dis Mt Sinai Sch Med One Gustave L Levy Pl New York NY 10029

HOLLAND, JAMES PHILIP, b Bowling Green, Ky, Dec 31, 34; m 67. ENDOCRINOLOGY. *Educ:* Ky State Col, BS, 56; Ind Univ, MS, 58, PhD(endocrinol), 61. *Prof Exp:* Fel endocrinol, Sch Med, Univ Wis, 61-62; asst prof zool, Howard Univ, 62-67; assoc prof, 67-74, prof zool & assoc dean, Grad Sch, 74-77, interim dean, 77-78, PROF BIOL, IND UNIV, BLOOMINGTON, 78- *Concurrent Pos:* Prin investr, NIH res grants & NSF res grants; mem, NSF Grad Fel Eval Panel, 83- *Mem:* AAAS; Am Soc Zool; Endocrine Soc; Soc Study Reproduction. *Res:* Reproduction physiology; interrelationships of the thyroid gland and the ovary. *Mailing Add:* Dept Biol Ind Univ Bloomington IN 47401

HOLLAND, JAMES READ, b Fulton, Ky, Aug 24, 30; m 56; c 4. PHYSICAL METALLURGY. *Educ:* Univ Ky, BS, 53, DEng(phys metall), 62; Univ Sheffield, MMet, 54. *Prof Exp:* Asst prof metall, Univ Indonesia, 56-58; br chief metall res, USAF Mat Lab, Wright-Patterson Air Force Base, Ohio, 60-61; res staff mem, Sandia Labs, 61-66, div supvr, 66-74; mgr reactor mat res, Westinghouse Res & Develop Ctr, 74-81; DIR, SCH MINES & ENERGY DEVELOP, UNIV ALA, 81- *Mem:* Am Soc Metals; Metall Soc; Am Nuclear Soc. *Res:* Anisotropy of crystalline solids; dislocation movement and generation; x-ray diffraction; propagation of stress waves in crystalline solids; plastic deformation; mechanical metallurgy; refractory metals; nuclear materials; radiation damage in solids; energy conversion; materials in energy systems. *Mailing Add:* Sch Mines & Energy Develop Univ Ala Box 6282 University AL 35486

HOLLAND, JOHN HENRY, b Ft Wayne, Ind, Feb 2, 29; m 56; c 3. APPLIED MATHEMATICS, COMPUTER SCIENCE. *Educ:* Mass Inst Technol, BS, 50; Univ Mich, MA, 54, PhD(commun sci), 59. *Prof Exp:* Assoc engr, Int Bus Mach Corp, 50-52, consult, 52-56; res assoc, Univ Mich, Ann Arbor, 56-59, assoc res mathematician & lectr psychol, 59-61, from asst prof to assoc prof commun sci, 61-67, actg dir logic of comput group, 65-66, actg chmn dept comput & commun sci, 71-72 & 74, prof, 67-80, assoc dir logic of comput group, 71-86, prof comput & commun sci, 80-86, PROF COMPUT SCI & ENG, UNIV MICH, ANN ARBOR, 86-, PROF PSYCHOL, 88- *Concurrent Pos:* Res assoc, Carnegie Inst, Washington, DC, 61-64; fel, Franklin Inst, 71; actg dir, Univ Mich, 77-78, consult, 64-; mem Steering Comt, Sci Adv Bd, Santa Fe Inst, 87-, dir, Adaptive Computation, 89-, dir, Univ Mich/Santa Fe Inst Res Prog, 90- *Honors & Awards:* Levy Medal, Franklin Inst. *Mem:* AAAS; Asn Comput Mach; Am Math Soc; Sigma Xi; Am Asn Artificial Intel; AAAS. *Res:* Logical and mathematical theory of computers and automata; theories and models of cognition; adaptive systems. *Mailing Add:* 3800 W Huron River Dr Ann Arbor MI 48103

HOLLAND, JOHN JOSEPH, b Pittsburgh, Pa, Nov 16, 29; m 60; c 2. MICROBIOLOGY. *Educ:* Loyola Univ, La, BS, 53; Univ Calif, Los Angeles, PhD(microbiol), 57. *Prof Exp:* From instr to asst prof bact, Univ Minn, 57-60; from asst prof to assoc prof microbiol, Univ Wash, 60-64; prof, Univ Calif, Irvine, 64-68; PROF BIOL, UNIV CALIF, SAN DIEGO, 68- *Honors & Awards:* Eli Lilly Award Microbiol, 63. *Mem:* AAAS; Soc Exp Biol & Med; Am Soc Microbiol. *Res:* Virology; cell biology. *Mailing Add:* Dept Biol C-016 Univ Calif San Diego Box 109 La Jolla CA 92093

HOLLAND, JOSHUA ZALMAN, b Chicago, Ill, June 23, 21; m 84; c 3. METEOROLOGY. *Educ:* Univ Chicago, BS, 41, Univ Wash, PhD(atmospheric sci), 68. *Prof Exp:* Instr meteorol, Univ Chicago, 42 & NY Univ, 42-43; weather officer, US Army Air Forces, 43-46; meteorologist in charge, Oak Ridge Off, US Weather Bur, Tenn, 48-53; meteorologist, Washington, DC, 53-55; exec secy, Adv Comt Reactor Safeguards, US AEC, 56-57, meteorologist, Environ Sci Br, Div Biol & Med, 57-59, chief fallout studies br, 59-69; dir, Barbados Oceanog & Meteorol Anal Proj, Nat Oceanog & Atmospheric Admin, US Dept Com, 69-71, dir, Ctr Exp Design & Data Anal, 71-78, dir, Environ Data & Info Serv, Ctr Environ Assessment Servs, 78-79; adj prof meteorol, 79-83, RES ASSOC, UNIV MD, 83- *Concurrent Pos:* Consult, Comt Meteorol Aspects Effects Atomic Radiation, Nat Acad Sci, 56-60, mem, Interdept Comt Community Air Pollution, 57-59, mem, Interdept Comt Atmospheric sci, 59-62, mem, Subcomt Mesometeorol, 61-62, mem, Ad Hoc Comt Int Progs Atmospheric Sci & Chmn Subcomt Atmospheric Chem, 62-63, mem, Adv Comt Civil Defense, 67-70, mem, Data Mgt Panel, US Comt Global Atmospheric Res Prog, 70-, mem, Adv Panel, Atlantic Trop Exp, Nat Acad Sci, 78-; chief scientist, Barbados Oceanog & Meteorol Exp, Sea-Air Interaction Prog, 69; mem Adv Comt Oceanic Meteorol Res, World Meteorol Orgn, 70-77; mem pub comn, Am Meteorol Soc, 80-88; mem Comt Geophys Data, Nat Res Coun, 84-87. *Honors & Awards:* Silver Medal, Dept Com, 56, Gold Medal, 74. *Mem:* Am Geophys Union; fel Am Meteorol Soc. *Res:* Atmospheric transport, dispersion and deposition of contaminants; atmospheric turbulence; micrometeorology; stack plume rise; air-sea exchange processes; environmental experiment design and data analysis. *Mailing Add:* Dept Meteorol Univ Md College Park MD 20742

HOLLAND, LEWIS, b Somerton, Ariz, July 3, 25; m 48; c 6. ANIMAL BREEDING, GENETICS. *Educ:* NMex State Univ, BS, 49; Colo State Univ, MS, 51; Iowa State Univ, PhD(animal breeding, genetics), 57. *Prof Exp:* Instr animal husb, Colo State Univ, 49-50; asst prof, Kans State Univ, 51-54 & 56-58; assoc prof, NMex State Univ, 59-69, prof animal sci, 69-81, assoc dean, Col Agr & Home Econ, 71-81, admin adv, NMex State Univ-Paraguay Aid, 76-81. *Concurrent Pos:* Asst, Iowa State Univ, 55. *Mem:* Am Soc Animal Sci; Am Genetics Asn. *Res:* Animal breeding and genetics of sheep and cattle. *Mailing Add:* Dept Animal & Range Sci Box 3AG NMex State Univ Las Cruces NM 88003

HOLLAND, LOUIS EDWARD, II, b Kansas City, Mo, Nov 15, 48; m 77; c 2. VIROLOGY. *Educ:* Baker Univ, BS, 70; Univ Calif, Irvine, PhD(molecular biol), 79. *Prof Exp:* Teaching fel virol, Univ Mich, 79-84; sr molecular biologist, Dept Biochem, Southern Res Inst, 84--88; SR VIROLOGIST LIFE SCI DEPT, ILL INST TECH RES INST, 88- *Mem:* Am Soc Microbiol; Am Soc Virol; AAAS; Soc Gen Microbiol; Sigma Xi; Int Soc Antiviral Res. *Res:* Gene mapping of herpes simplex viruses; mechanism of action for antiviral chemotherapeutic agents; pathogenesis of human immunodeficiency virus. *Mailing Add:* Life Sci Dept Ill Inst Tech Ten W 35th St Chicago IL 60616-3799

HOLLAND, LYMAN LYLE, b Portsmouth, Va, Apr 20, 40; m 68; c 2. CHEMICAL & ENVIRONMENTAL ENGINEERING. *Educ:* Univ Va, BChE, 63; Clemson Univ, MSChE, 65, PhD(chem eng, environ systs eng), 68. *Prof Exp:* SR RES ENGR, DACRON STAPLE RES & DEVELOP CTR, KINSTON, E I DU PONT DE NEMOURS & CO, INC, 69- *Mem:* Am Inst Chem Engrs. *Res:* Modeling photobiological processes using chemical engineering photochemical kinetics; effect of mixing on kinetic growth rates of Chlorella pyrenoidosa in parallel plate flow reactor; process development; waste recovery, recycle and utilization; polymer melt rheology. *Mailing Add:* 1805 Sunset Ave Kinston NC 28501

HOLLAND, MARJORIE MIRIAM, b Boston, Mass, Aug 18, 47. PLANT ECOLOGY, ENVIRONMENTAL SCIENCE. *Educ:* Conn Col, BA, 69; Smith Col, MA, 74; Univ Mass, PhD(bot), 77. *Prof Exp:* Teacher biol, Mountain Sch, Vershire, Vt, 69-70 & Dover-Sherborn Regional High Sch, Mass, 70-72; teaching fel, Smith Col, 72-76; vis lectr, Amherst Col, 76-77, vis asst prof, 77-78; exec dir environ sci, Water Supply Citizen's Adv Comt, Springfield, Mass, 78-80; asst prof, 80-88, ASSOC PROF, COL NEW ROCHELLE, 88- *Concurrent Pos:* Res assoc, Dept Biol Sci, Smith Col, 77-; ed, Lawrence Erdang Publ, 78-; consult, US Army CEngr, 79-80; consult, Secretariat of Unesco's MAB (Man & the Biosphere Prog), Paris, 86-88; dir, Ecol Soc Am, 87- *Mem:* AAAS; Ecol Soc Am; Sigma Xi; Bot Soc Am; Torrey Bot Club; Int Limnol Soc. *Res:* Plant systematics; wetlands ecology; riverine ecology; phytosociology; brackish tidal wetlands; natural history; water resource management; public policy development. *Mailing Add:* 1108 Lancaster Rd Silver Spring MD 20912

HOLLAND, MARY JEAN CAREY, b Dearborn, Mich, Feb 14, 42; m 68. PAIN MANAGEMENT, PHARMACOKINETICS. *Educ:* Vassar Col, AB, 63; NY Univ, MS, 69, PhD(biol), 71. *Prof Exp:* Res asst biophys, Parke-Davis & Co, 64-65; assoc microbiologist, Merck & Co, 65-67; instr biol, Lehman Col, City Univ New York, 70-71; NIH fel, 72-74, instr exp med, 75-79, asst prof med, Med Ctr, NY Univ, 79-82; from asst prof to assoc prof, 82-90, PROF BIOL, BARUCH COL, CITY UNIV NY, 91- *Concurrent Pos:* Fels, Fulbright, 63-64, Nat Found, March of Dimes, 74-75 & Arthritis Found, 76-79; adj fac mem biochem, Grad Prog, Sarah Lawrence Col, 77-79; young investr grant, NY Arthritis Found, 78-79; vis asst prof microbiol, NY Col Osteop Med, 79-82; res collabr, Chem Dept, Brookhaven Nat Lab, 81-; res asst prof psychiat, NY Univ Med Ctr, 82- *Mem:* Am Soc Microbiol; Am Soc Cell Biol; Soc Neurosci; Soc Math Biol; NY Acad Sci. *Res:* Mathematical models for receptor-mediated processes; pharmocokinetics and tissue distribution of opiates; opiate receptor binding kinetics; analgesia. *Mailing Add:* Dept Nat Sci Baruch Col 17 Lexington Ave New York NY 10010

HOLLAND, MONTE W, b De Kalb, NY, June 12, 38; m 66. PHYSICS. *Educ:* Union Col, NY, BS, 59; Northwestern Univ, PhD(physics), 63. *Prof Exp:* Lectr physics, State Univ NY Buffalo, 63-64, asst prof, 64-69, asst chmn dept, 67-69; assoc prof, 69-73, chmn dept, 69-73, PROF PHYSICS, SLIPPERY ROCK STATE COL, 72- *Mem:* Am Phys Soc; Am Asn Physics Teachers. *Res:* Particle physics; properties of hyperfragments; equipment of science instruction; uses of computers for instruction; basic physics textbook preparation. *Mailing Add:* RD 4 PO Box 222 Slippery Rock PA 16057

HOLLAND, NANCY H, b Paris, Ky; m 61. PEDIATRICS, NEPHROLOGY. *Educ:* Univ Denver, BS, 49; Univ Louisville, MD, 54; Am Bd Pediat, dipl, 59, cert pediat nephrology, 74. *Prof Exp:* From instr to asst prof, Sch Med, Univ Cincinnati, 58-64; from asst prof to assoc prof, 64-71, PROF PEDIAT, SCH MED, UNIV KY, 71- *Concurrent Pos:* Res assoc, Cincinnati Children's Hosp Res Found, 57-60, sr res asst, 60-62. *Mem:* Soc Pediat Res; Am Soc Pediat Nephrology; Am Pediat Soc; Am Acad Pediat; Int Soc Nephrology. *Res:* Pathogenesis of glomerulonephritis; role of B-oneC globulin and other complement components in etiology of glomerulonephritis; effectiveness of immunosuppressive drugs in treatment of glomerulonephritis; immunology. *Mailing Add:* Dept Pediat Univ Ky Col Med Lexington KY 40536

HOLLAND, NEAL STEWART, b Bracken, Sask, Oct 26, 29; US citizen. HORTICULTURE. *Educ:* NDak State Univ, BS, 51, MS, 60. *Prof Exp:* Teacher pub schs, Minn, 51-52; asst, NDak State Univ, 54-60, from asst prof to prof hort, 60-85, actg head, 81-83; RETIRED. *Honors & Awards:* Silver Medal, All-Am Selections Award, 66; R L Wodarz Award, 77. *Mem:* Am Pomol Soc; Int Lilac Soc. *Res:* Fruit culture and breeding; tomatoes, squash, and ornamentals breeding programs and genetic studies. *Mailing Add:* Rte 1 Box 36 Harwood ND 58042

HOLLAND, NICHOLAS DREW, b Washington, DC, Apr 24, 38; m 61; c 3. INVERTEBRATE ZOOLOGY. *Educ:* Carleton Col, AB, 60; Stanford Univ, PhD(biol), 64. *Prof Exp:* NSF fel zool, Naples Zool Sta, Italy, 64-66; from asst prof to assoc prof, 66-78, PROF MARINE BIOL & REGIONAL ED, MARINE ECOL PROGRESS SERIES, SCRIPPS INST OCEANOG, 78- *Res:* Physiology, anatomy, fine structure, development and natural history of echinoderms, especially crinoids. *Mailing Add:* Marine Biol Res Div A-002 Scripps Inst Oceanog Univ Calif San Diego La Jolla CA 92093

HOLLAND, PAUL VINCENT, b Toronto, Ont, Oct 29, 37; US citizen; m 62; c 4. INTERNAL MEDICINE. *Educ:* Univ Calif, Riverside, BA, 58; Univ Calif, Los Angeles, MD, 62; Am Bd Internal Med, dipl, 69; Am Bd Path, dipl, 73, cert blood banking. *Prof Exp:* Intern med, Univ Calif, Los Angeles, 62-63; staff physician blood bank, NIH, 63-66; asst res med, San Francisco Med Ctr, Univ Calif, 66-68; asst chief, 68-74, CHIEF BLOOD BANK, NIH, 74- *Concurrent Pos:* Clin instr med, Med Sch, George Washington Univ, 69-74, clin asst prof, 74-75, clin assoc prof, 75-; clin assoc prof path, Sch Med, Georgetown Univ, 74- *Mem:* Am Fedn Clin Res; Am Soc Hemat; NY Acad Sci; Am Asn Blood Banks. *Res:* Hepatitis, especially re Australia antigen and antibody; immunohematology; blood transfusion; component therapy. *Mailing Add:* 1625 Stockton Blvd Sacramento CA 95816-7089

HOLLAND, PAUL WILLIAM, b Tulsa, Okla, Apr 25, 40; m 61; c 2. STATISTICS, BIOSTATISTICS. *Educ:* Univ Mich, BA, 62; Stanford Univ, MS, 64, PhD(statist), 66. *Prof Exp:* Asst prof statist, Mich State Univ, 66; from asst prof to assoc prof statist, Harvard Univ, 66-72; sr res assoc, Nat Bur Econ Res, 72-75; DIR STATIST RES, EDUC TESTING SERV, 75- *Concurrent Pos:* Fac Res Grant, Social Sci Res Coun, 70-71; lectr statist, Harvard Univ, 72-75; mem, Panel Productivity Statist, Nat Res Coun, 77-78. *Mem:* Fel Am Statist Asn; fel Inst Math Statist; Biomet Soc; Am Sociol Asn; fel AAAS; Psychomet Soc (pres, 89-90). *Res:* Application of statistics and mathematical models to sociology, education and the behavioral sciences; problems of inference in studies of human populations. *Mailing Add:* Educ Testing Serv Rosedale Rd Princeton NJ 08541

HOLLAND, RAY W(ALTER), b Afton, Tenn, Feb 11, 24; m 46; c 2. MECHANICAL ENGINEERING. *Educ:* Duke Univ, BS, 47; Univ Tenn, MS, 52. *Prof Exp:* From instr to asst prof mech eng design, Duke Univ, 47-55; assoc prof, mech & aerospace eng, Univ Tenn, Knoxville, 55-69, prof, 69-89; RETIRED. *Concurrent Pos:* Indust consult, 52- *Mem:* Am Soc Mech Engrs; Soc Mfg Engrs; Am Soc Eng Educ; Sigma Xi; Soc Exp Mech. *Res:* Experimental stress analysis; machine design. *Mailing Add:* 2004 McClain Rd Knoxville TN 37912-4617

HOLLAND, REDUS FOY, b Frederick, Okla, Jan 8, 30; m 56; c 3. MOLECULAR SPECTROSCOPY. *Educ:* Panhandle State Col, BS, 51; Univ Okla, PhD(physics), 61. *Prof Exp:* RES PHYSICIST, LOS ALAMOS NAT LAB, UNIV CALIF, 61- *Mem:* Laser Inst Am; Optical Soc Am. *Res:* Emission spectra of atmospheric gases; probability of excitation by collisions with electrons and ions; long-lived states of diatomic ions; vibrational spectroscopy of polyatomic molecules; cryogenic solution spectroscopy; chemical kinetics. *Mailing Add:* 2774 Walnut St Los Alamos NM 87545

HOLLAND, ROBERT CAMPBELL, b Bushnell, Ill, Aug 16, 23; m 89; c 1. NEUROSCIENCES. *Educ:* Univ Wis, BS, 48, MS, 49, PhD(anat, zool), 55. *Prof Exp:* Instr histol & path, Dent Sch, Northwestern Univ, 49-51; res asst anat, Univ Wis, 51-54; from instr to asst prof neuroanat, Univ NDak, 54-60; assoc prof anat, Sch Med, Univ Ark, 60-66; vis prof & actg chmn dept, Mahidol Univ, Thailand, 66-75; fel anat, Med Sch, Univ Calif, Los Angeles, 76-77; prof & chmn, Dept Anat, 77-90, EMER PROF ANAT, MOREHOUSE SCH MED, 90- *Concurrent Pos:* Nat Found Infantile Paralysis fel, Univ Calif, Los Angeles, 57-58; mem field staff, Rockefeller Found, 66-77; vis prof anat, Med Sch, Univ Calif, Los Angeles, 76. *Mem:* AAAS; Am Asn Anatomists; Am Acad Neurol; Soc Exp Biol & Med; Soc Neurosci; Sigma Xi. *Res:* Neuroendocrinology; neuroanatomy; neurophysiology. *Mailing Add:* Dept Anat Morehouse Sch Med Atlanta GA 30310-1495

HOLLAND, ROBERT EMMETT, b Chicago, Ill, May 21, 20; m 49; c 3. EXPERIMENTAL NUCLEAR PHYSICS. *Educ:* Univ Iowa, BA, 42, MS, 44, PhD(physics), 50. *Prof Exp:* From res assoc to instr physics, Univ Iowa, 44-48; assoc physicist, Argonne Nat Lab, 50-68, sr physicist, 68-82, assoc ed, Appl Physics Lett, 83-89, ASSOC ED, J APPL PHYSICS, ARGONNE NAT LABS, ARGONNE, ILL, 83- *Concurrent Pos:* Fulbright res grant, Univ Helsinki, 61-62. *Mem:* Am Phys Soc. *Res:* Nuclear physics; bombardment by fast particles; lifetimes of nuclear states. *Mailing Add:* Argonne Nat Lab 9700 S Cass Ave Argonne IL 60439

HOLLAND, ROBERT FRANCIS, b Holley, NY, Sept 21, 08; m 30; c 4. FOOD SCIENCE. *Educ:* Cornell Univ, BS, 36, MS, 38, PhD(dairy indust), 40. *Prof Exp:* Instr dairy chem, Cornell Univ, 35-39; prof, NY Agr Exp Sta, Geneva, 39-41; dir chem res, Coop G L F Soil Bldg Serv, Inc, Ithaca, 41-45; prof dairy indust, 45-72, head depts dairy indust & food sci, 55-72, EMER PROF DAIRY INDUST, CORNELL UNIV, 73- *Mem:* Am Dairy Sci Asn; Inst Food Technologists. *Res:* Milk pasteurization; sanitary chemistry; dairy products marketing. *Mailing Add:* 114 Seneca Rd Trumansburg NY 14886

HOLLAND, RUSSELL SEDGWICK, b Westerly, RI, May 4, 29; m 57; c 2. PHOTOGRAPHIC SCIENCE. *Educ:* Brown Univ, BSc, 51; Princeton Univ, PhD(chem), 55. *Prof Exp:* Res chemist, Med Prod Dept, E I du Pont de Nemours & Co, Inc, 54-66, tech specialist, 66-77, sr tech specialist, 77-78, tech assoc, 78-90; RETIRED. *Mem:* Am Chem Soc; Soc Photog Sci & Eng. *Res:* Photographic chemistry; image evaluation; photographic film testing. *Mailing Add:* Dept Med Prod PO Box 80708 Wilmington DE 19880-0708

HOLLAND, SAMUEL S, JR, b Lawrence, Mass, June 29, 28; m 58; c 4. VALUATION THEORY, HERMITIAN FORMS. *Educ:* Mass Inst Technol, BS, 50; Univ Chicago, MS, 52; Harvard Univ, PhD(math), 61. *Prof Exp:* Staff scientist math physics, Tech Opers, Inc, 55-57; Nat Acad Sci-Nat Res Coun res assoc, 60-61; from asst prof to assoc prof math, Boston Col, 61-67; assoc prof, 67-71, PROF MATH, UNIV MASS, AMHERST, 71- *Concurrent Pos:* Consult, Tech Opers, Inc, 61-67; NSF res grant, 63- *Mem:* Am Math Soc. *Res:* Study of division rings with involution, their orderings, valuations, and the infinite dimensional hermitian forms they may support. *Mailing Add:* Dept Math & Statist Univ Mass Amherst MA 01003

HOLLAND, STEVEN WILLIAM, b Detroit, Mich, Dec 4, 51; m 75; c 2. ARTIFICIAL INTELLIGENCE, COMPUTER INTEGRATED MANUFACTURING. *Educ:* Gen Motors Inst, BA, 76; Stanford Univ, MS, 76. *Prof Exp:* Assoc sr res scientist, Gen Motors Res Labs, 75-77, staff res scientist, 77-79, sr staff res scientist, 79-82, asst dept head, 83-87, sect mgr, 87-91, DIR ROBOTICS, GEN MOTORS ADV ENG, 91- *Concurrent Pos:* Adj fac, Wayne state Univ, 77-81. *Honors & Awards:* Arch T Colwell Award, Soc Automotive Engrs, 80. *Mem:* Sr mem Inst Elec & Electronics Engrs; Am Asn Artificial Intel; Sigma Xi; Soc Mech Engrs. *Res:* Application of computer science and artificial intelligence to the problems and processes of a large manufacturing corporation; US patents. *Mailing Add:* 1670 Sumac Dr Rochester Hills MI 48309-2227

HOLLAND, WILBUR CHARLES, b Parkersburg, WVa, Feb 24, 35; m 55; c 5. MATHEMATICS. *Educ:* Tulane Univ, BS, 57, MS, 59, PhD(math), 61. *Prof Exp:* NATO fel, Univ Tübingen, 61-62; instr math, Univ Chicago, 62-64; from asst prof to prof math, Univ Wis-Madison, 64-72; chmn, Dept Math & Statist, 81-83, PROF MATH, BOWLING GREEN STATE UNIV, 78- *Concurrent Pos:* Vis prof, Simon Fraser Univ, 83-84. *Mem:* Am Math Soc; Math Asn Am. *Res:* Ordered algebraic structures. *Mailing Add:* Dept Math Bowling Green State Univ Bowling Green OH 43403

HOLLAND, WILLIAM FREDERICK, b Jersey City, NJ, Sept 30, 14; m 39; c 1. INDUSTRIAL & POLYMER CHEMISTRY. *Educ:* City Col New York, BS, 42; Polytech Inst Brooklyn, MS, 48. *Prof Exp:* Asst dir labs, Standard Varnish Works, NY, 43-50; tech dir org coatings, Benjamin Franklin Paint & Varnish Co, Pa, 50-53; vpres, Hanline Bros Inc, Baltimore, 53-67; res supvr chem div, PPG Industs, Inc, 67-71; chief chemist, Akron Paint & Varnish Co, 71-82; RETIRED. *Concurrent Pos:* Coatings Consult. *Mem:* Emer mem Am Chem Soc; Fedn Soc Coatings Technol. *Res:* Protective and decorative organic coatings and allied specialties; high polymer chemistry. *Mailing Add:* 1194 Highview Dr Wadsworth OH 44281

HOLLAND, WILLIAM JOHN, b Belleville, Ont, Jan 6, 20; m 45; c 2. ANALYTICAL CHEMISTRY. *Educ:* Queen's Univ, Ont, BSc, 45; Wayne State Univ, MSc, 53, PhD(chem), 56. *Prof Exp:* Chief chemist, R P Scherer Co, 46-59; fel org chem, Wayne State Univ, 59-60; from asst prof to assoc prof, 60-70, PROF ANALYTICAL CHEM, UNIV WINDSOR, 70- *Mem:* Chem Inst Can. *Res:* Synthesis of organic chelating agents and application to trace metal analysis and transition metal separations by solvent extraction. *Mailing Add:* 2293 Gladstone Ave Windsor ON N8W 2N8 Can

HOLLAND, WILLIAM ROBERT, b Van Nuys, Calif, July 31, 38; m 57; c 3. PHYSICAL OCEANOGRAPHY. *Educ:* Univ Calif, Los Angeles, AB, 60; Mass Inst Technol, MS, 61; Univ Calif, San Diego, PhD(oceanog), 66. *Prof Exp:* Fel oceanog, Cambridge Univ, 66-67; res oceanographer, Nat Oceanic & Atmospheric Admin, 67-74; RES OCEANOGR, NAT CTR ATMOSPHERIC RES, 74- *Mem:* AAAS; Am Geophys Union; Am Meterol Soc. *Res:* Numerical models of ocean circulation. *Mailing Add:* Nat Ctr Atmospheric Res PO Box 3000 Boulder CO 80307

HOLLAND-BEETON, RUTH ELIZABETH, b Weatherly, Pa, Apr 17, 36; m 66; c 2. HYDROBIOLOGY, ECOLOGY. *Educ:* St Olaf Col, BA, 58; Univ Mich, MS, 62. *Prof Exp:* Fishery res biologist, US Bur Com Fisheries, 63-66; res specialist, Ctr Great Lakes Studies, Univ Wis-Milwaukee, 66-76; RES INVESTR, DEPT ATMOSPHERIC & OCEANIC STUDIES, UNIV MICH, 76- *Concurrent Pos:* Consult, Atty Gen Off, Wis, 72-73; prin investr, Environ Protection Agency, 77-78, Sea Grant Prog, Wis, 70-76 & Sea Grant Col, Mich, 79-86; res assoc, dept oceanog, Ore State Univ, 82-83. *Mem:* fel AAAS; Int Asn Theoret & Appl Limnol; Am Soc Limnol & Oceanog; Fresh Biol Asn Gt Brit; Int Asn Great Lakes Res; Int Soc Diatom Res. *Res:* Ecology, growth rates, population fluxes and distribution of planktonic diatoms, especially of the Great Lakes region; effects of best management agricultural practices upon the aquatic habitat. *Mailing Add:* 2761 Oakcleft Ct Ann Arbor MI 48103

HOLLANDER, DAVID HUTZLER, pathology, for more information see previous edition

HOLLANDER, GERHARD LUDWIG, b Berlin, Ger, Feb 27, 22; nat US; m 57; c 3. SYSTEM ARCHITECTURE, COMPUTER SYSTEMS. *Educ:* Ill Inst Technol, BS, 47; Wash Univ, St Louis, MS, 48; Mass Inst Technol, EE, 53. *Prof Exp:* Radio buyer, Spiegel, Inc, 40-42; res engr, McDonnell Aircraft Co, 47; asst prof eng, St Louis Univ, 48-49; sr engr, Servo Lab, Raytheon Mfg Co, 49-51; res asst, Servomechanisms Lab, Mass Inst Technol, 52-54; sect head, data processing systs, Clevite Res Ctr, 54-57; sect mgr comput systs, Philco Corp, 57-60; mgr gen purpose comput dept, Hughes Aircraft Co, 60-61; PRES & TECH DIR, HOLLANDER ASSOCS, 61- *Concurrent Pos:* Consult, indust & govt, 53-; mem, Nat Joint Comput Comt, 59-61; mem,

Control Adv Comt, Am Automatic Control Coun, 60-62; dir, Am Fedn Info Processing Socs, 62-65; gen chmn, Joint Nat Conf Major Systs, 71; dir & chmn bd, Inst Elec & Electronics Engrs Winter Conv on Mil Systs, 75-79; gen chmn & chmn bd dir, Inst Elec & Electronics Engrs Conf Expert Systems, Westex, 85-; vpres & pres, Inst Elec & Electronics Engrs Computer Soc, 62-65, chmn, Los Angeles coun, Region 6 S, 71-73. *Honors & Awards:* Centennial Medal, Inst Elec & Electronics Engrs, 84- *Mem:* Opers Res Soc Am; fel Inst Elec & Electronics Engrs; Asn Comput Mach; Sigma Xi. *Res:* Computer system design and application; general methodology for decision making and for optimal structuring of large systems; system architecture and acquisition strategies for major systems. *Mailing Add:* Hollander Assocs PO Box 2276 Fullerton CA 92633-0276

HOLLANDER, JACK MARVIN, b Youngstown, Ohio, Apr 13, 27; m 48; c 3. ENERGY, RESOURCES. *Educ:* Ohio State Univ, BS, 48; Univ Calif, Berkeley, PhD(chem), 51. *Prof Exp:* Instr chem, Univ Calif, 51-53, coordr environ res, 71-73, asst dir, 71-73, dir energy & environ div, 73-76 & assoc dir, 73-79; assoc dir planning, Lawrence Berkeley Lab, 78-79, prof, 79-89, EMER PROF ENERGY & RESOURCES, UNIV CALIF, BERKELEY, 89- *Concurrent Pos:* Guggenheim fel, 58-59 & 65-66; mem, Subcomt Nuclear Struct, Nat Acad Sci-Nat Res Coun, 60-70; ed, Ann Rev Energy, 76-; mem & founding chair bd, Beijer Int Inst Energy & Human Ecol, Stockholm, 76-; exec dir, Comt Nuclear & Alternative Energy Systs, Nat Acad Sci, 76-77; sr staff scientist, Lawrence Berkeley Lab, 53-, assoc dir planning, 78-79; off res & grad studies, Ohio State Univ, Columbus, 89. *Mem:* AAAS; Am Phys Soc; Am Chem Soc; Royal Swed Acad Sci; World Acad Arts & Sci. *Res:* Nuclear spectroscopy and models; compilations of nuclear data; energy policy; energy conservation research. *Mailing Add:* Energy & Resources Group T-4 Univ Calif Berkeley CA 94720

HOLLANDER, JOSEPH LEE, b St Louis, Mo, Mar 8, 10; m 36, 74; c 2. RHEUMATOLOGY. *Educ:* Cornell Univ, AB, 32; Univ Pa, MD, 35. *Prof Exp:* Asst physician, Pa Hosp, Philadelphia, 37-46; instr med, 46-48, assoc, 48-51, assoc prof clin med, 53-58, assoc prof med, 58-62, chief arthritis sect, Univ Hosp, 46-72, prof, 62-78, EMER PROF MED, SCH MED, UNIV PA, 78- *Concurrent Pos:* Asst demonstr, Jefferson Med Col, 40-46; consult to Surgeon Gen, US Dept Army, 48-50, Valley Forge Gen Hosp, 50-52 & Childrens' Hosp, 51-; hon prof med, Univ Guadalajara, Mex & Univ Peruana Cayetano Heredia, Lima, Peru; mem, Comt of Honor, 15th Int Cong Rheumatology, Paris, 81. *Honors & Awards:* Harding Medal, Arthritis Found, 82; Master, Am Rheumatism Asn, 87. *Mem:* AMA; Am Rheumatism Asn; Asn Am Physicians; master Am Col Physicians. *Res:* Arthritis; clinical and basic research on joint physiology and action of cortisone-like hormones on joint tissues; effects of climate on arthritis; immunopathogenesis of rheumatoid arthritis. *Mailing Add:* 3400 Spruce Philadelphia PA 19104

HOLLANDER, JOSHUA, b New York, NY, Dec 28, 36; m 60; c 3. NEUROLOGY, BIOCHEMISTRY. *Educ:* Columbia Univ, BA, 56, MD, 60. *Prof Exp:* Intern med, Univ Hosp, Vanderbilt Univ, 60-61, asst resident, 61-62; resident neurol, Mass Gen Hosp, 62-63, actg resident, 64-65; clin & res assoc biochem, Geront Res Ctr, Nat Inst Child Health & Human Develop, 65-67; asst res biochemist, Ment Health Res Inst, Univ Mich, Ann Arbor, 67-69; asst prof, 69-77, ASSOC PROF NEUROL, SCH MED & DENT, UNIV ROCHESTER, 69-, SR ASSOC NEUROLOGIST, STRONG MEM HOSP, 77- *Concurrent Pos:* Clin fel neuropath, Mass Gen Hosp, 63-64; Nat Inst Neurol Dis & Blindness spec fel, 67-69; chief neurol serv, Rochester Gen Hosp, 69- *Mem:* AAAS; Am Acad Neurol; Geront Soc. *Res:* Biochemistry of aging nervous system; brain phospholipid metabolism-physiologic aspects. *Mailing Add:* 1425 Portland Ave Rochester NY 14621

HOLLANDER, LEONORE, b Ferguson, Mo, May 27, 06; div; c 3. CLINICAL BIOCHEMISTRY, SUPERVISION & ACCURACY CONTROL. *Educ:* Bryn Mawr Col, AB, 28; Univ Ill, MS, 29, PhD(physiol chem), 32. *Prof Exp:* Asst biochem, Cancer Res Inst, 31-33; grad scholar, Ger Polytech Inst, Prague, 33-35, Kaiser Hilhelm Inst, Heidelberg, 35-37; asst prof chem, Cedar Crest Col, Allentown, Pa, 47-49; res asst biochem, Inst Cancer Res, Fox Chase, Pa, 50-53; clin biochemist, St Luke's Hosp, Bethlehem, Pa, 53-62, Philadelphia Gen Hosp, 62-63; RETIRED. *Honors & Awards:* Meritorious Serv Award, Am Inst Chem. *Mem:* Fel Am Inst Chemists; fel AAAS; Am Inst Chem; Am Chem Soc; Am Asn Clin Chemists. *Res:* Salivary and pancreatic amylase, isolation and study of crystalline dextrine; elucidating bacterial polysaccharide complex from serratia marcescens, involving systemizing use of the anthrone reaction in identifying, distinguishing and quantifying sugars. *Mailing Add:* 4756 Sunshine Ave Santa Rosa CA 95405

HOLLANDER, MAX LEO, b Cologne, Ger, Dec 3, 23; nat US; m 48; c 3. CHEMISTRY. *Educ:* Rutgers Univ, BS, 44, PhD(chem), 47. *Prof Exp:* Instr chem, Rutgers Univ, 47; res chemist, Am Smelting & Refining Co, 47-66, sect leader, 66-70, supt chem res, 70-76, supt chem res, Asarco, Inc, 76-84; RETIRED. *Res:* Non-ferrous chemistry; electrochemistry; extractive metallurgy; water pollution control. *Mailing Add:* 740 Hart Dr Bridgewater NJ 08807

HOLLANDER, MILTON B(ERNARD), b Bayonne, NJ, Nov 29, 28; m 52; c 4. MECHANICAL ENGINEERING. *Educ:* Purdue Univ, BS, 51; Mass Inst Technol, MS, 53; Columbia Univ, PhD, 59. *Prof Exp:* Asst, Mass Inst Technol, 52-53; assoc proj engr, Bendix Aviation Corp, 53-55; chmn res dept & mgr develop task force, Am Mach & Foundry Co, 57-65, dir eng, AMFare Div, 65-67; vpres res & develop, Am Standard, Inc, 67-69, vpres corp res develop & technol, 69-72, vpres sci & technol, 68-72; VPRES SCI & TECHNOL, GULF & WESTERN INDUST, INC, 72- *Concurrent Pos:* Consult, Alpha Molykote Corp, 54- & electronics lab, Columbia Univ, 54-55; lectr, Univ Conn, 63; mem bd dirs, Am Nat Standards Inst, 71-76. *Mem:* Sigma Xi; Am Soc Mech Engrs; Soc Mfg Engrs; Nat Soc Prof Engrs; Am Inst Mining, Metall & Petrol Engrs. *Res:* Friction welding; automatic food equipment; automatic machinery; metal cutting; lubrication; temperature measurement; infrared measurements; friction and wear; oil field technology. *Mailing Add:* Gulf & Western Indust Inc One Gulf & Western Plaza New York NY 10023

HOLLANDER, MYLES, b Brooklyn, NY, Mar 21, 41; m 63; c 2. MATHEMATICAL STATISTICS, APPLIED STATISTICS. *Educ:* Carnegie Inst Technol, BS, 61; Stanford Univ, MS, 62, PhD(statist), 65. *Prof Exp:* From asst prof to prof, 65-78, chmn, dept, 78-81, PROF STATIST, FLA STATE UNIV, 78- *Concurrent Pos:* Vis prof biostatist, Stanford Univ, 72-73 & 81-82, Univ Wash, 89-90. *Mem:* AAAS; fel Am Statist Asn; Int Statist Inst; fel Inst Math Statist; Sigma Xi. *Res:* Nonparametric statistics; biostatistics; medical consulting. *Mailing Add:* Dept Statist Fla State Univ Tallahassee FL 32306

HOLLANDER, PHILIP B, b Chicago, Ill, May 4, 24; c 5. PHARMACOLOGY, BIOPHYSICS. *Educ:* Univ Southern Calif, MSc, 55, PhD(pharmacol), 60. *Prof Exp:* USPHS fel, 60-61; sr engr, Radio Corp Am, NJ, 61-64; asst prof pharmacol, Woman's Med Col Pa, 62-63; assoc prof, 64-69, dep chmn, 77-83, PROF PHARMACOL & BIOPHYS, COL MED, OHIO STATE UNIV, 69- *Concurrent Pos:* Mem fac, Jefferson Med Sch, 63-64; NIH develop award fel, 64-74. *Mem:* AAAS; Am Soc Pharmacol & Exp Therapeut; NY Acad Sci; Biomed Eng Soc; Am Heart Asn; Sigma Xi; Inst Elec & Electronics Engrs. *Res:* Electropharmacological principles of pharmacological activities as related to cardiac activity; information processing by biological systems; man-machine interrelationships for closed loop design analysis; substance abuse. *Mailing Add:* PO Box 8463 Columbus OH 43210-1239

HOLLANDER, VINCENT PAUL, b New York, NY, June 18, 17; m 52; c 4. INTERNAL MEDICINE. *Educ:* Univ Chicago, BS, 41, MS & PhD(biochem), 44; Northwestern Univ, MD, 47; Am Bd Internal Med, dipl, 54. *Prof Exp:* Asst resident med, Montefiore Hosp, 49-50; res fel, Sloan-Kettering Inst Cancer Res, 50-52; assoc prof internal med, Sch Med, Univ Va, 52-60, Am Cancer Soc prof internal med, 60-63; dir Res Inst Skeletomuscular Dis, Hosp Joint & Med Ctr, 63-80; PROF INTERNAL MED & RES PROF BIOCHEM, MT SINAI SCH MED, 69-, RES PROF SURG & NEOPLASTIC DIS, 80- *Concurrent Pos:* Clin asst, Mem Hosp Cancer & Allied Dis, 51-52; asst prof biochem, Sch Med, Univ Va, 52-63, physician & chief tumor clin, Univ Hosp, 52-58, coordr cancer prog, Univ, 52-63; mem endocrine study sect, Div Res Grants, NIH, 64-, mem cancer res training comt, Nat Cancer Inst, 69-; prof med & biochem, Am Cancer Soc, 77; attend physician Internal Med, Mt Sinai Hosp, 64-, attend physician Neoplastic dis, 82- *Mem:* AAAS; Am Soc Cancer Res; Endocrine Soc; Am Soc Biol Chem; Am Fedn Clin Res; Am Chem Soc. *Res:* Endocrinology; malignant disease; steroid chemistry. *Mailing Add:* Two Beekman Pl New York NY 10022

HOLLANDER, WALTER, JR, b Baltimore, Md, Aug 12, 22; m 55; c 2. INTERNAL MEDICINE, NEPHROLOGY. *Educ:* Haverford Col, BS, 44; Harvard Univ, MD, 50; Am Bd Internal Med, dipl, 60. *Prof Exp:* Intern & asst resident med, Presby Hosp, New York, 50-53; sr resident, Boston Vet Admin Hosp, 53-54; from instr to emer prof med, Sch Med, Univ NC, Chapel Hill, 56-88, dir, Gen Clin Res Unit, 60-66, asst to dean, 66-72; RETIRED. *Concurrent Pos:* USPHS fel, Sch Med, Univ NC, Chapel Hill, 54-56 & 72-73; Am Heart Asn res grant, 58-61; Markle scholar, 58-63; USPHS res grant, 61-66; asst, Sch Med, Boston Univ, 53-54 & Harvard Med Serv, Boston City Hosp, 53-54. *Mem:* AAAS; emer mem Am Fedn Clin Res; emer mem Am Soc Clin Invest; emer mem Am Soc Invest; Sigma Xi. *Res:* Renal and body fluid normal and abnormal function and structure. *Mailing Add:* 531 Dogwood Dr Chapel Hill NC 27516

HOLLANDER, WILLIAM, b Brooklyn, NY, May 21, 25; m 54; c 3. INTERNAL MEDICINE. *Educ:* NY Univ, BA, 45; Long Island Col Med, MD, 49. *Prof Exp:* Asst med, 51-55, instr, 55-57, assoc, 57-59, from asst prof to assoc prof, 59-69, from asst mem to assoc mem, Hosp, 57-65, vis physician, 57-65, PROF MED, SCH MED, BOSTON UNIV, 69-, DIR HYPERTENSION & ATHEROSCLEROSIS RES UNIT, UNIV HOSP, 65- *Mem:* AAAS; Am Heart Asn; Am Fedn Clin Res. *Res:* Metabolic and pharmacological studies in arterial hypertension and atherosclerosis. *Mailing Add:* Dept Med Physics & Biochem Sch Med Boston Univ 80 E Concord St Boston MA 02118

HOLLANDSWORTH, CLINTON E, b Pulaski, Va, Dec 7, 30. NUCLEAR PHYSICS. *Educ:* Va Polytech Inst, BS, 58, MS, 61; Duke Univ, PhD(nuclear physics), 63. *Prof Exp:* Asst nuclear physics, Duke Univ, 59-63, res assoc, 63-65; res physicist, US Army Nuclear Defense Lab, 65-71; RES PHYSICIST, US ARMY BALLISTIC RES LAB, 71- *Mem:* Am Phys Soc; Inst Elec & Electronics Engrs; Sigma Xi. *Res:* Experimental physics; fast neutron physics including elastic and inelastic scattering of neutrons; accelerator technology; physics of charged particle beams; pulsed power electromagnetics. *Mailing Add:* TBD-Ballistic Res Lab Aberdeen Proving Ground MD 21005-5066

HOLLBACH, NATASHA COFFIN, b Halifax, NS, Feb 4, 33; m 54; c 2. RADIOCHEMISTRY, CHEMICAL SAFETY. *Educ:* Dalhousie Univ, BSc, 53; McGill Univ, PhD, 57. *Prof Exp:* Chemist, Can Industs Ltd, Que, 57-58; asst prof chem, George Washington Univ, 58-61; spec lectr, Carleton Univ, 62-63; mem ed staff & ed off sci jour, Nat Res Coun Can, 63-66; mem fac, dept chem biochem, Algonquin Col Appl Arts & Technol, 67- 88; CONSULT, 88- *Honors & Awards:* Polysar Award, 89. *Mem:* Chem Inst Can. *Mailing Add:* Box 354 Almonte ON K0A 1A0 Can

HOLLDOBLER, BERTHOLD KARL, b Erling-Andechs, Ger, June 25, 36; m 80; c 3. BEHAVIORAL ECOLOGY, SOCIOBIOLOGY. *Educ:* Univ Wurzburg, Dr rer nat, 65; Univ Frankfurt, Dr habil, 69. *Hon Degrees:* Am, Harvard Univ. *Prof Exp:* Asst prof zool, Univ Frankfurt, 66-69, privatdozent on leave, 69-71, prof zool, 71-72; prof biol, Harvard Univ, 73-90, Alexander Agassiz prof zool, 82-90; ADJ PROF NEUROBIOL, UNIV ARIZ, TUCSON, 89-; PROF ZOOL, UNIV WÜRZBURG, GER, 89- *Concurrent Pos:* Res assoc zool, Harvard Univ, 69-71; co-ed, Psyche, 73- & Behav Ecol & Sociobiol, 76-; John Simon Guggenheim fel, 80; Psychobiol Panel, NSF, 84-87. *Honors & Awards:* Alexander von Humboldt Sr Scientist Award, 87; Leibniz Prize, 90. *Mem:* Am Acad Arts & Sci; Leopoldina Ger Acad Sci; fel

AAAS; Animal Behav Soc; Int Union Study Social Insects; Ger Zool Soc; Cambridge Entom Soc; Ecol Soc Am; Soc Study Evolution; Soc Am Naturalists; Int Soc Chem Ecol; Bavarian Acad Sci. *Res:* Behavioral ecology and sociobiology of insects; animal communication; chemical ecology. *Mailing Add:* Dept Zool Univ Wurzburg Rontgenring 10 Germany

HOLLE, MIGUEL, b Berlin, Ger, Jan 13, 37; Peru citizen; m 66; c 3. HORTICULTURE, PLANT BREEDING. *Educ:* Colo State Univ, BS, 58; Iowa State Univ, MS, 60, PhD(hort), 65. *Prof Exp:* From asst prof to prof hort, Univ Nac Agraria La Molina, Lima, Peru, 61-76; horticulturist, Res & Training Ctr, Turrialba, Costa Rica, 76-82; regional officer, Int Bd For Plant Genetic Resources, Cali, Colombia, 82-88; plant genetic resources prog, Lima, Peru, 88-89; DIR, ANDEAN CROP & ANIMAL RES PROJ, NAT AGR INST RES, PERU, 89- *Concurrent Pos:* Guggenheim Found fel, 74-75. *Mem:* Am Soc Hort Sci; Latin Am Asn Agr Sci; Int Soc Hort Sci. *Res:* Vegetable crops, especially germ plasm and field management studies within the cropping systems context. *Mailing Add:* Choquenhanca 851 Lima 27 Peru

HOLLE, PAUL AUGUST, b Decatur, Ind, July 5, 23; m 54; c 2. BIOLOGY. *Educ:* Valparaiso Univ, AB, 47; Univ Notre Dame, MS, 49, PhD(zool), 56. *Prof Exp:* Asst prof zool, Univ NH, 50-57; prof & chmn dept, 57-89, EMER PROF BIOL, WORCESTER STATE COL, 89- *Concurrent Pos:* Mem staff, Marine Biol Lab, Woods Hole, 50-51, Bermuda Marine Biol Labs, 51 & Mus Comp Zool, Harvard Univ, 53-55. *Mem:* AAAS; Soc Syst Zool; Nat Asn Biol Teachers; Am Malacol Union. *Res:* Malacology; marine invertebrates; science education; amphibian osteology. *Mailing Add:* 131 Holman St Shrewsbury MA 01545

HOLLEIN, HELEN CONWAY FARIS, b Ft Bragg, NC, Mar 21, 43; m 66; c 3. EDUCATIONAL ADMINISTRATION. *Educ:* Univ SC, BS, 65; NJ Inst Technol, MS, 79, DEng, 82. *Prof Exp:* Process engr, chem div, Exxon Res & Eng Co, 65-67; teacher chem & physics, Livingston High Sch, NJ, 67-69; substitute teacher math, Singapore Am High Sch, 70-71; teaching asst, dept chem eng & chem, NJ Inst Technol, 77-78, adj instr, 78-81; asst prof, 82-88, ASSOC PROF CHEM ENG, MANHATTAN COL, RIVERDALE, NY, 88-, DEPT CHAIR, 89- *Honors & Awards:* Ralph R Teetor Award, Soc Automotive Engrs, 84. *Mem:* Am Inst Chem Engrs; Soc Women Engrs; Am Chem Soc; Am Soc Eng Educ; Nat Soc Prof Engrs; Sigma Xi. *Res:* Adsorption and ion exchange; biotechnology; ultrafiltration; high-performance liquid chromatography. *Mailing Add:* Dept Chem Eng Manhattan Col Riverdale NY 10471

HOLLEMAN, KENDRICK ALFRED, b Normangee, Tex, Dec 11, 34; m 56; c 2. POULTRY SCIENCE. *Educ:* Tex A&M Col, BS, 58; Univ Nebr, Lincoln, MS, 62; Univ Mo, Columbis, PhD(poultry sci), 71. *Prof Exp:* Asst exten poultryman, Coop Exten Serv, Univ Nebr, Lincoln, 58-59; asst mgr turkey processing & mkt, Nebr Turkey Growers Coop Asn, 59-64; mgr small animal nutrit res, Small Animal Labs, Ralston Purina Co, Mo, 64-66; instr physiol res, Univ Mo-Columbia, 66-70; prof poultry sci & proj leader, Poultry Sci Exten, Clemson Univ, 70-80; PROG LEADER POULTRY SCI, EXTEN SERV, USDA, 81- *Honors & Awards:* Egg Sci Award, Am Egg Bd, 73. *Mem:* Poultry Sci Asn; World's Poultry Sci Asn; Sigma Xi. *Res:* Physiology of reproduction in avian species; environmental studies on domestic poultry; nutrition of domestic poultry; egg science and technology; PCB research. *Mailing Add:* Dept Poultry Sci Ore State Univ Dryden 208 Corvallis OR 97331-3402

HOLLEMAN, WILLIAM H, b Jamestown, Mich, Aug 18, 40; m 63; c 2. BIOCHEMISTRY. *Educ:* Hope Col, AB, 62; Mich State Univ, PhD(biochem), 66. *Prof Exp:* Res biochemist, Abbott Labs, 66-70, group leader, 70-77, sect head, antithrombosis, 77-80, head, Dept Virol & Biochem, 80-84, PROJ LEADER, CARDIOVASC RES, ABBOTT LABS, 84- *Concurrent Pos:* Lectr, Med Sch, Loyola Univ Chicago, 70-80. *Mem:* Am Soc Biol Chem; Am Soc Microbiol; Sigma Xi. *Res:* Biochemistry of macromolecules, ultracentrifugation; isolation and characterization of enzymes which have therapeutic importance; drug development for cardiovascular diseases; science administration. *Mailing Add:* Dept 473 Abbott Labs Abbott Park IL 60064

HOLLENBACH, EDWIN, b West Leesport, Pa, Nov 2, 18; m 44; c 1. MECHANICAL & ELECTRICAL ENGINEERING. *Educ:* Wyomissing Polytech, BSME, 39. *Prof Exp:* Designer, Birdsboro Steel, Steel Foundry & Mach Co, 39-41; eclipse-pioneer div, Bendix Corp, 41-45; instr math, Wyomissing Polytech, 45-46; design engr, Rheem Mfg Co, 46-49; proj engr, Air Prod, Inc, 49-50; sr engr, Scott Paper Co, 50-52; mgr serv, Burroughs Corp, 52-55; mgr serv sr res engr, Univac Corp, 55-56; vpres eng, Briggs Assocs, Inc, 56-59; exec vpres, Chem Serv, Inc, 60-69, pres, 69-80; vpres, Drexel Dynamics Corp, 59-80; RETIRED. *Mem:* Int Mat Mgt Soc; Am Chem Soc. *Res:* Design and development of automation and computer equipment; input-output automatic controls; general material handling equipment. *Mailing Add:* 30 Lyman Ave Woodbury NJ 08096-2837

HOLLENBAUGH, KENNETH MALCOLM, b Fostoria, Ohio, Sept 8, 34; m 61; c 3. ENVIRONMENTAL GEOLOGY. *Educ:* Bowling Green State Univ, BS, 57; Univ Idaho, MS, 59, PhD(geol), 68. *Prof Exp:* Teaching asst phys geol & paleont, Univ Idaho, 57-58, mine geologist, Idaho Bur Mines & Geol, 58-59; mine engr, Kaiser Cement & Gypsum Corp, 60-63, mine supt, 63-65; geologist, Kaiser Steel Corp, 65-68; from asst prof to assoc prof geol, 68-73, chmn dept, 71-77, PROF GEOL, BOISE STATE UNIV, 73-, DEAN GRAD COL, 75-, DIR, UNIV RES CTR, 81-, ASSOC EXEC VPRES, 81- *Concurrent Pos:* Pres, Kash Enterprises; vpres, Idaho Quartzite. *Mem:* Am Inst Mining, Metall & Petrol Engrs; Sigma Xi. *Res:* Research and consulting in environmental geology, geothermal geology and mineral economics. *Mailing Add:* Grad Col Boise State Univ Boise ID 83725

HOLLENBECK, CLARIE BEALL, b San Francisco, Calif, Apr 18, 47; m; c 4. EXPERIMENTAL BIOLOGY. *Educ:* Calif State Univ, BA, 71; Ore State Univ, PhD(nutrit biochem), 82. *Prof Exp:* Res asst, Univ Ore Health Sci Ctr, Portland, 79-81, Ore State Univ, Corvallis, 80-81, teaching asst, Dept Foods & Nutrit, 81-82, res asst, Nutrit Res Inst, 81-82; res scholar, 82-85, RES SCIENTIST, DEPT MED, STANFORD UNIV SCH MED, 85-; RES SCIENTIST, GERIAT EDUC & CLIN CTR, VET ADMIN MED CTR, PALO ALTO, CALIF, 85- *Concurrent Pos:* Nat res serv award, Nat Heart, Lung & Blood Inst, NIH, 83-85; mem, Prog Comt, Coun Nutrit Sci & Metab, Am Diabetes Asn, 84-88, chair, Educ Comt, 85-87, chair, Prof Educ Comt, 86-87, chair, Educ & Prog Comt, Coun Nutrit Sci & Metab, 86-88, vchair, Prof Educ Comt, 87-88, mem, 87-90, mem, Comt on Sci & Med Progs, 87-89, chair, Prof Educ Subcomt, 87-, vchair, Coun Nutrit Sci & Metab, 88-90, co-chmn, Task Force on Nutrit & Pregnancy, 88-91, chmn, Coun Nutrit Sci & Metab, 90-, mem, Med-Sci Comt, 90-; coordr diabetes res, Dept Med, Stanford Univ Sch Med, 85-86, assoc dir & res coordr, Gen Clin Res Ctr, 86-90, vis scholar, Dept Med, 86-87. *Mem:* Am Diabetes Asn; Am Fedn Clin Res; Am Heart Asn; Am Inst Nutrit; Am Soc Clin Nutrit; Sigma Xi. *Res:* Diabetes. *Mailing Add:* Div Endocrinol Geront & Metab Stanford Univ Med Ctr Rm S005 Stanford CA 94305-5103

HOLLENBECK, ROBERT GARY, b Oneida, NY, Oct 8, 49; c 2. PHARMACEUTICS, PHYSICAL PHARMACY. *Educ:* Albany Col Pharm, BS, 72; Purdue Univ, PhD(indust & phys pharm), 77. *Prof Exp:* Asst prof, 77-83, ASSOC PROF PHARMACEUT, SCH PHARM, UNIV MD, 83- *Mem:* Am Asn Pharmaceut Scientists; Am Pharmaceut Asn; Am Asn Col Pharm. *Res:* Application of physical-chemical principles to the design, development and evaluation of drug delivery systems. *Mailing Add:* 3405 Sylvan Lane Ellicott City MD 21043

HOLLENBERG, CHARLES H, b Winnipeg, Man, Sept 15, 30; m 56; c 1. MEDICINE. *Educ:* Univ Man, BSc, 50, BSc & MD, 55; FRCP(C), 59; FACP(US), 73. *Hon Degrees:* DSc, Univ Man, 83, McGill Univ, 85. *Prof Exp:* Asst med, Tufts Univ, 58-60; Markle scholar, 60; from lectr to prof, McGill Univ, 60-70; chmn dept, Univ Toronto, 70-81, Charles H Best prof med res, 81-83, vprovost, 83-89, PROF MED, UNIV TORONTO, 70-, DIR, BANTING & BEST DIABETES CTR, 81- *Concurrent Pos:* Physician-in-chief, Toronto Gen Hosp, 70-81, sr physician, 81- *Mem:* Fel Am Col Physicians; Endocrine Soc; Am Physiol Soc; Asn Am Physicians; Am Soc Clin Invest; Sigma Xi. *Res:* Endocrinology and lipid metabolism. *Mailing Add:* CCRW3-845 Toronto Gen Hosp 200 Elizabeth St Toronto ON M5G 2C4 Can

HOLLENBERG, DAVID HENRY, b Philadelphia, Pa, Feb 3, 46; m 67; c 3. CELLULOSE CHEMISTRY, HEMICELLULOSE CHEMISTRY. *Educ:* Wittenberg Univ, BS, 67; Univ Maine, PhD(chem), 72. *Prof Exp:* Fel, Sloan-Kettering Inst, 72-73, Am Cancer Soc, 73-74; sr develop chemist wood chem, St Regis Paper Co, 77-79, sr staff chemist, 79-82; mem staff, W R Grace & Co, 82-84; sr res assoc, James River Corp, 84-86, res fel, 86-91. *Mem:* Am Chem Soc; AAAS; Tech Asn Pulp & Paper Indust; Sigma Xi; NY Acad Sci. *Res:* Mechanisms of pulping and bleaching of wood fibers and natural products including tall oil, turpentine, polysaccharides and other natural polymers; chemical modification of cellulosic fibers; mechanisms of fiber-fiber bonding; papermaking chemistry surface and colloid science. *Mailing Add:* James River Corp 1915 Marathon Ave Neenah WI 54956

HOLLENBERG, J LELAND, b La Verne, Calif, Sept 17, 26; m 51; c 5. PHYSICAL CHEMISTRY. *Educ:* Univ Redlands, BA, 49, BS, 52; Univ Calif, Berkeley, MS, 52; Univ Southern Calif, PhD(chem), 62. *Prof Exp:* High sch instr chem & physics, 49-50 & 52-54; instr chem, Fullerton City Col, 54-59; staff consult, Chem Study, 61-63; PROF CHEM, UNIV REDLANDS, 63- *Concurrent Pos:* NSF fac fel, Univ Calif, Riverside, 71-72. *Mem:* Am Chem Soc. *Res:* Absolute infrared intensities of ions and molecules, especially in condensed phases; hydration numbers of small, biologically important molecules. *Mailing Add:* 31350 Vista Dr Redlands CA 92373

HOLLENBERG, JOEL WARREN, b New York, NY, Oct 20, 38; m 60, 74; c 2. TURBOMACHINERY, ALTERNATE ENERGY. *Educ:* Cooper Union, BME, 60; Stevens Inst Technol, MS, 62, PhD(mech eng), 78. *Prof Exp:* Res asst, Stevens Inst Technol, 60-61, res engr, Davidson Lab, 61-65; eng sect mgr, Diehl Div, Singer Co, 65-67; mem tech staff, Ingersoll Band Res, Inc, 67-71, prog mgr govt contracts, 71-73; instr, Stevens Inst Technol, 73-77; from asst prof to assoc prof, 77-82, PROF MECH ENG, COOPER UNION, 82- *Concurrent Pos:* Consult & expert witness for indust, 74-; prof in charge mech eng labs, Cooper Union, 81- *Mem:* Am Soc Mech Engrs; Am Soc Heating, Refrig & Air Conditioning Engrs; Am Soc Eng Educ; Sigma Xi. *Res:* Regenerative turbomachinery; transverse fans; unusual low speed aerodynamic problems; computer aided experimentation and alternative energy systems. *Mailing Add:* Cooper Union Cooper Square New York NY 10003

HOLLENBERG, MARTIN JAMES, b Winnipeg, Man, June 30, 34; m 59; c 2. ANATOMY. *Educ:* Univ Man, BSc & MD, 58; Wayne State Univ, MSc, 64, PhD(anat), 65. *Prof Exp:* From asst prof to assoc prof anat, Univ Western Ont, 65-71; prof anat, Univ BC, 71-75; prof morphol sci & head div, Fac Med, Univ Calgary, 75-78; dean med & prof anat, Univ Western Ont, 78-85; DEAN, FAC MED, UNIV BC, 85- *Mem:* Am Asn Anat; Can Asn Anat. *Res:* Structure of vertebrate eye. *Mailing Add:* Fac Med Univ BC 317-2194 Health Sci Mall Vancouver BC Z6T 1W5 Can

HOLLENBERG, MILTON, b New York, NY, July 29, 30; m 64; c 1. MEDICINE, CARDIOVASCULAR PHYSIOLOGY. *Educ:* Brooklyn Col, AB, 51; Cornell Univ, MD, 55. *Prof Exp:* Intern, Bellevue Hosp, New York, 55-56; resident, Med Ctr, Univ Calif, San Francisco, 58-61; fel physiol, Harvard Med Sch, 61-63; from instr to asst prof, Med Col, Cornell Univ, 63-68; assoc prof, 68-78, PROF, MED CTR, UNIV CALIF, SAN FRANCISCO, 78- *Concurrent Pos:* Health Res Coun NY career scientist, 63-68; chief cardiol, San Francisco Vet Admin Hosp, 68-77. *Mem:* AAAS;

Am Fedn Clin Res; Am Heart Asn; Vet Admin Cardiovasc Asn (pres, 84-85, vpres, 83-84, secy-treas, 82-83). *Res:* Exercise testing; uses of computers in science; cardiovascular pharmacology. *Mailing Add:* Vet Admin Hosp 111C-3 Univ Calif San Francisco 4150 Clement St San Francisco CA 94122

HOLLENBERG, MORLEY DONALD, b Winnipeg, Man, July 2, 42; m 65; c 2. ENDOCRINOLOGY. *Educ:* Univ Man, Can, BSc, 63, MSc, 64; Oxford Univ, PhD(pharmacol), 67; Johns Hopkins Med Sch, MD, 72. *Prof Exp:* Fel pharmacol, Oxford Univ, 67-68; intern internal med, Johns Hopkins Hosp, 71-72, fel pharmacol, Johns Hopkins Med Sch, 72-73, asst prof pharmacol, 73-79, asst prof med, 74-79; PROF & HEAD PHARMACOL DEPT, FAC MED, UNIV CALGARY, 79- *Concurrent Pos:* Investr med & pharmacol, Howard Hughes Med Inst, 74-79. *Mem:* Am Soc Pharmacol & Exp Therapeut; Am Soc Clin Invest; Can Soc Clin Invest. *Res:* Studies of the receptors and mechanisms of action of growth factors like insulin, somatomedins and epidermal growth factor-urogastrone. *Mailing Add:* Dept Pharmacol & Therapeut Univ Calgary Fac Med 3330 Hosp Dr NW Calgary AB T2N 4N1 Can

HOLLENBERG, PAUL FREDERICK, b Philadelphia, Pa, Sept 18, 42; m 67; c 2. TOXICOLOGY, ENZYMOLOGY. *Educ:* Wittenberg Univ, BS, 64; Univ Mich, MS, 66, PhD(biochem), 69. *Prof Exp:* Teaching asst biochem, Univ Mich, 64-69; asst prof biochem, Med Sch, Northwestern Univ, Chicago, 72-81, assoc prof path & pharmacol, 81-84, prof path & molecular biol, 84-87; PROF & CHMN PHARMACOL, WAYNE STATE UNIV SCH MED, 87- *Concurrent Pos:* Fel, Univ Mich, Ann Arbor, 69; fel, Univ Ill, Urbana-Champaign, 69-72; Schweppe Found Res Fel; assoc ed, Chem Res in Toxicol, Chem Path Study Sect, 87- *Mem:* AAAS; Am Chem Soc; NY Acad Sci; Am Asn Cancer Res; Am Soc Biol Chemists; Am Soc Pharmacol & Exp Therapeut; Soc Toxicol. *Res:* Enzyme mechanisms; role of enzyme structure in catalytic function; structure and properties of heme proteins; biological oxidations and peroxidations; mechanisms of carcinogensis and toxicity. *Mailing Add:* Dept Pharmacol Wayne State Univ Sch Med 540 E Canfield Detroit MI 48201

HOLLENDER, MARC HALE, b Chicago, Ill, Dec 19, 16; m 43; c 2. PSYCHIATRY, PSYCHOANALYSIS. *Educ:* Univ Ill, BS, 39, MD, 41. *Prof Exp:* Clin asst psychiat, Col Med, Univ Ill, 46-48, from clin instr to assoc prof, 48-56; prof, State Univ NY Upstate Med Ctr, 56-66; prof, Univ Pa, 66-70; prof & chmn dept, 70-82, EMER PROF PSYCHIAT, VANDERBILT UNIV, 82- *Concurrent Pos:* Consult, US Vet Admin, 52-56; mem staff, Inst Psychoanal, Chicago, 53-56; chmn dept psychiat, State Univ NY Upstate Med Ctr, 56-64; dir, Syracuse Psychiat Hosp, 57-64; dir, Am Bd Psychiat & Neurol, 72-81 (pres, 80). *Mem:* Am Psychosom Soc; fel Am Psychiat Asn; Am Psychoanal Asn. *Res:* Psychoanalysis; psychology of medical practice. *Mailing Add:* Dept Psychiat Vanderbilt Hosp Nashville TN 37232

HOLLENHORST, ROBERT WILLIAM, b St Cloud, Minn, Aug 12, 13; m 39; c 9. MEDICINE, OPHTHALMOLOGY. *Educ:* Univ Minn, BS, 38, MB, 40, MD, 41, MS, 48; Am Bd Ophthal, dipl. *Prof Exp:* Intern, Abbott Hosp, Minneapolis, 40; intern, Ancker Hosp, St Paul, 40-41; Mayo Found fel, Univ Minn, 46-48; from instr to prof, 49-79, consult, Mayo Clin, 49-79, EMER PROF OPHTHAL, MAYO GRAD SCH MED, 79- *Concurrent Pos:* State consult ophthal, Minn State Dept Pub Welfare, 63- *Honors & Awards:* Howe Medal, Am Ophthal Soc. *Mem:* Am Ophthal Soc; Am Acad Ophthal & Otolaryngol. *Res:* Neuro-ophthalmology with particular emphasis on intracranial vascular disease. *Mailing Add:* Mayo Clin 200 First St SW Rochester MN 55901

HOLLENSEN, RAYMOND HANS, b Madison Co, Nebr, Nov 5, 31; m 55; c 3. BRYOLOGY. *Educ:* Capital Univ, BA, 53; Univ Mich, MS, 58; Univ Cincinnati, PhD(bot), 62. *Prof Exp:* Asst bot, Univ Mich, 56-58 & Univ Cincinnati, 58-61; instr biol, Keuka Col, 61-63, actg chmn dept, 62-63, asst prof, 63-65; from asst prof to prof natural sci, Mich State Univ, 65-89, asst chairperson, 75-84, artg asst dir, Undergrad Univ Div, 87-90, PROF BOT & PLANT PATH, MICH STATE UNIV, 77-, ASSOC DIR, UNDERGRAD UNIV DIV, 90- *Concurrent Pos:* Curator, Cryptogamic Herbarium, Mich State Univ, 90- *Mem:* Am Bryol & Lichenological Soc. *Res:* Developmental morphology; tissue differentiation; systematics of the Hepaticae. *Mailing Add:* Undergrad Univ Div Mich State Univ 170 Bessey Hall East Lansing MI 48824-1033

HOLLER, ALBERT COCHRAN, b Erie, Pa, Mar 17, 21; m 46; c 5. CHEMISTRY. *Educ:* Univ Minn, BChem, 47. *Prof Exp:* Chief chemist, US Metal Prcd Co, Pa, 41-44; res fel chem, Off Naval Res Proj, Univ Minn, 47-49; chief anal chemist, Twin City Testing & Eng Lab, 49-52, dir chem div, 52-58, vpres chem div, 58-85; RETIRED. *Honors & Awards:* Thomas F Andrews Prize, 47. *Mem:* AAAS; Am Chem Soc; Am Inst Chemists; Sigma Xi; Nat Asn Corrosion Engrs; Royal Soc Chem London; Am Indust Hyg Assoc. *Res:* Visible and ultraviolet absorption spectrophotometry; chromatography; analytical methods for nitric acid esters; metallurgy of copper base alloys and their analysis; corrosion of metals; ferrous metals analysis; mineralogy; sampling of metals, alloys, vegetable oils and petroleum. *Mailing Add:* 3205 Wendhurst Ave NE Minneapolis MN 55418-1727

HOLLER, FLOYD JAMES, b Muncie, Ind, Aug 6, 46; m 67; c 3. ANALYTICAL CHEMISTRY. *Educ:* Ball State Univ, BS, 68, MS, 73; Mich State Univ, PhD(chem), 77. *Prof Exp:* Teacher chem & physics, Western Wayne Schs, 68-73; grad asst, Mich State Univ, 73-77; asst prof, 77-83, ASSOC PROF CHEM, UNIV KY, 83- *Mem:* Am Chem Soc. *Res:* Reaction rate methods; analytical instrumentation; mini and micro computer applications; science education. *Mailing Add:* Dept Chem Univ Ky Lexington KY 40506

HOLLER, JACOB WILLIAM, b Ft Edward, NY, Nov 27, 12; m 42; c 2. MEDICINE. *Educ:* Univ Rochester, MD, 41. *Prof Exp:* Buswell fel, Univ Rochester, 47-48; dir, Dorn Lab Med Res, Pa, 48-50; dir, clin labs & grad educ, Highland Hosp, 50-79, chief med, 56-70, dir clin serv, 70-82; from asst prof to prof, 68-82, EMER PROF MED, SCH MED, UNIV ROCHESTER, 82-; DIR, CLIN LABS & GRAD EDUC, HIGHLAND HOSP, 50-, DIR CLIN SERV, 70- *Concurrent Pos:* Consult, Genesee Hosp, Rochester, NY. *Mem:* Sigma Xi; Fel Am Col Physicians. *Res:* Endocrinology and metabolism. *Mailing Add:* 65 Greensward Cherry Hill NJ 08002

HOLLER, NICHOLAS ROBERT, b Plymouth, Ind, May 14, 39; m 67; c 2. WILDLIFE MANAGEMENT, MAMMALOGY. *Educ:* Univ Mo, AB, 61, AM, 63, PhD(zool), 73. *Prof Exp:* Staff specialist wildlife res, US Fish & Wildlife Serv, 70-73, supvr wildlife biologist animal damage, Patuxent Wildlife Res Ctr, 73-75, sr proj biologist, CFBC study, Fla Game & Fresh Water Fish Comn, 75-85, SUPV WILDLIFE BIOLOGIST ANIMAL DAMAGE, DENVER WILDLIFE RES CTR, US FISH & WILDLIFE SERV, 85-, LEADER, ALA COOP FISH & WILDLIFE RES UNIT, 85- *Concurrent Pos:* Assoc ed, J Wildlife Mgt, 80-81; ed, Wildlife Soc Bull, 90-91. *Mem:* Wildlife Soc; Am Soc Mammalogists; Soc Conserv Biol. *Res:* Wildlife damage control; mammalian ecology; social behavior; endangered species. *Mailing Add:* Ala Coop Fish & Wildlife Res Unit 331 Funchess Hall Auburn Univ AL 36849-5414

HOLLERAN, EUGENE MARTIN, b Pa, June 25, 22; m 47; c 8. PHYSICAL CHEMISTRY. *Educ:* Univ Scranton, BS, 43; Cath Univ Am, PhD(chem), 49. *Prof Exp:* Instr chem, Regis Col, 48-50; from asst prof to assoc prof, 50-54, chmn dept, 61-65 & 70-73, dir sci, 65-67, PROF CHEM, GRAD SCH, ST JOHN'S UNIV, NY, 54- *Mem:* AAAS; Am Chem Soc; Sigma Xi. *Res:* Kinetic theory of gases; correlation of fluid properties; equations of state. *Mailing Add:* Chem Dept St Johns Univ Jamaica NY 11432

HOLLEY, CHARLES ELMER, JR, geochemistry, thermochemistry, for more information see previous edition

HOLLEY, CHARLES H, b Pittsburgh, Pa, Apr 15, 19; c 4. POWER GENERATION & TECHNICAL MANAGEMENT. *Educ:* Duke Univ, BSEE, 41. *Prof Exp:* Mem staff, Gen Elec Co, 41-74, mgr TB-Generator Eng, 62-74, gen mgr, Elec Utility Syst Eng Dept, Energy Syst/Technol Div, 74-80, mgr turbine technol assessment oper, Turbine Bus Group, 80-83; RETIRED. *Concurrent Pos:* Consult, elec eng, power generation systs & tech mgt, 83- *Honors & Awards:* Tesla Award & Centennial Award, Inst Elec & Electronics Engrs. *Mem:* Nat Acad Eng; fel Inst Elec & Electronics Engrs. *Mailing Add:* 5603 Pipers Waite Sarasota FL 34235

HOLLEY, DANIEL CHARLES, b San Jose, Calif, Aug 1, 49. BIOLOGICAL SCIENCE. *Educ:* Cabrillo Col, AS, 69; Univ Calif, Davis, BS, 71, MS, 73, PhD(physiol), 76. *Prof Exp:* Staff res assoc I, Dept Surg, Sch Med, Univ Calif, Davis, 69-73, teaching asst, Dept Animal Physiol, 73-74, res & teaching asst, Dept Animal Sci, 71-76; instr physiol, Dept Physiol, La State Univ Med Sch, 76-78; staff mem, 78-86, PROF, DEPT BIOL SCI, SAN JOSE STATE UNIV, 86- *Concurrent Pos:* Consult, NIH; grants, San Jose State Univ, 80-82, NASA, 81-82, 82-86 & 82-85, NIH, 83-87; mem, People to People Aerospace Deleg, Peoples Repub China, 87. *Mem:* Sigma Xi; Fedn Am Soc Exp Biol; AAAS; Aerospace Med Asn; Am Asn Univ Prof; Am Inst Biol Sci. *Res:* Chronobiology, particularly physiological circadian rhythms; role of light as it relates to the circadian system; control of carbohydrate metabolism (endocrine aspects especially insulin); environmental physiology; space biology; numerous technical publications. *Mailing Add:* Dept Biol Sci San Jose State Univ San Jose CA 95912-0100

HOLLEY, FRIEDA KOSTER, b Albuquerque, NMex, Mar 12, 44; m 67. MATHEMATICS. *Educ:* Colo Col, BA, 65; Univ NMex, MA, 67, PhD(math), 70. *Prof Exp:* Lectr math, Ithaca Col, 68-69; asst prof, Col Notre Dame, 69-71; lectr asst prof level, Douglass Col, Rutgers Univ, 71-74; from asst prof to assoc prof, 75-83, PROF MATH, METROP STATE COL, 83- *Mem:* Math Asn Am; Am Math Soc; Asn Women Math; AAAS; Sigma Xi. *Res:* Topology, lattices. *Mailing Add:* 830 Columbia Pl Boulder CO 80303

HOLLEY, HOWARD LAMAR, b Marion, Ala, July 14, 14; m 46; c 5. MEDICINE. *Educ:* Univ SC, BS, 35; Med Col SC, MD, 41. *Hon Degrees:* DSc, 82. *Prof Exp:* Intern med, US Marine Hosp, Norfolk, Va, 41-42; resident, Jefferson-Hillman Hosp, Birmingham, Ala, 45-47; from instr to prof med, Sch Med, Univ Ala, Birmingham, 47-85, dir div rheumatology & chief sect rheumatic & allied dis, Dept Med, 55-70, dir univ tumor clin, 53-66, Anna Lois Waters prof rheumatol, 70-85; RETIRED. *Concurrent Pos:* Ed, Jefferson-Hillman Hosp Bull, 47-49; assoc ed, Southern Med J, 53-54; mem, Univ Ala Coun, Tuscaloosa, 55-58; mem arthritis training grants study comt, NIH, 61-65, chmn, 64-65; mem bd dirs & chmn adv comt, Ala Chapter, Arthritis Found; consult, Vet Admin Hosps, Birmingham & Tuscaloosa, Ala & Jefferson County Tuberc Sanit, Birmingham; archives cur, Dept Med, Sch Med, Univ Ala, Birmingham, 85- *Honors & Awards:* Seale Harris Res Award, 62; Hist Preserv Award, 78. *Mem:* Am Fedn Clin Res; master Am Col Physicians; Am Cancer Soc; assoc Royal Col Physicians; Am Osler Soc. *Res:* Rheumatoid arthritis and related diseases. *Mailing Add:* 4016 Old Leeds Circle Birmingham AL 35213

HOLLEY, RICHARD ANDREW, b Champaign, Ill, Sept 15, 43; m 67. MATHEMATICS. *Educ:* Univ NMex, BS, 65, MA, 66; Cornell Univ, PhD(math), 69. *Prof Exp:* Instr math, Cornell Univ, 69; Miller Inst fel, Univ Calif, Berkeley, 69-71; asst prof, Princeton Univ, 71-74; PROF MATH, UNIV COLO, BOULDER, 74- *Concurrent Pos:* Sloan Found fel, 75- *Mem:* Am Math Soc. *Res:* Infinite particle systems; random fields. *Mailing Add:* Dept Math Box 426 Univ Colo Boulder CO 80309

HOLLEY, RICHARD HOWARD, b Washington, DC, Nov 24, 43; m 66; c 3. CHEMICAL ENGINEERING, POLYMER SCIENCE. *Educ:* NC State Univ, BS, 65, PhD(chem eng), 69. *Prof Exp:* Ciba fel & res assoc, Swiss Fed Inst Technol, 69-70; res engr, Textiles, Deering Milliken Res Corp, Deering Milliken, Inc, Spartanburg, 70-77, dept head, 77-80, PLANT & BUS TECH MGR, DEERING MILLIKEN CO, 80- *Mem:* Am Chem Soc; Am Inst Chem Engrs. *Res:* Polymer membrane technology and science; textile coating and dyeing technology. *Mailing Add:* 729 Camellia Dr La Grange GA 30248

HOLLEY, ROBERT WILLIAM, b Urbana, Ill, Jan 28, 22; m 45; c 1. BIOCHEMISTRY. *Educ:* Univ Ill, AB, 42; Cornell Univ, PhD(org chem), 47. *Hon Degrees:* LittD, Keuka Col, 69; DSc, Univ Ill, 70. *Prof Exp:* Asst chem, Cornell Univ, 42-44, res chemist, Med Col, 44-46, Nat Res Coun fel chem, 46-47; Am Chem Soc fel, State Col Wash, 47-48; from asst prof to assoc prof org chem, NY Exp Sta, Cornell Univ, 48-57, res chemist, US Plant, Soil & Nutrit Lab, 57-64, prof, Cornell Univ, 64-69; AM CANCER SOC PROF MOLECULAR BIOL & RESIDENT FEL, SALK INST BIOL STUDIES, 68- *Concurrent Pos:* Guggenheim fel, Calif Inst Technol, 55-56; mem biochem study sect, NIH, 62-66; vis fel, Salk Inst Biol Studies & NSF sr fel, 66-67; adj prof, Univ Calif, San Diego, 69-; vis prof, Scripps Clin & Res Found, 66-67. *Honors & Awards:* Nobel Prize in Physiol or Med, 68; Lasker Award, 65; US Steel Found Award, Nat Acad Sci, 67. *Mem:* Nat Acad Sci; AAAS; Am Chem Soc; Am Soc Biochem & Molecular Biol; Am Acad Arts & Sci. *Res:* Control of growth of mammalian cells. *Mailing Add:* Salk Inst for Biol Studies PO Box 85800 San Diego CA 92186-5800

HOLLIBAUGH, WILLIAM CALVERT, b Corning, Iowa, May 13, 16; m 41; c 4. PETROLEUM PRODUCTS & LUBRICANTS. *Educ:* Southwestern Col, Kans, BA, 37. *Prof Exp:* Res asst, Mass Inst Technol, 37-40; from res engr to sect chief lubricants, Sun Oil Co, 40-57, sr res engr, Automotive Prod Staff, 57-64, mgr indust & automotive prod, Res & Develop Div, 64-70, mgr prod develop, Appl Res & Develop Dept, 70-77; tech secy, Comt Fuels & Lubricants, Am Nat Standards Inst, 81-84; RETIRED. *Concurrent Pos:* Consult fuels & lubricants, Res & Prod Mgt, 77-84; US deleg, Int Standards Orgn, 81-84; sci adv comt, South Western Col, 78-, chmn, 86-88; trustee, Southwestern Col, 86- *Mem:* Am Chem Soc; Soc Automotive Engrs; Am Soc Testing & Mat. *Res:* Lubricants; fuels. *Mailing Add:* 906 Wharton Ct Newton Square PA 19073-1052

HOLLIDAY, CHARLES WALTER, b Alexandria, Va, Sept 19, 46; m 71. EPITHELIAL TRANSPORT, INVERTEBRATE OSMOREGULATION. *Educ:* Marietta Col, BS, 68; Univ Ore, PhD(biol), 78. *Prof Exp:* Marine sci technician, US Coast Guard, 69-73; assoc res scientist, Mt Desert Island Biol Lab, Salisbury Cove, Maine, 78-81; asst prof-in-residence, biol sci group, Univ Conn, 81-82; asst prof, 82-86, ASSOC PROF BIOL, LAFAYETTE COL, 87- *Concurrent Pos:* Jones fac lectr, Lafayette Col, 87-88. *Mem:* AAAS; Am Physiol Soc; Am Soc Zoologists; Sigma Xi; Crustacean Soc; Western Soc Naturalists. *Res:* Osmoregulatory ion transport in crustacean and other invertebrate groups; organic anion and cation excretion in crustacean kidney; salt and water balance. *Mailing Add:* Dept Biol Lafayette Col Easton PA 18042-1778

HOLLIDAY, DALE VANCE, b Ennis, Tex, May 29, 40; m 62; c 3. UNDERWATER ACOUSTICS. *Educ:* Univ Tex, Austin, BA, 62, MS, 65; Univ Calif, San Diego, PhD(appl physics), 72. *Prof Exp:* Physicist, 62-66, dir, 66-70, sr scientist, 70-72, dir, San Diego Lab, 72-84, DIR RES, ANALYTIC & APPL RES DIV, TRACOR, INC, 84- *Mem:* Fel Acoust Soc Am; Am Soc Limnol & Oceanog. *Res:* Basic nature of underwater acoustics and its applications including antisubmarine warfare and the assessment of fisheries and other marine resources of a biological nature. *Mailing Add:* Tracor Inc 9150 Chesapeake Dr San Diego CA 92123

HOLLIDAY, GEORGE HAYES, b Toledo, Ohio, Oct 26, 22; m 44; c 3. GENERAL ENVIRONMENTAL SCIENCES. *Educ:* Univ Calif, BSME, 48; Univ Southern Calif, MSCE, 62, EME, 65; Univ Houston, PhD(civil eng), 70. *Prof Exp:* Mech engr drilling, Shell Oil Co, 48-52, sr mech engr prod, 53-57, asst chief engr prod, 57-65, environ engr, 71-85; res engr offshore, Shell Develop Co, 66-70; ENVIRON ENGR, HOLLIDAY ENVIRON ENG SERV INC, 86- *Concurrent Pos:* Distinguished lectr, Soc Petrol Engrs, 80-81; comt man, Safety & Environ Comt, Soc Petrol Engrs, 88- & Govt Affairs Comt, Am Acad Environ Engrs, 89- *Mem:* Fel Am Soc Mech Engrs; Am Petrol Inst. *Res:* Offshore civil engineering research on fatigue of leg connections and fracture mechanics of leg connections. *Mailing Add:* PO Box 1080 Tomball TX 77377

HOLLIDAY, MALCOLM A, b Staunton, Va, Jan 12, 24; m 46; c 5. PEDIATRICS. *Educ:* Univ Va, BA, 43, MD, 46. *Prof Exp:* Intern pediat, Children's Hosp, Boston, 46-47; asst resident, Vanderbilt Univ Hosp, Nashville, Tenn, 47-48; resident, Children's Hosp, Boston, 48-49; res fel, Harvard Med Sch, 49-50; res fel, Med Sch, Yale Univ, 50-51; asst prof, Sch Med, Ind Univ, 51-56; from asst prof to assoc prof, Sch Med, Univ Pittsburgh, 56-63; clin prof, 63-66, PROF PEDIAT & DIR, CHILDREN'S RENAL CTR, UNIV CALIF, SAN FRANCISCO, 66- *Concurrent Pos:* Physician-in-chief, Children's Hosp Med Ctr, Oakland, Calif, 63-66. *Mem:* Am Pediat Soc; Soc Pediat Res (vpres, 68); Am Acad Pediat; Am Soc Pediat Nephrology (pres, 74-75); Perinatal Res Soc. *Res:* Pediatric renal-electrolyte problem; nutrition in chronic renal disease; calorie balance and protein turnover; growth and kidney function; renal failure in children; problems in body fluids and osmolality. *Mailing Add:* Dept Pediat A-276 Univ Calif Children's Renal Ctr 400 Parnassus Ave San Francisco CA 94143

HOLLIEN, HARRY, b Brockton, Mass, July 16, 26; m 69; c 6. EXPERIMENTAL PHONETICS. *Educ:* Boston Univ, BS, 49, MEd, 51; Univ Iowa, MA, 53, PhD(exp phonetics), 55. *Prof Exp:* Dir speech & hearing prog, Baylor Univ, 55-58; asst prof logopedics, Univ Wichita, 58-62; res assoc, prof commun sci & assoc dir commun sci lab, 62-68, prof speech & dir commun sci lab, 68-75, DIR, INST ADVAN STUDY COMMUN PROCESSES, UNIV FLA, 75-, PROF CRIMINAL JUSTICE, 79-, PROF LING & ASSOC DIR LING, 89- *Concurrent Pos:* Guest prof, Paul Quinn Col, 56-58; fel, Northwestern Univ, 58; prin investr grants, NIH, 60-81, Univ Fla, 62-90, Off Naval Res, 65-74, Am Speech & Hearing Asn, 71, US Dept Com, 77-78, Voice Found, 78-80, Sea Grant, 79-83, Dreyfus Found, 80-, US Justice Dept, 89, Army Res Off, 81-84, USN, 85-; co-dir grants, Rehab Serv Admin, 63-68 & 70-72, NIH, 64- & NSF, 74-77; mem commun sci study sect, Div Res Grants, NIH, 63-67; pres, Hollien Assocs, 66-; res consult, Vet Admin Hosp, San Francisco, 67-; mem nerv & sensory systs res eval comt, Vet Admin, 69-74; vis scientist, Speech Transmission Lab, Royal Inst Technol, Sweden, 70; mem bd dirs, Div Sponsored Res, Univ Fla, 71-73 & 79-81; vis prof, Wroclaw Tech Univ, Poland, 74; chmn bd dirs, Wild Animal; sen fulbright prof, Univ Trier, Ger, 87. *Honors & Awards:* Manuel Garcia Int Award, Int Asn Logopedics & Phoniatrics, 71; Gould Int Res Prize, 78; Gutzmann Res Medal, Union Europ Phoniatrists & Int Asn Logopedics & Phoniatrics, 80; Kay Elemetrics Prize, Res Phonetics Int Soc Phonetic Sci, 87; John Hunt Award, Am Acad Forensic Sci, 88. *Mem:* Fel AAAS; fel Acoust Soc Am; fel Int Soc Phonetic Sci (secy gen, 75-89, exec vpres, 83- 89, pres, 89-); fel Am Speech & Hearing Asn; Marine Technol Soc; Am Asn Phonetic Sci (pres, 73-75); fel Am Acad Forensic Sci. *Res:* Diver communication; voice science; laryngeal function; psychoacoustics; underwater communication; forensic communication; author of 221 major books, articles and 161 minor publications. *Mailing Add:* Inst Advan Study Commun Proc 50 Dauer Hall Univ Fla Gainesville FL 32611

HOLLIES, NORMAN ROBERT STANLEY, physical chemistry; deceased, see previous edition for last biography

HOLLIMAN, ALBERT LOUIS, mechanical engineering, for more information see previous edition

HOLLIMAN, DAN CLARK, b Birmingham, Ala, Aug 25, 32; m 58; c 1. VERTEBRATE ZOOLOGY. *Educ:* Univ Ala, BS, 57, MS, 59, PhD(zool), 63. *Prof Exp:* From asst prof to assoc prof, 62-71, PROF BIOL, BIRMINGHAM-SOUTHERN COL, 71- *Res:* Taxonomy of mammals of the southeastern United States; ornithology in Alabama. *Mailing Add:* Dept Biol Birmingham-Southern Col Birmingham AL 35204

HOLLIMAN, RHODES BURNS, b Birmingham, Ala, Feb 28, 28; m 50; c 2. PARASITOLOGY. *Educ:* Howard Col, BS, 50; Univ Miami, MS, 53; Fla State Univ, PhD(zool), 60. *Prof Exp:* Asst invert anat, Univ Miami, 51-53; res assoc microbiol, Med Col Ala, 55-56; asst gen biol, Fla State Univ, 56-57, asst schistosomiasis, 57-59; asst prof zool, Jacksonville Univ, 60-61; instr gen biol, Fla State Univ, 61-62; assoc prof, 62-71, PROF ZOOL, VA POLYTECH INST & STATE UNIV, 71- *Concurrent Pos:* Fel trop med, La State Univ Med Sch, 73. *Mem:* Am Soc Parasitol; Am Soc Trop Med & Hyg; Paleopath Soc. *Res:* Infectious and parasitic diseases in tropical Americas; epidemiology of trichinosis; medical protozoology and medical helminthology. *Mailing Add:* Dept Biol Va Polytech Inst Blacksburg VA 24061

HOLLING, CRAWFORD STANLEY, b Theresa, NY, Dec 6, 30; Can citizen; m 53, 78. ZOOLOGY. *Educ:* Univ Toronto, BA, 52, MA, 54; Univ BC, PhD, 57. *Prof Exp:* Agr res officer predation, Forest Biol Div, Res Br, Can Dept Agr, 52-67; dir, Inst Resource Ecol, Inst Fisheries, Univ BC, 69-73, prof zool, 67-88; PROF ZOOL & EMER SCHOLAR ECOL SCI, UNIV FLA, 88- *Concurrent Pos:* Dir, Inst Appl Systs Anal, Laxenburg, Austria, 81-84; mem bd, Int Statist Ecol Prog. *Mem:* Ecol Soc Am; Entom Soc Can; Can Soc Zool; fel Royal Soc Can; Brit Ecol Soc; Japanese Ecol Soc; fel AAAS. *Res:* Systems ecology; population dynamics; policy analysis. *Mailing Add:* Zool Dept 223 Bartram Hall Univ Fla Gainesville FL 32661

HOLLING, HERBERT EDWARD, b Eng, Oct 8, 08; nat US. MEDICINE. *Educ:* Univ Sheffield, MSc & MD, 32. *Prof Exp:* Resident, Sheffield United Hosps, 33-36; mem sci staff, Brit Med Res Coun, 46-57; asst prof, 57-60, assoc prof, 60-80, EMER ASSOC PROF MED, SCH MED, UNIV PA, 80- *Concurrent Pos:* Rockefeller traveling fel, Harvard Univ, 36-37; mem staff, Guy's Hosp, 46-57; mem staff, Univ Hosp, Univ Pa, 57- *Mem:* Am Col Physicians; Am Heart Asn. *Res:* Diseases of the circulation. *Mailing Add:* Five E Locust St Oxford Manor No 6 Oxford PA 19363

HOLLINGDALE, MICHAEL RICHARD, b Newport, Gt Brit, Aug 27, 46. VACCINE DEVELOPMENT, EPIDEMIOLOGY. *Educ:* Univ Liverpool, UK, BSc, 67; Univ London, UK, PhD(microbiol), 71. *Prof Exp:* Sci officer microbiol, London Sch Hyg & Trop Med, 70-74; fel bacteriol, Sch Pub Health, Harvard Univ, 74-76; fel path, Johns Hopkins Univ, 76-78; res assoc parasitol, Rockefeller Univ, NY, 78-79; HEAD DEPT MALARIA, BIOMED RES INST, ROCKVILLE, MD, 79- *Concurrent Pos:* Prin investr, Agency Int Develop, 80- & WHO, 84-; mem, Study Sect Trop Med & Parasitol, NIH, 90- *Mem:* Am Soc Trop Med Hyg; Soc Protozool; Royal Soc Trop Med Hyg. *Res:* Development of anti-malarial vaccines and their effectiveness in human volunteer clinical trials by in vitro assays of protection; biochemical properties of malarial parasites, and their responses to drugs. *Mailing Add:* Biomed Res Inst 12111 Parklawn Dr Rockville MD 20852

HOLLINGER, F(REDERICK) BLAINE, b Hays, Kans, June 22, 35;; m 78; c 4. HEPATOLOGY, AIDS. *Educ:* Univ Kans, BA, 57, MD, 62. *Prof Exp:* Asst chief, Arbovirus Infection Unit, Ctr Dis Control, USPHS, Ga, 67-68; NIH spec fel, 68-70, from asst prof to assoc prof, 70-78, PROF VIROL & EPIDEMIOL, BAYLOR COL MED, 78-, PROF MED, 81- *Concurrent Pos:* Adv & consult prog projs, grants & contracts related to hepatitis & AIDS res, Nat Heart & Lung Inst, Nat Inst Allergy & Infectious Dis & Bur Biologics, Food & Drug Admin, 72-; consult ed, J Immunol, Gastro, Hepatology, New Eng J Med, Ann Internal Med, Intervirol & JAMA, 72-; consult, Expert Group Viral Hepatitis, Nat Libr Med, NIH, 78-84; assoc ed, J Inf Dis, 87-88; assoc ed, Hepatitis Knowledge-Base Proj, Mass Med Soc, 87-90; comn, Health Task Force AIDS, State of Tex, 86-; Am Asn Blood Banks Transfusion Transmitted Dis Comt, 85-; chmn, Int Symp Viral Hepatitis & Liver Dis, 87-90. *Mem:* Am Gastroenterol Asn; fel Infectious Dis Soc; Am Asn Study Liver Dis; Int Asn Study Liver Dis. *Res:* Characterization of hepatitis agents; develop of sensitive assays for hepatitis; prophylaxis of hepatitis; vaccine development; immunopathogenesis of viral hepatitis; epidemiol of hepatitis; AIDS research, virology quality assurance programs, isolation of agent, immunodiagnosis, immunopathogenesis and virology endpoints of therapy. *Mailing Add:* Baylor Col Med One Baylor Plaza Houston TX 77030

HOLLINGER, HENRY BOUGHTON, b Jacksonville, Fla, Feb 20, 33; m; c 3. PHYSICAL CHEMISTRY. *Educ:* Lebanon Valley Col, BS, 55; Univ Wis, PhD, 60. *Prof Exp:* Mem staff, Lebanon Valley Col, 59-61; asst prof, 61-64, assoc prof, 64-80, PROF PHYS CHEM, RENSSELAER POLYTECH INST, 80- *Mem:* Am Chem Soc; Am Phys Soc; Sigma Xi. *Res:* Kinetic theory of dense gases; nonequilibrium mechanics. *Mailing Add:* Dept Chem Rensselaer Polytech Inst Troy NY 12181

HOLLINGER, JAMES PIPPERT, b Elyria, Ohio, Oct 1, 33; m 60; c 2. REMOTE SENSING, PHYSICS. *Educ:* NMex Inst Mining & Technol, BS, 56; Univ Va, MS, 58, PhD(physics), 61. *Prof Exp:* Asst prof physics, George Washington Univ, 61-62; RES PHYSICIST, US NAVAL RES LAB, 63- *Mem:* Int Sci Radio Union. *Res:* Polarization of the radio wavelength radiation from discrete cosmic sources; remote microwave sensing of the earth's surface; millimeter wave passive imaging systems. *Mailing Add:* Code 4211 Naval Res Lab Washington DC 20375-5000

HOLLINGER, MANNFRED ALAN, b Chicago, Ill, June 28, 39; m 61; c 2. PHARMACOLOGY. *Educ:* NPark Col, BS, 61; Loyola Univ Chicago, MS, 65, PhD(pharmacol), 67. *Prof Exp:* Res pharmacologist, B Baxter Labs, 61-63; instr pharmacol, Med Sch, Stanford Univ, 67-69; from asst prof to assoc prof, 69-86, PROF PHARMACOL, SCH MED, UNIV CALIF, DAVIS, 86- *Concurrent Pos:* Fogarty Sr Int Fel, NIH, Burroughs-Wellcome Int Fel. *Mem:* Am Soc Pharmacol & Exp Therapeut; Soc Toxicol. *Res:* Reproductive biology; biochemistry of spermatogenesis; factors influencing gamete formation; drug-receptor interaction; pharmacology of fibrosis and pulmonary pharmacology and toxicology. *Mailing Add:* Dept Med Pharmacol & Toxicol Sch Med MSI Univ Calif Davis CA 95616

HOLLINGER, THOMAS GARBER, b Chicago, Ill, Oct 23, 42; m 65; c 3. REPRODUCTIVE BIOLOGY, DEVELOPMENTAL BIOLOGY. *Educ:* Northwestern Univ, BS, 66; Northern Ill Univ, MS, 69; Purdue Univ, PhD(zool), 74. *Prof Exp:* Investr biol, Oak Ridge Nat Labs, 74-76; asst prof, 78-81, ASSOC PROF ANAT, COL MED, UNIV FLA, 81- *Mem:* Am Asn Anat; Soc Develop Biol; Am Soc Cell Biol; Am Ageing Asn. *Res:* Yolk proteins in teleost fish oocytes; biological basis of meiosis and the initiation of cell division. *Mailing Add:* Dept Anat Box J-235 JHMHC Univ Fla Col Med Gainesville FL 32610-0235

HOLLINGSWORTH, CHARLES ALVIN, b Earl, Colo, June 22, 17; m 44; c 3. PHYSICAL CHEMISTRY. *Educ:* Western State Col Colo, BA, 41; Univ Iowa, PhD(phys chem), 46. *Prof Exp:* Res chemist, E I du Pont de Nemours & Co, 46-48; from instr to prof, 48-82, EMER PROF CHEM, UNIV PITTSBURGH, 82- *Mem:* Am Chem Soc. *Res:* Mathematical theoretical; rate processes; colloids; thermodynamics. *Mailing Add:* Dept Chem Univ Pittsburgh Pittsburgh PA 15260

HOLLINGSWORTH, CHARLES GLENN, b Erie, Pa, Aug 7, 46; m 70; c 2. EPIDEMIOLOGY, BIOSTATISTICS. *Educ:* Univ Pittsburgh, BS, 69, MSH, 70; Univ Mich, DrPH, 74. *Prof Exp:* Res asst asbestosis, Grad Sch Pub Health, Univ Pittsburgh, 70-71; biostatistician clin trials, Abbott Labs, 70-72; epidemiologist heart dis & clin trial, Sch Med, Univ Minn, 74-76; asst prof epidemiol, Sch Pub Health, Univ Mass, 76-78; CLIN APPLNS PROG DIABETES, NAT INST ARTHRITIS, METAB & DIGESTIVE DIS, NIH, 79- *Concurrent Pos:* Biomed res grant prostatic cancer, Sch Pub Health, Univ Mass, 78-79. *Mem:* Soc Epidemiol Res; Soc Clin Trials; Am Col Epidemiol. *Res:* Major chronic diseases; epidemiology of heart disease diabetes; clinical trials in heart disease diabetes and new pharmaceuticals. *Mailing Add:* 12504 Knightsbridge Ct Potomac MD 20850

HOLLINGSWORTH, CORNELIA ANN, b Carrollton, Ga, Mar 6, 57. MEAT SCIENCE, PROTEIN CHEMISTRY. *Educ:* Auburn Univ, BS, 79; Univ Nebr-Lincoln, MS, 81, PhD(animal sci), 84. *Prof Exp:* Res scientist, Armour Food Co-Con Agra, 84-88; MGR PROD DEVELOP, BIL MAR FOODS-SARA LEE CORP, 88- *Concurrent Pos:* Res & teaching asst, Univ Nebr-Lincoln, 79-84; chmn, Meat Indust Res Conf, 90-91; qual facilitator, Bil Mar Foods, 90-; mem, Nutrit Labeling Adv Comt, Am Meat Inst, USDA, 91-93. *Mem:* Inst Food Technologists; Am Meat Sci Asn. *Res:* Development of new products for a major meat processing company through application of chemistry, microbiology, sensory and textural evaluation. *Mailing Add:* 8300 96th Ave Zeeland MI 49464

HOLLINGSWORTH, DAVID S, CHEMICAL ENGINEERING. *Educ:* Lehigh Univ, BS, 48. *Prof Exp:* Chem engr, Res Ctr, Hercules Inc, Wilmington, Del, 48-49, Kalamazoo, Mich, 49-53, tech rep sales, New Orleans, La & Wilmington, Del, 53-61, asst sales mgr paper chem, 61-63, sales mgr specialty chem, 63-65, mgr specialty paper chem, 65-67, dir sales paper chem, Pine & Paper Chem Dept, 67-72, dir mkt, 72, asst gen mgr, 72-74, gen mgr, New Enterprise Dept, 74-75, gen mgr, Food & Fragrance Development Dept, 75-78, dir organics, Worldwide Bus Ctr, 78-79, vpres planning, 79-82, group vpres water-soluble polymers, 82-83, group vpres polypropylene, 83, div vpres mkt, 83-84, pres, Specialty Chem Co, 84-86, mem bd dirs, Hercules Inc, 82-90, vchmn, 86-90, chief exec officer, 87-90; RETIRED. *Concurrent Pos:* Mem bd dirs, Orbital Sci Corp II, Oryx Energy Co, Econ Strategy Inst & Chem Mfrs Asn; mem adv bd, Beckman Ctr for History of Chem; mem vis comt chem eng, Lehigh Univ. *Honors & Awards:* Nat Medal of Technol, 91. *Mem:* Am Inst Chem Engrs; distinguished mem Am Soc Metals Int. *Res:* Marketing of specialty chemicals. *Mailing Add:* Orbital Sci Corp 12500 Fair Lakes Circle Fairfax VA 22033

HOLLINGSWORTH, JACK W, b South Haven, Kans, Mar 3, 24; m 50; c 2. MATHEMATICS. *Educ:* Univ Kans, BS, 48, AB, 49; Univ Wis, MS, 51, PhD(math), 54. *Prof Exp:* Asst, Univ Kans, 48-49 & Univ Wis, 49-54; mathematician, Gen Elec Co, 54-57; assoc prof math, Rensselaer Polytechnic Inst, 57-61, supvr comput lab, 57-70, chmn interdisciplinary comt comput sci, 68-73, prof math, 61-79; dir, Sch Comput Sci & Technol, 80-82, prof comput sci, 79-86, PROF MATH ROCHESTER INST TECHNOL, 86- *Mem:* Am Math Soc; Soc Indust & Appl Math; Math Asn Am; Asn Comput Mach. *Res:* Numerical analysis; computing. *Mailing Add:* Dept Math Rochester Inst Technol Rochester NY 14623

HOLLINGSWORTH, JAMES W, b Mount Airy, NC, Jan 14, 26; m 51; c 3. INTERNAL MEDICINE. *Educ:* Duke Univ, MD, 48. *Prof Exp:* Instr internal med, Duke Univ, 51-52; res assoc, Yale Univ, 54-56, from asst prof to assoc prof, 56-68; prof med & chmn dept, Med Ctr, Univ Ky, 68-78; actg dean, 80-81, PROF MED & VCHMN DEPT, UNIV CALIF, SAN DIEGO, 78-; CHIEF MED SERV, SAN DIEGO VET ADMIN HOSP, 78- *Mem:* Am Col Physicians; Am Soc Clin Invest. *Res:* Rheumatology; hematology; immunology. *Mailing Add:* San Diego Med Ctr San Diego CA 92161

HOLLINGSWORTH, JOHN GRESSETT, b Decatur, Miss, July 15, 38; m 60; c 2. MATHEMATICS. *Educ:* Miss State Univ, BA, 59, MS, 61; Rice Univ, PhD(math), 67. *Prof Exp:* Assoc mathematician, Texaco Inc, 61-64; instr math, Rice Univ, 66-67; asst prof, 67-69, assoc prof, 72-81, PROF MATH, UNIV GA, 81- *Mem:* Am Math Soc. *Res:* Geometric topology; topology of manifolds. *Mailing Add:* Dept Math Univ Ga Athens GA 30602

HOLLINGSWORTH, JOSEPH PETTUS, b Mertens, Tex, Aug 20, 22; m 63; c 1. AGRICULTURAL ENGINEERING. *Educ:* Tex A&M Univ, BS, 43, MS, 61. *Prof Exp:* Res asst, Tex A&M Univ, 46-48; proj leader, USDA, 48-78; RETIRED. *Concurrent Pos:* Owner/operator, TV Sales & Serv Bus, 79-83; consult, Entom Consult Group, 84- *Mem:* Am Soc Agr Engrs; Entom Soc Am. *Res:* Measuring effects of near-ultraviolet and visible electromagnetic stimuli on phototactic responses of insects; developing equipment for using proven attractants in traps for survey and control applications. *Mailing Add:* 2204 Sharon Dr Bryan TX 77802-2438

HOLLINGSWORTH, RALPH GEORGE, b Wheeling, WVa, Mar 4, 47; m 66; c 2. USER INTERFACE DESIGN, SOFTWARE ENGINEERING. *Educ:* Univ Cincinnati, BSChE, 70; Univ Mich, MS, 71, PhD(bioeng), 74. *Prof Exp:* NSF Fel, Univ Mich, 74-75; asst prof physiol, Med Univ SC, 75-78; sr systs analyst, NCR, 78-81; ASSOC PROF COMPUT SCI & CHMN DEPT, MUSKINGUM COL, 81- *Concurrent Pos:* Consult analyst, SDRC, 84. *Mem:* Inst Elec & Electronics Engrs; Sigma Xi; Math Asn Am; Asn Comput Mach. *Res:* The application of expert system technology to the design of user interfaces, including the use of sound, color, N-dimensionality and cognitive style matching; applications neural networks. *Mailing Add:* Crows Nest 76147 Peoli Rd Port Washington OH 43837

HOLLINGWORTH, ROBERT MICHAEL, b Yorkshire, Eng, Oct 4, 39; m 61; c 2. TOXICOLOGY, PESTICIDE CHEMISTRY. *Educ:* Univ Reading, BSc, 62; Univ Calif, Riverside, PhD(insect toxicol), 66. *Prof Exp:* From asst prof to assoc prof, 66-75, prof insect toxicol, Purdue Univ, West Lafayette, 75-87; DIR PESTICIDE RES CTR & PROF ENTOM & ZOOL, MICH STATE UNIV, 87- *Concurrent Pos:* Vis prof, Stauffer Chem Co, 74-75; mem, Toxicol Study Sect, NIH, 76-80; Environ Protection Agency sci adv panel, Fifra, 82-84; chmn, Div Pestic Chem, Am Chem Soc, 84. *Mem:* AAAS; fel Am Chem Soc; Soc Toxicol; Entom Soc Am; Am Coun Sci Health. *Res:* Metabolism and mode of action of insecticides and related chemicals; selective toxicity, neurotoxicology, comparative neuropharmacology; pesticide resistance. *Mailing Add:* Pesticide Res Ctr MSU East Lansing MI 48824

HOLLINS, ROBERT EDWARD, b Chicago, Ill, Oct 19, 40. ENZYME KINETICS, INSTRUMENTAL TECHNIQUES. *Educ:* Roosevelt Univ, BS, 64; Northwestern Univ, PhD(phys chem), 70. *Prof Exp:* Sr res chemist, Sherwin-Williams Co, 69-72; asst prof chem, Roosevelt Univ, 72-77; ASSOC PROF CHEM, CHICAGO STATE UNIV, 77- *Mem:* Am Chem Soc; Am Asn Black Prof; Nat Orgn Black Chemists. *Res:* Observing the effects of lead and trace metals on the biochemical parameters in rats, specifically how they affect the metalloenzymes; instrumental analytical techniques as applied to the superoxide dismutase and the cytochrome oxioase enzymes. *Mailing Add:* Dept Phys Sci Chicago State Univ 95th & King Dr Chicago IL 60628-1599

HOLLINSHEAD, ARIEL CAHILL, b Allentown, Pa, Aug 24, 29; m 58; c 2. PHARMACOLOGY, VIROLOGY. *Educ:* Ohio Univ, AB, 51; George Washington Univ, MA, 55, PhD(pharmacol), 57. *Hon Degrees:* DSc, Ohio Univ, 77. *Prof Exp:* Res fel pharmacol, Med Sch, George Washington Univ, 57-58 & 59; asst prof pharmacol, George Washington Univ, 59-61, from asst prof to assoc prof med, 61-73, dir lab virus & cancer res, 64-89, PROF MED, MED SCH, GEORGE WASHINGTON UNIV, 74- *Concurrent Pos:* Res fel virol, Col Med, Baylor Univ, 58-59. *Mem:* Soc Exp Biol & Med; Am Soc Clin Oncol; Am Asn Cancer Res; Int Agency Res Cancer. *Res:* Chemotherapy of animal virus diseases and cancer; nucleoprotein chemistry of viruses; cancer immunogenetics; oncology; cancer immunochemotherapy and immunoprophylaxis; environmental carcinogens; first isolation, purification and identification of animal and human tumor-associated antigens; human-human hybridoma research, and development of epitopes for study and use in human cancers. *Mailing Add:* HT Virus & Cancer Res Inc 3637 Van Ness St NW Washington DC 20008

HOLLINSHEAD, MAY B, b New York, NY, Nov 28, 13; m 42; c 1. ANATOMY, EMBRYOLOGY. *Educ:* Hunter Col, AB, 36; Columbia Univ, PhD, 51. *Prof Exp:* Asst anat, Col Physicians & Surgeons, Columbia Univ, 49-51; instr, Col Med, NY Univ, 51-56; from asst prof to assoc prof, 56-72, PROF ANAT, COL MED & DENT NJ, NEWARK, 72- *Mem:* Am Asn Anat; AAAS; Sigma Xi. *Res:* Embryology; histochemistry of skeletal muscle; development of teeth and periodontal tissues in osteopetrotic mouse; effect of ascorbic acid on burns, frostbite and barbiturates; effect of aqueous humor on blood coaqulation. *Mailing Add:* Two Winthrop Pl Leonia NJ 07605

HOLLIS, BRUCE WARREN, b Elyria, Ohio, May 29, 51; m 80; c 4. METHODS DEVELOPMENT, CHROMATOGRAPHY METHODOLOGY. *Educ:* Ohio State Univ, BSc, 73, MSc, 76; Univ Guelph, PhD(exp nutrit), 79. *Prof Exp:* Teaching fel med, Case Western Reserve Univ, 79-82, asst prof nutrit, 82-86; ASSOC PROF PEDIAT, MED UNIV SC, 86-, ASSOC PROF BIOCHEM & MOLECULAR BIOL, 89- *Concurrent Pos:* Nat Res Serv Award, NIH, 80-82, Res Career Develop Award, 82-; acad consult, Immuno Nuclear Corp, 82- & Teltech Resource Network, 85-

Honors & Awards: Mead Johnson Award, Am Inst Nutrit, 91. *Mem:* Endocrine Soc; Am Soc Bone & Mineral Res; Am Inst Nutrit; NY Acad Sci; Sigma Xi. *Res:* Nutrient requirements of premature infants; development of new hormone analysis. *Mailing Add:* Dept Pediat Med Univ SC 171 Ashley Ave Charleston SC 29425

HOLLIS, CECIL GEORGE, b Eden, Ala, Nov 6, 24; c 3. MYCOLOGY. *Educ:* Univ Ala, PhD(bot), 54. *Prof Exp:* Asst biol, Univ Ala, 50-51; asst prof, Furman Univ, 51-52; jr res assoc plant physiol, Univ Ala, 52-53; prof biol, Ark Col, 54-57; microbiologist, Buckman Labs, Inc, Tenn, 57-70; prof biol, Memphis State Univ, 70-83; VPRES RES & DEVELOP, BUCKMAN LABS, 83- *Mem:* Mycol Soc Am; Am Soc Microbiol; Sigma Xi; Soc Indust Microbiol (pres, 88-89). *Res:* Taxonomy and physiology of fungi; industrial microbiology; biocides. *Mailing Add:* 1767 Poplar Estates Pkwy Germantown TN 38138

HOLLIS, DANIEL LESTER, JR, b Mexico City, Mex, Mar 7, 24; US citizen; m 56; c 2. NUCLEAR ENGINEERING. *Educ:* US Naval Acad, BS, 46; Univ Ala, MS, 58, MA, 59; Tex A&M Univ, PhD(nuclear eng), 67. *Prof Exp:* Instr, 59-61, asst prof, 61-62 & 65-67, Assoc prof, 67-76, PROF ELEC & NUCLEAR ENG, UNIV ALA, TUSCALOOSA, 76- *Mem:* Am Nuclear Soc; Am Soc Eng Educ; Sigma Xi; Inst Elec & Electronics Engrs. *Res:* Plasma-thermonuclear engineering; hydromagnetic models of magnetically confined plasmas; bremsstrahlung and space shielding; neutron activation and prompt gamma analyses. *Mailing Add:* Rte 3 Box 570 Cottondale AL 35453

HOLLIS, DONALD PIERCE, biophysics, for more information see previous edition

HOLLIS, GILBERT RAY, b Poplar Bluff, Mo, Nov 26, 39; m 61; c 3. ANIMAL NUTRITION. *Educ:* Univ Ark, BSA, 61, MS, 63; Purdue Univ, PhD(animal nutrit), 66. *Prof Exp:* Asst animal nutritionist, Univ Fla, 66-70; area swine specialist, Tex Agr Exten Serv, 70-77; PROF & EXTEN SWINE SPECIALIST, UNIV ILL, URBANA, 77- *Mem:* Am Soc Animal Sci. *Res:* Swine nutrition and management. *Mailing Add:* 1703 Stratford Dr Champaign IL 61821

HOLLIS, J(OHN) SEARCY, b Collier, Ga, Nov 6, 18; m 47; c 3. ELECTRICAL ENGINEERING. *Educ:* Ga Inst Technol, BEE, 50, MS, 56. *Prof Exp:* Res engr, Eng Exp Sta, Ga Inst Technol, 50-58, asst head, Radar Br, 58-59; vpres eng, Sci-Atlanta, Inc, 59-61, prin engr, 61-83; RETIRED. *Mem:* Inst Elec & Electronics Engrs; Sigma Xi. *Res:* Antennas; antenna measurements; microwave physics; radar; radiometry; propagation; electronic instrumentation; satellite communications. *Mailing Add:* 824 East 26th Ave New Smyrma Beach FL 32169

HOLLIS, JAN MICHAEL, b Martinsburg, WVa, June 5, 41; m 64; c 2. RADIO ASTRONOMY, COMPUTER SCIENCE. *Educ:* Duke Univ, AB, 63; Univ Va, MA, 72, PhD(astron), 76. *Prof Exp:* Comput systs analyst astron, Nat Radio Astron Observ, 73-79; astronr, 79-82, head sci opers br, 82-88, ASST CHIEF, SPACE DATA & COMPUT DIV, NASA GODDARD SPACE FLIGHT CTR, 88- *Honors & Awards:* Except Sci Achievement Medal, NASA, 90. *Mem:* Am Astron Soc; Int Astron Union. *Res:* Radio astronomy as it pertains to the physical conditions and chemical composition of the interstellar medium, comets and stars. *Mailing Add:* Code 930 NASA Goddard Space Flight Ctr Greenbelt MD 20771

HOLLIS, JOHN PERCY, JR, history & philosophy of science, for more information see previous edition

HOLLIS, MARK DEXTER, b Buena Vista, Ga, Sept 24, 08; m 27; c 2. ENGINEERING, ENVIRONMENTAL SCIENCES. *Educ:* Univ Ga, BSCE, 31, CE, 37. *Hon Degrees:* DSc, Univ Fla, 56. *Prof Exp:* Officer, USPHS, 31-47, asst surgeon gen, 47-61; dir environ health, UN World Health, Geneva, Switz, 61-73; VCHMN, FLA STATE ENVIRON REGULATION COMN, 74- *Concurrent Pos:* Mem interstate comn, Potomac River Basin, 50-61; chmn, Bd Nat Sanit Found, 52-60; mem interstate compact comn, Ohio River Basin, 54-58. *Honors & Awards:* Arthur Bedell Award & Charles A Emerson Medal, Water Pollution Control Fedn, 64; Distinguished Serv Medal, USPHS. *Mem:* Nat Acad Eng; Am Acad Environ Engrs; Am Soc Civil Engrs; Water Pollution Control Fedn (pres, 59-60). *Res:* Tropical diseases, especially malaria and typhus; pollution control, particularly air, water and land. *Mailing Add:* Carpenters Home Estates 1001 Carpenters Way No 1-520 Lakeland FL 33809

HOLLIS, RAYMOND HOWARD, reliability, for more information see previous edition

HOLLIS, THEODORE M, b Palo Alto, Calif, Dec 22, 39; m 63; c 3. PHYSIOLOGY. *Educ:* San Jose State Col, BA, 63; Ohio State Univ, MS, 67, PhD(physiol), 69. *Prof Exp:* Instr biol, Ohio Northern Univ, 66-69; from asst prof to assoc prof biol, 69-84, PROF PHYSIOL, PA STATE UNIV, 84- *Mem:* AAAS; Am Physiol Soc; Soc Exp Biol; Am Diabetic Asn; Asn Res in Vision & Ophthal. *Res:* Arterial metabolic changes associated with experimental atherogenesis, hypertension and diabetes; hemodynamics; vascular histamine. *Mailing Add:* Dept Biol 208 Mueller Lab Pa State Univ University Park PA 16802

HOLLIS, WALTER JESSE, b Bossier City, La, Mar 17, 21; m 45; c 3. INTERNAL MEDICINE, CARDIOLOGY. *Educ:* La State Univ, MD, 45. *Prof Exp:* Intern & resident internal med, Shreveport Charity Hosp, La, 45-48; clin assoc, Sch Med, La State Univ, 48; contract physician & chief med, USAF Hosp, Barksdale, 48-51; from instr to assoc prof med, Sch Med, La State Univ Med Ctr, New Orleans, 53-65, prof, 65-81; Disability Determination Serv, Dept Health & Human Resources, La, 81-89; RETIRED. *Concurrent Pos:* Clin assoc, La State Univ Sch Med, 48; pvt pract, 48-51; asst vis physician, Charity Hosp, 48, cardiologist, Heart Sta, 56-64; consult, Southeast La Hosp, Mandeville, 53-59. *Mem:* Ballistocardiographic Res Soc; fel Am Col Physicians. *Res:* Clinical cardiology; ballistocardiography. *Mailing Add:* 761 Glouster St Gretna LA 70056

HOLLIS, WILLIAM FREDERICK, b Cleveland, OH, May 25, 54; m 77; c 2. EXPERT SYSTEMS, COMPUTER BASED INFORMATION STORAGE & RETRIEVAL. *Educ:* Bowling Green State Univ, BS, 76; Kent State Univ, MLS, 79, PhD(instrnl technol). *Prof Exp:* Info specialist polymer sci, B F Goodrich Res & Develop Ctr, 79-82; instr libr & info sci, Col Wooster, 82-84; sr info specialist, 84, actg head, 85, HEAD INFO CTR POLYMER SCI, GENCORP RES, 86- *Concurrent Pos:* Instr, Stark Tech Col, 83-84. *Mem:* Am Chem Soc; Am Inst Physics; Am Soc Info Sci; Asn Educ Commun & Technol. *Res:* Information science, especially as it relates to the handling of science/technical information. *Mailing Add:* GenCorp Res 2990 Gilchrist Rd Akron OH 44305

HOLLISTER, ALAN SCUDDER, b Baltimore, Md, Feb 28, 47; m 70; c 2. CLINICAL PHARMACOLOGY, HYPERTENSION. *Educ:* Swarthmore Col, BA, 70; Univ NC, PhD(pharmacol), 76, MD, 77. *Prof Exp:* Intern & resident internal med, Univ Fla, 77-80; fel clin pharmacol, Vanderbilt Univ, 80-83, asst prof med, 83-90; ASSOC PROF MED, UNIV COLO, 90- *Concurrent Pos:* Clin assoc physician, Clin Res Ctr, Vanderbilt Univ, 83-85. *Mem:* Am Fedn Clin Res; Am Soc Pharmacol & Exp Therapeut; Am Heart Asn; Am Col Physicians; AAAS; Soc Neurosci. *Res:* Clinical pharmacology and pharmacological interactions in the control of blood pressure; adrenergic receptor mechanisms; atrial natriuretic factor; evaluation and treatment of hypertension and hypotension; autonomic pharmacology. *Mailing Add:* Div Clin Pharmacol Univ Colo Health Sci Ctr Denver CO 80262

HOLLISTER, CHARLES DAVIS, b Santa Barbara, Calif, Mar 18, 36. MARINE GEOLOGY, OCEANOGRAPHY. *Educ:* Ore State Univ, BS, 60; Columbia Univ, PhD(geol), 66. *Prof Exp:* Teaching asst geol, Ore State Univ, 60-61; oceanogr, US Dept Interior, 61; res asst submarine geol, Columbia Univ, 61-67; assoc scientist, 67-79, SCIENTIST & DEAN GRAD STUDIES, WOODS HOLE OCEANOG INST, 79- *Honors & Awards:* John Oliver La Gorce Medal, Nat Geog Soc. *Mem:* Fel AAAS; fel Geol Soc Am; Am Asn Petrol Geol; Sigma Xi. *Res:* Ocean bottom currents and their effects on the deep sea floor; sediment dynamics of the deep sea; sub-seabed disposal of radioactive waste; graduate education in oceanography. *Mailing Add:* Box 601 Falmouth MA 02540

HOLLISTER, CHARLOTTE ANN, b Santa Fe, NMex, Jan 2, 40; m 68; c 3. SYSTEM DESIGN, HUMAN ENGINEERING. *Educ:* Vassar Col, BA, 61; Yale Univ, PhD(chem), 65. *Prof Exp:* Asst res scientist, NY Univ, 65-67, asst prof chem, 67-68; prod planner, Gen Elec Co, 80-81; sr applications scientist, 69-80, sr scientist, 81-84, dept mgr, 84-89, PROG MGR, BOLT BERANEK & NEWMAN INC, CAMBRIDGE, 90- *Concurrent Pos:* Vis scientist, Dept Physics, Mass Inst Technol, 68-69; asst res scientist, NY Univ, 68-69. *Mem:* Am Chem Soc; Asn Comput Mach. *Res:* Design of software systems; application of computer technology, including quantum mechanical calculations and data retrieval methods, to relationship between structure and biological activity of pharmacologically interesting compounds. *Mailing Add:* Bolt Beranek & Newman Inc 10 Moulton St Cambridge MA 02238

HOLLISTER, LEO E, b Cincinnati, Ohio, Dec 3, 20; m 50; c 4. PHARMACOLOGY. *Educ:* Univ Cincinnati, BS, 40, MD, 43; Am Bd Internal Med, dipl, 51 & 74. *Prof Exp:* Chief med serv, 53-60, assoc chief staff, 60-70, SR MED INVESTR, VET ADMIN HOSP, 70-; MED DIR, HARRIS COUNTY PSYCH CTR. *Concurrent Pos:* Prof med, psychol & pharmacol, Sch Med, Stanford Univ, 70-86; prof psychol & pharmacol, Univ Tex Med Sch, Houston, 87-91. *Honors & Awards:* Menninger Award, Am Col Physicians; Hunter Award, Am Soc Clin Pharmacologists. *Mem:* Am Col Physicians; Am Soc Pharmacol & Exp Therapeut; Am Soc Clin Pharmacol & Therapeut (past pres); Int Col Neuropsychopharmacol (past pres); Am Col Neuropsychopharmacol (past pres). *Res:* Clinical psychopharmacology; drugs of abuse; adverse drug reactions. *Mailing Add:* Harris County Psych Ctr PO Box 20249 Houston TX 77225-0249

HOLLISTER, LINCOLN STEFFENS, b Rochester, Minn, Oct 16, 38; m 60; c 2. GEOLOGY, STRUCTURAL. *Educ:* Harvard Univ, AB, 61; Calif Inst Technol, PhD(geol & geochem), 66. *Prof Exp:* Asst prof geol, Univ Calif, Los Angeles, 66-69; from asst prof to assoc prof, 69-76, PROF GEOL, PRINCETON UNIV, 76- *Concurrent Pos:* Mem, Lunar Sample Rev Bd, 71-73; NATO fel, Ctr Res Petrol & Geochem, France, 72-73; assoc ed, Geol Soc Am, 82-, gastdozent, ETH, Switz, 87-88; vis geologist, Geol Surv Can, Vancouver, 81. *Mem:* Mineral Asn Can; Geol Asn Can; Am Geophys Union; Mineral Soc Am; Geol Soc Am. *Res:* Petrogenetic interpretations of compositional zoning in minerals from metamorphic and igneous rocks; origin of the Coast Range, British Columbia, batholithic complex; tectonics of the northeast Pacific; fluid inclusion studies. *Mailing Add:* Dept Geol & Geophys Sci Princeton Univ Princeton NJ 08544

HOLLISTER, VICTOR F, b Chicago, Ill, Mar 25, 25; Can citizen; c 4. MINING ENGINEERING. *Educ:* Univ Calif, Berkeley, BS, 48, MS,49. *Prof Exp:* Geologist, Am Smell & Refining Co, 49-62; mgr, Duval Corp, 62-80; CONSULT, 80- *Concurrent Pos:* Dir, Galactic Resources, Fairbanks Gold & SAm Gold. *Mem:* Soc Econ Geologists; Geol Asn Can; Assoc Inst Mining Engrs; Can Inst Mining Metall. *Res:* Author of twenty publications. *Mailing Add:* 8069 Philbert St Mission BC V2V 3W9 Can

HOLLISTER, WALTER M(ARK), b St Johnsbury, Vt, Nov 22, 30; m 61; c 3. ASTRONAUTICAL ENGINEERING. *Educ:* Middlebury Col, BA, 53; Mass Inst Technol, BS, 53, SM, 59, ScD(instrumentation), 63. *Prof Exp:* Eng asst, Sperry Gyroscope Co, 52, field serv engr, 53-54; teaching asst, 60-61, from instr to assoc prof, 62-82, PROF AERONAUT & ASTRONAUT, MASS INST TECHNOL, 82- *Res:* Technology in gyroscopic instrument theory; inertial guidance; instrumentation; air traffic control. *Mailing Add:* Dept Aeronaut Mass Inst Technol 77 Massachusetts Ave Cambridge MA 02139

HOLLOCHER, THOMAS CLYDE, JR, b Norristown, Pa, June 6, 31; m 53; c 3. REDOX BIOCHEMISTRY. *Educ:* Worcester Polytech Inst, BS, 53; Univ Rochester, PhD(biochem), 58. *Prof Exp:* Res assoc bot, Wash Univ, 58; from asst prof to assoc prof, 61-81, PROF BIOCHEM, BRANDEIS UNIV, 81- *Concurrent Pos:* Vis scientist, Karolinska Inst, Sweden, 68; vis prof, Univ Iceland, 75; Fulbright-Hays res fel, Waite Agr Inst, Adelaide, Australia, 80-81; traveling lectr, Peoples Repub China, 84; chmn environ studies prog, Brandeis Univ, 72-74. *Mem:* Am Soc Biol Chem; Am Soc Microbiol. *Res:* Mechanisms of action of oxidation reduction enzymes; denitrification; nitrification; inorganic chemistry of N-oxides. *Mailing Add:* Dept Biochem Brandeis Univ Waltham MA 02254

HOLLOMAN, JOHN L S, JR, b Washington, DC, Nov 22, 19; c 5. MEDICINE, HEALTH ADMINISTRATION. *Educ:* Va Union Univ, BS, 40; Univ Mich, MD, 43. *Hon Degrees:* DSc, Va Union Univ, 83. *Prof Exp:* Dir, Riverton Lab, 47-74; pres, New York City Health & Hosp Corp, 74-77; regional med officer, US Food & Drug Admin, 80-86; MED DIR, RYAN HEALTH CTR, 81- *Concurrent Pos:* Pvt med pract, NY, 47-; dir, Diag Lab, 69-74; chmn bd, Health Manpower Develop Corp, 69-79; dir, Health Ins Plan, 71-73, vpres, 73-74; prof staff, Ways & Means Health Subcomt, US Cong, 78-80; chmn, Clin Dirs Task Force on AIDS, Nat Asn Community Health Ctr, 87- *Mem:* Inst Med-Nat Acad Sci; AMA; Nat Med Asn (pres, 66-67); Am Pub Health Asn. *Res:* Clinical, pharmaceutical, social and administrative medicine. *Mailing Add:* 27-40 Ericsson St East Elmhurst NY 11369

HOLLOSZY, JOHN O, b Vienna, Austria, Jan 2, 33; US citizen; m 57. PHYSIOLOGY. *Educ:* Ore State Univ, 50-53; Wash Univ, MD, 57. *Prof Exp:* Intern med, Washington Univ Med Ctr, 57-58, from asst resident to chief resident, 58-61; from asst prof to assoc prof, 65-74, prof prev med & pub health & dir div appl physiol, 74-84, PROF MED & DIR SECT APPL PHYSIOL, SCH MED, WASH UNIV, 84- *Concurrent Pos:* Nat Inst Arthritis & Metab Dis spec res fel biochem, Sch Med, Wash Univ, 63-65; USPHS res grant, 65-91, res career develop award, 69-73. *Mem:* Am Inst Nutrit; Am Fedn Clin Res; Am Soc Clin Invest; Am Heart Asn; Am Physiol Soc; Am Soc Biol Chem; Geront Soc Am; Am Diabetes Asn. *Res:* Effects of exercise on sugar transport into muscle; serum lipids; cardiovascular function; enzymes involved in energy metabolism in muscle; body composition and food intake; aging health maintenance. *Mailing Add:* Dept Med Campus Box 8113 Wash Univ 4566 Scott Ave St Louis MO 63110

HOLLOWAY, CAROLINE T, b New York, NY, July 29, 37; m; c 2. LIPIDS, ENZYMES & PROTEINS. *Educ:* City Col NY, BS, 59; Duke Univ, PhD(biochem), 64. *Prof Exp:* Postdoctoral fel, NIMH, Duke Univ, 67; res asst prof, Med Ctr, Univ Va, 76-83; vis res scientist, E I Du Pont de Nemours & Co, Inc, 83-84; grants assoc, 84-85, head biol str sect, 85-90, DIR, OFF SCI POLICY, NAT CTR RES RESOURCES, NIH, 90- *Mem:* Am Soc Biochem & Molecular Biol; AAAS. *Res:* Develop planning, evaluation, legislation and science policy for intramural and extramural components of National Center Research Resources. *Mailing Add:* Off Sci Policy Nat Ctr Res Resources NIH Bldg 12A-4045 9000 Rockville Pike Bethesda MD 20892

HOLLOWAY, CLARKE L, b Atmore, Ala, May 2, 26; m 48; c 3. HISTOLOGY, GERONTOLOGY. *Educ:* Auburn Univ, DVM, 49, MS, 62; Iowa State Univ, PhD(anat), 69. *Prof Exp:* Pvt pract, 49-60; from instr to assoc prof anat, Auburn Univ, 60-67; assoc prof, Univ Ga, 67-68; prof anat & head dept, Sch Vet Med, Auburn Univ, 68-88; RETIRED. *Concurrent Pos:* Nat Inst Neurol Dis & Blindness spec fel, 64-68. *Mem:* Am Vet Med Asn; Am Asn Vet Anat. *Res:* Anatomy and histology of olfactory mechanism of the dog; changes with age in eye of domestic animals. *Mailing Add:* 426 Blake Ave Auburn AL 36830

HOLLOWAY, CLIVE EDWARD, b Bristol, Eng, Jan 28, 38; Can citizen. EDUCATION ADMINISTRATION, GENERAL ENVIRONMENTAL SCIENCES. *Educ:* Bristol Col Advan Technol, ARIC, 60; Univ Western Ont, MSc, 62, PhD(chem), 66. *Prof Exp:* Asst prof, 68-72, ASSOC PROF CHEM, YORK UNIV, 72- *Concurrent Pos:* Counr, Chem Inst Can, Toronto, 86-88; pres, Asn Chem Prof Ont, 87-89; dir, Natural Sci, York Univ, 89-; mem, Biol & Chem Defence Rev Comt, Defence Res Bd, Ottawa, Can, 90-; expert witness, Corp & Consumer Affairs Can, 90- *Mem:* Chem Inst Can. *Res:* Coordination complexes and organometallic compounds with emphasis on the transition elements and x-ray, electron and neutron diffraction stucture correlations. *Mailing Add:* Dept Chem York Univ 4700 Keele St North York ON M3J 1P3 Can

HOLLOWAY, DENNIS MICHAEL, surface physics; deceased, see previous edition for last biography

HOLLOWAY, FRANK A, b Houston, Tex, Jan 1, 40; m 67; c 3. BEHAVIORAL PHARMACOLOGY, BEHAVIORAL MEDICINE. *Educ:* Univ Houston, BS, 61, MA, 64, PhD(psychol), 66. *Prof Exp:* PROF DERMAT & INTERNAL MED, HEALTH SCI CTR, UNIV TEX, 84-; instr & asst prof med & biol psychol, 66-74, instr & assoc prof, 74-77, PROF BIOL PSYCHOL, DEPT PSYCHIAT & BEHAV SCI, HEALTH SCI CTR, UNIV OKLA, 77- *Mem:* Am Soc Clin Invest; Am Dermat Asn; Soc Neurosci; AAAS; Am Psychological Soc. *Res:* Psychopharmacology of learning and motivation; state-dependent learning; behavioral pharmacology of ethanol and drugs of abuse; biological rhythms & behavior. *Mailing Add:* Univ Okla Health Sci Ctr PO Box 26901 Oklahoma City OK 73190

HOLLOWAY, FREDERIC ANCRUM LORD, chemical energy; deceased, see previous edition for last biography

HOLLOWAY, G ALLEN, JR, b New York, NY, Oct 14, 38; m 87; c 2. ANGIOBIOLOGY, WOUND HEALING. *Educ:* Yale Univ, BA, 60; Harvard Univ, MD, 64. *Prof Exp:* Chief med, US Army Hosp, Camp Zama, Japan, 70-72; from asst prof to assoc prof bioeng, Ctr Bioeng, Univ Wash, Seattle, 75-88; DIR MED, VASCULAR LAB, MARICOPA MED CTR, 88- *Concurrent Pos:* Consult & lectr, mult orgn; renal fel, Pub Health Serv, 78-79, career develop award, 80-85. *Mem:* Biomed Eng Soc. *Res:* Vascular disease of both the macrovascular and microvascular systems as well as clinical aspects of wound healing. *Mailing Add:* Dept Surg Maricopa Med Ctr 2601 E Roosevelt Phoenix AZ 85008

HOLLOWAY, HARRY CHARLES, b Yukon, Okla, May 4, 33; m 55; c 3. NEUROPSYCHIATRY. *Educ:* Univ Okla, MD, 58; Am Bd Psychiat & Neurol, dipl, 66. *Prof Exp:* Chief Neurol & Psychiat Sect, 121st Evacuation Hosp, Korea, 62-63; res psychiatrist, Div Neuropsychiat, Walter Reed Army Inst Res, 63-65; chief Neuropsychiat Dept, USA Component, SEATO Med Res Lab, Thailand, 65-69; dep dir, Walter Reed Army Inst Res, 69-70, dir, Div Neuropsychiat, 70-77; PROF, DEPT PSYCHIAT, UNIFORM SERV UNIV HEALTH SCI, 77-, CHMN, 79- *Concurrent Pos:* Consult, Eighth US Army, 62-63; fac mem, Washington Sch Psychiat, 74-; consult, Community Ment Health Training Prog, 74-, mem, Res Study Sect & Res Adv Comt, Ill Dept Ment Health & Develop Disabilities, 75-; mem, Life Sci Comt, NASA, 75-; fac mem, Sch Med, Georgetown Univ, 76-79; mem, Leukemia Core Comt, Cancer & Leukemia Group B, 81-82. *Mem:* Fel Am Psychiat Asn; AAAS; Soc Neurosci; Acad Soc & Polit Sci. *Res:* Psychiatric epidemiology and psychopharmacology with emphasis on the impact of social and environmental factors upon the occurrence of psychiatric disease psychophysiology of cancer chemotherapy and drug abuse. *Mailing Add:* Dept Psychiat Rm B3066 4301 Jones Bridge Rd Bethesda MD 20874-4799

HOLLOWAY, HARRY LEE, JR, b York Co, Va, May 22, 26; m 48; c 4. PARASITOLOGY. *Educ:* Randolph-Macon Col, BS, 48; Univ Richmond, MA, 51; Univ Va, PhD, 56. *Prof Exp:* Asst, Univ Richmond, 50-51 & Univ Va, 51-53; from asst prof to prof biol, Roanoke Col, 53-69; prof biol & dean fac, Western Md Col, 69-71; chmn dept, 71-74, PROF BIOL, UNIV NDAK, 71- *Concurrent Pos:* Chmn dept biol, Roanoke Col, 59-69; US Antarctic Res Prog grants, 64-69, McMurdo Facil Antarctica, 64-65. *Mem:* Fel AAAS; Am Soc Parasitol; Am Micros Soc; Am Inst Biol Sci; Sigma Xi. *Res:* Morphology and taxonomy, development and life cycles, macroecology and zoogeography of animal parasites. *Mailing Add:* Dept Biol PO Box 8238 Univ Sta Grand Forks ND 58202

HOLLOWAY, JAMES ASHLEY, b Stanton, Tenn, Feb 7, 36. NEUROPHYSIOLOGY. *Educ:* San Jose State Univ, BA, 65; Univ Calif, PhD(physiol), 71. *Prof Exp:* Indust engr, Campbell Soup Co, 65-67, anal chemist, 67-68; asst prof, 71-73, assoc prof neurophysiol, 73-80, assoc prof physiol & biophys, 80-, PROF & CHMN, DEPT PHYSIOL & BIOPHYS, COL MED, HOWARD UNIV. *Concurrent Pos:* Comnr Parks & Recreation, State Calif, 70-71; consult, Am Cancer Soc, 70-71, Field Educ Publ, 72-73 & Nat Insts Gen Med Sci, 73-; fel, Johns Hopkins Univ, 71-72; NSF res grant, 72 & 73; Porter lectr, Meharry Med Col, 72- & Tuskegee Inst, 73-; NIH training grant, 74-78. *Mem:* AAAS; Am Physiol Soc; Soc Neurosci; Int Asn Study Pain. *Res:* Response characteristics of spinal cord dorsal horn cells to natural and electrical stimulation. *Mailing Add:* Dept Physiol Howard Univ Col Med Washington DC 20059

HOLLOWAY, JOHN LEITH, JR, b Hickory, NC, Dec 18, 27. DYNAMIC METEOROLOGY. *Educ:* Mass Inst Technol, BS, 52, MS, 53. *Prof Exp:* Res meteorologist, Univ Pa, 53-55 & Off Meteorol Res, US Weather Bur, 55-57; meteorologist, Geophys Fluid Dynamics Lab, Nat Oceanic & Atmospheric Admin, 57-83 & 89-90; RETIRED. *Honors & Awards:* Silver Medal, US Dept Com, 63. *Mem:* Am Meteorol Soc. *Res:* Statistical meteorology; simulation of climate and general circulation of atmosphere by numerical solution of meteorological equations on electronic computers; smoothing, filtering and spectral analysis of meteorological time series. *Mailing Add:* 10500 Rockville Pike No M10 Rockville MD 20852-3331

HOLLOWAY, JOHN REQUA, b Portland, Ore, Aug 2, 40; div; c 2. PETROLOGY, GEOCHEMISTRY. *Educ:* Univ Ore, BS, 63; Pa State Univ, PhD(geochem), 70. *Prof Exp:* From asst prof to assoc prof, 69-79, PROF CHEM, ARIZ STATE UNIV, 80- *Concurrent Pos:* Vis scientist, Geophys Lab, 75-76; vis prof, Stanford Univ, 79, Calif Inst Technol, 87; vis prof geol, Bristol, Eng; vis fel, Royal Soc, 90. *Mem:* Geol Soc Am; Mineral Soc Am; Am Geophys Union; Geochem Soc. *Res:* Experimental measurement and theoretical calculation of the stability of hydrous silicate phases. *Mailing Add:* Dept Chem Ariz State Univ Tempe AZ 85281

HOLLOWAY, JOHN THOMAS, b Cape Girardeau, Mo, June 19, 22; m 65; c 2. EXPERIMENTAL PHYSICS. *Educ:* Millikin Univ, AB, 43; Iowa State Univ, PhD(physics), 57. *Prof Exp:* With nuclear physics br, Off Naval Res, 46-53, actg br head, 51-52; asst, Iowa State Univ, 53-57; chief phys sci div, Off Dir Defense Res & Eng, 58-59, dept dir, Off Sci, 59-61; dept dir, Off Grants & Res Contracts, NASA, 61-65, actg dir, 65-67, with space applns div, 67-68; dir, Nat Hwy Safety Res Ctr, Fed Hwy Admin, US Dept Transp, 68-69; vpres res, Ins Inst for Hwy Safety, 69-72; assoc dir opers, Interdisciplinary Commun Prog, Smithsonian Inst, 72-77; consult hwy safety, biomed electronics & energy conserv, 77-78; sr staff officer, Radioactive Waste Mgt Bd, Nat Acad Sci-Nat Res Coun, 78-85; RETIRED. *Concurrent Pos:* Consult, radioactive waste mgt & hwy safety, 85- *Mem:* Am Phys Soc; Sigma Xi; AAAS. *Res:* Research administration and management; experimental nuclear physics; space science and engineering; highway safety; international population policy; radioactive waste management. *Mailing Add:* 2220 Cathedral Ave NW Washington DC 20008

HOLLOWAY, LELAND EDGAR, b Kansas City, Mo, Aug 19, 36; m 65; c 1. EXPERIMENTAL HIGH ENERGY PHYSICS. *Educ:* Mass Inst Technol, BS, 58; Univ Pa, PhD(physics), 62. *Prof Exp:* NSF fel, 62-63; res physicist, Lawrence Radiation Lab, 63-66 & Inst Nuclear Physics, Orsay, France, 66-67; from asst prof to assoc prof, 67-74, PROF PHYSICS, UNIV ILL, URBANA, 74- *Mem:* Fel Am Phys Soc. *Mailing Add:* Dept Physics Univ Ill 1110 W Green St Urbana IL 61801

HOLLOWAY, PAUL HOWARD, b Marion, Ind, Oct 31, 43; m 64; c 3. MATERIAL SCIENCE, SURFACE SCIENCE. *Educ:* Fla State Univ, BS, 65, MS, 66; Rensselaer Polytech Inst, PhD(mat eng), 72. *Prof Exp:* Phys metallurgist, Knolls Atomic Power Lab, Gen Elec Co, 66-69; staff mem mat eng, Sandia Labs, 72-78; assoc prof, 78-81, PROF MAT SCI & ENG, UNIV FLA, 81- *Honors & Awards:* E W Muller Award. *Mem:* Am Vacuum Soc; Am Soc Testing & Mat; Am Soc Metals; Am Inst Mining, Metall & Petrol Engrs. *Res:* Surface science; kinetics and thermodynamics of gas/solid interactions; metallurgical reactions in thin films; defect enhanced solid state atomic transport; electronic materials; compound semiconductors. *Mailing Add:* Dept Mat Sci & Eng Univ Fla Gainesville FL 32611

HOLLOWAY, PETER WILLIAM, b Manchester, UK, June 30, 38. MEMBRANE PROTEIN STRUCTURE. *Educ:* Univ Manchester, PhD(chem), 62. *Prof Exp:* PROF BIOCHEM, SCH MED, UNIV VA, 69- *Mem:* Am Biol Chemist; Biophys Soc. *Res:* Interaction of cytochrome b5 with membranes; fatty acid desaturation. *Mailing Add:* Dept Biochem Sch Med Univ VA Charlottesville VA 22908

HOLLOWAY, RALPH L, JR, b Philadelphia, Pa, Feb 6, 35; div; c 3. PHYSICAL ANTHROPOLOGY, NEUROSCIENCES. *Educ:* Univ NMex, BS, 59; Univ Calif, Berkeley, PhD(anthrop), 64. *Prof Exp:* From asst prof to assoc prof, 64-69, chmn, 79-81, PROF ANTHROP, COLUMBIA UNIV, 73- *Concurrent Pos:* NIMH res fel, 64; NSF res grant, 69-; vis prof, New Sch Social Res, 72-85; Guggenheim Found fel, 74-75; L S B Leakey Found Grant, 85. *Mem:* Fel AAAS; Am Anthrop Asn; Am Asn Phys Anthrop; fel NY Acad Sci; Soc Neurosci. *Res:* Evolution of human brain and behavior; comparative primate neuroanatomy; primate aggression and social evolution; fossil man; dendritic branching in cortex; fossil brain endocasts; cerebral asymmetrics; sexual dimorphism. *Mailing Add:* Dept Anthrop Columbia Univ New York NY 10027

HOLLOWAY, RICHARD GEORGE, b Norristown, Pa, Apr 19, 51. PALYNOLOGY, QUATERNARY PERIOD. *Educ:* Eastern NMex Univ, BA, 73; Tex A&M Univ, MS, 78, PhD(bot), 81. *Prof Exp:* Archaeol technician I, US Forest Serv, Sitka, Alaska, 79; palynologist, Res Found, State Univ NY, Binghamton, 81; res scientist, Anthrop Res Lab, Tex A&M Univ, 81-87; vis asst prof anthrop, Eastern NMex Univ, 87-89; RES ASSOC, BIOL DEPT, UNIV NMEX, 89- *Concurrent Pos:* Fel, Anthrop Prog, Tex A&M Univ, 81-87. *Mem:* AAAS; Am Asn Stratig Palynologists; Am Quaternary Asn; Electron Micros Soc Am; Int Asn Wood Anatomists; Int Asn Aerobiol; Am Bot Soc. *Res:* Quaternary palynology and quaternary plant ecology; palynological statistics; fossil pollen morphology and taxonomy; arctic and subarctic paleoecology; archaeological palynology; coprolite analysis; macro-botanical analysis of archaeological materials; prehistoric diets and nutrition; southeastern US paleoecology. *Mailing Add:* 9705 Toltec Rd NE Albuquerque NM 87111

HOLLOWAY, THOMAS THORNTON, b Dallas, Tex, Mar 6, 44; m 68. ENVIRONMENTAL QUALITY ANALYSIS. *Educ:* Rice Univ, BA, 66, PhD(chem), 69. *Prof Exp:* Res assoc chem, Johns Hopkins Univ, 69-71; Robert A Welch Found res fel, Tex Tech Univ, 71-73; assoc prof chem, William Jewell Col, 73-80; CHEMIST, US ENVIRON PROTECTION AGENCY, 80- *Concurrent Pos:* Mem, Exec Potential Prog, US Off Personnel Mgt, 91. *Mem:* Am Chem Soc; Sigma Xi. *Res:* Formal and applied perturbation theory to high order; theoretical study of molecular collisions which occur in the atmosphere; applications of computers in chemistry and chemical education; analysis of environmental quality. *Mailing Add:* 820 Northeast 73rd Pl Gladstone MO 64118

HOLLOWELL, JOSEPH GURNEY, JR, b Charleston, SC, Nov 19, 32; m 55; c 6. PEDIATRIC ENDOCRINOLOGY, GENETICS. *Educ:* Med Univ SC, MD, 56; Univ Calif, Berkeley, MPH, 77. *Prof Exp:* Intern, Hosp, Med Col SC, 56-57, resident pediat, 57-59; res fel pediat endocrinol & genetics, State Univ NY Upstate Ctr, 61-63; from instr to assoc prof pediat, Med Col Ga, 63-70; mem grad fac anthrop, 74-81, assoc prof pediat, 70-72, ASSOC PROF PEDIAT & PREV MED & DIR, CHILDREN'S REHAB UNIT, UNIV KANS MED CTR, 72- *Concurrent Pos:* Pediat consult & dir birth defects ctr, Gracewood State Sch & Hosp; US HEW grants, 64-67; dir health, Kans Dept Health & Environ. *Mem:* Am Soc Human Genetics; Soc Teachers Family Med; Ambulatory Pediat Asn; Sigma Xi; NY Acad Sci; Am Acad Pediat; Soc Develop Pediat. *Res:* Catecholamine metabolism; endocrinology, regulation of sodium transport; developmental aspects of transporting membranes; medical education; health care delivery; developmental aspects of mental retardation and disability; health concepts. *Mailing Add:* 3200 Lenox Rd NE-E 111 Atlanta GA 30324

HOLLSTEIN, ULRICH, b Berlin, Ger, July 7, 27; nat US; m 57; c 3. ORGANIC CHEMISTRY. *Educ:* Univ Amsterdam, BS, 48, MS, 53, PhD(chem), 56. *Prof Exp:* Nat Res Coun Can fel natural prod chem, 56-57; res chemist org radiochem, Inst Nuclear Res, Neth, 58-59; fel natural prod chem, Univ Calif, Berkeley, 60-63; from asst prof to assoc prof org chem, Eastern NMex Univ, 63-67; assoc prof org chem, 67-74, PROF CHEM, UNIV NMEX, 74- *Concurrent Pos:* Fulbright fel, Univ Tübingen, WGer, 73-74. *Mem:* Am Chem Soc; Royal Neth Chem Soc. *Res:* Natural products; alkaloids; ca-antagonists antibiotics; microbial metabolites; biosynthesis; interaction with polydeoxyribonucleotides; optical rotatory dispersion and circular dichroism; carbon-13 nuclear magnetic resonance; mass spectrometry. *Mailing Add:* Dept Chem Univ NMex Albuquerque NM 87131

HOLLUB, RAYMOND M(ATHEW), b Passaic, NJ, Mar 9, 28; m 52; c 2. AERONAUTICAL ENGINEERING. *Educ:* Univ Ala, BS, 55, MS, 58. *Prof Exp:* Assoc res engr, Lockheed Aircraft Corp, Ga, 55-56; instr eng mech, 56-58, asst prof aeronaut eng, 58-61, dir res comt proj, 59, dir NASA res contracts, 61-65, ASSOC PROF AEROSPACE ENG, UNIV ALA, 61-, DIR CONTINUING ENG EDUC, 72-, DIR OFF CAMPUS MASTERS DEGREE PROG, 75-, SPEC ASST CONTINUING ED, 80- *Concurrent Pos:* Consult, Lockheed Aircraft Corp, 58-61, Southern Serv Corp & Astro Space Labs, Ala; assoc dir, Am Soc Eng Educ-NASA fac fel prog, Marshall Space Flight Ctr & Univ Ala, 66, co-dir, 67; NSF-W Alton Jones fel, Ga Inst Technol, 67-70; dir contract consultation training, Occup Safety & Health Admin, US Dept Labor, 78- *Honors & Awards:* Frank Oppenheimer Award, Am Soc Eng Educ, 78. *Mem:* Am Soc Eng Educ; Am Inst Aeronaut & Astronaut; Nat Univ Exten Asn. *Res:* Mechanical properties of materials; structural analysis; heat transfer. *Mailing Add:* 44 Vestana Hills Northport AL 35476

HOLLWEG, JOSEPH VINCENT, b New York, NY, Mar 20, 44; m 76; c 2. SPACE PHYSICS, SOLAR PHYSICS. *Educ:* Mass Inst Technol, BS & MS, 65, PhD(plasma physics), 68. *Prof Exp:* Res fel space physics, Ctr Space Res, Mass Inst Technol, 67-68; vis scientist, Max Planck Inst Physics & Astrophys, 68-70; res assoc, Calif Inst Technol, 70-72; mem sci staff solar physics, Nat Ctr Atmospheric Res, 72-80; dir Space Sci Ctr, 82-83, PROF, DEPT PHYSICS, UNIV NH, 80- *Concurrent Pos:* Vis scientist, Max Planck Inst Aeronomy, 75-76; assoc ed, J Geophys Res, 82-85. *Mem:* Am Geophys Union; Am Astron Soc; Int Astron Union. *Res:* Physical processes of the solar and interplanetary plasma, with emphasis of mechanisms of energy and momentum balance. *Mailing Add:* Dept Physics Univ NH Durham NH 03824

HOLLY, FRANK JOSEPH, b Budapest, Hungary, Dec 3, 34; US citizen; m 69; c 4. OPTOMETRY, DENTISTRY. *Educ:* Cornell Univ, PhD(phys chem), 62. *Prof Exp:* Staff Scientist dent res, Procter & Gamble, 62-65; prof chem, Univ El Salvador, 65-66; staff scientist biomat, Thermo Electron Corp, 66-68; sr scientist cornea & retina, Eye Res Inst Retina Found, 68-78; prof ophthal & biochem, Health Sci Ctr, Tex Tech Univ, 78-90; DIR-OWNER, DAKRYON PHARMACEUT, 90- *Concurrent Pos:* Pres, Dry Eye Inst, 83- *Mem:* Am Chem Soc; Int Soc Contact Lens Res; Asn Res Vision & Ophthal; Int Soc Colloid & Interface Scientists; Int Soc Dakryology (pres, 84-90); hon fel Am Acad Optom; assoc fel Am Acad Ophthal. *Res:* Interface science, rheology, kinetics (physical chemistry) applied to biological systems, prosthesis, and pharmaceuticals; dental decay, biocompatibility of materials, lacrimal physiology, cell interaction, and bioadhesion; ocular surface disease. *Mailing Add:* PO Box 98069 Lubbock TX 79499

HOLLYFIELD, JOE G, b El Dorado, Ark, Aug 6, 38. DEVELOPMENTAL BIOLOGY. *Educ:* Hendrix Col, BA, 60; La State Univ, MS, 63; Univ Tex, Austin, PhD(zool), 66. *Prof Exp:* Asst prof bot & zool, Univ Tex, Austin, 66-67; asst prof ophthal res, Col Physicians & Surgeons, Columbia Univ, 69-75, assoc prof anat, 75-77; assoc prof, 77-80, PROF OPHTHAL, BAYLOR COL MED, 80- *Concurrent Pos:* Fel, Hubrecht Lab, Utrecht, Neth, 67-69; Olga K Weiss Scholars Award, Res Prevent Blindness Inc. *Honors & Awards:* Marjorie W Margohn Prize, Retina Res Found. *Mem:* Fel AAAS; Am Soc Zoologists; Asn Res Vision & Ophthal; Soc Develop Biol. *Res:* Development and cell biology of the retina and pigment epithelium. *Mailing Add:* Dept Ophthal Cullen Eye Inst Baylor Col Med One Baylor Plaza Houston TX 77030

HOLLYWOOD, JOHN M(ATTHEW), b Red Bank, NJ, Feb 4, 10; wid. ELECTRONICS. *Educ:* Mass Inst Technol, BS, 31, MS, 32. *Prof Exp:* Chief engr, Electron Res Labs, NY, 32-35; circuitry develop engr, Ken-Rad Tube & Lamp Corp, 35-36; electronics res engr, Columbia Broadcasting Syst, Inc, 36-42; consult, Am-Brit Lab, Harvard Univ-Radio Res Lab, Eng, 42-45, Naval Res Lab, 45-46 & Airborne Instruments Lab, 46-49; sr engr in charge adv res & develop, Columbia Broadcasting Syst, Inc, 49-54, coordr res, 54-58, sci aide to pres, 58-71; consult, Goldmark Commun Corp, 72-79; CONSULT, 79- *Mem:* Audio Eng Soc; fel Inst Elec & Electronics Engrs. *Res:* Circuit and information theory; monochrome; color television; audio; acoustics applications; stereo; radar; radio countermeasures; radar systems and displays; transistor applications. *Mailing Add:* 33 Peters Pl Red Bank NJ 07701

HOLM, DALE M, b Portland, Ore, Jan 23, 24; m 49; c 3. BIOPHYSICS, NUCLEAR PHYSICS. *Educ:* Lewis & Clark Col, BS, 49; Ore State Univ, MS, 52, PhD(physics), 55. *Prof Exp:* Res asst, Los Alamos Nat Lab, 52-55, staff mem, 55-73, group leader agr biosci & liaison between Los Alamos Nat Lab & USDA, 73-81, group leader, Biophys Instrumentation, 81-82; intel analyst, 82-87; RETIRED. *Concurrent Pos:* Consult, 87- *Mem:* AAAS. *Res:* Reactor design, construction, and diagnosis; gamma scanning of reactor fuel elements; activation analysis. *Mailing Add:* 264 Andanada Los Alamos NM 94087

HOLM, DAVID GARTH, b Shelley, Idaho, Aug 17, 50; m 74; c 2. HORTICULTURE, PLANT BREEDING. *Educ:* Univ Idaho, BS, 72, MS, 74; Univ Minn, PhD(hort), 77. *Prof Exp:* ASST PROF HORT, COLO STATE UNIV, 78- *Concurrent Pos:* Dir, 85-87, supt, Potato Asn Am, 82- *Mem:* Sigma Xi; Potato Asn Am; Am Soc Hort Sci; Europ Asn Potato Res. *Res:* Genotype-environment interactions of food crops; physiological and morphological characters associated with high yielding food crops. *Mailing Add:* San Luis Valley Res Ctr 0249 E Rd Nine N Center CO 81125

HOLM, DAVID GEORGE, b Rossland, BC, Apr 21, 35; m 57; c 2. GENETICS. *Educ:* Univ BC, BSc, 64; Univ Conn, PhD(genetics), 69. *Prof Exp:* Assayer, COMINCO, 55-61; from asst prof to assoc prof, 69-83, chmn, Biol Prog, 85-88, PROF, DEPT ZOOL, UNIV BC, 83- *Concurrent Pos:* Consult, biol course, Open Learning Inst, 80. *Mem:* AAAS; Genetics Soc Am; Can Genetics Soc; Zool Soc Can; Environ Mutagen Soc. *Res:* Meiotic behavior of chromosomes in Drosophila melanogaster; genetic properties of heterochromatin; maternal effect on position-effect variegation; radiation and chemical mutagenesis. *Mailing Add:* Dept Zool Univ BC 6270 Univ Blvd Vancouver BC V6T 1Z4 Can

HOLM, HARVEY WILLIAM, b Williston, NDak, Mar 3, 41; m 78; c 3. MICROBIOLOGY, BIOCHEMISTRY. *Educ:* Minot State Col, BS, 63; Univ NDak, MS, 65, PhD(microbiol), 69. *Prof Exp:* Res assoc microbiol, Univ Ga, 68-70; res microbiologist, Environ Res Lab, 70-80, CHIEF, ENVIRON SYSTS BR, US ENVIRON PROTECTION AGENCY, 80- *Mem:* Am Soc Microbiol; Am Inst Biol Sci; Am Soc Limnol & Oceanog. *Res:* Fate of pollutants in aquatic ecosystems; impact of pollutants on aquatic ecosystem structure and function. *Mailing Add:* Environ Res Lab US Environ Protection Agency Hatfield Marine Sci Ctr Newport OR 97365

HOLM, L(EROY) W(ALLACE), chemical engineering, petroleum engineering; deceased, see previous edition for last biography

HOLM, LEROY GEORGE, b Wyeville, Wis, Oct 2, 17; m 45; c 3. COMPILATION, BIOLOGY & DISTRIBUTION OF MAJOR WORLD WEED SPECIES. *Educ:* Wis State Univ, La Crosse, BS, 39; Univ Wis, PhD, 49. *Prof Exp:* From asst prof to prof hort, Univ Wis-Madison, 49-71; sr fel, East-West Ctr, Univ Hawaii, 71-72; COMPILATION OF WORLD WEED INVENTORY, 72- *Concurrent Pos:* Nat Acad Sci exchange prof, USSR, 63; mem staff, Div Plant Protection, UN Food & Agr Orgn, Rome, 65-66; consult weed sci, 60 countries, 72-; sr res scholar, East-West Ctr, Hawaii, 71-72. *Honors & Awards:* Distinguished Serv Award, Int Weed Sci Soc. *Mem:* Hon fel Weed Sci Soc Am; Europ Weed Res Soc; Int Weed Sci Soc; Asian-Pac Weed Sci Soc. *Res:* Distribution and biology of world weeds; survey of world weed vegetation. *Mailing Add:* 714 Miami Pass Madison WI 53711

HOLM, MYRON JAMES, b York, Nebr, May 31, 30; m 54; c 3. ORGANIC CHEMISTRY. *Educ:* Univ Nebr, AB, 52, MS, 56, PhD(chem), 58. *Prof Exp:* Res assoc chem, Ohio State Univ, 58-59; res chemist, Monsanto Co, 59-63, res specialist, 63-80, sr res specialist, 80-90; RETIRED. *Res:* Paper chemicals; photochemistry; antineoplastics; plant growth regulators; fungicides. *Mailing Add:* 9734 Mansfield Dr St Louis MO 63132

HOLM, RICHARD H, b Sept 24, 33; m 58; c 4. INORGANIC CHEMISTRY. *Educ:* Univ Mass, BS, 55; Mass Inst Technol, PhD(inorg chem), 59. *Hon Degrees:* DSc, Univ Mass, 79. *Prof Exp:* Prof chem, Mass Inst Technol, 67-75 & Stanford Univ, 75-80; PROF CHEM, HARVARD UNIV, 80- *Concurrent Pos:* Sloan fel, 64-67. *Honors & Awards:* Bailar Medal, 73; Inorg Chem Award, Am Chem Soc, 76; Harrison Howe Award, 77; Centenary Medal, Royal Chem Soc, 80; Dwyer Medal, Aust Chem Soc, 88; Distinguished Serv Inorg Chem Award, Am Chem Soc, 90. *Mem:* Nat Acad Sci; Am Chem Soc; Royal Soc Chem; Am Acad Arts & Sci. *Res:* Inorganic and bioinorganic chemistry. *Mailing Add:* Dept Chem Harvard Univ Cambridge MA 02138

HOLM, ROBERT E, b Lafayette, Ind, Feb 18, 40; m 64; c 2. AGRICULTURE. *Educ:* Purdue Univ, BS, 62, MS, 64, PhD(plant physiol & biochem), 69. *Prof Exp:* Sr res biochemist, Diamond Shamrock Corp, 69-75, res assoc, 75-76, sr res assoc, 76-78; head biol testing, Mobil Chem Co, 78-80, mgr crop chem res & develop, 80-81; DIR, AGROCHEM SCI DEPT, AGROCHEM DIV, RHONE-POULENC INC, 81- *Concurrent Pos:* Chmn, Plant Growth Regulator Working Group, 76-77; asst ed, J Plant Physiol, 80-82. *Honors & Awards:* Wilson Popenoe Award, Am Soc Hort Sci, 78. *Mem:* Plant Growth Regulator Soc Am (pres, 76-77); Weed Sci Soc Am; Coun Agr Sci & Technol; Am Soc Plant Physiologists. *Res:* Mode-of-action of plant hormones, regulators and herbicides; biological evaluation of herbicides and plant growth regulants; uptake, translocation and metabolism of agricultural chemicals. *Mailing Add:* Rhone-Poulenc Agr Co PO Box 12014 TW Alexander Dr Research Triangle Park NC 27709

HOLMAN, B LEONARD, b Sheboygan, Wis, June 26, 41; m 71; c 2. NUCLEAR MEDICINE, RADIOLOGY. *Educ:* Univ Wis, BS, 63; Wash Univ, MD, 66; Am Bd Radiol, cert diag radiol, 72, cert nuclear radiol, 79; Am Bd Nuclear Med, cert, 72. *Hon Degrees:* AM, Harvard Univ, 88. *Prof Exp:* Intern, Mt Zion Hosp Med Ctr, San Francisco, 66-67; resident & fel, Edward Mallinckrodt Inst Radiol, Wash Univ Med Sch, 67-70; from instr to prof, Harvard Med Sch, 70-88; dir Clin Nuclear Med Serv, Brigham & Women's Hosp, 80-88; PHILIP H COOK PROF RADIOL, HARVARD MED SCH, 88-; CHMN DEPT RADIOL, BRIGHAM & WOMEN'S HOSP, 88- *Concurrent Pos:* Mem, Radiopharmaceut Panel, US Pharmacopeia, 70-; radiologist nuclear med, Children's Hosp Med Ctr, 70-; consult, W Roxbury Vet Admin Hosp, 74-; radiologist nuclear med, Sidney Farber Cancer Ctr, 76-; chief clin nuclear med, Peter Bent Brigham Hosp, 75-80; assoc ed, Cardiovasc Radiol, 77-90; estab investr, Am Heart Asn, 77, fel, Coun Cardiovasc Radiol & Coun Circulation; mem, Nuclear Cardiol Rev Panel, Intersoc Comn Heart Dis Resources, 79-81 & Adv Comt Med Uses Isotopes, US Nuclear Regulatory Comn, 79-; ed, Diag Imaging Sect, Med Instrumentation, 80-85. *Honors & Awards:* Taplin Lectr Award, Soc Nuclear Med, 85; Blumgart Pioneer Award, Soc Nuclear Med, 86. *Mem:* Soc Nuclear Med (pres, 87-88); fel Am Col Cardiol; Am Heart Asn; fel Am Col Chest Physicians; fel Am Col Radiol; Am Bd Nuclear Med. *Res:* Cardiovascular nuclear medicine--the use of radioactive tracers to determine regional cardiac physiology in vivo and to diagnose acute myocardial infarction and alterations in regional myocardial blood flow; neuronuclear medicine; functional brain imaging spect. *Mailing Add:* Brigham & Women's Hosp 75 Francis St Boston MA 02115

HOLMAN, CRANSTON WILLIAM, b Pasadena, Calif, Jan 5, 07; m 28; c 2. SURGERY. *Educ:* Stanford Univ, AB, 27, MD, 31; Am Bd Surg, dipl; Am Bd Thoracic Surg, dipl. *Prof Exp:* Intern, Univ Hosp, Stanford Univ, 30; asst resident surg, Cincinnati Gen Hosp, 31-32; asst resident, NY Hosp, 32-35; dir, Second Surg Div, Bellvue Hosp, 51-63; instr, 35-36, assoc, 37-38, from asst prof to assoc prof, 38-58, prof, 58-88, EMER PROF CLIN SURG, MED COL, CORNELL UNIV, 88- *Concurrent Pos:* Resident surgeon, NY Hosp, 35-36, from asst to assoc attend surgeon, 37-53, attend surgeon, 53-, hon staff, Dept Cardiovasc Surg, 90; res fel, Med Col, Cornell Univ, 37; consult surgeon, Vet Admin Hosp, Montrose, NY, 50-, Manhattan Vet Admin Hosp, 54-68, NShore Hosp, Manhasset, Long Island, 54- & Hosp Spec Surg, New York, 55-60. *Honors & Awards:* Greenberg Award, 88. *Mem:* Soc Clin Surg; Soc Univ Surg; Soc Thoracic Surg; Am Surg Asn; Am Col Surg. *Res:* Surgery of gastrointestinal tract and chest; peptic ulcer; general and thoracic surgery; surgical pathology; clinical medicine; gross anatomy; gastric and thoracic conditions. *Mailing Add:* 200 E 66th St New York NY 10021

HOLMAN, GERALD HALL, b Winnipeg, Man, June 7, 29; m 53; c 4. PEDIATRICS, ENDOCRINOLOGY. *Educ:* Univ Man, BSc & MD, 53; FRCPS(C), 59. *Prof Exp:* Res fel physiol & med res, Univ Man, 50-52; rotating intern, Winnipeg Gen Hosp, 52-53; asst resident pediat, Winnipeg Children's Hosp, 53-54; from jr to sr asst resident, Harriet Lane Home, 54-56; NSF fel pediat endocrinol, Johns Hopkins Hosp, 56-58; asst prof pediat, Univ Hosp, Univ Sask, 58-61; assoc prof, Sch Med, Univ Kans, 61-64; prof & chmn dept, Med Col Ga, 64-69; chief dept pediat & head div, Univ Calgary, 69-73; prof pediat, Eastern Va Med Sch, 73-79, head dept, 73-75, dean & vpres acad affairs, 75-79; assoc dean & asst to pres, Tex Tech Health Sci Ctr, Amarillo, 79-84; MED DIR & MED DIR ADOLESCENT SUBSTANCE ABUSE, ST ANTHONY'S HOSPICE, AMARILLO, 85- *Concurrent Pos:* John & Mary Markle scholar acad med, 58-63; NIH res grants, 59-; consult, Proj Headstart, 65- *Honors & Awards:* Schwentker Res Award, Johns Hopkins Hosp, 56. *Res:* Developmental adrenal metabolism and growth disorders in newborns, infants and children; fragile X-syndrome; clinical genetics. *Mailing Add:* 2802 Travis Amarillo TX 79109

HOLMAN, GORDON DEAN, b Ft Lauderdale, Fla, July 15, 49; m 79; c 2. THEORETICAL PLASMA ASTROPHYSICS, SOLAR-STELLAR ASTROPHYSICS. *Educ:* Fla State Univ, BS, 71; Univ NC, Chapel Hill, MS, 73, PhD(astrophys), 77. *Prof Exp:* Teaching asst, dept physics & astron, Univ NC, Chapel Hill, 71-77; lectr astron, astron prog, Univ Md, College Park, 77-83, fel, Ctr Theoret Physics, 77-79, res assoc, astron prog, 79-83; res assoc, Nat Acad Sci & Nat Res Coun, 83-85, ASTROPHYSICIST, NASA, GODDARD SPACE FLIGHT CTR, GREENBELT, MD, 85- *Concurrent Pos:* Prin investr & co-prin investr, var NASA grants, 78- *Mem:* Am Astron Soc; Am Geophys Union; Int Astron Union. *Res:* Physical processes involved in solar and stellar flares and solar-stellar coronal heating; comet ion tail-solar wind interaction; radio and x-ray emission from clusters of galaxies; propagation of cosmic ray electrons. *Mailing Add:* Lab Astron & Solar Physics NASA-Goddard Space Flight Ctr Code 682 Greenbelt MD 20771

HOLMAN, HALSTED REID, b Cleveland, Ohio, Jan 17, 25; m 49, 84; c 4. INTERNAL MEDICINE, IMMUNOLOGY. *Educ:* Yale Univ, MD, 49. *Prof Exp:* Intern med, Montefiore Hosp, Bronx, NY, 52-53, resident, 53-55; asst physician immunol, Rockefeller Univ, 55-58, asst prof, 58-60; chmn dept, 60-71, PROF MED, SCH MED, STANFORD UNIV, 60- *Concurrent Pos:* Nat Res Coun fel, Carlsberg Labs, Copenhagen, Denmark, 49-50; dir, R W Johnson Clin Scholar Prog, Stanford Univ, 69-; dir, Stanford Multipurpose Arthritis Ctr, 77- *Honors & Awards:* Walter Bauer Mem Award, Arthritis Found, 64; Hewlett Award, 85. *Mem:* Am Soc Clin Invest (pres, 70-71); Asn Am Physicians; Asn Prof Med (treas, 67-70); AAAS. *Res:* Rheumatic disease; health services research. *Mailing Add:* Dept Family-Community Med Sch Med Stanford Univ HRP Rm 109 Stanford CA 94305

HOLMAN, J ALAN, b Indianapolis, Ind, Sept 24, 31. VERTEBRATE PALEONTOLOGY, HERPETOLOGY. *Educ:* Franklin Col, AB, 53; Univ Fla, MS, 57, PhD(biol), 61. *Prof Exp:* Asst prof biol, Howard Col, 60-61; assoc prof vert zool, Ill State Univ, 61-67; assoc prof, 66-69, PROF GEOL & ZOOL, MICH STATE UNIV, 69-, CUR VERT PALEONT, 67- *Mem:* Soc Study Amphibious Reptiles; Soc Vert Paleont; Herpet League. *Res:* Paleontology of Cenozoic amphibians and reptiles. *Mailing Add:* Dept Zool East Lansing MI 48824

HOLMAN, JACK PHILIP, b Dallas, Tex, July 11, 34; m 64. MECHANICAL ENGINEERING. *Educ:* Southern Methodist Univ, BS, 55, MS, 56; Okla State Univ, PhD(mech eng), 58. *Prof Exp:* Asst mech eng, Southern Methodist Univ, 55-56 & Okla State Univ, 56-58; task scientist fluid dynamics, USAF Aeronaut Res Lab, Wright-Patterson AFB, Ohio, 58-60; assoc prof, 60-66, chmn, dept civil &n mech eng, 73-78, PROF MECH ENG, SOUTHERN METHODIST UNIV, 66-, DIR, THERMAL & FLUID CTR, 67- *Concurrent Pos:* Consult ed, McGraw-Hill Book Co, 66 - *Honors & Awards:* Convair Award, 55-58; George Westinghouse Award, Am Soc Eng Educ, 72; James Harry Potter Gold Medal, 86; Worcester Reed Warner Gold Medal, Am Soc Mech Engrs, 87. *Mem:* Fel Am Soc Mech Engrs; Am Soc Eng Educ. *Res:* Heat transfer; thermodynamics; fluid dynamics; fluidized and vortex heat transfer; programmed learning methods; audio tutorial instruction techniques; air pollution technology. *Mailing Add:* Dept Civil & Mech Eng Southern Methodist Univ Dallas TX 75275

HOLMAN, JOHN ERVIN, JR, b St Louis, Mo, Nov 26, 33; m 54; c 3. VETERINARY PATHOLOGY. *Educ:* Univ Mo, BS, 56, DVM, 56; Ohio State Univ, PhD(vet path), 65. *Prof Exp:* Vet, Nat Heart Inst, 56-62; vet pathologist, Vet Path Div, Armed Forces Inst Path, 65-67; scientist adminr, Animal Resources Br, Div Res Resources, NIH, 67-71, chief, Lab Animal Med & Vivarial Sci Sect, 71-73, dir, Lab Animal Sci Prog, Animal Resources Br, 73-87; RETIRED. *Mem:* Am Vet Med Asn; Am Col Vet Path. *Res:* Comparative pathology; laboratory animal medicine. *Mailing Add:* 11807 Enid Dr Potomac MD 20854-3455

HOLMAN, RALPH THEODORE, b Minneapolis, Minn, Mar 4, 18; m 43; c 1. LIPIDS, POLYUNSATURATED ACIDS. *Educ:* Univ Minn, BS, 39, PhD(physiol chem), 44; Rutgers Univ, MS, 41. *Prof Exp:* Instr physiol chem, Med Sch, Univ Minn, 44-46; Nat Res Coun fel, Med Nobel Inst, Stockholm, 46-47; Am Scand Found fel, Inst Biochem, Univ Uppsala, 47; assoc prof biochem & nutrit, Tex A&M Univ, 48-51; assoc prof physiol chem, 51-56, prof biochem, Med Sch, 58-59, exec dir, 75-85, PROF BIOCHEM, HORMEL INST, UNIV MINN, 56-; PROF BIOCHEM, MAYO MED SCH, 77- *Concurrent Pos:* Consult nutrit study sect, NIH, 59-62; NIH spec fel, Univ Gothenburg, 62; organizer & pres, First Int Cong Essential Fatty Acids and Prostaglandins, 80; mem, Hormel Found, 79-86. *Honors & Awards:* Borden Award, Am Inst Nutrit, 66; Bailey Award, Am Oil Chem

Soc, 72 & Lipid Chem Award, 78; Fachini Prize, Italian Oil Chem Soc, 74. *Mem:* Nat Acad Sci; Am Chem Soc; Am Inst Nutrit; Am Oil Chem Soc (vpres, 73-74, pres, 74-75); Am Soc Biol Chemists. *Res:* Essential fatty acids; polyunsaturated fatty acids in disease; metabolism of unsaturated fatty acids; metabolism of hydrogenated fats, quantitative requirements for essential fatty acids; mass spectrometry of lipids; lipoxidase; quantitative chemical taxonomy. *Mailing Add:* Hormel Inst Univ Minn Austin MN 55912

HOLMAN, RICHARD BRUCE, b Washington, DC, Oct 26, 43; m 87; c 2. BEHAVIORAL NEUROCHEMISTRY. *Educ:* Colgate Univ, BA, 65; Purdue Univ, MSc, 67, PhD(psychol), 70. *Prof Exp:* Postdoctoral fel pharmacol, Inst Animal Physiol, Babraham, Cambridge, UK, 70-72; sr res assoc psychiat, Stanford Med Ctr, Stanford Univ, 72-79; lectr & reader, Physiol & Biochem, Univ Reading, UK, 79-88; PRIN SCIENTIST PHARMACOL, RECKITT & COLMAN PSYCHOPHARMACOL UNIT, DEPT PHARMACOL, UNIV BRISTOL, UK, 88-, SR RES FEL, 88- *Mem:* Am Soc Pharmacol & Exp Therapeut; Soc Neurosci; Int Brain Res Orgn; Int Soc Biomed Res Alcoholism. *Res:* Neurochemical basis of central nervous system disorders; monitoring changes in cerebral neurochemistry, in vivo microdialysis; behavioral modeling to investigate neurochemical regulation of behavior; HPLC techniques for monitoring changes in endogenous neurotransmitters. *Mailing Add:* Reckitt & Colman Psychopharmacol Unit Med Sch University Walk Bristol BS8 1RD England

HOLMAN, WAYNE J(AMES), JR, b Huntingdon, Tenn, Nov 12, 07; m 34; c 1. ELECTRICAL ENGINEERING. *Educ:* Ga Inst Technol, BS, 28; Yale Univ, MS, 30; Mass Inst Technol, MS, 39: NY Univ, PhD, 49. *Prof Exp:* Elec eng, Chicopee Mfg Corp Div, Johnson & Johnson, 40-46, vpres lumite div, 46-48, vpres & gen mgr, Chicopee Mills, Inc Div, 48-50, pres, 50-58, corp, 54-58, chmn br, mills & corp, 58-63, treas, Johnson & Johnson 63-67, dir, 51-73, chmn bd, Devro, Inc, 67-71, chmn bd, Jelco Labs, 67-73; RETIRED. *Concurrent Pos:* Mem adv coun & vis comt, Sch Indust Mgt, Mass Inst Technol, 54-62 & 80-; adv coun, Sch Eng, Yale Univ, 56-58; mem exec comt, Johnson & Johnson, 58-73. *Mem:* Inst Elec & Electronics Engrs; Sigma Xi. *Mailing Add:* One Johnson/Johnson Pl II St 200 New Brunswick NJ 08933

HOLMBERG, CHARLES ARTHUR, b Sayre, Okla, Nov 21, 44; m 64; c 3. VETERINARY PATHOLOGY. *Educ:* Okla State Univ, BS, 67, DVM, 71; Univ Calif, PhD(comp path), 75. *Prof Exp:* Assoc prof path, Tex A&M Univ, 71-; AT DEPT VET MED, UNIV CALIF, DAVIS. *Concurrent Pos:* Adj prof path, Univ Calif, Davis, 75-79. *Mem:* Am Col Vet Pathologists. *Mailing Add:* 23385 Rd 126 Tulare CA 93274

HOLME, THOMAS T(IMINGS), b Frankford, Pa, Mar 12, 13; m 36; c 3. INDUSTRIAL ENGINEERING. *Educ:* Lehigh Univ, BS, 35, MS, 40, IE, 48; Yale Univ, MA, 50. *Hon Degrees:* DEng, Lehigh Univ, 70. *Prof Exp:* Indust engr, E I du Pont de Nemours & Co, Del & Conn, 35-37; asst prof mech eng, Lehigh Univ, 37-41, assoc prof indust eng, 46-49, prof & head dept & dir curriculum, 49-50; chmn dept indust admin, Yale Univ, 54-63, prof, 50-73; exec dir, 53-81, EMER DIR, SIGMA XI, 81-; EMER PROF INDUST ENG, YALE UNIV, 73- *Concurrent Pos:* Fel, Trumbull Col, Yale Univ, dir Yale Coop Asn & Henry G Thompson Co; spec consult, Springfield Ord Dist, Ord Corps, US Army, 52-53, 56-57; consult, Hughes Aircraft Co, 58, 61 & 62, Southern New Eng Tel Co, 61, Hamilton Standard-United Aircraft, 63 & United Illum Co, 64; AAAS rep, exec comt, Sigma Xi. *Mem:* Fel AAAS; Am Soc Eng Educ; Sigma Xi; Am Inst Indust Engrs. *Res:* Engineering economy; industrial relations; education; operations research. *Mailing Add:* 773 Marlin Dr Fripp Island SC 29920

HOLMEN, REYNOLD EMANUEL, b Essex, Iowa, Oct 23, 16; m 42; c 3. POLYMER CHEMISTRY, ORGANIC CHEMISTRY. *Educ:* Augustana Col, Ill, AB, 36; Univ Mich, MS, 37, PhD(org chem), 49. *Prof Exp:* Res chemist, E I du Pont de Nemours & Co, Mich, 37-43 & Philadelphia, 43-46; res chemist, Cent Res Dept, Minn Mining & Mfg Co, 48-54, group supvr, 54-55, head tech info sect, 55-57, head inorg sect, 57-62, head org scouting group, 59-62, mgr gen res, Reflective Prod Div, 62-70, mgr gen res & vacuum coating, Spec Enterprises Dept, 70-73, mgr res & develop, Spec Enterprises Dept, 73-81, mgr, Life Sci Sector Develop Lab, 3M Co, 81-82, VPRES, R & D KEMSERCH, 85- *Concurrent Pos:* Consult, 82- *Mem:* AAAS; emer mem Am Chem Soc. *Res:* Synthetic resins; chemical specialties; elastomers; catalysis; vinyl monomers; boron compounds; coordination compounds; chemical binding; vacuum coating; electroluminescence; glass microbubbles; retroflection. *Mailing Add:* 2225 Lilac Lane White Bear Lake MN 55110

HOLMER, DONALD A, b Lead, SDak, Mar 29, 34; m 63; c 2. ORGANIC CHEMISTRY. *Educ:* SDak Sch Mines & Technol, BS, 56; Okla State Univ, PhD(org chem), 61. *Prof Exp:* Res chemist, Ethyl Corp, 61-68; res specialist, 68-81, SR SPECIALIST, MONSANTO TEXTILES CO, 81- *Mem:* Am Chem Soc. *Res:* Synthesis and characterization of high polymers; textile fibers. *Mailing Add:* 3705 Pompano Dr Pensacola FL 32514

HOLMER, RALPH CARROL, b Chicago, Ill, Nov 24, 16; m 39; c 3. EXPLORATION GEOPHYSICS. *Educ:* Colo Sch Mines, Geol Engr, 38, DSc(geophys), 54. *Prof Exp:* Geophysicist petrol explor, Mott-Smith Corp, 39-42; physicist, US Naval Bur Ord, 42-43; geophysicist petrol explor, Calif Co, 43-46; instr geophys, Colo Sch Mines, 46-50; geophysicist petrol explor, Exp Surv, Inc, 50-51; instr geophys, Colo Sch Mines, 51-52; chief geophysicist mineral explor, Kennecott Copper Corp, Colo, 52-64, dir explor serv, Utah, 64-72; PROF GEOPHYS, COLO SCH MINES, 72- *Mem:* Soc Explor Geophys; Am Inst Mining, Metall & Petrol Engr. *Res:* Application of geophysical and geochemical methods to mineral exploration and to regional and local structural geological problems; geologic research for application in mineral exploration; application of computers to exploration problems. *Mailing Add:* 1943 Foothill Rd Golden CO 80401

HOLMES, ALBERT WILLIAM, JR, b Chicago, Ill, Feb 3, 32; m 54; c 4. MEDICINE. *Educ:* Knox Col, Ill, BA, 52; Western Reserve Univ, MD, 56. *Prof Exp:* Intern, Presby Hosp, Chicago, 56-57, resident med, 57-59; resident, Rush-Presby-St Luke's Med Ctr, 61-62, dir hepatology, 66-75, assoc chmn dept med, 72-75; prof med & microbiol, Rush Med Col, 71-75, actg dean, Rush Grad Col, 73-74; prof med, Sch Med, Tex Tech Univ, 75-78, chmn dept, 75-83; PROF MED, UNIV ILL, COL MED AT PEORIA 85-, CHMN DEPT, 75-83. *Concurrent Pos:* Nat Inst Allergy & Infectious Dis spec fel med & microbiol, Presby-St Luke's Hosp, 62-66; prof, Univ Ill Col Med, 70-71. *Mem:* Am Asn Immunol; Am Soc Microbiol; Am Asn Study Liver Dis; Am Gastroenterol Asn. *Res:* Liver disease, especially viral hepatitis; virus inhibitors, interferon. *Mailing Add:* Dept Med Univ Ill Col Med Box 1649 Peoria IL 61656

HOLMES, ALVIN W(ILLIAM), metallurgy, electrical engineering; deceased, see previous edition for last biography

HOLMES, CALVIN VIRGIL, b Newhebron, Miss, Oct 21, 24; m 56; c 2. MATHEMATICS. *Educ:* Univ Miss, BA, 47, MA, 48; Univ Ill, MS, 52; Univ Kans, PhD, 55. *Prof Exp:* From instr to asst prof math, Murray State Col, 48-51; mathematician, Northrop Aircraft, Inc, 55-56; from asst prof to assoc prof, 56-64, PROF MATH, SAN DIEGO STATE UNIV, 64- *Mem:* Am Math Soc; Math Asn Am. *Res:* Group theory. *Mailing Add:* Dept Math San Diego State Univ San Diego CA 92182

HOLMES, CHARLES ROBERT, b Kemmerer, Wyo, Oct 11, 18; m 46; c 3. GEOPHYSICS. *Educ:* St Louis Univ, BS, 48, MS, 50; Pa State Univ, PhD, 58. *Prof Exp:* Geophysicist & instr physics & geophys, NMex Inst Mining & Technol, 49-52, res geophysicist, 54-56; asst, Pa State Univ, 52-54; res geophysicist, Atomic Energy Comn, 56-57& Birdwell, Inc, Pa, 57-59; from assoc prof to prof geophys, NMex Inst Mining & Technol, 59-89, sr geophysicist, 59-89; RETIRED. *Mem:* Fel AAAS; Am Geophys Union; Sigma Xi. *Res:* Atmospheric physics; thunderstorm acoustics; thunderstorm electrification; solid earth geophysics; electrical properties; tritium tracing and isotope exchange. *Mailing Add:* 1006 Lopezville Rd Socorro NM 87801

HOLMES, CLAYTON ERNEST, b Sechlerville, Wis, July 20, 04; m 31; c 2. POULTRY SCIENCE. *Educ:* Univ Wis, BS, 27, MS, 31, PhD(poultry genetics), 38. *Prof Exp:* Asst poultry husb, Pa State Col, 27-28; from instr to asst prof, Univ Wis, 28-41; assoc prof, Ore State Col, 41-46; mgr owner poultry breeding farm, 46-49; from assoc prof to prof, 49-70, EMER PROF POULTRY HUSB, VA POLYTECH INST & STATE UNIV, 70- *Mem:* Am Genetic Asn; Poultry Sci Asn. *Res:* Nutrition of poultry; control of intestinal parasites of poultry; genetics of poultry. *Mailing Add:* 2409 Bishop Rd NE Blacksburg VA 24060

HOLMES, CLIFFORD NEWTON, b Escanaba, Mich, Jan 1, 22; m 45; c 7. GEOLOGY. *Educ:* Univ Mich, BS, 43; Yale Univ, MS, 47; Univ Utah, PhD, 60. *Prof Exp:* Jr geologist, US Geol Surv, Kans & Calif, 43-45, asst geologist, Utah & La, 45-46, assoc geologist, Colo Plateau, 48-50; sr geologist, Phillips Petrol Co, 52, asst dir explor sect, 53, dir strategic minerals, 54-58, asst mgr mining & milling dept, 58-63, mgr minerals div, 63-71, VPRES & DIR, WESTERN HEMISPHERE & PHOSPHATE MINES, INC, PHILLIPS PETROL CO, 71-, ENERGY ADV, 72-, MGR MINERALS, PHILLIPS CHEM CO, 75- *Concurrent Pos:* Mem, Gov Energy Adv Coun, 75- *Mem:* Fel Geol Soc Am; Am Asn Petrol Geologists; Am Inst Mining, Metall & Petrol Eng; Am Petrol Inst; Soc Econ Geologists. *Res:* Mineral deposits. *Mailing Add:* 1431 Valley Rd Bartlesville OK 74003

HOLMES, CURTIS FRANK, b Baton Rouge, La, Feb 14, 43; m 70; c 1. ELECTROCHEMISTRY. *Educ:* La State Univ, BS, 65; Ind Univ, PhD(chem physics), 69. *Prof Exp:* Fel & instr chem, La State Univ, 71-72; prin chemist, Calspan Advan Technol, 73-76; VPRES TECHNOL, WILSON GREATBATCH LTD, 76- *Concurrent Pos:* Adj instr, State Univ NY, Buffalo, 83- *Mem:* Electrochem Soc; Am Chem Soc; Asn Advan Med Instrumentation; Nat Elec Mfrs Asn. *Res:* High energy density, high reliability lithium batteries for implantable devices and other speciality applications. *Mailing Add:* Wilson Greatbatch Ltd 10000 Wehrle Dr Clarence NY 14031

HOLMES, D BRAINERD, b New York, NY, May 24, 21; m. ELECTRICAL ENGINEERING. *Educ:* Cornell Univ, BSEE, 43. *Prof Exp:* Design engr, Western Elec Co, 45-53; mem staff, RCA Corp, 53-61, gen mgr major defense systs div, 61; dir manned space flight, NASA, 61-63; sr vpres, Raytheon Co, 63-69, exec vpres, 69-75, dir, 63-86, pres, 75-86; PRES, HOLMES ASSOC INC, 86- *Concurrent Pos:* Mem tech staff, Bell Tel Labs, 45-53; proj mgr Navy Talos land based missile syst develop, 54-57, Air Force Atlas launch control & checkout equip develop, 57 & Air Force ballistic missile early warning syst, 58-61. *Honors & Awards:* Paul T Johns Award, Arnold Air Soc. *Mem:* Nat Acad Eng; fel Inst Elec & Electronics Eng; fel Am Inst Aeronaut & Astronaut; Aerospace Indust Asn; Preparedness Asn. *Mailing Add:* 11301 Turtle Beach Rd North Palm Beach FL 33408

HOLMES, DALE ARTHUR, b Biwabik, Minn, Dec 31, 37; m 62; c 2. ELECTRICAL ENGINEERING, OPTICS. *Educ:* Purdue Univ, BS, 60; Carnegie Inst Technol, MS, 61, PhD(elec eng), 65; Univ Rochester, MS, 69. *Prof Exp:* Asst prof elec eng, Carnegie Inst Technol, 65-66; chief optical integration group, Laser Div, Air Force Weapons Lab, Kirtland AFB, NMex, 66-74; CHIEF SCIENTIST OPTICS, LASER PROGS, ROCKETDYNE DIV, ROCKWELL INT CORP, 74- *Honors & Awards:* Air Force Res & Develop Award, 71. *Mem:* Optical Soc Am; Inst Elec & Electronics Engrs. *Res:* Optical systems; high energy lasers. *Mailing Add:* 32850 Canyon Quail Trail Agua Dulce CA 91350

HOLMES, DAVID G, b Lethbridge, Can, Mar 29, 43; m 67; c 5. FOOD SCIENCE, BIOCHEMISTRY. *Educ:* Univ Alta, BS, 68; Brigham Young Univ, MS, 70; Utah State Univ, PhD(food sci), 73. *Prof Exp:* Teaching asst physiol, Brigham Young Univ, 68-70; biochemist & proj leader, Beatrice

Foods Res, Chicago, 73-75; group leader res & develop, Foremost Foods, 75-77, mgr prod & process develop, Foremost-Mckesson Res, 77-82, dir res & develop, 82-85; PRES, NAT FOOD LAB INC, 85- *Mem:* Am Dairy Sci Asn; Inst Food Technologists. *Res:* Sell contract research in sensory evaluation, analytical testing, new product and process development for the food industry. *Mailing Add:* 6363 Clark Ave Dublin CA 94568

HOLMES, DAVID KELLEY, solid state physics, for more information see previous edition

HOLMES, DAVID WILLIS, b Fremont, Ohio, Nov 24, 14; m 40; c 3. ORGANIC CHEMISTRY. *Educ:* Amherst Col, AB, 37; Univ Mich, MS, 38, PhD(org chem), 40. *Prof Exp:* Rackham fel, Univ Mich, 40-41; res chemist, E I du Pont de Nemours & Co, Inc, 41-42, supvr, Chambers Works, 42-49, tech supt, Louisville Works, 49-50, asst works mgr, 51-52, prod mgr, Org Chem Dept, 52-56, dir mfg, Elastomer Chem Dept, 57-66, dir sales, 66-77; RETIRED. *Concurrent Pos:* consult, 80- *Mem:* Am Chem Soc. *Res:* Synthetic hormone synthesis; synthetic rubber polymers and latices; synthesis of compounds related to female sex hormones. *Mailing Add:* 4619 Weldin Rd Wilmington DE 19803

HOLMES, DONALD EUGENE, b Wellington, Tex, Feb 17, 34; m 55; c 2. MEDICAL PHYSICS, MAGNETIC RESONANCE. *Educ:* Univ Okla, BS, 58; Calif State Col San Diego, MS, 60; Univ Calif, Los Angeles, PhD(biophys & nuclear med), 66. *Prof Exp:* Radiochemist, Gen Atomics, 59 & Northrop Aircraft-Norair Nuclear Sci, 60-61; chemist, Lab Nuclear Med, Univ Calif, Los Angeles, 61-66; Nat Acad Sci-James Picker Found fel, Lab Radiation Chem, Univ Newcastle-upon-Tyne, Eng, 66-67; Nat Acad Sci-James Picker Found fel, Dept Biophys, Univ Hawaii, 67-68, res biophysicist, 68-69; res biophysicist, Lawrence Berkeley Lab, Univ Calif, 69-71; PROF PHYSICS, CALIF STATE UNIV, FRESNO, 76- *Concurrent Pos:* Assoc prof physics, Calif State Univ, Fresno, 71-76; sabbatical leave med physics, Univ Calif, Los Angeles, 77-78. *Mem:* Health Physics Soc; Biophys Soc; Am Asn Ultrasound Med; Radiation Res Soc; Am Asn Physicists Med. *Res:* Biophysics of membranes; radiation biophysics of nucleic acids; proteins and nucleoproteins; radiation chemistry of aqueous systems; radiation physics of solids; magnetic resonance spectroscopy; physical chemistry of biochemical systems; radioisotope technology; medical physics. *Mailing Add:* Dept Chem Calif State Univ Fresno CA 93740

HOLMES, DOUGLAS BURNHAM, b Indianapolis, Ind, June16, 38; m 67; c 2. RHEOLOGY, HEAT TRANSFER. *Educ:* Rice Univ, BA, 60, MS, 62; T H Delft, Neth, DEng(appl physics), 67. *Prof Exp:* Res asst & lectr fluid mech, Tech Physics Lab, T H Delft, Neth, 63-68; res & develop engr, Res Labs, Polaroid Corp, 68-72; mgr camera opers, Polaroid, Europa, BV, 72-74; sr prin engr, Appl Tech Div, Polaroid Corp, 74-77, sr tech mgr, Battery Div, 77-84, sr mgr prod planning, 84-88; chief oper officer, Ecol Eng Assocs, 89; PRIN/CONSULT, MINERGY ASSOCS, 90- *Mem:* Am Solar Energy Soc; Brit Soc Rheology; Water Pollution Control Fedn; Koninkl Inst Eng; Sigma Xi. *Res:* Laminar, non-Newtonian fluid flow; heat transfer to such flows; heat transfer and fluid flow in solar heated dwellings. *Mailing Add:* Minergy House Four John Wilson Lane Lexington MA 02173-6033

HOLMES, EDWARD BRUCE, b Iola, Kans, Sept 29, 27; m 47. COMPARATIVE ANATOMY. *Educ:* Kans State Col Pittsburg, BS, 48; Univ Ill, MS, 50; Univ Kans, PhD(zool), 62. *Prof Exp:* Instr biol, Ark Polytech Col, 50-53 & Tex Col Arts & Indust, 53-56; asst prof, Westminster Col, 60-62 & St Olaf Col, 62-65; assoc prof, 65-72, PROF BIOL, WESTERN ILL UNIV, 72- *Mem:* Am Soc Zool. *Res:* Comparative vertebrate anatomy; avian myology; reptilian neuroanatomy; vertebrate heart. *Mailing Add:* Dept Biol Sci Western Ill Univ Adams St Macomb IL 61455

HOLMES, EDWARD LAWSON, food chemistry; deceased, see previous edition for last biography

HOLMES, EDWARD WARREN, b Winona, Miss, Jan 25, 41; m 67. BIOCHEMICAL GENETICS. *Educ:* Washington & Lee Univ, BS, 63; Univ Pa, MD, 67. *Prof Exp:* Intern med, Hosp Univ Pa, 67-68; res assoc nephrology, USPHS, 68-70; resident med, 70-74, fel metab & genetic dis, 71-73, chief resident med, 73-74, asst prof med, 74-77, assoc prof, 77-81, PROF MED, DUKE UNIV MED CTR, 81- ASST PROF BIOCHEM, 75- *Mem:* Am Soc Human Genetics; Am Fedn Clin Res; Am Soc Nephrology; Am Rheumatism Asn; AAAS. *Res:* Studies on the control of purine biosynthesis de novo; regulation of human PP-ribose-P amidotransferase and prime metabolism in muscle function. *Mailing Add:* Box 3034 Duke Univ Med Sch Durham NC 27710

HOLMES, FRANCIS W(ILLIAM), b Yonkers, NY, May 21, 29; m 53; c 3. PLANT PATHOLOGY, SHADE TREES. *Educ:* Oberlin Col, AB, 50; Cornell Univ, PhD(plant path), 54. *Prof Exp:* Asst plant path, Cornell Univ, 50-54; from asst prof to assoc prof, 54-70, dir, Shade Tree Labs, 73-88, Mass urban forestry coordr, 79-85, PROF, SHADE TREE LABS, UNIV MASS, AMHERST, 70- *Concurrent Pos:* NSF sr fel, Neth, 62-63 & Fulbright scholar, 62-63 & 70-71; Am Phil Soc grant, Neth, 84-85; sr res fel, Neth Agr Univ, 84-85; fel Int Agr Ctr, Wageningen, Neth, 84-85. *Honors & Awards:* Nat Arbor Day Fedn Award, 79. *Mem:* Am Phytopath Soc; corresp mem Royal Dutch Bot Soc; Can Phytopath Soc; Int Soc Arboriculture; Neth Soc Plant Path; Int Soc Plant Path. *Res:* Dutch elm disease; urban street- and shade-tree disease diagnosis; inheritance and mechanisms of pathogenicity and disease resistance; salt injuries to trees; phytopathological translations; shade tree pathology; biological control of tree diseases; urban forestry; extension service. *Mailing Add:* Shade Tree Labs Univ Mass Amherst MA 01003

HOLMES, GEORGE EDWARD, b Chicago, Ill, May 8, 37; m 67. MOLECULAR BIOLOGY. *Educ:* Wiley Col, BS, 60; Chicago State Univ, MS, 67; Univ Ariz, Med Ctr, PhD(molecular biol, biochem), 73. *Prof Exp:* Med technologist area hosps, Chicago, 61-67; teaching asst molecular genetics, Univ Calif, Davis, 67-68; res assoc hemoglobinopathies, Rockefeller Univ, 73-74; ASSOC PROF MICROBIOL, COL MED, HOWARD UNIV, 74- *Concurrent Pos:* Teacher high sch, Chicago Bd Educ, 64-67. *Mem:* Soc Protozoologists; Am Soc Microbiol; Am Soc Clin Pathologists; Am Soc Biol Chemists; Fedn Am Scientists; Am Soc Virologists. *Res:* Molecular basis of aging; apurinic/apyrimidinic lesions and DNA repair mechanisms; alterations of essential components of protein-synthesizing machinery by virus-specified proteins; intragenic complementation; protein-protein interaction. *Mailing Add:* Dept Microbiol Howard Univ Col Med Washington DC 20059

HOLMES, HELEN BEQUAERT, b Boston, Mass, Sept 6, 29; m 53; c 3. BIOETHICS, MEDICAL TECHNOLOGY ASSESSMENT. *Educ:* Oberlin Col, BA, 51; Cornell Univ, MS, 53; Univ Mass, PhD(zool), 70. *Prof Exp:* Tech asst parasitol, Col Vet Med, Cornell Univ, 53-54; teacher biol & chmn dept sci, Northampton Sch Girls, 65-67; teacher, Int Sch Beverweerd, Neth, 70-71; asst prof, Springfield Tech Community Col, 71-73; vis asst prof, Div Sci & Math, Eisenhower Col, 75-76; asst prof, Russell Sage Col, 76-78; vis lectr dept biol, Tufts Univ, 78-80; assoc fel, Inst Advan Studies Humanities, Univ Mass, Amherst, 88-90; scholar assoc, Womens Res Inst, 86-87, SCHOLAR ASSOC, OFF WOMENS RES, HARTFORD COL WOMEN, 90- *Concurrent Pos:* Teaching asst, Dept Zool, Univ Mass, 67-70; Adelia Field Johnston fel, Oberlin Col, 68-69; vis investr, Dept Biol, Amherst Col, 73-75; NSF & NEH grants, 78-80 & 82-83; reviewer, NSF-Ethics & Values Sci & Technol grant, 80 & 83; res assoc, Fedn Orgn Prof Women, 79-82; vis scholar, bioethics, Spelman Col, 82-83; vis scientist, Sci & Soc, Univ Groningen, Neth, 84-85; Fulbright scholar, Univ Waikato, NZ, 86; contractor, US Cong Technol Assessment, 87; guest ed, Hypatia Issue Feminist Ethics & Med, 89, ed, Issues Reproductive Technol I, 90-91; mem, Sci & Technol Task Force, Nat Women's Studies Asn. *Mem:* Sigma Xi; Inst Soc Ethics Life Sci; Am Philos Asn. *Res:* Population genetics of parasitic hymenoptera; technology assessment and ethical issues in reproductive medicine especially the new reproductive technlogies. *Mailing Add:* 24 Berkshire Terr Amherst MA 01002-1302

HOLMES, HOWARD FRANK, b Toledo, Ohio, May 21, 31; m 61; c 2. AQUEOUS CHEMISTRY. *Educ:* Tenn Tech Univ, BS, 50; Univ Tenn, PhD(chem), 58. *Prof Exp:* Tech asst, Oak Ridge Nat Lab, 55; asst, Univ Tenn, 55-58; CHEMIST, OAK RIDGE NAT LAB, 58- *Mem:* Sigma Xi; Am Chem Soc. *Res:* Surface chemistry; calorimetry; electrochemistry; electrokinetics; thermodynamics, solutions. *Mailing Add:* Chem Div Oak Ridge Nat Lab PO Box 2008 Oak Ridge TN 37831-6110

HOLMES, IVAN GREGORY, b Castana, Iowa, Mar 26, 35; m 56; c 2. ANALYTICAL CHEMISTRY, HISTORY OF SCIENCE. *Educ:* Ariz State Univ, BS, 57, MS, 61; Ore State Univ, PhD(chem), 68- *Prof Exp:* From instr to assoc prof chem, Andrews Univ, 60-72, assoc prof hist of sci, 70-72, dir freshman educ, 71-72; assoc acad dean, Col Arts & Sci, 74-77, dean col, 77-80, PROF CHEM, LOMA LINDA UNIV, 72- *Concurrent Pos:* consult, X-Ray & Nuclear Radiation Physics & Chem, Med & Gen Res Appln. *Mem:* AAAS; Am Chem Soc; Sigma Xi. *Res:* Analytical applications of metal chelates with emphasis on ultraviolet spectroscopy; x-ray fluorescence analysis of copper; unique instructional techniques using motion picture film; x-ray diffraction analysis of volcanic minerals. *Mailing Add:* Chem Dept Loma Linda Univ Riverside CA 92505

HOLMES, JAMES FREDERICK, b Billings, Mont, Sept 10, 37; m 59; c 3. APPLIED PHYSICS, ELECTRICAL ENGINEERING. *Educ:* Univ Wash, BSEE, 59, PhD(electromagnetic waves), 68; Univ Md, MSEE, 63. *Prof Exp:* Res engr, Boeing Co, 63-66; res assoc electromagnetic waves, Univ Wash, 66-68; asst prof elec eng, Ore State Univ, 69-74; assoc prof, 74-78, chmn dept physics & elec eng, 74-88, PROF APPL PHYSICS & ELEC ENG, ORE GRAD CTR, 78- *Mem:* Inst Elec & Electronics Engrs; fel Optical Soc Am; Am Soc Eng Educ. *Res:* Atmospheric optics; electrooptic and laser systems. *Mailing Add:* Dept Appl Physics & Elec Eng 19600 NW Von Neumann Dr Beaverton OR 97006-1999

HOLMES, JERRY DELL, b Mt Vernon, Tex, Nov 30, 35; m 57; c 3. ORGANIC CHEMISTRY. *Educ:* ETex State Col, BS, 56; Univ Tex, PhD(org chem), 64. *Prof Exp:* Jr chemist, Am Oil Co, Tex, 56-60; chemist, 63-68, sr chemist res & develop, 68-73, group leader org chem res, 73-74, div head, Res Div, 74-77, div dir, Develop Div, Tex Eastman Co, 77-80, asst to works mgr, 80-81, STAFF ASST, DEVELOP, EASTMAN CHEM DIV, TENN EASTMAN CO, 81- *Mem:* Am Chem Soc; Sigma Xi. *Res:* Product and process development and improvement in organic chemicals and polymers. *Mailing Add:* 202 Deerridge Ct Kingsport TN 37663-2934

HOLMES, JOHN CARL, b St Paul, Minn, Aug 30, 32; m 57; c 2. HELMINTH COMMUNITY ECOLOGY. *Educ:* Univ Minn, BA, 54; Rice Univ, MA, 57, PhD(biol), 59. *Prof Exp:* From asst prof to assoc prof, 59-71, PROF ZOOL, UNIV ALTA, 71- *Honors & Awards:* H B Ward Medal, Am Soc Parasitol, 72; R A Wardle Lectr, Can Soc Zoologists, 86. *Mem:* Ecol Soc Am; Am Soc Parasitologists (vpres, 80, pres, 91); Arctic Inst NAm; Wildlife Dis Asn; Can Soc Zool; Am Soc Naturalists. *Res:* Ecology of parasitic helminths, especially factors associated with the development and maintenance of communities of helminths. *Mailing Add:* Dept Zool Univ Alta Edmonton AB T6G 2E9 Can

HOLMES, JOHN LEONARD, b London, Eng, Nov 29, 31; m 58; c 2. PHYSICAL CHEMISTRY. *Educ:* Univ London, BSc, 54, PhD(chem), 57; DSc, 83. *Prof Exp:* Nat Res Coun Can fel, Ottawa, 58-60; Imp Chem Indust fel, Univ Edinburgh, 60-61, lectr chem, 61-62; from asst prof to assoc prof, 62-73, PROF CHEM, UNIV OTTAWA, 73- *Concurrent Pos:* NAm Ed, Org Mass Spectrometry, 77-; Overbeek vis prof, Univ Utrecht, 79; distinguished vis scientist, Univ Adelaide, 84; vis res fel, Australian Nat Univ, Canberra, 84. *Honors & Awards:* Barringer Res Award, Can Spectros Soc, 80; Medal, Chem Inst Can, 89. *Mem:* Fel Can Inst Chemists; Royal Soc Chem; Am Soc Mass Spectrometry; fel Royal Soc Can. *Res:* Kinetics of gas phase reactions; mechanisms and energetics of positive ion fragmentation; ion structures in the gas phase; ion beam chemistry and physics. *Mailing Add:* Dept Chem Univ Ottawa Ottawa ON K1N 6N5 Can

HOLMES, JOHN RICHARD, b Chula Vista, Calif, Sept 24, 17; m 51; c 3. ATOMIC SPECTROSCOPY. *Educ:* Univ Calif, AB, 38, AM, 41, PhD(physics), 42. *Prof Exp:* Asst physics, Univ Calif, 38-41, res physicist, Manhattan Proj, Radiation Lab, 42-45; from asst prof to prof & head dept, 45-63; chmn dept physics & astron, 63-72, PROF PHYSICS, UNIV HAWAII, HONOLULU, 63- *Concurrent Pos:* Fulbright lectr, Univ Madrid, 62-63; UNESCO Consult, Univ Buenos Aires, Arg, 70. *Mem:* AAAS; fel Am Phys Soc; fel Optical Soc Am; Sigma Xi. *Res:* Spectroscopy; optics of thin films; atomic transition probabilities. *Mailing Add:* Dept Physics & Astron Univ Hawaii Honolulu HI 96822

HOLMES, JOHN THOMAS, b Oak Park, Ill, Aug 10, 36; m 63; c 2. SOLAR ENERGY COMPONENT DEVELOPMENT. *Educ:* Univ Wis-Madison, BS, 58: Univ Calif, Berkeley, MS, 60. *Prof Exp:* Chem engr, Argonne Nat Lab, 60-76; mem tech staff, 76-85, DIV SUPV, SANDIA NAT LABS, ALBUQUERQUE, NMEX, 85- *Res:* Solar power-tower, component development and testing, systems analysis and system operation; nuclear reactor fuel cycle process development; coolant purity. *Mailing Add:* Sandia Nat Labs Orgn 7600 PO Box 5800 Albuquerque NM 87185

HOLMES, JOSEPH CHARLES, b Staunton, Va, Mar 28, 25; m 47; c 3. CHEMISTRY. *Educ:* Va Polytech Inst, BS, 48, MS, 49. *Prof Exp:* Supvr spec invests sect, Res & Develop Dept, Philip Morris, Inc, 49-60, sr supvr, 60-61, mgr chem develop, 61-64, dir res & develop, Am Safety Razor Co Div, Va, 64-66; sr res chemist, 66-69, supvr res, Spunbonded Prod Div, Textile Fibers Lab, Del, 69-70 & Old Hickory Res & Develop Lab, Tenn, 70-77; sr res chemist, E I du Pont de Nemours & Co, Inc, Richmond, Va, 77-85; RETIRED. *Mem:* Am Chem Soc; fel Am Inst Chemists. *Res:* Analytical and organic chemistry; spectroscopy; mass spectrometry; metallurgy; experimental design; research administration; polymer chemistry. *Mailing Add:* 2209 Lancashire Dr Richmond VA 23235

HOLMES, KATHRYN VOELKER, b Philadelphia, Pa, Sept 14, 40; m 62; c 2. VIROLOGY, CELL BIOLOGY. *Educ:* Radcliffe Col, AB, 62; Rockefeller Univ, PhD(virol), 68. *Prof Exp:* USPHS fel, Harvard Univ, 68-70; asst prof microbiol, Schs Med & Dent, Georgetown Univ, 70-72; asst prof microbiol, Southwestern Med Sch, Univ Tex Health Sci Ctr, Dallas, 72-76; assoc prof, 76-83, PROF PATH, UNIFORMED SERV UNIV HEALTH SCI, 83- *Concurrent Pos:* Consult, Molecular Anat Prog, NIH-Oak Ridge Nat Lab-Union Carbide Inc, 70-72. *Mem:* AAAS; Am Soc Microbiol; Sigma Xi; Electron Micros Soc Am; Am Soc Virol. *Res:* Virus-cell interactions, especially of moderate and cell fusing viruses; ultrastructure of virions and virus infected cells; pathogenesis of virus diseases; virus receptors. *Mailing Add:* Dept Path Uniformed Serv Univ Health Sci 4301 Jones Bridge Rd Bethesda MD 20814-4799

HOLMES, KENNETH ROBERT, b Lansing, Mich, Dec 14, 37; m 63; c 4. TISSUE BLOOD FLOW, BIOHEAT TRANSFER. *Educ:* Mich State Univ, BS, 59, MS, 66, PhD(physiol), 72. *Prof Exp:* Assoc engr, Martin-Marietta Corp, Baltimore, 59-62; elec engr, Mich Dept State Hwys, 62-66; instr anat, Mich State Univ, 71-72; asst prof physiol, Southern Ill Univ, Edwardsville, 72-75; asst prof, 75-83, ASSOC PROF ANAT, PHYSIOL & BIOENG & ASST HEAD DEPT, UNIV ILL, URBANA-CHAMPAIGN, 83-; PRES, PERFTEC INC, URBANA, ILL, 86- *Mem:* Am Physiol Soc; Sigma Xi; Am Asn Vet Anatomists. *Res:* Development of a new thermal method of measuring local tissue blood flow; effects of toxins on tissue blood flow; biophysical properties of tissues; bioheat transfer. *Mailing Add:* Dept Vet Biosci Uni Ill 2001 S Lincoln Urbana IL 61801

HOLMES, KING KENNARD, b St Paul, Minn, Sept 1, 37; c 3. MEDICINE. *Educ:* Harvard Col, AB, 59; Cornell Univ, MD, 63; Univ Hawaii, PhD(microbiol), 67; Am Bd Internal Med, dipl, 71, dipl infectious dis, 74; FRCP(E), 90. *Prof Exp:* Resident, Univ Wash, 67-68, chief resident, 68-69, from instr to assoc prof med, 69-78, vchmn, Dept Med, 84-89, PROF MED & ADJ PROF MICROBIOL & IMMUNOL, UNIV WASH, 78-, ADJ PROF EPIDEMIOL, 80-, DIR, CTR AIDS & SEXUALLY TRANSMITTED DIS, 89- *Concurrent Pos:* Epidemiologist, Div Prev Med, USN, Pearl Harbor, 65-67; head, Div Pulmonary Dis, USPHS Hosp, Seattle, 69-70, asst chief, Dept Med, 69-83, head, Div Infectious Dis, 70-83; dir, Sexually Transmitted Dis Clin, Harborview Med Ctr, 72-79, chief med, 84-89; mem, numerous adv comts, Nat Inst Allergy & Infectious Dis, NIH, USPHS, WHO & Nat Acad Sci, 74-; prin investr, NIH, Nat Cancer Inst, Nat Inst Allergy & Infectious Dis, Nat Inst Child Health & Human Develop, Centers Dis Control, 83- *Honors & Awards:* Squibb Award, Infectious Dis Soc Am, 78; Thomas Parran Award, Am Venereal Dis Asn, 83. *Mem:* Inst Med-Nat Acad Sci; Asn Am Physicians; fel Am Col Physicians; Am Epidemiol Soc; Am Fedn Clin Res; AMA. *Res:* Sexually transmitted diseases and acquired immune deficiency syndrome; etiology and natural history of cervical neoplasia; surveillance of gonorrhea. *Mailing Add:* Harborview Med Ctr Univ Wash 325 Ninth Seattle WA 98104

HOLMES, L(AWRENCE) B(RUCE), b Troy, NY, June 14, 32; m 55; c 4. ASTRONAUTICAL & MECHANICAL ENGINEERING. *Educ:* Rochester Inst Technol, BS, 60, MS, 64, PhD(mech & aerospace sci), 65. *Prof Exp:* Asst prof mech eng & astronaut sci, Northwestern Univ, 65-69; mem tech adv panel, Res Div, Gen Am Transp Corp, 69-70, MGR, THEORET & APPL RES DEPT, RES DIV, GEN AM TRANSP CORP, 70- *Res:* Plasma density ahead of pressure driven shock waves. *Mailing Add:* 1642 S Homan Chicago IL 60623

HOLMES, LARRY A, b Tulsa, Okla, Feb 7, 41; m 67; c 2. POLYMER CHEMISTRY & ENGINEERING, PLASTICS MARKETING. *Educ:* Univ Tex, BSChE, 63; Univ Wis, PhD(phys chem), 68. *Prof Exp:* Res & mkt specialist, Plastics Div, Exxon Chem Am, 67-76; vpres, Southwest Chem Serv, Suby M A Hanna, Inc, 76-85; pres, Polymix Consulting, 85-87; vpres, Chemtrusion, Inc, 87-90; PRES, ADVANTAGE POLYMER SYSTS, 90- *Mem:* Am Inst Chem Eng; Soc Plastics Eng. *Res:* Viscoelastic properties; rheology; free radical polymerization kinetics; polymer processing; polyolefin compounding; plastics marketing. *Mailing Add:* 3722 Nottingham Houston TX 77005

HOLMES, MARK LAWRENCE, b Tulsa, Okla, Sept 18, 38; m 74; c 2. GEOLOGY. *Educ:* Princeton Univ, BSE, 60; Univ Wash, MS, 67, PhD(oceanog), 75. *Prof Exp:* Sr oceanographer, Univ Wash, 67-75; GEOLOGIST, US GEOL SURV, 75- *Concurrent Pos:* Res assoc, Univ Wash, 75-81, affil asst prof, 81-82, affil assoc prof, 82- *Mem:* Geol Soc Am; Am Geophys Union; Soc Explor Geophysicist. *Res:* Structure and evolution of continental shelves on the west coast of North America, with particular emphasis on evaluating hydrocarbon potential and identifying geologic hazards to economic development. *Mailing Add:* US Geol Surv Univ Wash Sch Oceanog WB-10 Seattle WA 98195

HOLMES, NEAL JAY, b Mercer, Mo, Aug 2, 31; m 53; c 4. SCIENCE EDUCATION. *Educ:* Northeast Mo State Col, BS, 57; Washington Univ, MA, 62; Okla State Univ, EdD(sci educ), 67. *Prof Exp:* Pub sch teacher, 57-61; sci consult, Parkway Schs, 62-67; prof sci educ, 67-72, from assoc prof to prof chem, 72-87, head Dept Chem, 76-81, PROF SCI EDUC, CENT MO STATE UNIV, 87- *Concurrent Pos:* Consult, McGraw-Hill Publ Co, Mo, 63-66; panelist, Coop Col Sch Sci Proj Eval Comt, NSF, 68; mem, Teachers Indust & Environ Comt, 85- *Mem:* Nat Sci Teachers Asn. *Res:* Observational systems; systems to reliably determine the behavior of students, elementary and secondary, in science learning environments. *Mailing Add:* Dept Chem Cent Mo State Univ Warrensburg MO 64093

HOLMES, OWEN GORDON, b Swift Current, Sask, Apr 1, 29; m 55; c 3. CHEMISTRY. *Educ:* Univ Sask, BA, 48, MA, 50; Univ Calif, PhD(chem), 55. *Prof Exp:* Asst chem, Univ Calif, 51-52, res assoc, 54; asst prof, Regina Col, 55-61; assoc prof, Univ Sask, 61-65; assoc prof, 65-66, dean arts & sci, 66-71, PROF CHEM, UNIV LETHBRIDGE, 66-, ACAD VPRES, 72- *Mem:* Fel Chem Inst Can. *Res:* Inorganic chemistry; spectroscopy of material in condensed states; fused salt solutions. *Mailing Add:* 2708 S Parkside Dr Lethbridge AB T1K 0C5 Can

HOLMES, PAUL THAYER, b Sept 25, 35; US citizen; m 56; c 2. STATISTICS. *Educ:* Wash State Univ, BA, 57, MA, 59; Stanford Univ, PhD(statist), 66. *Prof Exp:* Asst prof statist, Purdue Univ, 63-67; assoc prof, Rutgers Univ, 67-69; assoc prof, 69-77, PROF MATH SCI, CLEMSON UNIV, 77- *Mem:* Am Statist Asn; Inst Math Statist. *Res:* Probability theory; Markov chains; mathematical statistics. *Mailing Add:* Dept Math Sci Clemson Univ 201 Sikes Hall Clemson SC 29634

HOLMES, PHILIP JOHN, b Lincolnshire, UK, May 24, 45; m 70; c 4. NONLINEAR MECHANICS, DYNAMICAL SYSTEMS. *Educ:* Oxford Univ, UK, BA, 67; Southampton Univ, UK, PhD(eng), 74. *Prof Exp:* Res asst eng, Southampton Univ, 70-74, res fel, 74-77; from asst prof to assoc prof theoret & appl mech, 77-84, dir, Ctr Appl Math, 81-86, PROF THEORET & APPL MECH & MATH, CORNELL UNIV, 84- *Concurrent Pos:* Prin investr, NSF, Air Force Off Sci Res & Off Naval Res grants, 78-; ed, Soc Indust & Appl Math J Appl Math, 84-89, Arch Rational Mech & Anal, 86-, J Nonlinear Sci, 89-; Aisenstadt chair, Ctr Math Res, Univ Montreal, 85-86; Fairchild scholar, Calif Inst Technol, 88-89. *Mem:* Am Math Soc; Soc Indust & Appl Math; Soc Natural Philos; Int Soc Interaction Math & Mech. *Res:* Mathematical theory of dynamical systems, with applications to the nonlinear dynamics of fluid and solid systems; nonlinear oscillations; Hamiltonian mechanics. *Mailing Add:* Dept Theoret & Appl Mech Cornell Univ Ithaca NY 14853-1503

HOLMES, R H LAVERGNE, organic chemistry; deceased, see previous edition for last biography

HOLMES, RANDALL KENT, b Muskegon, Mich, Nov 7, 40; m 62; c 2. INTERNAL MEDICINE, MICROBIOLOGY. *Educ:* Harvard Col, AB, 62; NY Univ, MD & PhD(microbiol), 68. *Prof Exp:* Intern med, Beth Israel Hosp, Boston, 68-69, resident, 69-70; res assoc, Lab Biochem & Metab, Nat Inst Arthritis & Metab Dis, 70-72; from instr to assoc prof internal med, Univ Tex Health Sci Ctr, Dallas, 72-76; PROF MICROBIOL & INTERNAL MED & CHMN DEPT MICROBIOL, UNIFORMED SERV UNIV OF HEALTH SCI, 76-, ASSOC DEAN ACAD AFFAIRS, 84- *Concurrent Pos:* NIH fel infectious dis, Univ Tex Health Sci Ctr, 72-73; mem, Vaccines Adv Comt, Nat Ctr Drugs & Biologics, 83-87, consult, 87-; mem, Microbiol Test Comt, Nat Bd Med Examrs, 84-86, chmn, 87-; mem, Bact Dis Rev Panel, Am Inst Biol Sci-Army Med Res & Develop Command, 85-, US Cholera Panel, NIH, 87-, Nat Bd Med Examrs, Comprehensive Part I Comt, 90- *Mem:* Am Soc Microbiol; Am Acad Microbiol; Am Soc Clin Invest; Infectious Dis Soc Am; Am Col Physicians. *Res:* Pathogenesis of bacterial infectious diseases. *Mailing Add:* Uniformed Serv Univ Health Sci Dept Microbiol 4301 Jones Bridge Rd Bethesda MD 20814-4799

HOLMES, RICHARD, b Victoria, BC, Oct 3, 12; nat US; m 36. BIOCHEMISTRY. *Educ:* Univ BC, BA, 35; Univ Toronto, PhD(biochem), 55. *Prof Exp:* Res assoc biochem, Ont Res Found, Can, 36-38 & 42-45; chief chemist fermentation, Chateau Gai, Ltd, 38-42; res assoc leather, Elkland Leather Co, 45-49; res assoc tissue cult, 55-78, RES CONSULT, ALFRED I DU PONT INST, NEMOURS FOUND, 78- *Concurrent Pos:* Res consult, Biochem, Vet Admin Med Ctr, Wilmington, 85- *Mem:* AAAS; Am Soc Biol Chem; Tissue Cult Asn; Am Chem Soc. *Res:* General metabolism of tissue cultures; nutritional requirements; protein metabolism; enzyme activity; relation of these factors to growth and function. *Mailing Add:* 213 Pine Cliff Dr Wilmington DE 19803

HOLMES, RICHARD BRUCE, b Washington, DC, Apr 5, 39; m 63; c 2. MATHEMATICAL ANALYSIS, SYSTEMS ANALYSIS. *Educ:* Harvard Univ, AB, 60; Mass Inst Technol, PhD(math), 64. *Prof Exp:* From instr to asst prof math, USAF Inst Technol, 64-67; from asst prof to prof math, Purdue Univ, 69-78; MEM TECH STAFF, LINCOLN LAB, MASS INST TECHNOL, LEXINGTON, 78- *Concurrent Pos:* Vis assoc prof, Univ Mass, 75-76; vis prof, Mass Inst Technol, 79; assoc ed, J Numerical Function Anal Appl, 78- *Mem:* Am Math Soc. *Res:* Functional analysis; approximation theory; optimization; system identification and modeling; signal processing and data analysis; probability density estimation; quantitative analysis and development of option investment strategies. *Mailing Add:* Lincoln Lab Mass Inst Technol Lexington MA 02173

HOLMES, RICHARD TURNER, b Monterey Park, Calif, Aug 7, 36; m 62; c 2. ANIMAL ECOLOGY, ANIMAL BEHAVIOR. *Educ:* Humboldt State Univ, BS, 59; Univ Calif, Berkeley, PhD(zool), 64. *Prof Exp:* Teaching asst zool, Univ Calif, Berkeley, 59-62, assoc, 62-64; lectr biol, Univ Calif, Santa Barbara, 64-65; asst prof, Tufts Univ, 65-67; from asst prof to assoc prof, 67-75, PROF BIOL, DARTMOUTH COL, 83-, CHMN DEPT, 83- *Concurrent Pos:* Res grants, Arctic Inst NAm, 66 & NSF, 67- *Honors & Awards:* Howell Award, Cooper Ornith Soc, 64, H R Painten Award, 67, 85. *Mem:* Animal Behav Soc; Ecol Soc Am; fel Am Ornith Union; Cooper Ornith Soc; fel AAAS. *Res:* Population and community ecology; structure and functioning of animal communities in forest and tundra ecosystems; behavioral ecology of birds; ecology of migrant birds in breeding and wintering areas. *Mailing Add:* Dept Biol Dartmouth Col Hanover NH 03755

HOLMES, ROBERT RICHARD, b Chicago, Ill, Aug 25, 28; m 56; c 3. INORGANIC CHEMISTRY. *Educ:* Ill Inst Technol, BS, 50; Purdue Univ, PhD(chem), 53. *Prof Exp:* From instr to assoc prof inorg chem, Carnegie Inst Technol, 53-62; mem tech staff, Bell Tel Labs, Inc, NJ, 62-66; PROF CHEM, UNIV MASS, AMHERST, 66- *Concurrent Pos:* Ed-in-chief, Phosphorus, Sulfur & Silicon; vis prof, Nat Inst Arthritis, Metab & Digestive Dis, NIH, Bethesda, MD, 73, Univ Braunschweig, Ger, 73, Louis Pasteur Univ, Strasbourg, France, 80 & Univ Calif, La Jolla, 87 & San Diego, 91; expert witness, organotin chem. *Mem:* Am Chem Soc. *Res:* Synthesis, nuclear magnetic resonance, single crystal x-ray and spectroscopic studies of main group compounds, particularly organo phosphorus, silicon, and tin compounds; stereochemistry and mechanisms of nucleophilic substitution reactions at phosphorus and silicon; formation of new classes of organotin cluster compounds; the development of an active site model for cyclic AMP action; abinitio studies of the comparative reactivity at pentacoordinate phosphorus and silicon centers; 170 publications and two invited ACS monographs. *Mailing Add:* Dept of Chem Univ of Mass Amherst MA 01003

HOLMES, ROBERT W, b Dover, NH, Sept 9, 25; m 49; c 3. BIOLOGICAL OCEANOGRAPHY, ECOLOGY. *Educ:* Haverford Col, BS, 49; Univ Calif, MS, 61; Univ Oslo, PhD, 61. *Prof Exp:* From jr res biologist to assoc res biologist, Scripps Inst Oceanog, Univ Calif, San Diego, 52-67; assoc prof biol, 67-74, PROF BIOL SCI, UNIV CALIF, SANTA BARBARA, 74-, DIR, MARINE SCI INST, 69- *Mem:* Am Soc Limnol & Oceanog; AAAS; Phycol Soc Am; Sigma Xi. *Res:* Diatom taxonomy, morphology and ecology; primary productivity of oceans. *Mailing Add:* 3749 Brenner Dr Santa Barbara CA 93105

HOLMES, ROGER ARNOLD, b Peekskill, NY, Aug 31, 31; m 55; c 3. ELECTRICAL ENGINEERING. *Educ:* USCG Acad, BSc, 53; Mass Inst Technol, SM, 58; Purdue Univ, PhD(elec eng), 62. *Prof Exp:* Asst elec eng, Mass Inst Technol, 56-58; from instr to assoc prof, Purdue Univ, 58-70; dean col eng, Univ SC, 70-77; DEAN ACAD AFFAIRS, GEN MOTORS INST, 78- *Concurrent Pos:* Jr scientist, Midwest Appl Sci Corp, 59-64; scientist, Adv Res Corp, Ga, 61-64; teaching consult, Western Elec Co, 62-63; dir res, Radiation Dynamics, Inc, NY, 63-64; consult, Nat Aeronaut & Space Admin, 73- *Mem:* Inst Elec & Electronics Engrs. *Res:* Remote sensing of earth resources; electronics. *Mailing Add:* 5590 Old Pond Dr Dublin OH 43017-3029

HOLMES, WILLIAM FARRAR, b St Louis, Mo, June 6, 32; m 79; c 2. BIOLOGICAL STRUCTURE, COMPUTER SCIENCE. *Educ:* Princeton Univ, AB, 53, Univ Pa, PhD(biophys), 60. *Prof Exp:* Res fel biophys, Johnson Found, Univ Pa, 60-64; res assoc comput sci, Wash Univ, 65-68, asst prof, 68-74, assoc prof biochem, 74-87; SR RES SCIENTIST, UNIV ARIZ, 87- *Mem:* Biophys Soc; Asn Comput Mach; Am Soc Mass Spectrometry. *Res:* Computer controlled mass spectrometry; computer analysis of biochemical self assembly. *Mailing Add:* Dept Radiation Oncol Univ Ariz Health Sci Ctr Tucson AZ 85724

HOLMES, WILLIAM LEIGHTON, biochemistry; deceased, see previous edition for last biography

HOLMES, WILLIAM NEIL, b New Mills, Eng, June 2, 27; m 55; c 4. ENDOCRINOLOGY. *Educ:* Univ Liverpool, Eng, BSc, 52, MSc, 54, PhD(zool), 57. *Hon Degrees:* DSc, Univ Liverpool, 68. *Prof Exp:* From asst prof to assoc prof zool, Univ BC, 57-64; assoc prof, 64-67, PROF PHYSIOL, UNIV CALIF, SANTA BARBARA, 67- *Concurrent Pos:* Guggenheim Found fel, 61-62; external examr undergrad degrees in zool, Univ Hong Kong, 76-79 & 85-88; vis prof, dept zool, Univ Hong Kong, 73 & 82-83; dir, exp animal res, Univ Calif, Santa Barbara, 68-82. *Mem:* Endocrine Soc; Am Physiol Soc; Am Soc Zool; Brit Soc Endocrinol; Brit Soc Exp Biol. *Res:* Endocrinology of salt and water balance in lower vertebrates; interactions between hormonal regulators and petroleum pollutants; ultrastructual characteristics of steroidogenic cells of the adrenal gland in relation to their embryological development and specific functional properties. *Mailing Add:* Dept Biol Sci Univ Calif Santa Barbara CA 93106

HOLMES, ZOE ANN, b Pittsburg, Kans, Feb 28, 42. FOOD QUALITY. *Educ:* Kans State Univ, Manhattan, BS, 64, MS, 65; Univ Tenn, Knoxville, PhD(food sci), 72. *Prof Exp:* Instr foods, Ore State Univ, 65-68 & Univ Tenn, Knoxville, 70; asst prof foods, Univ Tex, Austin, 72-74; from asst prof to assoc prof, 74-86, PROF FOODS, ORE STAT UNIV, 87- *Concurrent Pos:* Reviewer, Instrnl Sci Equip Prog, NSF, 78-81; sr consult, Nutrit Res & Develop Ctr, Borgor, Indonesia, 82. *Mem:* Inst Food Technologists; Am Asn Cereal Chemists; Am Chem Soc; Asn Develop Comput-Based Instrnl Systs; Am Meat Sci Asn; Am Home Econ Asn. *Res:* Food science. *Mailing Add:* Dept Nutrit & Food Mgt Milam 108 Ore State Univ Corvallis OR 97331-5103

HOLMGREN, ARTHUR HERMAN, b Midvale, Utah, Nov 18, 12; m 34; c 4. PLANT TAXONOMY, ECOLOGY. *Educ:* Univ Utah, BA, 36, MS, 42. *Prof Exp:* Chief party, US Forest Serv, 37-41; range mgr, Bur Land Mgt, 41-42; from asst prof to prof bot, Utah State Univ, 43-78, emer prof, 78-; RETIRED. *Mem:* AAAS; Bot Soc Am; Int Asn Plant Taxon; Am Soc Plant Taxon. *Res:* Flora of the intermountain region; taxonomy of flowering plants. *Mailing Add:* 1738 Country Club Dr Logan UT 84321

HOLMGREN, HARRY D, b Minneapolis, Minn, Apr 21, 28; m 49; c 4. NUCLEAR PHYSICS. *Educ:* Univ Minn, BS, 49, MA, 50, PhD, 54. *Prof Exp:* Physicist, US Naval Res Lab, 54-61; from assoc prof to prof nuclear physics, 61-74, PROF PHYSICS, UNIV MD, COLLEGE PARK, 74- *Mem:* Am Phys Soc. *Res:* The study of the mechanisms of nuclear interactions and the structure of nuclei; cyclotrons. *Mailing Add:* Dept Physics & Astron Univ Md College Park MD 20742

HOLMGREN, NOEL HERMAN, b Salt Lake City, Utah, Nov 18, 37; m 69. PLANT TAXONOMY. *Educ:* Utah State Univ, BS, 62; Columbia Univ, PhD(taxon), 68. *Prof Exp:* Asst prof bot, Ore State Univ, 67-68; res assoc, 68-69, assoc cur, 69-74, CUR, NY BOT GARDEN, 74- *Concurrent Pos:* Adj asst prof, Lehman Col, 74-; assoc ed, Brittonia, 75-77, ed, 77- *Mem:* AAAS; Am Soc Plant Taxon; Int Soc Plant Taxon; Bot Soc Am. *Res:* Floristic and biosystematic research; floristics of western United States; Penstemon and Castilleja; Scrophulariaceae of North and South America. *Mailing Add:* 304 Woodland Hills Rd White Plains NY 10603

HOLMGREN, PATRICIA KERN, b Athens, Ind, Jan 21, 40; m 69. PLANT TAXONOMY. *Educ:* Ind Univ, BS, 62; Univ Wash, NY Bot Garden, MS, 64, PhD(bot), 68. *Prof Exp:* Herbarium specialist, 68-69, assoc cur, 69-71, herbarium adminr, 71-73, herbarium supvr & adminr, Phanerogamic Herbarium, 73-75, head cur, 75-81, actg vpres bot sci, 88-91, ASST VPRES SCI & DIR HERBARIUM, NY BOT GARDEN, 81- *Concurrent Pos:* Assoc ed, Brittonia, 75-91, managing ed, 84-91, ed, News & Notes, Taxon, 83-88; copy ed, Intermountam Flora, 74-; bd dirs, Am Inst Biol Sci, Asn Systs Collections, 89-91. *Mem:* Am Soc Plant Taxon (treas, 85-88, pres-elect, 89, pres, 90); Int Asn Plant Taxon; Bot Soc Am (secy, 75-79, vpres, 80, pres, 81); Am Inst Biol Sci; Asn Systs Collections. *Res:* Floristics of western United States. *Mailing Add:* New York Bot Garden Bronx NY 10458-5126

HOLMGREN, PAUL, b Gary, Ind, Sept 25, 37; m 58; c 2. CELL BIOLOGY, ELECTRON MICROSCOPY. *Educ:* Ind Univ, BS, 60; Southern Methodist Univ, MA, 65; NTex State Univ, PhD(biol), 69. *Prof Exp:* High sch teacher, Ind, 60-65; asst prof biol, Pasadena Col, 65-66; teaching fel, NTex State Univ, 66-69; asst prof, 69-76, ASSOC PROF BIOL, NORTHERN ARIZ UNIV, 76- *Mem:* Am Soc Cell Biol. *Mailing Add:* Dept Biol Northern Ariz Univ Box 5640 Flagstaff AZ 86011-5640

HOLM-HANSEN, OSMUND, b Sandefjord, Norway, Sept 9, 28; nat US; m 63; c 1. PLANT PHYSIOLOGY. *Educ:* Harvard Univ, BA, 50; Univ Wis, PhD(plant physiol), 54. *Prof Exp:* Fulbright res award, Univ Oslo, 54-55; res biochemist, Radiation Lab, Univ Calif, 55-58; asst prof bot, Univ Wis, 58-62; RES BIOLOGIST, SCRIPPS INST OCEANOG, UNIV CALIF, SAN DIEGO, 62- *Res:* Intermediary metabolism of algae, especially photosynthesis, nitrogen-metabolism and microelement nutrition; biological and biochemical oceanography; nucleotide analyses. *Mailing Add:* 2336 Calle Chiquita La Jolla CA 92037

HOLM-KENNEDY, JAMES WILLIAM, b Los Angeles, Calif, Jan 2, 39. PHYSICS, ELECTRICAL ENGINEERING. *Educ:* Univ Calif, Riverside, AB, 60; Univ Minn, MS, 63, PhD(elec eng), 69. *Prof Exp:* Sta & dist mgr, Riverside Press-Enterprise Co, 55-59; lab asst & reader, Univ Calif, Riverside, 59-60; teaching asst physics, Univ Minn, 60-63; lectr elec eng, Univ Calif, Santa Barbara, 69; asst prof elec sci & eng, Univ Calif, Los Angeles, 69-77; mem fac, 77-80, PROF ELEC ENG, SCH ENG, UNIV HAWAII, 80- *Concurrent Pos:* Physicist, US Naval Ord Lab, 58-61. *Mem:* Am Inst Physics. *Res:* High field transport in semiconductors; semiconductor devices and device physics; solid state physics. *Mailing Add:* Dept Elec Eng-Hol 445 Univ Hawaii 2540 Dole St Honolulu HI 96822

HOLMLUND, CHESTER ERIC, b Worcester, Mass, Dec 14, 21; m 49; c 2. BIOCHEMISTRY. *Educ:* Worcester Polytech Inst, BS, 43, MS, 51; Univ Wis, PhD(biochem), 54. *Prof Exp:* Res chemist, E I du Pont de Nemours & Co, 46-47; adhesives chemist, US Envelope Co, 48-51; asst, Univ Wis, 51-54; res biochemist, Lederle Labs, Am Cyanamid Corp, 54-57, group leader, NY, 57-67; from assoc prof to prof biochem, 67-74, PROF CHEM, UNIV MD, COLLEGE PARK, 74- *Mem:* AAAS; Am Soc Microbiol; Am Soc Biol Chem; Am Oil Chemists Soc; Sigma Xi. *Res:* Biochemistry of microorganisms; mode of action of drugs on microorganisms and cell cultures; sterol metabolism in yeast and cell cultures; metabolic control mechanisms. *Mailing Add:* 16300 W Nine Mile Rd Apt 107 Southfield MI 48075-5909

HOLMQUIST, BARTON, b Los Angeles, Calif, Mar 12, 43; div; c 2. BIOLOGICAL CHEMISTRY. *Educ:* Univ Calif, Santa Barbara, BA, 65, PhD(chem), 68. *Prof Exp:* res assoc, 68-86, ASSOC PROF, DEPT BIOCHEM & MOLECULAR PHARM, HARVARD MED SCH, 86- *Mem:* Am Chem Soc; Fedn Am Soc Exp Biol. *Res:* Biophysical and mechanistic studies of enzymes and enzyme catalysis; kinetics and inhibition of enzymatic catalysis and enzyme activation accompanying chemical modification; magnetic circular dichroism of metalloproteins; biochemistry of alcoholism. *Mailing Add:* Ctr Biophys 75 Francis St Boston MA 02115

HOLMQUIST, HOWARD EMIL, organic chemistry, for more information see previous edition

HOLMQUIST, NELSON D, b Waterbury, Conn, Sept 27, 24; m 75; c 5. PATHOLOGY. *Educ:* Princeton Univ, AB, 47; Columbia Univ, MD, 51. *Prof Exp:* Intern surg, NY Hosp, 51-52, asst resident, 53-55; instr path, Med Col, Cornell Univ, 55-59; from asst prof to prof, 59-89, EMER PROF PATH, SCH MED, LA STATE UNIV MED CTR, NEW ORLEANS, 89- *Concurrent Pos:* Nat Cancer Inst trainee diag cytol, Med Col, Cornell Univ, 52-53; asst attend pathologist, NY Hosp, 58-59; vis pathologist, Charity Hosp La, New Orleans, 59-70, sr vis pathologist, 70-89. *Mem:* Am Soc Cytol; Am Soc Clin Path; AMA; Int Acad Cytol. *Res:* Diagnostic cytology; cancer research. *Mailing Add:* 12380 River Ridge Dr Folsom LA 70437

HOLMQUIST, WALTER RICHARD, b Kansas City, Mo, Dec 23, 34; m 68; c 2. BIOCHEMISTRY, MOLECULAR EVOLUTION. *Educ:* Washington & Lee Univ, BS, 57; Calif Inst Technol, BS, 61, PhD(chem), 66. *Prof Exp:* Lectr org chem & biochem, Univ Ife, Nigeria, 66-68; res fel biol, Harvard Univ, 68-70; asst res chemist, 70-74, assoc res chemist, 74-80, sr res fel, 80-82, RES CHEMIST, SPACE SCI LAB, UNIV CALIF, BERKELEY, 80- *Concurrent Pos:* Prin investr, Nat Heart & Lung Inst grant, Univ Calif, Berkeley, 71-73; Nat Sci Found grant, 77-81 & 83-87. *Mem:* AAAS; Am Chem Soc; Am Soc Biochem & Molecular Biol; Sigma Xi; NY Acad Sci; Soc Study Evolution; Am Inst Chemists. *Res:* Chemistry of heme- and electron transfer proteins; human hemoglobin, cytochrome c1; structure of thermostable bacterial spores; molecular evolution of proteins and nucleic acids; paleogenetics, methanobacteria. *Mailing Add:* 760 Mesa Way Richmond CA 94805-1743

HOLMSEN, THEODORE WAAGE, b Teaneck, NJ, Mar 1, 30; m 53; c 4. APPLIED PHYSIOLOGY. *Educ:* Rutgers Univ, BS & MS, 58; Univ Fla, PhD(plant physiol), 61. *Prof Exp:* Plant physiologist, Dow Chem Co, 61-63; fel, Lab Chem Biodyn, Univ Calif, Berkeley, 63-65; group leader herbicide res, Agr Res Ctr, Dow Chem Co, 65-67, res plant physiologist, 68-69, assoc scientist, 69-78, sr assoc scientist, 79-84, res scientist, 84-88; RETIRED. *Mem:* AAAS; Am Soc Agron; Crop Sci Soc; Plant Growth Regulator Soc Am. *Res:* Applied and basic phyto-pharmacology. *Mailing Add:* 20 Mt Scott Ct Clayton CA 94517

HOLMSTEDT, JAN OLLE VALTER, b Landskrona, Sweden, Mar 7, 33. DENTAL RESEARCH. *Educ:* Royal Dent Sch, Malmo, Sweden, DDS, 57; Univ Mich, MS, 66 & 68, PhD(anat), 71. *Prof Exp:* Instr orthod, Royal Dent Sch, Malmo, 61-64; teaching fel, Univ Mich, 65-66; asst prof anat, La State Univ Med Ctr, 71-76; vis scholar, Sch Dent, 76-77, ASSOC RES SCIENTIST GERONT, UNIV MICH, 77- *Mem:* Int Asn Dent Res; Am Soc Cell Biol; Am Asn Anatomists; Sigma Xi; AAAS. *Res:* Mechanisms of primary and secondary palate formation; cleft lip and palate formation; prenatal formation of eyelids; mechanisms of aging in oral mucosa; age-dependent effects of neutron and of irradiation on oral tissues. *Mailing Add:* Lab Cell Biol Rm 5212 Sch Dent Univ Mich Ann Arbor MI 48109

HOLMSTROM, FRANK ROSS, b Port Angeles, Wash, Dec 28, 36; m 61; c 2. ELECTRICAL ENGINEERING. *Educ:* Univ Washington, Seattle, BS, 58; Stanford Univ, MS, 60, PhD(elec eng), 65. *Prof Exp:* Res asst elec eng, Stanford Univ, 58-63; elec engr, NASA Electronics Res Ctr, 64-70; asst prof, 70-77, assoc prof, 77-81, PROF, UNIV LOWELL, 81-, chair elec eng, 81-84. *Concurrent Pos:* Elec engr, Transp Syst Ctr, Dept Transp, 70. *Mem:* AAAS; Inst Elec & Electronics Engrs. *Res:* Electronic aspects of rapid transit safety; semiconductor electronics. *Mailing Add:* Dept Elec Eng Univ Lowell One University Ave Lowell MA 01854

HOLMSTROM, FRED EDWARD, b Salt Lake City, Utah, Mar 31, 27; m 51; c 5. PHYSICS. *Educ:* Univ Utah, BA, 55, PhD(physics), 59. *Prof Exp:* Res physicist, Int Bus Mach Corp, 58-61; res specialist, Rocketdyne Div, NAm Aviation, Inc, Calif, 61-63; prof & chmn dept, 63-82, PROF PHYSICS, SAN JOSE STATE COL, 82- *Concurrent Pos:* Asst prof, San Jose State Col, 59-61 & San Fernando State Col, 61-63; indust consult, 64-; Esso Found grant, 68-69; NASA Consortium grant, 78-85. *Mem:* Am Phys Soc; Am Asn Physics Teachers. *Res:* Solid state, electron, ion, and MOS physics. *Mailing Add:* Dept Physics San Jose State Univ One Washington Square San Jose CA 95192-0106

HOLOB, GARY M, b Buffalo, NY, Feb 4, 47; m 70; c 1. CHEMICAL ENGINEERING, MATERIAL SCIENCE. *Educ:* State Univ NY Buffalo, BS, 69, MS, 70, PhD(chem eng), 75. *Prof Exp:* RES & DEVELOP ENGR SYNTHETIC RUBBERS, E I DU PONT DE NEMOURS & CO, INC, 73- *Mem:* Am Inst Chem Engrs. *Res:* Synthesis and refining of organic compounds; polymerization of synthetic rubbers; extrusion of non Newtonian materials; electrical conductivity of polymers. *Mailing Add:* 702 Westcliff Rd Wilmington DE 19803

HOLOIEN, MARTIN O, b Wolf Point, Mont, Nov 13, 28; m 48; c 4. GENERAL COMPUTER SCIENCES, APPLIED MATHEMATICS. *Educ:* Moorhead State Univ, BS, 51; NDak State Univ, MS, 59; Univ Minn, PhD(math educ), 70. *Prof Exp:* Sci instr math & sci, Hillcrest Lutheran Acad, 51-58; math instr col math, NDak State Univ, 58-60, asst prof math, 61-68; sci computer programmer, Boeing Co, 60-61; from asst prof to prof computer sci, Moorhead State Univ, 68-84; LECTR COMPUTER SCI, UNIV CALIF, SANTA BARBARA, 84- *Concurrent Pos:* Chmn, Minn Statewide Comt Educ Comput, 75-76; mem, Taxon Comt, Develop Computer Sci Taxon, US Off Educ, 76; mem, Spec Interest Group, Comp Sci Educ, Am Comput Mach & Spec Interest Group, Computers & Soc. *Mem:* Asn Comput Mach. *Res:* Author of four computer science textbooks. *Mailing Add:* 3810 Pueblo Ave Santa Barbara CA 93110

HOLONYAK, N(ICK), JR, b Ziegler, Ill, Nov 3, 28; m 55. SEMICONDUCTOR MATERIAL & DEVICES, APPLIED SOLID STATE PHYSICS. *Educ:* Univ Ill, BS, 50, MS, 51, PhD(elec eng), 54. *Prof Exp:* Mem tech staff, Transistor Develop Dept, Bell Tel Labs, NJ, 54-55; physicist & mgr, Advan Semiconductor Lab, Semiconductor Prod Dept, Gen Elec Co, 57-63; PROF ELEC ENG, UNIV ILL, URBANA, 63-, MEM, CTR ADVAN STUDY, 77- *Concurrent Pos:* Army instr, Univ Md, Yokohama, Japan, 56. *Honors & Awards:* Cordiner Award, Gen Elec Co, 62; Morris Liebmann Prize, Inst Elec & Electronics Engrs, 73; John Scott Medal, 75; Jack A Morton Award, Inst Elec & Electronics Engrs, 81, Edison Medal, 89; Solid State Sci & Technol Award, Electrochem Soc, 83; Monie A Ferst Award, Sigma Xi, 88; Welker Award 76; US Nat Medal Sci, 90. *Mem:* Nat Acad Sci; Nat Acad Eng; Electrochem Soc; fel Am Phys Soc; Math Asn Am; fel Inst Elec & Electronics Engrs; AAAS; Elec Chem Soc; fel Am Acad Arts & Sci; fel Optical Soc Am. *Res:* Solid state electronic devices; transistors; diodes; switches; negative resistance devices; semiconductor technology; semiconductor controlled rectifiers; semiconductor devices; III-V compounds; semiconductor lasers and light emitters; semiconductor materials. *Mailing Add:* Dept Elec & Comput Eng Univ Ill 1406 W Green St Urbana IL 61801

HOLOUBEK, VIKTOR, b Brno, Czech, Apr 19, 28; US citizen; m 64; c 2. BIOCHEMISTRY, MOLECULAR BIOLOGY. *Educ:* Charles Univ, Prague, PhD(physiol, biochem), 52. *Prof Exp:* Res fel chem, Fac Med, Charles Univ, Prague, 52-53; res fel, Cancer Res Inst, Bratislava, Czech, 53-54, assoc res biochemist, 54-60; res biochemist, Cancer Res Inst, Vienna, 60-61; asst res biochemist, Virus Lab, Univ Calif, Berkeley, 61-63; asst res biochemist, Cancer Res Inst, Med Ctr, Univ Calif, San Francisco, 63-68; from asst prof to assoc prof, 68-77, PROF BIOCHEM, UNIV TEX MED BR GALVESTON, 77- *Concurrent Pos:* Vis prof, Australia, 80-82. *Mem:* Am Soc Biol Chem; Am Soc Cell Biol; Am Asn Cancer Res; Sigma Xi. *Res:* Chemistry of histones and acidic nuclear proteins; regulation of gene expression, interaction of hormones, carcinogens and viruses with the cellular genome. *Mailing Add:* Dept Biochem Univ Tex Med Br Galveston TX 77550-2779

HOLOWATY, MICHAEL O, b Stanislav, Ukraine, Nov 21, 22; nat US; m 45; c 4. INORGANIC CHEMISTRY. *Educ:* Breslau Univ, PhD(chem), 44. *Prof Exp:* Asst, Breslau Univ, 43-44; res assoc, Univ Heidelberg, 45-48; res engr, Cleveland Cliffs Iron Co, Mich, 49-51; metallurgist, Inland Steel Co, 51-54, chief raw mat res, 54-57, chief res engr raw mat, 57-59, raw mat & reduction, 59-62, assoc mgr, Res Dept, 62-74, dir raw mat & mineral res, 74-76, sr adv, Res Dept, 76-91; RETIRED. *Concurrent Pos:* Mem, US Del Steel Indust Execs to USSR, 58. *Honors & Awards:* J E Johnson Award, Am Inst Mining, Metall & Petrol Eng, 56; J Metals Award, 57; Interprof Coop Award, Am Soc Tool & Mfg Eng, Soc Mfg Eng, 67; Joseph Becker Award, 81. *Mem:* Am Inst Mining, Metall & Petrol Eng. *Res:* Raw materials; beneficiation and agglomeration of iron ores; coal; preparation and coal chemistry; cooking processes; reduction of ferrous metals; heavy inorganic chemistry. *Mailing Add:* 12405 Clark St Crown Point IN 46307

HOLOWENKO, A(LFRED) R(ICHARD), b Boston, Mass, Dec 10, 16; m 44; c 2. MECHANICAL ENGINEERING. *Educ:* Harvard Univ, AB & MS, 38, 39. *Prof Exp:* Mech engr, Westinghouse Elec Corp, Pa, 39-42; asst prof mech eng, Pittsburgh, 42-44; instr, Rice Inst, 44-46; from asst prof to emer prof mech eng, Purdue Univ, 46-82; RETIRED. *Mem:* Am Soc Mech Engrs; Am Soc Eng Educ. *Res:* Dynamics of machinery; machine design. *Mailing Add:* 300 Chippewa West Lafayette IN 47906

HOLOWINSKY, ANDREW WOLODYMYR, b Ukraine, Oct 25, 36; US citizen; m 59; c 1. BIOCHEMISTRY, SCIENCE EDUCATION. *Educ:* La Salle Col, BA, 56; Univ Pa, PhD(bot), 61. *Prof Exp:* USPHS fel biol, Harvard Univ, 61-63, from instr to asst prof, 63-68; ASSOC PROF BIOL, BROWN UNIV, 68- *Concurrent Pos:* Vis sr res assoc, Brandeis Univ, 67-68, vis assoc prof photobiology, 76. *Mem:* Am Soc Plant Physiol; Sigma Xi; AAAS. *Res:* Plant growth and development; chloroplast development and light physiology. *Mailing Add:* Div Biol & Med Sci Brown Univ Providence RI 02912

HOLOWKA, DAVID ALLAN, b Rochester, NY, Aug 21, 48; m 79; c 3. MOLECULAR IMMUNOLOGY, CELL SURFACE RECEPTORS. *Educ:* St John Fisher Col, BS, 70; Tufts Univ, PhD(biochem), 75. *Prof Exp:* NIH fel biophys chem, dept chem, Cornell Univ, 75-77; Arthritis Found fel molecular immunol, Arthritis & Rheumatism Br, Nat Inst Arthritis, Metab & Digestive Dis, NIH, 77-80; res assoc, 80-86, SR RES ASSOC, DEPT CHEM, CORNELL UNIV, 86- *Concurrent Pos:* Reg ed, Molecular Immunol, 90- *Mem:* Am Asn Immunologists. *Res:* Biophysical and biochemical studies on the structure and function of cell surface receptors in the immune response, including spectroscopic studies on the receptor for immunoglobulin E. *Mailing Add:* Dept Chem Baker Lab Cornell Univ Ithaca NY 14853-1301

HOLPER, JACOB CHARLES, b Bosworth, Mo, May 20, 24; m 46; c 2. CANCER, IMMUNOLOGY. *Educ:* Univ Kans, AB, 49, MA, 51; Univ Mich, PhD(virol), 54. *Prof Exp:* Head, Infectious Dis Sect, Res Div, Abbott Labs, 54-64, dir, Infectious Dis & Parasitol Div Ill, 64-69, dir res & develop, Abbott Sci Prod Div, 69-72; dir, Cancer Res Br, Organon Teknika, 72-75, gen mgr, Biomed Res Div, 75-80, dir sci affairs, Bionetics Lab Prod, 80-89; RETIRED. *Concurrent Pos:* Mem, Virus Cancer Prog Sci Rev Comt B, Nat Cancer Inst, 74-78. *Mem:* Am Soc Microbiol; fel Am Acad Microbiol; Am Asn Immunol; Soc Exp Biol & Med. *Res:* Infectious diseases; antibiotics; chemotherapy of viruses; viral induced cancer; virus vaccines; biologic products research and development; immunodiagnostics. *Mailing Add:* 12 Windsor Dr Charleston SC 29407

HOLROYD, LOUIS VINCENT, b Vancouver, BC, Jan 22, 25; m 50; c 4. PHYSICS. *Educ:* Univ BC, BA, 45, MA, 47; Univ Notre Dame, PhD(physics), 50. *Prof Exp:* From asst prof to assoc prof, 50-61, chmn dept, 56-74, prof physics, 61-87, campus safety coordr, 77-87, EMER PROF PHYSICS, UNIV MO, COLUMBIA, 87- *Concurrent Pos:* Chmn, Hazardous Waste Task Force, Mo. *Mem:* Am Phys Soc; Am Asn Physics Teachers. *Res:* Electron spin resonance of ions and defects in crystals; improvement of high school teaching; enviromental science. *Mailing Add:* 2235 Bluff Blvd Columbia MO 65201-6101

HOLROYD, RICHARD ALLAN, b Jamestown, NY, Dec 31, 30; m 57; c 2. RADIATION CHEMISTRY. *Educ:* Col Wooster, BA, 52; Univ Rochester, PhD(phys chem), 56. *Prof Exp:* Asst prof eng, Univ Calif, Los Angeles-AID Prog Indonesia, 57-59; fel radiation chem, Mellon Inst, 59-64; res specialist, Atomics Int Div, NAm Rockwell Corp, Calif, 64-69; chemist, 69-89, SR CHEMIST, BROOKHAVEN NAT LAB, 89- *Concurrent Pos:* US sr scientist award, Humboldt Found, WGer, 75-76; assoc ed, Radiation Res Soc, 75-78. *Mem:* Radiation Res Soc; Am Chem Soc. *Res:* Photoionization and photodetachment processes in non-polar fluids; photolysis of aliphatic hydrocarbons; pulse radiolysis; electron mobility in fluids; reactions and energies of electrons in fluids. *Mailing Add:* Brookhaven Nat Lab 555 Upton NY 11973

HOLROYDE, CHRISTOPHER PETER, b Doncaster, Eng, Dec 22, 41. ONCOLOGY. *Educ:* Univ London, MB & BS, 66, MRCP, 68. *Prof Exp:* Asst lectr path, Univ London, 66-68; asst instr med, Univ Pa, 70-72; assoc clin oncol, Lankenau Hosp, 74-77, asst prof, 72-80, ASSOC PROF MED, THOMAS JEFFERSON UNIV, 80-, HEAD ONCOL RES, LANKENAU HOSP, 77-, ASSOC, DEPT INT MED, 77- *Concurrent Pos:* Sr investr, Eastern Coop Oncol Group, 75-; assoc, Dept Hemat, Lankenau Hosp, 78- *Mem:* Am Soc Clin Oncol. *Res:* Host-tumor metabolic relationships; chemotherapy of human malignancies. *Mailing Add:* Cancer Treatment Ctr Lankenau Hosp Philadelphia PA 19151

HOLSCLAW, DOUGLAS S, JR, b Tucson, Ariz, Nov 25, 34; m 77; c 2. PEDIATRICS, PULMONARY DISEASES. *Educ:* Univ Ariz, BA, 56; Columbia Univ, MD, 60; Inst Child Health, London, DCH, 64. *Prof Exp:* Intern med, Univ Chicago Clins, 60-61; resident, Children's Hosp, 61-63; registrar, Hosp Sick Children, 63-64; fel, Children's Hosp, 66-68; instr pediat, Harvard Med Sch, 68-70; from asst prof to assoc prof, 70-78, PROF PEDIAT, HAHNEMAN UNIV, 78- *Concurrent Pos:* Mem, Med Adv Coun, Cystic Fibrosis Found, 73-77; gov appointee, Adv Health Bd & Secy Health, 80-90; Pulmonary Acad Award, Nat Heart & Lung Inst, 74-77; Interim chmn, Dept Pediat, Hahnemann Univ, 88-90. *Mem:* Am Pediat Soc; Am Thoracic Soc; Am Acad Pediat. *Res:* Acute and chronic pediatric pulmonary disorders. *Mailing Add:* Dept Pediat Hahnemann Univ 230 N Broad St Philadelphia PA 19102

HOLSEN, JAMES N(OBLE), b Palo Alto, Calif, June 20, 24; m 50, 77; c 2. CHEMICAL ENGINEERING. *Educ:* Princeton Univ, BS, 48. *Hon Degrees:* DSc, Wash Univ, 54. *Prof Exp:* Chem engr, Olin Mathieson Chem Corp, 54-55; from asst prof to prof chem eng, Washington Univ, 55-73; prof chem eng, Univ Mo, Rolla, 73-76; SR TECH SPECIALIST, MCDONNELL DOUGLAS CORP, 77- *Concurrent Pos:* Vis prof mech eng, Kabul Univ, Afghanistan, 63-64 & 69-73, chem eng, Nat Tech Inst, Saigon, Vietnam, 73-74. *Mem:* Am Chem Soc; assoc fel Am Inst Aeronaut & Astronaut; Am Inst Chem Engrs; Am Soc Eng Educ; AAAS. *Res:* Mass transfer and chemical kinetics; shock tube studies in high temperature chemical kinetics; gas dynamics and re-entry thermodynamics; materials processing in space; satellite structural materials. *Mailing Add:* 419 E Argonne Dr Kirkwood MO 63122

HOLSEN, THOMAS MICHAEL, b Milwaukee, Wis, Oct 16, 59; m 82; c 2. ENVIRONMENTAL CHEMISTRY, DRY DEPOSITION. *Educ:* Univ Calif, Berkeley, BS, 83, MS, 85, PhD(civil & environ eng), 88. *Prof Exp:* Environ scientist, Kennedy-Jenks Engrs, 83-84; ASST PROF ENVIRON ENG, ILL INST TECHNOL, 88- *Mem:* Am Chem Soc; Asn Environ Eng Professors; Int Asn Water Pollution. *Res:* Environmental chemistry; fate and transport of organic chemicals in the environment; air toxics; dry deposition; sorption of organic chemicals on natural surfaces. *Mailing Add:* Dept Environ Eng Ill Inst Technol Chicago IL 60616

HOLSER, WILLIAM THOMAS, b Bakersfield, Calif, July 4, 20; m 54; c 3. MINERALOGY. *Educ:* Calif Inst Technol, BS, 42, MSc, 46; Columbia Univ, PhD(ore deposits), 50. *Prof Exp:* Asst geol, Calif Inst Technol, 45-46; lectr, Columbia Univ, 46-48; asst prof, Cornell Univ, 48-54; geologist, US Geol Surv, 48-54; prin geochemist, Battelle Mem Inst, 54-55; res geochemist, Inst Geophys, Univ Calif, Los Angeles, 55-58; sr res assoc mineral, Chevron Oil Field Res Co, 58-70; prof, 70-86, EMER PROF GEOL, UNIV ORE, 86- *Concurrent Pos:* Mem comt int tables, Int Union Crystallog, 63-65, 71-76; ed, Am Mineral, 66-72; Fulbright fel, 76; Alexander von Humboldt US sr scientist award, 76; Caswell Silver Res Prof, Univ NMex, 83; counr, Geochem Soc, 86-88; vis scientist, Cornell Univ, 89- *Mem:* Geol Soc Am; Am Geophys Union; hon life fel Mineral Soc Am; Geochem Soc; Soc Econ Geologists. *Res:* Mineralogy and geochemistry of bromine, sulfur, carbon, strontium and rare earths; geology of evaporites; stable isotope geochemistry and chemical history of the oceans and sediments; mass extinction events. *Mailing Add:* Dept Geol Sci Univ Ore Eugene OR 97403

HOLSHOUSER, DON F, b Dwight, Kans, Mar 23, 20; m 43; c 3. PHYSICAL ELECTRONICS. *Educ:* Kans State Col, BS, 42; Univ Ill, MS, 50, PhD(elec eng), 59. *Prof Exp:* Develop engr cathode ray tubes, Radio Corp of Am, Pa, 42-46; res assoc microwave tubes, Dept Elec Eng, Univ Ill, Urbana, 46-51, asst prof, 51-58, assoc prof & prof dir, 58-65, prof microwave & optical electronics, 65-82, EMER PROF, DEPT ELEC ENG, UNIV ILL, URBANA, 82- *Concurrent Pos:* Air Force Off Sci Res grants, 58-72; vis prof, Athens Tech, 64-65. *Mem:* Fel AAAS. *Res:* Modulation and demodulation of light at microwave frequencies; solar energy. *Mailing Add:* Two Light RR 5 Box 62 Cape Elizabeth ME 04107

HOLSHOUSER, W(ILLIAM) L(UTHER), b Blowing Rock, NC, Oct 23, 11; m 35; c 3. METALLURGY. *Educ:* Davidson Col, BS, 33. *Prof Exp:* Phys sci aide, Nat Bur Standards, 39-41, jr metallurgist, 41-42, asst metallurgist, 42-44, assoc metallurgist, 44-47, metallurgist, 47-57, aeronaut struct mat res engr, 57-59, phys metallurgist, 59-62; staff metallurgist, Civil Aeronaut Bd, 62-67; staff metallurgist, Nat Transp Safety Bd, 67-73, chief, Lab Serv staff, 71-73; RETIRED. *Concurrent Pos:* Conferee World Metall Cong, Chicago, 57; consult aircraft metall, 73-81. *Mem:* Am Soc Metals; Am Soc Testing & Mat. *Res:* Fatigue of metals; quality of aircraft welding; service failures and corrosion of metals; effects of prior fatigue stressing on the impact resistance of chromium molybdenum aircraft steel; evaluation of fatigue damage of steel by supplementary tension impact tests; investigation of aircraft accidents. *Mailing Add:* PO Box 1475 Banner Elk NC 28604

HOLSINGER, JOHN ROBERT, b Harrisonburg, Va, Apr 6, 34; m 70. INVERTEBRATE ZOOLOGY, BIOSPELEOLOGY. *Educ:* Va Polytech Inst, BS, 55; James Madison Univ, MS, 63; Univ Ky, PhD(biol), 66. *Prof Exp:* Instr, Fairfax Co Sec Sch Syst, 58-63; asst zool, Univ Ky, 63-66; asst prof biol & zool, ETenn State Univ, 66-68; from asst prof to assoc prof, 68-78, PROF BIOL, OLD DOMINION UNIV, 78- *Concurrent Pos:* Vis assoc cur, Dept Invert Zool, Smithsonian Inst, 72-73 & res assoc, 90-; eminent scholar, Old Dominion Univ, 90- *Mem:* AAAS; hon life mem Nat Speleol Soc; Soc Syst Zool; Crustacean Soc; Sigma Xi. *Res:* Systematics, ecology and zoogeography of freshwater amphipod crustaceans of North America, especially subterranean forms; zoogeography of cavernicolous invertebrates of the Appalachians; groundwater ecology. *Mailing Add:* Dept Biol Old Dominion Univ Norfolk VA 23529-0266

HOLSINGER, KENT EUGENE, b Oregon City, Ore, Oct 15, 56. THEORETICAL POPULATION GENETICS, PLANT POPULATION GENETICS. *Educ:* Col Idaho, BS, 78; Stanford Univ, PhD(biol sci), 82. *Prof Exp:* Res fel, Miller Inst Basic Res Sci, Univ Calif, Berkeley, 82-84; res assoc, Dept Biol Sci, Stanford Univ, 84-86; ASST PROF ECOL & EVOLUTIONARY BIOL, UNIV CONN, 86- *Concurrent Pos:* Assoc, dept genetics, Agr Exp Sta, Univ Calif, Davis, 85, adj lectr, 85. *Mem:* Soc Study Evolution; Am Soc Naturalists; Genetics Soc Am; Bot Soc Am; Am Soc Plant Taxonomists; Int Asn Plant Taxonomists. *Res:* Plant evolutionary biology, including systematics; theoretical population genetics, especially the evolution and consequences of self-fertilization in plants; comparative and experimental studies on genetic and phenotypic variation in natural populations of plants; philosophy of biology, especially theories of systematics and theories of explanation in the fields of organismal and evolutionary biology. *Mailing Add:* Dept Ecol & Evolutionary Biol Univ Conn U-43 75 N Eagleville Rd Storrs CT 06269-3043

HOLSINGER, VIRGINIA HARRIS, b Washington, DC. FOOD SCIENCES & TECHNOLOGY. *Educ:* Col William & Mary, BS, 58; Ohio State Univ, PhD(food sci & nutrit), 80. *Prof Exp:* Res chemist, 58-80, SUPVR RES CHEMIST, US DEPT AGR, 80- *Honors & Awards:* Col Rohland A Isker Award, 84; Indust Achievement Award (Team), Inst Food Technologists, 87. *Mem:* Am Chem Soc; Am Dairy Sci Asn; Inst Food Technologists. *Res:* Plan, supervise and execute research directed toward maintenance of quality and safety of processed dairy products and other foods. *Mailing Add:* Agr Res Serv US Dept Agr 600 E Mermaid Lane Philadelphia PA 19118

HOLSON, RALPH ROBERT, b Nashville, Tenn, July 26, 41; m 66; c 2. NEUROBIOLOGY OF NEOCORTEX, DEVELOPMENTAL PSYCHOBIOLOGY. *Educ:* Univ Calif, Berkeley, BA, 66; Univ Wash, PhD(psychol), 84. *Prof Exp:* Staff fel, 84-86, res psychologist, 86-88, ACTG CHIEF, PERINATAL & POSTNATAL EVAL BR, DIV REPRODUCTIVE & DEVELOP TOXICOL, NAT CTR TOXICOL RES, 88- *Mem:* Behav Teratology Soc; Soc Neurosci; Int Soc Develop Psychobiol; AAAS. *Res:* Effect of perinatal drug exposure on the developing dopamine system; effect of isolation rearing on brain and behavior; development and function of the pre-frontal cortex. *Mailing Add:* Nat Ctr Toxicol Res HFT-134 Nat Ctr Toxicol Res Dr Jefferson AR 72079

HOLST, EDWARD HARLAND, b Beaumont, Tex, Jan 25, 30; m 49; c 5. ORGANIC CHEMISTRY. *Educ:* NTex State Col, BA, 51, MS, 53; Pa State Univ, PhD(org chem), 55. *Prof Exp:* Asst, Pa State Univ, 54; res chemist, Texaco, Inc, 55-61, sr proj chemist, 61-67, asst supvr chem res, 67-74, supvr res, 74-87; CONSULT, 89. *Mem:* AAAS; Am Chem Soc. *Res:* Polymers; metal organic condensation and olefin metal alkyl catalysts; organic synthesis; fuel and lubricant additives process development; feasibility studies of commercial foreign joint ventures. *Mailing Add:* 2909 Lawrence Nederland TX 77627

HOLST, GERALD CARL, b Brooklyn, NY, Oct 29, 42. ELECTRO-OPTICAL SYSTEM ANALYSIS, THERMAL IMAGING. *Educ:* Brooklyn Polytech Inst, BSEE, 63; Univ Conn, MS, 65, PhD(mat sci), 68; George Washington Univ, MEngAd, 81. *Prof Exp:* Physicist, Am Optical Corp, 66-68; physicist, Frankford Arsenal, US Army, Philadelphia, 68-77, physicist, Chem Systs Lab, Aberdeen Proving Ground, Md, 77-84; SR TECH STAFF MEM, MARTIN MARIETTA, ORLANDO, FLA, 84- *Concurrent Pos:* Chmn, NATO Proj Group 16, 81-84; lectr, Martin Marietta Cont Educ Prog, 87-; adj prof, Univ Cent Fla, 88-; chmn, Conf Infrared Imaging Systs, Soc Photo-Optical Instrumentation Engrs, 90- *Mem:* Optical Soc Am; Soc Photo-Optical Instrumentation Engrs. *Res:* Knowledgeable in laser safety issues; researched single and multiple scattering by aerosols; measured image transmission and other optical phenomena through aerosols; fully conversant with infrared electro-optical system theory, modeling and testing; authored over 60 publications. *Mailing Add:* 2932 Cove Trail Maitland FL 32751

HOLST, JEWEL MAGEE, computer sciences; deceased, see previous edition for last biography

HOLST, TIMOTHY BAILEY, b Minneapolis, Minn, Mar 24, 51; m 74. STRUCTURAL GEOLOGY, TECTONICS. *Educ:* Univ Minn, BA, 73, PhD(geol), 77. *Prof Exp:* asst prof geol, Hope Col, 77-79; ASST PROF GEOL, UNIV MINN, DULUTH, 79- *Mem:* Geol Soc Am; Sigma Xi. *Res:* Methods of strain determination in rocks; regional joint patterns; joint formation; theory and modeling of fold development in rock; mechanism of overthrust faulting. *Mailing Add:* Dept Geol Univ Minn-Duluth Duluth MN 55812

HOLST, WILLIAM FREDERICK, US citizen. COMPUTER SCIENCES. *Educ:* Mass Inst Technol, BS & MS, 55. *Prof Exp:* Staff mem, MIT Lincoln Lab, 55-62; SR SCIENTIST, ARCON CORP, 62- *Concurrent Pos:* Dir develop, Union Spec Marcon Corp, 73-75. *Mem:* Inst Elec & Electronics Engrs; AAAS; fel Brit Interplanetary Soc. *Res:* Computer interactive graphics systems and civilian air traffic control. *Mailing Add:* Arcon Corp 260 Bear Hill Rd Waltham MA 02154

HOLSTEIN, ARTHUR G, b Mason City, Iowa, Oct 13, 11; m 37; c 2. BIOCHEMISTRY. *Educ:* Univ Ill, BS, 33. *Prof Exp:* Chief control chemist, Keokuk Electro Metals, 33-34 & anal & develop, Armour By-Prod, 34-37; tech sales adv, Gen Dye Stuffs Corp, 37-38; res chemist, Visking Corp, 38-42; ord officer, US Army, 42-46; chem eng & prod, Chicago Copper & Chem Co,

46-47; mgr, Chem Div, Pfanstiehl Chem Co, 47-54, pres, Pfanstiehl Labs, Inc, 54-75; RETIRED. *Concurrent Pos:* Mem sub-comt carbohydrates & comt specifications & criteria for biochem compounds & chmn subcomt carbohydrates, Nat Acad Sci-Nat Res Coun. *Honors & Awards:* Melville Wolfrom Award, Carbohydrate Div, Am Chem Soc. *Mem:* Am Chem Soc; Am Inst Chemists. *Res:* Cellulose viscose; carbohydrates. *Mailing Add:* One Laurens St Beaufort SC 29902

HOLSTEIN, BARRY RALPH, b Youngstown, Ohio, Nov 19, 43; m 66; c 2. THEORETICAL PHYSICS. *Educ:* Carnegie Inst Technol, BS, 65, MS, 67; Carnegie-Mellon Univ, PhD(physics), 69. *Prof Exp:* Instr physics, Princeton Univ, 69-71; from asst prof to assoc prof, 71-79, PROF PHYSICS, UNIV MASS, 79- *Concurrent Pos:* Vis fel, Princeton Univ, 75-76; prog officer theoret physics, NSF, 77-79; vis prof, Princeton Univ, 85. *Mem:* Fel Am Phys Soc; Am Asn Physics Teachers. *Res:* Theoretical physics, especially weak interaction theory; study of nonleptonic weak processes and of weak decays of nuclei. *Mailing Add:* Dept Physics & Astron Univ Mass Amherst MA 01003

HOLSTEIN, THOMAS JAMES, b Lansdowne, Pa, May 8, 43; m 65; c 2. DEVELOPMENTAL GENETICS. *Educ:* Providence Col, BS, 65; Brown Univ, MS, 67, PhD(biol), 69. *Prof Exp:* TEACHER BIOL, ROGER WILLIAMS COL, 69- *Concurrent Pos:* Vis investr, Brown Univ, 78- *Mem:* AAAS; Int Pigment Cell Soc. *Res:* Developmental genetics of the multiple forms of tyrosinase from mammalian melanocytes; chemically induced hypo and hypermelanosis. *Mailing Add:* 178 Poplar Dr Cranston RI 02920

HOLSTEN, JOHN ROBERT, b Tulsa, Okla, Feb 27, 31; m 51; c 3. ORGANIC CHEMISTRY, TEXTILE CHEMISTRY. *Educ:* Vanderbilt Univ, BA, 52, MA, 55, PhD(chem), 57. *Prof Exp:* Sr res chemist, Chemstrand Res Ctr, Inc, NC, 56-71; res specialist, M Lowenstein Corp, 71-75, dir chem res, 75-78, dir reg affairs, 78-83, chemist, 83-87; RES CHEMIST, SPRINGS INDUSTS INC, 87- *Mem:* Am Chem Soc; Am Asn Textile Chemists & Colorists; Am Oil Chemists Soc. *Res:* Condensation-type polymerizations; high temperature stable polymer systems; organic intermediates; acetylene chemistry; solvent effects; fabric flammability and flame retardants; textile chemical environmental and health problems; electroforming technology; textile printing technology. *Mailing Add:* Res & Develop Ctr Springs Industs Inc PO Box 70 Ft Mill SC 29715

HOLSTEN, RICHARD DAVID, b Wilmington, Del, Jan 26, 37; m 56; c 4. PLANT PHYSIOLOGY, BIOCHEMISTRY. *Educ:* Temple Univ, AB, 61; Cornell Univ, PhD(plant physiol), 65. *Prof Exp:* From instr to asst prof bot, Cornell Univ, 64-66; biologist, VCent Res Dept, Exp Sta, E I du Pont de Nemours, 66-71, res supvr, Instrument Prod Div, 71-74, explor res & develop supvr & biol Prod mgr, 71-77, mgr clin specialties, Instrument Prod Div, 77-79, supvr, Immunol Cent Res Dept, 79-81, mgr Biotechnol Prod Develop, 81-83, staff adminr molecular biol, Cent Res Dept, 83-87, res mgr, AG Biol, Agr Prod Dept, 87-88, PROD MGR, AG DIAGNOSTICS, AGR PROD DEPT, E I DU PONT DE NEMOURS, 87- *Mem:* AAAS; Sigma Xi. *Res:* Biochemistry of cell differentiation; biochemistry and physiology of nitrogen fixation; clinical chemistry; immunology; molecular biology. *Mailing Add:* Agr Prod Dept E I du Pont de Nemours & Co Inc Wilmington DE 19898

HOLSTIUS, ELVIN ALBERT, b Woonsocket, RI, Feb 9, 21; m 42; c 2. PHARMACY. *Educ:* RI Col Pharm, BS, 43; Purdue Univ, MS, 49, PhD(pharm chem), 50. *Prof Exp:* Pharm res chemist, Burroughs & Wellcome, NY, 45-47; sr chemist, Merck & Co, Inc, NJ, 50-53; assoc prof pharm, Univ Kansas City, 53-56, dir pharmaceut res, 54-56; dir pharm develop labs, Res Div, Geigy Chem Corp, 56-67; tech dir, Endo Labs, Inc, 67-68; DIR PHARMACEUT RES & DEVELOP, BURROUGHS WELLCOME CO, RESEARCH TRIANGLE PARK, 68- *Mem:* NY Acad Sci; Acad Pharmaceut Sci; Am Chem Soc; Am Pharmaceut Asn. *Res:* Physical characteristics which affect drugs and tableting. *Mailing Add:* Burroughs Wellcome Co PO Box 1887 Greenville NC 27834

HOLSTON, JAMES L, b Montgomery, Ala, Feb 9, 42; m 64; c 2. PUBLIC HEALTH. *Educ:* Huntingdon Col, BA, 64; Univ NC, MS, 70, PhD, 72. *Prof Exp:* Microbiologist, Clin Lab Admin, Ala Dept Pub Admin, 64-69, asst dir labs, 73-78, dir labs, 78-87; DOCTRITE PUB HEALTH, 87- *Mem:* Sigma Xi; Am Pub Health Asn. *Res:* Development of a neuraminidase hemagglutination assay for Hong Kong influenza; development and evaluation of an isolation and transport system for gonorrhoeae. *Mailing Add:* 4607 Kelly Ann Ct Montgomery AL 36109

HOLSWORTH, WILLIAM NORTON, wildlife ecology, for more information see previous edition

HOLSZTYNSKI, WLODZIMIERZ, b Nizny Tagil, USSR, Mar 3, 42; US citizen; m 62; c 4. MATHEMATICS, COMPUTER ARCHITECTURE. *Educ:* Univ Warsaw, MS, 62, PhD(math), 65. *Prof Exp:* Mem staff math, Inst Advan Studies, 73-74; lectr, Univ Southern Ill, 74-75; from assoc prof to prof math, Univ Western Ont, 75-79; mem sr prof staff, Martin Marietta Aerospace, 81-85; CONSULT, 85- *Concurrent Pos:* Fel, Lomonosoff Univ, 67-68; grant, Univ Western Ont, 75-76; Nat Res Coun Can grants, 76-77 & 77-80; res mathematician, Environ Res Inst Mich, 77-78, sr res mathematician, 78-81; vis prof, Bowling Green State Univ, 78 & 84; consult, 85-; Fel, Lomonosoff Univ, 67-68; res grants, NSF, 69-74. *Res:* Mathematics, especially topology and algebra; statistical mechanics; computer architecture, especially parallel processors, image processing, data compression; patents parallel processing-image processing including patents for Geometric Arithmetic Parallel Processor. *Mailing Add:* 14000 Short Hill Ct Saratoga CA 95070

HOLT, ALAN CRAIG, b Camp Lejeune, NC, Mar 16, 45; m 70; c 1. SPACE UTILIZATION & PAYLOAD INTEGRATION, ADVANCED SPACE SYSTEMS CONCEPTS. *Educ:* Iowa State Univ, BS, 67; Univ Houston, MS, 79. *Prof Exp:* Exp process specialist, Manned Spacecraft Ctr, NASA, 67-70, Skylab process & training specialist, 70-74, Spacelab crew opers specialist, Johnson Space Ctr, 74-78, Spacelab syst training supvr, 78-80, payload training group leader, 80-84, tech mgr, Space Sta Prog, 84-87, chief, Regt & Accom Br, NASA Hq, 87-89, DEP MGR, PAYLOAD INTEL OFF, SPACE STA FREEDOM PROG OFF, NASA HQ, 89- *Concurrent Pos:* Pres & dir, Holt Res & Develop, 83-86; prin investr, advan protective systs, 85-88; consult, Deep Space Propulsion/Interstellar Flight, 90-; lectr, nano systs & laser technol applications. *Mem:* Sr mem Am Inst Aeronaut & Astronaut; Am Astron Soc; Nat Classification Mgt Soc; Nat Space Soc. *Res:* Space station and space transport utilization and operations; advanced protection and propulsion systems; integration of new materials and nano/micro systems; advanced laser effects. *Mailing Add:* NASA Space Sta Freedom Prog Off Code MSU-1 10701 Parkridge Blvd Reston VA 22091

HOLT, BEN DANCE, b Tenn, June 3, 19; m 43; c 4. ATMOSPHERIC CHEMISTRY, MECHANISMS OF ATMOSPHERIC SULFATE FORMATION. *Educ:* Univ Tenn, BS, 41; Univ Chicago, SM, 51; Ill Inst Technol, PhD, 69. *Prof Exp:* Jr chemist, Tenn Valley Auth, 41-43; from jr chemist to assoc chemist, 44-72, CHEMIST, ARGONNE NAT LAB, 72- *Honors & Awards:* Mat Sci Award, US Dept Energy, 83. *Mem:* Am Chem Soc; Sigma Xi. *Res:* Gases in metals and oxides by high temperature; mass spectrometric techniques; environmental media by measurement of stable isotope changes; non-equilibrium gas-solid interactions. *Mailing Add:* Argonne Nat Lab Bldg 205 9700 S Cass Ave Argonne IL 60439

HOLT, CHARLENE POLAND, b Memphis, Tenn, Mar 2, 38; m 75; c 2. PEDIATRICS, ONCOLOGY. *Educ:* Fla Southern Col, BS, 59; Univ Miami, MD, 63; Am Bd Pediat, dipl. *Prof Exp:* Intern med pediat, Baptist Hosp, Memphis, 64, resident, 65; res fel pediat oncol, St Jude Children's Res Hosp, 66-68; asst prof pediat, Univ Tenn, 68-69; asst prof pediat, Med Ctr, Univ Colo, Denver, 69-75, assoc prof, 75-77; dir pediat oncol, Children's Hosp, 69-77; mem staff adolescent med & chief pediat oncol, Madigan Army Med Ctr, Tacoma Wash, 78-81, MEM, PEDIAT STAFF & CLIN ONCOL, GORGAS ARMY MED DEPT ACTIV, US ARMY MED CORP, REPUB PANAMA, 81- *Concurrent Pos:* Pediatrician chemother phase I-II liaison comt, Nat Cancer Inst; exec dir, Colo Regional Cancer Ctr, Inc, 74-75; pediat oncologist, Mountain States Tumor Inst, Idaho; dir, Intermountain Youth Cancer Ctr, Boise, Idaho; pvt practice, Weiser Idaho; assoc prof, Univ Wash, 78-81. *Mem:* Am Med Women's Asn; Am Asn Cancer Res; Am Soc Clin Oncol; Int Soc Lymphology; Int Soc Pediat Oncol. *Res:* Evaluation of new treatment modalities and chemotherapeutic agents in pediatric neoplasia. *Mailing Add:* 1500 Wallace Blvd Amarillo TX 79106

HOLT, CHARLES A(SBURY), b Staunton, Va, Feb 22, 21; m 47; c 4. ELECTRICAL ENGINEERING. *Educ:* US Mil Acad, BS, 43; Univ Ill, MS, 47, PhD(elec eng), 49. *Prof Exp:* Asst prof elec eng, US Mil Acad, 51-54; prof elec eng, Va Polytech Inst & State Univ, 54-85; RETIRED. *Mem:* Am Soc Eng Educ. *Res:* Electromagnetic waves; semiconductors; electric and magnetic phenomena in materials. *Mailing Add:* 1815 Winston Rd Charlottesville VA 22903

HOLT, CHARLES LEE ROY, JR, b Evanston, Ill, June 10, 24; m 46; c 1. GEOLOGY, STRATIGRAPHY & SEDIMENTATION. *Educ:* Univ Tex, BS, 48, MA, 50. *Prof Exp:* Geologist, Tex Bd Water Engrs, 48-49; geologist, US Geol Surv, Ga, 50-59, dist geologist, 59-66, dist chief, Wis, 66-75, prog & planning officer, Southeastern Region, Water Resource Div, 75-80; CONSULT, 80- *Concurrent Pos:* Mem US nat comt, Int Asn Hydrologists; pres, Hydroenological Asn, Hydrologists Am; consult groundwater, Amman Jordan, 80-84, Muscat, Oman, 84 & San Antonio, Tex, 85-88. *Mem:* Fel Geol Soc Am; Am Water Resources Asn; Am Water Works Asn; Am Geophys Union; Soil Conserv Soc Am. *Res:* Ground water geology; water resources; limestone hydrology. *Mailing Add:* PO Box 164 Port Aransas TX 78373

HOLT, CHARLES STEELE, b Chicago, Ill, June 18, 37; m 63; c 2. LIMNOLOGY, AQUATIC BIOLOGY. *Educ:* Beloit Col, BS, 59; Univ Wis-Madison, MS, 62; Univ Minn, PhD(fisheries sci), 65. *Prof Exp:* From asst prof to assoc prof, 65-76, PROF BIOL, BEMIDJI STATE UNIV, 76- *Concurrent Pos:* Consult, US Army CEngr, 73-74; vis scientist, Sci Mus Minn, St Paul, 76-77. *Mem:* Am Fisheries Soc; Am Soc Limnol & Oceanog. *Res:* Eutrophication of aquatic ecosystems; biotelemetry studies of walleye. *Mailing Add:* Dept Biol Bemidji State Univ Bemidji MN 56601

HOLT, DAVID LOWELL, b Greensville, SC, Apr 16, 61. PROCESS FLOWSHEET DEVELOPMENT. *Educ:* Clemson Univ, BS, 83, MS, 85. *Prof Exp:* Process engr, E I du Pont de Nemours & Co, 84-89; TECH MGR, WESTINGHOUSE SAVANNAH RIVER SITE, 89- *Mem:* Am Inst Chem Engrs. *Res:* Spent nuclear fuel processing. *Mailing Add:* 826 Riverfront Dr Augusta GA 30901

HOLT, DONALD ALEXANDER, b Joliet, Ill, Jan 29, 32; m 53; c 4. RESEARCH ADMINISTRATION, AG SYSTEMS. *Educ:* Univ Ill, Urbana, BS, 54, MS, 56; Purdue Univ, PhD(agron), 67. *Prof Exp:* Instr agron, Univ Ill, 56; self-employed, 56-63; from instr to assoc prof, 64-77, PROF AGRON, PURDUE UNIV, 77-; DIR, ILL AGR EXP STA & ASSOC DEAN, COL AGR, UNIV ILL. *Mem:* Fel Am Soc Agron; Crop Sci Soc Am. *Res:* Physiological response of plants to environmental changes; diurnal variation in plant constituency; hay management; large area yield prediction models; computer simulation of plant growth. *Mailing Add:* 1002 Ross Dr Urbana IL 61821-6632

HOLT, EDWARD C(HESTER), JR, b Philadelphia, Pa, Dec 23, 23; m 57; c 3. STRUCTURAL ENGINEERING. *Educ:* Mass Inst Technol, BS, 45, MS, 47; Pa State Univ, PhD(civil eng), 56. *Prof Exp:* Instr civil eng, Pa State Univ, 49-56; from asst prof to assoc prof, 56-84, PROF CIVIL ENG, RICE UNIV, 84- *Mem:* Am Soc Civil Engrs; Am Soc Eng Educ. *Res:* Structural engineering. *Mailing Add:* Dept Civil Eng Rice Univ Box 1892 Houston TX 77251-1892

HOLT, ETHAN CLEDDY, b Brilliant, Ala, Feb 6, 21; m 44; c 2. PLANT BREEDING, AGRONOMY. *Educ:* Auburn Univ, BS, 43 Purdue Univ, MS, 48, PhD(plant breeding), 50. *Prof Exp:* From asst prof to assoc prof grass breeding, 48-57, PROF FORAGE CROPS, TEX A&M UNIV, 57- *Honors & Awards:* Tex A&M Univ Distinguished Achievement Award Res, 70. *Mem:* Fel Am Soc Agron; Crop Sci Soc Am. *Res:* Grass breeding, including variety development, methodology, genetics, heritability and cytogenetics, especially warm-season grasses; forage physiology; pasture management. *Mailing Add:* 1110 Ashburn Ave E College Station TX 77840

HOLT, FREDERICK SHEPPARD, b Washington, DC, July 12, 20; m 49; c 3. MICROWAVE PHYSICS, MICROWAVE OPTICS. *Educ:* Kenyon Col, AB, 41; Mass Inst Technol, PhD(appl math), 50. *Prof Exp:* Res assoc theory group, Radiation Lab, Mass Inst Technol, 42-43, instr math, 47-50; mathematician, Microwave Physics Lab, Rome Air Develop Command, Air Force Cambridge Res Labs, 59-75, physicist, Electromagnetic Sci Div, 76-80; prof math, Tufts Univ, 70-85; RETIRED. *Concurrent Pos:* From asst prof to assoc prof, Tufts Univ, 55-69. *Mem:* Sigma Xi. *Res:* Design and analysis of antennas; microwave optics devices; 11 Air Force patent awards. *Mailing Add:* 46 Emerson Rd Winchester MA 01890

HOLT, HARVEY ALLEN, b San Jose, Calif, Oct 15, 43; m 64; c 2. FOREST ECOLOGY. *Educ:* Okla State Univ, BS, 65; Ore State Univ, MS, 67, PhD(forest ecol), 70. *Prof Exp:* Res asst forestry, Ore State Univ, 65-70; asst prof, Univ Ark, Fayetteville, 70-75; assoc prof, 75-83, PROF FORESTRY, PURDUE UNIV, 83- *Mem:* Soc Am Foresters; Weed Sci Soc Am; Coun Agr Sci & Technol; Int Soc Arboricult; Am Railway Eng Asn. *Res:* Effects of types and levels of vegetation on the desired species; rights-of-way and industrial vegetation management. *Mailing Add:* Dept Forestry & Natural Res Purdue Univ W Lafayette IN 47907

HOLT, HELEN KEIL, b West Palm Beach, Mar 23, 37; m 58; c 2. ATOMIC PHYSICS, LASERS. *Educ:* Barnard Col, BA, 58; Yale Univ, MS, 60, PhD(physics), 65. *Prof Exp:* Physicist, Ctr Absolute Phys Quantities, Nat Bur Standards, 65-86; CONSULT, 86- *Concurrent Pos:* Wood Wilson fel. *Mem:* Fel Am Phys Soc; Sigma Xi. *Mailing Add:* 6740 Melody Lane Bethesda MD 20817

HOLT, IMY VINCENT, b Billings, Okla, Apr 3, 19; m 46; c 1. BOTANY. *Educ:* NMex Col, BS, 50; Iowa State Col, MS, 52, PhD, 53. *Prof Exp:* Asst, Iowa State Col, 50-52, instr, 52; from asst prof to assoc prof bot, Okla State Univ, 53-60; from assoc prof to prof, 60-84, EMER PROF BIOL, WESTERN MICH UNIV, 84- *Mem:* Am Soc Agron; Sigma Xi. *Res:* Plant morphology; floral initiation. *Mailing Add:* Dept Biol Western Mich Univ Kalamazoo MI 49008

HOLT, JAMES ALLEN, b Antlers, Okla, May 24, 34; m 61; c 2. BIOCHEMISTRY. *Educ:* Okla Baptist Univ, AB, 56; Univ Ill, PhD(biochem), 61. *Prof Exp:* Res assoc biochem, Rockefeller Univ, 61-62; from asst prof to assoc prof chem, Okla Baptist Univ, 62-68; sr lectr & chmn dept, Fac Basic Sci, Njala Univ Col, Univ Sierra Leone, 66-67; asst prof chem, Williams Col, 68-71; prof chem, Univ Agr Malaysia, Peace Corps, 71-73; sci curric specialist, Maret Sch, 73-74; spec asst, Off Int Progs, 74-76, Prog Mgr, US-East Asia Coop Sci Prog, 76-81, Dep Dir, Div Int Prog, NSF, 81-84; DEAN, WARREN WILSON COL, 84- *Concurrent Pos:* Consult, Jersey Prod Res Co, 62. *Mem:* Sigma Xi. *Res:* Physical and chemical characterization of enzymes; mechanisms of enzyme catalysis; science education, policy and administration; management of international scientific activities; curriculum design and professional development in higher education; education with production. *Mailing Add:* 3016 Tilden St NW Apt 201 Washington DC 20008

HOLT, JAMES FRANKLIN, b Murray City, Ohio, June 14, 24; m 51. SOLAR PHOTOVOLTAICS, MAGNETOHYDRODYNAMICS. *Educ:* Ohio State Univ, BSc, 49, MSc, 51, PhD(physics), 62. *Prof Exp:* Electronics engr, Farnsworth Electronics Co, 51-56; res assoc plasma physics, Ohio State Univ, 58-62, assoc res supvr, 62-66; physicist, USAF Aero Propulsion Lab, Wright-Patterson AFB, 66-85; PHYSICIST, IAP RES, INC, 85- *Concurrent Pos:* Vis lectr, Ohio Acad Sci, 64-68. *Mem:* Am Phys Soc; Sigma Xi. *Res:* Magnetohydrodynamic power generation experiments and theory; gas lasers; electromagnetic heating and confinement of plasmas; thermonuclear fusion; television electronics; infrared radiation detection; solar V5 cells and solar power; EM accelerators. *Mailing Add:* 3795 Osborn Rd Medway OH 45341

HOLT, JOHN FLOYD, b Pittsburgh, Pa, Jan 20, 15; m 42; c 3. RADIOLOGY. *Educ:* Univ Pittsburgh, BS & MD, 38. *Prof Exp:* Asst resident, Univ Mich, 39-40, resident, 40-41, from instr to assoc prof, 41-53, prof, 53-84, EMER PROF RADIOL, MED CTR, UNIV MICH, ANN ARBOR, 84- *Honors & Awards:* Hon Lectr, Soc Pediat Radiol, Neuhauser LEC, 77; Gold Medal, Soc Pediat Radiol, 90. *Mem:* Int Skeletal Soc; fel Am Col Radiol; Soc Pediat Radiol (pres, 60-61); Asn Univ Radiol (pres, 56-57); Sigma Xi; Am Roentgenol Ray Soc (vpres, 65-66). *Res:* Clinical pediatric radiology; generalized bone dysplasias. *Mailing Add:* 250 Orchard Hill Dr Ann Arbor MI 48104

HOLT, JOHN GILBERT, b Buffalo, NY, Dec 16, 29; m 55; c 3. MICROBIOLOGY. *Educ:* Cornell Univ, BS, 52; Syracuse Univ, MS, 54; Purdue Univ, PhD(microbiol), 60. *Prof Exp:* Instr microbiol, Purdue Univ, 60; res assoc, Iowa State Univ, 60-62, USPHS fel, 62, from asst prof to prof microbiol, 63-90; PROF MICROBIOL, MICH STATE UNIV, 90- *Honors & Awards:* J Roger Porter Award, US Fed Cult Collections, 85. *Mem:* Am Soc Microbiol; Can Soc Microbiol; Sigma Xi. *Res:* Classification and nomenclature of the bacteria; ecology of soil bacteria. *Mailing Add:* 4351 Oakwood Dr Okemos MI 48864

HOLT, JOSEPH PAYNTER, SR, b Versailles, Ky, June 19, 11; m 40; c 2. PHYSIOLOGY. *Educ:* Univ Ky, BS, 32; Univ Louisville, MD, 36, MS, 37; Univ Chicago, PhD, 40. *Prof Exp:* Asst physiol, Med Sch, Univ Louisville, 34-36; intern, Marine Hosp, USPHS, New Orleans, 36-37; from instr to assoc prof, Med Sch, Univ Louisville, 37-45; asst clin prof indust med, NY Univ, 47-49, assoc prof, Post Grad Med Sch, 49-52; dir, Inst Med Res, 52-57, prof heart res, 52-77, EMER PROF HEART RES, SCH MED, UNIV LOUISVILLE, 77- *Concurrent Pos:* Med res dir, Standard Oil Co (NJ), 45-50. *Mem:* Am Physiol Soc; Biophys Soc; Am Fedn Clin Res; NY Acad Med; fel NY Acad Sci. *Res:* Physiology of the circulatory system; toxicology. *Mailing Add:* 9810 Sussex St Naples FL 33942-1624

HOLT, MATTHEW LESLIE, b Ellsworth, Iowa, June 19, 04; m 52. ELECTROCHEMISTRY. *Educ:* St Olaf Col, BA, 26; Univ Wis, MS, 28, PhD(inorg chem), 30. *Prof Exp:* Asst, 26-30, from instr to prof, 30-73, EMER PROF CHEM, UNIV WIS-MADISON, 73- *Concurrent Pos:* Instr, US Army Univ, France, 45-46; vis sr chemist, Argonne Nat Lab, 50. *Mem:* Am Chem Soc; Electrochemical Soc. *Res:* Inorganic chemistry; electrodeposition of tungsten alloys and less common metals and alloys. *Mailing Add:* 3502 Blackhawk Dr Madison WI 53705

HOLT, MAURICE, b Wildboarclough, Eng, May 16, 18; US citizen; m 42; c 5. APPLIED MATHEMATICS. *Educ:* Univ Manchester, BSc, 40, MSc, 44, PhD, 48. *Prof Exp:* Asst lectr appl math, Univ Liverpool, 48-49; lectr, Univ Sheffield, 49-51; prin sci off, Ministry of Supply, 52-55; vis lectr math, Harvard Univ, 55-56; assoc prof appl math, Brown Univ, 56-60; vis prof, 60-63, prof, 63-88, EMER PROF AERONAUT SCI, UNIV CALIF, BERKELEY, 88- *Concurrent Pos:* Consult, US Aerospace Co & US Res Labs. *Mem:* Fel Am Phys Soc; fel Am Inst Aeronaut & Astronaut. *Res:* Fluid dynamics; non-linear theory of compressible flow; hypersonic aerodynamics; shock wave phenomena; flutter and vibrations; numerical treatment of partial differential equations of hyperbolic and mixed type. *Mailing Add:* Div Mech Eng Univ Calif Berkeley CA 94720

HOLT, PERRY CECIL, b Overton County, Tenn, June 26, 12; m 42. INVERTEBRATE ZOOLOGY, SYSTEMATICS. *Educ:* Tenn Polytech Univ, BS, 42; Univ Va, MA, 48, PhD(biol), 51. *Prof Exp:* Instr biol, Univ Richmond, 48-50; from asst prof to assoc prof, ETenn State Col, 50-56; from asst prof to assoc prof, 56-65, prof zool, 65-78, dir, Ctr Systs Collections, 72-76, EMER PROF ZOOL, VA POLYTECH INST & STATE UNIV, 78- *Concurrent Pos:* Found Advan Educ fel, Univ Chicago, 55-56; vis res assoc, Smithsonian Inst, 67-68. *Mem:* Fel AAAS; Asn Southeastern Biologists (pres); Am Soc Zoologists; Am Micros Soc. *Res:* Morphology, taxonomy and zoogeography of the Branchiobdellida. *Mailing Add:* 1308 Crestview Dr Blacksburg VA 24061

HOLT, PETER STEPHEN, b Aberdeen, Md, Feb 19, 51; m 75. IMMUNOLOGY. *Educ:* Southern Ill Univ, BA, 73; Cath Univ Am, MS, 78; Univ Mo, PhD(microbiol), 85. *Prof Exp:* Sci asst microbiol, Walter Reed Army Inst Res, 75-78; chief bact, Letterman Army Med Ctr, 78-81; res assoc, Vet Toxicol & Entom Res Lab, 85-87, RES IMMUNOL, SOUTHEAST POULTRY RES LAB, AGR RES SERV, USDA, 87- *Mem:* Am Asn Immunologists; Am Soc Microbiol; Am Asn Vet Immunologists; Am Asn Avian Pathologists. *Res:* Stress effects on immunity in chickens; role of various components of chicken immune response in protecting against Salmonella enteritidis infection; development of intestinal mucosal immunity in the chicken. *Mailing Add:* 934 College Station Rd Athens GA 30605

HOLT, RICHARD A(RNOLD), b New York, NY, Sept 4, 42; m 78; c 3. LASER SPECTROSCOPY, ATOMIC BEAMS. *Educ:* Harvard Univ, AB, 64, AM, 66, PhD(physics), 73. *Prof Exp:* Fel physics, Brown Univ, 73-74; res scientist, 74-76, asst prof, 76-81, ASSOC PROF PHYSICS, UNIV WESTERN ONT, 82- *Concurrent Pos:* Assoc researcher, Spectrometry Lab, Univ Lyon, France, 82-83; vis scientist, Oxford Univ, 83. *Mem:* Am Phys Soc; Sigma Xi. *Res:* Laser and radiofrequency spectroscopy of neutral and ionic atomic and molecular beams; measurement of fine and hyperfine structure; quantum electrodynamic effects; measurement of excited atomic state lifetimes. *Mailing Add:* Dept Physics Univ Western Ont London ON N6A 3K7 Can

HOLT, RICHARD E(DWIN), b Deer Lodge, Mont, Feb 19, 23; m 44; c 2. METALLURGY. *Educ:* Univ Wash, BS, 47, MS, 57. *Prof Exp:* Partner, Joseph Holt & Co, 47-51; design engr, Boeing Airplane Co, 51-54; from instr to assoc prof mech eng, Univ Wash, 54-80, consult, appl physics lab, 58-80; RETIRED. *Concurrent Pos:* Consult, Leckenby Struct Steel Co, 55, Smith & Murray, 57 & R T Earley & Co, 57-58. *Mem:* Assoc mem Am Soc Mech Engrs; Am Welding Soc. *Res:* Welding metallurgy; flame straightening of metals; welding sequence to control residual stress; effect of pressure on metallurgical reactions. *Mailing Add:* 3531 NE 98th St Seattle WA 98115

HOLT, RICHARD THOMAS, b Derby, Eng, Dec 2, 38; Can citizen; m 69; c 2. PHYSICAL METALLURGY. *Educ:* Univ Sheffield, BMet, 60; Univ BC, PhD(metall), 68. *Prof Exp:* Fel mech eng, Univ Man, 68-70, asst prof civil eng, 70-72; RES OFFICER METALL, NAT RES COUN CAN, 73- *Concurrent Pos:* Secy subcomt aerospace mat & processes, Int Standards Orgn. *Mem:* Am Soc Metals; Asn Prof Engrs. *Res:* Metal and ceramic matrix composites reinforced with ceramic fibers; aluminum alloys for aerospace applications; hot isostatic pressing of metal and ceramic matrix composites; materials for orthopedic implants. *Mailing Add:* Inst Aerospace Res Nat Res Coun Ottawa ON K1A 0R6 Can

HOLT, ROBERT F, soil conservation, for more information see previous edition

HOLT, ROBERT LOUIS, b Blytheville, Ark, Sept 16, 21; m 42; c 1. ORGANIC CHEMISTRY, ANALYTICAL CHEMISTRY. *Educ:* Univ Miss, BS, 51; Univ NC, PhD, 59. *Prof Exp:* Sr chemist fuels res, Res & Tech Dept, Texaco Inc, 58-61; PROF CHEM, NORTHEAST LA UNIV, 61-, HEAD CHEM DEPT, 81-; MEM STAFF, GULF S RES INST, 75- *Mem:* Sigma Xi. *Res:* Instrumentation in organic analysis; pesticide analysis. *Mailing Add:* 913 Middleton Monroe LA 71204

HOLT, ROY JAMES, b Borger, Tex, Jan 22, 47; m 66; c 3. NUCLEAR PHYSICS. *Educ:* Southern Methodist Univ, BS, 69; Yale Univ, MPhil, 71, PhD(physics), 72. *Prof Exp:* Res staff physicist, Yale Univ, 72-74; asst scientist, 74-77, physicist, 77-88, SR PHYSICIST, ARGONNE NAT LAB, 88- *Concurrent Pos:* Mem bd dirs, Continuous Electron Beam Accelerator Facil, 83-84, chmn-elect, 89-; chmn, Gordon Res Conf Photonuclear Reactions, 90. *Mem:* Am Phys Soc. *Res:* Photonuclear reactions and electron scattering studies of nuclei; polarization phenomena; medium energy physics. *Mailing Add:* Physics Div Argonne Nat Lab Argonne IL 60439

HOLT, RUSH DEW, JR, b Weston, WVa, Oct 15, 48; m 85. SOLAR PHYSICS. *Educ:* Carleton Col, BA, 70; NY Univ, MA, PhD(physics), 81. *Prof Exp:* Acoust physicist, New York City Environ Protection Admin, 73; asst prof phys, Swarthmore Col, 80-88; SCI RES SPECIALIST US DEPT STATE, 87- *Concurrent Pos:* Consult, solar energy & environ planning, 75-; Cong scientist fel, Am Phys Soc, 82-83. *Mem:* Am Phys Soc; Sigma Xi; Am Asn Physics Teachers; AAAS. *Res:* Spectral line formation in the solar atmosphere; effect of magnetic fields in solar convection zone; stability of double-diffusive systems; development of solar pond energy systems. *Mailing Add:* 206 South Rd Lindamere Wilmington DE 19809

HOLT, SMITH LEWIS, JR, b Ponca City, Okla, Dec 8, 38; m 63; c 2. INORGANIC CHEMISTRY. *Educ:* Northwestern Univ, BSc, 61; Brown Univ, PhD(inorg chem), 65. *Prof Exp:* Res fel, Brown Univ, 65; NATO res fel, Univ Copenhagen, 65-66; asst prof inorg chem, Polytech Inst Brooklyn, 66-69; assoc prof chem, Univ Wyo, 69-73, prod, 73-78; prof chem & head dept, Univ Ga, 78-80; Okla secy educ, 86-87; PROF CHEM & DEAN ARTS & SCI, OKLA STATE UNIV, 80-86 & 88- *Concurrent Pos:* Fulbright grant & prof assoc, Univ Bordeaux, 74-75. *Mem:* Am Chem Soc; AAAS; Sigma Xi. *Res:* Structural inorganic chemistry; x-ray crystallography; inorganic synthesis; spectroscopy of transition metal ions; solid state chemistry. *Mailing Add:* Deans Off Arts & Sci Col Okla State Univ Stillwater OK 74078

HOLT, STEPHEN S, b New York, NY, May 17, 40; m 61; c 3. HIGH ENERGY ASTROPHYSICS, X-RAY ASTRONOMY. *Educ:* New York Univ, BS, 61, PhD(physics), 66. *Prof Exp:* Instr physics, NY Univ, 64-66; chief, High Energy Astrophys Progs, NASA, Washington, DC, 80-81, dir, Lab High Energy Astrophys, 83-90, ASTROPHYSICIST, GODDARD SPACE FLIGHT CTR, NASA, 66-, DIR, SPACE SCI, 90- *Concurrent Pos:* Lectr physics & astron, Univ Md, 67-88, adj prof astron, 88-; chmn, High Energy Astrophys Div, Am Astron Soc, 83, ed, Exp Astron (Sci J), 89- *Honors & Awards:* Except Sci Achievement Medal, NASA, 77 & 80, Outstanding Leadership Medal, 91. *Mem:* AAAS; fel Am Phys Soc; Am Geophys Union; Am Astron Soc; Int Astron Union; Sigma Xi. *Res:* Balloon, rocket and satellite-borne studies in cosmic ray physics and x-ray astronomy; space-borne scientific investigations, including the Einstein Observatory (the first X-ray astronomy telescope). *Mailing Add:* Code 600 Goddard Space Flight Ctr NASA Greenbelt MD 20771

HOLT, THOMAS MANNING, b Detroit, Mich, July 19, 43; m 63; c 2. NEUROCHEMISTRY. *Educ:* ECarolina Univ, BS, 66, MS, 68; La State Univ, PhD(anat), 74. *Prof Exp:* Instr biol, ECarolina Univ, 68-69; from instr to asst prof anat, Col Med, Univ SFla, 74-78; ASST PROF ANAT, COL MED, MED UNIV SC, 78- *Mem:* Am Asn Anatomists. *Res:* Plasticity of central nervous system catecholaminergic neurons. *Mailing Add:* Holt-Fernandez 12 Country Club Rd Shalimar FL 32579-1606

HOLT, VERNON EMERSON, b Miller, SDak, May 13, 30; m 56; c 4. HEAT TRANSFER, ENVIRONMENTAL CONTROL. *Educ:* SDak Sch Mines & Technol, BS, 51; NC State Univ, MS, 58; Purdue Univ, PhD, 61. *Prof Exp:* Mem lab staff, Gen Elec Co, Wash, 51-54; instr eng, NC State Univ, 56-58; mem tech staff, Bell Tel Labs, NJ, 60-65; asst prof eng mech, NC State Univ, 65-66, assoc prof & asst dean grad sch, 66-68; mem tech staff, Bell Labs, 68-90, supvr environ studies 72-86, design qual assurance, 86-90; RETIRED. *Concurrent Pos:* Consult, Comput Lab, Gen Elec Co, 65-; consult, Hayes Int Co, Ala, 65- *Mem:* Am Phys Soc; Sigma Xi. *Res:* Effect of environment on materials properties; phonon transport; superconductive devices; thermal radiation and emissivities; control of temperature, humidity and pollutants and effects on materials, devices and equipment; design and development quality assurance. *Mailing Add:* 45 Franklin Rd Mendhem NJ 07945

HOLT, WILLIAM HENRY, b San Antonio, Tex, Aug 5, 39; m 63; c 2. SOLID STATE PHYSICS. *Educ:* St Mary's Univ, Tex, BS, 60; Univ Tex, MA, 62, PhD(physics), 67. *Prof Exp:* Fel & lectr physics, Univ Man, 66-69; PHYSICIST, NAVAL SURFACE WARFARE CTR, 69- *Mem:* Am Phys Soc; Can Asn Physicists; Sigma Xi. *Res:* Shock waves in solids; positron annihilation. *Mailing Add:* 906 Carol Lane Falmouth VA 22405

HOLT, WILLIAM ROBERT, b Philadelphia, Pa, Feb 17, 30; m 52. MATHEMATICAL STATISTICS, MEDICAL STATISTICS. *Educ:* Pa State Univ, BS, 52, MF, 56; Ohio Wesleyan Univ, BA, 81. *Prof Exp:* Topog engr, US Geol Surv, 52; forester, Forest Serv, USDA, 52-56, entomologist, Southern Forest Exp Sta, 56-62, res entomologist, Cent States Forest Exp Sta, 62-66, res entomologist, Northeastern Forest Exp Sta, 66-69; biometrician, Merck Sharp & Dohme Res Labs, 69-72; biostatistician, Boehringer Ingelheim Ltd, 72-77; instr math, Miami Univ, 77; math statistician, La Statist Data Serv, Enterprise, Ala, 78-80; math statistician, US Army Aeromed Res Lab, Ft Rucker, Ala, 80-87; RETIRED. *Res:* Repeated measures designs for aeromedicial res; experimental design for aeromedical experiments; applied math modeling for aeromedical problems; sample surveys for aeromedical research. *Mailing Add:* Rte 1 Box 12 Coffee Springs AL 36318

HOLTAN, HEGGIE NORDAHL, b Stoughton, Wis, Apr 10, 09; m 38; c 3. AGRICULTURAL ENGINEERING, HYDROLOGY. *Educ:* Wis State Univ-La Crosse, BA, 33. *Prof Exp:* Lab aide, off res, soil conserv serv, US Dept Agr, Wis, 34-37, sci aide, DC, 37-39, eng aide, Edwardsville, Ill, 40-43, hydraul engr, Elmwood, Ill, 43-47 & Va, 47-54, hydraul engr cent tech unit, DC, 54-56, hydraul engr watershed hydrol, Agr Res Serv, 56-58, invests leader, 58-60, asst br chief, Northeast Br, 60-61, dir, Hydrograph Lab, 61-75; lectr agr eng, Univ Md, 75-81; CONSULT, 81- *Concurrent Pos:* Team mem preparation hydrol guide, Soil Conserv Serv, US Dept Agr, 54-56, mem subcomt hydrol, Interagency Comt Water Resources, 57-, chmn comt preparation field manual for res in agr hydrol, Agr Res Serv, 59-62, mem work group, subcomt on representative & exp basins & subcomt floods & their computation, UNESCO-Int Hydrol Decade, 67-69; mem work group, subcomt on representative & exp basins & subcomt floods & their computation, UNESCO-Int Hydrol Decade, 67-69; mem staff, Agr Exp Sta, Univ Md, 75-80. *Mem:* Soil Conserv Soc Am. *Res:* Watershed hydrology including precipitation and evaporation, soils treatment and vegetation, hydrogeology and hydrodynamics as to their effect on water quality and the hydrologic performance of agricultural watersheds; computer model for watershed hydrology including sediment and chemical transports. *Mailing Add:* 1734 27th Ave Vero Beach FL 32960

HOLTBY, KENNETH F, b Escanaba, Mich, May 18, 22; c 5. AERODYNAMICS. *Educ:* Calif Inst Technol, BS, 47; Mass Inst Technol, MS, 61. *Prof Exp:* Vpres 747 prog, Boeing Aircraft Co, 72-74, vpres 747 prog, 74-78, vpres new progs, 78-81, sr vpres, 81- 87; CONSULT, BOEING, 87- *Honors & Awards:* Design Award, Am Inst Aeronaut & Astronaut. *Mem:* Nat Acad Eng; fel Royal Aeronaut Soc; hon fel Am Inst Aeronaut & Astronaut. *Mailing Add:* 1616 Skyline Way No 1 Anacortes WA 98221

HOLTE, JAMES EDWARD, b Grand Forks, NDak, Apr 9, 31; m 55; c 3. ELECTRICAL ENGINEERING. *Educ:* Univ Minn, BS, 53, MS, 56, PhD(elec eng), 60. *Prof Exp:* From instr to asst prof, 55-63, dir continuing educ eng & sci, 64-69, ASSOC PROF ELEC ENG, UNIV MINN, MINNEAPOLIS, 63- *Mem:* Inst Elec & Electronics Engrs; Am Soc Eng Educ; Sigma Xi. *Res:* Electron gun design; computer-assisted problem solving; distributed system analysis and synthesis; network theory; continuing engineering education. *Mailing Add:* Elec Eng Dept 200 Union St SE Minneapolis MN 55455

HOLTE, KARL E, b Graettinger, Iowa, Apr 25, 31; div; c 3. PLANT TAXONOMY, PLANT ECOLOGY. *Educ:* Augustana Col, SDak, BA, 54; Northern Iowa Univ, MA, 61; Univ Iowa, PhD(bot), 66. *Prof Exp:* from asst prof to assoc prof, 65-83, PROF BOT, IDAHO STATE UNIV, 83- *Concurrent Pos:* Curator, Ray J Davis Herbarium; consult, govt agencies. *Mem:* Am Soc Plant Taxon; Bot Soc Am; Ecol Soc Am; Int Asn Plant Taxon. *Res:* Floristic studies in Idaho; threatened and endangered plants in Idaho and southeastern Oregon. *Mailing Add:* Dept Biol Idaho State Univ Pocatello ID 83209

HOLTEN, DAROLD DUANE, b Devils Lake, NDak, Sept 26, 35; m 62. BIOCHEMISTRY, ENZYMOLOGY. *Educ:* NDak State Univ, BS, 58, MS, 60; Univ NDak, PhD(biochem), 65. *Prof Exp:* Am Cancer Soc fel biochem, Oak Ridge Nat Lab, Tenn, 65-67; res scientist, Armour Pharmaceut Co, Ill, 67-68; from asst prof to assoc prof, 68-79, PROF BIOCHEM, UNIV CALIF, RIVERSIDE, 79- *Mem:* Am Inst Nutrit; Am Soc Biol Chemists; AAAS. *Res:* Mechanisms by which diet regulates the synthesis of lipogenic enzymes and apolipoproteins. *Mailing Add:* Dept Biochem Univ Calif Riverside CA 92521

HOLTEN, VIRGINIA ZEWE, b McKeesport, Pa, Mar 29, 38; m 62; c 1. BIOCHEMISTRY. *Educ:* Mt Mercy Col, Pa, BA, 60; Univ NDak, MS, 62, PhD(biochem), 65; Univ Glasgow, dipl mgt studies, 75. *Prof Exp:* Assoc scientist biochem & enzym, Oak Ridge Inst Nuclear Studies, 65 & biol div, Oak Ridge Nat Lab, 65-67; asst prof biol, Roosevelt Univ, 67-68; fel plant path, Univ Calif, Riverside, 68; instr chem, Riverside City Col, 69-78, asst dean, Div Natural Sci, 78-82; vpres, Victor Valley Col, 82-86; pres, Laffen Col, 86-90; PRES, SOLANO COL, 90- *Mem:* AAAS; Am Chem Soc; Sigma Xi. *Mailing Add:* Solano Col 4000 Suisun Valley Rd Suisun CA 94585

HOLTER, JAMES BURGESS, b New Bethlehem, Pa, Sept 7, 34; m 59; c 2. DAIRY SCIENCE. *Educ:* Pa State Univ, BS, 56, PhD(dairy sci), 63; Univ Md, MS, 59. *Prof Exp:* Assoc prof, 63-78, PROF ANIMAL SCI, UNIV NH, 78- *Concurrent Pos:* Consult, Dairy Nutrit, Venezuela & Can. *Mem:* Am Soc Animal Sci; Am Dairy Sci Asn; Am Inst Nutrit. *Res:* Efficiency of utilization of forages by dairy cattle; energy metabolism; computer dairy ration balancing; optimum supplementation in dairy rations. *Mailing Add:* Dept Animal Sci Univ NH Durham NH 03824

HOLTER, MARVIN ROSENKRANTZ, b Fairport, NY, July 4, 22; m 56; c 2. PHYSICS, MATHEMATICS. *Educ:* Univ Mich, BS, 49, MS, 51 & 58. *Prof Exp:* Res engr, Aeronaut Res Ctr, Univ Mich, 47-53, asst supvr proj Wizard, Missile Defense Systs, 53, supvr, 54, res engr, Willow Run Labs, 54-56; assoc prof eng mech, Univ Toledo, 56-57; mem tech opers & long range planning staffs, Lab Opers & Planning, Willow Run Labs, Univ Mich, 57-58, head sensory devices group infrared lab, 58-64, head infrared & optical sensor lab, Inst Sci & Technol, Willow Run Ctr, 64-70, prof, Sch Natural Resources, Univ, 68-70; chief earth observations div, NASA Manned Spacecraft Ctr, 70-72; dep dir, Willow Run Labs, Univ Mich, 72-73; exec vpres, 73-86, sr vpres, 86-89, CONSULT PHYSICIST, ENVIRON RES INST MICH, 89- *Concurrent Pos:* Mem comt remote sensing for agr purposes, Nat Acad Sci-Nat Res Coun, 61-70; mem div adv group, Aeronaut Syst Div, USAF, 64, mem, USAF Sci Adv Bd, 63-79, chmn electro optics panel, 75-77; mem, Joint US-USSR Working Group, Sensing of Environ, 71-77; mem space appl comt, NASA Space Progs Adv Coun, 71-77 & Defense Intel Agency Adv Comt, 78-81; mem, Nat Res Coun Comt Global Habitality, Nat Acad Sci, 85-86, US Army Sci Bd, 86- *Mem:* Explorers Club; Cosmos Club. *Res:* Infrared; optics; radar; digital computers; missile systems; large information handling and processing systems; reconnaissance; space systems. *Mailing Add:* Environ Res Inst Mich PO Box 8618 Ann Arbor MI 48107

HOLTER, NORMAN JEFFERIS, physics; deceased, see previous edition for last biography

HOLTFRETER, JOHANNES FRIEDRICH KARL, b Richtenberg, Ger, Jan 9, 01; m 59. EMBRYOLOGY. *Educ:* Univ Freiburg, PhD(zool), 24. *Hon Degrees:* DrNatSci, Univ Freiburg, 75. *Prof Exp:* Asst, Kaiser-Wilhelm Inst, Berlin, 28-33; assoc prof zool, Univ Munich, 33-38; Rockefeller fel, McGill Univ, 42-44, Guggenheim fel, 45-46; from assoc prof to prof, 46-69, Tracy N Harris prof, 66, EMER TRACY N HARRIS PROF ZOOL, UNIV ROCHESTER, 69- *Concurrent Pos:* Res worker, Marine Biol Sta, Naples & Helgoland, 25-26, Cambridge, 39-40; Rockefeller fel, Yale & McGill Univ, 36; vis prof, Montevideo, 49, Caracas, 54; res grants, Am Philos Soc, NSF & NIH, 46-73; vis prof & Fulbright fel, Univ Paris, 58; lectr, numerous foreign univs, 36-62. *Honors & Awards:* Jesup Lectr, Columbia Univ, 57. *Mem:* Nat Acad Sci; Am Acad Arts & Sci; hon mem Royal Swed Acad Sci; Am Asn Anatomists; Indian Soc Zoologists; Am Soc Naturalists; Soc Study Develop & Growth. *Res:* Experimental study of differentiation and organization in animals; embryonic induction; morphogenetic and cell movements; cell affinity and organelles; tissue culture. *Mailing Add:* Dept Biol Univ Rochester Rochester NY 14627

HOLTKAMP, DORSEY EMIL, b New Knoxville, Ohio, May 28, 19; m 42, 57, 84; c 7. MEDICAL SCIENCE, ENDOCRINOLOGY. *Educ:* Univ Colo, AB, 45, MS, 49, PhD(biochem), 51. *Prof Exp:* Asst chem & biol, Univ Colo, 45-46, biochem, Sch Med, 47-51; sr res scientist biochem sect, Smith, Kline & French Labs, 51-57, group leader, 57-58; head dept endocrinol, Merrell-Nat Labs Div, Richardson-Merrell, Inc, 58-70, group dir, Endocrine Clin Res, 70-81, group dir, Med Res Dept, Merrell Dow Pharmaceut Inc, Dow Chem Co, 81-87; INDEPENDENT CONSULT, 87- *Mem:* Am Chem Soc; Endocrine Soc; Am Soc Pharmacol & Exp Therapeut; Am Soc Clin Pharmacol & Therapeut; Am Fertil Soc; AAAS. *Res:* Homoiothermia control; manganese and thiamin nutrition; tumor, adjacent tissue metabolism; reticulo-endothelial system stimulation; endocrinodynamics; antihistaminic, anti-inflammatory, cholesterol-lowering, thyroidal, estrogenic, fertility-sterility drugs; reproduction; teratology; basic and clinical drug research and development. *Mailing Add:* 130 S Liberty-Keuter Rd Lebanon OH 45036

HOLTKAMP, FREDDY HENRY, b Needville, Tex, Jan 19, 34; m 57; c 3. ORGANIC CHEMISTRY. *Educ:* Tex Lutheran Col, BS, 57; Tex A&M Univ, MS, 64, PhD(chem), 69. *Prof Exp:* Chemist, Dow Chem Co, 59-61; asst prof, 68-74, Piedmont Univ Ctr fel, 70-74, assoc prof, 74-77, PROF CHEM, MARS HILL COL, 77-, CHMN DEPT, 74- *Mem:* AAAS; Am Chem Soc. *Res:* Organic synthesis of phosphonic acid analogs of naturally occurring heterocyclic amino acids and sulfur containing amino acids. *Mailing Add:* PO Box 147 Mars Hill NC 28754

HOLTMAN, DARLINGTON FRANK, b Randolph, Kans, Oct 12, 03; m 27; c 2. MEDICAL BACTERIOLOGY. *Educ:* Univ Kans, AB, 27; Univ Tenn, AM, 30; Ohio State Univ, PhD(bact), 37. *Prof Exp:* Asst bact, Univ Tenn, 27-29, instr, 29-32; asst, Ohio State Univ, 32-36, instr, 36-42; asst prof bact & hyg, Case Western Reserve Univ, 42-43; prof bact & head dept, Univ Tenn, Knoxville, 43-68, prof microbiol, 68-72; RETIRED. *Concurrent Pos:* Consult, Tenn Valley Authority, 44-48, Oak Ridge Nat Lab, 54-58 & Stewart Med Lab, Tenn, 72-; dipl, Am Bd Med Microbiol. *Mem:* AAAS; Soc Exp Biol & Med; Am Soc Microbiol; Radiation Res Soc; Am Acad Microbiol. *Res:* Heterophile antigenicity of bacteria; environment and susceptibility to poliomyelitis; antibiotics; fowl typhoid; cholera; pullorum disease; nitrogen, carbohydrate and lipid metabolism in host-parasite relationship; air and water pollution studies. *Mailing Add:* 7914 Gleason Rd No 1017 Knoxville TN 37919

HOLTMAN, MARK STEVEN, b Dayton, Ohio, Apr 3, 49; m 80; c 1. CHEMICAL APPLICATIONS OF COMPUTERS. *Educ:* Univ Dayton, BS, 71; Wright State Univ, MS, 73; Ohio State Univ, PhD(inorg chem), 79. *Prof Exp:* Sr res chemist, PPG Indust, Inc, 79-84; SYSTS DESIGNER, BP AMERICA, 85- *Mem:* Am Chem Soc. *Res:* Interfacing laboratory instrumentation/reactors to computers for data acquisition and process control; laboratory information management systems, software for chemical applications. *Mailing Add:* 970 Brookpoint Dr Macedonia OH 44056

HOLTMANN, WILFRIED, b Winnipeg, Man, June 18, 39; m 62; c 3. ANIMAL BREEDING. *Educ:* Univ Man, BSA, 61; Univ Guelph, MSA, 62; Univ Wis, PhD(dairy sci), 66. *Prof Exp:* Res scientist, Can Dept Agr, 66-67; asst prof animal sci, Laval Univ, 67-70, assoc prof, 70-76; MEM STAFF, SEMEN MKT, CTR D'INSEMINATION ARTIFICIELLE DU QUEBEC, CIAQ INC, 76- *Mem:* Am Soc Animal Sci; Can Soc Animal Sci. *Res:* Animal production, especially animal breeding; crossbreeding in swine and beef cattle; selection and improvement in dairy cattle. *Mailing Add:* Que AI Centre Box 518 St-Hyacinthe PQ J2S 7B8 Can

HOLTMYER, MARLIN DEAN, b Red Oak, Iowa, Feb 1, 37; m 60; c 2. PETROLEUM & POLYMER CHEMISTRY. *Educ:* Northwestern Mo State Col, BS, 62; Okla State Univ, PhD(phys chem), 67. *Prof Exp:* Sr res chemist, 66-70, group leader, 70-73, asst mgr, 73-78, RES ASSOC CHEM RES & DEVELOP, HALLIBURTON SERV, 78- *Mem:* Am Chem Soc. *Res:* Solution properties of polymers in both aqueous and nonaqueous media; chemistry of natural and synthetic polymers (solution properties), surfactants, emulsions, and their application in the petroleum well service industry. *Mailing Add:* 1219 N 12th Duncan OK 73533-3711

HOLTON, ADOLPHUS, b Norwood, Mass, June 1, 12; m 32; c 3. GEOLOGY. *Educ:* Northeastern Univ, BS, 36. *Prof Exp:* Inspector qual control, B F Sturtevant Co, Mass, 34-37; engr design & drafting, Tobe Deutschmann Corp, 38-40; real estate operator, 40-42; engr qual control, Bendix Aviation Corp, 42-43; design engr, Raytheon Mfg Co, 43-46; instr math, drawing & blueprint reading, Feener Tech Inst, 46-48; student financial aid dir, New England Col, 63-69, dir placement, 69-70, dir phys plant develop, 71-72, prof eng & geol, 48-78, res engr phys plant develop 72-78, emer prof, 86-87; RETIRED. *Concurrent Pos:* Engr, Eng Design Serv, Mass, 46-48; consult, 48-, consult engr, 78- *Mem:* Am Soc Mech Engrs. *Res:* Geology of New Hampshire, southern Rockies and gulf area of Mississippi. *Mailing Add:* 550 Mammoth Rd Londonderry NH 03053

HOLTON, GERALD, b Berlin, Ger, May 23, 22; nat US; m 47; c 2. PHYSICS, HISTORY OF SCIENCE. *Educ:* Wesleyan Univ, BA, 41, MA, 42; Harvard Univ, MA, 46, PhD(physics), 48. *Hon Degrees:* DSc, Grinnell Col, 67, Kenyon Col, 77, Bates Col, 79; LLD, Duke Univ, 81; LHD, Wesleyan Univ, 81. *Prof Exp:* instr, Brown Univ, 42-43; mem staff, Off Sci Res & Develop, 43-46, instr physics, 47-49, from asst prof to prof, 49-75, MALLINCKRODT PROF PHYSICS & PROF HIST OF SCI, HARVARD UNIV, 75- *Concurrent Pos:* Vis mem, Inst Advan Study, 64 & 67; trustee, Boston Mus Sci, 65-67, Sci Serv, 72-78, Wesleyan Univ, 75-89, Nat Humanities Ctr, 89-; mem bd dirs, AAAS, 67-71; mem, Comt Scholarly Commun with People's Repub China, Nat Acad Sci, 69-72, Nat Comt Hist & Philos Sci, 82-89; mem gov bd, Am Inst Physics, 69-74; mem, Adv Comt Ethical & Human Impact of Sci, NSF, 73-78, chmn, Adv Comt Sci & Eng Educ, 86-, adv comt, Directorate Sci & Eng Educ, 85-; mem, US Nat Comn, UNESCO, 75-80, US Nat Comn Excellence in Educ, 81-83; fel, Ctr Advan Study Behav Sci, Stanford Univ, 75-76; vis prof, Mass Inst Technol, 76-; Guggenheim fel, 80-81; mem, Coun Scholars, Libr Cong, 80-85 & 88-; mem adv bd, Radcliffe Inst Women, 82-85. *Honors & Awards:* George Sarton Mem lectr, AAAS, 62; R A Millikan Medal, Am Asn Physics Teachers, 67, Oersted Medal, 80; Herbert Spence lectr, Oxford Univ, 79; Jefferson lectr, 81; Presidential Citation Serv to Educ, 84; J D Bernal Prize, soc social Studies Sci, 89; Sarton Medal, Hist Sci Soc, 89; Gemant Award, Am Inst Physics, 89. *Mem:* AAAS (vpres, 61-62); fel Am Phys Soc; Hist Sci Soc(pres, 83-84); Am Asn Physics Teachers; fel Am Acad Arts & Sci; Int Acad Hist Sci (vpres, 81-). *Res:* High pressure phenomena; molecular physics; history and philosophy of physical science; science education. *Mailing Add:* 358 Jefferson Physics Lab Harvard Univ Cambridge MA 02138

HOLTON, JAMES R, b Spokane, Wash, Apr 16, 38; m 62; c 2. DYNAMIC METEOROLOGY. *Educ:* Harvard Univ, BA, 60; Mass Inst Technol, PhD(meteorol), 64. *Prof Exp:* NSF fel, Univ Stockholm, 64-65; from asst prof to assoc prof, 65-73, PROF ATMOSPHERIC SCI, UNIV WASH, 73- *Concurrent Pos:* Co-chief ed, J Atmospheric Sci, 78-83. *Honors & Awards:* Meisinger Award, Am Meteorol Soc, 73, Second Half Century Award, 82. *Mem:* Am Meteorol Soc; fel Am Geophys Union. *Res:* Stratospheric dynamics and tropical dynamics. *Mailing Add:* Dept Atmospheric Sci Univ Wash Seattle WA 98195

HOLTON, RAYMOND WILLIAM, b Riverside, Calif, Apr 30, 29; div; c 4. PLANT PHYSIOLOGY. *Educ:* Pomona Col, BA, 51; Univ Mich, MS, 54, PhD(bot), 58. *Prof Exp:* Chemist, US Naval Ord Test Sta, Calif, 51-52; instr bot, Flint Col, Univ Mich, 57-59, asst prof, 59-61; res scientist zool, Univ Tex, 61-62, USPHS trainee, 62-63; from asst prof to assoc prof, 63-65, actg head dept, 64-65, head dept, 65-84, co-chairperson biol consortium, 84-86, PROF BOT, UNIV TENN, KNOXVILLE, 65- *Concurrent Pos:* Sr Fulbright fel, Univ Durham, UK, 72-73; vis prof, Univ Groningen, Neth, 87. *Mem:* AAAS; Am Soc Plant Physiol; Bot Soc Am; Brit Phycol Soc; Phycol Soc Am. *Res:* Biochemistry and physiology of algae and fungi. *Mailing Add:* Dept Bot Univ Tenn Knoxville TN 37996-1100

HOLTON, ROBERT LAWRENCE, b Billings, Mont, Nov 1, 28; m 53; c 3. MARINE BIOLOGY, POLLUTION BIOLOGY. *Educ:* Univ Mont, BA, 50, ME, 58; Ore State Univ, MS, 62, PhD(biol oceanog), 68; Univ Minn, MS, 65. *Prof Exp:* Teacher high sch, Idaho, 50-51, Mont, 51-59, Wash, 59-60 & Ore, 61-64; assoc prof sci educ, Eastern Ore Col, 67-69; marine biologist, US AEC, 69-71; res assoc, 71-77, ASST PROF OCEANOG, SCH OCEANOG, ORE STATE UNIV, 77- *Concurrent Pos:* Res assoc, Ore Regional Primate Res Ctr, 62-63; consult, Nuclear Regulatory Comn, US Energy Res & Develop Admin, 73- *Mem:* Am Soc Limnol & Oceanog; Sigma Xi; Pac Estuarine Res Soc; Estuarine Res Fedn. *Res:* Effect of various pollutants on individuals, populations and communities of marine organisms; effects of pollutants at low sub-lethal levels on reproduction, behavior and productivity. *Mailing Add:* 8624 SW Miller Ct Tigard OR 97224-5409

HOLTON, WILLIAM COFFEEN, b Washington, DC, July 24, 30; m 54; c 3. SOLID STATE PHYSICS. *Educ:* Univ NC, BS, 52; Univ Ill, MS, 58, PhD(physics), 60. *Prof Exp:* Asst physics, Univ Ill, 56-60; mem tech staff, 60-65, mgr Quantum Electronics Br, Physics Res Lab, 65-74, dir Advan Components Lab,74-78, MGR RES DEVELOP ENG, TEX INSTRUMENTS INC, 78-; DIR MICROSTRUCT SCI, SEMICONDUCTOR RES CORP. *Mem:* Sr mem Inst Elec & Electronics Engrs; fel Am Phys Soc. *Res:* Electron paramagnetic resonance; luminescence and defects or imperfections in crystalline solids; quantum electronics; lasers; microwave devices; integrated circuits (LSI and VLSI); infrared physics; displays. *Mailing Add:* Semiconductor Res Corp 4501 Alexander Dr Suite 301 Research Triangle Park NC 27709

HOLTSLANDER, WILLIAM JOHN, b Moose Jaw, Sask, July 25, 37; m 60; c 3. PHYSICAL CHEMISTRY, CHEMICAL ENGINEERING. *Educ:* Univ Sask, BE & MSc, 60; Univ Alta, PhD(radiation chem), 67. *Prof Exp:* Res chemist, Chemcell Ltd, 60-61; teaching asst chem, Univ Alta, 61-66; PROF CHEM ENG, ATOMIC ENERGY CAN, LTD, 67- *Concurrent Pos:* Mgr, Int prog for the Nat Fusion Prog, Atomic Energy Can, Ltd; mem, Int Atomic Energy Agency Tech Comt on Tritium Technol; prog chmn, Third Trop Meeting on Tritium Technol, Toronto, 88. *Mem:* Fel Chem Inst Can; Can Soc Chem Eng. *Res:* Immobilization, storage and disposal of tritium containing wastes from nuclear power reactors; chemistry of aqueous sulfur containing systems; recovery of tritium from nuclear reactors; technology of handling concentrated tritium in fission and fusion systems; deuterium exchange reaction; recovery of tritium from nuclear reactors; technology of handling concentrated tritium. *Mailing Add:* Atomic Energy Can Ltd Chalk River Nuclear Labs Chalk River ON K0J 1J0 Can

HOLTUM, ALFRED G(ERARD), b Freeport, Ill, Aug 26, 18; m 41; c 6. ELECTRICAL ENGINEERING, MATHEMATICS. *Educ:* NY Univ, BA, 49; Rutgers Univ, MS, 52; Ill Inst Technol, PhD(elec eng), 72. *Prof Exp:* Electronic engr, Sig Corps Eng Labs, NJ, 45-54; chief, Radio Commun Div, Sig Commun Dept, Army Electronic Proving Ground, Ariz, 54-57; chief engr,

Andrew Corp, Calif, 57-61, dir antenna design, Orland Park, Ill, 61-66, dir new prod develop, 66-68, sr scientist, 68-70; mem tech staff, Justice Assocs, Bridgeview, Ill, 70-72; prin engr, Harris Corp, 72-75; consult, US Govt, 75-78; asst prof, Univ NC, Charlotte, 78-84; CONSULT, 84- *Concurrent Pos:* Adj prof, Fla Inst Technol, 72-74. *Mem:* Inst Elec & Electronics Engrs; Sigma Xi. *Res:* Antennas; electromagetic theory. *Mailing Add:* 300 Lakeview Ridge E Roswell GA 30076

HOLTY, DAVID WEBSTER, organic chemistry, finance, for more information see previous edition

HOLTZ, CARL FREDERICK, b Baltimore, Md, Sept 7, 40; m 61. PHOTOGRAPHIC & CLINICAL CHEMISTRY. *Educ:* Rensselaer Polytech Inst, BS, 61, PhD(org chem), 66. *Prof Exp:* Sr res chemist org synthesis, Eastman Kodak Co Res Labs, 65-68, res assoc, photog develop work, 68-74, lab head, 74-76, asst dir, Color Photog Div, 76-78, ASST DIR, BIOSCI DIV, EASTMAN KODAK CO RES LABS, 78- *Mem:* Am Chem Soc; Am Asn Clin Chem. *Res:* Administration of research and development work in the biosciences and clinical chemistry. *Mailing Add:* 13 Woodcliff Terr Fairport NY 14450-4208

HOLTZ, DAVID, b Los Angeles, Calif, Nov 11, 42; m 73; c 3. ADMINISTRATION, SOFTWARE DESIGN. *Educ:* Calif Inst Technol, BS, 64; Univ Calif, Berkeley, PhD(org chem), 68; Univ Calif, Los Angeles, MBA, 83. *Prof Exp:* Instr chem, Calif Inst Technol, 68-71; staff officer, Environ Studies Bd, Nat Acad Sci, 71-74; asst dir, Holcomb Res Inst, Butler Univ, 75-77; PRES, GLATT-HOLTZ INC, 78- *Concurrent Pos:* Adv, Comt Econ Develop, 71-74; mem, Eval Panel Air Qual, Nat Bur Standards, 71-74; cert pub acct, 85-; vpres finance, GNP Develop Corp, 85-86; pres, Medivest Corp, 86-87; mem bd dirs, Caltech Alumni Asn. *Mem:* Am Chem Soc; Am Inst Cert Pub Acct. *Res:* Entrepreneurship with companies based on new technologies; design and development of business oriented software. *Mailing Add:* 6842 Whitaker Ave Van Nuys CA 91406

HOLTZ, WESLEY G, b Riverside, Calif, June 5, 11; m 36; c 2. EARTHWORK FIELD & LABORATORY TESTING OF SOILS. *Educ:* Univ Calif, Berkeley, BS, 34. *Prof Exp:* Mat engr, US Bur Reclamation, Yama, Ariz, 34-40; supv mat engr, US Bur Reclamation, Altus, Okla, 40-43; chief, Soils Engr Br, US Bur Reclamation, Denver, Colo, 43-64, dep chief, Div Res, 64-69; CIVIL-GEOTECH ENGR CONSULT, WESLEY G HOLTZ, PE, 69- *Concurrent Pos:* Expert witness; mem, US Comt Int Comn Large Dams. *Honors & Awards:* Wellington Prize, Am Soc Civil Engrs, 56; Distinguished Serv, US Dept Interior, 66; Marburg Lectr, Am Soc Testing & Mat, 68. *Mem:* Fel Am Soc Civil Engrs; fel & hon mem Am Soc Testing & Mat. *Res:* Author or co-author of 52 published technical papers on civil and geotechnical engineering subjects; expansive clay soils; collapsible soils; spoon penetrating testing of foundations; earth dam materials; placement of earth dam materials; transmission line tower foundations; construction of pipelines; developer of soil-cement facings for earth dams; field project on disposal of retorted oil shale. *Mailing Add:* 3250 Moore St Wheat Ridge CO 80033

HOLTZBERG, FREDERIC, b New York, NY, Apr 12, 22; m 50; c 1. SYNTHETIC, INORGANIC & ORGANOMETALLIC CHEMISTRY. *Educ:* Brooklyn Col, BS, 47; Polytech Inst Brooklyn, PhD, 52. *Prof Exp:* Staff mem, Watson Sci Comput Lab, IBM Corp, 52-60, res staff mem, T J Watson Res Ctr, 60-74, mgr mat sci dept, 74-81, RES STAFF MEM, T J WATSON RES CTR, IBM CORP, 81- *Concurrent Pos:* Tech asst to spec asst to President for sci & technol, 59-60; consult, Off Sci & Technol, 60-61; adj prof, Univ Grenoble, 75-76. *Honors & Awards:* Int Prize New Mat, Am Phys Soc, 91. *Mem:* Am Chem Soc; fel Am Phys Soc; Am Crystallog Asn; fel Am Inst Chemists; Sigma Xi. *Res:* X-ray diffraction structure analysis; low temperature diffraction studies; crystal chemistry; phase equilibria in inorganic oxide systems; ferroelectric materials; physics and chemistry of rare earth semiconducting, ferromagnetic, magnetic intermediate valence and superconducting materials; high temperature superconductivity. *Mailing Add:* T J Watson Res Ctr IBM Corp PO Box 218 Yorktown Heights NY 10598

HOLTZCLAW, HENRY FULLER, JR, b Stillwater, Okla, July 30, 21; m 49; c 2. INORGANIC CHEMISTRY. *Educ:* Univ Kans, AB, 42; Univ Ill, MS, 46, PhD(inorg chem), 47. *Prof Exp:* Asst instr chem, Univ Kans, 42-43, instr math, 43; asst chem, Univ Ill, 43-44; shift foreman, Clinton Eng Works & Tenn Eastman Corp, Oak Ridge, 44-45; asst chem, Univ Ill, 45-46; from instr to prof chem, 47-67, found regents prof, 67-88, dean grad studies, 76-85, FOUND REGENTS EMER PROF CHEM, UNIV NEBR-LINCOLN, 88- *Concurrent Pos:* Ed-in-chief, Inorg Synthesis, Vol VIII. *Mem:* AAAS; Am Chem Soc; Sigma Xi. *Res:* Inorganic coordination compounds; polarography; metal chelates; inorganic polymers. *Mailing Add:* Univ Nebr 711 Hamilton Hall Lincoln NE 68588-0304

HOLTZER, ALFRED MELVIN, b Brooklyn, NY, Feb 22, 29; m 54, 69; c 2. BIOPHYSICAL CHEMISTRY. *Educ:* Wash Univ, AB, 50; Harvard Univ, PhD(chem), 54. *Prof Exp:* Instr chem, Yale Univ, 54-57; from asst prof to assoc prof, 57-65, PROF CHEM, WASH UNIV, 65- *Mem:* Am Chem Soc; Am Soc Biochem & Molecular Biol; Biophys Soc; Protein Soc. *Res:* Physical chemistry of macromolecules; protein folding; alpha-helical proteins. *Mailing Add:* Dept Chem Wash Univ Box 1134 St Louis MO 63130

HOLTZER, HOWARD, b New York, NY, Mar 8, 23; m 53; c 1. EMBRYOLOGY. *Educ:* Univ Chicago, PhD(zool), 52. *Prof Exp:* From assoc to assoc prof, 54-63, PROF ANAT, SCH MED, UNIV PA, 63-, DIR PROG CELL DIFFERENTIATION, 73- *Concurrent Pos:* USPHS fel, Col Physicians & Surgeons, Columbia Univ, 52-54; Fulbright & Guggenheim fels, Carlsberg Labs, Denmark & Sweden, 58-59; vis prof, Inst Molecular Biol, Cambridge, Eng, 70-71; consult, Nat Inst Child Health & Human Develop. *Mem:* Am Soc Zool; Am Asn Anat. *Res:* Differentiation of embryonic nervous system; influence of embryonic spinal cord on differentiation of cartilage and muscle; regeneration; tissue culture. *Mailing Add:* Dept Anat Univ Pa Sch Med Philadelphia PA 19104

HOLTZER, MARILYN EMERSON, b Belleville, Ill, July 22, 38; m 69; c 2. MACROMOLECULAR PHYSICAL CHEMISTRY, PROTEIN PHYSICAL CHEMISTRY. *Educ:* Wash Univ, AB, 60, AM, 63, PhD(chem), 66. *Prof Exp:* Res assoc, Wash Univ, 66-67; asst prof chem, Webster Col, 67-69; instr physics, John Burroughs Sch, 69-70; vis asst prof chem, Univ Mo, 71-72; instr chem, 73-80, res assoc, 80-90, RES ASST PROF, WASH UNIV, 90- *Mem:* Biophys Soc. *Res:* Stability of alpha-helical; two-chain coiled-coils; strength and specificity of interactions between alpha-helical chains in tropomyosin, tropomyosin segments, and paramyosin; protein folding kinetics in two-chain coiled coil molecules. *Mailing Add:* Dept Chem Wash Univ St Louis MO 63130

HOLTZMAN, ERIC, b New York, NY, May 25, 39; m 70; c 1. CELL BIOLOGY OF THE RETINA. *Educ:* Columbia Univ, BA, 59, MA, 61, PhD(cytochem), 64. *Prof Exp:* Res fel cytochem, Albert Einstein Col Med, 64-67; from asst prof to assoc prof, 67-75, chmn dept biol sci, 82-88, PROF BIOL SCI, COLUMBIA UNIV, 75- *Concurrent Pos:* NIH fel, 64-65; Am Cancer Soc fel, 65-67; lectr, Columbia Univ, 66-67; vis prof, Harvard Univ, 72; fel, AAAS, 90. *Mem:* Histochem Soc (pres, 74-75); Soc Neurosci; Am Soc Cell Biol; Asn Res Vision & Ophthal. *Res:* Intraneuronal transport and degradation; synaptic functioning in neurons and the retina; lysosomes, peroxisomes, golgi apparatus, and endoplasmic reticulum in nervous tissue; membrane circulation; intracellular calcium; acidified compartments. *Mailing Add:* Dept Biol Sci Columbia Univ Broadway & W 116th St New York NY 10027

HOLTZMAN, GOLDE IVAN, b Pittsburgh, Pa, Feb 10, 46; m 72; c 2. AGRO-ECOSYSTEM MODELING, STATISTICS. *Educ:* Univ Calif, Los Angeles, AB, 72; Univ Ariz, MA, 75; NC State Univ, PhD(biomath), 80. *Prof Exp:* asst prof, 80-86, ASSOC PROF STATIST, VA POLYTECH INST & STATE UNIV, 86- *Concurrent Pos:* Grad prog dir & actg dept head, Va Polytech Inst & State Univ. *Mem:* Am Statist Soc; Biomet Soc; AAAS; Sigma Xi. *Res:* Mathematical modeling of biological phenomena and the statistical analysis of experiments based on such models; statistical computing; environmental trend analysis. *Mailing Add:* Dept Statist Va Polytech Inst & State Univ Blacksburg VA 24061-0439

HOLTZMAN, JORDAN L, b Chicago, Ill, July 12, 33; m 58; c 3. PHARMACOLOGY. *Educ:* Univ Chicago, BA, 52, MS, 55, MD, 59, PhD(pharmacol), 64. *Prof Exp:* Pharmacologist, Lab Parasite Chemother, Nat Inst Allergy & Infectious Dis, 64-65; pharmacologist, Lab Chem Pharmacol, Nat Heart Inst, 65-67; sr scientist, Lab Pharmacol, Nat Cancer Inst, 67-71; CHIEF CLIN PHARMACOL SECT, VET ADMIN HOSP, 71-; PROF PHARMACOL & MED, MED SCH, UNIV MINN, MINNEAPOLIS, 71- *Concurrent Pos:* Vis lectr, Sch Med, George Washington Univ, 65. *Mem:* Am Soc Pharmacol & Exp Therapeut; Am Soc Biol Chemists; Am Soc Clin Pharmacol Therapeut; Soc Toxicol. *Res:* Investigations into the metabolic transformation of therapeutic and environmental agents; role of mechanism of toxicity of toxic metabolites. *Mailing Add:* Vet Admin Hosp Minneapolis MN 55417

HOLTZMAN, JULIAN CHARLES, b Staten Island, NY, Aug 14, 35; m 58. ELECTRICAL ENGINEERING, REMOTE SENSING. *Educ:* Polytech Inst Brooklyn, BS, 58; Univ Calif, Los Angeles, MS, 62; Cornell Univ, PhD(elec eng), 67. *Prof Exp:* Mem tech staff, Hughes Aircraft Co, 58-62; asst prof elec eng, San Jose State Col, 65-68; asst prof, 69-76, assoc prof, 76-80, PROF ELEC ENG & ACTG CHMN DEPT, CTR RES IN ENG SCI, UNIV KANS, 80- *Concurrent Pos:* Res scientist, Lockheed Missiles & Space Co, 66-67; consult, Appl Technol Inc, 67- *Mem:* Inst Elec & Electronics Engrs. *Res:* Communications theory; systems theory; radar systems; antennas; remote sensing. *Mailing Add:* Dept Elec & Comput Eng Univ Kans Learned Hall 2026 Learned Lawrence KS 66045

HOLTZMAN, NEIL ANTON, b Brooklyn, NY, Apr 8, 34; m 55; c 4. PEDIATRICS, MEDICAL GENETICS. *Educ:* Swarthmore Col, BA, 55; NY Univ, MD, 59; Univ Calif, Berkeley, MPH, 84; Am Bd Pediat, dipl, 65. *Prof Exp:* From intern to sr asst resident, Harriet Lane Home, Johns Hopkins Hosp, 59-62, USPHS fel biophys, Sch Med, Johns Hopkins Univ, 62-64, from asst prof to assoc prof, 64-84, PROF PEDIAT, SCH MED, JOHNS HOPKINS UNIV, 84-, PEDIATRICIAN, JOHNS HOPKINS HOSP, 64- *Concurrent Pos:* Fel pediat, Harriet Lane Home, Johns Hopkins Hosp, 59-64; USPHS res career develop award, 69-73; Joseph P Kennedy Jr Mem Found res scholar ment retardation, 72-74; resident fel, Nat Acad Sci, 74-75; dir genetics unit, Johns Hopkins Univ, 72-74; mem comt study inborn errors of metab, Nat Res Coun-Nat Acad Sci, 72-76; mem comn hereditary dis, State of Md, 74-83; mem comt genetics, Am Acad Pediat, 78- & chmn comn, 83-87; coordr, Hereditary Disorders Serv, Dept Health & Ment Hyg, State Md, 78-83; dir, Robert Wood Johnson Gen Pediat Acad Develop Prog, Johns Hopkins Univ, 79-83; fel epidemiol, Milbank Mem Fund, 83-84; sr analyst, US Cong Off Techol Assessment, 86-87; dept epidemiol, Sch Hyg, Johns Hopkins Univ, 81, dept health policy & mgt, 88. *Mem:* AAAS; fel Am Acad Pediat; Am Soc Human Genetics (secy, 83-); Am Soc Biochem & Molecular Biol; Am Pediat Soc; Soc Pediat Res. *Res:* Genetic screening and services; medical technology; diffusion of biotechnology. *Mailing Add:* 550 N Broadway Suite 301 Johns Hopkins Med Insts Baltimore MD 21205

HOLTZMAN, RICHARD BEVES, b Chicago, Ill, Sept 24, 27; m 53; c 3. RADIOCHEMISTRY, RADIOBIOLOGY. *Educ:* Univ Chicago, PhB, 45, BS, 46, MS, 50, PhD(chem), 53. *Prof Exp:* Res assoc geochem, Columbia Univ, 53-54; res physicist, Armour Res Found, Ill Inst Technol, 54 & 56-59; assoc chemist, Radiol Physics Div, Argonne Nat Lab, 59-74, chemist, Radiol & Environ Res Div, 74-84; RADIATION SPECIALIST, US NUCLEAR REGULATORY COMN, 84- *Concurrent Pos:* Mem, Nat Coun Radiation Protection & Measurements. *Mem:* Am Chem Soc; fel, Health Physics Soc; Sigma Xi. *Res:* Natural radioactivity in the environment; metabolism of radioactive and stable trace elements; reaction kinetics of radioactive tracers in biological and physical systems; dosimetry. *Mailing Add:* 6108 Carpenter Downers Grove IL 60516

HOLTZMAN, SAMUEL, b Mexico City, Mex, Feb 9, 55; m 85; c 2. INTELLIGENT DECISION SYSTEMS, DECISION ANALYSIS. *Educ:* Mass Inst Technol, SB, 77, SM & EE, 80; Stanford Univ, MS & PhD(eng-econ systs), 85. *Prof Exp:* Res asst comput sci, Mass Inst Technol, 77-80; res asst, 80-83, instr, 83-84, CONSULT ASST PROF ENG & ECON SYSTS, STANFORD UNIV, 85- *Concurrent Pos:* assoc, 81-86, sr assoc, 86-89, prin, Strategic Decisions Group, 89- *Mem:* Sigma Xi; Inst Elec & Electronics Engrs; Am Asn Artificial Intel; AAAS; Inst Mgt Sci; Asn Comput Math; Comput Scientists for Social Responsibility. *Res:* Decision analysis artificial intelligence; intelligent decision systems; medical decision making, medical ethics. *Mailing Add:* Strategic Decisions Group 3000 Sand Hill Rd Bldg 3 Suite 150 Menlo Park CA 94025-7127

HOLTZMAN, SEYMOUR, b Brooklyn, NY, June 22, 34; m 61; c 2. COMPARATIVE ENDOCRINOLOGY. *Educ:* City Col New York, BS, 55; City Univ New York, MA, 65, PhD(biol), 73. *Prof Exp:* Res asst histol, Cols Dent & Med, NY Univ, 58-59; res asst, Downstate Med Ctr, State Univ NY, 59-60; biologist endocrinol, Squibb Inst Med Res, 60-63; res assoc physiol, Inst Phys Med & Rehab, Med Ctr, NY Univ, 63-67; vis asst prof biol, Col VI, 73-74; NIH fel, Brookhaven Nat Lab, 74-75, res assoc radiobiol, 74-76, asst scientist, 76-78, assoc scientist carcinogenesis, Med Dept, 78-80; scientist, Off Res & Develop, US Environ Protection Agency, 80-82 & Brookhaven Nat Lab, 82-84; asst prof physiol, NY Col Oseopath Med, 85-87; res assoc, Biol Dept, Brooklyn Col, 88-90; SCIENTIST, DEPT APPL SCI, BROOKHAVEN NAT LAB, 90- *Concurrent Pos:* Adj asst prof biol, Brooklyn Col, 73-74; adj asst prof, Queensborough Community Col, 74-76, adj assoc prof, 84- *Mem:* Am Soc Zoologists; AAAS; Endocrine Soc; Sigma Xi; Am Asn Cancer Res; NY Acad Sci. *Res:* Health risk analysis of hazardous waste sites; comparative endocrinology of the pituitary gland in teleost fishes. *Mailing Add:* PO Box 902 Medford NY 11763-0902

HOLTZMANN, OLIVER VINCENT, b Highmore, SDak, June 26, 22; m 57; c 5. PLANT PATHOLOGY. *Educ:* Colo State Univ, BS, 50, MS, 52; Wash State Univ, PhD(plant path), 55. *Prof Exp:* Res asst plant path, Wash State Univ, 52-55; res asst prof, NC State Col, 55-56; asst plant pathologist, 56-66, from assoc prof to prof, 66-86, chmn dept, 70-83, & 85-86, EMER PROF PLANT PATH, UNIV HAWAII, 86- *Concurrent Pos:* Adv, Hawaii Turfgrass Asn; instr, Plant Quarantine. *Mem:* Am Phytopath Soc; Soc Nematol. *Res:* Phytonematology; diseases of turf grass; teaching of tropical plant pathology. *Mailing Add:* Dept Plant Path Univ Hawaii at Manoa 3190 Maile Way Honolulu HI 96822

HOLUB, BRUCE JOHN, b Sudbury, Ont, Nov 3, 44; m 72; c 2. NUTRITIONAL BIOCHEMISTRY. *Educ:* Univ Guelph, BSA, 67; Univ Toronto, MSc, 69, PhD(biochem, nutrit), 71. *Prof Exp:* Res chemist, Imperial Oil Ltd, 67; from asst prof to assoc prof, 73-83, PROF NUTRIT, UNIV GUELPH, 83- *Concurrent Pos:* Med Res Coun Can fel biol chem, Univ Mich, 71-73. *Honors & Awards:* Borden Award Nutrit. *Mem:* Am Oil Chemists Soc; Am Inst Nutrit; Can Biochem Soc; Nutrit Soc Can. *Res:* Metabolism of essential fatty acids; biological function of inositol; nutritional regulation of lipid metabolism. *Mailing Add:* Dept Nutrit Sci Univ Guelph Guelph ON N1G 2W1 Can

HOLUB, DONALD ARTHUR, b New York, NY, Nov 9, 28; m 50; c 3. INTERNAL MEDICINE. *Educ:* Columbia Univ, AB, 48, MD, 52. *Prof Exp:* Res fel med, Col Physicians & Surgeons, Columbia Univ, 57-59, from instr to assoc prof med, 59-75, PROF CLIN MED, COL PHYSICIANS & SURGEONS, COLUMBIA UNIV, 75- *Mem:* AAAS; Endocrine Soc; Harvey Soc; Am Fedn Clin Res. *Res:* Clinical endocrinology; adrenal, thyroid, gonadal and pituitary function and interrelationships; factors regulating the secretion of adrenocorticotropin by the pituitary gland. *Mailing Add:* Dept Med Columbia Univ Col Physicians & Surgeons 630 W 168th St New York NY 10032

HOLUB, FRED F, b Cleveland, Ohio, June 20, 21; m 49; c 1. CHEMISTRY. *Educ:* Western Reserve Univ, BA, Duke Univ, PhD(org chem). *Prof Exp:* RES SCIENTIST, RES & DEVELOP CTR, GEN ELEC CO, 56- *Concurrent Pos:* Consult mat sci & technol. *Mem:* Am Chem Soc; Sigma Xi. *Res:* Fluorination of organic compounds; synthesis and high temperature polymers; aromatic polyesters and nitrogen polymers; synthetic elastomers; chemical modification of polymers; composite structures; radiation chemistry; polymer blends. *Mailing Add:* 2263 Preisman Dr Schenectady NY 12309

HOLUB, JAMES ROBERT, b New Prague, Minn, Dec 5, 42; m 67; c 2. MATHEMATICS. *Educ:* Col St Thomas, BA, 64; La State Univ, MS, 66, PhD(math), 69. *Prof Exp:* From asst prof to assoc prof, 69-78, PROF MATH, VA POLYTECH INST & STATE UNIV, 78- *Mem:* Am Math Soc. *Res:* Functional analysis; operator, tensor product and basis theory. *Mailing Add:* Dept Math Va Polytech Inst Blacksburg VA 24061

HOLUBEC, ZENOWIE MICHAEL, b Ukraine, Nov 21, 37; US citizen; m 63; c 3. PETROLEUM CHEMISTRY, TRIBOLOGY. *Educ:* Case Western Reserve Univ, BA, 60; John Carroll Univ, MS, 64; Univ Ill, PhD(org chem), 68. *Prof Exp:* Res chemist, 60-63 & 68-70, proj leader, 70-71, res supvr, 71-72, proj mgr, 72-78, dir phys & anal chem, 78-86, SR PROJ COORDR, LUBRIZOL CORP, 86- *Mem:* Am Chem Soc; Am Soc Testing & Mat; Soc Automotive Engrs. *Res:* Tribology; synthesis; use and mode of action of additives for industrial, tractor transmission, gear, engine lubricants and railroad diesel lubricants; physical properties of lubricants; analysis of new and old lubricants. *Mailing Add:* 6908 Anthony Lane Cleveland OH 44130

HOLUBKA, JOSEPH WALTER, b Detroit, Mich, Jan 10, 50. ORGANIC CHEMISTRY, POLYMER CHEMISTRY. *Educ:* Univ Detroit, BS, 72; Wayne State Univ, PhD(chem), 77. *Prof Exp:* SR STAFF SCIENTIST POLYMER CHEM, FORD MOTOR CO, 77- *Mem:* Am Chem Soc. *Res:* Physical chemistry of polymer films, polymer synthesis and modification; surface analysis of polymer films. *Mailing Add:* 17902 Myron Livonia MI 48152-1997

HOLUJ, FRANK, b Poland, Oct 28, 27; Can citizen. SOLID STATE PHYSICS. *Educ:* Univ London, BSc, 54; McMaster Univ, MSc, 56, PhD(physics), 58. *Prof Exp:* Fel physics, McMaster Univ, 58-59; res officer gas masers, Appl Physics Div, Nat Res Coun Can, 59-61; from asst prof to assoc prof, 61-68, PROF PHYSICS, UNIV WINDSOR, 68- *Mem:* Can Asn Physicists; Am Phys Soc. *Res:* Nuclear and electron spin (magnetic) resonance in solids; optical study of solids using lasers; gamma-gamma correlation study in solids. *Mailing Add:* Dept Physics Univ Windsor Windsor ON N9B 3S4 Can

HOLUM, JOHN ROBERT, b Tracy, Minn, Aug 31, 28; m 50; c 3. ORGANIC CHEMISTRY. *Educ:* St Olaf Col, BA, 50; Univ Minn, Minneapolis, PhD(org chem), 54. *Prof Exp:* Chemist, Eastman Kodak Co, NY, 54; instr org chem, Augsburg Col, 57-58 & Pac Lutheran Col, 58-59; assoc prof, 59-64, PROF ORG CHEM, AUGSBURG COL, 64- *Concurrent Pos:* NSF grants, 59-60, 61-62 & 64-65; NSF sci fac fel, Calif Inst Technol, 62-63; consult, John Wiley & Sons, Inc, NY, 63-; consult, Control Data Corp, Minn, 66-67 & Plastics Inc, 67-69; sabbatical leave, Harvard Univ, 69-70, Univ Minn, 78-79. *Honors & Awards:* Minn Sect Col Chem Teaching Award, Am Chem Soc, 73. *Mem:* AAAS; Am Chem Soc. *Res:* The keto group as a neighboring group in chloroketones. *Mailing Add:* 3352 47th Ave S Minneapolis MN 55406

HOLVERSON, EDWIN LEROY, b Cottage Grove, Ore, Feb 1, 34; m 61; c 1. SOLID STATE PHYSICS. *Educ:* Humboldt State Col, AB, 58; Ariz State Univ, MS, 62, PhD(physics), 65. *Prof Exp:* Instr, Humboldt State Col, 57-58, AV technician, 58-59; asst physics, Ariz State Univ, 59-64, res fac assoc, 64-65; asst prof, Mont Col Mineral Sci & Technol, 65-67 & Midwestern Univ, 67-69; vis assoc prof, Am Univ Cairo, 69-70; from asst prof to assoc prof, 71-73, PROF PHYSICS, MIDWESTERN UNIV, 73- *Mem:* Am Phys Soc; Sigma Xi. *Res:* Pollution control utilizing electrokinetic techniques. *Mailing Add:* Dept Physics Midwestern Univ Wichita Falls TX 76308

HOLWAY, JAMES GARY, b Lake Placid, NY, Sept 28, 31; m 53; c 3. HUMAN ECOLOGY. *Educ:* State Univ NY Col Ed, Albany, AB, 59; Colo State Univ, MS, 61, PhD(bot sci), 62. *Prof Exp:* Prof biol, Yampa Valley Col, 62-63; from asst prof to assoc prof, 63-68, PROF BIOL, STATE UNIV NY COL ONEONTA, 68- *Res:* Life history of angiosperm alpine plants in Colorado; ecological study of forest-climate relationships in Adirondacks, NY; man-environment relationships. *Mailing Add:* Dept Biol State Univ NY Col Oneonta NY 13820

HOLWAY, LOWELL HOYT, JR, b St Louis, Mo, Oct 7, 31; m 55; c 1. PHYSICS, APPLIED MATHEMATICS. *Educ:* Dartmouth Col, AB, 53; Harvard Univ, AM, 55, PhD(appl math), 64. *Prof Exp:* Assoc res staff mem, Res Div, Raytheon Co, 55-58, sr engr, Govt Equip Div, 58-63, sr res scientist, 63-65, PRIN RES SCIENTIST, RES DIV, RAYTHEON CO, LEXINGTON, 65- *Mem:* Am Phys Soc; Inst Elec & Electronics Engrs; Am Nuclear Soc. *Res:* Kinetic theory of gases; shock waves; plasma physics; carrier transport in semiconductors; neutron diffusion; laser self-focusing; microwave breakdown, ionospheric and magnetospheric propagation; heat transfer; electron traps and deep levels in semiconductors; Impatt diodes. *Mailing Add:* Four Everett St Natick MA 01760

HOLWERDA, JAMES G, b Los Angeles, Calif, Dec 9, 25; m 55; c 2. GEOLOGY. *Educ:* Univ Southern Calif, BA, 48, MA, 51, PhD(geol), 58. *Prof Exp:* Proj geologist, Ralph M Parsons Co, Taiwan, 50-51; geologist, Johnston Int, Calif, 51-53; proj geologist, Ralph M Parsons Co, Iraq, 53-55, proj mgr, 55-57, proj mgr, Ethiopia, 58-64, VPRES, PARSONS CORP, 64-, VPRES, RALPH M PARSONS CO PTY LTD, 71- *Mem:* AAAS; Geol Soc Am; Asn Eng Geol; Am Asn Petrol Geol; Am Geophys Union; Sigma Xi. *Res:* Ground water and engineering geology; mineral exploration; project management; negotiations. *Mailing Add:* 1217 Paseo Del Mar San Pedro CA 90731

HOLWERDA, ROBERT ALAN, b Detroit, Mich, Sept 13, 47; m 71; c 3. INORGANIC CHEMISTRY. *Educ:* Stanford Univ, BS, 69; Calif Inst Technol, PhD(chem), 74. *Prof Exp:* From asst prof to assoc prof, 74-85, assoc dean, 85-87, PROF, COL ARTS & SCI, TEX TECH UNIV, 86- *Mem:* Am Chem Soc; Sigma Xi. *Res:* Reactivity and electronic structure of oxo-bridged polynuclear transition metal complexes; chemistry of metal stabilized organic free radical ligands; Kinetic stability of the copper (II) mercaptide sulphur bond; thermodynamic and kinetic aspects of palladium (II) carbon bond formation. *Mailing Add:* Dept Chem Tex Tech Univ Lubbock TX 79409

HOLY, NORMAN LEE, b Sturgis, Mich, June 3, 41; m 67; c 2. ORGANIC CHEMISTRY. *Educ:* Western Mich Univ, BA, 63; Purdue Univ, PhD(chem), 68. *Prof Exp:* Prof chem, Western Ky Univ, 67-84; SR SCIENTIST, ROHM & HAAS, 84- *Concurrent Pos:* Consult, Chem Systs, Inc. *Mem:* Am Chem Soc. *Res:* Homogeneous catalysis; surfactants; physical organic chemistry. *Mailing Add:* Four Lee Dr Maple Glen PA 19002

HOLYOKE, CALEB WILLIAM, JR, b York, PA, Sept 22, 43; m 66; c 2. AGRICHEMICALS. *Educ:* Wash Univ, BS, 67; Vanderbilt Univ, PhD(chem), 73. *Prof Exp:* Chemist, Petrolite Corp, 66-68; RES CHEMIST, E I DU PONT DE NEMOURS & CO, INC, 73- *Concurrent Pos:* Adj prof, Dept Entom & Appl Ecol, Univ Del, 81- *Mem:* Am Chem Soc; Sigma Xi; Soc Neurosci. *Res:* Toxicology; insecticide design and synthesis; structure-activity relationships. *Mailing Add:* 120 Country Club Dr Newark DE 19711

HOLYOKE, EDWARD AUGUSTUS, b Madrid, Nebr, Mar 10, 08; m 40, 81; c 2. ANATOMY, EMBRYOLOGY. *Educ:* Univ Nebr, BSc, 30, MA, 32, MD, 34, PhD(anat), 38. *Prof Exp:* From instr to assoc prof, 32-46, chmn dept, 60-73, prof anat, 46-83, EMER PROF, COL MED, UNIV NEBR, OMAHA, 83- *Mem:* AAAS; Am Asn Anatomists; Sigma Xi; Anatomical Soc Gt Brit & Ireland. *Res:* Experimental embryology; gross anatomy; hematology. *Mailing Add:* Col Med Univ Nebr Omaha NE 68105

HOLYOKE, THOMAS CAMPBELL, b Milwaukee, Wis, June 9, 22; m 47; c 3. ALGEBRA. *Educ:* Harvard Univ, AB, 43; Ohio State Univ, MA, 47, PhD(math), 50. *Prof Exp:* Instr math, Ohio State Univ, 49-50; instr, Northwestern Univ, 50-53, asst prof, 53-55; asst prof, Miami Univ, 55-58; assoc prof, Antioch Col, 58-63, prof math, 63-84, assoc acad dean, 77-80; RETIRED. *Concurrent Pos:* Vis mem, Dept Math, Air Develop Ctr, Wright-Patterson AFB, 56CF; dir, NSF Inst Teachers Math, Miami Univ, 57; NSF fel, Univ Calif, 57-58; chmn, Dept Math, Antioch Col, 61-64, dir, Sci Inst, 69-72; Fulbright-Hays lectr, Mindanao State Univ, 64-65, vis prof, 67-69; Fulbright-Hays lectr, prof math & head dept, Chancellor Col, Univ Malawi, 75-77; vis prof, Col Wooster, Ohio, 80-81; Chancellor Col, Univ Malawi, 75-77; Fulbright lect grant, Univ Botswana, 87-88. *Res:* Permutation groups; abstract algebra; elementary number theory; multiply transitive permutation groups. *Mailing Add:* 608 S High Yellow Springs OH 45387

HOLZ, GEORGE GILBERT, JR, protozoology, physiology; deceased, see previous edition for last biography

HOLZAPFEL, CHRISTINA MARIE, b Baltimore, Md, Jan 24, 42; m 70; c 1. BIOGEOGRAPHY. *Educ:* Goucher Col, BA, 64; Univ Mich, Ann Arbor, MS, 68, PhD(zool), 70. *Prof Exp:* NSF res asst biogeog plants, Canary Islands, Spain, 65-66; NSF res trainee biogeog insects, Univ Mich, Ann Arbor, 66-70, lectr exp biol, 70; res fel biogeog plants, Harvard Univ, 70-71; RES ASSOC BIOGEOG, UNIV ORE, 71- *Concurrent Pos:* Res assoc, Tall Timbers Res Sta, Tallahassee, Fla, 77-78; res fel, Imp Col, Silwood Park, Eng, 86; fel, Woods Hole Oceanog Inst, 63. *Mem:* Sigma Xi; Soc Study Evolution; Ecol Soc Am. *Res:* Population biology, life history patterns and predator-prey interactions among container-breeding mosquitoes. *Mailing Add:* Dept Biol Univ Ore Eugene OR 97403

HOLZBACH, R THOMAS, b Salem, Ohio, Aug 19, 29; m 56; c 3. HUMAN GALLSTONE PATHOGENESIS, CLINICAL GASTROENTEROLOGY & HEPATOLOGY. *Educ:* Georgetown Univ, BS, 51; Case Western Reserve Univ, MD, 55; Nat Bd Med Examiners, dipl, 56; Am Bd Internal Med, cert, 62, Med Subspecialty Gastroenterol, 64. *Prof Exp:* Internship & asst resident med, Univ Ill, Res & Educ Hosps, Chicago, 55-57; sr resident internal med, Cleveland Metrop Gen Hosp, Ohio, 59-60; fel gastroenterol, Case Western Reserve Univ, Cleveland, Ohio, 60-61; private pract gastroenterol, Cleveland, Ohio, 63-67; head, Gastrointestinal Res Unit & assoc physician, Div Med, St Luke's Hosp, Cleveland, Ohio, 67-73, dir, Div Gastroenterol, 70-73; HEAD, GASTROINTESTINAL RES UNIT, DEPT GASTROENTEROL, CLEVELAND CLIN FOUND, OHIO, 73- *Concurrent Pos:* Fel gastroenterol, Univ Calif Med Ctr, San Francisco, 61-62; instr med, Sch Med, Case Western Reserve Univ, Cleveland, Ohio, 61-64, clin instr, 64-71; asst physician, University Hosps, Cleveland, Ohio, 61-63; asst chief, Sect Gastroenterol, Vet Admin Med Ctr, Cleveland, Ohio, 61-63; sci dir, Res Projs Area, Cleveland Clin Found, 74-82; vis prof & lectr, numerous univs & hosps, foreign & US, 74-90; Alexander von Humboldt traveling fel, Univ Heidelberg, Ger, 78, Univ Munich, 82. *Mem:* Am Gastroenterol Asn; Am Asn Study Liver Dis; fel Am Col Physicians; Am Physiol Asn; Am Soc Biochem & Molecular Biol; AAAS; Sigma Xi; Am Soc Biol Chemists; Int Asn Study Liver Asn; Am Fedn Clin Res. *Res:* Discovered: glycopreteins secreted by liver are present in human bile; glycoproteins have opposing effects; an inhibitor form delays the growth of cholesterol crystals, an essential prelude to gallstone formation; promoter forms, of which there may be several, accelerate crystals formation and growth; characterizing these unique glycoproteins. *Mailing Add:* Gastrointestinal Res Unit Dept Gastroenterol Cleveland Clin Found 9500 Euclid Ave Cleveland OH 44195

HOLZBECHER, JIRI, b Prague, Czech, Nov 14, 43; Can citizen; m 68. PHYSICAL CHEMISTRY. *Educ:* Univ Chem Technol, Prague, MSc, 67; Dalhousie Univ, PhD(anal chem), 73. *Prof Exp:* Fel anal chem, Dalhousie Univ, 73-74; sr technologist clin chem, Grace Maternity Hosp, Halifax, 75-76; res assoc, Trace Anal Res Ctr, 76-79, PRIN SLOWPOKE OPERATOR ANALYTICAL CHEM, SLOWPOKE FACIL, DALHOUSIE UNIV, 79- *Mem:* Chem Inst Can. *Res:* Neutron activation analysis; infrared spectrometry; fluorimetry analysis. *Mailing Add:* Slowpoke Facil Life Sci Bldg Dalhousie Univ Halifax NS B3H 4J1 Can

HOLZBERLEIN, THOMAS M, b Oklahoma City, Okla, Apr 1, 26; m 51; c 2. ATOMIC PHYSICS, SOLAR MATERIALS. *Educ:* Okla State Univ, BS, 50, MS, 51; Univ Okla, PhD(physics), 63. *Prof Exp:* Instr, 51-54, assoc prof, 59-68, chmn dept, 64-73, Ken Smith Chmn Natural Sci Dept, 82, PROF PHYSICS, PRINCIPIA COL, 68-,. *Concurrent Pos:* Mem, Solar Energy Deleg, Peoples Repub China, 84. *Mem:* AAAS; Am Asn Physics Teachers; Int Solar Energy Soc; Am Underground Space Asn. *Res:* Tornado tracking by electronic means; solar energy; energy conservation; measurements of lifetimes of excited atomic states; passive solar teaching materials. *Mailing Add:* Dept Physics Principia Col Elsah IL 62028

HOLZER, ALFRED, b Vienna, Austria, Jan 23, 25; US citizen; m 50; c 2. NUCLEAR PHYSICS. *Educ:* Univ Mich, BS(physics) & BS(math), 50, MS, 51; Case Inst, PhD(physics), 60. *Prof Exp:* Chief spectroscopist, J H Herron Co, Ohio, 51-54; physicist, K Div, Univ Calif, 59-62, group leader, 62-66, from dep div leader to div leader, Plowshare Div, Lawrence Livermore Lab, 66-74, asst assoc dir, Energy & Resource Prog, 74-86; RETIRED. *Mem:* Am Phys Soc. *Res:* Shock waves in solids; explosion phenomena; energy & resources. *Mailing Add:* PO Box 7247 Carmel CA 93921

HOLZER, GUNTHER ULRICH, b Lahr, Ger, May 22, 42; m 81. GEOCHEMISTRY. *Educ:* Acad Chem & Physics, Isny, Ger, BS, 68; Univ Houston, Tex, PhD(org chem), 73. *Prof Exp:* Chem engr synthetic rubber, Farbenfabriken Bayer, Ger, 68-69; res asst peptide chem, Inst Org Chem, Univ Tubingen, Ger, 74; res assoc & vis asst prof chem & biochem, Dept Biophys Sci, Univ Houston, Tex, 75-80; from asst prof to assoc prof chem & geochem, Colo Sch Mines, 80-86; ASSOC PROF, SCH BIOL & CTR BIOTECHNOL GA INST TECHNOL, 86- *Concurrent Pos:* Consult, Solar Energy Res Inst, 81- *Mem:* Sigma Xi. *Res:* lipids in archaebacteria; role of microorganisms in petroleum formation; organic analysis of ancient and recent sediments by gas chromatography mass spectrometry; development of supercritical fluid chromatography. *Mailing Add:* Sch Biol Ctr Biotech Ga Inst Technol Atlanta GA 30332

HOLZER, ROBERT EDWARD, b Portland, Ore, Nov 21, 06; m 31; c 3. SPACE PHYSICS, PLASMA PHYSICS. *Educ:* Reed Inst, AB, 26, Univ Calif, Berkeley, MA, 28; PhD (physics), 30. *Hon Degrees:* DSc, Univ NMex, 89. *Prof Exp:* Instr physics, Univ Calif, 30-31; Nat Res Coun fel, Univ Chicago, 31-33; asst prof, Fenn Col, 33-34; instr, Univ Calif, 34-35; from instr to prof, Univ NMex, 35-46; prof & head dept, Pomona Col, 46-47; head dept planetary & space sci, 67-69, dean div phys sci, 73-74, prof, 47-74, EMER PROF GEOPHYS, UNIV CALIF, LOS ANGELES, 74- *Concurrent Pos:* Assoc dir war res proj, Univ NMex, 41-46; mem sci adv bd, USAF, 56-62; asst sci dir, Off Naval Res, London, 59-60. *Mem:* AAAS; fel Am Phys Soc; fel Am Geophys Union; Am Asn Physics Teachers. *Res:* Atmospheric electricity; electrical structure of thunderstorms; propagation of low audiofrequency atmospherics; experimental space physics; magnetospheric physics; auroral phenomena. *Mailing Add:* 10517 Wynton Dr Los Angeles CA 90024

HOLZER, SIEGFRIED MATHIAS, b Radstadt, Austria, Aug 24, 36; US citizen; m 63. STRUCTURAL DYNAMICS. *Educ:* Univ Kans, BSCE, 61, MSCE, 65; Univ Ill, Urbana, PhD(civil eng), 67. *Prof Exp:* Eng technician, Consult Firm Ferstl, Austria, 55-56; draftsman, Kansas City Struct Steel Co, Kas, 56-57; struct engr, J F Pritchard & Co, Mo, 62-64; asst prof civil eng, Univ Ill, Urbana, 67-72; from asst prof to assoc prof, 72-80, PROF CIVIL ENG, VA POLYTECH INST & STATE UNIV, 80- *Concurrent Pos:* NSF grant, 68-71; in chg Struct Eng Div, Dept Civil Eng, Va Polytech Inst & State Univ, 74- *Mem:* Am Soc Civil Engrs; Int Asn Shell & Spatial Struct; Sigma Xi. *Res:* Dynamic stability; noconservative and nonlinear structures; numerical methods in structural analysis. *Mailing Add:* Dept Civil Eng Va Polytech Inst Blacksburg VA 24061

HOLZER, THOMAS EDWARD, b Albuquerque, NMex, June 24, 44; m 68. SOLAR TERRESTRIAL PHYSICS, SPACE PLASMA PHYSICS. *Educ:* Pomona Col, BA, 65; Univ Calif, San Diego, PhD(appl physics), 70. *Prof Exp:* NATO fel space physics, Imp Col, Univ London, 70-71; Nat Res Coun res assoc, Aeronomy Lab, Nat Oceanic & Atmospheric Admin, Boulder, Colo, 71-73; scientist, 73-78, SR SCIENTIST, HIGH ALTITUDE OBSERV, NAT CTR ATMOSPHERIC RES, 78-; PROF ADJOINT, DEPT ASTROPHYS, PLANETARY & ATMOSPHERIC SCI, UNIV COLO, 81- *Concurrent Pos:* Lectr, Dept Astrophys, Planetary & Atmospheric Sci, Univ Colo, 72-81; mem, Heliospheric Hydromagnetics Subpanel, Space Plasma Physics Panel, Nat Acad Sci, 76-77, Comt Solar & Space Physics, 77-80, Advocacy Comt Study Physics of the Sun, 80-83, Ad Hoc Comt Role of Theory in Space Sci, 80-82, Study Major Directions Space Sci, Solar & Space Physics Task Group, Space Sci Bd, 84-86, Solar Panel, Astron & Astrophys Surv Comt, 89-90; mem, Solar Terrestrial Theory Panel, NASA, 78-79, Solar Cycle & Dynamics Mission Working Group, 78-79, Space Sci Steering Comt, Solar Terrestrial Theory Prog, Ad Hoc Adv Subcomt, 79-80, Solar Physics Working Group, Astron Surv Comt, 79-80, Solar Physics Mgt Opers Working Group, 79-80, Ad Hoc Comt, Prog Nex Solar Maximum, 84-86, Cosmic & Heliospheric Mgt Opers Working Group, Space Physics Div, 87-, Heliospheric Physics Working Group, 87-, Solar Probe Sci Working Team 88-; head, Coronal-Interplanetary Physics Sect, High Altitude Observ; mem steering comt, Off Interdisciplinary Earth Studies, Univ Corp Atmospheric Res, 86-88, Solar-Interplanetary Variability Study, Sci Comt Solar Terrestrial Physics, 87-, Solar-Terrestrial Energy Prog, Sci Comt Solar-Terrestrial Physics, 87-90; chmn, Heliospheric SR&T Rev Panel, Space Physics Div, NASA, 89, Solar Theory & Modelling SR&T Rev Panel, 90, Theory Working Group, 90-; chmn, Organizing Comt Workshop on Solar-Terrestrial Impacts on Global Change, 89-; mem, rev comt solar prog, NASA Goddard Space Flight Ctr, 90, rev comt, Inst Space & Terrestrial Sci, Ontario, Can, 90, theory panel, NASA Space Physics Study, 90, cosmic & heliospheric panel, 90; co-organizer, Interstellar Probe Workshop, 90; vis comt, Bartol Res Inst, Univ Del, 90- *Honors & Awards:* James B Macelwane Award, Am Geophys Union, 78. *Mem:* fel Am Geophys Union; Am Astron Soc; Int Astron Union; Int Union Radio Sci. *Res:* Physics of the sun; heliosphere; terrestrial magnetosphere, ionosphere and upper atmosphere. *Mailing Add:* High Altitude Observ Nat Ctr Atmospheric Res Box 3000 Boulder CO 80307

HOLZER, THOMAS LEQUEAR, b Lafayette, Ind, June 26, 44; m 68; c 2. GEOLOGY. *Educ:* Princeton Univ, BSE, 65; Stanford Univ, MS, 66, PhD(geol), 70. *Prof Exp:* Asst prof geol, Univ Conn, 70-75; geologist, 75-82, dep asst dir res, 82-84, GEOLOGIST, US GEOL SURV, 84- *Mem:* Geol Soc Am; Nat Water Well Asn; Am Geophys Union; Sigma Xi. *Res:* Rheology of fine grained sediments; land subsidence; ground water hydrology; earthquake liquefaction. *Mailing Add:* 151 Walter Hays Dr Palo Alto CA 94303

HOLZER, TIMOTHY J, b Minneapolis, Minn, Apr 13, 51. RETROVIROLOGY & HOST DEFENSE, MACROPHAGE-T CELL BIOLOGY. *Educ:* Univ Minn, BA, 73; Mankato State Univ, MA, 76; Univ Ill, Chicago, PhD(microbiol & immunol), 83. *Prof Exp:* Res assoc prev med, Univ Ill, Chicago, 83-84, instr med, 84-86, res asst prof, 86-87; immunologist, Transfusion Diag, 87-88, sr res immunol & microbiol, Exp Biol Res, 88-90, SR SCIENTIST, EXP BIOL RES, ABBOTT LABS, 90- *Concurrent Pos:* Adj asst prof, Univ Health Sci, Dept Microbiol & Immunol, Chicago Med Sch, 89- *Mem:* Am Asn Immunologists; AAAS; Am Soc Microbiol. *Res:* Investigation of host defense mechanisms in infectious diseases primarily focused on human retrovirus interactions with T-cells and monocytes; diagnostic assay development for AIDS and hepatitis fields. *Mailing Add:* Exp Biol Res Dept 90-D Abbott Labs 1400 Sheridan Rd North Chicago IL 60064

HOLZHEY, CHARLES STEVEN, b Malta, Mont, Apr 29, 36; m 66; c 2. SCIENCE ADMINISTRATION. *Educ:* Univ Idaho, BS, 58, MS, 64; Univ Calif, Riverside, PhD(soil sci), 68. *Prof Exp:* Soil scientist, Soil Conserv Serv, USDA, Mont, 58-68, res soil scientist, Soil Surv Lab, Plant Indust Sta, 68-72, head, Soil Surv Invest Unit, 72-75 & Nat Soil Surv Lab, 75-88, ASST DIR, SOIL SURV DIV, SOIL CONSERV SERV, USDA, 88- *Mem:* Am Soc Agron; Soil Sci Soc Am; Soils Conserv Soc Am; Nat Asn Conserv Dist; Sigma Xi. *Res:* Soil formation and classification; soil survey applications to land use and environmental concerns. *Mailing Add:* 430 Haverford Dr Lincoln NE 68510

HOLZINGER, JOSEPH ROSE, b Annville, Pa, Sept 30, 14; m 39; c 3. MATHEMATICS. *Educ:* Franklin & Marshall Col, BS, 35; Cornell Univ, MS, 48. *Prof Exp:* Teacher pub sch, Pa, 36-42; instr math, Wyomissing Polytech Inst, 45-46; prof math & astron, Franklin & Marshall Col, 48-80; RETIRED. *Concurrent Pos:* Dir Grundy Observ, Franklin & Marshall Col; NSF sci fac fel, 59-60. *Mem:* AAAS; Am Astron Soc; Math Asn Am. *Res:* Astronomy. *Mailing Add:* Dept Math & Astron Franklin & Marshall Col Lancaster PA 17604

HOLZINGER, THOMAS WALTER, b Nurnberg, Ger, Feb 23, 30; US citizen. FOOD SCIENCE. *Educ:* Univ Vt, BS, 53; Iowa State Univ, MS, 54. *Prof Exp:* Res asst biochem, Univ Vt, 54-55; asst mgr, NJ Dairy Labs, 55-61; res assoc, Milk & Ice Cream Div, 61-63, coordr res & qual control, 63-65, dir res, Dairy & Serv Div, NY, 66-68, dir qual assurance, Ohio, 68-73, assoc dir corp qual assurance, 73,76, dir qual assurance & compliance, 76-85, CORP TECH DIR, DAIRY & FOODS INT, BORDEN INC, 85- *Concurrent Pos:* Mem, Interstate Milk Shippers Conf, 66- & Milk Indust Found Tech Comt, 67-; dir, Am Cultured Dairy Prod Inst; chmn, Nat Dairy Bd, Sci Adv Comt; dir, Int Dairy Fedn, Am Dairy Sci Asn; tech comt mem, Nat Food Processors Asn. *Mem:* Am Dairy Sci Asn; Inst Food Technologists; NY Acad Sci; fel Am Inst Chemists; Am Soc Microbiol; Nat Food Processors Asn; Int Dairy Fedn. *Res:* Food technology; food chemistry; microbiology; food safety. *Mailing Add:* Borden Inc 960 Kingsmill Pkwy Columbus OH 43229

HOLZMAN, ALBERT G(EORGE), industrial engineering, operations research; deceased, see previous edition for last biography

HOLZMAN, GEORGE, b Los Angeles, Calif, Jan 25, 20; m 44; c 3. ORGANIC CHEMISTRY. *Educ:* Calif Inst Technol, BS, 42, PhD(org chem), 48. *Prof Exp:* Asst, Calif Inst Technol, 42-46; A D Little fel, Mass Inst Technol, 47-49; chemist, Shell Develop Co Div, Shell Oil Co, Calif, 49-53, res supvr, 53-62, spec technologist, 62-64, mgr, Aromatics Dept, 64, supt Woodriver Refining, 64-67, chief technologist, 67-68, mgr mfg technol, 68-70, refinery mgr, 70-71, gen mgr refining, 71-77, gen mgr logistics, 77-80; TRUSTEE & DIR, SOUTHWEST WASH HOSPS, 85-; DIR, HEALTH & WELFARE PLANNING COUN, 87- *Concurrent Pos:* Dir fuels & petrochem, Am Inst Chem Engrs, 70-71; chmn cood subcomt factors affecting petrol refining capacity, Nat Petrol Coun, 72-73. *Mem:* AAAS; Am Chem Soc; Am Inst Chem Engrs; The Chem Soc. *Res:* Petroleum process chemistry and processing. *Mailing Add:* PO Box 1909 Vancouver WA 98668

HOLZMAN, GEORGE ROBERT, b Long Island, NY, Dec 31, 19; m 44; c 2. PHYSICS, ELECTRICAL ENGINEERING. *Educ:* Rensselaer Polytech Inst, BEE, 42, PhD(physics), 51. *Prof Exp:* Radio engr radar, Naval Res Lab, 42; jr instr physics, Johns Hopkins Univ, 46; sr physicist, Behr-Manning Corp, 50-52; fel, Mellon Inst, 52-58; physicist, Int Bus Mach Corp, 58-59, mathematician programmer, 59-61, mgr phys sci, 61-63, mgr adv tech, 63-65, mgr new mkts develop, 65-69, planning mgr, Electromagnetic Compatibility Power Systs, 69-72, gas panel coordr, 72-76, electromagnetic compatibility mgr, ion implant, diffusion, IBM Corp, 76-84; RETIRED. *Mem:* Am Phys Soc. *Res:* Nuclear and electron spin resonance; electrical charging of small particles; study of surfaces, films and interfaces; ferroelectrics; laser communications; computer systems. *Mailing Add:* 166 Echo Dr Vernon CT 06066

HOLZMAN, GERALD BRUCE, b Los Angeles, Calif, Apr 22, 33; m 57; c 3. MEDICINE. *Educ:* Stanford Univ, BA, 54, MD, 57. *Prof Exp:* Intern, Los Angeles County Gen Hosp, 57-58; resident gynec & obstet, Johns Hopkins Hosp, Johns Hopkins Univ, 59, chief resident, 64-65, instr, Sch Med, 66-67; asst prof, Sch Med, Univ Calif, Los Angeles, 69-72; from assoc prof to prof, Dept Obstet, Gynec & Reproductive Biol, Col Human Med, Mich State Univ, 72-83, prof, Off Med Educ Res & Develop, 77-83, assoc chmn, 78-83; PROF & VCHMN, DEPT OBSTET-GYNEC, MED COL GA, 83- *Concurrent Pos:* Fel, Am Cancer Soc, 61-64, Joshi Macy Found, 65-66, John Hopkins Hosp, 66-67 & Off Med Educ Res & Develop, Mich State Univ, 72-73; private prac obstet & gynec, Baltimore, 65-68, Shelton Clin, Los Angeles, Calif, 68-69; assoc prof, Off Med Educ Res & Develop, Col Human Med, Mich State Univ, 73-77, actg asst dean for Lansing, 78; examr, Am Bd Obstet Gynec. *Mem:* Nat Bd Med Examiners; Am Col Obstetricians & Gynecologists; Asn Professors Gynec & Obstet. *Res:* Modeling medical decision making. *Mailing Add:* Dept Obstet-Gynec Med Col Ga 1120 Fifteenth St Augusta GA 30912

HOLZMAN, PHILIP S, b New York, NY, May 2, 22; m 46; c 3. PSYCHOLOGY. *Educ:* City Col New York, BA, 43; Univ Kans, PhD, 52. *Hon Degrees:* AM, Harvard Univ, 77. *Prof Exp:* Psychologist & dir, Res Training, Menninger Found, 50-68; prof Psychiat & Behav Sci, Univ Chicago, 68-77; ESTHER & SIDNEY R RABB PROF, PSYCHOL, HARVARD UNIV, 77- *Honors & Awards:* Stanley Dean Award, Am Col Psychiatrists; Lieber Prize, Nat Alliance Res Schizophrenia & Depression. *Mem:* Inst Med-Nat Acad Sci; Am Acad Arts & Sci. *Res:* Cognitive, psychophysiological and genetic aspects of mental disorders. *Mailing Add:* Harvard Univ 33 Kirkland St Cambridge MA 02138

HOLZMAN, ROBERT STEPHEN, b New York, NY, Apr 13, 40; m 63; c 2. IMMUNOLOGY, INFECTIOUS DISEASES. *Educ:* Rutgers Univ, BA, 61; Johns Hopkins Univ, MD, 65. *Prof Exp:* From intern to resident, Med Ctr, NY Univ, 65-70, fel immunol & infectious dis, 70-73, asst prof med, Med Sch, NY Univ & asst epidemiologist, Bellevue Hosp, 73-79, HOSP EPIDEMIOLOGIST, BELLEVUE HOSP, 79-, ASSOC PROF MED, NY UNIV. *Mem:* Am Fedn Clin Res; fel Am Col Physicians; fel Infectious Dis Soc Am. *Res:* Hospital acquired infections; delayed-type hypersensitivity; transfer factor; acquired immune deficiency syndrome. *Mailing Add:* Dept Med Med Ctr NY Univ 550 First Ave New York NY 10016

HOLZMANN, ERNEST G(UNTHER), b Ger, Nov 20, 21; US citizen; m 52; c 6. EXPERT SYSTEMS, OPERATIONS RESEARCH. *Educ:* Univ London, BSc, 47; Mass Inst Technol, MSEE, 51; Stanford Univ, MS, 66, PhD(elec eng), 69. *Prof Exp:* Engr process control, Shell Develop Co, 51-56; reactor safeguards, Atomic Power Equip Dept, Gen Elec Co, San Jose, 56-60, sr systs engr, Electronics Lab, Syracuse, 60-64; res assoc mat sci, Stanford Univ, 67-69; systs anal engr, Corp Res & Develop, Gen Elec Co, Schenectady, 69-83; mem tech staff, Anal Sci Corp, 83-85; mem tech staff, Sci Int Applns Corp, 85-88; SR ENG SPECIALIST, LORAL AEROSYS CORP, 88- *Concurrent Pos:* Adj prof, Dept Elec Eng, Union Col, 69-74 & Rensselaer Polytech Inst, 72-74. *Honors & Awards:* Region Award New Tech Concepts, Inst Elec & Electronics Engrs, 77 & 79. *Mem:* Inst Elec & Electronics Engrs. *Res:* Systems and operations analysis; expert systems; computer simulation for planning and design. *Mailing Add:* 2333 Barbour Rd Falls Church VA 22043

HOLZMANN, GERARD J, b Amsterdam, Neth, Nov 12, 51. RADIATION OF COMPUTER PROTOCOLS, DIGITAL IMAGE MANIPULATION. *Educ:* Delft Univ, BSc, 73, MSc, 76, PhD(tech sci), 79. *Prof Exp:* Fulbright fel oper systs, Univ Southern Calif, 78-80, AT&T Bell Labs, Murray Hill, 80-81; asst prof data commun, Delft Univ, 81-83; MEM TECH STAFF COMPUTER SCI, AT&T BELL LABS, MURRAY HILL, 83- *Concurrent Pos:* Vis lectr, Comput Sci Dept, Princeton Univ, NJ, 90-91. *Honors & Awards:* Babler Award, Royal Dutch Inst Engrs, 81. *Res:* Radiation methods for distributed systems. *Mailing Add:* AT&T Bell Labs Rm 2C-521 600 Mountain Ave Murray Hill NJ 07974-2070

HOLZMANN, RICHARD THOMAS, b New York, NY, Mar 24, 27; m 47; c 5. INORGANIC CHEMISTRY, PHYSICAL CHEMISTRY. *Educ:* St John's Univ, NY, BS, 51; Univ Del, MS, 55; Pa State Univ, PhD(inorg chem), 62; Georgetown Univ, LLB, 64. *Prof Exp:* Res chemist, Claymont Develop Lab, Gen Chem, 51-54 & Lawrence Radiation Lab, Univ Calif, 57-58; res supvr, Reaction Motors Div, Thiokol Chem Corp, 58-61; prog mgr, Advan Res Proj Agency, Washington, DC, 61-64, actg dir chem, 64; asst mgr chem & struct prod div, Von Karman Ctr, Aerojet-Gen Corp, 64-65, mgr chem prod div, 65-68; dir res, Quaker Chem Corp, 68-74; pres, Dynachem Corp, Santa Fe Springs, Calif, 74-79; MGR & DIR, CHRISTOPHER GROUP, LAGUNA HILLS, CALIF, 79- *Concurrent Pos:* Mem, Interagency Chem Rocket Propulsion Group, 62-64; consult, Advan Res Proj Agency, 64- *Mem:* Am Chem Soc; assoc fel Am Inst Aeronaut & Astronaut; Am Ord Asn; fel Am Inst Chemists; Am Inst Chem Engrs. *Res:* Hydride, fluorine and interhalogen chemistry; uranium recovery; phosphoric acid; microanalytical chemical surveillance; propellants and explosives; glass technology; composite structure; chemical specialties; lubricants; hydraulics; corrosion; textile and paper making auxiliaries; pollution chemistry and control. *Mailing Add:* Christopher Group Attn Matthew Holzmann 2601 S Oak St Santa Ana CA 92707

HOLZRICHTER, JOHN FREDERICK, b Chicago, Ill, June 8, 41; c 2. LASERS, OPTICS. *Educ:* Univ Wis, BS, 64; Stanford Univ, ME, 66, PhD(physics), 71. *Prof Exp:* Physicist, Res & Develop, Naval Res Lab, 71-73; solid state laser design leader, 73-74, dep group leader laser & plasma group, 74-75, assoc prog leader solid state lasers, 75-80, assoc prog leader, Fusion Lasers Prog, 80-81, leader Inertial Confinement Fusion, 81-84, leader adv solid state lasers, 84-87, DIR INST RES DEVELOP, LAWRENCE LIVERMORE NAT LAB, 87- *Concurrent Pos:* Consult, ILC Technol, 69-70; Fulbright fel, Sloan Found & Hertz Found. *Mem:* Am Phys Soc; Sigma Xi; Optical Soc Am; AAAS. *Res:* Design and construction of high power lasers for fusion research; scientific management for research, development and construction of large scientific projects. *Mailing Add:* 200 Hillcrest Rd Berkeley CA 94705

HOLZWARTH, GEORGE MICHAEL, b Dusseldorf, Ger, May 7, 37; US citizen; m 70; c 2. BIOPHYSICS, BIOPHYSICAL CHEMISTRY. *Educ:* Wesleyan Univ, BA, 59; Harvard Univ, PhD(biophys), 64. *Prof Exp:* Res assoc, Fla State Univ, 64-66; asst prof biophys & chem, Univ Chicago, 67-74; mem staff, Corp Res Lab, Exxon Res & Eng Co, 74-83; lectr, physics, 84-87, assoc prof, 88-89, PROF, WAKE FOREST UNIV, WINSTON-SALEM, NC, 89- *Mem:* Biophys Soc; Am Chem Soc; Am Phys Soc; Electrophoresis Soc. *Res:* Physical chemistry of water-soluble polymers; biophysical chemistry, especially solution conformation of polysaccharides, nucleic acids, and proteins; molecular spectroscopy, especially optical activity; polymer physics. *Mailing Add:* Dept Physics Wake Forest Univ PO Box 7507 Winston Salem NC 27109

HOLZWARTH, JAMES C, b Ft Wayne, Ind, June 20, 24; m 45; c 3. METALLURGICAL ENGINEERING, CORROSION. *Educ:* Purdue Univ, BS, 46, MS, 48. *Prof Exp:* Instr metall eng, Purdue Univ, 47-48; res metallurgist, Gen Motors Res Labs, 48-55, supvr, Metall Eng Dept, 55-60, asst head dept, 60-69, head dept, 69-73, tech dir, 73-87; RETIRED. *Mem:* Am Soc Metals; Am Soc Testing & Mat; Am Foundrymen's Soc; Soc Automotive Engrs; Sigma Xi. *Res:* Environmental simulation in corrosion of automobile bodies and components; corrosion product characterization; development of alloys for wear, fatigue and corrosion resistance; new technology in scrap metal reclamation and recycling. *Mailing Add:* PO Box 1317 Brackenridge CO 80424

HOLZWORTH, JEAN, b Port Chester, NY, Mar 26, 15. VETERINARY MEDICINE. *Educ:* Bryn Mawr Col, AB, 36, MA, 37, PhD(Latin), 40; State Univ NY Vet Col, Cornell Univ, DVM, 50. *Prof Exp:* Instr Latin, Mt Holyoke Col, 40-41, Bryn Mawr Col, 41-44 & Latin & Greek, Brearley Sch, NY, 45-46; intern, Angell Mem Animal Hosp, 50-51, resident, 51-52, clin staff vet med, 52-86; RETIRED. *Concurrent Pos:* Grants, USPHS, 56-57 & 64-66 & Am Philos Soc, 57; mem adv coun, Cornell Feline Health Ctr. *Mem:* Am Vet Med Asn. *Res:* Diseases of cats. *Mailing Add:* PO Box 2305 New Preston CT 06777

HOLZWORTH, ROBERT H, II, b Winston-Salem, NC, June 20, 50; m 70; c 2. SPACE SCIENCE, MAGNETOSPHERIC PHYSICS. *Educ:* Univ Colo, BA, 72; Univ Calif, Berkeley, MA, 74, PhD(physics), 77. *Prof Exp:* Res asst physics, Univ Calif, Berkeley, 74-77, asst res physicist space physics, Space Sci Lab, 78; mem tech staff space physics, Space Sci Lab, Aerospace Corp, 78-82; asst prof, 82-88, ASSOC PROF GEOPHYS & PHYSICS, UNIV WASH, SEATTLE, 88- *Concurrent Pos:* Prin Investr, NASA & NSF, 79- *Mem:* Am Geophys Union. *Res:* Ionospheric physics and atmospheric physics; particles and fields above tropopause; experimental work in electric fields. *Mailing Add:* Geophys Prog AK50 Univ Wash Seattle WA 98195

HOM, BEN LIN, b New York, NY, Feb 5, 36; m 83. HEMATOLOGY, BLOOD TRANSFUSION. *Educ:* Colby Col, BA, 57; NY Univ, MD, 61; Am Bd Path, cert anal & clin path, 69, cert hemat, 83 & cert blood banking, 84; Royal Col Pathologists, Australia, cert anat path-hemat, 85. *Prof Exp:* Intern med, Queen's Hosp, 61-62, resident path, 62-63; resident path, Univ Calif, Los Angeles, 63-64, New Eng Deaconess Hosp, Boston, 64-65 & Va Hosp, San Francisco, 68; res fel hemat, Bispeguerg Hosp, Copenhagen, 65-67; pathologist, Kaiser Hosp, Honolulu, 69-70; pathologist, Queen's Med Ctr, Honolulu, 70-84, fel hemat, 71-72; sr lectr, Chinese Univ, Hong Kong, 84-87; pathologist, Wagga Wagga Base Hosp, Australia, 87 & St John's Hosp, Springfield, Ill, 88; PATHOLOGIST, MERCY HOSP MED CTR, DES MOINES, IOWA, 88- *Concurrent Pos:* Consult hematologist, Prince of Wales Hosp, 84- *Mem:* Col Am Pathologists; Am Soc Clin Pathologists; Am Soc Hematologists; Int Soc Blood Transfusion; Am Asn Blood Banks. *Res:* Biology of lung cancer; physiology of plasma cells; immune thrombocytopenia; biology of blood coagulation. *Mailing Add:* Dept Path Mercy Hosp Med Ctr Sixth & University Des Moines IA 50314

HOM, FOO SONG, b Kwangtung, China, Sept 17, 29; US citizen; m 63; c 2. PHYSICAL PHARMACY. *Educ:* Philadelphia Col Pharm, BS, 54; Temple Univ, MS, 56; Univ Wis, PhD(pharm), 62. *Prof Exp:* Res asst chem, C P Hall Chem Co, 58-59; res assoc pharm develop, Parke, Davis & Co, 61-66; res scientist, Warner-Lambert Res Inst, 66-68; SR RES SCIENTIST, R P SCHERER CORP, 68- *Mem:* Am Pharmaceut Asn; Am Chem Soc; Acad Pharmaceut Sci. *Res:* Product developments; analytical developments, including spectrophotometry, gas chromatography, thin layer chromatography, titrimetry, and electrochemistry; dosage form design effect on bioavailability; gelatin properties; behavior of gelatin to oxygen permeation and light transmission and other physical parameters. *Mailing Add:* 3102 Blue Heron Clearwater FL 34695

HOMAN, CLARKE GILBERT, b Providence, RI, July 16, 31; m 54; c 4. SOLID STATE PHYSICS, PHYSICAL METALLURGY. *Educ:* Univ RI, BS, 58; Rensselaer Polytech Inst, MS, 65. *Prof Exp:* Teaching asst physics, Rensselaer Polytech Inst, 58-60; solid state physicist, US Army, 60-64, proj leader solid state physics, 64-70, sr proj leader, 70-73, group leader solid state physics, 73-85, chief mat eng, Benet Weapons Lab, Watervliet Arsenal, 85-87; RETIRED. *Concurrent Pos:* Adj prof, NY State Univ Albany, 76- *Honors & Awards:* Paul Siple Medal, USA, 79; Res & Develop Achievement Award, US Army, 81 & 85. *Mem:* Am Phys Soc; Asn Int Res Pressions & Temperature (corresp secy). *Res:* Solid state properties of materials at high pressure and cryogenic temperatures such as superconductivity, phase transitions, magnetic and electronic properties; diffusion and lattice defects; ion beam modification of materials. *Mailing Add:* 2084 Maple Ave Charleton NY 12019

HOMAN, ELTON RICHARD, b Glen Ridge, NJ, Oct 19, 32; m 54; c 4. TOXICOLOGY. *Educ:* Rutgers Univ, BSc, 54; Cornell Univ, PhD(econ entom), 62. *Prof Exp:* Fel, Kettering Lab, Med Ctr, Univ Cincinnati, 62-63, sr res assoc toxicol, 63-64; pharmacologist, Div Air Pollution, Taft Sanit Eng Ctr, USPHS, 64-65; pharmacologist, Lab Toxicol, Nat Cancer Inst, 65-74; pharmacologist, Off Toxic Substances, Environ Protection Agency, 74-75; fel chem hyg, Carnegie-Mellon Inst Res, 75-77; mgr, Union Carbide Corp, 77-79, assoc dir, Bushy Run Res Ctr, 80-85, assoc corp dir, Appl Toxicol, 85-87; SR TOXICOLOGIST, HERCULES, INC, 87- *Mem:* AAAS; Soc Toxicol; NY Acad Sci; Sigma Xi. *Res:* Toxicology of antineoplastic chemotherapeutic agents; regulatory toxicology; toxicology of household and industrial chemicals. *Mailing Add:* RD 1 Box 446 Lincoln Univ PA 19352

HOMAN, RUTH ELIZABETH, b Cincinnati, Ohio, Sept 24, 44. ANALYTICAL CHEMISTRY, BIOCHEMISTRY. *Educ:* Edgecliff Col, BA, 66; Xavier Univ, MEd, 72, MS, 75; St Thomas Inst, PhD(chem, biochem), 78. *Prof Exp:* Res chemist analytical, Lab Serv Sect, Air Qual & Emissions Data Proj, Nat Ctr Air Pollution Control, USPHS, 66-68; teacher sec physics, Williamsburg Sch Dist, 68-69; res chemist analytical chem, 69-79, SECT HEAD ANALYTICAL CHEM, MERRELL NAT LABS, MERRELL DOW PHARMACEUT, INC, 79- *Mem:* Am Chem Soc; NAm Thermal Analysis Soc; Soc Appl Spectros. *Res:* Characterization of organic compounds using differential scanning calorimetry, optical microscopy, infrared spectroscopy; organic elemental analysis. *Mailing Add:* Merrell Nat Labs 2110 E Galbraith Rd Cincinnati OH 45215

HOMANN, FREDERICK ANTHONY, b Philadelphia, Pa, July 3, 29. MATHEMATICS. *Educ:* St Louis Univ, AB, 53, PhL, 54, MS, 56; Univ Pa, PhD(math), 59; Woodstock Col, Md, STL, 63. *Prof Exp:* Instr math, Loyola Col, Md, 58-59 & 64-65; from asst prof to assoc prof, 65-70; ASSOC PROF MATH, ST JOSEPH'S UNIV, PA, 70- *Concurrent Pos:* Chmn dept math, Loyola Col, Md, 66-70, trustee, 70-77; rector Jesuit fac, St Joseph's Col, 71-74, actg chmn dept philos, 72-74. *Mem:* Soc Archit Historians. *Res:* Study of improper integrals arising in H Rademacher's analytic additive number theory; generalizations of algebraic and differential inequalities; history of Renaissance mathematics; history of calculus. *Mailing Add:* Dept Math St Joseph's Univ Philadelphia PA 19131

HOMANN, H ROBERT, b Hinsdale, Ill, Dec 11, 37; m 62. BIOCHEMISTRY. *Educ:* Beloit Col, BS, 59; Wash State Univ, PhD(biochem), 64. *Prof Exp:* Res fel, Wash State Univ, 64; from asst prof to assoc prof chem, 64-78, assoc acad dean, 69-73, NERE SUNDET PROF CHEM, CONCORDIA COL, MOORHEAD, MINN, 78-, ASSOC DEAN COL, 73- *Mem:* Am Chem Soc; Am Soc Microbiol. *Res:* Metabolism of facultative autotrophic bacteria. *Mailing Add:* Dept Chem Concordia Col Moorhead MN 56560

HOMANN, PETER H, b Wittenberge, Ger, Apr 3, 33; m 64; c 2. PLANT PHYSIOLOGY, BIOCHEMISTRY. *Educ:* Karlsruhe Tech Univ, Dipl chem, 59, Dr rer nat(chem, biochem), 62. *Prof Exp:* Res assoc, Inst Molecular Biophys, 62-66, from asst prof to assoc prof biol sci, 66-78, PROF PHOTOSYNTHESIS & PHOTOCHEM, INST MOLECULAR BIOPHYS, FLA STATE UNIV, 78- *Mem:* AAAS; Am Soc Plant Physiol; Am Soc Photobiol; Biophys Soc. *Res:* Photosynthesis; photobiology; plant biochemistry; organization of the protein complex of photosynthetic water oxidation, and its interaction with inorganic cofactors. *Mailing Add:* Inst Molecular Biophys Fla State Univ Tallahassee FL 32306-3015

HOMBERG, OTTO ALBERT, b New York, NY, Jan 10, 31; m 55; c 2. ORGANIC CHEMISTRY. *Educ:* Polytech Inst Brooklyn, BS, 52; Univ Md, MS, 60; Univ Cincinnati, PhD(chem), 65. *Prof Exp:* Res chemist, Gen Chem Div, Allied Chem Corp, 52-53; res chemist, US Indust Chem Div, Nat Distillers & Chem Corp, 57-59; sr chemist, Carlisle Chem Works, Inc, 59-64; res chemist, Cent Res Labs, GAF Corp, 65-67; res engr, Homer Res Labs, Bethlehem Steel Corp, 67-74, supvr chem, 74-77, res engr cokemaking, 77-83; CONSULT, 83- *Res:* Synthetic organic chemistry; organosulfur compounds; coal tar aromatics; heterocyclics; organometallics; process development; organophosphorus compounds; vapor phase oxidation; bio-oxidation; plastic additives; gas processing; environmental control. *Mailing Add:* Indian Purchase 5954 Green Point Rd E New Market MD 21631-9767

HOMBERGER, DOMINIQUE GABRIELLE, b Zurich, Switz, Apr 10, 48; m 85. FUNCTIONAL ANATOMY, EVOLUTIONARY BIOLOGY. *Educ:* Univ Zurich, Switz, MS, 72, PhD(zool), 76. *Prof Exp:* Asst, Zool Mus, Univ Zurich, Switz, 72-76; res assoc, dept biol sci, Columbia Univ, 77-79; asst prof, 79-84, ASSOC PROF COMP ANAT, LA STATE UNIV, 84- *Concurrent Pos:* Teacher, Sch Blind Adults, Zurich, 72-75; prin invest, sect syst biol, NSF, 84-87; Hon Mem, CSIRO-Wildlife & Ecol, Canberra, Australia, 87-88. *Mem:* Am Soc Zoologists; Am Asn Anatomists; Am Ornithologists Union; Brit Ornithologists Union; Ger Ornithologists Union; Soc Syst Zool. *Res:* Comparative, functional and ecological morphology of the avian feeding apparatus; feeding and drinking behavior of birds; biomechanical analysis of complex anatomical systems; systematics, phylogeny and biogeography of birds; theoretical foundations and methods of phylogenetic reconstruction and functional morphology. *Mailing Add:* Dept Zool & Physiol La State Univ Baton Rouge LA 70803

HOMBURGER, FREDDY, b St Gall, Switz, Feb 8, 16; nat US; m 39. MEDICINE, PATHOLOGY. *Educ:* Univ Geneva, MD, 41; Am Bd Toxicol, dipl, 82. *Prof Exp:* Res fel path, Sch Med, Yale Univ, 41-42; intern, New Haven Hosp, Conn, 42-43; intern med, Boston City Hosp, 43-44; res fel, Thorndike Mem Lab, Harvard Med Sch, 44-45; Teagle res fel, Sloan-Kettering Inst Cancer Res, New York, 45-46, chief dept clin invest, 46-48; res prof med, Med Sch, Tufts Univ, 48-57; pres & dir, Bio-Res Inst, Inc & Bio-Res Consults, Inc, 57-84; RES PROF PATH, SCH MED, BOSTON UNIV, 74- *Concurrent Pos:* Instr, Med Col, Cornell Univ, 46-48; dir cancer res & control unit, New Eng Med Ctr, 48-57; sci assoc, Jackson Lab, Maine, 52-59; visitor dept Ger lang & cultures, Harvard Univ, 65-71, 74-80 & Consul of Switz, Boston, 64-86; spec consult cancer control br, Nat Cancer Inst; consult adv comn serum albumins, Comn Plasma Fractionation & Related Processes; mem, Nat Adv Coun, Nat Hypertension Asn, Inc. *Honors & Awards:* Julius A Stratton Prize, Friends of Switz, 91. *Mem:* Endocrine Soc; Soc Exp Biol & Med; Am Asn Pathologists; Am Soc Pharmacol & Exp Therapeut; Toxicol Soc; Sci Acad Toxicol. *Res:* Protein metabolism; clinical investigation; experimental pathology of tumors; environmental carcinogenesis; toxicology; experimental pathology of Syrian hamster; cardiomyopathy. *Mailing Add:* 675 Massachusetts Ave Cambridge MA 02139

HOMBURGER, HENRY A, b Detroit, Mich, July 20, 40. PATHOLOGY, CLINICAL IMMUNOLOGY. *Educ:* Univ Mich, BS, 68, MD, 72; Am Bd Path, cert anat & clin path, 76. *Prof Exp:* Resident path, Univ Calif, San Francisco, 73-74; intern path, Mayo Grad Sch Med, 72-73, resident, 74-76, from asst prof to assoc prof, 77-87, PROF LAB MED, MAYO MED SCH & GRAD SCH MED, 87-, HEAD, SECT IMMUNOPATH, DIV PATH, DEPT LAB MED & PATH, 90-; DIR, SPECIFIC ANTIBODY LAB & CELLULAR CYTOTOXICITY LAB, SECT IMMUNOPATH, MAYO CLIN, 81- *Concurrent Pos:* Consult lab med, Sect Clin Chem & Regional Lab Serv, Mayo Clin, 76-81 & Sect Immunopath, 81-; resource consult, Dept Health & Human Serv, Food & Drug Admin, 91-; mem, var adv comts, Col Am Pathologists, Am Soc Clin Pathologists & Nat Comt Clin Lab Standards. *Mem:* Am Asn Immunologists; Sigma Xi; Col Am Pathologists; Am Soc Clin Pathologists; AAAS; Acad Clin Lab Physicians & Scientists; Am Fedn Clin Res; Am Acad Allergy & Immunol; AMA; Soc Anal Cytol. *Res:* Immunopathology. *Mailing Add:* Clin Immunol Mayo Clin Rm 160 Hilton Bldg Rochester MN 55905

HOMEIER, EDWIN H, JR, b Louisville, Ky, Dec 6, 37; m 59; c 3. PHYSICAL INORGANIC CHEMISTRY. *Educ:* Concordia Teachers Col, Ill, BS, 59; Univ Wis-Madison, PhD(inorg chem), 68. *Prof Exp:* Instr, High Sch, Chicago, 59-62; teaching asst chem, Univ Wis, 62-64; res chemist, UOP Inc, 67-80, res specialist, 80-83, assoc res scientist, 83-85, assoc anal scientist, 85-89, SR ANALYTICAL SPECIALIST, UOP INC, 89- *Concurrent Pos:* From asst prof to assoc prof chem, Concordia Teachers Col, Ill, 67-74; consult, Phys Sci for Nonsci Students Testing Prog, Rensselaer Polytech Inst, 69-70. *Mem:* Catalyst Soc NAm; Am Chem Soc. *Res:* Characterization of catalysts, feed and products for petroleum refinery and petrochemical applications; characterization of catalysts for pollution abatement processes; characterization of engineered materials; reduced states of the early transition metals. *Mailing Add:* Res Ctr 50 E Algonquin Rd Des Plaines IL 60017-5016

HOMER, EUGENE D(ANIEL), industrial engineering, for more information see previous edition

HOMER, GEORGE MOHN, b Cleveland, Ohio, Mar 28, 24; m 51; c 2. MEDICAL SCIENCE, TOXICOLOGY. *Educ:* Ohio State Univ, BA, 48, PhD(physiol chem), 58; Univ Colo, MS, 51; Nat Registry Clin Chem, cert, 69; Am Bd Clin Chem, dipl, 71. *Prof Exp:* Asst pharmacol, Wyeth Inst Med Res, 51-52; asst biochem & biophys, Dept Med, Ohio State Univ, 52-58; res chemist, Miami Valley Hosp, Akron, 58-63, asst dir res, Dept Res, 63-65; biochemist, Dept Path, Akron City Hosp, 65-80; instr chem & mem grad fac, Youngstown State Univ, 73-80; asst prof path, Northeastern Ohio Univs Col Med, 70-80; DIR CHEM, DR R G THOMAS & ASSOC, ELYRIA MEM HOSP, OHIO, 80- *Concurrent Pos:* Mem adj sci staff, med & dent Staff, Akron City Hosp, 74-80; health prof affil staff, Elyria Mem Hosp, 81-; inspector, forensic urine drug testing, Am Asn Clin Chem/CAP, 87- *Mem:* Nat Acad Clin Biochemists; Am Asn Clin Chemists; Sigma Xi. *Res:* Clinical chemistry techniques; TDM and toxicology pharmacokinetics. *Mailing Add:* 225 Westwind Dr No 26 Avon Lake OH 44012-2418

HOMER, LOUIS DAVID, b Washington, DC, Mar 21, 35; m 61; c 3. BIOMETRICS, PHYSIOLOGY. *Educ:* Columbia Univ, BA, 55; Med Col Va, PhD(physiol), 62, MD, 63. *Prof Exp:* Assoc biomet, Emory Univ, 63-64, asst prof, 63-69; assoc prof, Brown Univ, 69-71; MED OFFICER, NAVAL MED RES INST, NAT NAVAL MED CTR, 71- *Mem:* Am Physiol Soc; AAAS. *Res:* Mathematical models of capillary solute exchange. *Mailing Add:* Naval Med Res Inst Nat Naval Med Ctr Bethesda MD 20889-5055

HOMER, PAUL BRUCE, b Hempstead County, Ark, Feb 28, 39. PHYSICS, ENGINEERING. *Educ:* NMex Inst Mining & Technol, BS, 62; Univ Calif, Los Angeles, MS, 68; Mass Inst Technol, SM, 88. *Prof Exp:* Physicist weapons res & develop, Naval Ord Test Sta, Naval Weapons Ctr, 62-68, br head, 68-74, dir head weapons res & develop, 74-82, dept head electronic warfare, 82-87, dept head aircraft weapons integration, 87-88, DEPT HEAD ATTACK WEAPONS, NAVAL WEAPONS CTR, 88- *Concurrent Pos:* Task leader, TTCP Panel W-6, Generic Weapons Systs Effectiveness, 75-78; chmn, Joint Munitions Effectiveness manuals, Air-to-Surface Steering Comt, 75-79; Sloan fel, Mass Inst Technol, 87-88; panel mem, Adv Group Aerospace Res & Develop, NATO, 88- *Mem:* Res Soc Am; Sigma Xi. *Res:* Weapons systems research and development; warhead optimization; target vulnerability; fire control design analysis; combat assessment; military operations analysis; electronic warfare. *Mailing Add:* PO Box 2337 Ridgecrest CA 93555

HOMER, ROGER HARRY, b Long Beach, Calif, Dec 3, 24; m 52; c 5. MATHEMATICS. *Educ:* Univ Southern Calif, AB, 51; Univ Calif, Berkeley, PhD(math), 59. *Prof Exp:* From asst prof to assoc prof, 59-71, PROF MATH, IOWA STATE UNIV, 71- *Mem:* Am Math Soc. *Res:* Mathematical analysis; applied functional analysis. *Mailing Add:* 2212 Knapp St Ames IA 50010

HOMER, STEVEN ELLIOTT, b Chicago, Ill, Mar 27, 52; m 78; c 3. COMPLEXITY THEORY. *Educ:* Univ Calif Berkeley, BA, 73; Mass Inst Technol, PhD(math), 78. *Prof Exp:* Asst prof math, DePaul Univ, 78-81; asst prof, 81-85, ASSOC PROF COMPUTER SCI, BOSTON UNIV, 86- *Concurrent Pos:* Vis scholar, Mass Inst Technol, 79-80 & 84-88; Fulbright prof computer sci, Heidelburg Univ, 88-89. *Mem:* Sigmi Xi; Asn Symbolic Logic; Asn Comput Math. *Res:* Theory of computation; study combinatorial problems, trying to determine whether they are computable and if so, how difficult they are to solve. *Mailing Add:* Computer Sci Dept Boston Univ Boston MA 02215

HOMEYER, AUGUST HENRY, b Chicago, Ill, Apr 21, 08; m 34; c 2. CHEMISTRY. *Educ:* Wash Univ, BS, ChE, 30; Pa State Col, MS, 31, PhD(org chem), 33. *Prof Exp:* Chemist, Mallinckrodt Chem Works, 33-49, assoc dir res, 49-56, dir mkt res med chem, 52-60, dir foreign develop, 60-62, gen mgr, Int Div, 62-69, vpres, 69-73, consult, Mallinckrodt Inc, 73-83; RETIRED. *Concurrent Pos:* Mem, Comt Opium Scientists, UN, Geneva, 58. *Mem:* AAAS; Am Chem Soc; Am Inst Chemists. *Res:* Organic synthesis; alkaloids of opium; medicinal chemicals; compounds containing a neopentyl system; syntheses employing dialkyl carbonates. *Mailing Add:* 9033 Green Ridge Dr St Louis MO 63117

HOMMERSAND, MAX HOYT, b San Diego, Calif, July 10, 30. BOTANY. *Educ:* Univ Calif, BA, 54, PhD(bot), 58. *Prof Exp:* NSF res fel plant physiol, Harvard Univ, 57-59; from instr to assoc prof, 59-71, PROF BOT, UNIV NC, CHAPEL HILL, 71- *Mem:* AAAS; Int Phycol Soc; Bot Soc Am; Phycol Soc Am. *Res:* Algal morphology, physiology and systematics. *Mailing Add:* 304 Spruce St Chapel Hill NC 27514

HOMMES, FRITS A, b Bellingwolde, Neth, May 28, 34; m 58; c 2. BIOCHEMISTRY, DEVELOPMENTAL BIOCHEMISTRY. *Educ:* Univ Groningen, MSc, 58; Univ Nymegen, PhD(biochem), 60. *Prof Exp:* Res assoc chem, Univ Groningen, Neth, 56-58, res assoc biochem, Univ Nymegen, 60-61 & Univ Pa, 61-63; instr, Dept Biochem, Univ Nymegen, 65-66, head lab, Dept Pediat, Univ Groningen, 66-72, assoc prof, 72-79; PROF, DEPT BIOCHEM & MOLECULAR BIOL, MED COL GA, AUGUSTA, 79- *Concurrent Pos:* Fulbright fel, 61-63; lectr, ASBC, 65; consult, Dutch Health Coun, 74-79; chmn, Bioenergetics Study Sect, Netherlands, 74-76; lectr, Third Genetics Sem, Tokyo, 81. *Mem:* Europ Soc Pediat Res; Am Soc Human Genetics; Soc Study Inherited Metab Dis; AAAS; Am Soc Biol Chemists; Soc Inherited Metab Dis; Soc Pediat Res. *Res:* Molecular defects of in-born errors of metabolism; mechanism of brain dysfunction in amino acidemias; cerebral sulfate metabolism. *Mailing Add:* Dept Biochem & Molecular Biol Med Col Ga Augusta GA 30912-2100

HOMSEY, ROBERT JOHN, metal vapor lasers, program management, for more information see previous edition

HOMSHER, EARL EDWIN, II, b Painesville, Ohio, Feb 19, 42; m 65; c 1. PHYSIOLOGY. *Educ:* Marietta Col, BS, 64; Univ Pittsburgh, PhD(physiol), 69. *Prof Exp:* USPHS fel physiol, 69-70, asst prof, 70-75, assoc prof, 75-81, PROF PHYSIOL, UNIV CALIF, LOS ANGELES, 81-, ASSOC MEM, BRAIN RES INST, 70- *Concurrent Pos:* Assoc mem USPHS proj, Univ Calif, Los Angeles, 72-82. *Mem:* Am Physiol Soc; Biophys Soc. *Res:* Chemical and thermal energetics of excitation-contraction coupling and chemomechanical transduction in skeletal and cardiac muscle. *Mailing Add:* Dept Physiol Univ Calif Los Angeles Lewis Neuromuscular Res Ctr Los Angeles CA 90024

HOMSHER, PAUL JOHN, b Philadelphia, Pa, May 17, 31; m 59; c 3. REPRODUCTIVE BIOLOGY, ACAROLOGY. *Educ:* Pa State Univ, BS, 53, MS, 59, PhD(genetics), 67. *Prof Exp:* Instr bot, Forestry Sch, Pa State Univ, 59; from asst prof to assoc prof biol, 62-77, dir, Biomed Sci Prog, 79-82, PROF BIOL, OLD DOMINION UNIV, 77-, ASSOC DEAN, COL SCI, 85- *Concurrent Pos:* Asst prof dept biol & Old Dominion Univ Educ Found study grant, Ga Southern Col, 69-70; geneticist, Am Soc Eng Educ/NASA summer fel, 68, US Food & Drug Admin proj on miniature swine breeding, 79-86. *Mem:* Entom Soc Am; Am Genetic Asn; Acarology Soc Am; Sigma Xi. *Res:* Systematics and reproductive biology of ticks; ultrastructure of pheromone glands; analysis of pheromone biosynthesis; nucleic acid biochemistry of tick cells; viral host cell interactions in tick borne viruses. *Mailing Add:* Dean Off Col Sci Old Dominion Univ Norfolk VA 23529-0163

HOMSY, CHARLES ALBERT, b Boston, Mass, June 21, 32; m 56; c 3. BIOMEDICAL ENGINEERING. *Educ:* Mass Inst Technol, SB, 53, ScD(chem eng), 59. *Prof Exp:* Asst chem eng, Mass Inst Technol, 53-54, asst dir sch chem eng pract, 54-55, asst chem eng, 55-59; res engr, E I du Pont de Nemours & Co, 59-61, tech rep, 61-64, mkt rep, 64-66; coordr prosthetic mat & develop, 66-67, DIR PROSTHESIS RES LAB, METHODIST HOSP, HOUSTON, TEX, 66- *Concurrent Pos:* Fulbright scholar eng, Univ Sheffield, 57-58; res asst prof, Baylor Col Med, 69-77 & res assoc prof, 77-; pres, Vitek Inc. *Mem:* Orthop Res Soc; Soc Biomat; Am Inst Chem Eng; Am Chem Soc; Nat Asn Corrosion Eng; Am Soc Testing & Biomat; assoc & fel Am Acad Orthop Surgeons. *Res:* Industrial management; combustion similitude; process design for high polymer synthesis; prosthetic implant materials. *Mailing Add:* 11526 Raintree Circle Houston TX 77024

HON, DAVID NYOK-SAI, b Kluang Johor, Malaysia, June 19, 47; US citizen; m 72; c 2. WOOD CHEMISTRY, PULP & PAPER CHEMISTRY. *Educ:* Tokyo Univ Agr & Technol, BS, 72; Gunma Univ, MS, 74; Va Polytech Inst, PhD(wood chem) 77. *Prof Exp:* From asst prof to assoc prof res & teaching, Va Polytech Inst & State Univ, 77-84; assoc prof, 84-85, PROF RES & TEACHING, CLEMSON UNIV, 85-, DIR RES, 84- *Concurrent Pos:* Consult, Buckyle Cellulose, 77-78, Squibb, 83-84, Johnson Wax, 85-87, Carib-Med, 86-, Resource Chem & Off Technol Assessment (Congress), 88-89; ed, Am Chem Soc, 82 & Mercer Dekker, 90. *Mem:* Am Chem Soc; Tech Asn Pulp & Paper Indust; Forest Prods Res Soc; Soc Plastics Engrs. *Res:* Wood chemical modification and utilization; chemical modification of wood and polymer; weathering of wood; photochemistry of polymers; preservation of papers and woods with historic values. *Mailing Add:* Clemson Univ 128 Lehotsky Hall Clemson SC 29634-1003

HON, EDWARD HARRY GEE, b Canton, China, Jan 12, 17; m 48; c 3. OBSTETRICS & GYNECOLOGY. *Educ:* Union Col, Nebr, BS, 50; Col Med Evangelists, MD, 50. *Prof Exp:* From instr to asst prof obstet & gynec, Sch Med, Yale Univ, 55-60; prof, Col Med Evangelists, 60-64; from assoc prof to prof, Sch Med, Yale Univ, 64-69; prof obstet & gynec, Sch Med, Univ Southern Calif, 69-83; RETIRED. *Concurrent Pos:* Markle scholar, 55-60. *Mem:* Sigma Xi. *Res:* Evaluation of fetal responses to the stresses of labor and delivery; use of biophysical techniques for the study of uterine contractility. *Mailing Add:* 11 Bradbury Hills Rd Duarte CA 91010

HONAKER, CARL BOGGESS, b Oakvale, WVa, Sept 10, 26; m 90; c 1. INORGANIC CHEMISTRY. *Educ:* Concord Col, BS, 48; Univ Tenn, MS, 50, PhD(inorg chem), 57. *Hon Degrees:* DSc, Tenn Wesleyan Col, 78. *Prof Exp:* Asst chem, Univ Tenn, 48-50; pub sch teacher, SC, 50-51; prof chem & physics & chmn, dept phys sci, Tenn Wesleyan Col, 51-77; staff chemist, Res Dept, Rockwell Hanford Opers, Richland, 77-83; prin chemist, Bioorg Chem Dept, Midwest Res Inst, Kansas City, Mo, 83-85; RES ANALYST, BOEING TENN, INC, OAK RIDGE, TENN, 85- *Concurrent Pos:* Univ fel analytical chem, Univ Ariz, 60-61. *Mem:* Am Chem Soc; Sigma Xi; Royal Soc Chem. *Res:* Complex compounds; solvent extraction; emission spectroscopy; chromatography; low energy beta emitters. *Mailing Add:* 1329 Francis Station Knoxville TN 37909

HONDA, BARRY MARVIN, b Hamilton, Ont, July 5, 50. MOLECULAR BIOLOGY. *Educ:* McMaster Univ, BSc, 71; Univ BC, PhD(biochem), 75. *Prof Exp:* Fel molecular biol, MRC Lab, Cambridge, UK, 75-79; res assoc, Med Sch, Wash Univ, St Louis, 79-81; asst prof, 81-87, ASSOC PROF BIOL, SIMON FRASER UNIV, VANCOUVER, BC, 87- *Mem:* Can Biochem Soc. *Res:* Molecular studies of the structure, organization, packaging and expression of genes in higher organisms, aimed at understanding gene regulation during differentiation and development. *Mailing Add:* Dept Biol Sci Simon Fraser Univ Burnaby BC V5A 1S6 Can

HONDA, KAZUO, GASTROENTEROLOGY, CARDIOVASCULAR PHARMACOLOGY. *Educ:* Univ Tokyo, PhD(pharmacol), 85. *Prof Exp:* Pharmacologist, Mitsubishi Chem Indust, 74-79; SR PHARMACOLOGIST, YAMANOUCHI PHARMACEUT CO LTD, 79- *Concurrent Pos:* Postdoctoral, Stanford Univ, 88-89. *Res:* Cardiovascular, renal, gastrointestinal and receptor pharmacology. *Mailing Add:* Dept Pharmacol Cent Res Labs Yamanouchi Pharmaceut 1-1-8 Azusawa Itabashi-Ku Tokyo 174 Japan

HONDA, SHIGERU IRWIN, b Seattle, Wash, Nov 16, 27. CELL PHYSIOLOGY. *Educ:* Calif Inst Technol, BS, 50; Univ Wis, MS, 52, PhD(bot), 54. *Prof Exp:* Asst bot, Univ Wis, 50-53; res off, Commonwealth Sci & Indust Res Orgn, Div Food Preserv & Transport, Sydney, 54-56; chemist, US Plant, Soil & Nutrit Lab, USDA, 56-61; asst prof bot, Cornell Univ, 60-61; assoc res plant biochemist, Dept Bot & Plant Biochem, Univ Calif, Los Angeles, 62-67; assoc prof, 67-71, PROF BIOL, WRIGHT STATE UNIV, 71- *Concurrent Pos:* Fulbright grantee, Australia, 53-54; spec res fel, Div Gen Med Sci, USPHS, 61-62, res career develop award, 62-66; vis prof, Kasetsart Univ, Bangkok, 68 & Univ Chiengmai, 70; vis res plant physiologist, Univ Calif, Los Angeles, 71; NASA-Am Soc Eng Educ fel, Johnson Space Ctr, Houston, 73 & 74; vis plant physiologist, Commonwealth Sci & Indust Res Orgn, Div Hort Res, Adelaite, 85-86 & 88. *Honors & Awards:* Dipl Merit, NZ Sci Cong, 66. *Res:* Chloroplast replication and senescence. *Mailing Add:* Dept Biol Sci Wright State Univ Colonel Glenn Hwy Dayton OH 45435

HONDEGHEM, LUC M, b Jabbeke, Belg, Sept 22, 44; m; c 3. PHARMACOLOGY. *Educ:* Univ Louvain, Belg, MD, 70, MS, 71; Univ Calif, San Francisco, PhD(pharmacol), 73. *Prof Exp:* From asst prof to assoc prof pharmacol, Univ Calif, San Fransico, 73-85; PROF MED & PHARMACOL & STAHLMAN CHMN, CARDIOVASC RES PROG, VANDERBILT UNIV, 85- *Concurrent Pos:* Ed, Biophys J, 78. *Mem:* Am Heart Asn; Sigma Xi; Med Electronics & Data Soc; Am Soc Pharmacol & Exp Therapeut. *Res:* Ultrastructural and electrophysiological aspects of impulse transmission in cardiac tissue; mechanisms of cardiac arrhythmias; effects of antiarrhythmic drugs on normal and abnormal impulse transmission in the heart. *Mailing Add:* Dept Pharmacol - 260 Univ Calif San Francisco CA 94143

HONE, DANIEL W, b San Francisco, Calif, Jan 7, 37; m 58; c 3. THEORETICAL SOLID STATE PHYSICS. *Educ:* Univ Calif, Berkeley, AB, 58; Univ Ill, MS, 60, PhD(physics), 62. *Prof Exp:* Nat Acad Sci-Nat Res Coun fel, 62-63; asst prof physics, Univ Pa, 63-68; assoc prof, 68-74, PROF PHYSICS, UNIV CALIF, SANTA BARBARA, 74- *Concurrent Pos:* Exchange prof, Univ Paris, Orsay, 73; Nat Res Coun sr vis lectr, Oxford Univ, 74; res fel, Japan Soc Prom Sci, Tokyo, 82; chmn Physics Dept, Inst Theoret Physics, 82-86, dep dir, 86-87. *Mem:* Fel Am Phys Soc. *Res:* Theory of many body problems; magnetism; magnetic resonance theory; polymers. *Mailing Add:* Dept Physics Univ Calif Santa Barbara CA 93106

HONEA, FRANKLIN IVAN, b Hope, Ark, Aug 9, 31; m 59; c 10. CHEMICAL & MECHANICAL ENGINEERING. *Educ:* Univ Calif, Berkeley, BS, 55; Univ Southern Calif, MS, 62; Univ Denver, PhD(chem eng), 69. *Prof Exp:* Thermodynamicist, Aviation Div, NAm Rockwell Corp, Calif, 55-58; heat transfer engr, Northrop Corp, 58-61; res specialist, Space Div, NAm Rockwell Corp, 61-66; sr staff engr, Martin Marietta Corp, 66-68; adj prof chem eng, Univ Denver, 69; dept engr develop, Pantex Atomic Energy Comn Plant, Mason & Hanger-Silas Mason Co, Inc, 70-74; sr environ engr, Midwest Res Inst, Kansas City, 74-76; chem engr & proj mgr, Grand Forks Proj Off, US Dept Energy, NDak, 76-87; Consult Energy & Environ, 88-90; PROJ MGR, CTR RES SULFUR IN COAL, 90- *Mem:* Am Soc Testing & Mat; Am Inst Chem Engrs; Am Soc Mech Engrs; Sigma Xi. *Res:* Nationwide fossil energy project performance evaluations; technical and economic evaluation studies of energy-related processes; coal and peat combustion; environmental control systems. *Mailing Add:* 109 Edgewood Park Marion IL 62959-4847

HONECK, HENRY CHARLES, b Batavia, NY, Oct 4, 30; m 52; c 2. REACTOR PHYSICS, APPLIED MATHEMATICS. *Educ:* Rensselaer Polytech Inst, BS, 52; Mass Inst Technol, ScD(nuclear eng), 59. *Prof Exp:* Test engr reactor physics, Pratt & Whitney Aircraft, 52-56; scientist, Brookhaven Nat Lab, 59-65 & AEC, 65-67; res mgr, Comput Appln Div, Savannah River Lab, E I du Pont de Nemours & Co, 67-73, res fel, comput sci div, 73-80; PRES, COMPUT APPL TECHNOL, INC, 80- *Concurrent Pos:* Consult, Gen Atomic Div, Gen Dynamics Corp, 61-65. *Honors & Awards:* E O Lawrence Award, AEC, 74. *Mem:* Fel Am Nuclear Soc. *Res:* Neutron transport theory; neutron thermalization; reactor theory computer codes; data management; modular systems; computer science. *Mailing Add:* 621 Colleton Ave SE Aiken SC 29801-4607

HONES, MICHAEL J, b Rahway, NJ, Apr 18, 42; m 64. PARTICLE PHYSICS. *Educ:* Holy Cross Col, BS, 64; Univ Notre Dame, PhD(physics), 69. *Prof Exp:* ASST PROF PHYSICS, VILLANOVA UNIV, 69- *Mem:* Am Phys Soc. *Res:* Experimental high energy physics; meson and baryon spectroscopy. *Mailing Add:* Dept Physics Villanova Univ Villanova PA 19085

HONEY, RICHARD CHURCHILL, b Portland, Ore, Mar 9, 24; div; c 4. ELECTROOPTICS. *Educ:* Calif Inst Technol, BS, 45; Stanford Univ, PhD(elec eng), 53. *Prof Exp:* Asst, Electronics Res Labs, Stanford Univ, 47-52; sr res engr, Microwave Group, Electromagnetics Lab, Stanford Res Inst, 52-60, tech prog coordr, Electromagnetic Tech Lab, 60-64, mgr, 64-70, staff scientist, 70-84, sr staff scientist, Electromagnetic Sci Lab, 85-89, SR PRIN SCIENTIST, APPL ELECTROMAGNETICS & OPTICS LAB, SRI INT, 89- *Concurrent Pos:* Chmn working group lasers, Adv Group Electron Devices, Dir Defense Res & Eng, 69-75; mem, Army Sci Bd, 78-84. *Mem:* Fel Inst Elec & Electronics Engrs; Am Phys Soc; fel Optical Soc Am. *Res:* Lasers; microwave theory, antennae and measurements. *Mailing Add:* SRI Int Menlo Park CA 94025

HONEYCUTT, THOMAS LYNN, b Louisville, Ky, Aug 28, 42; m 63; c 1. OPERATIONS RESEARCH, COMPUTER SCIENCE. *Educ:* NC State Univ, BS, 65, MS, 67, PhD(eng), 69. *Prof Exp:* Sr analyst opers res, US Plywood-Champion Papers, 69-70; asst prof, 70-80, ADJ ASSOC PROF COMPUT SCI, NC STATE UNIV, 80-, ASSOC HEAD DEPT, 76- *Concurrent Pos:* Assoc consult, Mgt Decisions Develop Corp, 69- *Mem:* Opers Res Soc Am; Am Soc Agr Engrs. *Res:* Industrial plant location analysis. *Mailing Add:* Dept Comput Sci NC State Univ PO Box 5972 Raleigh NC 27650

HONEYMAN, MERTON SEYMOUR, b Hartford, Conn, Sept 27, 25; m 52; c 3. GENETICS. *Educ:* Univ Conn, BA, 50, MS, 51; Ohio State Univ, PhD(zool), 54. *Prof Exp:* Fel genetics, Yale Univ, 54-57; geneticist, Nat Cancer Inst, 57-59; geneticist, Chronic Dis Control Sect, Conn State Dept Health, 59-66, dir div res, Off Ment Retardation, 66-74, dir div health statist, 74-84, res scientist, 84-89; RETIRED. *Concurrent Pos:* Instr eve col, Teachers Col, Conn, 54-55; adj asst prof, Hillyer Col, 55-56 & Univ Hartford, 60-72; lectr dept epidemiol & pub health, Sch Med, Yale Univ, 60-69; mem, Expert Adv Comt Method of Twin Studies, WHO, Geneva, 65; consult, Swed Twin Registry, Karolinska Inst, 67 & Rheumatic Fever Clin, Hadassah Med Sch, Hebrew Univ, Israel, 67; lectr, Manchester Community Col, 71-73 & Tunxis Community Col, 73-75; consult, Dept Med, Hartford Hosp. *Mem:* Am Soc Human Genetics; Am Cancer Soc; Am Pub Health Asn; Am Heart Asn. *Res:* Human genetics; twin research; genetics of chronic disease; demography; genetic counseling. *Mailing Add:* 18 Sycamore Lane Avon CT 06001

HONEYMAN, PETER, computer theory, for more information see previous edition

HONEYWELL, WALLACE I(RVING), b North Platte, Nebr, Feb 6, 36; m 59; c 4. CHEMICAL ENGINEERING & PHYSICS. *Educ:* Stanford Univ, BS, 59; Calif Inst Technol, PhD(chem eng), 64. *Prof Exp:* NSF fel physics, State Univ Leiden, 64-65; asst prof chem eng, 65-67, assoc dean, Cullen Col Eng, 76-80, ASSOC PROF CHEM ENG, UNIV HOUSTON, 67-, DIR, ACAD ADV CTR, 80- *Mem:* Am Chem Soc; Am Inst Chem Engrs; Am Soc Eng Educ. *Res:* Magnetic field effects on transport properties of gases; non-equilibrium thermodynamics; high gradient magnetic separation processes; magnetothermal convection in gases. *Mailing Add:* Cullen Col Eng Univ Houston Houston TX 77004

HONG, BOR-SHYUE, OPHTHALMOLOGY. *Educ:* Colo State Univ, PhD(neurotoxins), 70. *Prof Exp:* ADJ ASSOC PROF OPHTHAL, TEX TECH UNIV HEALTH SCI CTR, 82- *Res:* structure and function of collagen cornea; calcium dependent neutral protease in eye lenses; catherization of vitamin A binding in tears. *Mailing Add:* Dept Ophthal & Visual Sci Tex Tech Univ Health Sci Ctr Lubbock TX 79430

HONG, CHUNG IL, b Koo-Sung, Korea, Dec 22, 38; m 64; c 3. MEDICINAL CHEMISTRY. *Educ:* Seoul Nat Univ, BS, 61; Univ NC, Chapel Hill, MS, 67, PhD(med chem), 68. *Prof Exp:* Instr biochem, Sch Pharm, Univ NC, 69; NSF grants & res scientist med chem, Gen Clin Res Ctr, 69-73, cancer res scientist, 73-74, sr res scientist, 74-80, CANCER RES SCIENTIST IV, DEPT NEUROSURG, ROSWELL PARK MEM INST, 80- *Mem:* Am Chem Soc; Am Pharmaceut Asn; Am Asn Cancer Res; NY Acad Sci. *Res:* Synthesis and biochemical studies of nucleic acid components and 3', 5'- cyclic nucleotides; kinetics of entry, distribution, and metabolism of antitumor agents in experimental animals; development of antitumor and antiviral agents by conjugation with natural carrier molecules. *Mailing Add:* Roswell Park Mem Inst 666 Elm St Buffalo NY 14263-0002

HONG, DONALD DAVID, b Kwangtung, China; US citizen. METHODS DEVELOPMENT. *Educ:* Univ Mich, BS, 62, MS, 65, PhD(pharm chem), 69. *Prof Exp:* Sr res chemist, Am Pharmaceut Asn Found, 69-70; assoc res chemist, Sterling-Winthrop Res Inst, 70-75; supvr, G D Searle Co, 76-78; scientist, Parke Davis & Co, 78-81; MGR, GLAXO, INC, 81- *Mem:* Am Pharmaceut Asn; Am Chem Soc. *Mailing Add:* Glaxo Inc PO Box 1217 Zebulon NC 27597-1217

HONG, JAU-SHYONG, b Nov 8, 43; m; c 2. NEUROPHARMACOLOGY. *Educ:* Univ Kans, PhD(pharmacol), 73. *Prof Exp:* HEAD, NEUROPHARMACOL SECT, LAB BEHAV & NEUROL TOXICOL, NAT INST ENVIRON HEALTH SCI, 80-; ASSOC PROF NEUROPHARMACOL, DUKE UNIV, 80- *Mem:* Am Soc Pharmacol & Exp Therapeut; Neurosci Soc. *Res:* Neuropeptides; neurotransmitters. *Mailing Add:* Lab Molecular & Integrative Neurosci Nat Inst Environ Health Sci NIH PO Box 12233 Research Triangle Park NC 27709

HONG, JEN SHIANG, b Mar 3, 39; m; c 2. GENETICS. *Educ:* Univ Calif, Berkeley, PhD(biochem), 69. *Prof Exp:* SR SCIENTIST, BOSTON BIOMED RES INST, 81- *Mem:* AAAS; Am Soc Biochem & Molecular Biol; Am Soc Microbiol. *Res:* Molecular genetics of membrane transport; gene regulation. *Mailing Add:* Dept Cell Physiol Boston Biomed Res Inst 20 Staniford St Boston MA 02114

HONG, KEELUNG, b Taiwan, Apr 8, 43; US citizen. MEMBRANE FUSION, ENDOCYTOSIS. *Educ:* Taiwan Cheng Kung Univ, BS, 65; Univ Tex, MS, 70; Univ Calif, Berkeley, PhD(chem), 75. *Prof Exp:* Postdoctoral res, Univ Calif, San Francisco, 75-76 & Univ Calif, Berkeley, 76-77; postdoctoral res, Stanford Univ, 77-78, res assoc res, 78-79; asst res biochemist, 79-91, ASSOC RES BIOCHEMIST, UNIV CALIF, SAN FRANCISCO, 91- *Mem:* AAAS; Am Chem Soc; Biophys Soc; Am Soc Cell Biol; NY Acad Sci; fel Am Inst Chemists. *Res:* Biomembrane structure and function; mechanism of membrane fusion; protein-lipid interactions; liposome as drug carrier; liposome-cell interactions; critique of social science research. *Mailing Add:* Univ Calif Box 0128 San Francisco CA 94143

HONG, KI C(HOONG), b Seoul, Korea, May 1, 36; m 63; c 4. CHEMICAL & PETROLEUM ENGINEERING. *Educ:* Iowa State Univ, BSc, 59, MSc, 61, PhD(chem eng), 62. *Prof Exp:* SR ENG ASSOC, CHEVRON OIL FIELD RES CO, LA HABRA, 62- *Concurrent Pos:* Lectr environ eng, Univ Calif, Irvine, 74-76; lectr mech eng, Calif State Univ, Fullerton, 83- *Mem:* Am Chem Soc; Am Inst Chem Engrs; Am Inst Mining, Metall & Petrol Engrs. *Res:* High analysis fertilizers; mathematical simulation; optimization; flow through porous media; thermal recovery of petroleum; water and air resources analysis; well completion methods; multiphase flow in pipes; reservoir engineering and analysis. *Mailing Add:* 7327 E Saddlehorn Way Orange CA 92669-4515

HONG, KUOCHIH, b Ping-Tong, Taiwan. PHYSICAL CHEMISTRY. *Educ:* Nat Taiwan Normal Univ, BS, 65; Nat Tsing Hua Univ, MS, 69; Univ Chicago, PhD(chem), 75. *Prof Exp:* Teacher sci, Taipei Girls' Mid Sch, 65-66; instr chem, Tamkung Col, 69-70; res assoc, Univ Chicago, 75-77; asst chemist, Brookhaven Nat Lab, 77-80; MEM STAFF, ENERGY CONVERSION DEVICES, 80- *Mem:* Am Chem Soc. *Res:* Molten salt; liquid metal and surface chemistry. *Mailing Add:* 1790 Rolling Woods Dr Troy MI 48098

HONG, PILL WHOON, b Ichon, Korea, Oct 10, 21; m 56; c 3. SURGERY. *Educ:* Yonsei Univ, Korea, MB, 42, DMSc(physiol), 60. *Prof Exp:* From asst prof to prof surg, Col Med, Yonsei Univ, Korea, 56-67; assoc prof, 67-70, PROF SURG, SCH MED, UNIV HAWAII, 70- *Concurrent Pos:* China Med Bd NY, Inc, travel fel, 60-66; Brit Nat Coun fel, 66. *Honors & Awards:* Acad Awards, Korean Med Asn, 62. *Mem:* Fel Am Col Surg; fel Am Col Chest Physicians; Korean Surg Soc (pres, 66). *Res:* Thoracic surgery. *Mailing Add:* 52-1001 Walker Hill AP Kwangjang Dong Sungdong KV Seoul Republic of Korea

HONG, RICHARD, b Danville, Ill, Jan 10, 29; m 52; c 4. PEDIATRICS, IMMUNOLOGY. *Educ:* Univ Ill, Urbana-Champaign, BS, 49; Univ Ill, Chicago, MD, 53. *Prof Exp:* Sr res assoc immunol, Univ Cincinnati, 60-65; from asst prof to prof pediat & immunol, Univ Minn, Minneapolis, 65-69; assoc dean clin affairs, 71-74, PROF PEDIAT & MICROBIOL, MED CTR, UNIV WIS-MADISON, 69- *Concurrent Pos:* USPHS fel, 60-62, res career develop award, 63-69; mem, Transplant Registry, 68-69. *Mem:* Am Soc Clin Invest; Am Asn Immunol; Soc Pediat Res; Am Pediat Soc. *Res:* Immunoglobulin structure and function; immunochemistry of hypogammaglobulinemia; treatment of immunological deficiency states; physiology of immunity. *Mailing Add:* Clin Sci Ctr Univ Wis 600 Highland Ave Madison WI 53792

HONG, SE JUNE, b Seoul, Korea, May 15, 44; m 68; c 1. COMPUTER & ELECTRICAL ENGINEERING. *Educ:* Seoul Nat Univ, BSc, 65; Univ Ill, Urbana, MS, 67, PhD(elec eng), 69. *Prof Exp:* Staff engr, 69-72, adv engr, 73-76, sr engr, 77, mem res staff, 78-82, SR MGR, IBM CORP, 82- *Concurrent Pos:* Distinguished vis, Inst Elec & Electronics Engrs, 71-74; vis assoc prof, Univ Ill, 74-75; vis prof, Korean Advan Inst Sci & Technol, 80. *Mem:* Fel Inst Elec & Electronics Engrs; Asn Comput Mach; Math Asn Am; Sigma Xi; Korean Scientists & Engrs in Am; Am Asn Artificial Intel; Asn Computational Ling. *Res:* Coding theory; fault tolerant computing; array logic; switching theory and minimization; test pattern generation; design automation; expert systems; artificial intelligence. *Mailing Add:* 1374 White Hill Rd Yorktown Heights NY 10598

HONG, SUK KI, b Kyonggi Do, Korea, Oct 16, 28; m 59; c 2. PHYSIOLOGY. *Educ:* Yonsei Univ, MD, 49; Univ Rochester, PhD(physiol), 56. *Hon Degrees:* DSc, Kyungpook Nat Grad Sch, Daegu, Korea, 83. *Prof Exp:* From instr to asst prof physiol, Univ Buffalo, 56-59; asst prof physiol, Yonsei Univ Korea, 59-62, assoc prof, 63-65, prof & chmn dept, 66-68; prof physiol, Univ Hawaii, 68-75, chmn dept, 71-75; asst prof, 62, assoc prof 65, PROF PHYSIOL, STATE UNIV NY BUFFALO, 75- *Concurrent Pos:* Consult, Off Surg Gen, Repub Korea Air Force, 60-68 & Korea Amateur Sports Asn, 62-68; deleg, Int Cong Physiol Sci, 68; consult, Tripler Army Med Ctr, Honolulu, 72-75; mem US-Japan coop prog natural resources, 72- *Honors & Awards:* Samil Cult Found Award, 63; Stover Link Award, Undersea Med Soc, 84; Stockton Kimball Award, SUNY at Buffalo, 87. *Mem:* Am Physiol Soc; Am Soc Nephrology; Undersea Med Soc; Int Soc Nephrology; Soc Exp Biol & Med. *Res:* Renal and membrane physiology; diving physiology; adaptation of man to cold environment. *Mailing Add:* Dept Physiol State Univ NY Buffalo NY 14214

HONG, WEN-HAI, b Tainan, Taiwan, Apr 11, 34; m 61; c 3. PHYSICAL PHARMACY. *Educ:* Nat Taiwan Univ, BS, 57; Univ Wis, MS, 66, PhD(pharm), 69. *Prof Exp:* Teaching asst pharm, Nat Taiwan Univ, 59-64, instr, 64-65; SR RES ASSOC, WARNER LAMBERT & PARKE DAVIS & CO, 69- *Mem:* Am Chem Soc. *Res:* Development of selective methods for analysis of investigational new dosage forms intended for use in humans. *Mailing Add:* Ten Sylvan Dr Pine Brook NJ 07058

HONIG, ARNOLD, b New York, NY, Feb 28, 28; m 47, 79; c 3. SEMICONDUCTORS, LOW TEMPERATURE PHYSICS. *Educ:* Cornell Univ, BA, 48; Columbia Univ, MA, 50, PhD(physics), 53. *Prof Exp:* Asst microwave spectros, Columbia Univ, 51-53; res physicist, Solid State Physics, Univ Calif, 53-54; res fel molecular physics, Ecole Normale Superieure, Paris, 54-56; from asst prof to assoc prof physics, 56-62, PROF PHYSICS, SYRACUSE UNIV, 62- *Concurrent Pos:* Vis prof, Hebrew Univ, 62, Comn Atomic Energy, Saclay, France, 65. *Honors & Awards:* Glover Mem Award, Dickinson Col, 66. *Mem:* Am Phys Soc; Fedn Am Scientists. *Res:* Spin-polarized nuclear fusion fuels; nuclear and electronic magnetic resonance in semiconductors and quantum solids at very low temperatures; luminescence and transport in semiconductors under conditions of high electron and hole spin-polarizations; bio-luminescence. *Mailing Add:* Dept Physics Syracuse Univ Syracuse NY 13244

HONIG, CARL ROBERT, b New York, NY, July 15, 25; m 47; c 2. TOXICOLOGY. *Educ:* Univ Rochester, BS, 45; Long Island Col Med, MD, 49; Am Bd Internal Med, dipl, 58. *Prof Exp:* Instr, Col Med, Univ Cincinnati, 52-53; asst, 54-57, from instr to assoc prof, 57-67, actg chmn dept, 68-69, PROF PHYSIOL, SCH MED & DENT, UNIV ROCHESTER, 67- *Honors & Awards:* Merit Award USPHS. *Mem:* Am Physiol Soc; Microcirculatory Soc; Int Soc Oxygen Transp to Tissue. *Res:* Circulation-metabolism coupling; oxygen transport; myoglobin microspectroscopy in vivo. *Mailing Add:* Dept Physiol Univ Rochester Sch Med & Dent Rochester NY 14642

HONIG, DAVID HERMAN, b Turtleford, Sask, Sept 9, 28; US citizen; m 67; c 3. FOOD CHEMISTRY. *Educ:* Bethany Col, BSc, 62; Kans State Univ, MSc, 65. *Prof Exp:* RES CHEMIST, NORTHERN REGIONAL RES CTR, SCI & EDUC ADMIN-FED RES, USDA, 65- *Mem:* Am Chem Soc; Am Asn Cereal Chemists. *Res:* Sources, causes and modification of flavor in soybeans and soybean products; chemical nature of soybeans and physiological effects of components; interactions of the components. *Mailing Add:* 1815 N University St Peoria IL 61604-3902

HONIG, GEORGE RAYMOND, PEDIATRIC HEMATOLOGY, ONCOLOGY. *Educ:* Univ Ill, MD, 61; George Washington Univ, PhD(biochem), 66. *Prof Exp:* PROF PEDIAT & HEAD DEPT, COL MED, UNIV ILL, 84- *Res:* Thalassemia; sickle cell anemia. *Mailing Add:* Dept Pediat Univ Ill Col Med 840 S Wood St Chicago IL 60612

HONIG, JOHN GERHART, b Vienna, Austria, Oct 30, 23; nat US; m 80; c 3. OPERATIONS RESEARCH, MANAGEMENT. *Educ:* Drew Univ, AB, 47; Univ Mich, MS, 48; Georgetown Univ, PhD(phys chem), 56. *Prof Exp:* Phys chemist, Nat Inst Cleaning & Dyeing, 48-50; phys chemist, Naval Res Lab, Washington, DC, 50-56; opers analyst, Opers Eval Group, Mass Inst Technol, 56-59; assoc proj leader, Weapons Syst Eval Group, Inst Defense Anal, 59-62; chief naval warfare tech, Honeywell Inc, 62-65; opers res officer, US Arms Control & Disarmament Agency, 65-66; sci adv, Dir Studies, Off Chief of Staff Army, 66-68, chief weapons requirements & anal, 68-74, asst dir systs rev, Off Res, Develop & Acquisitions, 74-78, chief of aircraft, missiles & electronics, Comptroller of the Army, 78-84; sr consult, Anal Serv Inc, 84-86, prin physicist, 86-88; CONSULT, 88- *Concurrent Pos:* Mem, Gov Sci Adv Coun, Md, 70- *Honors & Awards:* David Rist Prize, Mil Opers Res Soc, 71. *Mem:* Fel AAAS; Am Chem Soc; Mil Opers Res Soc (pres, 69-70); Inst Mgt Sci; Nat Coun Asns Policy Sci (pres, 77). *Res:* Military operations research; military and industrial long-range planning; research and development management; systems analysis; large molecular systems in nonaqueous media; non-Newtonian flow in nonaqueous media. *Mailing Add:* 7701 Glenmore Spring Way Bethesda MD 20817

HONIG, MILTON LESLIE, b New York, NY, Apr 10, 44;. ORGANIC CHEMISTRY. *Educ:* City Col New York, BS, 65; Polytech Inst Brooklyn, PhD(chem), 70. *Prof Exp:* Res chemist, Stauffer Chem Co, 70-77 & Fordham Univ Law Sch, 80; VPRES RES & DEVELOP, STANDARD LUBRICANTS INC, 77- *Concurrent Pos:* Adj lectr, Hunter Col, 72-73 & Barnard Col, 81-82. *Mem:* Am Chem Soc; The Chem Soc; Am Soc Lubrication Engrs. *Res:* Synthetic lubricant development; organophosphorus chemistry; flame retardant chemistry. *Mailing Add:* 3240 Henry Hudson Pkwy Bronx NY 10463-3212

HONIG, RICHARD EDWARD, b Gottingen, Ger, June 4, 17; nat US; m 43; c 2. MASS SPECTROMETRY. *Educ:* Robert Col, Istanbul, BS, 38; Mass Inst Technol, MS, 39, PhD(molecular physics), 44. *Prof Exp:* Asst physics, Mass Inst Technol, 39-40 & 41-44, res assoc, 44-46; sr physicist, Res Labs, Socony-Vacuum Oil Co, Inc, 46-50; tech staff mem, RCA Labs, 50-68, head mat characterization, 68-82, staff scientist, 82-87; RETIRED. *Concurrent Pos:* Instr, Bluffton Col, 40-41; res fel, Univ Brussels, 55-56; adj res prof, Rensselaer Polytech Inst. *Mem:* Fel Am Phys Soc; Am Soc Mass Spectrometry (pres, 70-72); Am Vacuum Soc. *Res:* Ion physics; mass spectrometry; solid state. *Mailing Add:* 3300 Darby Rd Pine No 7305 Haverford PA 19041-1095

HONIGBERG, BRONISLAW MARK, b Warsaw, Poland, May 14, 20; nat US; m 48; c 2. IMMUNOLOGY, MICROBIOLOGY. *Educ:* Univ Calif, AB, 43, MA, 46, PhD(zool), 50. *Prof Exp:* From instr to assoc prof, 50-61, PROF ZOOL, UNIV MASS, AMHERST, 61- *Concurrent Pos:* USPHS res grants, 55-, spec res fel, Lab Parasitic Dis, Nat Inst Allergy & Infectious Dis, 65-66; res assoc, Johns Hopkins Univ, 58-59; assoc ed, Trans Am Micros Soc, 66-71, ed, J Protozool, 71-80; trustee, Am Type Cult Collection, 67-72; mem, Int Comn Protozool, Int Union Biol Sci, 65-85, Trop Med & Parisitol Study Sect, Nat Inst Allergy & Infectious Dis, NIH, 73-77 & US Nat Comt, Int Union Biol Sci, 82-88; ed bd, Int J Parasitol, 71-74; USPHS, spec res fel, Ctr Trop Vet Med, Univ Edinburgh, UK, 73-74; managing ed NAm, Parasitol Res, Zeitschrift für Parasitenkd, 74-91; vpres & chmn sci prog comt, Int Cong Protozool, 77; bd reviewers, Acta Tropica, 77-80; vpres, Int Symposium Trichomoniasis, Poland, 81; fac fel award, Univ Mass, Amherst, 81-82; mem ed bd, J Protozool, 59-71 & 81-; Alexander von Humboldt US sr scientist award, 82; guest scientist, Int Ctr Insect Physiol Ecol, Nairobi, Kenya, 85; vis scientist, Int Lab Res Animal Dis, Nairobi, Kenya, 85-86. *Honors & Awards:* Chancellor's Lectr & Medalist, Univ Mass, 75; Gold Medal, Med Fac Comenius Univ, Czechoslovakia, 77; hon mem, Soc Protozool, 79; Constantin Janicki Medal, Polish Parasitol Soc, 84. *Mem:* Fel NY Acad Sci; fel AAAS; Soc Protozool (pres, 65-66); fel Royal Soc Trop Med & Hyg; Am Micros Soc (2nd vpres, 62-63, 1st vpres, 63-64, pres, 64-65); Am Soc Parasitol; hon mem Int Comn Protozool; corresp mem Belg Soc Trop Med; corresp mem Ger Soc Parasitol. *Res:* Protozoology; parasitology; cytology; pathogenicity and physiology of parasitic flagellates. *Mailing Add:* Dept Zool Morrill Sci Ctr Univ Mass Amherst MA 01003-0027

HONIGBERG, IRWIN LEON, b Brooklyn, NY, Jan 31, 30; m 58; c 4. MEDICINAL CHEMISTRY. *Educ:* Univ Conn, BS, 51; Univ NC, PhD(pharmaceut chem), 57. *Prof Exp:* Assoc prof pharmaceut chem, New Eng Col Pharm, 56-57; assoc chemist, Midwest Res Inst, 57-64; from asst prof to assoc prof, 64-79, PROF MED, SCH PHARM, UNIV GA, 79- *Mem:* AAAS; Am Chem Soc; Am Asn Pharmaceut Sci. *Res:* Pharmacy; liquid chromatography in pharmaceutical analysis; analysis of drugs in biological fluids; organic medicinals; mechanisms of antiallergy action. *Mailing Add:* Dept Med Chem Col Pharm Univ Ga Athens GA 30602

HONKALA, FRED SAUL, b Concord, NH, Nov 30, 19; m 51; c 3. GEOLOGY. *Educ:* Univ NH, BS, 40; Univ Mo, MA, 42; Univ Mich, PhD(geol), 49. *Prof Exp:* From instr to prof geol, Univ Mont, 48-68, chmn dept, 56-64, dean grad sch, 64-68, dir res, 66-68; dir, Advan Sci Educ Prog, NSF, 68-70; pres, Yankton Col, 70-72; dean fac, St Mary's Col, Md, 72-74; exec dir, Am Geol Inst, 74-78; consult, US Senate Select Comt on Indian Affairs, 79-81; GEOLOGIST, VA DIV MIN RESOURCES, 81- *Concurrent Pos:* Rep for Am Geol Inst, Nat Res Coun & US Nat Comt Geol, 74-78. *Mem:*

AAAS; Geol Soc Am; Am Asn Petrol Geologists; Sigma Xi. *Res:* Phosphate deposits of Western United States, including delineation of Centennial Mountains phosphate field, Montana-Idaho. *Mailing Add:* Rte 1 Box 312A Lexington Park MD 20653

HONKANEN, PENTTI A, b Brooklyn, NY, Nov 25, 32; m 54; c 5. COMPUTER SCIENCE, ELECTRICAL ENGINEERING. *Educ:* Univ Colo, BS(elec eng) & BS(appl math), 59; Syracuse Univ, MEE, 62; Pa State Univ, PhD(elec eng), 67. *Prof Exp:* Staff engr, IBM Corp, 59-64; res assoc, Ord Res Lab, Pa State Univ, 64-68, asst prof comput sci, 68-73, chmn grad affairs, 71-73; assoc prof, 73-, PROF INFO SYSTS, GA STATE UNIV, ATLANTA. *Mem:* Sigma Xi; Inst Elec & Electronics Engrs; Asn Comput Mach. *Res:* Database systems and data structures; computer simulation; analysis and simulations of nonlinear electrical networks on a digital computer; graph theory; numerical solution on nonlinear differential equations; systems simulation on a digital computer. *Mailing Add:* 2442 Cravey Dr NE Atlanta GA 30345

HONMA, SHIGEMI, b Haina, Hawaii, Feb 14, 20; m 45; c 2. HORTICULTURE. *Educ:* Cornell Univ, BA, 49; Univ Minn, PhD, 53. *Prof Exp:* Asst horticulturist, Univ Nebr, 53-55; from asst prof to assoc prof, 56-66, prof, 66-86, EMER PROF HORT, MICH STATE UNIV, 86- *Mem:* Fel Am Soc Hort Sci. *Res:* Plant geneticist and plant breeder. *Mailing Add:* Dept Hort Mich State Univ East Lansing MI 48824

HONNELL, MARTIAL A(LFRED), b Lyons, France, Oct 23, 10; nat US; m 38; c 2. ELECTRICAL ENGINEERING. *Educ:* Ga Inst Technol, BS, 34, MS, 40, EE, 45. *Prof Exp:* Radio engr, WGST, Atlanta, 34-35, 38-44 & Pan Am Airways, Inc, Fla, 36-37; from instr to prof elec eng, Ga Inst Technol, 37-53, instr, Civilian Pilot Training Prog, 40-43, supvr pre-radar sch, 42-43 & ultrahigh frequency course, Eng Sci & Mgt War Training, 42-44, fac res assoc, 44-53; vpres & chief engr, Measurements Corp Div, McGraw-Edison Co, NJ, 53-58; prof, 58-81, EMER PROF ELEC ENG & ADJ PROF, AUBURN UNIV, 81- *Concurrent Pos:* Consult, Redstone Arsenal Proj, NASA & Auburn Res Found, 58-; distinguised grad fac lectr, Auburn Univ, 76-77. *Honors & Awards:* M A Ferst Sigma Xi Res Award, Ga Tech, 51. *Mem:* Am Soc Eng Educ; Inst Elec & Electronics Engrs. *Res:* Instrumentation; electronics; antennas; lightning. *Mailing Add:* Dept Elec Eng Auburn Univ Auburn AL 36849

HONNELL, PIERRE M(ARCEL), b Paris, France, Jan 28, 08; US citizen; m 37; c 2. ELECTRICAL ENGINEERING. *Educ:* Agr & Mech Col Tex, BSc, 30, EE, 38; Mass Inst Technol, MSc, 39; Calif Inst Technol, MSc, 40; St Louis Univ, PhD(geophys), 50. *Prof Exp:* Radio operator stas WCAR & WOAI, Tex, 23-24; radio officer, US Merchant Marine, Tex Co & Gulf Refining Co, 26-28; mem tech staff, Bell Tel Labs, 30-33; geophysicist, Tex Co, 34-38; asst prof elec eng, Southern Methodist Univ, 40-41 & Univ Ill, 46; from assoc prof to prof, 46-76, EMER PROF ELEC ENG, WASH UNIV, 76- *Concurrent Pos:* Consult, McDonnell Aircraft Corp, Carter Carburetor Co & Martin Co, Colo. *Honors & Awards:* Marconi Premium, Brit Inst Radio Engrs, 56. *Mem:* Fel AAAS; fel Inst Elec & Electronics Engrs; Am Soc Eng Educ; Am Asn Physics Teachers; Int Asn Analogue Comput; fel Inst Radio Engrs. *Res:* Stability theory, electronic and mechanic networks; seismological instruments; matric computors over reals and integers; didactic applications of matric computors. *Mailing Add:* 3927 W Sharon Ave Phoenix AZ 85029

HONNOLD, VINCENT RICHARD, b Los Angeles, Calif, Feb 10, 24; m 48; c 5. SOLID STATE PHYSICS. *Educ:* Calif Inst Technol, BS, 48; Notre Dame Univ, PhD(physics), 54. *Prof Exp:* Physicist, Solid State Physics Br, US Naval Ord Test Sta, 53-57, head, 57-59; mem tech staff, Hughes Aircraft Co, 59-63, head basic mechanisms sect, Radiation Effects Res Dept, Fullerton Res & Develop, 63-69, mem staff, Vulnerability & Hardness Lab, TRW Systs, 70-77; mem staff, B-1 Div, Rockwell Int, 77-78; sr staff physicist, Hughes Aircraft Co, 78-86; RETIRED. *Concurrent Pos:* Instr, Univ Calif, Los Angeles, 55-56. *Mem:* Am Phys Soc; Sigma Xi. *Res:* High energy radiation effects; photoconductive and conduction processes in semiconductors and insulators; magnetic resonance. *Mailing Add:* 656 17th St Manhattan Beach CA 90266

HONRUBIA, VICENTE, b Valencia, Spain, July 13, 34; m 62; c 2. OTOLARYNGOLOGY, NEUROPHYSIOLOGY. *Educ:* Univ Valencia, MD, 57, dipl pub health, 58, DMSc, 59. *Hon Degrees:* Dr, Univ Valencia, Spain, 82. *Prof Exp:* Intern otolaryngol, Univ Valencia, 55-58; Marques de Urquijo Found fel, Univ Madrid, 58-60; NIH fel, Univ Chicago, 60-62 & Rockefeller Inst, 62-63; res assoc & guest investr neurophysiol, Rockefeller Univ, 63-64; asst prof physiol & otolaryngol, Vanderbilt Univ, 65-66, assoc prof otolaryngol & audiol & dir otolaryngol res, 66-68; assoc prof, 68-70, PROF OTOLARYNGOL & DIR RES, DIV HEAD & NECK SURG, UNIV CALIF, LOS ANGELES, 70- *Concurrent Pos:* Dir res, Hope for Hearing Found, 76- *Honors & Awards:* Ugo Foscolo Medallion, Univ Pavia, Italy, 85. *Mem:* Fel Acoust Soc Am; Am Physiol Soc; Collegium Oto-Rhino-Laryngologicum Amicitiae Sacrum; Asn Res Otolaryngol (pres, 77-78); Am Neurotology Soc; Soc Neurosci. *Res:* Physiopathology of the auditory and vestibular systems. *Mailing Add:* Div Head & Neck Surg Univ Calif Sch Med Los Angeles CA 90024

HONSAKER, JOHN LEONARD, b Pasadena, Calif, May 4, 34; m 56. ATMOSPHERIC PHYSICS, COMPUTER GRAPHICS. *Educ:* Calif Inst Technol, BS, 55, PhD(physics), 65. *Prof Exp:* Physicist & staff mem, Los Alamos Sci Lab, 57-58; instr physics, Univ Chicago, 64-67; mem fac, Inst Earth & Planetary Physics, Univ Alta, 67-90; RETIRED. *Mem:* AAAS; Can Cartog Asn. *Res:* Nuclear structure and reaction mechanisms; controlled thermonuclear reactions; climatic change; computer cartography; micrometeorology; erosion analysis; multidimensional scaling and geometry. *Mailing Add:* 20450 Township Rd 510 Sherwood Park AB T8G 1E5 Can

HONSBERG, WOLFGANG, b Ludwigshafen, Ger, Aug 17, 29; m 58; c 2. ORGANIC CHEMISTRY, POLYMER CHEMISTRY. *Educ:* Karlsruhe Tech Univ, dipl chem, 55, Dr rer nat, 58. *Prof Exp:* Res assoc radiation chem, Univ Chicago, 58-59, res assoc peroxides, Fla State Univ, 59-61, res assoc peptides, 62-63; SR TECH ASSOC, E I DU PONT DE NEMOURS & CO, 63- *Mem:* Am Chem Soc. *Res:* Organic polyvalent iodine compounds; radiation chemistry; peroxides; peptides; heterocycles; polymerization; emulsions; curing of elastomers. *Mailing Add:* Du Pont Polymers Exp Sta Bldg 269 Wilmington DE 19880-0269

HONSINGER, VERNON BERTRAM, ELECTRICAL ENGINEERING. *Educ:* Univ Mich, BS. *Prof Exp:* From engr to mgr develop eng, Allis Chalmers, Norwood, Ohio, 45-73; consult develop eng, Advan Technol Ctr & mgr, Elec Systs Prog, Allis Chalmers, Milwaukee, Wis, 73-76; elec engr, Corp Res & Develop, Gen Elec, 76-82; ELEC MACH CONSULT, 82- *Concurrent Pos:* Elec engr, USN; lectr, Rensselaer Polytech Inst, 80. *Mem:* Fel Inst Elec & Electronics Engrs. *Res:* Electrical machinery; computer programming; permanent magnet machines; reluctance machines; induction machines; actuators; author of over 25 technical publication; 50 internal and contractual technical reports. *Mailing Add:* Six Woodstead Rd Ballston Lake NY 12019

HONSTEAD, WILLIAM HENRY, b Waterville, Kans, May 21, 16; m 40; c 3. CHEMICAL ENGINEERING. *Educ:* Kans State Col, BS, 39, MS, 46; Iowa State Univ, PhD, 56. *Prof Exp:* Plant control chemist, Nat Aniline Div, Allied Chem & Dye Corp, NY, 39-43; from instr to prof chem eng, Kans State Univ, 43-83, head dept, 60-68, tech field rep, 68-70, dir, Kans Indust Exten Serv, 70-81, exec vpres, Res Found, 72-83; RETIRED. *Mem:* Am Chem Soc; Am Soc Eng Educ; Nat Soc Prof Engrs; Am Inst Chem Engrs. *Res:* Dehydration of vegetable materials, including grasses; industrial utilization of sorghum grains; heat transfer; fertilizer technology. *Mailing Add:* 2130 Meadowlark Rd Manhattan KS 66502

HONTZEAS, S, b Coroni, Greece, Aug 6, 23; Can citizen; m 57; c 1. NUCLEAR CHEMISTRY. *Educ:* Nat Univ Athens, BSc, 54; McGill Univ, PhD(nuclear chem), 62. *Prof Exp:* Dir pharmaceut, Profarm Labs, Greece, 54-56; res assoc nuclear chem, Mass Inst Technol, 61-63; res fel nuclear chem, McGill Univ, 63-65; from asst prof to assoc prof chem, 65-68, assoc prof physics, 71-74, PROF CHEM, UNIV REGINA, 74- *Concurrent Pos:* Consult, Pulp & Paper Res Inst, Que, 63-65; sabbatical leave, Physics Inst, Uppsala Univ & Atom-Fysik Stokholm, 71-72. *Mem:* Chem Inst Can; fel Can Asn Physics; Am Chem Soc; fel Asn Greek Chem. *Res:* Nuclear reactions; x-ray and laser spectroscopy; nuclear and atomic spectroscopy. *Mailing Add:* Dept Chem Univ Regina Regina SK S4S 0A2 Can

HOO, CHEONG SENG, b Malacca, Malaysia, Sept 15, 38; Can citizen; m 88. MATHEMATICS. *Educ:* Univ Auckland, BSc, 60, MSc, 61; Syracuse Univ, PhD(math), 64. *Prof Exp:* Instr math, Univ Ill, Urbana, 64-65; from asst prof to assoc prof, 65-74, PROF MATH, UNIV ALTA, 74- *Concurrent Pos:* Nat Res Coun Can res grant, 66-91. *Mem:* Am Math Soc; Math Asn Am; Can Math Cong; Southeast Asia Math Soc. *Res:* Algebraic topology; category theory and theory of topological semigroups; homotopy theory and the theory of H-spaces; lattice theory; topological semigroups, MV, BCI, and BCK algebras. *Mailing Add:* Dept Math Univ Alta Edmonton AB T6G 2E1 Can

HOO, JOE JIE, b Malang, Indonesia, July 7, 44; Can citizen; m 73. MEDICAL GENETICS, PEDIATRICS. *Educ:* Philipps Univ Marburg, WGer, MD, 72; FAAP(pediat)Am Acad Pediat, Cert, Am Bd Med Genetics, 82. *Prof Exp:* Sr res assoc, Inst Human Genetics, Univ Hamburg, WGer, 79-81, med geneticist, 81-82; asst prof pediat, Univ Calgary, 82-85; ASSOC PROF PEDIAT, UNIV ILL, CHICAGO, 85- *Concurrent Pos:* Assoc prof pediat, Southern Ill Univ, 86; consult geneticist, Univ Ill, Peoria, 86- *Mem:* Am Acad Pediat; Am Soc Human Genetics; Europ Soc Human Genetics; Int Soc Twin Studies; Human Genetics Soc Australasia. *Res:* Human karyotype-phenotype correlation; the origin and behavior of ring chromosome; usage of molecular genetic techniques in clinical cytogenetics; gene mapping. *Mailing Add:* Dept Pediat St Lukes Med Ctr 1750 W Harrison St Chicago IL 60612

HOOBER, JOHN KENNETH, b Lancaster, Pa, Sept 5, 38; m 60; c 2. BIOCHEMISTRY. *Educ:* Goshen Col, BA, 60; Univ Mich, PhD(biochem), 65. *Prof Exp:* USPHS fel cell biol, Rockefeller Univ, 66-68; asst prof biochem, Rutgers Med Sch, Col Med & Dent NJ, 68-71; assoc prof, 71-77, PROF BIOCHEM, MED SCH, TEMPLE UNIV, 77- *Mem:* AAAS; Am Soc Plant Physiologists; Am Soc Biol Chemists; NY Acad Sci; Am Soc Cell Biol; Am Soc Photobiol. *Res:* Structure and function of membrane proteins; photochemical induction of bacterial pili; biogenesis of cellular membranes. *Mailing Add:* Dept Biochem Temple Univ Med Sch Philadelphia PA 19140

HOOBLER, SIBLEY WORTH, b New York, NY, Apr 30, 11; m 40; c 2. MEDICINE. *Educ:* Princeton Univ, BA, 33; Johns Hopkins Univ, ScD, 37, MD, 38; Am Bd Internal Med, dipl. *Hon Degrees:* LHD, John Hopkins Univ, 85. *Prof Exp:* Asst resident, Peter Bent Brigham Hosp, Boston, 40-42; intern & asst resident, Univ Hosp, 38-40, from asst prof to prof internal med, Med Sch, 59-76; clin prof med, Sch Med, 76-82, EMER PROF, CASE WESTERN RESERVE UNIV, 82-; EMER PROF, UNIV MICH, ANN ARBOR, 76- *Concurrent Pos:* Res fel med, Harvard Med Sch, 40-42; dir hypertension unit, Univ Mich Hosp, Ann Arbor, 47-74, mem, 74-; mem coun high blood pressure res, Am Heart Asn; adj physician, Cleveland Clin, 76-82. *Mem:* Am Physiol Soc; Am Soc Clin Pharmacol & Therapeut; Am Soc Clin Invest; Soc Exp Biol & Med; Sigma Xi. *Res:* Internal medicine; cardiology; physiology. *Mailing Add:* 13515 Shaker Blvd Cleveland OH 44120

HOOD, CLAUDE IAN, b Purley, Eng, Jan 25, 25; nat US; m 52; c 2. PATHOLOGY. *Educ:* Univ Liverpool, MB, ChB, 48. *Prof Exp:* Casualty officer, Liverpool Stanley Hosp, 48-49; house physician, Kettering Gen Hosp, 49; resident med officer, St Andrews Hosp, 49-50; instr path, Cornell Univ, 50-56; from instr to assoc prof, 56-73, PROF PATH, COL MED, UNIV FLA, 73- *Concurrent Pos:* From asst resident to resident, NY Hosp, 50-56; asst med examr, Alachua County, Fla, 59-75; consult, Vet Admin Hosp, Gainesville, Fla, 74-78, chief lab serv, 78-88. *Mem:* Am Thoracic Soc; Brit Med Asn; Int Acad Path; Asn Res Vision & Ophthal. *Res:* Medical teaching. *Mailing Add:* Dept Path Univ Fla Col Med Box J275 Gainesville FL 32610

HOOD, DONALD C, b North Merrick, NY, June 2, 42; m 78. VISUAL SCIENCE. *Educ:* State Univ NY, BA, 65; Brown Univ, MSc, 68, PhD(psychol), 70. *Prof Exp:* From asst prof to assoc prof psychol, Columbia Univ, 69-78, chmn dept, 75-78, vpres arts & sci, 82-87, PROF PSYCHOL, COLUMBIA UNIV, 78-, JAMES F BENDER PROF, 90- *Concurrent Pos:* Mem, Comt Vision, Nat Acad Sci-Nat Res Coun, 87-91; trustee, Smith Col, 89- *Mem:* Asn Res Vision & Ophthal; Optical Soc Am; Am Asn Univ Professors; AAAS; Psychonomic Soc. *Res:* Behavioral and physiological structures of vision; behavioral economics; academic administration. *Mailing Add:* 450 Riverside Dr New York NY 10027

HOOD, DONALD WILBUR, b New Castle, Pa, July 12, 18; m 45; c 3. OCEANOGRAPHY. *Educ:* Pa State Col, BS, 40; Okla Agr & Mech Col, MS, 42; Agr & Mech Col Tex, PhD(biochem, nutrit), 50. *Prof Exp:* Chem engr, E I du Pont de Nemours & Co, 42-43; res chemist, Manhattan Proj, Univ Chicago, 44; chemist, Hanford Eng Works, 44-46; instr chem, Tex A&M Univ, 46-47, asst prof chem oceanog, 50-54; from assoc prof to prof, 54-65, prof marine sci & dir inst, 65-76, EMER PROF MARINE SCI, UNIV ALASKA, FAIRBANKS, 78-; OCEANOG CONSULT, 77- *Concurrent Pos:* NSF sr fel, 63-64; mem study sect environ sci & eng, NIH, 65-68 Int Comt Port & Coastal Eng under Arctic Conditions, 71, 73 & 75 & adv panel int decade ocean explor, NSF, 72-76; chmn, Gordon Conf Chem Oceanog, 72 & Alaska Comn Oceanog Advan Sci & Technol; consult, US Plywood/Champion Paper Co, Alyeska Pipeline Serv Co & US CEngr; vis prof, Seoul Nat Univ, 77-78. *Mem:* Am Chem Soc; Am Soc Limnol & Oceanog. *Res:* Chemical oceanography; organic chemistry of seawater; chemistry of carbon dioxide system; waste disposal and marine pollution. *Mailing Add:* PO Box 57 Friday Harbor WA 98250

HOOD, EDWARD E, JR, b Jonesville, NC, Sept 15, 30. NUCLEAR ENGINEERING. *Educ:* NC State Col, BS, 52, MS, 53. *Prof Exp:* Engr, 57-62, mgr supersonic trans eng proj, 62-67, vpres group exec, 72-73, vpres group exec power generation group, 73-77, VPRES GEN MGR, COM ENG DIV, GEN ELEC CO, 68-, VCHMN, 79- *Concurrent Pos:* mem, USAF, 52-56; mem bd, Nat Elec Mfrs Asn, Nat Action Coun Minorities Eng; chmn, bd gov, Aerospace Industs Asn, bd trustees, Rensselaer Polytech Inst, Steering Comt, Defense Indust Initiative Bus Ethics & Conduct & Adv Comt, Nat Security Telecommun. *Mem:* Nat Acad Eng; fel Am Inst Aeronaut & Astronaut; Aerospace Indust Asn. *Mailing Add:* 3135 Easton Turnpike Fairfield CT 06431

HOOD, HORACE EDWARD, b Sparta, Ill, Sept 16, 23; m 49; c 4. ORGANIC CHEMISTRY. *Educ:* Univ Ill, BS, 44; Univ Minn, PhD(org chem), 50. *Prof Exp:* Res chemist, Res Ctr, Hercules Powder Co, 50-63, res supvr, 64-66; chem res supvr, Mo Chem Works, Hercules, Inc, 67, tech supt, 67-70, res scientist, 71-77, res proj leader, Hercofina, 78-79, RES ASSOC, HERCULES, INC, 80- *Mem:* Am Chem Soc; NY Acad Sci. *Res:* Process development; hydrocarbon oxidation; peroxides; production technology. *Mailing Add:* Hercules Inc Res Ctr Wilmington DE 19894

HOOD, JAMES WARREN, b Houston, Tex, Nov, 18, 25; m 46; c 3. GEOLOGY. *Educ:* Univ Tex-Austin, BS, 48. *Prof Exp:* Ground water hydrologist, Water Resources Div, US Geol Surv, 48-84; RETIRED. *Concurrent Pos:* Consult & tech editing, 84-88. *Mem:* Geol Soc Am; Am Geophys Union. *Res:* Analysis of ground water systems; assisted in preparation of technical reports for publication, results given in 40 plus publications. *Mailing Add:* 1209 Princeton Ave Salt Lake City UT 84105-1913

HOOD, JERRY A, b McAlester, Okla, June 2, 34; m 57; c 4. ELECTRICAL ENGINEERING. *Educ:* Univ Okla, BSEE, 56; Univ NMex, MSEE, 64, PhD(elec eng), 67. *Prof Exp:* Staff mem, Sandia Corp, 56-65; res assoc solid state physics, Univ NMex, 65-66; staff mem, 66, div supvr, 67-69, DEPT MGR, SANDIA CORP, 69- *Mem:* Inst Elec & Electronics Engrs (secy, 62-). *Res:* Effects of nuclear radiation on electronic components and systems. *Mailing Add:* Sandia Labs Kirtland AFB Box 5800 Albuquerque NM 87185

HOOD, JOHN MACK, JR, b Alamosa, Colo, June 11, 25; m 46; c 3. PHYSICS. *Educ:* Univ Colo, BA, 49, MA, 51; Univ Reading, PhD, 69; Univ London, DIC, 67. *Prof Exp:* Instr physics, Univ Colo, 49-51; res engr, NAm Aviation Co, 51-53; assoc engr, Univ Calif, 53-55; supvry physicist, USN Electronics Lab, 55-75; LECTR NATURAL SCI, SAN DIEGO STATE UNIV, 75- *Concurrent Pos:* Consult, Galileo Electrooptics, 80- *Mem:* Optical Soc Am; Soc Photo-Optical Instrumentation Engrs; Sigma Xi. *Res:* Geophysics; optical instrument design; laser technology and systems; visual systems engineering; theory of coherent imaging systems; history of science. *Mailing Add:* 3802 Point Loma Ave San Diego State Univ San Diego CA 92106

HOOD, JOSEPH, b Commerce, Ga, Apr 14, 24; m 46; c 1. SOILS. *Educ:* Univ Ga, BSA, 47; Purdue Univ, MS, 49; Cornell Univ, PhD(soils), 54. *Prof Exp:* From asst prof to assoc prof, 49-59, PROF AGRON & SOILS, AUBURN UNIV, 59- *Mem:* Fel Am Soc Agron; Soil Sci Soc Am; Sigma Xi. *Res:* Ion interactions in soils and plants; potassium status of soils; fertilizer placement; soil classification and land use. *Mailing Add:* Dept Agron & Soils Auburn Univ Auburn AL 36830

HOOD, LAMARTINE FRAIN, b Johnstown, Pa, Feb 25, 37; m 60; c 3. FOOD CHEMISTRY. *Educ:* Pa State Univ, BS, 59, PhD(food sci), 68; Univ Minn, MS, 63. *Prof Exp:* From asst to prof, Food Sci & Dir Agr Exp Sta, Cornell Univ, 68-86; DEAN, COL AGR, PA STATE UNIV, 86- *Honors & Awards:* William F Geddis Mem Lectr, NW sect, Am Asn Cereal Chemists, 85. *Mem:* Am Asn Cereal Chemists; Inst Food Technologists. *Res:* Starch chemistry; comparative electron microscopic and chemical studies of the interactions among food ingredients during processing and storage; effect of processing on the nutritional quality of foods. *Mailing Add:* Pa State Univ University Park PA 16802

HOOD, LARRY LEE, b Noble, Ill, Mar 5, 44; m 78; c 1. PROTEIN CHEMISTRY, FOOD SCIENCE. *Educ:* Univ Ill, BS, 66; Mich State Univ, MS, 68, PhD(food sci), 73. *Prof Exp:* Res asst food microbiol & chem, Mich State Univ, 66-72; group leader biochem, Quaker Oats Co, 73-76; res assoc protein res, Int Tel & Tel Continental Baking Co, 77-80; MGR COLLAGEN CASING RES, DEVRO, INC, 80- *Mem:* AAAS; Sigma Xi; Inst Food Technologists; Am Chem Soc. *Res:* Physico-chemical characterization of plant proteins; chemical and enzymatic modification of proteins; functionality of proteins in food products; food microbiology. *Mailing Add:* ITT Continental Baking Co Checkerboard Sq St Louis MO 63164-0001

HOOD, LEROY E, b Missoula, Mont, Oct 10, 38; m 63; c 2. IMMUNOGENETICS, GENETICS. *Educ:* Calif Inst Technol, BS, 60, PhD(biochem), 68; Johns Hopkins Univ, MD, 64. *Hon Degrees:* DSc, Mt Sinai Sch Med, City Univ NY, Univ BC, Vancouver, Can, Univ Southern Calif; DHL, Johns Hopkins Univ. *Prof Exp:* USPHS fel, Calif Inst Technol, 64-67; sr investr immunol, Nat Cancer Inst, 67-70; from asst prof to prof, 70-77, chmn biol, 80-89, BOWLES PROF BIOL, CALIF INST TECHNOL, 77-; DIR, NSF CTR MOLECULAR BIOTECHNOL, 89- *Concurrent Pos:* Dreyfus teacher-scholar, 74. *Honors & Awards:* Lasker Award; Dickson Prize. *Mem:* Nat Acad Sci; AAAS; Am Asn Immunologists. *Res:* Genetics & evolution of multigene systems; genetics & evolution of immune recognition antibodies, T-cell receptors, MHC molecules; molecular biology of myelin protein; protein evolution; recombinant DNA; biotechnology. *Mailing Add:* Div Biol 139-74 Calif Inst Technol Pasadena CA 91125

HOOD, LONNIE LAMAR, b Marshall, Tex, June 13, 49; m 74; c 3. LUNAR GEOPHYSICS, LUNAR & METEORITIC PALEOMAGNETISM. *Educ:* Northeastern La Univ, BS, 71, MS, 73; Univ Calif, Los Angeles, PhD(geophys, 79. *Prof Exp:* Teaching asst, dept physics, Northeastern LA Univ, 71-73; res asst, Inst Geophys & Planetary Physics, Univ Calif, Los Angeles, 75-79; from res assoc to sr res assoc, 79-83, ASST RES SCIENTIST, LUNAR & PLANETARY LAB, UNIV ARIZ, 83- *Concurrent Pos:* Prin investr res grants, solar systs geophys, NASA, 81- *Mem:* Am Geophys Union; AAAS; Int Asn Geomagnetism & Aeronomy. *Res:* Synthesis of geophysical constraints on the nature and origin of the moon as derived from Apollo data; physics of outer planet radiation belts; effects of solar ultraviolet variability and particle precipitation on the stratosphere and mesosphere. *Mailing Add:* Lunar & Planetary Lab Univ Ariz Tucson AZ 85721

HOOD, RICHARD FRED, b Somerset, Ky, Dec 5, 26; m 56; c 5. PHYSICS. *Educ:* Univ Ky, BS, 52, MS, 57, PhD(physics), 66. *Prof Exp:* ASSOC PROF PHYSICS, FRANKLIN & MARSHALL COL, 63- *Mem:* Am Asn Physics Teachers; Am Phys Soc; Sigma Xi. *Res:* Nuclear spectroscopy; charged particle interactions. *Mailing Add:* Dept Physics Franklin & Marshall Col Lancaster PA 17604

HOOD, ROBERT L, b Shreveport, La, Dec 7, 29; m 56; c 3. INORGANIC CHEMISTRY. *Educ:* Centenary Col, BS, 50; Univ Tex, Austin, PhD(chem), 69. *Prof Exp:* Eng clerk, Gas Measurement Dept, Tex Eastern Transmission Corp, 50-54; dir, Gas Lab, Centenary Col, 54-76, asst prof, 68-72, lectr chem, 72-76; MGR LAB SERV, COUNTRY PRIDE FOODS, 76- *Mem:* AAAS; Am Chem Soc. *Res:* Reactions of the coordinated ethylenediamine ligand in his (ethylenediamine)-palladium (II) iodide. *Mailing Add:* PO Box 1997 El Dorado AR 71731

HOOD, ROBIN JAMES, b Detroit, Mich, Apr 1, 42; m 65; c 2. PHYSICAL CHEMISTRY, ANALYTICAL CHEMISTRY. *Educ:* Mich State Univ, BS, 66; Wayne State Univ, PhD(chem), 73. *Prof Exp:* Asst prof chem, Univ Nebr, 73-75; asst prof chem, Marygrove Col, 75-81; DIR, CENT INSTRUMENTATION FACIL, WAYNE STATE UNIV, 81- *Mem:* AAAS; Am Chem Soc; Am Phys Soc. *Res:* Solid state spectroscopy; microcomputers in undergraduate teaching laboratory. *Mailing Add:* 75 Chem Wayne State Univ Detroit MI 48202

HOOD, RODNEY TABER, b Kenmore, NY, Sept 29, 24; m 49; c 3. HISTORY OF MATHEMATICS. *Educ:* Oberlin Col, AB, 46; Univ Wis, MA, 47, PhD, 50; Colgate-Rochester Divinity Sch, BD, 53; Univ Chicago, MA, 65. *Prof Exp:* Instr math, Beloit Col, 49-50; asst Greek, Colgate-Rochester Divinity Sch, 52-53; asst math, Univ Chicago, 54-56; asst prof, Ohio Univ,56-60; from assoc prof to prof, 60-86, chmn dept, 62-72, 75-78, EMER PROF MATH, FRANKLIN COL, 87. *Honors & Awards:* Cert Meritorious Serv, Math Asn Am, 89. *Mem:* Math Asn Am; Am Math Soc; Sigma Xi. *Res:* Differential equations; geometry; history of mathematics; number theory. *Mailing Add:* 265 E South St Franklin IN 46131

HOOD, RONALD DAVID, b El Paso, Tex, July 5, 41; m 72; c 1. TERATOLOGY, REPRODUCTIVE TOXICOLOGY. *Educ:* Tex Tech Univ, BS, 63, MS, 65; Purdue Univ, PhD(physiol), 69. *Prof Exp:* From asst prof to assoc prof, 68-78, PROF BIOL, UNIV ALA, 78- *Concurrent Pos:* Consult, 78- *Mem:* Soc Study Reproduction; Teratol Soc; Am Asn Lab Animal Sci; Soc Toxicol. *Res:* Experimental teratology and reproductive toxicology: arsenic teratogenicity, including metabolism and distribution; mycotoxin teratogenicity; in vitro teratogenicity testing; contract research in teratology, reproductive toxicity; teratogenic mechanisms; teratogenicity screening systems. *Mailing Add:* Dept Biol Sci Univ Ala Tuscaloosa AL 35487-0344

HOOD, SAMUEL LOWRY, b Cambridge, Ohio, July 28, 18; m 50; c 1. BIOLOGY GENERAL, RADIATION BIOLOGY. *Educ:* Col Wooster, AB, 40; Cornell Univ, PhD(biochem), 49. *Prof Exp:* Biochemist soil fertil, Ohio Agr Exp Sta, 41; biochemist cereal chem, Fed Soft Wheat Lab, 42-43, USDA Nutrit Lab, Cornell Univ, 43-44 & 46-48; biochemist agr radiotracer chem,

Agr Res Prog, AEC, Oak Ridge, 48-53; biochemist & radiation biologist, Charles F Kettering Found, 53-63; sr scientist, Nuclear Sci & Eng Corp, 63-68; res assoc, Univ Pittsburgh, 68-69; prof, 69-83, EMER PROF BIOL, CALIF UNIV PA, 83- *Mem:* AAAS; Am Chem Soc; Radiation Res Soc; Am Inst Biol Sci; NY Acad Sci; Sigma Xi. *Res:* Mineral metabolism; radioisotope tracers; agricultural chemistry; cell culture; radiation biology, exobiology. *Mailing Add:* 2808 E North St No 34 Greenville SC 29615-1753

HOOD, WILLIAM BOYD, JR, b Sylacauga, Ala, Mar 25, 32. INTERNAL MEDICINE, CARDIOLOGY. *Educ:* Davidson Col, BS, 54; Harvard Med Sch, MD, 58; Am Bd Internal Med, dipl, 67. *Prof Exp:* House officer med, Peter Bent Brigham Hosp, 58-59, jr resident, 59-60, sr resident, 62-63; res assoc, Sch Pub Health, Harvard Univ, 65-67, assoc, Sch Med, 67-69, from asst prof to assoc prof, Harvard Med Sch, 69-71; assoc prof, 71-74, prof med, Sch Med, Boston Univ, 74-82; PROF, DEPT MED, CARDIOL UNIT, UNIV ROCHESTER MED CTR, 82- *Concurrent Pos:* NSF fels, Cardiol Lab, St Thomas' Hosp, London, Eng, 60-61 & Cardiol Lab, Peter Bent Brigham Hosp, Boston, 61-62; Mass Heart Asn, Am Heart Asn & NIH grants, 71- *Mem:* AAAS; Am Soc Clin Invest; Am Fedn Clin Res; Am Col Cardiol; Am Physiol Soc; Asn Am Physicians. *Res:* Investigation of normal and abnormal cardiovascular hemodynamics in man and animals; coronary artery disease and coronary ischemia. *Mailing Add:* Dept Med Cardiol Unit Univ Rochester Med Ctr Box 679 Rochester NY 14642

HOOD, WILLIAM CALVIN, b Kansas City, Mo, June 7, 37; m 56; c 3. PETROLEUM GEOLOGY, GEOCHEMISTRY. *Educ:* Univ Mo, BA, 59; Univ Mont, PhD(geol), 64. *Prof Exp:* Lectr geol, Southern Ill Univ, 62-63; asst prof, NC State Univ, 63-65 & Univ SFla, 65-68; from assoc prof to prof geol, Southern Ill Univ, Carbondale, 68-81; area geologist, 81-87, PROJ GEOLOGIST, AMOCO PROD CO, 87- *Mem:* Clay Minerals Soc; Geochem Soc; Am Inst Prof Geologists; Am Asn Petrol Geologists. *Res:* Mineralogy and geochemistry of surface and near surface geologic processes; origin and migration of hydrocarbons. *Mailing Add:* 18011 Ravenfield Dr Houston TX 77084

HOOFNAGLE, JAY H, b Washington, DC, July 31, 43; m; c 3. DIGESTIVE DISEASES. *Educ:* Univ Va, BA, 65; Yale Med Sch, MD, 77; Am Bd Internal Med, cert, 76; Am Bd Gastroenterol, cert, 79. *Prof Exp:* Intern, Internal Med, Univ Va Hosp, 70-71, resident, 71-72; staff assoc Hepatitis Br, Div Blood & Blood Prod, Bur Biologics, Food & Drug Admin, 72-74, actg dir, 74-76; asst chief res med, Vet Admin Hosp, Washington, DC, 75-76, res gastroenterol-hepatology, 76-78; clin dir, 86-88, SR INVESTR, LIVER DIS SECT, DIGESTIVE DIS BR, NAT INST DIABETES & DIGESTIVE & KIDNEY DIS, NIH, 78-, DIR, DIV DIGESTIVE DIS & NUTRIT, 88- *Concurrent Pos:* Mem, Clin Res Subpanel, Nat Inst Diabetes & Digestive & Kidney Dis, NIH, 80-, med bd, Clin Ctr, 86-88; chmn, Credentials Comt, Clin Ctr, NIH, 87-88, Digestive Dis, Interagency Coord Comt, 88- *Mem:* Am Asn Study Liver Dis (pres elect, 90-); Am Soc Clin Invest; AAAS; Int Soc Interferon Res; Am Gastroenterol Asn; Am Soc Parenteral & Enteral Nutrit. *Res:* Viral hepatitis, types A, B, C and D; epidemiology and serology of viral hepatitis; immunology of acute and chronic viral hepatitis; therapy of viral hepatitis; autoimmune liver disease including autoimmune chronic active hepatitis primary biliary cirrhosis and sclerosing cholangitis. *Mailing Add:* Nat Inst Diabetes Digestive & Kidney Dis Div Digestive Dis & Nutrit NIH Bldg 31 Rm 9A23 Bethesda MD 20892

HOOGENBOOM, GERRIT, b Monster, Neth, Dec 19, 55; m 84. COMPUTER SIMULATION, AGRICULTURAL METEOROLOGY. *Educ:* Agr Univ, Wageningen, Neth, BSc, 77, MSc, 81; Auburn Univ, PhD(agron & soils), 85. *Prof Exp:* Res asst, Auburn Univ, 81-85; postdoctoral assoc, Univ Fla, 85-89; ASST PROF, DEPT AGR ENG, UNIV GA, 89- *Concurrent Pos:* Vis scientist, Scottish Hort Res Inst, Dundee, 78-79 & Volcani Ctr, ARO, Israel; consult, Winrock Int, 90. *Honors & Awards:* Res Award, Sigma Xi, 86. *Mem:* Am Soc Agr Engrs; Am Soc Agron; Crop Sci Soc Am; Soil Sci Soc Am; Am Soc Plant Physiologists; Sigma Xi. *Res:* Effects of environment on crop growth, development and production using computer simulation; develop an automated environmental monitoring network for the state of Georgia. *Mailing Add:* Dept Biol & Agr Eng Univ Ga Griffin GA 30223-1797

HOOGHEEM, THOMAS JOHN, b Fulton, Ill, Dec 18, 51; m 74. ENVIRONMENTAL ENGINEERING, CIVIL ENGINEERING. *Educ:* Univ Ill, BS, 73, MS, 74. *Prof Exp:* Engr civil eng, Moline, Ill, 66-72; res asst water pollution, Univ Ill, 73-74; environ engr, Tenn Valley Authority, 74-76; environ specialist water, 76-82, sr environ specialist, 82-84, ENVIRON MGR, MONSANTO AGR PROD CO, MONSANTO CO, 85- *Mem:* Water Pollution Control Fedn; Am Inst Chem Engrs; Soc Environ Toxicol & Chem. *Res:* Water pollution control, especially desalination, removal of toxic organics and sampling of priority pollutants; chemical surveillance and monitoring systems; air pollution and control of hydrocarbons from tire manufcture and styrene-butadiene manufacture; dioxin and hazardous waste sampling; health and environmental aspects of herbicides. *Mailing Add:* 651 Charbray Dr St Louis MO 63011

HOOGSTRAAL, HARRY, b Chicago, Ill, Feb 24, 17; m 49. BIOLOGY. *Educ:* Univ Ill, BA, 38, MS, 42; Univ London, PhD(parasitol), 59, DSc(med zool), 71. *Hon Degrees:* Dr, Ain Shams Univ, Cairo, 78. *Prof Exp:* Asst entom & curator insects, Univ Ill, 38-42; head, Philippines Zool Exped, Chicago Natural Hist Mus & Philippines Govt, 46-47; med zoologist, Univ Calif, African Exped, 48-49; head, Sudan-Equatoria Subunit, 49-50, sci co-dir, Sudan-Upper Nile Subunit, 60-64, HEAD, MED ZOOL DEPT, NAVY MED RES UNIT-3, 50- *Concurrent Pos:* Vis lectr & external examr, Cairo & Alexandria Univs; res assoc, Chicago Natural Hist Mus; dir sci working party, EAfrica High Comn, 56; mem panel, Food & Agr Orgn, UN; consult, WHO, head, Collab Ctr Study Anthropod Vectors, 70-; mem, US Hemorrhagic Fever Deleg, USSR, 65 & 69; consult, Navy Med Res Unit-2, Taiwan & Indonesia, 70- *Honors & Awards:* Walter Reed Medal, Am Soc Trop Med Hyg, 78; Franklin lectr, Auburn Univ, 79. *Mem:* Am Soc Trop Med & Hyg; fel AAAS; Am Soc Parasitol; Am Soc Mammal; Am Soc Protozool; Sigma Xi. *Res:* Medical zoology and biology; exploration; relations between vertebrate hosts, ectoparasites and pathogens; interrelationships between vertebrate hosts; ectoparasites; pathogens (especially viral and rickettsial) causing human and animal diseases; biology, ecology and relationship to disease; epidemiology of leishmaniasis; bird migration in relation to dissemination of pathogens and vectors. *Mailing Add:* Navy Med Res Unit 3 FPO New York NY 09527

HOOGSTRATEN, JAN, b Winnipeg, Man, June 15, 17; m 46; c 4. PATHOLOGY. *Educ:* Univ Man, MD, 41; Cambridge Univ, PhD(med), 49; FRCPS(C), 81. *Prof Exp:* Asst pathologist, Royal Victoria Hosp, Montreal, Que, 44; pathologist, Royal Jubilee Hosp, 44-46 & Children's Hosp of Winnipeg, 46-49; dir path, Children's Hosp of Winnipeg, 49-82; from asst prof to assoc prof path, 51-80, prof path & pediat, 80-90, EMER PROF PATH & PEDIAT, UNIV MAN, 90- *Concurrent Pos:* Dir path, Grace Gen Hosp, 49-62, consult pathologist, 62-; examr path, Royal Col Physicians & Surgeons Can, 58-; consult, Can Tumour Registry, 60- *Mem:* Soc Hemat; Can Med Asn; Can Asn Path; Int Soc Hemat; Int Acad Path; Soc Pediat Path. *Res:* Pediatric pathology; leukemia; inherited metabolic diseases; muscular dystrophy; electron microscopy; virology. *Mailing Add:* 56 Queenston Winnipeg MB R3N 0W5 Can

HOOK, DEREK JOHN, b London, UK, Aug 2, 47. MICROBIAL BIOCHEMISTRY. *Educ:* Hull Univ, UK, BSc, 68; Dalhousie Univ, NS, PhD(biol), 72. *Prof Exp:* Fel biosynthesis, Purdue Univ, 72-74; biochemist pharmaceut, Raylo Chem, 74-77, BIOCHEMIST, BRISTOL-MYERS INDUST DIV, 78- *Mem:* Am Chem Soc; Soc Microbiol. *Res:* Natural product isolation and fermentations of microorganisms in connection with the development of new drugs and antibiotics; study of the biosynthesis of pharmacologically important compounds. *Mailing Add:* 24 Carriage Lane Roxbury CT 06783-1920

HOOK, DONAL D, b Cleveland, Okla, June 21, 33; m 56; c 3. FORESTRY, TREE PHYSIOLOGY. *Educ:* Utah State Univ, BS, 61, MS, 62; Univ Ga, PhD(tree physiol), 68. *Prof Exp:* Res asst silvicult & physiol, Utah State Univ, 60-61; res forester, Intermountain Forest & Exp Sta, 62; res asst physiol, Univ Minn, 62-63; asst silviculturist, Southeastern Forest Exp Sta, 63-64; from assoc silviculturist to silviculturist, Forest Serv, USDA, 65-68, from plant physiologist to prin plant physiologist, 68-70; assoc prof silvicult & physiol, Univ Ky, 70-73; prof & dir, Belle W Baruch Forest Sci Inst, 73-81, PROF, DEPT FORESTRY, CLEMSON UNIV, 81-, PROF & DIR, FOREST WETLANDS RES PROG, SOUTHEASTERN FOREST EXP STA, CLEMSON UNIV & USDA. *Concurrent Pos:* Mem, Southeastern Forest Exp Sta Ecol Res Comt, 64-68 & Southern Forest Environ Res Coun, 64-70; sabattical leave, Pac Northwest Res Sta, 84-85; chair, Appalachian Soc Am Foresters, 92. *Mem:* Soc Am Foresters; Soc Wetland Scientists. *Res:* Factors affecting the relative flood tolerance of tree species; aeration systems in trees; vegetative propagation of tree species; physiology and ecology of wetland species and ecosystems. *Mailing Add:* Dept Forestry Clemson Univ 2730 Savannah Hwy Charleston SC 29407

HOOK, EDWARD WATSON, JR, b Sumter, SC, Aug 10, 24; m 49; c 4. INTERNAL MEDICINE. *Educ:* Wofford Col, BS, 43; Emory Univ, MD, 49; Am Bd Internal Med, dipl, 58. *Hon Degrees:* Med Col Pa, DMSc, 86. *Prof Exp:* Intern med, Univ Hosps, Univ Minn, Minneapolis, 49-50; from jr asst resident to resident, Grady Mem Hosp, Atlanta, Ga, 50-55; instr, Sch Med, Emory Univ, 54-56; from instr to asst prof, Johns Hopkins Univ, 56-59; from assoc prof to prof, Cornell Univ, 59-69; chmn, Dept Med & physician-in-chief, Hosp, 69-90, HENRY B MULHOLLAND PROF MED, SCH MED, UNIV VA, 69- *Concurrent Pos:* Attend physician, New York Hosp, 59-69; mem, Allergy & Immunol Study Sect, Nat Inst Allergy & Infectious Dis, 66-70, numerous comts, Am Bd Internal Med & Am Col Physicians, 72-90, Bd Int Health, Nat Acad Sci, 84-87 & US deleg, US-Japan Coop Med Sci Prog, 84-; ed, Antimicrobial Agents & Chemother, 72-80, Am J Med, 74-79 & Current Opinion Infectious Dis, 88-90; Mary W Barr vis prof med, NY Hosp, Cornell Med Ctr, 76, Aaron Feder vis prof, Med Col, 88; M Glenn Koenig vis prof, Vanderbilt Univ, 76; vis prof, Tufts-New Eng Med Ctr, 78, Sch Med, Univ Louisville, 88 & La State Univ, 90. *Honors & Awards:* Helen & Payne Whitney Distinguished Lectr, North Shore Univ, 77; Paul S Rhoads Lectr, 77; Shaia Lectr, Med Col Va, 77; Robert H Moser Lectr, Am Col Physicians, 85; George Pedigo Lectr, Univ Louisville, 88; Marion D Hargrove, Sr Lectr, La State Univ, 90. *Mem:* Inst Med-Nat Acad Sci; fel Infectious Dis Soc Am (pres, 75-76); Am Soc Clin Invest; Asn Am Physicians; Am Clin & Climat Asn; fel Am Col Physicians (pres elect, 84-85, pres, 85-86); hon fel Royal Australian Col Physicians; Asn Professors Med (pres, 81-82); fel AAAS; Am Asn Immunologists; NY Acad Sci. *Res:* Infectious diseases with emphasis on pathogenesis; salmonella infections; activity of antibiotics. *Mailing Add:* Dept Med Univ Va Sch Med Charlottesville VA 22908

HOOK, ERNEST, b New York, NY. EPIDEMIOLOGY, TERATOLOGY. *Educ:* Oberlin Col, BA, 56; Univ Calif, Berkeley, MA, 58; NU Univ, MD, 62. *Prof Exp:* From intern to resident pediat, Univ Minn, 62-65; fel, Univ Wash, 65-68; from asst prof to prof, Albany Med Col, 68-78; chief human ecol sect, 69-74, chief epidemiol & human ecol, Birth Defects Inst, 74-81, CHIEF GENETICS SECT, BUR CHILD HEALTH, STATE DEPT HEALTH, NY, 81-, RES PHYSICIAN, 81-; PROF PEDIAT, ALBANY MED COL, 78- *Mem:* Am Soc Human Genetics; Soc Epidemiol Res; Soc Pediat Res; Teratology Soc; Environ Mutagen Soc; Am Statist Asn. *Res:* Epidemiology of birth defects; human population cytogenetics; human mutagenesis; teratology; pediatrics and medical genetics. *Mailing Add:* Albany Med Col 43 New Scotland Ave Albany NY 12208

HOOK, GARY EDWARD RAUMATI, b Wellington, NZ, Jan 31, 42; US citizen; m 65; c 4. PULMONARY BIOCHEMISTRY. *Educ:* Victoria Univ, Wellington, NZ, BSc, 63, MSc, 64, PhD(biochem), 68, DSc, 86. *Prof Exp:* Jr lectr biochem, Victoria Univ, Wellington, NZ, 67-68; lectr, Univ Wales, Gt Brit, 68-70; vis assoc, 70-74, sr staff fel, 74-79, SECT HEAD, BIOCHEM

PATH, NAT INST ENVIRON HEALTH SCI, 79- *Concurrent Pos:* Co-ed, Environ Health Perspectives, 73-; asst prof, dept med, Duke Univ, 76-; assoc prof, curric toxicol, Univ NC, 80-85, prof 85- *Honors & Awards:* Award of Merit, NIH, 87. *Mem:* Biochem Soc Gt Brit. *Res:* Regulation of pulmonary epithelial secretions; research aimed at elucidating mechanisms through which alveolar type II cells are activated and the subcellular events that lead to increased synthesis and hypersecretion of surfactant phosphospholids by those activated cells. *Mailing Add:* Nat Inst Environ Health Sci PO Box 12233 Research Triangle Park NC 27709

HOOK, JAMES EDWARD, b Washington, DC, Apr 24, 47; m 68; c 5. AGRONOMY, ENVIRONMENTAL SCIENCE. *Educ:* Pa State Univ, BS, 69, MS, 71, PhD(agron), 75. *Prof Exp:* Res asst agron, Pa State Univ, 71-75; res assoc, Mich State Univ, 75-78; asst prof, 78-85, ASSOC PROF AGRON, UNIV GA, 85- *Concurrent Pos:* Assoc ed, J Prod Agr. *Mem:* Am Soc Agron; Soil Sci Soc Am; Coun Agr Sci & Technol. *Res:* Soil-water-plant relations; irrigation management; waste water, sludge and solid waste treatment on land. *Mailing Add:* Dept Agron Coastal Plain Exp Sta PO Box 748 Tifton GA 31793-0748

HOOK, JERRY BRUCE, b Elk City, Okla, Sept 7, 37; m 85; c 2. TOXICOLOGY, DRUG DEVELOPMENT. *Educ:* Wash State Univ, BS & BPharm, 60; Univ Iowa, MS, 64, PhD(pharm), 66. *Hon Degrees:* DSc, John Jay Col Criminal Justice, NY. *Prof Exp:* Asst, Wash State Univ, 57-60; from asst prof to prof pharmacol, Mich State Univ, 66-85, prof toxicol, 78-83, dir, ctr environ toxicol, 80-83; vpres, Preclin Res & Develop, Smith Kline & French Labs, 83-87, vpres, Worldwide, 87-88. actg vpres, Res, 88-89, vpres, Develop, Res & Develop, 89-90, SR VPRES & DIR, DEVELOP, RES & DEVELOP, SMITHKLINE BEECHAM PHARMACEUT, 90- *Concurrent Pos:* Adj prof, Univ Mich, 77 & Philadelphia Col Pharm & Sci, 83-86; vis scientist, Imp Chem Industs, Alderley Park, Eng, 79-80; Burroughs-Wellcome vis prof, Univ NDak, 81; mem adv comt, Nat Toxicol Prog, 82-86, peer rev panel experts, 82-86, chmn, 84-86, adv bd, Toxicol Res & Training Ctr, John Jay Col Criminal Justice, NY, 86-; vis sci, Fedn Am Soc Exp Biol, Univ Puerto Rico Med Sch, 84, Herbert H Lehman Col of City Univ, Bronx, NY, 85, John Jay Col Criminal Justice, NY, 87, Calif State Univ, Long Beach, 88, Pembroke State Univ, NC, 89; adj prof pharmacol, Temple Univ, 86- *Mem:* Am Physiol Soc; Am Soc Pharmacol & Exp Therapeut; Int Soc Nephrology; Soc Pediat Res; Am Fedn Clin Res; AAAS; Soc Toxicol (vpres elect, 85-86, vpres, 86-87, pres, 87-88); Soc Exp Biol & Med. *Res:* Renal toxicology with emphasis on biomedical mechanisms of cellular damage; role of kidney in metabolic activation; nephrotoxicity in senescence; perinatal toxicology; developmental nephrology; developmental pharmacology. *Mailing Add:* Smithkline Beecham Pharmaceut 709 Swedeland Rd PO Box 1537 King of Prussia PA 19406-0937

HOOK, JOHN W, b Capon Bridge, WVa, June 11, 22; m 49; c 4. GEOLOGY, APPLIED, ECONOMIC & ENGINEERING. *Educ:* Univ Tenn, AB, 47. *Prof Exp:* Geologist, Am Zinc Co, 47-54 & Reynolds Metals Co, 54-74; consult, John W Hook & Assoc, 74-86; RETIRED. *Concurrent Pos:* Prin investr, Minerals availability Syst, Bauxite Pac NW State Ore & US Bur Mines, 74-76; geothermal consult, NW Natural Gas Co, 76-80 & Eugene Water & elect Bd, 78-81; mem, Geothermal Resources Coun; gen partner, Sea-Tac Geothermal Co, 79-; pres, Blue Lake Geothermal Co, 87. *Mem:* Fel Geol Soc Am; Soc Econ Geologists. *Res:* Geological structure and hydrothermal systems; development of the theory of solution breccias in the East Tennesse Zinc District which led to the discovery of Young Mine. *Mailing Add:* 7315 Battle Creek Rd SE Salem OR 97301

HOOK, MAGNUS AO, b St Malm, Sweden, Sept 10, 46. CONNECTIVE TISSUE. *Educ:* Univ Uppsala, Sweden, PhD(med chem), 74. *Prof Exp:* ASSOC PROF BIOCHEM, UNIV ALA, BIRMINGHAM, 80- *Mem:* Am Soc Cell Biol; Am Soc Biol Chemists. *Mailing Add:* Dept Biochem Univ Ala Birmingham 508 Basic Health Sci Bldg Birmingham AL 35294

HOOK, ROLLIN EARL, b Highland, Ind, Nov 26, 34; m 56; c 2. METALLURGICAL ENGINEERING. *Educ:* Purdue Univ, BS, 56; Ohio State Univ, MS, 59, PhD(metall eng), 66. *Prof Exp:* Metallurgist, Dow Chem Co, 56-57 & Aerospace Res Labs, Wright-Patterson AFB, 57-67; res engr, Armco Steel Corp, 67-68, sr res metallurgist, 68-70, res assoc, 70-75, sr staff metallurgist, 75-81, PRIN RES METALLURGIST, ARMCO INC, 81- *Concurrent Pos:* Vis prof, Ohio State Univ, 81. *Mem:* Fel, Am Soc Metals Int; Mat Res Soc. *Res:* Deep drawing sheet steels; alloy development; plastic deformation and fracture; yielding phenomena; strain aging behavior; iron and nickel aluminides. *Mailing Add:* Res Ctr Armco Inc Middletown OH 45043

HOOK, WILLIAM ARTHUR, b Washington, DC, Apr 24, 30; m 56; c 2. MICROBIOLOGY, IMMUNOLOGY. *Educ:* Univ Md, BS, 53, MS, 56, PhD(microbiol), 65. *Prof Exp:* Lab asst bact, Livestock Sanit Serv Lab, Md, 51-53, lab technician, 53-54; asst, Univ Md, 54-56; microbiologist serol, Walter Reed Army Inst Res, 56-66; microbiologist, Nat Cancer Inst, 66-68; RES MICROBIOLOGIST, LAB IMMUNOL, NAT INST DENT RES, 68- *Mem:* Am Asn Immunol; Am Soc Microbiol; fel Am Acad Microbiol. *Res:* Substances in natural resistance to infection; immunological mediators of inflammation; histamine release from basophils and mast cells. *Mailing Add:* Lab Microbiol & Immunol Nat Inst Dent Res Bethesda MD 20892

HOOKE, ROBERT, b Chattanooga, Tenn, Apr 8, 18; m 41; c 2. MATHEMATICS, STATISTICS. *Educ:* Univ NC, BA, 38, MA, 39; Princeton Univ, AM, 40, PhD(math), 42. *Prof Exp:* Instr math, NC State Univ, 41-43, asst prof, 43-46; assoc prof, Univ of the South, 46-51; mem staff, Opers Eval Group, US Navy, Mass Inst Technol, 51-52; res assoc, Princeton Univ, 52-55; res mathematician, Westinghouse Elec Corp, 54-56, mgr statist sect, 56-63, mgr math dept, 63-79; RETIRED. *Concurrent Pos:* Consult, 79-83. *Mem:* Fel AAAS; Opers Res Soc Am; fel Am Statist Asn. *Res:* Applied statistics; design of experiments; sampling from finite populations. *Mailing Add:* PO Box 1982 Pinehurst NC 28374

HOOKE, ROGER LEBARON, b Glenridge, NJ, Jan 3, 39; m 61; c 2. GEOLOGY. *Educ:* Harvard Univ, BA, 61; Calif Inst Technol, PhD(geol), 65. *Prof Exp:* Asst prof, 65-70, assoc prof geol, 70-79, PROF GEOL, UNIV MINN, MINNEAPOLIS, 79- *Concurrent Pos:* Guest researcher, Univ Uppsala, Sweden, 72-73 & Univ Stockholm, Sweden, 80-81; assoc dir res, Tarfala Field Sta, Sweden, 81- *Mem:* Fel Geol Soc Am; Glaciol Soc. *Res:* Processes on alluvial fans in arid regions; geomorphic evolution of the tongue of the ocean, Bahamas, using deep submersible Alvin; glaciological studies, Barnes Ice Cap, Baffin Island, Canada; distribution of shear stress and sediment in meandering rivers; glaciological studies on subpolar glaciers, Tarfala Field Station, Sweden. *Mailing Add:* Dept Geol & Geophys Univ Minn Minneapolis MN 55455

HOOKE, WILLIAM HINES, b Raleigh, NC, Apr 23, 43; m 76; c 2. ATMOSPHERIC PHYSICS. *Educ:* Swarthmore Col, BA, 64; Univ Chicago, SM, 66, PhD(geophys sci), 67. *Prof Exp:* Physicist, Inst Telecommun Sci, 67-70 & Wave Propagation Lab, 70-73, supvry physicist, Wave Propagation Lab, 73-83, DIR, ENVIRON SCI GROUP, NAT OCEANIC & ATMOSPHERIC ADMIN, US DEPT COM, 84- *Concurrent Pos:* Lectr astro-geophys, Univ Colo, Boulder, 69-74; fel, Coop Inst Res Environ Sci, 71-78; adjoint assoc prof astro-geophys, Univ Colo, Boulder, 74- *Mem:* Am Geophys Union; Int Sci Radio Union; fel Am Meteorol Soc. *Res:* Gravity-wave propagation and dynamics in the atmosphere; meso-scale and boundary-layer dynamics; upper atmosphere and ionospheric physics; remote sensing of the atmosphere. *Mailing Add:* 7813 Lee Ave Alexandria VA 22308-1006

HOOKER, ARTHUR LEE, b Lodi, Wis, Oct 12, 24; m 50; c 2. GENETICS, PHYTOPATHOLOGY. *Educ:* Univ Wis, BS, 48, MS, 49, PhD(plant path & agron), 52. *Prof Exp:* Asst plant path & agron, Univ Wis, 48-50, plant path, 51-52, assoc prof, 52; asst prof bot & plant path, Iowa State Col, 52-54; plant pathologist, Cereal Crops Sect, Field Crops Res Br, Agr Res Serv, USDA, 54-58; from asst prof to prof plant path & agron, Univ Ill, Urbana, 58-80; biosci dir, Pfizer Genetics, Inc, St Louis, 80-83, biosci dir, Dekalb-Pfizer Genetics, Dekalb, Ill, 83-86; SCI DIR, PLANT MOLECULAR BIOL CTR, NORTHERN ILL UNIV, DEKALB, ILL, 86- *Concurrent Pos:* Guggenheim Mem Found fel, 64-65; mem, Nat Plant Genetic Resources Bd, USDA, 81-87; mem, Maize Crop Adv Comt, USDA, 83-; plant breeder, Hughes Hybrids, Woodstock, Ill, 86- *Mem:* Fel AAAS; fel Am Phytopath Soc; fel Crop Sci Soc Am; fel Am Soc Agron. *Res:* Plant tissue culture; genetics and biology of plant parasite interactions; field crops diseases and breeding. *Mailing Add:* 39 W 749 Dearhaven St Charles IL 60175

HOOKER, MARK L, b Chadron, Nebr, Jan 23, 47; m 69; c 2. SOIL FERTILITY, IRRIGATION WATER MANAGEMENT. *Educ:* Univ Nebr, BS, 72, MS, 75, PhD(agron), 79. *Prof Exp:* Res assoc, Univ Nebr, 72-79; asst prof, 79-85, ASSOC PROF, KANS STATE UNIV, 85- *Mem:* Sigma Xi; Am Soc Agron. *Res:* Irrigated soil management including fertility, tillage and irrigation on corn, sorghum, wheat and soybeans. *Mailing Add:* 1106 Kingsbury Rd Garden City KS 67846

HOOKER, THOMAS M, JR, b Arlington, Ky, May 9, 36; m 58; c 3. BIOPHYSICAL CHEMISTRY. *Educ:* Univ Ky, AB, 58, BS, 61; Univ Mo-Kansas City, MS, 66; Duke Univ, PhD(biochem), 67. *Prof Exp:* NIH res assoc chem, Univ Ore, 66-69; from asst prof to assoc prof, 69-81, PROF CHEM, UNIV CALIF, SANTA BARBARA, 81- *Mem:* Am Chem Soc; Soc Appl Spectros. *Res:* Theoretical and experimental aspects of optical activity; molecular conformation in solution; conformation and biological activity of biopolymers; thermodynamics of molecular interactions of biological significance; biochemical applications of digital computers; Raman spectroscopy. *Mailing Add:* Dept Chem Univ Calif Santa Barbara CA 93106

HOOKER, WILLIAM JAMES, b Sycamore, Ill, Nov 30, 14; m 39; c 3. PLANT PATHOLOGY. *Educ:* Northern Ill State Teachers Col, BE, 37; Purdue Univ, MS, 39; Univ Wis, PhD(plant path), 42. *Prof Exp:* Instr, Univ Wis, 42-43; pathologist, Midwest Div, Calif Packing Corp, Ill, 43-44; asst & assoc prof bot & plant path, Iowa State Univ, 44-54; from assoc prof to prof bot & plant path, Mich State Univ, 55-79; virlogist, Int Potato Ctr, Lima, Peru, 79-84; RETIRED. *Mem:* Hom mem Potato Asn Am (secy, 55-57, pres, 59); fel Am Phytopath Soc. *Res:* Virology; diseases of potato; air pollution injury to plants. *Mailing Add:* 310 Oxford Rd East Lansing MI 48823

HOOKER, WILLIAM MEAD, b Takoma Park, Md, Sept 9, 42; m 64; c 3. MICROSCOPIC ANATOMY. *Educ:* Columbia Union Col, BA, 64; Loma Linda Univ, PhD(anat), 69. *Prof Exp:* asst prof, 69-77, ASSOC PROF ANAT & ASSOC DEAN, SCH MED, LOMA LINDA UNIV, 77- *Concurrent Pos:* Fel, Univ Iowa, 70-71. *Mem:* Am Asn Anatomists; Electron Micros Soc Am; Sigma Xi. *Res:* Quantitative ultrastructural descriptions of endocrine and other cell systems. *Mailing Add:* Dept Anat Loma Linda Univ Sch Med Loma Linda CA 92350

HOOKS, JAMES A, agronomy, plant breeding, for more information see previous edition

HOOKS, RONALD FRED, b San Antonio, Tex, Dec 2, 41; m 71; c 3. POMOLOGY, PLANT SCIENCE. *Educ:* Tex A&M Univ, BS, 64, MS, 66; Mich State Univ, PhD(pomol), 69. *Prof Exp:* Grad asst hort, Tex A&M Univ, 64-66; from asst prof to assoc prof, 69-75, PROF HORT, NMEX STATE UNIV, 75- *Mem:* Am Soc Hort Sci; Sigma Xi. *Res:* Fruit production; growth regulators; cultural practices; nutrition; chile pepper cultural practices; fertilizer and growth regulators. *Mailing Add:* 1036 Miller St SW Los Lunas NM 87031

HOOKS, WILLIAM GARY, b Asheville, NC, Oct 4, 27; m 51; c 4. STRATIGRAPHY. *Educ:* Univ NC, SB, 50, MS, 53, PhD, 61. *Prof Exp:* Asst geol, Univ NC, 50-54; from asst prof to assoc prof, Univ Ala, 54-66, head dept geol & geog, 69-77, actg head, 81-82, PROF GEOL, UNIV ALA, TUSCALOOSA, 77- *Concurrent Pos:* Consult, 55- *Mem:* Am Asn Petrol Geologists. *Res:* Geomorphology; environmental geology; sedimentation-stratigraphy. *Mailing Add:* Box 870338 Tuscaloosa AL 35487-0338

HOOL, JAMES N, b Streator, Ill, May 28, 38; m 66. INDUSTRIAL ENGINEERING, FOREST MANAGEMENT. *Educ:* Purdue Univ, BS, 60, MS, 62, PhD(forest mgt), 65. *Prof Exp:* Asst prof, 65-68, assoc prof, 68-79, PROF INDUST ENG, AUBURN UNIV, 79- *Mem:* Am Statist Asn; Am Inst Indust Engrs. *Res:* Experimental statistics; operations research; systems analysis; normal approximation of linear combinations. *Mailing Add:* Dept Indust Eng Auburn Univ Auburn AL 36849

HOOLEY, JOSEPH GILBERT, physical chemistry; deceased, see previous edition for last biography

HOOP, BERNARD, JR, b San Francisco, Calif, Feb 17, 39; m 65; c 2. PHYSICS IN MEDICINE AND THE LIFE SCIENCES. *Educ:* Stanford Univ, BS, 60; Univ Wis-Madison, MS, 62, PhD(physics), 66. *Prof Exp:* Res asst nuclear physics, Univ Wis-Madison, 62-65; res assoc, Physics Inst, Univ Basel, 65-67; asst appl physics, Mass Gen Hosp, 67-70, asst appl physicist, 70-73; assoc physicist, 73-87, PHYSICIST, MASS GEN HOSP, 87-; ASSOC PROF MED PHYSICS, HARVARD MED SCH, 87- *Concurrent Pos:* Affil fac, Health Sci & Technol, Harvard Univ & Mass Inst Technol, 79- *Mem:* Am Phys Soc; Am Physiol Soc. *Res:* Central nervous system control of respiration. *Mailing Add:* Pulmonary Unit Mass Gen Hosp Boston MA 02114

HOOPER, ALAN BACON, b Berkeley, Calif, Nov 30, 37. MICROBIAL PHYSIOLOGY. *Educ:* Oberlin Col, BA, 59; Johns Hopkins Univ, PhD(biol), 64. *Prof Exp:* Asst prof zool, Univ Minn, Minneapolis, 64-68; assoc prof, 68-75, PROF GENETICS & CELL BIOL, UNIV MINN, ST PAUL, 75- *Mem:* AAAS; Am Soc Microbiol; Am Soc Biol Chem; Am Soc Cell Biol; Am Soc Protozool. *Res:* Biochemistry of nitrification in microorganisms and of mitochondria biogenesis in Tetrahymena. *Mailing Add:* Dept Genetics & Cell Biol Univ Minn 250 Bioscience St Paul MN 55108

HOOPER, ANNE CAROLINE DODGE, b Groton, Mass, July 16, 26; m 52; c 3. PATHOLOGY. *Educ:* Wash Univ, AB, 47, MD, 52, Am Bd Path, cert, path anat, 58, forensic path, 60, clin path, 61. *Prof Exp:* Jackson-Johnson res fel pharmacol, Wash Univ, 49-50; intern, Virginia Mason Hosp, Seattle, 52-53; resident internal med, St Francis Hosp, Hartford, Conn, 53-54; resident path, New Britain Gen Hosp, Conn, 54-57 & Presby Hosp, Philadelphia, 57-58; forensic pathologist, Med Examr Off, Philadelphia, 58-60; consult radioisotopes, Vet Admin Hosp, Coatesville, Pa, 63, dir lab, 66; dir lab, Kerbs Hosp, St Albans, Vt, 66-71 & Williamson Appalachian Regional Hosp, Williamson, WVa, 71-74; pathologist, Beckley Appalachian Regional Hosp, 74-75, dir lab, 75-76; asst prof path, 77-78, ASSOC PROF PATH, WVA SCH OSTEOP MED, 78- *Concurrent Pos:* Dir lab, St Albans Hosp, Vt, 66-69; consult radioisotopes, Coatesville Hosp, Pa, 63. *Honors & Awards:* Borden Award, 52. *Mem:* Fel Am Soc Clin Pathologists; fel Am Acad Forensic Sci; Int Acad Path; Sigma Xi; Col Am Pathologists; AMA. *Mailing Add:* WVa Sch Osteop Med 400 N Lee St Lewisburg WV 24901

HOOPER, BILLY ERNEST, b Pawnee City, Nebr, June 22, 31; m 54; c 2. VETERINARY MEDICINE, EDUCATION ADMINISTRATION. *Educ:* Univ Mo, BS & DVM, 61; Purdue Univ, MS, 63, PhD(vet path), 65. *Prof Exp:* From asst prof to assoc prof vet path, Purdue Univ, 65-68; mem fac dept path, Sch Vet Med, Univ Mo-Columbia, 68, assoc prof, 68-71, chmn dept, 69-71; prof vet path, Col Vet Med, Univ Ga, 71-73; prof vet path, Sch Vet Med & assoc dean acad affairs, Purdue Univ, West Lafayette, 73-86; EXEC DIR, ASN AM VET MED COL, 86- *Mem:* Am Vet Med Asn; Am Col Vet Path. *Res:* Gastrointestinal pathology; virus induced enteric disease of swine. *Mailing Add:* 5893 Woodfield Estates Dr Alexandria VA 22310

HOOPER, CATHERINE EVELYN, b Brooklyn, NY, Nov 10, 39; m 74; c 2. HIGH ELECTRON MOBILITY TRANSFER TECHNOLOGY, GALLIUM ARSENIDE TECHNOLOGY. *Prof Exp:* Inspector, Amelco Semiconductor, 66-68; technician, Fairchild Res & Develop, 68-73; sr technician, Varian Cent Res, 73-84; SR DEVELOP ENGR, HUGHES RES LABS, 84- *Concurrent Pos:* Lectr, Grad Women Sci, Nat Speakers Bur, 90. *Mem:* Sigma Xi; Int Soc Optical Engrs; Am Vacuum Soc; Mat Res Soc. *Res:* Gallium arsenide field effect transistor; high electron mobility transistors for satellite and space applications. *Mailing Add:* Hughes Res Labs 3011 Malibu Canyon Rd Malibu CA 90265

HOOPER, CHARLES FREDERICK, JR, b Cambridge, Mass, June 3, 32; m 71; c 2. THEORETICAL ATOMIC & PLASMA PHYSICS. *Educ:* Dartmouth Col, AB, 54; Johns Hopkins Univ, PhD(physics), 63. *Prof Exp:* From asst prof to assoc prof, 63-76, PROF PHYSICS, UNIV FLA, 76-, dept chmn, 80-86. *Mem:* Fel Am Phys Soc. *Res:* Spectral line broadening in plasmas; electric microfield distributions; statistical physics; dense plasma physics. *Mailing Add:* Dept Physics Univ Fla Gainesville FL 32611

HOOPER, DONALD LLOYD, b St Stephen, NB, Can, Nov 18, 38; m 61; c 1. NUCLEAR MAGNETIC RESONANCE. *Educ:* Univ NB, BSc, 60, MSc, 61, PhD(nuclear magnetic resonance), 64. *Prof Exp:* NRC postdoctoral chem, Univ E Anglia, 64-66; asst prof, 66-73, ASSOC PROF CHEM, DALHOUSIE UNIV, 73-; MGR, NUCLEAR MAGNETIC RESONANCE, ATLANTIC REGION MAGNETIC RES CTR, 82- *Mem:* Chem Inst Can; Royal Soc Chem; Am Chem Soc. *Res:* Application of spectroscopic methods, primarily high field nuclear magnetic resonance to solution of problems in structure and dynamics of compounds. *Mailing Add:* Dept Chem Dalhousie Univ Halifax NS B3H 4J3 Can

HOOPER, EDWIN BICKFORD, JR, b Bremerton, Wash, June 18, 37; m 63; c 3. MAGNETIC FUSION ENERGY, MIRROR CONFINEMENT & HEATING. *Educ:* Mass Inst Technol, SB, 59, PhD(physics), 65. *Prof Exp:* Asst prof applied sci, Yale Univ, 66-70; PHYSICIST FUSION RES, LAWRENCE LIVERMORE NAT LAB, 70-, PHYSICS GROUP LEADER, MICROWAVE TOKAMAK EXP. *Mem:* Fel Am Phys Soc; AAAS. *Res:* Experimental and supporting theoretical research on magnetic fusion energy, with special attention to plasma physics and atomic physics issues; mirror machines and high energy neutral beams based on negative ions; Tokamak heating by electron cyclotron resonance. *Mailing Add:* L-637 Lawrence Livermore Nat Lab PO Box 5511 Livermore CA 94550

HOOPER, EMMET THURMAN, JR, b Phoenix, Ariz, Aug 19, 11; m 36; c 2. VERTEBRATE ZOOLOGY. *Educ:* Univ Calif, AB, 33, MA, 36, PhD(zool), 39. *Prof Exp:* Asst, US Bur Fisheries, 34; asst, Mus Vert Zool, Univ Calif, 34-36, asst zoologist, 36-38; from asst cur to cur, Mus Zool, 38-56, from asst prof to assoc prof, 46-58, prof & cur, 58-79, EMER PROF ZOOL & EMER CUR MAMMALS, UNIV MICH, 79- *Concurrent Pos:* Prog dir, NSF, 64-65, consult, 65-70; proj leader, Sea Otter Res Prog, Coastal Marine Studies, US Fish & Wildlife Serv, Univ Calif, Santa Cruz, 79-81, res assoc marine sci, 81- *Mem:* AAAS; Am Inst Biol Sci; Am Soc Mammal (corresp secy, 41-46, vpres, 58-62, pres, 62-64, hon mem, 76); Soc Study Evolution. *Res:* Mammalogy; systematics; anatomy; physiology; ecology. *Mailing Add:* 1101 Bear Valley Rd Aptos CA 95060

HOOPER, F(RANK) C(LEMENTS), b Toronto, Ont, Apr 10, 24; m 52; c 2. HEAT TRANSFER, HEAT ENGINEERING. *Educ:* Univ Toronto, BASc, 46; Univ London, DIC, 53. *Prof Exp:* Demonstr mech eng, Univ Toronto, 46-47, lectr, 47-55, from asst prof to assoc prof, 55-65, chmn, div eng sci, 77-82, prof mech eng, 65-89, chmn div eng sci, 88-89, EMER PROF, ENG, UNIV TORONTO, 89- *Concurrent Pos:* Exchange lectr, Univ London, 52-53; mem engines comt, Nat Res Coun Can, 52-54, heat transfer comt, 59-72, chmn, Assoc Comt on Heat Transfer, 66-72; consult, ERDA & Dept Energy, 74- & Sci Coun, 76-78, pres Hooper & Angus Assoc, 77-87; pres, Int Assembly Heat Transfer Conf, 78-82 & pres, Royal Can Inst, 81-82, pres, Can Soc Mech Eng, 87-88; consult eng univ lectr, 90- *Honors & Awards:* Eng Medal, Asn Prof Engrs Ont, 76; Can Cong Appl Mech Medal, 77; Silver Jubilee Medal, 77; E K Campbell Award, Am Soc Heating Refrig & Air Conditioning Engrs, 80; Stachiewic Heat Transfer Medal, 86- *Mem:* Am Soc Mech Engrs; fel Eng Inst Can; Can Soc Mech Engrs (pres, 87-88); Solar Energy Soc Can. *Res:* Conduction and radiation heat transfer; psychrometry; thermal storage; heat pumps; solar energy in space heating; distribution of the diffuse component of solar radiation; boiling and flashing of liquids. *Mailing Add:* Dept Mech Eng Univ Toronto Toronto ON M5S 1A4 Can

HOOPER, FRANK FINCHER, b Phoenix, Ariz, Jan 17, 18; m 43. ZOOLOGY. *Educ:* Univ Calif, BA, 39; Univ Minn, PhD(zool), 48. *Prof Exp:* Asst zool, Univ Minn, 40-42; instr, Univ Mich, 48-52, biologist, Inst Fisheries Res, 58-62; aquatic ecologist, Div Biol & Med, US AEC, 62-63; biologist-in-charge, Inst Fisheries Res, 65-66, PROF FISHERIES & ZOOL, UNIV MICH, ANN ARBOR, 66- *Mem:* Ecol Soc Am; Am Micros Soc (treas, 52-55); Am Fisheries Soc; Am Soc Zool; Am Soc Limnol & Oceanog (pres elect, 65, pres, 66). *Res:* Limnochemistry; aquatic food chains. *Mailing Add:* Dept Fisheries Sch Natural Resources Univ Mich 128 Dana Bldg Ann Arbor MI 48109-1115

HOOPER, GEORGE BATES, b Philadelphia, Pa, Nov 23, 24; m 53; c 4. BIOLOGY. *Educ:* Seton Hall Col, BS, 49; Princeton Univ, PhD(biol), 56. *Prof Exp:* Asst biol, Princeton Univ, 55-56, instr, 56-57; asst prof, Bard Col, 57-60; from asst prof to assoc prof biol, 60-67, chmn dept, 60-68, chmn div sci, 68-75, PROF BIOL, MARIST COL, NY, 68-, CHMN DIV SCI, 78- *Concurrent Pos:* Vis lectr, Princeton Univ, 64-65. *Mem:* AAAS; Am Soc Zoologists; Soc Study Evolution; Am Soc Naturalists. *Res:* Drosophila ecology, physiology and evolution. *Mailing Add:* 36 Circle Dr Hyde Park NY 12538

HOOPER, HENRY OLCOTT, b Washington, DC, Mar 9, 35; m 56; c 5. SOLID STATE PHYSICS. *Educ:* Univ Maine, ScB, 56; Brown Univ, ScM, 59, PhD(physics), 61. *Prof Exp:* Asst prof physics, Brown Univ, 61-64; from asst prof to prof, Wayne State Univ, 64-73; prof physics & chmn dept, Univ Maine, Orono, 73-80, dean grad sch, 77-80, act vpres acad affairs, 79-80; ASSOC VPRES ACAD AFFAIRS & DEAN GRAD COL, NORTHERN ARIZ UNIV, 81- *Mem:* AAAS; Am Phys Soc; Am Asn Physics Teachers. *Res:* Nuclear magnetic resonance; nuclear quadrupole resonance; amorphous magnetic materials. *Mailing Add:* Box 4085 Northern Ariz Univ Flagstaff AZ 86011

HOOPER, IRVIN P(LATT), b Lynn, Mass, June 5, 14; m 42; c 3. MECHANICAL ENGINEERING. *Educ:* Tufts Col, BS, 38; Univ Vt, MS, 49. *Hon Degrees:* DEng, Rose-Hulman Inst Technol, 79. *Prof Exp:* Asst prof mech eng, Rose Polytech Inst, 46-48 & Univ Vt, 48-49; from assoc prof to prof, 50-80, head dept, 57-65, bus mgr, 65-70, dir continuing educ,70-83, ROBERT SHATTUCK PROF MECH ENG, ROSE-HULMAN INST TECHNOL, 80- *Concurrent Pos:* Consult engr, USMC, 42-46. *Mem:* Am Soc Mech Engrs; Am Soc Eng Educ; Nat Soc Prof Engrs. *Res:* Analytical and experimental stress analysis; thermodynamics; heat transfer. *Mailing Add:* Dir Continuing Educ Rose-Hulman Inst Technol 5500 Wabash Ave Terre Haute IN 47803

HOOPER, IRVING R, b South Lyon, Mich, June 16, 21; m 44; c 3. MEDICINAL CHEMISTRY. *Educ:* Mich State Normal Col, AB, 41; Univ Ill, PhD(biochem), 44. *Prof Exp:* Sr chemist, Parke, Davis & Co, Detroit, 44-45; Res Found fel, Ohio State Univ, 45-46; proj leader, Bristol Labs Div, Bristol-Myers Co, 46-49, dir biochem res, 49-65, dir chem res, 65-72, assoc dir res, 72-77; res assoc, Marine Lab, Duke Univ, 78-87; CONSULT, 77- *Concurrent Pos:* Lectr, State Univ NY Upstate Med Ctr, 68-77. *Mem:* Am Chem Soc. *Res:* Isolation of natural products; structure determination; antibiotics; carbohydrate chemistry; immobilized enzymes; marine pheromones; anti-fouling surfaces. *Mailing Add:* Rte 1 Box 284 Beaufort NC 28516

HOOPER, JAMES R(IPLEY), b Boston, Mass, Mar 30, 15; m 40; c 5. SCIENCE EDUCATION. *Educ:* Harvard Univ, AB, 38, MS, 39, PhD(physics), 57. *Prof Exp:* Asst physics, Williams Col, 40-41; instr, Union Col, 41-42; instr & res assoc, Harvard Univ, 42-50; from instr to assoc prof elec eng, Case Western Reserve Univ, 50-75, dir spec progs, 58-60, assoc dean instr, 60-64, dean undergrad studies, 64-69, dean spec undergrad studies, 69-75; RETIRED. *Mem:* Am Phys Soc. *Res:* Innovation of undergraduate eductional programs. *Mailing Add:* Four Jerusalem Lane Cohasset MA 02025

HOOPER, JOHN WILLIAM, b Clarendon, Ark, June 9, 31; m 57; c 2. PHYSICAL ELECTRONICS, ACADEMIC ADMINISTRATION. *Educ:* Kans State Univ, BS(elec eng) & BS(bus admin), 54; Ga Inst Technol, MS, 55, PhD(elec eng), 61. *Prof Exp:* Instr math, Martin Br, Univ Tenn, 54; asst elec eng, Ga Inst Technol, 54-55; engr, Gen Elec Co, 55; res assoc, Army Ballistic Missile Agency, 55-57; asst elec eng, Ga Inst Technol, 57-58, from instr to prof, 58-71, regents' prof, 71-72; assoc vchancellor, Bd Regents Univ Syst Ga, 72-76, vchancellor, 76-79; regents prof & dir microelectronics res ctr, Ga Inst Technol, 79-86, actg vpres acad affairs, 86-87; RETIRED. *Concurrent Pos:* Consult, Ga Power Co, 57-65 & Oak Ridge Nat Lab, 65-79. *Mem:* Fel Am Phys Soc; Inst Elec & Electronics Engrs; Sigma Xi. *Res:* Atomic collisions; atomic cross section measurements; ion molecule reactions in gases at low and high pressures; charged and neutral beam detectors; solid state theory; ultra-high vacuum technology. *Mailing Add:* Rte 3 Box 185-H Jasper GA 30143

HOOPER, NIGEL, b Bristol, Eng, June 18, 61. BIOCHEMISTRY. *Educ:* Univ Salford, UK, BSc, 83; Univ Surrey, UK, MSc, 84; Univ Aston, UK, PhD(biochem), 87. *Prof Exp:* Researcher, Dept Biomed & Environ Health Sci, Sch Pub Health, Univ Calif, 87-89; RES INSTR, DIV MOLECULAR BIOL & BIOCHEM, SCH BASIC LIFE SCI, UNIV MO, 90- *Mem:* AAAS; Protein Soc; Am Soc Biochem & Molecular Biol. *Res:* Biomedical; molecular biology. *Mailing Add:* Sch Basic Life Sci Univ Mo 109BSB 5100 Rockhill Rd Kansas City MO 64110

HOOPER, ROBERT JOHN, b Wyoming, Pa, June 10, 31; m 60; c 1. INORGANIC CHEMISTRY, MOLECULAR SPECTROSCOPY. *Educ:* King's Col, Pa, BS, 53; Univ Notre Dame, MS, 59, PhD(inorg chem), 62. *Prof Exp:* Res assoc, Radiation Lab, Univ Notre Dame, 62-64; asst prof chem, Marist Col, 64-67; PROF CHEM, SOUTHEASTERN MASS UNIV, 67- *Mem:* Am Chem Soc; Soc Appl Spectros. *Res:* Spectroscopic studies of the structure and bonding in transition metal complexes. *Mailing Add:* Dept Chem Southeastern Mass Univ North Dartmouth MA 02747

HOOPER, WILLIAM JOHN, JR, b Buffalo, NY, July 15, 35; m 59; c 2. PHYSICS, ASTRONOMY. *Educ:* Rensselaer Polytech Inst, BS, 57, MS, 59; Boston Univ, EdD, 67. *Prof Exp:* Asst prof physics, Clinch Valley Col, 59-63; instr, Simmons Col, 63-64; assoc prof, Boston State Col, 64-67; assoc prof physics, 67-72, chmn dept phys sci, 67-74, PROF PHYSICS,CLINCH VALLEY CO,L 72-, CHMN NATURAL SCI DEPT, 86- *Mem:* US Metric Asn; Am Asn Physics Teachers. *Res:* Physics and physical science education. *Mailing Add:* Natural Sci Dept Univ Va Clinch Valley Col College Ave Wise VA 24293

HOOPES, JOHN A, b Berkeley, Calif, Mar 29, 36; m 59; c 3. FLUID MECHANICS, HYDRAULIC ENGINEERING. *Educ:* Univ Calif, Berkeley, BS, 58, MS, 60; Mass Inst Technol, PhD(civil eng), 65. *Prof Exp:* From asst prof to assoc prof, 68-72, PROF CIVIL ENG, UNIV WIS-MADISON, 72- *Honors & Awards:* Walter L Huber Civil Eng Res Prize, 72 & Karl Emil Hilgard Prize, 72, Am Soc Civil Engrs. *Mem:* Am Soc Civil Engrs; Am Geophys Union. *Res:* Transport and mixing of substances discharged into river, lake and groundwater flows; sediment erosion and transport; hydrology; wind and thermally driven circulations in impoundments. *Mailing Add:* Dept Civil Eng 2205 Eng Bldg Univ Wis 1415 Johnson Dr Madison WI 53706

HOOPES, JOHN EUGENE, b Boone, Iowa, Aug 8, 31; m 67; c 4. PLASTIC SURGERY. *Educ:* Rice Univ, AB, 53; Johns Hopkins Univ, MD, 57. *Prof Exp:* Asst prof surg, Sch Med, Johns Hopkins Univ, 64-68; assoc prof, Sch Med, Wash Univ, 68-70; PROF SURG, SCH MED, JOHNS HOPKINS UNIV, 70- *Concurrent Pos:* Am Cancer Soc fel, Sch Med, Johns Hopkins Univ, 64-66. *Res:* Wound healing. *Mailing Add:* Johns Hopkins Hosp Baltimore MD 21205

HOOPES, JOHN W(ALKER), JR, b Wilmington, Del, May 30, 22; m 47; c 2. CHEMICAL ENGINEERING. *Educ:* Bowdoin Col, BS, 43; Mass Inst Technol, BChE, 44; Columbia Univ, MS, 46, PhD(chem eng), 51. *Prof Exp:* Res engr, Sam Lab, Columbia Univ, 44, instr chem eng, 47-49; from assoc prof to asst prof, 50-55, mgr process eng sect, Chem Eng Dept, ICI Americas Inc, 57-58, sr chem engr, Atlas chem Industs, Inc, 55-57; mgr process eng sect, Chem Eng Dept, ICI Americas Inc, 57-58, from asst dir to dir chem eng, 58-81; RETIRED. *Concurrent Pos:* Co-ed, Advances Chem Eng, 56-81; vis prof, Widener Univ, 83-86, prof, 86-91. *Mem:* Am Chem Soc; Am Inst Chem Engrs; Sigma Xi; Am Soc Eng Educ. *Mailing Add:* Box 3992 Greenville Wilmington DE 19807

HOOPES, KEITH HALE, b Fairview, Wyo, July 23, 30; m 53; c 6. VETERINARY MEDICINE. *Educ:* Wash State Univ, DVM, 56; Utah State Univ, BS, 57. *Prof Exp:* Asst vet microbiol, Wash State Univ, 53-56; pvt pract, 56-57; from asst prof to assoc prof animal sci, 57-69, PROF ANIMAL SCI, BRIGHAM YOUNG UNIV, 69- *Concurrent Pos:* Res asst vet sci, Univ Nebr, 63-64. *Mem:* Am Vet Med Asn; Am Asn Sheep & Goat Practrs; Am Asn Bovine Practrs. *Res:* Pathology; physiology. *Mailing Add:* Dept Animal Sci Brigham Young Univ Provo UT 84602

HOOPES, LAURA LIVINGSTON MAYS, b Richmond, Va, Dec 1, 42; m 81; c 2. MOLECULAR BIOLOGY, GERONTOLOGY. *Educ:* Goucher Col, AB, 64; Yale Univ, PhD(biol), 68. *Prof Exp:* Fel microbiol, Scripps Clin & Res Found, 68-69; res assoc, Med Sch, Univ Colo, 69-73; from asst prof to assoc prof biol, 73-85, chair, 78-80 & 81-82, PROF BIOL, OCCIDENTAL COL, 85- *Concurrent Pos:* NSF fel, Yale Univ, 64-68; NIH fel, Scripps Clinic & Res Found, 68-69 & Med Sch, Univ Colo, 69-70; NIH res grant, Occidental Col, 74-80; mem Nat Bd, Am Aging Asn, 75-84; guest scientist, Nat Inst Aging, 80-81; res corp grants, 81-82 & 84-85; consult genetics & biochem, Addison-Wesley/Benjamin Cummings, 82, Merck undergrad res prog, 87; counr, Coun Undergrad Res, 85-, exec secy, 87-; NSF panelist, Col Sci Improv Prog, 86; vis assoc, Caltech Div Biol, 86-87 & 87-88; mem, Western Assoc Sch & Col accreditation team, 87-; co-auth & coordr, Pew Western Cluster grant, 87-90, mem vis comt, Wellesley Biol Dept, 88, mem Am Fedn Aging Res grants, 86-87 & 87-88. *Mem:* Am Aging Asn (pres, 76-78); Geront Soc; Am Soc Cell Biol; Genetics Soc Am; Am Soc Microbiol. *Res:* Molecular basis of changes during rodent aging; changes in DNA methylation during development and aging. *Mailing Add:* Dept Biol Occidental Col 1600 Campus Rd Los Angeles CA 90041

HOOPINGARNER, ROGER A, b Mich, May 27, 33; m 56; c 3. ENTOMOLOGY. *Educ:* Mich State Univ, BS, 55; Univ Wis, MS, 57, PhD(entom), 59. *Prof Exp:* Fel entom, Univ Wis, 59; from asst prof to assoc prof, 59-72, PROF ENTOM, MICH STATE UNIV, 72- *Mem:* AAAS; Int Bee Res Asn; Int Union Study Social Insects; Entom Soc Am. *Res:* Insect behavior; apiculture. *Mailing Add:* Dept Entom Mich State Univ East Lansing MI 48824-1115

HOOPS, RICHARD ALLEN, b Ft Wayne, Ind, Mar 3, 33; m 57; c 4. SPEECH PATHOLOGY, AUDIOLOGY. *Educ:* Oberlin Col, BA, 54; Univ Ill, MS, 56, PhD(audiol, speech path), 61. *Prof Exp:* Out-patient clinician speech path & audiol, Ball State Univ, 56-63; dir audiol unit, Univ Wis, 63-64; dir, Speech, Lang & Hearing Clin, 64-76, chmn dept Speech Path & Audiol, 76-89, PROF SPEECH PATH & AUDIOL, BALL STATE UNIV, 89- *Concurrent Pos:* Fulbright lectr, Belg, 71-72. *Mem:* Am Speech & Hearing Asn. *Mailing Add:* Dept Speech Path & Audiol Ball State Univ Muncie IN 47306

HOOPS, STEPHEN C, b Denver, Colo, Nov 15, 40; m 64; c 2. ORGANIC CHEMISTRY. *Educ:* Grinnell Col, BA, 62; Univ Kans, PhD(chem), 68. *Prof Exp:* Res chemist, Union Carbide Corp, NJ, 66-68; ASST PROF CHEM, PA STATE UNIV, 68- *Mem:* Am Chem Soc. *Res:* Reaction mechanisms; structure-reactivity correlations. *Mailing Add:* 524 Woodbind Ave Oakmont PA 15139-1226

HOORNBEEK, FRANK KENT, b Oak Park, Ill, Oct 25, 28; m 53; c 2. GENETICS. *Educ:* Ore State Univ, BS, 52, MS, 62, PhD(genetics), 64. *Prof Exp:* From asst to prof zool, Univ Nh, 64-88; RETIRED. *Mem:* AAAS; Am Genetics Asn; Genetics Soc Am. *Res:* Mammalian genetics; genetics of flatfish. *Mailing Add:* 311 NE Mistletoe Ct Corvallis OR 97330

HOORY, SHLOMO, b Baghdad, Iraq, June 1, 35; Israeli citizen; m 67; c 2. MEDICAL PHYSICS. *Educ:* Hebrew Univ, Jerusalem, MSc, 63, PhD(physics), 70. *Prof Exp:* Med physicist nuclear med, Med Ctr, Hadassah Univ & Tel-Hashamer, Jerusalem, 61-63; solar physicist-in-chg, Observ, Tel-Aviv Univ, Israel, 70-74; HEALTH PHYSICS & RADIATION SAFETY OFFICER NUCLEAR MED PHYSICS, LONG ISLAND JEWISH HILL-SIDE MED CTR, 74- *Honors & Awards:* Nuclear Med & Radiopharmacol Prize, Tel-Aviv Univ, 82; Res Award, Dept Physics, Hebrew Univ, Jerasulem, 70. *Mem:* Soc Nuclear Med; Am Asn Physicists Med; AAAS; Health Physics Soc. *Res:* Multilayer imaging reconstructions in the field of nuclear medicine; computerized radioisotope inventory and patient studies with radiopharmaceuticals; computerized approach of quality assurance in nuclear medicine; internal dosimetry evaluations using computerized models; new approach in evaluating bone mineral content using gamma camera. *Mailing Add:* Four E Mill Dr Great Neck NY 11021

HOOTMAN, HARRY EDWARD, b Oak Park, Ill, June 5, 33; m 63; c 3. NUCLEAR ENGINEERING, CHEMICAL ENGINEERING. *Educ:* Mich Col Mining & Technol, BS, 59, MS, 61. *Prof Exp:* Res assoc chem eng, Argonne Nat Lab, 60-62; reactor engr hydraul, Savannah River Plant, E I du Pont de Nemours & Co, Inc, 62-65, sr engr fuel design, Reactor Eng Div, Savannah River Lab, 65-68, staff physicist reactor physics, Theoret Physics Div, 68-74, res staff engr nuclear waste mgt, Environ Effects Div, 74-80, res staff engr, Isotope Separations Div, 80-85, res assoc, Nuclear Eng Div, 85-88; ADV ENGR, NEW PROD REACTOR DEPT, WESTINGHOUSE SAVANNAH RIVER CO, 88- *Mem:* Am Nuclear Soc; Nat Soc Prof Engrs; Am Phys Soc; Sigma Xi; Am Acad Environ Engrs. *Res:* Design and development of isotope production separation processing and disposal; incineration of nuclear wastes, shielding and criticality problems with nuclear wastes; production reactor design. *Mailing Add:* Savannah River Lab Westinghouse Savannah River Co Aiken SC 29808

HOOTS, FELIX R, b Nashville, Tenn, Sept 29, 47; m 72, 81; c 2. CELESTIAL MECHANICS, MATHEMATICS. *Educ:* Tenn Technol Univ, BS, 69, MS, 71; Auburn Univ, PhD(math), 76. *Prof Exp:* MATHEMATICIAN CELESTIAL MECH, AIR FORCE SPACE COMMAND, 74- *Mem:* Am Inst Aeronaut & Astronaut. *Res:* General perturbations math models for satellite orbit prediction under the influence of Earth and third body gravitation as well as atmospheric drag; rotational motion of satellites; satellite relative motion. *Mailing Add:* 4813 Chaparral Rd Colorado Springs CO 80917

HOOVER, CHARLES WILSON, JR, b Akron, Ohio, Oct 7, 25; m 53; c 3. MANUFACTURING ENGINEERING, ELECTRICAL ENGINEERING. *Educ:* Yale Univ, BE, 46, MS, 51, PhD(physics), 54; Mass Inst Technol, BS, 47. *Prof Exp:* Mem tech staff, Bell Tell Labs Inc, 54-55, supvr memory syst develop, 55-58, supvr mil res dept, 58-60, head mil electronics res dept, 60-64, dir mil electronics res lab, 64-69, exec dir mil res div, 69-70 & commun systs res div, 70-71, exec dir systs assembly technol div, 71-82, exec dir interconnections & power technol div, 82-88; PROF & DIR MFG ENG PROG, POLYTECH UNIV, 88- *Concurrent Pos:* Chmn, Comt Tech Educ, AT&T Res & Develop Community; pres & trustee, Summit Civic Found; adv, NAE Comt Career Long Educ Engrs; co-chmn, Comt Eng Design Theory & Methodology; mem, Mfg Studies Bd, Nat Res Coun, 90- *Mem:* Am Phys Soc; fel Inst Elec & Electronics Engrs. *Res:* Digital memory and logic; optics; pulse compression radar techniques; discrimination and reentry physics; defense system design; techniques for assembly and interconnection of large electronic systems; electronic power systems; computer aided design. *Mailing Add:* 87 Tanglewood Dr Summit NJ 07901

HOOVER, DALLAS GENE, b York, Pa, Apr 20, 51; m 77; c 1. BIOTECHNOLOGY, FOOD MICROBIOLOGY. *Educ:* Elizabethtown Col, BS, 73; Univ Del, MS, 77; Univ Minn, PhD(food sci & biochem), 81. *Prof Exp:* Res assoc toxicol, Univ Minn, 80-81; asst prof food microbiol, Drexel Univ, 81-82; res assoc toxicol, Cornell Univ, 82-84; ASST PROF FOOD MICROBIOL & BIOTECHNOL, UNIV DEL, 84- *Mem:* Am Soc Microbiol; Inst Food Technologists; Int Asn Milk, Food & Environ Sanitarians. *Res:* Physiology and genetics of bacteria in the genus Pediococcus. *Mailing Add:* 36 Stallion Dr Newark DE 19713

HOOVER, DONALD BARRY, b Sunbury, Pa, July 20, 50; m 71; c 2. AUTONOMIC, NEUROSCIENCE. *Educ:* Grove City Col, BS, 72; WVa Univ, PhD(pharmacol), 76. *Prof Exp:* Res asst, Dept Pharmacol, Med Ctr, WVa Univ, 72-76; res assoc pharmacol, Lab Clin Sci, NIMH, 76-78; from asst prof to assoc prof, 78-90, PROF PHARMACOL, JAMES H QUILLEN COL MED, E TENN STATE UNIV, 90- *Concurrent Pos:* Nat Inst Gen Med Sci fel, 76-78. *Mem:* Am Soc Pharmacol & Exp Therapeut; Soc Neurosci. *Res:* Functions and pharmacology of cholinergic neurons; role of neuropeptides in regulation of the heart and coronary circulation. *Mailing Add:* Dept Pharmacol Col Med E Tenn State Univ Johnson City TN 37614

HOOVER, DONALD BRUNTON, b Cleveland, Ohio, June 17, 30; m 54. GEOPHYSICS, ELECTRICAL ENGINEERING. *Educ:* Case Inst Technol, BS, 52; Univ Mich, MSE, 53; Colo Sch Mines, DSc(geophys), 66. *Prof Exp:* Res engr, Gulf Res & Develop Co, 56-58; geophysicist, Br Crustal Studies, 60-67, GEOPHYSICIST, BR REGIONAL GEOPHYS, US GEOL SURV, 67- *Mem:* AAAS; Inst Elec & Electronics Engrs; Soc Explor Geophys; Mineral Soc Am. *Res:* Geophysical instrumentation applied to solid earth problems, particularly seismic and electrical methods. *Mailing Add:* US Geol Surv Denver Fed Ctr Bldg 25 Denver CO 80225

HOOVER, FRED WAYNE, b North Manchester, Ind, Sept 26, 14; m 46; c 4. ORGANIC CHEMISTRY. *Educ:* Manchester Col, AB, 36; Purdue Univ, MS, 38, PhD(chem), 41. *Prof Exp:* Asst chem, Purdue Univ, 36-39; res chemist, Org Chem Dept, E I du Pont de Nemours & Co, Inc, 40-68, res assoc, 68-78; RETIRED. *Mem:* Royal Soc Chem; Am Chem Soc. *Res:* Finishes, plastics and organic syntheses; membrane technology; catalysis; photochemistry. *Mailing Add:* 206 W Pembrey Dr Wilmington DE 19803-2008

HOOVER, GARY MCCLELLAN, b Falls City, Nebr, Dec 10, 39; m 62; c 3. PHYSICS. *Educ:* Univ Nebr, Lincoln, BS, 61; Kans State Univ, MS, 65, PhD(physics), 68. *Prof Exp:* Engr, Atomic Energy Div, 61-63, res physicist, 68-78, SUPVR SEISMIC MEASUREMENTS RES, RES & DEVELOP, PHILLIPS PETROL CO, 78- *Concurrent Pos:* Res assoc, Dept Physics, Kans State Univ, 68. *Mem:* Optical Soc Am; Soc Explor Geophys; Sigma Xi. *Res:* Molecular physics; infrared spectroscopy; elastic and electromagnetic wave propagation; computer data processing. *Mailing Add:* 2919 Ridge Ct Bartlesville OK 74003

HOOVER, JAMES M(YRON), b Eagle Grove, Iowa, Oct 16, 28; m 50; c 2. GEOTECHNICAL ENGINEERING, HIGHWAY ENGINEERING. *Educ:* Iowa State Univ, BS, 53, MS, 56. *Prof Exp:* Surveyor & inspector, Eng Off, Pottawattamie County, Iowa, 48-50; draftsman, Safety & Traffic Dept, State Hwy Comn, Iowa, 50-53; asst, 53-55, from instr to prof, 55-90, EMER PROF CIVIL ENG, IOWA STATE UNIV, 90- *Concurrent Pos:* Consult; mem, Comt Earthwork, Comt Low-Volume Rds, Trans Res Bd, Nat Acad Sci-Nat Res Coun, Comt Stabilization Soils, Am Rd & Transp Builders Asn, Comt D-18 soils Eng Purposes, Am Soc Testing & Mat; chmn, Stabilization Sect, Group-2 Coun. *Mem:* Am Soc Civil Engrs; Sigma Xi; Am Soc Testing & Mat; Transp Res Bd. *Res:* Geotechnical, foundation and highway engineering; soil stabilization; dust and erosion control of soils. *Mailing Add:* Dept Civil Eng Iowa State Univ Ames IA 50011

HOOVER, JOHN RUSSEL EUGENE, b New Enterprise, Pa, Jan 3, 25; m 45; c 3. MEDICINAL CHEMISTRY. *Educ:* Juniata Col, BS, 47; Univ Pa, MS, 49, PhD(org chem), 53. *Prof Exp:* Asst instr org chem, Univ Pa, 47-51, res assoc, 53-56; sr scientist natural prod, Wyeth Inst Appl Biochem, Am Home Prod Corp, 51-53; sr med chemist, 56-59, group leader, 59-61, med chem sect head, 61-67, ASSOC DIR CHEM, SMITH KLINE & FRENCH LABS, PHILADELPHIA, 67- *Mem:* AAAS; Am Chem Soc; NY Acad Sci; Sigma Xi. *Res:* Medicinal chemistry related to microbial, viral and cancer chemotherapy; heterocycles; cage compounds; antibiotics; semisynthetic penicillins and cephalospores. *Mailing Add:* 624 Crescent Ave Glenside PA 19038

HOOVER, JOHN W(ESLEY), b Demopolis, Ala, Feb 24, 16; m 39; c 2. AEROSPACE ENGINEERING. *Educ:* Ala Polytech Inst, BS, 36; Ga Inst Technol, MS, 53. *Prof Exp:* Instr physics & english, Marion Inst, 36-37; engr hwy design, Ala State Hwy Dept, 37-41; dir academics, Southern Aviation Sch, 41-44; asst proj engr, Taylorcraft Aviation Corp, 44-45; design group engr, Globe Aircraft Corp, 45-46; assoc prof aeronaut eng, Univ Ala, 46-51; from assoc prof to prof, 51-85, interim head, 56-59, asst chmn dept, 72-72 EMER PROF, AEROSPACE ENG, UNIV FLA, 85- *Concurrent Pos:* Consult, Umbaugh Aircraft, 58-60, United Fruit Co, 59-61, US Naval Air Sta, JAX, 68 & Piper Aircraft Corp, 76-78; assoc & actg dir, Fla State Technol Appln Ctr, NASA, 77-80. *Mem:* Assoc fel Am Inst Aeronaut & Astronaut; Am Soc Eng Educ; Nat Soc Prof Engrs. *Res:* Aircraft design; structures; aerodynamic loads on ground structures. *Mailing Add:* 2107 NW Fourth Pl Gainesville FL 32603

HOOVER, L RONALD, b Martinsburg, Pa, July 23, 40; m 62; c 3. COMPUTER SCIENCE, ELECTRICAL ENGINEERING. *Educ:* Shippensburg State Col, BS, 62; Wash Univ, MA, 63; Univ Mo-Rolla, PhD(elec eng), 72. *Prof Exp:* Asst prof physics, Shippensburg State Col, 64-69; fel elec eng, Univ Mo-Rolla, 70-72; mem tech staff systs eng, Bell Tel Labs, 72-77; assoc prof comput sci, Shippensburg State Col, 77-80; DIR ENG & MFG COMPUT SYSTS, AMP INC, 80- *Concurrent Pos:* Consult comput systs & process appln, 77-80. *Mem:* Inst Elec & Electronics Engrs; Am Soc Physics; Asn Comput Mach. *Res:* Application of computer systems in new control and information system environments. *Mailing Add:* AMP Inc Bldg 161-075 PO Box 3608 Harrisburg PA 17105-3608

HOOVER, LINN, geology; deceased, see previous edition for last biography

HOOVER, LORETTA WHITE, b Stamford, Tex. DIETETICS, COMPUTER SCIENCES. *Educ:* NTex State Univ, BS, 62; Tex Tech Univ, MS, 69; Univ Mo-Columbia, PhD(food systs mgt), 73, MBA, 79. *Prof Exp:* Teacher home econ, Ozona Pub Schs, Tex, 62-67; vis instr food & nutrit, Tex Tech Univ, 69; from instr to assoc prof, 72-83, PROF FOOD SYSTS MGT, UNIV MO-COLUMBIA, 83- *Concurrent Pos:* Mem ed bd, J Am Dietetic Asn, 74-77, 83- *Mem:* Am Dietetic Asn; Decision Sci Inst; Sigma Xi; Foodserv Systs Mgt Educ Coun; Am Home Econ Asn. *Res:* Computer-assisted food management systems including inventory control, production control, food cost accounting, statistical forecasting, nutrient analyses, diet patterning, patient information systems, labor cost accounting, menu management, decision support systems and financial simulation models. *Mailing Add:* 601 Thilly Ave Columbia MO 65203

HOOVER, M FREDERICK, b Pittsburgh, Pa, Apr 11, 38; m 59; c 3. ORGANIC POLYMER CHEMISTRY, PAPER CHEMISTRY. *Educ:* Carnegie-Mellon Univ, BS, 60. *Prof Exp:* Chemist, Calgon Corp, 60-61, group leader org res, 61-67, mgr polymer res, 67-70, asst dir res, 70-74; DIR TECH OPERS, VENTRON CORP, 74- *Mem:* Tech Asn Pulp & Paper Indust; Am Chem Soc; Soc Plastics Engrs; Asn Res Dirs. *Res:* Water soluble polymers, structure, synthesis, production and applications; water treatment chemicals; corrosion inhibitors, biocides, dispersants; specialty chemicals for paper, oil and detergent industries; ion exchange resins; polyelectrolytes; metal hydrides; antimicrobials; catalysts. *Mailing Add:* 86 Rowley Bridge Rd Topsfield MA 01983-2308

HOOVER, MARK DOUGLAS, b Faribault, Minn, Aug 9, 48; m 70; c 2. AEROSOL SCIENCE, NUCLEAR ENGINEERING. *Educ:* Carnegie-Mellon Univ, BS, 70; Univ NMex, MS, 75, PhD(nuclear eng), 80. *Prof Exp:* Guest researcher, Inhalation Toxicol Res Inst, 75-77; vis scientist, Inst Aerobiol, Grafschaft, WGer, 77; AEROSOL SCIENTIST, INHALATION TOXICOL RES INST, LOVELACE BIOMED & ENVIRON RES INST, 77- *Concurrent Pos:* Mem, Biomed & Environ Effects Subpanel, Interagency Nuclear Safety Rev Panel, 88-; chmn, Environ Monitoring Working Group, Beryllium Monitoring Subcomt, Dept Energy-Dept Defense Beryllium Coordinating Comt, 90-; clin asst prof, Col Pharm, Univ NMex, 91- *Mem:* Am Nuclear Soc; Am Asn Aerosol Res. *Res:* Aerosol science and technology; inhalation toxicity associated with inhaled particles; sampling and characterization of aerosol sources; design and operation of aerosol instrumentation; measurement; exposure systems. *Mailing Add:* Inhalation Toxicol Res Inst PO Box 5890 Albuquerque NM 87185-5890

HOOVER, PETER REDFIELD, b Ann Arbor, Mich, Apr 29, 39; m 69; c 1. RESOURCES FOR PALEONTOLOGIC RESEARCH. *Educ:* Univ Pittsburgh, BS, 64; Ind Univ, MA, 66; Case-Western Reserve Univ, PhD(geol), 76. *Prof Exp:* Asst dir, 77-78, DIR, ED & CUR, PALEONT RES INST, 78- *Concurrent Pos:* Fel, Smithsonian Inst, 74 & Nat Res Coun, 75. *Mem:* Paleont Asn UK; Int Paleont Asn. *Res:* Permian and Triassic brachiopods; paleontology, paleoecology and functional morphology. *Mailing Add:* Paleont Res Inst 1259 Trumansburg Rd Ithaca NY 14850

HOOVER, RICHARD BRICE, b Sikeston, Mo, Jan 3, 43; m 70. X-RAY OPTICS, SOLAR PHYSICS. *Educ:* Henderson State Univ, BS, 64. *Prof Exp:* Instr physics, Univ Ark, 65-66; physicist optics, Appl Res Div, Astrionics Lab, 66-70, space scientist x-ray optics & solar physics, Solar Physics Br, Space Sci Lab, 70-76, space scientist x-ray astron, High Energy Astrophys Br, Space Sci Lab 76-80, ASTROPHYSICIST X-RAY TELESCOPES & SOLAR X-RAY ASTRON, SOLAR PHYSICS BR, SPACE-TERRESTRIAL PHYSICS DIV, SPACE SCI LAB, MARSHALL SPACE FLIGHT CTR, 80- *Concurrent Pos:* NSF grant math, Duke Univ, 64-65; consult gamma ray imaging systs, SCI Systs, Inc, Huntsville, Ala, 69-70; consult diatom community anal water pollution monitor, Environ Sci Lab, Teledyne Brown Eng, 70-71; dir, Environ Sci Lab, UNIDEV Inc, Huntsville, 70-72; consult diatom taxon, Henri Van Heurck Mus, Royal Soc Zool, Antwerp, Belg, 75-80; pres & chmn bd, Micromega, Inc, 83-; prin scientist, UHRXS Space Sta & MSSTA Rocket Exp. *Mem:* Fel Soc Photo-Optical Instrumentation Engrs. *Res:* X-ray optics, especially development of glancing incidence and multilayer x-ray telescope and microscope systems; solar physics, especially rapid changes in x-ray flare features; high energy astrophysics; micropaleontology, especially study of morphology and taxonomy of fossil & marine diatoms of the genera Entogonia and Actinoptychus; multiplate gamma ray collimator and multiplate focussing gamma ray collimator; develops advanced x-ray imaging systems; authored over 100 articles and 14 patents on x-ray telescopes and microscopes; international authority on x-ray optics and diatoms; produced diatom display for Smithsonian Museum of Natural History; author of a diatom article. *Mailing Add:* Marshall Space Flight Ctr Mail Code ES 52 Marshall Space Flight Ctr AL 35812

HOOVER, THOMAS BURDETT, b Wellsville, Pa, Jan 17, 20; m 45; c 2. ENVIRONMENTAL CHEMISTRY, ELECTROANALYTICAL CHEMISTRY. *Educ:* Pa State Univ, BS, 42, MS, 46, PhD(chem), 60; Am Inst Chemists, cert, 84. *Prof Exp:* Instr, Pa State Univ, 42-44; chemist, Union Carbide Nuclear Co Div, Union Carbide Corp, 48-52; sr chemist, Appl Sci Labs, Inc, 53-60; chemist, Nat Bur Standards, 60-70; res chemist, Environ Protection Agency, 70-85; RETIRED. *Mem:* AAAS; Am Inst Chem; Am Chem Soc; Am Water Works Asn. *Res:* Multielement analysis of water; electrochemical analysis. *Mailing Add:* 110 Richard Way Athens GA 30605

HOOVER, THOMAS EARL, b Temple, Tex, May 3, 41. ENERGY TECHNOLOGY. *Educ:* Univ Tex, BA, 62; Tex A&M Univ, PhD(marine chem), 66. *Prof Exp:* Sr res scientist, United Aircraft Res Labs, Conn, 67-70; DIR ENERGY TECHNOL, ADVAN TECHNOL, PARSONS, BRINCKERHOFF, QUADE & DOUGLAS, 70- *Concurrent Pos:* Environ consult; mem, Geothermal Resources Coun. *Mem:* AAAS; Am Geophys Union; Am Chem Soc; Am Soc Oceanog; Am Nuclear Soc. *Res:* Evaluation of new applications for advanced technology; gas exchange and heat exchange across liquid surface; use of computer modeling to evaluate river pollution; computer modeling of subway environment; long-term management of nuclear waste; ocean thermal energy conversion; geothermal energy direct-use application. *Mailing Add:* 93 Bedford St New York NY 10014

HOOVER, WILLIAM G(EORGE), b Pueblo, Colo, Nov 10, 06. ELECTRICAL ENGINEERING. *Educ:* Stanford Univ, AB, 28, EE, 29, PhD, 36. *Prof Exp:* With Westinghouse Elec & Mfg Co, 29-30, control engr, 30-31; from instr to prof elec eng, 32-59, LECTR, STANFORD UNIV, 59-; TECH DIR, GRANGER ASSOCS, 59- *Concurrent Pos:* Res engr, Pac Elec Mfg Corp, 36- *Mem:* Fel Inst Elec & Electronics Engrs. *Res:* Radio communication equipment; high-power radio frequency transformers; antennas; transmitters; aviation precipitation static discharge equipment. *Mailing Add:* 13820 La Paloma Rd Los Altos CA 94022

HOOVER, WILLIAM GRAHAM, b Boston, Mass, Apr 18, 36; div; c 2. CHEMICAL PHYSICS. *Educ:* Oberlin Col, AB, 58; Univ Mich, MS, 60, PhD(chem), 61. *Prof Exp:* Alfred P Sloan Found fel, Duke Univ, 61-62; instr, 66-73, PROF APPL SCI, UNIV CALIF, DAVIS, 74-; PHYSICIST, LAWRENCE LIVERMORE LAB, UNIV CALIF, 62- *Concurrent Pos:* Fulbright-Hays fel, Australian Nat Univ, 77-78; vis prof, Univ Australia, Univ Rome & Univ Vienna, 85, Keio Univ, Japan, 90-91. *Mem:* Fel Am Phys Soc. *Res:* Statistical mechanics; equilibrium and nonequilibrium molecular dynamics; solid phase cell models; irreversibility and time's arrow; phase diagrams; fracture; plasticity; transport and hydrodynamics. *Mailing Add:* PO Box 808 L-794 Livermore CA 94551

HOOVER, WILLIAM JAY, b Champaign, Ill, Mar 26, 28; m 50; c 3. FOOD SCIENCE. *Educ:* Univ Ill, BS, 50, MS, 54, PhD(food technol), 63. *Prof Exp:* Food technologist, Qm Food & Container Inst, 51-52; asst, Univ Ill, 52-53; asst to dir, Refrig Res Found, 53-56; mgr tech serv, Corn Indust Res Found, Inc, 56-62, admin vpres, 62-66; dir, Food & Feed Grain Inst & head, Dept Grain Sci & Indust, Kans State Univ, 66-76; PRES, AM INST BAKING, MANHATTAN, KANS, 76- *Mem:* Am Asn Cereal Chemists; Inst Food Technologists. *Res:* Research and development on grains, pulses and oilseeds and their products. *Mailing Add:* Am Inst Baking 1213 Bakers Way Manhattan KS 66502

HOOVER, WILLIAM L, b Joplin, Mo, June 20, 24; m 45; c 4. INORGANIC CHEMISTRY, SOIL CHEMISTRY. *Educ:* Tex A&M Univ, BS, 50, MS, 64, PhD(soil chem), 66. *Prof Exp:* Teacher, High Sch, Tex, 52-61; head fertilizer sect, Agr Anal Serv, 65-68, RES SCIENTIST, RES FOUND, TEX A&M UNIV, 63-, ACTG HEAD DEPT AGR CHEM, 66-, HEAD, AGR ANALYTICAL SERV, 68-, ASSOC PROF SOIL SCI, 74- *Concurrent Pos:* Staff mem, NSF Acad Yr Inst, Univ Tex, 57-58. *Mem:* AAAS; Am Chem Soc; Am Soc Agron; Soil Sci Soc Am. *Res:* Adsorption of organic chemicals on soil clays and the fixation of zinc and other micronutrients in soils; analytical methods by atomic absorption. *Mailing Add:* 1606 Feather Run College Station TX 77840

HOOVER, WILLIAM RINGO, metallurgy, materials science, for more information see previous edition

HOOZ, JOHN, b Brooklyn, NY, July 5, 35. ORGANIC CHEMISTRY. *Educ:* Brooklyn Col, BS, 56; Purdue Univ, PhD(org chem), 63. *Prof Exp:* NIH fel, Stanford Univ, 64-65; from asst prof to assoc prof, 65-75, PROF CHEM, UNIV ALTA, 75-, ASSOC DEAN SCI, 81- *Mem:* AAAS; Am Chem Soc; Royal Soc Chem; Chem Inst Can. *Res:* Exploratory synthetic organic chemistry; development of novel synthetic methods. *Mailing Add:* Dept Chem Univ Alta Edmonton AB T6G 2G2 Can

HOPCROFT, JOHN E(DWARD), b Seattle, Wash, Oct 7, 39; m 64; c 3. COMPUTER SCIENCES THEORY. *Educ:* Seattle Univ, BS, 61; Stanford Univ, MS, 62, PhD(elec eng), 64. *Hon Degrees:* DHH, Seattle Univ. *Prof Exp:* Asst prof theory of comput, Princeton Univ, 64-67; assoc prof, 67-71, PROF THEORY COMPUT, CORNELL UNIV, 72- *Concurrent Pos:* Assoc ed, J Comput & Systs Sci, 80-; mem, Comput & Technol Bd, Nat Res Coun, 80-; consult, IBM, 81- *Mem:* Asn Comput Mach; Inst Elec & Electronics Engrs; Soc Indust & Appl Math. *Res:* Theoretical computer science. *Mailing Add:* Dept Comput Sci Cornell Univ Ithaca NY 14853

HOPE, BRIAN BRADSHAW, b Preston, Eng, Apr 21, 36; Can citizen; m 64; c 2. CIVIL ENGINEERING, CONCRETE MATERIALS. *Educ:* Victoria Univ Manchester, BSc, 57; Queen's Univ, Ont, MSc, 59, PhD(civil eng), 62. *Prof Exp:* Structural specialist, H G Acres & Co Ltd, 61-65; asst prof civil eng, Univ Calgary, 65-67; assoc prof, 67-76, PROF CIVIL ENG, QUEEN'S UNIV, ONT, 76- *Honors & Awards:* Duggan Medal, Eng Inst Can, 60. *Mem:* Am Concrete Inst. *Res:* Properties of concrete and polymer impregnated concrete; concrete deterioration; corrosion of reinforcing steel; structural engineering. *Mailing Add:* Dept Civil Eng Queen's Univ Kingston ON K7L 3N6 Can

HOPE, ELIZABETH GREELEY, b Arlington, Mass, Jan 1, 43; m 85; c 5. SEISMOLOGY. *Educ:* Emmanuel Col, BS, 64. *Prof Exp:* Software analyst, Blue Cross-Blue Shield, 65; tech rep, 79-80, field supvr, 80-81, area mgr, 81-83, VPRES, PHILIP R BERGER & ASSOCS, 83- *Concurrent Pos:* Dir, Soc Explosives Engrs, 80-; ed, Leadline, Soc Explosives Engrs, 82-; mem, Div Pub Safety, Blasting Rev Bd, Commonwealth Mass, 85- *Mem:* Soc Explosives Engrs. *Res:* Theory of seismic vibrations; selective attenuation of wave components; correlation studies. *Mailing Add:* 7-5 Apple Ridge Rd Maynard MA 01754

HOPE, GEORGE MARION, b Waycross, Ga, Jan 24, 38; m 56; c 1. VISION, NEUROSCIENCE. *Educ:* Mercer Univ, AB, 65; Univ Fla, MA, 67, PhD(physiol psychol), 71. *Prof Exp:* Res asst vision, dept psychol, Univ Fla, 66-67 & dept ophthal, Wash Univ, 70-72; from instr to asst prof ophthal, Univ Louisville, 72-79, co-dir, Low Vision Clin, 73-79; ASSOC RES SCIENTIST, DEPT OPHTHAL, COL MED, UNIV FLA, 80-, DIR, LOW VISION SERV, EYE CTR, 80- & CTR LOW VISION, 81- *Concurrent Pos:* dir, Placement Serv, Asn Res Vision & Ophthal, 72-84; investr, numerous res contracts & grants, 74-; consult, Defense Div, Brunswick Corp, Deland, Fla, 80-81; mem, low vision consults comt, Div Blind Serv, State of Fla, 84- *Mem:* AAAS; Asn Res Vision & Ophthal; Sigma Xi. *Res:* Neuroanatomy; neurophysiology; function of the normal and abnormal vertebrate visual system. *Mailing Add:* Dept Ophthal Col Med Univ Fla Box J-284 JHMHC Gainesville FL 32610

HOPE, HAKON, b Forde, Norway, Dec 15, 30; m 57; c 3. X-RAY CRYSTALLOGRAPHY. *Educ:* Univ Oslo, Cand Mag, 54, Cand Real(chem), 58. *Prof Exp:* Res fel chem, Univ Oslo, 58-60, lectr, 61-65; from asst prof to assoc prof, 65-73, PROF CHEM, UNIV CALIF, DAVIS, 73- *Concurrent Pos:* Asst res chemist, Univ Calif, Los Angeles, 61-63; Fulbright travel grant, 61-63; NSF grants, 66-; US-Israel. *Mem:* Am Crystallog Asn; Norweg Chem Soc. *Res:* Determination of crystal structures; diffraction methods; biological macromolecules; low temperature methods; co-ed, Acta Cyst, 84. *Mailing Add:* Dept Chem Univ Calif Davis CA 95616

HOPE, HUGH JOHNSON, b Ottawa, Ont, Sept 12, 38; m 74; c 2. LOW TEMPERATURE TOLERANCE, WINTER SURVIVAL. *Educ:* Carleton Univ, BSc, 60, MSc, 63; Dalhousie Univ, PhD(plant biochem), 68. *Prof Exp:* Fel, Univ Calif, Riverside, 68-70; RES SCIENTIST, CAN DEPT AGR, 70- *Mem:* Phytochem Soc NAm; Am Soc Cryobiol; Can Soc Plant Physiologists; Am Soc Plant Physiologists; Am Soc Agron. *Res:* Identification and improvement of cold tolerance during germination and early growth in Zea mays (corn); protein metabolism as related to acquisition of cold tolerance and ice tolerance in winter wheat and forage plants; improvement of forage quality in timothy by selection for reduced protein losses. *Mailing Add:* Plant Res Ctr KW Neatby Bldg Agr Can Cent Exp Farm Ottawa ON K1A 0C6 Can

HOPE, LAWRENCE LATIMER, b New York, NY, Dec 28, 39; m 85; c 1. ELECTROLUMINESCENCE, ELECTRON DEVICES. *Educ:* Queens Col, BS, 61; Stevens Inst Technol, MS, 63, PhD(physics), 66; Mass Inst Technol, SM, 76. *Prof Exp:* Physicist, Gen Tel & Electronics Labs, Inc, Waltham, Mass, 66-76, PROG MGR, GTE PRODS CORP, 76- *Concurrent Pos:* Lectr, Hofstra Univ, 68-72. *Mem:* AAAS; Am Phys Soc; Sigma Xi; Inst Elec & Electronics Engrs; Soc Info Display. *Res:* Thin film electroluminescence; optical systems. *Mailing Add:* GTE Sylvania Lighting 60 Boston St Salem MA 01970

HOPE, RONALD RICHMOND, b Melbourne, Australia, Aug 12, 43; m 66; c 5. CARDIOLOGY. *Educ:* Univ Melbourne, MB & BS, 66; FRACP, 75; Am Bd Internal Med, dipl. *Prof Exp:* Jr resident med officer, Prince Henry's Hosp, Melbourne, 67-68, sr resident med officer, 68-69, registr gen med, 69-70; registr respiratory intensive care med & gen med, Dunedin Pub Hosp, NZ & asst lectr med, Otago Univ, 70-71, fel cardiol & lectr, Otago Univ, 71-72; house officer, Royal Children's Hosp, Melbourne, 72; fel pediat cardiol, Hosp Sick Children, Toronto, Ont, 72-73; fel adult cardiol, Mt Sinai Med Ctr, Miami Beach, Fla, 73-74; specialist adult & pediat cardiol, Waikato Hosp, Hamilton, NZ, 74-75; asst prof med, Sch Med, Univ Miami, 77-78; assoc prof, 78-80, CLIN ASSOC PROF MED, COL MED, UNIV OKLA, 80- *Mem:* Am Physiol Soc; Am Geriat Soc; Cardiac Soc Australasia. *Res:* Electrophysiology of ventricular arrhythmias in the context of myocardial infarction. *Mailing Add:* Dept Cardiol St Anthony's Hosp 826 NW 11th Oklahoma City OK 73103

HOPE, WILLIAM DUANE, b Ft Collins, Colo, June 7, 35; div; c 3. NEMATOLOGY, SYSTEMATIC ZOOLOGY. *Educ:* Colo State Univ, BS, 57, MS, 60; Univ Calif, PhD(nematol), 65. *Prof Exp:* Assoc cur, 64-72, chmn dept, 76-81, CUR DEPT INVERT ZOOL, MUS NATURAL HIST, SMITHSONIAN INST, 72- *Concurrent Pos:* Res Award, Smithsonian Inst, 66-67 & 78-79; fel parasitol, Univ Toronto, 67-68; chmn, Int Asn Meiobenthologists & ed Newsletter, 68-69; assoc ed, J Nematol, 70-75. *Mem:* Am Micros Soc; Int Asn Meiobenthologists; Soc Nematologists; Soc Syst Zool; Sigma Xi; Am Asn Zool Nomenclature (pres, 90-91). *Res:* Comparative and functional morphology, ultrastructure and systematics of marine nematodes. *Mailing Add:* Dept Invert Zool Nat Mus Natural Hist Washington DC 20560

HOPEN, HERBERT, b Madison, Wis, Jan 7, 34; m 59; c 1. HORTICULTURE. *Educ:* Univ Wis, BS, 56, MS, 59; Mich State Univ, PhD(hort), 62. *Prof Exp:* Asst prof agr, Univ Minn, 62-64; head veg crops div, 71-72, asst prof, 65-78, PROF HORT, UNIV ILL, URBANA, 78- *Concurrent Pos:* Vis assoc prof, Univ Ariz, 72-73; vis fel, Agr Univ Norway, 77. *Honors & Awards:* Campbell Award, Am Phytopath Soc, 80. *Mem:* Weed Sci Soc Am; Am Soc Hort Sci; Int Soc Hort Sci. *Res:* Chemical and biological weed control. *Mailing Add:* Dept Hort 385 Hort Univ Wis 1575 Linden Dr Madison WI 53706

HOPENFIELD, JORAM, b Warsaw, Poland, Jan 6, 34; US citizen; m 63; c 3. MECHANICAL ENGINEERING, TECHNICAL MANAGEMENT. *Educ:* Univ Calif, Los Angeles, BS, 60, MS, 62, PhD(mech eng), 68. *Prof Exp:* Mem tech staff mech eng, Atomics Int Div, NAm Rockwell Corp, 62-71; reactor engr, Div Reactor Develop & Technol, US AEC, 71-77; res engr, Magnetohydrodyn Div, Dept Energy, 77-82; NUCLEAR ENGR, NUCLEAR REGULATORY COMN, 82- *Concurrent Pos:* Lectr, Automotive Diesels, Montgomery Col, Md. *Honors & Awards:* Blackall Award, Am Soc Mech Engrs, 67. *Res:* Heat transfer; mass transfer; electrochemical machining; invented a method for the on-line detection of pipe thinning due to corrosion, patent pending. *Mailing Add:* 1724 Yale Pl Rockville MD 20850

HOPF, FREDERIC A, optics; deceased, see previous edition for last biography

HOPFENBERG, HAROLD BRUCE, b New York, NY, Aug 28, 38; m 59; c 2. CHEMICAL ENGINEERING, POLYMER SCIENCE. *Educ:* Mass Inst Technol, SB, 60, SM, 61, PhD(chem eng), 65. *Prof Exp:* From asst to prof chem eng, NC State Univ, 67-80, head chem eng dept, 80-87, assoc dean, Col Eng, 87-90, CAMILLE DREYFUS PROF, NC STATE UNIV, 80-, EXEC ASST TO CHANCELLOR, 90- *Concurrent Pos:* Grants, NSF & Army Res Off; consult, B Alza Corp, 74-, Am Can Co, 76-, Hoechst-Celanese, 81-, Hercules Inc, 83-, Duracell Inc, 84-, Grace Sierra Chem, 85-, Biosys, 86-; sr vis fel, Sci Res Coun, UK & vis fel, Clare Hall, Univ Cambridge, 77; res grants, Owens Ill, Inc, Alza Corp, E I du Pont Plastics Inst Am, Alcoa Corp & Hoechst-Celanese; vis prof, Univ Bologna, 82 & 84; vis prof, Appalachian State Univ, 85, Univ Naples, 89. *Honors & Awards:* Alcoa Found Eng Awards, 78 & 80; R J Reynolds Industs Award, 85. *Mem:* Am Inst Chem Engrs; Am Chem Soc. *Res:* Transport of small molecules in polymeric materials; controlled release; barrier plastics; migration and relaxation in glassy polymers; complex-biased separation processes. *Mailing Add:* Dept Chem Eng NC State Univ Box 7006 Raleigh NC 27695-7006

HOPFER, ROY L, b Guthrie, Okla, Mar 20, 44; m 74; c 3. MEDICAL MYCOLOGY, MEDICAL MICROBIOLOGY. *Educ:* Okla State Univ, BS, 67, PhD(microbiol), 72. *Prof Exp:* Fel med mycol, Skin & Cancer Hosp, Temple Univ, 72-74; asst microbiologist, 74-81, asst prof, 76-81, assoc microbiologist, assoc prof lab med microbiol & chief sect microbiol, Univ Tex M D Anderson Hosp & Tumor Inst, 81-87; ASSOC PROF, UNIV TEX GRAD SCH BIOMED SCI, 82-, UNIV TEX MED SCH, HOUSTON, 82- *Mem:* Am Soc Microbiol; Am Bd Med Microbiol; Med Mycol Soc Am; Sigma Xi. *Res:* Host-parasite relationships of fungi; infections in the immunosuppressed patient; diagnosis of systemic candidiasis in the cancer patient. *Mailing Add:* 202 Somerset Dr Chapel Hill NC 27514-2413

HOPFER, SAMUEL, b Rexingen, Ger, Nov 21, 14; nat US; m 42; c 4. PHYSICS. *Educ:* WVa Univ, BA, 44; Cornell Univ, MA, 46; Polytech Inst Brooklyn, PhD(physics), 54. *Prof Exp:* Mgr microwave res, PRD Electronics Co, 46-63; VPRES & DIR RES & DEVELOP, GEN MICROWAVE CORP, 63- *Concurrent Pos:* Adj prof, Polytech Inst Brooklyn, 56-65. *Mem:* Fel Inst Elec & Electronics Engrs. *Res:* Microwave broadband transmission systems, multimode coupling structures; instrumentation microwave power, radiation and impedance measurement; broadband solid-state switching, attenuation, modulation components; microstrip hybrids and microwave integrated circuit technology. *Mailing Add:* 17 Rechov Casuto Jerusalem Israel

HOPFER, ULRICH, b Burg, Ger, Apr 7, 39; m 63; c 2. CELL BIOLOGY. *Educ:* Univ Gottingen, MD, 66; Johns Hopkins Univ, PhD(biochem), 70. *Prof Exp:* Fel biochem, Mass Gen Hosp, Harvard Med Sch, 70-71, res assoc, 71; dozent, Swiss Fed Inst Technol, 72-74; from asst prof to assoc prof anat, 74-83, prof develop genetics & anat, 83-88, PROF PHYSIOL & BIOPHYS, CASE WESTERN RESERVE UNIV, 86- *Concurrent Pos:* Vis scientist, Max Planck Inst Biophys, Frankfurt, Fed Repub Ger, 80-81. *Honors & Awards:* Hoffmann-LaRoche Prize, Am Phys Soc, 83. *Mem:* AAAS; Am Soc Biol Chemists; Biophys Soc Am; Am Physiol Soc; Brit Biochem Soc; Soc Gen Physiol. *Res:* Structure and function of biomembranes; solute transport across plasma membranes; epithelial transport. *Mailing Add:* Dept Physiol & Biophys Sch Med 2119 Abington Rd Cleveland OH 44106

HOPFIELD, JOHN JOSEPH, b Chicago, Ill, July 15, 33; m 54; c 3. BIOPHYSICS, COMPUTER SCIENCE. *Educ:* Swarthmore Col, BA, 54; Cornell Univ, PhD(physics), 58. *Prof Exp:* Mem tech staff, AT&T Bell Tel Labs, NJ, 58-60 & 73-90; vis res physicist, Ecole Normale Superieure, Paris, 60-61; from asst prof to assoc prof physics, Univ Calif, Berkeley, 61-64; prof physics, Princeton Univ, 64-80; ROSCOE G DICKINSON PROF CHEM & BIOL, CALIF INST TECHNOL, 80- *Concurrent Pos:* Sloan Found fel, 62-64; Guggenheim fel, 69; mem, Neurosci Res Prog; MacArthur Prize fel, 83-88; trustee, Battelle Mem Inst, 83-, Harvey Mudd Col, 90- *Honors & Awards:* Oliver E Buckley Prize, Am Phys Soc, 69; Biophys Prize, Am Phys Soc, 85; Michelson-Morley Award, 88; Dudley Wright Prize, 89. *Mem:* Nat Acad Sci; Am Acad Arts & Sci; Biophys Soc; Am Phys Soc; Am Chem Soc; Am Philos Soc. *Res:* Electron transfer and chemical reactions in biological systems; accuracy and proofreading in biology; the physics of biological molecules and processes; collective aspects of neurobiology. *Mailing Add:* Crellin Lab 164-30 Calif Inst Technol Pasadena CA 91125

HOPFINGER, J ANTHONY, b Salem Ore, Sept 13, 51. CALCIUM NUTRITION, FRUIT QUALITY. *Educ:* Univ Calif, BS, 73, MS, 75; Wash State Univ, PhD(hort), 78. *Prof Exp:* Asst prof hort, Univ Mo, Columbia, 78-82, systs oper officer, analyst & programmer, Univ Mo Hosp & Clins, 82-84; SPECIALIST POMOLOGY EXTEN RES, AGR EXP STA, RUTGERS COOP EXTEN RUTGERS UNIV, NJ, 84- *Concurrent Pos:* Consult, Laura Exports USA, Inc, 87-, Blank, Rome, Cominsky & McCauley, Esqs, Philadelphia, 85-, Chubb Ins Co NAm, 86-; chmn, Nat Peach Coun, 88-89. *Mem:* Am Soc Hort Sci; Am Pomological Soc; Sigma Xi. *Res:* Whole plant nutrition (especially calcium) and its effect on post harvest fruit quality of apples and peaches; peach skin discoloration (inking); development of maturity standards for New Jersey peach industry; physiological disorders of apples and peaches. *Mailing Add:* Creamridge Fruit Res Ctr Rutgers Univ Creamridge NJ 08514

HOPKE, PHILIP KARL, b Sherman, Tex, Mar 22, 44; m 68; c 2. ENVIRONMENTAL CHEMISTRY, NUCLEAR CHEMISTRY. *Educ:* Trinity Col, Conn, BS, 65; Princeton Univ, MA, 67, PhD(chem), 69. *Prof Exp:* Res assoc nuclear chem, Mass Inst Technol, 69-70; asst prof chem, State Univ NY Col Fredonia, 70-74; from asst prof to prof, environ chem, Univ Ill, Urbana, 74- 89; ROBERT A PLANE PROF CHEM, CLARKSON UNIV, POTSDAM, 89- *Concurrent Pos:* Instr, Metrop Col, Boston Univ, 70; consult, Consultec Sci Inc, R J Lee Group, US Environ Protection Agency, Environ Sci Res Lab & Grants Prog, US Army Const Eng Res Lab; prin investr projs, US Environ Protection Agency, US Dept Energy, Electronic Power Res Inst, NSF, Environ Ont. *Mem:* AAAS; Am Chem Soc; Am Phys Soc; Air Pollution Control Asn; Am Asn Advan Aerosol Res; Geselshaft fur Aerosolforschung. *Res:* Chemical and physical behavior of radon and its decay products, development and application of methods of particle sizing for aerosol submicron particles; statistical analysis of enviromental data sets using factor regression, and cluster analysis techniques; studies of the transport and fate of toxic materials, including attachment to airborne particles; archaeological provenance studies using multi elemental analysis and pattern recognition methods. *Mailing Add:* Dept Chem Clarkson Univ Potsdam NY 13699-5810

HOPKE, PHILIP KARL, b Sherman, Tex, Mar 22, 44; m 68; c 2. RADON & RADON PROGENY BEHAVIOR, RECEPTOR MODELING. *Educ:* Trinity Col, Hartford, BS, 65; Princeton Univ, MA, 67, PhD(chem), 69. *Prof Exp:* Res assoc, Mass Inst Technol, 69-70; asst prof chem, State Univ Col, Fredonia, NY, 70-74; from asst prof to prof environ chem, Univ Ill, Urbana-Champaign, 75-89; ROBERT A PLANE PROF CHEM, CLARKSON UNIV, 89- *Concurrent Pos:* Vis asst prof chem, Univ Ill, Urbana-Champaign, 74-75; vis prof, Vrije Univ Brussels, Belg, 84-85; ed, Chemomet & Intel Lab Systs, 86-, assoc ed, Health Physics, 90-; chmn, Sci Rev Panel Air Chem & Physics, US Environ Protection Agency, 87-; mem, Comt Advan Assessing Human Exposure Airborne Pollutants, Nat Acad Sci, Nat Res Coun, 89-90, Comt Dosimetric Extrapolations Beir IV Risk Estimates Gen Pub, 89-91, Comt Assess Nat Inst Standards & Technol Lab Chem Sci & Technol, 91 & Task Force Air Toxics, Am Chem Soc, 89-; dir, Am Asn Aerosol Res, 89-94; chair, Tech Comt EM2, Air & Waste Mgt Asn, 90-92. *Mem:* Sigma Xi; fel Am Inst Chemists; Am Asn Aerosol Res; Am Chem Soc; Am Phys Soc; Air & Waste Mgt Asn; Health Physics Soc. *Res:* Atmospheric of radon and its progeny; measurement of environmental radioactivity; sampling, physical, and chemical characterization of airborne particles; multivariate statistical and pattern recognition analysis of environmental data sets; source-receptor relationships for airborne pollutants. *Mailing Add:* Dept Chem Clarkson Univ Potsdam NY 13676

HOPKIN, A(RTHUR) M(CMURRIN), b Burley, Idaho, Feb 25, 19; m 44; c 5. ELECTRICAL ENGINEERING. *Educ:* Ga Inst Technol, BS, 42; Northwestern Univ, MS, 47, PhD(elec eng), 50. *Prof Exp:* From instr to asst prof elec eng, Northwestern Univ, 46-54; assoc prof, 54-60, PROF ELEC ENG & COMPUT SCI, UNIV CALIF, BERKELEY, 60- *Mem:* Inst Elec & Electronics Engrs. *Res:* Rotating electrical machines; control systems and components. *Mailing Add:* Univ Calif 2200 University Ave Berkeley CA 94720

HOPKINS, ALAN KEITH, optical systems engineering, for more information see previous edition

HOPKINS, ALBERT LAFAYETTE, JR, b Chicago, Ill, May 6, 31; m 52; c 4. COMPUTER ENGINEERING. *Educ:* Harvard Univ, AB, 53, AM, 54, PhD, 57. *Prof Exp:* Instr control syst eng, Harvard Univ, 57-60; staff engr instrumentation lab, Mass Inst Technol, 60-70, assoc prof aeronaut & astronaut, 70-76; staff mem, C S Draper Lab, 76-81; pres, 81-84, DIR, TECHNOL DEPT, ITP BOSTON, INC, 84- *Honors & Awards:* Info Systs Award, Am Inst Aeronaut & Astronaut, 77. *Mem:* AAAS; Inst Elec & Electronics Engrs. *Res:* Fault-tolerant computing and control systems; computer integrated manufacturing systems. *Mailing Add:* 30 Spinelli Pl Cambridge MA 02138

HOPKINS, ALLEN JOHN, b Liverpool, Eng, Feb 8, 42; m 65; c 2. POLYMER CHEMISTRY. *Educ:* Liverpool Col Technol, BSc, 64; Univ Manchester Inst Sci & Technol, PhD(polymer chem), 68; Furman Univ, MBA, 76. *Prof Exp:* Res asst radiochem, UK Atomic Energy Authority, 64-65 & Unilever Corp, 65-66; res chemist fiber chem, Phillips Petrol Co, 68-70, res chemist fiber technol, Phillips Fibers Corp, 70-77; res chemist, Hoechst Fiber Industs, 77-83, GROUP LEADER PROCESS ASSISTANCE, HOECHST CELANESE CORP, 83-87. *Mem:* Am Asn Textile Chemists & Colorists. *Res:* Polymer adsorption studies; synthetic fiber technology particularly with respect to texturing of partially drawn polyester yarn. *Mailing Add:* 119 Continental Dr Greenville SC 29615

HOPKINS, AMOS LAWRENCE, b Boston, Mass, Jan 27, 26; m 49, 70; c 3. PHYSIOLOGY. *Educ:* Harvard Univ, BA, 48; Wash Univ, PhD(zool), 55. *Prof Exp:* From instr to asst prof, 55-66, ASSOC PROF ANAT, SCH MED, CASE WESTERN RESERVE UNIV, 66- *Mem:* NY Acad Sci; Soc Magnetic Resonance Med. *Res:* Membrane oxygenator; dielectric behavior of bound water; nuclear magnetic resonance of living muscle; nuclear magnetic resonance of human cerebrospinal fluid in vivo; oxygen-17 as a proton MRI contrast agent in cerebral blood flow studies. *Mailing Add:* Dept Anat Case West Reserve Univ Sch Med Cleveland OH 44106

HOPKINS, BETTY JO HENDERSON, b Harlan, Ky; m 57; c 3. RADIATION BIOLOGY, EXPERIMENTAL EMBRYOLOGY. *Educ:* Ky State Col, BS, 55; Univ Rochester, MS, 57, PhD(radiation biol), 61. *Prof Exp:* Res instr radiation biol, Univ Rochester, 61-65; from asst prof to assoc prof biol, 65-74, asst dean natural sci, 75-84, PROF BIOL, MONROE COMMUNITY COL, 74-, DEAN, NATURAL SCI, 84- *Concurrent Pos:* Fac res fel, State Univ NY, 71. *Mem:* AAAS; NY Acad Sci; Radiation Res Soc; Health Physics Soc; Am Inst Biol Sci; Am Chem Soc; Am Asn Higher Educ. *Res:* Retention, toxicity and pathologic effects of radiostrontium in the rat; radiation toxicity and oxygen effect mechanism in the chick embryo. *Mailing Add:* Dept Biol Monroe Community Col 1000 E Henrietta Rd Rochester NY 14623

HOPKINS, CARL DOUGLAS, b Rochester, NY, Apr 8, 44; m 70; c 2. ANIMAL BEHAVIOR. *Educ:* Bowdoin Col, AB, 66; Rockefeller Univ, PhD(behav sci), 72. *Prof Exp:* Fel neurosci, Univ Calif, San Diego, 72-73; asst prof ecol & behav biol, Univ Minn, Minneapolis, 73-78, assoc prof, 78-; AT NEUROL-BEHAV DEPT, CORNELL UNIV, ITHACA, NY. *Mem:* Sigma Xi; Animal Behav Soc; Am Physiol Soc; Am Soc Ichthyologists & Herpetologists. *Res:* Evolution of animal communication; electric communication among fish; sensory filtering and species recognition. *Mailing Add:* Neurobiol & Behav Dept Cornell Univ Ithaca NY 14853

HOPKINS, CHARLES B(EVERLEY), JR, b Birmingham, Ala, Aug 15, 22; m 48; c 4. CHEMICAL ENGINEERING. *Educ:* Ala Polytech Inst, BS, 43; Univ Wyo, MS, 54. *Prof Exp:* Anal chemist, Cities Serv Refining Corp, La, 43-46; pyrometrist, Tenn Coal, Iron & RR Co, Ala, 47-49; from chemist to chem engr, petrol & oil shale exp sta, US Bur Mines, Wyo, 49-56; res engr, Cities Serv Res & Develop, La, 56-60; chem engr, Chem Res & Develop Ctr, FMC Corp, 60-74, environ engr, 74-85; RETIRED. *Mem:* AAAS; Am Chem Soc. *Res:* Petroleum refining; catalytic processing; properties of porous solids; synthetic fuels; coal carbonization; reactivity of carbons; active carbon; water treating. *Mailing Add:* 11 Rosetree Lane Lawrenceville NJ 08648

HOPKINS, CLARENCE YARDLEY, b Kinmount, Ont, June 20, 03; wid; c 1. ORGANIC CHEMISTRY, PHYTOCHEMISTRY. *Educ:* Queen's Univ, Ont, BA, 24, MA, 26; NY Univ, PhD(org chem), 29. *Prof Exp:* Demonstr chem eng, Univ Toronto, 29-30; from asst res chemist to sr res chemist, Nat Res Coun Can, 30-68; RETIRED. *Concurrent Pos:* Lectr, Carleton Univ, 45-49; consult, 68-70; writer, 65-90. *Mem:* Am Oil Chem Soc; fel Chem Inst Can. *Res:* Protective coatings; oils and fats; novel fatty acids of seed oils; other natural products. *Mailing Add:* 180 Carleton St Ottawa ON K1M 0G7 Can

HOPKINS, COLIN RUSSELL, b Wales, June 4, 39; m 64; c 2. CELL BIOLOGY. *Educ:* Univ Wales, BSc, 61, PhD(biol), 64. *Prof Exp:* Lectr cell biol, Med Sch, Liverpoole Univ, 64-70, prof, 70-85; prof cell biol, Imp Col London, 86-90; PROF CELL BIOL, UNIV COL LONDON, 90-, DIR BIOL/MED, MRC LAB MOLECULAR CELL BIOL, 90- *Concurrent Pos:* Vis prof, Rockefeller Univ, 69-71. *Mem:* Am Soc Cell Biol; Brit Soc Cell Biol. *Res:* Cell biology; drug delivery; uptake and intracellular processing of cell surface receptors. *Mailing Add:* Lab Molecular Cell Biol Univ Col London Gowerst London W CIE 6BT England

HOPKINS, DANIEL T, b Peterborough, NH, Oct 3, 32; m 52; c 3. NUTRITIONAL BIOCHEMISTRY. *Educ:* Cornell Univ, BS, 56, MNS, 58, PhD(animal nutrit), 63. *Prof Exp:* From res asst to res assoc, Cornell Univ, 57-63; asst mgr broiler & roaster res, Ralston Purina Co, 63-65, coordr, Nutrit Labs, 65-66, mgr, Nutrit Biochem Lab, 66-73, dir, nutrit biochem lab, 73-83; vpres, Agr Nutrit Consult, Inc, 83-87; TECH DIR SPECIALTY BUS GROUP, PURINA MILLS, INC, 87- *Mem:* AAAS; Am Poultry Sci Asn; Am Inst Nutrit; Inst Food Technol; Sigma Xi; Am Dairy Sci Asn; Am Soc Animal Sci. *Res:* Lipid transport and metabolism; energy metabolism; protein quality of foods and feedstuffs. *Mailing Add:* 37 Lazy Ridge Ct St Charles MO 63303-7272

HOPKINS, DAVID ALAN, b Victoria, BC, Aug 31, 42; m 76; c 3. LIMBIC SYSTEM, AUTONOMIC NERVOUS SYSTEM. *Educ:* Univ Alberta, BSc, 64; McMaster Univ, MA, 67, PhD(psychol), 70. *Prof Exp:* Fel neuroanat, Erasmus Univ, Rotterdam, 70-72, asst prof, 72-77; assoc prof, 77-85, PROF NEUROANAT, DALHOUSIE UNIV, 85-, HEAD DEPT, 91- *Concurrent Pos:* Fel, Int Brain Res Orgn, 69. *Honors & Awards:* Murray L Barr Jr Scientist Award, Can Asn Anatomists, 83. *Mem:* Neurosci Soc; Am Asn Anatomists; Can Asn Anatomists; Europ Neurosci Asn. *Res:* Functional and anatomical organization of motor systems; limbic and basal ganglia connections with the brain stem; organization of the autonomic nervous system. *Mailing Add:* Dept Anat Dalhousie Univ Sir Charles Tupper Med Bldg Halifax NS B3H 4H7 Can

HOPKINS, DAVID MOODY, b Nashua, NH, Dec 26, 21; m 49, 57, 70; c 6. QUATERNARY GEOLOGY. *Educ:* Univ NH, BS, 42; Harvard Univ, MS, 48, PhD(geol), 55. *Prof Exp:* Jr geologist, US Geol Surv, 42-44, asst geologist, 46, assoc geologist, 47-48, geologist, 48-84; distinguished prof quaternary sci, 84-90, dir, Alaska Quaternary Ctr, 84-90, EMER PROF & DIR QUATERNARY SCI, UNIV ALASKA, FAIRBANKS, 90- *Concurrent Pos:* Mem, Int Geog Cong Comn Evolution of Slopes & Comn Periglacial Morphol, 59-63; vis prof, Stanford Univ, 61; mem, Alaska Glacial Map comn, US Geol Surv, 56-63, Quaternary Shorelines comn, Int Quaternary Asn, 65-, geol names comn, US Geol Surv, 66-82, Early Man Res comn, Nat Geog Soc, 75-80, vis comt, Univ Wash Quaternary Res Ctr, 81-82, adv bd, Univ Maine Ctr Early Man Studies, 85-, US Nat Comn Int Permafrost Asn, 86-90, sci adv comt, div polar progs, US Nat Sci Found, 86-88; panelist, Tundra Biome, Int Biol Prog, 68,; co-convenor, Burg-Wartenstein Conf Paleoecol of Beringia, 79; Nat Acad Sci exchange fel, USSR, 69. *Honors & Awards:* Kirk Bryan Award, Geol Soc Am, 68; Roal Fryxell Award, Soc Am Archeol, 88. *Mem:* Fel AAAS; fel Geol Soc Am; fel Arctic Inst NAm; Int Quaternary Asn (pres, 75-76); Soc Am Archaeol; Sigma Xi; Am Quaternary Asn (pres, 75-76). *Res:* Oceanography, geology, tectonics and paleogeography of Bering and Chukchi Seas and Bering land bridge; Cenozoic stratigraphy of Alaska; worldwide Pleistocene correlations; geomorphology of arctic and subarctic regions; geology of placers; paleocology of Beringia; offshore permafrost; Cenozoic stratigraphy of Alaska. *Mailing Add:* Alaska Quaternary Ctr Univ Alaska Fairbanks AK 99701

HOPKINS, DON CARLOS, b Charleston, Ill, Feb 18, 36; m 59; c 2. PHYSICS. *Educ:* Eastern Ill Univ, BS, 57; Univ Ill, MS, 59, PhD(physics), 62. *Prof Exp:* From asst prof to assoc prof physics, Univ Mo-Rolla, 62-68; head dept, 68-88, PROF PHYSICS, SDAK SCH MINES & TECHNOL, 68- *Mem:* AAAS; Am Phys Soc; Am Asn Physics Teachers; Sigma Xi. *Res:* Thermodynamic properties of superconductors; magnetic alloy systems at low temperatures; use of computers in undergraduate education. *Mailing Add:* Dept Physics SDak Sch Mines & Technol Rapid City SD 57701-3995

HOPKINS, DONALD LEE, b Sacramento, Ky, Mar 13, 43; m 67; c 3. PLANT PATHOLOGY. *Educ:* Western Ky Univ, BS, 65; Univ Ky, PhD(plant path), 68. *Prof Exp:* USDA res assoc, 68-69; asst plant pathologist, 69-74, assoc plant pathologist, 74-79, PLANT PATHOLOGIST, UNIV FLA, 79- *Mem:* AAAS; Sigma Xi; Am Phytopath Soc; Am Soc Microbiol. *Res:* Physiology of plant diseases and plant disease resistance; chemical control of fungus diseases; plant bacterial and virus diseases; Pierce's disease of grape; biological control of plant diseases. *Mailing Add:* CFREC Leesburg 5336 University Ave Leesburg FL 34748

HOPKINS, DONALD R, b Miami, Fla, Sept 25, 41. MEDICINE. *Educ:* Morehouse Col, BS, 62; Univ Chicago Sch Med, MD, 66; Harvard Univ, MPH, 70. *Hon Degrees:* DSc, Morehouse Col, 88. *Prof Exp:* Researcher, dept microbiol, Univ Chicago, 63-66; intern, San Francisco Gen Hosp, 66-67; med epidemiologist & dir, Sierra Leone Smallpox & Measles Prog, 67-69; resident pediat, Univ Chicago Hosp & Clin, 70-72; med officer, Off Prog Planning & Eval, Off Ctr Dir, Ctr Dis Control, 72-74; asst prof trop pub health, Harvard Sch Pub Health, 74-77; asst dir opers, Ctr Dis Control, 77-80, Int Health, 80-84, dep dir, 84-87; SR CONSULT, CARTER CTR, 87- *Concurrent Pos:* LSU fel trop med & parasitol, Gorgas Mem Lab, Panama City, 65; dep chief, Environ Health Serv Div, Bur State Serv, Ctr Dis Control, 74; dep adminr, Agency Toxic Substance & Dis Registry, 84-87; actg dir, Ctr Dis Control, 85; consult epidemiologist, Training Course Ethiopian Smallpox Eradication Prog, Peace Corps, WHO, 71, Yaws Prog, PAHO & WHO, Dominica, 72, Smallpox Prog, W Bengal, India, 73 & Anti-Yaws Campaign, Colombia, 75; consult to WHO hq staff, Expanded Immunization Prog, Geneva, 75; chmn, Interagency Working Group, Int Health Res Develop & Demo & spec adv, Int Health Res & African Regional Health Strategy, 78; mem, US Del World Health Assembly, 77-86, global adv group, Expanded Prog Immunization, WHO, 78-80, steering comt, Sci Working Group Epidemiol, Spec Prog Res & Training Trop Dis, 80-83, tech adv group, Diarrheal Dis Control, 83-84, vaccines & rel biol prod adv comt, Nat Cancer Drugs & Biol, Food & Drug Admin, 81-85; vis lectr, Harvard Sch Pub Health, 78-, Emory Univ Sch Med, Dept Community Health, Master Pub Health Prog, 84-, Stony Brook State Univ, 82-; mem, overseers comt vis sch pub health, Harvard Univ, 86-, gov bd, Nat Coun Int Health, 87-, bd dirs, Int Serv Asn Health & adv bd & Health of Pub, 86-, adv bd, Edna McConnell Clark Found & WHO Collab Ctr, Int Ctr Health Sci, Meharry Med Col, 87-90. *Honors & Awards:* One Millionth Dollar Award, Nat Med Fel Inc, 62; Joseph Mountin Lect Award, 81. *Mem:* Inst Med-Nat Acad Sci; Royal Acad Sci; Am Soc Trop Med & Hyg. *Res:* Author of over 75 publications in medicine. *Mailing Add:* Global 2000 Carter Ctr Atlanta GA 30307

HOPKINS, ELBERT ERSKINE, physical chemistry, for more information see previous edition

HOPKINS, ESTHER ARVILLA HARRISON, b Stamford, Conn, Sept 18, 26; m 53, 59; c 2. PHYSICAL CHEMISTRY. *Educ:* Boston Univ, AB, 47; Howard Univ, MS, 49; Yale Univ, PhD(phys chem), 67. *Hon Degrees:* JD, Suffolk Univ, 77. *Prof Exp:* Instr chem, Va State Col, 49-52; res asst biophys, New Eng Inst Med Res, 55-59; res chemist, Am Cyanamid Co, 59-61; scientist, 66-72, patent attorney, 73-79, sr proj admin, 79-86, TECHNOL LIAISON MGR, POLAROID CORP, 86- *Concurrent Pos:* Adv comt minority prog, Sci Educ Directorate, NSF. *Mem:* Am Chem Soc; AAAS. *Res:* Biophysical chemistry; metal catalysis of enzyme reactions; emulsion research and coating technology; fiber optics. *Mailing Add:* 1550 Worcester Rd Unit 524 W Framingham MA 01701-8968

HOPKINS, FREDERICK SHERMAN, JR, b Springfield, Mass, June 12, 22; m 46; c 4. FORESTRY, ECONOMICS & POLICY. *Educ:* Univ Mich, BSF, 46, BBA & MF, 47; State Univ NY, PhD(forest econ), 59. *Prof Exp:* Asst forester, Exp Sta & asst prof forestry, Univ Vt, 50-54; instr forest econ, State Univ NY Col Forestry, Syracuse Univ, 57-59; prof, 59-89, EMER PROF FORESTRY, IOWA STATE UNIV, 89- *Mem:* Soc Am Foresters. *Res:* Forestry economics. *Mailing Add:* Dept Forestry Iowa State Univ 249 Bessey Ames IA 50011

HOPKINS, GEORGE C, b Buffalo, NY, June 29, 35; m 54; c 3. POLYMER CHEMISTRY. *Educ:* State Univ NY Buffalo, BA, 61, PhD(org chem), 67. *Prof Exp:* Technician chem, 58-60, chemist, 60-68, proj coordr PVC & related polymers, 68-72, sect mgr, 72-75, ASSOC DIR RES & DEVELOP, HOOKER CHEM CORP, 75- *Mem:* Am Chem Soc; Soc Plastics Engrs. *Res:* Polymer and plastics synthesis, evaluation and process development. *Mailing Add:* 1730 Pinon Circle St Cloud FL 32769

HOPKINS, GEORGE H, JR, b San Saba, Tex, Jan 1, 33; c 2. ELECTRICAL ENGINEERING, GEOPHYSICS. *Educ:* Univ Tex, BS, 56, MS, 57, PhD, 65. *Prof Exp:* Instr elec eng, Univ Tex, 57-61, res engr, Elec Eng Res Lab, 61-65; chief engr, Geosci, Inc, Mass, 65-66, sr vpres, 66-68; pres, Geotronics Corp, 68-88; ASSOC PROF ELEC ENG, TEX A&I UNIV, 88- *Concurrent Pos:* Consult, Geosci, Inc, Mass, 64-65. *Mem:* Inst Elec & Electronics Engrs; Am Geophys Union; Soc Explor Geophys. *Res:* Electrical geoscience; magnetotellurics; instrumentation. *Mailing Add:* Dept Elec Eng Computer Sci Tex A&I Univ Campus Box 192 Kingsville TX 78363

HOPKINS, GEORGE ROBERT, b Pittsford, NY, July 1, 28; m 49; c 2. APPLIED PHYSICS, NUCLEAR ENGINEERING. *Educ:* Allegheny Col, BS, 48; Univ Rochester, MS, 51; Iowa State Univ, PhD(physics), 54. *Prof Exp:* Asst instrumentation, Univ Rochester, 48-50; asst physics, Iowa State Univ, 50-54; fel scientist, Westinghouse Elec Corp, 54-60; mgr, Triga Facil, Gen Atomic Co, 61-63, mgr res & develop, Peach Bottom Atomic Power Plant, 63-67, mgr eng & anal, Defense Systs Develop Div, 67-69, mgr eng & anal, Advan Reactors Div, 69-71, TECH MGR & SR STAFF ENGR FUSION ENG, GEN ATOMIC CO, SAN DIEGO, 71- *Concurrent Pos:* Lectr, Univ Pittsburgh, 56-60, adj mem, grad fac, 57-60. *Mem:* AAAS; Am Phys Soc; Am Nuclear Soc; Sigma Xi. *Res:* Fusion technology and engineering, structural ceramics, high heat flux components and plasma interactions with materials. *Mailing Add:* 281 Bear Hill Rd North Andover MA 01845-2115

HOPKINS, GEORGE WILLIAM, II, b Sewanee, Tenn, Mar 30, 47; m 70; c 1. OPTICAL ENGINEERING. *Educ:* Univ of the South, BA, 68; Univ Mass, MS, 70; Univ Ariz, PhD(optical sci), 76. *Prof Exp:* Res assoc, Univ Ariz, 76-77; develop engr, 77-80, 84-86 & 89-90, mem tech staff, 80-84, prod mgr, 86-89, MEM TECH STAFF, HEWLETT-PACKARD CO, 90- *Mem:* Optical Soc Am; Soc Photo-Optical Instrumentation Engrs. *Res:* Instrumentation using optical techniques; granted two United States patents. *Mailing Add:* Hewlett-Packard Co PO Box 10490 Palo Alto CA 94303-0969

HOPKINS, GORDON BRUCE, b Seattle, Wash; m 67. PATHOLOGY, NUCLEAR MEDICINE. *Educ:* Gonzaga Univ, BA, 62; Med Col Wis, MD, 66. *Prof Exp:* Intern path, Univ Wash, 66-67, resident, 67-70; resident path & nuclear med, Univ Ore, 70-71; ASST CLIN PROF, MED SCH, UNIV CALIF, SAN DIEGO, 72-; DIR DIV NUCLEAR MED, DEPT PATH, SCRIPPS MEM HOSP, 73- *Mem:* Col Am Path; Am Soc Clin Path; Am Col Nuclear Physicians; Soc Nuclear Med. *Res:* Clinical evaluation of diagnostic radiopharmaceuticals. *Mailing Add:* 988 Genessee Ave La Jolla CA 92037

HOPKINS, HARRY P, JR, b Norfolk, Va, July 10, 39; m; c 1. PHYSICAL CHEMISTRY. *Educ:* Univ Va, BA, 61; Carnegie Inst Technol, PhD(phys chem), 65. *Prof Exp:* AEC fel, Univ Pittsburgh, 65; Welch res fel, Rice Univ, 65-67; from asst prof to assoc prof chem, 67-74, PROF CHEM, GA STATE UNIV, 74- *Mem:* Am Chem Soc. *Res:* Calorimetry; solution chemistry; infrared and magnetic resonance spectroscopy. *Mailing Add:* Dept Chem Ga State Univ University Plaza Atlanta GA 30303

HOPKINS, HOMER THAWLEY, b Frederica, Del, July 27, 13; m 40; c 1. PLANT PHYSIOLOGY, INTERNATIONAL AGRICULTURAL DEVELOPMENT. *Educ:* Univ Del, BS, 35, Cornell Univ, MS, 39; Univ Md, PhD(plant physiol), 51. *Prof Exp:* Asst chemist, State Bd Agr, Del, 35-37; anal chemist, Forest Soils Lab, Cornell Univ, 37-38; soil conservationist, Soil Conserv Serv, USDA, 39-42, asst chemist, Bur Plant Indust, Agr Res Admin, 46-47, assoc soil scientist, 47-53, plant physiologist, AEC Proj, Agr Res Serv, 53-56, supvr chemist human nutrit, Agr Res Serv; chemist-in-charge invests membrane transport, Nutrit Div, US Food & Drug Admin, 61-69; biol sci adminr, Radiation Biol Lab, Smithsonian Inst, 69-70; staff officer, Food & Nutrit Bd, Nat Acad Sci-Nat Res Coun, Washington, DC, 70-71; consult tech & sci info, life & earth sci, 71-73; prog mgr environ sci & geochem, Hittman Assocs, Inc, 73-81; consult tech & sci info, life & health sci, 81-84; VOL COORDR INTL PROG, UNIV MD, INST APPL AGR, 84- *Concurrent Pos:* Asst adj prof, Univ Md Inst Appl Agr, 85- *Mem:* Fel AAAS; Am Soc Plant Physiol; Am Chem Soc; Am Inst Biol Sci; NY Acad Sci; Sigma Xi; Asn Int Agr & Exten Educ; Am Asn Adult & Continuing Educ. *Res:* Ion transport across living membranes; nutrient composition of national food supplies; phytotoxicity and persistence of chlorinated hydrocarbon insecticides in soils; soil biology; plant materials for the rehabilitation of lands disturbed by construction, mining and related activities; assessment of potential ecological and health effects of synthetic fuel technologies. *Mailing Add:* 4500 Elmwood Rd Beltsville MD 20705-2618

HOPKINS, HORACE H, JR, b Orange, NJ, Dec 24, 22; m 57; c 2. PHYSICAL CHEMISTRY, INDUSTRIAL ENGINEERING. *Educ:* Oberlin Col, BS, 43; Univ Calif, PhD(chem), 49. *Prof Exp:* Chemist, Metall Lab, Univ Chicago, 43-46; chemist, Gen Elec Co, 49-65; mgr, Puchem Lab, Isochem Co, 66-69; staff engr, Res & Develop Dept, Atlantic Richfield-Hanford Co, 69-76, mgr, Strategic Planning Dept, 74-75, staff engr res & develop, Rockwell Hanford Opers, 76-86, STAFF ENGR, WESTINGHOUSE HANFORD OPERS, ATLANTIC RICHFIELD-HANFORD CO, 86- *Mem:* Am Nuclear Soc; Am Chem Soc. *Res:* Radiochemistry; nuclear materials processing; inorganic chemistry; manufacturing engineering. *Mailing Add:* 25301 103rd Ave E Graham WA 98338

HOPKINS, JOHN CHAPMAN, b Palo Alto, Calif, June 30, 33; m 54; c 2. NUCLEAR PHYSICS. *Educ:* Univ Wash, BSc, 55, PhD(physics), 60. *Prof Exp:* Asst res, Dir Off, 71-72, assoc div leader, Weapons Div, 72; asst div leader, Field Test Div, 72-74, dep assoc dir test & verification, 79-84, DIV LEADER, FIELD TEST DIV, LOS ALAMOS NAT LAB, 74-, MEM STAFF, PHYSICS DIV, 60-, ASSOC DIR, 84- *Concurrent Pos:* Vis scientist, Nuclear Res Div, Atomic Weapons Res Estab, Eng, 63-64. *Mem:* Fel Am Phys Soc. *Res:* Beta decay; beta polarization; fission; neutron scattering. *Mailing Add:* 1251 41st St Los Alamos NM 87544

HOPKINS, JOHN RAYMOND, b Greenup, Ill, Nov 19, 44; m 66; c 2. GEOPHYSICS. *Educ:* Eastern Ill Univ, BS, 66; Iowa State Univ, PhD(physics), 72. *Prof Exp:* Res assoc physics, Okla State Univ, 72-74; res scientist, Continental Oil Co, 74-77, sr res scientist, 77-78, sr staff geophysicist, 78-80, ASST DIV EXPLOR MGR, CONOCO INC, DENVER, 74- *Mem:* Sigma Xi; Soc Explor Geophysicists. *Res:* Shear and compressional wave propagation and reflection properties and their applications in geophysics; Gulf Coast and Rocky Mountains hydrocarbon exploration. *Mailing Add:* Dir Admin Grinnell Col PO Box 805 Grinnell IA 50112

HOPKINS, JOHNS WILSON, b Darlington, Md, Apr 1, 33; m 55; c 1. MOLECULAR BIOLOGY. *Educ:* Haverford Col, BS, 55; Rockefeller Inst, PhD(biol), 60. *Prof Exp:* From instr to asst prof biol, Harvard Univ, 60-66; assoc prof, 66-69, chmn dept, 66-75, PROF BIOL, WASH UNIV, 69- *Res:* Cell biology of cultured plant cells; ion transport by plant cells; science education. *Mailing Add:* Dept Biol Wash Univ St Louis MO 63130

HOPKINS, LEMAC, economic entomology, for more information see previous edition

HOPKINS, LEON LORRAINE, b Ft Collins, Colo, Mar 7, 35; m 55; c 4. NUTRITION. *Educ:* Colo State Univ, BS, 57, MS, 59; Univ Wis, PhD(biochem, nutrit), 62. *Prof Exp:* Jr animal husbandman, Colo State Univ, 57-59; res scientist, NIH, 62-64; res biochemist, Div Nutrit, US Food & Drug Admin, 64-68, chief micronutrient res br, 68-69; asst to dir, Human Nutrit Res Div, Agr Res Serv, USDA, 69-72, asst dir, Colo-Wyo, 72-75; sect head, Joint Food & Agr Orgn-Int Atomic Energy Agency Div, Int Atomic Energy Agency, 75-78; chmn, Dept Food & Nutrit, Tex Tech Univ, 78-81, prof, 81-82; CONSULT, 82- *Mem:* Am Inst Nutrit; Am Soc Clin Nutrit; Sigma Xi. *Res:* Mineral nutrition. *Mailing Add:* Box 592 Norwood CO 81423

HOPKINS, M E, b Grove, Okla, Feb 4, 28; m 57; c 1. GEOLOGY. *Educ:* Univ Ark, BS, 50, MS, 51; Univ Ill, PhD(geol), 57. *Prof Exp:* Asst, Univ Ark, 50-51; asst, Coal Div, State Geol Surv, Ill, 51-53, asst geologist, 53-55; asst prof geol, Univ Tulsa, 55-63; assoc geologist, 63-68, actg head sect, 68-69, geologist & head coal sect, Ill Geol Surv, 69-75; consult coal geol, H Williamson, Inc, 75-81; dir geol, Peabody Coal Co, 81-83, DIR GEOL, PEABODY DEVELOP CO, 83- *Mem:* Geol Soc Am; Soc Econ Paleont & Mineral; Am Inst Mining, Metall & Petrol Engrs. *Res:* Coal mining geology; stratigraphy; sedimentation; field geology; subsurface geology; sedimentary petrology. *Mailing Add:* Peabody Develop Co PO Box 14222 St Louis MO 63178

HOPKINS, MANSELL HERBERT, JR, b Warrenton, Va, Oct 2, 27; m 52; c 3. ELECTRICAL ENGINEERING. *Educ:* Va Polytech Inst, BSEE, 53, MSEE, 58; Iowa State Univ, PhD(elec eng), 64. *Prof Exp:* Qual control engr, Lamp Div, Gen Elec Co, 53-55; asst prof, 55-58, ASSOC PROF ELEC ENG, VA POLYTECH INST & STATE UNIV, 58- *Concurrent Pos:* NSF Res fel, Iowa State Univ, 60-61. *Mem:* Am Soc Eng Educ; Inst Elec & Electronics Engrs. *Res:* Power systems; instrumentation and protection; controls. *Mailing Add:* Dept Elec Eng Va Polytech Inst & State Univ Blacksburg VA 24061

HOPKINS, NIGEL JOHN, b Indian Head, Sask, July 2, 22; m 45, 84; c 1. OPERATIONS RESEARCH. *Educ:* McGill Univ, BSc, 48, MSc, 49, PhD(nuclear physics), 52. *Prof Exp:* Res physicist, Atomic Energy Can Ltd, 52-53; staff mem, 53-68, dir maritime oper res, 68-70, dir land opers res, Can Forces Hq, 70-74, sr planning officer, 74-77, dir sci & tech intel, Nat Defense HQ, 77-85; PRES, H C CAMIRRAL LTD, 85- *Mem:* Can Oper Res Soc (pres, 72-73). *Res:* Operational research; systems analysis; planning of research and development; scientific intelligence. *Mailing Add:* 304-200 Rideau Terrace Ottawa ON K1M 0Z3 Can

HOPKINS, PAUL BRINK, b Indianapolis, Ind, June 13, 56. CHEMISTRY. *Educ:* Purdue Univ, BS, 77; Harvard Univ, PhD(chem), 82. *Prof Exp:* asst prof, 82-88, ASSOC PROF CHEM, DEPT CHEM, UNIV WASH, 88- *Concurrent Pos:* Searle scholar, 84-87; Sloan fel, 88-90. *Honors & Awards:* ICI Awardee, 88. *Mem:* Am Chem Soc; AAAS. *Res:* Organic and bio-organic chemistry. *Mailing Add:* Dept Chem Bldg 10 Univ Wash Seattle WA 98195

HOPKINS, PAUL DONALD, b Woodbury, NJ, Apr 1, 35; m 63; c 3. INORGANIC CHEMISTRY. *Educ:* Haverford Col, BS, 57; Univ Pittsburgh, PhD(inorg chem), 63. *Prof Exp:* Asst proj chemist, Am Oil Co, 63-65, proj chemist, 65-67, sr proj chemist, 67-74, res chemist, 74-81, sr res chemist, 81-89, RES ASSOC, AMOCO OIL CO, 89- *Mem:* Am Chem Soc; Chem Soc; Catalysis Soc; Clay Minerals Soc. *Res:* Transition metal, coordination and surface chemistry; catalytic conversions of petroleum; oxidation of liquid hydrocarbons; synthesis of heterogeneous catalysts and zeolites. *Mailing Add:* 1402 S 7th St St Charles IL 60174

HOPKINS, RICHARD ALLEN, b New Rochelle, NY, Apr 4, 43; m 70; c 2. ENGINEERING GEOPHYSICS. *Educ:* Murray State Univ, BA, 70; Univ Tenn, MS, 76. *Prof Exp:* mem staff eng geophys, Tenn Valley Authority, 72-88; CONSULT, 88- *Concurrent Pos:* Asst instr, Univ Tenn, 78; partner, H&H Geol Assocs, 76- *Mem:* Soc Explor Geophysicists; Am Geophys Union; Asn Eng Geologists; Nat Water Well Asn; Am Soc Testing & Mat. *Res:* Borehole and surface techniques as applied to foundation and earthquake and environmental investigations. *Mailing Add:* Marrich Inc 6000 Kaywood Rd SE Knoxville TN 37920-5905

HOPKINS, RICHARD H(ENRY), b Hartford, Conn, Dec 23, 40; m 68; c 1. MATERIALS SCIENCE, METALLURGY. *Educ:* Lehigh Univ, BS, 63, MS, 65, PhD(metall), 67. *Prof Exp:* Sr engr res labs, Westinghouse Elec Corp, 67-73, fel engr, Westinghouse Res & Develop Ctr, 73-77, prog mgr, 77-80, mgr, Silicon WEB Technol Develop, 84-87, mgr crystal sci, 80-87, mgr electro-optical mat, 88-90, MGR ELECTRONIC & PHOTONIC MAT, WESTINGHOUSE RES & DEVELOP CTR, 90- *Concurrent Pos:* Bd Dirs, Am Inst Mat Engrs, 82-; invited lectr, Int Conf Crystal Growth, 84; pres, Pittsburgh Chapter-Crystal Grower. *Mem:* AAAS; Am Soc Metals; Am Inst Mining, Metall & Petrol Engrs; Am Asn Crystal Growth. *Res:* Solidification and crystal perfection of metallic alloys; development of high melting inorganic materials for optical devices; growth and evaluation of synthetic crystals; crystal growth and perfection of metals, alloys, inorganic compounds and semiconductors; studies of composites, optical devices, and solar cells; ribbon crystal growth of silicon; development of acoustical crystals; studies of epitaxial ferrite films; silicon carbide crystal growth. *Mailing Add:* Westinghouse Res & Develop Ctr 1310 Bueluh Rd Pittsburgh PA 15235

HOPKINS, ROBERT CHARLES, b Pasadena, Calif, July 23, 37; m 65; c 2. DNA STRUCTURE, DNA FUNCTION. *Educ:* Univ Calif, Los Angeles, BS, 59; Harvard Univ, MA, 62, PhD(phys chem), 65. *Prof Exp:* Staff physicist, Shell Develop Co, 65-76; assoc prof & dir sci, 76-81, PROF CHEM & BIOPHYS, UNIV HOUSTON-CLEAR LAKE, 81- *Concurrent Pos:* Vis prof, Molecular Biol Inst, Univ Calif, Los Angeles, 82, dept chem, Univ Mich, 86. *Mem:* Am Phys Soc; Biophys Soc. *Res:* Theoretical studies of implications for a new family of double-helical DNA models including four-stranded structures; physical and biological investigations seeking evidence for such DNA forms. *Mailing Add:* Univ Houston-Clear Lake 2700 Bay Area Blvd Houston TX 77058

HOPKINS, ROBERT EARL, b Belmont, Mass, June 30, 15; m 39; c 6. PHYSICS. *Educ:* Mass Inst Technol, BS, 37; Univ Rochester, MS, 39, PhD(geomet optics), 45. *Prof Exp:* From asst to assoc prof physics, 38-51, dir, Inst Optics, 54-64 & 69-74, prof optics, 51-76, SCIENTIST, LAB FOR LASER ENERGETICS, UNIV ROCHESTER, 76-, PROF OPTICS, 80- *Concurrent Pos:* Researcher, Off Sci Res & Develop Contracts, 41-45; pres, Tropel, Inc, 68-75. *Mem:* Assoc Optical Soc Am (pres, 72-73); Am Phys Soc. *Res:* Optics for laser fusion systems; photolithographic optics for microelectronics. *Mailing Add:* 49 Reservoir Ave Rochester NY 14620-2753

HOPKINS, ROBERT WEST, b Springfield, Mass, May 26, 24; m 60; c 2. SURGERY. *Educ:* Harvard Med Sch, MD, 47. *Prof Exp:* Instr surg, Sch Med, Univ Pa, 55-58; from sr instr to assoc prof, Case Western Reserve Univ, 58-70; PROF MED SCI, BROWN UNIV, 70- *Concurrent Pos:* Asst surgeon, Pa Hosp, Philadelphia, 55-58; lower fel, Case Western Reserve Univ, 58-59; asst vis surgeon, Cleveland Metrop Gen Hosp, 55-65, assoc surgeon, 65-70; assoc surgeon-in-chief, Miriam Hosp, Providence, RI, 70-80, surgeon-in-chief, 80-86, sr surgeon, 86- *Mem:* Am Surg Asn; Soc Vascular Surg; Am Asn Surg Trauma; Sigma Xi. *Res:* Surgical physiology; physiologic and biochemical changes in circulatory shock in animals and in man. *Mailing Add:* 164 Summit Ave Providence RI 02906

HOPKINS, RONALD MURRAY, b Baltimore, Md, Jan 4, 42; m 87, 62. TOXICOLOGY, PHARMACY. *Educ:* Univ Md, BS, 63, MS, 67, PhD(pharmacol), 70. *Prof Exp:* Pharmacologist, Huntingdon Res Ctr, Inc, 65-69; instr pharmacol, Univ Md Sch Dent, 69-70; from res pharmacologist to sr res pharmacologist, Mallinckrodt, Inc, 70-76, asst pharm dir & toxicol, 76-80, from assoc dir to dir res, 80-84, dir res & develop, Opers Div, 84-88, bus dir, nuclear med prod, 88-90; VPRES RES & DEVELOP, ALLIANCE PHARMACEUT CORP, 90- *Mem:* Am Soc Pharmacol & Exp Therapeut; AAAS. *Res:* All phases of toxicological evaluation primarily involving diagnostic radiographic media and radiopharmaceuticals; cardiovascular and pulmonary pharmacology; industrial pharmaceutical formulation, primarily parenterals; medical device. *Mailing Add:* Alliance Pharmaceut Corp 3040 Sci Pk Rd San Diego CA 92121-1102

HOPKINS, THEODORE EMO, physical chemistry, computer science; deceased, see previous edition for last biography

HOPKINS, THEODORE LOUIS, b San Diego, Calif, June 12, 29; m 51, 71; c 3. ENTOMOLOGY. *Educ:* Ore State Univ, BS, 51, MS, 56; Kans State Univ, PhD(entom), 60. *Prof Exp:* Entomologist, USDA, 53-57; from instr to assoc prof, 58-70, PROF ENTOM, KANS STATE UNIV, 70- *Mem:* Entom Soc Am; Sigma Xi; Orthop Soc; AAAS. *Res:* Amino acid and catecholamine metabolism for insect cuticle tanning and hormonal regulation; grasshopper-plant biochemical interactions. *Mailing Add:* 5501 Turkeyfoot Lane Manhattan KS 66502

HOPKINS, THOMAS FRANKLIN, b Culpepper, Va, Dec 16, 24; m 47; c 4. REPRODUCTIVE ENDOCRINOLOGY. *Educ:* Calvin Coolidge Col, BS, 55; Mich State Univ, MS, 61; Boston Univ, PhD(physiol), 70. *Prof Exp:* Res asst, Worcester Found Exp Biol, 48-59, staff scientist, 61-70; assoc prof physiol, Univ Conn, 70-75; PROF BIOL & HEAD DEPT NATURAL SCI, UNIV MD, EASTERN SHORE, 75- *Mem:* AAAS; Am Soc Zool; Am Physiol Soc; Soc Study Reprod. *Res:* Toxic effects of heavy metals; physiology of aging. *Mailing Add:* Dept Acad Affairs Univ Md Eastern Shore Princess Anne MD 21853

HOPKINS, THOMAS R (TIM), b Urbana, Ill, Oct 12, 38; m 64; c 2. ENZYMOLOGY, PROTEIN CHEMISTRY. *Educ:* Univ Calif, Berkeley, BA, 61; Univ Utah, MS, 65, PhD(photodynamic action), 67. *Prof Exp:* USPHS fel phys chem, Univ Minn, Minneapolis, 67-69; sr lectr biochem, Med Sch, Univ Otago, NZ, 69-76; mgr biochem processes, Biotech Div, 76-88, PRES & CEO, BIOSPEC PROD INC, 81- *Concurrent Pos:* Consult, Biochem Instrumentation. *Honors & Awards:* IR 100 Award. *Mem:* Am Soc Photobiol; Am Soc Biol Chemists; Biophys Soc. *Res:* Protein biophysics and chemistry; excited and ground state energy transfer in biomolecules; biochemical instrumentation; industrial applications of enzymes and proteins. *Mailing Add:* Biospec Prod Inc PO Box 722 Bartlesville OK 74005-0722

HOPKINS, WILLIAM GEORGE, b Woodbury, NJ, Aug 15, 37; m 59; c 2. PLANT PHYSIOLOGY. *Educ:* Wesleyan Univ, AB, 59; Ind Univ, PhD(bot), 64. *Prof Exp:* Res assoc plant physiol, Brookhaven Nat Lab, 63-65; asst prof biol, Bryn Mawr Col, 65-69; from asst prof to assoc prof, 69-89, PROF BIOL, UNIV WESTERN ONT, 89- *Mem:* Am Soc Plant Physiol; Can Soc Plant Physiol; Scand Soc Plant Physiol. *Res:* Growth and development of higher plants; organization and development of photosynthetic membranes; bioenergetics. *Mailing Add:* Dept Plant Sci Univ Western Ont London ON N6A 5B7 Can

HOPKINS, WILLIAM STEPHEN, b Palo Alto, Calif, July 29, 31; m 64; c 2. PALYNOLOGY, GEOLOGY. *Educ:* Univ Wash, BSc, 54; Univ BC, MSc, 62, PhD(palynology, geol), 66. *Prof Exp:* Geologist, Humble Oil & Refining Co, 58-60; res scientist, Pan Am Petrol Res Ctr, Okla, 66-68; res scientist, Inst Sedimentary & Petrol Geol, Geol Surv Can, 68-84. *Concurrent Pos:* Consult & adv polynology & geol, San Juan Islands, Wash, 84-; founder & pres, Waldron Geol Consult, Inc. *Mem:* AAAS; fel Geol Soc Am; Am Asn Stratig Palynologists. *Res:* Palynology, paleoecology and past floral distributions in Tertiary and Cretaceous sedimentary rocks from British Columbia and Washington State. *Mailing Add:* PO Box 610 Eastsound WA 98245

HOPLA, CLUFF EARL, b Mapleton, Utah, Dec 28, 17; m 41; c 3. MEDICAL ENTOMOLOGY. *Educ:* Brigham Young Univ, BS, 41, MS, 47; Univ Kans, PhD(entom, bact), 50. *Prof Exp:* Res assoc, Univ Kans, 50-51; asst prof entom, 51-59, from assoc prof to prof zool, 59-69, GEORGE LYNN CROSS RES PROF ZOOL, UNIV OKLA, 69-, CHMN DEPT, 62-, CUR ENTOM, STOVALL MUS, 59- *Mem:* Am Soc Parasitol; Am Soc Trop Med & Hyg; Entom Soc Am; Soc Syst Zool. *Res:* Zoonoses; bionomics and taxonomy of Acarina, Culicidae and Siphonaptera. *Mailing Add:* Dept Zool Univ Okla 660 Parrington Oval Norman OK 73019

HOPP, WILLIAM BEECHER, b Terre Haute, Ind, Sept 4, 17; m 49; c 1. ZOOLOGY. *Educ:* Ind State Teachers Col, BS, 39; Purdue Univ, MS, 41, PhD(zool), 53. *Prof Exp:* Instr biol, Purdue Univ, 46-47; asst prof zool, Eastern Ky State Col, 47-55; from asst prof to assoc prof zool, Ind State Univ, 55-61, chmn div sci, 61-69, prof zool, 61-; RETIRED. *Mem:* AAAS; Am Soc Parasitol; Am Soc Ichthyol & Herpet; Sigma Xi. *Res:* Parasitology; herpetology. *Mailing Add:* Life Sci Div Ind State Univ Terre Haute IN 47809

HOPPE, DAVID MATTHEW, b West Concord, Minn, Aug 22, 42; m 69. VERTEBRATE BIOLOGY, EVOLUTIONARY ECOLOGY. *Educ:* Univ Minn, BA, 64, MS, 68; Colo State Univ, PhD(zool), 75. *Prof Exp:* Asst scientist genetics, Univ Minn, 64-66; asst prof biol, Mayville State Col, 68-72; instr zool, Colo State Univ, 73-74; PROF BIOL, UNIV MINN, MORRIS, 75- *Concurrent Pos:* Consult ecol, Houston Eng, Inc, 75- *Mem:* Sigma Xi; Soc Study Amphibians & Reptiles; Herpet League; Nat Wildlife Fedn. *Res:* Genetics and ecology of amphibian adaptations, particularly color polymorphism and various developmental phenomena; xenopus albinism. *Mailing Add:* Dept Math & Sci Univ Minn Morris MN 56267

HOPPE, JOHN CAMERON, b San Diego, Calif, Aug 18, 43; m 62; c 3. PHYSICS. *Educ:* Col William & Mary, BS, 64, MS, 67; Va Polytech Inst & State Univ, PhD(physics), 77. *Prof Exp:* Engr appl physics, Va Assoc Res Ctr, 64-65; electronics engr appl physics, 65-83, SECT HEAD, PHOTO-OPTICAL INST, LANGLEY RES CTR, NASA, 83- *Concurrent Pos:* Adj prof, St Leo Col, 80- 89. *Mem:* Am Phys Soc; Optical Soc Am; Inst Elec & Electronics Engrs; Soc Indust & Appl Math. *Res:* Applied optics; laser applications; scattering of electromagnetic radiation; spectroscopy; applications of laser light scattering methods to measure properties of gas flows; applications of video close-range photogrammetry; flow visualization technology. *Mailing Add:* NASA Langley Res Ctr Mail Stop 236 Hampton VA 23665

HOPPE, PETER CHRISTIAN, b Long Beach, Calif, Feb 16, 42; m 63; c 3. REPRODUCTIVE BIOLOGY. *Educ:* Calif State Polytech Univ, BS, 64; Kans State Univ, MS, 66, PhD(reprod physiol), 68. *Prof Exp:* Fel, 68-70, assoc staff scientist to staff scientist, 70-81, SR STAFF SCIENTIST, JACKSON LAB, 81- *Mem:* Soc Study Reprod. *Res:* Mammalian reproductive physiology; fertilization; parthenogenesis; differentiation; transgenic mice. *Mailing Add:* Jackson Lab Bar Harbor ME 04609

HOPPENJANS, DONALD WILLIAM, b Covington, Ky, June 21, 41; m; c 4. PHYSICAL INORGANIC CHEMISTRY. *Educ:* Bellarmine Col, AB, 63; Cath Univ Am, PhD(phys inorg chem), 68. *Prof Exp:* Fel, Univ Iowa, 67-69; res chemist, 69-73, tech supvr, 73, tech mgr, 74-78, prod mgr, 78-83, SR RES SUPVR, E I DU PONT DE NEMOURS & CO, INC, 83-; BIOMAT PROG CONSULT. *Mem:* Am Chem Soc. *Res:* Kinetics and mechanisms of the reactions of coordination compounds; fused salts; paint technology; T1O2 product development; biomaterials. *Mailing Add:* Cent Res Dept O-6028 E I du Pont de Nemours & Co Inc Dupont Bldg Wilmington DE 19898

HOPPER, ANITA KLEIN, b Chicago, Ill, Sept 24, 45; m 71. GENETICS, MOLECULAR BIOLOGY. *Educ:* Univ Ill, Chicago, BS, 67; Univ Ill, Urbana, PhD(biol), 72. *Prof Exp:* Fel genetics & molecular biol, Univ Wash, 72-75; asst prof genetics & molecular biol, Med Sch, Univ Mass, 75-78, assoc prof, 78-79; assoc prof, 79-87, PROF BIOCHEM, HERSHEY MED CTR, PA, 87- *Concurrent Pos:* Mem NSF panel genetic biol, 81-85, NIH genetics study sect, 85- *Mem:* Am Soc Microbiol; Am Soc Biol Chemists. *Res:* Molecular mechanisms of import of proteins into mitochondria and nuclei; synthesis and maturation of transfer RNA in yeast. *Mailing Add:* Dept Biol Chem Milton S Hershey Med Ctr Hershey PA 17033

HOPPER, ARTHUR FREDERICK, b Plainfield, NJ, Sept 7, 17; m 40; c 4. RADIATION BIOLOGY. *Educ:* Princeton Univ, AB, 38; Yale Univ, MS, 42; Northwestern Univ, PhD(zool), 48. *Prof Exp:* Instr zool, Northwestern Univ, 48; asst prof, Wayne Univ, 48-49; from asst prof to prof, 49-80, EMER PROF ZOOL, RUTGERS UNIV, NEW BRUNSWICK, 80- *Concurrent Pos:* Mem sci staff, Detroit Inst Cancer Res, 48-49; res collabr, Brookhaven Nat Lab, 63-80; vis prof, Univ Liege, 67-68. *Mem:* AAAS; Am Soc Zool; Soc Develop Biol; Radiation Res Soc; Sigma Xi. *Res:* Embryology and regeneration in fish and amphibians; cell population kinetics in the intestinal mucosa; radiation biology. *Mailing Add:* 231 Cocoanut Row Palm Beach FL 33480-4132

HOPPER, GRACE MURRAY, b New York, NY, Dec 9, 06; m 30. MATHEMATICS. *Educ:* Vassar Col, BA, 28; Yale Univ, MA, 30, PhD(math), 34. *Hon Degrees:* Various from US univ & cols, 72-81. *Prof Exp:* Asst math, Vassar Col, 31-34, from instr to assoc prof, 34-46; res fel eng sci & appl physics, Comput Lab, Harvard Univ, 46-49; staff mem, Univac Div, Sperry Rand Corp, 49-61, staff scientist, 61-67; prof lectr, George Washington Univ, 71-78; spec adv to comdr, Naval Data Automation Command, Dept Navy, 67-86; RETIRED. *Concurrent Pos:* Vassar Col fac fel, NY Univ, 41-42; adj prof, Moore Sch Elec Eng, Univ Pa, 63- *Honors & Awards:* Nat Medal of Technol, 91; Naval Ord Develop Award, 46; Harry Goode Mem Award, Am Fedn Info Processing Socs, 71; W Wallace McDowell Award, Inst Elec & Electronics Engrs, 79. *Mem:* Nat Acad Eng; fel AAAS; Franklin Inst; fel Inst Elec & Electronics Engrs; Asn Comput Mach. *Res:* Design, operation, preparation of problems for digital computers. *Mailing Add:* 1400 S Joyce St Arlington VA 22202

HOPPER, JACK R, b Highlands, Tex, May 12, 37; m 58; c 2. CHEMICAL ENGINEERING, CATALYSIS. *Educ:* Tex A&M Univ, BS, 59; Univ Del, MChE, 64; La State Univ, Baton Rouge, PhD(catalysis), 69. *Prof Exp:* Jr res engr, Humble Oil & Ref Co, 59-60, asst res engr, 60-61; res engr, Esso Res & Eng Co, 64-67; res assoc catalysis, La State Univ, Baton Rouge, 67-69; from asst prof to assoc prof chem eng, 69-75, PROF CHEM ENG, LAMAR UNIV, 75-, HEAD DEPT, 74- *Honors & Awards:* Dow Chem Co Award, Am Soc Eng Educ, 71. *Mem:* Fel Am Inst Chem Engrs; Am Chem Soc; Catalysis Soc; Am Soc Eng Educ. *Res:* Reaction engineering; catalysis, hazardous waste treatment and minimization. *Mailing Add:* Dept Chem Eng Lamar Univ Sta Beaumont TX 77710

HOPPER, JAMES ERNEST, b Madison, Wis, July 16, 42; m 71. CELL BIOLOGY, DEVELOPMENTAL BIOLOGY. *Educ:* Univ Wis-Madison, BS, 64, MS, 67, PhD(genetics), 70. *Prof Exp:* Comt Instnl Coop traveling scholar & res assoc plant physiol, Univ Ill, Urbana, 69-71; NIH res fel biochem, Univ Wash, 71-75; mem staff, Med Ctr, Univ Mass, Worcester, 75-; ASSOC PROF, DEPT BIOL CHEM, MILTON S HERSHEY MED CTR, PA STATE UNIV. *Res:* Molecular basis of regulatory mechanisms underlying metabolic differentiation and developmental transitions. *Mailing Add:* Dept Biol Chem Hershey Med Ctr, Pa State Univ Hershey PA 17033

HOPPER, JOHN HENRY, b Briggsville, Ark, May 17, 25; m 51; c 2. NUTRITION. *Educ:* Univ Ark, BSA, 49, MS, 51; Univ Ill, PhD(nutrit), 55. *Prof Exp:* Asst nutrit, Univ Ark, 50-51; instr animal nutrit, Ark Agr & Mech Col, 51-52; asst nutrit, Univ Ill, 52-55; asst prof, Univ Nebr, 55-56; head animal nutrit res sect, Armour & Co, 56-60, dir nutrit, 60-69, dir res, 69-70, vpres & dir food res, 70-76, sr vpres sci affairs, Kellog Co, 76-79; RETIRED. *Mem:* Am Chem Soc. *Res:* Vitamin, amino acid and protein metabolism; trace minerals, food composition; food metabolism; development of new foods. *Mailing Add:* PO Box 97 Hampton AR 71744

HOPPER, MICHAEL JAMES, b Los Angeles, Calif, Sept 23, 40; m 68; c 2. SPECTROSCOPY, ANALYTICAL CHEMISTRY. *Educ:* Univ Calif, Riverside, BA, 62; Univ Minn, PhD(phys chem), 67. *Prof Exp:* Teaching asst inorg chem, Univ Calif, Riverside, 60-62; gen teaching asst gen chem, Univ Minn, 62-64, phys chem, 64-66; mem staff, Celanese Res Co, NJ, 67-71, supvr, Anal & Phys Testing Dept, Celanar Res & Develop Div, 71-75, res & develop group leader appln & prod develop, Celanese Plastics Mat Co, 76-78, res & develop mgr film develop, 78-79; DEVELOP MGR RES & DEVELOP, FILM DIV, HOECHST-CELANESE CORP, 79- *Mem:* Am Chem Soc; Soc Appl Spectros; Coblentz Soc; Sigma Xi. *Res:* Ultraviolet spectroscopy; infrared spectroscopy of polymers; attenuated total reflectance spectroscopy of surfaces; thermal analysis; atomic absorption spectroscopy; gas chromatography; biaxially oriented polyester film technology; magnetic tape product development; dyed UV stabilized film development; process development; polarizing film technology. *Mailing Add:* Hoechst Celanese Corp Box 1400 Greer SC 29652

HOPPER, NORMAN WAYNE, b Ralls, Tex, Sept 1, 43; m 66; c 3. CROP PHYSIOLOGY, SEED PHYSIOLOGY. *Educ:* Tex Tech Univ, BS, 65, MS, 67; Iowa State Univ, PhD(agron), 70. *Prof Exp:* Asst prof agron, Ohio State Univ, 70-76; ASSOC PROF AGRON, TEX TECH UNIV, 76- *Mem:* Am Soc Agron; Crop Sci Soc Am. *Res:* Factors affecting seed quality; measurement of seed quality; seedling vigor studies. *Mailing Add:* Dept Plant & Soil Sci Tex Tech Univ Box 4169 Lubbock TX 79409

HOPPER, PAUL FREDERICK, b Troy, NY, Oct 26, 24; m 50; c 4. ORGANIC CHEMISTRY, TOXICOLOGY. *Educ:* Holy Cross Col, BS, 44; Univ Notre Dame, PhD(chem), 51. *Prof Exp:* Res chemist, Stamford Res Labs, Am Cyanamid Co, 49-52, group leader develop res, 52-53, tech rep, Govt Res Liaison, Washington, DC, 53-56, mgr consumer prod develop, Farm & Home Div, NY, 56-57, dir food indust develop, Agr Div, 57-60; prod mgr, Ethicon, Inc, NJ, 60-61; mgr prod develop, Devro Div, Johnson & Johnson, 61-63, vpres prod mgt, 64-70, vpres corp planning, 70-71; dir environ health sci, Gen Foods Corp, Tarrytown, 71-72, dir nutrit & health sci, 72-74, corp dir strategic tech planning, 74-78, corp dir sci affairs, White Plains, 78-88; ADJ PROF FOOD SCI, CORNELL UNIV, 88-; CONSULT, 88- *Concurrent Pos:* Pres, Flavor Mfr; adj prof, Cornell Univ. *Mem:* Am Chem Soc; fel Inst Food Technologists; fel Am Inst Chemists; AAAS; Coun Agr Sci & Technol. *Res:* Barbituates; methylene-dioxy compounds; cyanamide derivatives; cyanoethylation; s-triazines; organic nitrogen chemistry; synthetic organic chemistry; food science and technology; protein chemistry; food microbiology; food additives safety; nutrition; toxicology; biomedicine; research planning; food law. *Mailing Add:* 84 Londonderry Dr Greenwich CT 06830

HOPPER, ROBERT WILLIAM, b New York, NY. GLASS SCIENCE. *Educ:* Univ Wash, BS, 63; Harvard Univ, SM, 67; Mass Inst Technol, PhD(mat sci), 71. *Prof Exp:* Staff scientist metall, Manlabs, Inc, 65-67; mat engr, Microwave & Power Tube Div, Raytheon Co, 67-68; vpres res, Am Res & Instrument Corp, 71-72; asst prof polymers, Mass Inst Technol, 72-76; METALLURGIST, METALS & CERAMICS DIV, LAWRENCE LIVERMORE LAB, UNIV CALIF, 76- *Mem:* Am Ceramic Soc; Am Soc Metals; Soc Rheology; Am Phys Soc; AAAS. *Res:* Glasses, particularly inorganic, low molecular weight organic, polymeric and metallic; rheology of complex materials and its relationship to induced structural changes; interfaces; phase changes; mathematical modelling; low-density foams. *Mailing Add:* 430 Messina Dr Sacramento CA 95819

HOPPER, SAMUEL HERSEY, b Boston, Mass, July 4, 11; m 37; c 2. BACTERIOLOGY, PUBLIC HEALTH. *Educ:* Mass Inst Technol, BS, 33, MS, 34, PhD(bact), 37. *Prof Exp:* Res assoc, Mass Inst Technol, 37-39; asst prof bact & chem, Ga Inst Technol, 39-42; assoc prof chem, bact & sanit, Sch Pub Health, Univ NC, 42-44; mem staff sanit prog, War Shipping Admin, USPHS, Washington, DC, 44-45; assoc prof, 45-50, prof pub health & chmn dept, sch med, Ind Univ, Indianapolis, 50-81; RETIRED. *Concurrent Pos:* Vis prof, Butler Univ; consult, Ind State Bd Health; consult & dir scientist res corp, USPHS. *Mem:* AAAS; Am Soc Microbiol; fel Am Pub Health Asn; Sigma Xi; fel AAAS; Inst Food Tech; NY Acad Med. *Res:* Biochemistry of bacteria; bacteriology and pharmacology of wetting agents; flotation process for detection of toxins or bacteria; food microbiology. *Mailing Add:* 5064 Boardwalk Pl Indianapolis IN 46220-5382

HOPPER, SARAH PRIESTLY, b Yuma, Tenn, Dec 21, 25. BIOCHEMISTRY, PHARMACOLOGY. *Educ:* Union Univ, Tenn, BS, 56; Univ Pittsburgh, MS, 60; St Louis Univ, PhD(pharmacol), 63. *Prof Exp:* Med technician, Oak Ridge Hosp, Tenn, 46-49 & Md Gen Hosp, Baltimore, 50-55; staff fel, Nat Inst Arthritis, Metab & Digestive Dis, 63-65; res asst, Univ Pittsburgh, 56-59, grad teaching asst, 59-60, res assoc biochem, Fac Arts & Sci, 65-66, from asst prof to assoc prof pharmacol, 67-76, assoc head, Dept Pharmacol & Physiol, 69-81, PROF PHARMACOL & PHYSIOL, UNIV PITTSBURGH, 76-, ACTG CHMN, DEPT PHARMACOL & PHYSIOL, 88- *Concurrent Pos:* Recipient res grants, NSF, NIH & Univ Pittsburgh. *Mem:* Protein Soc; Am Chem Soc; AAAS; Am Soc Biochem & Molecular Biol. *Res:* Biochemical enzymology; characterization of the enzyme system for the reduction of ribonucleotides in animals. *Mailing Add:* Dept Pharmacol & Physiol Univ Pittsburgh Sch Dent Med Pittsburgh PA 15261

HOPPER, STEVEN PHILLIP, b Tucson, Ariz, June 14, 45; div; c 2. ORGANOMETALLIC CHEMISTRY. *Educ:* Wabash Col, BA, 67; Mass Inst Technol, PhD(inorg chem), 72. *Prof Exp:* Asst prof org chem, Wabash Col, 72-73; asst prof inorg chem, Vassar Col, 73-80; proj scientist, Union Carbide Corp, Tarrytown, NY, 80-84, res chemist, 84-86, mgr, Nat Tech Sales, 86-88, prog mgr, 88-90; MGR MKT, C P HALL, CHICAGO, 90- *Mem:* Am Chem Soc. *Res:* Preparative organometallic chemistry of the main group elements, particularly silicon and boron groups; organosilicon polymers. *Mailing Add:* 532 Turner Ave Glenellyn IL 60137

HOPPES, DALE DUBOIS, b Liberty, Ind, Sept 13, 28; m 50; c 4. NUCLEAR PHYSICS. *Educ:* Purdue Univ, BS, 50; Cath Univ Am, MS, 56, PhD(nuclear physics), 61. *Prof Exp:* LEADER, RADIOACTIVITY GROUP, NAT INST STANDARDS & TECHNOL, 50- *Mem:* Am Phys Soc. *Res:* Radioactivity standardization. *Mailing Add:* Ctr Radiation Res Nat Measurement Lab Dept Com Nat Inst Standards & Technol Gaithersburg MD 20899

HOPPIN, RICHARD ARTHUR, b St Paul, Minn, May 15, 21; m 47; c 4. GEOLOGY. *Educ:* Univ Minn, AB, 42, AM, 47; Calif Inst Technol, PhD(geol), 51. *Prof Exp:* Asst mineral, Univ Minn, 46-47; asst optical mineral & petrol, Calif Inst Technol, 47-49; from asst prof to assoc prof, 52-61, chmn dept, 74-83, PROF STRUCT, UNIV IOWA, 61- *Concurrent Pos:* Jr geologist, US Geol Surv, 47-48. *Mem:* AAAS; Nat Asn Geol Teachers; Geol Soc Am; Am Geophys Union; Am Asn Petrol Geologists; Am Asn Photogram Remote Sensing. *Res:* Structural geology; geotectonics; remote sensing. *Mailing Add:* Dept Geol Univ Iowa Iowa City IA 52242

HOPPING, RICHARD LEE, b Dayton, Ohio, July 26, 28; m 51; c 3. OPTOMETRY. *Educ:* Southern Col Optom, BS & OD, 52. *Hon Degrees:* DOS, Southern Col Optom, 72. *Prof Exp:* Optometrist, pvt pract, 52-73; PRES, SOUTHERN CALIF COL OPTOM, 73- *Concurrent Pos:* Chmn, Nat Acad Optom, 83-89; pres, Asn Schs & Cols Optom, 83-85; vchmn, 13th Dist Med Qual Rev Comt, State Calif Bd Med Qual Assurance, 85-; chmn, Nat Educ Summit Conf, Am Optom Asn/Asn Schs & Cols Optom, 91. *Honors & Awards:* Dr Raymond Myers Award, Am Optom Student Asn, 90. *Res:* Numerous articles written on vision care. *Mailing Add:* Southern Calif Col Optom 2575 Yorba Linda Blvd Fullerton CA 92631

HOPPMANN, WILLIAM HENRY, II, b Charleston, SC, Sept 23, 08; m 32; c 2. RHEOLOGY. *Educ:* Col Charleston, BS, 29; George Washington Univ, MA, 35; Columbia Univ, PhD(appl mech), 47. *Prof Exp:* Model tester, US Exp Basin, Washington, DC, 29-30, jr physicist, 31-33; eng aide, Bur Construct Repair, US Navy Dept, 33-34, asst naval architect, Bur Ships, 34-38; assoc mat engr, Mat Lab, NY Naval Shipyard, 38-41, mech engr, 41-43, sr mech engr & res dir, Div Noise Control, 43-46, consult appl mech, 47; from asst prof to assoc prof mech eng, Johns Hopkins Univ, 47-57; prof mech, Rensselaer Polytech Inst, 57-66; vis prof eng, 66-67, prof, 67-75, EMER PROF ENG, UNIV SC, 75- *Concurrent Pos:* Consult, US Army Materiel Systs Anal Activ, 75-; SC State Govt coordinator, Representing ASME, 87- *Honors & Awards:* Russell Award, Univ SC. *Mem:* AAAS; fel Am Soc Mech Engrs; Am Soc Rheol; Soc Natural Philos. *Res:* Fluid flow; mechanical shock and vibration; elasticity; impact; strength of materials; rheological mechanics; bioengineering, environmental engineering; theoretical and applied mechanics; bioengineering; engineering systems analysis; theory of models. *Mailing Add:* 3031 Duncan St Columbia SC 29205

HOPPONEN, JERRY DALE, b Fargo, NDak, Mar 26, 46; m 83; c 2. ELECTROMAGNETIC PROPAGATION. *Educ:* NDak State Univ, BS; Univ Ariz, MS; Univ Colo, PhD(math). *Prof Exp:* Mathematician, Inst Telecommun Sci, 72-77; STAFF ENGR, LOCKHEED MISSILES & SPACE CO, 77- *Mem:* Math Asn Am. *Res:* Simulation of electromagnetic wave propagation through the terrestrial atmosphere, with emphasis on efficient numerical methods, absorption by molecular oxygen and water vapor, and basic ionospheric phenomena; matrix theory over special algebraic structures. *Mailing Add:* 1514 Poppy Way Cupertino CA 95014

HOPPS, HOPE ELIZABETH BYRNE, microbiology; deceased, see previous edition for last biography

HOPPS, HOWARD CARL, b Schenectady, NY, Aug 14, 14; m 37, 68; c 3. GEOGRAPHIC PATHOLOGY. *Educ:* Univ Okla, BS, 35, MD, 37; Univ Chicago, PhD(path), 70. *Prof Exp:* Intern, Evanston Hosp, Ill, 37-39; asst path, Univ Chicago, 40-41, from instr to asst prof, 41-44; prof & chmn dept, Univ Okla, 44-56; prof path & chmn dept & pathologist-in-chief, Univ Hosp, Univ Tex Med Br, 57-63; vis prof anat, Northwestern Univ, 64; chief div geog path, Armed Forces Inst Path, 64-70; curators prof path, 70-82, EMER PROF, UNIV MO, COLUMBIA, 82-; ADJ PROF PATH, MED COL, OHIO, 82- *Concurrent Pos:* Chmn subcomt motion pictures, Intersoc Comt Res Potential Path, Inc, 54-55 & 60-64; mem, Path Study Sect, NIH, 54-59, Adv Comt Staph Infection & Influenza, 58, Gen Res Training Grants Comt, 58 & Res Career Award Comt, 59-64; Fulbright vis prof, Univ Otago, NZ, 55; mem path test comt, Nat Bd Med Exam, 56-60; mem adv comt personnel for res, Am Cancer Soc, 58-61, vchmn, 60-61, mem adv comt res pathogenesis of cancer, 61-65, chmn, 63-65; mem res adv coun, 67-73, chmn, 71-73; mem adv comt regulatory biol, NSF, 60-61; ed sect educ & res, Bull Col Am Path, 60-64 & Int Path, 64-70; NIH res training fel, Northwestern Univ, 64; registr, Am Registry Geog Path, 64-70; clin prof, Univ Md, 65-70; vis prof, Temple Univ, 67-70; consult geog path, NASA, 69-70 & WHO, 71; mem subcomt geochem environ related to health & dis, Nat Acad Sci, 69-75, co-chmn, 70-75; chmn, Nat Acad Sci Panel Aging & Geochem Environ, 75-; consult, Int Atomic Energy Agency, 77 & Water Resources Div, US Geol Surv, 79-82; vpres, Mo Acad Sci, 75-76, pres, 76-77; pres, Med Col Ohio Club, 88-89, pres elect, 90-91. *Honors & Awards:* Howard Taylor Ricketts Prize, 42. *Mem:* Am Asn Immunol; Am Soc Clin Path; Am Soc Tropical Med & Hyg; Int Acad Path; Soc Environ Geochem & Health (secy-treas, 71-78, pres elect, 78-79, pres, 79-80); Sigma Xi. *Res:* Geographic pathology; medical ecology; infectious diseases; parasitic diseases; trace elements; epidemiology;

inflammation; immunity and allergy; reticuloendothelial system; cardiovascular and renal disease; forensic pathology; medical education; graduate education in environmental sciences. *Mailing Add:* Med Col Ohio 5939 Swan Creek Dr Toledo OH 43614

HOPPS, JOHN ALEXANDER, b Winnipeg, Man, May 21, 19; m 43; c 3. BIOMEDICAL ENGINEERING. *Educ:* Univ Man, BScEE, 41. *Hon Degrees:* DSc, Univ Man, 76. *Prof Exp:* Jr assoc res officer elec eng, Nat Res Coun Can, 41-49, prin res officer med eng, 49-80, head med eng sect, 73-80; RETIRED. *Concurrent Pos:* Res assoc, Univ Toronto, 56- *Honors & Awards:* Medtronics Inc Can Award, 75; Gen A G L McNaughton Award, Inst Elec & Electronics Engrs, 86; Distinguished Scientist Award, NAm Soc Pacing & Electrophysiol, 85. *Mem:* Inst Elec & Electronics Engrs; fel Can Med & Biol Eng Soc (pres, 65-70); Int Fedn Med & Biol Engrs (pres, 71-73, secy gen, 76-85); Int Union Phys & Eng Sci Med (secy gen, 85-88). *Res:* Medical engineering, including standards, cardiac pacemaking and diagnostic instrumentation. *Mailing Add:* Nat Res Coun Can Rm 307 Bldg M-50 Ottawa ON K1A 0R8 Can

HOPSON, CLIFFORD ANDRAE, b Portland, Ore, Feb 2, 28; m 55; c 3. GEOLOGY. *Educ:* Stanford Univ, BS, 51; Johns Hopkins Univ, PhD(geol), 55. *Prof Exp:* Guest investr geophys lab, Carnegie Inst, 53-55; from asst prof to assoc prof geol, Johns Hopkins Univ, 55-64; assoc prof, 64-66, chmn dept, 66-69, PROF GEOL, UNIV CALIF, SANTA BARBARA, 66-; GEOLOGIST, US GEOL SURV, 52- *Concurrent Pos:* Mem, Earth Sci Adv Panel, NSF, 72-75, chmn, 74-75. *Mem:* AAAS; fel Geol Soc Am; Geochem Soc; Am Geophys Union. *Res:* Igneous and metamorphic petrology; volcanology; tectonics; ophiolites; geology of Cascade Mountains, Central Appalachian Piedmont, California coast ranges and Oman; diving scientist AMAR-78 project. *Mailing Add:* Dept Geol Univ Calif Santa Barbara CA 93106

HOPSON, JAMES A, b New Haven, Conn, Aug 27, 35; m 61; c 2. VERTEBRATE PALEONTOLOGY. *Educ:* Yale Univ, BS, 57; Univ Chicago, PhD(paleozool), 65. *Prof Exp:* Res asst vet paleont, Peabody Mus Natural Hist, Yale Univ, 63-65, res assoc geol, 65-67; asst prof anat, 67-73, ASSOC PROF ANAT, UNIV CHICAGO, 73-, MEM COMT EVOLUTIONARY BIOL, 80- *Concurrent Pos:* NSF res grants, 65- *Mem:* Soc Vert Paleont; Soc Study Evolution; Am Soc Naturalists; Am Soc Zool. *Res:* Functional anatomy and evolution of mammal-like reptiles; origin of mammals and evolution of Mesozoic mammals; vertebrate faunas of the Permian and Triassic; cranial morphology of dinosaurs. *Mailing Add:* Dept Anat Univ Chicago 1025 E 57th St Chicago IL 60637

HOPSON, JOHN WILBUR, JR, b San Antonio, Tex, Feb 3, 40; div; c 2. PHYSICS. *Educ:* Southwest Tex State Univ, BS, 60; Univ Tex, Austin, PhD(physics), 68. *Prof Exp:* Res staff mem physics, 68-76, leader shock wave physics group, 76-78, ALT DIV LEADER DYNAMIC TESTING, LOS ALAMOS SCI LAB, 78- *Mem:* Am Phys Soc. *Res:* Shock waves in condensed materials; solid state physics; dynamic mechanical properties of solids. *Mailing Add:* 1940 Camino Mora Los Alamos NM 87544

HOPTMAN, JULIAN, b New York, NY, Apr 23, 25; m 59; c 2. MICROBIOLOGY, ENVIRONMENTAL PHYSIOLOGY. *Educ:* City Col New York, BS, 48; Pa State Univ, MA, 50; George Washington Univ, PhD(physiol, viral pathogenesis), 63. *Prof Exp:* Res assoc, George Washington Univ, 59-63, asst res prof microbiol, 63-65; RES CONSULT, US GOVT, 51- *Mem:* AAAS; fel Aerospace Med Asn; Am Soc Microbiol; NY Acad Sci; Am Inst Biol Sci; Sigma Xi. *Res:* Environmental biology and medicine; aerospace and undersea medicine; public health; physiological aspects of immunity. *Mailing Add:* 6958 Duncraig Ct McLean VA 22101

HOPTON, FREDERICK JAMES, b Bristol, Eng, May 5, 36; m 60; c 4. ENVIRONMENTAL CHEMISTRY, ENVIRONMENTAL ENGINEERING. *Educ:* Bristol Univ, BSc, 58, PhD(phys chem), 62. *Prof Exp:* Fel, Ont Res Found, 62-63, assoc res scientist, 63-64; lectr, Bristol Univ, 64-66; res scientist, 66-71, sr res scientist, 71-76, prin scientist, 76-77, asst dir, 77-84, DIR, ONT RES FOUND, 84- *Mem:* Chem Inst Can; Can Soc Chem Engrs; Air Pollution Control Asn; Source Eval Soc; Occup Health Asn Ont. *Res:* Air pollution, specifically ambient air and stack sampling; research and development of pollution control methods; dispersion of pollutants; odor pollution; occupational and environmental health; risk assessment. *Mailing Add:* Ont Res Found Sheridan Park 2395 Speakman Dr Mississauga ON L5K 1B3 Can

HOPWOOD, LARRY EUGENE, b Frederick, Md, Dec 11, 45; m 83; c 2. RADIATION BIOLOGY, CELL BIOLOGY. *Educ:* Johns Hopkins Univ, BA, 67; Wash Univ, St Louis, PhD(molecular biol), 71. *Prof Exp:* Res instr radiol, Wash Univ Sch Med, 71-72; res assoc radiation & biol, Colo State Univ, 72-74, asst prof, 75-77; co-chmn biophysics grad prog, 80-84, ASSOC PROF RADIOL, MED COL WIS, 77-, ASSOC PROF PHARMACOL, 78-, ASSOC PROF RADIATION ONCOL, 82- *Concurrent Pos:* Res Career Develop Award, Nat Cancer Inst, 77-82; prin investr, Nat Cancer Inst grant, 78-82, 84-; grant, Am Cancer Soc, 79-82. *Mem:* Radiation Res Soc; AAAS; Am Soc Therapeut Radiologists; Am Soc Clin Oncologists; Am Asn Cancer Res. *Res:* Cell killing and mutation by combined hyperthermia and x-rays; membrane fluidity changes in mammalian cells; radiation response of human tumor cells; metal bound antineoplastic agents as radiosensitizers; molecular biology; resistance of tumors to drugs. *Mailing Add:* 2436 N 88th St Milwaukee WI 53226

HORACKOVA, MAGDA, b Braque, Czechoslovakia, Nov 5, 40. CELLULAR CARDIOLOGY. *Educ:* Acad Sci, PhD(physiol), 68. *Prof Exp:* Assoc prof, 78-83, PROF PHYSIOL & BIOPHYS, DALHOUSIE UNIV, 68- *Mailing Add:* Dept Physiol Dalhousie Univ Med Tupper Bldg Halifax NS B3H 4H7 Can

HORAI, KI-ITI, b Tokyo, Japan, Apr 25, 34. GEOPHYSICS, LUNAR SCIENCE. *Educ:* Univ Tokyo, BSci, 58, MSci, 60, DSci(geophys), 63. *Prof Exp:* Sakkokai Found fel geophys & asst, Earthquake Res Inst, Univ Tokyo, 63-66; res assoc, Mass Inst Technol, 66-70; vis scientist lunar sci, Lunar Sci Inst, 70-71; res assoc geophys, Lamont-Doherty Geol Observ, Columbia Univ, 71-82; prof geophys, Meteorol Col, 83-89, RES SCIENTIST, METEOROL RES INST, JAPAN METEOROL AGENCY, 89- *Honors & Awards:* Okada Prize, Oceanog Soc Japan, 68. *Mem:* Am Geophys Union; AAAS; Seismol Soc Japan. *Res:* Solid earth science; measurement and theoretical interpretation of lunar and terrestrial heat flow; measurement of physical properties of terrestrial materials and its application to the study of internal constitution of the earth. *Mailing Add:* Meteorol Inst Nagaminme 1-1 Tsukuba-shi Ibaraki-Ken 305 Japan

HORAK, DONALD L, b Wahoo, Nebr, July 13, 37; m 59; c 3. FISHERIES. *Educ:* Colo A&M Col, BS, 60; Colo State Univ, MS, 62, PhD(fish sci), 70. *Prof Exp:* FISH RES CHIEF, COLO DIV WILDLIFE, 62- *Mem:* Am Fisheries Soc; Wildlife Soc; Sigma Xi. *Res:* Rainbow trout, especially development of practical dry diets, culture techniques and disease control; forage species introductions; heavy metal studies; lake aeration. *Mailing Add:* 745 S Summit View Dr Ft Collins CO 80524

HORAK, JAMES ALBERT, b Plainfield, NJ, Oct 28, 31; m 54; c 3. MATERIALS SCIENCE, NUCLEAR REACTOR TECHNOLOGY. *Educ:* Univ Ill, Urbana, BS, 58; Northwestern Univ, MS, 63, PhD(mat sci), 66. *Prof Exp:* Asst metallurgist, Argonne Nat Lab, 58-68; staff scientist, Reactor Tech Div, Los Alamos Sci Lab, 68-69; assoc prof nuclear eng, Univ NMex, 69-74, prof, 74; SR RES SCIENTIST, OAK RIDGE NAT LAB, 74- *Concurrent Pos:* Consult, 68-70; staff mem, Reactor Studies Div, Sandia Labs, 69-71, staff scientist, 71-74; guest scientist, Los Alamos Sci Lab, 72-74; USA-USSR exchange team on fusion energy, US Dept Energy, 74-; US rep, working group on fast reactors, specialists meeting on mech properties of struct mat, Int Atomic Energy Agency, 83-; vchmn, Mat Sci & Technol Div, Am Nuclear Soc, 77-78, chmn, 78-79. *Mem:* Am Soc Testing & Mat; fel Am Nuclear Soc (secy-treas, 76-77). *Res:* Irradiation effects in metals, alloys and nuclear fuels; development of the materials to harness fission and fusion energy; development of improved fuels and structural materials for nuclear power reactors; development of refractory alloys for space nuclear power. *Mailing Add:* Metals & Ceramics Div Oak Ridge Nat Lab Oak Ridge TN 37831-6155

HORAK, MARTIN GEORGE, b Baltimore, Md, Dec 28, 36; m 64; c 3. APPLIED MATHEMATICS. *Educ:* Loyola Col, Md, BS, 58; Univ Notre Dame, MS, 60; Univ Md, College Park, PhD(math), 70. *Prof Exp:* Assoc mathematician, Appl Physics Lab, Johns Hopkins Univ, 60-62; asst prof math, Loyola Col, Md, 62-67; from asst prof to assoc prof, 67-73, PROF MATH, TOWSON STATE UNIV, 73-, CHMN, DEPT MATH, 73-76, 84- *Mem:* Math Asn Am. *Res:* Bochner p-integrable solution of nonlinear hyperbolic partial differential equations; periodic solutions of hyperbolic partial differential equations in the large. *Mailing Add:* Dept Math Towson State Univ Baltimore MD 21204

HORAK, VACLAV, b Prague, Czech, Dec 9, 22; US citizen; m 49; c 1. ELECTROCHEMISTRY, SURFACE CHEMISTRY. *Educ:* Charles Univ, Czech, ANDr, 48, CSc, 61. *Prof Exp:* Asst prof chem, Charles Univ, 48-68; PROF CHEM, GEORGETOWN UNIV, 68- *Concurrent Pos:* Vis scientist, NIH/Nat Health Inst Lab Chem Pharmacol, 65-66 & 68; vis prof, Univ Indust, Italy, 77, Univ Utah, 79 & 80 & Univ Belo Horizonte, Brazil, 78. *Mem:* Am Chem Soc; Royal Soc Chem; Czech Soc Art & Sci; Electrochem Soc. *Res:* Chemistry of organic sulfur compounds, guinones, pharmaceuticals; organic synthesis; electrosynthesis; electrochemistry; polymer surfaces modification and analysis; bioorganic chemistry of melanin; electron transfer; drug distribution. *Mailing Add:* 5508 Oakmont Ave Bethesda MD 20817-3528

HORAKOVA, ZDENKA, b Jindrichuv Hradec, Czech, Apr 6, 25; US citizen; m 49; c 1. PHARMACOLOGY, PHARMACY. *Educ:* Charles Univ, Prague, Magister Pharm, 49, RNDr, 52; Czechoslovakian Acad Sci, Prague, PhD(pharmacol), 61. *Prof Exp:* Teaching asst pharmacol, Med Fac Charles Univ, Prague, 49-50; res pharmacologist, Res Inst Pharm & Biochem, Prague, 50-68, exp therapeut br, Nat Heart, Lung & Blood Inst, NIH, Bethesda, 69-74, sect molecular pharmacol, Pulmonary Br, 74-77 & lab cellular metab, 77-78; toxicologist, Food Safety & Inspection Serv, 78-87, TOXICOLOGIST, FOREST SERV, USDA, 87- *Concurrent Pos:* Head pharmacol dept, Res Inst Pharm & Biochem, Prague, 58-68; vis guest, Lab Chem Pharmacol, Nat Heart Inst, 65-66. *Mem:* Am Soc Pharmacol & Exp Therapeut; Soc Exp Biol & Med; Inflammation Res Asn; Soc Toxicol; Asn Govt Toxicologists. *Res:* Inflammation, specifically mediators, cell involvement, role of the H receptors and the effect of drugs in inflammatory processes; the role of histamine in immediate hypertensitivity reactions in the stomach and the lungs; toxicology, particularly the screening of new drugs; toxicological evaluation of potentially toxic chemical residues such as drugs and pesticides in meat and poultry and environment, risk assessment evaluation; toxicology. *Mailing Add:* USDA Forest Serv PO Box 96090 Washington DC 20090-6090

HORAN, FRANCIS E, b Atchison, Kans, Mar 1, 14; m 46; c 4. PHYSICAL ORGANIC CHEMISTRY. *Educ:* St Benedict's Col, Kans, AB, 35; Columbia Univ, MA, 36, PhD(chem), 45. *Prof Exp:* Prof chem, St Francis Xavier, Can, 36-39; res assoc, Midwest Res Inst, 45-48; asst prof phys chem, Univ Ariz, 48; fel phys biochem, Med Col, Cornell Univ, 48-49; assoc prof chem, St Martin's Col, 49-52; res chemist, Huron Milling Co, 52-56; res chemist, Hercules Powder Co, 56-59; res mgr agr prod, 59-66, assoc dir res, 66-69, DIR, ARCHER DANIELS MIDLAND CO, 69- *Mem:* Am Chem Soc; Am Asn Cereal Chem. *Res:* Physico-chemical aspects of starch chemistry; industrial starch chemistry and utilization of grain sorghums; electrical conductivity of blood; streaming potentials of blood; ultrasonic investigations on high polymers; vegetable proteins; food and industrial grades; fermentation; research management. *Mailing Add:* 50 Malaga Way Hot Springs Village AR 71909

HORAN, MICHAEL, b Rochester, Minn, June 27, 45. CARDIOLOGY. *Educ:* Manhattan Col, BA, 67; Georgetown Univ, MD, 71; Johns Hopkins Univ, ScM, 78. *Prof Exp:* Attend physician, Baltimore USPHS Hosp, 77-81, asst chief ambulatory care, 77-81, dir, 81; instr med Johns Hopkins Univ, 77-81, dir, Johns Hopkins Hosp Hypertension Mgt Clin, 78-81; spec asst to dir, Div Heart & Vascular Dis, 81-82, chief Hypertension & Kidney Dis Br, 82-89, ASSOC DIR CARDIOL, NAT HEART, LUNG & BLOOD INST, 89-; ASST PROF, MED DEPT MED, JOHNS HOPKINS UNIV, 81- *Concurrent Pos:* Active staff, Johns Hopkins Hosp, 77-; sci adv, Nat High Blood Pressure Educ Prog, 82-; NIH coordr, US-USSR Sci Exchange Agreement in Hypertension, 82-; chmn, Nat Heart, Lung & Blood Inst Task Force Control of Blood Pressure in Children, 83-86, Res Activities Related to Blacks & other Minorities, 84-88; mem, Coun High Blood Pressure Res Educ Comt, Am Heart Asn, 89-, Nat Kidney & Urologic Dis Adv Bd, 89-, Nat Comn on Sleep Dis, 90-; sci adv, Nat Heart Attack Alert Prog, 91- *Mem:* Am Col Physicians; Am Pub Health Asn; Am Fedn Clin Res; Am Heart Asn. *Res:* Purification of thrombin and its amino acid sequence; primary care; hypertension; ischemic heart disease; ventricular ectopy; vascular biology and diseases. *Mailing Add:* Nat Heart Lung & Blood Inst Fed Bldg Rm 320 7550 Wisconsin Ave Bethesda MD 20892

HORAN, PAUL KARL, b Schenectady, NY, Sept 25, 42; m 63; c 2. PATHOLOGY, CELL BIOLOGY. *Educ:* State Univ NY, Albany, BS, 65; Pa State Univ, MS, 67, PhD(biophys), 70. *Prof Exp:* Emer lab technician, Albany Med Col Hosp, 63-65; guest researcher med physics, Donner Lab, Lawrence Radiation Lab, Berkeley, 68-69; exten lectr phys sci, Pa State Univ, 70; res fel cell biol, Donald S Walker Lab, Sloan-Kettering Inst Cancer Res, 71-72; staff mem, Los Alamos Sci Lab, 72-74; from asst prof to assoc prof path, Med Sch, Univ Rochester, 74-82; assoc dir immunol, Smith Kline & French, 82-88; CHMN BD, CHIEF SCI OFFICER & FOUNDER, ZYNAXIS CELL SCI, INC, 88- *Concurrent Pos:* NIH fel, Pa State Univ, 70-71; Nat Acad Sci nat sci travel award to IV Int Cong Radiation Biol, Evian, France, 70; lectr, Proj Newgate, Pa State Penitentiary, 70. *Mem:* AAAS; Soc Anal Cytol. *Res:* Studies of cell surface antigens on tumor cells using monoclonal antibodies; development of new methodologies in flow cytometry. *Mailing Add:* Zynaxis Cell Sci Inc 371 Phoenixville Pike Malvern PA 19355

HORBATSCH, MARKO M, b Goettingen, Repub Fed Ger, Mar 3, 54; m 82; c 2. SCATTERING THEORY, FIELD THEORY. *Educ:* JWV Goethe Universitat, Frankfurt, Vordiplm, 74, Dipl, 76, Dr phil nat(physics), 81. *Prof Exp:* Postdoctoral fel, Dept Physics, York Univ, 82-84; res assoc, Inst Physic theory, JWV Goethe Universitat, 84-85; res fel, NSERC, dept physics, 85-88; asst prof, 88-90, ASSOC PROF PHYSICS, DEPT PHYSICS, YORK UNIV, 90- *Res:* Theoretical description of scattering processes in atomic physics involving many particles by semiclassical time dependent field theory; quantum mechanical few body problem; atoms in intense laser fields. *Mailing Add:* Dept Physics York Univ 4700 Keele St Toronto ON M3J 1P3 Can

HORBETT, THOMAS ALAN, b Buffalo, NY, Aug 27, 43; m 81. BIOCHEMISTRY. *Educ:* State Univ NY, BSc, 65; Univ Wash, PhD(biochem), 70. *Prof Exp:* Res assoc bioeng, 71-75, from res asst prof to res prof, 75-88, PROF BIOENG & CHEM ENG, UNIV WASH, 88- *Concurrent Pos:* Fel, Univ Wash, 70-71 & 71-72, prin investr grants, 76-91; distinguished lectr, Controlled Drug Delivery, Col Pharm, Rutgers Univ, 89. *Honors & Awards:* Clemson Award for Basic Res, Soc Biomat, 89. *Mem:* AAAS; Soc Biomat; Am Chem Soc. *Res:* Protein-protein interactions; protein modification; interfacial proteins; cell adhesion; biomaterials; insulin delivery devices. *Mailing Add:* Dept Chem Eng Univ Wash BF-10 Seattle WA 98195

HORCH, KENNETH WILLIAM, b Cleveland, Ohio, 42; m; c 2. NEUROPHYSIOLOGY, SENSORY PHYSIOLOGY. *Educ:* Lehigh Univ, BS, 65; Yale Univ, MPhil, 68, PhD(biol), 71. *Prof Exp:* Instr biol, Purdue Univ, 70-71; fel physiol, 71-74, res instr, 73-75, asst prof physiol, Med Ctr, 75-81, ASSOC PROF PHYSIOL, UNIV UTAH SCH MED, 81-, ASSOC PROF BIOENG, 86- *Concurrent Pos:* Prin investr, NIH, NSF, 74-; vis prof, Univ Sci & Med, Grenoble, France, 86; dir neurol testing, Topical Testing Inc, 86- *Mem:* AAAS; NY Acad Sci; Soc Neurosci; Sigma Xi; Inst Elec & Electronics Engrs. *Res:* Somatosensory physiology; nerve regeneration and repair; neuroprosthetics. *Mailing Add:* Dept Bioeng Univ Utah 2480 MEB Salt Lake City UT 84112

HORD, CHARLES W, b Casper, Wyo, Feb 16, 37; m 59; c 2. PHYSICS, ATMOSPHERIC PHYSICS. *Educ:* Univ Colo, BA, 59, PhD(physics), 64. *Prof Exp:* Scientist, Rocky Flats Div, Dow Chem Co, 59-60; asst prof physics, Southern Colo State Col, 64-66; res physicist, Lab Atmospheric & Space Physics, 66-74, adj assoc prof, 74-76, adj prof, 76-80, PROF ASTRO-GEOPHYS, UNIV COLO, BOULDER, 80- *Concurrent Pos:* Planetary exploration missions; Mariner 6 and 7, 1969; Mariner 9, 1971; Voyager, 1978 and Pioneer Venus, 1978. *Mem:* Am Inst Physics. *Res:* Mariner Mars 1969 planetary exploration mission; ultraviolet spectrometer experiment on the Mars mission. *Mailing Add:* 980 Hartford Dr Boulder CO 80303

HORD, WILLIAM EUGENE, b Leola, SDak. ELECTRICAL ENGINEERING. *Educ:* Mo Sch Mines, BSc, 59; Univ Mo-Rolla, MSc, 63, PhD(elec eng), 66. *Prof Exp:* Assoc engr microwave, Sperry Gyroscope Co, 59-60; instr eng, Univ Mo-Rolla, 60-66; group engr microwave, Emerson Elec Co, 66-69; assoc prof eng, SOUTHERN ILL UNIV, EDWARDSVILLE, 74-chmn dept, 74-80,; VPRES, MICROWAVE APPL GROUP, SANTA MARIA, CALIF. *Concurrent Pos:* Consult, Emerson Elec Co, 69-, Monsanto Res Corp, 72. *Mem:* Sr mem Inst Elec & Electronics Engrs; Sigma Xi. *Res:* Propagation in ferrite-loaded waveguides; application of ferrite devices to phased array antennas. *Mailing Add:* 1509 Goldsmith Ct Santa Maria CA 93454

HORDON, ROBERT M, b New York, NY, July 10, 36; m 59; c 2. RESOURCE MANAGEMENT, ATMOSPHERIC SCIENCES. *Educ:* Brooklyn Col, BA, 59; Columbia Univ, MA, 65, PhD(geog), 70. *Prof Exp:* From instr to asst prof, 67-76, ASSOC PROF GEOG, RUTGERS UNIV, 76- *Mem:* Am Geophys Union; AAAS; Am Inst Hydrol; Am Water Resources Asn; Asn Am Geographers; Am Geog Soc. *Res:* Surface and ground water hydrology; ground water yield; wetlands; water quality; water demand; yield of consolidated rock areas. *Mailing Add:* Dept Geog Rutgers Univ New Brunswick NJ 08903

HORECKER, BERNARD LEONARD, b Chicago, Ill, Oct 31, 14; m 36; c 3. BIOCHEMISTRY. *Educ:* Univ Chicago, SB, 36, PhD(chem), 39. *Prof Exp:* Res assoc enzymes, Univ Chicago, 39-40; jr examr phys sci, US Civil Serv Comn, 40-41; biochemist, NIH, 41-53, chief, Biochem & Metab Lab, Nat Inst Arthritis & Metab Dis, 53-59; prof microbiol & chmn dept, Col Med, NY Univ-Bellevue Med Ctr, 59-63; prof molecular biol & chmn dept, Albert Einstein Col Med, 63-71, assoc dean sci affairs, 71-72; mem, Roche Inst Molecular Biol, 72-84, head, Lab Molecular Enzym, 77-84; prof, 84-89, EMER PROF BIOCHEM, MED COL, CORNELL UNIV, 89-, DEAN, GRAD SCH MED SCI, 84- *Concurrent Pos:* Vis prof, Univ Calif, Berkeley, 54, Univ Ill, 57, Univ Parana, Brazil, 60 & 63, Cornell Univ, Ithaca, NY, 65, Univ Rotterdam, 70, Indian Inst Sci, Bangalore, 71, Albert Einstein Col Med, 72-84; consult, Lederle Labs, 59-72; Fulbright scholar, 64; Career Develop Award Study Sect, NIH, 66-70; mem, comt personnel, Am Cancer Soc, 69-73, sci adv comt biochem & chem carcinogenesis, 74-78, coun res & clin invest awards, 84-88; mem, Biol Div Adv Comt, Oak Ridge Nat Lab, 76-80; adj mem, Roche Inst Molecular Biol, 84-89; assoc dean res & sponsored progs, Med Col, Cornell Univ, 84- *Honors & Awards:* Award Biol Chem, Wash Acad Sci, 54; Phillips Lectr, Haverford Col, 65; Reilly Lectr, Notre Dame Univ, 69; Merck Award, Am Soc Biol Chemists, 81; Neuberg Medal, Virchow-Pirquet Med Soc, 81. *Mem:* Nat Acad Sci; fel Am Acad Arts & Sci; Am Soc Biol Chemists (pres, 68-69); fel AAAS; Harvey Soc (vpres, 69-70, pres, 70-71); hon mem Brazil Acad Sci; Am Chem Soc; hon mem Swiss Biochem Soc; hon mem Japan Biochem Soc; hon mem Indian Sci Acad. *Res:* Isolation and characterization of respiratory enzymes; spectrophotometry of hemoglobin and derivatives; enzymology; carbohydrate metabolism; enzyme structure; mechanism of action regulation; thymic peptides. *Mailing Add:* Grad Sch Med Sci Cornell Univ 1300 York Ave New York NY 10021

HORECZY, JOSEPH THOMAS, b Falls City, Tex, Nov 2, 13; m 42; c 1. ORGANIC CHEMISTRY. *Educ:* Southwest Tex State Teachers Col, BS, 34; Univ Tex, Austin, MA, 39, PhD(org chem), 41. *Prof Exp:* Teacher, Pub Schs, Tex, 34-41; res chemist, Humble Oil & Refining Co, 41-48, sr res chemist, 48-54, sect head, 54-60, res assoc, 60-64, sr res assoc, 64-65, sr res assoc, Esso Res & Eng Co, 65-66, assoc sci adv, 66-67, sci adv, 67-75, chief scientist, plastic properties, 75-78, CONSULT, PLASTIC TECHNOL DIV, EXXON, CHEM CO, 78- *Concurrent Pos:* Staff Consult, S & B Engrs, Inc, Houston, TX. *Mem:* AAAS; Am Chem Soc; Am Inst Chem Eng; Soc Plastic Engr. *Res:* Synthetic organic chemicals from petroleum; aviation gasoline; motor gasoline; analytical chemistry; synthesis of new polymers; fibers from polymers; dyeability of fibers; flame retardancy of plastics and fibers; solids handling. *Mailing Add:* 105 Edgewood Dr Baytown TX 77520

HOREL, JAMES ALAN, b Eureka, Calif, Sept 11, 31; m 52; c 3. COGNITIVE NEUROSCIENCE. *Educ:* Humboldt State Col, AB, 58; Ohio State Univ, MA, 60, PhD(physiol psychol), 63. *Prof Exp:* Fel anat, Univ Fla, 62-63, Nat Inst Neurol Dis & Blindness fel, 63-64; from asst prof to assoc prof anat, Univ Fla, 64-69; assoc prof, 69-74, PROF ANAT, STATE UNIV NY UPSTATE MED CTR, 74- *Mem:* AAAS; Am Asn Anatomists; fel Am Psychol Asn; Soc Neurosci; fel Am Psychol Soc. *Res:* Neuroanatomy of behavior; neuroanatomy of learning and memory processes. *Mailing Add:* Dept Anat State Univ NY Health Sci Ctr Syracuse NY 13210

HORELICK, BRINDELL, b New York, NY, Sept 5, 32; m 65. GENERAL MATHEMATICS. *Educ:* Mass Inst Technol, BS, 52; Univ Chicago, MS, 53; Wesleyan Univ, PhD(math), 67. *Prof Exp:* Instr math, Villanova Univ, 59-60; from instr to asst prof, Lafayette Col, 60-67; assoc prof, State Univ NY Col Cortland, 67-68; ASST PROF MATH, UNIV MD, 68- *Mem:* Am Asn Univ Prof; Math Asn Am; Sigma Xi. *Res:* Topological dynamics. *Mailing Add:* 59 Glenwood Ave Cantonsville MD 21228

HOREN, DANIEL J, b New London, Conn, May 17, 28; m 59; c 2. NUCLEAR PHYSICS. *Educ:* Mass Inst Technol, SB, 52; Stanford Univ, PhD, 59. *Prof Exp:* AEC fel, Lawrence Radiation Lab, Univ Calif, 59-63; mem staff, USN Radiol Defense Lab, Calif, 63-67, head accelerator br, 67-69; dir nuclear data proj, 69-75, SR RES STAFF OAK RIDGE NAT LAB, 75- *Concurrent Pos:* Fel Sloan Found, 58-59; vis prof nuclear physics, Univ La Plata, Arg, 61, Inst Nuclear Sci, Univ Grenoble, 73-74; mem, US Nuclear Data Comt, 72-75. *Mem:* Fel Am Phys Soc; Sigma Xi; AAAS. *Res:* Low and intermediate energy nuclear physics; heavy-ion reactions. *Mailing Add:* Oak Ridge Nat Lab PO Box 2008 Oak Ridge TN 37831-6368

HORENSTEIN, EVELYN ANNE, b Philadelphia, Pa, July 7, 24. DEVELOPMENTAL BIOLOGY, CELL BIOLOGY. *Educ:* Univ Pa, AB, 45; Mich State Univ, PhD, 65. *Prof Exp:* Res technician, Philadelphia Serum Exchange, 45-46; asst bact enzymes, Wyeth Inst Appl Biochem, 49-53; res assoc mycol, Mich State Univ, 53-65, asst res prof, 65-66; grants assoc, 66-68, exec secy hemat study sect, 68-69, exec secy cell biol study sect, 69-77, health scientist admin, Div Heart & Vascular Dis, Nat Heart, Lung & Blood Inst, 77-81, EXEC SECY, CELL BIOL & PHYSIOL STUDY SECT, NIH, 81- *Concurrent Pos:* Res assoc mycol, Univ Pa, 53-65. *Mem:* Am Soc Cell Biol; Soc Develop Biol; AAAS; Sigma Xi. *Res:* Physiology and biochemistry of development. *Mailing Add:* Div Res Grants Nat Inst Health Bethesda MD 20892

HORENSTEIN, SIMON, b Providence, RI, Sept 8, 24; m 48; c 3. NEUROLOGY. *Educ:* Univ Ill, BS, 46, MD, 48. *Prof Exp:* Assoc neurol, Sch Med, Harvard Univ, 60-62; from asst prof to assoc prof, Case Western Reserve Univ, 62-70; PROF NEUROL & CHIEF DEPT, UNIV & CHIEF NEUROL, HOSPS, ST LOUIS UNIV, 70-, CHMN DEPT, 72-; ASSOC, VET ADMIN HOSP, 70-, CHIEF NEUROL, 72- *Concurrent Pos:* Nat Heart Inst fel, Boston City Hosp, 53-54; vis physician, Boston City Hosp, 54-62; neurologist, Mass Ment Health Ctr, 60-62; consult, Vet Admin Hosp, Cleveland, Ohio, 62-70, chief neurol, Highland View Hosp, 62-70; chmn stroke comt, Bistate Regional Med Prog, 71-; spec consult to dir, NIH, Neurol Sci Res Training Comt A, 71- *Mem:* Am Neurol Asn; fel Am Col Physicians; fel Am Acad Neurol (2nd vpres, 65-67); Asn Res Nerv & Ment Dis. *Res:* Behavioral consequences of visual fielo defects. *Mailing Add:* 3655 Vista Ave St Louis MO 63110-2594

HORGAN, CORNELIUS OLIVER, b Cork, Ireland, May 16, 44; US citizen; m 71; c 2. APPLIED MECHANICS, APPLIED MATHEMATICS. *Educ:* Nat Univ Ireland, BS, 64, MS, 65; Calif Inst Technol, PhD(appl mech), 70. *Hon Degrees:* DSc, Nat Univ Ireland, 83. *Prof Exp:* Lectr eng mech, Univ Mich, 70-72; sr res assoc appl math, Univ E Anglia, 72-74; assoc prof appl mech, Univ Houston, 74-78; vis assoc prof, Northwestern Univ, 77-78; from assoc prof to prof appl mech, Mich State Univ, 78-88; PROF APPL MATH, UNIV VA, 88- *Concurrent Pos:* Vis assoc appl mech, Calif Inst Technol, 79 & 81, vis prof, 84-85; vis prof appl math, Cornell Univ, 82. *Mem:* Am Soc Mech Engrs; Soc Indust & Appl Math; Am Math Soc; fel Am Acad Mech; Soc Eng Sci; Soc Natural Philos. *Res:* Continuum mechanics; partial differential equations arising in the physical sciences; linear and nonlinear elasticity. *Mailing Add:* Dept Appl Math Univ Va Charlottesville VA 22903-1226

HORGAN, JAMES D(ONALD), b Grand Rapids, Mich, May 21, 22; m 45; c 2. ELECTRICAL ENGINEERING. *Educ:* Marquette Univ, BEE, 47, MS, 51; Univ Wis, PhD, 57. *Prof Exp:* Mem staff, Radiation Labs, Mass Inst Technol, 43-45; from instr to assoc prof, 47-56, chmn dept elec eng, 56-61, prof elec eng, Marquette Univ, 57-87, clin prof eng med, Med Col Wis, 68-87; RETIRED. *Concurrent Pos:* Consult, Square D Co, Wis, 54-59, A C Spark Plug Div, Gen Motors Corp, 59- & Vet Admin Hosp, Wood, Wis, 63- *Mem:* Am Soc Eng Educ; Inst Elec & Electronics Engrs. *Res:* Biomedical engineering, particularly computer simulation of biological systems including human respiratory and circulatory systems. *Mailing Add:* Eng Sch Marquette Univ Milwaukee WI 53233

HORGAN, STEPHEN WILLIAM, b Springfield, Mass, Aug 23, 42; m 64; c 2. MEDICINAL CHEMISTRY. *Educ:* Xavier Univ, Ohio, BS, 64, MS, 66; Univ Cincinnati, PhD(org chem), 71. *Prof Exp:* Res asst med chem, 66-70, org res chemist, 70-74, proj leader, Chem Develop Dept, Merrell-Nat Labs, Richardson-Merrell, Inc, 74-79, sect head, Chem Develop Dept, Merrell Nat Labs, 79-81; RES ASSOC, CHEM DEVELOP DEPT, MERRELL DOW PHARMACEUT, 81- *Mem:* Am Chem Soc; Sigma Xi. *Res:* Synthetic medicinal agents; product process development including pilot plant scale-up. *Mailing Add:* Marion Merrell Dow 2110 Galbraith Rd Cincinnati OH 45215

HORGEN, PAUL ARTHUR, b Camp Rucker, Ala, July 8, 44; m 68; c 2. CELL BIOLOGY, MOLECULAR BIOLOGY. *Educ:* Univ Northern Iowa, BA, 66; Univ Iowa, MS, 68; State Univ NY & Syracuse Univ, PhD(bot), 71. *Prof Exp:* Res assoc bot & biochem, Univ Ga, 71-72; from asst prof to assoc prof, 72-79, PROF BOT, ERINDALE CAMPUS, UNIV TORONTO, 79-, CO-DIR, MUSHROOM RES GROUP, 83- *Concurrent Pos:* co-dir, NAm Poison Mushroom Res Ctr, 76-81; assoc chmn, dept bot, Univ Toronto, dir Ctr Plant Biotechnol. *Honors & Awards:* Res Award, Sigma Xi, 71. *Mem:* Int Plant Molecular Biol Soc; Mycol Soc Am; Can Soc Cell Biol; Can Genetics Soc; Can Soc Plant Molecular Biol. *Res:* Effects of steroid hormones on gene activation and gene product accumulation utilizing eukaryotic microbes as model systems; extra chromosomal elements in fungi; agaricus bisporus; biotechnology and genetic engineering of the commercial mushroom, Agaricus. *Mailing Add:* Dept Bot Univ Toronto Erindale Col Mississauga ON L5L 1C6 Can

HORGER, EDGAR OLIN, III, b Eutawville, SC, May 30, 37; m 60; c 3. OBSTETRICS. *Educ:* Furman Univ, BS, 59; Med Col SC, MD, 62, Am Bd Obstet & Gynec, cert, 70, cert maternal & fetal med, 74. *Prof Exp:* Intern med, Hosp, Med Col SC, 62-63, resident, obstet & gynec, 63-67; fel fetal physiol, Univ Pittsburgh, 67-68, asst prof obstet & gynec, 68-69; from asst prof to assoc prof, 69-76, PROF OBSTET & GYNEC, MED UNIV SC, 76-, PROF RADIOL, 78-, DIR MATERNAL & FETAL MED, 73- *Concurrent Pos:* Bd examiners, Am Bd Obstet & Gynec, 75-79 & 81-85; State Bd Med Examiners, SC, 85- *Mem:* AMA; Am Col Obstet & Gynec; Am Asn Obstet & Gynec; Am Gynec Soc; Soc Perinatal Obstetricians. *Res:* Assessment of fetal physiology and pathophysiology; detection of fetal abnormalities; management of maternal and fetal problems caused by complications of pregnancy; author or coauthor of over 80 publications. *Mailing Add:* Dept Obstetrics & Gynec Med Univ SC 171 Ashley Ave Charleston SC 29425

HORGER, LEWIS MILTON, b Dearborn, Mich, June 22, 27; m 52; c 3. PHYSIOLOGY. *Educ:* Univ Detroit, BS, 51; Purdue Univ, MS, 53, PhD(physiol), 56. *Prof Exp:* Res physiologist, Sherman Labs, Mich, 56-57; sr lit scientist, 57-60, group leader biochem, 60-66, assoc dir regulatory affairs, 69-71, dir res & develop, 71-80, VPRES, REGULATORY AFFAIRS, US PHARMACEUT PROD, SMITH KLINE & FRENCH LABS, 80- *Mem:* AAAS; Am Soc Zool; Endocrine Soc; Drug Info Asn. *Res:* Pharmaceutical research and development; federal drug regulations; data processing. *Mailing Add:* 7106 Fuller Circle Ft Worth TX 76133-6605

HORHOTA, STEPHEN THOMAS, b Buffalo, NY, June 9, 50; m 73; c 3. PHARMACY. *Educ:* Clarkson Col Technol, BS, 72; State Univ NY, Buffalo, PhD(pharmaceut), 78. *Prof Exp:* Pharmacist res & develop, Ayerst Labs, 72-74, group leader, 78-80; sr scientist res & develop, 80-85, SR PRIN SCIENTIST, BOEHRINGER INGELHEIM PHARMACEUT, 85- *Concurrent Pos:* Vis prof, Sch Pharm, State Univ NY Buffalo, 82-; adj prof, Sch Pharm, Univ RI, 84- *Mem:* Am Asn Pharmaceut Scientists. *Res:* Multivariate relations between properties of materials and methods of manufacture on features and attributes of pharmaceutical dosage forms. *Mailing Add:* Boehringer Ingelheim R&D Bldg PO Box 368 90 East Ridge Rd Ridgefield CT 06877

HORIE, YASUYUKI, b Tokyo, Japan, July 6, 37; m 64; c 1. PHYSICS. *Educ:* Int Christian Univ, Tokyo, BA, 61; Yale Univ, MS, 63; Wash State Univ, PhD(physics), 66. *Prof Exp:* Vis lectr math, Univ Strathclyde, 66-67; lectr, Manchester Col Sci & Technol, Eng, 67-68; assoc prof eng sci mech, Stanford Res Inst, 69-76, sr physicist, 77; assoc prof, 78-80, PROF CIVIL ENG & ASSOC MEM MAT SCI & ENG, NC STATE UNIV, 80- *Concurrent Pos:* Mem staff, US Army Res Off, 79-81, Sandia Nat Lab, 85, 86, US Army Mat Technol Lab, 87, US ARO, 87-88. *Mem:* Am Phys Soc; Mat Res Soc. *Res:* Physics of shock waves; nonlinear continuum physics; shock compression chemistry in materials synthesis. *Mailing Add:* Dept Civil Engr NC State Univ Raleigh NC 27650

HORING, SHELDON, b Brooklyn, NY, June 1, 36; m 61; c 2. ELECTRICAL ENGINEERING. *Educ:* City Col New York, BEE, 57; NY Univ, MEE, 59; Polytech Inst Brooklyn, PhD(elec eng), 61. *Prof Exp:* Instr elec eng, Polytech Inst Brooklyn, 60-61, asst prof, 61-62; mem tech staff, 57-60 & 62-72, head dept, 70-79, dir, 79-81, exec dir, Bell Tel Labs, 81-89; pres, CBIS Federal, 90, PRES, CBIS, 91- *Concurrent Pos:* Adj asst prof, Newark Col Eng, 62-63, adj assoc prof, 63-64; vis prof, Stevens Inst Technol, 64-68, assoc prof, 68-74. *Mem:* Inst Elec & Electronics Engrs. *Res:* Optimal control theory, particularly application and extension of the theory to terminal control problems which arise in guidance theory; teletraffic and congestion theory. *Mailing Add:* 65 Cornell Dr Livingston NJ 07039

HORITA, AKIRA, b Seattle, Wash, June 10, 28; m 54; c 2. PHARMACOLOGY. *Educ:* Univ Wash, AB, 50, MS, 51, PhD(pharmacol), 54. *Prof Exp:* From instr to assoc prof, 56-66, PROF PHARMACOL, MED SCH, UNIV WASH, 66-, PROF PSYCHIAT, 80- *Mem:* AAAS; Am Soc Pharmacol & Exp Therapeut; Sigma Xi. *Res:* Chemical pharmacology; biochemistry and pharmacology of the central nervous system and the autonomic nervous system. *Mailing Add:* 4201 NE 103rd Pl Seattle WA 98125

HORITA, ROBERT EIJI, b Vancouver, BC, Mar 15, 37; m 70; c 5. SPACE PHYSICS, GEOPHYSICS. *Educ:* Univ BC, BASc, 60, MASc, 62, PhD, 68. *Prof Exp:* Sci officer, Naval Res Estab, 62-66; res scientist, Commun Res Ctr, 68-69; Nat Res Coun Can Overseas Fel, Kyoto Univ, 69-70; from asst prof to assoc prof, 70-80, PROF PHYSICS, UNIV VICTORIA, 80- *Concurrent Pos:* Guest worker, Space Environ Lab, Nat Oceanic & Atmospheric Admin, 76-77; Exchange Prog Scientist, Nat Ctr Sci Res, France, 77. *Mem:* Can Asn Physicists; Am Geophys Union; Soc Terrestrial Magnetism & Elec Japan; Acoustic Soc Am; Inst Elec & Electronics Engrs. *Res:* Aeronomy; upper atmosphere physics; solar-terrestrial relationships; geomagnetic micropulsations. *Mailing Add:* Dept Physics Univ Victoria Box 1700 Victoria BC V8W 2Y2 Can

HORIUCHI, KENSUKE, b Tokyo, Japan, Sept 21, 33; m 62; c 3. MOLECULAR BIOLOGY. *Educ:* Univ Tokyo, BSc, 57, MSc, 59, PhD(biol), 62. *Prof Exp:* Fel microbiol, Yale Univ, 63-64; res assoc genetics, Rockefeller Inst, 64-66; asst biophys, Univ Tokyo, 66-69; res assoc genetics, 69-73, asst prof, 73-78, ASSOC PROF GENETICS, ROCKEFELLER UNIV, 78- *Concurrent Pos:* Fulbright travel grant, 63; Matsunaga Sci Found res grant, 67. *Mem:* Am Soc Microbiol. *Res:* Genetic and biochemical studies on small bacteriophages; restriction and modification enzymes. *Mailing Add:* 500 E 63rd St New York NY 10021

HORLICK, GARY, b Regina, Can, Apr 15, 44. ANALYTICAL CHEMISTRY. *Educ:* Univ Alta, BSc Hons, 65; Univ Ill, PhD(chem), 70. *Prof Exp:* PROF CHEM, UNIV ALTA, 69- *Honors & Awards:* Barringer Award, Spectros Soc Can, 77. *Res:* Analytical spectroscopy. *Mailing Add:* Dept Chem Univ Alta Edmonton AB T6G 2G2 Can

HORLICK, LOUIS, b Montreal, Que, Dec 2, 21; m 54; c 4. MEDICINE. *Educ:* McGill Univ, BSc, 44, MD & CM, 45, MSc, 52; FRCPS(C). *Prof Exp:* Demonstr med, McGill Univ, 50-53; lectr med & biochem, Univ Sask, 53-55, asst prof med & co-dir, Cardiopulmonary Lab, 55-57, assoc prof med, 57-62, head dept, 68-74, PROF MED, UNIV SASK, 62-, DIR , DEPT ELECTROCARDIOGRAPHY, 55- *Concurrent Pos:* Mem Coun Arteriosclerosis & Clin Cardiol, Am Heart Asn; mem Can Coun Hosp Accreditation. *Mem:* Am Col Physicians; Am Col Cardiol; Can Cardiovasc Soc; Can Med Asn; Can Soc Clin Invest; Can Soc Internal Med; Can Soc Arteriosclerosis. *Res:* Arteriosclerosis; clinical investigation; lipid metabolism; cardiac rehabilitation. *Mailing Add:* Dept Med Univ Hosp Saskatoon SK S7N 0X0 Can

HORMATS, ELLIS IRVING, b Albany, NY, Dec 10, 19; m 61; c 4. ELECTRON MICROSCOPY, APPLIED CHEMISTRY. *Educ:* Rensselaer Polytech Inst, BChE, 41, Univ Cincinnati, MS, 43; Cornell Univ, PhD(chem), 50. *Prof Exp:* Asst, Univ Cincinnati, 41-42; jr engr prod, Chem Warfare Serv, Edgewood Arsenal, 42-43, asst engr develop, Pinebluff Arsenal, 43-44; assoc scientist, Los Alamos Sci Lab, 45-46; chemist, Aerojet Gen Corp, 50-57; prin scientist, Basic Sci Lab, Stromberg-Carlson Co, 57-60; sr res staff mem, Gen Dynamics Electronics Div, 60-69; prin engr, Stromberg Carlson Co, 69-85; RETIRED. *Mem:* Am Chem Soc; fel Am Inst Chem; Sigma Xi. *Res:* Chemical kinetics; solid state chemistry and crystal growth; photochemistry; thin film techniques; electron microscopy; analytical chemistry; microelectronics. *Mailing Add:* 639 Lake Shores Dr Maitland FL 32751

HORN, ALFRED, b New York, NY, Feb 17, 18; m 45; c 2. ALGEBRA. *Educ:* City Col New York, BS, 38; NY Univ, MA, 41; Univ Calif, PhD(math), 46. *Prof Exp:* Asst math, Univ Calif, Berkeley, 41-42 & 45-46, instr, 46-47, mathematician, Radiation Lab, 42-45; from instr to prof, 47-88, EMER PROF MATH, UNIV CALIF, LOS ANGELES, 88- *Concurrent Pos:* Mem, Inst Advan Study, 52-53. *Res:* Lattice theory; universal algebra. *Mailing Add:* Dept Math Univ Calif Los Angeles CA 90024

HORN, ALLEN FREDERICK, JR, b Milwaukee, Wis, Apr 25, 29; m 54; c 3. FOREST MANAGEMENT. *Educ:* Mich State Univ, BS, 50, MS, 51; State Univ NY, PhD(forest mgt), 57; Syracuse Univ, LLB, 67. *Prof Exp:* Instr forestry, Mich State Univ, 51-53; asst prof & asst forester exp sta, Miss State Univ, 53-55; from asst prof to assoc prof, 57-69, PROF FOREST MGT, STATE UNIV NY COL ENVIRON SCI & FORESTRY, SYRACUSE UNIV, 69- *Concurrent Pos:* Consult, Wood Industs, Pa, 57 & Area Redevelop Admin, 63-64; forester, USDA Coop Stat Res Serv, 75-76. *Mem:* Soc Am Foresters. *Res:* Legal problems relating to the administration of natural resources; environmental law; business and financial aspects of forest management. *Mailing Add:* Fac Forestry SUNY Col Environ Sci & Forestry Syracuse NY 13210

HORN, CHRISTIAN FRIEDRICH, b Dresden, Germany, Dec 23, 27; US citizen; m 54; c 1. CHEMISTRY. *Educ:* Dresden Tech Univ, dipl chem, 51; Aachen Tech Univ, Dr rer nat(polymer chem), 58. *Prof Exp:* Res chemist synthetic fibers, Ger Acad Sci, 52-53; res chemist, Farbenwerke Hoechst, 53-54; res chemist, Union Carbide Chem Co, 54-57, group leader synthetic fibers, 58-61, res & develop mgr urethane polymers, 61-63, asst licensing mgr, 63-65; pres, Polymer Technol Inc, 65-71; managing dir & mem exec bd, Zimmer AG, 71-73; pres, Chrislon Corp, 73-74; vpres off opers, W R Grace & Co, 74-76, vpres, Corp Off Res & Tech Serv, 76-78, corp vpres & bd dirs, 78-81; CORP SR VPRES & HEAD INT DEVELOP, CHEMED CORP, 81-; PRES, GRACE VENTURES CORP, 83-, MEM, BD DIRS, W R GRACE & CO, 85- *Concurrent Pos:* Res chemist, Europ Res Assocs, Belg, 57-58; consult, 65-71. *Mem:* Am Chem Soc. *Res:* Chemical process engineering in synthetic fibers; plastics; fiber technology; synthesis of monomers; chemical processes; international development. *Mailing Add:* Grace Ventures Corp 20300 Stevens Creek Blvd No 330 Cupertino CA 95014

HORN, DAG, b Oslo, Norway, June 4, 50; m 76; c 2. NUCLEAR PHYSICS, HEAVY ION REACTIONS. *Educ:* Mass Inst Technol, SB, 72, PhD(nuclear physics), 76. *Prof Exp:* Res asst nuclear physics, Mass Inst Technol, 72-76; fel, Atomic Energy Can Ltd, Chalk River Nuclear Lab, 76-78; asst nuclear physicist, Brookhaven Nat Labs, 78-80; res fel, Australian Nat Univ, 80-81; RES OFFICER, CHALK RIVER LABS, ATOMIC ENERGY CAN, LTD, 81- *Concurrent Pos:* Nat Res Coun Can fel, 76-78; vis scientist, Lab Phys Corpuslaire, Caen, France, 89-90. *Mem:* Am Phys Soc; Can Asn Physicists; Sigma Xi. *Res:* Experimental nuclear physics with heavy ions; nuclear reaction studies and gamma ray spectroscopy of high spin states. *Mailing Add:* Chalk River Labs Chalk River ON K0J 1J0 Can

HORN, DAVID JACOBS, b Philadelphia, Pa, Feb 12, 43; m 66; c 2. ECOLOGY, ENTOMOLOGY. *Educ:* Harvard Univ, BA, 65; Cornell Univ MS, 67, PhD(entom), 69. *Prof Exp:* Asst entom, Cornell Univ, 65-69, asst ecol, 67-69; asst prof biol, Calif State Col, Hayward, 69-72; assoc prof, 72-78, PROF ENTOM, OHIO STATE UNIV, 78- *Mem:* AAAS; Am Inst Biol Sci; Ecol Soc Am; Entom Soc Am; Entom Soc Can. *Res:* Biological control of insects; predator-prey interrelationships; population and community ecology; human influences on insect ecology. *Mailing Add:* Dept Entom Ohio State Univ 1735 Neil Ave Columbus OH 43210

HORN, DAVID NICHOLAS, b London, Eng, Apr 17, 44; div; c 2. MULTIPOINT MULTIMEDIA COMMUNICATIONS SYSTEMS & NETWORK SERVICES, DISTRIBUTED & PARALLEL COMPUTING ARCHITECTURES & SYSTEMS. *Educ:* Beijing Univ, China, Grad cert, 69; Coun Nat Acad Awards, BSc(hons), 74; London Univ, Eng, PhD(elec eng), 81. *Prof Exp:* Tech asst test gear maintenance, Plessey Co Ltd, Eng, 69-71; trainee prod eng, Samwell & Hutton Ltd, Eng, 72-73; mem tech staff develop, 79-85, MEM TECH STAFF RES, AT&T BELL LABS, 85- *Concurrent Pos:* Lectr, Dept Elec Eng, Princeton Univ, 87. *Honors & Awards:* Inst Elec & Electronics Engrs Prize, 74. *Mem:* Inst Elec & Electronics Engrs; Inst Elec & Electronics Engrs Computer Soc; Inst Elec & Electronics Engrs Commun Soc. *Res:* Network architectures and bridging services to support multiparty, multimedia conferencing and computer-aided collaborative work; desktop multimedia conferencing systems; man-machine interfaces; distributed computing architectures and hardware. *Mailing Add:* AT&T Bell Labs Rm 4G602 Crawford's Corner Rd Holmdel NJ 07733-1988

HORN, DIANE, b Iowa City, Iowa. CELL BIOLOGY. *Educ:* Stanford Univ, BA, 65; Purdue Univ, PhD(biol sci), 71. *Prof Exp:* Res fel cell biol & genetics, Harvard Med Sch, 71-75; asst prof biol sci, Univ Southern Calif, 75-83; SR SCIENTIST ONCOGEN, SEATTLE, WASH, 83- *Concurrent Pos:* NIH fel, 71-73; res fel genetics, Clin Genetics Div, Children's Hosp Med Ctr, Boston, 71-75; fel, Workshops Huntington's Dis, 73-74. *Mem:* Am Soc Cell Biol; Sigma Xi; AAAS. *Res:* Control of gene expression in mammalian cells in tissue culture, particularly the regulation of growth and differentiated functions; biology of oncostatin M, monoclonal antibodies for cancer diagnosis and therapy. *Mailing Add:* Oncogen Div Bristol-Myers Squibb Pharmaceut Res Inst 3005 First Ave Seattle WA 98121

HORN, EDWARD GUSTAV, b Tucson, Ariz, Dec 18, 42; m 76; c 2. ECOLOGY, ENVIRONMENTAL SCIENCE. *Educ:* Duke Univ, BS, 64; Princeton Univ, PhD(biol), 69. *Prof Exp:* From asst prof to assoc prof biol, Russell Sage Col, 69-77, chmn dept, 71-76; res scientist III, NY State Dept Environ Conserv, 77-79, chief, Bur Environ Protection, 79-87; ENVIRON SCIENTIST VI, NY STATE DEPT HEALTH, 87- *Mem:* AAAS; Ecol Soc Am. *Res:* Role of food competition in community structure of cellular slime molds; ecology of aquatic macrophytes; fate of toxic substances in the environment; environmental remediation. *Mailing Add:* Two University Pl Albany NY 12233

HORN, EUGENE HAROLD, b Lancaster, Pa, Sept 27, 26; m 52; c 2. ENDOCRINOLOGY, ANATOMY. *Educ:* Franklin & Marshall Col, BS, 49; Rutgers Univ, PhD(endocrinol), 53. *Prof Exp:* Asst, Bur Biol Res, Rutgers Univ, 50-53; instr histol, neuroanat & gross anat, 53-60, assoc prof anat, 60-68, asst dean, 62-66, assoc dean, 66-79, PROF ANAT, ALBANY MED COL, 68-, INTERIM CHMN, 87- *Mem:* AAAS; Asn Am Med Cols; NY Acad Sci; Sigma Xi; Am Asn Anatomist; Asn Anat Chmn. *Res:* Reproductive maturation; nutrition; thyroid physiology, relaxin. *Mailing Add:* Dept Anat Albany Med Col Albany NY 12208

HORN, HARRY MOORE, b Brooklyn, NY, Jan 28, 31. CIVIL ENGINEERING. *Educ:* Polytech Inst Brooklyn, BCE, 53; Columbia Univ, MS, 56; Univ Ill, PhD(civil eng), 61. *Prof Exp:* Struct engr, Howard, Needles, Tammen & Bergendoff, NY, 53-55; soils engr, Tippetts, Abbett, McCarthy, Stratton, 56-57 & Haley & Aldrich, Mass, 60-62; asst prof civil eng, Mass Inst Technol, 62-65; assoc prof, Univ Ill, 65-67; PRIN, WOODWARD CLYDE CONSULTS, 67- *Honors & Awards:* C A Hogentogler Award, Am Soc Testing & Mat, 74. *Mem:* Am Soc Civil Engrs. *Res:* Frictional properties of minerals; performance of building foundations; foundation instrumentation. *Mailing Add:* Woodward-Clyde Consult 201 Willow Brook Blvd Wayne NJ 07470

HORN, HENRY STAINKEN, b Philadelphia, Pa, Nov 12, 41; m 63. POPULATION ECOLOGY, ANIMAL BEHAVIOR. *Educ:* Harvard Univ, BA, 62; Univ Wash, PhD(ecol), 66. *Prof Exp:* From asst prof to assoc prof, 66-78, PROF BIOL, PRINCETON UNIV, 78- *Mem:* AAAS; Ecol Soc Am; Animal Behav Soc; Lepidopterists Soc. *Res:* Spatial patterns of plants and animals; behavioral ecology of vertebrates and butterflies. *Mailing Add:* Dept Biol Princeton Univ Princeton NJ 08544

HORN, J(OHN) W(ILLIAM), b Martinsburg, WVa, Aug 6, 29; m 51; c 4. TRANSPORTATION, CIVIL ENGINEERING. *Educ:* WVa Univ, BS, 52; Mass Inst Technol, MS, 56. *Prof Exp:* Asst transp, Mass Inst Technol, 54-56; from asst prof to assoc prof, 56-69, tech dir hwy res, 58-74, admin coordr joint hwy res prog, 64-65, PROF CIVIL ENG, NC STATE UNIV, 69- *Concurrent Pos:* Res engr, Bruce Campbell & Assocs, Mass, 54-56; consult engr, 56-81; chmn bd, Kimley-Horn & Assocs, Prof Engrs, 66-84; prin & sr consult, Kimley-Horn & Assoc, Engrs, 84-88. *Mem:* Am Soc Civil Engrs; Inst Traffic Engrs; Am Pub Works Asn; Am Consult Engrs Coun; Sigma Xi. *Res:* Transportation planning, design and operations. *Mailing Add:* PO Box 33068 Raleigh NC 27606

HORN, KENNETH PORTER, b Ft Worth, Tex, Dec 10, 37; m 79. FLUID MECHANICS, SYSTEMS ENGINEERING. *Educ:* Rice Univ, BA, 60, MS, 62; Stanford Univ, PhD(aeronaut & astronaut), 66. *Prof Exp:* Aerospace engr, NASA Manned Spacecraft Ctr, 62; mem tech staff, Plasma Res Lab & Satellite Systs Div, Aerospace Corp, 66-75; PROG DIR, RAND CORP, 75-, DEPT HEAD, 88- *Mem:* Am Phys Soc; Am Inst Aeronaut & Astronaut. *Res:* High-temperature gas dynamics, including radiation-coupled flow and nonequilibrium processes; systems engineering, particularly space and missile systems. *Mailing Add:* Rand Corp 1700 Main St Santa Monica CA 90406

HORN, LEIF, b Tromso, Norway, Mar 15, 25; m 49, 79; c 3. BIOPHYSICS. *Educ:* Univ Oslo, Cand Phil, 47, PhD(physiol & biochem), 56. *Prof Exp:* Asst physiol, Univ Oslo, 49-50, asst, Inst Nutrit Res, 50-53; res assoc surg, Med Col Va, 57, from instr to asst prof biophys, 57-59; asst prof path, Sch Med, NY Univ, 59-62, asst prof exp surg, 62-64; assoc prof physiol, New York Med Col, 64-67; assoc prof, 67-68, PROF PHYSIOL, COL MED & DENT NJ, 68- *Concurrent Pos:* Res chemist, Aeromed Inst, Norway, 50; spec fel with Prof Benjamin W Zweifach, Nat Heart Inst, 59-61. *Honors & Awards:* Angiol Res Found-Purdue Frederick Co Achievement Award, 64-65. *Mem:* Microcirc Soc; Fedn Am Soc Exp Biol; Am Physiol Soc; Am Heart Asn. *Res:* Electrophysiology of vascular smooth muscle; vascular physiology and biophysics. *Mailing Add:* Dept Physiol UMDNJ NJ Med Sch 185 S Orange Ave Newark NJ 07103-2714

HORN, LYLE HENRY, b Two Rivers, Wis, Dec 26, 24; m 61; c 1. METEOROLOGY. *Educ:* Univ Wis, BS, 55, MS, 56, PhD(meteorol), 60. *Prof Exp:* From asst prof to assoc prof, 60-66, Chmn dept, 67-70, PROF METEOROL, 66-, CHMN DEPT, UNIV WIS-MADISON, 87- *Concurrent Pos:* Vis assoc prof, Dartmouth Col, 65-66. *Mem:* AAAS; Am Meteorol Soc; Sigma Xi. *Res:* Energetics of large-scale circulations; applications of satellite data to synoptic studies; climatology. *Mailing Add:* Dept Meteorol Univ Wis 1225 W Dayton St Madison WI 53706

HORN, LYLE WILLIAM, b St Paul, Minn, July 22, 43; m 70. BIOPHYSICS, CELL MEMBRANE TRANSPORT. *Educ:* Univ Colo, BS, 66; Johns Hopkins Univ, PhD(biomed eng), 73. *Prof Exp:* Res assoc mech, Johns Hopkins Univ, 73-74; asst prof physiol, Sch Med, Univ Md, 74-81; ASSOC PROF PHYSIOL & BIOPHYS, TEMPLE UNIV, 81- *Mem:* Biophys Soc; Soc Gen Physiol; AAAS. *Res:* Kinetics of cell membrane transport of amino acids; sodium-coupling; giant axon membranes; surfactants; muscle membranes. *Mailing Add:* Dept Physiol Temple Univ Sch Med 3223 N Broad St Philadelphia PA 19140

HORN, MICHAEL HASTINGS, b Tahlequah, Okla, Nov 14, 42. ICHTHYOLOGY, MARINE BIOLOGY. *Educ:* Northeastern State Col, BS, 63; Univ Okla, MS, 65; Harvard Univ, PhD(biol), 69. *Prof Exp:* Fel biol oceanog, Woods Hole Oceanog Inst, 68-69; NATO fel sci, Brit Mus, London, 70; from asst prof to assoc prof zool, 70-77, PROF BIOL, CALIF STATE UNIV, FULLERTON, 77- *Concurrent Pos:* Ed, Ichthyol Book Rev, 85-; vis res scientist, Dunstaff Nage Mar Res Lab, Oban, Scotland, 86. *Mem:* AAAS; Am Soc Ichthyol & Herpet; Am Fish Soc; Sigma Xi; Ecol Soc Am; Am Soc Zool. *Res:* Fish-plant interactions in coastal seas and tropical rainforests; physiological ecology of rocky shore fishes; feeding ecology and digestive physiology of fishes; ecology of coastal, harbor and bay-estuarine fishes. *Mailing Add:* Dept Biol Calif State Univ Fullerton CA 92634

HORN, MYRON K, b Miami, Fla, Jan 28, 30; m 55; c 3. PETROLEUM GEOLOGY. *Educ:* Univ Colo, BA, 52; Univ Houston, MS, 58; Rice Univ, PhD(geol), 64. *Prof Exp:* Sr res geologist, Pure Oil Co Res Ctr, 60-64; head dept geol res, Res Lab, 65-70, DIR RES, E&P RES LAB, CITIES SERV CO, 70- *Concurrent Pos:* Assoc ed, Am Asn Petrol Geologists, 75-78, ed, 79-; sci adv, Ocean Margin Drilling Proj, 81- *Mem:* AAAS; Sigma Xi; Soc Petrol Engrs; Am Petrol Inst; Am Asn Petrol Geologists. *Res:* Computer systems for well log interpretation; computer-derived retrieval systems; computer-derived geochemical balances; simulated neutron activation spectra; habitat of oil and gas on continental margins; production geology; remote sensing applications; composition of formation waters; global geology. *Mailing Add:* Cities Serv Oil & Gas Corp PO Box 3908 Tulsa OK 74102

HORN, ROGER ALAN, b Macon, Ga, Jan 19, 42; m 65; c 3. MATRIX ANALYSIS, COMPLEX VARIABLES. *Educ:* Cornell Univ, BA, 63; Stanford Univ, MS, 64, PhD(math), 67. *Prof Exp:* Asst prof math, Univ Santa Clara, 67-68; asst prof, Johns Hopkins Univ, 68-71; assoc prof, Univ Md, Baltimore County, 71-72; assoc prof, 72-75, chmn dept, 72-79, PROF MATH SCI, JOHNS HOPKINS UNIV, 75- *Concurrent Pos:* Alfred P Sloan Found res fel, 75-79. *Mem:* Am Math Soc; Soc Indust & Appl Math; Math Asn Am. *Res:* Analysis, complex variables; matrix analysis; probability. *Mailing Add:* Dept Math Sci Johns Hopkins Univ Baltimore MD 21218

HORN, SUSAN DADAKIS, b Cleveland, Ohio, Aug 30, 43; m 65; c 3. HEALTH SERVICES RESEARCH, MEDICAL STATISTICS. *Educ:* Cornell Univ, AB, 64; Stanford Univ, MS, 66, PhD(statist), 68. *Prof Exp:* Asst prof biostatist, Johns Hopkins Univ, 68-74, asst prof statist, 68-72, assoc prof math sci, 72-77, asst prof health care orgn, 73-77, assoc prof health policy & mgt, 77-86, PROF HEALTH POLICY & MGT, JOHNS HOPKINS UNIV, 86- *Concurrent Pos:* Statistician consult, var US hosps, 74- *Mem:* fel Am Statist Asn; Biomet Soc; Am Pub Health Asn. *Res:* Statistical methods; measuring patient's severity of illness for management, quality and prospective reimbursement; applications of statistics in public health. *Mailing Add:* Dept Health Policy & Mgt Johs Hopkins Univ 624 N Broadway Baltimore MD 21205

HORN, WILLIAM EVERETT, b Zanesville, Ohio, Aug 27, 28; m 54, 78. PHYSICS. *Educ:* Ohio State Univ, BSc, 50, MSc, 51; Johns Hopkins Univ, cert bus mgt, 60. *Prof Exp:* Res asst optical physics, Res Found, Ohio State Univ, 50-52; res engr, Res Labs, 52-54, sr engr, Aerospace Div, Md, 56-59, proj engr, 59-61, supvr physicist, 61-65, dir, Westinghouse Cambridge Lab, 65-69, MGR RES & DEVELOP PROGS, SYST DEVELOP DIV, WESTINGHOUSE ELEC CO, 69- *Mem:* Am Phys Soc; Optical Soc Am; sr mem Inst Elec & Electronics Engr; assoc fel Am Inst Aeronaut & Astronaut; Am Defense Preparedness Ord Asn; Sigma Xi. *Res:* Electromagnetic areas of laser techniques; infrared reconnaissance and aerospace applications; molecular electronics; microwaves; systems applications involving radar and optical guidance, control, stabilization and computation areas. *Mailing Add:* Two Thorndyke Carth Phoenix MD 21131

HORNA, OTAKAR ANTHONY, b Prague, Czech, Jan 8, 22; US citizen; m 57; c 1. COMMUNICATION ENGINEERING, MATHEMATICAL LOGIC. *Educ:* Czech Inst Technol, MSEE, 48; Inst Radiotechnol & Electronics, Prague, PhD(electronics), 62; Charles Univ, Prague, dipl math logic, 63. *Prof Exp:* Res asst measurement technol, Inst Theoret & Exp Mech, 48-55; res engr med electronics, Inst Med Electronics, Prague, 55; sr staff scientist comput, Inst Math Mach, Prague, 55-68; sr engr mil electronics, Multronics Corp, Rockville, Md, 68-69; sr staff scientist communs, Comsat Labs, Clarksburg, Md, 69-; AT SATELLITE BUS SYSTS. *Concurrent Pos:* Consult, Automotive Res Inst, Prague, 52-66; sci secy, Govt Adv Coun Comput, Prague, 63-65; consult, Turbine Test Dept, SKODA Works, Czech, 64-68; actg vchmn, Sci Coun, Res Inst Math Mach, Prague, 66-68; co-chmn, Govt Adv Bd Electronics Indust Planning, Prague, 67. *Honors & Awards:* Gold Medal, World Exhib, Brussels, 58 & Brno Int Fair & Exhib, Czech, 67; Res Award, COMSAT Labs, 76. *Mem:* Sr mem Inst Elec & Electronics Engrs; Czech Soc Arts & Sci Am. *Res:* Satellite communication; real-time signal processing; self-adaptive systems and filters; mathematical logic. *Mailing Add:* IBM 704 Quince Orchard Rd Gaithersburg MD 20787

HORNACK, FREDERICK MATHEW, b Philadelphia, Pa, June 10, 29; m 53; c 3. PHYSICAL CHEMISTRY. *Educ:* Lowell Technol Inst, BS, 50; Fla State Univ, PhD(chem), 55. *Prof Exp:* Asst, Fla State Univ, 50-55; res assoc chem, Univ Ark, 55-56; chemist, Robert A Taft Eng Ctr, USPHS, 56-58; asst prof chem, Univ Tampa, 58-60; asst prof, Va Polytech Inst, 60-64; PROF CHEM, UNIV NC, WILMINGTON, 64- *Honors & Awards:* Olney Prize, 50. *Mem:* Am Chem Soc. *Res:* Education in chemistry; solubility of gases. *Mailing Add:* Dept Chem Univ NC Wilmington NC 28403

HORNAK, THOMAS, b Bratislava, Czech, Oct 14, 24; US citizen; m 58; c 1. ELECTRONICS. *Educ:* Slovak Tech Univ, Bratislava, Czech, BSEE, 46, MSEE, 47; Czech Tech Univ, Prague, PhD(electronic eng), 66. *Prof Exp:* Sect mgr electronics, Tesla Radio Res Lab, Prague, 47-61; head adv, Comput Res Inst, Prague, 62-68; mem tech staff solid state circuits & optoelectronics, 68-73, DEPT MGR SOLID STATE CIRCUITS, OPTOELECTRONICS & COMMUN TECHNOL, HEWLETT PACKARD LABS, 73- *Concurrent Pos:* Vis prof, Tech Univ, Prague & Brno, Czech, 52-56. *Mem:* Fel Inst Elec & Electronics Engrs. *Res:* Advanced solid state circuits and optoelectronics and their applications to instrumentation, communication, computation and signal processing. *Mailing Add:* Hewlett Packard Co 1501 Page Mill Rd Palo Alto CA 94304

HORNBACK, JOSEPH MICHAEL, b Middletown, Ohio, Sept 16, 43; m 65; c 2. ORGANIC CHEMISTRY. *Educ:* Univ Notre Dame, BS, 65; Ohio State Univ, PhD(chem), 68. *Prof Exp:* Res assoc chem, Univ Wis, 68-70; asst prof, 70-75- ASSOC PROF CHEM, UNIV DENVER, 75-, ASSOC DEAN, 88- *Mem:* Am Chem Soc. *Res:* Photochemical reactions of organic compounds; synthesis of organic compounds. *Mailing Add:* Dept Chem Univ Denver Denver CO 80208

HORNBAKER, EDWIN DALE, b Louisburg, Kans, June 19, 29; m 57; c 2. SPECIALTY POLYMERS, CERAMICS. *Educ:* Univ Kans, BS, 51; Univ Va, MS, 53, PhD(chem), 55. *Prof Exp:* Sr res chemist, J T Baker Chem Co, 55-57; res chemist, Ethyl Corp, 57-62, from res assoc to sr res assoc, 62-64, supvr, 64-81, asst dir, 81-89, mat res dir, 89-90, SR RES ADV, ETHYL CORP, 90- *Mem:* Am Chem Soc. *Res:* Development of specialty polymers; high softening thermoplastics; specialty elastomers; water soluble polymers; ceramic precursor polymers; ceramics. *Mailing Add:* Ethyl Corp Tech Ctr 8000 GSRI Ave PO Box 14799 Baton Rouge LA 70898

HORNBECK, JOHN A, solid state physics; deceased, see previous edition for last biography

HORNBEIN, THOMAS F, b St Louis, Mo, Nov 6, 30; m 51; c 5. ANESTHESIOLOGY, PHYSIOLOGY. *Educ:* Univ Colo, BA, 52; Washington Univ, MD, 56; Am Bd Anesthesiol, dipl. *Prof Exp:* Instr anesthesiol, Sch Med, Washington Univ, 60-61; from asst prof to prof anesthesiol, physiol & biophys, 63-78, vchmn dept anesthesiol, 72-74, PROF & CHMN DEPT ANESTHESIOL, SCH MED, UNIV WASH, 78- *Concurrent Pos:* USPHS res fel, 59-61, res grant, 64-67, career develop award, 65-75. *Honors & Awards:* Hubbard Medal; George Norlin Award, Univ Colo, Denver, 70. *Mem:* Am Physiol Soc; Am Soc Anesthesiol; Asn Univ Anesthetists (pres, 75-76); Inst Med; Soc Acad Anesthesia Chmn. *Res:* Respiratory physiology; chemical regulation of ventilation; quantification of neural discharge from peripheral chemoreceptors in response to hypoxia, hypercapnia and acidosis; role of acidosis or carbon dioxide; ion regulation across blood-brain barrier. *Mailing Add:* Dept Anesthesiol RN-10 Univ Wash Sch Med Seattle WA 98195

HORNBERGER, CARL STANLEY, JR, b Chicago, Ill, June 27, 23; m 50; c 3. ORGANIC CHEMISTRY. *Educ:* NCent Col, Ill, AB, 47; Univ Ill, PhD(chem), 51; Am Bd Indust Hyg, cert. *Prof Exp:* Res fel, Univ Ill, 51-53; res chemist agr chem, E I Du Pont de Nemours & Co, Inc, 53-58, biochemist pharmaceut res, Stine Lab, 58-64, biochemist drug metab, 64-67, toxicologist, Haskell Lab, 67-74, chief dermalocular toxicol, 70-74, plant indust hygenist, 74-76, consult indust hyg, 76-80, sr info specialist, Haskell Lab, 80-82; CONSULT, 82- *Concurrent Pos:* Consult, indust hyg; adj prof toxicol, math modelling & on-line retrieval info (modem), Del State Col, Dover. *Mem:* Am Chem Soc; Am Indust Hyg Asn. *Res:* Biomolecular reduction of hindered keytones; lipoic acid, isolation, characterization and synthesis; agricultural chemicals; pharmaceuticals; drug metabolism; managerial methods for identification and assessment of chemical health hazards; setting standards for work place contaminants. *Mailing Add:* 11 N Cliffe Dr Wilmington DE 19809

HORNBERGER, GEORGE MILTON, b Ashland, Pa, June 22, 42; m 65; c 2. HYDROLOGY. *Educ:* Drexel Univ, BSCE, 65, MSCE, 67; Stanford Univ, PhD(hydrol), 70. *Prof Exp:* From asst prof to prof, 70-90, chmn dept, 79-83, DISTINGUISHED PROF ENVIRON SCI, UNIV VA, 90- *Concurrent Pos:* Vis fel, Centre Resource & Environ Studies, Australian Nat Univ, Canberra, 77-78; hon vis prof, Univ Lancaster, Eng, 84-85, mem, Bd Radioactive Waste Mgt, Nat Acad Sci, 86-, NAm ed, Hydrol Processes, 85-; vis prof, Stanford Univ, 90-91; vis scientist, US Geol Surv, 90-91. *Mem:* Am Geophys Union; Am Geol Inst; Sigma Xi. *Res:* Modelling of environmental systems with uncertainty; hydrogeochemical response of small cathments; ground water lake interaction; transport of microbes in groundwater. *Mailing Add:* Dept Environ Sci Clark Hall Univ Va Charlottesville VA 22903

HORNBROOK, K ROGER, b New Martinsville, WVa, Oct 23, 36; m 58; c 1. PHARMACOLOGY. *Educ:* WVa Univ, BS, 58; Univ Mich, PhD(pharmacol), 63. *Prof Exp:* Res assoc pharmacol, Univ Mich, 63; asst prof, Emory Univ, 65-72; assoc prof, 72-77, PROF PHARMACOL, UNIV OKLA HEALTH SCI CTR, OKLAHOMA CITY, 77- *Concurrent Pos:* Fel, Washington Univ, 63-65; Nat Inst Arthritis & Metab Dis spec fel, 67-70. *Mem:* Am Soc Pharmacol & Exp Therapeut; Soc Exp Biol & Med; Endocrine Soc. *Res:* Biochemical pharmacology; regulation of carbohydrate metabolism; mechanisms of hepatotoxicity. *Mailing Add:* Dept Pharmacol Univ Okla Health Sci Ctr Oklahoma City OK 73190

HORNBUCKLE, FRANKLIN L, b Birmingham, Ala, Jan 19, 41. ENGINEERING. *Educ:* Tenn State Univ, BSEE, 62; Rutgers Univ, MSEE, 71. *Prof Exp:* Engr, Radio Corp Am Space Div, 62-72; mgr spacecraft elec, 72-75, dir elec eng dept, 75-81, dir eng dept, 81-86, vpres eng dept, 86-88, VPRES SPACECRAFT ENG, FAIRCHILD SPACE CO, 88- *Mem:* Inst Elec & Electronics Engrs; Am Inst Aeronaut & Astronaut; Asn Naval Aviation; Armed Forces Commun & Electronics Asn. *Res:* Advanced satellite power systems concepts; high density solid state memory applications. *Mailing Add:* Spacecraft Eng Fairchild Space Co 20301 Century Blvd Germantown MD 20874

HORNBUCKLE, PHYLLIS ANN, b Mooresville, NC, Nov 18, 38. PSYCHOPHYSIOLOGY. *Educ:* Pfeiffer Col, BA, 59; Col William & Mary, MA, 66; Emory Univ, PhD(psychophysiol), 69. *Prof Exp:* ASST PROF PSYCHOL, MED COL VA, VA COMMONWEALTH UNIV, 68-, ASST PROF MED, 75- *Mem:* Sigma Xi. *Res:* Therapeutic and non-therapeutic chemical effects on behavior. *Mailing Add:* Dept Psychol Va Commonwealth Univ Box 2520 Richmond VA 23284

HORNE, ALEXANDER JOHN, b Doncaster, Eng. AQUATIC BIOLOGY. *Educ:* Univ Bristol, Eng, BSc, 64; Univ Dundee, Scotland, PhD(limnol & oceanog), 69. *Prof Exp:* Res asst nitrogen fixation, Dept Bot, Westfield Col, Univ London, 64-68; res fel nitrogen fixation, Dept Biol Sci, Univ Dundee, Scotland, 69-70; res fel limnol, Dept Ecol, Univ Calif, Davis, 70-71; assoc prof aquatic ecol, Dept Eng, 71-80, PROF SANITARY, ENVIRON, COASTAL & HYDRAULIC ENG, DEPT CIVIL ENG, UNIV CALIF, BERKELEY, 80-, ASSOC BIOLOGIST, SANIT ENG RES LAB, 71- *Concurrent Pos:* Consult, Lake County Flood Control & Water Conserv Dist, 70- & Calif Dept Water Resources, 73- *Mem:* Am Soc Limnol & Oceanog; Brit Phycol Soc; Int

Soc Limnol. *Res:* Aquatic ecology; environmental engineering; nitrogen-fixation; eutrophication, lakes, reservoirs, estuaries, rivers and coastal waters; pollution control; lake and estuarine management; oil pollution; geothermal pollution; sewage pollution. *Mailing Add:* Dept Civil Univ Calif Berkeley CA 94720

HORNE, FRANCIS R, b San Antonio, Tex, Dec 22, 39; c 3. COMPARATIVE PHYSIOLOGY. *Educ:* Tex Tech Col, BS, 62; Univ Wyo, MS, 64, PhD(physiol), 66. *Prof Exp:* NIH res fel, Inst Marine Sci, 66-67; from asst prof to assoc prof biol, 67-77, PROF BIOL, SOUTHWEST TEX STATE UNIV, 77- *Mem:* Am Soc Zool. *Res:* Comparative physiology and biochemistry of nitrogen excretion; respiratory pigments and ion regulation. *Mailing Add:* Dept Biol Southwest Tex State Univ San Marcos TX 78666

HORNE, FREDERICK HERBERT, b Kans City, Mo, Mar 11, 34; m 59; c 3. PHYSICAL CHEMISTRY, THEORETICAL CHEMISTRY. *Educ:* Harvard Univ, AB, 56; Univ Kans, PhD(phys chem), 62. *Prof Exp:* NSF fel, Stanford Univ, 62-63, instr chem, 63-64; from asst prof to assoc prof, Mich State Univ, 64-73, assoc chmn dept, 75-82, prof chem, 73-86, assoc dean, Col Natural Sci, 82-86; PROF CHEM & DEAN, COL SCI, ORE STATE UNIV, 86- *Concurrent Pos:* Vis scientist, Lawrence Livermore Lab, 71 & Odense Univ, Denmark, 79; vis prof chem, Arya Mehr Univ Technol, Iran, 75; Bd Sci & Technol Int Develop NRC, 88. *Mem:* Am Chem Soc; AAAS. *Res:* Chemical and nonequilibrium thermodynamics; nonequilibrium statistical mechanics; thermal diffusion; membrane transport; ion transport in liquids and solids; applied mathematics and statistics. *Mailing Add:* Col Sci Kidder Hall 128 Ore State Univ Corvallis OR 97331-4608

HORNE, G(ERALD) T(ERENCE), b New York, NY, June 24, 24; m 46; c 5. METALLURGY. *Educ:* Mont Sch Mines, BSc, 48; Carnegie Inst Technol, MS, 51, PhD(metall), 52; Univ Pittsburgh, MBA, 72. *Prof Exp:* Mem staff, Metals Res Lab, Carnegie Inst Technol, 52-64, asst prof, 52-60, assoc prof, Dept Metall Eng, 60-64; staff engr, Glass Res Ctr, PPG Indust, Inc, 64-65; asst mgr, 65-69; VPRES, INDUST TESTING LAB SERV CO, 69- *Honors & Awards:* Andrew Carnegie Lectr, 78. *Mem:* Am Soc Metals; Am Soc Testing & Mat; Am Inst Mining, Metall & Petrol Engrs; Am Soc Eng Educ; NY Acad Sci; Sigma Xi; Am Soc Mech Engr; AAAS. *Res:* Deformation of metals; fracture; fatigue of metals; diffusion in solids; mechanical metallurgy. *Mailing Add:* Ten Pine Creek Dr Pittsburgh PA 15238

HORNE, GREGORY STUART, b Minneapolis, Minn, June 11, 35; m 57; c 2. STRATIGRAPHY, MARINE GEOLOGY. *Educ:* Dartmouth Col, AB 57; Columbia Univ, PhD(geol), 68. *Prof Exp:* Geologist, Tidewater Oil Co, 57-64; geologist, Pan Am Petrol Corp, 65; assoc prof earth sci, 67-80, PROF EARTH & ENVIRON SCI, WESLEYAN UNIV, 80-; PRES, ESSEX MARINE LAB, 75- *Mem:* Geol Soc Am; Am Asn Petrol Geologists; Soc Econ Paleont & Mineral. *Res:* Petroleum geology; stratigraphy and structural geology of orogenic belts; sedimentology of chaotic deposits; paleoecology of Lower Paleozoic carbonate sequences; coastal processes and nearshore geological oceanography. *Mailing Add:* Dept Earth & Environ Sci Wesleyan Univ Middletown CT 06457

HORNE, JAMES GRADY, JR, b Ft Worth, Tex, Apr 6, 26; m 49; c 3. MATHEMATICS. *Educ:* Tulane Univ, BChE, 46, MS, 50, PhD, 56. *Prof Exp:* Asst math, Tulane Univ, 48-52, instr, 55; asst prof, Univ Ky, 56-59; from asst prof to assoc prof, 59-66, head dept, 69, PROF MATH, UNIV GA, 66- *Mem:* Am Math Soc; Math Asn Am. *Res:* Topological algebra; semigroups on manifolds; transformation groups. *Mailing Add:* 150 Bernice Dr Bogart GA 30622

HORNE, RALPH ALBERT, b Haverhill, Mass, Mar 10, 29. ENVIRONMENTAL CHEMISTRY. *Educ:* Mass Inst Technol, BS, 50; Univ Vt, MS, 52; Boston Univ, MS, 53; Columbia Univ, PhD(phys chem), 55, JD, Suffolk Univ Law Sch, 79- *Prof Exp:* Asst, Univ Vt, 50-52; asst, Columbia Univ, 53; res assoc chem, Brookhaven Nat Lab, 53-55; res assoc, Nuclear Sci Lab, Mass Inst Technol, 55-57; mem sr tech staff, Radio Corp Am, 57-59; sr scientist, Joseph Kaye & Co, Inc, 59-60; sr scientist, Arthur D Little, Inc, Mass, 60-68; assoc scientist, Woods Hole Oceanog Inst, 69-71; prin scientist, JBF Sci Corp, 71-72; mem sci staff, Arthur D Little, Inc, 72-78; sr scientist, GCA Tech Div, 78-80; sr scientist, Energy & Environ Eng Inc, 81-87; CONSULT, 87- *Mem:* AAAS. *Res:* Electron exchange reactions; structure of aqueous solutions; transport processes in aqueous solutions; chemical oceanography; high pressure; environmental law; history and philosophy of science; environmental chemistry; energy and resources. *Mailing Add:* Nine Wellington St Boston MA 02118

HORNE, ROBERT D, b Eufaula, Ala, June 17, 35; m 57; c 1. VETERINARY SURGERY. *Educ:* Auburn Univ, DVM, 59, MSc, 61; Am Col Vet Surg, dipl, 70. *Prof Exp:* From instr to assoc prof, 59-70, alumni assoc prof, 70, ALUMNI PROF SMALL ANIMAL SURG & MED, SCH VET MED, AUBURN UNIV, 70- *Concurrent Pos:* Orthop Res Soc grant, 63-64; Scott Rickey Grant, Orthopedic Res, 79-81; fac grant, Orthopedic Res, 78- *Honors & Awards:* Am Animal Hosp Asn Merit Award in Orthop, 73. *Mem:* Am Vet Med Asn; Am Asn Vet Clinicians; Animal Hosp Asn. *Res:* Thoracic, cardiovascular and orthopedic surgery. *Mailing Add:* Dept Animal Med Auburn Univ Auburn AL 36849

HORNE, ROLAND NICHOLAS, b London, Eng, Nov 27, 52; New Zealand citizen; m 78; c 1. GEOTHERMAL RESERVOIR ENGINEERING. *Educ:* Univ Auckland, New Zealand, BE, 72, PhD(appl mech), 75, DSc, 86. *Prof Exp:* Res fel, Univ Auckland, 74-76; actg asst prof chem eng, Stanford Univ, 76-77, actg asst prof petrol eng, 77-78; lectr appl mech, Univ Auckland, 78-79; asst prof, 80-84, ASSOC PROF PETROL ENG, STANFORD UNIV, 84- *Concurrent Pos:* Prin investr, Stanford Geothermal Prog, 81- *Mem:* Sigma Xi; Am Inst Mining, Metall & Petrol Engrs; Am Soc Mech Engrs. *Res:* Geothermal reservoir engineering; heat and mass transfer in porous media; pressure transient analysis of oil, gas and geothermal wells; hydrodynamic stability; reservoir simulation; optimisation of reservoir performance. *Mailing Add:* Dept Petrol Eng Stanford Univ Stanford CA 94305

HORNE, SAMUEL EMMETT, JR, b Jacksonville, Fla, July 26, 24; m 49; c 4. POLYMER CHEMISTRY, ORGANIC CHEMISTRY. *Educ:* Emory Univ, AB, 47, MA, 48, PhD(chem), 50. *Hon Degrees:* DSc, Emory Univ, 82. *Prof Exp:* Tech man, B F Goodrich Co, 50-54, sr tech man, 54-59, res assoc, 59-68, sr res assoc, res ctr, 68-82; sci adv, Rubber Tech Ctr, Polysar, Inc, Stow, Ohio, 82-84, prin scientist, 84-87; RETIRED. *Concurrent Pos:* Chmn, Gordon Conf Hydrocarbon Chem, 69; asst counr, Rubber Div, Am Chem Soc, 88- *Honors & Awards:* Pioneer Award, Am Inst Chemists, 74; Midgley Medal, Am Chem Soc, 78, & Goodyear Medal Rubber Div, 80. *Mem:* Am Chem Soc; NY Acad Sci; fel Am Inst Chemists; Sigma Xi; Tire Soc; AAAS. *Res:* Terpene research; natural products; polymerization, diene and vinyl; hydrocarbon reactions; stereospecific polymerization; organometallics; catalysis. *Mailing Add:* 10205 Echo Hill Dr Brecksville OH 44141

HORNEMANN, ULFERT, b Dresden, Ger, Apr 22, 39; m 70; c 4. GENETICS, MOLECULAR BIOLOGY. *Educ:* Hannover Tech Univ, BS, 61; Munich Tech Univ, MS, 64, PhD(org chem), 66. *Prof Exp:* From instr to prof med chem, Purdue Univ, 67-81; PROF PHARMACEUT BIOCHEM, SCH PHARM, UNIV WIS-MADISON, 81- *Concurrent Pos:* Vis scientist, Dept Genetics, John Inmes Inst, Norwich, England, 76-77. *Mem:* Ger Chem Soc; Am Chem Soc; Am Soc Microbiol; The Chem Soc; Am Soc Biol Chem. *Res:* Biosynthesis and chemistry of antibiotics; isolation and studies on regulation of enzymes of antibiotic biosynthesis; mode of action of antibiotics; biochemical genetics of antibiotic production; genetic engineering; DNA amplification in streptomycetes. *Mailing Add:* Sch Pharm Univ Wis 425 N Charter St Madison WI 53706

HORNER, ALAN ALFRED, b York, Eng, Feb 23, 34; m 58; c 3. BIOCHEMISTRY, PHYSIOLOGY. *Educ:* Univ Liverpool, BSc, 55, PhD(biochem), 58. *Prof Exp:* Biochemist, Blood Prods Lab, Lister Inst Prev Med, Eng, 58-59; sr biochemist, Res & Develop Labs, Can Packers Ltd, 59-63; from asst prof to assoc prof, Banting & Best Dept Med Res, 63-70, assoc prof, 70-80, PROF PHYSIOL, UNIV TORONTO, 80- *Mem:* Can Biochem Soc; Can Physiol Soc. *Res:* Biochemistry and physiology of heparin. *Mailing Add:* Dept Physiol Univ Toronto One Kings College Circle Toronto ON M5S 1A8 Can

HORNER, B ELIZABETH, b Merchantville, NJ, Apr 29, 17. ZOOLOGY. *Educ:* Rutgers Univ, BS, 38; Smith Col, MA, 40, Univ Mich, PhD(zool), 48. *Prof Exp:* Asst zool, 40-41, instr, 41-44 & 46-48, from asst prof to prof, 48-70, MYRA M SAMPSON PROF BIOL SCI, SMITH COL, 70- *Concurrent Pos:* Am Asn Univ Women fel, Univ Sydney, 54-55. *Mem:* AAAS; Ecol Soc Am; Am Soc Mammal; Soc Study Evolution; Animal Behav Soc. *Res:* Postnatal skeletal development of carnivor es and rodents; animal ecology; behavior and systematics of rodents and marsupials. *Mailing Add:* Dept Biol Sci Smith Col Northampton MA 01063

HORNER, CHESTER ELLSWORTH, b McMinnville, Ore, Mar 2, 25; m 53; c 3. PLANT PATHOLOGY. *Educ:* Walla Walla Col, BA, 50; Ore State Col, PhD(plant path), 54. *Prof Exp:* From asst prof to assoc prof, 54-65, leader hops & mint invests, USDA, 68-72, PROF BOT & PLANT PATH, ORE STATE UNIV, 65-, RES LEADER HOPS, MINT & FIELD CROPS, AGR RES SERV, USDA, 72- *Mem:* Am Phytopath Soc; Am Inst Biol Sci; Am Soc Agron; Crop Sci Soc Am. *Res:* Fungus diseases of plants; soil microbiology; breeding for disease resistance. *Mailing Add:* 1760 NW Menlo Dr Corvallis OR 97330

HORNER, DONALD RAY, b Beatrice, Nebr, Aug 8, 35; m 54; c 2. MATHEMATICAL ANALYSIS. *Educ:* Univ Tex, Arlington, BS, 61; NTex State Univ, MS, 62; NMex State Univ, PhD(math), 67. *Prof Exp:* Instr electronics, Philco Corp, 58-59; instr math, Univ Tex, El Paso, 62-66; assoc prof, 66-69, PROF MATH, EASTERN WASH UNIV, 69-, CHMN DEPT MATH & COMPUT SCI, 76- *Concurrent Pos:* Res mathematician, Schellenger Res Labs, 63-65; res assoc, Phys Sci Labs, 65-66; mathematician, Air Force Grant, 66. *Mem:* Math Asn Am; Am Math Soc. *Res:* Measure theory; locally convex spaces; general topology. *Mailing Add:* Dept Comput Sci Eastern Wash Univ Cheney WA 99004

HORNER, EARL STEWART, b Allison, Colo, Aug 22, 18; m 48; c 3. PLANT BREEDING. *Educ:* State Col Wash, BS, 40; Mich State Col, MS, 42; Cornell Univ, PhD(plant breeding), 50. *Prof Exp:* From asst prof to assoc prof, 50-64, PROF AGRON, UNIV FLA, 64- *Mem:* AAAS; Genetics Soc Am; fel Am Soc Agron; Sigma Xi. *Res:* Methods of corn breeding; genetics of heterosis in corn; alfalfa breeding. *Mailing Add:* Dept Agron 306 New Univ Fla Gainesville FL 32611

HORNER, GEORGE JOHN, b Sept 15, 23; US citizen; m 52; c 2. MEDICINE, CARDIOPULMONARY PHYSIOLOGY. *Educ:* Univ Sydney, MB & BS, 60. *Prof Exp:* Australian Heart Found res fel med & surg, Univ Sydney, 62-64; fel chest dis, Sch Med, Yale Univ, 64-65, Am Heart Asn res fel, 65-66; from instr to asst prof med, Yale Univ, 66-68, dir cardiopulmonary lab, 67-68; assoc prof med, Jefferson Med Col, 68-81; head, Respiratory Res Dept, 68-81, chief, Pulmonary Serv Lab, 72-81, CONSULT MED, LANKENAU HOSP, 81- *Concurrent Pos:* Clin prof med, Jefferson Med Col, 81-90, emer prof, 90- *Mem:* Am Fedn Clin Res; Am Thoracic Soc; Am Med Asn; fel Am Col Chest Physicians. *Res:* Experimental acute cor pulmonale in pulmonary embolism, incompatible blood transfusion and endotoxin shock; aging and pulmonary function; pulmonary gas exchange in patients with heart and/or lung disease. *Mailing Add:* 717 Old Eagle School Rd Strafford-Wayne PA 19087

HORNER, HARRY THEODORE, b Chicago, Ill, Jan 28, 36; m 61; c 3. BOTANY, CELL & DEVELOPMENTAL BIOLOGY. *Educ:* Northwestern Univ, BA, 59, MS, 61, PhD(biol), 64. *Prof Exp:* Asst biol, Northwestern Univ, 60-62; NIH fel, 62-66, from asst prof to assoc prof, 66-73, PROF BOT, IOWA STATE UNIV, 73- *Concurrent Pos:* Sigma Xi & Sci Res Soc Am grants, 62 & 65; grant, Coop State Res Serv, USDA, 71-73 & 82-84; NSF grant, 74; mem, Iowa Soybean Prom Bd, 81-85; Monsanto Res grant, 82-86; Iowa State

Univ res grant, 88-89. *Mem:* Am Micros Soc; Bot Soc Am; Am Inst Biol Sci; Sigma Xi; Electron Micros Soc Am. *Res:* Sporogenesis; cytoplasmic and nuclear sterility; plant secretion and calcification; bacterial leaf nodulation. *Mailing Add:* Dept Bot Iowa State Univ Ames IA 50011-1020

HORNER, JAMES M, b Phillipsburg, Mo, Feb 26, 35; m 57; c 2. MATHEMATICS. *Educ:* Univ Ala, BS, 61, MA, 62, PhD(math), 64. *Prof Exp:* Assoc sr res mathematician, Gen Motors Res Labs, 64-65; from asst prof to prof math, Univ Ala, Huntsville, 65-75, chmn dept, 66-68 & 70-72, dean fac, 73-77; vpres, provost & prof math, Ill State Univ, 75-79; pres, Cent Mo State Univ, 79-85, prof math, 79-90; DIR ACAD AFFAIRS, AIR FORCE INST TECHNOL, 90- *Mem:* Math Asn Am. *Res:* Functions of a complex variable, differential equations and special functions. *Mailing Add:* Air Force Inst Technol Wright Patterson AFB OH 45433-6583

HORNER, JAMES WILLIAM, JR, b Sioux Falls, SDak, Jan 24, 14; m 42, 53; c 3. CHEMICAL INFORMATION. *Educ:* Univ Minn, BCh, 35, PhD(org chem), 40. *Prof Exp:* Res chemist, Nat Starch Prod Inc, NY, 40-42 & Charles Bruning Co, Ill, 42-46; res chemist, Archer-Daniels-Midland Co, Minn, 46-58, group leader, 58-67; res chemist, Ashland Chem Co, Columbus, 67-80; RETIRED. *Mem:* Am Chem Soc; Spec Libr Asn. *Mailing Add:* 5800 St Cordix Ave No 310 Minneapolis MN 55422

HORNER, NORMAN V, b Brownwood, Tex, June 1, 42; m 63; c 2. ARACHNOLOGY, ENTOMOLOGY. *Educ:* NTex State Univ, BS, 65, MS, 67; Okla State Univ, PhD(entom), 71. *Prof Exp:* Instr, 67-69, asst prof, 71-74, assoc prof, 75-80, PROF BIOL, MIDWESTERN STATE UNIV, 81- *Mem:* Entom Soc Am; Am Arachnological Soc; Brit Arachnological Soc; Sigma Xi. *Res:* Life histories and taxonomy of local spider fauna. *Mailing Add:* Dept Biol Midwestern Univ Wichita Falls TX 76308

HORNER, RICHARD E(LMER), b Wrenshall, Minn, Oct 24, 17; m 41; c 2. AERONAUTICAL ENGINEERING. *Educ:* Univ Minn, BS, 40; Princeton Univ, MS, 47. *Prof Exp:* Tech dir & sr engr, Air Force Flight Test Ctr, 50-55; dep asst secy of Air Force, 55-57, asst secy, 57-59; assoc adminr, NASA, 59-60; gen mgr, Norair Div & sr vpres tech, Northrop Corp, 60-70; pres & dir, E F Johnson Co, 70-82, chmn bd & chief exec officer, 80-84; pres & chief exec officer, Western Union Personal Commun, Inc, 84-86; RETIRED. *Concurrent Pos:* Mem, Air Force Sci Adv Bd, 60-74, Nat Acad Sci Comt, 65-75 & Comn Govt Procurement, 69-73; chmn, Res & Technol Adv Coun, NASA, 71-75; dir, Medtronic, Inc, 75-87 & Northrop Corp, 77- *Mem:* Fel Am Inst Aeronaut & Astronaut; fel Am Astronaut Soc; Nat Space Asn. *Res:* Aircraft development; operational development. *Mailing Add:* 1581 Via Entrada del Lago San Marcos CA 92069

HORNER, SALLY MELVIN, b Fayetteville, NC, Nov 17, 35; div; c 2. INORGANIC CHEMISTRY. *Educ:* Univ NC, BS, 57, PhD(inorg chem), 61. *Prof Exp:* Res assoc chem, Univ NC, Chapel Hill, 61-62, instr, 62-63, res assoc, 63-67; from instr to asst prof, Meredith Col, 65-72, chmn dept, 72-77, assoc prof chem & chmn dept, 72-78, prof, 78; univ provost & dean, Col Arts & Sci, Univ Charleston, WVa, 78-81, actg dean, Col Health Sci, 80-81, vpres admin serv, 81-84, actg pres, 84; vchancellor admin & finance, 84-90, VCHANCELLOR, PLANNING & FISCAL AFFAIRS, UNIV SC, COASTAL CAROLINA COL, 90- *Concurrent Pos:* Consult, Res Triangle Inst, 66; sr vis res assoc chem, Duke Univ, 72; dir inst res, Meredith Col, 75-78, dir financial aid, 78; consult, Univ NC Gen Admin, 77, WVa Bd Regents, 83-86. *Mem:* AAAS; Am Chem Soc; Sigma Xi. *Res:* Coordination compounds of transition metals; spectra of transition metal ions; molecular orbital theory. *Mailing Add:* 608 D 35th Ave N Myrtle Beach SC 29578-2856

HORNER, THEODORE WRIGHT, b Clarksburg, WVa, Feb 29, 24; m 50; c 2. STATISTICS, GENETICS. *Educ:* NC State Col, BS, 49, MS, 51, PhD, 53. *Prof Exp:* Asst statistician, NC State Col, 51-53; asst prof statist, Iowa State Col, 53-57; sr opers res analyst, Gen Mills, Inc, 57-59; prin statistician, Booz-Allen Appl Res, Inc, 59-70; STATIST CONSULT, 70- *Concurrent Pos:* Statist consult, 70- *Mem:* Am Statist Asn; Biomet Soc; Inst Math Statist. *Res:* Mathematical theory of epistasis; operations research; industrial statistics; biometry; social statistics. *Mailing Add:* 659 Western Blvd Ext Suite 438 Cary NC 27511-4219

HORNER, WILLIAM HARRY, b Kenmore, NY, Sept 30, 23; m 62; c 4. BIOCHEMISTRY. *Educ:* Western Reserve Univ, MD, 47; Cornell Univ, PhD(biochem), 52. *Prof Exp:* From asst prof to assoc prof biochem, Med Col, State Univ NY Downstate Med Ctr, 52-59; assoc prof, 59-60, chmn dept, 60-81, PROF BIOCHEM, SCHS MED & DENT, GEORGETOWN UNIV, 60- *Concurrent Pos:* Prin investr res grants, 59-85. *Mem:* AAAS; Harvey Soc; Am Soc Biol Chemists; Sigma Xi. *Res:* Metabolism, amino acids and derivatives; metabolism of guanidine compounds; muscle disease. *Mailing Add:* Dept Biochem Georgetown Univ Washington DC 20007

HORNER, WILLIAM WESLEY, b Portsmouth, Ohio, Dec 26, 30; m 50; c 2. INORGANIC CHEMISTRY. *Educ:* Univ NC, BS, 57, MA, 61, PhD(inorg chem), 64. *Prof Exp:* Assoc prof, Elon Col, 66-71; assoc prof chem, Methodist Col, 71-79. *Concurrent Pos:* Pres, Triangle Chem Labs, Inc, 64-74. *Mem:* AAAS; Am Chem Soc; fel Am Inst Chemists. *Res:* Rhenium chemistry; anhydrous metal halides; halides and coordination compounds of transition metals; microanalysis; kinetics of hydrazine in liquid ammonia; synthetic methods in inorganic chemistry. *Mailing Add:* PO Box 335 Fayetteville NC 28302-0335

HORNEY, AMOS GRANT, b Hortonville, Ind, Mar 18, 07; m 37, 53. RESEARCH ADMINISTRATION, EDUCATIONAL ADMINISTRATION. *Educ:* Earlham Col, AB, 30; Ohio State Univ, MS, 32, MA, 33, PhD(org chem), 37. *Prof Exp:* Asst, Ohio State Univ, 30-32, asst to prof chem, 32-35, asst, 35-37; org res chemist, Ohio Chem & Mfg Co, 37-40, actg chief chemist, 40-41, chief chemist-in-charge res develop & control, 41-44; res assoc, Res Found, Ohio State Univ, 44-46; head dept chem & dean lib arts, Assoc Cols of Upper NY, 46-48; dep chief, Off Sci Intel, US Cent Intel Agency, Washington, DC, 48-49, sci liaison officer, 49-50; actg chief chem group, Flight Res Lab, Wright Air Develop Ctr, USAF, 50-51, chief chem div, Off Sci Res, 51-55, dir mat sci, 55-59, dir chem sci, 59-72; RETIRED. *Concurrent Pos:* Instr, Fenn Col, 39-42; consult chem & chem eng, 48-; adj prof chem, NMex State Univ, 75-81. *Mem:* Am Chem Soc; Am Phys Soc; Am Inst Chem Eng. *Res:* Unsaturated hydrocarbon synthesis; synthesis of new general anesthetics; sponge from synthetic latex; materials research; solid state chemistry; research, directing, planning, coordination and management; chemistry. *Mailing Add:* 13801 York Rd Apt N3 Cockeysville MD 21030

HORNFECK, ANTHONY J, electrical & electronics engineering; deceased, see previous edition for last biography

HORNG, SHI-JINN, b Taiwan, Repub of China, Oct 19, 57; m; c 2. PARALLEL ALGORITHM, PARALLEL SYSTEM. *Educ:* Nat Taiwan Inst Technol, Bachelor, 80; Nat Cent Univ, Master, 84; Nat Tsing-Hua Univ, PhD(computer sci), 89. *Prof Exp:* Assoc prof parallel prog, Inst Elec Eng, Nat Taiwan Inst Technol, 89-90; MEM TECH STAFF PARALLEL ALGORITHM, COMPUT SCI RES, AT&T BELL LABS, 90- *Res:* Reconfigurable array of processors. *Mailing Add:* 417 Morris Ave Apt 26 Summit NJ 07901

HORNG, WAYNE J, IDIOTYPE, GENETICS. *Educ:* Univ Tex, PhD(biochem), 74. *Prof Exp:* BIOCHEMIST IMMUNO-CHEM, ABBOTT LABS, 83- *Mailing Add:* Abbott Labs D-905 Abbott Park North Chicago IL 60064

HORNIBROOK, WALTER JOHN, plastics chemistry; deceased, see previous edition for last biography

HORNICK, RICHARD B, b Johnstown, Pa, Jan 27, 29; m 54; c 4. INTERNAL MEDICINE, INFECTIOUS DISEASES. *Educ:* Johns Hopkins Univ, AB, 51, MD, 55. *Prof Exp:* From instr to prof med, Sch Med, Univ Md, Baltimore, 71-79, dir div infectious dis, 63-79; chmn dept med, Sch Med, Univ Rochester, 79-85, assoc dean, Affil Hosps, 85-87; VPRES MED EDUC ORLANDO REG MED CTR, 87- *Concurrent Pos:* Mem epidemiol staff, USPH Lab, Mont, 60; consult, USPHS Hosp, 63-; mem comn epidemiol surv, US Armed Forces Epidemiol Bd, 65-86. *Honors & Awards:* Smadel Award, Infectious Dis Soc Am, 83; Bruce Award, Am Col Physicians, 86. *Mem:* Am Soc Microbiol; Infectious Dis Soc Am, (treas, 78-82, pres, 85-86); Am Fedn Clin Res; Am Soc Clin Invest; Am Col Physicians. *Res:* Value of immunoprophylactic agents in preventing infectious diseases in man; associated host defense mechanisms. *Mailing Add:* Orlando Regional Med Ctr 1414 Kuhl Ave Orlando FL 32806-2093

HORNICK, SHARON B, b Wilkes-Barre, Pa, 1948; m 67. HUMAN NUTRITION. *Educ:* Elizabethtown Col, BS, 70; Pa State Univ, MS, 74, PhD(soil chem), 76; Univ Md, PhD(human nutit), 86. *Prof Exp:* Res asst environ chem, Pa State Univ, 74-75; soil scientist, Md Environ Serv, 75-77 & US Environ Protection Agency, 77-78; SOIL SCIENTIST, USDA, 78- *Concurrent Pos:* Consult, Climax Molybdenum, 73-75; assoc ed, J Environ Quality, 80-82; instr nutrit, Univ Col, Univ Md, 88- *Honors & Awards:* Superior Serv Award, USDA, 77. *Mem:* Am Dietetic Asn; Am Soc Agron; Soc Nutrit Educ. *Res:* Effects of soil fertilizers and amendments, and management practices on the nutritional quality of crops with emphasis on macro- and micronutrient contents and ascorbic acid and beta-carotene levels. *Mailing Add:* Bldg 318 Rm 103 BARC-EAST USDA-ARS Beltsville MD 20705

HORNIG, DONALD FREDERICK, b Milwaukee, Wis, Mar 17, 20; m 43; c 4. CHEMISTRY. *Educ:* Harvard Univ, BS, 40, PhD(phys chem), 43. *Hon Degrees:* LLD, Temple Univ, 64, Notre Dame, 65, Boston Col, 66 & Dartmouth Col, 74; DHL, Yeshiva Univ, 65; ScD, Rensselaer Polytech Inst, 65, Univ Md, 65, Ripon Col, 66, Widener Col, 67, Univ Wis, 67, Univ Puget Sound, 68, Syracuse Univ, 68, Princeton Univ, 69, Seoul Nat Univ, Korea, 73, Univ Pa, 75 & Lycoming Col, 80; DEng, Worcester Polytech Inst, 67. *Prof Exp:* Res assoc, Woods Hole Oceanog Lab, 43-44; scientist & group leader, AEC, Los Alamos Sci Lab, Univ Calif, 44-46; from asst prof to prof chem, Brown Univ, 46-57, dir Metcalf Res Lab, 49-57, from assoc dean to actg dean grad sch, 52-54; Donner prof sci, Princeton Univ, 57-64, chmn dept chem, 58-64; spec asst sci & technol to President, Wash, DC, 64-69; vpres & dir, Eastman Kodak Co, NY, 69-70; pres, Brown Univ, 70-76; hon res assoc appl physics, Harvard Univ, 76-77, Alfred North Whitehead Prof Chem, 81-86, prof chem, Sch Pub Health, 77-90, dir interdisciplinary progs in health, 77-90, EMER PROF CHEM, HARVARD UNIV, 90- *Concurrent Pos:* Pres, Radiation Instruments Co, 46-48; Guggenheim & Fulbright fel, St John's Col, Oxford, 54-55; mem adv panel chem, NSF, 57-60 & Physics Adv Comt, Off Sci Res, USAF, 59-61; mem space sci bd, Nat Acad Sci, 58-64; mem bd dirs, W A Benjamin, Inc, 60-64; mem, President's Sci Adv Comt, 60-69, chmn, 64-69, consult-at-large, 69-72; vis prof, Calif Inst Technol, 62, mem, US deleg to negotiate space coop with USSR, 62-63; chmn, Fed Coun Sci & Technol, 64-69; dir off sci & technol, Exec Off President, 64-69; mem bd overseers, Harvard Univ, 64-70; prof, Univ Rochester, 69-70; dir, Overseas Develop Coun, 69-77; mem bd dirs, Upjohn Co, 71- & Westinghouse Elec Corp, 72-90; mem Adv Comt Health Sci & Tech Enterprise, Off Technol Assessment, 76-78, chmn, Adv Comt Effect Regulation Innovation, 79; chmn, Comt Instnl Arrangements Space Telescope Sci Inst, Nat Acad Sci, 76, mem Comt Satellite Power Syst, 79, Comt Post-Doctorals & Doctoral Res Staff, 79, vchmn, Bd Toxicol & Environ Health Hazards, 84-86; mem, Am Assembly Workshop, Improving Am Innovation, 83; mem, Twentieth Century Fund Task Force, Improving Am Innovation, 83; mem, Security Telecomm Policy Planning, 84-86, chmn, Bd Environ Studies & Toxicol, 86-88, Comn Life Sci, Nat Acad Sci, 87-; dir & treas, Overseas Develop Network, Inc, 85-; pres, Cambridge Water Bd, 86-; chmn, Bd Environ Studies & Toxicol, 86, Comn Life Sci, 87-; mem, comt to consider long term changes that may affect the habitability of the earth, NASA, 82. *Honors & Awards:* Eng Centennial Award, Widener Col, 67; Charles Lathrop Parsons Award, Am Chem Soc, 67; First Mellon Inst Award, Carnegie-Mellon Univ, 68; Moranjang Medal,

Korean Govt, 68. *Mem:* Nat Acad Sci; Am Chem Soc; fel Am Phys Soc; Am Acad Arts & Sci; Romanian Acad; Am Philos Soc. *Res:* Molecular spectroscopy; spectra of crystals; shock and detonation waves; fast reactions; theoretical chemistry. *Mailing Add:* Sch Pub Health Harvard Univ 665 Huntington Ave Boston MA 02115

HORNIG, HOWARD CHESTER, b Fond du Lac, Wis, July 14, 24; m 56; c 3. PHYSICAL CHEMISTRY, EXPLOSIVES. *Educ:* Univ Chicago, BS, 48, MS, 49, PhD(chem), 52. *Prof Exp:* Res chemist, 52-74, proj mgr, 74-76, group leader, 76-79, PROJ LEADER, LAWRENCE LIVERMORE NAT LAB, UNIV CALIF, 79- *Mem:* AAAS; Combustion Inst; Sigma Xi. *Res:* Energetic chemical systems, explosives, detonation, hydrodynamics, blasting and pyrotechnics; coal recovery; boron hydrides; high speed photography; radiochemical tracer techniques; kinetics of ionic reactions in solution; quality assurance. *Mailing Add:* 4763 E Wing Rd Castro Valley CA 94546

HORNIG, JAMES FREDERICK, b Milwaukee, Wis, Feb 22, 29; m 52; c 3. PHYSICAL CHEMISTRY. *Educ:* Harvard Univ, AB, 50; Univ Wis, MS, 52, PhD(chem), 54. *Prof Exp:* NSF fel, Univ Marburg, 54-55; res chemist, E I du Pont de Nemours & Co, 56-58; asst prof chem, Univ Calif, Riverside, 58-62; assoc prof, 62-66, dean grad sch, chmn sci div & assoc dean fac, 64-65, PROF CHEM, DARTMOUTH COL, 66-, ALBERT BRADLEY PROF SCI, 75- *Concurrent Pos:* Fulbright fel, Marburg, Ger, 69. *Mem:* Am Chem Soc; Am Phys Soc; Sigma Xi. *Res:* Energy transfer in organic crystals; environmental chemistry. *Mailing Add:* Dept Chem Dartmouth Col Hanover NH 03755

HORNING, DONALD O(URY), mechanical engineering, for more information see previous edition

HORNING, EVAN CHARLES, b Philadelphia, Pa, June 6, 16; m 42. ANALYTICAL BIOCHEMISTRY. *Educ:* Univ Pa, BS, 37; Univ Ill, PhD(org chem), 40. *Hon Degrees:* Dr, Karolinska Inst, Sweden, 73 & Univ Ghent, Belgium, 78. *Prof Exp:* Instr chem, Bryn Mawr Col, 40-41 & Univ Mich, 41-43; from asst prof to assoc prof, Univ Pa, 45-50; chief, Lab Chem Natural Prod, Nat Heart Inst, 50-61; dir, Lipid Res Ctr, Baylor Col Med, 61-66, chmn dept biochem, 62-66, prof chem, 61-86, dir Inst Lipid Res, 66-86, EMER PROF CHEM, BAYLOR COL MED, 86- *Concurrent Pos:* Res assoc comt med res, Nat Defense Res Comt, Univ Mich, 43-45; Guggenheim fel, 58. *Honors & Awards:* Bergman Award, Swedish Chem Soc, 73; Tswett Medal Chromatography, 75; Warner-Lambert Award, Am Asn Clin Chemists, 76; Scheele Award, Pharmaceut Soc Sweden, 76; Am Chem Soc Award Chromatography, 78; Tswett Comm Medal, Acad Sci USSR, 79; S S Dal Nogare Award, 80; Am Chem Soc Award, Mass Spectrometry, 90. *Mem:* Am Chem Soc; Brit Biochem Soc; Swiss Chem Soc; Sigma Xi; Am Soc Biochem & Molecular Biol; Am Soc Mass Spectrometry. *Res:* Lipid chemistry; atherosclerosis; drug metabolism; chromatography; mass spectrometry; analytical biochemistry. *Mailing Add:* 11610 Starwood Dr Houston TX 77024

HORNING, MARJORIE G, b Detroit, Mich, Aug 23, 17; m 42. PHARMACOLOGY. *Educ:* Goucher. *Hon Degrees:* DSc, Goucher Col, 77. Col, AB, 38; Univ Mich, MS, 40, PhD(biol chem), 43. *Prof Exp:* Res assoc pediat, Univ Mich Hosp, 44-45; res chemist, Univ Pa, 45-50; biochemist, Nat Heart Inst, 51-61; assoc prof, 61-69, PROF BIOCHEM, INST LIPID RES, BAYLOR COL MED, 69- *Honors & Awards:* Warner-Lambert Award, Am Asn Clin Chemists, 76; Garvan Medal, Am Chem Soc, 77. *Mem:* AAAS; Am Soc Pharmacol & Exp Therapeut; Am Chem Soc; NY Acad Sci. *Res:* Analytical biochemistry; drug metabolism; gas chromatography and mass spectrometry; toxicology. *Mailing Add:* Baylor Col Med One Baylor Plaza Rm 826E Houston TX 77030

HORNOR, SALLY GRAHAM, b Boston, Mass, June 10, 49; m 83; c 2. MICROBIAL ECOLOGY, AQUATIC ECOLOGY. *Educ:* Goucher Col, BA, 71; Univ Conn, MS, 74, PhD(ecol), 77. *Prof Exp:* Res assoc microbiol ecol, State Univ NY, Syracuse, 77-79; asst prof microbiol, Va Polytech Inst & State Univ, 79-85; ADJ PROF BIOL, ANNE ARUNDEL COMMUNITY COL, 86- *Mem:* AAAS; Am Soc Limnol & Oceanog; Am Soc Microbiol; Sigma Xi. *Res:* Energetics and comparative microbial decomposition processes in aquatic and terrestrial ecosystems; biochemical ecology of aquatic sediments and submerged soils. *Mailing Add:* 787 Mago Vista Rd Arnold MD 21012

HORNSBY, ARTHUR GRADY, b Marks, Miss, Feb 5, 40; m 66. SOIL PHYSICS, SOIL CHEMISTRY. *Educ:* Univ Ark, BS, 62, MS, 64; Okla State Univ, PhD(soil sci), 72. *Prof Exp:* Lab technician, Dept Water Sci & Eng, Univ Calif, Davis, 67-69; soil scientist, Robert S Kerr environ res lab, Environ Protection Agency, 71-; AT SOIL SCIENCE DEPT, UNIV FLA, GAINESVILLE, FLA. *Concurrent Pos:* Adj prof, Soil Sci Dept, Univ Fla, 81-, mem, Innovative Res Proj, 81-82. *Mem:* Am Soc Agron; Am Geophys Union; Soil Sci Soc Am; Sigma Xi. *Res:* Movement of water, salts and organics in soil systems; simulation techniques for water resource management; salinity control in irrigation return flows. *Mailing Add:* Soil Sci Dept Univ Fla 2169 McCarty Hall Gainesville FL 32611-0151

HORNSEY, EDWARD EUGENE, b Potosi, Mo, May 31, 37; m 59; c 2. ENGINEERING MECHANICS. *Educ:* Univ Mo, Rolla, BS, 59, MS, 61, PhD(mining eng), 67. *Prof Exp:* Instr eng mech, Univ Mo-Rolla, 60-62; mining methods res engr, Appl Physics Res Lab, US Bur Mines, Md, 62-63; from instr to asst prof, 63-75, ASSOC PROF ENG MECH, UNIV MO-ROLLA, 75-, ASSOC PROF BASIC ENG, 88- *Concurrent Pos:* Pvt consult. *Mem:* Nat Soc Prof Engrs; Am Soc Eng Educ. *Res:* Solid mechanics; mechanics of materials, dynamics, vibrations and stability; rock mechanics. *Mailing Add:* Dept Basic Eng Univ Mo Rolla MO 65401-0249

HORNSTEIN, IRWIN, b New York, NY, Jan 4, 17; m 42; c 2. NUTRITIONAL BIOCHEMISTRY. *Educ:* City Col New York, BChE, 37; Univ Md, MS, 51; Georgetown Univ, PhD, 60. *Prof Exp:* Chemist alloys, US Naval Gun Factory, 40-45; org chemist, Nat Bur Standards, 45-46; res chemist, Glenn L Martin Co, 46-48; res chemist, Agr Res Serv, USDA, 48-67, chief food qual & use, Lab Human Nutrit, Res Div, 67-69; mem staff, Aid US State Dept, 69-73, dep dir, Off Nutrit, 73-80; Mgr, Proj Sustain, 82-90, CONSULT, 90- *Concurrent Pos:* Co-ed, Food Rev Int, 84-; co-ed, Food Rev Int, 84- *Mem:* AAAS; Am Chem Soc; Inst Food Technol; Sigma Xi. *Res:* Analytical methods; insecticides; flavors; lipids; meats; nutrition; technology transfer. *Mailing Add:* 5920 Bryn Mawr Rd College Park MD 20740

HORNSTEIN, JOHN STANLEY, b New York, NY, Nov 18, 41; m 77. INFRARED RADIATION & SENSORS, DETECTION & ESTIMATION. *Educ:* Mass Inst Technol, BS, 63; Cornell Univ, PhD(physics), 75. *Prof Exp:* Staff scientist, Comput Sci Corp, 74-83; RES PHYSICIST, NAVAL RES LAB, 83- *Mem:* Am Phys Soc; Am Optical Soc; Inst Elec & Electronics Engrs; Am Math Soc; Math Asn Am. *Res:* Infrared sensors (especially focal plane arrays); infrared radiation in the atmosphere; detection and estimation (classical); atmospheric remote sensing; quantum optio, quantum noise, quantum detection, and estimation theory; functional analysis; differential geometry. *Mailing Add:* 10123 Hereford Pl Silver Spring MD 20901

HORNTVEDT, EARL W, b Detroit, Mich, Aug 21, 48; m 81; c 2. DESALINATION EQUIPMENT, PRODUCT & MARKET RESEARCH FLOW CONTROL. *Educ:* Western Mich Univ, BS, 78; Cardinal Stritch Col, MBA, 88. *Prof Exp:* Dir new prod develop, Aqua-Catem, Inc, 81-88; dir eng, Northland Stainless, 88-89; MGR NEW PROD DEVELOP, HAMILTON INDUSTS, 89- *Mem:* Am Soc Heating Refrig & Air Conditioning Engrs; Am Soc Testing & Mat. *Mailing Add:* 215 Cleveland Ave Manitowoc WI 54220

HORNUFF, LOTHAR EDWARD, JR, entomology, aquatic ecology; deceased, see previous edition for last biography

HORNUNG, DAVID EUGENE, b Latrobe, Pa, Apr 30, 45; m. PHYSIOLOGY. *Educ:* Geneva Col, BS, 67; Kent State Univ, MS, 69; State Univ NY Upstate Med Ctr, PhD(physiol), 75. *Prof Exp:* Teaching asst gen biol, Kent State Univ, 67-69; instr physiol, St Lawrence Univ, 69-73; teaching fel, State Univ NY Upstate Med Ctr, 73-75; from asst prof to assoc prof physiol, 75-83; res assoc prof, 80-85, RES PROF PHYSIOL, STATE UNIV NY UPSTATE MED CTR, 85-, CHMN DEPT BIOL, 86-; CHARLES A DANA PROF BIOL, ST LAWRENCE UNIV, 83- *Concurrent Pos:* Dir int educ, State Univ NY Med Ctr. *Mem:* AAAS; Am Physiol Soc; Am Chemoreception Scis; Europ Chemoreception Res Orgn. *Res:* Mechanism and functions of vertebrate olfactory and taste systems; initial distribution and subsequent resorption of odorant molecules from the olfactory sac. *Mailing Add:* Dept Biol St Lawrence Univ Canton NY 13617

HORNUNG, ERWIN WILLIAM, b Chicago, Ill, Mar 3, 19. PHYSICAL CHEMISTRY, CRYOGENICS. *Educ:* Univ Chicago, MS, 49; Univ Calif, PhD(chem), 54. *Prof Exp:* From asst res chemist to assoc res chemist, Univ Calif, Berkeley, 54-83; RETIRED. *Mem:* Sigma Xi. *Res:* Low temperature physical chemistry; magnetic properties of matter. *Mailing Add:* 975 Alvarado Berkeley CA 94705

HORNUNG, MARIA, immunology, for more information see previous edition

HORNYAK, WILLIAM FRANK, b Cleveland, Ohio, Aug 4, 22; m 50; c 2. NUCLEAR PHYSICS. *Educ:* City Col New York, BEE, 44; Calif Inst Technol, MS & PhD(physics), 49. *Prof Exp:* Jr engr, Radar Res, Westinghouse Mfg Co, 44; assoc physicist nuclear physics, Brookhaven Nat Lab, 49-53, physicist, 53-55; lectr, Princeton Univ, 55-56; assoc prof, 56-61, PROF PHYSICS & ASTRON, UNIV MD, COLLEGE PARK, 61- *Concurrent Pos:* Fel, NATO, Orsay, 62-63; guest prof, Univ Zurich, 66-67. *Mem:* Fel Am Phys Soc; Sigma Xi. *Res:* Beta ray spectroscopy; light element transmutations; fundamental scattering cross sections; nuclear radii; lifetime of nuclear excited states; cyclotron research. *Mailing Add:* Dept Physics Univ Md College Park MD 20742

HORNYIK, KARL, nuclear engineering; deceased, see previous edition for last biography

HORODYSKI, ROBERT JOSEPH, b Morristown, NJ, May 4, 43. STRATIGRAPHY, PALEONTOLOGY. *Educ:* Mass Inst Technol, BS, 65, MS, 68; Univ Calif, Los Angeles, PhD(geol), 73. *Prof Exp:* Res fel geol, Univ Calif, Los Angeles, 73-76; vis asst prof, Univ Notre Dame, 76-78; asst prof, 78-82, ASSOC PROF GEOL, TULANE UNIV, 82- *Mem:* Geol Soc Am; Soc Econ Paleontologists & Mineralogists; Int Asn Sedimentologists; Sigma Xi. *Res:* Precambrian geology and paleontology. *Mailing Add:* Dept Geol Tulane Univ New Orleans LA 70118

HOROSHKO, ROGER N, b Zahore, Byelorussia, June 8, 37; US citizen. NUCLEAR PHYSICS. *Educ:* City Col New York, BS, 59; Columbia Univ, MA, 61, PhD(physics), 66. *Prof Exp:* Res scientist, Columbia Univ, 65-66; fel physics, Bartol Res Found, Pa, 66-68; res assoc, Nuclear Structure Lab, Univ Rochester, 68-74; ASST PROF PHYSICS & ASTRON, LEHMAN COL, 74- *Mem:* Am Phys Soc. *Res:* Nuclear structure and nuclear reactions; determination of properties of nuclear states, their energies, spins, parities, electromagnetic transitions and interpretation in terms of nuclear structure; nuclear reaction mechanism studies. *Mailing Add:* 9-06 Parsons Blvd Malba NY 11352

HOROVITZ, ZOLA PHILLIP, b Pittsburgh, Pa, Oct 12, 34; m 58; c 2. PHARMACOLOGY. *Educ:* Univ Pittsburgh, BS, 55, MS, 58, PhD(pharmacol), 60. *Prof Exp:* Vis investr, Neuropsychiat Res Labs, Vet Admin, Pa, 58-59; res pharmacologist, Squibb Inst Med Res, 59-67, dir dept pharmacol, 67-72, assoc dir res, 72-78, vpres drug develop, 79- 85, vpres planning, 85-89, VPRES LICENSING, BRISTOL-MYERS SQUIBB, 90- *Concurrent Pos:* Vis prof, Sch Med, Rutgers Univ, 67-; fel, Am Found Pharmaceut Educ; mem sci adv bd, Princeton Univ, 75-; mem adv coun, Rutgers Sch Pharm, 77- *Honors & Awards:* A E Bennett Award, 65. *Mem:*

Fel AAAS; Am Pharmaceut Asn; Am Soc Pharmacol & Exp Therapeut; NY Acad Sci; assoc Brit Pharmacol Soc; Sigma Xi. *Res:* Neuropharmacology; neurophysiology; neurochemistry; behavioral and psychopharmacology; cardiovascular work. *Mailing Add:* 30 Philip Dr Princeton NJ 08540

HOROWICZ, PAUL, b New York, NY, June 17, 31; m 54; c 2. PHYSIOLOGY, BIOPHYSICS. *Educ:* NY Univ, AB, 51; Johns Hopkins Univ, PhD(biophys), 55. *Prof Exp:* Res fel physiol, Cambridge, 55-57; res assoc biophys, Johns Hopkins Univ, 57-59; asst prof physiol, Washington Univ, 59-61; assoc prof, Duke Univ, 61-69; PROF PHYSIOL & CHMN DEPT, SCH MED, UNIV ROCHESTER, 69- *Mem:* Biophys Soc; Am Physiol Soc; Soc Gen Physiol; NY Acad Sci; Sigma Xi. *Res:* Ionic transport across biological membranes; muscle physiology; neurophysiology; neurochemistry. *Mailing Add:* D-Physiol-Box 642 Univ Rochester 601 Elmwood Ave Rochester NY 14642

HOROWITZ, ALAN STANLEY, b Ashland, Ky, June 12, 30; m 75. PALEONTOLOGY, GEOLOGY. *Educ:* Washington & Lee Univ, AB, 52; Ohio State Univ, MS, 54; Ind Univ, PhD, 57. *Prof Exp:* Res geologist, Marathon Oil Co, 56-64; CUR PALEONT & PROF GEOL, IND UNIV, BLOOMINGTON, 64- *Concurrent Pos:* Guest prof, Aarhus Univ, Aarhus, Denmark, 71-72. *Mem:* Am Asn Petrol Geologists; Geol Soc Am; Paleont Soc; Soc Econ Paleontologists & Mineralogists; Soc Syst Zool. *Res:* Invertebrate paleontology; lower carboniferous bryozoans and echinoderms. *Mailing Add:* Dept Geol Ind Univ 1005 E Tenth St Bloomington IN 47405

HOROWITZ, CARL, b Lvov, Poland, Aug 10, 23; m 54; c 2. POLYMER CHEMISTRY, TEXTILE CHEMISTRY. *Educ:* Columbia Univ, BS, 50; Polytech Inst Brooklyn, MA & PhD, 63. *Prof Exp:* Vpres, Yardney Chem Corp, 51-63; PRES, POLYMER RES CORP, 63- *Honors & Awards:* John C Vaaler Awards, Chem Processing Mag, 64 & 72. *Mem:* Am Chem Soc; Am Inst Chem Engr. *Res:* Chemical grafting; fabric nonflammability; semipermeable membranes; silver zinc storage batteries. *Mailing Add:* 5607 Fillmore Ave Brooklyn NY 11234

HOROWITZ, ELLIS, b New York, NY, Feb 11, 44; m 68; c 3. COMPUTER SCIENCE. *Educ:* Brooklyn Col, BS, 64; Univ Wis, MS, 67, PhD(comput sci), 70. *Prof Exp:* PROF COMPUT SCI, UNIV SOUTHERN CALIF, 73- *Concurrent Pos:* Ed, Trans Math Software, 76-80, Comt ACM, 78-82. *Mem:* Soc Indust & Appl Math; Sigma Xi; Asn Comput Mach; Inst Elec & Electronics Engs. *Mailing Add:* Comput Sci Univ Southern Calif Los Angeles CA 90007

HOROWITZ, EMANUEL, b New York, NY, Mar 29, 23; m 50; c 4. MATERIALS SCIENCE & ENGINEERING. *Educ:* City Col New York, BS, 48; George Washington Univ, MS, 56, PhD, 63. *Prof Exp:* Chemist, Smithsonian Inst, 49-51; supvry chemist, Nat Bur Standards, 51-61; vis scientist, Univ Ill, 61-62; phys polymer chemist, Nat Bur Standards, 62-64, Dept Com sci & technol fel, 64-65, chief polymer characterization sect, 65-67, dep chief polymers div, 67-69, dep dir inst mat res, 69-76, dep dir resources & opers, Nat Meas Lab, 76-80; dir, Ctr Mat Res, Johns Hopkins Univ, 80-85, prof mat sci & eng, 80-88; CONSULT, MAT SCI & ENG, BIOMAT, SURG & DENT IMPLANTS, 88- *Concurrent Pos:* Chmn task group TC-38 (textiles), Int Orgn Standardization, 57-65, chmn task group polymer standard ref mat ISO/TC-61 (plastics) & mem, ISO/TC 150 (surg implants), ISO/TC 194 (biocompatibility), 65-80; leader, US Sci Team, Yugoslavia, 70, 72; chmn, Interagency Comn on Mat, 73-75; trustee, Inst Standards Res; prof, Johns Hopkins Univ, 88- *Honors & Awards:* Meritorious Awards, Nat Bur Standards, 61 & 65-68, Rosa Award, 72, Gold Medal, 75. *Mem:* Am Chem Soc; hon mem Am Soc Testing & Mat; NY Acad Sci; Cosmos Club; Fedn Mat Socs (pres, 83). *Res:* Synthesis and characterization of organic and coordination polymers; relation between structure and physical and chemical properties of materials; development of new or improved methods of chemical analysis and characterization of materials; properties and performance of surgical implant materials; biomaterials-polymers, metals and alloys, ceramics, composites, biosynthethics, biocompatibility. *Mailing Add:* 14100 N Gate Dr Silver Spring MD 20906

HOROWITZ, ESTHER, b New York, NY, Dec 17, 20. SPEECH PATHOLOGY, PSYCHOLOGY. *Educ:* Brooklyn Col, BA, 40; Univ Wis, MA, 49; Columbia Univ, PhD(psychol), 59. *Prof Exp:* Speech clinician, Queen's Col, 44-46; teacher, New York Schs, 46-50; assoc prof speech path, Hofstra Univ, 50-72, dir speech clin, 53-67, prof, 72-81; RETIRED. *Concurrent Pos:* Res grant speech path, Hofstra Univ, 61. *Mem:* Am Speech & Hearing Asn; Speech Asn Am. *Res:* Stuttering; aphasia. *Mailing Add:* 147-07 Charter Rd Jamaica NY 11435

HOROWITZ, GARY T, b Washington, DC, Apr 14, 55; m 80; c 2. GENERAL RELATIVITY, QUANTUM GRAVITY. *Educ:* Princeton Univ, BA, 76; Univ Chicago, PhD(physics), 79. *Prof Exp:* Fel, Univ Calif, Santa Barbara, 79-80; NATO fel, Oxford Univ, 80-81; mem, Inst Advan Study, 81-83; CHMN, DEPT PHYSICS, UNIV CALIF, SANTA BARBARA, 83- *Mem:* Sigma Xi. *Res:* Aspects of Einstein's theory of general relativity, approaches to quantum gravity, especially string theory, interested in nonperturbative formulations of the theory. *Mailing Add:* Dept Physics Univ Calif Santa Barbara CA 93106

HOROWITZ, HAROLD S, ceramic engineering, for more information see previous edition

HOROWITZ, HUGH H(ARRIS), b New York, NY, Sept 8, 28; m 53; c 3. ELECTROCHEMISTRY, ORGANIC CHEMISTRY. *Educ:* City Col New York, BS, 49; Columbia Univ, AM, 51, PhD(org chem), 53. *Prof Exp:* Chemist, Prod Res Div, Exxon Res & Eng Co, 52-55, proj leader lube oil additives, 55-61, proj leader & res assoc fuel cells, 61-66, proj leader & sr res assoc org electrochem, 66-71, proj leader & sr res assoc, Alsthom-Exxon Joint Fuel Cell Venture, 71-76, group leader, Corp Res Lab, 77-80, sr res assoc, Eng Mat Technol Div, 80-86, SR RES ASSOC, PRODS RES DIV, EXXON RES & ENG CO, 86- *Mem:* Am Chem Soc; Electrochem Soc. *Res:* Copolymerization; viscometry and rheology; hydrodynamic lubrication; automotive cold cranking; thermogravimetry; lubricant additives; catalysis; electrochemical oxidation of organics; fuel cells; electroanalytical chemistry; batteries; corrosion; antioxidants. *Mailing Add:* Exxon Res & Eng Co PO Box 51 Linden NJ 07036

HOROWITZ, ISAAC, b Safed, Israel, Dec 15, 20; nat US; m 45, 84; c 6. ELECTRICAL ENGINEERING. *Educ:* Univ Man, BSc, 44; Mass Inst Technol, SB, 48; Polytech Inst Brooklyn, MEE, 53, DEE, 56. *Prof Exp:* Asst prof elec eng, Polytech Inst Brooklyn, 56-58; sr scientist guid & controls div, Hughes Aircraft Corp, Calif, 58-66; prof elec eng, City Univ New York, 66-67; prof elec eng, Univ Colo, Boulder, 67-85; Cohen prof appl math, 69-85, EMER PROF APPL MATH, WEIZMANN INST SCI, ISRAEL, 85-; PROF ECE DEPT, UNIV CALIF, DAVIS, 85- *Concurrent Pos:* Guest lectr, Delft Technol Univ, 62 & Haifa Technion, 62; lectr, Calif Inst Technol, 64-; distinguished vis prof, Air Force Inst Technol, 83- *Honors & Awards:* Ann Award, Nat Electronics Conf, 56. *Mem:* Nat Electronics Conf; fel Inst Elec & Electronics Engrs. *Res:* Feedback theory; active network synthesis; adaptive systems. *Mailing Add:* Dept Elec Eng Univ Calif CB 425 Davis CA 95616

HOROWITZ, JACK, b Vienna, Austria, Nov 25, 31; nat US; m 61; c 2. BIOCHEMISTRY. *Educ:* City Col New York, BS, 52; Ind Univ, PhD(biochem), 57. *Prof Exp:* Asst chem, Ind Univ, 54-57; fel biochem, NSF, Columbia Univ, 57-59, res biochemist, 59-60, res assoc biochem, Col Physicians & Surgeons, 60-61; from asst prof to assoc prof, 61-71, chmn dept biochem & biophysics, 71-74, coordr molecular, cellular & develop biol prog, 77-80, PROF BIOCHEM, IOWA STATE UNIV, 71- *Concurrent Pos:* Vis scientist, Rockefeller Univ, 68, Mass Inst Technol, 90-91; vis prof, Yale Univ, 74-75; travel awards, Am Soc Biochem & Molecular Biol, 67, 73, 82. *Mem:* AAAS; Am Soc Biochem & Molecular Biol. *Res:* Nucleic acids; protein biosynthesis; nucleoproteins. *Mailing Add:* Dept Biochem & Biophys Iowa State Univ Ames IA 50011

HOROWITZ, JOHN M, b San Francisco, Calif, Sept 16, 34; m 70. HIBERNATION, HIPPOCAMPAL NETWORKS. *Educ:* Univ Calif, Berkeley, BS, 59, MS, 61, PhD(biophys), 68. *Prof Exp:* Res engr biophys, Univ Calif, Berkeley, 61-63; from asst prof to assoc prof, 69-78, PROF PHYSIOL, UNIV CALIF, DAVIS, 78- *Mem:* Soc Neurosci; Am Physiol Soc. *Res:* Neural control of temperature regulation and neural mechanisms associated with hibernation; hippocampal neural networks; coupling of signals from nerve cells to other cell types. *Mailing Add:* Dept Animal Physiol Univ Calif Davis CA 95616

HOROWITZ, LARRY LOWELL, b Flushing, NY, Apr 17, 49; m 78; c 2. ESTIMATION THEORY, ADAPTIVE ANTENNAS. *Educ:* Mass Inst Technol, SB, 72, SM, 72, PhD(elec eng), 74. *Prof Exp:* Engr, Appl Physics Lab, Johns Hopkins Univ, 74-75; staff mem, 75-83, asst group leader, 83-90, SR STAFF, LINCOLN LAB, MASS INST TECHNOL, 90- *Mem:* Inst Elec & Electronics Engrs; Sigma Xi. *Res:* Estimation theory and signal processing; practical superresolution techniques; adaptive antenna processing. *Mailing Add:* Lincoln Lab Mass Inst Technol 244 Wood St Rm D362 Lexington MA 02173

HOROWITZ, MARDI J, b Los Angeles, Calif. PSYCHIATRY, PSYCHOANALYSIS. *Educ:* Univ Calif, Los Angeles, BA, 55, MD, 58. *Prof Exp:* Intern, Univ Ore, 59; resident psychiat, Langley Porter Neuropsychiat Inst, 62; from instr to assoc prof, 69-74, PROF PSYCHIAT, SCH MED, UNIV CALIF, SAN FRANCISCO, 74- *Concurrent Pos:* NIMH res career award, Mt Zion Med Ctr, 64-72; dir, Prog Conscious & Unconscious Processes, John D & Catherine T MacArthur Found. *Honors & Awards:* Found Fund Prize; Royer Award; Strecker Award. *Mem:* Am Psychoanal Asn; Am Psychiat Asn; Am Psychosom Soc; Psychother Res Soc. *Res:* Visual imagery; abnormal thought; stress; character and character change. *Mailing Add:* Dept Psychiat Univ Calif San Francisco CA 94143

HOROWITZ, MARK CHARLES, b New York, NY, Aug 5, 50. IMMUNOBIOLOGY. *Educ:* Syracuse Univ, BS, 72; State Univ NY Upstate Med Ctr, PhD(immunol), 78. *Prof Exp:* Fel immunol, Howard Hughes Med Inst, Yale Univ, 77-78, res assoc, Dept Path, 78-80; RETIRED. *Mem:* Sigma Xi. *Res:* Immunoregulation mediated by different subsets of T and B cells; immunoregulatory defects in autoimmunity; control mechanisms of the immune response. *Mailing Add:* Path Dept Immunol Yale Univ Sch Med 310 Cedar St New Haven CT 06511

HOROWITZ, MARTIN I, b USA, May 17, 29; m 53; c 2. BIOCHEMISTRY. *Educ:* Brooklyn Col, MA, 52; Rutgers Univ, PhD(microbial chem), 57. *Prof Exp:* Asst nutrit chem, Sch Pub Health, Columbia Univ, 51-52; biochemist, Mt Sinai Hosp, 57-61; assoc prof, 61-69, PROF BIOCHEM, NY MED COL, 69- *Mem:* AAAS; Am Oil Chem Soc; Am Soc Biol Chemists; Am Chem Soc. *Res:* Biochemistry of mucins; digestive enzymes; glycolipids; nutrition; immunochemistry-blood group substances. *Mailing Add:* Dept Biochem Valhalla NY 10595

HOROWITZ, MYER GEORGE, b Boston, Mass, July 23, 24; m 50; c 3. BIOCHEMISTRY. *Educ:* Harvard Univ, BA, 45; Univ Ore, MA, 49, PhD(biochem), 52. *Prof Exp:* Asst biochemist, Univ Ore, 49-51; Nat Found Infantile Paralysis fel, Northwestern Univ, 52-54, USPHS fel, 54-55; CLIN CHEMIST, CLIN LAB, JEWISH HOSP, CINCINNATI, 55- *Mem:* AAAS; Am Chem Soc; Am Asn Clin Chem; Sigma Xi. *Res:* Mechanism of lactose biosynthesis; glycolytic processes in mammary tissue homogenates; interaction of dyes with proteins. *Mailing Add:* 1239 Avon Dr Cincinnati OH 45229

HOROWITZ, NORMAN HAROLD, b Pittsburgh, Pa, Mar 19, 15; wid; c 2. GENETICS, BIOCHEMISTRY. *Educ:* Univ Pittsburgh, BS, 36; Calif Inst Technol, PhD(biol), 39. *Prof Exp:* Fel, Nat Res Coun, Univ & Hopkins Marine Sta, Stanford Univ, 39-40; res assoc biol, Stanford Univ, 42-46; res fel biochem, 40-42, from assoc prof to prof biol, 47-82, mgr biosci sect, Jet Propulsion Lab, 65-70, chmn div, 77-80, EMER PROF BIOL, CALIF INST TECHNOL, 82- *Concurrent Pos:* Fulbright & Guggenheim fel, Genetics Lab, Univ Paris, 54-55; mem, Space Sci Bd, Nat Acad Sci, 71-75. *Honors & Awards:* Pub Serv Medal, NASA, 77. *Mem:* Nat Acad Sci; AAAS; Am Acad Arts & Sci; Am Soc Biol Chemists; Genetics Soc Am. *Res:* Biochemical genetics of Neurospora; enzymes; molecular evolution; exploration of Mars. *Mailing Add:* Div Biol Calif Inst Technol Pasadena CA 91125

HOROWITZ, PAUL MARTIN, b Jersey City, NJ, Oct 5, 39; m 65; c 3. BIOCHEMISTRY, PHYSICAL CHEMISTRY. *Educ:* NY Univ, BA, 62; Univ Chicago, PhD(biochem), 68. *Prof Exp:* AEC res fel spectros, Inst Molecular Biophys, Fla State Univ, 68-70; asst prof phys biochem, Dartmouth Col, 70-77; assoc prof, 77-84, PROF DEPT BIOCHEM, UNIV TEX HEALTH SCI CTR, 84- *Concurrent Pos:* Cottrell grant, 71; Petrol Res Fund grant, 71-74; adj asst prof, Dartmouth Med Sch, 73-; Erna & Jakob Michael vis prof, Weizman Inst Sci, Israel. *Mem:* AAAS. *Res:* Structure function relationships in biological macromolecules; enzymes and nucleic acids. *Mailing Add:* Dept Biochem 7703 Floyd Curl Dr San Antonio TX 78284

HOROWITZ, RICHARD E, b Vienna, Austria, May 17, 31; US citizen; div; c 1. PATHOLOGY, COMPUTER SCIENCES. *Educ:* Univ Calif, Los Angeles, AB, 53; Univ Calif, San Francisco, MD, 57; Am Bd Path, cert path anat & clin path, 63. *Prof Exp:* Intern, Los Angeles County Hosp, 58; supvr, Dept Germfree Res, Walter Reed Army Inst Res, 58-60; fel path, Mt Sinai Hosp, NY, 60-63; instr path, Col Physicians & Surgeons, Columbia Univ, 61-63; from asst prof to assoc prof, 63-72, CLIN PROF PATH, SCH MED, UNIV SOUTHERN CALIF, 72-; CLIN PROF PATH, SCH MED, UNIV CALIF, LOS ANGELES, 83- *Concurrent Pos:* Pathologist, Los Angeles County Gen Hosp, 63-66, asst dir labs, 66-68; consult pathologist, Vet Admin Hosp, Los Angeles, 64-, sr attend physician, Los Angeles County, Univ Southern Calif Med Ctr, 68-; assoc pathologist, St Joseph Med Ctr, 68-71, dir labs, 71-90, sr pathologist, 90-; mem, Comt Life Sci & Pub Policy, Nat Res Coun-Nat Acad Sci, 68-75; hmn, Verdugo Div, Am Heart Asn, 72-74; mem, Comt Computerized Lab Systs, Col Am Pathologists, 77-85, Calif Deleg to House of Delegates, 84-, bd dirs, Cap Found; mem comt Comput in Health Sci, Inst Elec & Electronics Engrs; mem bd dirs, Los Angeles Soc Pathologists, 82-87, pres, 86-; mem bd trustees, St Joseph Med Ctr Found, 82-, adv bd, 91-, chmn bd trustees, 91-; mem, Path Sci Adv Panel, Calif Med Asn, 85-89, chmn, 87-88; mem, finance comt, Am Soc Clin Pathologists, 85-88, vchair Res, Develop & Strategic Planning Comt, 86-90, bd dirs, 88- *Mem:* Col Am Path; Am Soc Clin Path (treas, 88-); AMA. *Res:* Pathologic anatomy; clinical pathology; immunopathology; germfree animal research; laboratory computer and information sciences. *Mailing Add:* St Joseph Med Ctr 501 S Buena Vista St Burbank CA 91505-4866

HOROWITZ, ROBERT MILLER, b Pittsburgh, Pa, Sept 5, 21; m 53; c 3. ORGANIC CHEMISTRY. *Educ:* Univ Calif, Los Angeles, AB, 43, PhD(chem), 49. *Prof Exp:* Jr chemist, Shell Develop Co, 43-46; res assoc, Smith, Kline & French Labs, 49-51; instr pharmacol, Univ Mich, 51-54; RES CHEMIST, FRUIT & VEG CHEM LAB, US DEPT AGR, 55- *Mem:* AAAS; Am Chem Soc; Sigma Xi. *Res:* Natural products chemistry; taste-structure relations; sweeteners; colchicine; flavonoids; siderophores. *Mailing Add:* US Dept Agr 263 S Chester Ave Pasadena CA 91106-3196

HOROWITZ, SAMUEL BORIS, b Perth Amboy, NJ, Aug 26, 27; m 55, 73; c 1. CELL PHYSIOLOGY, ZOOLOGY. *Educ:* Hunter Col, AB, 51; Univ Chicago, PhD(zool), 56. *Prof Exp:* Res assoc, Col Med, Univ Ill, 56-57; sr res fel basic res, Eastern Pa Psychiat Inst, 57-62; vis investr, Int Physiol & Med Biophys, Uppsala Univ, 62-63; assoc mem & co-head lab cellular biophys, Albert Einstein Med Ctr, 63-69, head, 69-72; chmn dept biol, 75-78, res scientist, 72-81, LAB CHIEF, MICH CANCER FOUND, 72-, CHMN DEPT PHYSIOL & BIOPHYS, 81- *Concurrent Pos:* Vis investr, Physiol Lab, Cambridge Univ, 71-72; adj prof dept biol, Wayne State Univ, 72-89; mem, Biophys & Biophysical Chem Study Sect, NIH, 78-79. *Mem:* Am Asn Cancer Res; AAAS; Am Soc Cell Biol; Biophys Soc; Soc Study Develop & Growth. *Res:* Intracellular transport; hormonal action at cellular level; solute transport; physics and chemistry of transport and solubility in proteinaceous systems; growth control. *Mailing Add:* Mich Cancer Found 110 E Warren Ave Detroit MI 48201

HOROWITZ, SIDNEY LESTER, b Jersey City, NJ, June 26, 21; m 50; c 4. ORTHODONTICS. *Educ:* Columbia Univ, BS, 42, cert, 49; NY Univ, DDS, 45. *Prof Exp:* Asst vis dentist, Children's Hosp, Philadelphia, 49-51; res asst dent, 51-54, from res assoc to asst clin prof, 54-69, PROF DENT & DIR DIV OROFACIAL DEVELOP, SCH DENT & ORAL SURG, COLUMBIA UNIV, 69- *Concurrent Pos:* Assoc attend orthodontist, Manhattan Eye, Ear & Throat Hosp, New York, 58-; consult, Roosevelt Hosp, 69- *Mem:* AAAS; Int Asn Dent Res; Sigma Xi. *Res:* Genetic variations in skull and dentition; orthodontic management in reconstructive plastic surgery of the face. *Mailing Add:* Columbia Univ Sch Dent & Oral Surg New York NY 10032

HOROWITZ, SYLVIA TEICH, b Brooklyn, NY; m 53; c 3. BIO-ORGANIC CHEMISTRY. *Educ:* Brooklyn Col, BA, 43; Columbia Univ, PhD(org chem), 49. *Prof Exp:* Res assoc org chem, Amherst Col, 49-50; res assoc steroid chem, Col Physicians & Surgeons, Columbia Univ, 50-53; res assoc biochem, Sch Med, Univ Mich, 53-55; VIS LECTR BIOCHEM, CALIF STATE UNIV, LOS ANGELES, 69- *Mem:* Am Chem Soc; Am Women Sci; Sigma Xi. *Res:* Metabolism of steroid hormones; synthesis of steroid conjugates; synthesis of analogues of penicillin. *Mailing Add:* Dept Chem Calif State Univ Los Angeles CA 90032

HORRES, ALAN DIXON, b Charleston, SC, Jan 12, 29; m 49; c 3. PHYSIOLOGY. *Educ:* Col Charleston, BS, 48; Syracuse Univ, MS, 50; Med Col SC, PhD(physiol), 55. *Prof Exp:* Asst prof biol, The Citadel, 55-56; instr physiol, Med Col SC, 56-58; from instr to asst prof physiol, 58-63, instr surg, 60-63, assoc surg, 64-65, dir pulmonary function lab, Emory Univ, 60-65, from asst prof to assoc prof physiol, 65-71, asst prof exp med, 65-67, PROF PHYSIOL, MED UNIV SC, 71- *Concurrent Pos:* Dir pulmonary function lab, Vet Admin Hosp, 65-67. *Mem:* Am Physiol Soc. *Res:* Respiratory control; cardiovascular physiology; neurophysiology. *Mailing Add:* Dept Physiol Med Univ SC 171 Ashely Ave Charleston SC 29425

HORRES, CHARLES RUSSELL, JR, b Charleston, SC, Aug 6, 45; m 67. PHYSIOLOGY, BIOMEDICAL ENGINEERING. *Educ:* Ga Inst Technol, BS, 68; Duke Univ, PhD(physiol & pharmacol), 75. *Prof Exp:* Res engr membrane develop, Chemstrand Res Ctr, Monsanto, 68-69; spec asst sanitary engr air pollution control, Nat Air Pollution Control Admin, 69-71; prin investr biochem & pharmacol, Becton-Dickenson Res Ctr, 75-77 & prin investr mat res, 77-78, dept mgr appl physiol, 78-83; vpres res & develop, Dancretec, Inc, 83-86; VPRES RES & DEVELOP, IVAC, INC, 86- *Concurrent Pos:* Adj res assoc physiol, Duke Univ, 76-78, adj asst prof, 78-82, adj assoc prof, 82-; lectr physiol, Univ NC, Chapel Hill, 78-80, adj asst prof, 80-83; adj asst prof, Univ NC, Chapel Hill, 80-83; vpres res & develop, Pancretec Inc, 83-86, IVAC Inc, 86- *Mem:* Biophys Soc; Am Soc Artificial Internal Organs; Cardiac Muscle Soc; Soc Gen Physiologists. *Res:* Membrane transport phenomena; cardiac cellular physiology. *Mailing Add:* Dept Physiol Duke Univ Med Ctr Box 3709 Durham NC 27706

HORRIDGE, PATRICIA EMILY, b Marshall, Tex, Jan 17, 37; c 1. TEXTILE CHEMISTRY. *Educ:* Univ Tex, Austin, BS, 58; Univ Houston, MS, 65; Tex Woman's Univ, PhD(textile sci), 69. *Prof Exp:* Asst prof textile & clothing, Baylor Univ, 66-68; res asst, Tex Woman's Univ, 68; asst prof, Fla State Univ, 69-71; assoc prof, Univ Southwestern La, 72; chairperson, Univ Ky, 73-76; chairperson textile & clothing, 76-85, chairperson merchandising, environ design & consumer econ, 85-90, PROF, TEX TECH UNIV, 90- *Concurrent Pos:* Numerous grants, Univs & Indust, 76-91. *Mem:* Int Fed Home Econ; fel Asn Col Prof Textiles & Clothing; Am Home Econ Asn. *Res:* Energy, especially determination of the effectiveness of window treatments as fuel savers; flammability of Cordelan/cotton blend fabrics. *Mailing Add:* Dept Merchandising Environ Design & Consumer Econ Tex Tech Univ Lubbock TX 79409-1162

HORRIGAN, FRANK ANTHONY, b Butte, Mont, Apr 19, 33; m 58, 75; c 4. THEORETICAL PHYSICS. *Educ:* Fordham Univ, BS, 54; Harvard Univ, AM, 55, PhD(physics), 61. *Prof Exp:* Sr res scientist, 61-65, prin res scientist, Res Div, Raytheon Co, Waltham, 65-74, mgr, Electrooptics Lab, 70-74; div mgr, Sci Appln, Inc, 74-80; tech mkt mgr, Equip Div, 80-87, prog mgr, AOSP, 81-85, TECH DIR ADVAN TECHNOL, EQUIP DEVELOP LABS, RAYTHEON, 85- *Concurrent Pos:* NATO fel, Saclay Nuclear Res Ctr, France, 62-63. *Mem:* Am Phys Soc. *Res:* Quantum mechanical many body theory; statistical mechanics; atomic and electron processes; plasma and laser physics; applied mathematics; fault tolerant computers; electro optic systems. *Mailing Add:* 283 Davis Rd Bedford MA 01730

HORRIGAN, PHILIP ARCHIBALD, b Brighton, Mass, Oct 8, 28; m 55; c 2. INORGANIC CHEMISTRY. *Educ:* Mass Inst Technol, BS, 48; Boston Col, MS, 50; Univ Ill, PhD(chem), 53. *Prof Exp:* Res chemist, Lever Bros Co, 50-51; mem staff tech sales, Rohm and Haas Co, 55-56; pres detergent mfg, Jet Prods, Inc, 57-62; from asst prof to assoc prof chem, 62-67, chmn dept, 67-80, PROF CHEM, SOUTHERN CONN STATE COL, 68- *Mem:* Am Chem Soc. *Res:* Manufacturing and marketing of chemical specialties. *Mailing Add:* Dept Chem Southern Conn State Univ 501 Crescent St New Haven CT 06515

HORRIGAN, ROBERT V(INCENT), b Brookline, Mass, June 8, 24; m 50; c 3. CHEMICAL & NUCLEAR ENGINEERING. *Educ:* Mass Inst Technol, BS, 44, MS, 47; Yale Univ, PhD(chem eng), 51. *Prof Exp:* Assoc engr, Brookhaven Nat Lab, 50-52; res investr, Titanium Alloy Mfg Div, Nat Lead Co, 52-57, chief develop div, 57-61; pres & tech dir, Transelco, Inc, 61-76, div mgr, Transelco Div, Ferro Corp, 76-79; RETIRED. *Concurrent Pos:* Consult, US AEC, 52-55. *Mem:* Am Chem Soc; Am Inst Chem Engrs; Am Ceramic Soc. *Res:* Industrial applications of titanium and zirconium products; concentration, disposal and utilization of radioactive wastes. *Mailing Add:* 3711 Country Lakes Circle Sarasota FL 33580

HORROBIN, DAVID FREDERICK, b Bolton, Eng, Oct 6, 39; m 65; c 2. ENDOCRINOLOGY, CARDIOVASCULAR PHYSIOLOGY. *Educ:* Oxford Univ, BA, 62, DPhil, 65, BM, BCh, 68. *Prof Exp:* Fel physiol, Magdalen Col, Oxford Univ, 63-68; prof med physiol & chmn dept, Nairobi Univ, Kenya, 69-72; reader physiol, Newcastle Univ, Eng, 72-75; dir endocrine pathophysiol, Clin Res Inst, 75-79; DIR, EFANOL RES INST, 79- *Mem:* Am Fedn Clin Res; Endocrine Soc; Royal Soc Med; Brit Physiol Soc; Brit Med Asn. *Res:* Actions of peptide hormones on renal, cardiovascular and immunological function; psychiatry; actions of hormones and prostaglandins on mental function. *Mailing Add:* Efanol Res Inst PO Box 818 Kentville NS B4N 4H8 Can

HORROCKS, LLOYD ALLEN, b Cincinnati, Ohio, July 13, 32; m 56; c 2. NEUROCHEMISTRY, BIOCHEMICAL NEUROPATHOLOGY. *Educ:* Ohio Wesleyan Univ, BA, 53; Ohio State Univ, MSc, 55, PhD(physiol chem), 60. *Prof Exp:* Res assoc neurochem, Cleveland Psychiat Inst, 60-68; from asst prof to assoc prof, 68-73, PROF PHYSIOL CHEM, COL MED, OHIO STATE UNIV, 73- *Concurrent Pos:* NIH fel, Univ Birmingham, 64-65; Josiah Macy Jr Found fac scholar, Ctr Neurochem, Nat Sci Res Ctr, Louis Pasteur Univ, Strasbourg, France, 74-75; mem rev B comt, Neurol Disorders Prog Proj, Nat Inst Health, 81-85; vis prof, Nat Res Coun Italy, NATO, 86. *Mem:* Soc Neurosci; Am Soc Biol Chemists; Am Soc Neurochem; Int Soc Neurochem; Brit Biochem Soc. *Res:* Brain lipid metabolism; brain function, membrane composition and structure; methods of lipid analysis; chromatography. *Mailing Add:* Dept Physiol Chem 214 Hamilton Hall Ohio State Univ 1645 Neil Ave Columbus OH 43210

HORROCKS, ROBERT H, b North Attleboro, Mass, Dec 23, 28; m 52; c 5. ORGANIC CHEMISTRY. *Educ:* Univ RI, BS, 51, MS, 61, PhD(chem), 64; Am Int Col, MBA, 57. *Prof Exp:* Plant chem engr, Plastics Div, Monsanto Co, 51 & 53-55, res chem engr, 55-61; asst, Univ RI, 61-63, res asst, 63-64; res assoc, Iowa State Univ, 64-65; asst prof, Univ Bridgeport, 65-72, assoc prof chem, 72-81, chmn, Dept Chem, 81-86, PROF CHEM, UNIV BRIDGEPORT, 81- *Concurrent Pos:* Res asst, Univ RI, 59-61; NSF grant, Boston Univ, 66-67; consult, Geigy Chem Co, 67. *Mem:* Am Chem Soc; Sigma Xi. *Res:* Free radical reactions, particularly hydrogen bromide additions to olefins and perester decompositions; both thermosetting and thermoplastic polymers; radical ions such as semidiones; mechanisms of phenazine formation. *Mailing Add:* 140 Salem Rd Stratford CT 06497

HORROCKS, RODNEY DWAIN, b Maeser, Utah, Oct 4, 38; m 60; c 5. AGRONOMY. *Educ:* Brigham Young Univ, BS, 62; Pa State Univ, MS, 64, PhD(agron), 67. *Prof Exp:* From asst prof to prof agron, Univ Mo-Columbia, 71-78; PROF DEPT AGRON & HORT, BRIGHAM YOUNG UNIV, 78-, CHMN DEPT, 81- *Concurrent Pos:* Vis prof, Dept Soils, Water & Eng, Univ Ariz, 75-76 & Dept Agron & Range Sci, Univ Calif, Davis, 89-90; USAID consult, Sierra Leone & Liberia, 76. *Mem:* Am Soc Agron; Crop Sci Soc Am; Sigma Xi; Am Inst Biol Sci. *Res:* Ecological modeling; forage crop production, management, physiology and ecology. *Mailing Add:* Dept Agron & Hort 273 Widtsoe Bldg Provo UT 84602

HORROCKS, WILLIAM DEWITT, JR, b Orange, NJ, Dec 7, 34; m 63; c 1. INORGANIC CHEMISTRY, BIOINORGANIC CHEMISTRY. *Educ:* Wesleyan Univ, BA, 56; Mass Inst Technol, PhD(inorg chem), 60. *Prof Exp:* From instr to asst prof chem, Princeton Univ, 60-69; assoc prof, 69-72, PROF CHEM, PA STATE UNIV, 72- *Concurrent Pos:* Vis lectr biol chem, Harvard Med Sch & Guggenheim fel, 74-75, vis prof, Mass Inst Technol, 88. *Mem:* Fel AAAS; Am Chem Soc; Am Soc Biochem & Molecular Biol; Biophys Soc. *Res:* Physical and inorganic chemistry, spectra and structure of transition metal compounds; physical inorganic and bioinorganic chemistry; transition metal and lanthanide coordination chemistry; paramagnetic nuclear magnetic resonance; Ligand field theory. *Mailing Add:* Dept Chem Pa State Univ University Park PA 16802

HORSBURGH, ROBERT LAURIE, Coronach, Sask, June 23, 31; m 55; c 3. ENTOMOLOGICAL EXTENSION. *Educ:* McGill Univ, BSc, 56; Pa State Univ, MS, 69, PhD(entom), 70. *Prof Exp:* Tree fruit entomologist, NS Dept Agr, 56-66 & 70-74; res assoc, Pa State Univ, 66-70; tree fruit entomologist, Va Polytech Inst & State Univ, 74-81; LAB SUPT & ENTOMOLOGIST, VA POLYTECH INST & STATE UNIV WINCHESTER FRUIT RES LAB, 81-, PROF ENTOM, 81- *Mem:* Etomologist Soc Am; Can Entom Soc. *Res:* Pest management of fruit pests with emphasis on biology of spiders, mirids, and anthocorids; efficacy of experimental pesticides on major pests and beneficial species on apple and peach. *Mailing Add:* Winchester Fruit Res Lab 2500 Valley Ave Winchester VA 22601

HORSCH, ROBERT BRUCE, b Pittsburgh, Pa, Sept 28, 52; m 72; c 3. PLANT SOMATIC CELL GENETICS, PLANT TRANSFORMATION. *Educ:* Univ Calif, Riverside, BSc, 74, PhD(genetics), 79. *Prof Exp:* Fel, Univ Sask, 79-81; sr res biologist, res specialist & res group leader, 81-85, sr res group leader, Corp Res Lab, 85-87, MGR, CROP TRANSFORMATION, MONSANTO AGR CO, 87- *Concurrent Pos:* Organizer for plant molecular biol course, Cold Spring, Harbor Labs, 85-88; adj prof biol, Washington Univ. *Mem:* Am Soc Plant Physiol; Tissue Cult Asn; Int Asn Plant Tissue Cult. *Res:* Somatic cell genetics of cultured plant cells including in vitro selection for mutants; gene-transfer techniques for genetic engineering of plants. *Mailing Add:* GG4J Monsanto Co 700 Chesterfield Village Pkwy St Louis MO 63198

HORSEMAN, NELSON DOUGLAS, b Dayton, Ohio, Sept 30, 51; m 75; c 1. ENDOCRINOLOGY, BIOLOGICAL RHYTHMS. *Educ:* Eastern Ky Univ, BS, 73, MS, 75; La State Univ, PhD(physiol), 78. *Prof Exp:* Res appointee, Argonne Nat Lab, 78-80; asst prof, 80-86, ASSOC PROF PHYSIOL, MARQUETTE UNIV, 86- *Concurrent Pos:* Lectr, Oakton Community Col, 79-80. *Mem:* Endocrine Soc; Am Physiol Soc; Sigma Xi; AAAS; Am Soc Zoologists. *Res:* Mechanisms of prolactin actions on gene expression and cell proliferation. *Mailing Add:* Dept Biol Marquette Univ Milwaukee WI 53233

HORSEMAN, NELSON DOUGLAS, b Dayton, Ohio, Sept 30, 51; m 75; c 2. PHYSIOLOGY. *Educ:* Eastern Ky Univ, BS, 73; La State Univ, PhD(physiol), 78. *Prof Exp:* Res assoc, La State Univ, 75-78; postdoctoral appointee, Argonne Nat Lab, 78-80; from asst prof to assoc prof physiol, Marquette Univ, 80-89; ASSOC PROF PHYSIOL, UNIV CINCINNATI, 89- *Concurrent Pos:* Dir, Biol & Biomed Res Inst, Marquette Univ, 87-89. *Mem:* Endocrine Soc; Am Soc Biochem & Molecular Biol; Sigma Xi. *Res:* Molecular biology of prolactin actions; transcription factor identification, receptor signal transduction and gene structure; function of annexin, lipocortin genes and proteins. *Mailing Add:* Dept Physiol Univ Cincinnati ML 576 Cincinnati OH 45267-0576

HORSFALL, JAMES GORDON, b Mountain Grove, Mo, Jan 9, 05; m 27; c 2. PLANT PATHOLOGY. *Educ:* Univ Ark, BS, 25; Cornell Univ, PhD(plant path), 29. *Hon Degrees:* DSc, Univ Vt, 58, Univ Turin, 64; LLD, Univ Ark, 69. *Prof Exp:* Asst plant path, Cornell Univ, 25-28, instr, 28-29; from asst prof to prof & chief res, NY Exp Sta, Geneva, 29-39, res assoc, 29-36; head dept, Plant Path & Bot, Conn Agr Exp Sta, 39-49, dir, 48-71, S W Johnson distinguished scientist, 72-75; RETIRED. *Concurrent Pos:* Prof, Bristol Univ, 36-37; mem crop protection comt, Nat Res Coun, 42-45, chem biol coordr, 46-55, chmn exec comt, 52-55; vpres, Int Cong Crop Protection, London, 49, pres, 79; lectr, Yale Univ, 50-64; trustee, Biol Abstr, 52; mem adv comt biol & med, AEC, 57-64; consult, President's Sci Adv Comt, 63-70, Environ Protection Agency, 75-78 & NASA, 74-77; mem, Nat Adv Comn Food & Fiber, 66-67; chmn, Gov Comt Environ Policy, Conn, 70; mem, Power Facil Eval Coun, Conn, 72- *Mem:* Nat Acad Sci; fel AAAS; Bot Soc Am; Am Phytopath Soc (vpres, 50, pres, 51); Am Acad Arts & Sci. *Mailing Add:* Conn Agr Exp Sta PO Box 1106 New Haven CT 06504

HORSFALL, WILLIAM ROBERT, b Mountain Grove, Mo, Jan 11, 08; m 30. ENTOMOLOGY. *Educ:* Univ Ark, BS, 28; Kans State Col, MS, 29; Cornell Univ, PhD(entom), 33. *Prof Exp:* Asst, Kans State Col, 28-29; asst, Cornell Univ, 30-33; prof biol, Agr & Mech Col, Monticello, 33-36; instr entom, Univ Ark, 36-37; asst prof entom & zool, Exp Sta, SDak State Col, 37-38; from asst prof to assoc prof entom, Univ Ark, 38-47; from asst prof to assoc prof, 47-55, prof, 55-76, EMER PROF ENTOM, UNIV ILL, URBANA, 76- *Concurrent Pos:* Asst entomologist, Exp Sta, Univ Ark, 38-45; lion mem, Am Mosquito Control Asn, 86. *Honors & Awards:* Award of Merit, Zool Soc Finland, 64; Harold Gray Award, Am Mosquito Control Asn, 70. *Mem:* Fel AAAS; fel Entom Soc Am; Am Mosquito Control Asn. *Res:* Bionomics of mosquitoes and ticks; medical entomology. *Mailing Add:* 320 Morrill Hall Univ Ill 505 S Goodwin Urbana IL 61801

HORSLEY, JOHN ANTHONY, b Scarborough, Eng, Mar 10, 43; m 69; c 2. SURFACE CHEMISTRY, LASER CHEMISTRY. *Educ:* Oxford Univ, BA, 64, DPhil, 68. *Prof Exp:* NATO fel, Ctr Appl Wave Mech, Paris, 67-68; res scientist, French Nat Sci Res Ctr, 68-75; res chemist, Exxon Res & Eng Co, 75-77, staff chemist, 77-80, sr staff chemist, 80-81, head, Surface Chem Group, 81-86; SR RES FEL, CATALYTICA, INC, 86- *Mem:* Am Chem Soc. *Res:* Quantum chemistry; surface chemistry; electronic structure of catalysts; metal clusters; x-ray near edge spectroscopy; laser chemistry; laser isotope separation. *Mailing Add:* 689 Torrington Dr Sunnyvale CA 94087-2446

HORSLEY, JOHN SHELTON, III, b Richmond, Va, Oct 3, 27; m 55; c 3. SURGERY. *Educ:* Univ Va, BA, 50, MD, 53; Am Bd Surg, cert, 60. *Prof Exp:* From intern to asst resident surg, Peter Bent Brigham Hosp, Boston, 53-58, chief resident, 59; clin instr surg, Med Col Va, 59-65; from asst prof to assoc prof, Med Ctr, Univ Va, 65-72, Am Cancer Soc prof clin oncol, 72-76, dir, McIntire Tumor Clin, 65-76, dir, Div Cancer Studies, 73-76, prof surg, 72-76, assoc dir, Cancer Ctr & dir, Joint Cancer Clin, 76-78; PROF SURG, DIV SURG ONCOL, MED COL VA, 76- *Concurrent Pos:* Nat Cancer Inst trainee, Col Physicians & Surgeons, Columbia Univ, Mem Ctr, New York, 55-56; Arthur Cabot teaching fel, Harvard Med Sch, 59; attend surgeon, St Elizabeth's Hosp & Sheltering Arms Hosp, Richmond, 59-65; surg consult, McGuire Vet Admin Hosp, Richmond, 59-65; mem courtesy staff, Richmond Mem Hosp, 59-65; clin cancer coordr, Univ Va, 65-, clin dir, Div Cancer Studies, 70-73; co-ed, Cancer Trends, 67-; consult, Salem Vet Admin Hosp, 68-76; asst chief surg serv, McGuire Vet Hosp, 78-80, chief, surg oncol, 80- *Honors & Awards:* Horsley Mem Award Merit, Am Cancer Soc, 73. *Mem:* Am Col Surg; Asn Acad Surg; James Ewing Soc; Int Soc Surg; Soc Surg Oncol; Am Surg Asn. *Res:* Cancer. *Mailing Add:* Div Surg Oncol Box 11 Med Col Va Richmond VA 23298

HORSMA, DAVID AUGUST, b Hubbell, Mich, Sept 3, 40; m 66; c 2. PHYSICAL CHEMISTRY. *Educ:* Mich Tech Univ, BS, 62; Univ Calif, Davis, PhD(chem), 66. *Prof Exp:* Res assoc chem, Univ Calif, Los Angeles, 66-67; asst prof, Calif Polytech State Univ, 66-68; group leader plastics appln, Rohm & Haas Co, 68-71; sr engr mat develop, Crown Zellerbach Corp, 71-72; mem staff, Raychem, 72-73, group leader polymer physics, 73-76, mgr electron tech, 76-78, tech mgr Telecom Div, Europe, 78-82, mkt develop mgr, 82-84, tech dir, 84-88; SR STAFF SCIENTIST, APS, 89- *Mem:* Am Chem Soc. *Res:* Physical and electrical properties of polymers and the applications of these materials to products. *Mailing Add:* 1141 Parkinson Ave Palo Alto CA 94301

HORST, G ROY, b Lancaster, Pa, July 13, 33; m 58; c 2. ZOOLOGY, HISTOLOGY. *Educ:* Wagner Col, BS, 59; Cornell Univ, PhD(zool), 67. *Prof Exp:* From instr to asst prof anat, Univ Ariz, 67-71; asst prof, Univ Vt, 71-75, USPHS grant, 72-75; prof biol & chmn dept, State Univ NY Col Potsdam, 75-80. *Concurrent Pos:* USPHS grant, Univ Ariz, 68-71; chmn, Ann NAm Symp Bat Res, 70-; mem exec comt, Desert Bighorn Sheep Coun, 71-72. *Mem:* AAAS; Am Asn Anatomists; Am Soc Mammalogists; Am Soc Zoologists; Sigma Xi. *Res:* Renal physiology and morphology in desert mammals; mammalian development osmoregulation and thermoregulation. *Mailing Add:* Dept Biol State Univ NY Col Potsdam Potsdam NY 13676

HORST, RALPH KENNETH, b Massillon, Ohio, June 22, 35; m 60, 69; c 4. PLANT PHYSIOLOGY, MICROBIOLOGY. *Educ:* Ohio Univ, BS, 57; Ohio State Univ, MS, 59, PhD(plant path), 61. *Prof Exp:* Supvr virus indexing, Yoder Bros, Inc, 62-63, dir lab plant path, 63-68; from asst prof to assoc prof, 68-80, PROF PLANT PATH, CORNELL UNIV, 80- *Mem:* Am Phytopath Soc; AAAS; Tissue Cult Asn; Int Soc Hort Sci; Int Soc Hort Path. *Res:* Virus indexing of mums, carnations, roses and meristem culture; characterization and investigations on viroid, virus and viruslike diseases affecting ornamental plants; epidemiology of Fusaria ornamentals. *Mailing Add:* Dept Plant Path Cornell Univ Ithaca NY 14853-5098

HORST, RALPH L, JR, b New Orleans, La, July 14, 25; m 54; c 3. CHEMICAL ENGINEERING, CORROSION. *Educ:* Columbia Univ, BS, 47, MS, 48. *Prof Exp:* Sales develop engr, Aluminum Co Am, 48-60, head petrol & chem sect, Sales Develop Div, 60-64, prod develop engr, 64-65, res engr, Alcoa Res Labs, 65-66, sr res engr, Alcoa Tech Ctr, 66-78, staff engr, 78-81, tech supvr, 81-84, ENG ASSOC, ALCOA LABS, 84- *Mem:* Sigma Xi; Nat Asn Corrosion Engrs. *Res:* Corrosion science; electrochemistry and metallurgy of corrosion processes; materials evaluation; aluminum alloy corrosion mechanisms; cathodic protection engineering. *Mailing Add:* 272 Fernledge Dr New Kensington PA 15068

HORST, RONALD LEE, b Waynesboro, Pa, Aug 16, 49; m 73; c 1. PHYSIOLOGY, BIOCHEMISTRY. *Educ:* WVa Univ, BS, 71; Univ Wis, MS, 72, PhD(dairy sci, biochem), 76. *Prof Exp:* Fel biochem, Univ Wis, 76-77; MEM STAFF, NAT ANIMAL DIS CTR, SCI & EDUC ADMIN, USDA, 78- *Mem:* Am Dairy Sci Asn; Am Soc Bone & Mineral Res; Am Inst Nutrit. *Res:* Vitamin D and calcium metabolism in the bovine. *Mailing Add:* Nat Animal Dis Ctr USDA ARS 2300 Dayton Ave PO Box 70 Ames IA 50010

HORSTMAN, ARDEN WILLIAM, b Manitowoc, Wis, Aug 23, 30; m 66, 78; c 2. GEOLOGY. *Educ:* Lawrence Univ, BS, 52; Univ Cincinnati, MS, 54; Univ Colo, PhD(geol), 66. *Prof Exp:* Geologist, Shell Oil Co, 54-56, Mene Grande Oil Co, Venezuela, Gulf Oil Corp, 57-59 & Belco Petrol Corp, 60; vis lectr geol, Univ Mont, 65 & Lawrence Univ, 66; asst prof, 66-71, ASSOC PROF GEOL, WESTERN CAROLINA UNIV, 71- *Mem:* Fel AAAS; Am Asn Petrol Geologists; Sigma Xi. *Res:* Stratigraphy and sedimentation. *Mailing Add:* PO Box 1551 Cullowhee NC 28723

HORSTMAN, DONALD H, b St Louis, Mo, Jan 4, 39. PULMONARY PHYSIOLOGY. *Educ:* Pa State Univ, PhD(physiol), 67. *Prof Exp:* RES PHYSIOLOGIST, US ENVIRON PROTECTION AGENCY, UNIV NC, 63- *Mailing Add:* US Environ Agency Health Effects Res Lab Human Studies Div MD-58 Research Triangle Park NC 27711

HORSTMANN, DOROTHY MILLICENT, b Spokane, Wash, July 2, 11. VIROLOGY, INFECTIOUS DISEASES. *Educ:* Univ Calif, AB, 36, MD, 40; Am Bd Internal Med, dipl, 47. *Hon Degrees:* DSc, Smith Col, 61; MA, Yale Univ, 61; DrMedSc, Woman's Med Col Pa, 63. *Prof Exp:* Instr prev med, Yale Univ, 43-44; instr med, Med Sch, Univ Calif, 44-45; from instr to assoc prof prev med, Yale Univ, 45-61, assoc prof pediat, 56-61, prof epidemiol & pediat, 61-69, J R Paul prof epidemiol & prof pediat, 69-82, EMER J R PAUL EPIDEMIOL, EMER PROF PEDIAT & SR RES SCIENTIST, EPIDEMIOL, SCH MED, YALE UNIV, 82- *Concurrent Pos:* Commonwealth Fund fel, Sch Med, Yale Univ, 42-43; NIH post-doctorate fel, Nat Inst Med Res, London, 47-48; distinguished alumni fac vis prof, Med Sch, Univ Calif, San Francisco, 79. *Honors & Awards:* James D Bruce Award, Am Col Physicians, 75; Maxwell Finland Award, Infectious Dis Soc Am; Thorvald Madsen Award, State Serum Inst, Copenhagen. *Mem:* Nat Acad Sci; Am Soc Clin Invest; Am Epidemiol Soc; Infectious Dis Soc Am (pres, 74-75); Asn Am Physicians; Pan-Am Med Asn; Am Pediat Soc; fel Am Acad Pediat; Europ Asn Virus Dis; master Am Col Physicians; hon mem Royal Soc Med. *Res:* Infectious disease; clinical virology; clinical epidemiology. *Mailing Add:* Dept Epidemiol & Pub Health Sch Med Yale Univ Box 3333 60 College St New Haven CT 06510

HORT, EUGENE VICTOR, b New York, NY, May 25, 21; m 43; c 3. ORGANIC CHEMISTRY. *Educ:* City Col New York, BS, 42; Polytech Inst Brooklyn, MS, 48, PhD(chem), 50. *Prof Exp:* Res chemist, Nopco Chem Co, 50-51; dir res, Marco Chem Co, 51-53; sr chemist, Air Reduction Res Labs, 53-55; asst mgr acetylene chem & polymer sect, Gen Aniline & Film Corp, 55-62, tech assoc, 62-63, mgr cent res lab, 63-68, sr tech assoc, GAF Corp, 68-72, mgr acetylene chem res & devel, cent res lab, 72-81, sr scientist, 81-86; CONSULT, 86- *Mem:* Am Chem Soc; Catalysis Soc. *Res:* Research management; synthetic organic chemistry; polymerization; catalysis. *Mailing Add:* 300 Winston Dr Apt 2318 Cliffside Park NJ 07010

HORTICK, HARVEY J, b Leyden Twp, May 28, 35; m 59; c 2. GENETICS, AGRICULTURAL CHEMISTRY. *Educ:* Univ Ill, BS, 57, MS, 59, PhD(hort & plant physiol), 62. *Prof Exp:* Plant physiologist veg crops res, Libby, McNeill & Libby, Inc, 62-65, asst dir, 65-67, assoc dir, 67-69 dir agr res, 69-78; DIR, FRUIT, VEG & TROP CROPS, AGR RES, NESTLE ENTERPRISES, INC, 78- *Concurrent Pos:* Rep, Agr Res Inst, 74- *Res:* Effect of environment on plant growth and development; management and technical implementation for fruit, vegetable and tropical crops (pesticides, nutrient fertility, plant genetics, agricultural engineering and cultural programs). *Mailing Add:* 12526 Lt Nichols Rd Fairfax VA 22033-2431

HORTMANN, ALFRED GUENTHER, b New York, NY, Mar 1, 37. ORGANIC CHEMISTRY. *Educ:* Mass Inst Technol, SB, 58; Harvard Univ, PhD(chem), 64. *Prof Exp:* NIH fel org chem, Harvard Univ, 64; from asst prof to assoc prof, 69-76, PROF ORG CHEM, WASH UNIV, 76- *Mem:* Am Chem Soc; AAAS. *Res:* Synthesis and structure determination in natural product area; photochemistry; synthesis of heterocyclic systems; C102 as an oxidant in organic synthesis. *Mailing Add:* Dept Chem Wash Univ St Louis MO 63130-4899

HORTON, AARON WESLEY, b Detroit, Mich, June 13, 19; m 41; c 5. ENVIRONMENTAL HEALTH. *Educ:* Yale Univ, BS, 40, MS, 47, PhD(phys/org chem), 48. *Prof Exp:* Sr res chemist, Socony-Vacuum Labs, 41-46; res engr, Res & Develop Dept, Franklin Inst, 48; from asst prof to prof indust health, Kettering Lab, Univ Cincinnati, 49-63; prof biochem & environ med, Med Sch, Ore Health Sci Univ, 62-76, actg head div environ med, 69-72, prof Pub Health, Prev Med & Biochem, 76-89, head Sect Chem Biol & Oncol, 76-89, EMER PROF, PUB HEALTH & PREV MED, ORE HEALTH SCI UNIV, 89- *Concurrent Pos:* Consult, indust and govt toxicol probs, 50-89; vis prof, Kettering Lab, Univ Cincinnati, 63-65 & Nat Inst Health & Med Res, Debrousse Hosp, Lyon, France, 72. *Mem:* Am Chem Soc; Sigma Xi. *Res:* Carcinogens, cocarcinogens and inhibitors in petroleum; physical models of biological membranes and their interactions with cocarcinogens and steroids; membrane-mediated control of latent cancer; etiologic factors in the geographical distribution of breast, lung and oesophageal cancer. *Mailing Add:* Dept Public Health Med Sch Ore Health Sci Univ Portland OR 97201

HORTON, BILLY D, b Guy, Ark, Jan 20, 30; m 62; c 1. PLANT PHYSIOLOGY, HORTICULTURE. *Educ:* Univ Ark, Fayetteville, BSA, 58, MS, 60; Univ Md, College Park, PhD(hort), 65. *Prof Exp:* Res assoc pollution, Univ Md, 65-66; PLANT PHYSIOLOGIST, AGR RES SERV, USDA, 66- *Mem:* Am Soc Hort Sci. *Res:* Irrigation and water relation studies, cultural practices and nutrition associated with peachtree short life and tree training for complete mechanization of peach production; tissue culture stone fruits. *Mailing Add:* Dept Agric USDA Rte 2 Box 45 Kearneysville WV 25430

HORTON, BILLY MITCHUSSON, b Bartlett, Tex, Dec 27, 18; m 41; c 2. CONTROL SYSTEMS, INNOVATIONS. *Educ:* Univ Tex, BA, 41; Univ Md, MS, 49. *Prof Exp:* Instr, Chanute Field, US Army, 41-42, Radar Officer, Signal Corp, 42-46; physicist, Naval Res Lab, 46-51; proj physicist, Nat Bur Standards, 51-53; prof mech & aero eng, Case Western Reserve Univ, 75-89; RETIRED. *Concurrent Pos:* Lectr, Univ Md, 54-58; consult, Chamberlain Mfg Corp, 75-79. *Honors & Awards:* Arnold O Beckman Award, Inst Soc Am, 60; John Scott Award, 66. *Mem:* Nat Acad Eng; fel Inst Elec & Electronics Engrs; AAAS. *Res:* Amplification processes: fluid, mechanical, parametric; frictional processes; correlation in signal processing; noise modulation in detection and ranging systems; high intensity radiation detectors. *Mailing Add:* 14250 Larchmere Blvd Shaker Heights OH 44120

HORTON, CHARLES ABELL, b Buffalo, NY, Mar 31, 18; m 47; c 2. ANALYTICAL CHEMISTRY. *Educ:* Cornell Univ, AB, 41; Univ Mich, MS, 46, PhD, 49. *Prof Exp:* Chemist, Mead Johnson Co, Trojan Powder Co & Bell Aircraft Corp, 41-44; res assoc, Manhattan Proj, Univ Rochester, 44-45; res assoc, Univ Mich, 46-49; res assoc nuclear div, Union Carbide Corp, 49-71; CONSULT, 71- *Concurrent Pos:* Sr Officer, Int Atomic Agency, Vienna, 66-68. *Mem:* Fel AAAS; fel Am Inst Chemists; Am Chem Soc; Sigma Xi. *Res:* Analytical chemistry, instrumental, metals and fluorides; analysis of biochemicals; uranium compounds; chemical documentation. *Mailing Add:* 384 East Dr Oak Ridge TN 37830

HORTON, CLAUDE WENDELL, JR, b Houston, Tex, Feb 3, 42; m 63; c 1. THEORETICAL PHYSICS. *Educ:* Univ Tex, Austin, BS, 63; Univ Calif, San Diego, MS, 65, PhD(physics), 67. *Prof Exp:* PROF PHYSICS, UNIV TEX, AUSTIN, 69- *Concurrent Pos:* Fel Inst Advan Study, Princeton, 67-69; consult, Lawrence Livermore Lab, 73-, Sloan fel, 75. *Mem:* Am Phys Soc; Am Geophys Union. *Res:* Stability theory of plasmas. *Mailing Add:* R L Moore Bldg 11-320 Univ Tex Austin TX 78712

HORTON, CLAUDE WENDELL, SR, b Cherryvale, Kans, Sept 23, 15; m 38; c 2. UNDERWATER ACOUSTICS. *Educ:* Rice Inst Technol, BA, 35, MA, 36; Univ Tex, PhD(physics), 48. *Prof Exp:* Asst seismologist, Shell Oil Co, 36-37, party chief, 38-43; asst math, Princeton Univ, 37-38; res assoc, Underwater Sound Lab, Harvard Univ, 43-45; asst prof, Univ Tex, 46-50, chmn dept, 57-63, prof physics, 50-76, prof geol, 65-76, res physicist, Defense Res Lab, 45-, EMER PROF, UNIV TEX, AUSTIN, 76- *Concurrent Pos:* Consult, 45-; mem corp, Woods Hole Oceanog Inst. *Honors & Awards:* Pioneers Underwater Acoust Medal, Acoust Soc Am, 80. *Mem:* Fel Am Phys Soc; Soc Explor Geophys; fel Acoust Soc Am; Am Geophys Union. *Res:* Interpretation of geophysical data; electromagnetic radiation problems; theory of underwater sound problems; propagation of sound in the ocean. *Mailing Add:* Dept Physics Univ Tex Austin TX 78712

HORTON, CLIFFORD E(DWARD), b Worcester, Mass, Oct 19, 22; m 45; c 2. PHYSICS, ELECTRICAL ENGINEERING. *Educ:* Purdue Univ, BS, 43. *Prof Exp:* Engr power tube develop, Victor Div, Radio Corp Am, 43-46; circuit develop, Gen Elec Co, 46-49, receiving tubes, 49-59, actg mgr circuit develop unit, 59-60, consult engr, 60-63; res scientist weapons effect, Kaman Sci Corp, 63-87; RETIRED. *Concurrent Pos:* Vis lectr, Ky Wesleyan Col, 48-49; instr eng graphics, Univ Colo, 74- *Res:* Physics of nuclear weapons effects; electronic receiving tubes. *Mailing Add:* 310 W Woodmen Rd Colorado Springs CO 80919

HORTON, DEREK, b Brit, Aug 31, 32; m 57; c 4. ORGANIC CHEMISTRY. *Educ:* Univ Birmingham, BSc, 54, PhD(chem), 57, DSc, 72. *Prof Exp:* Head sci dept, Sebright Sch, Eng, 57-59; res assoc, 59-60, assoc res supvr, 60-62, from asst prof to assoc prof, 62-69, PROF CHEM, OHIO STATE UNIV, 69- *Concurrent Pos:* Vis prof, Univ Paris, 70 & Univ Grenoble, 71, 72 & 75; chmn, Gordon Res Conf Carbohydrates, 71; mem Med Chem Study Sect, NIH; regional ed, Carbohydrate Res. *Honors & Awards:* C S Hudson Award, Am Chem Soc, 72. *Mem:* AAAS; Am Chem Soc; fel The Chem Soc; fel Am Inst Chemists. *Res:* Carbohydrate chemistry, carbohydrate antibiotics; synthesis, reactions, stereochemistry; mechanistic and biochemical aspects; bacterial antigens. *Mailing Add:* Dept Chem Ohio State Univ 120 W 18 Ave Columbus OH 43210-1106

HORTON, EDWARD S, MEDICINE. *Prof Exp:* PROF MED & CHMN, E L AMIDON, COL MED, UNIV VT, 89- *Mailing Add:* Dept Med Univ Vt Col Med Burlington VT 05405

HORTON, GLYN MICHAEL JOHN, b Brighton, Eng, Jan 30, 42; m 64; c 2. ANIMAL NUTRITION. *Educ:* Univ Natal, BSc, 67; Univ Alta, MSc, 70; Univ London, PhD(animal nutrit), 73. *Prof Exp:* Res fel beef prod, Univ London, 70-73; sr lectr animal nutrit, Fac Vet Sci, Univ Pretoria, 73-75; assoc prof animal sci, Univ Sask, 75-79; res scientist, Agr Res Ctr, Univ Fla, 79-81; dir, Hoffman-La Roche, 82-88; PROF, COOK COL, RUTGERS UNIV, 88- *Concurrent Pos:* Res assoc, Int Proj Food Irradiation, Karlsruhe, 73-75; consult, domestic & int, 75-76. *Mem:* Am Soc Animal Sci; Brit Soc Animal Prod; Can Soc Animal Sci; Nutrit Soc; Sigma Xi; Brit Grassland Soc. *Res:* Manipulation of rumen fermentation to improve feed utilization and livestock production; improving the nutritive value of low quality roughages for ruminants; nutritional consequences of gut parasites. *Mailing Add:* Dept Animal Sci Cook Col Rutgers Univ New Brunswick NJ 08903

HORTON, HORACE ROBERT, b St Louis, Mo, Aug 26, 35; m 59; c 4. BIOCHEMISTRY. *Educ:* Mo Sch Mines, BS, 56; Univ Mo, MS, 58, PhD(biochem), 62. *Prof Exp:* Asst instr biochem, Univ Mo, 58-59; Nat Acad Sci-Nat Res Coun fel, Brookhaven Nat Lab, 61-62, res assoc, 62-64; from asst prof to assoc prof, 64-72, PROF BIOCHEM, NC STATE UNIV, 72-, WILLIAM NEAL REYNOLDS PROF, 81- *Concurrent Pos:* Mem, Danforth Assoc, Carolinas Assoc, 68-; guest prof biochem, Univ Lund, 74. *Mem:* Am Soc Biol Chemists; Am Chem Soc. *Res:* Mechanisms of enzyme action; enzyme structure as related to catalytic activity; immobilized enzyme systems; formation of disulfide bonds in proteins and peptides. *Mailing Add:* Dept Biochem NC State Univ Box 7622 Raleigh NC 27695-7622

HORTON, HOWARD FRANKLIN, b Glendale, Calif, Sept 6, 26; m 54; c 4. FISH BIOLOGY. *Educ:* Calif State Polytech Col, BS, 53; Ore State Univ, MS, 55, PhD(fisheries), 63. *Prof Exp:* Fishery biologist, Calif Dept Fish & Game, 53 & Ore State Game Comn, 58; from instr to assoc prof, 58-69, marine adv prof leader, 80-87, PROF FISHERIES, ORE STATE UNIV, 69- *Concurrent Pos:* Dir, Pond Dynamics/Aquaculture CRSP, 87- *Honors & Awards:* Outstanding Publ Res Award, Western Agr Econ Asn; Prof Excellence Publ Res Award, Am Agr Econ Asn; Award Excellence, Soc Tech Publ. *Mem:* Am Fisheries Soc; Wildlife Soc; Am Inst Fishery Res Biol; Nat Shellfisheries Asn; Wildlife Soc. *Res:* Population dynamics; reproductive biology and ecology of fishes and invertebrates. *Mailing Add:* Dept Fisheries Ore State Univ Corvallis OR 97331

HORTON, JAMES CHARLES, plant pathology; deceased, see previous edition for last biography

HORTON, JAMES HEATHMAN, b Winston-Salem, NC, Jan 9, 31; m 54; c 5. TAXONOMY, BOTANY. *Educ:* Univ NC, AB, 52, MA, 58, PhD(bot), 61. *Prof Exp:* Instr bot, Univ NC, 60-61; assoc prof biol, 61-70, head dept, 67-74, PROF BIOL, WESTERN CAROLINA UNIV, 70- *Res:* Floristic research in Rowan County, North Carolina; taxonomy of Polygonella; floristics and descriptive ecology, Great Smoky Mountains. *Mailing Add:* 30 Old Longbranch Rd Cullowhee NC 28723

HORTON, JAMES WRIGHT, JR, b Anderson, SC, June 14, 50; m 73; c 2. STRUCTURAL GEOLOGY, PETROLOGY. *Educ:* Furman Univ, BA, 72; Univ NC, Chapel Hill, MS, 74, PhD(geol), 77. *Prof Exp:* Asst prof geol, Univ Southern Maine, 77-79; res assoc, 78-80, asst br chief, 84-85, RES GEOLOGIST, US GEOL SURV, 80- *Concurrent Pos:* Adj asst prof, Univ NC, Chapel Hill, 78-79; Penrose Conf Comt, Geol Soc Am, 83-85; Sci Adv Comt, US Geol Surv, Geol Div, 88-90; Assoc ed, Geol Soc Am Bull, 88-90, 91-, Geol Soc Am Spec Paper, 89- *Honors & Awards:* Spec Achievement Award, US Geol Surv, 86. *Mem:* Geol Soc Am; Am Geophys Union; AAAS. *Res:* Structural, metamorphic and igneous geology of the southern Appalachian Piedmont and Blue Ridge; regional tectonics; fault zones. *Mailing Add:* 928 Nat Ctr US Geol Surv Reston VA 22092

HORTON, JOHN, b Sheffield, Eng, June 25, 34; US citizen; c 3. ONCOLOGY. *Educ:* Univ Sheffield, MB & ChB, 57. *Prof Exp:* Intern med, Albany Hosp, NY, 57-58; sr house officer, Royal Hosp, Sheffield, Eng, 58-59; resident, Albany Med Ctr Hosp, 60-62, Nat Cancer Inst res fel, 62-63, from instr to assoc prof med, 64-74, PROF MED, ALBANY MED COL, 74- *Concurrent Pos:* Consult, NY State, 65-; dir, Natalie Warren Bryant Cancer Ctr, 75-77; ECOG grant; Cancer Educ grant. *Mem:* Am Asn Cancer Educ (pres, 81); Am Asn Cancer Res; Am Soc Clin Oncol; fel Am Col Physicians; fel Am Col Gastroenterol. *Res:* Clinical treatment trials, principally in relationship to cancer chemotherapy; cancer education; cancer prevention. *Mailing Add:* Div Med Oncol A52 47 New Scotland Ave Albany NY 12208

HORTON, JOHN EDWARD, b Brockton, Mass, Dec 30, 30; m 51; c 5. IMMUNOLOGY, PERIODONTOLOGY. *Educ:* Providence Col, BS, 52; Tufts Univ, DMD, 57; Baylor Univ, MSD, 65; George Washington Univ, MA, 78. *Prof Exp:* Vis scientist cellular immunol, Lab Microbiol & Immunol, Nat Inst Dent Res, 70-73; chief dept immunol, inst dent res, Walter Reed Army Med Ctr, 73-77; Assoc prof, Harvard Sch Dent Med, 80-, chmn dept periodont, 77-; PROF & CHMN, DEPT PERIODONT, COL DENT, OHIO STATE UNIV. *Concurrent Pos:* Asst prof lectr oral biol, Grad Sch Arts & Sci, George Washington Univ, 72-74, assoc prof lectr, 74-76, prof lectr, Sch Hyg & Pub Health, Johns Hopkins Univ, 75-76; consult, Vet Admin Hosp, West Roxbury, 78- *Mem:* AAAS; Int Asn Dent Res; Am Acad Periodont; Am Acad Oral Med; Am Acad Oral Path. *Res:* Investigations in cell-mediated immunological mechanisms as related to the pathogenesis of periodontal disease; specifically, actions of macrophages, lymphocytes and lymphokines in the host response to inflammation and bone resorption. *Mailing Add:* Det Periodont Ohio State Univ Col Dent 305 W 12th Ave Columbus OH 43210

HORTON, JOSEPH ARNO, JR, b Johnson City, Tenn, Feb 21, 51; m 78; c 3. TRANSMISSION ELECTRON MICROSCOPY, ORDERED ALLOYS. *Educ:* Tenn Technol Univ, BS, 73; Univ Va, MS, 75, PhD(mat sci), 79. *Prof Exp:* Res asst, Res Labs Eng Sci, Univ Va, 73-79; RES STAFF MEM, OAK RIDGE NAT LAB, MARTIN MARIETTA ENERGY SYSTS, 79- *Mem:* Am Soc Metals; Metall Soc; Electron Micros Soc Am; Sigma Xi; Assoc Inst Mech Engrs; Mat Res Soc. *Res:* Transmission electron microscopy microstructural studies of ordered intermetallic alloys. *Mailing Add:* Metals & Ceramics Div Bldg 5500-A113 Oak Ridge Nat Lab PO Box 2008 Oak Ridge TN 37831-6376

HORTON, JOSEPH WILLIAM, b Corpus Christi, Tex, Nov 9, 29; m 51; c 3. ORGANIC CHEMISTRY, ENVIRONMENTAL CHEMISTRY. *Educ:* Univ Portland, BS, 50, MS, 52; Mich State Univ, PhD(chem), 60. *Prof Exp:* Res assoc chem, Reed Col, 52; from instr to asst prof, 52-55, assoc prof, 58-71, PROF CHEM, UNIV WIS-SUPERIOR, 71- *Concurrent Pos:* Fel, Kans State Univ, 66. *Mem:* Am Chem Soc; Sigma Xi. *Res:* Nitrogen heterocyclics, especially in tetrazoles; tetrazole analogs of amino acids; pharmaceuticals; amino acids in peptides; organic mechanisms; trace metals in fuels, hydrosphere and biosphere; environmental quality of lakes and streams. *Mailing Add:* Dept Chem Univ Wis Superior WI 54880

HORTON, JURETA, b Ennis, Tex, Feb 11, 41. TRAUMA & SHOCK. *Educ:* Univ Tex, PhD(cardiophysiol), 81. *Prof Exp:* Asst prof, 86-90, RES INSTR SURG, UNIV TEX HEALTH SCI CTR, 80-, ASSOC PROF, 90- *Mailing Add:* Dept Gen Surgery Univ Tex Health Sci Ctr 5323 Harry Hines Blvd Dallas TX 75235

HORTON, KENNETH EDWIN, b Buffalo, NY, Aug 14, 32; m 58; c 2. PHYSICAL & NUCLEAR METALLURGY. *Educ:* Rensselaer Polytech Inst, BMetE, 54; Univ Wis, MS, 55. *Prof Exp:* Res engr, Atomics Int Div, NAm Aviation, Inc, 55-57; res lab analyst, Douglas Aircraft Co, 57-60; sr metallurgist, Adv Tech Labs Div, Am-Standard Corp, 60-65; metallurgist, US Atomic Energy Comn, 65-78; INT SPECIALIST, US DEPT ENERGY, 78- *Concurrent Pos:* Asst prof, Long Beach State Col, 59-60. *Mem:* Am Soc Metals; Am Inst Mining, Metall & Petrol Engrs. *Res:* Nuclear fuel elements; thermal stress fatigue; alloy development; liquid metals compatibilities; high temperature measurements. *Mailing Add:* US Dept Energy Energy Technol 1000 Independence SW Washington DC 20585

HORTON, M DUANE, b Salt Lake City, Utah, Nov 21, 35; m 60; c 2. CHEMICAL ENGINEERING. *Educ:* Univ Utah, BS, 57, PhD(chem eng), 61. *Prof Exp:* Chem engr, US Naval Ord Test Sta, 60-63; from asst prof to prof chem eng, Brighman Young Univ, 63-81; EXEC VPRES, MEGADIAMOND INDUST, 81- *Concurrent Pos:* Air Force Off Sci & Res grant, 65-74; consult, Hercules Powder Co, 65-75; co-founder, Megadiamond Corp; chmn bd, DBT Co. *Mem:* Am Inst Chem Engrs; Combustion Inst. *Res:* High pressure synthesis; combustion of pulverized coal. *Mailing Add:* 2900 Iroquois Dr Provo UT 84604

HORTON, MAURICE LEE, b Norman, Ind, May 23, 31; m 55; c 1. PLANT SCIENCE, SOIL PHYSICS. *Educ:* Purdue Univ, BS, 53, MS, 59; Iowa State Univ, PhD(soil physics), 62. *Prof Exp:* Res assoc agron, Iowa State Univ, 61-62; asst prof soil physics, Southern Ill Univ, 62-64; assoc prof soil physics, 64-74, prof soil & water resources, 74-78, PROF & HEAD PLANT SCI DEPT, SDAK STATE UNIV, 78- *Mem:* Soil Conserv Soc Am; Am Soc Agron; Soil Sci Soc Am. *Res:* Effects of compaction on soil properties; movement of air, heat, and water through soils; soil-plant-climate relationships; soil and water resources. *Mailing Add:* Dept Plant Sci SDak State Univ Box 2207a Brookings SD 57007

HORTON, OTIS HOWARD, b Des Moines, NMex, Nov 28, 17; m 46; c 2. DAIRY HUSBANDRY. *Educ:* NMex State Univ, BS, 41; Univ Mo, MA, 47; Univ Ill, PhD(nutrit prod), 54. *Prof Exp:* Asst dairying, Univ Mo, 46-47; asst prof dairy sci, Univ Ark, 47-50; asst prof, Univ Ill, 50-51 & 53-54; from asst prof to prof, Univ Ark, 54-70; PROF DAIRY SCI, ETEX STATE UNIV, 70- *Res:* Pasture quality; roughage utilization by young animals; baby calf nutrition and digestibility by ruminants. *Mailing Add:* Rte 1 Box 45 Campbell TX 75442

HORTON, PHILIP BISH, b Newark, Ohio, May 10, 26; m 58; c 2. SOLID STATE PHYSICS. *Educ:* Denison Univ, BS, 49; Univ Ark, MS, 54; La State Univ, PhD(physics), 64. *Prof Exp:* From asst prof to prof, 51-89, EMER PROF PHYSICS, PHILLIPS UNIV, 89- *Mem:* Am Asn Physics Teachers. *Res:* Galvanomagnetic and thermomagnetic transport phenomena in soilds; theoretical magnetotransport. *Mailing Add:* 117 Sammie Dr Hot Springs AR 71913

HORTON, RALPH M, b Salt Lake City, Utah, Oct 4, 34; m 61; c 6. METALLURGY. *Educ:* Univ Utah, BS, 57, PhD(metall), 61. *Prof Exp:* From asst prof metall to assoc prof, Metall Dept, Wash State Univ, 61-75; supvr, Mat Lab, Idaho Nat Eng Lab, 75-81; MGR METALL, DIAMOND TECHNOL CTR, NORTON-CHRISTENSEN INC, 82- *Concurrent Pos:* Vis prof, Hanford Eng Develop Labs, 63, 64 & 73; res trainee high pressure physics & chem, Brigham Young Univ, 71. *Mem:* Am Soc Metals; Am Powder Metall Inst. *Res:* Oxidation and corrosion; materials for coal gasification equipment; industrial diamond utilization. *Mailing Add:* 5184 Spring Clover Dr Murray UT 84123

HORTON, RICHARD, b New York, NY, May 3, 32; m; c 2. MEDICINE, PHYSIOLOGY. *Educ:* Univ Wash, BA, 54, MD, 58. *Prof Exp:* Intern med, Swed Hosp, Seattle, 58-59; resident, Vet Admin Hosp & Univ Calif, Los Angeles, 59-61; fel, Med Ctr, Univ Calif, San Francisco, 61-63, NIH spec res fel, Worcester Found Exp Biol, 63-65; asst prof med, Med Sch, Univ Calif, Los Angeles, 65-67; assoc prof med & physiol, Univ Ala, 67-68; assoc prof, 69-73, PROF MED & CHIEF SECT ENDOCRINOL, SCH MED, UNIV SOUTHERN CALIF, 73- *Concurrent Pos:* NIH career develop award, 65-67; Am Cancer Soc grant, 65-70; NIH grant, 65- *Mem:* AAAS; Am Fedn Clin Res; Endocrine Soc; NY Acad Sci; Am Soc Clin Invest; Am Asn Physicians. *Res:* Aldosterone pathophysiology and the study of sex hormone metabolism in man; androgen physiology; prostaglandens, resin and the kidney. *Mailing Add:* Dept Med Univ Southern Calif Sch Med 2025 Zonal Ave Los Angeles CA 90033

HORTON, RICHARD E, b Marshalltown, Iowa, Feb 5, 40; m 62; c 1. ELECTRICAL ENGINEERING. *Educ:* Iowa State Univ, BS, 62, MS, 63, PhD(elec eng), 67. *Prof Exp:* Instr electronics technol, 63-67, asst prof electronics technol & elec eng, 67-68, asst prof, 68-80, ASSOC PROF ELEC ENG, IOWA STATE UNIV, 80- *Mem:* Inst Elec & Electronics Engrs. *Res:* Estimation theory in random processes; simulation of random processes; data communications. *Mailing Add:* Dept Computer Eng Iowa State Univ Coover Hall Ames IA 50010

HORTON, RICHARD GREENFIELD, b Auburn, NY, May 29, 12; m 47; c 3. PHYSIOLOGY. *Educ:* Williams Col, AB, 34; Cornell Univ, MA, 35, PhD(physiol), 37. *Prof Exp:* Sterling fel, Med Sch, Yale Univ, 37-38; instr physiol, Univ Tenn, 38-39; asst prof, Mich State Col, 41-42; toxicologist, Chem Lab, Edgewood Arsenal, US Army, 42-72; CONSULT. *Concurrent Pos:* Consult toxicol, 72- *Mem:* Am Physiol Soc; Soc Toxicol; Sigma Xi. *Res:* Neuromuscular ph ysiology; toxicology; post-tetanic recovery of muscle. *Mailing Add:* 208 E Ring Factory Rd Bel Air MD 21014

HORTON, ROBERT, JR, b Ger, July 9, 54; US citizen; m 75; c 2. SOIL PHYSICS. *Educ:* Tex A&M Univ, BS, 75, MS, 77; NMex State Univ, PhD(soil physics), 81. *Prof Exp:* Trainee soil scientist, Soil Conserv Serv, 75; asst instr soil, Tex A&M Univ, 76, asst instr res, 76-77; asst instr res, NMex State Univ, 77-81; ASST PROF SOIL PHYSICS, IOWA STATE UNIV, 81- *Mem:* Soil Sci Soc Am; Am Soc Agron. *Res:* Heat and/or mass transport in soil; irrigation; drainage; pollution; waste disposal; nutrient uptake; soil temperature; tillage. *Mailing Add:* RR 3 Ames IA 50010

HORTON, ROBERT CARLTON, b Tonopah, Nev, July 25, 26; m 50; c 3. MINERAL ECONOMICS & POLICY. *Educ:* Univ Nev, BSc, 49. *Hon Degrees:* DSc, Univ Nev, 86. *Prof Exp:* Field asst geol, US Geol Surv, 49-50; geol consult, 51-53; mining engr econ geol, Nev Bur Mines, Univ Nev, Reno, 56-65, assoc dir, 65-66; pres, Nev Geoserv Inc, 66-76; regional geologist, Bendix Field Eng Corp, 76-77, dir, Geol Div, 77-81; dir, us bur mines, dept interior, 81-87; dir, Ctr Strategie Mat Res & Policy Study, 87-90, assoc dean,

88-90, EMER ASSOC DEAN, MACKAY SCH MINES, UNIV NEV, RENO, 90- *Concurrent Pos:* Mem, Nev Pollution Control Hearing Bd, 66-68 & Gov Adv Mining Bd, Nev, 66-73. *Mem:* Am Inst Mining, Metall & Petrol Engrs; Soc Econ Geologists; Mining & Metall Soc Am. *Res:* Mineral deposits of Nevada; development and utilization of geothermal resources; mineral economics and policy. *Mailing Add:* 654 W Riverview Circle Reno NV 89509

HORTON, ROBERT LOUIS, b Shreveport, La, Apr 28, 44; m 66; c 2. PHYSICAL CHEMISTRY, ENHANCED OIL RECOVERY. *Educ:* Rice Univ, BA, 66; Univ Calif, Berkeley, PhD(phys chem), 71. *Prof Exp:* Asst prof chem & physics, La State Univ, Alexandria, 71-75; res chemist, Phillips Petrol Co, 75-79, sr res chemist, Res & Develop Dept, 79-82; group leader II, BP Am, 82-85; enhanced oil recovery supvr, OXY USA, 85-90; SR RES ASSOC, TEXACO, 90- *Concurrent Pos:* Fel chem, Rice Univ, 73-75. *Mem:* Sigma Xi; Soc Petrol Engrs; Can Inst Mining; Gas Processors Asn; Soc Mining Engrs. *Res:* Enhanced oil recovery; alternative fuels; uranium extraction and refining; aqueous ionic/polyelectrolytic equilibria; reactions of non-aqueous polyelectrolytes; thermophysical properties of mixtures; phase equilibria; kinetics; gaseous ionic equilibria; ion-molecule reactions. *Mailing Add:* Texaco E&P Technol Div 3901 Briarpark Houston TX 77042-5301

HORTON, ROBERT LOUIS, b Miami, Fla, Sept 2, 20; m 41; c 1. ORGANIC CHEMISTRY. *Educ:* Cornell Univ, BA, 47, PhD(chem), 51. *Prof Exp:* Asst anal chem, Cornell Univ, 47-50; res chemist, Pharmaceut Res Dept, Am Cyanamid Co, 50-54, chief chemist, Pharmaceut Prod Dept, 54-56, group leader, Res & Develop Dept, 56-67, proj leader elastomers prod, 67-83; RETIRED. *Mem:* Am Chem Soc. *Res:* Antimetabolites; anticoagulants; tranquilizers; organic process research and development; antioxidents; ultraviolet adsorbers; synthetic polymers. *Mailing Add:* Box 132 Bound Brook NJ 08805

HORTON, ROGER FRANCIS, b Chislehurst, Eng, 43; m; c 3. PLANT PHYSIOLOGY. *Educ:* Univ Col Wales, BSc, 64; Oxford Univ, DPhil, 67. *Prof Exp:* Post-doctoral fel, 67-69, asst prof, 69-74, ASSOC PROF BOT, UNIV GUELPH, 74- *Concurrent Pos:* Vis scientist, Univ Cambridge, UK, 76. *Mem:* Can Soc Plant Physiologists; Am Soc Plant Physiologists; Int Plant Growth Substance Asn. *Res:* Plant growth regulators; plant biochemistry; stomatal mechanisms; growth correlations; senescence; abscission. *Mailing Add:* Dept Bot Univ Guelph Guelph ON N1G 2W1 Can

HORTON, THOMAS EDWARD, JR, b Houston, Tex, Jan 12, 35; m 63; c 2. FLUID MECHANICS, APPLIED MECHANICS. *Educ:* Univ Tex, BS, 57, PhD(eng mech), 64; Stanford Univ, MS, 58. *Prof Exp:* Jr mech engr, Shell Develop Co, 58-59; teaching asst eng mech, Univ Tex, 60-62; sr res engr, Jet Propulsion Lab, Calif Inst Technol, 62-66; assoc prof & res engr, 66-71, PROF & RES ENGR MECH ENG, UNIV MISS, 71- *Concurrent Pos:* Consult, Jet Propulsion Lab, Calif Inst Technol, 66-75, numerous govt & indust orgns, 68-; dir laser sci, US Army High Energy Laser Lab, 75-76; ed, Am Inst Aeronaut & Astronaut Progress Series, 81; chmn, Am Inst Aeronaut & Astronaut Thermophysic Tech Comt, 82-84; mem, Am Soc Mech Engrs Heat Transfer Div Tech Comt, 72-87. *Mem:* Am Soc Mech Engrs; Am Inst Aeronaut & Astronaut; Am Phys Soc; Sigma Xi (pres, 75). *Res:* Unsteady flow and acoustics, non-equilibrium phenomena, radiative transfer, and thermochemistry of high temperature gases. *Mailing Add:* Dept Mech Eng Univ Miss University MS 38677

HORTON, THOMAS ROSCOE, b Ft Pierce, Fla, Nov 17, 26; m 47; c 3. MATHEMATICS. *Educ:* Stetson Univ, BS, 49; Univ Fla, MS, 50, PhD(math), 54. *Hon Degrees:* LLD, Pace Univ, 76; DLitt, Univ Charleston, 80; DHum, Stetson Univ, 82. *Prof Exp:* Asst headmaster, Bolles Sch, Fla, 50-52; instr, Univ Fla, 53-54; appl sci res, IBM Corp, 54-55, mgr, Space Comput Ctr, 56, mgr spec systs mkt, 57-59, systs mgr intermediates systs, 59, mgr systs develop, Fed Systs Div, 59-64, vpres, Systs Develop Div, 64-67, dir, Univ Rels, 68-82; PRES, AM MGT ASN, 82- *Concurrent Pos:* Sr vpres, Nat Coun Philanthropy, 73-80; mem vis comt ling & philos, Mass Inst Technol, 71-82; trustee, Pace Univ, 75-, Am Grad Sch Int Mgt, 82-; chmn, Asn Gov Bds Univs & Cols, 82-83; dir, Mastery Educ Corp, 82-; dir, Am Precision Indust, 89-, Stanhome Corp, 91- *Mem:* AAAS; Am Math Soc; Math Asn Am; Inst Elec & Electronics Engrs. *Res:* Number theory, especially quadratic forms; high-speed computational techniques; computer systems design; technical management; education administration. *Mailing Add:* 165 Soundview Ave White Plains NY 10606

HORTON, WALTER JAMES, b Detroit, Mich, May 8, 13; m 39; c 4. SYNTHETIC ORGANIC CHEMISTRY. *Educ:* Wayne Univ, BS, 35; Univ Mich, MS, 40, PhD(chem), 42. *Prof Exp:* Anal chemist, Children's Fund, Univ Detroit, 35-37; instr anal & org chem, Wayne Univ, 37-40; res assoc explosives, Off Sci Res & Develop Proj, Univ Mich, 41-42; instr chem, Tulane Univ, 42-46; from asst prof to prof org chem, Univ Utah, 46-68; prof chem, Univ Ife, Nigeria, 68-71, head dept, 69-71; res prof, 71-88, EMER PROF CHEM, UNIV UTAH, 88- *Concurrent Pos:* Prof chem & head dept, Haile Sellassie Univ, 63-65. *Mem:* Am Chem Soc. *Res:* Synthesis of equilenin analogs, nitramines and polycyclic hydrocarbons-homoperinaphthenes; mechanisms of chemical reactions; Willgerodt reaction; synthesis of an analog of sex hormone equilenin; seven-membered ring compounds; synthesis of C-13-containing compounds of biochemical interest. *Mailing Add:* Dept Chem Univ Utah Salt Lake City UT 84112

HORTON, WILLIAM A, PEDIATRICS. *Prof Exp:* PROF MED & PEDIAT, UNIV TEX HEALTH SCI CTR, 89- *Mailing Add:* Dept Pediat Univ Tex Health Sci Ctr 6431 Fannin SHP 302 Houston TX 77030

HORVAT, ROBERT EMIL, b Chicago, Ill, Nov 4, 40. ENVIRONMENTAL EDUCATION, ENERGY EDUCATION. *Educ:* Valparaiso Univ, BS, 62; Univ Kans, MS, 69; Univ Wis, Madison, PhD(sci & environ educ), 74. *Prof Exp:* Instr chem, State Univ NY, Oneonta, 66-69; res asst, Univ Wis, Madison, 70-73; from asst prof to assoc prof, 73-89, CHMN, DEPT EARTH SCI & EDUC, STATE UNIV NY COL BUFFALO, 78-81 & 86-, PROF, 89- *Concurrent Pos:* Panel reviewer, NSF, 77-78; consult individualized sci instruct syst, 74-78 & proj for energy-enriched curriculum, 80-81; proj assoc, NY Sci, Technol Soc Educ Proj, 82-; coop sci educator, Chem Educ Prog Pub Understanding, 90- *Mem:* Asn Educ Teachers Sci; Nat Asn Sci Technol & Soc; Nat Sci Teachers Asn; AAAS. *Res:* Science, technology & society cirriculum development/dissemination; energy and environmental science curriculum development, kindergarten through 12th grade. *Mailing Add:* State Univ NY Col Buffalo 1300 Elmwood Buffalo NY 14222

HORVÁTH, CSABA GYULA, b Szolnok, Hungary, Jan 25, 30; US citizen; m 63; c 2. BIOTECHNOLOGY, SEPARATION SCIENCE. *Educ:* Budapest Tech Univ, Dipl chem eng, 52; Univ Frankfurt, DrPhil, 63. *Hon Degrees:* Dr, Budapest Tech Univ, 86. *Prof Exp:* Asst prof chem technol, Budapest Tech Univ, 52-56; res scientist, Farbwerke Hoechst AG, Ger, 56-60; res fel, Mass Gen Hosp & Harvard Univ, 63-64; res assoc, 64-70, lectr, 67-72, assoc prof, Sch Med, 70-75, from assoc prof to prof eng & appl sci, 72-81, PROF CHEM ENG, YALE UNIV, 81-, CHMN, 87- *Concurrent Pos:* US Sr Scientist, Humboldt Award, 82. *Honors & Awards:* S Dal Nogare Award, 78; M Tswett Award, 80 & Medal, 82; Chromatography Award, Am Chem Soc, 83. *Mem:* Am Chem Soc; Inst Food Technol; Am Inst Chem Engr; Deutsch Gesellschaft fuer Chemisches Apparatewesen; AAAS; NY Acad Sci; Sigma Xi; Am Ceramic Soc. *Res:* Biochemical separations; high performance liquid chromatography; biotechnology, enzyme reactors; separation science; biochemical engineering. *Mailing Add:* Dept Chem Eng Mason Lab Yale Univ PO Box 2159 Yale Sta New Haven CT 06520

HORVATH, DONALD JAMES, b Newark, NJ, Sept 8, 29; m 53; c 4. ANIMAL NUTRITION. *Educ:* Rutgers Univ, BS, 51, MS, 54; Cornell Univ, PhD, 57. *Prof Exp:* From asst prof to assoc prof, 57-69, PROF ANIMAL NUTRIT & PHYSIOL, WVA UNIV, 69- *Concurrent Pos:* Co-chmn subcomt, Goechem Environ Rel Health & Dis, Nat Acad Sci. *Mem:* Am Soc Animal Sci; Soc Environ Geochem & Health. *Res:* Swine nutrition; hypomagnesemic tetany of ruminants; minor element nutrition; environmental geochemistry and health. *Mailing Add:* Div Animal Sci WVa Univ Morgantown WV 26506

HORVATH, FRED ERNEST, b Pittsburgh, Pa, Nov 26, 24; m 87. NEUROPHYSIOLOGY. *Educ:* US Naval Acad, BS, 47; Univ Mich, PhD(physiol, psychol), 61. *Prof Exp:* Res asst psychol, Ment Health Res Inst, Univ Mich, 56-60; NIH fel neurophysiol, Univ Calif, Los Angeles, 60-61, res asst neurophysiol, 61-64; NIH spec fel, Paris, 64-66; from asst prof to assoc prof, 66-81, PROF PHYSIOL, NY MED COL, 81- *Mem:* AAAS; Am Physiol Soc; Soc Neurosci. *Res:* Caudate-thalamic-cortical spindles; termination of convulsant seizure activity; regulation of vasopressin release. *Mailing Add:* Dept Physiol NY Med Col Valhalla NY 10595

HORVATH, JOHN MICHAEL, b Budapest, Hungary, July 30, 24; US citizen; m 53; c 1. MATHEMATICAL ANALYSIS. *Educ:* Univ Budapest, PhD(math physics), 47. *Prof Exp:* Attache res math, Nat Ctr Sci Res, Paris, France, 48-51; prof math, Univ of the Andes, 51-57; from asst prof to assoc prof, 57-63, PROF MATH, UNIV MD, COLLEGE PARK, 63- *Concurrent Pos:* Vis prof, Univ Nancy, 63-64 & Univ Paris, 72-73. *Mem:* Am Math Soc; Math Asn Am; corresp mem Span Acad Sci. *Res:* Functional analysis, especially topological vector spaces, Schwartz distributors and their applications. *Mailing Add:* Dept Math Univ Md College Park MD 20742

HORVATH, KALMAN, b Győr, Hungary, Feb 13, 40; US citizen; m 64; c 2. QUALITY ASSURANCE MANAGEMENT, MICROBIOLOGY. *Educ:* Univ Calif, Los Angeles, BA, 63, MA, 65; Rice Univ, PhD(biol), 69. *Prof Exp:* Asst prof biol, Tex A&M Univ, 68-73; sr pharmacologist, Am McGaw Labs, 74-75; biol lab mgr, Cutter Labs, Inc, 75-76; QUAL ASSURANCE MICROBIOL MGR, CUTTER BIOL, MILES INC, 76- *Mem:* Am Soc Qual Control; Sigma Xi; Parenteral Drug Asn. *Res:* Resistance of microorganisms to industrial sterilization; environmental microbiology; environmental control for sterile drug manufacture; sterility, pyrogenicity, toxicity studies on MAb genetic engineered products MOAB-genetic engineered products; development of test methods in microbiology, biochemistry, immunology, chemistry. *Mailing Add:* 4484 Buckthorn Concord CA 94521

HORVATH, RALPH S(TEVE), b Scranton, Pa, Feb 9, 36; m 82; c 1. ELECTRONICS ENGINEERING, MICROCOMPUTER APPLICATIONS. *Educ:* Mich Technol Univ, BS, 60; NY Univ, MS, 62; Worcester Polytech Inst, PhD(elec eng), 68. *Prof Exp:* Mem tech staff, Bell Tel Labs, 60-62; from instr to asst prof, 62-72, ASSOC PROF ELEC ENG, MICH TECHNOL UNIV, 73- *Concurrent Pos:* Pacemaker consult, Medico Italia, Padua, Italy, 77-78; continuing educ consult, Gen Motors Inst, 78-; mem conf comt, Inst Life, Paris, 72-; lectr, Univ Md-Europe, 84-85; vis prof, Darling Downs Inst Advan Educ, Toowoomba, Queensland, Australia, 88. *Honors & Awards:* Ralph R Teetor Award, Soc Automotive Engrs, 81. *Mem:* Inst Elec & Electronics Engrs; Int Inst Arson Investigators; Sigma Xi; Int Asn Mini & Micro Comput. *Res:* Electronic circuits; computer applications in real time; neurophysiology; heart pacemaker design; continuing education (in-plant training) in microprocessor applications in industrial control. *Mailing Add:* Dept Elec Eng Mich Technol Univ Houghton MI 49931

HORVATH, STEVEN MICHAEL, b Cleveland, Ohio, Sept 15, 11; m 40; c 3. PHYSIOLOGY. *Educ:* Miami Univ, BS, 34, MS, 35; Ohio State Univ, MS, 39; Harvard Univ, PhD(physiol),42. *Prof Exp:* Asst physiol & zool, Miami Univ, 31-35; asst physiol, Col Med, Ohio State Univ, 35-37; instr physiol & zool, Miami Univ, 37-39; asst physiol, Fatigue Lab, Harvard Univ, 39-42; dir res & asst prof phys med, Grad Sch Med, Univ Pa, 46-48, assoc prof, 48-49; from assoc prof to prof physiol, Col Med, Univ Iowa, 49-51, actg dir, Inst Geront, 51-58; head dept physiol, Div Res, Lankenau Hosp, 58-61; prof physiol & biomed eng & dir, Inst Environ Stress, 61-86, PROF PHYSIOL & BIOMED ENG & DIR, NEUROSCI RES INST, UNIV CALIF, SANTA BARBARA, 86- *Concurrent Pos:* Asst, Woods Hole Marine Biol Lab, 36; res dir, Metrop State Hosp, Mass, 39-42; tutor, Harvard Univ, 40-42; assoc prof, Sch Med, Univ Pa, 58-61; consult, Off Surgeon, US Army, EPA & var industs. *Mem:* AAAS; Am Col Cardiol; Am Heart Asn; Am Physiol Soc; Am Pub Health Asn. *Res:* Environmental, stress, cardiovascular and respiratory physiology; gerontology; endocrinology; exercise biochemistry. *Mailing Add:* Neurosic Res Inst Univ Calif Santa Barbara CA 93106

HORVATH, WILLIAM JOHN, b New York, NY, Sept 13, 17; m 63; c 2. BIOPHYSICS. *Educ:* City Col New York, BS, 36; NY Univ, MS, 38, PhD(physics), 40. *Prof Exp:* Instr physics, NY Univ, 39; instr, Polytech Inst Brooklyn, 40; physicist, Bur Ord, US Dept Navy, 40-43, consult, Nat Defense Res Comt & Off Sci Res & Develop, US Navy, 43-45, analyst, Div Indust Co-op, Mass Inst Technol, 46-48, dept dir, Opers Eval Group, US Navy, 48-49, mem rev bd, Weapons Systs Eval Group, Joint Chiefs of Staff, 49-52, consult, 52-55; sect head dept med & biol physics, Airborne Instrument Lab, NY, 55-58; prof health systs, 74-88, RES SCIENTIST, MENT HEALTH RES INST, UNIV MICH, ANN ARBOR, 58-, EMER PROF HEALTH SYSTS, 88- *Concurrent Pos:* Mem health serv res study sect, NIH, 62-66, mem health care systs study sect, 68-70. *Honors & Awards:* Naval Ord Develop Award, 45; Presidential Cert of Merit, 47; Kimball Medal, Opers Res Soc, 77. *Mem:* Am Phys Soc; Opers Res Soc Am (vpres, 55); fel AAAS; fel Am Pub Health Asn; Royal Soc Health; Sigma Xi. *Res:* Nuclear physics; slow neutron absorption; electricity and magnetism; degaussing; cryogenics; properties of liquid helium; medical physics; mathematical models of behavior; health care research. *Mailing Add:* Ment Health Res Inst Univ Mich Ann Arbor MI 48109

HORVE, LESLIE A, b Decatur, Ill, Dec 5, 38; m 60; c 4. RHEOLOGY OF ELASTOMERS, OIL SEAL FUNCTION & DESIGN. *Educ:* Univ Ill, Urbana, BS, 60, MS, 64; Midwest Col Eng, Prof Degree Mech Eng, 71; Keller Grad Sch Mgt, MBA, 79. *Hon Degrees:* Dr Eng, Midwest Col Eng, 73. *Prof Exp:* Anal engr, Pratt & Whitney Aircraft, United Technologies, 64-67; appln engr, 67-69, mgr res & develop, 69-71, mgr prod eng, 72-75, vpres technol, OE Div, 75-83, vpres opers, Replacement Div, 83-85, vpres corp technol & qual assurance, 85-86, vpres technol, OE Group, Corp Staff, 86-90, SR VPRES, INT OPERS, CR INDUSTS, 90- *Concurrent Pos:* Instr, Midwest Col Eng, 69- & Judson Col, 70-75; chmn, Seals Tech Comt, Rubber Mfrs Asn, 82-83 & Soc Automotive Engrs, 82- *Honors & Awards:* Walter D Hobson Award, Am Soc Lubrication Engrs, 71. *Mem:* Soc Automotive Engrs; Rubber Mfrs Asn; Am Soc Lubrication Eng; Am Chem Soc. *Res:* Function of elastomeric sealing devices; rheology of elastomers. *Mailing Add:* 515 Rue Chamonix Deer Park IL 60010

HORVITZ, DANIEL GOODMAN, b New Bedford, Mass, Mar 4, 21; m 45; c 3. STATISTICS, SOCIAL SURVEYS. *Educ:* Mass State Col, BS, 43; Iowa State Col, PhD(statist), 53. *Prof Exp:* Assoc scientist, Manhattan Proj, Univ Calif, 46; asst statist, Iowa State Col, 46-49, instr, 49-51; asst prof biostatist, Grad Sch Pub Health, Univ Pittsburgh, 51-53; assoc prof statist, NC State Col, 53-56; vpres, A J Wood Res Corp, 56-60; vis prof statist, Univ Rangoon-Univ Chicago Proj, 60-62; group leader, Res Triangle Inst, 62-66, dep dir, Statist Res Div, 66-71, dir, Ctr Pop Res & Serv, 71-72; prof biostatist, Sch Pub Health, Univ NC, Chapel Hill, 73-74; vpres statist sci, 74-83, EXEC VPRES, RES TRIANGLE INST, 83- *Concurrent Pos:* Lectr, Villanova Univ, 58-60; adj prof statist, NC State Univ, 62-72; vis prof statist health sci, Univ NC, 63-64; consult, Nat Ctr Health Statist, 67-69 & 73; mem, Nat Defense Exec Reserve, 67-70; assoc ed, J Am Statist Asn, 69-71; mem adv comt to Bur Census, Am Statist Asn, 69-72 & 80-85; consult, Ctr Pop Res, NIH, 72-74; adj prof biostatist, Univ NC, 74-; chmn, panel statist family assistance & related progs, comt nat statist, Nat Res Coun, 80-82; mem, panel on qual control family assistance progs, Comt Nat Statist, Nat Res Coun, 86-87. *Mem:* Fel AAAS; fel Am Statist Asn (vpres, 85-87); Int Asn Surv Statist; Int Statist Inst. *Res:* Demographic simulation models; sample survey methods; measurement error research; social program evaluation; demographic measurement designs; health care expenditure surveys; randomized responses. *Mailing Add:* 3115 Eton Rd Raleigh NC 27608

HORVITZ, HOWARD ROBERT, b Chicago, Ill, May 8, 47. DEVELOPMENTAL GENETICS, BEHAVIORAL GENETICS. *Educ:* Mass Inst Technol, SB(math) & SB(econ), 68; Harvard Univ, MA, 72, PhD(biol), 74. *Prof Exp:* From asst prof to assoc prof, 78-86, PROF BIOL, 86-, INVESTR, HOWARD HUGHES MED INST, MASS INST TECHNOL, 88- *Concurrent Pos:* NIH res career develop award, 81-86; adv, Dept Biochem & Molecular Biol, Harvard Univ, 85-90 & Cold Spring Harbor Lab, 84-; mem, Sci Adv Bd, Hereditary Dis Found, 87- & Comt on Scholarly Commun with China, People's Rep of, US Nat Acad Sci, 87-; neurobiologist, Mass Gen Hosp, 89-, geneticist, 89-; lectr, Harvey Soc, 89; mem bd sci adv, Jane Coffin Childs Mem Fund Med Res, 89-; mem sci rev comt, Amyotrophic Lateral Sclerosis Asn, 90- *Honors & Awards:* Spencer Award in Neurobiol, Columbia Univ, 86; Warren Triennial Prize, Mass Gen Hosp, 86; US Steel Foundation Award in Molecular Biol, 88. *Mem:* Nat Acad Sci; Soc Develop Biol; Soc Nematologists; Soc Neurosci; Am Soc Cell Biol; Am Soc Microbiol; Am Asn Cancer Res; fel AAAS; Genetics Soc Am. *Res:* Developmental and behavioral genetics of the nematode caenorhabditis elegans, with particular emphasis on the cellular and genetic mechanisms utilized to control cell lineage and cell fate; author of numerous publications. *Mailing Add:* Dept Biol Rm 56-629 Mass Inst Technol 77 Mass Ave Cambridge MA 02139

HORWEDEL, CHARLES RICHARD, b Cleveland, Ohio, Aug 29, 02; m 31; c 2. PHYSICAL METALLURGY. *Educ:* Univ Dayton, BChE, 24; Univ Ala, MS, 25; Ohio State Univ, PhD, 29, MetEngr, 35. *Prof Exp:* Metall trainee, Thompson Prod Inc, 28-30; res laboratorian, Worcester, Mass Div, US Steel Corp, 30-36, asst dist metallurgist, Cleveland Div, 36-40, sr prod metallurgist, 40-43, div metall engr, 43-48, asst mgr stainless sales, 48-62; asst prof indust eng, Univ Dayton, 62-64, actg chmn mech eng & asst to dean sch eng, 64-66, dir grad progs, 66-68, admin asst to dean, 68-80; RETIRED. *Mem:* Am Soc Metals; Am Soc Eng Educ. *Res:* Stainless steel. *Mailing Add:* Nine Duncannen Ave Apt 3 Worcester MA 01604-5103

HORWITT, MAX KENNETH, b New York, NY, Mar 21, 08; m 33, 74; c 5. NUTRITION, BIOCHEMISTRY. *Educ:* Dartmouth Col, BA, 30; Yale Univ, PhD(physiol chem), 35; Am Bd Clin Chem, dipl; Am Bd Nutrit, dipl. *Prof Exp:* Asst physiol chem, Sch Med, Yale Univ, 32-35, Lead res fel, 35-37; assoc biochem, Univ Ill Col Med, 40-43, from asst prof to prof, 43-68; EMER PROF BIOCHEM, SCH MED, ST LOUIS UNIV, 77-; CONSULT NUTRIT SCI, 77- *Concurrent Pos:* Dir biochem res lab, Elgin State Hosp, 37-59, dir, L B Mendel Res Lab, 60-67; mem, Comts Food & Nutrit Bd, Nat Res Coun, 42-85, consult, 66-; mem study sect metab & nutrit, NIH, 55-58, mem study sect psychopharmacol, 59-63; mem WHO expert group vitamin requirements, Food & Agr Orgn, 65-67; consult human nutrit, USDA, 66-67; dir div res serv, Ill Dept Ment Health, 66-67; vis prof nutrit, Univ Hawaii, 66; consult med, Rush Med Sch, 66-84; field dir, Anemia & Malnutrit Res Ctr, Chiangmai Med Sch, Thailand, 68-69; mem, Comt Dietary Allowances, Nat Res Coun, 80-85; chmn, 81-83, mem, Inst Review Bd, St Louis Univ, 83-89. *Honors & Awards:* Osborne & Mendel Award, Am Inst Nutrit, 61. *Mem:* Fel AAAS; fel Am Soc Biol Chemists; fel Am Inst Nutrit; Am Soc Clin Nutrit; Soc Exp Biol & Med; fel Am Inst Chemists; fel Geront Soc. *Res:* Evaluation of human nutritional status and development of methods to establish nutritional requirements; effect of oral contraceptive agents on metabolic processes; effect of antioxidants on tissue lipids; vitamin E requirements. *Mailing Add:* Div Geriat Med Dept Internal Med St Louis Univ Sch Med St Louis MO 63104

HORWITZ, ALAN FREDRICK, b Minneapolis, Minn, Oct 26, 44; m 72; c 2. CELL BIOLOGY, DEVELOPMENTAL BIOLOGY. *Educ:* Univ Wis-Madison, BA, 66; Stanford Univ PhD(biophys), 69. *Hon Degrees:* MA, Univ Pa, 78. *Prof Exp:* NIH postdoctoral fel, Univ Calif, Berkeley, 70-73; res assoc cell biol, Bio Ctr, Univ Basel, 73-74; from asst prof to prof biochem, Univ Pa Med Sch, 74-87; PROF CELL BIOL & HEAD, DEPT CELL & STRUCT BIOL, UNIV ILL, URBANA-CHAMPAIGN, 87- *Concurrent Pos:* Mem, Pub Policy Comt, Am Soc Cell Biol, 87-; dir, Cell & Molecular Biol Training Prog, Univ Ill, 87-; ed, J Cell Biol, 88- & J Cell Sci, 90-; assoc ed, Develop Biol, 89- *Mem:* Am Soc Cell Biol; Soc Develop Biol; Am Soc Biol Chemists. *Res:* Cell adhesion molecules; membrane cytoskeletal linkages; muscle and neuronal development. *Mailing Add:* Dept Cell & Struct Biol Univ Ill Urbana IL 61801

HORWITZ, BARBARA ANN, b Chicago, Ill, Sept 26, 40; m 70. ANIMAL PHYSIOLOGY, CELL PHYSIOLOGY. *Educ:* Univ Fla, BS, 61, MS, 62; Emory Univ, PhD(physiol), 66. *Prof Exp:* Asst biol, Univ Fla, 59-61; res physiologist, Emory Univ, 66; USPHS res fel physiol, Sch Med, Univ Calif, Los Angeles, 66-67; USPHS res fel, 67-68, asst res phsiologist, 68-72, asst prof, 72-75, assoc prof & assoc physiologist, 75-78, PROF PHYSIOL & PHYSIOLOGIST, AGR EXP STA, UNIV CALIF, DAVIS 78- *Concurrent Pos:* Consult, Physiol Dept, Sch Med, Univ Calif, Los Angeles, 66-67, NSF, 80-83; consult, NSF, 80-83. *Mem:* Am Physiol Soc; Am Soc Zoologists; NY Acad Sci; Am Inst Nutrit; fel AAAS; Sigma Xi; Soc Exp Biol & Med; Am Asn Study of Obesity. *Res:* cellular physiology; bioenergetics; hormonal and neural control of cellular metabolism; thermogenesis; obesity. *Mailing Add:* Dept Animal Physiol Univ Calif Davis CA 95616

HORWITZ, DAVID LARRY, b Chicago, Ill, July 13, 42; m 65; c 2. ENDOCRINOLOGY, METABOLISM. *Educ:* Harvard Univ, BA, 63; Univ Chicago, MD, 67, PhD(physiol), 68; Lake Forest Grad Sch Mgt, MBA, 91. *Prof Exp:* Med officer, US Navy, 69-71; res fel endocrinol, Univ Chicago, 72-74, instr, 74-75, asst prof med endocrinol, 75-79; assoc prof med endocrinol, 79-90, CLIN PROF MED, UNIV ILL, 90- *Concurrent Pos:* Med dir, Baxter Healthcare Corp, 82- *Honors & Awards:* Res & Develop Award, Am Diabetes Asn, 74, 75. *Mem:* Fel Am Col Physicians; Am Fedn Clin Res; Am Diabetes Asn (pres, 87-); AAAS; Endocrine Soc. *Res:* Factors affecting the regulation of insulin secretion; clinical pharmacology of insulin; dietary treatment of diabetes mellitus; carbohydrate metabolism; exercise physiology. *Mailing Add:* Col Med M/C 787 Univ Ill 840 S Wood St Chicago IL 60612

HORWITZ, EARL PHILIP, b Cincinnati, Ohio, June 3, 30; m 55; c 2. SEPARATIONS CHEMISTRY. *Educ:* Univ Cincinnati, BS, 53; Univ Ill, MS, 55, PhD(chem), 57. *Prof Exp:* Res chemist, Dow Chem Co, 57-59; chemist, 59-80, SR CHEMIST, ARGONNE NAT LAB, 80- *Honors & Awards:* IR-100 Award, 87; Distinguished Assoc Award, US Dept Energy, 90. *Mem:* AAAS; Am Chem Soc; Sigma Xi. *Res:* Chemistry of the less familiar elements; chelate chemistry; liquid-liquid extraction; liquid chromatography; actinide chemistry; high speed-high efficiency liquid chromatography; nuclear fuel reprocessing; nuclear waste processing; radioanalytical chemistry; separation of kerogen and bitumen from oil shale; separation of maceral constituents of coal. *Mailing Add:* Argonne Nat Lab 9700 S Cass Ave Argonne IL 60439

HORWITZ, EDWIN M, b Indianapolis, Ind, June 29, 59. HEMATOLOGOY, ONCOLOGY. *Educ:* Ind Unvi, AB, 81, PhD, 85, MD, 88. *Prof Exp:* Assoc instr biol chem, Ind Univ, 81, res assoc, Dept Chem, 82-85, postdoctoral res assoc, Med Sci Prog, 85-86; intern, 88-89, PEDIAT RESIDENT, ST LOUIS CHILDREN'S HOSP, 89-; MED SCIENTIST, TRAINING PROG, DEPT PEDIAT, SCH MED, WASH UNIV, 88- *Concurrent Pos:* Assoc instr human anat, Ind Univ, 82 & physiol chem, 83, vis asst prof biochem, Med Sci Prog, 87. *Mem:* Am Acad Pediat; AMA; Am Soc Cell Biol. *Res:* Author of 16 technical publications. *Mailing Add:* Dept Pediat Children's Hosp Wash Univ 400 S Kingshighway Blvd St Louis MO 63110

HORWITZ, HARRY, b London, Eng, Mar 20, 27; US citizen; m 58; c 2. ONCOLOGY. *Educ:* Univ London, MB & BS, 50. *Prof Exp:* Intern, Univ London Teaching Hosps, 51-52; registr, St Bartholomew's Hosp, London, 54-58; sr registr, Addenbrookes Hosp, Cambridge, 58-60; from asst prof to assoc prof, 60-68, PROF RADIOL, COL MED, UNIV CINCINNATI, 68-, ASSOC DIR DEPT RADIOTHER, UNIV HOSP, 68- *Concurrent Pos:* Fel radiation ther, Mt Sinai Hosp, New York, 57-58; assoc dir, dept radiother & oncol, Jewish Hosp, Cincinnati. *Mem:* Brit Inst Radiol; Brit Med Asn; Am Asn Cancer Educ; Am Col Radiol; Radiol Soc NAm. *Res:* Clinical aspects of neoplastic diseases; radiation therapy; radiobiology; nuclear medicine; cancer therapy. *Mailing Add:* 7990 Ashley View Dr Cincinnati OH 45227

HORWITZ, JEROME PHILIP, b Detroit, Mich, Jan 16, 19; m 51; c 2. ORGANIC CHEMISTRY, ONCOLOGY. *Educ:* Univ Detroit, BS, 42, MS, 44; Univ Mich, PhD(chem), 48. *Hon Degrees:* Dr, Univ Detroit, 87. *Prof Exp:* US Army fel, Northwestern Univ, Evanston, 48-50; Parke Davis fel, Univ Mich, Ann Arbor, 50-51; asst prof chem, Ill Inst Technol, 51-55; res assoc org chem, Detroit Inst Cancer Res, 56-64; dir div chem, 65-69, sci dir, 70-73, DIR, DIV BASIC SCI, MICH CANCER FOUND, 73-; PROF CHEM, DIV ONCOL, SCH MED, WAYNE STATE UNIV, 71- *Mem:* Am Chem Soc; Am Asn Cancer Res. *Res:* Nucleic acid metabolism and new cancer chemotherapeutic agents. *Mailing Add:* PO Box 02188 Detroit MI 48202

HORWITZ, JOSEPH, b Petach-Tikva, Israel, Jan 10, 36; US citizen; m 64; c 2. BIOPHYSICS. *Educ:* Univ Calif, Los Angeles, BS, 65, PhD(biophysics), 70. *Prof Exp:* Fel, Lab Nuclear Med & Radiation Biol, 70-71, asst prof, 71-76, assoc prof, 76-80, PROF BIOPHYSICS, JULES STEIN EYE INST, SCH MED, UNIV CALIF, LOS ANGELES, 80- *Mem:* Am Chem Soc; Biophys Soc; Am Soc Photobiol; Sigma Xi. *Res:* Proteins, structure and function; vision on the molecular level. *Mailing Add:* 15451 Camarillo st Sherman Oaks CA 91423

HORWITZ, KATHRYN BLOCH, b Sosua, Dominican Repub; US citizen; c 2. MOLECULAR & CELL BIOLOGY, ENDORINOLOGY. *Educ:* NY Univ, MS, 66; Southwestern Med Sch, Dallas, PhD(med physiol), 75. *Prof Exp:* Fel, Univ Tex Health Sci Ctr, San Antonio, 78-79; from asst prof to assoc prof, 79-89, PROF MED, UNIV COLO MED CTR, 89- *Concurrent Pos:* Res career develop award, Nat Cancer Inst, 81-86. *Honors & Awards:* Wilson S Stone Award, M D Anderson Inst, 76. *Mem:* Am Fedn Clin Res; Endocrine Soc; Soc Cell Biol; Am Soc Cancer Res. *Res:* Steroid hormone action; estrogen, progesterone receptors; breast cancer; endocrinology. *Mailing Add:* Dept Med & Endocrinol Univ Colo Health Sci Ctr 4200 E Ninth Ave Box B-151 Denver CO 80262

HORWITZ, LAWRENCE D, b Nov 29, 39; m; c 2. CARDIOLOGY. *Educ:* Univ Rochester, BA, 61; Yale Univ, MD, 64; Am Bd Internal Med, dipl internal med & cardiovasc dis, 73. *Prof Exp:* Intern med, Bellevue Hosp, 64-65, asst resident, 65-66; med res officer, USAF Sch Aerospace Med, Brooks AFB, 66-68; Nat Heart Inst spec fel, Peter Bent Brigham Hosp & Harvard Med Sch, 68-70; asst prof med, Southwestern Med Sch, Univ Tex, Dallas, 70-73; from assoc prof to prof, Health Sci Ctr, Univ Tex, San Antonio, 73-79; head, Div Cardiol, 79-89, PROF MED, HEALTH SCI CTR, UNIV COLO, 79- *Concurrent Pos:* Attend physician, Parkland Mem Hosp, Dallas, 70-73, Bexar County Hosp, San Antonio, 73-79 & Univ Hosp, Denver, 79-; staff physician, Vet Admin Hosp, San Antonio, 73-79 & Vet Admin Med Ctr, Denver, 79-83; mem, Cardiovasc Study Comt, Vet Admin, 79-82, chmn, 82; mem, Clin & Circulation Coun, Am Heart Asn. *Mem:* Am Fedn Clin Res; Am Heart Asn; Am Physiol Soc; fel Am Col Cardiol; Am Soc Clin Invest; Am Asn Physicians. *Res:* Author of numerous technical publications. *Mailing Add:* Health Sci Ctr Univ Colo 4200 E Ninth Ave Denver CO 80262

HORWITZ, LAWRENCE PAUL, b New York, NY, Oct 14, 30; m 51; c 3. THEORETICAL PHYSICS. *Educ:* NY Univ, BS, 52; Harvard Univ, PhD(physics), 57. *Prof Exp:* Staff consult commun & electronics, Pickard & Burns, Inc, 54-55; physicist, Res Lab, Int Bus Machines Corp, 57-64; vis res assoc, Inst Theoret Physics, Univ Geneva, 64-66; from assoc prof to prof physics, Univ Denver, 66-71; PROF PHYSICS, TEL AVIV UNIV, 72- *Concurrent Pos:* Vis assoc prof, Tel Aviv Univ, 67-68; vis prof, Inst Theoret Physics, Univ Geneva, 71-72, Syracuse Univ, 80 & Eidgenossische Technische Hochschule, Zurich, Switzerland, 81. *Mem:* Am Phys Soc; Israeli Phys Soc; Swiss Phys Soc; Europ Phys Soc. *Res:* Numerical analysis; pattern recognition; advanced computer techniques; functional analysis and algebraic methods in quantum mechanics and particle physics; relativistic quantum theory; algebraic approach to non-Abelian gauge fields using modules over Clifford algebras; Steuckelberg-Feynman approach to relativistic quantum theory; unstable systems; relativistic statistical mechanics. *Mailing Add:* Sch Physics & Astron Tel Aviv Univ Ramat Aviv 69978 Israel

HORWITZ, MARCUS AARON, b Elmira, NY, May 3, 46; m 81; c 2. INFECTIOUS DISEASES, IMMUNOLOGY. *Educ:* Cornell Univ, BA, 68; Columbia Univ, MD, 72. *Prof Exp:* Internship med, Bronx Munic Hosp, 73, residency, 73-74; epidemic intel serv officer, Ctr Dis Control, 74-76; fel infectious dis, Albert Einstein Col Med, Affil Hosps, 76-77; NIH fel res, 77-80, asst prof res, Rockefeller Univ, 80-85; PROF MED, MICROBIOL & IMMUNOL, UNIV CALIF, LOS ANGELES, 85-, CHIEF, DIV INFECTIOUS DIS, 85- *Honors & Awards:* Alexander D Langmuir Award, Ctr Dis Control, 76. *Mem:* Am Fedn Clin Res; Am Soc Microbiol; Infectious Dis Soc Am. *Res:* Interaction between infectious agents and leukocytes; macrophage activation; intracellular parasitism; immunobiology of Legionnella pneumophila, Mycobacterium leprae and mycobacterium tuberculosis. *Mailing Add:* Div Infectious Dis Dept Med UCLA Sch Med Ctr Health Sci Rm 37-121 Los Angeles CA 90024

HORWITZ, MARSHALL SYDNEY, b Boston, Mass, Mar 26, 37; m 60; c 2. MICROBIOLOGY, PEDIATRICS. *Educ:* Harvard Univ, BA, 58; Tufts Univ, MD, 62. *Prof Exp:* From intern to resident pediat, New Eng Med Ctr, Boston, Mass, 62-65; instr, Med Sch, Emory Univ, 65-67; asst prof cell biol, 69-70, asst prof pediat, 69-73, asst prof microbiol & immunol, 70-73, assoc prof, 73-78, PROF PEDIAT, MICROBIOL & IMMUNOL, ALBERT EINSTEIN COL MED, 78-, DIR, DIV PEDIAT INFECTIOUS DIS, 81- *Concurrent Pos:* Virologist, Ctr Dis Control, US Dept Health, Educ & Welfare, 65-67; Am Cancer Soc fel, Albert Einstein Col Med, 67-69; career scientist, Health Res Coun, New York, 69-72; Nat Cancer Inst grant, 69-; NIH res career develop award, 72-77; Irma T Hirscl Career Develop Award, 76-81; mem study sect exp virol, NIH, 78-82, chmn, 85-87; mem, FDA adv comt on vaccines. *Mem:* Am Soc Microbiol; Harvey Soc; Soc Pediat Res; Am Soc Virol. *Res:* Virus replication; DNA synthesis; antiviral chemotherapy. *Mailing Add:* Dept Pediat Microbiol-Immunol Albert Einstein Col Med 1300 Morris Park Ave Bronx NY 10461

HORWITZ, NAHMIN, b Duluth, Minn, Oct 28, 27; m 49; c 4. PHYSICS. *Educ:* Western Reserve Univ, BS, 49; Univ Minn, MS, 51, PhD(physics), 54. *Prof Exp:* Physicist, Lawrence Radiation Lab, Univ Calif, 54-59; from asst prof to assoc prof, 59-65, PROF PHYSICS, SYRACUSE UNIV, 65- *Concurrent Pos:* Physicist, Rutherford High Energy Lab, Eng, 65-66; sci assoc, Europ Orgn Nuclear Res, Geneva, Switzerland, 75-76; Fulbright scholar & Lady Davis fel, Technion, Haifa, Israel, 82-83. *Mem:* Fel Am Phys Soc; Am Asn Physics Teachers. *Res:* Elementary particle physics. *Mailing Add:* Dept Physics Syracuse Univ Syracuse NY 13244

HORWITZ, ORVILLE, b Strafford, Pa, Nov 20, 09; m 34. MEDICINE. *Educ:* Harvard Univ, SB, 32; Johns Hopkins Univ, MD, 38. *Prof Exp:* From asst instr to instr pharmacol, Univ Pa, 40-52, from asst instr to instr med, 45-49, assoc, 49-53, from asst prof to assoc prof, 53-70, prof med, 70-80, PROF MED & PHARM, SCH MED, UNIV PA, 80-, SR WARD CHIEF, HOSP, 53- *Concurrent Pos:* Consult, US Vet Admin & US Dept Navy; trustee, Jackson Lab, 51-; chmn coun circulation & fel coun arteriosclerosis, Am Heart Asn, 65-67. *Mem:* Am Physiol Soc; AMA; Am Heart Asn; Am Clin & Climat Asn; fel Am Col Physicians; Sigma Xi. *Res:* Cardiovascular diseases; physiology; pathology and treatment. *Mailing Add:* Two Private Way Strafford PA 19087

HORWITZ, PAUL, b New York, NY, Dec 4, 38; m 64; c 3. SCIENCE EDUCATION. *Educ:* Harvard Univ, BA, 60; Columbia Univ, MA, 63; NY Univ, PhD(physics), 67. *Prof Exp:* SR SCIENTIST, BOLT BERANEK & NEWMAN, INC, 79- *Concurrent Pos:* Chmn, Forum Physics & Soc, 78-79. *Mem:* Am Phys Soc; Am Asn Physics Teachers. *Res:* Use of computers for teaching science and mathematics, particularly chaos theory, visual modeling, special relativity and quantum mechanics. *Mailing Add:* 32 Riverside Ave Concord MA 01742

HORWITZ, RALPH IRVING, b Philadelphia, Pa, June 25, 47; m 70. INTERNAL MEDICINE, CLINICAL EPIDEMIOLOGY. *Educ:* Albright Col, BA, 69; Pa State Univ, MD, 73. *Prof Exp:* ASST PROF MED, YALE UNIV, 78-, ASST DIR, ROBERT WOOD JOHNSON CLIN SCHOLAR PROG, 78- *Concurrent Pos:* Robert Wood Johnson clin scholar, 75-77. *Mem:* Am Fedn Clin Res; Soc Epidemiol Res; Am Col Physicians. *Res:* Quantification of prognosis and therapy in clinical care; clinical epidemiology; evaluation of diagnostic tests; methodology of case-control research. *Mailing Add:* C1 Scholars Prog 333 Cedar St New Haven CT 06510

HORWITZ, SUSAN BAND, b Cambridge, Mass; m 60; c 2. MOLECULAR PHARMACOLOGY, BIOCHEMISTRY. *Educ:* Bryn Mawr Col, BA, 58; Brandeis Univ, PhD(biochem), 63. *Prof Exp:* Fel pharmacol, Med Sch, Tufts Univ, 63-65; fel, Med Sch, Emory Univ, 65-67; from instr to asst prof pharmacol, 68-75, assoc prof, 75-80, PROF PHARMACOL & CELL BIOL, ALBERT EINSTEIN COL MED, 80-, CO-CHMN, DEPT MOLECULAR PHARMACOL, 85- *Mem:* Am Cancer Soc; Am Soc Pharmacol & Exp Therapeut; Am Chem Soc; Am Soc Microbiol; Am Soc Cell Biol. *Res:* Mechanism of action of chemotherapeutic agents which affect macromolecular synthesis in mammalian cells and their viruses; interaction of antitumor and antiviral drugs with nucleic acids. *Mailing Add:* Dept Pharmacol Albert Einstein Col Med 1300 Morris Park Ave Rm 248F Bronx NY 10461

HORWITZ, WILLIAM, b Gilbert, Minn, Feb 4, 18; m 40; c 3. FOOD CHEMISTRY. *Educ:* Univ Chicago, BS, 37; Univ Minn, MS, 38, PhD(phys chem), 47. *Prof Exp:* Asst chem, Univ Minn, 38-39; from chemist to chief chemist, Minn, 39-51, food & drug officer, 51-52, chief food res br, 52-63, asst to asst comnr sci resources, 63-67, asst dir bur sci, 67-70, dep dir, off sci, 70-81, SCI ADV, CTR FOOD SAFETY & APPLIED NUTRIT, FOOD & DRUG ADMIN, 81- *Concurrent Pos:* Instr, USDA Grad Sch, 51-61; sect ed foods, Chem Abstr, 55-; ed, Off Methods Anal, 55, 60, 65, 70, 75 & 80; US mem, adv & deleg, Food & Agr Orgn/WHO Codex Alimentarius Comts, 58- *Honors & Awards:* Harvey W Wiley Award, Asn Off Anal Chemists, 75. *Mem:* AAAS; Am Chem Soc; Asn Off Anal Chemists (secy, 52-79); Inst Food Technologists. *Res:* Methods of analysis for foods; research administration. *Mailing Add:* 9830 Cherry Tree Lane Silver Spring MD 20901-2732

HORWOOD, EDGAR M(ILLER), civil engineering; deceased, see previous edition for last biography

HORZ, FRIEDRICH, b Blaubeuren, Ger, Mar 23, 40. GEOLOGY, PETROGRAPHY. *Educ:* Univ Tubingen, BS, 62, PhD(petrog), 65. *Prof Exp:* Nat Res Coun resident assoc, Ames Res Ctr, NASA, Calif, 66-68; res fel, Calif Inst Technol, 68-69; vis scientist, Lunar Sci Inst, Houston, Tex, 69-70; SPACE TECHNOLOGIST, MANNED SPACECRAFT CTR, NASA, 71- *Concurrent Pos:* Mission sci trainer for Apollo 16 astronauts, NASA, 71- *Mem:* Fel Meteoritical Soc; Am Geophys Union; Ger Mineral Soc. *Res:* Impact cratering mechanics and experimental shock studies of geological materials; micrometeoroid impact and other small scale processes on the lunar surface. *Mailing Add:* 907 Layfair Pl Friendswood TX 77546

HORZEMPA, LEWIS MICHAEL, b Beverly, Mass, Sept 5, 49. ENVIRONMENTAL CHEMISTRY, GEOCHEMISTRY. *Educ:* Rensselaer Polytech Inst, BS, 71; Univ Mass, MS, 73; Univ Md, PhD(geochem), 77. *Prof Exp:* Asst prof earth sci, Western Conn State Col, 77-78; res assoc environ chem, Manhattan Col, 78-80; sr environ chemist, Envirosphere Co, NY, 80-; CHEMIST, EVASCO, BOSTON, MASS. *Mem:* Geochem Soc; Soil Sci Soc Am; Sigma Xi; Am Chem Soc. *Res:* Evaluation of the mobility and fate of chemicals in the environment; environmental chemical assessments and studies relating to hazardous waste disposal, site contamination, treatment technologies, combustion emissions, energy development and associated environmental health risks. *Mailing Add:* Evasco 211 Congress St 6th Floor Boston MA 02110

HORZEPA, JOHN PHILIP, b Bayonne, NJ, May 22, 43; m 85; c 3. PROCESS ENGINEERING, INTERNATIONAL PROJECT MANAGER. *Educ:* NJ Inst Technol, BS, 65, MS, 70. *Prof Exp:* Jr res engr, Engelhard Minerals & Chem Corp, 65-68, res engr, 68-72, sr res engr, 72-75, group leader, 75-78, mgr process develop catalyst & chem, minerals & chem div, 78-82, mgr div process eng, 83-85, proj mgr corp eng, 85-90; DIR ENG, NOBEL MET INDUST, 90- *Mem:* Am Inst Chem Engrs. *Res:* Catalysts, sorbents, and pigments, especially kaolin; precious metals. *Mailing Add:* 62 Johnson Rd Somerset NJ 08873

HOSAIN, FAZLE, b Murshidabad, India, Apr 1, 32; US citizen. NUCLEAR MEDICINE. *Educ:* Univ Calcutta, India, BSc Hons, 51, MSc, 53, AINP, 55, DPhil, 59. *Prof Exp:* Lectr physics, Chary Chandra Col, 55-56; Govt India nat res fel sci, Saha Inst Nuclear Physics, 59-61; Fulbright scholar & res fel hemat, Sch Med, Univ Wash, Seattle, 61-64; reader & in-chg, Dept Nuclear Med, Med Col, Banaras Hindu Univ, 64-66; fel radiol, Div Nuclear Med, Sch Med, Johns Hopkins Univ, 67-69, asst prof, Sch Hyg & Med, Div Nuclear Med & Radiation Health, 69-74, assoc prof environ health, Sch Hyg & Pub Health & asst prof radiol & radiation sci, Sch Med, 74-75; assoc prof nuclear med, 75-78, DIR RADIOPHYSICS, DEPT NUCLEAR MED, UNIV CONN HEALTH CTR, 75-, PROF NUCLEAR MED, 78- *Concurrent Pos:* Vis staff mem, Brookhaven Nat Lab, Mem Sloan-Kettering Cancer Ctr, NY & Naval Res Lab, Washington, DC. *Mem:* Soc Nuclear Med; Am Chem Soc; Am Physiol Soc. *Res:* Nuclear medicine; radiophysics and radiopharmacy; biomedical investigations with radionuclide tracers; author of more than 170 technical publications. *Mailing Add:* Dept Nuclear Med Univ Conn Health Ctr Farmington CT 06032

HOSAIN, MAHBUB UL, b Calcutta, W Bengal, India, Feb 4, 38; Can citizen; m 67; c 3. STRUCTURAL ENGINEERING. *Educ:* Univ Dacca, BSc, 60; Univ Man, Winnipeg, MSc, 63; Tech Univ of NS, Halifax, PhD (structure eng), 69. *Prof Exp:* Asst engr bldgs, Coun Sci & Indust Res, 60-61; engr class I struct design, Assoc Consult Engrs Ltd, 64-65; design engr bldgs & bridges, J Philip Vaughan & Assocs, Halifax, 67-68; from asst prof to assoc prof, 69-75, PROF STRUCT ENG, UNIV SASK, 79- *Concurrent Pos:* Commonwealth scholar, Univ Man, 61-63; postdoctoral fel, Univ Alta, Edmonton, 68-69; pres, Sask Eng Soc, 81-82; vis prof, Univ Ottawa, 78-79; comt, Can Standards Asn, 87- *Mem:* Assoc Am Soc Civil Engrs; fel Can Soc Civil Engrs; Eng Inst Can; Am Soc Engr Educ; Can Standards Asn. *Res:* Experimental investigation of steel and composite structures particularly stub-girders; computer aided design & analysis; finite element analysis and boundary element method; computer aided instruction. *Mailing Add:* Dept Civil Eng Univ Sask Saskatoon SK S7N 0W0 Can

HOSANSKY, NORMAN LEON, b NY, Feb 12, 24; m 49, 61; c 2. ORGANIC CHEMISTRY. *Educ:* Queens Col, NY, BS, 47; Columbia Univ, AM, 48; Rutgers Univ, PhD(chem), 53. *Prof Exp:* Prof asst antibiotics, E R Squibb & Sons, 48-50; res chemist natural prod, S B Penick & Co, 53-62; asst ed, 63-68, asst dept head org editing, 68-71, ORG CHEM MGR, CHEM ABSTR SERV, 71- *Mem:* Am Chem Soc. *Res:* Chemical documentation; antibiotics; steroids; alkaloids and other natural products; isolation and structural identification. *Mailing Add:* 181 S Chesterfield Rd Columbus OH 43209-1912

HOSE, RICHARD K, b Newark, NJ, March 23, 20. PETROLEUM GEOLOGY. *Educ:* Univ Ala, BS, 43. *Prof Exp:* Geologist, US Geol Surv, 44-80; sr geol assoc & adv, Atlantic Richfield Co, 80-85; CONSULT, 86- *Mem:* Fel Geol Soc Am; Am Asn Petrol Geologists; Sigma Xi; AAAS. *Res:* Geologic mapping, basin range province in Nevada and Utah; structural and stratigraphic studies in Wyoming. *Mailing Add:* 10335 Stonydale Dr Cupertino CA 95014

HOSEA, JOEL CARLTON, b Atlanta, Ga, Dec 12, 38; m 61; c 1. PLASMA PHYSICS. *Educ:* Auburn Univ, BS, 61; Stanford Univ, MS, 62, PhD(plasma physics), 66. *Prof Exp:* Res asst plasma physics, Stanford Univ, 61-66; NSF fel, Saclay Nuclear Res Ctr, France, 66-68; res fel, 68-69, MEM RES STAFF PLASMA PHYSICS, PRINCETON UNIV, 69- *Mem:* Am Phys Soc. *Res:* Interaction of electromagnetic energy with plasmas and solids; study of basic phenomena in all types of plasmas, especially in fusion oriented plasmas. *Mailing Add:* Plasma Physics Lab RF242 Princeton Univ Princeton NJ 08540

HOSEIN, ESAU ABBAS, b Trinidad, WI, Dec 4, 22. BIOCHEMISTRY. *Educ:* McGill Univ, BSc, 47, MSc, 50, PhD(biochem), 52. *Prof Exp:* Lectr, 52-58, from asst prof to assoc prof, 58-74, PROF BIOCHEM, McGILL UNIV, 74- *Mem:* Can Physiol Soc. *Res:* Epilepsy; myasthenia gravis; muscular dystrophy; neurohumoral transmitters in brain and muscle; neuropharmacology; molecular basis of action of narcotic drugs. *Mailing Add:* Dept Biochem McGill Univ 3655 Drummond St Montreal PQ H3G 1Y6 Can

HOSENEY, RUSSELL CARL, b Coffeyville, Kans, Dec 3, 34; m 56; c 3. CEREAL CHEMISTRY. *Educ:* Kans State Univ, BS, 57, MS, 60, PhD(cereal chem), 68. *Prof Exp:* Res chemist, Crops Res Div, Agr Res Serv, USDA, 56-70; res assoc cereal chem, 67-71, assoc prof grain sci, 71-75, PROF GRAIN SCI, KANS STATE UNIV, 75- *Mem:* Am Chem Soc; Am Asn Cereal Chemists; Inst Food Technologists. *Res:* Wheat and flour quality, including the role of proteins, carbohydrates and lipids. *Mailing Add:* Flour/Feed Mill Dept Kans State Univ Manhattan KS 66504

HOSEY, M MARLENE, PROTEIN PHOSPHORYLATION, MUSCARINIC RECEPTORS. *Educ:* Univ Ill, PhD(pharmacol), 74. *Prof Exp:* ASSOC PROF BIOCHEM, CHICAGO MED SCH, 78- *Res:* Signal transduction in muscle. *Mailing Add:* Dept Biol Chem & Structure Chicago Med Sch 3333 Green Bay Rd North Chicago IL 60064

HOSFORD, ROBERT MORGAN, JR, b Bremerton, Wash, Aug 18, 33; m 67. PLANT PATHOLOGY. *Educ:* Ore State Univ, BS, 56, MS, 59; Univ Ariz, PhD(plant path), 65. *Prof Exp:* Grad asst, Univ Ariz, 61-65; asst plant pathologist, Potato Invests Field Lab, Univ Fla, 65-67; from asst prof to assoc prof, 67-75, PROF PLANT PATH, NDAK STATE UNIV, 75- *Concurrent Pos:* Mem wheat comt, Nat Plant Dis Detection & Info Prog, 73- *Mem:* Am Phytopath Soc; Soc Nematol. *Res:* Identification, epidemiology, biology, control, physiology, genetics, cytogenetics of disease-causing organisms; root, foliage and seed diseases; field and horticultural crops; foliar diseases of wheat excluding rusts; biological control of leafy spurge; teaching & trouble shooting (nematology, diseases of horticultural crops). *Mailing Add:* Dept Plant Path NDak State Univ Fargo ND 58105

HOSFORD, WILLIAM FULLER, JR, b Orange, NJ, Mar 17, 28; m 54; c 4. METALLURGY & PHYSICAL METALLURGICAL ENGINEERING. *Educ:* Lehigh Univ, BS, 50; Yale Univ, ME, 51; Mass Inst Technol, ScD(metall), 59. *Prof Exp:* Asst prof metall, Mass Inst Technol, 59-63; assoc prof, 63-68, PROF METALL ENG, UNIV MICH, 68- *Mem:* Am Inst Mining, Metall & Petrol Engrs; fel Am Soc Metals Int. *Res:* Mechanical metallurgy; sheet metal forming. *Mailing Add:* Dept Mat Sci & Eng Univ Mich Dow Bldg Ann Arbor MI 48109

HOSHAW, ROBERT WILLIAM, b Lafayette, Ind, Dec 29, 21; m 47; c 3. PHYCOLOGY. *Educ:* Purdue Univ, BS, 42, MS, 48, PhD(bot), 50. *Prof Exp:* Asst bot, Purdue Univ, 46-49; from instr to assoc prof, 50-62, actg head dept, 65-66, PROF BOT, UNIV ARIZ, 62- *Concurrent Pos:* NSF fac fel, Ind Univ, 60-61; assoc ed bot, Trans Am Micros Soc, 75-78; chmn bd trustees, Phycol Soc Am, 80- *Mem:* AAAS; Bot Soc Am; Am Soc Plant Physiologists; Phycol Soc Am (vpres, 72, pres, 73); Am Inst Biol Sci; Int Phycological Soc; Am Micros Soc. *Res:* Systematic biology of the Zygnemataceae; role of ploidy in speciation and evolution in algae; morphology and cytology of Sirogonium. *Mailing Add:* Dept Ecol & Evolutionary Biol Univ Ariz Tucson AZ 85721

HOSHIKO, MICHAEL S, b Surrey, BC; US citizen; m 55; c 3. BIOFEEDBACK, RADIO COMMUNICATION. *Educ:* Heidelburg Col, BA, 48; Bowling Green State Univ, MA, 49; Purdue Univ, PhD(speech & hearing sci), 57. *Prof Exp:* Instr psychol, Univ Kans, 49-50; intern, Peoria State Hosp, 50 & Ill Inst Technol, 51; sch psychologist, Prep Sch, Montreal, 51-52; res psychologist, Univ Toronto Med Sch, 52-55; PROF COMMUN DIS, SOUTHERN ILL UNIV, 57-, COORDR, SPEECH & HEARING CLIN CTR, 86- *Concurrent Pos:* Prin investr, NIH, 61-66, State Ill Ment Health, 62-64 & Am Cancer Soc, 63-65; Am Speech-Lang-Hearing Asn int travel award, 62; res fel, Johns Hopkins Med Sch, 66-67; voiceprint consult, Southern Ill Univ, 75-; vis prof, MacMaster Med Sch, Hamilton, 81; site visitor, Am Speech-Lang-Hearing Asn, 84-, mem, Multicult Bd, 91- *Mem:* Fel Am Speech-Lang-Hearing Asn; Am Psychol Asn. *Res:* Electromyographic research respiratory muscles during speech; radio telemetry of physiological activity during speaking situations; radiophonic research communication without acoustic speech signals; skin potential as an indicator of hearing response voiceprint research. *Mailing Add:* Dept Commun Dis & Sci Southern Ill Univ Carbondale IL 62901

HOSHIKO, TOMUO, b BC, Can, Oct 5, 27; nat US; m 62; c 3. PHYSIOLOGY. *Educ:* Kent State Univ, BS, 49; Univ Minn, PhD(physiol), 53. *Prof Exp:* Asst physiol, Univ Minn, 49-53; instr, Univ Utah, 53-55; Am Heart Asn res fel zoophysiol, Copenhagen Univ, 55-56; from sr instr to assoc prof, 57-69, PROF PHYSIOL, SCH MED, CASE WESTERN RESERVE UNIV, 69- *Concurrent Pos:* Vis asst prof, Tokyo Med & Dent Univ, 64-65; vis prof, Dept Biol, Ill Inst Technol, 71-72; guest prof, Lab Physiol, Kath Univ, Leuven, Belgium, 80-81; mem prof adv comt, Ctr Biomed Ethics, Case Western Reserve Univ, 85- *Mem:* AAAS; Biophys Soc; Am Physiol Soc; Am Asn Univ Prof; Soc Gen Physiologists (secy, 73-75, pres, 81); Asn Comput Mach; Coun Sci Soc Pres; Am Sci Affil. *Res:* Ion transport through membranes; renal physiology; active transport. *Mailing Add:* Dept Physiol & Biophys Case Western Reserve Univ Sch Med Cleveland OH 44106

HOSICK, HOWARD LAWRENCE, b Champaign, Ill, Nov 1, 43; m 68; c 3. CELL BIOLOGY, CANCER. *Educ:* Univ Colo, Boulder, BA, 65; Univ Calif, Berkeley, PhD(zool), 70. *Prof Exp:* Fel molecular genetics, Karolinska Inst, Sweden, 70-72; asst res biochemist, Cancer Res Lab, Univ Calif, Berkeley, 72-73; from asst prof to assoc prof cell biol & cancer, 73-83, chmn, 83-87, PROF ZOOL & CELL BIOL, DEPT ZOOL, WASH STATE UNIV, 83-, CHMN, DEPT GENETICS & CELL BIOL, 87- *Concurrent Pos:* Courtesy prof, genetics prog, Wash State Univ, 74-, pharmacol & toxicol prog, 84-; vis scientist, Univ Reading, Eng, 79 & Aichi Cancer Res Inst, Nagoya, Japan, 86; res comt, Am Heart Asn, 83- *Honors & Awards:* Fac Develop Award, Shell Companies Found, 84. *Mem:* Am Soc Cell Biol; Int Asn Breast Cancer Res; Am Asn Cancer Res; Tissue Cult Asn. *Res:* Cell biol of growth regulation during normal development and tumorigenesis, particularly in the mammary gland; molecular basis of cell interaction in mammary gland and blood vessel wall; use and improvement of cell culture methods. *Mailing Add:* Dept Zool Wash State Univ Pullman WA 99164-4236

HOSKEN, WILLIAM H, b Zion, Ill, June 26, 37; c 3. COMPUTER SCIENCE. *Educ:* Antioch Col, AB, 60; Ohio State Univ, MA, 62; Purdue Univ, PhD(comput sci), 68. *Prof Exp:* Asst prof math & comput sci, Purdue Univ, 68; asst prof comput sci, Univ Nebr, 68-69 & Pa State Univ, 69-76; assoc prof, Univ NMex, 76-77; proj mgr, 77-80, MGR INFO SYSTS, UNIV CHICAGO, 80- *Mem:* Asn Comput Mach. *Res:* Logic; automata theory; formal systems; data processing systems. *Mailing Add:* 3022 Enoch Ave Zion IL 60099

HOSKER, RAYFORD PETER, JR, b Lynn, Mass, June 17, 43; m 66; c 2. AIR POLLUTION METEOROLOGY, MICROMETEOROLOGY. *Educ:* Boston Col, BS, 65; Univ Minn, MS, 67; Northwestern Univ, PhD(mech eng), 71. *Prof Exp:* Res fel, von Karman Inst Fluid Dynamics, NATO, Belgium, 70-71; phys scientist, 71-90, DIR, ATMOSPHERIC TURBULENCE & DIFFUSION DIV, ENVIRON RES LABS, NAT OCEANIC & ATMOSPHERIC ADMIN, US DEPT COM, 90- *Concurrent Pos:* Consult, Adv Comt Reactor Safeguards, US Nuclear Regulatory Comn, 82-89, Nat Coun Radiation Protection & Measurement, 84-89; field dir, US Dept Energy Atmospheric Studies Complex Terrain Prog, 85-90. *Mem:* AAAS; Am

Meteorol Soc; Air Pollution Control Asn; Am Soc Testing & Mat. *Res:* Transport and dispersion in complex terrain, including design and direction of large field experiments; theoretical and experimental studies of flow and dispersion near buildings; theory and monitoring of dry deposition of acidifying pollutants. *Mailing Add:* 108 Norton Rd Oak Ridge TN 37830

HOSKIN, FRANCIS CLIFFORD GEORGE, b Hanna, Alta, Oct 19, 22; nat US; m 45; c 1. BIOCHEMISTRY. *Educ:* Queen's Univ, Ont, BA, 50, MA, 51; Univ Sask, PhD(org chem), 53. *Prof Exp:* Asst, Queen's Univ, Ont, 49-51; asst, Univ Sask, 51-53; biochemist, Defence Res Bd, Suffield Exp Sta, Can, 53-57; res assoc & asst prof neurol, Col Physicians & Surgeons, Columbia Univ, 57-69; PROF BIOL, ILL INST TECHNOL, 69- *Concurrent Pos:* Res collabr, Brookhaven Nat Lab, 59-60; NIH spec fel, Leicester Univ, Eng, 67-68; mem, Marine Biol Lab Corps; vis prof biochem, Med Col, Rush Univ, 72-; mem, Chem Systs Labs Adv Panel, 81. *Mem:* AAAS; Am Soc Biol Chemists; Am Soc Neurochem. *Res:* Chemistry and biochemistry of phosphorus and sulfur compounds; biochemical basis of nerve function; neurochemistry; intermediary metabolism; quinones. *Mailing Add:* Dept Biol Ill Inst Technol Chicago IL 60616

HOSKIN, GEORGE PERRY, b Seattle, Wash, Oct 4, 41; m 67; c 2. PHYSIOLOGY, INVERTEBRATE PATHOLOGY. *Educ:* Univ Wash, BS, 63; Univ Hawaii, MS, 68; Lehigh Univ, PhD(biol), 72. *Prof Exp:* Instr biol, Lafayette Col, 71-72, asst prof, 72-80; ASSOC DIR, OFF SEAFOOD, FOOD & DRUG ADMIN, 80- *Concurrent Pos:* Consult, Brandt Water Testing Lab, 75-79. *Mem:* Sigma Xi; Am Soc Parasitologists; Soc Invert Path; Am Soc Zoologists; Asn Off Anal Chemists. *Res:* Molluscan physiology and pathobiology; invertebrate lipid metabolism; symbiosis among invertebrates, especially histo-pathological and biochemical alterations of host organisms in host-parasite relationships. *Mailing Add:* Food & Drug Admin HFF-503 200 C St SW Washington DC 20204

HOSKINS, BETH, AGING, PHARMACOLOGY & TOXICOLOGY. *Educ:* Univ Miss, PhD(pharmacol), 70. *Prof Exp:* ASSOC PROF PHARMACOL & TOXICOL, MED SCH, UNIV MISS, 72- *Mailing Add:* Dept Pharmacol Univ Miss Med Ctr Jackson MS 39216

HOSKINS, CORTEZ WILLIAM, b Long Beach, Calif, May 17, 30; m 54; c 3. PETROLEUM GEOLOGY, PALEOECOLOGY. *Educ:* Pomona Col, BA, 53; Claremont Col, MA, 54; Stanford Univ, PhD(geol), 57. *Prof Exp:* Res geologist paleont, Jersey Prod Res Co, 57-64; res geologist paleont prod res, Richfield Oil Corp, 65-66; res assoc, Unocal Sci & Technol Div, Union Oil Co Calif Res Ctr, 66-73, supvr geol & geochem, 73-74, mgr explor res, 74-90, MGR GEOL RES & STAFF CONSULT, EXPLOR RES, UNOCAL SCI & TECHNOL DIV, UNION OIL CO CALIF RES CTR, 90- *Mem:* Geol Soc Am; Soc Econ Paleontologists & Mineralogists; Am Asn Petrol Geologists; Sigma Xi. *Res:* Sedimentology; stratigraphy; paleontology. *Mailing Add:* Unocal Sci & Technol Div PO Box 76 Brea CA 92621

HOSKINS, DALE DOUGLAS, b Exeter, Calif, Feb 12, 28; m 53; c 2. BIOCHEMISTRY, REPRODUCTIVE PHYSIOLOGY. *Educ:* Ore State Univ, BS, 53, MS, 55; Univ Colo, PhD(biochem), 60. *Prof Exp:* Res asst biochem, Arctic Health Res Ctr, 55-57; instr dept pediat, Sch Med, Univ Wash, 60-61; from asst scientist to assoc scientist, Res Ctr, 61-69, from asst prof to assoc prof, 63-76, scientist, Res Ctr, 69-76, PROF BIOCHEM, UNIV ORE, 76-, SR SCIENTIST, ORE REGIONAL PRIMATE RES CTR, 76- *Concurrent Pos:* USPHS fel, 60-61 & spec res fel, Johns Hopkins Univ, 67-68; mem reproductive biol study sect, Div Res Grants, Nat Inst Child Health & Human Develop, 76-80; Fogarty int scholar, 81. *Mem:* Am Chem Soc; Am Soc Biol Chemists; Soc Study Reproduction. *Res:* Cellular respiration; mechanism of enzyme action; flavoproteins in electron transport; biochemistry of the male reproductive tract; enzymology of sperm maturation. *Mailing Add:* Dept Reproductive Biol & Behav Ore Reg Primate Res Ctr 505 NW 185th Ave Beaverton OR 97006

HOSKINS, DONALD MARTIN, b Lyons Fall, NY, May 22, 30; m 53; c 3. GEOLOGY. *Educ:* Union Col, NY, BS, 52; Univ Rochester, MS, 54; Bryn Mawr Col, PhD(geol), 60; Univ Pa, MGA, 74. *Prof Exp:* Geologist, 56-68, asst state geologist, Dept Environ Resources, 68-86, STATE GEOLOGIST, BUR TOPOG & GEOL SURV, PA GEOL SURV, 86- *Honors & Awards:* Ralph Digman Award, 90. *Mem:* Geol Soc Am; Paleont Soc. *Res:* Stratigraphy and structure of the Appalachian Mountain section of the Valley and Ridge province of Pennsylvania; paleontology of Silurian ostracodes; history of science. *Mailing Add:* Dept Environ Resources Bur Topog & Geol Surv Harrisburg PA 17105-2357

HOSKINS, EARL R, JR, b Chicago, Ill, May 2, 34; m 61; c 2. MINING ENGINEERING, GEOPHYSICS. *Educ:* SDak Sch Mines & Technol, BS, 56, MS, 64; Australian Nat Univ, PhD(geophysics), 68. *Prof Exp:* From instr to prof mining eng, SDak Sch Mines & Technol, 56-77, head dept mining eng, 73-77, head dept geophys, 83-88; TEX A&M UNIV, BROCKETT PROF GEOPHYSICS, GEOL & GEOG & PROF MINING ENG, 77-, DEP DEAN, COL GEOSCI. *Concurrent Pos:* Mem panel on earthquake modification, Nat Acad Sci, 69-75, mem comt on mech rope & cable, 72-75, mem, US Nat Comt on Rock Mech, 72-76 & 81- & mem panel on tectonophysics, 76-78; geophysicist, US Geol Surv, 69-75; vpres & mem bd dirs, Respec Inc, 70- *Mem:* Asn Eng Geologists; Am Inst Mining, Metall & Petrol Engrs; Geol Soc Am; Am Geophys Union; Soc Exp Stress Anal; Am Soc Photogram & Remote Sensing. *Res:* Rock mechanics; engineering geology; remote sensing; earthquakes; mined land reclamation; mineral economics. *Mailing Add:* Dept Geophysics Tex A&M Univ College Station TX 77843

HOSKINS, FREDERICK HALL, b Cincinnati, Ohio, May 17, 36; m 57; c 2. FOOD SCIENCE, NUTRITION. *Educ:* Univ Ariz, BS, 58, MS, 59; La State Univ, PhD(nutrit), 63. *Prof Exp:* Assoc animal sci & univ fel, La State Univ, Baton Rouge, 63-64, instr animal nutrit, 64-65, asst prof food sci & technol, 65-71, from assoc prof to prof food sci, 71-84; chair dept, 84-86, PROF FOOD SCI & HUMAN NUTRIT, WASH STATE UNIV, PULLMAN, 84- *Concurrent Pos:* Consult, Ethyl Corp, 65-66 & Int Food Technol Inc, 67-78; mem, Nutrit Res Coun; vis prof human nutrit, Univ Uppsala, Sweden, 78; mem bd dir, Nat Nutrit Consortium. *Mem:* Soc Animal Sci; Inst Food Technologists; Am Inst Nutrit; Soc Nutrit Educ; Asn Off Anal Chem. *Res:* Nutritional composition of meat; world food problems; nutrition as related to food processing; protein and amino acid relationships to rice and other cereal grains; nutritional interrelationships; food toxicology; therapeutic nutrition. *Mailing Add:* Dept Food Sci & Human Nutrit Washington State Univ Pullman WA 99164-6376

HOSKINS, HARTLEY, b Rochester, NY, Feb 15, 38; m 75; c 1. GEOPHYSICS, SCIENCE COMMUNICATION. *Educ:* Mass Inst Technol, BS, 59; Univ Chicago, MS, 61, PhD(geophys), 65. *Prof Exp:* Fel geophys sci, Univ Chicago, 65-66; lectr, Univ Ghana, 66-68; asst scientist, 68-72, RES ASSOC GEOL & GEOPHYS, WOODS HOLE OCEANOG INST, 72- *Concurrent Pos:* Coordr, Ocean Indust Prog, 77- *Mem:* Am Geophys Union; Soc Explor Geophys. *Res:* Oceanic rocks and sediments and their distribution; telecommunications; intellectual property; oceanography. *Mailing Add:* Dept Geol & Geophys Woods Hole Oceanog Inst Woods Hole MA 02543

HOSKINS, JOHN RICHARD, b Brewster, Wash, June 9, 19; m 46; c 3. MINING ENGINEERING. *Educ:* Univ Idaho, BS, 47; Univ Utah, PhD(mining eng), 62. *Prof Exp:* Proj engr, US Gypsum Co, 47-48, dept supt, 48-49; mining engr, Am Smelting & Refining Co, 49-50, shift boss, 50-51, foreman, 51-52; from instr to assoc prof mining eng, Col Mines, Univ Alaska, 52-59; dept asst, Univ Utah, 59-62; mining methods res engr, US Bur Mines, 62-67; PROF MINING ENG, UNIV IDAHO, 67-, HEAD DEPT, 69- *Concurrent Pos:* Consult, Prof Engrs Exam Mining Engrs, Alaska, 58-59 & US Bur Mines. *Mem:* Am Inst Mining, Metall & Petrol Engrs. *Res:* Applied research in explosives and rock mechanics problems dealing primarily with rock strength properties. *Mailing Add:* Dept Metall & Mat Univ Idaho Moscow ID 83843

HOSKINS, LEO CLARON, b Logan, Utah, July 27, 40; m 60; c 3. PHYSICAL CHEMISTRY. *Educ:* Utah State Univ, BS, 62; Mass Inst Technol, PhD(phys chem), 65. *Prof Exp:* From asst prof to assoc prof, 65-75, PROF CHEM, UNIV ALASKA, 75- *Mem:* Am Phys Soc; Sigma Xi. *Res:* Theory of resonanance raman spectroscopy of carotenoids; laser Raman studies in oceanography. *Mailing Add:* Dept Chem Univ Alaska Fairbanks AK 99701

HOSKINS, SAM WHITWORTH, JR, b Poplarville, Miss, Mar 29, 21; m 47; c 2. PERIODONTOLOGY. *Educ:* Atlanta-Southern Dent Col, DDS, 44; Northwestern Univ, Chicago, MSD(periodont), 50; Am Bd Periodont, dipl. *Prof Exp:* Dent officer, US Vet Admin, 46-48 & US Air Force, 50-70; prof periodont & coordr curric, Univ Tex Sci Ctr, San Antonio, 70-83; RETIRED. *Mem:* Am Dent Asn; Am Acad Periodont; fel Am Col Dent. *Res:* Periodontal therapy; mouth hygiene methods and practices including motivational factors. *Mailing Add:* 6243 Babcock San Antonio TX 78240

HOSLER, CHARLES FREDERICK, JR, b Flint, Mich, Sept 26, 39; m 62; c 2. BIOCHEMISTRY. *Educ:* Univ Mich, BS, 61; Univ Ill, MS, 63, PhD(biochem), 67. *Prof Exp:* Asst prof, 66-68, Bd Regents fac res grant, 66-78, admin dir nuclear radiation ctr, 68-70, assoc prof, 68-77, PROF CHEM, UNIV WIS-LA CROSSE, 77- *Mem:* Am Col Sports Med; Am Chem Soc. *Res:* Biochemistry related to exercise; physical fitness assessment; human performance; serum lipoprotein analysis. *Mailing Add:* Dept Chem Univ Wis La Crosse 1725 State St La Crosse WI 54601

HOSLER, CHARLES LUTHER, JR, b Honey Brook, Pa, June 3, 24; m 47, 71; c 4. METEOROLOGY. *Educ:* Pa State Univ, BS, 47, MS, 48 & PhD(meteorol), 51. *Prof Exp:* From instr to prof meteorol, Pa State Univ, 48-58, head dept, 60-65, dean, Col Mineral Industs, 65-66, Col Earth & Mineral Sci, 66-85, PROF METEOROL, PA STATE UNIV, UNIVERSITY PARK, 58-, ACTG EXEC VPRES & PROVOST, 90- *Concurrent Pos:* Hydrographer, Pa Dept Forests & Waters, 49-59; consult numerous pvt corps & fed agencies, 51-; mem, Nat Adv Comt on Oceans & Atmosphere, 72-75; mem comt atmospheric sci, Nat Acad Sci, 75-78; vpres, Res & Dean Grad Sch, 85-; mem Nat Sci Bd, 85. *Mem:* Nat Acad Eng; fel Am Meteorol Soc (pres, 76-77); Am Geophys Union. *Res:* Cloud physics; weather modification. *Mailing Add:* Pa State Univ 114 Kern Grad Bldg Univ Park PA 16802

HOSLER, E(ARL) RAMON, b Columbus, Ohio, Dec 24, 35; m 57; c 5. CHEMICAL ENGINEERING. *Educ:* Univ Dayton, BChE, 57; Univ Ill, MS, 59, PhD(chem eng), 61. *Prof Exp:* Engr thermal & hydraul res, Bettis Atomic Power Lab, Westinghouse Elec Corp, 61-63, sr engr, 63-73, eng mgr, 73-76, tech consult, 76-79; assoc prof, 79-85, PROF, UNIV CENTRAL FLA, 85- *Concurrent Pos:* Consult, Elec Power Res Inst; chmn, Heat Transfer & Energy Conversion Divs, Am Inst Chem Engrs, 81. *Mem:* Fel Am Inst Chem Engrs; Am Soc Mech Engrs; Am Soc Eng Educ. *Res:* Heat transfer and fluid mechanics, using high speed photography to discern basic mechanisms. *Mailing Add:* Univ Central Fla PO Box 25000 Orlando FL 32816-0993

HOSLER, JOHN FREDERICK, b Berwick, Pa, Apr 13, 20; m 44; c 3. ORGANIC CHEMISTRY. *Educ:* Pa State Col, BS, 42, MS, 48, PhD(chem), 51. *Prof Exp:* Chemist, Gen Chem Defense Corp, 42-43; res chemist, Calco Chem Div, Am Cyanamid Co, 51-54, group leader, 54-57, mgt pigment res, 57-58, asst to dir chem res, 58-59, mgr intermediate & explosives res, Org Chem Div, 59-61, mgr intermediates res & develop, NJ, 61-68; MGR DECORATIVE PROD RES & DEVELOP, FORMICA CORP, 68- *Mem:* Am Chem Soc. *Res:* Organic intermediates and plastic additives; high pressure laminates and plastics. *Mailing Add:* 1094 Little Shaglake Rd Gwinn MI 49841-9214

HOSLER, PETER, b Cleveland, Ohio, Mar 12, 27; m 48; c 5. PETROLEUM CHEMISTRY. *Educ:* Case Inst Technol, BS, 49; Univ Wis, MS, 51, PhD(biochem), 53. *Prof Exp:* Biochemist, Antibiotics Dept, Eli Lilly & Co, 53-61; RES ENGR, PROCESS DEVELOP, SUN OIL CO, 61- *Mem:* Am Chem Soc (secy, Div Microbial Chem, 65-67). *Res:* Development of industrial fermentation processes at pilot plant scale; applied studies of microbiol metabolism as related to environmental factors; petrochemicals; lubricants and industrial oils. *Mailing Add:* Res & Develop Div Sun Oil Co Marcus Hook PA 19061

HOSLEY, ROBERT JAMES, b Lima, NY, Apr 9, 23. MICROBIOLOGY. *Educ:* Univ Rochester, AB, 49; Univ Mich, MS, 51 PhD(bact), 55. *Prof Exp:* Bacteriologist, Eli Lilly & Co, 55-58, head dept poliomyelitis vaccine develop, 58-59, head dept tissue cult develop, 60-62, mgr biol develop, 62-64, staff asst, Biol-Pharmacol Res Div, 65-71, admin asst, Res Grants Admin, 71-79; RETIRED. *Mem:* Sigma Xi. *Res:* Biologics. *Mailing Add:* 515 Arnold St Bozeman MT 59715-6136

HOSMANE, NARAYAN SADASHIV, b Gokarn, Karnataka, India, June 30, 48; m 76; c 2. CARBORANES & METALLACARBORANES. *Educ:* Karnatak Univ, Dharwar, India, BSc, 68, MSc, 70; Edinburgh Univ, Scotland, PhD(inorg chem), 74. *Prof Exp:* Res asst organometallic chem, Queen's Univ Belfast, 74-75; res scientist pollution control, Lambeg Indust Res Asn, 75-76; res assoc inorg chem, Auburn Univ, 76-77 & Univ Va, 78-79; asst prof, Va Polytech Inst & State Univ, 79-82; from asst prof to assoc prof, 82-89, PROF INORG CHEM, SOUTHERN METHODIST UNIV, 89- *Concurrent Pos:* Vis prof, Ohio State Univ, 85-86; consult, Nat Cancer Inst, NIH, 90, Veritech, 90- *Honors & Awards:* Outstanding Res Award, Sigma Xi, 87. *Mem:* Fel Royal Soc Chem; Am Chem Soc; fel Am Inst Chemists; Sigma Xi. *Res:* Synthetic and structural investigations of cluster compounds, especially those of boron, carbon and transition metals in concert with a number of other elements including group 4 hetero atoms; organosilicon chemistry. *Mailing Add:* Dept Chem Southern Methodist Univ Dallas TX 75275

HOSMANE, RAMACHANDRA SADASHIV, b Gokarn, India, Dec 12, 44; US citizen; m 75; c 1. HETEROCYCLIC CHEMISTRY, NUCLEOSIDES & NUCLEOTIDES. *Educ:* Karnatak Univ, Dharwar, India, BSc, 66, MSc, 68; Univ SFla, MS, 76, PhD (chem), 78. *Prof Exp:* Res assoc chem, Univ Ill, Urbana, 78-82; asst prof, 82-86, ASSOC PROF CHEM, UNIV MD, BALTIMORE COUNTY, 86- *Concurrent Pos:* NIH new investr res award, 84. *Mem:* Sigma Xi; Am Chem Soc; affil mem Int union Pure & Appl Chem; Int Soc Heterocyclic Chem; AAAS. *Res:* Developing new and novel reagents for organic/biorganic synthesis; synthesis of medicinally significant analogues of natural products, heterocycles, and nucleosides; synthesis of polycyclic heterocycles and study of concepts of heteroaromaticity, stability, and reactivity; rearrangements in organic synthesis; cross-linking reagents for proteins and nucleic acids; author of more than 50 publications in international journals. *Mailing Add:* Dept Chem & Biochem Univ Md Baltimore MD 21228

HOSNER, JOHN FRANK, b Gillespie, Ill, Feb 25, 25; m 51; c 2. FORESTRY. *Educ:* Mich State Univ, BS, 48; Duke Univ, MF, 50; State Univ NY, PhD, 57. *Prof Exp:* Dist forester, State of Ill, 48-50; from instr to assoc prof forestry, Southern Ill Univ, 50-61; asst dean, Col Agr & Life Sci, 73-80, PROF FORESTRY, VA POLYTECH INST & STATE UNIV, 61-, DIR SCH FORESTRY & WILDLIFE RESOURCES, 69-, ASSOC DEAN, COL AGR & LIFE SCI, 80- *Concurrent Pos:* Vis prof, Col Forestry, State Univ NY, 58-59. *Mem:* Fel AAAS; fel Soc Am Foresters; Ecol Soc Am; Am Inst Biol Sci; Soil Conserv Soc Am; Sigma Xi. *Res:* Ecology and silviculture of piedmont and mountain forests. *Mailing Add:* Sch Forestry & Wildlife Resources Va Polytech Inst State Univ Blacksburg VA 24061

HOSNI, YASSER ALI, b Cairo, Egypt, July 30, 41; m 76; c 1. COMPUTER-AIDED MANUFACTURING, APPLIED OPERATIONS RESEARCH. *Educ:* Cairo Univ, BSc, 63, MSc, 67; Univ Ark, PhD(eng), 77. *Prof Exp:* Engr, Semaf Co, 53-64; chief eng, Prod Dept, Nasr Boiler & Mfg Co, 64-70; software specialist software design, NCR Co, 70-73; asst indust eng, Univ Ark, 73-77; ASSOC PROF INDUST ENG, UNIV CENT FLA, 77- *Concurrent Pos:* Consult, Morton Co, Fla Plumbing & Wakenhut, 76-81; prin investr, Dept Educ, 77-79, Fla Energy Off, 77-80 & Naval Training Equip Ctr, 78-82. *Mem:* Am Inst Indust Engrs; Inst Elec & Electronics Engrs; Am Soc Eng Educ; Am Soc Prof Engrs. *Res:* Software design and development; theory of programming; applied operations research; simulation; developing of industrial packages for decision making operations; energy conservation techniques. *Mailing Add:* Indust Eng & Mgt Systs Univ Cent Fla PO Box 25000 Orlando FL 32816

HOSNY, AHMED NABIL, mechanical & aerospace engineering, for more information see previous edition

HOSODA, JUNKO, BIOCHEMISTRY, MOLECULAR BIOLOGY. *Educ:* Univ Tokyo, PhD(biochem), 60. *Prof Exp:* STAFF SCIENTIST, LAWRENCE BERKELEY LAB, UNIV CALIF, BERKELEY, 77- *Res:* DNA replication, repair and recombination; protein interactions. *Mailing Add:* CMB Dept Bldg 934 Rm Eight Lawrence Berkeley Lab Univ Calif Berkeley CA 94720

HOSOKAWA, KEIICHI, b Minoh, Japan, Sept 19, 29; m 62; c 3. MOLECULAR CELL BIOLOGY, MOLECULAR VIROLOGY. *Educ:* Osaka Univ, MD, 55, PhD(physiol chem), 61. *Prof Exp:* Res fel biochem, Osaka Univ Dent Sch, 60-66; asst prof biochem, Univ Calif, Berkeley, 66-71; sr scientist, Worcester Found Exp Biol, 71-76; PROF BIOCHEM, KAWASAKI MED SCH, 76- *Concurrent Pos:* Res fel radiation med, Radiation Ctr Osaka Prefecture, 61-62; Miller res fel bact, Univ Calif, Berkeley, 63-65; proj assoc, Univ Wis, 65; Med Res Coun res fel biochem, Ont Cancer Inst, 65-66; vis scientist, City Hope Res Inst, 80; lectr, Dept Virol, Okayama Univ Med Sch, 80-82, Dept Biochem, 80- & Inst Cancer Res, 91-; vis sr scientist, Southern Ill Univ, 83; adj prof, Dept Nutrit Biochem & Chem Biopolymers, Kawasaki Col Allied Health Prof, 83-86; guest prof, Inst Genetics, Univ Cologne, 84-90. *Honors & Awards:* Sci Achievement Award, Sanyo Broadcasting Found Sci & Cult, 88. *Mem:* Am Soc Biochem & Molecular Biol; Am Soc Microbiol; Am Chem Soc; NY Acad Sci; Protein Soc. *Res:* Functionally active ribosome reconstituted from inactive subparticles and proteins and from RNAs and proteins; in vitro assembled adenovirus chromatin as a highly efficient probe for transfection. *Mailing Add:* 2-4-39 Kadotabunka-Machi Okayama 703 Japan

HOSPELHORN, VERNE D, biochemistry; deceased, see previous edition for last biography

HOSS, DONALD EARL, b Mexico, Mo, Dec 17, 36; m 60; c 2. PHYSIOLOGICAL ECOLOGY. *Educ:* Univ Mo-Columbia, BS, 58; NC State Univ, MS, 65, PhD(zool), 71. *Prof Exp:* Fishery biologist radiobiol, Bur Com Fisheries Radiobiol Lab, 58-69; fishery biologist, prog leader physiol ecol, 72-87, CHIEF, COASTAL & ESTUARINE ECOL DIV, NAT MARINE FISHERIES SERV, BEAUFORT LAB, 87- *Concurrent Pos:* Lab rep, Secy Interior Oil-Spill Comt, 70-71; adj fac mem, NC State Univ, 74-; vis investr, Dunstaffnage Marine Res Lab, Scottish Marine Biol Asn, Oban, 76-77. *Mem:* Am Fisheries Soc; Estuarine Res Fedn; Sigma Xi. *Res:* Effects of natural and man-induced changes on the survival and growth of fish; larval fish development and physiology. *Mailing Add:* SE Fisheries Ctr Nat Marine Fisheries Serv Nat Oceonog & Atmospheric Admin Beaufort NC 28516-9722

HOSS, WAYNE PAUL, b Paso Robles, Calif, Dec 11, 43; m 67. NEUROCHEMISTRY. *Educ:* Univ Idaho, BS, 66; Univ Nebr, PhD(chem), 71. *Prof Exp:* NSF teaching fel, Ctr Brain Res, Univ Rochester, 70-72, NIH fel, 72-75, from asst prof to assoc prof neurochem, 80-85; PROF MED CHEM, UNIV TOLEDO, 85-, CO-DIR, CTR DRUG DESIGN & DEVELOP & ASSOC VPRES, RES & DEVELOP, 89- *Concurrent Pos:* Vis assoc prof biochem, Nagoya City Univ, Japan, 77; NIMH res career develop award, 76-81; dir, Undergrad Neurosci Prog, Univ Rochester, 80- *Mem:* Am Soc Pharmacol & Exp Ther; Am Asn Cols Pharm; NY Acad Sci; Soc Neurosci; Am Chem Soc. *Res:* Biochemistry and pharmacology of cns receptors, including opioid, cholinergic and histaminergic; drugs of abuse; G-proteins; aging and dementia; development of new therapeutics. *Mailing Add:* Univ Toledo Col Pharm 2801 W Bancroft St Toledo OH 43606

HOSSAIN, SHAFI UL, b Calcutta, India, Feb 1, 26; US citizen; m 54; c 3. CHEMISTRY. *Educ:* Univ Calcutta, BSc, 44, MSc, 46; Univ Minn, MS, 52. *Prof Exp:* Res chemist, Abitibi Power & Paper Co Ltd, 53-58, sect supvr pioneering res sect, 58-64, res assoc, 64; res chemist, Res & Eng Div, 64-65, proj leader, 65-66, sr res scientist, 66-68, mgr, Tech Specialties Lab, Res & Eng Div, 68-76, res assoc pioneering res, 76-79, mgr explor res, 79-83, sr res assoc, 83-84, PRIN RES FEL, KIMBERLY CLARK CORP, 84- *Mem:* Am Chem Soc; Tech Asn Pulp & Paper Indust; Nat Asn Acad Sci. *Res:* Chemistry of pulp, paper and polymers; interaction between paper fibers and high polymers. *Mailing Add:* Kimberly-Clark Corp 2100 Winchester Rd Neenah WI 54956

HOSSAINI, ALI A, b Basra, Iraq, Sept 24, 28; nat US; m 61; c 4. CLINICAL PATHOLOGY. *Educ:* Am Univ Beirut, BA, 53; Tex Christian Univ, MS, 57; Ohio State Univ, PhD(path), 60. *Prof Exp:* Dir health educ, Am Point IV Prog, Iran, 53-55; asst instr clin path, Ohio State Univ, 59-60; dir, Ports Hosp, Basra, Iraq, 60-62; dir biochem, Med Ctr, Univ WVa, 62-63; from asst prof to assoc prof clin path, 63-76, PROF PATH, MED COL VA, 67-, DIR BLOOD BANK, 63-, DIR SCH BLOOD BANKING, 65- *Concurrent Pos:* Am Cancer Soc grant, 64-; consult, Vet Admin Hosp, 65- *Mem:* AAAS; Am Soc Clin Path; Am Thoracic. *Res:* Serologic differentiation between leukemic and normal leukocytes; diagnosis of leukemias serologically; attempts at finding antagonists destroying leukemic leukocytes; relationship between leukoagglutinins and febrile transfusion reactions. *Mailing Add:* Med Col Va Va Commonwealth Univ MCV Sta Box 451 Richmond VA 23298-0451

HOSSLER, FRED E, b Hamburg, Pa, May 7, 41; m 74; c 3. HISTOLOGY, CELL BIOLOGY. *Educ:* Muhlenberg Col, BS, 63; Pa State Univ, MS, 65; Univ Colo, PhD(path), 71. *Prof Exp:* Postdoctoral cell biol, Yale Univ, 71-74; from asst prof to assoc prof anat, histol & cell biol, La State Univ, 74-80; assoc prof, 80-89, PROF HISTOL & ELECTRON MICROS, ETENN STATE UNIV, 89- *Concurrent Pos:* Asst dean, Sch Grad Studies, ETenn State Univ, 87-89, actg dean & assoc vpres, 89-90. *Mem:* Am Asn Anatomists; Am Soc Cell Biol; Electron Micros Soc Am. *Res:* Sodium, potassium-astatine-pase-rich epithelia and their changes with changing osmotic conditions; microvascular of heart, lung, ion-transporting tissues as viewed with corrosion casting and scanning electron microscopy. *Mailing Add:* Dept Anat Col Med ETenn State Univ Box 19960A Johnson City TN 37614

HOSSNER, LLOYD RICHARD, b Ashton, Idaho, July 24, 36; m 58; c 3. SOIL CHEMISTRY, SOIL FERTILITY. *Educ:* Utah State Univ, BS, 58, MS, 61; Mich State Univ, PhD(soil chem), 65. *Prof Exp:* Res asst soil chem, Utah State Univ, 59-61; instr, Mont State Univ, 61-62; res asst, Mich State Univ, 62-65; res soil chemist, Int Minerals & Chem Corp, 65-68; from asst prof to assoc prof, 68-77, PROF SOIL CHEM, TEX A&M UNIV, 77- *Honors & Awards:* Super Achievement Award Res, Soil & Crop Sci, 89. *Mem:* AAAS; fel Am Soc Agron; fel Soil Sci Soc Am; Sigma Xi; Int Soil Sci Soc. *Res:* Applied soil chemistry; soil-fertilizer reaction products; waste management; surface mine reclamation; author of over 100 publications. *Mailing Add:* Dept Soil & Crop Sci Tex A&M Univ College Station TX 77843

HOSTER, DONALD PAUL, b Seneca Falls, NY, Aug 30, 41; m 82. ALGEBRA, INORGANIC CHEMISTRY. *Educ:* Union Col, NY, BS, 63; Univ Del, MS, 65, PhD(phys org chem), 68. *Prof Exp:* Asst prof chem, St Joseph Col, Md, 67-70; from asst prof to assoc prof chem, 70-89, coordr chem, Physics & Phys Sci, COMMMUNITY COL BALTIMORE, 86-, PROF CHEM, 89- *Concurrent Pos:* Lectr, Mt St Mary's Col, Md, 69-70; NSF res partic, 68; Summer res prog for fac, US Army, 90. *Honors & Awards:* NSF Equip Grant, 77. *Mem:* Am Chem Soc; Nat Sci Teachers Asn; Am Fedn

Teachers. *Res:* Gas phase reactions and pyrolysis of organic chlorides and acetates; deuterium exchange reactions; individualizing instruction; audio-visual aids; biochemistry for dental hygiene curriculum; computer applications in chemical education; chemistry for nonscientists; FTIR in chemical analysis. *Mailing Add:* 625 Parkwyrth Ave Baltimore MD 21218

HOSTERMAN, JOHN W, b East Lansing, Mich, March 24, 23. CLAY MINERALOGY. *Educ:* Pa State Univ, BS, 48, MS, 49. *Prof Exp:* RES GEOLOGIST, US GEOL SURVEY, 49- *Mem:* Fel Geol Soc Am; Clay Mineral Soc; Soc Mining Engrs. *Mailing Add:* US Geol Surv MS 928 Nat Ctr 2201 Sunrise Valley Dr Reston VA 22092

HOSTETLER, JEPTHA RAY, b Orrville, Ohio, June 23, 39; m 72; c 3. ANATOMY, ELECTRON MICROSCOPY. *Educ:* Goshen Col, BA, 62; Ohio State Univ, PhD(anat), 69. *Prof Exp:* NIH fel, Col Med, Tufts Univ, 68; instr anat, Univ Ky, 68-70; ASSOC PROF ANAT, OHIO STATE UNIV, 70- *Mem:* Am Asn Anatomists; Am Sci Affil. *Res:* Lymphopoiesis and lymph node histogenesis; electron microscopy and histochemistry of aneurytic vessels and of burns and scar formation. *Mailing Add:* Dept Prev Med Ohio State Univ 320 W Tenth Ave Columbus OH 43210

HOSTETLER, KARL YODER, b Goshen, Ind, Nov 17, 39; m 71; c 3. MEDICINE, BIOCHEMISTRY. *Educ:* De Pauw Univ, BA, 61; Western Reserve Univ, MD, 65. *Prof Exp:* from asst prof to assoc prof, 73-84, PROF MED, UNIV CALIF, SAN DIEGO, 84- *Concurrent Pos:* Fel, John Simon Guggenheim Found, 80 & Japan Soc Prom Sci, 86; founder & dir, Vical Inc, San Diego, 87- *Mem:* Am Soc Biochem & Molecular Biol; Am Soc Clin Invest; AAAS; Endocrine Soc; Western Asn Phys; Am Diabetes Asn. *Res:* Clinical endocrinology and metabolism; chemistry and biochemistry of phospholipids; mechanisms of drug-induced lipidosis; design, synthesis and evaluation of lipid prodrugs of antiviral and anticancer agents. *Mailing Add:* Dept Med (V-111G) Vet Admin Med Ctr Univ Calif San Diego San Diego CA 92161

HOSTETLER, ROBERT PAUL, b Holsopple, Pa, Apr 8, 37; m 60; c 5. MATHEMATICS EDUCATION. *Educ:* Eastern Mennonite Col, BS, 59; Pa State Univ, MA, 65, PhD(math educ), 70. *Prof Exp:* High sch teacher, Pa, 59-62; instr, 64-67, asst prof, 70-79, ASSOC PROF MATH, BEHREND COL, PA STATE UNIV, 79- *Mem:* Math Asn Am; Nat Coun Teachers Math. *Res:* Effective sequencing of mathematical content; mathematics as skill, art and process; calculus, consumer mathematics and precalculus. *Mailing Add:* Dept Math Behrend Col Pa State Univ Erie PA 16563

HOSTETLER, ROY IVAN, b Asotin, Wash, Sept 9, 16; m 41; c 3. VETERINARY MEDICINE. *Educ:* Wash State Univ, BS & DVM, 39. *Prof Exp:* Jr vet, USDA, 39-42; pvt pract, Wash, 46-58; supvr animal indust div, Wash State Dept Agr, 58-60; exten vet, Coop Exten Serv, Wash State Univ, 60-78; RETIRED. *Concurrent Pos:* Mem, Nat Mastitis Coun. *Mem:* Am Vet Med Asn; US Animal Health Asn. *Res:* Mastitis control; nitrate toxicity; selenium toxicity and industry contamination; internal parasites. *Mailing Add:* 2365 Valley View Court Clarkston WA 99403-1233

HOSTETTER, DONALD LEE, b Bartlett, Iowa, Jan 23, 41; m 83; c 3. INSECT PATHOLOGY. *Educ:* SDak State Univ, BS, 62, MS, 64. *Prof Exp:* Med entomologist, US Army, Ft Bragg, NC, 64-66; res technician, Agr Res, 66-70, res entomologist, Biol Control Insects Res Labs, 70-87; RES ENTOMOLOGIST, USDA-ARS GRASSHOPPER IPM PROJ, KIMBERLY, IDAHO, 87- *Mem:* Entom Soc Am; Soc Invert Path; Am Registry Prof Entomologists; Sigma Xi. *Res:* Insect baculoviruses and their environmental stability and persistence in agro-ecosystems; insect pathology and microbial control of insect pests of agricultural crop systems; conducting studies with insect pathogens in variety of ecosystems and developing protocols of use against major insect pests; population dynamics of rangeland grasshoppers and their parasites and predators. *Mailing Add:* USDA-ARS Snake River Conserv Res Ctr 3793 N 3600 E Kimberly ID 83341

HOSTETTLER, JOHN DAVISON, b Rockford, Ill, June 22, 41; m 65; c 2. PHYSICAL CHEMISTRY. *Educ:* Monmouth Col, BA, 62; Univ Wis-Madison, PhD(phys chem), 70. *Prof Exp:* asst prof, 70-77, ASSOC PROF CHEM, UNIV COLO, COLORADO SPRINGS, 77- *Mem:* AAAS; Am Chem Soc. *Res:* Fluorescence applied to polymers C-13 nuclear magnetic resonance; fluorescence applied to polymers; chemical education. *Mailing Add:* Dept Chem San Jose State Univ San Jose CA 95192

HOTCHIN, JOHN ELTON, b Sutton-On-Sea, Eng, Apr 7, 21; m 52; c 1. VIROLOGY. *Educ:* Univ London, MD & BS, 44, PhD, 52; FRCPath, 72. *Prof Exp:* House surgeon & physician, Kings Col Hosp, London, 43-44; mem res staff, Nat Insts Med Res, London, 48-55; asst prof bact & immunol, Univ BC, 55-57; asst dir, div labs & res, NY State Dept Health, Albany, 57-87; RETIRED. *Concurrent Pos:* Rockefeller travelling fel, Johns Hopkins Univ & Kerckhoff Lab, Calif Inst Technol, 52-54; res assoc, Western Div, Connaught Med Res Labs, 55-57; prof, Med Sch, Albany Hosp, 57-; ed, Infection & Immunity. *Mem:* Am Soc Microbiol; Am Asn Immunol; Infectious Dis Soc Am. *Res:* Pathogenesis of persistent and slow virus infections; virus induced behavioral disorders. *Mailing Add:* 18 Paxwood Rd Delmar NY 12054

HOTCHKISS, ARLAND TILLOTSON, plant morphology, for more information see previous edition

HOTCHKISS, DONALD K, b Eldora, Iowa, Dec 23, 28; m 52; c 3. APPLIED STATISTICS. *Educ:* Iowa State Univ, BS, 50, PhD(animal nutrit), 60. *Prof Exp:* County exten dir, Iowa State Univ, 54-56, asst dairy nutrit, 56-59; res statistician, Design & Anal Exp, Ralston Purina Co, 59-61; PROF & CONSULT STATIST, IOWA STATE UNIV, 61- *Concurrent Pos:* Vis prof, Nat Sch Agr, Mex, 67-68; vis lectr, Univ Costa Rica, 83. *Mem:* Biomet Soc; Am Statist Asn. *Res:* Nutrition of dairy cows; design of experiments in area of biological sciences. *Mailing Add:* Statist Lab Iowa State Univ Ames IA 50011

HOTCHKISS, HENRY, petroleum geology; deceased, see previous edition for last biography

HOTCHKISS, JULANE, b 1934; m; c 2. REPRODUCTIVE PHYSIOLOGY. *Educ:* Harvard Univ, PhD(physiol), 62. *Prof Exp:* from instr to res prof, Dept of Physiol, Univ Pittsburgh Sch of Med, 62-85; RES PROF PHYSIOL, MED SCH, UNIV TEX, 82- *Concurrent Pos:* Sect mem, Women Caucus of the Endocrine Soc, 83-86, pres, 88-90. *Mem:* Endocrine Soc; Am Physiol Soc; AAAS. *Res:* Reproductive endocrinology. *Mailing Add:* Dept Physiol & Cell Biol Med Sch Univ Tex PO Box 20708 Houston TX 77225

HOTCHKISS, ROLLIN DOUGLAS, b South Britain, Conn, Sept 8, 11; m 33, 67; c 2. BIOCHEMISTRY. *Educ:* Yale Univ, BS, 32, PhD(org chem), 35. *Hon Degrees:* ScD, Yale Univ, 62 & Univ Paris-South, 80. *Prof Exp:* Asst elem chem, Yale Univ, 32-34; fel chem, Rockefeller Inst, 35-36, asst, 36-37; Rockefeller Found fel, Carlsberg Lab, Copenhagen, 37-38; asst chem, Rockefeller Inst, 38-42, assoc path & microbiol, 42-50, assoc mem, 50-55, prof physiol, 55-82, EMER PROF PHYSIOL, ROCKEFELLER UNIV, 82-; RES PROF BIOL, STATE UNIV NY, ALBANY, 82- *Concurrent Pos:* Vis prof, Mass Inst Technol, 58, Univ Calif, Berkeley, 68 & Corpus Christi Col, Univ Cambridge, 70; mem sci adv bds, Roswell Park Mem Inst, Univ Buffalo, 58-65, Inst Cancer Res, 60-74, & Nat Inst Allergy & Infectious Dis, 61-64; mem adv bd, Biol Div, Oak Ridge Nat Lab, 62-70 & Nat Cancer Inst, 64-68; scholar, Fogarty Ctr, NIH, 71-72, UNESCO vis scholar, Biol Res Ctr, Szeged, 72-78; vis prof, Biol Sci Div, Univ Utah, 72 & 73, Univ Paris, Orsay, 75 & 79, Raine vis prof, Univ Perth, 80. *Honors & Awards:* Commercial Solvents Co Award, Am Soc Bact, 53; Dyer lectr, NIH, 61; Griffith lectr, Soc Gen Microbiol, 74. *Mem:* Nat Acad Sci; fel Am Soc Naturalists (vpres, 65); Harvey Soc (pres, 58-59); fel Am Acad Arts & Sci; Genetics Soc Am (pres, 72); fel AAAS; Hungarian Acad Sci; Royal Danish Acad Sci; Am Soc Cell Biol. *Res:* Bacterial metabolism; mode of action of antibacterial agents; bacterial genetics; fusion bacterial protoplasts; immunochemistry; protein chemistry; carbohydrates; nucleic acids; exonucleases; bacterial diploidy. *Mailing Add:* Two-Four Rolling Hills Condominium Lenox MA 01240

HOTCHKISS, SHARON K, b Eau Claire, Wis, Dec 1, 42; m 82. BIOLOGY. *Educ:* Univ Wis-Eau Claire, BS, 67; Univ Ky, PhD(microbiol), 73. *Prof Exp:* Instr cell biol, Med Sch, Univ Ky, 73-74; asst prof biol, Univ Wis-River Falls, 75-78; lectr biol, Northwestern Univ, 78-79; ASST PROF BIOL, CLARKSON COL, 79- *Mem:* AAAS; Am Soc Cell Biol; Asn Women Sci; Sigma Xi. *Res:* Genetics and developmental events associated with the circadian rhythm of ecolsion in Drosophila; cell division cycle. *Mailing Add:* 91 Maple St Potsdam NY 13676

HOTCHKISS, WILLIAM ROUSE, b Schenectady, NY, June 12, 37; m 65; c 6. GROUNDWATER MODELING, SNOW HYDROLOGY. *Educ:* Dartmouth Col, 55-59, Mont State Col, BS, 62, MS, 65; Colo State Univ, PhD(civil eng), 88. *Prof Exp:* Hydrologist, Water Resources Div, US Geol Surv, 65-78, proj leader, Northern Great Plains Regional Aquifier Syst Assessment, 78-82, supv hydrologist & chief Nat Training Ctr, 82-89, CENT REGION STAFF, US GEOL SURV, 89- *Honors & Awards:* Montgomery M Atwater Award, Nat Avalanche Found, 89. *Mem:* Am Inst Hydrol; Am Geophys Union; fel Geol Soc Am; Int Glacialogical Soc; Am Asn Avalanche Prof. *Res:* Glacial ablation studies in Montana & California; ground water studies in the San Joaquin & Sacramento Valleys of California; snow hydrology & the study of snow avalanches; author of 50 publications & reports. *Mailing Add:* US Geol Surv Cen Region WRD Mail Stop 406 Lakewood CO 80225-0046

HOTELLING, ERIC BELL, b Trenton, NJ, May 12, 23; m 56; c 2. ORGANIC CHEMISTRY. *Educ:* Rutgers Univ, BS, 48; Univ Mich, MS, 51, PhD(org chem), 53. *Prof Exp:* Res chemist, Merck & Co, Inc, 48-49; org chemist, Consol Coal Co, 53-58; res & develop group mgr, Am Mach & Foundry Co, 58-60; assoc dir res, Va Chem Inc, 60-63; lab mgr, Gen Foods Corp, 63-65; tech dir, Stein Hall & Co, Inc, NY, 65-66; sr vpres, Yoo-Hoo Beverage Co, 66-80, dir, 74-80; regional mkt dir, Beech-Nut Nutrit Corp, 80-82; DIR MKT & RES, STEENLAND LTD, 82- *Mem:* Fel Am Inst Chemists; Am Chem Soc; Am Mgt Asn; Am Inst Mgt. *Res:* Petrochemical derivatives; cellulose and amylose esters and ethers; fluorocarbons; electrodialysis membranes; food chemistry; instrumentation; coffee and tea processing; natural and synthetic gums; spiro-quaternary salts; morphine and cocaine derivatives; gelatin; proteins; high-protein beverages; research management. *Mailing Add:* 213 Harris Ave Croydon PA 19021

HOTH, DANIEL F, b Washington, DC, Apr 3, 46; m. MEDICAL RESEARCH. *Educ:* Franklin & Marshall Col, AB, 68; Georgetown Univ, MD, 72; Am Bd Internal Med, dipl, 76, dipl oncol, 77. *Prof Exp:* Intern med, Med Sch, Georgetown Univ, 72-73, jr & sr resident internal med, Georgetown Univ Med Hosp, 73-75, fel med oncol, 77-75, from instr to asst prof med, 77-80; head Drug Eval & Reporting Sect, Nat Cancer Inst, 80-81, chief Investigational Drug Br, Cancer Ther Eval Prog, Div Cancer Treatment, 81-87, actg assoc dir, 82-83; actg dir, 87-88, DIR, DIV AIDS, NAT INST ALLERGY & INFECTIOUS DIS, 88- *Concurrent Pos:* Jr fac clin fel, Am Cancer Soc, 78-80. *Mem:* Am Soc Clin Oncol; Am Col Physicians; Am Soc Clin Pharmacol & Therapeut; Soc Clin Trials; Infectious Dis Soc Am. *Res:* Internal medicine; AIDS. *Mailing Add:* Nat Inst Allergy & Infectious Dis Div AIDS 6003 Executive Blvd Rm 248 P Rockville MD 20892

HOTTA, SHOICHI STEVEN, b Stockton, Calif, Jan 8, 29; m 54; c 2. BIOCHEMISTRY, MEDICINE. *Educ:* Univ Calif, Berkeley, BA, 50, PhD(biochem), 53; Johns Hopkins Univ, MD, 58. *Prof Exp:* Jr asst physiologist biochem, Univ Calif, Berkeley, 53-54; asst prof, Med Col, Cornell Univ, 61-73; from assoc prof to prof biochem, Eastern Va Med Sch, 73-85; med officer, Food & Drug Admin, 85-89; MED OFFICER, PUB HEALTH SERV, 89- *Mem:* AAAS; Am Chem Soc; Soc Exp Biol & Med; NY Acad Sci; Am Soc Biol Chemists; Am Soc Clin Nutrit. *Res:* Cholesterol metabolism in diabetic and normal animals; pathways of glucose and glutamate metabolism in brain; roles of hexosemonophosphate shunt and glutathione in cells; biochemical basis for effects of compounds in brain; regulation of cholesterol metabolism. *Mailing Add:* 7001 Roundtree Rd Falls Church VA 22042-3912

HOTTA, YASUO, b Nagoya, Japan, Jan 12, 32; m 59; c 2. CELL BIOLOGY, GENETICS. *Educ:* Nagoya Univ, BSc, 54, MSc, 56, DSc(biol), 59. *Prof Exp:* Nat Res Coun Can res fel, 59-60; res assoc bot, Univ Ill, 60-65; assoc res biologist, 65-73, RES BIOLOGIST, UNIV CALIF, SAN DIEGO, 73- *Concurrent Pos:* Consult, Hirokawa Publ Co, Tokyo, Japan, 70- & Eisai Pharmaceut Co, Tokyo, 74-80; Oriental Nutrit Res Corp, San Diego, 83- *Mem:* Am Soc Cell Biol; Soc Develop Biol; Japan Soc Genetics. *Res:* Mechanisms and control of homologous chromosome pairing and crossing-over in meiosis; recombination, repair mechanisms and modification of DNA molecules in eukaryotic cells; analysis and control of chromosome pairing and chiasmata formation in meiosis. *Mailing Add:* Nagoya Univ Chikusa-Ku Nagoya 464 Japan

HOTTEL, HOYT C(LARKE), b Salem, Ind, Jan 15, 03; m 29; c 4. CHEMICAL ENGINEERING, COMBUSTION. *Educ:* Univ Ind, AB, 22; Mass Inst Technol, MS, 24. *Prof Exp:* Asst, 24-25, res assoc, 27-28, from asst prof to prof, 28-66, Carbon P Dubbs prof, 66-69, dir fuels res lab, 34-69, EMER PROF, MASS INST TECHNOL, 69- *Concurrent Pos:* Tech aide & sect chief, Fire Warfare Sect, Nat Defense Res Comt, 42-45; mem, Gas Turbine Subcomt, Nat Adv Comt Aerospace, 42-46, ad hoc comt, status jet propulsion, Whitman Comt, US Dept Defense, 43, Panel Coal Gasification Technol, Nat Acad Eng-Nat Res Coun, 71-73, rev comt, Task Force Energy, Nat Acad Eng, 74, Adv Group Arid Zone Probs Brazil, Nat Acad Sci, 74-75, ad hoc Panel Advan Power Cycles, Nat Res Coun, 75-78, Eval Panel Energy Conserv Prog, US Bur Standards, 76-80, Comt Chem Coal Utilization, Nat Res Coun, 76-80, Comt Assessment Indust Energy Conserv Prog Dept Energy, 80-82, Panel Fire Res, Assessment Nat Bur Standards, Prog, 85-; chmn, Thermal Panel, Armed Forces Spec Weapons Proj, US Dept Defense, 46-56; chmn, Am comt flame radiation, 53-74, hon chmn, 74-; chmn comt fire res, Nat Acad Sci-Nat Res Coun, 56-67, mem adv panel to bldg res div, US Bur Standards, 65-69, Ctr Fire Res, 85-91. *Honors & Awards:* King's Medal, Eng, 47; William H Walker Award, Am Inst Chem Eng, 47; Edgerton Medal, Combustion Inst, 60; Melchett Medal, Brit Inst Fuel, 60; Max Jakob Award, Am Soc Mech Engrs & Am Inst Chem Eng, 66; Farrington Daniels Award, Int Solar Energy Soc, 75; Royal Soc Esso Energy Award, 75; Founders Award, Nat Acad Eng, 80; Hottel Lectr, Mass Inst Technol, 85; Biennial Hottel Plenary Lectr, Combustion Inst, 85. *Mem:* Nat Acad Sci; Nat Acad Eng; Am Chem Soc; Combustion Inst (vpres, 53-64); Am Acad Arts & Sci. *Res:* Radiant heat transmission; optical methods of temperature measurement; flame propagation; combustion mechanisms; combustion in ramjets and turbines; radiative transfer; solar energy utilization; new energy technology. *Mailing Add:* 27 Cambridge St Winchester MA 01890

HOTTON, NICHOLAS, III, b Sault Ste Marie, Mich, Jan 28, 21; m 44; c 3. VERTEBRATE PALEONTOLOGY. *Educ:* Univ Chicago, BS, 47, PhD(paleozool), 50. *Prof Exp:* Asst paleont & zool, Univ Chicago, 47-50; asst instr zool, Univ Ill, 50-51; instr anat, Univ Kans Sch Med, 51-54, asst prof, 54-59; assoc cur to cur div vert paleont, 59-68, RES CUR, DEPT PALEONT, US NAT MUS, SMITHSONIAN INST, 68- *Mem:* Soc Vert Paleont; Sigma Xi. *Res:* Paleontology of lower tetrapods; cranial morphology; functional anatomy. *Mailing Add:* 101 Sheridan Avel Takoma Park MD 20912

HOTZ, HENRY PALMER, b Fayetteville, Ark, Oct 17, 25; m 52; c 3. INSTRUMENTAL ANALYSIS. *Educ:* Univ Ark, BS, 48; Wash Univ, PhD(physics), 53. *Prof Exp:* Asst prof physics, Auburn Univ, 53-58; asst prof Okla State Univ, 58-64; assoc prof, Marietta Col, 64-66; scientist-in-residence, US Naval Radiol Defense Lab, 66-67; assoc prof physics, Univ Mo-Rolla, 67-71; physicist, QantaþMetrix Corp, 71-74; sr scientist, Nuclear Equip Corp, 74-79; sr scientist, Wemco Div, Envirotech Corp, 79-82; sr scientist, Dohrmann Div, Xertex Corp, 82-86; SR SCIENTIST, DOHRMANN DIV, ROSEMOUNT ANALYSIS CORP, 86- *Concurrent Pos:* Consult, Air Force Missile Develop Ctr, Holloman AFB, 58-61 & US Naval Radiol Defense Lab, 64-65. *Mem:* AAAS; Am Phys Soc; Am Asn Physics Teachers; Comput Soc; Assn Comput Mach. *Res:* Design chemical analytical instruments. *Mailing Add:* 290 Stilt Ct Foster City CA 94404

HOTZ, PRESTON ENSLOW, b Sonoma, Calif, Mar 24, 13; m 40; c 3. GEOLOGY. *Educ:* Univ Calif, Berkeley, AB, 37, MA, 40; Princeton Univ, PhD, 49. *Prof Exp:* Recorder & mem field party, US Geol Surv, Ore, 38-41, geol field asst & jr geologist, Ore, 41, from jr geologist to geologist, Calif, NJ & Pa, 42-49, Washington, DC, 50 & Nev, 51-57, area supvr, Mineral Deposits Br, 58-59, geologist, 60-77; RETIRED. *Mem:* Fel Geol Soc Am. *Res:* Petrology of Triassic diabases of Pennsylvania; magnetite deposits of the Dillsburg district, Pennsylvania, and of New Jersey; structure and stratigraphy of northern Nevada; gold and nickel deposits of northern California and southwest Oregon; geology of northern Klamath Mountains, California. *Mailing Add:* 209 Blackburn Ave Menlo Park CA 94025

HOU, CHING-TSANG, b Taiwan, China, June 26, 35; m 61; c 3. BIOCHEMISTRY. *Educ:* Nat Taiwan Univ, BS, 58; Univ Tokyo, MS, 64, PhD(biochem), 67. *Prof Exp:* Chemist, Taiwan Sugar Coop, 60-62; res assoc biochem, Sch Pharm, Univ Wis-Madison, 68-69; microbiologist, Northern Regional Res Lab, USDA, 69-71; sr res biochemist, 71-79, RES ASSOC, EXXON RES & ENG CO, 79- *Mem:* Am Chem Soc; Am Soc Microbiol. *Res:* Biosynthesis of vitamin A; enzymatic synthesis of steryl glucoside; purification and properties of antibiotic lactonases; mycotoxins; oxygenases; microbial conversion of petroleum compounds; microbiology, biochemistry, and genetics of gaseous-hydrocarbon-utilizing microorganisms. *Mailing Add:* 705 W Deer Brook Dr Peoria IL 61615

HOU, GENE JEAN-WIN, b Yunlin, Taiwan, Feb 2, 52; US citizen; m 80; c 3. ENGINEERING DESIGN OPTIMIZATION, COMPUTATIONAL MECHANICS. *Educ:* Nat Cheng Kung Univ, Taiwan, BS, 74; Nat Taiwan Univ, MS, 76; Univ Iowa, PhD(mech eng), 83. *Prof Exp:* Asst prof, 83-90, ASSOC PROF MECH ENG, OLD DOMINION UNIV, 89- *Concurrent Pos:* Prin investr, NSF, NASA & local indust, 83-91; NSF presidential young investr, 87; liaison officer, Comt Design Theory & Methodology, Am Soc Mech Engrs, 88-89. *Honors & Awards:* Ralph R Teeter Educ Award, Eng Soc Advancing Mobility Land Sea Air & Space, 88. *Mem:* Am Soc Mech Engrs; Am Inst Aeronaut & Astronaut; Eng Soc Advancing Mobility Land Sea Air & Space. *Res:* Analysis tools supporting the engineering design machines; structural reanalysis techniques and design sensitivity analysis and optimization methodologies for multidisciplinary applications. *Mailing Add:* Mech Eng & Mech Off KDH Rm 241 Old Dominion Univ Norfolk VA 23529

HOU, KENNETH C, b Kiangsui, China, Apr 22, 29; m 66; c 2. RESEARCH & DEVELOPMENT ADMINISTRATION. *Educ:* Taiwan Univ, BS, 52; Univ Idaho, MS, 57; Univ Tex, PhD(phys chem), 62. *Prof Exp:* Res chemist, Celanese Res Co, 66-72; res engr, CUNO Div, AMF, 72-79, dir res, 79-85, VPRES RES & DEVELOP, CUNO INC, 85- *Mem:* Am Chem Soc. *Res:* Bioseparation and purification in large scale bioprocessing engineering; affinity chromatography; membrane filtration. *Mailing Add:* CUNO Inc 400 Research Pkwy Meriden CT 06450

HOU, ROGER HSIANG-DAH, b Changsha, China, Jan 6, 36; US citizen. ALGEBRA. *Educ:* Ind Univ, PhD(algebra), 65. *Prof Exp:* Asst prof math, Univ NH, 65-69; assoc prof math, Slippery Rock State Col, 69-85; RETIRED. *Concurrent Pos:* NSF res grant, 66-68. *Mem:* Am Math Soc; Math Asn Am. *Res:* Homological algebra. *Mailing Add:* 114 Williams Rd Butler PA 16001

HOUBOLT, JOHN C(ORNELIUS), b Altoona, Iowa, Apr 10, 19; m 49; c 3. CIVIL & AERONAUTICAL ENGINEERING. *Educ:* Univ Ill, BS, 40, MS, 42; Swiss Fed Inst Technol, PhD(tech sci), 58. *Hon Degrees:* Dr, Swiss Fed Inst Technol, 75. *Prof Exp:* Bridge engr, Ill Cent RR, 40; asst, Univ Ill, 40-42; aeronaut res engr, Langley Res Ctr, NASA, 42-48, exchange scientist, Eng, 49, asst chief, Dynamic Loads Div, 49-62, chief, Theoret Mech Div, 62-63; exec vpres & sr consult, Aeronaut Res Assocs, Princeton, 63-67, sr vpres & sr consult, 67-69; dir, Doweave, Inc, 69-73; CHIEF SCIENTIST, LANGLEY RES CTR, NASA, 73- *Concurrent Pos:* Jr city engr, Waukegan, Ill, 41; instr, Univ Va, 43-; mem div adv group, Air Force Sci Adv Bd; Dryden res lectr award, 71; consult, Air Force, Navy, Army & var com firms; assoc ed, Am Inst Aeronaut & Astronaut J Spacecraft & Rockets. *Honors & Awards:* Except Sci Achievement Award, NASA, 64; Am Inst Aeronaut & Astronaut, Award, 67. *Mem:* Nat Acad Eng; hon fel Am Inst Aeronaut & Astronaut. *Res:* Aeroelastic and structural problems of earthbound and space flight vehicles. *Mailing Add:* 51 Winster Fax Williamsburg VA 23185

HOUCHENS, DAVID PAUL, b Louisville, Ky, Jan 26, 37; m 63; c 3. IMMUNOLOGY, CHEMOTHERAPY. *Educ:* Stetson Univ, BS, 59; George Washington Univ, MS, 64, PhD(microbiol), 71. *Prof Exp:* Bacteriologist, Ft Detrick, Md, 60-62; supv technician immunol, Microbiol Assoc, Bethesda, 62-68; jr prof, Bionetics Res Labs, Kensington, 68-71; sr staff fel immunochemother, Nat Cancer Inst, 71-76; res immunologist, Battelle Mem Inst, 76-79, assoc sect mgr, 79-84, proj mgr, Columbus Div, 84-90; DIR, LAB & CLIN SCI, NEOPROBE CORP, 90- *Concurrent Pos:* Fel microbiol, George Washington Univ, 68-71; instr, Montgomery Col, 71-73; adj assoc prof, Ohio State Univ, 85- *Mem:* Am Asn Cancer Res; Am Soc Microbiol; Am Asn Immunologists; Drug Info Asn. *Res:* Immunotherapy and combine modality therapy of animal and human tumors; immunologic effects of chemotherapy; radioimmunologates for radioimmunodetection and radioimmunotherapy. *Mailing Add:* Neoprobe Corp 855 W Fifth Ave Columbus OH 43212

HOUCK, DAVID R, b Detroit, Mich, Apr 1, 56. BIOCHEMISTRY, ENDOCRINOLOGY. *Educ:* Alma Col, BS, 78; Purdue Univ, MS, 81; Ohio State Univ, PhD(chem), 86. *Prof Exp:* Biochemist, Dept Fermentation Microbiol, Merck & Co, Inc, 81-84, res biochemist, 84-86, sr biochemist, 86-87; postdoctoral fel isotope & struct chem group, Los Alamos Nat Lab, 87-89; SR RES INVESTR, DEPT NATURAL PROD, STERLING DRUG, INC, 89- *Mem:* Am Chem Soc; Am Soc Pharmacog; Am Soc Biochem & Molecular Biol. *Res:* Bioorganic chemistry; enzyme mechanisms; metabolism; nuclear magnetic resonance; author of 16 technical publications. *Mailing Add:* Dept Natural Prod Sterling Res Group 81 Columbia Turnpike Rensselaer NY 12144

HOUCK, FRANK SCANLAND, b Philadelphia, Pa, Aug 27, 30; m 55; c 3. INTERNATIONAL NUCLEAR SAFEGUARDS. *Educ:* Dickinson Col, BS, 52; Columbia Univ, MS, 57, PhD(chem), 59. *Prof Exp:* Opers analyst, Opers Eval Group, US Navy, 57-61 & Ctr for Naval Anal, Franklin Inst, 61-63; oper res officer, 63-72, phys sci officer, 72-81, SR SCIENTIST INT SAFEGUARDS, US ARMS CONTROL & DISARMAMENT AGENCY, 81- *Concurrent Pos:* Guest researcher, Mass Inst Technol, 60-62; standing adv group safeguards implementation, Int Atomic Energy Asn, 80- *Honors & Awards:* Meritorious Honor Awards, US Arms Control & Disarmament Agency, 67 & 80. *Res:* Operations research in naval and nuclear warfare; research and testing of inspection and verification for arms control; nuclear materials safeguards. *Mailing Add:* 320 21st St NW US AC & D Agency Washington DC 20451

HOUCK, JAMES RICHARD, b Mobile, Ala, Oct 5, 40; m 64; c 2. ASTROPHYSICS. *Educ:* Carnegie Inst Technol, BS, 62; Cornell Univ, PhD(physics), 67. *Prof Exp:* Res assoc astron, 67-69, asst prof, 69-74, assoc prof, 74-79, PROF ASTRON, CORNELL UNIV, 79- *Concurrent Pos:* Mem, Infrared Panel, Nat Acad Sci, 69-71, 75 & infrared satellite team, NASA, 75-; Sloan res fel, 71-75; mem, Bahcall Comt, Am Asn State Climatologists, 90. *Mem:* Int Astron Union; Am Astron Soc; AAAS. *Res:* Infrared studies of H II regions and ultraluminous galaxies. *Mailing Add:* 220 Space Sci Bldg Cornell Univ Ithaca NY 14853

HOUCK, JOHN CANDEE, b New York, NY, Feb 19, 31; m 53; c 6. IMMUNOBIOLOGY, CELL BIOLOGY. *Educ:* Columbia Univ, BA, 53; Univ Western Ont, MSc, 55, PhD(path chem), 56. *Prof Exp:* Dir, Surg Res Lab, Sch Med, Georgetown Univ, 58-60; dir biochem res lab, Children's Hosp, 59-76, sci dir res found, 69-76; DIR, VIRGINIA MASON RES CTR, 76- *Concurrent Pos:* Sr res fel surg res lab, Sch Med, Georgetown Univ, 57-58, from asst prof to prof lectr dept biochem, 58-76, asst prof dept pediat &

biochem, 67-69, prof pediat, 69-71, prof child health & develop, 71-76. *Honors & Awards:* A Cressy Mem Award, NY Acad Sci, 62; Am Dermat Asn Award, 65. *Mem:* AAAS; Am Soc Biol Chemists; Soc Exp Biol & Med; Am Rheumatism Asn; Am Soc Exp Path. *Res:* Biochemistry of inflammation; biochemical control of cell growth; chalones. *Mailing Add:* 5123 26th Ave Seattle WA 98105

HOUCK, LAURIE GERALD, b Tucson, Ariz, Aug 13, 28; m 58; c 2. PLANT PATHOLOGY, HORTICULTURE. *Educ:* Univ Ariz, BS, 52, MS, 54; Ore State Univ, PhD(plant path), 62. *Prof Exp:* Plant physiologist, Agr Res Serv, USDA, 54-55; asst horticulturist, Univ Ariz, 55-57; res asst plant path, Ore State Univ, 57-61; res plant pathologist, Mkt Qual Res, Riverside, 62-78; RES PLANT PATHOLOGIST, HORT CROPS RES STA, USDA, 78- *Concurrent Pos:* Instr, Calif State Polytech Univ, 63, lectr, 64 & 66. *Mem:* Am Phytopath Soc; Am Soc Hort Sci; Mycol Soc Am; Coun Agr Sci & Technol; Food Distrib Res Soc; Sigma Xi. *Res:* Post-harvest storing; chemical and nonchemical disease control; shipping, both domestic and export; marketing problems of citrus, avocados, dates and other subtropical fruits; herbicides for weed control; effect of maleic hydrazide upon respiration of roots; root rots and vascular diseases of strawberries and other small fruits of the Pacific Northwest. *Mailing Add:* Hort Crops Res Sta Agr Res Serv USDA 2021 S Peach Ave Fresno CA 93727

HOUCK, MARK HEDRICH, b Baltimore, Md, May 14, 51; m 72; c 3. SYSTEMS ENGINEERING, WATER RESOURCES ENGINEERING. *Educ:* Johns Hopkins Univ, BES, 72, PhD(eng), 76. *Prof Exp:* Res asst prof civil eng, Univ Wash, 75-77; from asst prof to assoc prof, 77-87, PROF CIVIL ENG, PURDUE UNIV, 87- *Concurrent Pos:* Assoc ed, Water Resources Res, 81-85; pres, Omtek Eng Inc, 84-; chmn, Water Resources Systs Comt, Am Soc Civil Engrs, 84; chmn, Expert Systs & Water Resources Comt, Am Soc Civil Engrs, 86-, Applns Emerging Technol Comt, 87-; vpres, Water Resources Mgt Inc, 88-89; Dr, Johns Hopkins Univ, 89-90. *Honors & Awards:* Huber Res Prize, Am Soc Civil Engrs, 88. *Mem:* Am Geophys Union; Am Soc Civil Engrs; Opers Res Soc Am; Sigma Xi; Inst Mgt Sci. *Res:* Engineering systems analysis with a focus on civil, water resources and environmental engineering, management, planning and design; operations research; civil engineering. *Mailing Add:* Sch Civil Eng Purdue Univ West Lafayette IN 47907

HOUDE, EDWARD DONALD, b Attleboro, Mass, Sept 4, 41; m 83; c 1. FISHERIES, BIOLOGICAL OCEANOGRAPHY. *Educ:* Univ Mass, BA, 63; Cornell Univ, MS, 65, PhD(fishery sci), 68. *Prof Exp:* Res biologist fishery sci, US Bur Com Fisheries, Fla, 68-70; res scientist, Div Biol & Living Resources, 70-71, asst prof fishery sci, Rosentiel Sch Marine & Atmospheric Sci, Univ Miami, 78-80; PROF, CHESAPEAKE BIOL LAB, UNIV MD, 80- *Concurrent Pos:* Prog dir, Biol Oceanog, NSF, 83-85. *Mem:* Am Fisheries Soc; Am Soc Limnol & Oceanog; AAAS; Oceanog Soc; Estuarine Res Fed. *Res:* Ecology and developmental biology of fish eggs and larvae; assessment of pelagic resources abundance; factors affecting distribution and abundance of eggs and larvae; feeding and energetics of fish larvae; experimental rearing of larval fishes in the laboratory. *Mailing Add:* Univ Md PO Box 38 Solomons MD 20688

HOUDE, RAYMOND WILFRED, b Claremont, NH, May 11, 16; m 50; c 2. PHARMACOLOGY, THERAPEUTICS. *Educ:* NY Univ, AB, 40, MD, 43. *Prof Exp:* Intern, Bellevue Hosp, 44; asst resident med, Mem Hosp, 47-48; from instr to asst prof, 50-65, ASSOC PROF MED, MED COL, CORNELL UNIV, 65-, PROF PHARMACOL, 80-; ASSOC MEM, SLOAN-KETTERING INST CANCER RES, 60- *Concurrent Pos:* Vis fel, Univ Mich, 48-49; res fel clin invest, Mem Sloan-Kettering Cancer Ctr, 48-50, asst attend physician, 55-65, attend physician, 65-; Nat Res Coun fel, Nat Cancer Inst, 49-51; mem res unit, USPHS Hosp, Ky, 49-50; asst attend physician, Med Serv, Mem Hosp, 50-55; asst vis physician, James Ewing Hosp, 50-55; asst, Sloan-Kettering Inst Cancer Res, 50-55, assoc, 55-60; assoc prof pharmacol, Cornell Univ, 67-80. *Mem:* AAAS; Am Col Neuropsychopharmacol; Am Soc Clin Pharmacol & Therapeut; Am Soc Pharmacol & Exp Therapeut; NY Acad Sci. *Res:* Clinical investigation of analgesic agents; psychophysiology of pain. *Mailing Add:* Sloan-Kettering Cancer Ctr 1275 York Ave New York NY 10021

HOUDE, ROBERT A, b Fall River, Mass, Apr 12, 31; m 57; c 2. COMMUNICATIONS SCIENCE. *Educ:* Northeastern Univ, BS, 54; Univ Rochester, MS, 65; Univ Mich, PhD(commun sci), 67. *Prof Exp:* Res staff mem, Gen Dynamics/Electronics, 56-68, mgr res dept, 68-70; PRES & DIR RES, CTR COMMUN RES INC, 70- *Mem:* Acoustical Soc Am; Sigma Xi; Am Speech & Hearing Asn; NY Acad Sci. *Res:* Speech acoustics. *Mailing Add:* Ctr for Commun Res 1895 Mount Hope Ave Rochester NY 14620

HOUGAS, ROBERT WAYNE, b Blythedale, Mo, June 17, 18; m 43. PLANT BREEDING, GENETICS. *Educ:* Univ Wis, BS, 42, PhD(genetics), 49. *Prof Exp:* From asst prof to assoc prof, 49-62, asst dir exp sta, 62-65, asst dean & dir col agr, 65-66, PROF GENETICS, UNIV WIS-MADISON, 62-, LEADER INTER-REGIONAL POTATO INTROD PROJ, 50-, ASSOC DEAN & DIR COL AGR, 66- *Mem:* Genetics Soc Am; Am Phytopath Soc; Potato Asn Am (secy, 52-55, vpres, 56, pres, 57); Am Genetic Asn. *Res:* Interspecific hybridization and genetics studies of tuber-bearing Solanum species; haploidy in Solanum tuberosum. *Mailing Add:* Hort Dept Univ Wis Linden Dr Madison WI 53706

HOUGEN, FRITHJOF W, b Oslo, Norway, Dec 22, 20. PLANT CHEMISTRY. *Educ:* Norway Inst Technol, Chem Engr, 48; Univ Cape Town, PhD(org chem), 55. *Prof Exp:* Sci asst, Nat Chem Res Lab, Coun Sci & Indust Res SAfrica, 48-49, asst res officer, 50-53, res officer, 54-55; fel, Prairie Regional Lab, Nat Res Coun Can, 56-58; res chemist, Univ Man, 58-67, assoc prof, 67-77, prof plant sci, 77-86; RETIRED. *Mem:* Chem Inst Can; Am Oil Chem Soc; Am Chem Soc; Sigma Xi. *Res:* Chemical composition of fats and lipids; development of chemical analytical methods for use in plant breeding research. *Mailing Add:* Dept Plant Sci Agr Blvd Univ Man Winnipeg MB R3T 2N2 Can

HOUGEN, JOEL O(LIVER), b Tacoma, Wash, Feb 26, 14; m 38; c 4. CHEMICAL ENGINEERING. *Educ:* Univ Wis, BS, 36; Univ Minn, MS, 46, PhD(chem eng), 48. *Prof Exp:* Res chem engr, Pan Am Ref Corp, Tex, 37-41; instr chem eng, Univ Minn, 41-44; process engr, Union Oil Co, Calif, 44-46; instr chem eng, Univ Ill, 46-48; from assoc prof to prof, Rensselaer Polytech Inst, 48-56; systs engr, Monsanto Chem Co, 56-66 & Monsanto Co, 66-67; independent consult process control & optimization, 67; vis prof chem, 67, prof, 67-80, EMER PROF CHEM, UNIV TEX, AUSTIN, 80- *Concurrent Pos:* Consult, Gen Elec Co, 49-56, Boeing Airplane Co, 54 & Amoco Chem Co, 76; mem sci adv bd, Environ Protection Agency, 76-78. *Honors & Awards:* Instrument Technol Award, Instrument Soc Am, 70. *Mem:* Am Soc Eng Educ; Am Inst Chem Engrs; Instrument Soc Am. *Res:* Process dynamics and control. *Mailing Add:* 1206 Falcon Ledge Dr Austin TX 78746

HOUGEN, JON T, b Sheboygan, Wis, Oct 23, 36. MOLECULAR SPECTROSCOPY. *Educ:* Univ Wis, BSc, 56; Harvard Univ, AM, 58, PhD(phys chem), 60. *Prof Exp:* Fel, Nat Res Coun Can, 60-62, assoc res officer, 62-66; physicist, 67-69, chief molecular spectros sect, 69-73, res scientist, 74-85, SR FEL, NAT INST STANDARDS & TECHNOL, 85- *Honors & Awards:* Coblentz Award, 68; Silver Medal, Nat Bur Standards, 74, Gold Medal, 80; Plyler Prize, 84; Lippincott Award, 84. *Mem:* Am Phys Soc; Optical Soc Am; Coblentz Soc. *Res:* Quantum mechanical and group theoretical problems in high-resolution molecular spectroscopy. *Mailing Add:* Molecular Physics Div Nat Inst Standards & Technol Gaithersburg MD 20899

HOUGH, ELDRED W, petroleum & chemical engineering; deceased, see previous edition for last biography

HOUGH, HUGH WALTER, b Hebron, Ind, Feb 10, 22; m 47; c 3. SOIL SCIENCE. *Educ:* Purdue Univ, BSA, 47, MS, 50; Mich State Univ, PhD(soil sci), 54. *Prof Exp:* Asst soils, Purdue Univ, 47-49; instr, Mich State Univ, 49-55; asst prof, 56-66, ASSOC PROF SOILS, UNIV WYO, 66- *Mem:* Soil Sci Soc Am; Am Soc Agron. *Res:* Soil structure; crops response to fertilizers; soil testing; effects of soil amendments. *Mailing Add:* 1703 E Ord St Laramie WY 82070

HOUGH, JANE LINSCOTT, LIPID BIOCHEMISTRY. *Educ:* Middlebury Col, BA, 72; Univ NH, PhD(biochem), 81. *Prof Exp:* fel lipid biochem, Div Nutrit Sci, Cornell Univ, 81-83; lectr chem, Colgate Univ, 83-85; res assoc, 85-87, RES SCIENTIST, MASONIC MED RES LAB, 88- *Mem:* Am Chem Soc; AAAS; NY Acad Sci; Sigma Xi. *Res:* Involvement of membrane dynamics in aging and disease development, in particular cancer and atherosclerosis; mechanism of enhancement of atherosclerosis by carcinogens. *Mailing Add:* RD 1 Box 61A Earlville NY 13332

HOUGH, LINDSAY B, b Paris, Ky, Oct 16, 50; m 79; c 2. NEUROPHARMACOLOGY, NEUROCHEMISTRY. *Educ:* Univ Ky, Lexington, BS, 73; Univ Mich, Ann Arbor, PhD(pharmacol), 78. *Prof Exp:* Postdoctoral fel pharmacol, Mt Sinai Med Sch, NY, 78-79, from asst prof to assoc prof, 80-84; assoc prof, 84-86, PROF PHARMACOL, ALBANY MED COL, 86- *Concurrent Pos:* Prin investr, Nat Inst Drug Abuse, 82- *Mem:* Am Soc Pharmacol; Soc Neurosci. *Res:* Exploration of the neurochemistry and neuropharmacology of histamine in the central nervous system. *Mailing Add:* Dept Pharmacol & Toxicol Albany Med Col Albany NY 12208

HOUGH, PAUL VAN CAMPEN, b Ellwood City, Pa, May 21, 25; m 45; c 2. MOLECULAR BIOLOGY, ELECTRON MICROSCOPY. *Educ:* Swarthmore Col, BA, 45; Cornell Univ, PhD(physics), 50. *Prof Exp:* Jr scientist electronics, Los Alamos Sci Lab, 45; from instr to assoc prof physics, Univ Mich, 49-60; from physicist to sr physicist, 61-74, SR BIOLOGIST, BROOKHAVEN NAT LAB 74- *Concurrent Pos:* Guggenheim fel, 59-60 & 73-74. *Mem:* AAAS; fel Am Phys Soc. *Res:* Molecular biology of transcription and replication in eukaryotes; macromolecular identification in electron microscopy. *Mailing Add:* Dept Biol Brookhaven Nat Lab Upton NY 11973

HOUGH, RALPH L, b Springfield, Ohio, Jan 13, 30; m 53; c 3. MATERIALS SCIENCE. *Educ:* Wittenberg Univ, BA, 54. *Prof Exp:* Engr, Equip Lab, Wright-Patterson AFB, 55-59, engr, Air Force Mat Lab, 59-63, res team leader reinforcements formation, 63-66; PRES, HOUGH LAB, RUSSELLS POINT, 66- *Concurrent Pos:* Consult, Space Gen Corp, Los Angeles, 66-67, Dow-Corning Corp, Midland & Univ Cincinnati, 67, Res Found, Ohio State Univ, 67-70 & Res Inst, Univ Dayton, 69; consult engr, Specialties Develop Corp, Hilliard, Ohio, 78- *Mem:* Sigma Xi; AAAS; Am Chem Soc; fel Am Inst Chem; Soc Aerospace Mat & Process Engrs. *Res:* Developing diffusion barriers for boron and silicon carbide filaments; development of large diameter carbonbased fibers; new coatings to enhance the strength of boron and other fibers currently being used as reinforcements; design, develop and install special halon fire suppression system for Air Force Avionics Lab Anachoic Chamber; design and developing microprosser controlled test stands for industrial materials industries; designing integrated computer systems for office or terminal use. *Mailing Add:* Hough Lab Box 81 Russells Point OH 43348-0081

HOUGH, RICHARD ANTON, b Houston, Tex, Sept 4, 42; m 69; c 2. LIMNOLOGY. *Educ:* Univ Ill, BS, 64; Univ Mich, MS, 66; Mich State Univ, PhD(bot), 73. *Prof Exp:* Oceanogr marine biol, US Naval Oceanogr Off, 66-69; asst prof, 73-80, assoc prof, 80-90, PROF BIOL, WAYNE STATE UNIV, 90-, ASSOC CHAIR, 90- *Mem:* Am Soc Limnol & Oceanogr; Ecol Soc Am; Int Soc Aquatic Vascular Plant Biologists; Int Soc Theoret & Appl Limnol; Sigma Xi. *Res:* Physiological ecology and productivity of aquatic macrophytes and algae. *Mailing Add:* Dept Biol Sci Wayne State Univ Detroit MI 48202

HOUGH, RICHARD R(ALSTON), b Trenton, NJ, Dec 13, 17; m 41; c 6. ELECTRICAL ENGINEERING. *Educ:* Princeton Univ, BSE, 39, EE, 40. *Hon Degrees:* DSc, Susquehanna Univ. *Prof Exp:* Instr elec eng, Princeton Univ, 39-40; mem tech staff, Bell Tel Labs, 40-51 & mil develop eng, 51-53, dir mil systs develop, 53-55 & mil electronics develop, 55-57, vpres, 57; asst chief engr, AT&T, 57-59; vpres opers, Ohio Bell Tell Co, 59-61; vpres eng, AT&T, 61-66, pres, AT&T-Long Lines, 61-78, exec vpres, AT&T, 78-82; RETIRED. *Concurrent Pos:* Mem tech adv panel electronics, Dept Defense, 53-57; dir, Am Can Co, Bell Tel Co, Can, 63-66 & Mt States Tel & Tel Co, 64-66; chmn bd, Bellcom, Inc, 62-66 & Bell Tel Labs, Inc, 65-66; civilian mem radar panel, Res & Develop Bd, 47-53; charter trustee, Princeton Univ, 60-88; dir, Commun Satellite Corp & Allegheny Corp; trustee, Wilson Col; mem adv bd, US Naval Postgrad Sch; defense sci bd, Dept Defense; dir, Midlantic Corp, Midlantic Nat Bank, Cyclops Indust. *Honors & Awards:* Alexander Graham Bell Medal, Inst Elect & Electronics Engrs. *Mem:* Nat Acad Eng; fel Inst Elec & Electronics Engrs; Sigma Xi. *Res:* Communications engineering. *Mailing Add:* Five Tenney Hill Rd Dunbarton NH 03301

HOUGH, WALTER ANDREW, b Philadelphia, Pa, May 3, 32; m 61; c 1. RESEARCH ADMINISTRATION, FORESTRY. *Educ:* Pa State Univ, BS, 55; Duke Univ, MF, 60, DF, 63. *Prof Exp:* Forester, US Forest Serv, 55-56, res forester, Southern Forest Fire Lab, 58-61; res assoc, Duke Univ, 63; res forester, Southern Forest Fire Lab, 63-67, proj leader fuel physics, 67-73, team leader mechanism of smoke generation, 73-75, staff specialist fire sci, Div Forest Fire & Atmospheric Sci Res, US Forest Serv, Washington, DC, 75-78, asst dir planning & appln, Southeastern Forest Exp Sta, Asheville, NC, 78-81, ASST DIR RES PROGS, SOUTHERN FOREST EXP STA, US FOREST SERV, NEW ORLEANS, LA, 81- *Mem:* Soc Am Foresters; Am Soc Plant Physiologists; Sigma Xi. *Res:* Use of prescribed fire in the forest; effect of forest fire on living vegetation; heat content and rate of energy release for forest fuels; influence of forest fuels on production and quality of forest fire smoke. *Mailing Add:* US Forest Serv Southern Stn 701 Loyola Ave New Orleans LA 70113

HOUGH-GOLDSTEIN, JUDITH ANNE, b Ann Arbor, Mich, Aug 27, 50; m 83; c 2. ENTOMOLOGY. *Educ:* Harvard Univ, BA, 72; Cornell Univ, MS, 77, PhD(entom), 81. *Prof Exp:* asst prof, 81-88, ASSOC PROF ENTOM, UNIV DEL, 88- *Mem:* Entom Soc Am; AAAS. *Res:* Insect/plant interactions; insect pest management. *Mailing Add:* Dept Entom & Applied Ecol Univ Del Newark DE 19717-1303

HOUGHTON, ALAN N, b Boston, Mass, Apr 12, 47; m 75; c 2. MEDICAL ONCOLOGY, TUMOR IMMUNOLOGY. *Educ:* Stanford Univ, BA, 70; Univ Conn, MD, 74. *Prof Exp:* ASSOC PROF MED & IMMUNOL, CORNELL UNIV MED CO, 88-; ASSOC MEM MED & IMMUNOL, MEM SLOAN-KETTERING CANCER CTR, 88-, CHIEF CLIN IMMUNOL SERV, 89- *Concurrent Pos:* Assoc/adv ed, Hybridoma, 86-, J Exp Med, 88- & Cancer Immunol Immunother, 90-; mem bd sci adv, Sterling Res Group, Eastman-Kodak, 87-; mem sci adv bd, Am Cancer Soc, 89-; mem exp ther study sect, Nat Cancer Inst, NIH, 90-; mem bd dirs, Soc Biol Ther, 90- *Honors & Awards:* Boyer Award, Mem Sloan-Kettering Cancer Ctr, 85. *Mem:* Am Soc Clin Invest; Soc Biol Ther; Am Asn Cancer Res; Am Soc Clin Oncol; Am Asn Immunologists. *Res:* Experimental cancer therapy with special emphasis on biologic treatments, including cytokines, monoclonal antibodies and cancer vaccines; pathogenesis and treatment of human melanoma; pigment cell biology; human tumor antigens. *Mailing Add:* Dept Immunol Mem Sloan-Kettering Cancer Ctr 1275 York Ave New York NY 10021

HOUGHTON, ARTHUR VINCENT, III, b Jacksonville, July 9, 26; wid; c 4. THERMODYNAMICS, COMBUSTION. *Educ:* Univ Ill, BS, 48, MS, 51; Purdue Univ, PhD(mech eng), 60. *Prof Exp:* Instr thermodyn & heat transfer, Univ Ark, 48-50; asst prof, Bradley Univ, 50-55; instr, Purdue Univ, 55-60; assoc prof thermodyn, 60-66, asst to dean grad sch, 67-68, asst dean res grad sch, 68-71, dir, Eric H Wang Civil Eng Res Facility, 71-72, PROF MECH ENG, UNIV NMEX, 66- *Concurrent Pos:* Engr, Caterpillar Tractor, 51 & Letoursneau, Westinghouse, 55; erector, Babcock-Wilcox Co, 53, field serv, 54; res engr, Schwitzer Corp, 56-57; consult, Air Force Off Sci Res, 60-63, Am Car & Foundry, ACF Industs, 61-63, Sandia Corp, 63-, Legal cases, 64-, Continental Casualty Co, 65-71, Mech Res Inst, 70-74 & Inst Cent Am Invests Technol Indust, 81-, Air Force Off Sci Res. *Mem:* Am Soc Mech Engrs; Sigma Xi. *Res:* Combustion; heat transfer; computers; research administration. *Mailing Add:* Star Rte 188 Tijeras NM 87059

HOUGHTON, CHARLES JOSEPH, b Castle Gate, Utah, June 26, 32; m 61; c 3. TOPOLOGY. *Educ:* Univ Utah, BA, 58; Ohio State Univ, MA, 61, PhD(math), 64. *Prof Exp:* Ballistics computer, Sign Missile Support Agency, White Sands Missile Range, NMex, 58-59; asst prof, 64-69, ASSOC PROF MATH, STATE UNIV NY BINGHAMTON, 69- *Mem:* Math Asn Am. *Mailing Add:* Dept Math State Univ NY Binghamton NY 13902-6000

HOUGHTON, DAVID DREW, b Philadelphia, Pa, Apr 26, 38; m 63; c 3. METEOROLOGY. *Educ:* Pa State Univ, BS, 59; Wash Univ, MS, 61, PhD(atmospheric sci), 63. *Prof Exp:* Res scientist dynamic meteorol, Nat Ctr Atmospheric Res, 63-68; from asst prof to assoc prof, 68-72, assoc chmn grad affairs, 70-72, chmn dept, 76-79, PROF METEOROL, UNIV WIS-MADISON, 72- *Concurrent Pos:* Prin investr & co-prin investr, NSF Grants; exchange scientist, Inst Atmospheric Physics, Moscow, USSR, 66; vis scientist, Courant Inst Math Sci, New York, 66; invited lectr, Nanjing Univ, People's Repub China, 80; vis sr scientist, Nat Meteorol Ctr, Washington, DC, 88 & Inst Atmospheric Physics, Beijing & Nanjing Univ, Nanjing, Peoples Repub China, 89. *Mem:* AAAS; fel Am Meteorol Soc. *Res:* General circulation; large-scale atmospheric dynamics; subsynoptic scale motions; initialization for numerical models; flow over mountains; gravity waves; short range prediction; large scale annual cycle dynamics; satellite meteorological data. *Mailing Add:* Dept Meteorol Univ Wis Madison WI 53706

HOUGHTON, JAMES RICHARD, b Morgantown, WVa, Aug 31, 36; m 61; c 2. MECHANICAL ENGINEERING, APPLIED MECHANICS. *Educ:* George Washington Univ, BME, 58; Univ Md, MSME, 65; Vanderbilt Univ, PhD(mech eng), 76. *Prof Exp:* Designer & supvr construct, Friends Africa Mission, 58-60; proj leader mech eng, Nat Bur Standards, 61-66; lectr mech eng, Univ Sierra Leone, 66-68 & Univ Nairobi, Kenya, 68-72; asst prof, Tenn State Univ, 72-77; PROF MECH ENG, TENN TECHNOL UNIV, 77- *Concurrent Pos:* Contractor, NASA, 73-77; consult, Energy Systs Div, Martin Marietta Corp, 74-; mem acoust emission working group; prod liabilities consult, TVA. *Mem:* Am Soc Mech Engrs; Instrument Soc Am. *Res:* Shock, vibrations, nondestructive testing and acoustic emission; signature analysis by mathematical model development for a system transfer function; computer aided machine design; modal analysis. *Mailing Add:* Dept Mech Eng Tenn Technol Univ Cookeville TN 38505

HOUGHTON, JANET ANNE, b Grantham, Eng, May 21, 52. PHARMACOLOGY. *Educ:* Univ Bradford, Eng, BPh Hons, 73; Univ London, PhD(biophys), 77. *Prof Exp:* Postdoctoral fel, Dept Biochem & Clin Pharmacol, St Jude Children's Res Hosp, 77-80, res assoc, 80-82, asst mem, 82-85, ASSOC MEM, DEPT BIOCHEM & CLIN PHARMACOL, ST JUDE CHILDREN'S RES HOSP, MEMPHIS, 85- *Mem:* Am Asn Cancer Res; Pharmaceut Soc Gt Brit. *Res:* Tumor models; human tumor xenografts; experimental chemotherapy; selectivity of drug action in vivo, drug resistance; 5-fluoropyrimidines; Vinca alkaloids; colon carcinoma and rhabdomyosarcoma as specific human diseases; tumor heterogeneity; folate metabolism; heterogeneity of thymidylate synthase; author of more than 120 technical publications. *Mailing Add:* Dept Biochem & Clin Pharmacol St Jude Children's Res Hosp 332 N Lauderdale Memphis TN 38101

HOUGHTON, JOHN M, b Knobnoster, Mo, July 18, 41; m 66; c 1. BIOTECHNOLOGY DEVELOPMENT & ASSESSMENT, NEW PRODUCT COMMERCIALIZATION. *Educ:* Southern Ill Univ, BS, 67; Univ Ill, MS, 70, PhD(agron), 73. *Prof Exp:* Sales rep, Ciba-Geigy Corp, 67-68; prod develop assoc & res group leader, Monsanto Agr Co, 73-78, mgr prod develop, Latin Am & US, 78-84, dir new prod, 84-88, dir new technol, 88-90; pres, J M Houghton & Assocs, 90-91; DIR, STEWART TECHNOL MGT, 91- *Concurrent Pos:* Partic, Comt Strategies Mgt Pesticides Resistant Pest Populations, 86. *Mem:* AAAS; Weed Sci Soc Am. *Mailing Add:* One Duddin Ct Ballwin MO 63021

HOUGHTON, KENNETH SINCLAIR, SR, b Hackensack, NJ, Aug 31, 40; m 63; c 4. INDOLE ALKALOIDS, TECHNICAL MANAGEMENT. *Educ:* St Peters Col, BS, 62; Seton Hall Univ, MS, 69, PhD(org chem), 72. *Prof Exp:* Chief chemist, Continental Chem Co, 66-68, plant mgr, 68-69; grad asst, Seton Hall Univ, 69-70; tech dir, Lightfoot Co, Philip Morris USA, 70-72, gen mgr, 72-74, mgr, Process Control & Qual Assurance, Park 500, 74-76 & Tech Serv, 76-77, gen mgr, Park 500, 77-80 & Richmond Processing Plants, 80-83, dir res & develop, Europe, Mid East, Africa, Philip Morris Int, 83-86, VPRES RES & DEVELOP, PHILIP MORRIS USA, 86- *Mem:* Am Chem Soc; Indust Res Inst; Sigma Xi. *Res:* Organic chemistry of natural products and flavors; extract and recombine processes; industrial waste water treatment; high-speed manufacturing techniques; tobacco chemistry and processing techniques; cigarette design, construction and product evaluation. *Mailing Add:* Philip Morris USA PO Box 26583 Richmond VA 23261

HOUGHTON, ROBERT W, b Philadelphia, Pa, Sept 4, 41; m 64; c 1. PHYSICAL OCEANOGRAPHY. *Educ:* Oberlin Col, Ohio, AB, 63; Univ Minn, PhD(physics), 68. *Prof Exp:* Res assoc physics, City Col, New York, 68-71; lectr, physics, statist & thermo, Univ Ghana, Legon, Accra, 71-75; science teacher, Friends Acad, NY, 75-76; res assoc, Dept Oceanog, Dalhousie Univ, 76-77; res scientist, 77-78, RES ASSOC, DEPT OCEANOG, LAMONT-DOHERTY GEOL OBSERV, COLUMBIA UNIV, 78- *Mem:* Am Phys Soc; Am Meteorol Soc; AAAS; Am Geophys Union. *Res:* Experimental physical oceanographic research on the thermal structure of the equatorial Atlantic Ocean; structure and exchange processes at the Middle Atlantic Bight coastal front. *Mailing Add:* Lamont-Doherty Geol Observ Rte 9W Palisades NY 10964

HOUGIE, CECIL, b Eng, Oct 29, 22; m 50; c 2. PATHOLOGY. *Educ:* Univ Manchester, MB, BS, 46. *Prof Exp:* Registr path, West London Hosp, 50-52, sr registr, 54-55; instr, Sch Med, Univ NC, 55-56; asst prof, Sch Med, Univ Va, 56-60; assoc prof, Sch Med, Univ Wash, 60-68; prof, 68-89, EMER PROF PATH, UNIV CALIF, SAN DIEGO, 89- *Concurrent Pos:* Mason res fel, Med Sch, West London Hosp, 52-54; established investr, Am Head Asn, 60-65. *Mem:* Soc Exp Biol & Med; Am Soc Exp Path. *Res:* Coagulation of blood and fibrinolytic enzymes. *Mailing Add:* Dept Path Univ Calif at San Diego La Jolla CA 92093

HOUGLAND, ARTHUR ELDON, b Omaha, Nebr, Jan 5, 35; m 61; c 3. MICROBIOLOGY, VIROLOGY. *Educ:* State Univ Iowa, BA, 58; Brigham Young Univ, MS, 61; Univ SDak, PhD(microbiol), 75. *Prof Exp:* Chief microbiol, Med Field Serv Sch, 62-64; res, Ft Detrick, 64-69; asst prof, 73-78, ASSOC PROF MICROBIOL, E TENN STATE UNIV, 79-, SPECIAL ASST TO DEAN, SCH PUB & ALLIED HEALTH, 90- *Concurrent Pos:* Am Cancer Soc fel, 69-71; Fraternal Order Eagles grant, 71-73; res consult, Oak Ridge Assoc Univ, 78- *Mem:* Am Soc Microbiol; Asn Mil Surgeons; Sigma Xi; Tissue Cult Asn. *Res:* Lipids of biological membranes; sulphydryl groups of cells and bacteria; cancer. *Mailing Add:* Health Sci ETenn State Univ PO Box 22690A Johnson City TN 37614

HOUH, CHORNG SHI, b Canton, China, Mar 23, 26; m 66; c 2. MATHEMATICS. *Educ:* Nat Taiwan Univ, BSc, 49; Tokyo Metrop Univ, PhD(math), 62. *Prof Exp:* Asst math, Nat Taiwan Univ, 49-56; from instr to asst prof, Univ Fla, 62-64; asst prof, Univ Man, 64-67; assoc prof, 67-75, PROF MATH, WAYNE STATE UNIV, 75- *Mem:* Am Math Soc; Math Soc Japan. *Res:* Connection theory of a Riemannian manifold; differential geometry on complex and almost complex spaces; differential geometry on submanifolds. *Mailing Add:* Dept Math 1150 FAB Wayne State Univ 5950 Cass Ave Detroit MI 48202

HOUK, ALBERT EDWARD HENNESSEE, b Glen Alpine, NC, May 20, 14; m 41; c 3. CHEMISTRY, BIOPHYSICS. *Educ:* Univ Minn, BS, 36, MS, 38; Columbia Univ, PhD(chem), 44. *Prof Exp:* Asst agr biochem, Univ Minn, 36-38; instr chem & agr chem, Purdue Univ, 38-39; from assoc to asst prof physiol chem, Chicago Med Sch, 44-45; admin asst & proj chief, Stand Brands, Inc, NY, 45-46; head nutrit res dept, H J Heinz Co, 46-49; asst dir lab serv, St Mary's Hosp, 55-56; chief chemist, Providence Hosp, 56; chemist, Food & Drug Admin, Rockville, 56-81; CONSULT, 81- *Concurrent Pos:* Fel, Yale Univ, 43-44. *Mem:* Sigma Xi; emer mem Am Chem Soc. *Res:* Drug manufacture and controls; clinical chemistry and biopharmaceutics; sanitary engineering and preventive medicine; nutrition; pesticides and food additives. *Mailing Add:* 3661 Jennings Chapel Rd PO Box 250 Woodbine MD 21797-7505

HOUK, CLIFFORD C, b Dayton, Ohio, July 7, 33; m 54; c 3. INORGANIC CHEMISTRY. *Educ:* Ohio Univ, BSEd, 55, MEd, 56; Mont State Univ, PhD(chem), 66. *Prof Exp:* Teacher, High Sch, Ohio, 56-62; asst chem, Mont State Univ, 62-63; asst prof, 66-71, assoc prof, 71-80, PROF CHEM, OHIO UNIV, 80-, DIR GEN CHEM, 66- *Mem:* AAAS; Am Chem Soc; Sigma Xi. *Res:* Synthesis of transition metal complexes; magnetic properties of transition metal compounds; chemical education. *Mailing Add:* Sch Health & SportSci Ohio Univ-Convocation Ctr Athens OH 45701-2979

HOUK, JAMES CHARLES, b Northville, Mich, June 3, 39; m 63; c 3. CNS NEUROPHYSIOLOGY. *Educ:* Mich Technol Univ, BS, 61; Mass Inst Technol, SM, 63; Harvard Univ, PhD(physiol), 66. *Prof Exp:* NIH fel, Fac Med, Toulouse, France, 66-67; from instr to asst prof physiol, Harvard Med Sch, 67-73; assoc prof physiol, Sch Med, Johns Hopkins Univ, 73-78; NATHAN SMITH DAVIS PROF PHYSIOL & CHMN DEPT, SCH MED, NORTHWESTERN UNIV, 78-, PROF BIOMED ENG, 79- *Concurrent Pos:* lectr, Dept Elec Eng, Mass Inst Technol, 71-73; adj assoc prof, Dept Physiol, Univ NC, 75. *Honors & Awards:* Javits Neurosci Investr Award, Nat Inst Neurol & Commun Disorders & Stroke, NIH. *Mem:* AAAS; Am Physiol Soc; Soc Neurosci; Sigma Xi; Inst Elec & Electronics Engrs; Europ Neurosci Asn; Int Neural Networks Soc. *Res:* Neural mechanisms controlling movement; cerebellar mechanisms for sensorimotor learning. *Mailing Add:* Dept Physiol 303 E Chicago Ave Chicago IL 60611

HOUK, KENDALL N, b Nashville, Tenn, Feb 27, 43; m 66, 90; c 1. THEORETICAL ORGANIC CHEMISTRY. *Educ:* Harvard Univ, AB, 64, MS, 66, PhD(chem), 68. *Prof Exp:* From asst prof to assoc prof chem, La State Univ, 68-80; prof chem, Univ Pittsburgh, 80-85; PROF CHEM, UNIV CALIF, LOS ANGELES, 86- *Concurrent Pos:* Vis prof, Princeton Univ, 75. *Honors & Awards:* Von Humboldt Sr Res Scientist Award, 82; James Flack Norris Award, Am Chem Soc, 91. *Mem:* Chem Soc; NAm Photochem Soc; Asn Quantum Biol. *Res:* Theoretical organic chemistry; photochemistry; cycloaddition reactions; asymmetric reagents; transition state modeling. *Mailing Add:* Dept Chem & Biochem Univ California Los Angeles CA 90024

HOUK, LARRY WAYNE, b Glasgow, Ky, Aug 25, 41; m 62; c 3. INORGANIC CHEMISTRY. *Educ:* Mid Tenn State Col, BS, 63; Univ Ga, PhD(inorg chem), 68. *Prof Exp:* From asst prof to assoc prof, 68-78, PROF DEPT CHEM, MEMPHIS STATE UNIV, 78- *Mem:* Am Chem Soc. *Res:* Low valence organophosphorus transition metal complexes. *Mailing Add:* Dept Chem Memphis State Univ Memphis TN 38152

HOUK, NANCY (MIA), b Potsdam, NY, July 18, 40. ASTRONOMY. *Educ:* Univ Mich, BS, 62; Case Western Reserve Univ, MS, 64, PhD(astron), 67. *Prof Exp:* Res assoc astron, Warner & Swasey Observ, Case Western Reserve Univ, 67-69; Netherlands Orgn Advan Pure Res grant & vis res assoc, Kapteyn Astron Lab, 70; res assoc astron, 71-74, from asst res scientist to assoc res scientist, 75-83, RES SCIENTIST ASTRON, UNIV MICH, ANN ARBOR, 84- *Mem:* Am Astron Soc; Int Astron Union; AAAS. *Res:* Spectral classification; M-type variable stars; stellar spectroscopy; galactic structure. *Mailing Add:* 1041 Dennison Bldg Dept Astron Univ Mich Ann Arbor MI 48109-1090

HOUK, RICHARD DUNCAN, b Hobart, Okla, June 13, 33; m 55; c 2. BOTANY, TAXONOMY. *Educ:* Southwest Mo State Col, BS, 59; Fla State Univ, PhD(bot), 66. *Prof Exp:* Instr biol, Southwest Mo State Col, 59-60; from asst prof to assoc prof, 64-73, asst vpres acad affairs, 74-75, vprovost, 75-80, PROF BIOL, WINTHROP COL, 75- *Mem:* Int Asn Plant Taxon; Nat Sci Teachers Asn; Am Soc Plant Taxonomists; Nat Asn Biol Teachers; Bot Soc Am. *Res:* Genecology of Sagittaria; methodology of biology teaching; conservation education. *Mailing Add:* Dept Biol Winthrop Col Rock Hill SC 29733

HOUK, ROBERT SAMUEL, b New Castle, Pa, Nov 23, 52; m 81; c 2. CHEMICAL INSTRUMENTATION. *Educ:* Slippery Rock State Col, BS, 74; Iowa State Univ, PhD(anal chem), 80. *Prof Exp:* Grad res asst, Ames Lab, US Dept Energy, 75-80, asst chemist, 80-81; asst prof, 81-87, ASSOC PROF ANALYTICAL CHEM, IOWA STATE UNIV, 87-; ASSOC CHEMIST, AMES LAB, US DEPT ENERGY, 81- *Concurrent Pos:* Prin investr, Plasma Mass Spectrometry Proj, Ames Lab, US Dept Energy, 80- *Honors & Awards:* Lester W Strock Medal, Soc Appl Spectros, 85. *Mem:* Am Chem Soc; Soc Appl Spectros; Am Soc Mass Spectrometry; Sigma Xi; AAAS. *Res:* Ionization principles and methods for mass spectrometry; applications and diagnostics of inductively coupled plasmas; trace elemental isotopic and organic analysis; mass spectrometry; high vacuum technology. *Mailing Add:* Dept Chem Iowa State Univ Ames IA 50011

HOUK, THOMAS WILLIAM, b Johnson City, Tenn, June 4, 44; m 73; c 3. BIOPHYSICS. *Educ:* NC State Univ, Raleigh, BS, 66; Univ Md, College Park, PhD(physics), 71. *Prof Exp:* USPHS fel, Univ Calif Med Ctr, Cardiovasc Res Inst, Med Sch, Univ Calif, San Francisco, 71-73; Robert A Welch fel, Dept Biomath, M D Anderson Hosp & Tumor Inst, Houston, 73-74; from asst prof to assoc prof, 74-82, PROF PHYSICS, MIAMI UNIV, OXFORD OHIO, 82-, CHMN DEPT, 85- *Mem:* Biophys Soc; Am Phys Soc; Am Asn Physics Teachers; Sigma Xi; AAAS. *Res:* Statistical physics; applications of the theory of stochastic processes in biology, medicine, and chemistry; diffusion coupled reaction theory; experimental biophysics; physical properties of macromolecules employing fluorescence spectroscopy, especially vertebrate striated muscle; effects of electromagnetic fields on biological systems and macromolecules. *Mailing Add:* Dept Physics Miami Univ Oxford OH 45056

HOULE, FRANCES ANNE, b Pasadena, Calif, Oct 22, 52; m 78; c 2. SURFACE CHEMISTRY. *Educ:* Univ Calif, Irvine, BA, 74, Calif Inst Technol, PhD(chem), 79. *Prof Exp:* Fel, Lawrence Berkeley Lab, Univ Calif, 79-80; MEM RES STAFF, RES DIV, INT BUS MACH CORP, 81- *Concurrent Pos:* assoc ed, J Vacuum Sci & Technol, 89-91; scholar trustee, Am Vacuum Soc, 90-92. *Mem:* Am Chem Soc; Am Phys Soc; Am Vacuum Soc. *Res:* Laser-induced chemical processes at surfaces; reaction kinetics and dynamics of gas-solid systems; laser etching, deposition and oxidation; chemistry of semiconductor etching. *Mailing Add:* IBM Almaden Res Center 650 Harry Rd K33/801 San Jose CA 95120-6099

HOULE, JOSEPH E, b Hartford, Conn, Oct 11, 30; m 54; c 6. MATHEMATICS, ACADEMIC ADMINISTRATION. *Educ:* Cath Univ Am, AB, 52, MA, 54, PhD(math), 59. *Prof Exp:* Asst math, Cath Univ Am, 52-53; from instr to assoc prof, Georgetown Univ, 53-62; assoc prof, Seton Hall Univ, 62-63; chmn dept, Pace Univ, NY, 63-70, dean, Dyson Col Arts & Sci, 71-90, vprovost, 87-90, DIR CTR APPL ETHICS, PACE UNIV, NY & WESTCHESTER CAMPUS, 90- *Concurrent Pos:* Prof, Pace Univ, 63- *Mem:* Math Asn Am. *Mailing Add:* 227 Garfield Pl South Orange NJ 07079-2108

HOULIHAN, JOHN FRANK, b Springfield, Ill, Sept 9, 42; m 65; c 2. PHOTOELECTROCHEMICAL MATERIALS. *Educ:* Ind Cent Univ, BA, 64; DePauw Univ, MA, 66; Pa State Univ, PhD(physics), 73. *Prof Exp:* Engr solid state, device group, Avco Electronic, 66-67; from instr to assoc prof physics, 67-82, PROF PHYSICS, PA STATE UNIV, 82- *Mem:* Am Phys Soc; Electrochem Soc; Am Asn Physics Teachers. *Res:* Materials preparation and characterization for use in photoelectrochemical device applications. *Mailing Add:* Dept Physics Pa State Univ Shenango Valley Campus Sharon PA 16146

HOULIHAN, RODNEY T, b Los Angeles, Calif, Dec 12, 26; m 48; c 4. PHYSIOLOGY, ENDOCRINOLOGY. *Educ:* Univ Redlands, BS, 57; Univ NMex, MSc, 58; Univ Calif, Davis, PhD(zool), 60. *Prof Exp:* Asst zool, Univ Calif, Davis, 58-60; instr biol, St Mary's Col, Calif, 60-61; from asst prof to assoc prof zool, Pa State Univ, 61-69; prof osteop med, Mich State Univ, 69-73; prof physiol, dean acad admin & dir res, 73-81, VPRES RES & DEVELOP, OKLA COL OSTEOP MED & SURG, 81- *Mem:* AAAS; Am Physiol Soc; Aerospace Med Soc; NY Acad Sci. *Res:* Adrenal gland function during chronic and acute stress; interrelations of steroids and catecholamine metabolism, hyperbaric medicine and oxygen toxicity, water and electrolyte metabolism. *Mailing Add:* Okla Col Osteop Med & Surg 1111 W 17th St Tulsa OK 74107-1805

HOULIHAN, WILLIAM J, b South Amboy, NJ, July 24, 30; m 59; c 3. ORGANIC CHEMISTRY, MEDICINAL CHEMISTRY. *Educ:* Seton Hall Univ, BS, 51; Rutgers Univ, PhD(org chem), 55. *Prof Exp:* Asst prof chem, St John's Univ, NY, 55-56; group leader, Universal Oil Prod Chem Co, 56-58; asst prof chem, Seton Hall Univ, 58-60; group leader, Universal Oil Prod Chem Co, 60-62; sect head, 62-73, sect dir chem develop & res, 73-84, DIR MED CHEM DEPT, SANDOZ RES INST, HANOVER, NJ, 84- *Concurrent Pos:* Lectr, Rutgers Univ, 61-62. *Mem:* Am Chem Soc. *Res:* Heterocyclic synthesis; synthesis of medicinal agents; organolithium reagents; antitumor agents. *Mailing Add:* Sandoz Inc Rte Ten East Hanover NJ 07936

HOUNSFIELD, G NEWBOLD, b 1919. X-RAY TECHNOLOGY. *Prof Exp:* SR STAFF SCIENTIST, CENT RES LABS, EMI, 77- *Honors & Awards:* Nobel Prize in Med, 79; Churchill Gold Medal, 76; Gairdner Found Award, 76; Medal, Inst Physics, 76. *Mailing Add:* 15 Crane Park Rd Whitton Twickenham Middlesex England

HOUNSHELL, DAVID A, b Denver, Colo, Oct 18, 50; m 73; c 3. HISTORY OF TECHNOLOGY. *Educ:* Southern Methodist Univ, BSEE, 72; Univ Del, MA, 75, PhD(hist), 78. *Prof Exp:* Asst prof hist, Harvey Mudd Col, 77-79; from asst prof to assoc prof, 79-88, PROF HIST, UNIV DEL, 88- *Concurrent Pos:* Cur technol, Hagley Mus & Libr, 79-86, sr scholar, 86-87; Marvin Bower fel, Harvard Bus Sch, 87-88, Ctr Advan Study, Univ Del, 91-92. *Honors & Awards:* Browder J Thompson Prize, Inst Elec & Electronics Engrs, 78; Dexter Prize, Soc Hist Technol, 87. *Mem:* AAAS; Inst Elec & Electronics Engrs; Soc Hist Technol; Orgn Am Historians; Am Hist Asn; Hist Sci Soc. *Res:* History of science, technology, and business in the United States, 1776-1990. *Mailing Add:* Dept Hist Univ Del Newark DE 19716

HOUNSHELL, WILLIAM DOUGLAS, b Atlanta, Ga, Aug 8, 51; m 73. ORGANIC CHEMISTRY. *Educ:* Calif Inst Technol, BS, 73; Princeton Univ, PhD(chem), 77. *Prof Exp:* Instr, Princeton Univ, 77-78; asst prof chem, Wake Forest Univ, 78-80; MEM STAFF, MOLECULAR DESIGN LTD, 80- *Mem:* Am Chem Soc; Sigma Xi. *Res:* Stereochemical analysis; computational chemistry; strained compounds; chemical information. *Mailing Add:* Molecular Design Ltd Mdl-AG Muhlebachweg 9 CH-4123 Ailschwil 2 Switzerland

HOUNSLOW, ARTHUR WILLIAM, b Reservoir, Australia, July 17, 33; m 70; c 3. ECOLOGY, HYDROGEOCHEMISTRY. *Educ:* Univ Melbourne, BSc, 60; Carleton Univ, MSc, 65, PhD(geol), 68. *Prof Exp:* Analyst, Imp Chem Industs Australia & NZ Ltd, 54-56; exp officer, Mineragraphic Invests Sect, Commonwealth Sci & Indust Res Orgn, Australia, 56-61; anal technician, Queen's Univ, Ont, 61-62; asst prof geol, Idaho State Univ, 69-74; sr proj mineralogist, Colo Sch Mines Res Inst, 74-81; geochemist, US Environ Protection Agency, Ada, Okla, 81-82; PROF GEOL, OKLA STATE UNIV,

STILLWATER, 82- *Mem:* Sigma Xi; Nat Waterwell Asn; The Geol Soc Am. *Res:* Prediction of rock/water interactions; materials analysis; determination of sources of pollutant as well as their rate of absorption retardation and degradation; hydrogeochemistry; pollutant attenuation in ground water systems. *Mailing Add:* Geol Dept 149 Phys Sci Okla State Univ Stillwater OK 74078

HOUPIS, CONSTANTINE H, b Lowell, Mass, June 16, 22. CONTROL THEORY, FLIGHT CONTROL APPLICATIONS. *Educ:* Univ Ill, BS, 47, MS, 48; Univ Wyo, PhD(control theory), 71. *Prof Exp:* Elec engr, Babcock & Willcox Corp, 49-51; instr elec engr, Wayne State Univ, 49-51; prin elec engr, Batelle Inst, 51-52; instr to assoc prof elec engr, 52-74, PROF ELEC ENGR, AIR FORCE INST TECHNOL & ENG, 74-, SR RES ASSOC, APPLY DYNAMICS DIRECTORATE, 80- *Mem:* Fel Inst Elec & Electronics Engrs Control Systs Soc; Am Soc Eng Educ; Sigma Xi. *Mailing Add:* Dept Electrical Eng Air Force Inst Technol & Eng Wright-Patterson AFB OH 45433

HOUPIS, JAMES LOUIS JOSEPH, b Binghamton, NY, Oct 11, 56; m 80; c 4. AIR POLLUTION EFFECTS & CLIMATE CHANGE EFFECTS ON FOREST SPECIES. *Educ:* Univ Calif, Berkeley, BA, 78, PhD(forest sci), 89; San Diego State Univ, MS, 84. *Prof Exp:* Res asst, Univ Calif, Berkeley, 80-86; TREE ECOPHYSIOLOGIST, LAWRENCE LIVERMORE NAT LAB, 86- *Concurrent Pos:* Teaching assoc silvicult, Univ Calif, Berkeley, 84, scientist, 86; prin investr, US Environ Protection Agency, US Forest Serv, 87-91, 88-89 & 88-91; adj prof, Calif State Univ, Chico, 91- *Mem:* Am Soc Plant Physiologists. *Res:* Comparison of physiological, morphological, biochemical and growth response of mature trees and seedlings to air pollution and climate change; author and co-author of over 50 publications. *Mailing Add:* Lawrence Livermore Nat Lab PO Box 5507 L-559 Livermore CA 94550

HOUPT, KATHERINE ALBRO, b Buffalo, NY, Jan 11, 39; m 62; c 2. VETERINARY PHYSIOLOGY, ANIMAL BEHAVIOR. *Educ:* Pa State Univ, BS, 60; Univ Pa, VMD, 63, PhD(biol), 72. *Prof Exp:* Instr physiol, Univ Pa, 63-64; res assoc, 73-75, asst prof, 75-80, PROF PHYSIOL, CORNELL UNIV, 80- *Concurrent Pos:* NIH fel, 63-64, spec fel, 73-75; vis scientist, Agr Res Coun, Inst Animal Physiol, Babraham, Cambridge, Eng, 78-79. *Mem:* Sigma Xi; Am Physiol Soc; Am Vet Med Asn; Animal Behav Soc; Am Soc Animal Sci. *Res:* Comparative control of ingestive behavior in swine, horses and laboratory rodents; dominance hierarchies in horses. *Mailing Add:* 1515 Trumansburg Rd Ithaca NY 14850

HOUPT, THOMAS RICHARD, b Roslyn, Pa, Oct 9, 25; m 62; c 2. ANIMAL PHYSIOLOGY. *Educ:* Univ Pa, VMD, 50, PhD(physiol), 58; Univ Ill, MS, 53. *Prof Exp:* Instr vet physiol & pharmacol, Col Vet Med, Univ Ill, 50-53; from instr to prof physiol, Sch Vet Med, Univ Pa, 53-71; PROF VET PHYSIOL, NY STATE COL VET MED, CORNELL UNIV, 71- *Concurrent Pos:* Res assoc, Duke Univ, 53-54; vis scientist, Agr Res Coun, Inst Animal Physiol, Babraham & vis scholar Corpus Christi Col, Cambridge Univ, Eng, 78-79 & 86-87; res career develop award, NIH, 62-67. *Mem:* AAAS; Am Physiol Soc; Am Soc Vet Physiologists & Pharmacologists; Am Soc Zoologists. *Res:* Comparative physiology of birds and mammals; environmental physiology; gastro-intestinal adaptations related to adverse nutritional conditions; nitrogen metabolism; controls of food and water intake. *Mailing Add:* Dept Physiol Vet Col Cornell Univ Ithaca NY 14853

HOURIGAN, WILLIAM R, b Lebanon, Ky, May 11, 29; c 1. SOIL CHEMISTRY. *Educ:* Univ Ky, BS, 52, MS, 56; Ohio State Univ, PhD(soil chem), 60. *Prof Exp:* Asst county agent, Univ Ky, 52-53 & 56-57; prof soil chem, 60-74, PROF AGR, WESTERN KY UNIV, 74-, ASSOC DEAN UNDERGRAD INSTR, 60-, DEAN COL APPL ARTS & HEALTH, 69- *Res:* Effect of aluminum in soils. *Mailing Add:* 9811 Nashville Rd Bowling Green KY 42101

HOURSTON, ALAN STEWART, b Toronto, Ont, July 9, 26; m 55; c 5. FISHERIES. *Educ:* Univ Toronto, BA, 47, MA, 49; Univ Calif, PhD(oceanog), 56. *Prof Exp:* Demonstr zool, Univ Toronto, 48-49; from asst scientist to res scientist, Fisheries Res Bd Can, 49-72, sci asst to dir, 59-62, asst dir, 62-63; prog coordr, Int Fisheries Br, Dept Fisheries & Oceans, 73, head herring prog, Pac Biol Sta, Resource Serv Br, 73-80; CONSULT, 81- *Res:* Population dynamics and ecology of Pacific and Atlantic herring and Pacific salmon; theoretical population studies; scientific editing. *Mailing Add:* 2948 Hammond Bay Rd Nanaimo BC V9T 1E2 Can

HOUSE, ARTHUR STEPHEN, b New York, NY, May 1, 21; m 43; c 2. SPEECH COMMUNICATION. *Educ:* City Col New York, BS, 42; Univ Denver, MA, 48; Univ Ill, PhD(speech), 51. *Prof Exp:* Instr speech sci, Univ Ill, 51-52; res assoc audition, Control Systs Lab, 52-53; staff mem speech commun, Acoustics Lab, Mass Inst Technol, 53-57; assoc prof audiol & speech path, Syracuse Univ, 57-59; staff mem, Res Lab Electronics, Mass Inst Technol, 59-64; prof audiol & speech sci, Purdue Univ, West Lafayette, 64-71; ADJ STAFF MEM, COMMUN RES DIV, INST DEFENSE ANALYSIS, 71- *Concurrent Pos:* Mem comt hearing, bioacoustics & biomech, Nat Acad Sci-Nat Res Coun, 65-66. *Mem:* Fel AAAS; fel NY Acad Sci; fel Am Speech & Hearing Asn; fel Acoustical Soc Am. *Res:* Experimental study of processes used to decode acoustic speech signals into sequences of discrete linguistic symbols, and to encode sequences of discrete linguistic symbols into acoustic signals. *Mailing Add:* 11 Thorngate Ct Princeton NJ 08540-3699

HOUSE, EDWARD HOLCOMBE, b Trenton, NJ, Sept 10, 29. PHYSICAL CHEMISTRY. *Educ:* Princeton Univ, AB, 50; Dartmouth Col, AM, 52; Univ Rochester, PhD(chem), 60. *Prof Exp:* Instr chem, Hobart & William Smith Cols, 55-59; asst prof, Univ NDak, 59-62; prof math & phys sci & chmn dept, Trenton Jr Col, NJ, 62-67; prof chem & chmn dept, 67-70, PROF CHEM, MERCER COUNTY COL, 70- *Mem:* AAAS; Am Chem Soc; Sigma Xi. *Res:* Reaction kinetics and catalysis; molecular spectroscopy; history and philosophy of science. *Mailing Add:* Dept Chem Box B Mercer County Col Trenton NJ 08690

HOUSE, EDWIN W, b Corning, NY, Oct 20, 39; m 64; c 2. ANIMAL PHYSIOLOGY, PATHOLOGY. *Educ:* Western Mont Col, BS, 60; Univ Mont, MS, 62; Univ NDak, PhD(physiol), 65. *Prof Exp:* asst prof zool, Univ Mont, 65-66; asst prof biol, 66-70, assoc prof, 70-75, PROF BIOL, IDAHO STATE UNIV, 75- *Concurrent Pos:* Dept chair, 81-86, Hosp Ethics Comt mem, 85-; chair, Idaho Higher Educ Res Coun. *Mem:* AAAS; Sigma Xi; Am Physiol Soc. *Res:* Newborn mammalian physiology; atherosclerosis; cardiovascular development and control; comparative mammalian physiology; cardiovascular control systems; comparative temperature regulation in mammals and reptiles; comparative anatomy and physiology of renal function in mammals. *Mailing Add:* Grad Studies & Res Idaho State Univ Pocatello ID 83201

HOUSE, GARY LAWRENCE, b St Marys, Pa, Dec 12, 47; m 70; c 3. CHEMICAL ENGINEERING, SOLID STATE PHYSICS. *Educ:* Grove City Col, BS, 73; Univ Ill, MS, 75, PhD(chem eng), 77. *Prof Exp:* Res chemist, 77-91, LAB HEAD, EASTMAN KODAK CO, 91- *Mem:* Am Inst Chem Engrs; Soc Imaging Sci & Technol. *Res:* Luminescence properties of inorganic semiconductors; photographic properties of silver halide imaging systems. *Mailing Add:* Eastman Kodak Co 1669 Lake Ave Bldg 59 Rochester NY 14650-1707

HOUSE, HERBERT OTIS, b Willoughby, Ohio, Dec 5, 29; m 80; c 2. ORGANIC CHEMISTRY. *Educ:* Miami Univ, BS, 50; Univ Ill, PhD(org chem), 53. *Prof Exp:* From instr to prof org chem, Mass Inst Technol, 53-71; PROF CHEM, GA INST TECHNOL, 71- *Concurrent Pos:* Consult, Union Carbide Chem Co; mem, Adv Bd, Org Reactions, Inc, 71-79 & Org Syntheses, 74-79; chmn comt on handling hazardous substances in labs, Nat Res Coun, 79-81. *Honors & Awards:* Award for Creative Work in Synthetic Org Chem, Am Chem Soc, 75 & Award Chem Health & Safety, 83. *Mem:* Am Chem Soc; The Chem Soc; Swiss Chem Soc. *Res:* Organic synthetic methods; chemistry of carbanions; stereochemistry of organic reactions; reactions involving electron transfer. *Mailing Add:* Dept Chem Ga Inst Technol Atlanta GA 30332

HOUSE, HOWARD LESLIE, b Brantford, Ont, Jan 20, 18; m 45; c 3. INSECT PHYSIOLOGY. *Educ:* Ont Agr Col, BSA, 43; Cornell Univ, PhD(entom), 48. *Prof Exp:* Agr scientist, Res Br, Can Dept Agr, 52-71, res officer, 71-79; RETIRED. *Mem:* Fel Entom Soc Am. *Res:* Insect nutrition and dietetics; food preference; biological control. *Mailing Add:* RR 1 Corbyville ON K0K 1V0 Can

HOUSE, JAMES EVAN, JR, b Benton, Ill, July 7, 36; m 55; c 4. SOLID STATE INORGANIC CHEMISTRY, CHEMICAL COMPUTATIONS. *Educ:* Southern Ill Univ, BS, 58, MA, 61; Univ Ill, Urbana, PhD(inorg chem), 70. *Prof Exp:* Instr chem, Southern Ill Univ, 60-61; fac asst phys sci, Univ Ill, Urbana, 62-63; asst prof chem, Western Ky Univ, 63-64; res chemist, A E Staley Co, 64-66; from asst prof to assoc prof, 66-70, asst dean fac, 70-72, assoc dean univ, 72-74, PROF CHEM, ILL STATE UNIV, 74- *Mem:* Am Chem Soc; Sigma Xi. *Res:* Thermal properties and mechanisms of reactions of coordination compounds; kinetics of solid state reactions; chemical computations and data analysis. *Mailing Add:* Dept Chem Ill State Univ Normal IL 61761

HOUSE, JAMES STEPHEN, b Philadelphia, Pa, Jan 27, 44; m 67; c 2. PSYCHOSOCIAL FACTORS ON HEALTH. *Educ:* Haverford Col, BA, 65; Univ Mich, PhD(social psychol), 72. *Prof Exp:* From instr to prof sociol, Duke Univ, 70-78; from adj asst prof to adj assoc prof epidemiol, Univ NC, 75-78; assoc res scientist epidemiol & assoc prof, Univ Mich, 78-83, assoc chair sociol, 81-84, chair sociol, 86-90, PROF SOCIOL & RES SCIENTIST, SURV RES CTR & INST GERONTOL, UNIV MICH, 83-, DIR, SURV RES CTR, 91- *Concurrent Pos:* Assoc ed, J Health & Social Behav, Am Sociol Asn, 80-82, prin investr, Nat Inst on Aging Proj grant, 85-, chair, Social Psychol Sect, 87-88. *Mem:* Am Sociol Asn; Am Psychol Asn; AAAS; Soc Epidemiol Res; Soc Psychol Study Social Issue. *Res:* Relation of social stress and social support to health, and the structural context and determinants of social stress and social support; health in middle and late life and in an action research project on occupational stress and health; productive activity; stress. *Mailing Add:* Inst Social Res Univ Mich Box 1248 Ann Arbor MI 48106

HOUSE, LELAND RICHMOND, b Richmond, Va, July 16, 08; div; c 4. OTOLARYNGOLOGY. *Educ:* Pac Union Col, BS, 32; Loma Linda Univ, MD, 34; Univ Pa, MSc, 45. *Prof Exp:* From instr to prof otolaryngol & head dept, Sch Med, Loma Linda Univ, 38-45 & 47-66; clin prof otolaryngol, Med Sch, Univ Southern Calif & chmn dept otolaryngol, Rancho Los Amigos Hosp, 66-72; RETIRED. *Concurrent Pos:* Mem attend staff, White Mem Hosp, 38-90, Calif Lutheran Hosp, 38-80, Glendale Adventist Med Ctr, 44-84 & Los Angeles County-Univ Southern Calif Med Ctr, 52-90; consult, St Francis Hosp, 44-70 & Cedars-Sinai Hosp, 56-90. *Mem:* Am Laryngol, Rhinol & Otol Soc; AMA; Am Acad Otolaryngol & Head & Neck Surg; Am Rhinologic Soc. *Res:* Problems of deafness and surgery on the ears for its correction; rhinological surgery for correction of nasal deformities and nasal disfunctions. *Mailing Add:* 1995 Cerco Alta Dr Monterey Park CA 91754

HOUSE, LEWIS LUNDBERG, b North Platte, Nebr, Jan 18, 33; m 54; c 3. ASTROPHYSICS, ATOMIC PHYSICS. *Educ:* Colo Sch Mines, geophys engr, 55; Rensselaer Polytech Inst, MS, 58; Univ Colo, Boulder, PhD(astrophys), 62. *Prof Exp:* Physicist, Hanford Atomic Prod Oper, Gen Elec Co, 55-56 & Knolls Atomic Power Lab, 56-58; res fel, 62-63, MEM SR RES STAFF SOLAR PHYSICS, HIGH ALTITUDE OBSERV, UNIV COLO, BOULDER, 63- *Concurrent Pos:* Lectr astrogeophys, Univ Colo, Boulder, 65-, lectr physics & astrophys, 68-; vis res appointment, Inst Astron, Univ Hawaii, 69. *Mem:* AAAS; Am Astron Soc; Int Astron Union; Am Asn Physics Teachers. *Res:* Transport of radiation in stellar atmospheres; quantum mechanical calculation of radiation-matter interaction coefficients in presence of magnetic fields; applications of transfer theory to the determination of magnetic field structures in the solar atmosphere; theory of origin and structure of stellar and solar atmospheres. *Mailing Add:* 2663 Grapewood Lane Boulder CO 80304

HOUSE, ROBERT W(ILLIAM), b Wellsville, Ohio, May 31, 27; m 48; c 5. SCIENCE POLICY, SYSTEMS ENGINEERING. *Educ:* Ohio Univ, BS, 49, MS, 52; Pa State Univ, PhD(elec eng), 59; Harvard Bus Sch, PMD, 66. *Prof Exp:* Mathematician, US Naval Proving Grounds, 50-51; electronic scientist, Wright-Patterson AFB, 51-54; sr mathematician, HRB-Singer, 54; from instr to asst prof elec eng, Pa State Univ, 54-59; consult, Systs Eng Div, Battelle-Columbus, 59-65, dir comput sci res, 65-66, assoc mgr, Systs & Electronics Dept, 66-70, mgr, Social & Systs Sci Dept, 70-74; pres exec, President's Exec Interchange Prog/AID, 74-75; prof technol & pub policy & dir, Vanderbilt Univ, 75-82, from assoc dean to dean grad sch, 81-84, dir mgt technol prog, 82-90, Orrin Henry Ingram Distinguished prof, 84-90, EMER ORRIN HENRY INGRAM DISTINGUISHED PROF & EMER PROF ELEC ENG, VANDERBILT UNIV, 90- *Concurrent Pos:* Adj prof, Ohio State Univ, 65-70; consult, Secy Sci & Technol, State of Sao Paulo, 75-78, Nat Ctr Productivity, 78, Brazilian Nat Alcohol Fuels Prog, 78-82 & US Biomass Energy Prog, 84-85. *Honors & Awards:* Centennial Medal & Commendation, Inst Elec & Electronics Engrs. *Mem:* Fel Inst Elec & Electronics Engrs; Eng Mgt Soc; Syst, Man & Cybernetics Soc. *Res:* Management of technology; systems analysis and synthesis; technological innovation process; technology and public policy; systems engineering; project management. *Mailing Add:* Mgt Technol Prog Vanderbilt Univ Box 6188 Sta B Nashville TN 37235

HOUSE, VERL LEE, b Wellsville, Mo, Apr 10, 19; m 48; c 3. GENETICS. *Educ:* Univ Calif, AB, 41, MA, 48, PhD(zool), 50. *Prof Exp:* Assoc zool, Col Agr, Univ Calif, 47-48; instr biol, Johns Hopkins Univ, 50-52, Sigma Xi res grant, 53, asst prof, 53-54; prof & chmn dept, Radford Univ, 54-58; assoc prof genetics, 58-67, prof, 67-81, EMER PROF GENETICS, OHIO STATE UNIV, 81- *Concurrent Pos:* NSF grant, Radford Univ, 54; grant, Ohio State Univ, 59-65, 69 & 71; assoc dir, Fel Off, Nat Acad Sci-Nat Res Coun, 66-67; vis prof genetics, NY State Univ-Cortland, 85-86; vis prof biol, Whitman Col, Walla Walla, WA, 87-88. *Mem:* Genetics Soc Am; Am Soc Zool; Am Soc Human Genetics. *Res:* Physiological genetics; physiological and developmental genetics of Drosophila; genetic control of venation development in the Drosophila wing. *Mailing Add:* 484 W 12th Ave Columbus OH 43210

HOUSE, WILLIAM BURTNER, b Kansas City, Mo, Aug 24, 18; m 42; c 6. BIOCHEMISTRY, NUTRITION. *Educ:* Univ Mo, BS, 42, AM, 49, PhD, 58. *Prof Exp:* Chemist, Hercules Powder Co, Del, 42-43, res chemist smokeless powder, 43-45, supvr & res chemist, 45-46; asst, Univ Mo, 46-52; nutritionist, Nat Alfalfa Dehydrating & Milling Co, 52-54; from assoc chemist to prin chemist, Midwest Res Inst, 54-60; dir res, Plan Foods, Inc, 60-61; dir biol, Sci Div, Midwest Res Inst, 63-77, prin adv, 81-82; RETIRED. *Concurrent Pos:* Tech dir, US Contract Mgt, Consumer Protection Dept, Saudi Arabia, 78-80. *Mem:* AAAS; Am Chem Soc; Sigma Xi; Am Inst Biol Sci. *Res:* Nutrition; inhalation toxicity; ecological assessment. *Mailing Add:* 5745 Grand Kansas City MO 64113

HOUSECROFT, CATHERINE ELIZABETH, b Bradford, Eng, Feb 23, 55. CLUSTER CHEMISTRY, METALLABORANE CHEMISTRY. *Educ:* Durham Univ, Eng, BSC, 76, PhD(inorg chem), 79. *Prof Exp:* Teacher chem, Oxford High Sch, Eng, 79-83; res assoc chem, Univ Notre Dame, 83-85; asst prof chem, Univ NH, 85-86; CONSULT, 86- *Mem:* Royal Soc Chem; Am Chem Soc. *Res:* Experimental and theoretical investigations of transition metal-boron containing cluster systems; changes in metal cluster reactivity brought about by the introduction of the main group element. *Mailing Add:* Dept Chem Univ NH Durham NH 03824

HOUSEHOLDER, ALSTON SCOTT, b Rockford, Ill, May 5, 04; m 26; c 2. MATHEMATICS. *Educ:* Northwestern Univ, BS, 25; Cornell Univ, MA, 27; Univ Chicago, PhD(math), 37. *Hon Degrees:* Dr, Munich Tech Inst, 65. *Prof Exp:* Instr math, Northwestern Univ, 26-28; tutor, Miss Harris Schs, Chicago, 29-30; from instr to asst prof, Washburn Col, 30-37; Rockefeller Found fel, Univ Chicago, 37-39, from res assoc to asst prof math biophys, 39-44; sr res psychophysiologist, Nat Defense Res Comt, Brown Univ, 44-45; math consult, Naval Res Lab, Washington, DC, 45-46; sr mathematician, Oak Ridge Nat Lab, 46-69; prof math, Univ Tenn, Knoxville, 64-74; CONSULT, CALIF INST TECHNOL, 74- *Concurrent Pos:* Instr, Northern Ill Col Optom, 40; mem div math, Nat Res Coun, 55-58 & 70-72. *Honors & Awards:* Harry Goode Mem Award, 69. *Mem:* Fel AAAS; fel Am Acad Arts & Sci; Soc Indust & Appl Math (vpres, 62-63, pres, 63-64); Asn Comput Mach (vpres, 52-54, pres, 54-56); Math Asn Am (vpres, 60); Sigma Xi. *Res:* Numerical analysis; programming for high speed digital computation. *Mailing Add:* 6235 Tapia Dr Malibu CA 90265

HOUSEHOLDER, JAMES EARL, b West Newton, Pa, Dec 26, 16; m 38; c 2. MATHEMATICS. *Educ:* Univ Ariz, BS, 52, MS, S3; Univ Colo, PhD(math), 59. *Prof Exp:* From asst prof to assoc prof, Humboldt State Univ, 59-69, prof math, 69-87; RETIRED. *Mem:* Am Math Soc; Math Asn Am. *Res:* Theory of numbers. *Mailing Add:* PO Box 4386 Arcata CA 95521

HOUSEHOLDER, MICHAEL K, b Chicago, Ill, May 31, 41; m 63; c 3. ENVIRONMENTAL ENGINEERING, HYDROMECHANICS. *Educ:* Valparaiso Univ, BSCE, 63, Purdue Univ, MSCE, 65, PhD(hydromech), 68. *Prof Exp:* Assoc prof civil eng, Youngstown State Univ, 68-77, prof, 77-80; ASSOC PROF CIVIL ENG, CALIF STATE UNIV, 80- *Mem:* Assoc mem Am Soc Civil Engrs; Am Soc Eng Educ. *Res:* Hydrology; fluid mechanics. *Mailing Add:* 410 N Leroux St Flagstaff AZ 86001

HOUSEHOLDER, WILLIAM ALLEN, b Newark, Nebr, May 19, 21; m 45; c 2. AGRICULTURE. *Educ:* Colo State Univ, BS, 49; Cornell Univ, MS, 57; Mich State Univ, PhD(agr educ), 65. *Prof Exp:* Teacher high sch, Colo, 49-52; tech adv agr & agr educ, Near East Found, Iran, 52-56; tech adv agr & community develop, US Agency Int Coop, Panama, 57-62; chmn dept, Dept Agr, Eastern Ky Univ, 65-77, prof agron, 65- 91; RETIRED. *Res:* Technical education in agriculture; human resource development; developing technical curriculums in agriculture. *Mailing Add:* Dept Agr Eastern Ky Univ Richmond KY 40475

HOUSEKNECHT, DAVID WAYNE, b Muncy, Pa, Mar 18, 51; m 73; c 2. SEDIMENTARY GEOLOGY. *Educ:* Pa State Univ, BS, 73, PhD(geol), 78; Southern Ill Univ, MS, 75. *Prof Exp:* ASST PROF GEOL, UNIV MO-COLUMBIA, 78- *Concurrent Pos:* Res geologist, US Bur Mines, 75- *Mem:* Soc Econ Paleontologists & Mineralogists; Int Asn Sedimentologists. *Res:* Sandstone petrogenesis; reconstruction of ancient depositional environments; statistical applications in sedimentary geology; coal geology. *Mailing Add:* Dept Geol Univ Mo 101 Geol Bldg Columbia MO 65211

HOUSEMAN, BARTON L, b Silver Spring, Md, Nov 12, 33; m 55; c 2. PHYSICAL CHEMISTRY. *Educ:* Calvin Col, AB, 55; Wayne State Univ, PhD(phys chem), 61. *Prof Exp:* Asst prof, 61-67, assoc prof, 67-73, PROF CHEM, GOUCHER COL, 73- *Concurrent Pos:* Danforth assoc, Danford Found, 64-81; consult, Nuclear Defense Lab, Edgewood Arsenal, 64-65; vis staff mem, Los Alamos Sci Lab, 66-82; res assoc, Univ Calif, Berkeley, 69-70; consult, Inorg Mat Res Div, Lawrence Radiation Lab, 69-70. *Mem:* Am Chem Soc. *Res:* Chemical education; solution thermodynamics; ion mobility spectrometry. *Mailing Add:* Dept Chem Goucher Col Baltimore MD 21204

HOUSEPIAN, EDGAR M, b New York, NY, Mar 18, 28; m 54; c 3. NEUROSURGERY, NEUROPHYSIOLOGY. *Educ:* Columbia Univ, BA, 49, MD, 53; Am Bd Neurol Surg, 61. *Prof Exp:* From instr to assoc prof neurol surg, 59-75, PROF CLIN NEUROL SURG, COL PHYSICIANS & SURGEONS, COLUMBIA UNIV, 75- *Concurrent Pos:* Parkinson's Dis Found res fel, 59-61; asst neurol surgeon, Columbia-Presby Med Ctr, 59-61, asst attend neurol surgeon, 61-64, assoc attend neurol surgeon, 64-; asst attend neurol surgeon, NY State Psychiat Inst, 61-; consult, Englewood Hosp, NJ, 64- & Greenwich Hosp, Conn, 65- *Honors & Awards:* Fulbright Traveling Fel, 68. *Mem:* Fel Am Col Surgeons; Am Asn Neurol Surgeons; Asn Res Nerv & Ment Dis; Soc Neurosci; Int Soc Res Stereoencephaltomy; Am Acad Neurosci; Res Soc Neurol Surg; Soc Neurosci. *Res:* Basic and applied electrophysiology; clinical research in brain tumor chemotherapy; transphenoidal microsurgery. *Mailing Add:* Dept Neurol Surg Columbia Univ 710 W 168th St New York NY 10032

HOUSER, HAROLD BYRON, b North Liberty, Ind, Nov 22, 21; m 44; c 5. EPIDEMIOLOGY. *Educ:* Ind Univ, AB, 42, MD, 44. *Prof Exp:* Resident internal med, Crile Vet Admin Hosp, Ohio, 47-49; res fel, Dept Prev Med, Western Reserve Univ, 49-52; from instr to asst prof med, Col Med, State Univ NY Upstate Med Ctr, 53-58; from asst prof to assoc prof prev med, 58-74, actg dir, dept biomet, 69-75, dir, 75-85, PROF EPIDEMIOL, CASE WESTERN RESERVE UNIV, 74-, ASSOC PROF MED, 58-, DIR DEPT EPIDEMIOL BIOSTAT, 85- *Concurrent Pos:* Arthritis & Rheumatism Found res fel, Col Med, State Univ NY Upstate Med Ctr, 53; dir lab, Housing & Illness, Armed Forces Epidemiol Bd, Sampson AFB, NY, 54-58; consult prev med, Surg Gen, US Army; dir, Dept Epidemiol Biostat, Case Western Reserve Univ, 85- *Honors & Awards:* Group Lasker Award. *Mem:* Infectious Dis Soc Am; Am Epidemiol Soc (pres, 91). *Res:* Long term illness; streptococcal diseases; rheumatic fever; nutrition epidemiology; AIDS epidemiology. *Mailing Add:* 2330 Delaware Dr Cleveland Heights OH 44106

HOUSER, JOHN J, b Abington, Pa, Feb 24, 39; m 68. PHOTOCHEMISTRY, CARBOCATIONS. *Educ:* Villanova Univ, BS, 60; Pa State Univ, PhD(chem), 64. *Prof Exp:* PROF CHEM, UNIV AKRON, 65- *Concurrent Pos:* Vis prof, Case Western Reserve Univ, 85. *Mem:* Am Chem Soc; Royal Soc Chem; Sigma Xi. *Res:* Photo dehalogenation; carbocation rearrangements; theoretical chemistry. *Mailing Add:* Dept Chem Univ Akron Akron OH 44325-3601

HOUSER, THOMAS J, b Chicago, Ill, Feb 9, 30; m 53; c 4. PHYSICAL CHEMISTRY. *Educ:* Ill Inst Technol, BS, 52, Univ Mich, MS, 54, PhD(phys chem), 57. *Prof Exp:* Chemist, Standard Oil Co, Ind, 57-58, asst proj chemist, 58-59; sr res engr, Rocketdyne Div, NAm Aviation, Inc, 59-62, prin scientist, 62-64; from asst prof to assoc prof phys chem, 64-76, PROF CHEM, WESTERN MICH UNIV, 76- *Mem:* Am Chem Soc; Combustion Inst. *Res:* Gas phase kinetics and mechanisms using flow systems; coal liquefaction mechanisms; supercritical fluid chemistry. *Mailing Add:* Dept Chem 5101 Mck Western Mich Univ Kalamazoo MI 49008

HOUSER, WESLEY G(RANT), electrical engineering, for more information see previous edition

HOUSEWEART, MARK WAYNE, forest entomology; deceased, see previous edition for last biography

HOUSEWRIGHT, RILEY DEE, b Wylie, Tex, Oct 17, 13; m 39, 69; c 1. TECHNICAL MANAGEMENT, SCIENCE ADMINISTRATION. *Educ:* NTex State Univ, BS, 34; Univ Tex, MA, 38; Univ Chicago, PhD(bact), 44; Am Bd Med Microbiol, dipl, 61. *Prof Exp:* Pub sch teacher, Tex, 34-36; instr biol & supvr sci teacher training, Southwest Tex State Col, 37-41; med consult, Fed Security Agency, 43-44; res & chief med microbiol div, Ft Detrick Md, 44-56, sci dir, 56-70; vpres & sci dir, Microbiol Assocs, Inc, Bethesda, Md, 70-75; prin staff officer, Nat Res Coun-Nat Acad Sci, Washington, DC, 75-81; exec dir, Am Soc Microbiol, Washington, DC, 81-84; RETIRED. *Concurrent Pos:* Am Mgt Asn fel, 57-58; mem, Fed Exec Prog, Brookings Inst, 60 & 69; Nat Acad Sci & US rep to Int Asn Microbiol Socs, Moscow, 66; consult, Nat Acad Sci, 67-68 & 71; mem panel regulatory biol & consult, NSF, 67-71; mem, Aspen Exec Prog, 68; Found Microbiol lectr, 68; mem, Gov Sci Adv Bd, Md, 68; consult, NIH, 68-70 & 81, Off Educ, Dept Health, Educ & Welfare, 73 & 75, NASA, 75, Leonard Wood Mem Found, 75-81 & US State Dept, 80-81; consult US Ctr for Dis Control, 76; US State Dept, 80-81; consult, Nat Can Inst, 86- *Honors & Awards:* Barnett L Cohen Award in Microbiol, 67. *Mem:* Fel AAAS; fel NY Acad Sci; fel Am Acad Microbiol; hon mem Am Soc Microbiol (pres, 66); Soc Exp Biol & Med; Cosmos Club. *Res:* Research and development management; chemotherapy; amino acid metabolism; nutrition and enzymology of pathogenic bacteria; bacterial toxins and antitoxins; holder of one US patent. *Mailing Add:* 147 Fairview Ave Frederick MD 21701

HOUSHOLDER, GLENN ETTA (THOLEN), b Granbury, Tex, Dec 29, 25; m 48. MEDICAL SCIENCES. *Educ:* Univ Houston, BS, 47; Univ Tex, Houston, PhD(pharmacol), 70. *Prof Exp:* From asst prof to assoc prof, 70-81, PROF PHARMACOL, DENT BR, UNIV TEX HEALTH SCI CTR, HOUSTON, 81- *Concurrent Pos:* Fel pharmacol, Univ Tex Dent Br Houston, 71-72. *Mem:* AAAS; Am Chem Soc; Am Asn Cancer Res; Soc Exp Biol & Med; NY Acad Sci; Am Asn Dent Schs. *Res:* Biological effects of dental materials; inflammation; hemostasis; inflammation mediators. *Mailing Add:* Dept Pharmacol Univ Tex Dent Br Houston TX 77030

HOUSKA, CHARLES ROBERT, b Cleveland, Ohio, May 16, 27; m 53; c 3. PHYSICS. *Educ:* Mass Inst Technol, SB, 51, SM, 54, ScD(metall), 57. *Prof Exp:* Mem res staff, Metall Dept, Mass Inst Technol, 57-59 & Union Carbide Res Inst, 59-63; from assoc prof to prof metall, 63-69, prof metals & ceramic eng & head dept, 69-71, PROF MAT ENG, VA POLYTECH INST & STATE UNIV, 71- *Mem:* Am Crystallog Asn; Am Inst Mining, Metall & Petrol Engrs; Sigma Xi. *Res:* Solid state diffusion and transformation kinetics; x-ray diffraction. *Mailing Add:* Dept Mat Eng Va Polytech Inst & State Univ Blacksburg VA 24061

HOUSLEY, ROBERT MELVIN, b Roseburg, Ore, June 24, 34; m 63; c 2. PLANETOLOGY, SURFACE SCIENCE. *Educ:* Reed Col, BA, 56; Univ Wash, PhD(physics), 64. *Prof Exp:* Res instr solid state physics, Univ Wash, 64; fel, State Univ Groningen, 64-65; STAFF MEM, ROCKWELL INT SCI CTR, 65- *Concurrent Pos:* Prin investr contract for lunar sample & meteorite studies, NASA; vis prof dept physics, Calif Inst Technol, 84-86. *Mem:* AAAS; Am Phys Soc; Am Geophys Union; Microbeam Anal Soc; Meteoritical Soc. *Res:* Lattice dynamics and spin interactions in solids via the Mossbauer effect; origin and evolution of the solar system; solar wind and micrometeorite interactions with the lunar surface; structure and origin of meteorites. *Mailing Add:* Rockwell Int Sci Ctr 1049 Camino Dos Rios Thousand Oaks CA 91360

HOUSLEY, THOMAS LEE, b Akron, Ohio, Feb 22, 42; m 66; c 2. PLANT PHYSIOLOGY. *Educ:* Taylor Univ, BS, 64; Univ Conn, MA, 69; Univ Ga, PhD(bot), 74. *Prof Exp:* Teacher life sci, Rocky Hill Jr High Sch, 64-66 & biol, Edwin O Smith High Sch, 66-68; fel agron, Univ Wis, 74-75; ASST PROF AGRON, PURDUE UNIV, 75- *Mem:* Am Soc Plant Physiologists; Crop Sci Soc Am; Am Soc Agron; Am Asn Advan Sci. *Res:* Long distance translocation; the interrelationship between sources of transported compounds and sinks, in crops; environmental factors that influence dry matter accumulation in corn, small grains and soybeans. *Mailing Add:* Dept Agron Purdue Univ West Lafayette IN 47907

HOUSMANS, PHILIPPE ROBERT H P, b Oostende, Belg, Aug 27, 53. ANESTHESIOLOGY. *Educ:* Univ Antwerp, Belg, BS, 74, MD, 78, PhD(physiol), 84. *Prof Exp:* Investr physiol, Nat Res Found, 83-84; SPEC CLIN FEL ANESTHESIOL, MAYO FOUND, 84-, ASST PROF, 85- *Concurrent Pos:* Consult, dept anesthesiol, Univ Antwerp, Belg, 83-84. *Mem:* Biophys Soc; NY Acad Sci; Int Anesthesia Res Soc; Am Soc Anesthesiologists; AAAS. *Res:* Physiology and pharmacology of cardiac muscle. *Mailing Add:* Dept Anesthesiol Mayo Clin Rochester MN 55905

HOUSMYER, CARL LEONIDAS, b Ind, Jan 24, 36; m 59; c 2. ANALYTICAL CHEMISTRY. *Educ:* Ohio State Univ, BS, 65, PhD(pharmaceut chem), 71. *Prof Exp:* Anal chemist, Warren-Teed Pharmaceut (Rohm & Haas), 65-67; sr res scientist anal chem, Pfizer, Inc, 71-78; sect head anal chem, 78-90, DEPT HEAD ANALYTICAL CHEM, MARION MERRELL DOW, INC, 90- *Mem:* Am Chem Soc; Am Asn Pharmaceut Sci; Am Asn Adv Sci. *Res:* High performance liquid chromatography; electroanalytical chemistry; general analytical methods for organic compounds. *Mailing Add:* PO Box 68470 Indianapolis IN 46268

HOUSNER, GEORGE W(ILLIAM), b Saginaw, Mich, Dec 9, 10. CIVIL ENGINEERING. *Educ:* Univ Mich, BS, 33; Calif Inst Technol, MS, 34, PhD(civil eng), 41. *Prof Exp:* Struct engr, Los Angeles, 34-39; group leader, Eng Seismol Res, Calif Inst Technol, 39-41; engr, US Eng Corps, 41-42; chief opers anal, US Air Force, Europe & Africa, 43-45; from asst prof to prof civil eng & appl mech, 45-74, C F Braun prof, 74-81, EMER C F BRAUN PROF ENG, CALIF INST TECHNOL, 81- *Concurrent Pos:* Consult, seismic eng, TransArabian Pipe Line, Lisbon, Portugal supension bridge, Calif Water Proj, San Francisco Bay Area Rapid Transit syst, Exxon Hondo drilling platform & Los Angeles Metro-Rail proj; mem or chmn numerous comts & councils, earthquake res & eng; lectr, Univ Nev, Reno, 84, Univ Ill, 84, Renssalaer Polytech Inst, 88, Carnegie-Mellon Univ, 90. *Honors & Awards:* Vincent Bendix Res Award, Am Soc Eng Educ, 67; Von Karman Medal, Am Soc Civil Eng, 74, Newmark Medal, 81; Seismol Soc Medal, 81; Nat Medal of Sci, 88. *Mem:* Nat Acad Sci; Nat Acad Eng; hon mem Int Asn Earthquake Eng (pres, 69-73); Seismol Soc Am (pres, 77-78); hon mem Earthquake Eng Res Inst (pres, 55-60); Indian Nat Sci Acad; fel AAAS; Am Soc Civil Engrs; Am Geophys Union; Am Concrete Inst. *Res:* Earthquake engineering; structural dynamics; author or co-author of over 15 books and 174 technical papers. *Mailing Add:* 211 Thomas Lab Calif Inst Technol Pasadena CA 91125

HOUSTIS, ELIAS N, b Portaria, Greece, Sept 12, 45. APPLIED MATHEMATICS, COMPUTER SCIENCE. *Educ:* Univ Athens, BS, 69; Purdue Univ, PhD(math), 74. *Prof Exp:* Asst prof comput sci, Purdue Univ, 75-78; asst prof appl math, 78-79, ASSOC PROF COMPUT SCI, UNIV SC, 79- *Mem:* Asn Comput Mach; Int Asn Math & Comput Simulation; Soc Indust & Appl Math. *Res:* Numerical analysis; mathematical software; modelling and performance evaluation of computer systems. *Mailing Add:* 142 Knox Dr West Lafayette IN 47906-2148

HOUSTON, ARTHUR HILLIER, b Quebec City, Que, May 7, 31; m 55; c 2. ENVIRONMENTAL PHYSIOLOGY. *Educ:* McMaster Univ, BSc, 54; Univ BC, MA, 56, PhD(comp physiol), 58. *Prof Exp:* Asst prof zool, Dalhousie Univ, 58-60; from asst prof to assoc prof, Univ Man, 60-64; assoc prof, Univ Wis-Milwaukee, 64-65; from assoc prof to prof biol & asst chmn dept, Marquette Univ, 65-71, adj assoc prof physiol, Med Sch, 68-71, mem biomed eng group, 68-71; chmn dept biol sci, 71-75, actg dean, Div Math & Sci, 81-82, PROF BIOL SCI, BROCK UNIV, 71-, DEAN, DIV MATH & SCI, 84-; CONSULT, MACLAREN PLANSEARCH INC, 82- *Concurrent Pos:* Mem panel lab instr, Comn Undergrad Educ Biol Sci, 68; appraisals comt, Ont Coun Grad Studies, 74-77 & 82-84; adj grad prof zool, Univ Toronto & Univ Guelph, 82-; sr univ adminr course, Univ Western Ont, 84; mem, res develop comt, NSERC, 84-87. *Mem:* Am Soc Zool; Soc Exp Biol & Med; Can Soc Zool; AAAS. *Res:* Respiratory and metabolic adaptations of fishes to natural and induced environmental variation; water electrolyte and acid-base regulation; erythropoiesis and leucopoiesis in fishes. *Mailing Add:* Brock Univ Ste Catherines ON L2S 3A1 Can

HOUSTON, BLAND BRYAN, JR, b Vero Beach, Fla, Sept 19, 26; m 51; c 2. SOLID STATE PHYSICS. *Educ:* Vanderbilt Univ, BA, 49, MS, 50; Univ Ill, PhD(physics), 55. *Prof Exp:* SOLID STATE PHYSICIST, WHITE OAK LAB, NAVAL SURFACE WEAPONS CTR, 55- *Mem:* Am Phys Soc. *Res:* Lead and tin chalcogenide semiconductors; study of high and low-field transport properties related to the use of these materials in devices, infrared and thermoelectric in particular. *Mailing Add:* 10711 Blossom Lane Silver Spring MD 20903

HOUSTON, CHARLES SNEAD, b New York, NY, Aug 24, 13; m 41; c 3. INTERNAL MEDICINE. *Educ:* Harvard Univ, AB, 35; Columbia Univ, MD, 39. *Prof Exp:* Intern med, Presby Hosp, New York, 39-41; internist & cardiologist, Exeter Clin, 47-56; med dir, Aspen Inst, 56-58, internist & cardiologist, Aspen Clin, 58-62; dir prog, Peace Corps, India, 62-65, spec asst to dir in chg vol dr progs, Washington, DC, 65-66; prof community med & chmn dept, 66-77, prof 77-80, EMER PROF EPIDEMIOL & ENVIRON HEALTH & MED, COL MED, UNIV VT, 80- *Concurrent Pos:* Consult, Vet Admin Hosp, Manchester, NH, 49-56; res assoc, Cleveland Clin, Ohio, 59; consult, Vet Admin Hosp, Denver, Colo, 59-62; mem adv bd, Water Pollution Control Admin, 66-69. *Mem:* Am Col Physicians; Am Col Cardiol. *Res:* High altitude physiology; cardiovascular physiology; mechanical replacement of damaged heart; environmental health; pollution abatement. *Mailing Add:* 77 Ledge Rd Burlington VT 05401

HOUSTON, CLARENCE STUART, b Williston, NDak, Sept 26, 27; Can citizen; m 51; c 4. RADIOLOGY. *Educ:* Univ Man, MD, 51; FRCPC, 64. *Hon Degrees:* DLitt, Univ Sask, 87. *Prof Exp:* Lectr, Univ Sask, 64-65, from asst prof to assoc prof, 65-69, asst dir dept diag radiol, Univ Hosp, 68-82, head dept med imaging, 82-87, PROF DIAG RADIOL, UNIV SASK, 69- *Concurrent Pos:* Ed, J Can Asn Radiologists, 76-81. *Honors & Awards:* Roland Michener Conserv Award, Can Wildlife Fedn, 86; Douglas H Pimlott Award. *Mem:* Radiol Soc NAm; Am Ornith Union; Coun Biol Eds; Can Asn Radiol; fel Sask Natural Hist Soc. *Res:* Pediatric radiology; congenital dislocation of hip; fetal dwarfism; hazards of maternal x-irradiation; raptor and colonial bird banding; history of science; history of medicine; history of natural history. *Mailing Add:* 863 University Dr Saskatoon SK S7N 0J8 Can

HOUSTON, CLYDE ERWIN, civil engineering; deceased, see previous edition for last biography

HOUSTON, DAVID ROYCE, b Worcester, Mass, Sept 18, 32; m 54; c 3. FOREST PATHOLOGY. *Educ:* Univ Mass, BS, 54; Yale Univ, MF, 55; Univ Wis, PhD(plant path), 61. *Prof Exp:* From instr to asst prof plant path, Univ Wis, 58-61; PLANT PATHOLOGIST, CTR BIOL CONTROL, NORTHEASTERN FOREST EXP STA, US FOREST SERV, 61- *Concurrent Pos:* Exchange scientist, Forestry Comn, Alice Holt Res Lab, Eng, 75-76; lectr, Sch Forestry & Environ Studies, Yale Univ, 84- *Mem:* Soc Am Foresters; Am Phytopath Soc. *Res:* Forest tree diseases; ecological aspects of dieback and decline diseases, with special attention to oak decline, beech bark disease and sapstreak disease of sugar maple. *Mailing Add:* Ctr Biol Control 51 Mill Pond Rd Hamden CT 06514

HOUSTON, FORREST GISH, b Devol, Okla, July 18, 16; m 42; c 3. BIOCHEMISTRY. *Educ:* Tex Tech Col, BS, 38; Univ Tex, MA, 42; Ohio State Univ, PhD(biochem), 47. *Prof Exp:* Instr biochem, Univ Tex, 42-46; asst chemist, Ky Agr Exp Sta, 47-53; ASSOC PROF BIOCHEM, UNIV TEX MED BR, GALVESTON, 53- *Mem:* AAAS; Am Chem Soc; Soc Exp Biol & Med. *Res:* Iron metabolism; metalloproteins; analytical methods in biochemistry. *Mailing Add:* 6713 E Bayou Dr Hitchcock TX 77563

HOUSTON, JACK E, b Arkansas City, Kans, Mar 11, 33; m 56; c 2. SURFACE PHYSICS. *Educ:* Okla State Univ, BS, 56, MS, 62, PhD(physics), 65. *Prof Exp:* Staff engr, Sandia Corp, NMex, 56-58; res engr, Jet Propulsion Lab, Calif, 59-60; res asst physics, Okla State Univ, 60-65; asst prof, SDak Sch Mines & Technol, 65-66; mem staff, 66-74, supvr surface physics div, 74-79, DISTINGUISHED MEM TECH STAFF, SANDIA NAT LABS, 85- *Mem:* Am Phys Soc; Am Vacuum Soc; Fel, Am Phy Soc. *Res:* Development and application of various electron spectroscopies as tools for studying the surface and its effect on the interaction of the solid with its environment. *Mailing Add:* Div 1114-Sandia Nat Lab PO Box 5800 Albuquerque NM 87185-5800

HOUSTON, JAMES GREY, b Farris, Okla, June 13, 38. ORGANIC CHEMISTRY, PHOTOCHEMISTRY. *Educ:* Okla State Univ, BS, 60; Ga Inst Technol, PhD(chem), 65. *Prof Exp:* Res chemist, Esso Res & Eng Corp, 65-66; head dept, 66-68, dean grad studies, 71-75, PROF CHEM, SUL ROSS STATE UNIV, 68- *Concurrent Pos:* Consult, Dow Chem Co, 67-69; vpres, Agrihol Corp, 80- *Mem:* Am Chem Soc; Sigma Xi. *Res:* Natural products; cresote bush and cedar oil; feasibility of feeding cresole bush to cattle; fermentation of carbohydrates. *Mailing Add:* Dept Chem Sul Ross State Univ Box 6080 Alpine TX 79832

HOUSTON, JAMES ROBERT, b Berkeley, Calif, Mar 31, 47; m 76. ENGINEERING MECHANICS, PHYSICS. *Educ:* Univ Calif, Berkeley, BA, 69; Univ Chicago, MS, 70; Univ Fla, MS, 74, PhD(eng mech), 78. *Prof Exp:* Res physicist weapons effects, US Army Corps Engrs, 70-72, res physicist hydraul, 72-79, res hydraul engr, 79-86, CHIEF, COASTAL ENG

RES CTR, WATERWAYS EXP STA, US ARMY CORPS ENGRS, 86- *Concurrent Pos:* Mem steering comt & chmn subcomt Tsunamis, Interagency Comt Seismic Safety Construct, Off President, 78- *Mem:* Am Soc Civil Engrs. *Res:* Numerical modeling of water waves; the phenomena modeled include tsuanmis, waves generated by seismic events, harbor seiching and interaction of waves, structures and sediment. *Mailing Add:* 22 Jil Marie Circle Vicksburg MS 39180

HOUSTON, JOHNNY LEE, b Sandersville, Ga, Nov 19, 41; m 69. MATHEMATICS. *Educ:* Morehouse Col, BA, 64; Atlanta Univ, MS, 66; Purdue Univ, PhD(math), 74. *Prof Exp:* Instr math, E E Smith High Sch, 64-65 & Stillman Col, 67-69; dir black cult ctr, Purdue Univ, 72-73; assoc prof, Savannah State Col, 74-75; ASSOC PROF & CHMN DEPT MATH, ATLANTA UNIV, 75-; SR RES PROF, DEPT MATH & COMPUTER SCI, ELIZABETH CITY STATE UNIV, 88- *Concurrent Pos:* Merrill grant, Atlanta Univ-Univ Strasbourg, 66-67; Spencer Found res grant, Atlanta Univ, 76-77; lectr statist, Atlanta Univ, 76-; lectr math, Math Asn Am-BAM prog, 77-; reviewer math, Zentralblatt Math, 77- *Mem:* Nat Asn Math; Am Math Soc; Math Asn Am; Nat Coun Teachers Math; AAAS. *Res:* Algebra, particularly group theory, matrices and generalized inverses; categorical algebra; graph theory, combinatorics and applications. *Mailing Add:* Dept Math & Computer Sci Elizabeth City State Univ Elizabeth City NC 27909

HOUSTON, L L, b Wichita, Kans, June 3, 40; m 62; c 2. BIOCHEMISTRY. *Educ:* Kans State Univ, BS, 62; Univ Wash, PhD(biochem), 67. *Prof Exp:* Res biochemist, Univ Calif, Berkeley, 67-69; asst prof, 69-73, assoc prof, 73-78, PROF BIOCHEM, UNIV KANS, 78-, PROF GENETICS, 80- *Concurrent Pos:* Res career develop award, NIH, 75-; Guest prof, Swiss Nat Sci Found, Univ Basel, 75-76. *Mem:* Am Soc Biol Chemists; Am Soc Microbiologists; AAAS. *Res:* Relation of structure of proteins to their function; mechanism of toxin action; multi- and bifunctional enzymes; microtubules; lectins; membrane biochemistry. *Mailing Add:* Dept Biochem Univ Kans Lawrence KS 66045

HOUSTON, MARSHALL LEE, b Poplar Bluff, Mo, Jan 27, 38; m 62; c 2. EMBRYOLOGY, TERATOLOGY. *Educ:* William Jewell Col, BA, 60; Kans State Univ, MS, 64, PhD(anat), 67. *Prof Exp:* Res assoc embryol, Southwest Found Res & Educ, Tex, 66-69; asst prof anat, Med Col Va, 69-70; asst prof anat, 70-73, ASSOC PROF ANAT & OBSTET & GYNEC, TEX HEALTH SCI CTR SAN ANTONIO, 73- *Mem:* Perinatal Res Soc; Am Asn Anatomists. *Res:* Comparative embryology and placentology of primates; anatomical and behavioral teratology of the brain; development of the hypothalamo-pituitary-adrenal axis. *Mailing Add:* Dept Anat, Obstet & Gynec Univ Tex Health Sci Ctr 7703 Floyd Curl Dr San Antonio TX 78284

HOUSTON, PAUL LYON, b Hartford, Conn, Jan 27, 47. CHEMICAL PHYSICS. *Educ:* Yale Univ, BS, 69; Mass Inst Technol, PhD(chem), 73. *Prof Exp:* Fel chem, Univ Calif, Berkeley, 73-75; from asst prof to assoc prof, 75-85, PROF CHEM, CORNELL UNIV, 85- *Concurrent Pos:* vis scientist, Max Planck Inst for Quantum Optus; Guggenheim fel, 87. *Honors & Awards:* Camille and Henry Dueyfus Teacher Scholar Award 80. *Mem:* Am Phys Soc; AAAS; Am Chem Soc. *Res:* Reaction kinetics of molecules in selected vibrational and electronic states; laser-induced chemical reactions, photodissociation and vibrational energy transfer; gas-surface interactions. *Mailing Add:* Dept Chem Baker Lab Cornell Univ Ithaca NY 14853

HOUSTON, ROBERT EDGAR, JR, b Detroit, Mich, Feb 17, 24; m 46; c 3. PHYSICS. *Educ:* Mich State Col, BS, 49, MS, 51; Pa State Univ, PhD, 57. *Prof Exp:* Instr physics, Mo Sch Mines, 51-53; from asst prof to assoc prof, 57-66, actg chmn dept, 65-66, chmn dept, 66-69, 72-73, PROF PHYSICS, UNIV NH, 66-, CHMN DEPT, 76- *Mem:* Am Phys Soc; Am Asn Physics Teachers; Am Geophys Union. *Res:* Ionospheric physics; tracking of earth satellites; electron distributions through rocket borne instrumentation. *Mailing Add:* Dept Physics Univ NH Durham NH 03824

HOUSTON, ROBERT S, b Monroe, NC, May 11, 23; m 52; c 1. ECONOMIC GEOLOGY, PETROLOGY. *Educ:* NC State Univ, BE, 48, MS, 50; Columbia Univ, PhD(econ geol), 54. *Prof Exp:* Geophysicist, Atlantic Refining Co, 48-49; from instr to assoc prof, 53-70, PROF GEOL & HEAD DEPT GEOL & GEOPHYSICS, UNIV WYO, 70- *Concurrent Pos:* US Geol Surv grants, 55-57 & 67-82; Geol Surv Wyo grant, 57-66; NSF grant, 67-82; Dept Energy grants, 75-81; NASA grant, 71-75; Consult, Los Alamos, 76-80; provost & vpres acad affairs, Univ Wyo, 86-88. *Mem:* Fel Geol Soc Am; Am Asn Petrol Geol; Soc Econ Geol; Soc Econ Paleont & Mineral; Am Geophys Union; Sigma Xi. *Res:* Study of layered mafic complexes; fossil marine and non-marine placers of the Rocky Mountain region; Precambrian geology of Wyoming province of Rocky Mountain region; uranium geology, remote sensing and Antarctic geology. *Mailing Add:* Dept Geol Univ Wyo Laramie WY 82071

HOUSTON, ROY SEAMANDS, b Lubbock, Tex, Dec 28, 42; m 61; c 4. MALACOLOGY, MARINE ECOLOGY. *Educ:* Univ Ariz, BS, 68, PhD(zool), 74; Univ of the Pac, MS, 70. *Prof Exp:* asst prof,74-80, ASSOC PROF MARINE BIOL, LOYOLA MARYMOUNT UNIV, 80- *Mem:* Malacol Soc London; Sigma Xi. *Res:* Functional morphology of reproductive and alimentary systems of marine molluscs. *Mailing Add:* Dept Biol Loyola Marymount Univ Los Angeles CA 90045

HOUSTON, SAMUEL ROBERT, b Los Angeles, Calif, May 20, 35; m 63; c 3. APPLIED STATISTICS. *Educ:* Univ Calif, Los Angeles, BA, 57; Calif State Univ, Los Angeles, MA, 61; Univ Ore, MS, 64; Colo State Col, PhD(appl statist), 67. *Prof Exp:* Instr math, Los Angeles City Sch Dist, 57-65; res fel, Bur Res, Colo State Col, 65-66; prog assoc, Grants Div, Charles F Kettering Found, 66-67, res specialist, Res & Develop Div, 67-68; from asst prof to assoc prof, 68-74, PROF MATH & APPL STATIST, UNIV NORTHERN COLO, 74-, CHMN DEPT, 75-78, 81-82, 89- *Concurrent Pos:* Scholar biostatist & psychol, Univ Calif, Los Angeles, 67-68; consult, CFK, Ltd, Colo, 68-73; pres, Multiple Linear Regression Spec Interest Group, Am Educ Res Asn; Nat Cancer Inst fel biomet, Sch Med, Yale Univ, 73-74; vis prof, Univ Wyo, 82-83; lectr & consult, Zagazig Univ, Egypt, 83; exec ed, J Exp Educ, 75-78; reviewer, Computing Reviews, 82-86; reviewer, Mult Linear Regression Viewpoints, 84-; vis prof, Univ Ga, 85-86. *Mem:* Am Educ Res Asn; Am Statist Asn; Sigma Xi. *Res:* Judgment analysis; policy capturing models; multidimensional policy models; evaluation models; management styles and stress; personality profile systems. *Mailing Add:* Math & Appl Statist Dept Univ Northern Colo Greeley CO 80639

HOUSTON, VERN LYNN, b Denver Colo, July 4, 47; m 70; c 3. REHABILITATION ENGINEERING. *Educ:* Univ Colo, BSAE, 69; Univ Calif, MSE, 72; Columbia Univ, MPh, 80, PhD(elec eng), 87. *Prof Exp:* Staff engr, Vet Admin Res Ctr, 74-77; mem tech staff, AT&T Bell Labs, 81-86; INSTR REHAB MED, MED CTR, NY UNIV, 86- *Mem:* NY Acad Med; Inst Elec & Electronics Engrs; Am Soc Mech Engrs; Soc Indust & Appl Math; Biomed Eng Soc; Int Soc Prosthetics & orthotics. *Res:* Research and development in prosthetics and orthotics computer-aided manufacturing; soft tissue viscoelastic finite element modeling and analysis; biomedical signal modeling and procssing with application in rehabilitation engineering. *Mailing Add:* 14 Erwin Park Montclair NJ 07042

HOUSTON, WALTER SCOTT, b Milwaukee, Wis, May 30, 12; m 38; c 3. ASTRONOMY, METEORITICS. *Educ:* Univ Wis, PhB, 35; Univ Ala, AM, 48. *Prof Exp:* Instr eng, Kans State Univ, 55-60; physics ed, Am Educ Publ, Inc, 60-66, sr sci ed, 66-74; RETIRED. *Concurrent Pos:* Civilian sr instr, US Army-Air Force Advan Navig Sch, Selman Field, La, 42-46; teacher astron courses, Univ Ala, Univ Cincinnati, Wash Univ, Kans State Univ & Univ Hartford; consult, Smithsonian Astrophys Observ, 61-71. *Honors & Awards:* Award, Astron League, 74; Leglie C Peltier Award, 84; Asteroid 3031 named Houston, Int Astron Union. *Mem:* Fel Meteoritical Soc; Am Asn Variable Star Observers; Am Meteor Soc; Asn Lunar & Planetary Observers; fel AAAS. *Res:* Recording of sudden enhancement of atmospherics; photometry of variable stars; fireball reduction; field exploration for meteorites; science writing in the physical sciences. *Mailing Add:* Boardman Rd East Haddam CT 06423

HOUSTON, WILLIAM BERNARD, JR, b Macon, Ga, July 3, 29; m 55; c 2. GEOMETRY. *Educ:* Ga Inst Technol, BEE, 52; Mass Inst Technol, PhD(math), 57. *Prof Exp:* Asst prof math, Morehouse Col, 57-59 & Carleton Col, 59-62; from asst prof to assoc prof, 62-70, PROF MATH, ANTIOCH COL, 70- *Concurrent Pos:* Vis assoc prof, Univ Col Cape Coast, Ghana, 66-67. *Mem:* Am Math Soc. *Res:* Connexions; functional analysis; calculus of variations; differential geometry. *Mailing Add:* Dept Math Antioch Univ Yellow Springs OH 45387

HOUSTON, WILLIE WALTER, JR, b Cedartown, Ga, Sept 14, 51. CELL BIOLOGY, HISTOLOGY. *Educ:* Morehouse Col, BS, 74; Atlanta Univ, MS, 76, PhD(biol), 81. *Prof Exp:* Lectr biol, Spelman Col, 79; instr, Ky State Univ, 79-80; PROF BIOL & CHMN, CENT STATE UNIV, 80- *Concurrent Pos:* Instr cell biol, Wright State Univ, 81. *Mem:* Nat Inst Sci; AAAS. *Res:* Changes in inferred proteins synthetic activity and its association with hyaluronic acid changes in chick brains development from 4 day-old embryos to 8 day-old embryos. *Mailing Add:* Dept Biol Cent State Univ Wilberforce OH 45384

HOUTMAN, THOMAS, JR, b Decatur, Mich, Apr 5, 18; m 42; c 2. ORGANIC CHEMISTRY. *Educ:* Hope Col, AB, 40; La State Univ, MS, 42. *Prof Exp:* Chemist, Dow Chem Co, 42-48, group leader, 48-50, asst dir, Strosackers Lab, 50-56, dir, 56-63, admin dir chem dept res, 63-68, admin dir prod dept res, 68-72, dir contract projs, 72-74, dir, Hydrocarbon Res Lab, 75-82, mgr, Mich Div Res & Develop Employee Relations, 82-85; RETIRED. *Concurrent Pos:* Mem, Grace Comn, 82; Cancer consult, Hope Col, 82. *Mem:* Am Chem Soc. *Res:* Synthesis of ultraviolet absorbers; chemistry of ethylene, propylene and butylene oxides; aliphatic amines; oxazolidionone; aromatic phenolic derivatives. *Mailing Add:* 765 Oakmont Lane Winter Haven FL 33884

HOUTS, GARNETTE EDWIN, b Chattanooga, Tenn, July 26, 36; m 65; c 3. VIROLOGY. *Educ:* Univ Tenn, Knoxville, BA, 58, MS, 64; Univ Tenn, Memphis, PhD(virol), 71. *Prof Exp:* Technician bact, Tenn State Health Dept, 58-59; technician biochem, Oak Ridge Nat Lab, 59-60, 62-64 & 65; res assoc, Med Ctr, Univ Ky, 65-68; res assoc oncornaviruses, St Joseph's Hosp, 71-73; assoc virologist oncornaviruses, Life Sci Res Lab, 73-83; FOUNDER & PRES, MOLECULAR GENETIC RESOURCES, INC, 83- *Res:* Physical, chemical and functional characterizations of reverse transcriptase enzyme from avian myeloblastosis virus; isolation of other proteins from AMV for use in screening drugs for AIDS treatment. *Mailing Add:* 6201 Johns Rd Suite Eight Tampa FL 33634

HOUTS, LARRY LEE, b Three Rivers, Mich, Oct 1, 42; m 73; c 2. DEVELOPMENTAL BIOLOGY, ANATOMY. *Educ:* Rockford Col, AB, 64; Fla State Univ, MS, 66; State Univ NY, Albany, PhD(biol), 77. *Prof Exp:* Instr biol, Winthrop Col, 66-69; res biologist, VA Hosp, 74-76; asst prof biol, Bethany Col, 76-81; asst prof biol, Wheeling Col, 81-82; intern, Garden City Hosp, 86-87, med resident, 88. *Concurrent Pos:* Consult allied health, East Cent Col Consortium, 77-81. *Mem:* Soc Develop Biol; Sigma Xi; Tissue Cult Asn; AAAS. *Res:* Aging of inductive tissue and control of anterioposterior polarity in the limb bud of the chick embryo; effects of various wavelengths of light on the estrous cycle of Mus musculus; zoology. *Mailing Add:* 56814 Feathers Ct Three Rivers MI 49093

HOUTS, PETER STEVENS, b Great Neck, NY, March 17, 33; m 60; c 2. SOCIAL PSYCHOLOGY, STRESS MANAGEMENT. *Educ:* Antioch Col, BA, 55; Univ Mich, PhD(soc psychol), 63. *Prof Exp:* Asst prof psychol, Goucher Col, 63-65; postdoctoral fel social psychol, psychait dept, Stanford Med Sch, 65-67; asst prof, 67-71, ASSOC PROF SOC PSYCHOL, PA STATE UNIV 71- *Mem:* Am Psychol Assoc; Am Pub Health Assoc; Am Assoc Clin Oncol; Am Assoc Univ Prof. *Res:* Psychological, social, and econimic impact of cancer on patients and their families; need for support services among cancer patients; use of cancer screening among persons at risk for cancer; long term psychological, social and economic impacts of the Three Mile Island Crisis. *Mailing Add:* Dept Behav Sci Col Med Pa State Univ PO Box 850 Hershey PA 17033

HOUTS, RONALD C(ARL), b Chicago, Ill, Sept 22, 37; m 60; c 2. DIGITAL SIGNAL PROCESSING, STATISTICAL COMMUNICATIONS. *Educ:* Univ Fla, BSE, 59, MSE, 60, PhD(elec eng), 63. *Prof Exp:* From asst prof to assoc prof, 66-73, PROF ELEC ENG, UNIV ALA, 73- *Concurrent Pos:* Researcher, Astrionics Labs, Telemetry Systs Br, Marshall Space Flight Ctr, Ala, 66-74; vis res prof, US Army Missile Res & Develop Command, 74-75, vis prof, US Mil Acad, West Point, 83-84; recipient grant, Electronics Div, US Army Res Off, 76-78; Consult, US Army Missile Command, 79-82 & 87, Universl Data Systs, 83-86. *Mem:* Sr mem Inst Elec & Electronics Engrs; Sigma Xi. *Res:* Statistical communications; data transmission; digital filtering; signal and system theory, especially digital signal processing techniques as applied to communications and radar; author of one publication. *Mailing Add:* Dept Elec Eng Univ Ala PO Box 870286 Tuscaloosa AL 35487-0286

HOUZE, RICHARD NEAL, b Atlanta, Ga, Oct 2, 38; m 60; c 2. CHEMICAL ENGINEERING. *Educ:* Ga Inst Technol, BChE, 60, Univ Houston, MS, 66, PhD(chem eng), 68. *Prof Exp:* Res engr, Esso Res & Eng Co, 60-63; NSF fel, Delft Technol Univ, 68-69; asst prof, 69-76, ASSOC PROF CHEM ENG, PURDUE UNIV, 76- *Mem:* Am Inst Chem Engrs. *Res:* Transport processes at gas-liquid interfaces; turbulent flow of fluids; transport processes in physiological systems and respiratory processes. *Mailing Add:* 734 1/2 Owen St West Lafayette IN 47905

HOVANEC, B(ERNARD) MICHAEL, b Riverside, Calif, May 26, 52; m 79; c 3. INDUCTIVITY COUPLED PLASMA MASS SPECTROMETRY, PARTICLE BEAM & LIQUID CHROMATOGRAPH MASS SPECTROMETRY. *Educ:* Univ Calif, Riverside, BA, 75. *Prof Exp:* Chemist, Indust Polymers, 76-77, Lever Bros, 77-78; sr chemist, W Coast Tech, 78-80; sr anal chemist, Rockwell Int, 80-86; SR STAFF CHEMIST, W COAST ANALYTICAL SERV, 86- *Mem:* Am Chem Soc; Am Soc Mass Spectrometry; Am Welding Soc. *Res:* Application of sample introduction methodologies to permit the determination of trace level metals in various pure solutions or compounds; development of appropriate automated sampling systems for hazardous chemicals; application of same for trace organics via particle beam interfaces into organic mass specs with on-line HPLC separation. *Mailing Add:* W Coast Anal Serv 9840 Alburtis Ave Santa Fe Springs CA 90670-5086

HOVANESIAN, JOSEPH DER, b Detroit, Mich, Aug 14, 30; m 60; c 2. ENGINEERING MECHANICS, MECHANICAL ENGINEERING. *Educ:* Mich State Univ, BS, 53, MA, 54, PhD, 58. *Prof Exp:* Res asst eng, Mich State Univ, 54-58; asst prof, Pa State Univ, 58-60; assoc prof eng mech, Wayne State Univ, 60-70; PROF ENG & CHMN, MECH ENG DEPT, OAKLAND UNIV, 70- *Honors & Awards:* Fel, Soc Exp Mech Anal; Shield of Strathelyde. *Mem:* Optical Soc Am; fel Soc Exp Mech Anal; Am Soc Mech Engrs. *Res:* Coherent optics and holography; photelasticity; theory of elasticity; experimental stress analysis; noise & vibration control; non-destructive testing. *Mailing Add:* Sch Eng Oakland Univ Rochester MI 48309-4401

HOVANESSIAN, SHAHEN ALEXANDER, b Teheran, Iran, Sept 6, 31; nat US; m 60; c 2. ELECTRONICS ENGINEERING, APPLIED MATHEMATICS. *Educ:* Univ Calif, Los Angeles, BS, 54, MS, 55, PhD(appl math, eng), 58. *Prof Exp:* Instr eng, Univ Calif, Los Angeles, 54-58; res engr, Standard Oil Co Calif, 58-61; LECTR ENG, UNIV CALIF, LOS ANGELES, 61-; SR SCIENTIST, HUGHES AIRCRAFT CO, 63-; CONSULT ENGR, 63- *Mem:* Am Soc Mech Engrs; Inst Elec & Electronics Engrs. *Res:* Radar detection and missile trajectory problems; signal processing and information theory studies; formulation of mathematical models and subsequent digital computer simulation; numerical solution of fluid flow and heat transfer problems; mathematical programming methods in economic studies; author of over 35 publications on radars and computers. *Mailing Add:* Aerospace Corp PO Box 92957 Los Angeles CA 90009

HOVDE, CHRISTIAN ARNESON, b Decorah, Iowa, Dec 25, 22; m 48. ANATOMY. *Educ:* St John's Col, AB, 48; Bucknell Univ, MA, 50; Columbia Univ, PhD(anat), 53. *Hon Degrees:* DD, Seabury-Western Theol Sem, 68. *Prof Exp:* Jr instr, Johns Hopkins Univ, 48-49; lab asst, Bucknell Univ, 49-50; asst neurol, Columbia Univ, 50-52, asst anat, 51-52, from instr to asst prof, 52-56; from asst prof to assoc prof, Seton Hall Col Med & Dent, 56-63; dir, Bishop Anderson Found, 63-88; prof & chmn, Dept Relig & Health, Col Sci, Rush Univ, 76-88; RETIRED. *Mem:* Assoc Asn Res Nerv & Ment Dis; Am Asn Anatomists; NY Acad Sci. *Res:* Physiology and disorders of the nervous system. *Mailing Add:* 2245 Club Hause Dr Lillian AL 36549

HOVDE, RUTH FRANCES, b Devils Lake, NDak, Mar 3, 17. MEDICAL TECHNOLOGY. *Educ:* Univ Minn, BS, 38, MS, 49. *Prof Exp:* Head technologist, Abbott Hosp, Minneapolis, 39-45; admin lab technologist, Univ Hosps, 45-46, from instr to assoc prof, 46-64, PROF MED TECHNOL, UNIV MINN, MINNEAPOLIS, 64- *Mem:* Am Soc Med Technol (pres-elect, 53, pres, 54). *Mailing Add:* 4033 20th Ave S Minneapolis MN 55407

HOVE, JOHN EDWARD, b Stoughton, Wis, July 19, 24; m 62; c 2. PHYSICS, MATERIALS SCIENCE. *Educ:* Purdue Univ, BS, 45; Univ Buffalo, MS, 51; Cornell Univ, PhD(physics), 53. *Prof Exp:* Engr aero, Cornell Aeronaut Lab, 46-50; group leader solid state physics, Atomic Int, NAA, 53-61; dir mat sci, Aerospace Corp, 61-67; vpres eng, Whittaker Corp, 67-73; MEM TECH STAFF, INST DEFENSE ANALYSES, 75- *Mem:* Assoc fel Am Inst Aeronaut & Astronaut; Am Phys Soc; Am Ceramic Soc. *Res:* Solid state physics; radiation damage; high temperature ceramics; metallurgy; composite materials; structures; aerodynamics. *Mailing Add:* 1600 S Eads St No 222N Arlington VA 22202

HOVEL, HAROLD JOHN, b Middletown, Ohio, July 6, 42; m 65; c 2. SOLID STATE PHYSICS, MATERIALS SCIENCE. *Educ:* Carnegie-Mellon Univ, BS, 64, MS, 65, PhD(elec eng), 68. *Prof Exp:* MEM RES STAFF, SOLID STATE DEVICE RES, IBM CORP, 68- *Mem:* Electrochem Soc; Am Phys Soc; Am Inst Physics; Inst Elec & Electronics Engrs. *Res:* Liquid phase epitaxy; solid state lasers; heterojunctions and heterojunction devices; vapor epitaxy; bistable switching and memory device; optical properties of thin films; investigations of trapping effects in solids. *Mailing Add:* Diane Ct Katonah NY 10536

HOVELAND, CARL SOREN, b Sand Creek, Wis, Oct 25, 27; m 51, 81. AGRONOMY. *Educ:* Univ Wis, BS, 50, MS, 52; Univ Fla, PhD(agron), 59. *Prof Exp:* Asst, Univ Wis, 50-52; asst agronomist, Exp Sta, Agr & Mech Col, Tex, 52-55; asst agron, Exp Sta, Univ Fla, 55-58, asst bot, 58-59; assoc agronomist, Auburn Univ, 59-68, prof agron, 68-81; PROF AGRON, UNIV GA, 81- *Concurrent Pos:* Staff scientist, Grassland Div, Dept Sci & Indust Res, NZ, 70 & 79; lectr, Univ Zagreb, Yugoslavia, 78 & Univ Rio Grande de Sul, Brazil, 79; consult, SAfrica Dept Agr, 79 & Partners Am, Ecuador, 83; vis prof, Univ Guelph, Ontario, Can, 84; pres, Am Forage & Grassland Coun, 83. *Honors & Awards:* Goddard lectr, Univ Tenn, 81; King Found lectr, Univ Ark, 83; Silver Medallion Award, Am Forage & Grassland Coun, 86. *Mem:* Fel Am Soc Agron; fel Crop Sci Soc Am; Weed Sci Soc Am; fel AAAS. *Res:* Pasture management and utilization; crop ecology; forage plant physiology; weed ecology; grazing management. *Mailing Add:* Dept Agron Univ Ga Athens GA 30602

HOVERMALE, JOHN BRUCE, b Martinsburg, WVa, Dec 7, 38; m 60; c 2. METEOROLOGY. *Educ:* Pa State Univ, BS, 60, MS, 62, PhD, 65. *Prof Exp:* Res meteorologist, US Weather Bur, 62-67; from asst prof to assoc prof, 67-74, ADJ PROF METEOROL, PA STATE UNIV, UNIVERSITY PARK, 74- *Concurrent Pos:* Consult, US Weather Bur, 67- *Mem:* Am Meteorol Soc. *Res:* Atmospheric dynamics on synoptic scales and mesoscales; utilization of numerical atmospheric models for theoretical studies and operational weather prediction. *Mailing Add:* 25224 Flanders Dr Carmel CA 93923-8306

HOVERSLAND, ARTHUR STANLEY, b Hallock, Minn, Apr 20, 22; m 46; c 4. REPRODUCTIVE PHYSIOLOGY, COMPARATIVE PHYSIOLOGY. *Educ:* Mont State Univ, BS, 51, MS, 58; Ore State Univ, PhD, 70. *Prof Exp:* From instr to assoc prof animal sci, Mont State Univ, 51-67; res assoc, Med Sch, Univ Ore, 68-70; assoc prof, Calif State Univ, Fresno, 71-73, prof animal sci & chmn dept, 73-78; EXEC SECY, HUMAN EMBRYOLOGY & DEVELOP STUDY SECT, NIH, 78- *Concurrent Pos:* Vis scientist, Univ of Calif, Davis, 77. *Mem:* Am Soc Animal Sci; Soc Study Fertil; Soc Study Reproduction; Am Physiol Soc; NY Acad Sci. *Res:* Comparative animal reproductive physiology, hemodynamic adjustments during pregnancy, fetal-maternal relationships; growth and maturation, respiratory and biochemical properties of mammalian blood; perinatal physiology and embryology; animal science. *Mailing Add:* NIH Westwood Bldg Rm 222 Bethesda MD 20892

HOVERSLAND, ROGER CARL, b Bozeman, Mont, June 10, 51; m 71; c 2. EMBRYO IMPLANTATION, SEX DEPENDENT TYPE II DIABETES. *Educ:* Calif State Univ, Fresno, BA, 74; Univ Ore, PhD(anat), 80. *Prof Exp:* NIH postdoctoral res, Dept Obstet & Gynec, Univ Kans, 80-82; asst prof, Dept Anat, Chicago Med Sch, 82-88; ASSOC PROF, DEPT ANAT, SCH MED, IND UNIV, 88- *Concurrent Pos:* Postdoctoral fel, individual nat res award, Nat Inst Child Health & Human Develop, 80-82; prin investr, NIH new investr award, Nat Inst Child Health & Human Develop, 82-87, Am Diabetes Found Feasibility grant, 86-88 & NIH Res Grant, 91- *Mem:* Sigma Xi; Soc Study Reproduction; Am Asn Anatomists. *Res:* Endocrine and maternal/fetal interactions of the immune regulatory responses in mice during embryo implanatation and pregnancy; hormone influences in genetic obesity and subsequent sex dependent diabetes in hypertensive rats. *Mailing Add:* Sch Med Ind Univ 2101 Coliseum Blvd E Ft Wayne IN 46805

HOVERSON, SIGMUND JOHN, b Seattle, Wash, Oct 18, 41; m 65; c 3. PHYSICS. *Educ:* Calif Inst Technol, BS, 63; Univ Wash, MS, 67; Tex A&M Univ, PhD(physics), 74. *Prof Exp:* Technician, Math Sci Northwest Inc, 72-75, scientist laser res & computer modeling, 75-; COMPUTER SYST ANALYSIS, KING COUNTY. *Mem:* Am Phys Soc. *Res:* Alternate energy source research especially those involving high power laser applications; computer modeling of physical systems. *Mailing Add:* 9642 NE 121st Ln Kirkland WA 98034

HOVEY, HARRY HENRY, JR, b Pittsfield, Mass, Aug 28, 30; m 57; c 2. ENVIRONMENTAL ENGINEERING, AIR POLLUTION. *Educ:* Rensselaer Polytech Inst, BCE, 52; Univ Minn, MPH, 60. *Prof Exp:* Hydraul engr, Fish & Wildlife Br, US Dept Interior, 56-57; sanit engr, Robert Wheeler Consult, 57-58; sr sanit engr, NY State Dept Environ Conserv, 58-65, assoc engr, 65-67, assoc dir, 67-76, dir, Div Air Resources, 76-88; PRIN CONSULT ENG, TRC ENVIRON CONSULTS INC, 88- *Concurrent Pos:* Mem, Clean Air Sci Adv Comt, US Environ Protection Agency, 78-82; mem, Int Air Qual Adv Bd & Int Joint Comt, 80-88. *Honors & Awards:* S Smith Griswald Award, Air Pollution Control Asn, 82. *Mem:* Air Pollution Control Asn (vpres, 77-79); Am Acad Environ Engrs; Nat Soc Prof Engrs. *Res:* Development of atmospheric pollution control regulations and ambient air quality standards; development of air pollution survey techniques. *Mailing Add:* 15 Sylvan Lane Troy NY 12180

HOVIN, ARNE WILLIAM, b Norway, Dec 30, 22; nat US; m 53; c 2. GENETICS, PLANT BREEDING. *Educ:* Norwegian Col Agr, BS, 49; Univ Calif, Los Angeles, PhD(plant genetics), 57. *Prof Exp:* Asst, Norwegian Col Agr, 49-52 & Univ Calif, Los Angeles, 53-57; res geneticist, Regional Pasture Res Lab, Agr Res Serv, USDA, 58-65, res leader grass & turf invests, 65-69; prof agron & plant genetics, Univ Minn, St Paul, 69-81; assoc dir, Mont Agr Exp Sta, Mont State Univ, Bozeman, 81-87; RETIRED. *Honors & Awards:* Merit Award, Am Forage & Grassland Coun. *Mem:* Am Soc Agron; Crop Sci Soc Am. *Res:* Grass breeding; quantitative and qualitative genetics. *Mailing Add:* 7734 Springhill Comm Rd Belgrade MT 58714-8421

HOVINGH, JACK, b Grand Rapids, Mich, May 5, 35. NUCLEAR ENGINEERING. *Educ:* Univ Mich, Ann Arbor, BSE(mech eng) & BSE(math), 58; Univ Calif, Berkeley, MS, 73. *Prof Exp:* NUCLEAR ENGR, UNIV CALIF, LAWRENCE LIVERMORE NAT LAB, 58- *Mem:* Am Nuclear Soc. *Res:* Design and analysis of both magnetically and inertially confined fusion reactors for energy applications. *Mailing Add:* 4250 Muirwood Dr Pleasanton CA 94588-4376

HOVIS, LOUIS SAMUEL, b Gastonia, NC, May 2, 26; m 49; c 2. CHEMISTRY, PHYSICS. *Educ:* NC State Col, BChE, 48, MS, 49; Univ Tenn, PhD(chem eng), 55. *Prof Exp:* Chem engr, Polychems Dept, E I du Pont de Nemours & Co, 49-52, res engr, Indust & Biochems Dept, 55-62; sr res engr, Chemstrand Res Ctr, Inc, Monsanto Co, 62-68; asst prof chem & math, 68-71, assoc prof chem & physics & chmn dept physics, 71-72, PROF NATURAL SCI & PHYSICS, CAMPBELL COL, NC, 72- *Mem:* AAAS; Am Inst Chem Engr; Am Chem Soc; Am Soc Qual Control; Am Asn Physics Teachers. *Res:* Heat and mass transfer in heterogeneous systems; statistics and computer applications in education; experimental design; reaction kinetics; experimental education. *Mailing Add:* 233 Marilyn Circle Cary NC 27513

HOVMAND, SVEND, b Nakskov, Denmark, Jan 3, 39; m 66; c 2. PARTICLE TECHNOLOGY, FLUIDIZATION OF PARTICLES. *Educ:* Denmark Tech Univ, MSci, 61; Cambridge Univ, Eng, PhD(chem eng), 68. *Prof Exp:* Postdoctoral fluidization, Cambridge Univ, Eng, 68-69; res engr, Haldor Topsoe, Denmark, 69-70; dir res, Niro Atomizer, Denmark, 71-77, vpres, Columbia, Md, 77-89; PRES, CROSSVILLE CERAMICS, TENN, 89- *Concurrent Pos:* Course dir, Indust Drying Technol, Ctr Prof Develop, NJ, 80-90; pres, Bowen Eng, NJ, 81-89 & Niro Ceramic, Inc, Md, 82-89. *Mem:* Am Chem Eng Soc; Am Ceramic Soc. *Res:* Industrial development of equipment for spray drying; fluid bed technology including chemical reactions; drying processes and particle agglomeration. *Mailing Add:* 3711 Spring Meadow Dr Ellicott City MD 21043

HOVNANIAN, H(RAIR) PHILIP, b Aleppo, Syria, Dec 17, 20; US citizen; m 48; c 3. BIOMEDICAL ENGINEERING, PHYSICS. *Educ:* Am Univ Beirut, AB, 42; Mass State Col Boston, MA, 50; State Univ NY, PE, 64; Univ Beverly Hills, PhD, 79; Chartered Physicist, UK, 84. *Prof Exp:* Instr physics, Am Univ Beirut, 42-47; teaching fel, Brown Univ, 47-49; sr engr, Western Elec Co, 51-52; asst chief engr electro-mech, Calidyne Co, 52-53; asst dir res, Boston Electronics Div, Norden-Ketay Corp, 53-56; dir physics, Neutronics Res Co, 56-58; dir med sci dept, Res & Advan Develop Div, Avco Corp, 58-66; consult & proj mgr biosci progs, NASA Hq, 66-67; dir biomed eng & biophys, Kollsman Instrument Corp, 67-69; vpres & corp dir biomed eng, Cavitron Corp, Cooper Labs, Inc, 69-85; CONSULT & CORP DIR MED DEVICE RES & DEVELOP, SONOKINETICS CO & VITAL SIGNS, INC, 89- *Concurrent Pos:* Prin investr, Nat Health Inst res grants, 59-63, co-prin investr, 63-66; vis lectr, Northeastern Univ, 62-68; res assoc surg res, Lahey Clin Found, 62-; exec secy biosci working group, NASA Hq, 66-67; guest lectr, Mass Inst Technol-Harvard Med Sch Study Group Biomed Eng, 68; mem Nat Acad Eng workshop on interaction between indust & biomed eng, 70; mem obstet & gynec devices panel, Food & Drug Admin. *Mem:* AAAS; Optical Soc Am; Inst Elec & Electronics Engrs; Sigma Xi; NY Acad Sci; fel Inst Physics London; fel Am Acad Dent Electrosurg; fel Am Soc Laser Med & Surg; Biomed Eng Soc; Int Microscopy Asn. *Res:* Fiber optic endoscopy; ultraviolet and fluorescence microscopy and microspectrophotometry; low frequency ultrasound surgery of tissue; diagnostic and therapeutic respiratory instrumentation; exobiology measurements; carbon dioxide, YAG and argon lasers in medicine and surgery; medical-dental instrumentation. *Mailing Add:* 3902 Manhattan Col Pkwy Suite 1B Bronx NY 10471

HOVORKA, JOHN, b New York, NY, Oct 17, 21; m 50; c 2. PHYSICS, SYSTEMS DESIGN & SYSTEMS SCIENCE. *Educ:* Queens Col, NY, BS, 42; Univ Ill, MS, 43; Mass Inst Technol, ScD, 61. *Prof Exp:* Physicist, Mass Inst Technol, 44-46 & Gulf Res & Develop Co, 46-48; instr physics, Univ Buffalo, 48-50; physicist, Instrumentation Lab, Mass Inst Technol, 50-63; prof & head dept, Hobart & William Smith Cols, 63-67; lectr aeronaut & astronaut, Mass Inst Technol, 67-88; chmn, Div Sci & Math, 69-81, PROF PHYSICS, CURRY COL, 69- *Concurrent Pos:* Consult, Inst Defense Anal, 62, Dept Army, 62, lunar astrnauts, Manned Spaceflight Ctr, NASA, 62-63, Off Secy Defense, 63 & Exp Astron Lab, Mass Inst Technol, 64-72; sr res assoc aeronaut & astronaut, Mass Inst Technol, 69-72; fac fel, US Dept Transp, 79-81. *Mem:* NY Acad Sci; Am Phys Soc; Sigma Xi. *Res:* Fire control; servomechanisms; infrared spectroscopy; inertial navigation; general relativity, experimental. *Mailing Add:* 674 Brush Hill Rd Milton MA 02186

HOVORKA, ROBERT BARTLETT, b Cleveland, Ohio, Mar 27, 36; m 57; c 3. CHEMICAL ENGINEERING. *Educ:* Case Western Reserve Univ, BS, 57, MS, 59, PhD(chem eng), 61. *Prof Exp:* Res engr chem eng, Houston Res Lab, 61-69, group leader petrochems, Wood River Refinery, 69-72, STAFF ENGR, SHELL OIL CO, 72- *Mem:* Am Inst Chem Engrs. *Res:* On-line computer control of petroleum processes; reaction kinetics as applied to the design and scale-up of chemical processes; process design and technical service related to production of petrochemicals. *Mailing Add:* 2634 Trotter Way Walnut Creek CA 94596

HOWALD, JEREMIAH MARK, b Pittsburgh, Pa, Nov 19, 27; m 53; c 5. ORGANIC CHEMISTRY. *Educ:* Oberlin Col, AB, 49; Cornell Univ, PhD(chem), 53. *Prof Exp:* Asst org chem, Cornell Univ, 49-52; chemist, Shell Chem Corp, 52-57; tech rep, Glaskyd, Inc, 57-65; tech dir, Glaskyd Dept, 65-73, CONSULT, PERRYSBURG LABS, AM CYANAMID CO, 73- *Mem:* Am Chem Soc. *Res:* Organic fluorine compounds; petrochemicals; plastics. *Mailing Add:* 13906 Roachton Rd Perrysburg OH 43551-1162

HOWALD, REED ANDERSON, b Pittsburgh, Pa, Nov 23, 30; m 61; c 3. THERMODYNAMICS & MATERIAL PROPERTIES. *Educ:* Oberlin Col, BA, 52; Univ Wis, PhD(phys chem), 55. *Prof Exp:* Asst, Univ Wis, 52-54; instr chem, Univ Calif, Los Angeles, 55-56 & Harvard Univ, 56-59; asst prof, Oberlin Col, 59-60 & St John's Univ, NY, 60-63; from asst prof to assoc prof, 63-69, PROF CHEM, MONT STATE UNIV, 69- *Mem:* Am Chem Soc. *Res:* Chemical kinetics; radiochemistry; geochemistry; phase equilibria; activity coefficient. *Mailing Add:* Dept Chem Mont State Univ Bozeman MT 59717

HOWARD, ARTHUR DAVID, geology, for more information see previous edition

HOWARD, AUGHTUM SMITH, b Almo, Ky, Nov 10, 06; c 2. MATHEMATICS. *Educ:* Georgetown Col, AB, 26; Univ Ky, MS, 38, PhD(geometry), 42. *Prof Exp:* Reader math, Georgetown Col, 25-26; lab technician, Parke, Davis Drug Co, Detroit, 26-27; pub sch teacher, Ky, 27-29; grad asst, Univ Ky, 37-41; from assoc prof math to prof math & physics, Ky Wesleyan Col, 42-58, chmn div sci & math, 57-58; from assoc prof to prof math, Eastern Ky Univ, 58-72; RETIRED. *Concurrent Pos:* Mem, Curric Study Comt, Comn Pub Educ, Ky, 61. *Mem:* Math Asn Am; Sigma Xi; Am Asn Univ Profs. *Res:* Linear second order partial differential equations with constant coefficients. *Mailing Add:* 206 Pembroke Dr Richmond KY 40475

HOWARD, BARBARA V, b East Orange, NJ, June 26, 41; m 62; c 3. BIOCHEMISTRY, CELL CULTURE. *Educ:* Bryn Mawr Col, BA, 63; Univ Pa, PhD(microbiol), 68. *Prof Exp:* Instr microbiol, Sch Med, Univ NC, Chapel Hill, 68-69; asst prof biochem, Sch Med, George Washington Univ, 69-72 & Sch Dent Med, Univ Pa, 72-75; asst prof biochem & physiol, Med Col Pa, 75-76; res biochemist, NIH, 76-80, assoc chief, Phoenix Clin Res Sect, Nat Inst Arthritis, Diabetes & Digestive & Kidney Dis, 80-88; PRES MEDLANTIC RES FOUND, 91- *Mem:* Am Soc Biol Chemists; AAAS; Tissue Cult Asn; fel Am Heart Asn. *Res:* Lipoproteins in plasma and in cultured cells. *Mailing Add:* 9120 Willowgate Lane Bethesda MD 20817

HOWARD, BERNARD EUFINGER, b Ludlow, Vt, Sept 22, 20; m 42. APPLIED MATHEMATICS, COMPUTER SCIENCE. *Educ:* Mass Inst Technol, SB, 44; Univ Ill, MS, 47, PhD(math), 51. *Prof Exp:* Mem staff, Radiation Lab, Mass Inst Technol, 42-45; asst math, Univ Ill, 45-49; sr mathematician, Inst Air Weapons Res, Univ Chicago, 51, asst to dir, 52-56, assoc dir, Inst Syst Res, 56-60, assoc dir, Appl Sci Labs, 58-60; dir, Sci Comput Ctr, Univ Miami, 61-64, co-investr, Positron Emission Tomography Ctr & Dept Neurol, Mt Sinai Med Ctr, 81-84, PROF MATH & COMPUT SCI, UNIV MIAMI, 60- *Concurrent Pos:* Exec secy, Air Force Adv Bd Simulation, 51-54; consult, Syst Res Labs, Inc, Dayton, Ohio, 63-67, actg dir math sci div, 65; consult, Electronics Assocs, Inc, Princeton, NJ, 64, Variety Children's Res Found, Miami, 64-66, Fla Power & Light Co, 68 & Shaw & Assocs, 64-75; vis fel, Dartmouth Col, 76; Am Soc Eng Educ & Off Naval Res fel, Naval Underwater Systs Ctr, New London, Conn, 80 & 81. *Mem:* Am Math Soc; Soc Indust & Appl Math; Asn Comput Mach; Am Phys Soc; Inst Elec & Electronics Engrs; Sigma Xi. *Res:* System analysis; simulation; plasma dynamics; electromagnetic wave propagation; guided missiles; weapon system evaluation; numerical analysis and computation; optimum curvature and torsion principles in highway routing; sociocybernetics; underwater sound propagation; position-emission tomography; artificial intelligence. *Mailing Add:* Dept Math & Comput Sci Univ Miami Coral Gables FL 33124

HOWARD, BETTY, b Warwickshire, Eng, Aug 5, 23; Can citizen; c 2. PHYSICS. *Educ:* Univ London, BSc, 44; Oxford Univ, PhD(physics), 52. *Prof Exp:* Sr sci officer physics, Atomic Energy Estab, Eng, 50-52; demonstr, 53-63, instr, 63-65, ASST PROF PHYSICS, UNIV BC, 65- *Mem:* Can Asn Physicists; Am Asn Physics Teachers. *Res:* Low temperature physics. *Mailing Add:* Dept Physics Univ BC 6224 Agricultural Rd Vancouver BC V6T 1Z1 Can

HOWARD, BRUCE DAVID, b Minneapolis, Minn, Oct 19, 37; m 64; c 1. BIOCHEMISTRY, MOLECULAR BIOLOGY. *Educ:* Univ Minn, BA, 59, MD, 62. *Prof Exp:* Res assoc biol, Mass Inst Technol, 62; Nat Cancer Inst fel, Purdue Univ, 62-64; fel biochem, Albert Einstein Col Med, 64-66; asst prof biol chem, 66-75, assoc prof, 75-81, PROF BIOL CHEM, MED SCH, UNIV CALIF, LOS ANGELES, 81- *Mem:* AAAS; Int Soc Neurochem; Soc Neurosci; Am Soc Biol Chemists. *Res:* Neurobiology; neurochemistry. *Mailing Add:* Dept Biol Chem Univ Calif Med Sch Los Angeles CA 90024

HOWARD, CARLETON JAMES, b Anacortes, Wash, July 20, 44; m 82; c 2. PHYSICAL CHEMISTRY, ATMOSPHERIC CHEMISTRY. *Educ:* Linfield Col, BA, 66; Univ Pittsburgh, PhD(chem), 71. *Prof Exp:* RES CHEMIST, AERONOMY LAB, NAT OCEANIC & ATMOSPHERIC ADMIN, 71- *Concurrent Pos:* NSF grad traineeship, Univ Pittsburgh, 66. *Mem:* Am Chem Soc; Am Phys Soc; Combustion Inst. *Res:* Gas kinetics of atoms, radicals, and ions with small molecules. *Mailing Add:* Aeronomy Lab Nat Oceanic & Atmopheric Admin R/E/AL2 325 Broadway Boulder CO 80303

HOWARD, CHARLES, b Evanston, Ill, Apr 2, 19; m 45; c 3. ANALYTICAL CHEMISTRY, INORGANIC CHEMISTRY. *Educ:* Univ Wis, BS, 40, PhD(chem), 43. *Prof Exp:* Chemist, Oscar Mayer & Co, Wis, 41-46, actg chief chemist, 46-50; dir prof control, Arbogast & Bastian, Inc, Pa, 51-54; asst supt, Valleydale Packers, Inc, Va, 55-60; supt, Roegelein Provision Co, Tex, 60-61; from instr to prof chem, San Antonio Col, 61-73, chmn dept, 67-73; prof chem, Univ Tex, San Antonio, 73-86; RETIRED. *Concurrent Pos:* Adj prof, St Mary's Univ, San Antonio. *Mem:* Am Chem Soc; Sigma Xi; Inst Food Technologists; Water Quality Asn. *Res:* Reduction of alumina; determination of structure of complex ions by radiotracer study; analytical chemistry of phosphorus compounds; ion selective electrodes education drinking water analysis. *Mailing Add:* 7039 San Pedro No 708 San Antonio TX 78216-6238

HOWARD, CHARLES FRANK, JR, b Colon, Panama, Dec 30, 32; US citizen; m 53, 75; c 4. BIOCHEMISTRY. *Educ:* Colo State Col, BA, 54, MA, 58; Univ Wis, MS, 61, PhD(biochem), 63. *Prof Exp:* Fel biochem, Brandeis Univ, 63-65; from asst scientist to assoc scientist, Ore Regional Primate Res Ctr, 65-74, scientist, Metab Dis Res Sect, 74-90, div head, 83-87; ASSOC DEAN, COL HEALTH & HUMAN SERV, WESTERN MICH UNIV, 90-

Concurrent Pos: Asst prof, 65-75, res prof biochem, dept biochem, Sch Med, Ore Health Sci Univ, Portland, 84-90; assoc prof physiol chem, Col Optom, Pac Univ, 73-79; affil scientist, Yerkes Regional Primate Res Ctr, Emory Univ, Atlanta, Ga, 79-; mem, Coun Arteriosclerosis, Am Heart Asn; collaborating scientist, Caribbean Primate Res Ctr, Univ Puerto Rico, 88- *Mem:* Am Inst Nutrit; Am Soc Biochemists & Molecular Biologists; Am Diabetes Asn; Am Soc Primatol; Soc Res Admin; Am Soc Allied Health Prof. *Res:* Academic research administration; diabetes; islet amyloid. *Mailing Add:* Col Health & Human Serv Western Mich Univ Kalamazoo MI 49008

HOWARD, CHARLES MARION, b Moundsville, WVa, Aug 28, 33; m 58; c 1. STRAWBERRY DISEASES, STRAWBERRY BREEDING. *Educ:* WVa Univ, BS, 60, MS, 62; Cornell Univ, PhD(plant path), 69. *Prof Exp:* PROF PLANT PATH, AGR RES & EDUC CTR, UNIV FLA, 67- *Mem:* Am Phytopath Soc. *Res:* Strawberry breeding; diseases of strawberries and vegetables. *Mailing Add:* 13138 Lewis Gallagher Rd Dover FL 33527

HOWARD, CLARENCE EDWARD, b Roseboro, NC, May 31, 29; m 55, 81; c 1. MINERALOGY, CRYSTALLOGRAPHY. *Educ:* Duke Univ, BS, 53; NC State Univ, MS, 55; La State Univ, PhD(geol), 63. *Prof Exp:* Teaching asst geol, NC State Univ, 53-55; mining engr, Tungsten Mining Corp, 55-57; teaching asst geol, La State Univ, 59-63; from asst prof to prof geol, Campbell Univ, NC, 63-76, chmn dept, 64-76; pres, Carolina Earth Resources Co, 76-81 & 83-84; dir, Div Soil & Water Conserv, NC Dept Natural Resources & Commun Develop, 81-82; vis prof geol, NC State Univ, 82-83; MGR EARTH & MINERAL SCI DEPT, CTR ENVIRON MEASUREMENTS, RES TRIANGLE INST, 84- *Concurrent Pos:* Pres, Carolina Earth Resources Co, 77-; dir, Div Soil & Water Conserv, NC Dept Nat Resources & Commun Develop. *Mem:* AAAS; fel Geol Soc Am; Soc Mining Eng; Asn Eng Geologists; Asn Prof Geol Scientists; Sigma Xi; Am Inst Prof Geologists. *Res:* Optical mineralogy; petrographic and statistical studies of sediments and sedimentary rocks. *Mailing Add:* PO Box 1386 Lillington NC 27546

HOWARD, DAVID K, b Glasgow, Scotland, Oct 15, 49; m 85; c 2. MOLECULAR BIOLOGY. *Educ:* Univ Glasgow, Scotland, BSc Hons, 71, PhD(biochem), 74. *Prof Exp:* Postdoctoral virol & oncol, Stanford Univ Sch Med, 74-75; sr res scientist, Meloy Labs, 76-80, prin res scientist, 80-81; scientist, IMC Corp, 81-83; mgr, 83-88, DIR, PITMAN-MOORE INC, 88- *Concurrent Pos:* Prin investr, Nat Cancer Inst, NIH. *Mem:* Sigma Xi; Am Soc Microbiol; Soc Gen Microbiol. *Res:* Developing new products for animal health industry; molecular biology; endocrinology. *Mailing Add:* Pitman-Moore Inc PO Box 207 Terre Haute IN 47808

HOWARD, DEAN DENTON, b Chatham, NJ, Jan 2, 27; m 50; c 2. MICROWAVE & MILLIMETER WAVE ANTENNA DEVELOPMENT. *Educ:* Purdue Univ, BS, 49; Univ Md, MS, 52. *Prof Exp:* Sect head microwaves, Naval Res Lab, 49-84; STAFF SCIENTIST ELECTRONICS, LOCUS INC, 84- *Concurrent Pos:* Lectr, George Washington Univ, 84- *Mem:* Fel Inst Elec & Electronics Engrs; Res Soc Am. *Res:* Monopulse tracking radar; system development, measurement and analysis of target caused errors and other sources of error; counter measures; development of error reduction techniques; computer modelling of monopulse radar. *Mailing Add:* Locus Inc 2560 Huntington Ave Alexandria VA 22303

HOWARD, DEXTER HERBERT, b Santa Monica, Calif, Apr 17, 27; m 51; c 2. MEDICAL MYCOLOGY. *Educ:* Univ Calif, Los Angeles, BA, 51, MA, 53, PhD(microbiol), 54. *Prof Exp:* Asst bact, 51-54, res microbiologist infectious diseases, Sch Med, 54-55, from instr to asst prof, 57-63, assoc prof microbiol & immunol, 63-71, PROF MICROBIOL & IMMUNOL, SCH MED, UNIV CALIF, LOS ANGELES, 71- *Concurrent Pos:* Consult, Univ Calif Hosp, Los Angeles; consult ed, Sabouraudia, 69-; ed, Infection & Immunity, 86- *Mem:* Am Soc Microbiol; Am Acad Microbiol; Mycol Soc Am; Int Soc Human & Animal Mycol. *Res:* Medical mycology, especially factors involved in the host-fungus relationship; dimorphism of fungi; genetics of zoopathogenic fungi; diagnostic mycology. *Mailing Add:* Microbiol 43-239 CHS Univ Calif 405 Hilgard Ave Los Angeles CA 90024

HOWARD, DONALD GRANT, b Richmond, Calif, Nov 28, 37; m 60; c 2. SOLID STATE PHYSICS. *Educ:* Univ Calif, Berkeley, AB, 59, PhD(physics), 64. *Prof Exp:* Res asst prof physics, Univ Wash, 63-65; asst prof, 65-69, assoc prof, 69-77, PROF PHYSICS, PORTLAND STATE UNIV, 77- *Mem:* Am Phys Soc; Sigma Xi. *Res:* Atoms and molecules adsorbed on surfaces; Mossbauer effect studies; phase transitions; lattice dynamics; micrometallurgy. *Mailing Add:* 356 SE 44th Ave Portland OR 97215

HOWARD, DONALD ROBERT, b Orange, NJ, Apr 9, 40; m 64; c 2. EDUCATION ADMINISTRATION, VETERINARY MEDICINE. *Educ:* Mich State Univ, BS, 63, DVM, 65; Tex A&M Univ, MS, 69; Univ Mo, PhD(neuroanat), 72. *Prof Exp:* Assoc prof surg, Tex A&M Univ, 66-72; res assoc anat, Univ Mo, 69-72; prof surg, Col Vet Med, Mich State Univ, 72-79, asst dean surg, 77-79; ASSOC DEAN SURG, COL VET MED, NC STATE UNIV, 80- *Concurrent Pos:* Fel admin, Univ Mich-Am Coun Educ Prog, 75-76; regent, Am Col Vet Surgeons, 75-80, chmn bd, 83-85. *Mem:* Am Vet Med Asn; Am Col Vet Surgeons; Am Asn Vet Clinicians. *Res:* Retinal geniculate projections in the dog; mapped the retinal projections to the lateral geniculate nucleus in the canine; developed cortical evoked potentials as normal study; surgical correcting procedure for cleft palate and esophageal rent repairs. *Mailing Add:* Col Vet Med 4700 Hillsbourough St Raleigh NC 27606

HOWARD, EDGAR, JR, b Westerly, RI, Aug 14, 22; m 53; c 3. ORGANOPHOSPHORUS CHEMISTRY. *Educ:* Brown Univ, ScB, 43; Univ Ill, PhD(org chem), 46. *Prof Exp:* Pittsburgh Plate Glass Co fel, Harvard Univ, 46-47; from instr to prof chem, Temple Univ, 47-86, chmn dept, 63-68, sr prof, 86-89, EMER PROF CHEM, TEMPLE UNIV, 89- *Mem:* Am Chem Soc; Sigma Xi. *Res:* Mechanism of organic reactions; synthesis and reactions of organophosphorus compounds. *Mailing Add:* 2323 Old Arch Rd Norristown PA 19401-2013

HOWARD, EDWARD GEORGE, JR, b Atco, NJ, Feb 17, 21; m 51; c 3. COMPOSITE STRUCTURES, FLUORO ORGANIC CHEMISTRY. *Educ:* Temple Univ, BA, 43, MA, 45; Indiana Univ, PhD(org chem), 48. *Prof Exp:* SCIENTIST, CENTRAL RES & DEVELOP DEPT, E I DUPONT DE NEMOURS CO, INC, 48- *Mem:* Am Chem Soc. *Res:* Free radical chemistry; flouro-organic chemistry; organic chemistry; catalysis; polymer chemistry; inorganic chemistry; numerous publications and patents. *Mailing Add:* 844 Old Public Rd Hockessin DE 19707

HOWARD, EUGENE FRANK, b Milwaukee, Wis, Aug 11, 38; m 61; c 2. CELL BIOLOGY. *Educ:* Univ Wis, BS, 60, MS, 62, PhD(molecular biol), 67. *Prof Exp:* Asst bot, Univ Wis, 60-62, asst zool, 62-67, res assoc, 67-68; asst in biol, M D Anderson Hosp, Houston, 69-71; asst prof, Grad Sch Biomed Sci, Univ Tex, Houston, 69-71; asst prof, 71-76, ASSOC PROF CELL & MOLECULAR BIOL, MED COL GA, 76- *Concurrent Pos:* Fel pharmacol, Col Med, Baylor Univ, 68-69. *Mem:* Am Soc Cell Biol; Am Asn Cancer Res; Tissue Cult Asn. *Res:* Cell proliferation; cancer biology; small nuclear RNA molecules. *Mailing Add:* Dept Cell & Molecular Biol Med Col Ga Augusta GA 30912

HOWARD, FORREST WILLIAM, b Denver, Colo, Feb 8, 36; m 66; c 2. ECOLOGY & CONTROL OF ARTHROPOD PESTS. *Educ:* NY State Col Forestry, BS, 61; La State Univ, MS, 71, PhD(entom), 75. *Prof Exp:* Jr forester surv, Fla Forest Serv, 58-59; forester mgt, Olinkraft, Brazil, 61-62 & US Forest Serv, 62-63; biologist wetland ecol, Fla Div Health, 65-70; asst prof, 76-81, ASSOC PROF ENTOM, FT LAUDERDALE RES & EDUC CTR, UNIV FLA, 81- *Mem:* Entom Soc Am; Palm Soc. *Res:* Insects and diseases of palms and other tropical and subtropical plants. *Mailing Add:* Fort Lauderdale Res & Educ Ctr Univ Fla 3205 College Ave Ft Lauderdale FL 33314

HOWARD, FRANK LESLIE, b Los Angeles, Calif, June 11, 03; m 24, 82; c 3. PLANT PATHOLOGY, MYCOLOGY. *Educ:* Ore State Col, BS, 25; Univ Iowa, PhD(mycol), 30. *Prof Exp:* Asst, Univ Iowa, 29-30; Nat Res Coun fel, Farlow Herbarium, Harvard Univ, 30-32; from instr to prof bot, 32-45, prof plant path, 45-70, head dept plant path & entom, 46-70, EMER PROF PLANT PATH, UNIV RI, 71- *Mem:* Fel Am Phytopath Soc; Mycol Soc Am; Bot Soc Am. *Res:* Biology of Myxomycetes; Rhode Island plant diseases; chemical soil fumigation; development of fungicides; plant chemotherapy; turf disease control. *Mailing Add:* 1686 Stuart Rd North Kingston RI 02852

HOWARD, FREDRIC TIMOTHY, b Ft Worth, Tex, May 17, 39; m 65. SPECIAL FUNCTIONS, COMBINATORICS. *Educ:* Vanderbilt Univ, BA, 61, MA, 63; Duke Univ, PhD(math), 66. *Prof Exp:* From asst prof to assoc prof, 66-78, PROF MATH, WAKE FOREST UNIV, 78- *Mem:* Am Math Soc; Math Asn Am; Fibonacci Soc. *Res:* Number theory. *Mailing Add:* Reynolds Station Wake Forest Univ Box 7311 Winston-Salem NC 27109

HOWARD, G(EORGE) MICHAEL, b Washington, DC, July 4, 35; m 59; c 3. CHEMICAL ENGINEERING. *Educ:* Univ Rochester, BS, 57; Yale Univ, MEng, 59; Univ Conn, PhD(chem eng), 67. *Prof Exp:* Develop engr, res & develop, Humphrey-Wilkinson, Inc, 59-61; from instr to assoc prof chem eng, 61-78, acting head dept, 69-70 & 76, assoc dean, Sch Eng, 74-88, PROF CHEM, UNIV CONN, 78- *Concurrent Pos:* Grant, Foxboro Corp, Chas Pfizer & Co, Olin Corp, 68-69. *Mem:* Am Inst Chem Engrs; Am Chem Soc; Am Soc Eng Educ. *Res:* Unsteady state behavior of processes; energy conservation; engineering education. *Mailing Add:* Dept Chem Eng U-139 191 Auditorium Univ Conn Storrs CT 06268

HOWARD, GILBERT THOREAU, b Torrington, Wyo, Sept 21, 41; m 63; c 2. OPERATIONS RESEARCH. *Educ:* Northwestern Univ, BS, 63; Johns Hopkins Univ, PhD(opers res), 67. *Prof Exp:* Prof opers res, 67-84, dir res admin, 84-91, ASSOC DEAN RES, NAVAL POSTGRAD SCH, 91- *Mem:* Opers Res Soc Am; Inst Mgt Sci. *Res:* Mathematical programming. *Mailing Add:* 1437 Deer Flat Rd Monterey CA 93940

HOWARD, GLENN WILLARD, JR, b Columbus, Ohio, Feb 16, 39; m 70; c 2. MOLECULAR BIOLOGY. *Educ:* Rice Univ, BA, 62; Purdue Univ, MS, 69, PhD(molecular biol), 70. *Prof Exp:* Asst prof biol, Adelphi Univ, 70-73; dir res, Environ 200 Ltd, 73-74, tech prod mgr, 74-76; MGR BIOL CONTROLS, PALL CORP, 76-, ASSOC DIR, SCIENTIFIC & LABORATORY SERVICES DEPT, 82- *Concurrent Pos:* Mem, Planning Bd City of Glen Cove, NY, 86-, Bd Dir, Glen Cove Chamber Com, 87- *Mem:* Am Soc Microbiol; Am Soc Testing & Mat. *Res:* Microbiology in regard to development of microbially and endotoxin retentive filters. *Mailing Add:* 11 Eastland Dr Glen Cove NY 11542

HOWARD, GUY ALLEN, US citizen. ENDOCRINOLOGY. *Educ:* Univ Wash, Seattle, BA, 65; Cent Wash Univ, MS, 67; Univ Ore, PhD(biochem), 70. *Prof Exp:* Postdoctoral fel, Dept Chem, Univ Ore, 70-71 & Dept Biol Chem, Sch Med, Univ Calif, Davis, 71-73; res assoc, Friedrich Miescher Inst, Basel, Switz, 73-76; from res asst prof to res assoc prof, Dept Med, Sch Med, Univ Wash, Seattle, 76-89, res assoc prof, Dept Oral Biol, Sch Dent & mem grad fac, 82-89, res prof, Dept Med, 89; RES PROF, DEPT MED & CHIEF, MINERAL METAB RES, DIV ENDOCRINOL, SCH MED, UNIV MIAMI, 89-, RES PROF, DEPT BIOCHEM & MOLECULAR BIOL, 90- *Concurrent Pos:* Res chemist, Vet Admin Med Ctr, Tacoma, Wash, 76-82, assoc res career scientist, 82-89; mem, Int Conf Calcium Regulating Hormones, Japan, 83 & France, 86, Prog Comt, Am Soc Bone & Mineral Res, 84, 88 & 90; sect head bone metab, Biochem Dept, Ayerst Labs Res, Inc, Princeton, 84-85; session chmn, Second Int Conf Basic & Clin Factors Influencing Bone Growth, 85; vis prof, Dept Orthop, Univ Zurich, Switz, 88; res career scientist, Vet Admin Med Ctr, Miami, 90- *Mem:* Am Fedn Clin Res; Soc Exp Biol & Med; Am Soc Bone & Mineral Res; Am Soc Biochem & Molecular Biol; Endocrine Soc; NY Acad Sci. *Res:* Author of more than 150 technical publications. *Mailing Add:* Dept Med Div Endocrinol Univ Miami PO Box 016960 D-503 Miami FL 33101

HOWARD, H(ENRY) TAYLOR, b Peoria, Ill, Apr 5, 32; m 55; c 3. ELECTRICAL ENGINEERING. *Educ:* Purdue Univ, EE, 51; Stanford Univ, BSEE, 55. *Prof Exp:* Res asst, Radio Propagation Lab, Stanford Univ, 55-56, res assoc, Radio Sci Lab, 59-67, sr res assoc, 67-76, Ctr Radar Astron, 71-76, res prof elec eng, 76-82, EMER PROF, STANFORD UNIV, 82-, DIR RES & DEVELOP & VCHMN, CHAPARRAL COMMUN, 80- *Concurrent Pos:* Leader, Celestial Mech & Radiosci Team, Mariner Venus-Mercury 73, NASA, prin investr, Apollo Bistatic Radar Exp, team leader, Radio Propagation Team, Galileo Mission to Jupiter. *Honors & Awards:* Medal for Except Sci Acheivement, NASA. *Mem:* AAAS; Am Astron Soc; Am Geophys Union; Int Union Radio Sci; fel Inst Elec & Electronics Engrs. *Res:* Radar astronomy; solar terrestrial relationships, interplanetary gas density and magnetic fields. *Mailing Add:* 215 Durand Bldg Stanford CA 94305

HOWARD, HAROLD HENRY, b New Britain, Conn, May 2, 28; m 53; c 2. BOTANY, LIMNOLOGY. *Educ:* Cornell Univ, BS, 53, MS, 56, PhD(limnol), 63. *Prof Exp:* Instr bot, Mont State Univ, 58-59; investr tundra lake biol, Arctic Res Lab, 59-60; instr bot, Mont State Univ, 60-61, instr biol, 61-63; from asst prof to assoc prof biol, Skidmore Col, 63-83; RETIRED. *Mem:* AAAS; Ecol Soc Am; Torrey Bot Club. *Res:* Floristic botany; studies of Cayuga Lake, New York, arctic tundra lakes and Adirondack mountain lakes; Adirondack vascular flora. *Mailing Add:* Rte 2 Box 202 Greenwich NY 12834

HOWARD, HARRIETTE ELLA PIERCE, b Texarkana, Ark, Dec 23, 54. MOLECULAR BIOCHEMISTRY, CELL BIOLOGY. *Educ:* Fisk Univ, BA, 75; Atlanta Univ, MS, 78, PhD(molecular biol & develop biochem), 81. *Prof Exp:* Res asst, Atlanta Univ, 76-81; ASST PROF BIOL, FAYETTEVILLE STATE UNIV, NC, 81- *Concurrent Pos:* Lab instr gen biol, Spelman Col, Atlanta, 79-81; prin investr, NIH, 82-86; consult, NSF, 83 & Bennett Col, Greensboro, NC, 84-85. *Mem:* Am Soc Cell Biol; NY Acad Sci; Fedn Am Soc Exp Biologists; Nat Asn Med Minority Educr. *Res:* Nucleic acid; the isolation and characterization of the total DNA of the bobwhite quail and pheasant; in-situ hybridization; gel electrophoresis; preparative and analytical centrifugation; reassociation kinetics. *Mailing Add:* PO Box 840382 Houston TX 77284

HOWARD, HENRY COBOURN, b Akron, Ohio, Sept 20, 28; m 64; c 2. MATHEMATICS, ORDINARY DIFFERENTIAL EQUATIONS. *Educ:* Col Wooster, BA, 50; Carnegie Inst Technol, MS, 55, PhD(math), 58. *Prof Exp:* Instr math, Carnegie Inst Technol, 57-58 & Univ Wis- Madison, 58-60; asst prof, Univ Wis-Milwaukee, 60-63; vis asst res prof, Fluid Dynamics Inst, Univ Md, 63-65; assoc prof, Univ Wis-Milwaukee, 65-67; assoc prof, 67-72, PROF MATH, UNIV KY, 72- *Concurrent Pos:* Vis prof math, Univ BC, 73-74. *Mem:* Am Math Soc; Math Asn Am. *Res:* Classical analysis; ordinary and partial differential equations. *Mailing Add:* Dept Math Univ Ky Lexington KY 40506

HOWARD, HILDEGARDE, b Washington, DC, Apr 3, 01; m 30. ORNITHOLOGY. *Educ:* Univ Calif, BA, 24, MA, 26, PhD(zool), 28. *Prof Exp:* Asst zool, Los Angeles, 24-25; res asst, Los Angeles County Mus, 24-25; teaching fel paleont & geol, Univ Calif, Berkeley, 25-27; res asst, Los Angeles County Mus, 28; avian paleontologist, Nat Hist Mus Los Angeles County, 29-38, cur avian paleont, 39-50, chief cur, Div Sci, 51-61, res assoc vert paleont, 61-73, emer chief cur, 74-90; RETIRED. *Concurrent Pos:* Guggenheim Found res fel, 62-63. *Honors & Awards:* Brewster Mem Award, Am Ornith Union, 53. *Mem:* Fel AAAS; hon mem Soc Vert Paleont; fel Geol Soc Am; hon mem Cooper Ornith Soc; fel Am Ornith Union. *Res:* Fossil birds, especially from southern California sites, Rancho La Brea Pleistocene and marine Pliocene and Miocene; cave deposits of western states; Pliocene and Pleistocene of Mexico. *Mailing Add:* 2045 Apt Q Via Mariposa E Laganahill CA 92653

HOWARD, IAN PORTEOUS, b Rochdale, Eng, July 20, 20; Brit & Can citizen; m 56; c 3. HUMAN VISION, VESTIBULAR SYSTEM. *Educ:* Univ Manchester, BSc, 52; Durham Univ, PhD(psychol), 66. *Prof Exp:* Res asst psychol, Univ Durham, 52-54, lectr, 54-65; assoc prof, NY Univ, 65-66; assoc prof, 66-68, chmn, 68-70, PROF PSYCHOL, YORK UNIV, 68- , DIR, LAB STUDY HUMAN PERFORMANCE SPACE, INST SPACE & TERRESTRIAL SCI, 88- *Mem:* Fel Can Psychol Soc. *Res:* Sensory-motor mechanisms responsible for human spatially coordinated behavior; visual and vestibular systems, eye movements and their interactions; psychophysical and objective recording procedures. *Mailing Add:* Human Performance Space Lab 103 Farquharson Bldg York Univ North York ON M3J 1P3 Can

HOWARD, IRIS ANNE, b Knoxville, Tenn, Sept 20, 53. SOLID STATE PHYSICS. *Educ:* Mass Inst Technol, SB, 75, PhD(physics), 81. *Prof Exp:* Res assoc, Webster Res Ctr, Xerox Corp, 81-; AT DEPT PHYSICS & ATRON, UNIV NMEX. *Mem:* Am Phys Soc; Asn Women Sci. *Res:* Transport properties of quasi-one-dimensional materials; soliton effects in quasi-one-dimensional crystals. *Mailing Add:* Dept Physics Queen Mary Col Mile End Rd London E1 4NS England

HOWARD, IRMGARD MATILDA KEELER, b Philadelphia, Pa, Jan 21, 41; m 69; c 4. CLINICAL CHEMISTRY, PROTEIN CHEMISTRY. *Educ:* Duke Univ, AB, 62, PhD(biochem), 70. *Prof Exp:* Asst prof, 70-74, ASSOC PROF CHEM, HOUGHTON COL, 74- *Concurrent Pos:* Res assoc biochem, Duke Univ, 80-81. *Mem:* Am Chem Soc; NY Acad Sci; Am Asn Clin Chem. *Res:* Pytohemagglutinin chemistry; molecular genetics; nutrition; consumer chemistry. *Mailing Add:* Houghton Col Houghton NY 14744-0128

HOWARD, JACK BENNY, b Monroe Co, Ky, Oct 16, 37. CHEMICAL ENGINEERING. *Educ:* Univ Ky, BS, 60, MS, 61; Pa State Univ, PhD(fuel tech), 65. *Prof Exp:* Ford Found res fel, 65-67, asst prof, 65-80, PROF CHEM ENG, MASS INST TECHNOL, 80- *Concurrent Pos:* Richard H Wilhelm lectr, Princeton Univ, 77. *Honors & Awards:* Henry H Storch Award, Am Chem Soc, 83; Silver Medal, Combustion Inst, 84; Oblad lectr, Univ Utah, 89. *Mem:* AAAS; Am Chem Soc; Am Inst Chem Engrs; Combustion Inst. *Res:* Combustion and pyrolysis of solid fuels; heterogeneous reaction kinetics; carbon formation in flames; energy technology; combustion generated pollution; coal gasification. *Mailing Add:* Dept Chem Eng Mass Inst Technol Cambridge MA 02139

HOWARD, JAMES ANTHONY, b Liverpool, Eng, Feb 9, 37; Can citizen; m 61; c 3. PHYSICAL CHEMISTRY, ORGANIC CHEMISTRY. *Educ:* Univ Birmingham, BSc, 58, PhD(chem), 61. *Prof Exp:* SR RES OFFICER CHEM, NAT RES COUN CAN, OTTAWA, 63- *Mem:* Chem Inst Can. *Res:* Kinetics and mechanistic studies of free-radical reactions in solution; metal vapor chemistry. *Mailing Add:* Steacie Inst Molecular Sci Nat Res Coun Ottawa ON K1A 0R9 Can

HOWARD, JAMES BRYANT, b Indianapolis, Ind, Apr 25, 42; m 64; c 1. BIOCHEMISTRY. *Educ:* De Pauw Univ, BA, 64; Univ Calif, Los Angeles, PhD(biochem), 68. *Prof Exp:* Asst prof biochem, 71-77, assoc prof, 77-82, PROF BIOCHEM MED, UNIV MINN, MINNEAPOLIS, 82- *Concurrent Pos:* Fel, Univ Calif, Berkeley, 69-71; vis prof, Dept Chem, Harvard Univ, 80-81. *Mem:* AAAS; Am Soc Biol Chemists; Am Chem Soc; Sigma Xi. *Res:* Protein chemistry; structure-function studies; sequencing methodology; enzymology; proteins of nitrogenase system, flavoproteins, and non-heme iron proteins; protein chemistry of blood protease inhibitors; protein chemistry of complement fixation. *Mailing Add:* 1610 Maple Knoll Dr St Paul MN 55113

HOWARD, JAMES DOLAN, b Chicago, Ill, Feb 26, 34. GEOLOGY. *Educ:* Drury Col, BA, 56; Univ Kans, MA, 62; Brigham Young Univ, PhD(geol), 66. *Prof Exp:* Asst prof geol, Marine Inst & Dept Geol, 65-71, assoc prof, 71-76, PROF GEOL, SKIDAWAY INST OCEANOG & DEPT GEOL, UNIV GA, 76- *Mem:* Geol Soc Am; Am Asn Petrol Geologists; Soc Econ Paleont & Mineral. *Res:* Stratigraphy; sedimentology; marine geology. *Mailing Add:* Skidaway Inst Oceanog Box 13687 Savannah GA 31406

HOWARD, JAMES HATTEN, III, b Jacksonville, Fla, June 11, 39; c 3. GEOCHEMISTRY, PLANETARY GEOLOGY. *Educ:* Duke Univ, BS, 61; Stanford Univ, PhD(geochem), 69. *Prof Exp:* Asst prof, 67-80, actg dept head geol, 81-83, ASSOC PROF GEOL, UNIV GA, 80- *Mem:* Geol Soc Am. *Res:* Geochemistry of selenium; geochemistry of earth-surface environments; geology of Mars; geology of Jovian moons. *Mailing Add:* Dept Geol GGS Univ Ga Athens GA 30602

HOWARD, JAMES LAWRENCE, b Glen Ellyn, Ill, Nov 30, 41; m 63; c 2. BEHAVIORAL PHARMACOLOGY, DRUG DEVELOPMENT. *Educ:* Univ NC, BA, 63; Tulane Univ, MS, 66, PhD(psychol), 68. *Prof Exp:* Asst prof psychol, Psychiat Dept, Univ NC, 68-74; SR RES SCIENTIST PSYCHOL, PHARMACOL DIV, BURROUGHS WELLCOME CO, 74- *Concurrent Pos:* Sci consult, NC Alcohol Res Authority, 75-; adj prof, Psychol Dept, NC State Univ, 78-; coun rep, Am Psychol Asn, 85-87. *Mem:* Fel Am Psychol Asn; fel Am Psychol Soc; Soc Stimulus Properties of Drugs (pres, 83-84); Soc Neurosci; Am Soc Pharmacol & Exp Therapeut; Behav Pharmacol Soc. *Res:* Behavioral pharmacology; drugs to treat psychiatric disorders, especially anxiety; animal models of human disorders and the effect of drugs on these models. *Mailing Add:* Div Pharmacol Burroughs Wellcome Co 3030 Cornwallis Rd Research Triangle Park NC 27709

HOWARD, JOHN, b Clitheroe, UK, Aug 26, 33; Can citizen; m 58; c 2. ORGANIC CHEMISTRY, INORGANIC CHEMISTRY. *Educ:* Leeds Univ, BSc, 54, PhD(org chem), 58. *Prof Exp:* Tech officer plastics technol, Imperial Chem Indust, 57-59; fel carbohydrate res, Univ Ottawa, 59-61 & penicillin res, Univ Alta, 61-62; res chemist, R&L Molecular Res, 63-65, ITT Rayonier, 65-69, Columbia Cellulose, 69-73, Econotech Serv, 73-75 & Multifibre Process Ltd, 75-77; res scientist gen chem, 77-80, SR RES CHEMIST, BC RES, 80- *Mem:* Chem Inst Can; Can Pulp & Paper Asn. *Res:* Wood, pulp and cellulose chemistry; process development and optimization. *Mailing Add:* BC Res 3650 Wesbrook Mall Vancouver BC V6S 2L2 Can

HOWARD, JOHN CHARLES, b Franklin, NY, Apr 19, 24; m 49; c 2. ORGANIC CHEMISTRY, BIOCHEMISTRY. *Educ:* Hobart Col, BS, 44; Cornell Univ, PhD(org chem), 53. *Prof Exp:* Sr res chemist, Stauffer Chem Co, 53-55; sr res chemist, Eaton Labs Div, Norwich Pharmacal Co, 55-57, unit leader & asst sect chief, 57-60; asst prof, 61-66, ASSOC PROF BIOCHEM, MED COL GA, 66- *Concurrent Pos:* USPHS grants, 61-63 & 64-67; NSF grants, 65 & 68. *Mem:* Am Chem Soc; The Chem Soc; Sigma Xi. *Res:* Heterocyclic compounds; bioorganic and physical organic chemistry. *Mailing Add:* 3008 Stratford Dr Augusta GA 30909-3530

HOWARD, JOHN HALL, b Norwalk, Conn, Aug 14, 34; m 57; c 3. GEOLOGY. *Educ:* Harvard Univ, AB, 56; Columbia Univ, PhD(geol), 61. *Prof Exp:* Res geologist, Shell Develop Co, 60-71; physicist, Lawrence Livermore Lab, Univ Calif, 71-77; geologist reservoir engr, Lawrence Berkeley Lab, 77-81; geologist, Lawrence Livermore Nat Lab, 81-82. *Concurrent Pos:* Continuing educ lectr, Am Asn Petrol Geologists, 75- *Mem:* Fel Geol Soc Am; Am Asn Petrol Geologists; Geothermal Resources Coun. *Res:* Structural geology; petroleum geology; geothermal resources; energy resources international. *Mailing Add:* 1614 Pebble Chase Katy TX 77450

HOWARD, JOHN MALONE, b Autaugaville, Ala, Aug 25, 19; m 43; c 6. SURGERY. *Educ:* Birmingham Southern Col, BS, 41; Univ Pa, MD, 44; Am Bd Surg, dipl; Am Bd Thoracic Surg, dipl. *Prof Exp:* Intern & resident surg, Hosp Univ Pa, 44-50, Am Cancer Soc fel, 49-50; asst prof surg, Col Med, Baylor Univ, 50-58; prof surg, Univ & surgeon, Hosp, Hahnemann Med Col, 58-74; PROF SURG, MED COL OHIO, 74- *Concurrent Pos:* Attend surgeon, Jefferson Davis, Methodist & Vet Admin Hosps, 50-58; consult, US Vet Admin, 55- & USPHS, 68-; surgeon-in-chief, Crozer-Chester Med Ctr, 71-73. *Mem:* Soc Vascular Surg; Soc Univ Surg; AMA; Am Surg Asn; Pan-Pac Surg Asn. *Res:* Physiology and surgery in the fields of the pancreas, trauma, and vascular disease. *Mailing Add:* Med Col Ohio CS 10008 PO Box 6190 Toledo OH 43699

HOWARD, JOHN NELSON, b Philadelphia, Pa, Feb 27, 21; m 50; c 4. PHYSICS. *Educ:* Univ Fla, BS, 43; Ohio State Univ, MS, 49, PhD(physics), 54. *Prof Exp:* Asst chemist, Dept Soils, Univ Fla, 43-44; assoc physicist, Spectros Lab, Nat Adv Comt Aeronaut, 44-46; res assoc physics, Ohio State Univ, 48-54; head infrared tech br, 54-60, chief optical physics lab, 60-64, chief scientist, Air Force Geophysics Lab, 64-82; RETIRED. *Concurrent Pos:* Ed, Appl Optics, 60-87, Optics News, 83-89. *Mem:* Fel Am Phys Soc; History Sci Soc; fel Optical Soc Am (vpres, 89, pres-elect, 90, pres, 91); Int Comn Optics (vpres & treas, 81-87). *Res:* Infrared physics; atmospheric optics; life and research of Lord Rayleigh. *Mailing Add:* Seven Norman Rd Newton Highlands MA 02161

HOWARD, JOHN WILLIAM, b Dayton, Md, Sept 1, 25; m 52; c 4. ANALYTICAL CHEMISTRY. *Educ:* Cath Univ Am, BS, 49; George Washington Univ, MS, 51; Georgetown Univ, PhD(chem), 55. *Prof Exp:* Chemist, Hazleton Labs, Inc, Va, 51-57 & Nat Cotton Coun Am, Washington, DC, 57-60; Food & Drug Admin, Washington, DC, 60-70; chief org & additives chem br, Bur Food, Food & Drug Admin, Washington, DC, 70-74, dir div chem & physics, 74-85; RETIRED. *Mem:* AAAS; Am Chem Soc. *Res:* Procedures for determining polycyclic aromatic hydrocarbons in petroleum products, smoked foods, and other food products; N-nitrosamines in foods; administration and research in development of analytical methods for determination of food additives, industrial chemicals, heavy metals, natural toxins, and others in foods and products of food additive significance. *Mailing Add:* HRC 32 Box 59C Middlebrook VA 24459

HOWARD, JOSEPH H, b Olustee, Okla, Jan 15, 31. AGRICULTURAL RESEARCH. *Educ:* Univ Okla, BS, 52, MS, 57. *Prof Exp:* Vocal music instr, Kiowa Pub Sch, Kans, 54-56; music librn, Univ Colo, 57-58, circulation librn, 58-59, assoc dir pub servs, 59-63; Peace Corp vol, Univ Malaya Libr & Malayan Teachers Col, 63-65; chief catalog dept, Wash Univ Librs, St Louis, Mo, 65-67; asst chief, Descriptive Cataloging Div, Libr Cong, 67-68, chief, 68-72, chief, Ser Records Div, 72-75, asst dir cataloging, Processing Dept, 75-76, asst librn, Processing Servs, 76-83; DIR, NAT AGR LIBR, USDA, 83- *Concurrent Pos:* Mem Joint Coun Food & Agr Sci, 83-, Am Libr Asn Historically Black Col & Univ Task Force, 87-, Nat Adv Comt, Libr Cong, 88-89; bd mem, Forest Press, 86-88; mem, bd regents, Nat Libr Med, gen admin bd, USDA grad sch, 88- *Honors & Awards:* Melvil Dewey Medal, Am Libr Asn, 85. *Mem:* Am Libr Asn; Spec Libr Asn; Nat Info Standards Orgn; Int Asn Agr Librn & Documentalists (pres, 90-). *Mailing Add:* Nat Agr Libr USDA 10301 Baltimore Blvd Rm 200 Beltsville MD 20705

HOWARD, JOSEPH H(ERMAN) G(REGG), b Grafton, Ont, Nov 13, 33; m 60; c 1. FLUID MECHANICS, TURBOMACHINERY. *Educ:* Queen's Univ, Ont, BSc, 56; Univ Birmingham, MSc, 58, PhD(turbomach), 60. *Prof Exp:* Anal engr, United Aircraft Can Ltd, 60-64; assoc prof, 64-71, PROF MECH ENG, UNIV WATERLOO, 71- *Concurrent Pos:* Lectr, Loyola Col, Montreal, 63-64; consult, Dominion Eng, 67-76, Creare, 73-83 & Concepts, 84-; vis res assoc, Creare, Hanover, NH, 73-74; vis res eng, Sundstrand Corp, 80-81; vis res assoc, Concepts ETI, Norwich, VT, 85-86. *Mem:* Am Soc Mech Engrs; Eng Inst Can. *Res:* Flow in passages of axial and centrifugal compressors and pumps; internal aerodynamics of ducts; computer-aided design methods for compressors and pumps. *Mailing Add:* Dept Mech Eng Univ Waterloo Waterloo ON N2L 3G1 Can

HOWARD, KEITH ARTHUR, b Price, Utah, Sept 12, 39; div; c 1. GEOLOGY. *Educ:* Univ Calif, Berkeley, BS, 61, MS, 62; Yale Univ, PhD(geol), 66. *Prof Exp:* GEOLOGIST, WESTERN REGIONAL GEOL BR, US GEOL SURV, 66- *Concurrent Pos:* Fulbright scholar, 88. *Mem:* Geol Soc Am; Am Geophys Union; Sigma Xi. *Res:* Structural geology; deformation and metamorphism in the western United States; planetology; volcanology; tectonics; volcanic structure and morphology. *Mailing Add:* US Geol Surv Menlo Park CA 94025

HOWARD, LORN LAMBIER, b Poplar Bluff, Mo, Nov 28, 17; m 41; c 3. MICROELECTRONIC SYSTEMS FOR NEURAL NETWORK ANALYSIS. *Educ:* Univ Ill, BSc, 47, MSc, 48; Mich State Univ, PhD(elec eng), 59. *Prof Exp:* Lab asst elec eng, Univ Ill, 43-44, res asst, physics, 46-47, res asst elec eng, 47-55; instr elec eng, Mich State Univ, 55-59; dir, biomed eng, 64-72 & 72-88, PROF ELEC ENG, SOUTHERN METHODIST UNIV, 59- *Concurrent Pos:* Consult & founder, Ill Res Group, 53-58; sci consult, Wadley Res Inst, Tex, 61-; sr res staff mem, Gen Dynamics/Electronics, NY, 63; clin prof biophys, Univ Tex Health Sci Ctr, Dallas, 64-; sr assoc engr, Collins Radio Co, Tex, 67; consult, numerous elec accident invests, 67-, Graham Magnetics, Inc, Tex, 68 & Armoloy, Inc, 70-71; consult biomed eng, Emarand, SA, Neuchatel, Switz, 71-77; sci exchange, Soviet Acad Sci, 90. *Mem:* Inst Elec & Electronics Engrs; Electron Micros Soc Am; Am Vacuum Soc; Sigma Xi. *Res:* Thin conducting films; microstructural analysis; simulation of biological systems; measurement of micropotentials in biological systems, especially neuronal networks. *Mailing Add:* Dept Elec Eng Southern Methodist Univ Dallas TX 75275

HOWARD, LOUIS NORBERG, b Chicago, Ill, Mar 12, 29; m 51; c 2. APPLIED MATHEMATICS, FLUID DYNAMICS. *Educ:* Swarthmore Col, AB, 50; Princeton Univ, MA, 52, PhD(math physics), 53. *Prof Exp:* Higgins lectr math, Princeton Univ, 54-55; res fel math & aeronaut, Calif Inst Technol, 55; from asst prof to prof, 55-84, EMER PROF MATH, MASS INST TECHNOL, 85-; PROF MATH, FLA STATE UNIV, 81- *Concurrent Pos:* Found fel, Guggenheim, 61. *Mem:* Nat Acad Sci; Am Phys Soc; Am Math Soc; Math Asn Am; Soc Indust & Appl Math; Soc Math Biol. *Res:* Fluid dynamics; chemical oscillations and waves; rotating stratified flow. *Mailing Add:* Dept Math Fla State Univ Tallahassee FL 32306

HOWARD, PAUL EDWARD, b St Louis, Mo, Aug 31, 43; m 65; c 2. SET THEORY, AXIOM OF CHOICE. *Educ:* Univ Mo, BA, 65; Univ Mich, PhD(math), 70. *Prof Exp:* From asst prof to assoc prof, 70-83, PROF MATH, EASTERN MICH UNIV, 83- *Mem:* Am Math Soc; Asn Symbolic Logic. *Res:* Zermelo-Fraenkel set theory; relationships between weakenings of the axiom of choice. *Mailing Add:* Dept Math Eastern Mich Univ Ypsilanti MI 48197

HOWARD, PHILIP HALL, b Worcester, Mass, Oct 7, 43; m 67, 82; c 4. ENVIRONMENTAL SCIENCES. *Educ:* Norwich Univ, BS, 65; Syracuse Univ, PhD(org chem), 70. *Prof Exp:* Res assoc environ res, 70-72, SR ENVIRON CHEMIST, CHEM HAZARD ASSESSMENT DIV, SYRACUSE RES CORP, 72- *Mem:* Am Chem Soc; AAAS; Soc Environ Toxicol & Chem. *Res:* Persistance of organic chemicals; biodegradation; environmental pollutants; commercial chemistry. *Mailing Add:* Syracuse Res Corp Merrill Lane Syracuse NY 13210

HOWARD, PHILLENORE DRUMMOND, b Detroit, Mich, Dec 10, 41; m 66. CLINICAL CHEMISTRY. *Educ:* Mich State Univ, BS, 63; Wash Univ, St Louis, PhD(biochem), 68. *Prof Exp:* Co-dir, Biochem Path Lab, Bexar Co Med Ctr, San Antonio, Tex, 68-72; ASST DIR, NMEX MED REF LAB, 76- *Concurrent Pos:* Instr biochem & path, Med Sch, Univ Tex, San Antonio, 68-72. *Mem:* Am Asn Clin Chemists; AAAS. *Mailing Add:* 200 Watson Blvd Stratford CT 06497

HOWARD, RICHARD ALDEN, b Stamford, Conn, July 1, 17; m 44; c 4. BOTANY. *Educ:* Miami Univ, AB, 38; Harvard Univ, AM, 40, PhD(biol), 42. *Hon Degrees:* DSc, Framingham State Col, 77. *Prof Exp:* Asst, Harvard Univ, 38-39, jr fel, Soc of Fels, 42 & 46-47; asst cur, NY Bot Garden, 47-48; asst prof biol, Harvard Univ, 48-53; prof bot & chmn dept, Univ Conn, 53-54; Arnold prof bot, Harvard Univ, dir, Arnold Arboretum, 54- 78, prof dendrol, 54-87; vpres, 88-90, EMER VPRES BOT SCI, NY BOT GARDEN, 91- *Concurrent Pos:* Chmn, Plant Records Ctr, 68-70; mem bot expeds, Caribbean Islands, Cent Am & Mex; Guggenheim fel, 78-79. *Mem:* Int Asn Bot Gardens (pres, 59-64); Bot Soc Am; Am Soc Plant Taxon (treas, 49-54); Am Acad Arts & Sci; Asn Trop Biol. *Res:* Floristic studies of the Caribbean Islands; monographic studies of the Icacinaceae, Magnoliaceae, Malvaceae, Coccoloba, and tropical American flowering plants; history of botanical gardens; flora of the lesser antilles. *Mailing Add:* Arnold Arboretum 22 Divinity Ave Cambridge MA 02138

HOWARD, RICHARD JAMES, b Appleton, Wis, May 29, 52; m 79; c 2. MYCOLOGY, CELL BIOLOGY. *Educ:* Univ Wis, Madison, BS, 74; Cornell Univ, MS, 77, PhD(plant path), 80. *Prof Exp:* NIH trainee, Sch Med, Wash Univ, 79-80, NSF fel, 80; res biologist, 81-84, sect res biologist, 84-85, PRIN INVESTR, E I DU PONT DE NEMOURS & CO, INC, 85- *Concurrent Pos:* Mem chmn, Mycol Soc Am, 89-91. *Honors & Awards:* Alexopoulos Prize, 90. *Mem:* Mycol Soc Am. *Res:* Cell biology of growth and morphogenesis in fungi. *Mailing Add:* Cent Res & Develop DuPont Exp Sta Wilmington DE 19880-0402

HOWARD, RICHARD JOHN, b Sprigg, WVa, Feb 9, 24; m 50; c 2. PHYSICS. *Educ:* Univ Ky, BS, 46, MS, 48; Ohio State Univ, PhD(physics), 59. *Prof Exp:* Instr physics, Univ Ky, 48 & 50; assoc prof math & physics, Morehead State Col, 51-52; asst prof physics, Univ Vt, 57-61; asst prof, 61-67, asst chmn dept, 69-70, ASSOC PROF PHYSICS, STATE UNIV NY BUFFALO, 67- *Concurrent Pos:* Sabbatical, Lab Infrared Spectros, Univ Bordeaux, 69 & dept physics, Univ SC, 77. *Mem:* Am Asn Physics Teachers. *Res:* Undercooling of liquids; supersaturation of vapors; molecular association in vapors; stereoscopy. *Mailing Add:* Dept Physics State Univ NY Amherst Campus Amherst NY 14260

HOWARD, ROBERT ADRIAN, b Los Angeles, Calif, Feb 23, 13; m 39, 71; c 7. PHYSICS. *Educ:* Calif Inst Technol, BS, 34, MS, 35; Wash Univ, PhD(physics),38. *Prof Exp:* Asst, Wash Univ, 35-38; res physicist, Carter Oil Co, Okla, 38-42; staff mem, Radiation Lab, Mass Inst Technol, 43-45; asst, Calif Inst Technol, 34-35, res physicist, Hydrodyn Lab, 46-47; from asst prof to assoc prof physics, 47-54, chmn dept phys & sch eng phys, 52-55, dir nuclear reactor lab, 58-60, chmn dept physics & astron, 64-66, prof physics, 55-76, EMER PROF PHYSICS, UNIV OKLA, 76- *Concurrent Pos:* Consult, Boeing Airplane Co, 52 & Oak Ridge Nat Lab, 54; mem tech staff, Thompson-Ramo-Wooldridge Corp, 55-57, consult, Space Technol Labs, 58-62; dir, Cent Inst Physics, Concepcion Univ, 62-64; vis res scientist, Inst High Energy Physics, Univ Heidelberg, 69-70. *Mem:* AAAS; Am Phys Soc; Am Asn Physics Teachers. *Res:* Geophysics; migration and accumulation of petroleum; design of microwave power and frequency measuring devices; x-rays; hydrodynamics; nuclear physics; high energy particle physics. *Mailing Add:* 711 W Timberdell Rd Norman OK 73072-6324

HOWARD, ROBERT BRUCE, b St Paul, Minn, Dec 25, 20; m 42; c 5. MEDICINE. *Educ:* Univ Minn, BA, 42, MB, 44, MD, 45, PhD(med), 52. *Prof Exp:* Fel, Univ Minn, Minneapolis, 45-46, 48, from instr to prof med, 48-82, dir dept continuation med educ, 52-57, from assoc dean to dean col med sci, 57-70, clin prof med, 82-87; RETIRED. *Concurrent Pos:* Mem nat adv coun health res facil, USPHS, 64-68; dir, residency internal med, Abbott-Northwestern Hosp, 71-82; ed-in-chief, Postgrad Med, 82-87. *Mem:* Fel Col Physicians; Asn Am Med Cols (treas, 66-68, vpres, 69, chmn, 70). *Res:* Medical education; hematology; iron metabolism. *Mailing Add:* 1338 Howard Rd Hudson WI 54016

HOWARD, ROBERT ERNEST, b Tulsa, Okla, Oct 29, 47; m 69; c 2. PHYSICAL CHEMISTRY, CHEMICAL PHYSICS. *Educ:* Cornell Col, BA, 69; Ind Univ, PhD(chem physics), 75. *Prof Exp:* Res assoc, IBM Res Labs, 75-76; fel chem physics, Univ Calif, Berkeley, 76-77; asst prof chem, Univ Tex, Permian Basin, 77-81; ASSOC PROF & CHMN CHEM DEPT, UNIV TULSA, 81- *Concurrent Pos:* NSF grant, 76-77 & 89-; Robert A Welch Found fel, 78-81; res corp grant, 88-90. *Mem:* Am Chem Soc; Am Phys Soc; Sigma Xi. *Res:* Theoretical chemical physics; ab initio molecular structure calculations; construction of potential energy surfaces; quasi-classical trajectory molecular dynamics. *Mailing Add:* Univ Tulsa Chem Dept 600 S College Tulsa OK 74107

HOWARD, ROBERT EUGENE, b St Louis, Mo, Mar 27, 37; m 66; c 2. MEDICAL INFORMATION SYSTEMS, PATHOLOGY. *Educ:* Wash Univ, AB, 59, MD, 65, PhD(pharmacol), 68. *Prof Exp:* From intern to resident path, Barnes Hosp, Wash Univ, 65-67, asst, Sch Med, 65-67; instr,

Med Ctr, Univ Colo, 67-68; from asst prof to assoc prof, Univ Tex Med Sch San Antonio, 68-72, dep chmn dept, 70-71; ASSOC PROF PATH, SCH MED, UNIV NMEX, 72-; chief clin chem, Path Serv, Bernalillo County Med Ctr, 72-79; chief clin chem, Path Serv, Univ NMex Hosp, 79-80; CONSULT, MED COMPUTER MGT & COMPUTER INFO SYSTS, 81- *Concurrent Pos:* USPHS trainee exp path, 66-67; Nat Inst Gen Med Sci res & Univ Colo Med Ctr fluid res grants, 67-68; co-dir biochem path labs, Bexar County Hosp Dist, Tex, 68-72, assoc chief path serv, 70-71; res subcontractor, Southwest Res Inst-Brooks Sch Aerospace Med, 70-71. *Mem:* Am Soc Clin Pathologists; Sigma Xi. *Res:* Computer-assisted interpretation of clinical data; clinical pathology and chemistry; computer-assisted medical decision making; endocrinology; laboratory management; computer utilization in pathology. *Mailing Add:* 588B Arapaho Lane Stratford CT 06497

HOWARD, ROBERT FRANKLIN, b Delaware, Ohio, Dec 30, 32; m 58. ASTRONOMY. *Educ:* Ohio Wesleyan Univ, BA, 54; Princeton Univ, PhD(astron), 57. *Prof Exp:* Carnegie fel, Mt Wilson & Palomar Observs, 57-59; asst prof astron, Univ Mass, 59-61; mem staff, Mt Wilson & Las Campanas Observ, 61-84; DIR, NAT SOLAR OBSERV, 85- *Mem:* Am Astron Soc; Int Astron Union. *Res:* Magnetic field of the sun; sunspots; solar rotation and velocity fields. *Mailing Add:* PO Box 26732 Tucson AZ 85726

HOWARD, ROBERT PALMER, history of medicine, internal medicine; deceased, see previous edition for last biography

HOWARD, ROBERT T(URNER), b Frederick, Okla, Sept 24, 20; m 47; c 5. MATERIALS SCIENCE, ENGINEERING. *Educ:* Mass Inst Technol, SB, 42, ScD(metall), 47. *Prof Exp:* Asst tool steels res, Mass Inst Technol, 43-46; mem staff, Los Alamos Sci Lab, Calif, 47-48; metallurgist & dir metall qual control lab, Black, Sivalls & Bryson, Inc, 48-52; sr staff engr & asst to dir eng, Bendix Aviation Corp, 52-56; assoc prof eng & chmn dept, Univ Kansas City, 56-60; prof, Univ Wichita, 60-65; vis prof metall & mat eng, Univ Kans, 65-66, prof, 66-68; adv metallurgist, IBM Corp, 68-74, sr engr, Gen Technol Div, 86-90; RETIRED. *Concurrent Pos:* Consult metallurgist, James W Weldon Lab, 52-60; NSF grant mat sci, Univ Wichita, 61-63; consult, IBM Corp, 65-68. *Mem:* Am Soc Eng Educ; Am Soc Metals; Am Welding Soc; Int Soc Hybrid Microelectronics; Inst Elec & Electronics Engrs; Mfg Technol Soc. *Res:* Diffusion in single and polycrystalline cobalt below 1000 degrees C; metallurgy of soldering; development and manufacture of hybrid microcircuits, reliability of solder joints in semiconductor packages; laser trimming of resistors; adhesive bonding; microcircuit welding; hybrid microelectronics; corrosion of microelectronic circuits, curing kinetics of encapsulants for semiconductor devices. *Mailing Add:* Two Forest Rd Essex Junction VT 05452

HOWARD, ROGER, b Bristol, Eng, Sept 11, 35; Can citizen. THEORETICAL PHYSICS. *Educ:* Univ Nottingham, BSc, 56, PhD(theoret physics), 60. *Prof Exp:* From instr to asst prof, 59-69, ASSOC PROF PHYSICS, UNIV BC, 69- *Mem:* Can Asn Physicists; Am Asn Physics Teachers. *Res:* Lattice dynamics, thermal properties and structure of solidified inert gases. *Mailing Add:* Dept Physics Univ BC 2075 Wesbrook Pl Vancouver BC V6T 1W5 Can

HOWARD, RONALD A(RTHUR), b New York, NY, Aug 27, 34; m 55; c 4. ENGINEERING. *Educ:* Mass Inst Technol, SB(elec eng) & SB(econ), 55, SMEE, 56, EE, 57, ScD(elec eng), 58. *Prof Exp:* From asst prof to assoc prof elec eng & indust mgt, Mass Inst Technol, 58-65; PROF ENG-ECON SYSTS, STANFORD UNIV, 65- *Concurrent Pos:* Consult, Gen Elec Co, Stanford Res Inst & Xerox Corp. *Mem:* Soc Indust & Appl Math; Opers Res Soc Am; Inst Elec & Electronics Engrs; Inst Mgt Sci; Brit Opers Res Soc; Sigma Xi. *Res:* Decision analysis and probabilistic systems; operations research; systems engineering; economic systems. *Mailing Add:* Eng-Econ Systs Dept Terman Eng Ctr Stanford Univ Stanford CA 94305

HOWARD, RONALD M, b Minneapolis, Minn, Aug 20, 20; m 43; c 1. BACTERIOLOGY, IMMUNOLOGY. *Educ:* Macalester Col, BA, 42; Univ Minn, MS, 50. *Prof Exp:* USAF, 43-66, chief res & develop br, Fitzsimons Army Hosp, 52-54, asst chief biophys br, Aeromed Lab, Wright Air Develop Ctr, Ohio, 54, asst prof chem, USAF Acad, 54-59, chief opers div & dir life sci, Hqs, Air Res & Develop Command, 59-60, dir life sci div, Off Aerospace Res, 60-64, dir res prog, Washington, DC, 65-66; staff scientist, Human Res & Biotechnol Div, NASA Hq, 66-68; dir, Inst Biol Sci, SDak State Univ, 68-82; RETIRED. *Mem:* AAAS; Am Soc Microbiol; Sigma Xi. *Res:* Bacteriology and immunology of tuberculosis; manned space flight; science education in elementary grades. *Mailing Add:* 1808 Derdall Dr Brookings SD 57006

HOWARD, RUFUS OLIVER, b Knoxville, Tenn, June 30, 29; m 55; c 7. MEDICINE, OPHTHALMOLOGY. *Educ:* Col William & Mary, BS, 49; Mass Inst Technol, SB, 49, PhD(phys properties of polymers), 53; Med Col Va, MD, 61. *Prof Exp:* Res chemist, E I du Pont de Nemours & Co, Inc, 54-57; res fel ophthal, 62-65, from asst prof to assoc prof, 66-76, CLIN PROF OPHTHAL, MED SCH, YALE UNIV, 76- *Mem:* Am Ophthal Soc. *Res:* Metabolism of ocular tissues. *Mailing Add:* Grove Hill Clin PC 300 Kensington Ave New Britain CT 06050

HOWARD, RUSSELL ALFRED, b Baltimore, Md, Aug 17, 41; m 66; c 2. SOLAR PHYSICS. *Educ:* Univ Md, BS, 64, PhD(chem physics), 69. *Prof Exp:* Nat Acad Sci-Nat Res Coun resident res assoc, 69-71, res physicist, 71-73, ASTROPHYSICIST, NAVAL RES LAB, 73- *Mem:* Am Geophys Union; Am Phys Soc; Am Astron Soc. *Res:* Structure and dynamics of solar corona; solar-terrestrial relationships; solar wind; atomic and molecular spectroscopy; upper atmospheric research. *Mailing Add:* 9936 Woodgrouse Ct Burke VA 22015

HOWARD, RUSSELL JOHN, b Melbourne, Australia, Aug 30, 50; m 73; c 2. MOLECULAR PARASITOLOGY, BIOCHEMISTRY. *Educ:* Univ Melbourne, BSc, 71, BSc, 72, PhD(biochem), 75. *Prof Exp:* Fel molecular parasitol, Walter & Eliza Hall Inst Med Res, Melbourne, Australia, 76-78; vis assoc malaria sect, Lab Parasitic Dis, Nat Inst Allergy & Infectious Dis, NIH, 79-81, res expert molecular parasitol, 81- 85, vis scientist, 85-87; SR STAFF SCIENTIST MOLECULAR PARASITOL, LAB INFECTIOUS DIS, DNAX RES INST MOLECULAR & CELLULAR BIOL, PALO ALTO, CALIF, 88- *Concurrent Pos:* Mem steering comt Sci Workshop Group Immunol Malaria, WHO, Geneva, 87-90. *Res:* Biochemical, immunochemical and molecular biological studies on malarial proteins inserted into the membrane of host erythrocytes; potential target antigens for an antimalarial vaccine or chemotherapeutic strategies. *Mailing Add:* DNAX 901 California Ave Palo Alto CA 94304-1104

HOWARD, SETHANNE, b Coronado, Calif, Feb 2, 44. TIDAL INTERACTIONS. *Educ:* Univ Calif, Davis, BS, 65; Rensselaer Polytech Inst, MS, 73; Ga State Univ, PhD, 89. *Prof Exp:* Astron asst, Lick Observ, 65-67; assoc res astron, Kitt Peak Nat Observ, 72-80; comput programmer analyst, Fleet Numerical Oceanog Ctr, Monterey, Calif, 80-84; adj fac, Ga State Univ, Atlanta, 84-89; post doctoral fel, Los Alamos Nat Lab, 89-90; CONSULT, NASA, 90- *Mem:* Am Astron Soc; AAAS; Planetary Soc. *Res:* Science and computer education; astrophysics; n-body simulations of tidally; interacting galaxies in three dimensions. *Mailing Add:* 1230 Main St Canton GA 30114

HOWARD, STEPHEN ARTHUR, b Bronx, NY, Dec 14, 41; m 65; c 2. PHARMACEUTICS. *Educ:* Columbia Univ, BS, 63; Univ Mich, Ann Arbor, MS, 66, PhD(pharmaceut chem), 68. *Prof Exp:* Teaching asst pharmacog, Sch Pharm, Univ Mich, 63-64; sr scientist prod develop, Wyeth Labs, Inc, 68-73; from asst prof to assoc prof pharm & pharmaceut, Sch Pharm, WVa Univ, 73-85; PRIN SCIENTIST, R W JOHNSON PHARMACEUT RES INST, 85- *Concurrent Pos:* Parenteral Drug Asn, res award, 79. *Mem:* Am Pharmaceut Asn; Acad Pharmaceut Sci; Sigma Xi. *Res:* Models for the dissolution of multisized powders; dissolution of pharmaceutical suspensions; cohesion of tablet ingredients; development of new dosage forms utillizing biodegradable microcapsules. *Mailing Add:* R W Johnson Pharmaceut Res Inst McKean & Welsh Rds Springhouse PA 19477

HOWARD, SUSAN CAROL PEARCY, b Enid, Okla, Jan 7, 54; m; c 2. GLYCOPROTEIN STRUCTURE-FUNCTION RELATIONSHIPS, ENZYMOLOGY. *Educ:* Okla State Univ, BS, 75; PhD(biochem), 80. *Prof Exp:* Fel, Johns Hopkins Univ Sch Med, 80-81; sr res biochem, 86-90, RES SPECIALIST, CORP RES LAB, MONSANTO CO, 90- *Mem:* Am Soc Cell Biol; Sigma Xi. *Res:* Development of antivirals targetting a critical protease; enzyme purification and development of highly specific protease inhibitors; fourteen publications, two patents and nine technical reports. *Mailing Add:* Monsanto Co 800 N Lindbergh Blvd St Louis MO 63167

HOWARD, THOMAS E(DWARD), b Longmont, Colo, Feb 25, 19; m 41; c 1. MINING ENGINEERING, GEOLOGY. *Educ:* Colo Sch Mines, EM, 41. *Prof Exp:* Asst engr, Anaconda Copper Co, 41-44; mining engr, St Louis Smelting & Ref Div, Nat Lead Co, 46-48; asst engr, Union Pac RR, 48-49; construct engr, US Army Corps Engrs, 49-51; mining engr, Spokane Field Off, US Bur Mines, 51-54, asst chief mining res & mineral develop, 54-59, br chief ceramic & fertilizer mat, 59-60, dir mining res, 60-70; dir res, Joy Mfg Co, 70-74, dir tech develop, Hard Rock Mining Div, 74-81; consult, 81-82; RETIRED. *Honors & Awards:* D C Jacking Award, Soc Mining Engrs, 75. *Res:* Applied scientific and engineering research, principally in earth sciences as related to earth excavation technology. *Mailing Add:* 4730 Nelson St Wheat Ridge CO 80033

HOWARD, THOMAS HYLAND, b Princeton, Ind, May 16, 45; m 76; c 1. VIROLOGY. *Educ:* Purdue Univ, DVM, 69; Colo State Univ, PhD(physiol), 82. *Prof Exp:* From asst vet to vet, Am Breeders Serv, 75-91; PVT VET CONSULT, 91- *Mem:* Am Vet Med Asn; Am Asn Bovine Practr; Soc for Theriogenology. *Res:* Virus diseases of cattle, particularly those affecting reproduction or transmitted via the reproductive process; artificial insemination of cattle. *Mailing Add:* N2197 Meilke Rd Poynette WI 53955

HOWARD, VOLNEY WARD, JR, b Catarina, Tex, Apr 9, 41; m 67; c 2. WILDLIFE RESEARCH. *Educ:* Tex A&M Univ, BS, 64; NMex State Univ, MS, 66; Univ Idaho, PhD(wildlife sci), 69. *Prof Exp:* PROF WILDLIFE SCI, NMEX STATE UNIV, 69- *Mem:* Wildlife Soc; Soc Range Mgt; Sigma Xi. *Res:* Big game; research with muledeer, whitetailed deer, pronghorned antelope, desert bighorn sheep, coyotes, Iranian and Siberian ibex and Persian gazelle, bobcats, elk (wapiti). *Mailing Add:* Dept Fishery & Wildlife Sci NMex State Univ Las Cruces NM 88003

HOWARD, W TERRY, b Pueblo, Colo, Apr 14, 36; m 60; c 3. ANIMAL NUTRITION, DAIRY HUSBANDRY. *Educ:* Univ Nebr, BS, 58, MS, 64; Purdue Univ, PhD(dairy mgt), 67. *Prof Exp:* Instr dairy sci, Univ Nebr, 59-64; from asst prof to assoc prof dairy sci, 71-80, PROF DAIRY SCI, UNIV WISMADISON, 80- *Mem:* Am Dairy Sci Asn. *Res:* Total mixed rations and computer ration formulation for dairy cattle and forage quality evaluation and use. *Mailing Add:* Dept Dairy Sci 266 An Sci Bldg Univ Wis Madison WI 53706

HOWARD, WALTER B(URKE), b Corpus Christi, Tex, Jan 22, 16; m 42; c 2. CHEMICAL ENGINEERING. *Educ:* Univ Tex, BA, 37, BS, 38, MS, 40, PhD(chem eng), 43. *Prof Exp:* From asst to sr chem engr, Bur Indust Chem, Tex, 39-52; from sr chem engr to scientist res dept, Plastics Div, Monsanto Co, 52-64, process safety mgr cent eng dept, 65-81; RETIRED. *Mem:* Am Chem Soc; Am Inst Chem Engrs; Combustion Inst. *Res:* Process hazard prevention; chemical technology; explosion phenomena as related to safety protection of plant equipment; acetylene manufacture from hydrocarbons; high temperature technology and fundamentals as applied to chemicals production; resonating electric circuits for large power supply; electric discharges in gases; engineering analysis. *Mailing Add:* 1415 Bopp Rd St Louis MO 63131

HOWARD, WALTER EGNER, b Woodland, Calif, Apr 9, 17; m 40; c 3. ECOLOGY, ANIMAL CONTROL. *Educ:* Univ Calif, AB, 39; Univ Mich, MS, 41, PhD(vert ecol), 47. *Prof Exp:* Asst lab vert biol, Univ Mich, 40-42; instr zool & jr zoologist, San Joaquin Exp Range, Col Agr, 47-49, lectr & asst zool, 49-54, from assoc specialist to specialist, 54-60, assoc vert ecologist, 60-64, prof wildlife biol & vert ecologist, Dept Animal Physiol, 64-73, prof, 74-87, EMER PROF, DEPT WILDLIFE & FISH BIOL, AGR EXP STA, UNIV CALIF, DAVIS, 87- *Concurrent Pos:* Fulbright res fel, NZ & Australia, 57; grant-in-aid, Animal Ecol Div, NZ Dept Sci Indust Res & Travel Grant, NZ Forest Serv, 62-63; UN consult, Food & Agr Orgn, Mex, 69, Argentina, 69 & 72, SKorea, 74 & 77, Peninsular Malaysia, 75; WHO consult, Lebanon, 69, Qatar, 69 & 71, Bahrain, 71, Egypt, 74 & Barbados, 79 & 80; consult & lectr, Nat Sci Coun, Taiwan, 80, 81 & 85, Chinese Acad Med, China & Mil Acad Med, Inst Zool Acad, Sinica, 81, 84, 85 & 88, Kuwait, 82, 83 & 85, Qatar, 84 & 85, India, 82, Finland, 82, Norway, 82, Denmark, 82, Peru, 83, Japan, 84 & Commonwealth N Marianas, 85, Brazil, 87, Italy, 87, NZ, 88, England, Scotland, 89. *Mem:* Am Soc Mammal; Wildlife Soc; Ecol Soc Am; Animal Behav Soc; Brit Ecol Soc; Soc Range Mgt. *Res:* Population dynamics and control; rodents, coyotes and other wild vertebrates, ecology and behavior. *Mailing Add:* Dept Wildlife & Fisheries Biol Univ Calif Davis CA 95616

HOWARD, WEBSTER EUGENE, b Winthrop, Mass, May 19, 34; m 57; c 5. SOLID STATE PHYSICS. *Educ:* Carnegie-Mellon Univ, BS, 55; Harvard Univ, AM, 56, PhD(physics), 62. *Prof Exp:* RES STAFF MEM APPL RES, IBM THOMAS J WATSON RES CTR, 61- *Honors & Awards:* Wetherill Medal, Franklin Inst, 81. *Mem:* Fel Am Phys Soc; Sigma Xi; fel Soc Info Display. *Res:* Semiconductor physics and devices, especially electroluminescence, surface transport, and thin film transistor devices. *Mailing Add:* IBM Res Ctr Box 704 Yorktown Heights NY 10598

HOWARD, WILLIAM EAGER, III, b Washington, DC, Aug 25, 32; m 57; c 2. ASTRONOMY, TECHNICAL MANAGEMENT. *Educ:* Rensselaer Polytech Inst, BS, 54; Harvard Univ, AM, 56, PhD(astron), 58. *Prof Exp:* Astronr, Harvard Radio Astron Sta, Tex, summer 58; from instr to assoc prof astron, Univ Mich, 59-64, res assoc radio astron, 59-61; assoc scientist, Nat Radio Astron Observ, Charlottesville, Va, 64-67, scientist, 67-77, asst to dir, 64-74, asst dir, Green Bank Opers, 74-77; dir div Astron Sci, NSF, 77-82; staff mem, Fed Govt, 82-84; tech dir, Naval Space Command, 85-91, DIR SPACE & STRATEGIC TECHNOL, US ARMY SECRETARIAT, 91- *Concurrent Pos:* Mem comn 40 radio astron, Int Astron Union, 61-; mem comn J. Int Sci Radio Union, 69-; chmn subcomt radio astron & comt radio frequencies, Nat Acad Sci, 71-76; mem US study group 2, Int Radio Consult Comt, 71-79, mem, US deleg, Geneva, 72-78. *Mem:* AAAS; Am Astron Soc (treas, 75-77); Sigma Xi; US Naval Inst; Nat Space Club. *Res:* Galactic and extra-galactic studies; optical and radio astronomy in the continuum and in the 21-centimeter line of neutral hydrogen; science administration. *Mailing Add:* 31 Woodlawn Terr Fredericksburg VA 22405-3358

HOWARD, WILLIAM GATES, JR, b Boston, Mass, Nov 6, 41. ELECTRICAL ENGINEERING, ELECTRONICS. *Educ:* Cornell Univ, BEE, 64, MSc, 65; Univ Calif, Berkeley, PhD(elec eng), 67. *Prof Exp:* Asst prof elec eng, Univ Calif, Berkeley, 67-70; mgr, Advan Linear Integrated Circuits Develop, 70-74, group opers mgr, Linear Integrated Circuits, 74-76, vpres & dir strategic opers, 76-79, vpres & dir technol, Planning Semiconductor Sect, 79-83, SR VPRES & DIR CORP RES DEVELOP, MOTOROLA, 83- *Concurrent Pos:* Sr fel, Nat Acad Eng, 87- *Mem:* Sr fel, Nat Acad Eng, 87; Nat Acad Eng; fel Inst Elec & Electronics Engrs; AAAS. *Res:* Electronic function realization in integrated circuit form. *Mailing Add:* 10642 E San Salvador Scottsdale AZ 85258

HOWARD, WILLIAM HENRY RICHARD, radiology, for more information see previous edition

HOWARD, WILLIAM JACK, b Kimball, Nebr, Aug 25, 22; m 46; c 2. ENGINEERING. *Educ:* NMex State Univ, BS, 46. *Hon Degrees:* Dr, NMex State Univ, 82. *Prof Exp:* Asst to secy defense atomic energy, Dept Defense, Sandia Nat Labs, 63-66, vpres, 66-73, exec vpres, 73-82; CONSULT, 82- *Concurrent Pos:* Deleg, Strategic Arms Limitation Talks, 76. *Honors & Awards:* Distinguished Pub Serv Medal, US Dept Defense. *Mem:* Nat Acad Eng. *Res:* Design and development of nuclear ordnance. *Mailing Add:* 920 McDuffie Circle NE Albuquerque NM 87110

HOWARD, WILLIAM WEAVER, b Ontario, Ore, Apr 25, 22; m 60; c 4. FIXED PROSTHODONTICS, OCCLUSION. *Educ:* Ore State Univ, BS, 47; Univ Ore, DMD, 50. *Prof Exp:* Pvt pract dent, 50-73; PROF & CHMN FIXED PROSTHODONTICS, SCH DENT, ORE HEALTH SCI UNIV, 73- *Concurrent Pos:* Ed, Gen Dent, Acad Gen Dent, 74-; dir, Dent Risk Mgt Found, 87- *Honors & Awards:* Geis Ed Award, Geis Found, 82 & 90; Golden Pen Award, Int Col Dentists, 81, Golden Scroll Award, 82, Golden Pencil Award, 86. *Mem:* Am Dent Asn; Am Acad Restorative Dent; Acad Gen Dent; fel Am Col Dentists; Am Asn Dent Ed; Int Col Dentists. *Res:* Clinical dentistry. *Mailing Add:* 611 SW Campus Dr Portland OR 97201

HOWARD, WILMONT FREDERICK, JR, b Brattleboro, Vt, Oct 27, 46; m 73; c 2. TECHNICAL SERVICE, PROCESS DEVELOPMENT. *Educ:* Univ Vt, BS, 69; St Michaels Col, MS, 72; Univ Va, MS, 74; Univ RI, PhD(inorg chem), 80. *Prof Exp:* Asst dir, Ctr Catalytic Sci & Technol, Univ Del, 80-84; res mgr, EMCA, subsid Rohm & Haas, 84-87; proj leader, Cermalloy Div, Heraeus Inc, 87-89; MGR CHEM RES, CYPRUS FOOTE MINERAL CO, 90- *Mem:* Am Chem Soc. *Res:* Production methods development of lithium chemicals, primarily inorganic. *Mailing Add:* Cyprus Foote Mineral Co Rte 100 Exton PA 19341

HOWARD-FLANDERS, PAUL, biophysics; deceased, see previous edition for last biography

HOWARD-LOCK, HELEN ELAINE, b Hamilton, Ont, Jan 5, 38; m 60; c 2. MOLECULAR SPECTROSCOPY. *Educ:* McMaster Univ, BSc, 59, PhD(chem physics), 68. *Prof Exp:* Nat Res Coun Can fel light scattering, Univ Toronto, 68-70; asst prof eng physics, 70-78, PROF SCIENTIST, INST FOR MAT RES, MCMASTER UNIV, 81-, ASSOC PROF PATH, 87- *Concurrent Pos:* Nat coordr remote sensing oil spills, Environ Can, 73-74; trustee, Halton Pub Sch, 73-81; vpres res, Castle Cement Co Ltd, Burlington, Ont, 77-82; treas, Howard Concrete & Mat Ltd, Burlington, Ont, 80-82; pres, Howard-Lock Assocs, Inc, 88. *Honors & Awards:* R Samuel McLaughlin Career Scientist Geront. *Mem:* Fel Chem Inst Can; Can Asn Physicists. *Res:* Molecular spectra and structures; properties of cement and other potential building materials; platinum anti-tumor agents; gold, D-penicillamine and non-steroidal anti-inflammatory anti-arthritic agents; structures of materials; medical research; chiral pharmacology. *Mailing Add:* Dept Chem McMaster Univ Hamilton ON L8S 4M1 Can

HOWARD-PEEBLES, PATRICIA NELL, b Lawton, Okla, Nov 24, 41; m 75. HUMAN GENETICS, CLINICAL CYTOGENETICS. *Educ:* Cent State Univ, Okla, BS, 63; Univ Tex, Austin, PhD(genetics), 69; Am Bd Med Genetics, dipl, 82. *Prof Exp:* Biochem technician, Biol Div, Oak Ridge Nat Lab, 64-66; instr res pediat, Univ Okla Health Sci Ctr, 71-72, cytotechnol, 71-72; from asst prof to assoc prof microbiol, Inst Genetics, Univ Southern Miss, & dir, Cytogenetics Lab, 73-80; assoc prof, Dept Pub Health & mem staff, Lab Med Genetics, Univ Ala, Birmingham, 80-81; assoc prof & dir, Cytogenetics Lab, Univ Tex Health Sci Ctr, Dallas, 81-85, prof, Dept Path, 85-87; CLIN CYTOGENETICIST & DIR, POSTNATAL LAB, GENETICS & IVF INST, FAIRFAX, VA, 87-; PROF, DEPT HUMAN GENETICS, MED COL VA, RICHMOND, 87- *Concurrent Pos:* Am Cancer Soc fel human genetics, Med Sch, Univ Mich, 69-70 & human genetics & develop, Col Physicians & Surgeons, Columbia Univ, 70-71; genetic consult, Ellisville State Sch, Miss, 73-80; attend staff, Dept Path, Parkland Mem Hosp, Dallas County Hosp Dist, 81-87; referee cytogeneticist, CAP Cytogenetics Survey, 89- *Mem:* Sigma Xi; Am Soc Human Genetics; Genetics Soc Am; AAAS. *Res:* Human cytogenetics, especially delineation of chromosomal disorders and mental retardation; X-linked mental retardation with fragile X, chromosomal fragile sites. *Mailing Add:* Genetics & IVF Inst 3020 Javier Rd Fairfax VA 22031

HOWARDS, STUART S, b Milwaukee, Wis, Mar 29, 37; m 66; c 2. PHYSIOLOGY, UROLOGY. *Educ:* Yale Univ, BA, 59; Columbia Univ, MD, 63. *Prof Exp:* Staff assoc renal physiol, NIH, 65-68; resident virol, Peter Bent Brigham Hosp, Boston, 68-71; PROF PHYSIOL, UNIV VA, 71- *Concurrent Pos:* Nat Inst Childhood Develop fel, 76-81; assoc ed, J Investigative Urol, 77-85, ed, 85-; co-ed, Year Book Urology, 79-; chmn exam comt, AVA/ABU, 87-92; mem, Am Bd Urol, 87-93. *Honors & Awards:* Gold Cystocope Award, Am Urol Asn; Scott Award, Am Urol Asn, 90. *Mem:* Soc Genitourinary Surgeons; Am Col Surgeons; Soc Study Reproduction; Am Soc Andrology; Am Soc Nephrology; Clin Soc Genitourinary Surgeons. *Res:* Reproductive biology; physiology of testis and epididymis. *Mailing Add:* Box 422 Univ Va Charlottesville VA 22901

HOWARTH, ALAN JACK, virology, for more information see previous edition

HOWARTH, BIRKETT, JR, b Sellersville, Pa, Dec 1, 36; m 61; c 4. REPRODUCTIVE PHYSIOLOGY. *Educ:* Del Valley Col, BS, 58; SDak State Univ, MS, 60; NC State Univ, PhD(reproductive physiol), 65. *Prof Exp:* Staff fel reproductive physiol, Intramural Res Prog, Develop Biol Br, Nat Inst Child Health & Human Develop, 65-68; from asst prof to assoc prof, 68-86, PROF POULTRY SCI, UNIV GA, 86- *Mem:* Soc Study Reprod; Am Poultry Sci Asn; World's Poultry Sci Asn. *Res:* Embryo transfer and in vitro culture techniques to investigate early development of mammalian embryos; effects of environment on mammalian spermatozoa and embryos; avian sperm physiology and sperm-egg interaction. *Mailing Add:* Dept Poultry Sci L-P Univ Ga Athens GA 30602

HOWARTH, FRANCIS GARD, b Worcester, Mass, Oct 23, 40; m 65; c 3. BIOSPELEOLOGY, MUSEUM COLLECTION MANAGEMENT. *Educ:* Univ Mass, Amherst, BS, 62; Cornell Univ, MS, 66; Univ Hawaii, Honolulu, PhD(entom), 74. *Prof Exp:* Vol agr exten, Int Vol Serv, Laos, 66-68; NIH trainee, B P Bishop Mus, 68-69; teaching asst entom, Univ Hawaii, Honolulu, 69-70; assoc entom, 70-74, ENTOMOLOGIST, B P BISHOP MUS, HONOLULU, 74- *Concurrent Pos:* Assoc ed, Int J Entom, 74-85; life sci ed, Nat Speleol Soc Bull, 75-85; prin investr, NSF grants, 76-77, 79-83 & 86-89; grad affil fac, dept entom, Univ Hawaii, 78-; chmn, Cave Species Specialist Group, Species Survival Comm, Int Union Conserv Nature & Natural Resources, 79-; mem, Sci Adv Comt, Cave Res Found, 83-; ecologist & co-dir, Explorer's Club, 85 & Chillagoe Caves Exped, 85; mem bd, Am Cave Conserv Asn, 84- *Mem:* Fel Nat Speleol Soc; Cave Res Found; Int Union Conserv Nature & Natural Resources; Am Cave Conserv Asn; Entom Soc Am; Int Soc Odonatology; Explorer's Club; AAAS. *Res:* Ecology and evolution of cave animals; tropical island ecology; biogeography; conservation; impact of alien species; biosystematics of aythropods, especially biting midges, Diptera, Ceratopogonidae. *Mailing Add:* B P Bishop Mus PO Box 19000-A Honolulu HI 96817-0916

HOWARTH, JOHN LEE, medical physics, biophysics, for more information see previous edition

HOWARTH, ROBERT W, b Boston, Mass, Feb 11, 52; m 87. BIOGEOCHEMISTRY, ECOSYSTEM BIOLOGY. *Educ:* Amherst Col, BA, 74; Mass Inst Technol & Woods Hole Oceanog Inst, PhD(oceanog), 79. *Prof Exp:* Fel, Marine Biol Lab, 79-80, asst scientist, 80-84, assoc scientist, 84-85; assoc prof ecol, 85-90, dir Ecosysts Res Ctr, 87-88, PROF ECOL, CORNELL UNIV, 91- *Concurrent Pos:* Adj prof, Univ Rhode Island, 82-; ed-in-chief, Biogeochemistry, 83-; consult, Nat Acad Sci Comt on Petroleum in Marine Environ, 81. Comt on Health & Ecol, 83; vis scientist, Inst Ecosystem Studies; NY Bot Garden, 85; mem, US Nat Comt SCOPE, 88-90;

mem, Comt on the Coastal Oceans, Nat Acad Sci, 89, Alternatives for Wastewater Disposal in Coastal Areas, 90; consult, state of Alaska on Exxon Valdez oil spill. *Mem:* Am Soc Limnol & Oceanog; Ecol Soc Am; Estuarine Res Fedn; AAAS; Sigma Xi. *Res:* Sulfur and molybdenum biogeochemistry; interactions of element cycles in aquatic ecosystems; environmental management and the effects of pollution on aquatic ecosystems and commercial fisheries; wetland ecosystems; microbial production and activity. *Mailing Add:* Sect Ecol & Systematics Corson Hall Cornell Univ Ithaca NY 14853

HOWARTH, RONALD EDWARD, b Wawota, Sask, Nov 20, 40; m 69; c 2. PLANT CHEMISTRY, NUTRITION. *Educ:* Univ Sask, BSA, 62, MSc, 65; Univ Calif, Davis, PhD(nutrit), 68. *Prof Exp:* Asst prof vet physiol & biochem, Univ Sask, 68-72; RES SCIENTIST, RES BR, AGR CAN, 72- *Mem:* Nutrit Today Soc; Phytochem Soc NAm; Crop Sci Soc Am; Can Soc Animal Sci; Agr Inst Can. *Res:* Ruminant bloat; plant biochemistry; animal and human nutrition. *Mailing Add:* Agr Can Res Sta 107 Sci Crescent Saskatoon SK S7N 0X2 Can

HOWATSON, JOHN, b Calgary, Alta, June 6, 20; nat US; m 45; c 2. X-RAY CRYSTALLOGRAPHY, SOLID STATE CHEMISTRY. *Educ:* Univ Wash, BSc, 43; Univ Wis, PhD(chem), 50. *Prof Exp:* Jr chemist, Corning Glass Works, 43-44; asst, SAM Labs, Columbia Univ, 44-45; jr physicist, Tenn-Eastman Co, 45-46; asst chem, Univ Wis, 46-47 & Wis Alumni Res Found, 47-49; res asst, Corning Glass Works, 49-55; instr chem, Colo Sch Mines, 55-56; from asst prof to assoc prof, 56-66, PROF CHEM, UNIV WYO, 66- *Mem:* Am Chem Soc; Sigma Xi. *Res:* Structure of small molecules; chemical reactions of oil shale minerals during thermal and hydrothermal processing; aqueous chemistry of aluminum. *Mailing Add:* Chem Dept-Phys Sci Bldg Univ Wyo Laramie WY 82071

HOWATT, WILLIAM FREDERICK, b Brewer, Maine, Aug 22, 29; m 61; c 4. PEDIATRICS, PHYSIOLOGY. *Educ:* Univ Maine, BS, 51; Boston Univ, MD, 55. *Prof Exp:* From instr to assoc prof, 59-75, PROF PEDIAT, UNIV MICH, ANN ARBOR, 75- *Concurrent Pos:* NIH fel, 63-64. *Mem:* Soc Pediat Res. *Res:* Pulmonary function in children; respiratory control of the neonate; pulmonary function changes in infants with congenital heart diesase undergoing surgical repair. *Mailing Add:* Dept Pediat D1209 0718 Univ Mich Ann Arbor MI 48109

HOWD, FRANK HAWVER, b Delmar, NY, Dec 2, 24; m 51; c 3. ECONOMIC GEOLOGY, ENVIRONMENTAL GEOLOGY. *Educ:* Univ Rochester, AB, 51, MS, 53; Wash State Univ, PhD(geol), 56. *Prof Exp:* Asst geol, Univ Rochester, 52-53; geologist, Div Mines & Geol, State Wash, 53; asst geol, Wash State Univ, 53-56; geologist, Bear Creek Mining Co, 56-59; asst prof, 59-65, ASSOC PROF GEOL, UNIV MAINE, ORONO, 65- *Concurrent Pos:* Consult, Kennecott Copper Corp, 60-70; consult, US Soil Conserv Serv, 74-75; consult, US Geol Survey, 79-83. *Mem:* Fel Geol Soc Am; Geochem Soc; Soc Econ Geologists. *Res:* Relationship of trace element dispersion and hydrothermal alteration to metallic mineral deposits; experimental replacement of rocks by sulfide minerals; geochemistry of stream sediments. *Mailing Add:* Dept Geol Boardman Hall Univ Maine Orono ME 04469

HOWD, ROBERT A, b McMinnville, Ore, Nov 19, 44; m 66; c 1. TOXICOLOGY, REGULATORY AFFAIRS & RISK ASSESSMENT. *Educ:* Linfield Col, BA, 66; Univ Wash, PhD(pharmacol), 73. *Prof Exp:* Biochem pharmacologist res, SRI Int, 75-88; toxicologist regulation, Toxic Substances Control Prog, 88-91, TOXICOLOGIST REGULATION, HAZARD IDENTIFICATION & RISK ASSESSMENT BR, CALIF DEPT HEALTH SERV, 91- *Concurrent Pos:* Postdoctoral fel, Lab Neuroendocrine Regulation, Mass Inst Technol, 73-75. *Mem:* Soc Toxicol; Am Soc Pharmacol & Exp Therapeut; Western Pharmacol Soc; AAAS. *Res:* Neurotoxicology; drug abuse; drug metabolism; risk assessment for solvents and pesticides; evaluation of dermal absorption of chemicals. *Mailing Add:* 2151 Berkeley Way Annex II PETS Berkeley CA 94704

HOWDEN, DAVID GORDON, b Scarborough, Eng, Aug 22, 37; m 62; c 1. METALLURGY, WELDING & FABRICATION. *Educ:* Univ Eng, BSc, 59, PhD(metall), 62. *Prof Exp:* Asst prof & researcher metall, Cent Tech Aeronaut, Sao Jose dos Campos, Brazil, 63-65; sr scientist welding metall, Dept Energy Mines & Resources, Can, 65-67; assoc mgr welding & fabric, Battelle Columbus Labs, 67-77; ASSOC PROF WELDING ENG, OHIO STATE UNIV, 77- *Concurrent Pos:* US Deleg, Int Inst Welding, 77-81. *Honors & Awards:* Oxygen of Brazil Prize, Asn Brazil de Metais, 75; James F Lincoln Award, Am Weld Soc, 82. *Mem:* Am Welding Soc; Am Soc Metals; Welding Res Coun. *Res:* Welding metallurgy; gas metal reactions in arc welding; health and safety in welding; welding process development; component failure analysis. *Mailing Add:* 3285 Wilson Rd Sunbury OH 43074

HOWDEN, HENRY FULLER, b Baltimore, Md, Aug 19, 25; m 49; c 3. ENTOMOLOGY. *Educ:* Univ Md, BS, 46, MS, 49; NC State Univ, PhD(entom), 53. *Prof Exp:* Asst prof entom, Univ Tenn, 53-57; res scientist, Entom Res Inst, Can Dept Agr, Ont, 57-68; PROF SYST ZOOL, CARLETON UNIV, 68- *Concurrent Pos:* Res partic, Oak Ridge Inst Nuclear Studies, 55; consult, US AEC, 56-58; Alexander Agassiz vis lectr, Mus Comp Zool, Harvard Univ, 67. *Mem:* Entom Soc Am; Asn Trop Biol; Soc Syst Zool; fel Entom Soc Can. *Res:* Taxonomy and biology of North American and world Scarabaeidae. *Mailing Add:* Dept Biol Carleton Univ Ottawa ON K1S 5B6 Can

HOWE, CALDERON, b Washington, DC, Mar 8, 16; m 44; c 4. MICROBIOLOGY. *Educ:* Yale Univ, AB, 38; Harvard Univ, MD, 42. *Prof Exp:* Chief res physician, Peter Bent Brigham Hosp, Boston, 47-48; assoc microbiol, Columbia Univ, 52-55, from asst prof to prof, Col Physicians & Surgeons, 55-71; asst physician, Presby Hosp, 52-57, asst attend physician, 57-64, assoc microbiologist, 64-71, attend microbiologist, 71; prof microbiol & chmn dept, Med Ctr, La State Univ, New Orleans, 72-82; RETIRED. *Concurrent Pos:* Assoc ed, J Virol, 74-, ed bd, Proc Soc Exp Biol & Med, 75. *Mem:* Am Soc Clin Invest; Am Soc Microbiol; Am Soc Cell Biol; Soc Exp Biol & Med; Am Asn Immunologists (secy-treas, 58-61). *Res:* Infectious diseases; immunology and immunochemistry; blood groups; virology; virus-erythrocyte, virus-cell interactions; immunochemical and immunoelectron microscopic investigations of membranes. *Mailing Add:* 90 Annandale Rd Newport RI 02840-6925

HOWE, CHIN CHEN, b Taipei, Taiwan, June 22, 33; US citizen; m 60; c 2. MOLECULAR BIOLOGY, DEVELOPMENTAL BIOLOGY. *Educ:* Univ Taiwan, BS, 56; Univ Pa, MS, 60, PhD(molecular biol), 72. *Prof Exp:* Fel virol, 73-76, STAFF DEVELOP BIOL, WISTAR INST ANAT & BIOL, 76 - *Concurrent Pos:* NIH fel, 75-76. *Mem:* Soc Develop Biol. *Res:* Gene expression during development and differentiation in mouse system; biochemical aspects of gene action and regulation during development. *Mailing Add:* Wistar Inst Anat & Biol 36th at Spruce Philadelphia PA 19104

HOWE, DAVID ALLAN, b Clark AFB, Philippine Isles, Apr 6, 49; m 73; c 2. HIGH RESOLUTION SPECTROSCOPY. *Educ:* Univ Colo, BA, 70. *Prof Exp:* PHYSICIST ATOMIC BEAM SPECTROS, DEPT COM, NAT INST STANDARDS & TECHNOL, 69- *Honors & Awards:* Gold Medal, Dept Com. *Mem:* Am Inst Physics. *Res:* Development of very high frequency stability using atomic and molecular spectroscopy techniques such as cesium beam and hydrogen maser devices; high-powered amplifier design. *Mailing Add:* Dept Com Nat Inst Standards & Technol Boulder CO 80303

HOWE, DAVID ALLEN, b Oct 29, 35; US citizen; m 61. PHYSICS. *Educ:* Ind Univ, BS, 58, PhD(physics), 62. *Prof Exp:* Staff mem physics, Los Alamos Sci Lab, 62-63; asst prof, 63-70, ASSOC PROF PHYSICS, TEX TECH UNIV, 70-; currently attache, US Embassy, Moscow. *Mem:* Am Phys Soc. *Res:* Beta decay; plasma physics. *Mailing Add:* US Embassy Moscow/ESO APO New York NY 09862

HOWE, DENNIS GEORGE, b Reading, Pa, Apr 8, 43; m 69. OPTICS. *Educ:* Cornell Univ, AB, 65; Univ Rochester, MS, 68, PhD(optics), 76. *Prof Exp:* Photog systs analyst & sr engr, Res & Eng Dept, Kodak Apparatus Div, 65-73, proj engr video imaging systs, Consumers Prod Dept, 73-75, sr physicist, 75-78, RES ASSOC OPTICAL, LASER & VIDEO SYSTS, EASTMAN KODAK CO RES LABS, 78- *Mem:* Optical Soc Am; Soc Info Display. *Res:* High density optical data storage and retrieval systems; electro-optical coherent and non-coherent light scanners for high resolution graphic arts, facsimile and video storage and display. *Mailing Add:* 14 Lodge Pole Rd Pittsford NY 14534

HOWE, EUGENE EVERETT, b Keats, Kans, June 6, 12; m 38; c 2. PHARMACEUTICAL CHEMISTRY. *Educ:* Kans State Col, BS, 36, MS, 37; Univ Ill, PhD(biochem), 40. *Prof Exp:* Asst chem, Kans State Col, 37; asst, Univ Ill, 38-40; res chemist, Merck & Co, Inc, 41-47, head process develop natural prod, 48-56, dir nutrit, Inst Therapeut Res, 57-67, dir exp biol, Calgon Consumers Products Res Lab, 67-74; chem res consult, 75-80; RETIRED. *Mem:* AAAS; Am Chem Soc; Am Soc Biol Chemists; Am Inst Nutrit; NY Acad Sci. *Res:* Supplementation of cereals; bile acid and cholesterol metabolism; dental therapeutics. *Mailing Add:* Rd 1 Box 53 Somerset NJ 08873-9764

HOWE, EVERETT D(UMSER), mechanical engineering; deceased, see previous edition for last biography

HOWE, GEOFFREY RICHARD, b Eng, Dec 2, 42; Can citizen; m 87; c 3. CANCER EPIDEMIOLOGY & BIOSTATISTICS. *Educ:* Univ London, BSc, 65; Univ Leicester, PhD(theoret chem, math modeling), 69. *Prof Exp:* PROF PREV MED & BIOSTATIST, UNIV TORONTO, 82- *Concurrent Pos:* Dir, Nat Cancer Inst of Can, Epidemiol Unit, 86- *Res:* Record linkage studies. *Mailing Add:* NCIC Epidemiol Unit Univ Toronto 12 Queen's Park Crescent W Third Floor Toronto ON M5S 1A8 Can

HOWE, GEORGE FRANKLIN, b Buffalo, NY, Nov 15, 31; m 55; c 4. PLANT PHYSIOLOGY. *Educ:* Wheaton Col, Ill, BS, 53; Ohio State Univ, MS, 56, PhD(plant physiol), 59. *Prof Exp:* Asst gen bot, Ohio State Univ, 55, asst plant physiol, 55-57; instr biol & bot, Westmont Col, 59-60, from asst prof to assoc prof biol, 60-68; PROF BIOL & CHMN DIV NATURAL SCI, THE MASTER'S COL, 68- *Concurrent Pos:* Partic, NSF Acad Yr Inst Radiation Biol, Cornell Univ, 65-66; ed, Creation Res Soc Quart, 69-74; dir, Grand Canyon Exp Sta, Creation Res Soc, 83- *Mem:* Creation Res Soc (pres, 77-82); Sigma Xi. *Res:* Relationship of the degree of stomatal opening to the time course of photosynthetic induction periods and photosynthetic rhythms; creation model of origins; plant physiology; botanical courses; radiation biology; post-fire regrowth of chaparral shrubs; pollination biology; micropaleontology; plant taxonomy. *Mailing Add:* Div Natural Sci The Master's Col Santa Clarita CA 91322

HOWE, GEORGE MARVEL, climatology, geography, for more information see previous edition

HOWE, GEORGE R, b Brattleboro, Vt, Apr 14, 33; m 55; c 4. REPRODUCTIVE PHYSIOLOGY. *Educ:* Univ Vt, BS; Pa State Univ, MS, 59; Univ Mass, PhD(physiol), 61. *Prof Exp:* Asst prof physiol & biophys, Col Med, Univ Vt, 61-68; asst prof physiol & biophys, 68-73, ASSOC PROF ANIMAL SCI, UNIV MASS, AMHERST, 73- *Concurrent Pos:* Grants, Pop Coun, 63 & 67, NIH, 63-65 & 69-76, Whitehall Found, 69-72, US Naval Res, 67-72 & Pop Coun, 72-74. *Mem:* Am Fertil Soc. *Res:* Equine reproduction; Anabolic steroids and reproduction. *Mailing Add:* Dept Vet & Animal Sci Univ Mass Amherst MA 01003

HOWE, HENRY BRANCH, JR, b Atlanta, Ga, Aug 5, 24; m 51; c 3. MICROBIOLOGY. *Educ:* Emory Univ, AB, 48, MA, 50; Univ Wis, PhD(genetics), 55. *Prof Exp:* Asst prof biol, Union Col, Ky, 54-57 & Wake Forest Univ, 57-59; from asst prof to assoc prof bact, Univ Ga, 59-70, prof microbiol, 70-90, assoc dean, Grad Sch, 81-90, EMER PROF MICROBIOL, UNIV GA, 90-, EMER ASSOC DEAN, GRAD SCH, 90- *Concurrent Pos:* NSF sci fac fel, Yale Univ, 59. *Honors & Awards:* Res Prize, Asn Southeastern Biologists, 67 & 71. *Mem:* Genetics Soc Am; Am Soc Microbiol; Mycol Soc Am; Am Ornith Union. *Res:* Microbial genetics; microbiology. *Mailing Add:* 130 Bishop Dr Athens GA 30606

HOWE, HENRY FRANKLIN, b Gardner, Mass, Dec 24, 46. ECOLOGY, EVOLUTIONARY BIOLOGY. *Educ:* Earlham Col, AB, 68; Univ Mich, AM, PhD(zool), 77. *Prof Exp:* Wildlife biologist, US Army, 69-71; instr biol, Phillips Acad, 71-72; asst prof zool, Univ Iowa, 78-82, assoc prof biol, 82-88; PROF BIOL, UNIV ILL, CHICAGO, 88- *Concurrent Pos:* Fel Smithsonian Trop Res Inst, 77-78. *Mem:* Soc Study Evolution; Brit Ecol Soc; Ecol Soc Am; Am Soc Naturalists; Asn Trop Biol; AAAS. *Res:* Evolutionary ecology; sex ratio, sexual selection, sexual dimorphism, and reproductive biology of birds; frugivory and the ecology of tropical seed-dispersal systems; reproductive ecology of plants; desert ecology; prairie ecology. *Mailing Add:* Biol Sci M/C 066 Univ Ill Box 4348 Chicago IL 60680

HOWE, HERBERT JAMES, b Baton Rouge, La, Jan 31, 31; m 56; c 2. STRATIGRAPHY. *Educ:* La State Univ, BS, 52; Columbia Univ, MA, 53, PhD(geol), 59. *Prof Exp:* Explor geologist, Humble Oil & Refining Co, 53-56; from asst prof to assoc prof geol, La State Univ, New Orleans, 58-67; assoc prof geosci, 67-88, EMER PROF, PURDUE UNIV, 88- *Mem:* Paleont Soc; Am Asn Petrol Geologists. *Res:* Stratigraphy; invertebrate paleontology. *Mailing Add:* Dept Earth & Atmospheric Sci Purdue Univ West Lafayette IN 47907

HOWE, JOHN A, b Uniopolis, Ohio, May 22, 27; m 53; c 1. VERTEBRATE PALEONTOLOGY. *Educ:* Bowling Green State Univ, BS, 53; Univ Nebr, MS, 56, PhD(geol), 61. *Prof Exp:* Instr & cur, Educ Serv, Univ Nebr State Mus, 58-65; from instr to asst prof, 65-70, ASSOC PROF GEOL, BOWLING GREEN STATE UNIV, 70- *Concurrent Pos:* Bowling Green State Univ res grant, 66- *Mem:* Paleont Soc; Am Quaternary Asn; Soc Vert Paleont; Sigma Xi. *Res:* Pleistocene vertebrates. *Mailing Add:* Dept Geol Bowling Green State Univ Bowling Green OH 43403

HOWE, JOHN A, b Ft Benning, Ga, Feb 20, 33; m 58; c 3. NAVAL OPERATIONS RESEARCH. *Educ:* Emory Univ, AB, 53, MS, 54; Harvard Univ, PhD(phys chem), 59. *Prof Exp:* Asst prof chem, Univ Calif, Berkeley, 59-62; mem staff, Bell Tel Labs, 62-65; opers analyst, Ctr Naval Anal, 65-84, OPERS ANALYST, NAVAL OCEAN SYST CTR, 84- *Mem:* Am Phys Soc; Opers Res Soc Am; Sigma Xi. *Res:* Molecular spectroscopy; chemical bonding theory; lasers; naval operations research; undersea warfare. *Mailing Add:* 425 Santa Dominga Solana Beach CA 92075

HOWE, JOHN P(ERRY), b Groton, NY, June 24, 10; m 41; c 4. NUCLEAR ENGINEERING. *Educ:* Hobart Col, BS, 33; Brown Univ, PhD(phys chem), 36. *Prof Exp:* Instr chem, Ohio State Univ, 36-38; asst prof, Brown Univ, 38-42; res assoc, Metall Lab, Univ Chicago, 42-44, assoc dir, 45; mgr metall, Knolls Atomic Power Lab, Gen Elec Co, 45-52, mgr res energy conversion & storage, Gen Elec Res Lab, 52-53; sect chief reactor mat, Atomics Int, 53-57, dir res dept, 57-61; Ford prof eng, Cornell Univ, 61-67, dir dept eng physics & mat sci, 62-65 & dept eng physics, 65; res staff mem, Sci & Technol Div, Inst Defense Anal, 67-68; asst lab dir, Gen Atomic Co, 68-69, assoc lab dir, 69-71, tech dir mat advan energy systs, 71-75; RETIRED. *Concurrent Pos:* Mem comt sr reviewers, Div Classification, AEC, 52-67, mem adv comt reactor safeguards, 61-63, mem bd atomic safety & licensing, 65-68; adv AEC deleg & vchmn session radiation effects, Geneva Conf Peaceful Uses Atomic Energy, Switz, 55, deleg, 58; mem rev comt fundamental res in mat, Nat Acad Sci-Dept Defense, 57-58; consult, Inst Defense Anal, 67-68 & Off Tech Assessment, US Congress, 75; mem sci adv group, Air Force Off Aerospace Res, 63-71; adj prof nuclear eng, Univ Calif, San Diego, 73-82. *Mem:* Fel AAAS; Am Chem Soc; fel Am Phys Soc; fel Am Nuclear Soc; fel Am Soc Metals; Sigma Xi. *Res:* Nuclear science and engineering; nuclear materials irradiation effects; energy policy. *Mailing Add:* 5725 Waverly Ave La Jolla CA 92037

HOWE, JULIETTE COUPAIN, b Woonsocket, RI, Aug 16, 44; m 79; c 2. NUTRITION. *Educ:* Univ RI, BS, 65; Univ Md, MS, 78, PhD(nutrit sci), 87. *Prof Exp:* Res biologist, Agr Res Serv, Protein Nutrit Lab, 77-79, res nutritionist, Sci & Educ Admin, High Nutrit, 79-80, RES CHEMIST, AGR RES SERV, BELTSVILLE HUMAN NUTRIT RES CTR, ENERGY & PROTEIN NUTRIT LAB, 81- *Mem:* Am Inst Nutrit; Am Chem Soc. *Res:* Effects of varying nutrient intake on energy metabolism in normal and hyperinsulinemic individuals; physiological bases of variation among individuals in resting energy expenditure; role of hormones in energy metabolism. *Mailing Add:* USDA ARS BHNRC EPNL Bldg 308 Rm 213 BARC-East 10300 Baltimore Ave Beltsville MD 20705

HOWE, KENNETH JESSE, b Rochester, NY, Nov 13, 23; m 50; c 4. PLANT PHYSIOLOGY. *Educ:* Univ Rochester, BA, 50, MS, 51; Cornell Univ, PhD(plant physiol), 56. *Prof Exp:* Asst gen bot & cytol, Univ Rochester, 48-51; asst gen bot, Cornell Univ, 52-54, instr, 54-56; asst prof plant physiol & gen bot, Univ Fla, 56-57; chmn dept biol sci, 65-78, from asst prof to assoc prof, 57-62, PROF GEN BOT & PLANT PHYSIOL, BRIDGEWATER STATE COL, 62-, DIV HEAD, 80- *Mem:* Am Soc Plant Physiol; Bot Soc Am; Sigma Xi. *Res:* Structure and function of plants, especially as affected by environment and growth regulators; production of essential oils in plants. *Mailing Add:* 47 Vernon St Bridgewater MA 02324

HOWE, KING LAU, b Shanghai, China, Sept 4, 28; US citizen; m 60; c 2. PHYSICAL ORGANIC CHEMISTRY, POLYMER CHEMISTRY. *Educ:* Dartmouth Col, AB, 52; Univ Wis, PhD(chem), 57. *Prof Exp:* From res chemist to sr res chemist, 57-68, supvr, 68, sr res chemist, 69-70, res assoc, 70-83, SR RES ASSOC, E I DU PONT DE NEMOURS & CO, INC, 83- *Mem:* Am Chem Soc. *Res:* Solvolysis reaction mechanism; stereochemistry of radical addition reaction; mechanism of base promoted dehydrohalogenation reaction; vinyl polymers; process and product development; ethylene copolymers and polymer blends; polyester engineering resins. *Mailing Add:* 414 Brentwood Dr Carrcroft Crest Wilmington DE 19803-4308

HOWE, MARSHALL ATHERTON, b Mt Kisco, NY, Dec 2, 44; m 87. ORNITHOLOGY. *Educ:* Alfred Univ, BA, 66; Univ Minn, PhD(zool), 72. *Prof Exp:* Zoologist, US Fish & Wildlife Serv, 72-75, chief Ornith, Bird Sect, Nat Fish & Wildlife Lab, 75-78, res zoologist, sect migratory nongame birds, 78-81, sect leader, 82-85, team leader, Br Migratory Bird Res, Patuxent Wildlife Res Ctr, 85-90, ASST CHIEF, OFF MIGRATORY BIRD MGT, US FISH & WILDLIFE SERV, 90- *Concurrent Pos:* Gen secy, Wader Study Group; treas, Pan-Am sect, Int Coun Bird Preserv. *Mem:* Am Ornith Union; Cooper Ornith Soc; Int Coun Bird Preserv; Int Asn Fish & Wildlife Agencies. *Res:* Behavior and ecology of birds, particularly social behavior, social organization and habitat utilization in shorebirds. *Mailing Add:* Off Migratory Bird Mgt US Fish & Wildlife Serv 4401 N Fairfax Dr Rm 634 Arlington VA 22011

HOWE, MARTHA MORGAN, b New York, NY, Sept 29, 45. MOLECULAR GENETICS, VIROLOGY. *Educ:* Bryn Mawr Col, AB, 66; Mass Inst Technol, PhD(microbiol), 72. *Prof Exp:* Europ Molecular Biol Orgn fel, Cent Nuclear Acad, Nat Res Coun Italy, 72; fel molecular biol, Cold Spring Harbor Lab, Cold Spring Harbor, NY, 72-74; fel Am Cancer Soc, 74; from asst prof to prof, Univ Wis-Madison, 75-84, Vilas prof bact, Col Agr & Life Sci, 84-86; VAN VLEET PROF VIROL, COL MED, UNIV TENN, MEMPHIS, 86- *Concurrent Pos:* NIH Res Career Develop Award, 78-83; assoc ed, Virol, 83-92; mem, rev panel, NSF, Howard Hughes Med Inst & Am Chem Soc; counr, Am Soc Microbiol, 89-91; dir, Genetics Soc Am, 89-91. *Honors & Awards:* Award, Eli Lilly & Co, 85, Am Acad Microbiol, 91. *Mem:* Am Soc Microbiol; Am Soc Biochem & Molecular Biol; Genetics Soc Am. *Res:* Analyses of the molecular mechanisms of genetic recombination and gene regulation, the study of these processes using bacteriophage Mu. *Mailing Add:* Dept Microbiol & Immunol 858 Madison Ave Memphis TN 38163

HOWE, NORMAN ELTON, JR, b San Antonio, Tex, Jan 16, 46; m 75. PHYSICAL ORGANIC CHEMISTRY. *Educ:* Univ Calif, Berkeley, BA, 68; Univ Calif, Los Angeles, PhD(org chem), 73. *Prof Exp:* Fel bio-org chem, Univ Chicago, 73-75; res chemist, 75-77; plant supt vitamins, 77-85, PROD MGR VITAMIN PLANT, BASF WYANDOTTE CORP, 85- *Mem:* Am Chem Soc. *Res:* High molecular weight block polymers. *Mailing Add:* 1529 Boxford Trenton MI 48183-1808

HOWE, RICHARD HILDRETH, b Providence, RI, Dec 25, 27; m 61; c 1. DURABILITY OF AGGREGATES & CONCRETE, PAVEMENT SKID RESISTANCE. *Educ:* Mass Inst Technol, SB, 51. *Prof Exp:* Petrol geologist, Calif Co, 53-54; grad student, Pa State Univ, 54-60, instr geol dept, 60-61; asst dist soils engr, 61-66, soils res engr, 66-69, concrete & aggregate testing engr, 69-87, STAFF SOILS ENGR, PA DEPT TRANS, 87- *Mem:* Fel Geol Soc Am; Am Soc Testing & Mat; Int Asn Eng Geol; Am Inst Prof Geologists; Asn Eng Geologists; Am Concrete Inst. *Res:* Factors affecting the level & seasonal variability of pavement skid resistance, especially those related to the properties of aggregates; development of procedures for evaluating the durability of aggregates & concrete. *Mailing Add:* 2911 Chestnut St Camp Hill PA 17011

HOWE, RICHARD SAMUEL, b Centralia, Ill, Aug 30, 36; m 68; c 2. WATER RESOURCE POLICY PLANNING MANAGEMENT, RECRUITING & RETENTING YOUNG WOMEN & MEN TO CAREERS IN ENGINEERING. *Educ:* Univ Ky, BS, 59; Mass Inst Technol, SM, 61; Univ Wis-Madison, MS(water res mgt), 68, MS(urban regional planning), 69, PhD(resource mgt), 71. *Prof Exp:* San engr, USPHS, 61-63; instr eng, Southern Ill Univ, 63-67; dir planning & res, City Chicago Dept Environ Control, 70-72; assoc prof environ sci, Ind Univ, 72-76; PROF, ENG & ENVIRON SCI, UNIV TEX, SAN ANTONIO, 76- *Concurrent Pos:* Prin, Design San Antonio, 88-; chmn bd, Hemisphere Inst Pub Serv, 89-; co-prin investr, Tex Water Develop Bd Proj, Univ Tex, San Antonio, 90-; prin investr, Comprehensive Regional Ctr Minorities, 90-; dir, Univ Tex, San Antonio, Alliance for Educ, 90-; dipl, Am Acad Environ Engrs. *Mem:* Am Acad Environ Engrs; Am Soc Civil Engrs; Nat Soc Prof Engrs; Am Inst Cert Planners; Am Soc Pub Admin; Water Pollution Control Fedn. *Res:* Decision support systems for resource management-water and solid waste; resource diplomacy; management of technology and technology transfer. *Mailing Add:* Div Eng Univ Tex 6900 N Loop 1604 West San Antonio TX 78249

HOWE, ROBERT C, b Baton Rouge, La, June 13, 39; m 88. REMOTE SENSING, MICROPALEONTOLOGY. *Educ:* La State Univ, BS, 61; Univ Wis, MS, 62, PhD(geol), 65. *Prof Exp:* Jr geologist, Humble Oil & Refining Co, summer 61; geologist, Atlantic Refining Co, summer 65; instr geol, Beloit Col, 67-68; from asst prof to assoc prof, 68-77, PROF GEOL, IND STATE UNIV, TERRE HAUTE, 77- *Mem:* AAAS; Am Asn Petrol Geol; Geol Soc Am; Sigma Xi; Am Soc Photogram & Remote Sensing. *Res:* Remote sensing, utilization of landsat, TIMS and other types of remote sensing data for geologic applications; nannofossils; ostracodes. *Mailing Add:* Dept Geog & Geol Ind State Univ Terre Haute IN 47809

HOWE, ROBERT GEORGE, b Sault Ste Marie, Ont, Apr 27, 26; m 52; c 3. ENTOMOLOGY. *Educ:* Ont Agr Col, BSA, 51; Cornell Univ, PhD(entom), 55. *Prof Exp:* Asst forest entom, Can Dept Agr, 48-49, asst med entom, 50; asst veg insects, NY Exp Sta, Geneva, 51-55; field specialist agr chem develop, Dow Chem Co, 55-67, mgr bioprod res & develop, Dow Chem Europe, Switz, 67-70, tech prod specialist, 70-78, prod develop mgr, 78-89; RETIRED. *Mem:* Entom Soc Am. *Res:* Insecticides; space, grain and soil fumigants; urban entomology row crops. *Mailing Add:* 714 Hollybrook Dr Midland MI 48642

HOWE, ROBERT H L, environmental science & engineering; deceased, see previous edition for last biography

HOWE, ROBERT JOHNSTON, b Providence, RI, May 2, 36. BEHAVIORAL BIOLOGY, VERTEBRATE BIOLOGY. *Educ:* Univ RI, BS, 67, MS, 71; Northern Ariz Univ, PhD(animal behav, zool), 77. *Prof Exp:* ASST PROF BIOL, SUFFOLK UNIV, 75- *Concurrent Pos:* Instr biol, Northeastern Univ, 76-77. *Mem:* Animal Behav Soc; Am Soc Mammalogists. *Res:* Social behavior of small mammals with an emphasis on scent marking and olfactory communication as it relates to reproduction. *Mailing Add:* PO Box 32 South Berlin MA 01549

HOWE, ROBERT KENNETH, b Kewanee, Ill, June 27, 39; m 64; c 1. ORGANIC CHEMISTRY. *Educ:* Univ Ill, BS, 61; Univ Calif, Los Angeles, PhD(org chem), 65. *Prof Exp:* MONSANTO SR FEL, NEW PROD DIV, MONSANTO AGR CO, 65- *Mem:* Am Chem Soc. *Res:* Metabolism studies of pesticides. *Mailing Add:* Monsanto Co 700 Chesterfield Village Pkwy Chesterfield MO 63198

HOWE, ROBERT T(HEODORE), b Cincinnati, Ohio, July 4, 20; m 46; c 3. CIVIL ENGINEERING. *Educ:* Univ Cincinnati, CE, 43, AM, 55; Purdue Univ, PhD(hwy eng), 59. *Prof Exp:* Inst math, Univ Cincinnati, 46-47, from inst to prof civil eng, 47-83, asst dean, 80-83; RETIRED. *Mem:* Am Soc Civil Engrs; Am Soc Eng Educ; Inst Transp Engrs. *Res:* Urban development and transportation. *Mailing Add:* 1516 Northview Ave Cincinnati OH 45223

HOWE, ROGER, b Chicago, Ill, May 23, 45; m 67; c 2. LIE THEORY, INVARIANT THEORY. *Educ:* Harvard Univ, BA, 66; Univ Calif, Berkeley, PhD(math), 69. *Prof Exp:* Asst prof math, State Univ NY Stony Brook, 69-74; PROF MATH, YALE UNIV, 74- *Concurrent Pos:* Mem, Inst Advan Study, 71-72; vis prof, Inst Theoret Math Res, Bonn, Ger, 73-74, Oxford Univ, Eng, 78, Tel Aviv Univ, Israel, 80, Univ Calif, San Diego, 81 & Rutgers Univ, 89-90; Guggenheim fel, 84-85; assoc ed, 85-87, ed, Res Announcements, Bull of AMS, 88-90; fel, Inst Advan Studies, Hebrew Univ, Jerusalem, 88. *Honors & Awards:* Lester R Ford Award, Math Asn Am. *Mem:* Am Math Soc; Math Asn Am. *Res:* Harmonic analysis and group representation theory, particularly of algebraic groups, and with emphasis on the theory of automorphic forms. *Mailing Add:* Dept Math Yale Univ Box 2155 Yale Sta New Haven CT 06520

HOWE, VIRGIL K, b King City, Mo, Apr 22, 31; m 56; c 2. FOREST PATHOLOGY. *Educ:* Northwest Mo State Col, BS, 56; Univ NMex, MS, 61; Iowa State Univ, PhD(forest path), 64. *Prof Exp:* High sch instr, Mo, 56-57; asst prof biol, Northwestern State Col, La, 64-68; from asst prof to prof biol, Western Ill Univ, 68-88, chmn dept, 78-80, assoc provost, 80-82 & 83-86, actg provost, 82-83 & 86-88; DEAN, COL HEALTH & LIFE SCI, FT HAYS STATE UNIV, 88- *Concurrent Pos:* Res assoc, Morton Arboretum, 70-82. *Mem:* Am Phytopath Soc; Mycol Soc Am; Int Soc Arboricult; Sigma Xi. *Res:* Forest mycology; mycorrhizae of forest and shade trees; urban tree problems. *Mailing Add:* Off Dean Col Health & Life Sci Ft Hays State Univ Hays KS 67601

HOWE, WALLACE BRADY, b Lexington, Mo, Aug 5, 26; m 48; c 1. GEOLOGY. *Educ:* Univ Mo, BA, 47, MA, 48; Univ Kans, PhD, 54. *Prof Exp:* Asst, Univ Mo, 47-48; asst instr geol, Univ Kans, 48-51; geologist, State Geol Surv, Mo, 51-60, sr geologist, 60-65, asst state geologist, 65-71, state geologist & dir, 71-74, dir & state geologist, Mo Div, Geol & Land Surv, 74-86, ASSOC DIR, 86- *Concurrent Pos:* Asst geologist, State Geol Surv, Kans, 48-50. *Mem:* Geol Soc Am; Am Asn Petrol Geol. *Res:* Stratigraphy; groundwater. *Mailing Add:* Mo Geol Surv PO Box 250 Rolla MO 65401

HOWE, WILLIAM EDWARD, b Jakarta, Indonesia, June 16, 48. MAMMALIAN CELL CULTURE, IMMUNOFLUORESCENCE. *Educ:* Univ Ariz, PhD(molecular biol), 76. *Prof Exp:* CELL BIOLOGIST, ALCON LABS, 76- *Mailing Add:* Alcon Labs 6201 S Freeway Ft Worth TX 76134

HOWE, WILLIAM JEFFREY, b Ottawa, Ont, July 15, 46; m 69; c 1. ORGANIC CHEMISTRY, COMPUTER SCIENCES. *Educ:* Carleton Univ, BSc, 68; Harvard Univ, PhD(chem), 72. *Prof Exp:* Fel org chem, Harvard Univ, 72-74; res scientist, 74-80, SR RES SCIENTIST, UPJOHN CO, 80- *Mem:* Am Chem Soc. *Res:* Computer handling of chemical structure information; computer graphics; drug design research. *Mailing Add:* Upjohn Co Unit 7247 Kalamazoo MI 49001-0199

HOWELL, ALVIN H(AROLD), b Sedgwick, Kans, Feb 5, 08; m 34; c 4. ELECTRICAL ENGINEERING, ELECTRONICS. *Educ:* Univ Kans, BS, 29; Mich Col Mining & Technol, MS, 34; Mass Inst Technol, ScD(elec eng), 38. *Prof Exp:* Test engr, Gen Elec Co, 29-31; instr elec eng, Mich Col Mining & Technol, 31-34; res asst, Mass Inst Technol, 37-39, res assoc, 39-40; asst prof elec eng, 40-41, assoc prof, 41-43, chmn dept, 41-70, prof & dir res, Balloon Instrumentation Lab, 43-78, EMER PROF ELEC ENG & EMER DIR RES, TUFTS UNIV, 78- *Concurrent Pos:* Prof, Radar Sch, Mass Inst Technol, 42-43; dir, Doble Eng Co, 60-, vpres, 61-63, chmn exec comt, 69-, chmn bd dirs, 79-; mem & chmn panel sci use of balloons, Nat Ctr Atmospheric Res, Univ Corp Atmospheric Res, 61-63; mem, Nat Res Coun. *Mem:* Am Soc Eng Educ; Am Phys Soc; Inst Elec & Electronics Engrs; Am Asn Univ Professors. *Res:* Instrumentation in the upper air research program for the Air Forces; microwave measurements; gas discharges; dieelectrics; electro-mechanical instrumentation; instrumentation for large plastic balloon systems; development of balloon-borne telescope for tracking planets and stars. *Mailing Add:* Dept Elec Eng Tufts Univ Bedford MA 02155

HOWELL, BARBARA FENNEMA, b Chicago, Ill, Dec 18, 24; m 46; c 3. PHYSICAL CHEMISTRY, POLYMER SCIENCE. *Educ:* Univ Mo, Columbia, PhD(phys chem), 64. *Prof Exp:* Instr chem, Kans State Univ, 46-49 & Univ Tex, Arlington, 57-61; asst prof, Kans State Teachers Col, 64-69; res assoc, Space Sci Res Ctr, Univ Mo, Rolla, 69-71; res chemist, Inorg Glass Sect, Nat Bur Standards, 71-72, Org Anal Res Div, 72-81, res chemist, Polymer Sci & Standards, 81-87; MAT ENGR, POLYMER COMPOSITES BR, DAVID TAYLOR RES CTR, 87- *Mem:* AAAS; Sigma Xi; Am Chem Soc. *Res:* Isotope exchange reactions; polymer morphology; electron microscopy; conducting polymers. *Mailing Add:* Code 2844 David Taylor Res Ctr Annapolis MD 21402

HOWELL, BARBARA JANE, b Waynesfield, Ohio, May 19, 27. PHYSIOLOGY. *Educ:* Kent State Univ, BS, 50; Univ Ill, MS, 54, PhD(physiol), 57. *Prof Exp:* Asst prof physiol, Mont State Univ, 57-58; from instr to assoc prof, 59-75, PROF PHYSIOL, STATE UNIV NY BUFFALO, 75- *Mem:* Am Physiol Soc; Am Soc Zool. *Res:* Respiratory research; comparative physiology. *Mailing Add:* Dept Physiol-Sherman Hl State Univ NY Sch Med & Dent Buffalo NY 14214

HOWELL, BARTON JOHN, b Salt Lake City, Utah, June 5, 17; m 42; c 6. OPTICS. *Educ:* Univ Utah, AB, 40, PhD(physics), 54; Univ Mich, MS, 44. *Prof Exp:* Dir, USAF Photog Sch, Utah, 43; asst contract, Off Sci Res & Develop, 45-46; instr AV educ & chg photog dept, Ind Univ, 46; instr physics, Univ Utah, 46-51; secy-treas, Universal Microfilming Corp, 51-55, res physicist, Explosives Res Dept, 55-56; prin engr, Sperry Utah Eng Lab, 56-57, eng sect head, 57-61; eng dept head, Sperry Gyroscope Co, 61-65; sect supvr, Sperry Support Facility, NASA, 65-71; optical physicist, Goddard Space Flight Ctr, NASA, 71-84, sect head optical design, 84-86; RETIRED. *Concurrent Pos:* Chmn patents panel, Appl Optics, 78- *Honors & Awards:* Second Prize, Int Photog in Sci Exhib, AAAS, 48. *Mem:* Optical Soc Am; Soc Photog Scientists & Engrs. *Res:* Infrared systems; optical analog computers; microfilming techniques; optical design; laser holography. *Mailing Add:* 287 E Sixth Ave Salt Lake City UT 84103-2705

HOWELL, BENJAMIN F, JR, b Princeton, NJ, June 12, 17; m 43; c 4. SEISMOLOGY. *Educ:* Princeton Univ, AB, 39; Calif Inst Technol, MSc, 42, PhD(geophys), 49. *Prof Exp:* Res engr, Div War Res, Univ Calif, 42-44 & United Geophys Co, 45-49; head, dept geophys & geochem, 49-63, asst dean, Grad Sch, 68-70, assoc dean, 70-82, prof, 49-85, EMER PROF GEOPHYS, PA STATE UNIV, UNIV PARK, 85- *Concurrent Pos:* Mem exec comt, Div Earth Sci, Nat Res Coun, 65-67. *Mem:* Geol Soc Am; Soc Explor Geophys; Seismol Soc Am (vpres, 60-63, pres, 63-64); Am Geophys Union (secy, 56-64). *Res:* Seismology; tectonics; geophysical surveying. *Mailing Add:* Pa State Univ 406 Deike Bldg University Park PA 16802

HOWELL, BOB A, b Jefferson, NC, Jan 14, 42; m 65; c 1. REACTION MECHANISM, POLYMER DEGRADATION & STABILIZATION. *Educ:* Berea Col, BA, 64; Ohio Univ, PhD(org chem), 71. *Prof Exp:* Assoc fel, Ohio Univ, 69-70, instr org chem, 70-71; assoc fel, Iowa State Univ, 71-74; asst prof, Univ Louisville, 74-76; from asst prof to assoc prof, 76-88, PROF ORG/POLYMER CHEM, CENT MICH UNIV, 88- *Concurrent Pos:* Res chemist, Tenn Corp, 63, E I Du Pont de Nemours & Co, 64; indust fel, Dow Chem Co, 85; consult, Dow Chem Co, 85-, vis scientist, 89. *Mem:* Am Chem Soc; Royal Soc Chem; AAAS; Sigma Xi. *Res:* Organic and polymer chemistry; carbocation rearrangements; acid-catalyzed aldehyde and oxirane rearrangements; polymer-suported organoplatinum antitumor agents; polymer degradation; polymer synthesis and characterization; chromium carbonyl compounds; hydrogenolysis. *Mailing Add:* Dept Chem Cent Mich Univ Mt Pleasant MI 48859

HOWELL, CHARLES FREDERICK, b Beardstown, Ill, July 14, 32; m 54; c 2. MEDICINAL CHEMISTRY. *Educ:* Univ Ill, BS, 54; Mass Inst Technol, PhD(org chem), 58. *Prof Exp:* Fulbright fel, Australian Nat Univ, Canberra, 58-59; res chemist, 59-68, SR RES CHEMIST, LEDERLE LABS DIV, AM CYANAMID CO, 68- *Mem:* Am Chem Soc; NY Acad Sci. *Res:* Nitrogenous heterocycles; psychotropic agents; peptides; metabolism of drugs; renin inhibitors. *Mailing Add:* 28 Drake Lane Upper Saddle River NJ 07458

HOWELL, CHARLES MAITLAND, b Thomasville, NC, Apr 14, 14; m 49; c 2. DERMATOLOGY, ALLERGY. *Educ:* Wake Forest Univ, BS, 35; Univ Pa, MD, 37; Am Bd Dermat, dipl, 54. *Prof Exp:* Intern med, Charity Hosp, New Orleans, 37-38; resident, Burlington County Hosp, NJ, 38-39; physician, Lawrenceville Sch, 39-42; asst resident path, NC Baptist Hosp, Winston-Salem, 47-48; asst resident & resident dermat, Presby Hosp, NY, 48-50; resident allergy, Roosevelt Hosp, NY, 50-51; prof dermat & allergy & assoc path, Bowman Gray Sch Med, Wake Forest Univ, 51-83; PVT PRACT, 83- *Mem:* Fel Am Acad Dermat; fel Am Acad Allergy; fel NY Acad Sci; Int Soc Trop Dermat. *Res:* Improved methods for identification and treatment for cutaneous allergies; topical and intralesional use of corticosteroids; topical and systematic control of acne. *Mailing Add:* 340 Pershing Ave Winston-Salem NC 27103

HOWELL, DANIEL BUNCE, b Mitchell, SDak, Aug 23, 37; m 62; c 2. INORGANIC CHEMISTRY. *Educ:* Yankton Col, BA, 59; Univ Nebr, MS, 62, PhD(phys chem), 65. *Prof Exp:* Great Lakes Col Asn teaching intern chem, Col Wooster, 64-65; asst prof, 65-67, assoc prof phys chem, 67-79, PROF CHEM, NEBR WESLEYAN UNIV, 79- *Mem:* AAAS; Am Chem Soc; Sigma Xi. *Res:* Molecular spectra and molecular structure; inorganic coordination chemistry. *Mailing Add:* Dept Chem Nebr Wesleyan Univ 50th St Paul Lincoln NE 68504

HOWELL, DAVID MCBRIER, b Erie, Pa, July 30, 33. VIROLOGY. *Educ:* Princeton Univ, AB, 55; Pa State Univ, MS, 63, PhD(microbiol), 66. *Prof Exp:* Res virologist, Virus & Rickettsia Div, US Army Biol Labs, Ft Detrick, 66-71; assoc viral oncol, Spec Animal Leukemia Ecol Segment, 71-72, staff scientist viral oncol, Off Prog Resources & Logistics, 72-77, spec asst to assoc dir viral oncol, 77-78, asst to dir div cancer cause & prev, 78-83, SPEC ASST TO DIR, DIV CANCER ETIOLOGY, NAT CANCER INST, NIH, 84- *Mem:* Am Soc Microbiol; AAAS; Tissue Cult Asn; Sigma Xi. *Res:* Viral oncology; arthropod-borne viruses. *Mailing Add:* Bldg 31 Rm 11A06 Nat Cancer Inst NIH Bethesda MD 20892

HOWELL, DAVID MOORE, b San Diego, Calif, July 7, 25; m 49; c 2. ORGANIC CHEMISTRY. *Educ:* Univ Calif, BS, 45; Univ Mich, MS, 47, PhD(chem), 52. *Prof Exp:* Asst, Univ Calif, 44-45; instr chem, Univ Pa, 49-51; from instr to asst prof, 51-59, ASSOC PROF CHEM, NORTHEASTERN UNIV, 59- *Mem:* Am Chem Soc; The Chem Soc; Sigma Xi. *Res:* Schmidt reaction; metal chelates; organic hydroxylamine derivatives; polymethine derivatives. *Mailing Add:* 110 Blake St Needham MA 02192-2205

HOWELL, DAVID SANDERS, b Montclair, NJ, Oct 4, 23; m 48; c 3. MEDICINE, PHYSIOLOGY. *Educ:* Bowdoin Col, AB, 44; Harvard Med Sch, MD, 47; Am Bd Internal Med, dipl, 58. *Prof Exp:* Intern, RI Hosp, 47-48, intern med res, 48-50; resident path, Hosp Univ Pa, 50-51; asst investr, Res Div, Nat Heart Inst, 51-54; vis fel med, Columbia Univ, 54-55; asst prof, 55-69, PROF MED, SCH MED, UNIV MIAMI, 69- *Concurrent Pos:* Markle Found scholar, 57-62; vis investr, Nobel Inst, Sweden, 61-62; med investr, Vet Admin Hosp, Miami, 69-77; consult, Coral Gables Vet Admin Hosp. *Mem:* Am Physiol Soc; fel Am Col Physicians; Am Soc Clin Invest; Asn Am Physicians; Am Soc Cell Biol. *Res:* Electrolyte physiology; cartilage and other connective tissues; rheumatic diseases. *Mailing Add:* Box 106960 D26 Univ Miami Sch Med Miami FL 33101

HOWELL, DILLON LEE, b New Bremen, Ohio, Aug 1, 27; m 48; c 2. ECOLOGY, GEOLOGY. *Educ:* Southwest Mo State Univ, BS, 51; Univ Mo, MA, 55, PhD(sci educ), 71. *Prof Exp:* Instr biol, Southwest Mo State Univ, 51-52; PROF BIOL & GEOL, STEPHENS COL, 55-, HEAD DEPT NATURAL SCI, 74- *Mem:* AAAS; Nat Asn Biol Teachers. *Res:* Prairie/forest ecotone; team learning in science education. *Mailing Add:* 700 Timberhill Rd Columbia MO 65201

HOWELL, DORIS AHLEE, b Brooklyn, NY, Dec 2, 23. PEDIATRICS, HEMATOLOGY. *Educ:* Park Col, BA, 44; McGill Univ, MD, CM, 49. *Prof Exp:* From intern to jr asst resident, Children's Mem Hosp, Montreal, 49-51; sr asst resident pediat, Duke Univ Hosp, 51-52; fel med, Children's Med Ctr, Harvard Med Sch, 52-53, res fel, 52-54, asst hematologist, 53-55, asst physician, 54-55, instr pediat, Harvard Med Sch, 54-55; from asst prof to assoc prof, Sch Med, Duke Univ, 55-63; prof & chmn dept, Med Col Pa, 63-73; dep dir instnl develop, Asn Am Med Cols, 73-74; chmn dept community med, 75-81, PROF PEDIAT, UNIV CALIF, SAN DIEGO, 74- *Concurrent Pos:* Pediat hematologist, Med Ctr, Duke Univ, 55-63; consult hemat, US Navy; vis prof pediat, Med Col Pa, 73-74; mem, Bd Maternal, Child & Family Health, Nat Acad Sci. *Mem:* Am Acad Pediat; Soc Pediat Res; Am Soc Hemat; AMA; NY Acad Sci. *Res:* Hematological disorders of children. *Mailing Add:* Dept Pediat T-002 Univ Calif San Diego La Jolla CA 92093

HOWELL, EDWARD TILLSON, b Dixon, Ill, Mar 29, 97; m 25; c 2. ORGANIC CHEMISTRY. *Educ:* Univ Ill, BS, 19, MS, 20. *Prof Exp:* Res chemist, Newport Co, Wis, 20-31; res chemist, 31-50, CHEMIST, E I DU PONT DE NEMOURS & CO, INC, 50-; DIR RES, BRANDYWINE PHOTO CHEM CO, INC, 67- *Mem:* Am Chem Soc; fel Photog Soc Am. *Res:* Spectrophotometry; synthetic organic dyes and intermediates. *Mailing Add:* 1131 McKinley Ave Woodland CA 95695-4630

HOWELL, ELIZABETH E, b Jersey City, NJ, Mar 16, 51. ENZYME KINETICS & MECHANISM, SITE-DIRECTED MUTAGENESIS. *Educ:* Muhlenburg Col, BS, 73; Lehigh Univ, MS, 80, PhD(chem), 82. *Prof Exp:* Postdoctoral, Agouron Inst, La Jolla, Calif, 82-85, res scientist, 85-88; ASST PROF BIOCHEM, UNIV TENN, KNOXVILLE, 88- *Concurrent Pos:* Prin investr, NIH, 85- *Mem:* Am Chem Soc; AAAS; Protein Soc; Fedn Am Socs Exp Biol. *Res:* Structure-function studies of E coli dihydrofolate reductase and R plasmid encoded R67 dihydrofolate reductase; site-directed mutagenesis and second site reversion techniques applied to study catalysis. *Mailing Add:* Dept Biochem Univ Tenn Knoxville TN 37996-0840

HOWELL, EMBRY MARTIN, b Bethesda, Md, Nov 18, 45; m 65; c 2. PUBLIC HEALTH & EPIDEMIOLOGY. *Educ:* Barnard Col, AB, 68; Univ NC, MSPH, 72. *Prof Exp:* Health planner, Health Systs Agency, NVa, 73-74; statistician, Nat Cap Med Found, 75-79; RES MGR SYSTEMETRICS, MCGRAW HILL INC, 80- *Mem:* Am Pub Health Asn; Asn Social Sci & Health; Am Eval Asn. *Res:* Health care financing and access to medical care, particularly for low income persons. *Mailing Add:* 2923 Macomb St NW Washington DC 20008

HOWELL, EVELYN ANNE, b Milwaukee, Wis, Feb 28, 47. PLANT ECOLOGY, RESTORATION. *Educ:* Carleton Col, BA, 69; Univ Wis, MS, 73, PhD(bot), 75. *Prof Exp:* Asst prof, 75-81, ASSOC PROF, UNIV WIS-MADISON, 82-, CHAIR, DEPT LANDSCAPE ARCHIT, 88- *Concurrent Pos:* Chair, Ctr Restoration Ecol Restoration & Mgt Notes; contrib ed, Univ Wis-Madison; mem, Comt to Rev Glen Canyon Environ Studies, Water Sci & Technol Bd, Nat Res Coun. *Mem:* Ecol Soc Am; AAAS; Am Inst Biol Sci. *Res:* Plant community analysis for purposes of preservation, management, and landscape design; impact analysis with emphasis on plant communities; restoration of natural plant communities, especially parairies and woodland. *Mailing Add:* Dept Landscape Archit Agr Hall Univ Wis Madison WI 53706

HOWELL, EVERETTE IRL, b Shelby, Miss, Jan 4, 14; m 43; c 3. PHYSICS. *Educ:* Miss Col, BA, 36; Vanderbilt Univ, MS, 37; Univ NC, PhD(physics), 40. *Prof Exp:* From assoc prof to prof phys sci, Belhaven Col, 40-48; prof physics & head dept, Miss State Univ, 48-79; RETIRED. *Mem:* Am Phys Soc; Am Asn Physics Teachers. *Res:* Electron scattering; production of secondary electrons by electrons of energy between .7 and 2.6 million electron volts. *Mailing Add:* PO Box 5384 Mississippi State MS 39762

HOWELL, FRANCIS CLARK, b Kansas City, Mo, Nov 27, 25; m 55; c 2. BIOLOGICAL ANTHROPOLOGY. *Educ:* Univ Chicago, PhB, 49, MA, 51, PhD(anthrop), 53. *Prof Exp:* Instr anat, Sch Med, Wash Univ, 53-55; from asst prof to prof anthrop, Univ Chicago, 55-70; PROF ANTHROP, UNIV CALIF, BERKELEY, 70- *Mem:* Nat Acad Sci; AAAS; Am Anthrop Asn; Am Asn Phys Anthrop; Soc Vert Paleont; Am Philos Soc; Am Acad Arts & Sci. *Res:* Human evolution; paleoanthropology. *Mailing Add:* Dept Anthrop Univ Calif Berkeley CA 94720

HOWELL, FRANCIS V, b Salt Lake City, Utah, Jan 12, 23; m 51; c 4. ORAL PATHOLOGY. *Educ:* Stanford Univ, AB, 48; Univ Pac, DDS, 50; Univ Ore, MS, 56. *Prof Exp:* From asst prof to assoc prof oral path & head dept, Univ Ore, 52-56; assoc clin prof path, Sch Med, Univ Southern Calif, 56-69; PROF ORAL MED, GRAD SCH, LOMA LINDA UNIV, 63- *Concurrent Pos:* Pvt pract, Calif, 56-; consult, US Navy Hosps, San Diego, 57-, Camp Pendleton, 60-72, Vet Admin Hosp, Long Beach, 61-74 & Vet Admin Hosp, San Diego, 74-; dir, Am Bd Oral Path, 69-; vis prof, Technol Univ Mex, 70-; coordr continuing educ, Sch Dent, Univ Baja Calif; chmn dept oral med, Scripps Clin & Res Found. *Mem:* AAAS; Am Dent Asn; Am Soc Clin Path; Am Acad Oral Path (pres, 68-69); fel Am Col Dent. *Res:* Diseases of the mouth, especially the tongue; fungus diseases; clinical pathological features of oral cancer. *Mailing Add:* PO Box 1965 La Jolla CA 92038

HOWELL, GOLDEN LEON, b Hamilton, Ala, July 10, 28. PLANT PHYSIOLOGY. *Educ:* Univ Ala, BS, 50, MS, 54, PhD(bot), 59. *Prof Exp:* Asst prof biol, High Point Col, 58-61; assoc prof, 61-69, PROF BIOL, MEMPHIS STATE UNIV, 69- *Mem:* AAAS; Am Soc Plant Physiol. *Res:* Cellular processes investigation by methods utilizing chromatography and radio-isotopes. *Mailing Add:* Box 11512 E Sta Memphis TN 38111

HOWELL, GORDON STANLEY, JR, b Mobile, Ala, Mar 7, 41; m 64; c 3. HORTICULTURE, PLANT PHYSIOLOGY. *Educ:* Miss State Univ, BS, 63, MS, 65; Univ Minn, St Paul, PhD(hort), 69. *Prof Exp:* Asst prof to assoc prof, 69-80, PROF HORT, MICH STATE UNIV, 80- *Mem:* Am Soc Hort Sci; Am Soc Enologists; Sigma Xi. *Res:* Cold hardiness in woody plants; pomology; viticulture; enology. *Mailing Add:* Dept Hort Mich State Univ A85 Plant & Soil Sci Bldg East Lansing MI 48824

HOWELL, GREGORY A, b Springfield, Mo, Feb 3, 43. CIVIL ENGINEERING, CONSTRUCTION MANAGEMENT. *Educ:* Stanford Univ, BS, 66, MS, 71. *Prof Exp:* Proj engr, Arcosanti Proj, 70-71 & Kingston Quarries, 72-75; PRES, TIMELAPSE, INC, 75-; PRES, HOWELL ASSOCS, 81- *Concurrent Pos:* Consult, Harrison Fraker, 73-75. *Mem:* Am Soc Civil Engrs; Am Inst Indust Engrs; Inst Transp Engrs. *Res:* Productivity improvement and measurement, management of the improvement process. *Mailing Add:* Howell Assocs 505 Aliso SE Albuquerque NM 87108

HOWELL, JAMES ARNOLD, b Murphysboro, Ill, Apr 19, 32; m 52; c 1. ANALYTICAL CHEMISTRY. *Educ:* Southern Ill Univ, BA, 59; Univ Ill, MS, 61; Wayne State Univ, PhD(anal chem), 64. *Prof Exp:* Asst prof, 64-69, assoc prof, 69-78, PROF ANALYTICAL CHEM, WESTERN MICH UNIV, 78- *Concurrent Pos:* Sci adv, US Food & Drug Admin, 76- *Mem:* Am Chem Soc; Soc Appl Spectros. *Res:* Ultraviolet and visible absorption spectroscopy; flame photometry and atomic absorption spectroscopy; chemical instrumentation. *Mailing Add:* 912 Weaver St Kalamazoo MI 49007-5539

HOWELL, JAMES LEVERT, b Gordo, Ala, June 16, 17; m 66. MATHEMATICS. *Educ:* Univ Ala, AB, 42; Yale Univ, MA, 48, PhD(math), 54. *Prof Exp:* Asst math, Lehigh Univ, 42-43; instr, Muhlenberg Col, 43-44 & Univ Ga, 45-46; asst instr, Yale Univ, 46-52; instr, Univ Del, 52-55; from asst to prof, 55-85, EMER PROF MATH, UNIV ALA, TUSCALOOSA, 85- *Concurrent Pos:* Instr, New Haven YMCA Jr Col, 50-52; consult & lectr, Vis Scientist Prog, Ala Acad Sci, 62-69. *Mem:* Math Asn Am. *Res:* Mathematical analysis; differential equations; complex variables. *Mailing Add:* 3411 Arcadia Dr Tuscaloosa AL 35404

HOWELL, JAMES MACGREGOR, b Quincy, Mass, Apr 10, 42; m 67; c 2. QUANTUM CHEMISTRY. *Educ:* Harvard Univ, AB, 64; Cornell Univ, MS, 70, PhD(chem), 71. *Prof Exp:* asst prof, 73-77, ASSOC PROF CHEM, BROOKLYN COL, 78- *Mem:* Am Chem Soc. *Res:* Quantum chemical investigation of the structure and reactions of organic and inorganic molecules. *Mailing Add:* Dept Chem Brooklyn Col Bedford Ave & Ave H Brooklyn NY 11210

HOWELL, JAMES MILTON, b Durham, NC, Nov 8, 44. PHYSICAL CHEMISTRY, SOLID STATE PHYSICS. *Educ:* Univ Miss, BS, 66; Univ NC, PhD(chem), 73. *Prof Exp:* Res assoc solid state physics, Univ NC, Chapel Hill, 72-77; RES CHEMIST, E I DU PONT DE NEMOURS, 78- *Mem:* Am Chem Soc. *Res:* Electrical effects in thin films, organic and metallic; electrical characteristics of polymeric fibers and films; surface chemistry and physics. *Mailing Add:* 108 S Baywood Lane Greenville NC 27834-6945

HOWELL, JERRY FONCE, b McDowell, Ky, Oct 18, 41; m 63. ECOLOGY, ICHTHYOLOGY. *Educ:* NC State Univ, BS, 63 & 64; Eastern Ky Univ, MS, 68; Univ Tenn, Knoxville, PhD(zool), 71. *Prof Exp:* Forester, US Forest Serv, Dept Agr, 64-66; prof & adminr environ studies, 72-80, prof & chmn, dept biol & environ sci, 80-86, PROF & ADMIN ENVIRON STUDIES, MOREHEAD STATE UNIV, 86- *Concurrent Pos:* HEW grant, 72-73 & 73-74; consult, S Miss Elec Power Asn, 76-78 & Gateway Area Develop Dist, 76-77; Area Health Educ Syst grant, 78, 79 & 80; NSF grant, 78-80; US Army Corps Engrs grants, 79 & 80; mem Southeastern Fisheries Coun. *Mem:* Sigma Xi; Wildlife Soc; Am Fisheries Soc. *Res:* Environmental education; ichthyology; land use; ecology; forestry; wildlife; environmental impact. *Mailing Add:* UPO 780 Morehead State Univ Morehead KY 40351

HOWELL, JO ANN SHAW, b Bonham, Tex, Sept 28, 45; c 2. COMPUTER SCIENCES, COMPUTER SECURITY. *Educ:* Univ Tex, Austin, BA, 67, MA, 69, PhD(comput sci), 71. *Prof Exp:* Res scientist, Ctr Numerical Anal & asst prof comput sci, Univ Tex, Austin, 71-72; J Willard Gibbs instr, Yale Univ, 72-73; STAFF MEM, LOS ALAMOS NAT LAB, UNIV CALIF, 73- *Concurrent Pos:* Secy, Spec Interest Group Symbolic & Algebraic Manipulation; mem, Bd Dirs, Special Interest Group Numerical Math, 73-74. *Mem:* Am Math Soc; Soc Indust & Appl Math; Asn Comput Mach; Sigma Xi. *Res:* Control systems; computer security. *Mailing Add:* Group N-4 Mail Stop E541 Los Alamos Nat Lab Los Alamos NM 87545

HOWELL, JOHN ANTHONY, b Hyde, Cheshire, Eng, Apr 26, 39; m 65; c 2. CHEMICAL ENGINEERING, BIOCHEMICAL ENGINEERING. *Educ:* Univ Cambridge, MA, 64; Univ Minn, Minneapolis, PhD(chem eng), 66. *Prof Exp:* Jr chem engr, Petrocarbon Develop Ltd, 61-63; res asst chem eng, Univ Minn, 63-66; asst prof, Fed Univ Rio de Janeiro, 66-67; from asst prof to assoc prof, State Univ NY Buffalo, 67-75; reader biochem eng, Univ Col Swansea, Wales, 75-85; PROF BIOCHEM ENG, UNIV BATH, ENG, 85- *Concurrent Pos:* Sr vis fel, Univ Col Swansea, Wales, 73-74; dir, Bio-Isolates Ltd, 84-87; vis prof, Univ Waterloo, Can, 81-82. *Mem:* Am Inst Chem Engrs; Int Asn Water Pollution Res; Brit Inst Chem Engrs; Royal Soc Arts. *Res:* Waste water treatment by biological and physical means; membrane separations; chromatographic separations; computer control of fermenters and downstream processing. *Mailing Add:* Sch Chem Eng Univ Bath Bath BA2 7AY England

HOWELL, JOHN EMORY, b Charles City, Iowa, Mar 14, 38; m 58; c 3. CATALYSIS, INORGANIC CHEMISTRY. *Educ:* Marion Col, BS, 59; Ariz State Univ, MNS, 64; Univ Iowa, PhD(chem), 70. *Prof Exp:* Instr chem, Miltonvale Wesleyan Col, 59-66; asst prof, 70-74, assoc prof, 74-78, PROF CHEM, UNIV SOUTHERN MISS, 78- *Mem:* Am Chem Soc; Sigma Xi. *Res:* Transition and rare earth metal catalyzed stereospecific polymerization of dienes is being studied to determine the role of the metal and the cocatalyst. *Mailing Add:* Southern Sta Box 5043 Univ Miss Hattiesburg MS 39401

HOWELL, JOHN FOSS, electronics, nuclear physics, for more information see previous edition

HOWELL, JOHN H(ANCOCK), b New York, NY, Feb 23, 13; m 38; c 3. CHEMICAL ENGINEERING. *Educ:* Mass Inst Technol, SB, 35, SM, 36. *Prof Exp:* Proj engr, Carbide & Carbon Chem Corp, 36-46, sr design engr, 47-56, proj leader, 57-60, asst dir, 60-63, assoc dir, 63-78; RETIRED. *Concurrent Pos:* Spec consult, US Dept Navy, 44. *Mem:* Am Inst Chem Engrs. *Res:* High pressure hydrogenation of organic materials. *Mailing Add:* 2105 Weberwood Dr Charleston WV 25303

HOWELL, JOHN N, b Franklin, NH, Jan 20, 40; m 61; c 2. PHYSIOLOGY. *Educ:* Kalamazoo Col, BA, 61; Univ Calif, Los Angeles, PhD(pharmacol), 68. *Prof Exp:* Asst prof biol sci, Calif State Polytech Col, 67-68; fel physiol, Univ Calif, Los Angeles, 68-70; from instr to asst prof physiol, Univ Pittsburgh, 70-77; ASSOC PROF ZOOL & BIOMED SCI, OHIO UNIV, 77- *Concurrent Pos:* Lectr, Univ Calif, Los Angeles, 69-70; vis prof, Berne Univ, Switz, 84-85; Danworth Found Fel, 61. *Mem:* Soc Gen Physiol; Biophys Soc; Am Physiol Soc; Soc Neurosci. *Res:* Cellular aspects of muscle physiology, especially excitation-contraction coupling; EMG and muscle pathophysiology. *Mailing Add:* Col Osteop Med Ohio Univ Athens OH 45701

HOWELL, JOHN REID, b Columbus, Ohio, June 13, 36; m 79; c 3. HEAT TRANSFER, ENERGY. *Educ:* Case Inst Technol, BSChE, 58, MSChE, 60, PhD(eng), 62. *Prof Exp:* Aerospace technologist heat transfer, NASA Lewis Res Ctr, 61-68; from assoc prof to prof mech eng, Univ Houston, 68-78; vis prof, Univ Tex, Austin, 78-79, prof mech eng, 79-83, E C H Bantel prof, 83-90, chmn mech eng, 87-90, DIR, CTR ENERGY STUDIES, UNIV TEX, AUSTIN, 88-, BAKER-HUGHES PROF, 90- *Concurrent Pos:* Assoc dir res, Univ Houston, 73-75; prog mgr heating & cooling, Solar Energy Lab, Univ Houston, 74-75, dir Energy Inst, 75-78; vis scholar, Japan Soc Prom of Sci, various Japanese Univs, 85. *Honors & Awards:* Ralph Coats Roe Award, Am Soc Eng Educ, 87; Thermophysics Award, Am Inst Aeronaut & Astronaut, 90. *Mem:* Fel Am Soc Mech Engrs; assoc fel Am Inst Aeronaut & Astronaut; Int Solar Energy Soc. *Res:* Thermal radiation heat transfer; solar energy; heat transfer in phase change; thermal radiation heat transfer; heat transfer in fusion devices. *Mailing Add:* Dept Mech Eng Univ Tex Austin TX 78712

HOWELL, LARRY JAMES, b West Frankfort, Ill, July 1, 43; m 65; c 2. STRUCTURAL DYNAMICS. *Educ:* Univ Ill, Urbana, BS, 66, MS, 68, PhD(aeronaut eng), 71. *Prof Exp:* Instr, Univ Ill, Urbana, 67-70; sr dynamics engr structural dynamics, Gen Dynamics, Convair, 70-72; assoc sr res engr, 72-75, staff res engr, 75-77, asst dept head eng mech, 77-82, DEPT HEAD ENG MECH, GEN MOTORS RES LABS, 82- *Concurrent Pos:* Lectr, Oakland Univ, 81- *Mem:* Am Inst Aeronaut & Astronaut; Soc Automotive Engrs; Am Soc Mech Engrs. *Res:* Structural analysis; structural dynamics (noise and vibration). *Mailing Add:* 256 Res Mech Bldg Gen Motors Tech Ctr Gen Motors Res Labs Warren MI 48090-9055

HOWELL, LEONARD RUDOLPH, JR, b Valdosta, Ga, May 19, 25; m 44; c 4. TOPOLOGY. *Educ:* Mercer Univ, AB, 48; Emory Univ, MS, 51; Fla State Univ, PhD(math), 65. *Prof Exp:* Instr math & chem, Emory Jr Col, 48-53; mathematician, Air Forces in Europe, USAF, 57-58, asst prof math, Air Force Acad, 58-62, mathematician, Air Force Hq, 65-69, from asst prof to assoc prof math, Air Force Inst Technol, 69-72; assoc prof math, Valdosta State Col, 72-81; RETIRED. *Mem:* Math Asn Am. *Res:* Point set topology. *Mailing Add:* Rte 1 Box 279 Quitman GA 31643

HOWELL, MARY GERTRUDE, b Wenona, Ill, May 25, 32; m 54; c 2. PHARMACEUTICAL CHEMISTRY. *Educ:* Univ Ill, BS, 54; Mass Inst Technol, PhD(org chem), 59. *Prof Exp:* Chemist, E I du Pont de Nemours & Co, 54; res chemist, Div Plant Indust, Commonwealth Sci & Indust Res Orgn, Canberra, Australia, 59; res lit scientist, 59-67, head lit serv, 67-70, dir tech info serv sect, 71-76, HEAD, TECH INFO SERV DEPT, CYANAMID MED RES DIV, LEDERLE LABS, AM CYANAMID CO, 77- *Concurrent Pos:* Mem, Nat Res Coun Comt Chem Info, Nat Acad Sci, 72-74 & Toxicol Info Prog Comt, 87- *Mem:* Am Chem Soc; Am Soc Info Sci. *Res:* Technical information and research; drug development. *Mailing Add:* Tech Info Serv Dept Cyanamid Med Res Div Pearl River NY 10965

HOWELL, NORMAN GARY, b Ithaca, NY, June 8, 49; m 71; c 1. ANALYTICAL CHEMISTRY, PHYSICAL CHEMISTRY. *Educ:* State Univ NY, Cortland, BS, 71; Cornell Univ, MS, 74, PhD(anal chem), 76. *Prof Exp:* Res chemist & group leader, 76-81, RES CHEMIST & SECT HEAD, CORP TECH DIV, PROCTER & GAMBLE, 81- *Mem:* Soc Appl Spectros; Sigma Xi. *Res:* Atomic spectroscopy for chemical analysis; X-ray fluorescence; molecular spectroscopy; digital image processing; analytical microscopy; their application to personal care products. *Mailing Add:* Miami Valley Labs Procter Gamble Co PO Box 39175 Cincinnati OH 45247

HOWELL, PAUL RAYMOND, b Coleford, Gloucestershire, Sept 9, 46; m 71; c 1. TRANSMISSION ELECTRON MICROSCOPY, CRYSTALLOGRAPHY. *Educ:* Univ Cambridge, BA, 69, MA, 73, PhD(metallurgy), 73. *Prof Exp:* Res fel, Univ Cambridge, 72-81; ASSOC PROF, PA STATE UNIV, 81- *Mailing Add:* Dept Mat Sci Pa State Univ Main Campus Univ Park PA 16802

HOWELL, PETER ADAM, b Des Moines, Iowa, Sept 11, 28; m 52; c 3. PHYSICAL INORGANIC CHEMISTRY. *Educ:* Calif Inst Technol, BS, 50; Univ Minn, PhD(chem), 55. *Prof Exp:* Asst, Univ Minn, 50-55; res chemist, Tonawanda Lab, Linde Co Div, Union Carbide Corp, 55-61; res chemist, 61-69, res specialist, 69-83, SR RES SPECIALIST, 3M CO, 83- *Concurrent Pos:* Adj assoc prof chem, Macalester Col, 68-69. *Mem:* Am Crystallog Asn; Am Ceramic Soc. *Res:* Crystal structures and chemistry of zeolite minerals; structures of boron compounds; clay mineralogy; x-ray structure determination; synthesis and properties of zeolites; polymer structure versus properties; chemical reactions in grinding processes; glass composition versus properties; particle properties & behavior. *Mailing Add:* 1896 Yorkshire Ave St Paul MN 55116

HOWELL, RALPH RODNEY, b Concord, NC, June 10, 31; m 60; c 3. PEDIATRICS, GENETICS. *Educ:* Davidson Col, BS, 53; Duke Univ, MD, 57. *Prof Exp:* Intern pediat, Sch Med, Duke Univ, 57-58, asst resident, 59-60; clin assoc metab dis, Nat Inst Arthritis & Metab Dis, 60-62, investr molecular biol, 62-64; assoc prof pediat, Sch Med, Johns Hopkins Univ, 64-72; DAVID R PARK PROF PEDIAT, UNIV TEX, HOUSTON, 72-; PEDIATRICIANS, UNIV CHILDREN'S HOSP HERMANN, HOUSTON, 72- *Concurrent Pos:* Res fel pediat & med, Med Ctr, Duke Univ, 58-59; Joseph P Kennedy, Jr sr res scholar, 64-72; pediatrician, Johns Hopkins Hosp, 64-72. *Mem:* Soc Pediat Res; Am Fedn Clin Res; Am Rheumatism Asn; Am Soc Human Genetics; Am Pediat Soc; Soc Inherited Metab Dis. *Res:* Inherited biochemical defects; cloning; isolation of human genes; genetic diagnosis and treatment. *Mailing Add:* Univ Miami R131 PO Box 016960 Miami FL 33101

HOWELL, ROBERT MAC ARTHUR, dentistry, oral pathology, for more information see previous edition

HOWELL, ROBERT RICHARD, b Chincoteague, Va, June 26, 52. SPECKLE INTERFEROMETRY, PLANETARY SCIENCES. *Educ:* Univ Mich, BSc, 74; Univ Ariz, PhD(planetary sci), 80. *Prof Exp:* Res asst, Lunar & Planetary Lab, Univ Ariz, 74-78, Steward Observ, 78-80; asst astronomer, Inst Astron, Univ Hawaii, 80-86; ASST PROF, PHYSICS & ASTRON DEPT, UNIV WYO, 86- *Mem:* Am Astron Sci; Astron Soc Pac; Am Geophys Union. *Res:* Infared speckle interferometry; infared observations of young and evolved stars; volcanism on Jupiter's satellite Io. *Mailing Add:* Physics & Astron Dept Box 3905 Univ Wyo Laramie WY 82071

HOWELL, ROBERT T, b Jamestown, NY, Jan 24, 27; m 57; c 3. BACTERIOLOGY, PUBLIC HEALTH & EPIDEMIOLOGY. *Educ:* Univ NMex, BS, 52; Univ NC, MPH, 67, DrPH(lab pract), 69. *Prof Exp:* Lab scientist bact, Div Pub Health Labs, NMex Dept Health, 51-66; dir, 69-81, ASST DIR, DIV PUB HEALTH LABS, ARK DEPT HEALTH, 81- *Concurrent Pos:* Pres, Southern Health Asn, 84. *Mem:* Am Soc Microbiol; Am Pub Health Asn; Sigma Xi; Asn Off Anal Chemists. *Res:* Differentiation of Clostridium botulinum and Clostridium sporogenes by culture, biochemical, and gas-liquid-chromatography of fatty acids. *Mailing Add:* 423 Springwood Dr Little Rock AR 72211

HOWELL, ROBERT WAYNE, b Houlka, Miss, Nov 26, 16; m 40; c 3. PLANT PHYSIOLOGY, AGRONOMY. *Educ:* Miss Col, BS, 49; Univ Wis, MS, 51, PhD(bot), 52. *Prof Exp:* Bus mgr, US Plant Soil & Nutrit Lab, NY, 40-43; asst treas, Pineapple Res Inst Hawaii, 46-47; plant physiologist, Regional Soybean Lab, USDA, 52-64, leader soybean invests, Agr Res Serv, 64-65, chief oilseed & indust crops res br, 66-71; prof agron & head dept, Univ Ill, Urbana, 71-82; RETIRED. *Concurrent Pos:* From asst prof to assoc prof, Univ Ill, 57-65; ed, J Crop Sci Soc Am, 68-70, ed-in chief, 71-73; consult, Food & Agr Orgn-China, 82-84. *Mem:* Fel AAAS; Crop Sci Soc Am; Am Soc Plant Physiol; fel Am Soc Agron; Scandinavian Soc Plant Physiol; hon mem Am Soybean Asn. *Res:* Mineral nutrition of soybeans; oil synthesis; energy considerations in agriculture. *Mailing Add:* 2012 S Cottage Grove Urbana IL 61801-6355

HOWELL, RONALD HUNTER, b Chicago, Ill, Oct 19, 35; m 56; c 3. MECHANICAL ENGINEERING, ENERGY ANALYSIS. *Educ:* Univ Ill, BS, 58, MS, 59, PhD, 67. *Prof Exp:* Instr mech eng, Univ Ill, 59-63, res assoc, 63-66; from asst prof to assoc prof, 66-73, PROF MECH ENG, UNIV MO, ROLLA, 73- *Concurrent Pos:* Consult, Union Elec Co , Am Soc Heating, Refrig & Air-Conditioning Engrs, 73-78 & Hussmann Refrig, 78- *Honors & Awards:* Teetor Award, Soc Automotive Engrs, 78. *Mem:* Am Soc Mech Engrs; Am Soc Heating, Refrig & Air Conditioning Engrs; Soc Automotive Engrs; Sigma Xi; Am Soc Engr Educ. *Res:* Measurement of drag forces in separated flows; dynamics and thermodynamics of fluid flow; turbulent wakes; emptying and filling processes; air curtain design; energy analysis and conversion; system simulation. *Mailing Add:* Dept Mech Eng Univ Mo Rolla MO 65401

HOWELL, STEPHEN HERBERT, b Davenport, Iowa, May 30, 41; m 64. PLANT MOLECULAR BIOLOGY. *Educ:* Grinnell Col, BA, 63; Johns Hopkins Univ, PhD(biol), 67. *Prof Exp:* USPHS fel biol, 67-69, res biologist, 69-70, actg asst prof biol, 70-71, from asst prof to prof biol, 71-88,; BOYCE SCHULZE DOWNEY SCIENTIST & PROG DIR IN MOLECULAR BIOL, BOYCE THOMPSON INST, 88- *Concurrent Pos:* Simon Guggenheim Fel, 76. *Mem:* AAAS. *Res:* Plant viruses; molecular aspects of plant development. *Mailing Add:* Boyce Thompson Inst Cornell Univ Ithaca NY 14853-1801

HOWELL, THOMAS HOWARD, b Atlanta, Ga, Oct 7, 45. DENTISTRY, PERIODONTAL DISEASE. *Educ:* Ga State Univ, BS, 67, MS, 69; Emory Univ, DDS, 73. *Prof Exp:* Fel periodont, 73-76, DIR UNDERGRAD PERIODONT DENT, HARVARD SCH DENT MED, 76-, ASSOC DEAN STUDENT AFFAIRS, 78- *Mem:* Am Acad Periodont; Am Asn Dent Schs. *Res:* periodontal disease. *Mailing Add:* Harvard Sch Dent Med 188 Longwood Ave Boston MA 02115

HOWELL, THOMAS RAYMOND, b New Orleans, La, June 17, 24; m 81; c 3. VERTEBRATE ZOOLOGY. *Educ:* La State Univ, BS, 46; Univ Calif, MA, 49, PhD(zool), 51. *Prof Exp:* From asst prof to prof zool, Univ Calif, Los Angeles, 51-86; RETIRED. *Honors & Awards:* Elliot Coues Award, Am Ornith Union, 85. *Mem:* AAAS; Cooper Ornith Soc; Wilson Ornith Soc; Am Ornith Union (pres, 82-84); Brit Ornith Union. *Res:* Ornithology; systematics, distribution, evolution, behavior and ecology. *Mailing Add:* PO Box 950 Gualala CA 95445

HOWELL, WALLACE EGBERT, b Central Valley, NY, Sept 14, 14; m 42; c 5. METEOROLOGY. *Educ:* Harvard Univ, AB, 36; Mass Inst Technol, ScD(meteorol), 48. *Prof Exp:* Chief meteorologist, Mid-Continent Airlines, Kans, 37-39; assoc meteorologist, Yankee Network Weather Serv, Mass, 39-40; asst regional forecaster, US Weather Bur, 40-41; field dir, Harvard-Mt Washington Icing Res Proj, Harvard Univ, 46-47, res fel meteorol, Blue Hill Observ, 47-51; pres, W E Howell Assocs, Inc, 51-67; vpres, E Bollay Assocs, Inc, 67-71; asst to chief, 71-84, SCI CONSULT, DIV ATMOSPHERIC RESOURCES MGT, BUR RECLAMATION, 84- *Concurrent Pos:* Actg dir, Mt Washington Observ, 48-53, pres, 58-67, trustee, 65-; consult meteorologist, 59-; adj prof, Northeastern Univ, 66-71. *Honors & Awards:* Am Meteorol Soc Award, 68. *Mem:* Fel Am Meteorol Soc; Am Geophys Union. *Res:* Cloud classification; physics of clouds; energy-environmental interactions; weather modification; weather-environment impacts. *Mailing Add:* 36 S Mt Vernon Country Club Rd Golden CO 80401

HOWELL, WILLIAM EDWIN, b Toronto, Ont, Mar 23, 23; m 48; c 6. ANIMAL BREEDING. *Educ:* Ont Agr Col, BSA, 49; Univ Minn, MSc, 50, PhD(animal breeding), 52. *Prof Exp:* Asst animal husb, Univ Minn, 49-52; from asst prof to prof animal sci, Univ Sask, 52-87; RETIRED. *Mem:* Am Soc Animal Sci; Genetics Soc Can; Can Soc Animal Sci (secy-treas, 54-59, pres, 59-60); fel Agr Inst Can. *Res:* Breeding for improvement of domestic animals; physiology of reproduction. *Mailing Add:* 38 Cantlon Crescent Saskatoon SK S7J 2T3 Can

HOWELLS, THOMAS ALFRED, b Girard, Ohio, Sept 21, 11; m 40; c 2. PAPER CHEMISTRY. *Educ:* Ohio Wesleyan Univ, BA, 33; Inst Paper Chem, Lawrence Univ, MS, 35, PhD(paper chem), 37. *Prof Exp:* Chemist, Munising Paper Co, 37-42; asst, Inst Paper Chem, 42-44, chief resins & plastics sect, 44-57, chief spec processes sect, 57-64, chmn tech sect, 64-69, dir continuing educ, 69-76; RETIRED. *Mem:* Tech Asn Pulp & Paper Indust. *Res:* Technology research and education in pulp and paper. *Mailing Add:* 13373 N Plaza Del Rio Blvd No 6658 Peoria AZ 85345

HOWELLS, WILLIAM WHITE, b New York, Nov 27, 08; m 29; c 2. PHYSICAL ANTHROPOLOGY. *Educ:* Harvard Univ, BS, 30, PhD(anthrop), 34. *Hon Degrees:* DSc, Beloit Col, 75 & Univ Witwatersrand, 85. *Prof Exp:* Res assoc phys anthrop, Am Mus Natural Hist, 32-43; from asst prof to prof anthrop, Univ Wis, 39-54; prof, 54-74, cur somatol, Peabody Mus, 55-74, EMER PROF ANTHROP, HARVARD UNIV, 74-, EMER CUR SOMATOL, PEABODY MUS, 74- *Concurrent Pos:* Lectr, Hunter Col, 37-39; ed, J Am Asn Phys Anthrop, 49-54. *Honors & Awards:* Viking Fund Medal, Wenner-Gren Found, 55; Broca Prix du Centenaire, Paris Soc Anthrop, 80. *Mem:* Nat Acad Sci; fel AAAS; fel Am Anthrop Asn (pres, 51); Am Asn Phys Anthrop; fel Am Acad Arts & Sci; fel Soc Antiquaries London; hon foreign assoc, Royal Soc S Africa. *Res:* Human paleontology; morphological population variation in man. *Mailing Add:* Peabody Mus Harvard Univ Cambridge MA 02138

HOWER, ARTHUR AARON, JR, b Royalton, Pa, Oct 21, 37; m 62; c 2. ENTOMOLOGY. *Educ:* Shippensburg State Col, BS, 59; Bucknell Univ, MS, 63; Pa State Univ, PhD(zool, entom), 67. *Prof Exp:* Teacher, Cent Dauphin Joint Sch Syst, 59-61; asst prof, 66-75, assoc prof entom, 75-80, PROF ENTOM, PA STATE UNIV, UNIVERSITY PARK, 80- *Mem:* Entom Soc Am; Entom Soc Can; Int Orgn Biol Control. *Res:* Ecology, biology and control of forage insects; radiation and parasites as a control mechanism for alfalfa weevil Hypera postica; impact of insecticides on parasite populations. *Mailing Add:* Dept Entom Pa State Univ University Park PA 16802

HOWER, CHARLES OLIVER, b Emmett, Idaho, May 13, 35; m 60; c 2. NUCLEAR CHEMISTRY, PHYSICAL CHEMISTRY. *Educ:* Whitman Col, BA, 56; Univ Wash, PhD(chem), 62. *Prof Exp:* Res assoc chem, Princeton Univ, 62-64; res assoc physics, Inst Nuclear Physics Res, 64-66; asst prof chem, Univ Idaho, 66-71; PROF CHEM, SOUTHWESTERN ORE COMMUNITY COL, 71-, PROF PHYS SCI, 80- *Mem:* AAAS. *Res:* Nuclear reaction cross sections; activation analysis; hot atom chemistry of halogen and nitrogen atoms. *Mailing Add:* Dept Life Sci Southwestern Ore Community Col Coos Bay OR 97420

HOWER, GLEN L, b Wenatchee, Wash, Feb 8, 34; m 56; c 3. ELECTRICAL ENGINEERING. *Educ:* Wash State Univ, BS, 56, MS, 61; Stanford Univ, PhD(elec eng), 63. *Prof Exp:* Mem staff, Gen Elec Co, 56-57; from instr to assoc prof, 57-73, PROF ELEC ENG, WASH STATE UNIV, 73- *Mem:* Am Geophys Union; Am Soc Eng Educ; Inst Elec & Electronics Engrs. *Res:* Radio wave propagation; electromagnetics. *Mailing Add:* Dept Elec Eng Wash State Univ Pullman WA 99164-2752

HOWER, JOHN, JR, geochemistry; deceased, see previous edition for last biography

HOWER, MEADE M, b Danielsville, Pa, July 6, 25; m 47; c 3. PHYSICS. *Educ:* Lehigh Univ, BS, 49, MS, 50. *Prof Exp:* Instr physics, Lehigh Univ, 49-50, mem tech staff, Bell Tel Labs, 50-54; mgr eng, Tung-Sol Elec Co, 54-60; mem tech staff, Bell Tel Labs, 60-64, supvr integ circuits develop, 64-87; RETIRED. *Mem:* Am Phys Soc; Sigma Xi. *Res:* Development of semiconductor devics and circuits; development and use of solid state devices. *Mailing Add:* 600 Snyder Rd Whitfield Reading PA 19609

HOWERTON, MURLIN T(HOMAS), chemical engineering; deceased, see previous edition for last biography

HOWERTON, ROBERT JAMES, b Hammond, Ind, Sept 27, 23; m 46. NUCLEAR PHYSICS, NUCLEAR ENGINEERING. *Educ:* Northwestern Univ, BS, 46, MS, 47. *Prof Exp:* Asst prof math, Regis Col, 48-54; nuclear engr, Phillips Petrol Co, 56-57; asst div leader, theoret physics div, 80-84, group leader phys data group, Theoret Physics Div, Lawrence Livermore Nat Lab, Univ Calif, 57-86, asst div leader, Comput Physics Div, 84-86; consult, Phys Data Systs, 86-91; LAB CONSULT, 91- *Concurrent Pos:* Consult, Am Inst Physics, 61-70 & Cent Intel Agency, 59-63. *Mem:* Fel Am Nuclear Soc. *Res:* Development of theoretical and empirical methods for providing nuclear and atomic data required for design of energy producing devices and shielding of radiation from such devices. *Mailing Add:* Lawrence Livermore Lab L298 Univ Calif Box 808 Livermore CA 94550

HOWERY, DARRYL GILMER, b Christiansburg, Va, June 19, 36; m 66; c 1. PHYSICAL CHEMISTRY. *Educ:* Roanoke Col, BS, 58; Univ NC, PhD(phys chem), 63. *Prof Exp:* Chemist, Univ Calif, Berkeley, 63-65; from asst prof to assoc prof, 65-75, PROF CHEM, BROOKLYN COL, 75- *Concurrent Pos:* Vis chemist & lectr, Kyoto Univ, 71-72, 79-80. *Mem:* Am Chem Soc. *Res:* Chemometrics, especially factor analysis. *Mailing Add:* Dept Chem Brooklyn Col Brooklyn NY 11210

HOWES, CECIL EDGAR, b Patten, Maine, Sept 23, 18; m 44; c 3. GENETICS, PHYSIOLOGY. *Educ:* Univ Maine, BS, 41; Cornell Univ, MS, 48, PhD(genetics physiol), 54. *Prof Exp:* Instr & asst poultry husb, Univ Maine, 46, from asst prof & asst poultry husbandman to prof & poultry husbandman, 48-58; head, Dept Poultry Sci, Va Polytech Inst & State Univ, 58-78, prof, 58-86, col liaison officer, 78-86; RETIRED. *Mem:* Poultry Sci Asn. *Res:* Avian genetics; genetic resistance to diseases; genetic and physiological correlations. *Mailing Add:* 714 Gracelyn Ct Blacksburg VA 24060

HOWES, HAROLD, JR, parasitology; deceased, see previous edition for last biography

HOWES, JOHN FRANCIS, b London, Eng, Jan 4, 43; m 69; c 2. PHARMACOLOGY, MEDICINAL CHEMISTRY. *Educ:* Univ London, BPharm, 64, PhD(pharmacol), 67. *Prof Exp:* Res assoc pharmacol, Sch Med, Univ NC, 67-69; sr scientist, Arthur D Little Inc, 69-70; dir, SISA Inc, 70-80, vpres pharmacol, 80-; dir, New Drug Develop, Key Pharmaceut Inc; VPRES DEVELOP, XENON VISION, INC. *Concurrent Pos:* Asst prof, Sch Pharm & Allied Health Sci, Northeastern Univ, 74- *Mem:* Am Soc Pharmacol & Exp Therapeut; Am Soc Clin Pharmacol Therapeut. *Res:* Pharmacology of centrally active drugs with reference to analgesics. *Mailing Add:* Xenon Vision Inc One Progress Blvd No 36 Alachua FL 32615

HOWES, ROBERT INGERSOLL, JR, b Santa Fe, NMex, Aug 3, 41; m 66; c 2. ANATOMY, DENTAL RESEARCH. *Educ:* Amherst Col, AB, 63; Columbia Univ, DDS, 67, PhD(anat), 71. *Prof Exp:* Asst prof, 71-78, ASSOC PROF ANAT, HEALTH SCI CTR, UNIV OKLA, 78- *Concurrent Pos:* Nat Inst Dent Res grant, 73- *Mem:* Am Asn Anatomists; Int Asn Dent Res; Soc Vert Paleont; AAAS; Sigma Xi. *Res:* Hard tissue research with interest in comparative odontology using experimental approaches to study tooth attachment, tooth development, and jaw regeneration in lower vertebrates. *Mailing Add:* 1012 Brookside Norman OK 73069

HOWES, RUTH HEGE, b Montpelier, Vt, Oct 18, 44; m 66; c 2. PHYSICS OF THE ARMS RACE, PHYSICS EDUCATION. *Educ:* Mt Holyoke Col, BA, 65; Columbia Univ, MA, 67, PhD(physics), 71. *Prof Exp:* Vis asst prof physics, Univ Okla, 71-72; adj instr, Okla City Univ, 72-76; PROF PHYSICS, BALL STATE UNIV, 76- *Concurrent Pos:* Foster fel, US Arms Control & Disarmament Agency, 84-85; educ comt, Forum Physics & Soc, Am Phys Soc, 85-88. *Mem:* Fel Am Phys Soc; Sigma Xi; AAAS. *Res:* Applications in physics that have impact on the society; status of energy technology in United States. *Mailing Add:* Dept Phys & Astron Ball State Univ Muncie IN 47306

HOWGATE, DAVID W, b Swedesboro, NJ, Oct 11, 32; m 64; c 3. LASER PHYSICS, PROJECT MANAGEMENT. *Educ:* Marshall Univ, BS, 54; WVa Univ, MS, 55; Univ Ala, PhD(physics), 67. *Prof Exp:* Physicist, Army Missile Command, Redstone Arsenal, 59-63, res physicist, 63-75, tech mgr, 75-84, consult, 84-88; PRES, LASER TOOLS, INC, HUNTSVILLE, ALA, 88- *Honors & Awards:* Paul A Siple Mem Award, 78. *Mem:* AAAS; Am Phys Soc; NY Acad Sci; Sigma Xi. *Res:* Mathematical physics; chemical lasers; computational physics; electron spin resonance and infrared diagnostics; gas and solid state lasers; nonlinear optics. *Mailing Add:* 7800 Smoke Rise SE Huntsville AL 35802

HOWICK, LESTER CARL, b Manchester, Iowa, Nov 2, 28; m 53. ANALYTICAL CHEMISTRY. *Educ:* Cornell Col, BA, 53; Univ Iowa, MS, 55, PhD(anal chem), 57. *Prof Exp:* From asst prof to assoc prof, 57-67, PROF CHEM, UNIV ARK, 67- *Mem:* Am Chem Soc; Sigma Xi. *Res:* Homogeneous precipitation; complex ions and the mechanism of their reactions. *Mailing Add:* Dept Chem Univ Ark Fayetteville AR 72701

HOWIE, DONALD LAVERN, b Monticello, Iowa, June 3, 24; m 47; c 4. INTERNAL MEDICINE, RESEARCH ADMINISTRATION. *Educ:* Univ Iowa, MD, 48; Am Bd Internal Med, dipl, 56; Harvard Univ, AMP, 65. *Prof Exp:* Med Corps, US Army, 44-, intern med, Denver Gen Hosp, 48-49, regimental surgeon, Seventh Inf Div, Hokkaido, Japan, 49-50, commanding officer, Hq Med Detachment, First Cavalry Div, Korea, 50-51, officer-in-chg hepatitis annex, 31st Sta Hosp, Kyoto & Tokyo, 51, resident med, Brooke Gen Hosp, Ft Sam Houston, 52-54, student mil med & allied sci, Walter Reed Army Inst Res, Walter Reed Army Med Ctr, 54-55, fel hemat, 57-58, chief med & prof serv, Ft Riley, Kans, 55-56, sr med specialist, Queen Alexandria Mil Hosp, London, 58-59, commanding officer, 46th Surg Hosp, Landstuhl, Ger & consult hemat, Surgeon, US Army, Europe, 59-60, dep dir div med & asst chief dept hemat, Walter Reed Army Inst Res, 60-62, chief med res br, US Army Med Res & Develop Command, Off of the Surgeon Gen, 62-63, chief plans, progs & funds div, 63-65, dep comdr, 65-67, chief life & sci div, Army Res Off of Chief of Res & Develop, Dept Army, 67-72, internist, Army Phys Disability Agency, Med Corps, US Army, 72-87; RETIRED. *Concurrent Pos:* Mem, NIH Study Sect Hemat, 61-65; student defense mgt, US Navy Postgrad Sch, 65; mem army res coun, Dept Army, 67-71. *Mem:* AMA; fel Royal Soc Med; fel Am Col Physicians; Asn Mil Surg US. *Res:* Research in laboratory and clinical hematology and oncology; bone marrow transplantation; 00088921now storage; research management. *Mailing Add:* 1051 Carnation Dr Rockville MD 20850

HOWITT, ANGUS JOSEPH, b Guelph, Ont, Feb 9, 19; m 55; c 1. ENTOMOLOGY. *Educ:* Ont Agr Col, BSA, 49; Mont State Col, MS, 51; Kans State Col, PhD(entom), 53. *Prof Exp:* Tech officer, Dom Parasite Lab, Belteville, 46-51, res officer, 52-53; asst entomologist, Western Wash Exp Sta, 53-58, assoc entomologist, 58-60; assoc prof entom, Mich State Univ, 60-66, prof, 66-; SUPVR, KALAMAZOO ORCHARD, TREVOR NICHOLS RES COMPLEX, MICH STATE UNIV. *Concurrent Pos:* Asst, Kans State Col, 52-53. *Mem:* Entom Soc Am; Entom Soc Can. *Res:* Vegetable and forage insect pests; biological control; insect resistance in crop plants; small fruit and tree fruit insect pests. *Mailing Add:* 166 Plant Biol Bot Mich State Univ East Lansing MI 48824

HOWKINS, STUART D, b London, Eng, Jan 24, 38; m 64; c 2. ACOUSTICS. *Educ:* Univ London, BSc, 59, MSc, 60, PhD(physics), 62. *Prof Exp:* Res assoc acoust, Univ Vt, 63-64; assoc physicist, IIT Res Inst, Ill, 64-69; mem res staff, Schlumberger-Doll Res Ctr, 69-75; mgr, Appln Lab, Krautkramer-Branson Inc, 76-79; SR SCIENTIST, DATA PROD CORP, 79- *Concurrent Pos:* Eve instr, Ill Inst Technol, 65-69. *Res:* High-amplitude sound propagation in liquids including experimental studies of attenuation of shock waves; cavitation phenomena including nonlinear bubble oscillations and cavitation erosion; ultrasonic nondestructive testing and transducer development; physics of impulse ink jet devices. *Mailing Add:* 1112 Fel Rd Data Prod Corp Brookfield CT 06804

HOWLAND, FRANK L, b Long Branch, NJ, July 19, 26; m 50; c 3. ENGINEERING, APPLIED MATHEMATICS. *Educ:* Rutgers Univ, BS, 50; Univ Ill, MS, 52, PhD(struct eng), 55. *Prof Exp:* Res assoc struct res, Univ Ill, 52-55; mem tech staff eng mech, 55-59, supvr, 59-65, head dept appl mech, 65-78, HEAD, DEPT FIBER & HYBRID LAB, BELL TEL LABS, 76-, HEAD, TECHNOL DEPT, 78- *Mem:* AAAS; Soc Exp Stress Anal; Inst Elec & Electronics Engrs; Int Soc Hybrid Microelectronics. *Res:* Mechanical engineering; electronic device assembly process and package design development; materials; fatigue; joining technology development; heat transfer; fracture mechanics. *Mailing Add:* 3076 Lindberg Ave Allentown PA 18103

HOWLAND, GEORGE RUSSELL, b Toronto, Can, July 15, 32; US citizen; m 57; c 3. MECHANICAL ENGINEERING. *Educ:* Univ Toronto, BASc, 56. *Prof Exp:* Performance engr, Lucas Rotax Co, Can, 56-59; performance analyst, Ford Motor Co, Mich, 59-62; res engr, Aerospace Div, Bendix Corp, 62-69; sr engr, Westinghouse Res & Develop Ctr, 69-82; CONSULT, 82- *Mem:* AAAS; assoc mem Am Soc Mech Engrs. *Res:* Solid state pneumatic and hydraulic amplifying and logic systems; analysis of gas turbine engines and subsystems; fluid flow and air moving equipment, including development of computer program analysis. *Mailing Add:* 46 Staymen Ct LaFayette IN 47905

HOWLAND, HOWARD CHASE, b Lafayette, Ind, May 26, 33; m 65; c 3. SENSORY PHYSIOLOGY, PHYSIOLOGICAL OPTICS. *Educ:* Univ Chicago, BA, 52; Tufts Univ, MS, 58; Cornell Univ, PhD, 68. *Prof Exp:* Instr biosci, State Univ NY, Stony Brook, 60-66, asst prof, 66-67; from asst prof to assoc prof, 68-85, PROF NEUROBIOL & BEHAV, CORNELL UNIV, 85- *Concurrent Pos:* Res inst guest, Max Planck Inst Physiol Behav, 60, 62-63 & 63-66, Wissenschaftliche asst, 72-73; vis researcher, Physiol Lab, Downing Site, Cambridge, 76-77; sr res fel, Inst Ophthal, Univ London, 76-77; vis scientist, Regional Primate Ctr, Univ Wash, 80. *Mem:* Am Physiol Soc; Soc Neurosci; Optical Soc Am; Asn Res Vision & Ophthal; AAAS. *Res:* Vertebrate vision, especially the development of focusing ability in infant humans and other vertebrates. *Mailing Add:* W-201 Seeley Mudd Hall Cornell Univ Ithaca NY 14853

HOWLAND, JAMES LUCIEN, b Toronto, Ont, Jan 24, 29; m 54; c 3. MATHEMATICAL ANALYSIS. *Educ:* Trinity Col, Can, BA, 51; Univ Toronto, MA, 52; Harvard Univ, PhD(math), 55. *Prof Exp:* Actg head data processing dept, Comput Devices Can, Ltd, 55-58; assoc prof, 58-69, prof math, Univ Ottawa, 69-88; RETIRED. *Mem:* Am Math Soc; Math Asn Am; London Math Soc. *Res:* Numerical analysis; integral equations of classical potential theory. *Mailing Add:* RR 3 Carp Ottawa ON K0A 1L0 Can

HOWLAND, JAMES SECORD, b Jacksonville, Fla, Dec 27, 37; m 63; c 2. MATHEMATICS. *Educ:* Univ Fla, BS, 59; Calif Inst Technol, MS, 62; Univ Calif, Berkeley, PhD, 66. *Prof Exp:* From asst prof to assoc prof, 66-75, PROF MATH, UNIV VA, 75- *Concurrent Pos:* Assoc ed, J Math Physics, 80-82, J Math Anal Appl, 87- *Res:* Functional analysis and mathematical physics. *Mailing Add:* Dept Math Math Astron Bldg Univ Va Cabell Dr Charlottesville VA 22903

HOWLAND, JOHN LAFOLLETTE, b Quincy, Mass, Dec 14, 35; m 61; c 2. BIOCHEMISTRY. *Educ:* Bowdoin Col, AB, 57; Harvard Univ, PhD(biol), 61. *Prof Exp:* USPHS fel biochem, Univ Amsterdam, 61-63; from asst prof to assoc prof, 63-71, PROF BIOL, BOWDOIN COL, 71-, JOSIAH LITTLE PROF NATURAL SCI, 75- *Concurrent Pos:* NSF res grant, 63-; USPHS res career develop award. *Mem:* Am Chem Soc; Am Soc Biol Chemists; Biophys Soc. *Res:* Oxidative enzymes, oxidative phosphorylation, membrane transport and genetic muscle disease. *Mailing Add:* Dept Biol Bowdoin Col Brunswick ME 04011

HOWLAND, JOSEPH E(MERY), b Providence, RI, Apr 2, 18; m 43, 69; c 2. PLANT PHYSIOLOGY. *Educ:* RI State Col, BS, 40; Mich State Univ, MS, 42; Cornell Univ, PhD(hort), 45. *Prof Exp:* Asst, Mich State Col, 40-42, Cornell Univ, 42-45; assoc ed, Better Homes & Gardens, 45-48; garden ed, House Beautiful, 48-54, asst to publisher, 53-54, dir spec projs, 55-56; asst to pres, O M Scott & Sons, 56-68; prof hort, Univ Nev, 69-73; vpres, George J Ball Inc, 73-78; pres, Pan Am Seed Co, 73-75 & Burgess Seed & Plant Co, 75-76; dir, Peto Seed, Peters-Wheeler Seed, Denholm Seed, 73-76, Carefree Garden Prods, 76-77; dir, Am Seed Inst, 80-83; prof hort, 78-85, PROF ADVERT & MKT, UNIV NEV, RENO, 80- *Concurrent Pos:* Vis prof, Victoria Agr Col, Melbourne, Australia, 85-; headmaster, Du Page Hort Sch, 74-78; China Daily Sem, Beijing, 83; Winrock & Rockefeller Found Seed Proj, Thailand, 85- *Honors & Awards:* Bradford Williams Medal, Am Soc Landscape Architects, 81. *Mem:* Am Soc Hort Sci; Am Soc Agron; Am Hort Soc; Am Mgt Asns; Garden Writers Asn Am; Sci Writers. *Res:* Ornamental phytopathology; plant nutrition; commercial greenhouse; lawns and commercial turf; horticultural marketing; psychographic marketing research. *Mailing Add:* MSS 6 Univ Nev Reno NV 89557

HOWLAND, LOUIS PHILIP, b Somerville, NJ, Dec 19, 29; m 56; c 5. SOLID STATE PHYSICS, THEORETICAL PHYSICS. *Educ:* Cornell Univ, BEngPhys, 52; Mass Inst Technol, PhD(physics), 57. *Prof Exp:* Mem staff physics res, Lincoln Lab, Mass Inst Technol, 56-58; from instr to asst prof physics, Dartmouth Col, 58-65; assoc prof, 65-74, chmn dept, 69-71 & 74-76, Paul Garrett fel, 71-74, fac chmn, 76-78, PROF PHYSICS, WHITMAN COL, 74- *Concurrent Pos:* Vis assoc prof, Univ Calif, Berkeley, 71-72. *Mem:* Am Phys Soc; Am Asn Physics Teachers; Sigma Xi. *Mailing Add:* 903 Woodlawn St Walla Walla WA 99362

HOWLAND, RICHARD A, mathematics, algebra; deceased, see previous edition for last biography

HOWLAND, RICHARD DAVID, b New York, NY, Oct 19, 42; m 65; c 7. PHARMACOLOGY. *Educ:* Drew Univ, BA, 64; Univ Calif, San Francisco, PhD(pharmacol), 70. *Prof Exp:* USPHS fel, Univ Calif, San Francisco, 70; fel, Roche Inst Molecular Biol, Nutley, NJ, 70-72; asst prof, 72-82, ASSOC PROF PHARMACOL, UNIV MED & DENT NJ, 82- *Concurrent Pos:* Assoc grad fac, Newark Col Arts & Sci, Rutgers Univ, 74- *Mem:* AAAS; Soc Neurosci; NY Acad Sci; Am Soc Pharmacol & Exp Therapeut. *Res:* Biochemical mechanisms of acrylamide neurotoxicity; neuron-specific enolase and its role in acrylamide neurotoxicity; neurotoxicology. *Mailing Add:* Dept Pharmacol Univ Med & Dent NJ 185 S Orange Ave Newark NJ 07103

HOWLAND, WILLARD J, b Neosho, Mo, Aug 28, 27; m 45; c 5. RADIOLOGY. *Educ:* Univ Kans, BS, 48, MD, 50; Univ Minn, MS, 58. *Hon Degrees:* DSc, Northeastern Ohio Univ, 90. *Prof Exp:* Intern, US Naval Hosp, Newport, RI, 50-51; gen pract med, Kans, 51-55; fel radiol, Mayo Clin, 55-58; assoc radiologist, Columbia Hosp, Milwaukee, Wis, 58-59; radiologist, Ohio Valley Gen Hosp, Wheeling, WVa, 59-67; prof radiol & dir diag radiol sect, Univ Tenn, 67-68; dir & chmn dept radiol, Aultman Hosp, 68-87; chmn coun radiol, Northeastern Ohio Univs Col Med, 74-87, dir integrated diag radiol, 76-87; CONSULT, 87- *Concurrent Pos:* NSF res grants radiol, 62-63 & 67-68. *Honors & Awards:* Silver Medal, DSRS, 82; Scroll of Appreciation, Radiol Soc NAm, 84. *Mem:* Fel Am Col Radiol; Roentgen Ray Soc; Radiol Soc NAm; Am Med Asn. *Res:* Diagnostic radiology; arteriography; radiobiology; radiology education. *Mailing Add:* 1405 Harbor Dr NW Canton OH 44708

HOWLAND, WILLIAM STAPLETON, b Savannah, Ga, July 21, 19; c 2. MEDICINE, ANESTHESIOLOGY. *Educ:* Univ Notre Dame, BS, 41; Columbia Univ, MD, 44; Am Bd Anesthesiol, dipl, 53. *Prof Exp:* Asst prof anesthesiol, Col Physicians & Surgeons, Columbia Univ, 53; assoc prof surg, Sloan-Kettering Div, Med Col, 54-55, assoc prof anesthesiol in surg, 55-68, PROF ANESTHESIOL IN SURG, CORNELL UNIV, 68-; ASSOC PROF EXP SURG, SLOAN-KETTERING INST, 54-, MEM STAFF, 76-, CHMN, DEPT CRITICAL CARE, 78- *Concurrent Pos:* Attend anesthesiologist & chmn dept anesthesiol, Mem Hosp, New York, 53-, dep chief med off, Dept Anesthesiol, 67-78, dep gen dir, 74-, vpres clin affairs, 77- *Mem:* Am Soc Anesthesiol. *Res:* Problems of hemorrhage and shock; post-operative care. *Mailing Add:* RR 3 Box 50A Brattleboro VT 05301

HOWLETT, ALLYN C, b Rowan Co, NC, June 21, 50; c 1. PHARMACOLOGY, PHYSIOLOGY. *Educ:* Pa State Univ, BS, 71; Rutgers Univ, PhD(pharmacol & toxicol), 76. *Prof Exp:* Postdoctoral fel, Dept Pharmacol, Sch Med, Univ Va, 76-79; from asst prof to prof, Dept Pharmacol, 79-90, PROF, DEPT PHARMACOL & PHYSIOL SCI, MED SCH, ST LOUIS UNIV, 90- *Concurrent Pos:* NIH nat res serv award, 76-78; Pharmaceut Mfrs Asn Found res starter award, 80 & fac develop award basic pharmacol, 81-83; mem, Neurol Sci Study Sect, NIH, 84-85 & Biochem Study Sect, Nat Inst Drug Abuse, 90-91; Nat Inst Neurol & Communicative Dis &

Stroke res career develop award, 85-89; prin investr, Nat Inst Drug Abuse, 89-92 & 90-93, NIH, 91-96. *Mem:* Am Soc Pharmacol & Exp Therapeut; Soc Neurosci. *Res:* Author of numerous technical publications. *Mailing Add:* Dept Pharmacol Med Sch St Louis Univ 1402 S Grand Blvd St Louis MO 63104-1028

HOWLETT, SUSAN ELLEN, b Montreal, Que, Sept 2, 57; m 84; c 2. CARDIOVASCULAR PHARMACOLOGY, CARDIAC PATHOPHYSIOLOGY. *Educ:* Concordia Univ, BSc, 79; Mem Univ, MSc, 82, PhD(physiol), 85. *Prof Exp:* Fel pharmacol, Univ Alta, 85-89; ASST PROF PHARMACOL, DALHOUSIE UNIV, 89- *Mem:* Soc Neurosci. *Res:* Role of calcium in the development of heart disease, especially in aging; techniques including voltage clamp of single cells and receptor binding studies to study calcium handling in the heart. *Mailing Add:* Dept Pharmacol Sir Charles Tupper Med Bldg Dalhousie Univ Halifax NS B3H 4H7 Can

HOWLEY, PETER MAXWELL, b New Brunswick, NJ, Oct 9, 46; m 69; c 3. PATHOLOGY. *Educ:* Princeton Univ, AB, 68; Rutgers Univ, MMS, 70; Harvard Univ, MD, 72. *Prof Exp:* Resident path, Mass Gen Hosp, 72-73; res assoc virol, Nat Inst Allergy & Infectious Dis, NIH, 73-75; resident, 75-77, sr investr, 77-79, sect chief path, 79-84, CHIEF, LAB TUMOR VIRO BIO, NAT CANCER INST, 84- *Concurrent Pos:* Ed, J Virol, 89- *Honors & Awards:* Warner Lambert-Parke Davis Award, Am Asn Pathologists, 83; Wallace Rowe Award, Nat Inst Allergy & Infectious Dis, 86; Meritorious Serv Award, USPHS, 89. *Mem:* Am Soc Microbiol; Am Soc Virol; Am Asn Pathologists; Am Soc Clin Invest. *Res:* Molecular virology and oncology with a particular interest in the molecular biology and genetic organization of the papilloma viruses; development of eukaryotic vectors for studying gene regulation; gene expression. *Mailing Add:* Lab Tumor Vir Biol Nat Cancer Inst Bldg 41 Rm C111 Bethesda MD 20892

HOWSE, HAROLD DARROW, b Poplarville, Miss, Nov 8, 28; m 60; c 2. CELL BIOLOGY. *Educ:* Univ Southern Miss, BS, 59, MS, 60; Tulane Univ, PhD(anat), 67. *Prof Exp:* Instr zool, Miss Col, 60; instr biol, Univ Southern Miss, 60-63; head sect micros, Gulf Coast Res Lab, 67-71, dir, 71-89; RETIRED. *Concurrent Pos:* Mem, Miss Marine Resources Coun, 71-; prof zool, Miss State Univ & prof biol, Univ Miss, 72-; prof biol, Univ Southern Miss, 73-; chmn adv comt, Miss Coastal Zone Mgt & Coastal Energy Impact Prog, 81. *Mem:* Am Asn Anatomists; AAAS; Am Micros Soc; Am Soc Cell Biol; Am Soc Zoologists; Electron Micros Soc Am; NY Acad Sci; Sigma Xi; Oceanog Soc. *Res:* Comparative histology, histochemistry and ultrastructure of the heart of marine invertebrates and vertebrates; histopathology of certain marine fishes and invertebrates. *Mailing Add:* Gulf Coast Res Lab PO Box 7000 Ocean Springs MS 39564

HOWSMON, JOHN ARTHUR, b Dayton, Ohio, July 4, 19; m 53. CHEMISTRY, RESEARCH ADMINISTRATION. *Educ:* Berea Col, AB, 41; Univ Ill, PhD, 44. *Prof Exp:* Res chemist, Am Viscose Corp, 44-57, mgr basic res, 57-59, mgr polyolefin dept, 59-62; assoc dir res & develop, Avisun Corp, Pa, 62-68, admin dir res & eng, 68-70; admin res & develop mgr, Amoco Chem Corp, 70-84; RETIRED. *Res:* Rayon fibers; relation of structure to properties of fibers; polyolefin plastics. *Mailing Add:* 23W389 Green Trails Dr Naperville IL 60540-9408

HOWTON, DAVID RONALD, b Tacoma, Wash, July 8, 20; m 73. BIOCHEMISTRY, RADIATION CHEMISTRY. *Educ:* Calif Inst Technol, BS, 42, PhD(org chem), 46. *Prof Exp:* Asst res biochemist, Atomic Energy Proj, Sch Med, 48-56, assoc res biochemist, 56-64, res biochemist, Lab Nuclear Med & Radiation Biol & prof physiol chem, Sch Med, Univ Calif, Los Angeles, 64-76; prof chem & chmn dept, Cols Med & Med Sci, King Faisal Univ, Dammam, Saudi Arabia, 76-78; PROF ALLIED MED SCI, JOHN A BURNS SCH MED, UNIV HAWAII, MANOA, 78- *Concurrent Pos:* Res Corp fel, Calif Inst Technol, 46-48; from clin instr to assoc prof physiol chem, Sch Med, Univ Calif, Los Angeles, 52-64, assoc prof biophys & nuclear med, 61-64, prof, 64-70. *Mem:* Am Chem Soc. *Res:* Radiation chemistry of lipids. *Mailing Add:* John A Burns Sch Med 1960 East-West Rd Honolulu HI 96822

HOY, CASEY WILLIAM, b Berea, Ohio, Jan 11, 54; m 82. INSECT SPATIAL DYNAMICS, PEST MANAGEMENT MODELING. *Educ:* Cornell Univ, BS, 81, PhD(entom), 88. *Prof Exp:* Pest mgt control consult, 81-83; grad res asst insect biol, Cornell Univ, 83-87; ASST PROF SYSTS ANALYTICAL INSECT PEST MGT, OHIO STATE UNIV, 87- *Mem:* Entom Soc Am; Sigma Xi. *Res:* Application of systems analysis to problems in vegetable pest management; intra-and interplant spatial dynamics of vegetable insect pests. *Mailing Add:* Dept Entom Ohio State Univ OARDC Madison Hill Wooster OH 44691

HOY, GILBERT RICHARD, b Cleveland, Ohio, June 17, 32; m 52, 81; c 4. GAMMA-RAY OPTICS. *Educ:* Davis & Elkins Col, BS, 54; Cornell Univ, MS, 58; Carnegie Inst Technol, PhD(physics), 63. *Prof Exp:* Scientist, Solid State Res Dept, Xerox Corp, 62-63; instr physics, Carnegie Inst Technol, 63-65; from asst prof to prof physics, 65-80, actg chair, Boston Univ, 79-80; prof & chmn, 80-89, PROF & EMINENT SCHOLAR, OLD DOMINION UNIV, 90- *Concurrent Pos:* Hon res assoc, Harvard Univ, 72; scientist, Centre Nuclear Studies, Grenoble, France, 76-77; consult, NIH, 78-81 & Inst Defense Anal, 85-90; collabr, Los Alamos Nat Lab, 87- *Mem:* Am Phys Soc. *Res:* Gamma-ray optics; application of Mossbauer spectroscopy to the study of solid state physics. *Mailing Add:* Physics Dept Old Dominion Univ Norfolk VA 23529

HOY, JAMES BENJAMIN, b Wauseon, Ohio, Apr 13, 35; m 61; c 1. ECOLOGY. *Educ:* Ohio Wesleyan Univ, BS, 57; Univ Mich, MS, 59; Univ Kans, PhD(entom), 66. *Prof Exp:* Res parasitologist, Univ Calif, Berkeley, 63-65; res entomologist, Agr Res Serv, USDA, 66-73; asst prof biol, State Univ NY Purchase & Mt Vernon, 74-76; mem res staff, Dept Entom Sci, Univ Calif, Berkeley, 76-86; ENTOMOLOGIST, FOREST SERV, USDA, BERKELEY, 87- *Concurrent Pos:* Field rep, Coun on Econ Priorities, 73-75. *Mem:* Entom Soc Am. *Res:* Animal behavior; medical and veterinary entomology; aquatic ecology; soil ecology. *Mailing Add:* 1004 Grizzly Peak Blvd Berkeley CA 94708

HOY, MARJORIE ANN, b Kansas City, Kans; m 61; c 1. INSECT ECOLOGY, GENETICS. *Educ:* Univ Kans, AB, 63; Univ Calif, MS, 66, PhD(entom), 72. *Prof Exp:* Res geneticist, Univ Calif, Berkeley, 64-66; lectr biol, Fresno State Col, 67-68; lab technician, Div Biol Control, Univ Calif, Berkeley, 68-70; lectr, Calif State Univ, Fresno, 73; res entomologist, Conn Agr Exp Sta, 73-75; res entomologist, Northeast Forest Exp Sta, US Forest Serv, USDA, 75-76; from asst prof to assoc prof, 76-82, PROF ENTOM SCI, UNIV CALIF, BERKELEY, 82- *Honors & Awards:* Bussart Mem Award, Entomol Soc Am, 86. *Mem:* Sigma Xi; Entom Soc Am; Int Orgn Biol Control; Acarology Soc Am; fel AAAS. *Res:* Genetics and biological control of insect and mite pests; studies of diapause in relation to insect and mite ecology; genetic selection of a pesticide-resistant predators and parasites for integrated pest management programs. *Mailing Add:* 201 Wellman Hall Univ Calif Berkeley CA 94720

HOY, ROBERT C, b New Orleans, La, Dec 2, 33; m 67. SOLID STATE PHYSICS, CRYSTALLOGRAPHY. *Educ:* Tulane Univ, BS, 54, MS, 60, PhD(physics), 65. *Prof Exp:* Res asst, Tulane Univ, 60-62, res assoc, 65-66; res assoc, Columbia Univ, 66; sr res engr, Space Div, Chrysler Corp, 66-67; asst prof physics, La State Univ, New Orleans, 67-73; PRES, GULF FOOD PROD CO, INC, 73- *Concurrent Pos:* Spec lectr, Univ New Orleans, 75- *Mem:* AAAS; Am Phys Soc. *Res:* Electron diffraction; surface physics; x-ray crystallography. *Mailing Add:* 1005 Andrews Ave Metairie LA 70005

HOY, RONALD RAYMOND, b Walla Walla, Wash, Jan 12, 39. BIOLOGY. *Educ:* Wash State Univ, BS, 62; Stanford Univ, PhD(biol), 68. *Prof Exp:* Asst prof biol, State Univ NY, Stony Brook, 71-73; asst prof, 73-77, ASSOC PROF BIOL, CORNELL UNIV, 77- *Concurrent Pos:* NIH fel, 68-70 & 78. *Mem:* Sigma Xi; Am Soc Zoologists; AAAS; Soc Neurosci. *Res:* Physiological and behavioral studies of insect audition; neuroethology; regeneration in the nervous system; developmental neurobiology. *Mailing Add:* Cornell Univ W 214 Mudd Hall Ithaca NY 14850

HOYE, THOMAS ROBERT, b New Castle, Pa, June 19, 50; m 73; c 3. ORGANIC CHEMISTRY. *Educ:* Bucknell Univ, BS & MS, 72; Harvard Univ, PhD(chem), 76. *Prof Exp:* From asst prof to assoc prof, 76-88, PROF CHEM, UNIV MINN, 88- *Concurrent Pos:* Adv comt mem, Res Corp Grants, 83-88; fel, Alfred P Sloan Found, 85-89. *Mem:* Am Chem Soc. *Res:* Synthetic organic chemistry; development and application of synthetic methodology to natural products total synthesis; synthetic applications of Fischer carbene complexes; stereochemistry; symmetry; NMR analysis; natural product structure determination. *Mailing Add:* Dept Chem Univ Minn 207 Pleasant St SE Minneapolis MN 55455

HOYER, BILL HENRIKSEN, b Seattle, Wash, Dec 4, 21; m 48; c 2. MICROBIOLOGY. *Educ:* Univ Wash, BS, 43, MS, 45; Univ Minn, PhD(microbiol), 48. *Prof Exp:* Instr, Univ Minn, 46-48; bacteriologist, NIH, Md, 48-50, Rocky Mt Lab, 50-64, sci dir lab biol viruses, 64-68; STAFF MEM DEPT TERRESTRIAL MAGNETISM, CARNEGIE INST WASH, 68- *Concurrent Pos:* Lectr, Mont State Univ, 52-64; mem, Human Develop & Aging Res & Training Comn, Nat Inst Child Health & Human Develop, 70-; sci dir, USPHS, 48-68. *Mem:* AAAS; Am Soc Microbiol; Sigma Xi. *Res:* Physiology of microorganisms and mammalian cells; chromatography and purification of viruses and viral components; DNA-DNA and DNA-RNA relationships of viruses, bacteria and animals; primate relationships; evolution. *Mailing Add:* 5901 Lone Oak Dr Bethesda MD 20814

HOYER, LEON WILLIAM, b Minneapolis, Minn, Mar 6, 36; m 60; c 3. BLOOD COAGULATION RESEARCH, HEMATOLOGY. *Educ:* Harvard Col, AB, 58; Univ Minn, MD, 62; Am Bd Internal Med, dipl, 70. *Prof Exp:* Asst prof med, Univ Rochester, Sch Med & Dent, 68-70; from assoc prof to prof, 70-85, CLIN PROF MED, SCH MED, UNIV CONN, 85-; NAT RES DIR, HOLLAND LAB, AM RED CROSS BLOOD SERV, 85- *Concurrent Pos:* Prin investr, NIH grants res factor VIII, 74-; mem, Hemat Study Sect, NIH, 76-80 & 87-91; vis scientist, Karolinska Inst, Stockholm, Sweden, 78-79; Chair, Med & Sci Adv Comn, Nat Hemophilia Found, 82-85; adj prof genetics, George Washington Univ, 88-; chmn, Van Willebrand factor subcomt, 87-90, Int Comt Haemostasis & Thrombosis, 89-93; dir, Spec Ctr Res Transfusion Med, 91- *Honors & Awards:* Murray Theilin Res Award, Nat Hemophilia Asn, 81; Hemophilia Res Award, French Hemophilia Asn, 83. *Mem:* Am Soc Hematol; Am Soc Clin Invest; Am Heart Asn Coun Thrombosis; Am Asn Immunologists; Int Soc Hematol. *Res:* Structure and function of coagulation on factor VIII; molecular basis of hemophilia A; immunochemical properties of human antibodies that inactivate factor VIII. *Mailing Add:* Holland Lab 15601 Crabbs Branch Way Rockville MD 20855

HOYER, WILMER ADOLF, physics, for more information see previous edition

HOYER-ELLEFSEN, SIGURD, b Olso, Norway, July 24, 32; US citizen; div; c 5. ELECTRICAL ENGINEERING. *Educ:* Mass Inst Technol, BSME, 56, MSME, 57, ScD, 62. *Prof Exp:* Sr engr, Potter Instrument Co, Inc, 57-69, vpres res & develop, 70-76; vpres res & develop, Smith-Corona, SCM Corp, 76-83; sr vpres, Servo Corp Am, 83-84; consult, 84-86; PRES & CHIEF EXEC OFFICER, LINTON ROOF TRUSS INC, 86- *Concurrent Pos:* Chief scientist & asst prog dir, Fairchild Industs, Inc, 64-70. *Mem:* Inst Elec & Electronics Engrs. *Mailing Add:* Linton Roof Truss Inc 1455 SW Fourth Ave Delray Beach FL 33444

HOYLE, HUGHES BAYNE, JR, mathematics; deceased, see previous edition for last biography

HOYT, CHARLES D, JR, b Indianapolis, Ind, Mar 17, 12; m 37; c 2. INDUSTRIAL ENGINEERING. *Educ:* Purdue Univ, BSChE, 35, MSIE, 60, PhD(indust eng), 62. *Prof Exp:* With Am Brass Co, Wis, 35-36; off mgr, Hoyt Mach Co, Ind, 36-37; off mgr, Charles D Hoyt Co, 37-45, pres & treas, 46-59; teaching asst indust eng, Purdue Univ, 60-62; assoc prof, 62-69, PROF INDUST ENG, ARIZ STATE UNIV, 69- *Mem:* Am Inst Indust Engrs; Am

Soc Eng Educ; Nat Soc Prof Engrs; Inst Mgt Sci. *Res:* Industrial productivity; manufacturing organization and management; rank-frequency statistics and its applications to organization, wage and salary administration, industrial marketing and manufacturing production; variable budget control of manufacturing operations. *Mailing Add:* Dept Indust Eng Ariz State Univ Tempe AZ 85287

HOYT, DONALD FRANK, b San Francisco, Calif, May 22, 45; m 69; c 3. PHYSIOLOGICAL ECOLOGY, ANIMAL ENERGETICS. *Educ:* Pomona Col, Calif, BA, 67; Univ Calif, Los Angeles, PhD(biol), 77. *Prof Exp:* Res assoc, Dept Physiol, State Univ NY, 77-78; res fel, Mus Comparative Zool, Harvard Univ, 78-79, preceptor, Dept Biol, 79-80; from assoc prof to asst prof, 80-88, PROF BIOL, CALIF STATE POLYTECH UNIV, 88- *Concurrent Pos:* Consult, A D Little Co, Mass, 80. *Mem:* Sigma Xi; Am Soc Zoologists; Am Ornithologists Union; Am Physiol Soc. *Res:* Physiological ecology of terrestrial vertebrates, in particular comparative aspects of energetics of locomotion in mammals and the relation between metabolic rate and growth in avian embryos; comparative aspects of energetics of locomotion in mammals; comparative physiology of detraining; the relation between metabolic rate and growth in avian embryos. *Mailing Add:* Dept Biol Sci Calif State Polytech Univ Pomona CA 91768

HOYT, EARLE B, JR, b New York, NY, July 14, 37; m 59; c 1. ORGANIC CHEMISTRY. *Educ:* Middlebury Col, BA, 59; Alliance Francaise, dipl, 60; Tufts Univ, PhD(chem), 66. *Prof Exp:* Instr chem, Northwestern Univ, 65-66; NIH fel, Cornell Univ, 66-67; asst prof, 67-74, ASSOC PROF CHEM, NORTHERN ARIZ UNIV, 67- *Mem:* AAAS; Am Chem Soc; Sigma Xi. *Res:* Organic synthesis, especially highly strained compounds and natural products. *Mailing Add:* 449 E Hutcheson Flagstaff AZ 86001

HOYT, HARRY CHARLES, b Grinnell, Iowa, June 20, 24; m 59; c 2. PHYSICS. *Educ:* Univ Colo, BS, 46; Calif Inst Technol, PhD(physics), 52. *Prof Exp:* Instr physics, Univ Colo, 47-48; res fel, Calif Inst Technol, 52-53; staff mem, Los Alamos Nat Lab, Univ Calif, 53-59, alt group leader, 59-68, asst div leader, Weapons Div, 68-70, assoc div leader, 70-71, alt div leader, Theoret Design Div, 71-72, asst dir, Weapon Planning, 72-76, asst dir, Weapon Planning & Coord, 76-79, asst dir policy, 79, assoc dir energy prog, 79-81, asst dir special prog, 81-86; RETIRED. *Concurrent Pos:* Consult, 75-76, 86- *Mem:* AAAS; Am Phys Soc; Opers Res Soc Am; Sigma Xi. *Res:* Nuclear spectroscopy; hydrodynamics; electricity and magnetism; numerical analysis. *Mailing Add:* 650 Barranca Rd Los Alamos NM 87544

HOYT, JACK W(ALLACE), b Chicago, Ill, Oct 19, 22; m 45; c 4. FLUID MECHANICS. *Educ:* Ill Inst Technol, BS, 44; Univ Calif, Los Angeles, MS, 52, PhD(eng), 62. *Prof Exp:* Aero-engr, Nat Adv Comt Aeronaut, 44-48; mech engr, Naval Ocean Systs Ctr, 48-79; prof mech engr, Rutgers Univ, 79-81; PROF MECH ENGR, SAN DIEGO STATE UNIV, 81- *Concurrent Pos:* Am Soc Mech Engrs Freeman scholar, 71; Benjamin Meaker vis prof, Univ Bristol, Eng, 87. *Mem:* Am Soc Mech Engrs; Soc Naval Architects & Marine Engrs; NY Acad Sci. *Res:* Fluid mechanics; drag-reducing polymer solutions; ocean engineering. *Mailing Add:* 4694 Lisann St San Diego CA 92117

HOYT, JOHN MANSON, polymer chemistry; deceased, see previous edition for last biography

HOYT, PHILIP M(UNRO), civil engineering, for more information see previous edition

HOYT, ROBERT DAN, b Conway, Ark, Sept 24, 41; m 62; c 5. ICHTHYOLOGY. *Educ:* Univ Cent Ark, BS, 63; Univ Ark, MS, 65; Univ Louisville, PhD(biol), 69. *Prof Exp:* Res asst fisheries, Univ Ark, 63-65; interpretive naturalist, Bernheim Forest Nature Mus, 65-66; asst prof, 69-74, assoc prof, 74-78, PROF BIOL, WESTERN KY UNIV, 78- *Concurrent Pos:* Actg dir, Univ Hons Prog, Western Ky Univ, 83-84, assoc dean, Ogden Sci Col, 85-86. *Mem:* Am Soc Ichthyol & Herpet. *Res:* Larval fish ecology; early anatomical development of fish. *Mailing Add:* Dept Biol Western Ky Univ Bowling Green KY 42101

HOYT, ROSALIE CHASE, b New York, NY, May 20, 14. PHYSICS. *Educ:* Columbia Univ, BA, 40; Bryn Mawr Col, MA, 41, PhD(physics), 45. *Prof Exp:* Instr physics, Univ Rochester, 45-48; from instr to assoc prof, 41-69, Marion Reilly prof, 69-82, EMER MARION REILLY PROF PHYSICS, BRYN MAWR COL, 82- *Mem:* Am Phys Soc. *Res:* Bioelectrics; bioelectric instrumentation; nerve models; counters and detectors for nuclear research. *Mailing Add:* Parker Head Rd Phippsburg ME 04562

HOYT, STANLEY CHARLES, b Oakland, Calif, Oct 4, 29; m 55; c 3. ACAROLOGY, ADMINISTRATION. *Educ:* Univ Calif, BS, 51, PhD, 58. *Prof Exp:* Asst entomologist, Wash State Univ, 57-63, assoc entomologist, 63-69, supt, 83-91, ENTOMOLOGIST, TREE FRUIT RES CTR, WASH STATE UNIV, 69- *Concurrent Pos:* Fulbright grant, 70-71; FAO consult, China, 87. *Honors & Awards:* Ciba-Geigy Award, Entom Soc Am, 73; C W Woodworth Award, Pac Br, Entom Soc Am, 89. *Mem:* Entom Soc Am; Acarological Soc Am. *Res:* Biology, ecology and control of insects affecting apples; host plant-pest interactions. *Mailing Add:* 351 NE 19th No 12 East Wenatchee WA 98802

HOYT, WILLIAM F, b Berkeley, Calif, June 17, 26; m 56; c 2. NEUROLOGY, OPHTHALMOLOGY. *Educ:* Univ Calif, San Francisco, AB, 47, MD, 50; Am Bd Ophthal, dipl, 58. *Prof Exp:* From asst prof to assoc prof, 58-70, PROF NEURO-OPHTHAL, SCH MED, UNIV CALIF, SAN FRANCISCO, 70- *Concurrent Pos:* Fulbright fel, Sch Med, Univ Vienna, 56-57; Heed fel, Johns Hopkins Hosp, 57-58; asst ed, Arch Ophthal, 62-65, assoc ed, 65- *Mem:* Soc Neurol Surg; Barany Soc. *Res:* Clinical teaching and laboratory investigation of problems related to anatomy, physiology and pathology of visual and oculomotor systems. *Mailing Add:* Dept Ophthal & Neurosurg Univ Calif San Francisco CA 94102

HOYT, WILLIAM LIND, b Nephi, Utah, Sept 8, 28; m 57; c 1. MATHEMATICS. *Educ:* Univ Utah, BA, 50, MS, 51; Univ Chicago, PhD(math), 58. *Prof Exp:* Instr math, Northwestern Univ, 56-57; instr, Hopkins, 58-59; asst prof, Brandeis Univ, 59; mem fac, Univ Ind, Bloomington, 59-67; ASSOC PROF MATH, RUTGERS UNIV, 67- *Mem:* Am Math Soc. *Res:* Algebraic geometry; algebraic number theory. *Mailing Add:* 153 N Eighth Ave Highland Park NJ 08904

HOYTE, ROBERT MIKELL, b New York, NY, Nov 8, 43; m 87. ORGANIC CHEMISTRY, BIOCHEMISTRY. *Educ:* Long Island Univ, BS, 64; Rutgers Univ, New Brunswick, MS, 67, PhD(chem), 68. *Prof Exp:* Asst chemist, Brookhaven Nat Lab, 68-71; asst prof chem, Medgar Evers Col, 71-72; from asst prof to prof, 72-90, DISTINGUISHED TEACHING PROF CHEM, COL OLD WESTBURY, STATE UNIV NY, 90- *Concurrent Pos:* Vis fel, Dept Obstet & Gynec, Yale Univ Sch Med, 80-81. *Mem:* AAAS; Am Chem Soc; Sigma Xi. *Res:* Synthetic organic chemistry; steroid biochemistry; synthesis of radiopharmaceuticals. *Mailing Add:* Col Old Westbury State Univ NY Old Westbury NY 11568-0210

HOYUMPA, ANASTACIO MANINGO, b Baybay, Leyte, Philippines, July 4, 37; m 63; c 4. INTERNAL MEDICINE, GASTROENTEROLOGY. *Educ:* Univ Santo Tomas, MD, 61. *Prof Exp:* Instr, Univ Cincinnati, 67-68, asst prof med, 68-72, actg assoc dir, 72; GE sect chief, Vet Admin Hosp, 72; from asst prof to assoc prof med, Vanderbilt Univ, 72-82; chief gastroenterol, Audie Murphy Mem Vet Admin Hosp, San Antonio, Tex, 82-87; PROF MED, UNIV TEX HEALTH SCI CTR, SAN ANTONIO, 82- *Concurrent Pos:* Fel, Univ Cincinnati, 65-67; consult, Longview State Hosp, 71-72; res & educ assoc, Vet Admin Hosp, 73-75, clin investr, 76-78, investr in alcoholism, 79-81. *Mem:* Am Soc Gastrointestinal Endoscopy; Am Asn Liver Dis; Am Gastroenterol Asn; Res Soc Alcoholism; Cent Soc Clin Res; Int Asn Study Liver. *Res:* Thiamine intestinal transport and mechanism of alcohol inhibition; effect of liver disease and alcohol on drug metabolism; fetal alcohol syndrome. *Mailing Add:* Dept Med Univ Tex Health Sci ctr 7703 Floyd Curl Dr San Antonio TX 78284

HOZUMI, NOBUMICHI, b Kitakata-City, Japan, Feb 25, 43. IMMUNOLOGY. *Educ:* Keio Univ, Japan, MD, 68, PhD(molecular biol), 72. *Prof Exp:* Lectr molecular biol, Dept Molecular Biol, Keio Univ, Japan, 72-75; mem immunol, Basel Inst Immunol, 75-78; asst prof molecular biol, Dept Med Biophysics, 79-84, assoc prof immunol, Dept Immunol, 84-88, PROF IMMUNOL, DEPT IMMUNOL, UNIV TORONTO, 88- *Concurrent Pos:* Reviewer, Nat Cancer Inst Can, 86-87. *Honors & Awards:* David Pressman Mem Award, 83; Boehringer Mannheim Can Prize, Can Biochem Soc, 84. *Mem:* Am Asn Immunol; Can Soc Immunologists. *Res:* Molecular mechanisms involved in B cell differentiation and immunological self-nonself recognition. *Mailing Add:* Samuel Lunenfeld Res Inst Mt Sinai Hosp 600 University Ave Toronto ON M5G 1X5 Can

HRABA, JOHN BURNETT, b New York, NY, Nov 21, 21; m 46; c 2. ELECTRICAL ENGINEERING. *Educ:* Univ NH, BS, 48; Yale Univ, MEng, 49; Univ Ill, PhD(elec eng), 55. *Prof Exp:* Prof elec eng, 49-73, assoc dean col technol, 49-68, dean instnl res & planning, 68-, systs dir planning & anal, 73-, EMER PROF ELEC & COMPUT ENG, UNIV NH. *Concurrent Pos:* Instr, Univ Ill, 55; asst prog dir eng sci, NSF, 59-60. *Mem:* Assoc Inst Elec & Electronics Engrs. *Res:* System electromagnetics. *Mailing Add:* Dept Elec & Comput Eng Univ NH Kingsbury Hall Durham NH 03824

HRAZDINA, GEZA, b Letenye, Hungary, Mar 16, 39; m 64; c 1. BIOCHEMISTRY. *Educ:* Swiss Fed Inst Technol, Dipl, 63, DSc(technol), 66. *Prof Exp:* From res assoc to assoc prof, 66-81, PROF BIOCHEM, NY STATE AGR EXP STA, CORNELL UNIV, 81- *Concurrent Pos:* Lectr fel, Alexander von Humboldt Stiftung, Ger, 74-75; vis prof, Tech Univ, Budapest, 79 & Univ Cologne, Ger, 81. *Mem:* Am Chem Soc; Phytochem Soc NAm (pres, 82-83); Phytochem Soc Europe; Am Soc Plant Physiol; Am Inst Biol Chem. *Res:* Enzyme chemistry; chemistry and biochemistry of plant natural products; cellular and subcellular localization in plant metabolism. *Mailing Add:* NY State Agr Exp Sta Cornell Univ Geneva NY 14456

HRBEK, GEORGE W(ILLIAM), b Oak Park, Ill, Dec 27, 27; m 53; c 3. ELECTRICAL ENGINEERING. *Educ:* Ill Inst Technol, BS, 53; Northwestern Univ, MS, 64. *Prof Exp:* Engr microwave tube res, Sperry Gyroscope Co, 53-54; engr, Zenith Radio Corp, 56-61, chief, Div Electron Device Res, 61-73, mgr, Electromech Res Dept, 73-80; dir res & develop, Ardev Co, Inc, Palo Alto, Calif, 80-; AT ZENITH RADIO CORP. *Mem:* Inst Elec & Electronics Engrs. *Res:* Acousto-optic devices and applications; ultrasonic delay lines. *Mailing Add:* Zenith Radio Corp 1000 Milwaukee Ave Glenview IL 60025

HRDINA, PAVEL DUSAN, b Uzhorod, Czech, Oct 3, 29; m 54; c 1. PSYCHIATRY, PHARMACOLOGY. *Educ:* Komensky Univ, MD, 55; Czech Acad Sci, PhD(pharmacol), 64. *Prof Exp:* Instr path, Med Sch, Komensky Univ, 53-54, asst prof med chem, 55-57, asst & assoc prof pharmacol, 58-68; assoc prof, 69-73, PROF PHARMACOL & PSYCHIAT, FAC MED, UNIV OTTAWA, 74- *Concurrent Pos:* Foreign res fel, Mario Negri Pharmacol Res Inst, Milan, 65-66; res fel, Med Col Va, 66-67; Med Res Coun Can vis scientist, Univ Ottawa, 68-69; vis prof, Synthelabo Lers, Paris, 79, & dept pharmacol, Melbourne Univ, 87; ed, J Psychiat & Neurosci. *Mem:* Am Soc Pharmacol & Exp Therapeut; Pharmacol Soc Can; Can Col Neuropsychopharmacol (vpres, 84-88, pres, 88-90); Int Soc Neurochem. *Res:* Pharmacology of antidepressant drugs; receptor binding; neuroscience. *Mailing Add:* Dept Pharmacol Univ Ottawa 451 Smyth Rd Ottawa ON K1H 8M5 Can

HRDY, SARAH BLAFFER, b Dallas, Tex, July 11, 46; m 72; c 3. PRIMATE BEHAVIOR, EVOLUTIONARY BIOLOGY. *Educ:* Radcliffe Col, BA, 69; Harvard Univ, PhD(behav biol), 75. *Prof Exp:* Instr anthrop, Univ Mass, 73; lectr biol anthrop, Harvard Univ, 75-76, fel biol, 77-78; ASSOC, PEABODY MUS, 79-; PROF, UNIV CALIF, DAVIS, 84- *Concurrent Pos:* Consult ed, Am J Primatol, Primates, Cult Anthrop; vis assoc prof anthrop, Rice Univ, 81-82. *Mem:* Nat Acad Sci. *Res:* Evolution of primate social behavior. *Mailing Add:* Dept Anthrop Univ Calif Davis CA 95616

HREN, JOHN J(OSEPH), b Milwaukee, Wis, Dec 3, 33; m 57; c 5. MATERIALS SCIENCE, SOLID STATE PHYSICS. *Educ:* Univ Wis, BS, 57; Univ Ill, MS, 60; Stanford Univ, PhD(mat sci), 62. *Prof Exp:* Patent exam, Metall Div, US Patent Off, 57-58; res physicist, Lawrence Radiation Lab, 61-62; NSF fels, Max Planck Inst, Stuttgart, Ger, 62-63 & Univ Cambridge, 63-64; from asst prof to assoc prof metall, Univ Fla, 64-72, prof mat sci & eng, 72-85; PROF HEAD MAT SCI & ENG, NC STATE UNIV, 85- *Concurrent Pos:* Vpres, Mat Consult, Inc; vis scientist, Div Tribophysics, Csiro, Melbourne, Australia, 70-71; consult, Bendix Corp, Mo, 71- & Sandia Labs, NM, 76-, Oak Ridge Nat Lab; vis prof mat sci, Vanderbilt Univ, 75-76; pres, Metamics Inc, Fla, 76-82; chmn, Mat Sci Prog, Nat Tech Univ; external adv & chmn, Nat Ctr Electron Micros; CONSULT, Oak Ridge Nat Lab. *Mem:* AAAS; Am Inst Physics; Am Inst Mining, Metall & Petrol Engrs; Am Soc Metals; Electron Microscopy Soc Am; Sigma Xi. *Res:* Field-ion microscopy; atom probe; atomic order; solute distribution and precipitation in solids; defects in crystals; radiation damage; analytical electron microscopy; electrosurgical devices. *Mailing Add:* Dept Mat Sci & Eng NC State Univ Raleigh NC 27695-7907

HRESHCHYSHYN, MYROSLAW M, b Kovel, Ukraine, Aug 30, 27; US citizen; m 58; c 4. OBSTETRICS & GYNECOLOGY. *Educ:* Univ Frankfurt, MD, 51. *Prof Exp:* Intern, St Joseph's Hosp, Yonkers, NY, 52-53; from asst resident to resident obstet & gynec, Cumberland Hosp, Brooklyn, 53-56; sr cancer res surgeon, Dept Gynec, Roswell Park Mem Inst, 58-65, assoc cancer res gynecologist, 65-71; from asst prof to assoc prof, 62-70, PROF OBSTET & GYNEC, STATE UNIV NY BUFFALO, 70-; ASSOC CHIEF CANCER RES GYNECOLOGIST, ROSWELL PARK MEM INST, 71-; AT CHILDREN'S HOSP, BUFFALO, NY. *Concurrent Pos:* Clin fel gynec cancer, Kings County Hosp, 56-57; fel chemother, Roswell Park Mem Inst, 57-59; chmn, Gynec Oncol Group, 71. *Mem:* AMA; Am Col Obstet & Gynec; Soc Gynec Oncol. *Res:* Gynecologic cancer; cancer chemotherapy; gonadotropins; endometriosis; osteoporosis. *Mailing Add:* Children's Hosp 219 Bryant St Buffalo NY 14222

HRIBAR, JOHN ANTHONY, b Pittsburgh, Pa, Jan 10, 34; m 67; c 4. CIVIL ENGINEERING. *Educ:* Carnegie Inst Technol, BS, 56, MS, 57, PhD(civil eng), 61. *Prof Exp:* From asst prof to assoc prof, 60-70, SR LECTR CIVIL ENG, CARNEGIE-MELLON UNIV, 70 -; VPRES, GAI CONSULTS, INC, 58- *Honors & Awards:* Collingwood Prize, Am Soc Civil Engrs, 66. *Mem:* Am Soc Civil Engrs; Am Soc Eng Educ; Am Concrete Inst; Am Soc Testing & Mat. *Res:* Soil mechanics and foundation engineering; numerical and finite element analysis of plates, shells and solids; experimental and theoretical stress analysis. *Mailing Add:* GAI Consults Inc Monroeville PA 15146

HRISKEVICH, MICHAEL EDWARD, b Timmins, Ont, Mar 7, 26; m 47; c 4. GEOLOGY. *Educ:* Queen's Univ (Ont), BSc, 47, MSc, 49; Princeton Univ, PhD(geol), 52. *Prof Exp:* Instr geol, Princeton Univ, 49-52; geologist subsurface, Stanolind Oil & Gas Co, 52-54, Triad Oil Co, 54-57 & Can Fina Oil, Ltd, 57-61; dist geologist, Atlantic Ref Co, 61-63; chief geologist, Banff Oil Co, 63-65, explor mgr, 65-70; explor mgr, Aquitaine Co, Can, 70-76, vpres, 76-80, sr vpres explor spec proj, 80-81; sr vpres, Canterra Energy Ltd, 81-83; PRES, BANAQU EXPLOR LTD, 83- *Concurrent Pos:* Bursary, Nat Res Coun Can. *Mem:* Am Asn Petrol Geol; Geol Soc Am; Can Soc Petrol Geologists. *Res:* Geology and petrology of basic rocks; subsurface stratigraphy. *Mailing Add:* 4103 14 A St SW Calgary AB T2T 3Y3 Can

HRIVNAK, BRUCE JOHN, b Johnstown, Pa, July 10, 49; m 73; c 3. EVOLVED STARS, BINARY STARS. *Educ:* Univ Pa, BA, 71, PhD(astron), 80. *Prof Exp:* Fel & instr, dept physics, Univ Calgary, 80-84; asst prof, 84-89, ASSOC PROF, PHYSICS DEPT, VALPARAISO UNIV, 89- *Concurrent Pos:* Vis res officer, Dominion Astrophys Observ, 84; prin investr, NASA, 87-, NSF, 89- *Mem:* Am Astron Soc; Sigma Xi; Int Astron Union; Am Sci Affil. *Res:* Observation and analysis binary stars; observation of evolved stars. *Mailing Add:* Dept Physics Valparaiso Univ Valparaiso IN 46383

HRKEL, EDWARD JAMES, b Des Moines, Iowa, July 3, 42; m 69; c 1. FUEL TECHNOLOGY, PETROLEUM ENGINEERING. *Educ:* Purdue Univ, BS, ME, 64; Mass Inst Technol, SM, 65, PhD(mech eng), 69. *Prof Exp:* Engr, Shell Develop Co, 69-70; PROF SPECIALIST, GETTY OIL CO, 70- *Mem:* Sigma Xi; Soc Petrol Engrs. *Res:* Numerical models for the prediction of performance in oil and gas reservoirs; fluid flow in Porous Media. *Mailing Add:* 8818 Carvel Houston TX 77036

HRONES, JOHN ANTHONY, b Boston, Mass, Sept 28, 12; m 38; c 4. MECHANICAL ENGINEERING. *Educ:* Mass Inst Technol, SB, 34, SM, 36, ScD(mech eng), 42. *Prof Exp:* Jr engr, Chase Brass & Cooper Co, Conn, 34; asst instr appl mech, Mass Inst Technol, 34-36, instr mech eng, 36-37; asst to factory mgr, Coldwell Lawnmower Co, NY, 37-39; from instr to prof mech eng, Mass Inst Technol, 39-57; vpres acad affairs, 57-64, provost, 64-77, EMER PROVOST, CASE WESTERN RESERVE UNIV, 77- *Concurrent Pos:* Consult mech eng, 40-; head mach design div, Mass Inst Technol, 46-57, dir dynamic anal & control lab, 50-57; mem bd trustees, Inst Defense Anal, 58-85; hon trustee, Cleveland Mus Natural Hist; pres, AIT Found, Inc, 68-; trustee, Asian Inst Technol, Bangkok. *Mem:* Nat Acad Engrs; Am Soc Mech Engrs; Am Soc Eng Educ; Am Acad Arts & Sci; Sigma Xi. *Res:* Automatic control; design and development of machinery. *Mailing Add:* 9397 Midnight Pass Rd Apt 306 Sarasota FL 34242

HROVAT, DAVORIN, b Zagreb, Yugoslavia, Jan 2, 49; m 74; c 2. CONTROL SYSTEMS, MODELING & SIMULATION. *Educ:* Univ Zagreb, Dipl Ing mech eng, 72; Univ Calif, Davis, MS, 76, PhD(mech eng), 79. *Prof Exp:* Asst prof mech eng, Wayne State Univ, 78-81; sr res engr, 81-85, prin res engr assoc control systs, 85- 89, PRIN STAFF ENG, SCI RES LAB, FORD MOTOR CO, 89- *Mem:* Sigma Xi; Am Soc Mech Engrs; Inst Elec & Electronics Engrs. *Res:* Modeling, analysis, and control of dynamic systems; vehicle dynamics; optimal mechanical structures; bond-graph modeling techniques; automotive power train modeling and control. *Mailing Add:* Sci Res Lab Rm E-1170 Ford Motor Co PO Box 2053 Dearborn MI 48121

HRUBAN, ZDENEK, b Czech, June 15, 21; nat US; m 55; c 3. PATHOLOGY. *Educ:* Univ Rostock, Ger, Cand Med, 44; Charles Univ, Prague, MUC, 48; Univ Chicago, MD, 56, PhD, 63. *Prof Exp:* USPHS res fel path, 57-60, Am Cancer Soc res fel, 60-63, from asst prof to prof, 63-91, EMER PROF PATH, UNIV CHICAGO, 91- *Concurrent Pos:* Lederle award, 64-67; res career develop award, 67-72; pres, coun higher educ, Am Asn Study liver Dis, 84-; bd dir, Czechoslovak Soc Arts & Sci Am, 74-; mem gac discussants, Chariles Louis Davis, DVM Found, 76. *Mem:* Sigma Xi; Electron Micros Soc Am. *Res:* Ultrastructural pathology; microbodies. *Mailing Add:* Dept Path Box 414 Univ Chicago 5841 S Maryland Ave Chicago IL 60637-1470

HRUBANT, HENRY EVERETT, b Aurora, Ill, Apr 5, 29; m 52; c 3. ANIMAL GENETICS. *Educ:* Univ Ill, BS, 50; Ohio State Univ, MS, 53, PhD(genetics), 57. *Prof Exp:* Res asst, Allied Chem & Dye Corp, 50-51; from asst prof to prof biol, Calif State Univ, Long Beach, 57-88; RETIRED. *Mem:* AAAS; Am Genetic Asn; Genetics Soc Am; Genetics Soc Can; Geront Soc. *Res:* Physiological genetics of metabolic disorders in mammals; genetic control of aging; mammalian cytogenetics. *Mailing Add:* PO Box 570 Talent OR 97540

HRUBESH, LAWRENCE WAYNE, b Eau Claire, Wis, Dec 6, 40; m 62; c 2. CHEMICAL INSTRUMENTATION. *Educ:* Wis State Univ-Eau Claire, BS, 65; Univ Wyo, MS, 67, PhD(physics), 70. *Prof Exp:* Electronic engr, 67-70, physicist, 70-74, dep div leader chem, 74-79, PROJ LEADER, LAWRENCE LIVERMORE NAT LAB, UNIV CALIF, 79- *Concurrent Pos:* Instr, Univ Calif, Berkeley, 68. *Mem:* Am Inst Physics; Optical Soc Am; Am Chem Soc. *Res:* Molecular spectroscopy, specifically in rotational & vibrational spectra, and in analytical instrumentation, including development of new systems and digital control automation; laser spectroscopy for analytical measurements; material science and corrosion; low density inorganic materials. *Mailing Add:* Lawrence Livermore Nat Lab L-325 PO Box 808 Livermore CA 94550

HRUBY, VICTOR J, b Valley City, NDak, Dec 24, 38; m 66; c 3. BIO-ORGANIC CHEMISTRY, BIOPHYSICS. *Educ:* Univ NDak, BS, 60, MS, 62; Cornell Univ, PhD(org chem), 65. *Hon Degrees:* Dr Honorus Causa, Free Univ Brussels, 89- *Prof Exp:* Teaching asst inorg & anal chem, Univ NDak, 60-61; teaching asst org chem, Cornell Univ, 62-63, instr biochem, Med Col, 65-67, res assoc bio-org chem, Cornell Univ, 67-68; from asst prof to assoc prof, 68-77, PROF CHEM, UNIV ARIZ, 77-, PROF BIOCHEM, 78-, REGENTS PROF, 89-; PROF, ARIZ RES LABS, 81- *Concurrent Pos:* Consult, Dow Chem Co, 73- & Nat Inst Health, 78-; Fulbright-Hays rector's lectr, Belg; mem, Physiol Chem Study Sect, NIH, 80-84, Biorg & Nat Prods Chem Study Sect, 85-89; ed, Int J Peptide & Protein Res, 88-; regents prof, Univ Ariz, 89- *Honors & Awards:* Javits Neurosci Award, 87. *Mem:* Fel AAAS; Am Chem Soc; fel Am Inst Chemists; Biophys Soc; Am Peptide Soc; fel NY Acad Sci; Am Soc Biol Chemists. *Res:* Synthesis, structure and properties of polypeptides, unusual amino acids and proteins; hormone and neurotransmitter chemistry; nuclear magnetic resonance as a structural tool for polypeptides; brain chemistry; use of stable isotopes in bio-organic and biophysical studies; brain chemistry; conformation-biological activity relationships; computer assisted molecular design. *Mailing Add:* Dept Chem Univ Ariz Tucson AZ 85721

HRUSCHKA, HOWARD WILBUR, b Brooklyn, NY, Sept 17, 15; m 40; c 3. HORTICULTURE, PLANT PHYSIOLOGY. *Educ:* Cornell Univ, BSA, 37. *Prof Exp:* Soil conservationist, Steuben Co, NY, Agr Res Serv, USDA, 38-44, plant physiologist, NY Mkt Lab, 46-54, Maine Potato Handling Res Ctr, Aroostook Farm, 54-57 & Qual, Maintenance & Improv Sect, Plant Indust Sta, 57-65, plant physiologist, 65-72, plant physiologist, Hort Crops Res Lab, Mkt Res Inst, 72-78; docent, Naturalist Ctr, Smithsonian Inst, Washington, DC, 78-80; COOPERATOR, PLANT INDUST STA, USDA, BELTSVILLE, MD, 80- *Concurrent Pos:* Civilian, Pub Serv, 44-46. *Honors & Awards:* Species of Miocene fossil porpoise named in honor, Rhabdosteus hruschkai. *Mem:* Am Soc Hort Sci; fel AAAS. *Res:* Handling, transportation and storage of horticultural crops; post harvest plant physiology paleobiological collection; nature study education; anthropology. *Mailing Add:* 9710 Wichita Ave College Park MD 20740

HRUSHESKY, WILLIAM JOHN MICHAEL, b Poughkeepsie, NY, Nov 9, 47; m 85; c 1. MEDICAL CHRONOBIOLOGY OF CANCER. *Educ:* Syracuse Univ, AB, 69; State Univ NY, Buffalo, MD, 73. *Prof Exp:* Intern med, Baltimore City Hosps, Johns Hopkins Univ, 73-74; assoc, Nat Cancer Inst, NIH, 74-76; resident med, Lab Med & Path, Univ Minn, 76-78, res specialist oncol, 78-79, from asst prof to assoc prof oncol, 79-89; ASSOC PROF MED, DIV MED ONCOL, ALBANY MED COL, 89- *Concurrent Pos:* Fel, dept internal med, Johns Hopkins Univ, 73-74; instr, dept internal med, George Washington Univ, 74-76 & oncol sect, Vet Admin, Wash, 75-76; staff mem, Georgetown Univ Hosp, 74-76; ed, Pract Appln Chronobiol to Cancer; prin investr grants, Nat Cancer Inst, 82-, Medtronic Corp, 85- & Nat Heart, Lung & Blood Inst, 86. *Mem:* Int Soc Chronobiol; Am Col Physicians; Am Soc Clin Oncol; Am Asn Cancer Res; NY Acad Sci; Am Fedn Clin Res. *Res:* Time structure of the universe; fundamental resonances of living things; development of time-based automated programmable drug delivery systems. *Mailing Add:* Div Med Oncol Albany Med Col 47 New Scotland Ave Albany NY 12208

HRUSKA, ANTONIN, b Prague, Czech, Apr 16, 34; Can citizen. SPACE PHYSICS, PLASMA PHYSICS. *Educ:* Charles Univ, Prague, dipl physics, 57; Czech Acad Sci, PhD(astrophysics), 60. *Prof Exp:* Res scientist astrophysics & geophysics, Czech Acad Sci, Prague, 57-67; vis scientist space physics, Univ BC, 67-69, Univ Alberta, 69-70 & Nat Ctr Atmospheric Res, 70-71; res officer space physics, Nat Res Coun Can, 71-90; RETIRED. *Mem:* Am Geophys Union; Europ Geophys Soc. *Res:* Structure of the magnetosphere and magnetotail; applications of plasma physics in the magnetosphere; instabilities in collisionless plasmas. *Mailing Add:* 400 Plumtree Cr Ottawa ON K1K 2N3 Can

HRUSKA, SAMUEL JOSEPH, b Detroit, Mich, Dec 13, 36; m 59; c 1. MATERIALS SCIENCE, METALLURGICAL ENGINEERING. *Educ:* Purdue Univ, BS, 59; Carnegie Inst Technol, MS, 62, PhD(metall eng), 63. *Prof Exp:* NSF fel physics, Bristol Univ, 63-64; from asst prof to assoc prof mat sci & metall eng, 64-74, PROF MAT ENG, PURDUE UNIV, 75- *Mem:* Metall Soc; Am Soc Eng Educ; Am Soc Metals. *Res:* Kinetics of phase transformations; formation of thin films; surface and interfacial phenomena; carburizing; brazing; cermets. *Mailing Add:* Dept Mat Sci Purdue Univ West Lafayette IN 47907

HRUTFIORD, BJORN F, b Blaine, Wash, Jan 31, 32; m 60; c 2. CHEMISTRY. *Educ:* Wash State Univ, BS, 54; Univ NC, PhD(org chem), 59. *Prof Exp:* Res instr chem eng, 59-65, asst prof wood chem, 65-73, assoc prof, 65-77, PROF WOOD CHEM & CHMN DIV PHYS SCI, COL FORESTRY, UNIV WASH, 77- *Mem:* Royal Soc Chem. *Res:* Lignin and extractives chemistry. *Mailing Add:* Col Forrest Resources Univ Wash AR Ten Seattle WA 98105

HRUZA, ZDENEK, b Prague, Czech, Oct 3, 26; US citizen; m 51; c 2. PHYSIOLOGY, GERONTOLOGY. *Educ:* Charles Univ, Prague, MD, 50; Czech Acad Sci, PhD(physiol), 55, ScD(physiol), 66. *Prof Exp:* Instr exp path, Med Sch, Charles Univ, Prague, 51-53; assoc scientist metab, Inst Human Nutrit, Prague, 53-56; head dept exp aging res, Czech Acad Sci, 56-66; RES PROF PATH, SCH MED, NY UNIV, 66- *Mem:* Am Physiol Soc; fel Geront Soc. *Res:* Aging; metabolism of lipids during aging; calcifications in aging; aging of collagen. *Mailing Add:* 499 Lakeshore Dr Hewitt NJ 07421

HRYCAK, PETER, b Przemysl, Poland, July 8, 23; US citizen; m 49; c 4. HEAT TRANSFER, FLUID MECHANICS. *Educ:* Univ Minn, BS, 54, MS, 55, PhD(mech eng), 60. *Prof Exp:* Teaching asst mech eng, Univ Minn, 54-55, instr, 55-60; mem tech staff, Bell Tel Labs, 60-65; sr proj engr, Curtiss-Wright Corp, 65; from assoc prof to prof mech eng, Newark Col Eng, 65-75; PROF MECH ENG, NJ INST TECHNOL, 75- *Concurrent Pos:* NASA res grant heat transfer from impinging jets, 67-68; NSF res grant, 82-85. *Mem:* Am Inst Aeronaut & Astronaut; Am Soc Mech Engrs; NY Acad Sci; Am Chem Soc; Ukrainian Engrs Soc Am (pres, 66-67); Am Geophys Union; sr mem Inst Environ Sci; Sigma Xi; Am Soc Eng Educ. *Res:* Heat transfer with a change of phase; heat transfer from impinging jets; effects of space environment on satellites; boundary layer theory applied to impinging jets; mechanics of transition from laminar to turbulent flow; problems related to thermal and chemical pollution; carbon dioxide balance of the atmosphere; principal designer of the telstar satellite thermal system. *Mailing Add:* 19 Roselle Ave Cranford NJ 07016

HRYNIUK, WILLIAM, b Central Patricia, Ont, Apr 29, 39; m 59; c 2. HEMATOLOGY, ONCOLOGY. *Educ:* Univ Man, MD, 61; FRCPS(C), 79. *Prof Exp:* Resident med, Wash Univ, 64-65; asst prof, Univ Man, 69-77, assoc prof med, 77-79; PROF MED, MCMASTER UNIV, HAMILTON, ONT, 79- *Concurrent Pos:* Clin fel hemat, Wash Univ, 65-66; Leukemia Soc Am fel clin pharmacol, Yale Univ, 66-68; investr, Nat Cancer Inst Can, 69-; attend physician & consult hematologist, Winnipeg Gen Hosp, 69-79; dir, Ont Cancer Treatment & Res Found & Hamilton Regional Ctr, Hamilton, Ont, 79- *Mem:* Can Soc Clin Invest; Can Hemat Soc; Am Soc Clin Oncol; Am Asn Cancer Res; Nat Cancer Inst. *Res:* Mechanisms of action of antifolates; chemotherapy of viral infections in man; fibrinolytic therapy; dose intensity in chemotherapy. *Mailing Add:* Ont Cancer Found Hamilton Clin 711 Concession St Hamilton ON L8V 1C3 Can

HSI, BARTHOLOMEW P, b Shanghai, China, Dec 10, 25; m 64; c 2. BIOMETRICS. *Educ:* Univ Minn, MA, 62, PhD(biostatist), 64. *Prof Exp:* Asst prof biomet, Case Western Reserve Univ, 64-70; assoc prof, 70-75, PROF BIOMET, UNIV TEX SCH PUB HEALTH, HOUSTON, 75 - *Concurrent Pos:* Consult, Vet Admin Coop Study Prog, 74 - *Mem:* Am Pub Health Asn; Am Soc Clin Pharmacol & Therapeut; Am Statist Asn; Biomet Soc. *Res:* Laboratory quality control; applied and theoretical statistics. *Mailing Add:* Dept Biomet Univ Tex Health Sci Ctr Houston TX 77225

HSI, DAVID CHING HENG, b Shanghai, China, May 17, 28; nat US; m 52; c 2. PLANT PATHOLOGY, GENETICS. *Educ:* St John's Univ, BS, 48; Univ Ga, MS, 49; Univ Minn, PhD, 51. *Prof Exp:* PROF, NMEX STATE UNIV, 68- *Concurrent Pos:* Oilseed consult, WPakistan, US Dept State, 70; vis scientist, NSF, Taiwan, Repub China, 79 & Argentina, 80; res fel, Dept Agr & Fisheries, Repub SAfrica, 81; vis scientist, Min Agr, Animal Husbantry & Fisheries, People's Repub China, 85. *Honors & Awards:* Distinguished Res Award, NMex State Univ, Col Agr, 71; Am Phytopath Soc; Am Peanut Res & Educ Asn; Sigma Xi; Distinguished Scientist, NMex Acad Sci. *Mem:* Am Peanut Res Educ Soc (pres, 81-82); Sigma Xi; NMex Acad Sci (pres, 83, 84); fel AAAS. *Res:* Peanut and sweet potato disease and improvement. *Mailing Add:* Agr Sci Ctr NMex State Univ 1036 Miller St SW Los Lunas NM 87031

HSI, RICHARD S P, b Shanghai, China, Sept 18, 33; US citizen; m 58; c 4. ORGANIC CHEMISTRY. *Educ:* Pomona Col, BA, 55; Mass Inst Technol, PhD(org chem), 58. *Prof Exp:* Res fel org chem, Mass Inst Technol, 58-59; RES ASSOC ORG CHEM, UPJOHN CO, 59- *Mem:* Am Chem Soc. *Res:* Organic synthesis; radioisotope labelling; drug metabolism; reaction mechanism; medicinal chemistry. *Mailing Add:* 2721 Parkwyn Dr Kalamazoo MI 49001

HSIA, HENRY TAO-SZE, b Peking, China, June 16, 23; US citizen; m 47; c 3. FLUID MECHANICS, ENERGY SOURCES. *Educ:* Chiao Tung Univ, BS, 44; Harvard Univ, MS, 48; Stanford Univ, Engr, 63, PhD(astronaut, aeronaut), 66. *Prof Exp:* Tech specialist, Nat Resources Comn, Repub of China, 45-47; design engr, Consol Edison Co NY, 48-50; sr design engr, Ebasco Serv Inc, 50-56; mech engr, atomic power equip dept, Gen Elec Co, 56-57, prog mgr, 76-83; res scientist, Lockheed Missile & Space Co, 57-62; sr staff scientist, United Technol Ctr, United Aircraft Corp, 62-73; sr staff engr, MB Assocs, 74-75; PRES, TECON SERV, 83- *Concurrent Pos:* Guest lectr, Stanford Univ, 67; consult, MB Assocs, 75-76, Bechtel Power Corp, 82-84, Sci & Technol Ctr, Jiangsu Prov, China, 85- *Mem:* Am Inst Aeronaut & Astronaut; Chinese Inst Engrs; Chinese Cult Asn (dir & exec secy, 66-); Am Nuclear Soc. *Res:* Propulsion; thermodynamics; heat transfer; nuclear reactor plant design; alternate energy source; rocket propulsion. *Mailing Add:* 865 Robb Rd Palo Alto CA 94306

HSIA, JACK JINN-GOE, b Anhwei, China, Oct 1, 37; US citizen; m 67; c 2. THERMAL PHYSICS, SPECTROPHOTOMETRY. *Educ:* Nat Taiwan Univ, BS, 59; Purdue Univ, MS, 64, PhD(thermal physics), 68. *Prof Exp:* Res asst thermal physics, Thermal Phys Properties Res Ctr, Purdue Univ, 68-69; physicist, 69-78, GROUP LEADER SPECTROPHOTOM, NAT BUR STANDARDS, 78- *Honors & Awards:* Bronze Medal, US Dept Com, 88. *Mem:* Optical Soc Am; Am Soc Mech Engrs; Am Soc Testing & Mat; Int Comn Illum; Inter-Soc Color Coun. *Res:* Experimental instrumentation and analytical method on reflection, scattering, transmission, retroreflection, densitometry, spectral fluorimetry and infrared methods. *Mailing Add:* Nat Inst Standards & Technol Rm B306 Bldg 220 Gaithersburg MD 20899

HSIA, JOHN S, b Shanghai, China, Dec 16, 38; US citizen; m 68; c 2. NUMBER THEORY, ABSTRACT ALGEBRA. *Educ:* Brown Univ, ScB & AB, 62; Mass Inst Technol, PhD(math), 66. *Prof Exp:* From asst prof to assoc prof math, 66-76, PROF MATH, OHIO STATE UNIV, 76- *Concurrent Pos:* Prin investr, NSF, 67-; vis asst prof math, Mass Inst Technol, 67; Alexander von Humboldt fel, WGer, 68-69; vis assoc prof math, Univ Calif, Santa Barbara, 75; hon res assoc, Harvard Univ, 78; distinguished lectr, Univ Southern Calif, 79, vis prof, 87; mem, Math Scis Res Inst, Berkeley, Calif, 86; ed, J Number Theory. *Mem:* Am Math Soc; Sigma Xi. *Res:* Arithmetic theory of integral quadratic forms over global fields; modular forms; combinatorial designs and coding theory. *Mailing Add:* Dept Math Ohio State Univ 231 W 18th Ave Columbus OH 43210

HSIA, MONG TSENG STEPHEN, b Shanghai, China, Sept 5, 46; m 71; c 2. BIOCHEMICAL TOXICOLOGY, CHEMICAL CARCINOGENESIS. *Educ:* Cheng Kung Univ, Taiwan, BS, 68; Univ Calif, San Diego, PhD(org chem), 74. *Prof Exp:* Res asst org chem, Univ Calif, San Diego, 69-74; from asst prof to prof toxicol, Dept Entom-Environ Toxicol Ctr, Univ Wis-Madison, 78-87; TOXICOLOGIST, HAZARDOUS SUBSTANCES EVAL GROUP, MITRE CORP, 87- *Concurrent Pos:* Prin investr, Am Cancer Soc grant, 76-77 & Nat Inst Environ Health Sci, 78-84; co-prin investr, Nat Cancer Inst grant, 77-80 & Nat Inst Environ Health Sci, 78-84; trainer, Nat Inst Environ Health Sci, 80-87; consult, Nat Sci Coun, Repub China, 80-82. *Honors & Awards:* Young Environ Health Scientist Award, Nat Inst Environ Health Sci, 78. *Mem:* AAAS; Soc Toxicol; Am Col Toxicol; Soc Environ Toxicol Chem. *Res:* Chemical carcinogenesis; mammalian toxicology of halogenated pesticides; insect growth regulators; risk assessment of synthetic chemicals; in vitro toxicity testing systems; hazardous substances evaluation. *Mailing Add:* 8313 Tuckermane Lane Potomac MD 20854-3747

HSIA, SUNG LAN, b China, Feb 6, 20; m 55; c 4. BIOCHEMISTRY. *Educ:* Cath Univ, Peiping, BS, 44; St Louis Univ, PhD(biochem), 52. *Prof Exp:* Fel biochem, St Louis Univ, 52-58, from sr instr to asst prof biochem, 58-63; assoc prof, 63-67, PROF DERMAT & BIOCHEM, SCH MED, UNIV MIAMI, 67- *Concurrent Pos:* NIH career develop award, 65-70. *Mem:* AAAS; Am Chem Soc; Am Soc Biol Chem; Soc Invest Dermat; Am Oil Chem Soc. *Res:* Chemistry and metabolism of sterols, bile acids and steroid hormones; biochemistry of the skin; atherosclerosis, serum cholesterol binding reserve. *Mailing Add:* Dept Dermat Rm R-117 Univ Miami Sch Med P O Box 016960 Miami FL 33101

HSIA, YUKUN, b Kunming, China, Feb 21, 41; US citizen; m 65; c 2. SOLID STATE ELECTRONICS, MICROELECTRONICS. *Educ:* Univ Calif, Berkeley, BS, 61; Univ Calif, Los Angeles, MS, 64, PhD(solid state electronics), 69. *Prof Exp:* Design leader circuits, Electronics Div, NCR, 61-67; mem, Tech Staff Infrared Mat, NAm Aviation, 67-68; prin investr nonvolatile memories, Guidance & Control Div, Litton Industs, 69-74; br chief advan microelectronics, McDonnell Douglas Corp, 74-82; dir eng mgr microprocessors, Fairchild-Schlumberger corp, 82-87; MGR MICROELECTRONICS & INFRARED TECHNOL, NORTHROP CORP, 87- *Concurrent Pos:* Mem standards comt, Inst Elec & Electronics Engrs, 78-88, prog chair, Computer Elements Workshop, 84; lectr, Univ Southern Calif, Los Angeles, 79-82, Santa Clara Univ, 82-87; vis scholar, Acad Sinica, 84. *Mem:* nonvolatile memories. *Res:* Silicon machine architecture and neural machines; Nine United States patents. *Mailing Add:* 1821 Park Skyline Rd Santa Ana CA 92705

HSIA, YU-PING, b Tungtai, Kiangsu, China, May 16, 36; m 61; c 2. PHYSICAL CHEMISTRY. *Educ:* Tunghai Univ, BS, 59; Univ Calif, Santa Barbara, MA, 63; Ill Inst Technol, PhD(phys chem), 67. *Prof Exp:* Asst prof chem, Univ Bridgeport, 67-68; from asst prof to assoc prof, 68-76, PROF CHEM, CALIF STATE POLYTECH UNIV, 76- *Concurrent Pos:* Vis assoc prof, Res Inst Chem, Tsing Hua Univ, Taiwan, 69-70; vis assoc, Calif Inst Technol, 71. *Mem:* Am Chem Soc; Chinese Chem Soc. *Res:* Molecular orbital calculations in inorganic compounds; coal chemistry; environmental chemistry. *Mailing Add:* Dept Chem Calif State Polytech Univ Pomona CA 91768

HSIANG, THOMAS Y, b Taiwan, China, Aug 4, 48; m; c 2. SOLID STATE ELECTRONICS, SUPERCONDUCTIVITY. *Educ:* Nat Taiwan Univ, BS, 70; Univ Calif, Berkeley, MA, 73, PhD(physics), 77. *Prof Exp:* Res assoc physics, Ames Lab, US Dept Energy, 77-79; asst prof, Ill Inst Technol, 79-81; ASST PROF ELEC ENG, UNIV ROCHESTER, 81- *Mem:* Am Phys Soc; Inst Elec & Electronics Engrs. *Mailing Add:* Dept Elec Eng Univ Rochester Rochester NY 14627

HSIAO, BENJAMIN S, b Taipei, Taiwan, Aug 12, 58; US citizen; m 87. STRUCTURE-PROPERTY-PROCESSING RELATIONS IN POLYMERS, POLYMER CRYSTALLIZATION & MORPHOLOGY. *Educ:* Nat Taiwan Univ, BS, 80; Univ Conn, MS, 85, PhD(polymer sci), 87. *Prof Exp:* Postdoctoral fel, Dept Polymer Sci & Eng, Univ Mass, 87-88; res engr, 88-90, SR ENGR, FIBERS RES & DEVELOP, DU PONT CO, 90- *Mem:* Am Chem Soc; Am Phys Soc; Soc Rheol; Soc Plastics Engrs; Mat Res Soc. *Res:* Polymer physics; structure-property-processing relations in polymers and advanced polymer composites. *Mailing Add:* Exp Sta Du Pont Co PO Box 80302 Wilmington DE 19880-0302

HSIAO, CHIH CHUN, b Peking, China, Oct 23, 19; m 53; c 3. ENGINEERING MECHANICS, PHYSICS. *Educ:* Yenching & Tsinghua Univs, BS, 41; Mass Inst Technol, SM, 44, PhD(physics, appl mech), 48. *Prof Exp:* From asst prof to assoc prof eng mech, Pa State Col, 47-53; prin res physicist, Jones & Laughlin Steel Corp, 53-55; assoc prof aeronaut & eng mech, 55-66, PROF AEROSPACE ENG & MECH, UNIV MINN, MINNEAPOLIS, 66- *Concurrent Pos:* Vis prof, Cambridge Univ, 68-69. *Mem:* Am Phys Soc. *Res:* Physics and mechanics of solids; fracture and strength of polymers; viscoelasticity; fracture mechanics; biomechanics. *Mailing Add:* Dept Aerospace Eng & Mech Univ Minn Minneapolis MN 55455

HSIAO, GEORGE CHIA-CHU, b Shanghai, China, Sept 9, 34; US citizen; m 63; c 2. APPLIED MATHEMATICS, CIVIL ENGINEERING. *Educ:* Nat Taiwan Univ, BS, 58; Carnegie Inst Technol, MS, 62; Carnegie-Mellon Univ, PhD(math), 69. *Prof Exp:* Instr math, Carnegie-Mellon Univ, 67-69; from asst prof to assoc prof, 69-77, PROF MATH, UNIV DEL, 77- *Concurrent Pos:* Guest prof math, T H Darmstadt, 75-76, Free Univ, Berlin, 79-80, 86; Alexander von Humboldt fel, Fed Repub Ger, 75-76, 83; vis prof, Univ Concepcion, Chile, 85. *Mem:* Sigma Xi; Soc Indust & Appl Math; Am Math Soc. *Res:* Partial differential equations; integral equations and applied mathematics. *Mailing Add:* Dept Math Sci Univ Del Newark DE 19716

HSIAO, HENRY SHIH-CHAN, b Chungking, China, Oct 12, 43; US citizen; c 3. BIOMEDICAL ENGINEERING, ENTOMOLOGY. *Educ:* Mass Inst Technol, BS, 65, EE, 67; Univ Calif, Berkeley, PhD(elec eng), 72. *Prof Exp:* Asst prof, 72-77, ASSOC PROF SURG & BIOMED ENG, UNIV NC, 77- *Concurrent Pos:* Asst prof neurobiol, Univ NC, 73-; NSF grant, 75-; consult, Environ Protection Agency & Northrup Serv, Inc, 77- *Mem:* Entomol Soc Am; Biomed Eng Soc. *Res:* Biomedical instrumentation; insect behavior. *Mailing Add:* Univ NC Chapel Hill Sch Med Univ NC Chapel Hill NC 27514

HSIAO, MU-YUE, b Changsha, Hunan, China, July 17, 33; nat US; m 62; c 3. COMPUTER SCIENCE. *Educ:* Nat Taiwan Univ, BS, 56; Univ Ill, MS, 60; Univ Fla, PhD(elec eng), 67. *Prof Exp:* Jr engr, Int Bus Mach Co, 60-61, assoc engr, 61-63, sr assoc engr, 63-65; res assoc elec eng, Univ Fla, 65-67; adv engr, 67-69, sr engr, 69-79, sr tech staff & mgr, 79-84, FEL, IBM CORP, 84- *Mem:* Fel Inst Elec & Electronics Engrs; NY Acad Sci. *Res:* Error detecting and correcting codes; switching theory and logic design; computer hardware error checking; fault detection and testing; computer maintainability. *Mailing Add:* IBM South Rd Box 390 Poughkeepsie NY 12602

HSIAO, SIDNEY CHIHTI, developmental biology; deceased, see previous edition for last biography

HSIAO, THEODORE CHING-TEH, b Peiping, China, Nov 28, 31; m 57; c 2. PLANT PHYSIOLOGY, ECOPHYSIOLOGY. *Educ:* Cornell Univ, BS, 55; Univ Conn, MSc, 60; Univ Ill, PhD(crop physiol, biochem), 64. *Prof Exp:* Jr chemist, Arrow Lacquer Corp, 55-56; anal chemist, S B Penick & Co, 58; res asst soil chem, Univ Conn, 58-60; asst res plant physiologist, Univ Calif, Los Angeles, 63-65; from asst prof & asst plant physiologist to assoc prof & assoc physiologist, 65-74, PROF WATER SCI & PLANT PHYSIOLOGIST, EXP STA, UNIV CALIF, DAVIS, 74- *Concurrent Pos:* Vis scientist, Int Rice Res Inst, Philippines, 78-79. *Mem:* Am Soc Plant Physiol; Am Soc Agron; Crop Sci Soc Am. *Res:* Plant-water relations and stress physiology; interaction of plants with hydroenvironment and the underlying physiological and physical processes; metabolism of nucleic acids in plants; biological regulation; water-plant relations. *Mailing Add:* Land Air & Water Resources Univ Calif Davis CA 95616

HSIAO, TING HUAN, b Hangchow, China, Feb 6, 36; m 61. INSECT PHYSIOLOGY. *Educ:* Taiwan Prov Col Agr, BSc, 57; Univ Minn, St Paul, MSc, 61; Univ Ill, Urbana, PhD(insect physiol), 66. *Prof Exp:* Res assoc insect physiol, Dept Entom, Univ Ill, Urbana, 66- 67; from asst prof to assoc prof, 67-79, PROF ENTOM, UTAH STATE UNIV, 79- *Concurrent Pos:* Vis prof, Entom Lab, State Agr Univ, Wageningen, Neth, 74- 75, 77, 78, 81 & 87. *Mem:* AAAS; Ecol Soc Am; Am Inst Biol Scientist; Entom Soc Am; Sigma Xi. *Res:* Physiological and ecological adaptations of insects, especially the chemical relationships in host finding and food selection of phytophagous insects; biotypes of insect pests; physiology, feeding behavior and ecological genetics. *Mailing Add:* Dept Biol Utah State Univ Logan UT 84322

HSIE, ABRAHAM WUHSIUNG, b Hsinwu, Taiwan, Mar 3, 40; m 62; c 2. GENETICS, TOXICOLOGY. *Educ:* Nat Taiwan Univ, Taipei, BS, 62; Ind Univ, Bloomington, MA, 65, PhD(microbiol), 68. *Prof Exp:* Res assoc, Nat Jewish Hosp, Denver, 68-69; NIH fel & instr biophys & genetics, Med Ctr, Univ Colo, 69-71, asst prof, 71-72; lectr, 72-76, group leader somatic cell genetics, 72-77, group leader, Mammalian Cell Genetic Toxicol, 77-84, Biol Div, Oak Ridge Nat Lab, 77-86, prof, Univ Tenn-Oak Ridge Grad Sch Biomed Sch, 76-86; PROF & ASSOC DIR, DIV ENVIRON TOXICOL, DEPT PREVENTIVE MED & COMMUNITY HEALTH, UNIV TX MED BR, GALVESTON TX, 86- *Honors & Awards:* Fel, Japan Soc Promotion Sci, 77; Distinguished Alumni Serv Award, Indiana Univ, Bloomington; Distinguished Vis Scientist, US Environ Protection Agency, 85-88. *Mem:* Am Soc Biol Chemists; Am Asn Cancer Res; Am Soc Cell Biol; Environ Mutagen Soc; Genetics Soc Am; Sigma Xi; Soc Risk Anal; Am Soc Microbiol; Am Soc Photobiol; Soc Toxicol; Radiation Res Soc; Japan Soc Prom Sci, 77. *Res:* Quantitative mammalian cell mutagenesis; environmental toxicology; Molecular epidemiology and biological desimetry growth regulation of mammalian cells. *Mailing Add:* Dept Prev Med Univ Tex Med Br Galveston TX 77550

HSIEH, CHUNG KUO, b Shanghai, China, May 29, 32; US citizen; m 64; c 2. THERMODYNAMICS, HEAT TRANSFER. *Educ:* Chinese Naval Col Technol, Taiwan, BSME, 54; Purdue Univ, MSME, 64, PhD(mech eng), 68. *Prof Exp:* From asst engr to engr, Chinese Naval Shipyards, Taiwan, 54-62; res asst thermophys properties res ctr, Purdue Univ, 62-66, res instr, 66-68; asst prof mech eng, SDak Sch Mines & Technol, 68-69; from asst prof to assoc prof, 69-79, PROF MECH ENG, UNIV FLA, 79- *Concurrent Pos:* Consult, Argonne Nat Lab, 76-; vis prof, Tientsin Univ, China, 80. *Mem:* Am Soc Mech Engrs; Am Inst Aeronaut & Astronaut. *Res:* Heat transfer, thermophysical properties; solar energy; energy conversion and conservation; engineering optics; infrared scanning; nondestructive testing; computer system modeling; thermodynamics; atmospheric science. *Mailing Add:* Dept Mech Eng Univ Fla Gainesville FL 32611

HSIEH, DENNIS P H, b Nantow, Taiwan, Feb 8, 37; m 60; c 3. ENVIRONMENTAL TOXICOLOGY, APPLIED MICROBIOLOGY. *Educ:* Nat Taiwan Univ, BA, 59; Mass Inst Technol, MS, 67, DSc(biochem eng), 69; Am Bd Toxicol, cert, 79. *Prof Exp:* Food technologist, Wei-Chuan Foods Inc, Taiwan, 61-64; lectr & asst biochem engr, 69-73, asst prof environ toxicol, 73-76, assoc prof, 76-80, PROF ENVIRON TOXICOL, UNIV CALIF, DAVIS, 80- *Concurrent Pos:* Consult, NIH, WHO, Stauffer Chem Co & City San Diego. *Mem:* Am Chem Soc; Am Soc Microbiol; Chinese Agr Chem Soc; Soc Toxicol. *Res:* Control of fungal toxigenicity and comparative toxicology of carcinogenic mycotoxins. *Mailing Add:* 726 Lake Terrace Circle Davis CA 95616

HSIEH, DIN-YU, b Nanking, China, Mar 25, 33; m 58; c 2. APPLIED MATHEMATICS, FLUID MECHANICS. *Educ:* Nat Taiwan Univ, BS, 54; Brown Univ, MSc, 57; Calif Inst Technol, PhD(eng sci), 60. *Prof Exp:* Res fel appl mech, Calif Inst Technol, 60-63, asst prof eng sci, 63-68; assoc prof, 68-78, PROF APPL MATH, BROWN UNIV, 78-; PROF MATH, HONG KONG UNIV SCI & TECHNOL, 90- *Mem:* Soc Indust & Appl Math; Am Phys Soc. *Res:* Two phase flows; hydrodynamical stabilities; asymptotic method on nonlinear problems; chaos. *Mailing Add:* Hong Kong Univ Sci Technol Clear Water Bay Rd Kowloon Hong Kong

HSIEH, HENRY LIEN, b Shanghai, China, Jan 17, 30; nat US; m 55; c 3. POLYMER CHEMISTRY. *Educ:* Univ Akron, BS, 54; Princeton Univ, MA, 56, PhD(chem), 57. *Prof Exp:* Asst, Princeton Univ, 54-56; explor res chemist, 57-59, group leader, 59-63, sect mgr, 63-79, sr res assoc, 79-83, SR SCIENTIST, PHILLIPS PETROL CO, 83- *Concurrent Pos:* Tech liaison, China. *Mem:* AAAS; Am Chem Soc; NY Acad Sci; Am Inst Chemists; Sigma Xi; Soc Plastics Eng. *Res:* Solution and stereospecific polymerizations; organo-metallic chemistry; rubber chemistry and technology; reaction mechanisms; polyolefino polymerization catalysts and processes; Ziegler-Natta catalysis; anionic polymerizations. *Mailing Add:* 1406 Meadow Lane Bartlesville OK 74006

HSIEH, HSUNG-CHENG, b Taichung, Taiwan, Feb 24, 29; US citizen; m 60; c 1. ELECTRICAL ENGINEERING, PLASMA PHYSICS. *Educ:* Dartmouth Col, AB, 54; Calif Inst Technol, MS, 55; Stanford Univ, EE, 57; Univ Calif, Berkeley, PhD(appl math), 60. *Prof Exp:* Res engr, Huggins Labs, Inc, Calif, 57; asst prof elec eng, Univ Wichita, 59-61; assoc prof, Univ Iowa, 61-64; vis assoc prof & scientist, Inst Sci & Technol, Univ Mich, Ann Arbor, 64-66, res engr & lectr, Electron Physics Lab, 66-68; assoc prof, 68-70, PROF ELEC ENG, IOWA STATE UNIV, 70- *Mem:* Am Geophys Union. *Res:* Electromagnetic field theory and its application; microwave electronic devices; electromagnetic wave interaction with plasmas; solid-state plasma phenomena. *Mailing Add:* Dept Elec Eng Iowa State Univ 201 Coover Ames IA 50011

HSIEH, HUI-KUANG, b Taiwan, Repub China, Feb 3, 44; m 72; c 2. STATISTICS. *Educ:* Fu-jen Univ, BS, 67; Nat Taiwan Univ, MA, 70; Columbia Univ, MS, 72; Univ Wis, PhD(statist), 76. *Prof Exp:* Statistician, Am Home Prod Corp, 71-73; ASSOC PROF STATIST, UNIV MASS, 76- *Concurrent Pos:* Vis expert, Inst Statist, Acad Sci, 82-83. *Mem:* Inst Math Statist; Am Statist Asn; Am Soc Qual Control. *Res:* Statistical inference; nonparametric statistics; multivariate analysis. *Mailing Add:* Dept Math & Statist Univ Mass Amherst MA 01003

HSIEH, JEN-SHU, health physics, medical physics, for more information see previous edition

HSIEH, JUI SHENG, b Chungking, China, Mar 5, 21; m 61; c 3. MECHANICAL ENGINEERING. *Educ:* Wuhan Univ, China, BS, 43; Univ Ky, MS, 50; Ohio State Univ, PhD(mech eng), 55. *Prof Exp:* From asst prof to assoc prof mech eng, Univ Bridgeport, 55-60; assoc prof, 60-65, PROF MECH ENG, NJ INST TECHNOL, 65- *Mem:* Am Soc Mech Engrs; Am Soc Eng Educ; Am Solar Energy Soc. *Res:* Thermodynamics; solar energy. *Mailing Add:* Dept Mech Eng NJ Inst Technol Newark NJ 07102

HSIEH, KE CHIANG, b Chungking, China, June 14, 40; m 68; c 2. COSMIC RAY PHYSICS. *Educ:* Wabash Col, BA, 63; Univ Chicago, PhD(physics), 69. *Prof Exp:* Res assoc cosmic ray physics, Enrico Fermi Inst, Univ Chicago, 69-71; asst prof, 71-76, ASSOC PROF PHYSICS, UNIV ARIZ, 76- *Mem:* AAAS; Am Phys Soc; Sigma Xi; Am Geophys Union. *Res:* Cosmic rays, energetic particles from the sun; space science; astrophysics. *Mailing Add:* Dept Physics Univ Ariz Tucson AZ 85721

HSIEH, MONICA C, m. NUTRITION, ENZYMOLOGY. *Educ:* Wash State Univ, PhD(nutrit), 75. *Prof Exp:* RES INSTR PEDIAT RES, CHILDREN'S HOSP RES FOUND CINCINNATI, 75- *Concurrent Pos:* Clin Lab. *Mem:* Am Inst Nutrit; Asn Clin Scientists. *Res:* Proteolytic defects of Cystic Fibrosis. *Mailing Add:* Children's Hosp Res Found IDR 526 Elland & Bethesda Ave Cincinnati OH 45229

HSIEH, PAUL YAO TONG, b Taichung, Taiwan, China, Oct 4, 27; US citizen; m 52; c 4. SURFACE CHEMISTRY, CHEMICAL ENGINEERING. *Educ:* Nat Taiwan Univ, BS, 50; Kans State Univ, MS, 57; Rensselaer Polytech Inst, PhD(chem eng), 59. *Prof Exp:* Asst, Nat Taiwan Univ, 50-54, instr, 54-56; sr res chemist, Nat Cash Register Co, Ohio, 59-65; res staff, Westvaco, 65-68; staff scientist, Olivetti Corp Am, 68-72; sect head, Hughes Aircraft Co, 73-75, dept mgr, 76-78, sr staff engr, 78-83, sr proj engr, 83-88, scientist & engr, 88-89; MEM TECH STAFF, ROCKWELL INT, 89- *Mem:* Am Chem Soc. *Res:* Colloid and surface chemistry; liquid crystals; display devices; microelectronics; infrared detector systems. *Mailing Add:* 3862 Banyan St Irvine CA 92714

HSIEH, PHILIP KWOK-YOUNG, ENZYME PURIFICATION, ENZYME KINETICS. *Educ:* Univ Colo, PhD(biol), 78. *Prof Exp:* SR SCIENTIST CHEM ENG, BIOPURE INC, 84- *Mailing Add:* Amgen Inc 1900 Oak Terrace Lane Thousand Oaks CA 91320

HSIEH, PO-FANG (PHILIP), b Tainan, Formosa, July 10, 34; m 61; c 2. MATHEMATICS. *Educ:* Nat Taiwan Univ, BSc, 57; Univ Minn, MSc, 61, PhD(math), 64. *Prof Exp:* From asst prof to assoc prof math, Western Mich Univ, 64-73; res mathematician, US Naval Res Lab, 70-71; PROF MATH, WESTERN MICH UNIV, 73- *Concurrent Pos:* NSF res grant, 67-71. *Honors & Awards:* Resl Award, US Naval Res Lab, 72. *Mem:* Am Math Soc; Math Asn Am; Soc Indust & Appl Math. *Res:* Asymptotic solutions and turning point problem of ordinary differential equations. *Mailing Add:* Dept Math & Statist Western Mich Univ Kalamazoo MI 49008

HSIEH, RICHARD KUOCHI, b China, June 7, 32; US citizen; m 60; c 2. PUBLIC HEALTH ADMINISTRATION, OPERATIONS RESEARCH. *Educ:* Johns Hopkins Univ, BES, 57, MS, 61, MPH, 62, DrPH, 66. *Prof Exp:* Indust engr, Johns Hopkins Hosp, 59-64; res assoc oper, Johns Hopkins Univ, 64-66; chief health serv res, USPHS Hosp, Baltimore, 66-80; admin health sci, extramural res prog, NCI, 81-86, DIR INT PROG, NLM, NIH, 86- *Concurrent Pos:* Sr consult, Social Security Admin, 65-66; consult, Pan Am Health Orgn, 70-71; mem task force, Nat Acad Eng, 70-72. *Mem:* Fel Am Med Informatics Asn; Oper Res Soc Am; fel Am Pub Health Asn. *Res:* Testing technological and innovative concepts in the delivery of medical care; analysis and evaluation of organizational and operational management problems in the hospitals and clinics. *Mailing Add:* 601 Stacy Ct Baltimore MD 21204

HSIEH, SHIH-YUNG, physics, engineering, for more information see previous edition

HSIEH, YOU-LO, b Taipei, Taiwan, Repub China, Feb 16, 53; m 80; c 2. POLYMER & FIBER CHEMISTRY SCIENCES. *Educ:* Fu-Jen Univ, Taipei, Repub China, BS, 75; Auburn Univ, MS, 77; Univ Md, College Park, PhD(textile chem), 81. *Prof Exp:* Asst prof, 81-87, ASSOC PROF TEXTILES, UNIV CALIF, DAVIS, 87- *Mem:* Am Chem Soc; Am Asn Textile Chemists & Colorists; Sigma Xi; Fiber Soc. *Res:* Modification of polymers and fibers to achieve functional properties; surface properties (hydrophilicity, hydrophobicity, reactivity, ionic nature, chemistry and morphology); bulk properties (hygroscopicity, mechanical properties, morphology, additives); fluid-substrate interaction and biomedical applications. *Mailing Add:* Div Textiles & Clothing Univ Calif Davis CA 95616

HSIEH, YU-NIAN, b Chungking City, China, Nov 30, 42; m 70; c 3. SOLID STATE PHYSICS. *Educ:* Nat Taiwan Univ, BS, 64; Univ Notre Dame, MS, 67; Univ Calif, Berkeley, PhD(physics), 73. *Prof Exp:* Res assoc chem physics, Univ Ill, 73-75; asst prof physics, Hunter Col, City Univ New York, 75-78; staff engr, 77-80, ADV ENGR ENGR, IBM CORP, 81- *Mem:* Am Phys Soc; AAAS; Sigma Xi; Inst Elec & Electronics Engrs. *Res:* Chemical physics; nuclear double resonance; semiconductor physics and devices; very large scale integration electronics. *Mailing Add:* 1083 Foxhurst Way San Jose CA 95120

HSIUNG, ANDREW K, b Feng-yang, China, Jan 4, 20; US citizen; m 57; c 4. ENVIRONMENTAL ENGINEERING, CIVIL ENGINEERING. *Educ:* Chiao-Tung Univ, BS, 43; Johns Hopkins Univ, MS, 60; Iowa State Univ, PhD(environ eng), 67. *Prof Exp:* Jr to sr engr civil eng, var govt agencies, Repub China, 43-64; res asst environ eng, Iowa State Univ, 64-67; res engr, 67-72, sr res engr environ eng, Neptune Microfloc Inc, 72-84; RETIRED. *Honors & Awards:* Rudolph Hering Medal, Am Soc Civil Engrs, 70. *Mem:* Am Waterworks Asn. *Res:* Water and wastewater treatment processes with special interest in liquid-solids separation. *Mailing Add:* 2226 Songbird Ct SE Salem OR 97306

HSIUNG, CHI-HUA WU, b Hangchow, China, Sept 26, 30; m 59; c 1. THEORETICAL CHEMISTRY. *Educ:* Univ Taiwan, BSc, 54; Univ Mich, MS, 57, PhD(chem), 62. *Prof Exp:* Res assoc chem, Univ Mich, 62-64; assoc prof physics, Hillsdale Col, Mich, 64-71; PROF HEALTH SCI, CALIF STATE UNIV, DOMINGUEZ HILLS, 72-, COORDR HEALTH SCI PROG, 84- *Concurrent Pos:* H H Rackham fel, Univ Mich, 62-63. *Mem:* Am Phys Soc; Sigma Xi. *Res:* Hot atom chemistry, particularly energy distribution and chemical nature of those highly energized atoms resulting from nuclear transformation; alternative energy sources; particularly passive solar energy devices. *Mailing Add:* Sch Health Calif State Univ Dominguez Hills CA 90747

HSIUNG, CHUAN CHIH, b Kiangsi, China, Feb 15, 16; nat US; m 42; c 1. MATHEMATICS. *Educ:* Nat Univ Chekiang, BS, 36; Mich State Univ, PhD(math), 48. *Prof Exp:* Asst math, Mich State Univ, 46-48; instr, Univ Wis, 48-50; lectr, Northwestern Univ, 50; res fel, Harvard Univ, 51-52; from asst prof to prof, 52-84, EMER PROF MATH, LEHIGH UNIV, 84- *Concurrent Pos:* Vis assoc prof, Math Res Ctr, US Army, Univ Wis, 59-60; vis specialist, Univ Calif, Berkeley, 62; founder & managing ed, J Differential Geom, 67-; ed, Bulletin Inst Math, Academia Sinica, Taiwan, 75-; ed adv math & ed, Series in Pure Math, Singapore, 81- *Mem:* Am Math Soc; Math Asn Am. *Res:* Differential geometry. *Mailing Add:* Dept Math Lehigh Univ Bethlehem PA 18015

HSIUNG, GUEH DJEN, b Hupeh, China, Sept 16, 18. VIROLOGY. *Educ:* Ginling Col, BS, 42; Mich State Col, MS, 48, PhD(bact), 51. *Hon Degrees:* DSc, Mich State Univ, 89. *Prof Exp:* Res asst microbiol, Sch Med, Yale Univ, 53-54, instr microbiol & prev med, 54-57, res assoc, 57-62, asst prof epidemiol & pub health, 62-65; assoc prof med, Sch Med, NY Univ, 65-67; lectr microbiol, 67-69, assoc prof, 69-74, PROF LAB MED, SCH MED, YALE UNIV, 74- *Concurrent Pos:* Dir, Virol Ref Lab, Vet Admin Med Ctr, W Haven, Conn, 84-89. *Honors & Awards:* Becton-Dickson Award in Clin Microbiol, 83. *Mem:* Am Soc Microbiol; Soc Exp Biol & Med; Tissue Cult Asn; Am Asn Immunol; fel Infectious Dis Soc. *Res:* Animal viruses; simian viruses; human viruses: characterization, pathogenesis, and epidemiology of virus infection; diagnostic virology; herpesvirus latency; animal model for cytomegalovirus infection; evaluation of antiviral agents. *Mailing Add:* Dept Lab Med Yale Univ Sch Med New Haven CT 06510

HSU, ANDREW C T, b Hangchow, China, July 25, 16; US citizen. CHEMICAL ENGINEERING, MATERIALS SCIENCE. *Educ:* Nanking Univ, BS, 38; Univ Wis, MS, 46; Univ Pa, PhD(chem eng), 53. *Prof Exp:* From asst prof to assoc prof chem eng, Ala Polytech Inst, 53-57; group leader res & develop, Air Prod, Inc, 57-60; res scientist, Lockheed Missiles & Space Co, 60-62; prof, 62-86, EMER PROF CHEM ENG, AUBURN UNIV, 86- *Concurrent Pos:* Fulbright vis prof, Univ Hong Kong, 64-65. *Mem:* Am Chem Soc; Am Inst Chem Engrs; Nat Soc Prof Engrs. *Res:* Mass transfer; engineering thermodynamics; chemical kinetics; adsorption; fuel cells; inorganic preparations; crystallization kinetics; glass reinforced composites. *Mailing Add:* PO Box 982 Auburn AL 36831

HSU, C(HIH) C(HI), b Shanghai, China, June 30, 23; m 55; c 3. ELECTRICAL ENGINEERING. *Educ:* Chiao Tung Univ, BSEE, 45; Univ Mich, MSE, 49; Ohio State Univ, PhD(elec eng), 51. *Prof Exp:* Jr engr, Tsingtao Power Co, 46; asst, Univ Mich, 49 & Ohio State Univ, 50-51; asst prof elec eng, Mich Col Mining & Technol, 52; proj engr, Bendix Radio Div, Bendix Aviation Corp, 53-57; from asst prof to assoc prof, 58-71, PROF ELEC ENG, UNIV WASH, 71- *Mem:* Inst Elec & Electronics Engrs. *Res:* Servomechanisms; sampled-data and digital control systems; optimal control; pattern recognition; systems optimization. *Mailing Add:* Dept Elec Eng Univ Wash Seattle WA 98195

HSU, CHAO KUANG, b Pingtung, China, Aug 12, 39; m 68; c 2. LABORATORY ANIMAL MEDICINE, PUBLIC HEALTH. *Educ:* Nat Taiwan Univ, DVM, 62; Univ Ill, MS, 67, PhD(vet med sci), 70; Johns Hopkins Univ, MPH, 71. *Prof Exp:* USPHS fel, Sch Hyg, Johns Hopkins Univ, 70-71, NIH spec res fel, Sch Med, 71-73, asst prof lab animal med, Sch Med & asst prof pathobiol, Sch Hyg & Pub Health, 73-76, asst prof comp med, 76-80; asst prof comp med, Sch Med, Univ Md, 80-; AT DEPT COMP MED, JOHNS HOPKINS UNIV. *Mem:* Am Asn Lab Animal Sci; Am Soc Trop Med & Hyg; Am Pub Health Asn. *Res:* IgE in parasitic infections in man and animals; diseases of laboratory animals. *Mailing Add:* 280 Stonegate Dr Devon PA 19333

HSU, CHARLES CHING-HSIANG, b Taiwan, 59; m 83; c 2. ELECTRICAL ENGINEERING. *Educ:* Tsing Hna Univ, Taiwan, BS, 81; Univ Ill, MS, 85, PhD(elec eng), 87. *Prof Exp:* Res asst, Univ Ill, 83-87; RES STAFF MEM, IBM, 87- *Mem:* Inst Elec & Electronic Engrs. *Res:* MOS theory; VLSI/Subprim CMOS design and technology; CMOS reliability; thin dielectrics for VLSI; non-volatile memory. *Mailing Add:* IBM 7-153 PO Box 218 Yorktown Heights NY 10598

HSU, CHARLES FU-JEN, b Kiangsu, China, May 9, 20; US citizen; m 47; c 4. NATURAL PRODUCTS CHEMISTRY. *Educ:* Nat Chung Cheng Univ, BS, 46; DePaul Univ, MS, 62. *Prof Exp:* Asst chem engr, Northeastern Cement Corp, China, 46-48; dir cellulose purification, Lin-Tow Co, 48-50; head teaching, Ga-Yi Indust Sch, 50-52; instr teaching & res, Taiwan Agr Col, 52-55; assoc prof, Taiwan Chung Hsin Univ, 55-59; res assoc cancer res, Ivy Cancer Res Fedn, Chicago, 61-64; res chemist, G D Searle & Co, 64-71, res chemist, Searle Labs, 71-85; RETIRED. *Concurrent Pos:* Adv, Taichung Technol Lab, Taiwan, 52-55 & Cent Mil Surv Sch, 55-56. *Mem:* Am Chem Soc; NY Acad Sci; Soc Indust Microbiol. *Res:* Microbial transformation of prostaglandins and steroids for search of diuretic and antiulcerogenic agents; pre-control sex of human embryo before foetal formation. *Mailing Add:* 4953 Elm St Skokie IL 60077

HSU, CHARLES TEH-CHING, b China, Dec 14, 35; US citizen. COMPUTER SCIENCE. *Educ:* Univ Wis-Madison, BS, 60, MS, 61. *Prof Exp:* Res scientist heat power, Am Standard, Inc, 61; mgr comput dept, Alcorn Combustion Co; int sales engr, Westinghouse Elec Co; mgr comput eng dept, Burns & Roe, Inc; mgr comput div, Stone & Webster Eng Co; prin computer engr, Kaiser Engrs, 74-84; gen mgr, Far E Oper, Digital Commun Assoc, 84-90; MIS MGR, SUN MOON STAR, 90- *Mem:* Am Soc Mech Engrs; Nat Coun Engr Examr. *Res:* Pioneered in the computerization of engineering applications. *Mailing Add:* 394 12th St Apt 9AM Oakland CA 94607

HSU, CHEN C, b Changhwa, Taiwan, June 29, 40; US citizen; m 66; c 2. CATALYSIS, SURFACE SCIENCE THERMODYNAMICS. *Educ:* Nat Taiwan Normal Univ, BS, 63; Brigham Young Univ, MS, 69; Univ Utah, PhD(phys chem), 72. *Prof Exp:* Res assoc phys chem, Univ Chicago, 72-74; comput syst analyst, Electronic Div, Bell & Howell, 74-75; chemist, Argonne Nat Lab, 75-82; actg br chief, 90, RES CHEMIST, US ARMY CHEM RES, DEVELOP & ENG CTR, 82- *Concurrent Pos:* Mem gas filter comt, US Army Chem Res, Develop & Eng Ctr, 83-85, chmn, Res Directorate, 89; adv, Nat Res Coun, 84-; tech consult, USAF, 87- *Mem:* Soc Appl Spectros; Am Chem Soc; Catalysis NAm; Sigma Xi. *Res:* Catalytic reaction kinetics and mechanisms; deactivation mechanisms of catalysts; surface active metals distribution and their oxidation states of oxidation catalysts; electrochemical means for toxic chemical sensing; surface chemical analysis. *Mailing Add:* 2409 Chatau Ct Fallston MD 21047

HSU, CHENG-TING, b Chekiang, China, Dec 1, 23; nat US; m 63; c 1. AEROSPACE ENGINEERING. *Educ:* Nat Southwestern Assoc Univ, China, BS, 44; Univ Minn, MS, 49, PhD, 54. *Prof Exp:* Engr aeronaut eng, Univ Minn, 52-53, assoc scientist, 53-54, sr engr, 54-56, res assoc, 56-58; assoc prof, 58-62, PROF AEROSPACE ENG, IOWA STATE UNIV, 62- *Mem:* Am Inst Aeronaut & Astronaut. *Res:* High temperature gas dynamics; tornado fluid dynamics. *Mailing Add:* Dept Aerospace Eng Iowa State Univ 304 Town Eng Bldg Ames IA 50011

HSU, CHEN-HSING, b Taiwan, China, April 6, 37; US citizen; m 66; c 1. INTERNAL MEDICINE, NEPHROLOGY. *Educ:* Nat Taiwan Univ, MD, 65. *Prof Exp:* From instr to asst prof, 72-78, ASSOC PROF MED, UNIV MICH, 78- *Mem:* Am Soc Nephrology; Int Soc Nephrology; Am Fedn Clin Res. *Res:* Vitamin D metabolism in hypertension and renal failure. *Mailing Add:* 3914 Taubman Med Ctr Univ Mich Ann Arbor MI 48109-0364

HSU, CHIEH-SU, b Soochow, Kiangsu,China, May 27, 22; m 53; c 2. ENGINEERING EDUCATION & RESEARCH. *Educ:* Nat Inst Technol, China, BS, 45; Stanford Univ, MS, 48, PhD(eng mech), 50. *Prof Exp:* Engr, Shanghai Naval Dockyard, 46-47; res asst, Stanford Univ, 48-51; proj engr, IBM Corp, 51-55; assoc prof eng mech, Univ Toledo, 55-58; PROF MECH ENG, UNIV CALIF, BERKELEY, 58- *Concurrent Pos:* Guggenheim fel, 64-65; chmn, Div Appl Mech, Univ Calif, 69-70; tech ed-in-chief, J Appl Mech, Am Soc Mech Engrs, 76-82; mem, US Nat Comt Theoret & Appl Mech, Nat Acad Sci, 85-89, sci adv bd, Humboldt Found, WGer, 85-; John Simon Guggenheim fel, 64-65; Miller res prof, Univ Calif, Berkeley, 73-74. *Honors & Awards:* Alexander von Humboldt Sr Scientist Award, 86. *Mem:* Sigma Xi; Acoust Soc Am; Am Soc Mech Engr; Acout Soc Am; Nat Acad Eng. *Res:* Research in the areas of elastic stability; nonlinear oscilations and systems analysis with emphasis on global analysis of nonlinear systems; elasticity; development of the cell to cell mapping methodology. *Mailing Add:* Dept Mech Eng Univ Calif Berkeley CA 94720

HSU, CHIN SHUNG, b Cha-Yi, Taiwan, Aug 10, 48; m 76; c 1. SYSTEM & CONTROL, NONLINEAR SYSTEMS. *Educ:* Nat Chiao-Tung Univ, BSEE, 70; Utah State Univ, MSEE, 74; Ore State Univ, PhD(elec eng), 79. *Prof Exp:* Vis asst prof systs & control, Wash State Univ, 78-79; ASST PROF SYSTS & CONTROL, WASH STATE UNIV, 80- *Concurrent Pos:* Vis asst prof systs & control, Cleveland State Univ, 78-80; asst engr, Univ Fla, 79; res asst, Ore State Univ, 73-78. *Mem:* Inst Elec & Electronics Engrs; Sigma Xi. *Res:* Analysis and control of large-scale systems; control and biomedical applications of bilinear systems; digital control systems. *Mailing Add:* Dept Elec Eng Wash State Univ Pullman WA 99164

HSU, CHIN-FEI, b Taipei, Taiwan, Oct 29, 47; m 74; c 2. WEATHER MODIFICATION, STATISTICS. *Educ:* Nat Taiwan Univ, BS, 72; Univ NC Chapel Hill, MS, 75, PhD(statist), 77. *Prof Exp:* Res asst, statist, Univ NC, Chapel Hill, 72-77; ASSOC PROF SCI, ILL STATE WATER SURV, 77- *Concurrent Pos:* Consult, Colo State Univ, 79-80; prin investr, Ill State Water Surv, 81- *Mem:* Am Statist Asn; Inst Math Statist; Am Meteorol Soc; Sigma Xi. *Res:* Statistical-physical techniques to evaluate weather modification operations. *Mailing Add:* 26 Parkway Pl Holmdel NJ 07733

HSU, CLEMENT C S, b Taiwan, China, Oct 9, 37; US citizen; m 65; c 2. CLINICAL INFECTIOUS DISEASES, HUMAN LYMPHOCYTE. *Educ:* Nat Taiwan Univ, MD, 63. *Prof Exp:* Intern med, Jersey City Med Ctr, 65-66; jr resident, Montefiore Hosp, NY, 66-67; sr resident, Boston City Hosp, 67-68; clin res fel, Div Liver & Nutrit, NJ Col Med & Dent, 68-69; res assoc path, Inst Cancer Res, Columbia Presby Med Ctr, NY, 69-70; clin res fel med genetics, dept pediat, Mt Sinai Sch Med, NY, 70-72; assoc med, 72-74, asst prof, 74-78, ASSOC PROF MED, INFECTIOUS DIS SECT, SCH MED, NORTHWESTERN UNIV, 78- *Concurrent Pos:* Vis attend physician, Northwestern Mem Hosp, Chicago, 77-; chief, Infectious Dis Sect, Columbus Hosp, Chicago, 77- *Honors & Awards:* Ann Med Res Award, Leukemia Res Found, Inc, 74, 75, 77. *Mem:* Am Asn Immunologists; Am Asn Cancer Res; Infectious Dis Soc Am; fel Am Col Physicians; AMA. *Res:* Cellular immunology and infectious diseases, specifically human lymphocyte activation in vitro, identification of lymphocyte surface molecules and geriatric clinical infectious diseases. *Mailing Add:* Tzu Chi Gen Hosp Shin-Shen S Rd Hualien Taiwan

HSU, DAVID KUEI-YU, b Shantung, China, Nov 30, 41; m 71. SOLID STATE PHYSICS. *Educ:* Nat Taiwan Univ, BS, 65; Wayne State Univ, PhD(physics), 71. *Prof Exp:* From instr to assoc prof physics, Colo State Univ, 78-84,; SR SCIENTIST & ADJ PROF AERIAL ENG & ENG MECH, AMES LAB & CTR FOR NONDESTRUCTIVE EVAL, IOWA STATE UNIV, 84- *Concurrent Pos:* Translr for Chinese physics & Chinese physics lasers. *Mem:* Am Phys Soc; Acoustical Soc Am; Sigma Xi. *Res:* Electromagnetic generation of ultrasonic waves in metals; nuclear acoustic resonance in metals, properties of metal hydrides and nondestructive evaluation; advanced composite materials. *Mailing Add:* 133 Appl Sci Complex II Iowa State Univ Ames IA 50011

HSU, DAVID SHIAO-YO, b Chunking, China, Nov 23, 44; m 70; c 1. CHEMICAL REACTION DYNAMICS, GAS-SURFACE REACTIONS. *Educ:* Univ Calif, Berkeley, BS, 67; Harvard Univ, MA, 69, PhD(chem physics), 74. *Prof Exp:* Sr res assoc, Brookhaven Nat Lab, 74-75; Nat Res Coun/Naval Res Lab res assoc, 75-77, res chemist, 77-91, SECT HEAD, NAVAL RES LAB, 91- *Concurrent Pos:* Prin investr, Naval Res Lab, 88- *Mem:* Am Chem Soc; fel Am Inst Chemists; Sigma Xi. *Res:* Molecular beam reaction dynamics; chemical kinetics; energy transfer; laser-induced gas phase and surface chemistry; high temperature combustion chemistry; gas-surface reaction dynamics; thin film nucleation; nanofabrication; authored 54 publications and 2 patents. *Mailing Add:* Chem Div Code 6114 Naval Res Lab Washington DC 20375-5000

HSU, DEH YUAN, b Shensi, China, Jan 1, 43; m 71; c 2. ENVIRONMENTAL ENGINEERING, WATER & WASTEWATER TREATMENT. *Educ:* Nat Taiwan Univ, BS, 64; Northwestern Univ, MS, 68, PhD(civil eng), 73. *Prof Exp:* Res assoc sanit eng, McGill Univ, 72-74; asst prof environ eng, Wayne State Univ, 75-78; PROCESS ENGR ENVIRON ENG, GREELEY & HANSEN ENGRS, CHICAGO, 78- *Mem:* Am Soc Civil Engrs; Am Water Works Asn; Water Pollution Control Fedn; Am Acad Environ Engrs. *Res:* Processes and techniques in water and wastewater treatment; industrial pollution control; treatment and disposal of water and wastewater treatment sludge; removal of toxic pollutants from wastewater by various processes. *Mailing Add:* Greeley & Hansen Engrs 222 S Riverside Plaza Chicago IL 60606

HSU, EDWARD CHING-SHENG, b Taiwan, May 9, 42; m 72; c 2. PHYSICAL CHEMISTRY. *Educ:* Cheng Kung Univ, Taiwan, BS, 64; Univ Chicago, PhD(phys chem), 70. *Prof Exp:* Fel appl physics, Yale Univ, 70-71; res assoc phys chem, Univ Chicago, 71-73; fel polymer sci, Univ Akron, 73-74; SR STAFF CHEMIST, EXXON RES & ENG CO, 74- *Mem:* Am Chem Soc. *Res:* Electrical properties of liquid and soild hydrocarbons and their derivatives. *Mailing Add:* Exxon Res & Eng Co PO Box 101 Florham Park NJ 07932

HSU, EN YUN, fluid mechanics, for more information see previous edition

HSU, FRANK HSIAO-HUA, b Hankow, China, Dec 26, 35; m 65; c 1. SOLID STATE PHYSICS. *Educ:* Nat Taiwan Univ, BS, 59; Columbia Univ, MS, 64, PhD(physics), 67. *Prof Exp:* Res scientist physics, Columbia Univ, 67-69; from asst prof to assoc prof, 69-81, PROF PHYSICS, GA STATE UNIV, 81- *Concurrent Pos:* Res Corp grant, 70; adv, Int Ctr Theoret Physics; consult, Int Adv Panel. *Mem:* Nat Acad Sci; Am Asn Physics Teachers; Am Phys Soc; Nat Sci Found. *Res:* Nuclear spectroscopy; positron annihilation in solids; radiation damage in solids. *Mailing Add:* Dept Physics Ga State Univ Atlanta GA 30303

HSU, H(WEI) P(IAO), b Miaoli, Taiwan, Jan 14, 30; m 53. ELECTRICAL ENGINEERING. *Educ:* Taiwan Univ, BSc, 52; Case Inst Technol, MSc, 59, PhD(elec eng), 61. *Prof Exp:* Asst prof, Univ Windsor, 61-63; assoc prof elec eng, Wayne State Univ, 63-72; dept res engr, Gen Motors Corp Res Labs, 72-83; prof & dept head elec eng, Univ Evansville, 83-86; PROF ELEC ENG, FAIRLEIGH DICKINSON UNIV, 86-, CHMN. *Concurrent Pos:* Consult, Gen Motors Corp Res Labs, 63-72. *Mem:* Inst Elec & Electronics Engrs. *Res:* Communication; electromagnetic compatability; signal analysis. *Mailing Add:* Elec Eng Dept Fairleigh Dickinson Univ 1000 River Rd Teaneck NJ 07666

HSU, HOWARD HUAI TA, b Taipei, Taiwan, Nov 13, 38; m 65; c 2. BIOCHEMISTRY, ENDOCRINOLOGY. *Educ:* Tunghai Univ, BS, 62; Univ NDak, MS, 66, PhD(biochem), 70. *Prof Exp:* Res fel, Cancer Res Inst, New Eng Deaconess Hosp, Boston, Mass, 69-71; res assoc biochem, Rockefeller Univ, 71-75; res assoc, State Univ NY, Downstate Med Ctr, 75-77; asst prof path, 77-81, ASSOC PROF PATH & ONCOL, UNIV KANS MED CTR, 81- *Concurrent Pos:* USPHS fel, 70-71. *Mem:* Harvey Soc; Sigma Xi; Am Soc Molecular Biol & Biochem. *Res:* Roles of thyroid hormones in cellular metabolism and calcification; mechanism of calcification of bone and cartilage; the biochemical relationship between cancerous and embryonic cells. *Mailing Add:* 6324 W 102 St Overland Park KS 66212

HSU, HSI FAN, parasitology, immunology; deceased, see previous edition for last biography

HSU, HSIEN-WEN, b Chia-Yi, Formosa, Apr 7, 28; m 61; c 2. CHEMICAL ENGINEERING. *Educ:* Nat Taiwan Univ, BS, 51; Kans State Univ, MS, 55; Univ Wis, PhD(chem eng), 59. *Prof Exp:* Fel thermodyn, sch mech eng, Purdue Univ, 59-61, asst prof fluid mech & heat transfer, 61-64; assoc prof chem & metall eng, 64-71, PROF CHEM & METALL ENG, UNIV TENN, KNOXVILLE, 71- *Concurrent Pos:* Consult, Northrop Space Labs, Ala, 65-67 & molecular anat prog, Chem Tech Div, Oak Ridge Nat Lab, 67-72 & 77- *Mem:* Am Inst Chem Engrs; Am Chem Soc; Japanese Soc Chem Engrs. *Res:* Viral coefficients and transport properties; heat and mass transfer; cavitation problems; optimization theory and applications; biophysical separation; zonal centrifugation; chromatography; fluidization. *Mailing Add:* Dept Chem Eng Univ Tenn Knoxville TN 37996-2200

HSU, HSIUNG, b Nantung, China, Jan 24, 20; m 55; c 3. APPLIED PHYSICS. *Educ:* Nat Wu-han Univ, China, BS, 41; Harvard Univ, MS, 46, PhD(appl physics), 50. *Prof Exp:* Engr, Int Broadcasting Sta, 41-45; res assoc, Harvard Univ, 50; sr physicist, Gen Elec Co, 50-62; assoc prof, 62-66, assoc supvr antenna lab, 62-66, PROF ELEC ENG, OHIO STATE UNIV, 66- *Mem:* Am Phys Soc; Inst Elec & Electronics Eng; Am Asn Physics Teachers. *Res:* Microwave physics; quantum electronics; electron tubes; nonlinear optics; electronic circuits; solid state devices. *Mailing Add:* Dept Elec Eng Ohio State Univ 2015 Neil Ave Columbus OH 43210

HSU, HSIU-SHENG, b Guangzhou, China, Oct 26, 31; nat US; m 58; c 4. MEDICAL MICROBIOLOGY, IMMUNOLOGY. *Educ:* McGill Univ, BSc, 55; Univ Pa, MS, 56, PhD(med microbiol), 59. *Prof Exp:* Fel med microbiol, Univ Pa, 59-62; fel, Med Sch, Johns Hopkins Univ, 63-64; from instr to asst prof, 64-69, ASSOC PROF MICROBIOL, MED COL VA, 69- *Concurrent Pos:* Res grants, Am Thoracic Soc, 65-68 & NIH, 65-73, 84-90. *Mem:* Reticuloendothelial Soc; Am Soc Microbiol; NY Acad Sci. *Res:* Host-parasite relationship, experimental pathology and immunology in bacterial infectious diseases; role of leucocytes in host defense against bacterial infections. *Mailing Add:* Dept Microbiol & Immunol Med Col Va Va Commonwealth Univ Richmond VA 23298-0678

HSU, IN-DING, applied mathematics, for more information see previous edition

HSU, JANG-YU, b Kadhsiung, Taiwan, Jan 20, 49. RADIO FREQUENCY, METABOLIC HEAT PRODUCTION. *Educ:* Nat Tsing Hua Univ, BS, 70; Univ Iowa, MS, 73; Princeton Univ, PhD(plasma physics), 77. *Prof Exp:* Sr scientist, Gen Atomics, 77-90; PRES, PHYSIONIX CORP, 90- *Mem:* Fel Am Phys Soc. *Res:* Stochastic heating, tokamak equilibrium profile and electron cyclotron resonance heating. *Mailing Add:* 520 Santa Carina Solana Beach CA 92075

HSU, JAY C, b Shanghai, China, July 18, 34; US citizen; m 61; c 3. SYSTEMS ENGINEERING. *Educ:* Cornell Univ, BEE, 57, MEE, 58, PhD(elec eng), 61. *Prof Exp:* Mem tech staff, 61-67, SUPVR, BELL TEL LABS, 67- *Mem:* Inst Elec & Electronics Engrs. *Res:* Feedback control theory, especially missile guidance and control; system engineering of operations support computer systems for the telecommunications network. *Mailing Add:* Bell Tel Labs Rm Ho 3j335 Crawford Rd Holmdel NJ 07733

HSU, JENG MEIN, b China, Nov 27, 20; nat US; m 54; c 4. BIOCHEMISTRY, NUTRITION. *Educ:* Nat Cent Univ, China, BS & DVM, 43; Wash State Univ, MS, 49, PhD(nutrit, biochem), 53. *Prof Exp:* Teacher, Agr Sch, China, 42-44; fel, Szechwan Exp Sta, Chengtu, China, 44-46; chief nutritionist, Nat Exp Sta, Shanghai, 46-47; asst poultry nutritionist, Wash State Univ, 50-53; res assoc biochem, Johns Hopkins Univ, 53-57; dir biochem res lab, St Joseph's Hosp, Elmira, NY, 57-58; assoc prof biochem, Johns Hopkins Univ, 58-74; PROF CHEM, UNIV SFLA, 75-; MEM STAFF, BIOCHEM RES PROJS, VET ADMIN CTR, BAY PINES, 75- *Concurrent Pos:* Fel, Nat Shanghai Demonstration Farm, 46-47; chief biochem res lab, Vet Admin Hosp, Baltimore, Md, 58-75; vis prof biochem & biophys sci, Johns Hopkins Univ, 75-76. *Mem:* AAAS; Am Chem Soc; Am Inst Nutrit; Soc Exp Biol & Med. *Res:* Zinc, vitamin B12, vitamin C and trace elements. *Mailing Add:* 13638 Pinecrest Dr Largo FL 34644

HSU, JOHN Y, b Nanking, China, Mar 17, 38; US citizen; m 65; c 2. COMPUTER SCIENCE. *Educ:* Nat Taiwan Univ, BS, 59; Univ Calif, Berkeley, MS, 64, PhD(comput sci), 69. *Prof Exp:* Proj engr various indust orgn, 64-67; comput architect, Varian Data Mach, 70; PROF COMPUT SCI, CALIF POLYTECH STATE UNIV, SAN LUIS OBISPO, 70- *Concurrent Pos:* Consult comput sci, Fed Elec Corp, Int Tel & Tel Corp, 71-, Illiac IV Proj, 73-, Control Data Corp, 81 & IBM Corp, 87-88. *Mem:* Inst Elec & Electronics Engrs; Asn Comput Mach. *Res:* Computer architecture, hardware and software. *Mailing Add:* 365 Mira Sol San Luis Obispo CA 93401

HSU, KATHARINE HAN KUANG, b Foochow, China, Feb 12, 14; m 41. PEDIATRICS. *Educ:* Yenching Univ, China, BS, 35; Peking Union Med Col, MD, 39. *Prof Exp:* From instr to asst prof pediat, Nat Chung Cheng Med Col, China, 42-47; researcher chemother tuberc, Henry Phipps Inst, Univ Pa, 49-50; sr physician, Children's Hosp, Mont Alto State Sanatorium, South Mountain, Pa, 50-53; from asst prof to prof pediat, Baylor Col Med, 53-79, dir, Pediat Chest Serv, 68-79; RETIRED. *Concurrent Pos:* Pediatrician-in-chg, Children's Tuberc Clin, Jefferson Davis Hosp, 53-68; dir tuberc control, Dept Pub Health, Houston, 65-68; consult, Tex Children's Hosp. *Mem:* Am Thoracic Soc; Am Col Chest Physicians. *Res:* Pediatric tuberculosis and respiratory diseases; pulmonary disease; chemotherapy, epidemiology and immunology of tuberculosis in children. *Mailing Add:* 9427 Denbury Way Houston TX 77025

HSU, KENNETH HSUEHCHIA, b Taiwan, Aug 21, 50. FOOD ENGINEERING. *Educ:* Kans State Univ, BS, 72, MS, 74, PhD(cereal sci), 78. *Prof Exp:* Chemist I prod develop, Durkee Food Co, SCM Corp, 78-79, sr chemist, 79-80; from asst prof to assoc prof, Dept Food Technol, Iowa State Univ, 80-87; SUPVR ENG, NABISCO BRANDS INC, 87- *Mem:* Am Asn Cereal Chemists; Am Inst Chem Engrs; Inst Food Technologists. *Res:* Determination of physical, chemical and functional properties of food systems, particularly as related to cereals and oilseeds; modeling of food processes and kinetics of food reactions. *Mailing Add:* Nabisco Brands Inc 200 DeForest Ave East Hanover NJ 07936

HSU, KENNETH JINGHWA, b Nanking, China, July 1, 29; m; c 4. GEOLOGY. *Educ:* Nat Cent Univ, BS, 48; Ohio State Univ, MA, 50; Univ Calif, Los Angeles, PhD(geol), 54. *Hon Degrees:* DSc, Nanjing Univ, 87. *Prof Exp:* Asst geol, Ohio State Univ, 48-49; asst geophys, Inst Geophys, Univ Calif, Los Angeles, 50-51, asst geol, Univ, 51-54; geologist, Shell Develop Co, 54-56, proj head, 57-62, res assoc, 63; assoc prof geol, Harpur Col, 63-64; assoc prof, Univ Calif, Riverside, 64-67; PROF GEOL, SWISS FED INST TECHNOL, 67-, CHMN, GEOL INST, 78- *Concurrent Pos:* Co-chief scientist, Mediter & Atlantic Cruises, Deep-Sea Drilling Proj; chmn, Mediter Panel, Atlantic Working Group, Joint Oceanog Inst Deep Earth Sampling; ed-in-chief, Sedimentol; assoc ed, J Sedimentary Petrol, Marine Geophys Res & Geophys Surv; Guggenheim Found fel, 72; chmn, Int Comn Marine Geol & Comt Global Change, Int Union Geol Sci; vis prof, Scripps Inst Oceanog, 72, Calif Inst Technol, 91. *Honors & Awards:* Wollaston Medal, Geol Soc London, 84; Twenhofel Medal, Soc Econ Paleontologists & Mineralogists, 84. *Mem:* Nat Acad Sci; Am Geophys Union; hon mem Soc Econ Paleontologists & Mineralogists; Swiss Geol Soc; Int Asn Sedimentology (pres, 78-82); hon fel Geol Soc Am. *Res:* Structural geology; sedimentation; petrology. *Mailing Add:* Geol Inst Swiss Fed Inst Technol Sonneggstrasse 5 Zurich Switzerland

HSU, KONRAD CHANG, b China, Aug 28, 01; nat US; m 51; c 7. IMMUNOBIOLOGY. *Educ:* St John's Univ, China, BS, 21; Columbia Univ, MA, 23, PhD(chem), 24. *Prof Exp:* Prof chem & dean col sci, Great China Univ, 24-26; with govt serv, China, 26-28; chemist & gen mgr, Chan Hwa & Co, 29-46; vpres, Sino Hawaiian Corp, 46-49; adv govt sugar orgn, Thailand, 50-51; consult sugar factories, Thailand, 51-54; from asst to assoc prof microbiol, 54-69, PROF MICROBIOL, COL PHYSICIANS & SURGEONS, COLUMBIA UNIV, 69- *Concurrent Pos:* Res consult microbiol, FDR Vet Hosp, Montrose, NY, 68-70. *Mem:* Am Asn Immunol; Am Asn Path; Soc Exp Biol & Med; Harvey Soc. *Res:* Diseases resulting from immune reactions utilizing fluorescein, ferritin and enzym labeled antibodies and antigens; immunologic studies of tumor virus antigens using such labeled antibodies. *Mailing Add:* 41-40 Union St Apt 12-R Flushing NY 11355

HSU, KWAN, b Kwang-Si, China, Mar 11, 13; US citizen. BIOPHYSICS. *Educ:* Shanghai Univ, BS, 36; Univ Minn, MS, 50; Univ Calif, Berkeley, PhD(biophys), 60. *Prof Exp:* Instr physics, Shanghai Univ, 37-47; Am Asn Univ Women int study grant, 47-48; asst physics, Univ Minn, 48-50; asst, Univ Iowa, 50- 52, Col Med, 52-53; assoc biophys, Univ Calif, Berkeley, 54-59; physicist, Vet Admin Hosp, Indianapolis, Ind, 60-61, asst chief biophys, 61-62; res assoc physics, Biophys Lab, Stanford Univ, 64; assoc prof, 64-74, prof biophys, 74-77, prof physics, 77-80, EMER PROF PHYSICS, PORTLAND STATE UNIV, 80- *Concurrent Pos:* Asst prof, Sch Med, Univ Ind, 60-62. *Mem:* Biophys soc; Radiation Res Soc; Soc Nuclear Med; Am Asn Physics Teachers; NY Acad Sci. *Res:* Cellular and radiation biophysics; health physics. *Mailing Add:* 4820 SW Barbur Blvd Portland OR 97201

HSU, LAURA HWEI-NIEN LING, b Kwei-Chow, China, Aug 22, 39; m 63; c 2. EVOLUTION. *Educ:* Acadia Univ, NS, Can, 61; Cornell Univ, MS, 64; Univ Miami, Coral Gables, PhD(biol), 74. *Prof Exp:* Supvr & instr microbiol, Mycol & Parasitol Labs, Evanston Hosp Asn, Evanston, Ill, 63-66; instr sci, Sch Nursing, St Francis Hosp, Evanston, 66-68; instr parasitol, Sch Med Technol, Med Col, Northwestern Univ, Chicago, 67-68; res asst, Inst Molecular & Cellular Evolution, Univ Miami, Coral Gables, Fla, 69-71, res asst prof, 74-; instr & coordr, Dept Biol, 80-84, DIR PROG, OFF CONTINUING STUDIES, RICE UNIV, 84- *Concurrent Pos:* Vis assoc prof, Nat Taiwan Univ, 78-79; vis fel plant path, Cornell Univ, Ithaca, NY, 82. *Mem:* Sigma Xi; Int Soc Study Origin Life; Am Soc Clin Pathologists; Am Soc Microbiologists; Asn Women Sci. *Res:* The origins and evolution of life on Earth; the development of metabolic systems and energy conversion in synthetic cell models; issues in continuing education. *Mailing Add:* 5034 Glenmeadow Meyerland Houston TX 77096

HSU, LIANG-CHI, b Formosa, Sept 1, 31; nat US; m 60; c 4. MINERALOGY, PETROLOGY. *Educ:* Univ Taiwan, BS, 56, MS, 61; Univ Calif, Los Angeles, PhD(geol), 66. *Prof Exp:* Instr geol, Univ Taiwan, 61-63; res assoc mineral & geochem, Pa State Univ, 66-68; asst prof geol & asst mineralogist, 69-72, assoc prof geol & assoc mineralogist, 72-78, PROF GEOL & MINERALOGIST, MACKAY SCH MINES, UNIV NEV, RENO, 78- *Concurrent Pos:* Res consult, Northrop Space Lab, Calif, 66; UN sr scientist, China, 82 & 87; tech peer rev, Nuclear Waste Isolation Prog, 80; fel, Ministry Economic Affairs, Li Found. *Mem:* Am Geophys Union; fel Geol Soc Am; fel Mineral Soc Am; Geochem Soc; Soc Econ Geologists; Asn Explor Geochem; Sigma Xi. *Res:* Hydrothermal investigations of minerals or mineral assemblages related to genesis of ores and rocks; determinative mineralogy with instrumental techniques in geochemical studies of ores and minerals. *Mailing Add:* Mackay Sch Mines Univ Nev Reno NV 89557

HSU, LINDA, b Kunming, China, May 5, 44. ANATOMY, DEVELOPMENTAL BIOLOGY. *Educ:* Pomona Col, BA, 66; Univ Mich, MS, 68, PhD(anat), 71. *Prof Exp:* Fel anat, Yale Univ, 71-72; from instr to asst prof, NY Med Col, 72-75; fel, Rutgers Med Sch, 76-77; asst prof anat, NJ Sch Osteop Med, 77-85; ASST PROF BIOL, SETON HALL UNIV, 85- *Concurrent Pos:* Res grant, Am Cancer Soc, 73-74; NIH grant, 78-81; Nat Oceanog Asn grant, 81-83. *Mem:* Tissue Cult Soc; Am Asn Anat; Neurosci Soc. *Res:* Nerve and muscle regeneration; amphibian limb regeneration; neurotrophic effects. *Mailing Add:* 1000 Sunset Rd Piscataway NJ 08854-5510

HSU, MING-TA, b Taipei, Taiwan, Jan 4, 44; US citizen; m 73; c 1. MICROBIOLOGY. *Educ:* Nat Taiwan Univ, BS, 66, MS, 68; Calif Inst Technol, PhD(chem biol), 73. *Prof Exp:* Instr chem, Nat Defense Med Ctr, Taiwan, 68-69; teaching asst, Calif Inst Technol, 69-71, res asst, 71-73; Leukemia Soc fel, Med Ctr, Stanford Univ, 73-75; from asst prof to assco prof virol, Rockefeller Univ, 75-85; PROF MICROBIOL, MT SINAI MED CTR, 85- *Concurrent Pos:* Irma T Hirschl Found Award, 78-83. *Honors & Awards:* Herbert Newby McCoy Award, Calif Inst Technol, 73. *Mem:* Am Soc Microbiol. *Res:* Higher order organization of viral as well as cellular DNA in chromosomes; mechanisms of DNA repair of DNA viruses in mammalian cells. *Mailing Add:* Dept Microbiol Mt Sinai Sch Med One Gustave Levy Plaza New York NY 10029

HSU, MING-TA SUNG, b Hopei, China, Aug 3, 37; m 72; c 1. ORGANIC CHEMISTRY. *Educ:* Nat Taiwan Univ, BS, 60; NMex Highlands Univ, MS, 63; Iowa State Univ, PhD(org chem), 67. *Prof Exp:* Fel chem, Synvar Res Inst, 67-68; spectroscopist, Stanford Univ, 68-69; res chemist, Appl Space Prods Inc, 69-70; res assoc, NASA Ames Res Ctr, 70-72; res chemist, San Jose State Univ, 72-82; RES CHEMIST, H C CHEM RES, 82- *Concurrent Pos:* NASA contract, 72- *Mem:* Am Chem Soc; Soc Advan Mat Process Eng. *Res:* Organic photochemistry; molecular complexes of psychoactive compounds; spectroscopy; characterization and thermal-oxidative degradation of polymers; polymer synthesis. *Mailing Add:* 1934 Cape Hilda Pl San Jose CA 95133

HSU, NAI-CHAO, b Formosa; c 3. MATHEMATICS, ABSTRACT ALGEBRA. *Educ:* Washington Univ, PhD(math), 60. *Prof Exp:* Assoc mathematician, Int Bus Mach Corp, 60-63; asst prof math, State Univ NY Buffalo, 63-66; assoc prof, 66-72, PROF MATH, EASTERN ILL UNIV, 72- *Mem:* Am Math Soc. *Res:* Homological algebra. *Mailing Add:* 1030 Colony Lane Charleston IL 61920

HSU, NELSON NAE-CHING, b Shanghai, China, May 20, 34; US citizen; m 57; c 4. POLYMER SCIENCE, TECHNICAL MANAGEMENT. *Educ:* Mass Inst Technol, SB, 57; Univ Akron, MS, 60, PhD(polymer sci), 66; Univ Conn, MBA, 70. *Prof Exp:* Res chemist, B F Goodrich Res Ctr, 57-65; group leader, Aerospace Prod Dept, 65-80, group leader, Eng Mat Dept, 80-82, MGR TECH SERV LAB, PROCESS CHEM DEPT, AM CYANAMID CO, 86- *Concurrent Pos:* Res chemist, Inst Polymer Sci, Univ Akron, 63-65. *Res:* Physics and chemistry of polymers; polymerization processes; structural adhesives; commercial development of chemicals and polymers; latices; surfactants; paper chemicals, specialty monomers, radiation curing. *Mailing Add:* 158 Thornwood Rd Stamford CT 06903

HSU, ROBERT YING, b China, Oct 10, 26; US citizen; m 62; c 2. BIOCHEMISTRY. *Educ:* Nanking Univ, BS, 50; Iowa State Univ, MS, 52; Cornell Univ, MS, 55; Univ Wis, PhD(biochem), 61. *Prof Exp:* Res scientist, Armour Pharmaceut Co, 61-63; fel biochem, Univ Wis, 63-66; asst prof, Rutgers Univ, 66-68; from asst prof to assoc prof, 68-77, PROF BIOCHEM, STATE UNIV NY HEALTH SCI CTR, 77- *Concurrent Pos:* Vis assoc prof, Nat Taiwan Univ, 62; travel award, Int Cong Biochem, Japan, 67; Brit Royal Soc vis scholar, Oxford Univ, 82. *Mem:* Am Soc Biol Chem & Molecular Biol; Am Chem Soc. *Res:* Enzymology of fatty acid biosynthesis; mechanism of enzyme action; purification and characterization of metabolically significant enzymes. *Mailing Add:* Dept Biochem & Molecular Biol State Univ NY Health Sci Ctr Syracuse NY 13210

HSU, SHAW LING, b Shanghai, China, July 14, 48; m 70; c 2. POLYMER PHYSICS, VIBRATIONAL SPECTROSCOPY. *Educ:* Rutgers Univ, BA, 70; Univ Mich, PhD(physics), 75. *Prof Exp:* Res assoc physics, Macromolecular Res Ctr, Univ Mich, 75-76; res chemist, chem physics, Allied Chem Corp, 76-78; from asst prof to assoc prof, 78-87, DIR, NSF MAT RES LAB, UNIV MASS, 85-, PROF POLYMER SCI & ENG, 87- *Mem:* Am Phys Soc; Am Chem Soc; Optical Soc Am. *Res:* Conformational analysis of synthetic and biological polymers by infrared and Raman spectroscopy; use of the longitudinal acoustic mode to study polymer morphology; structural analysis of polyacetlylene-dopant systems; vibrational studies of phase transitions in polymeric systems; deformation studies of polymers by time resolved fourier transform infrared spectroscopy; characterization of polymer microstructure and phase transformation mechanisms. *Mailing Add:* Dept Polymer Sci & Eng Univ Mass Amherst MA 01003

HSU, SHENG TENG, b Taiwan, Oct 15, 34; m 64; c 2. SOLID STATE ELECTRONICS. *Educ:* Nat Taiwan Univ, BS, 58; Chiao Tung Univ, MS, 60; Univ Minn, PhD(elec eng), 66. *Prof Exp:* Mem staff solid state devices, Fairchild Semiconductor, 66-70; asst prof elec eng, Univ Man, 70-72; mem staff, 72-85, FEL, RCA LABS, 85- *Mem:* Inst Elec & Electronics Engrs. *Res:* Solid state devices; IC technologies. *Mailing Add:* 94 President's Lane Quincy MA 02169

HSU, SHIH-ANG, b Hanchou, China, Sept 15, 36; nat US; m 63; c 3. METEOROLOGY. *Educ:* Nat Taiwan Univ, BS, 61; Univ Tex, Austin, MS, 67, PhD(meteorol), 69. *Prof Exp:* Meteorologist-in-chg, Yunlin Tidal Land Develop & Demonstration, Taiwan Sugar Corp, 61-63; meteorol officer, Ministry of Commun, Taipei, Taiwan, 63-65; res asst atmospheric sci group, Univ Tex, Austin, 65-67, res sci assoc, 67-69; from asst prof to assoc prof, 69-77, PROF COASTAL STUDIES INST & DEPT MARINE SCI, LA STATE UNIV, BATON ROUGE, 77- *Concurrent Pos:* Meteorol consult, govt & indust. *Mem:* Am Meteorol Soc; Am Geophys Union. *Res:* Coastal and marine meteorology; air-sea interaction. *Mailing Add:* Coastal Studies Inst La State Univ Baton Rouge LA 70803

HSU, SHU YING LI, b Peking, China, Aug 16, 20; m 54. PARASITOLOGY, IMMUNOLOGY. *Educ:* Nat Peking Norm Univ, BS, 38; Univ Iowa, PhD, 57. *Prof Exp:* Asst prof parasitol & head dept, Nat Shenyang Med Col, 47-48; instr, Mich State Univ, 52-53; res assoc, Sch Pub Health, Harvard Univ, 53-54; from asst prof to assoc prof, 57-73, PROF PREV MED, UNIV IOWA, 73- *Honors & Awards:* SC Res Award, Taiwan, 53. *Mem:* AAAS; Am Soc Parasitol; Am Soc Trop Med & Hyg; Am Asn Immunol; Soc Exp Biol & Med. *Res:* Parasitology and immunology, especially schistosomiasis. *Mailing Add:* State Univ Iowa Col Med Iowa City IA 52240

HSU, STEPHEN M, b Shanghai, China, Nov 20, 43; m 68; c 2. TRIBOLOGY, ADVANCED CERAMICS. *Educ:* Va Polytechnic Inst & State Univ, BS, 68; Pa State Univ, MS, 72, PhD(chem eng), 76. *Prof Exp:* Res engr, Amco Res Ctr, 74-78; res sci, Nat Bur Standards, 78-79, group leader tribology, 79-85; DIV CHIEF, CERAMICS, NAT INST STANDARDS & TECHNOL, 85- *Concurrent Pos:* Prog mgr, recycled oil progs, Nat Bur Standards, 78-83; bd dir, Soc Tribologist & Lub Engrs, 88-; adj prof, Pa State Univ, 83-; mem, Ceramics Subcomt, Comt Mat, 86-; mem, Nat Steering Comt, superconductivity for power transmission, 88-90; chmn, Gov Steering Comt, Comput Tribology Info Syst, 86-91, IEA Int Round Robin ceramics powders, Gordon Res Conf tribology & US chmn, VAMAS Int study wear mat; Diamond Shamrock fel; chmn, Am Soc Testing & Mat. *Honors & Awards:* Bronze Medal, Dept Com, 84, Silver Medal, 90. *Mem:* Soc Tribologist & Lubrication Engrs; Am Soc Mech Engrs; Am Soc Testing & Mat; Soc Automotive Engrs; Am Ceramics Soc; Sigma Xi. *Res:* Tribology: lubrication mechanisms and models, microstructural and environmental effects on wear of metals, ceramics, coatings; ceramics tribology: ceramic powders characterization; tribochemistry: interface chemical reactions and kinetics. *Mailing Add:* Nat Inst Standards & Technol Gaithersburg MD 20899

HSU, SU-MING, b In-Lin, Tawain, Jan 28, 49; US citizen; m 77; c 1. PATHOLOGY, MOLECULAR BIOLOGY. *Educ:* Nat Taiwan Univ, MD, 74. *Prof Exp:* Staff fel path, Nat Cancer Inst, 81-84; asst prof path, Univ Tex, Houston, 84-89; PROF PATH, UNIV ARK, LITTLE ROCK, 90- *Honors & Awards:* Young Man of Year, Taiwan Govt, 76. *Mem:* Am Soc Path. *Res:* The nature of Hodgkin's disease and lymphomas; development of new tests to facilitate the diagnosis of Hodgkin's disease. *Mailing Add:* Dept Path Univ Ark Slot 517 Little Rock AR 72205

HSU, SUSAN HU, b Jan 19, 43; US citizen; m 68; c 2. IMMUNOGENETICS. *Educ:* Nat Taiwan Univ, BS, 65; Univ Ill, MS, 68, PhD(genetics), 70. *Prof Exp:* Trainee immunogenetics, Div Med Genetics, Johns Hopkins Univ, 70-74, from asst prof to assoc prof med, 74-86; SCI DIR & DIR HISTOCOMPATIBILITY LAB, AM RED CROSS PA-JERSEY REGION, 87- *Mem:* Am Soc Human Genetics. *Res:* Genetic marker research; lymphocytes culture research. *Mailing Add:* Dept Res Histocompatibility Lab Am Red Cross Pa-Jersey Region 23rd & Chestnut Sts Philadelphia PA 19103

HSU, TSONG-HAN, b Chekiang, China, Oct 10, 22; US citizen; m 50; c 1. ORGANIC POLYMER CHEMISTRY. *Educ:* Amoy Univ, BS, 47; Auburn Univ, MS, 64, PhD(org chem), 68. *Prof Exp:* Sr technologist wood chem, Taiwan Forest Res Inst, 48-61; res chemist, Lawrence Ottinger Res Ctr, Champion Bldg Prod, 68-72; sr chemist org systs, RSA Corp, 72-76; polymer expert adhesive coatings, UN Indust Develop Orgn, 76-79; sr res chemist bldg prod, Jim Walter Corp, 80-82; SR RESIN CHEMIST PAINT & COATINGS, HILLYARD CHEM CO, 82- *Mem:* Am Chem Soc; Int Union Pure & Appl Chem. *Res:* Chemistry of thermosetting polymers such as phenolic, amino and epoxy resins and their uses for coatings, adhesives and molding compounds; high solids epoxy coatings technology; water-borne alkyl coatings development. *Mailing Add:* 1548 81st Ave N St Petersburg FL 33702

HSU, WALTER HAW, b Fu-Jian, China, July 10, 46; m 71; c 2. ENDOCRINE PHARMACOLOGY, VETERINARY PHARMACOLOGY. *Educ:* Nat Taiwan Univ, BVetMed, 69; Univ NC, PhD(pharmacol), 75. *Prof Exp:* Res assoc pharmacol, Purdue Univ, 75-77; from asst prof to assoc prof, 77-86, PROF VET PHARMACOL & PHYSIOL, IOWA STATE UNIV, 86- *Honors & Awards:* Ralston Purina Small Animal Res Award, 87. *Mem:* Am Vet Med Asn; Am Soc Pharmacol Exp Ther; Soc Exp Biol Med. *Res:* Adrenergic pharmacology; clinical pharmacology of veterinary drugs. *Mailing Add:* Dept Vet Physiol & Pharmacol Iowa State Univ Ames IA 50011

HSU, WEN-TAH, b Hwa-Lien, Formosa, Dec 29, 30; m 62; c 2. BIOCHEMISTRY, MICROBIOLOGY. *Educ:* Nat Taiwan Univ, BS, 54; Mich State Univ, MS, 58; Univ Ill, PhD(microbiol), 63. *Prof Exp:* Asst prof biochem, 66-81, ASSOC PROF BIOCHEM, BEN MAY LAB, UNIV CHICAGO, 81-, RES ASSOC BIOCHEM, FRANKLIN MCLEAN MEM RES INST, 62- *Mem:* Soc Am Biologists. *Res:* Molecular mechanism of action of polycyclic aromatic hydrocarbons. *Mailing Add:* Dept Molecular Genetics CLSC Univ Chicago Chicago IL 60637

HSU, WILLIAM YANG-HSING, b Chungking, China, Apr 8, 48; m 75; c 1. POLYMER PHYSICS, SOLID STATE PHYSICS. *Educ:* Chinese Univ Hong Kong, BSc, 70; Univ Calif, Berkeley, PhD(physics), 75. *Prof Exp:* Res assoc physics, Univ Ill, Urbana, 75-77; res staff mem physics, 77-80, group leader, 80-81, RES SUPVR, CENT RES & DEVELOP DEPT, EXP STA, E I DU PONT DE NEMOURS & CO, INC, WILMINGTON, 81- *Mem:* Am Phys Soc. *Res:* Properties of ionic and covalent polymers; percolation theory; properties of ionic conductors, metals and impurities in semiconductors; galvanomagnetic properties. *Mailing Add:* 102 Harvest Ct Autumn Hills Hockessin DE 19707

HSU, Y(OUN) C(HANG), b Kiangsu, China, Apr 4, 35; m 63; c 2. MECHANICS, MECHANICAL ENGINEERING. *Educ:* Cheng Kung Univ, Taiwan, BSME, 57; Univ Wash, MSME, 61; Rensselaer Polytech Inst, PhD(mech), 64. *Prof Exp:* Asst, Rensselaer Polytech Inst, 61-64; sr res engr mech & mech eng, Southwest Res Inst, 64-68; from asst prof to assoc prof, 67-77, PROF MECH ENG, UNIV NMEX, 77- *Mem:* AAAS; Soc Eng Sci; Am Soc Mech Engrs; Sigma Xi. *Res:* Applied mechanics; fracture mechanics; elasticity; thermal stress; composite body; lubrication. *Mailing Add:* Dept Mech Eng Univ NMex Albuquerque NM 87131

HSU, YU KAO, b Wukang, Hunan, China, Apr 24, 22; m 65; c 2. MATHEMATICS, AERONAUTICS. *Educ:* Nat Cent Univ, Nanking, BS, 48; Univ Md, MS, 59; Univ Ill, Urbana, MS, 62; Rensselaer Polytech Inst, PhD(mech, aeronaut), 66. *Prof Exp:* Res asst exterior ballistics, Ord Res Inst, 48-56; asst magnetogasdynamics & aerodyn, Univ Md, 56-59; asst aerodyn & plasmadynamics, Univ Ill, Urbana, 59-62; asst fluid mech & magnetogasdynamics, Rensselaer Polytech Inst, 62-66; asst prof aerodyn & magnetogasdynamics, WVa Univ, 66-71; chmn dept, 71-74, assoc prof, 71-83, PROF MATH, UNIV COL, UNIV MAINE, BANGOR, 83- *Concurrent Pos:* NSF fel, 75; fac fel, NASA, 87, 88 & 89. *Mem:* Am Inst Aeronaut & Astronaut; Am Math Soc; Math Asn Am; Soc Indust & Appl Math. *Res:* Exterior ballistics; magnetogasdynamics and aerodynamics; heat transfer and fluid mechanics. *Mailing Add:* Div Natural Sci & Math Univ Col Univ Maine Bangor ME 04401

HSU, YU-SHENG, b China, Oct 12, 45. MATHEMATICS, STATISTICS. *Educ:* Nat Tsing Hua Univ, BS, 68; Univ Wis, MS, 71; Purdue Univ, PhD(statist), 75. *Prof Exp:* Lectr math, Southern Ill Univ, 75-77; asst prof, Wright State Univ, 77-78; asst prof math, Eastern Mont Col, 78-79; asst prof, 79-84, ASSOC PROF MATH, GA STATE UNIV, 84- *Mem:* Inst Math Statist; Am Statist Asn. *Res:* Distribution theory of some multivariate test statistics; robustness studies of some multivariate tests; geostatistics. *Mailing Add:* Calif Polytech State Univ San Luis Obispo CA 93407

HSUAN, HULBERT C S, b Jiangsu, China, Nov 28, 39; m 63; c 2. PLASMA PHYSICS, ELECTRICAL ENGINEERING. *Educ:* Univ Taiwan, BS, 60; Univ Ill, MS, 63; Princeton Univ, PhD(plasma physics), 67. *Prof Exp:* Electronic engr, Makong Navy Yard, Taiwan, 60-61; from asst prof to prof elec eng, Univ Iowa, 66-75; FROM RES PHYSICIST TO PRIN RES PHYSICIST, PLASMA PHYSICS LAB, PRINCETON UNIV, 73- *Concurrent Pos:* Chinese scientists fel from HSU Found, 57-60; Ford Found Fel, Plasma Physics, 63-64, NSF grants, 65-75; exec comt mem, PSAC-IEEE-NPS, 78-80. *Mem:* Fel Am Phys Soc; Inst Elec & Electronics Engrs. *Res:* Controlled thermonuclear fusion research; kinetic theory; electromagnetic theory; electronics; basic plasma dynamics. *Mailing Add:* Plasma Physics Lab Princeton Univ Princeton NJ 08543

HSUEH, AARON JEN WANG, b Nanking, China, Mar 10, 48. ENDOCRINOLOGY, REPRODUCTIVE BIOLOGY. *Educ:* Nat Taiwan Univ, BS, 69; Purdue Univ, MS, 72; Baylor Col Med, PhD(cell biol), 75. *Prof Exp:* Fel reproduction, NIH, 75-76; asst prof, 76-81, ASSOC PROF REPRODUCTIVE MED, UNIV CALIF, SAN DIEGO, 81- *Concurrent Pos:* Res career develop award, NIH, 81- *Mem:* Endocrine Soc; Soc Study

Reproduction; Soc Gynec Invest; AAAS; Am Soc Cell Biol. *Res:* Mechanism of action of steroid and peptide hormones in the regulation of ovarian, testicular and pituitary functions. *Mailing Add:* Dept Reproductive Med T-002 Univ Calif San Diego Med Sch La Jolla CA 92093

HSUEH, ANDIE M, b July 22, 40; div; c 1. NUTRITION & FOOD SCIENCE. *Educ:* Tunghai Univ, Taiwan, BS, 59; Tex Woman's Univ, MS, 63; Johns Hopkins Univ, ScD(nutrit biochem), 70. *Prof Exp:* Res assoc, Dept Biochem, Sch Hyg & Pub Health, Johns Hopkins Univ, 70-72, res asst prof, 72-79; from asst prof to assoc prof, 79-89, PROF, DEPT NUTRIT & FOOD SCI, TEX WOMAN'S UNIV, 89- *Concurrent Pos:* Wellcome vis prof, 81; NIH res grant, 89-92. *Mem:* Am Inst Nutrit; Sigma Xi. *Res:* Author of numerous technical publications. *Mailing Add:* Dept Nutrit & Food Sci Col Health Sci Tex Woman's Univ PO Box 24134 Denton TX 76204

HSUEH, YA, b Kiangsu, China, Mar 19, 36; m 64; c 2. PHYSICAL OCEANOGRAPHY. *Educ:* Nat Taiwan Univ, BSc, 58; Johns Hopkins Univ, PhD(mech), 65. *Prof Exp:* Res assoc airsea interaction, Univ Wash, 65-67; from asst prof to assoc prof, 67-80, chmn dept, 82-85, PROF OCEANOG, FLA STATE UNIV, 80-, CHMN DEPT, 88- *Concurrent Pos:* Prin investr, Inst Naval Oceanog, 87-90; prog dir phys oceanog, NSF, 80-81; prin investr, Off Naval Res, 89- *Mem:* Am Geophys Union; Am Meteorol Soc. *Res:* Dynamic oceanography; coastal upwelling; boundary currents of oceans; theoretical models of the circulation on the continental shelf; meanders of ocean currents in a rotating tank. *Mailing Add:* Dept Oceanog Fla State Univ Tallahassee FL 32306

HSUI, ALBERT TONG-KWAN, b Canton, China, 45. GEOPHYSICS, MECHANICS. *Educ:* Univ Lowell, BS, 68; Cornell Univ, ME, 69, PhD(geol sci), 72. *Prof Exp:* Res assoc micrometeorol, Ames Res Ctr, NASA, 72-73; mem tech staff eng, Bell Labs, 73-75; res assoc geophys, Mass Inst Technol, 75-79; ASSOC PROF, DEPT GEOL, UNIV ILL, 80- *Concurrent Pos:* Lectr, State Seismol Bur, China, 81. *Mem:* Am Geophys Union; Sigma Xi; NY Acad Sci; AAAS; Soc Explor Geophys. *Res:* Geodynamics; basin evolution studies; borehole geophysics. *Mailing Add:* Dept Geol 245 NHB Univ Ill 1301 W Green St Urbana IL 61801

HTOO, MAUNG SHWE, b Yenangyaung, Burma, Aug 17, 27; US citizen; m 53; c 4. PHYSICAL CHEMISTRY. *Educ:* Univ Maine, BS, 52, MS, 54; Rensselaer Polytech Inst, PhD(phys chem), 61. *Prof Exp:* Sr res chem engr, Res Div, Int Paper Co, NY, 54-61; staff chemist supplies div, Int Bus Mach Corp, NY, 61-62, develop chemist, 62-65, adv chemist components div, Poughkeepsie, 65-67, proj mgr photosensitive processes, 67-69, mgr chem technol mfg res lab, 70-74, mgr prod technol mat res lab, 75-81, MGR TECH ASSURANCE LAB, MGR, IBM CORP, 82- *Concurrent Pos:* Mem fac dept chem eve div, Dutchess Community Col, 67-84; sr tech staff mem, IBM, 85. *Mem:* NY Acad Sci; distinguished mem, Soc Plastics Engrs; Am Chem Soc; fel Am Inst Chemists. *Res:* Photosensitive materials and processes; organic semiconductors; thin films; organic coatings. *Mailing Add:* Ten Rabbit Trail Rd Poughkeepsie NY 12603

HU, ALFRED SOY LAN, b Honolulu, Hawaii, Apr 1, 28; m 59; c 6. BIOCHEMISTRY. *Educ:* Univ Hawaii, BS, 50; Univ Ore, PhD(biol), 57. *Prof Exp:* Res assoc biochem, Univ Ore, 56-58; USPHS fel, Univ Wis, 58-59, Univ Res Found, 59-60; asst prof biochem, NMex Highlands Univ, 60-61; from asst prof to assoc prof, 61-74, PROF BIOCHEM, UNIV KY, 74- *Mem:* Am Soc Biol Chem. *Res:* Enzymology; cellular regulation; genetics. *Mailing Add:* Dept Biochem Univ Ky Lexington KY 40536-0084

HU, BAMBI, b Chongqing, Sichuan Prov, June 4, 45; US citizen; m 75; c 3. STATISTICAL MECHANICS, PARTICLES & FIELDS. *Educ:* Univ Calif, Berkeley, BA, 67; Cornell Univ, MS, 70, PhD(physics), 74. *Prof Exp:* Res assoc physics, Cornell Univ, 73-74, Centre d'Etudes Nucleaires de Saclay, 74-75, Ecole Polytechnique, France, 75-77 & Brown Univ, 77-78; from asst prof to assoc prof, 78-87, PROF PHYSICS, UNIV HOUSTON, 87- *Concurrent Pos:* Vis assoc prof, Univ Calif, Santa Cruz, 82-83; collabr, Los Alamos Nat Lab, 83-; ed, Int J Modern Physics B, 87-; adj prof, Beijing Normal Univ, Shandong Univ, Shandong Normal Univ, 88-; mem, Acad Adv Comt, Inst Physics, Academia Sinica, Taipei, 91-; div coordr, Overseas Chinese Physics Asn, 91-93. *Res:* Theoretical physics; nonlinear dynamics and chaos; phase transitions and critical phenomena; quantum field theory and particle physics. *Mailing Add:* Dept Physics Univ Houston Houston TX 77204-5504

HU, BEI-LOK BERNARD, b Chungking, China, Oct 4, 47; m 72; c 2. GENERAL RELATIVITY, COSMOLOGY. *Educ:* Univ Calif, Berkeley, BA, 67; Princeton Univ, MA, 69, PhD(physics), 72. *Prof Exp:* Res assoc physics, Princeton Univ, 72-73, & Stanford Univ, 73-74; mem natural sci, Inst Advan Study, Princeton, 73; res assoc physics & astron, Univ Md, 74-75; res mathematician, Univ Calif, Berkeley, 75-76; res astrophysicist, Inst Space Studies, NY, 76-77; res physicist, Univ Calif, Santa Barbara, 77-79; hon fel physics, Harvard Univ, 79-80; from asst prof to assoc prof, 80-88, PROF PHYSICS & ASTRON, UNIV MD, 88- *Concurrent Pos:* Coun, Chinese Soc & Gravitation Physics; bd, assoc mem Inst Advan Study, Princeton. *Mem:* Am Phys Soc; NY Acad Sci; Int Soc Gen Relativity & Gravitation; Chinese Soc Gravitation Physics & Relativistic Astrophys. *Res:* General relativity and cosmology; quantum field theory in curved spacetime; quantum theories of the early universe; quantum statistics; quantum gravity. *Mailing Add:* Dept Physics & Astron Univ Md College Park MD 20742

HU, CHENMING, b Beijing, China, July 12, 47; US citizen. INTEGRATED CIRCUIT TECHNOLOGY, SEMICONDUCTOR DEVICES. *Educ:* Nat Taiwan Univ, BS, 68, Univ Calif, Berkeley, MS, 70, PhD(Elec Eng), 73. *Prof Exp:* Asst prof elec eng, Mass Inst Technol, 73-76; PROF ELEC ENG, UNIV CALIF, BERKELEY, 76- *Concurrent Pos:* Mgr nonvolatile memory develop, Nat Semiconductor, 80-81; consult electronics indust, 75-; assoc ed, Inst Elec & Electronics Engrs, Trans Electron Devices, 86-88; hon prof, Beijing Univ, China. *Mem:* Fel Inst Elec & Electronics Engrs. *Res:* Advanced integrated circuit technologies; integrated circuit reliability physics; non-volatile memory; power devices. *Mailing Add:* Dept Elec Eng & Comp Sci Univ Calif Berkeley Berkeley CA 94720

HU, CHIA-REN, b Anhwei, China, May 25, 39; m 67. SUPERCONDUCTIVITY-SUPERFLUIDITY, LIGHT-SCATTERING. *Educ:* Nat Taiwan Univ, BS, 62; Univ Md, College Park, PhD(physics), 68. *Prof Exp:* Res assoc physics, Univ Ill, 68-69; asst prof physics, Univ Southern Calif, 69-76; assoc prof, 77-85, PROF PHYSICS, TEX A&M UNIV, 85- *Concurrent Pos:* Vis prof physics, Univ Southern Calif, 85-86. *Mem:* Am Phys Soc. *Res:* Quantum mechanical many body problem; theory of superconductivity magnetism and superfluid helium; theory of electromagnetic scattering. *Mailing Add:* Dept Physics Tex A&M Univ College Station TX 77843-4242

HU, CHING-YEH, b Chiang-su, China, Dec 15, 37; US citizen; m 65; c 2. HUMAN GENETICS, PLANT PHYSIOLOGY. *Educ:* Taichung Agr Col, BS, 60; WVa Univ, MS, 66, PhD(genetics), 68. *Prof Exp:* Asst prof genetics, Washington & Lee Univ, 68-69; asst prof genetics & bot, William Paterson Col NJ, 69-73, assoc prof, 73-78; vis res prof plain tissue cult, Academia Sinica, China, 78-79; PROF GENETICS & BOT, WILLIAM PATERSON COL NJ, 79- *Mem:* Am Soc Plant Physiologists; Holly Soc Am; Int Asn Plant Tissue Cult; Am Soybean Asn. *Res:* In vitro culture of glycine max, flex, taxus; crop genetic engineering. *Mailing Add:* William Paterson Col Wayne NJ 07470

HU, CHI-YU, b Szechwan, China, Feb 12, 33; c 4. NUCLEAR PHYSICS. *Educ:* Nat Taiwan Univ, BS, 55; Mass Inst Technol, PhD(physics), 62. *Prof Exp:* Res assoc physics, St John's Univ, NY, 62-63; from asst prof to assoc prof physics, 63-72, PROF PHYSICS, CALIF STATE UNIV, LONG BEACH, 72- *Concurrent Pos:* NSF res grant, 69-70, DOE 86-88; vis prof, NSF, 88-90. *Mem:* Am Phys Soc. *Res:* Collective properties of nuclei; pairing correlation and Hartree Fock approximation; nuclear three body problem; nuclear three and four body bound states; muon catayged fusion. *Mailing Add:* Dept Physics Calif State Univ Long Beach CA 90840

HU, FUNAN, b Shanghai, China, Sept 13, 19; US citizen; m 44. DERMATOLOGY, DERMATOPATHOLOGY. *Educ:* Nat Med Col Shanghai, MD, 42. *Prof Exp:* Resident, Med Br, Univ Tex, 47-48, univ fel, 50-53; assoc dermat, Henry Ford Hosp, Detroit, Mich, 53-65; PROF DERMAT, ORE HEALTH SCI UNIV, 65-, SCIENTIST, ORE REGIONAL PRIMATE RES CTR, 65- *Mem:* Am Soc Dermatopath; Am Acad Dermat; Soc Investigative Dermat; Am Asn Cancer Res; Am Soc Cell Biol. *Res:* Cutaneous biology; cytology; pigment cell biology; melanogenesis; pigmentary disorders. *Mailing Add:* Ore Regional Primate Res Ctr 4175 SW Crestwood Dr Portland OR 97225

HU, JING-SHAN, b Xinyang, Henan, Peoples Repub China, Oct 20, 63; m 87. DNA-PROTEIN INTERACTIONS, GENE REGULATION. *Educ:* Beijing Univ, BS, 83; Univ Tex, PhD(cell & molecular biol), 89. *Prof Exp:* POSTDOCTORAL RES FEL, DEPT GENETICS, HARVARD MED SCH, 89- *Concurrent Pos:* Postdoctoral fel, Leukemia Res Found, 90-91. *Mem:* Am Soc Cell Biol. *Res:* Regulation of tissue-specific gene expression by Helix-Loop Helix proteins. *Mailing Add:* Dept Molecular Biol Mass Gen Hosp Boston MA 02114

HU, JOHN NAN-HAI, b Kaifeng, China, Aug 14, 36; m 64; c 2. CHEMICAL ENGINEERING. *Educ:* Taiwan Univ, BS, 58; Univ Notre Dame, PhD(diffusion), 65. *Prof Exp:* CHEM ENGR, RES LABS, ETHYL CORP, 64- *Res:* Diffusion; refinery technology; economic evaluation. *Mailing Add:* Res Labs Box 14799 Ethyl Corp Baton Rouge LA 70898

HU, L(ING) W(EN), b Nanking, China, Jan 28, 22; m 53; c 2. ENGINEERING MECHANICS. *Educ:* Nat Cent Univ, China, BS, 43; Univ Wash, MS, 49; Pa State Univ, PhD(eng mech), 52. *Prof Exp:* Designer, Nat Cent Mach Works, China, 43-45; instr mech eng, Nat Cent Univ, 45-47; assoc prof, 49-62, PROF ENG MECH, PA STATE UNIV, 62- *Mem:* Soc Exp Stress Anal; Am Acad Mech. *Res:* Mechanical properties of metals; plasticity; experimental stress analysis; triaxial stress experiments and high pressure testing. *Mailing Add:* 728 Westerly Pkwy State College PA 16801

HU, MING-KUEI, b Anhwei, China, May 25, 18; m 46; c 1. ELECTRICAL ENGINEERING. *Educ:* Nat Cent Univ, China, BS, 41; Ore State Col, PhD(elec eng), 51. *Prof Exp:* Instr elec eng, Nat Cent Univ, China, 44-47; res asst prof, 51-56, sr res engr, 56-60, assoc prof, 60-63, PROF ELEC ENG, SYRACUSE UNIV, 63- *Mem:* Am Phys Soc; sr mem Inst Elec & Electronics Engrs; Asn Comput Mach. *Res:* Electromagnetic theory; antenna and wave propagation; high voltage; discharge phenomena; electronic computers; switching theory; artificial intelligence; theory of automata. *Mailing Add:* Col Eng Syracuse Univ Syracuse NY 13244

HU, PUNG NIEN, b Chenghai, China, Oct 29, 26; US citizen; m 58; c 3. PHYSICS. *Educ:* Cent Inst Technol, China, dipl elec eng, 47; Stevens Inst Technol, MS, 56, DSc(fluid dynamics), 58. *Prof Exp:* Engr, Chung Hwa Sci Co, China, 47-49; lectr physics, Taipei Inst Technol, Taiwan, 49-54; res asst, Stevens Inst Technol, 54-57, sr res engr, Davidson Lab, 57-59, staff res scientist, 60-62, div head flow phenomena, 64-65; asst prof eng sci, Pratt Inst, 59-60; vis res mem, Courant Inst Math Sci, NY Univ, 62-64; sr res scientist, Space Sci Inc, Waltham, 65-68, prin scientist & asst mgr physics dept, 68-71; consult, Aeronaut Res Assocs of Princeton 71-73; RES PROF & SR RES SCIENTIST, COURANT INST MATH, NY UNIV, 74- *Mem:* Am Phys Soc. *Res:* Plasma physics; fluid dynamics; kinetic theory of gases. *Mailing Add:* 212 Reynard Rd Bridgewater NJ 08807

HU, SHIU-LOK, b Hong Kong, Nov 10, 49; m 84; c 2. AIDS VACCINE RESEARCH, CANCER IMMUNOTHERAPY. *Educ:* Univ Calif, Berkeley, BA, 71; Univ Wis-Madison, PhD(molecular biol), 78. *Prof Exp:* Postdoctoral fel tumor virol, Cold Spring Harbor Lab, 78-81; sr scientist molecular biol, Molecular Genetics, Inc, 81-83, dir molecular biol, 83-84; sr scientist virol, 85-86, lab dir virol, 86-90, EXEC DIR VIROL, BRISTOL-MYERS SQUIBB PHARMACEUT RES INST, 91- *Concurrent Pos:* Res assoc prof, Microbiol Dept, Univ Wash, 88-; mem, Basic Sci II Subcomt, AIDS Res Rev Comt, Nat

Inst Allergy & Infectious Dis, NIH, 90-93. *Mem:* Am Soc Microbiol; Am Soc Virol; AAAS. *Res:* Virology and vaccine development; anti-tumor immunotherapy; recombinant DNA technology; control of gene expression. *Mailing Add:* 3005 First Ave Seattle WA 98121

HU, STEVE SENG-CHIU, b Yangchou City, China; US citizen; c 3. SYSTEMS RESEARCH. *Educ:* Chiao-Tung Univ, China, BS, 39; Rensselaer Polytech Inst, MS, 40; Mass Inst Technol, ScD(aero-space), 42. *Prof Exp:* Managing tech dir, China Aircraft/China Motor Prog, Douglas Aircraft Corp, 43-48 & Kelly Eng Corp, 49-54; systs engr & meteorol specialist, RCA Corp, 55-58; consult gas dynamics, Aerojet Gen Corp, 58-60; res scientist, Jet Propulsion Labs, Calif Inst Technol, 60-61; tech dir, Northrop Corp & Northrop Huntsville, Space Lab, 61-72; DIR, UNIV AM UNITED RES INST, 73-; DIR, CENTURY RES INST, 73- *Concurrent Pos:* Vis fel, Calif Inst Technol, 43-44; consult, Nuclear Reactor Prog, Tsin-Hwa Univ, Taiwan, 54-58; vis prof & lectr, Univ Southern Calif, 58-62 & 68-73; prof aerospace, Univ Ala, Huntsville, 63-68; prof mech eng, Auburn Univ, 63-66; vis prof, Chun-Shan Inst Technol, Taiwan, 68-69 & Calif State Univ, 68-72; ed, Proceedings Missiles & Aerospace Vehicle Sci, Am Astronaut Soc, 63-71; pres, Univ Am Found, 77- *Mem:* Am Astronaut Soc (vpres, 63-71); Am Inst Aeronaut & Astronaut; Nat Asn Tech Schs. *Res:* Applied mathematics, applied physics, and electro-mechanical engineering in the fields of propulsion, gas dynamics, navigation and guidance, environmental science and computer science. *Mailing Add:* 6491 Saddle Dr Long Beach CA 90815

HU, SUNG CHIAO, b Chekiang, China, Nov 4, 42; m 67; c 3. DIGITAL SYSTEMS DESIGN, COMPUTER CONTROL. *Educ:* Calif State Polytech Col, San Luis Obispo, BSEE, 65; Ore State Univ, MSEE, 67, PhD(elec eng), 70. *Prof Exp:* From asst prof to assoc prof elec eng, Cleveland State Univ, 70-80; PROF, SAN FRANCISCO STATE UNIV, 80- *Concurrent Pos:* Consult, numerous co. *Mem:* Inst Elec & Electronics Engrs; Instrument Soc Am; Am Soc Eng Educ. *Res:* Microcomputers; digital systems; logic design; switching theory; electric distribution automation; biomedical systems. *Mailing Add:* San Francisco State Univ 1600 Holloway Ave San Francisco CA 94132

HU, SZE-TSEN, mathematics, for more information see previous edition

HU, TE CHIANG, b Peking, China, Nov 28, 30; US citizen; m 60; c 3. COMBINATORIAL ALGORITHMS, OPERATIONS RESEARCH. *Educ:* Nat Taiwan Univ, BS, 53; Univ Ill, Urbana, MS, 56; Brown Univ, PhD(appl math), 60. *Prof Exp:* Res mathematician, IBM Res Ctr, 60-66; from assoc prof to prof math & comput sci, Univ Wis-Madison, 66-74; CHMN, COMPUT SCI DIV, UNIV CALIF, SAN DIEGO, 74- *Concurrent Pos:* Vis assoc prof, Univ Calif, Berkeley, 64-65; consult, Rand Corp, 64 & Off of Emergency Preparedness, Exec Off of President, 68-72. *Mem:* Soc Indust & Appl Math; Oper Res Soc Am; Inst Elec & Electronics Engrs; Asn Comput Mach. *Res:* Applied mathematics applied to computer science; mathematical programming; discrete optimization; combinatorial algorithms; very-large-scale-intergration design. *Mailing Add:* Comput Sci Dept Univ Calif San Diego La Jolla CA 92093

HU, WEI-SHOU, b Taiwan, Nov 5, 51; US citizen; m 76; c 2. BIOREACTOR DESIGN, ANIMAL CELL TECHNOLOGY. *Educ:* Nat Taiwan Univ, BS, 74; Mass Inst Technol, SM, 81, PhD(biochem eng), 83. *Prof Exp:* Asst prof, 83-89, ASSOC PROF CHEM ENG, UNIV MN, 89- *Concurrent Pos:* NSF presidential young investr, 85; vis fel, Int Ctr Biotechnol, Fac Eng, Osaka Univ, Japan, 89, vis prof, 91; co-chmn, 2nd Eng Found Conf Cell Cult Eng, Santa Barbara, Calif, 89; mem, grant selection panel, Nat Sci & Eng Res Coun Can, 90-91, deleg, US/Japan Joint Biotechnol Conf, Hawaii, 91. *Mem:* Am Chem Soc; Am Inst Chem Engrs; Am Soc Microbiologists. *Res:* Biochemical engineering with emphasis on reactor design and reaction kinetics of animal, plant and microbiological cell processes. *Mailing Add:* Dept Chem Eng & Mat Sci Univ Minn 421 Washington Ave SE Minneapolis MN 55455

HU, WILLIAM H(SUN), b Anhwei, China, Aug 17, 27; nat US; m 61; c 1. PHYSICAL METALLURGY, MATERIALS SCIENCE. *Educ:* Nat Chiao-Tung Univ, BS, 42; Lehigh Univ, MS, 47; Univ Notre Dame, PhD(metall), 51. *Prof Exp:* res metallurgist in charge x-ray diffraction lab, Inst Study Metals, Chicago, 51-55; res metallurgist, res labs, Westinghouse Elec Corp, 55-59; sr scientist, sect head & staff scientist, US Steel Corp, 59-75, res consult, 75-81, sr res consult, basic res div, 81-83; RES PROF, DEPT MAT SCI & ENG, UNIV PITTSBURGH, 84- *Concurrent Pos:* Vis prof, Clausthal Tech Univ, 67-68; founder & ed, Int J Texture, 72-82; sci consult, US Army, Dept Defense, 76-; vis scientist, Risø Nat Lab, Raskilde, Denmark, 89; vis res prof, Inst Metallphsik, Univ Göttingen, 87 & Inst Metalkd u Metallphysik, RWTH-Aachem, 88, Lab de Metallurgie Structurale, Univ Paris-Sud, Orsay, France, 90. *Honors & Awards:* Mathewson Gold Medal, Am Inst Mining, Metall & Petrol Engrs, 53; Distinguished Sr Scientist Award, Alexander von Humboldt Found, Ger, 87. *Mem:* Fel Am Soc Metals; Am Inst Mining, Metall & Petrol Engrs. *Res:* Structure and properties of materials, in particular, deformation and annealing textures; strengthening mechanisms and phase transformations; recovery, recrystallization and grain growth; nature, behavior and properties of grain boundaries. *Mailing Add:* Dept Mat Sci & Eng Univ Pittsburgh Pittsburgh PA 15261-0001

HUA, DUY HUU, b Saigon, Vietnam, June 30, 52; Taiwan citizen; m 76; c 2. ORGANIC CHEMISTRY. *Educ:* Kyoto Univ, Japan, BS, 76; Southern Ill Univ, PhD(chem), 80. *Prof Exp:* Asst chem, Harvard Univ, 80-82; ASST PROF ORG CHEM, KANS STATE UNIV, 82- *Mem:* Am Chem Soc. *Res:* Asymmetric synthesis and stereocontrolled total synthesis of natural products. *Mailing Add:* Dept Chem Kans State Univ Willard Hall Manhattan KS 66506

HUA, PING, b Nanchang, China, July 18, 63. MEDICAL IMAGING, COMPUTER VISION. *Educ:* Shanghai Jiao Tong Univ, Shanghai, China, BS, 82; Univ Wis-Madison, MS, 84, PhD(elec eng), 90. *Prof Exp:* Instr elec eng, Shanghai Jiao Tong Univ, 82-83; RES SCIENTIST, SIEMENS AG, 90- *Mem:* Inst Elec & Electronics Engrs. *Res:* Medical image processing; computer vision applications in medical imaging, especially in digital angiography; three dimensional reconstruction and visualization; computer-based medical instrumentation. *Mailing Add:* Res Dept Siemens Gammasonics Inc Hoffman Estates IL 60195

HUANG, ALICE SHIH-HOU, b Kiangsi, China, Mar 22, 39; US citizen; c 1. MICROBIOLOGY, VIROLOGY. *Educ:* Johns Hopkins Univ, AB, 61, MA, 63, PhD(microbiol), 66. *Hon Degrees:* DSc, Wheaton Col, 71, Mt Holyoke Col, 87. *Prof Exp:* Asst prof zool, Taiwan, 66; from asst prof to assoc prof, 71-79, prof microbiol & molecular genetics, Harvard Med Sch, 79-91; DEAN SCI, NEW YORK UNIV, 91- *Concurrent Pos:* USPHS fel biochem virol, Salk Inst Biol Studies, 67 & Mass Inst Technol, 68-70; mem ed bd, Intervirology, 73-; mem virol study sect, NIH, 73-77; dir, Lab Infectious Dis, Children's Hosp Med Ctr, Boston, 79- *Honors & Awards:* Eli Lilly Award in Microbiol & Immunol, 77. *Mem:* AAAS; Am Soc Biol Chem; Am Soc Microbiol; Infectious Dis Soc Am. *Res:* Replication of RNA animal viruses. *Mailing Add:* Box 500 1230 York Ave New York NY 10021-6399

HUANG, ANTHONY HWOON CHUNG, b China, Oct 10, 45; m 72; c 1. PLANT PHYSIOLOGY, MOLECULAR BIOLOGY. *Educ:* Nat Taiwan Univ, BS; Univ Calif, Santa Cruz, PhD(biol), 73. *Prof Exp:* From asst prof to prof, 73-88, Carolina res prof, Univ SC, 87-88; PROF BOT, UNIV CALIF, RIVERSIDE, 88- *Concurrent Pos:* Distinguished scholar travelling grant, Nat Acad Sci, 84. *Mem:* Am Soc Plant Physiologists. *Res:* Plant cell and molecular biology; metabolic compartmentation; seed maturation and germination; lipid metabolism; genetic engineering. *Mailing Add:* Dept Bot & Plant Sci Univ Calif Riverside CA 92521

HUANG, BARNEY K(UO-YEN), b Taiwan, China, Jan 27, 31; US citizen; m 64; c 3. BIOLOGICAL & AGRICULTURAL ENGINEERING. *Educ:* Nat Taiwan Univ, BS, 54; Univ Ill, MS, 60; Purdue Univ, PhD(agr eng), 63. *Prof Exp:* Instr & asst agr eng, Nat Taiwan Univ, 54-58; asst, Univ Ill, 58-60 & Purdue Univ, 60-62; from asst prof to assoc prof, 63-73, PROF BIOL & AGR ENG, NC STATE UNIV, 73- *Concurrent Pos:* NSF lectr human eng, NC State Univ; prin investr, NSF, Energy Res & Develop Asn, US Dept Agr, Dept Energy, HEW, Pub Health Serv, R J Reynolds Co, Am Can Co, NC Energy Inst & Dept Transp grants; vis prof, Tehran Univ, 75, Japan Adv Sci, 79, Korea Tobacco Res Inst, 80, Nat Taiwan Univ, 79-80 & Jilin Univ Technol, 87; consult, Coun Agr, 74-75, Taiwan Tobacco Bur, 76, Res Triangle Inst, 76-77, Union Carbide Corp, 77-79, UN, 80-81, Valmont Indust, 83-84 & World Bank, 91; hon prof, Shenyang Agr Univ, 87. *Honors & Awards:* Distinguished Leadership Award, ABI, 88. *Mem:* Am Soc Agr Engrs; Am Soc Eng Educ; Nat Geog Soc; AAAS; Sigma Xi. *Res:* Solar energy utilization in agriculture; biological servo-systems and human engineering; computer simulation and data acquisition; computer aided design and analysis of biological and physical systems; mechanics of vibrations, sounds and locomotion; agricultural power and machinery design; author & co-author of over 100 technical and research articles and books; patentee of over 20 United States, Canadian, Chinese and Japanese patents. *Mailing Add:* Dept Biol & Agr Eng NC State Univ 3332 Manor Ridge Dr Raleigh NC 27603-4845

HUANG, BESSIE PEI-HSI, genetics, biochemistry, for more information see previous edition

HUANG, C YUAN, b Tainan, Taiwan, Mar 14, 35; US citizen; m 61; c 3. TRANSPORT, OPTICAL. *Educ:* Nat Taiwan Univ, BS, 57; Harvard Univ, MS, 60, PhD(appl physics), 64. *Prof Exp:* Asst prof elec eng, Wash Univ, 65-66; assoc prof physics, Case Western Reserve Univ, 66-75; staff mem physics, Los Alamos Nat Lab, 75-87; SR STAFF SCIENTIST PHYSICS, LOCKHEED RES LAB, 87- *Concurrent Pos:* Adj prof, dept physics, Univ Houston, 87- *Honors & Awards:* Alan Berman Award, 85. *Mem:* Fel Am Phys Soc. *Res:* Superconductivity; magnetism; optics. *Mailing Add:* 097-40 B202 Lockheed Palo Alto Res Lab Palo Alto CA 94304-1191

HUANG, CHAOFU, b Taipei, Taiwan, Apr 3, 59. MICROELECTRONICS, IC PROCESSING. *Educ:* Nat Cheng-Kung Univ, BSEE, 81, MSEE, 83; Univ Calif, Santa Barbara, MSEE, 86. *Prof Exp:* SR ENGR RES & DEVELOP, ELECTRO-OPTEK CORP, 86- *Res:* Development of molecular beam epitaxy CMBEO processes for fabricating Insb infrared detector arrays; diamond film; high temperature superconductors; multiple quantum well infrared detector; surface emitting laser arrays. *Mailing Add:* 22818 Broadwell Ave Torrance CA 90502

HUANG, CHARLES T L, b Fukien, China, Oct 12, 38; m 70; c 1. ORGANIC CHEMISTRY, BIOCHEMISTRY. *Educ:* Nat Chung Hsing Univ, BSc, 60; La State Univ, PhD(chem), 70. *Prof Exp:* Res assoc biochem eng, La State Univ, 70-71; res assoc pediat, Baylor Col Med, 72-73, from instr to asst prof pediat, 73-82; SR RES CHEMIST, MERICHEM CO, 82- *Mem:* AAAS; Am Chem Soc; Inst Food Technologists; Am Asn Univ Prof; fel Am Inst Chemists. *Res:* Lipid metabolism; diarrheal disease; lipid chemistry; organic synthesis process development. *Mailing Add:* Merichem Co Res Ctr 1503 Cent Houston TX 77012

HUANG, CHARLES Y, b Nanking, China; US citizen. ENZYME KINETICS & MECHANISMS, CALCIUM REGULATION. *Educ:* Iowa State Univ, Ames, PhD(biochem), 68. *Prof Exp:* Pub Health Serv fel, Med Sch, Wash Univ, 68-70; staff fel, NIMH, NIH, 70-73; group leader, Dow Chem Co, 73-75; chemist I, Abbott Labs, 75-76; RES CHEMIST, LAB BIOCHEM, NAT HEART LUNG & BLOOD INST, NIH, 76- *Concurrent Pos:* Consult, Enzyme Kinetics Nomenclature, Int Union Biochem, 78-80, comt mem, Kinetics & Mechanisms Enzymes & Metab Interest Group, 83-; mem, Prog Comt, Am Soc Biochem & Molecular Biol, 81; guest lectr, Dept Chem, Georgetown Univ, 82; contract prof, Dept Cellular Biol, Univ Camerino, Italy, 83, 84 & 88; vis prof, Dept Biochem, Nanjing Univ, China, 85 & Inst Med Sci, Henan Med Univ, Zhengzhov, China, 85. *Mem:* Am Soc Biochem & Molecular Biol; Am Chem Soc; AAAS. *Res:* Enzyme regulatory mechanisms and kinetic behaviors, particularly enzyme systems involving calcium/calmodulin and phosphorylation/dephosphorylation; development of methods and theories applicable to enzyme research. *Mailing Add:* Bldg 3 Rm 218 NIH Bethesda MD 20892

HUANG, CHAU-TING, b Mar 28, 39; c 2. PHYSIOLOGY, PHARMACOLOGY. *Educ:* Univ Alta, Can, PhD(biochem toxicol), 70. *Prof Exp:* RES SCIENTIST, NORWICH EATON PHARMACEUT INC, 78- *Mem:* Am Physiol Soc. *Res:* Cardiac; hypertension. *Mailing Add:* Cardiac Dept Norwich Eaton Pharmaceut Inc PO Box 191 Norwich NY 13815

HUANG, CHE C, US citizen; m. RADIOCHEMISTRY. *Educ:* Tunghai Univ, BS, 69; State Univ NY, Binghamton, MA, 73; Univ Mich, PhD(med chem), 77. *Prof Exp:* Res fel nuclear med, Argonne Nat Lab, 77-80; sr scientist, 80-83, RES ASSOC, WARNER-LAMBERT/PARKE-DAVIS, 84- *Concurrent Pos:* Sr mgr, Warner-Lambert/Parke-Davis. *Mem:* Am Chem Soc; Int Isotope Soc. *Res:* Synthesis and design of radiolabeled drugs for metabolism and medical imaging; central nervous system drugs and steroid chemistry; development of new techniques for tritiation and radiohalogenation. *Mailing Add:* Warner-Lambert/Parke-Davis 2800 Plymouth Ann Arbor MI 48105

HUANG, CHENG-CHER, b Taipei, Taiwan, May 25, 47; m 72; c 2. PHYSICS. *Educ:* Nat Taiwan Univ, BS, 69; Univ Pa, PhD(physics), 75. *Prof Exp:* Res asst physics, Univ Pa, 72-75; res assoc, Univ Ill, 75-77; from asst prof to assoc prof, 77-87, PROF PHYSICS, UNIV MINN, 87- *Concurrent Pos:* Vis scientist, AT&T Bell Labs, 80, 81 & 85; consult, 3M Tech Res Lab, 83-85, vis prof, 84; vis prof, Chalmers Univ of Technol, Sweden, 84. *Mem:* Am Physics Soc. *Res:* Thermal and optical studies in phase transitions of various liquid-crystal mesophase. *Mailing Add:* Dept Physics Univ Minn Minneapolis MN 55455

HUANG, CHENG-CHUN, b Taipei, Taiwan, Feb 2, 38; m 67; c 2. OTOLARYNGOLOGY. *Educ:* Taiwan Univ, BS, 63; Vanderbilt Univ, MS, 66; Univ Iowa, Iowa City, PhD(biochem), 70. *Prof Exp:* Res assoc biochem, Univ Iowa, 70-71, assoc res scientist otolaryngol, 73-76, res scientist, 76-77; res assoc pharmacol, Yale Univ, 71-73; from asst prof to assoc prof, 77-89, PROF OTOLARYNGOL, COLUMBIA UNIV, 89- *Concurrent Pos:* Mem, Fac Coun, Col Physicians & Surgeons, Columbia Univ, 82- *Honors & Awards:* First Prize Award, Am Acad Ophthal & Otolaryngol, 78. *Mem:* Am Soc Biol Chemists; Asn Res Otolaryngol; AAAS; Centurions Deafness Res Asn; Chinese Agr Chem Soc. *Res:* Pathogenesis of bone destruction in chronic middle ear cyst (called cholesteatoma) and in inner ear (called otosclerosis). *Mailing Add:* Dept Otolaryngol Columbia Univ 630 W 168th St New York NY 10032

HUANG, CHESTER CHEN-CHIU, b Kiangsu, China, Jan 16, 27; m 61; c 3. BIOLOGY, CYTOGENETICS. *Educ:* Sun Yat-Sen Univ, BS, 49; SDak State Univ, MS, 61; State Univ NY Buffalo, PhD, 66. *Prof Exp:* Teacher agron, Prov Chi-yi Sr Agr Sch, Taiwan, 49-54 & Prov Col Agr, 55-58; res asst, SDak State Univ, 59-61; cancer res scientist, 61-66, sr cancer res scientist, 66-76, CANCER RES SCIENTIST V, ROSWELL PARK MEM INST, 77-; ASSOC PROF, STATE UNIV NY, BUFFALO, 74- *Concurrent Pos:* Asst prof, State Univ NY, Buffalo, 68-74. *Res:* Mammalian chromosome cytology; effects of viruses, carcinogens, radiation and other chemicals to chromosomes in vivo and vitro. *Mailing Add:* Dept Exp Biol 6th Floor Sci Bldg 666 Elm St Buffalo NY 14263

HUANG, CHI-LUNG DOMINIC, b Fujian, China, Oct 10, 30; US citizen; m 60; c 4. ENGINEERING MATHEMATICS, SYSTEM MODELING. *Educ:* Nat Taiwan Univ, BS, 55; Univ Ill, MS, 60; Yale Univ, DEngSc, 64. *Prof Exp:* Engr design, Taiwan Power Co, 55-59 & Amman & Whitney, 60-61; res asst res, Univ Ill, 59-60 & Yale Univ, 61-64; from asst prof to assoc prof, 64-74, PROF TEACHING & RES, KANS STATE UNIV, 74- *Concurrent Pos:* Vis prof, Chico-Tung Univ, China, 79; hon prof, Hua-Chico Univ, China, 86- *Mem:* Am Inst Aeronaut & Astronaut; Soc Eng Sci; Am Acad Mech; Am Soc Mech Engrs. *Res:* Continuum mechanics; mass and heat transfer; fluid mechanics; structural dynamics and machine design; engineering mathematics. *Mailing Add:* Dept Mech Eng Durland Kans State Univ KS 66506

HUANG, CHIN PAO, b Taiwan, China, Oct 4, 41; US citizen; m 71; c 2. AQUATIC CHEMISTRY. *Educ:* Nat Taiwan Univ, BS, 65; Harvard Univ, MS, 67, PhD(environ eng), 71. *Prof Exp:* Asst prof civil eng, Wayne State Univ, 71-74; from asst prof to assoc prof, 74-80, PROF CIVIL ENG, 80-, PROF MARINE CHEM, UNIV DEL, 87- *Concurrent Pos:* Mem, Water qual Comt, Am Geophys Union, 75-77, Newsletter Comt, Asn Environ Eng Profs, 76-77 & Standard Method Comt, Water Pollution Control Fedn, 79-; assoc ed, Am Soc Civil Engrs, 85-88, ed, 88-90. *Mem:* Am Chem Soc; Am Soc Civil Engrs; Water Pollution Control Fedn; Int Asn Colloid & Surface Scientists; Am Soc Limnol & Oceanog; Asn Environ Eng Profs. *Res:* Applied colloid and surface chemistry; industrial waste management; physical-chemical processes for water and wastewater treatment; chemistry and biology of natural waters. *Mailing Add:* Dept Civil Eng Univ Del Newark DE 19716

HUANG, CHING-HSIEN, b Tientsin, China, Oct 24, 35. BIOCHEMISTRY. *Educ:* Tunghai Univ, BS, 59; Johns Hopkins Univ, PhD(biochem), 65. *Prof Exp:* Fel phys chem, Max Planck Inst Phys Chem, Ger, 65-67; vis asst prof, 67-68, from asst prof to assoc prof, 68-77, PROF BIOCHEM, SCH MED, UNIV VA, 77- *Concurrent Pos:* Helen Hay Whitney Found res fels phys chem, 66-67 & biochem, 67-69. *Mem:* Am Soc Biol Chem; Biophys Soc. *Res:* Membrane biophysics and biochemistry; biophysical chemistry of phospholipids; molecular structure and mechanism of protein kinase. *Mailing Add:* Dept Biochem Sch Med Univ Va Charlottesville VA 22908

HUANG, CHING-RONG, b Hong Kong, Jan 19, 32; m 59; c 3. CHEMICAL & BIOMEDICAL ENGINEERING. *Educ:* Nat Taiwan Univ, BS, 54; Mass Inst Technol, SM, 58; Univ Mich, MS, 65, PhD(chem eng), 66. *Prof Exp:* Res engr, Am Polyplastics Lab, Mass, 58-62; asst prof chem eng, Newark Col Eng, 66-68, assoc prof, 68-78; PROF, NJ INST TECHNOL, 78-, ASST CHMN GRAD STUDIES, DEPT CHEM ENG, 81- *Concurrent Pos:* Hon prof biomed eng, Sichuan Med Univ, People's Repub China, 85. *Mem:* Am Inst Chem Engrs; Soc Rheology; Int Soc Biorheology. *Res:* Heterogeneous catalysis; applied mathematics; rheology. *Mailing Add:* Dept Chem Eng & Chem NJ Inst Technol Newark NJ 07102

HUANG, DENIS K, b China, May 14, 25; US citizen; m 57; c 1. CHEMICAL ENGINEERING, POLYMER CHEMISTRY. *Educ:* St John's Univ, China, BS, 44; Univ Calif, BS, 50; Univ Maine, MS, 51; Polytech Inst Brooklyn, DChE, 58. *Prof Exp:* Res engr, Single Serv Div, Int Paper Co, 58-62; sr chemist, Simoniz Div, Morton Int, Inc, 62-65; sr chemist, Laurel Res Lab, Westvaco Corp, 65-78; res consult, Fed Paper Bd, 78-90; ASSOC CONSULT, TECH CONSULTS INT, 90- *Concurrent Pos:* UN expert, India, 70; expert, Orgn Am States, 72; UNDP assignment, China, 83. *Mem:* Tech Asn Pulp & Paper Indust; Sigma Xi. *Res:* Pulp and paper emulsion and suspension polymerization; polymer application; pulp and paper technology; petroleum technology; industrial coating; coal slurries; forest products. *Mailing Add:* 3641 Nassau Dr Augusta GA 30909

HUANG, ENG-SHANG(CLARK), b Chia-Yi, Taiwan, Mar 17, 40; US citizen; m 65; c 2. VIROLOGY, CANCER RESEARCH. *Educ:* Nat Taiwan Univ, BS, 62, MS, 64; Univ NC, Chapel Hill, PhD(microbiol & immunol), 71. *Prof Exp:* Res fel virol, dept Microbiol & immunol, Univ NC, Chapel Hill, 71-73, vis asst prof, dept microbiol & immunol & dept med, 73-74, asst prof, dept med, 74-78, assoc prof virol & cancer res, dept med, Cancer Res Ctr, 78-85, PROF VIROL & CANCER RES, DEPT MED, CANCER RES CTR, UNIV NC, CHAPEL HILL, 86- *Concurrent Pos:* Res Career Develop Award, Nat Inst Allergy & Infectious Dis, NIH, 78-83; mem virol study sect, res & grant div, NIH, 79-83; head virol prog, Cancer Res Ctr, Univ NC, Chapel Hill, 79-; consult, Wu-han Inst Virol, Acad Sci, 86-; vis prof, IBMS, Academia Sinica, Taipei, 88; mem, Acquired Immune Deficiency Syndrom res rev comt, Nat Inst Allergy & Infectious Dis, NIH, 88-90. *Mem:* Am Soc Microbiol; AAAS; NY Acad Sci. *Res:* Molecular biology and pathobiology of human cytomegalovirus; rule of cytomegalovirus infection in sexually transmitted diseases; search for anti-herpes virus compund; the oncogenicity of cytomegalovirus; viral epidemiology; recombinant DNA research. *Mailing Add:* CB No 7295 Lineberger Cancer Res 237H Univ NC Chapel Hill NC 27599-7295

HUANG, EUGENE YUCHING, b Changsha, China, Nov 28, 17; nat US; m 55; c 6. CIVIL ENGINEERING. *Educ:* Univ Utah, BS, 50; Univ Mich, ScD(civil eng), 54. *Prof Exp:* Asst engr, Hunan Prov Hwy Admin, China, 39-40 & Hwy Admin, Ministry Commun, 40-41; from asst engr to assoc engr, Chinese Nat Hwy Admin, 41-48; from asst prof to assoc prof civil eng, Univ Ill, 54-63; prof, 63-84, EMER PROF TRANSP ENG, MICH TECHNOL UNIV, 84- *Concurrent Pos:* Mem, Transp Res Bd; fac res award, Mich Technol Univ, 67. *Mem:* AAAS; fel Am Soc Civil Engrs; Am Soc Eng Educ; Am Rwy Eng Asn; Am Soc Testing & Mat; Inst Mgt Sci. *Res:* Highway and transportation engineering; pavement analysis and design. *Mailing Add:* 400 Garnet St Houghton MI 49931-1420

HUANG, FAN-HSIUNG FRANK, b Pingtung, Taiwan, May 13, 40; m 69; c 3. EXPERIMENTAL SOLID STATE PHYSICS. *Educ:* Nat Taiwan Normal Univ, BS, 65; Rensselaer Polytech Inst, PhD(physics), 73. *Prof Exp:* Res assoc mat sci, Cornell Univ, NY, 73-77; PR SCIENTIST MECH PROPERTIES, WESTINGHOUSE HANFORD CO, 77- *Mem:* Am Soc Testing & Mat. *Res:* Surface diffusion and morphological instability of ceramics; the mechanical behavior of metals and alloys at elevated temperatures; fracture toughness of irradiated materials. *Mailing Add:* 531 Singletree Ct Richland WA 99352

HUANG, FRANCIS F, b Hong Kong, Aug 27, 22; US citizen; m 54; c 2. ENGINEERING THERMODYNAMICS. *Educ:* San Jose State Col, BS, 51; Stanford Univ, MS, 52; Columbia Univ, ME, 64. *Hon Degrees:* DSc, World Univ, 90. *Prof Exp:* Engr & job coordr, M W Kellogg Co, NY, 52-58; from asst prof to assoc prof, 58-67, chmn dept, 73-81, PROF MECH ENG, SAN JOSE STATE UNIV, 67- *Concurrent Pos:* Hon prof, Taiyuan Univ Technol, People's Repub China, 81- *Mem:* AAAS; Am Soc Mech Engrs; Am Soc Eng Educ; Am Inst Aeronaut & Astronaut; NY Acad Sci; Int Asn Sci & Tech Develop. *Res:* Engineering thermodynamics; intermolecular forces of gases through use of free expansion coefficients; energy systems design; second law analysis of energy processes. *Mailing Add:* Dept Mech Eng San Jose State Univ San Jose CA 95192

HUANG, H(SING) T(SUNG), b Malaya, Sept 9, 21; m 49; c 2. MICROBIOLOGY. *Educ:* Hong Kong Univ, BSc, 41; Oxford Univ, Eng, Dr Phil, 47. *Prof Exp:* Res fel chem, Univ Rochester, 47-48, Calif Inst technol, 48-51; res biochemist, Rohm & Haas Co, 51-55; res supvr biochem, Chas Pfizer Inc, 55-64; dir biol sci, Inst Minerals & Chem Corp, 67-73; dir tech serv, Wallerstein Div, Baxter Labs, 73-75; PROG DIR, ALTERNATIVE BIOL RESOURCES, NSF, 75- *Mem:* Am Chem Soc; Am Soc Biochem & Molecular Biol; Inst Food Technol; Soc Invert Pathol. *Res:* Organic synthesis and metabolism of amino acids, peptides and related compounds; fermentation and enzyme technology; microbial control of insect pests. *Mailing Add:* Apt 403 309 Yoakum Pkwy Alexandria VA 22304-3921

HUANG, HENRY HUNG-CHANG, b Taiwan, May 22, 39; Can citizen; m 67; c 2. PLANT PATHOLOGY, MYCOLOGY. *Educ:* Chung-Hsing Univ, BSc, 63; Univ Toronto, MSc, 69, PhD(plant path), 72. *Prof Exp:* Teaching asst, Chung Hsing Univ, 64-67; res scientist plant path, Agr Can Morden Res Sta, 74-81; RES SCIENTIST, AGR CAN LETHBRIDGE RES STA, 81- *Concurrent Pos:* Agr Can fel, Saskatoon Res Sta, 72-74; res grant, CSP Foods Ltd, 75-77; new crop develop fund, CSP Foods Ltd & Agr Can, 77-78, Canola Coun Can, 84; adj prof, Univ Man, 80-86. *Mem:* Am Phytopath Soc; Can Phytopath Soc; Plant Protection Soc Republic China. *Res:* Soilborne diseases such as sclerotinia diseases of sunflower, canola, safflower and field bean, and verticillium wilt of alfalfa; biological control of soilborne diseases; disease resistance and fungal cytology. *Mailing Add:* Res Sta Agr Can Lethbridge AB T1J 4B1 Can

HUANG, HUEY WEN, b Tokyo, Japan, Feb 22, 40; m; c 1. BIOPHYSICS, STATISTICAL PHYSICS. *Educ:* Nat Taiwan Univ, BS, 62; Cornell Univ, PhD(theoret physics), 67. *Prof Exp:* Res assoc theoret physics, Columbia Univ, 67-69; lectr & res assoc biophys, Yale Univ, 69-71; asst prof physics, Southern Ill Univ, 71-73; from asst prof to assoc prof, 73-82, PROF

PHYSICS, RICE UNIV, 82- *Concurrent Pos:* Vis prof biophys, Yale Univ, 87-88. *Mem:* Am Phys Soc; AAAS; Biophys Soc. *Res:* Protein dynamics by synchrotron radiation; statistical physics; biophysics. *Mailing Add:* Dept Physics Rice Univ PO Box 1892 Houston TX 77251

HUANG, JACK SHIH TA, b Nanchang, China, Aug 31, 33; US citizen; m 63; c 2. APPLIED PHYSICS. *Educ:* Univ Pa, BS, 55; Harvard Univ, MS, 58, PhD(appl physics), 63. *Prof Exp:* Mem tech staff, Bell Tel Labs, Inc, 63-67; asst prof elec eng, Univ Minn, 67-69; staff scientist, Solid State Electronics Ctr, Honey Inc, 69-75, mgr advan technol, 75-81, mgr advan technol planning, 81-; SECT HEAD, EXXON RES CO. *Mem:* Inst Elec & Electronics Engrs; Electrochem Soc. *Res:* Semiconductor devices and integrated circuits. *Mailing Add:* Exxon Res Co Clinton Township Rte 22E Annadale NY 08801

HUANG, JACOB WEN-KUANG, b Kuangehow, China, Nov 7, 35; m 67; c 2. OPTICAL PHYSICS, SOLID STATE PHYSICS. *Educ:* Nat Taiwan Univ, BS, 58; Johns Hopkins Univ, PhD(physics), 68. *Prof Exp:* From asst prof to assoc prof, 67-78, vchmn dept, 73-78, PROF PHYSICS, TOWSON STATE UNIV, 78- *Mem:* Am Phys Soc; Am Asn Physics Teachers; Sigma Xi; Am Sci Res Soc; Optical Soc Am. *Res:* Optics; spectroscopy; luminescence; lasers; holography; atomic physics; excited state absorption spectrum of ions in crystals and liquids; optical modulation spectroscopy of solids; physics education; holography; optical information processing. *Mailing Add:* Dept Physics Towson State Univ Baltimore MD 21204

HUANG, JAMIN, b Tapei, Taiwan, Aug 1, 51; m; c 2. PESTICIDAL CHEMISTRY. *Educ:* Nat Chung-Hsing Univ, BA, 73; Cornell Univ, MS, 78, PhD(org chem), 80. *Prof Exp:* Teaching asst org chem, Cornell Univ, 76-77, res asst, 77-80; sr chemist, Union Carbide Agr Prod Co, 80-83, proj scientist, 83-86; PROJ SCIENTIST, RHONE-POULENC AGR CO, 87- *Mem:* Am Chem Soc. *Res:* Natural product isolation; natural product total synthesis; exploration and syntheses of novel pesticides. *Mailing Add:* Rhone-Poulenc Agr Co PO Box 12014 Research Triangle Park NC 27709

HUANG, JENG-SHENG, b Taiwan, Repub China, May 11, 40; m 67; c 2. PLANT PATHOLOGY. *Educ:* Chung-Hsing Univ, Taiwan, BS, 62; Univ Mo-Columbia, MS, 69, PhD(plant path), 72. *Prof Exp:* Teaching asst plant path, Chung-Hsing Univ, Taiwan, 63-67; res asst plant path, Univ Mo-Columbia, 67-72, res microbiologist, 72-75; from asst prof to assoc prof plant path, 75-86, PROF PLANT PATH, NC STATE UNIV, 86- *Mem:* Am Phytopath Soc; AAAS; Am Soc Plant Physiologists. *Res:* Biochemistry and physiology of plant pathogenesis. *Mailing Add:* Dept Plant Path NC State Univ Box 7616 Raleigh NC 27695-7616

HUANG, JENNMING STEPHEN, b Changhua, Taiwan, July 30, 47; m; c 2. BATCH PROCESS AUTOMATION, COMPUTER INTEGRATED MAUFACTURING. *Educ:* Nat Taiwan Univ, Taipei, BS, 69; Syracuse Univ, NY, MS, 72, PhD(chem eng), 75. *Prof Exp:* Postdoctoral, Syracuse Univ, 75-76; res engr mixing & heat transfer, Int Flavors & Fragrances, Inc, 76-77, sr res engr separation technol, 78-79, proj engr batch process control, 79-81, mgr corp eng technol, mfg automation, 81-82, DIR CORP ENG TECHNOL, COMPUTER INTEGRATED MFG, INT FLAVORS & FRAGRANCES, INC, 83- *Honors & Awards:* Int Distinguished Leadership, Am Biog Inst, 90. *Mem:* Am Inst Chem Engrs. *Res:* Architecture and software of automation system for complex batch process manufacturing to capitalize advanced computer technology; assess the impact of automation systems on process performance, manpower issues and overall efficiency of production. *Mailing Add:* Three Magnolia Dr Marlboro NJ 07746

HUANG, JOHN S, b Chung-King, China, Feb 22, 40; m 67. CHEMICAL PHYSICS, SURFACE PHYSICS. *Educ:* Nat Taiwan Univ, BS, 62; Cornell Univ, MS, 66, PhD(physics), 69. *Prof Exp:* Asst prof physics, Rutgers Univ, New Brunswick, 69-76; res physicist, 76-81, res assoc, 81-88, SR STAFF PHYSICIST, EXXON RES & ENG CO, 88- *Concurrent Pos:* Adj prof, Rutgers Univ, 81-85. *Honors & Awards:* Max Planck Res Prize. *Mem:* Fel Am Phys Soc; NY Acad Sci; Am Chem Soc. *Res:* Critical phenomena in binary fluid mixtures; critical interface; dynamic light scattering; nucleation phenomena; microemulsions and colloid science. *Mailing Add:* Exxon Res & Eng Co Annandale NJ 08801

HUANG, JOHN WEN-CHIH, computer science, electrical engineering, for more information see previous edition

HUANG, JOSEPH CHI KAN, b Hang Chow, China, May 21, 38; m 67; c 3. OCEANOGRAPHY, METEOROLOGY. *Educ:* Chiao Tung Univ, BS, 58; Univ Mich, MS, 65, PhD(oceanog), 69. *Prof Exp:* Oceanogr, Scripps Inst Oceanog, Univ Calif, San Diego, 69-75; phys oceanogr & numerical modeler, Great Lakes Environ Res Lab, Nat Oceanic & Atmospheric Admin & assoc prof, Univ Mich, 75-81; SR SCIENTIST & PROJ MGR, NAT OCEANIC & ATMOSPHERIC ADMIN, 81- *Mem:* AAAS; Am Geophys Union; Am Meteorol Soc. *Res:* Large-scale air-sea interaction and numerical modeling of ocean and lake; topographic wave and ecosystem study. *Mailing Add:* Nat Atmospheric & Oceanic Admin 6010 Executive Blvd Rockville MD 20852

HUANG, JU-CHANG, b Kaohsiung, Taiwan, Jan 3, 41; m 65, 85; c 3. SANITARY ENGINEERING, CIVIL ENGINEERING. *Educ:* Nat Taiwan Univ, BS, 63; Univ Tex, Austin, MS, 66, PhD(civil eng), 67. *Prof Exp:* From asst prof to assoc prof, 67-70, PROF CIVIL ENG & DIR, ENVIRON RES CTR, UNIV MO, ROLLA, 75- *Concurrent Pos:* Prin investr, US Dept Interior grants, US Environ Protection Agency grants, NSF grants, Mo DNR grants & indust res grants, 68-90; consult, US & foreign govt agencies, WHO, industs & munic, 72- *Honors & Awards:* Walker L Huber Civil Eng Res Prize, Am Soc Civil Engrs, 79. *Mem:* Water Pollution Control Fedn; Am Soc Eng Educ; Am Soc Civil Engrs; Am Acad Environ Engrs; Asn Environ Eng Prof. *Res:* Water and wastewater treatment; water quality management; sludge handlings and disposals; reduction of metals concentrations in municipal waste treatment sludges; fixed films for anaerobic sludge digestion. *Mailing Add:* Dept Civil Eng Univ Mo Rolla MO 65401-0249

HUANG, JUNG SAN, CANCER, PROTEINS. *Educ:* Nat Taiwan Univ, Taipei, PhD(biochem), 72. *Prof Exp:* ASSOC PROF BIOCHEM, ST LOUIS UNIV, 84- *Mailing Add:* Dept Biochem St Louis Univ Med Sch 1402 S Grand Blvd St Louis MO 63104

HUANG, JUNG-CHANG, b Taiwan, Apr 7, 35; m 67; c 3. COMPUTER SCIENCE, ELECTRICAL ENGINEERING. *Educ:* Taipei Inst Technol, Taiwan, Dipl, 56; Kans State Univ, MS, 62; Univ Pa, PhD(elec eng), 69. *Prof Exp:* Engr telecommun, Chinese Govt Radio Admin, 58-60; electronics, Western Elec Co, 62-66; from asst to assoc prof, 69-80, PROF COMPUT SCI, UNIV HOUSTON, 80- *Concurrent Pos:* NSF res grant, 77-82. *Mem:* Inst Elec & Electronics Engrs; Asn Comput Mach. *Res:* Programming methodology; software engineering; real-time systems. *Mailing Add:* Dept Comput Sci Univ Houston Houston TX 77204-3475

HUANG, JUSTIN C, b Honolulu, Hawaii, Oct 10, 35; m 65; c 2. THEORETICAL PHYSICS. *Educ:* Univ Wash, Seattle, BS, 57; Mich State Univ, MS, 59, PhD(physics), 61. *Prof Exp:* Fulbright fel, Univ Tokyo, 61-62; asst prof, 62-69, ASSOC PROF PHYSICS, UNIV MO, COLUMBIA, 69- *Mem:* Am Phys Soc; Sigma Xi. *Res:* Quantum mechanics; scattering theory; theory of elementary particles and fields. *Mailing Add:* Dept Physics Univ Mo Columbia MO 65211

HUANG, KEE-CHANG, b Canton, China, July 22, 17; nat US; m 47; c 3. PHARMACOLOGY, PHYSIOLOGY. *Educ:* Dr Sun Yat-sen Univ, China, MD, 40; Columbia Univ, PhD, 53. *Prof Exp:* Res fel pharmacol, NIH, China, 40-46; instr, Nat Col Shanghai, 46-49; fel, Columbia Univ, 49-53; res assoc, 53-56, from asst prof to assoc prof, 56-63, PROF PHARMACOL, SCH MED, UNIV LOUISVILLE, 63- *Mem:* AAAS; Am Physiol Soc; Am Soc Pharmacol & Exp Therapeut; Soc Exp Biol & Med; Soc Clin Pharmacol & Chemother. *Res:* Renal and intestinal transport of electrolytes, sugars and amino acids. *Mailing Add:* Health Sci Ctr Univ Louisville Sch Med Louisville KY 40292

HUANG, KEH-NING, b Nanking, China, Dec 6, 47; nat US; m 69; c 2. THEORETICAL PHYSICS. *Educ:* Nat Cheng-Kung Univ, Taiwan, BS, 68; Yale Univ, PhD(physics), 74. *Prof Exp:* Fel physics, Univ Ore, 74-76; res assoc physics, Univ Nebr, 76-78; asst prof physics, Univ Notre Dame, 78-81; physicist, Argonne Nat Lab, 81-84; RES FEL, INST ATOMIC & MOLECULAR SCI, ACADEMIA SINICA, TAIWAN, 84-; PROF, NAT TAIWAN UNIV, 84- *Concurrent Pos:* Manuscript referee, Nuclear Physics, 76-, Phys Rev, 78- & J Optical Soc Am, 81-; co-prin investr, NSF grant, 79-81, prin investr, Res Corp grant, 80-82; assoc ed-in-chief, Chinese J Physics, 86-88, ed-in-chief, 88-89; exec dir, Chinese Phys Soc, 86-87, pres, 87-88. *Mem:* Am Phys Soc; Sigma Xi; Chinese Phys Soc. *Res:* Relativistic many-body theory; kinematics in scattering theory; structure and properties of exotic atoms; fusion related atomic spectroscopic data; interactions of atoms and ions with radiation. *Mailing Add:* Atomic & Molecular Sci Academia Sinica PO Box 23-166 Taipei 106 Taiwan

HUANG, KERSON, b Nan Ning, China, Mar 15, 28; nat US; m 56. THEORETICAL PHYSICS. *Educ:* Mass Inst Technol, BS, 50, PhD, 53. *Prof Exp:* Instr physics, Mass Inst Technol, 53-55; mem staff math, Inst Advan Study, 55-57; from asst prof to assoc prof, 57-66, PROF PHYSICS, MASS INST TECHNOL, 66- *Concurrent Pos:* Sloan fel, 61-64; Guggenheim fel, 65-66. *Mem:* Am Phys Soc; Am Acad Arts & Sci. *Res:* Statistical mechanics; theory of elementary particles. *Mailing Add:* Dept Physics Mass Inst Technol 77 Massachusetts Ave Cambridge MA 02139

HUANG, KUN-YEN, b Taiwan, Dec 11, 33; m 61; c 3. VIROLOGY, MEDICINE. *Educ:* Taiwan Univ, MD, 59; George Washington Univ, PhD(microbiol), 67. *Prof Exp:* Resident surg, Taiwan Univ Hosp, 61-62; teaching asst microbiol, Sch Med, George Washington Univ, 63-64; med officer, US Naval Med Res Inst, 64-67, Nat Acad Sci res assoc, 67-68; from asst prof to prof microbiol, Sch Med, George Washington Univ, 68-; DEAN, COL MED, NAT CHENG KUNG UNIV, TAIWAN, REPUB CHINA. *Concurrent Pos:* Res fel virol, US Naval Med Res Unit II, 62-63. *Mem:* AAAS; NY Acad Sci; Am Soc Microbiol. *Res:* Metabolic studies of Rickettsia quintana; interferon and macrophages; interferon and protozoan parasites; effects of abnormal gaseous environments on the resistance to infection. *Mailing Add:* Dean Col Med Nat Cheng Kung Univ One Ta-Hsueh Rd Tainan Taiwan

HUANG, KUO-PING, b Taichung, Taiwan, Jan 3, 42; US citizen; m 70; c 2. BIOCHEMISTRY. *Educ:* Nat Taiwan Univ, BS, 64; Univ Calif, Davis, PhD(biochem), 70. *Prof Exp:* Vis assoc, Nat Inst Arthritis, Metab & Digestive Dis, NIH, 72-73; sr staff fel, 73-77, RES CHEMIST, NAT INST CHILD HEALTH & HUMAN DEVELOP, NIH, 77- *Mem:* Am Soc Biol Chemists. *Res:* Regulation of glycogen metabolism in mammalian tissues and cells grown in culture. *Mailing Add:* Nat Inst Child Health & Human Develop NIH Bldg 10 Rm BlL400 Bethesda MD 20892

HUANG, LAURA CHI, b Nanking, China, Nov 17, 37; c 2. BIOCHEMISTRY. *Educ:* Nat Taiwan Univ, BS, 60; Univ Calif, Davis, PhD(comp biochem), 65. *Prof Exp:* Vis fel, NIH, 65-67; res chemist, Univ Calif, Los Angeles, 67-69; asst prof pharmacol, 69-74, ASSOC PROF PHARMACOL, UNIV VA, 74- *Concurrent Pos:* Res career develop award, 78-82. *Mem:* Am Soc Biol Chemists; Chinese Soc Biochemists. *Res:* Biological control mechanisms. *Mailing Add:* Dept Pharmacol Univ Va 1300 Jefferson Park Ave Charlottesville VA 22908

HUANG, LEAF, b Hunan, China, Sept 23, 46; m 72; c 2. BIOPHYSICS, BIOCHEMISTRY & CELL BIOLOGY. *Educ:* Nat Taiwan Univ, BS, 68; Mich State Univ, PhD(biophysics), 74. *Prof Exp:* Fel biophysics, Carnegie Inst Washington 74-76; from asst prof to assoc prof, 76-85, PROF BIOCHEM, UNIV TENN, KNOXVILLE, 85- *Concurrent Pos:* Prin investr, NIH grants; res career develop award, NIH, 81-86. *Mem:* Biophys Soc; Am Asn Cell Biologists; Am Soc Bio Chemists. *Res:* Structure and function of biological membranes; membrane fusion; membrane receptors; liposome-cell interactions; liposomes in immunology; liposome technology. *Mailing Add:* Dept Biochem Univ Tenn Knoxville TN 37996-0840

HUANG, LEO W, b Taiwan, China, Apr 10, 42; US citizen; m 68; c 1. HIGH TEMPERATURE CHEMISTRY, PHYSICAL ORGANIC CHEMISTRY. *Educ:* Nat Chung-Hsing Univ, BS, 64; Kans State Univ, MS, 68; Ill Inst Technol, PhD(phys chem), 73. *Prof Exp:* Res assoc, Univ Wis, Milwaukee, 73-74; ASST PROF CHEM, CITY COL CHICAGO, 74- *Mem:* Am Chem Soc. *Res:* Gas-surface interaction at high temperatures. *Mailing Add:* 445 Brookside Dr Wilmette IL 60091-3048

HUANG, LIANG HSIUNG, b I-Lan, Taiwan, July 16, 39; US citizen; m 70; c 2. TAXONOMY, FERMENTATION. *Educ:* Nat Taiwan Univ, Taipei, BS, 62; Univ Wis-Madison, MS, 66, PhD(bot), 71. *Prof Exp:* Res asst, Inst Enzyme Res, Univ Wis-Madison, 65-66; res fel, Ohio State Univ, Columbus, 72-73; res assoc, Univ Ga, Athens, 73-75; SR RES INVESTR, CENT RES, PFIZER INC, 75- *Concurrent Pos:* Ed, Newslett US Fedn Cult Collections, 87- *Mem:* Am Soc Microbiol; Mycol Soc Am; Soc Indust Microbiol; US Fedn Cult Collections. *Res:* Taxonomy of actinomycetes and soil fungi; strain development in antibiotic research; morphology of ascomycetes; isolation of microorganisms; culture collection. *Mailing Add:* Cent Res Pfizer Inc Eastern Point Rd Groton CT 06340

HUANG, MOU-TUAN, CANCER PREVENTION, CHEMOTHERAPY. *Educ:* Univ NC, PhD(biochem), 69. *Prof Exp:* Sr scientist, Hoffmann-La Roche Inc, 77-87; ASSOC PROF, DEPT BIOCHEM & PHARMACOL & DIR BIOCHEM RES LAB, RUTGERS UNIV, 87- *Res:* Drug metabolism. *Mailing Add:* Col Pharmacol Rutgers Univ Piscataway NJ 08855-0789

HUANG, NAI-CHIEN, b Nangtung, China, July 8, 32; US citizen; m 63; c 2. SOLID MECHANICS. *Educ:* Nat Taiwan Univ, BS, 53; Brown Univ, MS, 58; Harvard Univ, PhD(appl mech), 63. *Prof Exp:* Res assoc eng mech, Stanford Univ, 63-65; asst prof appl mech, Univ Calif, San Diego, 65-69; assoc prof, 69-83, PROF AEROSPACE & MECH ENG, UNIV NOTRE DAME, 83- *Concurrent Pos:* Vis assoc prof, Univ Wis-Madison, 74-75 & Mass Inst Technol, 79; consult, Amoco Prod Co, 78-86. *Mem:* Am Soc Mech Engrs; Am Inst Aeronaut & Astronaut; Soc Rheology; Sigma Xi; Am Fiber Soc; Am Soc Rheology; Am Acad Mech. *Res:* Viscoelasticity; stability; optimal design; textile mechanics; fracture mechanics; solid mechanics; structural mechanics; deformation and flow fields in an oil reservoir associated with hydraulic fracturing process; the speed of fatigue crack growth of metals based on the accumulative plastic work fracture criterion. *Mailing Add:* Dept Aerospace & Mech Eng Univ Notre Dame Notre Dame IN 46556

HUANG, P M, b Putze, Taiwan, Sept 2, 34; Can citizen; m 64; c 2. SOIL CHEMISTRY & MINERALOGY, ENVIRONMENTAL CHEMISTRY. *Educ:* Nat Chung Hsing Univ, BSA, 57; Univ Manitoba, MSc, 62; Univ Wis, Madison, PhD (soil sci), 66. *Prof Exp:* From asst prof to assoc prof, 65-78, PROF SOIL SCI, UNIV SASK, 78- *Concurrent Pos:* Nat vis prof, Nat Chung Hsing Univ, 75-76, head dept soil sci & dir soil res inst, 75-76; assoc ed, Soil Sci Soc Am J, 87- & ed, J Chinese Agr Chem Soc, 75-76; prin investr, Nat Sci Eng Res Coun Can, 78-, Found Agron Res, 82-83, Sask Agr Res Found, 77-82, Agr Can, 67-79 & Am Potash Inst, 67-69; mem, Rev Panel, US Dept Energy, 86-87 & Potash & Phosphate Inst Can, 84- *Mem:* Fel Am Soc Agron; fel Soil Sci Soc Am; fel Can Soc Soil Sci; Clay Minerals Soc; Can Asn Water Pollution & Res; Chinese Agr Chem Soc. *Res:* Solution and surface chemistry and mineralogy of soils and sediments; pedogenesis, dynamics and fate of nutrients and toxic pollutants in terrestrial and aquatic environments; food chain contamination and human health. *Mailing Add:* Dept Soil Sci Univ Sask Saskatoon SK S7N 0W0 Can

HUANG, PIEN-CHIEN, b Shanghai, China, July 13, 31; US citizen; m 65; c 2. BIOCHEMISTRY, GENETICS. *Educ:* Nat Taiwan Univ, BS, 53; Va Polytech Inst & State Univ, MS, 56; Ohio State Univ, PhD(genetics & biochem), 60. *Prof Exp:* NIH res fel biochem genetics & chem biol, Calif Inst Technol, 60-65; from asst prof to assoc prof, 65-76, PROF BIOCHEM, JOHNS HOPKINS UNIV, 76- *Concurrent Pos:* Abstractor, Chem Abstr Serv, 59-69; NIH res career develop award, 66-71; Am Cancer Soc & NSF fel, Fedn Europ Biochem Soc Sch Nucleic Acid Sequencing, 68; Med Res Coun vis scholar, Lab Molecular Biol, Post-Grad Med Sch, Univ Cambridge, 72; partic, NIH Cell Cult Workshop, Los Alamos, 73; NSC chair prof, Academia Sinica, Taipei, 87-88; Fogarty Sr Int fel, 87-88. *Honors & Awards:* NIH Res Career Develop Award. *Mem:* Genetics Soc Am; Biophys Soc Am; Am Soc Biol Chem; Am Soc Cell Biol; Sigma Xi; Academia Sinica. *Res:* Genetic chemistry; gene regulation. *Mailing Add:* Dept Biochem Johns Hopkins Univ Baltimore MD 21205

HUANG, ROBERT Y M, b Keelung, Formosa, Dec 23, 30; Can citizen; m 59; c 3. POLYMER CHEMISTRY, CHEMICAL ENGINEERING. *Educ:* Nat Taiwan Univ, BS, 53; Univ Toronto, MASc, 58, PhD(chem eng), 63. *Prof Exp:* Chem engr, Taiwan Fertilizer Co, Ltd, 53-56; res chemist, Int Cellulose Res Ltd & Can Int Paper Co, 62-65; from asst prof to assoc prof, 65-71, PROF CHEM ENG, UNIV WATERLOO, 71- *Mem:* Am Chem Soc; Fel Chem Inst Can. *Res:* Polymer science and engineering, especially properties of permselective polymer membranes, molecular weight distributions in polymerization, radiation-induced polymerization, applied cellulose chemistry; pervaporation membrane separation processes; synthesis and transport properties of thin film composite membranes for various separation processes. *Mailing Add:* Dept Chem Eng Univ Waterloo Waterloo ON N2L 3G1 Can

HUANG, RU-CHIH CHOW, b Nanking, China, Apr 2, 32; m 56; c 2. MOLECULAR BIOLOGY. *Educ:* Nat Taiwan Univ, BS, 53; Va Polytech Inst, MS, 56; Ohio State Univ, PhD, 60. *Prof Exp:* Asst plant physiol, Va Polytech Inst, 54-56; asst plant physiol, Ohio State Univ, 56-57, asst biochem, 57-58; abstracter biol, Chem Abstr Serv, 58-60; res fel, Calif Inst Technol, 60-65; from asst prof to assoc prof, 65-75, PROF BIOL, JOHNS HOPKINS UNIV, 75- *Mem:* Am Soc Biol Chemists; Biophys Soc; Chinese Acad Sci. *Res:* Biochemistry and enzymology of chromosomal nucleoproteins; gene action; differentiation and development. *Mailing Add:* Dept Biol Johns Hopkins Univ 34th & N Charles St Baltimore MD 21218

HUANG, SAMUEL J, b Canton, China, Mar 14, 37; m 68; c 1. ORGANIC CHEMISTRY, POLYMER CHEMISTRY. *Educ:* Nat Taiwan Univ, BS, 58; Polytech Inst Brooklyn, PhD(chem), 64. *Prof Exp:* Res fel, Univ Ill, 63-64; from asst prof to assoc prof, 67-78, PROF CHEM, UNIV CONN, 78- *Mem:* Am Chem Soc; The Chem Soc; Chinese Chem Soc. *Mailing Add:* IMS O-136 Univ Conn Storrs CT 06268

HUANG, SHAO-NAN, LIVER PATHOLOGY, VIRAL HEPITITIS. *Educ:* Nat Taiwan Univ, Tai Pei, MD, 54. *Prof Exp:* HEAD PATH, SUNNYBROOK MED CTR, 84- *Mailing Add:* Path Dept Univ Toronto Sunnybrook Med Ctr 2075 Bayview Ave Toronto ON M4N 3M5 Can

HUANG, SHAW-GUANG, b Taiwan, Aug 22, 49; US citizen; m 75; c 3. NUCLEAR MAGNETIC RESONANCE. *Educ:* Nat Taiwan Univ, BS, 71; Mich State Univ, PhD(phys chem), 77. *Prof Exp:* Res assoc, Univ Ill, Urbana, 77-78; head electronic shop, dept chem, Cornell Univ, 78, dir nuclear magnetic resonance facil, 78-82; DIR, MAGNETIC RESONANCE FACIL, HARVARD UNIV, 82- *Mem:* Am Chem Soc. *Res:* Molecular motions in molecules by nuclear magnetic resonance and other spectroscopic techniques; reaction kinetics; new nuclear magnetic resonance instrumental methodologies through both software and hardware designs. *Mailing Add:* Dept Chem Magnetic Resonance Lab Harvard Univ 12 Oxford St Cambridge MA 02138

HUANG, SHU-JEN WU, b Tainan, Taiwan, July 24, 45; US citizen; m 68; c 2. COAGULANTS, SURFACE ACTIVE AGENTS. *Educ:* Cheng Kung Univ, Taiwan, BS, 67; Univ Rochester, MS, 69. *Prof Exp:* Chief chemist, Universal Div, Lonza, Inc, 70-74; chemist, Hodag Chem, 74-75; assoc proj leader, Consumer Prod Div, A E Staley, 75; group leader, Org Chem Div, Richardson Co, 75-81; sr chemist, 81-83, group leader, 84-87, TECH DIR, NALCO CHEM CO, 88- *Mem:* Am Oil Chemists Soc; Am Water Works Asn; Soc Cosmetic Chemists; Soc Mfg Engrs. *Res:* Water clarification area; new coagulants development and testing; paint detackification; spray booth maintenance; waste treatment. *Mailing Add:* Nalco Chem Co One Nalco Ctr Naperville IL 60563-1198

HUANG, SUEI-RONG, b Tainan, Taiwan, Jan 1, 32; m 65; c 4. CHEMISTRY. *Educ:* Taiwan Univ, BS, 55; NMex Highlands Univ, MS, 60; Stevens Inst Technol, PhD(chem), 64. *Prof Exp:* From asst prof to assoc prof chem, 64-75, PROF CHEM, LONG ISLAND UNIV, 75- *Mem:* Am Chem Soc. *Res:* Radiation chemistry of inorganic salts; theoretical study of biological membrane structures; diffusion process of ions through biological membranes; analysis and elucidation of chemical components and their structures in ginseng; quantum mechanical calculations. *Mailing Add:* Dept Chem Long Island Univ Brooklyn NY 11201

HUANG, SUNG-CHENG, b Canton, China, Oct 26, 44; US citizen; m 71; c 2. TOMOGRAPHIC IMAGE RECONSTRUCTION. *Educ:* Nat Taiwan Univ, BS, 66; Wash Univ, MS, 69, DSc, 73. *Prof Exp:* Res assoc biomed comput, Biomed Comput Lab, Wash Univ, 74; proj engr, Picker Corp, 74-77; from asst prof to assoc prof, 77-86, PROF NUCLEAR MED & BIOMATH, DEPT RADIOL SCI, UNIV CALIF LOS ANGELES SCH MED, 86- *Concurrent Pos:* Prin investr, Lab Biomed & Environ Sci, Univ Calif Los Angeles, 77-; mem adv comt, Simulation Resource Facil, Univ Wash, 88- *Mem:* Soc Nuclear Med; Inst Elec & Electronics Engrs; AAAS. *Res:* Mathematical modeling of radionuclei tracer kinetics in positron emission tomography for quantitative assessment of blood perfusion; substrate utilization and receptor density in local brain-myocardial regions; tomographic image reconstruction in nuclear medicine and radiology; digital signal and image processing. *Mailing Add:* Div Nuclear Med & Biophys Dept Radiol Sci Univ Calif Sch Med Los Angeles CA 90024

HUANG, SYLVIA LEE, b Shanghai, China, July 14, 30; US citizen; m 57; c 3. BIOPHYSICS, BIOCHEMISTRY & MOLECULAR BIOLOGY. *Educ:* Nat Taiwan Univ, BS, 52; Univ Idaho, MS, 56; Univ Pittsburgh, PhD(biophys), 61. *Prof Exp:* Res asst biophys, Univ Pittsburgh, 58-61; res assoc, Sloan-Kettering Inst, 62-66; from res asst prof to res assoc prof, 69-71, ASSOC PROF BIOCHEM, SCH MED, NY UNIV, 71- *Concurrent Pos:* Instr, Sch Med, Cornell Univ, 64-66; prin investr, 72- *Mem:* Biophys Soc; Am Soc Biol Chemists; Harvey Soc. *Res:* Physical chemistry of nucleic acids; structure and function of macromolecules; regulation of gene expression during development and differentiation; erythropoietin and the control of erythropoiesis; mechanism of protein biosynthesis; mechanism of anti-HIV action of plant proteins and polycyclic compounds. *Mailing Add:* Dept Biochem NY Univ Sch Med New York NY 10016

HUANG, T(ZU) C(HUEN), b Shanghai, China; nat US; m 50; c 2. ENGINEERING MECHANICS. *Educ:* Nat Chiao-Tung Univ, China, BS, 46; Univ Ill, MS, 49, PhD(eng mech), 52. *Prof Exp:* Res stress analyst, Int Harvester Co, 52-55; asst prof mech eng, Ore State Col, 55-56; assoc prof eng mech, Univ Fla, 56-62; PROF ENG MECH, UNIV WIS-MADISON, 62- *Concurrent Pos:* Consult, NASA. *Mem:* Am Soc Mech Engrs; Am Soc Eng Educ; Am Soc Testing & Mat; Am Inst Aeronaut & Astronaut; Acoust Soc Am. *Res:* Vibration and dynamics; nonlinear mechanics and solid mechanics in elasticity; wave propagation; stress analysis. *Mailing Add:* Dept Eng Mech 2348 Eng Bldg Univ Wis 1415 Johnson Dr Madison WI 53706

HUANG, THOMAS SHI-TAO, b Shanghai, China, June 26, 36; m 59; c 4. ELECTRICAL ENGINEERING. *Educ:* Nat Taiwan Univ, BS, 56; Mass Inst Technol, MS, 60, ScD(elec eng), 63. *Prof Exp:* From instr elec eng to assoc prof, Mass Inst Technol, 63-72; prof elec eng, Purdue Univ, 72-80; PROF ELEC ENG, UNIV ILL, URBANA-CHAMPAIGN, 80- *Concurrent Pos:* Ford fel, 63-65; Alexander von Humboldt sr scientist award. *Honors & Awards:* Tech Achievement Award, Inst Elec & Electronics Engrs, 87. *Mem:* AAAS; fel Inst Elec & Electronics Engrs; fel Optical Soc Am. *Res:* Electrical communications; data processing; transmission and processing of pictorial information; design of efficient systems; picture quality evaluation. *Mailing Add:* Coordr Sci Lab Univ Ill 1101 W Springfield Ave Urbana IL 61801

HUANG, THOMAS TAO SHING, b China, July 3, 39; m 65; c 2. PHYSICAL CHEMISTRY. *Educ:* Taiwan Univ, BS, 61; East Tenn State Univ, MA, 64; Univ Ill, PhD(chem), 69. *Prof Exp:* Tech specialist, Great Lakes Res Co, 64; US AEC grant, Univ Ill, 68-69; asst prof phys chem, Univ Ky, 69-71; from asst prof to assoc prof, 71-82, PROF PHYS CHEM, E TENN STATE UNIV, 82-, CHMN DEPT, 79- *Concurrent Pos:* Vis prof, Beijing Norm Univ, Cent China Norm Univ & Tsinghua Univ. *Mem:* Am Chem Soc; Sigma Xi. *Res:* Theoretical and experimental studies of chemical kinetics; kinetic isotope effects; oscillatory reactions. *Mailing Add:* 812 W Maple St Johnson City TN 37601

HUANG, THOMAS TSUNG-TSE, b Taoyuan, Taiwan, Sept 1, 38; US citizen; m 62; c 2. FLUID DYNAMICS, ENGINEERING. *Educ:* Nat Taiwan Univ, BS, 61; State Univ Iowa, MS, 64; Cath Univ Am, PhD(fluid dynamics & appl physics), 69. *Prof Exp:* Res asst hydraul & mech, Iowa Inst Hydraul Res, State Univ Iowa, 62-63; assoc res scientist, Hydronautics, Inc, 63-68; NAVAL ARCHIT FLUID DYNAMICS & HYDRODYN, DAVID TAYLOR RES CTR, DEPT NAVY, 68- *Concurrent Pos:* Res prof, Johns Hopkins Univ, 88- *Mem:* Soc Naval Architects & Marine Engrs; Am Soc Mech Engrs. *Res:* Ship hydrodynamics; boundary layers; wakes; flow transition; drag reduction; cavitation; applied research in full scale ship performance prediction techniques; propeller/hull interaction. *Mailing Add:* David Taylor Res Ctr Code 1542 Bethesda MD 20084-5000

HUANG, TIAO-YUAN, b Kaohsiung, Taiwan, May 5, 49; US citizen; m 81; c 1. VERY LARGE SCALE INTEGRATION, SEMICONDUCTOR MEMORIES. *Educ:* Nat Cheng Kung Univ, Taiwan, BS, 71, MS, 73; Univ NMex, MS, 79, PhD(elec eng), 81. *Prof Exp:* Proj leader telemetry, Chung Shan Inst Sci & Technol, 75-77; mem tech staff, Semiconductor Process & Design Ctr, Tex Instruments, 81-83; mem res staff, Integrated Circuit Lab, Xerox Palo Alto Res Ctr, 84-86 & Electronics & Imaging Lab, 87-91; DEVICE MGR, INTEGRATED DEVICE TECHNOL, INC, 91- *Concurrent Pos:* Comt mem, Int Electron Devices Meeting, Inst Elec & Electronics Engrs, 90-91. *Mem:* Sr mem Inst Elec & Electronics Engrs; Electrochem Soc. *Res:* Integrated circuits; very large scale integration; novel device structures; lightly-doped drain transistors; thin film transistors; active matrix displays; static memories; eprom-eeprom; submicron cmos-bicmos technologies. *Mailing Add:* 2670 Seeley Ave San Jose CA 95134

HUANG, TSENG, b Kiangsu, China, Dec 9, 25; m 60; c 3. STRUCTURAL DYNAMICS, MECHANICS. *Educ:* Chiao Tung Univ, BS, 47; Univ Okla, MCE, 55; Univ Ill, PhD(struct mech), 60. *Prof Exp:* Civil engr, Keelung Harbour Bur, 47-54; bridge designer, Okla State Hwy Dept, 55-56; from res asst to res assoc struct dynamics, Univ Ill, 57-60, asst prof, 60-61; from assoc prof to prof mech, 61-69, PROF CIVIL ENG & ENG MECH, UNIV TEX, ARLINGTON, 69- *Concurrent Pos:* Consult, Jersey Prod Res Co, 62-64, Esso Prod Res Co, 64-65 & Maurer Engr Inc, 75-; mem, Struct Stability Res Coun; bd dir & secy, Offshore Mechs & Polar Eng, Coun Chmn, honors & awards comt; dir, Int Soc Offshore & Polar Engrs. *Honors & Awards:* Eugene W Jacobson Award, Am Soc Mech Engrs, 80; Achievement Award, Am Soc Mech Engrs, Offshore Mech & Arctic Eng Div, 88. *Mem:* Am Soc Civil Engrs; Am Soc Mech Engrs. *Res:* Structural analysis, eye mechanics; stability, vibration and design. *Mailing Add:* Dept Civil Eng Univ Tex PO Box 19308 Arlington TX 76019-0308

HUANG, WAI MUN, b China. MOLECULAR BIOLOGY. *Educ:* Johns Hopkins Unv, PhD(biophys chem), 67. *Prof Exp:* Fel biochem, Albert Einstein Col Med, 67-69; fel, Stanford Univ, 69-72; res assoc biol, Mass Inst Technol, 72-74; asst prof microbiol, 74-77, assoc prof cellular, viral & molecular biol, 77-90, PROF CELLULAR, VIRAL & MOLECULAR BIOL, UNIV UTAH, 90- *Concurrent Pos:* NIH career develop award, 75. *Mem:* Am Soc Microbiol; Fedn Am Soc Exp Biol. *Res:* DNA replication; nucleic acid enzymology; molecular biology of nucleic acids; protein-nucleic acid interactions. *Mailing Add:* Dept Cellular, Viral, & Molecular Biol Med Ctr Univ Utah Salt Lake City UT 84132

HUANG, WANN SHENG, b Szechuan, China, Oct 3, 39; m 64; c 2. ENHANCED OIL RECOVERY, STEAM FLOODING SIMULATION. *Educ:* Cheng Kung Univ, Taiwan, BS, 61; Univ SDak, MA, 63; Ill Inst Technol, PhD(chem eng), 68. *Prof Exp:* Sr chem engr, res chem engr, sr res chem engr, asst supvr, supvr, RES CONSULT, E P TECHNOL DIV TEXACO INC, 88- *Mem:* Am Inst Chem Engrs; Am Inst Mining, Metall & Petrol Engrs. *Res:* Telomer chemistry; mass transfer to and from bubbles and drops; oil production; thermal recovery of heavy crude oils; thermal and miscible flooding simulation. *Mailing Add:* E & P Technol Div Texaco Inc PO Box 425 Bellaire TX 77401

HUANG, WEI-FENG, b Shanghai, China, Jan 23, 30; m 60; c 4. PHYSICS. *Educ:* Univ Mich, BS, 58, MS, 59; Univ Va, PhD(nuclear physics), 63. *Prof Exp:* Res scientist, Univ Va, 62-64; from asst prof to assoc prof, 64-74, PROF PHYSICS, UNIV LOUISVILLE, 74- *Mem:* Am Phys Soc. *Res:* Positron annihilation studies of electronic structures of solids; superconductivity. *Mailing Add:* Dept Physics Univ Louisville Louisville KY 40292

HUANG, WEN HSING, b Taiwan, China, Dec 30, 36; US citizen; m 75. CLAY MINERALOGY, GEOCHEMISTRY. *Educ:* Nat Taiwan Univ, China, BS, 60; Wash Univ, St Louis, MA, 68; Univ Mo, Columbia, PhD(geol), 69. *Prof Exp:* Teaching asst, Wash Univ, 65-66; fel, Univ Mo, 69-70; from asst prof to assoc prof chem, Univ SFla, 70-76; assoc prof, Dept Geol, Tex A&M Univ, 76-81. *Concurrent Pos:* Wheeler fel, Wash Univ, St Louis, 64-65; Am Chem Soc Petrol Res Fund fel, Wash Univ, St Louis, 65-67; A P Green fel, Univ Mo, 68-70; NASA Res grant, NASA Apollo moon sample project, Univ Mo, 71; Fac Res Coun award, Univ San Francisco, 71-72, 72-73 & 74-75; Sigma Xi Res grant, 72; NSF grant, 72-74; Sarasota County grant, 74-75; Nat Oceanic & Atmospheric Agency sea grant, 75; assoc ed, Clay Minerals Soc, 74-76, mem publ comt; fel, Geol Soc Am, 75; adj assoc prof chem, Univ SFla, 75-76; With Dept Interior, Bur Land Mgt, 75-76. *Mem:* Clay Minerals Soc; Int Asn Study Clays; Int Asn Geochem & Cosmochem; Geol Soc Am; Sigma Xi. *Res:* Aqueous sedimentary geochemistry; exploration and production geochemistry of ores (uranium); aluminum extraction process; thermodynamics and kinetics of organo-chemical weathering processes; computer application in geology; low-temperature solution geochemistry; clay mineralogy, thermodynamics and kinetics of organochemical weathering processes; radioactivation analysis and computer application in geology. *Mailing Add:* 730 Ridge Dr Mclean VA 22101

HUANG, Y(EN) T(I), b Taipei, Formosa, Feb 4, 27; US citizen; m 58; c 1. CIVIL ENGINEERING, EARTHQUAKE ENGINEERING. *Educ:* Univ Taiwan, BSc, 50; Univ Toronto, MASc, 57; Columbia Univ, PhD(eng mech), 61. *Prof Exp:* Instr, Univ Taiwan, 50-55; res asst, Columbia Univ, 57-61; res staff mem, Sperry Rand Res Ctr, Mass, 61-63; sr res engr, Atlantic Ref Co, Tex, 63-64; proj geophysicist, Geotech div, Teledyne Indust, Inc, 64-66, sr proj geophysicist, Teledyne, Inc, 66-67; sr systs engr, Collins Radio Co, 67-70; pres, Y T Huang & Assocs, Inc, 70-; RETIRED. *Concurrent Pos:* Adj assoc prof, Univ Tex, Arlington, 77-79. *Mem:* NY Acad Sci; Seismol Soc Am; Am Soc Civil Engrs; Nat Soc Prof Engrs; Am Concrete Inst. *Res:* Structural engineering; elastic wave propagations; earthquake science; Geodetic domes. *Mailing Add:* 396 Foresesta Terr West Palm Beach FL 33415

HUBA, FRANCIS, b Peteri, Hungary, Apr 20, 20; US citizen; m 45; c 2. ORGANIC CHEMISTRY, ELECTROCHEMISTRY. *Educ:* Sch Commerce, Budapest, Hungary, BBA, 48; Polytech Univ, Budapest, MS, 55; Case Western Reserve Univ, PhD(electrochem), 77. *Prof Exp:* Asst vaccines, Serum & Health Inst, Budapest, 34-45; pesticides, Huba Bros, Inc, Budapest, 45-49; managing engr, Femtomegcikkmuvek Works, 52-56; sr res chemist, Hungary Acad Sci Chem Res Inst, 56-57; sr res & develop engr, 57-80, RES ASSOC, RES CTR, DIAMOND SHAMROCK CORP, PAINESVILLE, 80- *Concurrent Pos:* Res & develop chemist, Indust Org Res Inst, Budapest, 49-52. *Mem:* Am Chem Soc; Electrochem Soc. *Res:* Electrochemistry, including membranes, electrophoresis, plating and electrofluorination; organic chemistry, including acetylides, dyes and intermediates and fluoroorganics; biochemistry of pesticides, sanitizers, vaccines and serums. *Mailing Add:* 5521 80th St N Apt 312 St Petersburg FL 33709

HUBAY, CHARLES ALFRED, b Chagrin Falls, Ohio, Jan 23, 18; m 45; c 3. SURGERY. *Educ:* Western Reserve Univ, AB, 39, MD, 43; Am Bd Surg, dipl, 51. *Prof Exp:* Demonstr, 50, from instr to sr instr, 50-54, from asst prof to assoc prof, 54-65, PROF SURG, SCH MED, CASE WESTERN RESERVE UNIV & DIR GEN SURG, UNIV HOSP, 65- *Concurrent Pos:* Assoc surg, Highland View Hosp, 68- *Mem:* Soc Univ Surg; AMA; Am Heart Asn; Am Asn Surg of Trauma; Am Col Surg. *Res:* Tissue homotransplantation. *Mailing Add:* Dept Surg Case Western Reserve Univ 2065 Adelbert Rd Cleveland OH 44106

HUBBARD, ANN LOUISE, b Long Beach, Calif, May 30, 43; m 65; c 1. CELL BIOLOGY. *Educ:* Stanford Univ, AB, 65, MA, 66; Rockefeller Univ, NY, PhD(cell biol), 73. *Prof Exp:* Sec level teacher chem, Fairfax County Sch Syst, Groveton High Sch, 66-67; res assoc electrochem, Am Univ, Washington, DC, 67-68; fel, Sch Med, Yale Univ, 73-75, asst prof cell biol, Sect Cell Biol, 75-; assoc prof, 80-86, PROF, DEPT CELL BIOL & ANAT, JOHNS HOPKINS SCH MED, 86- *Concurrent Pos:* Fel Leukemia Soc Am, Inc, 73-75. *Mem:* Sigma Xi; Am Soc Cell Biol; fel Leukemia Soc Am. *Res:* Nature and biogenesis of protein constituents in the cellular membranes of eukaryotes; specific emphasis on membrane glycoproteins of the plasmalemma; receptor-mediated endocytosis. *Mailing Add:* Dept Cell Biol & Anat Johns Hopkins Sch Med 725 N Wolfe St Baltimore MD 21205

HUBBARD, ARTHUR T, b Alameda, Calif, Sept 17, 41; m 64; c 2. ANALYTICAL CHEMISTRY, ELECTROCHEMISTRY. *Educ:* Westmont Col, BA, 63; Calif Inst Technol, PhD(chem), 67. *Prof Exp:* Res fel chem, Calif Inst Technol, 66-67; prof chem, Univ Hawaii, 67-76; prof chem, Univ Calif, Santa Barbara, 76-86; PROF CHEM, UNIV CINCINNATI, 86- *Res:* Chemistry of surfaces studied by thin layer electrochemistry and electron techniques, such as low-energy electron diffraction, auger and electron energy-loss spectroscopy. *Mailing Add:* Surface Ctr Univ Cincinnati ML 172 Cincinnati OH 45221-0172

HUBBARD, BERTIE EARL, b Cameron, Ill, Aug 6, 28; m 52; c 2. MATHEMATICAL ANALYSIS. *Educ:* Western Ill Univ, BS, 49; Univ Iowa, MS, 52; Univ Md, PhD(math), 60. *Prof Exp:* Asst math, Univ Iowa, 49-52; mathematician, US Naval Ord Lab, 55-61; from res asst prof to res assoc prof, 61-67, RES PROF MATH, DEPT MATH & INST PHYS SCI & TECHNOL, UNIV MD, COLLEGE PARK, 67- *Mem:* Soc Indust & Appl Math; Am Math Soc. *Res:* Numerical solution of ordinary and partial differential equations. *Mailing Add:* Dept Math Univ Md College Park MD 20742

HUBBARD, COLIN D, b Ipswich, Eng, May 9, 39. PHYSICAL CHEMISTRY, BIOLOGICAL CHEMISTRY. *Educ:* Univ Sheffield, BSc, 61, PhD(chem), 64. *Prof Exp:* Res assoc chem, Mass Inst Technol, 64-65 & Cornell Univ, 65-66; res assoc biochem, Univ Calif, Berkeley, 66-67; from asst prof to assoc prof, 67-79, PROF CHEM, UNIV NH, 79- *Concurrent Pos:* Vis res fel, Univ Kent, Canterbury, 73-74, Univ Leicester, 80-81; vis prof, Univ Alta, Edmonton, 74; hon res fel, Univ Leicester, 87-88; sab fel, Univ Sevilla, 88. *Mem:* Am Soc Biol Chemists; Am Chem Soc; Royal Soc Chem. *Res:* Elementary steps in chemical reactions; application of rapid reaction techniques and high pressure kinetics to mechanism determination in inorganic, bioinorganic and biological systems and enzyme catalyzed processes; soluation. *Mailing Add:* Dept Chem Univ NH Durham NH 03824-3598

HUBBARD, DONALD, b Terra Ceia, Fla, Oct 4, 00. BIOCHEMISTRY, ASTRONOMY. *Educ:* Univ Fla, BS, 23, MS, 24; Am Univ, PhD(chem), 32. *Prof Exp:* Phys chemist res, Nat Bur Standards, 25-64; RETIRED. *Honors & Awards:* L J Daguerre Medal, Soc French Photog, 34. *Mem:* Am Chem Soc; Optical Soc Am; AAAS. *Res:* Photographic emulsion sensitivity; hypersentization of panchromatic emulsions; Donnan membrane heterogeneous equilibria; optical glass. *Mailing Add:* Three Fairfax Ct Apt 1 Chevy Chase MD 20015

HUBBARD, EDWARD LEONARD, b Phoenix, Ariz, July 7, 21; m 52; c 4. LINEAR ACCELERATORS, SYNCHROTRONS. *Educ:* Univ Calif, Los Angeles, AB, 43, AM, 48, PhD(physics), 51. *Prof Exp:* Asst physics, Univ Calif, Los Angeles, 43-44; physicist, Naval Ord Lab, 44-45; physicist, Lawrence Berkeley Lab, Univ Calif, 51-68; physicist, Fermi Nat Accelerator Lab, 68-74; mgr, Gen Atomics, San Diego, 74-86; CONSULT, 86-; CHIEF SCIENTIST, SAIC, SAN DIEGO, 90- *Mem:* Am Phys Soc. *Res:* Linear accelerators. *Mailing Add:* 5527 Chelsea Ave La Jolla CA 92037

HUBBARD, ERIC R, b Palm Springs, Calif, Oct 11, 43; c 4. PODIATRIC MEDICINE & SURGERY. *Educ:* Calif Col Podiatric Med, BS, 65, DPM, 68; Pepperdine Univ, MS, 77 Am Bd Pod Surg, dipl. *Prof Exp:* CLIN PROF PODIATRIC SURG, CALIF COL PODIATRIC MED, 72-; CLIN PROF PODIATRIC MED, UNIV SOUTHERN CALIF, LOS ANGELES CO HOSP, 77- *Concurrent Pos:* Consult, ins indust, 72- *Mem:* Am Podiatry Asn; fel Am Col Foot Orthopaedists; Am Podiatric Med Asn (pres, 88). *Res:* Silastic implants in foot surgery; phethesmograph and its relationship to vasoconstriction; the use of nizoral in onychomycosis. *Mailing Add:* 2333 Pacific Ave Long Beach CA 90806

HUBBARD, HARMON WILLIAM, b St Louis, Mo, Apr 15, 23. PHYSICS. *Educ:* Univ Ill, BS, 47; Univ Calif, PhD(physics), 52. *Prof Exp:* Physicist, Lawrence Radiation Lab, Univ Calif, 52-55; prof physics, Am Univ, Beirut, 56; physicist, Aeronutronic Systs, Inc, 57; phys scientist, Rand Corp, 58-71; PHYS SCIENTIST, R&D ASSOCS, 71- *Mem:* Am Phys Soc. *Res:* Theoretical physics and energy applications. *Mailing Add:* 12144 Travis St Los Angeles CA 90049

HUBBARD, HARVEY HART, b Swanton, Vt, June 17, 21; m 47; c 4. AEROACOUSTICS. *Educ:* Univ Vt, BS, 42. *Prof Exp:* Elec engr training, Westinghouse Elec Corp, 42; head atmosphere acoust br res, Nat Adv Comt for Aeronaut, 45-59; head acoust br res, NASA, 59-73, asst chief, Acoust & Noise Reduction Div, 73-80; SR RES ASSOC, COL WILLIAM & MARY, 81- *Honors & Awards:* Silver Medal for Noise Control, Acoust Soc Am, 78; Aeroacoust Award, Am Inst Aeronaut & Astronaut, 79. *Mem:* Assoc fel Am Inst Aeronaut & Astronaut; fel Acoust Soc Am; Inst Noise Control Eng. *Res:* Understanding of aircraft noise, its generation, propagation, and control, and its effects on people and structures. *Mailing Add:* 23 Elm Ave Newport News VA 23601

HUBBARD, JAMES STUART, b Ashland, Ala, Feb 3, 54; m; c 3. MATHEMATICS. *Educ:* Univ N Ala, BS, 76; Vanderbilt Univ, PhD(org chem), 80. *Prof Exp:* Res assoc org chem, Va Polytech Inst & State Univ, 80; process develop chemist, Parke-Davis, Div Warner-Lambert, 80-84; sr prod chemist, Koch Industries, Wichita, KS, 84-86; GEN PLANT MGR, KOCH CHEM CO, WHITEHALL, MICH, 86- *Mem:* Am Chem Soc. *Res:* Process research for drugs and drug intermediates; design and synthesis of specialty chemicals. *Mailing Add:* 718 Oakmere North Muskegon MI 49445

HUBBARD, JERRY S, b Tipton, Okla, Dec 10, 32; m 57; c 2. MICROBIOLOGY. *Educ:* Okla State Univ, BS, 58, MS, 60; Univ Tex, Austin, PhD(microbiol), 65. *Prof Exp:* Res fel, Lab Biochem, Nat Heart Inst, 65-66; res scientist, Jet Propulsion Lab, Calif Inst Technol, 66-68, group supvr microbiol, 68-73, sr biologist, Biol Div, Calif Inst Technol, 73-74; assoc prof, 74-79, PROF BIOL, GA INST TECHNOL, 79- *Concurrent Pos:* Assoc mem biol team, Viking 76 Mission, 73-77, mem surface sampler team, 74-77; Speakers bur, Am Inst Chem Engrs, 79-80. *Honors & Awards:* Newcomb-Cleveland prize, AAAS, 77. *Mem:* Am Soc Microbiol. *Res:* Microbial metabolism; bioconversions with immmobilized microorganisms; biodegradation, bioremediation, and dextoxification by soil microorganisms. *Mailing Add:* Dept Biol Ga Inst Technol 225 North Ave NW Atlanta GA 30332

HUBBARD, JOHN CASTLEMAN, b Louisville, Ky, Nov 2, 22; m 50; c 4. PATHOLOGY. *Educ:* Univ Ky, BS, 44; Univ Louisville, MD, 51. *Prof Exp:* Sr pathologist, Div Labs & Res, NY State Dept Health, 56-59; assoc cancer res pathologist, Roswell Park Mem Inst, 59-63; asst prof, 63-69, ASSOC PROF PATH, SCH MED, STATE UNIV NY-BUFFALO, 69- *Mem:* Electron Micros Soc Am. *Res:* Partial hepatectomy and other factors influencing recovery from experimental cirrhosis; primary tumor ultrastructure and induction; genesis of juxta glomerular cell granules; improved microscopy of mycoplasma. *Mailing Add:* Dept Path Rm 204 Farber Hall State Univ NY Health Sci Ctr 3435 Main St Buffalo NY 14214

HUBBARD, JOHN EDWARD, b Binghamton, NY, Apr 17, 33; m 58; c 3. HYDROLOGY, CLIMATOLOGY. *Educ:* Union Col, AB, 58; Brown Univ, MAT, 62; Colo State Univ, PhD(watershed sci), 68. *Prof Exp:* Asst prof phys sci, State Univ NY Agr & Tech Inst, Cobleskill, 61-63; from asst prof to assoc prof, 63-75, PROF EARTH SCI, STATE UNIV NY COL BROCKPORT, 75- *Mem:* Am Geophys Union; Am Meteor Soc; Am Water Res Asn. *Res:* Evapotranspiration; forest-soil-water relations; snow and suburban hydrology. *Mailing Add:* Dept Earth Sci State Univ NY Brockport NY 14420

HUBBARD, JOHN LEWIS, b Kingsport, Tenn, June 16, 47; m 78; c 2. ORGANIC CHEMISTRY, ORGANOMETALLIC CHEMISTRY. *Educ:* Univ NC, Chapel Hill, BS, 69; Purdue Univ, PhD(org chem), 76. *Prof Exp:* Vis asst prof, Purdue Univ, 77, res assoc, 77-78; from asst prof to assoc prof, 78-90, PROF CHEM, MARSHALL UNIV, 90- *Mem:* Am Chem Soc; Sigma Xi; Royal Soc Chem; AAAS. *Res:* New approaches to borohydrides; hydride transfer reactions involving boron and aluminum; novel syntheses of alkyaluminum compounds; hydride-induced carbonylation of organoboranes; synthesis of potential nephrotoxins. *Mailing Add:* Dept Chem Marshall Univ Huntington WV 25755-2520

HUBBARD, JOHN PATRICK, b Ft Bragg, NC, Feb 13, 35; c 2. CONSERVATION, ORNITHOLOGY. *Educ:* Western NMex Univ, BA, 61; Univ Mich, Ann Arbor, MS, 63, PhD(zool), 67. *Prof Exp:* Res cur, Smithsonian Inst, 66-68; Bailey-Law cur birds & mammals, Va Polytech Inst & State Univ, 69-71; cur New World birds, Del Mus Natural Hist, 71-74; SUPVR ENDANGERED SPECIES PROG, NMEX DEPT GAME & FISH, 74- *Mem:* Am Ornith Union; Wildlife Soc. *Res:* Avian systematics, distribution, ecology and paleohistory, particularly with reference to North America; biogeography; paleoecology; North American plant ecology and systematics; research and management of species that may be endangered or threatened in New Mexico and North America. *Mailing Add:* NMex Dept Game & Fish Santa Fe NM 87503

HUBBARD, JOHN W, b San Jose, Calif, June 8, 56; m 80. CARDIOVASCULAR PHARMACOLOGY, CLINICAL PHARMACOLOGY. *Educ:* Univ Santa Clara, BS, 78; Univ Tenn, PhD(physiol & physiol psychol), 83. *Prof Exp:* Teaching asst physiol psychol, Univ Tenn, 78-83; NIH postdoctoral fel pharmacol, Health Sci Ctr, Univ Tex, San Antonio, 83-85; sr res pharmacologist, Rorer-Revlon Health Care Corp, 85-86; sr res pharmacologist, 86-88, res assoc pharmacol, 88-89, ASST DIR CLIN PHARMACOL, HOECHST-ROUSSEL PHARMACEUT INC, 89- *Concurrent Pos:* Reviewer, Am J Physiol, 89-; adj assoc prof, Robert Wood Johnson Med Sch, 91. *Mem:* Am Heart Asn; Sigma Xi; Soc Neurosci; Am Physiol Soc; Am Col Clin Pharmacol; Drug Info Asn. *Res:* Cardiovascular and neuropharmacology; therapeutic agents for the treatment of hypertension, congestive heart failure, coronary artery disease, Alzheimer's disease and schizophrenia; role of enzyme adenylate cyclase in the regulation of myocardial contractility; pharmacokinetics and pharmacodynamics of new drugs in man. *Mailing Add:* Hoechst-Roussel Pharmaceut Inc Somerville NJ 08876

HUBBARD, LINCOLN BEALS, b Hawkesbury, Ont, Sept 8, 40; m 61; c 2. MEDICAL PHYSICS, RADIATION SAFETY. *Educ:* Univ NH, BS, 61; Mass Inst Technol, PhD(physics), 67. *Prof Exp:* Fel biol, Argonne Nat Lab, 66-68; asst prof math & physics, Knoxville Col, 68-70; asst prof physics, Furman Univ, 70-74; CHIEF PHYSICIST, MT SINAI HOSP MED CTR, CHICAGO, 74-75, 80-; PARTNER, FIELDS, GRIFFITH, HUBBARD & BROADBENT INC, 75- *Concurrent Pos:* Consult physics div, Oak Ridge Nat Lab, 68-73; consult, Vet Admin Hosp, Hines, Ill, 74-; chief physicist, Cook County Hosp, Chicago, 79-88; assoc prof, Rush Univ, 83- *Mem:* Am Phys Soc; Am Asn Physicists Med; Am Asn Physics Teachers; Am Col Radiol; Health Physics Soc. *Res:* Radiation dosimetry and radiological physics. *Mailing Add:* Mt Sinai Hosp Calif Ave & 15th St Chicago IL 60608

HUBBARD, M(ALCOLM) M(ACGREGOR), b New York, NY, Dec 12, 06; c 2. ENGINEERING. *Educ:* Mass Inst Technol, BS, 29. *Prof Exp:* Engr, New Eng Tel & Tel Co, 29-41; group leader, Eng Radiation Lab, Mass Inst Technol, 41-45, asst dir, Lab Nuclear Sci & Eng, 46-51, asst dir, Lincoln Lab, 51-55; pres, Hermes Electronics Co, 55-60; vpres & dir res, Itek Corp, 60-62; pres, M M Hubbard Assoc, Inc, 62-73; consult, 72-79; RETIRED. *Honors & Awards:* Pres Cert Merit. *Mem:* Am Phys Soc; fel Inst Elec & Electronics Engrs. *Res:* Communications; electronics; digital systems. *Mailing Add:* 72 Dane St Apt 1 Beverly MA 01915

HUBBARD, PAUL STANCYL, JR, b St Petersburg, Fla, July 15, 31; m 57; c 2. MAGNETIC RESONANCE, MOLECULAR PHYSICS. *Educ:* Univ Fla, BS, 53; Harvard Univ, AM, 54, PhD(physics), 58. *Prof Exp:* From asst prof to assoc prof physics, 58-68, assoc dean grad sch, 69-72, PROF PHYSICS, UNIV NC, CHAPEL HILL, 68- *Concurrent Pos:* Alfred P Sloan res fel, 62-65; NSF fel, Clarendon Lab, Oxford Univ, 64-65. *Mem:* Am Phys Soc; Sigma Xi. *Res:* Theoretical and experimental research in nuclear magnetic resonance and relaxation. *Mailing Add:* 1710 Audubon Ave Chapel Hill NC 27514

HUBBARD, PHILIP G(AMALIEL), b Macon, Mo, Mar 4, 21; m 43; c 5. HYDRAULICS, ELECTRICAL ENGINEERING. *Educ:* Univ Iowa, BS, 46, MS, 49, PhD, 54; St Ambrose Col, LHD, 69. *Prof Exp:* Res engr & instrumentation sect head, Inst Hydraul Res, 46-64, from instr to assoc prof mech & hydraul, 46-59, dean acad affairs, 66-71, PROF MECH & HYDRAUL, UNIV IOWA, 59-, PROF ENERGY ENG, 77-, VPROVOST, 69- *Mem:* Am Soc Eng Educ. *Res:* Instrumentation in hydraulic research. *Mailing Add:* 201 N First Ave Iowa City IA 52245

HUBBARD, RICHARD ALEXANDER, II, b Jersey City, NJ, July 18, 26; m 48; c 4. PHYSICAL INORGANIC CHEMISTRY. *Educ:* Villanova Col, BS, 48; Northeastern Univ, MS, 50; Fla State Univ, PhD(org chem), 53. *Prof Exp:* Instr chem, Rutgers Univ, 53, fel polymer chem, 53-55; chemist & sr technologist, Titanium Alloy Mfg Div, Nat Lead Co, 55-61; chmn dept chem, 61-73, prof chem, 61-78, MEM FAC, DEPT COMPUT & INFO SCI, NIAGARA UNIV, 78- *Concurrent Pos:* Consult, Titanium Alloy Mfg Div, 61- *Mem:* AAAS; Asn Comput Mfr; Inst Elec & Electronics Engrs. *Res:* Mechanisms of electron transfer in inorganic reactions; Hammett sigma function in polycyclic aromatic compounds; computer graphics. *Mailing Add:* Dept Comput Sci Niagara Univ Niagara University NY 14109

HUBBARD, RICHARD FOREST, b Kansas City, Mo, July 14, 48; m 73; c 2. PLASMA PHYSICS, SPACE PHYSICS. *Educ:* Univ Kans, BA, 70; Univ Iowa, MS, 72, MA, 75, PhD(physics), 75. *Prof Exp:* Res asst physics, Univ Iowa, 70-75; Nat Res Coun res assoc, NASA Goddard Space Flight Ctr, 76-77; res assoc, Inst Phys Sci & Technol, Univ Md, 77-79; res scientist physics, Jaycor, Inc, 79-85; RES PHYSICIST, NAVAL RES LAB, 85- *Concurrent Pos:* Consult, G T Devices, 81-88. *Mem:* Am Phys Soc; Am Geophys Union; AAAS; Planetary Soc; Fusion Power Assocs. *Res:* Stability and transport of intense charged particle beams in plasmas and neutral gases; instabilities and acceleration mechanisms in space plasmas; numerical simulation of plasmas. *Mailing Add:* Code 4790 Naval Res Lab Washington DC 20375-5000

HUBBARD, RICHARD W, b Battle Creek, Mich, Dec 24, 29; m 51; c 3. LABORATORY MEDICINE, BIOCHEMICAL NUTRITION. *Educ:* Pac Union Col, BA, 51; Purdue Univ, MS, 59, PhD(biochem), 61. *Prof Exp:* Analytical chemist, Willard Storage Battery Co, Calif, 51-53; clin lab technician, Los Angeles County Gen Hosp, 53-54; res asst med res, Res Inst, Cedars of Lebanon Hosp, Los Angeles, 54-55, res assoc, 55-56; asst biochem, Purdue Univ, 56-60; instr med sch, Univ Mich, 60-63; sr res chemist, Spinco Div, Beckman Instruments Inc, 63-67; biochemist, Stanford Res Inst, 67-70; clin chemist & asst prof biochem & med technol, 70-73, assoc prof biochem & med technol, Sch Med & chmn med technol, Sch Allied Health Professions, 73-87, ASSOC PROF PATH & NUTRIT, SCH MED & PUB HEALTH, LOMA LINDA UNIV, 87- *Concurrent Pos:* Prin investr, NIH res grant, 91-93. *Mem:* Am Asn Clin Chemists; Am Chem Soc; Am Soc Clin Path; Am Soc Med Technol; Am Inst Nutrit; Sigma Xi. *Res:* Amino acid metabolism; clinical chemistry; nutrition; physiological chemistry; dietary protein and insulin; serum protein metabolism and transport; chromatography of catecholamines. *Mailing Add:* Clin Lab Loma Linda Univ Med Ctr Loma Linda CA 92354

HUBBARD, ROBERT PHILLIP, b Portsmouth, Va, June 8, 43; m 66; c 2. BIOMECHANICS, ENGINEERING MECHANICS. *Educ:* Duke Univ, BSME, 65; Univ Ill, Urbana, MS, 67, PhD(theoret & appl mech), 70. *Prof Exp:* Asst biomech, Hwy Safety Res Inst, Univ Mich, 69-70, fel, 70-71; Sr res engr, Gen Motors Res Labs, 71-77; asst prof, 77-80, ASSOC PROF BIOMECH, MICH STATE UNIV, 80- *Concurrent Pos:* Consult, Franklin Res Inst, 78-, Hoover Universal, 81-, Motor Wheel Corp, 84-, Johnson Control, 84- & Ferno-Washington, 88-; pres, Biomechanical Design Inc, 87- *Mem:* Am Soc Mech Engrs; Am Acad Mech. *Res:* Human tolerance; head injury; tissue biomechanics; anthropometrics; biomechanics in design. *Mailing Add:* 160 Kenberry East Lansing MI 48823

HUBBARD, ROGER W, b Colchester, Vt, Feb 23, 39; m 62; c 2. ENVIRONMENTAL MEDICINE. *Educ:* Northeastern Univ, AB, 62; Brown Univ, PhD(biol), 66. *Prof Exp:* RES CHEMIST, HEAT RES DIV, US ARMY RES INST ENVIRON MED, NATICK DEVELOP CTR, 66- *Mem:* AAAS; Am Physiol Soc; Am Soc Biol Chemists; NY Acad Sci; Sigma Xi. *Res:* Conducts research on the biochemical and physiological mechanisms operative in the complex alterations that occur in animal models and humans when exposed to environmental extremes of heat and work. *Mailing Add:* RR 1 Box 113-23 Uxbridge MA 01569

HUBBARD, RUTH, b Vienna, Austria, Mar 3, 24; nat US; m 58; c 2. BIOLOGY. *Educ:* Radcliffe Col, PhD(biol), 50. *Prof Exp:* Res fel biol, Harvard Univ, 50-58, res assoc, 59-74, lectr, 68-73, prof, 73-90, EMER PROF BIOL, HARVARD UNIV, 90- *Concurrent Pos:* Guggenheim fel, Carlsberg Lab, Copenhagen Univ, 52-53; mem corp, Marine Biol Lab, Woods Hole, trustee, 73-78, emerita, 90-; vis prof, Mass Inst Technol, 72. *Honors & Awards:* Paul Karrer Medal, Swiss Chem Soc, 67. *Mem:* AAAS; Am Soc Biol Chemists; Biophys Soc; Soc Gen Physiol; Nat Women's Studies Asn. *Res:* Chemistry of vision; synthesis of visual pigments; health education of non-professionals; women's biology and health; sociology of science. *Mailing Add:* Biol Labs Harvard Univ Cambridge MA 02138

HUBBARD, T BRANNON, JR, surgery; deceased, see previous edition for last biography

HUBBARD, VAN SAXTON, b Middletown, NY, Mar 1, 45; m 74; c 2. GASTROENTEROLOGY, CLINICAL NUTRITION. *Educ:* Union Col, BS, 67; Med Col Va, MD & PhD(biochem), 74. *Prof Exp:* Resident pediat, Univ Minn Hosp, 74-76; fel pediat metab, 76-78, med officer, 78-83; asst prof, 81-86, ASSOC PROF CHILD HEALTH, GEORGE WASHINGTON UNIV, 86-; DIR NUTRIT SCI BR, NAT INST DIABETES, DIGESTIVE & KIDNEY DIS, NIH, 83- *Concurrent Pos:* Expert, Bur Drugs, Food & Drug Admin, 78-84; staff physician, Nat Naval Med Ctr, Bethesda, 79-; asst ed, Am J Clin Nutrit, 81-86; fel, Am Acad Pediat, 81- *Mem:* Am Inst Nutrit; Am Soc Parenteral & Enteral Nutrit; NAm Soc Pediat Gastroenterol; Soc Pediat Res; Sigma Xi; Am Soc Clin Nutrit. *Res:* Basic science and clinical applications relating to the study of clinical nutrition and cystic fibrosis emphasizing essential fatty acid metabolism. *Mailing Add:* NIH Westwood Bldg Rm 3A18B Bethesda MD 20892

HUBBARD, WILLARD DWIGHT, analytical chemistry; deceased, see previous edition for last biography

HUBBARD, WILLIAM BOGEL, b Liberty, Tex, Nov 14, 40; m 63; c 2. PLANETARY SCIENCES. *Educ:* Rice Univ, BA, 62; Univ Calif, Berkeley, PhD(astron), 67. *Prof Exp:* Res fel astrophys, Kellogg Radiation Lab, Calif Inst Technol, 67-68; asst prof astron, Univ Tex, Austin, 68-72; assoc prof planetary sci, 72-75, dir, Lunar & Planetary Lab & head, Dept Planetary Sci, 77-81, PROF PLANETARY SCI, UNIV ARIZ, 75-, mem, Div Planetary Sci Comt, 86-88. *Concurrent Pos:* Consult, NASA, 71-82, prin investr, 74-; mem, Div Planetary Sci Comt, Univ Ariz, 86-88. *Mem:* Am Astron Soc; Am Geophys Union; Int Astron Union; Sigma Xi; AAAS. *Res:* Interior structures of Jovian plants; occultation studies of planetary atmospheres and rings. *Mailing Add:* 2618 E Devon St Tucson AZ 85716

HUBBARD, WILLIAM JACK, b Ft Worth, Tex, June 10, 43; m 65; c 2. IMMUNOLOGY, BIOCHEMISTRY. *Educ:* Univ Tex, BA, 67; Incarnate Word Col, MS, 69; Iowa State Univ, PhD(zool), 73. *Prof Exp:* Res assoc immunol, Univ Ga, 73-77; FAC DEPT MICROBIOL & IMMUNOL, MED CTR, DUKE UNIV, 78- *Concurrent Pos:* Fel Cancer Res Inst, 78- *Res:* Immunology and physiological regulation of the immune system, especially plasma proteins and disease; immune evasion by pathogens as tumors and parasites; biochemistry and physiology of alpha 2 macroglobulin. *Mailing Add:* Innatus Found 81863 Dr Careon Blvd St 1 Indio CA 92201

HUBBARD, WILLIAM MARSHALL, b Houston, Tex, June 2, 35; m 55, 83; c 3. COMMUNICATION ENGINEERING, OPTICAL PHYSICS. *Educ:* Ga Inst Technol, BS, 57, PhD(physics), 63; Univ Ill, MS, 58. *Prof Exp:* Instr physics, Ga Inst Technol, 60-63; mem tech staff, Bell Labs, 63-83, DIST RES MGR, BELL COMMUN RES, 83- *Mem:* Am Phys Soc. *Res:* Theoretical nuclear physics, specifically orbital electron capture; millimeter wave communication; optical frequency communication; high-speed electronics & multiplexers; photonic switching and interconnects. *Mailing Add:* Bellcore NVC 3X-111 Box 7020 Red Bank NJ 07701-7020

HUBBARD, WILLIAM NEILL, JR, b Fairmont, NC, Oct 15, 19; m 45, 87; c 5. MEDICINE. *Educ:* Columbia Univ, BA, 41; NY Univ, MD, 44. *Hon Degrees:* ScD, Hillsdale Col, 67, Albany Med Col, 68, Hope Col, 79, Mich Tech Univ, 83 & Univ NC, 87. *Prof Exp:* From resident to chief resident, Bellevue Hosp, 44-50; from instr to asst prof med, Col Med, NY Univ, 50-59, from asst dean to assoc dean, 51-59; from assoc prof to prof internal med, Med Sch, Univ Mich, 59-70, dean, 59-70, dir med ctr, 69-70; vpres & gen mgr, Pharmaceut Div, Upjohn Co, 70-72, exec vpres, 72-74, pres, 74-84; RETIRED. *Concurrent Pos:* Consult, Nat Libr Med, 76-; mem panel educ consult, Comt Educ Health Admin, 73-; mem bd, Sci & Technol Int Develop, Nat Acad Sci, 78-80 & comt mem, Nat Res Coun, 75-; mem, Coun Sci & Technol Develop, 78-84; mem Nat Sci bd, NSF, 74-80, consult, 80-81; mem bd overseers Morehouse Col Med Sch, 76-81; mem bd dirs, Int Fert Res Prog, 81-90 & Upjohn Co, 68-, Family Health Int, 81-90 & Upjohn Co, 68-; mem bd trustees, Kellogg Found, 80-, Columbia Univ, 82-90; bd dir, Indust Technol Inst, 82-, vchmn, 86-; chmn, Coun Health Care Technol, 86-90; chmn vis comn, Univ Mich Med Ctr, 89- *Honors & Awards:* Award, Soc Health & Human Values, 76. *Mem:* Inst Med-Nat Acad Sci; Harvey Soc; AMA; fel Am Col Physicians; fel AAAS; Sigma Xi. *Res:* Medical education; administrative medicine. *Mailing Add:* 4630 Hickory Pt Hickory Corners MI 49060

HUBBART, JAMES E, b Walnut Hill, Ill, Mar 6, 25; m 48; c 3. AERODYNAMICS. *Educ:* Univ Notre Dame, BSAE, 47; Case Inst Technol, MSAE, 50. *Prof Exp:* Res scientist, Nat Adv Comt Aeronaut, 47-55; thermodyn engr, Lockheed Aircraft Corp, 55-57, res engr, 57-59, group leader, 59-60; from assoc prof to prof propulsion, Ga Inst Technol, 60-85; RETIRED. *Mem:* Am Inst Aeronaut & Astronaut; Am Helicopter Soc. *Res:* Propulsion; gas turbine cooling; air breathing propulsion system performance; aircraft engine inlets; missile base heating; turbulent wall jet; base and external burning propulsion; flow in fuselage-wing junction; helicopter rotor-body flow interactions. *Mailing Add:* Rte 2 Box 2517B Hartwell GA 30643

HUBBELL, D(EAN) S(TERLING), chemical engineering; deceased, see previous edition for last biography

HUBBELL, DAVID HEUSTON, b Albuquerque, NMex, Dec 8, 37; m 61; c 3. SOIL MICROBIOLOGY. *Educ:* Va Polytech Inst, BS, 60; NC State Univ, MS, 62, PhD(microbiol), 66. *Prof Exp:* Res assoc soil microbiol, Cornell Univ, 66-68, asst prof, 68-69; asst prof, 69-74, assoc prof, 74-78, PROF SOIL MICROBIOL, UNIV FLA, 78- *Concurrent Pos:* NIH res grant, 70-; Self-Help for Agr Res & Educ res grant, 71; NSF res grant, 71 & 75; Aid Int Develop res grant, 75-78; USDA res grant, 78-80; teaching & consult, Latin Am. *Mem:* Am Soc Microbiol; Am Soc Agron; fel Soil Sci Soc Am; Sigma Xi. *Res:* Mechanism of infection of legumes by Rhizobium; biochemical basis of Rhizobium-legume specificity; rhizosphere effects; plant-soil microflora relationships; grass-bacteria N-fixing associations. *Mailing Add:* Dept Soil Sci 2161 McCarty Hall Univ Fla Gainesville FL 32611

HUBBELL, DOUGLAS OSBORNE, b Norfolk, Va, Jan 6, 42; m 66; c 2. CHEMICAL ENGINEERING. *Educ:* Va Polytech Inst, BS, 64, MS, 66; Princeton Univ, PhD(chem eng), 69. *Prof Exp:* Develop engr, Hoechst Fibers Industs, 70-73, sr engr, 73-76, group leader fiber develop, 79-81, mgr chem technol, 81-87; CONSULT, QUALPRO, 87- *Concurrent Pos:* Fel Swiss Fed Inst Technol, 69-70. *Mem:* Fiber Soc; Am Soc Qual Control. *Res:* Process development in melt spinning and drawing; statistical experimental design; polymer processing. *Mailing Add:* 120 Rock Bridge Rd Spartanburg SC 29302-9678

HUBBELL, HARRY HOPKINS, JR, b Buffalo, NY, July 23, 14; m 41; c 1. PHYSICS, HEALTH PHYSICS. *Educ:* Williams Col, BA, 35; Lafayette Col, MS, 37; Princeton Univ, PhD(physics), 47; Am Bd Health Physics, cert, 60. *Prof Exp:* Asst physics, Lafayette Col, 36-37, Princeton Univ, 37-38, Williams Col, 38-40 & Wesleyan Univ, 40-41; asst physicist, Nat Bur Standards, 41-43; asst prof physics, Amherst Col, 43-44; instr, Princeton Univ, 44-45, asst, 45-46; asst prof, Williams Col, 46-47; assoc prof, Middlebury Col, 47-50; health physicist, Oak Ridge Nat Lab, 50-73; RETIRED. *Concurrent Pos:* Vis health physicist, Europ Orgn Nuclear Res, Geneva, 65-66; consult, Oak Ridge Nat Lab, 75- *Mem:* AAAS; Am Phys Soc; Health Physics Soc; Am Asn Physics Teachers. *Res:* Small angle x-ray scattering; radiation dosimetry and health physics; beta ray slowing down spectra; studies of radiation doses to survivors of atomic bombings of Hiroshima and Nagasaki; dosimetry of very high energy radiations; optical and electron beam studies of fiber and liquid surfaces. *Mailing Add:* 248 Outer Dr Oak Ridge TN 37830

HUBBELL, JOHN HOWARD, b Ann Arbor, Mich, Apr 9, 25; m 55; c 3. RADIATION PHYSICS. *Educ:* Univ Mich, Ann Arbor, BSE, 49, MS, 50. *Prof Exp:* Physicist, Nat Bur Standards, 50-51, dir, X-ray & Ionizing Radiation Data Ctr, Radiation Physics Div, Ctr Radiation Res, 63-81, staff mem, Photon & Charged Particle Data Ctr, Ionizing Radiation Div, 82-88; CONSULT, 88- *Concurrent Pos:* Consult, Int Comn on Radiol Units & Measurements, 65-; mem, Shielding Subcomt, Cross Sect Evaluation Working Group, 65-; chmn, ANS-6 Ad Hoc Comt on SI Units, Am Nuclear Soc, 74- & Gen Radiation Protection Sect, Health Physics Soc Standards Comt, 84- 90; secy, X-ray Attenuation Proj, Int Union Crystallography, 78-; consult, Particle Properties Data Proj, Europ Orgn Nuclear Res/Lawrence Berkeley Lab, 81-; exec coun, Int Radiation Physics Soc, 85-; consult, WHO, Int Atomic Energy Agency, 89-, Brookhaven Nat Lab, Upton, NY, 90-; ed, Appl Radiation & Isotopes. *Honors & Awards:* Paul C Aebersold Award, Soc Nuclear Med, 85; Radiation Indust Award, Am Nuclear Soc, 85; Prof Excellence Award, Am Nuclear Soc, 90. *Mem:* Fel Am Nuclear Soc; fel Health Physics Soc; Radiation Res Soc; Am Phys Soc; Soc Nuclear Med; Int Radiation Physics Soc. *Res:* Photon (x-ray, gamma-ray, bremsstrahlung) attenuation coefficients, cross sections, transport; atomic photoeffect; coherent, incoherent scattering; pair, triplet production; form factors; buildup factors; applied mathematics, especially distributed source calculations; scintillation counter response; x-ray crystallography; radiation gauging; x-ray fluorescence yields. *Mailing Add:* Nat Inst Standards & Technol Rm C-311 Radiation Phys Bldg Gaithersburg MD 20899

HUBBELL, STEPHEN PHILIP, b Gainesville, Fla, Feb 17, 42; m 76; c 2. ECOLOGY. *Educ:* Carleton Col, BA, 63; Univ Calif, Berkeley, PhD(zool), 69. *Prof Exp:* From asst prof to assoc prof zool, Univ Mich, 69-74; assoc prof, 74-80, PROF ZOOL, UNIV IOWA, 80-; BIOLOGIST, SMITHSONIAN INST, 83- *Concurrent Pos:* Vis distinguished prof, Univ Tex, 80; Guggenheim fel, 83. *Mem:* Ecol Soc Am; Asn Trop Biol; fel AAAS. *Res:* Tropical forest ecology; plant-animal interactions; theoretical population biology. *Mailing Add:* Dept Biol Princeton Univ Princeton NJ 08544

HUBBELL, WAYNE CHARLES, b New Orleans, La, Dec 14, 43; m 68; c 3. MAGNETISM, MATERIALS SCIENCE. *Educ:* La State Univ, New Orleans, BS, 66, MS, 68; Rice Univ, PhD(mat sci), 72. *Prof Exp:* Res assoc magnetism, Rice Univ, 72-73; mem tech staff magnetic bubbles, 73-80, SECT HEAD PLASMA CHARACTERIZATION, TEX INSTRUMENTS, INC, 80- *Mem:* Inst Elec & Electronics Engrs; Am Phys Soc. *Res:* Computer memories and other devices utilizing mobile cylindrical magnetic domains, magnetic bubbles, for data storage. *Mailing Add:* Qual Miero Systs One Magnum Pass Mobile AL 36609

HUBBEN, KLAUS, b Magdeburg, Ger, Jan 1, 30; US citizen; m 71; c 5. VETERINARY PATHOLOGY, TOXICOLOGY. *Educ:* Univ Pa, VMD, 53, MS, 58. *Prof Exp:* Instr path, Univ Pa, 56-61, asst prof, 61-67; dir, Safety Eval Dept, ICI Americas, Inc, 67-89; RETIRED. *Concurrent Pos:* Co-prin investr, NIH, 62-67. *Mem:* AAAS; Am Col Vet Path; Am Vet Med Asn; Am Col Vet Toxicol. *Res:* Pathology of canine heart disease; pathogenesis of bovine mastitis; pathology of avian sarcoma 13. *Mailing Add:* 2202 Brookline Rd Wilmington DE 19803

HUBBERT, MARION KING, geology, geophysics; deceased, see previous edition for last biography

HUBBLE, BILLY RAY, b South Bend, Ind, Sept 29, 44. PHYSICAL CHEMISTRY, ENVIRONMENTAL SCIENCES. *Educ:* Vincennes Univ, AS, 64; DePauw Univ, AB, 66; Kans State Univ, MS, 69, PhD(phys chem), 71. *Prof Exp:* Asst prof chem, Kans State Univ, 71 & Benedictine Col, 71-74; MEM STAFF, ENERGY & ENVIRON SYSTS DIV, ARGONNE NAT LAB, 74-82; NAVAL WEAPONS SUPPORT CTR, CRANE, IND, 82- *Mem:* Am Chem Soc; Am Phys Soc; Air Pollution Control Assoc; NAm Thermal Anal Soc. *Res:* High temperature, high pressure chemistry related to liquid state; molten salts; high temperature solid-gas reaction kinetics related to air pollution control processes; gas-liquid reaction kinetics related to air pollution control processes; environmental control technology. *Mailing Add:* RR 5 104 Laura Lane Bloomfield IN 47424

HUBBS, CLARK, b Ann Arbor, Mich, Mar 15, 21; m 49; c 3. ICHTHYOLOGY. *Educ:* Univ Mich, AB, 42; Stanford Univ, PhD, 51. *Prof Exp:* From instr to prof zool, Univ Tex, Austin, 49-88, chmn div biol sci, 74-76, chmn dept zool, 78-86, Clark Hubbs prof zool, 88-91; RETIRED. *Concurrent Pos:* Vis prof, Biol Sta. Univ Okla, 70-84; managing ed, Am Soc Ichthyologists & Herpetologists, 73-84, pres, 87; mem, Parks & Fish & Wildlife Adv Bd, US Dept Interior, 75-77. *Mem:* AAAS; Am Soc Ichthyologists & Herpetologists; Am Fisheries Soc; Ecol Soc Am; Am Soc Zool; Soc Syst Zoologists; Am Soc Naturalists; Am Inst Biol Sci. *Res:* Systematics; evolution, distribution and speciation of fishes; hybridization of fresh water fishes; environmental modification of freshwater fishes. *Mailing Add:* 5719 Marilyn Dr Austin TX 78712

HUBBS, ROBERT A(LLEN), b Sheridan, Wyo, Oct 22, 35; m 61; c 3. ELECTRICAL ENGINEERING. *Educ:* Stanford Univ, BS, 57, MS, 59, PhD(elec eng), 62. *Prof Exp:* Res specialist, Autonetics Div, NAm Aviation, Inc, 62-63, res supvr, 64-73, mgr radiation effects, 73-75, mgr systs eng, 75-78, chief scientist, 78-80, MGR SYSTS ANALYSIS, NAM ROCKWELL CORP, 80- *Res:* Analysis of linear and nonlinear control systems; stochastic processes; optimum filter theory; error analysis of inertial guidance systems. *Mailing Add:* Rockwell Int 3370 Miraloma Ave Anaheim CA 92803

HUBBUCH, THEODORE N(ORBERT), b Louisville, Ky, Aug 12, 02; m 42; c 2. CHEMICAL ENGINEERING, PHYSICAL CHEMISTRY. *Educ:* Univ Louisville, BS, 23; Harvard Univ, BS, 25, MS, 26. *Prof Exp:* Instr chem eng res, Mass Inst Technol, 26-28; chemist, Reynolds Metals Co, 28-29; sr res chemist, Girdler Corp, 29-30; engr tech sales, Am Potash & Chem Co, 31; consult engr, KenRad Radio Tube Corp, 33-38; proj leader indust anal, Tenn Valley Authority, 38-53; staff chem engr plant oper, US Army Chem Corps, 53-57; asst prof chem, 59-75, EMER PROF CHEM, UNIV N ALA, 75- *Concurrent Pos:* Coordr, Tenn Valley Authortiy res energy proj with Univ N Ala & Northwest Ala Coun Local Govts. *Mem:* Am Chem Soc. *Res:* Foods and gas processing; glass technology; electronics; fertilizer processing and production; nerve gas inter- mediates production; missile fuels. *Mailing Add:* 1702 Ingleside Dr Florence AL 35630

HUBBY, JOHN L, b Clovis, NMex, Mar 19, 32; m 52; c 3. BIOLOGICAL SCIENCES, GENETICS. *Educ:* Univ Tex, BS, 55, PhD(zool), 59. *Prof Exp:* NIH fel, Univ Chicago, 59-60, univ fel, 60-61, from instr to prof zool & biol, 60-88; RETIRED. *Mem:* AAAS; Genetics Soc Am; Soc Study Evolution; Am Soc Cell Biol; Am Soc Naturalists. *Res:* Physiological and evolutionary genetics. *Mailing Add:* 525 Camino Cabra Sante Fe NM 87501

HUBE, DOUGLAS PETER, b St Catharines, Ont, May 19, 41; m 65; c 2. ASTRONOMY. *Educ:* Univ Toronto, BSc, 64, MA, 65, PhD(astron), 68. *Prof Exp:* Lectr astron, Univ Toronto, 67-68; Nat Res Coun Can fel, Kitt Peak Nat Observ, Ariz, 68-69; asst prof physics & astron, 69-74, assoc prof physics, 74-82, PROF PHYSICS, UNIV ALTA, 82- *Honors & Awards:* Serv Award, Royal Astron Soc Can, 82. *Mem:* Am Astron Soc; Royal Astron Soc Can; Can Astron Soc; Int Astron Union; Brit Interplanetary Soc. *Res:* Stellar radial velocities; close binaries; peculiar A-type stars. *Mailing Add:* Dept Physics Univ Alta Edmonton AB T6G 2J1 Can

HUBEL, DAVID HUNTER, b Windsor, Ont, Feb 27, 26; US citizen; m 53; c 3. NEUROPHYSIOLOGY. *Educ:* McGill Univ, BSc, 47, MD, 51. *Hon Degrees:* AM, Harvard, 62; DSc, McGill Univ, 78, Univ Man, 83. *Prof Exp:* Rotating intern, Montreal Gen Hosp, 51-52; asst resident neurol, Montreal Neurol Inst, 52-53, fel EEG, 53-54; asst resident neurol, Johns Hopkins Hosp, 54-55; res fel neurophysiol, Walter Reed Army Inst Res, 55-58 & Wilmer Inst, Sch Med, Johns Hopkins Univ, 58-59; assoc neurophysiol & neuropharmacol, 59-60, from asst prof to assoc prof neurophysiol & neuropharmacol, 60-65, prof neurophysiol, 65-67, George Packer Berry prof neurobiol, 68-82, JOHN FRANKLIN ENDERS UNIV PROF, DEPT NEUROBIOL, HARVARD MED SCH, 82- *Concurrent Pos:* Sr fel, Harvard Soc Fels, 71-; mem, Bd Syndics, Harvard Univ Press, 79-83; Cecil H & Ida Green vis prof, Univ BC, Vancouver, 85. *Honors & Awards:* Nobel Prize Med & Physiol, 81; George H Bishop Lectr, Wash Univ, 64, James S McDonnell Lectr, Sch Med, 82; Bowditch Lectr, Am Physiol Soc, 66; Jessup Lectr, Columbia Univ, 70, Louisa Gross Horwitz Prize, 78; Jules C Stein Award, 71; Ferrier Lectr, Royal Soc, London, 72; James Arthur Lectr, Am Mus Natural Hist, 72; Lewis S Rosenstiel Award, Brandeis Univ, 72; Friedenwald Award, Asn Res Vision & Ophthal, 75; Harvey Lectr, Rockefeller Univ, 76; Grass Found Lectr, Soc Neurosci, 76; Karl Spencer Lashley Prize, Am Philos Soc, 77; Dickson Prize, Univ Pittsburgh, 79; Ledlie Prize, Harvard Univ, 80; David Marr Lectr, Cambridge Univ, Eng, 82; Paul Kayser Int Award Merit Retina Res, 89; Outstanding Sci Leadership Award, Nat Asn Biomed Res, 90. *Mem:* Nat Acad Sci; Ger Leopoldina Acad Scientists, EGer; fel Am Acad Arts & Sci; Am Physiol Soc; Soc Neurosci (pres, 88-89); Asn Res Vision & Ophthal; Am Philos Soc; foreign mem Royal Soc, London; hon mem Physiol Soc, UK; hon mem Am Neurol Soc. *Res:* Neurophysiology and neuroanatomy of the visual system of higher mammals. *Mailing Add:* Dept Neurobiol Harvard Med Sch 220 Longwood Ave Boston MA 02115

HUBEL, KENNETH ANDREW, b New York, NY, Nov 11, 27; m 57; c 3. MEDICINE, NEUROSCIENCES. *Educ:* Univ Rochester, AB, 50; Cornell Univ, MD, 54. *Prof Exp:* Intern med, State Univ NY Upstate Med Ctr, 54-55, resident, 55-56; assoc dir clin invest, Bristol Labs, NY, 56-59; resident med, State Univ NY Upstate Med Ctr, 59-60; Nat Found res fel physiol, George Washington Univ, 60-62; assoc, 62-63, from asst prof to assoc prof, 63-73, PROF MED, COL MED, UNIV IOWA, 73- *Concurrent Pos:* NIH spec res fel, Oxford Univ, 69-70; mem, Nat Inst Diabetes & Digestive Kidney Dis Study Sect, GMA-2, 84-88, chmn, 86-88. *Mem:* AAAS; Asn Am Med Cols; Am Fedn Clin Res; Am Gastroenterol Asn (secy, 78-83); Soc Exp Biol & Med; Am Physiol Soc. *Res:* Gastrointestinal absorption and secretion of drugs, electrolytes and water, particularly in the mechanisms of bicarbonate ion transport in the intestine and pancreas; autonomic control of intestinal function; neuro-endocrine control of epithelial ion transport. *Mailing Add:* Dept Med Univ Iowa Col Med Iowa City IA 52242

HUBER, BRIGITTE T, b Sins, Switz, Aug 21, 48; m 74; c 3. IMMUNOLOGY. *Educ:* Univ London, Eng, PhD(immunol), 74. *Prof Exp:* ASSOC PROF PATH, SCH MED, TUFTS UNIV, 83- *Mem:* Am Asn Immunologists. *Res:* Cellular and molecular biology of T lymphocytes; T cell receptor structure - function relationship; T lymphocyte subsets, growth factor production and control of expression; molecular characterization of B lympocyte differentiation antigens. *Mailing Add:* Dept Path Tufts Univ Sch Med 136 Harrison Boston MA 02111

HUBER, CALVIN, b Sheboygan Falls, Wis, June 7, 32; m 54; c 3. CHEMISTRY. *Educ:* Wheaton Col, BS, 54; Univ Wis, PhD(chem), 57. *Prof Exp:* From asst prof to assoc prof chem, Rockford Col, 57-61; from asst prof to assoc prof, 61-74, PROF CHEM, UNIV WIS-MILWAUKEE, 74- *Mem:* Am Chem Soc. *Res:* Electroanalytical chemistry; atomic absorption spectroscopy. *Mailing Add:* Dept Chem Univ Wis Milwaukee WI 53201

HUBER, CAROL (SAUNDERSON), b Winnipeg, Man, Apr 1, 37; m 66. CRYSTALLOGRAPHY. *Educ:* Univ Man, BSc, 59, MSc, 60; Oxford Univ, DPhil(chem crystallog), 63. *Prof Exp:* Nat Res Coun Can fel chem, 63-65, asst res off, 65-69, assoc res officer, 69-77, SR RES OFFICER, INST BIOL SCI, NAT RES COUN CAN, 78- *Concurrent Pos:* Secy, Can Nat Comt Crystallog, 78-85. *Mem:* Am Crystallog Asn. *Res:* X-ray crystallography; organic and protein structures; cysteine proteinases. *Mailing Add:* Inst Biol Sci Nat Res Coun Can Ottawa ON K1A 0R6 Can

HUBER, CLAYTON SHIRL, b Lapoint, Utah, Feb 28, 38; m 63; c 7. FOOD SCIENCE. *Educ:* Utah State Univ, BS, 62, MS, 64; Purdue Univ, PhD(food sci), 68. *Prof Exp:* Res asst, Utah State Univ, 62-63 & Purdue Univ, 65-68; assoc prin res scientist, Technol Inc, 68-71, mgr food sci sect, Life Sci Div, 71-75; res scientist, Am Potato Co, 75-76; chmn, Dept Food Sci & Nutrit, 76-88, DEAN COL BIOL & AGR, BRIGHAM YOUNG UNIV, 88- *Honors & Awards:* Snoopy Award, NASA, 70, Sci & Technol Award, 71; Cutler Award, 80; Signal Serv Award, 85. *Mem:* Inst Food Technologists; Am Dairy Sci Asn; Poultry Sci Asn; Sigma Xi. *Res:* Research and development of military feeding systems; space feeding program; chemistry of muscle and muscle proteins; food chemistry and food development; nutrition and biochemistry; dehydration of food. *Mailing Add:* 4380 N 189 E Provo UT 84604

HUBER, DAVID LAWRENCE, b New Brunswick, NJ, July 31, 37; m 62; c 4. THEORETICAL SOLID STATE PHYSICS. *Educ:* Princeton Univ, AB, 59; Harvard Univ, AM, 60, PhD(physics), 64. *Prof Exp:* Res assoc physics, Harvard Univ, 64; from instr to assoc prof, 64-69, PROF PHYSICS, UNIV WIS-MADISON, 69-, DIR, SYNCHROTRON RADIATION CTR, 85- *Concurrent Pos:* Sloan res fel, 65-67; Guggenheim fel, 72-73; Nat Asn State Univs & Land Grant Cols/CRPGE fel, 90-91. *Mem:* Fel Am Phys Soc; Sigma Xi; AAAS. *Res:* Theoretical studies of the magnetic and optical properties of solids; disordered systems. *Mailing Add:* Dept Physics Univ Wis Madison WI 53706

HUBER, DON MORGAN, b Mesa, Ariz, Mar 19, 35; m 59; c 11. PLANT PATHOLOGY, SOIL MICROBIOLOGY. *Educ:* Univ Idaho, BS, 57, MS, 59; Mich State Univ, PhD, 63. *Prof Exp:* Assoc plant pathologist, Univ Idaho, 63-71; assoc prof, 71-81, PROF PLANT PATH, PURDUE UNIV, 81- *Concurrent Pos:* Consult, 65-; chief exec officer, Decatl Mfg, 78- *Mem:* Am Phytopath Soc; Soc Prof Dispute Resolution; Sigma Xi. *Res:* Soilborne diseases; biochemistry of resistance; biological control; soil microflora; microbial interactions and activities in soil; nitrification; nitrification inhibitors. *Mailing Add:* Dept Bot & Plant Path Purdue Univ West Lafayette IN 47907

HUBER, DONALD JOHN, b Hamilton, Ohio, Sept 11, 51; m 86. POSTHARVEST PHYSIOLOGY. *Educ:* Miami Univ, BS, 73, MS, 75; Iowa State Univ, PhD(bot), 80. *Prof Exp:* Teaching asst plant physiol, Miami Univ, 73-75 & Iowa State Univ, 76-78; res asst, Iowa State Univ 78-80, assoc fel, 80; from asst prof to assoc prof, 81-90, PROF, UNIV FLA, 90- *Mem:* Am Soc Plant Physiologists; Am Soc Hort Sci; Brit Plant Growth Regulator Group. *Res:* Molecular basis of softening in horticultural crops; isolation, identification and characterization of cell wall enzymes and polysaccharides and how these change during ripening and softening. *Mailing Add:* Veg Crops Dept Univ Fla Gainesville FL 32611

HUBER, EDWARD ALLEN, b Alton, Ill, Aug 31, 29; m 54; c 4. ELECTRICAL ENGINEERING. *Educ:* Univ Ill, Urbana, BS & MS, 56, PhD(elec eng), 60. *Prof Exp:* Res assoc, coord sci lab, Univ Ill, 57-61; develop engr, Gen Tel & Electronics Labs, 61-63; sr eng specialist, Western Div, Sylvania Electronic Systs, 63-88; MEM STAFF, SRI INT, 88- *Concurrent Pos:* Lectr, Univ Santa Clara, 64-68. *Mem:* Inst Elec & Electronics Engrs. *Res:* Electronic signal analysis; software engineering. *Mailing Add:* 4243 Los Palos Ave Palo Alto CA 94306

HUBER, FLOYD MILTON, b Hinsdale, Ill, Dec 11, 37; m 76; c 4. MICROBIOLOGY. *Educ:* Ind State Univ, BS, 61; Univ Ill, MS, 64, PhD(plant path), 65. *Prof Exp:* SR RES SCIENTIST, ELI LILLY & CO, 65- *Mem:* Am Soc Microbiol. *Res:* Biosynthesis and control of antibiotic production. *Mailing Add:* 3314 W 100 S Danville IN 46122

HUBER, GARY LOUIS, b Spokane, Wash, Jan 30, 39; m; c 2. MEDICINE. *Educ:* Wash State Univ, BS, 61; Univ Wash, MD, 66, MS, 70. *Prof Exp:* Res trainee, Dept Biol Struct & Dept Anesthesiol, Univ Wash, Seattle, 63-64, student investr, Dept Anesthesiol, 64-66; teaching fel, Dept Med, Harvard Med Sch, 67-68, res fel, Dept Physiol, 68-70, instr med, 70-71, asst prof med, 71-80, dir, Smoking & Health Res Prog, 72-80; chief, Respiratory Dis Unit, Dept Med & dir, Student Res Prog, Beth Israel Hosp, 73-80; dir, Tobacco & Health Res Inst, Univ Ky, 80-81, prof med, Dept Med, Col Med, 80-85, prof anat, Col Med, Grad Sch, 80-82; clin prof, Dept Med, 85-86, physician in charge, Smoking Cessation Prog, 85-89, PROF MED, UNIV TEX HEALTH SCI CTR, TYLER, 85-, PROF MED, HOUSTON, 86- & DIR, NUTRIT UNIT, DEPT MED, TYLER, 89- *Concurrent Pos:* Clin fel, Div Respiratory Dis, Thorndike Mem Lab, Harvard Med Unit, 68-70; res consult path, Peter Bent Brigham Hosp, Boston, 68-71; asst dir, Dept Med Microbiol, Channing Lab & Dept Health & Hosp, Boston, physician in charge, Mycobacteria Ref Lab, 69-71 & dir, Clin Diag Lab, Div Chronic Dis, 69-71; vis physician, Pulmonary Med & Tuberculosis, Mattapan Chronic Dis Hosp & Long Island Chronic Dis Hosp, City of Boston, 69-71, diag clin, Tuberculosis Control Unit, Southend Health Unit, 70-74; dir, Respiratory Dis Clin, City of Boston, 70-73; chief, Div Respiratory Dis, Thorndike Mem Lab & Harvard Med Unit, Boston, 70-74, assoc physician, 71-74; consult, Mass Rehab Comn, Roxbury Div, 73-80, Univ Ky Med Ctr, Vet Admin Hosp, 82-85, Lincoln Trail Health Dept, Dept Human Resources, Ky, 83-85; physician-in-charge, Thoracic Clin, Beth Israel Hosp, 73-77, attending physician, Respiratory Intensive Care, 74-76, Pulmonary Function Lab, 76-77; assoc physician, Internal Med, Mount Auburn Hosp, Mass, 74-80, Dept Med, Beth Israel Hosp, 75-80; physician, Albert B Chandler Med Ctr, Univ Ky, 81-82, Div Pulmonary Dis, 81-83, Div Gen Internal Med, 82-85, Mountain Comprehensive Health Care, 83-85. *Mem:* AAAS; Am Asn Pathologists & Bacteriologists; fel Am Col Chest Physicians; Am Geriat Asn; Am Lung Asn; AMA; Am Physiol Soc; Am Thoracic Soc; NY Acad Sci. *Res:* Nutrition and nutritional research; modification of human behavior in appetitive dysfunctions; nutritional factors in the pathogenesis of cardiopulmonary diseases; general internal medicine; cardiopulmonary medicine; computer assisted literature analyses and metanalyses; interface of medicine, science and religious influences in Renaissance English literature. *Mailing Add:* Dept Med Univ Tex Box 2003 Tyler TX 75710

HUBER, HAROLD E, b Wauseon, Ohio, May 11, 25; m 64; c 10. PHARMACEUTICAL CHEMISTRY. *Educ:* Ohio State Univ, BS, 51, MS, 59, PhD(pharm), 62. *Prof Exp:* Pharmacist, Doast Rexall Drug, Ohio, 51-52; owner, Stryker Rexall Drug, 52-58; instr pharm, Ohio State Univ, 60-62; sr pharmaceut chemist, div Dow Chem Co, Merrell Nat Labs, Richardson-Merrell Inc, 62-74, GROUP LEADER, MERRELL-DOW PHARMACEUT INC, 74- *Concurrent Pos:* Dir Pharmaceut Res & Develop, Duramed Pham, Inc; consult. *Mem:* Am Pharmaceut Asn; Acad Pharmaceut Sci. *Res:* Pharmacology; pharmacy; physical-chemical characterizations of pharmaceutical compounds and dosage forms. *Mailing Add:* 9334 Wynnecrest Dr Cincinnati OH 45242

HUBER, IVAN, b Zagreb, Yugoslavia, Oct 15, 31; US citizen; m 61; c 2. BEHAVIOR-ETHOLOGY, GENETICS. *Educ:* Cornell Univ, AB, 54; Univ Kans, PhD(entom), 68. *Prof Exp:* Instr biol, Muhlenberg Col, 66-68; from asst prof to assoc prof, 68-84, PROF BIOL, FAIRLEIGH DICKINSON UNIV, 84- *Concurrent Pos:* Consult entom, biomed res & neurophysiol; pres, Periplaneta Inc. *Mem:* Soc Study Evolution; Soc Syst Zool; Entom Soc Am; Sigma Xi; Am Genetic Asn. *Res:* Cockroaches (Blattaria); phylogeny and biology; production of behavioral and learning mutants in cockroaches as useful models in neurobiology and biomedical research. *Mailing Add:* Dept Biol Fairleigh Dickinson Univ Madison NJ 07940-1006

HUBER, JOEL E, b Buffalo, NY, Aug 25, 36; m 60; c 3. ORGANIC CHEMISTRY. *Educ:* Canisius Col, BS, 58; Wayne State Univ, MS, 61, PhD(org chem), 63. *Prof Exp:* Asst chem, Wayne State Univ, 58-60, res assoc, 60-63; RES ASSOC, UPJOHN CO, 63- *Mem:* Am Chem Soc. *Res:* Physical organic chemistry; partial syntheses of steroids; synthesis of physiologically active compounds. *Mailing Add:* Upjohn Co 1500-91-2 7000 Portage Rd Kalamazoo MI 49001-0199

HUBER, JOHN TALMAGE, b Phoenix, Ariz, Sept 22, 31; m 55; c 3. ANIMAL NUTRITION, DAIRY SCIENCE. *Educ:* Ariz State Univ, BS, 56; Iowa State Univ, MS, 58, PhD(dairy nutrit), 60. *Prof Exp:* Res asst dairy nutrit, Iowa State Univ, 55-60; asst prof, Va Polytech Inst, 60-67; from assoc prof to prof dairy nutrit, Mich State Univ, 67-80, prof animal sci, 80-84; PROF ANIMAL SCI, UNIV ARIZ, 84- *Honors & Awards:* Outstanding Res in Dairy Nutrit Award, Am Feed Mfrs Asn, 68. *Mem:* Am Dairy Sci Asn; Am soc Animal Sci. *Res:* Digestion in the young calf; nutritional effects on milk composition; forage utilization studies in dairy cattle; mineral interactions in dairy cattle; milk residue and replacer studies. *Mailing Add:* Dept Animal Sci Univ Ariz Tucson AZ 85721

HUBER, JOSEPH WILLIAM, III, b Texas City, Tex, Aug 9, 44; m 70; c 1. ANALYTICAL ORGANIC, SOLVENT PURIFICATION. *Educ:* Univ Houston, BS, 67; Univ Miss, PhD(med chem), 74. *Prof Exp:* Res assoc, Univ Miss, 74; asst prof med chem, Sch Pharm, Northeast La Univ, 74-78; mkt chemist liquid chrom, Instrument Group, Varian Assoc, 78-80; SR CHEMIST, BURDICK & JACKSON LABS, INC, 80- *Mem:* Sigma Xi; Am Chem Soc. *Res:* Organophosphorus chemistry; amino phosphonic acids and peptides derived form aminophosphonic acids; reactions of organophosphorus compounds; liquid chromatography instrumentation and applications. *Mailing Add:* 12092 Sugar Pine Trail West Palm Beach FL 33414

HUBER, MELVIN LEFEVER, b Neffsville, Pa, Oct 11, 22; m 47; c 3. ORGANIC CHEMISTRY, INFORMATION SCIENCE. *Educ:* Franklin & Marshall Col, BS, 43; Univ Del, MS, 47, PhD(org chem), 50. *Prof Exp:* Res chemist, Eastern Lab, E I du Pont de Nemours & Co Inc, 50-53, group leader, 53-56, sect head, 56-61, supt lab serv, 61-63, tech asst, Cent Patent Index, 64-67, registr, Cent Chem Registry, 67-71, group supvr, Cent Report Index, 71-85; RETIRED. *Mem:* Am Chem Soc; Am Soc Info Sci. *Res:* Chemical structure codes; information retrieval systems; abnormal Grignard reactions; organic synthesis; process development; catalysis; rocket propellants. *Mailing Add:* 132 Colonial Ave Pitman NJ 08071

HUBER, NORMAN KING, b Duluth, Minn, Jan 14, 26; m 51; c 2. GEOLOGY. *Educ:* Franklin & Marshall Col, BS, 50; Northwestern Univ, MS, 52, PhD(geol), 56. *Prof Exp:* Asst, Franklin & Marshall Col, 48-50 & Northwestern Univ, 50-53; GEOLOGIST, US GEOL SURV, 52- *Honors & Awards:* Meritorious Serv Award, US Dept Interior, 89. *Mem:* AAAS; fel Geol Soc Am. *Res:* Geology of central Sierra Nevada, California, particularly Yosemite National Park. *Mailing Add:* US Geol Surv MS-975 345 Middlefield Rd Menlo Park CA 94025

HUBER, OREN JOHN, b Columbus, Ohio, Oct 9, 17; m 46. PHYSICS. *Educ:* Ohio State Univ, BSc, 39. *Prof Exp:* Res engr physics dept, Battelle Mem Inst, 39-44; radiographer, Jack & Heintz Co Inc, 44-45, exp engr, 46-48; res engr, Battelle Mem Inst, 48-59, proj leader metall dept, 59-60; metall engr, Livermore Labs, Sandia Corp, 60-65; asst dir metals eng inst, Am Soc Metals, 66-78; RETIRED. *Concurrent Pos:* Nat adv comt aeronaut, AEC, 45-46. *Res:* X-ray and electron diffraction; x-ray radiography; vacuum fusion; structural physical chemistry; physical metallurgy of non-ferrous and refractory metals. *Mailing Add:* 423 Royal Bonnet Ct Ft Myers FL 33908-1612

HUBER, PAUL W(ILLIAM), b Springfield, Ohio, Dec 17, 20; m 42; c 5. ELECTRICAL ENGINEERING, THERMODYNAMICS. *Educ:* Ohio Northern Univ, BSEE, 42. *Prof Exp:* Elec engr, Nat Adv Comt Aeronaut, 42-45, physicist, 46-48, aeronaut res scientist, 49-58; aeronaut res scientist, 58-62, head plasma appln sect & aerospace engr, 62-70, head flow field kinetics sect, 70-72, ASST HEAD HYPERSONIC PROPULSION BR, LANGLEY RES CTR, NASA, 72- *Mem:* AAAS; Am Inst Aeronaut & Astronaut. *Res:* Gas dynamics; plasma physics; chemical kinetics; combustion and electromagnetics. *Mailing Add:* Two Edgewood Dr Newport News VA 23666

HUBER, PETER WILLIAM, b Toronto, Ont, Nov 3, 52. THERMAL SCIENCES, FLUID MECHANICS. *Educ:* Mass Inst Technol, BS & MS, 74, PhD(mech eng), 76. *Prof Exp:* asst prof, 76-79, ASSOC PROF MECH ENG, MASS INST TECHNOL, 79- *Concurrent Pos:* Consult, Brookhaven Nat Lab, 76- *Mem:* Am Soc Mech Engrs; Can Soc Mech Engrs; Am Nuclear Soc. *Res:* Thermo-hydraulics; nuclear reactor hydrodynamics; two phase flows (solid-liquid and liquid-vapor). *Mailing Add:* 434 New Jersey Ave Wash DC 20003

HUBER, RAYMOND C, b Woodhaven, NY, Apr 14, 17; m 41; c 1. PHARMACY. *Educ:* City Col New York, BS, 38. *Prof Exp:* Res asst, Endo Labs, Inc, NY, 46-50; res scientist, Squibb Inst Med Res, 50-63, sr res investr, 63-78, proj leader, 79-80, consult, Pharmaceut Res & Develop Dept, 81-84; RETIRED. *Mem:* Am Chem Soc; Am Pharmaceut Asn. *Res:* New products for injection. *Mailing Add:* 1372 Crim Rd Bridgewater NJ 08807

HUBER, RICHARD V, b 1948. COMPUTER NETWORKS, OPERATING SYSTEMS. *Educ:* Mass Inst Technol, BS, 70; State Univ NY Stony Brook, MS, 72, PhD(comput sci), 75. *Prof Exp:* Asst prof comput sci, Tex A&M Univ, 75-77; mem tech staff, 77-80, SUPVR, BELL LABS, 80- *Mem:* Asn Comput Mach; Inst Elec & Electronics Engrs. *Res:* Computer software. *Mailing Add:* AT&T 307 Middletown-Lincroft Rd Lincroft NJ 07738

HUBER, ROBERT, b Feb 20, 37. BIOCHEMISTRY. *Prof Exp:* PROF & DIR DEPT, MAX-PLANCK INST BIOCHEM, GER. *Honors & Awards:* Nobel Prize in Chem, 88. *Mailing Add:* Max Planck Inst fur Biochem Am Klopferspitz 18A 8033 Martinsried Obb Germany

HUBER, ROBERT JOHN, b Payson, Utah, July 10, 35; m 57; c 2. SEMICONDUCTORS. *Educ:* Univ Utah, BS, 56, PhD(physics), 61. *Prof Exp:* Physicist & dir, Microelectronics Res & Develop Lab, Gen Instrument Corp, 66-71; dir microcircuit lab, Inst Biomed Eng, 71-76, res assoc prof surg, mat sci & eng & adj assoc prof physics, 76-78, PROF ELEC ENG, UNIV UTAH, 76- *Concurrent Pos:* Consult, Microelectronics Div, Gen Instrument Corp, 72-88, silicon integrated circuits, 88. *Mem:* Inst Elec & Electronics Engrs; Sigma Xi. *Res:* Semiconductor integrated circuits; large scale integrated circuits, semiconductor device physics; solid state chemical sensors. *Mailing Add:* 1145 E Millbrook Way Bountiful UT 84010

HUBER, ROGER THOMAS, b Duluth, Minn, Jan 25, 34; c 2. INSECT ECOLOGY, BIOMETEOROLOGY. *Educ:* Univ Del, BA, 61, MS, 63; Purdue Univ, PhD(entom), 69. *Prof Exp:* Instr entom, Purdue Univ, 65-69, asst prof, 69-74; assoc prof, 74-80, PROF ENTOM & PROG DIR, AGR & NATURAL RESOURCES ARIZ COOP EXTEN SERV, 81-, DEPT HEAD AGR, UNIV ARIZ, 88- *Concurrent Pos:* NSF Int Travel Grant, 69; rep, Int Soc Biometeorol, 69-75; pest mgt consult, Quatar, Mex, Ecuador & Peru, 77-85. *Mem:* Entom Soc Am; Int Soc Biometeorol. *Res:* Insect population ecology; insect pheromone research and utilization of pheromones for insect pest management; computer simulation and real-time forecasting of insect population phenology; pest management systems. *Mailing Add:* Dept Agr Educ Univ Ariz Forbes Bldg Rm 222A Tucson AZ 85721

HUBER, RUEBEN EUGENE, b Saskatchewan, Sask, Mar 24, 40; m 65; c 3. BIOCHEMISTRY. *Educ:* Univ Alta, BSc, 61, MSc, 62; Univ Calif, PhD(biochem), 66. *Prof Exp:* Fel biochem, Univ Aix-Marseille, 66-67; from asst prof to assoc prof, 67-79, PROF BIOCHEM & CHMN, DEPT BIOL SCI, DIV BIOCHEM, UNIV CALGARY, 79- *Mem:* Chem Inst Can; Am Chem Soc. *Res:* Protein and amino acid chemistry; enzymology and chemistry of disaccharidases. *Mailing Add:* Dept Biol Sci Univ Calgary Calgary AB T2N 1N4 Can

HUBER, SAMUEL G(EORGE), b Degraff, Ohio, Mar 5, 18; m 41; c 3. AGRICULTURAL ENGINEERING. *Educ:* Ohio State Univ, BS, 40, BAgrE, 41, MS, 58. *Prof Exp:* Agr engr, USDA, 41-42; res asst agr eng, Univ Ill, 42-43; exten engr, 46-54, instr agr eng, 56-58, assoc prof, 58-68, PROF AGR ENG, OHIO STATE UNIV, 68-, AGR ENGR, COOP EXTEN SERV, 75- *Concurrent Pos:* Consult, Ford Found, 64-65. *Mem:* Am Soc Agr Engrs; Am Soc Eng Educ. *Res:* Development of more efficient power systems for agricultural operations; management of farm machines. *Mailing Add:* 5815 Stoney Creek Ct Columbus OH 43210

HUBER, THOMAS LEE, b Brownstown, Ind, Feb 11, 35; m 62. PHYSIOLOGY, PHARMACOLOGY. *Educ:* Purdue Univ, BS, 57; Kans State Univ, MS, 58; Univ Ky, PhD(physiol, nutrit), 63. *Prof Exp:* Res assoc nutrit biochem, Univ Ill, 63-64, asst prof physiol & pharmacol, 64-66; assoc prof, 66-75, PROF PHYSIOL & PHARMACOL, UNIV GA, 75- *Concurrent Pos:* Mem, Conf Res Workers Animal Dis NAm. *Mem:* Am Soc Vet Physiol & Pharmacol; Soc Exp Biol & Med; World Asn Vet Physiol, Pharmacol & Biochem; Am Soc Animal Sci. *Res:* Digestive physiology of ruminants; lactic acid metabolism; renal physiology of ruminants. *Mailing Add:* Dept Physiol & Pharm Univ Ga Athens GA 30602

HUBER, THOMAS WAYNE, b Eddy, Tex, Sept 2, 42; m 63; c 3. MEDICAL MICROBIOLOGY, PUBLIC HEALTH. *Educ:* Univ Tex, Austin, BS, 64, PhD(microbiol), 68; Am Bd Med Microbiol, dipl & cert pub health & med lab microbiol, 73. *Prof Exp:* Fel, Nat Ctr Dis Control, Ga, 68-70; instr path, Med Sch, 70-72, asst prof, Health Sci Ctr, Univ Tex, San Antonio, 72-74; chief lab serv, Houston Health Dept, 74-81; adj asst prof microbiol & immunol, Baylor Col Med & adj asst prof med lab sci, Sch Allied Health Sci, Univ Tex, 77-80; MICROBIOLOGIST, LAB SERV, OLIN E TEAGUE VET CTR, TEMPLE, TEX, 81- *Concurrent Pos:* Assoc prof, Univ Tex Sch Pub Health, Houston, 74-81; asst prof path & microbiol & immunol, Baylor Col Med, 75-82; adj assoc prof, dept immunol, microbiol & path, Tex A&M Med Sch, 81-; Burroughs Wellcome vis prof, 90-91. *Mem:* Am Soc Microbiol; Am Acad Microbiol. *Res:* Induction and properties of spheroplasts, protoplasts and L-forms of Staphylococcus aureus; automation in microbiology; diagnostic and clinical microbiology; cultivation of L-forms, laboratory diagnosis of venereal diseases; occurrence and mechanisms of antibiotic resistance. *Mailing Add:* 4809 Arrowhead Temple TX 76502

HUBER, W(ILSON) FREDERICK, b Lebanon, Pa, May 6, 18; m 43; c 3. ORGANIC CHEMISTRY, SYNTHETIC ORGANIC & NATURAL PRODUCTS CHEMISTRY. *Educ:* Lebanon Valley Col, BS, 40; Univ Cincinnati, MS, 42, PhD(org chem), 43. *Prof Exp:* Chemist, Monsanto Chem Co, 43-46; Am Chem Soc fel, Univ Mich, 46-47; chemist, Procter & Gamble Co, Ohio, 47-56; vpres res & develop, Nease Chem Co, Inc, 56-62; dir prod develop, Martin-Marietta Chem, 63-65, dir res, 65-66, vpres opers, 67-71, sr vpres opers, Sodyeco Div, 71-80, sr vpres technol, 81-82; RETIRED. *Mem:* Am Chem Soc; Am Ord Soc. *Res:* Organic chemicals; fatty acids and their derivatives; organophosphorous compounds; dyestuffs. *Mailing Add:* 3738 Beresford Rd Charlotte NC 28211

HUBER, WAYNE CHARLES, b Shelby, Mont, Aug 2, 41; m 68; c 1. ENVIRONMENTAL ENGINEERING, HYDRODYNAMICS. *Educ:* Calif Inst Technol, BS, 63; Mass Inst Technol, MS, 65, PhD(civil eng), 69. *Prof Exp:* Res asst civil eng, Mass Inst Technol, 63-65 & 66-68; from asst prof to assoc prof , 68-78, PROF ENVIRON ENG SCI, UNIV FLA, 78- *Concurrent Pos:* Consult. *Honors & Awards:* Karl Hilgard Prize, Am Soc Civil Engrs, 73. *Mem:* Am Soc Civil Engrs; Am Geophys Union; Int Asn Hydraul Res; Am Water Resources Asn. *Res:* Diffusive processes in streams, estuaries, lakes and reservoirs; thermal structure of reservoirs; thermal pollution; water resources; urban hydrology; water pollution. *Mailing Add:* Dept Environ Eng Sci Univ Fla Gainesville FL 32611-2013

HUBER, WILLIAM RICHARD, III, b Indiana, Pa, Dec 23, 41; m 64; c 2. ELECTRICAL ENGINEERING, INTEGRATED CIRCUIT DESIGN. *Educ:* Univ Pittsburgh, BS, 62, DSc(elec eng), 69; Ohio State Univ, MS, 63. *Prof Exp:* Mem tech staff, Bell Tel Labs, Inc, 62-70, supvr, 70-81; MGR, GEN ELEC CO, 82- *Mem:* Sr mem Inst Elec & Electronics Engrs. *Res:* Semicustom integrated circuits; semiconductor memories; linear integrated circuits; computer-aided circuit design; integrated circuit processing; noise. *Mailing Add:* Gen Elec Co PO Box 13049 Research Triangle Park NC 27709

HUBER, WOLFGANG, biochemistry, for more information see previous edition

HUBER, WOLFGANG KARL, pharmacology, for more information see previous edition

HUBERMAN, BERNARDO ABEL, b Buenos Aires, Arg, Nov 7, 43; US citizen; m 70; c 2. NONLINEAR DYNAMICS, DISTRIBUTED COMPUTATION. *Educ:* Univ Buenos Aires, Licenciado, 66; Univ Pa, PhD(physics), 70. *Prof Exp:* mem res staff, 71-80, sr mem res staff physics, 80-84, RES FEL, XEROX PALO ALTO RES CTR, 85- *Concurrent Pos:* Lectr, Stanford Univ, 77-80 & consult prof, 80-; vis prof, Univ Paris, France, 81. *Mem:* Fel Japan Soc Prom Sci; fel Am Phys Soc. *Res:* Physics of complex systems; Nonlinear dynamics and distributed computation. *Mailing Add:* Xerox Palo Alto Res Ctr 3333 Coyote Hill Rd Palo Alto CA 94304

HUBERMAN, ELIEZER, b Lukow, Poland, Feb 8, 39; US citizen; m 67; c 2. CARCINOGENESIS, CELL DIFFERENTIATION. *Educ:* Tel-Aviv Univ, MSc, 64; Weizmann Inst Sci, PhD(genetics), 69. *Prof Exp:* Fel, McArdle Lab, Univ Wis, 69-71; sr scientist & assoc prof, dept genetics, Weizmann Inst Sci, 71-77; sr scientist & group leader, Biol Div, Oak Ridge Nat Lab, 76-81; dir & group leader, Biol & Med Res Div, Argonne Nat Lab, 81-; PROF MOLECULAR GENETICS & CELL BIOL, UNIV CHICAGO, 82-, PROF, RADIATION ONCOL, 84- *Concurrent Pos:* Vis assoc, Nat Cancer Inst, 68, vis scientist, 71, mem, Gene-Toxicol Comt, 79-81; chmn, Gene-Toxicol Comt, Environ Protection Agency, 79-81; mem, Comt Chem Environ Mutagens, Nat Acad Sci, 79-83; assoc ed, Cancer Res, 81-; mem, Health Sci Rev Comt, Nat Inst Environ Health Sci, 83-; fel Univ Tokyo, 86. *Mem:* Am Asn Cancer Res; Radio/Res Soc. *Res:* Mode of action of chemicals that initiate or promote tumor formation; analysis of chemically induced changes in gene expression (mutagenesis and differentiation) in cultured mammalian cells. *Mailing Add:* 424 Sunset Ave La Grange IL 60525

HUBERMAN, JOEL ANTHONY, b Washington, DC, Feb 13, 41; m 63; c 2. DNA REPLICATION, CHROMATIN STRUCTURE. *Educ:* Harvard Univ, BA, 63; Calif Inst Technol, PhD(biochem), 68. *Prof Exp:* From asst prof to assoc prof biol, Mass Inst Technol, 70-75; RES PROF CELLULAR & MOLECULAR BIOL, ROSWELL PARK CANCER INST, 75- *Concurrent Pos:* Ed, Chromosoma, 84- *Mem:* Am Soc Biol Chemists; Am Soc Cell Biol; Am Soc Microbiol; AAAS; Genetics Soc Am. *Res:* Identification and characterization of eukaryotic DNA replication origins; identification and characterization of proteins involved in initiation of DNA replication; Determination of the structural and DNA sequence requirements for eukaryotic DNA replication origins. *Mailing Add:* Molecular & Cellular Biol Dept Roswell Park Cancer Inst Buffalo NY 14263

HUBERMAN, MARSHALL NORMAN, b Brooklyn, NY, Jan 27, 32; m 67; c 2. ENGINEERING PHYSICS, ENERGY CONVERSION. *Educ:* Union Univ, NY, BS, 53; Univ Chicago, MS, 57, PhD(physics), 62. *Prof Exp:* Physicist, Atomics Int Div, NAm Aviation, Inc, 58; res asst nuclear physics, Enrico Fermi Inst Nuclear Studies, Univ Chicago, 59-62; sr physicist, Atomics Int Div, NAm Aviation, Inc, 62-64, res specialist energy conversion, 65-66; mem tech staff, 66-68, head colloid thruster tech sect, 68-74, head colloid & ion propulsion sect, 74-77, sr scientist, space & technol group, 77-88, ASST PROG MGR, TRW, 88- *Honors & Awards:* Assoc Fel, Am Inst Aeronaut & Astronaut. *Mem:* AAAS; Am Phys Soc; assoc fel Am Inst Aeronaut & Astronaut. *Res:* Low energy nuclear physics; physical electronics; work function variations on cathode surfaces; photovoltaic cells for energy conversion; electrical propulsion. *Mailing Add:* 3129 Haddington Dr Los Angeles CA 90064

HUBERT, HELEN BETTY, b New York, NY, Jan 22, 50; div. HEART DISEASE, ARTHRITIC DISEASES. *Educ:* Barnard Col, BA, 70; Yale Univ, MPH, 73, MA, 76, PhD(epidemiol), 78. *Prof Exp:* Assoc res, Sch Med, Yale Univ, 77-78; epidemiologist, div heart & vascular dis, Nat Heart, Lung & Blood Inst, 78-84; dir res, Gen Health Underwriting, Inc, 84-87; SR RES SCIENTIST, STANFORD UNIV SCH MED, 88- *Concurrent Pos:* Mem coun epidemiol, Am Heart Asn. *Mem:* Sigma Xi; Soc Epidemiol Res; fel Am Heart Asn; Am Col Epidemiol; Arthritis Health Prof Asn. *Res:* Understanding the relationship of health risks and behaviors to health care costs and utilization of services in population groups; design, conduct, and analysis of studies to elucidate the etiology of coronary heart disease and other cardiovascular diseases in human populations; design, conduct and analysis of studies to elucidate the etiology and prognosis of arthritic diseases and musculo skeletal disabilities. *Mailing Add:* 4142 Thain Way Palo Alto CA 94306

HUBERT, JAY MARVIN, b Denver, Colo, Apr 11, 44; m 68; c 3. CHEMICAL PHYSICS. *Educ:* Reed Col, BA, 66; Tex A&M Univ, MS, 68, PhD(physics), 70. *Prof Exp:* Res physicist, Chevron Res Co, 70-75, sr res physicist, 75-80, sr res assoc, Anal Div, 80-82, mgr, Tech Info Ctr, 82-84, TEAM LEADER, MOLECULAR IDENTIFICATIONS, CHEVRON RES CO, 85- *Concurrent Pos:* Comptrollers Dept, Chevron Corp, 84-85. *Mem:* AAAS; Am Phys Soc. *Res:* Low temperatures; liquids; superfluid helium; thermal analysis; physical properties of microporous catalysts; auger electron spectroscopy. *Mailing Add:* Chevron Res Co Analytical Unit 100 Chevron Way Richmond CA 94802-0627

HUBERT, JOHN FREDERICK, b Quincy, Mass, Nov 28, 30; m 55; c 3. GEOLOGY. *Educ:* Harvard Univ, AB, 52; Univ Colo, MS, 54; Pa State Univ, PhD(mineral, petrol), 58. *Prof Exp:* From asst prof to prof geol, Univ Mo, 58-70; PROF GEOL, UNIV MASS, AMHERST, 70- *Concurrent Pos:* Pres, eastern sect, Soc Econ Paleont & Mineral, 84-85; assoc ed Sedimentary Geol; vis prof, Desert Res Inst, Las Vegas, Nev, 84 & Utah Geol & Mineral Surv, Salt Lake City, Utah, 91. *Mem:* Fel Geol Soc Am; Soc Econ Paleont & Mineral; Am Asn Petrol Geol; Int Asn Sedimentol. *Res:* Sedimentology; sedimentary petrology; application of statistics to sedimentology. *Mailing Add:* Dept Geol Univ Mass Amherst MA 01003

HUBERTY, CARL J, b Lena, Wis, Nov 14, 34; m 66; c 4. APPLIED STATISTICS, EDUCATIONAL EVALUATION. *Educ:* Wis State Univ, BS, 56; Univ Wis, MS, 58; Univ Iowa, PhD(educ statist), 69. *Prof Exp:* Teacher math, Seymour High Sch, Wis, 56-57, Brown Deer High Sch, 58-59 & Oconto Falls High Sch, 59-61; teacher math, US Dept Army Schs, Orleans, France, 61-64; asst prof, Wis State Univ, 65-66; from asst prof to assoc prof, 69-79, PROF EDUC, UNIV GA, 79- *Concurrent Pos:* Consult projs, Bur Educ Handicapped. *Mem:* Am Statist Asn; Am Educ Res Asn; Nat Coun Measurement in Educ; Soc Multivariate Exp Psychol. *Res:* Applications of multivariate analysis; special education evaluation; research in applied statistical methodology; comparisons of statistical methods, techniques, indices; translation of theoretical developments for application; reviews of proposed methodologies; mostly in multivariate methods; educational evaluation methodology. *Mailing Add:* Dept Educ Psychol Aderhold Hall Univ Ga Athens GA 30602

HUBIN, WILBERT N, b Crosby, Minn, Apr 28, 38. AERONAUTICAL & ASTRONAUTICAL ENGINEERING. *Educ:* Wheaton Col, Ill, BS, 60; Univ Ill, Urbana, MS, 62, PhD(physics), 69. *Prof Exp:* From asst prof to assoc prof, 68-83, PROF PHYSICS, KENT STATE UNIV, 83- *Mem:* Am Phys Soc; Am Inst Aeronaut & Astronaut; Sigma Xi. *Res:* Dynamics of aircraft flight; digital electronic instrumentation. *Mailing Add:* Dept Physics Kent State Univ Kent OH 44242

HUBISZ, JOHN LAWRENCE, JR, b Salem, Mass, June 6, 38; m 74; c 4. ATOMIC & MOLECULAR PHYSICS. *Educ:* St Francis Xavier Univ, dipl eng, 58, BSc Hons, 59; Univ Tenn, MSc, 65, PhD(physics & space sci), 68; St Thomas Univ, BTh, 75. *Prof Exp:* Lectr physics, St Francis Xavier Univ, 59-68, asst prof, 68-71; INSTR, COL MAINLAND, 71- *Concurrent Pos:* Lectr, Univ Tenn, 63-65 & York Univ, 66-68; Shell merit fel, Stanford Univ, 69. *Mem:* Am Asn Physics Teachers; Am Phys Soc; Nat Coun Teachers Math; Sigma Xi. *Res:* Teaching of mathematics and physics; theoretical studies of diatomic molecules, especially those that are aeronomically and astronomically important. *Mailing Add:* 8017 College Ave Texas City TX 77591

HUBKA, WILLIAM FRANK, b Denver, Colo, June 18, 39; m 61; c 2. ENGINEERING SCIENCE. *Educ:* Univ Denver, BS, 62, MS, 65, PhD(mech eng), 72. *Prof Exp:* Engr, Martin Co, Colo, 62-64; res scientist, Kaman Sci Corp, Colo, 64-65; res engr, Univ Denver, 65-67; res scientist, Kaman Sci Corp, 67-72; div mgr, Gen Res Corp, Santa Barbara, 72-74; group opers mgr, Sci Applns, Inc, 74-89; SR STAFF SCIENTIST, SPACE SYST DIV, LOCKHEED CORP, 90- *Mem:* Am Soc Mech Engrs; Sigma Xi; NY Acad Sci. *Res:* Defense systems; energy systems; program management. *Mailing Add:* PO Box 4656 Mountain View CA 94040

HUBLER, GRAHAM KELDER, JR, b Suffern, NY, Feb 29, 44; m 69. APPLIED PHYSICS. *Educ:* Union Col, BS, 66; Rutgers Univ, PhD(physics), 72. *Prof Exp:* Nat Res Coun assoc, 72-74, RES SCIENTIST PHYSICS, NAVAL RES LAB, 74- *Mem:* Am Phys Soc; Mat Res Soc; Sigma Xi. *Res:* Study of the changes induced in the surfaces of metals and semiconductors by the implantation of ions, with regard to such altered surface properties as corrosion, wear, and optical properties. *Mailing Add:* Code 667 Naval Res Labs Washington DC 20375

HUBNER, KARL FRANZ, b Striegau, Ger, Jan 20, 34; US citizen; m 60; c 2. NUCLEAR MEDICINE, RADIATION BIOLOGY. *Educ:* Univ Heidelberg, MD, 59. *Prof Exp:* Resident clin invest, Med Div, Oak Ridge Assoc Univs, Tenn, 62-64, res assoc exp immunol, 67-70, sr clin staff mem, 71-73, dir outpatient nuclear med, Med & Health Sci Div, 75-84; DIR RADIOL RES, UNIV TENN MED CTR, KNOXVILLE, 84-, DIR NUCLEAR MED, 86- *Concurrent Pos:* Consult, Oak Ridge Hosp of Methodist Church, 77-; mem clin staff, Med & Health Sci Div, Oak Ridge Assoc Univ, Tenn, 71-82, dir, Radiation Emergency Assistance Ctr/Training Site, 77-82, chief clinician, 79-82, asst chmn & dir, Clin Nuclear Med Br, 82-84; dir, Multidisciplinary Breast Ctr, Univ Tenn Med Ctr, Knoxville, 85-; courtesy staff, internal med, E Tenn Baptist Hosp, Knoxville, 82-84. *Mem:* AMA; Soc Nuclear Med; Am Occup Med Asn; Int Asn Radiopharmacol; Soc Exp Biol & Med; Soc Magnetic Resonance; Radiol Soc NAm; Am Roentgen Ray Soc. *Res:* Radiopharmaceuticals; emission computerized tomography; radiobiology; treatment of radiation injury; diagnosis of neoplastic diseases. *Mailing Add:* Dept Radiol Univ Tenn Med Ctr 1924 Alcoa Hsy Knoxville TN 37920-6999

HUBRED, GALE L, b Alexandria, Minn, Jan 4, 39; div; c 2. ENVIRONMENTAL ENGINEERING. *Educ:* Univ Minn, BS, 62; Univ Hawaii, MS, 70; Univ Calif, Berkeley, PhD(metall), 73. *Prof Exp:* Supt, Dow Chem Co, 62-68; sr res engr, Kennecott Copper, 73-79 & Occidental Petrol, 79; SR SCIENTIST, CHEVRON RES CO, 79- *Concurrent Pos:* Chmn ocean technol, Am Inst Chem Engrs, 75-; vpres, Marine Technol Soc, 82- *Mem:* Fel Marine Technol Soc; Am Inst Chem Engrs; Am Inst Mining, Metall & Petrol Engrs; Water Pollution Control Fed. *Res:* Hydrometallurgical processes; manganese nodules; solvent extraction. *Mailing Add:* Chevron Res & Tech Co 100 Chevron Way Richmond CA 94802

HUBSCHMAN, JERRY HENRY, b Great Neck, NY, Feb 4, 29; m 53; c 3. INVERTEBRATE ZOOLOGY. *Educ:* Ohio State Univ, BS, 59, PhD(zool), 62. *Prof Exp:* Res biologist, Robert A Taft Sanit Eng Ctr, USPHS, 62-64; asst prof biol, 64-65, coordr, 65-67, assoc prof, 67-70, assoc provost, 74-78, PROF BIOL, WRIGHT STATE UNIV, 70- *Concurrent Pos:* Secy bd trustees, Wright State Univ, 75-78. *Mem:* Am Soc Parasitol; Societas Internationalis Limnologiae. *Res:* Growth and development of Crustacea; aquatic life cycles; parasitology; planktonic crustacea, aquatic benthos. *Mailing Add:* Dept Biol Sci Wright State Univ Dayton OH 45435

HUCHITAL, DANIEL H, b Brooklyn, NY, July 19, 40; m 60; c 2. INORGANIC CHEMISTRY. *Educ:* City Col New York, BSc, 61; Stanford Univ, PhD(chem), 65. *Prof Exp:* Res fel chem, State Univ NY Buffalo, 65-66; res assoc, Brookhaven Nat Lab, 66; from asst prof to assoc prof, 66-79, PROF CHEM, SETON HALL UNIV, 80- *Concurrent Pos:* Vis prof, Tex A&M, 72-73, Univ SCalif, 80-81. *Mem:* Am Chem Soc. *Res:* Coordination chemistry; mechanisms of electron-transfer reactions involving transition metal ion complexes. *Mailing Add:* Dept Chem Seton Hall Univ South Orange NJ 07079-2689

HUCHRA, JOHN P, b Jersey City, NJ, Dec 23, 48. ASTROPHYSICS. *Educ:* Mass Inst Technol, SB, 70; Calif Inst Technol, PhD(astron), 76. *Prof Exp:* Fel, Ctr Astrophys, 76-78, ASTRONR, SMITHSONIAN ASTROPHYS OBSERV, 78-; PROF ASTRON, HARVARD UNIV, 84-; ASSOC DIR, HARVARD-SMITHSONIAN CTR ASTROPHYS, 89- *Concurrent Pos:* Lectr, Harvard Univ, 79-; res grants, Space Telescope Working Group Galaxies & Clusters, NASA & Smithsonian Inst. *Honors & Awards:* Newcomb Cleveland Award, 90. *Mem:* Am Astron Soc; Sigma Xi; Am Inst Physics; Int Astron Union; Am Phys Soc; Royal Astron Soc; AAAS. *Res:* Extragalactic observational astronomy; galaxies and cosmology; gravitational lenses; quasars; globular starclusters. *Mailing Add:* Harvard-Smithsonian Ctr Astrophys 60 Garden St Cambridge MA 02138

HUCK, MORRIS GLEN, b Centralia, Ill, Mar 7, 37; m 66; c 1. PLANT SCIENCE, SYSTEM ANALYSIS. *Educ:* Univ Ill, Urbana, BS, 58, MS, 60; Mich State Univ, PhD(soil sci), 67. *Prof Exp:* Res soil scientist, Agr Res Serv, USDA & Auburn Univ, 67-85; AT DEPT AGRON, UNIV ILL, URBANA, 85- *Mem:* AAAS; Am Soc Agron; Am Soc Plant Physiol; Soil Sci Soc Am; Soc Comput Simulation; Sigma Xi. *Res:* Root metabolism; dynamic aspects of interactions between roots and shoots and between plant roots and their soil environment; computer simulation and operations analysis of plant growth processes; crop simulation and root system physiology. *Mailing Add:* S-216 Turner Hall 1102 S Goodwin Ave Urbana IL 61801

HUCKA, VLADIMIR JOSEPH, b Ostrava, Czech, Apr 22, 25; m 61; c 1. MINING ENGINEERING. *Educ:* Tech Univ Mines, MS, 62; Ostrava, Czech, PhD(mining), 66. *Prof Exp:* Engr, Inst Planning New Mines, Ostrava, 52-56; engr coal mine, 57-62; head rock mech, Sci Res Coal Inst, 63-68; sci officer, Coal Owner's Inst, Essen, 68; prof mining eng, Laval Univ, Que, 68-78; PROF MINING, UNIV UTAH, SALT LAKE CITY, 78- *Concurrent Pos:* Mining consult, Gaspe Copper Mines, Noranda Group, 74-77. *Mem:* Can Inst Mining & Metall; Soc Mining Engrs. *Res:* Longwall mining; methane drainage; robotics in mining; mine haulage systems. *Mailing Add:* 2282 E 7800 S Salt Lake City UT 84121

HUCKABA, CHARLES EDWIN, b Huntingdon, Tenn, Oct 20, 22; m 46; c 2. TECHNICAL MANAGEMENT, BIOENGINEERING. *Educ:* Vanderbilt Univ, BE, 44; Mass Inst Technol, MS, 47; Univ Cincinnati, PhD(chem eng), 53. *Prof Exp:* Asst chem eng, Mass Inst Technol, 44-45, mem res staff, USN Proj, 45-47; instr, Univ Cincinnati, 47-52; assoc prof, Lamar State Col, 53-55 & Univ Fla, 55-61; mem staff, E I du Pont de Nemours & Co, 61-62; assoc prof chem eng, Univ Fla, 62-63; prof & chmn dept, Drexel Univ, 63-67; vis prof, Columbia Univ, 67-68, sr res assoc, 68-69, mem-at-large, Fac Eng & Appl Sci & prof rehab med, Col Physicians & Surgeons, 69-74; sect head eng chem & energetics, NSF, 74-76; dir eng prog develop, Cooper Union, 76-80; PRES, CHARLES HUCKABA ASSOCS, 80- *Concurrent Pos:* Consult, Bethlehem Steel Co, 52-56, Thermal Res & Eng Corp, 65-67 & Foxboro Co, 70-71; adj prof clin eng, Med Sch, George Washington Univ, 74-78. *Honors & Awards:* Stephen L Tyler Award, Am Inst Chem Engrs, 72. *Mem:* Hon life mem Am Soc Eng Educ; fel Am Inst Chem Engrs; fel Am Inst Chemists; Sigma Xi. *Res:* Analysis of biochemical, genetic and chemical engineering systems; social implications of technology and public policy; impact of government regulatory activities upon industrial productivity; organization and financing of high-tech entrepreneurial companies. *Mailing Add:* Apartado de Correos 15 Marbella Malaga 29600 Spain

HUCKABA, JAMES ALBERT, b Charleston, Ill, Feb 6, 36; m 55; c 3. ALGEBRA. *Educ:* Univ Ill, BS, 60; Univ Iowa, MA, 64, PhD(math), 67. *Prof Exp:* From asst prof to assoc prof, 67-79, PROF MATH, UNIV MO-COLUMBIA, 79- *Concurrent Pos:* NSF grant, 69-71. *Mem:* Am Math Soc. *Res:* Commutative ring theory; ideal theory. *Mailing Add:* Dept Math Univ Mo Math Sci Bldg Columbia MO 65211

HUCKABAY, HOUSTON KELLER, b Shreveport, La, July 21, 32; m 54, 76; c 5. CHEMICAL ENGINEERING. *Educ:* La Polytech Inst, BSChE, 54; La State Univ, Baton Rouge, MSChE, 59, PhD(chem eng), 60. *Prof Exp:* Res chem engr, Crossett Co, 60-62; sr res engr, Forest Prod Div, Olin Mathieson Chem Corp, 62-63, res supvr paper converting, 63-64; assoc prof, 64-69, asst to dean, Col Eng, 67-71, dir eng grad studies, 67-74, PROF CHEM ENG, LA TECH UNIV, 69-, HEAD CHEM ENG, 86- *Concurrent Pos:* Consult, La Dept Revenue & Legal Firms; expert witness fires, explosions & chemically related accidents. *Mem:* Am Inst Chem Engrs; Am Chem Soc; Am Soc Eng Educ; Nat Fire Protection Asn; Nat Soc Prof Engrs. *Res:* Prediction of flame ignition characteristics; fire and chemical process safety; alternative energy sources; biotechnology; natural polymers. *Mailing Add:* Box 10348 Ruston LA 71272-0046

HUCKABAY, JOHN PORTER, b Paris, Tex, Sept 24, 28; m 51; c 2. BOTANY, BIOSYSTEMATICS. *Educ:* Southeastern State Col, BSEd, 55; Okla State Univ, MS, 60, PhD, 67. *Prof Exp:* Instr bot & biol, Cameron State Col, 58-62; instr biol, Okla State Univ, 64-66; from asst prof to assoc prof bot & cytol, 66-72, head dept biol, 72-77, PROF BIOL, SOUTHEAST MO STATE COL, 72- *Mem:* AAAS; Int Asn Plant Taxon. *Res:* Biosystematics of sorghum. *Mailing Add:* Dept Biol Southeast Mo State Univ Cape Girardeau MO 63701

HUCKABY, DALE ALAN, b Vicksburg, Miss, Sept 27, 44; m 70; c 2. PHYSICAL CHEMISTRY. *Educ:* La State Univ, Baton Rouge, BS, 66; Rice Univ, PhD(phys chem), 69. *Prof Exp:* From asst prof to assoc prof, 69-81, chmn, 87-89, PROF CHEM, TEX CHRISTIAN UNIV, 81- *Concurrent Pos:* Robert A Welch Found grant, 71-93; NSF grants, 73, 83, 84-85. *Mem:* Am Chem Soc; Sigma Xi. *Res:* Statistical mechanical calculations on model systems. *Mailing Add:* Dept Chem Tex Christian Univ Ft Worth TX 76129

HUCKABY, DAVID GEORGE, b Ponca City, Okla, Dec 8, 42; m 67; c 2. MAMMALOGY. *Educ:* La State Univ, BS, 63, MS, 67; Univ Mich, PhD(zool), 73. *Prof Exp:* Asst prof, 73-78, ASSOC PROF BIOL, CALIF STATE UNIV, LONG BEACH, 78- *Mem:* Am Soc Mammalogists; Soc Study Evolution; Am Inst Biol Sci; Soc Syst Zool; Am Soc Zoologists. *Res:* Taxonomy of muroid rodents; comparative anatomy of the reproductive system of rodents. *Mailing Add:* Dept Biol Calif State Univ Long Beach CA 90840-3702

HUCKE, DOROTHY MARIE, b Brooklyn, NY, Jan 14, 27. BIOSTATISTICS. *Educ:* St Joseph's Col Women, BA, 49; Columbia Univ, MS, 58; NY Univ, MS, 63. *Prof Exp:* Lab technician chem, Pfizer Inc, 49-50, lab supvr, Systs Planning & Tech Info, 50-52, statistician, 52-59, supvr statist servs, 60-67, mgr spec serv, 67-71, admin asst qual control mgt, 71-72, mgr systs & planning, 72-81, assoc dir, Systs, Planning & Tech Info, 81-87; RETIRED. *Res:* Quality control management; data processing; statistics; development of techniques that will focus on quality problems, increase efficiency, reduce cost and improve quality; computer sciences, general. *Mailing Add:* 85-35 66 Rd Rego Park NY 11374

HUCKE, EDWARD E, b Kansas City, Mo, Sept 1, 30; div; c 4. METALLURGY, CARBONS. *Educ:* Mass Inst Technol, SB, 51, SM, 52, ScD(metall), 54. *Prof Exp:* Res dir, LFM Mfg Co, Kans, 53-55; from asst prof to assoc prof, 55-61, PROF METALL ENG, UNIV MICH, ANN ARBOR, 61- *Concurrent Pos:* Vis prof, Max Planck Inst Metal Res, Stuttgart, WGer, 78. *Honors & Awards:* Lilliequist Award, Steel Founders Soc Am, 56; Bradley Stoughton Award, Am Soc Metals, 63. *Mem:* Fel Am Soc Metals; fel Am Inst Chem; Electrochem Soc; Am Inst Mining, Metall & Petrol Engrs; Brit Inst Metals. *Res:* Metallurgical thermodynamics; cemented carbides; carbons. *Mailing Add:* 3033 Lakehaven Ct Ann Arbor MI 48105-2501

HUCKER, HOWARD B(ENJAMIN), b St Louis, Mo, Mar 5, 26; m 58; c 4. DRUG METABOLISM. *Educ:* St Louis Univ, BS, 48; Univ Mo, PhD(org chem), 53; Med Col Va, ScD(pharmacol), 57. *Prof Exp:* Anal chemist, Shell Oil Co, 48-49; asst chem, Univ Mo, 49-53; Am Tobacco Co fel, Med Col Va, 53-56; res chemist, Nat Heart Inst, 56-60; sr res fel, Merck Inst Therapeut Res, 60-70, dir human drug metab, 70-76, sr investr, 76-84, sr scientist, 84-90; RETIRED. *Concurrent Pos:* Consult, Am Tobacco Co, 53-56; assoc ed, Drug Metab Disposition, 78-90. *Mem:* Sigma Xi; Am Chem Soc; Am Soc Pharmacol & Exp Therapeut. *Res:* Biochemical pharmacology; enzymology; analytical methods. *Mailing Add:* 915 Jenifer Rd Horsham PA 19044

HUDAK, MICHAEL J, b Johnson City, NY, Dec 4, 52. PATTERN CLASSIFICATION, ARTIFICIAL NEURAL SYSTEMS. *Educ:* State Univ NY, Binghamton, BA, 75, PhD(computer sci), 86; Northwestern Univ, MS, 77. *Prof Exp:* Instr, State Univ NY, Binghamton, 81-84; consult computer res, Digital Equip Corp, 85; RES SCIENTIST, SIEMENS CORP RES, 86- *Res:* Design and empirically study pattern classifiers inspired by neural systems. *Mailing Add:* 38 Oliver St Binghamton NY 13904

HUDAK, NORMAN JOHN, b Lorain, Ohio, Jan 24, 33; m 63; c 3. ORGANIC CHEMISTRY. *Educ:* DePauw Univ, BA, 54; Cornell Univ, PhD(chem), 59. *Prof Exp:* Instr chem, Oberlin Col, 58-60; asst prof, Haverford Col, 60-61; assoc prof, 61-65, chmn dept, 72-77 & 85-88, PROF CHEM, WILLAMETTE UNIV, 65- *Concurrent Pos:* NSF sci fac fel, Ore State Univ, 71-72. *Mem:* AAAS; Am Chem Soc; Royal Soc Chem. *Res:* Natural products; enzyme-catalyzed organic reactions; reduction reactions; carbocation rearrangements; organic photochemistry. *Mailing Add:* 1732 Toucan St SW Salem OR 97304

HUDAK, PAUL RAYMOND, b Baltimore, Md, July 15, 52; m 81; c 2. PARALLEL PROCESSING, FUNCTIONAL LANGUAGES PROGRAMMING. *Educ:* Vanderbilt Univ, BS, 73; Mass Inst Technol, MS, 74, Univ Utah, PhD(comput sci), 82. *Prof Exp:* Mem tech staff, Watkins Johnson Co, 74-77, head equip eng sect, 77-79; res assoc, univ res fel, Univ Utah, 79-82; asst prof, 82-85, ASSOC PROF COMPUT SCI, YALE UNIV, 85- *Concurrent Pos:* Collabr, Los Alamos Nat Lab, 83-; consult, Comput Technol Assocs, 83-; prin investr, NSF, DARPA & Dept Energy grants, 83- *Honors & Awards:* Presidential Young Investr Award, NSF, 85. *Mem:* Asn Comput Mach; Inst Elec & Electronics Engrs. *Res:* Formal analysis, design and implementation of very high-level programming languages; functional and logic programming languages targeted for parallel computers; developing semantic analysis methodologies and tools for optimizing. *Mailing Add:* PO Box 2158 Yale Sta New Haven CT 06520

HUDAK, WILLIAM JOHN, b Duquesne, Pa, Jan 3, 29; m 54; c 9. PHARMACOLOGY. *Educ:* Univ Pittsburgh, BS, 54, MS, 56, PhD(pharmacol), 59. *Prof Exp:* Head sect cardiovasc res, Richardson Merrell, Inc, 59-70, assoc group dir cardiovasc clin res, Merrell-Nat Labs, 70-80, asst to vpres res opers, 80-90, mgr res opers, 82-86, mgr med commun, Merrell-Dow Pharmaceut Inc, 87-90; RETIRED. *Mem:* Fel AAAS; Am Pharmaceut Asn; NY Acad Sci; Am Soc Pharmacol & Exp Therapeut; Am Heart Asn. *Res:* Cardiovascular clinical research. *Mailing Add:* 10476 Wintergreen Ct Sharonville OH 45241

HUDDLE, BENJAMIN PAUL, JR, b Ootacamund, India, May 25, 41; US citizen; m 64; c 2. PHYSICAL CHEMISTRY. *Educ:* Lenoir-Rhyne Col, BS, 63; Univ NC, Chapel Hill, PhD(phys chem), 68. *Prof Exp:* ASST PROF CHEM, ROANOKE COL, 68- *Mem:* Am Chem Soc; Am Phys Soc; Am Crystallog Asn; Sigma Xi. *Res:* Crystal structure of coordination compounds; structure and properties of phosphorous heterocycles and molecules with highly polar bonds. *Mailing Add:* Chem Dept Roanoke Col Salem VA 24153

HUDDLESTON, CHARLES MARTIN, b Dallas, Tex, Sept 27, 25; m 52. NUCLEAR PHYSICS. *Educ:* Northwestern Univ, BS, 48, MS, 49; Ind Univ, PhD(physics), 53. *Prof Exp:* Assoc physicist, Argonne Nat Lab, 53-61; dir physics & math div, US Naval Civil Eng Lab, 61-67; phys sci adminr, Naval Radiol Defense Lab, 67-69; res physicist & head shielding & dosimetry group, Naval Ord Lab, Naval Surface Weapons Ctr, 69-74, dir, Nuclear Radiation Br, 74-75, tech coord, Navy Charged Particle Beam Technol Prog & task mgr, Prog Strategic Defense Initiative Off & Tech Agt, Prog Defense Advan Res Projs Agency, 75-86; sr physicist, Booz Allen & Hamilton Inc, 87-89; CONSULT, 89- *Concurrent Pos:* Vis prof, Univ Ill, 68-69. *Mem:* Am Nuclear Soc; Sci Res Soc Am. *Res:* Radiation transport and energy deposition for neutrons, gamma rays, x-rays and electrons; nondestructive testing, underground nuclear weapons tests; radiation simulation; accelerators, changed particle beam technology. *Mailing Add:* Three Oyster Cove Dr Apt 3E Grasonville MD 21638

HUDDLESTON, ELLIS WRIGHT, b Knapp, Tex, Sept 10, 35; m 55; c 2. ECONOMIC ENTOMOLOGY, PESTICIDE APPLICATION. *Educ:* Tex Tech Col, BS, 56; Cornell Univ, MS, 58, PhD(entom), 60. *Prof Exp:* Asst entom, Cornell Univ, 56-59; from asst prof to prof, Tex Tech Univ, 60-75; head dept, 75-84, PROF ENTOM & PLANT PATH, NMEX STATE UNIV, 75- *Concurrent Pos:* Consult, USAID, Senegal, 86-88. *Mem:* Entom Soc Am. *Res:* Insect ecology; rangeland entomology; ecological effects of insecticides; insecticide residue sampling and decomposition; pesticide application technology. *Mailing Add:* Dept Entom & Plant Path NMex State Univ Las Cruces NM 88003

HUDDLESTON, GEORGE RICHMOND, JR, b Sandersville, Miss, Oct 7, 21; m 42, 77; c 1. PHYSICAL CHEMISTRY, CHEMICAL ENGINEERING. *Educ:* Miss State Col, BS, 43; La State Univ, MS, 49, PhD(chem), 60. *Prof Exp:* Control chemist, Copolymer Rubber & Chem Corp, 43-44, pilot plant shift chemist, 46-48, develop chemist, 48-50, supvr pilot plant, 50-54, supvr res, 54-58, mgr res lab, 58-63, res mgr, 63-67, process develop dir, 67-70; sr scientist, Independence Develop Ctr, B F Goodrich Chem Co, 70-75, res & develop assoc, Brecksville Res & Develop Ctr, 75-78, sr res & develop assoc, Avon Lake Tech Ctr, B F Goodrich Co, 78-83, res & develop fel, 83-86; RETIRED. *Mem:* Am Chem Soc. *Res:* Polymerization and copolymerization to elastomeric polymers and latices; fundamental colloidal and surface studies in latices and catalysts; reactions of elemental sulfur and organic materials; process development; polyvinyl chloride. *Mailing Add:* 4031 Woodstock Dr Lorain OH 44053-1568

HUDDLESTON, JAMES HERBERT, b Malone, NY, Mar 18, 42; m 62; c 4. SOIL MORPHOLOGY. *Educ:* Cornell Univ, BSc, 63, MS, 65; Iowa State Univ, PhD(soils), 69. *Prof Exp:* From asst prof to assoc prof ecosysts anal, Univ Wis-Green Bay, 69-76; ASSOC PROF SOIL SCI, ORE STATE UNIV, 76- *Concurrent Pos:* Wis Alumni Res Found res grant, 70-71; res partic, Eastern Deciduous Forest Biome, Int Biol Prog, Madison, Wis; dir NSF undergrad res participation grants, 74 & 75. *Mem:* AAAS; Am Soc Agron; Soil Conserv Soc Am. *Res:* Soil genesis and classification; soil geomorphology and landscape evolution; soil survey interpretation for agricultural and non-agricultural uses. *Mailing Add:* Dept Soil Sci Ore State Univ Corvallis OR 97331

HUDDLESTON, JOHN VINCENT, b Houston, Tex, Feb 1, 28; m 52; c 3. COMPUTER ENGINEERING. *Educ:* Columbia Univ, BS, 52, MS, 54, PhD(appl mech), 60. *Prof Exp:* Instr civil eng, Columbia Univ, 52-56; asst prof, Yale Univ, 56-62, assoc prof, 62-63, assoc prof eng & appl sci, 63-67; PROF ENG & APPL SCI, STATE UNIV NY, BUFFALO, 67- *Concurrent Pos:* Vis prof, Imp Col, London, 64-65, Univ Florence, 74, Univ Technol Malaysia, 75-76, Univ Malaya & MARA Inst Technol, 84-85; Fulbright fels, 75-76 & 84-85; consult, Custom Comput Serv, Inc, East Amherst, NY & Wan Mohamed & Khoo, Kuala Lumpur, Malaysia; prin investr, NSF, 76-79; educ consult, Inti Col, Kuala Lumpur, Malaysia. *Honors & Awards:* Robert Ridgway Award, Am Soc Civil Engrs, 52. *Mem:* Am Soc Civil Engrs; Am Soc Mech Engrs; Inst Elec & Electronics Engrs Comput Soc. *Res:* Dynamic stability; computer algorithms; population dynamics; mathematical modeling. *Mailing Add:* Dept Civil Eng State Univ NY Buffalo NY 14260

HUDDLESTON, PHILIP LEE, b 1947; US citizen; m 73. MATHEMATICAL PHYSICS, COMPUTATIONAL PHYSICS. *Educ:* Wash Univ, BS, 67; Boston Univ, MA, 69, PhD(physics), 74. *Prof Exp:* Asst prof physics & math, Edward Waters Col, 75-76; asst prof physics, Parks Col, St Louis Univ, 76-79; sr sci programmer/analyst, McDonnell Douglas Res Labs, 79-81, res scientist, 81-85, scientist radiation sci dept, 85-89, SR SCIENTIST, MCDONNEL DOUGLAS TECHNOLOGIES, 89- *Mem:* Inst Elec & Electronics Engrs; Am Phys Soc; AAAS; Sigma Xi. *Res:* Structure, properties and applications of Lie groups and Lie algebras; mathematical modeling of electromagnetic wave propagation through the magnetoplasma of a bumpy torus nuclear fusion device; classical electromagnetic scattering theory with applications to radar cross section predictions. *Mailing Add:* 5471 Kenrick Parke Dr St Louis MO 63119-5060

HUDDLESTON, ROBERT E, b Hugo, Okla, Mar 28, 39; m 87. NUMERICAL ANALYSIS. *Educ:* Tex Christian Univ, BA, 60, PhD, 66; Univ Ariz, MS, 62. *Prof Exp:* Fel & res assoc math, Inst Fluid Dynamics & Appl Math, Univ Md, 66-67; staff mem, Sandia Labs, Albuquerque, 67-69, numerical analyst, Sandia Nat Labs, Livermore, 69-76, supvr sci comput, 76-85; mgr, Computer Appl & Res, 85-87, dept head, Computer Sci Dept, 87-90, UNIV RELS, LAWRENCE LIVERMORE NAT LAB, 90- *Concurrent*

Pos: Vis lectr, Univ NMex, 67-69. *Mem:* Soc Indust & Appl Math; Spec Interest Group Numerical Math; Asn Comput Mach; Numerical Algorithms Group. *Res:* Numerical solutions of partial differential equations. *Mailing Add:* Univ Rels L-725 Lawrence Livermore Lab PO Box 808 Livermore CA 94550

HUDDLESTONE, RICHARD H, b Huntington Park, Calif, Dec 8, 26; m 52; c 4. THEORETICAL PHYSICS. *Educ:* Univ Calif, AB, 48, PhD(physics), 54. *Prof Exp:* Mem theoret group, Radiation Lab, Univ Calif, 50-54, jr res physicist, Dept Physics, 54-55; mem theoret physics sect, Phys Res Lab, Space Tech Labs, 55-60; head reentry & plasma electromagnetics dept, Aerospace Corp, 60-71, sr staff scientist, Lab Opers, 71-77, eng specialist, Info Processing Div, 77-; AT COMPUT SCI DEPT, CALIF STATE UNIV, CARSON. *Mem:* Am Phys Soc. *Res:* Quantum mechanics; low energy nuclear physics; plasma, thermonuclear, reentry and mathematical physics; applied mathematics. *Mailing Add:* Comput Sci Dept Calif State Univ Dominquez Hills Carson CA 90746

HUDECKI, MICHAEL STEPHEN, b Ft Bragg, NC, Nov 7, 43; m 73. MOLECULAR BIOLOGY, BIOCHEMISTRY. *Educ:* Niagara Univ, BS, 65, MS, 67; State Univ NY, Buffalo, MA, 70, PhD(biol), 73. *Hon Degrees:* DSc, Niagara Univ, 81. *Prof Exp:* Lectr biol, 76-77, res assoc, 77-78, res asst prof, 79-86, RES ASSOC PROF, DEPT BIOL SCI, STATE UNIV NY, BUFFALO, 87- *Concurrent Pos:* Muscular Dystrophy Asn fel, Syracuse Univ, 72-73 & State Univ NY, Buffalo, 74-76; NIH fel, State Univ NY, Buffalo, 77-79; Muscular Dystrophy Asn grant, 77-89, res career develop award, 80-85, grant, 80-83, NIH; consult, Off Serv Handicapped, State Univ NY, 77-, Proj Handicapped Sci, AAAS, 77- & NSF Phys Handicapped Sci Prog, 77-78; reviewer, biomed jour. *Mem:* AAAS; Soc Neurosci; NY Acad Sci; Fedn Sci & Handicapped; Am Physiol Soc. *Res:* Drug studies in muscular dystrophy. *Mailing Add:* Dept Biol Sci State Univ NY Buffalo NY 14260

HUDGENS, RICHARD WATTS, b Greenville, SC, Jan 10, 31; m 52, 82; c 4. PSYCHIATRY. *Educ:* Princeton Univ, BA, 52; Wash Univ, MD, 56. *Prof Exp:* From instr to prof clin psychiat, 63-89, from asst dean to assoc dean sch med, 64-74, PROF PSYCHIAT, WASH UNIV, 89- *Mem:* Am Psychiat Asn; AMA. *Res:* Affective disorders; suicide; psychiatric disorders in adolescence; transcultural psychiatry; medical education; history of psychiatry; psychiatric disorders. *Mailing Add:* 4940 Audubon Ave No 1117 St Louis MO 63110-1002

HUDGIN, DONALD EDWARD, b Greenville, SC, Aug 11, 17; m 43; c 3. ORGANIC CHEMISTRY. *Educ:* Clemson Col, BS, 38; Purdue Univ, MS, 40, PhD(org chem), 47. *Prof Exp:* Res chemist, Synthetic Detergents, Procter & Gamble Co, 47-52; res chemist, Mallinckrodt Chem Works, 52-55; sr res chemist, Celanese Corp Am, 55-56, group leader, 56-58, sect head, 58-60; dir res & develop, Cary Chem, Inc, 60-61; assoc dir res, Diamond Alkali Co, 61-66; sr staff adv, Esso Res & Eng Co, 66-67; res dir, Princeton Chem Res, Inc, 67, dir res & develop, 67-70, vpres, 70-80, dir res & develop, Princeton Polymer Labs, Inc, 70-89, PRES, PRINCETON POLYMER LABS, INC, 80- *Concurrent Pos:* Assoc ed, Int J Polymer Process Eng; adv coun, Polymer Processing Inst, 83-88; ser ed, Plastics Eng. *Mem:* AAAS; Am Chem Soc; fel Am Inst Chem; Soc Plastic Eng; Asn Res Dirs (pres, 73-74); Plastics Inst Am; Sigma Xi. *Res:* Vinyl and condensation polymers; thermoset resins; polymer properties; polyolefins. *Mailing Add:* 101 Westchester Circle Seneca SC 29678

HUDGINS, ARTHUR JUDSON, b Boise, Idaho, June 14, 20; m 55; c 2. APPLIED PHYSICS, PHYSICIST RECRUITMENT. *Educ:* Univ Calif, AB, 42, PhD(physics), 52. *Prof Exp:* Physicist, Lawrence Radiation Lab, Univ Calif, 42-45 & Tenn Eastman Corp, 45-46; physicist, Lawrence Livermore Nat Lab, Univ Calif, Livermore, 46-83. *Concurrent Pos:* Consult, 83- *Mem:* Am Phys Soc; AAAS. *Mailing Add:* 2294 N Livermore Ave Livermore CA 94550

HUDGINS, AUBREY C, JR, b Richmond, Va, Oct 15, 35; m 88; c 2. EXPERIMENTAL SOLID STATE PHYSICS, COMPUTER SIMULATION. *Educ:* Univ Richmond, BS, 59; WVa Univ, MS, 61, PhD(physics), 64. *Prof Exp:* Instr & res assoc physics, WVa Univ, 64-65; instr advan eng math, Univ Va Exten, 66-67; res physicist, Metall Res Div, Reynolds Metals Co, 65-71; asst prof physics, Randolph Macon Col, 72-75; instr physics, Va Commonwealth Univ, 75-77, asst prof info systs, 77-79; oper res analyst, 79-81, SUPVR OPER RES ANALYST GM-14, DEFENSE LOGISTICS AGENCY, DEFENSE OPER RES OFF, DEFENSE SUPPLY CTR, 81- *Concurrent Pos:* AEC grant, 64-65. *Mem:* Am Phys Soc; Am Asn Physics Teachers; Am Soc Metals. *Res:* Magnetism; heat transfer; computer simulation; ingot casting; ultrasonic nondestructive testing; experimental solid state research energy and systems analysis; operation research; data bank. *Mailing Add:* 4000 Cogbill Rd Richmond VA 23234

HUDGINS, PATRICIA MONTAGUE, b Buckhannon, WVa, Jan 31, 38; m 75; c 2. PHARMACOLOGY. *Educ:* WVa Univ, BS, 59, MS, 60, PhD(pharmacol), 66. *Prof Exp:* From instr to assoc prof pharmacol, Med Col Va, Va Commonwealth Univ, 66-75; assoc prof, 75-80, PROF PHYSIOL, KIRKSVILLE COL OSTEOP MED, 80- *Concurrent Pos:* Consult, Astra Pharmaceut Prod, Inc, 73-77. *Mem:* Am Soc Pharmacol & Exp Therapeut; Am Physiol Soc. *Res:* Biochemical pharmacology; active cation transport; ATPase. *Mailing Add:* Dept Physiol Kirksville Col Osteop Med 204 W Jefferson Kirksville MO 63501

HUDGINS, ROBERT R(OSS), b Toronto, Ont, Jan 25, 37; m 63; c 3. CHEMICAL ENGINEERING. *Educ:* Univ Toronto, BASc, 59, MASc, 60; Princeton Univ, PhD(chem eng), 64. *Prof Exp:* From asst prof to assoc prof, 64-75, PROF CHEM ENG, UNIV WATERLOO, 75- *Concurrent Pos:* Invited prof, Univ Sherbrooke, 71-72, Swiss Fed Inst Technol, Lausanne, 79. *Mem:* Fel Chem Inst Can; Sigma Xi. *Res:* Chemical reaction engineering; forced cycling of chemical reactors; influence of diluent gases in catalysis; sedimentation improvement in clarification; clarifiers. *Mailing Add:* Dept Chem Eng Univ Waterloo Waterloo ON N2L 3G1 Can

HUDIS, JEROME, b New York, NY, June 5, 25; m 53; c 2. NUCLEAR CHEMISTRY, INORGANIC CHEMISTRY. *Educ:* Wash Univ, BS, 47, PhD(chem), 52. *Prof Exp:* From assoc chemist to sr chemist, Brookhaven Nat Lab, 52-77, chmn, Chem Dept, 77-81, asst dir, 81-85; VPRES PROGRAMMATIC AFFAIRS, SECY & CONTROLLER, ASSOC UNIVS INC, 85- *Concurrent Pos:* Chmn, Div Nuclear Chem & Technol, Am Chem Soc, 81. *Mem:* Am Chem Soc. *Res:* Nuclear chemistry and reactions; x-ray photoelectron spectroscopy. *Mailing Add:* Assoc Univs Inc 1717 Massachusetts Ave Washington DC 20036

HUDLER, GEORGE WILLIAM, b Cloquet, Minn, Aug 19, 47. PLANT PATHOLOGY, FORESTRY. *Educ:* Univ Minn, BS, 70, MS, 73; Colo State Univ, PhD(plant path), 76. *Prof Exp:* ASST PROF PLANT PATH, CORNELL UNIV, 78- *Mem:* Am Phytopath Soc; Sigma Xi. *Res:* Diseases of forest and shade trees. *Mailing Add:* 315 Plant Sci Cornell Univ Ithaca NY 14853

HUDLESTON, PETER JOHN, b Osterley, Middlesex, UK, May 31, 44; m 71; c 2. MECHANICS FOLDING, STRAIN ANALYSIS. *Educ:* Imp Col, London Univ, BSc, 65, PhD(geol), 69. *Prof Exp:* Res fel, Univ Uppsala, Sweden, 69-70; from asst prof to assoc prof, 70-83, PROF & HEAD, DEPT GEOL & GEOPHYS, UNIV MINN, MINNEAPOLIS, 83- *Concurrent Pos:* Assoc ed, J Struct Geol; chmn, Struct Geol & Tech Div, Geol Soc Am, 89-90. *Mem:* Fel Geol Soc Am; Am Geophys Union; Am Asn Petrol Geologists; Geol Asn Can; Nat Asn Geol Teachers. *Res:* Analysis of deformation in rocks by field observation, numerical modeling and experiment with particular interest in the development of folds, foliation and crystallographic fabric. *Mailing Add:* Dept Geol & Geophys Univ Minn Minneapolis MN 55455

HUDLICKY, MILOS, b Prelouc, Czech, May 12, 19; m 46; c 2. ORGANIC CHEMISTRY, FLUORINE CHEMISTRY. *Educ:* Prague Tech Univ, PhD(org chem), 46. *Prof Exp:* Asst org chem, Prague Tech Univ, 45-54, assoc prof, 54-58; res chemist, Res Inst Pharm & Biochem, Prague, 58-68; vis prof, 68-69, from assoc prof to prof, 69-89, EMER PROF ORG CHEM, VA POLYTECH INST & STATE UNIV, 89- *Concurrent Pos:* Am Chem Soc & UNESCO fel, Ohio State Univ, 48. *Mem:* Am Chem Soc; The Chem Soc; Czech Chem Soc. *Res:* Chemistry of 1,3-dichloro-2-butene; chemistry of organic fluorine compounds. *Mailing Add:* Dept Chem Va Polytech Inst & State Univ Blacksburg VA 24061

HUDLICKY, TOMAS, b Dec 26, 49; US citizen; m; c 1. SYNTHETIC ORGANIC CHEMISTRY, NATURAL PRODUCTS. *Educ:* Va Polytech Inst & State Univ, BS, 73; Rice Univ, PhD(chem), 77. *Prof Exp:* Chemist, SPOFA Pharmaceut Works, 66-68 & Bio-Va Polytech Inst Antarctic Exped, 72; fel, Univ Geneva, 77-78; ASST PROF, ILL INST TECHNOL, 78-; PROF, DEPT CHEM, VA POLYTECH STATE UNIV, 88- *Concurrent Pos:* Sloan fel, 81-83, Fulbright fel, 84 & 85; NIH res career develop award, 84-89. *Honors & Awards:* NSF Antarctic Serv Award, 72. *Mem:* Am Chem Soc; Sigma Xi. *Res:* Synthesis of natural products, including alkaloids, terpenes, marine natural products; development of new reagents for organic synthesis; carbocyclic, heterocyclic annulation methodology; cyclopentanoid terpene synthesis; industrial consultant; pharmaceutical synthesis; biocatalysis; microbial oxidation of aromatics; carbohydrate synthesis; anti-viral agents; waste conversion. *Mailing Add:* Dept Chem Va Polytech State Univ Blacksburg VA 24061

HUDLOW, MICHAEL DALE, b Childress, Tex, Aug 13, 40; m 59; c 2. METEOROLOGY, HYDROLOGY. *Educ:* Tex A&I Univ, BS, 63; Tex A&M Univ, MS, 66, PhD(meteorol), 67. *Prof Exp:* Meteorol officer hydrometeorol, US Army Atmospheric Sci Lab, 67-69; hydrologist hydrometeorol, Barbados Oceanog & Meteorol Anal Proj Off, 69-71, hydrologist radar hydrol, Off Hydrol, 71-73, phys scientist radar meteorol, Ctr Exp Design & Data Anal, 73-78, ASST DIR, HYDROL RES LAB, NAT OCEANIC & ATMOSPHERIC ADMIN, 78- *Concurrent Pos:* Radar hydrol consult environ monitoring, Nat Oceanic & Atmospheric Admin, 71-73; mem various hydrometeorol comts & working groups, 68- *Mem:* Am Meteorol Soc; Am Geophys Union; Am Soc Civil Engrs. *Res:* All facets of the hydrologic cycle with particular emphasis on the use of remote sensing for measuring the hydrometeor components of the cycle. *Mailing Add:* 110 Elizabeth Ct Chester MD 21619

HUDNALL, PHILLIP MONTGOMERY, b Ward, WVa, Jan 7, 44; m 79. ORGANIC CHEMISTRY. *Educ:* Morris Harvey Col, BS, 67; Univ NC, PhD(org chem), 72. *Prof Exp:* Lab technician, Union Carbide Chem Co, 62-67, chemist, 67; chemist develop & control, 72-74, sr chemist intermediates dept staff, 74-77, sr chemist dyes dept staff, 77-78, sr chemist org chem div staff, 78-80, SR CHEMIST & DEVELOP COORDR, DEVELOP & CONTROL DEPT, TENN EASTMAN CO, 80- *Mem:* Am Chem Soc. *Res:* Process development. *Mailing Add:* PO Box 3031 Kingsport TN 37664-0031

HUDOCK, GEORGE ANTHONY, b Bridgetown Twp, Pa, Mar 28, 37; m 60; c 2. BIOCHEMICAL GENETICS. *Educ:* Harvard Univ, AB, 59, PhD(biol), 63. *Prof Exp:* NSF fel, Dartmouth Med Sch, 63-64, res assoc microbiol, 64-65; asst prof, 65-68, assoc prof zool, 68-80, ASSOC PROF BIOL, IND UNIV, BLOOMINGTON, 80- *Mem:* AAAS; Am Inst Biol Sci; Soc Protozool. *Res:* Biochemical aspects of genetics, regulation, chloroplast structure and function. *Mailing Add:* Dept Biol Ind Univ Bloomington IN 47405

HUDRLIK, ANNE MARIE, b Akron, Ohio, Aug 30, 41; m 67; c 2. SYNTHETIC ORGANIC CHEMISTRY. *Educ:* Ohio State Univ, BS, 63; Columbia Univ, MA, 64, PhD(chem), 67. *Prof Exp:* NIH fel, Stanford Univ, 68-69; fel org chem, Rutgers Univ, New Brunswick, 69-76; FEL ORG CHEM, HOWARD UNIV, 77- *Mem:* Am Chem Soc. *Res:* Organic synthesis. *Mailing Add:* Dept Chem Howard Univ Washington DC 20059

HUDRLIK, PAUL FREDERICK, b Portland, Ore, May 10, 41; m 67; c 2. SYNTHETIC ORGANIC CHEMISTRY, ORGANOSILICON CHEMISTRY. *Educ:* Ore State Univ, BS, 63; Columbia Univ, MA, 64, PhD(org chem), 68. *Prof Exp:* Fel chem, Stanford Univ, 68-69; asst prof, Sch Chem, Rutgers Univ, 69-76; assoc prof, 77-81, PROF CHEM, HOWARD UNIV, 81- *Mem:* AAAS; Royal Soc Chem; Am Chem Soc. *Res:* Development of new synthetic methods in organic chemistry; organosilicon chemistry. *Mailing Add:* Dept Chem Howard Univ Washington DC 20059

HUDSON, A SUE, neuromuscular morphology, carbonate effects, for more information see previous edition

HUDSON, ALAN P, b Batavia, NY, Dec 7, 48. GENETICS. *Educ:* Hamilton Col, BA, 71; City Univ NY, PhD(biol sci), 78. *Prof Exp:* AT RES SERV, VET ADMIN MED CTR, PHILADELPHIA, PA, 85-; ASST PROF, DEPT BIOL, MED COL PA, 88- *Mem:* Am Soc Microbiol; Sigma Xi; Am Soc Biol Chemists; Asn Res Vision & Ophthal. *Mailing Add:* Res Serv Vet Admin Med Ctr University & Woodland Aves Philadelphia PA 19104

HUDSON, ALBERT BERRY, b Washington, NC, Dec 2, 29; m 55; c 2. NATURAL PRODUCTS CHEMISTRY. *Educ:* Univ NC, BA, 56. *Prof Exp:* Chemist gas & oil, State of NC, 56-58; chemist tobacco, Lorillard Corp, 58-66, anal chemist foods, 66-69, res chemist fermentation, 69-72, supvr prod develop, 72-79, MGR, CIGARETTE PROD DEVELOP, LORILLARD CORP, 80- *Mem:* Am Chem Soc; Inst Food Technologists. *Res:* Separation and identification of chemical fractions and-or compounds which affect the organoleptic properties of various types of tobaccos. *Mailing Add:* 903 Ridgecrest Dr Greensboro NC 27410

HUDSON, ALICE PETERSON, b Sherburn, Minn, Aug 9, 42; m 63; c 2. SURFACE CHEMISTRY. *Educ:* Iowa State Univ, BS, 63; Fla Atlantic Univ, MS, 77. *Prof Exp:* Chemist, Pratt & Whitney Aircraft Co, 63-64; lab technician, SE Fla Tuberc Hosp, 64-65; chemist, 71-73, LAB DIR, SURFACE CHEMISTS FLA, INC, 73-, PRES, 83- *Concurrent Pos:* Dir & secy-treas bd dirs, Surface Chemists Fla, Inc, 73-83; secy bd dirs, Jessop, Prindle & Woodward, Inc, 75-81, Mining Reagents Corp, 81-83; pres, Sunrise Chem, Inc, 83-88; dir & treas, Surfactant Technol Corp, 84- *Mem:* Am Chem Soc; Am Oil Chemists Soc. *Res:* Mechanisms of halogen disinfection; studies of micelle-iodine interactions; applied surface chemistry in the areas of detergents; textile chemicals and cosmetics. *Mailing Add:* Surface Chemists Fla Inc 328 W 11th St Riviera Beach FL 33404

HUDSON, ALVIN MAYNARD, b Portland, Ore, June 28, 22. NUCLEAR PHYSICS. *Educ:* Stanford Univ, BS, 47, MS, 50, PhD(physics), 56. *Prof Exp:* From asst prof to assoc prof physics, 57-70, actg chmn dept, 58-61, chmn dept, 62-67, PROF PHYSICS, OCCIDENTAL COL, 70- *Concurrent Pos:* Mem staff, Sci Teaching Ctr, Mass Inst Technol, 64-65; mem area meeting adv comt, Phys Sci Study Comt, 65-69; mem advan placement exam comt physics, Col Entrance Exam Bd, Princeton, NJ, 70-, czmn, 74-78. *Mem:* AAAS; Am Asn Physics Teachers. *Res:* Relativity; science education. *Mailing Add:* 650 Crane Blvd Los Angeles CA 90065

HUDSON, ANNE LESTER, b Inverness, Miss, Jan 30, 32; m 60; c 2. MATHEMATICS. *Educ:* Hollins Col, AB, 53; Tulane Univ, MS, 58, PhD(math), 61. *Prof Exp:* NSF res fel math, 61-63; instr, Syracuse Univ, 63-71; PROF MATH, ARMSTRONG STATE COL, 71- *Mem:* Am Math Soc; Math Asn Am. *Res:* Structure of topological semigroups. *Mailing Add:* Dept Math & Computer Sci Armstrong State Col 11935 Abercom St Savannah GA 31419-1997

HUDSON, BILLY GERALD, b Pine Bluff, Ark, Oct 16, 41; m 68, 88; c 3. BIOCHEMISTRY. *Educ:* Henderson State Col, BS, 62; Univ Tenn, Memphis, MS, 64; Univ Iowa, PhD(biochem), 66. *Prof Exp:* Postdoctoral fel, US Army Res Inst Environ Med, 66-68; res fel biol chem, Harvard Med Sch, 68-70; from asst prof to assoc prof biochem, Okla State Univ, 70-74; assoc prof, 74-77, dean res, Col Health Sci, 80-86, PROF BIOCHEM, UNIV KANS MED CTR, 77- *Mem:* Am Chem Soc; Am Soc Biol Chem. *Res:* Pathogenesis of renal diseases; glycoprotein structure; structure and function of basement membranes. *Mailing Add:* Dept Biochem & Molecular Biol Univ Kans Med Ctr 39 Rainbow Blvd Kansas City KS 66103

HUDSON, BRUCE WILLIAM, b Minn, Oct 3, 28; m 51; c 4. IMMUNOLOGY, INFECTIOUS DISEASES. *Educ:* Univ Calif, BA, 54, PhD(entom), 57. *Prof Exp:* Entomologist, Kaiser Found Hosps, 57-60; biochemist, Bur State Serv, Nat Commun Dis Ctr, USPHS, Calif, 60-65, biochemist, Tech Br, Chem Sect, 65-66, supvry chemist ecologic invests prog, Zoonoses Sect, 66-74, staff mem, Immunochem Br, Vector Borne Dis Div, Ctr Dis Control, 74-82; RETIRED. *Concurrent Pos:* Mem, Int Comt Microbiol Stand; affil prof microbiol, Wild Animal Dis Ctr, Colo State Univ. *Mem:* AAAS; Wildlife Dis Asn; NY Acad Sci; Am Chem Soc. *Res:* Plague; biology-immunology; arbovirus; seroepidemiology. *Mailing Add:* 1709 Concord Dr Ft Collins CO 80526

HUDSON, CECIL IVAN, JR, b Atmore, Ala, Oct 21, 37; m 59; c 4. PHYSICS. *Educ:* Ga Inst Technol, BS, 58; Univ Va, PhD(physics), 62. *Prof Exp:* Physicist, Lawrence Livermore Lab, Univ Calif, 61-67, asst to assoc dir, Nuclear Design, 67-68, staff asst to assoc dir, Mil Applns, 68-69, group leader, 69-72; dep mgr, Systs Div, Sci Applns, Inc, 72-75, vpres, 75-76, corp vpres, 76-80; group vpres, Jaycor, 80-83; dir, Advan Technol & Policy Progs, Titan Corp, 83-87; vpres, Calif Res & Technol, 87-90; DIR TECHNOL PROGS, EXPERROFT, 90- *Concurrent Pos:* Mem reentry systs adv group, USAF Space & Missiles Systs Orgn, 69-70; mem reentry body comt, USN Strategic Systs Proj Off, 69-72; chmn futures planning coun, Episcopal Diocese Calif; chmn working group technol, arms control & foreign policy, Calif Sem Int Security & Foreign Policy. *Mem:* AAAS; Int Inst Strategic Studies; Am Inst Aeronaut & Astronaut; Am Defense Preparedness Asn; Asn; Am Phys Soc. *Res:* Experimental fast neutron physics; design of nuclear explosives; systems analysis; policy analysis; hypervelocity penetrators assumption-based truth maintenance systems. *Mailing Add:* 13005 Caminito Mar Villa Del Mar CA 92014

HUDSON, CHARLES MICHAEL, b Washington, DC, Mar 31, 35; m 63; c 4. MATERIALS ENGINEERING, ENGINEERING MECHANICS. *Educ:* Va Polytech Inst, BS, 58, MS, 65; NC State Univ, PhD(mat eng), 72. *Prof Exp:* Aerospace technologist fatigue res, 58-73, aerospace technologist fract mech, 73-75, ENG SUPVR PRESSURE SYSTS, NASA, 75- *Concurrent Pos:* Lectr fracture mechs, George Washington Univ; chmn, Comt E-24, Am Soc Testing & Mat. *Honors & Awards:* Medal for Except Eng Achievement, NASA, 84; Award of Merit, Am Soc Testing & Mat, 87. *Mem:* Am Soc Testing & Mat; Sigma Xi. *Res:* Recertification of existing high-pressure systems at Langley Research Center; fracture mechanics, fatigue, failure and stress analyses; and nondestructive examinations. *Mailing Add:* Mail Stop 437 NASA Langley Res Ctr Hampton VA 23665

HUDSON, DAVID FRANK, b Hudson, NY, Feb 23, 37; m 65; c 2. ATOMIC PHYSICS, GASEOUS ELECTRONICS. *Educ:* Union Col, BS, 59; Univ NH, MS, 68; Cath Univ Am, PhD(physics), 77. *Prof Exp:* Physicist, Naval Weapons Lab, 67-74, physicist, Naval Surface Weapons Ctr, 74-78, RES PHYSICIST, NAVAL SURFACE WARFARE CTR, 78- *Mem:* Am Phys Soc. *Res:* Collisions involving and between electronically excited atoms; recombination; three body processes; negative ions; spectroscopy; laser cooling. *Mailing Add:* F43 Naval Surface Warfare Ctr Silver Spring MD 20903-5000

HUDSON, DONALD E(LLIS), b Alma, Mich, Feb 25, 16. MECHANICAL ENGINEERING, APPLIED MECHANICS. *Educ:* Calif Inst Technol, BS, 38, MS, 39, PhD(mech eng), 42. *Prof Exp:* Geophys engr, Gen Petrol Corp, 37-41; from instr to assoc prof, Calif Inst Technol, 41-55, prof, 55-79, emer prof mech eng & applied mech, 81-87; prof & chmn civil eng, Univ Southern Calif, 81-87, Fred Champion prof eng, 82-88; RETIRED. *Concurrent Pos:* Ord res engr, Off Sci Res & Develop & Nat Defense Res Comt, Calif Inst Technol, 43-45. *Mem:* Nat Acad Eng; Am Soc Mech Engrs; Soc Exp Stress Anal; Earthquake Eng Res Inst; Seismol Soc Am (pres, 71-72). *Res:* Structural dynamics; earthquake engineering. *Mailing Add:* Div Eng & Appl Sci Thomas Lab 104-44 Calif Inst Technol Pasadena CA 91125

HUDSON, DONALD EDWIN, b Butte, Mont, July 17, 21; m 44; c 3. PHYSICS TEACHING. *Educ:* Univ Minn, BPhys, 42; Cornell Univ, PhD(physics), 50. *Prof Exp:* Asst physics, Univ Minn, 41-43, instr, 43-44; jr scientist, Los Alamos Sci Lab, 44-46; asst, Cornell Univ, 46-49; instr, Princeton Univ, 49-51; from asst prof to assoc prof, Iowa State Univ, 51-64; chmn dept, 64-67, prof, 64-83, EMER PROF PHYSICS, CALIF STATE UNIV, LOS ANGELES, 83- *Honors & Awards:* Sigma Xi. *Mem:* Am Phys Soc; Am Asn Physics Teachers. *Res:* Cosmic rays; chemical physics; solid state physics; physical electronics. *Mailing Add:* 1565 Washburn Rd Pasadena CA 91005

HUDSON, FRANK ALDEN, b Gallup, NMex, Dec 15, 23; m 60; c 2. ANIMAL SCIENCE. *Educ:* Ariz State Univ, BS, 52; NMex State Univ, MS, 53; Ore State Univ, PhD(genetics), 57. *Prof Exp:* Exp aide, Ore State Univ, 53-55; animal husbandman, Sheep & Fur Animal Res Br, Agr Res Serv, USDA, Md, 57-60; prof animal sci, Tex Tech Univ, 60-88; RETIRED. *Mem:* Am Soc Animal Sci; Soc Range Mgt. *Res:* Sheep breeding, feeding and management with emphasis on intensified production. *Mailing Add:* 3824 52nd St Lubbock TX 79413

HUDSON, FRANK M, b Clarksville, Ark, Sept 8, 35; m 57; c 2. MATHEMATICS. *Educ:* Univ Cent Ark BS, 57; Univ Tex, Austin, MA, 59, PhD(math), 65. *Prof Exp:* Grad asst math, Univ Tex, Austin, 56-59; instr, Western State Col Colo, 59-61; NSF fel, Harvard Univ, 61-62; spec instr, Univ Tex, Austin, 62-64; prof math & chmn dept, McMurry Col, 64-67; chmn dept, 67-83, PROF MATH, UNIV CENT ARK, 83- *Mem:* Math Asn Am; Nat Coun Teachers Math. *Res:* Integral transforms. *Mailing Add:* Dept Math Univ Cent Ark Conway AR 72032

HUDSON, FREDERICK MITCHELL, b Miami, Fla, Jan 27, 34; m 61, 76; c 5. ORGANIC CHEMISTRY, EXPLOSIVES. *Educ:* Davidson Col, BS, 55; Univ Tenn, Knoxville, PhD(org chem), 63. *Prof Exp:* Sr chemist, Amcel Propulsion Co, Celanese Corp Am, 61-65; sr chemist, Northrop Carolina Inc, Northrop Corp, 65-69, sr scientist, 69-71; sr scientist, 71-73, tech dir, Chemtronics Div, Airtronics, Inc, 73-76; dir res & develop, Plastifax, Inc, 76-80; vpres, Chemtronics, Inc, Div of Halliburton Co, 80-84; PRIN STAFF ENGR, ATLANTIC RES CORP, 85- *Concurrent Pos:* Consult, Hudson Assocs; mem, Am Defense-Preparedness Asn. *Mem:* Am Chem Soc; AAAS; Sigma Xi; Am Defense-Preparedness Asn. *Res:* Aromatic nitration and sulfonation; halogenation of aromatics and alkyl aromatics; rates of organic reactions in nonaqueous solvents; relation of structure and composition to physical and explosive properties; explosive manufacturing processes; insensitive high explosives. *Mailing Add:* 8194 Snowfall Dr Manassas VA 22111

HUDSON, GEORGE ELBERT, b Pittsburgh, Pa, Apr 25, 16; m 84; c 3. APPLIED PHYSICS, MATHEMATICS. *Educ:* George Washington Univ, BS, 38; Brown Univ, ScM, 40, PhD(physics), 42. *Prof Exp:* Asst physics, Brown Univ, 38-41; physicist, David Taylor Model Basin, Navy Dept, Washington DC, 42-46; from assoc prof to prof physics, NY Univ, 46-63; asst chief radio physics div, Radio Standards Lab, Nat Bur Standards, 63-67, consult, Time & Frequency Div, 67-70; sr res consult, Naval Ord Lab, 70-74; physicist, Naval Surface Weapons Ctr, White Oak Lab, 74-82; RETIRED. *Concurrent Pos:* Vis prof, Georgetown Univ, 43-44; in charge instrumentation group, Bikini Test, 46; assoc tech dir, Proj Squid, NY Univ, 46-48, tech dir, 48-50, dir gas & bubble oscillation projs, 53-56; dir res, Smyth Res Assocs, 56-57; consult, Naval Ord Lab, 57-59, Woods Hole Oceanog Inst, 58-60, Brookhaven Nat Lab, 58-61 & Avco Corp, 61-62; chmn int radio consult comt, US Time & Frequency Standards Study Group, 63-69. *Mem:* Fel AAAS; fel Am Phys Soc; Sigma Xi. *Res:* Time and relativity; stochastic processes; vortex flow; high energy laser beam propagation; relativistic particle beam propagation. *Mailing Add:* 1770 Avenida Del Mundo-1003 Coronado CA 92118-3044

HUDSON, HUGH T, b Jackson, Miss, Nov 20, 33; m 73; c 3. SOLID STATE PHYSICS, SCIENCE EDUCATION. *Educ:* Miss Col, BS, 58; Univ Va, MS, 60, PhD(physics), 62. *Prof Exp:* Asst prof physics, 62-67, ASSOC PROF PHYSICS, UNIV HOUSTON, 67- *Concurrent Pos:* Adj asst prof, Baylor Col Med, 65-80; resident coordr, Inst Int Educ, Bangladesh, 69-71. *Mem:* Am Phys Soc; Am Asn Physics Teachers; AAAS; Sigma Xi. *Res:* Growth of high quality metal single crystals; mechanical properties of single crystals; dislocations; response of skin to ultraviolet and visible light; identifying cognitive variables that influence the study of physics and developing support systems to reduce dropout in the introductory courses; computer assisted instruction. *Mailing Add:* Dept Physics Univ Houston Houston TX 77004

HUDSON, J(OHN) L, b Chicago, Ill, June 19, 37; m 63; c 3. CHEMICAL ENGINEERING. *Educ:* Univ Ill, BS, 59; Princeton Univ, MSE, 60; Northwestern Univ, PhD(chem eng), 62. *Prof Exp:* Fulbright fel, Univ Grenoble, 62-63; from asst prof to assoc prof chem & chem eng, Univ Ill, Urbana, 63-74; mgr div air pollution control, Ill Environ Protection Agency, 74-75; chmn dept, 75-85, prof chem eng, 75-85, Ctr Advan Studies, 85-86, WILLIS JOHNSON PROF CHEM ENG, UNIV VA, 88- *Concurrent Pos:* Fulbright fel, Univ Tubingen, 82-83. *Mem:* Am Inst Chem Engrs; Am Chem Soc; Air Pollution Control Asn. *Res:* Chemical reactors; air pollution. *Mailing Add:* Dept Chem Eng Thornton Hall Univ Va Charlottesville VA 22903

HUDSON, JACK WILLIAM, JR, b Denver, Colo, Oct 30, 26; m 47; c 2. COMPARATIVE PHYSIOLOGY. *Educ:* Occidental Col, BA, 48, MA, 50; Univ Calif, Los Angeles, PhD(zool), 60. *Prof Exp:* Instr zool, El Camino Col, 50-55; instr biol, Occidental Col, 55-59; res assoc vert zool, Univ Calif, Los Angeles, 60-62; from asst prof to assoc prof biol, Rice Univ, 62-67; assoc prof ecol & systs, 67-71, chmn sect, 71-74, prof ecol & systs, Cornell Univ, 71-78; chmn dept, 78-80, PROF BIOL, UNIV ALA, BIRMINGHAM, 78- *Concurrent Pos:* Consult, World Book Encycl, 71-; prog dir, NSF, 74-75; vis prof, Univ New South Wales, 74 & 78. *Mem:* AAAS; Am Physiol Soc; Ecol Soc Am; Am Soc Mammalogists; Am Soc Zoologists. *Res:* Comparative physiology of temperature regulation in birds and mammals including hibernation; cardiovascular physiology. *Mailing Add:* Dept Biol Univ Ala Camp 242 Birmingham AL 35294

HUDSON, JAMES BLOOMER, b New York, NY, Sept 23, 27; m 57; c 3. MEDICINE. *Educ:* Dartmouth Col, AB, 48; Boston Univ, MD, 52. *Prof Exp:* Intern med, Mass Mem Hosps, 52-53, from asst resident to chief resident, 53-56; asst, Sch Med, Boston Univ, 55-56, instr & asst dean, 56-58, assoc, 58-61; from asst prof to assoc prof, 61-68, PROF MED & VCHMN DEPT, MED COL GA, 68-, CHIEF NEPHROLOGY, 76- *Concurrent Pos:* Res fel infectious dis, Mass Mem Hosps, 54-55; Med Found Boston res fel, 58-61; USPHS spec res fel, 62-64; chmn cardiovasc res, Ga Heart Asn, 64- *Mem:* Am Fedn Clin Res; Sigma Xi. *Res:* Renal disease; electrolyte metabolism. *Mailing Add:* 2120 Gardner St Augusta GA 30904

HUDSON, JAMES GARY, b Mount Clemens, Mich, Aug 11, 46; m 70; c 4. CLOUD PHYSICS, AEROSOL SCIENCE. *Educ:* Western Mich Univ, BA, 68; Univ Mich, MS, 70; Univ Nev, Reno, PhD(atmos physics), 76. *Prof Exp:* From asst res prof to assoc res prof, 76-90, RES PROF, ATMOSPHERIC SCI CTR, DESERT RES INST, 90- *Mem:* Am Meterol Soc; Sigma Xi. *Res:* Instrumentation, laboratory and field measurements and climatology of cloud condensation nuclei; relationship between cloud condensation nuclei and cloud and fog droplet spectra; origin of atmospheric aerosol and atmospheric visibility. *Mailing Add:* 2055 Severn Dr Reno NV 89503

HUDSON, JOHN B(ALCH), b Plymouth, Mass, Dec 11, 34; m 57; c 3. SURFACE PHENOMENA. *Educ:* Rensselaer Polytech Inst, BChE, 56, MS, 58, PhD(metall), 60. *Prof Exp:* Phys chemist, Silicone Prod Dept, Gen Elec Co, 56, phys chemist, Res Lab, 57-60; sr res staff mem, Gen Dynamics/Electronics, 60-63; res assoc, 63-65, from asst prof to assoc prof, 65-72, PROF MAT ENG, RENSSELAER POLYTECH INST, 72- *Concurrent Pos:* Consult, Gen Elec Co, 64-68, Eastman Kodak Co, 64-70, Melpar, Inc, 65, Langley Res Ctr, NASA, 65-70, Aerovac Corp, 68-70, Gen Elec Co, 74-78, Watervliet Arsenal, 79-83, Mech Technol Inc, 78-79, Inficon Corp, 84-90, TPL Cordis, 87-88; vis prof physics, State Univ NY, Albany, 85-86. *Mem:* Am Inst Mining, Metall & Petrol Engrs; Am Vacuum Soc; Am Phys Soc; AAAS; Mat Res Soc. *Res:* Physics and chemistry of surfaces; ultrahigh vacuum instrumentation; heterogeneous catalysis; diffusion in solids; adsorption phenomena; environmental effects on materials; electronic materials processing; chemical vapor deposition. *Mailing Add:* Dept Mat Eng Rensselaer Polytech Inst Troy NY 12180

HUDSON, JOHN LESLIE, operations research, for more information see previous edition

HUDSON, LYNN DIANE, b Chicago, Ill, Aug 7, 53; m 77; c 1. GENETICS. *Educ:* Univ Wis-Madison, BS, 73; Univ Minn, PhD(genetics & cell biol), 77. *Prof Exp:* Fel genetics, Harvard Med Sch, 77-79; res assoc, Brown Univ, 79-87; AT NAT INST NEUROL & COMMUN DIS & STROKE, NIH, 87- *Mem:* Am Soc Cell Biol; Women Cell Biol; Am Soc Neurochem. *Res:* Regulation of brain-specific genes. *Mailing Add:* Bldg 36 Rm 5D04 NIH 9000 Rockville Pike Bethesda MD 20892

HUDSON, MARY KATHERINE, b Santa Monica, Calif, Jan 6, 49; m 84; c 2. SPACE PHYSICS, PLASMA PHYSICS. *Educ:* Univ Calif, Los Angeles, BS, 69, MS, 71, PhD(physics), 74. *Prof Exp:* Res physicist, Aerospace Corp, 69-71; res physicist space physics, Univ Calif, Berkeley, 74-84; ASSOC PROF PHYS & ASTRON DEPT, DARMOUTH COL. *Concurrent Pos:* Lectr, Univ Calif, Berkeley; vis asst prof physics, Mills Col; vis staff, LLNL & LANL; assoc ed, J Geophys Res. *Honors & Awards:* Macelwane Award, Am Geophys Union. *Mem:* Fel Am Geophys Union; Am Phys Soc. *Res:* Theoretical models of ionospheric plasma phenomena; E and F region irregularities; ionosphere-magnetosphere coupling and transport phenomena; ring current-plasmaparse interaction; other planetary magnetospheres. *Mailing Add:* Phys & Astron Dept Dartmouth Col Hanover NH 03755

HUDSON, PAGE, b Richmond, Va, Feb 14, 31; m 56; c 4. PATHOLOGY, FORENSIC MEDICINE. *Educ:* Univ Richmond, BA, 52; Med Col Va, MD, 56. *Prof Exp:* Asst path, Johns Hopkins Univ, 56-57, instr, 57-58; intern path, Johns Hopkins Hosp, 56-57, asst pathologist, 57-58; res fel, Dept Legal Med, Harvard Med Sch, 60-61; instr path, State Univ NY Downstate Med Ctr, 61-64; asst prof, Med Col Va, 64-65, from asst prof to assoc prof surg path, 65-68; chief med examr, State of NC, 68-86; chmn Div Forensic Path, 68-86, PROF PATH, UNIV NC, CHAPEL HILL, 71-; PROF PATH, E CAROLINA UNIV SCH MED, 86- *Concurrent Pos:* Residency in Anat, Clin & Forensic Path. *Mem:* Fel Am Acad Forensic Sci; AMA; US Can Acad Path; Nat Asn Med Examr. *Res:* Anatomic, surgical and forensic pathology; forensic toxicology. *Mailing Add:* Sch Med E Carolina Univ 205 W Church St Farmville NC 27828

HUDSON, PEGGY R, b Dayton, Wash, May 31, 47; m 69. PHYCOLOGY. *Educ:* Univ Wash, BS, 68, PhD(bot), 74. *Prof Exp:* Consult ultrastruct & phycol, Weyerhaeuser Co, 73-74; asst prof, 74-80, ASSOC PROF BIOL, SEATTLE UNIV, 80- *Mem:* Phycol Soc Am; Int Phycol Soc; Am Asn Univ Professors; Am Inst Biol Sci. *Res:* Ultrastructure of algae; algal structure and function, especially of seaweeds. *Mailing Add:* Dept Biol Seattle Univ Broadway & Madison Seattle WA 98122

HUDSON, RALPH P, b Wellingborough, Eng, Oct 14, 24; nat US; m 47; c 2. METROLOGY, LOW TEMPERATURE PHYSICS. *Educ:* Oxford Univ, BA, 44, MA & PhD, 49. *Prof Exp:* Sci officer, Brit Ministry Supply, 44-46; vis lectr physics, Purdue Univ, 49-50, asst prof, 50-51; mem cryogenics sect, Nat Bur Standards, 51-54, actg chief, 54-55, chief, 55-61, chief heat div, 61-78, dep dir, Ctr Absolute Phys Quantities, 78-80; MEM STAFF, INT BUR WEIGHTS & MEASURES, FRANCE, 80- *Concurrent Pos:* Guggenheim fel, 60; ed, Metrologia, 80- *Honors & Awards:* J P Wetherill Medal, Franklin Inst, 62; Stratton Award, Nat Bur Standards, 64, Condon Award, 76. *Mem:* Fel Franklin Inst; fel Am Phys Soc; Sigma Xi. *Res:* Temperature measurement and temperature physics; research near absolute zero: paramagnetism, liquid helium, superconductivity, nuclear orientation, thermometry and magnetic resonance. *Mailing Add:* Bur Int des Poids et Mesures Pavillon de Breteuil Sevres F-92312 France

HUDSON, REGGIE LESTER, b Newport News, Va, July 23, 52; m 82; c 1. PHYSICAL CHEMISTRY, MAGNETIC RESONANCE. *Educ:* Pfeiffer Col, AB, 74; Univ Tenn, PhD(chem), 78. *Prof Exp:* asst prof, 78-85, ASSOC PROF CHEM, ECKERD COL, 85- *Concurrent Pos:* Vis scientist, Univ Col, London, 85-86 & NASA-Goddard Space Flight Ctr, 87-88. *Mem:* Am Chem Soc; Sigma Xi; Royal Soc Chem. *Res:* Structure, bonding, and reactions of free radicals using electron spin resonance and optical spectroscopies; application of infrared, mass spectral, and radiation methods to problems in cometary chemistry. *Mailing Add:* Col Nat Sci Eckerd Col St Petersburg FL 33733

HUDSON, RICHARD DELANO, JR, b Newark, NJ, Dec 18, 24; m 73; c 4. OPTICAL PHYSICS. *Educ:* Univ Rochester, BS, 45, MS, 48. *Prof Exp:* Instr optics, Univ Rochester, 46-48; physicist, Am Optical Co, Mass, 48-51; proj engr, Am Cystoscope Co, NY, 51-52 & Olympic Develop Co, Conn, 52-55; group head, Infrared Labs, 56-59, sr staff engr, 59-63, sr scientist, 63-65, mgr advan tech educ, Aerospace Group, 65-66, mgr prof develop ctr, 66-70, corp dir Tech Educ, Hughes Aircraft Co, 70-83; PRES, ARJAY ASSOCS, 83- *Concurrent Pos:* Instr, Eve Div, Pierce Col, Calif, 56-60; mem exec comt, Infrared Info Symp, 62-72; consult adv group aerospace res & develop, Nato, 74-75. *Mem:* AAAS; fel Optical Soc Am; Inst Elec & Electronics Engrs; Am Soc Eng Educ; Soc Photog Scientists & Engrs. *Res:* Design, analysis and application of infrared and electromagnetic sensor systems; production of precision optics; problems of human vision under night time conditions; continuing education for technical personnel; high frequency communication techniques; astronomical photometry. *Mailing Add:* 4308 Alonzo Ave Encino CA 91316

HUDSON, ROBERT B, b Baltimore, Md, Oct 1, 20; m 42; c 4. CHEMICAL ENGINEERING, INORGANIC CHEMISTRY. *Educ:* Auburn Univ, BS, 42; St Louis Univ, MS, 59. *Prof Exp:* Tech trainee, Monsanto Chem Co, 46-47, res chem engr, 47-57, sr res chem engr, 57-60, res specialist, 60-62, res group leader, 62-64, sr res group leader, 64-70, mgr mfg technol, 70-78, sect mgr res & develop, 78-81, mgr res & develop, 81-82; RETIRED. *Honors & Awards:* Gaston Du Bois Award, 55. *Mem:* Am Chem Soc; Am Inst Chem Engrs. *Res:* Product and process development; phosphorus chemistry; process control system analysis and synthesis. *Mailing Add:* Ten Barberry St Louis MO 63122

HUDSON, ROBERT DOUGLAS, b Walton-on-Thames, Eng, Mar 11, 31; nat US; m 56; c 5. AERONOMY, ATOMIC SPECTROSCOPY. *Educ:* Univ Reading, BS, 56, PhD(ultraviolet spectros), 59. *Prof Exp:* Asst prof physics, Univ Southern Calif, 59-62; mem tech staff labs div, Aerospace Corp, 62-69; staff scientist, Sci Applns Directorate, Johnson Space Ctr, NASA, 69-74, mgr, Environ Effects Proj Off, 74-76, head, Stratospheric Chem & Physics Br, Goddard Space Flight Ctr, 76-90; CHMN & PROF, UNIV MD, 90- *Concurrent Pos:* Consult, Space Technol Labs, 60-61 & Aerospace Corp, 61-62. *Mem:* Am Geophys Union. *Res:* Measurements of vacuum ultraviolet absorption cross-sections of gases and vapors; applying these data to geophysical and astrophysical problems and to the formulation of rocket and spacecraft payloads; photo chemistry of the earth's and planetary atmospheres. *Mailing Add:* Dept Meteorol Univ Md College Park MD 20742

HUDSON, ROBERT FRANK, GEOLOGY. *Educ:* Colby Col, BA, 54, MS, 57; Univ Iowa, PhD, 65. *Prof Exp:* Explor & exploitation geologist, Texaco Inc, Houston, 66-78; exploitation geologist, La Land & Explor Co, Houston, 78-81; explor geologist, MCO Resources, Houston, 81-82; explor geologist, Donald C Slawson, Houston, 82-83; staff geologist, Sohio Petrol Co, Houston, 83-85; CONSULT, 85- *Concurrent Pos:* Explor geologist, Texaco Inc, Corpus Christi, 59-66. *Mem:* Am Asn Petrol Geologists; fel Geol Soc Am. *Res:* Prospect generation, prospect sales, evaluation of submittals & exploration research along the Texas Gulf Coast. *Mailing Add:* 6833 Hazen Houston TX 77074

HUDSON, ROBERT MCKIM, b Morristown, NJ, Oct 1, 26; m 52; c 2. PHYSICAL CHEMISTRY, INORGANIC CHEMISTRY. *Educ:* Yale Univ, BS, 47, PhD(chem), 50. *Prof Exp:* SR RES CONSULT SURFACE CHEM & GAS-METAL REACTIONS, LIGHT PROD DIV, RES TECH CTR, US STEEL CORP, MONROEVILLE, 50- *Mem:* Am Chem Soc; Sigma Xi; Iron & Steel Soc, Am Inst Mining Metall & Petrol Engrs. *Res:* Diffusion of electrolytes in aqueous solutions; gas-metal reactions during annealing of steel; hydrogen behavior in steel; reaction kinetics and mechanisms; thermodynamics; corrosion of steel and coated steel products; pickling of iron and steel; pickling inhibitors. *Mailing Add:* 1618 Williamsburg Pl Pittsburgh PA 15235

HUDSON, ROBERT Y(OUNG), b Algood, Tenn, Mar 13, 12; m 40; c 2. CIVIL ENGINEERING. *Educ:* Tenn Polytech Inst, BS, 36; Univ Iowa, BS, 39, CE, 52. *Prof Exp:* From eng aide to supv hydraul engr, Waterways Exp Sta, US Army, 37-63, chief, Waterways Br, 63-71, spec asst to chief, Hydraul Div, 71-72; RETIRED. *Concurrent Pos:* Consult, 72- *Honors & Awards:* Meritorious Civilian Serv Award, CEngr, US Army, 47 & 72; George W Goethals Medal, Am Mil Engrs, 71. *Res:* Hydraulic models; harbor wave action; breakwater design. *Mailing Add:* 205 Belva Dr Vicksburg MS 39180

HUDSON, ROY DAVAGE, b Hamilton Co, Tenn, June 30, 30; m 56; c 2. NEUROPHARMACOLOGY, HEALTH BUSINESS & MANAGEMENT. *Educ:* Livingstone Col, BS, 55; Univ Mich, MS, 57, PhD(pharmacol), 62. *Hon Degrees:* LLD, Lehigh Univ, 74 & Princeton Univ, 75. *Prof Exp:* Asst prof pharmacol, Med Sch Univ Mich, 61-66; assoc prof neurosci, Brown Univ, 66-70, assoc dean grad sch, 66-69; pres admin, Hampton Univ, 70-76; prof pharmacol, Univ Va, 72-74; vpres planning, Parke-Davis Co, 76-79; dir, Cent Nervous Syst Res, 81-87, vpres, PR&D Europe, 87-90, CORP VPRES, PUB RELS, UPJOHN CO, 90- *Concurrent Pos:* Chmn elect, Rhode Island Comn Econ Develop, 68-69; comnr, Norfolk Area Med Ctr Authority, Va, 70-74; mem, Adv Comt to the Dir, NIH, 74-77 & Off Technol Assessment Appln Sci & Technol, 76-77. *Mem:* AAAS; NY Acad Sci; Am Soc Pharmacol & Exp Therapeut; Am Col Neuropsychopharmacol; Am Asn Higher Educ. *Res:* Investigation of mechanics of action underlying motor pathologies; pyramidal and extrapyramidal motor dysfunctions occuring spontaneously and as a result of drug treatment; effects of centrally acting drugs on the electroencephalogram. *Mailing Add:* 7057 Oak Highlands Kalamazoo MI 49009

HUDSON, SIGMUND NYROP, b Memphis, Tenn, Sept 29, 36; m 60; c 2. MATHEMATICS, COMPUTER SCIENCE. *Educ:* Dartmouth Col, AB, 58; Tulane Univ, PhD(math), 63; Clarkson Univ, MS, 85. *Prof Exp:* Asst prof math, Syracuse Univ, 63-70; asst prof, Tulane Univ, 70-71; from assoc prof to prof math, Savannah State Col, 71-85; PROF MATH & COMPUTER SCI, ARMSTRONG STATE COL, 85- *Mem:* Am Math Soc; Math Asn Am; Asn Comput Mach; Computer Prof Social Responsibility. *Res:* Topological algebra. *Mailing Add:* Dept Math & Comp Sci Armstrong State Col 11935 Abercorm Savannah GA 31419-1997

HUDSON, STEVEN DAVID, b Syracuse, NY, Nov 19, 61; m 86; c 3. LIQUID CRYSTALS, ELECTRON MICROSCOPY. *Educ:* Cornell Univ, BS, 83; Univ Mass, PhD(polymer sci & eng), 90. *Prof Exp:* Process develop engr, Codenoll Tech Corp, 83-85; POSTDOCTORAL MEM TECH STAFF, AT&T BELL LABS, 90- *Mem:* Am Phys Soc; Mat Res Soc; Am Chem Soc. *Res:* Defect structure and interaction in liquid crystals; crystallization of polymers. *Mailing Add:* AT&T Bell Labs Rm 1B-216 600 Mountain Ave Murray Hill NJ 07974-0636

HUDSON, WILLIAM NATHANIEL, b Berkeley, Calif. MATHEMATICAL ANALYSIS, MATHEMATICAL STATISTICS. *Educ:* Univ Calif, Berkeley, AB, 60, MA, 63; Univ Calif, Irvine, PhD(math), 70. *Prof Exp:* Engr, Northrop Corp, 63-65; mem tech staff, Rockwell Corp, 65-70; lectr math, Univ Calif, Santa Barbara, 70-73; vis asst prof, Univ Utah, 73-75; assoc prof math, Bowling Green State Univ, 75-77; PROF, AUBURN UNIV, 78- *Concurrent Pos:* Vis assoc prof, Tulane Univ, 77-78; vis prof, Univ Ariz, 80-81, Univ NC, 84. *Mem:* Inst Math Statist. *Res:* Probability theory and stochastic processes; infinitely divisible distributions and processes with independent increments. *Mailing Add:* Dept Math Auburn Univ Auburn AL 36849

HUDSON, WILLIAM RONALD, b Temple, Tex, May 17, 33; m 58; c 3. CIVIL ENGINEERING. *Educ:* Tex A&M Univ, BS, 54, MS, 55; Univ Tex, Austin, PhD(civil eng), 65. *Prof Exp:* Civil engr, S J Buchanan & Assocs, 57-58; asst chief rigid pavement res br, Am Asn State Hwy Officials Rd Test, Nat Acad Sci, Ill, 58-61; supv designing res engr, Hwy Design Div, Tex Hwy Dept, 61-63; from instr to assoc prof, 63-73, asst dean col eng, 69-70, DeWitt C Greer prof, 81, PROF CIVIL ENG, UNIV TEX, AUSTIN, 73-, RES ENGR, CTR HWY RES, 63-, ASSOC DEAN ENG ADVAN PROGS, 70- *Concurrent Pos:* Mem pavement condition eval comt, Hwy Res Bd, Nat Acad Sci-Nat Res Coun, 63-, strength & deformation characteristics of pavement sect comt, 64-, award comt, 65, res needs subcomt, pavement condition eval comt, 65-66, subcomt, rigid pavement design comt, 65-69, theory of pavement design comt, 65-69, chmn, 69-, mem task force comt struct design of pavement systs, 70, adv comt struct design asphalt concrete pavement systs, 70, asst proj engr & consult, Proj 1-1, Nat Coop Hwy Res Prog, 63-64, mem adv sect, 67, surface eval subcomt, pavement condition eval comt, 67 & ad hoc comt interaction soils & design div, 67-; consult, Signal Oil & Gas Co, 65-66; prin investr, Mat Res & Develop Inc, Calif, 66- & C5A test pavement prog, Waterways Exp Sta & Ohio River Div Lab, US CEngr, 68; mem adv panel, Eric H Wany USAF Civil Eng Res Facil, 69; consult, Dow Chem Co & Shell Develop Co. *Honors & Awards:* James R Croes Medal, Am Soc Civil Engrs, 68. *Mem:* AAAS; Nat Soc Prof Engrs; Am Soc Civil Engrs; Am Concrete Inst; NY Acad Sci. *Mailing Add:* Dept Civil Eng Univ Tex 6-1 ECJ Hall Austin TX 78712

HUDSON, WILLIAM RUCKER, b Charlotte, NC, May 16, 25; m 47; c 3. SURGERY, OTOLARYNGOLOGY. *Educ:* Bowman Gray Sch Med, Wake Forest Univ, MD, 51. *Prof Exp:* NIH fel, Sch Med, Johns Hopkins Univ, 60-61; from asst prof to assoc prof, 61-65, PROF OTOLARYNGOL & CHIEF DIV, MED SCH, DUKE UNIV, 65- *Concurrent Pos:* Consult, Vet Admin Hosp, Durham, NC, 61- *Mem:* Am Acad Ophthal & Otolaryngol; Am Laryngol, Rhinol & Otolaryngol Soc; Am Laryngol Asn; Soc Univ Otolaryngol; Am Col Surg. *Res:* Otophysiology. *Mailing Add:* Div Otolaryngol Duke Univ Med Ctr Durham NC 27710

HUDSPETH, ALBERT JAMES, b Houston, Tex, Nov 9, 45; m 77; c 1. NEUROBIOLOGY, SENSORY TRANSDUCTION. *Educ:* Harvard Col, BA, 67; Harvard Univ, MA, 68, PhD(neurobiol), 73; Harvard Med Sch, MD, 74. *Prof Exp:* Asst prof biol, Calif Inst Technol, 75-78, assoc prof, 78-82, prof, 82-83; prof, Dept Physiol & Otolaryngol, Univ Calif, San Francisco, 85-88; PROF & CHAIR, DEPT CELL BIOL & NEUROSCI, UNIV TEX SOUTHWESTERN MED CTR, 89-, DIR, NEUROSCI PROG, 91- *Concurrent Pos:* Instr neurobiol, Cold Spring Harbor Lab, 71-72; vis res fel, Karolinska Hosp, Sweden, 74-75; res fel, Harvard Med Sch, 75; instr synapses, Cold Spring Harbor Lab, 78-82; assoc ed, J Neurosci, 81-; mem, Sci Prog, Univ Calif, San Francisco, 83-89, vchair, Dept Physiol, 86-89, dir, Cell Biol Prog, 86-89 & mem, Biophys Prog, 86-89. *Mem:* Nat Acad Sci; Asn Res Otolaryngol; Soc Neurosci; AAAS. *Res:* Transduction process whereby the sensory receptors of the inner ear, hair cells, convert acoustical and accelerational stimuli into electrical signals that the brain can interpret. *Mailing Add:* Dept Cell Biol Neurosci Southwestern Med Ctr Univ Tex 5323 Harry Hines Blvd Dallas TX 75235-9039

HUDSPETH, EMMETT LEROY, b Denton, Tex, Dec 3, 16; m 44; c 4. NUCLEAR PHYSICS. *Educ:* Rice Inst, AB, 37, AM, 38, PhD(nuclear physics), 40. *Prof Exp:* Res fel, Bartol Res Found, Franklin Inst, 40-41, asst dir, Found, 45-50; res assoc & mem staff radiation lab, Mass Inst Technol, 41-45; dir, Nuclear Physics Lab, 50-71, prof, 50-87, EMER PROF PHYSICS, UNIV TEX, AUSTIN, 87- *Concurrent Pos:* Mem comt undersea warfare, 47-48; sci adv comt radiobiol lab, Univ Tex & USAF, 54-58; bd dirs, Tex Nuclear Corp, 56-66; bd dirs, Nuclear-Chicago Corp, 61-66; bd dirs & consult, Medical Monitor Systems, Inc, 69-78. *Honors & Awards:* Emmett L Hudspeth Centennial Lectureship in Physics. *Mem:* AAAS; fel Am Phys Soc; Am Asn Physicists Med; Sigma Xi. *Res:* Nuclear physics with electrostatic machines; radar wave propogation; military operation and tactics; high voltage generators; energy levels of nuclei; neutron scattering; neutron therapy; vital signs measurements. *Mailing Add:* Dept Physics Univ Tex Austin TX 78712

HUEBERT, BARRY JOE, b Nebraska City, Nebr, Mar 13, 45; div; c 2. PHYSICAL CHEMISTRY, ANALYTICAL CHEMISTRY. *Educ:* Occidental Col, BS, 67; Northwestern Univ, MS, 68, PhD(chem), 71. *Prof Exp:* Instr chem, Northwestern Univ, 70-71; from asst prof to assoc prof chem, Colo Col, 71-86; dir athos chem prog, SRI Int, 86-87; PROF OCEANOG, OCEAN UNIV, RI, 87- *Concurrent Pos:* Sr fel, Nat Ctr Atmospheric Res, 77-78; mem, Atmospheric Chem & Climate Comt, IGAC Marine Aerosol & Gas Exchange, steering comt, Int Global Atmospheric Chem Prog, Comn Atmospheric Chem & Global Pollution. *Mem:* Am Chem Soc; Am Geophys Union. *Res:* Precipitation chemistry; global measurements of tropospheric particles and acidic gases; atmospheric chemistry and physics; atmospheric nitrogen chemistry; dry-deposition of acids; air/sea exchange climate change. *Mailing Add:* Ctr Atmospheric Chem Studies Grad Sch Ocean Univ Rhode Island Narragansett RI 02882-1197

HUEBNER, ALBERT LOUIS, b New York, NY, Feb 4, 31; m 50; c 3. APPLIED PHYSICS, BIOPHYSICS. *Educ:* Brooklyn Col, AB, 55; Univ Calif, Los Angeles, MA, 62. *Prof Exp:* Teaching asst, Columbia Univ, 55-57; from physicist to sr physicist, Rocketdyne Div, NAm Rockwell Corp, 57-66, prin scientist, 66-73; ASST PROF PHYSICS, CALIF STATE UNIV, 75- *Concurrent Pos:* Lectr, West Coast Univ, 62-65, sr lectr & coordr dept physics, 65-82; dir, Sci-Media Assocs, 73- *Mem:* AAAS; Am Phys Soc; Am Asn Physics Teachers; Am Inst Med Climat. *Res:* Electrified-fluid phenomena; solid state device physics; energy development and use analysis. *Mailing Add:* 20331 Mobile St Canoga Park CA 91306

HUEBNER, ERWIN, b July 14, 43; Can citizen; m 71. DEVELOPMENTAL BIOLOGY, REPRODUCTIVE BIOLOGY. *Educ:* Univ Alta, BSc, 65; Univ Mass, Amherst, PhD(zool), 70. *Prof Exp:* Fel insect physiol, MacDonald Col, McGill Univ, 70-72; from asst prof to assoc prof, 72-81, PROF ZOOL, UNIV MAN, 81- *Mem:* Can Soc Zool; Am Soc Develop Biol; Am Soc Cell Biol; Soc Exp Biol; Am Soc Zool; Can Soc Cell Biol; Sigma Xi. *Res:* The process of oogenesis, using comparative oogenesis to determine the morphological, cytochemical, bioelectrical and physiological parameters that occur and control this process. *Mailing Add:* Dept Zool Univ Man Winnipeg MB R3T 2N2 Can

HUEBNER, GEORGE J, b Detroit, Mich, Sept 8, 10; m. ENVIRONMENTAL RESEARCH. *Educ:* Univ Mich, BS, 32. *Hon Degrees:* DSc, Bucknell Univ, 69. *Prof Exp:* Lab engr, Eng Div, Chrysler Corp, 31-36, asst chief engr, Plymouth Div, 36-39, asst to dir res, Eng Div, 39-45, chief eng res, 45-52, exec eng, Missile Br Eng, 52-53, exec eng res eng staff, 53-64, dir res, 64-75; CHMN BD, ENVIRON RES INST MICH, ANN ARBOR, 75- *Honors & Awards:* L Ray Buckendale Award, Soc Automotive Engrs, 59. *Mem:* Nat Acad Eng; fel Soc Automotive Engrs (pres, 75); Sigma Xi. *Res:* Automotive gas turbine engine; metallurgy; chemistry; physics; author of various publications; granted over 40 patents. *Mailing Add:* Environ Res Inst Mich PO Box 134001 Ann Arbor MI 48113-4001

HUEBNER, GEORGE L(EE), JR, b Bay City, Tex, May 22, 18; m 41; c 4. ELECTRICAL ENGINEERING. *Educ:* Tex A&M Univ, BS, 46, MS, 51, PhD(physics, elec eng), 53. *Prof Exp:* Engr, Radio Intel Div, Fed Commun Comn, 42-43; res engr, Vector Mfg Co, Tex, 46-47 & Independent Explor Co, 47-50; res scientist, Tex A&M Res Found, 53-59; sect head, Electro Optics,

Tex Instruments, Dallas, 60-61; assoc prof, Tex A&M Univ, 61-70, prof meteorol, 71-; CHMN BD, ENVIRON RES INST MICH. *Concurrent Pos:* Consult, TRW Systs, Tex Instruments, Envicon Inc, Spectro Systs, Tri-Pak, Oceanografia & Tex Tech Univ. *Mem:* Nat Acad Eng; Am Meteorol Soc; Inst Elec & Electronics Engrs. *Res:* Electrical activity of thunderstorms, oceanographic instrumentation, remote sensing techniques, radar meteorology and analysis of weather systems with use of radar. *Mailing Add:* Environ Res Inst Mich PO Box 134001 Ann Arbor MI 48113-4001

HUEBNER, JAY STANLEY, b Bancroft, Kans, July 10, 39; m 61; c 3. BIOPHYSICS, ELECTRONICS. *Educ:* Kans State Univ, BSEE, 61; San Diego State Univ, MS, 65; Univ Calif, Riverside, PhD(physics), 71. *Prof Exp:* Assoc engr astronaut, Gen Dynamics Corp, 61-63; res proj supvr, Bourns Inc, 68-70; res assoc biophys, Mich State Univ, 71-72; from asst prof to assoc prof, 72-78, PROF NAT SCI, UNIV NFLA, 78-, DIR, CTR MEMBRANE PHYSICS, 83- *Concurrent Pos:* Fulbright scholar, 88. *Mem:* Nat Space Soc; Am Asn Physics Teachers; Am Soc Photobiol; Am Chem Soc. *Res:* Photoelectric effects in biological and artificial membranes and science teaching. *Mailing Add:* Dept Nat Sci Univ NFla Jacksonville FL 32216-6699

HUEBNER, JOHN STEPHEN, b Bryn Mawr, Pa, Sept 9, 40; m 62; c 2. PETROLOGY, MINERALOGY. *Educ:* Princeton Univ, BA, 62; Johns Hopkins Univ, PhD(geol), 67. *Prof Exp:* GEOLOGIST, US GEOL SURV, 67- *Concurrent Pos:* Lectr, George Washington Univ, 71; prin investr, NASA Lunar Sample Prog, 73-84; mem, Lunar & Planetary Rev Panel, 76-78; assoc ed, J Geophys Res, 77-79; counr, Min Soc Am, 85-88. *Honors & Awards:* Mineral Soc Am Award, 78. *Mem:* AAAS; fel Mineral Soc Am; Am Geophys Union; Geochem Soc (treas, 72-75); Am Geol Inst (secy-treas, 74-75); Sigma Xi. *Res:* Phase relations of rock forming minerals at elevated temperature, pressure; control of polycomponent fluids in experimental geochemistry; geochemistry of manganese; pyroxene phase relations; high temperature electrochemistry; diffusion in silicates. *Mailing Add:* US Geol Surv Stop 959 Reston VA 22092

HUEBNER, JUDITH DEE, b New York, NY, Mar 4, 47; m 71. FRESHWATER ECOLOGY. *Educ:* Queens Col, BA, 67; Univ Mass, Amherst, MS, 70, PhD(zool), 72. *Prof Exp:* Fel ecol res, Dept Zool, Univ Man, 73-74; instr, 75-81, ASST PROF BIOL, UNIV WINNIPEG, 81- *Concurrent Pos:* Researcher, Northwest Fisheries Ctr, Nat Oceanic & Atmospheric Admin, Woods Hole, Mass, 79-80. *Mem:* AAAS; Am Soc Zoologists; Ecol Soc Am; Can Soc Zoologists; Sigma Xi. *Res:* Ecological energetics and physiological ecology of large intertidal and fresh water molluscs; energetics, consumption and evacuation in inshore (marine) and freshwater fish; predator-prey interactions. *Mailing Add:* Dept Biol Univ Winnipeg Winnipeg MB R3B 2E9 Can

HUEBNER, ROBERT JOSEPH, b Cheviot, Ohio, Feb 23, 14; m 75; c 9. VIROLOGY. *Educ:* St Louis Univ, MD, 42. *Hon Degrees:* LLD, Univ Cincinnati, 65; DSc, Edgecliff Col, 70, Univ Parma, 70 & Cath Univ Louvain, 73. *Prof Exp:* Intern, USPHS Hosp, 42-43; med officer, NIH, 44-47, in chg res unit & Q fever lab, 47-50, chief virus & rickettsial dis, Nat Microbiol Inst, 50-56, chief lab viral dis, Nat Inst Allergy & Infectious Dis, 56-68, chief lab RNA tumor viruses, 68-76, sci coordr immunoprevention, 76-78, mem staff, Cellular & Molecular Biol Lab, Nat Cancer Inst, 78-82; RETIRED. *Honors & Awards:* Bailey K Ashford Award, Am Soc Trop Med, 49; James D Bruce Mem Award, Am Col Physicians, 64; Pasteur Medal, Pasteur Inst, 65; Nat Medal Sci, 69; Kimble Methodology Award, 70; Guido Lenghi Award, Nat Acad Lincei, Rome, 71; Founders Award Cancer Immunol, Cancer Res Inst Inc, 75. *Mem:* Nat Acad Sci; fel Am Pub Health Asn; fel AMA; fel NY Acad Sci. *Res:* Cancer viruses; medical research. *Mailing Add:* 12100 Whippoorwill Lane Rockville MD 20852

HUEBNER, RUSSELL HENRY, SR, b Indianapolis, Ind, Jan 24, 41; m 63; c 4. ELECTRON PHYSICS, RADIATION PHYSICS. *Educ:* Purdue Univ, BS, 62; Vanderbilt Univ, MS, 65; Univ Tenn, PhD(physics), 68; Univ Chicago, MBA, 83. *Prof Exp:* Health physics fel, Oak Ridge Nat Lab, 63-68; radiol physicist, Div Biol & Med, USAEC, 68-70; physicist, Radiol & Environ Res Div, 70-77, prog coodr biomed & environ res, 77-83, prof mgr, ERAB Sci Support, 83-86, assoc dir & off dir, Strategic Planning Group, 86-87, RES PROG MGR-ADMIN, ADVAN PHOTON SOURCE, ARGONNE, NAT LAB, 87- *Concurrent Pos:* Guest scientist, Electron & Optical Physics Sect, Optical Physics Div, Nat Bur Stand, 70; assoc ed, Radiation Res J, 78-81. *Mem:* Am Phys Soc; Health Physics Soc; Radiation Res Soc; Sigma Xi; AAAS. *Res:* Electron energy-loss spectroscopy of molecules and determination of oscillator strength distributions for use in radiation physics; atmospheric and environmental studies; synchrotron radiation. *Mailing Add:* 326 Carriage Hill Dr Naperville IL 60565

HUEBNER, WALTER F, b New York, NY, Feb 22, 28; m 57; c 4. ATOMIC PHYSICS, ASTROPHYSICS. *Educ:* Polytech Inst Brooklyn, BS, 52; Yale Univ, MS, 53, PhD(physics), 58. *Prof Exp:* Staff mem atomic physics, Los Alamos Sci Lab, Univ Calif, 57-68; vis staff mem, Max Planck Inst Astrophys, Munich, 68-70; staff mem, Los Alamos Nat Lab, Univ Calif, 70-72 & 88-91, alt group leader equation state & opacities, 72-76, group leader equation state & opacities, 77-88; INST SCIENTIST, SOUTHWEST RES INST, 87- *Concurrent Pos:* Mem staff, NASA Theoret Div, 59; Max Planck Soc grants, 64, 68-70 & 80; Fulbright grant, Nat Acad Sci, 68; consult, Int Astron Union, Comn 15, 70-73, Sci Appln Inc, 72-76 & Res Inst Swed Nat Defense, Stockholm, 72-80; alt group leader theoret astrophys, Los Alamos Nat Lab, Univ Calif, 80-81, dep group leader, 81-82; mem team photographed nucleus Halley's Comet, Giotto Spacecraft, 86; ed, Physics & Chem Comets. *Mem:* Am Phys Soc; Am Astron Soc; Ger Astron Soc; Int Astron Union; NY Acad Sci; Sigma Xi. *Res:* Calculation of opacity and equation of state data at all densities and temperatures for any material in the gaseous or plasma phase; research on the physics and chemistry of comets; identified first polymer (polyoxymethylene) in space. *Mailing Add:* SW Res Inst PO Drawer 28510 San Antonio TX 78228

HUEBSCH, IAN O, b New York, NY, Jan 18, 27. GEOPHYSICS, METEOROLOGY. *Educ:* Haverford Col, BA, 48; Univ Calif, Berkeley, MS, 65. *Prof Exp:* Physicist, Mil Eval Div, US Naval Radiol Defense Lab, 58-60, res physicist, 60-69; physicist, Physics Int Co, Calif, 69-71; MGR, EUCLID RES GROUP, 71- *Concurrent Pos:* Teacher physics, Merritt Col, Oakland, Calif, 80; teacher comput sci, Laney Col, Oakland, 81-84; mem cert comt, Nat Asn Environ Professionals, 79-86. *Honors & Awards:* Gold Medal Award Sci Achievement, US Naval Radiol Defense Lab, 65. *Mem:* Opers Res Soc Am; Am Geophys Union; Am Meteorol Soc. *Res:* Geophysical effects of nuclear explosions; statistical models of atmospheric phenomena; coagulation and diffusion of aerosols; fracture and comminution of rock; operations research; environmental impact statements. *Mailing Add:* Euclid Res Group 1760 Solano Ave Berkeley CA 94707

HUEBSCH, WILLIAM M, topology, for more information see previous edition

HUEBSCHMAN, EUGENE CARL, b Evanston, Ind, Oct 31, 19; m 46; c 4. SOLID STATE PHYSICS. *Educ:* Concordia Col, BS, 41; Purdue Univ, MS, 46; Univ Tex, PhD(physics), 56. *Prof Exp:* Prof physics, Concordia Col, 46-57; mgr adv anal math, Inertial Guidance Test Facility, Holloman AFB, 57-59; tech advisor, Weapons Guidance Labs, Wright Air Develop Ctr, Ohio, 59-60; mgr adv planning, Guided Missile Range, Pan Am World Airways, 60-64; mgr spec prods dept, Earth Sci Div, Teledyne, Inc, 65-66; pres, Nathaniel Hawthorne Col, 81-82; PROF ELEC ENG, SPACE INST, UNIV TENN, 66-; PRES, SYST ENG CORP, 68-; DIR ENG, CONCORDIA COL, WIS, 85- *Concurrent Pos:* Lectr, San Diego State Col, 58-59; lectr, Univ Calif, 59-60; head grad physics dept, Brevard Eng Col, 60-66. *Mem:* Am Phys Soc; Am Inst Aeronaut & Astronaut; Am Astronaut Soc; Marine Technol Soc. *Res:* Space technology; lunar firing window; track testing error model for inertial guidance system; missile guidance using solid state. *Mailing Add:* Syst Eng Corp 1100 Bragg Circle Tullahoma TN 37388

HUEBSCHMANN, JOHN W, b Baltimore, Md, July 8, 24; m 49; c 3. CHEMICAL ENGINEERING. *Educ:* Johns Hopkins Univ, BS, 53. *Prof Exp:* Asst chem engr, US Indust Chem, 46-51; chemist, Paul Jones & Co, 51-52; chief chemist Le Page's, Inc, 52-56; chemist, Ajex Adhesives, Inc, 56-57; sr develop chemist adhesives sect, Appln Res Div, A E Staley Mfg Co, Ill, 57-66; mgr, Indust Adhesives Div, Harad Chem Co, Ohio, 66-67; adhesive chemist, 67-77, sr chem engr, 77-80, MGR PROCESS INSTALLATION & EVAL, MOORE BUS FORMS INC, 80- *Mem:* Am Chem Soc; Tech Asn Pulp & Paper Indust. *Res:* Dextrinization of starch and cold water soluble adhesives; resin type adhesives; polymers. *Mailing Add:* 397 Howard Dr Youngstown NY 14174

HUEG, WILLIAM FREDERICK, JR, b New York, NY, Jan 12, 24; m 49, 78; c 7. AGRONOMY. *Educ:* Cornell Univ, BS, 48; Mich State Univ, MS, 54, PhD(farm crops, soils, agr econ), 59. *Prof Exp:* Asst county agr agent, NY Exten Serv, 48-50; instr crops & soils, State Univ NY Agr Tech Inst, Alfred, 50-55; instr farm crops, Mich State Univ, 55-57; assoc prof & exten agronomist, Univ Minn, 57-62, asst dir agr exp sta, 62-66, dir, 66-75, prof agron, Inst Agr, Forestry & Home Econ, 62-84, dep vpres & dean, 74-84; RETIRED. *Concurrent Pos:* Mem, Nat Sci Bd, 76-82; mgr, Bhella Holsteins Hammond, WI, 84- *Mem:* Fel Am Soc Agron; fel Crop Sci Soc Am; AAAS. *Res:* Forage physiology, utilization and seed production. *Mailing Add:* 1170 Dodd Rd St Paul MN 55118

HUEGE, FRED ROBERT, b New York, NY, Apr 3, 43; m 72; c 2. CLAY MINERALOGY, PAPER CHEMISTRY. *Educ:* Thiel Col, BA, 64; Pa State Univ, PhD(chem), 75. *Prof Exp:* Proj leader, 70-75, group leader, Phys Measurements Group, 74-75, group leader indust prod res, 75-78, mgr res pigments & extenders, 78-81, DIR RES & DEVELOP PIGMENTS & EXTENDERS, M&C DIV, ENGELHARD CORP, 81- *Mem:* Am Chem Soc; Clay Mineral Soc; Tech Asn Pulp & Paper. *Res:* Mineral based pigments for the paper, paint and other related industries; clay minerals such as kaolinite, attapulgite, talc, to develop new pigments and manufacturing processes. *Mailing Add:* NCH Corp 2730 Carl Rd Irving TX 75062-6453

HUELKE, DONALD FRED, b Chicago, Ill, Aug 20, 30; m 52; c 2. ANATOMY. *Educ:* Univ Ill, BS, 52, MS, 54, PhD(anat), 57. *Prof Exp:* From instr to assoc prof, 54-68, PROF ANAT, UNIV MICH, ANN ARBOR, 68- *Concurrent Pos:* Consult, Mich State Police Crime Lab, Ford Motor Co, Gen Motors Corp & Human Factors Test Devices Subcomt, Soc Automotive Engrs; mem, Nat Motor Vehicle Safety Adv Coun, 72-74. *Honors & Awards:* Automotive Safety Award, Med Tribune, 66. *Mem:* Int Asn Dent Res; Am Asn Anat; Am Asn Automotive Med (pres, 70); Am Trauma Soc; fel Am Acad Forensic Sci. *Res:* Effects of trauma on the human body. *Mailing Add:* Anat & Cell Biol 4643 Med Sci Two Univ Mich Med Sch 1301 Catherine Rd Ann Arbor MI 48109-0616

HUELSMAN, LAWRENCE PAUL, b Chicago, Ill, Jan 22, 26; c 1. ELECTRICAL ENGINEERING. *Educ:* Case Inst Technol, BSEE, 50; Univ Calif, MSEE, 56, PhD(elec eng), 60. *Prof Exp:* Sect chief in chg training opers, Western Elec Co, Inc, 51-56; assoc elec eng, Univ Calif, Berkeley, 56-60; assoc prof, 60-63, PROF ELEC ENG, UNIV ARIZ, 63- *Concurrent Pos:* Vis prof, Rice Univ, 64. *Mem:* Fel Inst Elec & Electronics Engrs; Sigma Xi. *Res:* Active resistor-capacitator circuit theory; linear vector space applications to network theory; computer software applications; computer-aided design; analysis of networks. *Mailing Add:* Dept Elec & Comput Eng Univ Ariz Tucson AZ 85721

HUENEMANN, RUTH L, b Waukon, Iowa, Feb 5, 10. NUTRITION. *Educ:* Univ Wis, BS, 38; Harvard Univ DSc, 53; Am Bd Nutrit, diploma. *Prof Exp:* Teacher pub schs, SDak, 28-35; staff dietitian, clins & hosps, Chicago, 39-40; nutritionist, Zoller Dent Clin, 40-43; assoc prof nutrit, Univ Tenn, 43-53; assoc prof nutrit, 53-67, prof, 67-80, EMER PROF PUB HEALTH NUTRIT, SCH PUB HEALTH, UNIV CALIF, BERKELEY, 80- *Concurrent Pos:* Nutrit consult, Inst Nutrit, Peru, 51-52; consult, Off Int Res, US Agency Int

Develop, 67 & WHO, Geneva, 74. *Mem:* Am Dietetic Asn; Am Home Econ; Am Pub Health Asn; Am Inst Nutrit; Am Bd Nutrit; Soc Nutrit Educ. *Res:* Community service; weight control. *Mailing Add:* Sch Pub Health Univ Calif Berkeley CA 94720

HUENING, WALTER C, JR, b Boston, Mass, Feb 10, 23; m 44, 88. INDUSTRIAL POWER SYSTEMS, ELECTRICAL ENGINEERING. *Educ:* Tufts Univ, BS, 44. *Prof Exp:* Metal rolling elec equip eng, 63-68; consult appln engr, Gen Elec Co, 68-89; RETIRED. *Honors & Awards:* Kaufmann Award, Inst Elec & Electronics Engrs, 88, Indust Applns Soc Achievement Award, 89. *Mem:* Inst Elec & Electronics Engrs. *Res:* Industrial electric power systems; calculation of short circuit currents. *Mailing Add:* 1229 Godfrey Lane Schenectady NY 12309

HUENNEKENS, FRANK MATTHEW, JR, b Galveston, Tex, Feb 16, 23; m 44; c 2. BIOCHEMISTRY. *Educ:* Univ Calif, BS, 43, PhD(chem), 48. *Prof Exp:* Asst prof enzyme chem, Univ Wis, 49-51; from asst prof to prof, Univ Wash, 51-62; mem & chmn dept biochem, 62-84, MEM, DIV BIOCHEM, SCRIPPS CLIN & RES FOUND, 84- *Concurrent Pos:* Williams-Waterman fel, Univ Wis, 48-49; adj prof, depts chem & biol, Univ Calif, San Diego, 67-; Outstanding Investr Grant, Nat Cancer Inst, 85. *Honors & Awards:* Paul Lewis Award, Am Chem Soc, 60. *Mem:* Fel, AAAS, 74; Am Chem Soc; Am Soc Biochem & Molecular Biol; Am Asn Cancer Res. *Res:* Mechanisms of enzyme action and membrane transport; cancer chemotherapy. *Mailing Add:* Div Biochem Dept & Molecular Exp Med Res Inst Scripps Clin 10666 N Torrey Pines Rd La Jolla CA 92037

HUENNEKENS, JOHN PATRICK, b Seattle, Wash, May 21, 52; m 78; c 2. LASER SPECTROSCOPY. *Educ:* Univ Calif, Berkeley, BA, 73, BA, 74; Univ Ill, Champaign-Urbana, MS, 76; Univ Colo, Boulder, PhD(physics), 82. *Prof Exp:* Res assoc, Joint Inst Lab Astrophys, Univ Colo, 82 & Princeton Univ, 82-84; asst prof, 84-89, ASSOC PROF PHYSICS, LEHIGH UNIV, 89- *Honors & Awards:* Presidential Young Investr Award, NSF, 85. *Mem:* Am Phys Soc. *Res:* Atomic collision physics; collisional line-broadening; collisional excitation transfer; radiation trapping in alkali vapors; development of tunable, near-infrared lasers based on excimer-like transitions in alkali molecules. *Mailing Add:* Dept Physics Lehigh Univ Bldg 16 Bethlehem PA 18015

HUERTA, MANUEL ANDRES, b Havana, Cuba, Nov 30, 43; m 67; c 2. STATISTICAL MECHANICS, PLASMA PHYSICS. *Educ:* Calif Inst Technol, Pasadena, BS, 65; Univ Miami, Coral Gables, MS, 67, PhD(physics), 70. *Prof Exp:* Instr physics, Univ Miami, 67-70; assoc res scientist plasma physics, Courant Inst Math Sci, NY Univ, New York, 70-72; from asst prof to assoc prof 72-83, chmn Dept Physics, 80- 85, PROF PHYSICS, UNIV MIAMI, CORAL GABLES, FLA, 83- *Concurrent Pos:* Consult magneto-fluid dynamics div, Courant Inst Math Sci, NY Univ, New York, 73-74; prog assoc, Div Int Progs, NSF, 77-78. *Mem:* Sigma Xi; Am Phys Soc. *Res:* Bifurcation phenomena; detonation waves; plasma physics; elecromagnetic launchers. *Mailing Add:* Dept Physics Univ Miami Coral Gables FL 33124

HUERTAS, JORGE, b Bogota, Colombia, Feb 26, 24; m 58; c 4. NEUROSURGERY, SPACE HUMAN NEUROBIOLOGY. *Educ:* Nat Univ Colombia, MD, 47. *Prof Exp:* Asst prof neurol & dir neurol res lab, Med Ctr, Georgetown Univ, 53-57; prof neurosurg & chmn dept neurol sci, Javeriana Univ, Colombia, 57-63; Ames Res Ctr, 63-64, chief neurobiol br, Environ Biol Div, NASA, 65-69; CHIEF NEUROSURG, KAISER HOSP, SANTA CLARA, 69- *Concurrent Pos:* Consult, Ames Res Ctr, NASA, 69-; lectr, Stanford Univ. *Mem:* Asn Am Med Cols; Am Acad Neurol; Am Asn Neurol Surg; NY Acad Sci; Am Inst Aeronaut & Astronaut; Am Bd Neurol Surgery. *Res:* Space neurobiology; epilepsy. *Mailing Add:* 900 Kiely Blvd Santa Clara CA 95051

HUESSY, HANS ROSENSTOCK, b Frankfurt, Ger, Aug 15, 21; US citizen; m 58; c 11. PSYCHIATRY. *Educ:* Dartmouth Col, BA, 42; Yale Univ, MD, 45; Univ Colo, MS, 51; Am Bd Psychiat & Neurol, dipl, cert psychiat, 52 & child psychiat, 63. *Prof Exp:* Intern pediat, Johns Hopkins Univ, 45-46; resident psychiat, USPHS Hosp, Tex, 47-48; staff psychiatrist, US Med Ctr Fed Prisoners, Springfield, Mo, 48-49; res psychiat, Univ Colo, 49-51, clin instr, 51-53; dir ment health progs, Saratoga, Warren & Washington Counties, NY, 53-58; clin instr, 59-64, from asst prof to assoc prof, 64-69, actg chmn dept, 67-70, PROF PSYCHIAT, UNIV VT, 69- *Concurrent Pos:* Regional psychiat consult, Pub Health Serv, Colo, 51-53; dir aftercare proj, NIMH, 58-67. *Honors & Awards:* Ment Health Sect Award, Am Pub Health Asn, 82. *Mem:* Fel Am Orthopsychiat Asn; fel Am Pub Health Asn; fel Am Psychiat Asn; fel Am Acad Child Psychiat; Am Psychopath Asn. *Res:* Community psychiatry; hyperkinetic syndrome; learning problems; epidemiology of behavior disorders. *Mailing Add:* Dept Psychiat Univ Vt Burlington VT 05405

HUESTIS, DAVID LEE, b St Paul, Minn, Dec 20, 46; m 68. SPECTROSCOPY, EXCITED STATE KINETICS. *Educ:* Macalester Col, Minn, BA, 68; Calif Inst Technol, MS, 69, PhD(chem), 73. *Prof Exp:* Fel, appl physics, Calif Inst Technol, 72; fel physicist, 73, physicist, 73-77, asst prog mgr, 77-80, PROG MGR, MOLECULAR PHYSICS LAB, SRI INT, 80- *Concurrent Pos:* Vis lectr, Dept Chem, Stanford Univ, 78. *Mem:* Am Phys Soc; Int Photochem Soc. *Res:* Theoretical description of molecular excited states and their interactions; scattering theory; electronic structure of solids and surfaces; experimental spectroscopic and kinetics studies of electronically excited states of molecules. *Mailing Add:* Chem Physics Lab SRI Int Menlo Park CA 94025

HUESTIS, DOUGLAS WILLIAM, b London, Ont, Mar 21, 25; US citizen; m 55; c 5. TRANSFUSION MEDICINE. *Educ:* McGill Univ, MD, 48; Am Bd Anat Path, dipl, 55; Am Bd Clin Path, dipl, 60. *Prof Exp:* Intern, Montreal Gen Hosp, 48; res fel exp cytol & biophys, Nobel Inst Cell Res, Karolinska Inst, Sweden, 49-50; sr house officer, Mayday Hosp, Croydon, Eng, 50-51; sr intern path, Montreal Gen Hosp, 51-52; demonstr path & asst prosector, Path Inst, McGill Univ, 52, demonstr path & asst surg path, 53; instr clin path, Ohio State Univ, 54-55; asst dir inst path, Western Pa Hosp, Pittsburgh, 55-60; assoc prof path, Chicago Med Sch, 60-66, prof clin path, 66-69; PROF PATH, COL MED, UNIV ARIZ, 69- *Concurrent Pos:* From clin instr to clin asst prof, Univ Pittsburgh, 55-60; dir blood ctr, Mt Sinai Med Res Found & clin pathologist, Mt Sinai Hosp, Chicago, Ill, 60-69; med dir, Southern Ariz Red Cross Blood Prog, 70-77, United Blood Serv Southern Ariz, 85-87. *Honors & Awards:* John Elliot Award, Am Asn Blood Banks, 75. *Mem:* Am Asn Blood Banks; Am Soc Hemat; Am Soc Clin Path; Int Soc Blood Transfusion. *Res:* Immunohematology, especially antigens and antibodies encountered in blood transfusion; clinical use of blood components; platelet and granulocyte transfusion; histocompatibility testing; therapeutic hemapheresis. *Mailing Add:* Dept Path Univ Ariz Col Med Tucson AZ 85724

HUESTIS, LAURENCE DEAN, b Roseville, Calif, July 19, 34; m 58; c 2. ORGANIC CHEMISTRY, ANALYTICAL CHEMISTRY. *Educ:* Univ Calif, Berkeley, BS, 56, Univ Calif, Davis, PhD(org chem), 60. *Prof Exp:* Res fel, Univ Minn, Minneapolis, 60-61; from asst prof to assoc prof chem, 61-73, chmn dept, 70-73, PROF CHEM, PAC LUTHERAN UNIV, 73- *Concurrent Pos:* Vis prof, Univ Nev, Reno, 82, Univ NMex, 89. *Mem:* Am Chem Soc; Mineral Soc Am. *Res:* Microanalytical techniques for cation analysis; Formaldehyde analysis by mass spectrometry; infrared spectra techniques for identification of minerals. *Mailing Add:* Dept Chem Pac Lutheran Univ Tacoma WA 98447

HUESTIS, STEPHEN PORTER, b Berkeley, Calif, Sep 11, 46; m 77. GEOPHYSICS. *Educ:* Harvey Mudd Col, BS, 68; Univ Calif, San Diego, MS, 69, PhD(earth sci), 76. *Prof Exp:* Lectr geophys, Eng-Geosci Group, Dept Mat Sci & Eng, Univ Calif, Berkeley, 76-77; asst res geophysicist, Inst Geophys & Planetary Physics, Univ Calif, San Diego, 77; asst prof, 77-83, ASSOC PROF GEOPHYS, DEPT GEOL, UNIV NMEX, 83- *Concurrent Pos:* Instr geosci, Geophys Sect, Sandia Nat Labs, 80-86. *Mem:* Am Geophys Union; Royal Astron Soc; Sigma Xi. *Res:* Geophysical inverse theory, with emphasis on problems in the interpretation of heat flow, gravity, and magnetic data. *Mailing Add:* Dept Geol Univ NMex Albuquerque NM 87131

HUETER, FRANCIS GORDON, b Baltimore, Md, June 29, 29; m 54; c 3. ANIMAL TOXICOLOGY. *Educ:* Univ Md, BS, 52, MS, 56, PhD(animal sci), 58. *Prof Exp:* Asst dairy husb, Univ Md, 54-58; asst prof, Ore State Univ, 58-61; chief physiol sect, Lab Med Biol Sci, Div Air Pollution, USPHS, Ohio, 61-67, chief health effects res prog, Nat Ctr Air Pollution Control, 67-70, asst dir div health effects res, Nat Air Pollution Control Admin, NC, 70-71; dir spec studies staff, Health Effects Res Lab, Nat Environ Res Ctr, Environ Protection Agency, 72-76, assoc dir, 76-89; RETIRED. *Mem:* AAAS; Nat Audubon Soc. *Res:* Animal science; biochemical and physiolgoical effects of inhaled toxicants on experimental animals and man. *Mailing Add:* 1227 Seaton Rd F61 Durham NC 27713

HUETER, THEODOR FRIEDRICH, b Baden-Baden, Ger, May 17, 17; nat US; m 49; c 1. ACOUSTICS, MATERIAL SCIENCES. *Educ:* Dresden Tech Univ, BS, 39; Munich Tech Univ, MS, 40, PhD, 48. *Prof Exp:* Supvr med electronics group, Siemens Reiniger Werke, Ger, 39, 47-49; res assoc physics, Mass Inst Technol, 50-56; mgr acoust dept, Submarine Sig Div, Raytheon Mfg Co, 56-59; mgr, Seattle Develop Lab, Honeywell Corp, 59-65, vpres & gen mgr, Marine Systs Div, 65-74, vpres ocean technol, 74-76, vpres technol & dir, Corp Technol Ctr, Minneapolis, 76-80, vpres control systs & tech adv Europ opers, 80-82; CONSULT TECH DEVELOP, 82- *Concurrent Pos:* Dir, Ultrasound Diag Unit, Mass Gen Hosp, 50-54; mem vis comt, Col Eng, Univ Wash; chmn, NMARC, Res & Develop Panel, Dept of Defense, 74-75; mem coun, Inst Technol, Univ Minn & adv bd, Carnegie-Mellon Res Inst, 76-80; mem, vis comt, Phys Dept, Univ Wash, 91- *Mem:* AAAS; fel Acoust Soc Am. *Res:* Wave propagation; signal processing; information and display systems; sonar transducers; ocean systems; ultrasonic attenuation in biological materials. *Mailing Add:* 5606 NE Keswick Dr Seattle WA 98105

HUETHER, CARL ALBERT, b Cincinnati, Ohio, Aug 22, 37; m 59; c 3. HUMAN POPULATION GENETICS. *Educ:* Ohio State Univ, BS, 59; NC State Univ, MS, 63; Univ Calif, Davis, PhD(evolutionary genetics), 66. *Prof Exp:* Asst prof genetics, 66-71, assoc prof biol, 71-81, PROF BIOL, UNIV CINCINNATI, 83- *Concurrent Pos:* Assoc dean hon, Col Arts & Sci, Univ Cincinnati, 74-77; Nat Serv Res fel, Div Human Genetics, Johns Hopkins Hosp, 77-78; mem adv comt, Pop Ref Bur Inc, Washington, DC, 78-87; vis scientist, CDC, Atlanta, 87, Univ BC, Univ Calif, Berkeley & Emory Univ, 89-90. *Mem:* AAAS; Am Soc Human Genetics; Soc Social Biol; Am Asn Univ Professors; Sigma Xi. *Res:* Demographic genetics; epidemiology and estimation of incidence of chromosomal aberrations, specifically Down's syndrome; probability as applied to genetic counseling; human genetics education. *Mailing Add:* Dept Biol Univ Cincinnati Cincinnati OH 45221

HUETTNER, DAVID JOSEPH, b New London, Wis, Feb 20, 38; m 61; c 4. PHYSICAL CHEMISTRY. *Educ:* St Norbert Col, BS, 60; Wash State Univ, PhD(chem), 66. *Prof Exp:* Res assoc chem, Univ Mass, 66-68; staff chemist appl res, IBM Corp, Endicott, NY, 68-75; asst prof & res assoc chem appl res, Inst Paper Chem, 75-78; sr res specialist, Akrosil Div, Hammermill Paper Co, 78-85; mat specialist, Dow Corning Corp, 85-89; SR RES CHEMIST, WACKER SILICONE, 89- *Concurrent Pos:* NSF & PRF grants, Univ Mass, 66-68. *Mem:* Am Chem Soc; Tech Asn Pulp & Paper Indust; Adhesion Soc; Radtech Int. *Res:* Chemistry and surface properties of silicone release coatings; radiation curing of silicone release coatings. *Mailing Add:* 516 Tilton Dr Tecumseh MI 49286-1640

HUEY, RAYMOND BRUNSON, b Bakersfield, Calif, Sept 14, 44. ANIMAL ECOLOGY, ANIMAL PHYSIOLOGICAL ECOLOGY. *Educ:* Univ Calif, Berkeley, AB, 66; Univ Tex, Austin, MA, 69; Harvard Univ, PhD(biol), 75. *Prof Exp:* Asst cur herpet, Mus Vert Zool, Univ Calif, Berkeley, 66-67; res assoc lizard ecol, Field Res, Kalahari, Univ Tex, 69-70; Richmond fel ecol, Harvard Univ, 71-74, res assoc lizard ecol, 74-75; Miller fel, Miller Inst Basic Res Sci, Univ Calif, Berkeley, 75-77; from asst prof to assoc prof, 77-84, assoc chmn dept zool, 88-91, actg chmn, 91-92, PROF, UNIV WASH, 84- *Mem:* Ecol Soc Am; Am Soc Ichthyologists & Herpetologists; Soc Study Evolution; Soc Study Amphibians & Reptiles; Sigma Xi; AAAS; Am Soc Nat. *Res:* Analyzing lizard thermoregulation and locomotion in ecological and evolutionary contexts; studies of lizard competition, lizard systematics, evolution of physiological ecology in lizards and fruitflies, and general vertebrate ecology. *Mailing Add:* Dept Zool NJ-15 Univ Wash Seattle WA 98195

HUEY, WILLIAM S, b Wichita Falls, Tex, Mar 26, 25; m 48. NATURAL RESOURCES MANAGEMENT. *Educ:* NMex State Univ, BS, 52. *Prof Exp:* Wildlife biologist, NMex Dept Game & Fish, 53-62, asst chief game mgt, 63, chief spec serv, 63-64, chief game mgt, 64-69, chief wildlife opers, 70, asst dir, 70-75, dir, 75-78; secy natural resources, NMex Natural Resources Dept, 78-82; RETIRED. *Honors & Awards:* George Allen Mem Award, Int Wild Waterfowl Asn, 62; Winchester Award, Game Conserv Int, 71; Spec Conserv Award, Nat Wildlife Fedn, 82; Oak Leaf Award, Nature Conservancy, 83. *Mem:* Am Forestry Asn; Wildlife Soc; Am Ornith Union; Whooping Crane Conserv Asn (pres, 71-75, 87-88); Int Asn Game, Fish & Conserv Comnrs. *Mailing Add:* PO Box 565 Santa Fe NM 87504

HUFF, ALBERT KEITH, b Burlington, Wash, Apr 16, 42; m 65; c 2. PLANT SCIENCE. *Educ:* Univ Wash, BS, 64; Colo State Univ, PhD(plant physiol), 74. *Prof Exp:* Chemist anal chem, Lab Radiation Biol, Univ Wash, 65-67; res assoc physiol, SDak State Univ, 74-77; ASST PROF PLANT SCI, UNIV ARIZ, 77- *Mem:* Am Soc Plant Physiologists; Am Chem Soc; AAAS; Sigma Xi; Am Inst Biol Sci. *Res:* Regulation of plant growth and development; plastid transformations; flower abortion; chlorophyll metabolism; plant productivity in desert climates; postharvest physiology and redox potentials in plant development. *Mailing Add:* 2925 E Loretta Tucson AZ 85716

HUFF, CHARLES WILLIAM, b Greenville Co, SC; June 6, 20; m 52; c 4. MATHEMATICS. *Educ:* Wofford Col, BS, 46; Univ SC, MS, 49; Univ Ga, PhD(math), 54. *Prof Exp:* Instr math, Pa State Univ, 48-49; instr, Univ SC, 49-50; from asst prof to assoc prof, Auburn Univ, 54-60; chmn dept, 60-68, prof, 60-85, EMER PROF MATH, WINTHROP COL, 85- *Mem:* Am Math Soc; Math Asn Am; Sigma Xi. *Res:* Linear algebra; exponential equation. *Mailing Add:* 113 Merrifield Dr Greenville SC 29615

HUFF, DALE DUANE, b Portland, Ore, Mar 22, 39; m 62; c 2. HYDROLOGY. *Educ:* Stanford Univ, BS, 61, MS, 64, PhD(hydrol), 68. *Prof Exp:* Actg asst prof nuclear hydrol, Stanford Univ, 67-68; from asst prof to assoc prof civil eng, Univ Wis-Madison, 68-74; staff hydrologist, Oak Ridge Nat Lab, 74-84, hydrol group leader, 84-88, sect head, environ eng & hydrol, 88-89, mgr, tech oversight & rev, environ restoration, 90, COORDR, GROUNDWATER PROG, OAK RIDGE NAT LAB, 91- *Concurrent Pos:* Radiochemist, Hazelton Nuclear Sci Corp; assoc, Adtech Consult, 68-73; consult, Oak Ridge Nat Lab, 72-74; mem steering comt, Int Joint Comn, PLUARG, 75-77. *Mem:* Sigma Xi; Am Water Well Asn; Am Geophys Union; Am Inst Hydrol. *Res:* Hydrologic simulation; land water interactions; hydrologic transport of trace materials; shallow-land disposal hydrology. *Mailing Add:* Oak Ridge Nat Lab Bldg 1505 PO Box 2008 Oak Ridge TN 37831-6036

HUFF, DENNIS KARL, b Des Moines, Iowa, Mar 2, 40. MICROBIOLOGY, IMMUNOLOGY. *Educ:* Drake Univ, BA, 61, MA, 63; Mich State Univ, PhD(microbiol), 66. *Prof Exp:* Asst prof biol, Univ Tulsa, 66-70; fel microbiol, Univ Ill Med Ctr, 70-73; MEM FAC, DEPT BIOL SCI, CALIF STATE UNIV, 73- *Mem:* Am Soc Microbiol; Int Soc Comp & Develop Immunol; Sigma Xi. *Res:* Immunity to protozoan infections; role of RNA in immune response. *Mailing Add:* Dept Biol Sci Calif State Univ Sacramento CA 95819

HUFF, GEORGE FRANKLIN, b Pittsburgh, Pa, Nov 4, 23; m 50; c 4. INDUSTRIAL CHEMISTRY. *Educ:* Yale Univ, BS, 44; Carnegie Inst Technol, DSc(chem), 49. *Prof Exp:* Res chemist, US Steel Co, 49-52; res chemist, Callery Chem Co, 52-53, mgr res dept, 53-56, dir res & develop, 56-60, vpres, 60-66; dir chem dept, Gulf Res & Develop Co, 66-68; world-wide coordr chem, Gulf Oil Corp, 68-70; vpres res & develop, Gulf Oil Chem Co, 70-80; PRES, ALT ENERGY ASSOCS INC, 80- *Mem:* Am Inst Chemists; Soc Chem Indust; Am Chem Soc. *Res:* Thermodynamics; thermochemistry. *Mailing Add:* Alt Energy Assocs Inc 155 N Dr Fox Chapel Pittsburgh PA 15238

HUFF, JAMES ELI, b Moscow, Idaho, Apr 7, 28; m; c 5. CHEMICAL ENGINEERING. *Educ:* Univ Idaho, BS, 50; Yale Univ, DEng, 57. *Prof Exp:* Trainee, Dow Chem Co, 55-56, chem engr, Phys Res Lab, 56-63, sr res engr, 63-68, tech expert, 68, process specialist, 68-71, sr process specialist, 71-80, assoc process consult, Mich Div Process Eng, 80-88; PVT CONSULT, 88- *Mem:* Fel Am Inst Chem Engrs; Sigma Xi. *Res:* Kinetics and reactor design; computer applications; phase equilibria; process development, analysis and control; emergency pressure relief systems. *Mailing Add:* 3628 Wood Duck Circle Stockton CA 95207

HUFF, JESSE WILLIAM, b Westmoreland Co, Pa, Dec 8, 16; m 44; c 5. BIOCHEMISTRY. *Educ:* Univ Pittsburgh, BS, 40; Duke Univ, PhD(biochem), 45. *Prof Exp:* Asst biochem, Med Sch, Duke Univ, 40-42, instr, 45-46; res biochemist, Sharp & Dohme Res Labs, 46-49, asst dir biochem res, 49-50, dir, 50-58, biochem, Merck Inst Therapeut Res, 58, dir biochem, Merck Sharp & Dohme Res Labs, 58-66; dir physiol chem, 66-71, asst dir exp path, 71-75, SR INVESTR, MERCK INST THERAPEUT RES, 75- *Concurrent Pos:* Biochemist, Off Sci Res & Develop contract, Duke Univ, 44-46. *Mem:* Am Soc Biol Chem; Am Heart Asn; NY Acad Sci. *Res:* Metabolism of nicotinic acid and of pyridoxine; isolation and identification of unknown biological compounds; nucleic acid and cholesterol metabolism; nutrition of experimental animals; lipid chemistry and metabolism; experimental atherosclerosis; human cardiovascular disease; lipoprotein metabolism. *Mailing Add:* PO Box 324 Fanwood NJ 07023

HUFF, KENNETH O, b Daleville, Ind, Dec 17, 26; m 57; c 4. PETROLEUM GEOLOGY, SUBSURFACE GEOLOGY. *Educ:* Indiana Univ, BS, 56. *Prof Exp:* Lab technician, Core Labs, Inc, 56-58, lab mgr & sales eng, 58-67, Rocky Mountain dist supv, 67-69; consult geologist, numerous energy co, 69-70; PRES, ADVENTURES, INC, 70- *Concurrent Pos:* Geol consult, numerous oil co, 70- *Mem:* Am Inst Prof Geologists; Am Asn Petrol Geologists; Soc Petrol Engrs. *Res:* Rotary drill rig design; tools for more efficient evaluation of subsurface rock samples for geological purposes. *Mailing Add:* 535 Lennox St Casper WY 82601-2144

HUFF, NORMAN THOMAS, b St Joseph, Mo, Jan 6, 40; m 64; c 2. GLASS FORMING, GLASS COMPOSITIONS. *Educ:* Mont State Univ, BS, 61; Carnegie-Mellon Univ, PhD(theoret chem), 64. *Prof Exp:* NSF fel, Indiana Univ, 64-66; res scientist, Owens-Ill Glass Co, 66-67, sr chemist, Owens-Ill, Inc, 67-85, SR SCIENTIST, OWENS CORNING FIBERGLASS, 85- *Concurrent Pos:* Vpres, Ultra Pure Chem, 78-82. *Mem:* Am Chem Soc; Am Ceramic Soc. *Res:* Computerized data retrieval systems; molecular structure of glasses; formulation of container and specialty glasses; glass container forming processes; glass melting reactions; continuous filament fiber glass forming processes. *Mailing Add:* 1684 Bryn Mawr Dr Newark OH 43055-1545

HUFF, THOMAS ALLEN, b Washington, DC, Oct 9, 35; m 59; c 4. INTERNAL MEDICINE. *Educ:* Southwestern at Memphis, AB, 57; Emory Univ, MD, 61. *Prof Exp:* From intern to resident, Grady Mem Hosp, 61-63; fel endocrinol, Med Ctr, Duke Univ, 65-68, assoc med, 68-70; from asst prof to assoc prof endocrinol & metab, 71-80, PROF MED, MED COL GA, 80- *Mem:* Am Diabetes Asn; Endocrine Soc; Am Fedn Clin Res. *Res:* Pharmacology of oral hypoglycemic compounds; physiology of diabetes syndromes associated with insulin resistance. *Mailing Add:* Dept Med Med Col Ga Augusta GA 30902

HUFF, WARREN D, b Omaha, Nebr, Apr 16, 37; m 70; c 1. GEOLOGY. *Educ:* Harvard Univ, AB, 59; Univ Cincinnati, PhD(geol), 63. *Prof Exp:* From asst prof to assoc prof geol, 63-69, PROF GEOL, UNIV CINCINNATI, 69- *Concurrent Pos:* NSF fac fel, 71-72. *Mem:* Geol Soc Am; Asn Int Study Clay; Soc Econ Paleont & Mineral; Clay Minerals Soc. *Res:* Diagenesis of clay minerals and characteristics of fine-grained sediments; stratigraphy and geochemistry of Paleozoic Bentonites. *Mailing Add:* Dept Geol Univ Cincinnati Cincinnati OH 45221-0013

HUFF, WILLIAM J, b Summerland, Miss, Mar 3, 19; m 58; c 1. MICROPALEONTOLOGY, STRATIGRAPHY. *Educ:* Univ Miss, LLB, 47, JD, 68; Miss State Univ, BS, 56; Rice Univ, MA, 57, PhD(geol), 60. *Prof Exp:* Assoc prof geol, Southern Miss Univ, 60-65; consult geologist & atty-at-law, 65-66; asst prof natural sci, Mich State Univ, 66-68; assoc prof geol, Univ SAla, 68-82; RETIRED. *Concurrent Pos:* Atty-at-law, law, Pascagoula, Miss, 75-82. *Mem:* Am Asn Petrol Geol; Soc Econ Paleont & Mineral; Paleont Soc; NY Acad Sci; Sigma Xi; Paleontological Res Inst. *Res:* Taxonomy, paleoecology and stratigraphic distribution of Ostracoda and Foraminifera. *Mailing Add:* 5917 Montfort Rd S Mobile AL 36608

HUFF, WILLIAM NATHAN, b Bryn Mawr, Pa, Dec 30, 12; m 38; c 3. MATHEMATICAL ANALYSIS. *Educ:* Haverford Col, AB, 35; Univ Pa, MA, 37, PhD(math), 47. *Prof Exp:* Asst math, Univ Pa, 37-40; instr, Northwestern Univ, 40-41; instr pvt sch, Pa, 41-43; instr army specialized training prog, Univ Nebr, 43-44; instr, Univ Rochester, 44-46; from asst prof to assoc prof math, Univ Okla, 46-60, chmn dept, 55-61, prof math, 61-82; RETIRED. *Mem:* Am Math Soc; Math Asn Am; Sigma Xi. *Res:* Special functions; differential equations. *Mailing Add:* 1600 Normandie Dr Norman OK 73072-6342

HUFFAKER, CARL BARTON, b Monticello, Ky, Sept 30, 14; m 38; c 4. INSECT ECOLOGY, BIOLOGICAL CONTROL. *Educ:* Univ Tenn, AB, 38, MS, 39; Ohio State Univ, PhD, 42. *Prof Exp:* Asst zool, Ohio State Univ, 40-41; asst entomologist, Univ Del, 41-43; assoc entomologist, Inst Inter-Am Affairs, Colombia, SAm, 43-44, entomologist, W Indies, 44-46; from asst entomologist to assoc entomologist, 46-57, etomologist & prof etom, 63-85, dir Int Ctr Biol Control, 70-85, EMER PROF BIOL, DIV BIOL CONTROL, UNIV CALIF, BERKELEY, 85- *Concurrent Pos:* NIH grants, 57-78; mem sci & cult exchange teams, USSR, 59, 74 & People's Republic of China, 75; Guggenheim fel, 63-64; NSF grants, 66-74; pres, Int Orgn Biol Control, 72-76 & 78-80; consult, UN Environ Prog, 75-; Rockefeller Found resident-scholar, Bellagio, 78. *Honors & Awards:* C W Woodworth Award, Entom Soc Am, 73; Louis E Levy Medal & Jour Premium Award, Franklin Inst, 76. *Mem:* Nat Acad Sci; AAAS; Entom Soc Am (pres, 81); Ecol Soc Am; Entom Soc Can; Japanese Soc Pop Ecol; Sigma Xi. *Res:* Population ecology, biological control, integrated control, natural balance and regulating mechanisms, roles of predators, parasites and environmental conditions in populations dynamics; integrated pest management, including use of experimental models. *Mailing Add:* Dept Entom Univ Calif Berkeley CA 94720

HUFFAKER, JAMES NEAL, b Atlanta, Ga, June 18, 37; m 73; c 4. PHYSICS. *Educ:* Univ Chicago, BA, 55; Emory Univ, MS, 59; Duke Univ, PhD(physics), 63. *Prof Exp:* Res assoc physics, Duke Univ, 62-63; asst prof, Univ Ala, Tuscaloosa, 63-67; ASSOC PROF PHYSICS, UNIV OKLA, 67- *Concurrent Pos:* Res grant, 65-68; fel, Japan Soc Prom Sci, Kobe Univ, 75-76; vis assoc prof, Dartmouth Col, 86-87. *Mem:* Am Phys Soc. *Res:* Theory of weak interactions, especially beta decay and mu capture; theory of nuclear structure; rovibrational states of diatomic molecules/quasiparticle surface states in superconductors. *Mailing Add:* Dept Physics Univ Okla Norman OK 73019

HUFFAKER, RAY C, b Murray, Utah, Dec 6, 29; m 52; c 4. BIOCHEMISTRY, PLANT PHYSIOLOGY. *Educ:* Brigham Young Univ, AB, 55; Univ Calif, Los Angeles, PhD(hort sci), 60. *Prof Exp:* Lab technician plant biochem, Univ Calif, Los Angeles, 56-60; asst agronomist, 60-65, assoc agronomist, 65-70, PROF PLANT BIOCHEM, UNIV CALIF, DAVIS, 70- *Mem:* Am Soc Plant Physiol; Am Soc Agron; Am Fedn Biol Chem; Scand Soc Plant Physiol; Sigma Xi. *Res:* Plant biochemistry; physiology. *Mailing Add:* Plant Growth Lab Univ Calif Davis CA 95616

HUFFINE, COY L(EE), b Knoxville, Tenn, Apr 2, 24; m 51; c 2. CHEMICAL ENGINEERING. *Educ:* Univ Tenn, BS, 45, MS, 47; Columbia Univ, PhD(chem eng), 53. *Prof Exp:* Engr metall & ceramic develop, Aircraft Nuclear Propulsion Dept, Gen Elec Co, 51-59, res ceramist, Res Lab, 59-60; proj mgr mat develop & fabrication, Res & Adv Develop Div, Avco Corp, 60-68; adv engr, Systs Prod Div, IBM Corp, 68-70, mgr prod technol, 71-80, mgr component technol, 81-86; RETIRED. *Mem:* AAAS; Am Inst Chem Engrs; Am Inst Mining, Metall & Petrol Engrs; NY Acad Sci; Nat Inst Ceramic Engrs; Sigma Xi. *Res:* Materials for nuclear reactors and for high temperature applications; powder processing technology; metal hydrides; ablative materials; composite materials; ferrites and magnetic recording components. *Mailing Add:* 2247 Fifth Ave NE Rochester MN 55906-4017

HUFFINE, WAYNE WINFIELD, b Grenville, NMex, July 21, 19; m 43. AGRONOMY. *Educ:* Okla Agr Mech Col, BS, 46, MS, 47; Purdue Univ, PhD(crop prod, soil fertil), 53. *Prof Exp:* Instr, 46-47, asst prof, 47-50, 53-55, assoc prof, 55-60, prof, 60-81, EMER PROF AGRON, OKLA STATE UNIV, 81- *Mem:* fel Am Soc Agron; Int Turfgrass Soc Crop Sci. *Res:* Turfgrass; selection, establishment and maintenance of roadside vegetation. *Mailing Add:* 1502 N Washington Stillwater OK 74075

HUFFINES, WILLIAM DAVIS, b Reidsville, NC, Jan 23, 27; m 57; c 4. MEDICINE, PATHOLOGY. *Educ:* Univ NC, BS, 51, MD, 55; Am Bd Path, dipl & cert anat & clin path, 63. *Prof Exp:* Intern, Osler Med Serv, Johns Hopkins Hosp, 55-56; asst resident path, NC Mem Hosp, 56-57; from instr to assoc prof path, 57-68, assoc dean basic sci, 71-80, PROF PATH, SCH MED, UNIV NC, CHAPEL HILL, 68- *Concurrent Pos:* Resident, Med Div, Oak Ridge Inst Nuclear Studies, 57-58; chief resident, NC Mem Hosp, 58-59; Markle scholar, 60-65. *Mem:* Sigma Xi. *Res:* Structural changes in renal diseases correlated with functional changes. *Mailing Add:* Dept Path Univ NC Sch Med Chapel Hill NC 27514

HUFFINGTON, NORRIS J(ACKSON), JR, b Baltimore, Md, July 23, 21; m 52; c 4. ENGINEERING MECHANICS, STRUCTURAL DYNAMICS. *Educ:* Johns Hopkins Univ, BE, 49, MSE, 51, DE(eng mech), 54. *Prof Exp:* Draftsman & designer, Glenn L Martin Co, 40-46; jr instr mech eng, Johns Hopkins Univ, 49-50, instr, 53-54; staff mem, Opers Res Off, 50-53; assoc prof appl mech, Va Polytech Inst, 54-56, prof eng mech, 56-58; chief dynamics res staff, Martin Co, 59-67; chief struct mech br, 67-73, chief target loading & response br, 73-80, RES MECH ENG, US ARMY BALLISTIC RES LAB, 80-; PROF, JOHNS HOPKINS UNIV, 86- *Concurrent Pos:* Prof, Univ Del, 74-82; fel, Ballistic Res Lab. *Mem:* Am Soc Mech Engrs; Am Acad Mech. *Res:* Structural dynamics; stiffened plate structures; dynamic properties of materials; terminal ballistics. *Mailing Add:* 3101 Rolling Green Dr Churchville MD 21028

HUFFINGTON, ROY MICHAEL, b Tomball, Tex, Oct 4, 17; m 45; c 2. OIL & GAS EXPLORATION & PRODUCTION. *Educ:* Southern Methodist Univ, BS, 38; Harvard Univ, MA, 41, PhD(geol), 42. *Hon Degrees:* LHD, Southern Methodist Univ, 90. *Prof Exp:* Instr geol, Harvard Univ, 42; from field geologist to sr geologist, Explor Geol Div, Humble Oil & Refining Co, 46-56; pres, Roy M Huffington, Inc, 56-83, chmn bd & dir, 58-90; AMBASSADOR, US EMBASSY, VIENNA, AUSTRIA, 90- *Concurrent Pos:* Dir, Independent Petrol Asn Am, 79-80 & Am Petrol Inst, 83-90. *Honors & Awards:* Petrol Indust Award, Am Soc Mech Engrs, 85; Gold Medallion Oil Pioneer Award, 85; Golden Plate Award, Am Acad Achievement, 86; Michel T Halbouty Human Needs Award, Am Asn Petrol Geologists, 91. *Mem:* Fel AAAS; fel Geol Soc Am; Am Asn Petrol Geologists; Am Inst Prof Geologists; Independent Petrol Asn Am. *Res:* International oil and gas exploration, production, engineering activities and gas liquefaction operations; author of articles on stratigraphic, structural and petroleum geology. *Mailing Add:* 1000 Louisiana St Suite 6700 PO Box 4337 Houston TX 77210-4337

HUFFMAN, ALLAN MURRAY, b Kennedy, Sask, Jan 6, 36; US citizen; m 65; c 5. ORGANIC CHEMISTRY. *Educ:* Univ Minn, BA, 59; Carnegie Inst Technol, MS, 63, PhD(org chem), 64. *Prof Exp:* Res chemist, E I du Pont de Nemours & Co, Del, 64-70; group leader res & develop, Am Color & Chem Corp, Inc, Charlotte, NC, 70-80; GROUP LEADER RES & DEVELOP, CROMPTON & KNOWLES CORP, NY, 80- *Mem:* Am Chem Soc; Am Asn Textile Chemists & Colorists. *Mailing Add:* PO Box 341 Reading PA 19603

HUFFMAN, DALE L, b Churchville, Va, July 23, 31; m 56; c 3. MEAT SCIENCE. *Educ:* Cornell Univ, BS, 59; Univ Fla, MS, 60, PhD(meats), 62. *Prof Exp:* Asst meats, Univ Fla, 59-62; meat scientist, Res & Develop Ctr, Swift & Co, 62-63; assoc prof, 63-72, PROF MEAT SCI, AUBURN UNIV, 72- *Concurrent Pos:* Indust fel, Armour & Co, 69-70. *Honors & Awards:* Meat Processing Award, Am Meat Sci Asn, 83, Signal Serv Award, 88; Harry L Rudnick Educr Award, Nat Asn Meat Purveyors, 84. *Mem:* Am Soc Animal Sci; Am Meat Sci Asn (pres-elect, 80-81, pres, 81-82); Inst Food Technol; Royal Soc Health; Sigma Xi. *Res:* Low-fat ground beef; restructured fresh meats; lean beef production. *Mailing Add:* 219 Deer Run Rd Auburn AL 36830

HUFFMAN, DAVID A(LBERT), b Alliance, Ohio, Aug 9, 25. ELECTRICAL ENGINEERING. *Educ:* Ohio State Univ, BEE, 44, MSE, 49; Mass Inst Technol, ScD, 53. *Prof Exp:* Instr elec eng & res assoc, Univ Res Found, Ohio State Univ, 46-50; from asst prof to assoc prof elec eng, Mass Inst Technol, 50-67; PROF INFO SCI, UNIV CALIF, SANTA CRUZ, 67- *Honors & Awards:* Levy Medal, Franklin Inst, 55; W Wallace McDowell Award, Inst Elec & Electronics Engrs, 74. *Mem:* Inst Elec & Electronics Engrs; Asn Comput Mach. *Res:* Switching theory; information theory; picture analysis; signal design. *Mailing Add:* Dept Comput & Info Sci Univ Calif Santa Cruz CA 95064

HUFFMAN, DAVID GEORGE, b Charleston, WVa, May 8, 41; div. PARASITOLOGY. *Educ:* WVa Univ, AB, 68; Marshall Univ, MS, 70; Univ NH, PhD(zool), 73. *Prof Exp:* Asst prof, 73-78, ASSOC PROF BIOL, SOUTHWEST TEX STATE UNIV, 78- *Mem:* Sigma Xi; Am Asn Parasitologists; Am Ornithologists Soc. *Res:* Freshwater parasitology, especially natural history, systematics, morphology and ecology of fish helminths. *Mailing Add:* Dept Biol Southwest Tex State Univ San Marcos TX 78666

HUFFMAN, DONALD MARION, b Pittsburg, Kans, Sept 28, 29; m 53. MYCOLOGY. *Educ:* Kans State Col, Pittsburg, BS, 51; Kans State Univ, MS, 52; Iowa State Univ, PhD(plant path), 55. *Prof Exp:* Asst prof plant path & bot, Kans State Univ, 55-57; from asst prof to assoc prof biol, Cent Col, Iowa, 57-61; NIH res fel mycol, Columbia Univ, 61-63; chmn, 63-76, PROF BIOL, DIV NATURAL SCI, CENT COL, IOWA, 63- *Concurrent Pos:* Univ Iowa vis scientist & lectr, 60-65. *Mem:* AAAS; Mycol Soc Am; Am Inst Biol Sci. *Res:* Higher fungi. *Mailing Add:* Dept Biol Cent Col Pella IA 50219

HUFFMAN, DONALD RAY, b Ft Worth, Tex, June 19, 35; m 63. SOLID STATE PHYSICS, ASTROPHYSICS. *Educ:* Tex A&M Univ, BS, 57; Rice Univ, MA, 59; Univ Calif, Riverside, PhD(physics), 66. *Prof Exp:* Res engr, Res Div, Humble Oil Co, 59-60; instr physics, Pepperdine Col, 60-62; NSF fel, Univ Frankfurt, 66-67; from asst prof to assoc prof, 67-75, PROF PHYSICS, UNIV ARIZ, 75- *Concurrent Pos:* Consult, Europ Space Agency, 75; vis scientist, Max Planck Inst Solid State Res, Stuttgart, WGer, 76. *Mem:* Am Phys Soc. *Res:* Low temperature physics of solids; thermal and optical studies of the antiferromagnetic phase transition; crystal field spectra; laboratory studies of likely interstellar and circumstellar dust particles. *Mailing Add:* Dept Physics Univ Ariz Tucson AZ 85721

HUFFMAN, ERNEST OTTO, b Hickory, NC, Aug 30, 11; m 35; c 3. PHYSICAL CHEMISTRY. *Educ:* Lenoir-Rhyne Col, BS, 32; Univ NC, MS, 34. *Prof Exp:* Control chemist, Tenn Eastman Corp, 35-36; res chemist, Tenn Valley Authority, 36-64, chief, Fundamental Res Br, 64-73, asst dir chem develop, 73-74, dir chem develop, 74-76; RETIRED. *Mem:* Am Chem Soc. *Res:* Fixation of atmospheric nitrogen; synthesis of phosphorous-nitrogen compounds; defluorination of phosphate rock in the molten state; rate of solution of phosphates; kinetics of heterogeneous processes; research administration; fertilizer science and production. *Mailing Add:* 909 N Pine St Florence AL 35630

HUFFMAN, FRED NORMAN, b Catawba Co, NC, Sept 25, 32; m 55; c 4. ENERGY CONVERSION, NUCLEAR ENGINEERING. *Educ:* Lenoir-Rhyne Col, BS, 54; Vanderbilt Univ, MS, 56; Johns Hopkins Univ, PhD(physics), 64. *Prof Exp:* Asst proj eng, Martin Marietta Corp, Md, 57-59, sr res scientist, 64-67, sect mgr biomed eng, 67-84, mgr direct energy conversion dept vpres, res & develop ctr, 74-85; THERMO ELECTRON CORP, WALTHAM, 86- *Concurrent Pos:* Nuclear consult, Gen Instrument Corp, 62-64. *Honors & Awards:* Fourth Hastings Lectr. *Mem:* Am Phys Soc; Inst Elec & Electronics Engrs; Am Ceramics Soc. *Res:* Energy conversion, particularly from nuclear sources; applied physics aspects of thermionic, thermoelectric and artificial heart development; vapor deposition; thermal insulation, heat pipe and high temperature technology. *Mailing Add:* 25 Colonial Rd Sudbury MA 01776

HUFFMAN, GEORGE GARRETT, petroleum geology; deceased, see previous edition for last biography

HUFFMAN, GEORGE WALLEN, b Renville Co, NDak, Dec 7, 21; m 42; c 3. BIOCHEMISTRY, ORGANIC CHEMISTRY. *Educ:* NDak State Univ, BS, 48, MS, 50; Univ Minn, Minneapolis, PhD(biochem), 56. *Prof Exp:* Chemist, Abbott Labs, 50-51; proj leader res, Quaker Oats Co, 56-61, group leader chem res, 61-64, cent dist sales mgr, 64-68, from com develop assoc to sr com develop assoc, Chem Div, 68-75, proj mgr bus develop, 75-77, mgr mkt develop, 77-84; RETIRED. *Mem:* AAAS; Am Chem Soc; Com Develop Asn. *Res:* Commercial development of organic chemicals; furan chemicals; carbohydrates and proteins. *Mailing Add:* 6502 Vermont Trail Crystal Lake IL 60012-3257

HUFFMAN, GERALD P, b Steubenville, Ohio, Sept 12, 38; m 61; c 3. SPECTROSCOPY & SPECTROMETRY, SOLID STATE PHYSICS. *Educ:* West Liberty State Col, BS, 60; Univ WVa, MS, 62, PhD(physics), 65. *Prof Exp:* Scientist, US Steel Corp Res Ctr, 65-75, assoc res consult, E C Bain Lab Fundamental Res, 75-80, res consult, 80-85; pres, Macroatom, Inc, 85-86; DIR, CONSORTIUM FOSSIL FUEL LIQUEFACTION SCI & RES PROF, UNIV KY, 86- *Mem:* Am Phys Soc; Am Chem Soc; Metall Soc; AAAS. *Res:* Coal structure; mineralogy; liquefaction; ash behavior; catalysis; XAFS spectroscopy; Mossbauer spectroscopy; electron microscopy. *Mailing Add:* Consortium Fossil Fuel Liquefaction Sci Univ Ky 233 Mining & Min Res Bldg Lexington KY 40506-0107

HUFFMAN, JACOB BRAINARD, b Maggie, Va, July 21, 19; m 48; c 2. FORESTRY, WOOD SCIENCE TECHNOLOGY. *Educ:* Va Polytech Inst, BS, 41; Duke Univ, MF, 47, DF(wood technol), 53. *Prof Exp:* Wood technologist, Tidewater Plywood Co, Ga, 47-49; instr, Sch Forestry, WVa Univ, 49-50; PROF FOREST PROD, SCH FOREST RESOURCES & CONSERV, UNIV FLA, 52- *Concurrent Pos:* Consult, 53- *Mem:* Forest Prod Res Soc; Soc Wood Sci & Technol; Soc Am Foresters; Am Wood Preservers' Asn; Hardwood Res Coun. *Res:* Wood properties, seasoning, gluing and preservation; biomass energy. *Mailing Add:* 528 NW 36th St Gainesville FL 32607

HUFFMAN, JOHN CURTIS, b Kokomo, Ind, Dec 9, 41; m 65; c 2. X-RAY CRYSTALLOGRAPHY. *Educ:* Ind Univ, BS, 64, MS, 68 & PhD(chem), 74. *Prof Exp:* Crystallographer, 74-80, SR SCIENTIST & DIR, IND UNIV MOLECULAR STRUCT CTR, 80- *Mem:* Am Crystallog Asn; Am Chem Soc; Sigma Xi; AAAS; Am Inst Physics. *Res:* Single crystal x-ray diffraction of small molecules; development of computers and techniques in crystallography; author of over 400 papers in field of structural chemistry. *Mailing Add:* Dept Chem Ind Univ Bloomington IN 47405

HUFFMAN, JOHN WILLIAM, obstetrics & gynecology; deceased, see previous edition for last biography

HUFFMAN, JOHN WILLIAM, JR, b Evanston, Ill, July 21, 32; m 54, 75; c 4. ORGANIC CHEMISTRY. *Educ:* Northwestern Univ, BS, 54; Harvard Univ, AM, 56, PhD(chem), 57. *Prof Exp:* Asst prof chem, Ga Inst Technol, 57-60; from asst prof to assoc prof, 60-67, PROF CHEM, CLEMSON UNIV, 67- *Concurrent Pos:* NIH career develop award, 65-70. *Mem:* Am Chem Soc. *Res:* Structure and synthesis of natural products; stereochemistry of carbocyclic systems. *Mailing Add:* Dept Chem Clemson Univ Clemson SC 29634-1905

HUFFMAN, K(ENNETH) ROBERT, b Akron, Ohio, Nov 13, 33; m 57; c 3. INFORMATION RETRIEVAL & CONSULTING. *Educ:* Oberlin Col, AB, 55; Univ Rochester, PhD(org chem), 59. *Prof Exp:* Res chemist, Am Cyanamid Co, 58-65, sr res chemist, 65-86; INFO CONSULT, 86- *Honors & Awards:* IR-100, Indust Res, Inc, 71. *Mem:* Am Chem Soc. *Res:* Synthetic organic chemistry; heterocyclic chemistry; organic photochemistry; polymer synthesis; bioabsorbable polymers; water soluble polymers; polymeric surfactants. *Mailing Add:* 24 Lolly Lane Stamford CT 06903

HUFFMAN, LOUIE CLARENCE, b Dundee, Tex, Dec 29, 26; m 54; c 3. MATHEMATICS. *Educ:* Midwestern Univ, BS, 52, ME, 54; Univ Tex, Austin, MA, 62, PhD(math), 69. *Prof Exp:* From instr to assoc prof math, 55-68, PROF MATH, MIDWESTERN STATE UNIV, 68-, CHMN DEPT, 69-, DIR DIV MATH SCI, 83- *Res:* Generalized discrete functions of more than one variable. *Mailing Add:* Div Math Sci Midwestern State Univ Wichita Falls TX 76308

HUFFMAN, ROBERT EUGENE, b Breckenridge, Tex, Apr 23, 31; m 56; c 3. PHYSICAL CHEMISTRY. *Educ:* Tex A&M Univ, BS, 53; Calif Inst Technol, PhD(chem, physics), 58. *Prof Exp:* Res chemist, Air Force Cambridge Res Labs, 58-67, supvry res chemist, 67-74, supvry res chemist, 75-86, chief, Ultraviolet Br, Air Force Geophys Lab, 86-89, SUPVRY PHYSICAL SCIENTIST, GEOPHYS DIRECTORATE, PHILLIPS LAB, HANSCOMB AFB, BEDFORD, MA, 89- *Concurrent Pos:* Dept Phys, Imp Col London, 67-68. *Mem:* Am Chem Soc; Am Phys Soc; Optical Soc Am; Am Geophys Union; Am Inst Aeronaut & Astronaut; Soc Photo-Optical Instrumentation Engrs. *Res:* Atmospheric ultraviolet radiation; vacuum ultraviolet spectroscopy; atmospheric chemistry and physics. *Mailing Add:* Phillips Lab Ionospheric Effects Div Hanscom AFB Bedford MA 01731

HUFFMAN, ROBERT WESLY, b North Lewisburg, Ohio, June 6, 32; m 56; c 2. ORGANIC CHEMISTRY. *Educ:* Ohio Univ, BS, 59; Ind Univ, Bloomington, PhD(org chem), 64. *Prof Exp:* Fel, directed by Prof Thomas C Bruice, Univ Calif, Santa Barbara, 64-67; asst prof, 67-74, ASSOC PROF CHEM, NORTHERN ARIZ UNIV, 74- *Mem:* Am Chem Soc; Sigma Xi. *Res:* Bio-organic reaction mechanisms such as mechanism of destruction of the amide bond in proteins and effects charge-transfer complexes on chemical reactions. *Mailing Add:* Northern Ariz Univ Box 5698 Flagstaff AZ 86011

HUFFMAN, RONALD DEAN, b Vandergrift, Pa, Dec 13, 37. NEUROPHARMACOLOGY, NEUROPHYSIOLOGY. *Educ:* Pa State Univ, BS, 59; Purdue Univ, MS, 61, PhD(pharmacol), 67. *Prof Exp:* Res assoc neurophysiol, Southwest Found Res & Educ, 66-67; instr physiol, Univ BC, 67-68; asst prof pharmacol & anat, Univ Tex Med Sch San Antonio, 68-73; ASSOC PROF PHARMACOL, UNIV TEX HEALTH SCI CTR SAN ANTONIO, 73- *Concurrent Pos:* Morrison Trust res support grants, San Antonio, 69-72 & 76-78; USPHS grants, 70-72, 74-76, 77-81 & 82-84; NSF grant 78-80. *Mem:* Soc Neurosci; Am Soc Pharmacol Exp Ther. *Res:* Physiology and pharmacology of synaptic transmission in the basal ganglia. *Mailing Add:* Dept Pharmacol Univ Tex Health Sci Ctr San Antonio TX 78284-7764

HUFFMAN, TOMMIE RAY, b Zincville, Okla, Mar 22, 29; m 53; c 4. ELECTRICAL ENGINEERING, SEMICONDUCTORS. *Educ:* Univ Denver, BS, 57. *Prof Exp:* Jr engr solid state diffusion, Radio Corp Am, 57-59; proj leader, Semiconductor Prod Sect, Motorola, Inc, 59-81, tech mgr, High Frequency & Optical Prod Div, 81-85, staff scientist, Adv Technol Discrete Group, 85-89; RETIRED. *Concurrent Pos:* Mem sci adv bd, Motorola, Inc, 72; Dan Noble fel. *Mem:* Int Environ Sci. *Res:* Pioneering work in solid state diffusion in germanium, silicon, gallium arsenide and cadmium sulfide; contamination control and electrostatic discharge (EDS) phenomenon. *Mailing Add:* 118 E Cairo Dr Tempe AZ 85282

HUFFORD, CHARLES DAVID, b Tiffin, Ohio, Oct 16, 44; m 68; c 2. PHARMACY, PHARMACOGNOSY. *Educ:* Ohio State Univ, BSPharm, 67, PhD(natural prod chem), 72. *Prof Exp:* Asst prof, 72-75, PROF PHARMACOG, SCH PHARM, UNIV MISS, 75-, CHMN, 87- *Concurrent Pos:* Nat Cancer Inst Fel, 74- *Mem:* Sigma Xi; Am Soc Pharmacog; Am Asn Pharmaceut Scientists; Am Chem Soc; Am Soc Microbiol; Am Asn Cols Pharm. *Res:* Natural products chemistry of bioactive constituents. *Mailing Add:* Dept Pharmacog Univ Miss Sch Pharm University MS 38677

HUFFORD, GEORGE (ALLEN), b San Francisco, Calif, June 1, 27; m 54; c 3. MATHEMATICS, ELECTROMAGNETISM. *Educ:* Calif Inst Technol, BS, 46; Univ Wash, MS, 48; Princeton Univ, PhD(math), 53. *Prof Exp:* Electronic scientist, Nat Bur Standards, 48-50; instr math, Princeton Univ, 53-55; instr, Stanford Univ, 55-58; asst prof, Univ Wash, 58-64; mathematician, Nat Bur Standards, 64-65; mathematician, Environ Sci Serv Admin, 65-70, Off Telecommun, 70-78, MATHEMATICIAN, NAT TELECOMMUN & INFO ADMIN, INST TELECOMMUN SCI, 78- *Mem:* Am Math Soc; Soc Indust & Appl Math; Inst Elec & Electronics Eng; Sigma Xi. *Res:* Radiowave propagation at VHF, UHF and millimeter wave frequencies. *Mailing Add:* Nat Telecommun & Info Admin Inst Telecommun Sci 325 Broadway Boulder CO 80303

HUFFORD, TERRY LEE, b Toledo, Ohio, Sept 19, 35; m 55; c 3. AQUATIC ECOLOGY. *Educ:* Bowling Green State Univ, BS, 61, MA, 62; Ohio State Univ, PhD(bot), 72. *Prof Exp:* Instr biol, Ind Cent Col, 62-63; asst prof biol, Cent Methodist Col, 63-64; asst prof biol, Grove City Col, 66-69; ASSOC PROF BOT, GEORGE WASHINGTON UNIV, 72- *Concurrent Pos:* Pres, Ctr Environ Educ & Res, 78-; chair, Four-Yr Col Sect, Nat Asn Biol Teachers, 90-91. *Mem:* Phycological Soc Am; Nat Asn Biol Teachers; Nat Sci Teachers Asn; Bot Soc Am; Am Inst Biol Sci; Am Micros Soc. *Res:* Taxonomy, morphology and anatomy of freshwater diatoms. *Mailing Add:* 7147 Parkview Ave Falls Church VA 22042

HUFHAM, JAMES BIRK, b Petersburg, Va, Jan 25, 35; m 57; c 2. BACTERIOLOGY, GENETICS. *Educ:* Univ Ala, BS, 57; Univ Nebr, Lincoln, MS, 59, PhD(microbiol), 68. *Prof Exp:* Res assoc microbiol, Sch Med, Wayne State Univ, 61-67; head res, Space-Defense Corp, 67-68; res microbiologist, Esso Res Lab, 68-69; asst prof, 69-77, ASSOC PROF LIFE SCI, UNIV MO, ROLLA, 78- *Concurrent Pos:* Consult, Space-Defense Corp, 66-67, Esso Res Lab, 69-70 & Anschustz Mining Corp, 79-81. *Mem:* Am Soc Microbiol; NY Acad Sci; Sigma Xi. *Res:* Carcinogenicity testing of chemicals using Ames Test; molecular genetics. *Mailing Add:* Dept Life Sci Univ Mo Box 249 Rolla MO 65401

HUFNAGEL, CHARLES ANTHONY, surgery; deceased, see previous edition for last biography

HUFNAGEL, LINDA ANN, b Teaneck, NJ, Nov 7, 39; m 84; c 2. DEVELOPMENTAL BIOLOGY OF CILIATED PROTISTAN CELLS, ULTRASTRUCTURAL CORRELATES OF BEHAVIOR. *Educ:* Univ Vt, BA, 61, MS, 63; Univ Pa, PhD(biol), 67. *Prof Exp:* Fel biol sci, Yale Univ, 67-69; res assoc biol, Columbia Univ, 69-70; res assoc path, Sch Med, Wayne State Univ, 71-73; lectr electron micros, 73-75, asst prof, 75-79, assoc prof, 79-86, PROF MICROBIOL, UNIV RI, 86- *Concurrent Pos:* NSF fel, Yale Univ, 67-69; NSF res grant, Univ RI, 76-78; Steps fel, Marine Biol Lab, Woods Hole, 78 & 79. *Mem:* AAAS; Am Soc Cell Biol; Soc Protozoologists; Electron Micros Soc Am; AAAS. *Res:* Morphogenesis and ciliogenesis in ciliated cells; structure and function of ciliary basal bodies; positioning of organelles in cells; membrane differentiation; mating interactions in protozoa; ultrastructure of epithelial conducting systems of Hydra. *Mailing Add:* Dept Microbiol Univ RI Kingston RI 02881

HUFNAGEL, ROBERT ERNEST, b Elizabeth, NJ, May 24, 32; m 57; c 3. OPTICS, ENGINEERING PHYSICS. *Educ:* Cornell Univ, BEngPhys, 55, PhD(eng physics), 59. *Prof Exp:* Sr staff physicist, Electro Optical Div, 59-61, mgr res br, 61-64, asst dir res, 64-66, assoc dir res, 66-67, dir res, 67-76, tech dir, 76-85, CHIEF SCIENTIST, PERKIN-ELMER CORP, 85- *Concurrent Pos:* Consult, Jason Comt, Inst Defense Anal, 63; assoc ed, J Optical Soc Am, 70-81. *Mem:* Inst Elec & Electronics Engrs; fel Optical Soc Am. *Res:* Atmospheric optics; optical and statistical systems analysis; analog and digital computers; communications theory; visual perception. *Mailing Add:* The Perkin-Elmer Group Mail Sta 882 100 Wooster Mts Rd Danbury CT 06810

HUFSCHMIDT, MAYNARD MICHAEL, b Catawba, Wis, Sept 28, 12; m 41; c 2. WATER RESOURCES MANAGEMENT, PUBLIC INVESTMENT THEORY. *Educ:* Univ Ill, BS, 39; Harvard Univ, MPA, 55, DPA, 64. *Prof Exp:* Planner, Ill State Planning Comn, 39-41; engr, US Nat Resources Planning Bd, 41-43; budget examr, US Bur Budget, 43-49; prog staff mem, US Dept Interior, 49-55; res assoc, Grad Sch Pub Admin, Harvard Univ, 55-65; prof planning & environ sci, Univ NC, Chapel Hill, 65-79; FEL & CONSULT, ENVIRON POLICY INST, EAST-WEST CTR, 79- *Concurrent Pos:* Consult, Resources for Future Inc, 55-56, 72-74, US Coun Econ Adv, 65-67, WHO, 70-71 & 76-78; chmn, panel, US Bur Budget, 61, & study group, Div Water Sci, UNESCO, 85-; adv, Nat Acad Sci, 67, 69 & 70; sr res fel, NSF, 70-71; actg dir, Environ Policy Inst, East-West Ctr, 85-86. *Mem:* Int Water Resource Asn. *Res:* Problems of water and related natural resources management in developing countries; the integration of bio-physical, social, economic, and institutional aspects of natural resource management. *Mailing Add:* 2525 Date St Apt 3001 Honolulu HI 96826

HUFSTEDLER, ROBERT SLOAN, b Idalou, Tex, July 20, 24; m 50; c 3. INORGANIC CHEMISTRY. *Educ:* Tex Tech Col, BS, 46, MS, 49; Colo Sch Mines, DSc(petrol refining eng), 58. *Prof Exp:* Instr chem, Tex Tech Col, 49-51; instr, Colo Sch Mines, 51-58, asst prof, 58-59; prof, Old Dominion Col, 59-68, chmn div natural sci, 61-63; PROF CHEM, VA WESLEYAN COL, 68-, CHMN DIV NATURAL SCI & MATH, 70- *Mem:* Am Chem Soc. *Res:* Transition-metal chemistry; organic chemistry; chemical engineering; petroleum refining. *Mailing Add:* General Delivery Lovettsville VA 22951-9999

HUFT, MICHAEL JOHN, b Highland Park, Mich, Jan 15, 49; m 72, 85; c 4. SYSTEMATICS, FLORISTICS. *Educ:* Univ Notre Dame, BS, 71; Univ Mich, PhD(bot), 79. *Prof Exp:* Res assoc & actg cur herbarium bot, dept bot & microbiol, Univ Okla, 80; curatorial trainee, 80-81, res botanist, 81-83, ASST CUR BOT, MO BOT GARDEN, 84- *Concurrent Pos:* Asst vis cur bot, dept bot, Field Mus Natural Hist, 81 - *Mem:* Am Soc Plant Taxonomists; Soc Econ Bot; Int Asn Plant Taxon. *Res:* Systematics of the Euphorbiaceae; floristics of the neotropics, especially Mexico and Central America. *Mailing Add:* Dept Bot Field Mus Natural Hist Roosevelt Rd at Lake Shore Dr Chicago IL 60605

HUG, CARL CASIMIR, JR, b Canton, Ohio, Dec 20, 36; m 56; c 5. PHARMACOLOGY, ANESTHESIOLOGY. *Educ:* Duquesne Univ, BS, 58; Univ Mich, PhD(pharmacol), 63, MD, 67; Am Bd Anesthesiol, dipl, 75. *Prof Exp:* From instr to assoc prof pharmacol, Med & Grad Schs, Univ Mich, Ann Arbor, 63-71; assoc prof pharmacol, 72-80, assoc prof anethesiol, 74-78, PROF ANESTHESIOL, SCH MED, EMORY UNIV, 78-, PROF PHARMACOL, 80- *Concurrent Pos:* Spec fel anesthesiol, Emory Univ, 71-73; ed, Anesthesiol, 79-88; dir, Am Bd Anesthesiol, 84- *Mem:* Am Soc Pharmacol & Exp Therapeut; Am Soc Clin Pharmacol & Therapeut; Am Soc

Anesthesiol; Int Anesthesia Res Soc; Soc Cardiovasc Anesthesiologists; Asn Univ Anesthesiologists (pres, 84-86). *Res:* Academic medicine, particularly pharmacology and anesthesiology; relationships between the biological disposition of drugs and their actions; opioids and drugs used in anesthesia and life support during cardiac and thoracic surgery and in intensive care. *Mailing Add:* Dept Anesthesiol Emory Univ Sch Med Atlanta GA 30322

HUG, DANIEL HARTZ, b Davenport, Iowa, Apr 9, 27; m 50; c 6. PHYSIOLOGICAL BACTERIOLOGY, PHOTOBIOLOGY. *Educ:* Iowa State Col, BS, 49, PhD(bact physiol), 56. *Prof Exp:* NIH fel, Univ Ill, 56-58; RES MICROBIOLOGIST, VET ADMIN HOSP, 58- *Concurrent Pos:* Res scientist internal med, Univ Iowa. *Mem:* Am Soc Biol Chemists; Am Soc Photobiol; Am Soc Microbiol; Am Chem Soc; Sigma Xi. *Res:* Enzymology; amino acid metabolism; histidine catabolism; photoactivation of urocanase. *Mailing Add:* 305 Fifth St Coralville IA 52240

HUG, GEORGE, b Zurich, Switz, June 2, 31; US citizen; m 63; c 5. PEDIATRICS, BIOCHEMISTRY. *Educ:* Univ Zurich, MD, 57; Am Bd Pediat, dipl, 64. *Prof Exp:* Asst pediat, Univ Children's Hosp, Bern, Switz, 57-58; intern pediat path, Children's Hosp Med Ctr, Boston, Mass, 60-61; resident, 61-63, from asst prof to assoc prof, 63-71, PROF PEDIAT, COL MED, UNIV CINCINNATI, 71- *Concurrent Pos:* Swiss Nat Found & Eli Lilly & Co fels biochem, Sch Med, Wash Univ, 58-60; attend pediatrician, Children's Hosp. *Mem:* Am Soc Clin Invest; Soc Pediat Res; Am Pediat Soc; Int Acad Path; Electron Micros Soc. *Res:* Inborn errors of metabolism; biochemical genetics. *Mailing Add:* Childrens Hosp Res Found Elland & Bethesda Ave Cincinnati OH 45229

HUG, VERENA, b Hinterwur, Switz, Nov 3, 41. ONCOLOGY. *Educ:* Univ Bern, Switz, BA, 66, MD, 72. *Prof Exp:* Instruct, 78-79, ASST PROF, M D ANDERSON CANCER CTR, 79- *Mem:* Am Asn Cancer Res; Am Soc Clin Oncol; Am Col Physicians; NY Acad Sci. *Res:* Tissue-specific hormones; breast tumors. *Mailing Add:* 8100 Cambridge Houston TX 77059

HUGELMAN, RODNEY D(ALE), b Bismarck, NDak, Aug 16, 34; m 55. AERONAUTICAL ENGINEERING. *Educ:* Ore State Univ, BS, 56, MS, 59; Okla State Univ, PhD(eng), 64. *Prof Exp:* Aerodynamicist, Gen Dynamics/Convair, 58; res scientist appl math, Aerospace Res Lab, Wright-Patterson AFB, 63-68; dir aerospace & eng mech, NDak State Univ, 68-72; exec consult res & develop, General Prog Corp, 72-74; dir res & develop, Warren Indust, 74-77; PROF GEN ENG, UNIV ILL, URBANA, 77- *Mem:* Am Inst Aeronaut & Astronaut. *Res:* Aerodynamics and applied mathematics, particularly fluidics, boundary layer flows and aircraft design; industrial controls; automation and robotics both fluidic and microprocessor controls. *Mailing Add:* Univ Ill 104 S Mathews Ave Rm 307 Transp Urbana IL 61801

HUGER, FRANCIS P, b Lexington, Va, Oct 25, 47; m 69. BIOCHEMICAL PHARMACOLOGY, NEUROSCIENCE. *Educ:* Va Military Inst, BS, 69; Med Col Va, PhD(pharmacol), 78. *Prof Exp:* Anal chemist, Va Dept Health, 69-70, Va Consol Lab, 72-74; grad student, Dept Pharmacol, Med Col Va, 74-78; res assoc, Uniformed Servs Univ, 79-80; SR RES BIOCHEMIST, HOECHST ROUSSEL PHARMACEUT, INC, 80- *Mem:* Soc Neurosci; Am Chem Soc. *Res:* Biochemical pharmacology and development of structure-activity relationships; regulation of synthesis and turnover of central neurotransmitters and associated receptor changes. *Mailing Add:* Hoechst Roussel Pharmaceuts Inc Dept Biochem Rte 202-206 N Somerville NJ 08876

HUGGANS, JAMES LEE, b Kahoka, Mo, Dec 30, 33; m 54; c 3. ENTOMOLOGY. *Educ:* Univ Mo, Columbia, BSF, 59, MS, 61; Iowa State Univ, PhD(entom), 68. *Prof Exp:* Res entomologist, USDA, 61-68; asst prof, 69-80, ASSOC PROF ENTOM & EXTEN ENTOMOLOGIST, UNIV MO, COLUMBIA, 80- *Mem:* Entom Soc Am. *Res:* Biology and biological control of alfalfa weevil; host plant resistance to European corn borer; parasites and predators of grasshoppers; field testing control measures on insect outbreaks. *Mailing Add:* 4551 E Hwy 163 Columbia MO 65201

HUGGETT, CLAYTON (MCKENNA), b Columbia Co, Wis, Mar 12, 17; m 46; c 3. FIRE SAFETY, COMBUSTION. *Educ:* Univ Wis, BS, 38; Univ Minn, MS, 43, PhD(phys chem), 45. *Prof Exp:* Res assoc, US Navy Proj, Univ Minn, 45-48; res chemist, Rohm & Haas Co, 48-49, resident dir res, Redstone Arsenal Res Div, 50-54, head high pressure lab, 54-59; dir res, Amcel Propulsion, Inc, 59-62; sr scientist, Atlantic Res Corp, 62-66, dir chem technol dept, 66-70; chief prog chem, Nat Bur Standards, 70-77, chief off extramural fire res, 77-80, dep dir, Ctr Fire Res, 80-85; RETIRED. *Mem:* Am Chem Soc; Combustion Inst. *Res:* Chemistry of propellants, explosives and incendiaries; polymer and acetylene chemistry; high pressure chemical processes; combustion hazards; spacecraft atmospheres; chemistry of combustion processes relating to fire safety. *Mailing Add:* 8301 Private Lane Annandale VA 22003

HUGGETT, RICHARD WILLIAM, JR, b Frazee, Minn, July 2, 30; div; c 4. PHYSICS. *Educ:* Concordia Col, AB, 51; Ind Univ, MS, 53, PhD(physics), 57. *Prof Exp:* Asst physics, Ind Univ, 51-57; from asst prof to assoc prof, 57-61, PROF PHYSICS, LA STATE UNIV, BATON ROUGE, 65- *Concurrent Pos:* Vis scientist, Max Planck Inst Extraterrestrial Physics, 69-70. *Mem:* Am Asn Physics Teachers. *Res:* Experimental nuclear physics; cosmic ray physics. *Mailing Add:* 10728 Leigh Ellen Ave Baton Rouge LA 70810

HUGGETT, ROBERT JAMES, b Newport News, Va, Apr 26, 42; m 63; c 2. MARINE GEOCHEMISTRY, ENVIRONMENTAL SCIENCES. *Educ:* Scripps Inst Oceanog, MS, 68; Va Inst Marine Sci, PhD(marine chem), 77. *Prof Exp:* Res technician chem, Dow Chem Co, 60-63; asst marine scientist, 68-71, sr marine scientist & chmn dept ecol pollution, 72-79, from asst prof to assoc prof, 77-81, chmn dept chem oceanog, 79- 86, PROF, VA INST MARINE SCI, 85- ASST DIR, 86- *Concurrent Pos:* Co-chmn permanent chem task force, Shellfish Sanit Br, US Food & Drug Admin, 73-77; mem eval panel, Off Air & Water Measurements, Nat Bur Standards, 74-77; lectr, Brookings Inst. *Honors & Awards:* Horsley Award Fundamental Res, Va Acad Sci. *Mem:* Soc Environ Toxicol & Chem; Am Chem Soc. *Res:* Biological availability of trace metals and organics in estuarine systems; toxic substances in aqueous systems; fete and effects of anthropogenic substances in agueous systems. *Mailing Add:* Sch Marine Sci Col Willaim & Mary Gloucester Point VA 23062

HUGGHINS, ERNEST JAY, b Bryan, Tex, Dec 25, 20; m 52; c 3. ANIMAL PARASITOLOGY. *Educ:* Baylor Univ, BA, 43; Tex A&M Univ, MS, 49; Univ Ill, PhD(zool), 52. *Prof Exp:* Instr microbiol, Univ Houston, 46-47; asst biol, Tex A&M Univ, 47-49; asst zool, Univ Ill, 49-51; from asst prof to prof entom & zool, 52-80, zoologist, Agr Exp Sta, 54-85, prof & head, dept biol, 81-85, EMER PROF BIOL, SDAK STATE UNIV, 85- *Concurrent Pos:* Consult, Off Naval Res, First Int Cong Parasitol, Rome, 64; Fulbright vis prof, Nat Univ Villarreal, Lima, Peru, 67; vis prof, Univ Okla Biol Sta, 60 & Black Hills Natural Sci Field Sta, 72-73; fel, Interam Prog, La State Univ, 63; partic, NATO Advan Study Inst on Animal Learning, WGer, 76; sabbatical leave, Kenya, Africa, 81. *Mem:* Fel AAAS; Am Soc Parasitol; Am Micros Soc; Am Soc Zool; Am Soc Mammal. *Res:* Wildlife parasitology; zoogeographical relationships of South American fish parasites; parasites of dairy cattle. *Mailing Add:* Agr Hall Box 2207 Brookings SD 57007

HUGGINS, CHARLES BRENTON, b Halifax, NS, Sept 22, 01; nat US; c 2. CANCER. *Educ:* Acadia Univ, BA, 20; Harvard Univ, MD, 24; Yale Univ, MSc, 47. *Hon Degrees:* Var degrees from colleges and univ. *Prof Exp:* Instr surg, Univ Mich, 26-27; from asst prof to prof, 28-36, dir, Ben May Lab Cancer Res, 51-69, WILLIAM B OGDEN DISTINGUISHED SERV PROF SURG, UNIV CHICAGO, 62- *Concurrent Pos:* Fel surg, Univ Chicago, 27-28; mem, Nat Adv Cancer Coun, USPHS, 46-48; Walker Prize, Royal Col Surgeons, 55-60; Mickle fel, Univ Toronto, 58; chancellor, Acadia Univ, 72-79. *Honors & Awards:* Nobel Prize in Physiol & Med, 66; Sheen Award, Am Med Asn, 60; Oscar B Hunter Award, Am Therapeut Soc & Valentine Prize, NY Acad Med, 62; Lasker Award, 63; Gold Medal, Rudolf Virchow Med Soc, 64; Passano Found Award, 65; Gold Medal, Worshipful Soc, Apothecaries London, 66; Gairdner Award, 66; Bigelow Medal, 67; Harvey Lectr, 46. *Mem:* Nat Acad Sci; Am Physiol Soc; Am Asn Cancer Res (pres, 48); Am Surg Asn; fel Am Acad Arts & Sci; Can Med Asn; Am Philos Soc; hon fel Royal Col Surgeons. *Res:* Calcium metabolism; sex hormones; experimental surgery; enzymes of blood; bone physiology. *Mailing Add:* Dept Surg Univ Chicago 5841 S Maryland Chicago IL 60637

HUGGINS, CHARLES EDWARD, surgery, cryobiology; deceased, see previous edition for last biography

HUGGINS, CHARLES MARION, physical chemistry, for more information see previous edition

HUGGINS, CLYDE GRIFFIN, b Watertown, Tenn, June 21, 22; m 44; c 3. BIOCHEMISTRY. *Educ:* Mid Tenn State Col, BS, 45; Univ Miss, BS, 48, MS, 50; Tulane Univ, PhD(biochem), 54. *Prof Exp:* Instr pharm, Univ Miss, 48-51; asst, Univ Tex, 51-52; asst prof physiol chem, Univ Miss, 54-55; from asst prof to assoc prof pharmacol, Med Ctr, Univ Kans, 55-61; from assoc prof to prof biochem, Sch Med, Tulane Univ, 61-72; chmn dept biochem, 72-73, asst dean basic med sci, 72-75, PROF BIOCHEM, COL MED, UNIV S ALA, 72-, ASSOC DEAN, 75- *Concurrent Pos:* Consult, Vet Admin, 64-72 & Nat Heart & Lung Inst, 68-72; coordr basic med sci, Tulane Univ, 68-72. *Mem:* AAAS; Am Soc Biol Chemists; Am Soc Pharmacol & Exp Therapeut; Asn Am Med Cols; Sigma Xi. *Res:* Metabolism of three-carbon compounds; biologically active polypeptides; phosphoinositides. *Mailing Add:* Asst Dean Col Med Univ SAla Mobile AL 36688

HUGGINS, ELISHA R, b Dover, NH, Apr 18, 34; m 58; c 2. THEORETICAL PHYSICS. *Educ:* Mass Inst Technol, BS, 55; Calif Inst Technol, PhD(physics), 62. *Prof Exp:* From asst prof to assoc prof physics, 61-74, PROF PHYSICS, DARTMOUTH COL, 74- *Concurrent Pos:* Consult, C A Rypinski Co, 58-59 & Benjamin Publ Co, 64-; vis scientist, Los Alamos Sci Lab, 66-67, consult, 67-; fel, US-Japan Coop NSF Prog, Kyoto Univ, 72. *Honors & Awards:* Am Asn Physics Teachers Award, 65. *Mem:* Am Phys Soc. *Res:* Theory of gravity; theory of superfluid helium; superfluid hydrodynamics. *Mailing Add:* Dept Physics Dartmouth Col Hanover NH 03755

HUGGINS, FRANK NORRIS, b Zephyr, Tex, Oct 25, 26; m 46. MATHEMATICS. *Educ:* Howard Payne Col, BA, 48; NTex State Col, MS, 50; Univ Tex, PhD, 67. *Prof Exp:* Teacher pub schs, Tex, 50-54; instr math, Agr & Mech Col, Tex, 54-57, asst prof, 57-61; spec instr, Univ Tex, 61-62; asst prof, Tex A&M Univ, 62-64; spec instr, Univ Tex, Austin, 64-67; from asst prof to assoc prof math, Univ Tex, Arlington, 67-84; RETIRED. *Concurrent Pos:* NSF fac fel, 59-60. *Mem:* Am Math Soc; Math Asn Am. *Res:* Point set theory; theory of functions of real and complex variables; pure mathematics; properties of real functions such as slope variation, generalized Lipschitz conditions and generalized convexity; theory of integration. *Mailing Add:* Rte 1 Box 92 Zephyr TX 76890

HUGGINS, JAMES ANTHONY, b Flint, Mich, Oct 1, 53; m 73; c 1. SORICID SYSTEMATICS, HOSPITAL-UNIVERSITY ALLIANCES. *Educ:* Ark State Univ, BSA, 75, MS, 77; Memphis State Univ, PhD(biol), 85. *Prof Exp:* Asst prof biol, Miss Indust Col, 80-82; instr anat & physiol, Memphis State Univ, 82-83; assoc prof anat & physiol, micro, Shelby State Community Col, 83-88; ASSOC PROF ANAT & PHYSIOL, MICRO, PATHOPHYSIOL, HUMAN GROSS ANAT, BAPTIST MEM HOSP, UNION UNIV, 88- *Concurrent Pos:* Chair, Div Sci, Baptist Mem Hosp, Union Univ, 88-; adj fac, Memphis State Univ, 90- *Mem:* Am Soc Mammalogists; Am Soc Syst Zoologists; Creation Res Soc; Bible Sci Asn. *Res:* Shrew systematics: morphometrics and chromosomal; hospital or health care association research with universities, joint partnerships, etc. *Mailing Add:* 5115 Battle Creek Memphis TN 38134

HUGGINS, LARRY FRANCIS, b Decatur, Ill, Nov 30, 37; m 56; c 3. AGRICULTURAL ENGINEERING. *Educ:* Univ Ill, Urbana, BS, 60, MS, 62; Purdue Univ, PhD(agr eng), 66. *Prof Exp:* From instr to assoc prof, 62-76, prof, 76-81, HEAD AGR ENGR, PURDUE UNIV, 81- *Concurrent Pos:* Nat dir prof devel, Am Soc Agr Engrs, 88-90. *Mem:* Am Soc Agr Engrs; Am Soc Eng Educ; Soil Conserv Soc Am; Sigma Xi; Nat Soc Prof Engrs. *Res:* Hydrology; water resources planning. *Mailing Add:* 2700 Covington West Lafayette IN 49706

HUGGINS, PATRICK JOHN, b Hertfordshire, Eng, Sept 25, 48. ASTRONOMY. *Educ:* Univ Cambridge, BA, 70, MA, 74, PhD(astron), 75. *Prof Exp:* Res assoc astrophys, Queen Mary Col, Univ London, 74-76; res fel, Bell Labs, 76-78; from asst prof to assoc prof, 78-88, PROF PHYSICS, NY UNIV, 88- *Concurrent Pos:* Fel, Alfred P Sloan Found, 81-85. *Mem:* Fel Royal Astron Soc; Am Astron Soc. *Res:* Interstellar medium. *Mailing Add:* Dept Physics NY Univ Wash Sq New York NY 10003

HUGGINS, ROBERT A(LAN), b Stanford, Calif, Mar 26, 29; m 51; c 3. MATERIALS SCIENCE. *Educ:* Amherst Col, BA, 50; Mass Inst Technol, SM, 52, ScD(metall), 54. *Prof Exp:* Instr metall, Mass Inst Technol, 53-54; asst prof, 54-58, actg exec head, Dept Metall Eng, 57-58, assoc prof metall eng, 58-62, dir, Ctr Mat Res, 61-80, PROF MAT SCI, STANFORD UNIV, 62- *Concurrent Pos:* NSF sr fel, Univ Gottingen, 65-66; dir mat sci, Advan Res Projs Agency, US Dept Defense, Washington, DC, 68-70; Case Centennial scholar, Case Western Reserve Univ, 80. *Honors & Awards:* Hardy Gold Medal, Am Inst Mining, Metall & Petrol Engrs, 57; Vincent Bendix Award, Am Soc Eng Educ, 78; Alexander von Humboldt Sr Scientist Award, Ger, 78. *Mem:* Am Soc Metals; Electrochem Soc; Am Inst Mining, Metall & Petrol Engrs; Am Phys Soc; Sigma Xi. *Res:* Solid state ionics; imperfections in crystals; solid-state kinetics; thermodynamics; solid-state electrochemistry; author or co-author of over 160 publications. *Mailing Add:* Dept Mat Sci & Eng Stanford Univ Stanford CA 94305

HUGGINS, SARA ESPE, b Denver, Colo, June 29, 13; m; c 2. ZOOLOGY, PHYSIOLOGY. *Educ:* Aurora Univ, AB, 34; Univ Ill, MS, 36; Case Western Reserve Univ, PhD(zool), 39. *Prof Exp:* Asst zool, Case Western Reserve Univ, 37-38; instr sci, Paine Col, 46-47; from asst prof to prof, 47-78, chmn dept, 52-64, EMER PROF BIOL, UNIV HOUSTON, 78- *Concurrent Pos:* NIH spec fel, Col Med, Baylor Univ, 67-69, vis prof, 70 & 71-; actg chmn, dept biol, Fac Sci, Mahidol Univ, Bangkok, Thailand, 72-73; vis prof, Univ Kebangsan, Malaysia, 76 & Fed Univ Pernambaco, 79-80 & 83. *Mem:* Am Physiol Soc; Soc Exp Biol & Med; Soc Trop Biol; Sigma Xi; Asn Women Sci. *Res:* Physiology of crocodilians; physiology of sloths. *Mailing Add:* Biol Dept Univ Houston Bellaire TX 77004

HUGGINS, W(ILLIAM) H(ERBERT), b Rupert, Idaho, Jan 11, 19. ELECTRICAL ENGINEERING. *Educ:* Ore State Col, MS, 42; Mass Inst Technol, ScD, 53. *Prof Exp:* Res assoc, Ore State Col, 41-43, instr elec eng, 43-44; res assoc, Harvard Univ, 44-46 & Mass Inst Technol, 49-54; prof, 54-86, chmn dept, 70-75, EMER PROF ELEC ENG, JOHNS HOPKINS UNIV, 86- *Concurrent Pos:* Actg chief, Comput Lab, Air Force Cambridge Res Ctr, 46-54; consult, Rand Corp, 54-63; ed, Circuit Theory J, Inst Elec & Electronics Engrs, 53-57. *Honors & Awards:* Thompson Award, Inst Radio Engrs, 48; Nat Electronics Conf Award, 55; Christian R & Mary F Lindback Award, 61; Ed Medal Award, Inst Elec & Electronics Engrs, 66; Western Elec Fund Award, Am Soc Eng Educ, 65. *Mem:* Nat Acad Eng; fel Acoust Soc Am; fel Inst Elec & Electronics Engrs; Am Soc Eng Educ. *Res:* Circuit theory; theory of hearing; signal theory; iconic communication. *Mailing Add:* One E University Pkwy Apt 1005 Baltimore MD 21218

HUGH, RUDOLPH, b Muskegon, Mich, Mar 3, 23; m 63. PATHOGENESIS, TAXONOMY. *Educ:* Mich State Univ, BS, 48; Loyola Univ Chicago, PhD, 54; Am Bd Med Microbiol, dipl, 64. *Prof Exp:* Instr bact, Cook County Sch Nursing, 50-51; dir labs, Dept Health, Ill, 53-54; asst res prof, 54-56, from asst prof to assoc prof, 57-68, PROF MICROBIOL, SCH MED, GEORGE WASHINGTON UNIV, 68- *Concurrent Pos:* Actg cur bact, Am Type Cult Collection, 61; vis assoc prof, Iowa State Univ, 63-64; mem fac, Naval Grad Dent Sch, 72-75. *Mem:* AAAS; Am Soc Microbiol; Soc Exp Biol & Med; fel Am Acad Microbiol; Brit Soc Gen Microbiol; Sigma Xi. *Res:* Systematic study of nonfermenting gram negative rods; flagellation of bacteria; nomenclature and taxonomy of Aeromonas, Pseudomonas, Alcaligenes, Acinetobacter and Vibrio. *Mailing Add:* Sch Med & Microbiol George Washington Univ 2300 I St NW Washington DC 20037

HUGHART, STANLEY PARLETT, b Spokane, Wash, Mar 17, 18; m 41; c 4. MATHEMATICS. *Educ:* Whitworth Col, BS, 40; Calif Inst Technol, PhD(math), 54. *Prof Exp:* Instr math, Princeton Univ, 43-44; from instr to asst prof, Univ Chicago, 44-54; from asst prof to assoc prof, 54-62, chmn dept, 66-69, prof math, 62-83, EMER PROF, CALIF STATE UNIV, SACRAMENTO, 83- *Mem:* Am Math Soc; Math Asn Am. *Res:* Group theory and homomorphisms; representations of dicatagories. *Mailing Add:* 5814 River Oak Way Carmichael CA 95608

HUGHEL, THOMAS J(OSIAH), b Anderson, Ind, Apr 9, 19; m 40; c 1. PHYSICAL METALLURGY. *Educ:* Purdue Univ, BS, 42, PhD(metall), 51. *Prof Exp:* Instr metall, Purdue Univ, 46-50, asst prof metall eng, 50-56; sr res metallurgist, Gen Motors Corp, 56-63, supvry res metallurgist, 63-77, staff res engr, 77-84; RETIRED. *Mem:* Am Soc Metals; Am Inst Mining, Metall & Petrol Engrs; Sigma Xi. *Res:* Alloy steels, x-ray diffraction, failure analysis, forensic metallurgy, hydrogen embrittlement. *Mailing Add:* 2012 Elmhurst Ave Royal Oak MI 48073

HUGHES, ABBIE ANGHARAD, b Whiston, UK, 1940; UK & Australian citizen; m 62; c 2. DESIGN & DEVELOPMENT OF EYE, RESEARCH ADMINISTRATION. *Educ:* Oxford Univ, BA Hons, 63, MA, 68; London Univ, DIC, 66; Edinburgh Univ, PhD(visual neurosci), 68, DSc, 82. *Prof Exp:* Asst lectr physiol, Edinburgh Univ Med Sch, 64-68; univ demonstr, Univ Lab Physiol, Oxford Univ, 68-72; res fel, John Curtin Sch Med Res, Australian Nat Univ, 72-74, sr res fel, 74-83; prof & dir, Nat Vision Res Inst Australia, 83-90; VIS FEL, OPTICAL SCI CTR, AUSTRALIAN NAT UNIV, 91- *Concurrent Pos:* Chmn, Royal Canberra Hosp Eye Bank Comt, 79-82; vis fel, Dept Appl Math, Australian Nat Univ, 82-88, mem, Visual Sci Ctr, 85-; hon secy, Nat Vision Res Found, Australia, 84- *Mem:* Fel Optical Soc Am; Am Acad Optom; Neurosci Soc; AAAS; Physiol Soc; Australian Physiol & Pharmacol Soc. *Res:* Comparative anatomy and physiology of ocular and visual function with attention to structure and function of optics and retina of the eye; evolution; development and engineering design. *Mailing Add:* 41 Outlook Dr Eaglemont 3084 Australia

HUGHES, ARTHUR D(OUGLAS), b Tacoma, Wash, June 16, 09; m 32; c 3. MECHANICAL ENGINEERING. *Educ:* State Col Wash, BS & MS, 32, ME, 53. *Prof Exp:* Instr mech eng, State Col Wash, 32-36 & Ore State Col, 38-40, asst prof, 40-43, assoc prof, 43-44; res engr, Gas Turbine Develop Div, Allis-Chalmers Mfg Co, Wis, 44-46; prof, 46-74, EMER PROF MECH ENG, ORE STATE UNIV, 74- *Concurrent Pos:* Chief proj engr, Birchfield Boiler, Inc, Wash, 63-64; pres, BTU Chasers, Inc, 74- *Mem:* Fel Am Soc Mech Engrs; Am Soc Heating, Refrig & Air-Conditioning Engrs; Am Soc Eng Educ; Am Inst Plant Engrs; Nat Soc Prof Engrs. *Res:* Gas turbines; heatpower equipment; air conditioning; peppermint oil distillation; energy audits. *Mailing Add:* 1162 Norsam Rd Gladwyne PA 19035

HUGHES, BENJAMIN G, b Lewistown, Ill, Aug 28, 37; m 60; c 4. INORGANIC CHEMISTRY, ANALYTICAL CHEMISTRY. *Educ:* Western Ill Univ, BS, 59; Iowa State Univ, MS, 62, PhD(inorg chem), 64. *Prof Exp:* Asst prof, 64-69, ASSOC PROF CHEM, WESTERN ILL UNIV, 69- *Mem:* Am Chem Soc. *Res:* Polynuclear halide compounds of niobium and tantalum. *Mailing Add:* 705 E Washington Macomb IL 61455-2433

HUGHES, BLYTH ALVIN, b Victoria, BC, Mar 10, 36; m; c 3. OCEAN WAVES, INTERNAL WAVES. *Educ:* Univ BC, BA, 56, MA, 60; Cambridge Univ, PhD(geophys fluid mech), 64. *Prof Exp:* Defense sci serv officer marine physics, Pac Naval Lab, Defense Res Bd Can, 56-89, defense Sci Ocean Physics, Defense Res Estab Pac, 69-91. *Res:* Surface and internal waves; remote sensing in oceanography; arctic acoustics; arctic ice morphology. *Mailing Add:* 760 Walema Ave Victoria BC V8Y 3B1 Can

HUGHES, BUDDY LEE, b Gastonia, NC, Apr 20, 42; m 64; c 3. POULTRY SCIENCE, REPRODUCTIVE PHYSIOLOGY. *Educ:* Clemson Univ, BS, 68; Ore State Univ, MS, 70, PhD(reprod physiol), 71. *Prof Exp:* Asst poultry sci, 65-68, assoc exten specialist, 71, asst prof poultry exten & res, 72-73, asst prof poultry res, 74-76, assoc prof, 76-81, PROF POULTRY RES & TEACHING, CLEMSON UNIV, 81- *Mem:* Poultry Sci Asn; Soc Study Reprod Physiol; Sigma Xi. *Res:* Reproductive performance of artificially inseminated, caged breeder chickens; reproductive characteristics of minor poultry groups; development of the guinea fowl for commerical meat production; housing environment and animal performance. *Mailing Add:* Dept Poultry Sci Clemson Univ 201 Sikes Hall Clemson SC 29634

HUGHES, CHARLES EDWARD, b Boston, Mass, Oct 25, 43; m 68; c 2. SOFTWARE TOOLS, SIMULATION. *Educ:* Northeastern Univ, BA, 66; Pa State Univ, MS, 68, PhD(comput sci), 70. *Prof Exp:* Programmer, Radio Corp Am, 61-66; from instr to asst prof, Pa State Univ, 68-74; from assoc prof to prof comput sci, Univ Tenn, 74-80; PROF COMPUT SCI, UNIV CENT FLA, 80- *Concurrent Pos:* Nat Bur Standards fel, 71-72; NSF res grant, 76-78, 79-82; vpres & co-founder, Gentleware Corp, 80-; Navy grant, 85-86, Fla high tech res grant & Army res grant, 88-89; Consult Nasa res grant, 88. *Mem:* Asn Comput Mach; Inst Elec & Electronics Engrs. *Res:* Software engineering; programming environments; simulation and training; computability theory. *Mailing Add:* 68 Greenfield Ave Ottawa ON K1S 0X7 Can

HUGHES, CHARLES JAMES, b Wallasey, Eng, Aug 28, 31; m 72; c 2. ENVIRONMENTAL SCIENCE. *Educ:* Oxford Univ, BA, 52, MA & DPhil(geol), 55. *Prof Exp:* Geologist, Anglo-Am Corp SAfrica, 56-60; sr petrologist, Geol Surv Ghana, 60-63; technical expert geol, UN, 64-65; teacher math, Oxford & Hants County Coun, 65-66; res assoc, Mem Univ, 66-67, from asst prof to prof geol, 67-88; RETIRED. *Mem:* Geol Asn Can. *Res:* Natural resources and the future. *Mailing Add:* 113 Ennisdale Newton West Kirby Wirrial Messey Side L48 94G England

HUGHES, DANIEL RICHARD, b Cincinnati, Ohio, Aug 7, 27; div; c 6. ALGEBRA, GEOMETRY. *Educ:* Univ Md, BS, 50, MA, 52; Univ Wis, PhD(math), 55. *Prof Exp:* Instr math, Univ Wis, 55; NSF fel, Ohio State Univ, 55-56, lectr, 56-57, asst prof, 57-58; res lectr, Univ Chicago, 58-60; from asst prof to assoc prof, Univ Mich, 60-64; reader, Westfield Col, 64-67, prof math, 67-84, head dept, 71-74, PROF MATH, QUEEN MARY COL, UNIV LONDON, 84- *Concurrent Pos:* Vis prof, Queen Mary Col, Univ London, 62-63, Univ Rome, 63-64 & 70-71 & Univ Perugia, 70-71. *Mem:* London Math Soc. *Res:* Finite projective planes; combinatorial analysis; finite group theory. *Mailing Add:* Dept Math Queen Mary Col Univ London 327 Mile End Rd London E1 4NS England

HUGHES, DAVID EDWARD, b Deansboro, NY, Nov 24, 22; m 63; c 5. VETERINARY MICROBIOLOGY. *Educ:* State Univ NY Vet Col, Cornell Univ, DVM, 51; Cornell Univ, MS, 52; Am Col Vet Microbiol, dipl, 67. *Prof Exp:* Res asst vet bact, Cornell Univ, 51-57; vet med off, Bact & Mycol Res Lab, Nat Animal Dis Ctr, NCent Region, Agr Res Serv, USDA, 57-79; RETIRED. *Concurrent Pos:* Exec secy, Leptospirosis Res Conf, 58-63; mem, Conf Res Workers Animal Dis. *Mem:* US Animal Health Asn; Am Vet Med Asn; Sigma Xi. *Res:* Infertility diseases of cattle, brucellosis and vibriosis; leptospirosis of animals; infectious bovine keratoconjunctivitis. *Mailing Add:* RR 4 Ames IA 50010

HUGHES, DAVID KNOX, b Lubbock, Tex, Dec 7, 40; m 62; c 1. MATHEMATICS. *Educ:* Abilene Christian Col, BA, 62; Univ Okla, MA, 64, PhD(math), 67. *Prof Exp:* Instr math, Univ Okla, 66-67; from asst prof to assoc prof, 67-77, PROF MATH, ABILENE CHRISTIAN COL, 77-, HEAD DEPT, 80- *Mem:* Soc Indust & Appl Math; Am Math Soc; Math Asn Am. *Res:* Differential equations; control theory. *Mailing Add:* Dept Math Abilene Christian Col Abilene TX 79699

HUGHES, DAVID WILLIAM, b Leicester, Eng, Aug 4, 33; Can citizen; m 57; c 2. ORGANIC CHEMISTRY. *Educ:* Univ Nottingham, BSc, 54, PhD(natural prod chem), 59. *Prof Exp:* Fel, Div Appl Biol, Nat Res Coun Can, 59-61; sr sci off, Nat Chem Lab, Eng, 61-64; res scientist, Can Food & Drug Directorate, 64-75, BUR DRUG RES, HEALTH PROTECTION BR, CAN, 75- *Mem:* Can Inst Chem. *Res:* Structural studies of compounds isolated from natural sources, especially moulds, bacteria and higher plants, analysis of medicinal plants; chemical methods of analysis of antibiotics. *Mailing Add:* Bur Drug Res Ottawa ON K1A 0L2 Can

HUGHES, EDWIN R, b Solano, NMex, May 27, 28; m 48; c 7. CHEMISTRY, MEDICINE. *Educ:* Eastern NMex Univ, BS, 51, MS, 52; Univ Utah, MD, 56; Am Bd Pediat, dipl. *Prof Exp:* Asst chem, Eastern NMex Univ, 51-52; asst pediat, Univ Utah, 53-56; intern, Univ Minn Hosp, 56-57; asst resident, Univ Utah, 57-58; resident pediat & path, Med Ctr, Univ Ark, 58-59; med res assoc, Brookhaven Nat Lab, 59-61; from asst prof to prof pediat & biochem, Med Ctr, Univ Ark, Little Rock, 61-72; prof, Med Ctr, WVa Univ, 72-73; prof pediat, 73-77, DIR GRAD STUDIES IN BASIC MED SCI, UNIV S ALA, 78- *Concurrent Pos:* Res collabr, Brookhaven Nat Lab, 61-62. *Mem:* Am Soc Exp Path; Endocrine Soc; Soc Pediat Res; Am Pediat Soc; Am Chem Soc; Sigma Xi. *Res:* Pediatrics; endocrinology; adrenal cortex. *Mailing Add:* MSB 2152 Univ South Ala Mobile AL 36688

HUGHES, EUGENE MORGAN, b Scottsbluff, Nebr, Apr 3, 34; m 54; c 3. MATHEMATICAL ANALYSIS. *Educ:* Chadron State Col, BS, 56; Kans State Univ, MS, 58; George Peabody Col, PhD(math), 68. *Prof Exp:* From instr to asst prof math & head dept, Chadron State Col, 57-62; asst to undergrad dean, George Peabody Col, 63-64, asst to pres, 64-65; from assoc prof to prof math, Chadron State Col, 65-70, dir res, 65-66 & Head Start Prog, 66, asst to pres, 66-68, asst dir Upward Bound, proj mgr regional training off, Proj Head Start & dir NSF In-Serv Insts Math & Sci, 66-70, dean admin, 68-70; dean col arts & sci, 70-71, provost univ arts & sci, 71-72, acad vpres, 72-79, PROF MATH, NORTHERN ARIZ UNIV, 70-, ACAD PRES, 79- *Concurrent Pos:* Vis lectr, Math Asn Am, 61-62; consult, Nebr State Dept Educ, 66-70; dir, Nebr State Cols Curriculum Develop Teacher Educ Proj, 67-69; co-dir, NCent Asn Cols & Sec Schs Workshop Teacher Educ, Univ Minn, 68-70, coordr asn, 68-72; chmn, State Elem & Sec Educ Act Adv Coun, Nebr, 69-70; chief Ariz Acad NSF fel, 63-64; bd dirs, Ariz Bank, 87-, adv bd, United Bank, 82-87, Flagstaff Chamber Com. *Mem:* Math Asn Am; Nat Educ Asn; Nat Coun Teachers Math. *Res:* Impact of selected experimental curriculum projects on commercially published elementary school mathematics textbooks; properties of groups and permutation groups; mathematics education. *Mailing Add:* Northern Ariz Univ Flagstaff AZ 86011

HUGHES, EVERETT C, b Wadena, Minn, Nov 22, 04; c 5. CHEMOTHERAPY, BIOENGINEERING. *Educ:* Carleton Col, AB, 27; Cornell Univ, PhD(chem), 30. *Prof Exp:* Res chemist, Standard Oil Co, Ohio, 30-43, chief chem & phys res div, 44-54, res mgr, 54-59, mgr res dept, 59-61, vpres res & develop, 61-69; RES ASSOC, ASST & ASSOC CLIN PROF, OTOLARYNGOL, SCH MED, UNIV SOUTHERN CALIF, 70- *Concurrent Pos:* Chmn mgt comn, Gordon Res Conf, 53-54; mem aviations fuels div, Coop Fuels Res Comt, Coord Res Coun. *Honors & Awards:* Pioneer Chemist Award, Am Inst Chemists, 71; Morley Award & Medal, Am Chem Soc, 74. *Mem:* Fel AAAS; Am Chem Soc; Am Inst Chemists; emer mem Soc Study Headache. *Res:* Causes and treatment of sensorineural hearing impairment and allergies; food allergy; immunology; electrochemical bioengineering. *Mailing Add:* 1200 N State St Rm 2P70 LAC-USC Med Ctr Los Angeles CA 90033

HUGHES, FRANCIS NORMAN, b Dresden, Ont, Jan 23, 08; m 35, 53; c 7. PHARMACY. *Educ:* Univ Toronto, PharmB, 29, MA, 44; Purdue Univ, BS, 40. *Hon Degrees:* LLD, Purdue Univ, 54, Dalhousie Univ, 73 & Univ Toronto, 80. *Prof Exp:* Asst prof pharm, Ont Col Pharm, 38-42, from assoc prof to prof mat med & pharm chem, 42-49, prof & asst dean, 49-52, dean, 52-53; dean fac pharm, 53-73, EMER DEAN & EMER PROF, FAC PHARM, UNIV TORONTO, 73- *Concurrent Pos:* Secy, Can Conf Pharmaceut Faculties, 44-50, chmn, 51-52; ed new pharmaceut sect, Can Pharmaceut J, 45-71; pres, Can Found Adv Pharm, 56-57; mem adv bd, Proprietary & Patent Med Act, 58-73; pres, Pharm Exam Bd Can, 64-66, registr-treas, 73-81; ed, Compendium Pharmaceut & Specialties, 60-67, co-ed, 67-71, consult ed, 71-79. *Honors & Awards:* Can Centennial Medal, Govt of Can, 67. *Mem:* Fel AAAS; Asn Faculties Pharm Can (pres, 70); Can Soc Hosp Pharmacists; Asn Deans Pharm Can (pres, 66-69); Can Pharmaceut Asn. *Mailing Add:* 74 Baxson Dr Aurora ON L4G 3P8 Can

HUGHES, GEORGE MUGGAH, b Sidney, NS, Aug 14, 29; m 57; c 2. HYDROGEOLOGY. *Educ:* Univ Alta, BSc, 51, MSc, 59; Univ Ill, PhD(geol), 62. *Prof Exp:* Geologist, Pan Am Petrol Corp, 52-57 & Ill State Geol Surv, 62-74; MEM STAFF, ONT MINISTRY ENVIRON, 74- *Mem:* Am Inst Hydrol. *Res:* Pleistocene geology and hydrogeology; groundwater pollution. *Mailing Add:* 135 St Clair Ave W Toronto ON M4V 1P5 Can

HUGHES, GILBERT C, b Homerville, Ga, Feb 28, 33; m 61; c 3. BOTANY, MYCOLOGY. *Educ:* Ga Southern Univ, BS, 53; Fla State Univ, MS, 57, PhD(bot, mycol), 60. *Prof Exp:* Nat Res Coun Can fel, Univ BC, 60-61; asst prof biol sci, Emporia Kans State Univ, 61-62; asst prof biol sci & marine bot, Univ Pac, 62-64, asst dir, Pac Marine Sta, 63-64; from asst prof to assoc prof, 64-76, PROF BOT & BIOL OCEANOG, UNIV BC, 76- *Mem:* Mycol Soc Japan; Brit Mycol Soc; Mycol Soc Am. *Res:* Development, distribution and taxonomy of marine fungi, especially lignicolous Ascomycetes and Fungi imperfecti of the Pacific Ocean; biology and host-parasite relationships of algicolous marine fungi and fungal symbionts of marine invertebrates. *Mailing Add:* Dept Bot & Dept Oceanog Univ BC Vancouver BC V6T 2B1 Can

HUGHES, GORDON FRIERSON, b Los Angeles, Calif, Sept 9, 37. APPLIED PHYSICS. *Educ:* Calif Inst Technol, BS, 59, MS, 60, PhD(elec eng), 64. *Prof Exp:* Mem res staff pattern recognition statist, Autonetics; prin scientist magnetic res & develop, Palo Alto Res Ctr, Xerox Corp, 69-; VPRES, EG&G ENERGY MEASUREMENTS. *Mem:* Inst Elec & Electronics Engrs; Sigma Xi. *Res:* Applied magnetism, mathematical analysis and theory of magnetic recording. *Mailing Add:* 390 Blackstone Dr Boulder Creek CA 95006

HUGHES, HAROLD K, b New York, Ny, Dec 31, 11; m 36; c 2. INDUSTRIAL PHYSICS. *Educ:* Columbia Univ, BA, 34, MA, 43, PhD, 48. *Prof Exp:* RETIRED. *Concurrent Pos:* Consult. *Mem:* Am Phys Soc; Am Chem Soc; Inst Elec & Electronics Engrs; Sigma Xi; Am Asn Physics Teachers. *Res:* Classical and quantum logic. *Mailing Add:* Hannawa Rd HCR 75 Box 37 Potsdam NY 13676-9231

HUGHES, HARRISON GILLIATT, b Princeton, Ind, Aug 14, 48; m. PLANT GENETICS, PLANT BREEDING. *Educ:* Eastern Ill Univ, BS, 69; Purdue Univ, PhD(plant genetics & breeding), 74. *Prof Exp:* Instr hort, Purdue Univ, 73-74; asst prof, State Univ NY, 74-76 & Univ RI, 76-77; ASSOC PROF HORT, COLO STATE UNIV, 77- *Concurrent Pos:* Vis prof tissue cult, Dept Agron, Univ Ky, 75, Dept Pomol, Univ Calif, 84. *Mem:* Int Asn Plant Tissue Cult; Am Soc Hort Sci; Tissue Cult Asn; Sigma Xi; Am Genetic Asn. *Res:* Tissue culture as applied to propagation and plant genetics and breeding; cryopreservation of plant parts; applied plant biotechnology. *Mailing Add:* 929 Timber Lane Ft Collins CO 80521

HUGHES, JAMES ARTHUR, b Sharon, Pa, June 10, 29; m 59; c 2. ASTRONOMY, ASTROMETRY. *Educ:* Columbia Univ, AB, 57, PhD(astron), 66. *Prof Exp:* Astronr, 59-77, dir transit circle div, 77-83, DIR ASTROMETRY DEPT, US NAVAL OBSERV, 84- *Concurrent Pos:* Observ dir, Yale-Columbia Southern Observ, 67-69; pres comt No 8, Int Astron Union; chmn, Working Group on Reference Frames. *Honors & Awards:* Berman Res Publ Award, NRL. *Mem:* Int Astron Union. *Res:* Fundamental (absolute) and differential astrometry, interferometry. *Mailing Add:* Dir Astrometry Dept US Naval Observ Washington DC 20392

HUGHES, JAMES GILLIAM, b Memphis, Tenn, Sept 11, 10; m 35; c 4. MEDICINE, PEDIATRICS. *Educ:* Southwestern at Memphis, AB, 32; Univ Tenn, Memphis, MD, 35. *Prof Exp:* Rotating intern, City of Memphis Hosps, 35-37; resident, Children's Mem Hosp, Chicago, 37-38 & City of Memphis Hosps, 38-39; instr pediat, Okla State Med Asn & Commonwealth Found NY, 40-42; from assoc prof to prof, 46-76, chmn dept, 60-76, EMER PROF PEDIAT, COL MED, UNIV TENN, MEMPHIS, 76- *Concurrent Pos:* Official examr, Am Bd Pediat, 51-; consult, Surgeon Gen, US Army, 54-, WHO, 55, 56 & 58, Rockefeller Found, 56 & AID-Alliance Progress, 63. *Honors & Awards:* Hon Prof, San Carlos Univ Guatemala, 56 & Univ Guadalajara, 63; Abraham Jacobi Award, 75. *Mem:* Am Acad Pediat (pres, 65-66); Am Pediat Soc; Soc Pediat Res; hon mem Arg, Brazilian, Colombian, Cuban, Mex, Panamanian, Uruguayan & Venezuelan Socs Pediat. *Res:* Studies on the causes of brain damage, prenatally, perinatally, and postnatally. *Mailing Add:* Le Bonheur Children's Hosp 848 Adams Ave Memphis TN 38103

HUGHES, JAMES MITCHELL, b Pittsburgh, Pa, Aug 11, 45; m 71; c 2. NOSOCOMIAL INFECTIONS, DIARRHEAL DISEASE. *Educ:* Stanford Univ, BA, 66, MD. *Prof Exp:* Chief, Water Related Dis Activ, Bacterial Dis Div, Bur Epidemiol, Ctrs Dis Control, 78-81, chief, Surveillance & Prev Br, Hosp Infections Prog, Ctr Infectious Dis, 81-83, dir, Hosp Infections Prog, 83-88, DEPT DIR, NAT CTR INFECTIOUS DIS, CRTS DIS CONTROL, ATLANTA, GA, 88-; STAFF PHYSICIAN, DEPT MED, DIV INFECTIOUS DIS, SCH MED, EMORY UNIV, 89- *Concurrent Pos:* Co-leader, Immunization & Infectious Dis Priority Area Work Group, Year 2000 Objectives for Nation, USPHS, 91- *Mem:* Fel Am Col Physicians; fel Infectious Dis Soc Am; AAAS; Am Soc Microbiol; Am Soc Trop Med & Hyg; Royal Soc Trop Med & Hyg. *Res:* Epidemiology, surveillance, prevention and control of infectious diseases in the United States and abroad, with emphasis on nosocomial infections, and diarrheal disease and antimicrobial resistance. *Mailing Add:* Nat Ctr Infectious Dis Ctrs Dis Control 1600 Clifton Rd NE Atlanta GA 30333

HUGHES, JAMES P, b Wilkinsburg, Pa, 1920. OCCUPATIONAL MEDICINE. *Educ:* Univ Pittsburgh, MD, 45. *Prof Exp:* Intern, St Francis Hosp, Pittsburgh, 45-46; res path, Univ Hosps, Cleveland, 48-49; fel indust med, Inst Indust Health, Kettering Lab, Cincinnati, 49-52; attend staff, Occup Med Sect, Clin Div Med, Ohio State Univ Hosp, 55-57; med dir, Kaiser Aluminum & Chem Corp, 57-82; ASSOC PROF OCCUP MED, UNIV CALIF, SAN FRANCISCO, 79-; MEM HEARING BD, BAY AREA AIR QUAL MGT DIST, CALIF, 89- *Concurrent Pos:* Asst prof indust med, Univ Cincinnati, 52-55; assoc prof prev med, Univ Ohio, 55-57; lectr med, Tulane Univ, 65-; exec vpres & dir, Kaiser Found Int, 64-74; vpres health affairs, Kaiser Industs Corp, 72-74. *Mem:* Inst Med-Nat Acad Sci; fel Am Col Occup Med; fel Am Acad Occup Med (pres, 76-77); fel Am Col Physicians; Am Indust Hyg Asn. *Res:* Author. *Mailing Add:* 124 Guilford Rd Piedmont CA 94611

HUGHES, JAMES SINCLAIR, b Woodbury, NJ, Oct 3, 34; m 64; c 2. BOOLEAN ALGEBRA IN NONLINEAR CIRCUITS, STATE MACHINE THEORY & DESIGN. *Educ:* Univ PR, BS, 60. *Prof Exp:* Electronic subcontractor, Computer Mechanisms, 67-68; electronic engr, KSM Welding Systs, Div Omark Industs, 68-73; physicist, Nuclear Res Corp, 73; electronic engr, Scriptomatic, 73-74 & 76-81; optoelectronic engr, Geom Data, 74-76; electronic engr, H L Yoh, 81-83; PRES, ARCTINURUS CO, INC, 83- *Concurrent Pos:* Lectr & author, Arctinurus Co, Inc, 74- *Res:* Star particle theory development: basic nature, structure of universe and unity-derived discreteness; proof unity is not remote; electric and magnetic charge lattices of three-dimensional space; time-cycles infer exacting sequences. *Mailing Add:* 20 Sullivan Ave Bellmawr NJ 08031-2332

HUGHES, JANICE S, b Bastrop, La, Sept 16, 37. FISHERIES. *Educ:* NE La Univ, BS, 59, MS, 66. *Prof Exp:* Biol aide, 59-61, fisheries biologist I, 61-70, fisheries biologist II, 70-73, fisheries biologist III, 73-83, DIST FISHERIES SUPVR, LA DEPT WILDLIFE & FISHERIES, 83- *Concurrent Pos:* Chmn, prog comt, Am Fisheries Soc, 76-77. *Mem:* Am Fisheries Soc (vpres, 80-82, pres, 83-84); Am Inst Fishery Res Biologists. *Res:* Lake management practices including the evaluation of stocking Florida largemouth bass in a new impoundment; techniques for producing striped bass and grass carp; the toxicity of various chemicals to several species of fish. *Mailing Add:* La Wildlife & Fisheries PO Box 4004 Monroe LA 71211-4004

HUGHES, JAY MELVIN, b Pueblo, Colo, Dec 1, 30; m 53; c 2. FOREST ECONOMICS, NON-MARKET VALUE ANALYSIS. *Educ:* Univ Colo, BA, 52; Colo State Univ, MF, 58; Mich State Univ, PhD(forestry), 64. *Prof Exp:* Forester, USDA Forest Serv, 58-59, res forester, 59-62, res forester & proj leader, 62-66, dir, Forest Resources Prog, Coop State Res Serv, 71-72, br chief, Nat Forest Surv, 72-74, dir, Forest Resources Econ Res, 74-77; prof regional anal forest mgt, Sch Forestry, Univ Minn, 66-71; dean & prof, Col Forestry & Natural Resources, Colo State Univ, 77-91; RETIRED. *Concurrent Pos:* Chmn, Working Group Recreation, Int Union Forest Res Orgn, 76-79, Forest Sci Bd, Soc Am Foresters, 78-80 & Fourth World Wilderness Cong, 83-87; mem, Panel Rev Natural Resources Progs Nepal, Nat Acad Sci, 82-, Nat Joint Coun Food & Agr Sci, 82-86, Nat Coop Forest Res Adv Coun, 83-87, Adv Bd Natural Resources Law Ctr, Univ Colo, 88-89 & bd trustees, Pan-Am Sch Agr, Honduras, 91- *Mem:* Soc Am Foresters; Am Agr Econ Asn. *Res:* Research administration; forest economics and marketing; feasibility studies for new forest products industries; economic impact analysis of changes in levels and kinds of industry; economic efficiency in forest management and product processing. *Mailing Add:* No 38 Las Casitas Las Cruces NM 88005

HUGHES, JOHN I(NGRAM), b Hudson, Wis, May 8, 19; m 43; c 2. ENGINEERING, CHEMISTRY. *Educ:* Univ Minn, BChE, 41. *Prof Exp:* Chem engr process develop, Ammonia Dept, 41-46, tech supt econ studies, 46-47, tech supt high pressure process develop, 47-48, asst prod supt nylon intermediates, Polychem Dept, 48-50, tech investr, Planning Div, 50-52, res supvr plastics res, 52-55, mgr mkt res plastics, 55-58, mgr mkt develop plastics, 58-61, mgr, Develop Dept, 61-63, NEW VENTURE MGR, E I DU PONT DE NEMOURS & CO, 63- *Mem:* Am Chem Soc; Am Inst Chem Engrs. *Mailing Add:* 1302 Copley Dr Wilmington DE 19803

HUGHES, JOHN LAWRENCE, b Evansville, Ind, June 9, 28; m 58; c 2. ORGANIC CHEMISTRY, MEDICINAL CHEMISTRY. *Educ:* Evansville Col, BS, 53; Univ Mich, MS, 60, PhD(pharmaceut chem), 62. *Prof Exp:* Chemist, Mead Johnson & Co, 56-58, sr scientist, 61-63; chemist, Union Carbide Chem Co, 63-65; res chemist, Armour Pharmaceut Co, 65-78; prin develop chem, Beckman Instruments, 78-85; assoc dir, Smith Kline & French Labs, 85-90; ASSOC DIR, PEPTIDES UNLIMITED, 90- *Mem:* Am Chem Soc; AAAS. *Res:* Synthesis of organic compounds with potentially useful biological properties; development of processes for the large-scale production of synthetic peptides; development of methods for the determination of composition and sequence of peptides and proteins. *Mailing Add:* Peptides Unlimited 1714 Cherrytree Lane Mountain View CA 94040

HUGHES, JOHN P, b Fresno, Calif, Mar 21, 22; c 2. VETERINARY MEDICINE. *Educ:* Kans State Col, DVM, 49; Am Col Theriogenologists, dipl. *Prof Exp:* Vet, private practice, Madera, Calif, 49-56; from asst prof to assoc prof, 56-67, chief ambulatory, 65-71, chmn dept reprod, 77-84, PROF CLIN SCI, UNIV CALIF, DAVIS, 67-, DIR, EQUINE RES LAB, & DIR SEROL, 81- *Concurrent Pos:* Vet, Agr Exp Sta, Univ Calif, Davis, 56-, chief equine reprod. Vet Med Teaching Hosp, 71- *Mem:* Am Vet Med Asn; Am Asn Equine Practitioners; Soc Study Reproduction; Am Asn Vet Clinicians; Am Col Theriogenologists; Soc Theriogenology. *Res:* Utero-ovarian relationships, ovarian tumors, infertility associated with chromosomal errors; estrous cycle of the mare; study of the normal physiology of the estrous cycle of the mare. *Mailing Add:* 1539 Brown Dr Davis CA 95616

HUGHES, JOHN RUSSELL, b Du Bois, Pa, Dec 19, 28; m 58; c 3. ELECTROENCEPHALOGRAPHY, NEUROPHYSIOLOGY. *Educ:* Franklin & Marshall Col, AB, 50; Oxford Univ, BA, 52, MA, 55; Harvard Univ, PhD(neurophysiol), 54; Northwestern Univ, Chicago, MD, 75. *Hon Degrees:* DM, Oxford Univ, 76. *Prof Exp:* Res neurophysiologist, Mass Inst Technol, 54; mem staff, Sect Cortical Integration, NIMH, 55-57; chief neurophysiol & EEG, Meyer Mem Hosp, State Univ NY, 57-63; from assoc prof to prof neurol & dir div EEG & neurophysiol, Northwestern Univ, Chicago, 64-77; PROF NEUROL & DIR EEG & EPILEPSY CLIN, MED CTR, UNIV ILL, 77- *Concurrent Pos:* Asst prof, Sch Med, Univ Buffalo, 57-63; chief EEG, Mercy Hosp, 57-63; consult, Gowanda State Hosp, NY & Craig Epileptic Colony, 57-63, Vet Admin Res Hosp & Evanston Hosp Assoc, 64-77, Nat Inst Neurol Dis Blind, 76-, Nat Inst Ment Health, 76- & Nat Cancer Inst, 76-; mem staff, Chicago Wesley Mem Hosp & Passavant Mem Hosp, 64-77, Mercy Ctr, Aurora, 64-, Copley Mem Hosp, 67-, Downey Vet Admin Hosp, 71-76, Community Hosp, Geneva & Delnor Hosp, St Charles, 72-, Univ Ill Hosp & West Side Vet Admin Hosp, 77-; full colonel, USAR, Med Corps, command surgeon, 86 USARCOM; USA rep, Int Brain Death Debate, BBC, London, 83; invited lectr, Int Conf EEG, London, 85. *Mem:* Soc Neurosci; Am Acad Neurol; Am EEG Soc; Am Epilepsy Soc; Am Med EEG Asn. *Res:* Electrophysiology of central nervous system. *Mailing Add:* Univ Ill Med Ctr 912 S Wood St Chicago IL 60612

HUGHES, JOHN RUSSELL, b Columbia, SC, June 7, 49. ADDICTION, BEHAVORIAL MEDICINE. *Educ:* Millsaps Col, BS, 71; Univ Miss, MD, 75. *Prof Exp:* Asst prof, psychiat, Univ Minn; PROF, PSYCHIAT, UNIV VT, 84- *Honors & Awards:* Res Scientist Develop Award, Nat Inst Drug Abuse, 84. *Res:* Human behavorial pharmacology. *Mailing Add:* Dept Psychiat Univ Vt 38 Fletcher Pl Burlington VT 05401

HUGHES, KAREN WOODBURY, b Madison, Wis, Aug 15, 40; div. PLANT GENETICS, TISSUE CULTURE. *Educ:* Univ Utah, BS, 62, MS, 64, PhD(genetics), 72. *Prof Exp:* Fel RNA develop, Univ Utah, 72-73; from asst prof to assoc prof, 73-84, PROF BOT, UNIV TENN, KNOXVILLE, 84-, HEAD DEPT, 85- *Concurrent Pos:* Assoc prog dir genetic biol, NSF, 80-81. *Mem:* Tissue Cult Asn; Int Asn Plant Molecular Biol. *Res:* induction of whole plants from single cells; transformation of higher plants and regulation of transgenic sequences. *Mailing Add:* Dept Bot Univ Tenn Knoxville TN 37996-1100

HUGHES, KENNETH E(UGENE), b Columbus, Ohio, May 15, 39; m 64; c 3. BIOMEDICAL ENGINEERING. *Prof Exp:* Technologist, 70-74, res scientist bioeng, Columbus Labs, 75-90, TECH LIAISON, MED PROD DEVELOP, BATTELLE MEM INST, 90- *Mem:* Soc Biomat; Am Soc Testing & Mat. *Res:* Biomedical device and equipment development; biomaterials and implantable prostheses for orthopedics and other surgery; bioengineering applications in space research; material specifications; processing for use in implantable prosthetics; orthopedic devices for bone and tendon repair; cardiovascular materials; instrumented probes and catheters; collagen biomaterials; disposable medical devices and diagnostic instuments. *Mailing Add:* Battelle Med Prod Develop Off 505 King Ave Columbus OH 43201-2693

HUGHES, KENNETH JAMES, b Glencoe, Okla, May 18, 21; m 46; c 2. CHEMICAL ENGINEERING, PHYSICAL CHEMISTRY. *Educ:* Okla State Univ, BS, 43. *Prof Exp:* Chem engr pilot plant fractionation, US Bur Mines, Bartlesville, 46-48, sr combustion engr, 48-55, chem engr automobile exhaust gas res, 55-59, supt mgr res ctr, 59-74; mgt officer & dir, Div Opers, Energy Tech Ctr, US Dept Energy, Bartlesville, 74-83, dir, Div Opers, NIPER, 83-85; RETIRED. *Concurrent Pos:* Dir admin, Emergency Petrol & Natural Gas Admin, 60- *Honors & Awards:* Outstanding Engr, Okla Soc Prof Engrs, 70 & Outstanding Engr in Mgt, 78. *Mem:* Nat Soc Prof Engrs; Am Chem Soc. *Res:* Fundamental investigation of heat release from combustion of liquid hydrocarbons; research dealt with thermodynamics and physical chemistry studies; application of gas chromatographic analysis of hydrocarbon combustion products. *Mailing Add:* 305 SE Rockwood Bartlesville OK 74006

HUGHES, KENNETH RUSSELL, b Winnipeg, Man, May 7, 33; m 63. PSYCHOPHYSIOLOGY. *Educ:* Univ Man, BA, 54, MA, 56; Univ Chicago, PhD(psychol), 61. *Prof Exp:* From asst prof to assoc prof, 61-74, actg chmn dept, 70- 74, PROF PHYSIOL, UNIV MAN, 74-, DEAN GRAD STUDIES. *Mem:* Am Physiol Asn; Can Physiol Soc. *Res:* Central nervous system and behavior; limbic system function and stress; environmental effects on behavior. *Mailing Add:* Dean Grad Studies Univ Man University Ctr Rm 500 Winnipeg MB R3T 2N2 Can

HUGHES, MALCOLM KENNETH, b Matlock, Eng, July 24, 43; c 2. PALEOCLIMATOLOGY, DENDROCHRONOLOGY. *Educ:* Univ Durham, Eng, BSc, 65, PhD(ecol), 70. *Prof Exp:* Amanuensis, Soil Biol Inst, Aarhus Univ, 68-69; res fel bot, Univ Durham, 69-71; lectr ecol, Liverpool Polytech, Eng, 71-73, sr lectr, 73-80, prin lectr, 80-82, reader, 82-86; PROF & DIR DENDROCHRONOLOGY, LAB TREE-RING RES, UNIV ARIZ, 86- *Concurrent Pos:* Mem, Terrestrial Life Sci Grants Comt, Natural Environ Res Coun, UK, 82-85, Combined Studies (Sci) Bd, Coun Nat Acad Awards, UK, 83-86, US Nat Comt for Int Union for Quaternary Res, 88-91 & Paleoclimat Adv Panel, Nat Oceanic & Atmospheric Admin, 89-; dir, Global Change Div, Ariz Res Labs, 89- *Mem:* Fel Royal Meteorol Soc; Am Geophys Union; Tree-Ring Soc. *Res:* Exploring climate of recent centuries using annual resolution natural records, primary tree rings, currently developing 3, 000 year drought history of California using annual rings of giant sequoia. *Mailing Add:* Lab Tree-Ring Res Univ Ariz Tucson AZ 85721

HUGHES, MARK, b Hutchinson, Minn, Aug 3, 31; m 54; c 7. ORGANIC CHEMISTRY. *Educ:* St John's Univ, Minn, BA, 53; Iowa State Col, PhD, 58. *Prof Exp:* From instr to assoc prof chem, 58-69, chmn dept, 67-77, PROF CHEM, ST JOHN'S UNIV, MINN, 69- *Concurrent Pos:* Sabbatical, 65-66. *Mem:* Am Chem Soc. *Mailing Add:* Dept Chem St John's Univ Collegeville MN 56321

HUGHES, MARYANNE ROBINSON, b Binghamton, NY, Dec 27, 30; m 61; c 3. AVIAN PHYSIOLOGY. *Educ:* Harpur Col, BA, 52; Duke Univ, MA, 56, PhD(zool), 62. *Prof Exp:* Res biochemist, Eaton Labs, Norwich Pharmacal Co, 52-54; asst zool, Duke Univ, 56-59; res assoc, Univ Wash, 60-61; res assoc, Univ of the Pac, 62-63, asst prof biol, 63-64; RES ASSOC ZOOL, UNIV BC, 64-, LECTR, 68- *Concurrent Pos:* NSF grant, 63-66; Nat Res Coun grant, 68- *Mem:* Am Soc Zool; Can Soc Zool. *Res:* Renal function in the octopus; renal and extrarenal salt and water excretion, body water and ion distribution, organ weight and water content in marine birds. *Mailing Add:* Dept Zool Univ BC 6270 University Blvd Vancouver BC V6T 1Z4 Can

HUGHES, MAYSIE J H, b Webster Groves, Mo, Aug 19, 18; wid; c 1. PHYSIOLOGY, PHARMACOLOGY. *Educ:* Wash Univ, AB, 43; St Louis Univ, PhD(pharmacol), 63. *Prof Exp:* Res assoc zool, Wash Univ, 62-64; from instr to asst prof pharmacol, Sch Med, St Louis Univ, 64-68, asst prof physiol, 68-72; assoc prof, Med Col Ga, 72-73; actg chmn dept, 74-75, assoc prof, 73-78, PROF PHYSIOL, SCH MED, TEX TECH UNIV, 78- *Mem:* AAAS; Am Physiol Soc; Am Soc Pharmacol & Exp Therapeut; Sigma Xi; Soc Neurosci. *Res:* Histamine action on heart muscle and its receptor. *Mailing Add:* Dept Physiol Tex Tech Univ PO Box 4569 Lubbock TX 79409

HUGHES, MICHAEL CHARLES, b Schenectady, NY, May 9, 42. ANALYTICAL CHEMISTRY. *Educ:* State Univ NY, Albany, BS, 65; Syracuse Univ, PhD(anal chem), 71. *Prof Exp:* Fel chem, Univ Mich, 74-75; asst prof anal chem, Lehigh Univ, 75-80; mem tech staff, 81-88, DISTINGUISHED MEM TECH STAFF, BELL TEL LABS, ALLENTOWN, PA, 88- *Mem:* Sigma Xi; AAAS; Am Chem Soc. *Res:* Electroanalytical chemistry; environmental trace analysis; analytical chemistry in the semiconductor industry. *Mailing Add:* 498 Rosewood Dr North Hampton PA 18067-9567

HUGHES, NORMAN, b Nashville, Tenn, Mar 26, 32; m 54; c 2. DEVELOPMENTAL GENETICS. *Educ:* Harding Col, BA, 54; Emory Univ, MS, 56, PhD(biol), 58. *Prof Exp:* Vis instr biol, Emory Univ, 58; prof, Lubbock Christian Col, 58-63; assoc prof, Harding Col, 63-68; prof, Christian Col S W, 68-70; PROF BIOL, PEPPERDINE UNIV, MALIBU, 70- *Mem:* Am Soc Zool; Sigma Xi. *Res:* Cytology and physiology of fertilization; vertebrates; developmental genetics. *Mailing Add:* Seaver Col Pepperdine Univ Malibu CA 90263

HUGHES, O RICHARD, b Washington, DC, May 14, 38; m 63; c 2. ADVANCED MATERIALS, POLYMER PROCESSING. *Educ:* DePaul Univ, BS, 60; Purdue Univ, MS, 64, PhD(phys chem), 66. *Prof Exp:* Res chemist, Celanese Res Co, 66-71, proj leader, Catalysis, 71-75, res assoc, 75-82, coordr univ res, 82-85, res assoc advan mat, 85-90, SR RES ASSOC ADVAN MAT, CELANESE RES CO, 90- *Concurrent Pos:* Pres, Chatham TWP Bd Health, 87- *Honors & Awards:* Shultheiss Lectr, Frankfurt, Ger, 89. *Mem:* Am Chem Soc; NY Acad Sci; Am Ceramic Soc; Mat Res Soc. *Res:* New structural polymers; synthesis and fabrication; polymer powder processing. *Mailing Add:* Hoechst Celanese R L Mitchell Tech Ctr 86 Morris Ave Summit NJ 07901

HUGHES, PATRICK HENRY, b Latrobe, Pa, May 28, 34; div; c 1. PSYCHIATRY. *Educ:* Univ Pittsburgh, AB, 56, MD, 60; Columbia Univ, MSc, 67. *Prof Exp:* Med internship, Med Ctr, Stanford Univ, 60-61; psychiat residency, Columbia Presby Med Ctr, 61-64; surgeon USPHS, Addiction Serv, NIMH Clin Res Ctr, Ft Worth, 64-66; chief prog consult, Ctr Studies Narcotic & Drug Abuse, NIMH, Md, 67-68; from asst prof to assoc prof psychiat, Univ Chicago, 68-77; SR MED OFFICER, DRUG DEPENDENCE PROG, DIV MENT HEALTH, WHO, SWITZ, 75-; prof lectr psychiat, Univ Chicago, 77-84; RETIRED. *Mem:* Am Psychiat Asn; Am Pub Health Asn. *Res:* Epidemiological and treatment evaluation research in drug dependence; mental health planning in developing countries. *Mailing Add:* 8028 Melvina Oak Lawn IL 60459

HUGHES, PATRICK RICHARD, b San Jose, Calif, March 17, 43; m 63; c 7. PLANT INSECT INTERACTIONS. *Educ:* San Jose State Col, BA, 64; Univ Calif, Davis, MS, 69, PhD(entom), 73. *Prof Exp:* Lab technician, Univ Calif, Davis, 64-66, res asst, 66-67, res entom, 67-68; res assoc, 69-72, asst, 73-77, assoc insect physiologist, 77-85, INSECT PHYSIOLOGIST, BOYCE THOMPSON INST PLANT RES, CORNELL UNIV, 85- *Concurrent Pos:* Teaching asst entom, Univ Calif, Davis, 70-71; guest prof, Forest Zool Inst, Univ Freiburg, WGer, 77; mem, Southern Regional Proj SY135, Develop Microbiol Pesticides, 79-; chmn, organizing comt, Am Registry Prof Entomologist, Northeastern Br, 82, NY Br, 81-82, nominating comt for sr examr, physiol & biochem specialty, 84; mem, Res Design Team to Brazil, Bean/Cowpea Collab Res Support Prog, 80; mem, USDA Competitive Res Grants Off Biol Stress Panel for Entomol, Nematol, 85; chmn nominating comt, subsect CE, Insect Path & Microbiol Control, Entom Soc Am, 86; mem, USDA competitive res grants, Off Biol Strero Panel Insect Pest Sci, 88; specialty mem, Behav, Prof Maintenance & Cert Comt, Am Registry Prof Entomologists, 88-91. *Mem:* Entom Soc Am; Am Registry Prof Entomologists; Soc Invert Path; Int Soc Chem Ecol; Am Soc Virol. *Res:* Response of insects and mites to environmentally-induced changes in host quality; quantitative relationships between insects and insect pathogens. *Mailing Add:* Boyce Thompson Inst Cornell Univ Tower Rd Ithaca NY 14853-0108

HUGHES, PETER C(ARLISLE), b St Catherines, Ont, June 5, 40. AEROSPACE SCIENCE. *Educ:* Univ Toronto, BASc, 62, MASc, 63, PhD(aerospace sci), 66. *Prof Exp:* From asst prof to assoc prof, 66-76, PROF AEROSPACE SCI, UNIV TORONTO, 76- *Concurrent Pos:* Consult, Can Defence Res Bd, 66, Aerospace Eng & Res Consults, 66-, Commun Res Ctr, 70- & Spar Aerospace Prod Ltd, 71- *Mem:* Am Inst Aeronaut & Astronaut; Can Aeronaut & Space Inst. *Res:* Spacecraft attitude dynamics, control; trajectory dynamics and control; flexible vehicle control; optimal control. *Mailing Add:* 42 Hamlyn Cres Toronto ON M9B 1Z1 Can

HUGHES, RAYMOND HARGETT, b Walla Walla, Wash, June 1, 27; m 52; c 4. PHYSICS. *Educ:* Whitman Col, BA, 49; Univ Wis, MS, 51, PhD, 54. *Prof Exp:* Asst, Univ Wis, 49-54; from asst prof to assoc prof, 54-65, PROF PHYSICS, UNIV ARK, FAYETTEVILLE, 65- *Mem:* Fel Am Phys Soc. *Res:* Spectra induced by ion and electron impact; laser generated ions; relativistic electron beams. *Mailing Add:* Dept Physics Univ Ark Fayetteville AR 72701

HUGHES, RICHARD DAVID, geology; deceased, see previous edition for last biography

HUGHES, RICHARD LAWRENCE, pharmacy; deceased, see previous edition for last biography

HUGHES, RICHARD ROBERTS, chemical engineering, applied mathematics; deceased, see previous edition for last biography

HUGHES, RICHARD V(AN VOORHEES), b Omaha, Nebr, Dec 27, 00; m 34; c 3. PETROLEUM ENGINEERING. *Educ:* Univ Nebr, BS, 25; Johns Hopkins Univ, PhD(geol), 33. *Prof Exp:* Jr geologist, State Geol Surv, Nebr, 25-26; geol draftsman, Lago Petrol Corp, Venezuela, 26-27, geologist, 27-29; asst chemist, Am Petrol Inst, 33; asst-in-charge acquisition surv, Cumberland Nat Forest, US Forest Serv, 33-35; geologist & asst chief geologist, Trop Oil Co, SAm, 35-40, chief geologist, 40-41; sr petrol engr & chief develop unit, Prod Div, Off Petrol Coordr, Petrol Admin for War, Washington, DC, 42-44; dir res, Pa Grade Crude Oil Asn, 44-49; vis lectr, prof petrol eng & head dept, Stanford Univ, 49-53; div reservoir engr & tech asst to vpres prod, Gulf Oil Corp, Tex, 53-59; prof, 59-67, emer prof petrol eng, Colo Sch Mines, 67-; RETIRED. *Concurrent Pos:* Sr petrol engr, Stanford Res Inst, 49-50; Int Exec Serv Corps adv, Univ Libya, 69-70; Fulbright fel, Univ Queensland, 70; petrol consult, 67-72. *Res:* Evaluation of oil properties. *Mailing Add:* PO Box 343 Americus GA 31709

HUGHES, ROBERT ALAN, b Milwaukee, Wis, June 27, 38; m 63; c 3. ENVIRONMENTAL CHEMISTRY, LIMNOLOGY. *Educ:* Univ Wis-Milwaukee, BS, 64; Univ Wis-Madison, MS, 68, PhD(water chem), 70. *Prof Exp:* Mgr environ res, Environ Systs Dept, Westinghouse Elec Corp, 70-71; prog supvr, Environ Assessment, Dept Sci Develop, Bechtel Corp, 71-72, group mgr environ sci, 72-74, group mgr environ monitoring, 74-76, proj mgr, 76-80, coal progs supvr, 80-81, CHEM WASTE TECHNOL MGR, BECHTEL NAT INC, 81- *Concurrent Pos:* Consult, Westinghouse Elec Corp, 70. *Mem:* Am Chem Soc; Am Soc Limnol & Oceanog; Am Water Works Asn; Int Asn Theoret & Appl Limnol; Int Asn Water Pollution Res; Sigma Xi. *Res:* Aquatic chemistry-biology interaction; environmental chemistry of organic nonelectrolytes; mitigation of environmental effects of large construction projects through advanced technology application; hazardous waste management and remedial action implementation. *Mailing Add:* Bechtel Environ Inc PO Box 3965 San Francisco CA 94119-3965

HUGHES, ROBERT CLARK, b Cleveland, Ohio, Nov 27, 40; m 67; c 3. MICROSENSOR SCIENCE. *Educ:* Carleton Col, BA, 63; Stanford Univ, PhD, 66. *Prof Exp:* Mem tech staff, 66-73, supvr physics of org solids, 73-78, SUPVR MICROSENSOR DIV, SANDIA NAT LAB, 84- *Mem:* Fel Am Phys Soc. *Res:* Physics of dielectrics; photoconductivity and photo detectors; solid state sensors including chemical and radiation sensors. *Mailing Add:* 1416 Mesilla NE Albuquerque NM 87110

HUGHES, ROBERT DAVID, b Chicago, Ill, Aug 17, 43; m 68; c 2. ORGANIC CHEMISTRY. *Educ:* Western Ill Univ, BS, 66; Univ Iowa, MS & PhD(org chem), 70. *Prof Exp:* Proj chemist, Am Oil Co, 70-79; PROJ CHEMIST, STANDARD OIL CO (IND), 79- *Mem:* Am Chem Soc. *Res:* Organic stereochemistry; reaction mechanisms; bicyclic compounds facilitated transport through membranes; synthetic fuels. *Mailing Add:* 518 Braemar Ave Naperville IL 60540-1373

HUGHES, ROBERT EDWARD, b New York, NY, May 24, 24; m 54; c 1. PHYSICAL CHEMISTRY, RESEARCH MANAGEMENT. *Educ:* Lehigh Univ, BS, 49; Cornell Univ, PhD(chem), 52. *Prof Exp:* Res asst, Bakelite Corp, 41-42, 46-47; instr Cornell Univ, 52-53; from asst prof to prof, Univ Pa, 53-64; dir, Mat Sci Ctr, Cornell Univ, 68-74, prof chem, 64-80; PRES, ASSOC UNIV INC, 80- *Concurrent Pos:* Consult, Rohm & Haas, 56-74, Sun Oil Co, 58-64 & 77-80 & Gen Motors, 78-80; sr NSF fel, Cambridge Univ, 67-68; ed, J Solid State Chem, 69-74; mem res & develop study group, US Comn Gov Proc, 71-72; mem solid state sci comt, Nat Res Coun-Nat Acad Sci, 67-74, chmn, 71-73; asst dir nat & int affairs, NSF, 74-75, astron, atmospheric, earth & ocean sci, 75-76 & actg asst dir sci, technol & int affairs, 75-76; head US deleg, 8th Antarctic Treaty Consult Meeting, Oslo, 75; assoc, Prog Sci, Technol & Soc, Cornell Univ, 77-80; mem nat mat adv bd, Nat Res Coun-Nat Acad Sci, 78-81, mem nat adv bd, Am Univ, 83-; vis prof, Oxford Univ, 79; mem, Res Coord Coun, Gas Res Inst, 85- *Mem:* Am Chem Soc; Am Crystallog Asn; AAAS; Am Astron Soc; Am Phys Soc; Sigma X. *Res:* X-ray crystallography; oligopeptide and antibiotic structures; boron and boride crystal chemistry; non-stoichiometric systems; physical chemistry of macromolecules. *Mailing Add:* Assoc Univ Inc 1717 Massachusetts Ave Washington DC 20036-2002

HUGHES, RUSSELL PROFIT, b Denbigh, Wales, Dec 23, 46. INORGANIC CHEMISTRY, ORGANOMETALLIC CHEMISTRY. *Educ:* Univ Manchester, Inst Sci & Technol, BSc, 67; Univ Toronto, PhD(chem), 72. *Prof Exp:* Res asst chem, Bristol Univ, 73-75; res assoc, McGill Univ, 75-76; asst prof, 76-82, assoc prof, 82-86, PROF CHEM, DARTMOUTH COL, 86- *Concurrent Pos:* Alfred P Sloan res fel, 80-84; Alexander von Humboldt fel, 83-84. *Mem:* Am Chem Soc. *Res:* Organotransition metal chemistry. *Mailing Add:* Dept Chem Dartmouth Col Hanover NH 03755-1477

HUGHES, STANLEY JOHN, b Llanelli, Wales, Sept 17, 18; m 58; c 3. MYCOLOGY. *Educ:* Univ Wales, BSc, 41, MSc, 43, DSc(mycol), 54. *Prof Exp:* Asst mycologist, Commonwealth Mycol Inst, Eng, 45-52; mycologist, 52-58, sr mycologist, 58-62, prin mycologist, 62-83, HON RES ASSOC, BIOSYST RES CTR, CENT EXP FARM, CAN DEPT AGR, 83- *Concurrent Pos:* Sr res fel, Dept Sci & Indust Res, NZ, 63. *Honors & Awards:* Jakob Eriksson Gold Medal, 69; George Lawson Medal, 81; Distinguished Mycologist Award, Mycol Soc Am, 85. *Mem:* Mycol Soc Am (pres, 75); mem Brit Mycol Soc (vpres, 87); Int Mycol Asn (vpres, 77-83); fel Royal Soc Can; Linnean Soc London. *Res:* Fungi imperfecti; classification; sooty molds. *Mailing Add:* Mycol Sect Biosysts Res Ctr Cent Exp Farm K W Neatby Bldg 960 Carling Ave Ottawa ON K1A 0C6 Can

HUGHES, STEPHEN EDWARD, b Newton, Mass, June 1, 53; m 84; c 2. TRANSPLANTATION, PHYSIOLOGY. *Educ:* Univ Miami, BS, 74; Univ Mass, Amherst, PhD(zool), 81. *Prof Exp:* Res & teaching asst biol, Univ Miami, 72-74; teaching assoc zool, Univ Mass, Amherst, 74-79; postdoctoral res assoc neurosci, Univ Mich, Ann Arbor, 80-85; ASST RES SCIENTIST NEUROBIOL, CENT INST DEAF, 85-; ASST PROF SPEECH & HEARING, WASH UNIV, 85- *Concurrent Pos:* Bd mem, St Louis Chap, Retinitis Pigmentosa Found, 88. *Mem:* AAAS; Asn Res Otolaryngol; Sigma Xi; Soc Neurosci. *Res:* Restoration of sensory function through the transplantation of sensory cells in the eye and ear; development of sensory systems. *Mailing Add:* Central Inst Deaf 818 S Euclid Ave St Louis MO 63110

HUGHES, THOMAS JOSEPH, b Brooklyn, NY, Aug 3, 43; m 72; c 3. ENGINEERING. *Educ:* Pratt Inst, BS, 65, MS, 67; Univ Calif, Berkeley, MS, 74, PhD(eng sci), 74. *Prof Exp:* Mech design engr, Grumman Aircraft Eng Co, 65-66; res & develop engr, Gen Dynamics Corp, 67-69; asst res engr, Univ Calif, Berkeley, 74-76, lectr, Dept Civil Eng, 75-76; asst prof struct mech, Calif Inst Technol, 76-78, assoc prof, 78-80; assoc prof, 80-82, PROF MECH ENG & CHMN, DIV APPL MECH, STANFORD UNIV, 83- *Concurrent Pos:* Ed, J Comput Methods Appl Mech & Eng. *Honors & Awards:* Walter Huber Res Prize, Am Soc Civil Engrs, 78; Melville Medal,

Am Soc Mech Engrs, 79. *Mem:* Am Soc Mech Engrs; Am Soc Civil Engrs; Sigma Xi; Am Inst Aeronaut & Astronaut. *Res:* Development of finite-element computer procedures for fluid, solid, soil and structural mechanics problems. *Mailing Add:* Div Appl Mech Stanford CA 94305

HUGHES, THOMAS ROGERS, JR, b Staten Island, NY, Oct 3, 31; m 67. PHYSICS. *Educ:* Yale Univ, BS, 53; Wash Univ, PhD(physics), 62. *Prof Exp:* INSTR PHYSICS, COL ALAMEDA, 70- *Mem:* Am Asn Physics Teachers; Int Solar Energy Soc; Astron Soc Pac; Am Solar Energy Soc. *Res:* Application of nuclear magnetic resonance techniques to the study of the metal ammonia system. *Mailing Add:* Dept Math Physics Chem & Biol Col Alameda 555 Atlantic Ave Alameda CA 94501

HUGHES, TRAVIS HUBERT, b Rapid City, SDak, Feb 21, 37; m 57; c 2. HYDROGEOLOGY, ENVIRONMENTAL GEOLOGY. *Educ:* Vanderbilt Univ, BA, 59, MS, 60; Univ Colo, PhD(geol), 67. *Prof Exp:* Geologist, Oman Construct Co, 60-62; from instr to prof geol, 66-82, chmn dept, UNIV ALA, 78-81; VPRES, P E LAMOREAUX & ASSOC, 82- *Concurrent Pos:* US rep, Conf Remobilization Ore Deposits, Sardinia, 68; sr staff scientist, Univ Ala Environ Inst Waste Mgt Studies, 84-88; comt Onshore Oil & Gas Leasing, Nat Acad Sci, 89. *Honors & Awards:* Waldemar Lindgren Citation Award, Soc Econ Geologists, 68; NASA Citation for Innovative Res, 78; Certificate of Merit, Am Inst Prof Geol, 83; Distinguished Achievement Earth Sci Award, Fedn of Lapidary & Mineral Socs, 79. *Mem:* Am Inst Prof Geologists (pres, 86); Nat Asn Geol Teachers; Geochem Soc; Geol Soc Am; Am Geol Inst; Asn Ground Water Scientists & Engrs. *Res:* Hydrogeology; geochemistry; environmental geology. *Mailing Add:* P E LaMoreaux & Assoc Box 2310 Tuscaloosa AL 35403

HUGHES, VERNON WILLARD, b Kankakee, Ill, May 28, 21; m 50; c 2. PHYSICS. *Educ:* Columbia Univ, AB, 41, PhD(physics), 50; Calif Inst Technol, MS, 42. *Hon Degrees:* Dr, Univ Heidelberg, 77. *Prof Exp:* Res assoc radiation lab, Mass Inst Technol, 42-46; instr & lectr physics, Columbia Univ, 49-52; asst prof, Univ Pa, 52-54; from asst prof to prof, 60-69, assoc chmn dept, 60-61, chmn, 61-66, Donner prof, 69-78, STERLING PROF PHYSICS, YALE UNIV, 78- *Concurrent Pos:* Assoc prof physics, Columbia Univ, 58-59, vis I I Rabi prof, 84, adj prof, 84-; trustee, Assoc Univs, Inc, 62-; mem, Naval Res Adv Comt, 68-74; counr-at-large, Am Phys Soc, 70-74; vis prof, Japan Soc Prom Soc Sci, 74; Alexander von Humboldt Sr US Scientist Award, 76; Guggenheim fel, 78-79; vis prof, Slac, Stanford Univ, 78-79, Col France, 81, Scuola Normale Superiore, Pisa, Italy, 82; consult, Los Alamos Sci Lab, Oak Ridge Nat Lab, NSF, Nat Res Coun, Dept Energy & others. *Honors & Awards:* Morris Loeb Lectr Physics, Harvard Univ, 72; Davisson-Germer Prize, Am Phys Soc, 78, Tom W Bonner Prize, 90; Rumford Prize, Am Acad Arts & Sci, 85. *Mem:* Nat Acad Sci; fel Am Acad Arts & Sci; fel Am Phys Soc; fel AAAS. *Res:* Radiofrequency and microwave spectroscopy of atoms and molecules; particle physics. *Mailing Add:* Dept Physics Yale Univ 217 Prospect St PO Box 6666 New Haven CT 06511-8167

HUGHES, VICTOR A, b Manchester, Eng, Mar 17, 25; m 58, 70; c 4. PHYSICS, ASTRONOMY. *Educ:* Univ Manchester, BSc, 45, MSc, 49, DSc, 77. *Prof Exp:* Mem staff radar & radio astron, Royal Radar Estab, Malvern, Eng, 52-61; prin sci officer, Radio Res Sta Slough, 61-63; prof physics & astron, 63-68, PROF PHYSICS, QUEEN'S UNIV (ONT), 68- *Concurrent Pos:* Chmn comn V, Can Nat Comt, Int Sci Radio Union, 65-70. *Mem:* Can Asn Physicists; Int Astron Union; fel Royal Astron Soc; Am Astron Soc; Can Astron Soc. *Res:* Galactic astronomy; interstellar medium. *Mailing Add:* Dept Physics Astron Group Queen's Univ Kingston ON K7L 3N6 Can

HUGHES, WALTER LEE, JR, b Trenton, NJ, Nov 19, 15; m 45; c 3. BIOCHEMISTRY. *Educ:* Mass Inst Technol, SB, 37, PhD(org chem), 41. *Prof Exp:* Fel protein chem, Harvard Univ, 40-42, res assoc phys chem, 42-46, assoc, 46-48, asst prof, 48-53; assoc prof, Johns Hopkins Univ, 53-55; head div microbiol, Med Res Ctr, Brookhaven Nat Lab, 55-58, head div biochem, 58-63; chmn dept, 63-76, PROF PHYSIOL, SCH MED, TUFTS UNIV, 63- *Concurrent Pos:* Guggenheim fel, Calif Inst Technol, 51-52; res collabr, Brookhaven Nat Lab. *Mem:* Am Acad Arts & Sci; Am Soc Biol Chemists; Fedn Am Scientists; Soc Gen Physiologists; Am Physiol Soc. *Res:* Proteins and nucleic acids; mechanisms of their synthesis and catabolism and their cellular localization; technics for studying metabolism; tritium-labeled nucleosides; chromium labeled proteins. *Mailing Add:* 54 Bellport Lane Bellport NY 11713

HUGHES, WALTER T, b Cleveland, Tenn, May 16, 30; m 57; c 3. PEDIATRICS. *Educ:* Tenn Polytech Inst, 48-50; Univ Tenn, MD, 54. *Prof Exp:* Intern, Gen Hosp, Knoxville, Tenn, 55; resident, Col Med, Univ Tenn, 57; staff mem, Walter Reed Army Med Ctr, Res & Develop Command, Ft Detrick, Md, 57-59; from instr to prof pediat, Sch Med, Univ Louisville, 61-69; prof pediat & microbiol, Univ Tenn Med Units, Memphis, 69-77; prof pediat & dir, Div Pediat Infectious Dis, Sch Med, Johns Hopkins Univ, 77-81, staff mem, Johns Hopkins Hosp, 77-81; CHMN, DIV INFECTIOUS DIS, ST JUDE CHILDREN'S RES HOSP, 81-; PROF PEDIAT, CTR HEALTH SCI, UNIV TENN, 81- *Concurrent Pos:* Consult, Ireland Army Hosp, US Army, Ft Knox, Ky, 63-69; lectr pediat, Sch Med, Johns Hopkins Univ, 81- *Mem:* Am Soc Microbiol; Soc Pediat Res; Infectious Dis Soc Am; Am Acad Pediat; Am Pediat Soc. *Res:* Infectious diseases in the immune-compromised host. *Mailing Add:* St Jude Children's Res Hosp 332 N Lauderdale Memphis TN 38101

HUGHES, WALTER WILLIAM, III, b Glendale, Calif, Apr 23, 51; m 72; c 2. INVERTEBRATE ZOOLOGY, INVERTEBRATE PALEONTOLOGY. *Educ:* Loma Linda Univ, BA, 73, PhD(biol), 78; Pac Union Col, MA, 74. *Prof Exp:* ASST PROF BIOL, ANDREWS UNIV, 78- *Concurrent Pos:* Res assoc, Univ Newcastle, Eng, 78-79. *Honors & Awards:* Edmund C Jaeger Award, 76. *Mem:* Sigma Xi; Paleontol Soc. *Res:* Invertebrate skeletal growth increments and their relation to paleontology, paleoecology and geophysics. *Mailing Add:* Dept Biol Andrews Univ Berrien Springs MI 49104

HUGHES, WILLIAM (LEWIS), b Rapid City, SDak, Dec 2, 26; c 4. ELECTRICAL ENGINEERING. *Educ:* SDak Sch Mines & Technol, BS, 49; Iowa State Univ, MS, 50, PhD(elec eng), 52. *Prof Exp:* TV studio & transmitter engr, WOI-TV, Ames, Iowa, 49-52; from instr to prof elec eng, Iowa State Univ, 49-60; head sch elec eng, 60-76, PROF ELEC ENG, OKLA STATE UNIV, 60-, DIR ENG ENERGY LAB, 75- *Concurrent Pos:* Mem, TV Allocations Study Orgn, 57-60; instigator & dir unconvention energy systs work, Okla State Univ, 62-; chmn study group XI-B, US Prep Comt, Consult Comt Int Radio, 66-73; nat chmn, Comn Elec Eng Dept Heads, 66; instigator & dir Themis invest elec characteristics severe storms, US Dept Defense, 67-72; US deleg & chmn Int Color TV Stand Group, Consult Comt Int Radio meeting, Palma, Spain, 68 & Geneva, 69-71; chmn, Nat Acad Sci Comt Energy Sources Develop Nations; tech adv, Ministry Elec, Arab Repub Egypt; consult renewable energy matters var industs, govt agencies & several foreign govt. *Mem:* Fel Inst Elec & Electronics Engrs; AAAS; Soc Motion Picture & TV Engrs; Am Soc Eng Educ; Solar Energy Soc. *Res:* Nonlinear systems; electromagnetics; color television systems; fuel cells; energy storage systems. *Mailing Add:* Im Em Corp 1016 E Airport Rd Stillwater OK 74075

HUGHES, WILLIAM BOND, organic geochemistry, organometallic chemistry, for more information see previous edition

HUGHES, WILLIAM CARROLL, b Albuquerque, NMex, Apr 5, 38; m 60; c 2. CIVIL ENGINEERING. *Educ:* Univ NMex, BS, 60, MS, 65, PhD(civil eng), 69; Univ Utah, BS, 61. *Prof Exp:* Asst engr, Calif State Dept Water Resources, 64-66; asst prof, 68-75, ASSOC PROF CIVIL ENG, UNIV COLO, DENVER, 75- *Concurrent Pos:* NSF initiation grant, 71-72. *Mem:* Am Soc Civil Engrs; Am Soc Eng Educ; Sigma Xi. *Res:* Hydrology and hydraulics. *Mailing Add:* 2318 Dennison Lane Boulder CO 80303

HUGHES, WILLIAM E, b Centreville, Miss, Sept 12, 32; m 60; c 3. SOLID STATE PHYSICS. *Educ:* Univ Southern Miss, BS, 58; Univ Ala, PhD(physics), 63. *Prof Exp:* Asst prof physics, Stetson Univ, 64-69; PROF PHYSICS & CHMN DEPT, UNIV SOUTHERN MISS, 69- *Res:* Electron spin resonance; nuclear resonance in solids. *Mailing Add:* Dept Physics Univ Southern Miss Box 5046 Hattiesburg MS 39406

HUGHES, WILLIAM F(RANK), b Ash, NC, Oct 20, 30; m 59; c 2. MECHANICAL ENGINEERING. *Educ:* Carnegie Inst Technol, BS, 52, MS, 53, PhD(mech eng), 55. *Prof Exp:* Assoc prof mech eng, 55-67, PROF MECH & ELEC ENG, CARNEGIE-MELLON UNIV, 67- *Concurrent Pos:* NSF fel, Cambridge Univ, 57-58. *Honors & Awards:* Fulbright lectr, Univ Sydney, 63. *Mem:* Am Soc Mech Engrs; Soc Automotive Engrs; Am Phys Soc; Am Geophys Union. *Res:* Fluid dynamics and hydrodynamic lubrication and friction; magnetohydrodynamics; nuclear and reactor engineering. *Mailing Add:* RD 3 Cambridge Springs PA 16403

HUGHES, WILLIAM TAYLOR, b Vidor, Tex, Nov 15, 36; m 65; c 1. BIOPHYSICS. *Educ:* Ind Univ, BS, 60, MA, 62; Northwestern Univ, PhD(astron), 67. *Prof Exp:* Astronr, Smithsonian Astrophys Observ Satellite Tracking Sta, Neth, WI, 58-59; instr astron, Univ Mo, 62-63; asst prof physics & astron, WVa State Col, 63-65; from asst prof to assoc prof, 67-75, chmn dept, 71-75, PROF PHYSICS & ASTRON, BOWDOIN COL, 75- *Concurrent Pos:* Consult, Southworth Planetarium, Univ Maine, 69-71; mem, Biosatellite eval panel, NASA, 70; mem staff, NATO Advan Study Inst Biophys, 71 & Cambridge Univ, 73-74. *Mem:* Fel Royal Astron Soc. *Res:* Astronomical spectroscopy and photometry; bioenergetics; membrane properties; DNA replication. *Mailing Add:* Dept Physics & Astron Bowdoin Col Brunswick ME 04011

HUGHETT, PAUL WILLIAM, b San Rafael, Calif, May 19, 50. IMAGE PROCESSING, COMPUTER GRAPHICS. *Educ:* Mass Inst Technol, BS, 86, MS, 86; Univ Calif, Berkeley, PhD(elec eng), 91. *Prof Exp:* Mem tech staff, Hewlett-Packard, 73-77; syst engr, Singer-Link Flight Simulation, 77-81; OWNER, HUGHETT RES, 81- *Concurrent Pos:* Instr, Univ Calif, Santa Cruz, 90- *Mem:* Prof & Tech Consults Asn (pres, 87-89); Soc Photo-Optical Instrumentation Engrs; Inst Elec & Electronics Engrs Computer Soc; Soc Motion Picture & TV Engrs; Asn Comput Mach. *Res:* Applications of image processing and computer graphics to scientific and engineering research; quantitative infrared thermography; ultrasound imaging; non-destructive testing; fringe pattern analysis. *Mailing Add:* Hughett Res PO Box 60 Palo Alto CA 94302

HUGILL, J(OHN) T(EMPLETON), chemical engineering, physical chemistry, for more information see previous edition

HUGLI, TONY EDWARD, b Logan, Ohio, June 26, 41; m 65; c 2. BIOCHEMISTRY, PROTEIN CHEMISTRY. *Educ:* Otterbein Col, BS, 63; Ind Univ, Bloomington, PhD(biochem), 68. *Prof Exp:* Res assoc protein chem, Rockefeller Univ, 68-72; assoc mem, 72-82, MEM, SCRIPPS CLIN & RES FOUND, 82- *Concurrent Pos:* Estab investigatorship, Am Heart Asn, 72-77, grant-in-aid, 74-77; exec ed, Anal Biochem, 85-; assoc ed, Immunopharmacol, 88-, Protein Sci, 91- *Mem:* Am Chem Soc; Am Soc Immunol; Am Soc Biol Chem; Am Soc Exp Pathol. *Res:* Structural comparison of dissolved and crystalline sperm whale metmyoglobin; active site studies of bovine pancreatic deoxyribonuclease A; method for determining tryptophan in proteins following alkaline hydrolysis; chemistry and biology of human serum complement components. *Mailing Add:* 10666 N Torrey Pines Rd La Jolla CA 92037

HUGO, NORMAN ELIOT, b Beverly, Mass, Sept 23, 33; m 59; c 5. PLASTIC SURGERY. *Educ:* Williams Col, AB, 55; Cornell Univ, MD, 59. *Hon Degrees:* DSc, Williams Col, 89. *Prof Exp:* Instr surg, Med Sch, Cornell Univ, 64-66; dir plastic surg lab, Ind Univ Med Ctr & asst prof surg, Med Sch, 66-67; assoc prof surg, Med Sch, Univ Chicago, 69-70; assoc prof surg, Northwestern Univ, 70-82; PROF SURG, COLUMBIA PRESBY MED CTR, CHMN, PLASTIC SURG, 82- *Concurrent Pos:* J J O'Neill res fel, New York Hosp-Cornell Med Ctr, 64-65; attend surgeon plastic surg, Ind Univ

Med Ctr, 66-67; asst chief plastic surg, Walter Reed Army Hosp, 67-69; attend surgeon & chief plastic surg, Michael Reese Hosp, 69-70; attend surgeon, Northwestern Mem Hosp, 70-82, Columbia-Presby Med Ctr; vchmn, Am Bd Plastic Surg, 88-89. *Mem:* Plastic Surg Res Coun; Am Asn Plastic Surg; Educ Found; Am Soc Plastic & Reconstruct Surgeons (pres, 88-89); Asn Acad Surgeons. *Res:* Wound healing, especially wound contraction and the role of the dermis in controlling myofibroblastic contraction; fetal wound healing. *Mailing Add:* Dept Surg Columbia Presbyterian Med Ctr 360 W 168th St New York NY 10032

HUGON, JEAN S, electron microscopy, gastroenterology, for more information see previous edition

HUGUENIN, GEORGE RICHARD, b East Stroudsburg, Pa, Nov 6, 37; m 75; c 1. ASTRONOMY, ASTROPHYSICS. *Educ:* Mass Inst Technol, SB, 59; Harvard Univ, PhD(astron), 63. *Prof Exp:* Asst prof astron, Harvard Univ, 63-68; from assoc prof to prof astron, Univ Mass, Amherst, 68-82; dir, Five Col Radio Astron Observ, 71-82; CHMN, MILLITECH CORP, 82- *Concurrent Pos:* Mem Comn J, US Nat Comt, Int Sci Radio Union, 64. *Honors & Awards:* Harvard Soc Fels; Bart J Bok Award. *Mem:* Int Astron Union; Am Astron Soc; Inst Elec & Electronics Engrs. *Res:* Radio astronomy investigating pulsars; the sun and other rapidly varying sources; millimeter wavelength spectroscopy of the interstellar medium; millimeter systems applications. *Mailing Add:* Millitech Corp S Deerfield Res Park Box 109 South Deerfield MA 01373

HUGULEY, CHARLES MASON, JR, b Macon, Ga, Mar 15, 18; m 47; c 1. INTERNAL MEDICINE, ONCOLOGY. *Educ:* Emory Univ, AB, 38; Wash Univ, MD, 42. *Prof Exp:* From intern to resident med, Barnes Hosp, 42-45; chief resident, Grady Mem Hosp, 45-46; Am Cancer Soc fel, Univ Utah, 46-47; from asst prof to prof med, 48-88, div dir hemat & oncol, 55-88, dir, Cancer Ctr, 72-82, Am Cancer Soc prof clin oncol, 83-88, EMER PROF MED, SCH MED, EMORY UNIV, 88- *Mem:* Am Soc Hemat; Int Soc Hemat; Am Asn Cancer Res; Am Soc Clin Oncol; Am Fedn Clin Res; Am Med Asn; Am Col Physicians. *Res:* Hematology, oncology. *Mailing Add:* 865 Clifton Rd Atlanta GA 30307

HUGUNIN, ALAN GODFREY, b Janesville, Wis, Apr 22, 45; m 67; c 2. FOOD SCIENCE, TECHNICAL MANAGEMENT. *Educ:* Univ Wis-Madison, BS, 67, MS, 70, PhD(food sci), 75; Univ Calif, Berkeley, MBA, 81. *Prof Exp:* Biosci asst serol, Walter Reed Army Inst Res, 70-71; proj leader food sci, Foremost Foods Co, Foremost-McKesson, Inc, 75-77, group leader & tech mgr indust prof, 77-85; dir prod develop, 85-87, SR DIR PROD TECHNOL, NAT FOOD PROCESSORS ASN, 87- *Concurrent Pos:* Consult. *Mem:* Inst Food Technologists. *Res:* Development of dairy-based ingredients and dehydrated vegetable products for use in food systems; formulation of dairy, bakery, confection and other processed food products with nutritional, functional and/or economic advantages; by-product utilization; protein functionality; enzyme applications in food systems; liqueurs; new business analysis and development. *Mailing Add:* Nat Food Lab Nat Food Processors Assoc 6363 Clark Ave Dublin CA 94568

HUGUS, Z ZIMMERMAN, JR, b Washington, DC, Aug 14, 23; m 47; c 4. INORGANIC CHEMISTRY. *Educ:* Williams Col, BA, 43; Univ Calif, Berkeley, PhD(chem), 49. *Prof Exp:* Res fel, Radiation Lab, Univ Calif, Berkeley, 49-52, instr chem, 50-52; asst prof, Univ Minn, 52-57, assoc prof, 57-63, prof & chief inorg chem, 63-67; dept head, 67-73, PROF CHEM, NC STATE UNIV, 67- *Mem:* Am Chem Soc; fel AAAS. *Res:* Application of nuclear quadrupole resonance measurements to the elucidation of the electronic structures of transition metal compounds. *Mailing Add:* Dept Chem NC State Univ Raleigh NC 27695-8204

HUH, OSCAR KARL, b Nov 29, 35; c 1. COASTAL OCEANOGRAPHY, GEOLOGY. *Educ:* Rutgers Univ, BA, 57; Pa State Univ, MS, 63, PhD(geol), 68. *Prof Exp:* Res asst stratig & sedimentol, Pa State Univ, 57-59, res geologist, 62-67; res oceanogr coastal & satellite oceanog, US Naval Oceanog Off, Washington, DC, 67-76; assoc prof oceanog & marine sci, 76-84, prof geol & geophys, 84-90, PROF OCEANOG & COASTAL SCI, COASTAL STUDIES INST, LA STATE UNIV, 91- *Concurrent Pos:* Substitute & guest lectr, Pa State Univ & Univ Md; NSF fel, 64; basic res grant, Off Naval Res, US Naval Oceanog Off, Coastal Oceanog, Sea of Japan, 71-72; consult, remote sensing oil spills, Tex, Kuwait, Greek coastal waters & Gulf Mex. *Mem:* Soc Econ Paleont & Mineral; AAAS; Am Geophys Union; fel Explorers Club. *Res:* Air-sea interactions during cold air outbreaks over continental shelf waters; oceanographic and hydrologic applications of satellite radiometric imagery; coastal and continental shelf oceanography and coastal sedimentary processes; oceanography of Sea of Japan, East China Sea, Yellow Sea, Bay of Bengal & Gulf Mex; stratigraphy of petrology of Mississippian Age Carbonate Strata, Overthrust Belt, Idaho-Montana; remote sensing of environment. *Mailing Add:* 615 Sunset Blvd Baton Rouge LA 70808

HUHEEY, JAMES EDWARD, b Cincinnati, Ohio, Aug 2, 35. INORGANIC CHEMISTRY, HERPETOLOGY. *Educ:* Univ Cincinnati, BS, 57; Univ Ill, MS, 59, PhD(inorg chem), 61. *Prof Exp:* Res fel chem, Univ Mich, 61; asst prof, Worcester Polytech Inst, 61-65; from asst prof to assoc prof, 65-75, PROF CHEM, UNIV MD, COLLEGE PARK, 75- *Concurrent Pos:* NSF grants, 65-67, 75-76 & 78-81; vis prof chem, Univ Calif, Los Angeles, 86, vis prof zool, Southern Ill Univ, Carbondale, 87, 89-90. *Mem:* Fel AAAS; Am Chem Soc; Ecol Soc Am; Soc Study Evolution; Soc Study Amphibians & Reptiles; Herpetologists League; fel Explorers Club. *Res:* Phosphorus, nitrogen and sulfur; Lewis acid-base interactions; orbital electronegativity, electronegativity equalization, group electronegativities and polar effects; zoology; mimicry; herpetology of Southern Appalachians, especially salamanders. *Mailing Add:* 6909 Carleton Terr College Park MD 20740

HUI, CHIU SHUEN, b Hong Kong, June 7, 42; m; c 1. PHYSIOLOGY, BIOPHYSICS. *Educ:* Univ Hong Kong, BS, 66; Mass Inst Technol, PhD(physics), 73. *Prof Exp:* Res assoc biophys, Ctr Theoret Studies, Miami, 73-74; fel physiol, Yale Med Sch, 74-79; from asst prof to assoc prof biol sci, Purdue Univ, 79-84; ASSOC PROF PHYSIOL & BIOPHYS, IND UNIV, 84- *Concurrent Pos:* Fel, NIH, 77, res career develop award, 83-84 & 85-89. *Mem:* Biophys Soc; Am Heart Asn. *Res:* Biophysics and physiology of excitable membrane in nerve and muscle; excitation-contraction coupling in muscle. *Mailing Add:* Dept Physiol & Biophys Ind Univ Med Ctr Indianapolis IN 46223

HUI, KOON-SEA, b Hong Kong, Sept 21, 48; m 74; c 1. NEUROCHEMISTRY, PSYCHIATRY. *Educ:* Chinese Univ Hong Kong, BSc, 71; Hong Kong Univ, MPhil, 74; Univ Sask, PhD(psychiat), 76. *Prof Exp:* Demonstr biol, Chinese Univ Hong Kong, 71-72; teaching asst pharmacol, Hong Kong Univ, 72-74; lab scientist II psychiat, Dept Health, Sask, 74-76; SR RES SCIENTIST, NY STATE RES INST NEUROCHEM & DRUG ADDICTION, 76- *Concurrent Pos:* Res assoc prof phychiat, NY Univ Med Ctr. *Mem:* AAAS; Am Soc Neurochem; Int Soc Neurochem; Am Soc Biol Chemists; NY Acad Sci. *Res:* Biochemistry of brain; brain protein and peptide turnover; neuropeptide breakdown and its regulation; inhibitors for neuropeptide breakdown. *Mailing Add:* Nathan S Kline Inst Psychiat Res Orangeburg NY 10962

HUI, SEK WEN, b Yunan, China, July 15, 35; m; c 2. MEMBRANE BIOPHYSICS, ELECTRON MICROSCOPY & DIFFRACTION. *Educ:* Univ Western Australia, BSc, 64; Monash Univ, Melboune, PhD(physics), 68. *Prof Exp:* Lectr phys sci, Flinders Univ S Australia, 68-69; res physicist, Carnegie-Mellon Univ, 70-71, res fel biophysics, 71-72; from asst prof to assoc prof, 76-84, PROF BIOPHYSICS, STATE UNIV NY, BUFFALO, 84-; HEAD, MEMBRANE BIOPHYSICS LAB, ROSWELL PARK MEM INST, 76- *Concurrent Pos:* Spec fel, NIH, 72; dir, Electron Micros Soc Am, 85-87; mem regional high voltage electron micros facil adv comt, NIH, 82-84, Biophys Chem Study Sect, 86-90; res scientist, Roswell Park Mem Inst, 72-; prin investr 6 grants, NIH & Am Cancer Soc, 74- *Mem:* Electron Micros Soc Am; Am Phys Soc; Biophys Soc; Am Soc Cell Biol; NY Acad Sci. *Res:* The effects of membrane lipids on the activities of biological membranes including transport, regeneration, fusion and electrical potential; physical chemistry of membrane lipids, molecular mechanism fusion and molecular organization in biomembranes; author of over 100 papers in professional journals and 20 chapters in scientific books. *Mailing Add:* Membrane Biophys Lab Roswell Park Mem Inst 666 Elm St Buffalo NY 14263

HUI, SIU LUI, b Hong Kong, Oct 12, 48. BIOMETRICS, BIOSTATISTICS. *Educ:* Univ Hong Kong, BSc, 70; Yale Univ, PhD(biomet), 79. *Prof Exp:* Asst prof biostatist, Med Ctr, Univ Ill, 79-81; asst prof to assoc prof, 81-88, PROF BIOSTATIST, SCH MED, IND UNIV, 88- *Concurrent Pos:* Prin investr, NIH, 80-83, 84-86, 87, mem study sect, 86-90, concensus panel, 90. *Mem:* Am Statist Asn; Biomet Soc. *Mailing Add:* Dept Med Ind Univ RR 135 702 Bamhill Dr Indianapolis IN 46202

HUIATT, TED W, MUSCLE DEVELOPMENT, MUSCLE PROTEIN ASSEMBLY. *Educ:* Iowa State Univ, PhD(biochem), 79. *Prof Exp:* ASSOC PROF BIOCHEM, IOWA STATE UNIV, 82- *Mailing Add:* Dept Animal Sci Muscle Biol Group Iowa State Univ Ames IA 50011

HUIE, ROBERT ELLIOTT, b Atlanta, Ga, Jan 24, 45; div; c 2. ATMOSPHERIC CHEMISTRY & PHYSICS. *Educ:* Macalester Col, BA, 66; Univ Md, Col Park, PhD(phys chem), 72. *Prof Exp:* RES CHEMIST, NAT INST STANDARDS & TECHNOL, 67- *Concurrent Pos:* Vis scientist phys chem, Cambridge Univ, 75-76. *Mem:* Am Chem Soc; Am Geophys Union. *Res:* Gas phase kinetics; solution kinetics; atmospheric chemistry; pulse radiolysis; flash photolysis. *Mailing Add:* Nat Inst Standards & Technol Gaithersburg MD 20899

HUIJING, FRANS, b Amsterdam, Neth, Jan 28, 36; m 63; c 2. CLINICAL BIOCHEMISTRY, GENETICS. *Educ:* Univ Amsterdam, BSc, 58, Drs, 61, DSc(biochem), 64. *Prof Exp:* Asst biochem, Univ Amsterdam, 61-64; res fel, Col Med Sci, Univ Minn, Minneapolis, 64-66; res scientist med enzymol, Univ Amsterdam, 66-68; assoc prof biochem & med, 68-73, prof biochem & assoc prof med, 73-86, PROF BIOCHEM & MOLECULAR BIOL, SCH MED, UNIV MIAMI, 86- *Mem:* Am Soc Biol Chemists; Soc Pediat Res. *Res:* Detection, genetics, biochemistry and nutritional management of inborn errors of metabolism, especially glycogen-storage diseases, lysosomes storage diseases; control of metabolism; computer-based medical education. *Mailing Add:* Dept Biochem & Molecular Biol PO Box 6129 Miami FL 33101

HUISINGH, DONALD, b Spokane, Wash, Mar 13, 37; m 67; c 2. PLANT PATHOLOGY, BIOCHEMISTRY. *Educ:* Univ Minn, BS, 61; Univ Wis, PhD(biochem, plant path), 65. *Prof Exp:* From asst prof to assoc prof plant path, 65-72, PROF UNIV STUDIES, NC STATE UNIV, 72- *Mem:* Am Soc Plant Physiol; Am Phytopath Soc; Am Soc Microbiol; Phytochem Soc NAm; Brit Soc Gen Microbiol. *Res:* Physiology of pathogenesis; biochemistry of differentiation; bacteriology; fungal physiology; bio-ecology of heavy metals as related to growth, development and disease resistance of plants; environmental science; alternative energy and aging. *Mailing Add:* Erasmus Studiecentrum Voor Milieukunde Erasmus Universiteit Postbus 1738 Rotterdam 3000 DR Netherlands

HUISJEN, MARTIN ALBERT, b Oak Park, Ill, May 30, 44; m 69; c 2. MAGNETIC RESONANCE, MICROWAVE ENGINEERING. *Educ:* Univ Ill, BS, 65; Cornell Univ, MS, 68, PhD(exp physics), 71. *Prof Exp:* Pulsed magnetic resonance, Varian Assocs, 71-73; mem tech staff microwave eng, Hughes Aircraft Co, 73-76; STAFF CONSULT MICROWAVE PHYSICIST, BALL AEROSPACE SYSTS DIV, 76- *Mem:* Sigma Xi; Am Phys Soc. *Res:* Magnetic resonance in solids and liquids, electron optics, radar and communications system engineering with emphasis on antenna systems; adaptive array antennas. *Mailing Add:* 2209 Juniper Ct Boulder CO 80304

HUISMAN, TITUS HENDRIK JAN, b Leeuwarden, Neth, Sept 1, 23; m 50; c 2. BIOCHEMISTRY. *Educ:* State Univ Groningen, MS, 46, PhD(chem), 48; State Univ Utrecht, DSc(biochem), 50. *Prof Exp:* Head biochem res lab, State Univ Groningen, 51-59; from assoc prof to prof biochem & path, 59-63, prof biochem, 63-64, REGENTS PROF CELL & MOLECULAR BIOL, MED COL GA, 64-, CHMN DEPT, 77- *Concurrent Pos:* Mem organizing comt colloquium, St Jans Hosp, Belg. *Honors & Awards:* D D Van Slyke Award, 71. *Mem:* Am Chem Soc; Am Soc Biol Chemists; Royal Neth Chem Soc; Europ Soc Hemat. *Res:* Inhomogeneity of hemoglobin and myoglobin types; functional differences of hemoglobin variants; genetic studies of human hemoglobin abnormalities. *Mailing Add:* Dept Cell & Molecular Biol Med Col Ga Augusta GA 30912-2100

HUITEMA, BRADLEY EUGENE, b Hammond, Ind, July 28, 38; m 61; c 2. ANALYSIS OF COVARIANCE, TIME-SERIES EXPERIMENTS. *Educ:* Southern Ill Univ, BA, 61; Western Mich Univ, MA, 62; Colo State Univ, PhD(psychol), 68. *Prof Exp:* Res psychologist, US Army Enlisted Eval Ctr, 62-64; res prof, res div, Ore State Univ, 67-68; PROF PSYCHOL, WESTERN MICH UNIV, 68- *Concurrent Pos:* vis prof, Univ Veracruz, Mex, 75; res design consult, 86-88; expert witness, Nabisco, 86-89; vis scholar, Univ Calif, San Diego, 88-89. *Mem:* Am Statist Asn; AAAS; Am Psychol Asn; Asn Behav Med; Asn Behav Anal; Am Educ Res Asn. *Res:* The design and analysis of single subject and group experiments; the effects of environment and behavior on atherosclerosis. *Mailing Add:* Dept Psychol Western Mich Kalamazoo MI 49008-5052

HUITINK, GERALDINE M, b Chicago, Ill, Oct 11, 42. ANALYTICAL CHEMISTRY. *Educ:* Mundelein Col, BS, 63; Iowa State Univ, MS, 65, PhD(anal chem), 67. *Prof Exp:* Asst prof chem, 67-73, ASSOC PROF CHEM, 73-, CHMN CHEM DEPT, IND UNIV SOUTH BEND, 82- *Mem:* Am Chem Soc; Sigma Xi; AAAS. *Res:* Synthesis and study of metallofluorochromic indicators. *Mailing Add:* Ind Univ PO Box 7111 South Bend IN 46634

HUITRIC, ALAIN CORENTIN, b Laurier, Man, July 28, 11; nat US. PHARMACEUTICAL CHEMISTRY. *Educ:* Loyola Univ, Calif, BS, 50; Univ Calif, MS, 52, PhD(pharmaceut chem), 54. *Prof Exp:* Asst prof, Univ San Francisco, 54-55; from asst prof to prof, 55-80, EMER PROF MED CHEM, UNIV WASH, 80- *Mem:* Am Chem Soc; Am Pharmaceut Asn; The Chem Soc. *Res:* Medicinal chemistry; organic synthesis; stereochemistry; configurational and conformational analysis by nuclear magnetic resonance; drug metabolism. *Mailing Add:* Dept Med Chem Sch Pharm Univ Wash Seattle WA 98195

HUIZENGA, JOHN ROBERT, b Fulton, Ill, Apr 21, 21; m 46; c 4. NUCLEAR CHEMISTRY, NUCLEAR PHYSICS. *Educ:* Calvin Col, AB, 44; Univ Ill, PhD(phys chem), 49. *Prof Exp:* Lab supvr, Tenn Eastman Co, 44-46; instr phys chem, Calvin Col, 46-47; assoc scientist, Argonne Nat Lab, 49-58, sr scientist, 58-67; prof chem & physics, chmn dept chem, 83-88, TRACY H HARRIS PROF CHEM & PHYSICS, UNIV ROCHESTER, 78- *Concurrent Pos:* Fulbright fel, Neth, 54-55; Guggenheim fel & vis prof, Univ Paris, 64-65; Guggenheim fel, Univ Calif, Berkeley & Niels Bohr Inst, Tech Univ Munich, 73-74. *Honors & Awards:* Ernest O Lawrence Mem Award, AEC, 66; Award for Nuclear Appln Chem, Am Chem Soc, 75. *Mem:* Nat Acad Sci; fel AAAS; fel Am Phys Soc; Am Chem Soc; Sigma Xi. *Res:* Heavy and light ion reactions; nuclear fission; nuclear level densities; nuclear properties of actinide elements; effect of environment on radioactive decay; muon-induced reactions in actinide elements. *Mailing Add:* Dept Chem Univ Rochester Rochester NY 14627

HUIZINGA, HARRY WILLIAM, b Rahway, NJ, Sept 4, 34; m 64; c 2. COMPARATIVE PATHOLOGY, PARASITOLOGY. *Educ:* Mich State Univ, BS, 56; Univ Md, MS, 61; Univ Conn, PhD(zool), 65. *Prof Exp:* Asst zool, Univ Md, 58-60; res technician, Patuxent Wildlife Res Ctr, US Fish & Wildlife Serv, Md, 60-62; asst & instr zool, Univ Conn, 62-63, lectr, 63-65; NIH trainee parasitol, Sch Pub Health, Univ NC, 65-67; from asst prof to assoc prof, 67-77, PROF PARASITOL, ILL STATE UNIV, 77- *Concurrent Pos:* Fel trop parasitol, La State Univ, 67; vis assoc prof path, Bull Fel, Ont Vet Col Guelph Univ, 74-75; vis assoc prof teaching, Peoria Med Sch, 72-87. *Mem:* Am Soc Trop Med & Hyg; Am Soc Parasitol; Wildlife Dis Asn; Sigma Xi. *Res:* Experimental host-parasite relationships of wildlife parasites. *Mailing Add:* Dept Biol Sci Ill State Univ Normal IL 61761

HUKILL, PETER BIGGS, b Lucerne, Switz, Feb 3, 27; US citizen; div; c 1. PATHOLOGY. *Educ:* Harvard Univ, AB, 47; Yale Univ, MD, 53. *Prof Exp:* From instr to assoc prof path, Sch Med, Yale Univ, 58-68; prof, Univ Ala, Birmingham, 68-69; prof path, Sch Med, Univ Conn & pathologist, Univ Hosps, 69-77; DIR LABS, CHARLOTTE HUNGERFORD HOSP, 77- *Concurrent Pos:* Asst attend path, Grace-New Haven Community Hosp, 59-61, attend, 61-68; clin prof path, Univ Conn, 77-; clin assoc prof, Yale Univ, 78- *Mem:* Col Am Pathologists; Am Soc Clin Pathologists; Int Acad Path; Am Soc Dermopath. *Res:* Histochemistry of human tumors; surgical pathology; dermatopathology. *Mailing Add:* Charlotte Hungerford Hosp Torrington CT 06790-0988

HULAN, HOWARD WINSTON, b Jeffrey's, Nfld, Oct 18, 41; m 64; c 4. NUTRITION, FOOD SCIENCE. *Educ:* McGill Univ, BSc, 65, MSc, 68; Univ Maine, PhD(nutrit), 71. *Prof Exp:* Asst nutrit, Univ Maine, 68-71; Nat Res Coun fel biochem, Carleton Univ, 71-72; nutritionist, Prod & Mkt Br, Kentville Res Sta Agr Can, 72-73, res scientist lipid metab, Res Br, Animal Res Inst, 73-77, sr scientist poultry sect, 77-88, prin res scientist, sect head/prog leader poultry, 88-90; PROF BIOCHEM, HEAD FOOD SCI, MEM UNIV, ST JOHN'S, NFLD, 90- *Concurrent Pos:* Consult nutrit & health, Prod & Mkt Br, Agr Can, 72-73; adj prof, Tech Univ Nova Scotia, Halifax, 88-91; prof & head dept poultry sci, Ore State Univ, Corvallis. *Honors & Awards:* Nutrit Res Award, Am Feed Indust, 87; Res Award, Am Broiler Coun, 88; Can Packers Medal Excellence Nutrit & Meat Sci Res, 88. *Mem:* Sigma Xi; Am Inst Nutrit; Poultry Sci Asn; Can Soc Nutrit Sci; World Poultry Sci Asn; Can Soc Animal Sci; Can Inst Food Sci & Technol. *Res:* Physiological, biochemical and pathological effects of feeding high levels of long chain monoenoic fatty acids on the rat, pig, chicken and monkey, with special emphasis on the myocardium; poultry nutrition-physiology; influence of strain and nutrition on leg abnormalities of chicken broilers and roasters; effect of nutrition on sudden death syndrome heart attack in chickens; nutrition-disease interrelationships; omega-3 fatty acids in broiler chickens and egg yolk lipids. *Mailing Add:* Food Sci Mem Univ St John's NF A1B 3X9 Can

HULBERT, LEWIS E(UGENE), b Somerton, Ohio, Nov 15, 24; m 48; c 3. ENGINEERING MECHANICS. *Educ:* Iowa State Col, BS, 47; Case Western Reserve Univ, MS, 51; Ohio State Univ, PhD, 63. *Prof Exp:* Asst chief metallurgist, True Temper Sporting Goods, 42-43 & 46; math consult, Res Lab, Am Soc Heating & Ventilating Engrs, 49; actuarial consult, Wyatt Co, 50-52; sr res engr, 52-67, chief, Adv Solid Mech Div, 67-77, RES LEADER TRANS & STRUCT DEPT, BATTELLE MEM INST, 77- *Concurrent Pos:* Vpres, Systs & Design; mem, Bd Res & Technol Develop. *Mem:* Fel Am Soc Mech Engrs; Sigma Xi. *Res:* Mechanics of deformable bodies; heat transfer; diffusion theory; theory of fibrous composites; biomechanics; numerical and computer solution of boundary value problems. *Mailing Add:* Battelle Mem Inst 505 King Ave Columbus OH 43201

HULBERT, LLOYD CLAIR, plant ecology; deceased, see previous edition for last biography

HULBERT, MATTHEW H, b Marlinton, WVa, Aug 25, 42; m 64; c 2. PHYSICAL CHEMISTRY, PHARMACEUTICAL MANUFACTURING PROCESSES. *Educ:* Washington & Lee Univ, BS, 64; Univ Wis, MS, 67, PhD(anal chem), 69. *Prof Exp:* Asst prof chem, Lehigh Univ, 69-75; Nat Res Coun res assoc, Atlantic Oceanogr & Meteorol Labs, 74-75; from asst prof to assoc prof chem, Conn Col, 75-81; sr res scientist, Int Minerals & Chem Corp, 81-87; SR RES SCIENTIST, PITMAN-MOORE INC, 87- *Concurrent Pos:* Consult, Pharmachem Corp, 70-74, Nat Oceanic & Atmospheric Admin, 75-79, USCG Res & Develop Ctr, 79-81; vis fel, Yale Univ, 77-78. *Mem:* Sigma Xi; Am Chem Soc; Parenteral Drug Asn; AAAS; Clay Minerals Soc. *Res:* Pharmaceutical process chemistry; marine sediments; resource management. *Mailing Add:* 45 Heritage Drive Terre Haute IN 47803

HULBERT, SAMUEL FOSTER, b Adams Center, NY, Apr 12, 36; m 60; c 3. BIOMEDICAL ENGINEERING, MATERIALS ENGINEERING. *Educ:* Alfred Univ, BS, 58, PhD(ceramic sci), 64. *Prof Exp:* Instr math & physics, Alfred Univ, 60-64; asst prof ceramic & metall eng, Clemson Univ, 64-68, assoc prof mat eng & head div interdisciplinary studies, 68-73, assoc dean eng res interdisciplinary progs, 70-73; prof surg & orthop & dean sch eng, Med Ctr, Tulane Univ, 73-76; PRES, ROSE-HULMAN INST TECHNOL, 76- *Concurrent Pos:* Chmn, Nat Conf Use of Ceramics in Surg Implants, 69; co-chmn, Nat Conf Mat for Implant Dent; co-chmn, Nat Conf Eng in Med-Bioceramics, 70. *Mem:* Am Ceramic Soc; Am Soc Artificial Internal Organs; Biomed Eng Soc; Nat Inst Ceramic Engr. *Mailing Add:* Rose-Hulman Inst Technol 5500 Wabash Ave Terre Haute IN 47803

HULBERT, THOMAS EUGENE, b Ft Edward, NY, Sept 1, 35; m 57; c 6. INDUSTRIAL & MANUFACTURING ENGINEERING. *Educ:* Rensselaer Polytech Inst, BMgtE, 57; Northeastern Univ, MS, 64. *Prof Exp:* Indust engr, Armstrong Corr Co, 57-60; sr indust engr, Raytheon Co, 60-63; assoc dir, Educ Resources, 67-69, from asst dean to actg dean eng, 69-81, INSTR & ASSOC PROF INDUST ENG, NORTHEASTERN UNIV, 63-, ASSOC DEAN & DIR ENG TECHNOL, 81- *Concurrent Pos:* Partner, Eng Mgt Assocs, 64-88; regional vpres & mem bd trustees, Inst Indust Engrs, 71-73. *Mem:* Inst Indust Engrs. *Res:* Design and analysis of manufacturing facilities to achieve maximum productivity; design and implementation of automated flexible manufacturing systems. *Mailing Add:* Sch Eng Technol 1205M Northeastern Univ Boston MA 02115

HULCHER, FRANK H, b Hampton, Va, Mar 12, 26; m 53; c 3. BIOCHEMISTRY. *Educ:* Va Polytech Inst, BS, 50, MS, 53, PhD(biochem), 57. *Prof Exp:* Instr bact, Va Polytech Inst, 53-54, asst biochem, 54-57; asst microbiol & fel, Yale Univ, 57-59, res assoc, 58; from instr to asst prof, 59-70, ASSOC PROF BIOCHEM, BOWMAN GRAY SCH MED, 70- *Concurrent Pos:* Vis scientist, Brookhaven Nat Lab, 60; problem-based tutor & evaluator Med Student Curriculum, Bowman Gray Sch Med, 87-91. *Honors & Awards:* Res Award, Sigma XI, 58. *Mem:* Fel AAAS; Am Chem Soc; Am Soc Microbiol; Am Soc Biol Chem; Sigma Xi. *Res:* Enzyme chemistry; biochemistry of myelin; microbial biochemistry; regulation of cholesterol metabolism, hydroxmethylglutaryl-coenzyme A reductase and cholesterol 7-alpha hydro xylase; role of isoprenoid and cholesterol synthesis in DNA replication; microbial hydroxamates and enzyme activation; control of plasma cholesterol by cortisol and cholesterol-7d hydroxylase. *Mailing Add:* Bowman Gray Sch Med/Biochem Wake Forest Univ 300 S Hawthorne Rd Winston-Salem NC 27103

HULET, CLARENCE VELOID, b Chinook, Mont, July 2, 24; m 52; c 8. ANIMAL SCIENCE. *Educ:* Brigham Young Univ, BS, 52; Univ Wis, MS, 53, PhD(reproductive physiol), 55. *Prof Exp:* Asst genetics, Univ Wis, 52-55; instr agr, Idaho State Col, 55-57; animal geneticist, Sheep Exp Sta, USDA, 57-62, animal physiologist, 62-76, supvry res physiologist, 76-90; RETIRED. *Concurrent Pos:* Fulbright-Hays sr res scholar, Ruakura Res Sta, Hamilton, NZ, 71-72. *Mem:* Am Soc Animal Sci. *Res:* Physiologic and genetics factors affecting reproduction in sheep, including semen studies; effects of nutrition on fertility; mating behavior; relationship between age at puberty and subsequent fertility; environmental and genetic factors affecting fertility in both the male and female ovine; selection for ovulation rate; effects of various hormones on reproductive phenomena; accelerated lambing. *Mailing Add:* 1905 E Powerhouse Rd Spanish Fork UT 84660

HULET, ERVIN KENNETH, b Baker, Ore, May 7, 26; m 49; c 2. ACTINIDE CHEMISTRY, EXPERIMENTAL NUCLEAR PHYSICS. *Educ:* Stanford Univ, BS, 49; Univ Calif, PhD(chem), 53. *Prof Exp:* Chemist nuclear chem, Lawrence Radiation Lab, Berkeley, 49-53, CHEMIST NUCLEAR CHEM, LAWRENCE LIVERMORE LAB, 53-, GROUP LEADER, 66- *Concurrent Pos:* Fulbright res scholar, Inst Atomic Energy, Univ Oslo, 62-63; co-discoverer of element 106; chmn & chmn-elect, Div Nuclear Chem & Technol, Am Chem Soc, 86-87. *Honors & Awards:* Wech Found Lectr, 90. *Mem:* Fel AAAS; fel Am Inst Chem; Am Phys Soc; Am Chem Soc. *Res:* Chemical and nuclear properties of the transplutonium elements. *Mailing Add:* L-232 Lawrence Livermore Lab Livermore CA 94550

HULET, RANDALL GARDNER, b Walnut Creek, Calif, Apr 27, 56; m 80; c 2. LASER COOLING OF ATOMS. *Educ:* Stanford Univ, BS, 78; Mass Inst Technol, PhD(physics), 84. *Prof Exp:* Postdoctoral assoc, Mass Inst Technol, 84-85; postdoctoral fel, NBS, 85-87; ASST PROF PHYSICS, RICE UNIV, 87- *Concurrent Pos:* Alfred P Sloan Found res fel, 89; NSF presidential young investr award, 89. *Mem:* Am Phys Soc. *Res:* Experimental atomic physics; quantum optics; laser cooling of atoms; interactions between ultra-cold atoms; laser and electro-optic-technology. *Mailing Add:* Dept Physics Rice Univ Houston TX 77251-1892

HULET, WILLIAM HENRY, b Minot, NDak, June 19, 25; m 69; c 1. INTERNAL MEDICINE. *Educ:* Minot State Col, BA, 47; Univ NDak, BS, 49; Univ Pa, MD, 51; Univ Miami, PhD(marine sci), 73; Am Bd Internal Med, dipl, 62. *Prof Exp:* Intern, George F Geisinger Mem Hosp, Danville Pa, 51-52; asst resident, Jackson Mem Hosp, Miami, Fla, 52-53; asst resident med, George F Geisinger Mem Hosp, Danville, Pa, 53-54, chief resident, 54-55; vis fel, Col Physicians & Surgeons, Columbia Univ, 55-56; NIH trainee, 55-57; fel, Sch Med, NY Univ, 56-58, instr physiol, 58-59; asst prof med, Sch Med, Univ Miami, 59-62, asst prof physiol, 60-63, from assoc prof to prof med & physiol, 63-73; prof med, physiol & biophys & chief Marine Med Div, Marine Biomed Inst, Univ Tex, Med Br Galveston, 73-83; ATTEND PHYSICIAN, JACKSON MEM HOSP, MIAMI, 61-; CONSULT, 62- *Concurrent Pos:* Asst physician, Presby Hosp, 55-56; clin asst vis physician, Bellevue Hosp, 56-59; Am Heart Asn res fel, 57-58; clin investr, Vet Admin Hosp, Coral Gables, Fla, 59-62, staff physician, 62-, assoc chief staff res, 62-63; NIH res career develop award, 63-; adj prof marine sci, Rosenstiel Sch Marine & Atmospheric Sci, Univ Miami, 77-83. *Honors & Awards:* C V Mosby Award, 51. *Mem:* AAAS; Am Fedn Clin Res; Am Physiol Soc; Am Col Physicians. *Res:* Marine science; diving medicine; renal physiology. *Mailing Add:* Rte 1 Box 78H Big Pine Key FL 33040

HULKA, BARBARA SORENSON, b Minneapolis, Minn, Mar 1, 31; m 54; c 3. EPIDEMIOLOGY. *Educ:* Radcliffe Col, BA, 52; Juilliard Sch Music, MS, 54; Columbia Col Physicians & Surgeons, MD, 59; Columbia Sch Pub Health & Admin Med, MPH, 61. *Prof Exp:* Asst pub health physician, Pa State Health Dept, 61-62; res instr, Dept Obstet & Gynec, Univ Pittsburgh, 62-65, res asst prof, 66-67; res assoc, Dept Prev Med, 67-68, asst prof epidemiol, Sch Pub Health, 67-71, asst prof med, Dept Family Med, 68-76, assoc prof epidemiol, Sch Pub Health, 72-76, assoc clin prof med, Dept Family Med & prof epidemiol, Sch Pub Health, Sch Med, 77-86, CHAIRPERSON DEPT EPIDEMIOL, SCH PUB HEALTH, UNIV NC, CHAPEL HILL, NC, 83-, CLIN PROF DEPT FAMILY MED, SCH MED, 86-, KENAN PROF, 87- *Concurrent Pos:* Reviewer, Health Manpower Prog & Res Proj, Nat Ctr Health Serv Res, 71-, Health Serv Develop grants study sect, 73-77, Comn Serv Fels Prog, 75-76, Cancer Control Prog, Nat Cancer Inst, 74- & Pop Res Prog, Nat Inst Child Health & Human Develop, 74-; fel, Health Care Resources Va, Nat Acad Sci, 74-75; consult var med insts & univs, 74-; aasoc clin prof epidemiol, community health sci, Med Ctr, Duke Univ, 76-82; comt toxic shock syndrome, Inst Med, Nat Acad Sci, 81-82; sci rev & eval bd, Health Serv Res Develop serv, Vet Admin, 83-85; breast cancer detection demonstration proj anal & publ comt, Nat Cancer Inst, 85; coord subcomt, Nat Acad Sci, 85, chair comt, 85-86; consult, Ortho Pharmaceut Corp, 88. *Honors & Awards:* Excellence Award, NC Pub Health Asn, 75. *Mem:* Soc Epidemiol Res (pres, 75-76); Am Pub Health Asn; Asn Teachers Prev Med; Am Col Prev Med; Int Epidemiol Asn. *Res:* Cancer epidemiology and health services research. *Mailing Add:* Dept Epidemiol Sch Pub Health Rosenau Hall 201 H Univ NC Chapel Hill NC 27514

HULKA, JAROSLAV FABIAN, b New York, NY, Sept 29, 30; c 3. OBSTETRICS & GYNECOLOGY. *Educ:* Harvard Univ, AB, 52; Columbia Univ, MD, 56. *Prof Exp:* Intern, Roosevelt Hosp, New York, 56-57; resident obstet & gynec, Sloane Hosp for Women, 57-60; Macy jr vis fel, Columbia-Presby Med Ctr, 60-61; asst prof, Univ Pittsburgh, 61-66, assoc mem microbiol, Grad Fac, 62-66, actg chmn dept obstet & gynec, Univ, 63-64; assoc prof obstet & gynec, Sch Med, 67-76, PROF MATERNAL & CHILD HEALTH, SCH PUB HEALTH, UNIV NC, CHAPEL HILL, 76-, PROF OBSTET & GYNEC, SCH MED, 76- *Mem:* Fel Am Col Obstet & Gynec; Am Fertil Soc; Am Asn Gynec Laparoscopists (pres, 80). *Res:* Antigenicity of the trophoblast; reproductive physiology; delivery of contraceptive service; sterilization techniques. *Mailing Add:* Dept Obstet Gynec Univ NC Sch Med Chapel Hill NC 27514

HULL, ALVIN C, JR, b Whitney, Idaho, Mar 25, 09; m 36; c 4. PLANT ECOLOGY. *Educ:* Utah State Univ, BS, 36, PhD, 59; Brigham Young Univ, MS, 40. *Prof Exp:* Range exam, Intermt Forest & Range Exp Sta, US Forest Serv, USDA, 36-41, forest ecologist, 42-47, range scientist, Rocky Mt Forest & Range Exp Sta, 48-51, asst, Div Range Mgt Res, Washington, DC, 51-55, coordr range seeding res, Western US, Agr Res Serv, 55-66, coordr range sci res, Western US, Agr Res Serv, 66-73; CONSULT DESERT LIVESTOCK, SKULL VALLEY CO, 73- *Concurrent Pos:* Range conserv work, Egypt & Israel, 52-53; range conserv res, Peru, 58. *Mem:* Fel Soc Range Mgt; Am Soc Agron; Sigma Xi. *Res:* Range seeding; range ecology; undesirable plant control; plant competition. *Mailing Add:* 321 N 400 E Logan UT 84321

HULL, ANDREW P, b New Britain, Conn, Jan 11, 20; m 42, 61; c 4. HEALTH PHYSICS. *Educ:* Cent Conn State Col, BS, 56; Vanderbilt Univ, MS, 61. *Prof Exp:* Res assoc meteorol, Travelers Weather Serv, 55-56; jr health physicist, Oak Ridge Nat Lab, 57-58; supvr health physics, Indust Reactor Lab, 58-61; assoc health physicist, 61-79, HEALTH PHYSICIST, BROOKHAVEN NAT LAB, 80- *Mem:* AAAS; Health Physics Soc; fel Am Pub Health Asn. *Res:* Detection and evaluation of low level radioactivity in the natural environment; evaluation of health risks from alternate energy sources. *Mailing Add:* Environ Protection Div Brookhaven Nat Lab Upton NY 11973

HULL, BRUCE LANSING, b Albany, NY, Feb 15, 41; m 65; c 2. VETERINARY MEDICINE, SURGERY. *Educ:* Cornell Univ, DVM, 65; Iowa State Univ, MS, 71. *Prof Exp:* Pvt pract, Delmar Animal Hosp, 65-66; post vet, US Army, 66-68; instr clin, Iowa State Univ, 68-71, from asst prof to assoc prof, 71-76; assoc prof clin, 76-81, PROF SURG, OHIO STATE UNIV, 81- *Mem:* Am Asn Vet Clinicians; Am Asn Bovine Practitioners; Am Vet Med Asn; Am Col Vet Surgeons. *Res:* Surgery and ultrasound of teat and udder; Stifle surgery; caprine arthritis encephalomylitis. *Mailing Add:* 1935 Coffey Rd Columbus OH 43210

HULL, CARL MAX, b Clinton, Ill, Nov 22, 08; m 34; c 4. ORGANIC CHEMISTRY. *Educ:* Univ Ill, BS, 30; Ohio State Univ, PhD(org chem), 35. *Prof Exp:* Res chemist, E I du Pont de Nemours & Co, Inc, 30-31; asst chem, Ohio State Univ, 31-35, res assoc, 43-46; res chemist, Standard Oil Co, Ind, 35-43; assoc prof chem & res head, Sampson Col, Assoc Cols Upper NY, 46-47, prof, 47-48, head, 48-50; prof chem & head dept, Champlain Col, 50-53; chmn div sci & math, 56-63, chmn dept chem, 65-68, prof, 53-74, EMER PROF CHEM, STATE UNIV NY, BINGHAMTON, 74- *Concurrent Pos:* Consult, Air Force Off Sci Res, 59-64. *Mem:* Fel AAAS; Am Chem Soc. *Res:* Organic sulfur compounds. *Mailing Add:* Dept Chem State Univ NY Vestal Pkwy E Binghamton NY 13901

HULL, CLARENCE JOSEPH, b Ft Madison, Iowa, July 14, 19; m 49; c 2. ORGANIC CHEMISTRY. *Educ:* Univ Iowa, BS, 41; Ind Univ, MA, 42, PhD(org chem), 44. *Prof Exp:* Res chemist, Am Cyanamid Co, Conn, 44-46; from instr to assoc prof gen org chem, Univ Detroit, 46-57; chmn sci dept, Moorhead State Col, 59-62; prof chem, Ind State Univ, Terre Haute, 62-; RETIRED. *Concurrent Pos:* Fel, Inst Int Educ, Zurich, 47; mem, Ky Contract Team, US Opers Mission, Indonesia, 57-61. *Mem:* AAAS; Am Chem Soc; Royal Soc Chem. *Res:* Indones; synthesis of isoflavones; homologues of vitamin B1; history and literature of chemistry. *Mailing Add:* 90 Deming Lane Terre Haute IN 47803

HULL, DALE O, b Oskaloosa, Iowa, Feb 26, 12; m 40; c 2. AGRICULTURAL ENGINEERING. *Educ:* Iowa State Univ, BS, 39, MS, 40. *Prof Exp:* Automotive engr, Standard Oil Co, Ind, 40-44, lubrication engr, 44-45; eng consult, Dale O Hull Assocs, 77-91; prof & exten agr engr, 45-77, EMER PROF AGR ENG EXTEN, IOWA STATE UNIV, 77- *Mem:* Am Soc Agr Engrs; Soc Automotive Engrs. *Res:* Automation of agriculture; farmstead engineering; agricultural land drainage; agricultural power and implements; farm and home safety; forensic engineering farm accidents. *Mailing Add:* Dale O Hull Assocs 2925 Ross Rd Ames IA 50010

HULL, DAVID G(EORGE), b Oak Park, Ill, Mar 27, 37; m 83; c 5. AEROSPACE ENGINEERING. *Educ:* Purdue Univ, BS, 59; Univ Wash, Seattle, Ms, 62; Rice Univ, PhD, 67. *Prof Exp:* Staff assoc aerospace res, Boeing Sci Res Labs, 59-64; res assoc, Aero-Astronaut Group, Rice Univ, 64-66; from asst prof to prof, 66-84, M J THOMPSON REGENTS PROF AEROSPACE ENG, UNIV TEX, AUSTIN, 85- *Concurrent Pos:* Assoc ed, J Optimization Theory & Applns. *Mem:* Am Inst Aeronaut & Astronaut. *Res:* Guidance and control; optimal control theory; flight mechanics. *Mailing Add:* Dept Aerospace Eng & Eng Mech Univ Tex Austin TX 78712

HULL, DAVID LEE, b Cleveland, Ohio, Aug 4, 39; m 63; c 2. APPLIED MATHEMATICS, STATISTICS. *Educ:* Ohio State Univ, BS, 61, PhD(math), 69; Univ Wis, MS, 64. *Prof Exp:* Teacher, Cleveland Bd Educ, 61-63; actuarial trainee, Nationwide Ins Co, 64-65; analyst, Ctr Naval Anal, 69-70; from asst prof to assoc prof, 70-80, PROF MATH, OHIO WESLEYAN UNIV, 80- *Concurrent Pos:* Consult statistician, Northeastern For Exp Sta, USDA, 70-71; consult, NSF, 73-85; assoc, Environ Anal Assocs, Inc, 73-88; independent comput consult, 78- *Mem:* Math Asn Am; Am Statist Asn. *Res:* Applications of mathematics and statistics to societal problems; methodology for evaluating federal programs in technology transfer; methodology for evaluating educational programs in entrepreneurship. *Mailing Add:* Dept Math Ohio Wesleyan Univ Delaware OH 43015

HULL, DONALD ALBERT, b Wallace, Idaho, July 9, 38; c 3. GEOLOGY, GEOTHERMAL ENERGY. *Educ:* Univ Idaho, BS, 60; McGill Univ, MS, 62; Univ Nev, PhD(geol), 70. *Prof Exp:* Dist mgr, Homestake Mining Co, 70-74; geothermal spec, 74-77, DIR, DEPT GEOL & MINERAL INDUSTS, STATE ORE, 77- *Mem:* Am Asn State Geologists; Can Inst Mining & Metall. *Res:* Geologic research; economic geology; mineral exploration; geologic hazards. *Mailing Add:* 3954 N Castle Ave Portland OR 97227

HULL, DONALD R(OBERT), b Minneapolis, Minn, Jan 31, 11; m 38; c 4. CHEMICAL ENGINEERING. *Educ:* Univ Minn, BChE, 34. *Prof Exp:* Engr, Textile Fibers Dept, E I du Pont de Nemours & Co, Inc, 34-36, res engr, 36-44, res supvr, 44-46, res mgr, 46-51, tech supt, 51-57, mgr eng develop, 57-69, res fel, 69-76; PVT CONSULT ENGR DYEING PROCESSES, NYLON PROCESSING, SMALL MED DEVICES, FIBER CONCEPTS INC, 76- *Mem:* AAAS; Am Chem Soc; Am Soc Mech Engrs. *Res:* Manufacturing and use of synthetic fibers. *Mailing Add:* 4925 Lancaster Pike Wilmington DE 19807

HULL, FREDERICK CHARLES, b Alliance, Ohio, Nov 9, 15; m 44; c 1. METALLURGY. *Educ:* Univ Mich, BS, 37; Carnegie Inst Technol, DSc(metall eng), 41. *Prof Exp:* Teaching asst, Carnegie Inst Technol, 37-39; res engr, Westinghouse Elec Corp, 41-57, adv engr, Res & Develop Ctr, 57-81; RETIRED. *Concurrent Pos:* Consult, 81-86. *Honors & Awards:* Charles B Dudley Medal, Am Soc Testing & Mat, 62; James F Lincoln Gold Medal, Am Welding Soc, 74. *Mem:* Fel Am Soc Metals Int; Am Inst Mining, Metall & Petrol Engrs; Am Soc Testing & Mat. *Res:* Grain size; high temperature alloys; stainless steels; hot cracking. *Mailing Add:* 109 Lavern St Pittsburgh PA 15235-1548

HULL, GEORGE, JR, b Indianola, Miss, Sept 30, 21; m 47; c 2. ENTOMOLOGY. *Educ:* Alcorn Agr & Mech Col, BS, 45; Tenn Agr & Indust State Univ, MS, 49; Ohio State Univ, PhD(entom), 57. *Prof Exp:* Asst instr biol, Tenn Agr & Indust State Univ, 48-49, from instr to prof, 49-64; prof biol sci & head dept, Grambling Col, 64-66; PROF BIOL & CHMN DEPT, FISK UNIV, 66-, DIR DIV NATURAL SCI & MATH, 77- *Mem:* AAAS; Entom Soc Am. *Res:* Biology and control of insects; insect pathology. *Mailing Add:* 4212 Enchanted Ct Nashville TN 37218

HULL, GORDON FERRIE, JR, b Hanover, NH, May 23, 12; m 37; c 5. PHYSICS. *Educ:* Dartmouth Col, AB, 33, AM, 34; Yale Univ, PhD(physics), 37. *Prof Exp:* Mem tech staff, Bell Tel Labs, Inc, NJ, 37-44; prof physics, Dartmouth Col, 44-55; sr staff mem, Lincoln Lab, Mass Inst Technol, 55-58; sr scientist, Sylvania Elec Prod, Inc, 58-63; dir res, Baird-Atomic, Inc, 63-65; PRES, HULL ASSOCS, SCI & EDUC CONSULTS, 65- *Mem:* Fel AAAS; fel Am Phys Soc. *Res:* Spectroscopy; electronics; microwaves; solid state physics; optics; astrophysics. *Mailing Add:* Two Jewett St Rockport MA 01966

HULL, HARRY H, b Beaver, Pa, Jan 30, 11; m 35; c 4. PHYSICAL CHEMISTRY. *Educ:* Purdue Univ, BS, 33; Univ Chicago, MS, 40. *Prof Exp:* Chem engr, Victor Chem Works, 34-43; res engr, R R Donnelley & Sons Co, 46-62; physicist, Graphic Arts Tech Found, 62-75; chem engr, C E Lummus Co, 75-77; RETIRED. *Mem:* AAAS; Soc Rheol; Am Chem Soc; Am Soc Qual Control. *Res:* Thermodynamics of rheology; printability of paper; Design of experiments. *Mailing Add:* 1710 Del Webb Blvd Sun City Center FL 33573-5010

HULL, HARVARD LESLIE, physics; deceased, see previous edition for last biography

HULL, HERBERT MITCHELL, b La Jolla, Calif, Aug 19, 19; m 50; c 2. PLANT PHYSIOLOGY. *Educ:* Univ Calif, BS, 46; Calif Inst Technol, PhD(plant physiol), 51. *Prof Exp:* prof, 66-85, EMER PROF WATERSHED MGT, UNIV ARIZ, 85- *Concurrent Pos:* Plant physiologist, Sci & Educ Admin-Agr Res, USDA, 52-78. *Mem:* Fel AAAS; Am Soc Plant Physiol; Bot Soc Am; Weed Sci Soc Am; Am Inst Biol Sci; Sigma Xi. *Res:* Absorption and translocation of organic substances; effect of air pollution upon plants; leaf and cuticle ultrastructure. *Mailing Add:* 4040 W Sweetwater Dr Tucson AZ 85745-9757

HULL, HUGH BODEN, b Greenfield, Ohio, Nov 1, 20; m 45; c 4. CARDIOVASCULAR PHYSIOLOGY. *Educ:* Ohio State Univ, BA, 42, MD, 44; Am Bd Internal Med, dipl, 55. *Prof Exp:* Intern, Good Samaritan Hosp, Phoenix, Ariz, 45-46; resident med, Ohio State Univ, 47-49; resident cardiol, White Cross Hosp, Columbus, 49-50; asst instr med & fel cardiol, Col Med, Ohio State Univ, 50-53, asst prof physiol, 52-59, asst prof med, 53-59; DIR CARDIOL DEPT, ST LUKE'S HOSP, PHOENIX, 59- *Concurrent Pos:* Actg chief res lab, Ohio Tuberc Hosp, Columbus, 51-52; mem med adv comt, Vet Admin Hosp, Phoenix; mem heart steering comt, Ariz Regional Med Prog. *Mem:* AMA; fel Am Col Chest Physicians. *Mailing Add:* 1800 E Van Buren Phoenix AZ 85006

HULL, JAMES CLARK, b Los Angeles, Calif, Nov 29, 45; m 71; c 2. PLANT ECOLOGY. *Educ:* Univ Calif, Santa Barbara, BA, 68, MA, 73, PhD(biol), 74. *Prof Exp:* Lectr bot, Univ Calif, Santa Barbara, 74-75; asst prof bot, Bishop's Univ, 75-76; ASSOC PROF BIOL, TOWSON STATE UNIV, 76- *Concurrent Pos:* Plant ecologist, Chesapeake Bay Ctr Environ Sci, Smithsonian Inst, 80; vis scholar, Stanford Univ, 84-85. *Mem:* Ecol Soc Am; Sigma Xi; Am Inst Biol Sci. *Res:* competitive and allelopathic interactions which result in dominance, stability and perturbations within vegetation; plant physiological adaptations for tolerance of adverse environmental conditions; serpentine vegetation; plant nutrient relations. *Mailing Add:* Dept Biol Sci Towson State Univ Towson MD 21204

HULL, JEROME, JR, b Canfield, Ohio, Dec 15, 30; m 57; c 4. AGRICULTURE, HORTICULTURE. *Educ:* Mich State Univ, BS, 52, PhD(agr), 58; Va Polytech Inst & State Univ, MS, 53. *Prof Exp:* Exten pomologist, Purdue Univ, 59-64; exten horticulturist, 64-88, PROF HORT, MICH STATE UNIV, 71- *Concurrent Pos:* Ed, Hoosier Hort, 69-74, Ann Report, Mich State Hort Soc, 71-; Stewart Green fel, 75; ext ed, Hortscience, 77-80. *Mem:* Am Soc Hort Sci. *Res:* Nutrition of deciduous fruit crops; chemical weed control and physiology of fruit crops. *Mailing Add:* Dept Hort Mich State Univ East Lansing MI 48824

HULL, JOHN LAURENCE, b Danville, Pa, Mar 29, 24; m 47, 76; c 3. POLYMER PROCESSING ENGINEERING. *Educ:* Mass Inst Technol, BS, 44. *Prof Exp:* Div officer, UN, China, 46-48, area officer, Ger, 48-50; design engr, Commonwealth Aircraft, Australia, 50-52; advert mgr, Fischer & Porter Co, Hatboro, Pa, 52-55; VPRES & VCHMN, HULL CORP, HATBORO, PA, 55- *Concurrent Pos:* Dir, Compex Corp, Hull Corp, Thinco, Inc, Barnes Corp, Finmac Corp, PCK Elastomerics, Tree Growers Inc, Kard Corp & Hull Int, Scotland, 55-; chmn, Publ Comt & Elec & Electronics Div, Soc Plastics Engrs, 55-; lectr, Ctr Prof Advan, 75-; pres, Hulltronics, Inc, 77-82. *Mem:* Sr mem Soc Plastics Engrs; Plastics Inst Am; Plastics Pioneers Asn. *Res:* Direct encapsulation of electronic components and semiconductor devices with plastic by transfer molding. *Mailing Add:* 3605 Edencroft Rd Huntingdon Valley PA 19006

HULL, JOHN R, b Des Moines, Iowa, Aug 15, 49; m 73; c 2. SOLAR ENERGY, SUPERCONDUCTIVITY. *Educ:* Iowa State Univ, BS, 71, PhD(physics), 79. *Prof Exp:* Fel, Iowa State Univ, 79-80; PHYSICIST, ARGONNE NAT LAB, 80- *Concurrent Pos:* Consult, UN; assoc ed, Solar Energy. *Mem:* Int Solar Energy Soc; Optical Soc Am; Inst Elec & Electronics Engrs; Am Soc Mech Engrs. *Res:* Solar pond physics; energy storage; electromagnetic levitation; natural convection; applications of superconductivity. *Mailing Add:* Argonne Nat Lab CT-308 Argonne IL 60439

HULL, JOSEPH A, b Portland, Kans, Jan 21, 25; m 46; c 6. ELECTRICAL ENGINEERING, ENGINEERING PHYSICS. *Educ:* Univ Kans, BS, 49; Univ NMex, MS, 57. *Prof Exp:* Asst cardiovasc res, Univ Kans, 49-52; res assoc high explosives, Los Alamos Sci Lab, 52-54, staff mem, 54-56; sr scientist, Avco Corp, 59-64, dept mgr space measurements, 64-67; br chief, Electronics Res Ctr, NASA, 67-70; assoc dir, 70-82, PHYSICIST, HIST TELECOMMUN SCI, NAT TELECOMMUN & INFO ADMIN, DEPT COM, 82- *Mem:* Am Inst Aeronaut & Astronaut; Sigma Xi; sr mem Inst Elec & Electronics Engrs. *Res:* Ballistic missile re-entry physics; non-nuclear vulnerability of missile and satellites; electronic and optical instrumentation; instrumentation aspects of medical research and technology; optical communications; fiber optics. *Mailing Add:* ITS NI 325 Broadway Boulder CO 80303

HULL, JOSEPH POYER DEYO, JR, b Tulsa, Okla, Jan 21, 31; m 62; c 2. PETROLEUM GEOLOGY. *Educ:* Hamilton Col, AB, 52; Columbia Univ, MA, 53, PhD(geol), 55. *Prof Exp:* Subsurface geologist petrol explor, Humble Oil & Refining Co, 55-58; staff geologist, Kerr-McGee Corp, 59-68, chief geologist, Can, 69-73, mgr Can explor, 73-75; pres, Impel Energy Corp, 75-80; pres, Page Petrol Inc, 80; vpres, Ensource, Inc, 81-84; vpres, Sabine Corp, 84-86; vpres, Wolf Energy Co, 87-90; CONSULT GEOLOGIST, 90- *Mem:* Geol Soc Am; Soc Econ Paleont & Mineral; Soc Econ Geol; Am Asn Petrol Geol. *Res:* Oil and gas; regional stratigraphy; sedimentary petrology; permian of Texas and New Mexico; geology of northern Canada. *Mailing Add:* 64 S Flora Way Golden CO 80401

HULL, LARRY ALLEN, b Gettysburg, Pa, April 23, 50; m 73; c 2. FRUIT ENTOMOLOGY, INTEGRATED PEST MANAGEMENT. *Educ:* Mt St Marys Col, BS, 72; Pa State Univ, PhD(entom), 77. *Prof Exp:* Res Asst, Pa State Univ, 72-74, grad res asst, 74-77, from instr to asst prof, 77-90, PROF, PA STATE UNIV, 90- *Concurrent Pos:* Prin investr, Cortium Integrated Pest Mgt through Tex A&M Univ, Environ Protection Agency, USDA, 78-82. *Mem:* Entom Soc Am. *Res:* Integrated pest-management programs for deciduous tree fruits, select natural enemies with resistance to pesticides for implementation of expanding pest management systems. *Mailing Add:* Pa State Univ PO Box 309 Biglerville PA 17307

HULL, MCALLISTER HOBART, JR, b Birmington, Ala, Sept 1, 23; m 46; c 2. THEORETICAL PHYSICS, NUCLEAR PHYSICS. *Educ:* Yale Univ, BS, 48, PhD(physics), 51. *Prof Exp:* Res technician theoret physics, Los Alamos Sci Lab, 46; asst res, Univ Wis, 46-47; asst res, Yale Univ, 47-51, from instr to assoc prof physics, 51-66; prof physics & chmn dept, Ore State Univ, 66-69; prof physics & astron & chmn dept, State Univ NY, Buffalo, 69-77, actg dean grad sch, 71-72, dean grad prof educ, 72-77; prof physics & provost, 77-85, counr to pres, 85-88, EMER PROF & PROVOST, DEPT PHYSICS & ASTRON, UNIV NMEX, 88- *Mem:* Fel Am Phys Soc; Am Asn Physics Teachers. *Res:* Nucleon scattering; analysis and tabulation of charged particle wave functions; interaction of radiation with matter; pion scattering; nuclear structure; planning and management in higher education. *Mailing Add:* Dept Physics & Astron Univ NMex Albuquerque NM 87131

HULL, MAURICE WALTER, b Clay Center, Kans, Mar 11, 22; m 45; c 3. VETERINARY PHYSIOLOGY. *Educ:* Kans State Univ, DVM, 45, MS, 61, PhD(physiol), 64. *Prof Exp:* Vet, Linden Animal Hosp, Mo, 45; owner, Clay Ctr Animal Hosp, Kans, 45-59; assoc vet physiologist, 62-71, PROF VET PHYSIOL, VET RES LAB, MONT STATE UNIV, 71- *Concurrent Pos:* Vis prof, Univ Alexandria, Egypt, 80-81. *Mem:* Am Soc Vet Physiol & Pharmacol; Am Vet Med Asn. *Res:* Ruminant physiology, especially bovine pulmonary emphysema and brisket disease; trace element nutrition. *Mailing Add:* 517 S 14th Bozeman MT 59715

HULL, MICHAEL NEILL, b Belfast, Northern Ireland, Nov 7, 42; US citizen; m 67; c 3. ELECTROCHEMISTRY. *Educ:* Queen's Univ, BSc, 64, PhD(electrochem), 67. *Prof Exp:* Res assoc electrochem, Duke Univ & US Army Electronics Command, 67-69; sr scientist electrochem, ESB, Inc, 69-78; sect mgr, IRDC, INCO, 78-82; MGR, INT PAPER, 82- *Concurrent Pos:* Lectr, Quality Improv Progs. *Mem:* Electrochem Soc; Tech Asn Pulp & Paper Indust. *Res:* Development of new battery systems and fundamental research in the electrochemistry of power systems; battery electrochemistry; pulping and bleaching. *Mailing Add:* 17 Arbor Lane Bardonia NY 10954

HULL, RICHARD JAMES, b Newport, RI, Nov 29, 34; m 66; c 2. PHYSIOLOGICAL ECOLOGY, PLANT NUTRITION. *Educ:* Univ RI, BS, 57, MS, 59; Univ Calif, Davis, PhD(bot), 64. *Prof Exp:* Asst prof plant physiol, Purdue Univ, 64-69; assoc prof, 69-78, PROF PLANT SCI, UNIV RI, 79- *Concurrent Pos:* Res assoc, Univ Calif, Davis, 77; chmn, Plant Sci Dept, Univ RI, 88- *Mem:* Am Soc Plant Physiol; Am Soc Hort Sci; Agron Soc Am; Crop Sci Soc Am; Soil Sci Soc Am. *Res:* Physiology of perennial plants; influence of environmental factors on the translocation and metabolism of photosynthetic products; stress physiology of turf, woody ornamental plants and tidal marsh halophytes; physiology of phanerogamic parasites. *Mailing Add:* Dept Plant Sci Univ RI Kingston RI 02881

HULL, ROBERT JOSEPH, b Salem, Mass, July 6, 36; m 58; c 6. PHYSICS. *Educ:* Mass Inst Technol, SB, 57, PhD(physics), 61. *Prof Exp:* Asst res physicist & lectr, Univ Calif, Berkeley, 63-64, physicist & lectr, 64-66; asst prof physics, State Univ NY, Buffalo, 66-69; STAFF MEM LINCOLN LAB, MASS INST TECHNOL, 69- *Mem:* Optical Soc Am; Am Phys Soc. *Res:* Optics; laser technology. *Mailing Add:* Lincoln Lab Mass Inst Technol PO Box 73 Lexington MA 02173

HULL, THOMAS EDWARD, b Winnipeg, Man, June 5, 22; m 44; c 2. COMPUTER SCIENCE. *Educ:* Univ Toronto, BA, 44, MA, 46, PhD(math), 49. *Prof Exp:* Instr math, Univ Toronto, 46-49; instr, Univ BC, 49-50, from asst prof to prof, 50-64; vis prof, 64-65, chmn dept comput sci, 68-77, PROF MATH & COMPUT SCI, UNIV TORONTO, 65- *Concurrent Pos:* Vis assoc prof, Calif Inst Technol, 58-59; ed, J Numerical Anal. *Mem:* Am Math Soc; Soc Indust & Appl Math; Math Asn Am; Asn Comput Mach; Royal Soc Can; Sigma Xi. *Res:* Numerical analysis. *Mailing Add:* Dept Comput Sci Univ Toronto Toronto ON M5S 1A7 Can

HULL, WILLIAM L(AVALDIN), b Bremerton, Wash, Apr 14, 13; m 40; c 2. MECHANICAL ENGINEERING. *Educ:* Univ Colo, BS, 34, ME, 46; Chrysler Inst Eng, MAE, 36; Purdue Univ, MS, 47. *Prof Exp:* Asst, Power Lab, Agr & Mech Col Tex, 34; exp engr, Chrysler Corp, 35-37; engr thermodyn, Purdue Univ, 37-40; engr heat power, Univ Colo, 40-42, from asst prof to assoc prof, 42-47; assoc prof heat power, Univ Ill, Urbana, 47-49, prof, 49-79; RETIRED. *Concurrent Pos:* Instr, Chrysler Inst Eng, 36-37; consult, Caterpillar Tractor Co, 55- & Int Harvester Co, 66- *Mem:* Soc Automotive Engrs; fel NY Acad Sci. *Res:* Internal combustion engines; gas turbines. *Mailing Add:* 101 W Windsor Rd No 4207 Urbana IL 61801

HULLAND, THOMAS JOHN, b Redcliff, Alta, Feb 2, 30; m 56; c 4. VETERINARY MEDICINE, PATHOLOGY. *Educ:* Univ Toronto, DVM, 54; Univ Edinburgh, PhD(vet path), 59; dipl, Am Col Vet Path, 63. *Prof Exp:* From asst to assoc prof, 54-64, head dept path, 64-69, assoc dean, 69-80, PROF VET PATH, ONT VET COL, UNIV GUELPH, 64- *Concurrent Pos:* Mem exam bd, Am Col Vet Path, 65-68, coun, 69-74, pres, 73; consult, Armed Forces Inst Path, 71. *Mem:* Can Vet Med Asn (pres, 71-72); Can Asn Vet Path; fel Royal Soc Med. *Res:* Veterinary pathology. *Mailing Add:* Dept Path Ont Vet Col Univ Guelph Guelph ON N1G 2W1 Can

HULLAR, THEODORE LEE, b Prescott, Wis, Mar 19, 35; m 58; c 2. ORGANIC CHEMISTRY, BIOCHEMISTRY. *Educ:* Univ Minn, BS, 57, PhD(biochem), 63. *Prof Exp:* NSF fel, State Univ NY, Buffalo, 63-64, asst prof, 64-69, assoc dean grad sch, 69-71, assoc prof med chem, 69-74; comnr environ qual, Erie County, NY, 74-75; dep comnr environ conserv, NY Dept Environ Conserv, 75-79; assoc dir res & adj prof, 79-81, DIR RES & PROF, COL AGR & LIFE SCI, CORNELL UNIV, 81-, DIR CORNELL UNIV AGR EXP STA, 81- *Mem:* AAAS; Am Chem Soc; The Chem Soc. *Res:* Chemistry and biochemistry of carbohydrates, nucleotides, pyridoxal phosphate; synthetic organic chemistry; mechanism of action and selective inhibition of enzymes; environmental toxicology; environmental policy; natural resources policy and management; agricultural research policy. *Mailing Add:* Chancellor's Office Univ Calif 16 College Park Davis CA 95616

HULLEY, CLAIR MONTROSE, b Cincinnati, Ohio, Mar 6, 25. ENGINEERING, COMPUTER SCIENCE. *Educ:* Univ Cincinnati, BS, 47, ME, 47. *Prof Exp:* Instr mech eng, Univ Cincinnati, 47-52, instr eng graphics, 52-57, from asst prof to assoc prof eng anal, 57-76, prof eng sci, 76-77, eng & computer graphic sci, 77-80, indust eng, 80-84, EMER PROF ENG & COMPUTER GRAPHIC SCI, UNIV CINCINNATI, 84- *Concurrent Pos:* Consult, Louvers & Dampers Corp, 73- & Design Dampers Corp, 77-78; consult prog graphics, Allis-Chalmers, 75- & NASA Tensgridity Dome Structs, 76-79. *Mem:* Am Soc Mech Engrs; Am Asn Univ Prof; Am Soc Elec Engrs. *Res:* Computer graphics; computer science and software; design improvements in agricultural and industrial tractors; thermodynamics; fuel saving devices. *Mailing Add:* 11560 Deerfield Rd Cincinnati OH 45242

HULLINGER, RONALD LORAL, b Des Moines, Iowa, Feb 4, 41; m 59; c 3. VETERINARY & MICROSCOPIC ANATOMY, IMMUNOLOGY. *Educ:* Iowa State Univ, BS, 64, DVM, 65, MS, 66, PhD(vet anat), 68. *Prof Exp:* Asst prof vet anat, Iowa State Univ, 68-69; from asst prof to assoc prof, 69-84, PROF VET ANAT, PURDUE UNIV, 85- *Concurrent Pos:* Vis prof, Cornell Univ, 74-75, Utrecht Univ, Holland, 76-77 & Univ Ill, 78-79, Free Univ Berlin, 84- *Mem:* Am Asn Vet Anat; Am Vet Med Asn; World Asn Vet Anat; Europ Asn Vet Anat; Asn Am Vet Med Educ; World Asn Vet Med Educ. *Res:* Normal microscopic and fine structural anatomy of the canine endocrine system and changes with age in that system; cell biology; development of endocrine system in mammals. *Mailing Add:* Dept Vet Anat Purdue Univ West Lafayette IN 47907

HULM, JOHN KENNETH, b Southport, Eng, July 4, 23; US citizen; m 48; c 5. SOLID STATE PHYSICS. *Educ:* Cambridge Univ, BA, 43, MA, 45, PhD(physics), 49. *Prof Exp:* Res fel physics, Union Carbide & Carbon Corp, Inst Study Metals, Univ Chicago, 49-51, asst prof, Univ, 51-54; adv physicist, Westinghouse Elec Corp, 54-56, mgr, Solid State Physics Dept, 56-60, assoc dir res, 60-64, dir cryogenic res, 64-69, res dir, 69-74; sci attache, US Dept State, US Embassy, London, 74-76; mgr chem res, 76-80, DIR CORP RES, WESTINGHOUSE RES LABS, 80- *Honors & Awards:* Wetherill Medal, Franklin Inst, Pa, 64; Int Prize New Mats, Am Phys Soc, 79. *Mem:* Nat Acad Sci; Nat Acad Eng; Brit Nuclear Energy Soc; fel Am Phys Soc; AAAS; Brit Inst Physics. *Res:* Cryogenics; superconductivity; semiconductors; ferroelectrics; magnetism; electrotechnology and the generation of high magnetic fields; energy technology. *Mailing Add:* Westinghouse Res Labs 1310 Beulah Rd Pittsburgh PA 15235

HULME, BERNIE LEE, b Chickasha, Okla, July 1, 39; m 60; c 1. NUMERICAL ANALYSIS, MATHEMATICAL MODELING. *Educ:* Univ Okla, BS, 62, MCE, 63; Harvard Univ, MS, 65, PhD(appl-math), 69. *Prof Exp:* Staff mem appl math, Sandia Nat Labs, 69-86; CONSULT APPL MATH, HULME MATH, 86- *Mem:* Soc Indust & Appl Math. *Res:* Numerical analysis, in particular the numerical solution of differential equations; applications of graph theory and Boolean algebra; operations research. *Mailing Add:* 3701 General Patch NE Albuquerque NM 87111

HULME, NORMAN ARTHUR, b Philadelphia, Pa, June 9, 24; m 50; c 3. PHARMACOLOGY. *Educ:* Philadelphia Col Pharm, BSc, 50; Johns Hopkins Univ, MS, 53; Univ Md, PhD(pharmacol), 54. *Prof Exp:* Clin pharmacologist, Sterling Winthrop Res Inst, 54-88; RETIRED. *Mem:* AAAS; Am Chem Soc; Brit Biochem Soc; NY Acad Sci; Am Soc Clin Pharmacol & Therapeut. *Res:* Mechanism of action of general anesthetics. *Mailing Add:* 50 Sunset Dr Delmar NY 12054-1126

HULSBOS, C(ORNIE) (LEONARD), b Given, Iowa, Aug 23, 20; wid; c 3. CIVIL ENGINEERING. *Educ:* Iowa State Univ, BS, 41, MS, 49, PhD(struct eng), 53. *Prof Exp:* Struct draftsman, Am Bridge Co, 41-46; instr theoret & appl mech, Iowa State Univ, 46-47, from instr to prof civil eng, 47-60; res prof, Lehigh Univ, 60-65; PROF & CHMN DEPT CIVIL ENG, UNIV NMEX, 65- *Concurrent Pos:* Mem concrete bridges comt, Transp Res Bd. *Mem:* Am Soc Civil Engrs; Am Soc Eng Educ; Am Concrete Inst; Nat Soc Prof Engrs. *Res:* Structural engineering. *Mailing Add:* 7608 Palo Duro NE Albuquerque NM 87110

HULSE, ARTHUR CHARLES, b Jersey City, NJ, Dec 20, 45; m 79; c 1. HERPETOLOGY, EVOLUTIONARY ECOLOGY. *Educ:* Bloomfield Col, BA, 67; Ariz State Univ, MS, 70, PhD(zool), 74. *Prof Exp:* Fel zool, Univ Tex, Austin, 73-75; from asst prof to assoc prof, 76-83, PROF BIOL, INDIANA UNIV OF PA, 83- *Concurrent Pos:* Res assoc, Carnegie Mus Natural Hist, 76- *Mem:* Soc Study Evolution; Soc Study Amphibian Reptile; Soc Syst Zoologists; Am Asn Ichthyologists & Herpetologists; Herpetologists League. *Res:* Ecological and systematic herpetology with major emphasis on evolution of life history strategies and community organization. *Mailing Add:* Dept Biol Indiana Univ of Pa Indiana PA 15705

HULSE, CHARLES O, b New York, NY, May 29, 28; m; c 2. CERAMICS. *Educ:* Rutgers Univ, BS, 51; Univ Calif, Berkeley, MS, 56, PhD(eng sci), 60. *Prof Exp:* RES SCIENTIST, UNITED TECHNOL CORP, 61- *Mem:* Am Ceramic Soc; Soc Exp Mech. *Res:* Mechanical behavior of materials; thin film sensors. *Mailing Add:* United Technol Corp Res Ctr Silver Lane East Hartford CT 06108

HULSE, RUSSELL ALAN, New York, NY, Nov 28, 50. FUSION PLASMA PHYSICS, COMPUTER MODELING. *Educ:* Cooper Union, BS, 70; Univ Mass, MS, 72, PhD(physics), 75. *Prof Exp:* Res assoc, Nat Radio Astron Observ, 75-77; mem tech staff, 77-80, staff res physicist II, 80-84, RES PHYSICIST, PLASMA PHYSICS LAB, PRINCETON UNIV, 84- *Mem:* Am Phys Soc; Am Astron Soc; Soc Indust & Appl Math. *Res:* Computer modeling of transport and atomic processes in tokamak controlled thermonuclear fusion plasmas; advanced computational environments. *Mailing Add:* Plasma Physics Lab Princeton Univ Princeton NJ 08543

HULSEY, J LEROY, b Sullian, Mo, Oct 6, 41; m 64; c 1. FINITE ELEMENT ANALYSIS, EXPERIMENTAL STRESS ANALYSIS. *Educ:* Mo Sch Mines & Metall, BS, 64; Univ Mo, Rolla, MS, 66, PhD(struct), 76. *Prof Exp:* Proj engr, Daily Assocs, Inc, US Army, 66-72; res asst, Univ Mo, Rolla, 72-76; asst prof, NC State Univ, 76-80; pres, Civil Eng & Appl Res, 79-85; asst dir, Inst Transp Res, 85-87; ASSOC PROF CIVIL ENG, UNIV ALASKA, FAIRBANKS, 87- *Concurrent Pos:* Mem, Gen Struct Comt, Transp Res Bd, 76-80 & Geotech Computer Appln & Numerical Methods Comt, Am Soc Civil Engrs, 76-79; pres, Appl Computer Serv, Inc, 79- *Mem:* Sigma Xi; Am Soc Civil Engrs; Am Concrete Inst. *Res:* Numerical methods and laboratory evaluation of bridge behavior for temperature stresses, load distribution for cable-stayed bridges, long term deflections of segmental bridges and life cycle performance of bridges. *Mailing Add:* Dept Civil Eng Univ Alaska Fairbanks AK 99775

HULSIZER, ROBERT INSLEE, JR, b East Orange, NJ, Nov 25, 19; m 41, 67; c 4. ELEMENTARY PARTICLE PHYSICS. *Educ:* Bates Col, BS, 40; Wesleyan Univ, MA, 42; Mass Inst Technol, PhD(physics), 48. *Prof Exp:* Asst physics, Wesleyan Univ, 40-42; mem staff, Radiation Lab, Mass Inst Technol, 42-46, res assoc physics, Lab Nuclear Sci, 46-49; from asst prof to prof, Univ Ill, 49-64; dir educ res ctr, Mass Inst Technol, 64-68, chmn fac, 77-79, prof physics & mem lab nuclear sci, 64-86, EMER PROF, MASS INST TECHNOL, 86- *Concurrent Pos:* Consult, Off Naval Res, 57-61, Xerox Corp, 68-72 & Phys Sci Study Comt, Educ Servs, Inc, 58-60. *Mem:* Fel Am Phys Soc; Am Asn Physics Teachers. *Res:* Acoustics of quartz bars; nature of primary cosmic radiation; use of digital computers in real-time data-handling; low temperature physics; beta decay; elementary particle reactions. *Mailing Add:* MIT Rm 24-410 Cambridge MA 02139

HULT, JOHN LUTHER, b Mulino, Ore, May 21, 16; m 43; c 2. ELECTRONICS, PHYSICS ENGINEERING. *Educ:* Ore State Col, BS, 39; Ohio State Univ, PhD(physics), 49. *Prof Exp:* Prin investr, Rand Corp, 49-72; CONSULT, 72- *Mem:* Am Phys Soc; World Future Soc. *Res:* Electronic warfare; communication satellites; remotely manned vehicles; Antarctic icebergs for fresh water and cooling. *Mailing Add:* 151 McAfee Court Thousand Oaks CA 91360

HULT, RICHARD LEE, b Grand Rapids, Mich, July 23, 45; m 69; c 3. BIOCHEMICAL PHARMACOLOGY, TOXICOLOGY. *Educ:* Ferris State Col, BS, 68; Ore State Univ, MS, 75, PhD(pharm & toxicol), 77. *Prof Exp:* Teaching asst, Ore State Univ, 71-76; asst prof, 76-82, ASSOC PROF PHARMACOL, FERRIS STATE UNIV, 82- *Mem:* Am Asn Cols Pharm. *Res:* Cancer chemotherapy; improvement of toxicity efficacy ratio; toxicity of para aminophenols. *Mailing Add:* Sch Pharm Ferris State Univ Big Rapids MI 49307

HULTGREN, FRANK ALEXANDER, b Lakewood, Ohio, Apr 13, 36; m 66. PHYSICAL METALLURGY. *Educ:* Case Inst Technol, BS, 58, PhD(metall), 62; Univ Mich, MSE, 60. *Prof Exp:* Res engr, Sci Lab, Ford Motor Co, 59-60; res asst, Case Inst Technol, 60-62, res assoc, 62; nuclear res sanlage, Jülich, Ger, 62-65; proj leader flat-rolled & stainless steels, Repub Steel Corp, 65-67, chief fundamental res sect, 67-73, chief phys metall &

formability, 73-75, develop metallurgist, 75-76, mgr process develop, 76-84; SR RES ASSOC, SOHIO METALL CO, 84- *Concurrent Pos:* Fulbright scholar, Ger, 62-63. *Mem:* Am Soc Metals; Am Inst Mining, Metall & Petrol Engrs; Am Iron & Steel Engrs. *Res:* Deformation; recrystallizations; mechanical metallurgy; metal physics; dislocation damping; internal friction; steel melting, refining and finishing. *Mailing Add:* 9551 230th St Lakeville MN 55044

HULTGREN, HERBERT NILS, b Santa Rosa, Calif, Aug 29, 17; m 48; c 3. MEDICINE. *Educ:* Stanford Univ, AB, 39, MD, 43. *Prof Exp:* Intern Prof Exp: Intern, San Francisco Hosp, 42-43; asst resident med, Stanford Hosp, 43-44, asst resident path, 46-47; fel, Thorndike Mem Lab, Harvard Med Sch, 47-48; from instr to assoc prof, 48-69, PROF MED, SCH MED, STANFORD UNIV, 69- *Concurrent Pos:* Markle Found scholar, 51-56. *Mem:* Am Soc Clin Invest; Am Fedn Clin Res; Am Col Physicians. *Res:* Cardiovascular diseases; valvular and congenital heart disease; phonocardiography; high altitude physiology; coronary artery disease. *Mailing Add:* Dept Med Stanford Univ Sch Med Vet Admin Med Ctr 3801 Miranda Ave C-111 Cardiol Palo Alto CA 94304

HULTGREN, RALPH RAYMOND, b Spokane, Wash, Sept 7, 05; m 33; c 3. PHYSICAL METALLURGY, THERMODYNAMICS. *Educ:* Univ Calif, BS, 28; Univ Utah, MS, 29; Calif Inst Technol, PhD(phys chem), 33. *Prof Exp:* Nat Res Coun fel chem, Mass Inst Technol, 33-35; instr metall, Harvard Univ, 35-38, asst prof, 38-41; from asst prof to prof metall, 41-72, asst dean col eng, 58-61, chmn dept mineral tech, 61-65, calorimetry conf, 65-66, EMER PROF METALL, UNIV CALIF, BERKELEY, 72- *Mem:* Am Soc Metals; Am Inst Mining, Metall & Petrol Engrs. *Res:* Crystal structure analysis by x-rays; chemical bonds; metallography; thermodynamics of alloys. *Mailing Add:* 1501 Le Roy Ave Berkeley CA 94708

HULTIN, HERBERT OSCAR, b Quincy, Mass, Jan 1, 34; m 58; c 5. FOOD SCIENCE. *Educ:* Mass Inst Technol, SB & SM, 56, PhD(food sci), 59. *Prof Exp:* Res asst food sci, Mass Inst Technol, 56-58, res assoc, 58-59; from asst prof to assoc prof, 59-69, PROF FOOD SCI, UNIV MASS, AMHERST, 69- *Concurrent Pos:* NIH fel, Inst Enzyme Res, Univ Wis, 62-63. *Honors & Awards:* Res Award, Inst Food Technol, 68; Earl P McFee Award, Atlantic Fisheries Technologists, 85. *Mem:* AAAS; Inst Food Technol; Am Chem Soc; Atlantic Fisheries Technol Conf. *Res:* Relationship of enzymic activity to subcellular environment; immobilized enzymes; enzymic control of postharvest metabolism; biochemistry of marine foods. *Mailing Add:* Dept Food Sci & Nutrit Univ Mass Marine Sta Box 7128 Lanesville Gloucester MA 01930

HULTQUIST, DONALD ELLIOTT, b Jamestown, NY, Sept 14, 34; m 61; c 5. BIOCHEMISTRY. *Educ:* Univ Rochester, BS, 56, PhD(biochem), 62. *Prof Exp:* NIH fel, Univ Minn, 62-63 & Univ Calif, Los Angeles, 63-64; from instr to assoc prof, 64-79, PROF BIOL CHEM, UNIV MICH, ANN ARBOR, 79- *Concurrent Pos:* Actg dir, Cellular & Molecular Biol Grad Prog, Univ Mich, 82. *Mem:* Am Chem Soc; Am Soc Biol Chemists. *Res:* Structure, properties and function of hemeproteins and proteases of erythroid cells. *Mailing Add:* 2185 Ayrshire Ann Arbor MI 48105

HULTQUIST, MARTIN EVERETT, b Laird, Colo, Oct 31, 10; m 34; c 2. ORGANIC CHEMISTRY. *Educ:* Univ Colo, BS, 31, MS, 33, PhD(chem), 35. *Prof Exp:* Asst org chem, Univ Colo, 31-35; res chemist, Calco Chem Div, Am Cyanamid Co, 35-44, chief chemist pharmaceut res, 44-46, asst dir, 46-50, res assoc, 50-54; res assoc, Arapahoe Chem Div, Syntex Corp, 54-63, assoc dir res, 63-78; RETIRED. *Concurrent Pos:* Chem consult, 78- *Honors & Awards:* Co-recipient Bruce M Cain Mem Award, Am Asn Cancer Res, 87. *Mem:* Am Chem Soc; fel NY Acad Sci. *Res:* Synthesis of sulfanilamide derivatives; folic acid and derivatives; synthesis of organic medicinal products. *Mailing Add:* 2847 Fourth St Boulder CO 80304

HULTQUIST, PAUL F(REDRICK), b Holdrege, Nebr, Mar 24, 20; m 46; c 2. ELECTRICAL ENGINEERING, APPLIED MATHEMATICS. *Educ:* Univ Colo, BA, 45, PhD(physics), 54. *Prof Exp:* Instr appl math, Univ Colo, 45-48 & 52-55, instr physics, 49-50 & 51-52; instr math, Univ Tex, 49; from asst prof to assoc prof appl math, Univ Colo, Colorado Springs, 55-63, prof appl math & elec eng, 65-67, prof elec eng & asst dean eng, 67-71, prof elec & computer eng, Denver, 71-88, EMER PROF ELEC ENG & COMPUT SCI, UNIV COLO, DENVER, 88-; CONSULT, 88- *Concurrent Pos:* Assoc res scientist, Lockheed Missile & Space Co, Calif, 56-57; exec dir, Numerical Anal Ctr, Univ Colo, 58-60 & 62-63; consult, Ball Bros Res Corp, Colo, 58, 59-60 & 65-, staff scientist, 60-61, sr staff scientist, 63-65; consult, Kaman Nuclear Div, Kaman Aircraft Corp, 65-70; consult, US Geol Surv, 76; lectr, Univ Nebr, Omah, 90- *Mem:* Soc Indust & Appl Math; Math Asn Am; Asn Comput Mach; Inst Elec & Electronics Engrs. *Res:* Numerical analysis and computers; systems engineering; aerospace systems analysis; energy systems. *Mailing Add:* 6803 N 68th Plaza No 414 Omaha NE 68152

HULTQUIST, ROBERT ALLAN, b Jamestown, NY, Nov 6, 29; m 56; c 5. MATHEMATICAL STATISTICS. *Educ:* Alfred Univ, BA, 51; Purdue Univ, MS, 53; Okla State Univ, PhD, 59. *Prof Exp:* Asst prof math, DePauw Univ, 59-60; from asst prof to assoc prof, Okla State Univ, 60-66; assoc prof, 66-71, PROF STATIST, PA STATE UNIV, 71- *Mem:* Am Statist Asn; Inst Math Statist. *Res:* Statistical methods; linear models; analysis of variance; distribution theory. *Mailing Add:* Dept Statist 315 Pond Bldg Pa State Univ University Park PA 16802

HULTS, MALCOM E, b Whitley Co, Ind, June 30, 26; m 52; c 3. PHYSICS. *Educ:* Manchester Col, BA, 49; State Univ NY Buffalo, MA, 52, PhD(physics), 64. *Prof Exp:* Asst physics, Univ Buffalo, 49-51; teacher high sch, Ind, 51-53; from instr to assoc prof, 53-67, PROF PHYSICS, BALL STATE UNIV, 67-, HEAD DEPT, 65- *Concurrent Pos:* Instr, State Univ NY, Buffalo, 60-61, res assoc, 61-63; vis prof, Univ Santa Maria, Brazil, 69. *Mem:* Optical Soc Am; Am Phys Soc; Am Asn Physics Teachers; Sigma Xi. *Res:* Atomic spectroscopy; solar eclipses. *Mailing Add:* Dept Physics Ball State Univ Muncie IN 47306

HULTSCH, ROLAND ARTHUR, b Columbus, Ind, June 18, 31; m. PHYSICS. *Educ:* Wabash Col, AB, 52; Univ Del, MS, 56; Iowa State Univ, PhD(physics), 61. *Prof Exp:* Asst prof, 61-67, ASSOC PROF PHYSICS, UNIV MO, COLUMBIA, 67- *Mem:* Am Phys Soc; Am Asn Physics Teachers. *Res:* Nuclear magnetic resonance. *Mailing Add:* 1848 Cliff Dr Columbia MO 65201

HULYALKAR, RAMCHANDRA K, b Mysore State, India, Jan 2, 29; m 57; c 2. POLYMER CHEMISTRY. *Educ:* Poona Univ, BSc, 51, MSc, 55; Queen's Univ, Ont, MSc, 61, PhD(org chem), 64. *Prof Exp:* Nat Res Coun Can fel, 64-66; sr res chemist, Polymer Corp Ltd, Can, 66-69; prin chemist, Dart Industries Inc, 69-78, sr res assoc, 78-80, prin res assoc, 80-83; res supvr, Spencer Kellog, 83-85; TECH DIR, ADCO CHEM CO, 85- *Mem:* Am Chem Soc; fel Am Inst Chemists; Fedn Socs Coating Technol. *Res:* Antifungal antibiotics; biosynthesis of rare sugars; synthesis of sugars; synthesis of carbon-14 labeled sugars; gas chromatography of sugar acids; structural examination of natural polysaccharides; synthesis of monomers and polymers; polymer characterization; modification of the polymers for tailor-made properties; rubber reinforced plastics; emulsion polymers and other water-based resins as applied to coating industry; paint formulation technology. *Mailing Add:* Arco Chem Co 49 Rutherford St PO Box 128 Newark NJ 07101

HUMAR, JAGMOHAN LAL, b Udaipur, Rajasthan, India, Sept 15, 37; Can citizen; m 60; c 3. STRUCTURAL DYNAMICS & EARTHQUAKE ENGINEERING, STRUCTURAL DESIGN & ANALYSIS. *Educ:* Banaras Hindu Univ, BSc, 58; Indian Inst Technol, MTech, 59; Carleton Univ, PhD(civil eng), 74. *Prof Exp:* Exec engr, Pub Works Dept, Govt India, 63-71, superintending engr, 71-75; from asst prof to assoc prof, 75-83, PROF CIVIL ENG, CARLETON UNIV, OTTAWA, 83-, CHMN DEPT, 90- *Concurrent Pos:* Spec consult, Adjeleian & Assocs, 88-89; chair, Computer Appln Div, Can Soc Civil Eng, 88- *Honors & Awards:* Gzowski Medal, Can Soc Civil Eng, 89. *Mem:* Fel Can Soc Civil Eng; Eng Inst Can; Indian Soc Earthquake Technol. *Res:* Dynamics of structures; response of structures to seismic ground motion; analysis of soil-structure interaction and dam-reservoir-foundation interaction under dynamics loading; dynamic response of bridges; computer-aided design including application of computers in civil engineering. *Mailing Add:* Dept Civil Eng Carleton Univ Ottawa ON K1S 5B6

HUMAYDAN, HASIB SHAHEEN, b Ain Anoub, Lebanon, Mar 6, 45; m 80; c 2. BREEDING FOR MULTIPLE DISEASE RESISTANCE IN PLANTS, TISSUE CULTURE. *Educ:* Am Univ Beirut, BS, 69, MS, 71; Univ Wis, Madison, PhD(plant path & plant genetics), 74. *Prof Exp:* Plant pathologist & Plant breeder, Joseph Harris Seed Co, Inc, 74-82, dir plant path & tissue cult, 82-84; eastern mgr res & develop, 84-85, VPRES RES & DEVELOP, HARRIS MORAN SEED CO, 85- *Mem:* Am Phytopath Soc; Am Soc Hort Sci; NY Acad Sci. *Res:* Breeding improved vegetable and flower cultivators with multiple disease resistance; traditional methods supplemented by tissue culture, electrophoresis, induced mutations, and protoplasm fusion. *Mailing Add:* c/o Harris Merran Seed Co 26239 Executive Pl Haywood CA 94545

HUMAYUN, MIR Z, b India, Feb 11, 49. MOLECULAR BIOLOGY MUTAGENESIS. *Educ:* Loyola Campus, Andhra Univ, India, BSc, 67; Madras Univ, India, MSc, 70; Indian Inst Sci, India, PhD(biochem), 75. *Prof Exp:* Res fel molecular biol, Biol Labs, Harvard Univ, 74-76; asst res scientist molecular biol, Med Ctr, NY Univ, 77-79; asst prof, 79-83, assoc prof molecular biol & microbiol, 83-89, PROF MICROBIOL & MOLECULAR GENETICS, NJ MED SCH, UNIV MED & DENT NJ, 89- *Concurrent Pos:* Res Career Develop Award, US Pub Health Serv, 84-89. *Mem:* Am Soc Biochem & Molecular Biol. *Res:* Molecular mechanisms by which damage to DNA leads to mutagenesis; mechanisms of mutagenicity, genotoxicity and carcinogenicity of the mycotoxin Alfatoxin B1 and the industrial chemical vinyl chloride. *Mailing Add:* Dept Microbiol NJ Med Sch 185 Bergen St MSB F607 Newark NJ 07103-2714

HUMBER, LESLIE GEORGE, b Kingston, Jamaica, Dec 19, 31; Can citizen; m 59; c 2. ORGANIC CHEMISTRY. *Educ:* Sir George Williams Univ, BSc, 53; Univ NB, PhD(org chem), 56. *Prof Exp:* Eli Lilly fel, Univ NB, 56-57; res chemist, Shawinigan Chem Ltd, 57; Sloan fel, Univ Rochester, 57-58; sr res chem, Wyeth-Ayerst Res, 58-67, res fel, 67-69, head, Mem Chem Sect, 69-71, assoc dir, 72-79, dir, chem dept, 80-83, DIR DRUG DESIGN & ANCILLARY SERV, WYETH-AYERST RES, 84- *Concurrent Pos:* Lectr, Concordia Univ, 59-; titular mem, Chem Med Sect, Int Union Pure & Appl Chem, 73-81. *Mem:* Am Chem Soc; Chem Inst Can. *Res:* Structure determination and synthesis of natural products; medicinal chemistry; design of enzyme inhibitors; drugs affecting the central nervous system and lipid metabolism; heterocyclic chemistry; antiinflammatory agents; dopaminergic agents; computer-assisted drug design. *Mailing Add:* Wyeth-Ayerst Res CN 8000 Princeton NJ 08543-8000

HUMBER, RICHARD ALAN, b San Francisco, Calif, May 9, 47; m 77; c 2. INSECT MYCOLOGY, TAXONOMY. *Educ:* Stanford Univ, AB, 69; Univ Washington, MS, 70, PhD(bot), 75. *Prof Exp:* Lectr bot, Univ Washington, 75-76; NIH fel, Univ Maine, 76-78; NIH fel, 78-79, res assoc, Boyce Thompson Inst, 79-82, RES SCIENTIST INSECT MYCOL, AGR RES SERV, CORNELL UNIV, USDA, 82-, ADJ PROF, PLANT PATH, 88- *Honors & Awards:* Grad Res Prize, Mycol Soc Am, 75. *Mem:* Mycol Soc Am; Soc Invert Path. *Res:* Basic and developmental biology; taxonomy and systematics of fungal pathogens of insects; entomophthorales (zygomycetes); curation of large collection of entomopathogenic fungal cultures. *Mailing Add:* USDA Insect Path Res Unit US Plant Soil & Nutrit Lab Tower Rd Ithaca NY 14853

HUMBERD, JESSE DAVID, b Roann, Ind, Dec 21, 21; m 42; c 2. MATHEMATICS, SCIENCE EDUCATION. *Educ:* Bryan Univ, BS, 43; Wittenberg Col, BA, 47; Ohio State Univ, MA, 50, PhD(math ed), 64; Grace Theol Sem, MDiv, 54. *Prof Exp:* Teacher pub schs, Ohio, 46-47 & 48-51; instr math & physics, Wittenberg Col, 47-48; PROF MATH & SCI, GRACE COL, 54- *Concurrent Pos:* Instr, Ohio State Univ, 58-59. *Mem:* Nat Coun Teachers Math; Nat Sci Teachers Asn. *Res:* Relations between religious and mathematical history; interrelations of sciences in general and liberal education. *Mailing Add:* 411 Kings Hwy Winona Lake IN 46590

HUMBERTSON, ALBERT O, JR, b Cumberland, Md, Sept 30, 33; m 56; c 3. ANATOMY, NEUROANATOMY. *Educ:* Marietta Col, BS, 56; Ohio State Univ, PhD(anat), 62. *Prof Exp:* From instr to assoc prof, 62-88, EMER ASSOC PROF ANAT, OHIO STATE UNIV, 89- *Mem:* Am Asn Anat; Soc Neurosci; Sigma Xi; Int Soc Develop Neurosci; AAAS. *Res:* Neuroanatomy; neuroembryology; neurohistochemistry. *Mailing Add:* Dept Anat Ohio State Univ Columbus OH 43210

HUMBURG, NEIL EDWARD, b LaCrosse, Kans, May 16, 33; m 61; c 2. AGRONOMY, WEED SCIENCE. *Educ:* Colo State Univ, BS, 55, MS, 65; Univ Wis-Madison, PhD(agron), 70. *Prof Exp:* Asst prof agron, Kans State Univ, 70-75; agronomist, Humburg Ranch, Inc, 75-76; asst prof weed sci, Univ Wyo, 76-83; AGR RESEARCHER & CONSULT, 83- *Concurrent Pos:* Consult, Osman-Omar, Inc, Agr Bank Saudi Arabia, 83, Colo pesticide use surv, 89-91. *Mem:* Weed Sci Soc Am; Am Soc Agron; Crop Sci Soc Am; Int Weed Sci Soc; Plant Growth Regulator Soc Am. *Res:* Weed control in agronomic crops. *Mailing Add:* 6615 Evers Blvd Cheyenne WY 82009-3209

HUME, ARTHUR SCOTT, b Columbia, Tenn, June 17, 28; m 55; c 3. PHARMACOLOGY, TOXICOLOGY. *Educ:* Univ Miss, BS, 58, MS, 61, PhD(pharmacol), 64. *Prof Exp:* Instr pharm, Univ Miss, 58-61; fel pharmacol, Sch Med, Vanderbilt Univ, 64-65, instr, 65-66; from asst prof to assoc prof, 66-86, PROF PHARMACOL, MED CTR, UNIV MISS, 86- *Concurrent Pos:* USPHS fel, 64-66; dir, Miss Crime Lab, 72-80. *Mem:* Am Soc Pharmacol & Exp Therapeut; Soc Toxicol; Sigma Xi. *Res:* Correlation of chemical structure; physicochemical properties and biological activities of drugs; toxicology of cyanide and carbon monoxide. *Mailing Add:* Dept Pharmacol Univ Miss Med Ctr 2500 N State St Jackson MS 39216

HUME, DAVID JOHN, b Milton, Ont, Aug 7, 40; m 64; c 2. CROP PHYSIOLOGY. *Educ:* Univ Toronto, BSA, 61, MSA, 63; Iowa State Univ, PhD(crop physiol), 66. *Prof Exp:* From asst prof to assoc prof, 66-79, PROF CROP SCI, UNIV GUELPH, 79- *Concurrent Pos:* Assoc prof crop sci, Univ Ghana, 72-74; mem, Nat Coun, Agr Inst Can, 84-86, chmn, Oilseeds Subcomt, East Can Expert Comt on Grain Breeding, 86- *Honors & Awards:* Outstanding Res Award, Can Soc Agron, 90. *Mem:* Am Soc Agron; Agr Inst Can; Crop Sci Soc Am; Can Soc Agron (pres, 84-85); Sigma Xi. *Res:* Physiology and management of soybeans and canola, including soybean nodulation, N2 fixation and crop production practices; spring and winter Canola production; canola seed constituent quality; canola variety evaluation. *Mailing Add:* Dept Crop Sci Univ Guelph Guelph ON N1G 2W1 Can

HUME, DAVID NEWTON, b Vancouver, BC, Dec 22, 17; nat US; m; c 2. ANALYTICAL CHEMISTRY. *Educ:* Univ Calif, Los Angeles, BA, 39, MA, 40; Univ Minn, PhD(phys chem), 43. *Prof Exp:* Res assoc plutonium proj, Metall Lab, Univ Chicago, 43; group leader, Clinton Labs, Oak Ridge, 43-44, sect chief, 45-46; asst prof chem, Univ Kans, 46-47; from asst prof to prof, 47-80, EMER PROF ANALYTICAL CHEM, MASS INST TECHNOL, 80- *Concurrent Pos:* Guggenheim fel, 54-55; NSF sr fel, 64-65; guest investr, Woods Hole Oceanog Inst, 71. *Honors & Awards:* Fisher Award, Am Chem Soc, 63. *Mem:* Am Chem Soc; Am Acad Arts & Sci. *Res:* Analytical chemistry; instrumental analysis; polarography; complex ions; environmental trace analysis; chemical oceanography; forensic science. *Mailing Add:* One Sylvan Rd Wellesley Hills MA 02181-3239

HUME, HAROLD FREDERICK, b Genesee Falls, NY; m 53; c 2. TEXTILE CHEMISTRY, GENERAL ENVIRONMENTAL SCIENCES. *Educ:* Houghton Col, BSc, 39. *Prof Exp:* Proj engr, Curtiss Airplane Co, NY, 39-45; res engr & res chemist, E I du Pont de Nemours & Co, Inc, 45-64, res assoc, 64-76; CONSULT, ENERGY CHEM & MGT AFFAIRS, BENCH MARK ASSOCS, 77- *Mem:* AAAS; Am Chem Soc; Fiber Soc. *Res:* Airplane design; food processing and preservation; textiles; physical chemistry; physics of fibers; housing development and design. *Mailing Add:* 2408 Cedar Ave Wilmington DE 19808

HUME, JAMES DAVID, b Fresno, Calif, Dec 17, 23; m 54; c 2. GEOLOGY. *Educ:* US Mil Acad, BS, 45; Univ Mich, BSE, 49, MS, 50, PhD(geol), 57. *Prof Exp:* Asst prof geol, Purdue Univ, 55-57; from asst prof to prof geol, Tufts Univ, 57-87, chmn dept, 69-77 & 81-87; RETIRED. *Concurrent Pos:* Investr, Arctic Inst NAm, Off Naval Res, 59-69. *Mem:* Geol Soc Am; Soc Econ Paleont & Mineral; Am Asn Petrol Geol; Nat Asn Geol Teachers; Int Asn Sedimentologists. *Res:* Sedimentary petrology; sedimentation; shoreline processes; archeological geology. *Mailing Add:* Dept Geol Tufts Univ Medford MA 02155

HUME, JAMES NAIRN PATTERSON, b Brooklyn, NY, Mar 17, 23; m 53; c 4. THEORETICAL PHYSICS. *Educ:* Univ Toronto, BA, 45, MA, 46, PhD(atomic spectra), 49. *Prof Exp:* Lab asst physics, Univ Toronto, 45-46, lectr math, 46-49; instr physics, Rutgers Univ, 49-50; from asst prof to assoc prof, 50-63, chmn dept, 75-80, prof, 63-88, EMER PROF COMPUT SCI, UNIV TORONTO, 88-; master, Massey Col, 81-88. *Mem:* Asn Comput Mach; Sigma Xi; fel Royal Soc Can. *Res:* Theoretical atomic spectra; high-speed digital computation; analysis of computer systems; programming languages. *Mailing Add:* Dept Comput Sci Univ Toronto Toronto ON M5S 1A4 Can

HUME, JOHN CHANDLER, b Brooklyn, NY, May 16, 11; m 33; c 3. PREVENTIVE MEDICINE, PUBLIC HEALTH. *Educ:* Princeton Univ, AB, 32; Vanderbilt Univ, MD, 36; Johns Hopkins Univ, MPH, 47, DrPH, 51. *Prof Exp:* Med intern, Vanderbilt Univ Hosp, 36-37; actg dir, Hardin County Dept Health, Tenn, 37-38; clinician & dir, Tri-County Demonstration Syphilis, Ga, 38-39; practicing physician, Ga, 39-42; dir venereal dis control, Wilmington City & New Hanover County Dept Health, NC, 42-44 & State Dept Health, WVa, 44-46; res assoc, Johns Hopkins Univ, 47-51, assoc prof, 51-55, asst dir, 48-55; chief health div, US Tech Coop Mission to India, Int Coop Admin, 55-61; assoc dean, Sch Hyg & Pub Health, 61-67, chmn dept pub health admin, 61-69, prof pub health admin, 61-81, dean, Sch Hyg & Pub Health, 67-77, EMER DEAN, JOHNS HOPKINS UNIV, 77-, EMER PROF HEALTH POLICY & MGT, 81- *Concurrent Pos:* Physician-in-chg Venereal Dis Div, Med Clin (PM), Johns Hopkins Hosp, 47-55; chief div venereal dis control, State Dept Health, Md, 52-55, mem residency rev comn prev med, 63-73; secy-treas, Am Bd Prev Med, 63-67, chmn, 67-74; mem, Am Bd Med Specialties, 64-72; mem, Nat Adv Coun Pub Health Training, 66-70; pres, Asn Schs Pub Health, 70-71; mem bd trustees & first vpres, Pan Am Health & Educ Found, 70-83, pres, 83-84; mem bd dirs, Am Asn World Health, US Comt, WHO, 70-81, pres, 72-74, vpres, 74-81; mem nat comn venereal dis, US Dept HEW, 71-72. *Honors & Awards:* Edward W Browning Award for Prev of Dis, Am Pub Health Asn, 77; William Freeman Snow Award for Distinguished Serv to Humanity, Am Social Health Asn, 85; Thomas A Parson Award for Long & Distinguished Contrib to Field of Veneral Dis Control, Am Venereal Dis Asn, 88. *Mem:* Fel Am Pub Health Asn; Int Health Soc Inc (pres, 77-78); Am Venereal Dis Asn (secy, 53-55, vpres, 64-67, pres, 69-70); Sigma Xi; Am Social Health Asn. *Res:* Public health administration; education for health professions; venereal and treponemal disease control; international health. *Mailing Add:* 317 Tuscany Rd Baltimore MD 21210

HUME, MERRIL WAYNE, b San Francisco, Calif, Apr 30, 39; m 63; c 3. STATISTICS. *Educ:* Col William & Mary, BS, 60; Va Polytech Inst, MS, 64, PhD(statist), 66. *Prof Exp:* Asst prof statist, Case Western Reserve Univ, 66-70; assoc prof, Univ Mo, Rolla, 70-74; RES SPECIALIST STATIST, ROCKWELL INT, 74- *Honors & Awards:* Harlan J Anderson Award, Am Soc Testing Mat. *Mem:* Am Statist Asn; fel Am Soc Testing & Mat; Inst Nuclear Mat Mgt; Nat Mgt Asn. *Res:* Rank correlation; assessment of transuranics in the environment; accountability and safeguards of nuclear material. *Mailing Add:* 7560 Johnson St Arvada CO 80005

HUME, MICHAEL, b Danbury, Conn, Sept 18, 24; m 58; c 3. SURGERY. *Educ:* Yale Univ, BA, 45; Columbia Univ, MD, 50; Am Bd Surg, dipl, 56. *Prof Exp:* Res fel surg, Sch Med, Yale Univ, 55-56; from instr to assoc prof, Yale Univ, 56-68; PROF SURG, SCH MED, TUFTS UNIV, 68-; SURGEON-IN-CHIEF & CHMN DEPT SURG, NEW ENG BAPTIST HOSP, 78- *Concurrent Pos:* Markle scholar, 59; Fulbright scholar, UK, 66. *Mem:* Sigma Xi; Am Col Surgeons; Soc Vascular Surg; Int Soc Thrombosis & Haemostasis; Int Cardiovasc Soc. *Res:* Detection of thrombosis and treatment of thromboembolism with thrombolytic agents. *Mailing Add:* Tufts Univ PO Box B30056 Medford MA 02155

HUME, WAYNE C, b Mankato, Minn, June 17, 36; m 62; c 2. ELECTRICAL ENGINEERING. *Educ:* Univ Minn, BEE, 59, MSEE, 61. *Prof Exp:* Assoc res eng, Res Lab, Univac Div, Sperry Rand Corp, Minn, 61-65; res scientist, Shock & Struct Dyn Lab, Kaman Nuclear Div, Kaman Sci Corp, 65-74; OWNER & PROD DESIGNER, VINTAGE REPRODUCTIONS, 74- *Concurrent Pos:* res grant, Colo Energy Res Inst, 78-79. *Mem:* Inst Elec & Electronics Engrs. *Res:* Pulse power engineering; high-speed photography; shock wave experiments; solar radiation. *Mailing Add:* 2606 Flintridge Dr Colorado Springs CO 80918

HUMENICK, MICHAEL JOHN, II, b Muskegon, Mich, Sept 24, 36. CIVIL & SANITARY ENGINEERING. *Educ:* Univ Mich, BS, 59, MS, 60; Univ Calif, Berkeley, PhD(civil eng), 70. *Prof Exp:* Asst res engr, Chevron Res Co, 60-61, assoc res engr, 61-63, trainee, Standard Oil Co Calif, 63-64, res engr, Chevron Res Co, 64-67; mgr water pollution control, Environ Qual Eng, Inc, 70; assoc prof civil eng, Univ Tex, Austin, 70-; AT DEPT CIVIL ENG, UNIV WYO. *Mem:* Am Soc Civil Engrs; Am Inst Chem Engrs. *Res:* Physical, chemical and biological treatment of water and wastewaters. *Mailing Add:* Dept Civil Eng Univ Wyo Box 3295 Laramie WY 82071

HUMENIK, FRANK JAMES, b Brooklyn, NY, May 26, 37; m 60; c 2. SANITARY ENGINEERING. *Educ:* Ohio State Univ, BSCE, 63, MS, 66, PhD(sanit eng), 69. *Prof Exp:* From asst prof to assoc prof, 69-78, PROF ENG, NC STATE UNIV, 78-, ASSOC HEAD DEPT BIOL & AGR ENG, 73- *Concurrent Pos:* Grants, Water Resources Res Inst, 69-71 & Environ Protection Agency, 71; consult, Environ Protection Agency Animal Waste Projs, 73-78, Nat Comn Water Qual, 75, animal waste projs, pvt eng firms, 76- & state non-point source projs, 82- *Honors & Awards:* Duggar Lectr, Auburn Univ, 77; Gunlogson Countryside Eng Award, Am Soc Agr Engrs, 78. *Mem:* Am Soc Agr Engrs; Water Pollution Control Fedn; Sigma Xi. *Res:* Animal waste management; wastewater and waste nutrient utilization; land application of waste; water quantity and quality. *Mailing Add:* Dept Biol & Agr Eng NC State Univ Raleigh NC 27650

HUMENIK, MICHAEL, JR, b Garfield, NJ, Nov 10, 24; m 48; c 3. CERAMICS, METALLURGY. *Educ:* Alfred Univ, BS, 49; Mass Inst Technol, ScD(ceramics), 52. *Prof Exp:* Asst, Mass Inst Technol, 49-52; res engr, Ford Motor Co, Dearborn, Mich, 52-54, supvr ceramic & powder metall, 54-62, mgr ceramics & glass res, 62-69, asst dir mfg res, 69-76, DIR MFG PROCESSES LAB RES, FORD MOTOR CO, 76- *Honors & Awards:* Ferro Enamels Award, 52. *Mem:* Am Ceramic Soc; Am Soc Metals; Am Inst Mining, Metall & Petrol Engrs. *Res:* Surface energy studies in ceramic and metal systems at elevated temperatures; sintering; powder metallurgy; cermets; glass. *Mailing Add:* 17097 Cambridge Allen Park MI 48101

HUMER, PHILIP WILSON, b Carlisle, Pa, Nov 8, 32; m 53, 81; c 3. ORGANIC CHEMISTRY. *Educ:* Dickinson Col, Pa, BS, 54; Pa State Univ, PhD(org chem), 64. *Prof Exp:* Res chemist surfactants & adhesives, Phillips Petrol Co, 64-69; sr res chemist paper additives & adhesives, Pa Indust Chem Corp, 69-73; sr res chemist adhesives, Hercules Inc, PICCO RESINS, 73-74; SR INFO SCIENTIST PATENT LIAISON, FMC CORP, 74- *Concurrent Pos:* Patent agt, 78- *Mem:* Am Chem Soc; AAAS. *Res:* Reaction mechanisms;

reactive intermediates, such as carbenes; stereochemistry of cycloadditions, surfactants, adhesives, paper additives, insecticides, nematicides, herbicides, fungicides, plant regulators and chemical processes. *Mailing Add:* Eight Glenwood South Gate Morrisville PA 19067-1022

HUMES, ARTHUR GROVER, b Seekonk, Mass, Jan 22, 16. PARASITOLOGY. *Educ:* Brown Univ, AB, 37; La State Univ, MS, 39; Univ Ill, PhD(invert zool, parasitol), 41. *Prof Exp:* Instr biol, Univ Buffalo, 41-42; asst prof zool, Univ Conn, 46-47; from asst prof to prof, 55-81, dir marine prog, 70-81, EMER PROF BIOL, BOSTON UNIV, 81- *Concurrent Pos:* Guggenheim fel, 54-55; instr, Univ NH, 51-52; res assoc, Mus Comp Zool, Harvard Univ, 50-51 & 60-65; chief scientist, Nosy Be, Madagascar, US Prog Biol, Int Indian Ocean Exped, 63-64; ed, J Crustacean Biol, 80- *Mem:* AAAS; Am Micros Soc (pres, 83); World Asn Copepodologists (pres, 90); Crustacean Soc; fel Am Acad Arts & Sci. *Res:* Morphology, taxonomy and life histories of free-living and parasitic copepods. *Mailing Add:* Boston Univ Marine Prog Marine Biol Lab Woods Hole MA 02543

HUMES, JOHN LEROY, b Uniontown, Pa, Feb 18, 36; m 63. ENDOCRINOLOGY, BIOCHEMISTRY. *Educ:* Ohio Wesleyan Univ, BA, 58; Rutgers Univ, PhD(zool), 74. *Prof Exp:* Res asst biochem, Univ Pittsburgh, 61-63; biochemist, 63-68, res biochemist, 68-73, sr res biochemist, 74-78, RES FEL, MERCK INST THERAPEUT RES, 78- *Concurrent Pos:* Assoc prof grad fac, Rutgers Univ, 74- *Mem:* Sigma Xi; Am Physiol Soc; NY Acad Sci; Am Asn Immunologists. *Res:* Role of intracellular and extracellular regulators of cell functions, especially the role of prostaglandins and cyclic-nucleotides in cell function. *Mailing Add:* 137 Hillside Ave Berkeley Heights NJ 07922

HUMES, PAUL EDWIN, b Colville, Wash, Nov 19, 42; m 65; c 2. ANIMAL SCIENCE, GENETICS. *Educ:* Wash State Univ, BS, 64; Ore State Univ, PhD(genetics), 68. *Prof Exp:* From asst prof to assoc prof, 68-77, PROF ANIMAL SCI, LA STATE UNIV, BATON ROUGE, 77- *Mem:* Am Soc Animal Sci; Genetics Soc Am; Sigma Xi. *Res:* Genetic evaluation of breeds and breed crosses of sheep, cattle and swine; genetic control of hormones associated with growth and reproduction; production efficiency and mature size relationships in beef cattle. *Mailing Add:* Dept Animal Sci La State Univ Baton Rouge LA 70803

HUMI, MAYER, b Bagdad, Iraq, Sept 29, 44; m 80; c 2. MATHEMATICAL PHYSICS. *Educ:* Hebrew Univ Jerusalem, BSc, 63, MSc, 64; Weizmann Inst Sci, PhD(math physics), 69. *Prof Exp:* Fel, Univ Toronto, 69-70, asst prof math, 70-71; from asst prof to prof, 71-87, SINCLAIR PROF MATH, WORCESTER POLYTECH INST, 88- *Concurrent Pos:* Asst prof, Clark Univ, 75-76; assoc prof, Ga Inst Technol, 78-79. *Mem:* Am Math Soc; Soc Indust & Appl Math; Int Asn Math Physics. *Res:* Lie groups theory and its applications to differential equations; relativity and cosmology; theoretical physics; turbulence; author of two books. *Mailing Add:* Dept Math Worcester Polytech Inst Worcester MA 01609

HUMIEC, FRANK S, JR, b Niagara Falls, NY, June 11, 33; div; c 2. INORGANIC CHEMISTRY. *Educ:* Niagara Univ, BS, 58; Fairleigh Dickinson Univ, MS, 68. *Prof Exp:* Anal technician, Union Carbide Corp, 51-53 & Hooker Electrochem Corp, 53-58; chemist, Mellon Inst, 58-63 & Witco Chem Soc, Inc, 63-68; res chemist, Hoffmann-La Roche Inc, 68-86; CHEMIST, AT&T BELL LAB, 86- *Mem:* Am Chem Soc. *Res:* Phosphorus chemistry; organic synthesis; instrumental analysis; inorganic polymers and complexes; amino acids; medicinal chemistry; isolation of natural products. *Mailing Add:* 614 Essex Pl Fair Lawn NJ 07410

HUMKE, PAUL DANIEL, b Milwaukee, Wis, Feb 16, 45; m 67; c 3. ANALYSIS & FUNCTIONAL ANALYSIS, TOPOLOGY. *Educ:* Univ Wis-Milwaukee, PhD(math), 72. *Prof Exp:* From asst prof to assoc prof, dept math, Western Ill Univ, 72-80; assoc prof, 80-85, PROF DEPT MATH, ST OLAF COL, 85- *Concurrent Pos:* Vis assoc prof, Univ Calif, Santa Barbara, 83; vis res scientist, Fulbright/Hungarian Ministry Educ, 87; vis prof, Univ Wis, Milwaukee, 88. *Mem:* Am Math Soc; Math Asn Am. *Res:* The exceptional behavior of real functions, fractals, set porosity, derivates and derivatives; dynamical systems. *Mailing Add:* Dept Math St Olaf Col Northfield MN 55057

HUMM, DOUGLAS GEORGE, b Montreal, Que, May 20, 17; m 47; c 2. PHYSIOLOGY. *Educ:* Yale Univ, BS, 39; Stanford Univ, PhD, 48. *Prof Exp:* Asst prof biol, Univ NMex, 48-51; assoc prof, 51-61, prof, 61-77, EMER PROF ZOOL, UNIV NC, CHAPEL HILL, 77- *Concurrent Pos:* Nat Inst Allergy & Infectious Dis spec fel, 64-65. *Mem:* AAAS; Am Soc Zool; Soc Gen Physiol; Soc Develop Biol. *Res:* Metabolism of fish melanomas; biochemistry of amphibian larval growth. *Mailing Add:* 1505 Smith Level Rd Chapel Hill NC 27514

HUMM, HAROLD JUDSON, b Lorain, Ohio, Feb 26, 12; m 36; c 3. MARINE BIOLOGY. *Educ:* Univ Miami, BS, 34; Duke Univ, MS, 42, PhD(bot), 45. *Prof Exp:* Res investr, Marine Lab, Duke Univ, 42-45, from asst dir to dir, 45-49; dir oceanog inst, Fla State Univ, 49-54; assoc prof bot, Duke Univ, 54-65; prof biol, Queens Col, NC, 65-67; prof, 67-82, dir, 67-73, EMER PROF, MARINE SCI INST, UNIV SFLA, 82- *Concurrent Pos:* Jacques Loeb assoc, Rockefeller Inst, 59-60; treas, Gulf Oceanog Develop Found, 68-; distinguished scholar, Marine Sci Inst, Univ SFla, 82- *Mem:* Int Phycol Soc; Phycol Soc Am. *Res:* Distribution, taxonomy and utilization of marine algae; marine bacteriology. *Mailing Add:* 13 E View Dr Brevard NC 28712

HUMMEL, DONALD GEORGE, b St Louis, Mo, Oct 26, 25; m 50; c 6. ORGANIC CHEMISTRY. *Educ:* St Louis Univ, BS, 48; Boston Col, MS, 50; Kans State Univ, PhD, 58. *Prof Exp:* Asst, Boston Col, 48-50 & Kans State Univ, 50-51; res chemist, Olin Mathieson Chem Corp, 53-57; res chemist, Org Chem Dept, Jackson Lab, 58-76, prod supvr, Chem, Dyes & Pigments Dept, 76-80, sr chemist, Chem & Pigments Dept, E I du Pont de Nemours & Co, Inc, 80-85; PRES, DONALD G HUMMEL, INC, 85- *Concurrent Pos:* Consult, 85- *Mem:* Am Chem Soc; Sigma Xi. *Res:* Explosives; fluorocarbon compounds; fluorocarbon chemistry; gasoline additives; surfactants; ethylene oxide chemistry; environmental consulting. *Mailing Add:* 603 Amberly Rd Wilmington DE 19803

HUMMEL, F(LOYD) A(LLEN), b Chicago, Ill, Sept 28, 15; m 38; c 4. CERAMIC CHEMISTRY. *Educ:* Univ Ill, BS, 37; Pa State Univ, MS, 48. *Prof Exp:* Res ceramist, Onondaga Pottery Co, Syracuse, NY, 37-41 & Corning Glass Works, Pa & NY, 41-44; res asst prof ceramics, Univ Ill, 44-45; from asst prof to prof, 48-79, head dept, 63-67, EMER PROF CERAMICS, PA STATE UNIV, 79- *Concurrent Pos:* Consult, Lamp Bus Div, Gen Elec Co, 53-75, Universal Dental Co, Philadelphia, 57-72, Ferro Corp, 62-77. *Honors & Awards:* Purdy Award, Am Ceramic Soc, 53, Meyer Award, 60, Bleininger Award, 83. *Mem:* Am Chem Soc; fel Am Ceramic Soc; fel Mineral Soc Am; Mineral Soc Can; Brit Ceramic Soc; Sigma Xi. *Res:* Crystal chemistry; phase equilibria and its application to luminescence; mechanical and thermal properties of solids; inorganic pigments. *Mailing Add:* 819 Fairway Rd State College PA 16803

HUMMEL, HANS ECKHARDT, b Heilbronn, Ger, Apr 30, 39; m 79. CHEMICAL ECOLOGY, CHEMICAL COMMUNICATION. *Educ:* Univ Stuttgart, BS, 60; Univ Munich, MS, 64; Univ Marburg, PhD(insect biochem), 68. *Prof Exp:* Fel res & teaching asst physiol chem, Philipps Univ, Marburg, 64-70; res entomologist, Div Toxicol & Physiol, Univ Calif, Riverside, 70-73; res assoc ecol chem, Dept Chem, State Univ NY, Syracuse, 73-74; ASST PROF ENTOM & CHEM ECOL, DEPT ENTOM, UNIV ILL, URBANA, 74- *Concurrent Pos:* Consult environ toxicol, 82- *Mem:* Entom Soc Am; Am Chem Soc; AAAS; Sigma Xi. *Res:* Chemical communication, inter and intraspecifically, among invertebrates, emphasis in insects, vertebrates, plants, by pheromones, allomones, kairomones, hormones; chemical ecology; bioassays; analytical techniques (chromatographic, spectrometric) for identifying submicrogram quantities of semiochemicals; environmental toxicology; integrated pest management. *Mailing Add:* Justus-Liebig-Univ Giessen Ludwigstrasse 21 D-6300 Giessen Germany

HUMMEL, HARRY HORNER, physics, for more information see previous edition

HUMMEL, JAMES ALEXANDER, b Santa Monica, Calif, Dec 14, 27; m 50; c 3. COMPLEX VARIABLES, UNIVALENT FUNCTIONS. *Educ:* Calif Inst Technol, BS, 49; Rice Univ, MA, 53, PhD(math), 55. *Prof Exp:* NSF fel, Stanford Univ, 55-57; from asst prof to assoc prof, 57-63, PROF MATH, UNIV MD, COLLEGE PARK, 63- *Mem:* AAAS; Am Math Soc; Math Asn Am; Sigma Xi. *Res:* Complex variables; univalent functions. *Mailing Add:* Dept Math Univ Md College Park MD 20742

HUMMEL, JOHN PHILIP, b Blue Earth, Minn, July 29, 31; m 53; c 4. NUCLEAR CHEMISTRY. *Educ:* Univ Rochester, BS, 53; Univ Calif, PhD(chem), 56. *Prof Exp:* From instr to assoc prof, 56-67, PROF CHEM & PHYSICS, UNIV ILL, URBANA, 67-, ASSOC HEAD DEPT CHEM, 71- *Mem:* Am Phys Soc. *Res:* Radiochemical studies of nuclear reactions induced by high energy x-rays; study of chemical effects on positron annihilation. *Mailing Add:* Dept Chem Univ Ill 505 S Mathews Ave Urbana IL 61801-3617

HUMMEL, JOHN RICHARD, b Kansas City, Mo, May 19, 51. ATMOSPHERIC SCIENCE. *Educ:* Pa State Univ, BS, 73, MS, 75; Univ Mich, PhD(atmospheric sci), 78. *Prof Exp:* Res asst meteorol, Ionosphere Res Lab, Pa State Univ, 73-75; res scientist, Univ Mich, 75-78; res scientist, 78-81, SR RES SCIENTIST, ATMOSPHERIC PHYSICS, GEN MOTORS RES LABS, 81- *Mem:* Am Geophys Union; Am Meteorol Soc; AAAS; Sigma Xi. *Res:* Climatology and radiative transfer. *Mailing Add:* 595 Great Elm Way Acton MA 01718

HUMMEL, JOHN WILLIAM, b Grantsville, Md, Nov 1, 40; m 64; c 4. CONSERVATION TILLAGE SYSTEMS, SOIL PHYSICAL PROPERTIES SENSORS. *Educ:* Univ Md, BS, 64, MS, 66; Univ Ill, PhD, 70. *Prof Exp:* from asst prof to assoc prof agr eng, Univ Md, Col Park, 69-76; AGR ENGR, AGR RES SERV, UNIV ILL, USDA, 76- *Mem:* Am Soc Agr Engrs. *Res:* Sensors for measurement of soil physical properties. *Mailing Add:* 376 Agr Engr Sci Bldg Univ Ill 1304 W Penn Ave Urbana IL 61801

HUMMEL, RICHARD LINE, b Dothan, Ala, Dec 27, 28; m 54; c 2. CHEMICAL ENGINEERING, PHYSICAL CHEMISTRY. *Educ:* Purdue Univ, BS, 50; Iowa State Univ, PhD(phys chem), 62. *Prof Exp:* Engr in training, E I du Pont de Nemours & Co, 51-52, engr, 52-55; asst quantum mech, Inst Atomic Res, Ames, Iowa, 55-59; instr chem eng, Univ Mich, 59-60; res assoc quantum mech, Iowa State Univ, 60-61; asst prof, 61-64, assoc prof, 64-76, PROF CHEM ENG, UNIV TORONTO, 76- *Mem:* Am Chem Soc; Am Inst Chem Engrs; Solar Energy Soc; Chem Inst Can. *Res:* Boiling catalysis to increase evaporation coefficients; mechanism of boiling; theoretical calculation of electronic spectra of catacondensed and pericondensed aromatic hydrocarbons. *Mailing Add:* Dept Chem Eng Univ Toronto Toronto ON M5V 1A4 Can

HUMMEL, ROBERT P, b Bellevue, Ky, Sept 17, 28; m 54; c 3. SURGERY. *Educ:* Xavier Univ, BS, 47; Univ Cincinnati, MD, 51. *Prof Exp:* From instr to assoc prof, 59-77, PROF SURG, UNIV CINCINNATI, 77-, VCHMN DEPT, 84- *Concurrent Pos:* Attend surgeon, C R Holmes Hosp, 60, Children's Hosp, 64 & Cincinnati Gen Hosp, 65; coordr, Trauma Ctr Res, 67. *Mem:* Am Col Surgeons; Am Surg Asn; Alimentary Tract; Am Burn Asn; Am Asn Surg Trauma. *Res:* Trauma; burns; surgical infections; gnotobiology. *Mailing Add:* Dept Surg Univ Cincinnati Med Ctr Cincinnati OH 45267-0558

HUMMEL, ROLF ERICH, b Sindelfingen, Ger, July 21, 34; m 61; c 3. PHYSICAL METALLURGY. *Educ:* Univ Stuttgart, BS, 56, dipl physics, 60, Dr rer nat(physics, metall), 63. *Prof Exp:* res assoc phys metall, Max Planck Inst Metals Res, 58-63; from asst prof to assoc prof, 64-74, PROF MAT SCI, UNIV FLA, 74- *Concurrent Pos:* Vis prof, Max Planck Inst Metals Res,

Stuttgart, 71-72, Solar Energy Res Inst, 80, Solid Optics Lab, Paris, 81 & Univ Kyoto, Japan, 88; vis scientist, Chinese Acad Sci, Bejing, 88. *Mem:* Am Inst Metall Eng; Am Phys Soc. *Res:* Optical properties of metals and alloys; corrosion optics; electrotransport in thin films; electronic materials; ion implantation; thin film technology; differential reflection spectroscopy. *Mailing Add:* Dept Mat Sci & Eng Univ Fla Gainesville FL 32611

HUMMEL, STEVEN G, b Tarrytown, NY, Jan 23, 58. CHLORIDE TRANSPORT GROWTH OF III-V COMPOUNDS, VAPOR LEVITATION EPITAXY. *Educ:* Rutgers Univ, BA, 80, MSEE, 86; Univ Southern Calif, PhD(elec eng), 91. *Prof Exp:* Sr staff technologist, AT&T Bell Labs, 80-83; mem tech staff, Bell Commun Res, 84-87; RES ASST, DEPT ELEC ENG, UNIV SOUTHERN CALIF, 87- *Concurrent Pos:* Consult, Lam Res, Inc, 88- *Mem:* Inst Elec & Electronics Engrs; Am Phys Soc; Am Asn Crystal Growth. *Res:* Metal organic chemical vapor deposition of III-V compounds; optical and electrical characterization techniques; in-situ growth probes. *Mailing Add:* Dept Elec Eng MC 0483 Univ Southern Calif SSC 502 Los Angeles CA 90089

HUMMELER, KLAUS, b Hamburg, Ger, Feb 23, 22; nat US; m 50; c 1. PATHOLOGY, VIROLOGY. *Educ:* Univ Hamburg, MD, 48. *Prof Exp:* Instr pediat & pub health, 51-52, assoc pediat & virol, 52-55, from asst prof to assoc prof, 55-70, PROF PEDIAT, UNIV PA, 70-, CHIEF DIV EXP PATH, 66-; MEM RES STAFF, CHILDREN'S HOSP, 50-; DIR, JOSEPH STOKES SR RES INST, 72- *Concurrent Pos:* Prof, Wistar Inst, 81- *Mem:* Fel AAAS; Am Asn Immunologists; Royal Soc Med; Path Soc Gt Brit & Ireland; Am Asn Path. *Res:* Virus immunology; cell biology; electron microscopy. *Mailing Add:* Childrens Hosp Philadelphia 34th St & Civic Ctr Blvd Philadelphia PA 19104

HUMMELS, DONALD RAY, b Morris, Ill, June 30, 36; m 57; c 5. ELECTRICAL ENGINEERING. *Educ:* Ariz State Univ, BS, 67, MS, 68, PhD(eng), 69. *Prof Exp:* Sr engr, Motorola, Inc, 57-70; asst prof, 70-75, assoc prof, 75-, PROF & HEAD ELEC ENG, KANS STATE UNIV. *Mem:* Inst Elec & Electronics Engrs. *Res:* Communication and information theory. *Mailing Add:* Dept Elec & Comput Eng Kans State Univ Durland Hall Manhattan KS 66506

HUMMER, DAVID GRAYBILL, b Manheim, Pa, Nov 4, 34; m 61; c 1. THEORETICAL ASTROPHYSICS, LOW ENERGY ATOMIC PHYSICS. *Educ:* Carnegie Inst Technol, BS, 57, MS, 59; Univ London, PhD(physics), 63. *Prof Exp:* Physicist, Gen Atomic Div, Gen Dynamics Corp, 57-59; vis fel, Joint Inst Lab Astrophys, Univ Colo, 63-64; lectr physics, Univ Col, Univ London, 64-66; chmn, Inst, 73-74 & 77-78, FEL, JOINT INST LAB ASTROPHYS, UNIV COLO, BOULDER, 66- *Concurrent Pos:* Mem, Interagency Coord Comt Astron, 75-77. *Honors & Awards:* Von Humboldt Sr Scientist Award, 84. *Mem:* Am Phys Soc; fel Royal Astron Soc; Int Astron Union; Am Astron Soc; Astron Soc of Pac. *Res:* Theory of spectral line shapes; radiative transfer; stellar atmospheres and accretion disks; experimental and theoretical atomic physics; numerical analysis; stellar opacities; equation of state. *Mailing Add:* Joint Inst Lab Astrophysics Campus Box 440 Univ Colo Boulder CO 80309

HUMMER, JAMES KNIGHT, b Titusville, Pa, June 13, 26. ORGANIC CHEMISTRY. *Educ:* Tufts Univ, BS, 46; Middlebury Col, MS, 50; Univ NC, PhD(chem), 56. *Prof Exp:* Instr chem, Col Wooster, 54-58; res fel, Univ Sydney, 58-59; mem steroid prog, Clark Univ, 59-60; asst prof chem, Bucknell Univ, 60-62; from assoc prof to prof chem, Lycoming Col, 62-88; RETIRED. *Concurrent Pos:* Teaching & res fel, Imp Col Sci & Technol, London, 70-71. *Mem:* Am Chem Soc; Royal Soc Chem. *Res:* Heterocyclics; steroids. *Mailing Add:* 2211 Fink Ave Williamsport PA 17701

HUMMERS, WILLIAM STRONG, JR, b Hackensack, NJ, Nov 1, 17; m 44; c 3. INORGANIC CHEMISTRY. *Educ:* Washington & Lee Univ, BS, 41; Univ NC, PhD(chem), 51. *Prof Exp:* Chemist paint & pigments, Nat Lead Co, 46-48; asst, Univ NC, 48-51; fel, Mellon Inst, 51-54; chemist molybdenum compounds, Climax Molybdenum Co, 54-58; assoc prof, 58-65, PROF CHEM, THE CITADEL, 65- *Res:* Anhydrous metal halides; organometallic complexes; clay; clay complexes; molybdenum compounds; catalyst. *Mailing Add:* 159 Sea Marsh Dr Johns Island SC 29455

HUMMERT, GEORGE THOMAS, b Pittsburgh, Pa, Oct 23, 38; m 62; c 2. ELECTROMAGNETIC LAUNCHERS, MAGNETOFLUIDYNAMICS. *Educ:* Carnegie-Mellon Univ, BS, 60, MS, 64, PhD(elec eng), 69. *Prof Exp:* Instrument engr, Hercules, Inc, 61-63, develop engr circuit design, 64-66; sr engr electrotechnol, Westinghouse Elec Corp, 68-80, mgr electromagnetics, Res & Develop Ctr, 80-89; DIR, ELECTROMAGNETIC APPLN, EEAD, MACNEAL SCHWENDLA CORP, 89- *Res:* Photoionization produced by gaseous discharges; application of minicomputers; electromagnetic analysis; numerical descriptions of liquid metal flow; computer graphics and sliding contacts; computer integrated manufacturing systems. *Mailing Add:* EEAD MacNeal Schwendla Corp 9076 N Deerbrook Trail Milwaukee WI 53223-2434

HUMMON, WILLIAM DALE, b Akron, Ohio, July 27, 32; m 58; c 2. POPULATION BIOLOGY, MEIOBENTHIC INVERTEBRATES. *Educ:* Univ Mont, BA, 55, BS, 60, MS, 61; Univ Mass, PhD(zool), 69. *Prof Exp:* Res trainee zool, Systs Ecol Prog, Marine Biol Lab, Woods Hole, 67-69; asst prof, 69-73, assoc prof, 73-79, PROF ZOOL, OHIO UNIV, 79- *Concurrent Pos:* Instr, Olympic Col, 61-65; NATO fel marine lab, Dept Agr & Fisheries Scotland, Aberdeen, 74-75, Ohio Univ fac fel, 78-79. *Mem:* Am Soc Zool; Am Micros Soc; Soc Study Evolution; Ecol Soc Am; fel AAAS; Am Soc Limnol Oceanog. *Res:* Evolutionay and quantitative ecology of meiobeuthic communities, with emphasis on spatio-temporal zoogeography and systematics of gastrotricha; interactions between science and religion, with emphasis on evolution and the challenge of creationism. *Mailing Add:* Dept Zool & Biomed Sci Ohio Univ Athens OH 45701

HUMPHERYS, ALLAN S(TRATFORD), b Idaho Falls, Idaho, Apr 4, 26; m 56; c 9. IRRIGATION, ON-FARM AUTOMATION OF IRRIGATION. *Educ:* Utah State Univ, BS, 54, MS, 60. *Prof Exp:* Res asst canal & reservoir lining res, Agr Exp Sta, Utah State Univ, 54-58; AGR ENGR IRRIG RES, AGR RES SERV, USDA, 58- *Concurrent Pos:* Mem, Water Supply & Conveyance Comt, Am Soc Agr Engrs, 72-; chmn, 84-87, mem, Am Soc Civil Engrs, Comt Control Systs Water Pipelines, 75-87; mem, Int Comt Irrig & Drain Working Group, Mechanized Irrig, 77-, chmn, 79-; consult, On-Farm Auto Irrig, Argentina, 85. *Mem:* Am Soc Agr Engrs; Sigma Xi; Am Soc Civil Engrs; Int Comn Irrig & Drainage; Irrigation Asn. *Res:* Control, conveyance and measurement of water; methods, hydraulic design and improvement of structures, facilities and systems for conveying, controlling and automatically applying irrigation water efficiently on farms. *Mailing Add:* Soil & Water Mgt Res USDA-ARS 3793 N 3600 E Kimberly ID 83341

HUMPHREY, ALBERT S, b Kansas City, Mo, June 2, 26; m 82; c 7. BUSINESS FAILURE RECONSTRUCTION, ORGANIZATIONAL POLITICS & STAGNATION. *Educ:* Univ Ill, BS, 46; Mass Inst Technol, MS, 48; Harvard Bus Sch, MBA, 55. *Prof Exp:* Staff engr ECoast Tech Serv, Esso Standard Oil Co, 48-50; off chief chem officer, US Army Chem Corp, Chief Chem & Protective Group, 50-54; asst to pres, Penberthy Instrument Co, 54-55; chief, Prod Planning, Boeing Airplane Co, 55-60; mgr Value Anal Prog Small Aircraft Div, Gen Elec, 60-61; mgr res & develop planning, P R Mallory & Co, Inc, 61-64; mgr, Mgt Admin Improv Gen Dynamics, 64-65; sr consult & dir, Int Exec Seminar in Bus Planning, Stanford Res Inst, Calif, 65-70; consult, Off Advan Res & Technol, NASA, 65-70; CHMN, BUS PLANNING & DEVELOP INC, 70- *Concurrent Pos:* Consult var US & foreign bus; dir, Sanbros Ltd, Eng, Tower Lysprodukter a/s, Norway, Petras Petrochemische Anwendungssysteme GmbH, Ger, Visual Enterprises Ltd, Eng, Long Life Herbal Classics Inc, NJ, Light Indust, Ltd, Eng; fac mem, Polytechnic Sch Bus & Mgt; vis prof, Newcastle upon Tyne. *Mem:* Sci Res Soc Am; Sigma Xi; Am Inst Chem Engrs. *Res:* Developed a method for inter-relating people in working parties which mimics the brain of the individual making it possible for a team of people from 2-21 to work together and think together at roughly the same speed as the individual working alone; application of the team work method to managing change in organizations and family units; creativity and product development programs. *Mailing Add:* Bus Planning & Develop Inc 34 Jellicoe House Osnaburgh St London NW-1 30H England

HUMPHREY, ARTHUR E(ARL), b Moscow, Idaho, Nov 9, 27; m 51; c 2. BIOCHEMICAL ENGINEERING, BIOTECHNOLOGY. *Educ:* Univ Idaho, BS, 48, MS, 50; Columbia Univ, PhD(chem eng), 53; Mass Inst Technol, MS, 60. *Hon Degrees:* DSc, Univ Idaho, 78. *Prof Exp:* Asst phys sci, Univ Idaho, 47-50 & Columbia Univ, 50-52; from asst prof to assoc prof chem eng, Univ Pa, 53-62, prof, 62-80, dean, Col Eng & Appl Sci, 77-80; provost, 80-86, prof chem eng, dir, Ctr Molecular Biosci & Biotech, Lehigh Univ, 86-91; PRES, AM INST CHEM ENG, 91- *Concurrent Pos:* Chem engr, Inst Coop Res, 54-58; NSF fel, Mass Inst Technol, 58-59; consult, Merck, Sharp & Dohme, Inc, 58-62, Sun Oil Co, 61-68, Merck Chem Co, 62-63, Int Minerals & Chem Co, 63-67, Fermentation Design, 66-, Squibb, 67- & Air Products, 71-86, & Hoffman LaRoche, 87-; Fulbright lectr, Tokyo, 63 & Univ New SWales, 70; chmn atmospheric regeneration panel, Space Sci Bd, Nat Acad Sci, 66-68, mem food sci panel, Comn Undergrad Educ Biol Sci, 66; lectr, Tunghai Univ, 67; mem ad hoc working group on single cell protein, Protein Adv Group, Food & Agr Orgn, WHO, UNICEF, 69-; mem eng adv bd, NSF, 70-; mem, Franklin Inst. *Honors & Awards:* Van Lannen Award, Am Chem Soc, 78; Founders Award, Am Inst Chem Engrs, 89. *Mem:* Nat Acad Eng; Soc Microbiol; Am Chem Soc; Am Soc Eng Educ; Am Inst Chem Engrs. *Res:* Enzyme engineering; air sterilization; media sterilization; kinetics of growth of cellular organisms; computerized fermenters; immobilized enzymes; recycle of wastes. *Mailing Add:* Lehigh Univ Bldg A MTC Bethlehem PA 18015

HUMPHREY, BINGHAM JOHNSON, b Proctor, Vt, Feb 9, 06; m 30; c 3. INDUSTRIAL ORGANIC CHEMISTRY. *Educ:* Univ Vt, BS, 27; Yale Univ, PhD(org chem), 30. *Hon Degrees:* LLD, Univ Vt, 80. *Prof Exp:* Res chemist, Firestone Tire & Rubber Co, 30-36, group leader, 36-42; tech dir, Conn Hard Rubber Co, 42-49; pres, Humphrey Chem Co, 49-72, chmn bd, 72-88; RETIRED. *Concurrent Pos:* Dir, Milfoam Corp, Milford Conn; trustee, Univ Vt, 68-74, chmn, 73-74. *Mem:* Am Chem Soc; Sigma Xi. *Res:* organic synthesis and manufacture. *Mailing Add:* PO Box 5142 Mt Carmel CT 06518

HUMPHREY, CHARLES HARVE, b Brewer, Maine, Sept 5, 25; m 48; c 2. PHYSICS. *Educ:* Univ Calif, Los Angeles, AB, 47, MA, 49, PhD(physics), 56. *Prof Exp:* Asst physics, Univ Calif, Los Angeles, 51-53; res scientist, Missile & Space Div, Lockheed Aircraft Corp, 54-58 & Aeronutronic Div, Ford Motor Co, 58-62; staff scientist, Palo Alto Res Lab, Lockheed Missiles & Space Co, Calif, 62-71; scientist, Sci Applns, Inc, 71-74; mem staff, R&D Assocs, Santa Monica, Calif, 74-77; MEM STAFF, VISIDYNE INC, 76- *Mem:* Am Phys Soc. *Res:* Nuclear physics; reactor and plasma physics; atomic physics; weapons effects physics. *Mailing Add:* 219 Follen Rd Lexington MA 02173

HUMPHREY, DONALD GLEN, b Ames, Iowa, Feb 28, 27; m 46; c 2. CYTOGENETICS. *Educ:* Univ Iowa, BS, 49; Univ Wash, MS, 50; Ore State Univ, PhD(zool), 56. *Prof Exp:* From instr to asst prof biol, Ore Col Educ, 50-57; instr, Ore State Univ, 54-55, from asst prof to prof & chmn dept, 57-70, dir honors prog, 65-70, asst dean fac, 69-70; dean div natural sci & math, 70-74, mem fac biol, 74-84, EMER MEM FAC BIOL, EVERGREEN STATE COL, 84- *Concurrent Pos:* Consult, Portland Curric Study, 58; dep dir Am Quintana Roo Expeds, 65 & 66; panel mem & consult, Comn Undergrad Ed Biol Sci, NSF; NSF fac fel hist & philos sci, Harvard Univ, 67-68 & Sci & Pub Policy, Kennedy Sch Govt, Harvard Univ, 80-81. *Mem:* Fel AAAS; Sigma Xi. *Res:* Chromosomes of vertebrates, Amphibia, Reptilia and Aves; evolution of genetic systems; science and public policy; electron microscopy. *Mailing Add:* 3026 43rd Ct NW Olympia WA 98502

HUMPHREY, DONALD R, b Tahlequah, Okla, July 31, 35; m 61; c 2. NEUROPHYSIOLOGY, NEUROANATOMY. *Educ:* San Jose State Univ, BA, 60; Univ Wash, PhD(physiol & biophys), 66. *Prof Exp:* USPHS staff assoc clin neurophysiol, Nat Inst Neurol Dis & Stroke, 66-68, staff neurophysiologist, Lab Neurol Control, 68-71; asst prof to assoc prof, 71-80, PROF PHYSIOL, SCH MED, EMORY UNIV, 80-, DIR LAB NEUROPHYSIOL, 71- *Concurrent Pos:* Mem, NIH Study Sect, 77-81; consult, Nat Inst Neurol Commun Disorders & Stroke, NIH, 80- *Honors & Awards:* Hans Berger Res Award, Am EEG & Clin Neurophysiol Soc, 68. *Mem:* Am Physiol Soc; Soc Neurosci; Biomed Eng Soc; Am Asn Anatomy. *Res:* Cerebral cortical mechanisms in the control of movement and posture; neuromuscular control systems. *Mailing Add:* Dept Physiol Emory Univ Sch Med Atlanta GA 30322

HUMPHREY, EDWARD WILLIAM, b Fargo, NDak, Dec 6, 26; m 50; c 2. SURGERY, PHYSIOLOGY. *Educ:* Univ Minn, BA, 48, BM, 51, MD, 52, PhD(physiol), 59. *Prof Exp:* From instr to assoc prof, 58-65, PROF SURG, UNIV MINN, MINNEAPOLIS, 65- *Concurrent Pos:* Staff surgeon, Vet Admin Hosp, 58-, chief surg serv, 62- *Honors & Awards:* Wangensteen Award, 82. *Mem:* Am Col Surgeons; Soc Univ Surgeons; Am Soc Cell Biol; Am Asn Thoracic Surg; Am Surg Asn; Am Physiol Soc. *Res:* Electrolyte flux in various tissues; therapy of cancer; pulmonary physiology. *Mailing Add:* Vet Admin Med Ctr Minneapolis MN 55417

HUMPHREY, FLOYD BERNARD, b Greeley, Colo, May 20, 25; m 55; c 4. MAGNETISM. *Educ:* Calif Inst Technol, BS, 50, PhD(chem), 56. *Prof Exp:* Mem tech staff, Bell Tel Labs, 55-60; res group supvr, Jet Propulsion Lab, Calif Inst Technol, 60-64, assoc prof elec eng, 64-71, prof elec eng & appl physics, 71-80; prof & head, dept elec eng, Carnegie-Mellon Univ, 80-87; RES PROF ELEC & COMPUTER & SYSTS ENG, BOSTON UNIV, 87- *Mem:* Am Phys Soc; fel Inst Elec & Electronics Engrs; Am Vacuum Soc; Sigma Xi. *Res:* Magnetism and magnetic computer devices; thin films. *Mailing Add:* Box 722 Little Run Meredith Neck Meredith NH 03253-0722

HUMPHREY, GEORGE LOUIS, b Belleville, WVa, Oct 5, 21; m 46; c 3. PHYSICAL CHEMISTRY. *Educ:* Marietta Col, AB, 43; Calif Inst Technol, MS, 48; Ore State Univ, PhD(phys chem), 50. *Prof Exp:* Phys chemist, Pac Exp Sta, US Bur Mines, 50-52; from asst prof to prof, 52-82, actg chem dept, 63-65, assoc chmn dept, 65-82, EMER PROF CHEM, WVA UNIV, 83- *Mem:* Am Chem Soc; Coblentz Soc. *Res:* Vibrational spectroscopy; normal coordinate and molecular orbital calculations; reactions in solid state. *Mailing Add:* 912 Riverview Dr Morgantown WV 26505

HUMPHREY, GORDON LAIRD, b Yuba City, Calif, Dec 2, 40; m 66; c 2. PHYSIOLOGY, PSYCHOLOGY. *Educ:* Willamette Univ, BA, 63; Univ Calif, Los Angeles, MA, 66, PhD(psychol), 68. *Prof Exp:* Actg asst prof psychol, Univ Calif, Los Angeles, 68, USPHS trainee, Brain Res Inst, 68-70, NIMH trainee, 70-71; asst prof physiol, 71-76, ASST PROF PHYSIOL IN OTOLARYNGOL, UNIV ILL MED CTR, 74-; SR RES BIOLOGIST, RES RESOURCES CTR, BENJAMIN GOLDBERG RES CTR, 76- *Mem:* AAAS; Soc Neurosci. *Res:* Neurophysiology of learning. *Mailing Add:* Res Resources Univ Ill 1940 W Taylor St Chicago IL 60612

HUMPHREY, HAROLD EDWARD BURTON, JR, b Lansing, Mich, Aug 5, 40; m 63; c 2. ENVIRONMENTAL HEALTH. *Educ:* Univ Mich, BS, 62, MS, 65; Mich State Univ, PhD(microbiol & pub health), 70. *Prof Exp:* Sci adv, Gov's Exec Off, State Mich, 69-70; asst prof environ sci, Mich State Univ, 70-71; ENVIRON EPIDEMIOLOGIST, MICH DEPT PUB HEALTH, 71- *Concurrent Pos:* Dir, Kelmik Corp, 71-; US rep, Health Effects Comt, Int Joint Comn, 75-, Res Adv Bd, 79-; adj prof, Col Med, Mich State Univ, 80-; lectr, Sch Pub Health, Univ Mich, 81- *Mem:* Am Soc Microbiol; AAAS; Am Chem Soc; Sigma Xi. *Res:* Chemical contaminants in the environment, detection and evaluation of the effects of exposure on human health; chemical solid-liquid separation processes. *Mailing Add:* 815 Stuart Ave East Lansing MI 48823

HUMPHREY, J RICHARD, b Clovis, NMex, Oct 21, 42; m 66; c 4. PROCESS DEVELOPMENT FOR TITANIUM INVESTMENT CASTING, METAL-MATERIAL SCIENCE. *Educ:* Univ Calif, Riverside, BA, 63. *Prof Exp:* Chemist anal chem, Atomics Int Div, Rockwell Int, 63-65; proj engr chem vapor deposition, San Fernando Labs, 65-67; solid state electrochemist, Mat & Device Develop, Gould Ionics, 67-69; tech dir titanium castings, Rem Metals Corp, 69-83; TECH VPRES TITANIUM CASTINGS, SELMET INC, 83- *Concurrent Pos:* Guest lectr, Linn-Benton Community Col, 70-71. *Mem:* Am Soc Metals; Metall Soc. *Res:* Electrochemical behavior of organo-titanium compounds, chemical vapor deposition process and solid state electrochemical devices; production of titanium castings, including thermoplastics, ceramics, melting processes, heat treatment and welding techniques; author of 14 technical papers; recipient of three patents. *Mailing Add:* Selmet Inc PO Box 689 Albany OR 97321

HUMPHREY, JIMMY LUTHER, b Fulbright, Tex, Oct 24, 36; m 78; c 2. SEPARATION PROCESSES, DISTILLATION EXTRACTION. *Educ:* Univ Tex, Arlington, AS, 61; Tex A&M Univ, BS, 63; Univ Tex, Austin, PhD(chem eng), 67. *Prof Exp:* Res supvr, Eastman Kodak Co, 67-69, sr res engr, 69-72, environ coordr, 73-74, prod mgr, 75, energy mgr, 76-79; proj mgr, Argonne Nat Lab, 79-82; assoc head, Separations Res Prog, Univ Tex, Austin, 82-86; PRES, J L HUMPHREY & ASSOCS, 86- *Concurrent Pos:* Prin investr & sr lectr, Univ Tex, Austin, 82-86; prog adv bd, US Dept Energy, 83-; indust consult, Alcoa, Gas Res Inst, Combustion Eng, Proctor & Gamble & Gen Foods, 84-; proj rev panel, NSF, 84-; consult, plate distillation subcomt, Am Inst Chem Engrs, 84-86; adv bd, Pub Utility Comn Tex, 84-86. *Mem:* Am Inst Chem Engrs; Am Chem Soc. *Res:* Fundamental and applied program in separation processes; development of mass transfer and hydraulic models for liquid-liquid and liquid-supercritical fluid contactors; adsorption; membranes. *Mailing Add:* 3605 Needles Dr Austin TX 78746

HUMPHREY, PHILIP STRONG, b Hibbing, Minn, Feb 26, 26; m 46; c 2. ORNITHOLOGY. *Educ:* Amherst Col, BA, 49; Univ Mich, MS, 51, PhD(ornith), 55. *Prof Exp:* Res asst, Mus Zool, Univ Mich, 49-55, res assoc, 55-57; asst prof zool, Yale Univ, 57-62, asst cur ornith, Peabody Mus, 57-62; cur div birds, US Nat Mus Natural Hist, Smithsonian Inst, 62-65, chmn dept vert zool, 65-67; chmn dept zool, 67-69, PROF & DIR MUS NATURAL HIST, UNIV KANS, 67- *Concurrent Pos:* Curatorial assoc, Peabody Mus, Yale Univ, res fel, Saybrook Col; res assoc, Fla State Mus; Guggenheim Mem fel; Rockefeller Found res award. *Mem:* Fel Am Ornith Union; fel AAAS; Soc Syst Zool. *Res:* Vertebrate ecology and epidemiology; tropical rain forest in Brazil; biological survey of the Central Pacific; ecology and distribution of birds in Patagonia and Tierra del Fuego. *Mailing Add:* Mus Natural Hist Dyche Hall Univ Kans Lawrence KS 66044-2454

HUMPHREY, RAY EICKEN, b La Harpe, Ill, Jan 27, 31; m 51; c 3. ANALYTICAL CHEMISTRY. *Educ:* Carthage Col, BA, 52; Western Ill State Col, MS, 55; Univ Iowa, PhD(anal chem), 58. *Prof Exp:* Anal res chemist, Ethyl Corp Res Labs, Mich, 58-61; assoc prof, 61-66, PROF CHEM, SAM HOUSTON STATE UNIV, 66- *Concurrent Pos:* Robert A Welch Found res grant, 63-71. *Mem:* Am Chem Soc. *Res:* Infrared spectra of charge-transfer complexes; determination of anions; ultraviolet study of the chemistry of mercury. *Mailing Add:* Dept Chem & Physics Sam Houston State Univ Huntsville TX 77341

HUMPHREY, RONALD DEVERE, b Denver, Colo, Mar 31, 38; m 60; c 2. MICROBIAL PHYSIOLOGY. *Educ:* Colo State Univ, BS, 60, MS, 63; Univ Tex, Austin, PhD(microbiol), 70. *Prof Exp:* PROF BIOL, PRAIRIE VIEW A&M UNIV, 70- *Concurrent Pos:* Fac res appointee, Argonne Nat Lab, Ill, 78 & Oak Ridge Nat Lab, 85. *Mem:* AAAS; Am Soc Microbiol; NY Acad Sci. *Res:* Microbial physiology; microbial genetics; microbial polysaccharide hydrolyzing enzymes; bacterial nitrogen fixation; bacteriophage enzymes. *Mailing Add:* Dept Biol Prairie View A&M Univ Prairie View TX 77446

HUMPHREY, RONALD MACK, b Abilene, Tex, Aug 13, 32; m 52; c 3. MICROBIOLOGY. *Educ:* Hardin-Simmons Univ, BA, 53; Univ Tex, MA, 55, PhD(bact), 58. *Prof Exp:* Res scientist bact, Univ, 54-58, res assoc radiobiol, Hosp, 58-63, assoc biologist, 63-69, PROF BIOPHYS, UNIV TEX M D ANDERSON HOSP & TUMOR INST, 69-, MEM GRAD FAC, BIOMED SCI, UNIV, 68-, ASHBELL SMITH CHAIR BIOPHYS, 80- *Mem:* AAAS; Radiation Res Soc. *Res:* Radiobiology; radiation biophysics; mutagenesis and carcinogenesis. *Mailing Add:* Univ Tex Syst Cancer Ctr Sci Park-Res Div Smithville TX 78957

HUMPHREY, S(IDNEY) BRUCE, inorganic chemistry; deceased, see previous edition for last biography

HUMPHREY, THOMAS MILTON, JR, b Inglewood, Calif, Mar 23, 24; m 50; c 2. GEOLOGY, NUCLEAR HYDROLOGY. *Educ:* Univ Ore, BS, 53, MS, 55. *Prof Exp:* Jr geologist, Int Petrol Colombia, 56-58; chief geol field party, Am Overseas Petrol Ltd, 58-60; engr metal extraction, Mountain Copper Co, 60-61; engr geol eng, Holmes & Narver, Inc, 61-62; sr geologist hydrol, Dept Conserv Natural Resources, 62-67; chief geol hydrol br, Atomic Energy Comn, 67-74; phys scientist geol & hydrol, 74-81, chief, environ br, Health Physics Div, Dept Energy, 81-86; CONSULT, DESERT RES INST & CALIF COASTAL COMN, 86- *Concurrent Pos:* Proj mgr, Radionuclide Migration Prog, 76, Tatum Dome Proj, 76, Community Monitoring Prog, 82. *Res:* The location, concentration and species of radionuclides in groundwater in a cavity and chimney resulting from an underground nuclear detonation and the rate of dispersion and movement in groundwater; Hydrologic monitoring programs for radiation at sites where nuclear tests have been conducted. *Mailing Add:* 1750 Patricks Point Dr Trinidad CA 95570-9706

HUMPHREY, WATTS S, b Battlecreek, Mich, Jul 4, 27; m 54; c 7. SYSTEM DESIGN, FINANCIAL MANAGEMENT. *Educ:* Univ Chicago, BS, 49, MBA, 51; Ill Inst Technol, MS, 50. *Prof Exp:* Elec engr, Fermi Inst, 49-51, Chicago Midway Labs, Univ Chicago, 51-53; mgr comput develop, Sylvania Elec Prods, 53-59; instr comput design, Northeastern Univ, 56-59; dir prog, IBM, 59-68, vpres, Systs Develop Div, 69, dir, Endicott Labs, 69-72, dir policy develop, 72-79, tech assessment, 79-82, prog qual, 82-86; DIR, SOFTWARE PROCESS PROG, SOFTWARE ENG INST, CARNEGIE MELLON UNIV, 86- *Concurrent Pos:* Chmn adv bd, Systs Res Inst, IBM, 73-82; reviewer, Inst Elec & Electronics Engrs Software & Compt, 84. *Mem:* Fel Inst Elec & Electronics Engrs; Asn Comput Mach. *Res:* Characterizing the software process, assessing the state of US software work, identifying key problems and issues, and initiating activities to address them, focus on process management, process definition and statistical control; research administration. *Mailing Add:* Software Eng Inst Carnegie Mellon Univ Pittsburgh PA 15213

HUMPHREYS, JACK BISHOP, b Knoxville, Tenn, Dec 17, 33; m 53; c 2. TRAFFIC ENGINEERING, HIGHWAY SAFETY. *Educ:* Univ Tenn, BS, 55, MS, 62; Tex A&M Univ, PhD(civil eng), 67. *Prof Exp:* Instr civil eng, Univ Tenn, 55-56 & 59-62; asst prof, Southwestern La Univ, 62-64; from asst prof to assoc prof, 66-77, PROF CIVIL ENG, UNIV TENN, KNOXVILLE, 77- *Concurrent Pos:* Supporting mem, Hwy Res Bd, Nat Acad Sci-Nat Res Coun. *Honors & Awards:* Teetor Ed Award, 68. *Mem:* Inst Transp Engrs; Am Soc Civil Engrs. *Res:* Highway safety as related to highway design, vehicle operation, driver competence and enforcement; temporary traffic control in construction and maintenance work zones; accident reconstruction. *Mailing Add:* Dept Civil Eng 112 Perkins Hall Univ Tenn Knoxville TN 37996

HUMPHREYS, JAN GORDON, b Jackson, Ohio, Mar 24, 41; m 64; c 8. ENTOMOLOGY. *Educ:* Ohio Univ, BS, 63, MS, 65; Va Polytech Inst, PhD(med entom), 69. *Prof Exp:* Prof zool, 69-80, PROF ZOOL & BIOL, INDIANA UNIV, PA, 80- *Concurrent Pos:* Mem bd dirs, Marine Sci Consort, 75-80. *Mem:* Sigma Xi; Soc Vector Ecologists. *Res:* Use of wildlife to monitor Lyme Disease; biology of squirrel fleas on the family Sciuridae; parasites of hyraxes in Southern Africa. *Mailing Add:* Dept Biol WEY 325 Indian Univ Pa Indiana PA 15705

HUMPHREYS, KENNETH K, b Pittsburgh, Pa, Jan 19, 38; m 61; c 4. COST ENGINEERING, PROJECT ENGINEERING. *Educ:* Carnegie Inst Technol, Pittsburgh, Pa, BS, 55; WVa Univ, Morgantown, MSE, 68; Kennedy-Western Univ, Agoura Hills, Calif, PhD(eng mgt), 90. *Prof Exp:* Res engr, Appl Res Lab, US Steel Corp, 59-65; from assoc dir to asst dir & cost engr, WVa State Bur Coal Res, 65-82; asst dean, prof & chmn mineral processing, Col Mineral & Energy Resources, WVa Univ, 70-81; EXEC DIR, AM ASN COST ENGRS, 71- *Concurrent Pos:* Consult, var indust firms & govt agencies, 65-81; dir, Am Asn Cost Engrs, 71, exec dir, 71-; adj prof, Col Mineral & Energy Resources, WVa Univ, 81- *Mem:* Fel Am Asn Cost Engrs; Int Cost Eng Coun (secy-treas, 76-); Am Inst Mining, Metall & Petrol Engrs; fel Asn Cost Engrs UK; Nat Soc Prof Engrs. *Res:* Cost and optimization engineering; coal utilization; coke production; fuel technology; ferrous metallurgy. *Mailing Add:* 305 Lebanon Ave Morgantown WV 26505

HUMPHREYS, MABEL GWENETH, b Vancouver, BC, Oct 22, 11; nat US. MATHEMATICS. *Educ:* Univ BC, BA, 32; Smith Col, AM, 33; Univ Chicago, PhD(math), 35. *Prof Exp:* Instr math & physics, Mt St Scholastica Col, 35-36; from instr to asst prof math, Newcomb Col, Tulane Univ, 36-49; from assoc prof to prof math & chmn dept, 50-73, Dana prof, 73-80, EMER DANA PROF MATH, RANDOLPH-MACON WOMAN'S COL, 80- *Concurrent Pos:* Fac fel, Fund Adv Ed, 55-56; NSF fac fel, 62-63. *Mem:* Am Math Soc; Math Asn Am; Can Math Soc. *Res:* Linear algebra; theory of numbers. *Mailing Add:* 1824 Clayton Ave Lynchburg VA 24503

HUMPHREYS, ROBERT EDWARD, b New York, NY, Dec 24, 42; m 68; c 2. IMMUNOCHEMISTRY. *Educ:* Yale Col, BS, 64; Yale Grad Sch, PhD(biochem), 69; Yale Med Sch, MD, 70. *Prof Exp:* Intern, US Naval Hosp, Bethesda, 70-71; med officer, US Naval Med Res Inst, 71-73; res fel biochem, Harvard Univ, 73-75; asst prof, 75-78, assoc prof, 78-86, PROF PHARMACOL & MED, UNIV MASS MED SCH. *Concurrent Pos:* Anna Fuller Fund fel, 73-74; USPHS fel, 74-75; Am Cancer Soc, Mass Div, cancer res scholar, 76-79. *Mem:* Am Asn Immunologists. *Res:* Regulation antigen processing and presentation; protein folding; vaccine design. *Mailing Add:* Univ Mass Med Sch 55 Lake Ave Worcester MA 01605-2397

HUMPHREYS, ROBERT WILLIAM RILEY, b Ottawa, Ont, April 25, 51; m 80. ADHESIVES CHEMISTRY, ORGANIC PHOTOCHEMISTRY. *Educ:* Univ Western Ont, BSc, 74, PhD(chem), 78. *Prof Exp:* Fel chem, Univ Utah, 79-80; res chemist Loctite Corp, 80-; AT LEVER RESEARCH INC, EDGEWATER, NJ. *Mem:* Am Chem Soc. *Res:* Peroxide and hydroperoxide chemistry; autoxidation; organic and organometallic photochemistry; free radical chemistry including polymarization; chemistry of charge-transfer complexes; polymer modification. *Mailing Add:* Loctite Corp 705 N Mountain Rd Newington CT 06111

HUMPHREYS, ROBERTA MARIE, b Indianapolis, Ind, May 20, 44; m 76; c 1. ASTRONOMY. *Educ:* Ind Univ, AB, 65; Univ Mich, MS, 67, PhD(astron), 69. *Prof Exp:* Res assoc astron, Dyer Observ, Vanderbilt Univ, 69-70 & Steward Observ, Univ Ariz, 70-72; from asst prof to assoc prof, 72-83, PROF ASTRON, UNIV MINN, 83- *Concurrent Pos:* Alfred P Sloan Found fel, 76-78; dir, Asn Univ Res Astron, 81-84; mem, NSF Astron Adv Comt, 81-84. *Honors & Awards:* George Taylor Award, 85; Humboldt Distinguished Sr Scientist Award, Fed Repub Ger. *Mem:* Am Astron Soc; Int Astron Union; fel AAAS; Sigma Xi. *Res:* Galactic structure; stellar spectroscopy; infrared sources. *Mailing Add:* Sch Physics & Astron Univ Minn Minneapolis MN 55455

HUMPHREYS, SUSIE HUNT, b Jackson, Miss, June 15, 39; div; c 1. CELL BIOLOGY, FOOD SCIENCE & TECHNOLOGY. *Educ:* Univ Chicago, BS, 60, MS, 62; Harvard Univ, PhD(biol), 69. *Prof Exp:* Res biologist, Univ Calif, San Diego, 66-69; trainee, Salk Inst Biol Studies, 69-71; asst researcher biol, Pac Biomed Res Ctr, Univ Hawaii, 72-79; sr staff fel, NIH, 79-83; SR RES SCIENTIST, KRAFT, INC, 83- *Mem:* Am Soc Cell Biol; Inst Food Technologists; Electron Micros Soc Am. *Mailing Add:* Dept Anat & Cell Biol Univ Ill 808 S Wood St Chicago IL 60612

HUMPHREYS, THOMAS ELDER, b Lakewood, Ohio, Feb 10, 24; m 54. PLANT BIOCHEMISTRY. *Educ:* Ohio State Univ, BS, 49, MS, 50; Univ Pa, PhD(bot), 54. *Prof Exp:* Jr res biochemist, Univ Calif, 53-55; from asst biochemist to assoc biochemist, 55-68, prof bot, 68-80, PROF, DEPT VEG CROPS, UNIV FLA, 80-, BIOCHEMIST, 68- *Mem:* Am Soc Plant Physiol; Sigma Xi. *Res:* Carbohydrate metabolism; sugar transport. *Mailing Add:* Vegetable Crops Dept Univ Fla Gainesville FL 32611-0514

HUMPHREYS, TOM DANIEL, b Arlington, Tenn, June 22, 36; div; c 1. DEVELOPMENTAL BIOLOGY. *Educ:* Univ Chicago, BS, 58, PhD(zool), 62. *Prof Exp:* NSF fel biol, Mass Inst Technol, 62-63, asst prof, 63-66; asst prof, Revelle Col, Univ Calif, San Diego, 66-71; assoc prof biochem, Kewalo Marine Lab Exp Biol, Pac Biomed Res Ctr, Univ Hawaii, 71-77; exec dir, 86-88, CHMN, CELL, MOLECULAR & NEUROSCI GRAD PROG, CANCER RES CTR, 88-; PROF GENETICS & MOLECULAR BIOL, KEWALO MARINE LAB & CANCER RES CTR, UNIV HAWAII, 77- *Concurrent Pos:* Dir embryol, Marine Biol Lab, Woods Hole, Ma, 75-79; dir basic sci, Cancer Res Ctr, Hawaii, 86- *Mem:* Biophys Soc; Soc Develop Biol; Am Soc Cell Biol; Marine Biol Lab Corp. *Res:* Mechanisms of cell adhesion; control of RNA and protein synthesis during embryonic development. *Mailing Add:* Kewalo Marine Lab Univ Hawaii 41 Ahui St Honolulu HI 96813

HUMPHREYS, WALLACE F, b Covington, Ky, Apr 9, 27; m 50; c 5. BIOCHEMISTRY, PHYSIOLOGY. *Educ:* Inst Divi Thomae, MS, 52, PhD(biochem), 55. *Prof Exp:* Instr, 52-54, assoc prof, 54-73, PROF BIOL, THOMAS MORE COL, 73- *Mem:* Am Inst Biol Sci. *Res:* Effects of cellular extracts on respiration and growth. *Mailing Add:* Dept Biol Thomas Moore Col 2771 Turkeyfoot Rd Ft Mitchell KY 41017

HUMPHREYS, WALTER JAMES, b Magnolia, Ark, July 5, 22; m 53; c 3. BIOLOGICAL STRUCTURE. *Educ:* Univ Calif, Berkeley, AB, 50, MA, 55, PhD(zool), 61. *Prof Exp:* Sr lab asst biochem, Univ Calif, Berkeley, 49-51; carpenter, Union Pac RR, 52-53; res asst zool, Univ Calif, Berkeley, 60-61, asst res zoologist, 61-67; assoc prof cell biol, Iowa State Univ, 67-68; prof zool, Univ Ga, 68-88; RETIRED. *Concurrent Pos:* Vis prof basic med sci, Univ NC, Chapel Hill, 81. *Mem:* AAAS; Soc Develop Biol; Electron Micros Soc Am; Am Soc Cell Biol. *Res:* Electron microscopy of gametes, zygotes and early embryos; reproductive biology; development of fractographic methods for studying interior substructures of tissues and cells by scanning electron microscopy and x-ray spectrometry. *Mailing Add:* Two C St St Augustine FL 32085

HUMPHRIES, ARTHUR LEE, JR, b Rock Hill, SC, 1928. SURGERY. *Educ:* The Citadel, BS, 48; Johns Hopkins Univ, MD, 52; Am Bd Surg, dipl, 62. *Prof Exp:* From instr to assoc prof, 60-70, PROF SURG, MED COL GA, 70- *Concurrent Pos:* NIH res grant, 61- *Mem:* Transplantation Soc; Am Soc Artificial Internal Organs; Soc Univ Surg. *Res:* Transplantation; storage of dog kidney by hypothermic perfusion. *Mailing Add:* Dept Surg Med Col Ga 1120 15th St Augusta GA 30912

HUMPHRIES, ASA ALAN, JR, b Anniston, Ala, Sept 6, 24; m 49, 72; c 4. CELL BIOLOGY, ZOOLOGY. *Educ:* Emory Univ, AB, 48, MS, 49; Princeton Univ, AM, 52, PhD(biol), 53. *Prof Exp:* Asst biol, Emory Univ, 48-49, instr, 49-50; asst instr, Princeton Univ, 50-52; instr anat, Univ Va, 53-54; from asst prof to prof biol, Emory Univ, 54-81, chmn dept, 74-81; prof biol & vpres, 81-83, DEAN COL, TRANSYLVANIA UNIV, 81-, EXEC VPRES, 83- *Concurrent Pos:* NSF sr fel, Free Univ Brussels, 62-63; NATO sr sci fel, Lab Molecular Embryol, Naples, 71; Physiol Lab, Univ of Cambridge, 78; scholar, Bellagio Study & Conf Ctr, Italy, 83. *Mem:* AAAS; Soc Develop Biol; Am Soc Cell Biol; Int Soc Develop Biol; Soc Study Fertility; Sigma Xi. *Res:* Embryology of amphibians; gametogenesis; control of cell division; heteroploidy; fertilization. *Mailing Add:* Transylvania Univ Lexington KY 40508

HUMPHRIES, ERVIN G(RIGG), b Shelby, NC, Apr 26, 36; m 58; c 2. AGRICULTURAL ENGINEERING. *Educ:* NC State Univ, BS, 58, MS, 60, PhD(agr eng), 64. *Prof Exp:* Res instr, 61-64, from asst prof to assoc prof, 64-73, PROF AGR ENG, NC STATE UNIV, 73- *Mem:* Am Soc Agr Engrs. *Res:* Investigations of the interaction of biological and engineering systems as related to production, harvesting and storage of food and fiber; physical properties of biological materials pertinent to engineering design. *Mailing Add:* Dept Agr Eng Box 7625 NC State Univ Raleigh NC 27695-7625

HUMPHRIES, J O'NEAL, b Columbia, SC, Oct 22, 31; m 54; c 3. MEDICINE. *Educ:* Duke Univ, AB, 52; Johns Hopkins Univ, MD, 56. *Prof Exp:* Res fel cardiol, Sch Med, Johns Hopkins Univ, 56-57 & 61-62, St George's Hosp, Univ London, 60-61; from instr to assoc prof, Johns Hopkins Univ, 64-76, assoc prof, 76-78, prof med, 78-79; PROF MED & CHMN, DEPT MED, SCH MED, UNIV SC, 79-, DEAN, SCH MED. *Concurrent Pos:* Consult, Union Mem Hosp, 64-; consult, Vet Admin Hosp, 69-; chmn Subspecialty Bd in Cardiovascular Dis, 76-79. *Mem:* Am Fedn Clin Res; Asn Univ Cardiologists; fel Am Col Physicians; fel Am Col Cardiol; Am Clin & Climat Asn. *Mailing Add:* Sch Med Univ SC Columbia SC 29208

HUMPHRIES, JACK THOMAS, b Middlesboro, Ky, May 7, 29; m 51; c 2. PHYSICS, NUCLEAR ENGINEERING. *Educ:* Univ Ky, BS, 51; US Air Force Inst Technol, MS, 58; Univ Fla, PhD(nuclear eng), 66. *Prof Exp:* Physicist, Mat Lab, Wright Air Develop Ctr, Ohio, 51-53; proj officer reactor shield mat, Mat Lab, Wright Air Develop Ctr, USAF, 53-56, instr physics, USAF Acad, 58-59, asst prof, 59-63, tenure prof, 66-72, head dept, 70-71; PROF NATURAL SCI & TECHNOL & ASST DEAN FACULTIES, UNIV NORTH FLA, 72- *Concurrent Pos:* Honorarium lectr, Univ Colo, Colorado Springs, 66-68; vis scholar, Dakota State Col, 71. *Mem:* Am Soc Eng Educ; Am Asn Physics Teachers. *Res:* Control of nuclear rocket engines; applications of nuclear radiation; reactor shielding materials; computers in physics instruction; attitudes of students toward science. *Mailing Add:* 936 Holly Lane Jacksonville FL 32207

HUMPHRIES, JAMES EDWARD, JR, b Shreveport, La, Aug 7, 40; m 63; c 1. TEACHING ADMINISTRATION, TECHNOLOGY APPLICATIONS. *Educ:* Sam Houston State Univ, BS, 65, MEd, 66; Univ Okla, PhD(biogeog), 71. *Prof Exp:* Instr geog, Univ Okla, 70-71; assoc prof, 71-82, PROF & HEAD EARTH SCI, E TEX STATE UNIV, 82- *Mem:* Asn Am Geogr; Asn Arid Land Studies; Am Geog Soc. *Res:* Distribution and adaptation of Mesquite (Prosopis glandulosa) in humid environments; ethnobotany of native species in arid and semi-arid regions; appropriate technologies in developing nations. *Mailing Add:* Dept Earth Sci ETex State Univ Commerce TX 75428

HUMPHRIES, LAURIE LEE, b Atlanta, Ga, Apr 26, 44; m 72; c 1. NUTRITION, MEDICINE. *Educ:* Emory Univ, BA, 66, MD, 73. *Prof Exp:* Resident adult psychiat, Emory Univ Sch Med, 73-75, fel child psychiat, 75-77, asst prof, 78-81; asst prof, 81-87, ASSOC PROF CHILD PSYCHIAT, COL MED, UNIV KY, 87- *Concurrent Pos:* Prin investr, zinc & eating dis, NIMH, 86-88, zinc deficiency in eating dis, McKnight Found, 86-90 & child & adolescent acad award, NIMH, 89- *Mem:* Am Acad Child & Adolescent Psychiat; Am Psychiat Asn; AMA; AAAS. *Res:* Study of nutritional factors, particulary zinc deficiency on anorexia nervosa and bulimia nervosa; medical complications of eating disorders. *Mailing Add:* Dept Psychiat Annex II Univ Ky 820 S Limestone St Lexington KY 40502

HUMPHRIES, ROBERT GORDON, b Edmonton, Alta, Jan 21, 45; m 72. METEOROLOGY. *Educ:* Univ Alta, BSc, 67, MSc, 69; McGill Univ, PhD(meteorol), 74. *Prof Exp:* Fel radar meteorol, Atmospheric Environ Serv, 73-74; assoc res officer radar meteorol, Alta Res Coun, 74-87, mgr, Precipitation Processes Sect; PROD MGR METEOROL SYSTS, MCDONALD-DETTWILER, 87- *Honors & Awards:* Dr Andrew Thomson

Prize, Can Meteorol Soc, 81. *Mem:* Am Meteorol Soc; Can Meteorol Soc; Inst Elec & Electronics Engrs. *Res:* Application of a polarization diversity radar to the study of rain and hail storms; relationship of rain drop size distributions to polarization parameters of forward and back scattered microwaves; application of weather radar to operational cloud seeding, hydrology and forecasting. *Mailing Add:* McDonald Dettwiler 13800 Commerce Pkwy Richmond BC V6V 2J3 Can

HUMPHRIS, ROBERT R, b Lexington, Va, Dec 30, 28; m 52; c 2. ELECTRICAL ENGINEERING. *Educ:* Univ Va, BSEE, 52, MSEE, 58, DSc(elec eng), 65. *Prof Exp:* Staff mem, Sandia Corp, 52-54; proj engr, Eng Exp Sta, 54-60, actg head instrumentation group, 60-62, res asst, 62-64, SR SCIENTIST, RES LABS ENG SCI, UNIV VA, 64-, RES PROF, 81- *Mem:* Inst Elec & Electronics Engrs; Am Vacuum Soc; Am Dowsing Soc. *Res:* Superconducting thin-film particle detector; supersonic wind tunnel balance using superconducting magnets; fluids research; turbomachinery vibrations; magnetic bearings. *Mailing Add:* Mech Eng Bldg Univ Va Charlottesville VA 22901

HUNDAL, MAHENDRA S(INGH), b Syana, India, Nov 25 34; nat US; m 64; c 1. MECHANICAL ENGINEERING. *Educ:* Osmania Univ, India, BE, 54; Univ Wis, MS, 62, PhD(mech eng), 64. *Prof Exp:* Design engr, Tata Iron & Steel Co, India, 55-60; asst prof eng, San Diego State Col, 64-67; assoc prof, 67-77, PROF MECH ENG, UNIV VT, 77- *Concurrent Pos:* Fulbright award, 87. *Honors & Awards:* R R Teetor Award, Soc Automotive Engrs, 67. *Mem:* Am Soc Mech Engrs; Inst Noise Control Engrs; Sigma Xi. *Res:* Design methods; computer-aided design; modeling of dynamic systems; industrial noise control. *Mailing Add:* Dept Mech Eng Votey Bldg Univ Vt Burlington VT 05405

HUNDERFUND, RICHARD C, b Haverstraw, NY, May 25, 29. MICROBIOLOGY, PARASITOLOGY. *Educ:* State Univ NY, BMS, 50; Fairleigh Dickinson Univ, BS, 59; Rutgers Univ, PhD(microbiol), 64. *Prof Exp:* Mem staff, Am Trading & Prod Corp, 50-57; asst microbiol, Fairleigh Dickinson Univ, 57-59; asst, 59-64, from asst prof to assoc prof microbiol, 65-74, chmn dept, 74-76, PROF BIOL, SAN FRANCISCO STATE UNIV, 74- *Concurrent Pos:* NSF instnl grant, 65-66; USPHS grant, 66-69. *Mem:* Nat Tuberc & Respiratory Dis Asn; Am Thoracic Soc; Am Soc Microbiol. *Res:* Growth stimulation of the mycobacteria; food microbiology; epidemiology of cryptococcus infections; pigeon genetics. *Mailing Add:* Dept Biol San Francisco State Univ 1600 Holloway Ave San Francisco CA 94132

HUNDERT, IRWIN, b New York, NY, Nov 9, 25; m 48; c 2. PETROLEUM REFING, ENVIRONMENTAL TRAINING. *Educ:* City Col New York, BChE, 45; NY Univ, MChE, 48. *Prof Exp:* Process & design engr, Heyden Chem Corp, 45-50; sr engr ref, supply & mkt, Chevron Corp, 50-83; VIS PROF CHEM ENG & ENVIRON SCI, NJ INST TECHNOL, 83- *Concurrent Pos:* Consult, Chevron Corp, Amerada Hess Corp, Englehardt Corp and Intercontinental Hotel; prin investr, Alt Teaching Mat/Energy Balances; assoc, various precol progs. *Mem:* Sigma Xi; Am Inst Chem Engrs. *Res:* Energy production; hazardous waste; precollege activities. *Mailing Add:* 48 Dutch Rd East Brunswick NJ 08816

HUNDERT, MURRAY BERNARD, b New York, NY, Sept 15, 19; m 48; c 2. POLYMER CHEMISTRY, ORGANIC CHEMISTRY. *Educ:* Brooklyn Col, BA, 40; NY Univ, MS, 43, PhD, 50. *Prof Exp:* Sr res chemist, US Indust Chems, Inc, 42-48; proj leader, Heyden-Newport Chem Corp, 48-50; tech dir, Crownoil Chem Co, Inc, 50-61 & Farnow, Inc, 61-62; from asst prof to prof chem, Fairleigh Dickinson Univ, 62-85, dep chmn, 63-65, chmn dept, 71-73 & 82-85; RETIRED. *Concurrent Pos:* Vis instr, NY Community Col, 57-62; consult; vis prof, Col Physicians & Surgeons, Columbia Univ, 81. *Mem:* AAAS; Am Chem Soc; Sigma Xi. *Res:* Polymer preparation and characterization; protective coatings, latexes, product development and synthetic organic chemistry. *Mailing Add:* 46 Wingate Dr Livingston NJ 07039-3518

HUNDHAUSEN, ARTHUR JAMES, b Wausau, Wis, Aug 2, 36; m 68; c 2. THEORETICAL PHYSICS, ASTROPHYSICS. *Educ:* Univ Wis, BS, 58, MS, 59, PhD(physics), 65. *Prof Exp:* Mem staff theoret div, Los Alamos Sci Lab, Univ Calif, 64-71; STAFF SCIENTIST, HIGH ALTITUDE OBSERV, 71- *Concurrent Pos:* Lectr, Univ Colo, 71- *Mem:* AAAS; Am Geophys Union; Am Astron Soc. *Res:* Observations and theoretical models of the interplanetary plasma and its interaction with the geomagnetic field; physical processes in tenuous plasmas. *Mailing Add:* 669 Linden Pask Dr Boulder CO 80304

HUNDHAUSEN, JOAN ROHRER, b Pittsburgh, Pa; m 67; c 3. MATHEMATICS. *Educ:* Duquesne Univ, BS, 58; Univ Wis, MS, 59; Carnegie-Mellon Univ, PhD(math), 67. *Prof Exp:* Instr math, Duquesne Univ, 59-61 & 62-63, asst prof, 64; staff mem, Los Alamos Sci Lab, Univ Calif, 67-71; adj asst prof, 74-75, asst prof, 75-80, ASSOC PROF MATH, COLO SCH MINES, 80- *Concurrent Pos:* Adj prof, Univ NMex, 68-71; mem, Consortium Coun Undergrad Math & Its Appl, 80-83, Comt Placement Exams, Math Asn Am, 86-89, subcomt symbolic computation, Comt Undergrad Prog Math, 89-; prin investr Curric Develop Prog Calculus, NSF, 88-91; prog chmn, Math Div, Am Soc Eng Educ, 91-92, div chmn, 92-93. *Mem:* Am Math Soc; Sigma Xi; Asn Women Math; Math Asn Am. *Res:* Discrete function theory; differential equations. *Mailing Add:* 669 Linden Park Dr Boulder CO 80304

HUNDLEY, JOHN GOWER, b New Orleans, La, Feb 19, 29; m 58; c 2. CHEMICAL ENGINEERING. *Educ:* Univ NDak, BS, 50; Univ Wis, MS, 51, PhD(chem eng), 53. *Prof Exp:* Engr res & develop, 53-57, asst proj engr, 57-59, proj engr, 59-61, proj engr res dept, Amoco Chem, 61-62, sr proj engr, 62-66, sr res engr, 66-67, group leader, 67-70, sect leader, Am Chem Corp, 70-85, RES ASSOC, AMOCO CHEM CO, STANDARD OIL CO, 85- *Mem:* Am Chem Soc. *Res:* Chemical process development; chemical plant design; process economics. *Mailing Add:* 4N159 Wild Rose Rd St Charles IL 60174

HUNDLEY, LOUIS REAMS, b Greenwood, Va, May 22, 26; m 53; c 1. PHYSIOLOGY. *Educ:* Va Mil Inst, BS, 50; Va Polytech Inst & State Univ, MS, 53, PhD(biol), 56. *Prof Exp:* Asst instr biol, Va Mil Inst, 50-51; wildlife biologist, Va Coop Wildlife Res Unit, 52-53; asst biol, Va Polytech Inst, 53-54 & Va Agr Exp Sta, 54-56; from asst prof to assoc prof, 56-65, actg head dept, 65-67, PROF BIOL, VA MIL INST, 65- *Mem:* AAAS; Am Physiol Soc; Am Inst Biol Sci. *Res:* Fat-free body weight; gross body composition; nutrients in deer browse. *Mailing Add:* Dept Biol Va Mil Inst Lexington VA 24450

HUNDLEY, RICHARD O'NEIL, b Joliet, Ill, May 27, 35; m 57; c 3. THEORETICAL PHYSICS. *Educ:* Calif Inst Technol, BS, 57, MS, 59, PhD(physics), 63. *Prof Exp:* Consult, Rand Corp, 60-62, phys scientist, 62-72; prod mgr, 72-83, VPRES, R&D ASSOCS, 83- *Concurrent Pos:* Mem, Army Sci Adv Panel, 72-78, Army Sci Bd, 78-80. *Mem:* AAAS; Am Phys Soc. *Res:* Quantum mechanics; quantum field theory; atomic and molecular physics; reentry physics; nuclear weapon effects; military systems research; laser physics; laser and satellite systems; electrooptic systems. *Mailing Add:* Rand Corp PO Box 2138 Santa Monica CA 90406

HUNEKE, HAROLD VERNON, b Arnett, Okla, Aug 26, 17; m 48; c 2. MATHEMATICS. *Educ:* Northwestern State Col, AB, 36; Univ Okla, MA, 41, PhD(math), 57. *Prof Exp:* Asst, Univ Okla, 39-42; head dept math, Northwestern State Col, Okla, 46-48; prof math, Univ Wichita, 51-60; assoc prof, Univ Okla, 60-67, prof math, 67-81; RETIRED. *Concurrent Pos:* Mem New Delhi staff, NSF, 69-71. *Mem:* Math Asn Am. *Res:* Changes in undergraduate and teacher education programs. *Mailing Add:* 517 S Flood Norman OK 73069

HUNEKE, JAMES THOMAS, materials science engineering, for more information see previous edition

HUNEKE, JOHN PHILIP, b Spokane, Wash, Apr 16, 42; m 65; c 4. MATHEMATICS. *Educ:* Pomona Col, BA, 64; Wesleyan Univ, PhD(math), 67. *Prof Exp:* Dunham Jackson res instr math, Univ Minn, Minneapolis, 67-69, asst prof, 69-70; from asst prof to assoc prof, 69-82, VCHMN, 80-, PROF MATH, OHIO STATE UNIV, 82- *Mem:* Am Math Soc; Math Asn Am. *Res:* Commuting functions; combinatorial topology; topological graph theory; topological dynamics. *Mailing Add:* Dept Math Ohio State Univ 231 W 18th Ave Columbus OH 43210

HUNEYCUTT, JAMES ERNEST, JR, b Gastonia, NC, Dec 14, 42; m 64. WEAPON SYSTEM ANALYSIS. *Educ:* Univ NC, Chapel Hill, BS, 63, MA, 65, PhD(math), 68. *Prof Exp:* Asst prof math, NC State Univ, 68-73, assoc prof, 73-78; sr software engr, Hadron Inc, 78-82; MATHEMATICIAN, APPL PHYSICS LAB, JOHNS HOPKINS UNIV, 82- *Mem:* Inst Elec & Electronics Engrs; Math Asn Am; Soc Indust Appl Math; Sigma Xi. *Res:* Linear error analysis; software development; vector measures and integration; probability; estimation; Kalman filtering. *Mailing Add:* 10840 Douglas Ave Wheaton MD 20902

HUNEYCUTT, MAEBURN BRUCE, b Troy, NC, Jan 7, 23; m 50; c 1. MYCOLOGY. *Educ:* Univ NC, BA, 46, MA, 49, PhD(mycol), 56. *Prof Exp:* From asst prof biol to assoc prof biol, 55-60, prof & chmn dept, 60-69, prof chem, 69-80, PROF BIOL, COL LIBERAL ARTS, 80-, DEAN, 69- *Mem:* Am Inst Biol Sci. *Res:* Morphology, taxonomy and physiology of lower aquatic fungi; forest litter fungi and wood decay fungi. *Mailing Add:* Dept Biol Univ Miss University MS 38677

HUNG, GEORGE WEN-CHI, b Penghu, Taiwan, Oct 14, 32; US citizen; m 63; c 2. ANALYTICAL CHEMISTRY, PHYSICAL CHEMISTRY. *Educ:* Tamkang Col Arts & Sci, Taiwan, 62; Auburn Univ, PhD(anal chem), 70. *Prof Exp:* Asst instr chem, Tamkang Col Arts & Sci, 62-65; res assoc, Univ Tenn Med Units, 70-72; res fel & NIH trainee, St Jude Children's Res Hosp, Memphis, 72-73; res assoc toxicol & biomat, Ctr Health Sci, Univ Tenn, 73-76; res scientist & head chemist, 76-77, dir & res scientist, Woodson-Tenent Labs, Nat Health Labs, Inc, Memphis, 78-83; ASSOC PROF CHEM DEPT, UNIV MONTEVALLO, AL, 84- *Mem:* Am Chem Soc; Chem Soc Japan; AAAS; Am Soc Testing & Mat; fel Am Inst Chemists. *Res:* Trace and ultratrace analysis of inorganic and organic compounds in biological and environmental systems by atomic absorption spectrophotometry and gas-liquid and high performance liquid chromatography. *Mailing Add:* 2428 Mountain Run Birmingham AL 35244

HUNG, JAMES CHEN, b Foochow, China, Feb 18, 29; nat US; m 58; c 3. ELECTRICAL ENGINEERING. *Educ:* Nat Taiwan Univ, BS, 53; NY Univ, MEE, 56, ScD(elec eng), 61. *Prof Exp:* Asst elec eng, NY Univ, 54-56, instr, 56-61; from asst prof to prof elec eng, 61-83, DISTINGUISHED SERV PROF, ELEC ENG DEPT, UNIV TENN, KNOXVILLE, 84-; VPRES, POLY-ANALYTICS, INC, KNOXVILLE, TN, 81- *Concurrent Pos:* Prin investr, NASA, 63-83, Army Res Off, US Army, 73-89; mem tech staff, IBM, Poughkeepsie, 67-68; eng consult, US Army Missile Command, 75-80; mem admin comt, Inst Elec & Electronics Engrs, 85-; consult investr, US Army, 89; hon prof, Nanjing Aeronaut Inst, China, 89; assoc ed, Inst Elec & Electronics Engrs Trans Indust Electronics, 85-90, ed, 91- *Mem:* Fel Inst Elec & Electronics Engrs. *Res:* System analysis, development and design; control engineering; system instrumentation, measurement, testing and evaluation; digital information and data processing; navigation and guidance systems; robotics and automation; 180 publications. *Mailing Add:* Dept Elec Eng Univ Tenn Knoxville TN 37996

HUNG, JAMES Y(UN-YANN), b Formosa, China, Oct 3, 35; m 62; c 2. ENGINEERING MECHANICS. *Educ:* Cheng Kung Univ, Taiwan, BS, 58; Univ Mass, MS, 61; Univ Wis, PhD(eng mech), 68. *Prof Exp:* Assoc prof math & physics, Inst Paper Chem, 67-77; sr appln engr, Voith Inc, 78-83; SR RES ASSOC, TEC SYSTS, 84- *Mem:* Am Soc Mech Engrs; Am Soc Eng Educ; Tech Asn Pulp & Paper Indust. *Res:* Applied mechanics and mathematics; computer simulations. *Mailing Add:* 3511 N Bracken Dr Appleton WI 54915-6748

HUNG, JOHN HUI-HSIUNG, b Taiwan, Republic of China, Mar 24, 37; US citizen; m 63; c 2. PHYSICAL CHEMISTRY. *Educ:* Nat Taiwan Univ, BS, 61; Clark Univ, PhD(phys chem), 68. *Prof Exp:* Sr res chemist, Norton Co, 68-69; SR RES ASSOC PHYS CHEM, INT PAPER CO, 69- *Mem:* Am Chem Soc. *Res:* Adhesive formulations; synthetic membranes; synthetic gauzes for medical applications and the antimicrobialization of medical devices and paper mill wastewater treatment. *Mailing Add:* 1301 15th St Ft Lee NJ 07024-1913

HUNG, JOHN WENCHUNG, electronics, mathematics, for more information see previous edition

HUNG, KUEN-SHAN, b Chia-Yi, Taiwan, Jan 13, 38; m 68; c 2. ANATOMY, HISTOLOGY. *Educ:* Nat Taiwan Univ, BS, 60; Univ Kans, PhD(anat), 69. *Prof Exp:* Teaching asst anat, Col Med, Nat Taiwan Univ, 61-64; teaching asst, Med Ctr, Univ Kans, 64-69; asst prof, Univ SDak, 69-71; asst prof path & anat, Univ Southern Calif, 71-74; from asst prof to assoc prof, 74-83, PROF ANAT, UNIV KANS MED CTR, 83- *Concurrent Pos:* Gen res support grant, Univ SDak, 69-71; Am Lung Asn res grant, 74-76; NIH, 76-81; affil, Am Heart Asn, Kansas, 81-87, 90-91. *Mem:* Electron Micros Soc Am; Am Asn Anatomists; Am Soc Cell Biol. *Res:* Innervation of lungs; hypoxia induced pulmonary hypertension; pulmonary endocrine cells. *Mailing Add:* Dept Anat & Cell Biol Univ Kans Col Health 39th St & Rainbow Blvd Kansas City KS 66103

HUNG, PAUL P, b Taipei, Formosa, Sept 30, 33; US citizen; m 56; c 3. VIROLOGY, MOLECULAR BIOLOGY. *Educ:* Millikin Univ, BS, 56; Purdue Univ, MS, 58, PhD(biochem), 60. *Prof Exp:* Res asst biochem, Purdue Univ, 56-60; sr biochemist, Abbott Labs, 60-70, assoc res fel, 70-73, res fel, 73-79, head molecular virol & biol, 74-81; gen mgr, Genetics Div, Bethesda Res Labs, Inc, 81-82; dir, Microbiol Div, Wyeth Labs, 82-88; ASST VPRES, BIOTECHNOL & MICROBIOL DIV, WYETH-AYERST RES, 88- *Concurrent Pos:* Adj asst prof, Stritch Sch Med, Loyola Univ, Chicago, 68-72; vis res fel, Sch Med, Stanford Univ, 69-70; adj assoc prof, Stritch Sch Med, Loyola Univ, Chicago, 72-76; adj prof, Northwestern Univ Med Sch, Chicago, 76- *Mem:* Am Chem Soc; Am Soc Biol Chem; Am Asn Cancer Res; Am Soc Microbiol; AAAS. *Res:* Nucleic acid; biosynthesis of antibiotics and carbohydrates; bacteriophages and oncogenic viruses; genetic engineering; vaccines; biopharmaceuticals. *Mailing Add:* Biotechnol & Microbiol Div Wyeth Ayerst Labs PO 8299 Philadelphia PA 19101

HUNG, RU J, b Taiwan, China, June 17, 34; US citizen; m 66; c 2. REMOTE SENSING, COMPUTATIONAL FLUID MECHANICS. *Educ:* Nat Taiwan Univ, BS, 57; Univ Osaka, MS, 66; Univ Mich, PhD(aerospace), 70. *Prof Exp:* Nat Acad Sci res assoc, Ames Res Ctr, NASA, 71-72; PROF MECH ENG, UNIV ALA, HUNTSVILLE, 72- *Concurrent Pos:* Prin investr, NASA, 72-, US Army Res Off, 76-78, 90-, Nat Acad Sci, 76-79 & NSF, 76- *Honors & Awards:* Nat Acad Sci Res Award; Res Award, Ministry Educ Japan. *Mem:* Assoc fel Am Inst Aeronaut & Astronaut; Am Geophys Union; Am Meteorol Soc; Nat Soc Prof Engrs; fel Royal Meteorol Soc. *Res:* Author or co-author of over 150 articles; space-based propulsion; cryogenic fluid management; fluid mechanics; heat transfer; computational fluid dynamics; propulsion; microgravity fluid mechanics; magneto hydrodynamics; plasma physics; numerical simulation of fog formation due to combustion related pollutants; satellite remote sensing; remote sensing of atmosphere; energy conversion. *Mailing Add:* Dept Mech Eng Univ Ala Huntsville AL 35899

HUNG, TONNEY H M, b Formosa, Nov 5, 32; m 65; c 1. MECHANICAL ENGINEERING. *Educ:* Cheng Kung Univ, Taiwan, BS, 55; Kans State Univ, MS, 61, PhD(mech eng), 64. *Prof Exp:* Design & systs control engr, Taiwan Power Co, 56-59; asst fluid mech, Cheng Kung Univ, 59; res asst mech eng, Kans State Univ, 60-64; sr res engr, Appl Sci Dept, Deere & Co, 64-68; team leader, Systs Anal Directorate, US Army Weapons Command, 68-71, Syst Res Div, Res Develop & Eng Directorate, 71-75, prog dir, US Army Reserve Command, 75-77, chief aircraft armament team, 77-80, chief systs eng br, Fir Control & Small Caliber Weapons Systs Lab, Armament Res & Develop Command, 80-, SCI ADV, US ARMY JAPAN/XI CORPS. *Concurrent Pos:* Lectr, Univ Iowa, 68-69; mem Simulation Coun. *Mem:* Am Soc Mech Engrs; Sigma Xi. *Res:* Heat transfer systems dynamics control; ordnance systems science and design. *Mailing Add:* Seven Nuko Terrace Randolph NJ 07869

HUNG, WILLIAM MO-WEI, b Chi-Kiang, China, Sept 17, 40; US citizen; m 68; c 2. POLYMER CHEMISTRY. *Educ:* Nat Chung-Hsing Univ, Taiwan, BS, 63; Univ Mass, Amherst, PhD(org chem), 70. *Prof Exp:* Res assoc, Ohio State Univ, 70-74; res chemist, Hilton-Davis Chem Co, 74-80, dir chem res, 80-84, sr res fel, 84-87; STAFF RES SCIENTIST, CIBA VISION CORP, 87- *Mem:* Am Chem Soc. *Res:* Organic synthesis, dye and intermediates; contact lens materials and selections. *Mailing Add:* 4062 Dover Ave Alpharetta GA 30201-1282

HUNG, YOU-TSAI JOSEPH, b Chunan, Taiwan, Aug 21, 32; m 61; c 3. IRRIGATION ENGINEERING. *Educ:* Nat Taiwan Univ, BS, 56; Mich State Univ, MS, 64, PhD(agr eng), 69. *Prof Exp:* Hydraul engr, Taiwan Soil & Water Conserv Bur, 56-58; asst instr agr eng, Nat Taiwan Univ, 58-63, instr, 64-66; res asst, Mich State Univ, 66-68; civil engr, Mich Dept State Hwy, 68-69; from asst prof to assoc prof, 69-77, PROF AGR ENG, CALIF STATE POLYTECH UNIV, POMONA, 77- *Concurrent Pos:* Consult, Agr Exp Sta, Hsinchu, Taiwan, 64-66, irrig & drainage comt, Taiwan Sugar Coop, 65-66 & Litton Systs Inc, Calif, 71; mem renovated water use study comt, Calif State Polytech Col, 69-; consult, Ministry Educ, Repub China, 73. *Mem:* Am Soc Agr Eng; Am Soc Eng Educ; Hydraul Eng Soc China; Chinese Soc Agr Eng. *Res:* Soil-water-plant relationships; irrigation and drainage systems design; erosion controls. *Mailing Add:* Dept Agr Eng Calif State Polytech Univ 3801 W Temple Ave Pomona CA 91768

HUNGATE, FRANK PORTER, b Cheney, Wash, June 6, 18; m 41; c 4. RADIOBIOLOGY. *Educ:* Univ Tex, AB, 40; Stanford Univ, PhD(biol), 46. *Prof Exp:* Instr zool, Univ Nev, 45; asst prof, Reed Col, 46-52; res scientist, Biol Sect, Gen Elec Co, 52-53, head plant nutrit & microbiol unit, 53-63; mgr radioecol oper, Hanford Labs, 63-65; mgr radioecol sect, Pac Northwest Lab, Battelle Mem Inst, 65-66, mgr plant physiol & agr sect, 66-68, staff scientist, Biol Dept, 68-82; RETIRED. *Concurrent Pos:* Instr, Univ Calif, Santa Barbara, 48; tech specialist, Int Atomic Energy Agency, Greece, 61-62; biol prog chmn, Joint Ctr, Grad Study, Wash, 67-73; affil assoc prof, Radiol Sci, Univ Wash, 75- *Mem:* AAAS; Genetics Soc Am; Radiation Res Soc; Soc Exp Biol & Med. *Res:* Radiation biology; genetics; transmutation effects; biochemical genetics in yeasts and molds; ion uptake and transport by plants; ecology; metabolism of actinides; genetic effects of microwaves; immunosuppression by blood irradiation; mutational effects from non-ionizing radiations; food irradiation. *Mailing Add:* Biol Dept Pac Northwest Labs Battelle Mem Inst Richland WA 99352

HUNGATE, ROBERT EDWARD, b Cheney, Wash, Mar 2, 06; m 33; c 3. BIOLOGY. *Educ:* Stanford Univ, AB, 29, PhD(biol), 35. *Hon Degrees:* Dr, Univ Göttingen, 89. *Prof Exp:* Actg instr biol, Stanford Univ, 30-33, instr, 33-35; from instr to asst prof zool, Univ Tex, 35-43, assoc prof & res assoc, Biochem Inst, 43-45; from assoc prof to prof bact, State Col Wash, 45-56; chmn dept, 56-62, prof bact, 56-73, EMER PROF BACT, UNIV CALIF, DAVIS, 73- *Concurrent Pos:* Guggenheim fel, 50; mem subcomt bloat, Nat Res Coun, 54; vis investr, EAfrica Vet Res Orgn, 57 & Univ Sheffield, 62-63; fac res lectr, Univ Calif, Davis, 67; Fulbright fel, NZ, 69. *Honors & Awards:* Fac Res Lectr, Univ Calif, Davis, 67. *Mem:* Am Soc Microbiol (pres, 71); Am Acad Microbiol. *Res:* Cohesion of water; carbohydrate and nitrogen nutrition of termites; biological decomposition of cellulose; nutrition of ruminant protozoa; biology and biochemistry of ruminant cellulose bacteria; growth of mammary cancers in eggs; rates of natural microbial processes; microbiology of acute indigestion and bloat in ruminants; microbiology of sludge; cultivation of strict anaerobes, synthesis in exergonic chemical reactions. *Mailing Add:* 801 Anderson Rd Davis CA 95616

HUNGER, HERBERT FERDINAND, b Vienna, Austria, Aug 25, 27; US citizen; m 55. ELECTROCHEMISTRY. *Educ:* Univ Vienna, PhD(chem), 55. *Prof Exp:* Res asst phys chem, Univ Vienna, 54-56; res chemist, US Army Electronics Res & Develop Command, Ft Monmouth, 56-86; RETIRED. *Concurrent Pos:* Mem, Fuel Cells Res & Develop Panel, President's Interdept Energy Study, 63. *Honors & Awards:* Dept of Army Res & Develop Achievement Award, 62. *Mem:* Nat Space Soc. *Res:* Fuel cells, hydrogen-oxygen, methanol, biochemical, ion-exchange membranes; direct production of electrical pulses from galvanic cells; mechanism of periodic electrode processes; solid electrolytes; ionic sensors; lithium nonaqueous electrolyte cells. *Mailing Add:* 18 Mark Dr Long Branch NJ 07740

HUNGER, ROBERT MARVIN, b Denver, Colo, Jan 14, 54; m 82; c 1. WHEAT PATHOLOGY & BREEDING FOR DISEASE RESISTANCE. *Educ:* Colo State Univ, BS, 76, MS, 78; Ore State Univ, PhD(plant path), 82. *Prof Exp:* asst prof, 82-87, ASSOC PROF PLANT PATH, OKLA STATE UNIV, 87- *Mem:* Am Phytopath Soc; Sigma Xi. *Res:* Epidemiological aspects of several fungal and viral pathogens of wheat; development of wheat lines with resistance to diseases caused by these pathogens. *Mailing Add:* Dept Plant Path 110 Noble Res Ctr Okla State Univ Stillwater OK 74078-9947

HUNGERFORD, ED VERNON, III, b New Orleans, La, Feb 18, 39; m 64; c 2. NUCLEAR PARTICLE PHYSICS. *Educ:* Ga Inst Technol, PhD(physics), 67. *Prof Exp:* Physicist, Oak Ridge Nat Lab, 67-69; asst prof physics, Rice Univ, 70-73; assoc prof, 73-79, dept chmn, 80-88, PROF PHYSICS, UNIV HOUSTON, 79- *Concurrent Pos:* Prin investr, Dept Energy. *Mem:* Fel Am Phys Soc; AAAS. *Res:* Hyperon nuclear interactions; anti nucleon-nucleon interactions; electroweak interactions. *Mailing Add:* Dept Physics Univ Houston Houston TX 77004

HUNGERFORD, GERALD FRED, b Anaheim, Calif, Feb 14, 23; m 55; c 1. ANATOMY. *Educ:* Univ Calif, AB, 44, MA, 47, PhD(anat), 51. *Prof Exp:* Asst, Univ Calif, 47-51, instr, Med Sch, 51-52; instr, Univ Southern Calif, 52-53; asst prof anat, Med Sch, George Washington Univ, 53-58; assoc prof, 58-61, PROF ANAT, SCH MED, UNIV SOUTHERN CALIF, 61- *Mem:* Soc Exp Biol & Med; Am Asn Anatomists. *Res:* Endocrine relationships to thoracic duct lymphocytes and eosinophiles; assay of ACTH hormone; pineal gland. *Mailing Add:* Dept Anat Univ Southern Calif Sch Med 2025 Zonal Ave Los Angeles CA 90033

HUNGERFORD, HERBERT EUGENE, b Hartford, Conn, Oct 3, 18; m 49. NUCLEAR ENGINEERING. *Educ:* Trinity Col, Conn, BS, 41; Univ Ala, MS, 50; Purdue Univ, PhD(nuclear eng), 65. *Prof Exp:* Instr, Brent Sch, Philippine Islands, 41 & Choate Sch, 45-46; head, Dept Physics, Marion Mil Inst, 46-48; physicist & head group, Oak Ridge Nat Lab, 50-54, mem opers team aircraft reactor exp, 54-55; head shielding & health physics sect, Atomic Power Develop Assocs, 55-62; from res asst to instr, 63-64, assoc prof, 64-68, PROF NUCLEAR ENG, PURDUE UNIV, 68- *Concurrent Pos:* Consult various orgn. *Mem:* Am Nuclear Soc; Am Phys Soc; Health Physics Soc; Am Asn Physics Teachers; Am Soc Eng Educ. *Res:* Fast reactor shield design; new shield materials; radiation streaming; deep penetration of neutrons; stochastic neutron transport. *Mailing Add:* 2104 4th Court SE Vero Beach FL 32960

HUNGERFORD, KENNETH EUGENE, b Madison, Wis, May 5, 16; m 40, 74; c 3. WILDLIFE ECOLOGY, MANAGEMENT. *Educ:* Univ Idaho, BS, 38; Univ Conn, MS, 40; Univ Mich, PhD(wildlife mgt), 52. *Prof Exp:* Proj supvr, US Fish & Wildlife Serv, 40; biologist, State Fish & Game Dept, Idaho, 40-42; instr & dir, Naval Training Sch, 42-45, instr forestry & wildlife mgt, 46-47, from asst prof to prof wildlife mgt, 48-78, chmn acad, 78-81, EMER PROF WILDLIFE RESOURCES, UNIV IDAHO, 78- *Concurrent Pos:* Consult, NZ Forest & Range Exp Sta, 69 & Nat Taiwan Univ, Taipei, China, 74; mem, Idaho Water Resource Bd, 81- *Honors & Awards:* Arthur S

Einarson Award, Wildlife Soc, 80. *Mem:* Wildlife Soc. *Res:* Ecological life history of the Idaho ruffed grouse; microenvironment, especially dew moisture; whitetailed deer in relation to habitat; environmental impact of animal populations on resource use; behavior of fossorial mammals. *Mailing Add:* Col Forest Wildlife & Range Sci Univ Idaho Moscow ID 83843

HUNGERFORD, THOMAS W, b Oak Park, Ill, Mar 21, 36; m 58; c 2. MATHEMATICS. *Educ:* Col Holy Cross, AB, 58; Univ Chicago, MS, 60, PhD(math), 63. *Prof Exp:* From instr to assoc prof math, Univ Wash, 63-80; PROF & CHMN, DEPT MATH, CLEVELAND STATE UNIV, 80- *Mem:* Am Math Soc; Math Asn Am. *Res:* Algebra. *Mailing Add:* Dept Math Cleveland State Univ Cleveland OH 44115

HUNKE, WILLIAM ALLEN, b Mandan, NDak, Aug 17, 50; m 72; c 2. PHYSICAL PHARMACY. *Educ:* NDak State Univ, BS, 73; Univ Iowa, PhD(pharmaceut), 78. *Prof Exp:* SR RES PHARMACIST, STERLING-WINTHROP RES INST, DIV STERLING DRUG INC, 79- *Mem:* Am Pharmaceut Asn; Acad Pharmaceut Sci. *Res:* Physical pharmacy and the application of physical-chemical principals and methods to drug substances and dosage forms. *Mailing Add:* 71 Dorchester St Selkirk NY 12158-9759

HUNKINS, KENNETH, b Lake Placid, NY, Mar 3, 28; m 60, 85; c 2. ARCTIC OCEAN, COASTAL & ESTUARINE STUDIES. *Educ:* Yale Univ, BS, 50; Stanford Univ, MS, 57, PhD, 60. *Prof Exp:* Res asst, Lamont Geol Observ, Columbia Univ, 57-60, res scientist, 60-64, sr res scientist, 64-89, ADJ PROF, COLUMBIA UNIV, 74-, SPEC RES SCIENTIST, LAMONT-DOHERTY GEOL OBSERV, 89- *Concurrent Pos:* Chmn oceanog panel, Comt Polar Res, Nat Acad Sci-Nat Res Coun, 67-72; mem, Oceanog Adv Comt to US Navy, 71-77. *Mem:* Am Geophys Union; Sigma Xi. *Res:* Physical oceanography and marine geophysics, with special interest in the Arctic Ocean. *Mailing Add:* Lamont-Doherty Geol Observ Palisades NY 10964

HUNN, JOSEPH BRUCE, b St Paul, Minn, Mar 12, 33; m 67; c 2. PHYSIOLOGY, FISHERIES. *Educ:* St John's Univ, Minn, BA, 55; Mich State Univ, MS, 57, PhD(physiol), 63. *Prof Exp:* NIH res fel fish dis, Eastern Fish Dis Lab, WVa, 63-65; fishery biologist, Fish Control Lab, US Fish & Wildlife Serv, LaCrosse, Wis, 65-77; supvry fishery biologist, Hammond Bay Biol Sta, Millersburg, Mich, 77-80; fishery biologist, Columbia Nat Fish Res Lab, Columbia, Mo, 80-85, asst chief biologist, 85-87; FISHERY BIOLOGIST, NAT FISH CONTAMINANT RES CTR, 87- *Concurrent Pos:* Adj prof, Sch Natural Resources, Univ Mo, Columbia, 90- *Mem:* AAAS; Am Fisheries Soc; Am Soc Zool; Japanese Soc Sci Fisheries. *Res:* aquatic toxicology. *Mailing Add:* US Fish & Wildlife Serv Rte 2 Columbia MO 65201

HUNNELL, JOHN WESLEY, b Chicago, Ill, July 23, 32; m 58; c 2. FOOD TECHNOLOGY. *Educ:* Univ Ill, BS, 53, MS, 54, PhD(food tech), 60. *Prof Exp:* Mgr com develop, Refrigerated Foods, Pillsbury Co, 60-62, sr scientist, 62-63; sr food technologist, Colgate-Palmolive Co, 63-65; mgr process & prod develop, 65-68, dir res & develop, 68-78, ASST VPRES, TECH SERVS, RIVIANA FOODS, INC, 78-, DIR RES, 80- *Mem:* Inst Food Technol; Am Asn Cereal Chemists; Am Oil Chemists Soc; Am Chem Soc; Sigma Xi. *Res:* Formulation, process and packaging development and commercialization of food products. *Mailing Add:* 12222 Pebblebrook Houston TX 77024

HUNNICUTT, RICHARD P(EARCE), b Asheville, NC, June 15, 26; m 54; c 4. METALLURGICAL ENGINEERING. *Educ:* Stanford Univ, BS, 51, MS, 52. *Prof Exp:* Res metallurgist, Res Labs, Gen Motors Corp, 52-55; head metall sect, Aerojet-Gen Corp, Div Gen Tire & Rubber Co, 55-57; head mat & processes, Eng Lab, Firestone Tire & Rubber Co, 57-58; supvr mat lab, Dalmo Victor Co, 57-62; VPRES, ANAMET LABS, INC, 62- *Mem:* Am Soc Metals; Electrochem Soc; Am Welding Soc; Am Soc Testing & Mat; Am Inst Aeronaut & Astronaut. *Res:* Development of high strength-high temperature structural materials, especially rocket engine applications; friction; lubrication; wear; fracture studies and failure analysis of materials. *Mailing Add:* Anamet Labs Inc 3400 Investment Blvd Hayward CA 94545-3811

HUNNINGHAKE, GARY W, b Seneca, Kans, July 10, 46; m 68; c 3. MEDICINE. *Educ:* St Benedicts Col, BS, 68; Kans Univ, MD, 72; Am Bd Internal Med, dipl pulmonary dis; Am Bd Allergy & Immunol, dipl. *Prof Exp:* Intern, Kans Univ Med Sch, 72-73, resident, Med Ctr, 73-74; clin assoc, Nat Inst Allergy & Infectious Dis, NIH, Bethesda, Md, 74-76, med officer, 76-77, sr investr, Pulmonary Br, Nat Heart, Lung & Blood Inst & consult, 77-81; assoc prof, 81-84, DIR PULMONARY DIS DIV, DEPT INTERNAL MED, UNIV IOWA COL MED & DEPT VET AFFAIRS CTR, IOWA CITY, 81-, PROF MED, 84-, DIR CRITICAL CARE PROG, 89- *Concurrent Pos:* Mem, Clin Res Comt, Nat Inst Allergy & Infectious Dis, NIH, 74-77, chmn, 76-77; mem, Clin Res Ctr Comt, Univ Iowa, 81-83; mem prog comt, Allergy & Immunol Assembly, Am Thoracic Soc, 81-90; mem, Pulmonary Allergy Drugs Adv Comt, Food & Drug Admin, 84-86; assoc ed, Chest, 86-90, Am Rev Respiratory Dis, 89-; numerous grants, NIH & Dept Vet Affairs, 86-95. *Mem:* Am Fed Clin Res (pres, 86-87); Am Soc Microbiol; Am Asn Immunologists; Am Soc Clin Invest; Asn Am Physicians; Am Acad Allergy; Am Col Chest Physicians; Am Thoracic Soc (secy-treas, 91-92); Am Col Physicians; Soc Critical Care Med. *Res:* Lung immunology and host defenses, interstitial lung diseases, asthma, viral lung disease. *Mailing Add:* Dept Internal Med Univ Iowa Col Med Iowa City IA 52242

HUNSAKER, DON, II, b Ft Worth, Tex, Apr 6, 30; m; m; c 4. PSYCHOBIOLOGY, ECOLOGY. *Educ:* Tex Tech Univ, BA, 52, MS, 57; Univ Tex, PhD, 60. *Prof Exp:* From asst prof to assoc prof, 64-67, PROF BIOL, SAN DIEGO STATE UNIV, 67- *Concurrent Pos:* Vis prof biol, Colo State Univ, 65-70, Univ Ariz, 70-73; coordr, US Peace Corps, Columbia prog ecol & nat parks, 70-72. *Mem:* Am Soc Ichthyol & Herpet; Herpetologists' League. *Res:* Behavior, speciation and vocalization of vertebrate; behavioral ecology of endangered speciese; fish and wildlife. *Mailing Add:* Dept Biol San Diego State Univ San Diego CA 92182

HUNSICKER, HAROLD YUNDT, b Frankfort, Ind, Dec 22, 14; m 39; c 3. METALLURGY & PHYSICAL METALLURGICAL ENGINEERING, MATERIALS SCIENCE ENGINEERING. *Educ:* Purdue Univ, BSChE, 36; Case Inst Tech, MSMetE, 39. *Prof Exp:* Engr phys metall, Aluminum Res Labs, Alcoa, Cleveland, 36-48, asst div chief phys metall, 48-58, div chief, Alcoa Labs, New Kensington, 58-60, div mgr, Alcoa Labs, Alcoa Ctr, 60-77, consult, 77-80; RETIRED. *Mem:* Fel Am Soc Metals; Am Inst Metall Engrs. *Res:* Developed aluminum alloys for automotive engine bearings, automotive, aerospace and architectural castings; advanced wrought products for automotive, aerospace, electrical and architectural industries; microstructural/property relationships for these aluminum products. *Mailing Add:* 508 Chester Dr New Kensington PA 15068-3304

HUNSINGER, BILL JO, b Roanoke, Ill, Feb 23, 39; m 60; c 2. ELECTRICAL ENGINEERING. *Educ:* Univ Ill, BS, 61, PhD(elec eng), 70; Bradley Univ, MS, 66. *Prof Exp:* Res engr elec eng, Gen Elec Co, 61-67; res engr, Magnavox Co, 67-74; ASSOC PROF ELEC ENG, UNIV ILL, 74- *Res:* Surface acoustic waves; ultrasonics; microwave devices; semiconductor devices; signal processing. *Mailing Add:* Electronic Decisions Inc 1776 E Washington Urbana IL 61801

HUNSLEY, ROGER EUGENE, b Pierre, SDak, Feb 21, 38; m 63; c 3. ANIMAL SCIENCE. *Educ:* SDak State Univ, BS, 59; NDak State Univ, MS, 61; Iowa State Univ, PhD(animal nutrit), 67. *Prof Exp:* Field rep, Am Hereford Asn, 61-63; instr animal sci, Iowa State Univ, 64-67; from asst prof to assoc prof, Purdue Univ, 67-76, prof animal sci, 76-83; EXEC SECY, AM SHORTHORN ASN, 83- *Concurrent Pos:* Tech adv comt mem, NAm Limousin Found, 74-75. *Mem:* Am Soc Animal Sci; Int Livestock Coaches Asn. *Res:* Management practices; nutritional regimes and genetics in relationship to their effects on growth and body composition of red meat animals; evaluation of red meat animals and their products. *Mailing Add:* American Shorthorn Assoc 8288 Hascall St Omaha NE 68124

HUNSPERGER, ROBERT G(EORGE), b Philadelphia, Pa, Mar 6, 40; m 58; c 1. APPLIED PHYSICS, ELECTRICAL ENGINEERING. *Educ:* Drexel Univ, BSEE, 62; Princeton Univ, MSE, 63; Cornell Univ, PhD(appl physics), 67. *Prof Exp:* Mem staff res ctr, Burroughs Corp, 58-62; asst electronic circuits, Princeton Univ, 62-63; asst semiconductor lasers, Cornell Univ, 63-67; mem tech staff, Hughes Res Labs, 67-76; PROF ELEC ENG, UNIV DEL, 76- *Concurrent Pos:* Instr, Dept Eng-Phys Sci Exten, Univ Calif, Los Angeles, 68-76; consult, Hughes Aircraft Co, 78-87, Martin Marietta Corp, 81-83, Westinghouse Corp, 82-84, E I du Pont de Nemours & Co, Inc, 84-, ISC Defense Systs, 85-87 & Nat Inst Standards & Tech, 88- *Mem:* Inst Elec & Electronics Engrs; AAAS; Int Platform Asn; Soc Photoinstrumentation Engrs; Optical Soc Am; AAAS. *Res:* Electrical and optical properties of semiconductors and semiconductor devices, particularly involving ion-implantation doping; integrated optics. *Mailing Add:* Dept Elec Eng Univ Del Newark DE 19716

HUNSTAD, NORMAN A(LLEN), b Pipestone, Minn, Aug 25, 24; m 50; c 3. MECHANICAL ENGINEERING. *Educ:* Univ Iowa, BS, 49. *Prof Exp:* Engr, Gen Motors Corp, 49-55, asst head, Fuels & Lubricants Dept, Res Lab, 55-87; PRES, BEECHCREEK CORP, 87- *Mem:* Fel Soc Automotive Engrs; fel Soc Tribologists & Lubrication Engrs; fel Am Soc Testing & Mat; Sigma Xi. *Res:* Automotive lubricants. *Mailing Add:* BeechCreek Corp 6281 Little Creek Rd Rochester Hills MI 48306

HUNSTON, DONALD LEE, b Springfield, Mo, July 27, 43. COMPOSITES, RHEOLOGY. *Educ:* Kent State Univ, BS, 65, PhD(chem), 69. *Prof Exp:* NIH fel, Northwestern Univ, 70-71; res chemist, Naval Res Lab, 71-80; SUPVRY PHYS SCIENTIST, NAT INST STANDARDS & TECH, 80- *Honors & Awards:* Am Inst Chem Award, 65; Silver Medal, Dept Com. *Mem:* Am Chem Soc; Soc Rheol; Am Inst Physics; Soc Advan Mat & Process Eng. *Res:* Relationship between molecular structure and the mechanical or chemical properties of macromolecules; fracture behavior of polymeric materials; molecular interactions of biological molecules; structure-property relations in polymer composites; fluid mechanics. *Mailing Add:* Polymer Div Nat Inst Standards & Tech Gaithersburg MD 20899

HUNSUCKER, ROBERT DUDLEY, b Portland, Ore, Mar 15, 30; m 56, 81; c 3. AERONOMY, RADIO PROPAGATION. *Educ:* Ore State Univ, BS, 54, MS, 58; Univ Colo, PhD(elec eng), 69. *Prof Exp:* Asst geophys, Univ Alaska, 58-59, from instr to asst prof, 60-64; physicist, Res Labs, Inst Telecommun Sci & Aeronomy, Environ Sci Serv Admin, Colo, 64-69, sr proj leader, 69-71; from assoc prof to prof geophys, 71-87, EMER PROF GEOPHYS & SR CONSULT, GEOPHYS INST, UNIV ALASKA, 88- *Concurrent Pos:* Mem, US Comm G & H Int Union Radio Sci, 67- & working groups incoherent scatter, ionospheric sounders & propagation, 75-; resident visitor, AT&T Bell Labs, 82-83; prin investr, NSF grants, 71-88; resident visitor, AT&T Bell Labs, 82-83. *Honors & Awards:* Achievement Award, Region Six Pac Coast, Inst Elec & Electronics Engrs, 88. *Mem:* Sr mem Inst Elec & Electronics Engrs; fel AAAS; Am Geophys Union; Sigma Xi. *Res:* Physics and radio propagation investigations of the high and middle latitude ionosphere using incoherent scatter radar and other techniques; author and co-author of over 60 papers and one book. *Mailing Add:* Geophys Inst Univ Alaska Fairbanks AK 99775-0800

HUNT, ALBERT MELVIN, analytical chemistry; deceased, see previous edition for last biography

HUNT, ANDREW DICKSON, b Staten Island, NY, Oct 1, 15; m 40; c 4. PEDIATRICS. *Educ:* Haverford Col, BS, 37; Cornell Univ, MD, 41; Am Bd Pediat, dipl, 48. *Hon Degrees:* LHD, Northern Mich Univ, 78. *Prof Exp:* Intern, Hosp Univ Pa, 41-42, resident pediat, 42-43; asst instr, Sch Med, Univ Pa, 46-48, instr pediat, 48-50, assoc, 50-52, asst prof, 52; asst prof clin pediat, Col Med, NY Univ, 52-55, asst prof pediat, 55-59; assoc prof pediat, Sch Med & dir ambulatory serv, Med Ctr, Stanford Univ, 59-64; prof pediat & dean col human med, Mich State Univ, 64-77, actg dir health serv prog, 77-78, coordr med humanities prog, 78-84; PROF PEDIAT, SCH MED, MERCER UNIV, 84-, ASSOC DEAN GLYNN-BRUNSWICK PROGS, 87- *Concurrent Pos:* Chief resident, Children's Hosp, Philadelphia, 46-47, asst dir clins, 47-50, asst physician, 48-50, dir diag clin, 48-52, dir clins, 50-52, sr physician, 51-52; dir pediat serv, Hunterdon Med Ctr, Flemington, NJ, 52-59; asst vis physician, Bellevue Hosp, NY, 52-59; chmn adv coun, Comprehensive State Health Planning Comn, 68-; fel adolescent med, Sch Med, Stanford Univ, 81-82; fel, Ctr Advan Study Behav Sci, 87-88. *Mem:* AAAS; Soc Pediat Res; Am Pediat Soc; AMA; Am Acad Pediat. *Res:* Infectious disease; application of medical knowledge to society; medical education; adolescent medicine. *Mailing Add:* 201 Hermitage Way St Simons Island GA 31522-1717

HUNT, ANGUS LAMAR, b Great Falls, Mont, Sept 18, 25; m 49; c 3. PHYSICS. *Educ:* Reed Col, BA, 50; Wash State Univ, PhD(physics), 59. *Prof Exp:* Physicist, Shell Develop Co, Calif, 56-59; PHYSICIST, LAWRENCE LIVERMORE NAT LAB, UNIV CALIF, 59- *Mem:* Am Phys Soc; Am Vacuum Soc. *Res:* Surface physics; vacuum technique; cryogenics; physical electronics; atomic beams; plasma physics; controlled thermonuclear research. *Mailing Add:* 1630 Las Trampas Rd Alamo CA 94507

HUNT, ANN HAMPTON, b Lexington, NC, Nov 4, 42. STRUCTURE ELUCIDATION. *Educ:* Univ NC, Greensboro, AB, 65; Duke Univ, PhD(phys chem), 70. *Prof Exp:* Res asst chem, Duke Univ, 68-69; from instr to asst prof, Converse Col, 69-71; fel biochem, Univ Tex M D Anderson Hosp & Tumor Inst, 71-73; res fel biol chem, Harvard Med Sch, 75-78; sr phys chemist, 78-82, RES SCIENTIST, LILLY RES LABS, ELI LILLY & CO, 83- *Concurrent Pos:* Proj dir, NSF res grant, 70-71; counr, Am Chem Soc, 85- *Mem:* AAAS; Am Chem Soc; Biophys Soc; Soc Magnetic Resonance Med. *Res:* Nuclear magnetic resonance of natural products; circular dichroism spectroscopy in structure elucidation; conformational studies involving drugs and biomolecules. *Mailing Add:* Lilly Res Labs Eli Lilly & Co Indianapolis IN 46285-1513

HUNT, ARLON JASON, b Council Bluffs, Iowa, Oct 8, 39; m 71; c 2. LIGHT SCATTERING, SOLAR ENERGY CONVERSION. *Educ:* Univ Minn, BA, 63; Univ Ariz, MS, 71, PhD(physics), 74. *Prof Exp:* Res assoc, Physics Dept, Univ Ariz, 68-74 & Optical Sci Ctr, 74-76; PHYSICIST, LAWRENCE BERKELEY LAB, 76- *Concurrent Pos:* Pres, Particle Technol Assoc, 79-; consult, 80-; vpres, Quantum Optics Inc, 86-; prin, Thermalux LP, 89- *Honors & Awards:* Fulbright res scholar, Africa, 85. *Mem:* Am Phys Soc; Am Optical Soc; Int Solar Energy Soc; Am Ceramic Soc; AAAS; Mat Res Soc. *Res:* Absorption and scattering of light by small particles; optical and physical properties of solids; solar energy conversion; instrumentation; microstructural materials-Aerogel. *Mailing Add:* 90-2024 Lawrence Berkeley Lab Univ Calif Berkeley CA 94720

HUNT, BOBBY RAY, b McAlester, Okla, Aug 24, 41; m 65; c 2. ELECTRICAL ENGINEERING, COMPUTER SCIENCE. *Educ:* Wichita State Univ, BSc, 64; Okla State Univ, MSc, 65; Univ Ariz, PhD(systs eng), 67. *Prof Exp:* Staff mem systs res, Sandia Lab, NMex, 67-68; staff mem comput sci, Los Alamos Sci Lab, Univ Calif, 68-72; adj prof elec eng & comput sci, Univ NMex, 72-75; chief scientist, Defense Systs Group, Sci Appln Int Corp, 85-89; PROF ELEC ENG, UNIV ARIZ, 75-85, 89- *Mem:* Fel Inst Elec & Electronics Engrs; fel Optical Soc Am. *Res:* Processing of images by digital computer; digital filtering; systems theory and applications in image processing. *Mailing Add:* Dept Elec & Computer Eng Univ Ariz Tucson AZ 85721

HUNT, C WARREN, b San Francisco, Calif, Dec 15, 24; Can citizen; m 50; c 4. GEOMORPHOLOGY, GLACIOLOGY. *Educ:* Calif Inst Technol, BS, 45. *Prof Exp:* Jr geologist, Stand Oil Co, Calif, 45; geologist, 46, Independent Explor Co, 47-48; Independent Explor Co, 47-48; CONSULT GEOLOGIST, 48- *Mem:* Am Asn Petrol Geol; Soc Econ Geol; Can Inst Mining & Metall; Am Inst Mining Engrs; Asn Explor Geochemists; Can Soc Petrol Geologists. *Res:* Oil, gas and mineral exploration; structural surface and subsurface geology; petroleum exploration; goldmine development; author of 1 book. *Mailing Add:* 1119 Sydenham Rd SW Calgary AB T2T 0T5 Can

HUNT, CARL E, PEDIATRICS. *Educ:* Yale Univ, Dr, 65; Am Bd Pediat, cert pediat neonatology. *Prof Exp:* PROF & VCHMN DEPT PEDIAT, NORTHWESTERN UNIV MED SCH & HEAD DIV NEONATOLOGY, CHILDREN'S MEM HOSP, 75- *Res:* Respiratory control; Sudden Infant Death Syndrome. *Mailing Add:* Dept Pediat Med Col Ohio 3000 Arlington Ave PO Box 10008 Toledo OH 43699-0008

HUNT, CARLTON CUYLER, b Waterbury, Conn, Aug 11, 18. NEUROPHYSIOLOGY. *Educ:* Columbia Univ, BA, 39; Cornell Univ, MD, 42. *Prof Exp:* Intern, New York Hosp, 42-43, asst resident med, 46; res fel pharmacol, Med Col, Cornell Univ, 46-48, instr, 48; Nat Res Coun sr fel neurol, Johns Hopkins Hosp, 48-51, asst prof physiol, Univ, 51-52; assoc, Rockefeller Inst, 52-55; prof physiol, Albert Einstein Col Med, 55-57; prof, Col Med, Univ Utah, 57-64; prof, Sch Med, Yale Univ, 64-67; PROF PHYSIOL, SCH MED, WASHINGTON UNIV, 67- *Concurrent Pos:* Hon res assoc, Univ Col London, 62-63; vis prof, Col de France, 83-86. *Mem:* Biophys Soc; Soc Neurosci. *Res:* Muscle sensory receptor function; spinal cord physiology. *Mailing Add:* Dept Neurol Wash Univ Sch Med St Louis MO 63110

HUNT, CHARLES E, b Riverside, Calif, Jan 10, 35; m 61; c 6. EXPERIMENTAL PATHOLOGY, NUTRITIONAL BIOCHEMISTRY. *Educ:* Wash State Univ, DVM, 59; Mass Inst Technol, PhD(nutrit, biochem), 67. *Prof Exp:* Intern, Rowley Mem Hosp, Springfield, 59-60; instr small animal med & surg, Auburn Univ, 60-62, Ala Heart Asn fel, 62-63; res assoc nutrit path, Mass Inst Technol, 63-66, asst prof, 66-68; assoc dir labs toxicol & path, Schick Pharmaceut, Inc, 68-69; assoc prof comp med & path, 69-74, PROF COMP MED, INST DENT RES, UNIV ALA, BIRMINGHAM, 74-, SR SCIENTIST, 69-, PROF & DIR NUTRIT PATH, DEPT NUTRIT SCI, 78- *Mem:* Am Col Vet Path; Int Acad Path; Sigma Xi. *Res:* Pathology and biochemistry of copper; effects of fluoride, strontium, molybdenum, boron and lithium on tooth and bone in rats; adult onset obesity in mice; experimental atherosclerosis and diabetes. *Mailing Add:* Path-Toxicol Res Upjohn Co 301 Henrietta St Kalamazoo MI 49001

HUNT, CHARLES EDMUND LAURENCE, b Norwich, Eng, Aug 8, 35; Can citizen; m 58; c 3. PHYSICAL METALLURGY, MECHANICAL ENGINEERING. *Educ:* Univ BC, BASc, 57; Univ Waterloo, MASc, 67, PhD(mech eng), 70. *Prof Exp:* Jr res engr, 57-59, opers engr, 59-65, develop engr, 65-67, RES ENGR, ATOMIC ENERGY CAN, LTD, 70- *Mem:* Eng Inst Can. *Res:* Phase transformations in alloys, specifically directed towards zirconium base alloys; physical and mechanical properties of metal alloys and the effects of radiation on properties; effects of high temperature transients on nuclear fuel sheathing; release mechanisms of fission products from uranium dioxides particularly under abnormal conditions such as high temperature and oxidation by air or steam. *Mailing Add:* Atomic Energy Res Co Chalk River Nuclear Labs Chalk River ON K0J 1J0 Can

HUNT, CHARLES MAXWELL, b Lima, Ohio, Feb 26, 11. ENVIRONMENTAL CHEMISTRY. *Educ:* George Washington Univ, BS, 34; Polytech Inst Brooklyn, MS, 40; Univ Md, PhD(chem), 59. *Prof Exp:* Chemist, Dept Res & Develop, US Rubber Co, 36-41 & Toxicol Res Div, Edgewood Arsenal, 41-44; chemist, Nat Bur Standards, 44-80; RETIRED. *Honors & Awards:* Bronze Award, Dept Com. *Mem:* Am Chem Soc; AAAS; Am Soc Heating Refrig & Air Conditioning Engrs. *Res:* Gas adsorption and surface chemistry; particle size measurement; infrared spectroscopy; building ventilation and indoor air quality. *Mailing Add:* 703 Mapleton Dr Rockville MD 20850

HUNT, DALE E, microbiology, for more information see previous edition

HUNT, DOMINIC JOSEPH, b Brooklyn, NY, Apr 30, 22; m 52; c 6. INORGANIC CHEMISTRY. *Educ:* St John's Univ (NY), BS, 42; Columbia Univ, AM, 47; St Louis Univ, PhD(chem), 58. *Prof Exp:* Chemist, Am Cyanamid Corp, 42-43; res physicist, Monsanto Chem Co, 48-54; res assoc, St Louis Univ, 54-56, instr chem, 56-57; from asst prof to assoc prof, John Carroll Univ, 57-70, chmn dept, 73-77, prof chem, 70-86; RETIRED. *Mem:* Am Chem Soc. *Res:* Coordination compounds. *Mailing Add:* 2569 Queenston Rd Cleveland OH 44118-4351

HUNT, DONALD F, b Hyannis, Mass, Apr 25, 41; m 65; c 2. ORGANIC CHEMISTRY. *Educ:* Univ Mass, BS, 62, PhD(chem), 67. *Prof Exp:* NIH trainee chem, Mass Inst Technol, 67-68; from asst prof to assoc prof, 68-78, PROF CHEM, UNIV VA, 78- *Honors & Awards:* Charles H Stone Award, Am Chem Soc, 90. *Mem:* Am Chem Soc; Protein Soc; Am Soc Mass Spectrometry. *Res:* Mass spectrometry; analytical biochemistry; development of new methods and instrumentation for the structural characterization of proteins, glycoproteins, oligosacharrides, and nucleic acids by mass spectrometry. *Mailing Add:* Dept Chem Univ Va Charlottesville VA 22901

HUNT, DONALD F(ULPER), b Philadelphia, Pa, May 10, 25; m 61; c 1. ELECTRICAL ENGINEERING. *Educ:* Univ Pa, BS, 48. *Prof Exp:* Instr elec eng, Univ Pa, 48-54; assoc prof, 63-66, PROF ELEC ENG, POLYTECH INST NEW YORK, 66- *Mem:* Inst Elec & Electronics Engrs; Am Soc Eng Educ; Am Asn Univ Profs. *Res:* Networks; solar energy; analog and digital computers; education. *Mailing Add:* Polytech Univ 333 Jay St Brooklyn NY 11201

HUNT, DONNELL RAY, b Danville, Ind, Aug 11, 26; m 51; c 2. AGRICULTURAL ENGINEERING. *Educ:* Purdue Univ, BS, 51; Iowa State Univ, MS, 54, PhD(agr eng), 58. *Prof Exp:* From instr to assoc prof agr eng, Iowa State Univ, 51-60; assoc prof, 60-68, PROF AGR ENG, UNIV ILL, URBANA, 68- *Concurrent Pos:* Consult, FMC Corp, 62, Massey-Ferguson Ltd, 63 & US Steel Corp, 65-66 & var law firms & small mfrs; lectr, Univ Col, Dublin, 68-69. *Mem:* Am Soc Agr Engrs; Am Soc Eng Educ; Sigma Xi. *Res:* Optimization of farm field machine operations; mechanization of crop production; alternative tractor fuels. *Mailing Add:* 405 Ira Urbana IL 61801

HUNT, EARLE RAYMOND, b Petersburg, Va, Mar 7, 36; m 56; c 4. PHYSICS. *Educ:* Rutgers Univ, BA, 58, PhD(physics), 62. *Prof Exp:* Res assoc physics, Duke Univ, 62-63, from instr to asst prof, 63-67; assoc prof, 67-75, PROF PHYSICS, OHIO UNIV, 75- *Mem:* Am Phys Soc. *Res:* Solid state physics; chaos. *Mailing Add:* Dept Physics Ohio Univ Athens OH 45701

HUNT, EDWARD EYRE, b Washington, DC, Mar 9, 22; m 52; c 4. PHYSICAL ANTHROPOLOGY. *Educ:* Harvard Univ, AB, 42, AM, 49, PhD, 51. *Prof Exp:* anthropologist, Harvard Univ, 52-54, Statist Lab, Peabody Mus, 53-54, lectr, 55-60, res fel, phys anthrop, 55-68, asst prof anthrop, 60-66; prof, Hunter Col, 66-69; prof, 69-85, EMER PROF ANTHROP & HEALTH EDUC, PA STATE UNIV, 85-; ADJ PROF ANTHROP, UNIV MASS, AMHERST, 86- *Concurrent Pos:* Physical anthropologist, Harvard Exped to Yap, Micronesia, 47-48; staff anthropologist, Forsyth Dent Infirmary for Children, 52-65; res assoc, Children's Med Ctr, 54-60; consult, Heart Dis Epidemiol Surv, Mass, 54-62; lectr, Sch Dent Med, Tufts Univ, 56-62; Fulbright vis lectr, Univ Melbourne, 56-57; mentor, Inst Advan Educ Dent Res, 63-65; mem study sect appl physiol, NIH, 65-69; adj prof, Grad Ctr, City Univ, NY, 69-74; film rev ed, Am J Phys Anthrop, 73-80. *Mem:* Fel AAAS; Am Asn Phys Anthrop (secy-treas, 58-60); Soc Study Human Biol. *Res:* Growth and aging of human body tissues; genetics of human body measurements and physique; population genetics and demography. *Mailing Add:* 77 Magnolia Ave Magnolia MA 01930

HUNT, EUGENE B, b Gary, SDak, Sept 21, 23. ELECTRICAL ENGINEERING. *Educ:* SDak State Univ, BS, 48; Kans State Univ, MS, 50; Purdue Univ, PhD, 55. *Prof Exp:* Instr elec eng, SDak State Univ, 48-49; asst prof, SDak Sch Mines & Technol, 50-53; res instr, Purdue Univ, 54-55; res specialist, Autonetics Div, NAm Aviation, Inc, 55-64; assoc prof elec eng,

Calif State Col, Long Beach, 64-66; chmn dept, Calif State Col, Fullerton, 66-79, prof elec eng, Calif State Univ, 80-90; RETIRED. *Mem:* Am Soc Eng Educ; Inst Elec & Electronics Engrs. *Res:* Electronics; automatic control; networks; computers; communications. *Mailing Add:* 26769 Potomac Dr Sun City CA 92381

HUNT, EVERETT CLAIR, b Stamford, Conn, Dec 28, 28; m 52; c 3. MARINE ENGINEERING, SHIP DESIGN. *Educ:* US Merchant Marine Acad, BS, 51; Rensselaer Polytech Inst, MS, 58; Northeastern Univ, MS, 72; Eurotech Univ, DSc, 88. *Prof Exp:* Eng officer, USN, 52-54; engr, Gen Elec Co, 54-65, proj mgr, 65-66, consult, 66-67, mgr eng, 67-69, mgr qual control, 69-75; dir, Sun Shipbldg Co, 75-79; prof eng, US Merchant Marine Acad, 79-84; PROF ENG & RES DIR, WEBB INST NAVAL ARCHIT, 84- *Concurrent Pos:* Adj prof, Union Col, 70-75 & Widener Univ, 78-79; chmn, Inst Marine Engrs, 80-83. *Honors & Awards:* Bronze Medal, US Dept Transp, 83. *Mem:* Soc Naval Architects & Marine Engrs; Inst Marine Engrs; Pan-Am Inst Naval Eng; Inst Indust Engrs. *Res:* Engineering marine power plants; ship propulsion systems; engineering economics; quality control; statistics. *Mailing Add:* 48 Cold Spring Hills Rd Huntington NY 11743

HUNT, FERN ENSMINGER, b Mt Perry, Ohio, May 11, 26; m 53. FOODS. *Educ:* Ohio State Univ, BS, 48, MS, 54, PhD(food, microbiol), 65. *Prof Exp:* Home economist, Ohio Edison Co, 48-52; asst food, Univ State Univ, 52-53, from instr to asst prof food & nutrit, Ohio Agr Res & Develop Ctr, 54-68, assoc prof food & equipment, 68-73, prof food & equipment, Ohio Agr Res & Develop Ctr, 73-80; RETIRED. *Mem:* AAAS; Am Soc Microbiol; Inst Food Technologists; Am Home Econ Asn; Sigma Xi. *Res:* Factors in cooking affecting organoleptic qualities of food; effect of freezing and thawing rates on microbial activity in frozen food; energy management in households. *Mailing Add:* 4692 Scenic Dr Columbus OH 43214

HUNT, GARY W, b Chicago, Ill, Dec 24, 42; m 68; c 1. X-RAY CRYSTALLOGRAPHY. *Educ:* East Tex State Univ, BS, 68; Univ Ark, PhD(inorg chem), 72. *Prof Exp:* Fel chem, Univ SC, 71-74; ASST PROF CHEM, PHYSICS & MATH, SHORTER COL, 74- *Mem:* Sigma Xi. *Res:* Computer-aided instruction; computerized drill and testing. *Mailing Add:* Dept Chem Shorter Col Shorter Ave Rome GA 30161

HUNT, GEORGE ALBERT, microbiology, for more information see previous edition

HUNT, GEORGE LESTER, JR, b Boston, Mass, Aug 10, 42; m 88. ECOLOGY. *Educ:* Harvard Univ, AB, 65, PhD(biol), 71. *Prof Exp:* PROF BEHAV ECOL & MARINE ORNITH, SCH BIOL SCI, UNIV CALIF, IRVINE, 71- *Mem:* AAAS; Am Ornith Union; Brit Ornith Union; Am Ecol Soc; Brit Ecol Soc; Am Soc Limnol Oceanog. *Res:* Ecology and reproductive biology of seabirds; coloniality and reproductive success; biological oceanography of seabirds; habitat selection; foraging behavior. *Mailing Add:* Dept Ecol & Evolutionary Biol Univ Calif Irvine CA 92717

HUNT, GRAHAM HUGH, b Melville, Sask, May 15, 30; m 56; c 3. ECONOMIC GEOLOGY. *Educ:* Univ Man, BS, 53; Univ Alta, MS, 58, PhD(geol), 61. *Prof Exp:* Geologist, Hudson Bay Mining & Smelting Co, 51-56, Shell Oil Co, 57-58 & Mobil Oil Co, 61-64; prof geol, Brandon Col, 64-65, Odessa Col, 65-67 & Eastern Ky Univ, 67-74; PROF GEOL & CHMN DEPT, UNIV LOUISVILLE, 74- *Concurrent Pos:* Consult geol map, Dept Mines, Can, 65-67. *Mem:* Nat Asn Geol Teachers. *Res:* Geologic research of Pre-Cambrian in western Canada; geology of Purcell rocks in British Columbia, Alberta and Montana; geology of Manitoba. *Mailing Add:* Dept Geol Univ Louisville Box 35260 Louisville KY 40232

HUNT, GUY MARION, JR, b Battle Creek, Mich, Aug 26, 15; m 41; c 3. NEUROANATOMY. *Educ:* Col Med Evangelists, MD, 42, MS, 59; Emmanuel Missionary Col, BS, 46; Am Bd Psychiat & Neurol, dipl, 55. *Prof Exp:* Rotating intern, Los Angeles County Gen Hosp, 41-42; from instr to asst prof, 45-46, assoc prof, 57-75, assoc prof neurol, 64-75, dir & electroencephlaographer, Neurodiag Lab, 64-81, chief, Neurol Sect, 73-81, PROF NEUROL & ANAT, MED CTR, LOMA LINDA UNIV, 75- *Concurrent Pos:* Resident & fel, White Mem Hosp, 50-52, consult neurologist, 52-; active sr neurologist & attend neurologist, Loma Linda Univ Hosp. *Mem:* AMA; Am Acad Neurol; Sigma Xi. *Res:* Electron microscopy of the nervous system; anatomy, physiology and pathology of the choroid plexus; convulsive disorders; diseases of the basal ganglia. *Mailing Add:* Anat Dept Loma Linda Univ Loma Linda CA 92354

HUNT, HAROLD RUSSELL, JR, b Evansville, Ind, Mar 22, 32; m 56; c 2. INORGANIC CHEMISTRY. *Educ:* Harvard Univ, AB, 53; Univ Chicago, PhD(chem), 57. *Prof Exp:* Asst prof, 57-64, ASSOC PROF CHEM, GA INST TECHNOL, 64- *Mem:* Am Chem Soc; Sigma Xi. *Res:* Complex ions; mechanisms of reactions. *Mailing Add:* Sch Chem Ga Inst Technol Atlanta GA 30332

HUNT, HEMAN DOWD, b Yakima, Wash, Feb 6, 19; m 41; c 3. PHOTOGRAPHIC CHEMISTRY. *Educ:* Col Puget Sound, BS, 49; Univ Wash, Seattle, PhD(phys chem), 52. *Prof Exp:* Sr chemist, Res Div, Photo-Prod Dept, E I du Pont de Nemours & Co, Inc, 52-65, res assoc, 65-80; RETIRED. *Mem:* Soc Photog Scientists & Engrs. *Res:* Vacuum ultraviolet spectroscopy of organic molecules; photographic optical sensitization mechanisms; new photosensitive systems; research and development of new photographic films. *Mailing Add:* 220 Shorewood Dr Webster NY 14580

HUNT, HOWARD BEEMAN, b Winthrop, NY, Sept 23, 02; m 30; c 2. RADIOLOGY. *Educ:* Univ Calif, AB, 22, MA, 26; Harvard Univ, MD, 27. *Prof Exp:* Intern & resident, Univ Mich Hosp, 27-29, instr radiol, Univ Mich, 29-30; asst prof, Univ Nebr Med Ctr, Omaha, 30-37, coordr cancer teaching, 50-65, prof radiol, 37-88, chmn dept, 30-88; RETIRED. *Mem:* AMA; Radiol Soc NAm (vpres, 55); Soc Nuclear Med; Am Cancer Soc; Am Radium Soc (pres, 52). *Res:* Radiotherapy; radiobiology; oncology. *Mailing Add:* 9315 Western Ave Omaha NE 68114

HUNT, HURSHELL HARVEY, b Wheeler, Tex, Apr 26, 30; m 59; c 4. BIOMETRICS. *Educ:* Panhandle State Col, BS, 53; Okla State Univ, MS, 59, PhD(math & statist), 68. *Prof Exp:* Asst prof math, Okla Cent State Univ, 59-65; asst math & statist, Okla State Univ, 65-68; asst prof statist, Tex A&M Univ, 68-74; ASSOC PROF BIOMET, DEPT BIOMET, MED UNIV SC, 74- *Mem:* Sigma Xi; Biomet Soc; Am Statist Asn. *Mailing Add:* Dept Biostatist Epidemiol & System Sci Med Univ SC Charleston SC 29425

HUNT, ISABELLE F, b Winnepeg, Man, 29. NUTRITION DURING PREGNANCY, DIET OSTEOPORSIS. *Educ:* Univ Calif, Los Angeles, DPH (pub health), 68. *Prof Exp:* PROF PUB HEALTH, UNIV CALIF, LOS ANGELES, 68- *Mem:* Am Inst Nutrit; Am Pub Health Asn; Am Dietetic Asn; Sigma Xi. *Mailing Add:* School Pub Health Univ Calif Los Angeles CA 90024

HUNT, JAMES CALVIN, b Lexington, NC, Sept 11, 25. INTERNAL MEDICINE, NEPHROLOGY. *Educ:* Catawba Col, Salisbury, AB, 49; Univ Minn, MSc, 58; Bowman Gray Sch Med, MD, 53. *Prof Exp:* Instr to prof med, Mayo Clin & Mayo Med Sch, Rochester, 58-78; PROF & DEAN MED, COL MED, UNIV TENN, 78-, CHANCELLOR & VPRES HEALTH AFFAIRS, 81- *Concurrent Pos:* Chmn nephrol, Mayo Clin & Mayo Med Sch, 63-72, assoc dean, 72-74, chmn med, 74-78. *Mem:* Am Soc Clin Pharmacol & Therapeut; Am Col Cardiol. *Res:* Renal failure; water and electrolyte metabolism; sodium influences on blood pressure regulation and control; mechanisms of renal hypertension; nutritional influences on hypertension control. *Mailing Add:* Univ Tenn Ctr Health Sci 343 S Goodwyn Memphis TN 38111

HUNT, JAMES HOWELL, b Memphis, Tenn, Oct 6, 44; div; c 3. EVOLUTIONARY ECOLOGY. *Educ:* NC State Univ, BS, 66; La State Univ, MS, 69; Univ Calif, Berkeley, PhD(zool), 73. *Prof Exp:* Res assoc biol, Harvard Univ, 73-74; asst prof, 74-81, ASSOC PROF BIOL, UNIV MO, ST LOUIS, 81- *Concurrent Pos:* Secy-treas, NAm Sect, Int Union Study Social Insects, 84-86, pres, 90; vis asst prof entomol, Univ Kans, 86. *Mem:* Entom Soc Am; Soc Study Evolution; Int Union Study Social Insects. *Res:* Laboratory and field studies on the evolution of sociality in hymenoptera, especially the role of trophic relationships among colony members, patterns of protein nutrition, and demography. *Mailing Add:* Dept Biol Univ of Mo St Louis MO 63121

HUNT, JAMES L, b Guelph, Ont, Feb 23, 33; m 55; c 3. PHYSICS. *Educ:* Queen's Univ, Ont, BA, 55; Univ Toronto, MA, 56, PhD(physics), 59. *Prof Exp:* Asst prof physics, Mem Univ, Nfld, 59-63; assoc prof, 63-70, PROF PHYSICS, UNIV GUELPH, 70- *Concurrent Pos:* Vis prof, Pa State Univ, 68-69. *Mem:* Can Asn Physicists; Am Asn Physics Teachers. *Res:* Molecular spectroscopy. *Mailing Add:* Dept Physics Univ Guelph Guelph ON N1G 2W1 Can

HUNT, JANET R, b Glendale, Ariz, Sept 23, 52; m 86; c 4. ZINC NUTRITION, IRON NUTRITION. *Educ:* Brigham Young Univ, BS, 73, MS, 75; Univ Minn, PhD(nutrit), 87. *Prof Exp:* Instr food & nutrit, Brigham Young Univ, 75-76; instr nutrit & dietetics, Univ Tex Health Sci Ctr, Dallas, 76-78; chief dietitian, 78-87, RES SCIENTIST, HUMAN NUTRIT RES CTR, AGR RES SERV, USDA, 88- *Concurrent Pos:* Adj instr, Univ NDak, 82-91; chmn-elect, Coun Res, Am Dietetic Asn, 91-92, chmn, 92-93. *Mem:* Am Dietetic Asn; Am Inst Nutrit; Am Soc Clin Nutrit; Sigma Xi. *Res:* Zinc and iron bioavailability from diets; effects of dietary protein on mineral and bone metabolism; iron deficiency and behavior; radioisotope tracers to measure trace element absorption and retention. *Mailing Add:* Human Nutrit Res Ctr USDA-Agr Res Serv Grand Forks ND 58202

HUNT, JERRY DONALD, b Hastings, Nebr, Nov 11, 38; div; c 2. ORGANIC CHEMISTRY, INORGANIC CHEMISTRY. *Educ:* Hastings Col, AB, 60; Iowa State Univ, MS, 63, PhD(org chem), 66. *Prof Exp:* Res scientist, 65-68, group leader, 68-77, assoc scientist chem, Firestone Tire & Rubber Co, 77-83; RES & DEVELOP ASSOC, GOODYEAR TIRE & RUBBER CO, 83- *Mem:* Am Chem Soc; Am Soc Testing & Mat. *Res:* Rubber chemistry; free radical chemistry; antioxidant and antiozonant chemistry. *Mailing Add:* 7355 Thatcher Ave NW North Canton OH 44720

HUNT, JOHN A, b Reading, Eng, June 22, 35; nat US; m 73; c 2. MOLECULAR GENETICS. *Educ:* Cambridge Univ, BA, 56, PhD(biochem), 60. *Prof Exp:* Fel biochem, Carlsberg Lab, Denmark, 59-60; mem sci staff, Nat Inst Med Res, Eng, 60-64; assoc prof, 64-70, PROF GENETICS, UNIV HAWAII, 70- *Mem:* Genetics Soc Am; Brit Biochem Soc. *Res:* Chemistry of genetic mutations affecting abnormal hemoglobins; biosynthesis of hemoglobin and transfer of genetic information in erythropoiesis; RNA synthesis; DNA sequence analysis in the evolution of Hawaiian Drosophila. *Mailing Add:* Dept Genetics Univ Hawaii 1960 E West Rd Honolulu HI 96822

HUNT, JOHN BAKER, b Pine Bluff, Ark, Oct 20, 33; m 59; c 3. INORGANIC CHEMISTRY. *Educ:* Tulane Univ, BS, 55; Georgetown Univ, MS, 60; Univ Chicago, PhD(chem), 62. *Prof Exp:* Instr chem, US Naval Acad, 57-59; from asst prof to assoc prof, 62-80, PROF CHEM, CATH UNIV AM, 80- *Mem:* Am Chem Soc. *Res:* Mechanisms of redox and substitution reactions of complex ions; structure and reactions of complex ions in fused salts. *Mailing Add:* Dept Chem Rm 340 NSF Washington DC 20550-0002

HUNT, JOHN MEACHAM, b Cleveland, Ohio, Dec 1, 18; m 47; c 2. PETROLEUM, GEOCHEMISTRY. *Educ:* Western Reserve Univ, AB, 41; Pa State Univ, MS, 43, PhD(org chem), 46. *Prof Exp:* Asst petrol ref, Pa State Univ, 43-46, hon asst to Dean F C Whitmore, 46, instr chem, 46-47; sr res chemist, Carter Oil Co, 48-56, sect head, 56-63; chmn dept chem, 64-74, sr res scientist, 74-84, EMER SCIENTIST, WOODS HOLE OCEANOG INST, 85- *Concurrent Pos:* Lectr, Oil & Gas Consults Int, Tulsa, Okla, 83- *Honors & Awards:* Treibs Medal, Geochem Soc, 82. *Mem:* Europ Asn Org Geochemists; Am Chem Soc; Am Asn Petrol Geol; Geochem Soc. *Res:*

Application of geochemical techniques to oil and gas exploration; origin, migration and accumulation of petroleum; identification of petroleum source rocks and correlation of crude oils. *Mailing Add:* Dept Chem Woods Hole Oceanog Inst Woods Hole MA 02543

HUNT, JOHN PHILIP, b Ann Arbor, Mich, Feb 2, 23; m 52; c 3. PHYSICAL CHEMISTRY, INORGANIC CHEMISTRY. *Educ:* Univ Mich, BS, 44; Univ Chicago, PhD(chem), 50. *Prof Exp:* Mem staff, Clinton Labs, Oak Ridge, Tenn, 44-46; res assoc, Univ Chicago, 50-51; asst prof chem, Cornell Univ, 51-55; from assoc prof to prof, 55-90, EMER PROF CHEM, WASH STATE UNIV, 90- *Mem:* Am Chem Soc; fel AAAS. *Res:* Inorganic reaction mechanisms. *Mailing Add:* N W 1325 Orion Dr Pullman WA 99163

HUNT, JOHN R, b Vancouver, BC, Apr 12, 28; m 53; c 3. POULTRY NUTRITION. *Educ:* Univ BC, BSA, 51; Wash State Univ, PhD(nutrit), 55. *Prof Exp:* Res asst nutrit, Wash State Univ, 51-55; res scientist, 56-74, head animal sci sect, Agassiz Res Sta, Can Dept Agr, 74-84, res scientist, 84-; RETIRED. *Mem:* Poultry Sci Asn; Nutrit Soc Can; Can Soc Animal Sci. *Res:* Measurement of egg shell strength; effect of nutrition on the strength of the hen's egg shell; factors effecting the acid-base balance in the laying hen; effect of early nutrition on the performance of the broiler breeder; effect of pesticides on shell formation; mineral metabolism; feed restriction of layers; distribution of egg weight; rabbit nutrition; sudden death syndrome in broilers. *Mailing Add:* PO Box 138 PO Box 1000 Agassiz BC V0M 1A0 Can

HUNT, JOHN WILFRED, b Regina Sask, May 30, 30; div; c 5. BIOPHYSICS, RADIATION CHEMISTRY. *Educ:* Univ Sask, BSc, 52, MSc, 53; McGill Univ, PhD(physics), 56. *Prof Exp:* Nat Cancer Inst Can fel, Univ London, 56-57; from asst prof to assoc prof, 59-71, PROF BIOPHYS, UNIV TORONTO, 71-, PHYSICIST, ONT CANCER INST, 57- *Concurrent Pos:* Gordon Richards Mem fel, 65-66, 82; mem, NAm Hyperthermia Group. *Honors & Awards:* Arturo Miolati Prize, 73. *Mem:* Biophys Soc; Am Inst Ultrasound Med. *Res:* Initial processes of radiation chemistry studied by nanosecond and picosecond pulse radiolysis; reactions and yields of free radicals and ions in liquids, particularly for biologically important molecules; diagnostic ultrasound techniques, including basic studies in high-resolution scanning and development of transvectal and breast scanners, Doppler; development of ultrasound devices designed to generate heating patterns in the body; clinical studies of hyperthermia. *Mailing Add:* Ont Cancer Inst Exp Therapeut & Dept Med Biol Univ Toronto 500 Sherbourne St Toronto ON M4X 1K9 Can

HUNT, KENNETH WHITTEN, b Westboro, Mass, May 29, 09; m 36; c 1. BOTANY, CONSERVATION. *Educ:* Mass State Col, BS, 30; Williams Col, Mass, AM, 32; Cornell Univ, PhD(plant morphol), 36. *Prof Exp:* Actg asst prof biol, St Lawrence Univ, 36-37; from assoc prof bot to prof bot & geol, Col Charleston, 37-46; from assoc prof to prof biol & dir, Glen Helen Dept, Antioch Col, 46-73; emer prof environ studies, Stockton State Col, 73-77; RETIRED. *Concurrent Pos:* Dir preforestry, Col Charleston, 37-46; Gen Educ Bd fel, Mt Lake Biol Sta, Univ Va, 43; Penrose Fund grant, Am Philos Soc, 45-46; fel, Yale Conserv Prog, 56-57. *Mem:* Nature Conservancy. *Res:* Origin of carpel; flora of Atlantic coastal plain; land and park management for conservation education and recreation. *Mailing Add:* Carol Woods Apt 1310 750 Weaver Dairy Rd Chapel Hill NC 27514

HUNT, LAWRENCE BARRIE, b Richmond, Ind, July 19, 32; m 55; c 2. ORNITHOLOGY, VERTEBRATE ECOLOGY. *Educ:* Miami Univ, BA, 55; Univ Wis-Madison, MS, 60, PhD(zool, wildlife ecol), 68. *Prof Exp:* Instr zool & bot, Univ Wis-Kenosha, 62-67; from assoc prof to prof zool, 67-89, EMER PROF, EASTERN ILL UNIV, 89- *Mem:* Am Ornith Union; Wilson Ornith Soc; Asn Field Ornithologists. *Res:* Effects of pesticides on vertebrate populations and ecosystem accumulation; avian breeding behavior and habitat preference. *Mailing Add:* 51 Heather Dr Charleston IL 61920

HUNT, LEE MCCAA, b Clio, Ala, Aug 18, 26; m 51; c 2. OCEANOGRAPHY, GEOLOGY. *Educ:* Am Univ, BS, 56. *Prof Exp:* Pres & gen mgr, Southern Iron Corp, 56-58; res asst oceanog, Tex A&M Univ, 58-60; tech asst, Mine Adv Comt, 60-62, exec secy, 62-74, exec secy, Army Countermine Adv Comt, 70-72 & Undersea Warfare Comt, 72-74, EXEC DIR, NAVAL STUDIES BD, NAT ACAD SCI, 74- *Mem:* Marine Technol Soc. *Res:* Underwater acoustics, magnetics and explosive phenomena; naval weapons and weapons countermeasures design; oceanographic environmental studies. *Mailing Add:* 7715 Lookout Ct Alexandria VA 22306

HUNT, LEON GIBSON, b Dallas, Tex, Oct 25, 31; m 54; c 4. MATHEMATICAL STATISTICS, OPERATIONS RESEARCH. *Educ:* La State Univ, BS, 52; Univ Denver, MA, 64. *Prof Exp:* Geologist, Marathon Oil Co, 54-62; sr engr, Martin-Marietta Corp, 62-64; mem prof staff, Ctr Naval Anal, 64-66; dir anti-submarine warfare studies, Off Asst Secy Defense, 66-67; mgr marine systs dept, Planning Res Corp, 67-68; pres, L G Hunt Assocs, 68-70; vpres & tech dir, Anal Sci Corp, 70-74; dir, Off Energy Statist, Fed Energy Admin, 74-75; CONSULT, 75- *Concurrent Pos:* Instr, Univ Denver, 64; consult, President's Comn Fed Statist, 70-71, US Comn RR Retirement, 71-72, Exec Off of the President, 73-, Nat Inst Drug Abuse, 74-, WHO, 74-75, UN, 75-, Nat Acad Sci, 78- & NIH, 80-; adj res prof, Antioch Col, 76-78. *Mem:* NY Acad Sci. *Res:* Stochastic processes; drug abuse; carcinogenesis; geostatistics; biomathematics; quantitative microbiology. *Mailing Add:* 4632-B S 36th St Arlington VA 22206

HUNT, LINDA MARGARET, b Lexington, NC, Oct 25, 47. DEVELOPMENTAL BIOLOGY. *Educ:* Univ NC, Greensboro, BA, 69, MA, 71; Univ Mich, MS, 73, PhD(zool), 75. *Prof Exp:* ASST PROF GEN BIOL, UNIV NOTRE DAME, 75- *Concurrent Pos:* Coordr gen biol, Univ Notre Dame, 75- *Mem:* AAAS; Entom Soc Am; Sigma Xi; Am Soc Zoologists; Am Inst Biol Sci. *Res:* Developmental biology of hemiptera with emphasis upon both histological and endocrinological studies; developmental biology. *Mailing Add:* 225 Galvin Life Sci Bldg Notre Dame IN 46556

HUNT, LINDSAY MCLAURIN, JR, b Oklahoma City, Okla, Sept 19, 39; m 60; c 2. ORAL BIOLOGY. *Educ:* Okla Univ, BA, 61; Baylor Univ, DDS, 65, PhD(physiol), 71. *Prof Exp:* Dent intern, St Anthony Hosp, Oklahoma City, 67-68; asst prof dent res & coordr basic health sci dent, Sch Dent, Emory Univ, 71-73, from assoc prof to prof oral biol & chmn dept, 73-85; PROF RESTORATIVE DENT & DEAN, SCH DENT, VA COMMONWEALTH UNIV, 85- *Mem:* AAAS; Int Asn Dent Res; Am Asn Dent Schs; Am Dent Asn; fel Am Col Dent. *Res:* Calcium metabolism; pain. *Mailing Add:* Sch Dent Va Commonwealth Univ Box 566 MCV Sta Richmond VA 23298

HUNT, LOIS TURPIN, b Norfolk, Va, Jan 14, 33; div; c 1. CYTOLOGY, MICROBIOLOGY. *Educ:* Univ Kans, BA, 58; Univ Wash, MS, 61; Univ Md, College Park, PhD(cytol), 68. *Prof Exp:* Instr zool, Univ Md, College Park, 64-66; jr res scientist, 66-68, SR RES SCIENTIST & PROTEIN DATA ED, ATLAS OF PROTEIN SEQUENCY & STRUCT STAFF, NAT BIOMED RES FOUND, 68- *Mem:* Soc Protozool; Coun Biol Educ; Soc Pharmacol & Environ Path; NY Acad Sci; Am Fisheries Soc; Sigma Xi. *Res:* Histopathology and protozoan parasites of freshwater fishes; biochemical evolution of organisms based on analyses of sequences of proteins; evolutionary relations of protein families. *Mailing Add:* 12317 Dalewood Dr Wheaton MD 20902

HUNT, MAHLON SEYMOUR, b Cairo, Ill, Apr 1, 25; m 53; c 4. GEODESY. *Educ:* Wash Univ, AB, 49, MA, 52. *Prof Exp:* Cartogr, Aeronaut Chart & Info Ctr, Mo, 51-54, geodesist, 57-59; cartogr, US Naval Observ, Washington, DC, 54-57; geodesist, Air Force Cambridge Res Labs, Mass, 59-68, chief, Lunar Laser Observ, Ariz, 68-72, geodesist, Air Force Geophys Lab, Mass, 72-76; RETIRED. *Res:* Astronomy; geodetic instrumentation and techniques. *Mailing Add:* 1749 Hillside Dr Springfield MO 65809-3303

HUNT, MICHAEL O'LEARY, b Louisville, Ky, Dec 9, 35; c 3. FOREST PRODUCTS. *Educ:* Univ Ky, BS, 57; Duke Univ, MF, 58; NC State Univ, PhD(wood sci), 70. *Prof Exp:* Prod engr, Wood Prod Div, Singer Co, 59-60; asst prof, 60-69, assoc prof, 69-79, PROF WOOD SCI, PURDUE UNIV, WEST LAFAYETTE, 79-, DIR, WOOD RES LAB, 82- *Concurrent Pos:* Consult eng wood composites, 62. *Honors & Awards:* Gottschalk Mem Award, Forest Prod Res Soc, 84. *Mem:* Soc Wood Sci & Technol; Forest Prod Res Soc; Am Soc Testing & Mat. *Res:* Use of wood and woodbase composite materials in structural systems. *Mailing Add:* Dept Forestry Purdue Univ FPRD Bldg West Lafayette IN 47907

HUNT, PAUL PAYSON, b Elkview, WVa, Dec 17, 30; m 55; c 3. PHYSICAL CHEMISTRY. *Educ:* Glenville State Col, BS & AB, 54; Univ Tenn, MS, 58, PhD(phys chem), 60. *Prof Exp:* Sr res chemist, Hercules Powder Co, Md, 59-65; assoc prof, 65-68, PROF CHEM, FROSTBURG STATE COL, 68-, CHMN DEPT, 66- *Mem:* Am Chem Soc; Sigma Xi. *Res:* Gas chromatography of hydrogen isotopes; high temperature kinetics of corrosive reactions; decomposition and stabilization studies of high-energy solid propellant ingredients. *Mailing Add:* Dept Chem Frostburg State Univ Frostburg MD 21532

HUNT, RICHARD HENRY, b Lexington, Ky, Jan 12, 25; m 52; c 2. ANALYTICAL CHEMISTRY. *Educ:* Univ Ky, BS, 45; Univ Wis, MS, 47, PhD(chem), 49. *Prof Exp:* Res chemist, Houston Res Lab, Shell Oil Co, 49-56, sr res chemist, 56, group leader, 56-69, supvr, 69-70, sr engr, Head Off, Mfg, Transp & Supplies & Mkt Res & Develop, Planning & Admin, 70-74, SR RES CHEMIST, WESTHOLLOW RES CTR, SHELL DEVELOP CO, 74- *Mem:* Am Chem Soc; Sigma Xi. *Res:* Stobbe condensation with cyclohexanone; organic synthesis; mass spectrometry; radiochemistry; organic analysis. *Mailing Add:* 10047 Wickersham Lane Houston TX 77042

HUNT, RICHARD LEE, b Lansing, Mich, Nov 1, 36; m 57; c 2. INORGANIC CHEMISTRY. *Educ:* Antioch Col, BS, 59; Univ Chicago, PhD(chem), 63. *Prof Exp:* NSF fel, Imp Col, Univ London, 63-65; from asst prof to assoc prof chem, Calif State Univ, Long Beach, 65-80; PROF CHEM, MOREHEAD STATE UNIV, 91- *Mem:* Am Chem Soc; Royal Soc Chem. *Res:* Kinetics of oxidations; electron microscopy; structure determination of organometallic compounds; nuclear quadruple resonance. *Mailing Add:* UPO 979 Morehead State Univ Morehead KY 40351

HUNT, RICHARD STANLEY, b Victoria, BC, Apr 22, 44; m 68; c 3. FOREST PATHOLOGY. *Educ:* Univ Victoria, BSc, 67; Univ Calif, Berkeley, PhD(plant path), 71. *Prof Exp:* Fel, Prairie Regional Lab, Nat Res Coun Can, 72-73; RES SCIENTIST FOREST PATH, CAN FORESTRY SERV, 73- *Concurrent Pos:* Nat Task Force, Future Forest Path, Can. *Mem:* Am Phytopath Soc; Can Phytopath Soc. *Res:* White pine blister rust; disease control through silviculture; hazard rating; genetic selection and biotechnology; phytogeography of conifers and their pathogens. *Mailing Add:* 506 W Burnside Pac Forest Ctr Victoria BC V8Z 1M5 Can

HUNT, ROBERT HARRY, b Ann Arbor, Mich, June 13, 32. PHYSICS. *Educ:* Univ Mich, BS, 54, MS, 55, PhD(molecular physics), 63. *Prof Exp:* Instr physics, Univ Mich, 63-64; from asst prof to assoc prof, 64-77, assoc chmn dept, 75-78, PROF PHYSICS, FLA STATE UNIV, 77- *Mem:* Am Phys Soc. *Res:* High resolution infrared spectroscopy; molecular structure. *Mailing Add:* Dept Physics Fla State Univ Tallahassee FL 32306

HUNT, ROBERT L, b Madison, Wis, June 3, 33; m 59; c 3. FISH BIOLOGY, AQUATIC ECOLOGY. *Educ:* Univ Wis-Madison, BS, 58, MS, 59. *Prof Exp:* Aquatic biologist, 59-67, COLD WATER GROUP LEADER TROUT RES, WIS DEPT NAT RESOURCES, 67- *Honors & Awards:* Nat Trout Unlimited Conserv Award, 81; Gulf Oil Conserv Award, 82. *Mem:* Am Fisheries Soc. *Res:* Life history and management of trout populations; ecology of trout streams; research to evaluate trout fishing regulations and trout habitat improvement; trout population dynamics. *Mailing Add:* Wis Dept Natural Resources 11084 Stratton Lake Rd Waupaca WI 54981

HUNT, ROBERT M, JR, b Cleveland, Ohio, July 18, 41; m 70; c 2. VERTEBRATE PALEONTOLOGY. *Educ:* Col Wooster, BA, 63; Univ NMex, MS, 65; Columbia Univ, PhD(biol), 71. *Prof Exp:* Instr anat, Columbia Univ, 70-73; asst prof, 73-75, ASSOC PROF GEOL, UNIV NEBR, 75- *Concurrent Pos:* Res assoc, Am Mus Natural Hist, NY, 73-79. *Mem:* Paleont Soc; Soc Vertebrate Paleont; Soc Study Evolution. *Res:* Biostratigraphy of Miocene mammals of North America; functional anatomy of mammals; evolution and biogeography of mammals. *Mailing Add:* Dept Geol Univ Nebr Lincoln NE 68588-0340

HUNT, ROBERT NELSON, b Edmonton, Alta, Sept 9, 46; m 68; c 2. MUNICIPAL ENGINEERING. *Educ:* Univ Alta, BS, 68, PhD(physics), 74. *Prof Exp:* Instr eng & comput, Col Arts, Sci & Technol & Univ WI, Jamaica, 72-73 & Camrose Lutheran Col, 73-74; chmn dept sci, 80-87, INSTR ENG, GRANDE PRAIRE REGIONAL COL, 74- *Concurrent Pos:* Civil engr, Beairsto, Stewart & Weir Eng, Ltd, 79-85; consult, Microelectronics Ctr & lectr, dept elec eng, Univ Man, 85, vis scientist, 85-86. *Mem:* Can Asn Physicists; AAAS. *Res:* Stable isotope studies; computer-aided control systems; computer-assisted design, manufacturing and engineering of printed circuit boards and very-large-scale integration gate arrays for custom design and semi-custom design; computer-aided design, manufacturing and engineering of chips and microelectronic devices. *Mailing Add:* Dept Sci Grande Prairie Regional Col 10726 106th Ave Grande Prairie AB T8V 4C4 Can

HUNT, ROBERT WELDON, b Portales, NMex, Nov 16, 35; m 55; c 10. MATHEMATICAL ANALYSIS, APPLIED MATHEMATICS. *Educ:* WTex State Univ, BS, 56; Univ Utah, MS, 58, PhD(math), 61. *Prof Exp:* Instr math, Univ Utah, 59-61; asst prof, Huntsville Br, Univ Ala, 61-62; assoc prof, Southern Ill Univ, 62-68 & Naval Postgrad Sch, 68-70; prof math & chmn dept, Calif State Col, Bakersfield, 70-77; PROF MATH, HUMBOLDT STATE UNIV, 77- *Concurrent Pos:* Consult, Marshall Space Flight Ctr, NASA, Ala, 62-67 & Douglas Aircraft Co, Calif, 64-65; vis prof, Calif State Univ, Los Angeles, 74-75 & Humboldt State Univ, 75-77. *Mem:* Am Math Soc; Math Asn Am; Sigma Xi. *Res:* Ordinary differential equations; oscillation properties and matrix forms; calculus of variations and its application to space and missile problems. *Mailing Add:* Humboldt State Univ Arcata CA 95521

HUNT, RONALD DUNCAN, b Los Angeles, Calif, Oct 9, 35. COMPARATIVE PATHOLOGY, RESEARCH ADMINISTRATION. *Educ:* Univ Calif, Davis, BS, 57, DVM, 59. *Prof Exp:* Mem, US Army Vet Corps Inst Path, , 59-61, captain, 61-63; pathologist-in-chg, Animal Res Ctr, Harvard Med Sch, 63-72, chmn div comp path, New Eng Regional Primate Res Ctr, 65-76, assoc prof comp path & assoc dir, Animal Res Ctr, 72-77, dir Animal Res Ctr, 79-88, DIR, NEW ENG REGIONAL PRIMATE RES CTR, HARVARD MED SCH, 76-, PROF COMP PATH, 77- *Concurrent Pos:* Lectr, Mass Inst Technol, 64-81; affil pathologist, Angell Mem Animal Hosp, 66-75. *Honors & Awards:* Res Awards, Am Asn Lab Animal Sci, 67, 69 & 73; Charles River Prize, Am Vet Med Asn, 83. *Mem:* Am Col Vet Path; Int Acad Path; Am Soc Exp Path; Am Vet Med Asn; NY Acad Sci. *Res:* Comparative pathology of human and animal diseases, especially diseases of primates; herpesvirus diseases and metabolic bone diseases; author or co-author of over 150 publications. *Mailing Add:* New Eng Regional Primate Res Ctr Southboro MA 01772

HUNT, ROY EDWARD, b New York, NY, Sept 4, 18; m 47; c 4. CHEMISTRY. *Educ:* Brown Univ, ScB, 40; Univ Rochester, PhD(chem), 47. *Prof Exp:* Res assoc, George Washington Univ, 42-45; res assoc, Gen Elec Co, 47-48, asst mgr, Tech Personnel Div, 48-53, mgr tech recruiting sci doctorates, 53-54, supvr tech personnel, Knolls Atomic Power Lab, 54-60; assoc, Exec Personnel Dept, Booz Allen & Hamilton, 60-68, managing assoc, 68-80; vpres, Spencer Stuart & Assocs, 80-83; RETIRED. *Mem:* Am Chem Soc. *Res:* Vapor pressures of binary liquid mixtures; photochemical investigations of ketones; fluorescence of ketone vapors; ultraviolet filters; burning rates of double-base powder; flame-thrower ballistics. *Mailing Add:* HC-01 Box 58 Kattskill Bay NY 12844-9702

HUNT, STEVEN CHARLES, b Salt Lake City, Utah, 53; m 77; c 4. GENETIC-EPIDEMIOLOGY. *Educ:* Univ Utah, BS, 76, PhD(med biophysics), 80. *Prof Exp:* Res instr, 80-82, res asst prof, 82-87, RES ASSOC PROF, MED, UNIV UTAH, 87- *Mem:* Am Pub Health Asn; Am Heart Asn. *Res:* Research on cause and prevention of early coronary disease and hypertension; separate genetic, common environment and sporadic causes; cellular ion transport and lipid abnormalities. *Mailing Add:* Res Park 410 Chipeta Way Rm 161 Salt Lake City UT 84108

HUNT, THOMAS KINTZING, b Boston, Mass, Oct 9, 37; m 64; c 4. ENERGY CONVERSION. *Educ:* Pomona Col, BA, 59; Calif Inst Technol, MS, 61, PhD(physics), 64. *Prof Exp:* Res fel physics, Calif Inst Technol, 64; staff scientist, Sci Lab, Ford Motor Co, 64-88; RES SCIENTIST & MGR, ENVIRON RES INST MICH, 89- *Concurrent Pos:* NSF fel, 59-61. *Mem:* Am Phys Soc; Am Asn Physics Teachers; AAAS; Sigma Xi. *Res:* Low temperature physics; superconductivity; energy conversion and storage. *Mailing Add:* 3420 Andover Rd Ann Arbor MI 48105

HUNT, THOMAS KNIGHT, b Chicago, Ill, Aug 6, 30; c 3. SURGERY. *Educ:* Harvard Univ, BS, 52, MD, 56. *Prof Exp:* Intern surg, Boston City Hosp, 56-57; resident, Med Sch, Univ Ore, 59-63, instr, 63-65; asst prof, 65-69, PROF SURG, UNIV CALIF, SAN FRANCISCO, 69-, PROF AMBULATORY & COMMUNITY MED, 71- *Concurrent Pos:* Res fel hyperbaric oxygen unit, Dept Surg, Western Infirmary, Univ Glasgow, 64; dir gen surg & dir surg outpatient dept, Univ Calif Hosp, 68- *Honors & Awards:* Mallinckrodt Award, Harvard Univ, 53; James IV Traveller, Alpha Omega Alpha, 74. *Mem:* Am Asn Surg of Trauma (vpres); Am Surg Asn; Soc Univ Surgeons; Am Col Surgeons; Am Trauma Soc (pres, 80-82). *Res:* Wound healing; tissue gas determination; general and endocrine surgery. *Mailing Add:* 839 HSE Univ Calif Med Ctr San Francisco CA 94143

HUNT, V DANIEL, b Baltimore, Md, Sept 30, 39. TECHNOLOGY ASSESSMENT, SYSTEM CONCEPT DEVELOPMENT. *Educ:* Univ Md, BSEE, 63, MBA, 68. *Prof Exp:* Reliability engr, Bendix Corp, 61-63; prob sponsor, Appl Physics Lab, John Hopkins Univ, 63-74; div mgr, TRW Inc, 80-83; PRES, TECHNOL RES CORP, 83- *Concurrent Pos:* Lectr, Nat Defense Univ, 90-91. *Mem:* Soc Mfg Engrs; Am Soc Qual Control. *Res:* system concept development, technology assesment, system integration, and technology transfer in high technology areas for defense systems; advanced manufacturing technology and applied artificial intelligence; author of numerous handbooks and sourcebooks in the fields of robotics and computer science. *Mailing Add:* Technol Res Corp 8328-A Traford Lane Springfield VA 22152

HUNT, WALTER ANDREW, ALCOHOLISM, DRUG ABUSE. *Educ:* WVa Univ, PhD(pharmacol), 71. *Prof Exp:* Chief Physiol-Psychol Div, Armed Forces Radiol Res Inst, 73-89; CHIEF, NEUROSCI & BEHAV RES BR, NAT INST ALCOHOL ABUSE & ALCOHOLISM, 89- *Res:* Neurochemistry. *Mailing Add:* Div Basic Res Rm 16C-05 Nat Inst Alcohol Abuse & Alcoholism Rockville MD 20857

HUNT, WILLIAM A(LFRED), b Whitefish, Mont, Aug 8, 30; m 56; c 4. HYDRAULIC ENGINEERING, FLUID MECHANICS. *Educ:* Mont State Univ, BS, 52; Univ Wis-Madison, MS, 53, PhD(hydraul eng), 60. *Prof Exp:* Res asst civil eng, Univ Wis-Madison, 52-53; instr, Mont State Univ, 55-58; res asst, Univ Wis-Madison, 58-60; from asst prof to assoc prof civil eng, 60-67, actg head, Dept Civil Eng & Eng Mech, 77-78, dir, Mont Water Resources Res Ctr, 78-82, PROF CIVIL ENG, MONT STATE UNIV, 67- *Concurrent Pos:* Eng consult, 60- *Honors & Awards:* Stephen D Bechtel Pipeline Engineering Award, Am Soc Civil Engrs, 78. *Mem:* Am Soc Civil Engrs; Am Water Works Asn. *Res:* Hydraulic transport of solids in pipelines; analysis and design of pipelines and distribution systems; hydraulic transients and control systems for water pipelines. *Mailing Add:* Dept Civil Eng & Eng Mech Mont State Univ Bozeman MT 59717

HUNT, WILLIAM B, JR, b Lexington, NC, Sept 27, 27; m 50; c 4. MEDICINE, IMMUNOLOGY. *Educ:* Wake Forest Col, BS, 48; Bowman Gray Sch Med, MD, 53; Am Bd Internal Med, dipl, 62, recert, 74; Am Bd Allergy, dipl, 66. *Prof Exp:* Instr med, NY Med Col, 59-60; res fel microbiol & immunol, 60-62, from instr to asst prof med, Sch Med, Univ Va, 60-67, assoc prof med & asst dean, 67-75, dir respiratory care unit, 69-75; DIR DEPT RESPIRATORY THER, CRAVEN COUNTY HOSP, NEW BERN, NC, 75-; CLIN ASSOC PROF MED, EASTERN CAROLINA SCH MED, 78- *Mem:* Fel Am Col Chest Physicians; fel Am Acad Allergy; Am Fedn Clin Res; fel Am Col Physicians. *Res:* Allergies; internal medicine; pulmonary disease. *Mailing Add:* PO Box 2157 New Bern NC 28560

HUNT, WILLIAM CECIL, b Conemaugh, Pa, June 21, 23; m 48; c 4. ORGANIC CHEMISTRY. *Educ:* Juniata Col, BS, 44; Univ Pa, MS, 49, PhD(org chem), 51. *Prof Exp:* Asst prof chem, Washington & Lee Univ, 50-51; fel protected metals, Mellon Inst, 51-53, sr fel, 53-57, admin fel, 57-59, head sci rels, 59-62; mgr lab facil, Celanese Corp Am, 62-64, mgr spec projs res dept, 64-65, vpres & tech dir, Devoe & Raynolds Co, Inc, Ky, 65-66, vpres & tech dir, Celanese Coatings Co, 66-67, vpres technol & mfg, 67-68, vpres, Celanese Res Co, NJ, 68-71; vpres corp technol, Great Lakes Chem Corp, 71-80; pres, Syngene Corp, 81-87; RETIRED. *Mem:* Indust Res Inst; Am Chem Soc; Soc Plastics Eng. *Res:* Polyester resins. *Mailing Add:* 305 Goodlette Rd No 401C Naples FL 33940

HUNT, WILLIAM DANIEL, b Jackson, Miss, Dec 21, 54. ACOUSTIC CHARGE TRANSPORT DEVICES, SURFACE ACOUSTIC WAVE DEVICES. *Educ:* Univ Ala, BS, 76; Mass Inst Technol, SM, 80; Univ Ill, Urbana-Champaign, PhD(elec eng), 87. *Prof Exp:* Engr, Govt Systs Div, Harris Corp, 76-78; staff engr, Bolt Beranek & Newman, Inc, 80-84; asst prof, 87-91, ASSOC PROF ELEC ENG, GA INST TECHNOL, 91- *Concurrent Pos:* Consult, Electronic Decisions Inc, 85-87 & United Technol Res Ctr, 88-90; NSF presidential young investr, 89. *Mem:* Acoust Soc Am; Inst Elec & Electronics Engrs; Am Phys Soc; Am Soc Eng Educ; Sigma Xi. *Res:* New types of acoustic charge transport devices; new substrate materials for acoustic charge transport devices; surface acoustic wave propagation in AlGaAs heterostructures and in II-IV compound semiconductors as they pertain to the performance of proposed acoustic charge transport devices. *Mailing Add:* Sch Elec Eng Ga Inst Technol Atlanta GA 30332-0250

HUNT, WILLIAM EDWARD, b Columbus, Ohio, Nov 26, 21; m 45; c 3. NEUROSURGERY. *Educ:* Ohio State Univ, BA, 43, MD, 45. *Prof Exp:* Resident gen surg, White Cross Hosp, Columbus, Ohio, 48-49; resident neurosurg, Barnes Hosp, St Louis, 49-50; fel, Washington Univ, 50-51; resident, Barnes Hosp, St Louis, 51-52; from instr to assoc prof, 52-64, PROF SURG & DIR, DIV NEUROL SURG, COL MED, OHIO STATE UNIV, 64- *Mem:* Cong Neurol Surg; Am Asn Neurol Surg; Am Col Surgeons; Neurosurg Soc Am; Soc Neurol Surg (secy); Sigma Xi. *Res:* Cerebral vascular disease and related fundamental problems; cerebral vascular physiology; sensory disorders and physiology; spinal cord injury. *Mailing Add:* Dept Neurosurg Ohio State Univ Columbus OH 43210

HUNTE, BERYL ELEANOR, b New York, NY. MATHEMATICS EDUCATION. *Educ:* Hunter Col, BA, 47; Columbia Univ, MA, 48; NY Univ, PhD, 65. *Prof Exp:* Instr math, Southern Univ, 48-51; teacher, Bloomfield High Sch, 51-57; teacher, Friends Sem, 57-62; asst prof, Rockland Community Col, 62-63; instr math & supvr pract teachers, NY Univ, 64; dean spec projs, Off Acad Affairs, City Univ NY, 88-89; asst prof & sr instr, Bor Manhattan Community Col, 64-66, assoc prof, 66-70, chmn dept, 66-67 & 70-73, actg dean students, 85-86, actg dean acad affairs, 87, PROF MATH, BOR MANHATTAN COMMUNITY COL, 70- *Concurrent Pos:* Deleg, Workshop Human Rels, PR Dept Educ, 65; speaker, Bur Math Educ, NY State Dept Educ, 67; consult, William H Sadlier, Inc, 67-70; mem, Mid States Eval Team, 71-73 & 77; math consult Title I progs, NY State Educ Dept, 72; test supvr, Am Col Testing, 73 & 74; oral examr, Dept Civil Serv, State NY,

75; NSF rev panelist, 76; tech rev panelist, Div Equal Educ Opportunities, HEW, 77; fac fel, Off Acad Affairs, Cent Off, City Univ NY, 80-81; rev panelist, Dept Higher Educ, State NJ, 84, 85 & 87; assoc, Univ Seminar Higher Educ, Columbia Univ, 89- *Mem:* NY Acad Sci; Am Math Soc. *Res:* Geometry; mathematics education; statistics and probability; educational opportunities; urban education. *Mailing Add:* 30 W 60th St New York NY 10023

HUNTEN, DONALD MOUNT, b Montreal, Que, Mar 1, 25; m 49; c 2. PLANETARY SCIENCES. *Educ:* Univ Western Ont, BSc, 46; McGill Univ, PhD(physics), 50. *Prof Exp:* Res assoc physics, Univ Sask, 50-52, from asst prof to assoc prof, 53-63; physicist, Kitt Peak Nat Observ, 63-77; prof, 77-88, REGENTS PROF PLANETARY SCI, LUNAR & PLANETARY LAB, UNIV ARIZ, 88- *Concurrent Pos:* Ed, Can Asn Physicists, 61-63; prof physics, Univ Sask, 64-66; Consult, NASA, 64-, vchmn sci steering groups, Pioneer-Venus Prog, 74-, mem, Space Sci Bd, 75-76 & 82-86, sci adv, Assoc Admin Space Sci. 76-77, mem, Adv Coun & Chmn, Space Sci Adv Comt, 78-; mem, Climatic Impact Comt, Nat Acad Sci, 72-75. *Honors & Awards:* Pub Serv Medals, NASA, 77 & 85; Space Sci Award, Am Inst Aeronaut & Astronaut, 80; Kuiper Prize Award, Am Astron Soc, 87. *Mem:* Nat Acad Sci; Am Geophys Union; Am Astron Soc; Am Phys Soc. *Res:* Spectroscopy; earth's upper atmosphere and planetary atmospheres; photoelectric and spectroscopic instrumentation; planetary entry probes. *Mailing Add:* Lunar & Planetary Lab Space Sci Bldg Univ Ariz Tucson AZ 85721

HUNTER, ALAN GRAHAM, b Pawtucket, RI, Apr 1, 34; m 55; c 2. ANIMAL PHYSIOLOGY. *Educ:* Univ RI, BS, 55, MS, 58; Mich State Univ, PhD(reprod physiol), 63. *Prof Exp:* Asst dairy sci, Univ RI, 55-58; res instr, Mich State Univ, 58-63; from asst prof to assoc prof, 63-70, PROF REPROD & LACTATION, UNIV MINN, ST PAUL, 70- *Concurrent Pos:* Vis Prof, Cambridge Univ, 85. *Mem:* AAAS; Am Soc Animal Sci; Am Dairy Sci Asn; Brit Soc Study Fertil; Sigma Xi. *Res:* Protein constituents of semen and female reproductive cells and fluids and their physiological role in reproduction and lactation. *Mailing Add:* Dept Animal Sci Univ Minn 495 An Sc/VM St Paul MN 55108

HUNTER, ALBERT SINCLAIR, b Worthington, Ind, Oct 21, 08; m 47; c 1. SOIL FERTILITY. *Educ:* Utah State Univ, BS, 38; Wash State Univ, MS, 40; Rutgers Univ, PhD(soils), 43. *Hon Degrees:* Comendador, Univ Federal de Santa Maria, 77. *Prof Exp:* Assoc soil scientist, Spec Guayule Res Proj, USDA, Calif, 43-45, soil technologist, US Plant, Soil & Nutrit Lab, NY, 45-48, sr agronomist, US Natural Rubber Res Sta, Calif, 48-49; prof soil sci, Ore State Univ & soil scientist, USDA, 49-57; prof, 57-74, EMER PROF SOIL FERTIL, PA STATE UNIV, UNIVERSITY PARK, 74- *Concurrent Pos:* Specialist, IRI Res Inst, Brazil, 65-66; prof, Grad Sch Agr Sci, Arg, 72-73; prof soil chem & soil fertility & grad course agron, Fed Univ Santa Maria, RS, Brazil, 74-77. *Mem:* Fel AAAS; fel Am Soc Agron; Soil Sci Soc Am. *Res:* Soil fertility; liming; crop production. *Mailing Add:* 1424 Harris St State College PA 16801

HUNTER, ALICE S (BAKER), b New York, NY, Sept 11, 23; m 54. EMBRYOLOGY. *Educ:* Queen's Col, NY, BS, 44; Columbia Univ, MA, 46, PhD(zool), 52. *Prof Exp:* Instr biol, Johns Hopkins Univ, 48-50; res assoc chem & physiol, Fla State Univ, 50-53; chemist, Fla State Dept Agr, 53-54; res assoc physiol, Col Med, Univ Ill, 54-55; asst prof biol, Roosevelt Univ, 55-57; prof, Univ Andes, Colombia, 57-64; prof, Univ Centro-Occidental, Venezuela, 64-70; assoc prof, 70-75, PROF BIOL SCI, UNIV OF PAC, 75- *Mem:* AAAS; Am Soc Zoologists; Entom Soc Am; Sigma Xi. *Res:* Taxonomy and ecology of Drosophila; pigmentation of frog skin; enzymes in erythrocytes; gametogenesis and development of amphibians; physiological embryology. *Mailing Add:* Dept Biol Sci Univ of the Pacific Stockton CA 95211

HUNTER, ARVEL HATCH, b Rigby, Idaho, Dec 19, 21; m 46; c 4. SOIL FERTILITY. *Educ:* Brigham Young Univ, BS, 54; Ohio State Univ, MS, 55; NC State Univ, PhD(soil sci), 59. *Prof Exp:* Res instr soil chem, NC State Univ, 55-59, asst prof soil fertil, 59-60; asst prof soil fertil, Wash State Univ, 60-61; regional agronomist, Calif Chem Co, 61-64; regional dir, NC State Univ-AID Contract, 64-67, dir control lab, 67-74, AID vis assoc prof soil sci, NC State Univ, 69-74; vpres, 74-87, PRES, AGRO SERV INT INC, 87- *Concurrent Pos:* Owner, Custom Lab Equip Co, 69- *Mem:* Am Soc Agron; Soil Sci Soc Am. *Res:* Agronomy; analytical chemistry. *Mailing Add:* PO Box 667 Agro Serv Int Inc Orange City FL 32763

HUNTER, BARRY B, b Turtle Creek, Pa, Apr 4, 39; m 63; c 3. MYCOLOGY, PLANT PATHOLOGY. *Educ:* California State Col, Pa, BS, 63, MEd, 69; Univ Minn, MS, 67; WVa Univ, PhD(plant path), 70. *Prof Exp:* Teacher high sch, 63-66; PROF BIOL, CALIFORNIA STATE COL, PA, 68- *Mem:* Am Phytopath Soc; Mycol Soc Am. *Res:* Soil microbial ecology of fungal pathogens as they relate to environmental, chemical and biological factors. *Mailing Add:* Dept Biol California State Col Third St California PA 15419

HUNTER, BYRON ALEXANDER, b Salt Lake City, Utah, Oct 15, 10; m 42; c 8. INDUSTRIAL ORGANIC CHEMISTRY, RUBBER CHEMISTRY. *Educ:* Univ Utah, AB, 33, AM, 38; Iowa State Univ, PhD(org chem), 41. *Prof Exp:* Asst chem, Iowa State Univ, 38-40; res chemist, Uniroyal Chem, Uniroyal Inc, 41-45, group leader org res, 45-48, sr res assoc, 58-75; RES ASSOC, BRIGHAM YOUNG UNIV, 77- *Honors & Awards:* Outstanding Achievement Award, Thermoplastics & Foams Div, Soc Plastics Engrs, 90; Thomas Midgley Award, Am Chem Soc, 90. *Mem:* Am Chem Soc; Sigma Xi. *Res:* Agricultural chemicals; rubber accelerators and antioxidants; stabilization of synthetic rubber; blowing agents, rubber and plastics; organic hydrazine compounds; high molecular weight aliphatic compounds of nitrogen and sulfur; cement chemicals; soil chemicals; recipient of 67 United States patents. *Mailing Add:* 352 E 426 N Alpine American Fork UT 84004

HUNTER, CHRISTOPHER, b Manchester, Eng, May 28, 34; m 61; c 4. APPLIED MATHEMATICS. *Educ:* Cambridge Univ, BA, 57, PhD(math), 60. *Prof Exp:* Res assoc math, Mass Inst Technol, 60-61, lectr, 61-62; res fel, Trinity Col, Cambridge Univ, 62-64; from asst prof to assoc prof, Mass Inst Technol, 64-70; dir appl math, 70-83, 89-91, PROF MATH, FLA STATE UNIV, 70- *Concurrent Pos:* Consult, Avco, Tyco Labs; Joint Inst Lab Astrophys fel, Univ Colo, 76-77. *Mem:* Am Astron Soc; Soc Indust & Appl Math; Int Astron Union; Royal Astron Soc London. *Res:* Dynamics of fluids and stellar systems; applied analysis. *Mailing Add:* Dept Math Fla State Univ Tallahassee FL 32306-3027

HUNTER, CYNTHIA L, b Wichita Falls, Tex, Mar 28, 54. CORAL REEF ECOLOGY, ECOTOURISM. *Educ:* Calif State Univ, BS, 75; Univ SFla, MS, 80; Univ Hawaii, PhD(zool), 88. *Prof Exp:* Res assoc, Conserv Consults Inc, 76-78, Smithsonian Inst, 78-80; staff biologist, Mote Marine Lab, Fla, 80-82; res asst, Dept Zool, 82-88, ASST RESEARCHER, HAWAII INST MARINE BIOL, UNIV HAWAII, 89- *Concurrent Pos:* Lectr, Scientists in Sch, Hawaii Dept Educ, 87-88; postdoctoral researcher, Univ Guam, 88-89; consult, Oceans Alive Inc, 89- *Mem:* Soc Study Evolution; Sigma Xi; Western Soc Naturalists; Pac Sci Asn. *Res:* Reproduction, genetics and ecology of corals; long term dynamics of coral reefs; biotoxicities for tropic pacific reef species. *Mailing Add:* Hawaii Inst Marine Biol PO Box 1346 Kaneohe HI 96744

HUNTER, DOUGLAS LYLE, b Belleville, Ont, Aug 29, 40; m 62; c 2. THEORETICAL PHYSICS. *Educ:* Univ Alta, BSc, 62, MSc, 64; Univ London, PhD(theoret physics), 67. *Prof Exp:* Nat Res Coun Can fel & lectr physics, Univ Alta, 67-69; asst prof math, St Francis Xavier Univ, 69-70; assoc mathematician, Brookhaven Nat Lab, 70-72; from asst prof to assoc prof, 72-84, chmn dept, 74-78 & 81-90, PROF PHYSICS, ST FRANCIS XAVIER UNIV, 84- *Concurrent Pos:* Vis assoc prof, Kings Col, Univ London, 78-79. *Res:* Statistical mechanical models of phase transitions; use of series expansion techniques to deduce behavior in critical region when exact solutions not available; methods of analyzing series expansions, percolation, cluster aggregation, polymers; self-avoiding walks. *Mailing Add:* Dept Physics St Francis Xavier Univ Antigonish NS B2G 1C0 Can

HUNTER, EDWIN THOMAS, nuclear chemistry, for more information see previous edition

HUNTER, ERIC, b Guisborough, Eng, Aug 18, 48. MOLECULAR VIROLOGY, MOLECULAR GENETICS. *Educ:* Univ Birmingham, Eng, BSc, 69; Imp Cancer Res Fund & Brunel Univ, Eng, PhD(tumor immunol), 72. *Prof Exp:* Fel microbiol, Sch Med, Univ Southern Calif, 72-76; from asst prof to assoc prof, 76-84, PROF MICROBIOL, UNIV ALA, BIRMINGHAM, 84-, ASSOC SCIENTIST, COMPREHENSIVE CANCER CTR, 76- *Concurrent Pos:* fel cancer immunol, Cancer Res Inst, NY; Fel, Daymon Runyon Mem Fund Cancer Res; Nat Cancer Inst Res Career Develop Award, 80. *Mem:* Am Soc Microbiol; Am Soc Biol Chemists; AAAS; Am Soc Virol. *Res:* Organization and expression of retroviral genes, synthesis and processing of products of such genes, particularly glycoproteins, and assembly of these products into virus particles. *Mailing Add:* Dept Microbiol Univ Ala Birmingham AL 35294

HUNTER, FRANCIS EDMUND, JR, b Alliance, Ohio, June 6, 16; m 40; c 2. BIOCHEMISTRY, PHARMACOLOGY. *Educ:* Mt Union Col, BS, 38; Univ Rochester, PhD(biochem), 41. *Prof Exp:* Asst pharmacol & biochem, Univ Rochester, 38-41; asst pharmacol, 41-42, from instr to assoc prof, 42-84, EMER PROF PHARMACOL, SCH MED, WASH UNIV, 84- *Mem:* Am Chem Soc; Am Soc Pharmacol & Exp Therapeut; Am Soc Biol Chem; Am Soc Cell Biol; Biophys Soc. *Res:* Phospholipids; sphingomyelin; barbiturates; oxidation of butyric and other fatty acids; oxidation of acetoacetic acid; tricarboxylic acid cycle; mechanisms of oxidative phosphorylation; mechanism of action of dinitrophenol; enzymatic activity and structural relationships in insolated mitochondria; glutathione-dependent enzymes. *Mailing Add:* 147 Timbercrest Rd Kirkwood MO 63122-1311

HUNTER, FRISSELL ROY, b Richmond, Va, Aug 28, 24; m 58; c 2. ZOOLOGY, COMPARATIVE PHYSIOLOGY. *Educ:* Va Union Univ, BS, 47; Howard Univ, MS, 52; Univ Iowa, PhD(zool), 57. *Prof Exp:* Prof biol, Southern Univ, 57-63 & Fla Agr & Mech Univ, 63-64; prof biol & chmn div natural sci & math, Allen Univ, 64-65; prof biol, Fayetteville State Col, 65-69 & Atlanta Univ, 69-71; prof biol, 71-86, HEAD, DEPT BIOL & LIFE SCI, SAVANNAH STATE COL, 86- *Mem:* Am Soc Zool; Sigma Xi; Am Inst Biol Sci; AAAS; Sigma Xi. *Res:* Protozoan enzymology; drug effects on enzyme systems. *Mailing Add:* Savannah State Col PO Box 20533 Savannah GA 31404

HUNTER, GEOFFREY, b Manchester, Eng, Apr 23, 34; m 71. QUANTUM CHEMISTRY. *Educ:* Royal Col Adv Technol, Eng, ARIC, 61; Univ Manchester, MSc, 62, PhD(theoret chem), 64. *Prof Exp:* Welch fel theoret chem, Rice Univ, 64-65; Nat Res Coun Can fel, 65-66, asst prof, 66-69, ASSOC PROF THEORET & PHYS CHEM, YORK UNIV, 69- *Mem:* The Chem Soc; assoc Royal Inst Chem. *Res:* Chemical and relativistic quantum mechanics; theory of electromagnetic interactions; hardware systems. *Mailing Add:* Dept Chem York Univ Downsview ON M3J 1P3 Can

HUNTER, GEORGE L K, b New York, NY, May 2, 22; m 46, 74; c 1. ORGANIC CHEMISTRY, ANALYTICAL CHEMISTRY. *Educ:* Univ Miami, BS, 51, MS, 52; Ohio State Univ, PhD(chem), 56. *Prof Exp:* Fel cool flame combustion, Ohio State Univ, 57-58; proj dir, Peninsular ChemRes, Inc, 58-60; proj dir, citrus flavor fruit & veg prod lab, Agr Res Serv, USDA, Fla, 61-65; proj dir, Minute Maid Co, 65-68; dir flavor res, Coca-Cola Co, 68-74, mgr appl flavor res & ingredients qual control, 74-87; RETIRED. *Mem:* AAAS; Am Chem Soc; Inst Food Technol. *Res:* Compositional studies of flavors in natural food products, particularly structural elucidation; analysis by thin layer, column, and gas chromatography; structural determinations by infrared, ultraviolet and mass spectroscopy; beverage and food formulation and processing; essential oil and flavor research and synthesis; flavor and essential oil quality control; agriculture and food chemistry. *Mailing Add:* 2408 Kings Lake Blvd Naples FL 33962

HUNTER, GEORGE TRUMAN, b Erie, Pa, May 10, 18; m 54; c 2. PHYSICS. *Educ:* Univ Tampa, BS, 39; Univ Fla, MS, 41; Univ Wis, PhD(physics), 49. *Hon Degrees:* DSc, Univ Tampa, 61. *Prof Exp:* Instr elec commun, Mass Inst Technol, 43-45; asst prof physics & electronics, US Naval Postgrad Sch, 45-49, assoc prof physics, 49-50; physicist, Int Bur Mach Corp, 50-53, asst to dir appl sci div, 53-55, ed consult, 56-57, consult exec develop, 57-59, proj coordr election processing activities, 59-61, mgr advan mgt control systs, 61, consult exec develop, 62, educ prog adminr, 62-68, consult, Univ Rels Dept, 68-71, prog adminr tech support, 71-72, sr educ indust mkt rep, Data Processing Div, 72-73, mgr APL, Syst Develop Div, 73-75; CONSULT, 75-; ADJ PROF PHYSICS, MIAMI UNIV, OXFORD OHIO, 86- *Mem:* Am Phys Soc; Asn Comput Mach. *Res:* Education; executive development; applications of computing machinery. *Mailing Add:* 1203 Dana Dr Oxford OH 45056-2513

HUNTER, GEORGE WILLIAM, b Jenkintown, Pa, Sept 29, 11; m 32; c 2. AGRICULTURAL BIOCHEMISTRY. *Educ:* Lincoln Univ, AB, 31; Columbia Univ, AM, 33; Pa State Col, PhD(agr biochem), 46. *Prof Exp:* Instr chem, Lincoln Univ, 31; head dept sci, Univ Tex, 33-34; assoc prof chem, Prairie View State Col, 34-35; prof sci, St Philips Jr Col, 35-36; assoc physiologist, City Col New York, 40-42; assoc prof chem & actg head dept, Hampton Inst, 42-48; prof & head dept, Md State Col, 48-51; prof, SC State Col, 51-56 & Agr & Tech Col NC, 56-59; head dept natural sci, SC State Col, 59-63, dean, Sch Arts & Sci, 63-77; acad dean, Allen Univ, 77-79; mem staff, SC State Col, 80-85; RETIRED. *Mem:* Fel AAAS; Am Chem Soc; Am Oil Chem Soc; Nat Inst Sci; fel Am Inst Chemists. *Res:* Agricultural chemistry; fat analysis; chemical education. *Mailing Add:* 1628 Belleville Rd Orangeburg SC 29115

HUNTER, GEORGE WILLIAM, III, medical parasitology, tropical medicine; deceased, see previous edition for last biography

HUNTER, GORDON EUGENE, b Sharon, Pa, July 17, 30; m 55; c 3. BOTANY. *Educ:* Miss Col, BS, 56; Wash Univ, MA, 60, PhD(bot), 63. *Prof Exp:* Assoc prof biol, Murray State Col, 63-67; chmn dept, 67-74, PROF BIOL, TENN TECHNOL UNIV, 74- *Concurrent Pos:* NSF res partic col teachers, 64-66. *Mem:* Am Inst Biol Sci; Sigma Xi. *Res:* Morphology and taxonomy of Mexican and Central America Saurauia; taxonomy of the Dilleniaceae of Panama; chromosome numbers, biochemical systematics and trichome morphology of North American Vernonia; floristics and plant ecology of central Tennessee. *Mailing Add:* Dept Biol Tenn Technol Univ Box 5063 Cookville TN 38501

HUNTER, HUGH WYLIE, b Greeley, Colo, Apr 16, 11; m 37; c 2. PHYSICS. *Educ:* Ind Univ, AB, 32, PhD(physics), 37. *Prof Exp:* Instr physics, Ind Univ, 37-38, res assoc, Univ Conn, 38-39, asst prof, 39-41; res assoc, Ind Univ, 41-42; asst prof, Univ Conn, 42-43; physicist, Div War Res, Univ Calif, 42-45, Manhattan Dist Proj, Calif Inst Technol, 45-46 & Naval Ord Test Sta, Pasadena, 46-47; asst prof physics, Univ Colo, 47-48; physicist, Tech Staff, US Naval Ord Test Sta, 48-51, assoc head res dept, 51-56, head propulsion dept, 56-59; sr scientist, Res Triangle Inst, NC, 59-65; head res dept, US Naval Weapons Ctr, 65-76; RETIRED. *Mem:* Assoc Am Phys Soc. *Res:* Spectroscopy; nuclear physics; underwater sound transmission; development of small mechanical devices; cyclotron construction and operation; solid propellants and propulsion. *Mailing Add:* 18192 Sencillo Dr Rancho Bernardo San Diego CA 92128

HUNTER, JAMES BRUCE, b Kansas City, Mo, Dec 7, 15; m 43; c 3. CHEMISTRY. *Educ:* Univ Ill, BS, 37; Univ Pa, PhD(chem eng), 42. *Prof Exp:* Res chemist, Atlantic Refining Co, 41-58, res supvr, 58-59; res chemist, J Bishop Co Platinum Works, Matthey Bishop Inc, 59-71, vpres res & develop, 71-78, vpres new technol develop, 78-80; pres, Contour Drawing Inc, 80-; RETIRED. *Concurrent Pos:* Consult, 80- *Mem:* Am Chem Soc. *Res:* Mass spectrometry; dewaxing; catalytic reactions; asphalt and light oil research; vaporization from aqueous solutions; evaluating failure of bituminous materials; pH of anaerobic corrosion; isomerization; cracking catalysts; additives; synthetic lubricants; permeation; development and fabrication, chemicals, catalysts and devices of platinum group metals. *Mailing Add:* 313 Echo Valley Lane Newtown Square PA 19073-2398

HUNTER, JAMES CHARLES, b Salt Lake City, Utah, Nov 1, 46; m 73; c 3. ELECTROCHEMISTRY, SOLID STATE CHEMISTRY. *Educ:* Calif Inst Technol, BS, 68; Univ Calif, Santa Barbara, PhD(chem), 74. *Prof Exp:* Electrochemist, 74-76, sr electrochemist, 76-79, staff electrochemist, 79-83, technol assoc, Battery Prod Div, Union Carbide Corp, 83-88, SR TECHNOL ASSOC, EVEREADY BATTERY CO, 88- *Mem:* Electrochem Soc. *Res:* Electrode processes; structural chemistry and electrochemical activity of manganese dioxide; use of complex impedance techniques to study electrochemical processes. *Mailing Add:* Westlake Technol Lab Eveready Battery Co PO Box 45035 Westlake OH 44145

HUNTER, JAMES EDWARD, b Groveton, NH, Feb 2, 35; m 56; c 1. PLANT PATHOLOGY, EPIDEMIOLOGY. *Educ:* Univ NH, BA, 61, PhD(bot), 64. *Prof Exp:* Asst prof microbiol, Calif State Polytech Col, 64-65; assoc plant pathologist, Agr Exp Sta, Univ Hawaii, 65-72; assoc prof & head dept, 72-82, PROF PLANT PATH, NY STATE AGR EXP STA, CORNELL UNIV, 82-, DIR, 90- *Mem:* Am Phytopath Soc; Am Inst Biol Sci; Am Soc Hort Sci. *Res:* Epidemiology of fungal and bacterial diseases of vegetables; control of plant diseases by chemicals and disease resistance. *Mailing Add:* Dept Plant Path Agr Exp Sta Cornell Univ Geneva NY 14456

HUNTER, JAMES EDWARD, b Philadelphia, Pa, May 4, 45; m 68; c 2. FOOD SCIENCE & TECHNOLOGY. *Educ:* Lehigh Univ, BS, 67; Univ Wis, MS, 69, PhD(biochem), 74. *Prof Exp:* Lab technician dermat, US Army, 69-71; NUTRITIONIST, PROCTER & GAMBLE CO, 74- *Concurrent Pos:* mem tech comt, Inst Shortening & Edible Oils, 81-; mem, Am Heart Asn, Oral Health Comt, Int Life Sci Inst-Nutrit Found, 85-89; mem bd dirs, Greater Cincinnati Nutr Coun, 86-90; assoc ed, J Am Oil Chem Soc, 87-89; mem, Subcomt Fatty Acids & Health, Int Life Sci Inst, Nutrit Found, 89-; mem, Human Nutrit Bd Sci Counselors, USDA, 90-; mem bd dirs, Ohio Valley Sect, Am Oil Chem Soc, 90- *Mem:* Am Chem Soc; Sigma Xi; Am Inst Nutrit; Am Oil Chemists; Int Life Sci Inst. *Res:* Availability of dietary trans fatty acids, omega-3 fatty acids from vegetable oils; iron availability in rats; effects of phytate; iron status and tissue ferritin levels; protein turnover; stability and induction of pyridoxal phosphate-dependent enzymes in vitamin B6 deficiency; effects of food processing conditions on nutritional quality; dietary fat and health; dietary trends. *Mailing Add:* Procter & Gamble Co 6071 Center Hill Rd Cincinnati OH 45224-1703

HUNTER, JAMES H, b Washburn, Maine, June 7, 31; m 55; c 4. AGRICULTURAL ENGINEERING. *Educ:* Univ Maine, BSAE, 53, MSAE, 57; Univ Mass, Amherst, PhD(food & agr eng), 77. *Prof Exp:* Asst agr eng, 55-57, asst prof, Exp Sta, 57-63, ASSOC PROF AGR ENG, UNIV MAINE, ORONO, 63- *Mem:* Am Soc Agr Engrs; Potato Asn Am. *Res:* Design, development and study of equipment; facilities and methods of handling potatoes. *Mailing Add:* Box 765 Presque Isle ME 04769

HUNTER, JAMES HARDIN, JR, b Lafayette, Ind, Mar 24, 38; m 60, 80; c 2. ASTRONOMY, HYDRODYNAMICS. *Educ:* Kalamazoo Col, BA, 60; Univ Calif, Los Angeles, PhD(astron), 64. *Prof Exp:* Res staff astronr, Yale Univ, 64-65, asst prof astron, 65-70; from assoc prof to prof, Univ SFla, 70-79; PROF ASTRON, UNIV FLA, 79-, PROF GRAD SCH, 74- *Mem:* Am Astron Soc; Int Astron Union. *Res:* Magnetohydrodynamics; star and galaxy formation; observational astronomy. *Mailing Add:* Dept Astron Univ Fla Gainesville FL 32611

HUNTER, JEHU CALLIS, b Washington, DC, Mar 11, 22; m 46; c 3. CELL PHYSIOLOGY. *Educ:* Howard Univ, BS, 43. *Prof Exp:* Asst zool, Howard Univ, 47-48; med biol aide, Nat Cancer Inst, 49-51, biologist, 53-62, res biologist, 62-65, scientist adminr, Nat Inst Child Health & Human Develop, 65-69, asst dir planning, 69-75, chief, Off Planning & Anal, 75-76, asst dir prog develop, Ctr Res Mothers & Children, 76-78, HEAD, PUB HEALTH ASSOC, 78- *Mem:* AAAS; Am Soc Cell Biol; Soc Develop Biol; Royal Soc Med. *Res:* Tumor metabolism; photosynthesis; cytochemistry; reproductive biology. *Mailing Add:* 7822 16th St NW Washington DC 20012

HUNTER, JERRY DON, b Abilene, Tex, Oct 31, 35. CELL BIOLOGY, CELL PHYSIOLOGY. *Educ:* Hardin-Simmons, BA, 58; Tex A&M Univ, MS, 60, PhD(zool), 66. *Prof Exp:* Res asst radioisotopes, Univ Tex M D Anderson Hosp & Tumor Inst, 58-59; instr biol, Howard Col, 60-62; asst prof, Muhlenberg Col, 65-66; asst prof zool, 66-70, ASSOC PROF BIOL SCI, UNIV TEX, EL PASO, 70- *Mem:* Fel AAAS; Sigma Xi. *Res:* Effects of ionizing radiations on cells and tissues; applications of autoradiographic techniques in study of reproductive cell differentiation and effects of chemical inhibitors on cellular DNA synthesis. *Mailing Add:* Dept Biol Sci Univ Tex El Paso TX 79968

HUNTER, JOHN EARL, b Galesville, Wis, Dec 12, 29; m 52; c 3. BACTERIOLOGY. *Educ:* Univ Wis, BS, 51, MS, 53, PhD(bact), 59. *Prof Exp:* Asst bact, Univ Wis, 51-53, 56-59; mem res staff, 59-65, SECT HEAD FOOD PROD DEVELOP, PROCTER & GAMBLE CO, 65- *Mem:* Am Soc Microbiol. *Res:* Microbial interactions; oral and food microbiology; food safety, science and technology. *Mailing Add:* 7615 View Place Dr Cincinnati OH 45224

HUNTER, JOHN ROE, b Los Angeles, Calif, Apr 11, 34; m 56; c 2. MARINE BIOLOGY, ICHTHYOLOGY. *Educ:* Univ Calif, Santa Barbara, BA, 56; Univ Wis, MS, 58, PhD(zool), 62. *Prof Exp:* Res fishery biologist, 62-88, DIV CHIEF, SOUTHWEST FISHERIES CTR, NAT MARINE FISHERIES SERV, 88- *Concurrent Pos:* Adj prof, Scripps Inst Oceanog, Univ Calif, San Diego; scholar in residence, Bellagio Study & Conf Ctr, Rockefeller Found, Italy, 81; pres, Am Inst Fishery Res Biol, 87-88. *Honors & Awards:* Gold Medal, US Dept Com, 80. *Mem:* Am Fishery Soc; Am Inst Fishery Res Biol. *Res:* Behavior and physiology of fishes in relation to the ecology and management of marine fish populations. *Mailing Add:* Southwest Fisheries Ctr Nat Marine Fisheries Serv PO Box 271 La Jolla CA 92037

HUNTER, JOHN STUART, b Holyoke, Mass, June 3, 23; m 52; c 3. STATISTICS, ENGINEERING STATISTICS. *Educ:* NC State Col, BS, 47, MS, 49, PhD(exp statist), 54. *Prof Exp:* Staff statistician, Am Cyanamid Co, NY, 54-57; res assoc statist, Tech Res Group, Princeton Univ, 57-59; mem staff, Army Math Res Ctr, Univ Wis, 59-61; assoc prof chem eng, Princeton Univ, 61-66; statistician in residence, Univ Wis, 66-67; prof, 68-84, EMER PROF STATIST & CIVIL ENG, PRINCETON UNIV, 85- *Concurrent Pos:* Founding ed, Technometrics, Am Statist Asn & Am Soc Qual Control, 59-64; mem staff, Nat Acad Sci, 75-76; mem, Comn Nat Statist, 77-81; chmn adv panel, Appl Math Ctr, Nat Bur Standards, 77-81; lectr, Nat Ctr Indust Sci & Technol, Dalian Inst Technol, People's Rep China, 81. *Honors & Awards:* Shewhart Medal, 70; Deming Medal, 86. *Mem:* AAAS; Royal Statist Soc; Am Inst Chem Engrs; Am Statist Asn; Inst Math Statist; Biometric Soc. *Res:* Experimental design in the engineering sciences, time series, production quality maintenance and improvement; development and application of multifactor experimental strategies for product design and manufacturing processes. *Mailing Add:* 503 Lake Dr Princeton NJ 08540

HUNTER, JOSEPH LAWRENCE, b New York, NY, May 16, 13. PHYSICS. *Educ:* Manhattan Col, BS, 34; Cath Univ Am, MS, 39. *Prof Exp:* Instr, Manhattan Community Col, 39-40; asst prof, 40-51, PROF PHYSICS, JOHN CARROLL UNIV, 51- *Mem:* Am Phys Soc; Am Math Soc. *Res:* High frequencies; supersonic waves. *Mailing Add:* 3938 E 149 St Cleveland OH 44128

HUNTER, JOSEPH VINCENT, b Brooklyn, NY, June 12, 25; m 57; c 5. ENVIRONMENTAL SCIENCES. *Educ:* St John's Univ, NY, BS, 47, MS, 49; Rutgers Univ, PhD(environ sci), 62. *Prof Exp:* Res chemist, Nopco Chem Co, NJ, 51-55; instr, 55-59, lectr, 59-62, from asst prof to assoc prof, 62-70, prof environ sci, 71-78, dir, Grad Prog, 79-86, CHMN, DEPT ENVIRON SCI, RUTGERS UNIV, 82-, DISTINGUISHED PROF, 83- *Concurrent Pos:* Am Chem Soc lectr tour speaker, 70 & 71. *Mem:* Am Chem Soc; Water Pollution Control Fedn; Am Water Works Asn; Am Pub Health Asn; fel Am Inst Chem. *Res:* Stream oxygen dynamics; composition of waste waters and treatment plant effluents; source, distribution and fate of toxic and carcinogenic substances; industrial waste water treatment. *Mailing Add:* 13 Patton Dr New Brunswick NJ 08816

HUNTER, KATHERINE MORTON, b Birmingham, Ala, Sept 16, 39; m 67; c 2. MICROBIOLOGY, MEDICAL TECHNOLOGY. *Educ:* Ala Col, BS, 59; Emory Univ, MT, 60; Vanderbilt Univ, MA, 62, PhD(biol), 68. *Prof Exp:* Asst microbiologist & teaching supvr, 67-71, educ coordr, 71-77, MICROBIOLOGIST, SCH MED TECHNOL, BAPTIST MED CTR, 71-, SPECIALIST MICROBIOLOGIST, 72-, PROG DIR, 77- *Concurrent Pos:* Dipl, Am Bd Med Microbiol, 77-; clin asst prof path, Univ Ala Med/Dent Sch, Birmingham, 77- *Mem:* Am Soc Microbiol; Am Soc Clin Path; Am Pub Health Asn. *Res:* Microbial nutrition and antibiotics; clinical microbiology; microbiology of hospital environment. *Mailing Add:* Lab Baptist Med Ctr 800 Montclair Rd Birmingham AL 35213

HUNTER, KENNETH W, JR, EXPERIMENTAL BIOLOGY. *Educ:* Ariz State Univ, BA, 72, MS, 73; Johns Hopkins Univ, ScD(immunol-parasitol), 77. *Prof Exp:* Res collab, Div Entom, Western Cotton Res Lab, Agr Res Serv, US Dept Agr, 73; John W Graham Fund Immunol Res grant, Johns Hopkins Univ, 75; vis res assoc, Malaria Res Proj, Agency Int Develop, Univ NMex, 76; postdoctoral fel, Infectious Dis Div, Uniformed Serv Univ Health Sci, F Edward Hebert Sch Med, 78-79, res asst prof, 79-82, assoc prof pediat & prev med/biometrics & dir pediat res, 82-86, adj assoc prof, 86-89; PROF BIOL & MICROBIOL, ASSOC VPRES RES & DEAN GRAD SCH, UNIV NEV, RENO, 89- *Concurrent Pos:* Founder & exec vpres, Antech Consult, Inc, 82-89; chief scientist, Westinghouse Bio-Anal Systs Co, 85-89; pres & chief exec officer, Biotronic Systs Corp, 86-89. *Mem:* AAAS; Am Asn Immunologists; Am Soc Trop Med & Hyg; Am Soc Clin Path; Asn Off Anal Chemists. *Res:* Gene regulation in leishmania; immunoregulatory functions in malaria; development of monoclonal antibodies for chemical haptens; somatic cell genetics-human hybridomas; biosensor and molecular electronics. *Mailing Add:* Univ Nev 239 Getchell Libr Reno NV 89557-0035

HUNTER, LARRY CLIFTON, mathematics, for more information see previous edition

HUNTER, LARRY RUSSEL, b Harvey, Ill, Apr 23, 53; m 82; c 4. PRECISION MEASUREMENTS, FUNDAMENTAL SYMMETRIES. *Educ:* Columbia Univ, BA, 74; Univ Calif, Berkeley, MA, 78, PhD(physics), 81. *Prof Exp:* Asst, Hertz Lab Spectros, 81-82; ASST PROF PHYSICS, AMHERST COL, 83- *Concurrent Pos:* Sloan Found fel; res award, Am Phys Soc. *Mem:* Am Phys Soc. *Res:* Atomic physics and fundamental symmetries; parity violation; time reversal violation. *Mailing Add:* Dept Physics Amherst Col Amherst MA 01002

HUNTER, LAWRENCE WILBERT, b London, Ont, July 15, 45; m 71; c 2. MATERIALS SCIENCE ENGINEERING. *Educ:* Carleton Univ, BSc, 67; Univ Wis-Madison, PhD(chem), 72. *Prof Exp:* Fel chem physics, Univ BC, 72-73; fel, 73-74, PRIN STAFF CHEMIST, APPL PHYSICS LAB, JOHNS HOPKINS UNIV, 74- *Mem:* Sigma Xi; Am Ceramic Soc; Combustion Inst. *Res:* Theory of moleclar collisions; kinetic theory of gases, polymer flammability and combustion; air-breating propulsion; advanced structural materials. *Mailing Add:* Appl Physics Lab Johns Hopkins Univ Johns Hopkins Rd Laurel MD 20810

HUNTER, LLOYD PHILIP, b Wooster, Ohio, Feb 11, 16; m 43; c 5. PHYSICS. *Educ:* Wooster Col, BA, 39; Mass Inst Technol, BS, 39; Carnegie Inst Technol, MS, 40, DSc(physics), 42. *Prof Exp:* Asst physics, Carnegie Inst Technol, 39-40; res physicist, Westinghouse Elec Corp, 42-49, mgr solid state electronics, 49-51; sr physicist, Int Bus Mach Corp, 51-63, mgr res dept, 53-57, resident mgr, Poughkeepsie Res Lab, 58-60; dir component res & develop, IBM Corp, 61-62; prof, 63-81, EMER PROF ELEC ENG, UNIV ROCHESTER, 81- *Concurrent Pos:* Res physicist, Radiation Lab, Univ Calif, 44-45 & Oak Ridge Nat Lab, 46-48; consult, Int Bus Mach Corp, 63- *Mem:* Fel Am Phys Soc; Fel Inst Elec & Electronics Engrs. *Res:* Development of magnetrons; photo and secondary electron emission; elasticity and physics of solids; development of calutron; nuclear fuel research; physics of semiconductors and transistors; diagnostic ultrasound; contact electrification. *Mailing Add:* Ten Schoolhouse Lane Rochester NY 14618

HUNTER, NORMAN ROBERT, b Odessa, Ont, Nov 9, 41; US citizen; m 64; c 2. CHEMISTRY CURRICULUM DEVELOPMENT. *Educ:* Carleton Univ, Ottawa, Ont, BSc, 65, MSc, 68; Univ NB, PhD(org chem), 70. *Prof Exp:* Postdoctoral fel, Dept Chem, Stanford Univ, 70-71 & Nat Res Coun Can, 71-72; sr med chemist, Merck-Frosst Labs, Montreal, Que, 72-74; from asst prof to assoc prof, 74-87, PROF ORG SYNTHESIS, DEPT CHEM, UNIV MAN, 87-, ASSOC HEAD, 85- *Res:* Development of synthetic methodology; selective oxidation processes; interconversion of fuels; oxidative processes in fatty acids; catalyst development; improvement of science teaching. *Mailing Add:* Dept Chem Univ Man Winnipeg MB R3T 2N2

HUNTER, NORMAN W, b Toledo, Ohio, June 10, 31; m 53; c 3. CHEMISTRY. *Educ:* Univ Toledo, BSE, 53, MSEd, 61, EdD, 68. *Prof Exp:* Teacher, Mason Consol Schs, Erie, Mich, 53-56 & Washington Local Schs, Toledo, Ohio, 56-65; from asst prof to assoc prof, 68-82, PROF CHEM, WESTERN KY UNIV, 82- *Concurrent Pos:* Instr, Univ Toledo, 59-67. *Mem:* Am Chem Soc. *Res:* History of science; computer-assisted instruction. *Mailing Add:* Dept Chem Western Ky Univ Bowling Green KY 42101

HUNTER, ORVILLE, JR, b Wellsville, Mo, Aug 20, 38; m 60; c 3. REFRACTORIES. *Educ:* Univ Mo, Rolla, BS, 60; Alfred Univ, PhD(ceramics), 64. *Prof Exp:* Res assoc ceramics, Alfred Univ, 64; res asst, US Army Mat Res Agency, Watertown Arsenal, Mass, 64-66; from asst prof to prof ceramic eng, Iowa State Univ, 66-84; VPRES RES, A P GREEN INDUST, INC, MEXICO, MO, 84- *Concurrent Pos:* Prof ceramic eng, Univ Mo, Rolla, 87. *Mem:* Am Ceramic Soc; Nat Inst Ceramic Engrs; Sigma Xi; Am Soc Testing & Mat. *Res:* High temperature mechanical and thermal properties; oxides; refractories. *Mailing Add:* 816 Bucks Run Columbia MO 65201-6109

HUNTER, OSCAR BENWOOD, JR, b Washington, DC, Oct 27, 15; m 41; c 7. MEDICINE. *Educ:* Cath Univ Am, BS, 36; Georgetown Univ, MD, 40. *Prof Exp:* Intern, Gallinger Munic Hosp, 40-41; fel, Mayo Clin, 41-42, asst, 42; asst chief lab serv, Borden Gen Hosp, Chickasha, Okla, US Army, 43, chief, 55th Gen Hosp, 44-45, assoc pathologist, Armed Forces Inst Path, 46; assoc pathologist, Doctor's Hosp, 47-51, dir dept path, 51-74, pres, Hosp, 65-73; CLIN PROF PATH, GEORGETOWN UNIV, 71- *Concurrent Pos:* Instr, Georgetown Univ, 46-59, lectr, 59-71; adj proj, Am Univ, 51-74; assoc pathologist, Sibley Mem Hosp, 51-; pres, Oscar B Hunter Mem Found Med Res & Educ. *Honors & Awards:* John Elliott Award, Am Asn Blood Banks, 70. *Mem:* Am Soc Clin Pharmacol & Therapeut (secy, Am Therapeut Soc, 52, pres-elect, Am Col Clin Pharmacol & Ther, 65-66); Am Soc Clin Path; Am Soc Hemat; Col Am Path (vpres, 65-67, pres, 67-69); Int Soc Hemat. *Res:* Clinical pathology and pathologic anatomy; hematology; neoplastic diseases; radioactive isotopes. *Mailing Add:* Dept Path Georgetown Univ 8218 Wisconsin Ave No 202 Bethesda MD 20814

HUNTER, PRESTON EUGENE, b Tonganoxie, Kans, Nov 4, 27; m 51; c 2. ENTOMOLOGY, ACAROLOGY. *Educ:* Univ Kans, AB, 51, PhD(entom), 56. *Prof Exp:* Res fel entom, Univ Minn, 56-59; from asst prof to assoc prof, 59-69, PROF ENTOM, UNIV GA, 69-, HEAD DEPT, 74- *Mem:* AAAS; Acarology Soc Am; Entom Soc Am; Sigma Xi. *Res:* Taxonomy and ecology. *Mailing Add:* Dept Entom Univ Ga Athens GA 30602

HUNTER, RALPH EUGENE, b Evansville, Ind, Jan 13, 35. MARINE & COASTAL GEOLOGY. *Educ:* Ind Univ, AB, 57; Johns Hopkins Univ, Phd(geol), 60. *Prof Exp:* Geologist, Bellaire Res Labs, Texaco, Inc, 60-63; asst geologist, Ill State Geol Surv, 64-67; GEOLOGIST, US GEOL SURV, 67- *Mem:* AAAS; Soc Econ Paleont & Mineral; Geol Soc Am; Int Asn Sedimentologists. *Res:* Sedimentology, especially sedimentary structures and petrography of sands and gravels; petrology of iron-rich rocks. *Mailing Add:* Off Marine Geol US Geol Surv 345 Middlefield Rd Menlo Park CA 94025

HUNTER, RAYMOND EUGENE, b Moultrie, Ga, Sept 4, 35; m 57; c 2. NUCLEONICS. *Educ:* Univ Ga, BS, 57, MS, 58; Fla State Univ, PhD(physics), 64. *Prof Exp:* Staff mem, Los Alamos Sci Lab, 65-66; prof physics & head dept, Valdosta State Col, 66-72; ASSOC DIV LEADER, LOS ALAMOS NAT LAB, 72 - *Mem:* Am Nuclear Soc. *Res:* Physics. *Mailing Add:* X-Div Los Alamos Nat Lab Los Alamos NM 87545

HUNTER, RICHARD EDMUND, b Jersey City, NJ, Jan 26, 23; m 46; c 3. PHYTOPATHOLOGY. *Educ:* Rutgers Univ, BS, 49; Okla State Univ, MS, 51, PhD, 68. *Prof Exp:* Asst plant pathologist, NMex State Univ, 51-55; agt, 58-65, res scientist, USDA, Stillwater, Okla, 65-72, Col Sta, Tex, 72-75, Byron, Ga, 75-79, location & res leader, W R Poage Pecan Field Sta, Brownwood, Tex, Agr Res Serv, 79-87; RETIRED. *Concurrent Pos:* From instr to assoc prof plant path, Okla State Univ, 58-72; regional ed, Pecan Quart, 77-79; mem Crop Adv Comt for Pecan & Hickory Germplasm, 84-87. *Res:* Etiology, epiphytology and control of pecan diseases; breeding disease resistance in pecans. *Mailing Add:* 3903 Glenwood Dr Brownwood TX 76801

HUNTER, ROBERT DOUGLAS, b Lafayette, Ind, July 13, 44; m. PHYSIOLOGICAL ECOLOGY. *Educ:* Marietta Col, AB, 66; Syracuse Univ, PhD(biol), 72. *Prof Exp:* Res asst neurol, Cornell Med Ctr, Bellevue Hosp, New York, 66; res asst physiol ecol, Syracuse Univ & Marine Biol Lab, 68-72; asst prof biol, 72-78, ASSOC PROF BIOL, OAKLAND UNIV, 79- *Concurrent Pos:* Hon res fel, Univ Glasgow, Scotland, 80; mem, Marine Biol Lab, Woods Hole; vis assoc prof biol, Simmons Col, 87-88. *Mem:* NAm Benthol Soc; Malacol Soc London; Int Asn Theoret & Appl Limnol; Ecol Soc Am; AAAS. *Res:* Lake acidification and molluscs: effects on development, juvenile survival, and shell erosion; ecology and control of the zebra mussel; effects of grazing by freshwater and marine invertebrates on attached microalgae. *Mailing Add:* Dept Biol Sci Oakland Univ Rochester MI 48309-4401

HUNTER, ROBERT L, b Chicago, Ill, Jan 27, 39; m 65; c 2. PATHOLOGY, IMMUNOLOGY. *Educ:* Harvard Univ, AB, 61; Univ Chicago, MS & MD, 65, PhD(path), 69. *Prof Exp:* Staff assoc immunol, Nat Inst Allergy & Infectious Dis, 67-69; from instr to asst prof path, Univ Chicago, 70-76, assoc prof path, 76-; PROF, DEPT PATH & LAB MED, EMORY UNIV HOSP, ATLANTA, GA. *Concurrent Pos:* Seymour Coman fel, Univ Chicago, 69-71; Am Cancer Soc fel, 70-72; Schweppe Found fel, 71-74. *Honors & Awards:* Third Award, SAMA-Mead Johnson Sci Forum, 65; Sheard Sanford Award, Am Asn Clin Pathologists, 65; Joseph A Capps Prize, Inst Med Chicago, 69; Hektoen Award, Chicago Path Soc, 70. *Mem:* AAAS; Reticuloendothelial Soc. *Res:* Elucidation of the factors which control the type of antibody or cellular immune response to a given antigenic stimulus, especially the localization of antigen in various types of phagocytic cells. *Mailing Add:* Dept Path & Lab Med Emory Univ Hosp 1364 Clifton Rd Atlanta GA 30322

HUNTER, ROBERT L, b Delaware, Ohio, June 29, 21; m 44; c 4. ANATOMY. *Educ:* Ohio Wesleyan Univ, BA, 43, MA, 49; Univ Mich, PhD(zool), 54. *Prof Exp:* Asst path, Sch Med, Washington Univ, 47-48; asst oncol, Sch Med, Univ Kans, 48-49; from instr to assoc prof anat, Med Sch, Univ Mich, 53-63; assoc prof, Sch Med, Stanford Univ, 63-66; chmn dept human anat, Sch Med, Univ Calif, Davis, 68-74, prof anat, 66-; RETIRED. *Concurrent Pos:* Vis res prof, Univ Recife, 61; USPHS career develop award,

61-63; vis prof anat, Univ Helsinki, 71. *Mem:* AAAS; Histochem Soc (pres, 68-69); Am Asn Anatomists; Sigma Xi. *Res:* Histology; cytology; enzymology; histochemistry; starch-gel and acrylamide-gel electrophoresis. *Mailing Add:* 809 Plum Lane Davis CA 95616

HUNTER, ROBERT P, b Newark, NJ, Jan 30, 33; m 56; c 4. MATHEMATICS. *Educ:* Univ Miami, BS, 54, MS, 56; La State Univ, PhD(math), 58. *Prof Exp:* Asst prof math, Univ Ga, 58-59 & 60-62; Sarah Moss Found fel, Oxford Univ, 59-60; assoc prof, 62-65, PROF MATH, PA STATE UNIV, 65- *Concurrent Pos:* Vis prof, Univ London, 68-69. *Res:* Algebra; topology. *Mailing Add:* Dept Math Pa State Univ University Park PA 16802

HUNTER, ROY, JR, b Birmingham, Ala, Jan 7, 30; m 65; c 1. DEVELOPMENTAL BIOLOGY. *Educ:* Morehouse Col, BS, 50; Univ Atlanta, MS, 53; Brown Univ, PhD(develop biol), 62. *Prof Exp:* Instr sci, Birmingham Baptist Col, 50-51; instr biol, Morgan State Col, 53-57; from asst prof to assoc prof, Morehouse Col, 61-64; assoc prof biol, Atlanta Univ, 64-68, prof, 64-70 & 73-81, chmn dept, 76-79; DIR, FAC DEVELOP FELS PROG, MOREHOUSE SCH MED, 83- *Concurrent Pos:* Univ Ctr Corp res grant, Atlanta Univ, 64-65; prof & chmn biol dept, Morgan State Col, 70-73. *Mem:* Am Inst Biol Sci; Am Soc Zoologists; Nat Inst Sci; Sigma Xi; NY Acad Sci. *Res:* Notochordal cells of amphibian embryos and larvae, including their involvement in fibrillogenesis; environmental teratogenesis. *Mailing Add:* Morehouse Sch Med 720 Westview Dr SW PO Box Eight Atlanta GA 30310

HUNTER, SAMUEL W, b Ireland, Nov 13, 21; nat US; m 44; c 6. THORACIC SURGERY. *Educ:* Cornell Univ, AB, 43; Univ Rochester, MD, 47; Univ Minn, MS, 56; Am Bd Surg, dipl, 55; Am Bd Thoracic Surg, dipl, 59. *Prof Exp:* Instr, 55-57, asst prof, 57-77, CLIN ASSOC PROF SURG, MED SCH, UNIV MINN, ST PAUL, 77- *Concurrent Pos:* Dir cardiac res lab, St Joseph's Hosp, 57-; mem comt hyperbaric oxygenation, Nat Acad Sci-Nat Res Coun. *Mem:* AAAS; fel Am Col Angiol; fel Am Col Surg; fel Am Col Chest Physicians. *Res:* Cardiac research, especially the atrioventricular block; use of the bipolar myocardial electrode; hyperbaric oxygenation; clinical application cardiac pacemakers; cardiac surgery of all phases, clinical and research. *Mailing Add:* 1175 Orchard Pl St Paul MN 55118

HUNTER, STANLEY DEAN, b Phoenix, Ariz, Feb 4, 54; m 83; c 2. ASTROPHYSICS. *Educ:* Univ Ariz BS(physics) & BS(math), 76; La State Univ, PhD(physics), 81. *Prof Exp:* Res assoc gamma-ray astrophys, Nat Res Coun, Goddard Space Flight Ctr, NASA, 81-83; res assoc x-ray astrophys, Inst Space & Astronaut Sci, Tokyo, 84; ASTROPHYSICIST GAMMA-RAY ASTROPHYS, NASA-GODDARD SPACE FLIGHT CTR, 85- *Mem:* Am Phys Soc. *Res:* High energy (30 mega electron volts to 100 giga electron volts) gamma rays produced by astrophysical processes; probe of the matter distribution in the galaxy; generation by energetic point sources; design and construction of satellite and balloon experiments for imaging detection of gamma rays. *Mailing Add:* Goddard Space Flight Ctr NASA Code 662 Greenbelt MD 20771

HUNTER, SUSAN JULIA, OSTEOCLAST STRUCTURE FUNCTION, BONE PROTEOGLYCANS. *Educ:* Pa State Univ, PhD(physiol), 80. *Prof Exp:* AT DEPT MOLECULAR BIOL, PA STATE UNIV. *Mailing Add:* Eight Cromwell Drive Orono ME 04473

HUNTER, THOMAS HARRISON, b Chicago, Ill, Oct 12, 13; m 43; c 5. INTERNAL MEDICINE. *Educ:* Harvard Univ, AB, 35, MD, 40. *Prof Exp:* Asst physician med, Presby Hosp, 41-42; instr, Col Physicians & Surgeons, Columbia Univ, 42-47; from asst to assoc dean, Sch Med, Wash Univ, 47-53; dean, Sch Med, Univ Va, 53-65, chancellor med affairs, 65-70, prof med & sci, 70-81; RETIRED. *Concurrent Pos:* Mem, Comt Infectious Dis & Chemother, Nat Res Coun; mem, Nat Arthritis & Metab Dis Coun, NIH. *Mem:* Am Acad Arts & Sci; AAAS; Am Soc Clin Invest; Asn Am Med Cols (pres, 60). *Res:* Infectious diseases; antibiotics and bacterial endocarditis; human ecology. *Mailing Add:* PO Box 212 Keswick VA 22947

HUNTER, TONY, b Ashford, Eng, Aug 23, 43; UK citizen. MOLECULAR BIOLOGY, VIROLOGY. *Educ:* Cambridge Univ, Eng, BA, 65, MA, 66, PhD(biochem), 69. *Prof Exp:* Res fel, Christ's Col, Cambridge, 68-71; res assoc, Salk Inst, San Diego, 71-73; res fel, Christ's Col, 73-75; from asst prof to assoc prof, 75-82, PROF, SALK INST, SAN DIEGO, 82- *Concurrent Pos:* Adj assoc prof biol, Univ Calif, San Diego, 79-83, adj prof, 83- *Honors & Awards:* Am Bus Found for Cancer Res Award, 88. *Mem:* Fel, Royal Soc London. *Res:* Molecular mechanisms of malignant cellular transformation by viral and tumor oncogenes; properties of oncogenic proteins; mechanisms of growth control of mammalian cells; regulation of protein function by phosphorylation; cancer. *Mailing Add:* Salk Inst PO Box 85800 San Diego CA 92186-5800

HUNTER, WANDA SANBORN, zoology, for more information see previous edition

HUNTER, WILLIAM G(ORDON), statistics, engineering; deceased, see previous edition for last biography

HUNTER, WILLIAM LESLIE, b East Liverpool, Ohio, May 2, 28; m 51; c 3. PLASTICS CHEMISTRY, SEMICONDUCTORS. *Educ:* Va Polytech Inst, BS, 50, MS, 57, PhD(phys chem), 60. *Prof Exp:* Res chemist, Basic Res Dept, Chemstrand Res Ctr, 59-62; asst prof phys & polymer chem, Va Polytech Inst, 62-68; mgr org mat sect, mat res lab, 68-75, MEM TECH STAFF, SEMICONDUCTOR PROD SECTOR, MOTOROLA, INC, 75- *Mem:* Soc Plastics Eng; Am Chem Soc; Sigma Xi. *Res:* Plastic encapsulants for semiconductors. *Mailing Add:* 8625 E Orange Blossom Lane Scottsdale AZ 85253

HUNTER, WILLIAM RAY, b Ancon, CZ, Oct 22, 24. PHYSICAL OPTICS. *Educ:* Univ Fla, BS, 48, MS, 49. *Prof Exp:* Physicist, Col Geophys Inst, Alaska, 49-50 & US Naval Mine Countermeasures Sta, Panama City, Fla, 50-52; physicist, US Naval Res Lab, 52-80; PHYSICIST, SACHS-FREEMAN ASSOC INC, 82- *Concurrent Pos:* Assoc ed, J Optical Soc Am, 74-77. *Honors & Awards:* Progress Medal, Photog Soc Am, 58. *Mem:* Fel Optical Soc Am; fel Am Phys Soc. *Res:* Optical properties in the vacuum ultraviolet; thin films; diffraction gratings photo emissive properties; vacuum ultraviolet detectors; particle radiation damage. *Mailing Add:* 6705 Caneel Ct Springfield VA 22152

HUNTER, WILLIAM SAM, b Amarillo, Tex, Sept 28, 40; m 72; c 2. PHYSIOLOGY. *Educ:* Univ Okla, BS, 62, MS, 65; Mich State Univ, PhD(physiol), 71. *Prof Exp:* Res asst, Civil Aeromed Inst, 64-65; res engr physiol, NAm Rockwell Corp, 65-67; consult, Ford Motor Co Safety Div, 67-68; asst prof physiol, Sch Med, St Louis Univ, 71-75; ASSOC PROF DEPT PHYSIOL, SCH MED, SOUTHERN ILL UNIV, CARBONDALE, 75- *Mem:* AAAS; Am Physiol Soc. *Res:* Thermoregulatory Physiology. *Mailing Add:* Dept Physiol Southern Ill Univ Med Sch Carbondale IL 62901-6512

HUNTER, WILLIAM STUART, b Orillia, Ont, May 5, 27; m 63; c 2. ORTHODONTICS, ANTHROPOLOGY. *Educ:* Univ Toronto, DDS, 50; Univ Mich, MS, 55, PhD(anthrop, orthod), 59. *Prof Exp:* From asst prof to assoc prof dent, Univ Mich, Ann Arbor, 61-68; PROF DENT & CHMN DIV ORTHOD, UNIV WESTERN ONT, 68- *Concurrent Pos:* Consult, Ctr Human Growth, Univ Mich, 68- *Mem:* Am Asn Orthod. *Res:* Genetics; growth; twins; clefting lip/palate. *Mailing Add:* Fac Dent Univ Western Ont London ON N6A 5C1 Can

HUNTER, WILLIAM WINSLOW, JR, b South Norfolk, Va, Mar 26, 30; m 51; c 3. INSTRUMENTATION. *Educ:* Univ Va, BEE, 58; Col William & Mary, MA, 66; Va Polytech Inst & State Univ, PhD(physics), 72. *Prof Exp:* Eng physicist, 63-69, SECT HEAD MEASUREMENT TECH & INSTRUMENTATION RES & DEVELOP, GAS PARAMETERS MEASUREMENT SECT, INSTRUMENT RES DIV, LANGLEY RES CTR, NASA, 69- *Mem:* Am Phys Soc. *Res:* Advanced measurement techniques and related instrumentation systems for wind tunnel flow field diagnostics. *Mailing Add:* Dept Radiol Ohio State Univ Columbus OH 43210

HUNTER, WOOD E, b Louisville, Ky, Sept 27, 40; m 59; c 6. ORGANIC CHEMISTRY. *Educ:* Bellarmine Col, BA, 62; Univ Notre Dame, PhD(org chem), 66. *Prof Exp:* Chemist, Nalco Chem Co, 66-67; res chemist, Celanese Coatings Co, 67-68; sr chemist, Nalco Chem Co, 68-77; res chemist, Buckman Labs, 77-78; SR RES FEL, CALGON CORP, 78- *Mem:* Am Chem Soc. *Res:* Emulsion polymerization; polyelectrolytes; organo-sulfur chemistry. *Mailing Add:* 637 Beverly Rd Mt Lebanon PA 15228-2098

HUNTING, ALFRED CURTIS, b Plainfield, NJ, Apr 30, 28; m 53; c 4. APPLIED PHYSICS. *Educ:* Swarthmore Col, BS, 49; Univ Mich, MS, 52, PhD(physics), 62. *Prof Exp:* Jr physicist, Brookhaven Nat Lab, 49-50; res asst, Eng Res Inst, Univ Mich, 50-54; res technologist, Field Res Lab, Mobil Oil Co, 54-57; sr res assoc, Res Inst, Univ Mich, 57-61; res specialist, Aerospace Div, Boeing Co, 61-69; sr engr, Equip Group, Tex Instruments Inc, 69-76; PRIN DEVELOP ENGR, HONEYWELL SEATTLE MARINE SYSTS CTR, 76- *Mem:* Am Phys Soc; Inst Elec & Electronics Engrs. *Res:* Modeling, analysis and simulation of engineering systems, including sonar systems and fiber-optic sensors; underwater acoustics; sonar signal processing. *Mailing Add:* 2357 Federal Ave E Seattle WA 98102

HUNTINGTON, CHARLES ELLSWORTH, b Boston, Mass, Dec 8, 19; m 56; c 4. ORNITHOLOGY. *Educ:* Yale Univ, BA, 42, PhD(zool), 52. *Prof Exp:* From instr to prof biol, Bowdoin Col, 53-86, dir sci sta, 53-89, chmn dept, 73-76; RETIRED. *Concurrent Pos:* NSF res grant, 58; Guggenheim mem fel, Oxford Univ, 63-64. *Mem:* AAAS; Soc Syst Zool; Am Ornith Union; Brit Ornith Union. *Res:* Population dynamics of sea birds. *Mailing Add:* RFD 2 Box 357 South Harpswell ME 04079

HUNTINGTON, DAVID HANS, b Westford, NY, Mar 19, 26; m 49; c 2. TECHNICAL COLLEGE ADMINISTRATION. *Educ:* Cornell Univ, BS, 46, MS, 48, PhD, 53. *Prof Exp:* Teacher pub sch, NY, 48-51; asst agr eng, Cornell Univ, 51-53; from asst prof to assoc prof, Univ Maine, 53-61, asst to dean col agr, 57-61, from asst dean to assoc dean, 61-64; pres, State Univ NY Agr & Tech Col Alfred, 64-86; RETIRED. *Mailing Add:* 5470 Jerico Hill Rd Alfred Station NY 14803

HUNTINGTON, HILLARD BELL, b Wilkes-Barre, Pa, Dec 21, 10; m 39; c 3. SOLID STATE PHYSICS. *Educ:* Princeton Univ, AB, 32, MA, 33, PhD(physics), 41. *Hon Degrees:* Dr hon causa, Univ Nancy I, France, 77. *Prof Exp:* Res assoc physics, Univ Pa, 39-41; instr, Wash Univ, 41-42; mem staff, Radiation Lab, Mass Inst Technol, 42-46; from asst prof to prof, 46-76, assoc head dept, 50-61, chmn dept, 61-69, EMER PROF PHYSICS & RES PROF, RENSSELAER POLYTECH INST, 76- *Concurrent Pos:* Liaison officer, US Off Naval Res, London, Eng, 54-55; vis prof, Yale Univ, 60-61 & Cornell Univ, 68-69; mem eval comt, Metall Div, Nat Bur Stand, 72-78; Fishback Fac Travel Grant, 88. *Mem:* Fel Am Phys Soc; Sigma Xi. *Res:* Theory of solids; megacycle ultrasonics; diffusion in metals; elastic constants; transport properties; electromigration; thermomigration; theoretical surface physics. *Mailing Add:* Dept Physics Rensselaer Polytech Inst Troy NY 12181

HUNTINGTON, ROBERT (WATKINSON), JR, pathology; deceased, see previous edition for last biography

HUNTLEY, DAVID, b San Mateo, Calif, Feb 28, 50; m 71. HYDROGEOLOGY, REMOTE SENSING. *Educ:* Univ Calif, Santa Barbara, BA, 72; Colo Sch Mines, PhD(geol eng), 76. *Prof Exp:* Res asst geol, Colo Sch Mines, 72-76; asst prof, Univ Conn, 76-78; asst prof, 78-80, ASSOC PROF GEOL, SAN DIEGO STATE UNIV, 81- *Mem:* Geol Soc Am; Am

Geophys Union; Nat Water Well Asn; Am Soc Photogram; Asn Eng Geologists. *Res:* Geologic and geophysical techniques in ground water hydrology; numerical modeling of ground water flow; remote sensing applications to hydrogeology; hydrology of geothermal systems; induced seismicity. *Mailing Add:* Dept Geol Sci 5300 Campanile Dr San Diego State Univ San Diego CA 92182-0763

HUNTLEY, JIMMY CHARLES, b Louisville, Miss, Nov 29, 46; m 68; c 2. FISH & WILDLIFE SCIENCE. *Educ:* Miss State Univ, BS, 69, MS, 77, PhD(biol sci), 80. *Prof Exp:* res wildlife biologist, Southern Forest Exp Sta, US Forest Serv, 78-83; WILDLIFE BIOLOGIST, SOUTHERN REGION, NAT FOREST SYST, 83- *Mem:* Wildlife Soc; Sigma Xi. *Res:* Effects of forestry management practices on wildlife habitat; wildlife habitat evaluation and classification; management of utility rights-of-way to reduce maintenance costs and improve wildlife habitat. *Mailing Add:* 1765 Highland Ave Montgomery AL 36107

HUNTLEY, ROBERT ROSS, b Wadesboro, NC, Sept 6, 26; m 76; c 5. PREVENTIVE MEDICINE, FAMILY MEDICINE HEALTH ADMINISTRATION. *Educ:* Davidson Col, BS, 47; Bowman Gray Sch Med, MD, 51. *Prof Exp:* From intern to resident, Univ Hosp, Univ Mich, 51-53; pvt pract, 53-58; resident & fel med, NC Mem Hosp, Univ NC, 59-62; assoc dir prog develop, Nat Ctr Health Serv Res & Develop, 68-70; prof & chmn dept, 70-89, EMER PROF COMMUNITY & FAMILY MED, SCH MED, GEORGETOWN UNIV, 89- *Concurrent Pos:* Exec co-dir health adv comt, Appalachian Regional Comn, 65-66; from instr to assoc prof med & prev med, Sch Med, Univ NC, 59-70, asst prof health admin, Sch Pub Health, 67-70; Milbank fac fel, 64-70; mem health serv res study sect, NIH, 66-68; mem test comt prev med & pub health, Nat Bd Med Examr, 70-74; trustee, Am Bd Prev Med, 74-78; chmn health care technol study sect, Nat Ctr Health Serv Res, 78-82; adj prof, Health Admin, Univ NC, Chapel Hill, 89- *Mem:* Asn Teachers Prev Med (pres, 74-75); Am Fedn Clin Res; fel Am Pub Health Asn; Asn Dept Family Med; Am Geriatrics Soc; Soc Teachers Family Med. *Res:* Organization and delivery of health services, quality, distribution of resources; clinical epidemiology and studies useful for planning the rational allocation of health resources. *Mailing Add:* PO Box 190 Chapel Hill NC 27514

HUNTOON, JACQUELINE E, b Encino, Calif, May 23, 59. FIELD GEOLOGY, NUMERICAL MODELLING GEOLOGIC PROCESSES. *Educ:* Univ Calif, Santa Cruz, BS, 81; Univ Utah, Salt Lake City, MS, 85. *Prof Exp:* Geol field asst, US Geol Serv, Menlo Park, Calif, 82-; exploration geol, Shell Offshore Inc, New Orleans, 87; consult geologist, Penn State Univ, 87-88; CONSULT, 88- *Mem:* Am Assoc Petrol Geologists; Geol Soc Am; Am Geophys Union. *Res:* Evolution of sedimentary basin-current research focuses on the interaction of thermal mechanical tectonic processes with sedimentologic processes that are controlled by climate, topography, pre-existing lithology and paleogeography. *Mailing Add:* Dept Geol Mich Tech Univ Houghton MI 96720

HUNTRESS, WESLEY THEODORE, JR, b Washington, DC, Apr 11, 42; m 73; c 1. ASTROCHEMISTRY. *Educ:* Brown Univ, BS, 64; Stanford Univ, PhD(chem physics), 68. *Prof Exp:* SR RES SCIENTIST, JET PROPULSION LAB, CALIF INST TECHNOL, 69- *Mem:* AAAS; Am Chem Soc; Am Astron Soc; Am Phys Soc. *Res:* Ion cyclotron magnetic resonance; mass spectrometry; ion-molecule reactions in the gas phase; photo-ionization, photoexcitation and photodissociation phenomena in the gasphase; the production and reactions of molecular species in planetary atmospheres; comets and interstellar clouds. *Mailing Add:* Jet Propulsion Lab Pasadena CA 91103

HUNTS, BARNEY DEAN, b Rehoboth, NMex, July 23, 36; m 55; c 4. ELECTRONICS. *Educ:* Univ Ariz, BSEE, 61; Univ Calif, Berkeley, MSEE, 62. *Prof Exp:* Electronic design engr, Gen Dynamics Corp, 62-66; dir microcircuits, Friden Res Lab, 66-69; mgr microelec res & develop, Singer Res Lab, Singer Co, 69-71, mgr microelec appl, Corp Res & Develop Lab, 71-74, dir, 74-77, sr dir advan technol, 77-; DIR, TECH DEVELOP, ALLIED ELECTRONIC COMPONENTS, MORRISTOWN, NJ; PRES, CIRCUIT TECHNOL INC. *Mem:* Inst Elec & Electronics Engrs; Am Phys Soc; Indust Res Inst; AAAS. *Res:* Solid-state devices; electro-optics; microelectronics; system architecture; logic design; circuit design. *Mailing Add:* Circuit Technol Inc 160 Smith St Farmingdale NY 11735

HUNTSBERGER, DAVID VERNON, b Harrisburg, Pa, July 22, 17; m 41; c 1. STATISTICS. *Educ:* Bethany Col, WVa, BS, 47; WVa Univ, MS, 48; Iowa State Col, PhD(statist), 54. *Prof Exp:* Instr, 50-53, from asst prof to assoc prof, 53-62, prof, 62-80, EMER PROF STATIST, IOWA STATE UNIV, 80- *Mem:* Fel Am Statist Asn; Int Math Statist; Int Asn Statist Phys Sci. *Res:* Preliminary tests of signficance; experimental design; engineering applications of statistics. *Mailing Add:* 1702 Maxwell Ames IA 50010

HUNTSBERGER, JAMES ROBERT, b Harrisburg, Pa, Feb 4, 21; m 45, 85; c 3. SURFACE CHEMISTRY, ADHESION. *Educ:* Bethany Col, BS, 42; WVa Univ, PhD(chem), 49. *Prof Exp:* Res assoc, Fabrics & Finishes Dept, E I du Pont de Nemours & Co, 48-65, res fel, 65-74, res fel, Plastics Dept, 74-78, res fel reverse osmosis, 78-82; RETIRED. *Concurrent Pos:* Chmn, Gordon Res Conf Adhesion, 64; assoc ed, J Adhesion, 71-84; consult, E I Du Pont de Nemours & Co, 82- *Mem:* Am Chem Soc; Sigma Xi; Adhesion Soc (vpres, 80-81). *Res:* Organic finishes; adhesion; dispersion; mechanical properties of polymers and composites; reverse osmosis. *Mailing Add:* 501 Field Lane Owls Nest Wilmington DE 19807

HUNTSMAN, GENE RAYMOND, b East St Louis, Ill, Aug 12, 40; m 63; c 1. FISH BIOLOGY. *Educ:* Cornell Univ, BS, 62; Iowa State Univ, MS, 64, PhD(fishery biol), 66. *Prof Exp:* NIH fel, Univ Miami, 66-67; FISHERY BIOLOGIST, NAT MARINE FISHERIES SERV, US DEPT COM, 67- *Concurrent Pos:* From adj asst prof to adj assoc prof, NC State Univ, 70-83, adj prof, 84-; mem, NC Coastal Resources Comn, 74-83, NC Coastal Resources Adv Coun, 83-87. *Mem:* Am Fisheries Soc (pres Marine Fishery Sect 87-88); Ecol Soc Am (pres, 88-89); Am Soc Limnol & Oceanog; Am Inst Fishery Res Biologists; Sigma Xi. *Res:* Population dynamics of coastal pelagic fishes; biology of deep water reef fishes; recreational fisheries. *Mailing Add:* Nat Marine Fisheries Serv Beaufort Lab Southeast Fisheries Ctr Beaufort NC 28516-9722

HUNTSMAN, LEE L, b Tacoma, Wash, June 11, 41. BIOMEDICAL ENGINEERING. *Educ:* Stanford Univ, BS, 63; Univ Pa, PhD(biomed eng), 68. *Prof Exp:* NIH fel, Biomed Eng, Univ Pa, 63-68; from res asst prof to res assoc prof, Ctr Bioeng & Dept Mech Eng, 68-81, actg dir, 80, DIR, CTR BIOMED ENG, COL ENG & SCH MED, UNIV WASH, SEATTLE, 80-, PROF, 81- *Concurrent Pos:* Mem bd dirs, Lawrence Med Systs, 80-85, Biomed Eng Soc, 84-87, Wash Res Found, 86-, Wash Exhib Sci & Technol, 87-; mem, Rocky Mountain Regional Heart Comt, Am Heart Asn, 81-85, Cardiovasc Study Sect, NIH, 83-85, Trinus adv bd, 87-; consult, Lawrence Med Systs, 80-89, CooperVision IOL, 83-85, Endosonics, 85-89, St Jude Med, 89-; grants, NIH prog proj, 89-94, W M Keck Found, 89-92 & Whitaker Found, 90-93. *Mem:* AAAS; Am Heart Asn; Biomed Eng Soc; Biophys Soc; Inst Elec & Electronics Engrs. *Res:* Bioengineering; heart; myocardial mechanics; ventricular performance; non-invasive measurements; ultrasound; doppler; technology transfer; numerous technical publications; granted three patents. *Mailing Add:* Ctr Bioeng Col Eng Sch Med Univ Wash Seattle WA 98195

HUNTSMAN, WILLIAM DUANE, b Dart, Ohio, Mar 1, 25; m 49; c 3. ORGANIC CHEMISTRY. *Educ:* Ohio Univ, BS, 47; Northwestern Univ, PhD(chem), 50. *Prof Exp:* Res assoc chem, Northwestern Univ, 50-51; from asst prof to assoc prof, 51-61, chmn dept, 63-69, prof, 61-69, DISTINGUISHED PROF CHEM, OHIO UNIV, 69- *Mem:* Am Chem Soc; The Chem Soc. *Res:* Thermal rearrangement reactions; catalytic hydrogenation; acid-catalyzed isomerization of hydrocarbons. *Mailing Add:* Dept Chem Clippinger Labs Ohio Univ Athens OH 45701

HUNTZICKER, JAMES JOHN, b St Clair, Mich, Oct 23, 41; m 67; c 2. AIR POLLUTION, ATMOSPHERIC CHEMISTRY. *Educ:* Univ Mich, BS, 63; Univ Calif, Berkeley, PhD(chem), 68. *Prof Exp:* Physicist, Lawrence Radiation Lab, Univ Calif, 68-69; guest prof physics, Free Univ Berlin, 69-70; actg asst prof, Indian Inst Technol, Kanpur, India, 71-72; res assoc environ health eng, Calif Inst Technol, 72-74, instr, 74; from asst prof to assoc prof environ sci, 74-85, chmn, dept environ sci, 76-83, chmn, dept chem, biol & environ sci, 84-86, PROF ENVIRON SCI, ORE GRAD CTR, 85- *Concurrent Pos:* Mem climatic impact assessment prog, Panel Aerosol Formation, Stratosphere Supersonic Transp Effluents, 73-75; consult, Environ Res & Technol Inc, 74-; exec vpres, Ore Grad Ctr, 86, pres, 87-, actg pres, 87-88. *Mem:* Am Chem Soc; Am Phys Soc. *Res:* Aerosol chemistry; air pollution chem. *Mailing Add:* Ore Grad Ctr 19600 NW Von Neumann Dr Beaverton OR 97006

HUNZEKER, HUBERT LAVON, b Pawnee City, Nebr, Nov 12, 20; m 43; c 4. MATHEMATICS, COMPUTER PROGRAMMING APPLICATIONS. *Educ:* Nebr State Teachers Col, Peru, AB, 48; Iowa State Col, MS, 50; Univ Mich, PhD(math), 59. *Prof Exp:* Actg instr math, Ohio Univ, 50-52; asst, Willow Run Res Ctr, Univ Mich, 53; from instr to asst prof math, DePauw Univ, 53-57; asst, Univ Mich, 57-58; asst prof, Univ Nebr, 58-62; from assoc prof to prof, Univ Omaha, 62-68, chmn dept, 62-68; head dept, 68-71, prof, 68-84, EMER PROF MATH, MICH TECHNOL UNIV, 84- *Concurrent Pos:* Consult, Res Inst, Univ Mich, 58-59, CAAM consult, 84- *Mem:* Am Math Soc; Math Asn Am; Sigma Xi; Asn Comput Mach. *Res:* Trend analysis of time series. *Mailing Add:* 858 12th Ave S Onalaska WI 54650-3410

HUNZICKER-DUNN, MARY, m. HORMONE ACTION, PROTEIN PHOSPHORATION. *Educ:* Univ Ill, PhD(physiol), 73. *Prof Exp:* ASSOC PROF, DEPT MOLECULAR BIOL, NORTHWESTERN UNIV, 79- *Concurrent Pos:* Res career develop award, Biochem Endocrinol Study Sect, NIH; bd dirs, Soc Study Reprod, 87-90. *Mem:* Am Soc Biol Chemists; Endocrine Soc. *Res:* Camp regulation; protein phosphorylation; molecular mechanisms by which the gonadotropic hormones luternizing hormone and follicle stimulating hormone regulate functions of ovarian follicles and corpira lutea; adenylyl cyclase enzyme and various protein kinases which mediate hormonal signals. *Mailing Add:* Molecular & Structural Biol Northwestern Univ Sch Med 303 E Chicago Ave Chicago IL 60611

HUNZIKER, HEINRICH ERWIN, b Aarau, Switz, Mar 16, 34; m 61; c 4. CHEMICAL PHYSICS. *Educ:* Swiss Fed Inst Technol, dipl, 58, Dr sc nat, 64. *Prof Exp:* Res asst, Swiss Fed Inst Technol, 64-65; res assoc, Brookhaven Nat Lab, 65-67; RES STAFF MEM, IBM RES LAB, 67- *Mem:* fel Am Phys Soc. *Res:* Molecular spectroscopy; gas phase photochemistry; kinetic spectroscopy of gas phase reactions; surface analysis. *Mailing Add:* Dept K33-802 IBM Almaden Res Ctr 650 Harry RD San Jose CA 95120-6099

HUNZIKER, RODNEY WILLIAM, b Everett, Wash, May 9, 22; m 48; c 1. ASTRONOMY. *Educ:* Univ Wash, BS, 44; San Francisco State Col, MA, 50; Univ Calif, Berkeley, MA, 64; PhD, 77. *Prof Exp:* Prof phys sci, Calif State Univ, Chico, 55-88, prof geol, 80-88; RETIRED. *Concurrent Pos:* NSF sci fac fel astron, Univ Calif, Berkeley, 60-62. *Res:* Astronomical spectroscopy; abundances in high-velocity Type A stars; science education; concept formation in the physical sciences. *Mailing Add:* PO Box 1671 Chico CA 95927

HUPE, DONALD JOHN, b Newark, NJ, Nov 27, 44; m 69; c 2. PARASITIC PROTOZOA, ENZYME INHIBITORS. *Educ:* Rutgers Univ, BA, 68; Dartmouth Univ, PhD(chem), 72. *Prof Exp:* Teaching fel biochem, Brandeis Univ, 72-74; asst prof chem & biochem, Univ Mich, 74-80; res fel, Merck Inst Therapeut Res, 80-86, sr res fel biochem, 86-90; DIR BIOCHEM, PARKE DAVIS PHARMACEUT, 90- *Concurrent Pos:* Vis assoc prof, Rutgers Univ, 81-82. *Mem:* Am Soc Biol Chem; Am Chem Soc. *Res:* Biochemistry and enzymology, particularly of parasitic protozoa; development of antiparasitic agents; study of mode of action of known antiparasitic agents; antimetastasis agents; nucleotide biosynthesis. *Mailing Add:* Parke Davis/Warner Lambert 2800 Plymouth Rd Ann Arbor MI 48105

HUPERT, JULIUS JAN MARIAN, b Tarnopol, Poland, May 6, 10; nat US; m 47. ELECTROPHYSICS. *Educ:* Polytech Inst, Warsaw, Poland, Dipl Ing, 33; Northwestern Univ, PhD(elec eng), 51. *Prof Exp:* Head, Transmission Dept, Bur Res, State Telecommun Estab, Poland, 33-39; dir res electronics, A R F Prod Inc, Ill, 47-57; lectr, 47-57, assoc prof, 57-58, prof, 58-78, EMER PROF PHYSICS, DEPAUL UNIV, 78- *Concurrent Pos:* Res consult, A R F Prod Inc. *Mem:* Am Asn Physics Teachers; sr mem Inst Elec & Electronics Engrs; fel Brit Inst Elec Eng. *Res:* Electromagnetism; modulation theory; circuit theory; electronic instrumentation. *Mailing Add:* 21 W 515 Clifton Ave Glen Ellyn IL 60137-4751

HUPKA, ARTHUR LEE, b South Bend, Ind, June 3, 40. PHARMACOLOGY. *Educ:* Butler Univ, BA, 64, MS, 66; Univ Utah, PhD(pharmacol), 71. *Prof Exp:* Fel, Dept Biochem, Univ Utah, 71-73; asst prof pharmacol, Univ Nebr Med Ctr, Omaha, 73-78; from asst prof to assoc prof pharmacol, Sch Health Sci, Grand Valley State Cols, 81-88; PROF & CHMN PHARMACOL, PONCE SCH MED, PR, 88- *Mem:* Res Soc Alcoholism; NY Acad Sci. *Res:* Drug metabolism; alcoholism; pharmacokinetics. *Mailing Add:* Dept Pharmacol Ponce Sch Med PO Box 7004 Ponce PR 00732

HUPP, EUGENE WESLEY, b Knox Co, Nebr, Feb 23, 33; m 57; c 3. PHYSIOLOGY. *Educ:* Univ Nebr, BS, 54, MS, 57; Mich State Univ, PhD, 59. *Prof Exp:* Asst & assoc scientist, Agr Res Lab, Atomic Energy Comn Proj, Tenn, 58-62; assoc prof, Biol Dept & Radiation Biol Lab, Tex A&M Univ, 62-65; assoc prof, 65-69, PROF BIOL, TEX WOMAN'S UNIV, 69- *Mem:* Radiation Res Soc; Brit Soc Study Fertil; Am Soc Animal Sci. *Res:* Effects of environmental agents on animals; reproductive physiology; laboratory animal management. *Mailing Add:* Box 23971/Biol Tex Woman's Univ Denton TX 76204

HUPPE, FRANCIS FROWIN, b Kansas City, Mo, Dec 11, 34; m 61; c 3. TECHNICAL MANAGEMENT, OPTICS. *Educ:* Rockhurst Col, BS, 56; Mass Inst Technol, MS, 59; Univ Wis, PhD(physics), 67. *Prof Exp:* Electronics engr, Bendix Aviation Corp, 59-60; instr math & physics, Rockhurst Col, 60-61; res physicist, 66-68, sr res physicist, 68, supvr, 68-69, res supvr, 69-71, div staff, 71-72, sr supvr, 72-74, mgr appl physics, 74-79, mgr instruments & control, 79-80, DIR, ENG PHYSICS LAB, E I DU PONT DE NEMOURS & CO, 81- *Mem:* Am Phys Soc; Sigma Xi. *Mailing Add:* E I du Pont de Nemours & Co Exp Sta Wilmington DE 19880-0304

HUPPERT, IRWIN NEIL, b Brooklyn, NY, Apr 18, 38; m 63; c 3. CHEMICAL ENGINEERING. *Educ:* Cooper Union, BSChE, 58; Newark Col Eng, MSChE, 64. *Prof Exp:* Res engr, Gen Dynamics Corp, 58-60; pilot plant engr, Merck & Co, 60-67; mgr process develop, Tenneco Inc, 67-82; mgr process eng & develop, Nuodex Inc, 82-85; MGR PROCESS ENG & DEVELOP, HULS AM, 85- *Mem:* Am Inst Chem Engrs. *Res:* Process development and scale-up of processes from laboratory to full scale production for synthetic organic chemicals. *Mailing Add:* Huls Am Inc PO Box 365 Piscataway NJ 08854-0365

HUR, J JAMES, b Greenville, Ohio, Aug 29, 20; m 44; c 4. CHEMICAL & MATERIALS ENGINEERING. *Educ:* Ohio State Univ, BChE, 42, MSc, 47. *Prof Exp:* Develop engr, Res & Develop Dept, Atlantic Richfield Co, 47-49, supv eng, 49-60, asst to vpres res, 60-65, planning dir, 65-69, mgr coordr & eval, 69-72, mgr prod supply, 72-73, mgr info systs & servs, 73-81, dir fed govt rels, 81-84; CONSULT, 84- *Concurrent Pos:* Assoc prof, Drexel Univ eve sch, 57-68. *Mem:* Am Inst Chem Engrs; Int Asn Energy Econ. *Res:* Energy sources, corrosion; process design; information systems and research administration; environmental engineering. *Mailing Add:* 109 Harrells Ct Williamsburg VA 23185-6542

HURA, GURDEEP SINGH, b Raipur, India, May 18, 50; m 76; c 2. PETRI NET MODELING SYSTEMS, DISTRIBUTED COMPUTING & COMPUTER NETWORKS. *Educ:* Jalalpur Univ, India, BE, 72; Univ Roorkee, India, ME, 75, PhD(computer eng), 84. *Prof Exp:* Lectr electronics, Regional Eng Col, Kurukshetra, India, 74-83, asst prof, 83-84; fel computer sci, Concordia Univ, Montreal, 85; ASST PROF COMPUTER SCI, WRIGHT STATE UNIV, DAYTON, OHIO, 85- *Concurrent Pos:* Consult, BDM Inc, Dayton, Ohio, 88-90; chmn, Dayton Chapter, Inst Elec & Electronics Engrs Computer Soc, 88-90, mem prog comt, Region 10 TENCON, 89-92 & Distrib Comput Syst, 90-92, chmn, Overseas Prog Comt, Region 10 TENCON, 93; guest ed, spec issue Microelectronics & Reliability Int J, 91; reviewer/panelist, NSF, 91. *Mem:* Sr mem Inst Elec & Electronics Engrs; Asn Comput Mach. *Res:* Published over 80 scholarly research papers in various international journals and proceedings; presented invited papers in the areas of modeling, computer networks, distributed systems, graph theory, real time design & software systems. *Mailing Add:* 1896 Hickory Ridge Dr Beavercreek OH 45432

HURAY, PAUL GORDON, b Knoxville, Tenn, Nov 3, 41; m 62; c 3. COMPUTER SCIENCES, INORGANIC CHEMISTRY. *Educ:* Univ Tenn, BS, 64, PhD(physics), 68. *Prof Exp:* Physicist, Solid State Div, Oak Ridge Nat Lab, 64 & Univ NC, Chapel Hill, 68-69; from asst prof to prof physics, 69-81, assoc dean lib arts, 81-84, dir, 84-85, sen policy analyst, White House Off Sci & Technol Policy, 85-87, DIR, THE SCIENCE ALLIANCE, UNIV TENN, 87- *Concurrent Pos:* Consult, Transuranium Res Lab, Chem Div, Oak Ridge Nat Lab, 70-; chmn, Comput Res & Appln Comt, Fed Coord Coun Sci Eng & Technol, 86-; mem acad adv bd, Indust Res Inst, 86-; mem, White House Initiative Sci Adv Bd, Hist Black Cols & Univs, 85-; prin investr, Dept Energy, grant, 79-90; consult, White House Sci Off, 87- *Mem:* Am Phys Soc; AAAS. *Res:* Electronic and magnetic properties of metals, alloys and compounds of transuranium elements; Mossbauer effect of iron, gold, uranium and neptunium isotopes; primary containment of transuranium elements for the purpose of nuclear waster disposal. *Mailing Add:* Physics Dept Univ Tenn Knoxville TN 37996

HURD, ALBERT EMERSON, b Montreal, Que. MATHEMATICS. *Educ:* Univ Toronto, BASc, 53, MA, 57; Stanford Univ, PhD(math), 62. *Prof Exp:* CLE Moore instr math, Mass Inst Technol, 61-63; asst prof, Univ Calif, Los Angeles, 63-71; assoc prof, 71-82, PROF MATH, UNIV VICTORIA, BC, 82- *Mem:* Can Math Soc. *Res:* Applications of nonstandard analysis to mathematical physics. *Mailing Add:* Dept Math Univ Victoria Victoria BC U8W 2V2 Can

HURD, COLIN MICHAEL, b Hull, Eng, Mar 9, 37; Can citizen; m 60; c 3. SOLID STATE PHYSICS, MATERIALS SCIENCE. *Educ:* Univ Hull, BSc, 58, PhD(physics), 61. *Prof Exp:* Fel chem, 61-63, sr res officer chem, 63-81, PRIN RES OFFICER, MICROSTRUCT, NAT RES COUN CAN, 81- *Concurrent Pos:* Vis scientist physics, CRTBT, Grenoble, France, 68-69 & Fac Sci, Orsay, France, 77-78; vis prof physics, Univ Geneva, Switz, 74. *Mem:* Am Phys Soc; fel Inst Physics. *Res:* Electron transport properties and magnetic properties of metals, alloys, semiconductors and heterostructures. *Mailing Add:* Inst Microstruct Sci Nat Res Coun Can Ottawa ON K1A 0R9 Can

HURD, CUTHBERT CORWIN, b Estherville, Iowa, Apr 5, 11; m 41; c 5. MATHEMATICS. *Educ:* Drake Univ, AB, 32; Iowa State Col, MS, 34; Univ Ill, PhD(math), 36. *Hon Degrees:* LLD, Drake Univ, 67. *Prof Exp:* From instr to asst prof math, Mich State Col, 36-42; dean, Allegheny Col, 45-47; tech res head, Union Carbide & Carbon Chem Corp, Oak Ridge, 47-49; dir appl sci dept, Int Bus Mach Corp, 49-53, dir appl sci div, 53-55, dir electronic data processing mach, 55-56, dir automation res, 56-60, dir indust process control systs, 60-61, spec asst vpres res, 61-62; chmn bd, Comput Usage Co, Inc, New York, NY & Palo Alto, Calif, 62-74; CHMN, CUTHBERT C HURD ASSOCS, 74-; chmn, HCC, Inc, 74-90; chmn, Picodyne Corp, 78-86; chmn, Quintus Comput Systs, 84-89; RETIRED. *Concurrent Pos:* Past mem comput comt, Nat Res Coun; mem Textile Res Inst; Nat Res Coun del, Int Sci Radio Union; pres exec coun, Found Instrumentation, Educ & Res; mem develop bd, Mass Inst Technol; chmn comput sci adv comt, Stanford Univ; comput hist proj, Smithsonian Inst; adv comt, Ctr Comput Sci & Technol, Inst Appl Technol, Nat Bur Stand. *Mem:* Fel AAAS; Inst Mgt Sci (vpres, 53-54); Opers Res Soc Am; Psychomet Soc; Am Math Soc. *Res:* Asymptotic solutions to differential equations; probability distribution theory of functions of indirect measurements; numerical analysis as applied to high-speed computing machines; logical design of computing machines. *Mailing Add:* 332 Westridge Dr Portola Valley CA 94025

HURD, ERIC R, b Columbus, Kans, July 5, 36; m; c 3. INTERNAL MEDICINE. *Educ:* Univ Tulsa, BS, 58; Univ Okla, MD, 62. *Prof Exp:* Intern, St John's Hosp, Tulsa, Okla, 62-63, resident internal med, 63-65; NIH fel, 65-67, from instr to prof, 67-81, CLIN PROF INTERNAL MED, SOUTHWESTERN MED SCH, UNIV TEX, DALLAS, 81-; ASSOC, BAYLOR ARTHRITIS CTR, BAYLOR UNIV MED CTR, 81- *Concurrent Pos:* Sr attend physician, Parkland Mem Hosp, Dallas, 67-; mem, Am Rheumatism Asn Coop Clin Comt, 68-74; dir, Arthritis Clin, John Peter Smith Hosp, Ft Worth, 72-; consult rheumatology, Vet Admin Hosp, 72-, attend physician, 73-, chief rheumatology, 81-; mem, Med & Sci Comt, Arthritis Found, 82-, chmn, Prof Educ Comt, 88- *Res:* Rheumatoid arthritis; author of numerous publications. *Mailing Add:* Dept Int Med Arthritis Ctrs Tex Baylor Univ Med Ctr 712 N Washington No 200 Dallas TX 75246

HURD, JAMES WILLIAM, b Grand Rapids, Minn, Jan 13, 51; m 79; c 2. LINEAR DYNAMICS, ACCELERATOR OPERATIONS. *Educ:* Univ Ala, Huntsville, BS, 74; Rutgers Univ, MS, 77; Univ NMex, MBA, 88. *Prof Exp:* Lab asst, Univ Ala, Huntsville, 71-72, res asst, 72-74; teaching res asst, dept physics, Rutgers Univ, 74-77, accelerator engr, Nuclear Physics Lab, 77-78; ACCELERATOR PHYSICIST, LOS ALAMOS NAT LAB, 78- *Mem:* Am Phys Soc. *Res:* Modeling and measurement of beam transports and accelerator systems; development of numerical and analytical models of transport elements; algorithms for accelerator operations. *Mailing Add:* Los Alamos Nat Lab MS-H812 MP-6 Los Alamos NM 87545

HURD, JEFFERY L, b Denver, Colo, July 31, 54. ELECTRON BEAM MICROANALYSIS, CRYSTAL GROWTH & CHARACTERIZATION. *Educ:* Colo Sch Mines, BS, 78; Univ Calif Berkeley, MS, 87. *Prof Exp:* Assoc scientist, Solar Energy Res Inst, 78-85; res asst, Lawrence Berkeley Lab, Calif, 85-88; STAFF ENGR, IBM E FISHKILL, HOPEWELL JUNCTION, NY, 88- *Concurrent Pos:* Teaching asst lab mat sci, Univ Calif, Berkeley, 86; peer Rev Panel mem, Dept Energy Nat Photovoltaics Prog, 89-90. *Mem:* Electrochem Soc; Am Asn Crystal Growth; Mat Res Soc; Microbeam Anal Soc. *Res:* Silicon sheet crystal growth and characterization for photovoltaics; hydrothermal crystal growth of copper-indium-selenium; medium temperature CVD of BN films for semiconductor applications; characterization of gradient freeze galium-arsenide; microbeam analysis of ceramics and semiconductors; high temperature thermal properties; author of various publications. *Mailing Add:* IBM E Fishkill Z/40E Rte 52 Hopewell Junction NY 12533-0999

HURD, JON RICKEY, b Columbus, Ohio, July 8, 45. NUCLEAR STRUCTURE & SPECIAL NUCLEAR MATERIAL SAFEGUARDS. *Educ:* Ohio State Univ, BSc, 68, MSc, 73; Fla State Univ, PhD(nuclear physics), 79. *Prof Exp:* Fel low energy nuclear physics, Univ Pa, 79-81; fel nuclear physics, Univ Va, 81-84; STAFF MEM, LOS ALAMOS NAT LAB, 84- *Mem:* Am Phys Soc; Am Asn Physics Teachers. *Res:* Low energy nuclear resonance phenomena, intermediate energy nuclear physics and safeguards for special nuclear materials. *Mailing Add:* 941 Calle Mejia No 822 Santa Fe NM 87501-1467

HURD, LAWRENCE EDWARD, b Syracuse, NY, Aug 13, 47; m 68; c 1. ENTOMOLOGY. *Educ:* Hiram Col, BA, 69; Syracuse Univ, PhD(ecol), 72. *Prof Exp:* Fel, Cornell Univ, 72-73; asst prof, 73-78, assoc prof, 78-86, PROF ECOL, UNIV DEL, 87- *Concurrent Pos:* Asst ed, Bull Royal Entom Soc London, 85-87. *Mem:* AAAS; Ecol Soc Am; fel Royal Ent Soc London. *Res:* Ecosystem theory; regulation of community diversity; population regulation in predators. *Mailing Add:* Biol Sci Dept Univ Del Newark DE 19711

HURD, MAGGIE PATRICIANNE, b Atlanta, Ga, Mar 6, 40. MYCOLOGY, ELECTRON MICROSCOPY. *Educ:* Spelman Col, BA, 62; Atlanta Univ, MS, 67, PhD(biol), 76. *Prof Exp:* Teacher sci, Atlanta Pub Sch, 62-75; ELECTRON MICROSCOPIST BIOL, FERNBANK SCI CTR, 75- *Concurrent Pos:* Consult sci, Spelman Col, 77; rev panelist, Sci Fac Prof Develop Prog, NSF, 77; adj prof, Ga State Univ, 77-; NSF guest lectr, Malcolm X Col, 81. *Mem:* Nat Sci Hon Soc; Educ Hon Soc; Electron Micros Soc Am; Am Inst Biol Sci; Nat Sci Teacher Asn. *Res:* Fungi including ceratocystic ulmi and giardia lamblia. *Mailing Add:* Fernbank Sci Ctr 156 Heaton Park Dr Atlanta GA 30307

HURD, PAUL DEHART, b Denver, Colo, Dec 25, 05; m 46; c 2. BIOLOGY, SCIENCE EDUCATION. *Educ:* Univ Northern Colo, AB, 29, MA, 32; Stanford Univ, EdD(sci educ), 49. *Hon Degrees:* DSc, Drake Univ, 74, Ball State Univ, 79, Univ Northern Colo, 80. *Prof Exp:* Teacher pub sch, Colo, 29-39; chmn dept sci, Menlo Jr Col, Calif, 39-51; prof, 51-71, EMER PROF SCI EDUC, STANFORD UNIV, 71- *Concurrent Pos:* Fac fel, Stanford Univ, Northeast Asia-US Forum Int Policy, 80- *Honors & Awards:* Apollo Award, NASA, 70. *Mem:* Nat Sci Teachers Asn; Nat Asn Biol Teachers; Nat Asn Res Sci Teaching; fel AAAS. *Res:* Biology curriculum; history of science teaching. *Mailing Add:* 549 Hilbar Lane Palo Alto CA 94303

HURD, RALPH EUGENE, b Columbus, Ohio, May 19, 50; m 71; c 2. NUCLEAR MAGNETIC RESONANCE. *Educ:* Calif State Col, San Bernadino, BA, 73; Univ Calif, Riverside, PhD(biochem), 78. *Prof Exp:* Res asst biochemist, Univ Calif, Riverside, 79-80; appl chemist, Nicolet Magnetics Corp, 80-83, MGR, RES & DEVELOP, GE NMR INST, 83- *Mem:* Am Chem Soc. *Res:* Application of nuclear magnetic resonance spectroscopy to biochemical problems, especially in the study of protein-nucleic acid interactions and structure. *Mailing Add:* G E NMR Inst 255 Fourier Ave Fremont CA 94539

HURD, RICHARD NELSON, b Evanston, Ill, Feb 25, 26; m 50; c 2. ORGANIC CHEMISTRY, PHOTOCHEMOTHERAPY. *Educ:* Univ Mich, BS, 46; Univ Minn, PhD(org chem), 56. *Prof Exp:* Group leader, Polyethylene Br, Res Div, Koppers Co, Inc, 56-57; res chemist, Org Res Dept, Mallinckrodt Chem Works, Inc, 57-64, group leader plastics additives res, Res Dept, Indust Chem Div, Mo, 64-66; group leader res dept, Com Solvents Corp, 66-69, sect leader, 69-71; mgr int sci affairs, G D Searle Int Co, 72-74, dir mfg & tech affairs, 74-77; vpres tech affairs, Elder Pharmaceut, 77-81; vpres res & develop, US Proprietary Drugs & Toiletries Div, Schering-Plough, Inc, 81-83; vpres sci affairs, Moleculon Inc, 83-88; VPRES REGULATORY AFFAIRS, PHARMACO, 89- *Mem:* AAAS; Am Chem Soc; Sigma Xi; NY Acad Sci; Am Acad Dermat; Am Pharm Asn; Am Asn Pharm Scientists; Soc Investigative Dermat. *Res:* Macrolide and natural product syntheses; non-steroidal estrogens; thioamide syntheses and reactions; structures and properties of organic chelates and complexes; stabilization of polymeric systems; photo-chemotherapy; photobiology; investigative dermatology. *Mailing Add:* 49 Austin Rd Sudbury MA 01776

HURD, ROBERT CHARLES, b Mt Vernon, Wash, Mar 7, 22; m 50; c 4. BACTERIOLOGY. *Educ:* Pac Univ, BS, 46; Wash State Univ, MS, 50, PhD(bact), 53. *Prof Exp:* Instr biol, Pac Univ, 46-48; res asst biol sci, State Col Wash, 48-53; instr microbiol, Gonzaga Univ, 53-57, head dept, 57-77, from asst prof to prof biol, 60-87; RETIRED. *Concurrent Pos:* Consult, bact. *Mem:* AAAS; Sigma Xi. *Res:* Pathogenic microbiology. *Mailing Add:* 3805 W Weile Ave Spokane WA 99208

HURD, SUZANNE SHELDON, b Elmira, NY, Dec 17, 39. BIOCHEMISTRY. *Educ:* Bates Col, BS, 61; Univ Wash, MS, 63, PhD(biochem), 67. *Prof Exp:* NIH fel biochem, Univ Calif, Berkeley, 67-69; scientist adminr, 69-81, DIR, DIV LUNG DIS, NAT HEART, LUNG & BLOOD INST, 77- *Mem:* Am Therapeut Soc. *Res:* Enzymology; muscle proteins; diseases of the heart and lung. *Mailing Add:* Dept Lung Dis Nat Heart Lung & Blood Inst Bethesda MD 02892

HURDIS, EVERETT CUSHING, b Providence, RI, Feb 5, 18; m 46; c 4. ORGANIC CHEMISTRY. *Educ:* Brown Univ, ScB, 39; Princeton Univ, MA, 41, PhD(phys chem), 43. *Prof Exp:* Res chemist, US Rubber Co, NJ, 42-58; mgr styrene polymers group, Koppers Co, Pa, 58-62; asst prof, 62-69, ASSOC PROF CHEM, TEX WOMAN'S UNIV, 69- *Concurrent Pos:* Lectr, Rutgers Univ, 48-49; instr, Fairleigh Dickinson Col, 50-52. *Mem:* Am Chem Soc; NY Acad Sci; fel Am Inst Chem. *Res:* Aromatic substitutions; redox indicators; polymer chemistry. *Mailing Add:* 621 E University Dr Denton TX 76201

HURDLE, BURTON GARRISON, b Roanoke, Va, Jan 2, 18; m 42; c 2. PHYSICS, ACOUSTICS. *Educ:* Roanoke Col, BS, 41; Va Polytech Inst, MS, 43; Open Univ, UK, PhD, 88. *Prof Exp:* Res physicist, Sound Div, Naval Res Lab, Washington, DC, 43-47; res physicist, Eng Res Assoc, Arlington, Va, 47-49; res physicist, Sound Div, Naval Res Lab, Washington, DC, 49-56, supvr res physicist & head, Acoust Scattering Sect, Acoust Div, 56-70, supvr res physicist & head, Propagation Br, 70-76; vis scientist, Admiralty Res Lab, Teddington, Eng, UK, 76-77; supvr res physicist & asst supt, 77-84, physicist & assoc supt, 84-89, SR SCIENTIST & CONSULT, ACOUST DIV, NAVAL RES LAB, WASHINGTON, DC, 89- *Mem:* Fel Acoust Soc Am; Sigma Xi. *Res:* Underwater acoustic propagation; scattering and noise; geophysics and physical oceanography of the ocean environment; author or co-author of over 25 publications. *Mailing Add:* 6222 Berkeley Rd Alexandria VA 22307

HURFORD, THOMAS ROWLAND, b Detroit, Mich, Nov 20, 41; m 64; c 2. POLYMER CHEMISTRY. *Educ:* Wayne State Univ, BS, 64, PhD(anal chem), 67. *Prof Exp:* Chemist, 67-69, tech rep polymer chem, 69-71, MKT REP PLASTICS, E I DU PONT DE NEMOURS & CO, INC, 72-, SALES & SALES MGT, 82- *Mem:* Am Chem Soc; Soc Plastics Eng. *Mailing Add:* 1001 Oriente Ave Wilmington DE 19807-2260

HURLBERT, BERNARD STUART, b Yarmouth, NS, Oct 11, 30; m 52, 87; c 3. LABORATORY DESIGN. *Educ:* Acadia Univ, BSc, 50 & 51; McGill Univ, MSc, 53; Univ NB, PhD(chem), 57. *Prof Exp:* Instr chem, Rochester Inst Technol, 57-59; res chemist, Wellcome Res Labs, 59-61, sr res chemist, 61-70, group leader, 70-80, sect head, 80-88; PRES, AQUA LIBRA ASSOCS, 88- *Concurrent Pos:* Consult, 88- *Res:* Organic chemistry; nuclear magnetic resonance; laboratory design. *Mailing Add:* Box 12222 Res Triangle Park NC 27709

HURLBERT, ROBERT BOSTON, b David City, Nebr, Aug 9, 26; m 55; c 4. BIOCHEMISTRY. *Educ:* Univ Nebr, BS, 47, MS, 49; Univ Wis, PhD(physiol chem), 52. *Prof Exp:* Am Cancer Soc fel, Kemiska Inst, Karolinska Inst, Sweden, 53-54; fel, Inst Cytophysiol, Copenhagen Univ, 54-55; instr biochem, McArdle Lab, Univ Wis, 55; from asst biochemist to assoc biochemist, 56-62, BIOCHEMIST, UNIV TEX M D ANDERSON HOSP & TUMOR INST HOUSTON, 62-, PROF BIOCHEM, 65-; PROF BIOCHEM, UNIV TEX GRAD SCH BIOMED SCI, 67- *Concurrent Pos:* From clin asst prof to clin assoc prof, Col Med, Baylor Univ, 56-67; assoc ed, Cancer Res, 68-80; mem adv comts, Am Cancer Soc, 66-71 & 79-82. *Mem:* Am Soc Biol Chemists; Am Asn Cancer Res; Am Soc Cell Biol. *Res:* Chemistry and metabolism of orotic acid; chromatography of ribonucleotides; biosynthesis of uridine nucleotides and cytidine nucleotides; synthesis of RNA and protein in isolated nuclei and nucleoli; chromatographic analysis of intra-cellular deoxynucleotides; mechanisms of action of antitumor drugs. *Mailing Add:* 8807 Manhattan Houston TX 77096

HURLBERT, STUART HARTLEY, b Savannah, Ga, July 16, 39; m 65; c 1. LIMNOLOGY, BIOSTATISTICS. *Educ:* Amherst Col, BA, 61; Cornell Univ, PhD(ecol), 68. *Prof Exp:* Instr zool, Cornell Univ, 66; res entomologist, 66-70, asst prof ecol & fel entom, Dept Entom, Univ Calif, Riverside, 70; from asst prof to assoc prof, 70-78, PROF BIOL, SAN DIEGO STATE UNIV, 78- *Concurrent Pos:* Vis prof, Univ Chile, 71-72. *Honors & Awards:* G W Snedecor Award, Am Statist Asn, 84. *Mem:* AAAS; Ecol Soc Am; Am Soc Limnol & Oceanog; Am Soc Nature. *Res:* Salamander migration; mathematical ecology; insecticides in aquatic ecosystems; predation and competition among freshwater zooplankters; limnology of the Central Andes; statistics; flamingo ecology. *Mailing Add:* Dept Biol San Diego State Univ San Diego CA 92182-0057

HURLBURT, DOUGLAS HERENDEEN, b Cleveland, Ohio, Apr 19, 41; m 68. COMMUNICATION SYSTEM DESIGN. *Educ:* Johns Hopkins Univ, BA, 62; McGill Univ, MSc, 64, PhD(eng), 72. *Prof Exp:* Demonstr physics, McGill Univ, 62-66, eng, 66-68; sci staff, Res & Develop Labs, RCA Ltd, Montreal, 69-76; staff scientist, Lincoln Lab, Mass Inst Technol, 76-90; SPEC ASST, STRATEGIC DEFENSE PROGS, DECISION SCI APPLICATIONS, INC, 90- *Concurrent Pos:* Consult, Oper Systs Develop, 82-84; prog mgr, Defense Advan Res Proj Agency/Systs Test Objectives, 88-90. *Honors & Awards:* Meritorious Serv Award, Defense Advan Res Projs Agency, 90. *Mem:* Inst Elec & Electronics Engrs. *Res:* Design and fabrication of surface acoustic wave devices including filters and acousto-electric devices; analysis and design of high-speed signal processors for ultrahigh frequency communication systems; design, development and analysis of waveforms and signal processors for advanced radar systems. *Mailing Add:* 12 Kenilworth St Newton MA 02158

HURLBURT, EVELYN MCCLELLAND, medical microbiology, for more information see previous edition

HURLBURT, H(ARVEY) ZEH, b Kellogg, Idaho, Sept 2, 21; m 43; c 8. CHEMICAL ENGINEERING. *Educ:* Univ Tex, BA, 42, BS, 43, MS, 47; Mass Inst Technol, ScD(chem eng), 50. *Prof Exp:* Process engr, synthetic rubber div, US Rubber Co, 43-46; lab instr, Univ Tex, 46-47; asst, Mass Inst Technol, 47-49; dir res, Consol Chem Industs Div, 50-67, mgr, Peiser Labs, Stauffer Chem Co, 67-85; CONSULT, 85- *Mem:* Am Chem Soc; Am Inst Chem Engrs; Sigma Xi. *Res:* Interphase resistance to mass transfer; inorganic chemical manufacture. *Mailing Add:* 7814 Santa Elena Houston TX 77017

HURLBUT, CORNELIUS SEARLE, JR, b Springfield, Mass, June 30, 06; m 32, 56; c 3. MINERALOGY, CRYSTALLOGRAPHY. *Educ:* Antioch Col, AB, 29; Harvard Univ, AM, 32, PhD(petrog), 33. *Prof Exp:* Instr petrog, 31-33, from instr to prof mineral, 33-72, EMER PROF MINERAL, HARVARD UNIV, 72- *Concurrent Pos:* Guggenheim fel, 55; vis prof, Boston Col, 72-77. *Honors & Awards:* Neil Miner Award, Nat Asn Geol Teachers, 66. *Mem:* AAAS; fel Am Mineral Soc (secy, 44-59, pres, 63); Nat Asn Geol Teachers; Am Acad Arts & Sci; fel Geol Soc Am; Gemmological Asn Gt Brit. *Res:* Gemology. *Mailing Add:* Dept Earth & Planatary Sci Harvard Univ Cambridge MA 02138

HURLBUT, FRANKLIN CHARLES, b Los Angeles, Calif, July 20, 20; m 42; c 2. PHYSICS. *Educ:* Univ Calif, Los Angeles, AB, 41; Univ Calif, Berkeley, PhD(physics), 54. *Prof Exp:* Physicist, Appl Res Lab, Univ Calif, Berkeley, 42-44; from lectr to assoc prof, 56-68, prof aeronaut sci, 56-91, EMER PROF AERONAUT SCI & MECH ENG, UNIV CALIF, BERKELEY, 91- *Concurrent Pos:* Consult, aerospace indust. *Mem:* Am Phys Soc. *Res:* Molecular and atomic interaction with solid surface; upper atmospheric aerodynamics. *Mailing Add:* 15 Tamalpais Rd Berkeley CA 94708

HURLBUTT, HENRY WINTHROP, zoology; deceased, see previous edition for last biography

HURLEY, FORREST REYBURN, b Greene Co, Ohio, Jan 11, 21; m 42; c 3. INORGANIC CHEMISTRY. *Educ:* Ohio State Univ, BChE, 42, PhD(chem), 54. *Prof Exp:* Control chemist, Gen Chem Co, 42-44; res chemist, Monsanto Chem Co, 46-49; asst chem, Ohio State Univ, 50-51; res chemist, Battelle Mem Inst, 51 & Davison Chem Co, 54; res supvr, W R Grace & Co, 54-66; dir res, Cement Div, Martin Marietta Corp, 66-77; DIR CEMENT & CONCRETE RES, W R GRACE & CO, 78- *Mem:* AAAS; Am Chem Soc; Am Ceramic Soc; Clay Minerals Soc; Am Soc Testing & Mat. *Res:* Chemistry of nitrogen, phosphorus, lanthanides, magnesia and lime; catalysis; inorganic colloids; portland cement and concrete; clay minerals; vermiculite; mica. *Mailing Add:* 144 Carrowmoor Ct Dublin OH 43017-1381

HURLEY, FRANCIS JOSEPH, b Waltham, Mass, May 18, 27. PARASITOLOGY. *Educ:* Univ Notre Dame, AB, 49; Cath Univ Am, MS, 58, PhD(parasitol), 60. *Prof Exp:* Head dept sci high sch, NY, 54-56; from asst prof to assoc prof, 60-71, PROF BIOL, STONEHILL COL, 71-, HEAD DEPT, 60- *Mem:* Am Soc Parasitol. *Res:* Nematode parasites of vertebrates; parasitic immunology. *Mailing Add:* Stonehill Col North Easton MA 02356

HURLEY, FRANK LEO, b Lowell, Mass, Oct 25, 44; m 80. BIOSTATISTICS. *Educ:* Georgetown Univ, BS, 66; Johns Hopkins Univ, PhD(biostatist), 70. *Prof Exp:* Lectr, Loyola Col, 67-70; asst prof biostatist, Sch Med, George Washington Univ, 70-73; SR PARTNER, BIOMET RES INST, 73- *Concurrent Pos:* Asst prof biostatist, Sch Med, Georgetown Univ, 71- *Mem:* Am Statist Asn; NY Acad Sci; Biomet Soc; AAAS. *Res:* Applications of stochastic models; statistical methods. *Mailing Add:* 4515 Potomac Ave NW Washington DC 20007

HURLEY, HARRY JAMES, b Philadelphia, Pa, Oct 10, 26; m 50; c 5. DERMATOLOGY. *Educ:* St Joseph's Col, Pa, 43-45; Jefferson Med Col, MD, 49; Univ Pa, DSc(med), 58. *Prof Exp:* Instr dermat, Sch Med, Boston Univ, 54-55; instr, Sch Med, Univ Pa, 56-57, instr, Grad Sch Med, 56-58, assoc, 57-59; prof, Hahnemann Med Col, 59-62; assoc prof, 62-68, PROF CLIN DERMAT, SCH MED, UNIV PA, 68- *Concurrent Pos:* Consult, US Vet Admin, 59-; mem bd trustees, Dermat Found, 73-75, pres, 75-; mem, Am Bd Dermat, 74- *Mem:* Soc Invest Dermat; AMA; Am Col Physicians; Am Acad Dermat; Am Dermat Asn. *Res:* Physiology of the skin; sweat gland and vascular activity; experimental granulomagenesis. *Mailing Add:* 39 Copley Rd Upper Darby PA 19082

HURLEY, IAN ROBERT, molecular dynamics, for more information see previous edition

HURLEY, JAMES EDGAR, b Morristown, Tenn, Sept 30, 29. MICROBIOLOGY. *Educ:* Tusculum Col, BA, 50; Vanderbilt Univ, PhD(biol), 63. *Prof Exp:* Instr biol, Mid Tenn State Col, 53-54; instr microbiol, US Naval Med Sch, Bethesda, Md, 55-58; chief microbiologist, Washington Hosp Ctr, DC, 58-61; assoc prof, 62-67, PROF BIOL, OKLA BAPTIST UNIV, 67-, CHMN DEPT, 68- *Concurrent Pos:* Danforth assoc, 70-; vis prof, Univ Tex Cancer Ctr, Houston, 76. *Mem:* AAAS; Am Soc Microbiol; NY Acad Sci. *Res:* Physiology of algae in mass culture; taxonomy and physiology of the Bacteriodaceae; physiology of Euglena and Paramecium bursaria. *Mailing Add:* Dept Nursing/Baccalaureate Okla Baptist Univ Shawnee OK 74801

HURLEY, JAMES FREDERICK, b Kansas City, Mo, Apr 3, 41; m 66; c 2. LIE ALGEBRAS, ALGEBRAIC GROUPS. *Educ:* Ariz State Univ,BA, 61, MA, 63; Univ Calif, Los Angeles, PHd(Lie algebras), 66. *Prof Exp:* Lectr math, Univ Calif, Riverside, 66-67, asst prof, 67-71; assoc prof, 71-82, PROF MATH, UNIV CONN, 82- *Concurrent Pos:* Fulbright-Hays vis assoc prof, Ateneo de Manila Univ & De La Salle Univ, Philippines, 73; vis prof, NSF Scientists & Engrs Econ Develop Prog, Univ Philippines, Quezon City, 75-76; vis fel, Inst Advan Studies, Australian Nat Univ Res Sch Phys Sci, Canberra, 76 & Tsukuba Univ, Japan, 80; comt employ & educ policy, Math Asn Am, 85-89; vis fel, Max-Planck-Inst für Math, 87; data comt, Am Math Soc, Math Asn Am, 90-; acad specialist, US Intel Agency, Philippine Sci Consortium, 89. *Mem:* Am Math Soc; Math Asn Am; AAAS; Asn Comput Math & Sci Teaching; Sigma Xi. *Res:* Lie algebras; Chevalley algebras; Chevalley groups over rings; classical linear algebraic groups over rings; simple groups of Lie type; Kac-Moody Lie algebras. *Mailing Add:* Dept Math U-9 Univ Conn 196 Auditorium Rd Rm 111 Storrs CT 06269-3009

HURLEY, JAMES P, b Detroit, Mich, Feb 13, 32. PHYSICS. *Educ:* NY Univ, AB, 55, PhD(physics), 61. *Prof Exp:* Res scientist, Courant Inst Math Sci, NY Univ, 61-63; asst prof, 63-71, ASSOC PROF PHYSICS, UNIV CALIF, DAVIS, 71- *Mem:* Am Phys Soc; Am Asn Physics Teachers. *Res:* Plasma physics; interaction of the solar wind with earth's magnetic field; plasma sheaths. *Mailing Add:* Dept Physics Univ Calif Davis CA 95616

HURLEY, JAMES R(OBERT), b Milwaukee, Wis, May 9, 31; m 53; c 4. CHEMICAL ENGINEERING, ELECTRICAL ENGINEERING. *Educ:* Univ Wis-Madison, BS, 53, MS, 61, PhD(chem eng), 63. *Prof Exp:* Res engr, Phillips Petrol Co, 53-57, develop engr, 57-58; eng analyst, Allis-Chalmers Mfg Co, 58-63, proj leader energy conversion, 63-64, group leader fuel cells, 64, sect mgr aerospace fuel cells, 64-67, dir mat res, 67; assoc prof mech eng, Marquette Univ, 69-74, lectr, 67-68; mgr, Advan Technol Power Systs Div, McGraw-Edison Co, 67-69, sr staff engr, 75-86; RETIRED. *Mem:* Inst Elec & Electronics Engrs. *Res:* Control theory; energy conversion; heat transfer. *Mailing Add:* McGraw-Edison Co 11131 Adams Rd Franksville WI 53126

HURLEY, LAURENCE HAROLD, b Birmingham, Eng, Jan 29, 44; US citizen; div; c 2. MICROBIAL BIOCHEMISTRY. *Educ:* Bath Univ, BPharm, 67; Purdue Univ, PhD(med chem), 70. *Prof Exp:* Fel org chem, Univ BC, 70-71; asst prof med chem, Col Pharm, Univ Md, 71-73; asst prof med chem, Col Pharm, Univ Ky, 73-77, assoc prof, 77-79, prof, 79-81; GEORGE HITCHING PROF, DRUG DESIGN, COL PHARM, UNIV TEX & ASST DIR, DRUG DYNAMICS INST, 81- *Concurrent Pos:* Chmn, NIH Bioorg & Nat Prod Study Sect; consult, Upjohn & Smith Kline & French Labs. *Honors & Awards:* Volwiler Award, 89; Outstanding Investr Award, Nat Cancer Inst. *Mem:* Am Chem Soc; Am Soc Pharmacog. *Res:* Mechanism of action of antitumor-antibiotics; drug-nucleic acid interactions; DNA damage and repair; molecular graphics and drug design. *Mailing Add:* Drug Dynamics Inst Col Pharm Univ Tex Austin TX 78712

HURLEY, MAUREEN, b Boston, Mass. MICROBIOLOGY, CELL BIOLOGY. *Educ:* Emmanuel Col, Mass, BA, 43; Cath Univ Am, MS, 55, PhD(biol), 59. *Prof Exp:* Teacher prep sch, Xavier Univ, 50-54; teacher chem & biol, Blessed Sacrament Col, 58-66; from asst prof to assoc prof, 66-70, PROF BIOL, XAVIER UNIV, LA, 71- *Concurrent Pos:* NSF fac fel, Med Sch, Johns Hopkins Univ, 69-70. *Res:* Medical microbiology; morphogenesis and differentiation in development. *Mailing Add:* Dept Biol Xavier Univ La New Orleans LA 70125

HURLEY, NEAL LILBURN, b Long Beach, Calif, Aug 3, 28; m 64; c 3. GEOPHYSICS, GEOLOGY. *Educ:* Calif Inst Technol, BS, 49; Stanford Univ, MS, 51, PhD(geol, geophys), 53. *Prof Exp:* Geologist, US Geol Surv, 49; res geologist, Continental Oil Co, 51; geophysicist explor, Richfield Oil Corp, 56-67; chief geophysicist, Union Pac RR Co, 67-70; explor mgr, Champlin Petrol Co, 70-74, chief geologist, 74-79, mgr explor serv, 79-88; RETIRED. *Mem:* Soc Explor Geophys; Am Asn Petrol Geol. *Res:* Mechanics of earth deformation; groundwater motion; exploration seismology. *Mailing Add:* 535 Regentview Houston TX 77079-6908

HURLEY, PATRICK MASON, b Hong Kong, China, Jan 12, 12; nat US; m 41; c 3. GEOCHRONOLOGY. *Educ:* Univ BC, BA & BASc, 34; Mass Inst Technol, PhD(econ geol), 40. *Prof Exp:* Mining engr & geologist, Gold Mines, BC, 34-37; field asst, Geol Surv Can, 38; res assoc, Mass Inst Technol, 40-42, Nat Defense Res Comt, 42-45 & Univ Wis, 46; from asst prof to assoc prof geol & geophys, 46-52, chmn fac, 58-60, prof, 53-81, EMER PROF GEOL & GEOPHYS, MASS INST TECHNOL, 81- *Concurrent Pos:* Mem comts, Nat Res Coun. *Mem:* Fel Geol Soc Am; fel Am Geophys Union; Geochem Soc; Am Acad Arts & Sci; hon mem Geol Soc London. *Res:* Geological age measurement and application of nuclear physics to geology; radiogenic isotopes in research on the development of the earth's crust. *Mailing Add:* 260 Seaview Ct Apt 611 Marco Island FL 33937

HURLEY, ROBERT EDWARD, b Cincinnati, Ohio, Oct 31, 11; m 41; c 4. CHEMICAL ENGINEERING, CERAMICS. *Educ:* Univ Cincinnati, BS, 35. *Prof Exp:* Analyst, Joseph E Seagram & Sons, Ind, 35-36, chief chemist, 40; chief chemist, Calvert Distilling Co, Md, 36-40; mem staff ord mil training, Training Div Ord, Aberdeen, Md & The Pentagon, 40-45; proj engr, Chem & Phys Lab, Radio Corp Am, NJ, 45-54, proj engr, Molded Prod Lab, 54-57, mgr ferrite eng, NJ, Ohio & Ind, 57-62, group leader, Ind, 62-68, MGR MAT ENG, CONSUMER ELECTRONICS DIV, RCA CORP, 68- *Mem:* Am Chem Soc; Am Ceramic Soc; fel Am Ceramic Soc. *Res:* Water chemistry and industrial water treatment; paper technology and dynamic properties of paper; sintering of pure oxides; technology of ferrite manufacture. *Mailing Add:* 324 E 45th St Indianapolis IN 46205

HURLEY, ROBERT JOSEPH, b Boston, Mass, July 11, 29; m 52; c 2. MARINE GEOLOGY, OCEANOGRAPHY. *Educ:* Tufts Univ, BS, 51; Scripps Inst Oceanog, Univ Calif, San Diego, MS, 57, PhD(oceanog), 60. *Prof Exp:* Res geologist oceanog, Scripps Inst Oceanog, Univ Calif, San Diego, 54-59; res oceanogr, Bell Tel Labs, NJ, 60-61; from asst prof to assoc prof oceanog, Univ Miami, 61-67, prof, 67-72, assoc dean grad studies, 69-72; dir, Moss Landing Marine Labs, 72-76; ENVIRON CONSULT, 76- *Concurrent Pos:* Dep dir off oceanol, Asst Secy Intergovt Comn Oceanog, UNESCO, Paris. *Mem:* Geol Soc Am; Am Geophys Union; Challenger Soc; Sigma Xi. *Res:* Morphology of the deep sea floor involving structure, geophysical investigations, sedimentary processes, ocean dynamics and the tectonic history of the ocean basins. *Mailing Add:* PO Box 305 Aptos CA 95003

HURLEY, WILLIAM CHARLES, b Milwaukee, Wis, Apr 24, 31; m 53; c 2. FOOD SCIENCE, BIOCHEMISTRY. *Educ:* Iowa State Univ, BS, 56; Univ Mo, MS, 57; Tex A&M Univ, PhD(food sci), 60. *Prof Exp:* Owner, Will-I-Will Poultry Farm, 45-52; dir qual control, Egg Processing, Des Moines Foods, 55; sect chief bakery, Res Div, Pet Milk Co, 60-64; assoc prof food sci & bakery mgt, Kans State Univ, 64-68; MGR BAKER MIX LAB, INT MULTIFOODS, INC, 68-; LAB MGR, KODAK, ST PAUL, MN. *Mem:* Poultry Sci Asn; Am Asn Cereal Chem; Am Soc Bakery Eng. *Res:* Flavor and enzymology of food; food problems associated with baking industry. *Mailing Add:* 4301 Queen Ave N Minneapolis MN 55412-1105

HURLEY, WILLIAM JORDAN, b Norwood, Mass, July 31, 43; m 70; c 2. OPERATIONS RESEARCH, MATHEMATICAL PHYSICS. *Educ:* Boston Col, BS, 65; Univ Rochester, PhD(physics), 71. *Prof Exp:* Res assoc math physics, Syracuse Univ, 70-72; res scientist, Univ Tex, Austin, 72-75; ANALYST OPER RES, CTR NAVAL ANALYSIS, UNIV ROCHESTER, 75- *Mem:* Am Phys Soc. *Res:* Mathematical modeling as applied to operations research, underwater acoustics and the physics of elementary particles. *Mailing Add:* 6332 Waterway Dr Falls Church VA 22044

HURLEY, WILLIAM JOSEPH, b Philadelphia, Pa, July 26, 40; m 63; c 4. PHOTOGRAPHIC EMULSION TECHNOLOGY, PLASTIC FILMS. *Educ:* Villanova Univ, BS, 62; Princeton Univ, PhD(chem), 67. *Prof Exp:* Res chemist magnetic tape, E I Du Pont De Nemours & Co, Inc, Wilmington, Del, 67-71, res chemist silver halide, Parlin, NJ, 71-73, res supvr silver halide chem, Parlin, NJ, 73-74 & Rochester, NY, 74-80; develop mgr magnetic mat, PD Magnetics, BV, Neth, 81-83; tech mgr magnetic mat, Wilmington, Del, 83-86, venture mgr, 86-88, tech mgr, high performance films, Circleville, Ohio, 88-89, LAB DIR, E I DU PONT DE NEMOURS & CO, INC, CIRCLEVILLE, OHIO, 89- *Mem:* Am Chem Soc. *Res:* Materials for magnetic and optical recording media; silver halide emulsions for x-ray and graphic arts films; coating technology; interferometric and raman spectroscopy; polyester and polyimide films; LAMINATES. *Mailing Add:* Du Pont Electronics E I Du Pont De Nemours & Co Inc Circleville OH 43113

HURLICH, MARSHALL GERALD, b Boston, Mass, Mar 26, 46; m 67; c 1. PHYSICAL ANTHROPOLOGY, EPIDEMIOLOGY. *Educ:* Univ Calif, Berkeley, BA, 68; State Univ NY, Buffalo, MA, 72, PhD(anthrop), 76; Univ Wash, Seattle, MPH, 88. *Prof Exp:* Asst prof anthrop, Univ Ga, 75-77; asst prof anthrop, Univ Wash, 77-85, res assoc, 85-87; RES ASSOC, FRED HUTCHINSON CANCER RES CTR, 87- *Concurrent Pos:* Res grants, NIH, NSF & Social Sci Res Ctr. *Mem:* Can Asn Phys Anthropologists; AAAS; Am Anthrop Asn; Am Asn Phys Anthropologists; Human Biol Coun; Am Pub Health Asn. *Res:* Human physiological and behavioral response to environmental stress, especially cold, heat and high altitude; causes of human morphological and physiological variability; biological and biomedical consequences of human migration; health and culture change among refugees from southeast Asia. *Mailing Add:* 10008 35th Ave Seattle WA 98125

HURN, R(ICHARD) W(ILSON), b Henrietta, Tex, Jan 13, 19; m 45; c 2. MECHANICAL ENGINEERING. *Educ:* Tex Tech Col, BS, 40; Univ Wis, MS, 47. *Prof Exp:* Petrol prod engr, Humble Oil Co, 40-48; supvr combustion res, Bartlesville Energy Res Ctr, 48-73, dir, Engine/Fuels Res, 73-79; PRES, R-H TECH SERV, 79- *Concurrent Pos:* Ord Officer, destroyers, USN, 42-46; consult alternative fuels, fuel behavior, engine fuel matching & engine fuel technol develop, fuel fires. *Honors & Awards:* Gold Medal, US Dept Interior, 67. *Mem:* Soc Automotive Engrs; Nat Soc Prof Engrs. *Res:* Combustion of hydrocarbon fuels; combustion reactions and engine emissions control; photochemical air pollution research; engine performance; engine/fuel interface; alternative fuels. *Mailing Add:* 800 SE Windingway Rd Bartlesville OK 74006

HURRELL, JOHN PATRICK, b Sanderstead, Eng, Sept 17, 38; US citizen; m 63; c 3. SOLID STATE PHYSICS. *Educ:* Oxford Univ, BA, 60, MA, 64, DPhil(physics), 64. *Prof Exp:* Postdoctoral fel, Clarendon Lab, Oxford Univ, 64-66; mem tech staff, Bell Tel Labs, 66-68; asst prof, Univ Southern Calif, 68-73; mem tech staff, 73-77, res scientist, 77-79, mgr device physics sect, 79-80, mgr galluim arsenide microelectronics, 80-84, MGR, SILICON MICROELECTRONICS SOLID-STATE ELECTRONICS DEPT, ELECTRONICS RES LAB, AEROSPACE CORP, 84- *Mem:* Am Phys Soc; Inst Physics Eng; Inst Elec & Electronics Engrs. *Res:* Light scattering, electro-optics, lasers and quantum electronics; resonance phenomena, atomic standards and magnetics; superconducting devices; silicon and gallium arsenide microelectronics. *Mailing Add:* 6903 Starstone Dr Palos Verdes Peninsula CA 90274

HURREN, WEILER R, b Salt Lake City, Utah, May 16 35; m 60; c 5. PLASMA PHYSICS, SOLID STATE PHYSICS. *Educ:* Utah State Univ, BS, 59; Univ SC, MS, 62; Brigham Young Univ, PhD(physics), 68. *Prof Exp:* Asst prof physics, Western Ill Univ, 68-72, assoc prof & chmn dept, 72-77; PROF PHYSICS & HEAD DEPT, THE CITADEL, 80- *Mem:* Am Phys Soc; Sigma Xi. *Res:* Magnetic resonance. *Mailing Add:* Dept Physics The Citadel Charleston SC 29409

HURSH, CHARLES RAYMOND, environmental sciences, for more information see previous edition

HURSH, JOHN BACHMAN, physiology, for more information see previous edition

HURSH, JOHN W(OODWORTH), b Duluth, Minn, Oct 8, 20; m 44; c 1. INERTIAL NAVIGATION, AUTOMATIC CONTROL & SATELLITE SYSTEMS. *Educ:* Harvard Univ, SB, 42, SM, 43, MES, 52. *Prof Exp:* Res engr, USM Corp, 46-48, proj leader, 48-51; mem staff, Instrumentation Lab, Mass Inst Technol, 51-52, asst proj engr, 52-54, proj engr, 54-55, asst dir, 55-63, assoc dir, 63-73; div leader, Charles Stark Draper Lab, Inc, 75-85; CONSULT, 85- *Honors & Awards:* Gold Medal Award, NY Acad Sci, 67. *Res:* Inertial navigation systems; administration of scientific activities; instrumentation and satellite systems. *Mailing Add:* 35 Newbury Park Needham MA 02192

HURSH, ROBERT W(ILLIAM), b Champaign, Ill, Mar 3, 16; m 47; c 2. MECHANICAL ENGINEERING. *Educ:* Univ Ill, BS, 37; Ohio State Univ, MS, 39. *Prof Exp:* Asst, Univ Ill, 37-38; fel, Ohio State, 38-39; eng trainee, B F Goodrich Co, Ohio, 40, physicist, 40-43, tire construct engr, 43-46; develop engr, Owens-Ill Glass Co, NJ, 46-49, Lukens Steel Co, Pa, 49-54, Gen Elec Co, Ohio, 54-58 & NY, 58-60; dir res & develop, Hansen Mfg Co, 60-63; head mech div, Electromech Res Ctr, Repub Steel Corp, 64-66; dir res & develop, Kinnear Corp, 67-70; indust hyg engr, Indust Comn Ohio, 70-77, mgr eng & indust hyg, 77-79, mgr tech res & eng, Div Safety & Hyg, 79-86; RETIRED. *Mem:* Am Soc Mech Engrs; Nat Soc Prof Engrs; Am Indust Hyg Asn. *Res:* Tires; process machinery; machinery products; electric insulation; fluid handling equipment; steel production and testing; industrial hygiene. *Mailing Add:* 983 Clubview Blvd Columbus OH 43235

HURST, DAVID CHARLES, b Bozeman, Mont, Aug 25, 28; m 54; c 2. STATISTICS. *Educ:* Mont State Col, BS, 50; NC State Col, MS, 57, PhD, 62. *Prof Exp:* Asst statist, NC State Col, 53-57; assoc prof, Va Polytech Inst & State Univ, 66; assoc prof, 66-70, prof statist & chmn dept biostatist, Sch Med, 70-82, PROF STATIST, UNIV ALA, BIRMINGHAM, 70- *Mem:* Am Statist Asn; Inst Math Statist; Biomet Soc. *Res:* Statistical techniques; design of experiments; nonlinear estimation; biostatistics; research consulting; biomedical modeling. *Mailing Add:* Dept Biostatist & Biomath Univ Ala Univ Sta Birmingham AL 35294

HURST, DONALD D, b Bronx, NY, July 8, 29; m 51; c 2. GENETICS. *Educ:* Univ Colo, AB, 50; Univ Mich, MS, 56, PhD(zool), 59. *Prof Exp:* From instr to assoc prof, 58-70, chmn dept, 71-80, PROF BIOL, BROOKLYN COL, 70- *Concurrent Pos:* USPHS fel genetics, Univ Calif, Berkeley, 69-70. *Mem:* AAAS; Genetics Soc Am; Am Soc Microbiol; Sigma Xi. *Res:* Microbial genetics; mechanism of gene conversion and meiotic recombination in Saccharomyces cerevisiae. *Mailing Add:* Dept Biol Brooklyn Col Brooklyn NY 11210

HURST, EDITH MARIE MACLENNAN, b Detroit, Mich, Aug 1, 26; m 58; c 3. NEUROANATOMY. *Educ:* Wayne State Univ, BS, 48; Univ Mich, MA, 49, PhD(anat), 56. *Prof Exp:* Instr anat, Hahnemann Med Col, 56-60, asst prof, 66-71; asst prof, Med Col Pa, 71-73, assoc prof, 73-76, PROF BIOL, EASTERN MICH UNIV, 76- *Mem:* Am Asn Anatomists. *Res:* Study of connections of cortical auditory areas by ablation experiments in monkey; histological and chemical studies of rodent hypothalamus. *Mailing Add:* Dept Biol Eastern Mich Univ Ypsilanti MI 48197

HURST, ELAINE H, b St Joseph, Mich, Mar 4, 20; m 46; c 2. PHYCOLOGY, AQUATIC ECOLOGY. *Educ:* Western Mich Univ, BS, 43, MA, 55; Mich State Univ, PhD(mycol), 65. *Prof Exp:* Teacher, Milford Consol Schs, 43-44; teacher, Richland Rural Agr Sch, 44-46; instr biol, Western Mich Univ, 55-63, asst prof, 63-66; assoc prof, Nazareth Col, Mich, 66-69, prof biol, 69-, head dept, 66- *Mem:* Am Micros Soc. *Res:* Growth of blue-green algae in culture and production of extra metabolites by these organisms; role of extra metabolites in ecology. *Mailing Add:* 3205 Hylle Ave Kalamazoo MI 49007

HURST, G SAMUEL, b Pineville, Ky, Oct 13, 27; m 48; c 2. RADIATION PHYSICS, PHOTOPHYSICS. *Educ:* Berea Col, AB, 47; Univ Ky, MS, 48; Univ Tenn, PhD(physics), 59. *Hon Degrees:* Dr, Berea Col. *Prof Exp:* Sr physicist, Oak Ridge Nat Lab, Tenn, 48-66; prof physics, Univ Ky, 66-70; sr physicist, Oak Ridge Nat Lab, 70-85; prof physics, Univ Tenn, 85-88; INDEPENDENT CONSULT, 88- *Concurrent Pos:* Vis res prof, Fla State Univ, 62-63; assoc prof, Univ Tenn, 63-; corp fel, Union Carbide & Martin Marietta Corps. *Honors & Awards:* IR-100 Touch Sensor & IR-100 One-Atom Detector Awards. *Mem:* Fel Am Phys Soc. *Res:* Laser and instrumentation applications; atomic and molecular physics; chemical kinetics-gas phase and photochemistry; atomic diffusion and mobility; one-atom detection; resonance ionization spectroscopy (RIS). *Mailing Add:* 10521 Res Dr Suite 300 Knoxville TN 37932-2567

HURST, HARRELL EMERSON, b Somerset, Ky, Mar 4, 49; m 74. MASS SPECTROMETRY, PHARMACOKINETICS. *Educ:* Univ Ky, BS, 72, MS, 76, PhD(toxicol), 78. *Prof Exp:* From instr to asst prof, 78-84, ASSOC PROF PHARMACOL & TOXICOL, SCH MED, UNIV LOUISVILLE, 84- *Concurrent Pos:* Dir, dept pharmacol & toxicol, Therapeut & Toxicol Lab, 82-; pres, Ohio Valley Regional Chap, Soc Toxicol, 86. *Mem:* Am Chem Soc; Am Soc Mass Spectrometry; Am Soc Pharmacol & Exp Therapeut; Soc Toxicol; Sigma Xi. *Res:* Analytical chemistry, pharmaco- and toxicokinetics, toxicological and pharmacological problems; techniques include gas chromatography-mass spectrometry, gas chromotography and high performance liquid chromotography for study of drug and chemical disposition in humans and animals. *Mailing Add:* Dept Pharmacol & Toxicol Sch Med Univ Louisville Louisville KY 40292

HURST, HOMER T(HEODORE), b Lone Rock, Ark, Oct 25, 19; m 45; c 4. AGRICULTURAL ENGINEERING. *Educ:* Ohio State Univ, BS, 50, MS, 51. *Prof Exp:* Instr agr eng, Ohio State Univ, 51-55; assoc prof, 55-60, PROF AGR ENG, VA POLYTECH INST & STATE UNIV, 60-, PROF ENVIRON & URBAN SYSTS, 77- *Mem:* Am Soc Agr Engrs. *Res:* Accelerated structural research with full-scale buildings or large sections of homes and service structures. *Mailing Add:* Dept Agr Eng Va Polytech Inst & State Univ Blacksburg VA 24061

HURST, JAMES KENDALL, b Maquoketa, Iowa, Oct 17, 40; m 66; c 2. BIOCHEMISTRY. *Educ:* Cornell Col, BA, 62; Stanford Univ, PhD(phys chem), 66. *Prof Exp:* NIH career develop awards, Cornell Univ, 66-67 & 68-69; from asst prof to assoc prof, 69-82, chmn dept, 78-82, PROF CHEM, ORE GRAD CTR, 78-, VCHAIR CHEM PROG, 82- *Concurrent Pos:* Consult, McDonell-Douglas Res Labs, St Louis 78-79; vis prof, Inst de Phys Chem, Lausanne, Switz, 80 & 85. *Mem:* Am Chem Soc. *Res:* Oxidation-reduction reactions of coordination compounds and biological particles; leukocyte microbicidal mechanisms. *Mailing Add:* Dept Chem Biol & Environ Sci Ore Grad Ctr Beaverton OR 97006-1999

HURST, JERRY G, b Belleview, Mo, Dec 8, 32; m 54; c 7. PHYSIOLOGY. *Educ:* Hardin-Simmons Univ, BA, 54; Okla State Univ, MS, 63, PhD(physiol), 67. *Prof Exp:* Instr, Buckner Acad, Tex, 54-56; teacher, Pampa Sch Dist, 59-60; asst prof, 64-69, ASSOC PROF PHYSIOL, OKLA STATE UNIV, 69- *Res:* Control of parathyroid hormone secretion rate; iodide metabolism in birds. *Mailing Add:* Dept Zool Okla State Univ Stillwater OK 74078

HURST, JOSEPHINE M, b Philadelphia, Pa, May 14, 38. HEMATOLOGY, ONCOLOGY. *Educ:* Immaculata Col, Pa, AB, 60; St John's Univ, NY, MS, 62, PhD(hemat-physiol), 65. *Prof Exp:* Teaching asst hemat, St John's Univ, 60-62; res asst hemat, NY Univ, 62-63; instr biol, St Johns Univ, 63-65; res assoc hemat, Holt Radium Inst, Christie Hosp, UK, 65-66; asst prof hemat, Grad Sch, St John's Univ, 66-71; res scientist hemat & oncol, NY Ocean Sci Lab, 71-72; RES ASSOC, HEMAT & ONCOL, GOODWIN INST CANCER RES, FLA, 79- *Concurrent Pos:* Res assoc hemat, Sch Med, Univ Bristol, UK, 66 & Med Dept, Brookhaven Nat Lab, 71-72; consult path & oncol, Cancer Lab, N Ridge Hosp, Fla, 82-83, actg head, 83-86; dir, SFla Biotherapeut Assocs, 87-88. *Mem:* Int Asn Breast Cancer Res. *Res:* Control of differentiation and proliferation of the hemopoietic cells; mechanism of tumor progression; drug sensitivity of human tumor cells. *Mailing Add:* Goodwin Inst Cancer Res 1850 69th Ave Plantation FL 33313

HURST, PEGGY MORISON, b Abington, Pa, May 15, 25. INORGANIC CHEMISTRY. *Educ:* Wilson Col, AB, 46; Univ Wis, MA, 48, PhD(chem), 56. *Prof Exp:* From instr to assoc prof, 55-75, PROF CHEM, BOWLING GREEN STATE UNIV, 75- *Mem:* AAAS; Am Chem Soc; Sigma Xi. *Res:* Peroxy compounds of transition elements; history of science. *Mailing Add:* Dept Chem Bowling Green State Univ Bowling Green OH 43403

HURST, REX LEROY, b Payson, Utah, Mar 10, 23; m 46; c 5. STATISTICS, COMPUTER SCIENCE. *Educ:* Utah State Univ, BS, 48, MS, 50; Cornell Univ, PhD, 52. *Prof Exp:* Asst prof appl statist, Utah State Univ, 52-55, assoc prof & head dept, 55-59; vis prof statist, Iowa State Univ, 59-60; PROF APPL STATIST & COMPUT SCI & HEAD DEPT, UTAH STATE UNIV, 60- *Mem:* Am Statist Asn; Biomet Soc. *Res:* Data analysis; experimental design; computer programming. *Mailing Add:* Dept Comput Sci Utah State Univ Logan UT 84322-4205

HURST, RICHARD WILLIAM, b Rockville Centre, NY, Aug 2, 48. GEOLOGY, GEOCHEMISTRY. *Educ:* State Univ NY Stony Brook, BS, 70; Univ Calif, Los Angeles, PhD(geol), 75. *Prof Exp:* Res asst geochem, Earth Sci, State Univ NY Stony Brook, 69-70; res asst geol, Univ Calif, Los Angeles, 72-75; asst res geol, Univ Calif, Santa Barbara, 75-77; ASST PROF GEOL, CALIF STATE UNIV, LOS ANGELES, 77- *Concurrent Pos:* Consult, Santa

Monica Mountains Task Force, 74-; Penrose Grant, Geol Soc Am, 74; instr geol, Santa Barbara City Col, 75-78. *Mem:* Meteoritical Soc; Geol Soc Am; Geochem Soc; Am Geophys Union. *Res:* Geochronology of early archaean gneisses of Labrador; archaean crustal evolution; evolution of the Santa Monica mountains; environmental geology. *Mailing Add:* Dept Geol Calif State Univ 5151 St Univ Dr Los Angeles CA 90032

HURST, ROBERT NELSON, b Ogden, Utah, May 21, 30; m 49; c 3. ENVIRONMENTAL PHYSIOLOGY, SCIENCE EDUCATION. *Educ:* Univ Utah, BS, 57; Harvard Univ, AMT, 58; Purdue Univ, MS, 64, PhD(physiol), 66. *Prof Exp:* Teacher, San Diego Unified Sch Dist, 58-63; from instr to asst prof biol, Purdue Univ, 63-67; assoc prof zool & educ, Univ Man, 67-69; asst prof, 69-75, ASSOC PROF BIOL, PURDUE UNIV, WEST LAFAYETTE, 75- *Concurrent Pos:* Area day consult, Biol Sci Curric Study, 65; vis scientist, Int Rice Res Inst, Los Banos, Philippines, 78-79. *Mem:* Am Inst Biol Sci; Nat Sci Teachers Asn; Int Audio-Tutorial Cong (pres, 72-73). *Res:* Seasonal behavior and bio-rhythmia of hybernating animals; simultaneous recording of physiological parameters; galvanic skin response; audio-tutorial instruction and learning theory. *Mailing Add:* Dept Biol Sci Purdue Univ West Lafayette IN 47907

HURST, ROBERT PHILIP, b Bozeman, Mont, May 30, 30; m 55; c 4. ATOMIC PHYSICS, MOLECULAR PHYSICS. *Educ:* Mont State Univ, BS, 52; Univ Tex, MA, 56, PhD(chem physics), 58. *Prof Exp:* Jr scientist, Gen Elec Co, 52-54; Teaching asst, 54-55, res asst, Univ Tex, Austin, 55-58; fel, Univ Wis-Madison, 58-59; fel, Univ Ill, Urbana, 59-60; asst prof physics & chem, Univ Okla, 60-62; PROF PHYSICS, STATE UNIV NY, BUFFALO, 62- *Concurrent Pos:* Vis prof, Univ Sussex, Eng, 68-69 & Univ Tex, 76. *Mem:* Am Phys Soc; NY Acad Sci. *Res:* Irridiated uranium crystal structures; calculations in atomic and molecular quantium mechanics; computer profiles and momentum space; non-linear electric polarizabilities. *Mailing Add:* Dept Physics & Astron State Univ NY Buffalo Amherst NY 14260

HURST, ROBERT R(OWE), b Washington, Iowa, Sept 17, 37; c 3. MEDICAL PHYSICS, RADIATION THERAPY PHYSICS. *Educ:* Pa State Univ, BS, 60, MS, 62, PhD(nuclear physics), 64. *Prof Exp:* Asst, Pa State Univ, 60-64; from asst prof to assoc prof physics, Univ Mo-Columbia, 64-83; DIR MED PHYSICS & CLIN ENG, BOONE HOSP CTR, COLUMBIA, MO, 83- *Mem:* Am Phys Soc; Am Asn Physicists in Med; Am Asn Physics Teachers; Sigma Xi; Soc Magnetic Resonance in Med. *Res:* Magnetic resonance imaging. *Mailing Add:* Dept Med Physics Boone Hosp Ctr 1600 E Broadway Columbia MO 65201

HURST, VERNON JAMES, b Glenmore, Ga, July 18, 23; m 50; c 2. GEOLOGY, CRYSTALLOGRAPHY. *Educ:* Univ Ga, BS, 51; Emory Univ, MS, 52; Johns Hopkins Univ, PhD(geol), 54. *Prof Exp:* Geologist, Dept Mines, Mining & Geol, Ga State Geol Surv, 53-61; prof geol & head dept, 61-66, chmn div phys sci, 66-69, RES PROF GEOL, UNIV GA, 69- *Mem:* Fel Geol Soc Am; Geochem Soc; fel Mineral Soc Am; Soc Econ Geol; French Soc Mineral & Crystallog. *Res:* Crystal growth; saprolitization. *Mailing Add:* Dept Geol-GGS 131 Univ Ga Athens GA 30602

HURST, WILBUR SCOTT, b Camden, NJ, May 20, 39; m 66; c 1. PHYSICS. *Educ:* Albion Col, BA, 61; Univ Pa, MS, 63; Pa State Univ, PhD(physics), 68. *Prof Exp:* PHYSICIST, NAT BUR STANDARDS, 68- *Mem:* Am Phys Soc. *Res:* Temperature metrology; stimulated raman spectroscopy. *Mailing Add:* Bldg 221 Rm B126 Nat Inst Standards & Technol Gaithersburg MD 20899

HURST, WILLIAM JEFFREY, b West Palm Beach, Fla, Jan 12, 48; m 69; c 2. ANALYTICAL CHEMISTRY. *Educ:* Ohio Univ, AB, 69; Youngstown State Univ, MS, 75; Columbia Pac Univ, PhD, 84. *Prof Exp:* Radiation safety officer, USAF, 69-72; instr environ sci, Muskingum Area Tech Col, Ohio, 72-73; assoc scientist, 76-77, scientist, 77-79, SR SCIENTIST, ANALTICAL GROUP, HERSHEY FOODS CORP, 79- *Concurrent Pos:* Instr, Chem Dept, Dickinson Col, 77-; adj asst prof chem, Lebanon Valley Col, 82-; tech abstrutor, Preston Pub, 81-; chem res officer, Air Force Wright Aeroneut Lab, Air Force Mat Lab, Wright Patterson AFB, 81-; instr, dept continuing educ, Hershey Med Ctr & clin asst prof comp med. *Mem:* Am Chem Soc; Inst Food Technologists; fel Am Inst Chemists; NY Acad Sci. *Res:* Analytical chemistry of compounds of nutritional or food safety emphasis and the use of high-performance liquid chromatography and its applicability to solve problems in that area. *Mailing Add:* PO Box 378 Mt Gretna PA 17064

HURT, ALFRED B(URMAN), JR, b NC, Sept 5, 20; m 49; c 1. SYNTHETIC TEXTILE FIBERS ENGINEERING, CHEMICAL ENGINEERING. *Educ:* NC State Col, BChE, 47, MSEE, 50. *Prof Exp:* Instr elec eng, NC State Col, 48-49; sr tel engr, Rural Elec Authority, NC, 49-50; chemist, E I du Pont de Nemours & Co, 41-44, engr, 52-55, sr engr, 55-60, sr res engr, 60-66, eng assoc, 66-74, design supvr eng, 74-78, proj engr, 78-79; RETIRED. *Concurrent Pos:* Captain, comndg officer, USN Res Unit, 76-78. *Mem:* Inst Elec & Electronics Engrs. *Res:* Automatic impedance measurement; industrial instrumentation; synthetic textile process development and equipment design. *Mailing Add:* Rte 1 Box 159 Crumpler NC 28617

HURT, H DAVID, b Del Norte, Colo, Jan 2, 41; m 62; c 2. NUTRITION. *Educ:* Colo State Univ, BS, 64; Univ Conn, MS, 67; Cornell Univ, PhD(nutrit), 70. *Prof Exp:* Asst dir nutrit res, Nat Dairy Coun, 69-70, assoc dir, 70-72; head, Div Nutrit Sci, Campbell Soup Co, 72-78; assoc dir biochem & nutrit, Del Monte Corp, 78-80; DIR NUTRIT, QUAKER OATS CO, 80- *Mem:* Am Inst Nutrit; Inst Food Technol; Am Dietetic Asn. *Res:* Protein quality evaluation; bioavailability of nutrients in foods; nutrient retention of processed foods. *Mailing Add:* Nat Live Stock & Mead Bd 444 N Michigan Ave Chicago IL 60611

HURT, JAMES EDWARD, b Newton, Kans, Feb 11, 35. PHYSICS. *Educ:* Okla State Univ, BS, 57, MS, 59, PhD(physics), 63. *Prof Exp:* Sr assoc physicist, 63-64, staff physicist, 64-68, ADV PHYSICIST, PROD DEVELOP LAB, IBM CORP, 68- *Mem:* AAAS; Am Phys Soc. *Res:* Solid state and surface physics. *Mailing Add:* 9906 E Colette Tucson AZ 85748-1907

HURT, JAMES JOSEPH, b Iowa City, Iowa, June 25, 39; m 61; c 3. OPERATIONS RESEARCH, APPLIED MATHEMATICS. *Educ:* Univ Iowa, BS, 61, MS, 63; Brown Univ, PhD(appl math), 67. *Prof Exp:* Asst prof math, Univ Iowa, 67-70; mathematician, Hq US Army Weapons Command & Rodman Labs, 72-76; sr staff mathematician, Deere & Co, 76-85; chief mathematician, Cognition, Inc, 85-89; SR PARTNER, HARDY, HURT & COIN, 89- *Mem:* Math Asn Am; Sigma Xi; Soc Indust & Appl Math; Asn Comput Mach; Am Math Soc. *Res:* Simulation of manufacturing process. *Mailing Add:* 324 Bell St Chagrin Falls OH 44022-2908

HURT, JOHN CALVIN, b Oak Hill, WVa, Apr 4, 41; m 63; c 1. CERAMICS, MATERIALS SCIENCE. *Educ:* Rutgers Univ, BS, 63, PhD(ceramic sci), 67. *Prof Exp:* Officer, US Army Ceramics Res & Develop, Army Mat & Mech Res Ctr, 67-69; asst prof mat eng, NC State Univ, 69-72; staff scientist, 72-74, ASSOC DIR MAT SCI, US ARMY RES OFF, 74- *Concurrent Pos:* Adj asst prof mat eng, NC State Univ, 72-77, adj assoc prof, 77- *Mem:* Sigma Xi; Am Ceramic Soc; Am Soc Metals. *Res:* Composite materials; transport phenomena; electrical, magnetic and optical properties of materials. *Mailing Add:* Mat Eng Dept NC State Univ Raleigh NC 27607

HURT, SUSAN SCHILT, b Buffalo, NY, Aug 6, 47; c 3. ENVIROMENTAL TOXICOLOGY, ENVIROMENTAL SCIENCES. *Educ:* Univ Mich, BSChem, 69; Univ Colo, PhD(org chem), 74; Am Bd Toxicol, dipl, 83. *Prof Exp:* Fel, Univ Colo, 74; sr chemist agr res, 75-80, prod tech mgr, Govt Rel Agr Chem, 80-81, PROG MGR TOXICOLOGY, ROHM AND HAAS CO, 81-, PROG MGR ECOTOXICOL, 86- *Mem:* Am Chem Soc; AAAS. *Res:* Design and synthesis of candidate herbicides; environmental/toxicological evaluation and risk assessment of pesticides and industrial chemicals, registration management. *Mailing Add:* 1528 Evans Rd RD 1 Ambler PA 19002-1211

HURT, VERNER C, b Yazoo City, Miss, Dec 19, 29; c 4. AGRICULTURAL ECONOMICS. *Educ:* Miss State Univ, BS, 50, MS, 55; Okla State Univ, PhD, 61. *Prof Exp:* Auditor, Miss Dept Agr & Com, 55-56; res asst, Agr Econ Dept, Okla State Univ, 58-60; asst prof & asst economist, Agr Econ Dept, Miss State Univ, 56-58, from assoc prof & economist to prof & economist, 60-71, asst dept head agr econ, 71-72, dept head, 72-87, DIR, AGR & FORESTRY EXP STA, MISS STATE UNIV, 87- *Concurrent Pos:* Economist & statist consult, Fla Citrus Comn; statist & computer programmer & consult, Southern Interstate Nuclear Bd; chmn, Econ Adv Comt Dairymen Inc & Alternative Publ Comt, Am Agr Econ Asn, 76; econ consult, Econ Res Serv, USDA & Nat Fertilizer Develop Ctr, Tenn Valley Authority; adv economist, Nat Boiler Mkt Asn. *Mem:* Am Agr Econ Asn; Southern Agr Econ Asn (pres, 73-74). *Mailing Add:* PO Drawer ES Mississippi State MS 39762

HURT, WAVERLY GLENN, b Richmond, Va, Sept 6, 38. GYNECOLOGY. *Educ:* Hampden-Sydney Col, BS, 60; Med Col Va, MD, 64. *Prof Exp:* PROF OBSTET & GYNEC, MED COL VA, 69- *Concurrent Pos:* Med dir, Ambulatory Surgery Ctr, Med Col Va, 81- *Mem:* Fel Am Col Obstetricians & Gynecologists. *Res:* Urogynecology; gynecologic surgery. *Mailing Add:* Dept Obstet & Gynec Box 34 Med Col Va Richmond VA 23298-0034

HURT, WILLIAM C, JR, b Tuskegee, Ala, Dec 27, 07; m 36; c 3. MECHANICAL ENGINEERING. *Educ:* Ala Polytech Inst, BS, 28, ME, 33. *Prof Exp:* Appln engr, Gen Elec Co, 31-36, mfg supt staff, 36-39, wage rate & planning, 39-40, design engr, Pittsfield Lab, 40-41, develop engr, 41-44 & 46-53, mgr & supvr adv process develop, 53-60, sr mech engr, mat eng div, 60-69, mat & tech lab, 69-70; RETIRED. *Concurrent Pos:* Eng Consult, Berkshire Community Col & Serv Corps Retired Execs. *Mem:* Am Soc Mech Engrs; Nat Soc Prof Engrs; Wire Asn Int. *Res:* Development of processes and material applications in area of electrical conductors; process developments; transformers and allied products. *Mailing Add:* 43 Arlington St Pittsfield MA 01201

HURT, WILLIAM CLARENCE, b Waynesboro, Miss, Nov 11, 22; m 43; c 4. PERIODONTICS, ORAL PATHOLOGY. *Educ:* Univ Miss, 40-42; Loyola Univ, La, DDS, 47; Am Bd Periodont, dipl, 62; Am Bd Oral Med, dipl, 82. *Prof Exp:* Resident periodont, Walter Reed Gen Hosp, 57-58; chief periodont serv, Brooke Gen Hosp, San Antonio, Tex, 63-69; prof periodont, Univ Tex Dent Br Houston, 69-72; prof periodont & chmn dept, 72-87, EMER PROF PERIODONT, BAYLOR COL DENT, 88- *Concurrent Pos:* Consult, Fourth US Army, San Antonio, 64-69, M D Anderson Hosp & Tumor Inst, 71-72, Brooke Gen Hosp, 72-81, Vet Admin Hosp, Dallas, 73-88 & Comn Dent Educ, Am Dent Asn, 73-88; ed, J Periodont, 79-88. *Honors & Awards:* Presidential Award, Am Acad Periodont, 89. *Mem:* Am Dent Asn; Int Asn Dent Res; fel Am Acad Oral Path; fel Am Col Dent; fel Am Acad Periodont. *Res:* Wound healing and bone regeneration; mucocutaneous disorders. *Mailing Add:* Dept Periodont Baylor Col Dent Dallas TX 75246

HURTER, ARTHUR P, b Chicago, Ill, Jan 29, 36; m; c 2. ENGINEERING ECONOMICS, LOGISTICS. *Educ:* Northwestern Univ, BS, MS & MA, 61, PhD(econ), 62. *Prof Exp:* Res assoc, Transp Ctr, Northwestern Univ, 62-65, assoc dir res, 65-69, from asst prof to assoc prof, 62-69, chmn dept, 70-90, PROF INDUST ENG & MGT SCI, NORTHWESTERN UNIV, 69- *Concurrent Pos:* Fac mem, Newspaper Mgt Ctr Univ, 90-; consult, Standard Oil Industs, Sears Roebuck & Ill Inst Natural Resources; chmn, Opers Res Soc Am; NSF res grants. *Mem:* Am Econ Asn; Inst Indust Engrs; Opers Res Soc Am; Inst Mgt Sci; Regional Sci Asn. *Res:* Author of over 60 books and papers on the integration of operations research and economics applied to facility location problems, plant and equipment investment and replacement, transportation, logistics, cost/benefit studies dealing with environment and energy, production/distribution problems-currently in newspaper industry. *Mailing Add:* 1505 W Norwood St Chicago IL 60660

HURTO, KIRK ALLEN, b Chicago, Ill, Nov 27, 51; m 75; c 2. WEED SCIENCE, TURFGRASS MANAGEMENT. *Educ:* Southern Ill Univ, BS, 74, MS, 75; Univ Ill, PhD(hort), 78. *Prof Exp:* Grad asst plant & soil, Southern Ill Univ, 74-75; grad asst turf, Univ Ill, Urbana, 75-78; asst prof plant & soil sci, Univ Mass, Amherst, 78-81; res scientist, 81-85, sr res scientist & group mgr, 86-90, DIR TECH SERV, CHEMLAWN SERV GROUP, 90- *Mem:* Am Soc Agron; Crop Sci Soc Am; Weed Sci Soc Am; Soil Sci Soc Am. *Res:* Turfgrass ecology and adaptation, weed competition and control, and herbicide activity in plants and soil environment. *Mailing Add:* ChemLawn Serv Corp 8275 N High St Columbus OH 43235

HURTT, WOODLAND, b Galena, Md, Feb 6, 32; m 59; c 2. PLANT PHYSIOLOGY. *Educ:* Univ Md, BS, 55, MS, 57; Mich State Univ, PhD(crop sci), 62. *Prof Exp:* Fel biochem, Mich State Univ, 61-63; plant physiologist, Vegetation Control Div, USDA, Ft Detrick, 63-74, plant physiologist, Weed Physiol Lab, Sci & Educ Admin-Agr Res, 74-87; SCIENTIST, DYNAMAC, INT, 87- *Mem:* AAAS; Am Soc Plant Physiol; Weed Sci Soc Am; Sigma Xi; Plant Growth Regulator Soc Am. *Res:* Weed seed germination; promotion, inhibition, and physiology as affected by chemicals and environmental variables. *Mailing Add:* 7302 Parkview Dr Frederick MD 21702

HURTUBISE, ROBERT JOHN, b Chicago, Ill, June 7, 41; m 65; c 3. ANALYTICAL CHEMISTRY. *Educ:* Xavier Univ, BS, 64, MS, 66; Ohio Univ, PhD(anal chem), 69. *Prof Exp:* Asst prof chem, Rockhurst Col, 69-71; supvr chem test, Pfizer Inc, 71-74; from asst prof to assoc prof, 74-83, PROF CHEM, UNIV WYO, 83- *Mem:* Am Chem Soc; Soc Appl Spectros; AAAS. *Res:* Luminescence analysis; chemical and physical separation methods; author of various publications. *Mailing Add:* Dept Chem Univ Wyo Laramie WY 82071

HURVICH, LEO MAURICE, b Malden, Mass, Sept 11, 10; m 48. PSYCHOPHYSIOLOGY. *Educ:* Harvard Col, BA, 32; Harvard Univ, MA, 34, PhD(psychol), 36. *Hon Degrees:* MA, Univ Pa, 72; DSc, State Univ NY, 89. *Prof Exp:* Asst psychol, Harvard Univ, 36-37, instr, 37-40, res asst, Grad Sch Bus Admin, 40-47; res psychologist vision, Eastman-Kodak Co, 47-57; prof & chmn dept psychol, Wash Square Col, NY Univ, 57-62; prof psychol, 62-79, EMER PROF PSYCHOL, INST NEUROL SCI, UNIV PA, 79- *Concurrent Pos:* Rep Am Psychol Asn, Optics Sect, Am Standards Asn, 53-54 & deleg, Inter Soc Color Coun, 55-79; Guggenheim fel, 64-65; chmn, Vision Tech Group, Optical Soc Am, 70-71, Troland Award Comt, Nat Acad Sci, 83-84; vis res prof, Columbia Univ, 71-72; vis prof psychol, Ctr Visual Sci, Univ Rochester, 74; mem, vis comt, Psychol Dept, Mass Inst Technol, 77-83; assoc ed, J Optical Soc Am, 80-83; fel, Ctr Advan Studies Behav Sci, 81-82; MEC-NAS Comt on Currency, 85-87. *Honors & Awards:* Howard Crosby Warren Award, Soc Exp Psychologists, 71; Distinguished Sci Contrib Award, Am Psychol Asn, 72; Godlove Award, Int Soc Color Coun, 73; Edgar D Tillyer Award, Optical Soc Am, 82; Deane B Judd-Int Color Asn Award, 85; Hermann von Helankoltz Prize, Cognitive Neurosci Inst, 87. *Mem:* Nat Acad Sci; Optical Soc Am; Soc Neurosci; Asn Res Vision & Ophthal; Soc Exp Psychologists; Am Acad Arts & Sci. *Res:* Sensory processes; psychophysics and psychophysiology of vision; color vision; neural coding. *Mailing Add:* 3815 Walnut St Philadelphia PA 19104-6196

HURVITZ, ARTHUR ISAAC, b Newton, Mass, Nov 29, 39; m 61; c 2. COMPARATIVE PATHOLOGY, IMMUNOPATHOLOGY. *Educ:* Mich State Univ, BS, 62, DVM, 64; Univ Calif, Davis, PhD(comp path), 67; Am Col Vet Path, dipl. *Prof Exp:* Res assoc comp path, Rockefeller Univ, 67-69, asst prof, 69-71; DIR RES & CHMN, DEPT PATH, ANIMAL MED CTR, NEW YORK, 71- *Concurrent Pos:* Adj asst prof immunol, Rockefeller Univ, 71-73; assoc scientist, Sloan-Kettering Inst Cancer Res, 71-73; adj asst prof, Col Physicians & Surgeons, Columbia Univ, 73-; consult, Off Dir, Bur Radiol Health, US Dept Health Educ & Welfare, 70-71. *Honors & Awards:* Small Animal Res Award, Ralston Purina. *Mem:* Am Vet Med Asn; Am Soc Exp Path; Int Acad Path; Am Col Vet Path. *Res:* Immunology; immunopathology; mechanisms of disease-correlating morphological with underlying molecular-biochemical alterations. *Mailing Add:* Animal Med Ctr 510 E 62nd St New York NY 10021

HURWICZ, LEONID, b Moscow, Russia, Aug 21, 17; nat US; m 44; c 4. ECONOMICS, STATISTICS. *Educ:* Univ Warsaw, LLM, 38. *Prof Exp:* Asst, Mass Inst Technol, 41; mem fac, Univ Chicago, 41-45; assoc prof econ & statist, Iowa State Col, 45-49; res prof econ & math statist, Univ Ill, 49-51; prof econ & math, 51-71, CURTIS L CARLSON PROF ECON & EMER REGENTS PROF, UNIV MINN, MINNEAPOLIS, 71- *Concurrent Pos:* Res assoc, Cowles Comn, 41-45, vis prof, 50; Guggenheim fel, 45-45; fel, Ctr Advan Study Behav Sci, 55-56 & 58-59; vis prof, Stanford Univ, 58-59; consult, Cowles Comn, 46-, Econ Comn for Europe, UN, 48, Rand Corp, 50-51 & US Bur Budget, 51-52; mem math div, Nat Res Coun, 54- *Honors & Awards:* Nat Medal Sci, 90. *Mem:* Nat Acad Sci; Am Statist Asn; fel Economet Soc. *Res:* Economic organization in resource allocation; decision-making under uncertainty; dynamic models. *Mailing Add:* 3710 Thomas Ave S Minneapolis MN 55410

HURWITZ, ALEXANDER, b Syracuse, NY, Oct 16, 37; m 69; c 2. COMPUTER SCIENCES. *Educ:* Univ Calif, Los Angeles, BA, 59, MA, 61, PhD(math), 65. *Prof Exp:* STAFF MEM COMPUT SCI, LOS ANGELES SCI CTR, IBM CORP, 65- *Mem:* Sigma Xi; Math Asn Am; Am Asn Artificial Intel; Asn Comput Mach. *Res:* Artificial intelligence; computer graphics. *Mailing Add:* 869 Santa Rita Ave Los Altos CA 94022

HURWITZ, CHARLES E(LLIOT), b Sioux City, Iowa, Nov 10, 37; m 59; c 3. ELECTRICAL ENGINEERING, SOLID STATE PHYSICS. *Educ:* Univ Mich, BSE, 59; Mass Inst Technol, SM, 60, PhD(elec eng), 63. *Prof Exp:* Staff mem, Lincoln Lab, Mass Inst Technol, 63-76, asst leader appl physics group, 76-78, assoc leader, 78-82; VPRES, LASERTRON, INC, 82- *Mem:* Am Phys Soc; Sigma Xi; Inst Elec & Electronics Engrs; sr mem Am Optical Soc. *Res:* Solid state devices; semiconductor lasers and detectors; guided-wave optics; electro-optical devices; fiber optics. *Mailing Add:* 42 Baskin Rd Lexington MA 02173

HURWITZ, DAVID ALLAN, b Lynn, Mass, Apr 17, 38; m 63; c 2. PHARMACOLOGY. *Educ:* Mass Col Pharm, BS, 61, MS, 64, PhD(pharmacol), 69. *Prof Exp:* Pharmacologist, US Army Res Inst Environ Med, 65-70; from asst prof to assoc prof pharmacol, Univ NMex, 70-75; radiopharmacist, Pharmaco Nuclear, Inc, 75-80, pres, 80-81; CONSULT. *Mem:* AAAS; Sigma Xi. *Res:* Behavioral pharmacology; biochemical correlates of behavior, namely the relationships between animal behavior, brain biogenic amines, drugs, and environmental stresses, and the effects of one upon the others. *Mailing Add:* 833 King George Blvd Cleveland OH 44121

HURWITZ, HENRY, JR, b New York, NY, Dec 25, 18; m 51; c 4. NUCLEONICS. *Educ:* Cornell Univ, AB, 38; Harvard Univ, PhD(quantum mech), 41. *Prof Exp:* Instr physics, Cornell Univ, 41-44; res assoc, Los Alamos Sci Lab, 44-46; from supvr theoret group to supvr advan naval reactor physics activ, Knolls Atomic Power Lab, Gen Elec Co, 46-56, mgr nucleonics & radiation, Res Lab, 57-65 & Res & Develop Ctr, 65-68, mgr theory & systs br, 68-72, physicist, Res & Develop Ctr, Gen Elec Co, 72-84; RETIRED. *Concurrent Pos:* GE Coolidge fel, Corp Res & Develop Ctr, Gen Elec Co, 75. *Honors & Awards:* E O Lawrence Mem Award, 61; Glenn Seaborg Medal, Am Nuclear Soc, 89. *Mem:* Fel AAAS; fel Am Phys Soc; fel Am Nuclear Soc; fel NY Acad Sci; sr mem Inst Elec & Electronics Engrs; Sigma Xi. *Res:* Nuclear energy; applied mathematics; physics. *Mailing Add:* 827 Jamaica Rd Schenectady NY 12309

HURWITZ, JAN KROSST, b Toronto, Ont, Apr 3, 24; m 48; c 2. ANALYTICAL CHEMISTRY. *Educ:* Univ Toronto, BASc, 46, MA, 47, PhD(physics), 50. *Prof Exp:* Physicist & sr sci officer, Mines Br, Dept Mines & Tech Surv, Govt Can, 48-57; physicist & assoc res consult, Res lab, US Steel Corp, 57-85. *Concurrent Pos:* Lectr, Univ Ottawa, 54-57 & Univ Pittsburgh, 63-70; consult & lectr, Pa State Univ, 87-88. *Mem:* Soc Appl Spectros (secy, 65-68, treas, 87-89); hon mem Spectros Soc Can (pres, 55-57); Iron & Steel Chemists Asn (secy-treas, 77, pres, 78 & 79); Am Soc Testing & Mat. *Res:* Emission spectroscopy; low pressure discharges; computer applications; development of certified chemical standard samples. *Mailing Add:* 109 Leslie Rd Monroeville PA 15146

HURWITZ, JERARD, b New York, NY, Nov 20, 28; wid; c 4. MICROBIOLOGY. *Educ:* Ind Univ, BA, 49; Western Reserve Univ, PhD(biochem), 53. *Prof Exp:* Res asst biochem, Western Reserve Univ, 49-50; Am Cancer Soc fel, Nat Med Res Coun, London, 53-54, NIH, 55-56; instr, Sch Med, Wash Univ, 56-58; asst prof, Sch Med, NY Univ, 58-60, assoc prof microbiol, 60-63; prof molecular biol, Albert Einstein Col Med, Bronx, NY, 63-65, prof & chmn, Dept Develop Biol & Cancer, 65-84; MEM STAFF & SLOAN KETTERING INST PROF, SLOAN KETTERING DIV, MED COL, CORNELL UNIV, 84- *Concurrent Pos:* Mem, Physiol Chem Study Sect, NIH, 64-68 & 81-84, Spec Viral Cancer Prog, 73-75; Charles Mickle fel award, Can, 67; Guggenheim fel award, Inst Pasteur, 68; mem study sect, Am Cancer Soc, 71-75; Fogarty scholar, 80; Louis & Bert Freeman Found prize res in biochem, NY Acad Sci, 82; Pfizer Sci award, 84-89; Guggenheim Mem Found fel, 86- *Honors & Awards:* Eli Lilly Award, 62; Brown-Hazen Lectr, 72. *Mem:* Nat Acad Sci; Am Soc Biol Chemists; Am Acad Arts & Sci; fel NY Acad Sci. *Res:* Enzyme purification; synthesis of coenzymes; studies of electron transport, cytochromes and flavin enzymes; chemistry of hydroxy amino acids; nucleic acids; author or co-author of over 260 publications. *Mailing Add:* Mem Sloan Kettering Cancer Ctr 1275 York Ave Box 97 New York NY 10021

HURWITZ, LEON, AUTONOMIC PHARMACOLOGY. *Educ:* Univ Rochester, PhD(pharmacol), 53. *Prof Exp:* PROF PHARMACOL & CHMN DEPT, SCH MED, UNIV NMEX, 72- *Res:* Physiology and pharmacology of smooth muscle; role of calcium in muscle contraction. *Mailing Add:* 8904 Harwood NE Albuquerque NM 87111

HURWITZ, MELVIN DAVID, b Medford, Mass, Dec 12, 17; m 43; c 3. ORGANIC CHEMISTRY. *Educ:* Harvard Univ, AB, 39; Univ Chicago, MS, 42; Cornell Univ, PhD(org chem), 48. *Prof Exp:* Res assoc, Billings Hosp, Chicago, 41-42; inspector powder & explosives, Elwood Ord Plant, 42; res assoc, Nat Defense Res Comt, 42-45; asst inorg chem, Cornell Univ, 45-47; res chemist, Rohm and Haas Co, 48-57, head textile prod develop lab, 57-61, head miscellaneous synthesis lab, 61-68, head textile chem lab, 68-73, mgr textile chem res dept, 73-77; prof, 77-86, EMER PROF, CLOTHING & TEXTILES DEPT, UNIV NC, GREENSBORO, 86- *Honors & Awards:* Olney Medalist, Am Asn Textile Chemists, 86. *Mem:* Am Chem Soc; Sigma Xi; Am Asn Univ Prof; Am Asn Textile Chemists & Colorists; Fed Soc Coatings Technol. *Res:* Vitamin analyses; catalytic reductions; explosives; nitrogenous resins; organic fluorine chemistry; high polymers; organic coatings; textile chemicals; textile processing; leather chemicals; flammability; dyestuffs. *Mailing Add:* 5000 Robert Andrews Rd Greensboro NC 27406

HURWITZ, SOLOMON, b New York, NY, Apr 11, 07; m 37; c 2. MATHEMATICS. *Educ:* City Col New York, BS, 27; Columbia Univ, MS, 29, PhD(math), 44. *Prof Exp:* Instr pub sch, NY, 27-42; instr math, Brooklyn Col, 42-46; prof, 46-72, EMER PROF MATH, CITY COL NEW YORK, 72- *Concurrent Pos:* Instr div gen studies, Columbia Univ, 46-48; vis prof, Grad Summer Sch Teachers, Wesleyan Univ, 61-73. *Mem:* Am Math Soc; Math Asn Am. *Res:* A class of Dirichlet series; class of functions suggested by the zeta of Riemann. *Mailing Add:* 2806 Bedford Ave Brooklyn NY 11210

HURYCH, ZDENEK, b Pelhrimov, Czech, July 19, 41; m 65; c 1. SOLID STATE PHYSICS, SURFACE PHYSICS. *Educ:* Charles Univ, Prague, MS, 63, PhD(physics), 67. *Prof Exp:* From asst prof to assoc prof, 68-77, PROF PHYSICS, NORTHERN ILL UNIV, 77- *Mem:* Sigma Xi; Am Phys Soc. *Res:* Photoemission spectroscopy of solids and solid surfaces using synchrotron radiation; optical properties of solids; surface physics; vacuum ultraviolet radiation; electronic structure of solids; layered solids; semiconductors; thin films. *Mailing Add:* Dept Physics Northern Ill Univ DeKalb IL 60115

HUSA, DONALD L, b Hastings, Nebr, Sept 13, 40. SOLID STATE LOW TEMPERATURE, LIGHT WAVE SYSTEMS ENGINEERING. *Educ:* Univ SC, BS, 62; Ohio State Univ, PhD(physics), 66. *Prof Exp:* Res assoc, Univ Pittsburgh, 67-70, res assoc, 70-79, res assoc prof, Stevens Inst Technol, 79-81; MEM TECH STAFF, AT&T BELL LABS, 81- *Mem:* Am Phys Soc; Inst Elec & Electronics Engrs. *Res:* Engineering of lightwave systems. *Mailing Add:* 22 Tudor Dr Ocean Township NJ 07712

HUSA, WILLIAM JOHN, JR, b Gainesville, Fla, Dec 16, 27. ANALYTICAL CHEMISTRY. *Educ:* Univ Fla, BS, 48, MS, 51, PhD, 53. *Prof Exp:* Prof chem, Southwest Mo State Col, 53-66; head phys sci div & chmn dept chem, 66-73, CHMN DIV NATURAL SCI & MATH & HEAD DEPT PHYS SCI, MID GA COL, 73- *Res:* Spectrophotometry; chemical education. *Mailing Add:* Dept Natural Sci Div Mid Ga Col Cochran GA 31014

HUSAIN, ANSAR, b Sitapur, India, Jan 15, 23; US citizen; m 60; c 6. ZOOLOGY, PARASITOLOGY. *Educ:* Aligarh Muslim Univ, India, BSc, 43, MSc, 45; Univ Liverpool, PhD(parasitol), 64. *Prof Exp:* Demonstr, Dept Zool, Aligarh Muslim Univ, India, 45-47; ed officer, Govt India & Govt Pakistan, 47-49; finance officer, Govt Pakistan, 49-50; lectr biol, Chittagong Col, Pakistan, 50-52; sci master several schs & cols, Eng, 52-58 & UN Int Sch, NY, 58-59; asst prof biol & math, State Univ NY Col New Paltz, 59-60; asst prof biol, Roanoke Col, 64-67; assoc prof biol, Albany State Col, 67-90; RETIRED. *Mem:* AAAS. *Res:* Entomology. *Mailing Add:* 2808 Newcomb Rd Albany GA 31705

HUSAIN, LIAQUAT, b Lucknow, India, July 12, 42; m 64; c 2. NUCLEAR CHEMISTRY. *Educ:* Univ Karachi, BS, 61, MS, 63; Univ Ark, PhD(nuclear chem), 68. *Prof Exp:* Asst lectr chem, Univ Karachi, 63-64; res assoc, Brookhaven Nat Lab, 68-70; lectr earth & space sci, State Univ NY Stony Brook, 70-75; sr res scientist, 75-80, DIR, LAB INORG & NUCLEAR CHEM, WADSWORTH CTR OF LABS & RES, NY STATE DEPT HEALTH, 80-; PROF, DEPT ENVIRON HEALTH & TOXICOL, STATE UNIV NY, ALBANY, 85- *Concurrent Pos:* Adv, Pakistan Inst Nuclear Sci & Technol, 74-75 & 81 & Pakistan Upper Atmosphere Res Comn, 85 & 87. *Mem:* Am Chem Soc; AAAS; Am Geophys Union. *Res:* Atmospheric chemstry and physics; aerosol composition and transport; chemical reactions in clouds; acid rain; origin of troposheric ozone; age determination of lunar rocks & meteorites; mechanisms of high-energy nuclear reactions; migration of radioactive isotopes from radioactive waste burial sites. *Mailing Add:* Wadsworth Ctr of Labs Res NY State Dept Health Empire State Plaza Albany NY 12201-0509

HUSAIN, SYED, b Hyderabad, India, Dec 9, 39; US citizen; m 68; c 3. PHARMACOLOGY, PHARMACY. *Educ:* Osmania Univ, BSc, 59; Univ Wis, BS, 65; Univ Mo, MS, 68, PhD(pharmacol), 71. *Prof Exp:* Res assoc pharmacol, Med Sch, Ind Univ, 71-74; biochem pharmacologist, SRI Int, 74-76; asst prof, 76-81, ASSOC PROF PHARMACOL, MED SCH, UNIV NDAK, 81- *Concurrent Pos:* Nat Inst Drug Abuse grant, 74-76 & 80-82; HEW grant, 74-76; Univ NDak Fac res grants, 76-77 & 78-79; NIH grant, 77-80. *Mem:* Int Soc Biochem Pharmacol; Sigma Xi; AAAS; Soc Exp Biol Med. *Res:* Disposition of anticoagulants and other drugs, their biliary and lymphatic excretion and metabolism; drug interactions with marijuana and cannabinoid effects on brain and testicular metabolism and function. *Mailing Add:* Dept Pharmacol Univ NDak Sch Med Grand Forks ND 58201

HUSAIN, SYED ALAMDAR, b Arrah, India, Feb 5, 31; nat US; m 63; c 5. MATHEMATICS. *Educ:* Aligarh Muslim Univ, India, BA, 50; Univ Dacca, MA, 52; Univ Chicago, MS, 56; Purdue Univ, PhD(math), 59. *Prof Exp:* Lectr math, Brindaban Col, Bangladesh53-54 & Rajshahi Univ, Bangladesh, 54-55; instr, Purdue Univ, 56-59; asst prof, Seattle Univ, 59-60 & Univ Sask, 60-61; from asst prof to assoc prof, Univ Idaho, 61-64; vis prof, Aegean Univ, Turkey, 64-65; assoc prof, 65-76, PROF MATH, UNIV WYO, 76- *Mem:* Am Math Soc. *Res:* Fourier analysis; divergent series and fixed point theorems. *Mailing Add:* Dept Math Univ Wyo Box 3036 Laramie WY 82071

HUSAIN, TAQDIR, b India, July 16, 29; m 59; c 3. PURE MATHEMATICS. *Educ:* Aligarh Muslim Univ, BA, 50, MA, 52; Syracuse Univ, PhD(math), 60. *Prof Exp:* Lectr math, Aligarh Muslim Univ, 52-53; res scholar, Tata Inst Fundamental Res, Bombay, India, 53-55; lectr, Forman Christian Col, Panjab, Pakistan, 55-57; asst, Syracuse Univ, 57-60, instr, 60-61; asst prof, Univ Ottawa, Can, 61-64; assoc prof, 64-67, PROF MATH & CHMN DEPT, MCMASTER UNIV, 67-73 & 79-82. *Concurrent Pos:* Assoc ed, Can J Math, 79-86. *Mem:* Am Math Soc; Math Asn Am; Can Math Soc; NY Acad Sci. *Res:* Functional analysis; topological vector spaces; topological groups and spaces; Banach algebras, summability methods, real and complex analysis; author of six books. *Mailing Add:* Dept Math & Statist McMaster Univ Hamilton ON L8S 4K1 CAN

HUSAINI, SAEED A, b Bhopal, India, July 1, 26; US citizen. FOOD SCIENCE. *Educ:* Aligarh Muslim Univ, BSc, 44; Univ Delhi, MSc, 46; Ohio State Univ, PhD(agr biochem), 50. *Prof Exp:* Fel agr biochem, Ohio State Univ, 51-52; asst biochem & food tech, Swed Inst Food Preservation Res, 53-55; res chemist, Rath Packing Co, Iowa, 55-57; RES CHEMIST, WESTRECO, INC, 57- *Mem:* Am Chem Soc; Inst Food Technol. *Res:* Agricultural biochemistry and food technology, particularly the preservation and research on meat and meat products, fish, coffee and tea. *Mailing Add:* 18149 Timber Lane Marysville OH 43040

HUSAR, RUDOLF BERTALAN, b Martonos, Yugoslavia, Oct, 29, 41; US citizen; m 67; c 2. MECHANICAL ENGINEERING. *Educ:* Tech Univ, WBerlin, dipl eng, 66; Univ Minn, PhD(mech eng), 71. *Prof Exp:* Design technician, W Hofer, Heating, Ventilation Consult, 62-63; res asst, Tech Univ, WBerlin, 63-66; res asst, Univ Minn, 66-69, res assoc & fel, 69-71; res fel, Calif Inst Technol, 71-73; assoc prof, 73-76, PROF MECH ENG, WASH UNIV, ST LOUIS, 73-, DIR CAPITA, 79- *Concurrent Pos:* Res fel, Univ Glasgow, Scotland, 65; vis prof, Meterol Inst, Univ Stockholm, 76; mem, Comt Sulfur Oxides, Nat Acad Sci, 78; Co-chmn, Interagency Comt Health & Environ Effects Advan Energy Technol, 78; mem, Coop Prog Develop & Appln Space Technol Air Pollution, Environ Protection Agency/NASA, 78; mem, Comn Atmospheric-Biosphere Interactions, Nat Acad Sci, 79-81. *Mem:* Am Meterol Asn; Air Pollution Control Asn; AAAS; Am Asn Aerosol Res; Ges Aerosolforschung. *Res:* Aerosol dynamics and plume studies of sulfur oxide kinetics; regional scale model of acid rain and optical effects; satellite imagery studies; historical trends of fuel use and concentrations for checks of the consistency in the systems's behaviour; visibility trends; environmental informatics. *Mailing Add:* CAPITA Box 1124 Wash Univ One Brookings Dr St Louis MO 63130

HUSBAND, DAVID DWIGHT, botany; deceased, see previous edition for last biography

HUSBAND, ROBERT MURRAY, b Russell, Man, Oct 5, 19; m 50; c 2. ORGANIC CHEMISTRY, CHEMICAL ENGINEERING. *Educ:* Univ Sask, BA & BE, 44; McGill Univ, PhD(chem), 47. *Prof Exp:* Bristol Labs res fel streptomycin res inst, Ohio State Univ, 47-50; asst prof cellulose industs, Toronto, 50-52; res assoc wood chem, State Univ NY Col Forestry, Syracuse Univ, 52-55, assoc prof pulp & paper technol, 55-57; chief chem sect, Res Ctr, Consol Paper Corp, Ltd, 57-61; mgr cent res, Riegel Paper Corp, 61-69; dir res, R T Vanderbilt Co, Inc, 69-72; PRES, POLYPULP TECHNOL, 73- *Concurrent Pos:* Proj leader, Empire State Paper Res Assocs, 52-57; expert adv, Cent Am Inst Indust Invest & Technol, 72-73. *Mem:* Am Chem Soc; Tech Asn Pulp & Paper Indust; Can Pulp & Paper Asn; NY Acad Sci. *Res:* Chemistry of wood and its components, carbohydrates and phenols; wood pulping and pulp bleaching; oxidation of organic compounds; surface chemistry; paper structure and rheology; chemistry of paper-making. *Mailing Add:* 5285 Madison Ave Trumbull CT 06611

HUSBAND, ROBERT W, b Hesperia, Mich, May 21, 31; m 55; c 3. ENTOMOLOGY. *Educ:* Univ Mich, AB, 53; Western Mich Univ, MA, 60; Mich State Univ, PhD(zool), 66. *Prof Exp:* Instr radar intercept tech, USAF Intercept Sch, 55-58; teacher pub schs, Mich, 59-60; teaching asst zool, Mich State Univ, 60-64; assoc prof biol, Adrian Col, 64-73, chmn dept 71-80, chmn div sci, 73-75 & 88-90, PROF BIOL, ADRIAN COL, 73- *Concurrent Pos:* Mem, Cong Acarologists, 64-; res assoc entom, Univ Ga, 71-72. *Mem:* Am Micros Soc (treas, 75-79); Entom Soc Am; Soc Syst Zool; Am Inst Biol Sci; Sigma Xi. *Res:* Mites associated with insects; bumblebees; taxonomy; ecology; zoogeography Acarina, Podapolipidae. *Mailing Add:* Dept Biol Adrian Col Adrian MI 49221

HUSBAND, THOMAS PAUL, b Flint, Mich, Jan 23, 50; m 70; c 4. WILDLIFE ECOLOGY. *Educ:* Univ Mich, Flint, BA, 72; Mich State Univ, MS, 74, PhD(wildlife ecol), 77. *Prof Exp:* ASSOC PROF NATURAL RESOURCES SCI, UNIV RI, 77- *Concurrent Pos:* Res asst, NSF, 74-76. *Mem:* Wildlife Soc; Soc Am Foresters; Am Asn Univ Professors; Am Soc Mammalogists. *Res:* Population dynamics of wildlife; bioenergetics of mammals; primary and secondary productivity of terrestrial ecosystems; faunal inventory of Brazilian Atlantic forest; population ecology, genetics and behavior of mountain goats. *Mailing Add:* Dept Natural Resources Sci Univ RI Kingston RI 02881

HUSBY, FREDRIC MARTIN, b Mt Vernon, Wash, Feb 16, 43; m 71; c 2. ANIMAL NUTRITION, RUMINANT NUTRITION. *Educ:* Wash State Univ, BS, 66, MS, 69, PhD(nutrit), 74. *Prof Exp:* Fel animal physiol & animal metab, Univ BC, 74-75; ASST PROF ANIMAL SCI & NUTRIT, AGR EXP STA, UNIV ALASKA, 75- *Mem:* Am Soc Animal Sci; Am Soc Dairy Sci. *Res:* Evaluation of feedstuffs, Alaska grains and animal-fish byproducts for use by swine; management and production systems for swine, beef and sheep in a northern environment; forage utilization. *Mailing Add:* Dept Animal Sci Univ Alaska 116 Bunnell Fairbanks AK 99701

HUSCH, LAWRENCE S, b Hollis, NY, Feb 17, 42; m 68; c 2. TOPOLOGY. *Educ:* St Francis Col, NY, BS, 63; Fla State Univ, MS, 64, PhD(math), 67. *Prof Exp:* Res asst math, Fla State Univ, 66-67; asst prof, Univ Ga, 67-69; from asst prof to assoc prof, Va Polytech Inst, 69-71; assoc prof, 71-75, PROF MATH, UNIV TENN, KNOXVILLE, 75- *Concurrent Pos:* Vis res scientist, Univ Zagreb, Yugoslavia, 76-77, Fulbright res scholar, 86; vis distinguished prof, Univ NC, Greensboro, 90-91. *Mem:* Am Math Soc. *Res:* Topology of manifolds; dynamics. *Mailing Add:* Dept Math Univ Tenn Knoxville TN 37916

HUSCHILT, JOHN, b Kitchener, Ont, June 1, 31; m 55; c 8. THEORETICAL PHYSICS. *Educ:* Univ Toronto, BA, 52, MA, 53; Wayne State Univ, PhD(theoret physics), 63. *Prof Exp:* Asst prof physics, 53-65, actg chmn dept, 58-59, ASSOC PROF PHYSICS, UNIV WINDSOR, 65- *Mem:* Am Asn Physics Teachers; Am Phys Soc; Can Asn Physicists. *Res:* Attempting to clarify basic interactions of elementary particles and interactions with radiation by investigating these in a consistent, classical, fully relativistic framework analytically and by means of numerical calculations. *Mailing Add:* Dept Physics Univ Windsor Windsor ON N9B 3P4 Can

HUSCHKE, RALPH ERNEST, b Utica, NY, Nov 28, 25; m 48; c 2. METEOROLOGY. *Educ:* Mass Inst Technol, BS, 46. *Prof Exp:* Prin asst, US Weather Bur, Conn, 48-51, field aide, Mass, 51-52; res staff mem meteorol, Div Indust Coop, Mass Inst Technol, 54-55; chief res prods br, Geophys Res Directorate, Air Force Cambridge Res Ctr, Mass, 56-59; eng specialist meteorol, Norair Div, Northrop Corp, Calif, 59-61; phys scientist, Rand Corp, 62-73, consult, 73-82; PRES, HUSCHKE ASSOCS, INC, 82- *Concurrent Pos:* Ed glossary meteorol, Am Meteorol Soc, Mass, 55-59, asst to dir ed film prog, 61-62; mem panel weather & climate modification, Nat Acad Sci, 65-66; assoc, Clin Consult Corp, 82- *Mem:* AAAS; Am Meteorol Soc; Am Geophys Union; Nat Coun Indust Meteorol. *Res:* Applied meteorological research and analysis; economic aspects of weather information; weather data processing; numerical simulation; military climatology; weather effects on systems and operations; forensic meteorology. *Mailing Add:* Six Admiralty Cross Coronado CA 92118

HUSEBY, ROBERT ARTHUR, medicine, for more information see previous edition

HUSEYIN, KONCAY, b Nicosia, Cyprus, Oct 7, 36; m 62; c 3. APPLIED MECHANICS, STABILITY THEORY. *Educ:* Istanbul Tech Univ, MSc, 60; Univ London, PhD(civil eng), 68, DSc(eng), 79. *Prof Exp:* Supvr civil eng, Tumpane Co, Inc, 61-64; asst prof, MidE Tech Univ, Ankara, 68-69; vis asst prof, 69-71, res assoc prof, solid mech, 71-72, assoc prof, systs design, 72-75, assoc chmn dept, 74-78, chmn systs design, 78-87, PROF SYSTS DESIGN, UNIV WATERLOO, 75- *Concurrent Pos:* NATO res fel, Alexander von Humboldt res fel & Nat Res Coun fel; vis prof, Inst Mech, Darmstadt Tech Univ, Ger, 78 & 83; vis prof, Inst Mech, Darmstadt Tech Univ, Ger, 87-88; ed, Dynamics & Stability of Systs. *Honors & Awards:* Fel CSME, 87. *Mem:* Am Acad Mech; Soc Natural Philos; Ger Asn Appl Math & Mech; Asn Mech Eng Can. *Res:* Elastic stability of conservative and non-conservative structural systems; general theory of stability of systems; bifurcation theory; nonlinear oscillations; catastrophe theory. *Mailing Add:* Dept Systs Design Univ Waterloo Waterloo ON N2L 3G1 Can

HUSHAK, LEROY J, b Belle Plaine, Iowa, Sept 15, 39; m 80; c 1. AGRICULTURAL ECONOMICS. *Educ:* Iowa State Univ, BS, 61; Univ Chicago, MA & PhD(econ), 68. *Prof Exp:* From asst prof to assoc prof, 68-78, PROF AGR ECON, OHIO STATE UNIV, 78- *Concurrent Pos:* Vis scholar, US Dept Agr, 84-85. *Mem:* Am Agr Econ Asn; Am Econ Asn; Am Fisheries Soc; Asn Environ & Resource Economists. *Res:* Economics of agricultural trade, regional economics, fishery economics. *Mailing Add:* 103 Agr Admin Bldg Ohio State Univ Columbus OH 43210

HUSK, GEORGE RONALD, b Waynesburg, Pa, Oct 19, 37. ORGANOMETALLIC CHEMISTRY, POLYMER CHEMISTRY. *Educ:* Waynesburg Col, BS, 59; Univ Mich, MS, 61, PhD(chem), 64. *Prof Exp:* Res assoc chem, Univ Wis, 64-66; asst prof, Villanova Univ, 66-71; actg chief org br, Chem Div, Army Res Off-Durham, 71-73, chief chem br, US Army Europ Res Off-London, England, 73-77, CHIEF CHEM SYNTHESIS & POLYMER CHEM BR, US ARMY RES OFF, 77- *Concurrent Pos:* Vis fac assoc, Duke Univ, 71-73 & Univ NC, 78-84; Vis res scientist, Univ Tex-Austin & adj prof, Southwest Tex State Univ, 84-85; secy, Army Fel, 84-85; consult, Univ Tex, 87-89; adj assoc prof, NC State Univ, 90- *Mem:* AAAS; Am Chem Soc; fel Am Inst Chemists. *Res:* Transition metal stabilized carbenes; study of unsaturated organoaluminum systems; oxidation of electron spin resonance studies of silyl-substituted anion radicals; stability of silyl-substituted organometallic compounds; phenyl migrations in lithiated hydrazines; chemistry and properties of polymides; gas transport through polymer films; main chain transition metal polymers. *Mailing Add:* US Army Res Off PO Box 12211 Res Triangle Park NC 27709

HUSKEY, GLEN E, b Hillsboro, Mo, Aug 9, 31; m 53; c 3. FOOD MICROBIOLOGY, BIOCHEMISTRY. *Educ:* Univ Mo, BS, 56, MS, 57, PhD(food microbiol), 66. *Prof Exp:* Prod trainee plant opers, Fresh Milk & Ice Cream Div, Carnation Co, Tex, 57-58; plant mgr, Int Div, Foremost Dairies, Inc, Calif, 59-64; asst prof dairy technol & exten specialist, Ohio State Univ, 66-67; prin scientist, Pet Inc, 67-69, asst mgr corp qual assurance, 70-71, mgr, 71-74, vpres processing & mfg, Dairy Div, 74-82. *Mem:* Am Soc Qual Control; Am Dairy Sci Asn; Inst Food Technologists. *Res:* Implementation of a total quality control program for the food industry and/or total production management in the dairy industry. *Mailing Add:* 25521 Via Impresso Valencia CA 91355

HUSKEY, HARRY D(OUGLAS), b Whittier, NC, Jan 19, 16; m 39; c 4. MATHEMATICS, ELECTRICAL ENGINEERING. *Educ:* Univ Idaho, BS, 37; Ohio State Univ, MS, 40, PhD(math), 43. *Prof Exp:* Asst math, Univ Ohio, 37-38; instr, Ohio State Univ, 41-43 & Univ Pa, 43-46; temp prin sci officer, Nat Phys Lab Eng, 47-48; temp prin officer, Nat Bur Standards, Washington, DC, 48-49, asst dir, inst numerical anal, Univ Calif, 49-54; from assoc prof to prof math & elec eng, Univ Calif, Berkeley, 54-68; dir Comput Ctr, 68-77, prof, 68-85, chmn dept, 76-79, EMER PROF INFO SCI, UNIV CALIF, SANTA CRUZ, 85- *Concurrent Pos:* Tech dir comput lab, Wayne Univ, 52-53; consult comput div, Bendix Corp, 54-63; Indian Inst Technol, India, 63-64 & 71; vis prof, Mass Inst Technol, 66; mem comput sci panel, NSF; Naval Res Adv Comt; consult comput for develop nations, UN; Von Humboldt award, 75. *Honors & Awards:* Pioneer Award, Inst Elec & Electronics Engrs Comput Soc, 82; Centennial Award, Inst Elec & Electronics Engrs, 84. *Mem:* Fel AAAS; fel Inst Elec & Electronics Engrs; fel Brit Comput Soc; Am Math Soc; Asn Comput Mach (vpres, 58-60, pres, 60-62). *Res:* Design and use of electronic digital computing machines and accessories; mathematical area of surfaces; solution of algebraic linear simultaneous equations. *Mailing Add:* Dept Info Sci Univ Calif Santa Cruz CA 95064

HUSKEY, ROBERT JOHN, b San Antonio, Tex, Dec 29, 38; m 60; c 2. GENETICS. *Educ:* Univ Okla, BS, 60, MS, 62; Calif Inst Technol, PhD(biol), 68. *Prof Exp:* Asst prof genetics, Syracuse Univ, 67-69; asst prof, 69-74, mem ctr advan studies, 69-71, ASSOC PROF BIOL, UNIV VA, 74-, ASSOC DEAN GRAD SCH ARTS & SCI, 83- *Mem:* Genetics Soc Am; Soc Develop Biol; AAAS; Sigma Xi. *Res:* Development genetics of algae; physiology and genetic recombination in bacteriophage lambda. *Mailing Add:* Dept Biol Univ Va Charlottesville VA 22903

HUSKINS, CHESTER WALKER, b Lincoln Co, NC, Jan 2, 21; m 53; c 1. ORGANIC CHEMISTRY. *Educ:* Davidson Col, BS, 48; Univ Fla, MS, 49. *Prof Exp:* Chemist, Fruit & Veg Lab, USDA, Fla, 49-52; chemist, Redstone Arsenal, US Army Missile Command, 52-59, chief org chem group, 59-62, explor res & develop, 62-70, prog mgr shoulder fired rocket propellant develop, Directorate Res, Develop & Eng Propellant Lab, 70-74, dir boron proj, Directorate Res, 74-80; RETIRED. *Concurrent Pos:* Tech projs officer, Weapons Data Exchange Agreements, Neth & Belg, 62-67; consult, 81- *Mem:* Am Chem Soc. *Res:* Solid rocket propellant development; propellant ingredient synthesis, characterization and evaluation; fluorine, polymer, ferrocene and carborane chemistry; synthesis of new boron compounds, development of pilot plant and manufacturing facilities for selected compounds. *Mailing Add:* 623 Owens Dr SE Huntsville AL 35801

HUSNI, ELIAS A, b Lebanon, Nov 26, 24; US citizen; c 3. CARDIOVASCULAR SURGERY. *Educ:* Am Univ Beirut, BA, 47, MD, 51; Am Bd Surg, dipl, 57. *Prof Exp:* Instr & asst surg, Western Reserve Univ, 57-58; dir surg educ, Huron Rd Hosp, 58-59, dir res, 66-, dir intensive care unit, 67-; AT DEPT SURG, OHIO COL PEDIAT MED, CLEVELAND. *Mem:* AMA; Am Col Surg; Am Col Angiol; Int Col Angiol; Int Cardiovasc Soc. *Res:* Peripheral vascular diseases. *Mailing Add:* 4947 Countryside Lane Lyndhurst OH 44124

HUSON, FREDERICK RUSSELL, b Sheridan, Wyo, Nov 22, 36; m 55. HIGH ENERGY PHYSICS. *Educ:* San Diego State Col, BA, 59; Univ Calif, Berkeley, PhD(physics), 64. *Prof Exp:* Res assoc physics, Polytech Sch, Paris, 64-65; res assoc, Accelerator Lineaire Labs, Orsay, 65-66; scientist, Brookhaven Nat Lab, 66-70; PHYSICIST, FERMI NAT ACCELERATOR LAB, 70- *Mem:* Am Phys Soc; Sigma Xi; AAAS. *Res:* Experimental high energy particle physics. *Mailing Add:* 55 Huntsmans Horn Circle The Woodlands TX 77380-1418

HUSS, GLENN I, b Haswell, Colo, May 10, 21; m 52; c 3. METEORITICS. *Educ:* Univ Denver, BA, 51, MA, 52. *Prof Exp:* Orgn & methods examr, US Army, Ft Carson, Colo, 52-55; mgr & preparator, Am Meteorite Mus, Ariz, 55-60; DIR, AM METEORITE LAB, 60- *Mem:* Int Asn Geochem & Cosmochem; AAAS; Geochem Soc; Am Astron Soc; Meteoritical Soc. *Res:* Location and collection of meteorites; studies of their aerial shaping and fracturing; terrestrial distribution and weathering; the relationship of external features to internal structures. *Mailing Add:* Am Meteorite Lab PO Box 2098 Denver CO 80201

HUSS, HARRY O(TTO), b Philadelphia, Pa, Mar 18, 12; m 44; c 2. VALUE & MECHANICAL ENGINEERING. *Educ:* Drexel Inst Technol, BS, 34; Univ Toledo, MS, 40. *Prof Exp:* design & test engr, Schutte & Koerting Co, 34-36; instr, Int Corresp Schs, 36-37; instr eng, Univ Toledo, 37-39; mech engr, Wright Field, Dayton, 39-41; mech engr, Chem Corps Technol Command, US Army Chem Ctr, 41-47, chief, Offense Materiel Sect, Develop Br, Res & Eng Div, Off Chief, 47-49, civilian chief, Develop Div, Hq Res & Eng Command, 49-53, asst dep, Biol Warfare & Civilian Dept, Chem Warfare, 53-54, dir prod eng, Eng Command, 65-71, chief, Defense Div, Mfg Technol Directorate, 71-73, dep dir, Technol Supt Dir, Edgewood Arsenal, 73-77; CONSULT, LARRY JONES & ASSOCS, 76- *Concurrent Pos:* Specialist on value analysis. *Mem:* Fel Am Soc Mech Engrs; Nat Soc Prof Engrs; Am Ord Asn; Soc Am Value Engrs. *Res:* Evaluation of agents and materiel against targets, including the mechanism of their action; insecticide dispersal equipment; staff supervision over offense and defense materiel development within the Chemical Corps; product design and engineering for mass production; product and industrial engineering of defense materiel. *Mailing Add:* 87675 Reed Rd Uhrichsville OH 44683

HUSS, RONALD JOHN, b Ft Atkinson, Wis, Dec 3, 53; m 77. MICROBIAL SCREENING & STRAIN IMPROVEMENT, FERMENTATION DEVELOPMENT. *Educ:* Univ Wis-Madison, BA, 75; Univ Ill, Champaign-Urbana, PhD(biochem), 85. *Prof Exp:* Group leader, 83-85, proj mgr, 86, DIR RES, BIO-TECH RESOURCES, INC, 86- *Concurrent Pos:* Lab mgr, Ladish Malting Co, 76-77. *Mem:* Am Chem Soc; Am Soc Plant Physiologists. *Res:* Microbial screening and strain improvement; fermentation development. *Mailing Add:* Bio Tech Resources Inc 1035 S Seventh St Manitowoc WI 54220

HUSSA, ROBERT OSCAR, b La Crosse, Wis, Mar 15, 41; m 63; c 2. BIOCHEMISTRY. *Educ:* Univ Wis-Madison, BS, 64; Univ Hawaii, PhD(biochem), 68. *Prof Exp:* Res assoc pharmacol, Sch Med, Ind Univ, 68-69, instr, 69; instr, 69-72, asst prof, 72-77, assoc prof biochem & assoc prof gynec & obstet, 78-, PROF, MED COL WIS. *Concurrent Pos:* Am Cancer Soc grant, Sch Med, Ind Univ, 68-69 & Med Col Wis, 70-71; Human Life Found grants, Med Col Wis, 70-72, Nat Cancer Inst res grant, 74-76 & 77-78. *Mem:* Endocrine Soc; Am Soc Biol Chemists; Am Chem Soc; Tissue Cult Asn. *Res:* Synthesis of polypeptide, protein, glycoprotein, and steroid hormones; glycogen metabolism. *Mailing Add:* Dept Gynec & Obstet Med Col Wis 8700 W Wisconsin Ave Milwaukee WI 53226

HUSSAIN, A K M FAZLE, b DHAKA, Bangladesh, Jan 20, 43; m 68; c 1. FLUID MECHANICS, TURBULENCE. *Educ:* E Pakistan Univ Eng & Technol, BSc, 63; Stanford Univ, MS, 66, PhD(mech eng), 70. *Prof Exp:* Lectr dept mech eng, E Pakistan Univ Eng & Technol, 63-65; vis asst prof dept mech, Johns Hopkins Univ, 69-71; from asst prof to assoc prof, 71-73, distinguished univ prof, 85-89, PROF MECH ENG, UNIV HOUSTON, 76-, CULLEN DISTINGUISHED PROF, 89- *Concurrent Pos:* Consult var co & adv bds of tech comt; Assoc ed, Turbulence in Liquids, Sci Press, 79- & Physics Fluids, Am Inst Physics, 80-84; Int Sci Comt, Asian Cong Fluid Mech, 79-; Freeman scholar, Am Soc Mech Engrs, 84, mem sci comt, Int Union Theoret & Appl Mech symp, topological fluid mech, Cambridge Univ, 89. *Honors & Awards:* Res Excellence Award, Univ Houston, 85. *Mem:* Fel Am Soc Mech Engrs; fel Am Phys Soc; assoc fel Am Inst Aeronaut & Astronaut. *Res:* Turbulence, experimental, numerical, and analytical studies of turbulent shear flows, mixing layers, jets and wakes, and turbulence control; hydrodynamic stability; chaos; aerodynamics; aeroacoustics; biofluid mechanics; flow visulization; digital image processing; vortex dynamics; vortex chaos; separation phenomena. *Mailing Add:* Dept Mech Eng Univ Houston Houston TX 77204-4792

HUSSAIN, MALEK GHOLOUM MALEK, b Kuwait City, Kuwait, Oct 7, 53; m 84; c 2. ULTRAWIDEBAND RADAR & COMMUNICATION TECHNOLOGY, DIGITAL COMMUNICATIONS. *Educ:* Cath Univ Am, BEE, 77, MEE, 79, PhD(elec eng), 83. *Prof Exp:* From asst prof to assoc prof elec eng, Kuwait Univ, 83-90; DEAN, DIV PROF & TECHNOL PROGS, INTER-AM UNIV PR, 91- *Concurrent Pos:* Consult, Signal Corps, Ministry Defense, Kuwait, 84-86, Gulf Coop Coun, Kuwait, 87-88; vis assoc prof, Univ Mich, Ann Arbor, 88-89. *Honors & Awards:* Sci Achievements Prize, Kuwait Found Advan Sci, 90. *Mem:* Sr mem Inst Elec & Electronics Engrs; Soc

Kuwaiti Engrs. *Res:* Ultrawideband, impulse, radar and communication technology; antenna theory and design; signal propagation and scattering; signal processing; radar system design and simulation; adaptive array processing; neural networks and systolic array processing. *Mailing Add:* 1522 Cavalieri St Urb Belisa Rio Piedras PR 00927

HUSSAIN, MOAYYED A, b Poona, India, Feb 25, 37; US citizen; m 66; c 2. FRACTURE MECHANICS, INTEGRAL EQUATIONS. *Educ:* Univ Poona, India, BE, 59; Renesselaer Polytech Inst, MS, 62, PhD(mech), 65. *Prof Exp:* Lectr mech, Col Eng, Univ Poona, India, 59-60; res asst aeronaut, Rensselaer Polytech Inst, 60-62; res engr appl math, US Army Watervliet Arsenal, 62-79; MATHEMATICIAN, CORP RES & DEVELOP, GEN ELEC CO, 79- *Concurrent Pos:* Vis scientist, Math Res Ctr, Univ Wis-Madison, 66-67; adj assoc prof, Union Col, 72-74. *Mem:* Soc Indust Appl Math; Am Soc Mech Engrs; Am Comput Mach; Sigma Xi; Inst Elec & Electronics Engrs. *Res:* Mathematical physics; mechanics fracture; vibration; electromagnetic phenomenon; integral equations; dual series relations; finite element methods; finite difference methods; array synthesis. *Mailing Add:* Bldg K-1 Rm KWC434A Corp Res-Develop Gen Elec Co PO Box 8 Schenectady NY 12301

HUSSAIN, NIHAD A, b Basrah, Iraq, Jan 9, 41; m 64; c 3. FORCED & NATURAL CONVECTION, RADIATION. *Educ:* Univ Baghdad, Iraq, BS, 62; Purdue Univ, MS, 65; Univ Notre Dame, PhD(mech eng), 69. *Prof Exp:* Grad asst heat transfer, Purdue Univ, 63-65; res assoc heat transfer, Univ Notre Dame, 65-69; from asst prof to prof heat transfer & fluid mech, San Diego State Univ, 69-81; ASSOC DEAN, COL ENG, SAN DIEGO STATE UNIV, 81- *Concurrent Pos:* Design engr, Al Damalogy Consult Eng, Baghdad, Iraq, 62-63; res fel, NASA-Lewis Res Ctr, 73-75, USAF-Wright Patterson AFB, 75-76; consult, Geosci, Inc, Solana Beach, Calif, 72-73, Burroughs Corp, Rancho Bernardo, Calif, 75-76, Boekamp, Inc, San Diego, Calif, 80-84, Langley Corp, San Diego, Calif, 81-86, Calif Energy Comm, 89-; chmn, K-19 comt, Heat Transfer Div, Am Soc Mech Engrs, 87-90, rep, Nat Heat Transfer Conf Coord Comn, 90-93. *Mem:* Sigma Xi; Am Soc Mech Eng; Am Soc Eng Educ. *Res:* Development of engineering correlations in the areas of combined free and forced convections in horizontal tubes; two phase buoyant plumes and in the areas of radiation in semi-transparent media. *Mailing Add:* 13026 Trigger St San Diego CA 92129

HUSSAIN, RIAZ, b Gujranwala, Pakistan, Jan 18, 37; US citizen; m 65, 78; c 1. PHYSICS. *Educ:* Forman Christian Col, Lahore, Pakistan, BSc, 54; Univ Punjab, Lahore, Pakistan, MSc, 56; Johns Hopkins Univ, Md, PhD(physics), 73; Univ Scranton, MBA, 80. *Prof Exp:* Lectr physics, Forman Christian Col, Lahore, W Pakistan, 56-60; from instr to assoc prof physics, 67-85, ASSOC PROF FINANCE, UNIV SCRANTON, 85- *Honors & Awards:* Fulbright Fel, 60. *Mem:* Am Asn Physics Teachers; Financial Analysts Fedn. *Res:* Atomic spectroscopy; optical instrumentation; teaching of physics; portfolio theory; financial analysis; management science. *Mailing Add:* 540 N Webster Ave Scranton PA 18510

HUSSAIN, SYED TASEER, b Lahore, Pakistan, Sept 18, 43; m 75. VERTEBRATE PALEONTOLOGY, FUNCTIONAL ANATOMY. *Educ:* Panjab Univ, Lahore, BS, 63, BS hon, 64, MS, 65; Utrecht Univ, Holland, PhD(paleont), 69. *Prof Exp:* Fel vertebrate paleont, Am Mus Natural Hist, New York, 70-72; instr geol, NY Univ, 71-72; from instr to assoc prof, 72-85, PROF ANAT, HOWARD UNIV, 85- *Concurrent Pos:* Res assoc, Smithsonian Inst, Washington, DC, 77-; prin investr & proj dir, Cenozoic Mammals & Biostrat, Pakistan, 76-; guest scientist, Chinese Acad Sci, 82. *Mem:* AAAS; Am Asn Anatomists; Soc Vert Paleont. *Res:* Mammalian evolution in Asia with particular interests in paleoecology, paleogeography, paleoclimates and climate changes and faunal migrations during the last sixty million years; science administration; educational administration; science education. *Mailing Add:* Dept Anat Col Med Howard Univ Washington DC 20059

HUSSAK, ROBERT EDWARD, b Perth Amboy, NJ, Feb 15, 61. POLYMER ENGINEERING, POLYMER CHEMISTRY. *Educ:* Bethany Col, BS, 83. *Prof Exp:* Mgr, Pigment Lab, STO Corp, 83-85, mgr plant qual assurance, 85-86, primary formulation chemist-Ger, 86-87, synthetic lab mgr, 87-90, CORP QUAL CONTROL/QUAL ASSURANCE GOVT REGULATIONS SAFETY MGR, STO CORP, 90- *Res:* Waterspace coatings, paints, sealers and allied coating used primarily in the exterior insulation market and concrete/cement restoration market; environmental control-production related safety; toxicology. *Mailing Add:* 63B Windy Lane North Clarendon VT 05759

HUSSAMY, SAMIR, b Tanta, Egypt, Mar 18, 35; US citizen; m 60; c 1. COLOR CHEMISTRY, TEXTILE CHEMISTRY. *Educ:* Leeds Univ, dipl, 56, MS, 60; ETH Zurich, Switz, DSc(Tech), 68. *Prof Exp:* Asst mgr dyeing & finishing, Polytex Co, Egypt, 60-61; res chemist color chem, ETH Zurich, Switz, 62-67; res & develop chemist dye synthesis, Clayton Aniline Co, UK, 67-70; dir chem res, Klopman Div, Burlington Industs, Inc, 70-87; sr scientist dyes & textile finishing, Inst Textile Technol, Charlottesville, Va, 87-88; mgr & formulator chemicals, Catawba-Charlab, Charlotte, NC, 88-90; PRES, INDUST CHEMICALS, HILTON INDUSTS, EVINGTON, VA, 90- *Mem:* Am Asn Textile Chemists & Colorists; Am Chem Soc; Swiss Asn Chemists & Colourists; Soc Dyers & Colourists. *Res:* Synthesis of dyes, pigments, polymers and textile auxiliaries; dispersion of dyes and pigments; preparation, dyeing, printing-discharge finishing of textile and industrial fabrics; chemical modification of polyester fibers. *Mailing Add:* 1008 White Pine Dr Lynchburg VA 24501

HUSSAR, DANIEL ALEXANDER, b Philadelphia, Pa, Feb 12, 41; m 67; c 3. PHARMACY. *Educ:* Philadelphia Col Pharm & Sci, BS, 62, MS, 64, PhD(pharm), 67. *Prof Exp:* From asst prof to prof, 66-75, dir dept, 71-75, dean fac, 75-84, REMINGTON PROF PHARM, PHILADELPHIA COL PHARM, 75- *Mem:* Am Pharmaceut Asn; Am Asn Col Pharm; Am Soc Hosp Pharmacists; Am Col Apothecaries; Drug Info Asn (treas, 72-74, pres, 77-78); Sigma Xi. *Res:* Aspects of antibiotic-metal ion interactions; protein-binding of drugs; drug interactions. *Mailing Add:* Philadelphia Col Pharm & Sci 43rd St-Kingsessing Mall Philadelphia PA 19104

HUSSEY, ARTHUR M, II, b Pittsburgh, Pa, Mar 9, 31; div; c 3. GEOLOGY, MINERALOGY-PETROLOGY. *Educ:* Pa State Univ, BS, 54; Univ Ill, PhD(geol), 61. *Prof Exp:* Vis asst prof geol, Purdue Univ, 60-61; from asst prof to assoc prof, 61-72, PROF GEOL, BOWDOIN COL, 72-, CHMN DEPT, 61- *Concurrent Pos:* Mem, educ comt, Geol Soc Am, 90-, chmn, NE Sect. *Mem:* AAAS; Geol Soc Am; Geol Asn Can; Sigma Xi; Am Geophys Union. *Res:* Structural geology; igneous and metamorphic petrology, especially of the Northern New England Appalachians. *Mailing Add:* Dept Geol Bowdoin Col Brunswick ME 04011

HUSSEY, CHARLES LOGAN, b San Diego, Calif, Dec 26, 47; m 71; c 1. ANALYTICAL CHEMISTRY, ELECTROCHEMISTRY. *Educ:* Univ Miss, BS, 71, PhD(anal chem), 74. *Prof Exp:* Res chemist & lectr, F J Seiler Res Lab, Air Force Systs Command, USAF Acad, 74-78; from asst prof to assoc prof, 78-87, PROF CHEM, UNIV MISS, 87- *Mem:* Am Chem Soc; Electrochem Soc; Sigma Xi; Soc Electroanal Chem. *Res:* Molten salt electrochemistry; electrochemical studies of transition metal solutes in haloaluminate melts; transference in molten haloaluminates; modified electrodes; electrochemical nitration. *Mailing Add:* Dept Chem Univ Miss University MS 38677

HUSSEY, CLARA VERONICA, b Milwaukee, Wis, Oct 24, 20. PATHOLOGY. *Educ:* Mt Mary Col, BS, 47; Marquette Univ, MS, 50, MD, 61. *Prof Exp:* Med technician, St Joseph's Hosp, Wis, 42-44; med technician, St Alban's Hosp, NY, 44-46; med technician, St Joseph's Hosp, Wis, 46-48; asst, Med Col Wis, 50-52, instr biochem, 52-64, from asst prof to prof path, 69-91; mem staff, Path Lab, Milwaukee County Gen Hosp, 64-91; RETIRED. *Concurrent Pos:* Intern, Med Ctr, Ind Univ, 62; mem coun thrombosis, Am Heart Asn. *Honors & Awards:* Franklin V Taylor Award, Soc Eng Psychol. *Mem:* Am Heart Asn; Int Soc Thrombosis & Haemostasis; Am Med Women's Asn. *Res:* Blood coagulation; study of hemorrhagic diseases regarding abnormalities and diagnostic tests. *Mailing Add:* 175 S Beaumontal 8700 W Wisconsin Ave Brookfield WI 53005

HUSSEY, EDWARD WALTER, b Petaluma, Calif, Sept 3, 38; c 4. ORGANIC CHEMISTRY. *Educ:* Univ Nev, Reno, BS, 61, PhD(org chem), 65. *Prof Exp:* Chemist, E I du Pont de Nemours & Co, Inc, 65-76, sr res chem, 76-83, res assoc, 83-90, SR RES ASSOC, E I DU PONT DE NEMOURS & CO, 90- *Mem:* Am Chem Soc. *Res:* Natural products; polymer science. *Mailing Add:* Rte 1 Box 297A Washington WV 26181

HUSSEY, HUGH HUDSON, medicine, for more information see previous edition

HUSSEY, KEITH MORGAN, b Rock Island, Ill, Dec 2, 08; m 37; c 2. GEOMORPHOLOGY. *Educ:* Augustana Col, AB, 36; La State Univ, MS, 39, PhD(geol), 40. *Prof Exp:* From instr to asst prof geol, Univ Houston, 40-46; assoc prof micropaleont, Univ Okla, 46-49; from assoc prof to prof geomorphol, Iowa State Univ, 49-79, head dept, 62-74; RETIRED. *Concurrent Pos:* Emer prof geomorphol, Iowa State Univ, 79- *Mem:* Fel Geol Soc Am; Am Asn Petrol Geol; Nat Asn Geol Teachers; fel Arctic Inst NAm; Am Geol Inst. *Res:* Photo interpretation; pediment development of Wyoming, Colorado and New Mexico piedmont areas; geomorphogeny of the Arctic Coatal Plain of Alaska; palynology of some Iowa coals. *Mailing Add:* 1910 Meadowlane Ave Ames IA 50010

HUSSEY, RICHARD SOMMERS, b Wheeling, WVa, Dec 18, 42; m 68; c 2. PLANT NEMATOLOGY. *Educ:* Miami Univ, AB, 65; Univ Md, MS, 68, PhD(plant path), 70. *Prof Exp:* Fel nematol, Dept Plant Path, NC State Univ, 70-73; chief nematologist, NC Dept Agr, 73-74; from asst prof to assoc prof, 74-82, PROF NEMATOL, DEPT PLANT PATH, UNIV GA, 82- *Mem:* Soc Nematologists; Sigma Xi; Am Phytopath Soc. *Res:* Host-parasite relationships involving plant-parasitic nematodes; nematode management; plant resistance; nematode secretions. *Mailing Add:* Dept Plant Path Univ Ga Athens GA 30602

HUSSEY, ROBERT GREGORY, b Shreveport, La, May 15, 35; m 60; c 4. FLUID DYNAMICS. *Educ:* Univ Notre Dame, BS, 57; La State Univ, PhD(physics), 62. *Prof Exp:* Asst physics, 57-62, from asst prof to assoc prof, 62-79, PROF PHYSICS, LA STATE UNIV, BATON ROUGE, 79-, ASSOC DEAN, COL BASIC SCI, 71- *Honors & Awards:* George B Pegram Award, Am Phys Soc, 82. *Mem:* Am Phys Soc; Am Asn Physics Teachers. *Res:* Flow of incompressible fluids; low Reynolds number flow. *Mailing Add:* Dept Physics & Astron La State Univ Baton Rouge LA 70803

HUSSIAN, RICHARD A, b Greensboro, NC, Aug 20, 52; m 79. AGING, BEHAVIOR MODIFICATION. *Educ:* Davidson Col, BA, 74; Univ NC-Greensboro, MA, 76, PhD(psychol), 78. *Prof Exp:* Psychologist, Evergreens, Inc, 78-80; psychologist, Guilford Co Mental Health, 80-82; PSYCHOLOGIST & UNIT DIR, TERRELL STATE HOSP, 82- *Concurrent Pos:* Ed, Int J Behav Geriat, 82-83; instr, Univ NC, Greensboro, 79-81, vis prof, 79-82; vis prof, Southwest Med Ctr, Dallas, Tex, 82- *Mem:* Am Psychol Asn; Asn Behav Anal; Geront Soc Am. *Res:* Effects of environmental changes on the behavior of patients with progressive dementia. *Mailing Add:* Terrell State Hosp PO Box 70 Terrell TX 75160

HUSSON, SAMIR S, b Nazareth, Israel, June 5, 34; nat US citizen; m 57; c 1. MANAGEMENT CONSULTING, SYSTEM & COMPONENT RELIABILITY. *Educ:* Greenville Col, BS, 57; Mich State Univ, BSEE, 58, MS, 59. *Prof Exp:* Develop engr, IBM, 60-69, staff, Syst Res Inst, 69-72, sr engr design Automation, 73-78, dir Tech Publ, 78-81, dir Quality Inst, 81-85; PRES, CONSULT, INT MGT SYST, 85- *Concurrent Pos:* Vis prof comput sci, Yale Univ, 72-73. *Mem:* Fel Inst Elec & Electronics Engrs; Comput Soc; Asn Comput Mach. *Res:* Micro programming system, concepts, organization and designs; microprogrammable systems architecture and implementation; numerous publications. *Mailing Add:* 9855 Bankside Dr Roswell GA 30076

HUSSONG, DONALD MACGREGOR, b Paterson, NJ, Aug 31, 42; c 3. MARINE GEOPHYSICS. *Educ:* Princeton Univ, BSE, 64; Univ Hawaii, MS, 67, PhD(geol, geophys), 72. *Prof Exp:* Res asst, Univ Hawaii, 66-68, jr geophysicist, 69-72, from asst geophysicist to assoc geophysicist, 72-82, geophysicist & prof, Hawaii Inst Geophys, 82-; SEAFLOOR SURV INT. *Concurrent Pos:* Coordr, Nazca Plate Proj, Int Decade Ocean Explor, Univ Hawaii, 72-79; tech adv, UN comt Coord Joint Prospecting, 74-75; chmn & mem, Active Margins Panel, Int Prog Ocean Drilling, Joint Oceanog Insts Deep Earth Sampling, 77-80; comt chmn, Ocean Crystal Dynamics Planning Comt, 78-80; mem, Oceanog Proposal Rev Panel, NSF, 80-81 & planning comt, Ocean Drilling Prog, 85-; prin investr, Sea MARC II proj, 81- *Mem:* Am Geophys Union. *Res:* Marine geophysics related to the tectonic evolution of the oceanic crust using seismic refraction experiments with ocean bottom seismometers; submarine morphology and tectonics using side-scan sonar and swath bathymetry mapping. *Mailing Add:* Seafloor Surv Int Pier 66 2201 Alaskan Way Seattle WA 98121

HUSSUNG, KARL FREDERICK, b Louisville, Ky, Feb 19, 31; m 53; c 4. ORGANIC CHEMISTRY. *Educ:* Murray State Col, BS, 53; Univ Louisville, PhD(chem), 57. *Prof Exp:* Asst prof, 57-59, assoc prof, 59-63, PROF CHEM, MURRAY STATE UNIV, 63- *Mem:* Sigma Xi; Am Chem Soc. *Res:* 2-pyrones, benzotriazoles and cancer chemotherapeutics. *Mailing Add:* Dept Chem Murray State Univ Murray KY 42071

HUST, JEROME GERHARDT, b Logan Co, NDak, May 21, 32; m 55; c 4. PHYSICS. *Educ:* SDak Sch Mines & Technol, BS, 58, MS, 60. *Prof Exp:* Physicist thermodyn, 61-66, PHYSICIST SOLID STATE, NAT BUR STANDARDS, 66- *Mem:* Am Soc Testing & Mat. *Res:* Thermophysical properties of solids; numerical data analysis; data evaluation and critical analysis; thermodynamic properties of fluids. *Mailing Add:* 222 Iroguois Dr Boulder CO 80303

HUSTED, JOHN E, b Lucasville, Ohio, Oct 12, 15; m 42; c 2. GEOLOGY, NATURAL RESOURCES. *Educ:* Hampden Sydney Col, BS, 39; Univ Va, MA, 42; Fla State Univ, PhD(geol), 70. *Prof Exp:* Sci teacher high sch, Va, 38-40; mem staff, US Geol Surv, 42, chemist, 42-44, geologist, 44-45; plant chemist, Consol Feldspar Corp, 45-46; instr geol & chem, Wash & Lee Univ, 46-48; jr geologist, Humble Oil & Ref Co, 48-49; assoc prof geol & head dept, Trinity Univ Tex, 49-51; res geologist, Va Iron Coal & Coke Co, 51-55; prin geologist, Battelle Mem Inst, 55-57; mem staff, Capital Univ, 58; res scientist, Eng Exp Sta, 58-63, head, Mineral Eng Group, 60-66, assoc prof geol, 63-67, res prof, 67-71, prof 74-84, dir, 80-84, PROF GEOL, ENG EXP STA, GA INST TECHNOL, 71-, EMER PROF MINERAL ENG, SCH CHEM ENG, 84-, EMER DIR, GA MINING & MINERAL RESOURCES INST, 84- *Concurrent Pos:* Consult, Va Iron Coal & Coke Co, 47-51, Slick-Morman Oil Co, 50-51, Transcontinental Oil Co, 52-55 & Battelle Mem Inst, 57-58; fel, NSF Sci Fac, 66-67; Civil Defense Exec Reserve, US Dept Interior, 69-90. *Mem:* Am Inst Mining, Metall & Petrol Eng; Geol Soc Am; Am Asn Petrol Geologists. *Res:* Mineral engineering; mineral and fuel resources; economic and structural geology; tectonophysics. *Mailing Add:* 2844 Woodthrush Dr Roanoke VA 24018-5043

HUSTED, RUSSELL FOREST, b Lafayette, Ind, Apr 4, 50; m 88; c 2. RENAL ION TRANSPORT. *Educ:* Colo State Univ, BS, 72; Univ Utah, PhD(pharmacol), 76. *Prof Exp:* Fel, Dept Med, Univ Iowa, 76-79, asst res scientist, 79-81; asst prof, Dept Med, Univ Conn, 81-82; ASST RES SCIENTIST, DEPT MED, UNIV IOWA, 82- *Mem:* Sigma Xi; Soc Gen Physiol; Am Physiol Soc; Am Soc Nephrology; NY Acad Sci; AAAS. *Res:* Papillary collecting duct of the kidney exerts the final control of urine composition; transport of sodium, chloride, and acid by cultured cells isolated from rat collecting ducts is studied to determine the effects of steroids and the role of the kidney in the development of hypertension. *Mailing Add:* 317 Med Labs Iowa City IA 52242

HUSTON, ERNEST LEE, b Uniontown, Pa, Mar 27, 40; m 76; c 2. MATERIALS SCIENCE, METALLURGY. *Educ:* Lafayette Col, BS, 62; Northwestern Univ, PhD(mat sci), 68. *Prof Exp:* Technician, Alcoa Res Labs, Aluminium Co Am, 60; res asst mat sci, Northwestern Univ, 63-68; res metallurgist, Paul D Merica Res Labs, Int Nickel Co, Inc, 68-72, energy systs sect mgr, Inco Res & Develop Ctr, 77-80; tech mgr, Ergenics Div, MPD Tech Corp, 30-84; pres, 84-86, MGR SPEC PROJS, ERGENICS INC, 86- *Concurrent Pos:* Consult, Tempel Steel Co Chicago, 67-68. *Mem:* AAAS; Am Soc Metals; Sigma Xi. *Res:* Phase equilibria; phase transformation; thermodynamics; magnetic alloys; magnetostriction; metal hydrides; photoelectrochemistry; getter alloys. *Mailing Add:* Box 38B-Maple Brook Tuxedo Park NY 10987

HUSTON, JEFFREY CHARLES, b Johnstown, Pa, Jan 30, 51; m 74; c 2. ENGINEERING MECHANICS, BIOMECHANICS. *Educ:* Ill Inst Technol, BS, 72; WVa Univ, MS, 73, PhD(mech eng), 75. *Prof Exp:* Mech engr res, Morgantown Energy Res Ctr, Energy Res Develop Admin, 75-76; from asst prof to assoc prof eng mech, 76-87, PROF ENG SCI & MECH, IOWA STATE UNIV, 87- *Concurrent Pos:* Lectr, asst prof, WVa Univ, 75-76. *Honors & Awards:* Teetor Award, Soc Automotive Engrs, 80; Michol Award, Am Soc Eng Educ, 84. *Mem:* Am Soc Mech Engrs; Am Soc Eng Educ; Soc Automotive Engrs; Sigma Xi; Nat Soc Prof Engrs. *Res:* Biodynamics of human body motion; vehicles dynamics/recreational vehicles. *Mailing Add:* 535 Valley West Ct West Des Moines IA 50265

HUSTON, JOHN LEWIS, b Lancaster, Ohio, Aug 19, 19. PHYSICAL CHEMISTRY, INORGANIC CHEMISTRY. *Educ:* Oberlin Col, AB, 42; Univ Calif, PhD(chem), 46. *Prof Exp:* Asst chem, Univ Calif, 42-46; from instr to asst prof, Ore State Col, 46-52; from asst prof to prof, 52-84, EMER PROF CHEM, LOYOLA UNIV CHICAGO, 84- *Concurrent Pos:* Consult, Argonne Nat Lab, 65-89. *Mem:* Am Chem Soc; Sigma Xi. *Res:* Inorganic tracers; nonaqueous solvents; kinetics in solution; isotopic exchange reactions; xenon and synthetic inorganic chemistry; organic and inorganic fluorine chemistry. *Mailing Add:* Dept Chem Loyola Univ 4401 Keeney St Skokie IL 60076

HUSTON, KEITH ARTHUR, b Cleveland, Ohio, Mar 18, 26; m 61. RESEARCH ADMINISTRATION. *Educ:* Univ Wis, BS, 49, MS, 50, PhD(dairy husb), 51. *Prof Exp:* Assoc prof exten dairy cattle breeding, Va Polytech, 51-54; assoc prof dairy cattle genetics, Kans State Univ, 54-63, prof dairy husb, 63-69, asst dir, Kans Agr Exp Sta, 69-71, prof & assoc dir, 71-75, adj prof vet path, 72; prof animal sci, Univ Minn, 75-79, dir, Agr Exp Sta, 75-79; dir-at-large, NCent Asn, State Agr Exp Sta, 79-89; EMER PROF, OHIO STATE UNIV, 89- *Concurrent Pos:* Adj prof dairy sci, Ohio State Univ, 79-89. *Mem:* AAAS; Am Soc Animal Sci; Am Dairy Sci Asn; Am Genetic Asn. *Res:* Dairy cattle breeding; pathological genetics. *Mailing Add:* 1689 Arthur Dr Wooster OH 44691

HUSTON, MERVYN JAMES, b Ashcroft, BC, Sept 4, 12; m 38; c 2. PHARMACOLOGY. *Educ:* Univ Alta, BSc, 37, MSc, 40; Univ Wash, Seattle, PhD(pharmacol), 43. *Hon Degrees:* LLD, Dalhousie Univ, 82; DSc, Univ Alta, 88. *Prof Exp:* From lectr to asst prof pharm, 39-46, actg dir sch pharm, 46-48, dir, 48-55, dean fac pharm, 55-78, EMER PROF, SCH PHARM, UNIV ALTA, 78- *Concurrent Pos:* Chmn, Can Conf Pharmaceut Facs, 48-49; mem, Can Drug Adv Comt, 64-70; sci ed, Can J Pharmaceut Sci, 65-70, ed-in-chief, 70-78; mem coun, Int Pharmaceut Fedn, 67-70; pres, Can Found Advan Pharm, 69-70. *Honors & Awards:* Squibb Pan-Am Pharmaceut Award, 71. *Mem:* Am Pharmaceut Asn; Can Pharmaceut Asn (pres, 68-69). *Res:* Effect of drugs on tissue respiration. *Mailing Add:* Fac Pharm Univ Alta Edmonton AB T6G 2N8 Can

HUSTON, NORMAN EARL, b Jefferson, Iowa, Jan 24, 19; m 43; c 3. NUCLEAR PHYSICS, TECHNICAL MANAGEMENT. *Educ:* Univ Calif, AB, 43; Univ Southern Calif, PhD(physics), 52. *Prof Exp:* Physicist, Radiation Lab, Univ Calif, 43-44; physicist, Atomics Int Div, NAm Aviation Corp, 47, res engr, 50-53, supt appl physics, 54-56, group leader instrumentation & control, 57-59, sr tech specialist, 60-61, chief tech oper, 61-62, dir radiation tech & instrumentation, 63-66; prof & dir Instrumentation Systs Ctr, 66-84, EMER PROF NUCLEAR ENG & CONSULT ENGR, UNIV WIS-MADISON, 84- *Concurrent Pos:* Dir, Ocean Eng Labs, 67-70; mem comt, Interplay Eng in Biol & Med, Nat Acad Eng, 68-72; dir, Adv Ctr Med Technol & Systs, Univ Wis-Madison, 72-78; UN Expert on Mission Singapore Inst Standards & Indust Res, 72; mem fact finding team on instrumentation req on mission to Nat Res Ctr, Cairo, Egypt, NSF, 74 & workshop team on sci instrumentation & biomed eng on mission to Rome, Italy & Cairo, Egypt, NSF, 75, prof leader mgt systs labs instruments, Nat Res Ctr, Cairo, 76, consult, 75-76 & prof leader, Egypt Sci & Technol Proj, Nat Res Ctr, Agency Int Develop, Cairo, 77; mem adv team to Saudi Arabian Nat Ctr Sci & Tech, NSF, 81; consult, World Bank & Indonesian Ministry Educ & Cult, 83. *Mem:* Fel AAAS; Am Phys Soc; Am Nuclear Soc; fel Instrument Soc Am (secy, 78, pres, 79); fel Inst Measurement & Control. *Res:* Reactor physics; instrumentation and control; radiation effects; radiation instrumentation; gaseous discharges; technical marketing; biomedical engineering; research management. *Mailing Add:* 4556 Winnequah Rd Monona WI 53716

HUSTON, ROBERT JAMES, b Riley County, Kans, Apr 30, 31; m 52; c 1. AERONAUTICAL ENGINEERING. *Educ:* Univ Kans, BS, 58, MS, 61; Stanford Univ, dipl bus, 72. *Prof Exp:* Res engr helicopters verticle takeoff & landing, Flight Res Br, Langley Res Ctr, NASA, 58-70, asst br head helicopters, 70-72, proj mgr, Rotor Systs Res Aircraft Proj Off, 72-76, asst chief helicopters, Flight Res Div, 76-78, proj mgr risk anal, Graphite Fibers Risk Anal Prog, 78-80, mgr Rotorcraft Res & Technol, 80-85 & 86-88, actg asst dir aeronaut, NASA Hq, 85-86, SPEC ASST TO DIR STRUCT, LANGLEY RES CTR, NASA, 89- *Honors & Awards:* NASA Except Serv Medal; Howard Hughes Award, Am Helicopter Soc. *Mem:* Am Helicopter Soc; Am Inst Aeronaut & Astronaut. *Res:* Flight dynamics of helicopter and vertical takeoff and landing aircraft. *Mailing Add:* Langley Res Ctr NASA Hampton VA 23665

HUSTON, RONALD L, b Central City, Pa, Aug 5, 37; m 56; c 3. DYNAMICS, BIOMECHANICS. *Educ:* Univ Pa, BS, 59, MS, 61, PhD(eng mech), 62. *Prof Exp:* From asst prof to prof, 62-70, head, Dept Eng Anal, 69-75, DIR INST APPL INTERDISCIPLINARY RES, UNIV CINCINNATI, 71-, PROF MECH, 77- *Concurrent Pos:* Div dir, NSF, 78-79; actg sr vpres & provost, Univ Cincinnati, 82-83. *Honors & Awards:* Teetor Award, Soc Automotive Engrs, 78. *Mem:* Soc Automotive Engrs; Am Soc Mech Engrs; Am Soc Eng Educ; Am Inst Aeronaut & Astronaut; Am Soc Biomech. *Res:* Multibody mechanics; accident reconstruction; crash-victim simulation; applied mathematics; finite element theory; robotics; multibody dynamics; gearing and transmissions. *Mailing Add:* Dept Mech Indust & Nuclear Eng Univ Cincinnati Cincinnati OH 45221-0072

HUSTRULID, WILLIAM A, b St Paul, Minn, Oct 31, 40; m 67; c 5. ROCK MECHANICS, MINERAL ENGINEERING. *Educ:* Univ Minn, BS, 62, MS, 65, PhD(rock mech), 68. *Prof Exp:* Res engr, Atlas Copco, Stockholm, Sweden, 65-66; from asst prof to assoc prof mining eng, Colo Sch Mines, 68-72; div dir, Gecamines, Repub Zaire, 72-74; assoc prof, Univ Utah, 74-76; dir mining res, Terra Tek, 76-77, PROF MINING ENG, COLO SCH MINES, 77- *Concurrent Pos:* Vis researcher, CSIRO, Australia,; vis prof, Univ Luleu, Sweden; consult, var mining co & other orgn. *Mem:* Am Inst Mining, Metall & Petrol Engrs; SAfrican Inst Mining & Metall; Sigma Xi. *Res:* Rock drilling and blasting; design of underground openings; nuclear waste disposal; surface and underground mining. *Mailing Add:* Dept Mining Colo Sch Mines Golden CO 80401

HUSZAR, GABOR, b Budapest, Hungary; m 65; c 2. BIOCHEMISTRY, MALE INFERTILITY. *Educ:* Med Univ Budapest, MD, 63. *Prof Exp:* Resident obstet & gynec, Univ Hosp, Med Univ Budapest, 63-66; res fel muscle biochem, Retina Found, Boston, Mass, 67-70, Boston Biomed Res Inst, 70-71 & biochem, Harvard Univ, 71-74; asst prof, 75-80, ASSOC PROF OBSTET, GYNECOL & PEDIAT, MED SCH, 81-, DIR, UTERINE PHYSIOL UNIT, 79-, DIR, SPERM PHYSIOL, YALE UNIV, 83- *Concurrent Pos:* Res fel neurol, Harvard Med Sch, 68-70; Mass Heart Asn res grant biochem, 70-71; Nat Inst Child Health & Human Develop spec fel, 71-74. *Mem:* AAAS; Am Chem Soc; Biophys Soc; Fedn Am Soc Biol Chemists; Soc Gynec Invest; Am Fertil Soc; Am Soc Androl. *Res:* Male fertility; myometrial contractility; biology and biochemistry of the uterine cervix; sperm fertility; human sperm physiology and biochemistry. *Mailing Add:* Dept Obstet & Gynec Med Sch Yale Univ 333 Cedar St New Haven CT 06510

HUT, PIET, b Sept 26, 52; Neth citizen. THEORETICAL ASTROPHYSICS, STELLAR DYNAMICS. *Educ:* Univ Utrecht, Neth, BSc, 74, MSc, 77; Univ Amsterdam, PhD(astron), 81. *Prof Exp:* Res & teaching asst, Inst Theoret Physics, Univ Utrecht, Neth, 77-78 & Astron Inst, Univ Amsterdam, Neth, 78-81; mem, 81-84, PROF, INST ADVAN STUDY, PRINCETON, 85- *Concurrent Pos:* Sloan Found fel, 85; asst prof, dept astron, Univ Calif, Berkeley, 85; sr vis fel, Japanese Soc Prom Sci, 845 & 89. *Mem:* Int Astron Union; Am Astron Soc; Dutch Astron Union; Royal Astron Soc. *Res:* Theoretical astrophysics, especially stellar dynamics; impacts of comets and asteroids and their relation to mass extinctions; astrophysical computer science. *Mailing Add:* Inst Advan Study Princeton NJ 08540

HUTCHCROFT, ALAN CHARLES, b Greeley, Colo, Mar 17, 41; m 65; c 2. ORGANIC CHEMISTRY. *Educ:* Kalamazoo Col, BA, 63; Univ Mich, MS, 65, PhD(chem), 69. *Prof Exp:* From asst prof to assoc prof, 67-80, prof chem & chmn dept, 80-87, GORDON H & VIOLET J BARTELS PROF, ROCKFORD COL, 84-, CHAIR DIV SCI, MATH & NURSING, 87- *Concurrent Pos:* Consult, Pierce Chem Co, 77- *Mem:* Sigma Xi; Am Chem Soc. *Res:* Ferrocene chemistry, including attempted synthesis and study of novel bridged ferrocene compounds; organocuprate chemistry. *Mailing Add:* Dept Chem Rockford Col Rockford IL 61108-2393

HUTCHCROFT, CHARLES DENNETT, b Mediapolis, Iowa, Aug 5, 18; m 47; c 3. AGRONOMY. *Educ:* Iowa State Univ, BS, 46, MS, 50, PhD, 55. *Prof Exp:* Assoc agron, Iowa State Univ, 46-53, from asst prof to prof agron, 53-80; RETIRED. *Concurrent Pos:* Secy-treas, Iowa Crop Improv Asn, 59- *Mem:* Am Soc Agron; Sigma Xi. *Res:* Seed certification; hybrid corn varietal testing; foundation seed increase and distribution. *Mailing Add:* 2222 Storm Ames IA 50010

HUTCHENS, JOHN OLIVER, b Noblesville, Ind, Nov 8, 14; m 39; c 3. TOXICOLOGY. *Educ:* Butler Univ, AB, 36; Johns Hopkins Univ, PhD(physiol), 39. *Prof Exp:* Nat Res Coun fel, Harvard Univ, 39-40; Johnston scholar, Johns Hopkins Univ, 40-41; from instr to prof physiol, Univ Chicago, 41-84, dir toxicity lab, 46-48 & 73-77, assoc dir, 70-73, chmn dept physiol, 46-58, prof pharmacol, 70-84; RETIRED. *Concurrent Pos:* Sci liaison officer, Off Naval Res, London, 54-55. *Mem:* AAAS; Am Physiol Soc; Soc Gen Physiologists; Brit Biochem Soc. *Res:* Cell metabolism; carbon and nitrogen metabolism; free energy efficiency of growth; biochemical mechanism of action of toxic chemicals; entropies of organic compounds; inhalation toxicology. *Mailing Add:* 5633 S Drexel Ave Chicago IL 60637

HUTCHENS, TYRA THORNTON, b Newberg, Ore, Nov 29, 21; m 42; c 3. CLINICAL PATHOLOGY, NUCLEAR MEDICINE. *Educ:* Univ Ore, BS, 43, MD, 45; Am Bd Path, dipl, 62; Am Bd Nuclear Med, dipl, 76. *Prof Exp:* Intern, Minn City Hosp, 45-46; AEC fel, Ore Health Sci Univ, 48-50, USPHS res fel, 51-53, assoc prof, 53-62, coordr allied health educ, 69-77, prof clin path & chmn dept, 62-87, EMER PROF CLIN PATH, ORE HEALTH SCI UNIV, 87- *Concurrent Pos:* Mem, Am Bd Nuclear Med, 71-77 & 82-84; pres, World Path Found, 87-89. *Mem:* AMA; Col Am Path (pres, 77-79); fel Am Soc Clin Path; Soc Nuclear Med; Sigma Xi; World Asn Socs Path (vpres, 85-87, pres, 89-); hon mem Asn Clin Pathologists; hon mem Ital Soc Lab MedMed. *Res:* clinical laboratory standards. *Mailing Add:* 15385 SW Petrel Lane Beaverton OR 97007

HUTCHEON, DUNCAN ELLIOT, b Kindersley, Sask, June 21, 22; nat US; m 46; c 4. PHARMACOLOGY, TOXICOLOGY. *Educ:* Univ Toronto, MD, 45, BSc, 47; Oxford Univ, DPhil(pharmacol), 50; Am Bd Internal Med, dipl. *Prof Exp:* Assoc prof pharmacol, Univ Sask, 50-53; sr pharmacologist, Pfizer Therapeut Inst, 53-56; resident med, Jersey City Med Ctr, 57-59; from asst prof to assoc prof pharmacol & from instr to asst prof med, 59-69, PROF PHARMACOL & PROF MED, UNIV MED & DENT NJ, 69-; PRES, PRINCETON INST ENVIRON MED, 80- *Concurrent Pos:* Ed, J Clin Pharmacol, 77-84, emer ed, 85- *Mem:* Am Soc Pharmacol & Exp Therapeut; Soc Exp Biol & Med; fel Am Col Physicians; fel Am Col Clin Pharmacol; Sigma Xi. *Res:* Effects of drugs and industrial chemicals on arrhythmia threshold of intact and isolated perfused hearts; clinical pharmacology of diuretics in hypertension and edema. *Mailing Add:* Col Med Dent NJ Newark NJ 07103

HUTCHERSON, JOSEPH WILLIAM, b Scooba, Miss, Sept 12, 40; m 64; c 1. ATOMIC PHYSICS, APPLIED MATHEMATICS. *Educ:* Univ Tenn, Chattanooga, BA, 62; VAnderbilt Univ, MS, 64, PhD(physics), 68. *Prof Exp:* Asst prof physics, Southern Missionary Col, 67-68 & Walla Walla Col, 68-69; asst prof math, 69-75, ASSOC PROF MATH, UNIV TENN, CHATTANOOGA, 75- *Mem:* Math Asn Am; Sigma Xi; Am Asn Univ Professors; Am Phys Soc. *Res:* Vacuum ultraviolet spectra of atoms and molecules. *Mailing Add:* Dept Math Univ Tenn Chattanooga TN 37401

HUTCHESON, DAVID PAUL, b Ft Worth, Tex, Nov 5, 41; m 66; c 2. ANIMAL NUTRITION, BIOSTATISTICS. *Educ:* Tex A&M Univ, BS, 63; Univ Mo-Columbia, MS, 67, PhD(animal husb), 70. *Prof Exp:* Asst prof vet physiol & pharmacol, 70-73, assoc prof animal sci, Univ Mo-Columbia, 73-, biostatistician Sinclair Res Farm, 70-; AT TEX AGR EXP STA, TEX A&M UNIV. *Mem:* AAAS; Am Asn Animal Sci; Am Soc Vet Physiol & Pharmacol. *Res:* Design and theoretical models of protein and energy metabolism. *Mailing Add:* Tex Agr Exp Sta Tex A&M Univ 6500 Amarillo Blvd W Amarillo TX 79106-1976

HUTCHESON, ELDRIDGE TILMON, III, b Atlanta, Tex, Dec 23, 42; m 66; c 1. BIOCHEMISTRY. *Educ:* Southwest Tex State Univ, BS, 66, MA, 68; Univ Tex, Austin, PhD(chem), 71; Am Bd Clin Chem, dipl, 77. *Prof Exp:* Clayton Found Biochem Inst fel, Univ Tex, Austin, 71; sr scientist & protein chemist, Flow Labs, Inc, 71-72; instr & NIH fel, Univ Tenn, Memphis, 72-73, asst prof biochem, Ctr Health Sci, 73-75; res chemist clin chem, Salem Vet Admin Med Ctr, 75-82; asst prof path, Sch Med, Univ Va, 77-82; CONSULT, 82- *Concurrent Pos:* Res chemist connective tissue, Memphis Vet Admin Hosp, 73-75; dir, Biolabs Med Lab, Roanoke, 77-82. *Mem:* Am Asn Clin Chem; Am Chem Soc. *Res:* Protein structure and function; enzymology; collagen structure and function; clinical chemistry; nutrition. *Mailing Add:* PO Box 4641 Roanoke VA 24015

HUTCHESON, HARVIE LEON, JR, b Tipton, Okla, Oct 6, 37; m 60; c 2. PLANT ECOLOGY. *Educ:* Okla State Univ, BS, 60, MS, 63; Univ Okla, PhD(plant ecol), 65. *Prof Exp:* From asst prof to assoc prof, 66-88, PROF BOT, SDAK STATE UNIV, 88- *Concurrent Pos:* NSF res grant-in-aid, Biol Sta, Univ Okla, 66. *Mem:* Ecol Soc Am; Soc Range Mgt. *Res:* Grassland ecosystem analysis; vegetation in relation to geological materials. *Mailing Add:* Dept Biol SDak State Univ Box 2207b Brookings SD 57007

HUTCHESON, J(OHN) A(LISTER), b Park River, NDak, Jan 21, 05; m 31; c 1. ELECTRICAL ENGINEERING. *Educ:* Univ NDak, BSc, 26. *Prof Exp:* Radio engr, Westinghouse Elec Corp, 26-36, sect engr, 36-40, mgr, Radio Eng Dept, 40-43, assoc dir, Res Labs, 43-48, dir res, 48-49, vpres res & develop, 49-55, eng, 55-62, planning atomic defense & space group, 62-65; chmn div eng & indust res, Nat Res Coun, 65-70; vchmn comt motor vehicle emissions, Nat Acad Sci, 70-73; RETIRED. *Mem:* Fel Am Phys Soc; fel Inst Elec & Electronics Engrs. *Res:* Electronics; vacuum tubes; radio circuits; apparatus for vibration testing; radio systems. *Mailing Add:* 946 Osage Rd Pittsburgh PA 15243

HUTCHESON, KERMIT, b Wrightsville, Ga, Dec 8, 29; m 50; c 3. BIOSTATISTICS. *Educ:* Ga Southern Col, BS, 53; Univ Miami, MS, 57; Va Polytech Inst, PhD(statist), 70. *Prof Exp:* Instr math, Miami Univ, 59-61; asst prof, Radford Col, 62-67; ASSOC PROF STATIST, UNIV GA, 67- *Mem:* Am Statist Asn; Sigma Xi. *Res:* Ecological diversity. *Mailing Add:* Dept Statist Univ Ga Athens GA 30602

HUTCHESON, PAUL HENRY, b Nashville, Tenn, Dec 7, 28; m 50; c 6. APPLIED MATHEMATICS, COMPUTER SCIENCE. *Educ:* David Lipscomb Col, BA, 50; Peabody Col, MA, 51; Univ Fla, PhD(math chem), 60. *Prof Exp:* Instr math & chem, Fla Col, 51-54; asst prof math, Univ Tampa, 54-57; from asst prof to assoc prof, 60-63, dir comput ctr, 63-78, PROF MATH, MID TENN STATE UNIV, 64- *Concurrent Pos:* Asst prof, Space Inst, Univ Tenn, 64-68; consult, ARO, Inc, Arnold Air Force Sta, 63-70. *Mem:* Am Math Soc; Math Asn Am; Asn Comput Mach. *Res:* Linear elasticity; reflector design; remote sensing; landsat processing. *Mailing Add:* Dept Math & Comput Sci Mid Tenn State Univ Murfreesboro TN 37132

HUTCHESON, THOMAS BARKSDALE, JR, agronomy; deceased, see previous edition for last biography

HUTCHIN, MAXINE E, b Kansas City, Mo, Dec 11, 22; m 46; c 2. PHYSIOLOGICAL CHEMISTRY, TOXICOLOGY. *Educ:* Univ Calif, Berkeley, AB, 43. *Prof Exp:* Jr chemist, Indust Lab, US Naval Dry Docks, Calif, 44; chemist, Insecticide Div, USDA, 44-46; res asst chem, S B Penick Co, NY, 46-47; sr lab technician, US Naval Hosp, Oakland, Calif, 47-49, res asst metab res unit, Off Naval Res, 49-50, res chemist, Naval Hosp, Oakland, 50-53, supvr, Clin Invest Ctr, 53-62, sr investr, US Naval Radiol Defense Lab, 62-69; asst toxicologist, San Mateo County Coroner's Lab, 70-74; CRIMINALIST, ALAMEDA COUNTY SHERIFF'S DEPT, CRIME LAB, 74- *Concurrent Pos:* Asst chemist, Permanente Metals Corp, Calif, 45. *Mem:* Am Acad Forensic Sci; Am Chem Soc. *Res:* Ion transport in terrestrial plants; physiological chemistry as applied to medical research; application of chemical methodology to radiobiology. *Mailing Add:* 5407 Greenridge Rd Castro Valley CA 94552-2621

HUTCHIN, RICHARD ARIEL, b Lafayette, Ind, May 9, 46; m; c 1. MATHEMATICAL PHYSICS. *Educ:* Princeton Univ, AB, 67; Stanford Univ, MS, 68, PhD(physics), 70. *Prof Exp:* Fel physics, State Univ NY, 70-72; staff scientist, Itek Corp, 72-74, prin scientist, 75, sect mgr electrooptics, 75-76, dept mgr advan anal, 76-77, chief scientist, 77-79; CONSULT, 80- *Concurrent Pos:* Treas, Novon, Inc, 81- *Honors & Awards:* Kusaka Physics Prize. *Mem:* Optical Soc Am; Sigma Xi; Math Asn Am. *Res:* Compensated imaging through turbulence; electro-optical reconnaisance; space optics; optical computing; active optics; active mirrors; pattern recognition; high energy laser beam control; speckle interferometry; optical tracking and acquisition; unconventional imaging; laser defense systems. *Mailing Add:* 554 S Helberta Ave Redondo Beach CA 90277

HUTCHINGS, BRIAN LAMAR, b South Jordan, Utah, June 11, 15; m 38; c 4. MICROBIAL BIOCHEMISTRY. *Educ:* Brigham Young Univ, BS, 38; Univ Wis, MS, 40, PhD(biochem), 42. *Prof Exp:* Asst, Univ Wis, 40-42; res chemist, Lederle Labs Div, Am Cyanamid Co, 42-49, head dept biochem, 49-52, asst dir chem & biol res, 52-56, dir biochem res, 56-65, sr res fel, 65-68; prof biol sci & chmn dept, 68-72, dean, Col Sci & Eng, 73-84, distinguished serv prof, 84- 85, EMER PROF, WRIGHT STATE UNIV, 85- *Mem:* Fel AAAS; Am Chem Soc; Am Soc Biol Chemists; fel NY Acad Sci; Am Soc Microbiol; Soc Exp Biol & Med. *Res:* Chemistry of vitamins, anti-malarials and antibiotics; microbial transport; biochemistry of membranes. *Mailing Add:* 2534 Grand Pt Cr Sandy UT 84092

HUTCHINGS, DONALD EDWARD, b Chicago, Ill, Jan 9, 34; m 60; c 3. BEHAVIORAL TERATOLOGY. *Educ:* Lake Forest Col, BA, 59; Univ Chicago, MA, 63, PhD(psychol), 65. *Prof Exp:* From res scientist to sr res scientist, NY State Psychiat Inst, 63-74; instr psychiat, 66-68, res assoc, 68-72, ASST PROF MED PSYCHOL, DEPT PSYCHIAT, COLUMBIA

UNIV, 72-, ASST PROF MED PSYCHOL, DEPT PEDIAT, 75-; RES SCIENTIST, NY STATE PSYCHIAT INST, 74- *Concurrent Pos:* Adj asst prof, Dept Psychol, Columbia Univ, 66- & Barnard Col, 71- *Mem:* Am Psychol Asn; Teratol Soc; AAAS; Int Soc Develop Psychobiol; Sigma Xi. *Res:* Developmental behavior toxicology; effects of prenatal drug exposure on behavior of the offspring. *Mailing Add:* 722 W 168th St New York NY 10032

HUTCHINGS, JOHN BARRIE, b Johannesburg, SAfrica, July 18, 41. ASTRONOMY. *Educ:* Univ Witwatersrand, BSc, 62, BSc(Hons), 63, MSc, 64; Cambridge Univ, PhD(astron), 67. *Prof Exp:* Fel astron, 67-69, RES SCIENTIST, NAT RES COUN CAN, 69- *Honors & Awards:* Beals Award, 82. *Mem:* Fel Royal Astron Soc; Am Astron Soc; Int Astron Union; Can Astron Soc. *Res:* Optical and radio imaging and spectroscopy of quasi-stellar objects and active galaxies; optical and ultraviolet astronomy research into early type stars with particular interest in evolution, mass loss, rotation, masses and dimensions; x-ray sources and cataclysmic variables; space instrumentation including Hubble Space Telescope. *Mailing Add:* Dom Astrophys Observ 5071 W Saanich Rd Victoria BC V8X 4M6 Can

HUTCHINGS, L(EROI) E(ARL), b Northfield, Minn, Aug 3, 20; m 49; c 2. CHEMICAL ENGINEERING. *Educ:* Univ Mich, BSE, 43, MS, 44; Northwestern Univ, PhD(chem eng), 48. *Prof Exp:* Process control engr, US Rubber Co, 43; res assoc, Univ Mich, 44-45; div dir process & prod res, Pure Oil Co, 48-61; assoc dir res, Great Lakes Res Corp, 61-65, mgr tech serv, Great Lakes Carbon Corp, 65-69; sr coordr eng res & develop, Universal Oil Prod Co, Des Plaines, 69-82; RETIRED. *Mem:* Am Chem Soc. *Res:* Petroleum coke; unit operations; hydrotreating. *Mailing Add:* 1002 W Gregory Mt Prospect IL 60056

HUTCHINGS, WILLIAM FRANK, b Rochester, NY, July 13, 38; m 64; c 3. MECHANICAL ENGINEERING. *Educ:* Case Western Reserve Univ, BS, 60; Univ Rochester, MS, 69. *Prof Exp:* Design engr, Cornell Aeronaut Lab, 61-66; supv engr, Consol Vacuum Corp, 66-70; proj engr, Genera Signal Corp, 70-71, mgr design eng, 71-76, mgr prod standards & develop, 76-82, DIR RES & PROD DEVELOP, LIGHTNIN, GENERA SIGNAL CORP, 82- *Mem:* Am Soc Mech Engrs; Am Soc Metals. *Res:* Fluid dynamics of mixing along with the dynamics of mixing machinery; development of fluid mixers. *Mailing Add:* 19 Cannock Dr Fairport NY 14450

HUTCHINS, GROVER MACGREGOR, b Baltimore, Md, Aug 17, 32; m 56; c 3. ANATOMIC PATHOLOGY, AUTOPSY PATHOLOGY. *Educ:* Johns Hopkins Univ, BA, 57, MD, 61; Am Bd Path, cert anat, 72 & pediat, 90. *Prof Exp:* Resident path, Johns Hopkins Hosp, 64-65; res fel, Scripps Clin & Res Found, 65-66; from instr to assoc prof, 66-83, prof path, JOHNS HOPKINS UNIV SCH MED, 83- *Concurrent Pos:* Assoc dir autopsy path, 67-76, dir, 76- *Mem:* Col Am Pathologists; Am Heart Asn; Int Acad Pathologists; Soc Cardiovasc Path; Soc Pediat Path; Teratol Soc. *Res:* Studies in human cardiovascular, pulmonary and pediatric pathology. *Mailing Add:* Dept Path Johns Hopkins Hosp 600 N Wolfe St Baltimore MD 21205

HUTCHINS, HASTINGS HAROLD, SR, b Philadelphia, Pa, Oct 1, 22; m 45; c 3. PHARMACY. *Educ:* Fordham Univ, BS, 43; Rutgers Univ, MS, 49; Purdue Univ, PhD(pharmaceut chem), 52. *Prof Exp:* Instr, Chem Lab, Rutgers Univ, 47-49; sr pharmaceut chemist, Merck & Co, Inc, 52-54, head, Tech Serv Lab, 54-56; sr res scientist, Johnson & Johnson, New Brunswick 56-62, group leader, 62-65, dir pharmaceut develop, 65-69, dir, Baby & Toiletries Prod Develop, 70-87; RETIRED. *Mem:* Am Chem Soc; Pharmaceut Mfrs Asn; Am Pharmaceut Asn; fel Am Inst Chem. *Res:* Anti-oxidants for vitamin A stabilization; separation of radioactive sulfonamides via ion exchange resins; stabilized cosmetic, ethical and proprietary pharmaceutical preparations; tabletting; physical pharmacy and aerosol technology; parenterals and veterinary drug dosage forms; in vitro evaluations of antacids, basic studies on particle size, dielectric constants and zeta potential. *Mailing Add:* 15 Hillwood Ave Edison NJ 08820

HUTCHINS, JOHN R(ICHARD), III, b Perth Amboy, NJ, May 6, 34; m 58; c 4. CERAMICS, PHYSICAL CHEMISTRY. *Educ:* Rensselaer Polytech Inst, BS, 55. *Hon Degrees:* ScD, Mass Inst Technol, 59 & Rutgers Univ, 84. *Prof Exp:* Res asst ceramics, Mass Inst Technol, 55-58; res physicist, Knox Labs, Inc, NJ, 59-60; sr ceramist, Tech Staff Div, Corning Glass Works, NY, 60-61, res ceramist, 61-62, res assoc ceramics, 62-66, mgr mat res, 66-67, mgr surface chem res, Tech Staffs Div, 67-69, dir bio-org technol, 69-72, dir appl physics & biol, 72-73, vpres res & develop, 73-80, dir, 73-85, sr vpres, 80-85; exec vpres staffs, 85-88, EXEC VPRES, DIR TECHNOL & NEW BUS DEVELOP, SIECOR CORP, HICKORY NC, 88- *Concurrent Pos:* Dir, Cormedics Corp, NJ, Diag Res, Inc, NY, 71-72 & Genecor Inc, 82-85; mem, NY State Adv Coun on High Technol, 78-; mem, Nat Mat Adv Bd, 78-80; mem, Res & Develop Coun, Am Mgt Asn, Dir Indust Res & Technol Coun ITG, Electronic Indust Asn. *Mem:* AAAS; Am Chem Soc; Sigma Xi; fel Am Ceramic Soc; Brit Soc Glass Technol. *Res:* Gases in glass; ion selective electrodes; immobilized enzymes; chromatography; immunochemistry; optical waveguides; medical instrumentation; hypodermic syringe manufacture; solid state reactions; dissolution; glass ceramics, photosensitive glasses; refractories; specialty glasses and ceramics. *Mailing Add:* 1058 25th Ave Dr NW Hickory NC 28601

HUTCHINS, MARYGAIL KINZER, b Los Angeles, Calif, Jan 17, 40; m 62; c 2. SURFACE CHEMISTRY, SPECTRAL ANALYSIS. *Educ:* Mt St Mary's Col, BA, 61; St Joseph's, MS, 77; Temple Univ, PhD(chem), 81. *Prof Exp:* Technician, Douglas Aircraft Co, 58-59; lit chemist, Richfield Res, Anaheim, 61-62; lit chemist, Thermophys Properties Res Ctr, 62-64; sci teacher, Cent Catholic High Sch, Ind, 64-65; spectros technician, Purdue Univ, 65-66; nuclear magnetic resonance technician, Univ Notre Dame, 66-68; substitute teacher, Philadelphia Bd Educ, 68-69; instr, Dept Chem, Temple Univ, 69-71; technician, Drexel Univ, 71-72; lit chemist, Inst Sci Info, 72-73; teaching asst, St Joseph's Univ, 74-76; teaching asst & res asst, Temple Univ & Fels Res Found, 76-79, Pub Health Serv trainee & res assoc, 79-80, grad res assoc, 80-81; prod develop chemist, LNP Eng Plastics Corp, 81-86; ANALYTICAL CHEMIST, ICI ADVAN MAT, 86- *Mem:* Am Inst Chem; Am Chem Soc; Sigma Xi; Fiber Soc; Soc Advan Mat & Process Eng. *Res:* Compositing of thermoplastics with fiber; nature of any coating required to better these composites; surface of and bulk polymer analyses. *Mailing Add:* ICI Advan Mat 475 Creamery Way Exton PA 19341

HUTCHINS, PHILLIP MICHAEL, b Winston-Salem, NC, Aug 15, 42; m 64; c 3. PHYSIOLOGY, BIOMEDICAL ENGINEERING. *Educ:* NC State Univ, BSEE, 64; Bowman Gray Sch Med, MS, 66, PhD(physiol), 69. *Prof Exp:* Assoc med, Harvard Med Sch, 69-70; asst prof to assoc prof, 70-81, PROF PHYSIOL, BOWMAN GRAY SCH MED, 81-, DIR DEPT BIOMED ENG, 70- *Concurrent Pos:* Res assoc, Peter Bent Brigham Hosp, 69-70; NASA & NC Heart Asn grants, 71-; Nat Heart & Lung Inst grant, 71- *Mem:* Am Heart Asn; Am Physiol Soc; Microcirc Soc; Biomed Eng Soc; Planetary Soc; Am Soc Gravitational & Space Biol. *Res:* Microcirculation and effects of oxygen; hypertension; hemorrhagic shock; pharmacology of the microcirculation; computers; medical instrumentation; gravitational (space) physiology; chronic cardiovascular monitoring. *Mailing Add:* Microcirc Lab Bowman Gray Sch Med Winston-Salem NC 27103

HUTCHINS, ROBERT OWEN, SR, b Danville, Ill, Sept 25, 39; m 62; c 2. ORGANIC CHEMISTRY. *Educ:* Univ Calif, Berkeley, BS, 61; Calif State Col, Long Beach, MA, 62; Purdue Univ, PhD(chem), 67. *Prof Exp:* NSF res fels, Univ Notre Dame, 66-68; from asst prof to assoc prof, 68-79, PROF ORG CHEM, DREXEL UNIV, 79-, HEAD DEPT, 86- *Honors & Awards:* Drexel Res Award, 82; Res Award, Am Chem Soc, Philadelphia Sect, 87. *Mem:* AAAS; Am Chem Soc; Chem. *Res:* New synthetic techniques; conformational analysis and stereochemistry. *Mailing Add:* Dept Chem Drexel Univ Philadelphia PA 19104

HUTCHINS, SAMUEL F, electrical engineering, for more information see previous edition

HUTCHINS, WILLIAM R(EAGH), b New York, NY, Mar 23, 19; m 45; c 3. ELECTRONICS ENGINEERING, COMPUTER SYSTEMS. *Educ:* Columbia Univ, BA, 39, BS, 40, MS(EE), 41. *Prof Exp:* Eng asst to Maj Edwin H Armstrong, Columbia Univ, 39-46; mgr & chief engr, E Anthony & Sons, Mass, 46-48; mgr, adv develop dept, missile systs div, Raytheon Co, 48-58; chief, ballistic missile defense br, ARPA, 58-59; sci liaison officer, Raytheon Co, 59-60; dir, eastern tech opers, Aerospace Corp, 61-65; vpres adv develop, Nat Co, Inc, 65-66; systs consult, Sanders Assocs, Inc, 66-69, mgr, S-3A, 69-71, mgr, Antisubmarine Warfare Systs, 71-72, tech dir, Systs Div, 72-73, mgr, Air Systs, 74-79, mgr, VHSIC, 79-81, mgr, Adv Electronic Warfare, 81-85; PRES & CONSULT, FUTURE CONCEPTS, 85- *Concurrent Pos:* Mem proj Lamplight, Off Naval Res, Exec Off President, 44-45; security resources panel, 47; consult, NY Eye & Ear Infirmary, 48-50; dir, mission area study on integrated air surveillance, USAF Systs Command, 74-75. *Mem:* Acoust Soc Am; Am Inst Aeronaut & Astronaut; Inst Elec & Electronics Engrs. *Res:* Frequency modulation-continuous wave radar; acoustics, missile and defense systems; audio; audiometry; radio and microwaves. *Mailing Add:* Future Concepts Inc 5605 Brisbane Dr Chapel Hill NC 27514

HUTCHINSON, BENNETT B, b Honolulu, Hawaii, Aug 7, 42; m 70; c 1. SPECTROCHEMISTRY. *Educ:* Abilene Christian Col, BS, 63; Univ Tex, Austin, MA, 65; Ill Inst Technol, PhD(chem), 70. *Prof Exp:* From instr to assoc prof, 65-76, PROF CHEM, ABILENE CHRISTIAN UNIV, 76- *Mem:* Am Chem Soc; Soc Appl Spectros; AAAS. *Res:* Spectroscopic study of transition metal complexes, their preparation and bonding, and the role of these metals in biological systems. *Mailing Add:* Abilene Christian Univ Box 8208 Abilene TX 79601

HUTCHINSON, CHARLES E(DGAR), b Parkersburg, WVa, Dec 18, 35; m 60; c 2. VLSI, COMPUTER AIDED DESIGN. *Educ:* Ill Inst Technol, BSEE, 57; Stanford Univ, MSEE, 61, PhD(elec eng), 63. *Prof Exp:* Res specialist, Autonetics Div, NAm Aviation, Inc, Calif, 63-65; prof elec eng, Univ Mass, Amherst, 65-84; PROF & DEAN, THAYER SCH ENG, DARTMOUTH COL. *Concurrent Pos:* Lectr, Univ Calif, Los Angeles, 64-65. *Mem:* Inst Elec & Electronics Engrs; Am Soc Eng Educ. *Res:* Automatic control systems; stochastic processes; inertial navigation systems; computer processing. *Mailing Add:* Thayer Sch Eng Dartmouth Col Hanover NH 03755

HUTCHINSON, CHARLES F, b Riverside, Calif, Sept 22, 46; m 66; c 2. REMOTE SENSING, NATURAL RESOURCE INVENTORY & MONITORING. *Educ:* Univ Calif, Riverside, AB, 72, MA, 74, PhD(geog), 78. *Prof Exp:* Geographer, Earth Resources Observ Syst, US Geol Surv, 76-78; tech staff, Jet Propulsion Lab, Calif Inst Technol, 78-80; DIR, ARIZ REMOTE SENSING CTR, ASSOC DIR, OFF ARID LANDS STUDIES & CHMN, ARID LANDS RESOURCE SCI DOCTORAL PROG, UNIV ARIZ, 80-, ASST PROF GEOG, 81- *Concurrent Pos:* Secy, US Man & Biosphere Desert Zone Directorate, 80-; consult, Food & Agr Orgn, UN, People's Repub China, 84 & 85; Gilbert F White fel, Resources Future, 88-89. *Mem:* Am Soc Photogram; Sigma Xi; Asn Am Geographers; AAAS. *Res:* Development and application of remote sensing techniques, both manual and automated; techniques for land resource inventory; economic development of the arid zone. *Mailing Add:* 2027 E Ninth St Tucson AZ 85719

HUTCHINSON, CHARLES S, JR, b Topeka, Kans, Oct 17, 30; m 56; c 2. GEOLOGY. *Educ:* Principia Col, BA, 52. *Prof Exp:* Ed-in-chief & corp secy, Burgess Pub Co, 55-65; ed-in-chief, prof books, Reinhold Book Corp, 65-68; ed-in-chief, prof books, Van Nostrand Reinhold Co, 68-70; pres & co-founder, Dowden, Hutchinson & Ross Inc, 70-83; publ geol, physics, math & statist, Van Nostrand Reinhold Co, 84-86; PUBL, GEOSCI PRESS, INC, 88- *Concurrent Pos:* Consult, Hutchinson Assoc, 87- *Mem:* Fel Geol Soc Am; Am Soc Petrol Geologists; Am Geophys Union; Soc Econ Paleontologists & Mineralogists. *Mailing Add:* 1040 Hyland Circle Prescott AZ 86303

HUTCHINSON, DONALD PATRICK, b Laurel, Miss, Jan 15, 47; m 67; c 2. PLASMA & LASER PHYSICS. *Educ:* Univ Miss, BS, 68; Mass Inst Technol, SM, 70, ScD(appl plasma physics), 74. *Prof Exp:* PHYSICIST, OAK RIDGE NAT LAB, 74- *Mem:* Am Phys Soc; Am Optical Soc. *Res:* Development of CW and pulsed submillimeter lasers for use as plasma diagnostic tools. *Mailing Add:* 909 Woodsmoke Circle Knoxville TN 37922-1634

HUTCHINSON, ERIC, b Morton, Eng, Dec 25, 20; nat US; m 42. COLLOID CHEMISTRY. *Educ:* Cambridge Univ, BA, 41, MA, 44, PhD(colloid chem), 45. *Prof Exp:* Lectr phys chem, Univ Sheffield, 45-46; Bristol Myers fel colloid chem, Stanford Univ, 46-48; asst prof phys chem, Fordham Univ, 48-49; from asst prof to assoc prof, 49-59, assoc exec head, dept chem, 58-62, PROF PHYS CHEM, STANFORD UNIV, 59-, ACAD SECY, 74- *Concurrent Pos:* Fulbright lectr, Yokohama Nat Univ, 59-60; vis prof, Univ Sussex, 67. *Mem:* Am Chem Soc; The Chem Soc. *Res:* Properties of interfacial monolayers; adsorption on surfaces; thermodynamics of colloidal electrolytes; microcalorimetry; science and public policy; studies of government support of scientific and technological research and development. *Mailing Add:* Dept Chem Stanford Univ Stanford CA 94305-1684

HUTCHINSON, FRANKLIN, b Brooklyn, NY, Feb 29, 20; m 44; c 4. BIOPHYSICS, MOLECULAR BIOLOGY. *Educ:* Mass Inst Technol, BS, 42; Yale Univ, PhD(physics), 48. *Prof Exp:* Instr radiol & physics, 48-51, asst prof physics, 51-57, assoc prof biophys, 57-60, chmn dept, 60-63 & 67-69, chmn Dept Molecular Biophys & Chem, 73-76, PROF BIOPHYS, YALE UNIV, 60- *Concurrent Pos:* Guggenheim fel, King's Col, Univ London, 63-64; consult radiation physics, Yale-New Haven Hosp. *Mem:* AAAS; Biophys Soc; Am Phys Soc; Radiation Res Soc. *Res:* Damage to DNA by physical and chemical agents; repair of DNA damage and the biological consequences of damage and of errors in repair; mutagenesis. *Mailing Add:* Dept Molecular Biophys & Biochem Box 6666 Yale Univ New Haven CT 06511

HUTCHINSON, FREDERICK EDWARD, b Atkinson, Maine, June 1, 30; m 52; c 2. SOIL CHEMISTRY. *Educ:* Univ Maine, BS, 53, MS, 58; Pa State Univ, PhD, 67. *Prof Exp:* Instr soil chem, 58, from asst prof to assoc prof soil fertil, 58-67, dean col life sci & agr, 72-75, PROF SOIL FERTIL, UNIV MAINE, ORONO, 67-, CHMN DEPT PLANT & SOIL SCI, 71-, VPRES RES & PUB SERV, 75- *Mem:* Am Soc Agron; Soil Conserv Soc Am; Am Soc Hort Sci. *Res:* Nutrition of snap beans, sweet corn and peas; relationship of soil fertility to plant growth; lime requirement of specific soil types; chemical characteristics of selected soil types. *Mailing Add:* Col Agr Ohio State Univ 2120 Fyffe Rd Columbus OH 43210

HUTCHINSON, GEORGE ALLEN, b Brooklyn, NY, Apr 24, 36; m 69; c 3. ALGEBRA. *Educ:* Columbia Univ, BA, 58, MA, 60, PhD(math), 67. *Prof Exp:* Jr mathematician res & develop lab, US Army Signal Corps, NJ, 58; mem tech staff, Int Tel & Tel Commun Systs, Inc, 62-64; asst prof math, Fairleigh Dickinson Univ, 64-65; RES MATHEMATICIAN, NIH, 67- *Mem:* Am Math Soc; Math Asn Am. *Res:* Algebra, especially category theory, lattice theory and ring and modules; computer science. *Mailing Add:* Rm 3045 Bldg 12A NIH Bethesda MD 20892

HUTCHINSON, GEORGE EVELYN, zoology, ecology; deceased, see previous edition for last biography

HUTCHINSON, GEORGE KEATING, b Belfast, Maine, Nov 20, 32; m 58; c 5. COMPUTER SCIENCE, MANAGEMENT SCIENCES. *Educ:* Univ Maine, BS, 55; Carnegie Inst, MS, 56; Stanford Univ, PhD(mgt sci), 64. *Prof Exp:* Teaching asst mech drawing & anal geom, Carnegie Inst, 55-56; design engr, Lockheed Missiles & Space Co, 57-58, eng opers analyst, 58, from math analyst to sr math analyst, 58-62, proj leader & math engr, 62-66; asst prof indust eng & dir comput ctr, Tex Tech Univ, 66-68; assoc prof mgt info systs & dir comput ctr, 68-71, PROF MGT INFO SYSTS, UNIV WIS-MILWAUKEE, 71-; CONSULT, HUTCHINSON ASSOCS, LTD. *Concurrent Pos:* Res asst, Western Data Processing Ctr, Stanford Univ, 59-60; consult var nat & int orgn, 68-; mem comt, Nat Acad Sci; US rep, Int Fedn Info Processing. *Mem:* AAAS; Asn Comput Mach; Oper Res Soc Am; Inst Elec & Electronics Engrs; Inst Mgt Sci. *Res:* Development and application of generalized simulation models to solve complex systems problems; management information systems; automation of manufacturing and production systems, CAM and CIM. *Mailing Add:* Hutchinson Assocs Ltd 3404 Colette Ct Mequon WI 53092

HUTCHINSON, GORDON LEE, b Yuma, Colo, Feb 6, 43; m 62; c 3. MICROMETEOROLOGY. *Educ:* Colo State Univ, BS, 65, MS, 68; Univ Ill, PhD(soil sci), 73. *Prof Exp:* SOIL SCIENTIST, AGR RES SERV, USDA, 65- *Concurrent Pos:* Assoc ed, Soil Sci Soc Am J, 91- *Honors & Awards:* Emil Truog Soil Sci Award, Soil Sci Soc Am, 74. *Mem:* Am Soc Agron; Soil Sci Soc Am; Am Soc Plant Physiologists; AAAS; Coun Agr Sci & Technol. *Res:* Aerodynamic exchange of gaseous nitrogen compounds between the atmosphere and soil, water and crop surfaces; implications of the exchange to crop physiology; ecology and the environment. *Mailing Add:* Agr Res Serv USDA PO Box E Ft Collins CO 80522

HUTCHINSON, HAROLD DAVID, b Moneymore, Northern Ireland, Apr 15, 31; US citizen; m 57; c 2. ANIMAL NUTRITION. *Educ:* Queen's Univ Belfast, BAgr, 53; Univ Ill, MS, 55, PhD(animal sci), 57. *Prof Exp:* Res asst animal sci, Univ Ill, 53-56, instr vet physiol, 56-57; swine nutritionist, 57-64, head explor res, 64-67, head accessory prod develop & control coord, Res Dept, 67-71, dir res, 71-77, V PRES & DIR RES, MOORMAN MFG CO, 77- *Mem:* AAAS; Am Soc Animal Sci. *Res:* Factors influencing amino acids requirements of swine; methods of controlling anemia in baby pigs; effects of feeding moldy grains to swine. *Mailing Add:* 2901 Curved Creek Rd Quincy IL 62301

HUTCHINSON, HAROLD LEE, petroleum & chemical engineering; deceased, see previous edition for last biography

HUTCHINSON, JAMES HERBERT, JR, b Jackson, Miss, Aug 7, 33; m 62; c 2. ORGANIC CHEMISTRY. *Educ:* Univ Southern Miss, BS, 55; Univ Iowa, MS, 60; Auburn Univ, PhD(org chem), 68. *Prof Exp:* Res chemist, Ethyl Corp, La, 57-58; develop chemist, Geigy Chem Corp, Ala, 61-62; US Army Med Res & Develop Command, Dept Army grant, Ind Univ, Bloomington, 67-69; PROF CHEM, MID TENN STATE UNIV, 69- *Mem:* Am Chem Soc; Sigma Xi. *Res:* Heterocyclic chemistry; synthesis of physiologically active compounds. *Mailing Add:* Mid Tenn State Univ PO Box 196 Murfreesboro TN 37132

HUTCHINSON, JAMES R(ICHARD), b San Francisco, Calif, June 1, 32; m 56; c 2. ENGINEERING MECHANICS. *Educ:* Stanford Univ, BS, 54, PhD(eng mech), 63; Univ Pittsburgh, MLitt, 58. *Prof Exp:* Engr, atomic power div, Westinghouse Elec Corp, 54-58; res specialist, Lockheed Missiles & Space Co, 58-64; from asst prof eng to assoc prof civil eng, 64-73, PROF CIVIL ENG, UNIV CALIF, DAVIS, 73- *Mem:* Am Soc Mech Engrs; Am Acad Mech. *Res:* Vibrations; structural dynamics; development of exact solutions for the vibration of solids such as cylinders, rectangular parallelepipeds, and hollow cylinders; comparison of these solutions with previous approximate beam plate and shell theories; research into the application of Boundary Element and Boundary Collocation Methods to vibration problems. *Mailing Add:* Dept Civil Eng Univ Calif Davis CA 95616

HUTCHINSON, JOAN PRINCE, b Philadelphia, Pa, Apr 19, 45; m 75. MATHEMATICS, GRAPH THEORY. *Educ:* Smith Col, BA, 67; Univ Pa, MA, 71, PhD(math), 73. *Prof Exp:* John Wesley Young res instr math, Dartmouth Col, 73-75; asst prof, Tufts Univ, 75-76; asst prof math, Smith Col, 76-90; PROF MATH, MACALESTER COL, 90- *Concurrent Pos:* NSF res grant, 77-79, 84-88; Benedict distinguished vis prof math, Carleton Col, 78-79; vis prof women, NSF, 89-90. *Mem:* Am Math Soc; Math Asn Am; Asn Women Math; Soc Indust & Appl Math. *Res:* Chromatic and topological graph theory; combinatorial analysis; graph algorithms; computer science theory. *Mailing Add:* Dept Math Macalester Col St Paul MN 55105

HUTCHINSON, JOHN W(ENDLE), b Williamstown, Ky, Dec 1, 27; m 50; c 7. CIVIL ENGINEERING. *Educ:* Univ Ill, BS, 51, MS, 54, PhD(civil eng), 61. *Prof Exp:* Engr in training, Stephens-Adamson Co, Ill, 51; from instr to assoc prof civil eng, Univ Ill, 54-64; assoc prof, 64-67, PROF CIVIL ENG, UNIV KY, 67- *Concurrent Pos:* mem, Transp Res Bd, Nat Acad Sci-Nat Res Coun; sci adv, Int Ctr Transp Studies, Amalfi, Italy. *Honors & Awards:* Teetor Award, Soc Automotive Engrs, 79. *Mem:* Nat Acad Eng; Am Soc Civil Engrs; Int Ctr Transp Studies; Am Asn Automotive Med; Inst Transp Engrs; Am Soc Eng Educ. *Res:* Highway materials, geometrics, safety and administration; traffic planning; accident reconstruction; vehicle dynamics. *Mailing Add:* Dept Civil Eng Univ Ky Lexington KY 40506

HUTCHINSON, KENNETH A, b Dugger, Ind, Sept 8, 32. ANALYTICAL CHEMISTRY, CHEMICAL ENGINEERING. *Educ:* Wayne State Univ, BS, 55, MS, 56, PhD(anal chem), 60. *Prof Exp:* Instr anal chem, Wayne State Univ, 58-60; res chem engr, 60-69, engr, Water Systs Eng Div, 69-74, ENGR, ANALTICAL CHEM, DETROIT EDISON CO, 74- *Concurrent Pos:* Res compounder org chem, US Rubber Co, 59-60. *Mem:* Am Chem Soc; Air Pollution Control Asn; Sigma Xi; Am Soc Testing & Mat. *Res:* Gas chromatography; infrared spectrophotometry; gas analysis; rubber; paint; plastics; electrical insulation; research engineering of liquid alkali metals; asbestos analysis; scanning electron microscopy; x-ray diffraction spectrometry. *Mailing Add:* Detroit Edison Co 2000 Second Ave H-9 WSC Detroit MI 48226

HUTCHINSON, RICHARD ALLEN, b Wayne, Mich, Jan 6, 42; m 67; c 3. PHARMACY. *Educ:* Univ Mich, BS, 66, Pharm D, 67. *Prof Exp:* Resident clin pharm, Univ Mich, 68; instr clin pract, Wayne State Univ, 68-69; asst prof clin pharm, State Univ NY Buffalo, 69-73; assoc prof, 73-78, PROF & HEAD CLIN PHARM, UNIV ILL, 78- *Concurrent Pos:* Dir pharmaceut serv, Sinai Hosp, Detroit, Mich, 68-69; chief clin pract, E J Meyer, 69-70; chief clin pract, Buffalo Gen Hosp, 69-70, dir pharmaceut servs, 70-73. *Mem:* Am Soc Hosp Pharmacists; fel Am Col Clin Pharmacol; Am Asn Col Pharm. *Res:* Establishment and audit of minimum standards of clinical practice. *Mailing Add:* Dept Pharm Pract MC 886 Univ Ill Col Pharm Rm 244 833 S Wood St Chicago IL 60612

HUTCHINSON, RICHARD WILLIAM, b London, Ont, Nov 17, 28; m 51; c 4. GEOLOGY. *Educ:* Univ Western Ont, BSc, 50; Univ Wis, MS, 51, PhD(geol), 54. *Prof Exp:* Geologist, Am Metal Climax, Inc, 54-64; from assoc prof to prof econ geol & mineral deposits, Univ Western Ont, 64-83; CHARLES F FOGARTY PROF ECON GEOL, COLO SCH MINES, 83- *Concurrent Pos:* Nat Res Coun Can res grant, 64-83; consult geologist, Callahan Mining Corp, NY, 64-70, Chevron Resources, 72-84, Utah Int, 84-; Geol Surv Can res grant, 65-69; NATO Res Coun res grant, 67-69, Nat Sci & Eng Res coun, Can Res grant, 64-83; coun, Geol Soc Am, 88-90. *Honors & Awards:* Barlow Gold Medal, Can Inst Mining & Metall, 72 & 79; Duncan Derry Gold Medal, Mineral Deposits Div Geol Assoc Can, 83; Silver Medal, Soc Econ Geol, 85. *Mem:* Soc Econ Geologists (pres, 83); Geol Soc Am; Can Inst Mining & Metall; Geol Asn Can; Soc Geol Appl; Prospectors & Developers Asn Can; Geol Soc SAfrica. *Res:* Descriptive geology on pegmatites and deposits of rare-element minerals in pegmatites; genesis and tectonic relationships of massive sulfide base metal ore deposits; origin of K-bearing marine evaporite deposits; metallogenic evolution of ore deposits; genesis of stratiform sulfide-rich tin lodes; geology and genesis of Precambrian gold deposits. *Mailing Add:* Dept Geol & Geol Eng Colo Sch Mines Golden CO 80401

HUTCHINSON, ROBERT MASKIELL, b Trenton, NJ, Dec 24, 28; m 48; c 8. GEOLOGY. *Educ:* Princeton Univ, AB, 41; Univ Mich, AM, 48; Univ Tex, PhD(geol), 53. *Prof Exp:* Geol field asst, Nfld Geol Surv, 39; civil engr, Eng Dept, CEngr, US War Dept, Trinidad, BWI, 41-42; geologist, Strategic Minerals Div, US Geol Surv, 42-45, groundwater div, 45; instr geol, Univ Tex, 48-53; asst prof, Kans State Col, 53-56; from asst prof to assoc prof, 56-74,

PROF GEOL, COLO SCH MINES, 74- *Concurrent Pos:* Geol Soc Am res grant, 52-55; NSF grant, 60-63. *Mem:* Geol Soc Am; Am Inst Prof Geologists; AAAS. *Res:* Igneous, metamorphic and structural petrology; economic geology, genesis and occurrence of metalliferous ore deposits associated with batholiths; time-space, time-span studies of Precambrian basement rocks of central Texas and Colorado Rocky Mountains; erosion rates in the Colorado Rocky Mountains during Precambrian time; economic geology of precious and base metal ore deposits; nuclear waste repository siting in Precambrian crystalline basement rocks; structural analysis of heavy oil reservoirs in overthrust belts. *Mailing Add:* Dept Geol Colo Sch Mines Golden CO 80401

HUTCHINSON, THOMAS C, b Sunderland, UK, Feb 18, 39; div; c 1. BIOLOGY, AGRICULTURE. *Educ:* Victoria Univ Manchester, BSc, 60; Univ Sheffield, PhD(bot), 66. *Prof Exp:* Sir James Knott fel, Univ Newcastle, 64-67; from asst prof to assoc prof bot, Univ Toronto, 67-71, assoc exec, Inst Environ Sci & Eng, 71-73, assoc dir, 73-76, chmn dept, 76-82, grad coordr, 83-87, prof bot, 74-91, prof forestry, 78-91; PROF BIOL & ENVIRON RESOURCE STUDIES, TRENT UNIV, 91- *Concurrent Pos:* Chmn, Can Water Pollution Res Coun, 74-75; consult, WHO, 74-; chmn heavy metal subcomt, Nat Res Coun, 77- & Nat Sci & Eng Coun, 81-82. *Honors & Awards:* George Lawson Medal, Can Bot Asn, 83. *Mem:* Brit Ecol Soc; Ecol Soc Am; Can Bot Soc; Am Soc Agron; Arctic Inst NAm; fel Royal Soc Can. *Res:* Heavy metal toxicity; oil spill phytotoxicity; air pollution; acid rain phenomena; smelter metal studies on lead, arsenic, copper and nickel; tailings revegetation, metal tolerant grasses; algae; forest decline; climatic change on boreal forest. *Mailing Add:* Dept Biol Univ Trent Peterborough ON K9J 7B8 Can

HUTCHINSON, THOMAS EUGENE, b York, SC, Aug 1, 36; c 2. BIOLOGICAL MICROANALYSIS, CELL-ION TRANSPORT. *Educ:* Clemson Univ, BS, 58, MS, 59; Univ Va, PhD(physics), 63. *Prof Exp:* Teaching asst, Univ Va, 60-61; res fel, AEC, 62; sr scientist, 3M Co, 63-66, res specialist, 66-67; assoc prof chem eng & mat sci, 67-74, Univ Minn, prof, 74-76; prof biomed & chem eng, Univ Wash, 76-82, assoc dean eng res, 82; PROF, SCH ENG & APPL SCI, UNIV VA, 82- *Concurrent Pos:* Consult for var orgns including 3M Co, RCA, North Star Res & others, 67-73; chmn, Gordon Res Conf, 70; vis prof, Cavendish Lab, Cambridge, Eng, 71; sr res fel, Univ Glasgow, Scotland, 74-; chmn, Battelle Conf Microprobe Anal, 80. *Mem:* Electron Micros Soc Am; Am Vacuum Soc. *Res:* Microprobe analysis of biological tissue-an application of physics tools to solution of questions of ion transport in excitable cells. *Mailing Add:* Dept Eng & Appl Sci Univ Va Thornton Hall Charlottesville VA 22901

HUTCHINSON, WILLIAM BURKE, b Seattle, Wash, Sept 6, 09; m 39; c 5. SURGICAL CANCER. *Educ:* Univ Wash, BSc, 31; McGill Univ, MD, 36. *Hon Degrees:* DHH, Seattle Univ, 82. *Prof Exp:* Clin prof, 58-83, EMER PROF SURG, UNIV WASH, 83-; FOUNDER, PRES & DIR, PAC NORTHWEST RES FOUND, 55- *Concurrent Pos:* Pvt pract surgeon, 41-; founder & dir, Fred Hutchinson Cancer Res Ctr, 70-81, pres, 70-85, emer pres, 85- *Mem:* Hon mem Soc Surg Oncol; Am Col Surgeons. *Res:* Surgical oncology; early identification of clinical cancer with new technologies in the basic sciences. *Mailing Add:* Pac Northwest Res Found 720 Broadway Seattle WA 98122

HUTCHISON, CLYDE ALLEN, JR, b Alliance, Ohio, May 5, 13; m 37; c 3. PHYSICAL CHEMISTRY. *Educ:* Cedarville Col, AB, 33; Ohio State Univ, PhD(chem), 37. *Hon Degrees:* DSc, Cedarville Col, 53. *Prof Exp:* Nat Res Coun fel, dept chem, Columbia Univ, 37-38, res assoc, dept chem, 38-39 & SAM Labs, 43-45; asst prof chem, Univ Buffalo, 39-45; asst, Metall Lab, 45-46, from asst prof to prof, 45-63, chmn dept, 59-62, Carl William Eisendrath prof, 63-69, Carl William Eisendrath distinguished serv prof, 69-83, CARL WILLIAM EISENDRATH DISTINGUISHED SERV PROF CHEM EMER, ENRICO FERMI INST & DEPT CHEM, UNIV CHICAGO, 83- *Concurrent Pos:* Res assoc, Manhattan Dist Proj, dept physics, Univ Va, 42-43; consult, Argonne Nat Lab, 46- & Los Alamos Sci Lab, 53-62; ed, J Chem Physics, 53-59; Guggenheim fel, Oxford Univ, 55-56 & 72-73; mem chem panel, NSF, 60-63; mem vis comt, Brookhaven Nat Lab, 60-63 & Oak Ridge Nat Lab, 63-66; mem bd dirs, Ohio State Univ Res Found, 63-68; consult, Nat Bur Stand, 63-69; mem chem eval panel, USAF Off Sci Res, 66-70; mem vis comt, Argonne Nat Lab, 70- & Mass Inst Technol, 71-; Japan Soc Prom Sci vis prof, 75; mem, Eval Panel, Phys Chem Div, Nat Bur Standards, 77-80; Eastman prof, Univ Oxford, 81-82. *Honors & Awards:* Peter Debye Award, Am Chem Soc, 72. *Mem:* Nat Acad Sci; Am Chem Soc; fel Am Acad Arts & Sci; Am Phys Soc. *Res:* Separation and tracer studies of isotopes; magnetic susceptibilities of heavy elements; electron paramagnetic resonance absorption; paramagnetism of triplet state organic molecules; organic crystals; application of electron magnetic resonance to protein structure determination. *Mailing Add:* Searle Lab Univ Chicago 5735 Ellis Ave Chicago IL 60637

HUTCHISON, CLYDE ALLEN, III, b New York, NY, Nov 26, 38; div; c 3. MOLECULAR BIOLOGY, GENETICS. *Educ:* Yale Univ, BS, 60; Calif Inst Technol, PhD(biophys), 69. *Prof Exp:* From asst prof to prof microbiol, 68-83, KENAN PROF MICROBIOL, SCH MED, UNIV NC, CHAPEL HILL, 83- *Concurrent Pos:* NIH grant, 69- & res career develop award, 73-78; vis scientist, MRC lab molecular biol, Cambridge, Eng, 75-76 & 87-88. *Honors & Awards:* NIH Merit Award, 87. *Res:* Mammalian genome organization; hemoglobin genes; long interspersed sequences (retroposons); directed mutagenesis using chemically synthesized DNA; DNA sequence analysis. *Mailing Add:* Dept Microbiol & Immunol Univ NC CB No 7290 FLOB Chapel Hill NC 27599

HUTCHISON, DAVID ALLAN, b Milwaukee, Wis, Oct 6, 47; m 74. FUEL & LUBRICANT CHEMISTRY. *Educ:* Univ Wis-Milwaukee, BS, 69; Purdue Univ, PhD(org chem), 75. *Prof Exp:* Chemist res & develop, Explor Res-Lubrizol Corp, 75-79; sr chemist, Specialty Prod, Arco Petrol Prod Co, 79-81, mgr, Fuel Develop, 81-82, mgr engine oils res & develop, 82-85; SR RES CHEMIST, AMOCO PETROL ADDITIVES CO, 85- *Mem:* Am Chem Soc; Am Soc Lubricant Eng; Soc Automotive Eng. *Res:* Fuel development; motor oils; industrial lubricants; lubricant additive development; railroad engine oil formulation. *Mailing Add:* Amoco Res Ctr PO Box 3011 Mail Sta J8 Naperville IL 60566-7011

HUTCHISON, DAVID M, b Chicago, Ill, Aug 3, 35; m 58; c 3. GEOLOGY. *Educ:* Beloit Col, BS, 57; Univ Mont, MS, 59; WVa Univ, PhD(geol), 68. *Prof Exp:* Asst geol, Mich State Univ, 59-62; instr, Flint Community Jr Col, 62-65; asst, WVa Univ, 65-66; asst prof, 67-71, assoc prof, 71-78, PROF GEOL, HARTWICK COL, 78- *Mem:* Geol Soc Am. *Res:* Clay mineralogy; correlation of volcanic ash; provenance study of sand in sand dunes. *Mailing Add:* Dept Geol Hartwick Col Oneonta NY 13820

HUTCHISON, DORRIS JEANNETTE, b Carrsville, Ky, Oct 31, 18. MICROBIOLOGY. *Educ:* Western Ky State Col, BS, 40; Univ Ky, MS, 43; Rutgers Univ, PhD(microbiol), 49. *Prof Exp:* Res asst, Univ Ky, 40-42; instr biol, Russell Sage Col, 42-44; instr plant sci, Vassar Col, 44-46; res fel & asst, Rutgers Univ, 46-48, res assoc, 48-49; instr bot, Wellesly Col, 49-51; asst, Sloan-Kettering Inst Cancer Res, 51-56, assoc, 56-60, sect head, drug resistance, 56-72, assoc mem, 60-69, actg chief, div exp chemother, 65-66, chief, div drug resistance, 67-72, co-head, lab exp tumor ther, 73-74, head, lab drug resistance & cyto-regulation, 73-83, coordr, field of educ, 75-81; instr bact, Cornell Univ, New York, NY, 52-53, res assoc, 53-54, from asst prof to prof microbiol, 54-90, chmn biol unit, 68-74, assoc dir, 74-87, EMER PROF MICROBIOL, SLOAN-KETTERING DIV, GRAD SCH MED SCI, CORNELL UNIV, NEW YORK, NY, 90- *Concurrent Pos:* Fac fel, Vassar Col, 46; USPHS fel, 51-53; Philippe Found fel, 59; assoc dean, Grad Sch Med Sci, NY & asst dean, Grad Sch, Cornell Univ, Ithaca, 78-87; mem, Sloan-Kettering Inst Cancer Res, 69-90, Sloan-Kettering Cancer Ctr, NY, 84-90, emer mem, 90-; nat deleg dir, Am Cancer Soc, 86- *Mem:* AAAS; Am Soc Microbiol; Am Asn Cancer Res; fel Am Acad Microbiol; fel NY Acad Sci; Am Cancer Soc; Am Inst Nutrit; Genetics Soc Am; Soc Cryobiol. *Res:* Biochemistry; microbial antagonism; antibiotics, especially with tuberculosis; antimetabolites; mechanisms of resistance to antimetabolites in bacteria, mouse and human leukemia; enzymatic formation of citrovorum factor; pteridine metabolism. *Mailing Add:* Southgate Bronxville NY 10708

HUTCHISON, GEORGE B, b Lexington, Ky, Oct 18, 22. EPIDEMIOLOGY. *Educ:* Harvard Univ, AB, 43, MD, 51, MPH, 60. *Prof Exp:* Fel med, Lahey Clin, 52-55; asst med dir, Equitable Life Assurance Soc US, 55-56 & Health Ins Plan Gtr NY, 56-57; dir res, City Health Dept, NY, 57-59; from asst prof to assoc prof epidemiol, Sch Pub Health, Harvard Univ, 60-66; epidemiologist, Michael Reese Hosp, Chicago, 66-71; prof epidemiol, Sch Pub Health, Harvard Univ, 72-88; RETIRED. *Mem:* Fel Am Pub Health Asn; Am Statist Asn; World Med Asn; Asn Teachers Prev Med; NY Acad Med. *Res:* Epidemiologic study of effects of ionizing radiation and study of causes of cancer and other chronic diseases. *Mailing Add:* 115 St Francis Ct No 96 Louisville KY 40205

HUTCHISON, JAMES A, b Salina, Kans, Jan 28, 30; m 53; c 2. MYCOLOGY. *Educ:* Kans Wesleyan Univ, BA, 58; Univ Kans, MA, 60, PhD(bot), 63. *Prof Exp:* Asst prof biol, Southwest Col, Kans, 63-65; asst prof bot, Ark State Col, 65-68, assoc prof, 68-71, PROF BOT, DIV BIOL, ARK STATE UNIV, 71- *Mem:* Mycol Soc Am; Med Mycol Soc Ams. *Res:* Zygomycetes, fungal agents causing disease in insects and/or fungi inhabiting the intestine of the ectotherms. *Mailing Add:* Box 262 Ark State Univ State University AR 72467

HUTCHISON, JAMES ROBERT, b Smithville, Ohio, July 23, 40; m 63; c 4. PHYSICAL INORGANIC CHEMISTRY. *Educ:* Wittenberg Univ, BS, 62; Princeton Univ, PhD(chem), 68. *Prof Exp:* From instr to asst prof chem, Swarthmore Col, 66-73; from asst prof to prof, 73-90, TOWSLEY PROF CHEM, ALMA COL, 90-, CHMN DEPT, 78- *Concurrent Pos:* Res assoc, Mass Inst Technol, 69-70; adj prof chem, Mich State Univ, 79-81; SERAPHIM fel, NSF-Proj SERAPHIM, Eastern Mich Univ, 86-87. *Mem:* Am Chem Soc; Sigma Xi. *Res:* Nuclear magnetic resonance of paramagnetic systems; kinetics and mechanisms of rearrangement reactions of metal complexes; nuclear magnetic resonance of metal nuclides; instructional uses of computers. *Mailing Add:* Dept Chem Alma Col Alma MI 48801

HUTCHISON, JEANNE S, b Newburgh, NY, Nov 13, 42; wid; c 1. MATHEMATICS. *Educ:* Creighton Univ, BS, 64; Univ Calif, Los Angeles, MA, 67, PhD(math), 70. *Prof Exp:* ASST PROF MATH, UNIV ALA, BIRMINGHAM, 70- *Concurrent Pos:* Assoc dean nat sci & math, Univ Ala, Birmingham, 77-81. *Mem:* Am Math Soc; Math Asn Am. *Res:* Convexity. *Mailing Add:* Dept Math Univ Ala Univ Sta Birmingham AL 35294

HUTCHISON, JOHN HOWARD, b Chicago, Ill, Dec 29, 39. VERTEBRATE PALEONTOLOGY. *Educ:* Univ Fla, MS, 62; Univ Ore, MA, 64; Univ Calif, Berkeley, PhD(paleont), 76. *Prof Exp:* Sr mus paleontologist, 66-84, asst res, 84-86, ASSOC RES PALEONTOLOGIST, UNIV CALIF MUS PALEONT, 86-, ADJ PROF, INTEGRATIVE BIOL, 90- *Mem:* Soc Vert Paleont; Am Soc Mammalogists; Soc Syst Zool; Am Soc Ichthyologists & Herpetologists; Soc Study Amphibians & Reptiles. *Res:* Systematics, morphology and ecology of recent and fossil moles, Talpidae, Mammalia; systematics, morphology and ecology of North American fossil turtles. *Mailing Add:* Mus Paleont Univ Calif Berkeley CA 94720

HUTCHISON, JOHN JOSEPH, b New York, NY, Nov 16, 39; m 64; c 2. ORGANIC CHEMISTRY, POLYMER CHEMISTRY. *Educ:* Polytech Inst Brooklyn, BChE, 61, PhD(chem), 65. *Prof Exp:* Sr res scientist, Jet Propulsion Lab, NASA, 66-68; Humboldt Found res fel, Univ Heidelberg, 68-70; CHEMIST, BASF AG, 70- *Mem:* Am Chem Soc. *Res:* Organic chemistry of polymers, particularly polyurethanes. *Mailing Add:* BASF AG Polyurethene Ludwigshafen 6700 Germany

HUTCHISON, KENNETH JAMES, b Portadown, Northern Ireland, June 23, 36; Can citizen; m 62; c 3. MEDICAL PHYSIOLOGY, NON-INVASIVE VASCULAR DIAGNOSIS. *Educ:* Queen's Univ Belfast, MB, BCh & BAO, 61, MD, 66. *Prof Exp:* Tutor physiol, Queen's Univ Belfast, 62-66, lectr, 66-69; assoc prof, 69-76, PROF PHYSIOL, UNIV ALTA, 76-, SCI & RES ASSOC STAFF, UNIV ALTA HOSP, 84- *Concurrent Pos:* Med Res Coun Can fel, Univ Western Ont, 67-68; lectr, Northern Ireland Hosps Authority, 69; acting chmn physiol, Univ Alta, 77-78; vis scientist, Dept Surg, Univ Wash, 79-80. *Mem:* Brit Physiol Soc; Can Physiol Soc (treas, 85-88); fel Am Col Angiology. *Res:* Endothelial response to shear stress; arterial hemodynamics; peripheral circulation; hemodynamics and atherosclerosis; non-invasive diagnosis of peripheral arterial disease. *Mailing Add:* Dept Physiol Univ Alta Edmonton AB T6G 2H7 Can

HUTCHISON, NANCY JEAN, b Dallas, Tex, June 21, 50. GENETICS. *Educ:* Univ Tex, BA, 77; Univ Calif Irvine, PhD(develop & cell biol), 82. *Prof Exp:* Damon Runyon-Walter Winchell Cancer Fund fel, 82, NIH postdoctoral fel, 83-84, STAFF SCIENTIST, DEPT GENETICS, FRED HUTCHINSON CANCER RES CTR, 84- *Concurrent Pos:* Sci adv Fred Hutchinson Cancer Res Ctr Image Anal & chair Users Comt, 88-90; mem, Clin Sci Study Sect, Subcomt 1, NIH. *Mem:* Am Soc Cell Biol; Asn Women Sci; Am Soc Microbiol; AAAA. *Res:* Developmental and cell biology; genetics; numerous technical publications. *Mailing Add:* Dept Genetics AC136 Fred Hutchinson Cancer Res Ctr 1124 Columbia St Seattle WA 98104

HUTCHISON, ROBERT B, b Freeport, Ill, June 14, 35; m 58; c 3. ORGANIC CHEMISTRY. *Educ:* Kent State Univ, BS, 57; Univ Calif, Berkeley, PhD(org chem), 60. *Prof Exp:* Res chemist, Miami Valley Labs, Procter & Gamble Co, 60-63; asst prof chem, Bowling Green State Univ, 63-67; res mgr, 67-74, dir cent res, 74-78, dir corp res develop, 78-79, VPRES RES & DEVELOP, HENKEL CORP, EMERY GROUP, 79- *Concurrent Pos:* USPHS grant, 65-69. *Mem:* Am Chem Soc; Soc Cosmetic Chem. *Res:* Structure studies of naturally occurring compounds; reaction study of compounds leading to electron deficient nitrogen species. *Mailing Add:* Henkel Corp Emery Group 4900 Este Ave Cincinnati OH 45232

HUTCHISON, THOMAS SHERRET, b Scotland, Aug 12, 21; m 46; c 2. PHYSICS. *Educ:* St Andrews Univ, BSc, 42, PhD(physics), 48. *Hon Degrees:* DSc, Queen's Univ, 85. *Prof Exp:* Exp off, Admiralty Res Estab, 42-45; lectr, St Andrews Col, 48-49; sr sci off, Atomic Energy Res Estab, Eng, 49-50; assoc prof physics, 50-54, head dept, 54-62, dean grad studies & res, 62-80, chmn, Div Sci, 72-80, PROF PHYSICS, ROYAL MIL COL CAN, 54- *Honors & Awards:* Medal, Can Metal Physics Conf, 54. *Mem:* Fel Brit Inst Physics; fel Am Phys Soc; fel Royal Soc Edinburgh. *Res:* X-ray diffraction; solid state physics. *Mailing Add:* Sibbit Rd RR 1 Kingston ON K7L 4V1 Can

HUTCHISON, VICTOR HOBBS, b Blakely, Ga, June 15, 31; m 52; c 4. PHYSIOLOGICAL ECOLOGY, ZOOLOGY. *Educ:* NGa Col, BS, 52; Duke Univ, MA, 56, PhD(zool), 59. *Prof Exp:* Asst zool, Duke Univ, 54-57, instr, 58-59; from instr to assoc prof, Univ RI, 59-67, dir inst environ biol, 66-70; prof zool & chmn dept, 70-81, GEORGE LYNN CROSS RES PROF, UNIV OKLA, 79- *Concurrent Pos:* Guggenheim fel, 65-66. *Mem:* Fel AAAS; Ecol Soc Am; Am Soc Ichthyologists & Herpetologists (pres, 88); Am Soc Zoologists; Am Physiol Soc; Soc Study Amphibians & Reptiles. *Res:* Physiological ecology; comparative physiology; metabolism and gas exchange in lower vertebrates; temperature relations; herpetology. *Mailing Add:* Dept Zool Univ Okla Norman OK 73019

HUTCHISON, WILLIAM FORREST, b Lakeland, Fla, Oct 7, 25; m 51; c 5. PARASITOLOGY, PREVENTIVE MEDICINE. *Educ:* Emory Univ, BA, 49, MS, 52; Tulane Univ, PhD(parasitol), 58. *Prof Exp:* Instr parasitol, Tulane Univ, 53-55; from asst prof to prof, 55-90, EMER PROF PREV MED, SCH MED, UNIV MISS, 90- *Concurrent Pos:* Consult, Vet Admin Hosp, Jackson, Miss, 55-90. *Mem:* Am Soc Parasitol; Am Soc Trop Med & Hyg; Int Filariasis Asn. *Res:* Epidemiology and immunologic diagnosis of hydatid disease; metabolism of filarial worms. *Mailing Add:* Dept Prev Med Univ Miss Med Ctr Jackson MS 39216

HUTCHISON, WILLIAM MARWICK, b Cleveland, Ohio, Oct 18, 19; m 59; c 3. ANALYTICAL CHEMISTRY. *Educ:* Ohio State Univ, BS, 50. *Prof Exp:* Chemist, Harshaw Chem Co, 40-46, anal chemist, 50-63, mgr anal serv, 63-67, sr anal res assoc, Anal Res Dept, 67-88, CONSULT, N & T HARSHAW, 88- *Mem:* Am Chem Soc. *Res:* Inorganic analytical chemistry. *Mailing Add:* 1224 Quilliams Rd Cleveland Heights OH 44121

HUTCHISON, WILLIAM WATT, geology; deceased, see previous edition for last biography

HUTH, EDWARD J, b Philadelphia, Pa, May 15, 23; m 57; c 2. MEDICINE. *Educ:* Wesleyan Univ, BA, 45; Univ Pa, MD, 47. *Prof Exp:* From intern to resident, Univ Pa Hosp, 47-51, asst instr pharmacol, Sch Med, Univ, 48-49, assoc med, 51-58, asst prof, 58-61; from asst ed to ed, 61-90, EMER ED, BOOK REVIEW ED, ANN INTERNAL MED, AM COL PHYSICIANS, 90- *Concurrent Pos:* From adj asst prof to adj assoc prof, Sch Med, Univ Pa, 65-74, adj prof med, 74- *Mem:* AMA; master Am Col Physicians; Am Fedn Clin Res; Coun Biol Ed; fel Royal Col Physicians (Eng); Soc Gen Internal Med; Soc Scholarly Publ. *Res:* Medical editing. *Mailing Add:* 1124 Morris Ave Bryn Mawr PA 19010

HUTHNANCE, EDWARD DENNIS, JR, b Macon, Ga, Aug 31, 42. ALGEBRA, COMPUTER SCIENCE. *Educ:* Ga Inst Technol, BS, 64, MS, 66, PhD(math), 69. *Prof Exp:* Asst math, Ga Inst Technol, 64-66, instr, 68-69; asst prof, Newberry Col, 69-71, assoc prof, 71-, head dept, 69-; AT MATH DEPT, MIDWESTERN STATE UNIV, WICHITA FALLS. *Mem:* Am Math Soc; Math Asn Am; Sigma Xi. *Res:* Modern, abstract algebra; generalizations of group theory; loop theory. *Mailing Add:* Dept Math Bloomsberg State Univ Bloomsburg PA 17815

HUTJENS, MICHAEL FRANCIS, b Green Bay, Wis, Nov 21, 45; m 68; c 4. ANIMAL SCIENCE, NUTRITION. *Educ:* Univ Wis-Madison, BS, 67, MS, 69, PhD(dairy sci, nutrit), 71. *Prof Exp:* From asst prof to prof animal sci, Univ Minn, 71-79; PROF DAIRY SCI & PROGRAM LEADER, UNIV ILL, URBANA, 79- *Honors & Awards:* Midwest Young Exten Award, Am Asn Animal Sci, 84; DeLavel Exten Award, Am Dairy Sci Asn, 85. *Mem:* Am Dairy Sci Asn; Am Soc Animal Sci. *Res:* Dairy cattle nutrition, especially lipid and protein metabolism; mastitis detection with practical application. *Mailing Add:* Dairy Sci 315 Ar Sci Lab Univ Ill Urbana IL 61801

HUTNER, SEYMOUR HERBERT, b Brooklyn, NY, Oct 28, 11; m 37; c 1. MICROBIOLOGY. *Educ:* City Col New York, BS, 32; Cornell Univ, PhD(microbiol), 37. *Hon Degrees:* DSc, St Francis Col, Brooklyn, NY, 86. *Prof Exp:* Res assoc physics, Mass Inst Technol, 35-36; technician, Labs & Res Div, State Health Dept, NY, 38-41; mem staff biochem & microbiol, 41-65, assoc res dir, Haskins Labs, 65-77; Haskins adj prof biol, 70-78, EMER PROF IN RESIDENCE, PACE UNIV, 78-, EMER STAFF MEM, HASKINS LABS. *Concurrent Pos:* Vis prof, Inst Microbiol, Univ Brazil, 63-; adj prof, Fordham Univ, 64-68; vis prof, Univ Brasilia, 70. *Mem:* Soc Protozool (pres, 61-62); Am Soc Microbiol; Am Chem Soc; Phycol Soc Am; Tissue Cult Asn; Biochem Soc. *Res:* Nutritional physiology of bacteria, algae, fungi and protozoa; chemotherapy of trypanosomatid diseases; biopterin metabolism; polyamine metabolism; cell-injury repair. *Mailing Add:* Haskins Labs Pace Univ 41 Park Row at Pace Plaza New York NY 10038-1502

HUTNIK, RUSSELL JAMES, b Register, Pa, Feb 9, 24; m 55; c 3. FOREST ECOLOGY. *Educ:* Pa State Univ, BS, 50; Yale Univ, MF, 52; Duke Univ, PhD(forestry), 64. *Prof Exp:* Res forester, Northeastern Forest Exp Sta, Pa, 50-57; from asst prof to prof, 57-86, EMER PROF FORESTRY, PA STATE UNIV, 86- *Concurrent Pos:* Consult elec utilities, USDA, 64- *Mem:* Soc Am Foresters; Ecol Soc Am; Am Inst Biol Sci; Sigma Xi; Am Soc Surface Mining & Reclamation. *Res:* Ecosystem dynamics; reclamation of disturbed land; impact of air pollution on forest ecosystems; vegetation management on right-of-way. *Mailing Add:* 1658 Princeton Dr State College PA 16803

HUTSON, ANDREW RHODES, solid state physics; deceased, see previous edition for last biography

HUTSON, RICHARD LEE, b Pittsburg, Kans, Nov 25, 36; m 58; c 2. ELEMENTARY PARTICLE PHYSICS. *Educ:* Univ Colo, BA, 60, PhD(nuclear physics), 67; Mass Inst Technol, SB, 63. *Prof Exp:* Appointee, 67-69, MEM STAFF NUCLEAR PHYSICS, LOS ALAMOS SCI LAB, 69- *Mem:* Am Asn Physicists Med. *Res:* Study of medical applications of particle beams and accelerator technology. *Mailing Add:* Los Alamos Nat Lab MP-5 MS M838 Los Alamos NM 87545

HUTT, FREDERICK BRUCE, b Guelph, Ont, Aug 20, 1897; m 30; c 3. GENETIC RESISTANCE TO DISEASE IN DOMESTIC ANIMALS. *Educ:* Univ Toronto, BSA, 23; Univ Wis, MS, 25; Univ Man, MA, 27; Univ Edinburgh, PhD(genetics), 29, DSc, 39. *Hon Degrees:* DSc, Univ Agr, Brno, Czechoslovakia, 65, Univ Guelph, 74. *Prof Exp:* Lectr poultry husb, Univ Man, 23-27; from asst prof to prof poultry & genetics, Univ Minn, 28-34; prof poultry & head dept, Cornell Univ, 34-40, prof zool & chmn dept, 39-44, prof animal genetics, 40-65, EMER PROF ANIMAL GENETICS, CORNELL UNIV, 65- *Concurrent Pos:* Borden fel, 47; consult poultry breeding, Kimber Farms, Calif, 54-56, F&G Sykes, Warminster, Eng, 57-72; vis lectr, NC Univ, 56, Ore State Univ & Va Polytech Univ, 65; hon mem, Int Congress Genetics Appl to Livestock Prod, Madrid, 74. *Honors & Awards:* Res Award, Poultry Sci Asn, 29; Borden Award, 45; G Scott Robinson Mem Lectr, Queen's Univ, Belfast, Northern Ireland, UK, 59; Halpin Mem Lectr, Univ Wis, 78; Poultry Hall of Fame, 80. *Mem:* Emer mem & fel AAAS; Genetic Soc Am; Am Genetic Asn; fel Poultry Sci Asn (pres, 32-33); Am Soc Human Genetics; hon fel Royal Soc Edinburgh, 75. *Res:* Spotted genes in chromosome map of domestic fowl; breeding fowls genetically resistant to pullorum disease and Marek's disease; lethal genes in dogs, cattle, and horses; animal genetics; author four books. *Mailing Add:* 411 Butternut Dr Newfield NY 14867

HUTT, MARLAND PAUL, JR, organic chemistry, for more information see previous edition

HUTT, MARTIN P, b New York, NY, Sept 27, 26; m 49; c 4. INTERNAL MEDICINE. *Educ:* NY Univ, MD, 49; Am Bd Internal Med, dipl, 58. *Prof Exp:* Instr med, Med Sch, Northwestern Univ, 55-57; from instr to asst prof, 57-80, PROF MED, SCH MED, UNIV COLO, DENVER, 80- *Concurrent Pos:* Consult, Vet Admin Hosp, Denver, 59- & Fitzsimmons Army Hosp, 65- *Mem:* Am Soc Nephrology. *Res:* Clinical nephrology. *Mailing Add:* 4200 E Ninth Ave Denver CO 80220

HUTT, RANDY, b New York, NY, Dec 11, 47; m 84; c 2. IMMUNOLOGY, MOLECULAR BIOLOGY. *Educ:* State Univ NY, Buffalo, BA, 68; Univ Fla, MS, 72; Pa State Univ, PhD(microbiol), 77. *Prof Exp:* Lab technologist indust bacteriol, West Chem Co, 68-69; lab technologist clin bacteriol, Queen's Gen Hosp, 70; res scientist virol, Southwest Found Res & Educ, 77-79; mgr immunol & virol, Res & Develop Labs, Gibco Div, Dexter Corp, 79-80; mgr biol control, Schering Corp, 80-83, qual assurance coordr, interferon mfg proj team, 83-84, tech serv proj coordr, 84-85; dept head, tech support, 86-88, DEPT HEAD, COMPOUNDING, FILLING & PACKAGING, STERILE PROD DIV, BURROUGHS WELLCOME CO, 88- *Mem:* Am Soc Microbiol; Pharmaceut Mfg Asn; Parenteral Drug Asn; Mensa. *Res:* Basic virology research in vitro and in vivo; development of diagnostic tests; immunology; molecular virology; cell kinetics. *Mailing Add:* Burroughs Wellcome Co PO Box 1887 Greenville NC 27835-1887

HUTTA, PAUL JOHN, b Coaldale, Pa, Jan 20, 46; m 69; c 1. CHEMICAL PHYSICS. *Educ:* Pa State Univ, BS, 67; Lehigh Univ, MS, 69, PhD(physics), 74. *Prof Exp:* Lectr, 75-76, res assoc phys chem, Tex A&M Univ, 76-78; RES SCIENTIST, BETZ LABS, 78- *Mem:* Am Phys Soc. *Res:* Study of transition metal complexes in aluminosiliates utilizing electron paramagnetic resonance, and infrared visible ultra violet spectroscopy with the intention of developing heterogeneous analogs to homogeneous analogs to catalysts. *Mailing Add:* 1224 Oakleaf Lane Warminster PA 18974

HUTTENLOCHER, DIETRICH F, b Oberlahnstein, Ger, June 27, 28; US citizen; m 61; c 2. ORGANIC CHEMISTRY, LUBRICATION ENGINEERING. *Educ:* Univ Buffalo, BA, 53; Univ Cincinnati, PhD(org chem), 58. *Prof Exp:* Teaching asst, Univ Cincinnati, 53-55; chemist, Texaco Res Labs, NY, 58-60 & SKF Industs, Pa, 60-63; chemist, 63-74, res assoc, 74-78, PROG MGR HERMETIC SYSTS, MAJOR APPLIANCE LABS, GEN ELEC CO, 78- *Mem:* Am Chem Soc; Am Soc Lubrication Eng; Am Soc Heat, Refrig & Air-Conditioning Eng. *Res:* Synthetic organic chemistry; lubricants and lubrication technology; chemistry of sealed refrigeration systems. *Mailing Add:* 32 Sterling Rd Louisville KY 40222-3524

HUTTER, EDWIN CHRISTIAN, electrooptics; deceased, see previous edition for last biography

HUTTER, GEORGE FREDERICK, b Glen Ridge, NJ, Jan 7, 43. ORGANIC & POLYMER CHEMISTRY. *Educ:* Rutgers Univ, BA, 71; Seton Hall Univ, MS, 74, PhD, 85. *Prof Exp:* Chemist, Inmont Corp/BASF, 71-78, sr chemist, 78-86; RES CHEM, WESTVACO, 86- *Mem:* Assoc mem Sigma Xi; Am Chem Soc. *Res:* Synthesis of polymers and resins used as coatings and ink vehicles; study of the interactions of vehicle and dispersant molecules with the surfaces of pigment particles. *Mailing Add:* 968 Orange Grove Rd Charleston SC 29407

HUTTER, ROBERT V P, b Yonkers, NY, May 25, 29; m 55; c 3. PATHOLOGY, ONCOLOGY. *Educ:* Syracuse Univ, BA, 50; State Univ NY, MD, 54. *Hon Degrees:* MA, Yale Univ, 69. *Prof Exp:* Intern, Yale Med Ctr, 54-55, resident, 55-56; resident path, Mem Ctr, NY, 56-58; asst attend pathologist, Mem Hosp, NY, 60-64, asst attend cytologist, 62-64, assoc attend pathologist, 65-68, attend pathologist, 67-68; chief path serv, Yale-New Haven Hosp, 68; prof path & dir anat path & cytopath, Sch Med, Yale Univ, 68-70; prof path & chmn dept, Col Med NJ, 70-73; dir, 70-73, CHMN, DEPT PATH, ST BARNABAS MED CTR, 73- *Concurrent Pos:* Asst vis pathologist, James Ewing Hosp, New York, 60-64, assoc vis pathologist, 65, attend pathologist, 67-68; consult, NY State Div Labs, 60, Mary Swift Tumor Clin & Registry, Butte, Mont, 64, West Haven Vet Admin Hosp, Conn, 67, St Peter's Gen Hosp, New Brunswick, NJ, 68, Lyndon B Johnson Trop Med Ctr, Am Samoa, Laurel Heights Hosp, Shelton, Conn, Windham Community Mem Hosp, Willimantic, 69 & Waterbury Hosp, 70; consult pathologist & cytopathologist, USPHS, NY, 61-68; prof clin oncol, Am Cancer Soc, 71-73. *Mem:* AAAS; AMA; Am Acad Forensic Sci; fel Col Am Path; Am Soc Cytol; hon fel Am Col Radiol; Am Cancer Soc (pres, 81-82); Soc Surg Oncol (pres, 86-87). *Mailing Add:* St Barnabas Med Ctr Old Short Hills Rd Livingston NJ 07039

HUTTERER, FERENC, b Budapest, Hungary, Jan 27, 29; US citizen; m 53; c 1. CLINICAL BIOCHEMISTRY, CLINICAL PATHOLOGY. *Educ:* Univ Szeged, MD, 53. *Prof Exp:* Prof clin chem, Mt Sinai Sch Med, 68-75; PROF BIOCHEM & MOLECULAR PATH & CHMN DEPT, COL MED, NORTHEASTERN OHIO UNIV, 75- *Concurrent Pos:* Dir dept chem, Mt Sinai Hosp, 68-75. *Mem:* Fedn Am Soc Exp Biol; Asn Study Liver Dis; NY Acad Sci; AAAS. *Res:* Microsomal electron transfer system and its relation to human diseases, especially cholestasis and atherosclerosis; molecular pathology. *Mailing Add:* Northeastern Ohio Univ Col Med Rootstown OH 44272

HUTT-FLETCHER, LINDSEY MARION, b Cardiff, South Wales, July 2, 47; nat US. PATHOLOGY. *Educ:* Liverpool Univ, Eng, BSc, 68; Univ London, PhD(viral immunol), 73. *Prof Exp:* Res asst, Dept Virol, Lister Inst Prev Med, Univ London, Eng, 68-73; res assoc, Dept Bact & Immunol & Cancer Res Ctr, Univ NC, Chapel Hill, 76-77, res asst prof, 77-82; from asst prof to assoc prof, 82-90, PROF DEPT COMP & EXP PATH, COL VET MED, DEPT IMMUNOL & MED MICROBIOL DEPT PATH & LAB MED, COL MED, UNIV FLA, GAINESVILLE, 90- *Res:* Comparative and experimental pathology. *Mailing Add:* Dept Comp & Exp Path Univ Fla Box J-145 JHMHC Gainesville FL 32610

HUTTLIN, GEORGE ANTHONY, b Abington, Pa, Mar 19, 47. PHOTOCONDUCTIVITY, PULSED-POWER. *Educ:* La Salle Col, Philadelphia, BA, 69; Univ Notre Dame, MS, 72, PhD(nuclear physics), 75. *Prof Exp:* PHYSICIST PULSED POWER, DEPT ARMY, HARRY DIAMOND LABS, 75- *Mem:* Am Phys Soc; Am Asn Physics Teachers; Inst Elec & Electronics Engrs. *Res:* Numerical and analytical modeling; design and diagnostics of multi-terawatt, pulsed-power machines; nuclear structure studies through nuclear reactions induced by polarized deuteron beams; polarization of nuclear reaction; application of photoconductivity to the control of high- powered microwave and pulsed-power. *Mailing Add:* 10500 Truxton Rd Adelphi MD 20783

HUTTO, FRANCIS BAIRD, JR, b Savannah, Ga, June 25, 26; m 52; c 6. FILTRATION & SEPARATION, CHEMICAL MICROSCOPY. *Educ:* Clemson Univ, BS, 48; Cornell Univ, MS, 50, PhD(chem), 52. *Prof Exp:* Sr res chemist, Res & Eng Ctr, Johns-Manville Corp, 52-61, from sect chief to sr sect chief, 61-69, res mgr, 69-72, sr res assoc, Res Ctr, Johns-Manville Prod Corp, 72-76; DIR RES & DEVELOP, PABCO INSULATION DIV, FIBREBOARD CORP, 76- *Mem:* Am Chem Soc; fel Am Inst Chem Engrs; Filtration Soc. *Res:* Calcium silicate chemistry as related to industrial insulation, fire proofing and fillers; filter and filtration. *Mailing Add:* 676 Peony Dr Grand Junction CO 81501

HUTTON, ELAINE MYRTLE, b Toronto, Ont, Aug 26, 40; m 65; c 2. MEDICAL GENETICS, PRENATAL DIAGNOSIS. *Educ:* Univ Toronto, BSc, 63, MA, 66, PhD(human genetics), 70. *Prof Exp:* STAFF GENETICIST, DEPT GENETICS, HOSP SICK CHILDREN, 80-; ASST PROF PEDIAT, OBSTET & GYNEC, UNIV TORONTO, 90- *Concurrent Pos:* Lectr, Dept Obstet & Gynec, Univ Toronto, 80-90. *Mem:* Am Soc Human Genetics; fel Can Col Med Geneticists. *Res:* Prenatal genetic diagnosis; gene carrier screening; prenatal genetic screening tests. *Mailing Add:* Dept Genetics Hosp Sick Children 555 University Ave Toronto ON M4G 1V2 Can

HUTTON, HAROLD M, b Prince Albert, Sask, June 30, 37; m 63; c 2. PHYSICAL CHEMISTRY, NUCLEAR MAGNETIC RESONANCE. *Educ:* Brandon Col, BSc, 58, Univ Man, MSc, 61, PhD(chem), 63. *Prof Exp:* Res assoc chem, Johns Hopkins Univ, 63-64; asst prof, Univ Man, 64-65; from asst prof to assoc prof, Brandon Univ, 65-70; assoc prof, 70-77, PROF CHEM, UNIV WINNIPEG, 77-, CHMN DEPT, 83- *Concurrent Pos:* Adj prof, Univ Man, 74- *Mem:* Chem Inst Can. *Res:* Proton and carbon-13 magnetic resonance of organic compounds for the determination of physical properties. *Mailing Add:* Dept Chem Univ Winnipeg Winnipeg MB R3B 2E9 Can

HUTTON, JAMES ROBERT, b Lilbourn, Mo, Mar 22, 43; m 64; c 1. MOLECULAR BIOLOGY. *Educ:* Univ Mo-St Louis, BS, 69; Univ Ill, MS, 71, PhD(biochem), 73. *Prof Exp:* Res fel biol, Harvard Med Sch, 73-75; asst prof biol, Marquette Univ, 75-; MGR, PURE CULT FERMENTATION FAC, UNIVERSAL BIOVENTURES. *Concurrent Pos:* Am Cancer Soc fel, 73-75. *Mem:* Am Chem Soc; AAAS; Sigma Xi. *Res:* Organization and structure of DNA in chromosomes; molecular biology and control of gene expression; factors affecting polynucleotide-polynucleotide interactions. *Mailing Add:* Universal Bioventures 6143 N 60th St Milwaukee WI 53218

HUTTON, JOHN JAMES, JR, b Ashland, Ky, July 24, 36; m 64; c 3. EDUCATIONAL ADMINISTRATION. *Educ:* Harvard Univ, AB, 58, MD, 64. *Prof Exp:* Intern med, Mass Gen Hosp, 64-65; staff assoc biochem, Nat Heart Inst, 65-67; resident & instr med, Univ Ky, 67-68; sect chief mammalian genetics, Roche Inst Molecular Biol, 68-71; from assoc prof to prof med, Sch Med, Univ Ky, 71-80; prof med, Sch Med, Univ Tex, San Antonio, 81-84; chief Heniat, Vet Admin Hosp, San Antonio, 81-84; prof, Children's Hosp Med Ctr, 84-87, DEAN & PROF MED & PEDIAT, UNIV CINCINNATI, 87- *Concurrent Pos:* Vis investr, Jackson Lab, 68-69; adj asst prof, Columbia Univ, 69-71; chief med serv, Vet Admin Hosp, Lexington, 74-80. *Mem:* AAAS; Am Soc Clin Invest; Am Soc Biol Chemists; Am Soc Hemat; Am Fedn Clin Res; Asn Am Physicians. *Res:* Mammalian molecular genetics. *Mailing Add:* Ped/054 ch-ldr 729 Univ Cincinnati Cincinnati OH 45221

HUTTON, KENNETH EARL, b Chicago, Ill, Apr 17, 28. PHYSIOLOGY. *Educ:* Kalamazoo Col, AB, 51; Purdue Univ, MS, 52, PhD, 55. *Prof Exp:* Instr zool, Tulane Univ, 55-58; from asst prof to assoc prof, 58-66, PROF BIOL, SAN JOSE STATE UNIV, 66- *Mem:* AAAS; Am Soc Zoologists; Sigma Xi. *Res:* Reptilian blood chemistry; blood and immunochemistry of fish. *Mailing Add:* Dept Biol Sci San Jose State Col San Jose CA 95192

HUTTON, LUCREDA ANN, b Logansport, Ind, Jan 5, 26; m 50; c 2. MATHEMATICS, MATHEMATICS EDUCATION. *Educ:* Butler Univ, BA, 67; Purdue Univ, MS, 72; Ind Univ, Bloomington, EdD(math educ), 76. *Prof Exp:* Asst prof, 76-81, ASSOC PROF MATH, IND UNIV-PURDUE UNIV, INDIANAPOLIS, 81- *Concurrent Pos:* Dir res proj, Math Div, Indianapolis Pub Schs, 76-77, consult, 78-; math consult, Optical Apprenticeship Prog, Workmate, Inc, 77-78. *Mem:* Math Asn Am; Nat Coun Teachers Math; Am Math Asn Two Year Cols; Am Math Soc. *Res:* Effects of calculators and micro-computers in the teaching of mathematics; mathematics anxiety. *Mailing Add:* Dept Math Ind Univ Purdue Univ Indianapolis IN 46202

HUTTON, THOMAS WATKINS, b Milwaukee, Wis, June 1, 27; m 54; c 3. ORGANIC CHEMISTRY. *Educ:* Brown Univ, ScB, 50; Univ Wash, PhD, 54. *Prof Exp:* Asst, Univ Calif, 54-56; res scientist, Weyerhaeuser Co, 56-61; chemist, Rohm & Haas Co, Philadelphia, 61-80; CONSULT, 80- *Mem:* Am Chem Soc. *Res:* Natural products; effect of ultraviolet light on organic compounds; photolysis of organic compounds; synthetic tanning agents; organic synthesis and coatings; adhesives; caulks. *Mailing Add:* 153 Woodview Dr Doylestown PA 18901-2923

HUTTON, WILBERT, JR, b Denver, Colo, July 30, 27; m 53; c 2. INORGANIC CHEMISTRY, SCIENCE EDUCATION. *Educ:* Univ Denver, BS, 50; Mich State Univ, PhD(chem), 59. *Prof Exp:* Asst, Mich State Univ, 50-59; assoc prof chem, Bowling Green State Univ, 59-68; prof, 68-90, EMER PROF CHEM, IOWA STATE UNIV, 90- *Concurrent Pos:* Kettering vis lectr, Univ Ill, 63-64. *Mem:* Am Chem Soc; Nat Sci Teachers Asn; Soc Col Sci Teachers; Sigma Xi. *Res:* Nonaqueous solvents; chemical education; ion-association products in solution. *Mailing Add:* Dept Chem Iowa State Univ Ames IA 50011

HUTZENLAUB, JOHN F, b Derby, Conn, Oct 8, 18; m 46; c 2. PHYSICS, MATHEMATICS. *Educ:* Rensselaer Polytech Inst, BS, 39; Mass Inst Technol, PhD(physics), 43. *Prof Exp:* Staff mem, Div Indust Coop, Mass Inst Technol, 42-44, proj engr, Instrumentation Lab, 44-50; tech dir exp eng, AC Spark Plug Div, Gen Motors Corp, 50-52; group leader res & develop, Lincoln Lab, Mass Inst Technol, 52-62, assoc div head eng, 62-63, div head eng, 63-80; CONSULT ENGR, 80- *Mem:* NY Acad Sci. *Res:* Experimental solid state physics; instrumentation; inertial navigation systems; automatic control systems; structural design. *Mailing Add:* 37 Leslie Rd Winchester MA 01890

HUTZINGER, OTTO, b Vienna, Austria, Mar 14, 33; Can citizen; m 60; c 3. ENVIRONMENTAL CHEMISTRY. *Educ:* Fed Inst Chem, Vienna, Ing, 56; Univ Sask, MS, 63, PhD(org chem), 65. *Prof Exp:* Chemist, Ebewe Pharmaceuts, Austria, 53-58; res scientist, Synthesis of Drugs, Psychiat Res Unit, Univ Hosp, Univ Sask, 60-65; res biochemist metabolism, Univ Calif, Davis, 65-67; scientist, Indoles, Pesticides & Pollutants, Atlantic Regional Lab, Nat Res Coun Can, 67-74; dir, Lab Environ & Toxicol Chem, Univ Amsterdam, 74-; CHMN ECOL CHEN & GEOCHEM DEPT, UNIV BAYREUTH, WGER. *Concurrent Pos:* Lectr, Dalhousie Univ, 70-; dir, Lab Environ Chem, Univ Amsterdam, Holland, 74- *Honors & Awards:* Sr Res Award, Alexander von Humboldt Found, Ger, 73; Frank R Blood Award, Soc Toxicol, 74. *Mem:* Am Chem Soc; Chem Inst Can; Dutch Chem Soc; Soc Environ Toxicol & Chem; Sigma Xi. *Res:* Chemistry, metabolism and environmental breakdown of pesticides and pollutants; development of new analytical procedures. *Mailing Add:* Ecol Chem & Geochem Dept Univ Bayreuth Postfach 3008 Bayreuth D-8580 Germany

HUTZLER, JOHN R, b New York, NY, Aug 8, 29; m 52; c 4. ELECTRICAL INSULATION, DIELECTRICS. *Educ:* Fordham Univ, BS, 51; Rensselaer Polytech Inst, MS, 54. *Prof Exp:* Supvr chem, Gen Elec, 51-59; sr engr, Western Elec, 63-64; prod mgr, Union Carbide, 64-70; dir res & develop, Aerovox Inc, subsid RTE Corp, 71-; PRES, HIGH ENERGY, PARKESBURG, PA. *Concurrent Pos:* Mem, Int Conf Large High Tension Elec Systs. *Mem:* Inst Elec & Electronics Engrs; Illuminating Engrs Soc. *Res:* Lower cost, smaller size capacitors for electrical and electronic applications. *Mailing Add:* High Energy Lower Valley Rd Parkesburg PA 19365

HUTZLER, LEROY, III, b Richmond, Va, Oct 18, 21; m 44; c 4. CHEMICAL, SAFETY & ENVIRONMENTAL ENGINEERING. *Educ:* Va Polytech Inst, BS, 42. *Prof Exp:* Chem engr, process develop & serv, Gen Elec Co, 42-47, chem engr & process supvr, 47-52, supvr, chem eng unit, 52-53, supvr, process eng lab, 53-61, supvr, adv mfg eng, 61-65, adv mfg engr mat & process, 65-68, mgr, process & equip develop, 68-72, environ engr, 72-83; CONSULT, HUTZLER CONSULT SERV, 84- *Mem:* Am Indust Hyg Asn; Am Acad Indust Hyg; Am Soc Safety Engrs. *Res:* Drying and vacuum treating processes for insulating materials; painting and metal finishing; wire manufacturing and insulating processes; plastics molding; industrial hygiene; environmental health. *Mailing Add:* 11 Biltmore Ave Providence RI 02908-3513

HUWE, DARRELL O, b Lemmon, SDak, Sept 22, 32; m 64; c 3. PARTICLE PHYSICS. *Educ:* SDak Sch Mines & Technol, BS, 54; Univ Calif, Berkeley, PhD(physics), 64. *Prof Exp:* Res assoc physics, Univ Colo, 64-66; asst prof, 66-70, ASSOC PROF PHYSICS, OHIO UNIV, 70- *Concurrent Pos:* Vis assoc prof, Am Univ Cairo, 75-77. *Mem:* Am Phys Soc; Sigma Xi. *Res:* Study of strong interactions among elementary particles using hydrogen bubble chamber. *Mailing Add:* Dept Physics Ohio Univ Athens OH 45701

HUXLEY, HUGH E, b 1924; m 66; c 1. MOLECULAR BIOLOGY, PHYSIOLOGY. *Educ:* Univ Cambridge, MA, 48, PhD(molecular biol), 52. *Prof Exp:* Dep dir, MRC Lab Molecular Biol, Cambridge, UK, 77-87; PROF BIOL, BRANDEIS UNIV, 87-, DIR, ROSENSTIEL BASIC MED SCI RES CTR, 88- *Honors & Awards:* Int Award, Gairdner Found, 75; Baly Medal, Royal Col Physicians, 75; Royal Medal, Royal Soc Eng, 77; Albert Einstein World Award of Sci, 87; Benjamin Franklin Medal, 90. *Mem:* Nat Acad Sci; Royal Soc London; Am Acad Arts & Sci. *Res:* Structural biology, especially mechanism of muscle contraction. *Mailing Add:* 349 Nashawtuc Rd Concord MA 01742

HUXSOLL, DAVID LESLIE, b Aurora, Ind, July 18, 36; m 59; c 2. VETERINARY MEDICINE. *Educ:* Univ Ill, Urbana, BS, 59, DVM, 61; Univ Notre Dame, PhD(microbiol), 65. *Prof Exp:* Lab officer, Dept Vet Microbiol, Div Vet Med, Walter Reed Army Inst Res, 61-62, asst chief dept, 64-67, dep dir div, 66-67, chief dept vet med, 9th Army Med Lab, Vietnam, 67-68, chief dept vet diag serv, div vet med, Vet Corps, Walter Reed Army Inst Res, 68-, dep dir div vet med, 72-; US Army Med Res Inst Infectious Dis; ASSOC DEAN RES & ADVAN STUDIES, SCH VET MED LA STATE UNIV, BATON ROUGE, 90- *Mem:* Am Vet Med Asn; Am Soc Microbiol; Wildlife Dis Asn; US Animal Health Asn. *Res:* Virus diseases of laboratory animals; infectious diseases of dogs; zoonotic diseases. *Mailing Add:* Sch Vet Med La State Univ Five Stadium Dr Baton Rouge LA 70803-8402

HUXTABLE, RYAN JAMES, b Bristol, Eng, Sept 20, 43; US citizen; m 68. PHARMACOLOGY. *Educ:* Bristol Univ, BSc, 64, PhD(chem), 68. *Prof Exp:* Tech officer chem, Mond Div, Imperial Chem Indust, 67-68; res assoc chem, Univ Ill, 68-69; res assoc biochem, Duke Univ, 69-70; asst prof, 70-75, assoc prof, 75-79, PROF PHARMACOL, COL MED, UNIV ARIZ, 79- *Concurrent Pos:* Vis prof, Univ Montreal, 77, Univ Mex, 80 & Univ Florence, 81. *Mem:* Am Soc Pharmacol & Exp Therapeut; AAAS. *Res:* Pharmacology and biochemistry of taurine, with emphasis on involvement in epilepsy and cardiac dysfunctioning; cardiopulmonary toxicology of pyrrolizidine alkaloids. *Mailing Add:* Dept Pharmacol Univ Ariz Health Sci Ctr Tucson AZ 85724

HUYER, ADRIANA (JANE), b Giessendam, Neth, May 19, 45; Can citizen. PHYSICAL OCEANOGRAPHY. *Educ:* Univ Toronto, BSc, 67; Ore State Univ, MS, 71, PhD(oceanog), 74. *Prof Exp:* Sci officer phys oceanog, Marine Sci Directorate, Environ Can, 67-74, res scientist, 74-75; res assoc phys oceanog, 75-79, from asst prof to assoc prof, 79-85, PROF OCEANOG, ORE STATE UNIV, 85- *Mem:* Am Geophys Union; AAAS; Am Meteorol Soc; Can Meteorol & Oceanog Soc. *Res:* Coastal upwelling; circulation over continental shelves, distributions of physical properties in the ocean. *Mailing Add:* Col Oceanog Ore State Univ Corvallis OR 97331

HUYSER, EARL STANLEY, b Holland, Mich, May 27, 29; m 52; c 4. ORGANIC CHEMISTRY. *Educ:* Hope Col, AB, 51; Univ Chicago, PhD(org chem), 54. *Prof Exp:* Res fel, Univ Chicago, 54-55; res chemist, Dow Chem Co, 57-59; from asst prof to assoc prof, 59-66, PROF ORG CHEM, UNIV KANS, 66- *Concurrent Pos:* Vis prof, State Univ Grottingen, 64-65; consult, Dow Chem Co & Dowell Schlumberger, Inc; vis scientist, Univ Wash, 81. *Mem:* Am Chem Soc. *Res:* Chemistry of free radical reactions; kinetic analysis of chain reactions; halogenation reactions; biochemical redox reactions; theoretical chemistry. *Mailing Add:* Dept Chem Univ Kans Lawrence KS 66044

HUZINAGA, SIGERU, b Chanchun, Manchuria, May 23, 26; m 54; c 2. THEORETICAL CHEMISTRY, PHYSICS. *Educ:* Kyushu Univ, BSc, 48; Kyoto Univ, DSc(physics), 59. *Prof Exp:* Assoc prof physics, Kyushu Univ, 58-65; PROF CHEM, UNIV ALTA, 69- *Mem:* Phys Soc Japan; Am Phys Soc. *Res:* Atomic and molecular structures. *Mailing Add:* Dept Chem Univ Alta Edmonton AB T6G 2J1 Can

HVATUM, HEIN, b Tonsberg, Norway, Apr 17, 23; m 47; c 2. ELECTRONICS. *Educ:* Chalmers Tech Univ, Sweden, MSc, 54, PhD, 58. *Prof Exp:* Asst electronics, res lab electronics, Chalmers Tech Univ, Sweden, 48-54, assoc prof, 54-58 & 60-61; res assoc, 58-60, asst dir, 65-72, SCIENTIST, NAT RADIO ASTRON OBSERV, 61-, ASSOC DIR, 72- *Mem:* Am Astron Soc; Inst Elec & Electronics Engrs. *Res:* Development and applications of electronics in radio astronomy. *Mailing Add:* 104 Lancaster Ct Charlottesville VA 22901

HWA, JESSE CHIA HSI, b Hankow, China, July 12, 24; nat US; m 49; c 3. POLYMER CHEMISTRY. *Educ:* St John's Univ China, BS & BS Hons, 45; Univ Ill, MS, 47, PhD(chem), 49. *Prof Exp:* Chemist, China Biol & Chem Co, 45-46; res chemist, Rohm and Haas Co, 49-55, group leader, 55-63; sr group leader, Stauffer Chem Co, 63-64, head org sect, 64-65, mgr polymer sect, 65-67, mgr polymer res dept, 67-71, mgr res planning, 71-77, dir develop, 77-82, corp proj dir, 82-84, tech dir electronic mat, 84-85; PRES, HWA INT, INC, 85- *Concurrent Pos:* Mem adv bd, Polymer Res Inst, Univ Conn, 71-82 & Mat Res Inst, Univ Mass, 81-83; mem comm macromolecular chem, Nat Res Coun, Nat Acad Sci, 72-77; tech comt Nat Asn Mfg, 75-78; mem bd dirs, Sci & Ed Soc, 82-; sr adv, Ministry Light Indust, China, 85-; mem bd dirs, AI Technol Inc, 85, Int Multitrade Corp, 87- *Mem:* Am Chem Soc; Soc Plastics Eng; Chinese Am Chem Soc (pres, 81-85); Soc Plastics Indust. *Res:* Industrial organic intermediates; surfactants; monomer and polymer synthesis; ion exchange resins; emulsion polymers; stereospecific polymers; polyvinyl chloride; surface chemistry; polymer physics; electronic materials. *Mailing Add:* HWA Int Inc 54 Westhill Circle Stamford CT 06902

HWA, RUDOLPH CHIA-CHAO, b Shanghai, China, Nov 1, 31; nat US; m 62, 84; c 4. HIGH ENERGY NUCLEAR PHYSICS. *Educ:* Univ Ill, BS, 52, MS, 53, PhD(elec eng), 57; Brown Univ, PhD(physics), 62. *Prof Exp:* Res asst prof, Univ Ill, 57-58; asst physics, Brown Univ, 58-62; res physicist, Lawrence Radiation Lab, Univ Calif, Berkeley, 62-64; mem, Inst Advan Study, 64-66; asst prof, Inst Theoret Physics, State Univ NY Stony Brook, 66-71; assoc prof physics & res assoc, Inst Theoret Sci, 71-74; dir, Inst Theoret Sci, 73-77 & 79-84, PROF PHYSICS, UNIV ORE, 74- *Concurrent Pos:* Ed, Int J Mod Physics & Mod Physics Lett, 85- *Mem:* Am Phys Soc. *Res:* High energy physics; scattering and production processes; symmetries; structure of hadrons; quantum chromodynamics; quark-gluon plasma; heavy-ion collisions. *Mailing Add:* Inst Theoret Sci Univ Ore Eugene OR 97403

HWANG, BRUCE YOU-HUEI, b Taiwan, Jan 6, 36; US citizen; m 66; c 3. ORGANIC CHEMISTRY, MEDICINAL CHEMISTRY. *Educ:* Nat Taiwan Univ, BS, 58; ETenn State Univ, MA, 61; Ohio State Univ, PhD, 65. *Prof Exp:* Sr med chemist, 65-69, assoc sr investr, 69-72, SR INVESTR, SMITH KLINE & FRENCH LABS, 72- *Mem:* AAAS; Am Chem Soc; NY Acad Sci; fel Am Inst Chemists; Int Soc Study Xenobiotics. *Res:* Organic chemistry; drug metabolism; natural products; structure elucidation. *Mailing Add:* 210 Lindbergh Ave Broomall PA 19008-2607

HWANG, CHERNG-JIA, b Changhua, Taiwan, Oct 1, 37; US citizen; m 65; c 2. SEMICONDUCTOR LASERS, FIBER OPTICS. *Educ:* Nat Taiwan Univ, BS, 60; Univ Wash, MS, 64, PhD(elec eng), 66. *Prof Exp:* Mem tech staff, Bell Telephone Labs, Murray Hill, 66-73; proj leader, Hewlett Packard Labs, HP Co, 73-77; pres & dir, Gen Optronics Corp, 77-88; vpres & dir, Laser Diode, Inc, Morgan Crucible Co, 88-90; PRES & DIR, PHOTON IMAGING CORP, 90- *Concurrent Pos:* Chief exec officer, pres & dir, Appl Optronics Corp, 91- *Mem:* Am Phys Soc; Am Optical Soc; Soc Photo-Optical Instrumentation Engrs. *Res:* Application of super radiant diode array and fiber optic bundle in the development of high resolution printers; high power semiconductor laser fabrication and application; quantum well laser structure; strained-layer laser structure. *Mailing Add:* 170 Hill Hollow Rd Watchung NJ 07060

HWANG, CHING-LAI, b Taiwan, Formosa, Jan 22, 29; m 54; c 3. MECHANICAL & INDUSTRIAL ENGINEERING. *Educ:* Univ Taiwan, BS, 52; Kans State Univ, MS, 60, PhD(mech eng), 62. *Prof Exp:* Instr, Ta-Tung Inst Technol, 54-58; asst prof eng & physics, Washburn Univ, Topeka, 62-65; from asst prof to assoc prof indust eng, 64-73, PROF INDUST ENG, KANS STATE UNIV, 73- *Concurrent Pos:* Guest prof, Tech Univ Denmark, 74-75. *Mem:* Am Soc Mech Engrs; sr mem Am Inst Indust Engrs; Japan Asn Automatic Control Engrs. *Res:* Heat and mass transfer; decision theory; numerical methods optimization techniques; operation research. *Mailing Add:* Dept Indust Eng Kans State Univ Manhattan KS 66506

HWANG, FRANK K, b Shanghai, China, Aug 24, 40; m 63; c 2. STATISTICS, DISCRETE MATHEMATICS. *Educ:* Nat Taiwan Univ, BA, 60; City Univ New York, MBA, 64; NC State Univ, MES, 66, PhD(statist), 68. *Prof Exp:* MEM TECH STAFF STATIST, BELL LABS, AT&T CO, 67- *Concurrent Pos:* Vis assoc prof statist, Nat Tsing Hua Univ, Taiwan, 70-71. *Mem:* Inst Elec & Electronics Engrs Commun Soc; Soc Indust & Appl Math. *Res:* Combinatorial theory; applied probability; operations research; computational geometry. *Mailing Add:* Math & Statist Ctr Bell Labs Ctr 121 600 Mountain Ave Murray Hill NJ 07974

HWANG, HU HSIEN, b China, Dec 1, 23; US citizen; m 64. ELECTRICAL ENGINEERING. *Educ:* Chiao Tung Univ, BSc, 49; Lehigh Univ, MS, 57, PhD(elec eng), 59. *Prof Exp:* Instr elec eng, Lehigh Univ, 57-60; asst prof, Univ Windsor, 60-64; chmn dept math, Chung Chi Col, Hong Kong, 61-62; assoc prof elec eng, Univ Windsor, 64-66; PROF ELEC ENG, UNIV HAWAII, 66- *Mem:* Inst Elec & Electronics Engrs. *Res:* Electromechanical energy conversion and systems analysis. *Mailing Add:* Dept Elec Eng Hol 44 Univ Hawaii Manoa Honolulu HI 96822

HWANG, JOHN DZEN, b Shanghai, China, Sept 8, 41; US citizen; m 67; c 3. TELECOMMUNICATIONS, SYSTEMS ENGINEERING. *Educ:* Univ Calif, Berkeley, BS, 64; Ore State Univ, MA, 66, PhD(math), 68; Harvard Bus Sch, PMD, 81. *Prof Exp:* Asst math, Ore State Univ, 65-68; opers researcher, Hq, US Army Materiel Command, 68-70, mathematician, Weapons

Command, Ill, 70-71, res scientist, US Army Air Mobility Res & Develop Lab, Ames Res Ctr, Moffett Field, Calif, 71-75; prog mgr, Defense Commun Agency, Washington, DC, 75-82; DIR SYSTS ENG, FED EMERGENCY MGT AGENCY, WASHINGTON DC, 82- *Concurrent Pos:* Asst comput physics, Ore State Univ, Eng Exp Sta, 67-68; systs anal specialist, M60A1E2 tank ad hoc panel, US Army Weapons Command, Ill, 69; vis prof, US Army Logistics Mgt Ctr, Va, 69-70; mem, Army Math Steering Comt, 71-75; chmn comt retail comput facil issues, Army Materiel Command Sci Comput Coun, 74-75 & subcomt opers res, Army Math Steering Comt, 75; vchmn, Army Materiel Command Sci & Eng Comput Coun, 75; consult to mgr, Nat Commun Syst, 78-82; dir, Fed Emergency Mgt Agency, 81; fel, Harvard Univ, 81; adj prof, George Mason Univ, Va, 78-83; adj prof, Am Univ, Washington, DC, 83-90; adj prof, Marymount Univ, Va, 91- *Mem:* Inst Elec & Electronics Engrs; NY Acad Sci; Opers Res Soc Am. *Res:* Information systems and telecommunications; defense materiel acquisition, particularly research and development management, decision risk analysis, military systems analysis; applied mathematics, particularly numerical algorithms for computational physics and mathematical systems theory. *Mailing Add:* 1200 Meadow Green Lane McLean VA 22102

HWANG, KAI, b Lan-Chou, China, Feb 28, 43; m 71; c 2. COMPUTER & ELECTRICAL ENGINEERING. *Educ:* Nat Taiwan Univ, BS, 66; Univ Hawaii, MS, 69; Univ Calif, Berkeley, PhD(elec eng), 72. *Prof Exp:* From asst prof to assoc prof elec eng, Purdue Univ, 81-85; PROF ELEC ENG & COMPUTER SCI, UNIV SOUTHERN CALIF, 85- *Mem:* Inst Elec & Electronics Engrs; Asn Comput Mach. *Res:* Computer architecture; parallel processing arithmetic systems; advanced automation; information processing and computer applications. *Mailing Add:* Dept Elec Eng & Computer Sci Univ Southern Calif Unversity Park Los Angeles CA 90089-2562

HWANG, LI-SAN, b Chikiang, China, Aug 12, 35; m 62; c 2. PHYSICAL OCEANOGRAPHY, ENGINEERING & RESEARCH MANAGEMENT. *Educ:* Nat Univ Taiwan, BS, 58; Mich State Univ, MS, 62; Calif Inst Technol, PhD(civil eng), 65. *Prof Exp:* Mem tech staff, Nat Eng Sci Co Calif, 65-66, mem sr staff, 66-67, sr scientist, 67-71, assoc dir eng div, 71-72, dir eng div, 72-74, vpres eng div, 74-77, sr vpres eng div, 77-88, CHMN & PRES, TETRA-TECH, INC, 88- *Concurrent Pos:* Chmn & chief exec officer, Edward H Richardson Assocs, 81- *Mem:* Am Soc Civil Engrs; Am Geophys Union; Int Asn Hydraul Res; Am Nuclear Soc. *Res:* Harbor oscillation due to both periodic and dispersive waves; problem of water wave propagation in shallow water; problem of wave run-up; tsunami generation, propagation and coastal effects; container ships problems; scale model studies of harbor problems. *Mailing Add:* 1955 Kinclair Dr Pasadena CA 91107

HWANG, NEDDY H C, b Hunan, China, Dec 28, 34; m 64; c 2. FLUID MECHANICS. *Educ:* Cheng Kung Univ, Taiwan, BSc, 57; Univ Calif, Berkeley, MS, 61; Colo State Univ, PhD(fluid mech), 66. *Prof Exp:* Asst prof fluid mech, 64-68, assoc prof civil eng, 68-74, PROF BIOMECH, UNIV HOUSTON, 74- *Concurrent Pos:* Adj assoc prof, Baylor Col Med, 70-; eng consult, Houston Lighting & Power Co & Driscoll Found Children's Hosp; vis scientist, Boeing Aircraft Co & Shell Pipeline Res & Develop Lab. *Res:* Wind generated waves; turbulent mixing; wake turbulence; clear air turbulence; biological fluid dynamics and its simulations. *Mailing Add:* Dept Civil Eng Univ Houston 4800 Calhoun Rd Houston TX 77004

HWANG, S STEVE, b Taichung, Taiwan, Oct 28, 38; US citizen; m 72; c 3. FLUID MECHANICS, HEAT TRANSFER. *Educ:* Nat Taiwan Univ, BS, 62; Univ Rochester, MS, 65, PhD(mech & aerospace sci), 67. *Prof Exp:* Assoc scientist, Xerox, 67-70, scientist, 70-74, proj mgr & sr scientist, 74-84, SR MEM RES STAFF & PROJ MGR, WEBSTER RES CTR, XEROX, 84- *Mem:* Am Soc Mech Eng; Sigma Xi; Soc Imaging Sci & Technol. *Res:* Novel color imaging systems including analysis, hardware development and integration. *Mailing Add:* 11 Valley Green Circle Penfield NY 14526

HWANG, SAN-BAO, b Kaohsiung, Taiwan, Sept 8, 46; US citizen; m 70; c 2. BIOCHEMISTRY, BIOPHYSICS. *Educ:* Nat Taiwan Univ, BS, 69; Univ Calif, Berkeley, PhD(biophys), 76. *Prof Exp:* Postdoctoral physiol, Univ Calif, San Francisco, 76-78, res asst physiologist physiol, 78-79; sr res biophysicist, Merck Sharp & Dohme Res Lab, 79-83, res fel, 84-88, sr res fel, 88-91; DIR, CYTOMED, INC, 91- *Mem:* Biophys Soc; Am Soc Biochem & Molecular Biol; NY Acad Sci; AAAS. *Res:* Characterization of platelet-activating factor receptors; identification of receptor antagonists of platelet-activating factor; identification of various receptor subtypes and intracellular receptors of platelet-activating factor. *Mailing Add:* CytoMed Inc One Mountain Rd Framingham MA 01701

HWANG, SHY-SHUNG, b Taichung, Taiwan, Oct 28, 38; m 72; c 3. FLUID MECHANICS, ENGINEERING SCIENCE. *Educ:* Nat Taiwan Univ, BS, 62; Univ Rochester, MS, 66, PhD(fluid mech), 67. *Prof Exp:* Assoc scientist, 67-70, scientist, 70-74, proj mgr & sr scientist, 74-84, PROJ MGR & SR MEM RES STAFF, XEROX CORP, 84- *Mem:* Am Soc Mech Engrs; Sigma Xi; Soc Photographic Scientists & Engrs. *Res:* Photoelectrophoresis; engineering research; electronic printing. *Mailing Add:* Xerox Corp Xerox Sq Bldg 114-210 Rochester NY 14644

HWANG, SUN-TAK, b Taijon, Korea, June 24, 35; m 63; c 2. CHEMICAL ENGINEERING. *Educ:* Seoul Nat Univ, BS, 58; Univ Iowa, MS, 62, PhD(chem eng), 65. *Prof Exp:* Instr fluid flow, Univ Iowa, 64-65; res engr, Field Res Lab, Mobil Oil Corp, 65-66; from asst prof to assoc prof thermodyn transport processes, 66-73, PROF MAT ENG, UNIV IOWA, 73- *Mem:* Am Inst Chem Engrs; Am Chem Soc. *Res:* Membrane separations. *Mailing Add:* Dept of Chem Eng Mail Location 171 Univ Cincinnati Cincinnati OH 45221-0171

HWANG, WILLIAM GAONG, b Shanghai, China, June 17, 49; US citizen; m 78; c 2. MILSATCOM SYSTEMS. *Educ:* Calif Inst Technol, BS, 71; Cornell Univ MS, 74, PhD(appl math), 78. *Prof Exp:* Eng aide, Calif Dept Water Resources, 67, Calif Inst Technol, 68-69 & Sacramento Munic Utility Dist, 69-70; teaching asst eng, Cornell Univ, 71-76; mem staff, BDM Corp, 78-82; asst vpres, MIA-Com Linkabit, 82-88; ASST, PRES, SAIC, 88- *Mem:* Inst Elec & Electronics Engrs; Am Math Soc; Soc Indust & Appl Math; Armed Forces Commun & Electronics Asn. *Res:* Autoregressive and moving average time series; robust estimation of order of process; covariance analysis and discriminant analysis; system identification; control theory; command, control, and communications systems analysis; military satellite communications analyses. *Mailing Add:* SAIC 8619 Westwood Center Dr Vienna VA 22182

HWANG, YU-TANG, b Tainan, Taiwan, May 20, 30; m 61; c 3. POLYMERIZATION PROCESSES, POLYOLEFIN CATALYSIS. *Educ:* Univ Taiwan, BS, 54; Univ Mich, MS, 57, PhD(chem eng), 61. *Prof Exp:* Sr res engr, Pure Oil Co, Ill, 61-63, res scientist, 63-65; res assoc, Hawshaw Chem Co, 65-69; res assoc, 69-75, asst mgr, tech sect, 75-85, TECH PROJ MGR, QUANTUM CHEM, USI DIV, 85- *Mem:* Am Chem Soc; Am Inst Chem Engrs; Sigma Xi. *Res:* Industrial catalysis; process research and development; polyethylene. *Mailing Add:* USI Div Quantum Chem Corp 8935 N Tabler Rd Morris IL 60450-9153

HWU, MARK CHUNG-KONG, b Canton, China, Nov 10, 26; m 60; c 4. CHEMICAL ENGINEERING. *Educ:* Nat Taiwan Univ, BS, 51; Kans State Col, MS, 54; Univ Cincinnati, PhD(chem eng), 59. *Prof Exp:* Asst phys chem, Univ Cincinnati, 58-59, res assoc, 59-60, res assoc polymer characterization, 62; chief chem engr, Que Lithium Corp, Can, 60-61, tech adv, 61; asst prof chem eng, Univ NB, 61; chem engr, Chem Div, 62-67, proj scientist, 67-84, ENG SCIENTIST, CHEM DIV & SR DEVELOP SCIENTIST, POLYOLEFINS DIV, UNION CARBIDE CORP, 84- *Concurrent Pos:* Fel, Univ Cincinnati, 59-60; vis lectr, WVa Inst Technol, 64 & 66. *Res:* Polymerization of ethylene and associated fields; homopolymers and copolymers in high-pressure processes and low-pressure fluid bed processes; chemical engineering kinetics; high pressure and high temperature processing; process modelling, control and optimization; computer control of polyethylene processes. *Mailing Add:* Res & Develop Dept PO Box 670 Bound Brook NJ 08805

HYAMS, HENRY C, physics, electronics, for more information see previous edition

HYATT, ABRAHAM, b Ukraine, July 15, 10; US citizen; m 38; c 2. AEROSPACE ENGINEERING. *Educ:* Ga Inst Technol, BS, 33. *Prof Exp:* Sr stress analyst, Glenn L Martin Co, 36-39; chief struct engr, McDonnell Aircraft Corp, 39-44; head airframe design res, Res Div, Bur Aeronaut, USN, 46-48, asst dir aeronaut systs, 48-54, dir, Res Div, 54-56, chief scientist, 56-58; asst dir propulsion systs, NASA, 58-60, dep dir launch vehicle systs, 60, dir prog planning, 60-63; consult space systs, Douglas Aircraft Corp & IBM Corp, 63-64; Hunsaker prof aeronaut & astronaut, Mass Inst Technol, 64-65; exec dir bus develop, N Am Rockwell Corp, 65-76; CONSULT, PLANNING & MGT RES & DEVELOP PROGS IN AEROSPACE. *Concurrent Pos:* USMC, 44-46; mem, Polaris sci adv bd, Navy Dept, 54-58; adv panel aeronaut, Dept Defense, 56-58; comt aircraft, missile & spacecraft aerodynamics, Nat Adv Comt Aeronaut, 56-58, spec comt space technol, 57-58; panel high altitude detection, President's Sci Adv Comt, 58; comt long range planning, Fed Coun Sci & Technol, 60 & USAF Sci Adv Bd, 60-63. *Honors & Awards:* Distinguished Civilian Serv Award, USN. *Mem:* AAAS; fel Am Inst Aeronaut & Astronaut; Am Nuclear Soc; Int Acad Astron. *Res:* Research and optimization of engineering systems principally involving aeronautical systems and manned space systems. *Mailing Add:* 18038 Bluesail Dr Pacific Palisades CA 90272

HYATT, ASHER ANGEL, b London, Eng, July 31, 30; US citizen; m 57; c 2. SCIENCE ADMINISTRATION, ORGANIC CHEMISTRY. *Educ:* Univ London, BSc, 51, PhD(org chem), 54. *Prof Exp:* Demonstr, Queen Mary Col, Univ London, 52-54; res assoc, Mass Inst Technol, 54-55; res chemist, Brit Thomson-Houston, Eng, 55-56; dir, Nucleus Chem Labs, London, 56-57; res chemist, Ionics, Inc, Mass, 57-58; sr org chemist, Monsanto Res Corp, 58-64; head org res, Collab Res, Inc, 64-65, sr scientist, 65-66; exec secy, Med & Org Chem Fels Comt, 66-69, Med Chem Study Sect, 69-78, CHIEF, BIOMED SCI REV SECT, REFERRAL & REV BR, DIV RES GRANTS, NIH, 78- *Mem:* AAAS; Am Chem Soc; Royal Inst Chem. *Res:* Research administration and management; peer review. *Mailing Add:* NIH Westwood Bldg 336 Bethesda MD 20892

HYATT, DAVID ERNEST, b Syracuse, NY, Nov 25, 42; m 65; c 2. INORGANIC CHEMISTRY, ORGANOMETALLIC CHEMISTRY. *Educ:* Colgate Univ, AB, 64; Univ Ill, MS, 66, PhD(chem), 68. *Prof Exp:* Res chemist, Eastern Res Ctr, Stauffer Chem Co, 68-70; sr res chemist, Vulcan Mat Co, Colo, 70-72; sr proj engr & environ scientist, Colo Sch Mines Res Inst, 72-75, mgr, Anal Serv Lab, 75-77; sr proj engr, Hazen Res, 77-81; res mgr, Chem & Metal Industs, 81-90; SR RES CHEMIST, ADA, 90- *Mem:* Am Chem Soc; Am Inst Mining & Metall Engrs; Am Soc Testing & Mat. *Res:* Boron hydride and carborane synthesis; homogeneous and heterogeneous catalysis systems; pollution abatement; minerals processing. *Mailing Add:* 6735 S Downing Circle E Littleton CO 80122-1344

HYATT, EDMOND PRESTON, b Joliet, Ill, Nov 15, 23; c 10. CERAMIC ENGINEERING. *Educ:* Univ Mo-Rolla, BS, 49, MS, 50; Univ Utah, PhD(ceramic eng), 56. *Prof Exp:* Instr, Brigham Young Univ, 50-54; co-founder, Electro-Ceramics Inc, 54, vpres opers, 57-63; mgr tech ceramics, Centralab Electronics Div, 64-70; CERAMIC ENG CONSULT, 70- *Mem:* Fel Am Ceramic Soc; Inst Ceramic Engrs; Int Soc Hybrid Microelectronics. *Mailing Add:* 31 E 400 S Orem UT 84058

HYATT, GEORGE, JR, b Toledo, Ohio, Dec 14, 14; m 38; c 3. DAIRY HUSBANDRY. *Educ:* Mich State Col, BS, 37; Rutgers Univ, MS, 41; Univ Wis, PhD, 61. *Prof Exp:* Milk inspector, Detroit Health Dept, 37-38; asst exten dairyman, WVa Univ, 41-43 & Univ Md, 43-46; prof dairy husb, WVa Univ, 46-51; farm mgr & ed adv, Hoard's Dairyman, Wis, 51-52; head, Dairy

Exten, 52-58, head, Dept Animal Sci, 58-61, assoc dir, Exten Serv, 61-63, assoc dean, Sch Agr & Life Sci, 72-78, emer prof animal sci, Agr Exten Serv & Prof Adult Educ, NC State Univ, 78-84, dir, Agr Exten Serv, 63-78; RETIRED. *Mem:* Am Dairy Sci Asn; Am Soc Animal Sci. *Res:* Agricultural administration; some bases for coordination of cooperative extension programs with research and resident instruction in selected landgrant institutions. *Mailing Add:* 1419 Lutz Ave Raleigh NC 27607

HYATT, JOHN ANTHONY, b Harlan, Ky, Aug 21, 48; m 71; c 1. SYNTHETIC ORGANIC CHEMISTRY. *Educ:* Wake Forest Univ, BS, 70; Ohio State Univ, PhD(chem), 73. *Prof Exp:* Res fel chem, Harvard Univ, 73-74; SR RES ASSOC, EASTMAN CHEM DIV, EASTMAN KODAK CO, 75- *Concurrent Pos:* Vis prof chem, ETenn State Univ, 82-83. *Mem:* Am Chem Soc; Sigma Xi. *Res:* Synthetic chemistry of complex organic compounds; heterocyclic chemistry; synthetic photochemistry; photographic chemistry; chemistry of wood-derived natural products; cellulose chemistry; lignin chemistry. *Mailing Add:* Res Labs Eastman Chem Div Eastman Kodak Co Kingsport TN 37662

HYATT, RAYMOND R, JR, b Pawtucket, RI, June 8, 50; m 73; c 2. MULTIVARIATE STATISTICS, COMPUTER GRAPHICS. *Educ:* Brown Univ, ScB, 80; Mass Inst Technol, MS, 82. *Prof Exp:* Res asst math, Mass Inst Technol, 80-81; instr, Northeastern Univ, 82; chmn, Dept Computer Sci, Daniel Webster Col, 83-85, instr, 83-90; SR ANALYST, FED CONTRACT RES, ABT ASSOCS, 90- *Concurrent Pos:* Sr res methodologist, Ctr Appl Social Res, Northeastern Univ, 82- *Mem:* Am Math Soc; Sigma Xi; Asn Comput Mach. *Res:* The application of advanced methods in computer science and mathematics to the anlysis of very large data bases such as are found in sociology, criminology and economics. *Mailing Add:* PO Box 493 Belmont MA 02178

HYATT, ROBERT ELIOT, b Trenton, NJ, June 2, 25; m 52; c 2. PHYSIOLOGY, INTERNAL MEDICINE. *Educ:* Univ Rochester, AB, 47, MD, 50; Am Bd Internal Med, dipl. *Prof Exp:* Resident internal med, Sch Med, Wash Univ, 50-52, fel cardiol, 52-53; clin assoc, Nat Heart Inst, 53-55, investr respiratory physiol, 55-58; assoc chief med & dir cardiopulmonary lab, Beckley Mem Hosp, 58-62; consult & asst prof physiol, Mayo Clin, 62-67, assoc prof physiol, Mayo Grad Sch Med, Univ Minn, 67-71, assoc prof med, 71-73, prof med & physiol, 73-87; RETIRED. *Concurrent Pos:* Intern, Barnes Hosp, St Louis, 50-51, asst resident, 51-52. *Mem:* Am Physiol Soc; Am Thoracic Soc; Am Fedn Clin Res. *Res:* Physiology of ventilatory mechanics; lung growth and development. *Mailing Add:* 1912 Lake Ridge Dr Vandalia IL 62471

HYATT, ROBERT MONROE, b Orlando, Fla, Jul 14, 43; m 76; c 3. SYSTEMS ENGINEERING. *Educ:* Univ Fla, BSEE, 65, Univ Dallas, MBA, 74. *Prof Exp:* Design eng electro-optics, Tex Instruments, 69-74; region mgr eng, Teccor, 74-75, Portescap, 75-77; gen mgr, Amco Enterprises, 77-78; VPRES ENGR MKT, TRANSICOIL INC, 78- *Mem:* Inst Elec & Electronics Engrs; Am Helicopter Soc; Army Aviation Am; Soc Automotive Eng. *Res:* Electro magnetic device development; precision drive trains for position system. *Mailing Add:* PO Box 419 Worchester PA 19490-0419

HYBL, ALBERT, b Iowa City, Iowa, Jan 8, 32; m 62; c 2. BIOPHYSICS, CRYSTALLOGRAPHY. *Educ:* Coe Col, BA, 54; Calif Inst Technol, PhD(chem), 61. *Prof Exp:* Fel chem, Iowa State Univ, 61-64; asst prof, 64-71, ASSOC PROF BIOPHYS, MED SCH, UNIV MD, BALTIMORE CITY, 71- *Mem:* AAAS; Am Crystallog Asn; Am Chem Soc; Biophys Soc; Sigma Xi. *Res:* Biological structures; molecular structure of organic compounds; structure of membranes and carbohydrates. *Mailing Add:* Dept Biophys Univ Md Med Sch Baltimore MD 21201

HYDE, BEAL BAKER, b Dallas, Tex, June 26, 23; m 47; c 3. GENETICS, MOLECULAR BIOLOGY. *Educ:* Harvard Univ, AB, 48, Am, 50, PhD(biol), 52. *Prof Exp:* Res assoc bot, Ind Univ, 51-54; assoc prof plant sci, Univ Okla, 54-61; NSF res fel, Calif Inst Technol, 61-64; vis assoc prof bot, Univ Tex, 64-65; chmn dept, 65-77, PROF BOT, UNIV VT, 65- *Concurrent Pos:* Mem staff, Inst Genetics, Copenhagen Univ, 71-72 & Plant Breeding Inst, Cambridge, Eng, 78-79. *Mem:* AAAS; Genetics Soc Am; Am Soc Cell Biologists; Sigma Xi. *Res:* Plant cytogenetics and cytochemistry; ultrastructure of nucleus. *Mailing Add:* Dept Bot Marsh Life Sci Bldg Univ Vt Burlington VT 05405

HYDE, DALLAS MELVIN, b Mar 12, 45. ANATOMY. *Educ:* Univ Calif, Irvine, BA, 67, Whittier Col, MS, 72, Univ Calif, Davis, PhD(anat), 76. *Prof Exp:* Captain, USMC Res, 67-70; staff res assoc II, Sch Vet Med, Univ Calif, Davis, 72-74, Walter Foster Found fel, 74-76; asst prof, Dept Metab, Col Vet Med, Univ Fla, Gainesville, 76-79; from asst prof to assoc prof, 79-88, PROF & CHMN DEPT ANAT, SCH VET MED, UNIV CALIF, DAVIS, 88- *Concurrent Pos:* Mem, Grad Studies & Res Comt, Col Vet Med, Univ Fla, 76-78, Electron Micros Comt, 76-79, Student Advisement Comt, 76-79, Admis Comt, 77-79, Off Vet Med Educ Comt, 77-79; mem, Curric Comt, Grad Group Anat, Univ Calif, 80-81, mem, Health Sci Libr, Sch Vet Med, 80-81, chmn, Curric Comt, 81-83, grad adv, 83-85; Nat Res Serv Award sr fel, NIH, Nat Jewish Hosp, Denver, Colo, 85-86. *Mem:* Am Asn Anatomists; AAAS; Am Asn Vet Anatomists; Am Asn Pathologists; Am Thoracic Soc; Electron Micros Soc Am; Int Soc Stereology. *Res:* Morphometric and morphologic evaluation of the mechanisms of pulmonary inflammation in response to inhale environmental pollutants and fibrogenic chemicals. *Mailing Add:* Dept Anat Univ Calif Davis CA 95616

HYDE, EARL K, b Rossburn, Man, Aug 9, 20; US citizen; m 49; c 4. NUCLEAR CHEMISTRY, NUCLEAR PHYSICS. *Educ:* Univ Chicago, BS, 41, PhD(chem), 46. *Prof Exp:* Res assoc inorg war res proj, Univ Chicago, 42-44, jr chemist, Metall Lab, Manhattan Dist, 44-46; assoc chemist, Argonne Nat Lab, 46-49; chemist, Nuclear Chem Div, Lawrence Berkeley Lab, Univ Calif, 49-72, dep head, 71-73, dep dir, 73-83, assoc dir-at-large, 84-86, SR STAFF SCIENTIST, LAWRENCE BERKELEY LAB, UNIV CALIF, 73-, EMER DEP DIR, 86- *Mem:* Am Chem Soc; fel Am Phys Soc. *Res:* Chemistry of heavy elements; mechanisms of interaction of giga electron volt particles with complex nuclei studied by physical techniques; identification of new alpha emitting isotopes of short half life; radioactive isotopes produced by high energy spallation or fission reactions. *Mailing Add:* 852 The Arlington Berkeley CA 94707

HYDE, GEOFFREY, b Toronto, Ont, Apr 10, 30; US citizen; m 53; c 3. SATELLITE COMMUNICATIONS, RADIOPHYSICS. *Educ:* Univ Toronto, BSc, 53, MSc, 59; Univ Pa, PhD(elec eng), 67. *Prof Exp:* Engr power equip, Can Gen Elec Co, 53-55; engr antennas, Sinclair Radio Labs, Ltd, 55-57; engr antennas & radar, Avro Aircraft Ltd, 58-59; engr antennas & electromagnetic theory, Radio Corp Am, 59-65, engr radar systs, 67-68; sr staff scientist antennas & electromagnetic theory, 68-74, mgr propagation studies, 74-80, sr staff scientist & coordr, Corp Res Develop Prog, 80-84, ASST TO DIR, COMSAT LABS, 84- *Concurrent Pos:* Mem US Study Group 5, Int Radio Consult Comt, Int Telecommun Union, 74-, mem US deleg, Switz, 76, US deleg, France, 77 & subcomt chmn, US Study Group 5F, 78-; David Sarnoff fel, 65-67. *Mem:* Fel Inst Elec & Electronics Engrs; Int Scientists Radio Union; Am Inst Aeronaut & Astronaut. *Res:* Propagation of electromagnetic waves in non-ionized media; microwave antenna research and development; related areas of satellite communications systems and radio meteorology; related areas of electromagnetic theory; communications engineering and systems; radio engineering; radio physics. *Mailing Add:* Comsat Labs 22300 Comsat Dr Rm 0217 Clarksburg MD 20871-0310

HYDE, HENRY VAN ZILE, medicine; deceased, see previous edition for last biography

HYDE, JAMES STEWART, b Mitchell, SDak, May 20, 32; m 59; c 3. ELECTRON PARAMAGNETIC RESONANCE SPECTROSCOPY. *Educ:* Mass Inst Technol, SB, 54, PhD(physics), 59. *Hon Degrees:* Dr, Jagiellonian Univ, Krakow, Poland, 89. *Prof Exp:* Asst, Lab Insulation Res, Mass Inst Tech, 55-59; res physicist, Varian Assocs, 59-64, mgr, Electron Paramagnetic Resonance Res, 64-75; PROF RADIOL & RADIOL & BIOPHYS, MED COL WIS, 75-, DIR NAT BIOMED ELECTRON SPIN RESONANCE CTR, 80- *Concurrent Pos:* Adj prof biomed eng, Marquette Univ, 81- *Mem:* Fel Am Phys Soc; Biophys Soc; Soc Magnetic Resonance Med; Soc Magnetic Resonance Imaging; Radiol Soc NAm. *Res:* Electron paramagnetic resonance; electron nuclear double resonance; free radicals in liquids; spin-labels; spin resonance instrumentation; dynamics of macromolecular complexes; oxygen transport; physico-chemical properties of melanin; in-vivo magnetic resonance imaging and spectroscopy. *Mailing Add:* Biomed Electron Spin Resonance Ctr Med Col Wis 8701 Watertown Plank Rd Milwaukee WI 53222

HYDE, JOHN WELFORD, b London, Eng, Jan 13, 10; US citizen; m 37; c 2. GEOLOGY. *Educ:* Univ London, Eng, BSc, 36, ChE, 41. *Prof Exp:* Mem staff, Brit Petrol Co, 37-63, sr vpres, BP NAm Inc, 63-70; vpres, Pace Co Consults & Engrs, 70-80; pres, Hyde Assocs, 80- *Concurrent Pos:* Pres, Heat Transfer Res Inc, 65-70. *Mem:* Fel Inst Petrol; Brit Inst Chem Engrs; Am Inst Chem Engrs. *Res:* Petroleum and chemical engineering; awarded 22 patents. *Mailing Add:* 26 Winthrop Dr Riverside CT 06878

HYDE, KENDELL HEMAN, b Afton, Wyo. MATHEMATICS. *Educ:* Univ Wyo, BS, 59; Univ Utah, MS, 62, PhD(math), 69. *Prof Exp:* Asst prof math, Church Col Hawaii, 62-66; asst prof, 69-71, assoc prof, 71-78, PROF MATH, WEBER STATE COL, 78- *Mem:* Math Asn Am; Asn Educ Data Systs. *Res:* Group theory. *Mailing Add:* Weber State Col 1702 Ogden UT 84408

HYDE, KENNETH E, b Pittsburgh, Pa, July 26, 41; c 1. INORGANIC CHEMISTRY. *Educ:* Carnegie-Mellon Univ, BS, 63; Univ Md, PhD(inorg chem), 69. *Prof Exp:* Instr chem, Univ Md, 67-68; from asst prof to assoc prof, 68-78, PROF CHEM, STATE UNIV NY COL OSWEGO, 78- *Concurrent Pos:* Grant-in-aids, State Univ NY Res Found, 69-71, 73-74, 77-78, Petrol Res Fund, 73-76 & Cottrell res grant, 76-78; vis assoc prof, State Univ NY Buffalo, 74-76; vis scientist, Inst Phys Chem, Univ Frankfurt, applications prog, Gen Elec Corp, Syracuse, 81-85; vis fac res partic, Oak Ridge Nat Lab, 90-91. *Mem:* Am Chem Soc; Sigma Xi. *Res:* Magnetic and thermal properties of inorganic complex compounds; kinetics and mechanisms of inorganic complexes in aqueous solutions; applications of computer systems to on-line data acquisition; connecting serial instruments to microcomputers. *Mailing Add:* Dept Chem State Univ NY Col Oswego NY 13126

HYDE, KENNETH MARTIN, b Warren, Ohio, Aug 8, 43; m 63; c 3. BIOLOGY, WILDLIFE MANAGEMENT. *Educ:* Hiram Col, BA, 68; La State Univ, MS, 70, PhD(entom), 72. *Prof Exp:* Assoc prof, 73-80, prof biol, 80-, chmn dept, 73- DEPT CHEM, PT LOMA NOZARENE COL. *Concurrent Pos:* Mem comt on pesticides, San Diego Qual Life Bd, 73-75 & Environ Health Sect, Comprehensive Health Planning Asn, Imperial, Riverside & San Diego Counties, 74. *Mem:* Sigma Xi. *Res:* Effects of chronic exposure of environmental contaminants on nervous systems of endemic mammals as monitored through the electroencephalogram. *Mailing Add:* Dept Chem Pt Loma Nozarene Col 3900 Lomaland Dr San Diego CA 92106

HYDE, PAUL MARTIN, b San Francisco, Calif, Jan 27, 23; m 56; c 3. BIOCHEMISTRY, ANALYTICAL CHEMISTRY. *Educ:* Univ San Francisco, BS, 47; Univ Calif, MS, 50; St Louis Univ, PhD(biochem), 53. *Prof Exp:* Asst pharmaceut chem, Col Pharm, Univ Calif, 47-50; res assoc, Univ Wash, 53-55, res instr, 55-57; from asst prof to assoc prof, 57-69, PROF BIOCHEM, SCH MED, LA STATE UNIV MED CTR, NEW ORLEANS, 69- *Concurrent Pos:* Res assoc, Urban Maes Res Found, La State Univ New Orleans, 57-61; consult, Xavier Univ, 72-79, Lab Specialists, Inc, New Orleans, 80- *Mem:* Soc Exp Biol & Med; Endocrine Soc; Am Soc Biol Chemists. *Res:* Metabolism of androgens, adrenal steroids, bile acids and drugs; pharmacology. *Mailing Add:* Dept Biochem La State Univ Med Ctr 1901 Perdido St New Orleans LA 70112

HYDE, RICHARD MOOREHEAD, b Pierre, SDak, Feb 11, 33; m 53; c 3. MICROBIOLOGY, IMMUNOLOGY. *Educ:* Univ SDak, BA, 55, MA, 56; Univ Minn, PhD(immunol), 62. *Prof Exp:* Asst prof microbiol, San Francisco State Col, 60-62; from instr to asst prof, Univ Mo, 62-65; from asst prof to assoc prof, 65-72, PROF MICROBIOL, UNIV OKLA, 72-, VCHMN, 85- *Concurrent Pos:* Res assoc, Sch Dent, Univ of the Pac, 61-62; USPHS spec fel, Scripps Clin & Res Found, 68-69; spec fel, Univ Ill & Univ Dundee, 84. *Mem:* AAAS; Am Asn Immunologists; Am Soc Microbiol; Soc Exp Biol & Med; Reticuloendothelial Soc. *Res:* innate immunity; medical education. *Mailing Add:* Dept Microbiol Univ Okla Health Sci Ctr PO Box 26901 Oklahoma City OK 73190

HYDE, RICHARD WITHERINGTON, b Plainfield, NJ, Sept 6, 29; m 57; c 4. RESPIRATORY PHYSIOLOGY. *Educ:* Yale Univ, BA, 53; Columbia Univ, MD, 57. *Prof Exp:* Intern & resident, Hosp Univ Pa, 57-60, asst prof, 65-69; PROF MED, RADIATION BIOL & BIOPHYS, UNIV ROCHESTER, 69- *Concurrent Pos:* USPHS res fel physiol, Hosp Univ Pa, 60-64; estab investr, Am Heart Asn, 67-72. *Mem:* AAAS; Am Fedn Clin Res; Am Physiol Soc; Am Soc Clin Invest; Am Thoracic Soc; Am Heart Asn. *Res:* Respiratory gas exchange; aerosols; lung metabolism; pulmonary edema. *Mailing Add:* Pulmonary Unit Box 692 Univ Rochester Med Ctr Rochester NY 14642

HYDE, WALTER LEWIS, b Minneapolis, Minn, May 30, 19; m 41; c 5. PHYSICS, ACADEMIC ADMINISTRATION. *Educ:* Harvard Univ, BS, 41, AM, 43, PhD(physics), 49. *Prof Exp:* Res physicist, Polaroid Corp, 43-46 & Baird Assoc, 48-50; sci liaison officer, Off Naval Res, London, 50-53; res physicist, Am Optical Co, 53-58, asst dir res, 58-60; dir develop, J W Fecker Div, 60-63; prof optics, Univ Rochester, 63-68; provost, NY Univ, Univ Heights Campus, 68-71, spec asst to pres, Univ, 71-72; exec dir, Conn Conf Independent Cols, 72-79 & Conn State Tech Cols, 79-84; RETIRED. *Concurrent Pos:* Mem, US Nat Comt, Int Comn Optics, 56-, vpres, 62-; dir, Inst Optics, 65-68. *Mem:* Optical Soc Am (pres, 70). *Res:* Optical instruments; polarized light; education policy. *Mailing Add:* 337 English Neighborhood Rd Woodstock CT 06281

HYDE, WILLIAM W, b Twin Falls, Idaho, Jan 19, 46; m 63; c 2. FORCE ENERGY ENGINEERING, RENEWABLE ENERGY ENGINEERING. *Educ:* Gen Motors Res Inst, BSME, 68; Univ Calif, BSEE, 71. *Prof Exp:* Eng mgr, Dept Defense, 71-75, NASA, 75-79; ENG MGR, AUTOMOTIVE RES CORP, 79- *Concurrent Pos:* Consult, Dept Defense, NASA, Gen Motors, Ford, Chrysler, DOW, Gen Dynamics, Gen Elec, Defense Advan Res Projs Agency, IBM & others, 79- *Mem:* Soc Automotive Engrs; Inst Elec & Electronics Engrs. *Res:* Advanced electrical power generator from natural forces and sources such as the electric field, magnetic field, gravitational field, solar, wind, nuclear; advanced research in automotive hydrogen propulsion systems using the above mentioned forces; issued 20 patents. *Mailing Add:* 1685 Whitney Idaho Falls ID 83402-1768

HYDER, DONALD N, b Crossville, Tenn, Apr 24, 21; m 47; c 3. RANGE SCIENCE. *Educ:* Univ Idaho, BS, 47; Utah State Univ, MS, 49; Ore State Univ, PhD, 61. *Prof Exp:* Asst range mgt, Utah State Univ, 47-48; range conservationist, Bur Land Mgt, US Dept Interior, Ore, 49-56; res agronomist, Sci & Educ Admin-Agr Res, USDA, 56-61, range scientist, 61-73, leader forage & range res, 73-76; retired, 76-78; RANGE CONSULT, 78- *Honors & Awards:* Outstanding Achievement Award, Soc Range Mgt, 73. *Mem:* Soc Range Mgt. *Res:* Range improvement; plant morphology; range and management. *Mailing Add:* 1008 E Elizabeth Ft Collins CO 80524

HYDER, MONTE LEE, b Maryville, Tenn, June 16, 36; m 64; c 3. NUCLEAR CHEMISTRY, ANALYTICAL CHEMISTRY. *Educ:* Rice Univ, BS, 58; Univ Calif, Berkeley, PhD(chem), 62. *Prof Exp:* Res chemist, Savannah River Lab, E I du Pont de Nemours & Co, Inc, 62-67, res supvr, 67-76, res mgr, 76-84, sr res assoc, 84-89; SR ADV SCIENTIST, WESTINGHOUSE SAVANNAH RIVER CO, 89- *Mem:* Am Chem Soc; Sigma Xi. *Res:* Radiation chemistry of aqueous solutions; nuclear fuel reprocessing; chemistry of uranium and transuranium elements; nuclear reactor safety. *Mailing Add:* Savannah River Lab Aiken SC 29808

HYDER, SYED S, computer systems, for more information see previous edition

HYDOCK, JOSEPH J, b Plains, Pa, Apr 13, 18; m 45; c 4. CHEMISTRY. *Educ:* Univ Scranton, BS, 40; Univ Notre Dame, PhD(chem), 50. *Prof Exp:* Res assoc, Nat Defense Res Comt, 42-43; res chemist, Sinclair Res Labs, 49-54; sr res engr, Autonetics Div, NAm Aviation, Inc, 54-65; mem tech staff, Hughes Aircraft Co, Calif, 65-66 & Autonetics Div, NAm Rockwell Corp, 67-69; consult, Radkowski Assocs, 69-73; pres, QAS Corp, 73-74; SR CORROSION ENGR, FLUOR ENGRS, INC, 74- *Concurrent Pos:* Consult design groups on corrosion. *Mem:* Am Chem Soc; Nat Asn Corrosion Engrs. *Res:* Lubrication; corrosion; radiation; materials; liquid metals; dielectric coolants. *Mailing Add:* 11832 Peacock Ct Garden Grove CA 92641-2590

HYER, PAUL VINCENT, marine science, for more information see previous edition

HYERS, DONALD HOLMES, b Los Angeles, Calif, Apr 1, 13; m 40, 87; c 2. MATHEMATICS. *Educ:* Univ Calif, Los Angeles, AB, 33, MA, 34; Calif Inst Technol, PhD(math), 37. *Prof Exp:* Asst math, Calif Inst Technol, 36-37; fel mech eng, Nat Defense Res Comt, 42-44; instr, Univ Wis, 37-42; from assoc prof to prof math, 44-78, head dept, 45-50, EMER PROF MATH, UNIV SOUTHERN CALIF, 78- *Mem:* Am Math Soc; Math Asn Am; Soc Indust & Appl Math. *Res:* Functional analysis; nonlinear integral equations; applications to hydrodynamics. *Mailing Add:* 3114 Pine Ave Long Beach CA 90807-5049

HYERS, THOMAS MORGAN, b Jacksonville, Fla, June 16, 43; c 2. PULMONOLOGY, CRITICAL CARE MEDICINE. *Educ:* Duke Univ, MD, 68; Am Bd Internal Med, dipl, 74. *Prof Exp:* Chief resident & instr med, Vet Admin Med Ctr, Univ Wash, Seattle, 74-75; fel pulmonary dis, Univ Colo, Denver, 75-76, res fel, Cardiovasc Pulmonary Res Lab, Health Sci Ctr, 76-77, asst prof med, 77-82; asst prof med & dir, Div Pulmonary, 82-85, PROF MED, ST LOUIS UNIV MED CTR, 85- *Concurrent Pos:* Res assoc med, Vet Admin Med Ctr, St Louis, 79-82; dir, Nat Heart, Lung & Blood Inst, NIH grant, Scor Adult Respiratory Failure, 83-93; mem, Coun Thrombosis & Cardiopulmonary Dis, Am Heart Asn. *Mem:* Am Physiol Soc; Int Soc Thrombosis & Haemostasis; Am Fedn Clin Res; Am Heart Asn; Am Thoracic Soc; fel Am Col Physicians. *Res:* Pathogenesis of the adult respiratory distress syndrome; relationship of adult respiratory distress syndrome to multiple organ injury; interactions of protease inhibitors and various cytokines in modulating neutrophil activation; role of growth factors in the recovery of the acutely injured lung. *Mailing Add:* Div Pulmonary & Occup Med St Louis Univ Med Ctr PO Box 15250 St Louis MO 63110-0250

HYGH, EARL HAMPTON, b Atlanta, Ga, Apr 19, 33; m 53; c 4. FACILITIES MAINTENANCE MANAGEMENT. *Educ:* Univ Calif, Riverside, BA, 60, MA, 62, PhD(physics), 64. *Prof Exp:* Physicist, US Naval Ord Lab, Calif, 60-61; res asst solid state physics, Univ Calif, Riverside, 61-64; asst prof solid state theory, Univ Utah, 64-73; chemist, Naval Air Sta, Lemoore, Calif, 73-77, environmentalist & natural resources specialist, Pub Works Dept, 77-79, training div dir, Civilian Personnel Dept, 79-83, dir maintenance & utilities, Pub Works Dept, 83-88. *Concurrent Pos:* Solid state physicist, US Naval Ord Lab, 63-64. *Mem:* Am Phys Soc. *Res:* Theoretical investigation of energy band theory; theory of ideal and imperfect insulating crystals; application of granular activated carbon to removal of organic contamination of water. *Mailing Add:* Civilian Personnel Naval Air Sta Lemoore CA 93246-0001

HYLAND, JOHN R(OTH), b Cleveland, Ohio, Aug 25, 25; div; c 4. GEOLOGICAL & CIVIL ENGINEERING. *Educ:* Colo Sch Mines, GeolE, 50. *Prof Exp:* Head eng geol sect, Ohio Dept Natural Resoucres, 51-54, asst chief shore erosion, 54-59, chief shore erosion & secy water ways comt, 59-61, asst chief div water, 61-63; consult, Found, G K Jewell Assocs, Ohio, 63-64; proj dir, Monongahela River Eng Proj, USPHS, then Fed Water Pollution Control Admin, US Dept Interior, 64-68, staff engr spec control projs, 69, actg chief, Spec Control Prog Br, 69-71, chief indust wastes eval sect, Environ Protection Agency, 71-75; consult, 76-78; div engr, Ohio Div Reclamation, 78-80; CONSULT, 80- *Mem:* Am Soc Civil Engrs; Water Pollution Control Fedn. *Res:* Wave energy and bottom deposits of Lake Erie; waterborne pollutants from mining and oil; automated monitoring for organic pollutants in water. *Mailing Add:* 5812 Linworth Rd Columbus OH 43235

HYLAND, KERWIN ELLSWORTH, JR, b York, Pa, Apr 7, 24; c 4. ENTOMOLOGY, PARASITOLOGY. *Educ:* Pa State Univ, BS, 47; Tulane Univ, MS, 49; Duke Univ, PhD(zool), 53. *Prof Exp:* Asst zool, Duke Univ, 49-51; sci instr, Christchurch Sch, Va, 51-53; from instr to assoc prof, 53-66, chmn dept, 65-68, PROF ZOOL, UNIV RI, 66- *Concurrent Pos:* Fulbright res scholar, Prince Leopold Inst Trop Med, Belg, 60-61; vis prof, Univ Lovanium, Kinshasha, 64; vis prof entom, Mich State Univ, 58-60; USPHS spec res fel, Prince Leopold Inst Trop Med, Belg, 68-69, sr res prof, 69-; guest prof, Univ Instelling, Antwerp, Belg, 77 & 78. *Mem:* Am Soc Parasitol; Entom Soc Am; Wildlife Dis Asn; Acarological Soc Am; Soc Vector Ecologists; French Acarological Soc; Am Soc Trop Med & Hyg. *Res:* Acarology; medical entomology; taxonomy; life histories and host-parasite relations of parasite mites and ticks; epidemiology of Lyme disease and honeybee mites. *Mailing Add:* Dept Zool Univ RI Kingston RI 02881-0816

HYLANDER, DAVID PETER, b Chicago, Ill, Nov 2, 24; m 53; c 5. ORGANIC CHEMISTRY. *Educ:* NCent Col Ill, BA, 49; Univ Ill, MS, 50. *Prof Exp:* From asst res chemist to assoc res chemist, Parke Davis & Co, 50-68, res chemist, 68-76; scientist, Warner-Lambert Co, 76-86; RETIRED. *Mem:* Am Chem Soc. *Res:* Synthetic organic, peptide and heterocyclic chemistry; bench scale to pilot plant development of synthe. *Mailing Add:* 219 Country Club Rd Holland MI 49423

HYLANDER, WILLIAM LEROY, b Chicago, Ill, Mar 5, 38; m 72. PRIMATOLOGY, DENTAL RESEARCH. *Educ:* Univ Ill, BS, 61, DDS, 63; Univ Chicago, AM, 69, PhD(anthrop), 72. *Prof Exp:* Dent intern, Vet Admin Hosp, Ann Arbor, Mich, 63-64; pvt pract dent, Plymouth, 64-66; asst anat, 71-72, asst prof anat & anthrop, 72-74, ASSOC PROF ANAT & ANTHROP, DUKE UNIV, 74-; ASST PROF ORTHOD, UNIV NC, 75- *Concurrent Pos:* Duke Univ Med Ctr res award, 75; NIH Res Career Develop Award, 76-81, NIH grant, 76-80; NSF grant, 76-79; assoc ed, J Phys Anthrop, 76-80. *Mem:* Int Asn Dent Res; Am Asn Phys Anthropologists; Am Soc Zoologists; Sigma Xi. *Res:* Analyzing the biomechanics of the primate craniofacial region utilizing histological, biometrical, radiographical, electromyographical and strain gage techniques. *Mailing Add:* Dept Anat Duke Univ Durham NC 27706

HYLIN, JOHN WALTER, b Brooklyn, NY, Jan 28, 29; m 54; c 3. PLANT BIOCHEMISTRY, PESTICIDE SCIENCE. *Educ:* Marietta Col, AB, 50; Purdue Univ, MS, 53; Columbia Univ, PhD(bot), 57. *Prof Exp:* Asst, Delafield Hosp & Columbia Univ, 53-54; res assoc biochem, Univ Tenn, 57-58, instr, 58-59; from asst prof to prof, 59-85, EMER PROF AGR BIOCHEM, UNIV HAWAII, 85- *Concurrent Pos:* Ed-in-chief, Bull Environ Contamination & Toxicol, 65-75, assoc ed, 75-; Fulbright res fel, Denmark, 72-73; chmn dept, Univ Hawaii, 85-; IAEA expert, Korea, 87, Algeria, 90. *Mem:* AAAS; Am Chem Soc; Sigma Xi; Am Inst Chem. *Res:* Pesticide metabolism; biosynthesis and metabolism of alkaloids and naturally occurring sulfur compounds; phytochemistry. *Mailing Add:* PO Box 6323 Incline Village NV 89450-6323

HYLTON, ALVIN ROY, b Los Angeles, Calif, Feb 19, 24. ENVIRONMENTAL SCIENCES. *Educ:* Iowa State Univ, BS, 50; Johns Hopkins Univ, ScD, 65; dipl, US Army Command & Gen Staff Col, 65. *Prof Exp:* Res entomologist, US Army, Ft Detrick, Md, 62-64, asst dir biol res, 64-65, res & develop coordr, Army Res Off, 65-66, chem officer, First Infantry Div, Vietnam, 66-67, staff officer, Pentagon, Washington, DC, 68-69; sr biologist, Midwest Res Inst, 69-71, sr environ biologist, 71-72, head ecol assessments, 72-75, dir Denver opers, 75-76, dir prog develop, 77-84; TEACHER ENVIRON SCI, LAS VEGAS COMMUNITY COL, SOUTHERN NEV, UNIV NEV, LAS VEGAS, 88- *Concurrent Pos:* Rep for Chief, Res & Develop US Army to Armed Forces Pest Cent Bd, 65-66; chmn, Mid Continent Res & Develop Coun, 82-83; dir, Sales & Mgt Execs, 80-82. *Honors & Awards:* Enterprise Award, Midwest Res Inst Coun Prin Scientists, 74. *Mem:* Entom Soc Am; Am Inst Biol Sci; Sigma Xi; Am Forestry Asn; Ecol Soc Am. *Res:* Insect pathology and physiology, particularly aging, fecundity and longevity in disease vectors; chemical and biological research and development programs; use and effects of herbicides; environmental studies and impact statements. *Mailing Add:* 501 Kennedy Dr Las Vegas NV 89110

HYMAN, ABRAHAM, b Brooklyn, NY, Mar 8, 24; m 55; c 3. ELECTRONICS, ELECTRICAL ENGINEERING. *Educ:* Polytech Inst Brooklyn, BEE, 45; Newark Col Eng, MSEE, 54. *Prof Exp:* Develop engr, Int Tel & Tel Lab, 45-48, supvry elec engr, Fed Aviation Agency, 48-55; chief elec eng br, Med Equip Res & Develop Lab, 55-64, head electronics lab, USN Training Device Ctr, 64-66; tech adminstr electronics, US AEC, Brookhaven Nat Lab, 66-71; supvry indust hygienist, Occup Safety & Health Admin, US Dept Labor, 71-81, regional indust hygienist, 81-84; SAFETY & HEALTH MGR, UNISYS CORP, 84- *Concurrent Pos:* Consult, JFD Electronics Corp, 52-53, Sterling Electronic Corp, 53-55, Dakon Corp, 57-59, Int Electronics Corp, 59-60, Taffett Electronics Corp, 62-63 & Poison Control Ctr, Nassau County Med Ctr, NY; lectr, dept elec sci, State Univ NY, Stony Brook, adj asst prof, dept allied health professions; adj prof, York Col, City Univ New York, 74-78 & Staten Island Col, City Univ New York, 83-; mem, bd dirs, Am Lung Asn; environ consult, Staten Island Col, City Univ. *Mem:* Inst Elec & Electronics Engrs; Sigma Xi; dipl Am Acad Environ Engrs. *Res:* Bio-medical engineering instrumentation, automation, transducers; radar, sonar and scoring trainers; signal anaylisis and synthesis; environmental and industrial hygiene; noise reduction; measurement of toxic substances; ventilation and engineering improvement; nine US patents. *Mailing Add:* 142 Claudy Lane New Hyde Park NY 11040

HYMAN, ARTHUR BERNARD, b London, Eng, Aug 22, 05; US citizen; m 30; c 2. MEDICINE. *Educ:* Univ London, MB, BS, 28; Am Bd Dermat, dipl, 45. *Prof Exp:* PROF CLIN DERMAT, SCH MED & IN-CHARGE TEACHING DERMATOPATH, POSTGRAD MED SCH, NY UNIV, 59- *Concurrent Pos:* Dermatopathologist, Beth Israel Hosp, 56-; chief serv skin & cancer unit & attend dermatologist, Univ Hosp, 58-; attend dermatologist, Vet Admin Hosp, 58- *Mem:* AMA; Am Acad Dermat. *Res:* Dermatology; dermatopathology. *Mailing Add:* 205 W 86th St New York NY 10024

HYMAN, BRADLEY CLARK, b San Diego, Calif, May 14, 52; m 75; c 2. MOLECULAR EVOLUTION. *Educ:* Univ Calif, San Diego, BA, 74; Univ Calif, Los Angeles, PhD(biol), 80. *Prof Exp:* NIH fel, lab molecular biol, Univ Wis Madison, 80-83; ASST PROF BIOL, DEPT BIOL, UNIV CALIF, RIVERSIDE, 83- *Mem:* Am Soc Cell Biol. *Res:* Molecular genetics of yeast nuclear and mitochondrial DNA replication; mitochondrial genome diversity in non-free-living, pathogenic nematodes. *Mailing Add:* Dept Biol Univ Calif 900 Univ Ave Riverside CA 92521

HYMAN, EDWARD SIDNEY, b New Orleans, La, Jan 22, 25; m 56; c 4. INTERNAL MEDICINE, PHYSIOLOGY. *Educ:* La State Univ, BS, 44; Johns Hopkins Univ, MD, 46. *Prof Exp:* Intern internal med, Barnes Hosp, Wash Univ, 46-47; fel med, Sch Med, Stanford Univ, 49-50, resident, Univ Hosps, 50-51; resident, Peter Bent Brigham Hosp, Harvard Univ, 51-53; clin instr, La State Univ, 53-55; INVESTR, TOURO RES INST, 58- *Concurrent Pos:* Teaching fel, Harvard Med Sch, 52-53; dir artificial kidney, Charity Hosp, 53-55; NIH & Am Heart Asn res grants, 55-80. *Mem:* Am Physiol Soc; Biophys Soc; Am Soc Artificial Internal Organs; fel Am Col Physicians; Am Fedn Clin Res; AAAS; Am Soc Microbiol. *Res:* Renal physiology; salt metabolism; physical chemistry of electrolytes; artificial kidney and heart; salt retaining hormone; delivery of medical care; bacteriuria. *Mailing Add:* Touro Res Inst New Orleans LA 70115

HYMAN, HOWARD ALLAN, physics, for more information see previous edition

HYMAN, JAMES MACKLIN, b Lakeland, Fla, Mar 20, 50; m 72; c 2. NUMERICAL ANALYSIS, MATHEMATICAL PHYSICS. *Educ:* Tulane Univ, BS, 72; Courant Inst Math Sci, NY Univ, MS, 73, PhD(math), 76. *Prof Exp:* Res staff mem math, Courant Inst Math Sci, 76-77; staff mem numerical anal, 77-83, assoc chmn, Ctr Nonlinear Studies, 83-84, GROUP LEADER APPL MATH, LOS ALAMOS NAT LAB, 85- *Mem:* Soc Indust & Appl Math. *Res:* Numerical methods and software for the approximate solution of partial differential equations. *Mailing Add:* Los Alamos Sci Lab Mail Stop B284 Los Alamos NM 87545

HYMAN, LLOYD GEORGE, b New York, Mar 18, 28; m 58; c 4. EXPERIMENTAL HIGH ENERGY PHYSICS. *Educ:* Mass Inst Technol, BS, 53, PhD(physics), 59. *Prof Exp:* Res fel, Harvard Univ, 59-61; from asst scientist to assoc scientist, 61-71, SR SCIENTIST, ARGONNE NAT LAB, 71- *Concurrent Pos:* Vis prof, Univ Tel-Aviv, Ramat-Aviv, Israel, 73-74. *Mem:* Fel Am Phys Soc; Fedn Am Sci. *Res:* High energy physics; neutrino interactions at high energies; Interactions in colliding beams; superconducting magnet design, construction, instrumentation and operation. *Mailing Add:* 303 N Ashland Ave LaGrange Park IL 60525

HYMAN, MELVIN, b US, June 20, 27. SPEECH PATHOLOGY. *Educ:* Brooklyn Col, BA, 49; Ohio State Univ, MA, 50, PhD(voice sci), 53. *Prof Exp:* EMER PROF PATH & AUDIOL, BOWLING GREEN STATE UNIV, 77-, CO-DIR, HYMAN SPEECH, LANG & HEARING CTR, 85- *Concurrent Pos:* Consult, New Voice Club Northwest Ohio, 54-; chmn bd trustees, Northwest Ohio Heart Asn, 77-; mem, Comt Stroke Rehab, Am Heart Asn, 62-; mem, Diag Team, Toledo Cleft Palate Ctr, 55-; mem bd trustees, Am Cancer Soc. *Mem:* Fel Am Speech & Hearing Asn. *Res:* Voice problems, particularly those concerning contact ulcers, nodules, largyngectomy, cerebral palsy, cleft palate, aphasia, language retardation, stuttering and other areas of speech pathology and related psychology. *Mailing Add:* Hyman Speech Lang & Hearing Ctr 5950 Airport Hwy Suite 17 Toledo OH 43615

HYMAN, RICHARD W, b San Francisco, Calif, Oct 22, 41. MICROBIOLOGY, BIOCHEMISTRY. *Educ:* Univ Calif, Berkeley, BS, 62; Cornell Univ, MS, 64; Calif Inst Technol, PhD(chem), 70. *Prof Exp:* Res assoc radiol, Med Sch, Yale Univ, 70-73; from asst prof to assoc prof, 73-82, PROF MICROBIOL, PA STATE UNIV, 82- *Mem:* Biophys Soc; Am Soc Biochem & Molecular Biol; Am Soc Microbiol; Am Soc Virol; Protein Soc. *Res:* Herpesviruses; electron microscopy. *Mailing Add:* Dept ICDB Syntex Res 3401 Hillview Ave Palo Alto CA 94304

HYMAN, SEYMOUR C, b New York, NY, June 3, 19; m 45; c 2. CHEMICAL ENGINEERING. *Educ:* City Col New York, BChE, 39; Va Polytech Inst, MS, 40; Columbia Univ, PhD(chem eng), 50. *Hon Degrees:* LLD, William Paterson Col, 85. *Prof Exp:* Chem engr, Ashland Oil Ref Co, 40-42 & US War Dept, NJ, 42- 47; from asst prof to prof & assoc dean sch technol, City Univ New York, 47-66, vchancellor campus planning & develop, 66-69, dep chancellor, 69-77, actg chancellor, 71; pres, William Paterson Col NJ, 77-85; RETIRED. *Mem:* Am Soc Mech Engrs; Am Inst Chem Engrs. *Res:* Heat transfer; fluid flow. *Mailing Add:* 6904 W Country Club Dr Sarasota FL 34243-3501

HYMANS, WILLIAM E, b Detroit, Mich, Mar 30, 40; m 66; c 2. ORGANIC CHEMISTRY. *Educ:* Cornell Univ, BA, 62; Ohio State Univ, PhD(org chem), 67. *Prof Exp:* Res chemist, Chem Res & Develop Ctr, FMC Corp, 67-78, regist specialist, 78-81, mgr tech serv, 81-85, prod mgr, 85-87, REGIST MGR, FMC CORP, 87- *Res:* Synthetic organic chemistry. *Mailing Add:* FMC Corp PO Box 8 Princeton NJ 08543

HYMER, WESLEY C, b Ironwood, Mich, July 22, 35; m 61; c 4. ENDOCRINOLOGY, CELL BIOLOGY. *Educ:* Univ Wis, BA, 57, MS, 59, PhD(zool, biochem), 62. *Prof Exp:* Res fel, Lab Biochem, Nat Cancer Inst, 62-64, staff fel, 64-65; asst prof zool, 65-68, assoc prof biol, 68-73, PROF BIOCHEM & BIOPHYS, PA STATE UNIV, 73- *Concurrent Pos:* NIH res career develop award, 69-74. *Res:* Hypothalamic regulation of the anterior pituitary; cytophysiology of separated adenohypophysial cells. *Mailing Add:* Dept Biochem & Biophys 401 Althouse Lab Pa State Univ University Park PA 16802

HYMOWITZ, THEODORE, b New York, NY, Feb 16, 34; m; c 3. GENETICS, PLANT BREEDING. *Educ:* Cornell Univ, BS, 55; Univ Ariz, MS, 57; Okla State Univ, PhD(plant breeding, genetics), 63. *Prof Exp:* Fulbright scholar, Indian Agr Res Inst, New Delhi, 62-63; agronomist, IRI Res Inst, Campinas, Brazil, 64-66; from asst prof to assoc prof, 67-76, PROF PLANT GENETICS, UNIV ILL, URBANA, 76- *Concurrent Pos:* Partic, AID-Univ Ill coord soybean res prog, G B Pant Univ Agr & Technol, India, 67 & Nat Acad Sci-Romanian Acad Fac Exchange Prog, 69; Food & Agr Orgn consult soybean prod, Yugoslavia & Hungary, 73; consult biotechnol, Indonesia, 91. *Honors & Awards:* Frank N Meyer Medal, 88; Funk Award, 90. *Mem:* Fel AAAS; fel Am Soc Agron; Genetics Soc Am; Soc Econ Botanists; fel Crop Sci Soc Am; fel Linnean Soc London. *Res:* Plant introduction of chemurgic and forage crops; genetics and origin of Cyamopsis tetragonoloba; biosystematics and genetics of Glycine; origin of cultivated plants. *Mailing Add:* Dept Agron Univ Ill 1102 S Goodwin Urbana IL 61801

HYNDMAN, ARNOLD GENE, b Los Angeles, Calif, Oct 16, 52; c 4. NEUROBIOLOGY. *Educ:* Princeton Univ, AB, 74; Univ Calif, Los Angeles, PhD(biol), 78. *Prof Exp:* Teacher biol, Upward Bound Prog, Univ Calif, Los Angeles, 75-77, lectr, 78; fel, Ohio State Univ, 78-79; res fel, Univ Calif, San Diego, 79-81; asst prof, 81-90, ASSOC PROF BIOL SCI, RUTGERS UNIV, 90- *Concurrent Pos:* Dir minority adv prog teaching & res, Rutgers Univ, 83-90, assoc provost, 90- *Mem:* Soc Neurosci; Asn Res Vision & Ophthal; Sigma Xi. *Res:* Development of visual cells; intact retinal and purified monolayer cultures are used in the analysis of retinal cells; analysis includes the study of cellmorphology, survival, transmitter development and cellular metabolism. *Mailing Add:* Dept Biol Sci Rutgers Univ Piscataway NJ 08855

HYNDMAN, DONALD WILLIAM, b Vancouver, BC, Apr 15, 36; m 60; c 2. PETROLOGY, TECTONICS. *Educ:* Univ BC, BASc, 59; Univ Calif, Berkeley, PhD(geol), 64. *Prof Exp:* From asst prof to assoc prof, 64-72, chmn, 75-77, PROF GEOL, UNIV MONT, 72- *Concurrent Pos:* Prin investr, NSF, 67-72, 78-80 & 83-86 & Nat Park Serv, 79-82; vis prof, Stanford, 77-78. *Mem:* Fel Geol Soc Am; fel Mineral Soc Am; Mineral Asn Can; fel Geol Asn Can; Sigma Xi; Am Geol Inst. *Res:* Igneous and metamorphic petrology; petrography; geochemistry; tectonics of Northern Rockies; granites of Idaho batholith; high-grade regional metamorphism and partial melting; alkaline igneous rocks; books on petrology and roadside geology. *Mailing Add:* Dept Geol Univ Mont Missoula MT 59812

HYNDMAN, HARRY LESTER, b Springfield, Ill, Sept 23, 40; m 65; c 3. ORGANIC & ANALYTICAL CHEMISTRY. *Educ:* Univ Ill, Urbana, BS, 62; Calif Inst Technol, PhD(chem), 68. *Prof Exp:* Sr res chemist, 68-75, res specialist, 75-77, sr res specialist, 77-81, res group leader, 81-82, SR RES GROUP LEADER, MONSANTO CO, 83- *Concurrent Pos:* Secy, Regulatory Rev Comt, Soc Qual Assurance, 89-90, chmn, 90- *Mem:* Am Chem Soc; Chemometrics Soc; Soc Qual Assurance. *Res:* Agriculture chemical residue analysis; computer applications in chemistry; quality assurance. *Mailing Add:* 9310 Old Bonhomme Rd Olivette St Louis MO 63132

HYNDMAN, JOHN ROBERT, b Kendallville, Ind, July 3, 20; m 46; c 5. PHYSICAL CHEMISTRY. *Educ:* Ind Univ, AB, 41; Univ Wis, PhD(chem), 50. *Prof Exp:* Asst phys chem, Univ Wis, 47-49; chemist, Redstone Arsenal Res Div, Rohm and Haas Co, 50-52, group leader, 52-59, sr scientist, Bristol Res Lab, 59-62; chemist, Goodyear Tire & Rubber Co, 62-65, sect head, Res Div, 65-78, prin engr, 78-84; RETIRED. *Mem:* Am Chem Soc; Soc Plastics Eng. *Res:* Polymer properties and processing. *Mailing Add:* 1952 State Rte 44 RD 1 Atwater OH 44201-9373

HYNDMAN, ROY D, b Vancouver, BC, Feb 20, 40; m 69. GEOPHYSICS, OCEANOGRAPHY. *Educ:* Univ BC, BASc, 62, MASc, 64; Australian Nat Univ, PhD(geophys), 67. *Prof Exp:* Asst prof, 67-71; vis assoc prof, Univ BC, 71-72; assoc prof, Dalhousie Univ, 72-74; vis prof, Inst Phys Glove, Paris, 74-75; RES SCIENTIST, EARTH PHYSICS BR, DEPT ENERGY MINES RESOURCES, PAC GEOSCI CENTRE, 75- *Concurrent Pos:* Mem subcomts geomagnetism, seismol & physics of earth's interior, Nat Res Coun Can Assoc Comt & Int Heat Flow Comt; assoc ed, J Geophys Res & Marine Geophys Res; chmn, Joint Oceanogr Inst Deep Earth Sampling Downhole Measurement Panel. *Mem:* Can Geophys Union; Geol Asn Can; Am Geophys Union. *Res:* Continental and oceanic heat flow marine seismology; temperatures in the earth; magnetotellurics; general oceanic geophyiscal measurements; physical properties of rocks; marine geophysics. *Mailing Add:* 8320 Alec Saanichton BC V0S 1M0 Can

HYNE, JAMES BISSETT, b Dundee, Scotland, Nov 23, 29; m 58. CHEMISTRY. *Educ:* St Andrews Col, BSc, 50, PhD(chem), 54. *Prof Exp:* Fel chem kinetics, Nat Res Coun Can, 54-56; instr phys org chem, Yale Univ, 56-59; asst prof, Dartmouth Col, 59-60; assoc prof, 60-63, First Head Chem, 63-66, Dean Fac Grad Studies, 66-89, PROF PHYS ORG CHEM, UNIV CALGARY, 65- *Concurrent Pos:* Res dir, Alta Sulphur Res, Ltd, 64-; consult, Oil, Gas & Sulphur, 78- *Honors & Awards:* R S Jane Mem lectr, 77; Queen Elizabeth II Jubilee Medal, 77. *Mem:* Am Chem Soc; The Chem Soc; fel Chem Inst Can; Can Res Mgt Asn. *Res:* Fundamental and applied sulphur research; study of the chemical reaction of water with components of heavy oils during steam stimulated enhanced recovery. *Mailing Add:* Dept Chem Univ Calgary Calgary AB T2N 1N4 Can

HYNE, NORMAN JOHN, b Berwyn, Ill, Nov 17, 39; c 2. PETROLEUM GEOLOGY. *Educ:* Pomona Col, BA, 61; Fla State Univ, MS, 65; Univ Southern Calif, PhD(marine geol), 69. *Prof Exp:* From asst prof to assoc prof earth sci, 69-79, head dept, 72-79, PROF PETROL GEOL, UNIV TULSA, 79-; PRES, NJH ENERGY, 85- *Mem:* Geol Soc Am; Soc Econ Paleontologists & Mineralogists; Am Asn Petrol Geol; Soc Petrol Engrs. *Res:* Sedimentation; petroleum geology. *Mailing Add:* Continuing Educ Univ Tulsa Tulsa OK 74104

HYNEK, ROBERT JAMES, b Phillips, Wis, Sept 7, 27; m 52; c 5. ANALYTICAL CHEMISTRY. *Educ:* Carroll Col, Wis, BS, 51; Marquette Univ, MS, 59. *Prof Exp:* Chemist, Allis-Chalmers Mfg Co, Milwaukee, 51-52, group leader, sect leader & res chemist, Metals & Ultra High Purity Gases, 52-70; supvr pollution control lab, Autorol Corp, Milwaukee, 70-72, mgr pilot plant prog, 72-76, mgr process develop, 76-78, mgr process verification, 78-82; SUPVR PROCESS ENG, ENVIREX, INC, 82- *Mem:* Am Chem Soc; Water Pollution Control Fedn. *Res:* Water and waste analysis; gas chromatography; elemental analysis. *Mailing Add:* 1120 Parkmoor Dr Brookfield WI 53005

HYNES, HUGH BERNARD NOEL, b Devizes, Eng, Dec 20, 17; Can citizen; m 42; c 4. FRESHWATER ECOLOGY. *Educ:* Univ London, BSc & ARCS, 38, PhD(biol), 41, DSc, 58. *Hon Degrees:* DSc, Univ Waterloo, 87. *Prof Exp:* Field asst entom, Brit Ministry Agr, 41; entomologist locust control, Brit Colonial Agr Serv, Kenya, Ethiopia & Somalia, 42-46; lectr zool, Univ Liverpool, 47-58, sr lectr, 58-64; vis prof, Ind Univ, 62-63, Monash Univ, Australia, 71-72; prof, 64-83, EMER PROF BIOL, UNIV WATERLOO, 83- *Concurrent Pos:* Sabbatical leave, Monash Univ, Australia, 71-72 & Univ Tasmania & Adelaide Univ, Australia, 78-79; vis prof, Addis Ababa Univ, Ethiopia, 73-74, Univ Louisville, 85. *Honors & Awards:* Can Centenniel Medal, 67; Hilary Jolly Award, Australian Limnol Soc, 84. *Mem:* Royal Soc Can; Freshwater Biol Asn; Australian Limnol Soc; Int Soc Limnol; Can Soc Zool; NAm Benthol Soc. *Res:* Ecology of the desert locust and aquatic invertebrates, especially running water and pollution; Plecoptera; Amphipoda. *Mailing Add:* Dept Biol Univ Waterloo Waterloo ON N2L 3G1 Can

HYNES, JOHN BARRY, b Orange, NJ, Oct 24, 36; m 60; c 2. ORGANIC CHEMISTRY. *Educ:* Colgate Univ, AB, 58; Duke Univ, PhD(chem), 61. *Prof Exp:* Res assoc chem, Duke Univ, 61-62; pres, Hynes Chem Res Corp, 62-68; assoc prof, 68-72, prof pharmaceut chem, 72-80, PROF PHARMACEUT CHEM & ASSOC PROF BIOCHEM, COL PHARM, MED UNIV SC, 80- *Mem:* Am Chem Soc; Am Soc Biol Chemists. *Res:* Synthesis of potential antagonists of folic acid metabolism; synthesis and evaluation of new potential chemotherapeutic agents, especially those targeted for the treatment of malaria, leprosy, fungal infections and cancer. *Mailing Add:* Dept Pharm Sci Med Univ SC 171 Ashley Ave Charleston SC 29425

HYNES, JOHN EDWARD, b New Orleans, La, Sept 25, 40; m 67; c 2. PHYSICS. *Educ:* La State Univ, BS, 62; Fla State Univ, MS, 66, PhD(solid state physics), 69. *Prof Exp:* Res assoc solid state physics, Fla State Univ, 68-69; sr physicist magnetic physics, Pitney Bowes, 69-77; vpres res & develop, EMI Data-Malco Plastics Inc, 77-87, vpres & gen mgr, Malco Security Magnetics, Owings Mill, 87-; CONSULT. *Res:* Magnetic tape; magnetic materials; security systems encoding; code analysis; date entry devices. *Mailing Add:* 6417 Dear Park Rd Reisterstown MD 21136

HYNES, JOHN THOMAS, b Brooklyn, NY, Sept 6, 33; m 57; c 7. FOOD BIOCHEMISTRY. *Educ:* Manhattan Col, BS, 56. *Prof Exp:* Chemist, Kraft Inc, 56-59, res chemist, 59-68, sr chemist, 68-77, group leader, 77-85, sr res scientist, 81-84, sr group leader, 85-89, RES PRIN, RES CTR, KRAFT USA, 89- *Mem:* AAAS; Dairy Sci Asn; Inst Food Technol. *Res:* Chemistry of edible proteins; food enzyme control and utilization; effect of heat on protein denaturation; cheese research and development. *Mailing Add:* 840 Rolling Pass Glenview IL 60025

HYNES, MARTIN DENNIS, III, b Albany, NY, Dec 23, 49; m 82; c 2. PHARMACOLOGY, PSYCHOPHARMACOLOGY. *Educ:* Providence Col, BA, 72; Univ RI, MS, 75, PhD(pharmacol), 78. *Prof Exp:* Asst pharmacol, Univ RI, 72-77; fel pharmacol, Roche Inst Molecular Biol, 77-79; sr pharmacologist, 79-84, head cent nervous syst & endocrine res, 84-86, mgr pharmaceut proj, 86-87, DIR CLIN RES, ELI LILLY, JAPAN, KK, 87- *Mem:* AAAS; Sigma Xi; Soc Neurosci; NY Acad Sci; Am Soc Pharmacol & Exp Therapeut. *Res:* Major interest in the area of the central nervous system; endocrine and clinical research. *Mailing Add:* Eli Lilly & Co Lilly Corp Ctr Indianapolis IN 46285

HYNES, RICHARD OLDING, b Nairobi, Kenya, Nov 29, 44; Brit citizen; m 66; c 2. CELL BIOLOGY, DEVELOPMENTAL BIOLOGY. *Educ:* Univ Cambridge, BA, 66, MA, 70; Mass Inst Technol, PhD(biol), 71. *Prof Exp:* Res fel, Imperial Cancer Res Fund, London, 71-74; from asst prof to assoc prof, 75-83, assoc head, 85-89, PROF BIOL, MASS INST TECHNOL, 83-, HEAD, 89- *Concurrent Pos:* Mem, Cell Biol Study Sect, NIH, 78-82; assoc ed, Cell, 78-, Develop Biol, 78-85, BBA Rev Cancer, 80-86 & J Cell Biol, 84-85, Develop, 87-90, Proc Roy Soc B, 90-; hon res fel, Univ Col, London, 82-83; Howard Hughes Med Inst investr, 88-; Guggenheim fel, 82-83. *Honors & Awards:* Harvey lectr, 86. *Mem:* fel AAAS; Am Soc Cell Biol; Soc Develop Biol; Royal Soc. *Res:* Molecular basis of cell adhesion and migration; structure-function relations. *Mailing Add:* Ctr Cancer Res Mass Inst Technol E17-227 Cambridge MA 02139

HYNES, THOMAS VINCENT, b New Haven, Conn, Aug 11, 38; m 68; c 3. SOLID STATE PHYSICS, MATERIALS SCIENCE. *Educ:* St Joseph's Col, BS, 59; St Louis Univ, PhD(physics), 68. *Prof Exp:* Physicist, E I du Pont de Nemours & Co, 60; physicist & Nat Acad Sci fel physics, US Army Mat & Mech Res Ctr, 68-70; Orgn Am States vis prof, Cath Univ Rio de Janeiro, 70-71, assoc prof, 71-73; Nat Acad Sci res fel, US Army, 73-74, staff physicist, Mat & Mech Res Ctr, 74-84, phys sci adminr, Mat Command, 85, chief, Ceramics Res Div, 85-90, dir, Technol Integration & Mgt Div, Mat Technol Lab, 86-90, MGR OPTICAL & ELECTRO OPTICAL MAT, MAT LAB, US ARMY, 90- *Mem:* Am Phys Soc; AAAS; Sigma Xi; Am Ceramics Soc. *Res:* Electromagnetic properties of structural materials and structures; response of materials to ultra-short pulses of electromagnetic radiation; quantum mechanical properties of systems with nondiagonal density matrices; structure of glasses. *Mailing Add:* PO Box 126 Waverly MA 02179

HYSLOP, NEWTON EVERETT, JR, b Newton, Mass, Oct 14, 35; m 57; c 2. INFECTIOUS DISEASES, CLINICAL IMMUNOLOGY. *Educ:* Harvard Col, AB, 57; Harvard Med Sch, MD, 61. *Prof Exp:* Intern & resident med, Mass Gen Hosp, 61-63; res assoc immunol, Lab Immunol, Nat Inst Allergy & Infectious Dis, NIH, 63-65; sr resident, Peter Bent Brigham Hosp, 65-66; clin & res fel infectious dis, Mass Gen Hosp, 66-68; vis scientist immunochem, Oxford Univ, UK, 68-69; instr, 69-71, asst prof med, Harvard Med Sch & Mass Gen Hosp, 71-85; PROF MED & CHIEF, INFECTIOUS DIS SECT, TULANE UNIV MED CTR, 84- *Concurrent Pos:* Investr, Howard Hughes Med Inst, 71-75; prin investr, Tulane-LSU AIDS Treatment Eval/Clin Trials Unit, 87- *Mem:* Am Asn Immunologists; fel Am Col Physicians; Am Soc Microbiol; fel Infectious Dis Soc Am. *Res:* Clinical aspects and chemotherapy of AIDS; immunochemistry and pathophysiology of hypersensitivity reactions to beta-lactam antibiotics (penicillins and cephalosporins) and other antimicrobials; factors regulating normal and abnormal chemotactic responses of human leukocytes. *Mailing Add:* Infectious Dis Sect Tulane Univ Med Ctr 1430 Tulane Ave New Orleans LA 70112

HYSLOP, PAUL A, b London, Eng, July 12, 52. BIOCHEMISTRY. *Educ:* Univ South Hampton, London, BMS, 76, PhD(physiol & biochem), 81. *Prof Exp:* Researcher, Scripps Clin & Res Found, La Jolla, Calif, 82-88; AT LILLY RES LABS, ELI LILLY & CO, 88- *Mem:* Am Soc Biochem & Molecular Biol. *Mailing Add:* Lilly Res Labs Eli Lilly & Co MC907 28-1 Indianapolis IN 46285

HYSON, ARCHIBALD MILLER, chemistry; deceased, see previous edition for last biography

HYUN, KUN SUP, b Seoul, Korea, Feb 25, 37; m 63; c 3. CHEMICAL ENGINEERING. *Educ:* Seoul Nat Univ, BS, 59; Univ Mo-Columbia, MS, 62, PhD(chem eng), 66. *Prof Exp:* Res asst chem eng, Univ Mo-Columbia, 61-65; chem engr, 66, res engr, 66-71, res specialist II, Styrene Molding Polymers Res & Develop, 71-76, res specialist II, 76-79, res assoc, 79-84, ASSOC SCIENTIST, SARAN & CONVERTED PROD, RES DOW CHEM USA, 84- *Concurrent Pos:* Abstr, Chem Abstract Serv, 67-76. *Mem:* Am Inst Chem Engrs; Am Chem Soc; Sigma Xi; Soc Rheology; Soc Plastics Engrs. *Res:* Polymer processing; melt rheology; characterization of polymers; emulsion polymerization; latex stability; emulsion finishing. *Mailing Add:* 613 Nakoma Dr Midland MI 48640

HYZER, WILLIAM GORDON, b Janesville, Wis, Mar 25, 25; m 49; c 3. CLOSE-RANGE PHOTOGRAMMETRY, HUMAN VISIBILITY. *Educ:* Univ Minn, BEE, 46; Univ Wis, BS, 48. *Prof Exp:* Chief physicist, Parker Pen Co, 48-53; CONSULT, 53- *Concurrent Pos:* Contrib ed, Photomethods mag, 56-; lectr, Univ Wis, 60-; columnist, Res & Develop mag, 70-78 & Optical Eng J, 76-79; US nat deleg, Int Cong High Speed Photog & Photonics, 78-86; vis prof, Xian Inst Optics & Precision mech, People's Repub China, 79. *Honors & Awards:* DuPont Gold Medal, Soc Motion Picture & TV Engrs, 69;

Coleman Award, Brit Inst Physics, 80; Hon Master Photog, Prof Photogr Am, 81. *Mem:* Soc Motion Pictures & TV Engrs (vpres, 66-69); fel Soc Photo Optical Instrumentation Engrs; fel Am Acad Forensic Sci; Illum Eng Soc; Prof Photogr Am; NAm Photonics Asn (pres, 80-82). *Res:* Developed and publicized techniques of high speed photography and videography as they apply to scientific and industrial research; developed photographic techniques for forensic applications. *Mailing Add:* 136 S Garfield Ave Janesville WI 53545

I

I, TING-PO, b Yunnan, China, Feb 20, 41; m 69; c 2. PHYSICAL CHEMISTRY, ANALYTICAL CHEMISTRY. *Educ:* Cheng-Kung Univ, Taiwan, BSc, 65; State Univ NY Buffalo, PhD(phys chem), 72. *Prof Exp:* Res assoc phys org chem, Brandeis Univ, 72-74; sr res chemist, 74-79, sr res investr, 79-83, PRIN RES INVESTR, FERMENTATION PROCESS RES & DEVELOP, CENT RES, PFIZER INC, 83- *Mem:* Am Chem Soc; Am Inst Chemists; NY Acad Sci. *Res:* Aqueous and non-aqueous solution chemistry; purification and separation science, including solvent extraction, chromatography, crystal growth, membrane separation and supercritical fluid extraction; high performance liquid chromatography and thin layer chromatography; analytical process automation; bio-engineering produced compound recovery; continuous flow analysis; protein and peptide chemistry; biotechnology down stream processing; protein engineering. *Mailing Add:* Fermentation Process Res & Develop Pfizer Inc Groton CT 06340

IACOBUCCI, GUILLERMO ARTURO, b Buenos Aires, Arg, May 11, 27; m 52; c 2. BIO-ORGANIC CHEMISTRY. *Educ:* Univ Buenos Aires, MChS, 50, PhD(chem), 52. *Prof Exp:* Res chemist org chem, Res Labs, E R Squibb & Sons, Arg, 52-57; Guggenheim fel, Harvard Univ, 58-59; prof phytochem, Fac Pharm & Biochem, Univ Buenos Aires, 60-61; sr res chemist org chem, Squibb Inst Med Res, 62-67; sr res scientist, Fundamental Res Dept, Coca-Cola Co, 67-68 & Corp Res Dept, 68-70, head, Biochem Sect, 70-74, asst dir, Corp Res & Develop Dept, 74-87, MGR, BIOCHEM & BASIC ORG CHEM GROUP, COCA-COLA CO, 88- *Concurrent Pos:* Adj prof chem, Emory Univ, 75- *Mem:* AAAS; Am Chem Soc; Am Soc Pharmacog; Sigma Xi; Int Union Pure & Appl Chem; Am Inst Chemists. *Res:* Chemistry of natural products. *Mailing Add:* Corp Res & Develop Dept Coca-Cola Co PO Drawer 1734 Atlanta GA 30301

IACOCCA, LEE A, b Allentown, Pa, Oct 15, 24. SCIENCE ADMINISTRATION. *Educ:* Lehigh Univ, BS, 45; Princeton Univ, MSME, 46. *Hon Degrees:* Var from Univs. *Prof Exp:* From mgt trainee to Pres & chief oper officer, Ford Motor Co, 46-78; PRES & CHIEF OPERATING OFFICER, CHRYSLER CORP, 78-, CHMN BD DIRS & CHIEF EXEC OFFICER CHRYSLER MOTORS CORP, CHRYSLER FINANCIAL CORP & CHRYSLER TECHNOL, 79- *Concurrent Pos:* Chmn comt, Corp Support Joslin Diabetes Found; founder, Iacocca Inst Am Enterprise, Lehigh Univ; chmn, Iacocca Found; co-chmn, Gov Mich Comn Jobs & Econ Develop; emer chmn, Statue Liberty-Ellis Island Found. *Mem:* Nat Acad Eng; Soc Automotive Engr. *Res:* Advancement of diabetes research. *Mailing Add:* Chrysler Corp 1200 Chrysler Dr Highland Park MI 48288

IACONO, JAMES M, b Chicago, Ill, Dec 11, 25; m 50; c 4. BIOCHEMISTRY, NUTRITION. *Educ:* Loyola Univ, Chicago, Ill, BS, 50; Univ Ill, Urbana, MS, 52, PhD(nutrit biochem), 54. *Prof Exp:* Res asst agr, Univ Ill, 50-53; asst chief, Physiol Div, US Army Med Nutrit Lab, Fitzsimmons Army Hosp, Denver, Colo, 54-57; from asst prof to assoc prof exp med & biol chem, Col Med, Univ Cincinnati, 58-70; chief, Lipid Nutrit Lab, Nutrit Inst, 70-75, dep asst, nat prog staff, 75-77, assoc adminr, Off Human Nutrit, 78-82, S&E, USDA, DIR WESTERN HUMAN NUTRIT RES CTR, AGR RES SERV, USDA, 82- *Concurrent Pos:* Vis prof, Inst Pharmacol, Med Sch, Univ Milan, Italy, 66-67; mem, Coun Arteriosclerosis & Coun Thrombosis, Am Heart Asn. *Mem:* AAAS; Am Chem Soc; Am Inst Nutrit; Am Soc Clin Nutrit; Am Oil Chemists Soc; fel Am Heart Asn; fel Am Inst Chemists. *Res:* Nutritional biochemistry; nutrient requirements; role of dietary fats in cardiovascular disease. *Mailing Add:* 480-1 Point Pacific Dr Daly City CA 94014

IADECOLA, COSTANTINO, b Aquino, Italy, Mar 15, 53; m 87; c 1. NEUROSCIENCES. *Educ:* Univ Rome, MD & PhD(physiol), 77. *Prof Exp:* Instr physiol, Univ Rome, 78-79; instr neurol, Cornell Univ Med Col, 82-83, asst prof neurobiol, 84-86; neurol residency, 86-90; ASST PROF NEUROL, UNIV MINN, 90- *Honors & Awards:* McHenry Award, Hist Neurol, 90. *Mem:* Am Physiol Soc; Neurosci Soc; Am Acad Neurol; Soc Cerebral Blood Flow. *Res:* Regulation of blood circulation to the brain with respect to intrinsic neural pathways and neuro transmitters, in health and in the disease state. *Mailing Add:* Dept Neurol Univ Minn Box 295 420 Delaware St SE Minneapolis MN 55455

IAFRATE, GERALD JOSEPH, b Brooklyn, NY, Apr 8, 41; c 1. PHYSICS OF NANOELECTRONICS, QUANTUM TRANSPORT. *Educ:* Long Island Univ, BS, 63; Fordham Univ, MS, 65; Polytech Inst Brooklyn, PhD(physics), 70. *Prof Exp:* Physicist, Inst Explor Res, 65-68; res phys scientist, Electronics Technol & Devices Lab, 68-85, prin scientist, 85-89; DIR, US ARMY RES OFF, 89- *Concurrent Pos:* Adj prof, Monmouth Col, 65-82, Ocean County Col, 77-84, adj prof physics, Georgian Ct Col, 80-90; vis prof elec eng, NC State Univ, 86-, Physics Dept, Duke Univ, 90- *Mem:* Sr mem Inst Elec & Electronics Engrs; fel Am Phys Soc; Am Asn Physics Teachers. *Res:* Physics of small dimensions; conceiving, planning and conducting avante-garde theoretical research in areas of ultra-small, ultra-fast device physics leading to the realization of ultra-high information processing systems for applications scenarios. *Mailing Add:* US Army Res Off PO Box 12211 Research Triangle Park NC 27709-2211

IAMMARINO, RICHARD MICHAEL, b Cleveland, Ohio, Aug 17, 26; m 52; c 4. PATHOLOGY. *Educ:* John Carroll Univ, BS, 49; Loyola Univ Chicago, MD, 53. *Prof Exp:* Fel, Med Ctr, Univ Kans, 56-58; fel, Western Reserve Univ, 58-59; dir labs, St Alexis Hosp, Cleveland, Ohio, 59-63; assoc prof, Univ Pittsburgh, 63-79; PROF PATH, WVA UNIV, 79- *Concurrent Pos:* Sigma Xi fel, Univ Pittsburgh, 63-64. *Mem:* Am Asn Clin Chemists; Col Am Path; Acad Clin Lab Physicians & Scientists. *Res:* Clinical laboratory medicine, particularly plasma protein disorders, lipoproteins, acute phase protein reactants and alpha 1-antitrypsin; electrophoresis methodology. *Mailing Add:* Clin Labs Rm 2288 WVa Univ Morgantown WV 26506

IAMPIETRO, P(ATSY) F, b Middleboro, Mass, Jan 5, 25; m 54; c 3. PHYSIOLOGY. *Educ:* Univ Mass, BS, 49, MA, 51; Univ Rochester, PhD(physiol), 54. *Prof Exp:* Chief environ physiol sect, Qm Res & Eng Ctr, 54-60; chief physiol lab, Civil Aeromed Inst, Fed Aviation Admin, 60-73; dir life sci, 73-82, dir, Air Force Off Sci Res, Far East, Tokyo, 83-85; RETIRED. *Concurrent Pos:* Adj prof zool & prof res physiol, Univ Okla, 62-73; consult, Nat Inst Environ Health Sci, 67-71; mem adv group aerospace res & develop, NATO, 75-82. *Honors & Awards:* Hitchcock Award, Aerospace Med Asn, 74. *Mem:* Fel Aerospace Med Asn; Am Phys Soc; Sigma Xi; Am Polar Soc; Soc Exp Biol & Med. *Res:* Temperature regulation; acclimatization to heat and cold; tolerance to extreme environments; water and electrolytes; drugs. *Mailing Add:* 3803 Barrington Dr No 18-A San Antonio TX 78217-4101

IANDOLO, JOHN JOSEPH, b Chicago, Ill, Sept 26, 38; m 60; c 3. BACTERIAL PHYSIOLOGY, FOOD MICROBIOLOGY. *Educ:* Loyola Univ Ill, BS, 61; Univ Ill, Urbana, MS, 63, PhD(microbiol), 65. *Prof Exp:* Assoc prof biol, 67-80, prof path, 90, PROF BIOL, KANS STATE UNIV, 80- *Concurrent Pos:* Mem consult bact & mycol study sect, NIH, 75-79; ed, Appl & Environ Microbiol. *Mem:* Am Soc Microbiol; Am Acad Microbiol. *Res:* Genetics and biosynthetic regulation of the staphylococcal enterotoxins. *Mailing Add:* Dept Path VCS Bldg Kans State Univ Manhattan KS 66502

IANNARONE, MICHAEL, bacteriology; deceased, see previous edition for last biography

IANNICELLI, JOSEPH, b New York, NY, Aug 5, 29; m 58; c 3. ORGANIC CHEMISTRY, NATURAL RESOURCES. *Educ:* Mass Inst Technol, SB, 51, PhD(org chem), 55. *Prof Exp:* Chemist, Explosives Dept, E I du Pont de Nemours & Co, 51, res chemist, Dacros Res Lab, Textile Fibers Dept, 55-56, chemist, Tech Lab, Org Chem Dept, 56, Carothers Res Lab, 57-58, Pioneering Res Lab, Textile Fibers Dept, 58-60; res dir, J M Huber Corp, 60-68, asst tech dir, 68-69, tech dir, 69-71; PRES & BD CHMN, AQUAFINE CORP, 71- *Concurrent Pos:* Consult & co-investr, NSF Proj Magnetic Beneficiation, 74-; consult, Proj Magnetic Filtration, Environ Protection Res Inst, 80. *Mem:* Am Inst Mining, Metall & Petrol Engr; fel Am Inst Chemists; Tech Asn Pulp & Paper Indust; Clay Minerals Soc; Am Chem Soc. *Res:* Synthetic penicillin analogs and intermediates; synthesis of fiber forming polymers; polyesters; irradiation chemistry; pyrolysis of organic compounds; mineral beneficiation; pigment modification; mining equipment; water purification systems; spray and flash dryers; high intensity magnetic separation of industrial minerals, metallic ores and coal. *Mailing Add:* Aquafine Corp 157 Darien Hwy Brunswick GA 31520

IANUZZO, C DAVID, Concord, NH, Oct 15, 38; c 4. MUSCLE PHYSIOLOGY & BIOCHEMISTRY, EXERCISE PHYSIOLOGY. *Educ:* Springfield Col, BS, 66; Wash State Univ, MS, 68, PhD(exercise physiol), 71. *Prof Exp:* Teaching asst, Wash State Univ, 66-67, res asst electron micros, 67-68, res asst exercise physiol, 68-71; asst prof biol & health sci, Boston Univ, 71-75; assoc prof, 75-80, PROF BIOL & PHYS EDUC, YORK UNIV, 81-; ADJ PROF, DEPT SURG, UNIV TORONTO, 87- *Concurrent Pos:* Res grants, Atkinson Charitable Found, 78-79, Health & Welfare, Can, 78-79 & 84-85, Nat Sci & Eng Res Coun, 78-85 & Ont & Can Heart Found, 84-85; consult, Civil Aviation Med, Health & Welfare, Can, 78-85; founder & pres, Bio-Clin Res Assays Ltd, 88- *Mem:* Can Soc Cell Biol; Am Physiol Soc; fel Am Col Sports Med; Sigma Xi. *Res:* Biochemical characteristics and adaptability of skeletal and cardiac muscle cells. *Mailing Add:* Farquharson Life Sci Bldg York Univ 4700 Keele St Downsview ON M3J 1P3 Can

IATROPOULOS, MICHAEL JOHN, b Athens, Greece, Nov 8, 38; US citizen; m 66; c 2. COMPARATIVE & EXPERIMENTAL PATHOLOGY. *Educ:* Univ Tubingen, MD, 64, DrMed, 65. *Prof Exp:* Res assoc cytol, Brown Univ, 66-69, instr path, 69-71; resident path, Univ Mo, Columbia, 71-72; spec fel, Albany Med Col, 72-74, from asst prof to assoc prof toxicol, 74-78; head, Exp Path Dept, Med Res Div, Am Cyanamid Corp, 78-89; HEAD REGULATORY PATH, AM HEALTH FOUND, 89- *Concurrent Pos:* Adj prof, NMex State Univ, 75-78 & Old Dominion Univ, Va, 81-; assoc ed, J Toxicol Path. *Mem:* Sigma Xi; Am Inst Biol Sci; Soc Toxicol Pathologists; Soc Toxicol; NY Acad Sci; Int Soc Xenobiotics. *Res:* Tumor induction potential and safety evaluation of drugs, vaccines, medical devices, cosmetics, food additives, agricultural and chemical products; toxicology. *Mailing Add:* Am Health Found One Dana Rd Valhalla NY 10595

IATROU, KOSTAS, b Athens, Greece, June 10, 46. EUKARYOTIC GENE REGULATION, RNA METABOLISM. *Educ:* Univ Thessaloniki, Greece, BSc, 70; Univ Calgary, Alta, Can, PhD(med sci), 77. *Prof Exp:* Res fel molecular biol, dept cellular & develop biol, Harvard Univ, 77-80, res assoc, 80-81; from asst prof to assoc prof, 81-91, PROF MOLECULAR BIOL, DEPT MED BIOCHEM, UNIV CALGARY, 91- *Concurrent Pos:* Med Res Coun Can Scholar, 81-86, scientist, 86-91; Alberta Heritage Found Med Res Scholar, 81-91; chmn, Molecular Develop Biol Res Group, Fac Med, Univ Calgary, 89-; mem bd dirs, Insect Biotech, Can, 90- *Mem:* Soc Develop Biol; Am Soc Cell Biol; Int Soc Develop Biologists; NY Acad Sci; Int Sericult Comn. *Res:* Regulation of gene expression during oogenesis; evolution of eukaryotic multigene families; molecular biology of insect viruses. *Mailing Add:* Dept Med Biochem Fac Med Univ Calgary 3330 Hospital Dr NW Calgary AB T2N 4N1 Can

IBANEZ, MANUEL LUIS, b Worcester, Mass, Sept 23, 35; m 70; c 4. BIOCHEMISTRY, MICROBIOLOGY. *Educ:* Wilmington Col, BS, 57; Pa State Univ, MS, 59, PhD(microbiol), 61. *Prof Exp:* Asst prof biol, Bucknell Univ, 61-62; res fel biochem, Sch Med, Univ Calif, Los Angeles, 62; sr biochemist plant tech, Interam Inst Agr Sci, Orgn Am States, 62-65; assoc prof biol, 65-77, chmn dept, 65-70, health sci coordr, 71-76, assoc dean grad sch, 78-82, assoc vchancellor acad affairs, 82-83, actg vchancellor acad affairs, 83-85, PROF BIOL, UNIV NEW ORLEANS, 77-, VCHANCELLOR ACAD AFFAIRS & PROVOST, 85- *Concurrent Pos:* NSF coop fel, 59-60. *Mem:* Am Soc Microbiol. *Res:* Bacterial photosynthesis; cacao seed physiology; phospholipids in chromatophore of Rhodospirillum rubrum. *Mailing Add:* Dept Biol Sci Univ New Orleans Lake Front New Orleans LA 70148

IBANEZ, MICHAEL LOUIS, b Havana, Cuba, Nov 9, 16; nat US; m 44; c 3. PATHOLOGY. *Educ:* Col Havana, Cuba, BA, 37; Havana Med Sch, MD, 48; Am Bd Path, cert path anat. *Prof Exp:* Intern, McKennan Hosp, 49-50; resident path, Hermann Hosp, 50; resident, Robert B Green Mem Hosp, 50-53; resident, 53-56, asst res pathologist, 56-59, asst pathologist, 59-64, assoc pathologist, Cancer Ctr, 64-74, assoc prof path, 64-75, pathologist, Cancer Ctr, Univ Tex M D Anderson Hosp & Tumor Inst, 74-, prof path, 75-; RETIRED. *Concurrent Pos:* Instr exp path, Univ Tex Postgrad Sch Med Houston, 57-59, asst prof path, 59-64; mem fac, Div Continuing Educ, Univ Tex Grad Sch Biomed Sci Houston, 64- *Mem:* Am Asn Path & Bact; Int Acad Path; Col Am Pathologists; Am Soc Clin Path; Am Thyroid Asn. *Res:* Pathologic anatomy of cancer; pathology of the thyroid gland. *Mailing Add:* 7510 Hopewell Houston TX 77071

IBELE, WARREN EDWARD, b New Orleans, La, Aug 17, 24; m 47; c 4. MECHANICAL ENGINEERING. *Educ:* Tulane Univ, BS, 44; Univ Minn, MS, 47, PhD(mech eng), 53. *Prof Exp:* Asst prof mech eng, 53-56, assoc prof mech eng & thermodyn, 56-59, assoc dean, 67-75, dean, Grad Sch, 75-83, PROF MECH ENG, UNIV MINN, MINNEAPOLIS, 59- *Concurrent Pos:* Field eng, Babcock & Wilcox Co, 47, Pratt & Whitney Aircraft Div, United Aircraft Corp, 57-58; eng educ & accreditation comt, Engrs Coun Prof Develop; chmn adv comt transit, Twin Cities Area Metrop Transit Comn. *Mem:* Fel Am Soc Mech Engrs; Am Soc Eng Educ; Sigma Xi. *Res:* Thermodynamics; power; fluid mechanics; heat transfer; transport properties. *Mailing Add:* Dept Mech Eng Univ Minn Church St SE Minneapolis MN 55455

IBEN, ICKO, JR, b Champaign, Ill, June 27, 31; m 56; c 4. ASTROPHYSICS. *Educ:* Harvard Univ, AB, 53; Univ Ill, MS, 54, PhD(physics), 58. *Prof Exp:* Asst prof, Williams Col, 58-61; sr res fel, Calif Inst Technol, 61-64; from assoc prof to prof, Mass Inst Technol, 64-72; head dept astron, 72-84, prof, 72-89, DISTINGUISHED PROF ASTRON & PHYSICS, UNIV ILL, URBANA CHAMPAIGN, 90- *Concurrent Pos:* Vis prof, Harvard Univ, 66, 68 & 70; vis fel joint inst lab astrophys, Univ Colo, Boulder, 71-72; Eberly Family Chair Astron & Physics, Pa State Univ, State Col, 89-90; vis prof physics & astron & mem, Inst Astron, Univ Hawaii, Manoa, 77; mem adv panel, Astron Sect of NSF, 72-74; univ scholar, Univ Ill, 85-88; vis comt, Nat Astron Optical Observ, 79-82; Telescope Allocation Comt, Cerro Tololo Inter Am Observ, 82-88; vis prof, Dept Earth Sci Astron, Col Arts & Sci, Univ Tokyo, 84; vis scientist, Astron Coun, USSR Acad Scis, 85; vis prof, Astron Dept, Univ Bologna, Italy, 86-; Guggenheim fel, 85-86, US & USSR exchange scholar, 85, sr vis fel Australian Nat Univ, 86, Univ Sussex, Eng ,86. *Honors & Awards:* George Darwin Lectr, Royal Astron Soc, 84; Henry Norris Russell Lectr, Am Astron Soc, 89; Eddington Medal, Royal Astron Soc, 90. *Mem:* Nat Acad Sci; Int Astron Union; Am Astron Soc; fel Japan Soc Prom Sci. *Res:* Stellar structure; evolution; pulsation; binary star evolution. *Mailing Add:* Dept Astron Univ Ill Urbana IL 61801

IBER, FRANK LYNN, b Eaton, Ohio, Oct 28, 28; m 53; c 3. MEDICINE. *Educ:* Miami Univ, BA & MA, 49; Johns Hopkins Univ, MD, 53. *Prof Exp:* Intern med, Johns Hopkins Hosp, 53-54, asst resident, 56-58, from instr to assoc prof, Johns Hopkins Univ, 58-68; prof, Tufts Univ, 68-73; PROF MED, SCH MED, UNIV MD, BALTIMORE, 73-; CHIEF, GASTROENTEROL DIV, HINES VET ADMIN HOSP, LOYOLA UNIV, 86- *Concurrent Pos:* Markle scholar; asst & USPHS fel, Postgrad Med Sch, Univ London, 58-59, resident, 59-60; consult, Liver Res Unit, Univ Calcutta; chief med, Lemuel Shattuck Hosp, 68-73; chief gastroenterol div, Baltimore Vet Admin Hosp & Univ Md Hosp, 73-85. *Mem:* Am Gastroenterol Soc; Am Asn Study Liver Dis; Am Physiol Soc; Am Soc Clin Nutrit. *Res:* Liver disease; protein disorders; biochemistry of human disease; alcoholism; human nutrition; aging nutrition. *Mailing Add:* Vet Admin Med Ctr Hines IL 60141

IBERALL, ARTHUR SAUL, b New York, NY, June 12, 18; m 40; c 4. PHYSICS. *Educ:* City Col New York, BS, 40. *Prof Exp:* Gen physicist instrumentation & measurement, Nat Bur Standards, 41-53; res dir instruments & aircraft accessories, Aro Equip Corp, 53-54; chief physicist res & develop, Rand Develop Corp, 54-64; pres, Gen Tech Serv, Inc, 64-81; VIS SCHOLAR, UNIV CALIF, LOS ANGELES, 81- *Concurrent Pos:* Consult, 53- *Mem:* Am Soc Mech Eng; NY Acad Sci; Am Phys Soc; Biomed Eng Soc. *Res:* Hydrodynamics; elasticity; molecular physics; biophysics; hydrology; control theory; instrumentation; physics of complex systems including behavior and social systems; physics of space; scientific systems analysis, physical, biological and social. *Mailing Add:* 4675 Willis Ave Apt 106 Sherman Oaks CA 91403

IBERS, JAMES ARTHUR, b Los Angeles, Calif, June 9, 30; m 51; c 2. INORGANIC CHEMISTRY. *Educ:* Calif Inst Technol, BS, 51, PhD(chem), 54. *Prof Exp:* NSF fel, Commonwealth Sci & Indust Res Orgn, Australia, 54-55; chemist, Shell Develop Co, 55-61 & Brookhaven Nat Lab, 61-64; PROF CHEM, NORTHWESTERN UNIV, 65-, MORRISON PROF, 86- *Honors & Awards:* Am Chem Soc Award in Inorg Chem, 78. *Mem:* Nat Acad Sci; Am Chem Soc. *Res:* Coordination, solid state and bionorganic chemistry. *Mailing Add:* Dept Chem Northwestern Univ Evanston IL 60208

IBRAHIM, A MAHAMMAD, b Veppanapalli, India, Apr 12, 53; US citizen; m 82; c 2. MATERIALS ENGINEERING, POLYMERS & COMPOSITES. *Educ:* Univ Madras, BSc, 74, MSc, 76; Indian Inst Technol, PhD(polymer chem), 81. *Prof Exp:* Postdoctoral polymer eng & sci, Univ Wash, 81-82, res assoc, Polymeric Composites Lab, 82-85; scientist, 85-88, SR SCIENTIST & GROUP LEADER, MARTIN MARIETTA LABS, 88- *Concurrent Pos:* Mem, gov bd, Int Thermal Expansion Forum, 86-; prin investr electronic packaging, Martin Marietta Labs, 88-90; lectr, Stevens Inst Technol, 88. *Mem:* Soc Advan Mat & Processing Eng; Int Electronic Packaging Soc; Mat Res Soc; Am Chem Soc; Soc Plastics Engrs. *Res:* Processing-structure-property relationships of polymeric materials and advanced composites, thermoset and thermoplastic; composites for electronic, space aerospace and underwater applications; rigid-rod polymers and molecular composites; electronic packaging technologies; surface mount technology; multichip module and chip-on-board. *Mailing Add:* 8001 Mayfield Ave Baltimore MD 21227

IBRAHIM, ADLY N, b Egypt, Jan 4, 17; m 43; c 2. MICROBIOLOGY. *Educ:* Cairo Univ, DVM, 39, dipl bact, 60; Univ Miami, MS, 64; Univ Pittsburgh, DSc(virol), 67. *Prof Exp:* Vet, 40-51; med rep, Parke, Davis & Co, Egypt, 51-61; teaching asst microbiol, Univ Miami, 61-64; fel virol, Grad Sch Pub Health, Univ Pittsburgh, 64-68; res virologist, Gulf South Res Inst, 68-70; assoc prof, 70-73, PROF MICROBIOL, GA STATE UNIV, 73- *Mem:* Am Soc Microbiol; Am Soc Trop Med & Hyg; Tissue Cult Asn. *Res:* Degradation of hyaluronic acid by certain strains of Hemophilus influenzae; application of immunodiffusion methods to the antigenic analysis of dengue viruses; tumor immunology with emphasis on immunodiagnosis of human cancer. *Mailing Add:* 946 Bridgegate Dr NE Marietta GA 30068

IBRAHIM, BAKY BADIE, b Assiut, Egypt, Sept 18, 47; US citizen; m 78; c 2. FLUID MECHANICS, HEAT TRANSFER. *Educ:* Assiut Univ, BS, 68; Kans State Univ, MS, 71 & PhD(mech eng), 75. *Prof Exp:* Instr mech eng, Helwan Inst Technol, 68-69 & Kans State Univ, 71-75; asst prof, Bradley Univ, 75-76; sr engr, 76-86, FEL ENGR, WESTINGHOUSE ELEC CORP, 86- *Mem:* Am Soc Mech Engrs. *Res:* Propose, design and conduct tests in the fluid mechanics area simulating different nuclear reactor components; fluid flow, fluid-solid interaction and fluid transients and instabilities. *Mailing Add:* 11496 Drop Rd North Huntingdon PA 15642

IBRAHIM, MEDHAT AHMED HELMY, b Alexandria, Egypt, Apr 25, 39; m 69. POWER SYSTEMS, SYSTEMS ENGINEERING. *Educ:* Cairo Univ, BSE, 61; Univ Mich, MSE, 65, MA, 68, PhD(elec eng), 69. *Prof Exp:* Instr elec eng, Cairo Univ, 61-63; asst res eng, Radiation Lab & asst lectr, Univ Mich, 65-69; lectr math, Eastern Mich Univ, 70; asst prof elec eng, Mich Technol Univ, 70-74; vis asst prof, Purdue Univ, 74-76; sr engr, Bechtel Power Corp, 76-80; PROF ELEC ENG, CALIF STATE UNIV, FRESNO, 80- *Concurrent Pos:* Vis scholar, Univ Mich, 78-79. *Mem:* Sr mem Inst Elec & Electronics Engrs; Am Soc Eng Educ. *Res:* Systems; control; computer applications. *Mailing Add:* Dept Elec Eng Calif State Univ Fresno CA 93612

IBRAHIM, MICHEL A, b Egypt, Jan 28, 34; US citizen; m 62; c 4. EPIDEMIOLOGY. *Educ:* Cairo Univ, MD, 57; Univ NC, MPH, 61, PhD(epidemiol), 64. *Prof Exp:* Hosp dir, Walaga Hosp, Egypt, 59-60; univ fel, Univ NC, 61-62, USPHS fel, 62-64; from asst to assoc prof med, State Univ NY Buffalo, 65-71; assoc prof, 71-73, chmn epidemiol, 75-82, PROF EPIDEMIOL, SCH PUB HEALTH, UNIV NC, CHAPEL HILL, 73-, DEAN, SCH PUB HEALTH, 82- *Concurrent Pos:* Dep comnr health, Erie County Dept Health, Buffalo, 68-71; chmn ed bd, Am J Pub Health, 74-; consult, Nat Inst Neurol & Commun Dis & Stroke, 77-; assoc ed, Am J Epidemiol. *Mem:* Fel Am Pub Health Asn; fel Am Heart Asn; Am Epidemiol Soc; Fel Am Col Epidemiol. *Res:* Epidemiology of cardiovascular and cerebrovascular diseases; group psychotherapy and prognosis of coronary heart disease; community-based intervention studies for hypertension; health service planning and evaluation. *Mailing Add:* 201H Sch Pub Health Univ NC Chapel Hill NC 27514

IBRAHIM, RAGAI KAMEL, b Cairo, Egypt, Aug 3, 29; Can citizen; m 62. PLANT BIOCHEMISTRY, PLANT TISSUE CULTURE. *Educ:* Cairo Univ, BSc, 49; Univ Alexandria, MSc, 58; McGill Univ, PhD(plant biochem), 61. *Prof Exp:* Demonstr bot, Univ Alexandria, 50-58, asst lectr, 58-61, lectr, 61-65, asst prof, 65-66; prof assoc biol, Macdonald Col, McGill Univ, 66-67; from asst prof to assoc prof biol, Sir George Williams Univ, 67-77; PROF BIOL, CONCORDIA UNIV, 77- *Concurrent Pos:* Nat Res Coun Can oper grant, 67-; deleg, Int Conf Plant Tissue Cult, Calgary, Can, 78; Govt Que res grant, 71-78. *Mem:* Can Soc Plant Physiol; Am Soc Plant Physiol; Phytochem Soc NAm; Int Asn Plant Tissue Cult. *Res:* Enzymology of flavonoid compounds in intact plants and in tissue cultures; isolation, purification, property studies and subcellular localization of the enzymes involved in polymethylated flavone synthesis. *Mailing Add:* Dept Biol Sci Concordia Univ 1455 Demaisonneuve Blvd W Montreal PQ H3G 1M8 Can

IBSEN, KENNETH HOWARD, b New York, NY, Feb 4, 31; m 58; c 2. BIOCHEMISTRY. *Educ:* Univ Calif, Los Angeles, BS, 54, PhD(physiol chem), 59. *Prof Exp:* Res biochemist, Sepulveda Vet Admin Hosp, Calif, 59-61; asst res physiol chemist, Univ Calif, Los Angeles, 61-64; asst prof, 63-71, ASSOC PROF BIOCHEM, SCH BIOL SCI & CALIF COL MED, UNIV CALIF, IRVINE, 71-, ASST DEAN, 88- *Mem:* AAAS; Am Soc Biol Chem; Am Chem Soc. *Res:* Regulation of energy metabolism; properties of regulatory enzymes. *Mailing Add:* Dept Biol Chem Univ Calif Col Med Irvine CA 92717

IBSER, HOMER WESLEY, b Ft Worth, Tex, Sept 17, 20. PHYSICS. *Educ:* Nebr Wesleyan Univ, AB, 41; Univ Wis, MS, 48, PhD, 54. *Prof Exp:* Jr physicist, Manhattan Proj, 42-46; res assoc physics, Univ Wis, 54-55; asst prof, Northwestern State Col La, 55-58; assoc prof, Lincoln Univ Mo, 58-59; from asst prof to assoc prof, 59-68, PROF PHYSICS, CALIF STATE UNIV, SACRAMENTO, 68- *Mem:* AAAS; Am Asn Physics Teachers. *Res:* Thermodynamics; nuclear physics; nuclear energy controversy. *Mailing Add:* Dept Physics Calif State Univ Sacramento CA 95819

ICE, RODNEY D, b Ft Lewis, Wash, Apr 24, 37; m 58; c 3. PHARMACY, NUCLEAR MEDICINE. *Educ:* Univ Wash, BS, 59; Purdue Univ, MS, 65, PhD(bionucleonics), 67; Am Bd Health Physics, cert health physicist, 72. *Prof Exp:* Asst prof radiochem & radiation safety officer, Temple Univ, 67-69, assoc prof radiochem, 69-70; from assoc prof to prof pharm, Univ Mich, Ann Arbor, 70-76, dir radiopharmaceut serv, 70-75; prof pharm & dean, Univ Okla, Oklahoma City, 76-83; VPRES, EAGLE-PICHER INDUSTS, 85- *Concurrent Pos:* Consult, Philadelphia Tech Inst, 68-69, Philadelphia Pub Health Dept, 69-74 & Vet Admin Hosps, 72-74; chmn, Govt Adv Comt Ionizing Radiation, 73-74; US Pharmacopeia Comt of Rev, 81-90. *Mem:* AAAS; Am Pharmaceut Asn; Am Sci Affil; Health Physics Soc; Soc Nuclear Med; USP. *Res:* Radiopharmaceutical development, dosimetry and design. *Mailing Add:* PO Box 3042 Edmond OK 73083

ICERMAN, LARRY, b Muncie, Ind, Sept 22, 45. MANAGEMENT CONSULTING, PUBLIC POLICY. *Educ:* Mass Inst Technol, BS, 67; Univ Calif, San Diego, MS, 68, PhD(eng sci), 76; San Diego State Univ, MBA, 76. *Prof Exp:* Asst prof technol & human affairs, Washington Univ, 76-79, assoc prof, 79-80; dir, NMex Energy Res & Develop Inst, Santa Fe, NMex State Univ, 80-89; PRES, ICERMAN & ASSOCS, 89- *Concurrent Pos:* Bd dirs, NMex Entrepreneurs Asn, 90-, Trade, 90-, RhoMed Inc, 90- *Honors & Awards:* Spec Recognition Award for Energy Innovation, US Dept Energy, 85, Award for Energy Innovation, 86, 88. *Mem:* Am Inst Aeronaut & Astronaut; AAAS; Int Solar Energy Soc; Int Asn Hydrogen Energy; Am Chem Soc; Am Mgt Asn. *Res:* Technology, science, and policy; geothermal, solar, and wind energy; environmental impact of energy use; energy conservation; energy analysis and economics. *Mailing Add:* 2999 Calle Cerrada Santa Fe NM 87505

ICHIDA, ALLAN A, b Seattle, Wash, Aug 26, 29; m 62; c 1. MYCOLOGY, BACTERIOLOGY. *Educ:* Ohio Wesleyan Univ, BA, 53; Univ Tenn, MS, 55; Univ Wis, PhD(bot), 60. *Prof Exp:* Res asst mycol, Univ Wis, 59-60, res assoc, 60-61; from asst prof to assoc prof, 61-71, PROF BOT & BACT & CHMN DEPT, OHIO WESLEYAN UNIV, 71- *Concurrent Pos:* Sci fac fel, NSF, 66-67; microbiologist, Agr Res Serv, USDA, 70-71. *Mem:* Am Soc Microbiol; Mycol Soc Am; Sigma Xi. *Res:* Ecology and taxonomy of aspergilli; ultrastructure of fungi. *Mailing Add:* Dept Bot & Bact Ohio Wesleyan Univ Delaware OH 43015

ICHIKAWA, SHUICHI, b Tokyo, Japan, Aug 9, 43; m 72; c 4. CHEMICAL & EXPERIMENTAL HYPERTENSION, CARDIOLOGY. *Educ:* Gunma Univ, MD, 69, PhD(hypertension), 78. *Prof Exp:* Postdoctoral fel hypertension, Univ Mo, 75-77; asst prof internal med, 2nd Dept Internal Med, Sch Med, Gunma Univ, 83-89; PRES, CARDIOVASC HOSP CENT JAPAN, 89- *Concurrent Pos:* Invited asst prof, 2nd Dept Internal Med, Sch Med, Gunma Univ, 89-, Col Med Care & Technol, 89- *Mem:* Am Physiol Soc; Soc Exp Biol & Med; Endocrine Soc. *Res:* Vascular reactivity and the involvement in the pathogenesis of hypertension; renin-angiotesin system. *Mailing Add:* Cardiovasc Hosp Cent Japan Hokkitsu-Mura Sete-Gun 377 Japan

ICHIKI, ALBERT TATSUO, b Lahaina, Hawaii, Sept 21, 36. IMMUNOLOGY. *Educ:* Purdue Univ, BS, 58, MS, 61; Univ Calif, Los Angeles, PhD(microbiol), 69. *Prof Exp:* Grad res microbiol, Univ Calif, Los Angeles, 66-69; fel immunol, John Curtin Sch Med Res, Australia, 69-71; res assoc, 71-72, res asst prof immunol, 72-78, asst prof, 78-80, ASSOC PROF MED BIOL, DEPT MED BIOL, UNIV TENN MEM RES CTR, 80- *Concurrent Pos:* NIH fel, USPHS, 69. *Mem:* Am Soc Hemat; AAAS; Am Asn Cancer Res; Sigma Xi; Int Soc Hematol. *Res:* Immunological evaluation of colorectal cancer and malignant melanoma patients; surface receptors on chronic myelogenous leukemia cells; isolation and characterization of colorectal tumor and malignant melanoma associated antigens. *Mailing Add:* Univ Tenn Mem Res Ctr 1924 Alcoa Hwy Knoxville TN 37920

ICHINOSE, HERBERT, b Koloa, Hawaii, July 25, 31; m 55; c 4. PATHOLOGY. *Educ:* Tulane Univ, BS, 53, MD, 57, dipl path, 62. *Prof Exp:* Rotating intern med path, Charity Hosp, New Orleans, 57-58; from instr to prof, Sch Med, Tulane Univ, 58-80, clin prof path, 80-85; MED DIR, DRAMATOPATH INC, 85- *Concurrent Pos:* Nat Cancer Inst trainee, 58-62; jr attend pathologist, Charity Hosp, New Orleans, 58-63, vis pathologist, 63-69, sr vis pathologist, 69-; NIH grant, 62-63; Am Cancer Soc advan clin fel, 62-65. *Honors & Awards:* Mellon Award, 64. *Res:* Study of neoplasia; chemical carcinogenesis; dermatopathology. *Mailing Add:* 10555 Lake Forest Blvd Suite 5-A New Orleans LA 70127

ICHIYE, TAKASHI, b Kobe, Japan, Oct 1, 21; m 52; c 2. PHYSICAL OCEANOGRAPHY. *Educ:* Univ Tokyo, BS, 44, DSc(geophys), 53. *Prof Exp:* Res assoc phys oceanog, Kobe Marine Observ, 45-55; assoc to chief oceanog sect, Japan Meteorol Agency, Tokyo, 55-57; Rockefeller Found fels & vis scientist, Woods Hole Oceanog Inst, 57-58; res assoc oceanog inst, Fla State Univ, 58-59, asst prof oceanog, 59-63; res scientist, Lamont Geol Observ, Columbia Univ, 63-64, sr res scientist, 65-68; PROF OCEANOG, TEX A&M UNIV, 68- *Concurrent Pos:* Chmn Emer, Japan & East China Seas Study Prog, UNESCO, 81-; over sea ed, La Mer, Tokyo, 83-; vis lectr, Acad Sinica, People's Repub China, 85, summer res fel, Am Soc Educ Eng, 86, res fel, Japan Soc Prom Sci, 87. *Mem:* Am Meteorol Soc; Am Geophys Union; Oceanog Soc Japan; French-Japanese Oceanog Soc; Am Soc Limnol & Oceanog; AAAS. *Res:* Dynamical studies of oceanic circulation; turbulence and diffusion in oceans; shallow water physical oceanography; oceanic long waves, pollutants and sediment transport. *Mailing Add:* Dept Oceanog Tex A&M Univ College Station TX 77843

ICHNIOWSKI, CASIMIR THADDEUS, b Baltimore, Md, Mar 4, 09; m 51; c 4. PHARMACOLOGY. *Educ:* Univ Md, BS, 30, MS, 32, PhD(pharmacol), 36. *Prof Exp:* Asst pharmacol, Univ Md, 30-36; asst toxicologist, Chem Warfare Serv, Edgewood Arsenal, 36-38; pharmacologist, Warner Inst Therapeut Res, NY, 38-46 & Wyeth, Inc, Pa, 46-51; asst dean, Sch Pharm, Univ Md, Baltimore City, 68-74, Emerson prof pharmacol, 51-74; RETIRED. *Mem:* AAAS; Am Pharmaceut Asn. *Res:* Bioassays; stability of pharmaceuticals; absorption; drug combinations. *Mailing Add:* 625 Woodbine Ave Baltimore MD 21204

ICHNIOWSKI, THADDEUS CASIMIR, b Baltimore, Md, June 1, 33; m 59; c 4. CHEMISTRY. *Educ:* Wash Col, BS, 55; Purdue Univ, MS, 60, PhD(chem), 62. *Prof Exp:* From asst prof to assoc prof, 61-66, PROF CHEM, ILL STATE UNIV, 66- & COORDR COOP EDUC, CHEM DEPT, 80- *Concurrent Pos:* Vis prof, Dow Chem Co, 80-81; dir, Univ Co-op Prog, Ill State Univ, 87. *Mem:* Am Chem Soc; Coop Educ Asn. *Res:* Synthetic inorganic chemistry; inorganic polymers; cooperative education in chemistry; water soluble polymers. *Mailing Add:* Dept Chem Ill State Univ Normal IL 61761-6901

ICKE, VINCENT, b Utrecht, Neth, July 23, 46; m 72; c 1. ASTROPHYSICS, HYDRODYNAMICS. *Educ:* State Univ Utrecht, BSc, 66, MSc, 69; Univ Leiden, PhD(astrophys), 72. *Prof Exp:* Res asst astrophys, State Univ Leiden, 69-72, Univ Sussex, 72-73, Cambridge Univ, 73-76 & Calif Inst Technol, 76-78; asst prof astrophys, Univ Minn, 78-83; AT DEPT ASTRON, UNIV LEYDEN, 83- *Concurrent Pos:* Vis lectr, Univ Milano, 75 & State Univ Amsterdam, 76. *Mem:* Dutch Astron Soc; Royal Astron Soc; Am Astron Soc; Int Astron Union; Europ Phys Soc. *Res:* Cosmic hydrodynamics; cosmology, galactic structure, galaxy formation; star formation; relativistic hydrodynamics. *Mailing Add:* Univ Leiden Postbus 9513 Leiden 2300 RA Netherlands

ICKES, WILLIAM K, b Salt Lake City, Utah, Feb 4, 26; m 46; c 4. SPEECH PATHOLOGY, AUDIOLOGY. *Educ:* Univ Utah, BS, 48, MS, 49; Southern Ill Univ, PhD(audiol), 60. *Prof Exp:* Audiologist, Detroit Hearing Ctr, 50-52 & Mich Asn Better Hearing, 52-54; exec dir, Des Moines Hearing & Speech Ctr, 54-62; dir hearing & speech clin, 62-69, chmn dept speech & theatre arts, 69-76, PROF SPEECH PATH & AUDIOL, TEX TECH UNIV, 64- *Mem:* Fel Am Speech & Hearing Asn; Nat Rehab Asn. *Res:* Establishing psychophysical thresholds, auditory; stuttering as a learned phenomenon; noise-induced hearing loss; industrial audiology. *Mailing Add:* Dept Speech Path & Audiol Tex Tech Univ Lubbock TX 79409-4266

IDDINGS, CARL KENNETH, b New York, NY, June 28, 33. THEORETICAL PHYSICS. *Educ:* Harvard Univ, AB, 55; Calif Inst Technol, PhD(physics), 60. *Prof Exp:* Res assoc theoret physics, Enrico Fermi Inst Nuclear Studies, Univ Chicago, 60-62; asst prof physics, Stanford Univ, 62-65; assoc prof, 65-70, PROF PHYSICS, UNIV COLO, BOULDER, 70- *Concurrent Pos:* Consult, Phys Res Lab, Space Tech Labs, 60. *Mem:* Am Phys Soc; Sigma Xi. *Res:* Elementary particle physics. *Mailing Add:* Dept Physics Campus Box 390 Univ Colo Boulder CO 80309-0390

IDDINGS, FRANK ALLEN, b Abilene, Kans, Jan 20, 33; m 54; c 4. NUCLEAR SCIENCE, NONDESTRUCTIVE TESTING. *Educ:* Midwestern Univ, BS, 54; Univ Okla, MS, 56, PhD(anal chem), 59. *Prof Exp:* Chemist anal develop, Sabine River Works, E I du Pont de Nemours & Co, 55-56; instr chem, Univ Okla, 56-59; chemist, Esso Res Labs, La, 59-64; from asst prof to assoc prof, 64-75, prof nuclear sci, 75-85, asst dir, Nuclear Sci Ctr, La State Univ, Baton Rouge, 83-85; DIR, NONDESTRUCTIVE RES & TESTING INFO ANAL CTR, SOUTHWEST RES INST, SAN ANTONIO, TEX, 85- *Concurrent Pos:* Tech expert, Int Atomic Energy Agency, 71 & 83. *Honors & Awards:* Bausch-Lomb Sci Award, 51; Tutorial Award, Am Soc Nondestructive Testing. *Mem:* Am Nuclear Soc; fel Am Soc Nondestructive Testing; Health Physics Soc; Sigma Xi; Am Welding Soc. *Res:* Industrial radioisotope applications; liquid scintillation spectrometry; neutron activation analysis; nondestructive testing; neutron radiography; environmental monitoring. *Mailing Add:* 1635 Rob Roy Lane San Antonio TX 78251

IDE, CARL HEINZ, b New York, NY, Nov 15, 28; m 62; c 2. OPHTHALMOLOGY. *Educ:* NY Univ, BA, 52; Univ Hamburg, MD, 59. *Prof Exp:* From instr to assoc prof, 64-72, actg chief dept, 66-67, PROF OPHTHAL, SCH MED, UNIV MO, COLUMBIA, 72- *Concurrent Pos:* Assoc ophthal, Mo Crippled Children's Serv, 65- *Mem:* AMA; Asn Res Vision & Ophthal; Am Col Surg; Ger Ophthal Soc. *Res:* Retinal surgery; clinical and experimental evaluation of grafts with preserved sclera; congenital abnormalities of the eye. *Mailing Add:* Univ Mo Inst Ophthal Columbia MO 65212

IDE, HIROYUKI, b Fukuoka, Japan, Mar 22, 33; m 56; c 2. DRUG DEVELOPMENT IN JAPAN, RESEARCH & SCIENCE ADMINISTRATION. *Educ:* Kyushu Univ, PhD(drug metab), 65. *Prof Exp:* Res fel oncol, Sch Med, Tufts Univ, 65-67, instr, 67-68; assoc prof hyg chem, Sch Pharmacol Sci, Fukuoka Univ, 68-71; dir drug develop, Hisamitsu Pharmaceut Co Ltd, 71-74, gen mgr, 74-80; pres, Contract Labs, Kyudo Co Ltd, 80-82; PRES CONTRACT LABS, PANAPHARM LABS CO LTD, 85- *Concurrent Pos:* Pres, Ide Off, 82- *Honors & Awards:* Prize Sci Technol, Japanese Govt, 77, 79 & 90. *Res:* Drug metabolism; pathology; toxicology. *Mailing Add:* Panapharm Lab Co Ltd Kumamoto 869-04 Japan

IDE, ROGER HENRY, b Port Clinton, Ohio, July 8, 37. NUCLEAR CHEMISTRY, RESEARCH ADMINISTRATION. *Educ:* Wabash Col, BA, 59; Univ Calif, Los Angeles, PhD(nuclear chem), 64. *Prof Exp:* Chemist, Lawrence Livermore Lab, Univ Calif, 64-69, asst div leader, 69-72, test group dir, 72-76, asst assoc dir, 76-80, dep assoc dir, 80-87; DEPT ENERGY TECH TEAM LEADER NUCLEAR TEST NEGOTIATIONS WITH USSR, 87- *Res:* Interaction of artificial radionuclides with the environment; high energy fission yields. *Mailing Add:* 4627 Almond Circle Livermore CA 94550

IDELL-WENGER, JANE ARLENE, b Halifax, Pa. BIOCHEMISTRY, PHYSIOLOGY. *Educ:* Elizabethtown Col, BS, 65; Univ Minn, MS, 68, PhD(biochem), 70. *Prof Exp:* NIH fel pharmacol, Mayo Clin & Grad Sch Med, 70; USPHS fel, Univ Mich, 71, NIH fel biochem, 72-73; ASST PROF PHYSIOL, COL MED, MILTON S HERSHEY MED CTR, PA STATE UNIV, 74-, ASST PROF MED, 81- *Mem:* Am Physiol Soc; Biophys Soc; Asn Women in Sci; Int Asn Women Bioscientists; Am Heart Asn. *Res:* Myocardial and renal fatty acid metabolism in normal and ischemic conditions. *Mailing Add:* 111 Esheman Rd Lancaster PA 17601

IDELSOHN, SERGIO RODOLFO, b Parana, Arg, Nov 15, 47; m 71; c 3. NUMERICAL METHODS. *Educ:* Tech Sch, Parana, Arg, Dipl Tech, 65; Univ Rosario, Arg, Dipl Eng, 70; Univ Liege, Belg, PhD(comp mech), 74. *Prof Exp:* Sr scientist comp mech, Univ Liege, Belg, 71-74; dir res, Regional Res Ctr, Santa Fe, 75-80; assoc prof, 75-80, PROF APPL MECH, UNIV ROSARIO, ARG, 80-; PROF COMP MECH, UNIV LITORAL, ARG, 89- *Concurrent Pos:* Prin res comp mech, Conicet, Arg, 81-; chmn, Comp Mech Lab, 81-; ed, Mecamica Computacional, 85; pres, Arg Asn Comp Mech, 85-; consult, several Arg Industs, 85-; vis prof, Inst Advan Study, Princeton, 87-88 & Univ Paris VI, France, 89-90. *Mem:* Int Asn Computational Mech; Am Soc Mech Eng. *Res:* Numerical methods in engineering; computational mechanics; fluid mechanics; aerodynamics; structural dynamics problems. *Mailing Add:* Guemes 3450 Santa Fe 3000 Argentina

IDELSON, MARTIN, b Staten Island, NY, Aug 23, 28; m 57; c 2. ORGANIC CHEMISTRY, PHOTOGRAPHIC CHEMISTRY. *Educ:* Polytech Inst Brooklyn, BS, 52, PhD, 55. *Prof Exp:* With Children's Med Ctr, Boston, 54-57; scientist, Polaroid Corp, 57-65, sr scientist, 65-66, res group leader, 66-68, sr res group leader, 68-69, asst mgr, 69-70, res assoc, 70-75, res fel, Res Div, 75-85, dir dye chem, 80-85; DEPT MGR, SHIPLEY CO, INC, 85- *Concurrent Pos:* Instr, Northeastern Univ, 56-64. *Mem:* AAAS; Am Chem Soc; fel Soc Photog Scientists & Engrs. *Res:* Photographic chemistry; polymers; dyes. *Mailing Add:* 1603 Commonwealth Ave W Newton Boston MA 02165

IDEN, CHARLES R, b NJ, 42. ANALYTICAL CHEMISTRY. *Educ:* Lafayette Col, BA, 64; Johns Hopkins Univ, PhD(phys chem), 71. *Prof Exp:* Fel, Johns Hopkins Univ, 71-72; res asst, Brookhaven Nat Lab, 72-74; instr, 77-80, RES ASSOC PHARMACOL, STATE UNIV NY, STONY BROOK, 74-, ASST PROF, 80- *Concurrent Pos:* Consult, Brookhaven Nat Lab, 77-81 & County Suffolk, NY, 79-; vis scientist, Brookhaven Nat Lab, 82- *Mem:* Am Chem Soc; Am Soc Mass Spectrometry; AAAS. *Res:* Biomedical applications of mass spectrometry. *Mailing Add:* Dept Pharmacol State Univ NY Stony Brook NY 11794

IDLER, DAVID RICHARD, b Winnipeg, Man, Mar 13, 23; m 56; c 2. BIOCHEMISTRY. *Educ:* Univ BC, BA, 49, MA, 50; Univ Wis, PhD(biochem), 53. *Hon Degrees:* Univ Guelph, DSc, 87. *Prof Exp:* Assoc chemist, Fisheries Res Bd Can, 53-55, investr-in-chg, 55-59, asst dir, chem dept, Vancouver Lab, 59 -61, dir & investr-in-chg steroid biochem, Halifax Lab, 61-69, dir res, Atlantic Region, 69-71; DIR & PROF BIOCHEM, 71-87, PATON RES PROF, MARINE SCI RES LAB, MEM UNIV NFLD, 87- *Concurrent Pos:* Chmn, Fisheries Res Bd Can, 72-75; mem grants selection comt animal biol, NSERC, 76-80, chmn, 78-79, adv, 79-80, chmn comt selection univ res fels, 84-85; mem grants comt, Can Nat Sportsmen's Fund, 81-84; mem Fisheries & Environ Adv Coun to Nfld Govt, 81- *Honors & Awards:* Fry Medal, Can Soc Zool, 86. *Mem:* Am Chem Soc; Can Soc Zool (vpres, 85, pres, 87-88); Can Biochem Soc; AAAS; NY Acad Sci; Am Zool Soc; Endocrine Soc; Europ Soc Comp Endocrinol; fel Royal Soc Can. *Res:* the isolation, structure and biological activities of natural productions in particular hormones involved in the regulation of fish reproduction; 250 scientific publications. *Mailing Add:* Marine Sci Res Lab Mem Univ Nfld St John's NF A1C 5S7 Can

IDOL, JAMES DANIEL, JR, b Harrisonville, Mo, Aug 7, 28. INDUSTRIAL CHEMISTRY, ORGANIC POLYMER CHEMISTRY. *Educ:* William Jewell Col, AB, 49; Purdue Univ, MS, 52, PhD(chem), 55. *Hon Degrees:* Purdue Univ, DSc, 80. *Prof Exp:* Proj assoc, Standard Oil Co, Ohio, 55-56, proj leader, 56-60, res assoc, 60-63, sect supvr, 63-65, res supvr, 65-68, res mgr, 68-77; vpres venture res & develop, Ashland Chem Co, 77-88; indust rel dir, Ohio State Univ, 88; DIR & PMMI PROF, CTR PACKAGING SCI & ENG, RUTGERS UNIV, 88- *Concurrent Pos:* Mem consult comt chem eng, Okla State Univ, 74-76, mem adv comt, Dept Chem, 78-88, Ohio State Univ, Purdue Univ & Tex A&M Univ, 79-81; mem & fel lectr, Am Inst Chemists, 80, mem bd dir, 81-83; chmn bd, Inst Chemists, 77-78; mem res & develop coun & packaging coun, Am Mgt Asn; mem Nat Res Coun Toxic Waste Handling & Energy Conserv. *Honors & Awards:* Modern Pioneer Award, Nat Asn Mfrs, 65; Chem Pioneer Award, Am Inst Chemists, 68; FG Ciapetti Lectr, 88; Perkin Medal, Soc Chem Indust, 79. *Mem:* Nat Acad Eng; Am Chem Soc; Soc Plastic Engrs; fel Am Inst Chemists; Am Inst Chem Engrs; fel AAAS. *Res:* Vapor phase oxidation and catalysis; organic halogen compounds; monomer synthesis; organic process chemistry; separation processes; polymer synthesis and physical properties; plastics processing; barrier properties of polymeric materials. *Mailing Add:* Ctr Packaging Sci & Eng Rutgers Univ Piscataway NJ 08855

IDOUX, JOHN PAUL, b Houston, Tex, Feb 5, 41; m 66; c 4. ORGANIC CHEMISTRY. *Educ:* Univ St Thomas, Tex, BA, 62; Tex A&M Univ, MS, 65, PhD(chem), 66. *Prof Exp:* Teaching & res assoc, Tex A&M Univ, 63-66; vis res assoc chem, Ohio State Univ, 66-67; asst prof, Northeast La State Col, 67-70; from asst prof to assoc prof, Univ Cent Fla, 70-77, asst dean, Col Natural Sci, 77-80, prof chem, 77-84, assoc dean, Col Arts & Sci, 80-84; dean, Col Arts & Sci & prof chem, 84-90, EXEC VPRES, LAMAR UNIV, 90- *Mem:* AAAS; Am Chem Soc; Chem Soc; Sigma Xi. *Res:* Effect of structure on properties of organic compounds; model peptide bond chemistry; polymer supported catalyst systems; organofluoro chemistry; phase transfer catalysis. *Mailing Add:* Lamar Univ PO Box 10002 Beaumont TX 77710

IDOWU, ELAYNE ARRINGTON, b Homestead, Pa, Feb 3, 40; m 64; c 2. ALGEBRA, PURE MATHEMATICS. *Educ:* Univ Pittsburgh, BS, 61; Univ Dayton, MS, 68; Univ Cincinnati, PhD(math), 74. *Prof Exp:* Aerospace engr, Foreign Technol Div, Wright-Patterson AFB, 62-66 & Parachute Br, 67-68; instr math, Univ Cincinnati, 73-74; ASST PROF MATH, UNIV PITTSBURGH, 74- *Mem:* Am Math Soc; Math Asn Am; Nat Asn Mathematicians. *Res:* Finite group theory, particularly formation theory and automorphisms of finite groups. *Mailing Add:* Dept Math & Statist Univ Pittsburgh 301 Zachary Hall Pittsburgh PA 15260

IDRISS, IZZAT M, CIVIL ENGINEERING. *Educ:* Rensselaer Polytech Inst, BCE, 58; Calif Inst Technol, MS, 59; Univ Calif, Berkeley, PhD(civil eng), 66. *Prof Exp:* Field engr, Moran, Proctor, Meuser & Rutledge, 58; from field engr to sr engr, Danes & Moore, 59-66 & 68-69; res engr, Dept Civil Eng, Univ Calif, Berkeley, 66-75; from proj engr to prin, vpres & dir, Woodward-Clyde Consults, Oakland & San Francisco, Calif, 69-82, managing prin, Orange County, Los Angeles & Santa Barbara area offices & vpres, 82-87, sr consult prin & vpres, Oakland, Calif, 87-89; PROF CIVIL ENG & DIR, CTR GEOTECH MODELING, DEPT CIVIL ENG, UNIV CALIF, DAVIS, 89- *Concurrent Pos:* Consult, several architect engrs & other firms, 66-69; lectr, Dept Civil Eng, Univ Calif, Berkeley, 66-75; consult prof, Dept Civil Eng, Stanford Univ, 78-82; adj prof, Dept Civil Eng, Univ Calif, Los Angeles, 84-86; mem, External Rev Panel, US Geol Surv, 90. *Honors & Awards:* Thomas A Middlebrooks Award, Am Soc Civil Engrs, 71, J James Croes Medal, 72, Walter L Huber Civil Res Prize, 75 & Norman Medal, 77. *Mem:* Nat Acad Eng; Sigma Xi; Am Soc Civil Engrs; Earthquake Eng Res Inst; Seismol Soc Am. *Res:* Geotechnical earthquake engineering; soil mechanics and foundation engineering; earthfill and rockfill dam engineering; probabilistic applications to geotechnical problems; numerical modeling; author of various publications. *Mailing Add:* Dept Civil Eng Univ Calif 2097 Bainer Hall Davis CA 95616

IDSO, SHERWOOD B, b Thief River Falls, Minn, June 12, 42; m 63; c 7. CLIMATOLOGY, REMOTE SENSING. *Educ:* Univ Minn, BPhysics, 64, MS, 66, PhD(soil sci), 67. *Prof Exp:* RES PHYSICIST, AGR RES SERV, USDA, 67- *Concurrent Pos:* Adj prof geog, Arizona State Univ, 80-, adj prof geol, 81- & adj prof bot & microbiol, 84- *Honors & Awards:* Arthur S Flemming Award, 77. *Mem:* Am Meteorol Soc; Am Geophys Union; AAAS; Royal Meteorol Soc; Sigma Xi; Am Soc Agron. *Res:* Global climate, specializing in effects of carbon dioxide; remote sensing to detect plant water stress; schedule irrigations and predicting crop yields. *Mailing Add:* US Water Conserv Lab 4331 E Broadway Rd Phoenix AZ 85040

IDSON, BERNARD, chemistry; deceased, see previous edition for last biography

IDZIAK, EDMUND STEFAN, b Montreal, Que, Sept 23, 35; m 63; c 4. FOOD MICROBIOLOGY. *Educ:* McGill Univ, BSc, 56, MSc, 57; Delft Univ Technol, DSc(microbiol), 62. *Prof Exp:* Bacteriologist, Food & Drug Directorate, Dept Nat Health & Welfare, 62-65; from asst prof to assoc prof, 65-78, dir, Sch Food Sci, 75-78, PROF MICROBIOL, MACDONALD CAMPUS, MCGILL UNIV, 78- *Concurrent Pos:* Mem adv comt food safety assessment, Minister Nat Health & Welfare, 73-76; consult, Cent Health Lab, WHO, Cairo, Egypt, 81, Amalgamated Dairies Ltd, PEI Salmonella in Cheese Outbreak, 84, Minister, Nat Health & Welfare, Can, Eval Food Safety, Qual & Nutrit Prog, 85-86, Cent Health Lab, WHO, Mogadishu, Somalia, 87; mem, Food Safety Assessment Rev, Can, 85; food microbiol exchange expert, Nankai Univ, China, 85. *Mem:* Inst Food Technologists; Am Soc Microbiol; Can Inst Food Technol; Can Soc Microbiol; Sigma Xi; Que Asn Microbiol; Asn Microbiol. *Res:* interactions between spoilage and food poisoning microorganisms; microbiology of meat spoilage. *Mailing Add:* Dept Microbiol Macdonald Campus McGill Univ 21111 Lakeshore Rd Ste Anne de Bellevue PQ H9X 1C0 Can

IDZKOWSKY, HENRY JOSEPH, b Pittsburgh, Pa, Mar 19, 08; m 37; c 2. ENDOCRINOLOGY, EMBRYOLOGY. *Educ:* Univ Pittsburgh, BS, 32, MS, 33, PhD(biol), 36. *Prof Exp:* Asst, Univ Pittsburgh, 33-36, teaching fel zool & Buhl Found fel, 36-37; prof biol & head dept, St Francis Col, Pa, 37-45; from asst prof to assoc prof, 45-58, prof biol, 58-74, head dept, 68-74, EMER PROF BIOL, UNIV PITTSBURGH, JOHNSTOWN CAMPUS, 74- *Mem:* Sigma Xi. *Res:* Biological effects of vitamin C; experimental embryology; host-donor relationships; endocrinology; gonad-adrenal relationships; larval transplants to adult Amphibia; ovarian implant; energy transfer relationships in freshwater ecosystems. *Mailing Add:* 1324 Christopher Johnstown PA 15905

IEYOUB, KALIL PHILLIP, b Lake Charles, La, Aug 21, 35; m 59; c 4. ORGANIC CHEMISTRY. *Educ:* McNeese State Col, BS, 58; La State Univ, MS, 65, PhD(org chem), 67. *Prof Exp:* Lab instr, 59-62, from asst prof to assoc prof, 66-70, prof chem, 70-80, head dept, 80-87, dean, Col Sci, 87-88, VPRES ADMIN & STUDENT AFFAIRS, MCNEESE STATE UNIV, 88- *Concurrent Pos:* Consult. *Mem:* Am Chem Soc. *Res:* Chemistry of tertiary amides; study of the syntheses, reactions, and stereochemistry of these compounds; acid rain studies; geo-pressured, gas-thermal water studies; environmental studies. *Mailing Add:* 2020 Charvais Dr Lake Charles LA 70601-6089

IEZZI, ROBERT ALDO, b Chester, Pa, Jan 13, 44; m 73; c 1. METAL & ORGANIC COATINGS, PACKAGING. *Educ:* Widener Univ, BS, 65; Kent State Univ, MS, 69; Lehigh Univ, PhD(metall), 79. *Prof Exp:* Res engr, corrosion, Res Ctr, Repub Steel Corp, 65-74; res engr, 74-80, supvr org coatings, Homer Res Labs, Bethlehem Steel Corp, 81-85; DIR, PACKAGING TECHNOL, CAMPBELL SOUP CO, 85- *Mem:* Am Soc Metals; Packaging Inst; Nat Metal Decorators Asn. *Res:* Process and product development of coated steel products, including metal and organic coatings; corrosion, cans. *Mailing Add:* 33 Fox Sparrow Turn Tabernacle NJ 08088-9010

IFFLAND, DON CHARLES, b Blissfield, Mich, Nov 26, 21; m 44; c 2. ORGANIC CHEMISTRY. *Educ:* Adrian Col, BS, 43; Purdue Univ, MS, 46, PhD(org chem), 47. *Hon Degrees:* DSc, Adrian Col, 73. *Prof Exp:* From asst prof to assoc prof chem, WVa Univ, 47-56; fel, Purdue Univ, 54-55; assoc prof, 56-59, chmn dept chem, 68-78, PROF CHEM, WESTERN MICH UNIV, 59- *Mem:* Am Chem Soc; Royal Soc Chem; Sigma Xi. *Res:* Aromatic and aliphatic amines; aliphatic nitro compounds; reaction of oximes and hydrazones; configuration of substituted diphenyls. *Mailing Add:* 3430 Northview Dr Kalamazoo MI 49007

IFFT, EDWARD M, b Grove City, Pa, July 19, 37; m 67. PHYSICS. *Educ:* Antioch Col, BS, 60; Ohio State Univ, PhD(physics), 67. *Prof Exp:* Res assoc low temperature physics, Ohio State Univ, 65-67; phys sci officer, US Arms Control & Disarmament Agency, 67-73; dep dir, Off Arms Control & Disarmament, US State Dept, 73-78; chief, Int Prog Policy Office, US Nat Aeronaut & Space Admin, 78-81; dep state dept rep, US Deleg to the Strategic Arms Reduction Talks, 81-84, SR STATE DEPT REP, NUCLEAR & SPACE TALKS USSR, GENEVA, 85- *Concurrent Pos:* Mem US deleg, Strategic Arms Limitations Talks with Soviet Union, 69-78; exec secy, US-USSR agreement on space coop, 78-81. *Mem:* Am Phys Soc; Int Inst Strategic Studies. *Res:* Arms control; properties of liquid mixtures of helium-3 and helium-4 at low temperatures. *Mailing Add:* 6825 Wheatley Ct Falls Church VA 22042

IFJU, GEZA, b Szeged, Hungary, Jan 26, 31; US citizen; m 60; c 7. FORESTRY. *Educ:* Univ BC, BSF, 59, PhD(wood sci), 63; Yale Univ, MF, 60. *Hon Degrees:* Dr, Univ Forestry, Hungary, 90. *Prof Exp:* Asst prof, 64-71, PROF FORESTRY, VA POLYTECH INST & STATE UNIV, 71-, HEAD DEPT, 77- *Concurrent Pos:* Asst specialist & fel forest prod, Forest Prod Lab, Univ Calif, 63-64. *Honors & Awards:* Wood Award, 64; Marwardt Award, Forestry Prod Res Soc, 84. *Mem:* Forest Prod Res Soc; Tech Asn Pulp & Paper Indust; Soc Wood Sci & Technol; Can Pulp & Paper Asn. *Res:* Wood science and technology; relationship between chemical constitution and strength behavior of wood; wood anatomy and physical properties of cell wall of woody plants. *Mailing Add:* Sch Forestry & Wildlife Va Polytech Inst & State Univ Blacksburg VA 24061-0323

IGEL, HOWARD JOSEPH, b Omaha, Nebr, May 10, 34; m 59; c 4. PATHOLOGY, VIROLOGY. *Educ:* Creighton Univ, BS, 55, MD, 59. *Prof Exp:* Intern, St Joseph's Hosp, Omaha, Nebr, 59-60; resident path, Western Reserve Univ, 60-64; res pathologist-virologist, Nat Cancer Inst, 64-68; dir infectious dis & virol, 68-72, dir labs, 72-75, PROF & CHMN PATH & LAB MED, AKRON CHILDREN'S HOSP, 75- *Concurrent Pos:* Nat Cancer Inst res fel, 61-64; assoc clin prof path, Case Western Reserve Univ; adj prof biol, Akron Univ & Kent State Univ; Nat Adv Panel, Food & Drug Admin, 82- *Mem:* AAAS; Am Asn Path; Am Soc Microbiol; Am Soc Clin Path. *Res:* Mouse leukemia and sarcoma viruses; virus-chemical cocarcinogenesis; in vitro systems of carcinogenesis; in vitro transformation of human cells; in vitro growth of human skin for grafting burn patients. *Mailing Add:* Dept Path Akron Childrens Hosp Med Ctr 281 Locust St Akron OH 44308

IGLAR, ALBERT FRANCIS, JR, b New Kensington, Pa, July 17, 39; m 64; c 2. ENVIRONMENTAL HEALTH, SANITARY ENGINEERING. *Educ:* Carnegie-Mellon Univ, BS, 61; Univ Minn, Minneapolis, MPH, 66, PhD(environ health), 70. *Prof Exp:* Sanit engr I, Regional Off, Pa Dept Health, Meadville, 61-62, sanit engr II, Sewerage Sect, Cent Off, Harrisburg, 62-65; res fel environ health, Univ Minn, Minneapolis, 66-70; asst prof, 70-73, assoc prof, 73-81, PROF ENVIRON HEALTH, ETENN STATE UNIV, 81- *Mem:* Nat Environ Health Asn; Air Pollution Control Asn; Am Indust Hyg Asn. *Res:* Institutional solid waste management; treatment of coal gasification wastewater; land disposal of low level radiological waste. *Mailing Add:* 605 Pine Ridge Rd Johnson City TN 37601

IGLEHART, DONALD LEE, b Baltimore, Md, May 11, 33; m 61; c 2. OPERATIONS RESEARCH, MATHEMATICAL STATISTICS. *Educ:* Cornell Univ, BEngPhys, 56; Stanford Univ, MS, 59, PhD(math statist), 61. *Prof Exp:* Asst, Appl Physics Lab, Johns Hopkins Univ, 56-58; assoc prof, Cornell Univ, 61-67; PROF OPERS RES, STANFORD UNIV, 67, CHMN, 85- *Concurrent Pos:* Nat Acad Sci-Nat Res Coun res fel, 61-62. *Mem:* Opers Res Soc Am; fel Inst Math Statist. *Res:* Mathematical theory of inventory; Markov processes; limit theorems in queueing theory; simulation methodology; weak convergence of probability measures. *Mailing Add:* 833 Tolman Dr Palo Alto CA 94305

IGLESIA, ENRIQUE, b Havana, Cuba, Aug 27, 54; US citizen; m 78; c 3. HETEROGENEOUS CATALYSIS, RESEARCH MANAGEMENT. *Educ:* Princeton Univ, BSE, 77; Stanford Univ, MS, 79, PhD(chem eng), 82. *Prof Exp:* PRIN INVESTR CATALYSIS RES & DEVELOP, CORP RES LABS, EXXON RES & ENG, 81-, SECT HEAD, 88- *Concurrent Pos:* Consult prof chem eng, Stanford Univ, 89- *Mem:* Am Chem Soc; Am Inst Chem Engrs; Mat Res Soc; NAm Catalysis Soc. *Res:* Heterogeneous catalysis and its applications to petrochemical processing; mechanisms of surface reactions and engineering of catalytic materials. *Mailing Add:* Exxon Res & Eng Co Rte 22 E Annandale NJ 08801

IGLEWICZ, BORIS, b Omsk, Russia, Oct 11, 39; US citizen; m 73; c 2. STATISTICS. *Educ:* Wayne State Univ, BS, 62, MA, 63; Va Polytech Inst, PhD(statist), 67. *Prof Exp:* Instr math, Mich Technol Univ, 63-64; asst prof statist, Case Western Reserve Univ, 67-69; assoc prof, 69-74, dir PhD prog, 70-75, chmn dept, 78-82, PROF STATIST, TEMPLE UNIV, 74- *Concurrent Pos:* Fel Harvard Univ, 78; vis prof, Harvard Univ, 84-85. *Mem:* Inst Math Statist; Biomet Soc; Am Statist Asn; Int Statist Inst; sr mem Am Soc Qual Control; fel Royal Statist Soc. *Res:* Quality Control; clinical trial models; survey sampling; robust methods; sequential analysis. *Mailing Add:* Dept Statist Speakman Hall Temple Univ Philadelphia PA 19122

IGLEWICZ, RAJA, b Nov 27, 45; US citizen; m 73; c 2. ENVIRONMENTAL HEALTH & SCIENCES. *Educ:* Temple Univ, BS, 75, MS, 80. *Prof Exp:* Indust hygienist, NJ Dept Health, 80-84; consult, Gen Elec Co, 84-85; sup & indust hygienist, 85-87, octg prog mgr, 88-90, RES SCIENTIST, NJ DEPT HEALTH, 87-88 & 91- *Mem:* Am Indust Hyg Asn; Am Pub Health Asn. *Res:* Health hazards of elevated levels of carbon monoxide in ambulances and vehicles; AIDS issues in occupational settings, and the control of occupational health hazards in industrial settings. *Mailing Add:* 1912 Rolling Lane Cherry Hill NJ 08003

IGLEWSKI, BARBARA HOTHAM, b Freeport, Pa, Mar 23, 38; m 65; c 2. MICROBIOLOGY. *Educ:* Allegheny Col, BS, 60; Pa State Univ, MS, 62, PhD(microbiol), 64. *Prof Exp:* Fel, Pa State Univ, 64-65, Univ Colo Med Ctr, 65-66 & Pub Health Res Inst, NY, 66-68; instr, 68-69, asst prof, 69-73, assoc prof, 73-79, PROF MICROBIOL, ORE HEALTH SCI UNIV, 79- *Concurrent Pos:* Sr fel, Walter Reed Army Inst Res, 76-77; mem, Bacterial & Mycotic Dis Study Sect, NIH, 79-83, Res & Training Comt, Nat Cystic Fibrosis Found, 81-84 & Vaccine & Related Biol Prod Adv Comt, 81-82. *Mem:* Am Soc Microbiol. *Res:* Bacterial toxins and pathogenesis of gram negative bacteria, including Pseudomonas aeruginosa and Legionella pneumophila. *Mailing Add:* Dept Microbiol Univ Rochester PO Box 672 601 Elmwood Ave Rochester NY 14642

IGLEWSKI, WALLACE, b Cleveland, Ohio, Aug 17, 38; m 65; c 2. VIROLOGY, MOLECULAR BIOLOGY. *Educ:* Western Reserve Univ, BA, 61; Pa State Univ, MS, 63, PhD(microbiol), 65. *Prof Exp:* Fel virol, Med Sch, Univ Colo, 65-66; fel, Pub Health Res Inst City of New York, Inc, 66-68; from asst prof to prof microbiol, Med Sch, Univ Ore, 68-86; PROF, DEPT MICROBIOL-IMMUNOL, MED SCH, UNIV ROCHESTER 86- *Mem:* Am Soc Microbiol; Am Soc Virol. *Res:* Replication of RNA viruses, microbial toxins, somatic cell genetics, especially the translation and transcription of viral ribonucleic acid and the effects of virus multiplication on cellular metabolism; control of translation in eucaryotes; mechanism of intoxication of animal cells; ADP-Ribosylation in eucaryotic and procaryotic organisms. *Mailing Add:* Dept Microbiol & Immunol Univ Rochester Med Sch Rochester NY 14642

IGNARRO, LOUIS JOSEPH, b Brooklyn, NY, May 31, 41; c 1. PHARMACOLOGY, CELL BIOLOGY. *Educ:* Columbia Univ, BS, 62; Univ Minn, PhD(pharmacol), 66. *Prof Exp:* NIH fel, 66-68; res scientist pharmacol, Ciba-Geigy Corp, 68-72; asst prof to assoc prof, 72-78, prof pharmacol, Med Sch, Tulane Univ, 79-86; PROF PHARMACOL, SCH MED, UNIV CALIF, 86-, ASST DEAN RES, 91- *Concurrent Pos:* Merck res grant, Tulane Univ, 73-74; Arthritis Found grant, 74-75; Nat Inst Arthritis, Metab & Digestive Dis grant, 74-77, 78-81 & 82-85; USPHS res career develop award, 75-80. *Honors & Awards:* Res Award, Pharmaceut Mfrs Asn Found, 73; Res Award, Edward G Schlieder Found Educ, 73; Smith, Kline & French Award. *Mem:* Am Soc Pharmacol & Exp Therapeut; Soc Exp Biol & Med; Am Rheumatism Asn; Am Heart Asn; Am Soc Cell Biol; Sigma Xi; Am Soc Biochem & Molecular Biol; Am Physiol Soc. *Res:* Inflammation and arthritis; cyclic nucleotide research; bioregulation of human cell function; hormonal control mechanisms; free radicals and enzyme activation; nitric oxide metabolics; regulation of vascular and platelet function. *Mailing Add:* Dept Pharmacol Univ Calif Los Angeles CA 90024

IGNATIEV, ALEX, b Wehingen, Ger, Feb 14, 45; US citizen; m 67; c 2. SURFACE PHYSICS. *Educ:* Univ Wis, BS, 66; Cornell Univ, PhD(mat sci), 72. *Prof Exp:* Fel mat sci, State Univ NY Stony Brook, 71-73; from asst prof to assoc prof physics, 74-83, PROF PHYSICS & CHEM, UNIV HOUSTON, 83-, ASSOC DIR, MAGNETIC INFO RES LAB, 84- *Concurrent Pos:* Mem, Energy Lab, Univ Houston, 75-; lectr physics, Aarhus Univ, Denmark, 77-78; Fulbright sr scholar, 83; assoc dir, Space Vacuum Epitaxy Ctr, 86-88, dir, 88- *Mem:* Am Phys Soc; Am Vacuum Soc; Am Chem Soc; Int Solar Energy Soc. *Res:* Structure and properties of surfaces, including catalysis, two dimensional phase transitions, surface image, small particle structures, solar energy absorbing and reflecting coatings; epitaxial growth of thin film semiconductors by MBE/CBE, thin film high Tc superconductor materials. *Mailing Add:* Dept Physics Univ Houston Univ Park Houston TX 77004

IGNIZIO, JAMES PAUL, b Akron, Ohio, Oct 28, 39; m 61; c 2. OPERATIONS RESEARCH. *Educ:* Univ Akron, BSEE, 62; Univ Ala, MSE, 68; Va Polytech Inst & State Univ, PhD(opers res), 71. *Prof Exp:* Lead engr, Commun Systs, NAm Aviation, 62-63; proj engr, Space Craft, Inc, 63-65; proj engr, Boeing Co, 65-66; flight test engr, NAm Rockwell Corp, 66-67; asst prof, Dept Indust Eng, Univ Ala, Huntsville, 67-74; PROF, DEPT INDUST ENG, PA STATE UNIV, 74- *Concurrent Pos:* Consult, Gen Res Corp, 70-71, 79-82, Litton Industs, 71, US Army Ballistic Missile Defense Agency, 72-74 & Teledyne Brown Eng, 72- *Honors & Awards:* First Hartford Prize, 80. *Mem:* Am Inst Indust Engrs; Opers Res Soc Am; Inst Mgt Sci; Am Soc Eng Educ. *Res:* Application of operations research to energy, space and military systems; research in multiple objective optimization, mathematical modeling and simulation, site deployment, resource allocation and cost analysis; export systems and artificial intelligence. *Mailing Add:* Dept Ind Eng Univ Houston 4800 Calhoun Houston TX 77204

IGNOFFO, CARLO MICHAEL, b Chicago Heights, Ill, Aug 24, 28; m 49. ENTOMOLOGY, INVERTEBRATE PATHOLOGY. *Educ:* Northern Ill Univ, BS, 50; Univ Minn, MS, 54, PhD(entom), 56. *Prof Exp:* Assoc prof entom, Iowa Wesleyan Col, 57-59; res entomologist, Agr Res Serv, USDA, 59-65; dir entom, Bioferm Div, Int Minerals & Chem Corp, 65-67, assoc dir entom, Indust Chem Div, 67-71; DIR BIOL CONTROL INSECTS RES LAB, ENTOM RES DIV, AGR RES SERV, USDA, 71- *Concurrent Pos:* Mem, Int Comt Nomenclature Invert Viruses & Int Comt Standardization & Assay of Insect Pathogens. *Mem:* AAAS; Entom Soc Am; Soc Invert Path; Am Inst Biol Sci; Int Orgn Biol Control. *Res:* Insect pathogens and their effects on invertebrates; viruses, fungi and bacteria infecting insects. *Mailing Add:* 2905 W Rollins Rd Apt B-6 Columbia MO 65203

IGO, GEORGE (JEROME), b Greeley, Colo, Sept 2, 25; m 53; c 1. PHYSICS. *Educ:* Harvard Univ, AB, 49; Univ Calif, PhD(physics), 53. *Prof Exp:* Res assoc, Radiation Lab, Univ Calif, 52-53; res assoc, Sloane Physics Lab, Yale Univ, 53-54; assoc physicist, Brookhaven Nat Lab, 54-56; assoc physics, Stanford Univ, 56-58; Fulbright fel, Univ Heidelberg, 58-59; staff scientist, Lawrence Radiation Lab, Univ Calif, 60-64; dir cyclotron inst, Tex A&M Univ, 64; mem vis staff, Los Alamos Sci Lab, 65-69; chmn dept, 77-80, PROF PHYSICS, UNIV CALIF, LOS ANGELES, 69- *Concurrent Pos:* Sr scientist award, Alexander von Humboldt-Stiftung, 91-93. *Mem:* Am Phys Soc. *Res:* Nuclear physics; isotope shift; optical model. *Mailing Add:* Dept Physics 3174 Knudsen Hall Univ Calif Los Angeles CA 90024

IGUSA, JUN-ICHI, b Japan, Jan 30, 24; m 48; c 3. MATHEMATICS. *Educ:* Tokyo Univ, BS, 45; Kyoto Univ, PhD, 53. *Prof Exp:* Asst math, Inst Statist Sci, 47-48; asst prof, Tokyo Univ Educ, 48-49 & Kyoto Univ, 49-53; res assoc, Harvard Univ, 53-55; from asst prof to assoc prof, 55-61, PROF MATH, JOHNS HOPKINS UNIV, 61- *Concurrent Pos:* Mem, Inst Advan Study, Princeton, 59-60 & 70-71; guest, Inst Advan Study Sci, Paris, 64, Tata Inst, Bombay, 78; ed, Am J Math; vis prof, Harvard Univ, 81-82. *Mem:* Am Math Soc; Japan Math Soc. *Res:* Algebraic geometry connected with number theory. *Mailing Add:* Dept Math Johns Hopkins Univ 3400 N Charles St Baltimore MD 21218

IH, CHARLES CHUNG-SEN, b Hankow, China, May 15, 33; US citizen; c 2. ELECTRO-OPTICAL DEVICES, ELECTRO-OPTICAL SYSTEMS. *Educ:* Nat Taiwan Univ, BSEE, 56; Lehigh Univ, MSEE, 59; Univ Pa, PhD(physics), 66. *Prof Exp:* Engr, Sperry Rand Univac, 59-63; mem tech staff, RCA Lab, 67-71; dir, Galor Lab, CBS Lab, 71-75; PROF ELECTRO-OPTIC DEVICES & SYST, DEPT ELEC ENG, UNIV DEL, 75- *Concurrent Pos:* Consult, Perkin Elmer Corp, 75, Naval Air Develop Ctr, Warminister, Pa, 80-81, IBM & Sperry Univac, 81; mem fac, IBM Res Ctr, Yorktown Heights, NY, 81. *Mem:* Optical Soc Am; Inst Elec & Electronics Engrs; Am Inst Physics; Soc Photo-Optical Instrumentation Engrs. *Res:* Holography applications related to color image preservation; holographic scanners and fiber optical communications. *Mailing Add:* Dept Elec Eng Univ Del Newark DE 19711

IHA, FRANKLIN TAKASHI, b Honolulu, Hawaii, Dec 29, 37; m 71. MATHEMATICAL ANALYSIS. *Educ:* Univ Hawaii, BA, 61, MA, 63; Univ Calif, Los Angeles, PhD(math), 69. *Prof Exp:* From instr to asst prof math, Univ Hawaii, 63-74; lectr, 74-79, INSTR, DIV MATH & NATURAL SCI, LEEWARD COMMUNITY COL, 79- *Mem:* Am Math Soc; Math Asn Am. *Res:* Spectral theory of partial differential operators. *Mailing Add:* 1667 Lime St Honolulu HI 96819

IHAS, GARY GENE, b Hillsdale, Mich, Sept 5, 45; m 66; c 3. ULTRA-LOW TEMPERATURE PHYSICS. *Educ:* Calif Inst Technol, BA & BS, 67; Univ Mich, MA, 70, PhD(physics), 71. *Prof Exp:* Res physicist low temp, Kernforschungsanlage, Juelich, WGer, 71-73; vis asst prof physics, Ohio State Univ, 73-75, res assoc physics, 75-77; from asst prof to assoc prof, 77-85, PROF PHYSICS, UNIV FLA, 85- *Concurrent Pos:* Alfred P Sloan Found fel, 77-82. *Mem:* Am Phys Soc. *Res:* Surface effects in liquid helium; superfluid helium three; critical phenomena; impurities and excitations in liquid helium; nuclear demagnetization refrigeration. *Mailing Add:* Dept Physics Univ Fla 201 Williamson Hall Gainesville FL 32611-2085

IHDE, AARON JOHN, b Neenah, Wis, Dec 31, 09; wid; c 2. CHEMISTRY, HISTORY OF SCIENCE. *Educ:* Univ Wis, BS, 31, MS, 39, PhD(food chem), 41. *Prof Exp:* Res chemist, Blue Valley Creamery Co, Ill, 31-38; asst chem, Univ Wis, 39-41; instr, Butler Univ, 41-42; from instr to prof, 42-80, chmn integrated lib studies, 63-70, EMER PROF CHEM, HIST SCI & INTEGRATED LIB STUDIES, UNIV WIS-MADISON, 80- *Concurrent Pos:* Carnegie fel, Harvard Univ, 51-52; consult, Wis Food Stands Adv Comt, 55-67. *Honors & Awards:* Dexter Award, Am Chem Soc. *Mem:* AAAS; Am Chem Soc; Hist Sci Soc; Soc Hist Technol. *Res:* Metals, vitamins, and synthetic additives in foods; history of chemistry; biochemistry; social implications of science; science education. *Mailing Add:* Dept Chem Univ Wis 9369 Chem Bldg Madison WI 53706

IHDE, DANIEL CARLYLE, b Parsons, Kans, July 10, 43; m 68; c 2. INTERNAL MEDICINE, MEDICAL ONCOLOGY. *Educ:* Eastern NMex Univ, BS, 64; Stanford Univ, MD, 69. *Prof Exp:* Intern & resident internal med, NY Hosp, 69-71; resident & fel internal med & med oncol, Mem Hosp, NY, 71-73; fel, Bethesda, Md, 73-75, sr investr, Vet Admin Med Oncol Br, Vet Admin Med Ctr, Washington, DC, 75-81, HEAD, CLIN INVEST MED ONCOL, NAT CANCER INST-NAVY MED ONCOL BR, NAVAL HOSP, BETHESDA, MD, 81- *Concurrent Pos:* Prof med, Uniformed Serv Univ Health Sci, 85- *Honors & Awards:* Commendation Medal, USPHS, 84. *Mem:* Am Col Physicians; Am Soc Clin Oncol; Am Asn Cancer Res; Int Asn Study Lung Cancer. *Res:* Clinical investigations in cancer chemotherapy; staging and the natural history of human cancer; clinical-cell biologic correlation, particularly lung cancer mycosis fungoides and prostatic cancer. *Mailing Add:* 3404 Kenilworth Dr Chevy Chase MD 20815

IHLER, GARRET MARTIN, b Milwaukee, Wis, Nov 4, 39; m 70; c 2. MOLECULAR BIOLOGY, HEMATOLOGY. *Educ:* Calif Inst Technol, BS, 61; Harvard Univ, PhD(biochem), 67; Univ Pittsburgh, MD, 76. *Prof Exp:* NIH res fel biochem, Harvard Med Sch, 67-69; from asst prof to assoc prof biochem, Sch Med, Univ Pittsburgh, 71-76; PROF MED BIOCHEM & HEAD DEPT, COL MED, TEX A&M UNIV, 76- *Mem:* Am Chem Soc; Am Soc Microbiol; Acad Clin Lab Physicians & Scientists; Am Soc Cell Biol; Am Soc Biol Chemists. *Res:* Enzyme therapy using red cells as carrier; Gaucher's disease; Enzymology, especially RNA polymerase; macromolecules, especially DNA replication; protein synthesis; bacterial and phage genetics. *Mailing Add:* Dept Med Biochem Tex A&M Univ Col Med College Station TX 77843

IHNAT, MILAN, b Montreal, Que, Oct 1, 41; m 70; c 2. ANALYTICAL CHEMISTRY. *Educ:* McGill Univ, BSc, 62, PhD(phys chem), 67. *Prof Exp:* Res assoc biophys chem, Columbia Univ, 66-69; res asst, McGill Univ, 69-71; res scientist anal chem, Chem & Biol Res Inst, 71-86, LAND RESOURCE RES CENTRE, RES BR AGR CAN, 86- *Concurrent Pos:* Asst prof, McGill Univ, 70-71. *Mem:* Am Chem Soc; Chem Inst Can; fel Asn Off Analytical Chemists; Spectros Soc Can. *Res:* Quantitative analytical methodology for the determination of macro and trace concentrations of metals and other inorganic constituents in biological materials; analytical atomic absorption spectroscopy; development of biological reference materials for chemical composition; atomic mass spectrometry. *Mailing Add:* Land Resource Res Centre Res Br Agr Can Ottawa ON K1A 0C6 Can

IHNDRIS, RAYMOND WILL, b Sterling, Ind, Apr 1, 20; m 42; c 2. ORGANIC CHEMISTRY. *Educ:* Rollins Col, BSc, 55. *Prof Exp:* Chemist, Bur Entom & Plant Quarantine, USDA, 46-55, chemist, Pesticid Chem Res Br, Entom Res Div, 56-60; head, Sci Records Sect, Res Commun Br, Cancer Chemother Nat Serv Ctr, NIH, 60-67, sr info chemist, Toxicol Info Prog, Nat Libr Med, 67-76, info chem consult, 76-81; RETIRED. *Mem:* Am Chem Soc; Am Soc Info Sci. *Res:* Organic chemical abstracts system of nomenclature; synthetic organic insecticides and attractants; formulation of insecticides, repellants and attractants; information storage and retrieval of organic compounds by structural fragmentation. *Mailing Add:* 682 Granville Dr Winter Park FL 32789

IHRIG, JUDSON LA MOURE, b Santa Maria, Calif, Nov 5, 25; m 50; c 2. PHYSICAL CHEMISTRY. *Educ:* Haverford Col, BS, 49; Princeton Univ, AM, 51, PhD(phys chem), 52. *Prof Exp:* Asst instr, Princeton Univ, 49-52; from asst prof to assoc prof, 52-72, dir lib studies prog, 73-79, prof chem & chmn dept, 81-86, DIR HON PROG, UNIV HAWAII, 87- *Mem:* Am Chem Soc; Sigma Xi. *Res:* Transport properties of aqueous solutions; magnetochemistry; inhibition of chain reactions. *Mailing Add:* 386 Wailupe Circle Honolulu HI 96821

IHRKE, CHARLES ALBERT, b Oshkosh, Wis, June 9, 38; m 60; c 2. GENETICS, PLANT BREEDING. *Educ:* Wis State Univ, BS, 60; Univ Nebr, Omaha, MS, 66; Ore State Univ, PhD(plant genetics), 69. *Prof Exp:* Instr biol, Onarga Mil Sch, 60-64; asst prof genetics, 69-75, assoc prof pop dynamics, 75-80, ASSOC PROF HUMAN BIOL, UNIV WIS-GREEN BAY, 81-, CHMN DEPT, 84- *Mem:* Am Soc Agron; Crop Sci Soc Am; Genetics Soc Am; Am Inst Biol Sci; Am Genetics Asn. *Res:* Investigations of environmental effects on genetic recombination; incidence of genetic disorders in some small United States populations. *Mailing Add:* 2201 Sunrise Ct Green Bay WI 54302

IHRMAN, KRYN GEORGE, b Kalamazoo, Mich, Nov 5, 30; m 64. ORGANIC CHEMISTRY. *Educ:* Kalamazoo Col, BA, 52; Ohio State Univ, PhD(org chem), 57. *Prof Exp:* RES CHEMIST, ETHYL CORP, 57- *Mem:* Am Chem Soc. *Res:* Organo-metallic synthesis involving transition elements; organic synthesis by organo-metallic catalysis; synthesis of polymer additives; synthesis of organic intermediates. *Mailing Add:* 3119 Valcour Aime Baton Rouge LA 70820

IHSSEN, PETER EDOWALD, b Bremen, Ger, Jan 12, 39; m 64. GENETICS, FISH BIOLOGY. *Educ:* Univ Man, BSc, 63; Univ Toronto, MSc, 66, PhD(zool), 71. *Prof Exp:* RES SCIENTIST FISHERIES GENETICS, FISHERIES BR, ONT MINISTRY NATURAL RESOURCES, 71- *Mem:* Can Soc Zoologists; Genetics Soc Can; Am Fisheries Soc. *Res:* Application of quantitative and population genetics principles to fisheries problems; conservation of biodiversity of fishes; physiological and biochemical genetics of fish. *Mailing Add:* Ministry Natural Resources Box 5000 Maple ON L6A 1S9 Can

IIJIMA, SUMIO, solid state physics, crystallography, for more information see previous edition

IJAMS, CHARLES CARROLL, b Jackson, Tenn, Dec 23, 13; m 47; c 1. CHEMICAL PHYSICS. *Educ:* Union Univ, Tenn, AB, 36; Vanderbilt Univ, MS, 37, PhD(phys chem), 41. *Prof Exp:* Asst chemist, Wolf Creek Ord Plant, Tenn, 41-42; chief chemist, Gulf Ord Plant, 42-43; asst prof chem, Union Univ, Tenn, 47-48; from assoc prof to prof physics & head dept, Memphis State Univ, 48-77; RETIRED. *Concurrent Pos:* Consult, USN Air Tech Training Command. *Mem:* Am Asn Physics Teachers. *Res:* Spreading of monomolecular films on water; solubility of gases in organic liquids; maintenance of sonar equipment. *Mailing Add:* 249 N Rose Rd Memphis TN 38117

IJAZ, LUBNA RAZIA, b Lahore, Pakistan, Mar 27, 40; US citizen; m 60; c 5. TECHNICAL MANAGEMENT, INDUSTRIAL PRODUCTION. *Educ:* Peshawar Univ, BS, 57; Punjab Univ, MS, 60; Va Polytech Inst & State Univ, PhD(physics educ), 75. *Prof Exp:* Asst prof physics, Va Polytech Inst & State Univ, Blacksburg, 75-79; pres, Solar Energy Educ & Res Corp, 80-83; CHMN, INT SOLAR & ELECTRONIC INDUST LTD, PAKISTAN, 83-, MANAGING DIR, 90- *Concurrent Pos:* Vis solar scientist & lectr, Int Summer Col Physics & Contemp Needs, Nathiagali, 76 & 77; adj prof elec eng, Va Polytech Inst & State Univ, 79-83. *Honors & Awards:* VERA Res Merit Award, 75, VERA Merit Award, 76. *Mem:* AAAS; Nat Sci Teachers Asn; Inst Elec & Electronics Engrs; Am Asn Physics Teachers. *Res:* Solar cell materials; thin film solar cells for photovoltaic energy conversion; developing solar photovoltaic module production facility and transfering electronic chip technology to Pakistan; US patent on solar cell. *Mailing Add:* 2236 Archer Rd Shawsville VA 24162

IJAZ, MUJADDID A, b Lahore, Pakistan, Oct 1, 37; US citizen; m 60; c 5. NUCLEAR PHYSICS, ELEMENTARY PARTICLE PHYSICS. *Educ:* Govt Col, Lahore, BSc, 57; Punjab Univ, Pakistan, MS, 59; Fla State Univ, MS, 62; Ohio Univ, PhD(physics), 64. *Prof Exp:* Assoc head physics, 64-78, PROF PHYSICS, VA POLYTECH INST & STATE UNIV, 78- *Concurrent Pos:* Res scientist, Univ Isotope Separator Proj, Oak Ridge Nat Lab, 69-; vis prof, Univ Petrol & Minerals, Saudi Arabia, 79-81, Int Ctr Theoret Physics, Trieste, Italy, 85. *Mem:* Am Phys Soc. *Res:* Neutron deficient isotopes of rare earth alpha emitters; multiple pion production reactions at high energies. *Mailing Add:* Dept Physics Va Polytech Inst & State Univ Blacksburg VA 24061

IKA, PRASAD VENKATA, b Eluru, Andhra Pradesh, India, Apr 8, 58; m 87; c 2. PHYSICAL & MECHANICAL PROPERTIES, STRUCTURE-PROPERTY CORRELATIONS. *Educ:* Indian Inst Technol, Delhi, MS, 81; State Univ NY, MS, 83, PhD(polymer chem), 85. *Prof Exp:* Res engr indust res, Polymer Prod Develop, 85-88, sr res engr, 88-90, RES SUPVR, SUPV POLYMER RES, NORTON CO, 90- *Mem:* Am Chem Soc; Soc Plastics

Engrs; AAAS; Sigma Xi. *Res:* Product development research of polymer based abrasive products; structure-property correlationships of polymers in-general and thermosets-in-particular; polymer characterization tools to probe into polymer wear characteristics and degradation. *Mailing Add:* 14 Thestland Dr Shrewsbury MA 01545

IKAWA, HIDEO, fluid-aerodynamics, endo-exo atmospheric flight mechanics, for more information see previous edition

IKAWA, MIYOSHI, b Venice, Calif, Feb 14, 19; m 50; c 2. ALGAL & FUNGAL TOXINS, MARINE TOXINS. *Educ:* Calif Inst Technol, BS, 41; Univ Wis, MS, 45, PhD(biochem), 48. *Prof Exp:* Res fel chem, Calif Inst Technol, 48-52; res scientist, Biochem Inst, Univ Tex, 52-56; from asst res biochemist to assoc res biochemist, Univ Calif, 56-63; prof, 63-87, EMER PROF BIOCHEM & ADJ PROF ZOOL, UNIV NH, 87- *Mem:* Fel AAAS; Am Chem Soc; Int Soc Toxinol; Japan Asn Mycotoxicol. *Res:* Algal and fungal toxins; microbial. *Mailing Add:* Dept Biochem Univ NH Durham NH 03824

IKE, ALBERT FRANCIS, b East Orange, NJ, July 18, 32; m 56; c 2. FOREST SOILS, ACADEMIC ADMINISTRATION. *Educ:* Rutgers Univ, BS, 54; Cornell Univ, MS, 57; NC State Univ, PhD(soil sci), 69. *Prof Exp:* Soil Scientist soil surv, Monogahela Nat Forest, US Forest Serv, 59-60; res forester soils & tree nutrit, Southeastern Forest Exp Sta, US Forest Serv, Asheville, NC, 60-70; assoc prof outdoor recreation, 70-74, ASSOC DIR, UNIV GA INST COMMUNITY & AREA DEVELOP, 74- *Concurrent Pos:* Consult, Conserv Found, 73-74, US Army CEngr, 79, Nat Park Serv, 82. *Mem:* Soc Am Foresters; Sigma Xi. *Res:* Coastal Zone Management Planning, including carrying capacity estimates for Cumberland Island National Seashore; soil-site relations; soil variability; tree nutrition; land management planning and decision making; forest site evaluation; carrying capacity; natural resource policy analysis. *Mailing Add:* 240 Kings Rd Athens GA 30606

IKEDA, GEORGE J, b Pahoa, Hawaii, Jan 9, 35; m 62. BIOCHEMISTRY. *Educ:* Univ Hawaii, BA, 57; Ore State Univ, MS, 60, PhD(biochem), 67. *Prof Exp:* Res asst phys & anal chem, Upjohn Co, 60-64; pharmacologist, Dept Drug Metab, Abbott Labs, 67-75; pharmacologist, Div Toxicol, Bur Foods, 75-84, SUPVRY PHARMACOLOGIST, DIV TOXICOL STUDIES, CTR FOOD SAFETY & APPL NUTRIT, FOOD & DRUG ADMIN, 85- *Mem:* Am Chem Soc; Phytochem Soc NAm; AAAS; Sigma Xi; fel Am Inst Chemists; Asn Off Anal Chemists; Int Soc Study Xenobiotics. *Res:* Drug metabolism; intermediary metabolism; plant biochemistry; radiotracer methodology; pharmacokinetics. *Mailing Add:* Metab Br HFF-169 Food & Drug Admin 8501 Muirkirk Rd Laurel MD 20708

IKEDA, RICHARD MASAYOSHI, b Long Beach, Calif, Feb 21, 34; m 55; c 2. POLYMER SCIENCE. *Educ:* Juniata Col, BS, 55; Univ Ill, PhD(phys chem), 58. *Prof Exp:* NSF fel, Harvard Univ, 58-59; res chemist, 59-67, staff scientist, Film Dept, 67-73, res assoc, 73-76, RES ASSOC, CENT RES & DEVELOP DEPT, E I DU PONT DE NEMOURS & CO, INC, 76- *Mem:* Am Chem Soc; Am Phys Soc. *Res:* Chemical kinetics; polymer physics. *Mailing Add:* 39 Cedarwood Lane Rd 1 Chadds Ford PA 19317

IKEDA, ROBERT MITSURU, b Tracy, Calif, Feb 4, 25; m 51; c 3. AGRICULTURAL CHEMISTRY. *Educ:* Univ Calif, BS, 50, PhD(agr chem), 55. *Prof Exp:* Sr lab technician, Univ Calif, 50-55, jr res food technologist, 55-57; chemist, Fruit & Veg Chem Lab, USDA, 57-62; sr chemist, 62-67, mgr anal div, 67-69, mgr chem & biol div, 69-72, assoc prin scientist, 72-76, PRIN SCIENTIST, PHILIP MORRIS, INC, 76- *Mem:* Am Chem Soc; Inst Food Technol; Sigma Xi. *Res:* Isolation and identification of flavor constituents. *Mailing Add:* 1915 Southcliff Rd Richmond VA 23225

IKEDA, TATSUYA, b Tokyo, Japan, Apr 1, 40; m 69; c 3. MOLECULAR SPECTROSCOPY. *Educ:* Univ Tokyo, BE, 63; Univ Wis-Milwaukee, MS, 68; Mass Inst Technol, PhD(phys chem), 72. *Hon Degrees:* JD, Suffolk Univ, 81. *Prof Exp:* Mem staff petrol chem, Res & Develop Dept, Tonen Sekiyu Kagaku Co, 63-66; fel, Rice Univ, 72-74; fel, Univ Calif, Santa Barbara, 74-75; tech adv polymer chem, 74-81; PATENT ATTY, HOECHST-CELANESE CORP, 81- *Mem:* Sigma Xi. *Res:* Microstructures of polymers; application of tunable dye laser and fixed-frequency laser to the study of electronic spectra of small molecules. *Mailing Add:* Six Blackberry Lane Whitehouse Station NJ 08889-9692

IKEDA-SAITO, MASAO, STRUCTURE & FUNCTION OF METALLO PROTEINS. *Educ:* Osaka Univ, PhD(biophys), 78. *Prof Exp:* Asst prof molecular spectros, Univ Pa, 81-89; ASSOC PROF, DEPT PHYSIOL/ BIOPHYS, CASE WESTERN RESERVE UNIV, 89- *Mailing Add:* Dept Physiol/Biophys Rm E559 Case Western Reserve Univ 2109 Adelbert Rd Cleveland OH 44106

IKEHARA, YUKIO, b Kagoshima, Japan, Apr 11, 39; m 70; c 3. MOLECULAR CELL BIOLOGY. *Educ:* Kyushu Univ, Japan, MD, 64, PhD(biochem), 69. *Prof Exp:* Postdoctoral biochem, Sch Med, Univ Ill, 69-71, Univ Wis, 71-72; assoc prof, Fukuoka Women's Univ, 72-73; assoc prof, Kyushu Univ Sch Pharm Sci, 73-78; PROF BIOCHEM, SCH MED, FUKUOKA UNIV, 78- *Concurrent Pos:* Vis prof, Sch Med, NY Univ, 80. *Mem:* Am Soc Cell Biol. *Res:* Intracellular transport and modification of proteins; biogenesis and function of the Golgi complex; membrane anchoring of proteins; structure and function of ectoenzymes. *Mailing Add:* Dept Biochem Fukuoka Univ Sch Med 34 Nanakuma Jonan-ku Fukuoka 814-01 Japan

IKENBERRY, DENNIS L, b Glendale, Calif, Oct 14, 39; m 64. PHYSICS. *Educ:* Occidental Col, BA, 61; Univ Calif, Riverside, PhD(physics), 65. *Prof Exp:* From asst prof to assoc prof, 65-77, PROF PHYSICS, CALIF STATE COL, SAN BERNARDINO, 77- *Concurrent Pos:* Res leave, Univ Utah, 69-71. *Mem:* Am Phys Soc. *Res:* Theoretical chemical physics; biophysics; air pollution and exercise. *Mailing Add:* Dept Computer Sci Calif State Univ San Bernardino CA 92407

IKENBERRY, GILFORD JOHN, JR, b Fargo, NDak, Dec 26, 29; m 51; c 3. PLANT ANATOMY, PLANT DEVELOPMENT. *Educ:* McPherson Col, BS, 52; Okla State Univ, MS, 56; Iowa State Univ, PhD(plant morphol), 59. *Prof Exp:* Instr bot, Iowa State Univ, 57-59; asst prof, Mich State Univ, 59-61; assoc prof biol, 61-64, PROF BIOL, MCPHERSON COL, 64- *Concurrent Pos:* NSF sci fac fel biol, Yale Univ, 67-68. *Mem:* AAAS; Bot Soc Am; Am Soc Cell Biol; Am Inst Biol Sci; Sigma Xi. *Res:* Developmental anatomy of angiosperms; plant morphogenesis; somatic cell culture. *Mailing Add:* Dept Biol Sci McPherson Col McPherson KS 67460

IKENBERRY, LUTHER CURTIS, b Franklin Co, Va, Jan 8, 17; m 43; c 3. ANALYTICAL CHEMISTRY. *Educ:* Bridgewater Col, AB, 37; Va Polytech Inst & State Univ, MS, 40. *Prof Exp:* Analyst chem, Middletown Works, 40-41, from chemist to sr chemist, Res Ctr, 41-68, supvr res chemist, 68-74, PRIN RES CHEMIST, RES CTR, ARMCO INC, 74- *Honors & Awards:* Lundell-Bright Award, Am Soc Testing & Mat, 80. *Mem:* Am Chem Soc; Am Soc Testing & Mat. *Res:* Analytical methods using gas chromatography for trace impurities in the environment; laboratory automation of analytical chemical equipment; gas chromatography and mass spectrometry analysis of organic pollutants in air and water. *Mailing Add:* 314 N Marshall Rd Middletown OH 45042

IKENBERRY, RICHARD W, b Des Moines, Iowa, June 17, 37. PLANT PATHOLOGY. *Educ:* Iowa State Univ, BS, 60, PhD(plant path), 64. *Prof Exp:* Instr bot & plant path, Iowa State Univ, 64-65; from asst prof to assoc prof, 65-73, PROF BIOL, KEARNEY STATE COL, 73- *Mailing Add:* Dept Biol Kearney State Col Kearney NE 68849-0531

IKENBERRY, ROY DEWAYNE, b Raton, NMex, Apr 21, 40; m 60; c 2. VERTEBRATE PHYSIOLOGY, HEMATOLOGY. *Educ:* Eastern NMex Univ, BS, 62; Tex Tech Univ, MS, 64; Univ Okla, PhD(zool), 69. *Prof Exp:* Instr biol, Eastern NMex Univ, 64-66; asst prof, 69-77, ASSOC PROF BIOL, E TENN STATE UNIV, 77- *Mem:* Sigma Xi. *Res:* Biosystematics of blood serum; reproductive studies of mammalian species; cardiovascular physiology of fishes and amphibians; mammalian pheromones. *Mailing Add:* Five Sheffield Ct Johnson City TN 37604

IKEZI, HIROYUKI, b Kochi, Japan, Apr 5, 37; US citizen; m 62; c 3. HIGH POWER SOLITON GENERATION. *Educ:* Univ Electrocommun, BS, 60; Tokyo Univ Educ, MS, 62; Nagoya Univ, PhD(physics), 68. *Prof Exp:* Res assoc plasma physics, Inst Plasma Physics, Nagoya Univ, 62-68; asst prof physics, physics dept, Univ Calif Los Angeles, 68-70; assoc prof plasma physics, Inst Plasma Physics, Nagoya Univ, 70-75; mem tech staff physics, Bell Lab, 75-81; SR TECH ADV APPL PHYSICS, GEN ATOMICS, 81- *Concurrent Pos:* Consult, Gen Atomics, 69; vis scientist, Culham Lab, UK Atomic Energy Authority, 71. *Honors & Awards:* Nishina Prize, Nishina Mem Found, Japan, 69. *Mem:* Am Phys Soc. *Res:* Nonlinear plasma waves such as solitons, trapped particle effects and echoes; two dimensional electron system by using electrons on liquid helium; development of high-power microwave system employing soliton generation. *Mailing Add:* Gen Atomics PO Box 85608 San Diego CA 92138

IKONNE, JUSTUS UZOMA, b Aba, Nigeria, May 24, 49. BIOCHEMISTRY. *Educ:* Univ Ghana, BSc, 70; Univ London, PhD(biochem), 73. *Prof Exp:* Res specialist biochem genetics, Univ Minn, 73-78; SR SCIENTIST BIOCHEM, CORNING GLASS WORKS, CORNING, NY, 78- *Mem:* Sigma Xi. *Res:* Lysosomal enzymes and their relevance to genetic diseases; enzyme systems currently used in the identification and diagnosis of myocardial infarctions; preparation and characterization of enzyme labels for enzyme immunoassays. *Mailing Add:* Box 1133 Elmira NY 14902

IKUMA, HIROSHI, b Nishinomiya-Shi, Japan, Jan 23, 32; m 66; c 2. PLANT PHYSIOLOGY. *Educ:* Kobe Univ, BSc, 56; Harvard Univ, AM, 58, PhD(biol), 62. *Prof Exp:* Res fel plant biochem, Johnson Res Found, Univ Pa, 62-65; from asst prof to assoc prof bot, 65-75, chmn dept cellular & molecular biol, 76-78, PROF BIOL SCI, UNIV MICH, ANN ARBOR, 75- *Mem:* Am Soc Plant Physiologists. *Res:* Respiratory properties of higher plant mitochondria; cellular regulation of plant growth and developmental processes; germination processes of onoclea fern spores. *Mailing Add:* Dept Cell & Molecular Biol Univ Mich Div Biol Sci Ann Arbor MI 48109-1048

ILARDI, JOSEPH MICHAEL, b New York, NY, Dec 11, 39; m 63; c 3. PHYSICAL CHEMISTRY. *Educ:* Fordham Univ, BS, 61, MS, 63, PhD(chem), 66. *Prof Exp:* Res chemist, Res Labs, 66-74, mgr phys chem sect, 74-78, MGR INORG PROCESS DEVELOP, FMC CORP, PRINCETON, 78- *Mem:* Am Chem Soc; Sigma Xi. *Res:* Nuclear and radiation chemistry of chelate compounds; high temperature studies of fluorapatite melt systems; crystallographic studies of system sodium carbonate, carbon dioxide, water. *Mailing Add:* 111 Skyline Dr Sparta NJ 07871

ILER, RALPH KINGSLEY, chemistry; deceased, see previous edition for last biography

ILIC, MARIJA, b Zajecar, Yugoslavia, Feb 11, 51; US citizen; m; c 2. ELECTRICAL POWER SYSTEMS. *Educ:* Univ Belgrade, dipl elec eng, 74, MEE, 77; Washington Univ, St Louis, MSc, 79, DSc, 80. *Prof Exp:* Asst prof elec eng, Drexel Univ, 81-82, Cornell Univ, 82-84; from asst prof to assoc prof elec eng, Univ Ill, Urbana, 86-89; SR RES ENGR, MASS INST TECHNOL, 89- *Concurrent Pos:* Vis assoc prof elec eng, Mass Inst Technol, 87-89; Presidential Young Investr Award, NSF, 84-89. *Mem:* Inst Elec & Electronics Engrs. *Res:* Modeling and control design for interconnected electric power systems; impact of high technology on the operation of power systems and electromechanical and electronic devices; power electronics based motion control. *Mailing Add:* Bldg 10-059 Mass Inst Technol 77 Massachusetts Ave Cambridge MA 02139

ILLANGASEKARE, TISSA H, b Kanay, Sri Lanka, Feb 19, 49; Us citizen; m 77; c 2. GEOHYDROLOGY, NUMERICAL MODELING & COMPUTATIONAL METHODS. *Educ:* Univ Ceylon, Sri Lanka, BSc Hons, 71; Asian Inst Technol, MEng,74; Colo State Univ, Ft Collins, PhD(civil Eng), 78. *Prof Exp:* Instr civil eng, Univ Ceylon, 71-72; res assoc, Asian Inst Technol, 74; asst prof, Colo State Univ, 78-83 & La State Univ, 83-86; assoc prof, 86-90, PROF CIVIL ENG, UNIV COLO, BOULDER, 90- *Concurrent Pos:* Prin investr, NSF, Dept Energy,Environ Protection Agency, US Dept Interior & other orgn, 79-; mem, Publ Comt, Am Water Resources Asn, 85- *Mem:* Am Geophys Union; Am Soc Civil Engrs; Nat Water Well Asn; Am Water Resources Asn. *Res:* Transport and flow through porous media as applied to chemical and waste transport in ground water Ö656 global hydrology and greenhouse effect; global hydrology and greenhouse effect; flow in snow; dam stability; mathematical and numerical modeling of hydrolic systems. *Mailing Add:* 3867 Campo Ct Boulder CO 80301

ILLG, PAUL LOUIS, zoology, for more information see previous edition

ILLIAN, CARL RICHARD, b Trenton, NJ, Dec 31, 41; m 63; c 4. ANALYTICAL CHEMISTRY, PHARMACEUTICALS. *Educ:* Rutgers Univ, BS, 63, MS, 67; Univ Kans, PhD(analytical pharm chem), 71. *Prof Exp:* Sr chemist, Ciba-Geigy Corp, 71-77; mem staff, Pharm Div, Pennwalt Corp, 77-88, dir qual control, Pharm Div, 80-88; VPRES TECH AFFAIRS, GENETIC INST, 89- *Mem:* AAAS. *Res:* Methods development for analysis of pharmaceuticals; high speed liquid chromatography; stability of pharmaceutical systems; quality control pharmaceutical systems. *Mailing Add:* Six Meadowbrook Rd Derry NH 03038

ILLICK, J(OHN) ROWLAND, b Nanchang, China, Feb 22, 19; US citizen; m 43; c 4. THIRD WORLD DEVELOPMENT, CARTOGRAPHY & REMOTE SENSING. *Educ:* Syracuse Univ, BA, 40, MA, 41; Harvard Univ, AM, 43, PhD(geog), 54. *Prof Exp:* Prof geog, Am Univ Beirut, Lebanon, 58-60; prof, Col Petrol & Minerals, Dhahran, Saudi Arabia, 69-71; environmentalist, USAID, WAfrica, 78-80; from instr to prof, 46-86, EMER PROF GEOG, MIDDLEBURY COL, 86- *Concurrent Pos:* Prof geog, numerous univs, 56-85. *Mem:* Asn Am Geographers; Am Geog Soc; Nat Coun Geog Educ. *Res:* Regional geographic studies of China, Saudi Arabia, Lebanon, Mauritania, Vermont and Southeast Asia; environmental studies. *Mailing Add:* 16 Springside Rd Middlebury VT 05753

ILLINGER, JOYCE LEFEVER, b Philadelphia, Pa, Mar 5, 37; div; c 1. POLYMER SCIENCE, PHYSICAL CHEMISTRY. *Educ:* Pa State Univ, BS, 59; Tufts Univ, MS, 66; Univ Mass, Amherst, PhD(polymer sci), 76. *Prof Exp:* Teacher chem & biol, Frankford High Sch, Philadelphia, 59-62; teacher chem, Boy's Cent High Sch, Philadelphia, 62-63; asst, Tufts Univ, 63-66; res chemist surface chem, Cabot Corp, 66-68; res chemist polymers, Army Mat & Mech Res Ctr, US Army Europe, 69-81, chief, Mgt Info Systs, 81-85, res coordr, Mat & Technol Lab & Chem Res, Develop & Eng Ctr, 85-89, SCI ADV TO CMNDG GEN, V CORPS, US ARMY EUROPE, 89- *Mem:* Am Chem Soc; AAAS; Sigma Xi. *Res:* Studies of the interaction of water with polymeric materials; transport behavior, effects on transition temperatures, effects on mechanical properties and durability of materials; structure property of relationships in polymers. *Mailing Add:* 74 Pine Hill Circle Waltham MA 02154

ILLINGER, KARL HEINZ, b Nuernberg, Ger, May 4, 34. PHYSICAL CHEMISTRY. *Educ:* Univ Pa, AB, 56; Princeton Univ, MS, 58, PhD(phys chem), 60. *Prof Exp:* Res assoc chem, Princeton Univ, 59-60; from instr to asst prof, 60-67, chmn dept, 82-86, ASSOC PROF CHEM, TUFTS UNIV, 67- *Mem:* Am Phys Soc; Am Chem Soc; NY Acad Sci. *Res:* Intermolecular forces and collisional perturbation of molecular spectra; experimental microwave and millimeter-wave spectroscopy; interaction between electromagnetic radiation and biological systems. *Mailing Add:* Dept Chem Tufts Univ Medford MA 02155

ILLINGWORTH, GEORGE ERNEST, b Somerville, Mass, Mar 11, 35; m 56; c 3. PHYSICAL ORGANIC CHEMISTRY, INDUSTRIAL CHEMISTRY. *Educ:* Loyola Univ, Los Angeles, BS, 56; Univ Calif, Los Angeles, PhD(org chem), 63. *Prof Exp:* Res chemist, 62-74, mgr petrochem, 74-75, DIR BIOCHEM & CHEM RES, UNIVERSAL OIL PROD, INC, 75- *Mem:* Am Chem Soc; Am Inst Chem; Sigma Xi. *Res:* Elucidation of mechanisms involved in free radical reactions; addition reactions; hydrocarbon oxidation processes; petrochemicals; surfactants; fragrance and specialty chemicals; rubber and petroleum additives, water treatment and purification systems; immobilized enzymes. *Mailing Add:* 807 W White Oak St Arlington Heights IL 60005

ILLINGWORTH, KEITH, b Bradford, Eng, Nov 16, 32; m 56; c 3. FOREST GENETICS. *Educ:* Aberdeen Univ, BSc, 53, Hons, 54. *Prof Exp:* Forest ecologist, 55-58, regional res forester, Prince Rupert Dist, 58-60, Nelson Dist, 60-65 & Vancouver Dist, 65-67, forest geneticist, 67-79, mgr Tree Improv Sect & Res Sta, 79-85, DIR, RES BR, BC MINISTRY FORESTS, 85- *Mem:* Can Tree Improv Asn (exec secy, 71-75). *Res:* Silviculture; geographic variation in Pseudotsuga menziesii, Picea sitchensis and Pinus contorta; gene pool conservation; tree breeding; propagation; seed orchards. *Mailing Add:* 2835 Lincoln Rd Victoria BC V8R 6E8 Can

ILLIS, ALEXANDER, b Hungary, Jan 16, 17; nat Can; m 44; c 1. CHEMICAL METALLURGY. *Educ:* St Francis Xavier Univ, BSc, 42. *Prof Exp:* Anal chemist, Int Nickel Co Can Ltd, 42-45, res chemist, 46-50, group leader, 50-64, sr res scientist, 64-66, tech asst to dir, 67-70, res mgr, 71-74, dir process technol, 74-82; RETIRED. *Mem:* Am Chem Soc; Chem Inst Can; Can Inst Mining & Metall; Am Inst Mining, Metall & Petrol Engr. *Res:* Extractive metallurgy of nickel, copper, cobalt and associated metals. *Mailing Add:* 3303 Don Mills Rd Apt 2802 Willowdale ON M2J 4T6 Can

ILLMAN, WILLIAM IRWIN, b Chatham, Ont, Apr 10, 21; m 45; c 3. MYCOLOGY. *Educ:* Univ Western Ont, BA, 43, MSc, 46, PhD, 61. *Prof Exp:* Jr res officer, Div Appl Biol, Can Nat Res Coun, 44-46 & 48-49; FROM LECTR TO ADJ PROF BOT, CARLETON UNIV, 49- *Mem:* Mycol Soc Am; Can Bot Asn. *Res:* Mold deterioration of material; mold fermentations; sexuality and nutrition of fungi in culture; wood pathology; taxonomy of wood inhabiting fungi; morphology of plants and the taxonomy of fungi and algae. *Mailing Add:* ELBA Carleton Univ Ottawa ON K1S 5B6 Can

ILMET, IVOR, b Tartu, Estonia, Mar 29, 30; US citizen; m 56; c 2. PHYSICAL CHEMISTRY, ANALYTICAL CHEMISTRY. *Educ:* NY Univ, AB, 56, PhD(phys chem), 61. *Prof Exp:* From instr to asst prof chem, Univ Conn, 61-67; assoc prof chem, State Univ NY Col Buffalo, 67-80. *Mem:* AAAS; Am Chem Soc; Sigma Xi. *Res:* Structure and chemistry of organic charge-transfer complexes; solvent effects in spectrophotometry. *Mailing Add:* 475 Casey Rd E Amherst NY 14051

ILNICKI, RICHARD DEMETRY, b Proctor, Vt, Sept 1, 28; m 55; c 3. WEED SCIENCE. *Educ:* Rutgers Univ, BS, 49, MS, 51; Ohio State Univ, PhD(agron, plant physiol), 55. *Prof Exp:* Asst farm crops, Rutgers Univ, 50-51; asst plant genetics, Conn Agr Exp Sta, 51; res asst agron, Ohio Agr Exp Sta, 51-55; RES PROF WEED SCI, RUTGERS UNIV, NEW BRUNSWICK, 58- *Concurrent Pos:* Vpres, Northeastern Weed Control Conf, 65-66, pres, 66. *Mem:* Am Soc Agron; Weed Sci Soc Am (secy, 73-74); Crops Sci Soc Am. *Res:* Weed control in agronomic and horticultural crops; life cycle studies of economically important weeds; uptake, translocation and degradation of herbicides in plants; use of adjuvants to increase herbicide efficiency; weed/crop competition. *Mailing Add:* Dept Soils & Crops Rutgers Univ New Brunswick NJ 08903

ILOFF, PHILLIP MURRAY, JR, b State College, Pa, Jan 8, 21; m 54; c 2. ORGANIC CHEMISTRY, INORGANIC CHEMISTRY. *Educ:* Stanford Univ, BS, 48, PhD(org chem), 57. *Prof Exp:* Res chemist, Univ Calif, Berkeley, 52-56; res chemist, Callery Chem Co, Pa, 56-59; res chemist, Aerojet-Gen Corp Div, Gen Tire & Rubber Co, 59-62; asst prof chem, Whittier Col, 62- 70; assoc prof chem, Piedmont Col, 70-74, prof chem & physics, 74-87, chmn, Div Nat Sci, 83-87; RETIRED. *Mem:* Am Chem Soc; Sigma Xi. *Res:* Terpenes; turpentines; organo-boron compounds; azo compounds; free radicals; boro-hydrides; synthesis of high energy fuels and propellants. *Mailing Add:* 991 S Todd St Jupiter FL 33458

ILTEN, DAVID FREDERICK, b Marshalltown, Iowa, July 24, 38; m 68; c 3. PHYSICAL CHEMISTRY. *Educ:* Yale Univ, BA, 60; Univ Calif, Berkeley, PhD(chem), 64. *Prof Exp:* Alexander von Humboldt fel, Inst Phys Chem, Frankfurt, 64-66, Ger Res Asn res fel, 66-68; chemist, Fishkill Labs, IBM Corp, 68-72; res & teaching fel, Tech Univ Berlin, Ger, 72-74; sr res assoc, 74-78, SYST ANALYST CONTROL DATA, UNIV REGENSBURG, 78- *Mem:* NY Acad Sci; Am Chem Soc; Electrochem Soc; AAAS; Am Phys Soc. *Res:* Photochemistry of charge-transfer complexes; electron paramagnetic resonance; theoretical physical chemistry; biophysical chemistry. *Mailing Add:* CDC Stresemannala 30 Frankfurt 6 70 Germany

ILTIS, DONALD RICHARD, b Leon, Iowa, Mar 28, 36; m 60; c 2. MATHEMATICS. *Educ:* SDak Sch Mines & Tech, BS, 58; Stanford Univ, MS, 62; Univ Ore, PhD(math), 66. *Prof Exp:* Programmer, Gen Elec Co, 59-62; fel, Univ Toronto, 66-67; asst prof math, Univ NC, Chapel Hill, 67-72; from assoc prof to prof math, Willamette Univ, 72-81. *Mem:* Am Math Soc. *Res:* Abstract harmonic analysis; Banach algebras. *Mailing Add:* Williamette Univ Salem OR 97301

ILTIS, HUGH HELLMUT, b Brno, Moravia, Czech, Apr 7, 25; nat US; div; c 4. SYSTEMATIC BOTANY, BIOGEOGRAPHY. *Educ:* Univ Tenn, AB, 48; Wash Univ, St Louis, MA, 50, PhD(syst bot), 52. *Prof Exp:* From instr to asst prof, Univ Ark, 52-55; from asst prof to assoc prof, 55-67, PROF BOT, UNIV WIS-MADISON, 67-, CUR HERBARIUM, 55-, DIR, 68- *Concurrent Pos:* Expeditions, Costa Rica, 49, 89, Mex, 60, 71-88, Peru, 62-63, Hawaii, 67, Equador, 77, USSR, 79, Venezuela, 91. *Honors & Awards:* Presidential Citation, Mex, 88. *Mem:* Fel AAAS; Am Inst Biol Sci; Am Soc Plant Taxon; Int Asn Plant Taxon; Bot Soc Am; Soc Study Species Biol; Bot Soc Mex; fel Linnaean Soc London; Ecol Soc Am; Soc Study Evolution; New Eng Bot Club. *Res:* Taxonomy and evolution of New World Capparidaceae; origins of agriculture and preservation of genetic diversity in cultivated plants, especially corn; human ecology, especially human needs and adaptations for natural environment, diversity and beauty; taxonomy and evolution of Zea mays and the teosintes; flora of Wisconsin; flora of Sierra de Manantlán, Jalisco, Mexico. *Mailing Add:* Dept Bot Birge Hall Univ Wis Madison WI 53706

ILTIS, WILFRED GREGOR, b Brno, Czech, Apr 20, 23; m 48. BIOLOGY, MOSQUITOES. *Educ:* Univ Minn, BA, 57; Univ Calif, Davis, PhD(entomol), 66. *Prof Exp:* ASSOC PROF BIOL, SAN JOSE STATE UNIV, 68- *Concurrent Pos:* Fel, Sch Pub Health, Harvard Univ, 67-68. *Mem:* Ecol Soc Am; AAAS; Soc Vector Ecol; Am Mosquito Control Asn. *Res:* Biosystematics and physiological ecology of Culicidae, especialy the Culex pipiens complex; effect of social and political decisions on ecosystems and human survival. *Mailing Add:* Dept Biol Sci San Jose State Univ San Jose CA 95192

IM, JANG HI, b Seoul, Korea, June, 42; US citizen; m 73; c 2. MATERIALS SCIENCE, FRACTURE MECHANICS. *Educ:* Seoul Nat Univ, BS, 64; Mass Inst Technol, MSME, 71, ScD, 76. *Prof Exp:* Design engr machine design, Doerfer Eng & Design, 67-70; sr res engr polymer physics, Dow Chem Co, 76-79, proj leader, 79-83, res leader, microlayer coextrusion, 83-86, res assoc, polymer processing, 86-91, ASSOC SCIENTIST, HIGH PERFORMANCE FIBER & COMPOSITES, DOW CHEM CO, 91- *Concurrent Pos:* Adj prof, Univ Ill, Chicago, 89- *Mem:* Sigma Xi; Korean Scientist & Engrs Asn Am; Am Phys Soc; Mat Res Soc; Soc Advan Mat & Process Eng. *Res:* Plastics extrusion; polymer structure to processing relationship; fracture mechanics of reinforced composite materials; microlayer coextrusion; liquid crystalline fibers and composites. *Mailing Add:* Cent Res Advan Composites Lab 1702 Bldg Dow Chem Co Midland MI 48674

IM, UN KYUNG, b Seoul, Korea, June 27, 34; m 60; c 2. CHEMICAL ENGINEERING. *Educ:* Univ Mo-Columbia, BS, 57; Univ Kans, MS, 59, PhD(chem eng), 70. *Prof Exp:* Res engr, Butler Mfg Co, 57-58; process engr, Chungju Fertilizer Co, Korea, 59-60; chem engr, US Agency Int Develop, 60-67; res engr, 70-75, sr res engr, 75-78, res assoc, 78-81, SR PROJ MGR, AMOCO CHEM CORP, 81- *Mem:* Am Inst Chem Engrs; Am Chem Soc. *Res:* Thermodynamic properties of hydrocarbons, particularly phase behavior of multicomponent systems; transport properties of pure and mixed hydrocarbons; mathematical modeling of reaction systems; olefins technology. *Mailing Add:* Amoco Chem 2800 Farm Rd PO Box 568 Texas City TX 77592-0568

IMAEDA, TAMOTSU, b Nogoya, Japan, Nov 9, 27; m 55; c 2. MYCOBACTERIOLOGY, LEPROLOGY. *Educ:* Third Nat Col, BS, 49; Kyoto Univ, MD, 53, PhD(cytol), 59. *Prof Exp:* Res fel, leprosy res lab, Sch Med, Kyoto Univ, 53-54, asst dermat, Univ, 54-59; contract investr microbiol, Venezuelan Inst Sci Invest, 59-61, from assoc investr to investr, 61-70; from asst prof to assoc prof, 70-75, PROF MICROBIOL, NJ MED SCH, UNIV MED & DENT NJ, 75- *Honors & Awards:* Sakurane Prize, Japanese Leprosy Asn, 63. *Mem:* Am Soc Microbiol; Int Leprosy Asn. *Res:* Genetic characteristics of mycobacteria. *Mailing Add:* Dept Microbiol NJ Med Sch Col Med & Dent NJ 185 S Orange Ave Newark NJ 07103-2757

IMAGAWA, DAVID TADASHI, b Calif, Mar 24, 22; m 55; c 2. VIROLOGY. *Educ:* Macalester Col, BA, 44; Univ Minn, MS, 46, PhD(bact, immunol), 50; Am Bd Med Microbiol, dipl. *Prof Exp:* Instr bact & immunol, Univ Minn, 50-52; from asst prof to assoc prof, 52-76, PROF PEDIAT MICROBIOL & IMMUNOL, SCH MED, UNIV CALIF, LOS ANGELES, 76- *Concurrent Pos:* Consult, Hyland Labs, Seifuen-Imai Inst, Hyogo-Ken, Japan; vis prof, Nat Inst Animal Health, Tokyo, 63-64; dir, Pediat Res Labs, Harbor Gen Hosp, 66-76; co-dir, Ctr Child Health Res, 76-88. *Mem:* Am Soc Microbiol; Soc Exp Biol & Med; Am Asn Pathologists; Am Soc Virol; Sigma Xi. *Res:* Viruses and tumors; leukemia viruses; relationships of measles, distemper and rinderpest viruses; tissue culture; immunology; biology of human immunodeficiency virus (HIV). *Mailing Add:* Dept Pediat Harbor-Univ Calif Los Angeles Med Ctr Torrance CA 90509

IMAI, HIDESHIGE, pathology, anatomy; deceased, see previous edition for last biography

IMBER, MURRAY, b New York, NY. MECHANICAL ENGINEERING. *Educ:* Univ Ill, BS, 51; Columbia Univ, MS, 53, DEngSc, 58. *Prof Exp:* PROF MECH ENG, POLYTECH INST NY, 58- *Mem:* Am Soc Mech Engrs; Am Soc Eng Educ; assoc fel Am Inst Aeronaut & Astronaut; Sigma Xi. *Res:* Non-linear heat transfer in solids; temperature measurements; determination of golf system and sports equipment characteristics. *Mailing Add:* Dept Mech Eng 333 Jay St Brooklyn NY 11201

IMBERSKI, RICHARD BERNARD, b Amsterdam, NY, Nov 5, 35; m 67; c 2. DEVELOPMENTAL GENETICS. *Educ:* Univ Rochester, BS, 59, PhD(biol), 66. *Prof Exp:* Teaching asst biol, Univ Rochester, 60-63, res trainee, 63-65; res assoc, Johns Hopkins Univ, 65-67; asst prof zool, 67-73, ASSOC PROF ZOOL, UNIV MD, 73- *Concurrent Pos:* NIH trainee, 65-67; sr fel sci, NATO, 70; vis investr, Univ Nijmegen, 70 & 73-74, Univ Leiden, 77. *Mem:* AAAS; Genetics Soc Am; Am Soc Zool; Soc Develop Biol. *Res:* Developmental genetics; genetics of hormone production. *Mailing Add:* Dept Zool Univ Md College Park MD 20742

IMBODEN, JOHN BASKERVILLE, b Morrilton, Ark, Sept 17, 25. PSYCHIATRY. *Educ:* Univ Notre Dame, 44-46; Johns Hopkins Univ, MD, 50. *Prof Exp:* Psychiatrist in chief, Sinai Hosp Baltimore, 69-90; ASSOC PROF PSYCHIAT, JOHNS HOPKINS UNIV, 63-; PVT PRACT, 90- *Concurrent Pos:* Instr med, Johns Hopkins Univ, 58- *Mem:* AMA; Am Psychiat Asn; Am Psychoanal Asn. *Res:* Psychosomatic medicine. *Mailing Add:* 111 W Lake Ave Baltimore MD 21210

IMBRIE, JOHN, b Penn Yan, NY, July 4, 25; m 47; c 2. INVERTEBRATE PALEONTOLOGY, PALEOCLIMATOLOGY. *Educ:* Princeton Univ, BA, 48; Yale Univ, MS, 49, PhD(geol), 51. *Hon Degrees:* DSc, Univ Edinburgh, Scotland, 89. *Prof Exp:* Asst prof geol, Univ Kans, 51-52; from asst prof to prof, Columbia Univ, 52-67, chmn dept, 66-67; prof geol, 67-76, HENRY L DOHERTY PROF OCEANOG, DEPT GEOL SCI, BROWN UNIV, 76-, EMER HENRY L DOHERTY PROF, 90- *Concurrent Pos:* Vis res assoc, Lamont-Doherty Geol Observ, Columbia Univ, 71-; co-prin investr or prin investr, NSF, 71-; adj prof oceanog, Univ RI, 76-; MacArthur Prize fel, 81-85; mem, several permanent fed comts sci & sci policy & several ad hoc comts & panels. *Honors & Awards:* William Smith Lectr, Geol Soc, London, 84; Maurice Ewing Medal in Geophys, Am Geophys Union-USN, 86; Leopold von Buch Medal, Dutch Geol Asn, 90. *Mem:* Nat Acad Sci; fel Geol Soc Am; Am Philos Soc; Am Acad Arts & Sci. *Res:* Paleoecology; biometrics; history of climate aimed at identifying the main mechanisms of climatic variability on time scales ranging from a month to 100,000 years. *Mailing Add:* Dept Geol Sci Brown Univ Providence RI 02912

IMBRIE, JOHN Z, b Englewood, NJ, May 16, 56; m 85. MATHEMATICAL PHYSICS, DISORDERED SYSTEMS. *Educ:* Harvard Univ, AB, 78, AM, 79, PhD(physics), 80. *Prof Exp:* Jr fel, Harvard Soc Fel, 81-84, asst prof physics, 84-86, ASSOC PROF PHYSICS & MATH, HARVARD UNIV, 86- *Concurrent Pos:* Vis mem, Courant Inst Math Sci, 86; res fel, Alfred P Sloan Found, 86-; Presidential Young Investr award, 88- *Mem:* Am Phys Soc; Am Math Soc; Int Asn Math Physics. *Res:* Mathematical physics; phase transitions in quantum field theory and statistical mechanics; expansion methods and renormalization group methods; disordered systems. *Mailing Add:* Dept Math & Physics Harvard Univ Cambridge MA 02138

IMBRUCE, RICHARD PETER, b New York, NY, Aug 27, 42; c 1. PULMONARY PHYSIOLOGY. *Educ:* St Peter's Col, BS, 63; NY Univ, MS, 68, PhD(biol), 71. *Prof Exp:* Dir clin physiol, Sect Chest Dis, Norwalk Hosp, 70-; assoc prof respiratory ther, Univ Bridgeport, 78-; PRES, DATAMED CORP. *Concurrent Pos:* Chmn subcomt on humidifiers & nebulizers, Am Nat Standards Inst. *Mem:* Am Asn Respiratory Ther; Am Thoracic Soc. *Res:* Computerized respiratory cardio-pulmonary monitoring; effect of pharmacological agents on airway dynamics. *Mailing Add:* Pneumedics Inc 291 Pepe's Farm Rd Milford CT 06460

IMEL, ARTHUR MADISON, b San Francisco, Calif, June 30, 32; m 57; c 2. ORGANIC CHEMISTRY. *Educ:* Willamette Univ, BS, 54; Ore State Univ, MS, 58, PhD(org chem), 60. *Prof Exp:* Teaching fel chem, Ore State Univ, 54-59; sr res chem, Richmond Res Ctr, Stauffer Chem Co, 59-63; mgr res, Cent Processing Co, 63-64; prof, 64-68, HEAD DEPT CHEM, NORTHWEST NAZARENE COL, 68- *Concurrent Pos:* Guest prof, Col Idaho, 74 & 81. *Mem:* Am Chem Soc. *Res:* Organic chemistry, especially structure and mechanism; photochemical reactions; structure biological activity relationships; organic phosphorus compounds. *Mailing Add:* Northwest Nazarene Col Nampa ID 83651

IMHOF, WILLIAM LOWELL, b Oakland, Calif, Aug 17, 29; m 53. SPACE PHYSICS, NUCLEAR PHYSICS. *Educ:* Univ Calif, Berkeley, BA, 51, MA, 53, PhD(nuclear physics), 56. *Prof Exp:* Res asst phys res, Radiation Lab, Univ Calif, Berkeley, 52-56; SR STAFF SCIENTIST PHYS RES, LOCKHEED MISSILES & SPACE CO, PALO ALTO, 56- *Mem:* Sigma Xi. *Res:* Meson production by photons and protons; neutron and charged particle reactions; electrons and protons trapped in the earth's magnetic field. *Mailing Add:* 1130 Westfield Dr Menlo Park CA 94025

IMHOFF, DONALD WILBUR, b West Salem, Ohio, Dec 8, 39; m; c 2. ANALYTICAL CHEMISTRY, SPECTROSCOPY. *Educ:* Manchester Col, BS, 61; Ohio State Univ, MS, 64, PhD(analytical chem), 66. *Prof Exp:* RES ASSOC NUCLEAR MAGNETIC RESONANCE & INTERNAL REFLECTION SPECTROS, ETHYL CORP, 66- *Mem:* Am Chem Soc. *Res:* Research and methods development utilizing nuclear magnetic resonance and internal reflection spectroscopies in support of the overall research and development programs. *Mailing Add:* 3048 Woodland Ridge Blvd Baton Rouge LA 70816

IMHOFF, JOHN LEONARD, b Baltimore, Md, Feb 9, 23; m 48; c 3. INDUSTRIAL ENGINEERING. *Educ:* Duke Univ, BS, 45; Univ Minn, MS, 47; Okla State Univ, PhD, 71. *Prof Exp:* Eng asst, Crosse & Blackwell Co, 40 & Am Rolling Mill Co, 41-43; asst prof indust eng, Univ Minn, 47-52; prof indust eng & head dept, 52-80, DIR, PRODUCTIVITY CTR, UNIV ARK, FAYETTEVILLE, 80- *Concurrent Pos:* Consult engr, 49- & US Army Ord Mgt Training Hq, Pentagon, 53; partic, AEC nuclear eng inst, Univ Calif, 58; adv nuclear eng inst, Argonne Nat Lab, 59; sci fel, Stanford Univ, 60-61; nat chmn, Nat Coun Indust Eng Dept Heads, 66-67; guid chmn for Ark, Engrs Coun Prof Develop, 71. *Mem:* AAAS; Am Soc Eng Educ; Am Soc Qual Control; Nat Soc Prof Engrs; fel Am Inst Indust Engrs; Sigma Xi. *Res:* Plant design; organization and controls; statistical analysis; operations research; data processing; engineering and nuclear materials; manufacturing processes. *Mailing Add:* Dept Indust Eng Univ Ark Fayetteville AR 72701

IMHOFF, MICHAEL ANDREW, b Los Angeles, Calif, Dec 4, 42; m 65; c 2. ORGANIC CHEMISTRY. *Educ:* Univ Calif, Riverside, BA, 64; Univ Colo, Boulder, PhD(chem), 69. *Prof Exp:* Res assoc & fel, NIH, 69-70; from asst prof to assoc prof, 70-79, PROF CHEM, AUSTIN COL, 79- *Concurrent Pos:* Res grants, Robert A Welch Found, 72-81 & 86-88; vis prof, Ind Univ, 80. *Mem:* Am Chem Soc; Sigma Xi. *Res:* Carbonium ion rearrangements, reactivities and stabilities; neighboring group participation in carbonium ion and free radical reactions; isotope effects. *Mailing Add:* Dept Chem Austin Col Sherman TX 75091

IMIG, CHARLES JOSEPH, b Waterloo, Iowa, Oct 14, 22; m 44; c 6. PHYSIOLOGY. *Educ:* Coe Col, AB, 44; Univ Iowa, BS, 48, PhD(physiol), 51. *Prof Exp:* Asst, 46-51, instr, 51-52, res assoc, 52-54, asst prof, 54-58, ASSOC PROF PHYSIOL, UNIV IOWA, 58- *Concurrent Pos:* Lederle med fac award, 55-57; Arthritis & Rheumatism Found fel, 52-55. *Honors & Awards:* Sci Exhibit Award, Am Cong Phys Med, 49 & 53. *Mem:* Am Physiol Soc; Soc Exp Biol & Med. *Res:* Peripheral blood flow; microwaves; vascular muscle energetics. *Mailing Add:* Dept Physiol Univ Iowa Iowa City IA 52240

IMIG, THOMAS JACOB, b Omaha, Nebr, Jan 17, 45; m 84; c 2. NEUROBIOLOGY. *Educ:* Pomona Col, BA, 67; Univ Calif, Irvine, PhD(biol), 72. *Prof Exp:* PROF PHYSIOL, MED CTR, UNIV KANS, 85- *Mem:* Soc Neurosci; Asn Res Otolaryngol; Int Brain Res Orgn. *Res:* Structure and function of the mammalian auditory forebrain. *Mailing Add:* Dept Physiol Med Ctr Univ Kans Kansas City KS 66103

IMLAY, RICHARD LARRY, b Lockport, NY, Sept 9, 40; m 78; c 1. HIGH ENERGY PHYSICS. *Educ:* Univ Md, BS, 62; Princeton Univ, PhD(physics), 67. *Prof Exp:* Instr physics & res assoc, Cornell Univ, 67-71; res assoc, Univ Wis-Madison, 71-75; asst prof physics, Rutgers Univ, 75-79; assoc prof, 79-83, PROF PHYSICS, LA STATE UNIV, BATON ROUGE, 83- *Mem:* Am Phys Soc. *Res:* Photoproduction; weak decays; weak interactions. *Mailing Add:* Dept Physics & Astron La State Univ Baton Rouge LA 70803-4001

IMLE, ERNEST PAUL, b Marshall, Ill, Oct 15, 10; m 47; c 4. PLANT PATHOLOGY. *Educ:* Purdue Univ, BSc, 33, MSc, 36; Cornell Univ, PhD(plant path), 42. *Prof Exp:* Asst bot, Purdue Univ, 34-36, asst botanist, 36-37; asst plant path, Cornell Univ, 37-41, instr, 42; assoc pathologist, Bur Plant Indust, Soils & Agr Eng, USDA, 42-44, pathologist, 45, sr pathologist, prin agriculturist & dir, Regional Rubber Exp Sta, Turrialba, Costa Rica, 45-54, botanist, Plant Introd Sect, 55-57; dir res, Am Cocoa Res Inst, 57-71; ASST DIR, INT PROGS DIV, AGR RES SERV, USDA, 71- *Concurrent Pos:* Plant pathologist & res fel, Boyce Thompson Inst, 37-41; res adv,

Ministry of Agr, San Salvador, 70-71. *Mem:* AAAS; Phytopath Soc; Mycol Soc; Soc Hort Sci; Soc Econ Bot. *Res:* Crop improvement and diseases of tropical crops; research and training needs in tropical agriculture; plant introduction; quarantine and germ plasm problems. *Mailing Add:* 10802 Bornedale Dr Adelphi MD 20783

IMMEDIATA, TONY MICHAEL, b Riverside, NJ, June 1, 13; m 49; c 1. ORGANIC CHEMISTRY, POLYMER CHEMISTRY. *Educ:* Philadelphia Col Pharm, BSc, 36; Univ Pa, MS, 37, PhD(chem), 40. *Prof Exp:* Res chemist, Sharp & Dohme Inc, 41-45; res chemist, Int Resistance Co, 45-49, develop chemist, 49-55, chief chem engr, 55-62, sr chem consult, 62-68; sr chem consult, TRW Inc, Philadelphia, 68-70, mgr org & inorg chem, 70-78; RETIRED. *Concurrent Pos:* Consult, 78- *Mem:* Am Chem Soc. *Res:* Development on materials, mainly polymers and plastics, for electronic components and hybrid circuits; organic research, synthesis of medicinals. *Mailing Add:* 401 Bellefonte Ave Wilmington DE 19809

IMMERGUT, EDMUND H(EINZ), b Vienna, Austria, Mar 23, 28; US citizen; m 55; c 3. SCIENCE PUBLISHING, POLYMER CHEMISTRY. *Educ:* Univ Calif, Berkeley, BS, 49; Polytech Inst Brooklyn, MS, 51, PhD(polymer chem), 54. *Prof Exp:* Res assoc cellulose chem, Inst Phys Chem, Sweden, 54-55; fel, Mass Inst Technol, 55-57; mgr explor res, Dunlop Res Ctr, Ont, Can, 57-59; ed chem, Intersci Publ Div, John Wiley & Sons, New York, NY, 59-65, head bk div, 65-69; vpres & ed-in-chief, Gordon & Breach Sci Publs Inc, 69-75; ed & dir prof educ book dept, Sci Am, 75-80; ED DIR, HANSER PUBL, 80-, CONSULT ED, VCH PUBL, 80- *Concurrent Pos:* Consult, Dunlop Res Ctr, Ont, Can, 59-79; adj prof, Polytech Inst Brooklyn, 62-74. *Honors & Awards:* Fel, NY Acad Sci, Polytech Univ. *Mem:* AAAS; Am Chem Soc; Soc Plastics Engrs; Mat Res Soc; Am Phys Soc. *Res:* Cellulose; fibers; elastomers; polymer chemistry. *Mailing Add:* Two Sidney Pl Brooklyn NY 11201

IMMING, HARRY S(TANLEY), b Newark, NJ, Jan 2, 18; m 49. MECHANICAL ENGINEERING. *Educ:* Univ Mich, BS, 42; Stevens Inst Technol, MS, 58. *Prof Exp:* Aeronaut scientist, Nat Adv Comt Aeronaut, 42-48; assoc marine engr, US Naval Boiler & Turbine Lab, 48-50; asst proj engr, Curtiss-Wright Corp, 50-56; design engr, Walter Kidde & Co, 56-57; asst prof mech eng, Rutgers Univ, 57-60; equip design engr, Foster Wheeler Corp, 60-67; sr engr, C F Braun & Co, 67-70; sr engr, Ebasco Serv, Inc, Enserch Corp, 70-81; RETIRED. *Res:* Heat transfer and fluid dynamics; stress analysis as related to steam raising equipment for nuclear energy power plants. *Mailing Add:* 110 Midvale Terr Westfield NJ 07090

IMONDI, ANTHONY ROCCO, b Providence, RI, Aug 21, 40; m 63; c 4. PHARMACOLOGY, BIOCHEMISTRY. *Educ:* Univ RI, BS, 62; Univ Maine, MS, 64, PhD(animal nutrit), 66. *Prof Exp:* Res assoc biochem, Med Col, Cornell Univ, 66-69; sr pharmacologist, Warren-Teed Pharmaceut, Rohm & Haas Co, 69-74, proj leader pre-clin res, 74-77; mgr pharmacol, Adria Labs, 77-83, mgr strategic planning oncol, 83-84, dir res, 85-89, DIR PRECLIN DEVELOP, ADRIA LABS, 90- *Concurrent Pos:* USPHS res fel biochem, Walker Lab, Sloan-Kettering Inst Cancer Res, 67-69. *Mem:* AAAS; Am Soc Pharmacol & Exp Therapeut; Soc Exp Biol & Med. *Res:* Evaluation of preclinical pharmacology and toxicology for human drugs. *Mailing Add:* Adria Labs Research Park PO Box 16529 Columbus OH 43216

IMPAGLIAZZO, JOHN, b Brooklyn, NY, July 25, 41. COMPUTER MODELING. *Educ:* St John's Univ, BS, 64; State Univ NY, Stony Brook, MS, 66; Adelphi Univ, MS, 78, PhD(math), 83. *Prof Exp:* ASSOC PROF COMPUTER SCI, HOFSTRA UNIV, 85-, CHMN DEPT, 87- *Concurrent Pos:* Mem, Two Yr Col Comput Curric Task Force, Asn Comput Mach, 86-, chair, 89-91; chair, Accreditation Comt Comput Sci, Asn Comput Mach, 91- *Mem:* Asn Comput Mach. *Mailing Add:* Dept Computer Sci Hofstra Univ Hempstead NY 11550

IMPARATO, ANTHONY MICHAEL, b New York, NY, July 29, 22; m 43; c 2. MEDICINE. *Educ:* Columbia Univ, AB, 44; NY Univ, MD, 46; Am Bd Surg, dipl, 57 & 83. *Prof Exp:* Instr anat, Col Med, NY Univ, 49-50, clin surg, 53-54 & surg, 54-56, from asst prof to assoc prof, 59-71, PROF SURG, POSTGRAD MED SCH, NY UNIV, 71-, DIR, DIV VASCULAR SURG, 75- *Concurrent Pos:* From asst attend surgeon to assoc attend surgeon, NY Univ Hosp, 56-68, attend surgeon, 69-; from asst vis surgeon to assoc vis surgeon, 4th Surg Div, Bellevue Hosp, 56-68, vis surgeon, 69-; co-prin investr joint study Extracranial Arterial Occlusion, 62-72; consult, Paterson Gen Hosp, NJ, 59- & Norwalk Hosp, Conn, 76-; fel, Coun Cerebrovasc Dis, Am Heart Asn, 70; consult surg, Manhattan Vet Admin Hosp, 72- & Lenox Hill Hosp. *Mem:* Int Cardiovasc Soc; Soc Clin Vascular Surg; fel Am Col Cardiol; fel Am Surg Asn; fel Am Col Surgeons; Soc Vascular Surg (pres, 84-85); Am Surg Asn; hon mem Australasian Col Surg; James IV Asn Surg (treas). *Res:* Gastric physiology, including autonomic innervation and intrinsic circulation; hemodynamic factors in atherosclerosis; toxicity of vascular contrast media; hemodynamic factors in atherosclerosis and postoperative intimal hyperplasia; evaluation of surgical procedures on cerebrovascular insufficiency states; data monitoring committee, North American symptomatic caroted endorterestomy trial and VA cooperative study 167, etiologic importance in development of stroke; evolution of caroted bifurcation atherosclerotic plague. *Mailing Add:* 530 First Ave New York NY 10016

IMPASTATO, FRED JOHN, b Palermo, Sicily, Sept 20, 29; US citizen; m 56; c 6. ORGANIC CHEMISTRY. *Educ:* Univ Ill, BS; Fla State Univ, PhD(org chem), 59. *Prof Exp:* Control chemist, Div Hwys, State Ill, 51-53; chemist chem res & develop, Ethyl Corp, 59-61, chemist polymers group, 61-62, chemist in charge absorption spectros lab, 62-67, supvr instrumental anal group, Res & Develop Serv, 67-75, supvr indust chem res, 75-79, sr com develop assoc, 79-84, mgr, Fine Chem, 84-89, CHEM RES & DEVELOP DIR, ETHYL CORP, 89- *Mem:* Sigma Xi; Am Chem Soc. *Res:* Reaction mechanisms; cyclopropanes; organotransition metal compounds; nuclear magnetic resonance spectroscopy studies of organometallic compounds. *Mailing Add:* Ethyl Corp PO Box 14799 Baton Rouge LA 70899

IMPERATO, PASCAL JAMES, b New York, NY, Jan 13, 37; m 77; c 3. INTERNAL MEDICINE, PREVENTIVE MEDICINE. *Educ:* St John's Univ, NY, BS, 58; State Univ NY Downstate Med Ctr, MD, 62; Tulane Univ, MPH & TM, 66. *Hon Degrees:* DSc, St John's Univ, 77. *Prof Exp:* From intern to resident internal med, Long Island Col Hosp, New York, 62-65; Tulane Univ fel, Univ Valle, Colombia, 65; Glorney-Raisbeck fel, Tulane Univ, 65-66; med epidemiologist, Smallpox Eradication-Measles Control Prog, Ctr Dis Control, USPHS, Mali, 66-72; dir bur infectious dis control, New York City Dept Health, prin epidemiologist & dir immunization prog, 72-74, dir residency training prog pub health, 74-77, first dep health comnr, 74-77, comnr health, 77-78; PROF PREV MED & COMMUNITY HEALTH & CHMN DEPT, HEALTH SCI CTR, BROOKLYN, STATE UNIV NY, 78- *Concurrent Pos:* Trustee, Martin & Osa Johnson Safari Museum, 64-, Milton Helpern Libr Legal Med, 77-; consult, US Agency Int Develop, 74, Nat Res Coun, 85; lectr commun med, Mt Sinai Sch Med, 74-; asst attend physician, Dept Med, NY Hosp, 74-78; asst clin prof med, Med Col, Cornell Univ, 74-78, from asst clin prof to assoc clin prof pub health, 74-78, adj prof pub health, 78-; chmn, NY City Swine Influenza Immunization Task Force, 76-77, Bd Dirs NY City Health & Hosps Corp, 77-78, NY City Bd Health, 77-78, Exec Comt NY City Health Systs Agency & NY City Inter-Agency Health Coun, 77-78; mem bd dirs, Community Coun Greater New York, 77-78, Pub Health Res Inst, 77-78, Int Med Res Found, 78-79 & NY Heart Asn, 83-84; attend physician, Dept Med, State Univ Hosp, 78-, Kings County Hosp, 78-; mem adv bd, Physicians Social Responsibility, 83-84; dep ed, NY State J Med, 83-85, ed, 86-; mem adv coun, NY Tech Col, 84-87; sr Fulbright Fel, Yemen Arab Repub, 85; ed, J Community Health, 85-; mem, NY State Bd Med, 85-, vchmn, 91-; mem, Bd Zoning & Appeals, Plandome Heights, NY, 86-90, trustee, 90-; chmn, Metrop NY City Area Syphilis Task Force, 90-91. *Honors & Awards:* Meritorious Honor Award & Medal, US Agency Int Develop, 70, US Dept State, 71; Frank Babbot Award, Downstate Med Ctr, State Univ NY, 80; Spec Serv Award, Smallpox Eradication, USPHS, 87. *Mem:* Am Soc Trop Med & Hyg; Am Col Physicians; Am Col Prev Med; African Studies Asn; Am Col Epidemiol. *Res:* International health, preventive medicine; smallpox; measles; malaria; health care administration, health risks; African history and traditional African medicine; acquired immunodeficiency syndrome; medical editing. *Mailing Add:* Dept Prev Med State Univ NY Health Sci Ctr Brooklyn Box 43 450 Clarkson Ave Brooklyn NY 11203

IMPERIAL, GEORGE ROMERO, b Dumaguete City, Philippines, Apr 8, 29; m 56; c 3. SURFACE CHEMISTRY, POLYMER CHEMISTRY. *Educ:* Silliman Univ, Philippines, BSCh, 52; Pa State Univ, MS, 57, PhD(fuel sci), 62. *Prof Exp:* Chemist, Petrol Lab, Caltex, Inc, Manila, 53-55; res asst fuel sci, Pa State Univ, 57-61; mem tech staff, Appl Chem Group, Bell Tel Labs, Pa, 61-64; actg head dept chem, Silliman Univ, Philippines, 64-65; res assoc fuel sci, Pa State Univ, 65-66; SCIENTIST, RES & DEVELOP CTR, XEROX CORP, 66- *Mem:* AAAS; Am Chem Soc; Am Phys Soc. *Res:* Surface and semiconductor chemistry; properties of carbons and graphites; protective coats; imaging processes; electrical and surface properties of polymers; polymer/elastomer properties and structure/adhesion. *Mailing Add:* Highland Mills Highland Falls NY 10903-9527

IMPINK, ALBERT J(OSEPH), JR, b Reading, Pa, Sept 4, 31; m 53; c 4. NUCLEAR ENGINEERING. *Educ:* Villanova Univ, BME, 53; Mass Inst Technol, SM, 61, PhD(nuclear eng), 63. *Prof Exp:* Staff mem, Sandia Corp, 53-54; sr engr, Westinghouse Elec Co, 62-65, fel eng, 65-69; from asst prof to assoc prof nuclear eng, 69-78, prof, 78-80, ADJ PROF NUCLEAR ENG, CARNEGIE-MELLON UNIV, 80- *Concurrent Pos:* Consult, Westinghouse Elec Corp, 69- *Mem:* Am Nuclear Soc; Sigma Xi. *Res:* Nuclear power reactors, reactor physics testing and operations; thermonuclear power reactors, feasibility and initial design. *Mailing Add:* 18 Bel Aire Rd Belmont PA 15626

IMPRAIM, CHAKA CETEWAYO, b Sunyani, Ghana, Sept 30, 51; m 80; c 1. MOLECULAR CYTOGENETICS. *Educ:* Univ Sci & Technol, Kumasi, Ghana, BSc, 75; Univ London, PhD(biochem), 80. *Prof Exp:* Res fel, 81-, ASST RES SCIENTIST, DEPT CYTOGENETICS & CYTOL, CITY OF HOPE NAT MED CTR, DUARTE, CALIF. *Res:* Application of recombinant DNA technology to the study of human genetic diseases. *Mailing Add:* Life Technol Inc 8717 Grovemont Circle Gaithersburg MD 20877

IMSANDE, JOHN, b Grass Range, Mont, June 14, 31; m 56; c 2. GENETICS, PLANT PHYSIOLOGY. *Educ:* Univ Mont, BA, 53; Mont State Univ, MS, 56; Duke Univ, PhD(biochem), 60. *Prof Exp:* USPHS fel, Univ Calif, Berkeley, 60-61; lectr & vis fel, Princeton Univ, 61-62; from asst prof to assoc prof biol, Case Western Reserve Univ, 62-69; assoc prof genetics & biochem, 69-73, PROF GENETICS, IOWA STATE UNIV, 73- *Concurrent Pos:* Vis prof, Univ Queensland, Australia, 86-87. *Mem:* Am Soc Agron. *Res:* Genetics and biochemistry of dinitrogen fixation. *Mailing Add:* Dept Agron Iowa State Univ Ames IA 50011

INADA, HITOSHI, b Nara, Japan, Mar 10, 37; m 70; c 2. ELECTRICAL ENGINEERING. *Educ:* Tokyo Elec Eng Col, BS, 60; Northwestern Univ, MS, 66, PhD(elec eng), 69. *Prof Exp:* Res engr, Toa Electronics Ltd, Tokyo, 60-64; res asst, Northwestern Univ, 64-69; asst prof elec eng, Univ Ill, Chicago, 69-75; MEM STAFF, MASS INST TECHNOL LINCOLN LAB, 75- *Concurrent Pos:* NSF res grant, Univ Ill, Chicago, 70-71; mem US Nat Comt, Int Union Radio Sci, Wash, 71. *Honors & Awards:* Award, Inst Elec & Electronics Engrs, 71. *Mem:* Inst Elec & Electronics Engrs; Inst Electronics & Commun Eng Japan; Optical Soc Am. *Res:* Electromagnetic wave scattering; radar systems; optical wave propagation through atmosphere; imaging technique. *Mailing Add:* Mass Inst Technol Lincoln Lab PO Box 73 Lexington MA 02173-0073

INAGAMI, TADASHI, b Kobe, Japan, Feb 20, 31; m 61; c 2. BIOCHEMISTRY. *Educ:* Kyoto Univ, BS, 53, MS, 60, DAgrSc, 63; Yale Univ, MS, 55, PhD(chem), 58. *Prof Exp:* Res asst chem, Yale Univ, 57-59; instr, Hanazono Univ, Japan, 61-62; instr biochem, Sch Med, Nagoya City Univ, 62; res asst chem, Yale Univ, 62-63, res assoc molecular biophys, 63-66; from asst prof to assoc prof biochem, 66-74, PROF BIOCHEM, SCH MED, VANDERBILT UNIV, 75- *Honors & Awards:* CIBA Award, Am Heart Asn, 85; SPA Award, Belgian Nat Sci Found, 86. *Mem:* Am Soc Biol Chem; Am Chem Soc; Am Heart Asn; Japanese Biochem Soc; Endocrine Soc; Am Soc Cell Biol; Neurosci Soc; Am Soc Hypertension; AAAS; Soc Exp Biol Med. *Res:* Biochemistry and hypertension; protein and peptide chemistry; cell biology. *Mailing Add:* Dept Biochem Vanderbilt Univ Sch Med Nashville TN 37232

INAMINE, EDWARD S(EIYU), b Honolulu, Hawaii, Nov 18, 26; m 52; c 4. MICROBIOLOGY. *Educ:* Univ Hawaii, BA, 50; Wash State Univ, MS, 53, PhD(chem), 55. *Prof Exp:* Res assoc biochem, Col Med, Cornell Univ, 55-57; res chemist, Merck Sharp & Dohme Res Labs, NJ, 57-63; res biochemist, Western Regional Res Lab, Calif, 63-65; sr res chemist, Merck Sharp & Dohme Res Lab, 65-69, res fel, 69-73, sr res fel, 73-81, dir, 81-90; RETIRED. *Mem:* Am Chem Soc. *Res:* Natural products; microbial metabolism. *Mailing Add:* 376 Russel Ave Rahway NJ 07065

INANA, GEORGE, b Tokyo, Japan, Sept 14, 47; US citizen. OPHTHALMOLOGY, HUMAN GENETICS. *Educ:* Johns Hopkins Univ, BA, 70; Univ Chicago, PhD(biochem), 77, MD, 78. *Prof Exp:* Intern, dept path, Stanford Univ Hosp, 78-79, resident, 79-80; res assoc, Lab Molecular Genetics, Nat Inst Child Health & Human Develop, NIH, 80-82, Lab Molecular & Develop Biol, Nat Eye Inst, 82-83, Lab Ophthalmic Path, 83-85, chief, Sect Molecular Path, Lab Mech Ocular Dis, 85-89. *Mem:* AAAS; Asn Res Vision & Ophthal. *Res:* Molecular genetic investigations of human hereditary ocular diseases, including gyrate atrophy and retinoblastoma. *Mailing Add:* Bascom Palmer Eye Inst 1638 NW Tenth Ave Miami FL 33136

INCARDONA, ANTONINO L, b Johnstown, Pa, Mar 2, 36; m 60; c 4. VIROLOGY, NUCLEOPROTEIN STRUCTURE. *Educ:* Georgetown Univ, BS, 58; Univ Wis-Madison, MS, 60, PhD(biochem, phys chem), 62. *Prof Exp:* Mellon independent biophys fel, Mellon Inst, 62-65; asst prof chem, Fla State Univ, 65-73; ASSOC PROF VIROL, CTR HEALTH SCI, UNIV TENN, MEMPHIS, 73- *Concurrent Pos:* USPHS spec fel biol, Univ Calif, San Diego, 75-76. *Mem:* Am Soc Microbiol; Sigma Xi; Am Soc Biochem & Molecular Biol. *Res:* Characterization of protein-protein and protein-nucleic acid interactions and elucidating their role in the control of virus uncoating and assembly; theory of ultracentrifugation. *Mailing Add:* Dept Microbiol & Immunol Univ Tenn Memphis TN 38163

INCE, A NEJAT, b Bodrum, Turkey, Nov 16, 28; m 52; c 3. COMMUNICATIONS SATELLITES & EARTH STATIONS, REMOTE SENSING. *Educ:* Birmingham Univ, UK, BSc, 52; Cambridge Univ, UK, PhD(electronics), 55. *Prof Exp:* Div chief commun, Shape Tech Ctr, The Hague, Holland, 61-78; asst vpres advan develop, Western Union Tel Co, NJ, 78-79; dir res & develop, Pa Ctr Advan Studies, Cambridge, 79-81; prof telecoms, Istanbul Tech Univ, Turkey, 81-82; chief engr systs, Marconi Space & Defence Systs Ltd, UK, 82-83; secy gen res & develop, Turkish Sci Coun, 83-86; dept dir gen commun, Proj Off, PTT, Turkey, 87-90; PROF, TECH UNIV ISTANBUL, 84-; PRES RES, MARMARA RES CTR, TURKISH SCI COUN, 90- *Concurrent Pos:* Mem, Avionics Panel, NATO Adv Group Aerospace Res & Develop, 68-, adv bd, Am Int Open Un, St Louis, Mo, 75-80, Europ Sci Found, 84-86 & NATO Sci Comt, Brussels, 85-87; adv, Europ Comn, Luxemberg, 79-81; bd mem, Aselsan Electronics Co, Ankara, 84-; pres space, Turkish Space Sci & Technol Comn, 90- *Honors & Awards:* Int Commun Award, Inst Elec & Electronics Engrs, 79; Sci Award, Turkish Sci Coun, 82. *Mem:* Fel Inst Elec & Electronics Engrs; Int Acad Astronaut. *Res:* Satellite communications systems including propagation, modulation, multiple access, on-board processing, and system control; design and planning of automatically switched, stored-program controlled integrated services, digital networks; digital speech processing. *Mailing Add:* Burumcuk Sok 7/10 Cankaya Ankara Turkey

INCE, SIMON, b Istanbul, Turkey, Nov 6, 21; Can citizen; m 61. HYDRAULIC & WATER RESOURCES ENGINEERING, HYDROLOGY. *Educ:* Robert Col, Istanbul, BSc, 43; Univ Iowa, MS, 48, PhD(mech, hydraul), 53. *Prof Exp:* Instr math & civil eng, Robert Col, Istanbul, 43-46; asst to chief engr, Istanbul Off, Braithwaite & Co, 46-47; res asst hydraul, inst hydraul res, Univ Iowa, 48-50, res assoc mech & hydraul, 50-53; Nat Res Coun Can, 57-61, sr res officer, 61-63, head hydraul sect, 63-71; PROF CIVIL ENG, HYDROL & WATER RESOURCES, UNIV ARIZ, 71- *Concurrent Pos:* Secy, assoc comt, waves & littoral drift, Nat Res Coun Can, 60-69; mem working group ice in navig waters, Can Comt Oceanog, 61-71; vis lectr, Univ Ariz, 63-64, vis prof, 65 & 66. *Mem:* AAAS; Am Water Resources Asn; Int Asn Hydraul Res; Am Soc Civil Engrs; Am Geophys Union; NY Acad Sci. *Res:* Hydrology and water resources development; coastal engineering; arid lands resources; history of science. *Mailing Add:* Dept Hydrol & Water Resources Bldg 11 Univ Ariz Tucson AZ 85721

INCE, WILLIAM J(OHN), b London, Eng, Jan 19, 33; m 56, 87; c 4. ELECTRICAL ENGINEERING, SOLID STATE PHYSICS. *Educ:* Victoria Univ, Manchester, BSc, 55; Mass Inst Technol, SM, 65, PhD(elec eng), 69. *Prof Exp:* Engr, E M I Electronics Ltd, 55-59; sr engr, Raytheon Co, 59-60; asst group leader, Lab, 60-80, asst prof elec eng, Inst, 69-72, assoc group leader, 80-83, GROUP LEADER, LINCOLN LAB, MASS INST TECHNOL, 83- *Mem:* Fel Brit Inst Physics; Inst Elec & Electronics Engrs. *Res:* Theoretical and experimental investigations of microwave circuits and components; microwave magnetics; interactions between electromagnetic energy and magnetic insulators; radar systems; communications systems. *Mailing Add:* 19 Taft Ave Lexington MA 02173-4129

INCH, WILLIAM RODGER, b Port Hope, Ont, Feb 21, 28; m 54; c 3. RADIOBIOLOGY. *Educ:* Queen's Univ, BSc, 50; Univ Western Ont, PhD(biophys), 54. *Prof Exp:* Analyst & asst, Eldorado Mining & Ref Co, 46-47; asst, Nat Res Coun Can, 50; Brit Empire exchange fel radiobiol, Mt Vernon Hosp, London, Eng, 56-64; assoc prof biophys, 64-72, PROF RADIATION ONCOL, UNIV WESTERN ONT, 72-, PROF BIOPHYS, 74-; RADIOBIOLOGIST, ONT CANCER FOUND, 64- *Concurrent Pos:* Assoc, Inst Physics, Gt Brit, 56. *Mem:* Biophys Soc; Radiation Res Soc; Am Asn Cancer Res; Can Asn Physicists; Can Asn Med Physics. *Res:* Effect of radiation and cytotoxic chemicals on malignant mammalian cells; cardiotoxicity monitored by MRS-31p. *Mailing Add:* Ont Cancer Found London Clinic Victoria Hosp London ON N6A 4G5 Can

INCHIOSA, MARIO ANTHONY, JR, b Weehawken, NJ, Jan 9, 29; m 55, 77; c 3. PHARMACOLOGY, TOXICOLOGY. *Educ:* Rutgers Univ, BS, 50, MS, 53; Univ Ill, PhD(physiol), 56. *Prof Exp:* Asst, Rutgers Univ, 51-53; asst, Univ Ill, 53-56; resident res assoc pharmacol, Argonne Nat Lab, 56-58; sr res scientist physiol neuro-endocrine res unit, NY State Ment Hyg Dept, Willowbrook State Sch, 58-60; res assoc med, Harvard Med Sch, 60-66; assoc prof, 66-76, assoc dean acad affairs, 85-88, PROF PHARMACOL, NY MED COL, 76-, RES PROF ANESTHESIOL, 80-, DIR MD/PHD PROG, 81-, VCHMN PHARMACOL, 83- *Concurrent Pos:* Assoc, Beth Israel Hosp, Boston, 60-66; prof pharmacol extraodinario, Univ Guadalajara, Autoomous, 84-85; vis prof pharmacol, Univ Lille, France, 86-88. *Mem:* Am Col Clin Pharmacol; Am Chem Soc; Am Col Toxicol; Am Soc Pharmacol & Exp Therapeut; Int Soc Study Xenobiotics. *Res:* Biochemical and physiological importance of oxidation products of epinephrine in relation to involuntary muscle contraction; biochemistry of cardiac muscle proteins; doxorubicin cardiotoxicity. *Mailing Add:* Dept Pharmacol NY Med Col Valhalla NY 10595

INCROPERA, FRANK P, b Lawrence, Mass, May 12, 39; m 61; c 3. HEAT TRANSFER. *Educ:* Mass Inst Technol, SB, 61; Stanford Univ, MS, 62, PhD(mech eng), 66. *Prof Exp:* Heat transfer specialist, Lockheed Missiles & Space Co, 62-64; from asst prof to assoc prof, 66-73, PROF MECH ENG, PURDUE UNIV, 73-, HEAD, SCH MECH ENG, 89- *Concurrent Pos:* Prin investr heat transfer res grants, NSF & energy conversion grants, Dept Energy; consult, Alcoa, 3M & John Wiley & Sons. *Honors & Awards:* Roe Award, Am Soc Eng Educ, 82, Westinghouse Award, 83; Alexander von Humboldt Stiftung, Fed Repub Ger, 87; Heat Transfer Mem Award, Am Soc Mech Eng, 88, Melville Medal, 88. *Mem:* Am Soc Mech Engrs; Am Soc Eng Educ. *Res:* Heat transfer; electronic equipment cooling; manufacturing and material processing. *Mailing Add:* Sch Mech Eng Purdue Univ West Lafayette IN 47907

INDECK, RONALD S, b Minneapolis, Minn, Dec 3, 58; m 80; c 3. ELECTRICAL ENGINEERING. *Educ:* Univ Minn, BS, 80, MS, 84, PhD(elec eng), 87. *Prof Exp:* Res fel, Res Inst Elec Commun, Tohoku Univ, Japan, 87-88; ASST PROF, DEPT ELEC ENG, WASH UNIV, 88- *Concurrent Pos:* Mem, Local Prog Comt, Intermag Conf, Inst Elec & Electronics Engrs, 84-85, Secy, Magnetics Group, Twin Cities Sect, 86, vchmn, 87, chmn, Local Prog Comt, Intermag Conf, 88-92, vchmn, St Louis Combined Group, 89, session chmn, Intermag Conf, 90, chmn, St Louis Sect, Combined Chap, Inst Elec & Electronics Engrs, 90-, mem, Magnetics Soc Admin Comt, 91- *Honors & Awards:* Presidential Young Investr Award, NSF & Int Exchange Award; Centennial Young Engr Key to the Future Award, Inst Elec & Electronics Engrs; Bausch & Lomb Hon Sci Award. *Mem:* Am Phys Soc; Inst Elec & Electronics Engrs. *Res:* Fundamental applied magnetic recording research; transducers and limitations of ultra-high density magnetic recording systems; author of 16 publications. *Mailing Add:* Dept Elec Eng Wash Univ St Louis MO 63130-4899

INDELICATO, JOSEPH MICHAEL, b Cleveland, Ohio, Oct 20, 41; m 64; c 2. ORGANIC CHEMISTRY. *Educ:* St Louis Univ, BS, 63, PhD(org chem), 69. *Prof Exp:* Chemist, Anheuser Busch, Inc, 64-66; sr pharmaceut chemist, 70-75, res scientist, 75-81, SR RES SCIENTIST, ELI LILLY & CO, 81- *Mem:* Am Chem Soc; Am Asn Pharmaceut Scientists. *Res:* Solution chemistry of lactam antibiotics; drug development; dosage form design; stability; pharmaceutical chemistry. *Mailing Add:* Lilly Res Labs Eli Lilly & Co Corp Ctr Indianapolis IN 46285

INDERBITZEN, ANTON LOUIS, JR, b Sacramento, Calif, Dec 9, 35; m 59; c 4. MARINE GEOLOGY, POLAR EARTH SCIENCE. *Educ:* Stanford Univ, BS, 57; Univ Southern Calif, MA, 60; Stanford Univ, PhD(geol, geotechnol), 70. *Prof Exp:* Eng geologist, Maurseth & Howe Found Engrs, 58-61; scientist, Lockheed-Calif Co, 61-62, sr scientist, 62-67, sr scientist oceanog, Lockheed Missiles & Space Co, 67-72; dir marine opers, Univ Del, 72-77; sr staff scientist, MAR, Inc, 77-78; prog mgr appl geophys sci, 78-80, actg chief scientist, Ocean Drilling Progs, 80-81, prog dir, Sci Drilling, 81-83, assoc chief scientist, 83-86, HEAD ANTARCTIC STAFF, POLAR PROGS, NSF, 86- *Concurrent Pos:* Instr, Loyola Univ, Calif, 59-60, San Diego City Col, 63-67 & Univ Calif, Los Angeles, 67-69 & 71; Outer Continental Shelf Environ Assessment Prog Coord Panel, Dept Com, Nat Oceanic & Atmospheric Admin, 79-80, & Outer Continental Shelf Mining Policy Task Force, 79-81, Dept Interior, 79-81; US-Japan Coop Prog in Natural Resources Marine Mining Panel, 82-; USCG User Coun Working Group on Icebreaker Design, 84-86; US Deleg & NSF rep to the Antarctic Minerals Regime Negotiations, 87- *Mem:* Fel Geol Soc Am; Marine Technol Soc; AAAS; 05490066xs Union. *Res:* Mass physical properties of marine sediments; sea floor stability; marine mining; instrumentation for sea floor exploration. *Mailing Add:* US Geol Surv 104 Nat Ctr Reston VA 22092

INDICTOR, NORMAN, b Philadelphia, Pa, June 23, 32; m 59; c 1. ORGANIC CHEMISTRY, POLYMER CHEMISTRY. *Educ:* Univ Pa, AB, 53; Columbia Univ, MA, 54, PhD(chem), 58. *Prof Exp:* Res chemist, Interchem Corp, NY, 58-59; res assoc chem, Princeton Univ, 59-61; res chemist, FMC Corp, NJ, 61-63; from instr to assoc prof, 63-71, PROF CHEM, BROOKLYN COL, 72- *Concurrent Pos:* Vis scientist, Inst Fine Arts, NY Univ, 69-70, res assoc, 70- *Mem:* Am Chem Soc. *Res:* Reaction mechanisms; technical aspects of the conservation of art objects. *Mailing Add:* Dept Chem Brooklyn Col Brooklyn NY 11210

INDUSI, JOSEPH PAUL, b Ossining, NY, Apr 12, 42; m 78; c 1. NUCLEAR SAFEGUARDS & ARMS CONTROL VERIFICATION. *Educ:* Univ Bridgeport, BS, 65; State Univ NY, Stony Brook, MS, 69, PhD(appl math), 71. *Prof Exp:* Systs consult, Burndy Corp, 72-73; SCI STAFF, BROOKHAVEN NAT LAB, 73-, HEAD TECH SUPPORT ORGN, 86- *Concurrent Pos:* Consult, security prog for energy systs; dep chmn, Dept Nuclear Energy, Brookhaven Nat Lab, 90- *Mem:* Inst Nuclear Mat Mgt; Am Defense Preparedness Asn. *Res:* Development and application of mathematical models and optimization methods for analysis of material measurement, accountability systems, and physical protection systems; assessment methods of safeguard systems; development of arms control verification systems. *Mailing Add:* Bldg 197 Brookhaven Nat Lab Upton NY 11973

INFANGER, ANN, b Newark, NJ, Dec 20, 33. MITOCHONDRIAL GENETICS. *Educ:* Seton Hill Col, BA, 55; Cornell Univ, PhD(genetics), 63. *Prof Exp:* Instr, 56-59, assoc prof, 63-71, PROF BIOL, SETON HILL COL, 72- *Concurrent Pos:* Res grant, Gen Med Div, NIH, 63-69; dir, NSF grant coop prog col biol educ, 69-73. *Mem:* AAAS; Genetics Soc Am; Sigma Xi; Am Inst Biol Sci. *Res:* Cytoplasmic inheritance in Neurospora; mitochondrial genetics of Neurospora; interaction of normal and SG-1 strains. *Mailing Add:* Dept Biol Seton Hill Col Greensburg PA 15601

INFANTE, ANTHONY A, b Philadelphia, Pa, June 29, 38; m 64; c 4. BIOCHEMISTRY. *Educ:* Temple Univ, BA, 59; Univ Pa, PhD(biochem), 63. *Prof Exp:* NIH res fel, 63-64; res assoc biochem, Inst Cancer Res, 64-67; asst prof biol, 67-72, assoc prof, 72-78, PROF BIOL, WESLEYAN UNIV, 78- *Concurrent Pos:* Am Cancer Soc faculty res award, 75-80. *Mem:* Am Soc Biol Chemists; Soc Develop Biol. *Res:* Metabolism of ribonucleic acids and regulation of protein synthesis during embryonic development; DNA polymerases and regulation of DNA synthesis in eucaryotic cells; chemistry of muscle contraction; heat shock response in sea urchin embroys. *Mailing Add:* Dept Molecular Biol Wesleyan Univ Middletown CT 06457

INFANTE, ETTORE F, b Modena, Italy, Aug 20, 38; m 62. APPLIED MATHEMATICS. *Educ:* Univ Tex, BA, 58, BS, 59, PhD, 62. *Prof Exp:* Asst prof mech eng, Univ Tex, Austin, 62-64; from asst prof to prof appl math, Brown Univ, 64-84; PROF MATH & DEAN, INST TECHNOL, UNIV MINN, 84- *Concurrent Pos:* Consult, Humble Oil & Refining Co, Tex, 64-65, NSF, 80-81; dir, Div Math & Comput Sci, NSF, 81-84. *Mem:* Soc Indust & Appl Math; Am Soc Mech Eng; Inst Elec & Electronics Eng; Am Math Soc; Sigma Xi. *Res:* Differential equations; stability theory; nonlinear control systems; elasticity and plasticity of porous media; vibrations; mathematical macroeconomics. *Mailing Add:* Inst Tech Dean's Off 107 Walten Univ Minn 117 Pleasant St Minneapolis MN 55455

INFANTE, GABRIEL A, b Habana, Cuba, Nov 3, 45; US citizen; m 69; c 3. RADIATION CHEMISTRY, CANCER. *Educ:* Cath Univ PR, BS, 67; Univ PR, Mayaguez, MS, 69; Tex A&M Univ, PhD(chem), 73. *Prof Exp:* Instr chem, Cath Univ PR, 69-71; res fel, Tex A&M Univ, 71-73; fel, Carnegie-Mellon Univ, 73-74; asst prof, 74-76, assoc prof, 76-81, PROF CHEM, CATH UNIV PR, 81- *Concurrent Pos:* Dir biomed res prog, Cath Univ PR, 77-; mem, Adv Comt, Biomed Res Symp, 80, Radiation Study Sect, NIH, 80-84 & Liason Comt, Univ PR-NSF Resource Ctr Sci & Eng, 80- *Mem:* Am Chem Soc; Latin Am Chemists Fedn. *Res:* Radiosensitization characteristics of newly synthesized chemical compounds and the chemical basis of their actions; radiation chemical investigations and in vitro and in vivo tests of potential radio-chemotherapeutic agents. *Mailing Add:* Dept Chem Cath Univ PR Ponce PR 00732

INFANTE, RONALD PETER, b Newark, NJ, Sept 10, 40; m 67; c 3. DIFFERENCE ALGEBRA. *Educ:* Rutgers Univ, AB, 61; Yale Univ, MA, 63, PhD(math), 73. *Prof Exp:* Instr, Newark Col Eng, 66-67; ASSOC PROF MATH, SETON HALL UNIV, 67- *Mem:* Am Math Soc; Math Asn Am. *Res:* Transcendental Galois theory of difference field extensions. *Mailing Add:* One Arnold Dr Randolph NJ 07869

INFELD, MARTIN HOWARD, b New York, NY, Aug 3, 40; m 65; c 2. PHARMACEUTICAL DOSAGE FORM DEVELOPMENT, NEW DRUG DELIVERY SYSTEMS. *Educ:* Columbia Univ, BS, 62; Univ Wis, MS, 67, PhD(pharm), 69. *Prof Exp:* Sr chemist, 69-74, sr scientist, 74-80, tech fel, 80-84, res investr, 84-87, RES LEADER, HOFFMANN-LA ROCHE INC, 87- *Mem:* Am Asn Pharmaceut Scientists; Sigma Xi. *Res:* Development of new, stable and pharmaceutical dosage forms and development of new drug delivery systems. *Mailing Add:* Six Tuers Pl Upper Montclair NJ 07043

ING, HARRY, b China, July 5, 40; Can citizen; m 65; c 4. HEALTH PHYSICS, NUCLEAR PHYSICS. *Educ:* Univ Toronto, BASc, 65, MSc, 67, PhD(nuclear physics), 69. *Prof Exp:* res officer, Neutron Dosimetry, Atomic Energy Can Ltd, 69-88; PRES, BUBBLE TECHNOL INDUST, 88- *Concurrent Pos:* Vis scientist, Stanford Univ, 79-80; ed, J Health Physics, J Radiation Protection Dosimetry; adv mem, Nat Coun Radiation Protection & Measurements. *Mem:* Health Phys Soc; Can Asn Physicists. *Res:* Interaction of neutrons with nuclei; neutron spectra determination, experimentally and using Monte Carlo calculation; moderation of fission neutrons by ordinary materials; bubble damage polymer detector, space dosimetry. *Mailing Add:* Bubble Technol Indust Hwy 17 Chalk River ON K0J 1J0 Can

ING, SAMUEL W(EI-HSING), JR, b Shanghai, China, Sept 26, 32; US citizen; m 58; c 2. CHEMICAL ENGINEERING, PHYSICAL CHEMISTRY. *Educ:* Mass Inst Technol, SB, 53, SM, 54, ScD(chem eng), 59. *Prof Exp:* Asst prof chem eng, Bucknell Univ, 58-59; sr engr, Raytheon Co, 59-60; physicist, adv semiconductor lab, Gen Elec Co, 60-62, electronics lab, 62-64; sr scientist res labs, 64-71, prin scientist, 71-77, LAB MGR, XEROX CORP, 77- *Mem:* Fel Am Inst Chem; Am Chem Soc; Soc Photog Scientists & Engrs; Sigma Xi. *Res:* Materials, chemistry and physics of semiconductors and devices; electrophotographic sciences. *Mailing Add:* 740 John Glenn Blvd Webster NY 14580

INGALLS, JAMES WARREN, JR, b Barre, Vt, July 31, 19; m 44; c 3. LIMNOLOGY & LAKE BASINS. *Educ:* Univ Maine, BS, 42; NY Univ, MS, 49, PhD(parasitol), 53. *Prof Exp:* Asst zool, Univ Maine, 42; asst biol, NY Univ, 47-50; instr microbiol, Long Island Univ, 51, from asst prof to prof pharmacol, 52-82, chmn dept, 63-77, EMER PROF PHARMACOL, LONG ISLAND UNIV, 82- *Concurrent Pos:* Mem comn schistosomiasis, US Army, 45-46; vis asst prof, Einstein Col Med, 58-67; consult, Air Reduction Co, 66-71. *Mem:* AAAS; Sigma Xi; fel NY Acad Sci. *Res:* Practical political protection of lakes and lakeshore and forest residents. *Mailing Add:* Box 570 North Hudson NY 12855

INGALLS, JESSE RAY, b Randolph, Vt, Apr 4, 36; m 63; c 3. ANIMAL SCIENCE. *Educ:* Univ Vt, BS, 58, MS, 60; Mich State Univ, PhD(dairy nutrit), 65. *Prof Exp:* From asst prof to assoc prof, 64-74, PROF ANIMAL SCI, UNIV MAN, 74- *Mem:* Am Soc Animal Sci; Am Dairy Sci Asn; Can Soc Animal Prod; Am Forage & Grassland Coun. *Res:* Dairy nutrition; forage utilization; use of rape seed meal in dairy rations. *Mailing Add:* Dept Animal Sci Univ Man Winnipeg MB R3T 2N2 Can

INGALLS, PAUL D, b Hood River, Ore, May 18, 44. NUCLEAR PHYSICS, UNDERWATER ACOUSTICS. *Educ:* Univ Wash, BS, 66; Princeton Univ, PhD(nuclear physics), 71. *Prof Exp:* Res assoc, Univ Colo, 71-73; res fel, Calif Inst Technol, 73-76; vis asst prof physics, Univ Ore, 76-78; PHYSICIST, APPL PHYSICS LAB, UNIV WASH, 78- *Mem:* Am Phys Soc; Acoust Soc Am. *Mailing Add:* Appl Phys Lab Univ Wash, HN-10 Seattle WA 98195

INGALLS, ROBERT L, b Spokane, Wash, June 15, 34; m 61; c 3. PHYSICS. *Educ:* Univ Wash, BS, 56; Carnegie Inst Technol, MS, 60, PhD(physics), 62. *Prof Exp:* Instr physics, Carnegie Inst Technol, 61-63; res assoc, Univ Ill, 63-65, res asst prof, 65-66; from asst prof to assoc prof, 66-74, PROF PHYSICS, UNIV WASH, 74- *Concurrent Pos:* Vis scholar, State Univ Groningen, 72-73. *Mem:* Am Phys Soc; Fedn Am Sci. *Res:* Synchrotron radiation; solid state physics; high pressure physics; Mossbauer effect. *Mailing Add:* Dept Physics FM-15 Univ Wash Seattle WA 98195

INGALLS, THEODORE HUNT, epidemiology; deceased, see previous edition for last biography

INGALLS, WILLIAM LISLE, b Franklin Co, Ohio, Sept 28, 18; m 46; c 2. VETERINARY PATHOLOGY, VETERINARY MEDICINE. *Educ:* Ohio State Univ, DVM, 42, MS, 47. *Prof Exp:* Asst pathologist, State Lab, Ohio, 42-45; assoc animal pathologist, Exp Sta, Va Polytech Inst, 45-47; from instr to asst prof vet path, Ohio State Univ, 47-52; vpres, Columbus Serum Co, 52-74, secy, Col Vet Med, 77-78, assoc prof, 74-76, PROF VET PREV MED, COL VET MED, OHIO STATE UNIV, 76-, PROF VET COOP EXT, 76-, PROF ANIMAL SCI, 78- *Concurrent Pos:* Prof vet sci, Ohio Agr Res & Develop Ctr, Wooster, 76-, prof animal sci, 78-; health comnr, Pickaway Co, Circleville, Ohio, 88. *Mem:* Fel Vet Med Asn. *Res:* Bovine mastitis, bacteriology, pathology and trichomoniasis; protozoology; avian pathology; Newcastle disease; sulfaquinoxaline-swine-diseases. *Mailing Add:* Col Vet Med Ohio State Univ 1900 Coffey Rd Columbus OH 43210

INGALSBE, DAVID WEEDEN, b Yakima, Wash, May 29, 27; m 51; c 3. FOOD SCIENCE, NATURAL PRODUCTS CHEMISTRY. *Educ:* Wash State Univ, BS, 51. *Prof Exp:* Res asst, Ore State Univ, 51-54; res chemist, USDA, 54-67; vpres & tech dir, Ermey Vineyards, 67-71; owner, D W Ingalsbe Co, 72-73; PRES, INGALSBE ENTERPRISES, INC, 74- *Mem:* Inst Food Technol. *Res:* Postharvest physiology of tree fruits, grapes and berries; anthocyanin pigments in fruits; chemistry of essential oils; measurement of quality of processed fruit and vegetable products. *Mailing Add:* 909 S 41st Ave Yakima WA 98908

INGARD, KARL UNO, b Gothenburg, Sweden, Feb 24, 21; nat US; m 48; c 4. FLUID DYNAMICS. *Educ:* Chalmers Univ Technol, Sweden, EE, 44, Techn lic, 48; Mass Inst Technol, PhD(physics), 50. *Prof Exp:* Res engr, Nat Lab Defense, Stockholm, 45-46; dir acoust lab, Chalmers Univ Technol, Sweden, 46-52; from asst prof to assoc prof, Mass Inst Technol, 52-66, prof physics, 66-91, prof aeronaut & astronaut, 71-91, EMER PROF PHYSICS & AERONAUT & ASTRONAUT, MASS INST TECHNOL, 91- *Concurrent Pos:* Armstrong Cork fel, Mass Inst Technol, 50; Guggenheim fel, 59. *Honors & Awards:* Biennial Award, Acoust Soc Am, 54; Gustaf Dalen Medal, Sweden, 70; John Ericsson Medal, Am Soc Swed Engr, 72. *Mem:* Nat Acad Eng; fel Acoust Soc Am; fel Am Phys Soc; Inst Noise Control Eng. *Res:* Plasma physics; acoustics. *Mailing Add:* 22 Captains Way Gerrish Island Kittery Point ME 03908

INGE, WALTER HERNDON, JR, b Mobile, Ala, July 24, 33; m 56; c 2. ENVIRONMENTAL PHYSIOLOGY, OCCUPATIONAL HEALTH. *Educ:* Univ Ala, BS, 55; Univ Calif, Berkeley, MA, 65, PhD(physiol), 72. *Prof Exp:* Instr physiol, USAF Acad,Colo, 65-66; opers analyst, Weapons Effectiveness Test Div, Armament Test Ctr, Air Force, Eglin AFB, Fla, 70-71; proj officer biol & behav sci, Europ Off of Aerospace Res, London, 71-73, chief, Eng & Biotech Div, 73-75; res physiologist environ physiol, Aerospace Med Res Lab, Air Force, Ohio, 75-77; asst prof, Dept Health Sci, Sargent Col, Boston Univ, 77-83; prof, Emory Univ, 83-87; prof, St Ga Sch Med, 87-88; MED WRITER, 88- *Concurrent Pos:* US Rep, Working Party, Aerospace Physiol & Med & Working Party Effects Ionizing Radiations, Coun Europe, Strasbourg, France, 72-75; informal liaison, NASA Europ Space Res Orgn, 72-75; mem bd dirs, Nat Safety Training Found, Inc, 78-; consult occup health & safety, Western Elec Co, 78- *Mem:* Aerospace Med Asn; Am Col Sports Med; Am Heart Asn; Am Inst Biol Sci; Med Electronics & Data Soc. *Res:* Environmental physiology such as extremes of temperature, altitude, pressure, acceleration; occupational hazards and disabilities; exercise physiology; cardiopulmonary physiology. *Mailing Add:* 753-7 Houston Mill Rd NE Atlanta GA 30329

INGELS, FRANKLIN M(URANYI), b New York, NY, Aug 7, 37; m 55; c 3. ELECTRICAL ENGINEERING, MATHEMATICS. *Educ:* Univ Kans, BS, 60, MS, 62; Miss State Univ, PhD(elec eng), 67. *Prof Exp:* Engr, Oread Electronics Lab, Inc, 60-62; group engr, Dynatronics, Inc, 62-64; from instr to assoc prof, 64-76, PROF ELEC ENG, MISS STATE UNIV, 76- *Concurrent Pos:* Consult, Battelle Mem Inst, 77-78, 84-, FWG assoc, 89, SCI Systs, 80, 81, Gen Dynamics, 81; consult, Inst Defense Anal, 67, staff consult, Los Alamos Sci Lab, 67-; staff consult & sci adv, Mil Asst Command, Vietnam, 68; army consult, 85- *Mem:* Inst Elec & Electronics Engrs. *Res:* Communications and electronic systems; information theory and radar system analysis. *Mailing Add:* Drawer EE Miss State Univ Mississippi State MS 39762

INGELS, NEIL BARTON, JR, b Evanston, Ill, July 15, 37; m 59; c 2. BIOMEDICAL ENGINEERING, CARDIOVASCULAR PHYSIOLOGY. *Educ:* Univ Ark, BSEE, 59; Santa Clara Univ, MSEE, 63; Stanford Univ, PhD(biomed eng), 66. *Prof Exp:* Systs designer guided missiles, Sperry Utah Eng Lab, 59-60; systs designer space satellites, Lockheed Missile & Space Co, 60-62; sr res assoc cardiovasc physiol, Palo Alto Med Res Found, 62-78; CONSULT ASST PROF MED, STANFORD UNIV MED CTR, CALIF, 79-, LECTR, DEPT ELEC ENG, 80- *Concurrent Pos:* Consult, Stanford Univ Sch Med, 74-, Alza Res, 78- *Mem:* Biomed Eng Soc; Am Heart Asn; Am Physiol Soc; AAAS. *Res:* Basic and clinical studies of heart function with special emphasis on cardiac muscle dynamics. *Mailing Add:* Palo Alto Med Res Found 860 Bryant St Palo Alto CA 94301

INGENITO, ALPHONSE J, b Harrison, NJ, Oct 8, 32; m 62; c 4. PHARMACOLOGY. *Educ:* St John's Univ, BS, 53; Rutgers Univ, MS, 60; NJ Med Col, PhD(pharmacol), 65. *Prof Exp:* From instr to assoc prof pharmacol, Albany Med Col, 65-76; assoc prof 76-78, PROF PHARMACOL, E CAROLINA UNIV, 78- *Mem:* AAAS; Am Soc Pharmacol & Exp Therapeut; Am Col Clin Pharmacol. *Res:* Actions of anti-hypertensive and other drugs on the central and reflex neural control of the cardiovascular system; role of endogenous opioid peptides in experimental hypertension. *Mailing Add:* Dept Pharmacol Sch Med ECarolina Univ Greenville NC 27858

INGENITO, FRANK LEO, b Brooklyn, NY, Aug 15, 32; m 59. UNDERWATER ACOUSTICS. *Educ:* Univ Rochester, BS, 54; Brown Univ, PhD(physics), 67. *Prof Exp:* Physicist, Pratt & Whitney Div, United Aircraft Co, 59-61; res assoc physics, Mich State Univ, 66-68; RES PHYSICIST, US NAVAL RES LAB, 68- *Mem:* Acoust Soc Am; Am Phys Soc. *Res:* Acoustic signal and ambient noise characteristics in shallow water. *Mailing Add:* 5031 Fulton St NW Washington DC 20016

INGERSOL, ROBERT HARDING, b Stillwater, Okla, Mar 25, 21; m 48; c 2. MAMMALOGY, ECOLOGY. *Educ:* Okla State Univ, BS, 49, MS, 53, PhD(zool), 68. *Prof Exp:* Teacher sr high sch, Okla, 53-59; asst prof biol sci, Okla State Univ, 59-66; assoc prof biol, Drury Col, 68-88; RETIRED. *Mem:* AAAS; Am Soc Mammal; Am Inst Biol Sci; EAfrican Wildlife Soc. *Res:* Ethiopian mammals and their ecological circumstances. *Mailing Add:* 436 Lakeview Terr Springfield MO 65807

INGERSOLL, ANDREW PERRY, b Chicago, Ill, Jan 2, 40; m 61; c 5. PLANETARY ATMOSPHERES. *Educ:* Amherst Col, BA, 60; Harvard Univ, AM, 61, PhD(atmos physics), 65. *Prof Exp:* Res fel atmos physics, Harvard Univ, 65-66; from asst prof to assoc prof planetary sci, 66-76, PROF PLANETARY SCI, CALIF INST TECHNOL, 76- *Concurrent Pos:* mem summer study prog, Geophys Fluid Dynamics, Woods Hole Oceanog Inst, 65, 70-73, 76, 80; chmn Div Planetary Sci, Am Astron Soc, 89-90; mem Spacecraft exp teams, Pioneer 10/11, Pioneer Venus, Nimbus 7, Voyager, Galileo, Mars Observer, Cassini. *Honors & Awards:* Except Sci Achievement Award, NASA, 81. *Mem:* Am Astron Soc; fel Am Geophys Union; fel AAAS. *Res:* Dynamics of planetary atmospheres; solar system exploration by means of deep space probes; climates of earth and planets; geophysical fluid dynamics. *Mailing Add:* Div Geol & Planetary Sci Calif Inst Technol Pasadena CA 91125

INGERSOLL, EDWIN MARVIN, b Minn, Dec 23, 19; m 48; c 4. ZOOLOGY. *Educ:* Bemidji State Teachers Col, BS, 42; Univ Minn, MS, 49, PhD(zool), 54. *Prof Exp:* Asst prof, Miami Univ, 51-70, asst chmn dept zool & physiol, 66-70, prof zool, 70-85, asst chmn dept zool, 75-85; RETIRED. *Mem:* Soc Parasitol. *Res:* Parasitology. *Mailing Add:* 502 Glenview Dr Oxford OH 45056

INGERSOLL, HENRY GILBERT, b Chestertown, Md, Sept 11, 15; m 43; c 5. PHYSICAL CHEMISTRY. *Educ:* Washington Col, Md, BS, 35; Univ Md, MS, 37; Mass Inst Technol, PhD(phys chem), 40. *Prof Exp:* Res chemist, E I du Pont de Nemours & Co, Inc, 40-51, sr res chemist, 51-74, res assoc, 74-79; RETIRED. *Mem:* Am Chem Soc. *Res:* Compressibility of hydrocarbons; physical structure of high polymers; mechanism of film and filament formation; heterogeneous catalysis. *Mailing Add:* 633 Horseshoe Hill RD 3 Box 288 Hockessin DE 19707-9803

INGERSOLL, RAYMOND VAIL, b New York, NY, June 17, 47; m 72; c 3. SEDIMENTARY GEOLOGY, TECTONICS. *Educ:* Harvard Univ, AB, 69; Stanford Univ, MS, 74, PhD(geol), 76. *Prof Exp:* Teacher phys sci, Putney Sch, Vt, 69-72; from asst prof to assoc prof geol, Univ NMex, 76-82; dept earth & space sci, 82-85, PROF GEOL, UNIV CALIF, 85- *Concurrent Pos:* Assoc ed, J Sedimentary Petrol, 84-88, Geol Soc Am Bulletin, 84- *Honors & Awards:* A I Levorsen Award, Am Asn Petrol Geologists, 78. *Mem:* Fel Geol Soc Am; Int Asn Sedimentologists; Soc Sedimentary Geol; Sigma Xi; Am Geophys Union. *Res:* Sandstone petrology; turbidite sedimentation; basin analysis; plate tectonics. *Mailing Add:* Dept Earth & Space Sci Univ Calif Los Angeles CA 90024-1567

INGERSON, FRED EARL, b Barstow, Tex, Oct 28, 06; m 30, 83; c 1. GEOLOGY, GEOCHEMISTRY. *Educ:* Simmons Univ, Tex, AB, 28, MA, 31; Yale Univ, PhD(geol), 34. *Hon Degrees:* ScD, Hardin-Simmons Univ, 42. *Prof Exp:* Instr chem, Simmons Univ, Tex, 30-31; instr mineral, Yale Univ, 32-34; Sterling fel, Innsbruck Univ, 34-35; from asst phys chemist to phys chemist, Carnegie Inst Geophys Lab, 35-43, petrologist, 43-47; geologist & chief br geochem & petrol, US Geol Surv, 47-58; prof, 58-77, assoc dean grad sch, 61-64, EMER PROF GEOL, UNIV TEX, AUSTIN, 77- *Concurrent Pos:* Mem Yale Univ exped, Nfld, 33; mem fac, USDA Grad Sch, 40-44; ed, Geochem & Cosmochem Acta, 50-63; mem adv comt geophys, Off Naval Res, 50-53; mem div earth sci, Nat Res Coun, 51-54, div chem & chem technol, 51-57, chmn comt earth sci Fulbright awards, 53-56; mem comn geochem, NSF, 53-55; ed, Trans Geokhimija, 56-61; secy, Int Comn Geochem, 57-60, vpres, 60-63; sr ed, Int Geol Rev, 59-62; consult, Exxon Res, 59-78; hon fel, Comn Ore-Forming Fluids, Inst Asn Genesis Ore Deposits. *Honors & Awards:* Day Medal, Geol Soc Am, 55; Ingerson distinguished lect, 85. *Mem:* Fel Mineral Soc Am (treas, 41-58); hon fel Geochem Soc (pres, 55-57); Soc Econ Geol; fel Am Geophys Union; hon fel Int Asn Geochem & Cosmochem (pres, 65-72). *Res:* Quartz deposits; artificial quartz; hydrothermal silicate systems; petrofabric analysis; apparatus for accurate orientation of thin sections; universal compass for measuring lineations directly; geologic thermometry; isotope geology; tektites; geochemistry of sedimentary rocks; origin of ore deposits. *Mailing Add:* Dept Geol Sci Univ Tex Austin TX 78713-7909

INGHAM, HERBERT SMITH, JR, b Los Angeles, Calif, Nov 15, 31; m 60; c 2. SOLID STATE PHYSICS. *Educ:* Rensselaer Polytech Inst, BS, 53; Carnegie Inst Technol, MS, 56, PhD(physics), 59; Hofstra Univ, JD, 82. *Prof Exp:* Engr, Photocircuits, Inc, 53; res asst, Carnegie Inst Technol, 53-58; assoc physicist, Res Ctr, Int Bus Mach Corp, 58-60; mgr res & develop, Metco, Inc, 60-79, adminr patents & contracts, 79-83; SR PATENT ATTY, PERKIN-ELMER, 83- *Mem:* Am Bar Asn. *Res:* Nuclear radiation effects in solids; ionic conductivity in crystals; vapor deposition, crystal growth; plasma and combustion flames, flame-spray materials and process; coating properties. *Mailing Add:* 38 Milmohr Ct Northport NY 11768

INGHAM, KENNETH CULVER, b Ann Arbor, Mich, Feb 14, 42; m 63; c 2. PHYSICAL BIOCHEMISTRY. *Educ:* Eastern Mich Univ, BS, 64; Univ Colo, PhD(chem), 70. *Prof Exp:* Res assoc biophys, Mich State Univ, 70-72; staff fel, Nat Inst Arthritis, Metab & Digestive Dis, 72-75; res scientist, 75-80, sr res scientist, Blood Res Lab, 80-85, HEAD, BIOCHEM LAB, BIOMED RES & DEVELOP LABS, AM NAT RD CROSS, 85- *Concurrent Pos:* Res career develop award, Nat Inst Health, 77-82. *Mem:* Biophys Soc; Am Phys Soc; Am Soc Biol Chemists; Protein Soc; Int Soc Biorecognition Technol. *Res:* Purification, structure and function of human plasma proteins; fluorescense probs of protein-protein interactions. *Mailing Add:* 11308 Rokeby Ave PO Box 58 Garrett Park MD 20766

INGHAM, KENNETH R, b Cambridge, Mass, Aug 26, 38; m 58; c 3. PHYSICS, COMPUTER SCIENCE. *Educ:* Boston Univ, BA, 60; Brandeis Univ, MA, 63, PhD(physics & astrophys), 67. *Prof Exp:* Staff engr, Res Lab Electronics, Mass Inst Technol, 63-67, res assoc, 67-71, res affil, 71-77; dir, Arts Bur, Protestant Guild for Blind, 71-73; founder & pres, ASI Teleprocessing, 69-78, chmn bd, 78-81; pres, founder, chief exec officer & treas, Jupiter Technol, Inc, 81-86, chmn bd & treas, 86-89; consult, Intel Corp, 89; CHMN BD & CHIEF EXEC OFFICER, ALLMEDIA SOLUTIONS, INC, 89- *Concurrent Pos:* Sr consult, Mitre Corp, 67-69; consult, Comput Notation Braille Proj, Bur Educ Handicapped, Fla State Univ; consult, Mass Inst Technol, 72-73; consult, Inst Corp & Govt Strategy, 82-83; mem bd, Lexington Waldorf Sch; trustee, Carroll Ctr Blind, Stone Soup Arts Trust Inc, Christian Community, Inc; dir, Rudolf Steiner Libr Blind, Inc. *Mem:* Inst Elec & Electronics Engrs. *Res:* Speech analysis, generation, synthesis and recognition; computer use in artificial intelligent language translation especially punctiliographics such as Braille, word processing and text, data base formatting; packet switching network analysis; communications, telecommunication and data; Information Service Data Network standards; author of numerous professional and business publications. *Mailing Add:* 111 Gibbs St Newton Centre MA 02159

INGHAM, MERTON CHARLES, b Stockton, Calif, Jan 9, 30; m 55; c 3. PHYSICAL OCEANOGRAPHY, BIOLOGICAL OCEANOGRAPHY. *Educ:* Ore State Univ, BS, 53, MS, 59, PhD(oceanog), 65. *Prof Exp:* Teacher jr high sch, Ore, 55-56 & sr high sch, 56-58 & 59-62; res oceanogr, Trop Atlantic Biol Lab, Bur Com Fisheries, 65-69; dir oceanog unit, USCG, 69-72; chief, Atlantic Environ Group, 72-85, CHIEF PHYS OCEANOG BR, NAT MARINE FISHERIES SERV, NAT OCEANIC & ATMOSPHERIC ADMIN, 85- *Res:* Fishery oceanography; descriptive physical oceanography. *Mailing Add:* Phys Oceanog Br NOAA/NMFS 28 Tarzwell Dr Narragansett RI 02882

INGHAM, ROBERT KELLY, b Bristol, Va, Sept 26, 26; m 52. ORGANIC CHEMISTRY. *Educ:* King Col, AB, 47; Iowa State Univ, PhD(chem), 52. *Prof Exp:* Res assoc chem, Iowa State Univ, 53; from asst prof to assoc prof, 53-63, PROF CHEM, OHIO UNIV, 63- *Mem:* Am Chem Soc; Royal Soc Chem. *Res:* Chemistry of organophosphorus compounds; heterocyclic compounds; organometallic chemistry; biological action; chemical constitution. *Mailing Add:* Dept Chem Ohio Univ Athens OH 45701-2978

INGHAM, STEVEN CHARLES, b Portland, Ore, Mar 6, 61; m 88. MICROBIOLOGY. *Educ:* Cornell Univ, BS, 83, MS, 85, PhD(food sci), 88. *Prof Exp:* Food technologist, US Army Natick Res & Develop Ctr, 85; asst prof, Dept Food Sci, La State Univ, 88-89; ASST PROF, DEPT APPL MICROBIOL & FOOD SCI, UNIV SASK, 89- *Mem:* Inst Food Technologists; Int Asn Milk Food & Environ Sanitarians; Can Inst Food Sci & Technol. *Res:* Microbiological safety ramifications of modern food processing and packaging techniques; anaerobic microbiological techniques for evaluation of microbiological safety of foods. *Mailing Add:* Dept Appl Microbiol & Food Sci Univ Sask Saskatoon SK S7N 0W0 Can

INGHRAM, MARK GORDON, b Livingston, Mont, Nov 13, 19; m 46; c 2. PHYSICS. *Educ:* Olivet Col, BA, 39; Univ Chicago, PhD(physics), 47. *Prof Exp:* Jr physicist, Univ Minn, 42; physicist, Manhattan Proj, Columbia Univ, 42-45; sr physicist, Argonne Nat Lab, 45-49; from instr to prof physics, 47-69, chmn dept, 59-70, assoc dean, Div Phys Sci, 64-71 & 81-85, Samuel K Allison Distinguished Serv Prof physics, 69-85, master & assoc dean col, 81-85, EMER PROF PHYSICS, UNIV CHICAGO, 85- *Concurrent Pos:* Mem, Comt Sci & Pub Policy, Nat Acad Sci, 66-69. *Honors & Awards:* Smith Medal, Nat Acad Sci, 57. *Mem:* Nat Acad Sci; fel Am Phys Soc; Am Acad Art & Sci. *Res:* Nuclear physics, geophysics and chemical physics as studied using mass spectrometric techniques. *Mailing Add:* 3077 Lakeshore Dr Holland MI 49424

INGLE, DONALD LEE, b Kendrick, Idaho, Sept 4, 36; m 66; c 3. ANIMAL NUTRITION. *Educ:* Univ Idaho, BS, 58, MS, 68; Univ Ill, PhD(ruminant nutrit), 71. *Prof Exp:* Ext agr agent, Bonner County, Idaho, 59-61; ext agr agent, Boundary County, Idaho, 61-65; sr res nutritionist, Am Cyanamid Co, 72-76, group leader nutrit & physiol, 76-81, mgr, Animal Indust Res, Agr Res Div, Nutrit & Physiol Res Dept, 81-82, mgr, Animal Indust Develop, 82-89, MGR INT REGULATORY AFFAIRS, ANIMAL INDUST DEVELOP, AM CYANAMID CO, 89- *Concurrent Pos:* Res assoc, Dairy Sci Dept, Univ Ill, 71-72; co-adj fac human nutrit, Div Nursing, Trenton State Col, 73-78. *Mem:* Am Soc Animal Sci; Am Dairy Sci Asn; AAAS; Nutrit Today Soc; Am Inst Nutrit; Am Meat Sci Asn. *Res:* Improving growth and body composition of farm animals; mechanisms controlling fat deposition and lipid metabolism. *Mailing Add:* Dept Nutrit & Physiol Res Am Cyanamid Co PO Box 400 Princeton NJ 08543-0400

INGLE, GEORGE WILLIAM, b Lynbrook, NY, May 11, 17; m 46; c 4. PHYSICAL CHEMISTRY, RESEARCH ADMINISTRATION. *Educ:* Colgate Univ, AB, 38; Inst Paper Chem, MS, 40. *Prof Exp:* Control chemist, Plastic Div, Monsanto Co, 40-42, color physicist, 42-46, operating supt, Plant Labs, 46-49, group leader color res, 49-54, sect leader polystyrene process & prod develop, Res Dept, 54-60, asst dir res, 60-64, mgr res, Property & Standards Sect, 64-67, mgr res prod property develop, 67-68, dir tech liaison, 68-75; asst tech dir, Mfr Chemists Asn, 75-78; dir asn liaison nat & int, Chem Mfrs Asn, 78-85; RETIRED. *Concurrent Pos:* Chmn, Int Standardization Orgn Tech Comt 61, 74-77. *Mem:* Am Soc Testing & Mat; Am Chem Soc. *Res:* Colorimetry; coloring of plastics; process and product development of polystyrene plastics; toxicity of plastics for food packaging; plastics waste management; industrial solid waste management; toxicity; toxic substances control legislation and regulation, United States and International. *Mailing Add:* 125 Mountain Rd Hampden MA 01036

INGLE, JAMES CHESNEY, JR, b Los Angeles, Calif, Nov 6, 35; m 58; c 1. MICROPALEONTOLOGY, MARINE GEOLOGY. *Educ:* Univ Southern Calif, BS, 59, MS, 62, PhD(geol), 66. *Prof Exp:* Asst, Univ Southern Calif, 59-60; jr geologist, Shell Oil Co, 60; res assoc, dept geol & Allan Hancock Found, Univ Southern Calif, 61-67; from asst prof to prof, 68-84, chmn dept, 82-86, W H PECK PROF EARTH SCI, STANFORD UNIV, 84- *Concurrent Pos:* Res assoc, Los Angeles Co Mus, 63-; vis scholar, Inst Geol & Paleont, Tohoku Univ, Japan, 66-67; scientist, Leg 18, Deep-Sea Drilling Proj, 71, co-chief scientist, Leg 31, 73; geologist, US Geol Surv, 76-81; distinguished lectr, Am Asn Petrol Geologists, 86; co-chief scientist, Leg 128, Ocean Drilling Prog, 89. *Honors & Awards:* Lewis G Weeks lectr, Univ Wis, 86; A I Levorsen Award, Am Asn Petrol Geol, 88. *Mem:* AAAS; fel Geol Soc Am; Soc Econ Paleont & Mineral; Am Geophys Union. *Res:* Marine stratigraphy; biostratigraphy; marine geology; Cenozoic foraminiferal biostratigraphy and paleoceanography of the Pacific region; Cenozoic geology of the Pacific coast of North America. *Mailing Add:* Dept Geol Sch Earth Sci Stanford Univ Stanford CA 94305-2115

INGLE, JAMES DAVIS, b Louisiana, Mo, Sept 21, 13; m 38; c 1. FOOD SCIENCE. *Educ:* Ottawa Univ, BS, 32; Univ Kans, MA, 34, PhD(phys chem), 37. *Prof Exp:* Asst instr chem, Univ Kans, 32-37; res chemist, Swift & Co, 37-42, head dairy res div, 46-52, asst chief chemist, 53-55; tech dir, Food Mat Corp, 55-65, vpres res & develop, 65-78. *Mem:* AAAS; Am Oil Chem Soc; Am Chem Soc; Am Dairy Sci Asn; Am Asn Cereal Chem. *Res:* Glass electrode in sugar solutions; vapor densities of liquids near their boiling points; dairy products; food flavors. *Mailing Add:* 408 Meadowgreen Dr Santa Rosa CA 95405

INGLE, JAMES DAVIS, JR, b Chicago, Ill, Oct 9, 46; m 70; c 1. CHEMICAL INSTRUMENTATION, TRACE ANALYSIS. *Educ:* Univ Ill, Urbana-Champaign, BS, 68, Mich State Univ, PhD(anal chem), 71. *Prof Exp:* PROF CHEM, ORE STATE UNIV, 71- *Mem:* Am Chem Soc; Soc Appl Spectros. *Res:* Instrumentation for the determination of trace amounts of inorganic and organic species with emphasis on environmental and clinical samples. *Mailing Add:* Dept Chem Ore State Univ Corvallis OR 97331

INGLE, JOHN IDE, b Colville, Wash, Jan 19, 19; m 40; c 3. DENTISTRY. *Educ:* Northwestern Univ, DDS, 42; Univ Mich, Ann Arbor, MSD, 48; Am Bd Periodont, dipl, 51; Am Bd Endodont, dipl, 63. *Prof Exp:* Asst, Northwestern Univ, 42-43; from asst prof to prof periodont & endodont, Sch Dent, Univ Wash, 48-55, from actg exec officer to exec officer dept, 53-64; prof periodont & endodont & dean, Sch Dent, Univ Southern Calif, 64-72; sr prof assoc, Inst Med-Nat Acad Sci, 73-78; PRES, PALM SPRINGS SEMINARS, INC, 78- *Concurrent Pos:* Mem dent educ rev comt, NIH, 70; mem adv comt on dent health, Off Secy HEW, 70-72; liaison comt on dent educ & lic, Am Asn Dent Schs, 70-73; vis lectr, Univ Calif, Los Angeles, 80- & Loma Linda Univ, 81- *Mem:* Int Asn Dent Res; Am Asn Endodont (past pres); Am Dent Asn; Am Acad Periodont; fel Am Col Dent. *Res:* Periodontics; endodontics. *Mailing Add:* 255 N El Cielo Suite 114 B Palm Springs CA 92262

INGLE, L MORRIS, b Covina, Calif, July 28, 29; m 60; c 3. PLANT PHYSIOLOGY, HORTICULTURE. *Educ:* Univ Calif, Santa Barbara, AB, 51; Univ Calif, Davis, MA, 56; Purdue Univ, PhD(plant physiol), 60. *Prof Exp:* Asst plant physiologist, United Fruit Co, 59-62, assoc biochemist, 62-63; from asst prof to assoc prof, 63-71, PROF HORT, WVA UNIV, 71- *Mem:* Am Soc Plant Physiol; Am Soc Hort; Sigma Xi. *Res:* Post-harvest physiology of fruits; plant growth regulators; carbohydrate and amino acid metabolism. *Mailing Add:* G-164 Agr Sci Bldg WVa Univ Morgantown WV 26506-6108

INGLE, MORTON BLAKEMAN, b Carlsbad, NMex, Apr 25, 42. MICROBIOLOGY, BIOCHEMISTRY. *Educ:* Ft Lewis Col, BS, 64; Colo State Univ, MS, 66, PhD(microbiol), 68. *Prof Exp:* Vpres res & develop, Miles Labs, Inc, 68-80; vpres res & develop, Int Minerals & Chem Corp, 80-85, vpres & chief admin officer, 85-87, sr vpres, chief admin & tech officer, 87-89, pres, Pitman-Moore Inc, 88-89, PRES & CHIEF OPERATING OFFICER, IMCERA, 89-, CHIEF EXEC OFFICER, 90- *Mem:* Am Soc Microbiol; Soc Indust Microbiol. *Res:* Regulation of protein biosynthesis; mutation mechanisms. *Mailing Add:* Int Minerals & Chem Corp 2315 Sanders Rd Northbrook IL 60062

INGLEDEW, WILLIAM MICHAEL, b Vancouver, BC, Mar 8, 42; m 66; c 2. BREWERY & GASOHOL MICROBIOLOGY, FERMENTATIONS. *Educ:* Univ BC, BSc, 65, PhD(microbiol), 69. *Prof Exp:* Nat Res Coun Can Postdoctorate fel, cellular chem dept, Coun Sci Res, Madrid, Spain, 69-79; from asst prof to assoc prof agr microbiol, 70-79, PROF APPL MICROBIOL & FOOD SCI, UNIV SASK, 79-, ASSOC MEM, DEPT MICROBIOL, SCH MED, 78- *Concurrent Pos:* Nat Res Coun Can sr indust fel, Molson Res Lab, Montreal, 78-79; assoc Exp Station Viticulture & Enology, Univ Calif, Davis, 83-84; ed-in-chief, Am Soc Brewing Chemists J, 88- *Mem:* Can Soc Microbiol; Am Soc Brewing Chemists; Can Inst Food Technol; Master Brewers' Asn Am; Am Soc Microbiol; Soc Ind Microbiologists. *Res:* Brewery yeast management; brewing and gasohol; production of alcohols from starch through biotechnology; heat resistance of microbes; utilization of spent industrial microbes; fuel alcohol; yeast nutrition. *Mailing Add:* Dept Appl Microbiol & Food Sci Univ Sask Saskatoon SK S7N 0W0 Can

INGLES, CHARLES JAMES, b Halifax, NS, Jan 31, 42; m 66; c 1. BIOCHEMISTRY, MOLECULAR BIOLOGY. *Educ:* Univ Toronto, BSc, 64; Univ BC, PhD(biochem), 68. *Prof Exp:* Nat Res Coun Can fel biochem, Beatson Inst Cancer Res, Glasgow, 68-70; res biochemist, Univ Calif, San Francisco, 70-71; from asst prof to assoc prof biochem, 71-81, ASSOC PROF MED RES, UNIV TORONTO, 81- *Mem:* Can Biochem Soc. *Res:* Molecular biology of cellular differentiation; control of gene expression in eukaryotes. *Mailing Add:* Banting & Best Dept Med Res Univ Toronto 112 College St Toronto ON M5G 1L6 Can

INGLESSIS, CRITON GEORGE S, b Piraeus, Greece, Apr 23, 30; US citizen; m 76; c 3. ORGANIC CHEMISTRY. *Educ:* Colgate Univ, BA, 53; Clark Univ, PhD(chem), 58. *Prof Exp:* Res chemist, Vineland Chem Co, 59-60; PRES, FRINTON LABS, 60- *Mem:* AAAS; Am Chem Soc. *Res:* Synthetic organic chemistry; liquid crystals; herbicides. *Mailing Add:* PO Box 2310 South Vineland NJ 08360-1310

INGLETT, GEORGE EVERETT, b Waltonville, Ill, Aug 3, 28; m 54; c 2. AGRICULTURAL CHEMISTRY, FOOD CHEMISTRY. *Educ:* Univ Ill, BS, 49; Univ Iowa, PhD(biochem), 52. *Prof Exp:* Asst biochemist, Univ Ill, 52-54; asst scientist cancer res, Nat Cancer Inst, 54-55; sr asst scientist air & water pollution control, Robert A Taft Sanit Eng Ctr, USPHS, 55-56; res chemist, Corn Prod Co, 56-60; sr res chemist, Griffith Labs, 60-63; sr food technologist, Inst Minerals & Chem Corp, 63-64, supvr natural prod chem, 64-65, mgr natural prod & food technol, 65-67; CHIEF CEREAL SCI & FOODS LAB, NORTHERN REGIONAL RES CTR, AGR RES SERV, USDA, 67- *Concurrent Pos:* Adj prof food sci, Univ Ill, 77- *Honors & Awards:* Philadelphia Sect Award, Inst Food Tech, 81; Advan Appln Agr & Food Chem Award, Agr & Food Chem Div, Am Chem Soc, 83, Distinguished Serv Award, 85. *Mem:* Am Chem Soc; Inst Food Technol; Am Asn Cereal Chem. *Res:* Chemistry of carbohydrates, flavorings, proteins, sweeteners and plant constituents; biochemistry of foods, cereal chemistry. *Mailing Add:* Cereal Sci & Foods Ctr Agr Res Serv USDA Peoria IL 61604

INGLIS, JACK MORTON, b Houston, Tex, Dec 31, 23; m 44; c 5. ECOLOGY. *Educ:* Tex A&M Univ, BS, 50, MS, 52, PhD(zool), 67. *Prof Exp:* Proj leader wildlife res, Tex Game & Fish Comn, 52-54; res biologist, Tex Agr Exp Sta, 54-58; from instr to assoc prof, 58-76, PROF WILDLIFE & FISHERIES SCI, TEX A&M UNIV, 76- *Concurrent Pos:* Scientist, Seregenti Res Inst, 71-72; ecol consult, Bechtel Power Corp, 73-78; vis prof, Univ Dar es Salaam, Tanzania, 79; consult wildlife, Tex Munic Power Agency, 81. *Mem:* Ecol Soc Am; Am Soc Mammal; Sigma Xi; Wildlife Soc; Animal Behav Soc. *Res:* Integration of wildlife management into ranching systems; habitat relationships and movements of white tailed deer; sociology of white-tailed deer; radioecology; wildlife conservation in developing countries. *Mailing Add:* 607 Jersey College Station TX 77840

INGLIS, JAMES, b Cleveland, Ohio, June 19, 45; m 67; c 2. STATISTICS, MATHEMATICAL STATISTICS. *Educ:* Amherst Col, BA, 67; Stanford Univ, MS, 68, PhD(statist), 73. *Prof Exp:* Mathematician, US Army Engr Div, Huntsville, 69-70; asst prof statist & biostatist, Univ Rochester, 72-78; mem tech staff, 78-81, Bell Lab, 78-81, supvr, 81-87; DEPT HEAD, AT&T BELL LABS, 87- *Mem:* Am Statist Asn; Inst Math Statist. *Res:* Data analysis; reliability. *Mailing Add:* AT&T Bell Labs 600 Mountain Ave Rm 7D-518 Murray Hill NJ 07974-2070

INGOGLIA, NICHOLAS ANDREW, NEUROSCIENCE. *Educ:* NY Univ, PhD(biol), 69. *Prof Exp:* PROF PHYSIOL & NEUROSCI, NJ MED SCH, 81- *Mailing Add:* 382 Woodland Pl South Orange NJ 07079-2447

INGOLD, DONALD ALFRED, b Columbus, Nebr, Dec 12, 34; m 58; c 4. ETHOLOGY, ORNITHOLOGY. *Educ:* Univ Nebr, Lincoln, BS, 58, MS, 61; Univ Wyo, PhD(zool), 69. *Prof Exp:* Asst prof biol, Sterling Col, 62-66; ASSOC PROF BIOL, E TEX STATE UNIV, 69- *Concurrent Pos:* Res fac grant, ETex State Univ, 71 & 75-78; consult, US Army corps Engr, 74-76. *Mem:* AAAS; Am Soc Mammal; Sigma Xi. *Res:* Avian ecology; ornithology; animal behavior. *Mailing Add:* 212 Brookhaven Commerce TX 75428

INGOLD, KEITH USHERWOOD, b Leeds, Eng, May 31, 29; Can citizen; m 56; c 3. PHYSICAL ORGANIC CHEMISTRY. *Educ:* Univ London, BSc, 49; Oxford Univ, DPhil(phys chem), 51. *Hon Degrees:* DSc, Univ Guelph, Ont, 85, Univ St Andrews, Scotland, 89; LLD, Mt Allison Univ, Sackville, NB, 87. *Prof Exp:* Nat Res Coun fel chem, 51-53; Defense Res Bd fel, Univ BC, 53-55; assoc dir, div chem, 77-90, RES OFFICER, NAT RES COUN, 55-, DIST SCIENTIST, STEACIE INST MOLECULAR SCI, 91- *Concurrent Pos:* Vis scientist, Chevron Res Corp, Calif, 66, Exxon Res & Develop Co, NJ, 72, Univ Adelaide, Australia, Univ Sci & Med Grenole & Univ Bordeaux, France; vis prof, Univ Col, Univ London, 69 & 72, Univ Western Ont, 75, Univ Bologna, 75 & Univ St Andrews, 77; vis lectr, Japan Soc Prom Sci, 82; adj prof, Dept Chem, Brunel Univ, UK, 83-, Dept Chem & Biochem, Univ Guelph, Ont, Can, 85-87; vis fel, Australian Nat Univ, Canberra, 87; mem, Frontiers Free Radicals & Radical Traps, Tex A&M Univ, 89. *Honors & Awards:* Award Petrol Chem, Am Chem Soc, 68; Frontiers Chem Lectr Series, Case Western Reserve Univ, 69; Frank Burnett Dains Mem Lectr, Univ Kans, 69; Queen's Silver Jubilee Medal, 77; Award Kinetics & Mech, Am Chem Soc, 78, Pauling Award, 88; Medal, Chem Inst Can, 81; Centennial Medal, Royal Soc Can, 82; Syntex Award Phys Org Chem, Chem Inst Can, 83; J A McRae Mem Lectr Chem, Queens Univ, Kingston, Ont, Can, 80; C I L Lect, Acadia Univ, Wolfville, Nova Scotia, Can, 87; Imp Oil Lectr, Univ Western Ont, London, Ont, 87; Douglas Hill Mem Lect, Duke Univ, 87; Rayson Huang Lect, Univ Hong Kong, Japan, 88; Alfred Bader Award in Organic Chem, Can Soc Chem, 89; Sir Christopher Ingold Lectr Award, Royal Soc Chem, UK, 89. *Mem:* Am Chem Soc; fel Chem Inst Can; Brit Chem Soc; fel Royal Soc Can; fel Royal Soc London; Can Soc Chem (vpres, 85-87, pres, 87-88). *Res:* Kinetics and mechanisms of free radical reactions in solution. *Mailing Add:* Steacie Inst Molecular Sci Nat Res Coun Ottawa ON K1A 0R6 Can

INGRAHAM, JOHN CHARLES, b Seattle, Wash, Nov 15, 36. PLASMA PHYSICS. *Educ:* Mass Inst Technol, SB, 58, PhD(physics), 63. *Prof Exp:* From instr to asst prof physics, Mass Inst Technol, 63-68; sci specialist, Edgerton, Germeshausen & Grier, Inc, 68-70; MEM STAFF, LOS ALAMOS SCI LAB, 70- *Mem:* Am Phys Soc; Sigma Xi. *Res:* Plasma physics; microwave techniques; atomic and molecular physics. *Mailing Add:* 304 Venado Los Alamos NM 87544

INGRAHAM, JOHN LYMAN, b Berkeley, Calif, Sept 22, 24; m 50; c 2. MICROBIOLOGY. *Educ:* Univ Calif, BS, 47, PhD(microbiol), 51. *Prof Exp:* Res scientist microbiol, Stine Lab, E I du Pont de Nemours & Co, 51-56; chemist, Western Regional Lab, USDA, 56-58; asst prof enol, 58-62, from assoc prof to prof, 62-89, EMER PROF BACT, UNIV CALIF, DAVIS, 89- *Concurrent Pos:* Guggenheim fel, 65-66. *Mem:* AAAS; Am Soc Microbiol; Sigma Xi. *Res:* Microbial physiology and genetics. *Mailing Add:* 8560 Holmes Ln Winters CA 95694

INGRAHAM, JOSEPH STERLING, b Grand Rapids, Minn, Nov 13, 20; div; c 2. IMMUNOLOGY. *Educ:* Univ Minn, BA, 43; Univ Chicago, SM, 47, PhD(biochem), 50. *Prof Exp:* Res chemist, Armour & Co, 43-46; instr biochem, Univ Chicago, 50-54; from asst prof to assoc prof, 54-69, prof microbiol, 69-87, prof immunol, 80-87, EMER PROF, IND UNIV, INDIANAPOLIS, 69-, PROF IMMUNOL, 87- *Concurrent Pos:* Vis asst prof microbiol, Sch Med, Univ Colo, 61; vis scientist, Pasteur Inst, Paris, 62 & 77. *Mem:* Am Chem Soc; Am Soc Microbiol; Am Asn Immunol; Soc Exp Biol & Med; AAAS; French Soc Immunol. *Res:* Fate of radioactive labeled antigens in vivo; formation of antibody homologous to simple hapten groups; antibody formation by individual cells; allotypy of antibodies; antigen-antibody reactions. *Mailing Add:* Dept Microbiol/Immunol Ind Univ Med Ctr 635 Barnhill Dr Indianapolis IN 46223

INGRAHAM, LLOYD LEWIS, b Berkeley, Calif, Jan 24, 20; m 47; c 2. PHYSICAL CHEMISTRY, ORGANIC CHEMISTRY. *Educ:* Univ Calif, BS, 42; Univ Calif, Los Angeles, PhD(chem), 49. *Prof Exp:* Chemist, Assoc Oil Co, 42-44; chemist, Western Regional Res Lab, USDA, 49-58; asst prof enzyme chem, 58-59, assoc prof biophys, 59-65, PROF BIOPHYS, UNIV CALIF, DAVIS, 65- *Concurrent Pos:* Guggenheim fel, Harvard Univ, 55-56; NSF sr fel, Copenhagen Univ, 64-65. *Mem:* Sigma Xi. *Res:* Biochemical kinetics and mechanisms. *Mailing Add:* PO Box 202 Woodland CA 95695

INGRAHAM, RICHARD LEE, b Des Moines, Iowa, Aug 29, 23; m 51, 55; c 2. THEORETICAL PHYSICS. *Educ:* Harvard Univ, BS, 47, MA, 50, PhD(physics), 52. *Prof Exp:* Asst, Inst Advan Study, 52-54; instr math, Univ Conn, 54-55; asst prof physics, Johns Hopkins Univ 55-56; res assoc, Univ Md, 56-57; assoc prof, NMex State Univ, 57-59; assoc prof, Tech Inst Aeronaut, Brazil, 59-60; res prof physics, NMex State Univ, 60-; AT DEPT BIOL SCI, SAN JOSE STATE UNIV. *Concurrent Pos:* Sheldon prize traveling fel from Harvard Col, 48-49; Fulbright fel, Paris, 50-51; consult, US Naval Res Lab, 56-57. *Mem:* Sigma Xi. *Res:* Quantum field theory; group theory and elementary particles; conformal relativity. *Mailing Add:* Dept Biol Sci San Jose State Univ San Jose CA 95192

INGRAHAM, THOMAS ROBERT, b Sydney, NS, Dec 7, 20; m 49. PHYSICAL CHEMISTRY, AIR POLLUTION. *Educ:* Dalhousie Univ, BSc, 43, MSc, 45; McGill Univ, PhD(chem), 47. *Prof Exp:* Prof chem, Loyola Col, Concordia Univ, 47-51; res scientist metall, Mines Br, Mines & Tech Surv, 53-54; head res, Mines Br, Energy, Mines & Resources, 54-72; dir technol, Environ Protection, 72-78; dir progs, Natural Sci & Eng Res Coun, 78-84; CONSULT, 84- *Concurrent Pos:* Fel, Univ Toronto, 51-53. *Mem:* Fel Chem Inst Can; fel Metall Soc-Am Inst Mining, Metall & Petrol Engrs; Air Pollution Control Asn; NAm Thermal Anal Soc (pres, 72). *Res:* Metallurgical thermodynamics; heterogeneous chemical kinetics; air pollution control technology. *Mailing Add:* Seven Opeongo Rd Ottawa ON K1S 4K9 Can

INGRAM, ALVIN JOHN, b Jackson, Tenn, Mar 31, 14; m 43; c 3. ORTHOPEDIC SURGERY. *Educ:* Univ Tenn, BS & MD, 39, MS, 47. *Prof Exp:* Intern, Univ Mich Hosp, Ann Arbor, 39-40, asst resident surg, 40-41; fel orthop surg, Campbell Clin, Memphis, 41-42 & 46-47, mem staff, 47-67, dep chief of staff, 67-69; assoc prof, 60-71, prof orthop surg & chmn dept, 71-78, EMER PROF & CHMN, COL MED, UNIV TENN, 78-; consult, Orthop Surg, Richards Med Co, 84-90; RETIRED. *Concurrent Pos:* Med dir, Crippled Children's Hosp, 48-61, chief of staff, 61-71; med dir, Les Passes Cerebral Palsy Treatment Ctr, 53-56; mem staff, Baptist Mem Hosp, exec comt med staff, 69-70, chmn orthop dept, 70-72, pres med staff, 73; chief of staff, Campbell Clin, 70-77, emer chief, 77-; mem Am Bd Orthop Surg, pres, 76-78; emer staff mem, St Joseph Hosp, LeBonheur Children's Hosp & Methodist Hosp. *Mem:* Inst Med-Nat Acad Sci; Am Acad Orthop Surgeons; Am Orthop Asn (pres, 73); Int Soc Orthop & Traumatology; Am Acad Cerebral Palsy (pres, 58); Am Bd Orthop Surg (pres, 75 & 76). *Mailing Add:* 190 Belle Mead Lane Memphis TN 38117

INGRAM, ALVIN RICHARD, b Enfield, NH, May 16, 18; m 42; c 2. ORGANIC POLYMER CHEMISTRY. *Educ:* Univ NH, BS, 40; Northeastern Univ, MS, 42; Univ Pittsburgh, PhD(org chem), 55. *Prof Exp:* Chemist, Gen Chem Defense Corp, WVa, 42-43; fel, Mellon Inst, 43-44; res proj leader, Johnson & Johnson, NJ, 44-48; fel, Mellon Inst, 48-53; chemist, tech coordr & group leader, Koppers Co, Inc, 53-58, mgr, Expandable Polymers Res Group, Koppers Co & Sinclair-Koppers Co, 58-73; mgr, Dylite Polymer Res Group, Arco Polymers Co & Arco Chem Co, 74-84; RETIRED. *Concurrent Pos:* Res consult, Arco Chem Co, 84-90. *Mem:* Am Chem Soc; Soc Plastics Indust. *Res:* Synthesis and application of thermoplastic and thermoset polymers; polystyrene bead foams; amino acids; plaster of paris; orthopedic bandages; organic reactions. *Mailing Add:* 439 Cardinal Lane West Chester PA 19382-7801

INGRAM, DAVID CHRISTOPHER, b Nottingham, Eng, Sept 9, 53; m 79; c 4. ION IMPLANTATION, ION BEAM EQUIPMENT. *Educ:* Salford Univ, BS, 75, MS, 76, PhD(elec eng), 80. *Prof Exp:* Res fel, Atomic Energy Res Estab Harwell, 79-82; sr scientist, Universal Energy Systs, 82-87; chief scientist, Whickam Ion Beams Systs Ltd, 87-89, gen mgr, 89; ASSOC PROF PHYSICS, OHIO UNIV, 89- *Concurrent Pos:* Vis res fel, Univ Durham, 87-90; consult, 89- *Mem:* Mat Res Soc; Am Soc Metals; Am Phys Soc; Am Vacuum Soc; Inst Physics; Bohmische Phys Soc. *Res:* Application of ion beams to materials for modification and analysis; deposition of diamond-like carbon and diamond coatings. *Mailing Add:* Dept Physics & Astron Clippinger Res Lab Ohio Univ Athens OH 45701

INGRAM, FORREST DUANE, b Lafayette Co, Wis, Jan 17, 38; m 65; c 2. NUCLEAR PHYSICS, BIOPHYSICS. *Educ:* Wis State Col, Platteville, BS, 59; Univ Iowa, MS, 61, PhD(physics), 68. *Prof Exp:* From instr to asst prof physics, Wis State Univ, Platteville, 62-65; instr physiol & biophys, Univ Iowa, 68-80, res scientist, Cardiovasc Ctr, 75-80; ASST PROF PEDIAT, BAYLOR COL MED, HOUSTON, 80- *Concurrent Pos:* Lectr-consult, Oak Ridge Inst Nuclear Studies, 66-69; vis asst prof internal med, Univ Tex Health Sci Ctr Dallas, 75-; vis scientist, Univ Nijmegeu, Neth, 79. *Mem:* Microbeam Anal Soc; NY Acad Sci; Am Asn Physics Teachers; Sigma Xi; fel Royal Micros Soc; Am Phys Soc. *Res:* Electron microprobe studies of the distributions of soluble ions in soft biological tissue; studies of the final state Coulomb interactions resulting from three body breakup reactions involving charged reaction products. *Mailing Add:* Dept Phys Sci Rock Valley Col 3301 N Mulford Rockford IL 61101

INGRAM, GERALD E(UGENE), b Phoenix, Ariz, Mar 8, 28; m 54; c 3. CIVIL ENGINEERING. *Educ:* Univ Ariz, BSCE, 49; Purdue Univ, MSCE, 57, PhD(civil eng), 61. *Prof Exp:* Asst county engr, Pinal County Hwy Dept, Ariz, 49-50, county engr, 52-55; instr civil eng, Purdue Univ, 55-60; assoc prof & chmn dept, Univ Denver, 60-63; mgr advan res & develop, Arinc Res Corp, 63-64; mem prof staff systs anal, ctr advan studies, Tempo-Gen Elec Co, 64-71; mgr reliability progs, Adcon Corp, 71-74; sr engr reliability eng, Advan Reactor Systs Dept, Gen Elec Co, 75-85. *Concurrent Pos:* Instr, Engrs Sch, Ft Belvoir, Va, 50-52; consult, Hq USAF Opers Anal Off, 61-65; consult engr, 74-75 & 85- *Mem:* Am Soc Civil Engrs; Nat Soc Prof Engrs; Am Astron Soc; Am Concrete Inst; Soc Exp Stress Anal. *Res:* Probabilistic approaches to analysis and design of structural and other type systems; basic technology associated with systems effectiveness analysis with emphasis on reliability; efficient techniques for safety reliability and risk analysis of nuclear power plant systems. *Mailing Add:* 13575 Howen Dr Saratoga CA 95070-5406

INGRAM, GLENN R, b Terry, Mont, Apr 25, 28; m 81; c 2. COMPUTER SCIENCE. *Educ:* Mont State Col, BS, 52, MS, 54; Wash State Univ, PhD(math), 62. *Prof Exp:* From instr to assoc prof math, Mont State Col, 52-65, dir comput ctr, 62-65; engr, Mont Hwy Dept, 54-55; asst comput analyst, Wash State Univ, 57-62; assoc prog dir comput sci, NSF, Washington, DC, 65-66, prog dir comput facs, 66-69, actg head off comput activities, 69-70; assoc prof comput sci & dir comput ctr, Wash State Univ, 70-73; chief tech appl div, Nat Inst Educ, 73-75; chief comput serv & technol, Energy Res & Develop Admin, 75-78; ASSOC DIR COMPUT, CENT AM MISSION, NAT BUR STANDARDS, 78- *Concurrent Pos:* Consult, NSF; mem steering comt, Comput in Undergrad Curricula. *Mem:* Asn Comput Mach; Soc Indust & Appl Math; Inst Elec & Electronics Engrs. *Res:* Applied mathematics; numerical analysis. *Mailing Add:* 18417 Kingshill Rd Germantown MD 20874

INGRAM, JOHN (WILLIAM), JR, b Buena Park, Calif, Aug 15, 24; m 51; c 3. TAXONOMIC BOTANY. *Educ:* Univ Southern Calif, AB, 49, MS, 52; Univ Calif, PhD(bot), 56. *Prof Exp:* Herbarium technician, Univ Calif, 53-55, asst bot, 55-56; from asst prof to assoc prof bot, L H Bailey Hortorium, Cornell Univ, 57-90; RETIRED. *Mem:* Am Soc Plant Taxon; Int Asn Plant Taxon. *Res:* Taxonomy of cultivated plants; taxonomy and morphology of genus Argythamnia. *Mailing Add:* 23893 Green Haven Lane Ramona CA 92065

INGRAM, JORDAN MILES, b Ottawa, Ont, Aug 24, 36; m 63; c 3. BIOCHEMISTRY. *Educ:* McGill Univ, BSc, 59, MSc, 61; Mich State Univ, PhD(biochem), 65. *Prof Exp:* Res scientist, Can Fed Govt, 65-68; from asst prof to assoc prof, 72-82, PROF MICROBIOL, MACDONALD COL, MCGILL UNIV, 82- *Concurrent Pos:* Assoc ed, Can J Biochem, 76- *Mem:* Can Soc Microbiol. *Res:* Enzymology from a structure and reaction mechanism point of view; antibiotic resistance. *Mailing Add:* Dept Microbiol Macdonald Col Ste Anne de Bellevue PQ H9X 1C0 Can

INGRAM, LONNIE O'NEAL, b Greenwood, SC, Dec 30, 47; m 68; c 3. MICROBIAL PHYSIOLOGY, INDUSTRIAL MICROBIOLOGY. *Educ:* Univ SC, BS, 69; Univ Tex, Austin, PhD(bot), 71. *Prof Exp:* Fel microbiol physiol, Oak Ridge Nat Lab, Biol Div, 71-72; asst prof, 72-76, assoc prof, 76-81, PROF MICROBIOL, UNIV FLA, 82- *Mem:* Soc Indust Microbiol; Am Soc Microbiol; Sigma Xi. *Res:* Relationship of lipids in the cell membrane to cellular functions; the effects of alcohols and other lipophilic agents on membranes; role of membranes in alcohol tolerance; production of alcohol by zymomonas and yeasts; regulations of glycolysis; molecular biology of alcohol production; glycolytic enzymes. *Mailing Add:* Dept Microbiol Univ Fla 2047 McCarty Hall Gainesville FL 32611

INGRAM, MARYLOU, b Ashtabula, Ohio, June 14, 20; div. MEDICAL RESEARCH. *Educ:* Western Reserve Univ, BA, 42, MS, 43; Univ Rochester, MD, 47. *Prof Exp:* Asst biol, Western Reserve Univ, 42-43; from instr to asst prof radiation biol, Sch Med & Dent, Univ Rochester, 46-55, assoc prof radiation biol & biophys, 59-71; med scientist & res assoc bio-med eng, Jet Propulsion Lab, Calif Inst Technol, 71-75; staff med scientist, Health Div, Los Alamos Sci Lab, Univ Calif, 75-77; DIR, INST CELL ANALYSIS & PROF MED & BIOMED ENG, UNIV MIAMI SCH MED, 77- *Concurrent Pos:* From asst to sect head, Atomic Energy Proj, Univ Rochester, 46-71, proj physician, 47-65, instr med, Sch Med & Dent, 53-61, sr instr, 61-71; consult, Nat Cancer Inst, 57-60 & Armed Forces Radiobiol Res Inst, 64-; NIH resident res collabr, Jet Propulsion Lab, Calif Inst Technol, 71-72, John A Hartford Found grant, 72-75; res collabr, Brookhaven Nat Lab, 70-74; clin prof path, Sch Med, Univ Southern Calif, 71-75; mem, Nat Coun Radiation Protection & Measurements, 73-79 & Environ Radiation Mgt & Control Adv Comt, 79- *Mem:* Soc Exp Biol & Med; Radiation Res Soc; Am Inst Biol Sci; Anal Cytol Soc; Am Soc Hemat; Sigma Xi. *Res:* Analytical cytology; mammalian and human hematology; biological effects of radiation clinical medicine. *Mailing Add:* 371 Patrician Way Pasadena CA 91105

INGRAM, PETER, b London, Eng, Nov 14, 38; US citizen; m 59; c 3. BIOPHYSICS. *Educ:* Univ Southampton, BSc, 59, PhD(physics), 63. *Prof Exp:* SR PHYSICIST, DREYFUS LAB, RES TRIANGLE INST, 63- *Concurrent Pos:* Adj asst prof path, Duke Univ, 81- *Honors & Awards:* Cecil Hall Award, Ecectron Micros Soc Am, 89. *Mem:* Electron Micros Soc Am; Microbeam Analysis Soc. *Res:* Analytical electron microscopy; morphology of natural and synthetic polymers, ultrastructure of biological tissue; scanning and transmission electron microscopy and x-ray analysis, especially of submicron cellular inclusions, intracellular elemental distributions and quantitative analytical and imaging techniques. *Mailing Add:* PO Box 12194 Research Triangle Park NC 27709

INGRAM, RICHARD GRANT, b Warrington, Eng, Feb 15, 45; Can citizen; m 70. PHYSICAL OCEANOGRAPHY. *Educ:* McGill Univ, BSc, 65, MSc, 67; Mass Inst Technol, PhD(oceanog), 71. *Prof Exp:* From asst prof to assoc prof phys oceanog, Inst Oceanog, 72- 87, chmn Ctr, 79-84, PROF METEOROL, MCGILL UNIV, 87- *Concurrent Pos:* Exec coun, Inter-Univ Res Group Oceanog, Que, 71- *Mem:* Am Geophys Union; Am Meteorol Soc; Can Meteorol Oceanog Soc. *Res:* Estuarine dynamics; coastal oceanography; mixing and entrainment processes in under-ice plumes and coastal waters. *Mailing Add:* Dept Meteorol McGill Univ 805 Sherbrooke St W Montreal PQ H3A 2K6 Can

INGRAM, ROLAND HARRISON, b Birmingham, Ala, March 10, 35; m 61; c 1. INTERNAL MEDICINE, PULMONARY DISEASES. *Educ:* Univ Ala, BS, 57; Yale Univ, MD, 60. *Hon Degrees:* AM, Harvard Univ, 80. *Prof Exp:* Dir, Respiratory Div, Sch Med, Emory Univ, 67-73; DIR, RESPIRATORY DIV, MED PULMONARY DIS, BRIGHAM & WOMEN'S HOSP, 73-; PARKER B FRANCIS PROF MED, HARVARD MED SCH, 79- *Concurrent Pos:* Dir respiratory div, Beth Israel Hosp, 80-85. *Mem:* Asn Am Physicians; Am Soc Clin Invest; Am Physiol Soc; Am Thoracic Soc (pres, 82-83). *Res:* Respiratory and cardiopulmonary physiology; pulmonary and critical care medicine problems. *Mailing Add:* Dept Med/Harvard Sch Med Brigham & Women Hosp 45 Francis St Boston MA 02115

INGRAM, ROY LEE, b Mamers, NC, Mar 12, 21; m 44; c 2. GEOLOGY. *Educ:* Univ NC, BS, 41; Univ Okla, MS, 43; Univ Wis, PhD(geol), 48. *Prof Exp:* From asst prof to assoc prof, Univ NC, Chapel Hill, 47-57, chmn dept, 57-64 & 74-79, prof geol, 57-91; RETIRED. *Mem:* Geol Soc Am; Soc Econ Paleont & Mineral; Am Asn Petrol Geol; Clay Minerals Soc; Nat Asn Geol Teachers. *Res:* Sedimentation of ancient and modern sediments; marine geology; clay mineralogy. *Mailing Add:* 601 Oteys Rd Chapel Hill NC 27514

INGRAM, SAMMY WALKER, JR, b Easonville, Ala, Nov 17, 33; m 60; c 2. ORGANIC CHEMISTRY. *Educ:* Jacksonville State Col, BS, 55; Univ Tex, PhD(org chem, math), 59. *Prof Exp:* Asst prof chem, Jacksonville State Col, 59-62, assoc prof sci & head dept chem, 62-64; asst prof chem, Univ Ala, 64-65; prof, Troy State Univ, 65; AT DEPT CHEM, LIVINGSTON UNIV. *Mem:* Am Chem Soc. *Res:* Isomerization of the 2-bromo-2-butenes; isomerization of haloalkenes by means of transition element complex ions; adamantine and its compounds; natural products. *Mailing Add:* 1712 Woodwind Lane Austin TX 78758

INGRAM, VERNON MARTIN, b Breslau, Ger, May 19, 24; m 50; c 2. BIOCHEMISTRY. *Educ:* Univ London, BSc, 43 & 45, PhD, 49, DSc(biochem), 61. *Prof Exp:* Asst lectr chem, Birkbeck Col, Univ London, 47-50; fel, Rockefeller Inst, 50-51; Coxe fel, Yale Univ, 51-52; mem sci staff, Med Res Coun, Cavendish Lab, Cambridge Univ, 52-58; assoc vis prof, 58-59, assoc prof, 59-61, PROF BIOCHEM, MASS INST TECHNOL, 61- *Concurrent Pos:* Lectr, Col Physicians & Surgeons, Columbia Univ, 61-73. *Honors & Awards:* William Allen Award, Amer Soc Human Genetics, 67. *Mem:* Am Acad Arts & Sci; Am Chem Soc; Brit Biochem Soc; The Chem Soc; fel Royal Soc. *Res:* Human and animal hemoglobins, their chemistry and the control of their biosynthesis; development of erythropoiesis in embryos; DNA methylation; biochemical mechanisms in Alzheimers disease; molecular basis of mental retardation in Downs syndrome. *Mailing Add:* Rm 56-601 Dept Biol Mass Inst Technol Cambridge MA 02139

INGRAM, WILLIAM THOMAS, b McKenzie, Tenn, Nov 26, 37; m 58; c 3. MATHEMATICS. *Educ:* Bethel Col, Tenn, BA, 59; La State Univ, MS, 61; Auburn Univ, PhD(math), 64. *Prof Exp:* Instr math, Auburn Univ, 61-63; from instr to prof math, Univ Houston, 64-89; PROF MATH & CHMN MATH & STAT, UNIV MO, ROLLA, 89- *Mem:* Am Math Soc; Math Asn Am. *Res:* Point set topology, particularly problems concerned with tree-like continua. *Mailing Add:* Dept Math & Statist Univ Mo Rolla MO 65401

INGRATTA, FRANK JERRY, b Chatham, Ont, Can, Aug 30, 49. COMMERCIAL MUSHROOM PRODUCTION. *Educ:* Univ Guelph, BSc, 71, MSc, 74; Univ Toronto, PhD(fungal physiol), 84. *Prof Exp:* Exten horticulturist, Ont Ministry Agr & Food, 72-75, res scientist, 75-83, chief res scientist, Hort Res Inst Ont, 83-90. *Mem:* Can Soc Hort Sci; Am Soc Hort Sci; Int Soc Hort Sci; Agr Inst Can; Can Mushroom Growers Asn. *Res:* Commercial production technologies for Agaricus bisporus mushrooms and greenhouse vegetable crop production; crop management; energy conservation. *Mailing Add:* 69 Monticello Crescent Guelph ON N1H 4P4 Can

INGRUBER, OTTO VINCENT, b Gosau, Austria, Jan 3, 19; nat Can; m 44; c 4. CHEMISTRY, ELECTROCHEMISTRY. *Educ:* Vienna Tech Univ, MSc, 46; Graz Tech Univ, PhD, 69. *Prof Exp:* Qual control chemist, Ger Air Ministry, 41-45; asst to chief chemist, Austrian Alpine Montan Co, 47-52; assoc scientist chem processes, Pulp & Paper Res Inst Can, 52-60; res chemist, Indust Cellulose Res Ltd, Can Int Paper Res Ltd, 60-67, group leader pulping res, 67-70, sr group leader process develop, 70-73, res assoc, 73-80, sr res assoc, 80-83; RETIRED. *Concurrent Pos:* Independent consult. *Honors & Awards:* Weldon Medal, Can Pulp & Paper Asn, 73 & 82. *Mem:* Tech Asn Pulp & Paper Indust; Can Pulp & Paper Asn. *Res:* Chemical pulping research, methods, control, automation and equipment, energy, economy; recovery of chemicals; byproduct utilization; chemistry of hot solutions under pressure; electrochemical measurements and control in systems at elevated pressure and temperature; alk sulfite pulping with redox catalysis. *Mailing Add:* RR 1 Vankleek Hill ON K0B 1R0 Can

INGUVA, RAMARAO, b Chinamuttevi, India, Dec 6, 41; m 64; c 1. CONDENSED MATTER PHYSICS, STATISTICAL MECHANICS. *Educ:* Osmania Univ, India, BSc, 60, MSc, 62; Univ Colo, PhD(physics), 69. *Prof Exp:* Res assoc theoret physics, Stanford Univ, 69-70; fel, Tata Inst Fundamental Res, Bombay, India, 70-78; asst prof, 78-85, ASSOC PROF PHYSICS, UNIV WYO, 85-; NAT RES COUN ASSOC, MICOM REDSTONE ARSENAL, 87- *Concurrent Pos:* Consult, McAdams, Roux & O'Connor & Co, Denver, 81; vis scientist, Explor Res Div, Conoco Inc, 85-86. *Mem:* Sigma Xi; NY Acad Sci; Am Phys Soc. *Res:* Ground state properties of quantum fluids; dielectric properties of heterogeneous layered media such as oil shales; microwave heating of oil shales; critical phenomena; information theory and statistical mechanics; nonlinear optics. *Mailing Add:* Dept Physics Univ Wyo Laramie WY 82071

INGVOLDSTAD, D(ONALD) F(ANNON), metallurgical engineering; deceased, see previous edition for last biography

INGWALL, JOANNE S, b Syracuse, NY, Oct 23, 41; m 63. BIOENERGETICS. *Educ:* Le Moyne Col, NY, BS, 63; Cornell Univ, PhD(biophys chem), 68. *Prof Exp:* Nat Heart & Lung Inst trainee, Cardiovasc Res Inst, Univ Calif, San Francisco, 68-69 & 71; fel, Stanford Res Inst, 69-71; clin asst prof physiol, Sch Dent, Univ Pac, 72-73; asst res biochemist cardiol, Univ Calif, San Diego, 73-74, asst prof in residence med cardiol, 74-76; asst prof physiol, Dept Med, Harvard Med Sch & mem assoc staff, Peter Bent Brighamn Hosp, 77-83; ASSOC PROF PHYSIOL & BIOPHYSICS, DEPT MED, HARVARD MED SCH & DIR, NMR LAB/BIOCHEMIST, BRIGHAM & WOMEN'S HOSP, 84- *Concurrent Pos:* Muscular Dystrophy Asn Am fel, Sch Dent, Univ Pac, 72-73. *Honors & Awards:* Louis N Katz Basic Sci Res Award, Am Heart Asn, 72. *Mem:* Biophys Soc; Am Heart Asn; Fedn Am Socs Exp Biol; Am Soc Cell Biol; Soc Magnetic Res Med. *Res:* Muscle biochemistry, especially regulation of cardiac energy metabolism. *Mailing Add:* Brigham & Women's Hosp 221 Longwood Ave Boston MA 02115

INGWALSON, RAYMOND WESLEY, b Rockford, Ill, Mar 13, 12; m 39; c 1. INDUSTRIAL ORGANIC CHEMISTRY. *Educ:* Carthage Col, BA, 35; Univ Fla, MS, 48, PhD(chem), 52. *Prof Exp:* Inspector, Vol Ord Works, 42-43; chemist, Chem Div, W F & Johns Barnes Co, 43-46; res chemist, Tenn Prod & Chem Corp, 52-63; res supvr, Velsicol Chem Corp, 63-74; OWNER, PRO CHEM CO, 74-, CONSULT, 77- *Concurrent Pos:* Consult, Velsicol Chem Corp, 75-; vis assoc prof, Fla Technol Univ, 75-76, adj assoc prof, 76-77. *Mem:* Am Chem Soc. *Res:* Photochlorination and oxidation of hydrocarbons; synthesis of biological chemicals and organic intermediates; specialty chemicals-processes: plasticiers, synthetic lubricants, ultraviolet absorbers, esterification, oxidation, chlorination. *Mailing Add:* 686 Barrington Circle Winter Springs FL 32708-6115

INHABER, HERBERT, b Montreal, Que, Can, Jan 25, 41; m 87. RISK ANALYSIS, ENVIRONMENTAL QUALITY INDICES. *Educ:* McGill Univ, BSc, 62; Univ Ill, MS, 64; Univ Okla, PhD(physics & math), 71. *Prof Exp:* Assoc physicist, US Steel Corp, 64-65; sci adv sci policy, Sci Coun Can, 71-72; policy analyst, Can Dept Environ, 72-77; sci adv, Atomic Energy

Control Bd, Can, 77-80; sr staff mem, Oak Ridge Nat Lab, 80-84; pres, Risk Concepts Inc, 84-87; PRES & CHIEF SCIENTIST, LIGHT FANTASTIC INC, 86-; EXEC SCIENTIST, NUS CORP, GAITHERSBURG, MD, 87- *Concurrent Pos:* Sessional lectr physics, Carleton Univ, 74-80; vis lectr hist sci, Yale Univ, 75-76; columnist, Oak Ridger, 81-; mem, Comt Technol Safety, Scientists & Engrs for Secure Energy, 83-; consult, Technol Energy Corp, 84 & Tenn Valley Authority & Martin Marietta Energy Systs, 85. *Mem:* AAAS; Air Pollution Control Asn; Health Physics Soc; Am Nuclear Soc; Soc Risk Anal. *Res:* Risk analysis of energy systems; measurement and analysis of environmental quality; probabilistic risk analysis in nuclear energy and other fields; determining the bottom line or lines of technological effects on human and environmental well-being. *Mailing Add:* 62 Clarendon Pl Buffalo NY 14209-1008

INHORN, STANLEY L, b Philadelphia, Pa, Aug 1, 28; m 54; c 3. PATHOLOGY. *Educ:* Western Reserve Univ, BS, 49; Columbia Univ, MD, 53. *Prof Exp:* Intern med, 53-54, resident path, 56-60, from instr to assoc prof, 59-69, asst dir Wis State Lab Hyg, 60-66, dir 66-79, chmn dept, 78-81, PROF PREV MED & PATH, UNIV WIS, MADISON, 69-, MED DIR WIS STATE LAB HYG, 79- *Concurrent Pos:* Consult, Vet Admin Hosp, 62-; NIH res grant, 63-66; Children's Bur res grant cytogenetics, 66- *Honors & Awards:* Papanicolaou Award, 81. *Mem:* Am Asn Path; Am Soc Clin Path; Am Soc Cytol; Am Soc Human Genetics; fel Am Pub Health Asn. *Res:* Human cytogenetics of congenital malformations; cancer cytology; clinical laboratory improvement programs. *Mailing Add:* Dept Path Univ Wis Sch Clin Ctr Madison WI 53792

INIGO, RAFAEL MADRIGAL, b Madrid, Spain, June 18, 32; m 61; c 2. ELECTRICAL ENGINEERING. *Educ:* Valparaiso Tech Univ, Chile, Ing Elec, 57; Univ Va, MS, 65, DSc, 66. *Prof Exp:* Asst prof elec eng, Valparaiso Tech Univ, 61-63, prof, 66-68; from assoc prof to prof elec eng, Va Mil Inst , 68-78; assoc prof, 79-86, PROF ELEC ENG, UNIV VA, 86- *Concurrent Pos:* Invited prof, Univ Navarra, Spain, 74-75; vis prof, Univ Va, 78-79; Helen Wessel fel. *Honors & Awards:* Halliburton Educ Found Award, 78. *Mem:* Inst Elec & Electronics Engrs; Sigma Xi. *Res:* Application of machine vision and neural networks to control, guidance & robotics. *Mailing Add:* Dept Elec Eng Univ Va Charlottesville VA 22901

INKLEY, SCOTT RUSSELL, b Cleveland, Ohio, Mar 8, 21; m 43; c 4. INTERNAL MEDICINE. *Educ:* Western Reserve Univ, MD, 45. *Prof Exp:* Instr med, Sch Med, 51-54, sr instr, 54-55, asst prof, 55-66, assoc clin prof, Sch Med & physician, Univ Hosps, 66-71, assoc prof, 71-75, PROF MED, SCH MED & PHYSICIAN, UNIV HOSPS, CASE WESTERN RESERVE UNIV. *Concurrent Pos:* Physician in chg outpatient dept, Case Western Reserve Univ, 62-; dir pulmonary function lab & dir dept inhalation ther, Univ Hosps Cleveland, 65-77, chief staff, 78-82. pres & chief exec officer, 82-86. *Mem:* Fel Am Col Physicians; fel Am Col Chest Physicians; Soc Nuclear Med; Am Fedn Clin Res; Am Thoracic Soc; Sigma Xi. *Res:* Cardio-pulmonary disease. *Mailing Add:* County Line Rd Chagrin Falls OH 44022

INMAN, CHARLES GORDON, b New York, NY, Apr 12, 29; m 54; c 3. ORGANIC AND INORGANIC PIGMENT CHEMISTRY. *Educ:* Rensselaer Polytech Inst, BS, 51; Mass Inst Technol, MS, 53. *Prof Exp:* Res chemist, Durez Plastics, Inc, 53-54 & Hooker Chem Corp, 56-57; chemist imp dept, Hercules, Inc, 57-65, res supvr coatings & specialty prod dept, 65-73, asst supt res & develop, Coatings & Specialty Prod Dept, 73-79; RES MGR, PIGMENTS DEPT, CIBA-GEIGY CORP, 79- *Mem:* Am Chem Soc. *Res:* Azo and chelate pigments. *Mailing Add:* RR 1 Box 1372 Ft Edward NY 12828-9733

INMAN, DOUGLAS LAMAR, b Guam, Marianas Islands, July 7, 20; US citizen; m 46; c 3. COASTAL OCEANOGRAPHY, GEOTECHNICAL SCIENCE. *Educ:* San Diego State Col, BA, 42; Univ Calif, MS, 48, PhD(oceanog), 53. *Prof Exp:* Asst phys geol, San Diego State Col, 40-42, instr, 64; asst, 47-48, assoc marine geologist, 58-53, asst prof marine geol, 53-57, assoc prof, 57-65, prof oceanog, 65-80, DIR, CTR COASTAL STUDIES, SCRIPPS INST OCEANOG, UNIV CALIF, SAN DIEGO, 80- *Concurrent Pos:* Guggenheim fel, 61. *Honors & Awards:* Int Coastal Eng Award, Am Soc Civil Engrs, 88; Ocean Sci Educ Award, Off Naval Res, 89. *Mem:* Fel Geol Soc Am; Am Asn Petrol Geol; Am Geophys Union; AAAS. *Res:* Coastal oceanography; nearshore sediment transport; waves; effect of waves on beaches; beach and nearshore processes. *Mailing Add:* Scripps Inst Oceanog Univ Calif San Diego La Jolla CA 92093

INMAN, FRANKLIN POPE, JR, immunochemistry, departmental administration; deceased, see previous edition for last biography

INMAN, FRED WINSTON, b Mountain Home, Ark, Mar 30, 31; m 55; c 2. NUCLEAR PHYSICS. *Educ:* Univ Calif, AB, 53, MA, 55, PhD(physics), 57. *Prof Exp:* Prof physics & chmn dept, Howard Payne Col, 57-63, chmn div sci & math, 63-64; assoc prof physics, Univ Pac, 64-67; chmn dept, 67-75, PROF PHYSICS, MANKATO STATE UNIV, 67- *Mailing Add:* Dept Physics & Elec Eng Mankato State Univ Mankato MN 56001

INMAN, JOHN KEITH, b St Louis, Mo, May 21, 28; m 54; c 3. PROTEIN CHEMISTRY, IMMUNOLOGY. *Educ:* Calif Inst Technol, BS, 50; Harvard Univ, PhD(biochem), 56. *Prof Exp:* Res biochemist labs, Mich Dept Health, 56-60, dir, Div Biochem, Ortho Pharmaceut Corp, Div Johnson & Johnson, 60-63; NIH spec fel, Sch Med, Johns Hopkins Univ, 63-65; RES BIOCHEMIST, LAB IMMUNOL, NAT INST ALLERGY & INFECTIOUS DIS, 65- *Honors & Awards:* Director's Award, NIH, 78. *Mem:* AAAS; Fedn Am Soc Exp Biol; Am Asn Immunol; Am Chem Soc. *Res:* Synthesis of immunomodulators; antibody specificity; immunochemistry. *Mailing Add:* Lab Immunol Rm 11N252 Bldg 10 Nat Inst Allergy Infectious Dis Bethesda MD 20892

INMAN, ROBERT DAVIES, b Toronto, Canada, Mar 6, 49; m 71; c 3. RHEUMATOLOGY AND IMMUNOLOGY. *Educ:* Yale Univ, BA, 71; McMaster Univ, MD, 74. *Prof Exp:* Asst prof, Cornell Univ, 79-83; ASSOC PROF, TORONTO WESTERN HOSP, 83- *Concurrent Pos:* Fel, Cornell Univ, 77-79. *Mem:* NY Acad Sci; Am Asn Immunol; Am Rheumat Asn. *Res:* Infection, autoimmunity and immunogenetics. *Mailing Add:* Dept Med Rheumat Dis Unit Toronto Western Hosp 399 Bathurst St Toronto ON M5T 2S8 Can

INMAN, ROSS, b Adelaide, Australia, Nov 4, 31; m 57; c 3. DNA REPLICATION, ELECTRON MICROSCOPY. *Educ:* Univ Adelaide, BSc, 56, Hons, 57, PhD(molecular biol), 60. *Prof Exp:* Fel molecular biol, Med Sch, Stanford Univ, 60-64; res fel, Univ Adelaide, 64-67; assoc prof biophys, 67-69, PROF BIOPHYS & BIOCHEM, UNIV WIS-MADISON, 70- *Concurrent Pos:* NIH grant, 67-; Am Cancer Soc grant, 70-; consult, US & Australian Govt, 82- *Res:* Structure and function of nucleic acids. *Mailing Add:* Biophys Lab Univ Wis Madison WI 53706

INN, EDWARD CHANG YUL, physics, for more information see previous edition

INNANEN, KIMMO A, b Kirkland Lake, Ont, Mar 12, 37; m 64; c 3. ASTROPHYSICS. *Educ:* Univ Toronto, BASc, 59, PhD(astron), 64; Univ Waterloo, MSc, 60. *Prof Exp:* Lectr astron, Western Ont Univ, 63-64, asst prof, 64-66; from asst prof to assoc prof, 66-73, actg chmn dept, 84-85, PROF PHYSICS, YORK UNIV, 73-, DEAN PURE & APPL SCI, 86- *Concurrent Pos:* Chmn bd, Inst Space & Terrestrial Sci, Innovation York, 87- 88. *Mem:* Am Astron Soc; Royal Astron Soc Can; Can Astron Soc; Asn Prof Engr Ont. *Res:* Galactic structure; stellar and planetary dynamics. *Mailing Add:* Dept Physics York Univ North York ON M3J 1R1 Can

INNERARITY, TOM L, b Durant, Okla, Oct 29, 43. LIPOPROTEIN METABOLISM. *Educ:* Univ Okla, PhD(biochem), 70. *Prof Exp:* SR SCIENTIST BIOCHEM, GLADSTONE FOUND LAB CARDIOVASC DIS, UNIV CALIF, SAN FRANCISCO, 79- *Mailing Add:* Gladstone Found Lab Cardiovasc Dis Univ Calif 2550 23rd St PO Box 40608 San Francisco CA 94140

INNERS, JON DAVID, b York, Pa, July 1, 42; m 68; c 3. GEOLOGY. *Educ:* Susquehanna Univ, AB, 64; Univ Mass, Amherst, PhD(geol), 75. *Prof Exp:* Soils engr, Pa Dept Transp, Dist 6-0, St Davids, Pa, 69-70 & 72-73; geologist soil conserv, USDA, Amherst, 71-72; geologist, 73-79, ASSOC STATE GEOLOGIST, BUR TOPOG & GEOL SURV, PA DEPT ENVIRON RESOURCES, HARRISBURG, 89- *Mem:* Geol Soc Am; Asn Eng Geol; Soc Mining Engrs; Hist Earth Sci Soc; Int Asn Eng Geol; Soc Indust Archeol. *Res:* Areal geology of the central Appalachians; geotechnical evaluation of bedrock and surficial mapping units; economic geology of anthracite coal. *Mailing Add:* Bur Topog & Geol Surv Pa Dept Environ Resources PO Box 2357 Harrisburg PA 17105

INNES, DAVID LYN, b Cleveland, Ohio, Dec 19, 41; m 64; c 2. PHYSIOLOGY, GASTROENTEROLOGY. *Educ:* Ohio Wesleyan Univ, BA, 64; Univ Cincinnati, MS, 66; Ohio State Univ, PhD(physiol), 69. *Prof Exp:* Instr physiol, Ohio State Univ, 69-70; from asst prof to assoc prof physiol, Temple Univ, 70-80; dir res, 80-82, PROF PHYSIOL, SCH MED, MERCER UNIV, 80-, DIR OFF PROTECTION RES RISKS & ASST PROVOST MED AFFAIRS, 85- *Mem:* AAAS; Am Soc Zool; Sigma Xi; Am Physiol Soc; Am Gastroenterol Asn; Am Asn Lab Animal Sci. *Res:* Gastrointestinal and neural endocrinology; nuclei within the brain controlling gastrointestinal and/or metabolic-endocrine functions of the body and their mechanisms of action; mechanisms of gastric stress ulceration. *Mailing Add:* Sch Med Mercer Univ Macon GA 31207

INNES, IAN ROME, pharmacology, for more information see previous edition

INNES, JOHN EDWIN, b Philadelphia, Pa, Apr 3, 38; m 60; c 2. INDUSTRIAL ORGANIC CHEMISTRY. *Educ:* Ursinus Col, BS, 60; Univ Del, PhD(org chem), 64; Ohio Univ, MBA, 85. *Prof Exp:* Chemist, 64-72, group leader org chem, 72-76, prod supt, 76-77, tech dir, Marietta Plant, 77-88, PLANT MGR, AM CYANAMID CO, 88- *Mem:* Am Chem Soc; Sigma Xi. *Mailing Add:* Am Cynamid Co 1405 Greene St Marietta OH 45750

INNES, KENNETH KEITH, b Fayette, Mo, June 6, 28; m 52; c 2. CHEMICAL PHYSICS, MOLECULAR PHYSICS. *Educ:* Cent Col, Mo, AB, 47; Brown Univ, MS, 49; Univ Wash, Seattle, PhD(phys chem), 51. *Prof Exp:* Nat Res Coun Can fel, 51-53; asst prof chem, Univ Okla, 53-55; assoc prof, Vanderbilt Univ, 55-62, prof, 62-69; prof chem, 69-79, actg vpres, Acad Affairs, 74-75, DISTINGUISHED PROF, STATE UNIV NY BINGHAMTON, 79- *Concurrent Pos:* Guggenheim fel, 61-62; mem adv panel chem, NSF, 65-69, sr fel, 67-68; mem petrol res fund adv bd, Am Chem Soc, 73-75; vis prof, Univ Toronto, 75-76. *Mem:* Am Chem Soc; fel Am Phys Soc. *Res:* Molecular spectroscopy and structure. *Mailing Add:* Dept Chem State Univ NY Binghamton NY 13901

INNES, WALTER RUNDLE, b Stamford, Conn, Dec 29, 45; m 66; c 2. PARTICLE PHYSICS. *Educ:* Calif Inst Technol, BS, 67; Univ Calif, San Diego, MS, 69, PhD(physics), 74. *Prof Exp:* Res assoc particle physics, Fermi Nat Accelerator Lab, 74-78; MEM STAFF, STANFORD LINEAR ACCELERATOR CTR, 79- *Mem:* Am Phys Soc. *Res:* Hyperon beta decay; dihadron, dilepton, and upsilon production; electron positron colliding beam physics; search for fractional charge states; studies of Z bosons. *Mailing Add:* Stanford Linear Accelerator Ctr PO Box 4349 Stanford CA 94305

INNES, WILLIAM BEVERIDGE, b Cambria, Calif, Mar 8, 13; m 38; c 4. CATALYSIS, OXIDES OF NITROGEN. *Educ:* Univ Calif, BS, 37; Univ Iowa, MS, 39, PhD(phys chem), 41. *Prof Exp:* Asst, Univ Iowa, 39-41; res chemist, Champion Paper & Fibre Co, Ohio, 41-43 & Am Cyanamid Co,

43-44; res scientist, SAM Labs, Columbia, 44-45; sr chemist, Am Cyanamid Co, 45-55, group leader, 55-58, res assoc, 58-64; PRES, PURAD INC, 64- *Mem:* Air Pollution Control Asn; Am Chem Soc. *Res:* Surface phenomena; adsorption and pore structure; catalysis, vehicle exhaust treatment; thermocatalytic detectors, instrumentation; oxides of nitrogen, photochemical smog. *Mailing Add:* Purad Inc 724 Kilbourne Dr Upland CA 91786

INNIS, GEORGE SETH, b Victoria, Tex, Jan 7, 37; m 58; c 2. MATHEMATICS, SYSTEM ANALYSIS. *Educ:* Univ Tex, BA, 58, MA, 61, PhD(math), 62. *Prof Exp:* Res asst physics & math, Defense Res Lab, Univ Tex, 57-58, res scientist, 58-60 & 62, asst univ, 58-59, spec instr, 60-62; Nat Acad Sci-Nat Res Coun fel, Harvard Univ, 62-63; asst prof math, Rice Univ, 63-64; asst prof math, Univ Tex, 64-67, res scientist physics & math, Defense Res Lab, 64-67; staff mem, Los Alamos Sci Lab, 67-68; assoc prof math & dir comput serv, Tex Tech Univ, 68-71; assoc prof math & dir systs anal, Nat Resource Ecol Lab, Colo State Univ, 71-73; assoc prof, Utah State Univ, 73-76, prof fisheries & wildlife & adj prof math, 76-, dept head, 80-; AT COLO STATE UNIV; CONSULT, 90- *Concurrent Pos:* Res mathematician, Tracor Corp, 60-61; consult, Antioch Col, 70-72; consult, NSF, Washington, 70- & W F Sigler & Assocs, Logan, Utah, 74- *Mem:* AAAS; Wildlife Soc; Sigma Xi; Ecol Soc Am; Soc Computer Simulation. *Res:* Complex variables; signal processing; acoustics; applications of quantitative techniques to ecological problems including the development of simulation languages and mathematical theories addressed to specific needs. *Mailing Add:* 14111 Red Mulberry Woods San Antonio TX 78249

INNISS, DARYL, b St Thomas, VI, June 15, 61; US citizen; m 83; c 2. REACTION MECHANISM, SURFACE ANALYSIS. *Educ:* Princeton Univ, AB, 83; Univ Calif, Los Angeles, PhD(chem), 88. *Prof Exp:* MEM TECH STAFF, AT&T BELL LABS, 88- *Mem:* Am Ceramic Soc; Nat Orgn Prof Advan Black Chemists & Chem Engrs. *Res:* Chemical reaction theory is applied to the fracture of silica fibers; the contribution of the chemical environment to the failure mechanism is explored. *Mailing Add:* AT&T Bell Labs Rm 6D-213 600 Mountain Ave Murray Hill NJ 07974

INNISS, WILLIAM EDGAR, b Toronto, Ont, Feb 25, 36; m 58; c 2. MICROBIOLOGY. *Educ:* Univ Toronto, BSA, 58, MSA, 59; Mich State Univ, PhD(microbiol), 61. *Prof Exp:* Microbiologist, Biochem Res Lab, Dow Chem Co, Mich, 61-63; from asst prof to assoc prof, 63-80, PROF BIOL, UNIV WATERLOO, 80-, ASSOC DEPT CHMN, 89- *Concurrent Pos:* Vis prof, Univ Calif, Davis, 73. *Mem:* Am Soc Microbiol; Can Soc Microbiol. *Res:* Microbial physiology; environmental and physiological activities of psychrotrophic and psychrophilic microorganisms; microbial activity in cold environments; effect of toxicants on microorganisms; macromolecular synthesis and metabolism. *Mailing Add:* Dept Biol Univ Waterloo Waterloo ON N2L 3G1 Can

INOKUTI, MITIO, b Tokyo, Japan, July 6, 33; m 60; c 1. MOLECULAR PHYSICS. *Educ:* Univ Tokyo, BS, 56, MS, 58, PhD(appl physics), 62. *Prof Exp:* Instr math physics, Univ Tokyo, 60-62; res assoc, Dept Chem & Mat Res Ctr, Northwestern Univ, 62-63; resident res assoc, 63-65, physicist, 65-73, SR PHYSICIST, ENVIRON RES DIV, ARGONNE NAT LAB, 73- *Concurrent Pos:* Vis fel, Joint Inst Lab Astrophys, Univ Colo, 69-70; mem comt average energy required produce & ion pair, Int Comn Radiation Units & Measurements, 73-79, vchmn comt stopping power, 76-; mem gen comt, Int Conf Physics Electronics & Atomic Collisions, 73-77, adj officer exec comt, 77-79; assoc ed, Radiation Res, 76-79; vis prof, Inst Space & Aeronaut Sci, Univ Tokyo, Japan, 78-79, Odense Univ, Denmark & Tokyo Inst Technol, Japan, 89; counr, Radiation Res Soc, 78-81; corresp, Comments Atomic & Molecular Physics, 82-; mem, Int Comn Radiation Units & Measurements, 85- *Mem:* Fel Am Phys Soc; Radiation Res Soc; fel Brit Inst Physics & Phys Soc; Phys Soc Japan; Int Radiation Physics Soc; Sigma Xi. *Res:* Theoretical problems concerning action of ionizing radiations on molecular substances, especially primary elementary processes. *Mailing Add:* Argonne Nat Lab 9700 S Cass Ave Argonne IL 60439

INOMATA, AKIRA, b Tochigi, Japan, May 13, 31; m 58; c 2. THEORETICAL PHYSICS. *Educ:* Kyushu Univ, BS, 56, MS, 58; Rensselaer Polytech Inst, PhD(physics), 64. *Prof Exp:* Res physicist, Benet Res Labs, 64-67; from asst prof to assoc prof, 67-84, PROF PHYSICS, STATE UNIV NY ALBANY, 84- *Concurrent Pos:* Vis lectr, State Univ NY, 66-67; adj assoc prof, Rensselaer Polytech Inst, 78; vis prof, Univ Munich, 81. *Mem:* Am Phys Soc; Phys Soc Japan. *Res:* Neutrinos in general relativity; geometric models for hadrons; de Sittar and conformal symmetries; gauge theory; path integrals; strong-gravity; monopoles. *Mailing Add:* Dept Physics State Univ NY Albany NY 12222

INOUE, MICHAEL SHIGERU, b Tokyo, Japan, June 27, 36; m 65; c 5. INDUSTRIAL ENGINEERING. *Educ:* Univ Dayton, BS, 59; Ore State Univ, MS, 64, PhD(indust eng), 67. *Prof Exp:* Res engr, appl res lab, Black & Decker Mfg Co, 61-62, sr res engr, 62-64; teaching asst physics, indust eng & elec eng, 64-65, from instr to prof indust eng, Ore State Univ, 64-82; pres, Productive Resources, Inc, 77-82; asst to the pres, Kyocera Northwest, Inc, 82-83, mgr consumer prod, 83-84, mgr corp planning & admin, 84, vpres admin & secy, 84-87, VPRES TECHNOL & PLANNING, KYOCERA INT, INC, 87- *Concurrent Pos:* Prin investr res proj, Nat Marine Fisheries Serv, 68-70, Educ Coord Coun, 70-71 & Sea-Grant Activities, 71-75; consult, minicomput mgt syst, 73-82, microprocessor-based syst design, 79, educ consult, qual circle implementation, 80-, Forest Prod Exportation, 81-; vis prof, Kyoto Univ, Japan, 74, Monterrey Inst Technol, Mexico, 76, Latvia Acad Sci, USSR, 79, Costa Rica Inst Technol, 82; invited speaker, Ore Productivity Conf, 85. *Mem:* Am Inst Indust Engrs; Data Processing Mgt Asn; Opers Res Soc Am; Am Soc Eng Educ. *Res:* Systems analysis; systems design and systems science; computer sciences; operations research; management science; data processing; computer simulation languages and models; organizational productivity and technology; sea-food harvesting and processing techniques; corporate planning and decision processes. *Mailing Add:* 8611 Balboa Ave San Diego CA 92123-1580

INOUE, RIICHI, fluid power engineering, oil hydraulic engineering, for more information see previous edition

INOUE, SHINYA, b London, Eng, Jan 5, 21; m 52; c 5. CELL BIOLOGY, MICROSCOPY. *Educ:* Tokyo Imp Univ, Rigakushi, 44; Princeton Univ, MA, 50, PhD(biol), 51. *Hon Degrees:* MA, Dartmouth Col, 62. *Prof Exp:* Instr micros & submicros anat, Sch Med, Univ Wash, Seattle, 51-53; asst prof biol, Tokyo Metrop Univ, 53-54; res assoc, Univ Rochester, 54-55, asst prof & fel, Am Cancer Soc, 55-58, assoc prof, 58-59; prof cytol & chmn dept, Dartmouth Med Sch, 59-65, John Laporte Given prof, 65-66; prof biol, Univ Pa, 66-82, grad group chmn, 77-79; sr scientist, 82-87, DISTINGUISHED SCIENTIST, MARINE BIOL LAB, 87- *Concurrent Pos:* Consult, Am Optical Co, 54-60; scholar in cancer res, Am Cancer Soc, 55-58; chmn dept anat, Dartmouth Med Sch, 59-63; instr, Woods Hole Marine Biol Labs, 62-65, 80-, mem bd trustees, 73-79 & 81-84; mem rev panels, NIH & NSF; John Simon Guggenheim fel, 71-72; pres, Universal Imaging Corp, 84-87; adj prof, Univ Pa, 82-89. *Honors & Awards:* Lewis S Rosenstiel Award for Distinguished Work in Basic Med Res, 88; Brown-Hazen Award for Outstanding Contrib to Basic Life Sci, 88. *Mem:* Fel AAAS; Soc Gen Physiol (pres, 69-70); fel Am Acad Arts & Sci; Am Soc Cell Biol; Biophys Soc; Hon fel Royal Micros Soc, 88. *Res:* Submicroscopic structure in relation to cell function and mechanisms of cell division; experimental cytology; biophysics; optics; video microscopy. *Mailing Add:* Marine Biol Lab Woods Hole MA 02543

INOUYE, DAVID WILLIAM, b Philadelphia, Pa, Jan 7, 50; m 69; c 2. ECOLOGY. *Educ:* Swarthmore Col, BA, 71; Univ NC, Chapel Hill, PhD(zool), 76. *Prof Exp:* Dir, Univ Colo Mountain Res Sta, 88-90; asst prof, 76-81, ASSOC PROF ZOOL, UNIV MD, 81- *Concurrent Pos:* NATO fel ecol, Bot Inst, Univ Wien, 77-78; mem bd trustees, Rocky Mt Biol Lab, 78-89. *Mem:* Ecol Soc Am; Soc Study Evolution; Animal Behav Soc; Sigma Xi; AAAS. *Res:* Plant-animal interactions; pollination ecology; plant-ant mutualisms; plant population biology. *Mailing Add:* Dept Zool Univ Md College Park MD 20742

INOUYE, MASAYORI, MEMBRANE BIOCHEMISTRY. *Educ:* Osaka Univ, Japan, PhD(biochem), 63. *Prof Exp:* Prof biochem & chmn dept, State Univ NY, Stonybrook, 81-87; PROF BIOCHEM & CHMN DEPT, UNIV MED & DENT, NJ, 87- *Res:* Regulation of gene expression. *Mailing Add:* Dept Biochem Univ Med & Dent NJ Robert Wood Johnson Med School 675 Hoes Lane Piscataway NJ 08854-5635

INSALATA, NINO F, b Brooklyn, NY, Aug 6, 26; m 49; c 1. MICROBIOLOGY, BACTERIOLOGY. *Educ:* St John's Univ, BSc, 48, MSc, 50. *Prof Exp:* Chem bacteriologist, Nat Distillers Prod Corp, 49-53; assoc bacteriologist, Gen Foods Res Ctr, 53-60, sr microbiologist, 60-67, sect head, Post Div, 67-73, lab mgr, 71-73, res assoc, Res Ctr, 73-75, AREA MGR MICROBIOL, CENT RES, GEN FOODS CORP, 75- *Concurrent Pos:* Mem subcomt sampling & methodology, USPHS, 64-75; assoc referee standardization microbiol test methods, Asn Off Anal Chemists, 64-75; prin investr, USPHS Contract, 66-67; US Food & Drug Admin res contract, 69-71; mem comn, Microbiol of Foods, Nat Acad Sci, Nat Res Coun; assoc referee, Am Asn Cereal Chemists & Am Off Anal Chemists; mem, Food Hyg Comn, Codex Alimentarius; mem subcomt microbiol standards for foods, Am Pub Health Asn; mem USDA salmonella comn, Am Asn Vet Lab Diagnosticians; consult, US Food & Drug Admin, USDA, Communicable Dis Ctr, USPHS, Am Pub Health Asn, Nat Fisheries Inst, Nat Food Protection Comt & pharmaceut & cosmetics industs; mem, Nat Referral Ctr, Libr Cong; co-chmn, Interagency Comn Compendium Microbiological Methods for Foods, Am Pub Health Asn, 71-75; mem subcomt, Microbiol Food Protection, Am Pub Health Asn, 74-75; tech advr, US Food & Drug Admin, Int Standards Orgn, Berlin, 80 & Paris, 81. *Honors & Awards:* Res Achievement Award, Gen Foods Res Comt, 73. *Mem:* Am Soc Microbiol; Soc Indust Microbiol; NY Acad Sci; Inst Food Technol; Int Asn Milk, Food & Environ Sanit. *Res:* Development of rapid microbiological test methods; incidence and survival studies with C1 Botulinum; methodology of recovery of Salmonellae, Staphylococci, Streptococci and fecal enterocci and development of fluorescent-antibody methods; environmental health aspects of industrial bioengineering and good manufacturing practices. *Mailing Add:* Four Sleator Dr Ossining NY 10562

INSCOE, MAY NILSON, b Geneva, Ill, May 16, 25; m 54; c 4. INSECT PHEROMONES. *Educ:* Wheaton Col, Ill, BS; Northwestern Univ, MS, 48, PhD(org chem), 51. *Prof Exp:* Teacher chem & physics, Tarsus Koleji, Turkey, 49-50; instr phys chem, Wellesley Col, 51-52; asst prof chem, Am Col Girls, Turkey, 52-54; chemist, Nat Bur Standards, Washington, DC, 54-66; CHEMIST, BELTSVILLE AGR RES CTR, AGR RES SERV, USDA, 66- *Mem:* Am Chem Soc. *Res:* Insect pheromones and other semiochemicals, attractants and repellents. *Mailing Add:* 10007 Thornwood Rd Kensington MD 20895-4228

INSEL, ARNOLD J, b Danbury, Conn, Jan 30, 40; m 69; c 2. MATHEMATICS. *Educ:* Univ Fla, BA, 60, MA, 62; Univ Calif, Berkeley, PhD(math), 69. *Prof Exp:* From asst prof to assoc prof, 69-87, PROF MATH, ILL STATE UNIV, 87- *Mem:* Am Math Soc; Math Asn Am. *Res:* Harmonic analysis; topological groups; linear algebra. *Mailing Add:* Dept Math Ill State Univ Normal IL 61761

INSEL, PAUL ANTHONY, b New York, NY, Nov 22, 45; m 77; c 2. RECEPTOR BIOLOGY & PHARMACOLOGY. *Educ:* Univ Mich, MD, 68. *Prof Exp:* Intern, Harvard Unit, Boston City Hosp, 68-70; clin assoc med officer, Nat Inst Child Health & Human Develop Geront Res Ctr, NIH, 70-74; res fel, Cardiovasc Res Inst, Univ Calif, San Francisco, 74-77, asst prof med, 77-78; from asst prof med, to assoc prof med, 78-87, PROF PHARMACOL & MED, UNIV CALIF, SAN DIEGO, 87- *Concurrent Pos:* Asst med, Johns Hopkins Univ, 71-74; assoc, Cardiovasc Res Inst, Univ Calif, San Francisco, 77-78; estab investr, Am Heart Asn, 77-82; master res, Inserm, France, 81, 86; mem US-France coop cancer res, NIH, 81; mem, Pharmacol Study Sect, NIH, 82-86. *Mem:* Am Soc Clin Invest; Am Soc Biol Chemists; Am Soc Pharmacol & Exp Therapeut; Am Soc Cell Biol; Endocrine Soc; Coun High Blood Pressure Res. *Res:* Mechanisms of action of adrenergic receptors and regulation of those receptors by target cells; cardiovascular disorders. *Mailing Add:* Dept Pharmacol & Med MO36 Univ Calif San Diego La Jolla CA 92093

INSELBERG, ALFRED, b Athens, Greece, Oct 22, 36. COMPUTER SCIENCE, AUTOMATIC CONFLICT RESOLUTION FOR AIR TRAFFIC CONTROL. *Educ:* Univ Ill, BS, 58, MS, 59, PhD(appl math), 65. *Prof Exp:* Res asst prof, Biol Comput Lab, Univ Ill, 65-66; sr sci staff mem, 66-87, SR CORP TECH STAFF MEM, IBM SCI CTR, LOS ANGELES, 87- *Concurrent Pos:* Lectr math & comput sci, Univ Calif, Los Angeles, 68-71; sr lectr appl math, Israel Inst Technol, 71-73; assoc prof math & comput sci, Ben Gurion Univ of the Negev, Beersheva, Israel, 77-83; Adj full prof, Dept Comp Sci, Univ Calif, Los Angeles, 85- & Univ Southern Calif, 87- *Honors & Awards:* Cochlear Model Award Oustanding Corp Contrib, IBM Corp, 75. *Mem:* Soc Indust & Appl Math; Sigma Xi; Asn Comput Mach. *Res:* Visualization of multivariate (multi-dimensional) relations X Data; conflict resolution algorithms for air traffic control; parallel coordinators; patents for multivariade information displays, conflict resolution for ATC. *Mailing Add:* IBM Sci Ctr 2525 Colorado Ave Santa Monica CA 90904

INSELBERG, EDGAR, b Athens, Greece, June 15, 30; nat US; m 56; c 1. PHOTOSYNTHESIS. *Educ:* Cornell Univ, BS, 53; Univ Ill, MS, 54, PhD, 56. *Prof Exp:* Res asst, Univ Ill, 53-56; dir res, Na-Churs Plant Food Co, Ohio, 56-61; res assoc photosynthesis, Univ Pittsburgh, 61-63, fel space prog, 63-65; res scientist, Volcani Inst, Israel, 65-66; ASSOC PROF BIOL, WESTERN MICH UNIV, 66- *Concurrent Pos:* Consult, 61- *Mem:* Am Soc Agron; Am Soc Plant Physiol; Am Inst Biol Sci; Am Soc Photobiol. *Res:* Earshoot development in corn; methodology and statistics of radioassay; photosynthesis; flash spectroscopy of algae. *Mailing Add:* Dept Biol Sci Western Mich Univ Kalamazoo MI 49008-5050

INSKEEP, EMMETT KEITH, b Petersburg, WVa, Jan 11, 38; m 60; c 2. REPRODUCTIVE PHYSIOLOGY, ENDOCRINOLOGY. *Educ:* WVa Univ, BS, 59; Univ Wis, MS, 60, PhD(endocrinol), 64. *Prof Exp:* From asst prof to assoc prof, 64-73, PROF ANIMAL PHYSIOL, WVA UNIV, 73- *Concurrent Pos:* Sect ed, Am Soc Animal Sci J, 72-74. *Honors & Awards:* Nat Asn Animal Breeders Res Award, 81; Am Soc Animal Sci Physiol & Endocrinol Award, 87. *Mem:* Am Soc Animal Sci; Endocrine Soc; Soc Study Reproduction; Brit Soc Study Fertil. *Res:* Control of ovulation in mammals; prostaglandins in uterine and ovarian tissues in relation to luteal function and life span; follicular maturation and ovulation in mammals. *Mailing Add:* Div Animal & Vet Sci WVa Univ Morgantown WV 26506-6108

INSKEEP, GEORGE ESLER, b Wilmington, Del, Dec 25, 18; m 55. CHEMISTRY. *Educ:* Pa State Col, BS, 40; Univ Ill, MS, 41, PhD(org chem), 43. *Prof Exp:* Asst chem, Univ Ill, 41-43, spec asst, US Off Rubber Reserve contract, 43-45; res chemist, Firestone Plastics Co, 45-50 & E I du Pont de Nemours & Co, 50-61; res chemist, Philip Morris, Inc, 61-66, asst patent off, 66-84; RETIRED. *Mem:* Am Chem Soc; Sci Res Soc Am; Am Inst Chemists. *Res:* High polymer chemistry; plastics characterization; smoke chemistry. *Mailing Add:* Rte 1, Box 2460 Kilmarnock VA 22482

INSKEEP, RICHARD GUY, b East Liberty, Ohio, Mar 11, 23; m 51. PHYSICAL CHEMISTRY. *Educ:* Miami Univ, Fla, AB, 44; Univ Ill, MS, 47, PhD(phys chem), 49. *Prof Exp:* Res engr, Dept Fuels, Battelle Mem Inst, 45-46; asst phys chem, Univ Ill, 46-48; fel chem, Univ Minn, 49-51; instr, Brown Univ, 51-53; from asst prof to assoc prof, Univ Vt, 53-61; assoc prof, Univ Hawaii, 61-65, chmn dept, 62-71, prof, 65-85, EMER PROF CHEM, UNIV HAWAII, 85- *Concurrent Pos:* NSF sci fac fel, Tech Univ Denmark, 58-59; vis prof, Univ BC, 74-75; vis scholar, Stanford Univ, 82-83. *Mem:* Am Chem Soc. *Res:* Molecular spectroscopy; complex ions; resonance Raman spectroscopy. *Mailing Add:* 2545 The Mall Univ Hawaii Honolulu HI 96822

INSKIP, ERVIN BASIL, b Nebo, Ill, Oct 27, 41; m 65; c 2. PROCESS CHEMISTRY, ANALYTICAL CHEMISTRY. *Educ:* Southern Ill Univ, BS, 63. *Prof Exp:* Chemist analytic chem, 63-65, sr chemist, 65-70, supvr tantalum process technol, 70-74, group leader, res & develop process chem, 74-77, MGR RES & DEVELOP, PROD DEVELOP, MALLINCKRODT INC, 77- *Mem:* Chem Soc. *Res:* Analytical chemistry in chemical process control; niobium and tantalum benefication, iodine chemistry, iodine in disinfection of water; derivatives of p-aminophenol-manufacturing methods and applications; monomer stabilization. *Mailing Add:* 213 W Second St Trenton IL 62293

INSKIP, HAROLD KIRKWOOD, b Buffalo, NY, Nov 7, 22; m 49; c 3. ORGANIC CHEMISTRY. *Educ:* Yale Univ, BS, 44, MS, 48, PhD(org chem), 52. *Prof Exp:* Asst & instr org chem, Univ Ill, 50-52; from res chemist to res supvr, 52-61, staff scientist, 61-66, RES ASSOC, ELECTROCHEM DEPT, E I DU PONT DE NEMOURS & CO, 66- *Mem:* Am Chem Soc. *Res:* Vinyl polymerization. *Mailing Add:* 38 Bridle Brook Lane Newark DE 19711

INSLEY, EARL GLENDON, physical chemistry; deceased, see previous edition for last biography

INSLEY, ROBERT H(ITESHEW), b Washington, DC, June 20, 23; m 47; c 3. CERAMICS. *Educ:* Hamilton Col, BA, 49; Pa State Col, MS, 52. *Prof Exp:* Asst mineral, Pa State Col, 49-52; petrographer, ceramic div, 52-58, sr res engr, 58-68, mgr ceramic res, 68-73, asst dir, res & eng, 73-75, DIR, RES & DEVELOP, CERAMIC DIV, CHAMPION SPARK PLUG CO, 75- *Honors & Awards:* Ross Coffin Purdy Award, Am Ceramic Soc, 66; Award of Merit, Am Soc Testing & Mat, 83. *Mem:* Fel Am Ceramic Soc; Am Soc Mineralogists; Sigma Xi; German Ceramic Soc; fel Am Soc Testing & Mat. *Res:* Petrography of ceramic raw materials and finished products; high temperature investigations of oxides; semiconductors; glass to metal seals. *Mailing Add:* 855 Pine Hill Dr Bloomfield Hills MI 48013

INTAGLIETTA, MARCOS, b Buenos Aires, Argentina, Aug 10, 35. BIOENGINEERING. *Educ:* Univ Calif, Berkeley, BS, 57; Calif Inst Technol, MS, 58, PhD(appl mech), 63. *Prof Exp:* Estab investr, Los Angeles County Heart Asn, Calif Inst Technol, 64-66; from asst prof to assoc prof bioeng, 66-76, PROF BIOENG, UNIV CALIF, SAN DIEGO, 76- *Concurrent Pos:* Dir, Int Inst Microcirculation; Hoffman La-Roche Int fel. *Mem:* Microcirculatory Soc (pres, 85-86); Am Physiol Soc; Soc Biorheology. *Res:* Transport phenomena; instrumentation at the microscopic level; microcirculatory physiology. *Mailing Add:* AMES-Bioeng 0412 Univ Calif San Diego La Jolla CA 92093-0412

INTEMANN, GERALD WILLIAM, b North Bergen, NJ, Jan 12, 43; c 3. PARTICLE PHYSICS. *Educ:* Stevens Inst Technol, BS, 64, MS, 66, PhD(particle physics), 68. *Prof Exp:* Asst prof physics, State Univ NY Binghamton, 68-72; asst prof physics, Seton Hall Univ, 72-77, assoc prof, 77-80; PROF PHYSICS & HEAD DEPT, UNIV NORTHERN IOWA, 80- *Concurrent Pos:* Vis physicist, Argonne Nat Lab, 83. *Mem:* Am Phys Soc; Am Asn Physics Teachers; Sigma Xi; AAAS. *Res:* Theoretical investigation of multi-quark states; phenomenological studies of strong and electromagnetic decays of hadrons which are bound states of heavy quarks. *Mailing Add:* Dept Physics Univ Northern Iowa Cedar Falls IA 50614

INTEMANN, ROBERT LOUIS, b North Bergen, NJ, Feb 23, 38; m 64; c 1. THEORETICAL PHYSICS. *Educ:* Stevens Inst Technol, BE, 59, MS, 61, PhD(physics), 64. *Prof Exp:* From asst prof to assoc prof physics, Temple Univ, 64-84, asst dean, Col Lib Arts, 71-81, chmn dept, 85-90, PROF PHYSICS, TEMPLE UNIV, 84- *Mem:* AAAS; Am Phys Soc; Am Asn Physics Teachers; Sigma Xi. *Res:* Theoretical atomic physics; inner shell processes in atoms; atomic collisions; radiation theory. *Mailing Add:* Dept Physics Temple Univ Philadelphia PA 19122

INTERRANTE, LEONARD V, b Brooklyn, NY, Apr 6, 39; m 59; c 2. INORGANIC CHEMISTRY, MATERIALS CHEMISTRY. *Educ:* Univ Calif, Riverside, AB, 60; Univ Ill, PhD(inorg chem), 64. *Prof Exp:* NSF fel, Univ Col, Univ London, 63-64; asst prof inorg chem, Univ Calif, Berkeley, 64-68; inorg chemist, Corp Res & Develop, Gen Elec Co, 68-85; AT DEPT CHEM, RENSSELAER POLYTECH INST, 85- *Concurrent Pos:* Consult, indust, Heineman-Butterworths Publ Co; ed, Am Chem Soc Jour, "Chem of Mats". *Mem:* Am Chem Soc; Royal Soc Chem; fel AAAS; Am Ceramic Soc; Mats Res Soc. *Res:* Synthesis and solid state properties of coordination compounds; synthesis, electrical and magnetic properties of donor-acceptor complexes; organometallic chem vapor deposition; development of organometallic precursors to ceramic materials. *Mailing Add:* Dept Chem Rensselaer Polytech Inst Troy NY 12181

INTRES, RICHARD, MOLECULAR BIOLOGY. *Educ:* Windham Col, BA, 78; Wesleyan Univ, PhD, 86. *Prof Exp:* Asst staff scientist, May & Baker Ltd, Dagenham, Essex, Eng, 78-79; res scientist, Norwich-Eaton Pharmaceut, Inc, 79-81; Peterson fel biochem, Wesleyan Univ, 83, res asst, 83-86; res assoc, Dept Biochem & Biophys, Howard Hughes Med Inst, Univ Calif, San Francisco, 86-87; asst scientist res & develop, Bionique Lab, Inc, Saranac Lake, NY, 87-89; RES SCIENTIST, W ALTON JONES CELL SCI CTR, LAKE PLACID, NY, 89- *Mem:* Am Soc Cell Biol; Asn Res Vision & Ophthal. *Res:* Tissue culture; cell culture; cell separation; electron microscopy; immunofluorescence. *Mailing Add:* W Alton Jones Cell Sci Ctr Ten Old Barn Rd Lake Placid NY 12946

INTRILIGATOR, DEVRIE SHAPIRO, b New York, NY; m 63; c 3. SPACE PHYSICS. *Educ:* Mass Inst Technol, SB, 62, SM, 64; Univ Calif, Los Angeles, PhD(planetary & space physics), 67. *Prof Exp:* Res asst cosmic ray group, Physics Dept, Mass Inst Technol, 60; consult physicist, Inst Physics, Univ Milan, 61; physicist, Cosmic Ray Br, Air Force Cambridge Res Labs, 62-63; asst res geophysicist, Inst Geophys & Planetary Physics, Univ Calif, Los Angeles, 67; Nat Acad Sci-Nat Res Coun resident res assoc, Space Sci Div, Ames Res Ctr, NASA, 67-69; res fel physics, Calif Inst Technol, 69-74; mem staff, Stauffer Hall Sci, Univ Southern Calif, 74-77, asst prof physics, 77-79; SR RES PHYSICIST, CARMEL RES CTR, 79- *Concurrent Pos:* Dir, Space Plasma Lab, 80- *Mem:* Am Geophys Union; Am Phys Soc. *Res:* High energy nuclear physics; plasma physics; astrophysics. *Mailing Add:* Carmel Res Ctr PO Box 1732 Santa Monica CA 90406

INTURRISI, CHARLES E, b Waterbury, Conn, Apr 15, 41; m 66; c 1. PHARMACOLOGY. *Educ:* Univ Conn, BS, 62; Tulane Univ, MS, 65, PhD(pharmacol), 67. *Prof Exp:* Res assoc chem pharmacol, Nat Heart Inst, 67-69; from asst prof to assoc prof, 69-78, PROF PHARMACOL, MED COL, CORNELL UNIV, 78- *Honors & Awards:* Litchfield lectr, Oxford Univ, 82. *Mem:* AAAS; NY Acad Sci; Am Soc Pharmacol & Exp Therapeut. *Res:* Biochemical pharmacology; relationship of disposition of narcotics and antagonists to pharmacologic effects and development of tolerance. *Mailing Add:* Dept Pharmacol Cornell Univ Med Col New York NY 10021

INUI, THOMAS S, b Baltimore, Md, July 10, 43. MEDICINE. *Educ:* Haverford Col, BA, 65; Johns Hopkins Univ, MD, 69, ScM, 73; Am Bd Internal Med, dipl, 72. *Prof Exp:* Intern, Johns Hopkins Hosp, 69-70, asst resident med, 70-71, sr asst resident, 71-72, chief resident, 73-74, instr, Dept Med Care Orgn, Sch Hyg & Pub Health, Johns Hopkins Univ, 73-76; chief med, USPHS Indian Hosp, Albuquerque, NMex, 74-76, physician-in-chief, 74-76; asst prof, 76-80, assoc prof, 80-85, PROF MED & HEALTH SERV, UNIV WASH, 85-, HEAD, DIV GEN INTERNAL MED, 86- *Concurrent Pos:* Carnegie-Commonwealth Clin scholar, Johns Hopkins Univ, 71-73, instr, Dept Med, Sch Med, 73-74; dir, Robert Wood Johnson Clin Scholars Prog, Univ Wash, 77-; chief, Med Comprehensive Care Unit, Seattle Vet Admin Med Ctr, 76-86, dir, Health Serv Res & Develop Ctr, 76-82, co-dir, Health Serv Res Training Prog, 79-86; assoc mem, Hastings Ctr, 81-; nat coun mem, Soc Gen Internal Med, 83-86; chief, Gen Internal Med Sect, Harborview Med Ctr, Seattle, 86-; mem, Med Res Serv Coop Studies Eval Comt, Vet Admin, 88-91; assoc ed, J Gen Internal Med; correspondent, Comt Human Rights, Nat Acad Sci. *Mem:* Inst Med-Nat Acad Sci; Am Pub Health Asn; Soc Gen Internal Med (pres-elect, 86-89); Am Fedn Clin Res; fel Am Col Physicians. *Res:* Author or co-author of over 70 publications. *Mailing Add:* Div Gen Internal Med Harborview Med Ctr 325 Ninth Ave Seattle WA 98104

IOACHIM, HARRY L, b Bucharest, Rumania, Oct 22, 24; US citizen; m 54; c 2. PATHOLOGY, ONCOLOGY. *Educ:* Cultura-Lyceum, Bucharest, Rumania, BS, 44; Fac Med, Univ Bucharest, MD, 51. *Prof Exp:* Chief pathologist, First Surg Clin, Fac Med, Univ Bucharest, 54-61, asst prof path, Dept Path, 56-61; chief res cancer, Inst Cancer, Paris, France, 61-62; from asst prof to assoc prof, 62-72, CLIN PROF PATH, COL PHYSICIANS & SURGEONS, COLUMBIA UNIV, 72-; PROF PATH, MED COL, CORNELL UNIV, 90- *Concurrent Pos:* Health res career award, City New York Health Res Coun, 66; attend-on-staff path, Lenox Hill Hosp, 68-; grant cancer res, Nat Cancer Inst, 67, 70, & 73-, & Am Cancer Soc, 82-; ed, Pathobiol Ann, Raven Press, 70-80; dir dept path, Med Col, Cornell Univ, 82- *Mem:* Int Acad Path; NY Acad Sci; Am Asn Cancer Res; Am Asn Immunol; Harvey Soc. *Res:* Cancer immunology, lung cancer, ovarian cancer; cancer pathology; etiology and mechanisms of leukemia; author 3 books on pathology and AIDS. *Mailing Add:* Dept Path Lenox Hill Hosp 100 E 77th St New York NY 10021

IODICE, ARTHUR ALFONSO, b Rome, NY, Nov 7, 28. BIOCHEMISTRY. *Educ:* Columbia Col, AB, 50; State Univ NY Upstate Med Ctr, PhD(biochem), 58. *Prof Exp:* Jane Coffin Childs Mem Fund fel biochem, Univ Calif, Berkeley, 58-60, res assoc, 60-62; res assoc, Biochem Div, Inst Muscle Dis, 62-65, asst mem, 65-69, assoc mem, 69-74; RES SCIENTIST, MASONIC MED RES LAB, 75- *Mem:* AAAS; NY Acad Sci; Am Heart Asn. *Res:* Roles of the lysosomal and neutral proteases in protein turnover and in muscular dystrophy; intracellular uptake of quinidine and other antiarrhythmic agents by cardiac tissues. *Mailing Add:* Masonic Med Res Lab Utica NY 13501-1787

IONA, MARIO, b Berlin, Ger, June 17, 17; nat US; m 49; c 2. COSMIC RAY PHYSICS. *Educ:* Univ Vienna, PhD(physics), 39. *Prof Exp:* Int Student Serv fel, Univ Uppsala, 39-41; fel, Univ Chicago, 41-42, from res asst physics to instr, 41-46; from asst prof to prof physics, 46-85, coordr, inter-univ high altitude labs, 48-62, coordr, high altitude lab, 62-82, EMER PROF PHYSICS, UNIV DENVER, 85- *Concurrent Pos:* Asst coord, NSF Summer Inst High Sch Teachers, 58, coord phys, In-serv Inst High Sch Teachers, 58-59; consult, Denver Pub Schs, 62-65, 76- & summer inst col teachers, Univ Saugar, India, 66; consult, Denver Res Inst 55-77; consult, Jefferson County Pub Schs, 73 & Adams County Sch Dist R12, 85; consult, J Sci & Children, Nat Sci Teachers Asn, 75-, J Sci Scope, 82-, J Sci Teacher, Physics Teacher, Am Asn Physics Teachers, 83- *Honors & Awards:* Robert A Milliken Lectr Award, Am Asn Physics Teachers, 86. *Mem:* Fel AAAS; Am Phys Soc; Am Asn Physics Teachers; Nat Sci Teachers Asn. *Res:* Nuclear physics; cosmic rays; electronics; distribution of lattice vibrations of the KCL crystal; science education. *Mailing Add:* Dept Physics Univ Denver Denver CO 80208-0202

IONESCU, DAN, b Oravita, Romania, July 23, 43; Can citizen; m 69; c 2. COMPUTERS, ELECTRICAL ENGINEERING. *Educ:* Polytech Inst Bucharest, Dipl Eng, 66, DrSci, 82; Univ Timisoara, Dipl Math, 82. *Prof Exp:* ASST PROF COMPUTERS CONTROL, UNIV OTTAWA, 85- *Mem:* Sr mem Inst Elec & Electronics Engrs; Soc Indust & Appl Math; Am Math Soc; NY Acad Sci. *Res:* Multivariable control; robust control; artificial intelligence in control; expert systems for process control; learning in control. *Mailing Add:* Dept Elec Eng Univ Ottawa 770 King Edward Ottawa ON K1N 6N5 Can

IONESCU, LAVINEL G, b Varset, Yugoslavia, May 19, 43; US citizen. PHYSICAL CHEMISTRY, BIOCHEMISTRY. *Educ:* Univ NMex, BS, 64, MS, 66; NMex State Univ, PhD(chem), 70. *Prof Exp:* NSF fel, NMex State Univ, 71; asst prof, NMex Highlands Univ, 72-75; asst prof chem, Univ Detroit, 75-78; PROF PHYS CHEM, FED UNIV SANTA CATARINA, BRAZIL, 78- *Concurrent Pos:* USPHS fel, Univ Calif, Santa Barbara, 71-72. *Mem:* AAAS; Sigma Xi; Am Chem Soc. *Res:* Thermodynamics; kinetics; application of physical chemical principles in the elucidation of basic biologic processes; noble gases, micelles, membranes, respiratory pigments, clathrates and radioactive exchange. *Mailing Add:* Inst De Quimica Univ Fed Do Rio Grande Dosul Porto Alegre RS 90000 Brazil

IONESCU TULCEA, CASSIUS, b Bucarest, Roumania, Oct 14, 23; US citizen. STATISTICS, MATHEMATICAL ANALYSIS. *Educ:* Univ Bucarest, MS, 46; Yale Univ, PhD(math), 59. *Prof Exp:* Instr math, Univ Bucarest, 46-51, from asst prof to assoc prof, 52-57; res assoc, Yale Univ, 57-59, vis lectr, 59-61; assoc prof, Univ Pa, 61-64; prof, Univ Ill, Urbana, 64-66; PROF MATH, NORTHWESTERN UNIV, 66- *Honors & Awards:* Prize, Roumanian Acad Sci, 57. *Res:* Research monographs on probability and statistics, lifting theory and Hilbert Spaces; set theory and topology; game theory. *Mailing Add:* Dept Math Northwestern Univ Evanston IL 60208

IORILLO, ANTHONY J, b Southington, Conn, Feb 12, 38; m; c 4. AVIATION ENGINEERING. *Educ:* Calif Inst Technol, ME, 59, MS, 60. *Prof Exp:* SR VPRES & GROUP PRES SPACE & COMMUN GROUP, HUGHES AIRCRAFT CO. *Concurrent Pos:* Fulbright scholar, Politecnico, Turin, Italy, 61; mem, Air Force Sci Adv Bd & Defense Commun Agency Sci Adv Group. *Honors & Awards:* Lawrence A Hyland Patent Award, 70; Spacecraft Design Award, Am Inst Aeronaut & Astronaut, 71. *Mem:* Nat Acad Eng. *Mailing Add:* Hughes Aircraft Co PO Box 92919 Bldg S64 MS A400 Los Angeles CA 90009

IORNS, TERRY VERN, b Kenosha, Wis, Apr 15, 44. ANALYTICAL CHEMISTRY. *Educ:* Northwestern Univ, BA, 66; Univ Wis, PhD(chem), 70. *Prof Exp:* Fel chem, Univ Calif, Los Angeles, 70-71; instr chem, Syracuse Univ, 71-72; res chemist, 72-77, supvr chem methods, Phillips Petrol Co, 77-87, VPRES, PROD PLANNING AXIOM SYSTS INC, 87- *Mem:* Am Chem Soc; Sigma Xi. *Res:* Automation of chemical methods of analysis including wet chemical and combustion methods; application of computers to chemical analysis. *Mailing Add:* Coopers & Lybrand One Sylvan Way Parsippany NJ 07054

IOSSIFIDES, IOULIOS A, pathology, hematopathology, for more information see previous edition

IOVINO, ANTHONY JOSEPH, b Brooklyn, NY, Apr 25, 25; m 49; c 1. PHYSIOLOGY. *Educ:* St John's Univ, NY, BS, 46, MS, 52; NY Univ, PhD(biol), 59. *Prof Exp:* Instr biol, St John's Univ, NY, 49-56; PROF BIOL, LONG ISLAND UNIV, 56-, DIR DIV SCI, 77- *Mem:* AAAS; NY Acad Sci; Sigma Xi. *Res:* Endocrinology. *Mailing Add:* 170 Powerhouse Rd Long Island Univ Roslyn Heights NY 11577

IP, CLEMENT CHEUNG-YUNG, b Hong Kong, Aug 21, 47; m 72. BIOCHEMISTRY. *Educ:* McGill Univ, BSc, 69; Univ Wis-Madison, PhD(biochem), 73. *Prof Exp:* Res assoc endocrinol, State Univ NY Upstate Med Ctr, 73-75; CANCER RES SCIENTIST, BREAST CANCER, ROSWELL PARK MEM INST, 75- *Mem:* Am Asn Cancer Res. *Res:* Nutritional modification of mammary carcinogenesis. *Mailing Add:* Dept Breast Surg Cell & Virus Roswell Park Mem Inst 666 Elm St Buffalo NY 14263

IP, MARGOT MORRIS, b Toronto, Ont, Mar 10, 43; m 72. ENDOCRINOLOGY, CANCER RESEARCH. *Educ:* Univ Toronto, BA, 65; Harvard Sch Pub Health, MS, 67; Univ Wis-Madison, PhD(biochem), 72. *Prof Exp:* Fel nutrit biochem, Univ Wis-Madison, 72-73; res assoc endocrinol, Upstate Med Ctr, Syracuse, 73-75; CANCER RES SCIENTIST, ROSWELL PARK MEM INST, 75- *Mem:* Am Asn Cancer Res; Am Inst Nutrit; AAAS; Endocrine Soc. *Res:* Endocrinology of neoplastic tissues. *Mailing Add:* Dept Exp Therapeut Roswell Park Cancer Inst Buffalo NY 14263

IP, STEPHEN H, b China, Feb 9, 49. BIOPHYSICS. *Educ:* Bridgewater Col, BA, 70; Univ Va, PhD(biochem & biophys), 76. *Prof Exp:* NIH postdoctoral fel hemat & oncol, Med Sch Univ Pa, assoc scientist, 76-79; group leader cancer biol, Ortho Diag Systs, Inc, Cambridge, Mass, Johnson & Johnson subsid, mgr technol develop, 79-84; dir res, T Cell Sci, Inc, vpres & corp officer technol develop, 84-89; PRES & CHIEF OPER OFFICER, CYTOMED INC, FRAMINGHAM, MASS, 89- *Concurrent Pos:* Adj vis assoc prof path, Med Sch, Columbia Univ, 82-84. *Mem:* Am Asn Immunologists; Am Asn Exp Hemat; Am Chem Soc; Am Asn Clin Chemists. *Res:* Bone marrow transplantation; blood transfusion and immuno-hematology; author of numerous scientific articles. *Mailing Add:* CytoMed One Mountain Rd Framingham MA 01701

IP, WALLACE, b Hong Kong, Dec 29, 48; US citizen; m; c 1. CYTOSKELETON STRUCTURE & FUNCTION, ELECTRON MICROSCOPY. *Educ:* Ill Inst Technol, PhD(cell biol), 77. *Prof Exp:* ASSOC PROF ANAT & CELL BIOL, SCH MED, UNIV CINCINNATI, 85- *Concurrent Pos:* Estab investr, Am Heart Asn. *Mem:* Am Soc Cell Biol. *Res:* Structure and assembly of intermediate filaments. *Mailing Add:* Dept Anat & Cell Biol Col Med Univ Cincinnati 231 Bethesda Ave Cincinnati OH 45267-0521

IPPEN, ERICH PETER, b Fountain Hill, Pa, Mar 29, 40; m 66; c 2. QUANTUM ELECTRONICS. *Educ:* Mass Inst Technol, SB, 62; Univ Calif, Berkeley, MS, 65, PhD(elec eng), 68. *Prof Exp:* Mem tech staff, Bell Labs, Holmdel, NJ, 68-80; prof, dept elec eng & comput sci, 80-87, ELIHU THOMSON PROF ELEC ENG, MASS INST TECHNOL, 87- *Concurrent Pos:* Vis prof, Dept Elec Eng & Comput Sci, Mass Inst Technol, 78; NSF spec rev panels, 80, 85 & 87; mem adv comt, Lasers & Electro Optics Soc, Inst Elec & Electronics Engrs, 82-85, chmn, Fel Eval Comt, 85-86; mem prog comt, Conf Picosecond Electronics & Opto-electronics, Optical Soc Am, 85 & 87; Humboldt award, 86; mem, Air Force Studies Bd, Nat Acad Sci, 86-, Comt Optical Data Collection, 87, chmn, Sect Eng, 90-; mem, Lincoln Lab Adv Bd, Mass Inst Technol, 90- *Honors & Awards:* R W Wood Prize, Optical Soc Am, 81; Edward Longstreth Medal, Franklin Inst, 82; Morris Leeds Awards, Inst Elec & Electronics Engrs, 83; R V Pole Lectr, Optical Soc Am, 88; Walter Schottky Lectr, RWTH Aachen, Ger, 89; Harold E Edgerton Award, Int Soc Optical Eng, 89. *Mem:* Nat Acad Sci; Nat Acad Eng; fel Am Acad Arts & Sci; fel Inst Elec & ELectronics Engrs; fel Optical Soc Am; fel Am Phys Soc. *Res:* Development of sub-picosecond and femtosecond optical techniques; application to studies of ultrafast processes in materials; extension of picosecond techniques to semiconductor optical electronics and signal processing in optical waveguides; author of over 100 publications. *Mailing Add:* Dept Elec Eng Bldg 36 Rm 319 Mass Inst Technol 77 Massachusetts Ave Cambridge MA 02139

IPPEN-IHLER, KARIN ANN, b Fountain Hill, Pa, Mar 13, 42; m 70; c 2. BACTERIAL GENETICS. *Educ:* Wellesley Col, BA, 63; Univ Calif, Berkeley, PhD(biochem), 67. *Prof Exp:* Harold Ernst fel bact & immunol, Harvard Med Sch, 67-68, Am Cancer Soc fel, 68-69; Am Cancer Soc fel, Med Res Coun Molecular Genetics Unit, Univ Edinburgh, 69-70; res asst prof, Univ Pittsburgh, 70-71, asst prof microbiol, 71-77; assoc prof, 77-84, PROF MED MICROBIOL, COL MED, TEX A&M UNIV, 84- *Concurrent Pos:* Mem, Microbiol Genetics Study Sect, NIH, 78-82, NIH Cellular & Molecular Basis Dis Rev Comt, 85-89; ed bd, J Bacteriol, 79- *Mem:* AAAS; Am Soc Microbiol. *Res:* Genetics and biochemistry of the conjugative transfer systems expressed by bacterial plasmids. *Mailing Add:* Dept Med Microbiol Tex A&M Univ Col Med College Station TX 77843

IPSER, JAMES REID, b New Orleans, La, July 13, 42; m 65; c 2. THEORETICAL ASTROPHYSICS. *Educ:* Loyola Univ, La, BS, 64; Calif Inst Technol, MS, 67, PhD(physics), 69. *Prof Exp:* Fel physics, Calif Inst Technol, 69-70; actg asst prof astron, Univ Wash, 70-71; AT PHYSICS DEPT, UNIV FLA. *Mem:* Am Astron Soc; Am Phys Soc; Int Astron Union. *Res:* Theoretical astrophysics and applications of general relativity to astrophysical problems. *Mailing Add:* Dept Physics Univ Fla Williamson Hall Gainesville FL 32611

IQBAL, ZAFAR, b Lucknow, India, July 12, 46. NEUROCHEMISTRY, NEUROSCIENCES. *Educ:* Univ Lucknow, BS, 61, MS, 63; All-India Inst Med Sci, PhD(biochem), 71. *Prof Exp:* Jr res fel biochem, Coun Sci & Indust Res, India, 63-66; res scholar, Directorate-Gen Health Serv, India, 66-67; res fel, Coun Sci & Indust Res, 67-68; asst res officer, Indian Coun Med Res,

68-71; res assoc & investr physiol, Sch Med, Ind Univ, Indianapolis, 72-77, asst prop med biophys, 77-79, asst prof biochem, 79-82; ASSOC PROF NEUROL, NORTHWESTERN UNIV SCH MED, 82- *Concurrent Pos:* Mem, Inst Neurosci, Northwestern Univ, Ctr Develop Biol & Amyolateral Sclerosis Res Ctr; res grants, NIH, 73-77 & 88-91, Muscular Dystrophy Asn, 75-77, Am Cancer Soc, 79-80, NSF, 81 & 84, Juv Diabetes Found, 81, Am Diabetes Asn, 80; FIDIA Res Found Award, 87. *Mem:* Int Brain Res Orgn; Int Soc Neurochem; Soc Neurosci; Am Soc Neurochem; Biophys Soc; Am Physiol Soc; Soc Exp Biol & Med; Sigma Xi. *Res:* Chemical exploration of developing nervous system; characterization of enzymes, proteins and polypeptides associated with the axoplasmic transport system; calmodulin associated neurobiological processes; role of polyamines in neuronal signal transduction; contributor or editor of several books. *Mailing Add:* Northwestern Univ Sch Med PO Box 11538 Chicago IL 60611-0538

IRALU, VICHAZELHU, microbiology, bacteriology; deceased, see previous edition for last biography

IRANI, KEKI B, b India. ELECTRONICS. *Educ:* Univ Bombay, BE, 47; Univ Mich, MSE, 49, PhD, 53. *Prof Exp:* Res & develop engr, Philips Telecommun Industs, Neth, 50-56; from asst prof to assoc prof elec eng, Univ Kans, 56-60; assoc res engr, Inst Sci & Technol, 61-62, from asst prof to assoc prof elec eng, 62-67, PROF ELEC ENG, UNIV MICH, ANN ARBOR, 67- *Mem:* Sr mem Inst Elec & Electronics Engrs; Sigma Xi. *Res:* Computers; large-scale systems. *Mailing Add:* 2785 Parkridge Dr Ann Arbor MI 48103-1732

IRANI, N F, b Haifa, Israel, July 20, 34; m 60; c 2. PHYSICAL CHEMISTRY. *Educ:* Am Univ Beirut, BS, 55, MS, 60; Univ NC, Chapel Hill, PhD(phys chem), 66. *Prof Exp:* Chmn, Sci Sect, Nat Col Chouifat, Lebanon, 55-60; instr chem, Am Univ Beirut, 60-61; sr res chemist, Westvaco Corp, SC, 66-68; SR RES CHEMIST, RES LABS, EASTMAN KODAK CO, 68- *Mem:* Am Chem Soc. *Res:* Adsorption of dispersants on colloidal particles; solid liquid interface; characterization of polymers involved in adsorption on a colloidal substrate. *Mailing Add:* Ten Enfield Dr Pittsford NY 14534

IRANI, RIYAD RAY, b Beirut, Lebanon, Jan 15, 35; nat US; m 56; c 3. PHYSICAL CHEMISTRY. *Educ:* Am Univ Beirut, BS, 53; Univ Southern Calif, PhD(phys chem), 57. *Prof Exp:* Res chemist, Monsanto Chem Co, 57-60, group leader, 60-63, sr res group leader, Monsanto Co, Mo, 63-67; assoc dir, T R Evans Res Ctr, Diamond Shamrock Corp, 67-69, dir res, 69-73; vpres res & develop, Olin Corp, 73-74, sr vpres, 74-76, pres & chief operating officer, 78-80, exec vpres, Chem Group, 77-83; CHMN, PRES & CHIEF EXEC OFFICER, OCCIDENTAL PETROL CORP, 83- *Mem:* Am Chem Soc; Sigma Xi. *Res:* Ionic solutions; theory of liquids; physical properties of powders; complex ion formation; thermodynamics; phosphorous compounds chemistry. *Mailing Add:* Occidental Petrol Corp 10889 Wilshire Blvd Los Angeles CA 90024

IRAUSQUIN, HILTJE, b Netherlands, Jan 14, 37; US citizen; div; c 2. BIOCHEMISTRY, NUTRITION. *Educ:* Univ Amsterdam, BS, 58, MS, 64, PhD(biochem), 69. *Prof Exp:* Surv dir nutrit, Cent Health Lab, Curacao, Neth Antilles, 64-67; res asst, Royal Trop Inst, Amsterdam, 69; res assoc cell biol, Children's Hosp Nat Med Ctr, Washington, DC, 69-72; biomed indexer, Nat Libr Med, Bethesda, Md, 75-77; res assoc, Univ Md, 77-78; sr staff fel, 78-81, rev toxicologist, 81-85, GROUP LEADER, FOOD & DRUG ADMIN, 85- *Mem:* AAAS; NY Acad Sci; Soc Toxicol. *Res:* Metabolism of artificial sweeteners and their interactions with glucose homeostasis; regulatory mechanism of enzyme activity; safety of food additives; database management and priority ranking; toxicology. *Mailing Add:* 200 C St SW HFF 156 HFF 268 Washington DC 20204

IRBY, BOBBY NEWELL, b Meridian, Miss, Mar 17, 32; m 53; c 1. SCIENCE EDUCATION. *Educ:* Univ Wash, Seattle, BA, 57; Univ Miss, MS, 62, DEd(sci educ), 67. *Prof Exp:* Teacher private sch, Miss, 57-58; head dept sci, Pub Sch, 59-64; instr chem, Northeast La State Col, 64-65; from asst prof to assoc prof, 67-72, PROF SCI EDUC, UNIV SMISS, 72-, HEAD DEPT, 69- *Concurrent Pos:* Prin investr, Miss-Ala sea grant, Marine Educ Curric Proj, 79, 80 & 81; Dir, Dept Educ Inserv Marine Educ Proj, Miss, 81. *Mem:* AAAS; Am Chem Soc; Nat Sci Teachers Asn; fel Am Inst Chem. *Res:* Vapor pressures of the methylacetylenes; effect of ion hydrate size and shape on the surface interaction between metal ions and highly porous materials; science curriculum development and teacher education. *Mailing Add:* Dept Sci Educ Univ SMiss Southern Sta Box 5087 Hattiesburg MS 39406

IRBY, WILLIAM ROBERT, b Blackstone, Va, May 31, 23; m 50; c 3. MEDICINE. *Educ:* Hampden-Sydney Col, AB, 43; Med Col Va, MD, 48. *Prof Exp:* Trainee arthritis, Nat Inst Arthritis & Metab Dis, 53-54; assoc med, 56-58, from asst prof to assoc prof, 58-70, PROF MED, MED COL VA, VA COMMONWEALTH UNIV, 70- *Concurrent Pos:* Arthritis med consult, USPHS, 64-72; arthritis prog consult, Arthritis Found, 66-72, chmn educ comt, 67-72. *Mem:* AMA; Am Rheumatism Asn; Nat Soc Clin Rheumatol; fel Am Col Physicians. *Res:* Bone and joint changes observed in patients undergoing renal transplantation. *Mailing Add:* Dept Med Va Commonwealth Univ Box 163 Richmond VA 23298

IRELAND, CAROL BEARD, b Southington, Conn, July 6, 48; m 73. PHYSICAL CHEMISTRY. *Educ:* Earlham Col, AB, 70; Univ Ill, Urbana-Champaign, PhD(chem), 74. *Prof Exp:* Res chemist, Exp Sta, 75-90, SUPVR TECHNOL SERV, E I DU PONT DE NEMOURS & CO, INC, 90- *Mem:* Am Chem Soc; Sigma Xi. *Res:* Use of physical chemical techniques to characterize and improve elastomeric materials; gas chromatography and gas chromatography/mass spectrometry methods development. *Mailing Add:* 106 Hobson Dr Hockessin DE 19707-2105

IRELAND, GORDON ALEXANDER, b Belfast, Northern Ireland, Feb 13, 44; US citizen. CLINICAL PHARMACY. *Educ:* Univ Md, BS, 73; Univ Minn, PharmD, 76. *Prof Exp:* Asst pharm, 73-74, ASST PROF CLIN PHARM, SCH PHARM, UNIV MD, 76-; CLIN PHARMACIST, BALTIMORE VET ADMIN MED CTR, 76- *Concurrent Pos:* Consult, Shangri-La Nursing Home, 78-; pharmacist consult, McGilvray's Pharm, 78-80. *Mem:* Am Pharmaceut Asn; Am Asn Cols Pharm; Am Soc Hosp Pharmacists. *Res:* Pharmacokinetics of medications especially in the patient with liver disease. *Mailing Add:* 35 Chestnut Hill St Louis MO 63119

IRELAND, HERBERT O(RIN), b Buckley, Ill, June 12, 19; m 41; c 3. CIVIL & GEOTECHNICAL ENGINEERING. *Educ:* Univ Ill, BS, 41, MS, 47, PhD(civil eng), 55. *Prof Exp:* From asst to prof civil eng, 46-79, EMER PROF CIVIL ENG, UNIV ILL, 79- *Concurrent Pos:* Consult, geotech probs; mem, US Nat Coun Soil Mech & Found Eng. *Honors & Awards:* Fourth Across Can Lectr, Nat Res Coun Can, 67. *Mem:* Fel Am Soc Civil Engrs; fel Geol Soc Am; Am Rwy Eng Asn; Sigma Xi. *Res:* Soils and foundations; foundation and retaining wall behavior. *Mailing Add:* RR 1 Box 185C Gilman IL 60938

IRELAND, JAMES, b Clarksville, Tenn, June 30, 47; m 67; c 2. REPRODUCTIVE BIOLOGY, ENDOCRINOLOGY. *Educ:* Austin Peay State Univ, BS, 69; Univ Tenn, PhD(animal sci), 75. *Prof Exp:* PROF ANIMAL SCI & PHYSIOL, MICH STATE UNIV, 77- *Concurrent Pos:* NIH & Ford Found fel, 76-77 & Univ Mich postdoctoral fel, 77; prin investr, Nat Inst Child Health & Human Develop, 79-91; NSF & USDA grants, 79-91; invited lectr, twenty univs incl Univ Wis, WVa Univ, Yale Univ, Dublin, Nebr Univ & Ohio State Univ, 85; NIH sr fel, Yale Sch Med, 87; dir, Ctr Animal Prod & Toxicol, 90-91. *Honors & Awards:* Young Investr Award, Soc Study Reproduction, 77. *Mem:* Soc Study Reproduction; Endocr Soc. *Res:* Hormonal regulation of ovarian follicular development; published over 100 scientific articles. *Mailing Add:* Dept Animal Sci Mich State Univ East Lansing MI 48824

IRELAND, ROBERT ELLSWORTH, b Cincinnati, Ohio, Apr 12, 29; div; c 2. ORGANIC CHEMISTRY. *Educ:* Univ Wis, PhD(org chem), 54. *Prof Exp:* NSF fel & res chemist, Univ Calif, Los Angeles, 54-56; instr chem, Univ Mich, 56-59, from asst prof to assoc prof, 59-65; PROF ORG CHEM, CALIF INST TECHNOL, 65-, PROF & CHMN CHEM DEPT, 87- *Concurrent Pos:* Consult, Merrell Nat Labs, 72-; mem adv ed bd, J of Am Chem Soc, 72- & Org Synthesis, 75- *Mem:* Am Chem Soc; The Chem Soc; Swiss Chem Soc. *Res:* Synthetic organic chemistry relative to natural products. *Mailing Add:* Chem Dept Univ Va Charlottesville VA 22901

IRENE, EUGENE ARTHUR, b Brooklyn, NY, Oct 22, 41; m 64; c 2. SOLID STATE CHEMISTRY. *Educ:* Manhattan Col, BS, 63; Rensselaer Polytech Inst, PhD(solid state chem), 72. *Prof Exp:* Proj scientist, Rocket Propulsion Lab, USAF, 63-66, instr electronics, Electronics Training Ctr, 66-68; commun officer, Defense Commun Agency, Dept of Defense, 68-69; res staff mem solid state chem, T J Watson Res Ctr, IBM Corp, 72-82; AT DEPT CHEM, UNIV NC, 82- *Honors & Awards:* Callinan Award, Electrochem Soc, 88. *Mem:* Am Chem Soc; Electrochem Soc. *Res:* Thin films, particularly dielectrics for field effect transistor applications, including preparation, mechanical properties, optical properties, electrical properties, electron microscopy and ellipsometry for structure of thin films and oxidation kinetics. *Mailing Add:* Dept Chem CB3290 Univ NC Chapel Hill NC 27599-3290

IRESON, W(ILLIAM) GRANT, b North Tazewell, Va, Dec 23, 15; m 38; c 2. INDUSTRIAL ENGINEERING. *Educ:* Va Polytech Inst, BS, 37, MS, 43. *Prof Exp:* Indust engr, Wayne Mfg Corp, 37-41; from instr to assoc prof & acting head dept, indust eng, Va Polytech Inst, 41-48; prof, Ill Inst Technol, 48-51; exec head, 54-75, prof, 51-81, EMER PROF INDUST ENG, STANFORD UNIV, 81- *Honors & Awards:* E L Grant Award, Am Soc Qual Control, 75; Frank & Lillian Gilbreth Award, Am Inst Indust Engrs, 80; Wellington Award, Am Soc Eng Educ, 81; Austin J Bonis Award, Am Soc Qual Control, 87. *Mem:* Am Soc Eng Educ; fel Am Soc Qual Control; fel Am Inst Indust Engrs. *Res:* Plant layout and factory planning; materials handling; statistical quality control and reliability; sampling inspection by variables; engineering economy; development of small industries and engineering education programs in developing countries; engineering economic analysis. *Mailing Add:* 735 Alvarado Ct Stanford CA 94305

IREY, NELSON SUMNER, b Lewisburgh, Pa, July 18, 11; m 40; c 5. PATHOLOGY. *Educ:* Univ Pittsburgh, BS, 35, MD, 38. *Prof Exp:* Chief, Dept Path, 97th Gen Hosp, Frankfurt, WGer, 50-54, Letterman Gen Hosp, 54-60 & Walter Reed Gen Hosp, 61-65; pathologist, Registry Tissue Reactions to Drugs, 65-78, CHMN, DEPT ENVIRON & DRUG-INDUCED PATHOL, ARMED FORCES INST PATH, 78- *Concurrent Pos:* Clin prof path, Sch Med, George Washington Univ, 67-; prof lectr, Dept Forensic Sci, Grad Sch Arts & Sci, George Washington Univ, 75-76. *Mem:* Fel Am Soc Clin Pathologists; fel Am Col Physicians; fel Col Am Pathologists; Soc Pharmacol & Environ Pathologists (pres, 74-75); Int Acad Pathologists. *Res:* Adverse reactions to drugs. *Mailing Add:* Armed Forces Inst Path Washington DC 20306

IREY, RICHARD KENNETH, b Hackensack, NJ, Dec 11, 36; m; c 3. MECHANICAL ENGINEERING. *Educ:* Rose Polytech Inst, BSME, 58; Purdue Univ, MSME, 62, PhD(mech eng), 64. *Prof Exp:* Test & res engr, E I du Pont de Nemours & Co, Inc, 58-60; from asst prof to assoc prof mech eng, 64-69, PROF MECH ENG, UNIV FLA, 69- *Mem:* Am Soc Mech Engrs. *Res:* Cryogenic heat transfer; thermophysical properties; statistical thermodynamics; radiation heat transfer. *Mailing Add:* Dept Mech Eng Univ Toledo 2801 W Bancroft Toledo OH 43606

IRFAN, MUHAMMAD, b Meerut, India, Jan 7, 33; Can citizen; m 68; c 2. RADIATION PHYSICS, MEDICAL PHYSICS. *Educ:* Univ Punjab, Pakistan, BSc, 52; Univ Dacca, MSc, 54; Glasgow Univ, PhD(physics), 62. *Prof Exp:* Demonstr physics, Univ Karachi, 55-57, lectr, 58; attache res, Univ Montreal, 62-64; from asst prof to assoc prof, 64-81, PROF PHYSICS, MEM UNIV NFLD, 81- *Concurrent Pos:* Nat Res Coun Can fel, 62-63, radiation control officer & chmn radiation control comt, Mem Univ, Nfld, 69-74, 75-78; consult, UN Develop Prog, Pakistan, 81. *Mem:* Am Phys Soc; Int Radiation Physics Soc; Can Asn Physicists; Can Radiation Protection Asn.

Res: Nuclear reactions, especially those induced by fast neutrons; scintillation counters for gamma rays; measurements of radioactivity in water samples, in air, and like problems; neutron activation analysis; beta dosimetry (medical physics). *Mailing Add:* Dept Physics Mem Univ Nfld St John's NF A1B 3X7 Can

IRGENS, ROAR L, b Trondheim, Norway, Oct 15, 30; US citizen; m 58; c 2. MICROBIOLOGY. *Educ:* Univ Ill, BS, 57, PhD(microbiol), 63. *Prof Exp:* Res asst waste treatment, Univ Ill, 63-65; chemist, Minn State Dept Health, 65-66; from asst prof to assoc prof res teaching, 66-77, PROF LIFE SCI, SOUTHWEST MO STATE COL, 77- *Concurrent Pos:* Res grant, Southwest Mo State Col, 67-68. *Res:* Microbiology of waste treatment. *Mailing Add:* Dept Life Sci Southwest Mo State Col 901 S National Springfield MO 65804

IRGOLIC, KURT JOHANN, b Hartberg, Austria, Sept 28, 38; m 64; c 1. RESEARCH ADMINISTRATION, SYNTHETIC FUELS. *Educ:* Univ Graz, PhD(inorg anal chem), 64. *Prof Exp:* Fel chem, 64-66, from asst prof to assoc prof, 66-77, assoc dir energy, Ctr Energy & Mineral Resources, 75-86, PROF INORG CHEM, TEX A&M UNIV, 77- *Concurrent Pos:* Prog coordr res, Off Univ Res, Tex A&M Univ, 73-75. *Mem:* Am Chem Soc. *Res:* Synthetic chemistry of organic compounds of arsenic, selenium and tellurium; metal ion extraction with organic arsenic compounds; isolation and characterization of arsenic and releniune compounds from biological systems; mass spectrometry of organometallic compounds; element specific detectors for high pressure liquid chromatography; inductively coupled argon plasma emission spectrometry; trace element and trace element compound determinations in environmental materials. *Mailing Add:* Inst Analytical Chem Universitsplatz Karl Franzens Univ Graz A 8010 Graz Austria

IRGON, JOSEPH, b Polonnoe, Russia, Dec 30, 19; US citizen; m 48; c 3. ENERGETICS, CATALYSIS. *Educ:* Northeastern Univ, BS, 43; Mass Inst Technol, PhD(phys chem), 48. *Prof Exp:* Res assoc chem eng, Chem Warfare Serv Develop Lab, Mass Inst Technol, 43-45; proj leader phys chem, Cent Labs, Gen Foods Corp, 48-52; dept head space sci, Reaction Motors, 52-56; vpres res & develop space & ocean sci, Fulton-Irgon Corp, 56-62; pres ocean sci & technol, Proteus, Inc, 62-69; vpres res & develop ocean sci, Ocean Recovery Systs, 69-73; prin scientist radiant energy sci, 73-80, PRIN SCIENTIST & CONSULT ENERGETICS, J IRGON & ASSOCS, 80- *Concurrent Pos:* Coffin fel, Mass Inst Technol, 46-48; consult energy, Stauffer Chem Corp, 56-59, Allied Chem Corp, 58-59, McGraw Edison Co, 63-65, Union Carbide Corp, 63-66; US Navy, 64-66 & Pakistan Govt, 75-77; consult dir res, Energy Technol Inc, 79-83. *Mem:* Am Chem Soc; Am Soc Mech Engrs. *Res:* Development and/or extension of basic principles large-scale collection and conversion of radiant energy; integrated waste-to-energy systems; computer assisted catalysis of oxidation/reduction processes; human chemical warfare test participation and current long term evaluation of effects. *Mailing Add:* 144 Emmans Rd Flanders NJ 07836

IRIBARNE, JULIO VICTOR, b Buenos Aires, Argentina, Nov 11, 16; m 57; c 1. PHYSICAL CHEMISTRY. *Educ:* Univ Buenos Aires, DChem, 42. *Prof Exp:* Mem staff phys & inorg chem, Univ Buenos Aires, 45-53; indust consult & chemist, 53-55; assoc prof phys chem, Univ Buenos Aires, 56-57, head inst atmospheric physics, 57-66, prof meteorol, 58-66; from assoc prof to prof, 66-82, EMER PROF PHYSICS, UNIV TORONTO, 82- *Concurrent Pos:* Fel, Inst Phys Res Rio de Janeiro, Brazil, 53- & Imp Col, London, 65-66. *Mem:* AAAS; Am Meteorol Soc; Can Meteorol Soc; Can Asn Physicists; Am Geophys Union. *Res:* Cloud physics; atmospheric electricity; aerosol physics; atmospheric chemistry. *Mailing Add:* 29 Banstock Dr Willowdale ON M2K 2H5 Can

IRICK, GETHER, JR, b Stone, Ky, Jan 29, 36; m 80; c 5. PHOTOCHEMISTRY, CATALYSIS. *Educ:* Eastern Ky State Col, BS, 57; Univ Louisville, PhD(org chem), 60. *Prof Exp:* From chemist to sr res chemist, 60-70, res assoc, 70-78, SR RES ASSOC, TENN EASTMAN CO, 78- *Mem:* AAAS; Am Chem Soc; Sigma Xi; Am Asn Textile Chem & Colorists. *Res:* Synthesis and photochemistry of dyes and stabilizers for synthetic polymers; catalysts for synthesis gas conversions. *Mailing Add:* Tenn Eastman Co Box 1972 Kingsport TN 37662

IRICK, PAUL EUGENE, b Greenville, Ohio, Nov 4, 18; m 58; c 5. APPLIED STATISTICS, INFORMATION SCIENCE. *Educ:* Purdue Univ, BS, 40, MS, 45, PhD(statist), 50. *Prof Exp:* Assoc prof statist, Purdue Univ, 54-56; res statistician hwy engr, Nat Res Coun, 56-57, asst dir spec projs hwy res, Transp Res Bd, 68-82; RETIRED. *Concurrent Pos:* Vpres, Engr Index, Inc, 70-74, pres, 74-76. *Mem:* Am Statist Asn. *Res:* Scientific and technical information systems. *Mailing Add:* 484 Windmill Pt Rd Hampton VA 23664

IRIE, REIKO FURUSE, b Japan, Sept 11, 40; m 68; c 1. ONCOLOGY, CANCER. *Educ:* Nat Ochanomizu Univ, Japan, BS, 63; Nat Niigata Univ, Japan, MD, 72. *Prof Exp:* Researcher immunol, Virol Div, Nat Cancer Inst, Japan, 66-71; res oncol cancer immunol, 71-77, from asst prof to assoc prof, 77-84, PROF SURG, SCH MED, UNIV CALIF, LOS ANGELES, 84- *Mem:* Am Asn Cancer Res. *Res:* Immunobiology of cancer; immunotherapy and immunodiagnosis of human cancer. *Mailing Add:* Sch Med Louis Factor Bldg 9th Floor Univ Calif Los Angeles CA 90024

IRISH, DONALD EDWARD, b Uxbridge, Ont, June 14, 32; m 56; c 4. PHYSICAL CHEMISTRY, VIBRATIONAL SPECTROSCOPY. *Educ:* Univ Western Ont, BSc, 55; McMaster Univ, MSc, 56; Univ Chicago, PhD(chem), 62. *Prof Exp:* Teacher tech sch, Ont, 56-57; from lectr to asst prof, 57-65, assoc prof, 65-71, chmn dept, 77-83, PROF CHEM, UNIV WATERLOO, 71-, EXEC DIR, FAC SCI FOUND, 85- *Concurrent Pos:* Vis fac, Bell Tel Labs, Inc, 63; guest mem staff, Sch Chem, Univ Newcastle, 70-71; exchange scientist, Nat Res Coun Can-Nat Sci Res Ctr, France, 76; vis prof, Univ Karlsruhe, WGer, 83, Dept Sci & Indus Res, Lower Hutt, NZ, 85 & Univ Queensland, Australia, 85. *Honors & Awards:* Union Carbide Award for Chem Educ, Chem Inst Can, 90. *Mem:* Chem Inst Can; Royal Soc Chem; Electrochem Soc; Spectros Soc Can. *Res:* Raman and infrared spectroscopy; constitution and processes in electrolyte solutions; high temperature aqueous chemistry; raman and infrared at electrode surfaces. *Mailing Add:* Dept Chem Univ Waterloo Waterloo ON N2L 3G1 Can

IRISH, JAMES DAVID, b Bay City, Mich, Dec 7, 43; m 65. PHYSICAL OCEANOGRAPHY. *Educ:* Antioch Col, BS, 67; Univ Calif, San Diego, MS, 69, PhD(oceanog), 71. *Prof Exp:* Res asst tides, Scripps Inst Oceanog, Univ Calif, San Diego, 67-71, assoc marine instrumentation, 71-72; res asst prof internal waves, dept oceanog, Univ Wash, 72-79, oceanographer, Appl Physics Lab, 74-79; res scientist, Univ NH, 79-83, res assoc prof, Dept Earth Sci, 83-91, Dept Ocean Eng, 85-91; RES SPECIALIST, WOODS HOLE OCEANOG INST, 91- *Mem:* Marine Technol Soc; Am Geophys Union; Oceanog Soc. *Res:* Tides, tidal dissipation and internal tides; internal wave behavior and the effect on acoustic propagation, wind driven shelf circulation; marine instrumentation and satellite data telemetry. *Mailing Add:* 307 Smith Lab Woods Hole Oceanog Inst Woods Hole MA 02543

IRISH, JAMES MCCREDIE, III, b Portland, Maine, Sept 30, 43. RENAL PHYSIOLOGY. *Educ:* State Univ NY, New Paltz, BS, 70; Univ Ariz, PhD(physiol), 75. *Prof Exp:* ASST PROF PHYSIOL, MED CTR, WVA UNIV, 78- *Concurrent Pos:* Nat Kidney Found renal fel, Med Ctr, Univ Kans, 75-76 & USPHS renal fel, 76-78. *Mem:* Am Physiol Soc; Am Soc Nephrol. *Res:* Organic anion and cation transport mechanisms in the kidney. *Mailing Add:* 2862 University Ave Morgantown WV 26505

IRITANI, W M, b Denver, Colo, Apr 29, 23; m 54; c 3. HORTICULTURE, PLANT PHYSIOLOGY. *Educ:* Univ Minn, BS, 51; Univ Idaho, MS, 53; Univ Ill, PhD(hort), 58. *Prof Exp:* Assoc hort, Univ Idaho, 58-68; PROF & HORTICULTURIST, WASH STATE UNIV, 68- *Concurrent Pos:* Fulbright grant to Japan; sabbatical, Neth 81, Scotland, 87. *Mem:* Am Soc Hort Sci; Potato Asn Am (pres, 84-85); Europ Potato Asn; Asian Potato Asn. *Res:* Physiology and post-harvest problems of potatoes. *Mailing Add:* Dept Hort Wash State Univ Pullman WA 59164-6414

IRMITER, THEODORE FERER, b Meadville, Pa, July 9, 22; m 45; c 2. FOOD CHEMISTRY. *Educ:* Kent State Univ, BS, 43; Ohio State Univ, MSc, 49, PhD(dairy chem), 51. *Prof Exp:* Res chemist, B F Goodrich Co, 46-47; res chemist, Chr Hansen's Lab, Inc, 51-54, lab dir, 54-58; assoc dir res & develop, Salada-Shirriff-Horsey, Ltd, 58-61; assoc prof foods & nutrit, Mich State Univ, 61-70; dir sch, 70-80, PROF FOODS, KENT STATE UNIV, 70- *Mem:* Inst Food Technol; Sigma Xi. *Res:* Teaching; food preparation, experimental foods, meal management and food ecology, cultural foods; rate of heat transfer in meats, sensory evaluation. *Mailing Add:* 2070 Skyview Lane Kent OH 44240

IRONS, EDGAR T(OWAR), b Detroit, Mich, Oct 11, 36. COMPUTER SCIENCE. *Educ:* Princeton Univ, BSE, 58; Calif Inst Technol, MS, 59. *Prof Exp:* Mem tech staff comput res, Inst Defense Anal, 61-69; from assoc prof to prof comput sci, Yale Univ, 69-80; dir res, Interactive Syst Corp, 80-84; PRES, SLATER TOWER LTD, 84- *Concurrent Pos:* Vis lectr, Princeton Univ, 65-66. *Mem:* Asn Comput Mach. *Res:* Communication between man and digital computers; computer languages; compilers; operating systems. *Mailing Add:* Slater Tower Ltd 586 Longs Peak Rd Longs Peak Rte Estes Park CO 80517

IRONS, MARGARET JEAN, b London, Ont, May 24, 51; m 76; c 2. CELL BIOLOGY, RETINA. *Educ:* Univ Toronto, BSc, 74; McGill Univ, MSc, 76, PhD(anat), 80. *Prof Exp:* Asst prof anat, Univ Toronto, 82-88; ASSOC PROF ANAT, MCMASTER UNIV, 88- *Mem:* Am Soc Cell Biol; Asn Res Vision & Ophthalmol; Int Soc Eye Res. *Res:* Interactions of photoreceptors and retinal pigment epithelial (RPE) cells in normal retinas and those with inherited degenerative conditions. *Mailing Add:* Dept Anat Fac Health Sci McMaster Univ 1200 Main St W Hamilton ON L8N 3Z5 Can

IRONS, RICHARD DAVIS, b Oakland, Calif, Sept 30, 47; m 79; c 3. TOXICOLOGY, CELL BIOLOGY. *Educ:* Univ Pac, BA, 68; Univ Calif, MT, 70; Univ Rochester, PhD(toxicol), 75; Am Bd Toxicol, cert, 80. *Prof Exp:* Fel path, Strong Mem Hosp, Rochester, NY, 74-76; PATHOLOGIST IMMUNO PATH, CHEM INDUST INST TOXICOL, 76- *Concurrent Pos:* Adj prof, Duke Univ Sch Med; mem, Ad Hoc Study Sect Immunol & Toxicol, NIH, 79, 81-82, permanent mem toxicol, 83-87, chmn, Toxicol Study Sect, 86-87, mem, EPA Health Effects Rev Panel, 87- *Mem:* AAAS; Am Asn Pathologists; Fedn Am Soc Exp Biol; Soc Toxicol; Reticuloendothelial Soc. *Res:* Myelotoxicity and immunopathology; molecular aspects of the effects of chemical agents on the cytoskeleton; regulation of cell growth and differentiation; leukemogenesis. *Mailing Add:* Dept Molecular Toxicol Univ Colo Health Sci Ctr C-235 4200 E Ninth Ave Denver CO 80262

IRR, JOSEPH DAVID, b Pittsburgh, Pa, Sept 19, 34. GENETICS. *Educ:* Univ Pittsburgh, BS, 62; Univ Calif, Santa Barbara, PhD(microbial genetics), 67. *Prof Exp:* NIH fel, Univ Wash, 67-69; from asst prof to assoc prof biol, Marquette Univ, 69-75; NSF fel, 70-74, Peter M Stanka Found fel, 71-74, March of Dimes fel, 74-75, NIH fel, 75-77; res fel, Harvard Med Sch-Mass Gen Hosp, 74-77; res molecular biologist, DuPont Merck Pharmaceut, 77-80, sr res genetic toxicologist, Haskell Lab, 81-84, sr res geneticist & cent res & develop biotechnol, 84-86, RES ASSOC, MED PROD DEPT, DUPONT MERCK PHARMACEUT, 77-; PROF, UNIV DELAWARE, 81- *Concurrent Pos:* Vis prof, Univ Hamburg, WGer, 76; vis scientist, Mass Inst Technol, 77; adj assoc prof genetics, Univ Delaware, 78-81. *Mem:* Genetics Soc Am; Am Soc Microbiol; Environmental Mutagen Soc; Genetic Toxicol Asn. *Res:* Genetic control mechanisms; gene expression in human cells, cellular therapy. *Mailing Add:* E I du Pont de Nemours Co Inc Glenolden Lab Glenolden PA 19036

IRSA, ADOLPH PETER, b New York, NY, Dec 9, 22; m 53; c 1. MASS SPECTROMETRY, PHYSICAL CHEMISTRY. *Educ:* City Col New York, BS, 47; Adelphi Col, MS, 55. *Prof Exp:* Chemist mass spectrometry, Hydrocarbon Res Inc, 47-49; chemist mass spectrometry, Brookhaven Nat Lab, 49-87; RETIRED. *Mem:* Am Chem Soc; Am Soc Mass Spectrometry. *Res:* Isotope effects; biomedical mass spectrometry; ion molecule reactions; analytical mass spectrometry. *Mailing Add:* 79 Morton Blvd Plainview NY 11803

IRVIN, HOWARD BROWNLEE, b Pittsburgh, Pa, Oct 21, 19; m 48; c 1. CHEMICAL ENGINEERING, ECONOMICS. *Educ:* Pa State Col, BS, 42, MS, 46; Purdue Univ, PhD(chem eng), 49. *Prof Exp:* Res asst fluids, Pa State Col, 42-46; engr, 48-66, sect mgr process design, 66-72, SR DESIGN ENGR, PHILLIPS PETROL CO, 72- *Mem:* Am Inst Chem Engrs; Am Chem Soc. *Res:* Process design and evaluation of plants for manufacture of polymers, petrochemicals and fertilizers. *Mailing Add:* 1927 Polaris Dr Bartlesville OK 74006-6113

IRVIN, HOWARD H, b Munich, Ger, Nov 19, 18; nat US; m 43; c 2. POLYMER CHEMISTRY. *Educ:* Rose Polytech Inst, BS, 43. *Hon Degrees:* DEng, Rose Hubean Inst, 87. *Prof Exp:* Metal inspector, Inland Steel Co, Ind, 43; asst to dir res, Marbon Chem Div, Borg-Warner Chem, 59-65, exec vpres, 65-69, pres, Marbon Int, 69-74, vpres eastern hemisphere, 74-79, vpres external technol, 80-81; PRES, HOWARD H IRVIN & ASSOCS, 82- *Mem:* Am Chem Soc; Am Soc Testing & Mat; Soc Plastics Indust; Tech Asn Pulp & Paper Indust. *Res:* Development of synthetic rubbers and special resins used in rubber compounding; electrical insulation materials; adhesives; adhesive for elastomers; special resins; administration; plastics and paint resins. *Mailing Add:* 175 N Harbor Dr Chicago IL 60601-7344

IRVIN, JAMES DUARD, b Grand Island, Nebr, Sept 24, 42; m 63; c 2. BIOCHEMISTRY. *Educ:* Gonzaga Univ, Wash, BS, 65; Mont State Univ, PhD(chem), 70. *Prof Exp:* Res assoc biochem, Univ Tex, Austin, 70-73; asst prof, 73-78, assoc prof, 78-83, PROF CHEM, SOUTHWEST TEX STATE UNIV, 83- *Concurrent Pos:* Grants, Res Corp, 74-75, Robert A Welch Found, 74-84 & Pub Health Serv, 78-81, 86-88. *Mem:* Am Chem Soc; Am Soc Biochem & Molecular Biol. *Res:* Eukaryotic protein synthesis; antiviral and toxic proteins; ribosomes structure and function. *Mailing Add:* Dept Chem Southwest Tex State Univ San Marcos TX 78666

IRVINE, CYNTHIA EMBERSON, b Washington, DC, Aug 14, 48; m 71; c 2. STELLAR SPECTROSCOPY, PRE-COLLEGE SCIENCE ENRICHMENT. *Educ:* Rice Univ, BA, 70; Case Western Reserve Univ, PhD(astron), 75. *Prof Exp:* Res assoc, US Naval Postgrad Sch, 75-81; pres, 82-86, dir educ, 85-86, RES SCIENTIST, MONTEREY INST RES ASTRON, 72-; PROJ LEADER, GEMINI COMPUT, 87- *Concurrent Pos:* Consult, Calif Capitol, 84-85; instr, Monterey Peninsula Col, 85-86; consult & thesis adv, US Naval Postgrad Sch, 85-86. *Mem:* Am Astron Soc; Optical Soc Am; Inst Elec & Electronics Engrs. *Res:* Design and implementation of trusted computer system; design of applications to run on trusted systems. *Mailing Add:* Gemini Computers Inc PO Box 222417 Carmel CA 93922

IRVINE, DONALD GRANT, b Victoria, BC, Oct 28, 30; m 54; c 3. GEOGRAPHIC & GEOCHEMICAL TOXICOLOGY, ALTERNATIVE & COMPUTERIZED BIOASSAYS DEVELOPMENT. *Educ:* Univ BC, BA, 52, MA, 54; Univ Sask, PhD(biol psychiat), 81. *Prof Exp:* Res asst zool, Univ BC, 53-54; res asst physiol, Dept Physiol, Univ Sask, 54-58; res scientist, Psychiat Res Div, Sask Health, 68-84; res scientist, 83-90, SR RES SCIENTIST, TOXICOL RES CTR, UNIV SASK, 90- *Concurrent Pos:* Prin investr, var projs, Nat Health, Can, 67-71, Med Res Coun Can, 70-73, Schizophrenia Biol Res Found, 71 & Wildlife Toxicol Fund, 86-90; res assoc psychiat, Univ Sask, 69-72; adj prof toxicol, Fac Grad Studies & Res, Univ Sask, 85-, actg dir, Toxicol Res Ctr, 88-89. *Mem:* Soc Toxicol Can; Can Biochem Soc; Can Soc Clin Chemists; Air & Waste Mgt Asn; Am Soc Testing & Mat; World Asn Theoret Chemists. *Res:* Geographic, geochemical and hydrogeochemical factors in toxicoses and related diseases of man, animals and plants; use of minature organisms in developing alternative bioassays for toxicity; application of computers and automation to bioassays. *Mailing Add:* Toxicol Res Ctr Univ Sask Saskaton SK S7N 0W0 Can

IRVINE, DONALD MCLEAN, b Toronto, Ont, Mar 22, 20; m 58; c 4. DAIRY SCIENCE. *Educ:* Ont Agr Col, BSA, 42; Univ Wis, MSc, 50, PhD(dairy sci), 56. *Prof Exp:* Lectr dairy eng, Ont Agr Col, 46-47; instr, Univ Wis, 48-55; head dept, Ont Agr Col, Univ Guelph, 55-66, prof dairy sci, 55-85; RETIRED. *Concurrent Pos:* Consult. *Honors & Awards:* Pfizer Award, Am Dairy Sci Asn. *Mem:* Am Dairy Sci Asn; Int Food Technol; Nat Dairy Coun; Can Inst Food Technol. *Res:* Cheese mechanization; cheese food products; cheese varieties; enzymes. *Mailing Add:* 107 College Ave W Guelph ON N1G 1S3 Can

IRVINE, GEORGE NORMAN, b Calgary, Alta, Apr 6, 22; m 45; c 2. PHYSICAL BIOCHEMISTRY. *Educ:* Univ Man, BSc, 43; McGill Univ, PhD(chem), 49. *Prof Exp:* Res chemist, Can Grain Comn, 45-63, dir, Grain Res Lab, 63-78; CONSULT, 79- *Concurrent Pos:* Adj prof, Univ Man. *Honors & Awards:* William F Geddes Mem Award, Am Asn Cereal Chem, 78; Neumann Medal, Arbeitsgemeinschaft Getreideforsch, Ger, 78. *Mem:* Am Asn Cereal Chem; fel Chem Inst Can. *Res:* Chemical kinetics; enzymes; wheat pigments; quality factors in durum and bread wheat; milling and baking technology; flour quality. *Mailing Add:* 994 Cottontree Close Victoria BC V8X 4E9 Can

IRVINE, JAMES BOSWORTH, b Lexington, Ky, Apr 15, 14; m 37; c 2. PHYSICAL CHEMISTRY. *Educ:* Univ Ky, BS, 37. *Prof Exp:* Anal chemist, Naval Stores, Hercules Powder Co, 37-38; anal & develop chemist, Textile Auxiliaries, O F Zurn Co, 38-41; textile chemist, Collins & Aikman Corp, 41-45; chemist textile auxiliaries, Quaker Chem Prods Corp, 45-47, group leader, Customer Serv & Textile Auxiliaries, 47-52, dir, New Prod Develop, 52-58, Textile process Develop, 58-60 & Tech Sales Develop, 60-64, process engr, Quaker Chem Corp, 64-82; PRES, ROCKREATION, INC, 82- *Mem:* Am Chem Soc; Am Asn Textile Chem & Colorists; Friends Mineral. *Res:* Textile auxiliaries and processes; physical sciences. *Mailing Add:* 3508 Starmount Dr Greensboro NC 27403

IRVINE, JAMES ESTILL, b Charlottesville, Va, Feb 2, 28; m 51; c 3. PLANT PHYSIOLOGY. *Educ:* Univ Miami, Fla, BS, 51, MS, 52; Univ Va, PhD(biol), 57. *Prof Exp:* Asst prof biol, Bridgewater Col, 53-54; instr, Univ Va, 55-56; plant physiologist & dir, US Sugarcane Lav, Sci & Educ Admin-Agr Res, USDA, 57-; AT RES EXT CTR, TEXAS A&M AGR CTR. *Mem:* Am Soc Plant Physiol; Am Soc Agron; Asn Trop Biol; Int Soc Sugarcane Technol. *Res:* Photosynthetic efficiency; leaf physiology; cold tolerance; pre- and post-harvest changes in sugarcane quality; tropical agriculture. *Mailing Add:* Agr Res Extension Ctr Tex A&M 2415 E Hwy 83 Weslaco TX 78596

IRVINE, JOHN WITHERS, JR, radiochemistry; deceased, see previous edition for last biography

IRVINE, MERLE M, b San Francisco, Calif, Jan 5, 24; m 45; c 4. PHYSICS. *Educ:* Mont State Univ, BS, 50; Lehigh Univ, MS, 52, PhD(physics), 55. *Prof Exp:* Instr physics, Lehigh Univ, 52-55; mem tech staff, Bell Labs, 55-61, supvr, 61-63, dept head, 63-71, dir, 71-84; RETIRED. *Mem:* Am Asn Physics Teachers; Asn Comput Mach; Am Phys Soc; Inst Elec & Electronics Engrs. *Res:* Cathode sputtering in glow discharges; digital computers; electromagnetic theory; plasma physics; data management systems. *Mailing Add:* 61 Cheshire Sq Little Silver NJ 07739

IRVINE, STUART JAMES CURZON, b Broadstairs, Kent, UK, May 14, 53; m 74; c 3. CRYSTAL GROWTH, INFRARED DETECTORS. *Educ:* Loughborough Univ Technol, UK, BSc, 74; Univ Birmingham, UK, PhD(metall & mat sci), 78. *Prof Exp:* Res fel, Dept Phys, Univ Birmingham, UK, 77-78; higher sci officer, Royal Signals & Radar Estab, Ministry of Defence, UK, 78-80, sr sci officer, 80-84, prin sci officer, 84- 90; ASST MGR ARRAY PROD RES, SCI CTR, ROCKWELL INT, 90- *Concurrent Pos:* Tutor physics, Open Univ, 79-82; guest ed, J Crystal Growth, 81-85. *Mem:* Am Asn Crystal Growth; Mat Res Soc; Inst Phys. *Res:* Growth of narrow band gap II-VI semiconductors using Metal Organic Vapor Phase Epitaty; ultra violet and laser stimulation of epitatial growth processes. *Mailing Add:* 1049 Camino Dos Rios PO Box 1085 Thousand Oaks CA 91358

IRVINE, T NEIL, b Manitoba, Can, Jan 5, 33; m 62; c 1. GEOLOGY. *Educ:* Univ Man, BSc, 53, MSc, 56; Calif Inst Technol, PhD(geol), 59. *Prof Exp:* Asst prof geol, McMaster Univ, 59-62; petrologist, Geol Surv Can, 62-72; PETROLOGIST, GEOPHYS LAB, WASHINGTON, DC, 72- *Mem:* Geol Soc Am; Am Geophys Union. *Res:* Petrology and geochemistry of ultramafic rocks. *Mailing Add:* 5251 Broad Branch Rd NW Washington DC 20015

IRVINE, THOMAS FRANCIS, b Northmont, NJ, June 25, 22; m 66; c 2. HEAT TRANSFER, FLUID MECHANICS. *Educ:* Pa State Univ, BS, 46; Univ Minn, MS, 51, PhD(mech eng), 56. *Prof Exp:* From instr to assoc prof mech eng, Univ Minn, 50-59; prof, NC State Univ, 59-61; dean, 61-72, PROF MECH ENG, STATE UNIV NY STONY BROOK, 72- *Concurrent Pos:* Ed transl jours: Heat Transfer-Soviet Res, 69-, Heat Transfer-Japanese Res, 72- & Previews Heat & Mass Transfer, 74-; co-ed, Advances in Heat Transfer, 60; vis prof, Tech Univ Munich, 68, Univ Florence, 85; vis scientist, Boris Kidric' Inst, Belgrade, 72. *Mem:* Fel Am Soc Mech Engrs; fel AAAS; fel Int Ctr Heat & Mass Transfer. *Mailing Add:* Dept Mech Eng State Univ NY Stony Brook NY 11794

IRVINE, WILLIAM MICHAEL, b Los Angeles, Calif, Aug 31, 36; m; c 4. PLANETARY SCIENCE, RADIO ASTRONOMY. *Educ:* Pomona Col, BA, 57; Harvard Univ, MA, 58, PhD(physics), 61. *Prof Exp:* Physicist, Smithsonian Astrophys Observ & res fel & lectr, Harvard Col Observ, 62-66; assoc prof astron & physics, 66-69, head astron prog & chmn, Five Col Astron Dept, Univ Mass, Amherst, Hampshire, Smith & Mt Holyoke Cols, 66-78, PROF ASTRON, UNIV MASS, AMHERST, 69- *Concurrent Pos:* NATO fel astron, 61-62; vchmn, Northeast Radio Observ Corp, 68-73, mem bd trustees, 67-79; sr fel, Int Res & Exchanges Bd, Sweden, 73-74; chmn, Div Planetary Sci, Am Astron Soc, 73-74; assoc ed, Icarus, 76-; vis prof, Kanazawa Inst Technol, Japan, 77; vis prof & chmn sci comn, Onsala Space Observ, Chalmers Univ Technol, Sweden, 79-81; vis prof, Japanese Nat Astron Observ, 90. *Mem:* Am Phys Soc; Am Astron Soc; Am Geophys Union; Int Astron Union; Int Sci Radio Union; Int Soc Study Origin Life. *Res:* Light scattering and radiative transfer in planetary atmospheres and surfaces; spectral line radio astronomy. *Mailing Add:* Dept Physics & Astron Univ Mass Amherst MA 01003

IRVING, CHARLES CLAYTON, b Memphis, Tenn, Oct 12, 32; m 54; c 5. BIOCHEMISTRY. *Educ:* Memphis State Univ, BS, 53; Univ Tenn, MS, 55, PhD(chem), 57. *Prof Exp:* Asst chem, Univ Tenn, 56-57; res fel physiol chem, Univ Minn, 57-59; from asst prof to assoc prof biochem, 60-73, PROF BIOCHEM, MED UNITS, UNIV TENN, MEMPHIS, 73-, PROF UROL, 76-; RES BIOCHEMIST, VET ADMIN HOSP, 60- *Mem:* AAAS; Am Chem Soc; Am Soc Biol Chem; Am Asn Cancer Res. *Res:* Chemical carcinogenesis; drug metabolism. *Mailing Add:* Cancer Res Lab Vet Admin Hosp 1030 Jefferson Ave Memphis TN 38104

IRVING, EDWARD, b Colne, Eng, May 27, 27; m 57; c 4. PALEOMAGNETISM. *Educ:* Cambridge Univ, BA, 50, MA, 53, ScD(nat sci), 65. *Hon Degrees:* DSc, Carleton Univ, 79, Mem Univ Nfld, 86. *Prof Exp:* From res fel to sr fel geophys, Australian Nat Univ, 54-64; sr sci officer geomagnetism, Dominion Observ, 64-66; prof geophys, Univ Leeds, 66-67; res scientist geomagnetism, Earth Physics Br, Dept Energy Mines & Resources, Can, 67-81; RES SCIENTIST, PAC GEOSCI CTR, 81- *Concurrent Pos:* Adj prof, Carleton Univ, Ottawa, 75-77 & Univ Victoria, 85- *Honors & Awards:* Gondwanaland Medal, Mining, Geol & Metallog Inst India, 65; Logan Medal, Geol Asn Can, 75; Walter Bucher Medal, Am Geophys Union, 79; Wilson Medal, Can Geophys Union, 84. *Mem:* Fel Royal Astron Soc; fel Am Geophys Union; fel Royal Soc Can. *Res:* Paleomagnetism and its application to geological and geophysical problems. *Mailing Add:* Pac Geosci Ctr 9860 W Saanich Rd Box 6000 Sidney BC V8L 4B2 Can

IRVING, FRANK DUNHAM, b Plainfield, NJ, July 30, 23; m 48; c 3. FORESTRY. *Educ:* Rutgers Univ, BS, 48; Univ Minn, BS, 49, MF, 50, PhD(forestry), 60. *Prof Exp:* Dist game mgr, Wis Conserv Dept, 50-55; from instr to assoc prof forestry, 55-66, PROF FORESTRY, UNIV MINN, ST PAUL, 66- *Mem:* Soc Am Foresters; Wildlife Soc; Sigma Xi. *Res:* Southeastern Minnesota hardwood management; patterns of administrative organization in forestry and wildlife management; techniques of prescribed burning. *Mailing Add:* Col Forestry Univ Minn St Paul MN 55108

IRVING, GEORGE WASHINGTON, JR, b Caribou, Maine, Nov 20, 10; m 38; c 2. BIOCHEMICAL PHARMACOLOGY. *Educ:* George Washington Univ, BS, 33, MS, 35, PhD(biochem), 39. *Prof Exp:* Lab asst, Nat Bur Standards, 27-28; lab asst, Bur Chem, USDA, 28-35, jr chemist, Bur Entom & Plant Quarantine, 35; res fel biochem, George Washington Univ Med Sch, 36-38; res fel biochem, Cornell Univ Med Col, 38-39; asst chem, Rockefeller Inst, 39-42; biochemist southern regional res lab, Bur Agr & Indust Chem, USDA, La, 42-43, sr biochemist, 43-44, sr biochemist, Agr Res Ctr, Md, 44-46, prin chemist & head div biol active compounds, 46-47, asst chief, agr & indust chem, Washington, DC, 47-53, chief biol sci br, Mkt Res Div, Agr Mkt Serv, 53-54, dep adminstr, Agr Res Serv, 54-64, from assoc adminstr to administr, 64-71; consult, 71-72; res assoc, Fedn Am Soc Exp Biol, Bethesda, Md, 72-77; sr res, 78-88, exec vpres agr res, Inst Bethesda, Md, 82-84; CONSULT, 78- *Concurrent Pos:* Lectr, USDA Grad Sch, 46-52 & med sch, George Wash Univ, 47-54; trustee, The Nutrit Found, 56; mem, Expert Comt Food Additives, Joint Food & Agr Orgn, WHO, UN, Geneva, Switz, 76-77; chmn bd, The Nutrit Found, 83-84; Asn Ed Hexagon, 84. *Honors & Awards:* Honor Award, Am Inst Chem, 69; Honor Award, Am Leather Chem Asn, 69. *Mem:* AAAS (vpres, 62); Am Chem Soc; Am Soc Biochem & Molecular Biol; Inst Food Technol; Sigma Xi. *Res:* Biochemistry of pituitary hormones; plant and animal proteolytic enzymes; chemistry of plant proteins; chemistry of antibiotics and plant growth regulators; toxicological evaluation of food additives. *Mailing Add:* 4601 N Park Ave Apt 613 Chevy Chase MD 20815

IRVING, JAMES P, b New York, NY, Apr 28, 36; m 63; c 3. CHEMICAL ENGINEERING. *Educ:* Univ Notre Dame, BS, 57; Univ Northwestern, MS, 59; Yale Univ, DEng(chem eng), 66. *Prof Exp:* Res engr, Jet Propulsion Lab, Calif Inst Technol, 59-61; res engr, Chevron Res Co, 65-68, prod eng, 72-73, sr res engr, Chevron Oil Field Res Co, 68-71, sr eng assoc, 71-72, prod eng, 72-73; mgr systs & eng serv, Standard Oil Co Calif, 73-75, mgr oil recovery, 75-78, mgr planning, Mgt Planning & Develop Staff, 78-80, mgr reservoir eng, 80-81, Western Region coordr prod, Chevron USA, 81-83, vpres prod res, Chevron Oil & Fuel, Lahbra Co, pres, 83-85, GEN MGR PROD, SOUTHERN REGION, CHEVRON USA, 87- *Mem:* Am Inst Chem Engrs. *Res:* Assisted recovery techniques. *Mailing Add:* 575 Market 21st F San Francisco CA 94105

IRVING, JAMES TUTIN, b Christchurch, NZ, May 3, 02; US citizen; m 37. BIOCHEMISTRY. *Educ:* Cambridge Univ, BA, 23, PhD(biochem), 27, MD, 31. *Hon Degrees:* AM, Harvard Univ, 61. *Prof Exp:* Prof physiol, Univ Cape Town, 39-53; prof odontol, Univ Witwatersrand, 54-59; prof anat, 59-61, prof physiol, 61-68, vis lectr oral biol, Sch Dent Med, Harvard Univ, 68-77; emer prof, Nat Inst Aging, NIH, 78-81; RETIRED. *Concurrent Pos:* Consult, Warner Lambert Res Inst, 69-72; ed, Archives Oral Biol, 62-87. *Mem:* Am Physiol Soc; Int Asn Dent Res; Soc Exp Biol & Med; Biochem Soc; hon mem NY Acad Sci. *Res:* Bone and tooth formation; calcium and phosphorus metabolism. *Mailing Add:* Forsyth Dental Ctr 140 Fenway Boston MA 02115

IRVING, PATRICIA MARIE, b Kenosha, Wis, May 28, 50. PHYSIOLOGICAL ECOLOGY. *Educ:* Dominican Col, BS, 72; Univ Wis-Milwaukee, MS, 75, PhD(bot), 79. *Prof Exp:* Res assoc, Argonne Nat Lab, 74-79, asst ecologist, 79-86, ecologist, 86-88; assoc, 88-90, DIR NAT ACID PRECIPITATION ASSESSMENT PROG, WHITE HOUSE COUN ENVIRON QUAL, 91- *Concurrent Pos:* Instr, Milwaukee Area Tech Col, 76; ecologist, M H Gabriel & Assocs, 78-79; adj prof, Northern Ill Univ, 81-83; US rep UN-ECE Conv Transboundry Air Pollution Task Force Terrestrial Effects, 87-; US rep Int Meeting Acid Rain Coord, 89- *Honors & Awards:* Wolf Vishniac Mem Excellence Award, 85. *Mem:* Ecol Soc Am; Fedn Am Scientists; AAAS. *Res:* Causes, effects and control strategy options for pollution; effects of environmental stress on ecosystems; environmental economics. *Mailing Add:* NAPAP Coun Environ Qual 722 Jackson Pl NW Washington DC 20503

IRWIN, ARTHUR S(AMUEL), b South Bend, Ind, Oct 31, 12; m 38; c 3. ENGINEERING MECHANICS. *Educ:* Univ Mich, BS, 35. *Prof Exp:* Design engr, Bendix Prods Div, Bendix Aviation Corp, 35-37; engr, Marlin-Rockwell Corp, 37-40; sr mech develop engr, Bell Aircraft Corp, 40-46; chief engr, Warren Ricketts & Sons, 46-47; asst chief engr & dir res, Marlin-Rockwell Co, 47-57, dir res, Marlin-Rockwell Co Div, TRW Inc, 57-69, mgr sales res, 62-69, mgr res & develop, 69-77; CONSULT, 77- *Concurrent Pos:* Mem subcomt lubrication & wear, Nat Adv Comt Aeronaut, 55. *Mem:* Assoc fel Am Inst Aeronaut & Astronaut; Am Ord Asn; Am Soc Mech Engrs; fel Am Soc Lubrication Engrs; Jet Pioneers Asn US. *Res:* Bearings; lubricants and mechanisms. *Mailing Add:* Driftwood-RFD 1 Bemus Point NY 14712

IRWIN, CAROL LEE, biochemistry, for more information see previous edition

IRWIN, CHARLES EDWIN, JR, b Medford, Mass, Dec 15, 45; m 79; c 1. PEDIATRICS, ADOLESCENT MEDICINE. *Educ:* Hobart Col, BS, 67; Dartmouth Med Sch, BMS, 69; Univ Calif, San Francisco, MD, 71. *Prof Exp:* DIR ADOLESCENT MED & PROF PEDIAT, SCH MED, UNIV CALIF, SAN FRANCISCO, 77- *Concurrent Pos:* Clin scholar, Robert Wood Johnson Found, Univ Calif, San Francisco, 74-77; dir, Interdisciplinary Adolescent Health Training Prog, Health & Human Serv training grant, 77-; assoc dir, San Francisco Unified Sch Dist Adolescent Sch Prog, 76- *Honors & Awards:* Res award, Soc Adolescent Med, 83; Nat Youth Law Ctr Award, 88. *Mem:* Soc Adolescent Med; Ambulatory Pediat Asn; Am Acad Pediat; Am Pub Health Asn; Am Venereal Dis Asn; Soc Pediat Res; Soc Res Child Develop. *Res:* Health behaviors of adolescents; compliance and self care in adolescents; adolescent reproductive health problems; risk-taking behaviors in adolescence. *Mailing Add:* Dept Pediat Univ Calif San Francisco CA 94143

IRWIN, DAVID, b Leeds, Eng, Oct 17, 45; US citizen; m 69; c 2. ONCOLOGY, HEALTH SCIENCES ADMINISTRATION. *Educ:* Univ Leeds, BSc, 69, PhD, 74. *Prof Exp:* Postdoctorate fel surg, Harvard Med Sch, 73-75, instr, 75-79; asst biochemist, Mass Gen Hosp, 75-79; sr res scientist, George Washington Univ Med Ctr, 79-83; spec expert genetics, Nat Inst Gen Med Sci, NIH, 83-84, exec secy molecular biol, Div Res Grants, 84-85, health sci admin grant rev, Nat Cancer Inst, 85-89, chief, 89- *Concurrent Pos:* Lectr biol, Boston Univ, 73-75; vis prof, Univ Mass, Boston, 77-79, St George's Med Sch, Grenada, 81-83; vis asst prof, Northern Va Community Co, 84-85; freelance ed, Cambridge Sci Abstr, 85-87. *Mem:* Asn Cancer Res; Biochem Soc; Am Soc Cell Biol; AAAS. *Res:* Molecular biology and regulation of cellular growth in normal and neoplastic tissues. *Mailing Add:* Res Pro Sect Grants Revi Br Nat Cancer Inst NIH Westwood Bldg Rm 818 Bethesda MD 20892

IRWIN, GEORGE RANKIN, b El Paso, Tex, Feb 26, 07; m 33; c 4. PHYSICS, MECHANICAL ENGINEERING. *Educ:* Knox Col, AB, 30; Univ Ill, MS, 33, PhD, 37. *Hon Degrees:* DEng, Lehigh Univ, 77. *Prof Exp:* Physicist, US Naval Res Lab, 37-67; prof mech, Lehigh Univ, 67-72; PROF MECH ENG, UNIV MD, COLLEGE PARK, 72- *Concurrent Pos:* Vis prof, Univ Ill, 61 & 62. *Honors & Awards:* Timoshenko Medal, Am Soc Mech Engrs; Gold Medal, Am Soc Mech. *Mem:* Nat Acad Eng; Am Soc Mech; Am Soc Mech Engrs; hon mem Am Soc Testing & Mat; Soc Exp Mech. *Res:* Fracture mechanics. *Mailing Add:* 7306 Edmonton Ave College Park MD 20740

IRWIN, GLENN WARD, JR, b Roachdale, Ind, July 18, 20; m 43; c 3. ENDOCRINOLOGY, INTERNAL MEDICINE. *Educ:* Ind Univ, BS, 42, MD, 44. *Hon Degrees:* LLD, Ind Univ, 86, Marian Col, 87. *Prof Exp:* From instr to assoc prof, Ind Univ-Purdue Univ, Indianapolis, 50-61, dean, Sch Med, 65-73, chancellor, Univ, 73-74, vpres, Univ, 74-86, PROF MED, SCH MED, IND UNIV-PURDUE UNIV, INDIANAPOLIS, 61- *Mem:* Fel Am Col Physicians; Endocrine Soc; Am Diabetes Asn; Am Thyroid Asn; Am Fedn Clin Res; Sigma Xi. *Mailing Add:* Ind Univ 1120 South Dr Indianapolis IN 46202

IRWIN, HOWARD SAMUEL, b Louisville, Ky, Mar 28, 28; div; c 2. SYSTEMATIC BOTANY, PLANT ECOLOGY. *Educ:* Univ Puget Sound, BA, 50, BEd, 52; Univ Tex, PhD(trop bot), 60. *Hon Degrees:* DSc, Fordham Univ, 77. *Prof Exp:* Fulbright English & biol, Queen's Col, Georgetown, Guyana, 52-56; res assoc trop bot, 60-63, assoc cur, 63-66, cur & herbarium adminr, 66-68, head cur, 68-71, exec dir inst mgt, 71-72, exec vpres, 72-73, PRES, NY BOT GARDEN, 73-; PROF BOT, COLUMBIA UNIV, 68- *Concurrent Pos:* Adj prof bot, City Univ New York, 71-; NSF grantee, 64- *Mem:* Fel NY Acad Sci; Sigma Xi; Asn Trop Biol (secy-treas, 63-64); Am Soc Plant Taxonomists (pres, 71-72); Asn Systs Collections (pres, 73-75). *Res:* Systematics of neotropical Cassia, Mimosa; flora of planalto do Brasil. *Mailing Add:* 31 Grandview St Huntington NY 11743

IRWIN, JAMES JOSEPH, b Hamilton, Ont, July 1, 58. NOBLE GAS ISOTOPE GEOCHEMISTRY, NUCLEAR PROCESSES IN ROCKS. *Educ:* McGill Univ, BS, 80; Univ Calif, Berkeley, PhD(geol), 86. *Prof Exp:* Res asst geol, Dept Geol, 81-86, POSTDOCTORAL FEL, DEPT PHYSICS, UNIV CALIF, BERKELEY, 86- *Mem:* Am Geophys Union. *Res:* Origin and significance of fluid inclusions in ancient hydrothermal systems as evidenced from halogen chemistry and abundance of naturally occurring isotopes of AR, KR and XO. *Mailing Add:* Dept Physics Univ Calif Berkeley CA 94704

IRWIN, JOHN (HENRY) BARROWS, b Princeton, NJ, July 7, 09; m 36; c 4. ASTRONOMY. *Educ:* Univ Calif, BS, 33, PhD(astron), 46. *Prof Exp:* Jr astronomer, US Naval Observ, Washington, DC, 37-39; asst, Lick Observ, Univ Calif, 41-42; asst physics, Off Sci Res & Develop Proj, Calif Inst Technol, 42-44, res assoc, 44-45; physicist, US Naval Ord Test Sta, Calif, 45; asst prof astron, Univ Pa, 46-48; assoc prof, Ind Univ, 48-51, prof, 51-64; staff assoc, Carnegie Inst, 64-67; vis res prof, Steward Observ, Univ Ariz, 67-68; vis prof, Univ Calif, Los Angeles, 68-70; assoc prof astron, Kean Col, 71-77; RETIRED. *Concurrent Pos:* Guggenheim fel, 54-55. *Mem:* AAAS; Am Astron Soc; Royal Astron Soc. *Res:* Photoelectric photometry; binary star orbits; cepheid variables; history of astronomy. *Mailing Add:* 2744 N Tyndall Ave Tucson AZ 85719

IRWIN, JOHN CHARLES, b Rossburn, Man, June 3, 35; m 61; c 4. PHYSICS. *Educ:* Univ BC, BASc, 58, PhD(physics), 65. *Prof Exp:* From asst prof to assoc prof, 65-78, chmn, 80-88, PROF PHYSICS, SIMON FRASER UNIV, 78- *Concurrent Pos:* Nat Res Coun Can grant, 65-; mem bd mgt, BC Res Coun, 74-77. *Mem:* Can Asn Physicists; Am Phys Soc; Mat Res Soc. *Res:* Roman scattering studies of solids: dielectrics, semiconductors, metals and superconductors. *Mailing Add:* Dept Physics Simon Fraser Univ Burnaby BC V5A 1S6 Can

IRWIN, JOHN DAVID, b Minneapolis, Minn, Aug 9, 39; m; c 3. NOISE & VIBRATION CONTROL. *Educ:* Auburn Univ, BS, 61; Univ Tenn, MS, 62, PhD(elec eng), 67. *Prof Exp:* Supvr & mem tech staff, Bell Telephone Labs, 67-69; PROF & DEPT HEAD, ELEC ENG, AUBURN UNIV, 69- *Honors & Awards:* Centennial Medal, Inst & Electronics Engrs, 84. *Mem:* Fel Inst Elec & Electronics Engrs; Indust Electronics Soc (pres 80-81). *Res:* Author or co-author of books on industrial noise, computer logic & engineering circuits. *Mailing Add:* Dept Elec Eng Auburn Univ Auburn AL 36849-5201

IRWIN, JOHN MCCORMICK, b Peking, China, Nov 11, 29; US citizen; m 56; c 4. MATHEMATICS. *Educ:* Purdue Univ, BS, 53; Univ Kans, MA, 56, PhD(math), 60. *Prof Exp:* From asst prof to assoc prof math, NMex State Univ, 60-65; PROF MATH, WAYNE STATE UNIV, 65- *Concurrent Pos:* Co-recipient, NSF res grant, 61-64. *Mem:* Am Math Soc; Math Asn Am. *Res:* Algebra, specifically infinite Abelian groups & differential equations; applied math. *Mailing Add:* Dept Math Wayne State Univ Detroit MI 48202

IRWIN, LAFAYETTE K(EY), b Tenn, June 2, 22; m 48; c 2. MECHANICAL METROLOGY, APPLIED MECHANICS. *Educ:* Univ Ala, BS, 49. *Prof Exp:* Jr engr mech design, Goslin-Birmingham Mfg Co, 47-48; mech engr, Nat Bur Standards, 49-59, chief engr mech sect, 59-69, chief mech div, 69-78; PRES, L K IRWIN, INC, 78- *Concurrent Pos:* Convener, Symbols & Terminology Tech Comt 164, Int Standardization Orgn, 76-89; adj prof mech eng, Univ SC, 78-79; consult & secretariat, Int Orgn Legal Meterol, 78-79; ed, SC Engr, 88- *Honors & Awards:* Templin Award, Am Soc Testing & Mat, 57. *Mem:* Am Soc Mech Engrs; fel Am Soc Testing & Mat. *Res:* Experimental stress analysis; static and dynamic mechanical properties of materials; physical test methods; mechanical fasteners; fluid flow. *Mailing Add:* 510 Greene St Camden SC 29020

IRWIN, LOUIS NEAL, b Big Spring, Tex, Jan 8, 43; m 67; c 2. NEUROBIOLOGY, NEUROCHEMISTRY. *Educ:* Tex Tech Univ, BA, 65; Univ Kans, PhD(biochem, physiol), 69. *Prof Exp:* NIH trainee, Parsons State Hosp, Kans, 69-70; asst prof biol, Col Pharmaceut Sci, Columbia Univ, 70-73; asst prof physiol, Sch Med, Wayne State Univ, 73-76; assoc biochemist, E K Shriver Ctr, 76-79; PROF BIOL, SIMMONS COL, 80- *Concurrent Pos:* Staff scientist, Neurosci Res Prog, Mass Inst Technol, 77-78. *Mem:* AAAS; Soc Neurosci; Am Soc Neurochem; Sigma Xi; Int Soc Neurochem. *Res:* Molecular neurobiology; biochemical correlates of brain organization and function; developmental neurochemistry. *Mailing Add:* 420 Lowell Ave Newton MA 02160

IRWIN, LYNNE HOWARD, b Los Angeles, Calif, July 15, 41; m 65; c 3. HIGHWAY ENGINEERING, PAVEMENT DESIGN & EVALUATIONS. *Educ:* Univ Calif, BS, 65, MS, 66; Tex A&M Univ, PhD(civil eng), 73. *Prof Exp:* Asst prof eng, Calif State Univ, 66-69; res assoc, Tex A&M Univ, 69-72; ASSOC PROF ENG, CORNELL UNIV, 73- *Concurrent Pos:* NSF fel, 69-71; hwy engr, US Forest Serv, 79-80; res engr, Cold Regions Res & Eng Lab, US Army Corps Engrs, 80. *Mem:* Asn Soc Civil Engrs; Asn Asphalt Pavement Technologists; Am Rd & Transportation Builders Asn. *Res:* Structural evaluation of pavements using nondestructive deflection testing; recycling of pavements, surfaces and bases, using chemical stabilization; development of computer programs for calculation of moduli of elasticity of pavement layers. *Mailing Add:* 104 Riley-Robb Hall Cornell Univ Ithaca NY 14853

IRWIN, MICHAEL EDWARD, b Los Angeles, Calif, Aug 10, 40; m 71. ENTOMOLOGY, SYSTEMATICS. *Educ:* Univ Calif, Davis, BS, 63; Univ Calif, Riverside, PhD(entom), 71. *Prof Exp:* Res entomologist surv entom, Univ Calif-Univ Chile, 66-67; asst spec entom, Univ Calif, Riverside, 70-71; sr prof officer entom, Nat Mus, Pietermaritzburg, SAfrica, 71-74; asst prof entom, 74-78, assoc prof, 79-89, PROF ENTOM & PLANT PATH, UNIV ILL & ILL NAT HIST SURV, 84-; HEAD, DEPT AG ENTOM, UNIV ILL, 90- *Concurrent Pos:* Mem, Int Soybean Prog; Travel Award, Entom Soc Am, 80; prog mgr, Plant/Pests Ineractions, USDA Nat Res Initiative Competetive Grants, 90; dir, Ctr Econ Entom, Ill Nat Hist Surv, 90- *Mem:* Entom Soc Am (secy, 78); Can Entom Soc; Am Inst Biol Sci; AAAS; Int Asn Ecol; Sigma Xi. *Res:* Epidemiol of plant viruses by insect vectors; long distance movement of insects; biosystematics of Therevidae (Insecta Diptera). *Mailing Add:* 1108 S Busey St Urbana IL 61801

IRWIN, PETER ANTHONY, b Thetford, Eng, Aug 10, 45; Can citizen; m 70; c 2. WIND ENGINEERING, FLUID MECHANICS. *Educ:* Southampton Univ, UK, BSc, 67, MSc, 69; McGill Univ, PhD(fluid mech), 74. *Prof Exp:* Sci officer aerodyn, Royal Aircraft Estab, Farnborough, UK, 68-71; res officer wind eng, Nat Res Coun Can, 74-80; dir tech serv, Morrison Hershfield Ltd, 80-86; PRIN, ROWAN WILLIAMS DAVIES & IRWIN, INC, 86- *Honors & Awards:* Two Merit Awards, Asn Consult Engrs Can. *Mem:* Can Soc Civil Engrs; Eng Inst Can; Am Soc Civil Engrs. *Res:* Response of buildings, bridges and structures to wind; aerodynamics; fluid mechanics; turbulence; aeroelasticity; airborne pollution studies. *Mailing Add:* Rowan Williams Davies & Irwin Inc 650 Woodlawn Rd W Guelph ON N1G 2N9 Can

IRWIN, PHILIP GEORGE, b Duquesne, Pa, Nov 20, 34; m 61; c 6. PHYSICAL ORGANIC CHEMISTRY. *Educ:* Duquesne Univ, BS, 56; Purdue Univ, MS, 58; Pa State Univ, PhD(org chem), 62. *Prof Exp:* Res chemist, Gulf Res & Develop Co, 62-65; appl res, US Steel Corp, 65-69; group leader synthetic resin res, Picco Resins, Hercules Inc, 69-80; tech dir, Ameron Ind Coatings, 80; tech mgr, Ga-Pac Corp, 82-83; PRES, PRESTIGE CHEMICALS INC, 83- *Concurrent Pos:* Consult. *Mem:* Am Chem Soc. *Res:* Resin and chemical synthesis; chemical modification of polymers; new product and process research; polymer and chemical research in adhesives, printing inks, toners, paper size, film, coatings. *Mailing Add:* RD 1 Box 1034 Ruffsdale PA 15679-9626

IRWIN, RICHARD LESLIE, b Fullerton, Nebr, Sept 3, 17; m 43; c 3. PHARMACOLOGY. *Educ:* Univ Denver, BA, 49; Univ Colo, PhD(pharmacol), 53. *Prof Exp:* Chief clin neuropharmacol, Nat Inst Neurol Dis & Stroke, 53-71, asst dir intramural res, 71-75, LAB DIR, NAT INST NEUROL & COMMUN DIS & STROKE, BETHESDA, MD, 75- *Res:* Respiratory and renal physiology; curariform drugs; neuropharmacology; pharmacodynamics; enzyme inhibition; muscle physiology. *Mailing Add:* 105 Upton St Rockville MD 20850

IRWIN, RICHARD STEPHEN, b New London, Conn, Nov 15, 42; m 69; c 4. COUGH, ASTHMA. *Educ:* Tufts Univ, BS, 64; Tufts Sch Med, MD, 68; dipl, Nat Bd Med Examiners, 69, Am Bd Internal Med, 72, Am Bd Pulmonary Med, 74, Am Bd Critical Care Med, 87. *Prof Exp:* Internship med, Tufts New England Med Ctr, 68-69, residency, 69-70; post doctorate fel, cardio response physiol, Columbia Presby Hosp, 70-72; asst prof med, Brown Univ, 74-79; assoc prof med, 79-82, PROF MED, UNIV MASS MED SCH, 82- *Concurrent Pos:* Dir, div Pulmonary & Critical Care Med, 79- *Mem:* Am Thoracic Soc; Am Col Chest Physicians; Am Col Physicians; Am Fedn Clin Res; Nat Assoc Med Dirs Respiratory Care. *Res:* The pathophysiology, pathogenesis, diagnosis and treatment of cough and the diagnosis and treatment of asthma. *Mailing Add:* Univ Mass Med Ctr 55 Lake Ave N Worcester MA 01655

IRWIN, ROBERT COOK, b Hastings, Nebr, Oct 25, 29; m 60; c 1. MATHEMATICS, COMPUTER SCIENCES. *Educ:* Univ Nebr, BS, 51; Univ Calif, MA, 58; Univ Ariz, PhD(math), 63. *Prof Exp:* Engr, Int Bus Mach Corp, 54-56; mem staff, Lincoln Lab, Mass Inst Technol, 58-59; mem tech staff, Mitre Corp, Mass, 59-67; sr res mathematician, Dikewood Corp, NMex, 67-71; MEM TECH STAFF, MITRE CORP, 71- *Concurrent Pos:* Lectr, Univ Colo, 64-67. *Mem:* Math Asn Am; Am Math Soc. *Res:* Systems science; numerical analysis; computer science. *Mailing Add:* Mitre Corp 1259 Lake Plaza Dr Colorado Springs CO 80906

IRWIN, WILLIAM EDWARD, b Reading, Pa, Sept 26, 26; m 50; c 3. PLASTICS CHEMISTRY. *Educ:* Kutztown State Col, BS, 48; Columbia Univ, MA, 54; Pa State Univ, PhD(org chem), 62. *Prof Exp:* Chemist prod develop, Armstrong World Industs, Inc, 62-64, res supvr, 64-69, sr res scientist, 69-73, unit mgr, 73-82, res assoc, 82-89; RETIRED. *Mem:* Am Chem Soc; fel Am Soc Testing & Mat. *Res:* Chemical and physical property studies of plastic materials. *Mailing Add:* 3139 Parker Dr Lancaster PA 17601

IRWIN, WILLIAM ELLIOT, b Ferndale, Mich, Nov 1, 28; m 48; c 2. BIOCHEMISTRY, FOOD TECHNOLOGY. *Educ:* Western Mich Univ, BS, 50; Univ Notre Dame, MS, 65, PhD(biochem), 70. *Prof Exp:* Chem buyer, R P Scherer Corp, 50-55; tech serv mgr, Miles Labs, Inc, 55-65, dir prod develop, 55-89; USA REP, PALATNIT GMBH, 89- *Concurrent Pos:* Dir, Food Protein Coun, 71-73. *Mem:* AAAS; Am Chem Soc; Am Asn Cereal Chemists; Inst Food Technologists. *Res:* Development of economical, biologically efficient, protein foods; protein, carbohydrate enzyme mechanisms. *Mailing Add:* Irwin Serv 1501 Dogwood Dr Elkhart IN 46514-4329

ISA, ABDALLAH MOHAMMAD, b Bassa, Palestine, June 15, 38; m 62; c 3. MICROBIOLOGY, IMMUNOLOGY. *Educ:* Am Univ Beirut, BS, 60; Univ Calif, Berkeley, MA, 66; Univ Calif, San Francisco, PhD(microbiol), 68. *Prof Exp:* Asst, Univ Calif, Berkeley, 63-65; asst, Univ Calif, San Francisco, 66-68, Fight for Sight, Inc fel, Med Ctr, 68-69; asst prof, 69-73, ASSOC PROF IMMUNOL, MEHARRY MED COL, 73- *Concurrent Pos:* Fight for Sight, Inc grant, Meharry Med Col, 69-72; Res Corp grant, 70-73; spec asst, Off Exec Vpres Near Eastern Affairs, Meharry Med Col. *Mem:* AAAS; Europ Dialysis & Transplant Soc; Transplantation Soc; Am Asn Immunologists. *Res:* Immunology of trachoma and other Chalamydia agents; transplantation immunology, especially the rejection reaction and the role of the thymocyte and bone marrow-derived cells in rejection; cellular immunology; role of the suppressor cell in tumor immunity. *Mailing Add:* Dept Biol Sci Tenn State Univ 3500 J A Merritt Blvd Nashville TN 37209

ISAAC, PETER ASHLEY HAMMOND, b Grimsby, Lincolnshire, Eng, Jan 25, 45; m 72; c 2. PULP & PAPER TECHNOLOGY. *Educ:* Cambridge Univ, BA, 67, MA, 71; Carnegie-Mellon Univ, PhD(org chem), 72; Rutgers Univ, MBA, 76. *Prof Exp:* Res chemist, 72-74, group leader paper prod res, Minerals & Chem Div, Engelhard Minerals & Chem Corp, 74-76; group leader, 76-79, planning mgr, 79-81, sales mgr, 81-82, opers mgr, 82-83, bus develop mgr, 83-84, mkt mgr elastomer, 84-86, mkt mgr, Urethane Chem, 86-87, MGR BUS DEVEL, AM CYANAMID CO, 87- *Concurrent Pos:* Chmn, Mkt Comt, Polyurethane Mfrs Asn. *Mem:* Am Chem Soc. *Res:* Application of inorganic materials in paper coating and filling; mineralogy, synthetic inorganic chemistry, colloid and surface chemistry; polymers and plastics. *Mailing Add:* 21 Oakley Ave Summit NJ 07901

ISAAC, RICHARD EUGENE, b New York, NY, Jan 2, 34; m 63; c 1. MATHEMATICS. *Educ:* Cornell Univ, BA, 55; Univ Calif, Berkeley, PhD(math), 59. *Prof Exp:* Fulbright fel math, Poincare Inst, Paris, 59-60; instr, Fordham Univ, 61; res assoc, Yeshiva Univ, 61-62; asst prof, Hunter Col, 62-69; PROF MATH, LEHMAN COL, 69- *Concurrent Pos:* Vis asst prof, Cornell Univ, 64-65. *Mem:* Am Math Soc. *Res:* Probability theory and related fields; analysis. *Mailing Add:* Dept Math Herbert H Lehman Col Bronx NY 10468

ISAAC, ROBERT A, b Georgetown, SC, Feb 19, 36; m 66; c 3. ANALYTICAL CHEMISTRY. *Educ:* Col Charleston, BS, 58; Clemson Univ, MS, 62, PhD(anal chem), 66. *Prof Exp:* Teacher high sch, SC, 58-60; chemist, Savannah River Lab, E I DuPont de Nemours & Co, Inc, 62-63; head anal chem dept, Tenn Corp, NJ, 66-68; dir, Soil & Plant Anal Lab, Coop Exten Serv, 80-87, DIR AGR SERV LABS, UNIV GA, 87- *Concurrent Pos:* Assoc referee, Asn Off Anal Chemists, 66-; anal chemist, Univ Ga, 68- *Mem:* Am Soc Agron; Am Chem Soc; fel Am Inst Chem; Sigma Xi; Soc Appl Spectroscopy. *Res:* Analytical methodology for the analysis of major and micronutrients in soils and plant tissue; use of atomic absorption and plasma emission spectroscopy along with automated analytical techniques. *Mailing Add:* Soil & Plant Analysis Lab Univ Ga 2400 College Station Rd Athens GA 30605

ISAAC, WALTER, b Cleveland, Ohio, June 13, 27; m 49; c 2. PSYCHOPHYSIOLOGY. *Educ:* Western Reserve Univ, BS, 49; Ohio State Univ, MA, 50, PhD(psychol), 53. *Prof Exp:* Res instr physiol, Sch Med, Univ Wash, 54-56; asst res psychol, Univ Calif, Los Angeles, 56-57; from asst prof to prof psychol, Emory Univ, 57-68; PROF PSYCHOL, UNIV GA, 68- *Mem:* Psychonomic Soc; fel AAAS; Sigma Xi; Soc Behav Med; Am Asn Lab Animal Sci; Int Neuropsychol Soc. *Res:* Changes in arousal levels with aging; effects of pharmacological stimulants and depressants on behavior; functions of the cerebral cortex. *Mailing Add:* Dept Psychol Univ Ga Athens GA 30602

ISAACKS, RUSSELL ERNEST, b Humble, Tex, July 25, 35; m 54; c 2. BIOCHEMISTRY, NUTRITION. *Educ:* McNeese State Col, BS, 57; Tex A&M Univ, MS, 59; PhD(biochem), 61. *Prof Exp:* Res asst nutrit, Tex A&M Univ, 56-61; scientist, USPHS, HEW, Md, 61-64; res chem, res serv, Vet Admin Hosp, Dallas, 64-65; from asst prof to res assoc prof biochem, 65-83, RES PROF, MED SCH, UNIV MIAMI, 83-; RES CHEM RES SERV, VET ADMIN HOSP, MIAMI, 65- *Mem:* AAAS; Sigma Xi; Poultry Sci Asn; Am Chem Soc; Am Physiol Soc; Am Zool. *Res:* Metabolism and function of organic phosphates in oxygen transport of red blood cells of birds, reptiles, amphibians, and fishes; membrane transport. *Mailing Add:* 8965 SW 115th Terr Miami FL 33156

ISAACS, CHARLES EDWARD, b Brooklyn, NY, Oct 26, 49; m 77; c 2. IMMUNOLOGY OF HUMAN MILK, ANTIVIRAL COMPOUNDS IN BLOOD. *Educ:* State Univ NY, Stony Brook, BS, 70; Rutgers Univ, PhD(microbiol), 77. *Prof Exp:* Postdoctoral fel, Roche Inst Molecular Biol, 77-78; RES SCIENTIST, NY STATE INST BASIC RES, 78- *Concurrent Pos:* Adj asst prof, Dept Pediat, Mt Sinai Sch Med, 78-; prin investr, Nat Inst Heart, Lung & Blood, 88- *Mem:* Am Inst Nutrit; AAAS; NY Acad Sci; Int Soc Res Human Lactation. *Res:* Antiviral and antibacterial in humans. *Mailing Add:* Dept Develop Biochem NY State Inst Basic Res 1050 Forest Hill Rd Staten Island NY 10314

ISAACS, GERALD W, b Crawfordsville, Ind, Sept 3, 27; m 48; c 4. AGRICULTURAL ENGINEERING. *Educ:* Purdue Univ, BSEE, 47, MSEE, 49; Mich State Univ, PhD(agr eng), 54. *Prof Exp:* Asst, Purdue Univ, 47, instr, 48-52; asst, Mich State Univ, 52-54; from asst prof to prof agr eng, Purdue Univ, 54-81, head dept, 64-81; PROF & CHMN, DEPT AGR ENG, UNIV FLA, 81- *Concurrent Pos:* Consult, var farm equip mfrs, 58- & US Agency Int Develop, 66-; guest prof, Hohenheim Agr Univ, 63 & 74. *Honors & Awards:* Silver Medal, Max Eyth Gesellschaft, 66. *Mem:* Fel Am Soc Agr Engrs (pres, 82-83); Am Soc Eng Educ; Nat Soc Prof Engrs. *Res:* Grain storage; grain drying; solar energy utilization. *Mailing Add:* Dept Agr Eng Univ Fla Gainesville FL 32611

ISAACS, GODFREY LEONARD, b Cape Town, SAfrica, Feb 9, 24. MATHEMATICAL ANALYSIS. *Educ:* Univ Cape Town, BSc, 44, MSc, 45; London Univ, PhD(math), 50. *Prof Exp:* Lectr math, Univ Natal, 46-47 & Birkbeck Col, London, 48-49; lectr, Univ Witwatersrand, 52-55, sr lectr, 55-59, prof, 60-67; vis prof, State Univ NY Stony Brook, 67-68; from assoc prof to prof, 68-86, EMER PROF MATH, LEHMAN COL, CITY UNIV NY, 86- *Concurrent Pos:* Nuffield fel res assoc, Univ Col, London, 59-60; Carnegie Corp Traveling grant, 66; fac res fel, City Univ New York, 70. *Mem:* SAfrican Math Asn; Am Math Soc; Math Asn Am; NY Acad Sci. *Res:* Summability; integration; Fourier series; set theory. *Mailing Add:* 3111 N Ocean Dr Apt 1206 Hollywood FL 33019

ISAACS, HUGH SOLOMON, b Johannesburg, SAfrica, Aug 13, 36; US citizen; m 59; c 2. MATERIAL SCIENCE, ELECTROCHEMISTRY. *Educ:* Univ Witwatersrand, BSc, 58; Imp Col Sci Technol, DIC, 61; Univ London, PhD(metal corrosion), 64. *Prof Exp:* Vis sci oxidation metals, UK Atomic Energy Authority, Harwell, 63-64; res scientist metall, Atomic Energy Bd, SAfrica, 64-67; asst metallurgist, Brookhaven Nat Lab, 67-72; assoc metallurgist, Oak Ridge Nat Lab, 72-74; METALLURGIST, BROOKHAVEN NAT LAB, 74- *Concurrent Pos:* Adj prof, Univ NY, 77-80. *Honors & Awards:* H E Armstrong Medal & Prize, Imperial Col Sci & Technol, 64; Sam Tour Award, Am Soc Testing & Mat, 83. *Mem:* Electrochem Soc; Nat Asn Corosion Engrs. *Res:* Corrosion, and electrochemical kinetics at solid and liquid electrolyte interfaces. *Mailing Add:* Brookhaven Nat Lab Upton NY 11973

ISAACS, I(RVING) MARTIN, b New York, NY, Apr 14, 40. FINITE GROUPS, GROUP REPRESENTATIONS. *Educ:* Polytech Inst Brooklyn, BS, 60; Harvard Univ, AM, 61, PhD(math), 64. *Prof Exp:* Instr math, Univ Chicago, 66-68, vis asst prof, 68-69; assoc prof, 69-71, PROF MATH, UNIV WIS-MADISON, 71- *Concurrent Pos:* Sloan res fel, 71-73; vis prof, Univ Calif, Berkeley, 73-74 & 89. *Mem:* Am Math Soc; Math Asn Am. *Res:* Theory of finite groups and their characters. *Mailing Add:* Dept Math Univ Wis Madison WI 53706

ISAACS, LESLIE LASZLO, b Berehovo, Czech, Aug 5, 33; US citizen; m 62; c 2. EXPERIMENTAL SOLID STATE PHYSICS, THERMODYNAMIC PROPERTIES. *Educ:* Columbia Univ, BSc, 55; Mass Inst Technol, PhD(phys chem), 60. *Prof Exp:* Fel metal physics, Mellon Inst, 60-65; mem metall div, Argonne Nat Lab, 65-71; vis fac chem, Univ Wash, Seattle, 72-74; ASSOC PROF CHEM ENG, CITY COL NEW YORK, 74-, ASSOC DEAN SCH ENG, 88- *Mem:* Am Phys Soc; Am Inst Chem Eng; Am Chem Soc; Mat Res Soc. *Res:* Properties of materials; low temperature physics; thermodynamics of coal chars; catalysis; thermodynamics, materials science. *Mailing Add:* Dept Chem Eng City Col Convent Ave & 138th Sts New York NY 10031

ISAACS, PHILIP KLEIN, b New York, NY, Oct 18, 27; m 53; c 3. POLYMER CHEMISTRY. *Educ:* Bard Col, BA, 48; Columbia Univ, MA, 50; Univ Cincinnati, PhD, 51. *Prof Exp:* Res chemist, Dewey & Almy Div, W R Grace & Co, 51-58, group leader, 58-62, res assoc, Res Div, 63-66; res assoc, Machteshim Chem Co, Israel, 66-68; head textile finishing dept, Israel Fibers Inst, 68-74; vis scientist, Weizmann Inst Sci, 74-75; chief technologist, Off Chief Scientist, Ministry Com & Indust, Israel, 75-86; RES CHEMIST, ISRAEL FIBER INST, 87- *Concurrent Pos:* Sr lectr, Sch Appl Sci, Hebrew Univ, 69-75; UN fel, 71; vis scientist, Casali Inst, Hebrew Univ, 81-86. *Mem:* Am Chem Soc. *Res:* Polymer modification; latex applications; flame retardancy of textiles; surface chemistry; crosslinking of polymers; plastic coatings for foods; halogenated polymers; adhesives. *Mailing Add:* 2 Dov Kimche St Jerusalem 92549 Israel

ISAACS, TAMI YVETTE, b Elizabeth, NJ, Feb 17, 52; m; c 3. PHOTOCHEMISTRY, PHYSICAL ORGANIC CHEMISTRY. *Educ:* Rennselaer Polytech Inst, BS, 74; Johns Hopkins Univ, MA, 76, PhD(org chem), 80. *Prof Exp:* Lectr chem, Essex Community Col, Baltimore County, 76-79; asst prof, Rider Col, Lawrenceville, NJ, 80; ASST PROF CHEM, COL NOTRE DAME, MD, 80- *Concurrent Pos:* Adj prof, Towson State Univ, Md, 81-82; researcher, Johns Hopkins Univ, 81-83. *Mem:* Am Chem Soc; Sigma Xi. *Res:* Organic photochemical reactions, including mechanisms and using various methods to increase control of the products formed. *Mailing Add:* 6006 Berkeley Ave Baltimore MD 21209

ISAACSON, ALLEN, b Brooklyn, NY, Jan 15, 32; m 59; c 2. PHYSIOLOGY, BIOPHYSICS. *Educ:* City Col New York, BS, 53; Harvard Univ, AM, 54; NY Univ, PhD(biol), 62. *Prof Exp:* Jr engr, Amperex Electronics Corp, 55-56; proj engr, Kollsman Instrument Corp, 56-58; res asst biophys, Sloan-Kettering Inst Cancer Res, 58-60; res assoc physiol, Inst Muscle Dis, 60-62, asst mem, 62-69; ASSOC PROF BIOL, WILLIAM PATERSON COL, NJ, 69- *Concurrent Pos:* Pt-time instr biol sci, Fairleigh-Dickenson Univ, 67-71; NSF res partic, biochem & biophys, Oregon State Univ, 70, 71 & 75, Neurol Dept, Columbia Univ, 82; post-doctoral res, Inst Muscle Dis, Muscular Dystrophy Asn, 72 & 73 & Physiol Dept, Downstate Med Ctr, State Univ NY, 74. *Mem:* Fel AAAS; Am Physiol Soc; Soc Gen Physiol; Biophys Soc; Nat Asn Biol Teachers. *Res:* Excitation-contraction coupling of muscle; kinetics of calcium and zinc movements in skeletal muscle; radiation safety; radioisotope applications; caffeine and related drug effects on the integrity of membranes of muscle. *Mailing Add:* Dept Biol William Paterson Col 300 Pompton Rd Wayne NJ 07470

ISAACSON, DAVID, b New York, NY, Oct 22, 48; m 71; c 2. NUMERICAL METHODS, SCIENTIFIC COMPUTING. *Educ:* NY Univ, BA, 70, MS, 72, PhD(math), 76. *Prof Exp:* Lectr math, NY Univ, 74-75; lectr, Rutgers Univ, 75-76, asst prof, 76-80; ASSOC PROF MATH, RENSSELAER POLYTECH INST, 80- *Mem:* Am Math Soc; Math Asn Am; Soc Indust Appl Math; Sigma Xi; Inst Elec & Electronics Engrs. *Res:* Numerical and analytical methods for approximating the spectral properties of Schrodinger; energy; operators; mathematical problems in cardiology. *Mailing Add:* 15 Riviera Dr Latham NY 12110

ISAACSON, DENNIS LEE, b Los Angeles, Calif, May 5, 42; m 66; c 2. ENTOMOLOGY, REMOTE SENSING. *Educ:* Portland State Univ, BS, 69; Ore State Univ, MS, 72, MAg, 75. *Prof Exp:* Entomologist biol control weeds, Ore Dept Agr, 74-78; Sr analyst, Environ Remote Sensing Applns Lab, Ore State Univ, 78-90; PROG MGR, NOXIUS WEED CONTROL, ORE DEPT AGR, 90- *Concurrent Pos:* Proj leader, Pac NW Regional Comn, 74-75 & 75-76; contract adminr, Bur Land Mgt, US Dept Interior, 75-76; World Bank consult, Gadhah Mada Univ, Indonesia, 87. *Mem:* Am Soc Photography and Remote Sensing. *Res:* Vegetation inventories by application of remote sensing techniques; distribution and abundance of plant species; insect-plant interactions with special interest in biological weed control. *Mailing Add:* 303 NW 31st St Corvallis OR 97330

ISAACSON, EUGENE, b Brooklyn, NY, June 14, 19; m 47; c 2. MATHEMATICS. *Educ:* City Col New York, BS, 39; NY Univ, MS, 41, PhD(math), 49. *Prof Exp:* Comput operator, US Bur Standards, 43; asst, 44, from asst prof to assoc prof, 49-58, actg chief, AEC Comput & Appl Math Ctr, 52-53, assoc dir, 58-65, chmn, Comput Ctr, 63-73, PROF MATH, NY UNIV, 58- *Concurrent Pos:* Chmn ed comt, J Math of Computation, 66-74, mem bd assoc ed, 75-; vis prof & actg chmn dept comput sci, City Col New York, 70-71; consult, US Army Corps Eng; managing ed, Soc Indust Appl Math J Numerical Anal, 80-86; prin investr, NASA grant, 83-; trustee, Univ Space Res Asn, 84- *Mem:* Am Math Soc; Math Asn Am; Soc Indust & Appl Math. *Res:* Applied mathematics; waterwaves; numerical analysis; climatology; meteorology. *Mailing Add:* Courant Inst Math Sci New York Univ 251 Mercer St New York NY 10012

ISAACSON, EUGENE I, b Wilson, Wis, June 30, 33; m 53; c 3. PHARMACEUTICAL CHEMISTRY, ORGANIC CHEMISTRY. *Educ:* Univ Minn, BS, 56, PhD(pharmaceut chem), 63. *Prof Exp:* Asst prof pharmaceut chem, Col Pharm, Univ Tex, Austin, 63-69; assoc prof, 69-74, PROF PHARMACEUT CHEM, COL PHARM, IDAHO STATE UNIV, 74- *Mem:* Am Pharmaceut Asn; Acad Pharmaceut Sci; Am Asn Cols Pharm; NY Acad Sci; fel Am Inst Chemists. *Res:* Synthesis and structure-activity-relationships among cholinergic and anti-cholinergic agents; psychopharmacologic agents; anticonvulsants. *Mailing Add:* Col Pharm Idaho State Univ Pocatello ID 83201

ISAACSON, HENRY VERSCHAY, b Chicago, Ill, Nov 11, 39; m 66; c 2. ORGANIC POLYMER CHEMISTRY. *Educ:* Univ Ill, Urbana, BS, 61; Univ Minn, Minneapolis, PhD(org chem), 69. *Prof Exp:* Chemist, Sinclair Res Labs, 61-65; sr res chemist, 69-76, RES ASSOC, RES LABS, EASTMAN KODAK CO, 77- *Mem:* Am Chem Soc. *Res:* Synthesis of polymers; emulsion polymerization; electrophotographic toners; modification of polymers. *Mailing Add:* 1260 Holley Rd Webster NY 14580

ISAACSON, LAVAR KING, b Provo, Utah, July 15, 34; m 61; c 3. MECHANICAL ENGINEERING, THERMODYNAMICS. *Educ:* Univ Utah, BS, 56, PhD(mech eng), 62. *Prof Exp:* Engr, Sperry Utah Co, 57; instr mech eng, 57-58, from asst prof to assoc prof, 61-70, PROF MECH ENG, UNIV UTAH, 70- *Concurrent Pos:* Res specialist, Space & Info Systs Div, NAm Aviation, Inc, 63-64; proj mgr, Elec Power Res Inst, 74-75, prog mgr, 75-76. *Mem:* Inst Aeronaut & Astronaut; Am Soc Eng Educ; Sigma Xi. *Res:* Reactive boundary layers; combustion gas dynamics; nonequilibrium thermodynamics; turbulent flow. *Mailing Add:* 4016 S Mercury Dr Salt Lake City UT 84124

ISAACSON, MICHAEL SAUL, b Chicago, Ill, July 4, 42; m 65; c 2. ELECTRON PHYSICS, ELECTRON MICROSCOPY. *Educ:* Univ Ill, BS, 65; Univ Chicago, SM, 66, PhD(physics), 71. *Prof Exp:* Staff scientist, Brookhaven Nat Lab, 70-73; asst prof physics, Univ Chicago & Enrico Fermi Inst, 73-; AT DEPT PHYSICS, CORNELL UNIV. *Concurrent Pos:* Consult, Coates & Welter Instrument Corp, 71-73 & Siemens A G, 76; adv, Annual Scanning Microscope Symp, Ill Inst Technol Res Inst, 74-77; ed adv, J Ultramicros, 75- & Electron Micros Soc Am Bull, 77-; Alfred P Sloan res fel, 75-; mem organizing comt, 9th Int Cong Electron Micros, Toronto, 1978, 75-; adv, SEM, Inc, 78- *Honors & Awards:* Burton Award, Electron Micros Soc Am, 76. *Mem:* Radiation Res Soc; Electron Micros Soc Am; Am Asn Physics Teachers; Biophys Soc. *Res:* Development of new advanced electron optical systems and techniques, and the illucidation of new principles for use in increasing the limits of ultrastructure research by means of electrons. *Mailing Add:* Dept Physics 210 Clark Hall Cornell Univ Main Campus Ithaca NY 14853

ISAACSON, PETER EDWIN, b Seattle, Wash, Mar 29, 46; m 79; c 2. BIOSTRATIGRAPHY, PALEOECOLOGY. *Educ:* Univ Colo, BA, 68; Ore State Univ, PhD(geol), 74. *Prof Exp:* Adj asst prof geol, Univ Mass, Amherst, 74-78; res assoc, Amherst Col, 74-78; asst prof, 78-83, PROF GEOL, COL MINES, UNIV IDAHO, 83- *Concurrent Pos:* Vis asst prof, Franklin & Marshall Col, 75; exchange scientist, Eastern Europe Prog, Nat Acad Sci, 82, 90. *Honors & Awards:* Fulbright Res Fel, Czechoslovakia, 87. *Mem:* Asn Paleontol Agr; Geol Soc Am; Soc Econ Paleontologists & Mineralogists; Int Paleontol Union. *Res:* Biostratigraphic and paleo-depositional modelling of Paleozoic-age rock sequences. *Mailing Add:* Dept Geol Univ Idaho Moscow ID 83843

ISAACSON, ROBERT B, b New York, NY, Jan 14, 36; m 75. POLYMER CHEMISTRY, ORGANIC CHEMISTRY. *Educ:* City Col New York, BS, 56; Univ Md, PhD(org chem), 61. *Prof Exp:* Proj leader chem res, Polymer Res & Develop, Esso Res & Eng Co, 61-65; group leader new prod, Plastics Res & Develop, 65-68, tech mgr, 68-71, tech mgr resins Res & Develop, 71-74, tech dir, 74-78, DIR PLANNING & TECHNOL ASSESSMENT, RES & DEVELOP, CELANESE RES CO, 78- *Concurrent Pos:* Res corp fel, 59-60. *Mem:* Sigma Xi; Am Chem Soc; AAAS; Soc Plastics Engrs. *Res:* Management of industrial research; applied polymer science; plastics research and development; materials science; structure and property relationships of polymeric materials. *Mailing Add:* 14 Post Lane Livingston NJ 07039-4907

ISAACSON, ROBERT JOHN, b Flushing, NY, July 12, 32. ORTHODONTICS, ANATOMY. *Educ:* Univ Minn, Minneapolis, BS, 54, DDS, 56, MSD, 61, PhD(anat), 62. *Prof Exp:* From asst prof to prof dent, Sch Dent, Univ Minn, Minneapolis, 62-77, chmn orthod div, 65-77; prof & chmn dept Growth & Develop, Sch Dent, Univ Calif, San Francisco, 77-87; PROF & CHMN DEPT ORTHOD, VA COMMONWEALTH UNIV, 87- *Concurrent Pos:* Mem grad fac, assoc mem dent & anat & dir res training prog, Univ Minn-Minneapolis, 63-77, lectr, Sch Dent, 64-77. *Mem:* Int Asn Dent Res. *Res:* Quantitation of loads in relation to skeletal changes in orthodontics, teratogenesis and cleft lip and palate; facial skeletal growth and dental occlusion. *Mailing Add:* Dept Orthod Va Commonwealth Univ Richmond VA 23298-0566

ISAACSON, ROBERT LEE, b Detroit, Mich, Sept 26, 28. PHYSIOLOGICAL PSYCHOLOGY. *Educ:* Univ Mich, BA, 50, MA, 54, PhD(psychol), 58. *Prof Exp:* Instr psychol, Univ Mich, Ann Arbor, 58-60, from asst prof to prof, 60-68; prof psychol & neurosci, Univ Fla, 68-77, grad res prof, 77-78; DISTINGUISHED PROF PSYCHOL, STATE UNIV NY, BINGHAMTON, 78- *Concurrent Pos:* Dir, NSF Res Participation Prog for Col Teachers Psychol, 61-64; dir, Ctr Neurobiol Sci, 70-78; mem biol sci training review comt, NIMH, 70-74; mem Neurol A Study Sect, NIH, 76-80; dir, Ctr Neurobehav Sci, 78- *Mem:* Am Physiol Soc; Soc Neurosci. *Res:* Effects of brain lesions and hormones on behavior; limbic system, drug actions and behavior; calcium channel antogonists and recovery from brain damage. *Mailing Add:* Dept Psychol State Univ NY Binghamton NY 13902-6000

ISAACSON, STANLEY LEONARD, b Baltimore, Md, Jan 31, 27; m 52; c 3. MATHEMATICAL STATISTICS. *Educ:* Johns Hopkins Univ, BA, 45, MA, 47; Columbia Univ, PhD(math statist), 50. *Prof Exp:* Jr instr math, Johns Hopkins Univ, 45-47; asst, Off Naval Res Contract, Columbia Univ, 49-50; asst prof statist, Iowa State Col, 50-55; sr statistician, Westinghouse Elec Corp, Pa, 55-56; sales mgr, 56-64, vpres, 64-80, pres, Gendler Stone Prod Co, 80-86, GEN MGR, GENDLER AGGREGATES CO, 86- *Concurrent Pos:* Vis prof, Stanford Univ, 53-54; lectr, Drake Univ, 56-59; corp secy, 87-90, dir, CIM Tech, Inc, 87- *Mem:* Am Statist Asn; Inst Math Statist. *Res:* Statistical decision theory; industrial application of statistics. *Mailing Add:* 4706 Lakeview Dr Des Moines IA 50311

ISAAK, DALE DARWIN, b Bismarck, NDak, July 2, 48; m 70; c 2. CELLULAR IMMUNOLOGY, VIRAL ONCOLOGY. *Educ:* Eastern Mont Col, BS, 70; Mont State Univ, MS, 73, PhD(microbiol), 76. *Prof Exp:* Cancer res scientist I viral oncol, Roswell Park Mem Inst, 76; res assoc cellular immunol, Harvard Sch Public Health, 76-79; ASST PROF MED MICROBIOL & IMMUNOL, KIRKSVILLE COL OSTEOP MED, 79- *Concurrent Pos:* Adj asst prof, Northeast Mo State Univ; prin investr res grant, NIH, 79-82, Am Osteop Asn, 79-83 & Nat Sci Found, 80-83. *Mem:* Am Soc Microbiol; Reticuloendo Thelial Soc; Sigma Xi. *Res:* Elucidatioin of processes involved in lymphoyte differentiation into subpopulation; the role of major histocompatibility complex-coached proteins in lymphoyte interactions; epigenetic factors involved in malignant transformation of lymphoytes by leukemia viruses. *Mailing Add:* Dept Molecular Biol Univ Wyo Laramie WY 82071

ISAAK, ROBERT D(EETS), b Spokane, Wash, Jan 1, 21; m 44; c 3. ELECTRICAL ENGINEERING. *Educ:* Univ Colo, BS, 43, MS, 49. *Prof Exp:* Test engr, Gen Elec Co, 43-44; instr elec eng, Univ Colo, 47-52; sect head underwater acoust, US Naval Electronics Lab, 52-64, div head, Sonic Sonar & Countermeasures, 64-68; dir eng, Honeywell Marine Systs Ctr, 68-79, advan prog mgr, 81-85, chief sci consult, 85-86, sci tech adv, Honeywell-Elac, Kiel, WGer, 79-81,; infrared radar, US Navy Aircorp, 44-47; RETIRED. *Mem:* Inst Elec & Electronics Engrs. *Res:* Acoustic communication between submarines; active and passive means of detecting submarines; digital computers and digital signal processing techniques as applied to sonar devices; acoustic warfare techniques; numerous patents in areas of signal processing, communications and ocean engineering. *Mailing Add:* 7111 35th Ave NW Seattle WA 98117

ISACHSEN, YNGVAR WILLIAM, b Oslo, Norway, Mar 16, 20; nat US; m 44; c 3. GEOLOGY. *Educ:* Syracuse Univ, BA, 42; Washington Univ, St Louis, MA, 49; Cornell Univ, PhD(geol), 53. *Prof Exp:* Instr geol, Lafayette Col, 49-51; dist geologist Colo plateau, AEC, 53-57; radioactive minerals expert, UN Tech Asst Mission, Turkey, 57; prof geol, State Univ NY Col Plattsburgh, 57-58; assoc geologist, 58-74, PRIN GEOLOGIST, NY STATE GEOL SURVEY, 74- *Concurrent Pos:* Adj assoc prof, Rensselaer Polytech Inst, 64-67, adj prof, 67-; adj prof, State Univ NY Albany, 75-86. *Honors & Awards:* Photog Interpretation Award, Am Soc Photogram, 74. *Mem:* Fel AAAS; fel Geol Soc Am; Soc Econ Geol; Am Geophys Union; Int Basement Tectonics Asn. *Res:* Metamorphic, igneous, structural, economic geology and neotectonics of New York including remote sensing of natural resources; geology of uranium in Colorado Plateau, Colombia and Turkey. *Mailing Add:* New York Geol Surv State Educ Dept Albany NY 12234

ISACKS, BRYAN L, b New Orleans, La, July 25, 36; m 58; c 3. SEISMOLOGY. *Educ:* Columbia Univ, AB, 58, PhD(seismol), 65. *Prof Exp:* Res asst seismol, Lamont Geol Observ, Columbia Univ, 62-65, res assoc, 65-68; res geophysicist, Earth Sci Labs, Environ Sci Serv Admin Res Labs, US Dept Com, 68-71; ASSOC PROF GEOL SCI, CORNELL UNIV, 71- *Mem:* Am Geophys Union; Soc Explor Geophys; Seismol Soc Am. *Res:* Analysis of seismicity; seismicity of island arc regions; deep-focus earthquakes. *Mailing Add:* Dept Geol Sci Cornell Univ Ithaca NY 14853

ISADA, NELSON M, b Dao, Philippines, July 29, 23; m 52; c 3. ENGINEERING, STRUCTURAL DYNAMICS. *Educ:* Univ Mich, BS, 50, MS, 52, PhD(eng), 56. *Prof Exp:* Engr, Smith, Hinchman & Grylls, Inc, Mich, 55-56, Giffels & Vallet, Inc, 56-57 & Harley-Ellington-Day, Inc, 57; asst prof graphics & mech, Syracuse Univ, 57-59; assoc prof shock & vibration, 59-77, PROF SHOCK & VIBRATION, STATE UNIV NY, BUFFALO, 77- *Concurrent Pos:* Consult, Konski Engrs, 57-59, Nat Gypsum Res Lab, 60-61, Bell Aerosysts Co, 63-64, Electronic Assocs, Inc, 65-66 & lawyers in Niagara Frontier, 66-; res assoc, Cornell Aeronaut Lab, 63- *Mem:* Am Soc Civil Engrs. *Res:* Vehicle ride; crash mechanics; shock; vibration; structural dynamics; ultrasonics; earthquake engineering. *Mailing Add:* Dept Eng & Aerospace State Univ NY 608 Furnas Bldg Buffalo NY 14260

ISAKOFF, SHELDON ERWIN, b Brooklyn, NY, May 25, 25; m 46; c 1. MATERIALS SCIENCE ENGINEERING. *Educ:* Columbia Univ, BS, 45, MS, 47, PhD(chem eng), 52. *Prof Exp:* Res engr, Eng Res Lab, E I du Pont de Nemours & Co, Inc, 51-54, res supvr, 55-58, res mgr appl physics, 58-60, asst dir, Mech Res Lab, 61-62, dir, Eng Mat Lab, 63-69, dir, Eng Physics Lab, 70-73, asst dir eng res & develop, 74, dir eng res & develop, 75-90; RETIRED. *Concurrent Pos:* Dir, Am Inst Chem Eng, 77-79; adv, NSF, Univ Pa, Columbia Univ. *Honors & Awards:* Inst Lectr Award, Am Inst Chem Eng, 84; Reilly Lectr, Notre Dame Univ, 89. *Mem:* Nat Acad Eng; fel Am Inst Chem Engrs (vpres, 89, pres); Am Chem Soc; fel AAAS; Am Inst Chem Eng. *Res:* Heat transfer; fluid mechanics; process dynamics and process control; polymer processing; engineering materials. *Mailing Add:* PO Box 6090 Newark DE 19714-6090

ISAKS, MARTIN, b Riga, Latvia, Sept 24, 35. PHYSICAL ORGANIC CHEMISTRY. *Educ:* Purdue Univ, BS, 57; Iowa State Univ, MS, 60; Univ Cincinnati, PhD(chem), 63. *Prof Exp:* Res fel org chem, Ga Inst Technol, 63-64; res assoc, Brown Univ, 64-65; asst prof, 65-71, ASSOC PROF ORG CHEM, UNIV LOWELL, 71- *Concurrent Pos:* Vis prof, Univ Toronto, 82-84. *Mem:* AAAS; Am Chem Soc; The Chem Soc; Sigma Xi. *Res:* Carbenes; oxidation mechanisms; acidity functions; Hammett equation; cyclopentadienones; reaction mechanisms; photochemistry. *Mailing Add:* Dept Chem Univ Lowell Lowell MA 01854

ISAMAN, FRANCIS, b Lewiston, Idaho, Jan 18, 30; m 56; c 1. PHYSICS. *Educ:* Univ Utah, BA, 51; San Fernando Valley State Col, MS, 68. *Prof Exp:* Res engr, Hughes Aircraft Co, 54-57; asst to mgr, Electronics Div, Statham Instruments Inc, 57-60; pres, Pac Telemetry Systs, Inc, 60-62; mgr, Instrument Div, Scionics Corp, 62-64; res engr, Rocketdyne Div, NAm Aviation, Inc, 64-65; PRES, CELTIC INDUSTS, INC, VAN NUYS, 65- *Res:* Electronics and materials science. *Mailing Add:* Celtic Industs Inc 14654 Keswick St Van Nuys CA 91405

ISARD, HAROLD JOSEPH, b Philadelphia, Pa, Nov 22, 10; m 34; c 2. MEDICINE, RADIOLOGY. *Educ:* Temple Univ, MD, 34. *Prof Exp:* Chmn div radiol, 50-76, EMER CHMN DIV RADIOL, ALBERT EINSTEIN MED CTR, 76- CLIN PROF RADIOL, SCH MED, TEMPLE UNIV, 65- *Concurrent Pos:* Consult, Vet Admin Hosp, Coatesville, Pa, 47- *Mem:* NY Acad Sci; fel Am Acad Thermology (pres, 75-76); Am Roentgen Ray Soc; fel Am Col Radiol; Radiol Soc N Am. *Res:* Clinical radiology; mammography; diaphanography; thermography. *Mailing Add:* Albert Einstein Med Ctr York & Tabor Rds Philadelphia PA 19141

ISBELL, ARTHUR FURMAN, b Lubbock, Tex, Feb 12, 17; m 42; c 3. ORGANIC CHEMISTRY. *Educ:* Baylor Univ, BA, 37; Univ Tex, MA, 41, PhD(chem), 43. *Prof Exp:* Lab asst chem, Baylor Univ, 36-37; anal chemist, First Tex Chem Mfg Co, 37-39; instr chem & quiz-master, Univ Tex, 39-42; res chemist, Gen Mills, Inc, Minn, 43-50; chief chemist, Buckman Labs, Inc, Tenn, 50-51; sr res chemist, Monsanto Chem Co, Ala, 51-53; from asst prof to assoc prof chem, Tex A&M Univ, 53-61, prof, 61-77; RETIRED. *Mem:* Am Chem Soc; Am Inst Chemist; Am Asn Univ Profs; Sigma Xi. *Res:* Synthesis of anticonvulsants; amino acids; quinoline derivatives; surface active agents; organo-phosphorus compounds; indexing of scientific information. *Mailing Add:* 800 Delma Circle Bryan TX 77802

ISBELL, HORACE SMITH, b Denver, Colo, Nov 13, 98; m 30. CARBOHYDRATE CHEMISTRY. *Educ:* Univ Denver, BS, 20, MA, 23; Univ Md, PhD(org chem), 26. *Prof Exp:* Asst chemist, Am Smelting & Refining Co, 20-21; asst chemist, Bur Animal Indust, USDA, 23-25; res chemist, Nat Bur Standards, 27-57, chief org chem sect, 57-68; SR RES SCIENTIST, DEPT CHEM, AM UNIV, 68- *Honors & Awards:* Hillebrand Prize, Am Chem Soc, 52, Hudson Award, 54; W N Haworth Medal, Chem Soc London, 73. *Mem:* Am Chem Soc; NY Acad Sci. *Res:* Sugars and sugar derivatives; polyhydroxy acids; mutarotation reactions; aldo condensations; effect of neighboring groups and electronic structures on the course of chemical reactions; C14 and tritium-labeled carbohydrates; peroxy-radical reactions of carbohydrates. *Mailing Add:* 3401 38th St NW Apt 216 Washington DC 20016

ISBELL, JOHN ROLFE, b Portland, Ore, Oct 27, 30; m 60; c 3. MATHEMATICS. *Educ:* Univ Chicago, BS, 51; Princeton Univ, PhD(math), 54. *Prof Exp:* Fel, NSF Inst Adv Study, 56-57; asst prof math, Univ Wash, Seattle, 57-59, from assoc prof to prof, 59-65; prof, Case Western Reserve Univ, 65-69; PROF MATH, STATE UNIV NY BUFFALO, 69- *Mem:* Am Math Soc. *Res:* Categories; topology; theory of games. *Mailing Add:* Dept Math Diefendorf Hall State Univ NY Buffalo NY 14214-3093

ISBELL, RAYMOND EUGENE, b Colbert Co, Ala, Jan 13, 32; m 53; c 2. ORGANIC CHEMISTRY. *Educ:* Florence State Col, BS, 53; Univ Ala, MS, 57, PhD(phys chem), 59. *Prof Exp:* Lab instr, Univ Ala, 56-57; chemist, Interior Ballistics Res, Rohm & Haas Co, 58-60 & Chem Process Res, 60; res chemist, Fundamental Res Br, Tenn Valley Authority, 60-65, proj leader, 65; assoc prof, 65-68, prof, Florence State Univ, 68-74, coord chem dept, 71, PROF CHEM, UNIV NALA, 74-, HEAD DEPT, 73- *Mem:* Am Chem Soc. *Res:* Mechanism of organic reactions. *Mailing Add:* 101 Hiram St Sheffield AL 35660-1739

ISBERG, CLIFFORD A, b Tomahawk, Wis, May 25, 35; m 57; c 5. ELECTRICAL ENGINEERING, COMPUTER SCIENCE. *Educ:* Univ Alaska, BS, 57; Stanford Univ, MS, 58, PhD(elec eng), 65. *Prof Exp:* Adv engr, IBM Corp, 59-70; sr res engr, Comput Synectics, 70; consult, Comput Performance, 70-71; PROD PLANNER COMPUT, MEMOREX CORP, SANTA CLARA, 71- *Concurrent Pos:* Sr res engr, SRI Int, 70- *Mem:* AAAS; Asn Comput Mach; Inst Elec & Electronics Engrs; Sigma Xi. *Res:* Computer systems performance; automata theory; metamathematics; formal linguistics; simulation. *Mailing Add:* 1521 Upper Ellen Rd Los Gatos CA 95030

ISBIN, HERBERT S(TANFORD), b Seattle, Wash, Dec 8, 19; m 48; c 4. CHEMICAL ENGINEERING. *Educ:* Univ Wash, BS, 40, MS, 41; Mass Inst Technol, ScD, 47. *Prof Exp:* Chem engr, Md Res Labs, 43-45; res assoc chem eng, Mass Inst Technol, 45-47; chem engr, Hanford Works, Gen Elec Co, 47-50; prof, 50-83, EMER PROF CHEM ENG, UNIV MINN, MINNEAPOLIS, 83- *Concurrent Pos:* Mem, Adv Comt Reactor Safeguards, Nuclear Regulatory Comn. *Mem:* AAAS; Am Inst Chem Engrs; Am Soc Eng Educ; Am Chem Soc. *Res:* Two-phase flow; nuclear safety; radiation chemistry. *Mailing Add:* 2815 Monterey Pkwy Minneapolis MN 55416

ISBISTER, ROGER JOHN, b Waterville, Maine, May 18, 42; m 64; c 2. POLYMER CHEMISTRY. *Educ:* Colby Col, BA, 64; Univ Colo, Boulder, PhD(org chem), 68. *Prof Exp:* Res chemist, 68-74, GROUP LEADER PACKAGING POLYMERS, MORTON-NORWICH PROD, INC, 74- *Mem:* Am Chem Soc. *Res:* Small ring organic nitrogen compounds; polymer latices; barrier films and foils; gel permeation chromatography. *Mailing Add:* 14817 Dogwood Lane Woodstock IL 60098-9756

ISBRANDT, LESTER REINHARDT, b Chicago, Ill, Jan 29, 46; m 67; c 1. PHYSICAL CHEMISTRY, NUCLEAR MAGNETIC RESONANCE. *Educ:* Northern Ill Univ, BS, 67; Mich State Univ, PhD(phys chem), 72. *Prof Exp:* Staff chemist, Gulf Res & Develop Co, 72-73; staff chemist, 73-77, ANALYTICAL SECT HEAD, PROCTER & GAMBLE CO, 77- *Concurrent Pos:* Lectr, Xavier Univ, 75. *Mem:* Am Chem Soc. *Res:* Application of nuclear magnetic resonance spectroscopy to organic and inorganic chemical systems for molecular structure identification and characterization. *Mailing Add:* Drackett Co 5020 Spring Grove Ave Cincinnati OH 45232-1926

ISEBRANDS, JUDSON G, b Los Angeles, Calif, Oct 20, 43; m 65; c 2. WOOD SCIENCE & TECHNOLOGY, PHYSIOLOGY OF PLANTS & TREES. *Educ:* Iowa State Univ, BS, 65, PhD(wood sci & technol), 69. *Prof Exp:* TREE PHYSIOLOGIST, FORESTRY SCI LAB, NCENT FOREST EXP STA, US FOREST SERV, 69- *Mem:* Soc Am Foresters; Tech Asn Pulp & Paper Indust; Int Asn Wood Anat; Soc Wood Sci & Technol; Bot Soc Am. *Res:* Physiology of forest trees; application of statistics to forestry research. *Mailing Add:* NCent Forest Exp Sta US Forest Serv PO Box 898 Rhinelander WI 54501

ISELER, GERALD WILLIAM, b Port Hope, Mich, May 23, 38; m 60; c 3. MATERIALS SCIENCE. *Educ:* Northwestern Univ, BS, 61, PhD(sci eng), 66. *Prof Exp:* MEM RES STAFF MAT SCI, MASS INST TECHNOL LINCOLN LAB, 66- *Mem:* Am Phys Soc. *Res:* Luminescence due to isoelectronic centers in II-VI compounds; donor levels in compound semiconductors; crystal growth of nonlinear optical materials and indium phosphide and their optical and electrical properties. *Mailing Add:* 26 State St Chelmsford MA 01824

ISELIN, DONALD G, b Racine, Wis, Sept 5, 22. CIVIL ENGINEERING. *Educ:* US Naval Acad, BS, 45; Rensselaer Polytech Inst, MSC, 48. *Prof Exp:* Vpres, Raymond Kaiser Engrs, 81-84, sr vpres, 85; CONSULT, 85- *Mem:* Nat Acad Eng; fel, Am Soc Military Engrs (pres, 78-79). *Mailing Add:* 2695 Sycamore Canyon Rd Santa Barbara CA 93108

ISELY, DUANE, b Bentonville, Ark, Oct 24, 18; m 40; c 2. PLANT TAXONOMY, ECONOMIC BOTANY. *Educ:* Univ Ark, BA, 38, MS, 39; Cornell Univ, PhD(econ & taxon bot), 42. *Prof Exp:* Asst, Univ Ark, 38-39; asst bot, Cornell Univ, 40-42, instr, 42-43; sr seed analyst, Ala Dept Agr, 43-44, exten assoc, 44-46, from asst prof to prof, 46-81, DISTINGUISHED PROF, IOWA STATE UNIV, 81- *Concurrent Pos:* Ed, Iowa State J Res, 78-87. *Honors & Awards:* Award, Asn Off Seed Anal, 65. *Mem:* Soc Econ Bot; Asn Off Seed Anal (pres, 53); Int Asn Plant Taxonomists; Am Soc Plant Taxon. *Res:* Leguminosae of the United States, a taxonomic summary; plant taxon; economic utilization of legumes; history of botany. *Mailing Add:* Dept Bot Iowa State Univ Ames IA 50011

ISENBERG, ALLEN (CHARLES), b Philadelphia, Pa, Aug 18, 38; m 67. PHARMACEUTICAL CHEMISTRY, INFORMATION SCIENCE. *Educ:* Temple Univ, BSc, 60; Univ Wis, MSc, 62, PhD(pharmaceut chem), 65. *Prof Exp:* Res asst pharmaceut chem, Univ Wis, 60-65; res assoc med chem, Univ Calif, 65-66; vis asst lectr pharmaceut chem, Univ Strathclyde, 66-67; asst prof pharmacog & pharmaceut chem, Temple Univ, 67-70; assoc ed, 70-73, info scientist, 73-78, sr assoc ed, 78-87, SR ED, CHEM ABSTR SERV, 87- *Mem:* Am Chem Soc. *Res:* Chemical information science; chemical nomenclature. *Mailing Add:* Chem Abstr Serv Columbus OH 43210

ISENBERG, GEORGE RAYMOND, JR, b Altoona, Pa, Apr 12, 29; m 62. BIOLOGY. *Educ:* Lock Haven State Col, BS, 54; Pa State Univ, MEd, 59, DEd, 66. *Prof Exp:* Teacher high schs, Pa, 54-58; instr, 60-61, from asst prof to assoc prof, 61-69, PROF BIOL, STATE UNIV NY COL POTSDAM, 69- *Mem:* NY Acad Sci; Sigma Xi. *Res:* Microscopy of albatross nasal gland; freshwater mussels; physiological effects of high altitudes; ultrastructural studies on cell communications. *Mailing Add:* Dept Biol State Univ NY Col Potsdam NY 13676

ISENBERG, HENRY DAVID, b Giessen, Ger, Mar 9, 22; nat US; m 48; c 2. MEDICAL MICROBIOLOGY. *Educ:* City Col New York, BS, 47; Brooklyn Col, MA, 51; St John's Univ, NY, PhD, 59; Am Bd Med Microbiol, dipl, 68. *Prof Exp:* Asst dir labs, Labs of Dr A Angrist, 47-54; asst prof orthop surg, State Univ NY Downstate Med Ctr, 62-65, clin assoc prof, 65-71; prof clin path, Sch Basic Sci, State Univ NY Stony Brook, 71-89; clin prof microbiol & immunol, Sch Med, Univ SFla, 82-87; microbiologist, Long Island Jewish Med Ctr, 54-69, attend microbiologist, 69-77, PROF LAB MED, ALBERT EINSTEIN COL MED, 87- , CHIEF MICROBIOL, LONG ISLAND JEWISH MED CTR, 77- *Concurrent Pos:* Res assoc, Sloan-Kettering Inst Cancer Res, 59-63; consult, Health Care Facil Serv, Health Servs & Ment Health Admin, 68-73; prof lectr orthop surg, State Univ NY Downstate Med Ctr, 71-; ed, J Clin Microbiol, Am Soc Microbiol, 75-79, Critical Reviews Microbiol, 78-82 & ed-in-chief, J Clin Microbiol, 79-89; consult, Biomed Opers & Res Br, NASA, 90- *Honors & Awards:* Becton-Dickinson Award, Am Soc Microbiol, 79, Alexander Sunnenwirth Mem Lectr, 89; Kimble Award, 80. *Mem:* Fel NY Acad Sci; fel Am Inst Chem; fel Asn Clin Sci; Am Bd Med Microbiol; fel Am Acad Microbiol; assoc fel NY Acad Med; fel Infectious Dis Soc. *Res:* Microbial metabolism; host-parasite relationships; mineral deposition; antibiotics; automation; nosocomial disease. *Mailing Add:* Div Microbiol Long Island Jewish Med Ctr Albert Einstein Col Med New Hyde Park NY 11042-1433

ISENBERG, IRVING HARRY, b Buffalo, NY, Sept 11, 09; m 41; c 3. WOOD ANATOMY. *Educ:* State Univ NY, Syracuse, BS, 31, MS, 32; Univ Calif, Berkeley, PhD(plant biochem), 36. *Hon Degrees:* MS, Lawrence Univ, 73. *Prof Exp:* Asst, State Univ NY Col Environ Sci & Forestry, Syracuse, 31-32; Pac SW Forest & Range Exp Sta, US Forest Serv, Univ Calif, 32-34, tech asst forestry, 35-37; tech assoc, Inst Paper Chem, 37-38, res asst wood technol, 38-41, res assoc, 41-55, res assoc wood technol & fiber micros, 55-73, group leader wood technol, 41-46, group leader wood technol & fiber micros, 46-53, instr wood technol, 38-55, instr fiber micros, 46-55, assoc prof wood technol & fiber micros, 55-73, EMER ASSOC PROF WOOD TECHNOL & FIBER MICROS, INST PAPER CHEM, 73- *Concurrent Pos:* Consult, 73-83; instr, Fox Valley Tech Inst, 75-76; vis prof, Inst Paper Chem, 79. *Mem:* Soc Am Foresters; Soc Wood Sci & Technol; Tech Asn Pulp & Paper Indust; Int Asn Wood Anat. *Res:* Pulp and paper microscopy; wood structure and fiber identification; wood chemistry; wastewater treatment. *Mailing Add:* 529 N Linwood Ave Appleton WI 54914-3360

ISENBERG, LIONEL, b Detroit, Mich, Feb 26, 25; m 49; c 3. ENERGY SYSTEMS ENGINEERING, LIQUID SYNTHETIC FUELS. *Educ:* Univ Calif, Los Angeles, BS(chem) & BS(elec eng), 50; Alexander Hamilton Inst, MBA, 70. *Prof Exp:* Plant mgr chem, Prudential Chem Co, 50-55; asst div mgr struct mat, Aerojet-Gen Corp, 55-63; mgr tech transfer, Rockwell Int, 63-71; mgr energy prog, Fairchild Industs, 76-80; PRES, IR ASSOCS, 70-; TECH MGR SYST ENG, CALIF TECH/JPL, 80- *Concurrent Pos:* Consult, Proj Gold Eagle, USAF, 65. *Honors & Awards:* Polaris Citation, Navy, 63; Presidential Award for Innovation, 71 & 72. *Mem:* Assoc fel Am Inst Aeronaut & Astronaut; Am Inst Chem Engrs; Am Chem Soc; fel Am Inst Chemists; AAAS; Sigma Xi. *Res:* Heat transfer and insulation systems, especially nuclear cryogenic and ultra high temperatures; synthetic fuels, liquid natural gas transport and storage; ablative systems; acoustic control; chemical process; netted computer systems; space nuclear power; author of two books, over 180 publications and over 50 patents. *Mailing Add:* 1205 Sunbird Ave La Habra CA 90631

ISENBERG, NORBERT, b Saarbruecken, Ger, June 17, 23; nat US; c 4. ORGANIC CHEMISTRY. *Educ:* Columbia Univ, BA, 48, MA, 50; Rensselaer Polytech Inst, PhD(org chem), 63. *Prof Exp:* Asst chem, Columbia Univ, 49-50; from instr to assoc prof, Skidmore Col, 50-64, dir NSF undergrad res prog, 60-63; from asst prof to assoc prof, Univ Wis-Parkside, 64-69, actg chmn div sci, 68-69, chmn, 69-71, 73-76, prof chem, 69-90, EMER PROF CHEM, UNIV WIS-PARKSIDE, 90- *Concurrent Pos:* Fel oncol, Univ Wis, 63-64; vis prof, Univ Wis-Madison, 71-72. *Mem:* Fel AAAS; Am Chem Soc; Sigma Xi. *Res:* Organic sulfur compounds; heterocyclic compounds; synthetic organic chemistry; precipitation with thioacetamide; undergraduate research projects. *Mailing Add:* 4118 Pennington Lane Racine WI 53403

ISENECKER, LAWRENCE ELMER, b Cleveland, Ohio, Feb 8, 24. MATHEMATICS. *Educ:* Xavier Univ, Ohio, LittB, 46; West Baden Col, PhL, 50, STL, 58; Cath Univ Am, MS, 54, PhD(math), 62. *Prof Exp:* Teacher math, Loyola Acad, Ill, 50-52; asst prof, 63-68, ASSOC PROF MATH, XAVIER UNIV, OHIO, 69- *Mem:* Am Math Soc; Math Asn Am. *Res:* Mathematical analysis; foundations of mathematics. *Mailing Add:* Dept Math Xavier Univ Cincinnati OH 45207

ISENHOUR, THOMAS LEE, b Statesville, NC, Jan 29, 39; m 60; c 2. ANALYTICAL CHEMISTRY. *Educ:* Univ NC, BS, 61; Cornell Univ, PhD(anal chem), 65. *Prof Exp:* Asst prof chem, Univ Wash, 65-69; assoc prof, Univ NC, Chapel Hill, 69-74, chmn dept, 75, prof chem, 74-; dean, Chem Dept, Utah State Univ, 84-87; DEAN, KANS STATE UNIV, 87- *Concurrent Pos:* Sloan res fel, 71; I M Kolthoff vis prof, Hebrew Univ, 81. *Honors &*

Awards: Am Award Anal Chem, 83. *Mem:* Am Chem Soc; Pattern Recognition Soc. *Res:* Computerized chemical information processing, search and retrieval system, molecular structure encoding; computerized learning machines: application to pattern recognition and interpretation in mass, infrared, and gamma-ray spectra; metal chelate mass spectrometry. *Mailing Add:* Col Arts & Sci Kans State Univ Manhattan KS 66506

ISENOR, NEIL R, b Dutch Settlement, NS, Jan 6, 32; c 3. PHYSICS. *Educ:* Acadia Univ, BSc, 54; McMaster Univ, MSc, 55, PhD(physics), 59. *Prof Exp:* Lectr physics, Univ NB, 55-56; asst prof, Bishop's Univ, Can, 59-61; from asst prof to assoc prof, 61-73, assoc dean, 76-77, chmn dept, 77-82, PROF PHYSICS, UNIV WATERLOO, 73- *Concurrent Pos:* Nat Res Coun Can res grant, 62-; vis res assoc, Univ Rochester, 65; vis res fel, Univ New S Wales, 82-83, Macquarie Univ, 88-89. *Mem:* Am Phys Soc; Can Asn Physicists; Optical Soc Am. *Res:* Laser interactions with matter; quantum optics. *Mailing Add:* Dept Physics Univ Waterloo Waterloo ON N2L 3G1 Can

ISENSEE, ALLAN ROBERT, b Sparta, Wis, Dec 25, 39; m 64; c 2. SOIL SCIENCE, PLANT PHYSIOLOGY. *Educ:* Wis State Univ, Stevens Point, BS, 62; Univ Wis, Madison, MS, 65, PhD(soils), 68. *Prof Exp:* Res asst soils, Univ Wis, Madison, 62-67; PLANT PHYSIOLOGIST, BELTSVILLE AGR RES SERV, USDA, 67- *Mem:* Am Soc Agron; Weed Sci Soc Am; Am Chem Soc; Sigma Xi. *Res:* Persistence and absorption of pesticides in soil; influence of soil and pesticide properties on plant uptake of pesticides; development and use of model ecosystems to determine the fate and behavior of pesticides in the aquatic environment. *Mailing Add:* US Dept Agr Beltsville Agr Res Ctr Bldg 050 Headhouse One Beltsville MD 20705

ISENSEE, ROBERT WILLIAM, b Portland, Ore, Nov 2, 19; m 43; c 2. ORGANIC CHEMISTRY. *Educ:* Reed Col, BA, 41; Ore State Col, MA, 43, PhD(org chem), 48. *Prof Exp:* Res chemist, Hercules Powder Co, Del, 43-44, indust chemist, Va, 44-45; instr chem, Ore State Univ, 47-48; From asst prof to prof, San Diego State Univ, 48-82, chmn dept, 58-61; RETIRED. *Mem:* Am Chem Soc. *Res:* Synthesis of organic heterocycles; catalytic hydrocarbon oxidation; fine chemicals; explosives; synthesis of a new amino alcohol derived from quinazoline; acid-base equilibria studies. *Mailing Add:* 5036 Art San Diego CA 92115

ISERI, LLOYD T, b Los Angeles, Calif, Aug 31, 17; m 40; c 2. MEDICINE. *Educ:* Univ Calif, BA, 39; Wayne State Univ, MD, 44; Am Bd Internal Med & Cardiol, dipl. *Prof Exp:* Resident, Receiving Hosp, Detroit, 45-48; instr internal med, Col Med, Wayne State Univ, 48-51, res assoc, 49-51, from asst prof to assoc prof med, 51-57; assoc clin prof, Col Med Evangelists, 57-63; assoc prof, Calif Col Med, 63-70; asst chmn dept, 70-72, prof, 70-88, EMER PROF MED, UNIV CALIF, IRVINE-CALIF COL MED, 88- *Concurrent Pos:* Fel, Receiving Hosp, Detroit, 45-48, jr assoc, 49-; staff physician, Yales Mem Clin, 51-52; attend staff, Vet Admin Hosp, Dearborn, Mich, 52-53, consult, 53-; consult med clin, Vet Admin Regional Off, 53-; chief cardiovasc res, Rancho Los Amigos Hosp, Downey, 57-67; dir med serv, Col Med, Univ Calif, Irvine, 69-70, chmn div cardiol, 69-74. *Mem:* Fel Am Col Physicians; fel Am Col Cardiol; Am Fedn Clin Res. *Res:* Fluid and electrolyte metabolism in cardiovascular renal diseases; magnesium in cardiovasc diseases. *Mailing Add:* Dept Med Univ Calif Irvine CA 92717

ISERI, OSCAR AKIO, b Thomas, Wash, Aug 23, 27; m 61; c 3. PATHOLOGY. *Educ:* Antioch Col, BS, 52; Harvard Univ, MD, 56; Am Bd Path, dipl, 64. *Prof Exp:* Intern internal med, King County Hosp, Seattle, Wash, 56-57; asst resident, Univ Wash, 57-58, trainee exp path, 58-61, res instr path, 61-62; USPHS res fel, Mallory Inst Path, Boston City Hosp, 62-64, asst pathologist, Inst, 64-69; from asst prof to assoc prof path, Tufts Univ, 64-74; PROF PATH, SCH MED, UNIV MD, BALTIMORE, 74-; STAFF PATHOLOGIST, VET ADMIN HOSP, 74- *Concurrent Pos:* Lectr, Harvard Med Sch, 65-74; asst physician path, Boston City Hosp, 65-69, assoc vis physician, 69-74; assoc pathologist, Mallory Inst Path, 69-74. *Mem:* AAAS; Am Asn Pathologists; Int Soc Stereology; Electron Micros Soc Am; Am Asn Study Liver Dis; Int Acad Path. *Res:* Ultrastructural and chemical pathology; liver and gastro-intestinal pathology. *Mailing Add:* Vet Admin Med Ctr 3900 Loch Raven Blvd Baltimore MD 21218

ISERSON, KENNETH VICTOR, b Washington, DC, Apr 8, 49; m 73. EMERGENCY MEDICINE, BIOETHICS. *Educ:* Univ Md Col Park, BS, 71; Univ Md, Baltimore, MD, 75; Univ Phoenix, Tucson, MBA, 87. *Prof Exp:* Clin assoc prof emergency med, Scott & White Clin, Tex A&M Col Med, 80-81; ASSOC PROF EMERGENCY MED, UNIV ARIZ COL MED, 81- *Concurrent Pos:* Bd dirs, Wilderness Med Soc & Ariz Bioethics Network, 87-; chmn practice mgt comt, Am Col Emergency Physicians, 87-89; mem, Coun Med Sch, Am Med Asn, 88-; bioethics fel, Univ Chicago, 90-91. *Mem:* Am Col Emergency Physicians; Am Med Asn; Am Acad Med Dirs; Soc Acad Emergency Med; Wilderness Med Soc; Soc Health & Human Values; Soc Bioethics Consult. *Res:* Evaluation of current and new methods of emergency vascular access; health policy and bioethics; rapid blood warming. *Mailing Add:* Dept Surg Emergency Med Sect 1501 N Campbell Ave Tucson AZ 85724

ISETT, ROBERT DAVID, b Jenkintown, Pa, Feb 23, 42; m 84; c 3. CLINICAL PSYCHOLOGY, SOCIAL GERONTOLOGY. *Educ:* La Salle Col, BA, 65; Villanova Univ, MA, 67; Peabody-Vanderbilt Univ, PhD(clin psychol), 73. *Prof Exp:* Psychologist, State Off Ment Retardation, Pa, 72-74; proj dir, dept psychiat, Women's Med Col, 74-76; dir res, Woodhaven Ctr, 76-82, dir psychol, 77-78, dir res, Inst Aging, Temple Univ, 82-86; DIR, INSTNL RES & PLANNING, BUCKS COUNTY COMMUNITY COL, 86- *Concurrent Pos:* Adj assoc prof educ psychol, Temple Univ, 77-; assoc ed, Eval Health Professions, 81-; mem, Pa Develop Disabilities Coun, 83-85; prin investr, minority aging proj, Philadelphia AAA, 84-86 & Pa Dept Aging, 86-; co-dir, Psychol Serv Assocs, 85-; mem, Pa Bd Psychologist Examr, 85-87. *Mem:* Am Psychol Asn. *Res:* Minority aging issues; aging and developmental disabilities; respite care services to the elderly; services to families caring for relatives with Alzheimers disease; mental retardation. *Mailing Add:* Bucks County Community Col Newtown PA 18940

ISGUR, BENJAMIN, b Boston, Mass, June 19, 11; m 38; c 2. SOIL CONSERVATION. *Educ:* Mass State Col, BS, 33, MS, 35, PhD(agron), 40. *Prof Exp:* Instr agron, Mass State Col, 35-38; agronomist, 38-42, soil conservationist, 42-45, dist conservationist, 45-57, dep state conservationist, Mass, 55-57, STATE CONSERVATIONIST, MASS, SOIL CONSERV SERV, USDA, 57- *Concurrent Pos:* Asst dir & agr adv, Dominican Repub Settlement Asn, 42-43; adj prof resources planning, Univ Mass, Amherst, 71- *Mem:* Fel AAAS; fel Soil Conserv Soc Am. *Res:* Crops; soils; soil colloids; vitamins in relation to fertilizers; vitamins in plants related to maturity; plant oxidants; mineral balance in plants; environmental quality planning of the natural resources base; environmental quality indices. *Mailing Add:* 79 Maynard Rd Northampton MA 01060

ISGUR, NATHAN, b Houston, Tex, May 25, 47. THEORETICAL SUBATOMIC PHYSICS. *Educ:* Calif Inst Technol, BS, 68; Univ Toronto, PhD(physics), 74. *Prof Exp:* Fel physics, 74-76, from asst prof to assoc prof, 76-82, PROF PHYSICS, UNIV TORONTO, 82-; THEORY GROUP LEADER, CONTINUOUS ELECTRON BEAM ACCELERATOR FACIL, 90- *Honors & Awards:* E W R Steacie Prize, Natural Sci Eng Res Coun, Can,; Rutherford Medal, Royal Soc Can, 89. *Mem:* Can Asn Physicists; Am Inst Physics; Can Inst Particle Physics; fel Royal Soc Can; fel Am Phys Soc. *Res:* Quantum chromodynamics; weak interactions; theories of elementary interactions. *Mailing Add:* Continuous Electron Beam Accelerator Facil 1200 Jefferson Ave Newport News VA 23606

ISH, CARL JACKSON, b Mansfield, Ohio, Mar 20, 19; m 42; c 2. CHEMISTRY. *Educ:* Ashland Col, BA, 42; Ohio State Univ, MA, 48, PhD, 55. *Prof Exp:* Process engr, US Rubber Co, 42-43; sr chemist, Crown Cent Petrol Corp, 43-45; asst dept chem, Ohio State Univ, 45-47; res engr, Battelle Mem Inst, 48-52, prin chemist, 52-58; assoc ed, 58-68, spec projs mgr, 68-70, asst to ed, 70-71, MANAGING ED, ED PROCESSING, CHEM ABSTR SERV, OHIO STATE UNIV, 71- *Mem:* AAAS; Am Chem Soc; Sigma Xi. *Res:* Physical chemistry of membranes; crystal chemistry; thermal dissociation of inorganic compounds; high purity boron; documentation. *Mailing Add:* 221 Ceramic Dr Columbus OH 43214

ISHAM, ELMER REX, b Bovina, Tex, July 26, 35; m 56; c 2. PLASMA PHYSICS, SOLID STATE PHYSICS. *Educ:* Tex A&M Univ, BS, 58, MS, 61, PhD(physics), 65. *Prof Exp:* Instr physics, Tex A&M Univ, 61-64, instr math, 64-65; from asst prof to assoc prof, 65-75, asst pres, 77-79, PROF PHYSICS, SAM HOUSTON STATE UNIV, 75-, DIR RES, 67-, ASST TO PRES, 79- *Mem:* Am Phys Soc. *Res:* Theoretical plasma physics; luminescence of solids. *Mailing Add:* Dept Physics Sam Houston State Univ Huntsville TX 77341

ISHAQ, KHALID SULAIMAN, b Basra, Iraq, Jan 16, 33; US citizen; m 68. ORGANIC CHEMISTRY. *Educ:* Am Univ Beirut, BS, 56; Univ Minn, Minneapolis, PhD(med chem), 69. *Prof Exp:* Fel lipids res, 68-70, instr, 70-73, asst prof, 73-80, ASSOC PROF MED CHEM, UNIV NC, CHAPEL HILL, 80- *Mem:* Am Chem Soc; Sigma Xi; Am Asn Pharmaceut Scientists. *Res:* Lipid chemistry; anti-tumor agents. *Mailing Add:* Sch Pharm Univ NC Chapel Hill NC 27514

ISHERWOOD, DANA JOAN, b Lewiston, NY; m 70. GEOCHEMISTRY. *Educ:* San Francisco State Univ, BS, 64; Univ Colo, PhD(geol), 75. *Prof Exp:* geochemist, 77-86, SR ANALYST, LAWRENCE LIVERMORE LAB, 86- *Concurrent Pos:* Sci cong fel, Am Geophys Soc, 85-86. *Mem:* AAAS; Am Geophys Soc; Soc Women Geographers; Grad Women Sci. *Res:* Geochemistry of radionuclides in natural systems; soil geochemistry; rock weathering processes; groundwater pollution. *Mailing Add:* Lawrence Livermore Lab PO Box 808 L-1 Livermore CA 94550

ISHERWOOD, WILLIAM FRANK, b Stoneham, Mass, Apr, 30, 41; m 70. GEOTHERMAL RESOURCES, CONTAMINANT HYDROGEOLOGY. *Educ:* Princeton Univ, AB, 63; Univ Utah, MS, 67; Univ Colo PhD(geol sci), 75; Golden Gate Univ, MBA, 82. *Prof Exp:* Geophysicist, Geophys Polar Res Ctr, Univ Wis, 65-67, Stanford Res Inst, 67-70, Nat Oceanic & Atmospheric Admin, 71-73 & US Geol Surv, 73-77; chief geothermal eval, US Geol Surv, 77-79, dep conserv mgr, geothermal conserv div, 79-83; sr geophysicist, Geothermex Inc, 83-87; PROJ LEADER, LIVERMORE NAT LAB, 87- *Concurrent Pos:* Bd dirs, Geothermal Resources Coun, 81. *Mem:* Am Geophys Union; Soc Explor Geophysicists; Nat Water Well Asn. *Res:* Using and integrating geophysical methods for evaluation and understanding of geothermal systems; interpretation of reservoir dynamics from time-changes in the gravitational field. *Mailing Add:* 37 La Encinal Orinda CA 94563

ISHIDA, HATSUO, b Kosai, Japan, Dec 16, 48; m 75; c 2. SURFACE SPECTROSCOPY, COMPOSITE INTERFACE. *Educ:* Doshisha Univ, Japan, BS, 71, MS, 73; Case Western Reserve Univ, PhD(macromolecular sci), 76. *Prof Exp:* Postdoctoral res assoc, Case Western Reserve Univ, 77-78, sr res assoc, 79, asst prof phys chem, 79-83, assoc prof, 84-87, PROF MACROMOLECULAR SCI, CASE WESTERN RESERVE UNIV, 88- *Concurrent Pos:* Dir, CR Newphor Polymer Composite Processing, Case Western Reserve Univ, 80-; gen chmn, Int Conf Composite Interfaces, 83- *Mem:* Am Chem Soc; Soc Appl Spectros; Japan Soc Polymer Sci; Processing Soc; Am Composite Soc; Soc Adv Mat Processing Engrs. *Res:* Molecular characterization techniques to investigate structure of composite interfaces, coatings on metals, and structural changes during processing of composites; synthesize and characterize organic ferro-magnetic polymers. *Mailing Add:* Dept Macromolecular Sci Case Western Reserve Univ 10900 Euclid Ave Cleveland OH 44106-1712

ISHIDA, TAKANOBU, b Kyoto, Japan, Mar 22, 31; m 64; c 2. ISOTOPE EFFECTS, STABLE ISOTOPE SEPARATION. *Educ:* Kyoto Univ, BS, 53, MS, 55; NY Univ, MS, 58; Mass Inst Technol, PhD(nuclear eng), 64. *Prof Exp:* Res assoc chem, Brookhaven Nat Lab, 64-66, Belfer Grad Sch Sci, Yeshiva Univ, 66-68; from asst prof to assoc prof chem, Brooklyn Col, 68-74, prof chem, 74-79; AT CHEM DEPT, STATE UNIV NY, STONY BROOK,

74- *Concurrent Pos:* Res collabr, Brookhaven Nat Lab, 66-; res grant, AEC, Energy Res & Develop Admin & Dept of Energy, 71-; vis assoc prof, Univ Rochester, 73-74, vis prof, 74-79; vis scholar, Max Planck Inst Chem, Mainz, 86. *Mem:* Am Chem Soc; Am Phys Soc. *Res:* Separation of stable isotopes; statistical mechanics of isotope effects and condensed phase. *Mailing Add:* Chem Dept State Univ NY Stony Brook NY 11794-3400

ISHIDA, YUKISATO, b Tokyo, Japan, July 7, 48; m 74; c 2. BIOCHEMISTRY, NEUROSCIENCES. *Educ:* Toyama Univ, BS, 71; Univ Tokyo, MS, 74, PhD(pharmacol), 83. *Prof Exp:* Asst prof vet pharmacol, Fac Agr, Univ Tokyo, 75-77; researcher pharmacol, 77-85, SR RESEARCHER PHYSIOL, MITSUBISHI KASEI INST LIFE SCI, 85- *Concurrent Pos:* Vis asst prof, Dept Physiol Biophys, Col Med, Univ Cincinnati, 84-86; vis scientist, Inst Cell Biol, Swiss Fedn Inst Technol, ETH, Hönggerberg, 89. *Mem:* Corresp mem Am Physiol Soc; fel Japan Pharmacol Soc; fel Japan Physiol Soc; Japan Soc Smooth Muscle Res; Japan Soc Vet Med. *Res:* Energy metabolism of smooth muscle; phosphagen content and tension of muscle; identification of mitochrondrial creatine kinase in smooth muscle; cardiovascular physiology and pharmacology. *Mailing Add:* Mitsubishi Kasei Inst Life Sci Machida Tokyo 194 Japan

ISHIHARA, KOHEI, b Kyoto, Japan, Apr 9, 41; m 65; c 2. ENGINEERING. *Educ:* Kyoto Univ, BS, 65; Okla State Univ, MS, 67, PhD(chem eng), 71. *Prof Exp:* Res engr polymer extrusion, Toyobo Co, 71-72; res engr heat transfer, Chicago Bridge & Iron Co, 72-75; RES ENGR HEAT TRANSFER, HEAT TRANSFER RES, INC, 75- *Mem:* Am Inst Chem Engrs. *Res:* Condensation and convection heat transfer; single and two-phase flow; numerical analysis of fluid flow. *Mailing Add:* 1835 Vistillas Rd Altadena CA 91001

ISHIHARA, TERUO (TERRY), b Ogden, Utah, Apr 9, 27; m 52; c 5. MECHANICAL & AEROSPACE ENGINEERING. *Educ:* Wash State Univ, BS, 49; Univ Calif, Berkeley, teaching cert, 51; San Jose State Col, MS, 58; Univ Ariz, MS, 66, PhD(mech eng), 69. *Prof Exp:* Jr mech engr, US Naval Air Missile Test Ctr, Calif, 51-53; res analyst, Northrop Aircraft Corp, 53-54; teacher math, jr high sch, Calif, 54-55; mech engr, Lawrence Radiation Lab, 55-59; instr eng, San Jose City Col, 59-61; instr math, Monterey Peninsula Col, 61-63; asst prof mech eng, Univ Mo, Columbia, 66-70; assoc prof mech eng, Rose-Hulman Inst Technol, 70-80; PROF MECH ENG & TECHNOL, SAGINAW VALLEY STATE COL, 80- *Mem:* AAAS; Am Soc Eng Educ; Am Soc Mech Engrs; Am Asn Univ Profs. *Res:* Applied mathematics; applications of optimization theory; educational methods in the teaching of engineering; numerical analysis; digital computer applications; engineering mechanics; thermodynamics; heat transfer. *Mailing Add:* Dept Eng & Technol Saginaw Valley State Col Univ Ctr MI 48710

ISHII, DOUGLAS NOBUO, b Santa Anita, Calif, July 30, 42; m 82; c 3. MOLECULAR BASIS OF NERVE DEVELOPMENT & REGENERATION. *Educ:* Univ Calif, Berkeley, BA, 67; Stanford Univ PhD(pharmacol), 74. *Prof Exp:* Instr pharmacol, Univ Pac, San Francisco, 68-70; postdoctoral fel neurobiol, 74-76; from asst prof to assoc prof pharmacol, Columbia Univ, New York City, 76-85; assoc prof, 86-89, PROF PHYSIOL & BIOCHEM, COLO STATE UNIV, FT COLLINS, 89- *Concurrent Pos:* Res career develop award, NIH, 78; prin investr grants, Nat Inst Neurol & Communicative Dis & Stroke, NIH, 78-, ad hoc mem, neurol B study sect, 81, site rev team, Cancer Therapeut Prog Rev Comt, Nat Cancer Inst, 83 & 86, spec reviewer, neurol C study sect, 87, mem, Special Prog Proj Rev Comt, Nat Inst Child Health & Human Develop, 87, spec reviewer, Exp Cardiovasc Sci Study Sect, 89; ad hoc consult, Develop Neurosci Panel, NSF, 85, 88 & 89. *Mem:* Soc Neurosci; Am Soc Pharmacol & Exp Therapeut; AAAS; Am Diabetes Asn; Tissue Cult Asn. *Res:* Neurobiology and molecular mechanism of action of insulin and insulin-like growth factors; role of insulin growth factors in nerve development and regeneration; new theory for pathogenesis of diabetic neuropathy has been developed and is under test. *Mailing Add:* Physiol Dept Colo State Univ Ft Collins CO 80523

ISHII, T(HOMAS) KORYU, b Japan, Mar 18, 27; m 58; c 4. ELECTRICAL ENGINEERING, MICROWAVES. *Educ:* Nihon Univ, Tokyo, BS, 50, DEng, 61; Univ Wis, MS, 57, PhD(elec eng), 59. *Prof Exp:* Elec engr, Japan Naval Res Lab, 45 & Japan Broadcasting Corp, 49; instr elec eng, Nihon Univ, Tokyo, 50-56; from asst prof to assoc prof, 59-64, PROF ELEC ENG, MARQUETTE UNIV, 64- *Honors & Awards:* Inst Elec & Electronics Engrs TC Burnam Awards, 69, Centennial Awards, 84. *Mem:* Inst Elec & Electronics Engrs. *Res:* Microwave applications and electronics; millimeter circuit; millimeter wave amplification, detection and generation; masers and lasers; industrial sensors. *Mailing Add:* Dept Elec Eng Marquette Univ Milwaukee WI 53233

ISHIKAWA, HIROSHI, b Tokyo, Japan, Nov 11, 41; m 68; c 2. DIFFERENTIATION. *Educ:* Jikei Univ, BMed, 68, DMed, 72. *Prof Exp:* Ast anat, Dept Anat, Sch Med, Showa Univ, 72-73, lectr, 73-74; asst prof, Dept Anat, Sch Med, Tohoku Univ, 74-80; PROF ANAT, DEPT ANAT, SCH MED, JIKEI UNIV, 80- *Concurrent Pos:* Med examr, Tokyo Med Examiner's Off, 72-76 & 80- *Honors & Awards:* Prize of Japan Endocrine Soc, Japan Endocrine Soc, 74. *Res:* Functional and morphological studies on the differentiation of anterior pituitary cells in in vivo and in vitro. *Mailing Add:* Dept Anat Sch Med Jikei Univ 3-25-8 Nishishinbashi Minato-ku Tokyo 105 Japan

ISHIKAWA, SADAMU, b Osaka, Japan, July 6, 32; m 61; c 1. MEDICINE, PHYSIOLOGY. *Educ:* Kyoto Med Col, MD, 57. *Prof Exp:* From asst prof to assoc prof path, Univ Man, 67-71; physician in charge pulmonary function, Cent Blood Gas & Lung Physiol Labs, Boston City Hosp, 71-73; ASSOC PROF MED, SCH MED, TUFTS UNIV, 71-; DIR, PULMONARY LAB, ST ELIZABETH HOSP, 81- *Concurrent Pos:* Dir, Pulmonary Function Lab, Lemuel Shattuck Hosp, 73-80. *Mem:* Am Thoracic Soc; fel Am Col Chest Physicians. *Res:* Ecology of man, adaptation of lung and heart to abnormal environment; air pollution and emphysema; structure-function of aging lung. *Mailing Add:* 736 Cambridge St Boston MA 02135

ISHIMARU, AKIRA, b Fukuoka, Japan, Mar 16, 28; US citizen; m 56; c 4. ELECTRICAL ENGINEERING. *Educ:* Univ Tokyo, BS, 51; Univ Wash, PhD(elec eng), 58. *Prof Exp:* From asst prof to assoc prof, 58-65, PROF ELEC ENG, UNIV WASH, 65- *Concurrent Pos:* Vis assoc prof, Univ Calif, Berkeley, 63-64; consult, Boeing Co, Wash, 59- & Jet Propulsion Lab, 64-; ed, Radio Sci, Am Geophys Union, 79-82; mem comn, Int Sci Radio Union; Inst Elec & Electronics Engrs; ed, Waves In Random Media, Inst Physics, UK. *Honors & Awards:* Achievement Award, Inst Elec & Electronics Engrs, 68; Centennial Medal, Inst Elec & Electronics Engrs, 84. *Mem:* Inst Elec & Electronics Engrs. *Res:* Antenna pattern synthesis; propagation and antennas; plasmas; waves in random and turbulent media; bioengineering applications; remote sensing; space communications; ultrasound imaging. *Mailing Add:* Dept Elec Eng FT-10 Univ Wash Seattle WA 98195

ISHIZAKA, KIMISHIGE, b Tokyo, Japan, Dec 3, 25; m 49; c 1. IMMUNOLOGY, IMMUNOCHEMISTRY. *Educ:* Univ Tokyo, MD, 48, DrMedSci(immunol), 54. *Prof Exp:* Res mem, NIH, Tokyo, 50-53, chief div immunoserol, 53-62; from asst prof to assoc prof microbiol, Univ Colo, Denver, 62-70, chief div serol, Children's Asthma Res Inst & Hosp, 62-63, dir dept basic sci, 63-70; prof med & microbiol, Sch Med, Johns Hopkins Univ, 70-81, dir, Sub-dept Immunol & prof immunol & med, 81-89; SCI DIR, LA JOLLA INST ALLERGY & IMMUNOL, CALIF, 89- *Concurrent Pos:* Res fel chem, Calif Inst Technol, 57-59; res fel microbiol, Johns Hopkins Univ, 59; mem adv comt immunol, WHO. *Honors & Awards:* Passano Found Award; Paul Ehrlich & Ludwig-Darmstaedter Prize, WGer; Int Award, Gairdner Found, Can; Emperor's Award, Japan; Borden Award, Asn Am Med Cols; Achievement Award, Am Col Physicians. *Mem:* Foreign assoc Nat Acad Sci; Am Asn Immunol; hon fel Am Acad Allergy; Soc Exp Biol & Med; fel AAAS. *Res:* Molecular bases of hypersensitivity reactions; immunochemical and physicochemical properties of human antibodies, especially reaginic antibodies; regulation of the immunoglobulin-E response. *Mailing Add:* 11149 N Torrey Pines Rd La Jolla CA 92037

ISHIZAKA, TERUKO, b Sept 28, 26; c 1. IMMEDIATE HYPERSENSITIVITY, ALLERGIES. *Educ:* Tokyo Women's Med Sch, MD, 49; Univ Tokyo, PhD, 55. *Prof Exp:* From assoc prof to prof microbiol, 70-81, from assoc prof to prof med, 72-81, PROF IMMUNOL & MED, JOHNS HOPKINS UNIV, 79- *Concurrent Pos:* Res mem, Dept Serol, NIH, Tokyo, 51-57, Dept Immunoserol, 59-62; res fel, div chem & chem eng, Calif Inst Technol, 57-59 & dept microbiol, Sch Hyg & Pub Health, Johns Hopkins Univ, 59; res immunologist, Children's Asthma Res Inst & Hosp, Denver, 62-70. *Honors & Awards:* Passano Found Award, USA, 72; Gairdner Found Int Award, Can, 73; First Sci Achievement Award, Inst Asn Allergology, 73; Borden Award, Asn Am Med Cols, 79. *Mem:* Japanese Soc Bacteriol; Japanese Soc Allergy; Am Asn Immunologists; hon mem Am Acad Allergy; AAAS; Am Soc Exp Path; Int Col Allergol; hon mem Span Soc Allergy & Clin Immunol; hon mem Venezuelan Soc Allergy & Immunol. *Res:* Immediate hypersensitivity, allergies. *Mailing Add:* La Jolla Inst Allergy/Immunol 11149 N Torrey Pines Rd La Jolla CA 92037

ISHLER, NORMAN HAMILTON, food science; deceased, see previous edition for last biography

ISIED, STEPHAN SALEH, b Jerusalem, Jordan, Aug 4, 46. INORGANIC CHEMISTRY. *Educ:* Am Univ Beirut, BS, 67, MS, 69; Stanford Univ, PhD(inorg chem), 74. *Prof Exp:* Teaching asst inorg & phys chem, Am Univ Beirut, 68-69; teaching & res asst inorg & anal chem, Stanford Univ, 70-73; postdoctoral bioinorg chem, Univ Calif, Berkeley, 74-75; from asst prof to assoc prof, 75-83, PROF INORG CHEM, RUTGERS UNIV, 83- *Honors & Awards:* Camille & Henry Dreyfus Award; Rutgers Excellence in Res Award. *Mem:* Am Chem Soc. *Res:* Intramolecular electron transfer in electron transfer proteins; the role of amino acids and peptides in electron mediation; reactivity of coordinated small molecules. *Mailing Add:* Dept Chem Rutgers Univ New Brunswick NJ 08903

ISIHARA, AKIRA, b Kofu, Japan; US citizen. CONDENSED MATTER PHYSICS. *Educ:* Univ Tokyo, DSC. *Prof Exp:* Res assoc statist physics, Univ Tokyo, 52-55; res assoc statist physics, Univ Md, 55-58; from assoc prof to prof statist physics, Polytech Inst NY, 58-64; prof statist physics, State Univ NY, Buffalo, 64-90; VCHMN & PRES, JIDELO, BARDSTOWN, INC, 91- *Concurrent Pos:* Consult, Oak Ridge Nat Lab, 62-69, NAm Rockwell, 61-65 & Boeing Sci Lab, 60-63; vis prof, Vander Waals Lab, Univ Amsterdam, 77, Free Univ Brussels, 71 & Univ Rochester, 70-71; vis scholar, Harvard Univ, 87. *Mem:* Fel Am Phys Soc. *Res:* Transport and many body phenomena in 2D electron systems; polymer physics; condensed matter physics; many body theory. *Mailing Add:* 901 Withrow Ct Bardstown KY 40004

ISKANDER, FELIB YOUSSEF, b Assiout, Egypt, Sept 19, 49; m 72; c 2. TRACE ELEMENT MEASUREMENT, RADIOCHEMISTRY. *Educ:* Cairo Univ, Egypt, BS, 71, MS, 73; Washington State Univ, MS, 82, PhD(chem), 83. *Prof Exp:* Radio chemist, 83-85, res assoc, 85-90, MGR NUCLEAR ANALYTICAL SERV, UNIV TEX, 90- *Concurrent Pos:* Instr analytical chem, fac pharmacy, Cairo Univ, 71-78; teaching asst, dept chem, Wash State Univ, 79-80, res asst, radio chem, Nuclear Radiation Ctr, Wash State Univ, 80-83. *Mem:* Am Chem Soc; Am Nuclear Soc; Egyptian Pharmaceut Soc; NY Acad Sci. *Res:* Innovative approaches to solve problems associated with trace element measurement in food, geological deposites and environment samples using radioanalytical method of analysis. *Mailing Add:* Dept Mech Eng Univ Tex Nuclear Eng Teaching Lab Austin TX 78712

ISKANDER, SHAFIK KAMEL, b Cairo, Egypt, Dec 15, 34; US citizen; m 62; c 2. FRACTURE MECHANICS, FINITE ELEMENTS. *Educ:* Cairo Univ, Egypt, BMechEng, 56, MSc, 67; Univ Tenn, PhD(eng mech), 72. *Prof Exp:* Head tech off eng, Gen Indust Co, GIMCO, Egypt, 58-62; head mat testing unit eng, Nat Inst Standards, 62-68; instr eng sci & mech, Col Eng, Univ Tenn, 69-72; engr, Nuclear Div, 72-77, head solid mech sect, Comput Sci Div, 77-86, ENGR, METALS & CERAMICS DIV, OAK RIDGE NAT LAB, TENN, 86- *Concurrent Pos:* Consult, govt & pvt industs, 58-68; instr, Mech

Eng Dept, Cairo Univ, 58-68; researcher, Bundesanstalt fur Mat Res, West Berlin, Ger, 62-64; assoc prof, Eng Sci & Mech Dept, Univ Tenn, 72- *Mem:* Am Soc Mech Engrs; Soc Exp Stress Anal; Sigma Xi. *Res:* Computational mechanics; finite elements; boundary integral methods; fracture mechanics; analytical and experimental. *Mailing Add:* 103 Berwick Dr Oak Ridge TN 37830

ISLAM, MIR NAZRUL, b Barisal, Bangladesh, Jan 1, 47; m 70; c 2. FOOD SCIENCE, NUTRITION. *Educ:* Dacca Univ, BSc, 66, MSc, 67; La State Univ, PhD(food sci), 72. *Prof Exp:* Lectr biochem & nutrit, Dacca Univ, Bangladesh, 67-69; asst prof food & nutrit, Southern Univ, La, 72-74; asst prof, 74-79, ASSOC PROF FOOD SCI & NUTRIT, UNIV DEL, 79- *Concurrent Pos:* Consult, Int Paper Co, Tuxedo, NY, 76-77, Shoregood Corp, Federalsburg, Md, 78- & Container Corp Am, Oaks, Pa, 81-; distinguished vis prof, Univ Panama, 80-81. *Mem:* Inst Food Technologists; Am Asn Cereal Chemists; Nutrit Today Soc; Am Dietetic Asn. *Res:* Post-harvest preservation; product development and utilization of halophyte crops. *Mailing Add:* Dept Food Sci & Human Univ Del Newark DE 19716

ISLAM, MUHAMMAD MUNIRUL, b Chittagong, Bangladesh, May 18, 36; m 62; c 1. THEORETICAL HIGH ENERGY PHYSICS. *Educ:* Univ Dacca, BSc, 56, MSc, 57; Univ London, PhD(theoret physics), 61. *Prof Exp:* Res asst theoret physics, Imp Col, Univ London, 61-62; res assoc high energy physics, Brown Univ, 62-65, asst prof res, 65-67; from asst prof to assoc prof, 67-75, prof physics, Univ Conn, 75-; AT PHYS DEPT, STATE UNIV NY, POTSDAM. *Concurrent Pos:* Vis scientist, Saclay Nuclear Res Ctr, France, 74-75. *Mem:* Am Phys Soc. *Res:* Scattering of elementary particles, their interactions and structure. *Mailing Add:* Dept Phys U-46 Univ Conn Main Campus 2152 Hillside Rd Storrs CT 06268

ISLAM, NURUL, b Bogra, Bangladesh, Dec 13, 39; m 71. PLANT PHYSIOLOGY, AGRONOMY. *Educ:* Univ Dacca, BAgr, 60, MAgr, 61; Tuskegee Inst, MAgr, 65; WVa Univ, PhD(plant physiol), 69. *Prof Exp:* Res asst bot, EPakistan Agr Res Inst, Dacca, 61-62; instr agr, Pakistan-Japan Training Inst, 63; annotator, WVa Univ, 68-69; planner, Div Water Resources, WVa Dept Natural Resources, 69-75, chief planning, 75-81; AT NORTHWEST AREA ELEC, USDA RURAL ELECTRIFICATION ASN. *Mem:* Am Soc Agron. *Res:* Environmental physiology of crops with respect to day length, temperature, and plant growth substances. *Mailing Add:* Northwest Area Elec USDA Rural Electrification Asn 14th & Independence SW Washington DC 20250

ISLEIB, DONALD RICHARD, b Paterson, NJ, June 2, 27; m 51; c 3. PLANT PHYSIOLOGY. *Educ:* Rutgers Univ, BS, 51, MS, 52; Iowa State Univ, PhD, 54. *Prof Exp:* Asst chem weed control & plant physiol, Rutgers Univ, 51-52 & Iowa State Univ, 52-54; assoc prof farm crops, Mich State Univ, 54-61; agr res mgr, Frito-Lay, Inc, 61-65 & Int Minerals & Chem Corp, 65-71; from sci adv to dir, Mich Dept Agr, 71-74, chief dep dir, 74-79; exec secy, Mich Toxic Substance Control Comn, 80, DIR, BEAN/COWPEA COLLABORATIVE RES SUPPORT PROG, MICH STATE UNIV, 81- *Mem:* Potato Asn Am (pres, 66); Am Soc Agr; Crop Sci Soc. *Res:* Special emphasis on crop plants and chemicals used in crop production. *Mailing Add:* 101 Soil Sci Crop/Soil Sci Mich State Univ East Lansing MI 48824

ISLER, GENE A, b Marion, Ohio, Apr 13, 40; m 68; c 2. ANIMAL SCIENCE. *Educ:* Ohio State Univ, BS, 62, MS, 66, PhD(animal sci), 69. *Prof Exp:* EXTEN SPECIALIST ANIMAL SCI, OHIO STATE UNIV, 69- *Mem:* Am Soc Animal Sci. *Res:* Swine genetics; swine performance testing; adjustment factors for growth; swine ultrasonics. *Mailing Add:* Dept Animal Sci Ohio State Univ 2029 Fyffe Rd Columbus OH 43210

ISLER, HENRI GUSTAVE, histology, biochemistry; deceased, see previous edition for last biography

ISLER, RALPH CHARLES, b Pittsburgh, Pa, Apr 23, 33; m 58; c 3. ATOMIC SPECTROSCOPY, PLASMA PHYSICS. *Educ:* Univ Pittsburgh, BS, 55; Johns Hopkins Univ, PhD(physics), 64. *Prof Exp:* Instr physics, Johns Hopkins Univ, 62-64; res assoc, Columbia Univ, 64-66; from asst prof to prof physics, Univ Fla, 66-78; RES STAFF MEM, OAK RIDGE NAT LAB, 76- *Mem:* Fel Am Phys Soc. *Res:* Spectroscopic research as applied to plasmas; rocket studies of planetary atmospheres and low energy ion-atom and ion-molecule collisions; level-crossing spectroscopy. *Mailing Add:* Fusion Energy Div Oak Ridge Nat Lab PO Box 2009 Oak Ridge TN 37831-8072

ISLES, DAVID FREDERICK, b Rahway, NJ, Sept 23, 35; m 64. MATHEMATICS. *Educ:* Princeton Univ, AB, 57; Mass Inst Technol, PhD(math logic), 64. *Prof Exp:* Asst prof, 63-69, ASSOC PROF MATH, TUFTS UNIV, 69- *Mem:* Am Math Soc. *Res:* Mathematical logic. *Mailing Add:* Dept Math Tufts Univ Medford MA 02155

ISLEY, JAMES DON, b Kingsport, Tenn, May 7, 28; m 54; c 2. RESEARCH ADMINISTRATION. *Educ:* Va Polytech Inst, BS, 51. *Prof Exp:* First lieutenant artillery, US Army, 51-53; jr engr, 51, engr archit, 53-58, mech engr, 53-60, sr indust engr, 60-65, sr group engr, 66-68, sr supv engr, mat handling engr, eng div, 69-83, MGR ENG SERV, EASTMAN CHEM DIV, TENN EASTMAN CO, EASTMAN KODAK, 83- *Concurrent Pos:* Chmn task group, Chemists Mfg Asn, 69-72; chmn, Am Soc Mech Engrs, MH14 Comt, 82-85. *Mem:* Am Inst Indust Engrs. *Res:* Design of pilot plants; expansion and development of research facilities. *Mailing Add:* 1713 Longview St Kingsport TN 37660

ISMAIL, AMIN RASHID, b Bombay, India, Mar 18, 58; m 90. PERSONAL COMPUTERS - HARDWARE & SOFTWARE, MICROPROCESSOR CONTROLLERS. *Educ:* Univ Dayton, BS, 78, MS, 81. *Prof Exp:* From instr to asst prof, 78-90, ASSOC PROF ELECTRONICS, UNIV DAYTON, 90- *Concurrent Pos:* Consult, Sch Eng, Univ Dayton, 77-85, instr, Spec Progs, 77-89; dir, Microcon Int, 81-87; instr, Kettering Col Med Arts, 81, Air Force Inst Technol, 82; consult, Milmar Century Corp, 90- *Mem:* Am Soc Eng Educ; Inst Elec & Electronics Engrs; Asn Comput Mach. *Res:* Conducted research and development of an aircraft relay life-test system for the United States Air Force; co-author of three books on microprocessors and digital electronics. *Mailing Add:* 5640 Waterloo Rd Dayton OH 45459-1830

ISMAIL, MOURAD E H, b Cairo, Egypt, Apr 27, 44; Can & Egyptian citizen; m 69. ORTHOGONAL POLYNOMIALS. *Educ:* Cairo Univ, BSc, 64, Univ Alta, MSc, 69, PhD(math), 74. *Prof Exp:* Demonstr math, Dept Math, Fac Sci, Cairo Univ, 64-68; asst scientist, Math Dept & Res Ctr, Univ Wis-Madison, 74-75; fel & vis lectr, Univ Toronto, 75-76; asst prof appl math, McMaster Univ, 76-78; from asst prof to prof math, Ariz State Univ, 78-88; PROF MATH, UNIV SFLA, 87- *Mem:* Am Math Soc; Soc Indust & Appl Math; Math Asn Am. *Res:* Orthogonal polynomials and their applications to spectra of Jacobi matrices; birth and death processes and queueing theory; special functions and infinite divisibility problems. *Mailing Add:* Dept Math Univ SFla Tampa FL 33620

ISOM, BILLY GENE, b Eagleville, Tenn, May 9, 32; m 51, 72; c 3. MALACOLOGY, AQUATIC TOXICOLOGY. *Educ:* Middle Tenn State Univ, BS, 56, MA, 57;. *Prof Exp:* Biologist pollution, Tenn State Pollution Control Bd, 57-63; aquatic biologist/malacologist, Fisheries & Wildlife Br, Tenn Valley Authority, 63-66, aquatic biologist, water Qual & EcolBr, 66-79 & prog mgr aquatic environ res, Fisheries & Aquatic Eco Br, 80-86; DIR, AQUATIC TOXICOL & ECOL, ECKENFELDER INC, NASHVILLE, TENN, 86- *Concurrent Pos:* Univ Mich Scholar, 68-69; lectr, aquatic toxicol; reviewer, Student Originated Res, NSF & Women in Sci Proposals; chmn, Standard methods water & wastewater, Am Water Works Asn, 75, 80, 85 & 90; chmn, Subcomt Biol Field Methods, Comt on Water, Am Soc Testing Mat,80-85 & Biol Field Testing Comt Environ Fate Effects, 84-87. *Mem:* Am Malacol Union; Am Fisheries Soc; Sigma Xi; Am Soc Testing Mat; NAM Benthological Soic (pres, 71); Nat Mgt Asn (treas, 83-84); Water Pollution Control Fedn. *Res:* Aquatic environmental effects of reservoir releases from hydroelectric projects (low dissolved oxygen); discharges from fossil and nuclear plants; new coal gasification technologies; acid precipitation; biofouling control in power plants, and freshwater mussel research; limnology; aquatic toxicology; consulting; author of numerous papers and publications; one patent. *Mailing Add:* Eckenfelder Inc 227 French Landing Dr Nashville TN 37228

ISOM, GARY E, b Twin Falls, Idaho, June 21, 46; m 67; c 3. PHARMACOLOGY. *Educ:* Idaho State Univ, BS, 69; Wash State Univ, PhD(pharmacol), 73. *Prof Exp:* Res asst pharmacol, Wash State Univ, 69-70; fel pharmacol, Am Found Pharmaceut Educ, 70-71; partic molecular pharmacol, NIH & Am Soc Pharmacol & Exp Therapeut, 71; fel pharmacol, Burroughs-Welcome Co, 71-73; clin clerkship, Wash State Univ, 73; from asst prof to assoc prof pharmacol, Idaho State Univ, 73-80; assoc prof, 80-83, PROF TOXICOL, SCH PHARM & PHARMACEUT SCI, PURDUE UNIV, 83- *Concurrent Pos:* Pharmaceut Mfgrs Asn res grant, 73-75; mem, US Pharmacopeia Conv, 75-80; regist pharmacist, Idaho; prin investr, NIH grants, 78- *Mem:* Am Soc Pharmacol Exp Therapeuts; Sigma Xi; Soc Toxicol; Am Asn Col Pharm; Drug Info Asn. *Res:* Study of the biochemical mechanisms associated with morphine physical dependence and tolerance; neurotoxicology of cyanide; pharmacology of atrial natriuretic factor receptor. *Mailing Add:* Dept Pharmacol & Toxicol Sch Pharm Purdue Univ West LaFayette IN 47905

ISOM, HARRIET C, b 1947; m 70; c 1. DIFFERENTIATION, VIRUS TRANSFORMATION. *Educ:* Bryn Mawr Col, AB, 69; Univ Ill, MS, 71, PhD, 73. *Prof Exp:* NIH fel microbiol, Univ Pa Sch Med, 73-74, Pa plan scholar, 74-76, assoc, 76; from asst prof to assoc prof, 76-87, PROF MICROBIOL, PA STATE UNIV COL MED, 87- *Concurrent Pos:* Assoc mem grad fac, Pa State Univ Col Med, 76-82, mem grad fac, 82- *Mem:* Am Soc Microbiol; AAAS; Sigma Xi; NY Acad Sci; Am Asn Cancer Res; Am Soc Virol; Tissue Cult Asn; Am Soc Cell Biol; Am Soc Biol Chemists. *Res:* Viral transformation of differentiated cells to determine the virus genes involved in converting, in particular normal adult mondividing liver cells to replicating tumorogenic cells; effects of virus infection and transformation on expression of liver proteins. *Mailing Add:* 98 Woodbine Hershey PA 17033

ISOM, JOHN B, b Nashville, Tenn, June 16, 25; m 47; c 3. PEDIATRICS, NEUROLOGY. *Educ:* Vanderbilt Univ, BA, 50, MD, 54; Am Bd Pediat, dipl. *Prof Exp:* Intern, Vanderbilt Univ Hosp, 54-55; resident, Univ Iowa Hosp, 55-56; resident, St Louis Children's Hosp, 56-57; mem staff, Mease Diag & Treatment Ctr, Dunedin, Fla, 57-58; instr pediat, Vanderbilt Univ Hosp, 58-59, USPHS spec fel neurol, 59-60; USPHS fel pediat neurol, Mass Gen Hosp, 60-61, USPHS fel neuropath, 61-62; res fel biochem, Harvard Med Sch, 62-63; assoc prof pediat, 63-70, PROF PEDIAT & ASSOC PROF NEUROL, MED SCH, UNIV ORE, 70- *Concurrent Pos:* Clin fel neurol & res fel neurosurg, Mass Gen Hosp, 62-63. *Mem:* AMA; Am Acad Neurol. *Res:* Pediatric neurology; water and electrolyte metabolism in isolated central nervous tissue; reading disability in children. *Mailing Add:* Dept Pediat Ore Health Sci Univ 3181 SW S Jackson Park Rd Portland OR 97201

ISOM, MORRIS P, b Miami, Fla, Mar 12, 28; m 56; c 3. AERONAUTICAL ENGINEERING, APPLIED MATHEMATICS. *Educ:* Harvard Col, BA, 52; Mass Inst Technol, MS, 56; Princeton Univ, PhD(aerospace & mech sci), 64. *Prof Exp:* Res engr, Mass Inst Technol, 53-55; asst proj eng fluid mech, Wright Aeronaut Div, Curtis-Wright Corp, 57-58; res assoc, Princeton Univ, 58; res analyst strategic & defense systs, Hudson Inst, 62-64; assoc prof aeronaut & astronaut, NY Univ, 64-, asst chmn dept, 67-; ASSOC PROF AEROSPACE ENG, POLYTECH INST NEW YORK. *Concurrent Pos:* Consult, Hudson Inst, 64-66, Sikorsky Aircraft, 67-69 & US Army, 70- *Res:* Aerodynamics. *Mailing Add:* Dept Aerospace Eng Polytech Inst New York 333 Jay St Brooklyn NY 11201

ISOM, WILLIAM HOWARD, b Hurricane, Utah, Oct 30, 17; m 48; c 3. AGRONOMY. *Educ:* Utah State Univ, BS, 40, MS, 51; Cornell Univ, PhD(plant breeding, genetics), 55. *Prof Exp:* Res agronomist, Agr Res Serv, USDA, 55-60; farm adv, 60-62, exten agronomist, 62-73, AGRICULTURIST, COOP EXTEN, UNIV CALIF, RIVERSIDE, 73-, LECTR PLANT SCI, UNIV, 74- *Mem:* Am Soc Agron; Sigma Xi. *Res:* Flax breeding and improvement; field crops; dry edible legumes; agricultural extension. *Mailing Add:* 4138 Batchelor Hall Dept Bot & Plant Sci Univ Calif Riverside CA 92521

ISON-FRANKLIN, ELEANOR LUTIA, b Dublin, Ga, Dec 24, 29; m 65; c 1. PHYSIOLOGY, ENDOCRINOLOGY. *Educ:* Spelman Col, Atlanta, AB, 48; Univ Wis-Madison, MS, 51, PhD(zool), 57. *Prof Exp:* Instr biol, Spelman Col, 48-49 & 51-53; asst zool, Univ Wis, 50-51, res asst zool, 55-57; instr biol, Spelman Col, 51-53; asst prof physiol, Sch Vet Med, Tuskegee Inst, 57-60, assoc prof physiol, Carver Found, 60-63; from asst prof to assoc prof physiol, 63-71, assoc dean gen admin, 70-72, assoc dean acad affairs, 72-80, PROF PHYSIOL, COL MED, HOWARD UNIV, 71-, DIR CARDIOVASC RES LAB, 80- *Concurrent Pos:* Porter physiol lectr, Am Physiol Soc, 67-74; vis prof Harvard Med Sch, 77-78; co-prin investr, NASA grant NAG 2-250, 83-86. *Mem:* Am Physiol Soc; Am Heart Asn; Am Soc Hypertension; Sigma Xi; Fedn Am Soc for Exp Biol. *Res:* Cardiovascular physiology; cardiac dynamics and hypertension. *Mailing Add:* Howard Univ Col Med 520 W St NW Washington DC 20059

ISPHORDING, WAYNE CARTER, b Willow Grove, Pa, Sept 26, 37; m 59; c 3. GEOCHEMISTRY, MINERALOGY. *Educ:* Univ Fla, BS, 62, MS, 63; Rutgers Univ, PhD(geol), 66. *Prof Exp:* Assoc prof, 66-78, PROF CRYSTALLOG & MINERAL, UNIV SALA, 78- *Concurrent Pos:* Grants, Army Res Off, NSF, Nat Oceanic Atmospheric Admin. *Mem:* Geol Soc Am; Am Asn Petrol Geol; Soc Econ Paleont & Mineral; Am Inst Prof Geologists. *Res:* Application of computer techniques to sediment analysis; engineering properties of tropical soils; paleogeographic studies of the upper Tertiary of eastern United States; geochemistry of Laterites in Yucatan; genesis and petrology of palygorskite-sepiolite clays; heavy metal chemistry of estuarine sediments. *Mailing Add:* Dept Geol Univ SAla 307 University Dr Mobile AL 36688

ISQUITH, IRWIN R, b Brooklyn, NY, Jan 23, 42; m 64; c 2. PROTOZOOLOGY, ECOLOGY. *Educ:* Brooklyn Col, BS, 62; NY Univ, MS, 64, PhD(biol), 66. *Prof Exp:* Asst cur, Acad Nat Sci, Philadelphia, 66-68; from asst prof to assoc prof, 68-76, PROF BIOL, FAIRLEIGH DICKINSON UNIV, 76-, CHMN, DEPT BIOL SCI, 82- *Mem:* AAAS; Am Soc Protozool; NY Acad Sci; Sigma Xi. *Res:* Taxonomy, genetics and evolution of protozoa, especially the ciliate, Blepharisma; ecology; biomagnetics. *Mailing Add:* Fairleigh Dickinson Univ Dept Biol 1000 River Rd Teaneck NJ 07666

ISRAEL, HAROLD L, b Boston, Mass, Nov 16, 09; m 38, 75; c 4. PULMONARY DISEASES. *Educ:* Amherst Col, AB, 30; Jefferson Med Col, MD, 34; Univ Pa, MPH, 42. *Prof Exp:* Assoc prof med, Grad Sch Med, Univ Pa, 50-59; prof, 59-79, EMER PROF MED, JEFFERSON MED COL, 80- *Concurrent Pos:* Mem, Int Comt Sarcoidosis, 50- *Mem:* Sigma Xi; hon mem Brit Thoracic Soc. *Res:* Sarcoidosis; Wegner's granulomatosis; immunologic lung disease. *Mailing Add:* 1015 Walnut St Rm 804 Philadelphia PA 19107

ISRAEL, HARRY, III, b Dayton, Ohio, Sept 30, 34; m 60; c 2. DENTISTRY, PERIODONTICS. *Educ:* Univ Mich, Ann Arbor, BA, 56; Western Reserve Univ, DDS, 60; Univ Pa, MSc, 69; Univ Ala, Birmingham, PhD(anat), 71. *Prof Exp:* NIH fels, Boston Univ Hosps, 61-62; NIH fels, Philadelphia Gen Hosp-Univ Pa, 62-63; res assoc growth & genetics, Fels Res Inst, 65-68, sr investr, 68-71, chief dent res sect, 71-76; DIR DENT RES, CHILDREN'S MED CTR, DAYTON, 76- *Concurrent Pos:* NIH fel, Univ Ala, 69-70; assoc clin prof, Wright State Univ, Sch Med, 76- *Mem:* Am Dent Asn; Am Acad Periodont; AAAS; Am Inst Nutrit; Int Asn Dent Res; Am Soc Clin Nutrit. *Res:* Growth and aging in the human cranio-facial skeleton; nutriton and oral-facial development; cranio-facial congenital malformation; periodontal disease. *Mailing Add:* Dent Res Sect One Childrens Plaza Dayton OH 45404

ISRAEL, HERBERT WILLIAM, b Chicago, Ill, July 17, 31; m 53; c 3. PLANT CYTOLOGY, PLANT PATHOLOGY. *Educ:* Concordia Teachers Col, Ill, BS, 53; Univ Wis, MS, 59; Univ Fla, PhD(bot, zool), 62. *Prof Exp:* Chmn dept sci, High Sch, Ill, 53-60; Turtox fel, Univ Fla, 62-63; Nat Inst Gen Med Sci fel, 63-64; res assoc, 64-70, sr res assoc cell physiol, 70-72, SR RES ASSOC PLANT PATH, CORNELL UNIV, 72- *Concurrent Pos:* Partic, Academic Year Inst, NSF, 59; Hatch New York res grants, Coop State Res Serv, USDA, 74, 77, 79 & 83, spec prog grant, 73; ed adv bd, Protoplasma, 73-81; NSF res grant, 76, 79 & 81; res grant, Comp Res Grant Off, Sci Educ Admin, USDA, 79, 81 & 89; assoc ed, Phytopathology, 80-83; bd regents, Concordia Univ, 86-92. *Mem:* AAAS; Sigma Xi; Am Phytopath Soc. *Res:* Cellular ultrastructure; virology; pathogen-host, primary interactions; developmental morphology; embryology. *Mailing Add:* Room 334 Plant Sci Bldg Cornell Univ Ithaca NY 14853

ISRAEL, JAY ELLIOT, computer sciences, applied mathematics, for more information see previous edition

ISRAEL, MARTIN HENRY, b Chicago, Ill, Jan 12, 41; m 65; c 2. COSMIC-RAY ASTROPHYSICS. *Educ:* Univ Chicago, SB, 62; Calif Inst Technol, PhD(physics), 69. *Prof Exp:* From asst prof to assoc prof, 68-75, PROF PHYSICS, WASH UNIV, ST LOUIS, 75-, DEAN FAC ARTS & SCI, 88- *Concurrent Pos:* Alfred P Sloan Found fel, 70-72; prin investr, Heavy Nuclei Exp on High Energy Astron Observ-3, 71-; mem, exec comt, Div Cosmic Phys, Am Phys Soc, 75-77 & 79-82, balloon study comt, Nat Acad Sci, 75, space astron and astrophys comt, 76-79, mgt opers group high energy astrophys div, Am Astron Soc, 82-84, sub-panel for cosmic rays, space & astrophys plasmas & physics surv comt, Nat Acad Sci, 83-85 & space & earth sci adv comt, NASA, 85-88; chmn, exec comt, Bevalac Users Group, 75, div cosmic physics, Am Physics Soc, 80-81, chmn, nat org comt, 19th Int Cosmic Ray Conf, 82-85, particle astrophys superconducting magnet definition team, NASA, 85-88; assoc dir, McDonnell Ctr Space Sci, Wash Univ, St Louis, 82-87, actg dean fac arts & sci, 87-88. *Honors & Awards:* Except Sci Achievement Award, NASA, 80. *Mem:* Fel Am Phys Soc; Am Astron Soc; Am Asn Univ Professors; AAAS. *Res:* Cosmic ray physics; very heavy nuclei; electrons; observations using electronic detectors on satellites and high-altitude balloons. *Mailing Add:* Fac Arts & Sci Campus Box 1094 Wash Univ St Louis MO 63130-4899

ISRAEL, STANLEY C, b Brooklyn, NY, Dec 30, 42; m 66; c 1. PYROLYSIS MASS SPECTROMETRY, REACTION MECHANISMS. *Educ:* Parsons Col, BS, 65; Lowell Technol Inst, PhD(chem), 70. *Prof Exp:* From instr to assoc prof, 68-80, PROF CHEM, UNIV LOWELL, 80- *Concurrent Pos:* Assoc ed, Fire Res; consult, Optimers Co & Polymer Technol Corp; vis prof, Univ Utah, 75-77, dir chem res, Flammability Res Ctr, Univ Utah, 76-77; prin investr numerous grants & contracts; mem, Am Chem Soc, div polymer chem (treas, 79-84, chmn, 89), div org chem, div chem educ; consult, chem indust, legal profession. *Honors & Awards:* A L Lipschitz Award, 87; Phoenix Award, Am Chem Soc, 89. *Mem:* AAAS; Am Chem Soc; NY Acad Sci; Am Asn Univ Prof; Am Soc Mass Spectrometry; Sigma Xi. *Res:* Pioneering the techniques of Direct-Pyrolysis-Chemical Ionization Mass Spectrometry for the study of the reactions and mechanisms of thermal decomposition of polymeric materials; identification and characterization of materials by high temperature pyrolysis; surface characterization of fine fibers and lens like curved surfaces by Laser Contact Angle Gonionmetry; author of over 70 papers and patents. *Mailing Add:* Dept Chem Univ Lowell One University Ave Lowell MA 01854

ISRAEL, WERNER, b Berlin, Ger, Oct 4, 31; m 58; c 2. THEORETICAL PHYSICS. *Educ:* Univ Cape Town, BSc, 51, MSc, 54; Trinity Col, Dublin, PhD(math), 60. *Hon Degrees:* DSc, Queens Univ, Kingston, Ont, 87. *Prof Exp:* Lectr math, Univ Cape Town, 55-56; res scholar theoret physics, Dublin Inst Advan Studies, 56-58; from asst prof to prof math, 58-72, PROF PHYSICS, UNIV ALTA, 72-; PROV ALTA FEL, CAN INST ADVAN RES, 86- *Concurrent Pos:* Vis prof, Dublin Inst Advan Studies, 66-68; mem, Int Comt Gen Relativity & Gravitation, 71-80; Sherman Fairchild distinguished scholar, Calif Inst Technol, 74-75; sr visitor, Dept Appl Math & Theoret Physics, Univ Cambridge, 75-76; maitre de recherche assoc, Inst Henri Poincare, Paris, 76-77; vis prof, Univ Berne, 80, Univ Kyoto, 86; mem, Comn Math Physics, Int Union Pure & Appl Physics, 84-; vis fel, Gonville & Caius Col, Cambridge, 85. *Honors & Awards:* Medal for Achievement in Physics, Can Asn Physicists, 81; Killam Mem Prize, Can Coun, 84. *Mem:* Int Astron Union; fel Royal Soc Can; Can Asn Physicists; Can Astron Soc; fel, Royal Soc (London). *Res:* General relativity; cosmology; statistical mechanics. *Mailing Add:* Dept Physics Univ Alta Edmonton AB T6G 2J1 Can

ISRAEL, YEDY, b Temuco, Chile, Sept 19, 39; m 62; c 2. PHARMACOLOGY. *Educ:* Univ Chile, biochemist, 62; Univ Toronto, PhD(biochem, pharmacol), 65. *Prof Exp:* NIH int fel biochem, Nat Heart Inst, 65-66; from asst prof to prof, Univ Chile, 66-69; PROF PHARMACOL, SCH MED, UNIV TORONTO, 70- *Concurrent Pos:* Head, Dept Biochem, Addiction Res Found, Ont, 81- *Honors & Awards:* Award, Med Soc Chile, 70; Jellinek Award, 80. *Mem:* Am Soc Pharmacol; Soc Neurosci; Pharmacol Soc Can. *Res:* Mechanisms of psychotropic drug action; alcoholic liver disease; alcoholism. *Mailing Add:* Dept Pharmacol Univ Toronto Toronto ON M5S 1A8 Can

ISRAELS, LYONEL GARRY, b Regina, Sask, July 31, 26; m 50; c 2. HEMATOLOGY, ONCOLOGY. *Educ:* Univ Sask, BA, 46; Univ Man, MD, 49, MSc, 50; FRCP(C), 54. *Prof Exp:* From lectr to asst prof biochem, 54-59, asst prof med, 59-64, assoc prof internal med, 64-66, actg chmn internal med, 78-79, distinguished prof, 83, PROF INTERNAL MED, UNIV MAN, 66-; EXEC DIR, MAN CANCER TREATMENT & RES FOUND, 73- *Concurrent Pos:* Mem bd, Nat Cancer Inst Can, 67-78, mem, Adv Res Comt, 83-87; dir, Man Inst Cell Biol, 69-73; mem, Med Res Coun Can, 73-75; chmn, Man Health Res Coun, 80-87. *Mem:* Nat Cancer Inst Can (pres, 76-78); Can Soc Hemat (pres, 74); Can Soc Clin Invest (pres, 68); Am Soc Clin Invest. *Res:* Chemotherapy of malignant tumors; biological effects of alkylating agents, bilirubin synthesis and heme metabolism. *Mailing Add:* 100 Olivia St Winnipeg MB R3E 0V9 Can

ISRAELSEN, C EARL, b Hyrum, Utah, Apr 21, 28; m 53; c 8. HYDROLOGY, EROSION CONTROL. *Educ:* Utah State Univ, BS, 59; Univ Ariz, PhD(hydrol), 68. *Prof Exp:* Asst res engr, Eng Exp Sta, 60-63, res engr, Utah Water Res Lab, 66-68, assoc prof hydrol, Univ, 68-70, dir interam ctr integral develop land & water resources, Utah State Univ-Orgn Am States, Venezuela, 70-72, PROF HYDROL WATER RESOURCES, UTAH WATER RES LAB, UTAH STATE UNIV, 73-, PROF CIVIL & ENVIRON ENG, 83- *Concurrent Pos:* Training adv irrig system mgt proj, Pakistan, 86-88. *Res:* Physical hydraulic modelling; evaporation reduction by use of monolayers; measurement of streamflow by capacitance methods; weather modification in mountainous areas using silver iodide; saline water; erosion control during construction; development of water education materials for public schools; development of state water atlas. *Mailing Add:* Utah Water Res Lab Utah State Univ Logan UT 84322-8200

ISRAELSTAM, GERALD FRANK, b Johannesburg, SAfrica, Jan 19, 29; Can citizen; m 53; c 1. PLANT PHYSIOLOGY. *Educ:* Univ Witwatersrand, BSc, 51, BSc, 53; Univ London, Eng, PhD(bot), 63. *Prof Exp:* Lectr bot, Univ Col, Ft Hare, SAfrica, 57-60; lectr biol, Medway Col Technol, Eng, 60-61; lectr plant physiol, Birmingham Univ, Eng, 63-64; ASSOC PROF PLANT PHYSIOL, SCARBOROUGH COL, UNIV TORONTO, 64- *Mem:* Can Soc Plant Physiol; Soc Exp Bot. *Res:* Mechanism of action of plant growth substances. *Mailing Add:* Div Life Sci West Hall Scarborough Col 1265 Military Trail Toronto ON M1C 1A4 Can

ISRAILI, ZAFAR HASAN, b Sambhal, India, July 2, 34; US citizen; m 70; c 3. HYPERTENSION, MEDICINAL CHEMISTRY. *Educ:* Aligarh M Univ, India, BSc, 51, MSc, 53; Univ Kans, PhD(med chem), 68. *Prof Exp:* Lectr chem, Aligarh M Univ, 53-54, sr res scholar chem, 54-57; res asst & sci officer radiation chem, Atomic Energy Comn, India, 57-61; sr res chemist, Inst Pharmceut Chem, Alza Corp, Kans, 69-70; asst prof med & chem, 70-75, assoc prof chem, 74-78, ASSOC PROF MED, EMORY UNIV, ATLANTA, 75-, PROF CHEM, 78-, res pharmacologist, Vet Admin Med Ctr, 79-87. *Concurrent Pos:* Fel, Dept Med Chem, Univ Kans, 68-69; mem sci staff, Grady Hosp, Atlanta, 74-; assoc ed, Drug Metab Rev, 74-; res grant, Merck, Sharpe, & Dohme, 77-; mem med res serv, Vet Admin Hosp, 78-87; res grants, Nat Cancer Inst, 78-81 & Nat Inst Aging, 80-82, Vet Admin, 79-87. *Mem:* Am Asn Cancer Res; Am Soc Clin Pharmacol & Therapeut; Am Soc Pharmacol & Exp Therapeut; Soc Exp Biol & Med; Am Chem Soc; Interam Soc Clin Pharmacol Therapeut; Am Aging Soc; Am Heart Asn; Int Soc Study Xenobiotics; Int Soc Hypertension Blacks. *Res:* Drug metabolism; clinical pharmacology and toxicology; hypertension; medicinal chemistry; analytical biochemical methods; aging; cancer chemotherapy. *Mailing Add:* 3567 Cloudland Dr Stone Mountain GA 30083

ISSAQ, HALEEM JERIES, b Haifa, Palestine, Feb 19, 36; US citizen; m 75; c 1. ANALYTICAL CHEMISTRY. *Educ:* Robert Col, BS, 65; Technion-Israel Inst Technol, MS, 68; Georgetown Univ, PhD(anal chem), 72. *Prof Exp:* SCIENTIST ANALYTICAL CHEM, FREDERICK CANCER RES & DEVELOP CTR, 72- *Concurrent Pos:* Assoc ed, J Liquid Chromatography; adv bd, CRC Critical Rev Anal Chem. *Mem:* Am Chem Soc; Soc Appl Spectros; Sigma Xi. *Res:* Role of trace metals in cancer; detection of differences of trace metals between cancerous and non-cancerous serum and tissue; detection and identification of trace chemicals by thin layer and liquid chromatography; solvent effects and selectivity optimization in liquid chromatography; methods development; capillary zone electrophoresis application in the biomedical field. *Mailing Add:* Frederick Cancer Res Ctr PO Box B Frederick MD 21702

ISSEKUTZ, BELA, JR, b Kolozsvar, Hungary, Dec 24, 12; nat Can; m 41; c 2. PHYSIOLOGY. *Educ:* Univ Szeged, MD, 36. *Hon Degrees:* Dr habil, Univ Budapest, 42. *Prof Exp:* Instr pharmacol, Univ Szeged, 36-38; from asst prof to assoc prof pharmacol, Univ Budapest, 38-45; prof physiol & head dept, Univ Szeged, 45-56; vis scientist physiol, Max Planck Inst, Heidelberg, WGer, 56-57; vis scientist & res grant, Smith Kline & French Lab, Pa, 57-58; head dept biochem, Res Div, Lankenau Hous, Philadelphia, 59-61, head dept physiol, 61-67; prof, 67-83, EMER PROF PHYSIOL, DALHOUSIE UNIV, 83- *Concurrent Pos:* Hoffman-LaRoche fel physiol, Univ Basel, 37; Hungarian State fel pharmacol, Univ Berlin, 38-39 & Univ Gottingen, 43; vis prof physiol & head dept, Univ Greifswald, 55-56. *Mem:* Am Physiol Soc; Can Physiol Soc. *Res:* Endocrine and nervous regulation of metabolism; mode of action of thyroxine and insulin; exercise metabolism; obesity. *Mailing Add:* 50 Nightingale Dr Halifax NS B3M 1V4 Can

ISSEL, CHARLES JOHN, b San Francisco, Calif, Mar 25, 43; m 64; c 2. VETERINARY VIROLOGY, EPIDEMIOLOGY. *Educ:* Univ Calif, Berkeley, 65; Univ Calif, Davis, DVM, 69; Univ Wis, Madison, MS, 71, PhD(vet sci), 73. *Prof Exp:* Vet med officer, USDA, Plum Island, 73-74; from asst prof to assoc prof vet viral, La State Univ, 74-81, prof, 81-90; WRIGHT-MARKEY PROF EQUINE INFECTIOUS DIS, UNIV, KY, 90- *Mem:* AAAS; Sigma Xi; Wildlife Dis Asn; Am Soc Microbiol; Am Soc Virol. *Res:* Epidemiology diagnosis and control of viral diseases of horses and wildlife. *Mailing Add:* Dept Vet Sci Univ Ky Gluck Equine Res Ctr Lexington KY 40546

ISSELBACHER, KURT JULIUS, b Wirges, Ger, Sept 12, 25; nat US; m 55; c 4. MEDICINE, BIOCHEMISTRY. *Educ:* Harvard Univ, AB, 46; Harvard Med Sch, MD, 50. *Prof Exp:* Intern med, Mass Gen Hosp, 50-51, from asst resident to resident, 51-53; clin instr, Sch Med, Johns Hopkins Univ, 53-54 & Sch Med, George Wash Univ, 54-56; instr, 56-58, from asst prof to prof, 60-72, MALLINCKRODT PROF MED, HARVARD MED SCH, 72-, ASSOC, 58-, CHMN EXEC COMT, DEPTS MED, 68- *Concurrent Pos:* Fel, Harvard Med Sch, 51-53; clin investr, Nat Inst Arthritis & Metab Dis, 53-56; asst, Mass Gen Hosp, 56-58, chief, Gastroenterol Unit & Res Lab, 57-, from asst physician to assoc physician, 58-66, physician, 66-, chmn comt res, 67-70; chmn cancer comt, Harvard Univ, 72; dir, Cancer Ctr, Mass Gen Hosp, 86; ed-in-chief, Harrison's Prin Internal Med, 90- *Mem:* Nat Acad Sci; Am Gastroenterol Asn (pres, 74-75); Soc Exp Biol & Med; Am Soc Biol Chem; Am Acad Arts & Sci. *Res:* Biochemical basis for normal and disturbed function of the intestinal tract and liver; cellular changes in malignancy. *Mailing Add:* Dept Med Mass Gen Hosp Boston MA 02114

ISSENBERG, PHILLIP, food science, toxicology; deceased, see previous edition for last biography

ISSEROFF, HADAR, b Newark, NJ, Dec 24, 38; m 60; c 2. IMMUNOLOGY, ZOOLOGY. *Educ:* Brooklyn Col, BS, 60; Purdue Univ, MS, 63, PhD(microbiol), 66. *Prof Exp:* Teaching assoc biol, Purdue Univ, 61-63; NIH fel physiol of parasites, Rice Univ, 66-68; from asst prof to assoc prof, 68-78, PROF BIOL, STATE UNIV NY COL BUFFALO, 78- *Concurrent Pos:* Fel & grant, State Univ NY Res Found, State Univ NY Col Buffalo, 69 & 70; res grants, United Health Found Western NY, 70-72, NIH, 71-74, 76-83 & Edna McConnell Clark Found, 79-81; vis prof, Med, Sch, Univ Tel Aviv, 75; vis scientist & prof, Roswell Park Mem Inst, 83, 87-88 & 90-91. *Mem:* Fel AAAS; Am Physiol Soc; Am Soc Parasitol; Am Soc Zool; NY Acad Sci; Sigma Xi. *Res:* Symbiology; proline in fascioliasis and schistosommiasis; proline as a regulator of bile duct morphogenesis; electron microscopy of larval trematodes; pheromones regulating population growth in molluscs; molecular basis of immuo-endocrine interactions in schistosomiasis; suppression of the immune response by schistosomiasis at the genomic level. *Mailing Add:* 58 Redwood Terr Williamsville NY 14221

ISSEROW, SAUL, b Berlin, Ger, Mar 8, 22; US citizen; m 54; c 2. MATERIALS SCIENCE, MATERIALS ENGINEERING. *Educ:* Brooklyn Col, BA, 41; Pa State Univ, MS, 45, PhD(phys chem), 50. *Prof Exp:* Assoc chem, Armour Res Found, 49-52; chemist, Mass Inst Technol, 52-54; chemist, Nuclear Metals Inc, 54-56, group leader chem metall, 56-59, proj mgr, 59-63, mgr process develop, 63-66, mgr mat res, 66-68, sr scientist, 68-70; head develop dept, Saphikon Div, Tyco Labs, Inc, 70-71; METALLURGIST, ARMY MAT TECHNOL LAB, 71- *Mem:* Am Chem Soc; Am Soc Metals; Am Soc Testing & Mat; Am Powder Metall Inst; Nat Asn Corsosion engrs. *Res:* Powder metallurgy; chemical thermodynamics; titanium metallurgy; materials for nuclear reactors, notably metallic fuel elements; high temperature materials; aircraft and ordnance materials; corrosion. *Mailing Add:* Army Mat Technol Lab Watertown MA 02172-0001

ISSITT, PETER DAVID, b London, Eng, Jan 29, 33. IMMUNOHEMATOLOGY. *Educ:* Inst Med Lab Sci, AIMLS, 55, FIMLS, 57; Inst Biol, LIBiol, 68, MIBiol, 78, MRCPath, 77, FIBiol, 85, CB, 85. *Prof Exp:* Supvr immunohemat, St Mary's Hosp, London, Eng, 58-62; supvr hemat & immunohemat, Peace Mem Hosp, Watford, Eng, 62-64; res fel immunohemat, New York Blood Ctr, 64-68; gen mgr, Spectra Biol Div, Becton-Dickinson Co, 68-71; asst prof res surg immunohemat, Paul I Hoxworth Blood Ctr, Univ Cincinnati, 71-77, dir labs, 71-81, assoc prof, 77-81; SCI DIR, SOUTH FLA BLOOD SERV, MIAMI, 81-; ADJ ASSOC PROF MED, UNIV MIAMI, 81- *Concurrent Pos:* Examnr, New York City Bd Health-Immunohemat, 65-68; sr sci consult, Spectra Biol Div, B-D Diag, 75-80; assoc ed, J Transfusion, Am Asn Blood Banks, 76-81 & 84- *Honors & Awards:* Ivor Dunsford Mem Award, Am Asn Blood Banks, 74; John Weaver King lectr, Cleveland Clin, 81; Jean Stubbins Mem lectr, Galveston, 82. *Mem:* Am Asn Blood Banks; Inst Med Lab Sci; Brit Inst Biol; Int Soc Blood Transfusion; Royal Col Pathologists; Brit Blood Transfusion Soc. *Res:* Investigation of human blood group systems as they pertain to genetics, immunology and biochemistry; investigation of hemolytic disorders caused by blood group autoantibodies; quantitation of human blood group antigens; investigation of polymorphisms of human blood groups. *Mailing Add:* Transfusion Serv Duke Univ Med Ctr PO Box 2928 Durham NC 27710

ISTOCK, CONRAD ALAN, b Aug 31, 36; m 61; c 2. ANIMAL ECOLOGY, POPULATION GENETICS. *Educ:* Wayne State Univ, AB, 59; Univ Mich, MA, 61, PhD(zool), 64. *Prof Exp:* Asst prof zool, Univ Ill, 64-65; from asst prof to prof biol, Univ Rochester, 65-84; PROF & HEAD DEPT ECOL & EVOLUTIONARY BIOL, UNIV ARIZ, 84- *Concurrent Pos:* Vis prof, Biol Sta, Univ Mich, 74, 75, 78 & 80. *Mem:* AAAS; Ecol Soc Am; Am Soc Naturalists (secy 83-86); Gen Soc Am; Soc Study Evolution. *Res:* Population ecology; intraspecific and interspecific competition in animal populations; insect and microbiol population genetics; evolutionary theory. *Mailing Add:* Dept Ecol & Evolutionary Biol Univ Ariz Tucson AZ 85721

ISTRE, CLIFTON O, JR, b Jennings, La, Sept 17, 32; m 56; c 3. AUDIOLOGY. *Educ:* Univ Ala, MA, 56; Ind Univ, PhD(hearing), 65. *Prof Exp:* Asst prof, 65-71, ASSOC PROF OTOLARYNGOL, MED SCH, TULANE UNIV, 71- *Concurrent Pos:* Consult, USPHS & USAF. *Mem:* Acoust Soc Am; Am Speech & Hearing Asn; assoc fel Am Acad Ophthal & Otolaryngol. *Res:* Noise and human hearing; electronystagmorphy and hearing. *Mailing Add:* 3812 Ridgelake Dr Metairie LA 70002

ITABASHI, HIDEO HENRY, b Los Angeles, Calif, July 7, 26; m 52; c 2. NEUROPATHOLOGY, NEUROLOGY. *Educ:* Boston Univ, AB, 49, MD, 54; Am Bd Path, dipl & cert neuropath, 66. *Prof Exp:* Intern, Univ Mich Hosp, 54-55, resident neurol, 55-58; assoc res neurologist, Univ Calif, San Francisco, 58-60, clin instr psychiat & neurol, Med Ctr, 60-64, asst clin prof, 64-65; asst prof neurol, Med Sch, Univ Mich, Ann Arbor, 65-68, asst prof path, 66-68, assoc prof neurol & path, 68-71; assoc prof, 71-75, PROF PATH & NEUROL, SCH MED, UNIV CALIF, LOS ANGELES, 75- *Concurrent Pos:* Nat Inst Neurol Dis & Blindness spec fel neuropath, 58-60; electroencephalographer, State Dept Ment Hyg, Sonoma State Hosp, Calif, 59-60; asst neuropathologist, Langley Porter Neuropsychiat Inst, Calif, 60-65; consult neuropath, San Francisco Gen Hosp, 64-65; consult neurol, Ypsilanti State Hosp, Mich, 66-71; consult, Vet Admin Hosp, Ann Arbor, 69-71; attend, Wadsworth Gen Hosp Vet Admin, Los Angeles; consult path, Sepulveda Vet Admin Hosp, 77-; consult neuropath, Dept Chief Med Examiner-Coroner, Los Angeles County, 77- *Mem:* Am Asn Neuropath; Am Acad Neurol; Am Acad Forensic Sci; Nat Asn Med Exam. *Res:* Electron microscopy of human muscle and degenerative, viral disorders of the central nervous system; neuropathology of dementias and amnestic disorders and head and neck trauma. *Mailing Add:* Dept Path Harbor-Univ Calif Los Angeles Med Ctr Torrance CA 90509

ITAKURA, KEIICHI, b Tokyo, Japan, Feb 18, 42; m 70; c 2. MOLECULAR BIOLOGY. *Educ:* Tokyo Col Pharm, BS, 65, PhD(pharm sci), 70. *Prof Exp:* Fel chem, Nat Res Coun Can, 71-74; sr scientist, Calif Inst Technol, 74-76; assoc res scientist, 76-78, SR RES SCIENTIST BIOL, CITY OF HOPE NAT MED CTR, 78- *Concurrent Pos:* Vis scientist, Nat Res Coun Can, 74-75; co-prin investr, Genentech, Inc, 76-; prin investr, Calif Inst Technol, 77-, vis assoc, 78-; prin investr, NIH, 78-81. *Res:* Chemical synthesis of nucleic acids for the study of molecular biology. *Mailing Add:* Dept Molecular Genetics City Hope Nat Med Ctr 1500 E Duarte Rd Duarte CA 91010

ITANO, HARVEY AKIO, b Sacramento, Calif, Nov 3, 20; m 49; c 3. BIOCHEMISTRY. *Educ:* Univ Calif, BS, 42; St Louis Univ, MD, 45; Calif Inst Technol, PhD(chem), 50. *Hon Degrees:* DSc, St Louis Univ, 87. *Prof Exp:* Intern, City of Detroit Receiving Hosp, 45-46; Am Chem Soc fel, Calif Inst Technol, 46-48, NIH fel, 48-50; sr asst surgeon, Nat Cancer Inst, 50-54, surgeon, 54-56, sr surgeon, Nat Inst Arthritis & Metab Dis, 56-58, med dir, 58-70, chief sect chem genetics, Lab Molecular Biol, 62-70; prof dept path, 70-88, EMER PROF DEPT PATH, SCH MED, UNIV CALIF, SAN DIEGO, 88- *Concurrent Pos:* Res fel chem, Calif Inst Technol, 50-52, sr res fel, 52-54; Minot lectr, AMA, 55; vis prof, Osaka Univ, 61-62; Harrington lectr, Univ Buffalo, 63; vis prof, Univ Chicago, 65, Univ Calif, San Francisco,

67; fel, Japan Soc Prom Sci, Okayama Univ, 83-84. *Honors & Awards:* Lilly Award Biol Chem, Am Chem Soc, 54. *Mem:* Nat Acad Sci; Am Soc Biol Chem; Am Soc Hemat; AAAS; Int Soc Hemat; Am Chem Soc. *Res:* Abnormal hemoglobins; chemical modification of amino acids and proteins; heme degradation; oxidative degradation of hemoglobin; chemically induced hemolytic anemia; heme ligands. *Mailing Add:* Dept Path Sch Med D-006 Univ Calif at San Diego La Jolla CA 92093

ITANO, WAYNE MASAO, b Pasadena, Calif, June 1, 51; m 83; c 1. ATOMIC & MOLECULAR PHYSICS. *Educ:* Yale Univ, BS, 73; Harvard Univ, MA, 75, PhD(physics), 79. *Prof Exp:* PHYSICIST, TIME & FREQUENCY DIV, NAT INST STANDARDS & TECHNOL, 79- *Concurrent Pos:* Secy-treas, Laser Sci Topical Group, Am Phys Soc, 90-; sci & technol agency fel, Commun Res Lab, Tokyo, 90. *Honors & Awards:* Gold Medal, Dept Com, 85; Stratton Award, Nat Inst Standards & Technol, 89. *Mem:* fel Am Phys Soc; Optical Soc Am. *Res:* Atomic and molecular hyperfine structure and g-factors, radio frequency and optical spectroscopy of stored ions, molecular beam radio frequency spectroscopy; radiation-pressure cooling of atoms and ions, atomic frequency standards. *Mailing Add:* Nat Inst Standards & Technol 325 Broadway St Boulder CO 80303-3328

ITEN, LAURIE ELAINE, b Los Angeles, Calif, Jan 5, 47; m 77. DEVELOPMENTAL BIOLOGY. *Educ:* Univ Calif, Irvine, BSc, 71, PhD(biol), 75. *Prof Exp:* Fel develop biol, Ctr Pathobiol, Univ Calif, Irvine, 75-76 & Dept Avian Sci, Univ Calif, Davis, 76-77; ASST PROF BIOL, PURDUE UNIV, WEST LAFAYETTE, 77- *Concurrent Pos:* Nat Inst Child Health & Human Develop fel, 75-76; Am Cancer Soc fel, 76-77; prin investr, NSF res grant, 77-83. *Mem:* Int Soc Develop Biologists; Soc Develop Biol; Am Soc Zoologists; AAAS. *Res:* Pattern formation during animal development. *Mailing Add:* Dept Biol Sci Purdue Univ West Lafayette IN 47907

ITIABA, KIBE, b Jamaica, WI, Oct 1, 31; m 64; c 3. BIOCHEMISTRY, PHYSIOLOGY. *Educ:* Univ Toronto, BSc, 62, PhD(physiol, biochem), 66; Univ Leicester, MSc, 63. *Prof Exp:* Res assoc physiol & biochem, Banting & Best Inst Med Res, Univ Toronto, 63-66; lectr physiol, Med Sch, Makerere Univ, Uganda, 66-68; BIOCHEMIST, DEPT EXP MED, CLIN, MCGILL UNIV & ROYAL VICTORIA HOSP, MONTREAL, 68- *Concurrent Pos:* Asst prof dept exp med, McGill Univ, 68- *Mem:* Can Biochem Soc; Brit Biochem Soc; Can Soc Clin Chemists. *Res:* Purine phosphorinosyl transferases in erythrocyte and fibroblast cell extracts. *Mailing Add:* Dept Biochem Royal Victoria Hosp Montreal PQ H3A 1A1 Can

ITIL, TURAN M, b Bursa, Turkey, Aug 12, 24; m 55; c 2. PSYCHIATRY. *Educ:* Istanbul Univ, MS, 43, MD, 48; Univ Erlangen, Venia Legendi, 63; Ger Bd Psychiat & Neurol, dipl, 60. *Prof Exp:* Intern med, Istanbul Univ Hosps, 47-48; intern, Seferihisar Mil Hosp & Workers Hosp, Istanbul, 49; resident surg & internal med, Neuropsychiat Clin, Univ Tubingen, 49-50, resident neurol, 50-51, resident psychiat, 51-52, asst in neurol, Col Med & resident child psychiat, 52-53; resident neuropath, Neuropsychiat Clin, Univ Erlangen, 53, instr psychiat, Col Med & res assoc neurol & psychiat, 53-55, instr psychiat & EEG, neuropsychiatrist, dir out-patient dept & chief EEG dept, 55-62; assoc prof psychiat & neurol, Sch Med, Univ Mo-Columbia, 62-64, from assoc prof to prof psychiat, 64-73, from asst chmn to assoc chmn dept, 68-73; RES PROF & DIR DIV BIOL PSYCHIAT, NEW YORK MED COL, 73- *Concurrent Pos:* Assoc chmn, Mo Inst Psychiat, 62-63, sr psychiatrist & prin res scientist, 63-65, chief sect EEG & clin neurophysiol, 65-73, chief sect psychopharmacol, 67-73; co-investr, USPHS Res Grants, 63-66 & 70-75, proj dir, 63-68 & 66-68, prin investr, 71-75; pvt docent, Univ Erlangen, 63-; consult, Clin EEG Lab, St Louis State Hosp, 64-65 & 69; docent, Istanbul Univ, 67-; mem, Am Schizophrenia Found; prin scientist, Int Asn Psychiat Res, 73-; res psychiatrist, Vet Admin Hosp, Montrose, NY, 74- *Mem:* Am Col Neuropsychopharmacol; Am Soc Med Psychiat; Acad Psychosom Med; Am Psychiat Asn; Am Psychopath Asn. *Res:* Electroencephalography; psychopharmacology; neurology; drugs for psychiatric patients using computer-analyzed model; electrophysiological markers to predict the outcome of psychiatric patients during drug treatment; biological diagnosis of psychiatric syndromes, treatment of therapy resistant depression and schizophrenia. *Mailing Add:* 150 White Plains Rd Tarrytown NY 10591

ITKIN, IRVING HERBERT, b New York, NY, Aug 1, 17; m 42; c 3. ALLERGY. *Educ:* Ind Univ, Bloomington, AB, 35, MD, 39. *Prof Exp:* Intern, Kings County Hosp, Brooklyn, 39-41; resident, Mt Sinai Hosp, New York, 41-42; attend & chief allergy clin, Worcester Mem Hosp, 47-59; chief dept asthma-allergy, Nat Jewish Hosp Denver, 59-69; dir, Div Clin Immunol, Hahnemann Univ, 69-82, prof med, 69-90; RETIRED. *Concurrent Pos:* Asst, Mass Gen Hosp, 49-59; asst clin prof, Univ Colo, 61-66, asst prof, 66-69; vpres, Sect Allergy, AMA, 64-65. *Mem:* AMA; fel Am Col Physicians; fel Am Acad Allergy & Immunol. *Res:* Inhalation challenge in asthma; Candida Albicans as an allergen; chemical mediators of hypersensitivity; exercise in asthma; role of infection and antibiotics in asthma. *Mailing Add:* 11 Lochwick Rd Palm Beach Gardens FL 33418-3702

ITO, JUNETSU, MICROBIOL GENETICS, BIOTECHNOLOGY. *Educ:* Kyoto Univ, Japan, PhD(biochem), 67. *Prof Exp:* PROF MED MICROBIOL, COL MED, UNIV ARIZ, 83- *Mailing Add:* Dept Microbiol & Immunol Col Med Univ Ariz, 1501 N Campbell Ave Tucson AZ 85724

ITO, KEITH A, b Sebastopol, Calif, Mar 21, 39. BACTERIOLOGY, FOOD SCIENCE. *Educ:* Univ Calif, Berkeley, BA, 61. *Prof Exp:* DIR MICROBIOL & PROCESSING, NFPA, 75-, ASST MGR, WESTERN RES LAB, 79-, VPRES, 81- *Mem:* Am Soc Microbiol; Soc Appl Bacteriol; Inst Food Technologists. *Res:* Growth and resistance of clostridium botulinum, food spoilage; thermal processing of foods. *Mailing Add:* Western Res Lab 6363 Clark Ave Dublin CA 94568-3097

ITO, PHILIP J, b Kapaa, Kauai, Hawaii, Mar 11, 32; m 60; c 2. HORTICULTURE & PLANT BREEDING, TROPICAL FRUITS & NUTS. *Educ:* Univ Hawaii, BS, 58; Univ Minn, PhD(veg breeding), 64. *Prof Exp:* Assoc horticulturist, 64-80, PROF & HORTICULTURIST, UNIV HAWAII, HILO, 80- *Concurrent Pos:* Consult, tropical fruits & nuts. *Mem:* Am Soc Hort Sci. *Res:* Tropical and subtropical fruit and nut breeding and culture. *Mailing Add:* Exp Sta Univ Hawaii 461 W Lanikaula St Hilo HI 96720

ITO, SUSUMU, b Stockton, Calif, July 27, 19; m 48; c 4. ELECTRON MICROSCOPY, CYTOLOGY. *Educ:* Fenn Col, BS, 50; Western Reserve Univ, MS, 51, PhD, 54. *Hon Degrees:* MS, Harvard Univ, 66. *Prof Exp:* Asst zool, Western Reserve Univ, 52-54; USPHS fels, 54-58; instr anat, Med Col, Cornell Univ, 58-60; assoc, 60-64, from asst prof to assoc prof, 64-69, PROF ANAT, HARVARD MED SCH, 69- *Mem:* Am Soc Cell Biol; Electron Micros Soc; Tissue Cult Asn; Am Gastroenterol Asn; Am Asn Anatomists. *Res:* Electron microscopy of biological material; cell biology; correlation of fine structure with function; gastronintestinal tract; gastric mucosal repair; carbonic anhydrase localization. *Mailing Add:* Dept Anat Harvard Med Sch 220 Long Wood Ave Boston MA 02115

ITO, TAKERU, b Tokyo, Japan, May 10, 28; US citizen; m 59; c 3. BIOCHEMISTRY. *Educ:* Trinity Univ, Tex, BS, 54; Okla State Univ, MS, 56; Univ Calif, Berkeley, PhD(biochem), 58. *Prof Exp:* Asst prof physiol chem & res assoc nutrit, Ohio State Univ, 62-67; PROF BIOL, E CAROLINA UNIV, 67- *Concurrent Pos:* Univ fel, Univ Pa, 58-60; USPHS fel, Osaka Univ, 61-62. *Res:* Microbial iron metabolism; essential fatty acids; oxidative phosphorylation in animal and plant mitochondria and in microorganisms. *Mailing Add:* Dept Biol East Carolina Univ Greenville NC 27858-4353

ITO, Y(ASUO) MARVIN, b Los Angeles, Calif, July 4, 40; m 65; c 2. ENGINEERING MECHANICS. *Educ:* Univ Calif, Los Angeles, BS, 63, MS, 65, PhD(eng), 68. *Prof Exp:* Res asst eng, 63-65, res engr, 65-68, ASST PROF ENG, UNIV CALIF, LOS ANGELES, 68- *Concurrent Pos:* NSF grants, 69-72; res fel, Harvard, 71-72. *Mem:* Am Acad Mech; Am Soc Mech Engrs. *Res:* Solid mechanics, especially physical theories of inelastic solids and composites and computational approaches. *Mailing Add:* 11260 Overland Ave Unit 15a Culver City CA 90230

ITOGA, STEPHEN YUKIO, b Honolulu, Hawaii. COMPUTER SCIENCE. *Educ:* Cornell Univ, BS, 65, ME, 66; Univ Calif, Los Angeles, PhD(syst sci), 73. *Prof Exp:* Mem tech staff, TRW Systs, TRW Inc, 66-75; ASST PROF COMPUT SCI, UNIV HAWAII, 75- *Concurrent Pos:* Mem staff, Fairchild Test Systs, 81-82. *Mem:* Asn Comput Mach; Soc Indust & Appl Math; Inst Elec & Electronics Engrs. *Res:* Programming methodology; software systems. *Mailing Add:* Dept Info & Comput Sci 2565 Mall Univ Hawaii Honolulu HI 96822

ITOH, TATSUO, b Tokyo, Japan, May 5, 40; m 69; c 2. MICROWAVES, MILLIMETER WAVES. *Educ:* Yokohama Nat Univ, BS, 64, MS, 66; Univ Ill, PhD(elec eng), 69. *Prof Exp:* Res assoc fel elec eng, Univ Ill, 69-71, res asst prof, 71-76; sr res engr, Stanford Res Inst, 76-77; assoc prof elec eng, Univ Ky, 77-78; assoc prof elec eng, Univ Tex, Austin, 78-81, prof, 81-90; PROF ELEC ENG, UNIV CALIF, LOS ANGELES, 91- *Concurrent Pos:* Consult, Hughes Aircraft Co, 79- *Mem:* Int Sci Radio Union; fel Inst Elec & Electronics Engrs. *Res:* Millimeter wave circuit; field theory. *Mailing Add:* Dept Elec Eng Univ Calif Los Angeles CA 90024-1594

ITON, LENNOX ELROY, b St Vincent, WI, Jan 3, 49; UK citizen; m 80. PHYSICAL CHEMISTRY, SURFACE SCIENCE. *Educ:* McGill Univ, BSc, 70; Princeton Univ, PhD(chem), 76. *Prof Exp:* Fel solid state sci, 75-77, ASST CHEMIST SOLID STATE SCI, ARGONNE NAT LAB, US DEPT ENERGY, UNIV CHICAGO, 78- *Mem:* Am Chem Soc; AAAS; Sigma Xi; Am Phys Soc. *Res:* Electron paramagnetic resonance; nuclear magnetic resonance; x-ray spectroscopic absorption techniques for electronic, dynamic, structural and chemical characterizations of surfaces; adsorbates and heterogeneous catalysts. *Mailing Add:* Argonne Nat Lab Sci & Tech Div 9700 S Cass Ave Argonne IL 60439

ITTEL, STEVEN DALE, b Hamilton, Ohio, Nov 8, 46; m 68; c 2. INORGANIC CHEMISTRY. *Educ:* Miami Univ, Ohio, BS, 68; Northwestern Univ, Evanston, PhD(inorg chem), 74. *Prof Exp:* Chemist, Dept Health & Human Serv, USPHS, 68-70; chemist, 74-79, RES SUPVR, E I DU PONT DE NEMOURS & CO, INC, 79- *Mem:* Am Chem Soc; Catalysis Soc. *Res:* Homogeneous catalysis; olefin polymerization; activation of carbon; metal vapor synthesis; use of phosphorus ligands to control organometallic reactions through steric and electronic effects. *Mailing Add:* Cent Res & Develop Exp Sta E I Du Pont de Nemours & Co Inc Wilmington DE 19880-0328

ITTER, STUART, biochemistry, for more information see previous edition

ITURRIAN, WILLIAM BEN, b Hudson, Wyo, May 17, 39; m 65; c 4. NEUROPHARMACOLOGY. *Educ:* Univ Wyo, BS, 62; Ore State Univ, PhD(pharmacol), 68. *Prof Exp:* Asst pharmacol, Wash State Univ, 62-63; asst, Ore State Univ, 64-66, res assoc, 66-67; asst prof, 67-71, ASSOC PROF PHARMACOL, UNIV GA, 71- *Mem:* Soc Neurosci; Int Soc Develop Psychobiol; Am Soc Exp Pharmacol & Therapeut. *Res:* Behavioral pharmacology; response of the immature nervous system to drugs or noise. *Mailing Add:* Dept Pharmacol Univ Ga Col Pharm Athens GA 30602

ITZKAN, IRVING, b Brooklyn, NY, Dec 4, 29; m 57; c 2. PHYSICS, ELECTRICAL ENGINEERING. *Educ:* Cornell Univ, BEngPhys, 52; Columbia Univ, MSEE, 61; NY Univ, PhD(physics), 69. *Prof Exp:* Engr, Sperry Rand Corp, 57-69; PRIN RES SCIENTIST, AVCO-EVERETT RES LABS, 69- *Mem:* Inst Elec & Electronics Engrs; Am Phys Soc; Optical Soc Am. *Res:* Lasers and modern optics. *Mailing Add:* Res Sci MIT 6-014 Spectro Lab 77 Massachusetts Ave Cambridge MA 02139

ITZKOWITZ, GERALD LEE, b Brooklyn, NY, June 2, 38; m 67; c 2. MATHEMATICS. *Educ:* Mass Inst Technol, BS, 60; Univ Rochester, MA, 63, PhD(math), 65. *Prof Exp:* Asst math, Univ Rochester, 60-64; lectr, State Univ NY, Buffalo, 64-65, asst prof, 65-71; asst prof, 71-77, ASSOC PROF MATH, QUEENS COL, NY, 78- *Concurrent Pos:* Assoc investr, NSF grant, 66-67; instr, NSF, Inst High Sch Teachers, 86-87. *Mem:* Math Asn Am; Am Math Soc. *Res:* Harmonic analysis on extensions of Haar measure and other invariant integrals on groups; embedding theorems for topological groups, density character; functional equations; integration theory; uniform spaces and uniformities on topological groups. *Mailing Add:* Dept Math Queens Col Flushing NY 11367

IULIUCCI, JOHN DOMENIC, b Camden, NJ. TOXICOLOGY. *Educ:* Temple Univ, BS, 67, MS, 70, PhD(pharmacol), 73; Am Bd Toxicol, dipl. *Prof Exp:* Scientist toxicol, Warner-Lambert Co, 72-75; supvr, Adria Labs Inc, 75-77, mgr toxicol, 77-84; dir preclin, Clin & Regulatory Affairs, 84-86, DIR TOXICOL & PHARMACOL, CENTOCOR, INC, 86- *Mem:* Environ Mutagen Soc; Soc Toxicol; Am Col Toxicol; Teratology Soc. *Res:* Safety evaluation of new investigational drugs intended for human use; preclinical acute, subacute and chronic toxicity; carcinogenic potential; teratology; reproduction toxicology; mutagenicity. *Mailing Add:* Centocor Inc 244 Great Valley Pkwy Malvern PA 19355

IUVONE, PAUL MICHAEL, b New York, NY, Sept 4, 51; m 79; c 3. NEUROCHEMISTRY, NEUROPHARMACOLOGY. *Educ:* Univ Fla, BS, 72, PhD(neurosci), 76. *Prof Exp:* Fel pharmacol, NIMH, 76-78; from asst prof to assoc prof pharmacol, Emory Univ, 78-90, asst prof ophthal, 80-86, dir, Pharmacol Grad Prog, 85-88, ASSOC PROF OPHTHAL, SCH MED, EMORY UNIV, 86-, PROF PHARMACOL, 90- *Honors & Awards:* Nat Res Serv Award, NIMH, 76. *Mem:* Am Soc Neurochem; Int Soc Neurochem; Am Soc Pharmacol Exp Therapeut; Asn Res Vision Ophthamol; AAAS; Soc Neurosci. *Res:* Cellular and systems approaches to neurotransmitter function; retinal cell biology and neurotransmitters; neurotransmitter receptors and second messengers; dopamine in brain function and dysfunction. *Mailing Add:* Dept Pharmacol Emory Univ Atlanta GA 30322

IVAN, MICHAEL, b Falkusovce, Slovakia, Aug 1, 38; Can citizen; m 64; c 2. ANIMAL NUTRITION, ANIMAL PHYSIOLOGY. *Educ:* Univ Nitra, Ing, 62, DSc, 91; Univ Alta, MSc, 71; Univ New Eng, PhD(animal nutrit), 74. *Prof Exp:* Nutritionist, animal nutrit, UKSUP, 62-64; chemist, org synthesis, Delmar Chem, 65-67; RES SCIENTIST, ANIMAL NUTRIT, ANIMAL RES CTR, 75- *Honors & Awards:* Medal for Excellence in Nutrit & Meat Sci, Can Packers. *Mem:* Brit Nutrit Soc; Am Soc Animal Sci; Can Soc Animal Sci; Am Dairy Sci Asn. *Res:* Rumen fermentation-protozoa. *Mailing Add:* Animal Res Ctr Agr Can Ottawa ON K1A 0C6 Can

IVANETICH, RICHARD JOHN, b San Francisco, Calif, Feb 12, 41; m 66; c 2. OPERATIONS RESEARCH, INFORMATION SYSYTEMS. *Educ:* Univ Calif, Berkeley, BS, 63; Harvard Univ, PhD(physics), 69. *Prof Exp:* Asst prof physics, Harvard Univ, 69-74; res staff mem, 75-84, asst dir, Syst Eval Div, 84-90, DIR, COMPUTER & SOFTWARE ENG DIV, INST DEFENSE ANALYSIS, 90- *Mem:* Am Phys Soc. *Res:* Defense systems, operations, and policy analysis; primarily concerned with computer and information systems strategic and theater nuclear systems; command, control and communication systems and procedures; crisis management; theoretical high energy physics. *Mailing Add:* 1801 N Beauregard St Alexandria VA 22311

IVANKOVICH, ANTHONY D, b Delrljaca, Yugoslavia, Mar 25, 39. ANESTHESIOLOGY. *Educ:* Univ Zagreb Med Sch, Yugoslavia, MD, 63; Am Bd Anesthesiol, dipl, 71. *Prof Exp:* Instr anesthesiol, Univ Chicago, Pritzker Sch Med, 69; asst prof, Loyola Univ, Stritch Sch Med, 70-71; prof, Univ Ill, Lincoln Sch Med, 75-80; PROF & CHMN ANESTHIOL, RUSH MED COL, RUSH-PRESBY ST LUKE'S MED CTR, 80- *Concurrent Pos:* Attend anesthesiologist, Loyola Univ Hosp, 70-71; consult, Suburban Tuberc Sanitorium, Hinsdale, Ill, 70-71; attend & dir anesthesia res, Michael Reese Med Ctr, 71-74; chief, Operating Rm Serv, 801st Gen Hosp, US Army Res, 71-73, chief surg, 73-74 & assoc chief, 74-76; lectr anesthesiol, Loyola Univ, Stritch Med Sch, 71-81; chmn, Ill Masonic Med Ctr, 74-80; consult, Shriner's Hosp Crippled Children, 77-82. *Mem:* AMA; Am Soc Anesthesiologists; Int Anesthesia Res Soc; Am Heart Asn; Sigma Xi; Am Col Chest Physicians; Am Pain Soc. *Res:* Neurosurgical and cardiovascular clinical anesthesia, pain treatment, cerebral blood flow, pharmacology of digitalis and anti-hypertensive drugs, patient monitoring and safety in the operating room. *Mailing Add:* Dept Anesthesiol Rush-Presby St Luke's Med Ctr 1753 W Congress Pkwy Chicago IL 60612

IVANSO, EUGENE V, b Braddock, Pa, Dec 8, 08; m 38, 52; c 1. PHYSICAL METALLURGY. *Educ:* Case Inst Technol, BS, 31, MS, 32. *Prof Exp:* Molder apprentice, Johnston & Jennings Co, 26-29; instr math, Case Inst Technol, 31-32; field engr & mgr elec sound equip, Cleveland Sound Equip Co, 32-34; res metallurgist alloy develop, Brush Beryllium Co, 34-36; phys metallurgist, Bundy Tubing Co, 36-40; chief metallurgist, corrosion & foundry, Wyandotte Chem Corp, 40-44; mgr alloy sales & field eng nickel alloys, Steel Sales Corp, 44-52; vpres & dir, Commercial Lab, Detroit Testing Lab, Inc, 52-55, dir & consult, 55-65; assoc dir, Appl Mgt & Tech Ctr, Wayne State Univ, 65-74; RETIRED. *Concurrent Pos:* Vpres, March Corp, 55-65. *Res:* Constitution of beryllium-copper alloys; internal stresses in aluminum alloys; desulfurization of cast iron; investigations in mechanism of galvanic corrosion; surface tension of molten metals; diffusion of metals; high temperature and room temperature corrosion of all types alloys and effects of properties; educational up-dating of industrial and professional personnel in science, technology and management; administration. *Mailing Add:* 911 Lakepointe Ave Grosse Pointe Park MI 48230

IVARSON, KARL C, soil microbiology, for more information see previous edition

IVASH, EUGENE V, b Windsor, Ont, July 24, 25; nat US; m 53; c 3. THEORETICAL PHYSICS. *Educ:* Univ Mich, BS, 45, MS, 47, PhD(physics), 52. *Prof Exp:* Asst physics, Univ Mich, 47-52; from asst prof to assoc prof, 52-66, PROF PHYSICS, UNIV TEX, AUSTIN, 66-, RES SCIENTIST, CTR NUCLEAR STUDIES, 67- *Concurrent Pos:* Res scientist, Nuclear Physics Lab, 52-67; res partic, Oak Ridge Nat Lab, 55-57, consult, 57-58; tech adv, Chulalongkorn Univ, Bankok, 58-59; lectr, Cambridge Univ, 59; vis scientist, Lawrence Radiation Lab, Univ Calif, 61; consult, Gen Atomic Div, Gen Dynamics, 63. *Mem:* Am Phys Soc; Am Asn Physics Teachers. *Res:* Nuclear physics; quantum mechanics; microwave spectroscopy; plasma physics. *Mailing Add:* Dept Physics Univ Tex Austin TX 78712

IVASHKIV, EUGENE, b Ukraine, Mar 21, 23; US citizen; m 53; c 2. PHARMACEUTICS, CHEMISTRY. *Educ:* Ukrainian Polytech Inst, Ger, BSForestry, 50; Columbia Univ, BSChem, 57; Polytech Inst Brooklyn, BSChemEng, 59; Newark Col Eng, MSChemEng, 63. *Prof Exp:* SR RES SCIENTIST, SQUIBB INST MED RES, 57-, RES FEL, 63-, SR RES LEADER, 89- *Honors & Awards:* Quality/Productivity Innovation Award, Squibb, 87. *Mem:* Am Inst Chem Engrs; NY Acad Sci; Ukrainian Engrs Soc Am (pres, 72-74); Shevchenko Sci Soc. *Res:* Development of new methods for following the biosynthesis of antibiotics, microbial and enzymatic conversion of steroids, for drugs in dosage forms, residual methods, bioavailibity of drugs, metabolites; extraction, separation and purification, kinetic studies and characterization of new products; modern instrumentation, automation. *Mailing Add:* Squibb Inst Med Res New Brunswick NJ 08903

IVATT, RAYMOND JOHN, b Hong Kong, Jan 1, 49; Brit citizen; m 72; c 2. VASCULAR CELL BIOLOGY, GROWTH FACTOR BIOLOGY. *Educ:* London Univ, BSc, 70, PhD(biochem), 75. *Prof Exp:* Postdoctorate tumor biol, ETH-Zurich, 72-75; postdoctorate biochem, Princeton Univ, 75-78; res assoc tumor biol, Cancer Ctr, Mass Inst Technol, 78-80; asst prof tumor biol, M D Anderson Hosp, 81-87; sr scientist, proj mgr, Cetus Corp, 87-91; DIR BIOL RES, BERLEX LABS, 91- *Concurrent Pos:* Lectr biochem, Princeton Univ, 77-78; consult, Cetus Corp, 80-87; adj asst prof, Univ Tex Cancer Ctr, 87-90. *Mem:* Am Soc Cell Biol; Am Heart Asn. *Res:* Development of ethical drugs to treat acute and chronic vascular disorders. *Mailing Add:* Berlex Labs Inc 110 E Hanover Ave Cedar Knolls NJ 07927

IVATURI, RAO VENKATA KRISHNA, b Madras, India, Jan 20, 60; m 88. NUTRITION EDUCATION, INTERNATIONAL NUTRITION. *Educ:* Andhra Pradesh Agr Univ, BS, 81; Kans State Univ, MS, 83; Univ Nebr, PhD(nutrit), 86. *Prof Exp:* Coordr nutrit, Univ Nebr, Lincoln, 85-86; ASST PROF NUTRIT, IND STATE UNIV, 86- *Concurrent Pos:* Nutrit consult, WCent Ind Econ Develop Dist Inc, 86-; sem leader, Weight Mgt Clin, Oper Weight Loss, Adult Fitness Prog, Ind State Univ, 86-, coordr-World Food Day, World Food Day Teleconf, 86 & 88 & prin investr res projs, 86-; Fullbright-Hays partic, Fullbright-Hays: Group to Brazil, 91. *Mem:* Am Inst Nutrit; Fedn Am Socs Exp Biol; Sigma Xi; Soc Nutrit Educ. *Res:* Dietary sugars and their interactions with minerals in humans; nutrition education research with senior citizens; investigating factors that affect food consumption in young adults; nutrition education research in the developing countries. *Mailing Add:* Dept Home Econ Ind State Univ Terre Haute IN 47809

IVENS, MARY SUE, b Maryville, Tenn, Aug 23, 29. MEDICAL MYCOLOGY, INFECTIOUS DISEASES. *Educ:* Tenn State Univ, BS, 49; Tulane Univ Sch Med, MS, 63; La State Univ Med Ctr, PhD(microbiol & mycol), 66. *Prof Exp:* Res asst nuclear med, Oak Ridge Nat Lab, 49-51; dir, Microbiol Lab, Lewis-Gale Hosp, Roanoke, Va, 53-57; res mycologist, Ctr Dis Control, Atlanta, Ga, 57-60; res assoc med, Med Ctr, La State Univ, 63-66, instr med, 66-72, instr microbiol, 68-72; ASSOC PROF NATURAL SCI, DILLARD UNIV, 72-; CLIN PROF MED, LA STATE UNIV MED CTR, 72- *Concurrent Pos:* Res asst, Baroness Erlanger Hosp, Chatanooga, Tenn, 52-53; dir, Mycol Lab, Med Ctr, La State Univ, 63-72, lectr, Sch Dentistry, 68-70; consult, Conf Ctr Mycotic Sera, WHO, 69, Med Mycol, Charity Hosp, New Orleans, 66-; prin investr NSF grants, Med Ctr, La State Univ, 68-72, clin prof med, 72-; prin investr NIH grants, Dillard Univ, 76-; Macy fel, Marine Biol Lab, Woods Hole, Mass, 78. *Mem:* Int Soc Human & Animal Mycol; Med Mycol Soc Am; Am Soc Microbiol; AAAS; Sigma Xi; Infectious Dis Soc Am. *Res:* Immunology of the systemic mycoses; biochemical characterization of fungal skin testing antigens; mechanisms of infectious diseases. *Mailing Add:* Div Natural Sci Dillard Univ New Orleans LA 70122

IVERS, DREW RUSSELL, b Vincennes, Ind, Aug 12, 46; m 68; c 3. RESEARCH ADMINISTRATION. *Educ:* Purdue Univ, BS, 68; Iowa State Univ, PhD, 74. *Prof Exp:* Res assoc, Iowa State Univ, Ames, 72-74; res mgr, Cargill, Inc, 74-76; DIR PLANT RES, LAND O'LAKES, INC, 88- *Concurrent Pos:* Mem bd, Nat Soybean Variety Rev, 77-86 & Nat Plant Genetic Resources Bd, Washington, DC, 82-88; chmn, Basic Soy Res Comt, Am Seed Trade Asn, 87-89. *Mem:* Crop Sci Soc Am; Am Seed Trade Asn; Nat Coun Com Plant Breeders. *Res:* Soybean breeding methods; awarded 12 patents. *Mailing Add:* Land O'Lakes Res Farm RR 2 Webster City IA 50595

IVERSEN, EDWIN SEVERIN, b Ferndale, Mich, Dec 4, 22; m 58; c 2. FISH BIOLOGY. *Educ:* Univ Wash, BS, 49, MS, 53; Tex A&M Univ, PhD(biol oceanog), 61. *Prof Exp:* Fishery res biologist, Fishery Res Inst, Univ Wash, 48-52 & US Fish & Wildlife Serv, Hawaii, 53-56; assoc prof fisheries biol, 56-74, PROF BIOL & LIVING RESOURCES, ROSENSTIEL SCH MARINE & ATMOSPHERIC SCI, UNIV MIAMI, 74- *Mem:* Am Fisheries Soc; Am Inst Fishery Res Biol; World Mariculture Soc. *Res:* General marine fishery biology; population dynamics; marine parasites; mariculture. *Mailing Add:* Dept Marine Sci Univ Miami Coral Gables FL 33124

IVERSEN, GUDMUND R, b Norway, Sept 14, 34; US citizen; m 62, 74; c 4. APPLIED STATISTICS. *Educ:* Univ Mich, AM, 60 & 61; Harvard Univ, PhD(statist), 69. *Prof Exp:* Instr math, Eastern Mich Univ, 61-62; res assoc sociol, Univ Oslo, 62-64; fac assoc, Inst Social Res & asst prof sociol, Univ Mich, 69-72, assoc prof, 72-77, PROF STATIST & CONSULT STATISTICIAN, CTR SOCIAL & POLICY STUDIES, SWARTHMORE COL, 72- *Mem:* Am Statist Asn; Am Sociol Asn; Biomet Soc. *Res:* Sociological methodology; application of statistics to the social sciences. *Mailing Add:* Dept Math Swarthmore Col Swarthmore PA 19081

IVERSEN, JAMES D(ELANO), b Omaha, Nebr, Apr 1, 33; m 60; c 2. AEROSPACE ENGINEERING. *Educ:* Iowa State Univ, BS, 56, MS, 58, PhD, 64. *Prof Exp:* From instr to assoc prof, 56-70, PROF AEROSPACE ENG, IOWA STATE UNIV, 70- *Concurrent Pos:* Opers analyst, US Air Force, NASA; chmn, Dept Aerospace Eng, Iowa State Univ, 87-90. *Mem:* Am Inst Aeronaut & Astronaut; Am Soc Eng Educ; Am Geophys Union; Am Soc Civil Eng. *Res:* Aerodynamics; micrometeorology; aeolian geology. *Mailing Add:* Dept Aerospace Eng & Eng Mech Iowa State Univ Ames IA 50011

IVERSON, A EVAN, b Stanley, NDak. COMPUTER VISION, IMAGE EXPLOITATION. *Educ:* Univ NMex, BS; Univ Ariz, MS, PhD(appl math). *Prof Exp:* Scientist, Los Alamos Nat Lab, 77-90; SR SCIENTIST, SCI APPLICATIONS INT CORP, 90- *Mem:* Soc Indust & Appl Math; Appl Computational Electromagnetics Soc. *Res:* Mathematical modeling of physical systems, numerical analysis, signal and image processing and computer vision; over 20 publications in these areas. *Mailing Add:* SAIC 5151 E Broadway Suite 900 Tucson AZ 85711

IVERSON, F KENNETH, b Downers Grove, Ill, Sept 18, 25. STEEL PRODUCTION. *Educ:* Cornell Univ, BS; Purdue Univ, MS. *Hon Degrees:* Dr, Univ Nebr & Purdue Univ. *Prof Exp:* Vpres & gen mgr, Vulcraft Div, Nucor Corp, 62-63, group vpres, 63-65, pres, 65-84, CHMN & CHIEF EXEC OFFICER, NUCOR CORP, 84- *Honors & Awards:* Nat Medal of Technol, 91. *Res:* Low-cost steel production; author of numerous technical and business articles. *Mailing Add:* Nucor Corp 4425 Randolph Rd Charlotte NC 20211

IVERSON, JOHN BURTON, b Omaha, Nebr, Oct 4, 49; m 71; c 2. VERTEBRATE BIOLOGY, HERPETOLOGY. *Educ:* Hastings Col, BA, 71; Univ Fla, MS, 74, PhD(zool), 77. *Prof Exp:* ADJ CUR HERPET, FLA STATE MUS, UNIV FLA, 77-; ASST PROF BIOL, EARLHAM COL, 78- *Concurrent Pos:* Prin investr, Sigma Xi grant, 77-78, Theodore Roosevelt Mem Fund, 77-78 & Am Philos Soc, 78, NSF, 80-82 & 84, US Fish & Wildlife Serv Grant, 81-83 & 85; asst investr, BLM grant, 77-78. *Mem:* Am Soc Ichthyologists & Herpetologists; Herpetologists League; AAAS; Ecol Soc Am; Soc Studies Amphibians & Reptiles. *Res:* Herpetology, specifically systematics and ecology of freshwater turtles; ecology and behavior of large herbivorous Iguanine lizards in the American tropics; reproductive strategies of turtles. *Mailing Add:* Dept Biol Earlham Col Richmond IN 47374

IVERSON, KENNETH EUGENE, b Camrose, Alta, Dec 17, 20; m 46; c 4. APPLIED MATHEMATICS. *Educ:* Queen's Univ, Can, BA, 50; Harvard Univ, AM, 51, PhD(appl math), 54. *Prof Exp:* Instr appl math, Harvard Univ, 54-55, asst prof, 55-60; I P Sharp Assoc, Toronto, Can, 80-87; res staff mem, Sci Ctr, IBM Corp, 60-71, IBM FEL, 71- *Honors & Awards:* Harry Goode Award, Am Fedn Info Processing Soc, 75; Tueing Award, Asn Comput Mach, 79. *Mem:* Nat Acad Eng; Inst Elec & Electronics Engrs. *Res:* Automatic computers and programming. *Mailing Add:* 70 Erskine Ave Toronto ON M4P 1Y2 Can

IVERSON, LAURA HIMES, b Michigan City, Ind, Apr 25, 60; m 82; c 1. HEALTH POLICY ANALYSIS. *Educ:* Carleton Col, BA, 82; Univ Southern Calif, MSG & MPA, 85. *Prof Exp:* Res assoc, 85-87, SR RES ASSOC HEALTH CARE, INTERSTUDY CTR AGING & LONG TERM CARE, 87- *Mem:* sigma Xi; Geront Soc Am; Am Pub Health Asn; Am Pub Aging. *Res:* Improved financial planning and long term health care services to the elderly; author of numerous publication and articles related to long term care financing , medicaire HMOs and case management. *Mailing Add:* 2464 Elm Dr White Bear Lake MN 55110

IVERSON, RAY MADS, b Tremonton, Utah, Nov 3, 27; m 53; c 3. BIOLOGY. *Educ:* Reed Col, BA, 51; Stanford Univ, PhD, 57. *Prof Exp:* Asst prof zool, Univ Miami, 58-64, assoc prof, 64-66, prof, 66-72; PROF BIOL SCI, FLA ATLANTIC UNIV, 72- *Mem:* AAAS; Sigma Xi. *Res:* Cellular physiology. *Mailing Add:* Dept Biol Sci Fla Atlantic Univ Boca Raton FL 33432

IVERSON, STUART LEROY, b Albert Lea, Minn, Oct 12, 39; m 62; c 2. MAMMALIAN ECOLOGY, RADIATION ECOLOGY. *Educ:* St Olaf Col, BA, 61; Univ NDak, MA, 63; Univ BC, PhD(zool), 67. *Prof Exp:* Group leader ecol, Environ Res Br, Whiteshell Labs, Atomic Energy Can Ltd, 67-77, head, Ecol Res Sect, 77-81, Environ Res Br, 81-83, sr adv exec vpres, Head Off, Res Co, 83-84, HEAD RADIATION, APPL RES BR, WHITESHELL LABS, ATOMIC ENERGY CAN LTD, 85- *Mem:* Ecol Soc Am; Am Soc Mammal. *Res:* Use of radiation for food preservation, crosslinking and other industrial processes; mammalian population ecology and physiological ecology; long term environmental implications of methods of radioactive waste management. *Mailing Add:* 122 Burrows Rd Pinawa MB R0E 1L0 Can

IVES, DAVID HOMER, b Rockford, Ill, Apr 6, 33; m 56; c 2. ENZYMOLOGY, BIOCHEMISTRY. *Educ:* Cornell Col, AB, 55; Univ Minn, PhD(physiol chem), 60. *Prof Exp:* From asst prof to assoc prof, 62-69, PROF BIOCHEM, OHIO STATE UNIV, 69- *Concurrent Pos:* Nat Cancer Inst fel, Univ Wis, 60-62; NIH & NSF res grants. *Mem:* Am Soc Biol Chem; Am Sci Affiliation; Am Chem Soc. *Res:* Control mechanisms regulating the synthesis of nucleic acids precursors in animal, plant and bacterial cells. *Mailing Add:* Dept Biochem Col Biol Sci Ohio State Univ 484 W 12th Ave Columbus OH 43210

IVES, JEFFREY LEE, b Torrington, Conn, Feb 6, 51; m 74; c 1. ORGANIC CHEMISTRY, MEDICINAL CHEMISTRY. *Educ:* Colgate Univ, BA, 73; Yale Univ, MS, MPhil & PhD, 78. *Prof Exp:* MED CHEMIST, PFIZER, INC, 78- *Mem:* Am Chem Soc. *Res:* Synthesis of pharmacologically active agents involving the design and discovery of new organic reactions and transformations. *Mailing Add:* 80 Goose Hill Dr CHester CT 06412-1229

IVES, JOHN (JACK) DAVID, b Grimsby, Eng, Oct 15, 31; Can citizen; m 54; c 4. PHYSICAL GEOGRAPHY. *Educ:* Univ Nottingham, BA, 53; McGill Univ, PhD(geog), 56. *Prof Exp:* Asst cur geog, Sub-Arctic Res Lab, McGill Univ, 55, dir geog, Res Lab & asst prof, Univ, 57-60; asst to dir & chief div phys geog, Geog Br, Can Dept Mines & Tech Surv, 60-63, dir, 64-67; dir inst arctic & alpine res, Univ Colo, Boulder, 67-79, prof geog, 67-89; PROF GEOG & CHAIRPERSON, DEPT GEOG, UNIV CALIF, DAVIS, 89- *Concurrent Pos:* Mem subcomt glaciers, Nat Res Coun, 60-67; mem Can comt, Int Geog Union, 61-63, secy-treas, 64-67; chmn, Nat Adv Comt Geog Res, 65-67; alpine site coordr, Tundra Biome, Int Biol Prog, 69-73; chmn, US Directorate Proj, Human Impacts on Mountain Ecosysts, 74-, Int Geog Union Comn, Mountain Geoecol, UN Univ Proj, Highland-Lowland Interactive Syst, 74-90, UN Univ Mountain Ecol & Sustainable Develop, 90-; John Simon Guggenheim Mem fel, 76-77; vis prof, Bern Univ, Switz, 76-77; founder & ed, J Arctic & Alpine Res, 69-81 & J Mountain Res & Develop, 81-; chmn, Int Geog Union Comm Mountain Geoecol, 88- *Mem:* Glaciol Soc; Asn Am Geographers; Int Mountain Soc (pres, 80-); Arctic Inst NAm. *Res:* Geomorphology with emphasis on processes of deglaciation and effects of erosion by glacier melt-water; investigation of history of recession of the last major ice sheets; glaciology; coordination of interdisciplinary studies on mountain geoecology; permafrost studies in alpine areas; natural hazards in mountain regions; renewable natural resources in mountain regions, Nepal, China, Thailand, South America and USSR. *Mailing Add:* Dept Geog Hart Hall Univ Calif Davis CA 95616

IVES, MICHAEL BRIAN, b Bournemouth, Eng, Sept 30, 34; m 61. MATERIALS SCIENCE, CORROSION. *Educ:* Bristol Univ, BSc, 57, PhD(physics), 59. *Prof Exp:* Res metall engr, Carnegie Inst Technol, 58-61; assoc dean eng, 79-88, from asst prof to assoc prof, 61-72, PROF MAT SCI & ENG, MCMASTER UNIV, 72- *Concurrent Pos:* Sabbatical leave, Univ Milan, 67-68; Alexander Von Humboldt Found teaching fel, vis prof, Max Planck Inst Iron Res, Dusseldorf, 75, Japan Soc Promotion of Sci, 79, Univ Erlangen, Nurnberg, 83 & EPFL, Switz, 89-90. *Mem:* Am Soc Metals Int; Nat Asn Corrosion Engrs. *Res:* Localized corrosion of metals and alloys, dissolution of crystals. *Mailing Add:* Dept Mat Sci & Eng McMaster Univ 1280 Main St W Hamilton ON L8S 4M1 Can

IVES, NORTON C(ONRAD), b Rolfe, Iowa, Mar 6, 17; m 42; c 7. AGRICULTURAL ENGINEERING. *Educ:* Iowa State Univ, BS, 38, MS, 39, PhD, 59. *Prof Exp:* Exten agr engr, Univ Minn, 39-44 & Iowa State Univ, 44-45; chief dept agr eng, Inter-Am Inst Agr Sci, Costa Rica, 45-53; agr engr, USDA, Iowa State Univ, 53-61; CONSULT AGR ENGR, 61- *Mem:* AAAS; Am Soc Agr Engrs; Soil Conserv Soc Am; Sigma Xi. *Res:* Farm structures; grain drying and storage investigations; crop processing; farmstead engineering. *Mailing Add:* Rte 1 Rolfe IA 50581

IVES, PHILIP TRUMAN, b Amherst, Mass, Aug 15, 09; m 40; c 4. GENETICS. *Educ:* Amherst Col, AB, 32, AM, 34; Calif Inst Technol, PhD, 38. *Prof Exp:* Asst, 41-45, res assoc, 45-75, EMER RES ASSOC BIOL, AMHERST COL, 75- *Mem:* AAAS; Genetics Soc Am; Ecol Soc Am; Am Soc Zool; Soc Study Evolution. *Res:* Population genetics of Drosophila; spontaneous and induced mutagenesis. *Mailing Add:* Dept Biol Amherst Col Box 2237 Amherst MA 01002

IVES, ROBERT SOUTHWICK, b Salem, Mass, July 7, 13; m 40; c 7. CHEMISTRY. *Educ:* Univ Maine, BS, 33. *Prof Exp:* Chemist, Lever Bros Co, 37-41 & F C Huyck & Sons Co, 41-43; res chemist, Sylvania Elec Corp, 43-47; lab dir, Esselen Res Corp, 47-52; pres, Ives Lab, 52-58; dir res, Ludlow Papers, Inc, 58-60; TREAS, CONTOUR CHEM CO, 60- *Mem:* Am Chem Soc; Tech Asn Pulp & Paper Indust; Am Asn Textile Chemists & Colorists. *Res:* Research and development of plastics; adhesives; functional paper coatings and textiles. *Mailing Add:* 144 Wheeler St Gloucester MA 01930-1649

IVESON, HERBERT TODD, b Mirror, Alta, Sept 28, 15; m 42; c 3. CHEMISTRY. *Educ:* Univ Ill, BS, 37, MS, 40. *Prof Exp:* From res chemist to sr chemist, Glidden Co, 40-51, indust consult, 51-58; mgr lecithin prod, Cent Soya Co, Inc, 58-62, div prod mgr, 62-64, div prod adminr, 64-65, mgr indust applns, 65-70; qual assurance mgr, Glidden-Durkee Div, SCM Corp, Joliet, 70-81; RETIRED. *Mem:* Am Chem Soc; Am Oil Chemists' Soc; Sigma Xi. *Res:* Design of radio frequency induction heating units; ionization from hot filaments; development of products from fats, oils and lipids; lecithin product showing increased efficiency in chocolate viscosity reduction; method of producing a non-break bleached oil; edible emulsifiers. *Mailing Add:* 219 Clinton St Elmhurst IL 60126

IVETT, REGINALD WILLIAM, b Stockton, NY, Oct 27, 15; m 43. INDUSTRIAL CHEMISTRY. *Educ:* Allegheny Col, AB, 36; Purdue Univ, MS, 39, PhD(phys chem), 41. *Prof Exp:* Res chemist, Hercules Powder Co, 41-43, res supvr, 43-49, res mgr, 49-63, dir develop, Fibers Dept, 63-68; patent coordr, Hercules, Inc, 68-83; RETIRED. *Mem:* Am Chem Soc. *Res:* Fiber science; polymer applications; research administration; rosin and terpene chemicals. *Mailing Add:* 536 Kerfoot Farm Rd Woodbrook Wilmington DE 19803

IVEY, DON LOUIS, b Ft Worth, Tex, Nov 17, 35. CIVIL ENGINEERING. *Educ:* Lamar State Col, BS, 60; Tex A&M Univ, MS, 62, PhD(struct eng), 64. *Prof Exp:* Asst res eng, Tex Transport Inst & resident engr, Struct Res Lab, 63-69; asst, 60-63, PROF CIVIL ENG, TEX A&M UNIV, 71-, ASST DIR, TEX TRANSP INST, 76-; PRES, SCI INQUIRY INC, 76- *Concurrent Pos:* Res engr & head, Hwy Safety Res Ctr & Safety Div, Tex Transp Inst, 71-76. *Honors & Awards:* Kummer lectr, Am Soc Testing & Mat, 79. *Mem:* Am

Concrete Inst; Am Soc Testing & Mat. *Res:* Highway safety engineering; concrete technology; reinforced lightweight structural concrete design practices; vehicle handling and stability; collision dynamics; tire pavement friction; highway geometrics; highway safety structures. *Mailing Add:* CE/TTI Bldg Suite 802 Tex A&M Univ College Station TX 77843-3135

IVEY, DONALD GLENN, b Clanwilliam, Man, Feb 6, 22; m 44; c 3. PHYSICS, POLYMER PHYSICS. *Educ:* Univ BC, BA, 44, MA, 46; Univ Notre Dame, PhD(physics), 49. *Prof Exp:* Demonstr math & physics, Univ BC, 43-46; res assoc physics, Univ Notre Dame, 46-49; from asst prof to assoc prof, 49-63, prin, New Col, 63-74, vpres, Instnl Rels, 80-84, prof, 63-87, EMER PROF PHYSICS, UNIV TORONTO, 87- *Concurrent Pos:* Secy & treas, Can High Polymer Forum, 53-55, chmn, 58. *Mem:* Am Phys Soc; Am Asn Physics Teachers; Can Asn Physicists. *Res:* Physical properties of solid polymers. *Mailing Add:* Dept Physics Univ Toronto Toronto ON M5S 1A7 Can

IVEY, E(DWIN) H(ARRY), JR, b Galveston, Tex, Apr 19, 21; m 84; c 6. PETROCHEMICALS MANUFACTURE, PROJECT MANAGEMENT. *Educ:* Agr & Mech Col, Tex, BSChE, 41, MSChE, 47. *Prof Exp:* Instr chem, Agr & Mech Col, Tex, 42-43, instr chem eng, 46-47; res & develop engr, Houdry Process Corp, 47-50; res & develop engr, Dow Chem Co, 50-52, asst supt, Light Hydrocarbons Dept, 53-55, supt, 55-66, proj mgr, Dow Chem Nederland, 66-67, hydrocarbons tech mgr, Dow Chem Europe, Switz, 67-68; consult & proj mgr, P R Olefins Co, 68-70; gen mgr, 70-72, CONSULT & PRES, E H IVEY & CO, INC, 72- *Mem:* Am Chem Soc; Am Inst Chem Engrs. *Res:* Project management; petrochemical manufacture in Europe and North America. *Mailing Add:* 1000 Country Pl No 137 Houston TX 77079-4719

IVEY, ELIZABETH SPENCER, b Schenectady, NY, Apr 21, 35; m 57, 82; c 5. ACOUSTICS. *Educ:* Simmons Col, BS, 57; Harvard Univ, MAT, 59; Univ Mass, PhD(mech eng acoust), 76. *Prof Exp:* Prof physics, Simmons Col, 58-59, Bucknell Univ, 60-63 & Colo State Univ, 64-68; assoc dean fac, 82-85, Louise Wolff Kahn Chair, 85, PROF PHYSICS, SMITH COL, 69-, CHAIR DEPT, 83- *Concurrent Pos:* Vis prof, Yale Univ, 82. *Mem:* Acoust Soc Am; AAAS; Am Asn Physics Teachers. *Res:* Study of noise propagation in urban and suburban situations; sound attenuation by building-size barriers; diffraction, absorption, and scattering processes; ground vibrations due to urban rail transportation systems; attenuation of loud sounds by helmets. *Mailing Add:* Dept Physics Smith Col Northampton MA 01063

IVEY, HENRY FRANKLIN, b Augusta, Ga, June 16, 21; m 48; c 2. LUMINESCENCE. *Educ:* Univ Ga, AB, 40, MS, 41; Mass Inst Technol, PhD(physics), 44. *Prof Exp:* Asst physics, Univ Ga, 40-41; mem staff, Radiation Lab, Mass Inst Technol, 42-45; sr engr, Nat Union Radio Corp, NJ, 45-46; res engr, Lamp Div, Westinghouse Elec Corp, 46-52, adv engr, 53-56, res sect mgr, 56-61, res eng consult, 61-63, adv scientist, Res Labs, 63-69, mgr optical physics, 69-74, adv scientist technol assessment, Res Labs, 74-86; RETIRED. *Mem:* Am Phys Soc; hon mem Electrochem Soc; Optical Soc Am; fel Inst Elec & Electronics Engrs. *Res:* Color centers in alkali halides; cathode ray tube screens; thermionic electron emission; phosphors; space charge; electroluminescence; laser materials; lamps. *Mailing Add:* 9259-B Jamison Ave Philadelphia PA 19115

IVEY, JERRY LEE, b Laredo, Tex, Oct 14, 42; m 64; c 3. THEORETICAL SOLID STATE PHYSICS. *Educ:* McMurry Col, BA, 65; Purdue Univ, MS, 67, PhD(physics), 71. *Prof Exp:* Sr res physicist mat res, Monsanto Res Corp, Mound Lab, 74-; AT EG&G IDAHO INC. *Concurrent Pos:* Res assoc, Purdue Univ, 71-72; res assoc, Aerospace Res Lab, Wright-Patterson AFB, Ohio, 72-74. *Mem:* Am Phys Soc. *Res:* Properties of materials, hydrogen in metals. *Mailing Add:* 3209 Tipperrary Lane Idaho Falls ID 83404

IVEY, MARVIN, b Orlando, Fla, Jan 17, 32; m 53; c 2. PHYSICAL GEOLOGY. *Educ:* Univ Fla, BSE, 53, MEd, 57, EdD(educ & geol), 61; Stetson Univ, JD, 78. *Prof Exp:* High sch teacher, Fla, 53-54; asst geol, Univ Fla, 56-59; instr geol, St Petersburg Jr Col, 59-64, chmn dept natural sci, 64-75 & 77-84, asst div dir sci, 75-77, prof geol, 84-91; RETIRED. *Mem:* Fel AAAS; Geol Soc Am; Nat Asn Geol Teachers; Nat Asn Res Sci Teaching; Nat Sci Teachers Asn; Am Inst Prof Geologists. *Res:* Local investigations into ground water conditions and associated problems of erosion. *Mailing Add:* 14452 Hillview Dr Largo FL 34644

IVEY, MICHAEL HAMILTON, b Auburn, Ala, Jan 24, 30; m 52; c 2. MEDICAL PARASITOLOGY. *Educ:* Ala Polytech Inst, BS, 51; Univ NC, MSPH, 53, PhD, 56. *Prof Exp:* Asst prof microbiol, Univ Mo, 56-62; fel, Tulane Univ, 62-63; assoc prof prev med & pub health, Sch Med, from assoc prof to prof lab practice & chmn dept parasitol & lab practice, Sch Health, 67-72, PROF MICROBIOL & IMMUNOL, HEALTH SCI CTR, UNIV OKLA, 73- *Concurrent Pos:* Fel, China Med Bd, 57; consult to Saigon Med Sch, AMA Proj, 68-75. *Mem:* AAAS; Am Soc Microbiol; Am Soc Parasitol; Am Soc Trop Med & Hyg. *Res:* Host-parasite relationships; immunology; pneumocystis. *Mailing Add:* Dept Microbiol & Immunol Univ Okla Health Sci Ctr PO Box 26901 Oklahoma City OK 73190

IVEY, ROBERT CHARLES, b Portland, Ore, Aug 3, 43; m 67; c 2. CHEMICAL PHYSICS, PETROLEUM ENGINEERING. *Educ:* Abilene Christian Univ, BS, 65; Univ Tex, Austin, PhD(phys chem), 69. *Prof Exp:* Lectr chem & comput sci, Univ Tex, El Paso, 68-69; asst prof, 69-77, chmn dept physics & dir observ, 71-81, PROF PHYSICS, ABILENE CHRISTIAN UNIV, 77-; VPRES & GEOPHYSICIST, LAJET GEOPHYS, INC, 81- *Concurrent Pos:* Welch Found grant electron diffraction res, 72-81; co-owner, CSA Software Co; consult & vpres res & develop, GeoNuclear Consults, Inc; consult independent oil & gas explor. *Mem:* Am Asn Physics Teachers; Am Inst Mech Engrs. *Res:* Electromagnetic geophysics, nuclear well logging, petroleum formation evaluation research; electron diffraction of gases; applications of group theory in quantum mechanics; scattering theory; determination of the structures of gas phase radicals. *Mailing Add:* PO Box 5198 Abilene TX 79608

IVEY, WILLIAM DIXON, embryology, for more information see previous edition

IVIE, GLEN WAYNE, pesticide chemistry, phytochemistry; deceased, see previous edition for last biography

IVINS, RICHARD O(RVILLE), b Chicago, Ill, Nov 1, 34; m 72; c 4. CHEMICAL ENGINEERING. *Educ:* Northwestern Univ, BS, 58, MS, 62; Univ Chicago, MBA, 79. *Prof Exp:* Res assoc, Argonne Nat Lab, 58-59, from asst chem engr to assoc chem engr, 59-67, group leader reactor safety, 67-68, sect mgr, 68-70, tech dir, FEFP Proj, 70, proj mgr treat facil improv, 71-72, prog mgr treat fast reactor safety tests, 72-74, assoc mgr, energy Storage Progs & mgr, Lithium/Metal Sulfide Battery Proj, 74-76, dep dir & actg dir, Coal Progs, 76-78, dep for commercialization, 78-80; dir, Off Indust Interaction & Technol Transfer, 80-85; pres, TDM Corp, 85-87; pres, Midcor, 87-89; PRES, ROI ASSOCS, 89- *Concurrent Pos:* Lectr, Proj Mgt Systs. *Mem:* Am Inst Chem Engrs; fel Am Inst Chemists. *Res:* Coal technology; electrochemical systems; fast reactor safety; vapor explosions between reactor fuels and sodium; core debris heat transfer; in-reactor testing. *Mailing Add:* 13457 S Redberry Circle Plainfield IL 60544-9366

IVORY, JOHN EDWARD, b Buffalo, NY, May 8, 29; m 57; c 3. PHYSICS. *Educ:* Canisius Col, BS, 50; Univ Notre Dame, MS, 52, PhD(physics), 54. *Prof Exp:* Physicist aerodyn, Bell Aircraft Corp, 54-55; asst prof physics, Col St Thomas, 55-56, chmn dept, 56-60; res physicist thermoelec, US Naval Res Lab, 60-68; physist, Off Naval Res, Chicago Br, 68-81; CONSULT, NAVAL RES LAB, 87- *Concurrent Pos:* Lectr, Univ Md, 61-68; adj prof, Harper & Triton Col, 81-; adj prof, Col DuPage, 84-85; assoc prof, Ind Univ, 86-87; vis assoc prof, Ind Univ, 88. *Mem:* Am Phys Soc; Sigma Xi. *Mailing Add:* 302 S Dwyer Arlington Heights IL 60005-1642

IVORY, THOMAS MARTIN, III, b Corpus Christi, Tex, Aug 24, 43; m 72; c 1. BIOLOGY, MICROBIOLOGY. *Educ:* Univ Utah, BA, 65, MS, 67, PhD(biol), 73. *Prof Exp:* Res asst limnol, Univ Utah, 65-67; environ health officer, USN Med Serv Corp, 67-70; res asst limnol, Univ Utah, 70-73; proj mgr ecol, 73-81, MGR WESTERN REGION, ESE, NUS CORP, 81- *Mem:* Am Soc Limnol & Oceanog. *Res:* Limnology; primary productivity of fresh water lakes; phyxology; environmental microbiology; in stream flow requirements of aquatic organisms; environmental permit studies for mining projects. *Mailing Add:* 6263 S Niagara Way Englewood CO 80111

IWAMOTO, REYNOLD TOSHIAKI, b Honolulu, Hawaii, Nov 20, 28; m 51; c 3. ANALYTICAL CHEMISTRY. *Educ:* Univ Hawaii, BS, 50, MS, 52; Harvard Univ, PhD(chem), 56. *Prof Exp:* Instr chem, Princeton Univ, 55-56; from asst prof to assoc prof, 56-65, PROF CHEM, UNIV KANS, 65- *Mem:* Am Chem Soc. *Res:* Electrochemical, spectral and magnetic resonance studies of the nature and behavior of inorganic species in nonaqueous media; porphyrin redox chemistry. *Mailing Add:* Dept Chem Univ Kans Lawrence KS 66045

IWAMOTO, TOMIO, b Los Angeles, Calif, July 16, 39; m 71; c 3. ICHTHYOLOGY, SYSTEMATICS. *Educ:* Univ Calif, Los Angeles, AB, 61; Univ Miami, MS, 68, PhD(marine sci), 72. *Prof Exp:* Biologist, Pascagoula Fishery Sta, US Bur Comm Fisheries, 62-65; instr fisheries, Ore State Univ, 71-72; CUR ICHTHYOL, CALIF ACAD SCI, 72- *Concurrent Pos:* Prin investr, NSF grant, 75-90; consult, Environ Mgt & Res, 78- *Mem:* Am Soc Ichthyologists & Herpetologists; Ichthyol Soc Japan; Indian Soc Ichthyol; Systematic Zool; Sigma Xi. *Res:* Biology of deep-sea fishes, particularly grenadier family Macrouridae. *Mailing Add:* Calif Acad Sci Golden Gate Park San Francisco CA 94118

IWAN, DEANN COLLEEN, astrophysics, physics; deceased, see previous edition for last biography

IWANIEC, TADEUSZ, b Elblag, Poland, Oct 9, 47; m 71; c 1. QUASICONFORMAL MAPPINGS, PARTIAL DIFFERENTIAL EQUATIONS. *Educ:* Univ Warsaw, Habilitation, 79, PhD(math), 75. *Prof Exp:* From asst to assoc prof Univ Warsaw,71-81; assoc prof math, Polish Acad Sci, 81-83; PROF, MATH, SYRACUSE UNIV, 86- *Concurrent Pos:* Vis prof math, Univ Lomonoff, Moscow, 78-79, Univ Bonn, 79, Univ Piza, 81, Univ Helsinki, 82, Univ Mich, Ann Arbor, 83-84, Univ Tex, Austin, 84-85, Courant Inst, 85-86. *Mem:* Am Math Soc. *Res:* Quasiconformal analysis and related problems in non-linear elliptic equations, complex functions and differential geometry; complex functions and differntial geometry; harmonic analysis. *Mailing Add:* Syracuse Univ 200 Carnegie Syracuse NY 13210

IWASA, KUNIHIKO, b Tokushima, Japan, Nov 26, 44; US citizen; m 73; c 3. BIOPHYSICS. *Educ:* Osaka City Univ, BSc, 67; Nagoya Univ, MSc, 69, PhD(physics), 74. *Prof Exp:* Killam fel chem, Dalhousie Univ, 75-76; res assoc chem, Ind Univ, 76-77; vis fel, Univ Ljubljana, 77-78; res assoc chem eng, Rice Univ, 78-79; chemist biophys, NIMH, 79-83; biophysicist, Nat Inst Neurol Commun Dis & Stroke, 83-85; physicist, Ctr Drugs & Biol, Fed Drug Admin, 85-86; BIOPHYSICIST, NAT INST NEUROL COMM DIS & STROKE, NIH, 86- *Mem:* Am Chem Soc; Biophys Soc. *Res:* Excitable membrane; thermodynamical study of polyelectrolytes. *Mailing Add:* NIH Bldg Nine Rm 1E124 Bethesda MD 20892

IWASA, YUKIKAZU, b Kyoto, Japan, Feb 15, 38; m 69; c 2. ELECTRICAL ENGINEERING. *Educ:* Mass Inst Technol, SB, SM(mech eng) & SM(elec eng), 62, EE, 64, PhD(elec eng), 67. *Prof Exp:* SR SCIENTIST SUPERCONDUCTIVITY, FRANCIS BITTER NAT MAGNET LAB, MASS INST TECHNOL, 64- *Honors & Awards:* Oyama Mem Award, 84; M Hetényi Award, 87; Cryog & The Brit Cryog Coun Prize, 90. *Mem:* AAAS. *Res:* Applied superconductivity; materials, magnets and devices. *Mailing Add:* Francis Bitter Nat Magnet Lab Rm NW17-201 Mass Inst Technol Cambridge MA 02139

IWASAKI, IWAO, b Tokyo, Japan, Feb 6, 29; m 72; c 3. METALLURGY, CHEMISTRY. *Educ:* Univ Minn, BS, 51, MS, 53; Mass Inst Technol, ScD(metall), 57. *Hon Degrees:* DEng, Tohoku Univ, Japan, 61. *Prof Exp:* Asst prof metall, Univ Minn, 57-59; group leader, Fuji Iron & Steel Co, Japan,

59-63; assoc prof, 63-66, PROF METALL, UNIV MINN, MINNEAPOLIS, 66- *Honors & Awards:* A M Gaudin Award & A F Taggart Award, Soc Mining Engrs of Am Inst Mining, Metall & Petrol Engrs, 81; R H Richard Award, Am Inst Mining, Metall & Petrol Engrs, 86. *Mem:* Am Inst Mining, Metall & Petrol Engrs; Mining Inst Japan; Iron & Steel Inst Japan; Electrochem Soc Japan. *Res:* Mineral processing, ore flotation, flocculation, hydrometallurgy, iron ore agglomeration and reduction; solid waste treatment; grinding media wear; chlorination and segregation roasting; water treatment methods. *Mailing Add:* Mineral Resources Res Ctr Univ Minn Minneapolis MN 55455

IWASAWA, KENKICHI, b Kiriu, Japan, Sept 11, 17; m 41; c 3. MATHEMATICS. *Educ:* Univ Tokyo, DSc(math), 45. *Prof Exp:* Asst prof math, Univ Tokyo, 49-52; asst prof, Mass Inst Technol, 52-54, assoc prof, 55-57; mem, Inst Advan Study, 57-58; prof math, Mass Inst Technol, 58-67; PROF MATH, PRINCETON UNIV, 67- *Concurrent Pos:* Mem, Inst Advan Study, 50-52, 57-58, 66-67, 71, 79 & 85. *Honors & Awards:* Cole Prize, Am Math Soc, 62. *Mem:* Am Math Soc; Am Acad Arts & Sci. *Res:* Algebra and number theory. *Mailing Add:* 1-34-5 Hieashigaoka Meguro-Ku Tokyo 152 Japan

IWIG, MARK MICHAEL, b Topeka, Kans, Feb 6, 51; m 73; c 3. WHEAT & CORN BREEDING. *Educ:* Baker Univ, BS, 73; Purdue Univ, MS, 75, PhD(plant breeding & genetics), 77. *Prof Exp:* Res sta mgr, 77-86, dir, Dept Cereal Seed Res, 86-89, DIR, DEPT NAM CORN BREEDING, PIONEER HI-BRED INT, 89- *Mem:* Agron Soc Am; Crop Sci Soc Am. *Res:* Development of new improved cultivars of Soft Red Winter Wheat, Hard Red Winter Wheat, Hard Red Spring Wheat and new improved corn hybrids of North America. *Mailing Add:* Dept NAm Corn Breeding 7301 NW 62nd Ave PO Box 85 Johnston IA 50131

IYENGAR, DORESWAMY RAGHAVACHAR, b Nanjangud, India, July 3, 30; m 66; c 2. PHYSICAL CHEMISTRY, SURFACE CHEMISTRY. *Educ:* Univ Mysore, BSc, 51; Univ Madras, MSc, 54; Univ Miami, PhD(chem), 63. *Prof Exp:* Lectr chem, Univ Mysore, 51-56; Du Pont fel, Lehigh Univ, 62-64, US Army Signal Corps fel, 64-65; phys scientist, Frankford Arsenal, Pa, 65-66 & Army Mat Res Agency, Mass 66-67; res assoc prof chem, Lehigh Univ, 67-70; sr res chemist, Cent Inorg Res Lab, Sherwin-Williams Co, 70-71, staff scientist & leader, Pigments Lab, 71-75; prof consult, 75; res assoc, Dispersions Pigments Div, 75-80, sr res assoc & head, Dispersions Characterization, Basf Wyandotte Corp, 81-86; tech mgr, 86-90, CONSULT, FLINT INK CORP, 90- *Concurrent Pos:* Prof chem, Roosevelt Univ, 72-75. *Mem:* Am Chem Soc; fel Royal Soc Chem; fel Am Inst Chem; fel Tech Surface Coatings. *Res:* Heterogeneous catalysis; surface chemistry of metals; oxides; semiconductors and pigments; adsorption and chemisorption; neutron inelastic scattering from surfaces; adsorbed species by electron spin resonance; corrosion and stress corrision; studies on pigments and coatings; dispersions and emulsions; printing inks and lithography. *Mailing Add:* 1347 N Silo Ridge Dr Ann Arbor MI 48108

IYENGAR, RAJA M, biochemistry; deceased, see previous edition for last biography

IYER, HARIHARAIYER MAHADEVA, b Kerala, India, June 21, 31; US citizen; m 72; c 2. GEOPHYSICS, SEISMOLOGY. *Educ:* Univ Kerala, BSc, 51, MSc, 53; Univ London, PhD(oceanog), 59. *Prof Exp:* Sci officer oceanog, Indian Naval Phys Lab, Cochin, 53-60; Sverdrup Mem fel inst geophys & planetary physics, Univ Calif, San Diego, 60-61; Gassiot fel seismol, Brit Meteorol Off, Bracknell, 61-63; sci officer seismol, Bhabha Atomic Res Ctr, Bombay, 63-67; geophysicist, 67-80, SUPVRY GEOPHYSICIST, US GEOL SURV, 80- *Concurrent Pos:* Consult inst geophys & planetary physics, Univ Calif, San Diego, 62 & 64; mem, Dept Energy Consortium Active & Passive Seismic Methods for Geothermal Explor, 77; USA-India exchange scientist, 78; UN expert, seismology, 81; consult, Regional Govt Azores, 83; fel, Indo-US Subcomt Educ & Cult, Nat Geophys Inst, Hyderabad, India, 85-86; adv, Wadan Inst, Himalayan Geol, India, 90; mem, Working Group Microseisms, Int Asn Seismology & Physics Earth's. *Honors & Awards:* Krishnan Mem Gold Medal, Indian Geophys Union, 66. *Mem:* Fel Asn Explor Geophysicists India; Int Asn Seismology & Physics Earth's Interior; Seismological Soc Am; Am Geophys Union. *Res:* Structure and evolution of the earth's crust and mantle; mapping the heterogeneous structure of the earth's crust and upper mantle using seismic tomography; imaging magma chambers using seismic tomography; physics and evolution of volcanic and geothermal systems; geothermal exploration using seismic techniques; microseismic, seismic and earthquake prediction; induced seismicity associated with dams and reservoirs; author of 70 publications including six in books. *Mailing Add:* US Geol Surv MS 977 345 Middlefield Rd Menlo Park CA 94025

IYER, RAJUL V, b Bombay, India, Jan 14, 30; m 56; c 2. MICROBIOLOGY. *Educ:* Univ Bombay, BSc, 49, MSc, 52, PhD(chem, bot, biochem), 53. *Prof Exp:* Indian Coun Agr Res grant soil microbiol, Indian Inst Sci, Bangalore, 54-56; res asst intermediary metab, Univ Ill, Urbana, 56-57; Nat Inst Sci India res fel microbiol, Gujarat Univ, India, 58-59; res assoc photobiol, Univ Rochester, 60-62; from asst prof to assoc prof, 66-88, ADJ PROF MICROBIOL, UNIV OTTAWA, 88- *Concurrent Pos:* Med Res Coun Can res grant, 67-; sabbatical, Univ Leicester, 75. *Res:* Infectious drug resistance in gram-negative enteric organisms; R plasmid induced membrane changes in E coli. *Mailing Add:* Dept Microbiol & Immunol Univ Ottawa Ottawa ON K1N 6N5 Can

IYER, RAM R, b Pallavaram, India, Apr 17, 53; m 82; c 2. PIPE STRESS ANALYSIS, FINITE ELEMENT ANALYSIS. *Educ:* Bangalore Univ, India, BE, 74; Indian Inst Sci, ME, 76; Brooklyn Polytech, DE, 78; Drexel Univ, MBA, 86. *Prof Exp:* Res assoc, Ctr Regional Technol, 77-78; design engr, SunTech, Inc, 78-83; sr staff engr, 83-87, sr proj engr, 87-89, MGR, SUN REFINING & MKT CO, 89- *Mem:* Am Soc Civil Engrs; Am Soc Mech Engrs. *Res:* Pipe stress analysis; pipe support design; earthquake resistant design of concrete structures; design of petroleum and petrochemical plants; vibration and fatigue evaluation of rotating equipment and finite element analysis of mechanical and structural components. *Mailing Add:* Seven Bon Air Dr Marlton NJ 08053

IYER, RAVI, b Kumbakonam, India, July 13, 58; m 85; c 1. PASSIVATION & INSULATOR RESEARCH ON III-V SEMICONDUCTORS, OPTICAL CHARACTERIZATION & KINETIC STUDIES. *Educ:* Regional Eng Col, BTech, 79; Colo State Univ, MS, 84, PhD(elec eng), 91. *Prof Exp:* Res asst solar energy, 81-84, res asst semiconductors, 85-87, res assoc, 87-89, POSTDOCTORAL SEMICONDUCTORS, COLO STATE UNIV, 89- *Res:* Development of a suitable insulator on InP for MIS technology; kinetics of SiO_2 deposition from silane and plasma O_2; passivation of III-V semiconductors and characterization using photoluminescence and raman spectroscopy. *Mailing Add:* Dept Elec Eng Colo State Univ Ft Collins CO 80521

IYPE, PULLOLICKAL THOMAS, b Kottayam, India. CELL BIOLOGY, CANCER. *Educ:* Univ Madras, BSc, 55; Univ Baroda, India, MSc, 58, PhD(zool), 61. *Prof Exp:* Demonstr zool, Univ Baroda, 58-61; sci officer biochem, Regional Res Lab, India, 61-65; Damon Runyon fel carcinogenesis, McArdle Lab, Madison, Wis, 65-68; sci officer, Regional Res Lab, India, 68-69; scientist, Paterson Lab, Christie Hosp, Eng, 69-77; sect head carcinogenesis, Frederick Res Ctr, Nat Cancer Inst, 77-85; PRES & OWNER, BIOL RES FAC & FACIL, 85- *Concurrent Pos:* Lectr, Univ Manchester, 72-77. *Mem:* Am Asn Cancer Res; Brit Asn Cancer Res; Brit Soc Cell Biol; AAAS. *Res:* Cellular aspects of chemical carcinogenesis using model systems to analyze the early events, especially the control of cell proliferation, during malignant transformation of epithelial cells. *Mailing Add:* Biol Res Facility Inc 10075 Tyler Pl Ijamsville MD 21754

IZANT, ROBERT JAMES, JR, b Cleveland, Ohio, Feb 4, 21; m 47; c 3. PEDIATRIC SURGERY. *Educ:* Amherst Col, BA, 43; Western Reserve Univ, MD, 46. *Prof Exp:* From intern to resident gen surg, Univ Hosps, Cleveland, 46-52; resident pediat surg, Boston Childrens Hosp, 52-55; asst prof, Ohio State Univ, 55-58; from asst prof to assoc prof, PROF PEDIAT SURG, CASE WESTERN RESERVE UNIV, 72- *Honors & Awards:* Belle Sherwin Award. *Mem:* Am Col Surg; Am Pediat Surg Asn (pres, 87-88); Am Asn Surg Trauma; Am Burn Asn; Am Acad Pediat; Sigma Xi. *Res:* Congenital malformations, malignancies and trauma in infants and children. *Mailing Add:* Rainbow Babies & Childrens Hosp Univ Hosps Cleveland OH 44106-5000

IZATT, JERALD RAY, b Preston, Idaho, Sept 22, 28; m 51; c 5. LASERS, SPECTROSCOPY. *Educ:* Univ Utah, BS, 52; Johns Hopkins Univ, PhD(physics), 60. *Prof Exp:* Analyst, digital comput prog, Douglas Aircraft Co, Inc, 52-53; engr weapon syst anal, Westinghouse Elec Corp, 54-55; res scientist, Electro-Optical Systs Anal, Northrop Space Labs, 60-61; from asst prof to prof physics, NMex State Univ, 61-70; vis prof physics, Laval Univ, 70-71, prof, 71-81; PROF PHYSICS, UNIV ALA, 81-, ADJ PROF MAT SCI, 88- *Concurrent Pos:* Consult, 61-; vis scientist, Cambridge Res Labs, USAF, 67-68, Max Planck Inst Solid State Physics, 79-80; adj prof, NMex State Univ, 70-75; vis scholar, Univ Utah, 89-90. *Mem:* Optical Soc Am; Sigma Xi. *Res:* Infrared and submillimeter lasers; molecular spectroscopy; optical techniques for industrial inspection. *Mailing Add:* Dept Physics Univ Ala Tuscaloosa AL 35487-0324

IZATT, REED MCNEIL, b Logan, Utah, Oct 10, 26; m 49; c 6. INORGANIC CHEMISTRY, METAL SEPARATIONS. *Educ:* Utah State Univ, BS, 51; Pa State Univ, PhD, 54. *Prof Exp:* Fel chem, Mellon Inst, 54-56; from asst prof to assoc prof, 56-64, PROF CHEM, BRIGHAM YOUNG UNIV, 64- *Mem:* AAAS; Am Chem Soc; The Chem Soc; Sigma Xi. *Res:* Thermodynamics of coordination compounds in aqueous solutions; selective transport of cations across liquid membranes using macrocycles; heats of mixing at elevated temperatures and pressures. *Mailing Add:* Dept Chem Brigham Young Univ Provo UT 84602-1022

IZEN, JOSEPH N, b Brooklyn, NY, Nov 14, 56. EXPERIMENTAL ELEMENTARY PARTICLE PHYSICS. *Educ:* Cooper Union, BS, 77; Harvard Univ, AM, 78, PhD(physics), 82. *Prof Exp:* Res assoc, Univ Wis, 82-85; ASST PROF, UNIV ILL, 86- *Mem:* Am Phys Soc. *Res:* Experiments observing the high energy annihilation of electrons and positions. *Mailing Add:* 437 Loomis Lab Univ Ill 1110 W Green St Urbana IL 61801

IZENOUR, GEORGE C(HARLES), b New Brighton, Pa, July 24, 12; m 37; c 1. ELECTRICAL ENGINEERING. *Educ:* Wittenberg Univ, AB, 34, MA, 36. *Hon Degrees:* DFA, Wittenberg Univ, 60. *Prof Exp:* Lighting dir, Fed Theatre, 37-39; Rockefeller Found res fel & founder, Electromech Lab, Drama Sch, Yale Univ, 39-43; res engr, Airborne Instruments Lab, Off Sci Res & Develop, Columbia Univ, 43-46; dir, Electomech Lab, Drama Sch, 46-77, from assoc prof to prof theatre eng, 50-77, EMER PROF THEATRE ENG, YALE UNIV, 77- *Concurrent Pos:* Consult, US State Dept, Brazil, 57; grantee award, Ford Found, 60, grant, 67. *Mem:* AAAS. *Res:* Electromechanical-electronic aspects of control systems as they affect the theatre and television. *Mailing Add:* 432-4771 16 Flying Point Rd Stoney Creek CT 06405

IZOD, THOMAS PAUL JOHN, b London, UK, Apr 12, 45; US citizen; m 69; c 3. FIBER FINISHING, FLUORO-CHEMICAL RESEARCH. *Educ:* London Univ, UK, BSc, 66; Oxford Univ, DPhil(phys chem), 69. *Prof Exp:* Teaching fel, res, Harvard Univ, Cambridge, Mass, 69-71; res assoc phys chem, St Andrews Univ, Scotland , 71-74; group leader res, Union Carbide Corp, Tarrytown, NJ, 74-78; technol mgr, res & develop, Waters Assoc, Millipore Corp, 78-82; sr res group leader, res & develop, Polaroid Corp, 82-85; MGR EXPLOR RES, RES & DEVELOP, ALLIED CORP, 85- *Mem:* Am Chem Soc; Royal Chem Soc. *Res:* Chemical surface modification of polymers and inorganic surfaces; characterization of these surfaces such as adsorption catalysis; techniques for making small (colloidal) particles and mechanisms for stabilizing particles in aqueous and non-aqueous media. *Mailing Add:* Allied Corp Columbia Rd Morristown NJ 07960-4658

IZQUIERDO, RICARDO, b Spain, June 8, 62; Span & Can citizen. LASER PROCESSING. *Educ:* Ecole Polytech, BIng, 86, MScA, 88. *Prof Exp:* Asst engr, Mitel Corp, 86; RES ASSOC LASER PROCESSING, ECOLE POLYTECH, 88- *Res:* Laser deposit of metals at different wavelengths for application in microelectronics. *Mailing Add:* Eng Physics Dept Ecole Polytech Montreal PQ H3C 3A7 Can

IZUI, SHOZO, b Kanazawa, Japan, Jan 22, 46; m 77; c 2. AUTOIMMUNOLOGY, IMMUNOPATHOLOGY. *Educ:* Univ Tokyo, MD, 70. *Hon Degrees:* Dr, Univ Geneva, 83. *Prof Exp:* Res assoc immunopath, Scripps Clin & Res Found, USA, 77-78, asst Mb I immunopath, 78-79, Mb II immunopath, 79-81; res fel immunol, WHO/IRIC, 73-74, asst immunol, Div Hemat, 74-77, chg res immunol, Div Hemat, 81-84, Dept Path, 84-87, PROF ADJ IMMUNOL, DEPT PATH, FAC MED, UNIV GENEVA, 87- *Mem:* Am Asn Immunologists. *Res:* Genetic, cellular and molecular investigation on immunopathogenesis of spontaneous murine model of autoimmune diseases, especially systemic lupus erythematosus. *Mailing Add:* Dept Path Geneva One rue Michel Servet 1211 Geneva 4 Switzerland

IZUNO, TAKUMI, agronomy, for more information see previous edition

IZYDORE, ROBERT ANDREW, b McKeesport, Pa, July 13, 43; div; c 2. ORGANIC & PHARMACEUTICAL CHEMISTRY. *Educ:* Pa State Univ, BS, 65; Duquesne Univ, PhD(org chem), 69. *Prof Exp:* US Army Res Off-Durham Res Assoc, Duke Univ, 69-71; assoc prof, 71-80, PROF CHEM, NC CENT UNIV, 80- *Concurrent Pos:* Instr, Pa State Univ, McKeesport, 68-69. *Mem:* Am Chem Soc. *Res:* Synthesis and chemistry of azo compounds and their cyclo-addition products; synthesis of pharmacologically active compounds. *Mailing Add:* 311 Hubbard Chem Bldg NC Cent Univ Durham NC 27707

IZZO, JOSEPH ANTHONY, JR, b Rochester, NY, Feb 7, 17; m 58; c 1. MATHEMATICS. *Educ:* Columbia Univ, PhD, 57. *Prof Exp:* Asst prof math, Sampson & Champlain Cols, 46-50; instr, Hunter Col, 50-56; from asst prof to assoc prof, 56-71, PROF MATH, UNIV VT, 71- *Concurrent Pos:* Instr, Columbia Univ, 51-56. *Mem:* Am Math Soc; Math Asn Am. *Res:* Statistics and probability; foundations of geometry; training of teachers of mathematics. *Mailing Add:* Dept Math Univ Vt Burlington VT 05405

IZZO, PATRICK THOMAS, b Beverly, Mass, May 10, 18; m 53; c 2. CHEMISTRY, PHARMACEUTICAL CHEMISTRY. *Educ:* Fordham Univ, BS, 42, MS, 44; Columbia Univ, PhD(chem), 47. *Prof Exp:* Fel, Mass Inst Technol, 47-48; res chemist, Lederle Labs, Am Cyanamid Co, 48-91; RETIRED. *Concurrent Pos:* Instr, Rockland Community Col, Suffern, NY, 72- *Mem:* Am Chem Soc. *Res:* Organic chemistry; anti-infective agents; central nervous system agents. *Mailing Add:* Five Rollins Ave Pearl River NY 10965